Students: Campbell offers many tools to help you succeed

✓ Access your text anywhere with **the Pearson eText**

✓ Strengthen your knowledge in **the Mastering Biology Study Area**

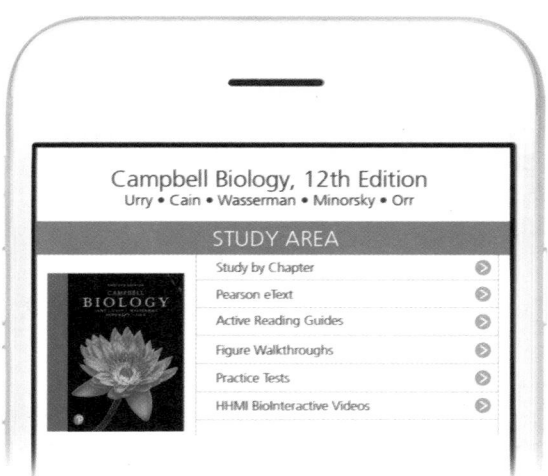

✓ Prepare for your exams with **the Dynamic Study Modules**

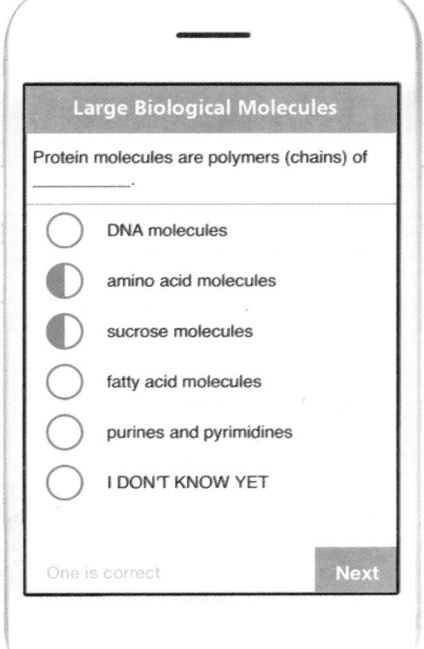

✓ Use the tools in *Campbell Biology* and **Mastering Biology** to create a **Study Plan**. Your Study Plan might include:

Bio Study Plan

☐ Review the syllabus, assignments, and notes from my instructor
☐ Read the chapter and...
    ☐ Use the Study Tip
    ☐ Fill out the Active Reading Guide
       (download from the Study Area)
    ☐ Watch the Figure Walkthroughs, Videos, and Animations
       (in the eText or Study Area)
    ☐ Answer the questions in the chapter
☐ Do the assignments
☐ Study for the test!
    ☐ Review lecture notes and assignments
    ☐ Read the Summary
    ☐ Answer the questions at the end of the chapter
    ☐ Use the Dynamic Study Modules
       (in Mastering Biology)
    ☐ Take the Practice Test
       (in the eText or Study Area)

BIOLOGY

TWELFTH EDITION

Lisa A. Urry
MILLS COLLEGE, OAKLAND, CALIFORNIA

Michael L. Cain
NEW MEXICO STATE UNIVERSITY

Steven A. Wasserman
UNIVERSITY OF CALIFORNIA, SAN DIEGO

Peter V. Minorsky
MERCY COLLEGE, DOBBS FERRY, NEW YORK

Rebecca B. Orr
COLLIN COLLEGE, PLANO, TEXAS

Pearson

Director, Global Higher Ed Content Management and Strategy, Science & Health Sciences: *Jeanne Zalesky*
Manager, Higher Ed Global Content Strategy, Life Sciences: *Joshua Frost*
Associate Content Analyst: *Chelsea Noack*
Editorial Assistant: *Ashley Fallon*
Director, Higher Ed Product Management, Life Sciences: *Michael Gillespie*
Product Manager: *Rebecca Berardy Schwartz*
Managing Producer: *Michael Early*
Senior Content Producer: *Lori Newman*
Director, Content Development & Partner Relationships: *Ginnie Simione Jutson*
Supervising Editors: *Beth N. Winickoff, Pat Burner*
Senior Developmental Editors: *John Burner, Mary Ann Murray, Hilair Chism, Andrew Recher, Mary Hill*
Specialist, Instructional Design and Development: *Sarah Young-Dualan*
Senior Content Developer, Mastering Biology: *Sarah Jensen*

Project Manager: *Katie Cook*
Content Producers, Mastering Biology: *Kaitlin Smith, Ashley Gordon*
Supervising Media Producer: *Tod Regan*
Media Producer: *Ziki Dekel*
Full-Service Vendor: *Integra Software Services, Inc.*
Design Manager: *Mark Ong*
Cover & Interior Designer: *Jeff Puda*
Illustrators: *Lachina Creative*
Rights & Permissions Project Manager: *Matt Perry, SPi Global*
Rights & Permissions Manager: *Ben Ferrini*
Photo Researcher: *Maureen Spuhler*
Product and Solutions Specialist: *Kelly Galli*
Senior Product Marketing Manager: *Alysun Estes*
Manufacturing Buyer: *Stacey Weinberger, LSC Communications*
Cover Photo Credit: *Robert Rohrbaugh/Rohrbaugh Photography*

Library of Congress Cataloging-in-Publication Data
Names: Urry, Lisa A., author. | Cain, Michael L. (Michael Lee), author. | Wasserman, Steven Alexander, author. | Minorsky, Peter V., author. | Orr, Rebecca B., author. | Campbell, Neil A., Biology.
Title: Campbell biology / Lisa A. Urry, Michael L. Cain, Steven A. Wasserman, Peter V. Minorsky, Rebecca B. Orr, Neil A. Campbell.
Description: Twelfth edition. | New York, NY : Pearson, 2020. | Includes index.
Identifiers: LCCN 2019039139 | ISBN 9780135188743 (hardcover) | ISBN 9780135988046 (ebook)
Subjects: LCSH: Biology.
Classification: LCC QH308.2 .C34 2020 | DDC 570—dc23
LC record available at https://lccn.loc.gov/2019039139

1 2019

ISBN 10: 0-135-18874-1; ISBN 13: 978-0-135-18874-3 (Rental Edition)
ISBN 10: 0-136-62344-1; ISBN 13: 978-0-136-62344-1 (Instructor Review Copy)
www.pearson.com

Setting the Standard for Excellence, Accuracy, and Innovation

Campbell Biology, **12th Edition,** delivers an authoritative, accurate, current, and pedagogically innovative experience that helps students make connections so they learn and understand biology. This edition presents new, engaging visual and digital resources that meet demonstrated student needs.

TWELFTH EDITION

CAMPBELL
BIOLOGY
URRY • CAIN • WASSERMAN
MINORSKY • ORR

A New Visual Experience for Every Chapter

NEW! Chapter Openers introduce each chapter and feature a question answered with a clear, simple image to help students visualize and remember concepts as they move through each chapter. Each opener includes a Study Tip and highlights of interactive media in Mastering Biology.

17 Gene Expression: From Gene to Protein

KEY CONCEPTS

17.1 Genes specify proteins via transcription and translation *p. 336*

17.2 Transcription is the DNA-directed synthesis of RNA: *A Closer Look p. 342*

17.3 Eukaryotic cells modify RNA after transcription *p. 345*

17.4 Translation is the RNA-directed synthesis of a polypeptide: *A Closer Look p. 347*

17.5 Mutations of one or a few nucleotides can affect protein structure and function *p. 357*

Study Tip

Make a visual study guide: Sketch the process shown below, and add labels and details as you read the chapter. (In this exercise, assume all processes take place in a eukaryotic cell.)

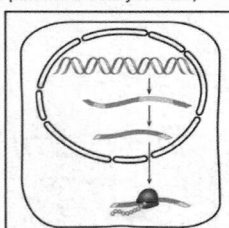

◉ Go to Mastering Biology

For Students (in eText and Study Area)
• Get Ready for Chapter 17
• BioFlix® Animation: Protein Synthesis
• Figure 17.27 Walkthrough: Types of Small-Scale Mutations that Affect mRNA Sequence

For Instructors to Assign (in Item Library)
• BioFlix® Tutorial: Protein Synthesis (1 of 3): Overview
• Tutorial: CRISPR: A Revolution in Genome Editing

Ready-to-Go Teaching Module (in Instructor Resources)
• Gene Expression: Mutations (Concept 17.5)

Figure 17.1 A population of albino donkeys grazes on vegetation on the hillsides of Asinara, an Italian island. Several centuries ago, a recessive mutation that disables pigment synthesis arose in the DNA of one donkey and was passed down through the generations. Inbreeding has resulted in a large number of homozygous albino donkeys living on the island today.

How can one change in DNA result in such a dramatic change in appearance?

Proteins are the link between genotype and phenotype. Gene expression is the process by which DNA directs the synthesis of proteins:

Genotype: DNA — Gene for pigmentation

TRANSCRIPTION

mRNA

TRANSLATION

Protein: Enzyme required for pigment synthesis

Phenotype: Brown pigment ✓

Genotype: DNA — Gene for pigmentation with mutation

mRNA

Protein: Enzyme required for pigment synthesis does not work

Phenotype: No pigment ✗

NEW! A Visual Overview helps students start with the big picture.

335

39 Plant Responses to Internal and External Signals

KEY CONCEPTS

39.1 Signal transduction pathways link signal reception to response *p. 843*

39.2 Plants use chemicals to communicate *p. 845*

39.3 Responses to light are critical for plant success *p. 855*

39.4 Plants respond to a wide variety of stimuli other than light *p. 861*

39.5 Plants respond to attacks by pathogens and herbivores *p. 866*

Study Tip

Make a table: As you read the chapter, add specific examples for each of the general categories of responses shown in the diagram.

Factor	Example of plant response
Light	Seed germination in response to red light

→ Go to Mastering Biology

For Students (in eText and Study Area)
• Get Ready for Chapter 39
• Video: Gravitropism
• Video: *Mimosa* leaves

For Instructors to Assign (in Item Library)
• Activity: Leaf Abscission
• Activity: Plant Hormones

Figure 39.1 Sunflowers track the sun from east to west each day. After sunset, they reverse direction, facing the direction of the next sunrise. By facing the hot sun during the day, the floral heads become warmer and release greater amounts of chemicals that attract pollinators. Light is just one of the many factors to which a plant responds.

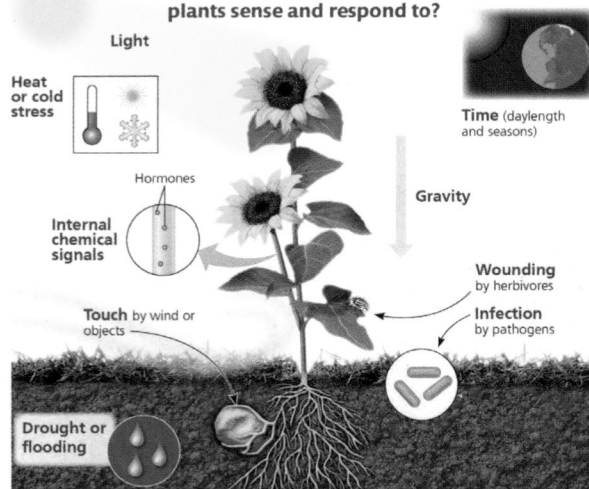

What are some factors that plants sense and respond to?

Light

Heat or cold stress

Time (daylength and seasons)

Internal chemical signals

Hormones

Gravity

Touch by wind or objects

Wounding by herbivores

Infection by pathogens

Drought or flooding

Pearson eText for *Campbell Biology:*

HHMI BioInteractive Video Jennifer Doudna: CRISPR Provides a Tool for Changing Genes

Over the past 15 years, biologists have developed a powerful new technique for gene editing, called the **CRISPR-Cas9** system, that is transforming the field of

BioFlix Animation Mechanics of breathing

Expanding the thoracic cavity during inhalation involves the animal's rib muscles and the **diaphragm**, a sheet of skeletal muscle that forms the bottom wall of the cavity. Contracting the rib

A Whole New Reading Experience

NEW! **The Pearson eText** is a simple-to-use, mobile-optimized, personalized reading experience. It allows students to easily highlight, take notes, and review vocabulary all in one place—even when offline. **Pearson eText for *Campbell Biology*** also includes **Get Ready for This Chapter Questions, Practice Tests, Figure Walkthroughs,** and **500 videos and animations.**

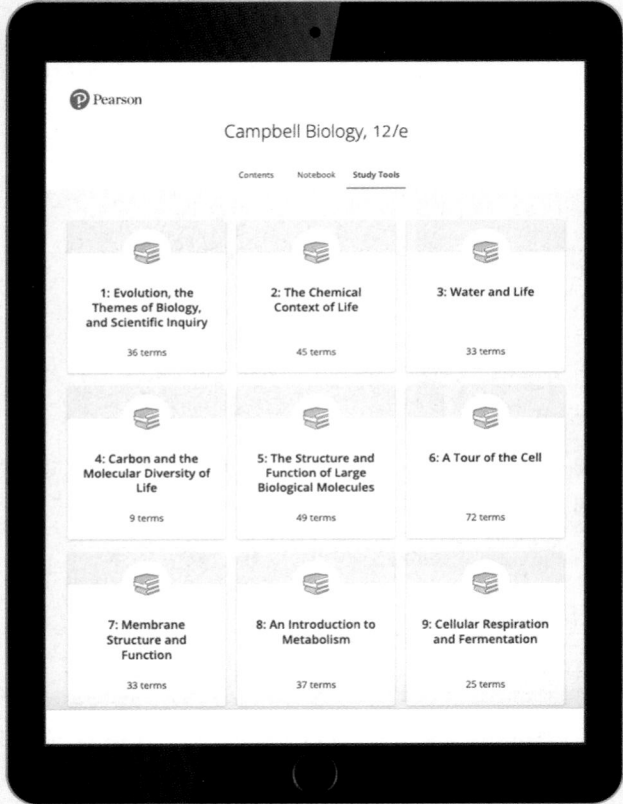

The Pearson eText app is available for download in the app store for approved devices.

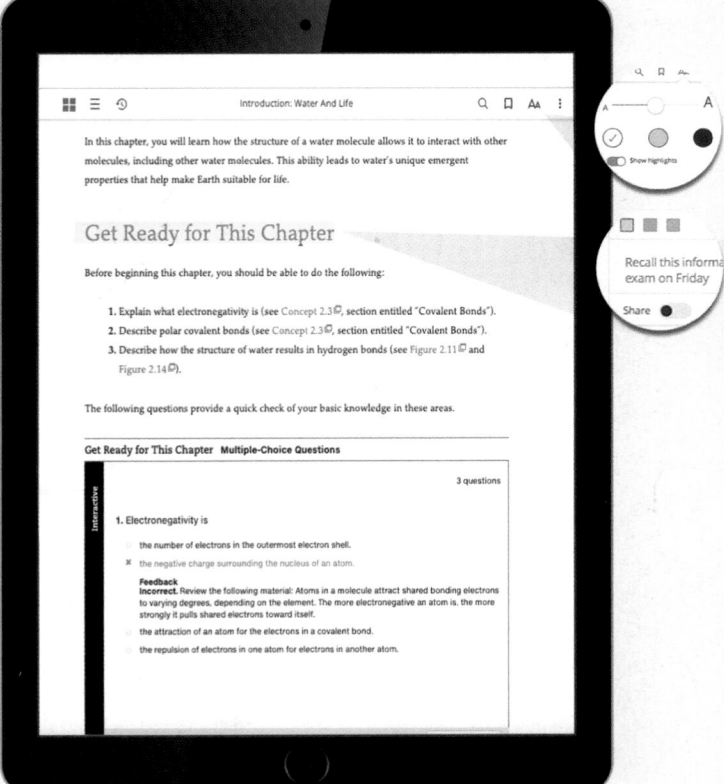

Bringing Innovative Art to Life

NEW! An expanded collection of Figure Walkthroughs guide students through key figures with narrated explanations and figure mark-ups that reinforce important points. **These are embedded in the eText and available for assignment in Mastering Biology**.

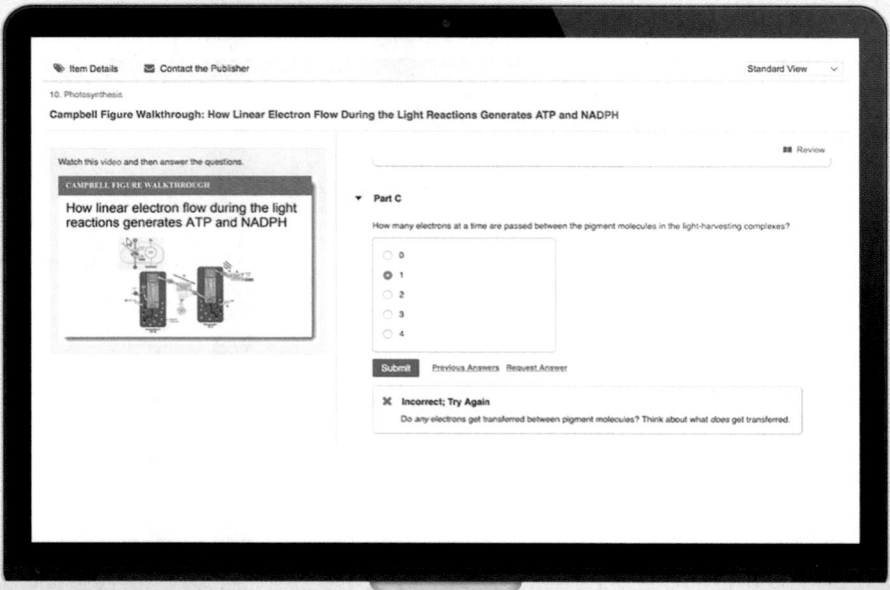

Giving Students the Tools They Need to Succeed

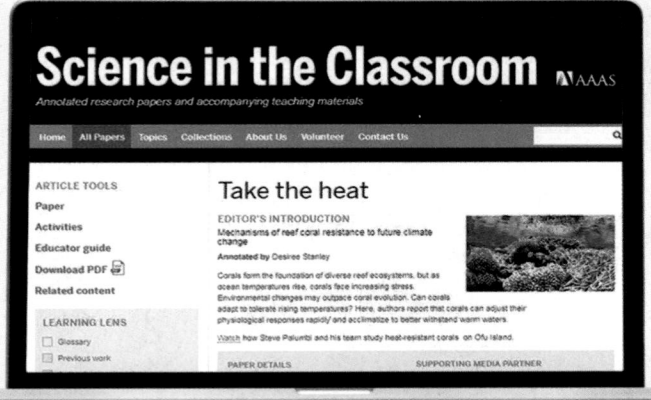

Science in the Classroom AAAS
Annotated research papers and accompanying teaching materials

Home All Papers Topics Collections About Us Volunteer Contact Us

ARTICLE TOOLS
Paper
Activities
Educator guide
Download PDF
Related content

LEARNING LENS
☐ Glossary
☐ Previous work

Take the heat

EDITOR'S INTRODUCTION
Mechanisms of reef coral resistance to future climate change
Annotated by Desiree Stanley

Corals form the foundation of diverse reef ecosystems, but as ocean temperatures rise, corals face increasing stress. Environmental changes may outpace coral evolution. Can corals adapt to tolerate rising temperatures? Here, authors report that corals can adjust their physiological responses rapidly and acclimatize to better withstand warm waters.

Watch how Steve Palumbi and his team study heat-resistant corals on Ofu Island.

PAPER DETAILS   SUPPORTING MEDIA PARTNER

NEW! Science in the Classroom presents annotated journal articles from the **American Association for the Advancement of Science (AAAS)** and makes reading and understanding primary literature easier for students. The articles include assessments in Mastering Biology, allowing instructors to assign the journal articles.

NEW! Active Reading Guides support students in actively reading their biology text. Students can download the worksheets from the Study Area in Mastering Biology.

35. On these diagrams of plant and animal cells, label each organelle and give a brief statement of its function.

Concept 6.6 The cytoskeleton is a network of fibers that organizes structures and activities in the cell

36. What is the cytoskeleton?

37. What are the three roles of the cytoskeleton?

38. There are three main types of fibers that make up the cytoskeleton. Name them.

39. *Microtubules* are hollow rods made of a globular protein called tubulin. Each tubulin protein is a dimer made of two subunits. These are easily assembled and disassembled. What are four functions of microtubules?

Make Connections Across Multiple Concepts

Make Connections Figures pull together content from different chapters, providing a visual representation of "big picture" relationships.

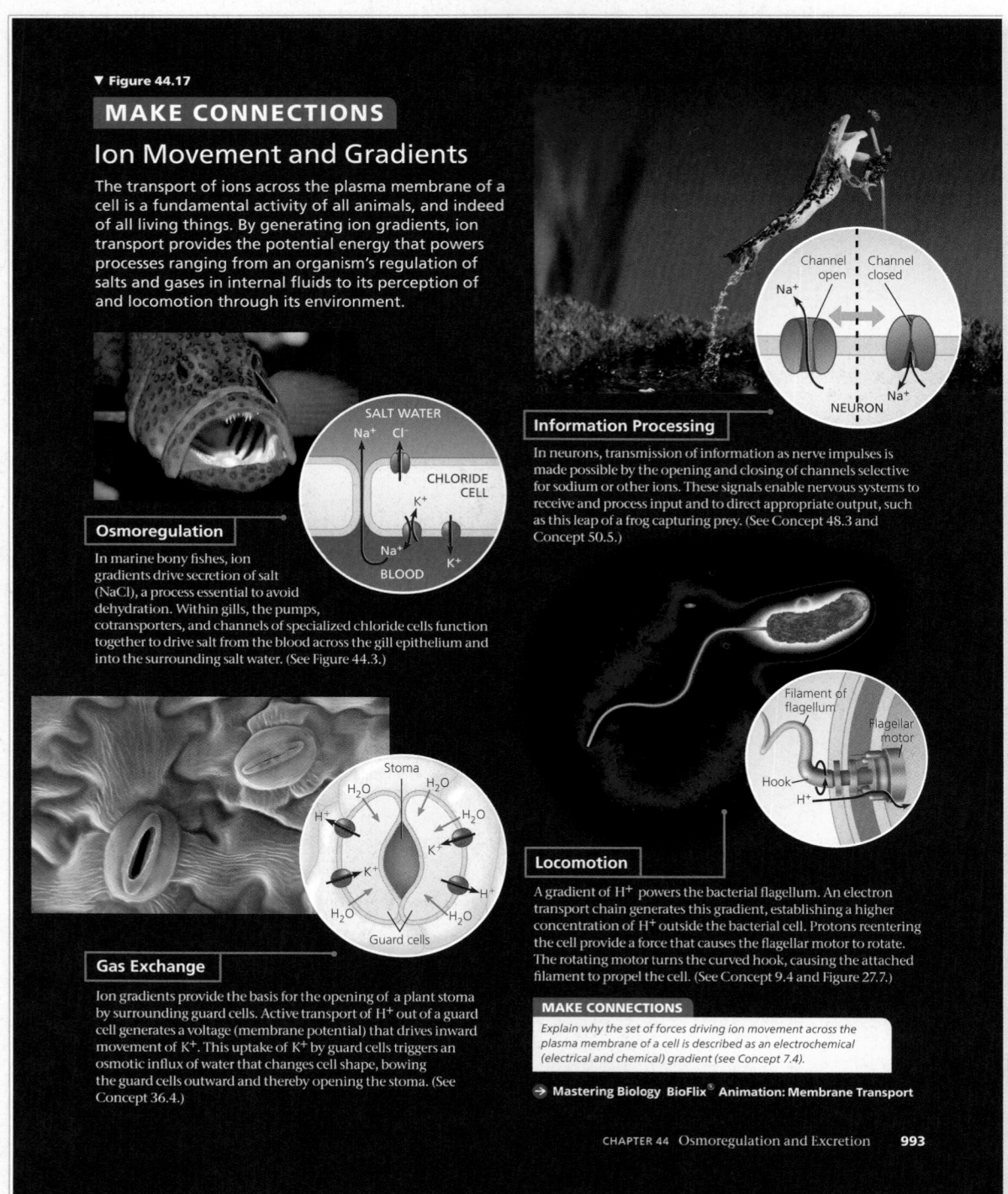

▼ **Figure 44.17**

MAKE CONNECTIONS

Ion Movement and Gradients

The transport of ions across the plasma membrane of a cell is a fundamental activity of all animals, and indeed of all living things. By generating ion gradients, ion transport provides the potential energy that powers processes ranging from an organism's regulation of salts and gases in internal fluids to its perception of and locomotion through its environment.

Osmoregulation

In marine bony fishes, ion gradients drive secretion of salt (NaCl), a process essential to avoid dehydration. Within gills, the pumps, cotransporters, and channels of specialized chloride cells function together to drive salt from the blood across the gill epithelium and into the surrounding salt water. (See Figure 44.3.)

Gas Exchange

Ion gradients provide the basis for the opening of a plant stoma by surrounding guard cells. Active transport of H^+ out of a guard cell generates a voltage (membrane potential) that drives inward movement of K^+. This uptake of K^+ by guard cells triggers an osmotic influx of water that changes cell shape, bowing the guard cells outward and thereby opening the stoma. (See Concept 36.4.)

Information Processing

In neurons, transmission of information as nerve impulses is made possible by the opening and closing of channels selective for sodium or other ions. These signals enable nervous systems to receive and process input and to direct appropriate output, such as this leap of a frog capturing prey. (See Concept 48.3 and Concept 50.5.)

Locomotion

A gradient of H^+ powers the bacterial flagellum. An electron transport chain generates this gradient, establishing a higher concentration of H^+ outside the bacterial cell. Protons reentering the cell provide a force that causes the flagellar motor to rotate. The rotating motor turns the curved hook, causing the attached filament to propel the cell. (See Concept 9.4 and Figure 27.7.)

MAKE CONNECTIONS

Explain why the set of forces driving ion movement across the plasma membrane of a cell is described as an electrochemical (electrical and chemical) gradient (see Concept 7.4).

→ **Mastering Biology BioFlix® Animation: Membrane Transport**

CHAPTER 44 Osmoregulation and Excretion   **993**

x

Make Connections
Questions
in every chapter ask students to relate content to material presented earlier in the course.

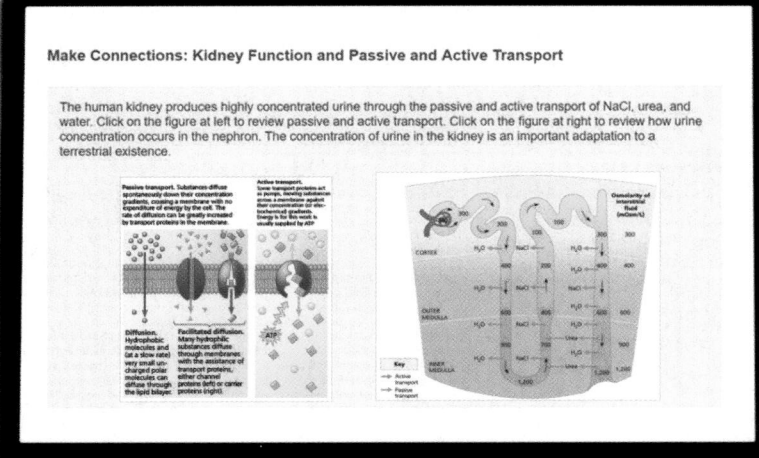

Make Connections: Kidney Function and Passive and Active Transport

The human kidney produces highly concentrated urine through the passive and active transport of NaCl, urea, and water. Click on the figure at left to review passive and active transport. Click on the figure at right to review how urine concentration occurs in the nephron. The concentration of urine in the kidney is an important adaptation to a terrestrial existence.

Make Connections
Tutorials
connect content from two different chapters using art from the book. Make Connections Tutorials are assignable and automatically graded in Mastering Biology and include answer-specific feedback for students.

Develop Scientific Skills

Scientific Skills Exercise

Analyzing Polypeptide Sequence Data

Are Rhesus Monkeys or Gibbons More Closely Related to Humans? In this exercise, you will look at amino acid sequence data for the β polypeptide chain of hemoglobin, often called β-globin. You will then interpret the data to hypothesize whether the monkey or the gibbon is more closely related to humans.

How Such Experiments Are Done Researchers can isolate the polypeptide of interest from an organism and then determine the amino acid sequence. More frequently, the DNA of the relevant gene is sequenced, and the amino acid sequence of the polypeptide is deduced from the DNA sequence of its gene.

Data from the Experiments In the data below, the letters give the sequence of the 146 amino acids in β-globin from humans,

rhesus monkeys, and gibbons. Because a complete sequence would not fit on one line here, the sequences are divided into three segments: amino acids 1–50, 51–100, and 101–146. The sequences for the three different species are aligned so that you can compare them easily. For example, you can see that for all three species, the first amino acid is V (valine) and the 146th amino acid is H (histidine).

INTERPRET THE DATA

1. Scan the monkey and gibbon sequences, letter by letter, circling any amino acids that do not match the human sequence. **(a)** How many amino acids differ between the monkey and the human sequences? **(b)** Between the gibbon and human?

2. For each nonhuman species, what percent of its amino acids are identical to the human sequence of β-globin?

3. Based on these data alone, state a hypothesis for which of these two species is more closely related to humans. What is your reasoning?

4. What other evidence could you use to support your hypothesis?

⊙ **Instructors:** A version of this Scientific Skills Exercise can be assigned in **Mastering Biology**.

Species	Alignment of Amino Acid Sequences of β-globin				
Human	1 VHLTPEEKSA	VTALWGKVNV	DEVGGEALGR	LLVVYPWTQR	FFESFGDLST
Monkey	1 VHLTPEEKNA	VTTLWGKVNV	DEVGGEALGR	LLLVYPWTQR	FFESFGDLSS
Gibbon	1 VHLTPEEKSA	VTALWGKVNV	DEVGGEALGR	LLVVYPWTQR	FFESFGDLST
Human	51 PDAVMGNPKV	KAHGKKVLGA	FSDGLAHLDN	LKGTFATLSE	LHCDKLHVDP
Monkey	51 PDAVMGNPKV	KAHGKKVLGA	FSDGLNHLDN	LKGTFAQLSE	LHCDKLHVDP
Gibbon	51 PDAVMGNPKV	KAHGKKVLGA	FSDGLAHLDN	LKGTFAQLSE	LHCDKLHVDP
Human	101 ENFRLLGNVL	VCVLAHHFGK	EFTPPVQAAY	QKVVAGVANA	LAHKYH
Monkey	101 ENFRLLGNVL	VCVLAHHFGK	EFTPPVQAAY	QKVVAGVANA	LAHKYH
Gibbon	101 ENFRLLGNVL	VCVLAHHFGK	EFTPPVQAAY	QKVVAGVANA	LAHKYH

Data from Human: http://www.ncbi.nlm.nih.gov/protein/AAA21113.1; rhesus monkey: http://www.ncbi. nlm.nih.gov/protein/122634; gibbon: http://www.ncbi.nlm.nih.gov/protein/122616

Scientific Skills Exercises in every chapter of the text use real data to build key skills needed for biology, including data analysis, graphing, experimental design, and math skills. Each exercise is also available as an automatically graded assignment in Mastering Biology with answer-specific feedback for students.

Problem-Solving Exercises

Problem-Solving Exercises guide students in applying scientific skills and interpreting real data in the context of solving a real-world problem. A version of each Problem-Solving Exercise can also be assigned in Mastering Biology.

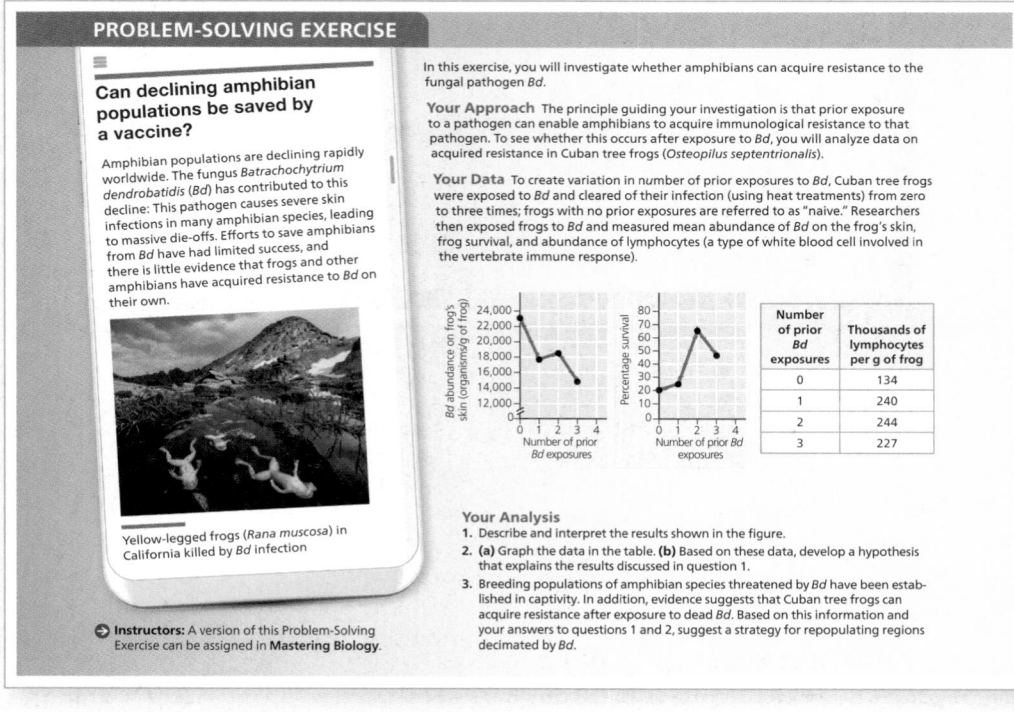

PROBLEM-SOLVING EXERCISE

Can declining amphibian populations be saved by a vaccine?

Amphibian populations are declining rapidly worldwide. The fungus *Batrachochytrium dendrobatidis* (*Bd*) has contributed to this decline: This pathogen causes severe skin infections in many amphibian species, leading to massive die-offs. Efforts to save amphibians from *Bd* have had limited success, and there is little evidence that frogs and other amphibians have acquired resistance to *Bd* on their own.

Yellow-legged frogs (*Rana muscosa*) in California killed by *Bd* infection

⊙ **Instructors:** A version of this Problem-Solving Exercise can be assigned in **Mastering Biology**.

In this exercise, you will investigate whether amphibians can acquire resistance to the fungal pathogen *Bd*.

Your Approach The principle guiding your investigation is that prior exposure to a pathogen can enable amphibians to acquire immunological resistance to that pathogen. To see whether this occurs after exposure to *Bd*, you will analyze data on acquired resistance in Cuban tree frogs (*Osteopilus septentrionalis*).

Your Data To create variation in number of prior exposures to *Bd*, Cuban tree frogs were exposed to *Bd* and cleared of their infection (using heat treatments) from zero to three times; frogs with no prior exposures are referred to as "naive." Researchers then exposed frogs to *Bd* and measured mean abundance of *Bd* on the frog's skin, frog survival, and abundance of lymphocytes (a type of white blood cell involved in the vertebrate immune response).

Number of prior *Bd* exposures	Thousands of lymphocytes per g of frog
0	134
1	240
2	244
3	227

Your Analysis

1. Describe and interpret the results shown in the figure.

2. **(a)** Graph the data in the table. **(b)** Based on these data, develop a hypothesis that explains the results discussed in question 1.

3. Breeding populations of amphibian species threatened by *Bd* have been established in captivity. In addition, evidence suggests that Cuban tree frogs can acquire resistance after exposure to dead *Bd*. Based on this information and your answers to questions 1 and 2, suggest a strategy for repopulating regions decimated by *Bd*.

Innovation in Assessment

Dynamic Study Modules use the latest developments in cognitive science to help students study by adapting to their performance in real time. Students build confidence and understanding, enabling them to participate and perform better, both in and out of class. Available on smartphones, tablets, and computers.

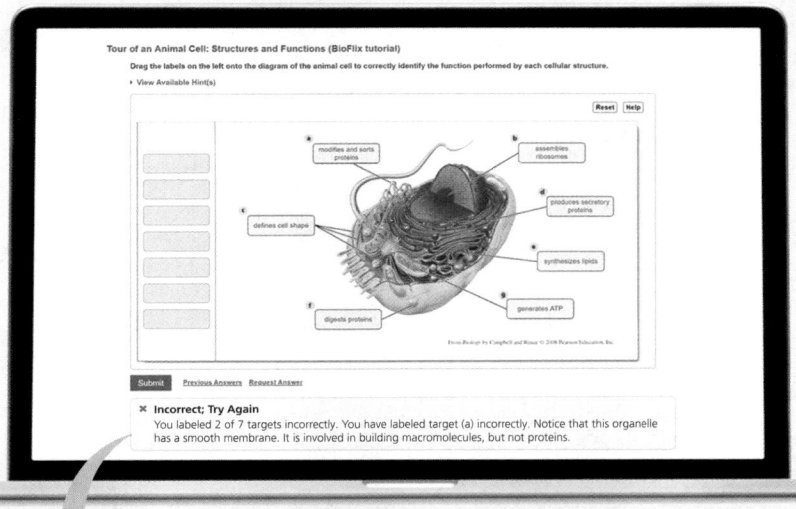

✗ Incorrect; Try Again
You labeled 2 of 7 targets incorrectly. You have labeled target (a) incorrectly. Notice that this organelle has a smooth membrane. It is involved in building macromolecules, but not proteins.

Wrong-Answer Feedback Using data gathered from all of the students using the program, **Mastering Biology** offers wrong-answer feedback that is specific to each student. Rather than simply providing feedback of the "right/wrong/try again" variety, Mastering Biology guides students toward the correct final answer without giving the answer away.

"I wouldn't have passed my class without Mastering Biology. The feedback doesn't just tell me I'm wrong, it gave me a paragraph of feedback on why I was wrong and how I could better understand it."

—Student, University of Texas at Arlington

UPDATED! Test Bank questions have been analyzed and revised with student success in mind. Revisions account for how students read, analyze, and engage with the content.

Innovation in Instructor Resources

NEW! 5 new Ready-to-Go Teaching Modules expand the number of modules to 15. These instructor resources are designed to make use of teaching tools before, during, and after class, including new ideas for in-class activities. The modules incorporate the best that the text, **Mastering Biology**, and **Learning Catalytics** have to offer and can be accessed through the Instructor Resources area of Mastering Biology.

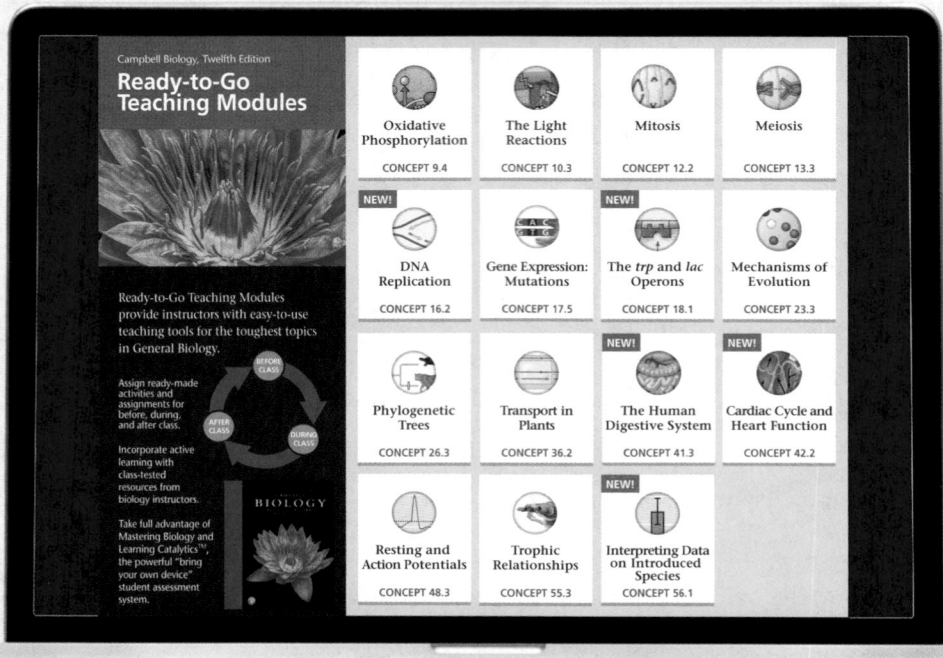

NEW! Early Alerts in **Mastering Biology** help instructors know when students may be struggling in the course. This insight enables instructors to provide personalized communication and support at the moment students need it so they can stay—and succeed—in the course.

About the Authors

The author team's contributions reflect their biological expertise as researchers and their teaching sensibilities gained from years of experience as instructors at diverse institutions. They are also experienced textbook authors, having written *Campbell Biology in Focus* in addition to *Campbell Biology*.

Lisa A. Urry (Chapter 1 and Units 1–3) is Professor of Biology at Mills College. After earning a B.A. at Tufts University, she completed her Ph.D. at the Massachusetts Institute of Technology (MIT). Lisa has conducted research on gene expression during embryonic and larval development in sea urchins. Deeply committed to promoting opportunities in science for women and underrepresented minorities, she has taught courses ranging from introductory and developmental biology to an immersive course on the U.S./Mexico border.

Michael L. Cain (Units 4, 5, and 8) is an ecologist and evolutionary biologist who is now writing full-time. Michael earned an A.B. from Bowdoin College, an M.Sc. from Brown University, and a Ph.D. from Cornell University. As a faculty member at New Mexico State University, he taught introductory biology, ecology, evolution, botany, and conservation biology. Michael is the author of dozens of scientific papers on topics that include foraging behavior in insects and plants, long-distance seed dispersal, and speciation in crickets. He is also a coauthor of an ecology textbook.

Steven A. Wasserman (Unit 7) is Professor of Biology at the University of California, San Diego (UCSD). He earned an A.B. from Harvard University and a Ph.D. from MIT. Working on the fruit fly *Drosophila*, Steve has done research on developmental biology, reproduction, and immunity. Having taught genetics, development, and physiology to undergraduate, graduate, and medical students, he now focuses on introductory biology, for which he has been honored with UCSD's Distinguished Teaching Award.

Peter V. Minorsky (Unit 6) is Professor of Biology at Mercy College in New York, where he teaches introductory biology, ecology, and botany. He received his A.B. from Vassar College and his Ph.D. from Cornell University. Peter taught at Kenyon College, Union College, Western Connecticut State University, and Vassar College; he is also the science writer for the journal *Plant Physiology*. His research interests concern how plants sense environmental change. Peter received the 2008 Award for Teaching Excellence at Mercy College.

Rebecca B. Orr (Ready-to-Go Teaching Modules, Interactive Visual Activities, eText Media Integration) is Professor of Biology at Collin College in Plano, Texas, where she teaches introductory biology. She earned her B.S. from Texas A&M University and her Ph.D. from University of Texas Southwestern Medical Center at Dallas. Rebecca has a passion for investigating strategies that result in more effective learning and retention, and she is a certified Team-Based Learning Collaborative Trainer Consultant. She enjoys focusing on the creation of learning opportunities that both engage and challenge students.

Neil A. Campbell (1946–2004) earned his M.A. from the University of California, Los Angeles, and his Ph.D. from the University of California, Riverside. His research focused on desert and coastal plants. Neil's 30 years of teaching included introductory biology courses at Cornell University, Pomona College, and San Bernardino Valley College, where he received the college's first Outstanding Professor Award in 1986. For many years he was also a visiting scholar at UC Riverside. Neil was the founding author of *Campbell Biology*.

To Jane, our coauthor, mentor, and friend. Enjoy your retirement! LAU, MLC, SAW, and PVM

Preface

We are honored to present the Twelfth Edition of *Campbell Biology*. For the last three decades, *Campbell Biology* has been the leading college text in the biological sciences. It has been translated into 19 languages and has provided millions of students with a solid foundation in college-level biology. This success is a testament not only to Neil Campbell's original vision but also to the dedication of hundreds of reviewers (listed on pages xxviii–xxxi), who, together with editors, artists, and contributors, have shaped and inspired this work.

Our goals for the Twelfth Edition include:

- **supporting students** with new visual presentations of content and new study tools
- **supporting instructors** by providing new teaching modules with tools and materials for introducing, teaching, and assessing important and often challenging topics
- **integrating text and media** to engage, guide, and inform students in an active process of inquiry and learning

Our starting point, as always, is our commitment to crafting text and visuals that are accurate, are current, and reflect our passion for teaching biology.

New to This Edition

Here we provide an overview of the new features that we have developed for the Twelfth Edition; we invite you to explore pages iii–xiv for more information and examples.

- **NEW! Chapter Openers Re-envisioned.** Catalyzed by feedback from students and instructors, informed by data analytics, and building on the results of science education research, we have redesigned the opening of every chapter of the text. The result is more visual, more interactive, and more engaging. In place of an opening narrative, the first page of each chapter is organized around three new elements that provide students with the specific tools and approaches needed to achieve the learning objectives of that chapter:
  - **NEW! Visual Overview.** Centered on a basic biological question related to the opening photo and legend, the Visual Overview illustrates a core idea of the chapter with straightforward art and text. Students get an immediate sense of what the chapter is about and what kinds of thinking will underlie its exploration.
  - **NEW! Study Tip.** Just as the Visual Overview introduces students to *what* they will learn, the study tip offers guidance in *how* to learn. It encourages students to learn actively through such proven strategies as drawing a flow chart, labeling a diagram, or making a table. Each tip provides an effective strategy for tackling important content in the chapter.

- **NEW! Highlights of Digital Resources**. In conversations with users of the textbook, we often encounter a limited awareness of the digital tools the text provides to facilitate instruction and learning. We therefore created *Go to Mastering Biology*, a chapter opener section where we highlight some of the tutorials, animations, and other interactives available for students to explore on their own or for instructors to assign. These resources include Get Ready for This Chapter questions, Figure Walkthroughs, HHMI BioInteractive videos, Ready-to-Go Teaching Modules, and more.

- **NEW! Updated Content.** As in each new edition of *Campbell Biology*, the Twelfth Edition incorporates new content, summarized on pages xviii–xx. Content updates reflect rapid, ongoing changes in knowledge about climate change, genomics, gene-editing technology (CRISPR), evolutionary biology, microbiome-based therapies, and more. In addition, Unit 7 includes a new section on "Biological Sex, Gender Identity, and Sexual Orientation in Human Sexuality," which provides instructors and students with a thoughtful, clear, and current introduction to topics of tremendous relevance to biology, to student lives, and to current public discourse and events.

- **NEW! Active Reading Guides.** These worksheets provide students with self-assessment activities to complete as they read each chapter. Students can download the Active Reading Guides from the Mastering Biology Study Area.

- **5 NEW! Ready-to-Go Teaching Modules.** The Ready-to-Go Teaching Modules provide instructors with active learning exercises and questions to use in class, plus Mastering Biology assignments that can be assigned before and after class. A total of 15 modules are now available in the Instructor Resources area of Mastering Biology.

Pearson eText

Students using the Pearson eText will reap all the benefits of the new text features, while also benefiting from the following new and existing interactive resources, which are integrated directly into the online text:

- **NEW!** An expanded collection of the popular **Figure Walkthroughs** guide students through key figures with narrated explanations and figure mark-ups that reinforce important points.

- **NEW!** Links to the **AAAS Science in the Classroom** website provide research papers from *Science* with annotations to help students understand the papers. These links are included at the end of each appropriate chapter.

- **EXPANDED! 500 animations and videos** bring biology to life. These include new resources from **HHMI BioInteractive** that engage students in topics from CRISPR to coral reefs.
- **Get Ready for This Chapter** questions provide a quick check of student understanding of the background information needed to learn a new chapter's content, with feedback to bolster their preparation.
- **Vocabulary Self-Quizzes** and **Practice Tests** at the end of each chapter provide opportunities for students to test their understanding.
- Links to **Interviews** from all editions of *Campbell Biology* are included in the chapter where they are most relevant. The interviews show students the human side of science by featuring diverse scientists talking about how they became interested in biology and what inspires them.

For more information, see pages vi–ix.

Mastering Biology

Mastering Biology provides valuable resources for instructors to assign homework and for students to study on their own:

- **Assignments.** Mastering Biology is the most widely used online assessment and tutorial program for biology, providing an extensive library of thousands of tutorials and questions that are graded automatically.
  - **NEW! Early Alerts** give instructors a quick way to monitor students' progress and provide feedback, even before the first test.
  - **NEW! AAAS Science in the Classroom** journal articles can be assigned with automatically graded questions.
  - Hundreds of self-paced **tutorials** provide individualized coaching with specific hints and feedback on the most difficult topics in the course.
  - Optional **Adaptive Follow-up Assignments** provide additional questions tailored to each student's needs.
- **Pearson eText.** The Pearson eText, described above, can be directly accessed from Mastering Biology.
- **Dynamic Study Modules.** These popular review tools can be assigned, or students can use them for self-study.
- **Study Area.** Media references in the printed book direct students to the wealth of online self-study resources available to them in the Mastering Biology Study Area, including Active Reading Guides, Figure Walkthroughs, videos, animations, Get Ready for This Chapter, Practice Tests, Cumulative Test, and more.
- **Instructor Resources.** This area of Mastering Biology provides one-stop shopping for Ready-to-Go Teaching Modules, PowerPoints, Clicker Questions, animations, videos, the Test Bank, and more.

For more information, see pages xiii–xiv and xxiv–xxv and visit www.masteringbiology.com.

Our Hallmark Features

Teachers of general biology face a daunting challenge: to help students acquire a conceptual framework for organizing an ever-expanding amount of information. The hallmark features of *Campbell Biology* provide such a framework, while promoting a deeper understanding of biology and the process of science. As such, they are well-aligned with the core competencies outlined by the **Vision and Change** national conferences. Furthermore, the core concepts defined by Vision and Change have close parallels in the unifying themes that are introduced in Chapter 1 and integrated throughout the book.

Chief among the themes of both Vision and Change and *Campbell Biology* is **evolution.** Each chapter of this text includes at least one Evolution section that explicitly focuses on evolutionary aspects of the chapter material, and each chapter ends with an Evolution Connection Question and a Write About a Theme Question.

To help students distinguish "the forest from the trees," each chapter is organized around a framework of three to seven carefully chosen **Key Concepts**. The text, Concept Check Questions, Summary of Key Concepts, and Mastering Biology resources all reinforce these main ideas and essential facts.

Because text and illustrations are equally important for learning biology, **integration of text and figures** has been a hallmark of *Campbell Biology* since the First Edition. The new Visual Overviews, together with our popular Visualizing Figures, Exploring Figures, and Make Connections Figures, epitomize this approach.

To encourage **active reading** of the text, *Campbell Biology* includes numerous opportunities for students to stop and think about what they are reading, often by putting pencil to paper to draw a sketch, annotate a figure, or graph data. Answering these questions requires students to write or draw as well as think and thus helps develop the core competency of communicating science.

Finally, *Campbell Biology* has always featured **scientific inquiry**. The inquiry activities provide students practice in applying the process of science and using quantitative reasoning, addressing core competencies from Vision and Change.

Our Partnership with Instructors and Students

The real test of any textbook is how well it helps instructors teach and students learn. We welcome comments from both students and instructors. Please address your suggestions to:

Lisa Urry (Chapter 1 and Units 1–3): lurry@mills.edu
Michael Cain (Units 4, 5, and 8): mlcain@nmsu.edu
Peter Minorsky (Unit 6): pminorsky@mercy.edu
Steven Wasserman (Unit 7): stevenw@ucsd.edu
Rebecca Orr (Media): rorr@collin.edu

Highlights of New Content

This section highlights selected new content in *Campbell Biology*, Twelfth Edition. In addition to the content updates noted here, every chapter has a **new Visual Overview** on the chapter opening page.

Unit 1 THE CHEMISTRY OF LIFE

In Unit 1, new content engages students in learning foundational chemistry. Chapter 2 includes a new micrograph of the tiny hairs on a gecko's foot that allow it to walk up a wall. The opening photo for Chapter 3 features a ringed seal, a species endangered by the melting of Arctic sea ice due to climate change. Chapter 3 also has added coverage on the discovery of a large subsurface reservoir of liquid water on Mars and the first CO_2 enhancement study done on an unconfined natural coral reef (both reported in 2018). Chapter 4 now includes the discovery of carbon-based compounds on Mars reported by NASA in 2018. In Chapter 5, the technique of cryo-electron microscopy is introduced, due to its increasing importance in the determination of molecular structure.

Unit 2 THE CELL

Our main goal for this unit was to make the material more accessible, inviting, and exciting to students. Chapter 6 includes a new text description of cryo-electron microscopy (cryo-EM) and a new cryo-EM image in Figure 6.3. Art has been added to Figure 6.17 to illustrate the dynamic nature of mitochondrial networks. Chapter 7 begins with a new chapter-opening image showing neurotransmitter release during exocytosis. **Figure 8.1** includes a new photo of bioluminescent click beetle larvae on the outside of a termite mound and a new Visual Overview that illustrates how the laws of thermodynamics apply to metabolic reactions like bioluminescence.

Chapter 9 includes new information on human brown fat usage, the role of fermentation during the production of chocolate, and recent research on the role of lactate in mammalian metabolism. Chapter

▼ Figure 8.1

Figure 8.1 The green glowing spots on the outside of this Brazilian termite mound are larvae of the click beetle, *Pyrophorus nyctophanus*. These larvae convert the energy stored in organic molecules to light, a process called bioluminescence, which attracts termites that the larvae eat. Bioluminescence and other metabolic activities in a cell are energy transformations that are subject to physical laws.

How do the laws of thermodynamics relate to biological processes?

The first law of thermodynamics: *Energy can be transferred and transformed, but it cannot be created or destroyed.* For example:

Light energy from the sun
↓ Transformed by plants to
Chemical energy in organic molecules in plants
↓ Transformed by termites to
Chemical energy in organic molecules in termites
↓ Transformed by click beetle larvae to
Light energy produced by bioluminescence

The second law of thermodynamics: *Every energy transfer or transformation increases the entropy (disorder) of the universe. For example, during every energy transformation, some energy is converted to thermal energy and released as heat:*

Heat — Plant using light to make organic molecules
Heat — Termite digesting plant and making new molecules
Heat — Click beetle larva digesting termite and emitting light

10 begins with a new concept that puts photosynthesis into a big-picture ecological context. Chapter 10 also includes a discussion of the 2018 discovery of a new form of chlorophyll found in cyanobacteria that can carry out photosynthesis using far-red light. In Chapter 11, the relevance of synaptic signaling is underscored by mentioning that it is a target for treatment of depression, anxiety, and PTSD. In Chapter 12, the cell cycle figure (Figure 12.6) now includes cell images and labels describing the events of each phase.

Unit 3 GENETICS

Chapters 13–17 incorporate changes that help students grasp the more abstract concepts of genetics and their chromosomal and molecular underpinnings. For example, a new Concept Check 13.2 question asks students about shoes as an analogy for chromosomes. In Chapter 14, the classic idea of a single gene determining hair or eye color, or even earlobe attachment, is discussed as an oversimplification. Also, the "Fetal Testing" section has been updated to reflect current practices in obstetrics. Chapter 15 now includes new information on "three-parent" babies. In Concept 16.3, the text and Figure 16.23 have been extensively revised to reflect recent models of the structure and organization of interphase chromatin, as well as how chromosomes condense during preparation for mitosis. Chapter 17 now describes the mutation responsible for the albino phenotype of the Asinara donkeys featured in the chapter-opening photo. To make it easier to cover CRISPR, a new section has been added to Concept 17.5 describing the CRISPR-Cas9 system, including Figure 17.28, "Gene editing using the CRISPR-Cas9 system" (formerly Figure 20.14).

Chapters 18–21 are extensively updated, driven by exciting new discoveries based on DNA sequencing and gene-editing technology. In Chapter 18, the coverage of epigenetic inheritance has been enhanced and updated, including the new **Figure 18.8**. Also in Chapter 18, a description of topologically associated domains has been added, along with an update on the 4D Nucleome Network. In Chapter 19, the topic of emerging viral diseases has been updated extensively and reorganized to clearly differentiate influenza viruses that are emerging from those that cause seasonal flu. Other Chapter 19 updates include

▼ Figure 18.8 **Examples of epigenetic inheritance.**

(a) Effects of maternal diet on genetically identical mice.

(b) The Dutch Hunger Winter.

information on vaccine programs, mentioning a large measles outbreak in 2019 that correlated with lower vaccination rates in that region. Information has also been added on improvement of treatment regimes for HIV. Chapter 20 has been extensively updated, including addition of two new subsections, "Personal Genome Analysis" and "Personalized Medicine," with new information on direct-to-consumer genome analysis. Other updates include the first cloning of a primate, stem cell treatment of age-related macular degeneration, CRISPR correction of the sickle-cell disease allele in mice, and a report of gene editing of fertilized human eggs that resulted in live births. Chapter 21 updates include results of the Cancer Genome Atlas Project, a newly discovered function of retrotransposon transcription, and new information on the *FOXP2* gene.

Unit 4   MECHANISMS OF EVOLUTION

The revision of Unit 4 uses an evidence-based approach to strengthen how we help students understand key evolutionary concepts. For example, new text in Concept 24.3 describes how hybrids can become reproductively isolated from both parent species, leading to the formation of a new species. Evidence supporting this new material comes from a 2018 study on the descendants of hybrids between two species of Galápagos finches and provides an example of how scientists can observe the formation of a new species in nature. In Concept 25.2, the discussion of fossils as a form of scientific evidence is supported by a new figure (Figure 25.5) that highlights five different types of fossils and how they are formed. The unit also features new material that connects evolutionary concepts and societal issues. For example, in Chapter 23, new text and a new figure **(Figure 23.19)** describe how some snowshoe hare populations have not adapted to ongoing climate change, causing them to be poorly camouflaged in early winter and leading to increased mortality. Additional changes include a new section of text in Chapter 22 and a figure (Figure 22.22) describing biogeographical evidence for evolution in a group of freshwater fishes that cannot survive in salt water, yet live in regions separated by wide stretches of ocean. In Chapter 25, a new figure (Figure 25.11) provides fossil evidence of an enormous change in the evolutionary history of life: the first appearance of large, multicellular eukaryotes.

▼ **Figure 23.19 Lack of variation in a population can limit adaptation.**

Unit 5   THE EVOLUTIONARY HISTORY OF BIOLOGICAL DIVERSITY

In keeping with our goal of developing students' skills in interpreting visual representations in biology, we have added a new Visualizing Figure, Figure 32.8, "Visualizing Animal Body Symmetry and Axes." New Visual Skills Questions provide practice on topics such as interpreting phylogenetic trees and using graphs to infer how rapidly antibiotic resistance evolves in bacteria. Chapter 31 has been significantly revised to account for new fossil discoveries and updates to the phylogenetic tree of fungi (Figure 31.10). Chapter 34 has been updated with recent genomic data and fossil discoveries indicating that Neanderthals and Denisovans are more closely related to each other than to humans and that they interbred with each other (and with humans), including two new figures (Figures 34.51 and 34.52b). In Chapter 29, a new figure (Figure 29.1) provides a visual overview of major steps in the colonization of land by plants, and revisions to text in Concept 29.1 strengthen our description of derived traits of plants that facilitated life on land. Chapter 27 includes a new section of text that describes the rise of antibiotic resistance and multidrug resistance and discusses novel approaches in the search for new antibiotics. This new material is supported by two new figures, **Figure 27.22** and Figure 27.23. Other updates include the revision of many phylogenies to reflect recent phylogenomic data; a new Inquiry Figure (Figure 28.26) on the root of the eukaryotic tree; and new text describing the 2017 discovery of 315,000-year-old fossils of a hominin that had facial features like those of humans, while the back of its skull was elongated, as in earlier species.

▶ **Figure 27.22 The rise of antibiotic resistance.**

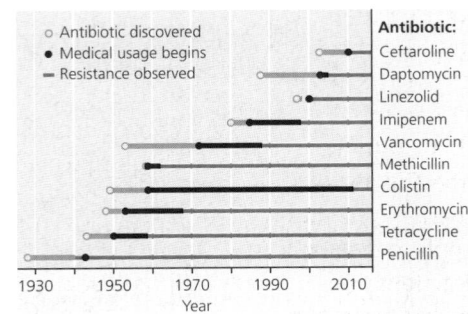

Unit 6   PLANT FORM AND FUNCTION

In Chapter 35, greater emphasis is placed on how structure fits function in vascular plants by way of a new Visual Overview. In Chapter 36, a new Visual Skills Question provides a quantitative exercise in estimating stomatal density. Chapter 37 begins with an emphasis on the importance of crop fertilization in feeding the world. To increase student engagement, renewed emphasis is placed on the link between the nutrition of plants and the nutrition of the organisms, including humans, that feed on them. Table 37.1 concerning plant essential elements has been expanded to include micronutrients as well as macronutrients. In Concept 37.2, a new subsection titled "Global Climate Change and Food Quality" discusses new evidence that global climate change may be negatively impacting the nutritional mineral content of crops. In Chapter 38, the discussion of genetic engineering and agriculture has been enhanced by a discussion of biofortification and by updates concerning "Golden Rice." Chapter 39 includes new updates on the location of the IAA receptor in plant cells and the role of abscisic acid in bud dormancy. The introduction to Concept 39.2 has been revised to emphasize that plants use many classes of chemicals in addition to the classic hormones to communicate information.

Unit 7   ANIMAL FORM AND FUNCTION

The Unit 7 revisions feature pedagogical innovations coupled with updates for currency. A striking new underwater image of Emperor penguins (Figure 40.1) opens the unit and highlights the contributions of form, function, and behavior to homeostasis in general as well as to the specific topic of thermoregulation. The artwork used to introduce and explore homeostasis throughout the unit (Figures 40.8, 40.17, 41.23, 42.28, 44.19, 44.21, and 45.18) has been improved and refined to provide a clear and consistent presentation of the role of perturbation in triggering a response. In Chapter 43, the introduction of the adaptive immune response has been shifted to later in the chapter, allowing students to build on the features of innate immunity before tackling the more demanding topic of the adaptive response. In Chapter 46, a new section of text in Concept 46.4 provides a clear and current introduction to "Biological Sex, Gender Identity, and Sexual Orientation in Human Sexuality." In Chapter 48, the structural overview of neurons is now completed before the introduction of information processing. A new illustration, **Figure 49.8,** provides a concise visual comparison of sympathetic and parasympathetic neurons with each other and with motor neurons of the CNS. In addition, in-depth consideration of glia is now provided in Concept 49.1, where it is more logically integrated into the overview of nervous systems. At the end of the unit, an eye-catching photograph of the male frigatebird's courtship display (Figure 51.1) introduces the topic of animal behavior. Among the content updates that enhance currency and student engagement throughout the unit are discussions of phage

therapy and fecal transplantation, state-of-the-art treatments that both rely on microbiome data, and chronic traumatic encephalopathy (CTE), as well as the latest findings on dinosaur locomotion (Concept 40.1), the awarding of a Nobel Prize in 2017 in the field of circadian rhythms (Concept 40.2), and reference to the ongoing public health crisis of opioid addiction in the context of considering the brain's reward system (Concept 49.5).

Unit 8   ECOLOGY

Complementary goals of the Unit 8 revision were to strengthen our coverage of core concepts while also increasing our coverage of how human actions affect ecological communities. Revisions include a new section of text and a new figure (Figure 52.7) on how plants (and deforestation) can affect the local or regional climate; a new section of text in Concept 55.1 that summarizes how ecosystems work; new text and a new figure (Figure 52.25) illustrating how rapid evolution can cause rapid ecological change; new material in Concept 55.2 on eutrophication and how it can cause the formation of large "dead zones" in aquatic ecosystems; and new text and a new figure (Figure 54.22) on how the abundance of organisms at each trophic level can be controlled by bottom-up or top-down control. A new figure **(Figure 56.23)** shows the extent of the record-breaking 2017 dead zone in the Gulf of Mexico and the watershed that contributes to its nutrient load. In addition, Concept 56.1 includes a new section that describes attempts to use cloning to resurrect species lost to extinction, while Concept 56.4 includes a new section of text and two new figures (Figure 56.27 and 56.28) on plastic waste, a major and growing environmental problem. In keeping with our book-wide goal of expanding our coverage of climate change, Chapter 56 has a new Scientific Skills Exercise in which students interpret changes in atmospheric CO_2 concentrations. Chapter 55 describes how climate warming is causing large regions of tundra in Alaska to release more CO_2 than they absorb (thereby contributing to further climate warming); a new figure (Figure 56.32) describes human and natural factors that contribute to rising global temperatures; and a new section of text in Concept 56.4 describes how global climate change models are developed and why they are valuable.

► **Figure 49.8 Comparison of pathways in the motor and autonomic nervous systems.**

Key to neurotransmitters • Acetylcholine   • Norepinephrine

▼ **Figure 56.23 A dead zone arising from nitrogen pollution in the Mississippi basin.**

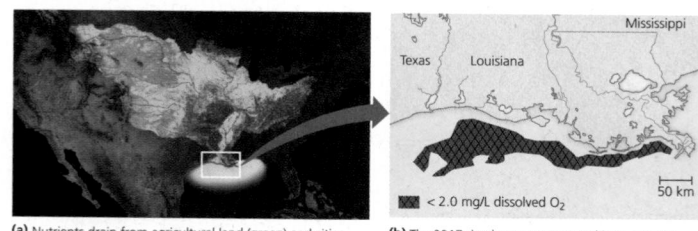

(a) Nutrients drain from agricultural land (green) and cities (red) through the vast Mississippi watershed to the Gulf of Mexico.

(b) The 2017 dead zone, represented here, was the largest yet measured. It occupied 22,730 km² (8,776 mi²), an area slightly larger than New Jersey.

Skills Exercises

Scientific Skills Exercises

1 Interpreting a Pair of Bar Graphs 23

2 Calibrating a Standard Radioactive Isotope Decay Curve and Interpreting Data 33

3 Interpreting a Scatter Plot with a Regression Line 54

4 Working with Moles and Molar Ratios 58

5 Analyzing Polypeptide Sequence Data 89

6 Using a Scale Bar to Calculate Volume and Surface Area of a Cell 99

7 Interpreting a Scatter Plot with Two Sets of Data 136

8 Making a Line Graph and Calculating a Slope 157

9 Making a Bar Graph and Evaluating a Hypothesis 179

10 Making Scatter Plots with Regression Lines 205

11 Using Experiments to Test a Model*

12 Interpreting Histograms 250

13 Making a Line Graph and Converting Between Units of Data 264

14 Making a Histogram and Analyzing a Distribution Pattern 283

15 Using the Chi-Square (x^2) Test 304

16 Working with Data in a Table 318

17 Interpreting a Sequence Logo 351

18 Analyzing DNA Deletion Experiments 376

19 Analyzing a Sequence-Based Phylogenetic Tree to Understand Viral Evolution 411

20 Analyzing Quantitative and Spatial Gene Expression Data*

21 Reading an Amino Acid Sequence Identity Table 458

22 Making and Testing Predictions 483

23 Using the Hardy-Weinberg Equation to Interpret Data and Make Predictions 493

24 Identifying Independent and Dependent Variables, Making a Scatter Plot, and Interpreting Data 513

25 Estimating Quantitative Data from a Graph and Developing Hypotheses 538

26 Using Protein Sequence Data to Test an Evolutionary Hypothesis 570

27 Calculating and Interpreting Means and Standard Errors 590

Making a Bar Graph and Interpreting the Data*

28 Interpreting Comparisons of Genetic Sequences 595

29 Making Bar Graphs and Interpreting Data 628

30 Using Natural Logarithms to Interpret Data 639

31 Interpreting Genomic Data and Generating Hypotheses 657

Synthesizing Information from Multiple Data Sets*

32 Calculating and Interpreting Correlation Coefficients 678

33 Understanding Experimental Design and Interpreting Data 700

34 Determining the Equation of a Regression Line 751

35 Using Bar Graphs to Interpret Data 762

36 Calculating and Interpreting Temperature Coefficients 790

37 Making Observations 812

38 Using Positive and Negative Correlations to Interpret Data 834

39 Interpreting Experimental Results from a Bar Graph 864

40 Interpreting Pie Charts 892

41 Interpreting Data from an Experiment with Genetic Mutants 918

42 Making and Interpreting Histograms 938

43 Comparing Two Variables on a Common x-Axis 972

44 Describing and Interpreting Quantitative Data 981

45 Designing a Controlled Experiment 1014

46 Making Inferences and Designing an Experiment 1030

47 Interpreting a Change in Slope 1049

48 Interpreting Data Values Expressed in Scientific Notation 1082

49 Designing an Experiment using Genetic Mutants 1095

50 Interpreting a Graph with Log Scales 1136

51 Testing a Hypothesis with a Quantitative Model 1150

52 Making a Bar Graph and a Line Graph to Interpret Data 1186

53 Using the Logistic Equation to Model Population Growth 1200

54 Making a Bar Graph and a Scatter Plot 1217

55 Interpreting Quantitative Data 1247

56 Graphing Data and Evaluating Evidence 1279

Problem-Solving Exercises

5 Are you a victim of fish fraud? 89

11 Can a skin wound turn deadly? 214

17 Are insulin mutations the cause of three infants' neonatal diabetes? 359

24 Is hybridization promoting insecticide resistance in mosquitoes that transmit malaria? 518

34 Can declining amphibian populations be saved by a vaccine? 733

39 How will climate change affect crop productivity? 863

45 Is thyroid regulation normal in this patient? 1010

55 Can an insect outbreak threaten a forest's ability to absorb CO_2 from the atmosphere? 1245

*Available only in Mastering Biology. All other Scientific Skills Exercises are in the print book, eText, and Mastering Biology.

Featured Figures

Visualizing Figures

5.16 Visualizing Proteins 79

6.32 Visualizing the Scale of the Molecular Machinery in a Cell 122

16.7 Visualizing DNA 319

25.8 Visualizing the Scale of Geologic Time 532

26.5 Visualizing Phylogenetic Relationships 556

32.8 Visualizing Animal Body Symmetry and Axes 680

35.11 Visualizing Primary and Secondary Growth 767

47.8 Visualizing Gastrulation 1050

55.13 Visualizing Biogeochemical Cycles 1249

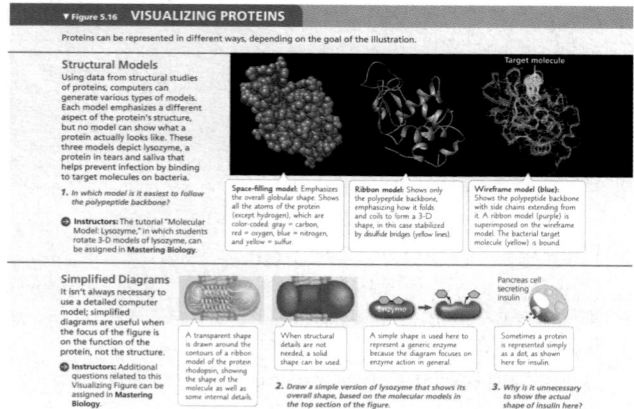

Make Connections Figures

5.26 Contributions of Genomics and Proteomics to Biology 88

10.22 The Working Cell 208

18.27 Genomics, Cell Signaling, and Cancer 392

23.18 The Sickle-Cell Allele 502

33.8 Maximizing Surface Area 695

37.9 Mutualism Across Kingdoms and Domains 813

39.27 Levels of Plant Defenses Against Herbivores 868

40.23 Life Challenges and Solutions in Plants and Animals 894

44.17 Ion Movement and Gradients 993

55.19 The Working Ecosystem 1256

56.31 Climate Change Has Effects at All Levels of Biological Organization 1280

Exploring Figures

1.3 Levels of Biological Organization 4

5.18 Levels of Protein Structure 80

6.3 Microscopy 95

6.8 Eukaryotic Cells 100

6.30 Cell Junctions in Animal Tissues 120

7.21 Endocytosis in Animal Cells 140

11.8 Cell-Surface Transmembrane Receptors 218

12.7 Mitosis in an Animal Cell 238

13.8 Meiosis in an Animal Cell 260

16.23 Chromatin Packing in a Eukaryotic Chromosome 330

24.3 Reproductive Barriers 508

25.7 The Origin of Mammals 531

27.17 Bacterial Diversity 584

28.5 Protistan Diversity 598

29.5 Alternation of Generations 620

29.13 Bryophyte Diversity 626

29.19 Seedless Vascular Plant Diversity 632

30.7 Gymnosperm Diversity 642

30.17 Angiosperm Diversity 650

33.42 Insect Diversity 712

34.42 Mammalian Diversity 745

35.10 Examples of Differentiated Plant Cells 764

41.5 Four Main Feeding Mechanisms of Animals 903

44.12 The Mammalian Excretory System 986

46.11 Human Gametogenesis 1028

49.11 The Organization of the Human Brain 1092

50.10 The Structure of the Human Ear 1113

50.17 The Structure of the Human Eye 1118

53.17 Mechanisms of Density-Dependent Regulation 1204

55.14 Water and Nutrient Cycling 1250

55.17 Restoration Ecology Worldwide 1254

Inquiry Figures

1.25 Does camouflage affect predation rates on two populations of mice? 21

4.2 Can organic molecules form under conditions estimated to simulate those on the early Earth? 57

7.4 Do membrane proteins move? 128

†10.9 Which wavelengths of light are most effective in driving photosynthesis? 194

12.9 At which end do kinetochore microtubules shorten during anaphase? 241

12.14 Do molecular signals in the cytoplasm regulate the cell cycle? 245

14.3 When F_1 hybrid pea plants self- or cross-pollinate, which traits appear in the F_2 generation? 271

14.8 Do the alleles for one character segregate into gametes dependently or independently of the alleles for a different character? 276

†15.3 In a cross between a wild-type female fruit fly and a mutant white-eyed male, what color eyes will the F_1 and F_2 offspring have? 296

15.9   How does linkage between two genes affect inheritance of characters? 301

16.2   Can a genetic trait be transferred between different bacterial strains? 315

16.4   Is protein or DNA the genetic material of phage T2? 316

*†**16.12**   Does DNA replication follow the conservative, semiconservative, or dispersive model? 322

17.3   Do individual genes specify the enzymes that function in a biochemical pathway? 338

18.22   Could Bicoid be a morphogen that determines the anterior end of a fruit fly? 387

19.2   What causes tobacco mosaic disease? 399

20.16   Can the nucleus from a differentiated animal cell direct development of an organism? 429

20.21   Can a fully differentiated human cell be "deprogrammed" to become a stem cell? 432

21.18   What is the function of a gene (*FOXP2*) that may be involved in language acquisition? 462

22.13   Can a change in a population's food source result in evolution by natural selection? 477

*†**23.16**   Do females select mates based on traits indicative of "good genes"? 500

24.7   Can divergence of allopatric populations lead to reproductive isolation? 512

24.12   Does sexual selection in cichlids result in reproductive isolation? 515

24.19   How does hybridization lead to speciation in sunflowers? 521

25.27   What causes the loss of spines in lake stickleback fish? 546

26.6   What is the species identity of food being sold as whale meat? 557

28.26   What is the root of the eukaryotic tree? 612

29.14   Can bryophytes reduce the rate at which key nutrients are lost from soils? 627

31.22   Do fungal endophytes benefit a woody plant? 667

33.29   Did the arthropod body plan result from new *Hox* genes? 706

36.18   Does phloem sap contain more sugar near sources than near sinks? 801

37.10   How variable are the compositions of bacterial communities inside and outside of roots? 814

39.5   What part of a grass coleoptile senses light, and how is the signal transmitted? 847

39.6   What causes polar movement of auxin from shoot tip to base? 848

39.16   How does the order of red and far-red illumination affect seed germination? 857

40.16   How does a Burmese python generate heat while incubating eggs? 888

40.22   What happens to the circadian clock during hibernation? 893

*†**41.4**   Can diet influence the frequency of neural tube defects? 902

42.25   What causes respiratory distress syndrome? 944

44.20   Can aquaporin mutations cause diabetes? 995

46.8   Why is sperm usage biased when female fruit flies mate twice? 1024

†**47.3**   Does the distribution of Ca^{2+} in an egg correlate with formation of the fertilization envelope? 1045

47.23   How does distribution of the gray crescent affect the developmental potential of the first two daughter cells? 1061

47.24   Can the dorsal lip of the blastopore induce cells in another part of the amphibian embryo to change their developmental fate? 1062

47.26   What role does the zone of polarizing activity (ZPA) play in limb pattern formation in vertebrates? 1063

51.8   Does a digger wasp use landmarks to find her nest? 1145

51.24   Are differences in migratory orientation within a species genetically determined? 1157

53.13   How does caring for offspring affect parental survival in kestrels? 1201

†**54.3**   Can a species' niche be influenced by competition? 1216

54.20   Is *Pisaster ochraceus* a keystone species? 1226

55.6   Which nutrient limits phytoplankton production along the coast of Long Island? 1243

55.12   How does temperature affect litter decomposition in an ecosystem? 1248

*†**56.12**   What caused the drastic decline of the Illinois greater prairie chicken population? 1267

Research Method Figures

5.21   X-Ray Crystallography 83

6.4   Cell Fractionation 96

10.8   Determining an Absorption Spectrum 193

13.3   Preparing a Karyotype 256

14.2   Crossing Pea Plants 270

14.7   The Testcross 275

15.11   Constructing a Linkage Map 305

20.3   Sequencing by Synthesis: Next-Generation Sequencing 417

20.7   The Polymerase Chain Reaction (PCR) 421

20.11   RT-PCR Analysis of the Expression of Single Genes 425

26.15   Applying Parsimony to a Problem in Molecular Systematics 563

35.21   Using Dendrochronology to Study Climate 773

37.7   Hydroponic Culture 810

48.6   Intracellular Recording 1072

53.2   Determining Population Size Using the Mark-Recapture Method 1191

54.14   Determining Microbial Diversity Using Molecular Tools 1223

*The Inquiry Figure, original research paper, and a worksheet to guide you through the paper are provided in *Inquiry in Action: Interpreting Scientific Papers*, Fourth Edition.

†A related Experimental Inquiry Tutorial can be assigned in Mastering Biology.

Student and Lab Supplements

For Students

NEW! *Active Reading Guides,* Twelfth Edition

by Fred and Theresa Holtzclaw, The Webb School of Knoxville

These worksheets provide students with self-assessment activities to complete as they read each chapter. Students can download the Active Reading Guides from the Mastering Biology Study Area.

Study Guide, Eleventh Edition

by Martha R. Taylor, Ithaca, New York, and Michael Pollock, Mount Royal University

978-0-134-44377-5/0-134-44377-2

This popular study aid provides concept maps, chapter summaries, word roots, and a variety of interactive activities including multiple-choice, short-answer essay, art labeling, and graph interpretation questions.

Study Card, Eleventh Edition

978-0-134-48648-2/0-134-48648-X

This quick-reference card provides students with an overview of the entire field of biology, helping them see the connections between topics.

Inquiry in Action: Interpreting Scientific Papers, Fourth Edition*

by Ruth V. Buskirk, University of Texas at Austin, and Christopher M. Gillen, Kenyon College

978-0-134-47861-6/0-134-47861-4

This guide helps students learn how to read and understand primary research articles. Part A presents complete articles accompanied by questions that help students analyze the article. Related Inquiry Figures are included in the supplement. Part B covers every part of a research paper, explaining the aim of the sections and how the paper works as a whole.

Practicing Biology: A Student Workbook, Sixth Edition*

by Jean Heitz and Cynthia Giffen, University of Wisconsin, Madison

978-0-134-48603-1/0-134-48603-X

This workbook offers a variety of activities to suit different learning styles. Activities such as modeling and concept mapping allow students to visualize and understand biological processes. Other activities focus on basic skills, such as reading and drawing graphs.

Biological Inquiry: A Workbook of Investigative Cases, Fifth Edition*

by Margaret Waterman, Southeast Missouri State University, and Ethel Stanley, BioQUEST Curriculum Consortium and Beloit College

978-0-134-48646-8/0-134-48646-3

This workbook offers ten investigative cases. Each case study requires students to synthesize information from multiple chapters of the text and apply that knowledge to a real-world scenario as they pose hypotheses, gather new information, analyze evidence, graph data, and draw conclusions. A link to a student website is in the Study Area in Mastering Biology.

Spanish Glossary for Biology

By Laura P. Zanello, University of California, Riverside 978-0-32183498-0/0-321-83498-4

Into the Jungle: Great Adventures in the Search for Evolution

by Sean B. Carroll, University of Wisconsin, Madison

978-0-32155671-4/0-321-55671-2

Get Ready for Biology

by Lori K. Garrett, Parkland College

978-0-32150057-1/0-321-50057-1

A Short Guide to Writing About Biology, Ninth Edition

by Jan A. Pechenik, Tufts University

978-0-13414373-6/0-134-14373-6

An Introduction to Chemistry for Biology Students, Ninth Edition

by George I. Sackheim, University of Illinois, Chicago

978-0-805-39571-6/0-805-39571-7

* An Instructor Guide is available for downloading through the Instructor Resources area of Mastering Biology.

For Lab

Investigating Biology Laboratory Manual, Ninth Edition

by Judith Giles Morgan, Emory University, and M. Eloise Brown Carter, Oxford College of Emory University

978-0-13447346-8/0-134-47346-9

With its distinctive investigative approach to learning, this best-selling laboratory manual is now more engaging than ever, with full-color art and photos throughout. The lab manual encourages students to participate in the process of science and develop creative and critical-reasoning skills.

Annotated Instructor's Edition for Investigating Biology Laboratory Manual, Ninth Edition

978-0-13451922-7/0-134-51922-1

This Annotated Instructor's Edition features teaching information including margin notes with suggestions for lab procedures and answers to questions from the Student Edition. Also included is a detailed Teaching Plan at the end of each lab with specific suggestions for organizing labs, including estimated time allotments for each part

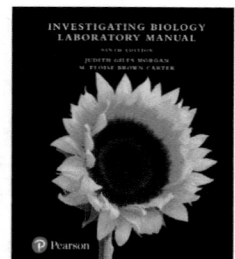

of the lab and suggestions for encouraging independent thinking and collaborative discussions.

Preparation Guide for Investigating Biology, Ninth Edition

978-013451801-5/0-134-51801-2

This guide contains materials lists, suggested vendors, instructions for preparing solutions and constructing materials, schedules for planning advance preparation, and more. It is available for downloading through the Instructor Resources area of Mastering Biology.

Pearson Collections Custom Library

This library gives instructors the power to create custom lab manuals using Pearson content as well as original materials. Learn more at www.pearsonhighered.com/collections.

NEW! Mastering Biology LabBench

The LabBench pre-labs feature 13 online tutorials in Mastering Biology that will both prepare students for their lab work and reinforce key biological principles.

Instructor Resources

The Instructor Resources Area of Mastering Biology

- **5 NEW! Ready-to-Go Teaching Modules** help instructors efficiently make use of the available teaching tools for the toughest topics. Before-class assignments, in-class activities, and after-class assignments are provided for ease of use. Instructors can incorporate active learning into their course with the suggested activity ideas and clicker questions or Learning Catalytics Questions. A total of 15 modules are now available.

- **Lecture Presentations in PowerPoint®** for each chapter with lecture notes, editable figures (art and photos with enlarged, customizable labels), tables, and links to animations and videos:

- **Accessible Lecture Presentations in PowerPoint** with alt text for every image; students can access alt text with a screen reader if needed

- **Editable Images in PowerPoint** (all art and photos from the text) and all tables from the text in PowerPoint with enlarged, customizable labels

- **Labeled and Unlabeled JPEG Images,** including art, photos from the text, and extra photos

- **Clicker Questions in PowerPoint,** which can be used to stimulate effective classroom discussions (for use with or without clickers):

- **EXPANDED! 500 Instructor Animations and Videos,** including BioFlix 3-D Animations, HHMI BioInteractive Animations and Videos, BBC Videos, and much more

- **Test Bank** questions in TestGen® software and Microsoft® Word. This extensively revised resource contains over 4,500 questions, including scenario-based questions and art, graph, and data interpretation questions. **NEW!** Every image in the Test Bank has alt text, which students can access with a screen reader if needed.

- **NEW! Statistics Worksheets for Biology**

- **Instructor Answers** to Scientific Skills Exercises, Problem-Solving Exercises, Interpret the Data Questions, and Essay Questions; includes rubric and tips for grading short-answer essays

- **Instructor Guides for Supplements:** Instructor Guide for *Active Reading Guides*; Instructor Guide for *Practicing Biology: A Student Workbook*; Instructor Guide for *Biological Inquiry: A Workbook of Investigative Cases*; Answer Key for *Inquiry in Action: Interpreting Scientific Papers*; *Investigating Biology Lab Prep Guide*; and *Investigating Biology* Lab Data Tables

Learning Catalytics™

Learning Catalytics allows students to use their smartphone, tablet, or laptop to respond to questions in class. For more information, visit learningcatalytics.com.

Learning Management Systems

Integration with various learning management systems is available for Mastering Biology. Contact your sales representative for details.

Interviews

Unit 1   THE CHEMISTRY OF LIFE   27

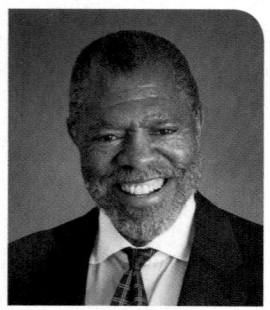

Kenneth Olden
National Center for
Environmental Assessment

Unit 2   THE CELL   92

Diana Bautista
University of California,
Berkeley

Unit 3   GENETICS   253

Francisco Mojica
University of Alicante,
Spain

Unit 4   MECHANISMS OF EVOLUTION   467

Cassandra Extavour
Harvard University

Unit 5   THE EVOLUTIONARY HISTORY OF BIOLOGICAL DIVERSITY   552

Penny Chisholm
Massachusetts Institute of
Technology

Unit 6   PLANT FORM AND FUNCTION   757

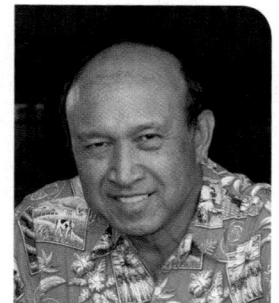

Dennis Gonsalves
Agricultural Research
Center, Hilo, Hawaii

Unit 7   ANIMAL FORM AND FUNCTION   872

Steffanie Strathdee
University of California,
San Diego

Unit 8   ECOLOGY   1163

Chelsea Rochman
University of Toronto

Acknowledgments

The authors wish to express their gratitude to the global community of instructors, researchers, students, and publishing professionals who have contributed to the Twelfth Edition of *Campbell Biology*.

As authors of this text, we are mindful of the daunting challenge of keeping up to date in all areas of our rapidly expanding subject. We are grateful to the many scientists who helped shape this text by discussing their research fields with us, answering specific questions in their areas of expertise, and sharing their ideas about biology education. We are especially grateful to the following, listed alphabetically: Graham Alexander, Elizabeth Atkinson, Kristian Axelsen, Ron Bassar, Christopher Benz, David Booth, George Brooks, Abby Dernberg, Jean DeSaix, Alex Engel, Rachel Kramer Green, Fred Holtzclaw, Theresa Holtzclaw, Tim James, Kathy Jones, Azarias Karamanlidis, Gary Karpen, Joe Montoya, Laurie Nemzer, Kevin Peterson, T. K. Reddy, David Reznick, Thomas Schneider, Alastair Simpson, Martin Smith, Steven Swoap, and John Taylor. In addition, the biologists listed on pages xxviii–xxxi provided detailed reviews, helping us ensure the text's scientific accuracy and improve its pedagogical effectiveness. Thanks also to Mary Camuso and Ann Sinclair for contributing a creative Study Tip for their fellow students.

Thanks also to the other professors and students, from all over the world, who contacted the authors directly with useful suggestions. We alone bear the responsibility for any errors that remain, but the dedication of our consultants, reviewers, and other correspondents makes us confident in the accuracy and effectiveness of this text.

Interviews with prominent scientists have been a hallmark of *Campbell Biology* since its inception, and conducting these interviews was again one of the great pleasures of revising the book. To open the eight units of this edition, we are proud to include interviews with Kenneth Olden, Diana Bautista, Francisco Mojica, Cassandra Extavour, Penny Chisholm, Dennis Gonsalves, Steffanie Strathdee, and Chelsea Rochman.

Mastering Biology and the other electronic accompaniments for this text are invaluable teaching and learning aids. We are grateful to the contributors for the Ready-to-Go Teaching Modules: Chad Brassil, Ruth Buskirk, Eileen Gregory, Angela Hodgson, Molly Jacobs, Bridgette Kirkpatrick, Maureen Leupold, Jennifer Metzler, Karen Resendes, Justin Shaffer, Allison Silveus, Jered Studinski, Cynthia Surmacz, Sara Tallarovic, and Carole Twichell. We would also like to extend our sincere appreciation to Carolyn Wetzel for her hard work on the Figure Walkthroughs. And our gratitude goes to Bryan Jennings and Roberta Batorsky for their work on the Reading Questions. Thanks also to Ann Brokaw and Bob Cooper for their contributions to the AAAS Science in the Classroom activities; we also appreciate the support of Beth Reudi, Shelby Lake, and Lydia Kaprelian from AAAS.

The value of *Campbell Biology* as a learning tool is greatly enhanced by the supplementary materials that have been created for instructors and students. We recognize that the dedicated authors of these materials are essentially writing mini (and not so mini) books. We appreciate the hard work and creativity of all the authors listed, with their creations, on pages xxiv–xxv. We are also grateful to Kathleen Fitzpatrick and Nicole Tunbridge (PowerPoint® Lecture Presentations); Roberta Batorsky, Douglas Darnowski, James Langeland, and David Knochel (Clicker Questions); Sonish Azam, Ford Lux, Karen Bernd, Janet Lanza, Chris Romero, Marshall Sundberg, Justin Shaffer, Ed Zalisko, and David Knochel (Test Bank).

Campbell Biology results from an unusually strong synergy between a team of scientists and a team of publishing professionals.

Our editorial team at Pearson Education again demonstrated unmatched talents, commitment, and pedagogical insights. Josh Frost, our Manager of Higher Ed Global Content Strategy for Life Sciences, brought publishing savvy, intelligence, and a much-appreciated level head to leading the whole team. The clarity and effectiveness of every page owe much to our extraordinary Supervising Editors Beth Winickoff and Pat Burner, who worked with a top-notch team of Senior Developmental Editors in John Burner, Mary Ann Murray, Hilair Chism, Andrew Recher, and Mary Hill. Our unsurpassed Director of Content Development Ginnie Simione Jutson and Courseware Portfolio Management Director Beth Wilbur were indispensable in moving the project in the right direction. We also want to thank Robin Heyden for organizing the annual Biology Leadership Conferences and keeping us in touch with the world of AP Biology. We also extend our thanks to Ashley Fallon, Editorial Assistant, Chelsea Noack, Associate Content Analyst, and Rebecca Berardy Schwartz, Product Manager.

You would not have this beautiful text if not for the work of the production team: Director, Content Production & Digital Studio Erin Gregg; Managing Producer Michael Early; Senior Content Producer Lori Newman; Photo Researcher Maureen Spuhler; Copy Editor Joanna Dinsmore; Proofreader Pete Shanks; Rights & Permissions Manager Ben Ferrini; Rights & Permissions Project Manager Matt Perry; Senior Project Manager Margaret McConnell and the rest of the staff at Integra Software Services, Inc.; Art Production Manager Rebecca Marshall, Artist Kitty Auble, and the rest of the staff at Lachina Creative; Design Manager Mark Ong; Text and Cover Designer Jeff Puda; and Manufacturing Buyer Stacey Weinberger. We also thank those who worked on the text's supplements: Project Manager Shiny Rajesh and her team at Integra Software Services.

For creating the wonderful package of electronic media that accompanies the text, we are grateful to Senior Content Developer Sarah Jensen; Content Producers Kaitlin Smith and Ashley Gordon; Director, Production & Digital Studio Katie Foley; Director, Production & Digital Studio Laura Tommasi; Supervising Media Producer Tod Regan; Specialist, Instructional Design and Development Sarah Young-Dualan; Digital Program Manager, Science, Caroline Ayres; Project Manager Katie Cook; Media Producer Ziki Dekel; Manager, Creative Technology Greg Davis; and Senior Learning Tools Strategist Kassi Foley.

For their important roles in marketing the text and media, we thank Alysun Estes, Kelly Galli, Jane Campbell, Brad Parkins, and Stacey Abraham. For their enthusiasm, encouragement, and support, we are grateful to Jeanne Zalesky, Director, Global Higher Ed Content Management and Strategy, Science & Health Sciences; Michael Gillespie, Director, Higher Ed Product Management, Life Sciences; Adam Jaworski, VP Product Management Higher Ed, Science; and Paul Corey, SVP Global Content Strategy, Higher Ed.

The Pearson sales team, which represents *Campbell Biology* on campus, is an essential link to the users of the text. They tell us what you like and don't like about the text, communicate the features of the text, and provide prompt service. We thank them for their hard work and professionalism. For representing our text to our international audience, we thank our sales and marketing partners throughout the world. They are all strong allies in biology education.

Finally, we wish to thank our families and friends for their encouragement and patience throughout this long project. Our special thanks to Lily and Alex (L.A.U.); Debra and Hannah (M.L.C.); Aaron, Sophie, Noah, and Gabriele (S.A.W.); Natalie (P.V.M.); and Jim, Abby, Dan, and Emily (R.B.O). Thanks to Jane Reece, now retired, for her generosity and thoughtfulness throughout her many years as a Campbell author. And, as always, thanks to Rochelle, Allison, Jason, McKay, and Gus.

Lisa A. Urry, Michael L. Cain, Steven A. Wasserman, Peter V. Minorsky, and Rebecca B. Orr

Reviewers

Twelfth Edition Reviewers

Sheena Abernathy, *College of the Mainland*; James Arnone, *William Paterson University*; Josh Auld, *West Chester University*; Gemma Bartha, *Springfield College*; Louise Beard, *University of Essex*; Marin Beaupre, *Massasoit Community College*; Kevin Bennett, *University of Hawai'i*; Kelsie Bernot, *North Carolina A&T State University*; Christine Bezotte, *Elmira College*; Chris Bloch, *Bridgewater State University*; Aiwei Borengasser, *University of Arkansas – Pulaski Tech*; Robert Borgon, *University of Central Florida*; Nicole Bournias-Vardiabasis, *California State University, San Bernardino*; George Brooks, *University of California, Berkeley*; Michael Buoni, *Delaware Technical Community College*; Kelcey Burris, *Union High School*; Elena Cainas, *Broward College*; Mickael Cariveau, *Mount Olive College*; Billy Carver, *Lees-McRae College*; Anne Casper, *Eastern Michigan University*; Bruce Chase, *University of Nebraska, Omaha*; Amanda Chau, *Blinn College*; Katie Clark, *University of California, Riverside*; Catharina Coenen, *Allegheny College*; Curt Coffman, *Vincennes University*; Juliet Collins, *University of Wisconsin, Madison*; Bob Cooper, *Pennsbury High School*; Robin Cotter, *Phoenix College*; Marilyn Cruz-Alvarez, *Florida Gulf Coast University*; Noelle Cutter, *Molloy College*; Deborah Dardis, *Southeastern Louisiana University*; Farahad Dastoor, *University of Maine, Orono*; Andrew David, *Clarkson University*; Jeremiah Davie, *D'Youville College*; Brian Deis, *University of Hawai'i* Jean DeSaix, *University of North Carolina at Chapel Hill* Kelly Dubois, *Calvin College*; Cynthia Eayre, *Fresno City College*; Arri Eisen, *Emory University*; Lisa Elfring, *University of Arizona*; Kurt Elliott, *Northwest Vista College*; Ana Esther Escandon, *Los Angeles Harbor College*; Linda Fergusson-Kolmes, *Portland Community College*; April Fong, *Portland Community College*; Robert Fowler, *San Jose State University*; Brittany Gasper, *Florida Southern College*; Carri Gerber, *The Ohio State University*; Marina Gerson, *California State University, Stanislaus*; Brian Gibbens, *University of Minnesota*; Sara Gremillion, *Georgia Southern University*; Ron Gross, *Community College of Allegheny County*; Melissa Gutierrez, *University of Southern Mississippi*; Gokhan Hacisalihoglu, *Florida A&M University*; Monica Hall-Woods, *St. Charles Community College*; Catherine Hartkorn, *New Mexico State University*; Valerie Haywood, *Case Western Reserve University*; Maryann Herman, *St. John Fisher College*; Alexander Heyl, *Adelphi University*; Laura Hill, *University of Vermont*; Anne-Marie Hoskinson, *South Dakota State University*; Katriina Ilves, *Pace University*; James Jacob, *Tompkins Cortland Community College*; Darrel James, *Fred C. Beyer High School*; Jerry Johnson, *Corban University*; Greg Jones, *Santa Fe College*; Kathryn Jones, *Howard Community College*; Seth Jones, *University of Kentucky*; Steven Karafit, *University of Central Arkansas*; Lori Kayes, *Oregon State University*; Ben Kolber, *Duquesne University*; Catherine Konopka, *John Carroll University*; Bill Kroll, *Loyola University, Chicago*; MaryLynne LaMantia, *Golden West College*; Neil Lamb, *HudsonAlpha Institute for Biotechnology*; Michelle LaPorte, *St. Louis Community College*; Neil Lax, *Duquesne University*; John Lepri, *Carrboro High School*; Jani Lewis, *State University of New York at Geneseo*; Eddie Lunsford, *Southwestern Community College*; Alyssa MacDonald, *Leeward Community College*; Charles Mallery, *University of Miami*; Marlee Marsh, *Columbia College*; Nicole McDaniels, *Herkimer College*; Mike Meighan, *University of California, Berkeley*; Jennifer Metzler, *Ball State University*; Grace Ju Miller, *Indiana Wesleyan University*; Terry Miller, *Central Carolina Community College*; Shamone Mizenmayer, *Central High School*; Cam Muir, *University of Hawai'i at Hilo*; Heather Murdock, *San Francisco State University*; Madhavan Narayanan, *Mercy College*; Jennifer Nauen, *University of Delaware*; Karen Neal, *J. Sargeant Reynolds Community College, Richmond*; Leonore Neary, *Joliet Junior College*; Shanna Nifoussi, *University of Superior*; Jennifer Ortiz, *North Hills School District*; Fernanda Oyarzun, *Universidad Católica de la Santísima Concepción, Chile*; Stephanie Pandolfi, *Wayne State University*; John Plunket, *Horry-Georgetown Technical College*; Elena Pravosudova, *University of Nevada, Reno*; Pushpa Ramakrishna, *Chandler-Gilbert Community College*; Sami Raut, *University of Alabama, Birmingham*; Robert Reavis, *Glendale Community College*; Linda Rehfuss, *Bucks County Community College*; Deborah Rhoden, *Snead State Community College*; Linda Richardson, *Blinn College*; Brian Ring, *Valdosta State University*; Rob Ruliffson, *Minneapolis Community and Technical College*; Judy Schoonmaker, *Colorado School of Mines*; David Schultz, *University of Louisville*; David Schwartz, *Houston Community College*; Duane Sears, *University of California, Santa Barbara*; J. Michael Sellers, *University of Southern Mississippi*; Pramila Sen, *Houston Community College*; Jyotsna Sharma, *University of Texas San Antonio*; Joan Sharp, *Simon Fraser University*; Marcia Shofner, *University of Maryland, College Park*; Linda Sigismondi, *University of Rio Grande*; Davida Smyth, *Mercy College*; Helen Snodgrass, *YES Prep North Forest*; Ayodotun Sodipe, *Texas Southern University*; Kathy Sparace, *Tri-County Technical College*; Patricia Steinke, *San Jacinto College Central*; Elizabeth Sudduth, *Georgia Gwinnett College*; Aaron Sullivan, *Houghton College*; Yvonne Sun, *University of Dayton*; Andrea Swei, *San Francisco State University*; Greg Thurmon, *Central Methodist University*; Stephanie Toering-Peters, *Wartburg College*; Monica Togna, *Drexel University*; Gail Tompkins, *Wake Technical Community College*; Tara Turley-Stoulig, *Southeastern Louisiana University*; Bishnu Twanabasu, *Weatherford College*; Erin Ventresca, *Albright College*; Wei Wan, *Texas A&M University*; Alan Wasmoen, *Metropolitan Community College, Nebraska*; Fred Wasserman, *Boston University*; Vicki Watson, *University of Montana*; Bill Wesley, *Mars Area High School*; Clay White, *Lone Star College*; Lisa Whitenack, *Allegheny College*; Larry Wimmers, *Towson University*; Heather Woodson, *Gaston College*; Shelly Wu, *Texas Christian University*; Mary Wuerth, *Tamalpais High School*; John Yoder, *University of Alabama*; Alyson Zeamer, *University of Texas San Antonio*.

Reviewers of Previous Editions

Steve Abedon, *Ohio State University*; Kenneth Able, *State University of New York, Albany*; Thomas Adams, *Michigan State University*; Martin Adamson, *University of British Columbia*; Dominique Adriaens, *Ghent University*; Ann Aguanno, *Marymount Manhattan College*; Shylaja Akkaraju, *Bronx Community College of CUNY*; Marc Albrecht, *University of Nebraska*; John Alcock, *Arizona State University*; Eric Alcorn, *Acadia University*; George R. Aliaga, *Tarrant County College*; Philip Allman, *Florida Gulf Coast College*; Rodney Allrich, *Purdue University*; Richard Almon, *State University of New York, Buffalo*; Bonnie Amos, *Angelo State University*; Katherine Anderson, *University of California, Berkeley*; Richard J. Andren, *Montgomery County Community College*; Estry Ang, *University of Pittsburgh, Greensburg*; Jeff Appling, *Clemson University*; J. David Archibald, *San Diego State University*; David Armstrong, *University of Colorado, Boulder*; Howard J. Arnott, *University of Texas, Arlington*; Mary Ashley, *University of Illinois, Chicago*; Angela S. Aspbury, *Texas State University*; Robert Atherton, *University of Wyoming*; Karl Aufderheide, *Texas A&M University*; Leigh Auleb, *San Francisco State University*; Terry Austin, *Temple College*; P. Stephen Baenziger, *University of Nebraska*; Brian Bagatto, *University of Akron*; Ellen Baker, *Santa Monica College*; Katherine Baker, *Millersville University*; Virginia Baker, *Chipola College*; Teri Balser, *University of Wisconsin, Madison*; William Barklow, *Framingham State College*; Susan Barman, *Michigan State University*; Steven Barnhart, *Santa Rosa Junior College*; Jim Barron, *Montana State University Billings*; Andrew Barton, *University of Maine Farmington*; Rebecca A. Bartow, *Western Kentucky University*; Ron Basmajian, *Merced College*; David Bass, *University of Central Oklahoma*; Stephen Bauer, *Belmont Abbey College;* Bonnie Baxter, *Westminster College*; Tim Beagley, *Salt Lake Community College*; Margaret E. Beard, *College of the Holy Cross*; Tom Beatty, *University of British Columbia*; Chris Beck, *Emory University*; Wayne Becker, *University of Wisconsin, Madison*; Patricia Bedinger, *Colorado State University*; Jane Beiswenger, *University of Wyoming*; Anne Bekoff, *University of Colorado, Boulder*; Marc Bekoff, *University of Colorado, Boulder*; Tania Beliz, *College of San Mateo*; Adrianne Bendich, *Hoffman-La Roche, Inc.*; Marilee Benore, *University of Michigan, Dearborn*; Barbara Bentley, *State University of New York, Stony Brook*; Darwin Berg, *University of California, San Diego*; Werner Bergen, *Michigan State University*; Gerald Bergstrom, *University of Wisconsin, Milwaukee*; Anna W. Berkovitz, *Purdue University*; Aimee Bernard, *University of Colorado Denver*; Dorothy Berner, *Temple University*; Annalisa Berta, *San Diego State University*; Paulette Bierzychudek, *Pomona College*; Charles Biggers, *Memphis State University*; Teresa Bilinski, *St. Edward's University*; Kenneth Birnbaum, *New York University*; Sarah Bissonnette, *University of California, Berkeley*; Catherine Black, *Idaho State University*; Michael W. Black, *California Polytechnic State University, San Luis Obispo*; William Blaker, *Furman University*; Robert Blanchard, *University of New Hampshire*; Andrew R. Blaustein, *Oregon State University*; Judy Bluemer, *Morton College*; Edward Blumenthal, *Marquette University*; Robert Blystone, *Trinity University*; Robert Boley, *University of Texas, Arlington*; Jason E. Bond, *East Carolina University*; Eric Bonde, *University of Colorado, Boulder*; Cornelius Bondzi, *Hampton University*; Richard Boohar, *University of Nebraska, Omaha*; Carey L. Booth, *Reed College*; Allan Bornstein, *Southeast Missouri State University*; David Bos, *Purdue University*; Oliver Bossdorf, *State University of New York, Stony Book*; James L. Botsford, *New Mexico State University*; Lisa Boucher, *University of Nebraska, Omaha*; Jeffery Bowen, *Bridgewater State University*; J. Michael Bowes, *Humboldt State University*; Richard Bowker, *Alma College*; Robert Bowker, *Glendale Community College, Arizona*; Scott Bowling, *Auburn University*; Barbara Bowman, *Mills College*; Barry Bowman, *University of California, Santa Cruz*; Deric Bownds, *University of Wisconsin, Madison*; Robert Boyd, *Auburn University*; Sunny Boyd, *University of Notre Dame*; Jerry Brand, *University of Texas, Austin*; Edward Braun, *Iowa State University*; Theodore A. Bremner, *Howard University*; James Brenneman, *University of Evansville*; Charles H. Brenner, *Berkeley, California*; Lawrence Brewer, *University of Kentucky*; Donald P. Briskin, *University of Illinois, Urbana*; Paul Broady, *University of Canterbury*; Chad Brommer, *Emory University*; Judith L. Bronstein, *University of Arizona*; David Broussard, *Lycoming College*; Danny Brower, *University of Arizona*; Carole Browne, *Wake Forest University*; Beverly Brown, *Nazareth College*; Mark Browning, *Purdue University*; David Bruck, *San Jose State University*; Robb T. Brumfield, *Louisiana State University*; Herbert Bruneau, *Oklahoma State University*; Gary Brusca, *Humboldt State University*; Richard C. Brusca, *University of Arizona, Arizona-Sonora Desert Museum*; Alan H. Brush, *University of Connecticut, Storrs*; Howard Buhse, *University of Illinois, Chicago*; Arthur Buikema, *Virginia Tech*; Beth Burch, *Huntington University*; Tessa Burch, *University of Tennessee*; Al Burchsted, *College of Staten Island*; Warren Burggren, *University of North Texas*; Meg Burke, *University of North Dakota*; Edwin Burling, *De Anza College*; Dale Burnside, *Lenoir-Rhyne University*; William Busa, *Johns Hopkins University*; Jorge Busciglio, *University of California, Irvine*; John Bushnell, *University of Colorado*; Linda Butler, *University of Texas, Austin*; David Byres, *Florida Community College, Jacksonville*; Patrick Cafferty, *Emory University*; Guy A. Caldwell, *University of Alabama*; Jane Caldwell, *West Virginia University*; Kim A. Caldwell, *University of Alabama*; Ragan Callaway, *The University of Montana*; Kenneth M. Cameron, *University of Wisconsin, Madison*; R. Andrew Cameron, *California Institute of Technology*; Alison Campbell, *University of Waikato*; Iain Campbell, *University of Pittsburgh*; Michael Campbell, *Penn State University*; Patrick Canary, *Northland Pioneer College*; W. Zacheus Cande, *University of California, Berkeley*; Deborah Canington, *University of California, Davis*; Robert E. Cannon, *University of North Carolina, Greensboro*; Frank Cantelmo, *St. John's University*; John Capeheart, *University of Houston, Downtown*; Gregory Capelli, *College of William and Mary*; Cheryl Keller Capone, *Pennsylvania State University*;

Richard Cardullo, *University of California, Riverside*; Nina Caris, *Texas A&M University*; Mickael Cariveau, *Mount Olive College*; Jeffrey Carmichael, *University of North Dakota*; Robert Carroll, *East Carolina University*; Laura L. Carruth, *Georgia State University*; J. Aaron Cassill, *University of Texas, San Antonio*; Karen I. Champ, *Central Florida Community College*; David Champlin, *University of Southern Maine*; Brad Chandler, *Palo Alto College*; Wei-Jen Chang, *Hamilton College*; Bruce Chase, *University of Nebraska, Omaha*; P. Bryant Chase, *Florida State University*; Doug Cheeseman, *De Anza College*; Shepley Chen, *University of Illinois, Chicago*; Giovina Chinchar, *Tougaloo College*; Joseph P. Chinnici, *Virginia Commonwealth University*; Jung H. Choi, *Georgia Institute of Technology*; Steve Christensen, *Brigham Young University, Idaho*; Geoffrey Church, *Fairfield University*; Henry Claman, *University of Colorado Health Science Center*; Anne Clark, *Binghamton University*; Greg Clark, *University of Texas*; Patricia J. Clark, *Indiana University-Purdue University, Indianapolis*; Ross C. Clark, *Eastern Kentucky University*; Lynwood Clemens, *Michigan State University*; Janice J. Clymer, *San Diego Mesa College*; Reggie Cobb, *Nashville Community College*; William P. Coffman, *University of Pittsburgh*; Austin Randy Cohen, *California State University, Northridge*; Bill Cohen, *University of Kentucky*; J. John Cohen, *University of Colorado Health Science Center*; James T. Colbert, *Iowa State University*; Sean Coleman, *University of the Ozarks*; Jan Colpaert, *Hasselt University*; Robert Colvin, *Ohio University*; Jay Comeaux, *McNeese State University*; David Cone, *Saint Mary's University*; Erin Connolly, *University of South Carolina*; Elizabeth Connor, *University of Massachusetts*; Joanne Conover, *University of Connecticut*; Ron Cooper, *University of California, Los Angeles*; Gregory Copenhaver, *University of North Carolina, Chapel Hill*; John Corliss, *University of Maryland*; James T. Costa, *Western Carolina University*; Stuart J. Coward, *University of Georgia*; Charles Creutz, *University of Toledo*; Bruce Criley, *Illinois Wesleyan University*; Norma Criley, *Illinois Wesleyan University*; Joe W. Crim, *University of Georgia*; Greg Crowther, *University of Washington*; Karen Curto, *University of Pittsburgh*; William Cushwa, *Clark College*; Anne Cusic, *University of Alabama, Birmingham*; Richard Cyr, *Pennsylvania State University*; Curtis Daehler, *University of Hawaii at Manoa*; Marymegan Daly, *The Ohio State University*; W. Marshall Darley, *University of Georgia*; Douglas Darnowski, *Indiana University Southeast*; Cynthia Dassler, *The Ohio State University*; Shannon Datwyler, *California State University, Sacramento*; Marianne Dauwalder, *University of Texas, Austin*; Larry Davenport, *Samford University*; Bonnie J. Davis, *San Francisco State University*; Jerry Davis, *University of Wisconsin, La Crosse*; Michael A. Davis, *Central Connecticut State University*; Thomas Davis, *University of New Hampshire*; Melissa Deadmond, *Truckee Meadows Community College*; John Dearn, *University of Canberra*; Maria E. de Bellard, *California State University, Northridge*; Teresa DeGolier, *Bethel College*; James Dekloe, *University of California, Santa Cruz*; Eugene Delay, *University of Vermont*; Patricia A. DeLeon, *University of Delaware*; Veronique Delesalle, *Gettysburg College*; T. Delevoryas, *University of Texas, Austin*; Roger Del Moral, *University of Washington*; Charles F. Delwiche, *University of Maryland*; Diane C. DeNagel, *Northwestern University*; William L. Dentler, *University of Kansas*; Jennifer Derkits, *J. Sergeant Reynolds Community College*; Daniel DerVartanian, *University of Georgia*; Jean DeSaix, *University of North Carolina, Chapel Hill*; Janet De Souza-Hart, *Massachusetts College of Pharmacy & Health Sciences*; Biao Ding, *Ohio State University*; Michael Dini, *Texas Tech University*; Kevin Dixon, *Florida State University*; Andrew Dobson, *Princeton University*; Stanley Dodson, *University of Wisconsin, Madison*; Jason Douglas, *Angelina College*; Mark Drapeau, *University of California, Irvine*; John Drees, *Temple University School of Medicine*; Charles Drewes, *Iowa State University*; Marvin Druger, *Syracuse University*; Gary Dudley, *University of Georgia*; David Dunbar, *Cabrini College*; Susan Dunford, *University of Cincinnati*; Kathryn A. Durham, *Lorain Community College*; Betsey Dyer, *Wheaton College*; Robert Eaton, *University of Colorado*; Robert S. Edgar, *University of California, Santa Cruz*; Anna Edlund, *Lafayette College*; Douglas J. Eernisse, *California State University, Fullerton*; Betty J. Eidemiller, *Lamar University*; Brad Elder, *Doane College*; Curt Elderkin, *College of New Jersey*; William D. Eldred, *Boston University*; Michelle Elekonich, *University of Nevada, Las Vegas*; George Ellmore, *Tufts University*; Mary Ellard-Ivey, *Pacific Lutheran University*; Kurt Elliott, *North West Vista College*; Norman Ellstrand, *University of California, Riverside*; Johnny El-Rady, *University of South Florida*; Bert Ely, *University of South Carolina*; Dennis Emery, *Iowa State University*; John Endler, *University of California, Santa Barbara*; Rob Erdman, *Florida Gulf Coast College*; Dale Erskine, *Lebanon Valley College*; Margaret T. Erskine, *Lansing Community College*; Susan Erster, *Stony Brook University*; Gerald Esch, *Wake Forest University*; Frederick B. Essig, *University of South Florida*; Mary Eubanks, *Duke University*; David Evans, *University of Florida*; Robert C. Evans, *Rutgers University, Camden*; Sharon Eversman, *Montana State University*; Olukemi Fadayomi, *Ferris State University*; Lincoln Fairchild, *Ohio State University*; Peter Fajer, *Florida State University*; Bruce Fall, *University of Minnesota*; Sam Fan, *Bradley University*; Lynn Fancher, *College of DuPage*; Ellen H. Fanning, *Vanderbilt University*; Paul Farnsworth, *University of New Mexico*; Larry Farrell, *Idaho State University*; Jerry F. Feldman, *University of California, Santa Cruz*; Lewis Feldman, *University of California, Berkeley*; Myriam Alhadeff Feldman, *Cascadia Community College*; Eugene Fenster, *Longview Community College*; Linda Fergusson-Kolmes, *Portland Community College, Sylvania Campus*; Russell Fernald, *University of Oregon*; Danilo Fernando, *SUNY College of Environmental Science and Forestry, Syracuse*; Rebecca Ferrell, *Metropolitan State College of Denver*; Christina Fieber, *Horry-Georgetown Technical College*; Melissa Fierke, *SUNY College of Environmental Science and Forestry*; Kim Finer, *Kent State University*; Milton Fingerman, *Tulane University*; Barbara Finney, *Regis College*; Teresa Fischer, *Indian River Community College*; Frank Fish, *West Chester University*; David Fisher, *University of Hawaii, Manoa*; Jonathan S. Fisher, *St. Louis University*; Steven Fisher, *University of California, Santa Barbara*; David Fitch, *New York University*; Kirk Fitzhugh, *Natural History Museum of Los Angeles County*; Lloyd Fitzpatrick, *University of North Texas*; William Fixsen, *Harvard University*; T. Fleming, *Bradley University*; Abraham Flexer, *Manuscript Consultant, Boulder, Colorado*; Mark Flood, *Fairmont State University*; Margaret Folsom, *Methodist College*; Kerry Foresman, *University of Montana*; Norma Fowler, *University of Texas, Austin*; Robert G. Fowler,

San Jose State University; David Fox, *University of Tennessee, Knoxville*; Carl Frankel, *Pennsylvania State University, Hazleton*; Stewart Frankel, *University of Hartford*; Robert Franklin, *College of Charleston*; James Franzen, *University of Pittsburgh*; Art Fredeen, *University of Northern British Columbia*; Kim Fredericks, *Viterbo University*; Bill Freedman, *Dalhousie University*; Matt Friedman, *University of Chicago*; Otto Friesen, *University of Virginia*; Frank Frisch, *Chapman University*; Virginia Fry, *Monterey Peninsula College*; Bernard Frye, *University of Texas, Arlington*; Jed Fuhrman, *University of Southern California*; Alice Fulton, *University of Iowa*; Chandler Fulton, *Brandeis University*; Sara Fultz, *Stanford University*; Berdell Funke, *North Dakota State University*; Anne Funkhouser, *University of the Pacific*; Zofia E. Gagnon, *Marist College*; Michael Gaines, *University of Miami*; Cynthia M. Galloway, *Texas A&M University, Kingsville*; Arthur W. Galston, *Yale University*; Stephen Gammie, *University of Wisconsin, Madison*; Carl Gans, *University of Michigan*; John Gapter, *University of Northern Colorado*; Andrea Gargas, *University of Wisconsin, Madison*; Lauren Garner, *California Polytechnic State University, San Luis Obispo*; Reginald Garrett, *University of Virginia*; Craig Gatto, *Illinois State University*; Kristen Genet, *Anoka Ramsey Community College*; Patricia Gensel, *University of North Carolina*; Chris George, *California Polytechnic State University, San Luis Obispo*; Robert George, *University of Wyoming*; J. Whitfield Gibbons, *University of Georgia*; J. Phil Gibson, *University of Oklahoma*; Frank Gilliam, *Marshall University*; Eric Gillock, *Fort Hayes State University*; Simon Gilroy, *University of Wisconsin, Madison*; Edwin Ginés-Candelaria, *Miami Dade College*; Alan D. Gishlick, *Gustavus Adolphus College*; Todd Gleeson, *University of Colorado*; Jessica Gleffe, *University of California, Irvine*; John Glendinning, *Barnard College*; David Glenn-Lewin, *Wichita State University*; William Glider, *University of Nebraska*; Tricia Glidewell, *Marist School*; Elizabeth A. Godrick, *Boston University*; Jim Goetze, *Laredo Community College*; Lynda Goff, *University of California, Santa Cruz*; Elliott Goldstein, *Arizona State University*; Paul Goldstein, *University of Texas, El Paso*; Sandra Gollnick, *State University of New York, Buffalo*; Roy Golsteyn, *University of Lethbridge*; Anne Good, *University of California, Berkeley*; Judith Goodenough, *University of Massachusetts, Amherst*; Wayne Goodey, *University of British Columbia*; Barbara E. Goodman, *University of South Dakota*; Robert Goodman, *University of Wisconsin, Madison*; Ester Goudsmit, *Oakland University*; Linda Graham, *University of Wisconsin, Madison*; Robert Grammer, *Belmont University*; Joseph Graves, *Arizona State University*; Eileen Gregory, *Rollins College*; Phyllis Griffard, *University of Houston, Downtown*; A. J. F. Griffiths, *University of British Columbia*; Bradley Griggs, *Piedmont Technical College*; William Grimes, *University of Arizona*; David Grise, *Texas A&M University, Corpus Christi*; Mark Gromko, *Bowling Green State University*; Serine Gropper, *Auburn University*; Katherine L. Gross, *Ohio State University*; Gary Gussin, *University of Iowa*; Edward Gruberg, *Temple University*; Carla Guthridge, *Cameron University*; Mark Guyer, *National Human Genome Research Institute*; Ruth Levy Guyer, *Bethesda, Maryland*; Carla Haas, *Pennsylvania State University*; R. Wayne Habermehl, *Montgomery County Community College*; Pryce Pete Haddix, *Auburn University*; Mac Hadley, *University of Arizona*; Joel Hagen, *Radford University*; Jack P. Hailman, *University of Wisconsin*; Leah Haimo, *University of California, Riverside*; Ken Halanych, *Auburn University*; Jody Hall, *Brown University*; Heather Hallen-Adams, *University of Nebraska, Lincoln*; Douglas Hallett, *Northern Arizona University*; Rebecca Halyard, *Clayton State College*; Devney Hamilton, *Stanford University* (student); E. William Hamilton, *Washington and Lee University*; Matthew B. Hamilton, *Georgetown University*; Sam Hammer, *Boston University*; Penny Hanchey-Bauer, *Colorado State University*; William F. Hanna, *Massasoit Community College*; Dennis Haney, *Furman University*; Laszlo Hanzely, *Northern Illinois University*; Jeff Hardin, *University of Wisconsin, Madison*; Jean Hardwick, *Ithaca College*; Luke Harmon, *University of Idaho*; Lisa Harper, *University of California, Berkeley*; Deborah Harris, *Case Western Reserve University*; Jeanne M. Harris, *University of Vermont*; Richard Harrison, *Cornell University*; Stephanie Harvey, *Georgia Southwestern State University*; Carla Hass, *Pennsylvania State University*; Chris Haufler, *University of Kansas*; Bernard A. Hauser, *University of Florida*; Chris Haynes, *Shelton State Community College*; Evan B. Hazard, *Bemidji State University* (emeritus); H. D. Heath, *California State University, East Bay*; George Hechtel, *State University of New York, Stony Brook*; S. Blair Hedges, *Pennsylvania State University*; Brian Hedlund, *University of Nevada, Las Vegas*; David Heins, *Tulane University*; Jean Heitz, *University of Wisconsin, Madison*; Andreas Hejnol, *Sars International Centre for Marine Molecular Biology*; John D. Helmann, *Cornell University*; Colin Henderson, *University of Montana*; Susan Hengeveld, *Indiana University*; Michelle Henricks, *University of California, Los Angeles*; Caroll Henry, *Chicago State University*; Frank Heppner, *University of Rhode Island*; Albert Herrera, *University of Southern California*; Scott Herrick, *Missouri Western State College*; Ira Herskowitz, *University of California, San Francisco*; Paul E. Hertz, *Barnard College*; Chris Hess, *Butler University*; David Hibbett, *Clark University*; R. James Hickey, *Miami University*; Karen Hicks, *Kenyon College*; Kendra Hill, *San Diego State University*; William Hillenius, *College of Charleston*; Kenneth Hillers, *California Polytechnic State University, San Luis Obispo*; Ralph Hinegardner, *University of California, Santa Cruz*; William Hines, *Foothill College*; Robert Hinrichsen, *Indiana University of Pennsylvania*; Helmut Hirsch, *State University of New York, Albany*; Tuan-hua David Ho, *Washington University*; Carl Hoagstrom, *Ohio Northern University*; Elizabeth Hobson, *New Mexico State University*; Jason Hodin, *Stanford University*; James Hoffman, *University of Vermont*; A. Scott Holaday, *Texas Tech University*; Mark Holbrook, *University of Iowa*; N. Michele Holbrook, *Harvard University*; James Holland, *Indiana University, Bloomington*; Charles Holliday, *Lafayette College*; Lubbock Karl Holte, *Idaho State University*; Alan R. Holyoak, *Brigham Young University, Idaho*; Laura Hoopes, *Occidental College*; Nancy Hopkins, *Massachusetts Institute of Technology*; Sandra Horikami, *Daytona Beach Community College*; Kathy Hornberger, *Widener University*; Pius F. Horner, *San Bernardino Valley College*; Becky Houck, *University of Portland*; Margaret Houk, *Ripon College*; Laura Houston, *Northeast Lakeview College*; Daniel J. Howard, *New Mexico State University*; Ronald R. Hoy, *Cornell University*; Sandra Hsu, *Skyline College*; Sara Huang, *Los Angeles Valley College*; Cristin Hulslander, *University of Oregon*; Donald Humphrey, *Emory University School of Medicine*;

Catherine Hurlbut, *Florida State College, Jacksonville*; Diane Husic, *Moravian College*; Robert J. Huskey, *University of Virginia*; Steven Hutcheson, *University of Maryland, College Park*; Linda L. Hyde, *Gordon College*; Bradley Hyman, *University of California, Riverside*; Jeffrey Ihara, *Mira Costa College*; Mark Iked, *San Bernardino Valley College*; Cheryl Ingram-Smith, *Clemson University*; Erin Irish, *University of Iowa*; Sally Irwin, *University of Hawaii, Maui College*; Harry Itagaki, *Kenyon College*; Alice Jacklet, *State University of New York, Albany*; John Jackson, *North Hennepin Community College*; Thomas Jacobs, *University of Illinois*; Kathy Jacobson, *Grinnell College*; Mark Jaffe, *Nova Southeastern University*; John C. Jahoda, *Bridgewater State College*; Douglas Jensen, *Converse College*; Jamie Jensen, *Brigham Young University*; Dan Johnson, *East Tennessee State University*; Lance Johnson, *Midland Lutheran College*; Lee Johnson, *The Ohio State University*; Randall Johnson, *University of California, San Diego*; Roishene Johnson, *Bossier Parish Community College*; Stephen Johnson, *William Penn University*; Wayne Johnson, *Ohio State University*; Kenneth C. Jones, *California State University, Northridge*; Russell Jones, *University of California, Berkeley*; Cheryl Jorcyk, *Boise State University*; Chad Jordan, *North Carolina State University*; Ann Jorgensen, *University of Hawaii*; Alan Journet, *Southeast Missouri State University*; Walter Judd, *University of Florida*; Ari Jumpponen, *Kansas State University*; Thomas W. Jurik, *Iowa State University*; Caroline M. Kane, *University of California, Berkeley*; Doug Kane, *Defiance College*; Thomas C. Kane, *University of Cincinnati*; The-Hui Kao, *Pennsylvania State University*; Tamos Kapros, *University of Missouri*; Kasey Karen, *Georgia College & State University*; E. L. Karlstrom, *University of Puget Sound*; David Kass, *Eastern Michigan University*; Jennifer Katcher, *Pima Community College*; Laura A. Katz, *Smith College*; Judy Kaufman, *Monroe Community College*; Maureen Kearney, *Field Museum of Natural History*; Eric G. Keeling, *Cary Institute of Ecosystem Studies*; Patrick Keeling, *University of British Columbia*; Thomas Keller, *Florida State University*; Elizabeth A. Kellogg, *University of Missouri, St. Louis*; Paul Kenrick, *Natural History Museum, London*; Norm Kenkel, *University of Manitoba*; Chris Kennedy, *Simon Fraser University*; George Khoury, *National Cancer Institute*; Stephen T. Kilpatrick, *University of Pittsburgh at Johnstown*; Rebecca T. Kimball, *University of Florida*; Shannon King, *North Dakota State University*; Mark Kirk, *University of Missouri, Columbia*; Robert Kitchin, *University of Wyoming*; Hillar Klandorf, *West Virginia University*; Attila O. Klein, *Brandeis University*; Karen M. Klein, *Northampton Community College*; Daniel Klionsky, *University of Michigan*; Mark Knauss, *Georgia Highlands College*; Janice Knepper, *Villanova University*; Charles Knight, *California Polytechnic State University*; Jennifer Knight, *University of Colorado*; Ned Knight, *Linfield College*; Roger Koeppe, *University of Arkansas*; David Kohl, *University of California, Santa Barbara*; Greg Kopf, *University of Pennsylvania School of Medicine*; Thomas Koppenheffer, *Trinity University*; Peter Kourtev, *Central Michigan University*; Margareta Krabbe, *Uppsala University*; Jacob Krans, *Western New England University*; Anselm Kratochwil, *Universität Osnabrück*; Eliot Krause, *Seton Hall University*; Deborah M. Kristan, *California State University, San Marcos*; Steven Kristoff, *Ivy Tech Community College*; Dubear Kroening, *University of Wisconsin*; William Kroll, *Loyola University, Chicago*; Janis Kuby, *San Francisco State University*; Barbara Kuemerle, *Case Western Reserve University*; Justin P. Kumar, *Indiana University*; Rukmani Kuppuswami, *Laredo Community College*; David Kurijaka, *Ohio University*; Lee Kurtz, *Georgia Gwinnett College*; Michael P. Labare, *United States Military Academy, West Point*; Marc-André Lachance, *University of Western Ontario*; J. A. Lackey, *State University of New York, Oswego*; Elaine Lai, *Brandeis University*; Mohamed Lakrim, *Kingsborough Community College*; Ellen Lamb, *University of North Carolina, Greensboro*; William Lamberts, *College of St Benedict and St John's University*; William L'Amoreaux, *College of Staten Island*; Lynn Lamoreux, *Texas A&M University*; Carmine A. Lanciani, *University of Florida*; Kenneth Lang, *Humboldt State University*; Jim Langeland, *Kalamazoo College*; Dominic Lannutti, *El Paso Community College*; Allan Larson, *Washington University*; Grace Lasker, *Lake Washington Institute of Technology*; John Latto, *University of California, Santa Barbara*; Diane K. Lavett, *State University of New York, Cortland, and Emory University*; Charles Leavell, *Fullerton College*; C. S. Lee, *University of Texas*; Daewoo Lee, *Ohio University*; Tali D. Lee, *University of Wisconsin, Eau Claire*; Hugh Lefcort, *Gonzaga University*; Robert Leonard, *University of California, Riverside*; Michael R. Leonardo, *Coe College*; John Lepri, *University of North Carolina, Greensboro*; Donald Levin, *University of Texas, Austin*; Joseph Levine, *Boston College*; Mike Levine, *University of California, Berkeley*; Alcinda Lewis, *University of Colorado, Boulder*; Bill Lewis, *Shoreline Community College*; Jani Lewis, *State University of New York*; John Lewis, *Loma Linda University*; Lorraine Lica, *California State University, East Bay*; Harvey Liftin, *Broward Community College*; Harvey Lillywhite, *University of Florida, Gainesville*; Graeme Lindbeck, *Valencia Community College*; Clark Lindgren, *Grinnell College*; Eric W. Linton, *Central Michigan University*; Diana Lipscomb, *George Washington University*; Christopher Little, *The University of Texas, Pan American*; Kevin D. Livingstone, *Trinity University*; Andrea Lloyd, *Middlebury College*; Tatyana Lobova, *Old Dominion University*; Sam Loker, *University of New Mexico*; David Longstreth, *Louisiana State University*; Christopher A. Loretz, *State University of New York, Buffalo*; Donald Lovett, *College of New Jersey*; Jane Lubchenco, *Oregon State University*; Douglas B. Luckie, *Michigan State University*; Hannah Lui, *University of California, Irvine*; Margaret A. Lynch, *Tufts University*; Steven Lynch, *Louisiana State University, Shreveport*; Lisa Lyons, *Florida State University*; Richard Machemer Jr., *St. John Fisher College*; Elizabeth Machunis-Masuoka, *University of Virginia*; James MacMahon, *Utah State University*; Nancy Magill, *Indiana University*; Christine R. Maher, *University of Southern Maine*; Linda Maier, *University of Alabama, Huntsville*; Jose Maldonado, *El Paso Community College*; Richard Malkin, *University of California, Berkeley*; Charles Mallery, *University of Miami*; Keith Malmos, *Valencia Community College, East Campus*; Cindy Malone, *California State University, Northridge*; Mark Maloney, *University of South Mississippi*; Carol Mapes, *Kutztown University of Pennsylvania*; William Margolin, *University of Texas Medical School*; Lynn Margulis, *Boston University*; Julia Marrs, *Barnard College* (student); Kathleen A. Marrs, *Indiana University-Purdue University, Indianapolis*; Edith Marsh, *Angelo State University*; Diane L. Marshall, *University of New Mexico*; Mary Martin,

Northern Michigan University; Karl Mattox, *Miami University of Ohio*; Joyce Maxwell, *California State University, Northridge*; Jeffrey D. May, *Marshall University*; Mike Mayfield, *Ball State University*; Kamau Mbuthia, *Bowling Green State University*; Lee McClenaghan, *San Diego State University*; Richard McCracken, *Purdue University*; Andrew McCubbin, *Washington State University*; Kerry McDonald, *University of Missouri, Columbia*; Tanya McGhee, *Craven Community College*; Jacqueline McLaughlin, *Pennsylvania State University, Lehigh Valley*; Neal McReynolds, *Texas A&M International*; Darcy Medica, *Pennsylvania State University*; Lisa Marie Meffert, *Rice University*; Susan Meiers, *Western Illinois University*; Michael Meighan, *University of California, Berkeley*; Scott Meissner, *Cornell University*; Paul Melchior, *North Hennepin Community College*; Phillip Meneely, *Haverford College*; John Merrill, *Michigan State University*; Brian Metscher, *University of California, Irvine*; Jenny Metzler, *Ball State University*; Ralph Meyer, *University of Cincinnati*; James Mickle, *North Carolina State University*; Jan Mikesell, *Gettysburg College*; Roger Milkman, *University of Iowa*; Grace Miller, *Indiana Wesleyan University*; Helen Miller, *Oklahoma State University*; John Miller, *University of California, Berkeley*; Jonathan Miller, *Edmonds Community College*; Kenneth R. Miller, *Brown University*; Mill Miller, *Wright State University*; Alex Mills, *University of Windsor*; Sarah Milton, *Florida Atlantic University*; Eli Minkoff, *Bates College*; John E. Minnich, *University of Wisconsin, Milwaukee*; Subhash Minocha, *University of New Hampshire*; Michael J. Misamore, *Texas Christian University*; Kenneth Mitchell, *Tulane University School of Medicine*; Ivona Mladenovic, *Simon Fraser University*; Alan Molumby, *University of Illinois, Chicago*; Nicholas Money, *Miami University*; Russell Monson, *University of Colorado, Boulder*; Joseph P. Montoya, *Georgia Institute of Technology*; Frank Moore, *Oregon State University*; Janice Moore, *Colorado State University*; Linda Moore, *Georgia Military College*; Randy Moore, *Wright State University*; William Moore, *Wayne State University*; Carl Moos, *Veterans Administration Hospital, Albany, New York*; Linda Martin Morris, *University of Washington*; Michael Mote, *Temple University*; Alex Motten, *Duke University*; Jeanette Mowery, *Madison Area Technical College*; Deborah Mowshowitz, *Columbia University*; Rita Moyes, *Texas A&M, College Station*; Darrel L. Murray, *University of Illinois, Chicago*; Courtney Murren, *College of Charleston*; John Mutchmor, *Iowa State University*; Elliot Myerowitz, *California Institute of Technology*; Barbara Nash, *Mercy College*; Gavin Naylor, *Iowa State University*; John Neess, *University of Wisconsin, Madison*; Ross Nehm, *Ohio State University*; Tom Neils, *Grand Rapids Community College*; Kimberlyn Nelson, *Pennsylvania State University*; Raymond Neubauer, *University of Texas, Austin*; Todd Newbury, *University of California, Santa Cruz*; James Newcomb, *New England College*; Jacalyn Newman, *University of Pittsburgh*; Harvey Nichols, *University of Colorado, Boulder*; Deborah Nickerson, *University of South Florida*; Bette Nicotri, *University of Washington*; Caroline Niederman, *Tomball College*; Eric Nielsen, *University of Michigan*; Maria Nieto, *California State University, East Bay*; Anders Nilsson, *University of Umeå*; Greg Nishiyama, *College of the Canyons*; Charles R. Noback, *College of Physicians and Surgeons, Columbia University*; Jane Noble-Harvey, *Delaware University*; Mary C. Nolan, *Irvine Valley College*; Kathleen Nolta, *University of Michigan*; Peter Nonacs, *University of California, Los Angeles*; Mohamed A. F. Noor, *Duke University*; Shawn Nordell, *St. Louis University*; Richard S. Norman, *University of Michigan, Dearborn* (emeritus); David O. Norris, *University of Colorado, Boulder*; Steven Norris, *California State University, Channel Islands*; Gretchen North, *Occidental College*; Cynthia Norton, *University of Maine, Augusta*; Steve Norton, *East Carolina University*; Steve Nowicki, *Duke University*; Bette H. Nybakken, *Hartnell College*; Brian O'Conner, *University of Massachusetts, Amherst*; Gerard O'Donovan, *University of North Texas*; Eugene Odum, *University of Georgia*; Mark P. Oemke, *Alma College*; Linda Ogren, *University of California, Santa Cruz*; Patricia O'Hern, *Emory University*; Olabisi Ojo, *Southern University at New Orleans*; Nathan O. Okia, *Auburn University, Montgomery*; Jeanette Oliver, *St. Louis Community College, Florissant Valley*; Gary P. Olivetti, *University of Vermont*; Margaret Olney, *St. Martin's College*; John Olsen, *Rhodes College*; Laura J. Olsen, *University of Michigan*; Sharman O'Neill, *University of California, Davis*; Wan Ooi, *Houston Community College*; Aharon Oren, *The Hebrew University*; John Oross, *University of California, Riverside*; Rebecca Orr, *Collin College*; Catherine Ortega, *Fort Lewis College*; Charissa Osborne, *Butler University*; Gay Ostarello, *Diablo Valley College*; Henry R. Owen, *Eastern Illinois University*; Thomas G. Owens, *Cornell University*; Penny Padgett, *University of North Carolina, Chapel Hill*; Kevin Padian, *University of California, Berkeley*; Dianna Padilla, *State University of New York, Stony Brook*; Anthony T. Paganini, *Michigan State University*; Fatimata Pale, *Thiel College*; Barry Palevitz, *University of Georgia*; Michael A. Palladino, *Monmouth University*; Matt Palmtag, *Florida Gulf Coast University*; Stephanie Pandolfi, *Michigan State University*; Daniel Papaj, *University of Arizona*; Peter Pappas, *County College of Morris*; Nathalie Pardigon, *Institut Pasteur*; Bulah Parker, *North Carolina State University*; Stanton Parmeter, *Chemeketa Community College*; Susan Parrish, *McDaniel College*; Cindy Paszkowski, *University of Alberta*; Robert Patterson, *San Francisco State University*; Ronald Patterson, *Michigan State University*; Crellin Pauling, *San Francisco State University*; Kay Pauling, *Foothill Community College*; Daniel Pavuk, *Bowling Green State University*; Debra Pearce, *Northern Kentucky University*; Patricia Pearson, *Western Kentucky University*; Andrew Pease, *Stevenson University*; Nancy Pelaez, *Purdue University*; Shelley Penrod, *North Harris College*; Imara Y. Perera, *North Carolina State University*; Beverly Perry, *Houston Community College*; Irene Perry, *University of Texas of the Permian Basin*; Roger Persell, *Hunter College*; Eric Peters, *Chicago State University*; Larry Peterson, *University of Guelph*; David Pfennig, *University of North Carolina, Chapel Hill*; Mark Pilgrim, *College of Coastal Georgia*; David S. Pilliod, *California Polytechnic State University, San Luis Obispo*; Vera M. Piper, *Shenandoah University*; Deb Pires, *University of California, Los Angeles*; J. Chris Pires, *University of Missouri, Columbia*; Jarmila Pittermann, *University of California, Santa Cruz*; Bob Pittman, *Michigan State University*; James Platt, *University of Denver*; Martin Poenie, *University of Texas, Austin*; Scott Poethig, *University of Pennsylvania*; Crima Pogge, *San Francisco Community College*; Michael Pollock, *Mount Royal University*; Roberta Pollock, *Occidental College*; Jeffrey Pommerville, *Texas A&M University*; Therese M. Poole, *Georgia State University*; Angela R. Porta, *Kean University*; Jason Porter,

University of the Sciences, Philadelphia; Warren Porter, *University of Wisconsin*; Daniel Potter, *University of California, Davis*; Donald Potts, *University of California, Santa Cruz*; Robert Powell, *Avila University*; Andy Pratt, *University of Canterbury*; David Pratt, *University of California, Davis*; Halina Presley, *University of Illinois, Chicago*; Eileen Preston, *Tarrant Community College Northwest*; Mary V. Price, *University of California, Riverside*; Mitch Price, *Pennsylvania State University*; Steven Price, *Virginia Commonwealth University*; Terrell Pritts, *University of Arkansas, Little Rock*; Rong Sun Pu, *Kean University*; Rebecca Pyles, *East Tennessee State University*; Scott Quackenbush, *Florida International University*; Ralph Quatrano, *Oregon State University*; Peter Quinby, *University of Pittsburgh*; Val Raghavan, *Ohio State University*; Deanna Raineri, *University of Illinois, Champaign-Urbana*; David Randall, *City University Hong Kong*; Talitha Rajah, *Indiana University Southeast*; Charles Ralph, *Colorado State University*; Pushpa Ramakrishna, *Chandler-Gilbert Community College*; Thomas Rand, *Saint Mary's University*; Monica Ranes-Goldberg, *University of California, Berkeley*; Samiksha Raut, *University of Alabama at Birmingham*; Robert S. Rawding, *Gannon University*; Robert H. Reavis, *Glendale Community College*; Kurt Redborg, *Coe College*; Ahnya Redman, *Pennsylvania State University*; Brian Reeder, *Morehead State University*; Bruce Reid, *Kean University*; David Reid, *Blackburn College*; C. Gary Reiness, *Lewis & Clark College*; Charles Remington, *Yale University*; Erin Rempala, *San Diego Mesa College*; David Reznick, *University of California, Riverside*; Fred Rhoades, *Western Washington State University*; Douglas Rhoads, *University of Arkansas*; Eric Ribbens, *Western Illinois University*; Christina Richards, *New York University*; Sarah Richart, *Azusa Pacific University*; Wayne Rickoll, *University of Puget Sound*; Christopher Riegle, *Irvine Valley College*; Loren Rieseberg, *University of British Columbia*; Bruce B. Riley, *Texas A&M University*; Todd Rimkus, *Marymount University*; John Rinehart, *Eastern Oregon University*; Donna Ritch, *Pennsylvania State University*; Carol Rivin, *Oregon State University East*; Laurel Roberts, *University of Pittsburgh*; Diane Robins, *University of Michigan*; Kenneth Robinson, *Purdue University*; Thomas Rodella, *Merced College*; Luis Rodriguez, *San Antonio College*; Deb Roess, *Colorado State University*; Heather Roffey, *Marianopolis College*; Rodney Rogers, *Drake University*; Suzanne Rogers, *Seton Hill University*; William Roosenburg, *Ohio University*; Kara Rosch, *Blinn College*; Mike Rosenzweig, *Virginia Polytechnic Institute and State University*; Wayne Rosing, *Middle Tennessee State University*; Thomas Rost, *University of California, Davis*; Stephen I. Rothstein, *University of California, Santa Barbara*; John Ruben, *Oregon State University*; Albert Ruesink, *Indiana University*; Patricia Rugaber, *College of Coastal Georgia*; Scott Russell, *University of Oklahoma*; Jodi Rymer, *College of the Holy Cross*; Neil Sabine, *Indiana University*; Tyson Sacco, *Cornell University*; Glenn-Peter Saetre, *University of Oslo*; Rowan F. Sage, *University of Toronto*; Tammy Lynn Sage, *University of Toronto*; Sanga Saha, *Harold Washington College*; Don Sakaguchi, *Iowa State University*; Walter Sakai, *Santa Monica College*; Per Salvesen, *University of Bergen*; Mark F. Sanders, *University of California, Davis*; Kathleen Sandman, *Ohio State University*; Davison Sangweme, *University of North Georgia*; Louis Santiago, *University of California, Riverside*; Ted Sargent, *University of Massachusetts, Amherst*; K. Sathasivan, *University of Texas, Austin*; Gary Saunders, *University of New Brunswick*; Thomas R. Sawicki, *Spartanburg Community College*; Inder Saxena, *University of Texas, Austin*; Karin Scarpinato, *Georgia Southern University*; Carl Schaefer, *University of Connecticut*; Andrew Schaffner, *Cal Poly San Luis Obispo*; Maynard H. Schaus, *Virginia Wesleyan College*; Renate Scheibe, *University of Osnabrück*; Cara Schillington, *Eastern Michigan University*; David Schimpf, *University of Minnesota, Duluth*; William H. Schlesinger, *Duke University*; Mark Schlissel, *University of California, Berkeley*; Christopher J. Schneider, *Boston University*; Thomas W. Schoener, *University of California, Davis*; Robert Schorr, *Colorado State University*; Patricia M. Schulte, *University of British Columbia*; Karen S. Schumaker, *University of Arizona*; Brenda Schumpert, *Valencia Community College*; David J. Schwartz, *Houston Community College*; Carrie Schwarz, *Western Washington University*; Christa Schwintzer, *University of Maine*; Erik P. Scully, *Towson State University*; Robert W. Seagull, *Hofstra University*; Edna Seaman, *Northeastern University*; Duane Sears, *University of California, Santa Barbara*; Brent Selinger, *University of Lethbridge*; Orono Shukdeb Sen, *Bethune-Cookman College*; Wendy Sera, *Seton Hill University*; Alison M. Shakarian, *Salve Regina University*; Timothy E. Shannon, *Francis Marion University*; Victoria C. Sharpe, *Blinn College*; Elaine Shea, *Loyola College, Maryland*; Stephen Sheckler, *Virginia Polytechnic Institute and State University*; Robin L. Sherman, *Nova Southeastern University*; Richard Sherwin, *University of Pittsburgh*; Alison Sherwood, *University of Hawaii at Manoa*; Lisa Shimeld, *Crafton Hills College*; James Shinkle, *Trinity University*; Barbara Shipes, *Hampton University*; Brian Shmaefsky, *Lone Star College*; Richard M. Showman, *University of South Carolina*; Eric Shows, *Jones County Junior College*; Peter Shugarman, *University of Southern California*; Alice Shuttey, *DeKalb Community College*; James Sidie, *Ursinus College*; Daniel Simberloff, *Florida State University*; Rebecca Simmons, *University of North Dakota*; Anne Simon, *University of Maryland, College Park*; Robert Simons, *University of California, Los Angeles*; Alastair Simpson, *Dalhousie University*; Susan Singer, *Carleton College*; Sedonia Sipes, *Southern Illinois University, Carbondale*; John Skillman, *California State University, San Bernardino*; Roger Sloboda, *Dartmouth University*; John Smarrelli, *Le Moyne College*; Andrew T. Smith, *Arizona State University*; Kelly Smith, *University of North Florida*; Nancy Smith-Huerta, *Miami Ohio University*; John Smol, *Queen's University*; Andrew J. Snope, *Essex Community College*; Mitchell Sogin, *Woods Hole Marine Biological Laboratory*; Doug Soltis, *University of Florida, Gainesville*; Julio G. Soto, *San Jose State University*; Susan Sovonick-Dunford, *University of Cincinnati*; Rebecca Sperry, *Salt Lake Community College*; Frederick W. Spiegel, *University of Arkansas*; Clint Springer, *Saint Joseph's University*; John Stachowicz, *University of California, Davis*; Joel Stafstrom, *Northern Illinois University*; Alam Stam, *Capital University*; Amanda Starnes, *Emory University*; Karen Steudel, *University of Wisconsin*; Barbara Stewart, *Swarthmore College*; Gail A. Stewart, *Camden County College*; Cecil Still, *Rutgers University, New Brunswick*; Margery Stinson, *Southwestern College*; James Stockand, *University of Texas Health Science Center, San Antonio*; John Stolz, *California Institute of Technology*; Judy Stone, *Colby College*; Richard D. Storey, *Colorado College*; Stephen Strand, *University of California, Los Angeles*; Eric Strauss, *University of Massachusetts, Boston*; Antony Stretton, *University of Wisconsin, Madison*; Russell Stullken, *Augusta College*; Mark Sturtevant, *Oakland University, Flint*; John Sullivan, *Southern Oregon State University*; Gerald Summers, *University of Missouri*; Judith Sumner, *Assumption College*; Marshall D. Sundberg, *Emporia State University*; Cynthia Surmacz, *Bloomsburg University*; Lucinda Swatzell, *Southeast Missouri State University*; Daryl Sweeney, *University of Illinois, Champaign-Urbana*; Diane Sweeney, *Punahou School*; Samuel S. Sweet, *University of California, Santa Barbara*; Janice Swenson, *University of North Florida*; Michael A. Sypes, *Pennsylvania State University*; Lincoln Taiz, *University of California, Santa Cruz*; David Tam, *University of North Texas*; Yves Tan, *Cabrillo College*; Samuel Tarsitano, *Southwest Texas State University*; David Tauck, *Santa Clara University*; Emily Taylor, *California Polytechnic State University, San Luis Obispo*; James Taylor, *University of New Hampshire*; John W. Taylor, *University of California, Berkeley*; Kristen Taylor, *Salt Lake Community College*; Martha R. Taylor, *Cornell University*; Franklyn Tan Te, *Miami Dade College*; Thomas Terry, *University of Connecticut*; Roger Thibault, *Bowling Green State University*; Kent Thomas, *Wichita State University*; Rebecca Thomas, *College of St. Joseph*; William Thomas, *Colby-Sawyer College*; Cyril Thong, *Simon Fraser University*; John Thornton, *Oklahoma State University*; Robert Thornton, *University of California, Davis*; William Thwaites, *Tillamook Bay Community College*; Stephen Timme, *Pittsburg State University*; Mike Toliver, *Eureka College*; Eric Toolson, *University of New Mexico*; Leslie Towill, *Arizona State University*; James Traniello, *Boston University*; Paul Q. Trombley, *Florida State University*; Nancy J. Trun, *Duquesne University*; Constantine Tsoukas, *San Diego State University*; Marsha Turell, *Houston Community College*; Victoria Turgeon, *Furman University*; Robert Tuveson, *University of Illinois, Urbana*; Maura G. Tyrrell, *Stonehill College*; Catherine Uekert, *Northern Arizona University*; Claudia Uhde-Stone, *California State University, East Bay*; Gordon Uno, *University of Oklahoma*; Lisa A. Urry, *Mills College*; Saba Valadkhan, *Center for RNA Molecular Biology*; James W. Valentine, *University of California, Santa Barbara*; Joseph Vanable, *Purdue University*; Theodore Van Bruggen, *University of South Dakota*; Kathryn VandenBosch, *Texas A&M University*; Gerald Van Dyke, *North Carolina State University*; Brandi Van Roo, *Framingham State College*; Moira Van Staaden, *Bowling Green State University*; Martin Vaughan, *Indiana University-Purdue University Indianapolis*; Sarah VanVickle-Chavez, *Washington University, St. Louis*; William Velhagen, *New York University*; Steven D. Verhey, *Central Washington University*; Kathleen Verville, *Washington College*; Sara Via, *University of Maryland*; Meena Vijayaraghavan, *Tulane University*; Frank Visco, *Orange Coast College*; Laurie Vitt, *University of California, Los Angeles*; Neal Voelz, *St. Cloud State University*; Thomas J. Volk, *University of Wisconsin, La Crosse*; Leif Asbjørn Vøllestad, *University of Oslo*; Amy Volmer, *Swarthmore College*; Janice Voltzow, *University of Scranton*; Margaret Voss, *Penn State Erie*; Susan D. Waaland, *University of Washington*; Charles Wade, *C.S. Mott Community College*; William Wade, *Dartmouth Medical College*; John Waggoner, *Loyola Marymount University*; Jyoti Wagle, *Houston Community College*; Edward Wagner, *University of California, Irvine*; D. Alexander Wait, *Southwest Missouri State University*; Claire Walczak, *Indiana University*; Jerry Waldvogel, *Clemson University*; Dan Walker, *San Jose State University*; Robert Lee Wallace, *Ripon College*; Jeffrey Walters, *North Carolina State University*; Linda Walters, *University of Central Florida*; James Wandersee, *Louisiana State University*; James T. Warren Jr., *Pennsylvania State University*; Nickolas M. Waser, *University of California, Riverside*; Fred Wasserman, *Boston University*; Margaret Waterman, *University of Pittsburgh*; Charles Webber, *Loyola University of Chicago*; Peter Webster, *University of Massachusetts, Amherst*; Terry Webster, *University of Connecticut, Storrs*; Beth Wee, *Tulane University*; James Wee, *Loyola University, New Orleans*; Andrea Weeks, *George Mason University*; John Weishampel, *University of Central Florida*; Peter Wejksnora, *University of Wisconsin, Milwaukee*; Charles Wellman, *Sheffield University*; Kentwood Wells, *University of Connecticut*; David J. Westenberg, *University of Missouri, Rolla*; Richard Wetts, *University of California, Irvine*; Christopher Whipps, *State University of New York College of Environmental Science and Forestry*; Jessica White-Phillip, *Our Lady of the Lake University*; Matt White, *Ohio University*; Philip White, *James Hutton Institute*; Susan Whittemore, *Keene State College*; Murray Wiegand, *University of Winnipeg*; Ernest H. Williams, *Hamilton College*; Kathy Williams, *San Diego State University*; Kimberly Williams, *Kansas State University*; Stephen Williams, *Glendale Community College*; Elizabeth Willott, *University of Arizona*; Christopher Wills, *University of California, San Diego*; Paul Wilson, *California State University, Northridge*; Fred Wilt, *University of California, Berkeley*; Peter Wimberger, *University of Puget Sound*; Robert Winning, *Eastern Michigan University*; E. William Wischusen, *Louisiana State University*; Clarence Wolfe, *Northern Virginia Community College*; Vickie L. Wolfe, *Marshall University*; Janet Wolkenstein, *Hudson Valley Community College*; Robert T. Woodland, *University of Massachusetts Medical School*; Joseph Woodring, *Louisiana State University*; Denise Woodward, *Pennsylvania State University*; Patrick Woolley, *East Central College*; Sarah E. Wyatt, *Ohio University*; Grace Wyngaard, *James Madison University*; Shuhai Xiao, *Virginia Polytechnic Institute,* Ramin Yadegari, *University of Arizona*; Paul Yancey, *Whitman College*; Philip Yant, *University of Michigan*; Linda Yasui, *Northern Illinois University*; Anne D. Yoder, *Duke University*; Hideo Yonenaka, *San Francisco State University*; Robert Yost, *Indiana University-Purdue University Indianapolis*; Tia Young, *Pennsylvania State University*; Gina M. Zainelli, *Loyola University, Chicago*; Edward Zalisko, *Blackburn College*; Nina Zanetti, *Siena College*; Sam Zeveloff, *Weber State University*; Zai Ming Zhao, *University of Texas, Austin*; John Zimmerman, *Kansas State University*; Miriam Zolan, *Indiana University*; Theresa Zucchero, *Methodist University*; Uko Zylstra, *Calvin College*.

Detailed Contents

1 Evolution, the Themes of Biology, and Scientific Inquiry 2

CONCEPT 1.1 The study of life reveals unifying themes 3
Theme: New Properties Emerge at Successive Levels of Biological Organization 4
Theme: Life's Processes Involve the Expression and Transmission of Genetic Information 6
Theme: Life Requires the Transfer and Transformation of Energy and Matter 9
Theme: From Molecules to Ecosystems, Interactions Are Important in Biological Systems 9

CONCEPT 1.2 The Core Theme: Evolution accounts for the unity and diversity of life 11
Classifying the Diversity of Life 12
Charles Darwin and the Theory of Natural Selection 14
The Tree of Life 15

CONCEPT 1.3 In studying nature, scientists form and test hypotheses 16
Exploration and Observation 17
Gathering and Analyzing Data 17
Forming and Testing Hypotheses 17
The Flexibility of the Scientific Process 18
A Case Study in Scientific Inquiry: Investigating Coat Coloration in Mouse Populations 20
Variables and Controls in Experiments 20
Theories in Science 21

CONCEPT 1.4 Science benefits from a cooperative approach and diverse viewpoints 22
Building on the Work of Others 22
Science, Technology, and Society 23
The Value of Diverse Viewpoints in Science 24

Unit 1   The Chemistry of Life   27

Interview: Kenneth Olden 27

2 The Chemical Context of Life 28

CONCEPT 2.1 Matter consists of chemical elements in pure form and in combinations called compounds 29
Elements and Compounds 29
The Elements of Life 29
Case Study: Evolution of Tolerance to Toxic Elements 30

CONCEPT 2.2 An element's properties depend on the structure of its atoms 30
Subatomic Particles 30
Atomic Number and Atomic Mass 31
Isotopes 31
The Energy Levels of Electrons 32
Electron Distribution and Chemical Properties 34
Electron Orbitals 35

CONCEPT 2.3 The formation and function of molecules and ionic compounds depend on chemical bonding between atoms 36
Covalent Bonds 36
Ionic Bonds 37
Weak Chemical Interactions 38
Molecular Shape and Function 39

CONCEPT 2.4 Chemical reactions make and break chemical bonds 40

3 Water and Life 44

CONCEPT 3.1 Polar covalent bonds in water molecules result in hydrogen bonding 45

CONCEPT 3.2 Four emergent properties of water contribute to Earth's suitability for life 45
Cohesion of Water Molecules 45
Moderation of Temperature by Water 46
Floating of Ice on Liquid Water 48
Water: The Solvent of Life 49
Possible Evolution of Life on Other Planets 50

CONCEPT 3.3 Acidic and basic conditions affect living organisms 51
Acids and Bases 51
The pH Scale 51
Buffers 52
Acidification: A Threat to Our Oceans 53

4 Carbon and the Molecular Diversity of Life 56

CONCEPT 4.1 Organic chemistry is key to the origin of life 57

CONCEPT 4.2 Carbon atoms can form diverse molecules by bonding to four other atoms 58
The Formation of Bonds with Carbon 58
Molecular Diversity Arising from Variation in Carbon Skeletons 60

CONCEPT 4.3 A few chemical groups are key to molecular function 62
The Chemical Groups Most Important in the Processes of Life 62
ATP: An Important Source of Energy for Cellular Processes 64
The Chemical Elements of Life: *A Review* 64

5 The Structure and Function of Large Biological Molecules 66

CONCEPT 5.1 Macromolecules are polymers, built from monomers 67
The Synthesis and Breakdown of Polymers 67
The Diversity of Polymers 67

CONCEPT 5.2 Carbohydrates serve as fuel and building material 68
Sugars 68
Polysaccharides 70

CONCEPT 5.3 Lipids are a diverse group of hydrophobic molecules 72
Fats 72
Phospholipids 74
Steroids 75

CONCEPT 5.4 Proteins include a diversity of structures, resulting in a wide range of functions 75
Amino Acids (Monomers) 75
Polypeptides (Amino Acid Polymers) 78
Protein Structure and Function 78

CONCEPT 5.5 Nucleic acids store, transmit, and help express hereditary information 84
The Roles of Nucleic Acids 84
The Components of Nucleic Acids 84
Nucleotide Polymers 85
The Structures of DNA and RNA Molecules 86

CONCEPT 5.6 Genomics and proteomics have transformed biological inquiry and applications 86
DNA and Proteins as Tape Measures of Evolution 87

Unit 2 The Cell 92

Interview: Diana Bautista 92

6 A Tour of the Cell 93

CONCEPT 6.1 Biologists use microscopes and biochemistry to study cells 94
Microscopy 94
Cell Fractionation 96

CONCEPT 6.2 Eukaryotic cells have internal membranes that compartmentalize their functions 97
Comparing Prokaryotic and Eukaryotic Cells 97
A Panoramic View of the Eukaryotic Cell 99

CONCEPT 6.3 The eukaryotic cell's genetic instructions are housed in the nucleus and carried out by the ribosomes 102
The Nucleus: Information Central 102
Ribosomes: Protein Factories 102

CONCEPT 6.4 The endomembrane system regulates protein traffic and performs metabolic functions 104
The Endoplasmic Reticulum: Biosynthetic Factory 104
The Golgi Apparatus: Shipping and Receiving Center 105
Lysosomes: Digestive Compartments 107
Vacuoles: Diverse Maintenance Compartments 108
The Endomembrane System: *A Review* 108

CONCEPT 6.5 Mitochondria and chloroplasts change energy from one form to another 109
The Evolutionary Origins of Mitochondria and Chloroplasts 109
Mitochondria: Chemical Energy Conversion 110
Chloroplasts: Capture of Light Energy 110
Peroxisomes: Oxidation 112

CONCEPT 6.6 The cytoskeleton is a network of fibers that organizes structures and activities in the cell 112
Roles of the Cytoskeleton: Support and Motility 112
Components of the Cytoskeleton 113

CONCEPT 6.7 Extracellular components and connections between cells help coordinate cellular activities 118
Cell Walls of Plants 118
The Extracellular Matrix (ECM) of Animal Cells 118
Cell Junctions 119

CONCEPT 6.8 A cell is greater than the sum of its parts 121

7 Membrane Structure and Function 126

CONCEPT 7.1 Cellular membranes are fluid mosaics of lipids and proteins 127
The Fluidity of Membranes 128
Evolution of Differences in Membrane Lipid Composition 129
Membrane Proteins and Their Functions 129
The Role of Membrane Carbohydrates in Cell-Cell Recognition 130
Synthesis and Sidedness of Membranes 131

CONCEPT 7.2 Membrane structure results in selective permeability 131
The Permeability of the Lipid Bilayer 132
Transport Proteins 132

CONCEPT 7.3 Passive transport is diffusion of a substance across a membrane with no energy investment 132
Effects of Osmosis on Water Balance 133
Facilitated Diffusion: Passive Transport Aided by Proteins 135

CONCEPT 7.4 Active transport uses energy to move solutes against their gradients 136
The Need for Energy in Active Transport 136
How Ion Pumps Maintain Membrane Potential 137
Cotransport: Coupled Transport by a Membrane Protein 138

CONCEPT 7.5 Bulk transport across the plasma membrane occurs by exocytosis and endocytosis 139
Exocytosis 139
Endocytosis 139

8 An Introduction to Metabolism 143

CONCEPT 8.1 An organism's metabolism transforms matter and energy 144
Metabolic Pathways 144
Forms of Energy 144
The Laws of Energy Transformation 145

CONCEPT 8.2 The free-energy change of a reaction tells us whether or not the reaction occurs spontaneously 147
Free-Energy Change, ΔG 147
Free Energy, Stability, and Equilibrium 147
Free Energy and Metabolism 148

CONCEPT 8.3 ATP powers cellular work by coupling exergonic reactions to endergonic reactions 150
The Structure and Hydrolysis of ATP 150
How ATP Provides Energy That Performs Work 151
The Regeneration of ATP 153

CONCEPT 8.4 Enzymes speed up metabolic reactions by lowering energy barriers 153
The Activation Energy Barrier 153
How Enzymes Speed Up Reactions 154
Substrate Specificity of Enzymes 155
Catalysis in the Enzyme's Active Site 156
Effects of Local Conditions on Enzyme Activity 157
The Evolution of Enzymes 159

CONCEPT 8.5 Regulation of enzyme activity helps control metabolism 159
Allosteric Regulation of Enzymes 160
Localization of Enzymes Within the Cell 161

9 Cellular Respiration and Fermentation 164

CONCEPT 9.1 Catabolic pathways yield energy by oxidizing organic fuels 165
Catabolic Pathways and Production of ATP 165
Redox Reactions: Oxidation and Reduction 165
The Stages of Cellular Respiration: *A Preview* 168

CONCEPT 9.2 Glycolysis harvests chemical energy by oxidizing glucose to pyruvate 170

CONCEPT 9.3 After pyruvate is oxidized, the citric acid cycle completes the energy-yielding oxidation of organic molecules 171
Oxidation of Pyruvate to Acetyl CoA 171
The Citric Acid Cycle 172

CONCEPT 9.4 During oxidative phosphorylation, chemiosmosis couples electron transport to ATP synthesis 174
The Pathway of Electron Transport 174
Chemiosmosis: The Energy-Coupling Mechanism 175
An Accounting of ATP Production by Cellular Respiration 177

CONCEPT 9.5 Fermentation and anaerobic respiration enable cells to produce ATP without the use of oxygen 179
Types of Fermentation 180
Comparing Fermentation with Anaerobic and Aerobic Respiration 181
The Evolutionary Significance of Glycolysis 182

CONCEPT 9.6 Glycolysis and the citric acid cycle connect to many other metabolic pathways 182
The Versatility of Catabolism 182
Biosynthesis (Anabolic Pathways) 183
Regulation of Cellular Respiration via Feedback Mechanisms 183

10 Photosynthesis 187

CONCEPT 10.1 Photosynthesis feeds the biosphere 188

CONCEPT 10.2 Photosynthesis converts light energy to the chemical energy of food 189
Chloroplasts: The Sites of Photosynthesis in Plants 189
Tracking Atoms Through Photosynthesis 189
The Two Stages of Photosynthesis: *A Preview* 191

CONCEPT 10.3 The light reactions convert solar energy to the chemical energy of ATP and NADPH 192
The Nature of Sunlight 192
Photosynthetic Pigments: The Light Receptors 192
Excitation of Chlorophyll by Light 195
A Photosystem: A Reaction-Center Complex Associated with Light-Harvesting Complexes 195
Linear Electron Flow 197
Cyclic Electron Flow 198
A Comparison of Chemiosmosis in Chloroplasts and Mitochondria 199

CONCEPT 10.4 The Calvin cycle uses the chemical energy of ATP and NADPH to reduce CO_2 to sugar 201

CONCEPT 10.5 Alternative mechanisms of carbon fixation have evolved in hot, arid climates 203
Photorespiration: An Evolutionary Relic? 203
C_4 Plants 203
CAM Plants 205

CONCEPT 10.6 Photosynthesis is essential for life on Earth: a review 206

11 Cell Communication 212

CONCEPT 11.1 External signals are converted to responses within the cell 213
Evolution of Cell Signaling 213
Local and Long-Distance Signaling 215
The Three Stages of Cell Signaling: *A Preview* 216

CONCEPT 11.2 Signal reception: A signaling molecule binds to a receptor, causing it to change shape 217
Receptors in the Plasma Membrane 217
Intracellular Receptors 220

CONCEPT 11.3 Signal transduction: Cascades of molecular interactions transmit signals from receptors to relay molecules in the cell 221
Signal Transduction Pathways 221
Protein Phosphorylation and Dephosphorylation 222
Small Molecules and Ions as Second Messengers 223

CONCEPT 11.4 Cellular response: Cell signaling leads to regulation of transcription or cytoplasmic activities 226
Nuclear and Cytoplasmic Responses 226
Regulation of the Response 226

CONCEPT 11.5 Apoptosis requires integration of multiple cell-signaling pathways 229
Apoptosis in the Soil Worm *Caenorhabditis elegans* 230
Apoptotic Pathways and the Signals That Trigger Them 230

12 The Cell Cycle 234

CONCEPT 12.1 Most cell division results in genetically identical daughter cells 235
Key Roles of Cell Division 235
Cellular Organization of the Genetic Material 235
Distribution of Chromosomes During Eukaryotic Cell Division 236

CONCEPT 12.2 The mitotic phase alternates with interphase in the cell cycle 237
Phases of the Cell Cycle 237
The Mitotic Spindle: *A Closer Look* 240
Cytokinesis: *A Closer Look* 241
Binary Fission in Bacteria 242
The Evolution of Mitosis 243

CONCEPT 12.3 The eukaryotic cell cycle is regulated by a molecular control system 244
The Cell Cycle Control System 244
Loss of Cell Cycle Controls in Cancer Cells 248

Unit 3   Genetics   253

Interview: Francisco Mojica 253

13 Meiosis and Sexual Life Cycles 254

CONCEPT 13.1 Offspring acquire genes from parents by inheriting chromosomes 255
Inheritance of Genes 255
Comparison of Asexual and Sexual Reproduction 255

CONCEPT 13.2 Fertilization and meiosis alternate in sexual life cycles 256
Sets of Chromosomes in Human Cells 256
Behavior of Chromosome Sets in the Human Life Cycle 257
The Variety of Sexual Life Cycles 258

CONCEPT 13.3 Meiosis reduces the number of chromosome sets from diploid to haploid 259
The Stages of Meiosis 259
Crossing Over and Synapsis During Prophase I 262
A Comparison of Mitosis and Meiosis 262

CONCEPT 13.4 Genetic variation produced in sexual life cycles contributes to evolution 265
Origins of Genetic Variation Among Offspring 265
The Evolutionary Significance of Genetic Variation Within Populations 266

14 Mendel and the Gene Idea 269

CONCEPT 14.1 Mendel used the scientific approach to identify two laws of inheritance 270
Mendel's Experimental, Quantitative Approach 270
The Law of Segregation 271
The Law of Independent Assortment 274

CONCEPT 14.2 Probability laws govern Mendelian inheritance 276
The Multiplication and Addition Rules Applied to Monohybrid Crosses 277
Solving Complex Genetics Problems with the Rules of Probability 277

CONCEPT 14.3 Inheritance patterns are often more complex than predicted by simple Mendelian genetics 278
Extending Mendelian Genetics for a Single Gene 278
Extending Mendelian Genetics for Two or More Genes 281
Nature and Nurture: The Environmental Impact on Phenotype 282
A Mendelian View of Heredity and Variation 282

CONCEPT 14.4 Many human traits follow Mendelian patterns of inheritance 284
Pedigree Analysis 284
Recessively Inherited Disorders 285
Dominantly Inherited Disorders 287
Multifactorial Disorders 287
Genetic Testing and Counseling 287

15 The Chromosomal Basis of Inheritance 294

CONCEPT 15.1 Mendelian inheritance has its physical basis in the behavior of chromosomes 295
  Morgan's Choice of Experimental Organism 295
  Correlating Behavior of a Gene's Alleles with Behavior of a
    Chromosome Pair: *Scientific Inquiry* 295

CONCEPT 15.2 Sex-linked genes exhibit unique patterns of inheritance 298
  The Chromosomal Basis of Sex 298
  Inheritance of X-Linked Genes 299
  X Inactivation in Female Mammals 300

CONCEPT 15.3 Linked genes tend to be inherited together because they are located near each other on the same chromosome 301
  How Linkage Affects Inheritance 301
  Genetic Recombination and Linkage 302
  Mapping the Distance Between Genes Using Recombination Data:
    Scientific Inquiry 305

CONCEPT 15.4 Alterations of chromosome number or structure cause some genetic disorders 306
  Abnormal Chromosome Number 307
  Alterations of Chromosome Structure 307
  Human Conditions Due to Chromosomal Alterations 308

CONCEPT 15.5 Some inheritance patterns are exceptions to standard Mendelian inheritance 310
  Genomic Imprinting 310
  Inheritance of Organelle Genes 311

16 The Molecular Basis of Inheritance 314

CONCEPT 16.1 DNA is the genetic material 315
  The Search for the Genetic Material: *Scientific Inquiry* 315
  Building a Structural Model of DNA 317

CONCEPT 16.2 Many proteins work together in DNA replication and repair 320
  The Basic Principle: Base Pairing to a Template Strand 321
  DNA Replication: *A Closer Look* 322
  Proofreading and Repairing DNA 327
  Evolutionary Significance of Altered DNA Nucleotides 328
  Replicating the Ends of DNA Molecules 328

CONCEPT 16.3 A chromosome consists of a DNA molecule packed together with proteins 330

17 Gene Expression: From Gene to Protein 335

CONCEPT 17.1 Genes specify proteins via transcription and translation 336
  Evidence from Studying Metabolic Defects 336
  Basic Principles of Transcription and Translation 337
  The Genetic Code 340

CONCEPT 17.2 Transcription is the DNA-directed synthesis of RNA: *A Closer Look* 342
  Molecular Components of Transcription 342
  Synthesis of an RNA Transcript 342

CONCEPT 17.3 Eukaryotic cells modify RNA after transcription 345
  Alteration of mRNA Ends 345
  Split Genes and RNA Splicing 345

CONCEPT 17.4 Translation is the RNA-directed synthesis of a polypeptide: *A Closer Look* 347
  Molecular Components of Translation 348
  Building a Polypeptide 350
  Completing and Targeting the Functional Protein 352
  Making Multiple Polypeptides in Bacteria and Eukaryotes 355

CONCEPT 17.5 Mutations of one or a few nucleotides can affect protein structure and function 357
  Types of Small-Scale Mutations 357
  New Mutations and Mutagens 360
  Using CRISPR to Edit Genes and Correct Disease-Causing
    Mutations 360
  What Is a Gene? *Revisiting the Question* 361

18 Regulation of Gene Expression 365

CONCEPT 18.1 Bacteria often respond to environmental change by regulating transcription 366
  Operons: The Basic Concept 366
  Repressible and Inducible Operons: Two Types of Negative Gene
    Regulation 368
  Positive Gene Regulation 369

CONCEPT 18. 2 Eukaryotic gene expression is regulated at many stages 370
  Differential Gene Expression 370
  Regulation of Chromatin Structure 371
  Regulation of Transcription Initiation 373
  Mechanisms of Post-transcriptional Regulation 377

CONCEPT 18.3 Noncoding RNAs play multiple roles in controlling gene expression 379
  Effects on mRNAs by MicroRNAs and Small Interfering RNAs 379
  Chromatin Remodeling and Effects on Transcription by
    ncRNAs 380

CONCEPT 18.4 A program of differential gene expression leads to the different cell types in a multicellular organism 381
  A Genetic Program for Embryonic Development 381
  Cytoplasmic Determinants and Inductive Signals 382
  Sequential Regulation of Gene Expression During Cellular
    Differentiation 383
  Pattern Formation: Setting Up the Body Plan 384

CONCEPT 18.5 Cancer results from genetic changes that affect cell cycle control 388

Types of Genes Associated with Cancer 388
Interference with Normal Cell-Signaling Pathways 389
The Multistep Model of Cancer Development 391
Inherited Predisposition and Environmental Factors Contributing to Cancer 394
The Role of Viruses in Cancer 394

19 Viruses 398

CONCEPT 19.1 A virus consists of a nucleic acid surrounded by a protein coat 399

The Discovery of Viruses: *Scientific Inquiry* 399
Structure of Viruses 399

CONCEPT 19.2 Viruses replicate only in host cells 401

General Features of Viral Replicative Cycles 401
Replicative Cycles of Phages 402
Replicative Cycles of Animal Viruses 404
Evolution of Viruses 406

CONCEPT 19.3 Viruses and prions are formidable pathogens in animals and plants 408

Viral Diseases in Animals 408
Emerging Viral Diseases 409
Viral Diseases in Plants 412
Prions: Proteins as Infectious Agents 412

20 DNA Tools and Biotechnology 415

CONCEPT 20.1 DNA sequencing and DNA cloning are valuable tools for genetic engineering and biological inquiry 416

DNA Sequencing 416
Making Multiple Copies of a Gene or Other DNA Segment 418
Using Restriction Enzymes to Make a Recombinant DNA Plasmid 419
Amplifying DNA: The Polymerase Chain Reaction (PCR) and Its Use in DNA Cloning 420
Expressing Cloned Eukaryotic Genes 422

CONCEPT 20.2 Biologists use DNA technology to study gene expression and function 423

Analyzing Gene Expression 423
Determining Gene Function 426

CONCEPT 20.3 Cloned organisms and stem cells are useful for basic research and other applications 428

Cloning Plants: Single-Cell Cultures 428
Cloning Animals: Nuclear Transplantation 428
Stem Cells of Animals 430

CONCEPT 20.4 The practical applications of DNA-based biotechnology affect our lives in many ways 433

Medical Applications 433
Forensic Evidence and Genetic Profiles 436
Environmental Cleanup 437
Agricultural Applications 437
Safety and Ethical Questions Raised by DNA Technology 438

21 Genomes and Their Evolution 442

CONCEPT 21.1 The Human Genome Project fostered development of faster, less expensive sequencing techniques 443

CONCEPT 21.2 Scientists use bioinformatics to analyze genomes and their functions 444

Centralized Resources for Analyzing Genome Sequences 444
Identifying Protein-Coding Genes and Understanding Their Functions 445
Understanding Genes and Gene Expression at the Systems Level 446

CONCEPT 21.3 Genomes vary in size, number of genes, and gene density 448

Genome Size 448
Number of Genes 449
Gene Density and Noncoding DNA 449

CONCEPT 21.4 Multicellular eukaryotes have a lot of noncoding DNA and many multigene families 450

Transposable Elements and Related Sequences 451
Other Repetitive DNA, Including Simple Sequence DNA 452
Genes and Multigene Families 452

CONCEPT 21.5 Duplication, rearrangement, and mutation of DNA contribute to genome evolution 454

Duplication of Entire Chromosome Sets 454
Alterations of Chromosome Structure 454
Duplication and Divergence of Gene-Sized Regions of DNA 455
Rearrangements of Parts of Genes: Exon Duplication and Exon Shuffling 456
How Transposable Elements Contribute to Genome Evolution 459

CONCEPT 21.6 Comparing genome sequences provides clues to evolution and development 459

Comparing Genomes 459
Widespread Conservation of Developmental Genes Among Animals 463

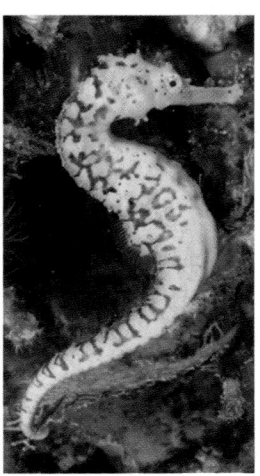

Unit 4 Mechanisms of Evolution 467

Interview: Cassandra Extavour 467

22 Descent with Modification: A Darwinian View of Life 468

CONCEPT 22.1 The Darwinian revolution challenged traditional views of a young Earth inhabited by unchanging species 469
Endless Forms Most Beautiful 469
Scala Naturae and Classification of Species 470
Ideas About Change over Time 470
Lamarck's Hypothesis of Evolution 471

CONCEPT 22.2 Descent with modification by natural selection explains the adaptations of organisms and the unity and diversity of life 471
Darwin's Research 471
Ideas from *The Origin of Species* 473
Key Features of Natural Selection 476

CONCEPT 22.3 Evolution is supported by an overwhelming amount of scientific evidence 476
Direct Observations of Evolutionary Change 477
Homology 479
The Fossil Record 481
Biogeography 482
What Is Theoretical About Darwin's View of Life? 483

23 The Evolution of Populations 486

CONCEPT 23.1 Genetic variation makes evolution possible 487
Genetic Variation 487
Sources of Genetic Variation 488

CONCEPT 23.2 The Hardy-Weinberg equation can be used to test whether a population is evolving 489
Gene Pools and Allele Frequencies 490
The Hardy-Weinberg Equation 490

CONCEPT 23.3 Natural selection, genetic drift, and gene flow can alter allele frequencies in a population 493
Natural Selection 494
Genetic Drift 494
Gene Flow 496

CONCEPT 23.4 Natural selection is the only mechanism that consistently causes adaptive evolution 497
Natural Selection: *A Closer Look* 497
The Key Role of Natural Selection in Adaptive Evolution 498

Sexual Selection 499
Balancing Selection 500
Why Natural Selection Cannot Fashion Perfect Organisms 501

24 The Origin of Species 506

CONCEPT 24.1 The biological species concept emphasizes reproductive isolation 507
The Biological Species Concept 507
Other Definitions of Species 510

CONCEPT 24.2 Speciation can take place with or without geographic separation 511
Allopatric ("Other Country") Speciation 511
Sympatric ("Same Country") Speciation 513
Allopatric and Sympatric Speciation: *A Review* 516

CONCEPT 24.3 Hybrid zones reveal factors that cause reproductive isolation 516
Patterns Within Hybrid Zones 516
Hybrid Zones and Environmental Change 517
Hybrid Zones over Time 518

CONCEPT 24.4 Speciation can occur rapidly or slowly and can result from changes in few or many genes 520
The Time Course of Speciation 520
Studying the Genetics of Speciation 522
From Speciation to Macroevolution 523

25 The History of Life on Earth 525

CONCEPT 25.1 Conditions on early Earth made the origin of life possible 526
Synthesis of Organic Compounds on Early Earth 526
Abiotic Synthesis of Macromolecules 527
Protocells 527
Self-Replicating RNA 528

CONCEPT 25.2 The fossil record documents the history of life 528
The Fossil Record 529
How Rocks and Fossils Are Dated 529
The Origin of New Groups of Organisms 530

CONCEPT 25.3 Key events in life's history include the origins of unicellular and multicellular organisms and the colonization of land 532
The First Single-Celled Organisms 533
The Origin of Multicellularity 535
The Colonization of Land 536

CONCEPT 25.4 The rise and fall of groups of organisms reflect differences in speciation and extinction rates 537
Plate Tectonics 538
Mass Extinctions 540
Adaptive Radiations 542

CONCEPT 25.5 Major changes in body form can result from changes in the sequences and regulation of developmental genes 544
Effects of Developmental Genes 544
The Evolution of Development 545

CONCEPT 25.6 Evolution is not goal oriented 547
Evolutionary Novelties 547
Evolutionary Trends 548

Unit 5   The Evolutionary History of Biological Diversity   552

Interview: Penny Chisholm 552

26 Phylogeny and the Tree of Life   553

CONCEPT 26.1 Phylogenies show evolutionary relationships 554
- Binomial Nomenclature 554
- Hierarchical Classification 554
- Linking Classification and Phylogeny 555
- What We Can and Cannot Learn from Phylogenetic Trees 555
- Applying Phylogenies 557

CONCEPT 26.2 Phylogenies are inferred from morphological and molecular data 558
- Morphological and Molecular Homologies 558
- Sorting Homology from Analogy 558
- Evaluating Molecular Homologies 559

CONCEPT 26.3 Shared characters are used to construct phylogenetic trees 559
- Cladistics 559
- Phylogenetic Trees with Proportional Branch Lengths 561
- Maximum Parsimony and Maximum Likelihood 562
- Phylogenetic Trees as Hypotheses 564

CONCEPT 26.4 An organism's evolutionary history is documented in its genome 565
- Gene Duplications and Gene Families 565
- Genome Evolution 566

CONCEPT 26.5 Molecular clocks help track evolutionary time 566
- Molecular Clocks 566
- Applying a Molecular Clock: Dating the Origin of HIV 567

CONCEPT 26.6 Our understanding of the tree of life continues to change based on new data 568
- From Two Kingdoms to Three Domains 568
- The Important Role of Horizontal Gene Transfer 568

27 Bacteria and Archaea   573

CONCEPT 27.1 Structural and functional adaptations contribute to prokaryotic success 574
- Cell-Surface Structures 574
- Motility 576
- Internal Organization and DNA 577
- Reproduction 577

CONCEPT 27.2 Rapid reproduction, mutation, and genetic recombination promote genetic diversity in prokaryotes 578
- Rapid Reproduction and Mutation 578
- Genetic Recombination 579

CONCEPT 27.3 Diverse nutritional and metabolic adaptations have evolved in prokaryotes 581
- The Role of Oxygen in Metabolism 582
- Nitrogen Metabolism 582
- Metabolic Cooperation 582

CONCEPT 27.4 Prokaryotes have radiated into a diverse set of lineages 583
- An Overview of Prokaryotic Diversity 583
- Bacteria 583
- Archaea 585

CONCEPT 27.5 Prokaryotes play crucial roles in the biosphere 586
- Chemical Recycling 586
- Ecological Interactions 587

CONCEPT 27.6 Prokaryotes have both beneficial and harmful impacts on humans 587
- Mutualistic Bacteria 587
- Pathogenic Bacteria 588
- Antibiotic Resistance 588
- Prokaryotes in Research and Technology 589

28 Protists   593

CONCEPT 28.1 Most eukaryotes are single-celled organisms 594
- Structural and Functional Diversity in Protists 594
- Endosymbiosis in Eukaryotic Evolution 594
- Four Supergroups of Eukaryotes 597

CONCEPT 28.2 Excavates include protists with modified mitochondria and protists with unique flagella 597
- Diplomonads and Parabasalids 600
- Euglenozoans 600

CONCEPT 28.3 SAR is a highly diverse group of protists defined by DNA similarities 601
- Stramenopiles 602
- Alveolates 604
- Rhizarians 606

CONCEPT 28.4 Red algae and green algae are the closest relatives of plants 609
- Red Algae 609
- Green Algae 610

CONCEPT 28.5 Unikonts include protists that are closely related to fungi and animals 611
- Amoebozoans 612
- Opisthokonts 613

CONCEPT 28.6 Protists play key roles in ecological communities 614
- Symbiotic Protists 614
- Photosynthetic Protists 614

29 Plant Diversity I: How Plants Colonized Land 618

CONCEPT 29.1 Plants evolved from green algae 619
Evidence of Algal Ancestry 619
Adaptations Enabling the Move to Land 619
Derived Traits of Plants 621
The Origin and Diversification of Plants 621

CONCEPT 29.2 Mosses and other nonvascular plants have life cycles dominated by gametophytes 623
Bryophyte Gametophytes 624
Bryophyte Sporophytes 625
The Ecological and Economic Importance of Mosses 627

CONCEPT 29.3 Ferns and other seedless vascular plants were the first plants to grow tall 629
Origins and Traits of Vascular Plants 629
Classification of Seedless Vascular Plants 631
The Significance of Seedless Vascular Plants 633

30 Plant Diversity II: The Evolution of Seed Plants 636

CONCEPT 30.1 Seeds and pollen grains are key adaptations for life on land 637
Advantages of Reduced Gametophytes 637
Heterospory: The Rule Among Seed Plants 638
Ovules and Production of Eggs 638
Pollen and Production of Sperm 638
The Evolutionary Advantage of Seeds 639

CONCEPT 30.2 Gymnosperms bear "naked" seeds, typically on cones 640
The Life Cycle of a Pine 640
Early Seed Plants and the Rise of Gymnosperms 641
Gymnosperm Diversity 641

CONCEPT 30.3 The reproductive adaptations of angiosperms include flowers and fruits 644

Characteristics of Angiosperms 644
Angiosperm Evolution 647
Angiosperm Diversity 649

CONCEPT 30.4 Human welfare depends on seed plants 651
Products from Seed Plants 651
Threats to Plant Diversity 651

31 Fungi 654

CONCEPT 31.1 Fungi are heterotrophs that feed by absorption 655
Nutrition and Ecology 655
Body Structure 655
Specialized Hyphae in Mycorrhizal Fungi 656

CONCEPT 31.2 Fungi produce spores through sexual or asexual life cycles 657
Sexual Reproduction 658
Asexual Reproduction 658

CONCEPT 31.3 The ancestor of fungi was an aquatic, single-celled, flagellated protist 659
The Origin of Fungi 659
The Move to Land 660

CONCEPT 31.4 Fungi have radiated into a diverse set of lineages 660
Cryptomycetes and Microsporidians 661
Zoopagomycetes 662
Mucoromycetes 663
Ascomycetes 663
Basidiomycetes 665

CONCEPT 31.5 Fungi play key roles in nutrient cycling, ecological interactions, and human welfare 667
Fungi as Decomposers 667
Fungi as Mutualists 667
Practical Uses of Fungi 670

32 An Overview of Animal Diversity 673

CONCEPT 32.1 Animals are multicellular, heterotrophic eukaryotes with tissues that develop from embryonic layers 674
Nutritional Mode 674
Cell Structure and Specialization 674
Reproduction and Development 674

CONCEPT 32.2 The history of animals spans more than half a billion years 675
Steps in the Origin of Multicellular Animals 675
Neoproterozoic Era (1 Billion–541 Million Years Ago) 676
Paleozoic Era (541–252 Million Years Ago) 677
Mesozoic Era (252–66 Million Years Ago) 679
Cenozoic Era (66 Million Years Ago to the Present) 679

CONCEPT 32.3 Animals can be characterized by body plans 679
Symmetry 679
Tissues 679
Body Cavities 680
Protostome and Deuterostome Development 681

CONCEPT 32.4 Views of animal phylogeny continue to be shaped by new molecular and morphological data 682
The Diversification of Animals 682
Future Directions in Animal Systematics 684

33 An Introduction to Invertebrates 686

CONCEPT 33.1 Sponges are basal animals that lack tissues 690

CONCEPT 33.2 Cnidarians are an ancient phylum of eumetazoans 691
Medusozoans 692
Anthozoans 693

CONCEPT 33.3 Lophotrochozoans, a clade identified by molecular data, have the widest range of animal body forms 694
Flatworms 694
Rotifers and Acanthocephalans 697
Lophophorates: Ectoprocts and Brachiopods 698
Molluscs 699
Annelids 703

CONCEPT 33.4 Ecdysozoans are the most species-rich animal group 705
Nematodes 705
Arthropods 706

CONCEPT 33.5 Echinoderms and chordates are deuterostomes 713
Echinoderms 713
Chordates 715

34 The Origin and Evolution of Vertebrates 718

CONCEPT 34.1 Chordates have a notochord and a dorsal, hollow nerve cord 719
Derived Characters of Chordates 719
Lancelets 720
Tunicates 721
Early Chordate Evolution 722

CONCEPT 34.2 Vertebrates are chordates that have a backbone 722
Derived Characters of Vertebrates 722
Hagfishes and Lampreys 723
Early Vertebrate Evolution 724

CONCEPT 34.3 Gnathostomes are vertebrates that have jaws 725
Derived Characters of Gnathostomes 725
Fossil Gnathostomes 726
Chondrichthyans (Sharks, Rays, and Their Relatives) 726
Ray-Finned Fishes and Lobe-Fins 728

CONCEPT 34.4 Tetrapods are gnathostomes that have limbs 730
Derived Characters of Tetrapods 730
The Origin of Tetrapods 731
Amphibians 731

CONCEPT 34.5 Amniotes are tetrapods that have a terrestrially adapted egg 734
Derived Characters of Amniotes 734
Early Amniotes 735
Reptiles 735

CONCEPT 34.6 Mammals are amniotes that have hair and produce milk 741
Derived Characters of Mammals 741
Early Evolution of Mammals 741
Monotremes 742
Marsupials 743
Eutherians (Placental Mammals) 744

CONCEPT 34.7 Humans are mammals that have a large brain and bipedal locomotion 748
Derived Characters of Humans 748
The Earliest Hominins 748
Australopiths 749
Bipedalism 750
Tool Use 750
Early *Homo* 750
Neanderthals 752
Homo sapiens 753

Unit 6 Plant Form and Function 757

Interview: Dennis Gonsalves 757

35 Vascular Plant Structure, Growth, and Development 758

CONCEPT 35.1 Plants have a hierarchical organization consisting of organs, tissues, and cells 759
Vascular Plant Organs: Roots, Stems, and Leaves 759
Dermal, Vascular, and Ground Tissues 762
Common Types of Plant Cells 763

CONCEPT 35.2 Different meristems generate new cells for primary and secondary growth 766

CONCEPT 35.3 Primary growth lengthens roots and shoots 768
Primary Growth of Roots 768
Primary Growth of Shoots 769

CONCEPT 35.4 Secondary growth increases the diameter of stems and roots in woody plants 772
The Vascular Cambium and Secondary Vascular Tissue 773
The Cork Cambium and the Production of Periderm 774
Evolution of Secondary Growth 774

CONCEPT 35.5 Growth, morphogenesis, and cell differentiation produce the plant body 775
Model Organisms: Revolutionizing the Study of Plants 776
Growth: Cell Division and Cell Expansion 776
Morphogenesis and Pattern Formation 777
Gene Expression and the Control of Cell Differentiation 778
Shifts in Development: Phase Changes 778
Genetic Control of Flowering 779

36 Resource Acquisition and Transport in Vascular Plants 784

CONCEPT 36.1 Adaptations for acquiring resources were key steps in the evolution of vascular plants 785
Shoot Architecture and Light Capture 785
Root Architecture and Acquisition of Water and Minerals 787

CONCEPT 36.2 Different mechanisms transport substances over short or long distances 787
The Apoplast and Symplast: Transport Continuums 787
Short-Distance Transport of Solutes Across Plasma Membranes 788
Short-Distance Transport of Water Across Plasma Membranes 788
Long-Distance Transport:The Role of Bulk Flow 791

CONCEPT 36.3 Transpiration drives the transport of water and minerals from roots to shoots via the xylem 792
Absorption of Water and Minerals by Root Cells 792
Transport of Water and Minerals into the Xylem 792
Bulk Flow Transport via the Xylem 792
Xylem Sap Ascent by Bulk Flow: *A Review* 796

CONCEPT 36.4 The rate of transpiration is regulated by stomata 796
Stomata: Major Pathways for Water Loss 796
Mechanisms of Stomatal Opening and Closing 797
Stimuli for Stomatal Opening and Closing 798
Effects of Transpiration on Wilting and Leaf Temperature 798
Adaptations That Reduce Evaporative Water Loss 798

CONCEPT 36.5 Sugars are transported from sources to sinks via the phloem 799
Movement from Sugar Sources to Sugar Sinks 799
Bulk Flow by Positive Pressure: The Mechanism of Translocation in Angiosperms 800

CONCEPT 36.6 The symplast is highly dynamic 801
Changes in Plasmodesmatal Number and Pore Size 802
Phloem: An Information Superhighway 802
Electrical Signaling in the Phloem 802

37 Soil and Plant Nutrition 805

CONCEPT 37.1 Soil contains a living, complex ecosystem 806
Soil Texture 806
Topsoil Composition 806
Soil Conservation and Sustainable Agriculture 807

CONCEPT 37.2 Plant roots absorb many types of essential elements from the soil 809
Essential Elements 810
Symptoms of Mineral Deficiency 810
Global Climate Change and Food Quality 812

CONCEPT 37.3 Plant nutrition often involves relationships with other organisms 812
Bacteria and Plant Nutrition 814
Fungi and Plant Nutrition 817
Epiphytes, Parasitic Plants, and Carnivorous Plants 818

38 Angiosperm Reproduction and Biotechnology 822

CONCEPT 38.1 Flowers, double fertilization, and fruits are key features of the angiosperm life cycle 823
Flower Structure and Function 823
Methods of Pollination 825
The Angiosperm Life Cycle: An Overview 826
Development of Female Gametophytes (Embryo Sacs) 826
Development of Male Gametophytes in Pollen Grains 826
Seed Development and Structure 828
Sporophyte Development from Seed to Mature Plant 829
Fruit Structure and Function 830

CONCEPT 38.2 Flowering plants reproduce sexually, asexually, or both 833
Mechanisms of Asexual Reproduction 833
Advantages and Disadvantages of Asexual and Sexual Reproduction 833
Mechanisms That Prevent Self-Fertilization 834
Totipotency, Vegetative Reproduction, and Tissue Culture 835

CONCEPT 38.3 People modify crops by breeding and genetic engineering 836
Plant Breeding 837
Plant Biotechnology and Genetic Engineering 837
The Debate over Plant Biotechnology 839

39 Plant Responses to Internal and External Signals 842

CONCEPT 39.1 Signal transduction pathways link signal reception to response 843
Reception 844
Transduction 844
Response 845

CONCEPT CHECK 39.2 Plants use chemicals to communicate 845
General Characteristics of Plant Hormones 846
A Survey of Plant Hormones 847

CONCEPT 39.3 Responses to light are critical for plant success 855
Blue-Light Photoreceptors 855
Phytochrome Photoreceptors 856
Biological Clocks and Circadian Rhythms 857
The Effect of Light on the Biological Clock 858
Photoperiodism and Responses to Seasons 859

CONCEPT 39.4 Plants respond to a wide variety of stimuli other than light 861
Gravity 861
Mechanical Stimuli 861
Environmental Stresses 862

CONCEPT 39.5 Plants respond to attacks by pathogens and herbivores 866
Defenses Against Pathogens 866
Defenses Against Herbivores 867

Unit 7   Animal Form and Function   872

Interview: Steffanie Strathdee 872

40 Basic Principles of Animal Form and Function 873

CONCEPT 40.1 Animal form and function are correlated at all levels of organization 874
Evolution of Animal Size and Shape 874
Exchange with the Environment 874
Hierarchical Organization of Body Plans 876
Coordination and Control 880

CONCEPT 40.2 Feedback control maintains the internal environment in many animals 881
Regulating and Conforming 881
Homeostasis 881

CONCEPT 40.3 Homeostatic processes for thermoregulation involve form, function, and behavior 884
Endothermy and Ectothermy 884
Variation in Body Temperature 884
Balancing Heat Loss and Gain 885
Acclimatization in Thermoregulation 888
Physiological Thermostats and Fever 888

CONCEPT 40.4 Energy requirements are related to animal size, activity, and environment 889
Energy Allocation and Use 889
Quantifying Energy Use 890
Minimum Metabolic Rate and Thermoregulation 890
Influences on Metabolic Rate 891
Torpor and Energy Conservation 892

41 Animal Nutrition 898

CONCEPT 41.1 An animal's diet must supply chemical energy, organic building blocks, and essential nutrients 899
Essential Nutrients 899
Variation in Diet 901
Dietary Deficiencies 901
Assessing Nutritional Needs 902

CONCEPT 41.2 Food processing involves ingestion, digestion, absorption, and elimination 902
Digestive Compartments 904

CONCEPT 41.3 Organs specialized for sequential stages of food processing form the mammalian digestive system 905
The Oral Cavity, Pharynx, and Esophagus 905
Digestion in the Stomach 907
Digestion in the Small Intestine 908
Absorption in the Small Intestine 909
Processing in the Large Intestine 910

CONCEPT 41.4 Evolutionary adaptations of vertebrate digestive systems correlate with diet 911
Dental Adaptations 911
Stomach and Intestinal Adaptations 912
Mutualistic Adaptations 912

CONCEPT 41.5 Feedback circuits regulate digestion, energy storage, and appetite 915
Regulation of Digestion 915
Regulation of Energy Storage 915
Regulation of Appetite and Consumption 917

42 Circulation and Gas Exchange 921

CONCEPT 42.1 Circulatory systems link exchange surfaces with cells throughout the body 922
Gastrovascular Cavities 922
Open and Closed Circulatory Systems 923
Organization of Vertebrate Circulatory Systems 924

CONCEPT 42.2 Coordinated cycles of heart contraction drive double circulation in mammals 926
Mammalian Circulation 926
The Mammalian Heart: *A Closer Look* 926
Maintaining the Heart's Rhythmic Beat 928

CONCEPT 42.3 Patterns of blood pressure and flow reflect the structure and arrangement of blood vessels 929
Blood Vessel Structure and Function 929
Blood Flow Velocity 930
Blood Pressure 930
Capillary Function 932
Fluid Return by the Lymphatic System 933

CONCEPT 42.4 Blood components function in exchange, transport, and defense 934
Blood Composition and Function 934
Cardiovascular Disease 937

CONCEPT 42.5 Gas exchange occurs across specialized respiratory surfaces 939
Partial Pressure Gradients in Gas Exchange 939
Respiratory Media 939
Respiratory Surfaces 940
Gills in Aquatic Animals 940
Tracheal Systems in Insects 941
Lungs 942

CONCEPT 42.6 Breathing ventilates the lungs 944
How an Amphibian Breathes 944
How a Bird Breathes 944
How a Mammal Breathes 945
Control of Breathing in Humans 946

CONCEPT 42.7 Adaptations for gas exchange include pigments that bind and transport gases 947
Coordination of Circulation and Gas Exchange 947
Respiratory Pigments 947
Respiratory Adaptations of Diving Mammals 949

43 The Immune System 952

CONCEPT 43.1 In innate immunity, recognition and response rely on traits common to groups of pathogens 953
Innate Immunity of Invertebrates 953
Innate Immunity of Vertebrates 954
Evasion of Innate Immunity by Pathogens 957

CONCEPT 43.2 In adaptive immunity, receptors provide pathogen-specific recognition 957
Antigens as the Trigger for Adaptive Immunity 958
Antigen Recognition by B Cells and Antibodies 958
Antigen Recognition by T Cells 959
B Cell and T Cell Development 960

CONCEPT 43.3 Adaptive immunity defends against infection of body fluids and body cells 963
Helper T Cells: Activating Adaptive Immunity 963
B Cells and Antibodies: A Response to Extracellular Pathogens 964
Cytotoxic T Cells: A Response to Infected Host Cells 966
Summary of the Humoral and Cell-Mediated Immune Responses 967
Immunization 968
Active and Passive Immunity 968
Antibodies as Tools 969
Immune Rejection 969

CONCEPT 43.4 Disruptions in immune system function can elicit or exacerbate disease 970
Exaggerated, Self-Directed, and Diminished Immune Responses 970
Evolutionary Adaptations of Pathogens That Underlie Immune System Avoidance 971
Cancer and Immunity 974

44 Osmoregulation and Excretion 977

CONCEPT 44.1 Osmoregulation balances the uptake and loss of water and solutes 978
Osmosis and Osmolarity 978
Osmoregulatory Challenges and Mechanisms 978
Energetics of Osmoregulation 980
Transport Epithelia in Osmoregulation 981

CONCEPT 44.2 An animal's nitrogenous wastes reflect its phylogeny and habitat 982
Forms of Nitrogenous Waste 982
The Influence of Evolution and Environment on Nitrogenous Wastes 983

CONCEPT 44.3 Diverse excretory systems are variations on a tubular theme 983
Survey of Excretory Systems 984

CONCEPT 44.4 The nephron is organized for stepwise processing of blood filtrate 987
From Blood Filtrate to Urine: *A Closer Look* 987
Solute Gradients and Water Conservation 989
Adaptations of the Vertebrate Kidney to Diverse Environments 991

CONCEPT 44.5 Hormonal circuits link kidney function, water balance, and blood pressure 994
Homeostatic Regulation of the Kidney 994

45 Hormones and the Endocrine System 999

CONCEPT 45.1 Hormones and other signaling molecules bind to target receptors, triggering specific response pathways 1000
Intercellular Information Flow 1000
Chemical Classes of Hormones 1001
Cellular Hormone Response Pathways 1002
Endocrine Tissues and Organs 1003

CONCEPT 45.2 Feedback regulation and coordination with the nervous system are common in hormone pathways 1004
Simple Endocrine Pathways 1004
Simple Neuroendocrine Pathways 1005
Feedback Regulation 1005
Coordination of the Endocrine and Nervous Systems 1006
Thyroid Regulation: A Hormone Cascade Pathway 1008
Hormonal Regulation of Growth 1009

CONCEPT 45.3 Endocrine glands respond to diverse stimuli in regulating homeostasis, development, and behavior 1011
Parathyroid Hormone and Vitamin D: Control of Blood Calcium 1011
Adrenal Hormones: Response to Stress 1012
Sex Hormones 1014
Hormones and Biological Rhythms 1015
Evolution of Hormone Function 1015

46 Animal Reproduction 1019

CONCEPT 46.1 Both asexual and sexual reproduction occur in the animal kingdom 1020
Mechanisms of Asexual Reproduction 1020
Variation in Patterns of Sexual Reproduction 1020
Reproductive Cycles 1021
Sexual Reproduction: An Evolutionary Enigma 1021

CONCEPT 46.2 Fertilization depends on mechanisms that bring together sperm and eggs of the same species 1022
Ensuring the Survival of Offspring 1023
Gamete Production and Delivery 1023

CONCEPT 46.3 Reproductive organs produce and transport gametes 1025
Human Male Reproductive Anatomy 1025
Human Female Reproductive Anatomy 1026
Gametogenesis 1027

CONCEPT 46.4 The interplay of tropic and sex hormones regulates reproduction in mammals 1030
Biological Sex, Gender Identity, and Sexual Orientation in Human Sexuality 1031
Hormonal Control of the Male Reproductive System 1031
Hormonal Control of Female Reproductive Cycles 1032
Human Sexual Response 1034

CONCEPT 46.5 In placental mammals, an embryo develops fully within the mother's uterus 1034
Conception, Embryonic Development, and Birth 1034
Maternal Immune Tolerance of the Embryo and Fetus 1037
Contraception and Abortion 1037
Modern Reproductive Technologies 1039

47 Animal Development 1043

CONCEPT 47.1 Fertilization and cleavage initiate embryonic development 1044
Fertilization 1044
Cleavage 1046

CONCEPT 47.2 Morphogenesis in animals involves specific changes in cell shape, position, and survival 1049
Gastrulation 1049
Developmental Adaptations of Amniotes 1053
Organogenesis 1054
The Cytoskeleton in Morphogenesis 1056

CONCEPT 47.3 Cytoplasmic determinants and inductive signals regulate cell fate 1057
Fate Mapping 1058
Axis Formation 1059
Restricting Developmental Potential 1060
Cell Fate Determination and Pattern Formation by Inductive Signals 1061
Cilia and Cell Fate 1064

48 Neurons, Synapses, and Signaling 1067

CONCEPT 48.1 Neuron structure and organization reflect function in information transfer 1068
Neuron Structure and Function 1068
Introduction to Information Processing 1068

CONCEPT 48.2 Ion pumps and ion channels establish the resting potential of a neuron 1069
Formation of the Resting Potential 1070
Modeling the Resting Potential 1071

CONCEPT 48.3 Action potentials are the signals conducted by axons 1072
Hyperpolarization and Depolarization 1072
Graded Potentials and Action Potentials 1073
Generation of Action Potentials: *A Closer Look* 1073
Conduction of Action Potentials 1075

CONCEPT 48.4 Neurons communicate with other cells at synapses 1077
Generation of Postsynaptic Potentials 1078
Summation of Postsynaptic Potentials 1079
Termination of Neurotransmitter Signaling 1079
Modulated Signaling at Synapses 1080
Neurotransmitters 1080

49 Nervous Systems 1085

CONCEPT 49.1 Nervous systems consist of circuits of neurons and supporting cells 1086
Organization of the Vertebrate Nervous System 1087
The Peripheral Nervous System 1088
Glia 1090

CONCEPT 49.2 The vertebrate brain is regionally specialized 1091
Arousal and Sleep 1094
Biological Clock Regulation 1094
Emotions 1095
Functional Imaging of the Brain 1096

CONCEPT 49.3 The cerebral cortex controls voluntary movement and cognitive functions 1096
Information Processing 1097
Language and Speech 1098
Lateralization of Cortical Function 1098
Frontal Lobe Function 1098
Evolution of Cognition in Vertebrates 1098

CONCEPT 49.4 Changes in synaptic connections underlie memory and learning 1099
Neuronal Plasticity 1100
Memory and Learning 1100
Long-Term Potentiation 1101

CONCEPT 49.5 Many nervous system disorders can now be explained in molecular terms 1102
Schizophrenia 1102
Depression 1102
The Brain's Reward System and Drug Addiction 1103
Alzheimer's Disease 1103
Parkinson's Disease 1104
Future Directions in Brain Research 1104

50 Sensory and Motor Mechanisms 1107

CONCEPT 50.1 Sensory receptors transduce stimulus energy and transmit signals to the central nervous system 1108
- Sensory Reception and Transduction 1108
- Transmission 1109
- Perception 1109
- Amplification and Adaptation 1109
- Types of Sensory Receptors 1110

CONCEPT 50.2 In hearing and equilibrium, mechanoreceptors detect moving fluid or settling particles 1112
- Sensing of Gravity and Sound in Invertebrates 1112
- Hearing and Equilibrium in Mammals 1112
- Hearing and Equilibrium in Other Vertebrates 1116

CONCEPT 50.3 The diverse visual receptors of animals depend on light-absorbing pigments 1117
- Evolution of Visual Perception 1117
- The Vertebrate Visual System 1119

CONCEPT 50.4 The senses of taste and smell rely on similar sets of sensory receptors 1123
- Taste in Mammals 1123
- Smell in Humans 1124

CONCEPT 50.5 The physical interaction of protein filaments is required for muscle function 1125
- Vertebrate Skeletal Muscle 1126
- Other Types of Muscle 1131

CONCEPT 50.6 Skeletal systems transform muscle contraction into locomotion 1132
- Types of Skeletal Systems 1132
- Types of Locomotion 1135

51 Animal Behavior 1139

CONCEPT 51.1 Discrete sensory inputs can stimulate both simple and complex behaviors 1140
- Fixed Action Patterns 1140
- Migration 1140
- Behavioral Rhythms 1141
- Animal Signals and Communication 1141

CONCEPT 51.2 Learning establishes specific links between experience and behavior 1143
- Experience and Behavior 1143
- Learning 1144

CONCEPT 51.3 Selection for individual survival and reproductive success can explain diverse behaviors 1148
- Evolution of Foraging Behavior 1148
- Mating Behavior and Mate Choice 1149

CONCEPT 51.4 Genetic analyses and the concept of inclusive fitness provide a basis for studying the evolution of behavior 1154
- Genetic Basis of Behavior 1155
- Genetic Variation and the Evolution of Behavior 1155
- Altruism 1156
- Inclusive Fitness 1157
- Evolution and Human Culture 1159

Unit 8   Ecology   1163

Interview: Chelsea Rochman 1163

52 An Introduction to Ecology and the Biosphere 1164

CONCEPT 52.1 Earth's climate varies by latitude and season and is changing rapidly 1167
- Global Climate Patterns 1167
- Regional and Local Effects on Climate 1167
- Effects of Vegetation on Climate 1169
- Microclimate 1169
- Global Climate Change 1170

CONCEPT 52.2 The distribution of terrestrial biomes is controlled by climate and disturbance 1171
- Climate and Terrestrial Biomes 1171
- General Features of Terrestrial Biomes 1172
- Disturbance and Terrestrial Biomes 1172

CONCEPT 52.3 Aquatic biomes are diverse and dynamic systems that cover most of Earth 1177
- Zonation in Aquatic Biomes 1177

CONCEPT 52.4 Interactions between organisms and the environment limit the distribution of species 1178
- Dispersal and Distribution 1183
- Biotic Factors 1184
- Abiotic Factors 1184

CONCEPT 52.5 Ecological change and evolution affect one another over long and short periods of time 1187

53 Population Ecology 1190

CONCEPT 53.1 Biotic and abiotic factors affect population density, dispersion, and demographics 1191
- Density and Dispersion 1191
- Demographics 1193

CONCEPT 53.2 The exponential model describes population growth in an idealized, unlimited environment 1196
- Changes in Population Size 1196
- Exponential Growth 1196

CONCEPT 53.3 The logistic model describes how a population grows more slowly as it nears its carrying capacity 1197
- The Logistic Growth Model 1198
- The Logistic Model and Real Populations 1199

CONCEPT 53.4 Life history traits are products of natural selection 1200
- Diversity of Life Histories 1200
- "Trade-offs" and Life Histories 1201

CONCEPT 53.5 Density-dependent factors regulate population growth 1202
- Population Change and Population Density 1202
- Mechanisms of Density-Dependent Population Regulation 1203
- Population Dynamics 1205

CONCEPT 53.6 The human population is no longer growing exponentially but is still increasing extremely rapidly 1207
  The Global Human Population 1207
  Global Carrying Capacity 1209

54 Community Ecology 1214

CONCEPT 54.1 Interactions between species can help, harm, or have no effect on the individuals involved 1215
  Competition 1215
  Exploitation 1217
  Positive Interactions 1220

CONCEPT 54.2 Diversity and trophic structure characterize biological communities 1222
  Species Diversity 1222
  Diversity and Community Stability 1223
  Trophic Structure 1223
  Species with a Large Impact 1225
  Bottom-Up and Top-Down Controls 1226

CONCEPT 54.3 Disturbance influences species diversity and composition 1228
  Characterizing Disturbance 1228
  Ecological Succession 1229
  Human Disturbance 1231

CONCEPT 54.4 Biogeographic factors affect community diversity 1231
  Latitudinal Gradients 1232
  Area Effects 1232
  Island Equilibrium Model 1232

CONCEPT 54.5 Pathogens alter community structure locally and globally 1234
  Effects on Community Structure 1234
  Community Ecology and Zoonotic Diseases 1234

55 Ecosystems and Restoration Ecology 1238

CONCEPT 55.1 Physical laws govern energy flow and chemical cycling in ecosystems 1239
  Energy Flow and Chemical Cycling 1239
  Conservation of Energy 1239
  Conservation of Mass 1239
  Energy, Mass, and Trophic Levels 1240

CONCEPT 55.2 Energy and other limiting factors control primary production in ecosystems 1241
  Ecosystem Energy Budgets 1241
  Primary Production in Aquatic Ecosystems 1242
  Primary Production in Terrestrial Ecosystems 1243

CONCEPT 55.3 Energy transfer between trophic levels is typically only 10% efficient 1246
  Production Efficiency 1246
  Trophic Efficiency and Ecological Pyramids 1246

CONCEPT 55.4 Biological and geochemical processes cycle nutrients and water in ecosystems 1248
  Decomposition and Nutrient Cycling Rates 1248
  Biogeochemical Cycles 1249
  Case Study: Nutrient Cycling in the Hubbard Brook Experimental Forest 1252

CONCEPT 55.5 Restoration ecologists return degraded ecosystems to a more natural state 1253
  Bioremediation 1253
  Biological Augmentation 1255
  Ecosystems: *A Review* 1255

56 Conservation Biology and Global Change 1260

CONCEPT 56.1 Human activities threaten earth's biodiversity 1261
  Three Levels of Biodiversity 1261
  Biodiversity and Human Welfare 1262
  Threats to Biodiversity 1263
  Can Extinct Species Be Resurrected? 1266

CONCEPT 56.2 Population conservation focuses on population size, genetic diversity, and critical habitat 1266
  Extinction Risks in Small Populations 1266
  Critical Habitat 1269
  Weighing Conflicting Demands 1270

CONCEPT 56.3 Landscape and regional conservation help sustain biodiversity 1270
  Landscape Structure and Biodiversity 1270
  Establishing Protected Areas 1272
  Urban Ecology 1273

CONCEPT 56.4 Earth is changing rapidly as a result of human actions 1274
  Nutrient Enrichment 1274
  Toxins in the Environment 1275
  Greenhouse Gases and Climate Change 1278
  Depletion of Atmospheric Ozone 1283

CONCEPT 56.5 Sustainable development can improve human lives while conserving biodiversity 1284
  Sustainable Development 1284
  The Future of the Biosphere 1285

APPENDIX A Answers A-1

APPENDIX B Classification of Life B-1

APPENDIX C A Comparison of the Light Microscope and the Electron Microscope C-1

APPENDIX D Scientific Skills Review D-1

CREDITS   CR-1

GLOSSARY   G-1

INDEX   I-1

1 Evolution, the Themes of Biology, and Scientific Inquiry

KEY CONCEPTS

1.1 The study of life reveals unifying themes *p. 3*

1.2 The Core Theme: Evolution accounts for the unity and diversity of life *p. 11*

1.3 In studying nature, scientists form and test hypotheses *p. 16*

1.4 Science benefits from a cooperative approach and diverse viewpoints *p. 22*

Study Tip

Make a table: List the five unifying themes of biology across the top. Enter at least three examples of each theme as you read this chapter. One example is filled in for you. To help you focus on these big ideas, continue adding examples throughout your study of biology.

Evolution	Organization			
Beach mouse's coat color matches its sandy habitat.				

→ Go to Mastering Biology

For Students (in eText and Study Area)
- Get Ready for Chapter 1
- Figure 1.8 Walkthrough: Gene Expression: Cells Use Information Encoded in a Gene to Synthesize a Functional Protein
- Video: Galápagos Biodiversity by Peter and Rosemary Grant

For Instructors to Assign (in Item Library)
- Scientific Skills Exercise: Interpreting a Pair of Bar Graphs
- Tutorial: The Scientific Method

Figure 1.1 The light, dappled color of this beach mouse (*Peromyscus polionotus*) allows it to blend into its habitat—brilliant white sand dunes dotted with sparse clumps of beach grass along the Florida seashore. Mice of the same species that inhabit nearby inland areas are much darker, blending with the soil and vegetation where they live.

How do these mice illustrate the unifying themes of biology?

Beach mouse

As a result of **evolution** through natural selection over long periods of time, the fur colors of these two populations of mice resemble their surroundings, providing protection from predators.

Inland mouse

Structure fits function at all levels of a mouse's **organization**.

Organ (heart)

Tissue

Cell

Molecule

Genetic **information** encoded in DNA determines a mouse's fur colors.

Gene for brown fur

Gene for white fur

Energy flows one way from the sun to plants to a mouse; **matter** cycles between a mouse and its environment.

Flow of energy

Cycling of matter

A plant being eaten by a mouse and a mouse being preyed upon by a hawk are **interactions** within a system.

The study of life reveals unifying themes

At the most fundamental level, we may ask: What is life? Even a child realizes that a dog or a plant is alive, while a rock or a car is not. Yet the phenomenon we call life defies a simple definition. We recognize life by what living things do. **Figure 1.2** highlights some of the properties and processes we associate with life.

Biology, the scientific study of life, is a subject of enormous scope, and exciting new biological discoveries are being made every day. How can you organize into a comprehensible framework all the information you'll encounter as you study biology? Focusing on a few big ideas will help. Here are five unifying themes—ways of thinking about life that will still be useful decades from now.

- Organization
- Information
- Energy and Matter
- Interactions
- Evolution

In this section and the next, we'll briefly explore each theme.

▼ **Figure 1.2 Some properties of life.**

▼ **Order.** This close-up of a sunflower illustrates the highly ordered structure that characterizes life.

▲ **Evolutionary adaptation.** The overall appearance of this pygmy sea horse camouflages the animal in its environment. Such adaptations evolve over countless generations by the reproductive success of those individuals with heritable traits that are best suited to their environments.

▲ **Regulation.** The regulation of blood flow through the blood vessels of this jackrabbit's ears helps maintain a constant body temperature by adjusting heat exchange with the surrounding air.

▲ **Energy processing.** This butterfly obtains fuel in the form of nectar from flowers. The butterfly will use chemical energy stored in its food to power flight and other work.

⊝ **Mastering Biology**
Animation: Signs of Life
Video: Sea Horse Camouflage

▲ **Growth and development.** Inherited information carried by genes controls the pattern of growth and development of organisms, such as this oak seedling.

▲ **Response to the environment.** The Venus flytrap on the left closed its trap rapidly in response to the environmental stimulus of a grasshopper landing on the open trap.

▼ **Reproduction.** Organisms (living things) reproduce their own kind.

Theme: New Properties Emerge at Successive Levels of Biological Organization

ORGANIZATION The study of life on Earth extends from the microscopic scale of the molecules and cells that make up organisms to the global scale of the entire living planet. As biologists, we can divide this enormous range into different levels of biological organization. In **Figure 1.3**, we zoom in from space to take a closer and closer look at life in a mountain meadow. This journey, depicted as a series of numbered steps, highlights the hierarchy of biological organization.

Zooming in at ever-finer resolution illustrates the principle that underlies *reductionism*, an approach that reduces complex systems to simpler components that are more manageable to study. Reductionism is a powerful strategy in biology. For example, by studying the molecular structure of DNA that had been extracted from cells, James Watson and Francis Crick inferred the chemical basis of biological inheritance. Despite its importance, reductionism provides an incomplete view of life on Earth, as you'll see next.

Emergent Properties

Let's reexamine Figure 1.3, beginning this time at the molecular level and then zooming out. This approach allows us to see novel properties emerge at each level that are absent

---

▼ **Figure 1.3** **Exploring Levels of Biological Organization**

◀ **1 The Biosphere**

Even from space, we can see signs of Earth's life—in the mosaic of greens indicating forests, for example. We can also see the **biosphere**, which consists of all life on Earth and all the places where life exists: most regions of land, most bodies of water, the atmosphere to an altitude of several kilometers, and even sediments far below the ocean floor.

◀ **2 Ecosystems**

Our first scale change brings us to a North American mountain meadow, which is an example of an ecosystem, as are a tropical forest, grassland, desert, and coral reef. An **ecosystem** consists of all the living things in a particular area, along with all the nonliving components of the environment with which life interacts, such as soil, water, atmospheric gases, and light.

▶ **3 Communities**

The array of organisms inhabiting a particular ecosystem is called a biological **community**. The community in our meadow ecosystem includes many kinds of plants, various animals, mushrooms and other fungi, and enormous numbers of diverse microorganisms, such as bacteria, that are too small to see without a microscope. Each of these forms of life belongs to a *species*—a group whose members can only reproduce with other members of the group.

▶ **4 Populations**

A **population** consists of all the individuals of a species living within the bounds of a specified area that interbreed with each other. For example, our meadow includes a population of lupines (some of which are shown here) and a population of mule deer. A community is therefore the set of populations that inhabit a particular area.

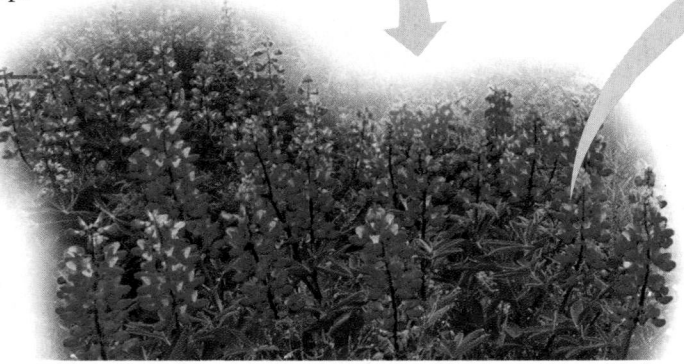

▲ **5 Organisms**

Individual living things are called **organisms**. Each plant in the meadow is an organism, and so is each animal, fungus, and bacterium.

from the preceding one. These **emergent properties** are due to the arrangement and interactions of parts as complexity increases. For example, although photosynthesis occurs in an intact chloroplast, it will not take place if chlorophyll and other chloroplast molecules are simply mixed in a test tube. The coordinated processes of photosynthesis require a specific organization of these molecules in the chloroplast. Isolated components of living systems—the objects of study in a reductionist approach—lack a number of significant properties that emerge at higher levels of organization.

Emergent properties are not unique to life. A box of bicycle parts won't transport you anywhere, but if they are arranged in a certain way, you can pedal to your chosen destination. Compared with such nonliving examples, however, biological systems are far more complex, making the emergent properties of life especially challenging to study.

To fully explore emergent properties, biologists today complement reductionism with **systems biology**, the exploration of a biological system by analyzing the interactions among its parts. In this context, a single leaf cell can be considered a system, as can a frog, an ant colony, or a desert ecosystem. By examining and modeling the dynamic behavior of an integrated network of components, systems biology enables us to pose new kinds of questions. For example, how

▼ 6 Organs

The structural hierarchy of life continues to unfold as we explore the architecture of a complex organism. This lupine leaf (consisting of six leaflets) is an example of an **organ**, a body part that is made up of multiple tissues and has specific functions in the body. Leaves, stems, and roots are the major organs of plants. Within an organ, each tissue has a distinct arrangement and contributes particular properties to organ function.

▼ 7 Tissues

Viewing the tissues of a leaf requires a microscope. Each **tissue** is a group of cells that work together, performing a specialized function. The leaf shown here has been cut on an angle. The honeycombed tissue in the interior of

the leaf (left side of photo) is the main location of photosynthesis, the process that converts light energy to the chemical energy of sugar. The jigsaw puzzle–like "skin" on the surface of the leaf (right side of photo) is a tissue called epidermis. The pores through the epidermis allow entry of the gas CO_2, a raw material for sugar production.

50 μm

▶ 8 Cells

Cell   10 μm

The **cell** is life's fundamental unit of structure and function. Some organisms consist of a single cell, which performs all the functions of life. Other organisms are multicellular and feature a division of labor among specialized cells. Here we see a magnified view of a cell in a leaf tissue. This cell is about 40 micrometers (μm) across—about 500 of them would reach across a small coin. Within these tiny cells are even smaller green structures called chloroplasts, which are responsible for photosynthesis.

▼ 9 Organelles

Chloroplasts are examples of **organelles**, the various functional components present in cells. The image below, taken by a powerful microscope, shows a single chloroplast.

Chloroplast   1 μm

▼ 10 Molecules

Our last scale change drops us into a chloroplast for a view of life at the molecular level. A **molecule** is a chemical structure consisting of two or more units called atoms, represented as balls in this computer graphic of a chlorophyll molecule. Chlorophyll is

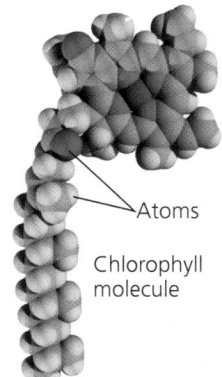

the pigment that makes a leaf green, and it absorbs sunlight during photosynthesis. Within each chloroplast, millions of chlorophyll molecules are organized into systems that convert light energy to the chemical energy of food.

Atoms

Chlorophyll molecule

do networks of molecular interactions in our bodies generate our 24-hour cycle of wakefulness and sleep? At a larger scale, how does a gradual increase in atmospheric carbon dioxide alter ecosystems and the entire biosphere? Systems biology can be used to study life at all levels.

Structure and Function

At each level of the biological hierarchy, we find a correlation between structure and function. Consider the leaf in Figure 1.3: Its broad, flat shape maximizes the capture of sunlight by chloroplasts. Because such correlations of structure and function are common in all living things, analyzing a biological structure gives us clues about what it does and how it works. For example, the hummingbird's anatomy allows its wings to rotate at the shoulder, so hummingbirds have the ability, unique among birds, to fly backward

or hover in place. While hovering, the birds can extend their long, slender beaks into flowers and feed on nectar. The elegant match of form and function in the structures of life is explained by natural selection, which we'll explore shortly.

The Cell: An Organism's Basic Unit of Structure and Function

The cell is the smallest unit of organization that can perform all activities required for life. The so-called Cell Theory was first developed in the 1800s, based on the observations of many scientists. The theory states that all living organisms are made of cells, which are the basic unit of life. In fact, the actions of organisms are all based on the activities of cells. For instance, the movement of your eyes as you read this sentence results from the activities of muscle and nerve cells. Even a process that occurs on a global scale, such as the recycling of carbon atoms, is the product of cellular functions, including the photosynthetic activity of chloroplasts in leaf cells.

All cells share certain characteristics. For instance, every cell is enclosed by a membrane that regulates the passage of materials between the cell and its surroundings. Nevertheless, we distinguish two main forms of cells: prokaryotic and eukaryotic. Prokaryotic cells are found in two groups of single-celled microorganisms, bacteria (singular, *bacterium*) and archaea (singular, *archaean*). All other forms of life, including plants and animals, are composed of eukaryotic cells.

A **eukaryotic cell** contains membrane-enclosed organelles **(Figure 1.4)**. Some organelles, such as the DNA-containing nucleus, are found in the cells of all eukaryotes; other organelles are specific to particular cell types. For example, the chloroplast in Figure 1.3 is an organelle found only in eukaryotic cells that carry out photosynthesis. In contrast to eukaryotic cells, a **prokaryotic cell** lacks a nucleus or other membrane-enclosed organelles. Furthermore, prokaryotic cells are generally smaller than eukaryotic cells, as shown in Figure 1.4.

Theme: Life's Processes Involve the Expression and Transmission of Genetic Information

INFORMATION Within cells, structures called chromosomes contain genetic material in the form of **DNA (deoxyribonucleic acid)**. In cells that are preparing to

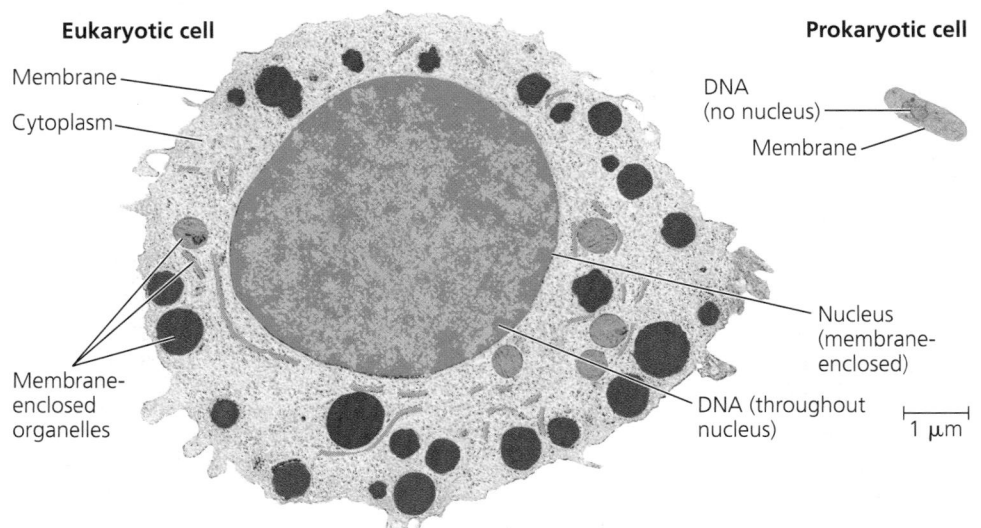

Eukaryotic cell

Membrane

Cytoplasm

Membrane-enclosed organelles

Prokaryotic cell

DNA (no nucleus)

Membrane

Nucleus (membrane-enclosed)

DNA (throughout nucleus)

1 μm

◄ **Figure 1.4 Contrasting eukaryotic and prokaryotic cells in size and complexity.** The cells are shown to scale here; to see a larger magnification of a prokaryotic cell, see Figure 6.5.

VISUAL SKILLS *Measure the scale bar, the length of the prokaryotic cell, and the diameter of the eukaryotic cell. Knowing that this scale bar represents 1 μm, calculate the length of the prokaryotic cell and the diameter of the eukaryotic cell in μm.*

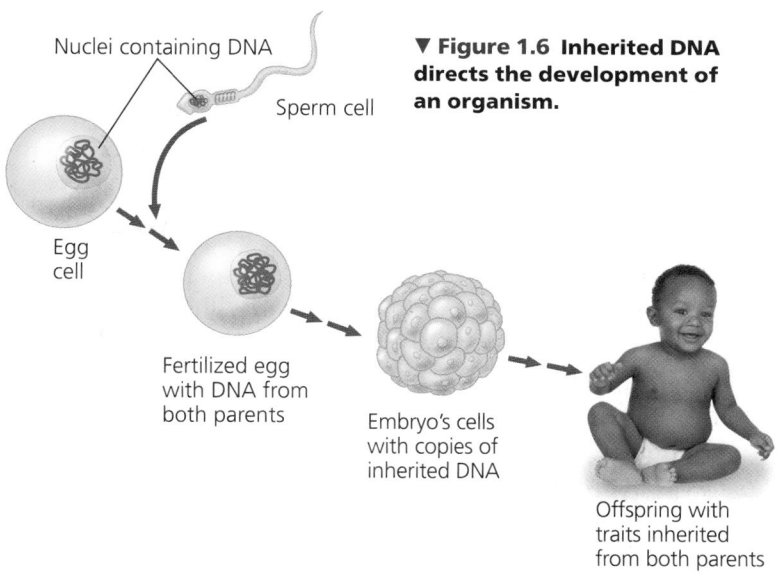

Nuclei containing DNA

Sperm cell

Egg cell

▼ Figure 1.6 Inherited DNA directs the development of an organism.

Fertilized egg with DNA from both parents

Embryo's cells with copies of inherited DNA

Offspring with traits inherited from both parents

divide, the chromosomes may be made visible using a dye that appears blue when bound to the DNA **(Figure 1.5)**.

DNA, the Genetic Material

Each chromosome contains one very long DNA molecule with hundreds or thousands of **genes**, each a section of the DNA of the chromosome. Transmitted from parents to offspring, genes are the units of inheritance. They encode the information necessary to build all of the molecules synthesized within a cell, which in turn establish that cell's identity and function. You began as a single cell stocked with DNA inherited from your parents. The replication of that DNA prior to each cell division transmitted copies of the DNA to what eventually became the trillions of cells of your body. As the cells grew and divided, the genetic information encoded by the DNA directed your development **(Figure 1.6)**.

The molecular structure of DNA accounts for its ability to store information. A DNA molecule is made up of two long chains, called strands, arranged in a double helix. Each chain is made up of four kinds of chemical building blocks called nucleotides, abbreviated A, T, C, and G **(Figure 1.7)**. Specific sequences of these four nucleotides encode the information in genes. The way DNA encodes information is analogous to how we arrange the letters of the alphabet into words and phrases with specific meanings. The word *rat*, for example, evokes a rodent; the words *tar* and *art*, which contain the same letters, mean very different things. We can think of nucleotides as a four-letter alphabet.

For many genes, the sequence provides the blueprint for making a protein. For instance, a given bacterial gene may specify a particular protein (such as an enzyme) required to break down a certain sugar molecule, while one particular human gene may denote an enzyme, and another gene a different protein (an antibody, perhaps) that helps fight off infection. Overall, proteins are major players in building and maintaining the cell and carrying out its activities.

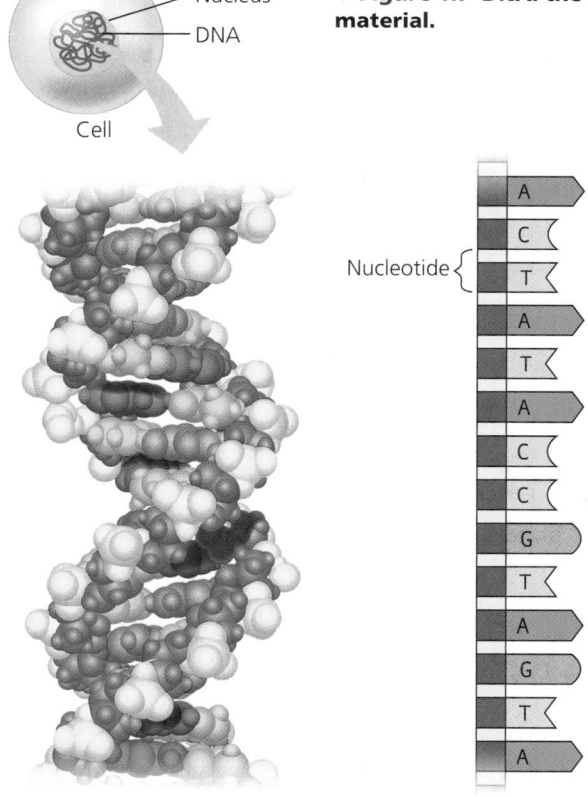

Nucleus

DNA

Cell

Nucleotide

▼ Figure 1.7 DNA: the genetic material.

(a) DNA double helix. This model shows the atoms in a segment of DNA. Made up of two long chains (strands) of building blocks called nucleotides, a DNA molecule takes the three-dimensional form of a double helix.

(b) Single strand of DNA. These geometric shapes and letters are simple symbols for the nucleotides in a small section of one strand of a DNA molecule. Genetic information is encoded in specific sequences of the four types of nucleotides. Their names are abbreviated A, T, C, and G.

→ **Mastering Biology** Animation: Heritable Information: DNA

Protein-encoding genes control protein production indirectly, using a related molecule called RNA as an intermediary. The sequence of nucleotides along a gene is transcribed into mRNA, which is then translated into a linked series of protein building blocks called amino acids. Once completed, the amino acid chain forms a specific protein with a unique shape and function. The entire process by which the information in a gene directs the manufacture of a cellular product is called **gene expression (Figure 1.8)**.

In carrying out gene expression, all forms of life employ essentially the same genetic code: A particular sequence of nucleotides means the same thing in one organism as it does in another. Differences between organisms reflect differences between their nucleotide sequences rather than between their genetic codes. This universality of the genetic code is a strong piece of evidence that all life is related. Comparing the sequences in several species for a gene that codes for a particular protein can provide valuable information both about the protein and about the relationship of the species to each other.

Molecules of mRNA, like the one in Figure 1.8, are translated into proteins, but other cellular RNAs function differently. For example, we have known for decades that some types of RNA are actually components of the cellular machinery that manufactures proteins. In the last few decades, scientists have discovered new classes of RNA that play other roles in the cell, such as regulating the function of protein-coding genes. Genes specify these RNAs as well, and their production is also referred to as gene expression. By carrying the instructions for making proteins and RNAs and by replicating with each cell division, DNA ensures faithful inheritance of genetic information from generation to generation.

Genomics: Large-Scale Analysis of DNA Sequences

The entire "library" of genetic instructions that an organism inherits is called its **genome**. A typical human cell has two similar sets of chromosomes, and each set has approximately 3 billion nucleotide pairs of DNA. If the one-letter abbreviations for the nucleotides of a set were written in letters the size of those you are now reading, the genomic text would fill about 700 biology textbooks.

Since the early 1990s, the pace at which researchers can determine the sequence of a genome has accelerated at an astounding rate, enabled by a revolution in technology. The genome sequence—the entire sequence of nucleotides for a representative member of a species—is now known for humans and many other animals, as well as numerous plants, fungi, bacteria, and archaea. To make sense of the deluge of data from genome-sequencing projects and the growing catalog of known gene functions, scientists are applying a systems biology approach at the cellular and molecular levels. Rather than investigating a single gene at

▼ **Figure 1.8 Gene expression: Cells use information encoded in a gene to synthesize a functional protein.**

(a) The lens of the eye (behind the pupil) is able to focus light because lens cells are tightly packed with transparent proteins called crystallin. How do lens cells make crystallin proteins?

Lens cell

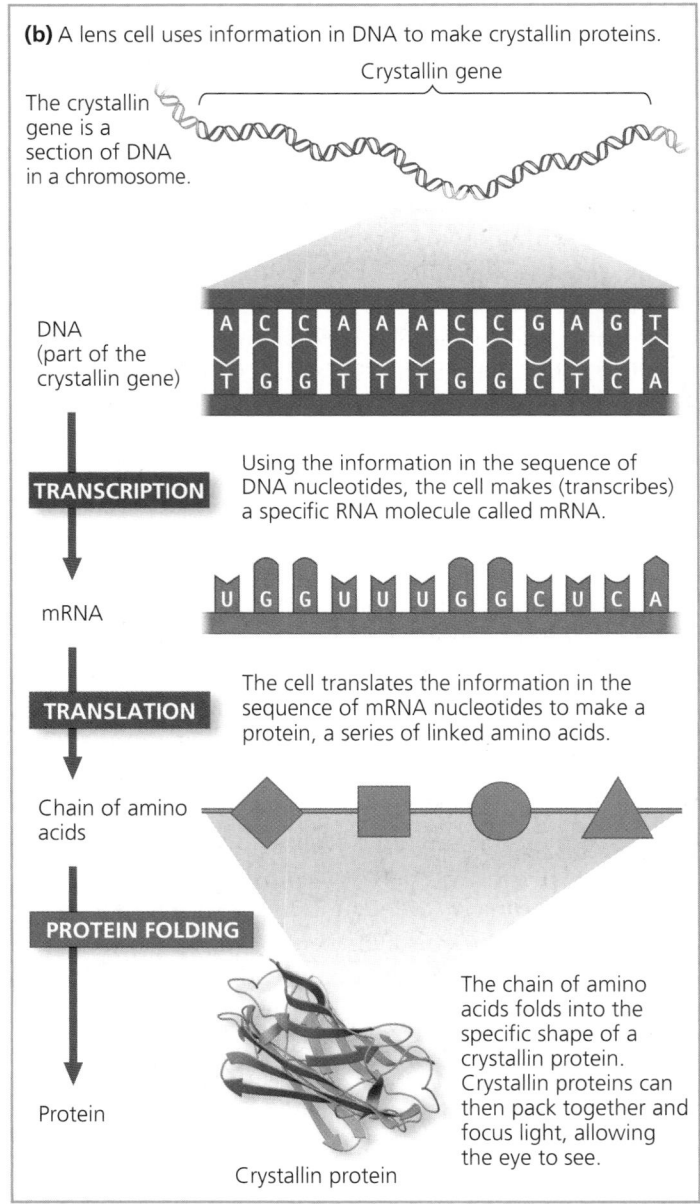

(b) A lens cell uses information in DNA to make crystallin proteins.

Crystallin gene

The crystallin gene is a section of DNA in a chromosome.

DNA (part of the crystallin gene)

A C C A A A C C G A G T
T G G T T T G G C T C A

TRANSCRIPTION

Using the information in the sequence of DNA nucleotides, the cell makes (transcribes) a specific RNA molecule called mRNA.

mRNA

U G G U U U G G C U C A

TRANSLATION

The cell translates the information in the sequence of mRNA nucleotides to make a protein, a series of linked amino acids.

Chain of amino acids

PROTEIN FOLDING

Protein

The chain of amino acids folds into the specific shape of a crystallin protein. Crystallin proteins can then pack together and focus light, allowing the eye to see.

Crystallin protein

→ **Mastering Biology** Figure Walkthrough

a time, researchers study whole sets of genes (or other DNA) in one or more species—an approach called **genomics**. Likewise, the term **proteomics** refers to the study of sets of proteins and their properties. (The entire set of proteins expressed by a given cell, tissue, or organism is called a **proteome**.)

Three important research developments have made the genomic and proteomic approaches possible. One is "high-throughput" technology, tools that can analyze many biological samples very rapidly. The second major development is **bioinformatics**, the use of computational tools to store, organize, and analyze the huge volume of data that results from high-throughput methods. The third development is the formation of interdisciplinary research teams—groups of diverse specialists that may include computer scientists, mathematicians, engineers, chemists, physicists, and, of course, biologists from a variety of fields. Researchers in such teams aim to learn how the activities of all the proteins and RNAs encoded by the DNA are coordinated in cells and in whole organisms.

Theme: Life Requires the Transfer and Transformation of Energy and Matter

ENERGY AND MATTER Moving, growing, reproducing, and the various cellular activities of life are work, and work requires energy. The input of energy, primarily from the sun, and the transformation of energy from one form to another make life possible **(Figure 1.9)**. When a plant's leaves absorb sunlight in the process of photosynthesis, molecules within the leaves convert the energy of sunlight to the chemical

energy of food, such as sugars. The chemical energy in the food molecules is then passed along from plants and other photosynthetic organisms (**producers**) to consumers. A **consumer** is an organism that feeds on other organisms or their remains.

When an organism uses chemical energy to perform work, such as muscle contraction or cell division, some of that energy is lost to the surroundings as heat. As a result, energy *flows through* an ecosystem in one direction, usually entering as light and exiting as heat. In contrast, chemicals *cycle within* an ecosystem, where they are used and then recycled (see Figure 1.9). Chemicals that a plant absorbs from the air or soil may be incorporated into the plant's body and then passed to an animal that eats the plant. Eventually, these chemicals will be returned to the environment by decomposers such as bacteria and fungi that break down waste products, leaf litter, and the bodies of dead organisms. The chemicals are then available to be taken up by plants again, thereby completing the cycle.

Theme: From Molecules to Ecosystems, Interactions Are Important in Biological Systems

INTERACTIONS At any level of the biological hierarchy, interactions between the components of the system ensure smooth integration of all the parts, such that they function as a whole. This holds true equally well for molecules in a cell and the components of an ecosystem; we'll look at both as examples.

▶ **Figure 1.9 Energy flow and chemical cycling.** There is a one-way flow of energy in an ecosystem: During photosynthesis, plants convert energy from sunlight to chemical energy (stored in food molecules such as sugars), which is used by plants and other organisms to do work and is eventually lost from the ecosystem as heat. In contrast, chemicals cycle between organisms and the physical environment.

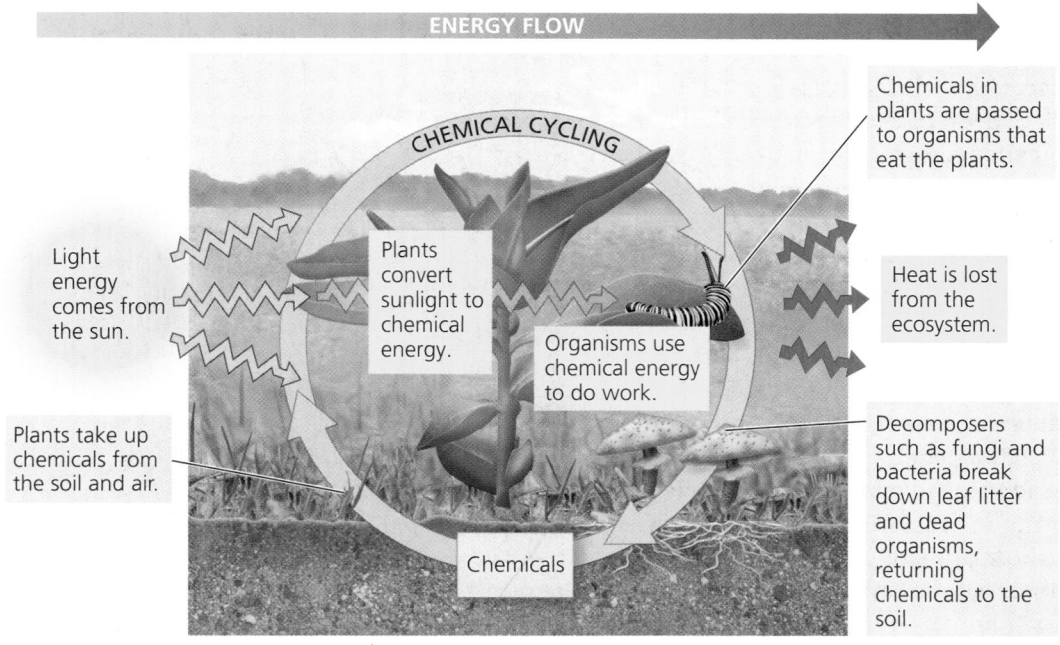

Molecules: Interactions Within Organisms

At lower levels of organization, the interactions between components that make up living organisms—organs, tissues, cells, and molecules—are crucial to their smooth operation. Consider the regulation of blood sugar level, for instance. Cells in the body must match the supply of fuel (sugar) to demand, regulating the opposing processes of sugar breakdown and storage. The key is the ability of many biological processes to self-regulate by a mechanism called feedback.

In **feedback regulation**, the output or product of a process regulates that very process. The most common form of regulation in living systems is *negative feedback*, a loop in which the response reduces the initial stimulus. As seen in the example of insulin signaling **(Figure 1.10)**, after a meal the level of the sugar glucose in your blood rises, which stimulates cells of the pancreas to secrete insulin. Insulin, in turn, causes body cells to take up glucose and liver cells to store it, thus decreasing the blood glucose level. This eliminates the stimulus for insulin secretion, shutting off the pathway. Thus, the output of the process (insulin) negatively regulates that process.

Though less common than processes regulated by negative feedback, there are also many biological processes regulated by *positive feedback*, in which an end product *speeds up* its own production. The clotting of your blood in response to injury is an example. When a blood vessel is damaged, structures in the blood called platelets begin to aggregate at the site. Positive feedback occurs as chemicals released by the platelets attract *more* platelets. The platelet pileup then initiates a complex process that seals the wound with a clot.

Ecosystems: An Organism's Interactions with Other Organisms and the Physical Environment

At the ecosystem level, every organism interacts with other organisms. For instance, an acacia tree interacts with soil microorganisms associated with its roots, insects that live on it, and animals that eat its leaves and fruit **(Figure 1.11)**. Interactions between organisms include those that are mutually beneficial (as when "cleaner fish" eat small parasites on a turtle) and those in which one species benefits and the other is harmed (as when a lion kills and eats a zebra). In some interactions between species, both are harmed—for example, when two

▼ **Figure 1.10 Feedback regulation.** The human body regulates the use and storage of glucose, a major cellular fuel. This figure shows negative feedback: The response to insulin reduces the initial stimulus.

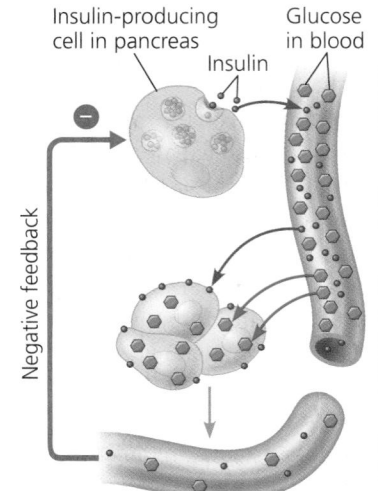

❶ A high blood glucose level stimulates cells in the pancreas to secrete insulin into the blood.

❷ Insulin circulates in the blood throughout the body.

❸ Insulin binds to body cells, causing them to take up glucose and liver cells to store glucose. This lowers the glucose level in the blood.

❹ The lowered blood glucose level does not stimulate secretion of insulin.

VISUAL SKILLS *In this example, what is the response to insulin? What is the initial stimulus that is reduced by the response?*

plants compete for a soil resource that is in short supply. Interactions among organisms help regulate the functioning of the ecosystem as a whole.

Each organism also interacts continuously with physical factors in its environment. The leaves of a tree, for example, absorb light from the sun, take in carbon dioxide from the air,

▼ **Figure 1.11 Interactions of an African acacia tree with other organisms and the physical environment.**

Sunlight

Leaves absorb light energy from the sun.

Leaves take in carbon dioxide from the air and release oxygen.

CO_2

O_2

Leaves fall to the ground and are decomposed by organisms that return minerals to the soil.

Water and minerals in the soil are taken up by the tree through its roots.

Animals eat leaves and fruit from the tree, returning nutrients and minerals to the soil in their waste products.

and release oxygen to the air (see Figure 1.11). The environment is also affected by organisms. For instance, in addition to taking up water and minerals from the soil, the roots of a plant break up rocks as they grow, contributing to the formation of soil. On a global scale, plants and other photosynthetic organisms have generated all the oxygen in the atmosphere.

Like other organisms, we humans interact with our environment. Our interactions sometimes have dire consequences: For example, over the past 150 years, humans have greatly increased the burning of fossil fuels (coal, oil, and gas). This practice releases large amounts of carbon dioxide (CO_2) and other gases into the atmosphere, causing heat to be trapped close to Earth's surface (see Figure 56.29). Scientists calculate that the CO_2 added to the atmosphere by human activities has increased the average temperature of the planet by about 1°C since 1900. At the current rates that CO_2 and other gases are being added to the atmosphere, global models predict an additional rise of at least 3°C before the end of this century.

This ongoing global warming is a major aspect of **climate change**, a directional change to the global climate that lasts for three decades or more (as opposed to short-term changes in the weather). But global warming is not the only way the climate is changing: Wind and precipitation patterns are also shifting, and extreme weather events such as storms and droughts are occurring more often. Climate change has already affected organisms and their habitats all over the planet. For example, polar bears have lost much of the ice platform from which they hunt, leading to food shortages and increased mortality rates. As habitats deteriorate, hundreds of plant and animal species are shifting their ranges to more suitable locations—but for some, there is insufficient suitable habitat, or they may not be able to migrate quickly enough. As a result, the populations of many species are shrinking in size or even disappearing **(Figure 1.12)**. (For more examples of how climate change is affecting life on Earth, see Make Connections Figure 56.30.)

The loss of populations due to climate change can ultimately result in extinction, the permanent loss of a species.

▶ **Figure 1.12 Threatened by global warming.** A warmer environment causes lizards in the genus *Sceloporus* to spend more time in refuges from the heat, reducing time for foraging. Their food intake drops, decreasing reproductive success. Surveys of 200 *Sceloporus* populations in Mexico show that 12% of these populations have disappeared since 1975.

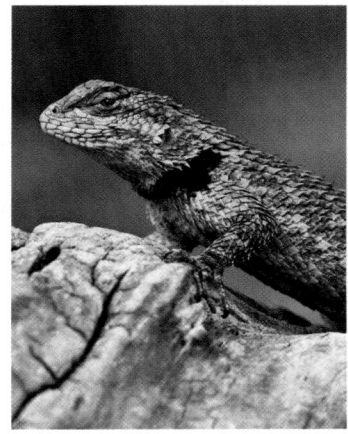

As we'll explore in greater detail in Concept 56.4, the consequences of these changes for humans and other organisms may be profound.

Having considered four of the unifying themes (organization, information, energy and matter, and interactions), let's now turn to evolution. There is consensus among biologists that evolution is the core theme of biology, and it is discussed in detail in the next section.

CONCEPT CHECK 1.1

1. Starting with the molecular level in Figure 1.3, write a sentence that includes components from the previous (lower) level of biological organization, for example: "A molecule consists of *atoms* bonded together." Continue with organelles, moving up the biological hierarchy.

2. Identify the theme or themes exemplified by (a) the sharp quills of a porcupine, (b) the development of a multicellular organism from a single fertilized egg, and (c) a hummingbird using sugar to power its flight.

3. **WHAT IF?** For each theme discussed in this section, give an example not mentioned in the text.

For suggested answers, see Appendix A.

CONCEPT 1.2

The Core Theme: Evolution accounts for the unity and diversity of life

EVOLUTION An understanding of evolution helps us to make sense of everything we know about life on Earth. As the fossil record clearly shows, life has been evolving for billions of years, resulting in a vast diversity of past and present organisms. But along with the diversity there is also unity, in the form of shared features. For example, while sea horses, jackrabbits, hummingbirds, and giraffes all look very different, their skeletons are organized in the same basic way.

The scientific explanation for the unity and diversity of organisms is **evolution**: a process of biological change in which species accumulate differences from their ancestors as they adapt to different environments over time. Thus, we can account for differences between two species (diversity) with the idea that certain heritable changes occurred after the two species diverged from their common ancestor. However, they also share certain traits (unity) simply because they have descended from a common ancestor. An abundance of evidence of different types supports the occurrence of evolution and the mechanisms that describe how it takes place, which we'll explore in detail in Chapters 22–25. To quote one of the founders of modern evolutionary theory, Theodosius Dobzhansky, "Nothing in biology makes sense except in the light of evolution." To understand this statement, we need to examine how biologists think about the vast diversity of life on the planet.

Classifying the Diversity of Life

Diversity is a hallmark of life. Biologists have so far identified and named about 1.8 million species of organisms. Each species is given a two-part name: The first part is the name of the genus (plural, *genera*) to which the species belongs, and the second part is unique to the species within the genus. (For example, *Homo sapiens* is the name of our species.)

To date, known species include at least 100,000 species of fungi, 290,000 plant species, 57,000 vertebrate species (animals with backbones), and 1 million insect species (more than half of all known forms of life)—not to mention the myriad types of single-celled organisms. Researchers identify thousands of additional species each year. Estimates of the total number of species range from about 10 million to over 100 million. Whatever the actual number, the enormous variety of life gives biology a very broad scope. Biologists face a major challenge in attempting to make sense of this variety.

The Three Domains of Life

Humans tend to group diverse items according to their similarities and relationships to each other. Consequently, biologists have long used careful comparisons of structure, function, and other obvious features to classify forms of life into groups. In the last few decades, new methods of assessing species relationships, such as comparisons of DNA sequences, have led to a reevaluation of the classification of life. Although this reevaluation is ongoing, biologists currently place all organisms into three groups called domains: Bacteria, Archaea, and Eukarya **(Figure 1.13)**.

Two of the three domains—**Bacteria** and **Archaea**—consist of single-celled prokaryotic organisms. All the eukaryotes (organisms with eukaryotic cells) are in domain **Eukarya**. This domain includes four subgroups: kingdom Plantae, kingdom Fungi, kingdom Animalia, and the protists. The three kingdoms are distinguished partly by their modes of nutrition: Plants produce their own sugars and other food

▼ **Figure 1.13 The three domains of life.**

(a) Domain Bacteria

Bacteria are the most diverse and widespread prokaryotes and are now classified into multiple kingdoms. Each rod-shaped structure in this photo is a bacterial cell.

(b) Domain Archaea

Domain Archaea includes multiple kingdoms. Some of the prokaryotes known as **archaea** live in Earth's extreme environments, such as salty lakes and boiling hot springs. Each round structure in this photo is an archaeal cell.

(c) Domain Eukarya

◀ **Kingdom Animalia** consists of multicellular eukaryotes that ingest other organisms.

▲ **Kingdom Plantae** (plants) consists of multicellular eukaryotes that carry out photosynthesis, the conversion of light energy to the chemical energy in food. Most plant species live on land.

▶ **Kingdom Fungi** is characterized in part by the nutritional mode of its members (such as this mushroom), which absorb nutrients from outside their bodies.

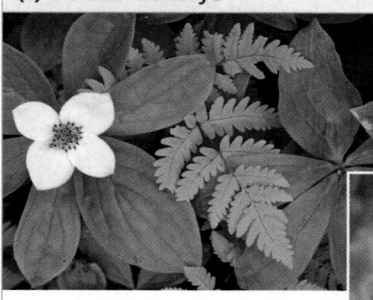

▶ **Protists** are mostly unicellular eukaryotes and some relatively simple multicellular relatives. Pictured here is an assortment of protists inhabiting pond water. Scientists are currently debating how to classify protists in a way that accurately reflects their evolutionary relationships.

molecules by photosynthesis, fungi absorb nutrients in dissolved form from their surroundings, and animals obtain food by eating and digesting other organisms. Animalia is, of course, the kingdom to which we belong.

The most numerous and diverse eukaryotes are the protists, which are mostly single-celled organisms. Although protists were once placed in a single kingdom, they are now classified into several groups. One major reason for this change is the recent DNA evidence showing that some protists are less closely related to other protists than they are to plants, animals, or fungi.

Unity in the Diversity of Life

As diverse as life is, there is also remarkable unity among forms of life. Consider, for example, the similar skeletons of different animals and the universal genetic language of DNA (the genetic code), both mentioned earlier. In fact, similarities between organisms are evident at all levels of the biological hierarchy. For example, unity is obvious in many features of cell structure, even among distantly related organisms **(Figure 1.14)**.

How can we account for life's dual nature of unity and diversity? The process of evolution, explained next, illuminates both the similarities and differences in the world of life. It also introduces another important dimension of biology: the passage of time. The history of life, as documented by fossils and other evidence, is the saga of an ever-changing Earth billions of years old, inhabited by an evolving cast of living forms **(Figure 1.15)**.

▼ **Figure 1.14 An example of unity underlying the diversity of life: the architecture of cilia in eukaryotes.** Cilia (singular, *cilium*) are extensions of cells that function in locomotion. They occur in eukaryotes as diverse as *Paramecium* (found in pond water) and humans. Even organisms so different share a common architecture for their cilia, which have an elaborate system of tubules that is striking in cross-sectional views.

▼ **Figure 1.15 Studying the history of life.** Researchers in South Africa reconstruct skeletons of *Homo naledi*, an extinct relative of *Homo sapiens*. The fossils were discovered in an underground cave that may have been a burial chamber.

Cross section of a cilium, as viewed with an electron microscope

15 μm

0.1 μm

5 μm

Cilia of *Paramecium*. The cilia of the single-celled *Paramecium* propel the organism through pond water.

Cilia of windpipe cells. The cells that line the human windpipe are equipped with cilia that help keep the lungs clean by sweeping a film of debris-trapping mucus upward.

▼ **Figure 1.16 Charles Darwin.** The portrait shows Darwin in about 1840, well before the 1859 publication of his revolutionary book, commonly referred to as *The Origin of Species*.

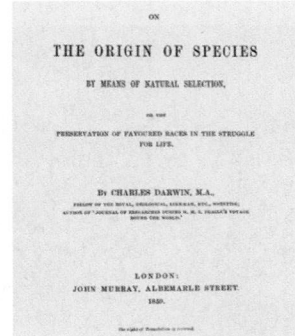

Charles Darwin and the Theory of Natural Selection

An evolutionary view of life came into sharp focus in November 1859, when Charles Darwin published one of the most important and influential books ever written, *On the Origin of Species by Means of Natural Selection* **(Figure 1.16)**. *The Origin of Species* articulated two main points. The first point was that, as species adapt to different environments over time, they accumulate differences from their ancestors. Darwin called this process "descent with modification." This insightful phrase captured the duality of life's unity and diversity—unity in the kinship among species that descended from common ancestors and diversity in the modifications that evolved as species branched from their common ancestors **(Figure 1.17)**. Darwin's second main point was his proposal that "natural selection" is a primary cause of descent with modification.

Darwin developed his theory of natural selection from observations that by themselves were neither new nor profound. However, although others had described the pieces of the puzzle, it was Darwin who saw how they fit together. He started with the following three observations from nature: First, individuals in a population vary in their traits, many of which seem to be heritable (passed on from parents to offspring). Second, a population can produce far more offspring than can survive to produce offspring of their own. With more individuals than the environment is able to support, competition is inevitable. Third, species generally are suited to their environments—in other words, they are adapted to their circumstances. For instance, a common adaptation among birds that eat hard seeds is an especially strong beak.

By making inferences from these three observations, Darwin developed a scientific explanation for how evolution occurs. He reasoned that individuals with inherited traits that are better suited to the local environment are more likely to survive and reproduce than less well-suited individuals.

▼ **Figure 1.17 Unity and diversity among birds.** These four birds are variations on a common body plan. For example, each has feathers, a beak, and wings. However, these common features are highly specialized for the birds' diverse lifestyles.

▼ **Red-tailed hawk (*Buteo borealis*)**

▲ **European robin (*Erithacus rubecula*)**

▼ **American flamingo (*Phoenicopterus ruber*)**

▲ **Gentoo penguin (*Pygoscelis papua*)**

Over many generations, a higher and higher proportion of individuals in a population will have the advantageous traits. Evolution occurs as the unequal reproductive success of individuals ultimately leads to adaptation to their environment, as long as the environment remains the same.

Darwin called this mechanism of evolutionary adaptation **natural selection** because the natural environment consistently "selects" for the propagation of certain traits among naturally occurring variant traits in the population. The example in **Figure 1.18** illustrates the ability of natural selection to "edit" an insect population's heritable variations in coloration. We see the products of natural selection in the exquisite adaptations of various organisms to the special circumstances of their way of life and their environment. The wings of the bat shown in **Figure 1.19** are an excellent example of adaptation.

▼ Figure 1.18 Natural selection. This imaginary beetle population has colonized a locale where the soil has been blackened by a recent brush fire. Initially, the population varies extensively in the inherited coloration of the individuals, from very light gray to charcoal. For hungry birds that prey on the beetles, it is easiest to spot the beetles that are lightest in color.

1 Population with varied inherited traits

2 Elimination of individuals with certain traits that make them more visible

3 Reproduction of survivors

4 Increased frequency of traits that enhance survival and reproductive success

DRAW IT *Over time, the soil will gradually become lighter in color. Draw another step to show how the soil, when lightened to medium color, would affect natural selection. Write a caption for this new step 5. Then explain how the population would change over time as the soil becomes lighter.*

 Mastering Biology HHMI Video: The Making of The Fittest: Natural Selection and Adaptation (Rock Pocket Mouse)

▲ Figure 1.19 Evolutionary adaptation. Bats, the only mammals capable of active flight, have wings with webbing between extended "fingers." Darwin proposed that such adaptations are refined over time by natural selection.

The Tree of Life

Take another look at the skeletal architecture of the bat's wings in Figure 1.19. These wings are not like those of feathered birds; the bat is a mammal. The bones, joints, nerves, and blood vessels in the bat's forelimbs, though adapted for flight, are very similar to those in the human arm, the foreleg of a horse, and the flipper of a whale. Indeed, all mammalian forelimbs are anatomical variations of a common architecture. According to the Darwinian concept of descent with modification, the shared anatomy of mammalian limbs reflects inheritance of the limb structure from a common ancestor—the "prototype" mammal from which all other mammals descended. The diversity of mammalian forelimbs results from modification by natural selection operating over millions of years in different environmental contexts. Fossils and other evidence corroborate anatomical unity in supporting this view of mammalian descent from a common ancestor.

Darwin proposed that natural selection, by its cumulative effects over long periods of time, could cause an ancestral species to give rise to two or more descendant species. This could occur, for example, if one population of organisms became fragmented into several subpopulations isolated in different environments. In these separate arenas of natural selection, one species could gradually radiate into multiple species as the geographically isolated populations adapted over many generations to different environmental conditions.

The Galápagos finches are a famous example of the process of radiation of new species from a common ancestor. Darwin collected specimens of these birds during his 1835 visit to the remote Galápagos Islands, 900 kilometers (km) off the Pacific coast of South America. These relatively young volcanic islands are home to many species of plants and animals found nowhere else in the world, though many Galápagos organisms are clearly related to species on the South American mainland. The Galápagos finches are thought to have descended from an ancestral finch species that reached the archipelago from South America or the Caribbean. Over time, the Galápagos finches diversified from their ancestor as populations became adapted to different food sources on their particular islands. Years after Darwin collected the finches, researchers began to sort out their evolutionary relationships, first from anatomical and geographic data and more recently with the help of DNA sequence comparisons.

Mastering Biology Video: Galápagos Biodiversity by Peter and Rosemary Grant

Biologists' diagrams of evolutionary relationships generally take treelike forms, though the trees are often turned

► **Figure 1.20 Descent with modification: adaptive radiation of finches on the Galápagos Islands.** This "tree" illustrates a current hypothesis for the evolutionary relationships of finches on the Galápagos. Note the various beaks, which are adapted to particular food sources. For example, heavier, thicker beaks are better at cracking seeds, while the more slender beaks are better at grasping insects.

➲ **Mastering Biology** HHMI Video: The Origin of Species: The Beak of the Finch

hhmi BioInteractive

ANCESTRAL FINCH

Green warbler finch
Certhidea olivacea
(insect-eater)

Vegetarian finch
Platyspiza crassirostris
(fruit-eater)

Woodpecker finch
Camarhynchus pallidus
(insect-eater)

Small tree finch
Camarhynchus parvulus
(insect-eater)

Common cactus finch
Geospiza scandens
(cactus-eater)

Large ground finch
Geospiza magnirostris
(seed-eater)

Branch point: where lineages diverge from a common ancestor

Common ancestor of finches in the genera *Camarhynchus* and *Geospiza*

sideways as in **Figure 1.20**. Tree diagrams make sense: Just as an individual has a genealogy that can be diagrammed as a family tree, each species is one twig of a branching tree of life extending back in time through ancestral species more and more remote. Species that are very similar, such as the Galápagos finches, share a relatively recent common ancestor. Through an ancestor that lived much further back in time, finches are related to sparrows, hawks, penguins, and all other birds. Furthermore, finches and other birds are related to us through a common ancestor even more ancient. Trace life back far enough, and we reach the early prokaryotes that inhabited Earth over 3.5 billion years ago. We can recognize their vestiges in our own cells—in the universal genetic code, for example. Indeed, all of life is connected through its long evolutionary history.

CONCEPT CHECK 1.2

1. Explain why "editing" is a metaphor for how natural selection acts on a population's heritable variation.

2. Referring to Figure 1.20, provide a possible explanation for how, over a very long time, the green warbler finch came to have a slender beak.

3. **DRAW IT** The three domains you learned about in Concept 1.2 can be represented in the tree of life as the three main branches, with three subbranches on the eukaryotic branch being the kingdoms Plantae, Fungi, and Animalia. What if fungi and animals are more closely related to each other than either of these kingdoms is to plants—as recent evidence strongly suggests? Draw a simple branching pattern that symbolizes the proposed relationship between these three eukaryotic kingdoms.

For suggested answers, see Appendix A.

CONCEPT 1.3

In studying nature, scientists form and test hypotheses

Science is a way of knowing—an approach to understanding the natural world. It developed out of our curiosity about ourselves, other life-forms, our planet, and the universe. The word *science* is derived from a Latin verb meaning "to know." Striving to understand seems to be one of our basic urges.

At the heart of science is **inquiry**, the search for information and explanations of natural phenomena. There is no formula for successful scientific inquiry, no single scientific method that researchers must rigidly follow. As in all quests, science includes elements of challenge, adventure, and luck, along with careful planning, reasoning, creativity, patience, and the persistence to overcome setbacks. Such diverse elements of inquiry make science far less structured than most people realize. That said, it is possible to highlight certain characteristics that help to distinguish science from other ways of describing and explaining nature.

Scientists use a process of inquiry that includes making observations, forming logical, testable explanations (*hypotheses*), and testing them. The process is necessarily repetitive: In testing a hypothesis, more observations may inspire revision of the original hypothesis or formation of a new one, thus leading to further testing. In this way, scientists circle closer and closer to their best estimation of the laws governing nature.

Exploration and Observation

Biology, like other sciences, begins with careful observations. In gathering information, biologists often use tools such as microscopes, precision thermometers, or high-speed cameras that extend their senses or facilitate careful measurement. Observations can reveal valuable information about the natural world. For example, a series of detailed observations have shaped our understanding of cell structure, and another set of observations is currently expanding our databases of genome sequences from diverse species and databases of genes whose expression is altered in various diseases.

In exploring nature, biologists also rely heavily on the scientific literature, the published contributions of fellow scientists. By reading about and understanding past studies, scientists can build on the foundation of existing knowledge, focusing their investigations on observations that are original and on hypotheses that are consistent with previous findings. Identifying publications relevant to a new line of research is now easier than at any point in the past, thanks to indexed and searchable electronic databases.

Gathering and Analyzing Data

Recorded observations are called **data**. Put another way, data are items of information on which scientific inquiry is based. The term *data* implies numbers to many people. But some data are *qualitative*, often in the form of recorded descriptions rather than numerical measurements. For example, Jane Goodall spent decades recording her observations of chimpanzee behavior during field research in a Tanzanian jungle **(Figure 1.21)**. In her studies, Goodall also enriched the field of animal behavior with volumes of *quantitative* data, such as the frequency and duration of specific behaviors for different members of a group of chimpanzees in a variety of situations. Quantitative data are generally expressed as numerical measurements and often organized into tables and graphs. Scientists analyze their data using a type of mathematics called *statistics* to test whether their results are significant or merely due to random fluctuations. All results presented in this text have been shown to be statistically significant.

Collecting and analyzing observations can lead to important conclusions based on a type of logic called **inductive reasoning**. Through induction, we derive generalizations from a large number of specific observations. "The sun always rises in the east" is one example. Another biological example is the generalization "All organisms are made of cells," which was based on two centuries of microscopic observations made by biologists examining cells in diverse biological specimens. Careful observations and data analyses, along with generalizations reached by induction, are fundamental to our understanding of nature.

▼ **Figure 1.21 Jane Goodall collecting qualitative data on chimpanzee behavior.** Goodall recorded her observations in field notebooks, often with sketches of the animals' behavior.

→ **Mastering Biology**
Interview with Jane Goodall:
Living with chimpanzees

Forming and Testing Hypotheses

Our innate curiosity often stimulates us to pose questions about the natural basis for the phenomena we observe in the world. What caused the different chimpanzee behaviors observed in the wild? What explains the variation in coat color among the mice of a single species, shown in Figure 1.1? Answering such questions usually involves forming and testing logical explanations—that is, hypotheses.

In science, a **hypothesis** is an explanation, based on observations and assumptions, that leads to a testable prediction. Said another way, a hypothesis is an explanation on trial. The hypothesis is usually a rational accounting for a set of observations, based on the available data and guided by inductive reasoning. A scientific hypothesis must lead to predictions that can be tested by making additional observations or by performing experiments. An **experiment** is a scientific test, carried out under controlled conditions.

We all make observations and develop questions and hypotheses in solving everyday problems. Let's say, for example, that your desk lamp is plugged in and turned on but the bulb isn't lit. That's an observation. The question is obvious: Why doesn't the lamp work? Two reasonable hypotheses based on your experience are that (1) the bulb is burnt out or (2) the bulb is not screwed in properly. Each of these alternative hypotheses leads to predictions you can test with experiments. For example, the burnt-out bulb hypothesis predicts that replacing the bulb will fix the problem. **Figure 1.22** diagrams this informal inquiry. Figuring things out in this way by trial and error is a hypothesis-based approach.

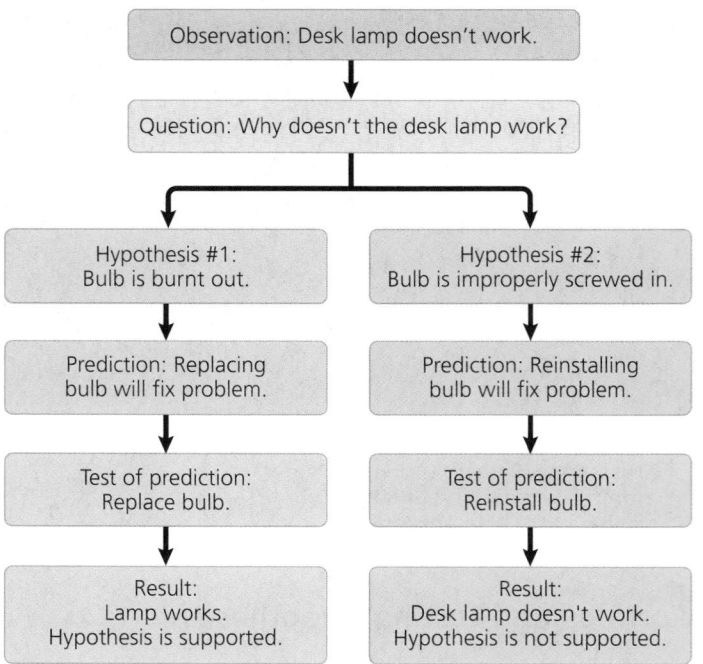

▼ **Figure 1.22 A simplified view of the scientific process.** The idealized process sometimes called the "scientific method" is shown in this flow chart, which illustrates hypothesis testing for a desk lamp that doesn't work.

Observation: Desk lamp doesn't work.

Question: Why doesn't the desk lamp work?

Hypothesis #1:
Bulb is burnt out.

Hypothesis #2:
Bulb is improperly screwed in.

Prediction: Replacing bulb will fix problem.

Prediction: Reinstalling bulb will fix problem.

Test of prediction:
Replace bulb.

Test of prediction:
Reinstall bulb.

Result:
Lamp works.
Hypothesis is supported.

Result:
Desk lamp doesn't work.
Hypothesis is not supported.

Deductive Reasoning

A type of logic called deduction is also built into the use of hypotheses in science. While induction entails reasoning from a set of specific observations to reach a general conclusion, **deductive reasoning** involves logic that flows in the opposite direction, from the general to the specific. From general premises, we extrapolate to the specific results we should expect if the premises are true. In the scientific process, deductions usually take the form of predictions of results that will be found if a particular hypothesis (premise) is correct. We then test the hypothesis by carrying out experiments or observations to see whether or not the results are as predicted. This deductive testing takes the form of "*If . . . then*" logic. In the case of the desk lamp example: *If* the burnt-out bulb hypothesis is correct, *then* the lamp should work if you replace the bulb with a new one.

We can use the desk lamp example to illustrate two other key points about the use of hypotheses in science. First, one can always devise additional hypotheses to explain a set of observations. For instance, another hypothesis to explain our nonworking desk lamp is that the wall socket is faulty. Although you could design an experiment to test this hypothesis, you can never test all possible hypotheses. Second, we can never *prove* that a hypothesis is true. Suppose that replacing the bulb fixed the lamp. The burnt-out bulb hypothesis would be the

most likely explanation, but testing supports that hypothesis *not* by proving that it is correct, but rather by failing to prove it incorrect. For example, even if replacing the bulb fixed the desk lamp, it might have been because there was a temporary power outage that just happened to end while the bulb was being changed.

Although a hypothesis can never be proved beyond all doubt, testing it in various ways can significantly increase our confidence in its validity. Often, rounds of hypothesis formulation and testing lead to a scientific consensus—the shared conclusion of many scientists that a particular hypothesis explains the known data well and stands up to experimental testing.

Questions That Can and Cannot Be Addressed by Science

Scientific inquiry is a powerful way to learn about nature, but there are limitations to the kinds of questions it can answer. A scientific hypothesis must be *testable*; there must be some observation or experiment that could reveal if such an idea is likely to be true or false. The hypothesis that a burnt-out bulb is the sole reason the lamp doesn't work would not be supported if replacing the bulb with a new one didn't fix the lamp.

Not all hypotheses meet the criteria of science: You wouldn't be able to test the hypothesis that invisible ghosts are fooling with your desk lamp! Because science only deals with natural, testable explanations for natural phenomena, it can neither support nor contradict the invisible ghost hypothesis, nor whether spirits or elves cause storms, rainbows, or illnesses. Such supernatural explanations are simply outside the bounds of science, as are religious matters, which are issues of personal faith. Science and religion are not mutually exclusive or contradictory; they are simply concerned with different issues.

The Flexibility of the Scientific Process

The way that researchers answer questions about the natural and physical world is often idealized as the *scientific method*. However, very few scientific inquiries adhere rigidly to the sequence of steps that are typically used to describe this approach. For example, a scientist may start to design an experiment, but then backtrack after realizing that more preliminary observations are necessary. In other cases, observations remain too puzzling to prompt well-defined questions until further study provides a new context in which to view those observations. For example, scientists could not unravel the details of how genes encode proteins until *after* the discovery of the structure of DNA (an event that took place in 1953).

A more realistic model of the scientific process is shown in **Figure 1.23**. The focus of this model, shown in the central circle in the figure, is the forming and testing of hypotheses. This core set of activities is the reason that science does so well in explaining phenomena in the natural world. These activities, however, are shaped by exploration and discovery (the upper circle in Figure 1.23) and influenced by interactions with other scientists and with society more generally (lower circles). For example, the community of scientists influences which hypotheses are tested, how test results are interpreted, and what value is placed on the findings. Similarly, societal needs—such as the push to cure cancer or understand the process of climate change—may help shape what research projects are funded and how extensively the results are discussed.

Now that we have highlighted the key features of scientific inquiry—making observations and forming and testing hypotheses—you should be able to recognize these features in a case study of actual scientific research.

▼ **Figure 1.23 The process of science: a realistic model.** In reality, the process of science is not linear, but instead involves backtracking, repetitions, and feedback between different parts of the process. This illustration is based on a model (How Science Works) from the website Understanding Science (www.understandingscience.org).

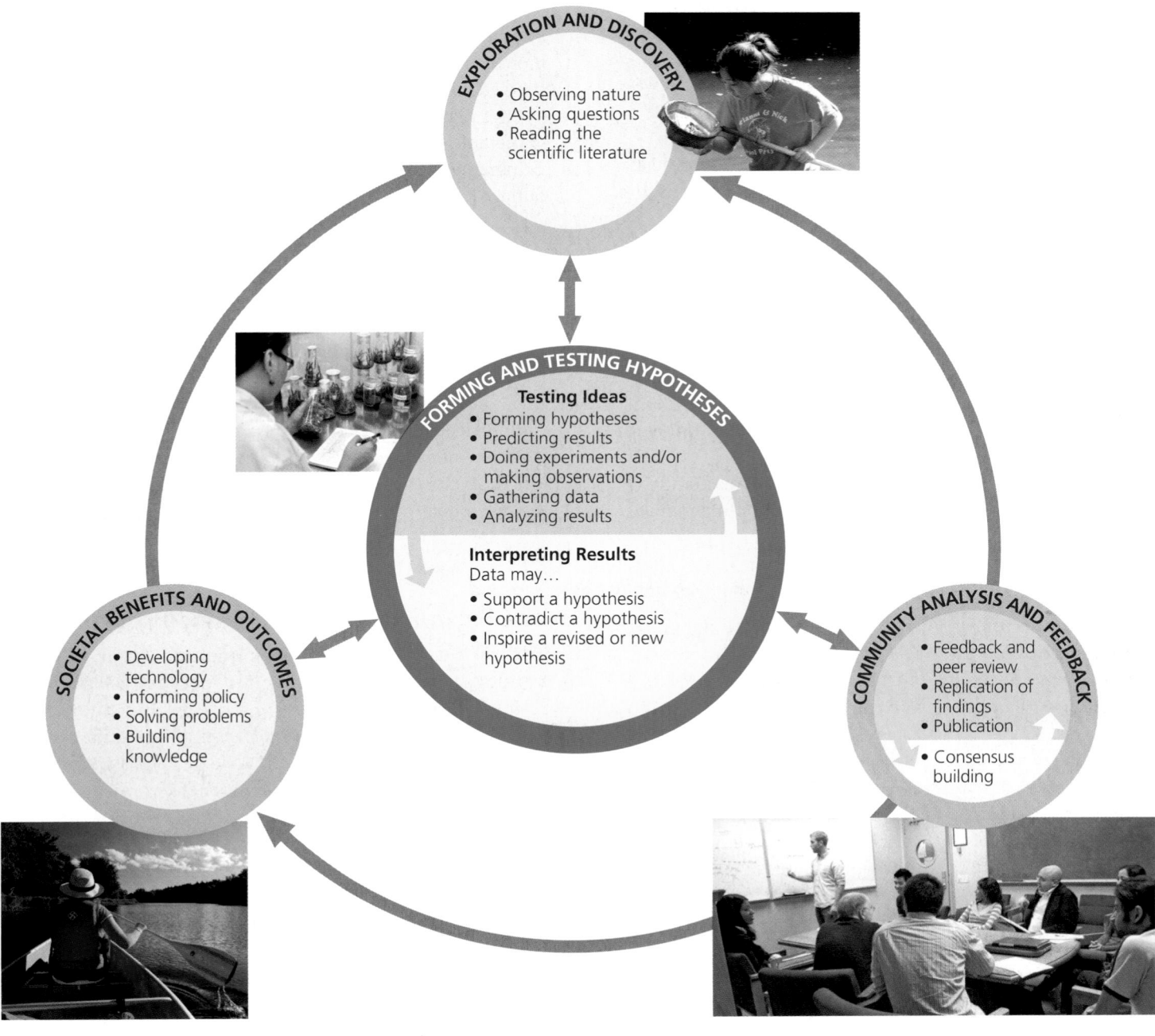

EXPLORATION AND DISCOVERY
- Observing nature
- Asking questions
- Reading the scientific literature

FORMING AND TESTING HYPOTHESES

Testing Ideas
- Forming hypotheses
- Predicting results
- Doing experiments and/or making observations
- Gathering data
- Analyzing results

Interpreting Results
Data may…
- Support a hypothesis
- Contradict a hypothesis
- Inspire a revised or new hypothesis

SOCIETAL BENEFITS AND OUTCOMES
- Developing technology
- Informing policy
- Solving problems
- Building knowledge

COMMUNITY ANALYSIS AND FEEDBACK
- Feedback and peer review
- Replication of findings
- Publication
- Consensus building

A Case Study in Scientific Inquiry: Investigating Coat Coloration in Mouse Populations

Our case study begins with a set of observations and inductive generalizations. Color patterns of animals vary widely in nature, sometimes even among members of the same species. What accounts for such variation? As you may recall, the two mice depicted at the beginning of this chapter are members of the same species (*Peromyscus polionotus*), but they have different coat (fur) color patterns and reside in different environments. The beach mouse lives along the Florida seashore, a habitat of brilliant white sand dunes with sparse clumps of beach grass. The inland mouse lives on darker, more fertile soil farther inland **(Figure 1.24)**. Even a brief glance at the photographs in Figure 1.24 reveals a striking match of mouse coloration to its habitat. The natural predators of these mice, including hawks, owls, foxes, and coyotes, are all visual hunters (they use their sense of sight to look for prey). It was logical, therefore, for Francis Bertody Sumner, a naturalist studying populations of these mice in the 1920s, to form the hypothesis that their coloration patterns had evolved as adaptations that camouflage the mice in their native environments, protecting them from predation.

As obvious as the camouflage hypothesis may seem, it still required testing. In 2010, biologist Hopi Hoekstra of Harvard University and a group of her students headed to Florida to test the prediction that mice with coloration that did not match their habitat would be preyed on more heavily than the native, well-matched mice. **Figure 1.25** summarizes this field experiment, introducing a format we will use throughout the book to walk through other examples of biological inquiry.

The researchers built hundreds of models of mice and spray-painted them to resemble either beach or inland mice, so that the models differed only in their color patterns. The researchers placed equal numbers of these model mice randomly in both habitats and left them overnight. The mouse models resembling the native mice in the habitat were the *control* group (for instance, light-colored mouse models in the beach habitat), while the mouse models with the non-native coloration were the *experimental* group (for example, darker models in the beach habitat). The following morning, the team counted and recorded signs of predation events, which ranged from bites and gouge marks on some models to the outright disappearance of others. Judging by the shape of the predators' bites and the tracks surrounding the experimental sites, the predators appeared to be split fairly evenly between mammals (such as foxes and coyotes) and birds (such as owls, herons, and hawks).

For each environment, the researchers then calculated the percentage of predation events that targeted camouflaged models. The results were clear-cut: Camouflaged models showed much lower predation rates than those lacking camouflage in both the beach habitat (where light mice were less vulnerable) and the inland habitat (where dark mice were less vulnerable). The data thus fit the key prediction of the camouflage hypothesis.

Variables and Controls in Experiments

In carrying out an experiment, a researcher often manipulates one factor in a system and observes the effects of this change. The mouse camouflage experiment described in Figure 1.25 is an example of a **controlled experiment**, one that is designed to compare an experimental group (the non-camouflaged mice models, in this case) with a control group (the camouflaged models). Both the factor that is manipulated and the factor that is subsequently measured are **variables**—a feature or quantity that varies in an experiment.

▼ **Figure 1.24 Different coloration in beach and inland populations of *Peromyscus polionotus*.**

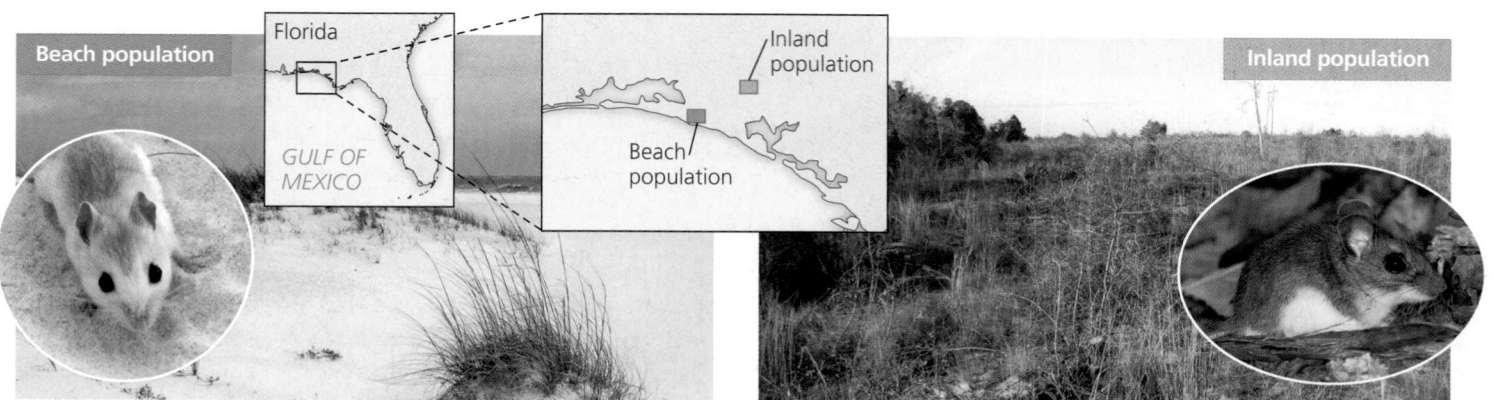

Beach mice live on sparsely vegetated sand dunes along the coast. The light tan, dappled fur on their backs causes them to blend into their surroundings, providing camouflage.

Members of the same species living about 30 km inland have dark fur on their backs, camouflaging them against the dark ground of their habitat.

▼ Figure 1.25 Inquiry

Does camouflage affect predation rates on two populations of mice?

Experiment Hopi Hoekstra and colleagues tested the hypothesis that coat color-ation provides camouflage that protects beach and inland populations of *Peromyscus polionotus* mice from predation in their habitats. The researchers spray-painted mouse models with light or dark color patterns that matched those of the beach and inland mice and placed models with each of the patterns in both habitats. The next morning, they counted damaged or missing models.

Results For each habitat, the researchers calculated the percentage of attacked mod-els that were camouflaged or non-camouflaged. In both habitats, the models whose pattern did not match their surroundings suffered much higher "predation" than did the camouflaged models.

Conclusion The results are consistent with the researchers' prediction: that mouse models with camouflage coloration would be attacked less often than non-camouflaged mouse models. Thus, the experiment supports the camouflage hypothesis.

Data from S. N. Vignieri, J. G. Larson, and H. E. Hoekstra, The selective advantage of crypsis in mice, *Evolution* 64:2153–2158 (2010).

INTERPRET THE DATA *The bars indicate the percentage of the attacked models that were either light or dark. Assume 100 mouse models were attacked in each habitat. For the beach habitat, how many were light models? Dark models? Answer the same questions for the inland habitat. Do the results of the experiment support the camouflage hypothesis? Explain.*

> **Mastering Biology**
> Interview with Hopi Hoekstra: Investigating the genetics and natural selection of mouse coat color

In our example, the color of the mouse model was the **independent variable**—the factor being manipulated by the researchers. The **dependent variable** is the factor being measured that is predicted to be affected by the independent variable; in this case, the researchers measured the amount of predation in response to variation in color of the mouse model. Note also that the experimental and control groups differ in only one independent variable: color.

As a result, the researchers could rule out other factors as causes of the more frequent attacks on the non-camouflaged mice—such as different numbers of predators or different tem-peratures in the different test areas. The clever experimental design left coloration as the only factor that could account for the low predation rate on models camouflaged with respect to the surrounding environment.

A common misconception is that the term *controlled experiment* means that scientists control all features of the experimental environment. But that's impossible in field research and can be very difficult even in highly regulated laboratory environments. Researchers usually "control" unwanted variables not by *eliminating* them through envi-ronmental regulation, but by *canceling out* their effects by using control groups.

> **Mastering Biology** Animation:
> **Introduction to Experimental Design**

Theories in Science

"It's just a theory!" Our everyday use of the term *theory* often implies an untested speculation. But the term *theory* has a dif-ferent meaning in science. What is a sci-entific theory, and how is it different from a hypothesis or from mere speculation?

First, a scientific **theory** is much broader in scope than a hypothesis. *This* is a hypothesis: "Coat coloration well-matched to their habitat is an adaptation that protects mice from predators." But *this* is a theory: "Evolutionary adaptations arise by natural selection." This theory proposes that natural selection is the evo-lutionary mechanism that accounts for an enormous variety of adaptations, of which coat color in mice is but one example.

Second, a theory is general enough to spin off many new, testable hypoth-eses. For example, the theory of natural selection motivated two researchers at Princeton University, Peter and Rosemary Grant, to test the specific hypothesis that the beaks of Galápagos finches evolve in response to changes in the types of available food. (Their results supported their hypothesis; see Figure 23.2.)

And third, compared to any one hypothesis, a theory is gen-erally supported by a much greater body of evidence. The the-ory of natural selection has been supported by a vast quantity of evidence, with more being found every day, and has not been contradicted by any scientific data. Those theories that become widely adopted in science (such as the theory of natural selec-tion and the theory of gravity) explain a great diversity of obser-vations and are supported by a vast accumulation of evidence.

Finally, scientists will sometimes modify or even reject a previously supported theory if new research consistently produces results that don't fit. For example, biologists once lumped bacteria and archaea together as a kingdom of prokaryotes. When new methods for comparing cells and molecules could be used to test such relationships, the evidence led scientists to reject the theory that bacteria and archaea are members of the same kingdom. If there is "truth" in science, it is at best conditional, based on the weight of available evidence.

CONCEPT CHECK 1.3

1. What qualitative observation led to the quantitative study in Figure 1.25?
2. Contrast inductive reasoning with deductive reasoning.
3. Why is natural selection called a theory?
4. **WHAT IF?** In the deserts of New Mexico, the soils are mostly sandy, with occasional regions of black rock derived from lava flows that occurred about 1,000 years ago. Mice are found in both sandy and rocky areas, and owls are known predators. What might you expect about coat color in these two mouse populations? Explain. How would you use this ecosystem to further test the camouflage hypothesis?

For suggested answers, see Appendix A.

CONCEPT 1.4

Science benefits from a cooperative approach and diverse viewpoints

Movies and cartoons sometimes portray scientists as loners in white lab coats, working in isolated labs. In reality, science is an intensely social activity. Most scientists work in teams, which often include both graduate and undergraduate students. And to succeed in science, it helps to be a good communicator. Research results have no impact until shared with a community of peers through seminars, publications, and websites. And, in fact, research papers aren't published until they are vetted by colleagues in what is called the "peer review" process. The examples of scientific inquiry described in this book, for instance, have all been published in peer-reviewed journals.

Building on the Work of Others

The great scientist Isaac Newton once said: "To explain all nature is too difficult a task for any one man or even for any one age. 'Tis much better to do a little with certainty, and leave the rest for others that come after you." Anyone who becomes a scientist, driven by curiosity about how nature works, is sure to benefit greatly from the rich storehouse of discoveries by others who have come before. In fact, Hopi Hoekstra's experiment benefited from the work of another

researcher, D. W. Kaufman, 40 years earlier. You can study the design of Kaufman's experiment and interpret the results in the **Scientific Skills Exercise**.

Scientific results are continually scrutinized through the repetition of observations and experiments. Scientists working in the same research field often check one another's claims by attempting to confirm observations or repeat experiments. If scientific colleagues cannot repeat experimental findings, this failure may reflect some underlying weakness in the original claim, which will then have to be revised. In this sense, science polices itself. Integrity and adherence to high professional standards in reporting results are central to the scientific endeavor, since the validity of experimental data is key to designing further lines of inquiry.

It is not unusual for several scientists to converge on the same research question. Some scientists enjoy the challenge of being first with an important discovery or key experiment, while others derive more satisfaction from cooperating with fellow scientists working on the same problem.

Cooperation is facilitated when scientists use the same organism. Often, it is a widely used **model organism**—a species that is easy to grow in the lab and lends itself particularly well to the questions being investigated. Because all species are evolutionarily related, such an organism may be viewed as a model for understanding the biology of other species and their diseases. Genetic studies of the fruit fly *Drosophila melanogaster*, for example, have taught us a lot about how genes work in other species, even humans. Some other popular model organisms are the mustard plant *Arabidopsis thaliana*, the soil worm *Caenorhabditis elegans*, the zebrafish *Danio rerio*, the mouse *Mus musculus*, and the bacterium *Escherichia coli*. As you read through this book, note the many contributions that these and other model organisms have made to the study of life.

Biologists may approach interesting questions from different angles. Some biologists focus on ecosystems, while others study natural phenomena at the level of organisms or cells. This text is divided into units that look at biology at different levels and investigate problems through different approaches. Yet any given problem can be addressed from many perspectives, which in fact complement each other. For example, Hoekstra not only carried out field studies showing that coat coloration can affect predation rates but also did lab studies that uncovered at least one genetic mutation that underlies the differences between beach and inland mouse coloration. Her lab includes biologists specializing in different biological levels, allowing links to be made between the evolutionary adaptations she focuses on and their molecular basis in DNA sequences.

As a biology student, you can benefit from making connections between the different levels of biology. You can develop this skill by noticing when certain topics crop up again and again in different units. One such topic is sickle-cell disease, a well-understood genetic condition that is prevalent among

Interpreting a Pair of Bar Graphs

How much does camouflage affect predation on mice by owls with and without moonlight? D. W. Kaufman hypothesized that the extent to which the coat color of a mouse contrasted with the color of its surroundings would affect the rate of nighttime predation by owls. He also hypothesized that the contrast would be affected by the amount of moonlight. In this exercise, you will analyze data from his studies of owl predation on mice that tested these hypotheses.

How the Experiment Was Done Pairs of mice (*Peromyscus polionotus*) with different coat colors, one light brown and one dark brown, were released simultaneously into an enclosure that contained a hungry owl. The researcher recorded the color of the mouse that was first caught by the owl. If the owl did not catch either mouse within 15 minutes, the test was recorded as a zero. The release trials were repeated multiple times in enclosures with either a dark-colored soil surface or a light-colored soil surface. The presence or absence of moonlight during each assay was recorded.

Data from the Experiment

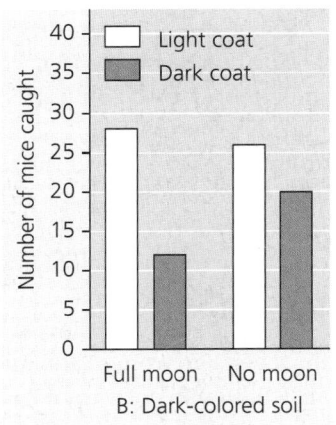

Data from D. W. Kaufman, Adaptive coloration in *Peromyscus polionotus*: Experimental selection by owls, *Journal of Mammalogy* 55:271–283 (1974).

INTERPRET THE DATA

1. First, make sure you understand how the graphs are set up. Graph A shows data from the light-colored soil enclosure and graph B from the dark-colored enclosure, but in all other respects the graphs are the same. **(a)** There is more than one independent variable in these graphs. What are the independent variables, the variables that were tested by the researcher? Which axis of the graphs has the independent variables? **(b)** What is the dependent variable, the response to the variables being tested? Which axis of the graphs has the dependent variable?

2. **(a)** How many dark brown mice were caught in the light-colored soil enclosure on a moonlit night? **(b)** How many dark brown mice were caught in the dark-colored soil enclosure on a moonlit night? **(c)** On a moonlit night, would a dark brown mouse be more likely to escape predation by owls on dark- or light-colored soil? Explain your answer.

3. **(a)** Is a dark brown mouse on dark-colored soil more likely to escape predation under a full moon or with no moon? **(b)** What about a light brown mouse on light-colored soil? Explain.

4. **(a)** Under which conditions would a dark brown mouse be most likely to escape predation at night? **(b)** A light brown mouse?

5. **(a)** What combination of independent variables led to the highest predation level in enclosures with light-colored soil? **(b)** What combination of independent variables led to the highest predation level in enclosures with dark-colored soil?

6. Thinking about your answers to question 5, provide a simple statement describing conditions that are especially deadly for either color of mouse.

7. Combining the data from both graphs, estimate the number of mice caught in moonlight versus no-moonlight conditions. Which condition is optimal for predation by the owl? Explain.

➡ **Instructors:** A version of this Scientific Skills Exercise can be assigned in **Mastering Biology**.

---

native inhabitants of Africa and other warm regions and their descendants. Sickle-cell disease will appear in several units of the text, each time addressed at a new level. In addition, Make Connections figures connect the content in different chapters, and Make Connections questions ask you to make the connections yourselves. We hope these features will help you integrate the material you're learning and enhance your enjoyment of biology by encouraging you to keep the big picture in mind.

Science, Technology, and Society

The research community is part of society at large, and the relationship of science to society becomes clearer when we add technology to the picture (see Figure 1.23).

Though science and technology sometimes employ similar inquiry patterns, their basic goals differ. The goal of science is to understand natural phenomena, while that of **technology** is to *apply* scientific knowledge for some specific purpose. Because scientists put new technology to work in their research, science and technology are interdependent.

The potent combination of science and technology can have dramatic effects on society. Sometimes, the applications of basic research that turn out to be the most beneficial come out of the blue, from completely unanticipated observations in the course of scientific exploration. For example, discovery of the structure of DNA by Watson and Crick in 1953 and subsequent achievements in DNA science led to the technologies of DNA manipulation that are transforming applied fields

▼ Figure 1.26 DNA technology and forensics. Since 1992, the Innocence Project has used forensic analysis of DNA samples from crime scenes to exonerate over 360 wrongly convicted prisoners. Most had served many years in prison. To read about the four people shown here who were found to be innocent, go to the Innocence Project website.

Lynn DeJac

John White

Michael Morton

Clemente Aguirre-Jarquin

such as medicine, agriculture, and forensics **(Figure 1.26)**. Perhaps Watson and Crick envisioned that their discovery would someday lead to important applications, but it is unlikely that they could have predicted exactly what all those applications would be.

The directions that technology takes depend less on the curiosity that drives basic science than on the current needs and wants of people and on the social environment of the times. Debates about technology center more on "*should* we do it" than "*can* we do it." With advances in technology come difficult choices. For example, under what circumstances is it acceptable to use DNA technology to find out if particular people have genes for hereditary diseases? Should such tests always be voluntary, or are there circumstances when genetic testing should be mandatory? Should insurance companies or employers have access to the information, as they do for many other types of personal health data? These questions are becoming much more urgent as the sequencing of individual genomes becomes quicker and cheaper.

Ethical issues raised by such questions have as much to do with politics, economics, and cultural values as with science and technology. All citizens—not only professional scientists—have a responsibility to be informed about how science works and about the potential benefits and risks of technology. The relationship between science, technology, and society increases the significance and value of any biology course.

The Value of Diverse Viewpoints in Science

Many of the technological innovations with the most profound impact on human society originated in settlements along trade routes, where a rich mix of different cultures ignited new ideas. For example, the printing press, which helped spread knowledge to all social classes, was invented by the German Johannes Gutenberg around 1440. This invention relied on several innovations from China, including paper and ink. Paper traveled along trade routes from China to Baghdad, where technology was developed for its mass production. This technology then migrated to Europe, as did water-based ink from China, which was modified by Gutenberg to become oil-based ink. We have the cross-fertilization of diverse cultures to thank for the printing press, and the same can be said for other important inventions.

Along similar lines, science benefits from a diversity of backgrounds and viewpoints among its practitioners. But just how diverse a population are scientists in relation to gender, race, ethnicity, and other attributes?

The scientific community reflects the cultural standards and behaviors of the society around it. It is therefore not surprising that until recently, women, people of color, and other underrepresented groups have faced huge obstacles in their pursuit to become professional scientists in many countries around the world. Over the past 50 years, changing attitudes about career choices have increased the proportion of women in biology and some other sciences, so that now women constitute roughly half of undergraduate biology majors and biology Ph.D. students.

The pace has been slow at higher levels in the profession, however, and women and many racial and ethnic groups are still significantly underrepresented in many branches of science and technology. This lack of diversity hampers the progress of science. The more voices that are heard at the table, the more robust, valuable, and productive the scientific interchange will be. The authors of this text welcome all students to the community of biologists, wishing you the joys and satisfactions of this exciting field of science.

CONCEPT CHECK 1.4

1. How does science differ from technology?
2. **MAKE CONNECTIONS** The gene that causes sickle-cell disease is present in a higher percentage of residents of sub-Saharan Africa than among those of African descent living in the United States. Even though this gene causes sickle-cell disease, it also provides some protection from malaria, a serious disease that is widespread in sub-Saharan Africa but absent in the United States. Discuss an evolutionary process that could account for the different percentages of the sickle-cell gene among residents of the two regions. (See Concept 1.2.)

For suggested answers, see Appendix A.

1 Chapter Review

→ Go to **Mastering Biology** for Assignments, the eText, the Study Area, and Dynamic Study Modules.

SUMMARY OF KEY CONCEPTS

→ To review key terms, go to the **Vocabulary Self-Quiz** in the **Mastering Biology** eText or Study Area, or go to goo.gl/zkjz9t.

CONCEPT 1.1

The study of life reveals unifying themes (pp. 3–11)

Organization Theme: New Properties Emerge at Successive Levels of Biological Organization

- The hierarchy of life unfolds as follows: biosphere > ecosystem > community > population > organism > organ system > organ > tissue > cell > organelle > molecule > atom. With each step from atoms to the biosphere, new **emergent properties** result from interactions among components at the lower levels. In an approach called reductionism, complex systems are broken down to simpler components that are more manageable to study. In **systems biology**, scientists attempt to model the dynamic behavior of whole biological systems by studying the interactions among the system's parts.
- Structure and function are correlated at all levels of biological organization. The cell, an organism's basic unit of structure and function, is the lowest level that can perform all activities required for life. Cells are either prokaryotic or eukaryotic. **Eukaryotic cells** have a DNA-containing nucleus and other membrane-enclosed organelles. **Prokaryotic cells** lack such organelles.

Information Theme: Life's Processes Involve the Expression and Transmission of Genetic Information

- Genetic information is encoded in the nucleotide sequences of **DNA**. It is DNA that transmits heritable information from parents to offspring. DNA sequences (called **genes**) program a cell's protein production by being transcribed into mRNA and then translated into specific proteins, a process called **gene expression**. Gene expression also produces RNAs that are not translated into protein but serve other important functions. **Genomics** is the large-scale analysis of the DNA sequences of a species (its **genome**) as well as the comparison of genomes between species. **Bioinformatics** uses computational tools to deal with huge volumes of sequence data.

Energy and Matter Theme: Life Requires the Transfer and Transformation of Energy and Matter

- Energy flows through an ecosystem. All organisms must perform work, which requires energy. **Producers** convert energy from sunlight to chemical energy, some of which is used by them and by **consumers** to do work, and is eventually lost from the ecosystem as heat. Chemicals cycle between organisms and the environment.

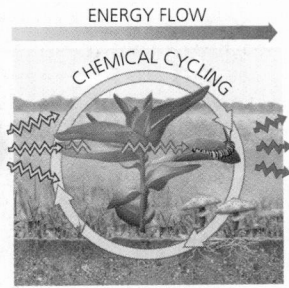
ENERGY FLOW
CHEMICAL CYCLING

Interactions Theme: From Molecules to Ecosystems, Interactions Are Important in Biological Systems

- In **feedback regulation**, a process is regulated by its output or end product. In negative feedback, accumulation of the end product slows its production. In positive feedback, an end product speeds up its own production.

- Organisms interact continuously with physical factors. Plants take up nutrients from the soil and chemicals from the air and use energy from the sun.

? *Thinking about the muscles and nerves in your hand, how does the activity of text messaging reflect the four unifying themes of biology described in this section?*

CONCEPT 1.2

The Core Theme: Evolution accounts for the unity and diversity of life (pp. 11–16)

- **Evolution**, the process of change that has transformed life on Earth, accounts for the unity and diversity of life. It also explains evolutionary adaptation—the match of organisms to their environments.
- Biologists classify species according to a system of broader and broader groups. Domain **Bacteria** and domain **Archaea** consist of prokaryotes. Domain **Eukarya**, the eukaryotes, includes various groups of protists and the kingdoms Plantae, Fungi, and Animalia. As diverse as life is, there is also evidence of remarkable unity, revealed in the similarities between different species.
- Darwin proposed **natural selection** as the mechanism for evolutionary adaptation of populations to their environments. Natural selection is the evolutionary process that occurs when a population is exposed to environmental factors that consistently cause individuals with certain heritable traits to have greater reproductive success than do individuals with other heritable traits.

Population of organisms

Hereditary variations

Overproduction of offspring and competition

Environmental factors

Differences in reproductive success of individuals

Evolution of adaptations in the population

- Each species is one twig of a branching tree of life extending back in time through more and more remote ancestral species. All of life is connected through its long evolutionary history.

? *How could natural selection have led to the evolution of adaptations such as camouflaging coat color in beach mice?*

In studying nature, scientists form and test hypotheses
(pp. 16–22)

- In scientific **inquiry**, scientists collect **data** and use **inductive reasoning** to draw a general conclusion, which can be developed into a testable **hypothesis**. **Deductive reasoning** uses predictions to test hypotheses. Hypotheses must be testable; science can address neither the possibility of supernatural phenomena nor religious beliefs. Hypotheses can be tested by conducting **experiments** or, when that is not possible, by making observations. In the process of science, the core activity is testing ideas. This endeavor is influenced by exploration and discovery, community analysis and feedback, and societal outcomes.
- **Controlled experiments** are designed to demonstrate the effect of one **variable** by testing control groups and experimental groups that differ in only that one variable.
- A scientific **theory** is broad in scope, generates new hypotheses, and is supported by a large body of evidence.

? *What are the roles of gathering and interpreting data?*

Science benefits from a cooperative approach and diverse viewpoints *(pp. 22–24)*

- Science is a social activity. The work of each scientist builds on the work of others who have come before. Scientists must be able to repeat each other's results, and integrity is key. Biologists approach questions at different levels; their approaches complement each other.
- **Technology** consists of any method or device that applies scientific knowledge for some specific purpose that affects society. The impact of basic research is not always immediately obvious.
- Diversity among scientists promotes progress in science.

? *Explain why different approaches and diverse backgrounds among scientists are important.*

TEST YOUR UNDERSTANDING

→ For more multiple-choice questions, go to the **Practice Test** in the **Mastering Biology** eText or Study Area, or go to goo.gl/GruWRg.

Levels 1-2: Remembering/Understanding

1. All the organisms on your campus make up
   - (A) an ecosystem.
   - (B) a community.
   - (C) a population.
   - (D) a taxonomic domain.

2. Systems biology is mainly an attempt to
   - (A) analyze genomes from different species.
   - (B) simplify complex problems by reducing the system into smaller, less complex units.
   - (C) understand the behavior of entire biological systems by studying interactions among its component parts.
   - (D) build high-throughput machines to rapidly acquire data.

3. Which of these best demonstrates unity among organisms?
   - (A) emergent properties
   - (B) descent with modification
   - (C) the structure and function of DNA
   - (D) natural selection

4. A controlled experiment is one that
   - (A) proceeds slowly so a scientist can make careful records.
   - (B) tests experimental and control groups in parallel.
   - (C) is repeated many times to make sure the results are accurate.
   - (D) keeps all variables constant.

5. Which of the following statements best distinguishes hypotheses from theories in science?
   - (A) Theories are hypotheses that have been proved.
   - (B) Hypotheses are guesses; theories are correct answers.
   - (C) Hypotheses usually are relatively narrow in scope; theories have broad explanatory power.
   - (D) Theories are proved true; hypotheses are often contradicted by experimental results.

Levels 3-4: Applying/Analyzing

6. Which of the following is an example of qualitative data?
   - (A) The fish swam in a zigzag motion.
   - (B) The contents of the stomach are mixed every 20 seconds.
   - (C) The temperature decreased from 20°C to 15°C.
   - (D) The six pairs of robins hatched an average of three chicks each.

7. Which sentence best describes the logic of scientific inquiry?
   - (A) If I generate a testable hypothesis, tests and observations will support it.
   - (B) If my prediction is correct, it will lead to a testable hypothesis.
   - (C) If my observations are accurate, they will support my hypothesis.
   - (D) If my prediction turns out to be correct, my hypothesis is supported.

8. **DRAW IT** Draw a biological hierarchy similar to the one in Figure 1.3 but using a coral reef as the ecosystem, a fish as the organism, its stomach as the organ, and DNA as the molecule. Include all levels in the hierarchy.

Levels 5-6: Evaluating/Creating

9. **EVOLUTION CONNECTION** A typical prokaryotic cell has about 3,000 genes in its DNA, while a human cell has about 21,300 genes. About 1,000 of these genes are present in both types of cells. Explain how such different organisms could have this same subset of 1,000 genes. What sorts of functions might these shared genes have?

10. **SCIENTIFIC INQUIRY** Based on the results of the mouse coloration case study, suggest another hypothesis researchers might use to study the role of predators in natural selection.

11. **SCIENTIFIC INQUIRY** Scientists search the scientific literature using electronic databases such as PubMed, a free online database maintained by the National Center for Biotechnology Information. Use PubMed to find the abstract of an article that Hopi Hoekstra published in 2017 or later.

12. **WRITE ABOUT A THEME: EVOLUTION** In a short essay (100–150 words), discuss Darwin's view of how natural selection resulted in both unity and diversity of life. Include in your discussion some of his evidence. (For help in writing essays, see "Writing Tips and Rubrics" in the Study Area of Mastering Biology under "Additional Resources.")

13. **SYNTHESIZE YOUR KNOWLEDGE**

Can you pick out the mossy leaf-tailed gecko lying against the tree trunk in this photo? How is the appearance of the gecko a benefit in terms of survival? Given what you learned about evolution, natural selection, and genetic information in this chapter, describe how the gecko's coloration might have evolved.

For selected answers, see Appendix A.

Dr. Kenneth Olden recently retired from a long and distinguished career in medical research and public health, including serving as Director of the National Institute of Environmental Health Sciences from 1991 to 2005, as founding Dean of the School of Public Health at the City University of New York from 2008 to 2012, and as Director of the National Center for Environmental Assessment from 2012 to 2016. He has published over 220 research papers and has received many honors and awards, among them the three most distinguished awards in public health. Dr. Olden grew up in Parrottsville, Tennessee, the son of a sharecropper. He remembers walking up a long hill to high school every morning and daydreaming about helping the poor people—both black and white—in the neighborhoods he'd walk through, wanting to make a difference.

AN INTERVIEW WITH

Kenneth Olden

What got you interested in biology?

I was always cerebral, I always liked to read and think. For me, role models were important. At that time, I knew about only two professions that blacks were in: medicine and teaching. There was one black physician in town—unusual for a rural town. My high school principal—he was black—would tell us, "By golly, you could be anything you want to be!" I paid attention and listened. He helped me apply to Knoxville College, and I decided I would be a physician, so I majored in biology and minored in chemistry. Then, in my senior year, my professor at Knoxville—he was interested in diversity—got me into a research program at the University of Tennessee, which was not integrated at that time—blacks couldn't attend. But I was allowed to do research on tapeworms, irradiate them and examine their chromosomes, and I was allowed to attend the seminars. I was fascinated by the research,

▼ Dr. Olden established Children's Environmental Health and Disease Prevention Research Centers.

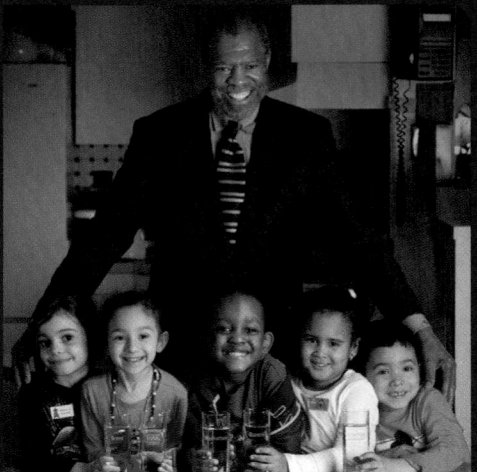

I was just turned on—finally, I realized this is what I'd really like to do.

Can you tell me about how you got into cancer research?

After my Ph.D. and my postdoctoral research at Harvard, I realized I wanted to work on animal cells, so I joined Ira Pastan's group at the National Cancer Institute at the National Institutes of Health, where I eventually got my own lab. Together with Ken Yamada, I was working on a protein called fibronectin, which was present on the outside surface of normal cells but not cancer cells. Fibronectin is a glycoprotein—it has carbohydrates (sugars) attached to it. At the time, there was a hypothesis that the carbohydrates were necessary for fibronectin to be exported from the cell, and we decided to test that hypothesis using a drug called tunicamycin that prevented carbohydrate attachment. We showed that carbohydrates weren't necessary for export, but instead they were important for stabilizing the protein's structure. That turned out to be one of the most cited papers for 1978; it was huge.

In 1991, you became Director of the National Institute of Environmental Health Sciences. What were your goals and accomplishments there?

When I interviewed for the position, I told the Director of the NIH, "My first priority would be to make the Institute responsive to the needs of the American people." She immediately offered me the job—and that changed my life. It gave me the opportunity to think big and to address

"One of us from rural America had to make it—and I thought, 'Why not me?'"

a lot of issues I felt weren't being dealt with, kind of what I'd been dreaming of. Environmental health research at that time focused on chemical carcinogenesis, and I wanted to expand that focus to social and behavioral issues also, as well as genetics. Over my time there, I engaged communities in identifying areas of concern for our research, such as disproportionate exposure to chemicals in certain neighborhoods. I founded the Environmental Genome Project, which used a novel genomic approach to determine susceptibility to toxins. I also expanded Environmental Health Centers around the country, developing the Breast Cancer and the Environment Research Program and Children's Environmental Health and Disease Prevention Research Centers. Children are really important to me—they are especially susceptible to environmental toxins, and we needed to address that.

What is your advice to an undergraduate considering a career in biology?

Most people, I think, will figure out what is the right thing to do, but often it takes a lot of courage to do the right thing. When I accepted the Sackler Prize, I talked about walking to high school and realizing that government was making a lot of decisions that affected rural America without ever bothering to consult rural Americans. In order to change that, one of us from rural America had to make it—and I thought, "Why not me?" In being awarded the prize, for creating community-based participatory research, it looks like I actually succeeded in what I set out to do: to get the public health decision-makers to pay attention to the needs of the poor.

2 The Chemical Context of Life

KEY CONCEPTS

2.1 Matter consists of chemical elements in pure form and in combinations called compounds *p. 29*

2.2 An element's properties depend on the structure of its atoms *p. 30*

2.3 The formation and function of molecules and ionic compounds depend on chemical bonding between atoms *p. 36*

2.4 Chemical reactions make and break chemical bonds *p. 40*

Study Tip

Make a table: As you read the chapter, make a summary table like the following. Add more rows as you proceed.

Property	Element (atom)			
	C	H	O	N
Atomic number				
# Electrons				
# Neutrons				
Mass number				
Electron distribution diagram				
# Valence electrons				

→ Go to Mastering Biology

For Students (in eText and Study Area)
- Get Ready for Chapter 2
- Figure 2.6 Walkthrough: Energy Levels of an Atom's Electrons
- Animation: Introduction to Chemical Bonds

For Instructors to Assign (in Item Library)
- Chemistry Review: Atoms and Molecules: Covalent Bonds
- Chemistry Review: Atoms and Molecules: Electronegativity

Figure 2.1 Wood ants (*Formica rufa*) use chemistry to ward off enemies. When threatened from above, they shoot volleys of formic acid from their abdomens into the air. The acid bombards and stings potential predators, such as hungry birds.

What determines the properties of a compound such as formic acid?

A compound is made of atoms joined by bonds. Formic acid (CH_2O_2) consists of carbon (C), hydrogen (H), and oxygen (O).

Formic acid

The number of protons (⊕) determines an atom's identity. Oxygen has 8 protons.

Oxygen atom

An atom's electron (⊖) distribution determines its ability to form bonds. Oxygen has space for 2 more electrons, so it can form 2 bonds.

A compound's properties depend on its atoms and how they are bonded together.

In formic acid, this O attracts H's electron, releasing H^+ and making this compound an acid, which stings.

Matter consists of chemical elements in pure form and in combinations called compounds

Organisms are composed of **matter**, which is anything that takes up space and has mass. Matter exists in many forms. Rocks, metals, oils, gases, and living organisms are a few examples of what seems to be an endless assortment of matter.

Elements and Compounds

Matter is made up of elements. An **element** is a substance that cannot be broken down to other substances by chemical reactions. Today, chemists recognize 92 elements occurring in nature; gold, copper, carbon, and oxygen are examples. Each element has a symbol, usually the first letter or two of its name. Some symbols are derived from Latin or German; for instance, the symbol for sodium is Na, from the Latin word *natrium*.

A **compound** is a substance consisting of two or more different elements combined in a fixed ratio. Table salt, for example, is sodium chloride (NaCl), a compound composed of the elements sodium (Na) and chlorine (Cl) in a 1:1 ratio. Pure sodium is a metal, and pure chlorine is a poisonous gas. When chemically combined, however, sodium and chlorine form an edible compound. Water (H_2O), another compound, consists of the elements hydrogen (H) and oxygen (O) in a 2:1 ratio. These are simple examples of organized matter having emergent properties: A compound has characteristics different from those of its elements **(Figure 2.2)**.

The Elements of Life

Of the 92 natural elements, about 20–25% are **essential elements** that an organism needs to live a healthy life and reproduce. The essential elements are similar among organisms, but there is some variation—for example, humans need 25 elements, but plants need only 17.

▼ **Figure 2.2 The emergent properties of a compound.** The metal sodium combines with the poisonous gas chlorine, forming the edible compound sodium chloride, or table salt.

| **Na** | + | **Cl** | → | **NaCl** |
| Sodium | | Chlorine (gas) | | Sodium chloride |

Just four elements—oxygen (O), carbon (C), hydrogen (H), and nitrogen (N)—make up approximately 96% of living matter. Calcium (Ca), phosphorus (P), potassium (K), sulfur (S), and a few other elements account for most of the remaining 4% or so of an organism's mass. **Trace elements** are required by an organism in only minute quantities. Some trace elements, such as iron (Fe), are needed by all forms of life; others are required only by certain species. For example, in vertebrates (animals with backbones), the element iodine (I) is an essential ingredient of a hormone produced by the thyroid gland. A daily intake of only 0.15 milligram (mg) of iodine is adequate for normal activity of the human thyroid. An iodine deficiency in the diet causes the thyroid gland to grow to abnormal size, a condition called goiter. Consuming seafood or iodized salt reduces the incidence of goiter. Relative amounts of all the elements in the human body are listed in **Table 2.1**.

Some naturally occurring elements are toxic to organisms. In humans, for instance, the element arsenic has been linked to numerous diseases and can be lethal. In some areas of the world, arsenic occurs naturally and can make its way into the groundwater. As a result of using water from drilled wells in southern Asia, millions of people have been inadvertently exposed to arsenic-laden water. Efforts are under way to reduce arsenic levels in their water supply.

> ⊙ **Mastering Biology** Interview with Kenneth Olden: **Assessing susceptibility to environmental toxins using genomics (see the interview before Chapter 2)**

Table 2.1 Elements in the Human Body

Element	Symbol	Percentage of Body Mass (including water)	
Oxygen	O	65.0%	
Carbon	C	18.5%	
Hydrogen	H	9.5%	96.3%
Nitrogen	N	3.3%	
Calcium	Ca	1.5%	
Phosphorus	P	1.0%	
Potassium	K	0.4%	
Sulfur	S	0.3%	3.7%
Sodium	Na	0.2%	
Chlorine	Cl	0.2%	
Magnesium	Mg	0.1%	

Trace elements (less than 0.01% of mass): Boron (B), chromium (Cr), cobalt (Co), copper (Cu), fluorine (F), iodine (I), iron (Fe), manganese (Mn), molybdenum (Mo), selenium (Se), silicon (Si), tin (Sn), vanadium (V), zinc (Zn)

INTERPRET THE DATA *Given the makeup of the human body, what compound do you think accounts for the high percentage of oxygen?*

Case Study: Evolution of Tolerance to Toxic Elements

EVOLUTION Some species have become adapted to environments containing elements that are usually toxic; an example is serpentine plant communities. Serpentine is a jade-like mineral that contains elevated concentrations of elements such as chromium, nickel, and cobalt. Although most plants cannot survive in soil that forms from serpentine rock, a small number of plant species have adaptations that allow them to do so **(Figure 2.3)**. Presumably, variants of ancestral, nonserpentine species arose that could survive in serpentine soils, and subsequent natural selection resulted in the distinctive array of species we see in these areas today. Serpentine-adapted plants are of great interest to researchers because studying them can teach us so much about natural selection and evolutionary adaptations on a local scale.

▼ **Figure 2.3 Serpentine plant community.** These plants are growing on serpentine soil, which contains elements that are usually toxic to plants. The insets show a close-up of serpentine rock and one of the plants, a Tiburon Mariposa lily (*Calochortus tiburonensis*). This specially adapted species is found only on this one hill in Tiburon, a peninsula that juts into San Francisco Bay.

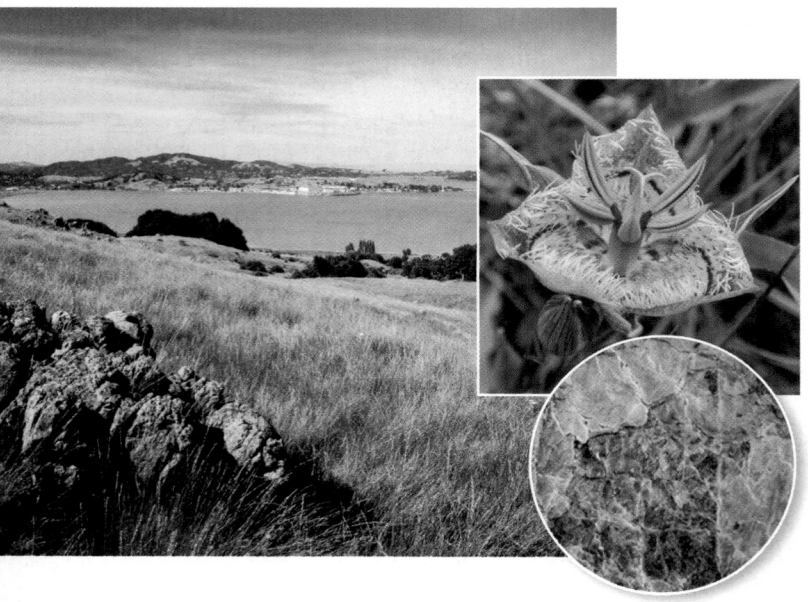

CONCEPT CHECK 2.1

1. **MAKE CONNECTIONS** Explain how table salt has emergent properties. (See Concept 1.1.)

2. Is a trace element an essential element? Explain.

3. **WHAT IF?** In humans, iron is a trace element required for the proper functioning of hemoglobin, the molecule that carries oxygen in red blood cells. What might be the effects of an iron deficiency?

4. **MAKE CONNECTIONS** Explain how natural selection might have played a role in the evolution of species that are tolerant of serpentine soils. (Review Concept 1.2.)

For suggested answers, see Appendix A.

An element's properties depend on the structure of its atoms

Each element consists of a certain type of atom that is different from the atoms of any other element. An **atom** is the smallest unit of matter that still retains the properties of an element. Atoms are so small that it would take about a million of them to stretch across the period printed at the end of this sentence. We symbolize atoms with the same abbreviation used for the element that is made up of those atoms. For example, the symbol C stands for both the element carbon and a single carbon atom.

Subatomic Particles

Although the atom is the smallest unit having the properties of an element, these tiny bits of matter are composed of even smaller parts, called *subatomic particles*. Using high-energy collisions, physicists have produced more than 100 types of particles from the atom, but only three kinds of particles are relevant here: **neutrons**, **protons**, and **electrons**. Protons and electrons are electrically charged. Each proton has one unit of positive charge, and each electron has one unit of negative charge. A neutron, as its name implies, is electrically neutral.

Protons and neutrons are packed together tightly in a dense core, or **atomic nucleus**, at the center of an atom; protons give the nucleus a positive charge. The rapidly moving electrons form a "cloud" of negative charge around the nucleus, and it is the attraction between opposite charges that keeps the electrons in the vicinity of the nucleus. **Figure 2.4** shows two commonly used models of the structure of the helium atom as an example.

▼ **Figure 2.4 Simplified models of a helium (He) atom.** The helium nucleus consists of 2 neutrons (brown) and 2 protons (pink). Two electrons (yellow) exist outside the nucleus. These models are not to scale; they greatly overestimate the size of the nucleus in relation to the electron cloud.

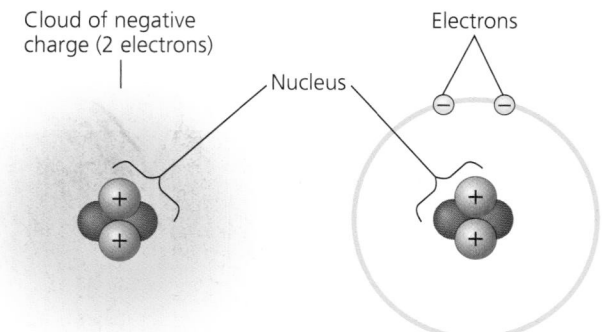

Cloud of negative charge (2 electrons)

Electrons

Nucleus

(a) This model represents the two electrons as a cloud of negative charge, a result of their motion around the nucleus.

(b) In this more simplified model, the electrons are shown as two small yellow spheres on a circle around the nucleus.

The neutron and proton are almost identical in mass, each about 1.7×10^{-24} gram (g). Grams and other conventional units are not very useful for describing the mass of objects that are so minuscule. Thus, for atoms and subatomic particles (and for molecules, too), we use a unit of measurement called the **dalton**, in honor of John Dalton, the British scientist who helped develop atomic theory around 1800. (The dalton is the same as the *atomic mass unit*, or *amu*, a unit you may have encountered elsewhere.) Neutrons and protons have masses close to 1 dalton. Because the mass of an electron is only about 1/2,000 that of a neutron or proton, we can ignore electrons when computing the total mass of an atom.

Atomic Number and Atomic Mass

Atoms of the various elements differ in their number of subatomic particles. All atoms of a particular element have the same number of protons in their nuclei. This number of protons, which is unique to that element, is called the **atomic number** and is written as a subscript to the left of the symbol for the element. The abbreviation $_2\text{He}$, for example, tells us that an atom of the element helium has 2 protons in its nucleus. Unless otherwise indicated, an atom is neutral in electrical charge, which means that its protons must be balanced by an equal number of electrons. Therefore, the atomic number tells us the number of protons and also the number of electrons in an electrically neutral atom.

We can deduce the number of neutrons from a second quantity, the **mass number**, which is the total number of protons and neutrons in the nucleus of an atom. The mass number is written as a superscript to the left of an element's symbol. For example, we can use this shorthand to write an atom of helium as $_2^4\text{He}$. Because the atomic number indicates how many protons there are, we can determine the number of neutrons by subtracting the atomic number from the mass number. In our example, the helium atom $_2^4\text{He}$ has 2 neutrons. For sodium (Na):

Mass number = number of protons + neutrons
= 23 for sodium

$_{11}^{23}\text{Na}$

Atomic number = number of protons
= number of electrons in a neutral atom
= 11 for sodium

Number of neutrons = mass number − atomic number
= 23 − 11 = 12 for sodium

The simplest atom is hydrogen $_1^1\text{H}$, which has no neutrons; it consists of a single proton with a single electron.

Because the contribution of electrons to mass is negligible, almost all of an atom's mass is concentrated in its nucleus. Neutrons and protons each have a mass very close to

1 dalton, so the mass number is close to, but slightly different from, the total mass of an atom, called its **atomic mass**. For example, the mass number of sodium ($_{11}^{23}\text{Na}$) is 23, but its atomic mass is 22.9898 daltons; the difference is explained below.

> **Mastering Biology** Animation: Atomic Number and Atomic Mass

Isotopes

All atoms of a given element have the same number of protons, but some atoms have more neutrons than other atoms of the same element and therefore have greater mass. These different atomic forms of the same element are called **isotopes** of the element. In nature, an element may occur as a mixture of its isotopes. As an example, the element carbon, which has the atomic number 6, has three naturally occurring isotopes. The most common isotope is carbon-12, $_6^{12}\text{C}$, which accounts for about 99% of the carbon in nature. The isotope $_6^{12}\text{C}$ has 6 neutrons. Most of the remaining 1% of carbon consists of atoms of the isotope $_6^{13}\text{C}$, with 7 neutrons. A third, even rarer isotope, $_6^{14}\text{C}$, has 8 neutrons. Notice that all three isotopes of carbon have 6 protons; otherwise, they would not be carbon. Although the isotopes of an element have slightly different masses, they behave identically in chemical reactions. (For an element with more than one naturally occurring isotope, the atomic mass is an average of those isotopes, weighted by their abundance. Thus, carbon has an atomic mass of 12.01 daltons.)

Both ^{12}C and ^{13}C are stable isotopes, meaning that their nuclei do not have a tendency to lose subatomic particles, a process called decay. The isotope ^{14}C, however, is unstable, or radioactive. A **radioactive isotope** is one in which the nucleus decays spontaneously, giving off particles and energy. When the radioactive decay leads to a change in the number of protons, it transforms the atom to an atom of a different element. For example, when a carbon-14 (^{14}C) atom decays, a neutron decays into a proton, transforming the atom into a nitrogen (^{14}N) atom. Radioactive isotopes have many useful applications in biology.

Radioactive Tracers

Radioactive isotopes are often used as diagnostic tools in medicine. Cells can use radioactive atoms just as they would use nonradioactive isotopes of the same element. The radioactive isotopes are incorporated into biologically active molecules, which are then used as tracers to track atoms during metabolism, the chemical processes of an organism. For example, certain kidney disorders are diagnosed by injecting small doses of radioactively labeled substances into the blood and then analyzing the tracer molecules excreted in the urine. Radioactive tracers are also used in combination with sophisticated imaging instruments, such as PET scanners that

► **Figure 2.5 A PET scan, a medical use for radioactive isotopes.** PET, an acronym for positron-emission tomography, detects locations of intense chemical activity in the body. The bright yellow spot marks an area with an elevated level of radioactively labeled glucose, which in turn indicates high metabolic activity, a hallmark of cancerous tissue.

Cancerous throat tissue

can monitor growth and metabolism of cancers in the body **(Figure 2.5)**.

Although radioactive isotopes are very useful in biological research and medicine, radiation from decaying isotopes also poses a hazard to life by damaging cellular molecules. The severity of this damage depends on the type and amount of radiation an organism absorbs. One of the most serious environmental threats is radioactive fallout from nuclear accidents. The doses of most isotopes used in medical diagnosis, however, are relatively safe.

Radiometric Dating

EVOLUTION Researchers measure radioactive decay in fossils to date these relics of past life. Fossils provide a large body of evidence for evolution, documenting differences between organisms from the past and those living at present and giving us insight into species that have disappeared over time. While the layering of fossil beds establishes that deeper fossils are older than more shallow ones, the actual age (in years) of the fossils in each layer cannot be determined by position alone. This is where radioactive isotopes come in.

A "parent" isotope decays into its "daughter" isotope at a fixed rate, expressed as the **half-life** of the isotope—the time it takes for 50% of the parent isotope to decay. Each radioactive isotope has a characteristic half-life that is not affected by temperature, pressure, or any other environmental variable. Using a process called **radiometric dating**, scientists measure the ratio of different isotopes and calculate how many half-lives (in years) have passed since an organism was fossilized or a rock was formed. Half-life values range from very short for some isotopes, measured in seconds or days, to extremely long—uranium-238 has a half-life of 4.5 billion years! Each isotope can best "measure" a particular range of years: Uranium-238 was used to determine that moon rocks are approximately 4.5 billion years old, similar to the estimated age of Earth. In the **Scientific Skills Exercise**, you can work with data from an experiment that used carbon-14 to determine the age of an important fossil. (Figure 25.6 explains more about radiometric dating of fossils.)

The Energy Levels of Electrons

The simplified models of the atom in Figure 2.4 greatly exaggerate the size of the nucleus relative to that of the whole atom. If an atom of helium were the size of a typical football stadium, the nucleus would be the size of a pencil eraser in the center of the field. Moreover, the electrons would be like two tiny gnats buzzing around the stadium. Atoms are mostly empty space. When two atoms approach each other during a chemical reaction, their nuclei do not come close enough to interact. Of the three subatomic particles we have discussed, only electrons are directly involved in chemical reactions.

An atom's electrons vary in the amount of energy they possess. **Energy** is defined as the capacity to cause change—for instance, by doing work. **Potential energy** is the energy that matter possesses because of its location or structure. For example, water in a reservoir on a hill has potential energy because of its altitude. When the gates of the reservoir's dam are opened and the water runs downhill, the energy can be used to do work, such as moving the blades of turbines to generate electricity. Because energy has been expended, the water has less energy at the bottom of the hill than it did in the reservoir. Matter has a natural tendency to move toward the lowest possible state of potential energy; in our example, the water runs downhill. To restore the potential energy of a reservoir, work must be done to elevate the water against gravity.

The electrons of an atom have potential energy due to their distance from the nucleus **(Figure 2.6)**. The negatively charged electrons are attracted to the positively charged nucleus.

▼ **Figure 2.6 Energy levels of an atom's electrons.** Electrons exist only at fixed levels of potential energy called electron shells.

(a) A ball bouncing down a flight of stairs can come to rest only on each step, not between steps. Similarly, an electron can exist only at certain energy levels, not between levels.

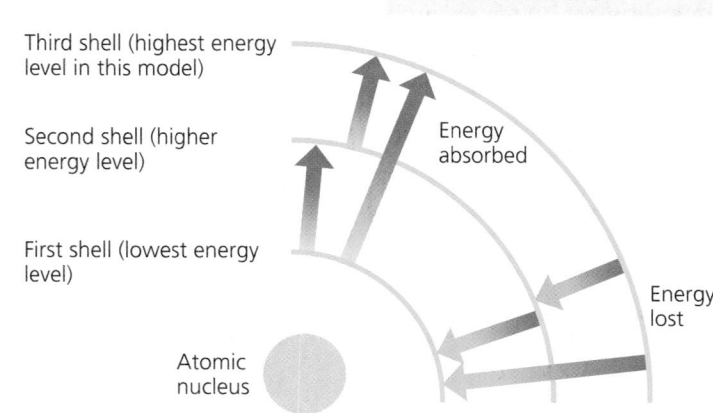

Third shell (highest energy level in this model)

Second shell (higher energy level)

First shell (lowest energy level)

Energy absorbed

Energy lost

Atomic nucleus

(b) An electron can move from one shell to another only if the energy it gains or loses is exactly equal to the difference in energy between the energy levels of the two shells. Arrows in this model indicate some of the stepwise changes in potential energy that are possible.

⊙ **Mastering Biology Figure Walkthrough**

Calibrating a Standard Radioactive Isotope Decay Curve and Interpreting Data

How Long Might Neanderthals Have Co-Existed with Modern Humans (*Homo sapiens*)? Neanderthals (*Homo neanderthalensis*) were living in Europe by 350,000 years ago and may have coexisted with early *Homo sapiens* in parts of Eurasia for hundreds or thousands of years before Neanderthals became extinct. Researchers sought to more accurately determine the extent of their overlap by pinning down the latest date Neanderthals still lived in the area. They used carbon-14 dating to determine the age of a Neanderthal fossil from the most recent (uppermost) archeological layer containing Neanderthal bones. In this exercise you will calibrate a standard carbon-14 decay curve and use it to determine the age of this Neanderthal fossil. The age will help you approximate the last time the two species may have coexisted at the site where this fossil was collected.

How the Experiment Was Done Carbon-14 (^{14}C) is a radioactive isotope of carbon that decays to ^{14}N at a constant rate. ^{14}C is present in the atmosphere in small amounts at a constant ratio with both ^{13}C and ^{12}C, two other isotopes of carbon. When carbon is taken up from the atmosphere by a plant during photosynthesis, ^{12}C, ^{13}C, and ^{14}C isotopes are incorporated into the plant in the same proportions in which they were present in the atmosphere. These proportions remain the same in the tissues of an animal that eats the plant. While an organism is alive, the ^{14}C in its body constantly decays to ^{14}N but is constantly replaced by new carbon from the environment. Once an organism dies, it stops taking in new ^{14}C but the ^{14}C in its tissues continues to decay, while the ^{12}C in its tissues remains the same because it is not radioactive and does not decay. Thus, scientists can calculate how long the pool of original ^{14}C has been decaying in a fossil by measuring the ratio of ^{14}C to ^{12}C and comparing it to the ratio of ^{14}C to ^{12}C present originally in the atmosphere. The fraction of ^{14}C in a fossil compared to the original fraction of ^{14}C can be converted to years because we know that the half-life of ^{14}C is 5,730 years—in other words, half of the ^{14}C in a fossil decays every 5,730 years.

Data from the Experiment The researchers found that the Neanderthal fossil had approximately 0.0078 (or, in scientific notation, 7.8×10^{-3}) as much ^{14}C as the atmosphere. The following questions will guide you through translating this fraction into the age of the fossil.

INTERPRET THE DATA

1. The graph shows a standard curve of radioactive isotope decay. The line shows the fraction of the radioactive isotope over time (before the present) in units of half-lives. Recall that a half-life is the amount of time it takes for half of the radioactive isotope to decay. Labeling each data point with the corresponding fractions will help orient you to this graph. Draw an arrow to the data point for half-life = 1 and write the fraction of ^{14}C that will remain after one half-life. Calculate the fraction of ^{14}C remaining at each half-life and write the fractions

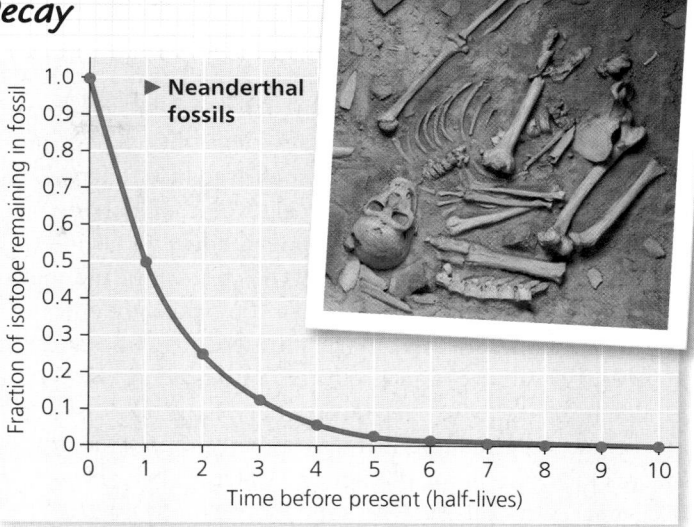

Data from R. Pinhasi et al., Revised age of late Neanderthal occupation and the end of the Middle Paleolithic in the northern Caucasus, *Proceedings of the National Academy of Sciences USA* 147:8611–8616 (2011). doi 10.1073/pnas.1018938108

on the graph near arrows pointing to the data points. Convert each fraction to a decimal number and round off to a maximum of three significant digits (zeros at the beginning of the number do not count as significant digits). Also write each decimal number in scientific notation.

2. Recall that ^{14}C has a half-life of 5,730 years. To calibrate the *x*-axis for ^{14}C decay, write the time before present in years below each half-life.

3. The researchers found that the Neanderthal fossil had approximately 0.0078 as much ^{14}C as found originally in the atmosphere. **(a)** Using the numbers on your graph, determine how many half-lives have passed since the Neanderthal died. **(b)** Using your ^{14}C calibration on the *x*-axis, what is the approximate age of the Neanderthal fossil in years (round off to the nearest thousand)? **(c)** Approximately when did Neanderthals become extinct according to this study? **(d)** The researchers cite evidence that modern humans (*H. sapiens*) became established in the same region as the last Neanderthals approximately 39,000–42,000 years ago. What does this suggest about possible overlap of Neanderthals and modern humans?

4. Carbon-14 dating works for fossils up to about 75,000 years old; fossils older than that contain too little ^{14}C to be detected. Most dinosaurs went extinct 65.5 million years ago. **(a)** Can ^{14}C be used to·date dinosaur bones? Explain. **(b)** Radioactive uranium-235 has a half-life of 704 million years. If it was incorporated into dinosaur bones, could it be used to date the dinosaur fossils? Explain.

> **Instructors**: A version of this Scientific Skills Exercise can be assigned in **Mastering Biology.**

It takes work to move a given electron farther away from the nucleus, so the more distant an electron is from the nucleus, the greater its potential energy. Unlike the continuous flow of water downhill, changes in the potential energy of electrons can occur only in steps of fixed amounts. An electron having a certain amount of energy is something like a ball on a staircase (see Figure 2.6a). The ball can have different amounts of potential energy, depending on which step it is on, but it cannot spend much time between the steps. Similarly, an electron's potential energy is determined by its energy level. An electron can exist only at certain energy levels, not between them.

An electron's energy level is correlated with its average distance from the nucleus. Electrons are found in different **electron shells**, each with a characteristic average distance and energy level. In diagrams, shells can be represented by concentric circles, as they are in Figure 2.6b. The first shell is closest to the nucleus, and electrons in this shell have the lowest potential energy. Electrons in the second shell have more energy, and electrons in the third shell even more energy. An electron can move from one shell to another, but only by absorbing or losing an amount of energy equal to the difference in potential energy between its position in the old shell and that in the new shell. When an electron absorbs energy, it moves to a shell farther out from the nucleus. For example, light energy can excite an electron to a higher energy level. (Indeed, this is the first step taken when plants harness the energy of sunlight for photosynthesis, the process that produces food from carbon dioxide and water.) When an electron loses energy, it "falls back" to a shell closer to the nucleus, and the lost energy is usually released to the environment as visible light or ultraviolet radiation.

Electron Distribution and Chemical Properties

The chemical behavior of an atom is determined by the distribution of electrons in the atom's electron shells. Beginning with hydrogen, the simplest atom, we can imagine building the atoms of the other elements by adding 1 proton and 1 electron at a time (along with an appropriate number of neutrons). **Figure 2.7**, a modified version of what is called the *periodic table of the elements*, shows this distribution of electrons for the first 18 elements, from hydrogen ($_1$H) to argon ($_{18}$Ar). The elements are arranged in three rows, or *periods*, corresponding to the number of electron shells in their atoms. The left-to-right sequence of elements in each row corresponds to the sequential addition of electrons and protons. (See the complete periodic table at the back of the book.)

Hydrogen's 1 electron and helium's 2 electrons are located in the first shell. Electrons, like all matter, tend to exist in the lowest available state of potential energy. In an atom, this state is in the first shell. However, the first shell can hold no more than 2 electrons; thus, hydrogen and helium are the

▼ **Figure 2.7 Electron distribution diagrams for the first 18 elements in the periodic table.** In a standard periodic table (see the back of the book), information for each element is presented as shown for helium in the inset. In the diagrams in this table, electrons are represented as yellow dots and electron shells as concentric circles. These diagrams are a convenient way to picture the distribution of an atom's electrons among its electron shells, but these simplified models do not accurately represent the shape of the atom or the location of its electrons. The elements are arranged in rows, each representing the filling of an electron shell. As electrons are added, they occupy the lowest available shell.

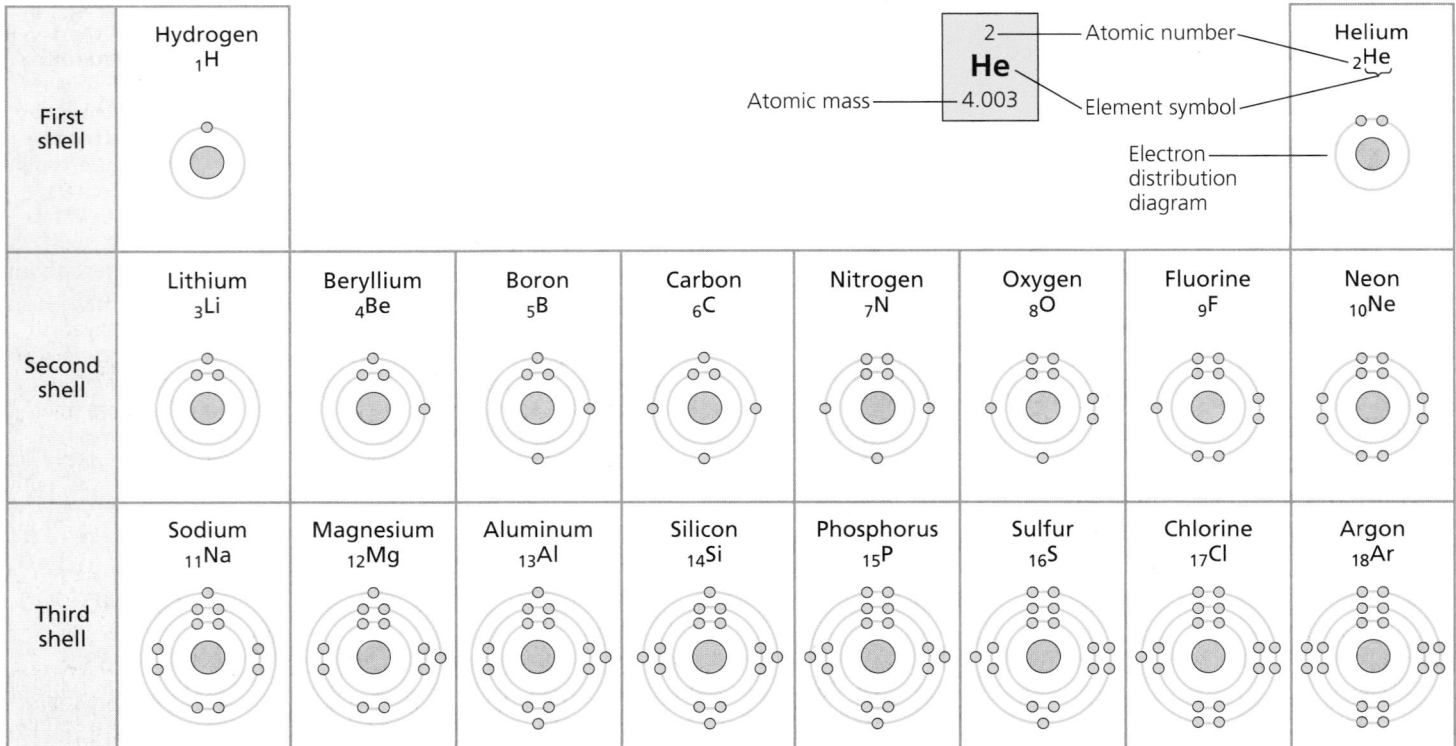

VISUAL SKILLS *What is the atomic number of magnesium? How many protons and electrons does it have? How many electron shells? How many valence electrons?*

⊘ **Mastering Biology** Animation: **Electron Distribution Diagrams** only elements in the first row of the table. In an atom with more than 2 electrons, the additional electrons must occupy higher shells because the first shell is full. The next element, lithium, has 3 electrons. Two of these electrons fill the first shell, while the third electron occupies the second shell. The second shell holds a maximum of 8 electrons. Neon, at the end of the second row, has 8 electrons in the second shell, giving it a total of 10 electrons.

The chemical behavior of an atom depends mostly on the number of electrons in its *outermost* shell. We call those outer electrons **valence electrons** and the outermost electron shell the **valence shell**. In the case of lithium, there is only 1 valence electron, and the second shell is the valence shell. Atoms with the same number of electrons in their valence shells exhibit similar chemical behavior. For example, fluorine (F) and chlorine (Cl) both have 7 valence electrons, and both form compounds when combined with the element sodium (Na): Sodium fluoride (NaF) is commonly added to toothpaste to prevent tooth decay, and, as described earlier, NaCl is table salt (see Figure 2.2). An atom with a completed valence shell is unreactive; that is, it will not interact readily with other atoms. At the far right of the periodic table are helium, neon, and argon, the only three elements shown in Figure 2.7 that have full valence shells. These elements are said to be *inert*, meaning chemically unreactive. All the other atoms in Figure 2.7 are chemically reactive because they have incomplete valence shells.

Electron Orbitals

In the early 1900s, the electron shells of an atom were visualized as concentric paths of electrons orbiting the nucleus, somewhat like planets orbiting the sun. It is still convenient to use two-dimensional concentric-circle diagrams, as in Figure 2.7, to symbolize three-dimensional electron shells. However, you need to remember that each concentric circle represents only the *average* distance between an electron in that shell and the nucleus. Accordingly, the concentric-circle diagrams do not give a real picture of an atom. In reality, we can never know the exact location of an electron. What we can do instead is describe the space in which an electron spends most of its time. The three-dimensional space where an electron is found 90% of the time is called an **orbital**.

Each electron shell contains electrons at a particular energy level, distributed among a specific number of orbitals of distinctive shapes and orientations. **Figure 2.8** shows the orbitals of neon as an example, with its electron distribution diagram for reference. You can think of an orbital as a component of an electron shell. The first electron shell has only one spherical *s* orbital (called 1*s*), but the second shell has four orbitals: one large spherical *s* orbital (called 2*s*) and three dumbbell-shaped *p* orbitals (called 2*p* orbitals). (The third shell and other higher electron shells also have *s* and *p* orbitals, as well as orbitals of more complex shapes.)

▼ **Figure 2.8 Electron orbitals.**

(a) Electron distribution diagram. An electron distribution diagram is shown here for a neon atom, which has a total of 10 electrons. Each concentric circle represents an electron shell, which can be subdivided into electron orbitals.

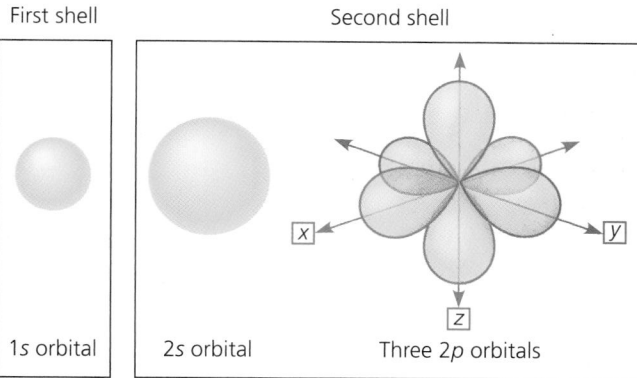

(b) Separate electron orbitals. The three-dimensional shapes represent electron orbitals—the volumes of space where the electrons of an atom are most likely to be found. Each orbital holds a maximum of 2 electrons. The first electron shell, on the left, has one spherical (*s*) orbital, designated 1*s*. The second shell, on the right, has one larger *s* orbital (designated 2*s* for the second shell) plus three dumbbell-shaped orbitals called *p* orbitals (2*p* for the second shell). The three 2*p* orbitals lie at right angles to one another along imaginary *x*-, *y*-, and *z*-axes of the atom. Each 2*p* orbital is outlined here in a different color.

1*s*, 2*s*, and 2*p* orbitals

(c) Superimposed electron orbitals. To reveal the complete picture of the electron orbitals of neon, we superimpose the 1*s* orbital of the first shell and the 2*s* and three 2*p* orbitals of the second shell.

No more than 2 electrons can occupy a single orbital. The first electron shell can therefore accommodate up to 2 electrons in its *s* orbital. The lone electron of a hydrogen atom occupies the 1*s* orbital, as do the 2 electrons of a helium atom. The four orbitals of the second electron shell can hold up to 8 electrons, 2 in each orbital. Electrons in each of the four orbitals in the second shell have nearly the same energy, but they move in different volumes of space.

The reactivity of an atom arises from the presence of unpaired electrons in one or more orbitals of the atom's valence shell. As you will see in the next section, atoms interact in a way that completes their valence shells. When they do so, it is the *unpaired* electrons that are involved.

CONCEPT 2.3

The formation and function of molecules and ionic compounds depend on chemical bonding between atoms

Now that we have looked at the structure of atoms, we can move up the hierarchy of organization and see how atoms combine to form molecules and ionic compounds. Atoms with incomplete valence shells can interact with certain other atoms in such a way that each partner atom completes its valence shell: The atoms either share or transfer valence electrons. These interactions usually result in atoms staying close together, held by attractions called **chemical bonds**. The strongest kinds of chemical bonds are covalent bonds in molecules and ionic bonds in dry ionic compounds. (Ionic bonds in aqueous, or water-based, solutions are weak interactions, as we will see later.)

→ **Mastering Biology Animation: Introduction to Chemical Bonds**

Covalent Bonds

A **covalent bond** is the sharing of a pair of valence electrons by two atoms. For example, let's consider what happens when two hydrogen atoms approach each other. Recall that hydrogen has 1 valence electron in the first shell, but the shell's capacity is 2 electrons. When the two hydrogen atoms come close enough for their 1s orbitals to overlap, they can share their electrons **(Figure 2.9)**. Each hydrogen atom is now associated with 2 electrons in what amounts to a completed valence shell. Two or more atoms held together by covalent bonds constitute a **molecule**, in this case a hydrogen molecule.

▼ **Figure 2.9 Formation of a covalent bond.**

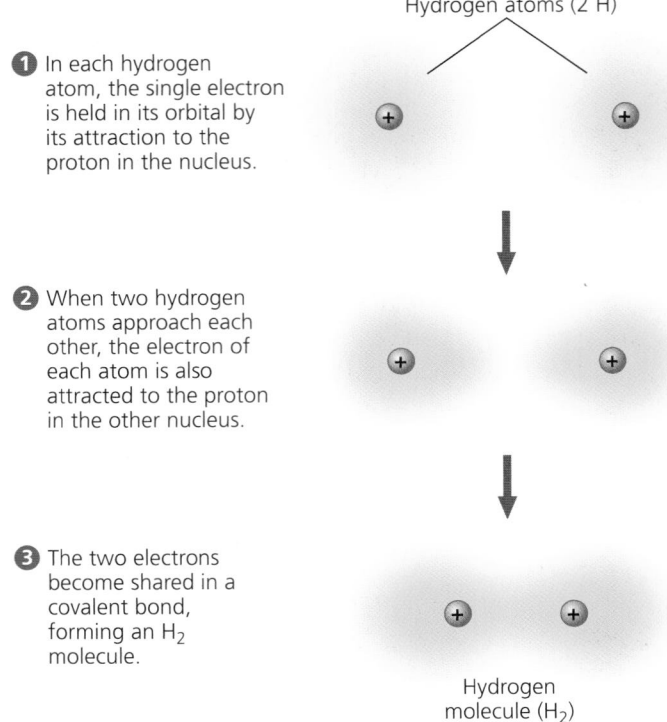

Hydrogen atoms (2 H)

❶ In each hydrogen atom, the single electron is held in its orbital by its attraction to the proton in the nucleus.

❷ When two hydrogen atoms approach each other, the electron of each atom is also attracted to the proton in the other nucleus.

❸ The two electrons become shared in a covalent bond, forming an H_2 molecule.

Hydrogen molecule (H_2)

Figure 2.10a shows several ways of representing a hydrogen molecule. Its *molecular formula*, H_2, simply indicates that the molecule consists of two atoms of hydrogen. Electron sharing can be depicted by an electron distribution diagram or by a *Lewis dot structure*, in which element symbols are surrounded by dots that represent the valence electrons (H:H). We can also use a *structural formula*, H—H, where the line represents a **single bond**, a pair of shared electrons. A *space-filling model* comes closest to representing the actual shape of the molecule. (You may also be familiar with ball-and-stick models, which are shown in Figure 2.15.)

Oxygen has 6 electrons in its second electron shell and therefore needs 2 more electrons to complete its valence shell. Two oxygen atoms form a molecule by sharing *two* pairs of valence electrons **(Figure 2.10b)**. The atoms are thus joined by what is called a **double bond** (O=O).

Each atom that can share valence electrons has a bonding capacity corresponding to the number of covalent bonds the atom can form. When the bonds form, they give the atom a full complement of electrons in the valence shell. The bonding capacity of oxygen, for example, is 2. This bonding capacity is called the atom's **valence** and usually equals the number of electrons required to complete the atom's outermost (valence) shell. See if you can determine the valences of hydrogen, oxygen, nitrogen, and carbon by studying the electron distribution diagrams in Figure 2.7. You can see that the valence of hydrogen is 1; oxygen, 2; nitrogen, 3; and carbon, 4. The situation is more complicated for phosphorus, in the third row of the periodic table, which can have a valence of 3 or 5 depending on the combination of single and double bonds it makes.

▼ **Figure 2.10 Covalent bonding in four molecules.** The number of electrons required to complete an atom's valence shell generally determines how many covalent bonds that atom will form. This figure shows several ways of indicating covalent bonds.

Name and Molecular Formula	Electron Distribution Diagram	Lewis Dot Structure and Structural Formula	Space-Filling Model
(a) Hydrogen (H_2). Two hydrogen atoms share one pair of electrons, forming a single bond.		H:H H—H	
(b) Oxygen (O_2). Two oxygen atoms share two pairs of electrons, forming a double bond.		:Ö::Ö: O=O	
(c) Water (H_2O). Two hydrogen atoms and one oxygen atom are joined by single bonds, forming a molecule of water.		:Ö:H H O—H | H	
(d) Methane (CH_4). Four hydrogen atoms can satisfy the valence of one carbon atom, forming methane.		H H:C:H H H | H—C—H | H	

→ **Mastering Biology** Animation: Covalent Bonds

The molecules H_2 and O_2 are pure elements rather than compounds because a compound is a combination of two or more *different* elements. Water, with the molecular formula H_2O, is a compound. Two atoms of hydrogen are needed to satisfy the valence of one oxygen atom. **Figure 2.10c** shows the structure of a water molecule. (Water is so important to life that Chapter 3 is devoted to its structure and behavior.)

Methane, the main component of natural gas, is a compound with the molecular formula CH_4. It takes four hydrogen atoms, each with a valence of 1, to complete the valence shell of a carbon atom, with its valence of 4 **(Figure 2.10d)**. (We'll look at other carbon compounds in Chapter 4.)

Atoms in a molecule attract shared bonding electrons to varying degrees, depending on the element. The attraction of a particular atom for the electrons of a covalent bond is called its **electronegativity**. The more electronegative an atom is, the more strongly it pulls shared electrons toward itself. In a covalent bond between two atoms of the same element, the

▼ **Figure 2.11 Polar covalent bonds in a water molecule.**

Because oxygen (O) is more electronegative than hydrogen (H), shared electrons are pulled more toward oxygen. Thus, the covalent bonds in water are polar.

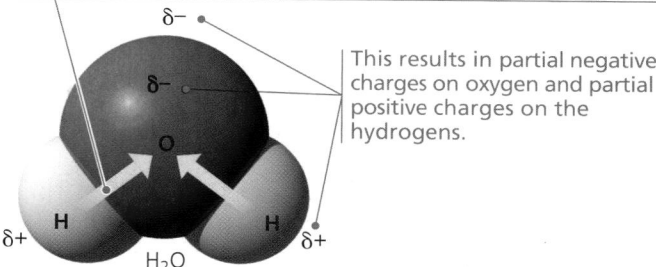

This results in partial negative charges on oxygen and partial positive charges on the hydrogens.

H_2O

→ **Mastering Biology** Animation: Nonpolar and Polar Molecules

electrons are shared equally because the two atoms have the same electronegativity—the tug-of-war is at a standoff. Such a bond is called a **nonpolar covalent bond**. For example, the single bond of H_2 is nonpolar, as is the double bond of O_2. However, when an atom is bonded to a more electronegative atom, the electrons of the bond are not shared equally. This type of bond is called a **polar covalent bond**. Such bonds vary in their polarity, depending on the relative electronegativity of the two atoms. For example, the bonds between the oxygen and hydrogen atoms of a water molecule are quite polar **(Figure 2.11)**.

Oxygen is one of the most electronegative elements, attracting shared electrons much more strongly than hydrogen does. In a covalent bond between oxygen and hydrogen, the electrons spend more time near the oxygen nucleus than near the hydrogen nucleus. Because electrons have a negative charge and are pulled toward oxygen in a water molecule, the oxygen atom has partial negative charges (indicated by the Greek letter δ with a minus sign, δ−, or "delta minus"), and the hydrogen atoms have partial positive charges (δ+, or "delta plus"). In contrast, the individual bonds of methane (CH_4) are much less polar because the electronegativities of carbon and hydrogen are quite similar.

Ionic Bonds

In some cases, two atoms are so unequal in their attraction for valence electrons that the more electronegative atom strips an electron completely away from its partner. The two resulting oppositely charged atoms (or molecules) are called **ions**. A positively charged ion is called a **cation**, while a negatively charged ion is called an **anion**. (It may help you to think of the *t* in *cation* as a plus sign, and of *anion* as "a negative ion.") Because of their opposite charges, cations and anions attract each other; this attraction is called an **ionic bond**. Note that the transfer of an electron is not, by itself, the formation of a bond; rather, it allows a bond to form because it results in two ions of opposite charge. Any two ions of opposite charge can form an ionic bond. The ions do not need to have acquired their charge by an electron transfer with each other.

▼ **Figure 2.12 Electron transfer and ionic bonding.** The attraction between oppositely charged atoms, or ions, is an ionic bond. An ionic bond can form between any two oppositely charged ions, even if they have not been formed by transfer of an electron from one to the other.

❶ The single valence electron of a sodium atom is transferred to join the 7 valence electrons of a chlorine atom.

❷ Each resulting ion has a completed valence shell. An ionic bond can form between the oppositely charged ions.

Na
Sodium atom

Cl
Chlorine atom

Na⁺
Sodium ion
(a cation)

Cl⁻
Chloride ion
(an anion)

Sodium chloride (NaCl)

➡ **Mastering Biology Animation: Formation of Ions and Ionic Bonds**

This is what happens when an atom of sodium ($_{11}$Na) encounters an atom of chlorine ($_{17}$Cl) **(Figure 2.12)**. A sodium atom has a total of 11 electrons, with its single valence electron in the third electron shell. A chlorine atom has a total of 17 electrons, with 7 electrons in its valence shell. When these two atoms meet, the lone valence electron of sodium is transferred to the chlorine atom, and both atoms end up with their valence shells complete. (Because sodium no longer has an electron in the third shell, the second shell is now the valence shell.) The electron transfer between the two atoms moves one unit of negative charge from sodium to chlorine. Sodium, now with 11 protons but only 10 electrons, has a net electrical charge of 1+; the sodium atom has become a cation. Conversely, the chlorine atom, having gained an extra electron, now has 17 protons and 18 electrons, giving it a net electrical charge of 1−; it has become a chloride ion—an anion.

Compounds formed by ionic bonds are called **ionic compounds**, or **salts**. We know the ionic compound sodium chloride (NaCl) as table salt **(Figure 2.13)**. Salts are often found in nature as crystals of various sizes and shapes. Each salt crystal is an aggregate of vast numbers of cations

and anions bonded by their electrical attraction and arranged in a three-dimensional lattice. Unlike a covalent compound, which consists of molecules having a definite size and number of atoms, an ionic compound does not consist of molecules. The formula for an ionic compound, such as NaCl, indicates only the ratio of elements in a crystal of the salt. "NaCl" by itself is not a molecule.

Not all salts have equal numbers of cations and anions. For example, the ionic compound magnesium chloride ($MgCl_2$) has two chloride ions for each magnesium ion. Magnesium ($_{12}$Mg) must lose 2 outer electrons if the atom is to have a complete valence shell, so it has a tendency to become a cation with a net charge of 2+(Mg^{2+}). One magnesium cation can therefore form ionic bonds with two chloride anions (Cl^-).

The term *ion* also applies to entire molecules that are electrically charged. In the salt ammonium chloride (NH_4Cl), for instance, the anion is a single chloride ion (Cl^-), but the cation is ammonium (NH_4^+), a nitrogen atom covalently bonded to four hydrogen atoms. The whole ammonium ion has an electrical charge of 1+ because it has given up 1 electron and thus is 1 electron short.

Environment affects the strength of ionic bonds. In a dry salt crystal, the bonds are so strong that it takes a hammer and chisel to break enough of them to crack the crystal in two. If the same salt crystal is dissolved in water, however, the ionic bonds are much weaker because each ion is partially shielded by its interactions with water molecules. Most drugs are manufactured as salts because they are quite stable when dry but can dissociate (come apart) easily in water. (In Concept 3.2, you will learn how water dissolves salts.)

Weak Chemical Interactions

In organisms, most of the strongest chemical bonds are covalent bonds, which link atoms to form a cell's molecules. But weaker interactions within and between molecules are also indispensable, contributing greatly to the emergent properties of life. Many large biological molecules are held in their functional form by weak interactions. In addition, when two molecules in the cell make contact, they may adhere temporarily by weak interactions. The reversibility of weak interactions can be an advantage: Two molecules can come together, affect one another in some way, and then separate.

Several types of weak chemical interactions are important in organisms. One is the ionic bond as it exists between ions dissociated in water, which we just discussed. Hydrogen bonds and van der Waals interactions are also crucial to life.

▼ **Figure 2.13 A sodium chloride (NaCl) crystal.** The sodium ions (Na⁺) and chloride ions (Cl⁻) are held together by ionic bonds. The formula NaCl tells us that the ratio of Na⁺ to Cl⁻ is 1:1.

Na⁺
Cl⁻

Hydrogen Bonds

Among weak chemical interactions, hydrogen bonds are so central to the chemistry of life that they deserve special attention. When a hydrogen atom is covalently bonded to an electronegative atom, the hydrogen atom has a partial positive charge that allows it to be attracted to a different electronegative atom with a partial negative charge nearby. This noncovalent attraction between a hydrogen and an electronegative atom is called a **hydrogen bond**. In living cells, the electronegative partners are usually oxygen or nitrogen atoms. **Figure 2.14** shows hydrogen bonding between water (H_2O) and ammonia (NH_3).

▼ **Figure 2.14 A hydrogen bond.**

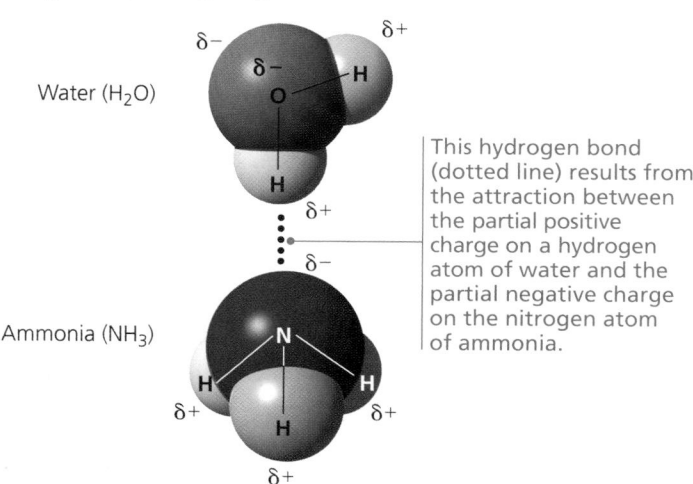

Water (H_2O)

Ammonia (NH_3)

This hydrogen bond (dotted line) results from the attraction between the partial positive charge on a hydrogen atom of water and the partial negative charge on the nitrogen atom of ammonia.

DRAW IT *Draw one water molecule hydrogen bonded to four other water molecules around it. Use simple outlines of space-filling models. Draw the partial charges on the water molecules and use dots for the hydrogen bonds.*

→ **Mastering Biology Animation: Hydrogen Bonds**

Van der Waals Interactions

Even a molecule with nonpolar covalent bonds may have positively and negatively charged regions. Electrons are not always evenly distributed; at any instant, they may accumulate by chance in one part of a molecule or another. The results are ever-changing regions of positive and negative charge that enable all atoms and molecules to stick to one another. These **van der Waals interactions** are individually weak and occur only when atoms and molecules are very close together. When many such interactions occur simultaneously, however, they can be powerful: Van der Waals interactions allow the gecko lizard shown here to walk straight up a wall! The anatomy of the gecko's foot—including toes with hundreds of thousands of tiny hairs, each with

multiple projections—maximizes surface contact with the wall. The van der Waals interactions between the foot molecules and the molecules of the wall's surface are so numerous that despite their individual weakness, together they can support the gecko's body weight.

Van der Waals interactions, hydrogen bonds, ionic bonds in water, and other weak interactions may form not only between molecules but also between parts of a large molecule, such as a protein or nucleic acid. The cumulative effect of weak interactions is to reinforce the three-dimensional shape of the molecule. (You will learn more about the very important biological roles of weak interactions in Figures 5.18 and 5.24.)

Molecular Shape and Function

A molecule has a characteristic size and shape, which are key to its function in the living cell. A molecule consisting of two atoms, such as H_2 or O_2, is always linear, but most molecules with more than two atoms have more complicated shapes. These shapes are determined by the positions of the atoms' orbitals **(Figure 2.15)**. When an atom forms covalent bonds, the orbitals in its valence shell undergo rearrangement.

▼ **Figure 2.15 Molecular shapes due to hybrid orbitals.**

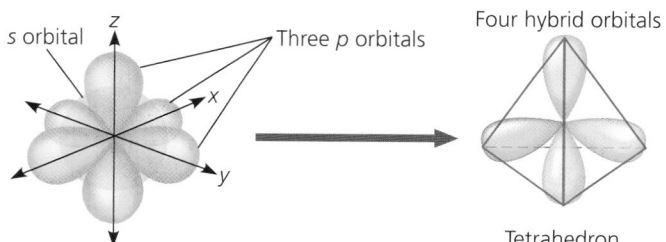

s orbital   Three *p* orbitals   Four hybrid orbitals

Tetrahedron

(a) Hybridization of orbitals. The single *s* and three *p* orbitals of a valence shell involved in covalent bonding combine to form four teardrop-shaped hybrid orbitals. These orbitals extend to the four corners of an imaginary tetrahedron (outlined in pink).

Space-Filling Model	Ball-and-Stick Model	**Hybrid-Orbital Model** (with ball-and-stick model superimposed)
Water (H_2O)	104.5°	Lone (unbonded) pairs of electrons
Methane (CH_4)		

(b) Molecular-shape models. Three models representing molecular shape are shown for water and methane. The positions of the hybrid orbitals determine the shapes of the molecules.

For atoms with valence electrons in both *s* and *p* orbitals (review Figure 2.8), the single *s* and three *p* orbitals form four new hybrid orbitals shaped like identical teardrops extending from the region of the atomic nucleus, as shown in Figure 2.15a. If we connect the larger ends of the teardrops with lines, we have the outline of a geometric shape called a tetrahedron, a pyramid with a triangular base.

For water molecules (H_2O), two of the hybrid orbitals in the oxygen's valence shell are shared with hydrogens. The other two hybrid orbitals are occupied by lone (unbonded) pairs of electrons (see Figure 2.15b). The result is a molecule shaped roughly like a V, with its two covalent bonds at an angle of 104.5°.

The methane molecule (CH_4) has the shape of a completed tetrahedron because all four hybrid orbitals of the carbon atom are shared with hydrogen atoms (see Figure 2.15b). The carbon nucleus is at the center, with its four covalent bonds radiating to hydrogen nuclei at the corners of the tetrahedron. Larger molecules containing multiple carbon atoms, including many of the molecules that make up living matter, have more complex overall shapes. However, the tetrahedral shape of a carbon atom bonded to four other atoms is often a repeating motif within such molecules.

Molecular shape is crucial: It determines how biological molecules recognize and respond to one another with specificity. Biological molecules often bind temporarily to each other by forming weak interactions, but only if their shapes are complementary. Consider the effects of opiates, drugs such as morphine and heroin derived from opium. Opiates relieve pain and alter mood by weakly binding to specific receptor molecules on the surfaces of brain cells. Why would brain cells carry receptors for opiates, compounds that are not *endogenous*, made by the body? In 1975, this question was answered by the discovery of endorphins (or "endogenous morphines"). Endorphins are signaling molecules made by the pituitary gland that bind to the receptors, relieving pain and producing euphoria during times of stress, such as intense exercise. Opiates have shapes similar to endorphins and can bind to endorphin receptors in the brain. That is why opiates and endorphins have similar effects **(Figure 2.16)**. The role of molecular shape in brain chemistry illustrates how biological organization leads to a match between structure and function, one of biology's unifying themes.

CONCEPT CHECK 2.3

1. Why does the structure H—C=C—H fail to make sense chemically?
2. What holds the atoms together in a crystal of magnesium chloride ($MgCl_2$)?
3. **WHAT IF?** If you were a pharmaceutical researcher, why would you want to learn the three-dimensional shapes of naturally occurring signaling molecules?

For suggested answers, see Appendix A.

▼ **Figure 2.16 A molecular mimic.** Morphine affects pain perception and emotional state by mimicking the brain's natural endorphins.

(a) **Structures of endorphin and morphine.** The boxed portion of the endorphin molecule (left) binds to receptor molecules on target cells in the brain. The boxed portion of the morphine molecule (right) is a close match.

(b) **Binding to endorphin receptors.** Both endorphin and morphine can bind to endorphin receptors on the surface of a brain cell.

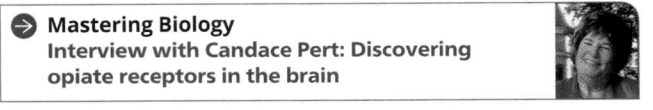

→ **Mastering Biology**
   Interview with Candace Pert: Discovering opiate receptors in the brain

CONCEPT 2.4

Chemical reactions make and break chemical bonds

The making and breaking of chemical bonds, leading to changes in the composition of matter, are called **chemical reactions**. An example is the reaction between hydrogen and oxygen molecules that forms water:

$2 H_2$ + O_2 → $2 H_2O$
Reactants   Chemical   Products
              reaction

This reaction breaks the covalent bonds of H_2 and O_2 and forms the new bonds of H_2O. When we write the equation for a chemical reaction, we use an arrow to indicate the conversion of the starting materials, called the **reactants**, to the resulting materials, or **products**. The coefficients indicate the number of molecules involved; for example, the coefficient 2 before the H_2 means that the reaction starts with two molecules of hydrogen. Notice that all atoms of the reactants must be accounted for in the products. Matter is conserved in a chemical reaction: Reactions cannot create or destroy atoms but can only rearrange (redistribute) the electrons among them.

Photosynthesis, which takes place within the cells of green plant tissues, is an important biological example of how chemical reactions rearrange matter. Humans and other animals ultimately depend on photosynthesis for food and oxygen, and this process is at the foundation of life in almost all ecosystems.

The following summarizes photosynthesis:

The raw materials of photosynthesis are carbon dioxide (CO_2) and water (H_2O), which land plants absorb from the air and soil, respectively. Within the plant cells, sunlight powers the conversion of these ingredients to a sugar called glucose ($C_6H_{12}O_6$) and oxygen molecules (O_2), a by-product that can be seen when released by a water plant **(Figure 2.17)**. Although photosynthesis is actually a sequence of many chemical reactions, we still end up with the same number and types of atoms that we had when we started. Matter has simply been rearranged, with an input of energy provided by sunlight.

All chemical reactions are theoretically reversible, with the products of the forward reaction becoming the reactants for the reverse reaction. For example, hydrogen and nitrogen molecules can combine to form ammonia, but ammonia can also decompose to regenerate hydrogen and nitrogen:

$$3\,H_2 + N_2 \rightleftharpoons 2\,NH_3$$

The two opposite-headed arrows indicate that the reaction is reversible.

One of the factors affecting the rate of a reaction is the concentration of reactants. The greater the concentration of reactant molecules, the more frequently they collide with one

▶ **Figure 2.17**
Photosynthesis: a solar-powered rearrangement of matter. *Elodea*, a freshwater plant, produces sugar by rearranging the atoms of carbon dioxide and water in the chemical process known as photosynthesis, which is powered by sunlight. Much of the sugar is then converted to other food molecules. Oxygen gas (O_2) is a by-product of photosynthesis; notice the bubbles of O_2 gas escaping from the leaves submerged in water.

DRAW IT *Add labels and arrows on the photo showing the reactants and products of photosynthesis as it takes place in a leaf.*

another and have an opportunity to react and form products. The same holds true for products. As products accumulate, collisions resulting in the reverse reaction become more frequent. Eventually, the forward and reverse reactions occur at the same rate, and the relative concentrations of products and reactants stop changing. The point at which the reactions offset one another exactly is called **chemical equilibrium**. This is a dynamic equilibrium; reactions are still going on in both directions, but with no net effect on the concentrations of reactants and products. Equilibrium does *not* mean that the reactants and products are equal in concentration, but only that their concentrations have stabilized at a particular ratio. The reaction involving ammonia reaches equilibrium when ammonia decomposes as rapidly as it forms. In some chemical reactions, the equilibrium point may lie so far to the right that these reactions go essentially to completion; that is, virtually all the reactants are converted to products.

We will return to the subject of chemical reactions after more detailed study of the various types of molecules that are important to life. In the next chapter, we focus on water, the substance in which all the chemical processes of organisms occur.

CONCEPT CHECK 2.4

1. **MAKE CONNECTIONS** Consider the reaction between hydrogen and oxygen that forms water, shown with ball-and-stick models at the beginning of Concept 2.4. After studying Figure 2.10, draw and label the Lewis dot structures representing this reaction.

2. Which type of chemical reaction, if any, occurs faster at equilibrium: the formation of products from reactants or that of reactants from products?

3. **WHAT IF?** Write an equation that uses the products of photosynthesis as reactants and the reactants of photosynthesis as products. Add energy as another product. This new equation describes a process that occurs in your cells. Describe this equation in words. How does this equation relate to breathing?

For suggested answers, see Appendix A.

2 Chapter Review

→ Go to **Mastering Biology** for Assignments, the eText, the Study Area, and Dynamic Study Modules.

SUMMARY OF KEY CONCEPTS

→ To review key terms, go to the **Vocabulary Self-Quiz** in the **Mastering Biology** eText or Study Area, or go to goo.gl/zkjz9t.

CONCEPT 2.1

Matter consists of chemical elements in pure form and in combinations called compounds *(pp. 29–30)*

- **Elements** cannot be broken down chemically to other substances. A **compound** contains two or more different elements in a fixed ratio. Oxygen, carbon, hydrogen, and nitrogen make up approximately 96% of living matter.

? *Compare an element and a compound.*

CONCEPT 2.2

An element's properties depend on the structure of its atoms *(pp. 30–36)*

- An **atom**, the smallest unit of an element, has the following components:

Nucleus

Protons (+ charge) determine element

Neutrons (no charge) determine isotope

Electrons (– charge) form negative cloud and determine chemical behavior

Atom

- An electrically neutral atom has equal numbers of electrons and protons; the number of protons determines the **atomic number**. The **atomic mass** is measured in **daltons** and is roughly equal to the **mass number**, the sum of protons plus neutrons. **Isotopes** of an element differ from each other in neutron number and therefore mass. Unstable isotopes give off particles and energy as radioactivity.
- In an atom, electrons occupy specific **electron shells**; the electrons in a shell have a characteristic energy level. Electron distribution in shells determines the chemical behavior of an atom. An atom that has an incomplete outer shell, the **valence shell**, is reactive.
- Electrons exist in **orbitals**, three-dimensional spaces with specific shapes that are components of electron shells.

DRAW IT *Draw the electron distribution diagrams for neon ($_{10}$Ne) and argon ($_{18}$Ar). Use these diagrams to explain why these elements are chemically unreactive.*

CONCEPT 2.3

The formation and function of molecules and ionic compounds depend on chemical bonding between atoms *(pp. 36–40)*

- **Chemical bonds** form when atoms interact and complete their valence shells. **Covalent bonds** form when pairs of electrons are shared:

H· + H· ⟶ H:H
Single covalent bond

:Ö· + ·Ö: ⟶ Ö::Ö
Double covalent bond

- **Molecules** consist of two or more covalently bonded atoms. The attraction of an atom for the electrons of a covalent bond is its **electronegativity**. If both atoms are the same, they have the same electronegativity and share a **nonpolar covalent bond**. Electrons of a **polar covalent bond** are pulled closer to the more electronegative atom, such as the oxygen in H_2O.
- An **ion** forms when an atom or molecule gains or loses an electron and becomes charged. An **ionic bond** is the attraction between two oppositely charged ions:

Ionic bond

Electron transfer forms ions

Na
Sodium atom

Cl
Chlorine atom

Na⁺
Sodium ion (a cation)

Cl⁻
Chloride ion (an anion)

- Weak interactions reinforce the shapes of large molecules and help molecules adhere to each other. A **hydrogen bond** is an attraction between a hydrogen atom carrying a partial positive charge ($\delta+$) and an electronegative atom carrying a partial negative charge ($\delta-$). **Van der Waals interactions** occur between transiently positive and negative regions of molecules.
- A molecule's shape is determined by the positions of its atoms' valence orbitals. Covalent bonds result in hybrid orbitals, which are responsible for the shapes of H_2O, CH_4, and many more complex biological molecules. Molecular shape is usually the basis for the recognition of one biological molecule by another.

? *In terms of electron sharing between atoms, compare nonpolar covalent bonds, polar covalent bonds, and the formation of ions.*

CONCEPT 2.4

Chemical reactions make and break chemical bonds *(pp. 40–41)*

- **Chemical reactions** change **reactants** into **products** while conserving matter. All chemical reactions are theoretically reversible. **Chemical equilibrium** is reached when the forward and reverse reaction rates are equal.

? *What would happen to the concentration of products if more reactants were added to a reaction that was in chemical equilibrium? How would this addition affect the equilibrium?*

TEST YOUR UNDERSTANDING

→ For more multiple-choice questions, go to the **Practice Test** in the **Mastering Biology** eText or Study Area, or go to goo.gl/GruWRg.

Levels 1-2: Remembering/Understanding

1. Compared with ^{31}P, the radioactive isotope ^{32}P has
   (A) a different atomic number.
   (B) one more proton.
   (C) one more electron.
   (D) one more neutron.

2. In the term *trace element*, the adjective *trace* means that
   (A) the element is required in very small amounts.
   (B) the element can be used as a label to trace atoms through an organism's metabolism.
   (C) the element is very rare on Earth.
   (D) the element enhances health but is not essential for the organism's long-term survival.

3. The reactivity of an atom arises from
   (A) the average distance of the outermost electron shell from the nucleus.
   (B) the existence of unpaired electrons in the valence shell.
   (C) the sum of the potential energies of all the electron shells.
   (D) the potential energy of the valence shell.

4. Which statement is true of all atoms that are anions?
   (A) The atom has more electrons than protons.
   (B) The atom has more protons than electrons.
   (C) The atom has fewer protons than does a neutral atom of the same element.
   (D) The atom has more neutrons than protons.

5. Which of the following statements correctly describes any chemical reaction that has reached equilibrium?
   (A) The concentrations of products and reactants are equal.
   (B) The reaction is now irreversible.
   (C) Both forward and reverse reactions have halted.
   (D) The rates of the forward and reverse reactions are equal.

Levels 3-4: Applying/Analyzing

6. We can represent atoms by listing the number of protons, neutrons, and electrons—for example, $2p^+$, $2n^0$, $2e^-$ for helium. Which of the following represents the ^{18}O isotope of oxygen?
   (A) $7p^+$, $2n^0$, $9e^-$
   (B) $8p^+$, $10n^0$, $8e^-$
   (C) $9p^+$, $9n^0$, $9e^-$
   (D) $10p^+$, $8n^0$, $9e^-$

7. The atomic number of sulfur is 16. Sulfur combines with hydrogen by covalent bonding to form a compound, hydrogen sulfide. Based on the number of valence electrons in a sulfur atom, predict the molecular formula of the compound.
   (A) HS   (C) H_2S
   (B) HS_2 (D) H_4S

8. What coefficients must be placed in the following blanks so that all atoms are accounted for in the products?

   $$C_6H_{12}O_6 \rightarrow \underline{\quad} C_2H_6O + \underline{\quad} CO_2$$

   (A) 2; 1   (C) 1; 3
   (B) 3; 1   (D) 2; 2

9. **DRAW IT** Draw Lewis dot structures for each hypothetical molecule shown below, using the correct number of valence electrons for each atom. Determine which molecule makes sense because each atom has a complete valence shell and each bond has the correct number of electrons. Explain what makes the other molecule nonsensical, considering the number of bonds each type of atom can make.

Levels 5-6: Evaluating/Creating

10. **EVOLUTION CONNECTION** The percentages of naturally occurring elements making up the human body (see Table 2.1) are similar to the percentages of these elements found in other organisms. How could you account for this similarity among organisms?

11. **SCIENTIFIC INQUIRY** Female luna moths (*Actias luna*) attract males by emitting chemical signals that spread through the air. A male hundreds of meters away can detect these molecules and fly toward their source. The sensory organs responsible for this behavior are the comblike antennae visible in the photograph shown here. Each filament of an antenna is equipped with thousands of receptor cells that detect the sex attractant. Based on what you learned in this chapter, propose a hypothesis to account for the ability of the male moth to detect a specific molecule in the presence of many other molecules in the air. What predictions does your hypothesis make? Design an experiment to test one of these predictions.

12. **WRITE ABOUT A THEME: ORGANIZATION** While waiting at an airport, Neil Campbell once overheard this claim: "It's paranoid and ignorant to worry about industry or agriculture contaminating the environment with their chemical wastes. After all, this stuff is just made of the same atoms that were already present in our environment." Drawing on your knowledge of electron distribution, bonding, and emergent properties (see Concept 1.1), write a short essay (100–150 words) countering this argument.

13. **SYNTHESIZE YOUR KNOWLEDGE**

This bombardier beetle is spraying a boiling hot liquid that contains irritating chemicals, used as a defense mechanism against its enemies. The beetle stores two sets of chemicals separately in its glands. Using what you learned about chemistry in this chapter, propose a possible explanation for why the beetle is not harmed by the chemicals it stores and what causes the explosive discharge.

For selected answers, see Appendix A.

3 Water and Life

KEY CONCEPTS

3.1 Polar covalent bonds in water molecules result in hydrogen bonding *p. 45*

3.2 Four emergent properties of water contribute to Earth's suitability for life *p. 45*

3.3 Acidic and basic conditions affect living organisms *p. 51*

Study Tip

Make a visual study guide: Draw a diagram and write a caption that explains how the structure of water supports life for each of the following properties of water:

The Properties of Water	
Cohesion of water molecules	Moderation of temperature
Floating of ice	The solvent of life

→ Go to Mastering Biology

For Students (in eText and Study Area)
- Get Ready for Chapter 3
- BioFlix® Animation: Adhesion and Cohesion in Plants
- Animation: Acids, Bases, and pH

For Instructors to Assign (in Item Library)
- Chemistry Review–Atoms and Molecules: Polar Attractions
- Chemistry Review–Water: Properties of Water

Figure 3.1 Ringed seals (*Phoca hispida*) depend on Arctic sea ice as a platform from which to hunt for fish in the water below. As Earth warms from climate change, the melting of sea ice is a threat to species that live on, under, and around the floating ice.

How does water's structure allow its solid form (ice) to float on liquid water?

Water (H_2O) is a polar molecule: At one end, the O has partial negative charges ($\delta-$) because O pulls electrons toward itself. At the other end, the H atoms have partial positive charges ($\delta+$).

Weak attractions between oppositely charged regions of water molecules, called hydrogen bonds, allow water molecules to bond to each other.

In liquid water, the hydrogen bonds constantly break and re-form. As a result, **the water molecules can slip closer together.**

In ice, the hydrogen bonds are stable and **the water molecules are farther apart.** Therefore, ice is less dense than liquid water, so it floats.

Floating ice insulates the water below, enabling survival of aquatic life. Water also has other life-supporting properties, as you'll see.

Polar covalent bonds in water molecules result in hydrogen bonding

Water is so familiar to us that it is easy to overlook its many extraordinary qualities. Following the theme of emergent properties, we can trace water's unique behavior to the structure and interactions of its molecules.

Studied on its own, the water molecule is deceptively simple. It is shaped like a wide V, with its two hydrogen atoms joined to the oxygen atom by single covalent bonds. Oxygen is more electronegative than hydrogen, so the electrons of the covalent bonds spend more time closer to oxygen than to hydrogen; these are **polar covalent bonds** (see Figure 2.11). This unequal sharing of electrons and water's V-like shape make it a **polar molecule**, meaning that its overall charge is unevenly distributed. In water, the oxygen of the molecule has partial negative charges ($\delta-$), and the hydrogens have partial positive charges ($\delta+$).

The properties of water arise from attractions between oppositely charged atoms of different water molecules: The partially positive hydrogen of one molecule is attracted to the partially negative oxygen of a nearby molecule. The two molecules are thus held together by a hydrogen bond **(Figure 3.2)**. When water is in its liquid form, its hydrogen bonds are very fragile, each only about 1/20 as strong as a covalent bond. The hydrogen bonds form, break, and re-form with great frequency. Each lasts only a few trillionths of a second, but the molecules are constantly forming new hydrogen bonds with a succession of partners. Therefore, at any instant, most of the water molecules

are hydrogen-bonded to their neighbors. The extraordinary properties of water emerge from this hydrogen bonding, which organizes water molecules into a higher level of structural order.

1. **MAKE CONNECTIONS** What is electronegativity, and how does it affect interactions between water molecules? (Review Figure 2.11.)
2. **VISUAL SKILLS** Look at Figure 3.2 and explain why the central water molecule can hydrogen-bond to other water molecules.
3. Why is it unlikely that two neighboring water molecules would be arranged like this?
4. **WHAT IF?** What would be the effect on the properties of the water molecule if oxygen and hydrogen had equal electronegativity?

For suggested answers, see Appendix A.

Four emergent properties of water contribute to Earth's suitability for life

We will examine four emergent properties of water that contribute to Earth's suitability as an environment for life: cohesive behavior, ability to moderate temperature, expansion upon freezing, and versatility as a solvent.

Cohesion of Water Molecules

Water molecules stay close to each other as a result of hydrogen bonding. Although the arrangement of molecules in a sample of liquid water is constantly changing, at any given moment many of the molecules are linked by multiple hydrogen bonds. These linkages make water more structured than most other liquids. Collectively, the hydrogen bonds hold the substance together, a phenomenon called **cohesion**.

One result of cohesion due to hydrogen bonding is high **surface tension**, a measure of how difficult it is to stretch or break the surface of a liquid. At the air-water interface is an ordered arrangement of water molecules, hydrogen-bonded to one another and to the water below, but not to the air above. This asymmetry gives water an unusually high surface tension, making it behave as though it were coated with an invisible film. The spider in **Figure 3.3** takes advantage of the surface tension of water to walk across a pond without breaking

▼ **Figure 3.2 Hydrogen bonds between water molecules.**

The charged regions in a water molecule are due to its polar covalent bonds.

Regions of neighboring water molecules with opposite partial charges are attracted to each other, forming hydrogen bonds.

Each water molecule can hydrogen-bond to several others; these associations are constantly changing.

Oxygen can form two hydrogen bonds.

DRAW IT *Draw partial charges on the water molecule at the far left, and draw three more water molecules hydrogen-bonded to it.*

➔ **Mastering Biology** Animation: Polarity of Water

▶ **Figure 3.3 Walking on water.** The high surface tension of water, resulting from the collective strength of its hydrogen bonds, allows this raft spider to walk on the surface of a pond.

the surface, and some plants can float on water as well. You can observe the surface tension of water by slightly overfilling a drinking glass; the water will stand above the rim.

Cohesion also contributes to the transport of water and dissolved nutrients against gravity in plants **(Figure 3.4)**. Water from the roots reaches the leaves through a network of water-conducting cells. As water evaporates from a leaf, hydrogen bonds cause water molecules leaving the veins to tug on molecules farther down, and the upward pull is transmitted through the water-conducting cells all the way to the roots. **Adhesion**, the clinging of one substance to another, also plays a role. Adhesion of water by hydrogen bonds to the molecules of cell walls helps counter the downward pull of gravity (see Figure 3.4).

▼ **Figure 3.4 Water transport in plants.** Evaporation from leaves pulls water upward from the roots through water-conducting cells. Because of the properties of cohesion and adhesion, the tallest trees can transport water more than 100 m upward—approximately one-quarter the height of the Empire State Building in New York City.

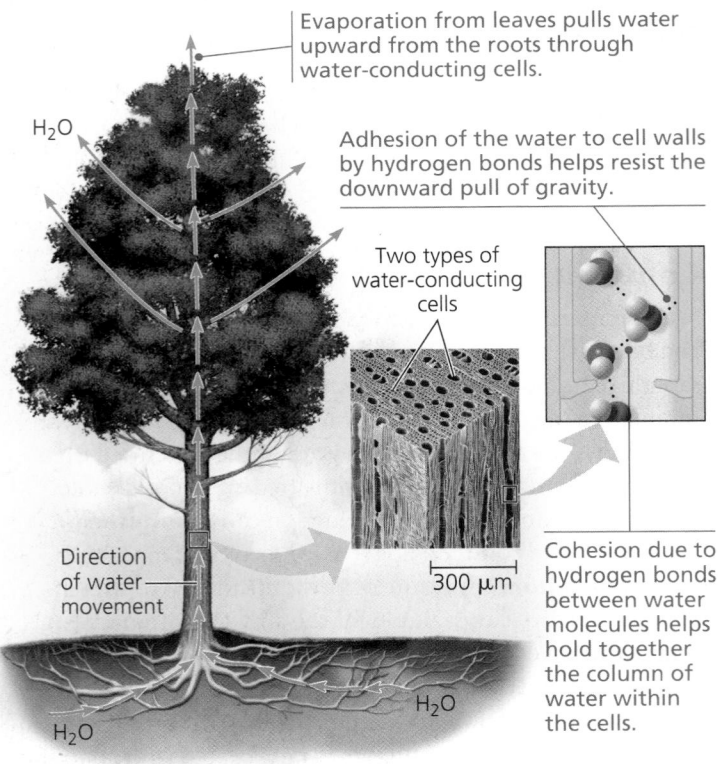

Evaporation from leaves pulls water upward from the roots through water-conducting cells.

H_2O

Adhesion of the water to cell walls by hydrogen bonds helps resist the downward pull of gravity.

Two types of water-conducting cells

Direction of water movement

300 μm

H_2O

H_2O

Cohesion due to hydrogen bonds between water molecules helps hold together the column of water within the cells.

Mastering Biology BioFlix® Animation: Adhesion and Cohesion in Plants • Animation: Cohesion of Water

Moderation of Temperature by Water

Water moderates air temperature by absorbing heat from air that is warmer and releasing stored heat to air that is cooler. Water is effective as a heat bank because it can absorb or release a relatively large amount of heat with only a slight change in its own temperature. To understand this capability of water, let's first look at temperature and heat.

Temperature and Heat

Anything that moves has **kinetic energy**, the energy of motion. Atoms and molecules have kinetic energy because they are always moving, although not necessarily in any particular direction. The faster a molecule moves, the greater its kinetic energy. The kinetic energy associated with the random movement of atoms or molecules is called **thermal energy**. Thermal energy is related to temperature, but they are not the same thing. **Temperature** represents the *average* kinetic energy of the molecules in a body of matter, regardless of volume, whereas the thermal energy of a body of matter reflects the *total* kinetic energy, and thus depends on the matter's volume. When water is heated in a coffeemaker, the average speed of the molecules increases, and the thermometer records this as a rise in temperature of the liquid. The total amount of thermal energy also increases in this case. Note, however, that although the pot of coffee has a much higher temperature than, say, the water in a swimming pool, the swimming pool contains more thermal energy because of its much greater volume.

Whenever two objects of different temperature are brought together, thermal energy passes from the warmer to the cooler object until the two are the same temperature. Molecules in the cooler object speed up at the expense of the thermal energy of the warmer object. An ice cube cools a drink not by adding coldness to the liquid but by absorbing thermal energy from the liquid as the ice itself melts. Thermal energy in transfer from one body of matter to another is defined as **heat**.

One convenient unit of heat used in this book is the **calorie (cal)**. A calorie is the amount of heat it takes to raise the temperature of 1 g of water by 1°C. Conversely, a calorie is also the amount of heat that 1 g of water releases when it cools by 1°C. A **kilocalorie (kcal)**, 1,000 cal, is the quantity of heat required to raise the temperature of 1 kilogram (kg) of water by 1°C. (The "Calories" on food packages are actually kilocalories.) Another energy unit used in this book is the **joule (J)**. One joule equals 0.239 cal; one calorie equals 4.184 J.

Water's High Specific Heat

The ability of water to stabilize temperature stems from its relatively high specific heat. The **specific heat** of a substance is defined as the amount of heat that must be absorbed or lost for 1 g of that substance to change its temperature by 1°C. We already know water's specific heat because we have defined a calorie as the amount of heat that causes 1 g of water to change its temperature by 1°C. Therefore, the specific heat of water is 1 calorie per gram and per degree Celsius, abbreviated as 1 cal/(g · °C). Compared with most other substances, water has an unusually high specific heat. For example, ethyl alcohol, the type of alcohol in alcoholic beverages, has a specific heat of 0.6 cal/(g · °C); that is, only 0.6 cal is required to raise the temperature of 1 g of ethyl alcohol by 1°C.

Because of the high specific heat of water relative to other materials, water will change its temperature less than other liquids when it absorbs or loses a given amount of heat. The reason you can burn your fingers by touching the side of an iron pot on the stove when the water in the pot is still luke-warm is that the specific heat of water is ten times greater than that of iron. In other words, the same amount of heat will raise the temperature of 1 g of the iron much faster than it will raise the temperature of 1 g of the water. Specific heat can be thought of as a measure of how well a substance resists changing its temperature when it absorbs or releases heat. Water resists changing its temperature; when it does change its temperature, it absorbs or loses a relatively large quantity of heat for each degree of change.

We can trace water's high specific heat, like many of its other properties, to hydrogen bonding. Heat must be absorbed in order to break hydrogen bonds; by the same token, heat is released when hydrogen bonds form. A calorie of heat causes a relatively small change in the temperature of water because much of the heat is used to disrupt hydrogen bonds before the water molecules can begin moving faster. And when the temperature of water drops slightly, many additional hydrogen bonds form, releasing a considerable amount of energy in the form of heat.

What is the relevance of water's high specific heat to life on Earth? A large body of water can absorb and store a huge amount of heat from the sun in the daytime and during summer while warming up only a few degrees. At night and during winter, the gradually cooling water can warm the air. This capability of water serves to moderate air temperatures in coastal areas **(Figure 3.5)**. The high specific heat of water also tends to stabilize ocean temperatures, creating a favorable environment for marine life. Thus, because of its high specific heat, the water that covers most of Earth keeps temperature fluctuations on land and in water within limits that permit life. Also, because organisms are made primarily of water, they are better able to resist changes in their own temperature than if they were made of a liquid with a lower specific heat.

▼ **Figure 3.5 Temperatures for the Pacific Ocean and Southern California on an August day.**

☐	70s (°F)
☐	80s
▨	90s
▨	100s

INTERPRET THE DATA *Explain the pattern of temperatures shown in this diagram.*

Evaporative Cooling

Molecules of any liquid stay close together because they are attracted to one another. Molecules moving fast enough to overcome these attractions can depart the liquid and enter the air as a gas (vapor). This transformation from a liquid to a gas is called vaporization, or evaporation. Recall that the speed of molecular movement varies and that temperature is the *average* kinetic energy of molecules. Even at low temperatures, the speediest molecules can escape into the air. Some evaporation occurs at any temperature; a glass of water at room temperature, for example, will eventually evaporate completely. If a liquid is heated, the average kinetic energy of molecules increases and the liquid evaporates more rapidly.

Heat of vaporization is the quantity of heat a liquid must absorb for 1 g of it to be converted from the liquid to the gaseous state. For the same reason that water has a high specific heat, it also has a high heat of vaporization relative to most other liquids. To evaporate 1 g of water at 25°C, about 580 cal of heat is needed—nearly double the amount needed to vaporize a gram of alcohol or ammonia. Water's high heat of vaporization is another emergent property resulting from the strength of its hydrogen bonds, which must be broken before the molecules can exit from the liquid in the form of water vapor.

The high amount of energy required to vaporize water has a wide range of effects. On a global scale, for example, it helps moderate Earth's climate. A considerable amount of solar heat absorbed by tropical seas is consumed during the evaporation of surface water. Then, as moist tropical air circulates poleward, it releases heat as it condenses and forms rain. On an organismal level, water's high heat of vaporization accounts for the severity of steam burns. These burns are caused by the heat energy released (during formation of hydrogen bonds) when steam condenses into liquid on the skin.

As a liquid evaporates, the surface of the liquid that remains behind cools down (its temperature decreases). This **evaporative cooling** occurs because the "hottest" molecules, those with the greatest kinetic energy, are the most likely to leave as gas. It is as if the 100 fastest runners at a college transferred to another school; the average speed of the remaining students would decline.

Evaporative cooling of water contributes to the stability of temperature in lakes and ponds and also provides a mechanism that prevents terrestrial organisms from overheating. For example, evaporation of water from the leaves of a plant helps keep the tissues in the leaves from becoming too warm in the sunlight. Evaporation of sweat from human skin dissipates body heat and helps prevent overheating on a hot day or when excess heat is generated by strenuous activity. High humidity on a hot day increases discomfort because the high concentration of water vapor in the air inhibits the evaporation of sweat from the body. Animals without sweat glands, such as elephants, may spray water on themselves to cool down **(Figure 3.6)**.

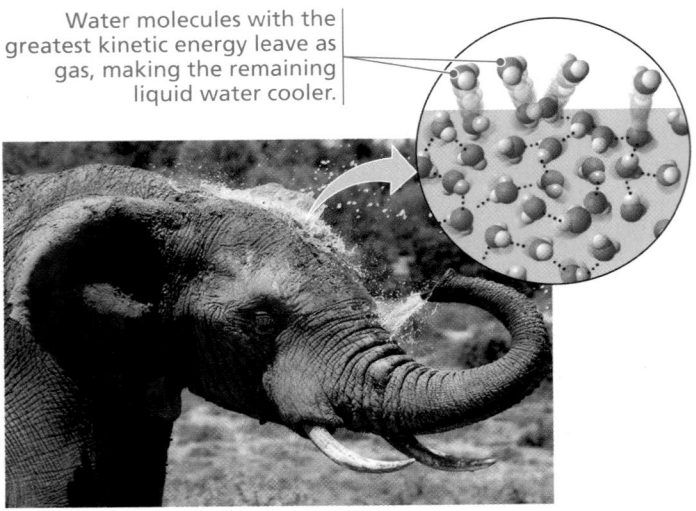

▼ **Figure 3.6 Evaporative cooling.** In hot weather, an elephant sprays water from its trunk onto its head. Evaporation of this water cools the elephant down.

Water molecules with the greatest kinetic energy leave as gas, making the remaining liquid water cooler.

Floating of Ice on Liquid Water

Water is one of the few substances that are less dense as a solid than as a liquid. As a result, ice floats on liquid water. While other materials contract and become denser when they solidify, water expands. The cause of this exotic behavior is, once again, hydrogen bonding. At temperatures above 4°C, water behaves like other liquids, expanding as it warms and contracting as it cools. As the temperature falls from 4°C to 0°C, water begins to freeze because more and more of its molecules are moving too slowly to break hydrogen bonds. At 0°C, the molecules become locked into a crystalline lattice, each water molecule hydrogen-bonded to four partners (see Figure 3.1). The hydrogen bonds keep the molecules at "arm's length," far enough apart to make ice about 10% less dense (10% fewer molecules in the same volume) than liquid water at 4°C. When ice absorbs enough heat for its temperature to rise above 0°C, hydrogen bonds between molecules are disrupted. As the crystal collapses, the ice melts and molecules have fewer hydrogen bonds, allowing them to slip closer together. Water reaches its greatest density at 4°C and then begins to expand as the molecules move faster. Even in liquid water, many of the molecules are

connected by hydrogen bonds, though only transiently: The hydrogen bonds are constantly breaking and re-forming.

The ability of ice to float due to its lower density is an important factor in the suitability of the environment for life. If ice sank, then eventually ponds, lakes, and even oceans could freeze solid, making life as we know it impossible on Earth. During summer, only the upper few inches of the ocean would thaw. Instead, when a deep body of water cools, the ice floats, insulating the liquid water below. This prevents it from freezing and allows life to exist under the frozen surface, as shown in Figure 3.1. Besides insulating the water below, ice also provides a solid habitat for some animals, such as polar bears and seals.

Many scientists are worried that these bodies of ice are at risk of disappearing. Global warming, which is caused by carbon dioxide and other "greenhouse" gases in the atmosphere (see Figure 56.30), is having a profound effect on icy environments around the globe. In the Arctic, the average air temperature has risen 2.2°C just since 1961. This temperature increase has affected the seasonal balance between Arctic sea ice and liquid water, causing ice to form later in the year, to melt earlier, and to cover a smaller area. The rate at which glaciers and Arctic sea ice are disappearing is posing an extreme challenge to animals that depend on ice for their survival **(Figure 3.7)**.

▼ **Figure 3.7 Effects of climate change on the Arctic.** Warmer temperatures in the Arctic cause more sea ice to melt in the summer. The loss of ice disrupts the ecosystem, affecting many species. (Map data is from the National Snow and Ice Data Center.)

Warm water and more light result in more phytoplankton, which are eaten by other organisms. Harmful algal blooms are also a threat.

Populations of bowhead whales and some fish species may be increasing because more plankton is available to eat.

Less ice reduces feeding opportunities for polar bears, who hunt from the ice.

Black guillemots in Alaska cannot fly from their nests on land to their fishing grounds at the edge of the ice, which is now too far from land; young birds are starving.

Loss of floating ice as habitat has caused a decline in Pacific walrus populations due to overcrowding and deadly stampedes on land.

Russia

Bering Strait

Alaska

Canada

Arctic Ocean

Extent of ice sheet in Sept. 2019

North Pole

Median extent of ice sheet in Sept., 1981–2010

Greenland

> **Mastering Biology**
> **Interview with Susan Solomon: Understanding climate change**

Water: The Solvent of Life

A sugar cube placed in a glass of water will dissolve. In time, the glass will contain a uniform mixture of sugar and water; the concentration of dissolved sugar will be the same everywhere in the mixture. A liquid that is a completely homogeneous mixture of two or more substances is called a **solution**. The dissolving agent of a solution is the **solvent**, and the substance that is dissolved is the **solute**. In this case, water is the solvent and sugar is the solute. An **aqueous solution** is one in which the solute is dissolved in water; water is the solvent.

Water is a very versatile solvent, a quality we can trace to the polarity of the water molecule. Suppose, for example, that a spoonful of table salt, the ionic compound sodium chloride (NaCl), is placed in water **(Figure 3.8)**. At the surface of each crystal (grain) of salt, the sodium and chloride ions are exposed to the solvent. These ions and regions of the water molecules are attracted to each other due to their opposite charges. The oxygens of the water molecules have regions of partial negative charge that are attracted to sodium cations. The hydrogen regions are partially positively charged and are attracted to chloride anions. As a result, water molecules surround the individual sodium and chloride ions, separating and shielding them from one another. The sphere of water molecules around each dissolved ion is called a **hydration shell**. Working inward from the surface of each salt crystal, water eventually dissolves all the ions. The result is a solution of two solutes, sodium cations and chloride anions, mixed homogeneously with water, the solvent. Other ionic compounds also dissolve in water. Seawater, for instance, contains a great variety of dissolved ions, as do living cells.

A compound does not need to be ionic to dissolve in water; many compounds made up of nonionic polar molecules, such as the sugar in the sugar cube mentioned earlier, are also

▼ **Figure 3.8 Table salt dissolving in water.** A sphere of water molecules, called a hydration shell, surrounds each solute ion.

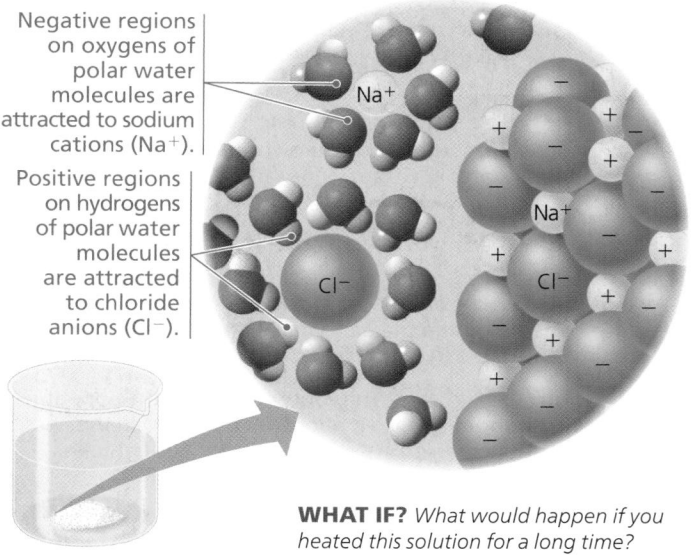

Negative regions on oxygens of polar water molecules are attracted to sodium cations (Na⁺).

Positive regions on hydrogens of polar water molecules are attracted to chloride anions (Cl⁻).

WHAT IF? *What would happen if you heated this solution for a long time?*

▼ **Figure 3.9 A water-soluble protein.** Human lysozyme is a protein found in tears and saliva that has antibacterial action (see Figure 5.16). This model shows the lysozyme molecule (purple) in an aqueous environment. Ionic and polar regions on the protein's surface attract the partially charged regions on water molecules.

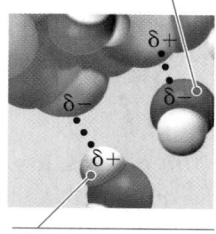

This oxygen is attracted to a partial positive charge on the lysozyme molecule.

This hydrogen is attracted to a partial negative charge on the lysozyme molecule.

water-soluble. Such compounds dissolve when water molecules surround each of the solute molecules, forming hydrogen bonds with them. Even molecules as large as proteins can dissolve in water if they have ionic and polar regions on their surface **(Figure 3.9)**. Many different kinds of polar compounds are dissolved (along with ions) in the water of such biological fluids as blood, the sap of plants, and the liquid within all cells. Water is the solvent of life.

Hydrophilic and Hydrophobic Substances

Any substance that has an affinity for water is said to be **hydrophilic** (from the Greek *hydro*, water, and *philos*, loving). In some cases, substances can be hydrophilic without actually dissolving. For example, some molecules in cells are so large that they do not dissolve. Another example of a hydrophilic substance that does not dissolve is cotton, a plant product. Cotton consists of giant molecules of cellulose, a compound with numerous regions of partial positive and partial negative charges that can form hydrogen bonds with water. Water adheres to the cellulose fibers. Thus, a cotton towel does a great job of drying the body, yet it does not dissolve in the washing machine. Cellulose is also present in the walls of water-conducting cells in a plant; you read earlier how the adhesion of water to these hydrophilic walls helps water move up the plant against gravity.

There are, of course, substances that do not have an affinity for water. Substances that are nonionic and nonpolar (or otherwise cannot form hydrogen bonds) actually seem to repel water; these substances are said to be **hydrophobic** (from the Greek *phobos*, fearing). An example from the kitchen is vegetable oil, which, as you know, does not mix stably with water-based substances such as vinegar. The hydrophobic behavior of the oil molecules results from a high number of relatively nonpolar covalent bonds, in this case bonds between carbon and hydrogen, which share electrons almost equally. Hydrophobic molecules related to oils are major ingredients of cell membranes. (Imagine what would happen to a cell if its membrane dissolved!)

Solute Concentration in Aqueous Solutions

Most of the chemical reactions in organisms involve solutes dissolved in water. To understand such reactions, we must know how many atoms and molecules are involved and calculate the concentration of solutes in an aqueous solution (the number of solute molecules in a volume of solution).

When carrying out experiments, we use mass to calculate the number of molecules. We must first calculate the **molecular mass**, which is the sum of the masses of all the atoms in a molecule. As an example, let's calculate the molecular mass of table sugar (sucrose), $C_{12}H_{22}O_{11}$, by multiplying the number of atoms by the atomic mass of each element (see the periodic table at the back of the book). In round numbers of daltons, the mass of a carbon atom is 12, the mass of a hydrogen atom is 1, and the mass of an oxygen atom is 16. Thus, sucrose has a molecular mass of $(12 \times 12) + (22 \times 1) + (11 \times 16) = 342$ daltons. Because we can't weigh out small numbers of molecules, we usually measure substances in units called moles. Just as a dozen always means 12 objects, a **mole (mol)** represents an exact number of objects: 6.02×10^{23}, which is called Avogadro's number. Because of the way in which Avogadro's number and the unit *dalton* were originally defined, there are 6.02×10^{23} daltons in 1 g. Once we determine the molecular mass of a molecule such as sucrose, we can use the same number (342), but with the unit *gram*, to represent the mass of 6.02×10^{23} molecules of sucrose, or 1 mol of sucrose (sometimes called the *molar mass*). To obtain 1 mol of sucrose in the lab, therefore, we weigh out 342 g.

The practical advantage of measuring a quantity of chemicals in moles is that a mole of one substance has exactly the same number of molecules as a mole of any other substance. If the molecular mass of substance A is 342 daltons and that of substance B is 10 daltons, then 342 g of A will have the same number of molecules as 10 g of B. A mole of ethyl alcohol (C_2H_6O) also contains 6.02×10^{23} molecules, but its mass is only 46 g because the mass of a molecule of ethyl alcohol is less than that of a molecule of sucrose. Measuring in moles makes it convenient for scientists working in the laboratory to combine substances in fixed ratios of molecules.

How would we make a liter (L) of solution consisting of 1 mol of sucrose dissolved in water? We would measure out 342 g of sucrose and then gradually add water, while stirring, until the sugar was completely dissolved. We would then add enough water to bring the total volume of the solution up to 1 L. At that point, we would have a 1-molar (1 *M*) solution of sucrose. **Molarity**—the number of moles of solute per liter of solution—is the unit of concentration most often used by biologists for aqueous solutions.

Water's capacity as a versatile solvent complements the other properties discussed in this chapter. Since these remarkable properties allow water to support life on Earth so well, scientists who seek life elsewhere in the universe look for water as a sign that a planet might sustain life.

→ **Mastering Biology** MP3 Tutor: The Properties of Water

Possible Evolution of Life on Other Planets

EVOLUTION Biologists who look for life elsewhere in the universe (known as *astrobiologists*) have concentrated their search on planets that might have water. More than 800 planets have been found outside our solar system, with evidence for the presence of water vapor on a few. In our own solar system, Mars has been a focus of study. Like Earth, Mars has an ice cap at both poles. Images from spacecraft sent to Mars showed that ice is present just under the surface of Mars and that enough water vapor exists in its atmosphere for frost to form. In 2015, scientists found evidence of water flowing on Mars **(Figure 3.10)**, and a study using radar in 2018 concluded there is a large reservoir of liquid water one mile below surface ice. Drilling below the surface may be the next step in the search for signs of life on Mars. If any life-forms or fossils are found, their study will shed light on the process of evolution from an entirely new perspective.

▼ **Figure 3.10 Evidence for liquid water on Mars.** Water appears to have helped form these dark streaks that run downhill on Mars during the summer. NASA scientists also found evidence of hydrated salts, indicating water is present. (This digitally treated photograph was taken by the Mars Reconnaissance Orbiter.)

Dark streaks

CONCEPT CHECK 3.2

1. Describe how properties of water contribute to the upward movement of water in a tree.
2. Explain the saying "It's not the heat; it's the humidity."
3. How can the freezing of water crack boulders?
4. **WHAT IF?** A water strider (an insect that can walk on water) has legs that are coated with a hydrophobic substance. What might be the benefit? What would happen if the substance were hydrophilic?
5. **INTERPRET THE DATA** The concentration of the appetite-regulating hormone ghrelin is about 1.3×10^{-10} *M* in the blood of a fasting person. How many molecules of ghrelin are in 1 L of blood?

For selected answers, see Appendix A.

Acidic and basic conditions affect living organisms

Occasionally, a hydrogen atom participating in a hydrogen bond between two water molecules shifts from one molecule to the other. When this happens, the hydrogen atom leaves its electron behind, and what is actually transferred is a **hydrogen ion** (H^+), a single proton with a charge of 1+. The water molecule that lost a proton is now a **hydroxide ion** (OH^-), which has a charge of 1−. The proton binds to the other water molecule, making that molecule a **hydronium ion** (H_3O^+). We can picture the chemical reaction as follows:

$2\ H_2O$ Hydronium ion (H_3O^+)   Hydroxide ion (OH^-)

→ **Mastering Biology** Animation: Dissociation of Water Molecules

By convention, H^+ (the hydrogen ion) is used to represent H_3O^+ (the hydronium ion), and we follow that practice in this book. Keep in mind, though, that H^+ does not exist on its own in an aqueous solution. It is always associated with a water molecule in the form of H_3O^+.

As indicated by the double arrows, this is a reversible reaction that reaches a state of dynamic equilibrium when water molecules dissociate at the same rate that they are being reformed from H^+ and OH^-. At this equilibrium point, the concentration of water molecules greatly exceeds the concentrations of H^+ and OH^-. In pure water, only one water molecule in every 554 million is dissociated; the concentration of H^+ and of OH^- in pure water is therefore $10^{-7}\ M$ (at 25°C). This means there is only one ten-millionth of a mole of hydrogen ions per liter of pure water and an equal number of hydroxide ions. (Even so, this is a huge number—over 60,000 *trillion*—of each ion in a liter of pure water.)

Although the dissociation of water is reversible and statistically rare, it is exceedingly important in the chemistry of life. H^+ and OH^- are very reactive. Changes in their concentrations can drastically affect a cell's proteins and other complex molecules. As we have seen, the concentrations of H^+ and OH^- are equal in pure water, but adding certain kinds of solutes, called acids and bases, disrupts this balance. Biologists use something called the pH scale to describe how acidic or basic (the opposite of acidic) a solution is. In the remainder of this chapter, you will learn about acids, bases, and pH and why changes in pH can adversely affect organisms.

Acids and Bases

What would cause an aqueous solution to have an imbalance in H^+ and OH^- concentrations? When acids dissolve in water, they donate additional H^+ to the solution. An **acid** is a substance that increases the hydrogen ion concentration of a solution. For example, when hydrochloric acid (HCl) is added to water, hydrogen ions dissociate from chloride ions:

$$HCl \rightarrow H^+ + Cl^-$$

This source of H^+ (dissociation of water is the other source) results in an acidic solution—one having more H^+ than OH^-.

A substance that reduces the hydrogen ion concentration of a solution is called a **base**. Some bases reduce the H^+ concentration directly by accepting hydrogen ions. Ammonia (NH_3), for instance, acts as a base when the unshared electron pair in nitrogen's valence shell attracts a hydrogen ion from the solution, resulting in an ammonium ion (NH_4^+):

$$NH_3 + H^+ \rightleftharpoons NH_4^+$$

Other bases reduce the H^+ concentration indirectly by dissociating to form hydroxide ions, which combine with hydrogen ions and form water. One such base is sodium hydroxide (NaOH), which in water dissociates into its ions:

$$NaOH \rightarrow Na^+ + OH^-$$

In either case, the base reduces the H^+ concentration. Solutions with a higher concentration of OH^- than H^+ are known as basic solutions. A solution in which the H^+ and OH^- concentrations are equal is said to be neutral.

Notice that single arrows were used in the reactions for HCl and NaOH. These compounds dissociate completely when mixed with water, so hydrochloric acid is called a strong acid and sodium hydroxide a strong base. In contrast, ammonia is a weak base. The double arrows in the reaction for ammonia indicate that the binding and release of hydrogen ions are reversible reactions, although at equilibrium there will be a fixed ratio of NH_4^+ to NH_3.

Weak acids are acids that reversibly release and accept back hydrogen ions. An example is carbonic acid:

$$\underset{\text{Carbonic acid}}{H_2CO_3} \quad \rightleftharpoons \quad \underset{\text{Bicarbonate ion}}{HCO_3^-} \quad + \quad \underset{\text{Hydrogen ion}}{H^+}$$

Here the equilibrium so favors the reaction in the left direction that when carbonic acid is added to pure water, only 1% of the molecules are dissociated at any particular time. Still, that is enough to shift the balance of H^+ and OH^- from neutrality.

The pH Scale

In any aqueous solution at 25°C, the product of the H^+ and OH^- concentrations is constant at 10^{-14}. This can be written

$$[H^+][OH^-] = 10^{-14}$$

(The brackets indicate molar concentration.) As previously mentioned, in a neutral solution at 25°C, $[H^+] = 10^{-7}$ and $[OH^-] = 10^{-7}$. Therefore, the product of $[H^+]$ and $[OH^-]$ in a neutral solution at 25°C is 10^{-14}. If enough acid is added to a solution to increase $[H^+]$ to 10^{-5} M, then $[OH^-]$ will decline by an equivalent factor to 10^{-9} M (note that $10^{-5} \times 10^{-9} = 10^{-14}$). This constant relationship expresses the behavior of acids and bases in an aqueous solution. An acid not only adds hydrogen ions to a solution, but also removes hydroxide ions because of the tendency for H^+ to combine with OH^-, forming water. A base has the opposite effect, increasing OH^- concentration but also reducing H^+ concentration by the formation of water. If enough of a base is added to raise the OH^- concentration to 10^{-4} M, it will cause the H^+ concentration to drop to 10^{-10} M. Whenever we know the concentration of either H^+ or OH^- in an aqueous solution, we can deduce the concentration of the other ion.

The pH scale **(Figure 3.11)** is a simple numerical method for expressing the range of H^+ concentrations. The H^+

▼ **Figure 3.11 The pH scale and pH values of some aqueous solutions.**

pH Scale

0

1 Battery acid

2 Gastric juice (in stomach)
   Lemon juice

3 Vinegar, wine, cola
   Formic acid (from ants)

4 Tomato juice
   Beer
5 Black coffee

   Rainwater
6 Urine

   Saliva
7 **Pure water**
   Human blood, tears

8 Seawater
   Inside small intestine

9

10
   Milk of magnesia
11
   Household ammonia
12

   Household
13 bleach
   Oven cleaner
14

Acidic solution — Increasingly **Acidic** $[H^+] > [OH^-]$

Neutral solution — **Neutral** $[H^+] = [OH^-]$

Basic solution — Increasingly **Basic** $[H^+] < [OH^-]$

BLEACH

⮕ **Mastering Biology** Animation: Acids, Bases, and pH

concentrations of solutions can vary by a factor of 100 trillion or more. Instead of using moles per liter, the pH scale compresses the range of H^+ concentrations by employing logarithms. The **pH** of a solution is defined as the negative logarithm (base 10) of the H^+ concentration:

$$pH = -\log [H^+]$$

For a neutral aqueous solution, $[H^+]$ is 10^{-7} M, giving us

$$-\log 10^{-7} = -(-7) = 7$$

Notice that pH *decreases* as H^+ concentration *increases* (see Figure 3.11). Notice, too, that although the pH scale is based on H^+ concentration, it also implies OH^- concentration. A solution of pH 10 has a hydrogen ion concentration of 10^{-10} M and a hydroxide ion concentration of 10^{-4} M.

The pH of a neutral aqueous solution at 25°C is 7, the midpoint of the pH scale. A pH value less than 7 denotes an acidic solution; the lower the number, the more acidic the solution. The pH for basic solutions is above 7. Most biological fluids, such as blood and saliva, are within the range of pH 6–8. There are a few exceptions, however, including the strongly acidic digestive juice of the human stomach (gastric juice), which has a pH of about 2.

Remember that each pH unit represents a tenfold difference in H^+ and OH^- concentrations. It is this mathematical feature that makes the pH scale so compact. A solution of pH 3 is not twice as acidic as a solution of pH 6, but 1,000 times ($10 \times 10 \times 10$) more acidic. When the pH of a solution changes slightly, the actual concentrations of H^+ and OH^- in the solution change substantially.

Buffers

The internal pH of most living cells is close to 7. Even a slight change in pH can be harmful because the chemical processes of the cell are very sensitive to the concentrations of hydrogen and hydroxide ions. The pH of human blood is very close to 7.4, which is slightly basic. A person cannot survive for more than a few minutes if the blood pH drops to 7 or rises to 7.8, and a chemical system exists in the blood that maintains a stable pH. If 0.01 mol of a strong acid is added to a liter of pure water, the pH drops from 7.0 to 2.0. If the same amount of acid is added to a liter of blood, however, the pH decrease is only from 7.4 to 7.3. Why does the addition of acid have so much less of an effect on the pH of blood than it does on the pH of water?

The presence of substances called buffers allows biological fluids to maintain a relatively constant pH despite the addition of acids or bases. A **buffer** is a substance that minimizes changes in the concentrations of H^+ and OH^- in a solution. It does so by accepting hydrogen ions from the solution when they are in excess and donating hydrogen ions to the solution when they have been depleted. Most buffer solutions contain a weak acid and its corresponding base, which combine reversibly with hydrogen ions.

Several buffers contribute to pH stability in human blood and many other biological solutions. One of these is carbonic acid (H_2CO_3), which is formed when CO_2 reacts with water in blood plasma. As mentioned earlier, carbonic acid dissociates to yield a bicarbonate ion (HCO_3^-) and a hydrogen ion (H^+):

$$H_2CO_3 \; \underset{\text{Response to a drop in pH}}{\overset{\text{Response to a rise in pH}}{\rightleftharpoons}} \; HCO_3^- \; + \; H^+$$

H_2CO_3		HCO_3^-	+	H^+
H^+ donor (acid)		H^+ acceptor (base)		Hydrogen ion

The chemical equilibrium between carbonic acid and bicarbonate acts as a pH regulator, the reaction shifting left or right as other processes in the solution add or remove hydrogen ions. If the H^+ concentration in blood begins to fall (that is, if pH rises), the reaction proceeds to the right and more carbonic acid dissociates, replenishing hydrogen ions. But when the H^+ concentration in blood begins to rise (when pH drops), the reaction proceeds to the left, with HCO_3^- (the base) removing the hydrogen ions from the solution and forming H_2CO_3. Thus, the carbonic acid–bicarbonate buffering system consists of an acid and a base in equilibrium with each other. Most other buffers are also acid-base pairs.

Acidification: A Threat to Our Oceans

Among the many threats to water quality posed by human activities is the burning of fossil fuels, which releases CO_2 into the atmosphere. The resulting increase in atmospheric CO_2 levels has caused global warming and other aspects of climate change (see Concept 56.4). In addition, about 25% of human-generated CO_2 is absorbed by the oceans. In spite of the huge volume of water in the oceans, scientists worry that the absorption of so much CO_2 will harm marine ecosystems.

Recent data have shown that such fears are well founded. When CO_2 dissolves in seawater, it reacts with water to form carbonic acid, which lowers ocean pH. This process, known as **ocean acidification**, alters the delicate balance of conditions for life in the oceans **(Figure 3.12)**. Based on measurements of the CO_2 level in air bubbles trapped in ice over thousands of years, scientists calculate that the pH of the oceans is 0.1 pH unit lower (more acidic) now than at any time in the past 420,000 years. Recent studies predict that it will drop another 0.3–0.5 pH unit by the end of this century.

As seawater acidifies, the extra hydrogen ions combine with carbonate ions (CO_3^{2-}) to form bicarbonate ions (HCO_3^-), thereby reducing the carbonate ion concentration (see Figure 3.12). Scientists predict that ocean acidification will cause the carbonate ion concentration to decrease by 40% by the year 2100. This is of great concern because carbonate ions are required for calcification, the production of calcium carbonate ($CaCO_3$) by many marine organisms, including reef-building corals and animals that build shells. The **Scientific Skills Exercise** allows you to work with data from an experiment examining the effect of carbonate ion concentration on coral reefs, using

▼ **Figure 3.12 Atmospheric CO_2 from human activities and its fate in the ocean.**

Some carbon dioxide (CO_2) in the atmosphere dissolves in the ocean, where it reacts with water to form carbonic acid (H_2CO_3).

$$CO_2 + H_2O \rightarrow H_2CO_3$$

Carbonic acid dissociates into hydrogen ions (H^+) and bicarbonate ions (HCO_3^-).

$$H_2CO_3 \rightarrow H^+ + HCO_3^-$$

The added H^+ combines with carbonate ions (CO_3^{2-}), forming more HCO_3^-.

$$H^+ + CO_3^{2-} \rightarrow HCO_3^-$$

Less CO_3^{2-} is available for calcification — the formation of calcium carbonate ($CaCO_3$)— by marine organisms such as corals.

$$CO_3^{2-} + Ca^{2+} \rightarrow CaCO_3$$

VISUAL SKILLS *Summarize the effect of adding excess CO_2 to the oceans on the calcification process in the final equation.*

an artificial system. In 2018, researchers carried out the first CO_2 enhancement study on an unconfined natural coral reef, observing that addition of CO_2 suppressed calcification and concluding that ocean acidification is likely to cause "profound, ecosystem-wide changes in coral reefs." Coral reefs are sensitive ecosystems that act as havens for a great diversity of marine life. The disappearance of coral reef ecosystems would be a tragic loss of biological diversity.

If there is any reason for optimism about the future quality of water resources on our planet, it is that we have made progress in learning about the delicate chemical balances in oceans, lakes, and rivers. Continued progress can come only from the actions of informed individuals, like yourselves, who are concerned about environmental quality. This requires understanding the crucial role that water plays in the suitability of the environment for continued life on Earth.

CONCEPT CHECK 3.3

1. Compared with a basic solution at pH 9, the same volume of an acidic solution at pH 4 has _____ times as many hydrogen ions (H^+).

2. HCl is a strong acid that dissociates in water: $HCl \rightarrow H^+ + Cl^-$. What is the pH of 0.01 M HCl?

3. Acetic acid (CH_3COOH) can be a buffer, similar to carbonic acid. Write the dissociation reaction, identifying the acid, base, H^+ acceptor, and H^+ donor.

4. **WHAT IF?** Given a liter of pure water and a liter solution of acetic acid, what would happen to the pH, in general, if you added 0.01 mol of a strong acid to each? Use the reaction from question 3 to explain the result.

For suggested answers, see Appendix A.

Interpreting a Scatter Plot with a Regression Line

How Does the Carbonate Ion Concentration of Seawater Affect the Calcification Rate of a Coral Reef? Scientists predict that acidification of the ocean due to higher levels of atmospheric CO_2 will lower the concentration of dissolved carbonate ions, which living corals use to build calcium carbonate reef structures. In this exercise, you will analyze data from a controlled experiment that examined the effect of carbonate ion concentration ($[CO_3^{2-}]$) on calcium carbonate deposition, a process called calcification.

How the Experiment Was Done For several years, scientists conducted research on ocean acidification using a large coral reef aquarium at Biosphere 2 in Arizona. They measured the rate of calcification by the reef organisms and examined how the calcification rate changed with differing amounts of dissolved carbonate ions in the seawater.

Data from the Experiment The black data points in the graph form a scatter plot. The red line, known as a linear regression line, is the best-fitting straight line for these points.

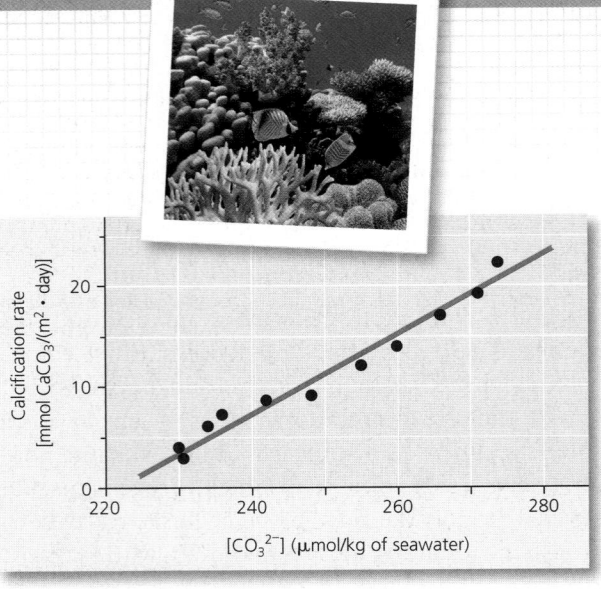

Data from C. Langdon et al., Effect of calcium carbonate saturation state on the calcification rate of an experimental coral reef, *Global Biogeochemical Cycles* 14:639–654 (2000).

INTERPRET THE DATA

1. When presented with a graph of experimental data, the first step in analysis is to determine what each axis represents. **(a)** In words, what is shown on the *x*-axis? (Include the units.) **(b)** What is on the *y*-axis? **(c)** Which variable is the independent variable—the one that was *manipulated* by the researchers? **(d)** Which is the dependent variable—the one that responded to or depended on the treatment, which was *measured* by the researchers? (For additional information about graphs, see the Scientific Skills Review in Appendix D.)

2. Based on the data shown in the graph, describe in words the relationship between carbonate ion concentration and calcification rate.

3. **(a)** If the seawater carbonate ion concentration is 270 μmol/kg, estimate the rate of calcification and how many days it would take 1 square meter of reef to accumulate 30 mmol of calcium carbonate ($CaCO_3$). **(b)** If the seawater carbonate ion concentration is 250 μmol/kg, what is the approximate rate of calcification, and approximately how many days would it take 1 square meter of reef to accumulate 30 mmol of calcium carbonate? **(c)** If the carbonate ion concentration decreases, how does the calcification rate change, and how does that affect the time it takes coral to grow?

4. **(a)** Which step of the process in Figure 3.12 is measured in this experiment? **(b)** Are the results of this experiment consistent with the hypothesis that increased atmospheric $[CO_2]$ will slow the growth of coral reefs? Why or why not?

→ **Instructors:** A version of this Scientific Skills Exercise can be assigned in **Mastering Biology**.

3 Chapter Review

→ Go to **Mastering Biology** for Assignments, the eText, the Study Area, and Dynamic Study Modules.

SUMMARY OF KEY CONCEPTS

→ To review key terms, go to the **Vocabulary Self-Quiz** in the **Mastering Biology** eText or Study Area, or go to goo.gl/zkjz9t.

CONCEPT 3.1

Polar covalent bonds in water molecules result in hydrogen bonding (p. 45)

- Water is a **polar molecule**. A hydrogen bond forms when a partially negatively charged region on the oxygen of one water molecule is attracted to the partially positively charged hydrogen of

a nearby water molecule. Hydrogen bonding between water molecules is the basis for water's properties.

DRAW IT *Label a hydrogen bond and a polar covalent bond in the diagram of five water molecules. Is a hydrogen bond a covalent bond? Explain.*

CONCEPT 3.2

Four emergent properties of water contribute to Earth's suitability for life (pp. 45–50)

- Hydrogen bonding keeps water molecules close to each other, giving water **cohesion**. Hydrogen bonding is also responsible for water's **surface tension**.
- Water has a high **specific heat**: Heat is absorbed when hydrogen bonds break and is released when hydrogen bonds form. This helps keep **temperatures** relatively steady, within limits that permit life. **Evaporative cooling** is based on water's high **heat**

of vaporization. The evaporative loss of the most energetic water molecules cools a surface.

- Ice floats because it is less dense than liquid water. This property allows life to exist under the frozen surfaces of lakes and polar seas.
- Water is an unusually versatile **solvent** because its polar molecules are attracted to ions and polar substances that can form hydrogen bonds. **Hydrophilic** substances have an affinity for water; **hydrophobic** substances do not. **Molarity**, the number of moles of **solute** per liter of **solution**, is used as a measure of solute concentration in solutions. A **mole** is a certain number of molecules of a substance. The mass of a mole of a substance in grams is the same as the **molecular mass** in daltons.
- The emergent properties of water support life on Earth and may contribute to the potential for life to have evolved on other planets.

? *Describe how different types of solutes dissolve in water. Explain what a solution is.*

CONCEPT 3.3

Acidic and basic conditions affect living organisms *(pp. 51–54)*

- A water molecule can transfer an H^+ to another water molecule to form H_3O^+ (represented simply by H^+) and OH^-.
- The concentration of H^+ is expressed as **pH**; $pH = -\log [H^+]$. A **buffer** consists of an acid-base pair that combines reversibly with hydrogen ions, allowing it to resist pH changes.
- The burning of fossil fuels increases the amount of CO_2 in the atmosphere. Some CO_2 dissolves in the oceans, causing **ocean acidification**, which has potentially grave consequences for marine organisms that rely on calcification.

Acidic
$[H^+] > [OH^-]$ — 0

Acids donate H^+ in aqueous solutions.

Neutral
$[H^+] = [OH^-]$ — 7

Basic
$[H^+] < [OH^-]$ — 14

Bases donate OH^- or accept H^+ in aqueous solutions.

? *Explain what happens to the concentration of hydrogen ions in an aqueous solution when you add a base and cause the concentration of OH^- to rise to 10^{-3}. What is the pH of this solution?*

TEST YOUR UNDERSTANDING

➡ For more multiple-choice questions, go to the **Practice Test** in the **Mastering Biology** eText or Study Area, or go to goo.gl/GruWRg.

Levels 1-2: Remembering/Understanding

1. Which of the following is a hydrophobic material?
   (A) paper   (C) wax
   (B) table salt   (D) sugar

2. We can be sure that a mole of table sugar and a mole of vitamin C are equal in their
   (A) mass.   (C) number of atoms.
   (B) volume.   (D) number of molecules.

3. Measurements show that the pH of a particular lake is 4.0. What is the hydrogen ion concentration of the lake?
   (A) $4.0\,M$ (C) $10^{-4}\,M$
   (B) $10^{-10}\,M$ (D) $10^4\,M$

4. What is the *hydroxide* ion concentration of the lake described in question 3?
   (A) $10^{-10}\,M$ (C) $10^{-7}\,M$
   (B) $10^{-4}\,M$ (D) $10.0\,M$

Levels 3-4: Applying/Analyzing

5. A slice of pizza has 500 kcal. If we could burn the pizza and use all the heat to warm a 50-L container of cold water, what would be the approximate increase in the temperature of the water? (Note: A liter of cold water weighs about 1 kg.)
   (A) 50°C   (C) 100°C
   (B) 5°C   (D) 10°C

6. **DRAW IT** Draw the hydration shells that form around a potassium ion and a chloride ion when potassium chloride (KCl) dissolves. Label the positive, negative, and partial charges.

Levels 5-6: Evaluating/Creating

7. Right before a predicted overnight freeze, farmers spray water on crops to protect the plants. Use the properties of water to explain how this method works. Be sure to mention why hydrogen bonds are responsible for this phenomenon.

8. **MAKE CONNECTIONS** What do climate change (see Concepts 1.1 and 3.2) and ocean acidification have in common?

9. **EVOLUTION CONNECTION** This chapter explains how the emergent properties of water contribute to the suitability of the environment for life. Until fairly recently, scientists assumed that other physical requirements for life included a moderate range of temperature, pH, atmospheric pressure, and salinity, as well as low levels of toxic chemicals. That view has changed with the discovery of organisms known as extremophiles, which flourish in hot, acidic sulfur springs, around hydrothermal vents deep in the ocean, and in soils with high levels of toxic metals. Why would astrobiologists study extremophiles? What does the existence of life in such extreme environments say about the possibility of life on other planets?

10. **SCIENTIFIC INQUIRY** Design a controlled experiment to test the hypothesis that water acidification caused by acidic rain would inhibit the growth of *Elodea*, a freshwater plant (see Figure 2.17).

11. **WRITE ABOUT A THEME: ORGANIZATION** Several emergent properties of water contribute to the suitability of the environment for life. In a short essay (100–150 words), describe how the ability of water to function as a versatile solvent arises from the structure of water molecules.

12. **SYNTHESIZE YOUR KNOWLEDGE**

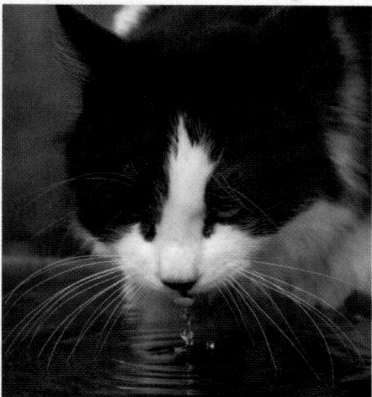

How do cats drink? Scientists using high-speed video have shown that cats use an interesting technique to drink aqueous substances like water and milk. Four times a second, the cat touches the tip of its tongue to the water and draws a column of water up into its mouth (as you can see in the photo), which then shuts before gravity can pull the water back down. Describe how the properties of water allow cats to drink in this fashion, including how water's molecular structure contributes to the process.

For selected answers, see Appendix A.

Explore Scientific Papers with Science in the Classroom ▲AAAS
How are coral reefs responding to climate change?
Go to "Take the Heat" at www.scienceintheclassroom.org.
➡ **Instructors:** Questions can be assigned in **Mastering Biology**.

4 Carbon and the Molecular Diversity of Life

KEY CONCEPTS

4.1 Organic chemistry is key to the origin of life *p. 57*

4.2 Carbon atoms can form diverse molecules by bonding to four other atoms *p. 58*

4.3 A few chemical groups are key to molecular function *p. 62*

Study Tip

Label chemical groups: After you have read through Figure 4.9, look through Chapters 4 and 5 for molecules that have the chemical groups shown in that figure. Circle and label the chemical groups you find, as in the following example:

Figure 4.1 The Qinling golden snub-nosed monkeys and other living organisms in this mountainous forest in southwest China are made up of chemicals based mostly on the element carbon. Of all chemical elements, carbon is unparalleled in its ability to form molecules that are large, complex, and varied, making possible the diversity of organisms that have evolved on Earth.

→ Go to Mastering Biology

For Students (in eText and Study Area)
- Get Ready for Chapter 4
- Animation: Diversity of Carbon-Based Molecules
- Animation: Functional Groups

For Instructors to Assign (in Item Library)
- Activity: Isomers
- Tutorial: Carbon Bonding and Functional Groups

What makes carbon the basis for all biological molecules?

Carbon can form four bonds, and therefore can bond to up to four other atoms or groups of atoms.

Carbon can bond to other carbons, resulting in carbon skeletons. Carbon also commonly bonds to

 H hydrogen,

 O oxygen, and

 N nitrogen.

The properties of a carbon-containing molecule depend on the arrangement of its **carbon skeleton** and on its **chemical groups.**

Carbon skeleton

Chemical groups

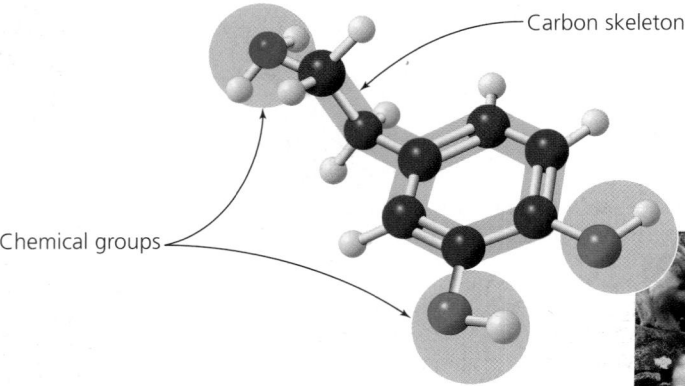

The signaling molecule shown here, **dopamine,** has many functions, including promoting mother-infant bonding.

Organic chemistry is key to the origin of life

For historical reasons, compounds containing carbon are said to be organic, and their study is called **organic chemistry**. Organic compounds range from simple molecules, such as methane (CH_4), to colossal ones, such as proteins, with thousands of atoms.

EVOLUTION In 1953, Stanley Miller, a graduate student of Harold Urey at the University of Chicago, designed an experiment on the abiotic (nonliving) synthesis of organic compounds to investigate the origin of life. Study **Figure 4.2** to learn about his classic experiment. From his results, Miller concluded that complex organic molecules could arise spontaneously under conditions thought at that time to have existed on early Earth. You can work with the data from a related experiment in the **Scientific Skills Exercise**. These experiments support the idea that abiotic synthesis of organic compounds, perhaps near volcanoes, could have been an early stage in the origin of life (see Figure 25.2).

In Concept 3.2, you learned about evidence for the presence of water on Mars. Even more exciting, in 2018, NASA reported that the rover *Curiosity* had found carbon-based compounds on Mars in a crater where a lake once existed. While these compounds might have been brought to Mars on a meteorite or formed by geologic processes, an intriguing possibility is that they might have been the relics of life-forms that once existed on that planet.

The overall percentages of the major elements of life—C, H, O, N, S, and P—are quite uniform from one organism to another, reflecting the common evolutionary origin of all life. Because of carbon's ability to form four bonds, however, this limited assortment of atomic building blocks can be used to build an inexhaustible variety of organic molecules. Different species of organisms, and different individuals within a species, are distinguished by variations in the types of organic molecules they make. In a sense, the great diversity of living organisms we see on the planet (and in fossil remains) is made possible by the unique chemical versatility of the carbon atom.

→ **Mastering Biology**
Interview with Stanley Miller:
Investigating the origin of life

CONCEPT CHECK 4.1

1. **VISUAL SKILLS** See Figure 4.2. Miller carried out a control experiment without discharging sparks and found no organic compounds. What might explain this result?

For suggested answers, see Appendix A.

▼ **Figure 4.2** Inquiry

Can organic molecules form under conditions estimated to simulate those on the early Earth?

Experiment In 1953, Stanley Miller set up a closed system to mimic conditions thought at that time to have existed on the early Earth. A flask of water simulated the primeval sea. The water was heated so that some vaporized and moved into a second, higher flask containing the "atmosphere"—a mixture of gases. Sparks were discharged in the synthetic atmosphere to mimic lightning.

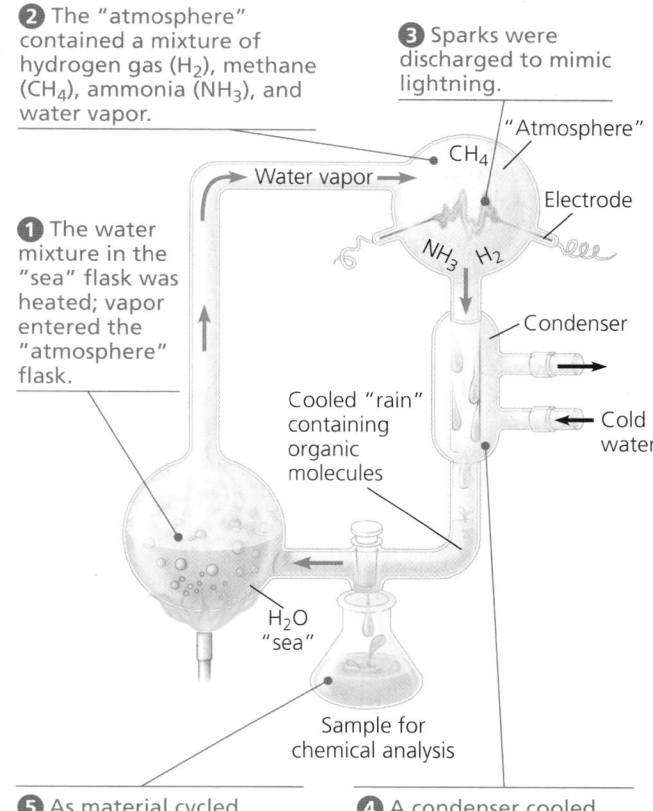

2 The "atmosphere" contained a mixture of hydrogen gas (H_2), methane (CH_4), ammonia (NH_3), and water vapor.

3 Sparks were discharged to mimic lightning.

"Atmosphere"

CH_4

Water vapor

Electrode

1 The water mixture in the "sea" flask was heated; vapor entered the "atmosphere" flask.

NH_3 H_2

Condenser

Cooled "rain" containing organic molecules

Cold water

H_2O "sea"

Sample for chemical analysis

5 As material cycled through the apparatus, Miller periodically collected samples for analysis.

4 A condenser cooled the atmosphere, raining water and dissolved molecules into the sea flask.

Results Miller identified a variety of organic molecules that are common in organisms. These included simple compounds, such as formaldehyde (CH_2O) and hydrogen cyanide (HCN), and more complex molecules, such as amino acids and long chains of carbon and hydrogen known as hydrocarbons.

Conclusion Organic molecules, a first step in the origin of life, may have been synthesized abiotically on the early Earth. Although later evidence indicated that the early-Earth atmosphere was different from the "atmosphere" used by Miller in this experiment, recent experiments using the revised list of chemicals also produced organic molecules. (We will explore this hypothesis in more detail in Concept 25.1.)

Data from S. L. Miller, A production of amino acids under possible primitive Earth conditions, *Science* 117:528–529 (1953).

WHAT IF? *If Miller had increased the concentration of NH_3 in his experiment, how might the relative amounts of the products HCN and CH_2O have differed?*

Working with Moles and Molar Ratios

Could the First Biological Molecules Have Formed Near Volcanoes on Early Earth? In 2007, Jeffrey Bada, a former graduate student of Stanley Miller, discovered some vials of samples that had never been analyzed from an experiment performed by Miller in 1958. In that experiment, Miller used hydrogen sulfide gas (H_2S) as one of the gases in the reactant mixture. Since H_2S is released by volcanoes, the H_2S experiment was designed to mimic conditions near volcanoes on early Earth. In 2011, Bada and colleagues published the results of their analysis of these "lost" samples. In this exercise, you will make calculations using the molar ratios of reactants and products from the H_2S experiment.

How the Experiment Was Done According to his laboratory notebook, Miller used the same apparatus as in his original experiment (see Figure 4.2), but the mixture of gaseous reactants included methane (CH_4), carbon dioxide (CO_2), hydrogen sulfide (H_2S), and ammonia (NH_3). After three days of simulated volcanic activity, he collected samples of the liquid, partially purified the chemicals, and sealed the samples in sterile vials. In 2011, Bada's research team used modern analytical methods to analyze the products in the vials for the presence of amino acids, the building blocks of proteins.

Data from the Experiment The table below shows 4 of the 23 amino acids detected in the 2011 analysis of the samples from Miller's 1958 H_2S experiment.

Product Compound	Molecular Formula	Molar Ratio (Relative to Glycine)
Glycine	$C_2H_5NO_2$	1.0
Serine	$C_3H_7NO_3$	3.0×10^{-2}
Methionine	$C_5H_{11}NO_2S$	1.8×10^{-3}
Alanine	$C_3H_7NO_2$	1.1

Data from E. T. Parker et al., Primordial synthesis of amines and amino acids in a 1958 Miller H_2S-rich spark discharge experiment, *Proceedings of the National Academy of Sciences USA* 108:5526-5531 (2011). www.pnas.org/cgi/doi/10.1073/pnas.1019191108.

INTERPRET THE DATA

1. A *mole* is the number of particles of a substance with a mass equivalent to its molecular (or atomic) mass in daltons. There are 6.02×10^{23} molecules (or atoms) in 1.0 mole (Avogadro's number; see Concept 3.2). The data table shows the "molar ratios" of some of the products from the Miller H_2S experiment. In a molar ratio, each unitless value is expressed relative to a standard for that experiment. Here, the standard is the number

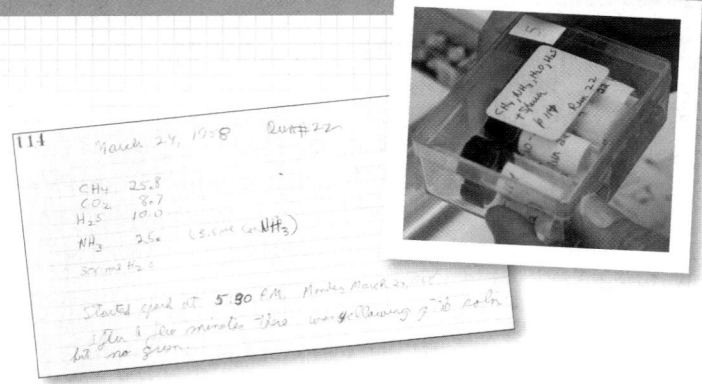

▲ **Some of Stanley Miller's notes from his 1958 hydrogen sulfide (H_2S) experiment along with his original vials.**

of moles of the amino acid glycine, which is set to a value of 1.0. For instance, serine has a molar ratio of 3.0×10^{-2}, meaning that for every mole of glycine, there is 3.0×10^{-2} mole of serine. **(a)** Give the molar ratio of methionine to glycine and explain what it means. **(b)** How many molecules of glycine are present in 1.0 mole? **(c)** For every 1.0 mole of glycine in the sample, how many molecules of methionine are present? (Recall that to multiply two numbers with exponents, you add their exponents; to divide them, you subtract the exponent in the denominator from that in the numerator.)

2. **(a)** Which amino acid is present in higher amounts than glycine? **(b)** How many more molecules of that amino acid are present than the number of molecules in 1.0 mole of glycine?

3. The synthesis of products is limited by the amount of reactants. **(a)** If one mole each of CH_4, NH_3, H_2S, and CO_2 is added to 1 liter of water ($= 55.5$ moles of H_2O) in a flask, how many moles of hydrogen, carbon, oxygen, nitrogen, and sulfur are in the flask? **(b)** Looking at the molecular formula in the table, how many moles of each element would be needed to make 1.0 mole of glycine? **(c)** What is the maximum number of moles of glycine that could be made in that flask, with the specified ingredients, if no other molecules were made? Explain. **(d)** If serine or methionine were made individually, which element(s) would be used up first for each? How much of each product could be made?

4. The earlier published experiment carried out by Miller did not include H_2S in the reactants (see Figure 4.2). Which of the compounds shown in the data table can be made in the H_2S experiment but could not be made in the earlier experiment?

→ **Instructors:** A version of this Scientific Skills Exercise can be assigned in **Mastering Biology**.

CONCEPT 4.2

Carbon atoms can form diverse molecules by bonding to four other atoms

The key to an atom's chemical characteristics is its electron configuration. This configuration determines the kinds and number of bonds an atom will form with other atoms. Recall that it is the valence electrons, those in the outermost shell, that are available to form bonds with other atoms.

The Formation of Bonds with Carbon

Carbon has 6 electrons, with 2 in the first electron shell and 4 in the second shell; thus, it has 4 valence electrons in a shell that can hold up to 8 electrons. A carbon atom usually completes its valence shell by sharing its 4 electrons with other atoms so that 8 electrons are present. Each pair of shared electrons constitutes a covalent bond (see Figure 2.10d). In organic molecules, carbon usually forms single or double covalent bonds. Each carbon atom acts as an intersection point from which a molecule can branch off in as many as four directions. This enables carbon to form large, complex molecules.

▼ Figure 4.3 The shapes of three simple organic molecules.

Molecule and Molecular Shape	Molecular Formula	Structural Formula	Ball-and-Stick Model (molecular shape in pink)	Space-Filling Model
(a) Methane. When a carbon atom has four single bonds to other atoms, the molecule is tetrahedral.	CH_4	H | H — C — H | H		
(b) Ethane. A molecule may have more than one tetrahedral group of single-bonded atoms. (Ethane consists of two such groups.)	C_2H_6	H H | | H — C — C — H | | H H		

(c) Ethene (ethylene). When two carbon atoms are joined by a double bond, all atoms attached to those carbons are in the same plane, and the molecule is flat.	C_2H_4	H H
		  \ /
		   C = C
		  / \
		 H H		

When a carbon atom forms four single covalent bonds, the arrangement of its four hybrid orbitals causes the bonds to angle toward the corners of an imaginary tetrahedron. The bond angles in methane (CH_4) are 109.5° **(Figure 4.3a)**, and they are roughly the same in any group of atoms where carbon has four single bonds. For example, ethane (C_2H_6) is shaped like two overlapping tetrahedrons **(Figure 4.3b)**. In molecules with more carbons, every grouping of a carbon bonded to four other atoms has a tetrahedral shape. But when two carbon atoms are joined by a double bond, as in ethene (C_2H_4), the bonds from both carbons are all in the same plane, so the atoms joined to those carbons are in the same plane as well **(Figure 4.3c)**. We find it convenient to write molecules as structural formulas, as if the molecules being represented are two-dimensional, but keep in mind that molecules are three-dimensional and that the shape of a molecule is central to its function.

The number of electrons required to fill the valence shell of an atom is generally equal to the atom's **valence**, the number of covalent bonds it can form. **Figure 4.4** shows the valences of carbon and its most frequent bonding partners—hydrogen, oxygen, and nitrogen. These are the four main atoms in organic molecules.

The electron configuration of carbon gives it covalent compatibility with many different elements. Let's consider how valence and the rules of covalent bonding apply to carbon atoms with partners other than hydrogen. We'll look at two examples, the simple molecules carbon dioxide and urea.

▼ Figure 4.4 **Valences of the major elements of organic molecules.** Valence, the number of covalent bonds an atom can form, is generally equal to the number of electrons required to fill the valence shell. (Sodium, phosphorus, and chlorine are exceptions.)

	Hydrogen	Oxygen	Nitrogen	Carbon
Lewis dot structure showing existing valence electrons	H·	·Ö:	·N̈·	·C̈·
Electron distribution diagram with red circles showing electrons needed to fill the valence shell	(H)	(O)	(N)	(C)
Number of electrons needed to fill the valence shell	1	2	3	4
Valence: Number of bonds the element can form	1	2	3	4

MAKE CONNECTIONS *Draw the Lewis dot structures for sodium, silicon, phosphorus, sulfur, and chlorine. (Refer to Figure 2.7.)*

In the carbon dioxide molecule (CO_2), a single carbon atom is joined to two atoms of oxygen by double covalent bonds. The structural formula for CO_2 is shown here:

$$O = C = O$$

Each line in a structural formula represents a pair of shared electrons. Thus, the two double bonds in CO_2 have the same number of shared electrons as four single bonds. The arrangement completes the valence shells of all atoms in the molecule:

Because CO_2 is a very simple molecule and lacks hydrogen, it is often considered inorganic, even though it contains carbon. Whether we call CO_2 organic or inorganic, however, it is clearly important to the living world as the source of carbon, via photosynthetic organisms, for all organic molecules in organisms (see Concept 2.4).

Urea, $CO(NH_2)_2$, is an organic compound found in urine. Again, each atom has the required number of covalent bonds. In this case, one carbon atom participates in both single and double bonds.

Urea

Urea and carbon dioxide are molecules with only one carbon atom. But a carbon atom can also use one or more valence electrons to form covalent bonds to other carbon atoms, linking the atoms into chains, as shown here for C_3H_8:

Molecular Diversity Arising from Variation in Carbon Skeletons

Carbon chains form the basis of most organic molecules. Carbon skeletons vary in length and may be straight, branched, or arranged in closed rings **(Figure 4.5)**. Some carbon chains have double bonds, which vary in number and location. Such variation in carbon chains is one important source of the molecular complexity and diversity that characterize living matter. In addition, the skeletons of biological molecules often include atoms of other elements, like oxygen and phosphorus; such atoms can also be bonded to carbons of the skeleton.

Hydrocarbons

All of the molecules that are shown in Figures 4.3 and 4.5 are **hydrocarbons**, organic molecules consisting of only carbon and hydrogen. Atoms of hydrogen are attached to the carbon skeleton wherever electrons are available for covalent bonding. Hydrocarbons are the major components of petroleum, which is called a fossil fuel because it consists of the partially decomposed remains of organisms that lived millions of years ago.

> **Mastering Biology Interview with Deborah Gordon: Studying How Ants Use Hydrocarbons to Communicate**

▼ **Figure 4.5 Four ways that carbon skeletons can vary.**

(a) Length

Ethane   Propane

Carbon skeletons vary in length.

(b) Branching

Butane   2-Methylpropane (commonly called isobutane)

Skeletons may be unbranched or branched.

(c) Double bond position

1-Butene   2-Butene

The skeleton may have double bonds, which can vary in location.

(d) Presence of rings

Cyclohexane   Benzene

Some carbon skeletons are arranged in rings. In the abbreviated structural formula for each compound (to its right), each corner represents a carbon and its attached hydrogens.

> **Mastering Biology Animation: Diversity of Carbon-Based Molecules**

Although hydrocarbons are not prevalent in most living organisms, some of a cell's organic molecules have regions consisting of only carbon and hydrogen. For example, the molecules known as fats have long hydrocarbon tails attached to a nonhydrocarbon component **(Figure 4.6)**. Neither petroleum nor fat dissolves in water; both are hydrophobic compounds because the great majority of their bonds are relatively nonpolar carbon-to-hydrogen linkages. Another characteristic of hydrocarbons is that they can undergo reactions that release a relatively large amount of energy. The gasoline that fuels a car consists of hydrocarbons, and the hydrocarbon tails of fats serve as stored fuel for plant embryos (seeds) and animals.

▼ Figure 4.6 The role of hydrocarbons in fats.

(a) Mammalian adipose cells stockpile fat molecules as a fuel reserve. This colorized micrograph shows part of a human adipose cell with many fat droplets, each containing a large number of fat molecules. **(b)** A fat molecule consists of a small, nonhydrocarbon component joined to three hydrocarbon tails that account for the hydrophobic behavior of fats. The tails can be broken down to provide energy. (Black = carbon; gray = hydrogen; red = oxygen.)

— Nucleus

— Fat droplets

10 μm

(a) Part of a human adipose cell   **(b) A fat molecule**

MAKE CONNECTIONS *How do the tails account for the hydrophobic nature of fats? (See Concept 3.2.)*

Isomers

Variation in the architecture of organic molecules can be seen in **isomers**, compounds that have the same numbers of atoms of the same elements but different structures and hence different properties. We will examine three types of isomers: structural isomers, *cis-trans* isomers, and enantiomers.

Structural isomers differ in the covalent arrangements of their atoms. Compare, for example, the two five-carbon compounds in **Figure 4.7a**. Both have the molecular formula C_5H_{12}, but they differ in the covalent arrangement of their carbon skeletons. The skeleton is straight in one compound but branched in the other. The number of possible isomers increases tremendously as carbon skeletons increase in size. There are only three forms of C_5H_{12} (two of which are shown in Figure 4.7a), but there are 18 variants of C_8H_{18} and 366,319 possible structural isomers of $C_{20}H_{42}$. Structural isomers may also differ in the location of double bonds.

In *cis-trans* **isomers** (also known as *geometric isomers*), carbons have covalent bonds to the same atoms, but these atoms differ in their spatial arrangements due to the inflexibility of double bonds. Single bonds allow the atoms they join to rotate freely about the bond axis without changing the compound. In contrast, double bonds do not permit such rotation. If a double bond joins two carbon atoms, and each C also has two different atoms (or groups of atoms) attached to it, then two distinct *cis-trans* isomers are possible. Consider a simple molecule with two double-bonded carbons, each of which has an H and an X attached to it **(Figure 4.7b)**. The arrangement with both Xs on the same side of the double bond is called a *cis isomer,* and that

▼ Figure 4.7 Three types of isomers. Isomers are compounds that have the same molecular formula but different structures.

(a) Structural isomers

Pentane   2-Methylbutane

Structural isomers differ in the arrangement of covalent bonding partners, as shown in these two isomers of C_5H_{12}.

(b) *Cis-trans* isomers (also known as geometric isomers)

cis isomer: The two Xs are on the same side.

trans isomer: The two Xs are on opposite sides.

Cis-trans isomers differ in arrangement about a double bond. In these diagrams, X represents an atom or group of atoms attached to a double-bonded carbon.

(c) Enantiomers

L isomer   D isomer

Enantiomers differ in spatial arrangement around an asymmetric carbon, resulting in molecules that are mirror images, like left and right hands. The two isomers here are designated the L and D isomers from the Latin for "left" and "right" (*levo* and *dextro*). Enantiomers cannot be superimposed on each other.

DRAW IT *There are three structural isomers of C_5H_{12}; draw the one not shown in (a).*

→ **Mastering Biology Animation: Isomers**

with the Xs on opposite sides is called a *trans isomer*. The subtle difference in shape between such isomers can have a dramatic effect on the biological activities of organic molecules. For example, the biochemistry of vision involves a light-induced change of retinal, a chemical compound in the eye, from the *cis* isomer to the *trans* isomer (see Figure 50.17). Another example involves *trans* fats, harmful fats formed during food processing that are discussed in Concept 5.3.

Enantiomers are isomers that are mirror images of each other and that differ in shape due to the presence of an *asymmetric carbon*, one that is attached to four different atoms or groups of atoms. (See the middle carbon in the ball-and-stick models shown in **Figure 4.7c**.) The four groups can

▼ **Figure 4.8 The pharmacological importance of enantiomers.** Ibuprofen and albuterol are drugs whose enantiomers have different effects. (S and R are used here to distinguish between enantiomers, rather than D and L as in Figure 4.7c.) Ibuprofen is commonly sold as a mixture of the two enantiomers; the S enantiomer is 100 times more effective than the R form. Albuterol is synthesized and sold only as the R form of that drug; the S form counteracts the active R form.

Drug	Effects	Effective Enantiomer	Ineffective Enantiomer
Ibuprofen	Reduces inflammation and pain	S-Ibuprofen	R-Ibuprofen
Albuterol	Relaxes bronchial (airway) muscles, improving airflow in asthma patients	R-Albuterol	S-Albuterol

be arranged in space around the asymmetric carbon in two different ways that are mirror images. Enantiomers are, in a way, left-handed and right-handed versions of the molecule. Just as your right hand won't fit into a left-handed glove, a "right-handed" molecule won't fit into the same space as the "left-handed" version. Usually, only one isomer is biologically active because only that form can bind to specific molecules in an organism.

The concept of enantiomers is important in the pharmaceutical industry because the two enantiomers of a drug may not be equally effective, as is the case for both ibuprofen and the asthma medication albuterol **(Figure 4.8)**. Methamphetamine also occurs in two enantiomers that have very different effects. One enantiomer is the highly addictive stimulant drug known as "crank," sold illegally in the street drug trade. The other has a much weaker effect and is the active ingredient in an over-the-counter vapor inhaler for treatment of nasal congestion. The differing effects of enantiomers in the body demonstrate that organisms are sensitive to even the subtlest variations in molecular architecture. Once again, we see that molecules have emergent properties that depend on the specific arrangement of their atoms.

CONCEPT CHECK 4.2

1. **DRAW IT** (a) Draw a structural formula for C_2H_4. (b) Draw the *trans* isomer of $C_2H_2Cl_2$.
2. **VISUAL SKILLS** Which two pairs of molecules in Figure 4.5 are isomers? For each pair, identify the type of isomer.
3. How are gasoline and fat chemically similar?
4. **VISUAL SKILLS** See Figures 4.5a and 4.7. Can propane (C_3H_8) form isomers? Explain.

For suggested answers, see Appendix A.

A few chemical groups are key to molecular function

The distinctive properties of an organic molecule depend not only on the arrangement of its mostly carbon skeleton but also on the various chemical groups attached to that skeleton. These groups may participate in chemical reactions or may contribute to function indirectly by their effects on molecular shape; they help give each molecule its unique properties.

The Chemical Groups Most Important in the Processes of Life

Consider the differences between estradiol (a type of estrogen) and testosterone. These compounds are female and male sex hormones, respectively, in humans and other vertebrates. Both are steroids, organic molecules with a common carbon skeleton in the form of four fused rings. They differ only in the chemical groups attached to the rings (shown here in abbreviated form, where each corner represents a carbon and its attached hydrogens); the distinctions in molecular architecture are shaded in blue:

The different actions of these two molecules on many targets throughout the body are the basis of sexual characteristics, producing the contrasting features of male and female vertebrates. In this case, the chemical groups are important because they affect molecular shape, contributing to function.

In other cases, chemical groups are directly involved in chemical reactions; such groups are known as **functional groups**. Each has certain properties, such as shape and charge, that cause it to participate in chemical reactions in a characteristic way.

The seven chemical groups most important in biological processes are the hydroxyl, carbonyl, carboxyl, amino, sulfhydryl, phosphate, and methyl groups. The first six groups can be chemically reactive; of these six, all except the sulfhydryl group are also hydrophilic and thus increase the solubility of organic compounds in water. The methyl group is not reactive, but instead often serves as a recognizable tag on biological molecules. Study **Figure 4.9** to become familiar with these biologically important chemical groups. As shown at the right of the figure, the carboxyl group and the amino group are ionized at normal cellular pH.

▼ **Figure 4.9 Some biologically important chemical groups.**

Chemical Group	Group Properties and Compound Name	Examples
Hydroxyl group (—OH) (may be written HO—)	Is polar due to electronegative oxygen. Forms hydrogen bonds with water, helping dissolve compounds such as sugars. Compound name: **Alcohol** (specific name usually ends in -*ol*)	**Ethanol**, the alcohol present in alcoholic beverages
Carbonyl group (⟩C＝O)	Sugars with ketone groups are called ketoses; those with aldehydes are called aldoses. Compound name: **Ketone** (carbonyl group is within a carbon skeleton) or **aldehyde** (carbonyl group is at the end of a carbon skeleton)	**Acetone**, the simplest ketone **Propanal**, an aldehyde
Carboxyl group (—COOH)	Acts as an acid (can donate H^+) because the covalent bond between oxygen and hydrogen is so polar. Compound name: **Carboxylic acid**, or **organic acid**	**Acetic acid**, which gives vinegar its sour taste Ionized form of —COOH (carboxylate ion), found in cells
Amino group (—NH_2)	Acts as a base; can pick up an H^+ from the surrounding solution (water, in living organisms). Compound name: **Amine**	**Glycine**, an amino acid (note its carboxyl group) Ionized form of —NH_2, found in cells
Sulfhydryl group (—SH) (may be written HS—)	Two —SH groups can react, forming a "cross-link" that helps stabilize protein structure. Hair protein cross-links maintain the straightness or curliness of hair; in hair salons, "permanent" treatments break cross-links, then re-form them while the hair is in the desired shape. Compound name: **Thiol**	**Cysteine**, a sulfur-containing amino acid
Phosphate group (—OPO_3^{2-})	Contributes negative charge (1– when positioned inside a chain of phosphates; 2– when at the end). When attached, confers on a molecule the ability to react with water, releasing energy. Compound name: **Organic phosphate**	**Glycerol phosphate**, which takes part in many important chemical reactions in cells
Methyl group (—CH_3)	Affects the expression of genes when bonded to DNA or to proteins that bind to DNA. Affects the shape and function of male and female sex hormones. Compound name: **Methylated compound**	**5-Methylcytosine**: Cytosine, a component of DNA, has been modified by addition of a methyl group.

⊙ **Mastering Biology Animation: Functional Groups**

ATP: An Important Source of Energy for Cellular Processes

The "Phosphate group" row in Figure 4.9 shows a simple example of an organic phosphate molecule. A more complicated organic phosphate, **adenosine triphosphate**, or **ATP**, is worth mentioning here because its function in the cell is so important. ATP consists of an organic molecule called adenosine attached to a string of three phosphate groups:

Where three phosphates are present in series, as in ATP, one phosphate may be split off as a result of a reaction with water. This inorganic phosphate ion, $HOPO_3^{2-}$, is often abbreviated $\textcircled{P}_i$ in this book, and a phosphate group in an organic molecule is often written as $\textcircled{P}$. Having lost one phosphate, ATP becomes adenosine *di*phosphate, or ADP. Although ATP is sometimes said to store energy, it is more accurate to think of it as storing the potential to react with water or other molecules. Overall, the process releases energy that can be used by the cell. You'll learn more about this in Concept 8.3.

ATP   ADP   Inorganic phosphate

The Chemical Elements of Life: *A Review*

Living matter, as you have learned, consists mainly of carbon, oxygen, hydrogen, and nitrogen, with smaller amounts of sulfur and phosphorus. These elements all form strong covalent bonds, an essential characteristic in the architecture of complex organic molecules. Of all these elements, carbon is the virtuoso of the covalent bond. The versatility of carbon makes possible the great diversity of organic molecules, each with particular properties that emerge from the unique arrangement of its mostly carbon skeleton and the chemical groups attached to that skeleton. This variation at the molecular level provides the foundation for the rich biological diversity found on our planet.

CONCEPT CHECK 4.3

1. **VISUAL SKILLS** What does the term *amino acid* signify about the structure of such a molecule? See Figure 4.9.
2. What chemical change occurs to ATP when it reacts with water and releases energy?
3. **DRAW IT** Suppose you had an organic molecule such as cysteine (see Figure 4.9, sulfhydryl group example), and you chemically removed the —NH₂ group and replaced it with —COOH. Draw this structure. How would this change the chemical properties of the molecule? Is the central carbon asymmetric before the change? After?

For suggested answers, see Appendix A.

---

4 Chapter Review

→ Go to **MasteringBiology** for Assignments, the eText, the Study Area, and Dynamic Study Modules.

SUMMARY OF KEY CONCEPTS

→ To review key terms, go to the **Vocabulary Self-Quiz** in the **Mastering Biology** eText or Study Area, or go to goo.gl/zkjz9t.

CONCEPT 4.1

Organic chemistry is key to the origin of life *(pp. 57–58)*

- **Organic** compounds, once thought to arise only within living organisms, were finally synthesized in the laboratory.
- Living matter is made mostly of carbon, oxygen, hydrogen, and nitrogen. Biological diversity results from carbon's ability to form a huge number of molecules with particular shapes and properties.

❓ *How did Stanley Miller's experiments support the idea that, even at life's origins, physical and chemical laws govern the processes of life?*

CONCEPT 4.2

Carbon atoms can form diverse molecules by bonding to four other atoms *(pp. 58–62)*

- Carbon, with a valence of 4, can bond to various other atoms, including O, H, and N. Carbon can also bond to other carbon

atoms, forming the carbon skeletons of organic compounds. These skeletons vary in length and shape and have bonding sites for atoms of other elements.
- **Hydrocarbons** consist of carbon and hydrogen.
- **Isomers** are compounds that have the same molecular formula but different structures and therefore different properties. Three types of isomers are **structural isomers**, *cis-trans* isomers, and **enantiomers**.

VISUAL SKILLS *Refer back to Figure 4.9. What type of isomers are acetone and propanal? How many asymmetric carbons are present in acetic acid, glycine, and glycerol phosphate? Can these three molecules exist as forms that are enantiomers?*

CONCEPT 4.3

A few chemical groups are key to molecular function *(pp. 62–64)*

- Chemical groups attached to the carbon skeletons of organic molecules participate in chemical reactions (**functional groups**) or contribute to function by affecting molecular shape (see Figure 4.9).
- **ATP (adenosine triphosphate)** consists of adenosine attached to three phosphate groups. ATP can react with water or other molecules, forming ADP (adenosine diphosphate) and inorganic phosphate. This reaction releases energy that can be used by the cell.

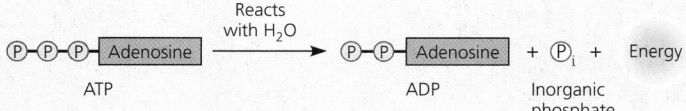

ATP   ADP   Inorganic phosphate

? *In what ways does a methyl group differ chemically from the other six important chemical groups shown in Figure 4.9?*

TEST YOUR UNDERSTANDING

→ For more multiple-choice questions, go to the **Practice Test** in the **Mastering Biology** eText or Study Area, or go to goo.gl/GruWRg.

Levels 1-2: Remembering/Understanding

1. Organic chemistry is currently defined as
   (A) the study of compounds made only by living cells.
   (B) the study of carbon compounds.
   (C) the study of natural (as opposed to synthetic) compounds.
   (D) the study of hydrocarbons.

2. **VISUAL SKILLS** Which functional group is present in this molecule?
   (A) sulfhydryl
   (B) carboxyl
   (C) methyl
   (D) phosphate

3. **MAKE CONNECTIONS** Which chemical group is most likely to be responsible for an organic molecule behaving as a base (see Concept 3.3)?
   (A) hydroxyl   (C) amino
   (B) carbonyl   (D) phosphate

Levels 3-4: Applying/Analyzing

4. **VISUAL SKILLS** Visualize the structural formula of each of the following hydrocarbons. Which hydrocarbon has a double bond in its carbon skeleton?
   (A) C_3H_8 (C) C_2H_4
   (B) C_2H_6 (D) C_2H_2

5. **VISUAL SKILLS** Choose the term that correctly describes the relationship between these two sugar molecules.
   (A) structural isomers
   (B) *cis-trans* isomers
   (C) enantiomers
   (D) isotopes

6. **VISUAL SKILLS** Identify the asymmetric carbon in this molecule.

7. Which action could produce a carbonyl group?
   (A) the replacement of the —OH of a carboxyl group with hydrogen
   (B) the addition of a thiol to a hydroxyl
   (C) the addition of a hydroxyl to a phosphate
   (D) the replacement of the nitrogen of an amine with oxygen

8. **VISUAL SKILLS** Which of the molecules shown in question 5 has an asymmetric carbon? Which carbon is asymmetric?

Levels 5-6: Evaluating/Creating

9. **EVOLUTION CONNECTION • DRAW IT** Some scientists think that life elsewhere in the universe might be based on the element silicon, rather than on carbon, as on Earth. Look at the electron distribution diagram for silicon in Figure 2.7 and draw the Lewis dot structure for silicon. What properties does silicon share with carbon that would make silicon-based life more likely than, say, neon-based life or aluminum-based life?

10. **SCIENTIFIC INQUIRY** Fifty years ago, pregnant women who were prescribed thalidomide for morning sickness gave birth to children with birth defects. Thalidomide is a mixture of two enantiomers; one reduces morning sickness, but the other causes severe birth defects. Today, the FDA has approved this drug for non-pregnant individuals with Hansen's disease (leprosy) or newly diagnosed multiple myeloma, a blood and bone marrow cancer. The beneficial enantiomer can be synthesized and given to patients, but over time, both the beneficial *and* the harmful enantiomer can be detected in the body. Propose a possible explanation for the presence of the harmful enantiomer.

11. **WRITE ABOUT A THEME: ORGANIZATION** In 1918, an epidemic of sleeping sickness caused an unusual rigid paralysis in some survivors, similar to symptoms of advanced Parkinson's disease. Years later, L-dopa (below, left), a chemical used to treat Parkinson's disease, was given to some of these patients. L-dopa was remarkably effective at eliminating the paralysis, at least temporarily. However, its enantiomer, D-dopa (right), was subsequently shown to have no effect at all, as is the case for Parkinson's disease. In a short essay (100–150 words), discuss how the effectiveness of one enantiomer and not the other illustrates the theme of structure and function.

L-dopa   **D-dopa**

12. **SYNTHESIZE YOUR KNOWLEDGE**

Explain how the chemical structure of the carbon atom accounts for the differences between the male and female lions seen in the photo.

For selected answers, see Appendix A.

5 The Structure and Function of Large Biological Molecules

KEY CONCEPTS

5.1 Macromolecules are polymers, built from monomers *p. 67*

5.2 Carbohydrates serve as fuel and building material *p. 68*

5.3 Lipids are a diverse group of hydrophobic molecules *p. 72*

5.4 Proteins include a diversity of structures, resulting in a wide range of functions *p. 75*

5.5 Nucleic acids store, transmit, and help express hereditary information *p. 84*

5.6 Genomics and proteomics have transformed biological inquiry and applications *p. 86*

Study Tip

Make a visual study guide: For each class of biological molecules, draw two examples and list their structural similarities and their functions.

Important Biological Molecules	
Carbohydrates	Proteins
Nucleic acids	Lipids

→ Go to Mastering Biology

For Students (in eText and Study Area)
- Animation: Making and Breaking Polymers
- Figure 5.11 Walkthrough: The Structure of a Phospholipid
- BioFlix® Animation: Gene Expression

For Instructors to Assign (in Item Library)
- Molecular Model: Lysozyme
- Tutorial: Nucleic Acid Structure

Figure 5.1 Alcohol dehydrogenase, a protein that breaks down alcohol in the body, is shown here as a molecular model. The form of this protein that an individual possesses affects how well that person tolerates drinking alcohol. Proteins are one class of large molecules, or macromolecules.

What are the structures and functions of the four important classes of biological molecules?

Three classes are macromolecules that are polymers (long chains of monomer subunits).

Monomer

Polymer

Carbohydrates are a source of energy and provide structural support.

Glucose

Carbohydrate (starch)

Proteins have a wide range of functions, such as catalyzing reactions and transporting substances into and out of cells.

Amino acid

Protein (alcohol dehydrogenase)

Nucleic acids store genetic information and function in gene expression.

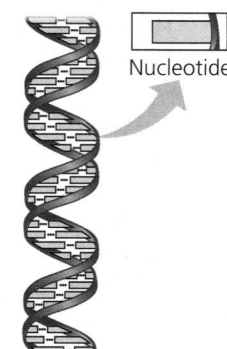

Nucleotide

Nucleic acid (DNA)

The fourth class, lipids, are not polymers or macromolecules.

Lipids are a group of diverse molecules that do not mix well with water. Key functions include providing energy, making up cell membranes, and acting as hormones.

Lipid (phospholipid)

Macromolecules are polymers, built from monomers

Large carbohydrates, proteins, and nucleic acids, also known as **macromolecules** for their huge size, are chain-like molecules called polymers (from the Greek *polys*, many, and *meros*, part). A **polymer** is a long molecule consisting of many similar or identical building blocks linked by covalent bonds, much as a train consists of a chain of boxcars. The repeating units that serve as the building blocks of a polymer are smaller molecules called **monomers** (from the Greek *monos*, single). In addition to forming polymers, some monomers have functions of their own.

The Synthesis and Breakdown of Polymers

Although each class of polymer is made up of a different type of monomer, the chemical mechanisms by which cells make polymers (polymerization) and break them down are similar for all classes of large biological molecules. In cells, these processes are facilitated by **enzymes**, specialized macromolecules (usually proteins) that speed up chemical reactions. The reaction that connects a monomer to another monomer or a polymer is a *condensation reaction*, a reaction in which two molecules are covalently bonded to each other with the loss of a small molecule. If a water molecule is lost, it is known as a **dehydration reaction**. For example, carbohydrate and protein polymers are synthesized by dehydration reactions. Each reactant contributes part of the water molecule that is released during the reaction: One provides a hydroxyl group (—OH), while the other provides a hydrogen (—H) **(Figure 5.2a)**. This reaction is repeated as monomers are added to the chain one by one, lengthening the polymer.

Polymers are disassembled to monomers by **hydrolysis**, a process that is essentially the reverse of the dehydration reaction **(Figure 5.2b)**. Hydrolysis means water breakage (from the Greek *hydro*, water, and *lysis*, break). The bond between monomers is broken by the addition of a water molecule, with a hydrogen from water attaching to one monomer and the hydroxyl group attaching to the other. An example of hydrolysis within our bodies is the process of digestion. The bulk of the organic material in our food is in the form of polymers that are much too large to enter our cells. Within the digestive tract, various enzymes attack the polymers, speeding up hydrolysis. Released monomers are then absorbed into the bloodstream for distribution to all body cells. Those cells can then use dehydration reactions to assemble the monomers into new, different polymers that can perform specific functions required by the cell. (Dehydration reactions and hydrolysis can also be involved in the formation and breakdown of molecules that are not polymers, such as some lipids.)

The Diversity of Polymers

A cell has thousands of different macromolecules; the collection varies from one type of cell to another. The inherited

▼ Figure 5.2 **The synthesis and breakdown of carbohydrate and protein polymers.**

(a) Dehydration reaction: synthesizing a polymer

Short polymer | Unlinked monomer

Dehydration removes a water molecule, forming a new bond.

H_2O

Longer polymer

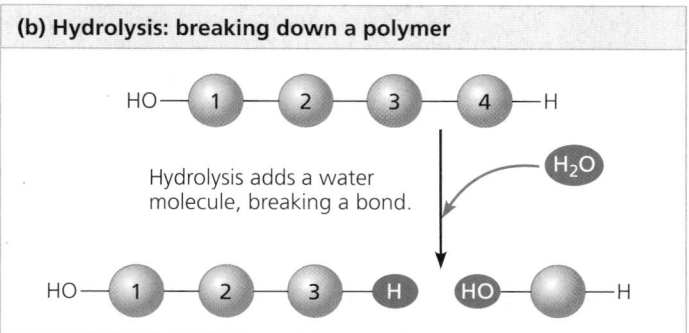

(b) Hydrolysis: breaking down a polymer

Hydrolysis adds a water molecule, breaking a bond.

H_2O

→ **Mastering Biology** Animation: Making and Breaking Polymers

differences between close relatives, such as human siblings, reflect small variations in polymers, particularly DNA and proteins. Molecular differences between unrelated individuals are more extensive, and those between species greater still. The diversity of macromolecules in the living world is vast, and the possible variety is effectively limitless.

What is the basis for such diversity in life's polymers? These molecules are constructed from only 40 to 50 common monomers and some others that occur rarely. Building a huge variety of polymers from such a limited number of monomers is analogous to constructing hundreds of thousands of words from only 26 letters of the alphabet. The key is arrangement—the particular linear sequence that the units follow. However, this analogy falls far short of describing the great diversity of macromolecules because most biological polymers have many more monomers than the number of letters in even the longest word. Proteins, for example, are built from 20 kinds of amino acids arranged in chains that are typically hundreds of amino acids long. The molecular logic of life is simple but elegant: Small molecules common to all organisms act as building blocks that are ordered into unique macromolecules.

Despite this immense diversity, molecular structure and function can still be grouped roughly by class. Let's examine each of the four major classes of large biological molecules. For each class, the large molecules have emergent properties not found in their individual components.

1. What are the four main classes of large biological molecules? Which class does not consist of polymers?

2. How many molecules of water are needed to completely hydrolyze a polymer that is ten monomers long?

3. **WHAT IF?** If you eat a piece of fish, what reactions must occur for the amino acid monomers in the protein of the fish to be converted to new proteins in your body?

For suggested answers, see Appendix A.

CONCEPT 5.2

Carbohydrates serve as fuel and building material

Carbohydrates include sugars and polymers of sugars. The simplest carbohydrates are the monosaccharides, or simple sugars; these are the monomers from which more complex carbohydrates are built. Disaccharides are double sugars, consisting of two monosaccharides joined by a covalent bond. Carbohydrate macromolecules are polymers called polysaccharides, composed of many sugar building blocks.

⊖ **Mastering Biology** Animation: Carbohydrates

Sugars

Monosaccharides (from the Greek *monos*, single, and *sacchar*, sugar) generally have molecular formulas that are some multiple of the unit CH_2O. Glucose ($C_6H_{12}O_6$), the most common monosaccharide, is of central importance in the chemistry of life. In the structure of glucose, we can see the trademarks of a monosaccharide: The molecule has a carbonyl group, $>C=O$, and multiple hydroxyl groups, —OH **(Figure 5.3)**. Depending on the location of the carbonyl group, a monosaccharide is either an aldose (aldehyde sugar) or a ketose (ketone sugar). Glucose, for example, is an aldose; fructose, an isomer of glucose, is a ketose. (Most names for sugars end in *-ose*.) Another criterion for classifying monosaccharides is the size of the carbon skeleton, which ranges from three to seven carbons long. Glucose, fructose, and other sugars that have six carbons are called hexoses. Trioses (three-carbon sugars) and pentoses (five-carbon sugars) are also common.

Still another source of diversity for simple sugars is in the way their parts are arranged spatially around asymmetric carbons. (Recall that an asymmetric carbon is a carbon attached to four different atoms or groups of atoms.) Glucose and galactose, for example, differ only in the placement of parts around one asymmetric carbon (see the purple boxes in Figure 5.3). What seems like a small difference is significant enough to give the two sugars distinctive shapes and binding activities, thus different behaviors.

Although it is convenient to draw glucose with a linear carbon skeleton, this representation is not completely accurate. In aqueous solutions, glucose molecules, as well as most other

▼ **Figure 5.3 The structure and classification of some monosaccharides.** Sugars vary in the location of their carbonyl groups (orange), the length of their carbon skeletons, and the way their parts are arranged spatially around asymmetric carbons (compare, for example, the purple portions of glucose and galactose).

MAKE CONNECTIONS *In the 1970s, a process was developed that converts the glucose in corn syrup to its sweeter-tasting isomer, fructose. High-fructose corn syrup, a common ingredient in soft drinks and processed food, is a mixture of glucose and fructose. What type of isomers are glucose and fructose? (See Figure 4.7.)*

⊖ **Mastering Biology** Animation: Monosaccharides

▼ Figure 5.4 Linear and ring forms of glucose.

(a) Linear and ring forms. Chemical equilibrium between the linear and ring structures greatly favors the formation of rings. The carbons of the sugar are numbered 1 to 6, as shown. To form the glucose ring, carbon 1 (magenta) bonds to the oxygen (blue) attached to carbon 5.

(b) Abbreviated ring structure. Each unlabeled corner represents a carbon. The ring's thicker edge indicates that you are looking at the ring edge-on; the components attached to the ring lie above or below the plane of the ring.

DRAW IT *Start with the linear form of fructose (see Figure 5.3) and draw the formation of the fructose ring in two steps, as shown in (a). First, number the carbons starting at the top of the linear structure. Then draw the molecule in a ringlike orientation, attaching carbon 5 via its oxygen to carbon 2. Compare the number of carbons in the ring portions of fructose and glucose.*

five- and six-carbon sugars, form rings, because they are the most stable form of these sugars under physiological conditions **(Figure 5.4)**.

Monosaccharides, particularly glucose, are major nutrients for cells. In the process known as cellular respiration, cells extract energy from glucose molecules by breaking them down in a series of reactions. Not only are monosaccharides a major fuel for cellular work, but their carbon skeletons also serve as raw material for the synthesis of other types of small organic molecules, such as amino acids and fatty acids. Monosaccharides that are not immediately used in these ways are generally incorporated as monomers into disaccharides or polysaccharides, discussed next.

A **disaccharide** consists of two monosaccharides joined by a **glycosidic linkage**, a covalent bond formed between two monosaccharides by a dehydration reaction (*glyco* refers to carbohydrate). For example, maltose is a disaccharide formed by the linking of two molecules of glucose **(Figure 5.5a)**. Also known as malt sugar, maltose is an ingredient used in brewing beer. The most prevalent disaccharide is sucrose, or table sugar. Its two monomers are glucose and fructose **(Figure 5.5b)**. Plants generally transport carbohydrates from leaves to roots and other nonphotosynthetic organs in the form of sucrose. Lactose, the sugar present in milk, is another disaccharide, in this case a glucose molecule joined

▼ Figure 5.5 Examples of disaccharide synthesis.

(a) Dehydration reaction in the synthesis of maltose. The bonding of two glucose units forms maltose. The 1–4 glycosidic linkage joins the number 1 carbon of one glucose to the number 4 carbon of the second glucose. Joining the glucose monomers in a different way would result in a different disaccharide.

(b) Dehydration reaction in the synthesis of sucrose. Sucrose is a disaccharide formed from glucose and fructose. Notice that fructose forms a five-sided ring, though it is a hexose like glucose.

DRAW IT *Referring to Figures 5.3 and 5.4, number the carbons in each sugar in this figure. How does the name of each linkage relate to the numbers?*

→ **Mastering Biology Animation: Synthesis of Sucrose**

CHAPTER 5 The Structure and Function of Large Biological Molecules   **69** to a galactose molecule. Disaccharides must be broken down into monosaccharides to be used for energy by organisms. Lactose intolerance is a common condition in humans who lack lactase, the enzyme that breaks down lactose. The sugar is instead broken down by intestinal bacteria, causing formation of gas and subsequent cramping. The problem may be avoided by taking the enzyme lactase when eating or drinking dairy products or consuming dairy products that have already been treated with lactase to break down the lactose.

Polysaccharides

Polysaccharides are macromolecules, polymers with a few hundred to a few thousand monosaccharides joined by glycosidic linkages. Some polysaccharides serve as storage material, hydrolyzed as needed to provide monosaccharides for cells.

Other polysaccharides serve as building material for structures that protect the cell or the whole organism. The architecture and function of a polysaccharide are determined by its monosaccharides and by the positions of its glycosidic linkages.

Storage Polysaccharides

Both plants and animals store sugars for later use in the form of storage polysaccharides **(Figure 5.6)**. Plants store **starch**, a polymer of glucose monomers, as granules within cellular structures known as plastids. (Plastids include chloroplasts.) Synthesizing starch enables the plant to stockpile surplus glucose. Because glucose is a major cellular fuel, starch represents stored energy. The sugar can later be withdrawn by the plant from this carbohydrate "bank" by hydrolysis, which breaks the bonds between the glucose monomers. Most animals, including humans, also have enzymes that can hydrolyze plant starch, making glucose

▼ **Figure 5.6 Polysaccharides of plants and animals. (a)** Starch stored in plant cells, **(b)** glycogen stored in muscle cells, and **(c)** structural cellulose fibers in plant cell walls are all polysaccharides composed entirely of glucose monomers (green hexagons). In starch and glycogen, the polymer chains tend to form helices in unbranched regions because of the angle of the 1–4 linkages between glucose molecules. There are two kinds of starch: amylose and amylopectin. Cellulose, with a different kind of glucose linkage, is always unbranched.

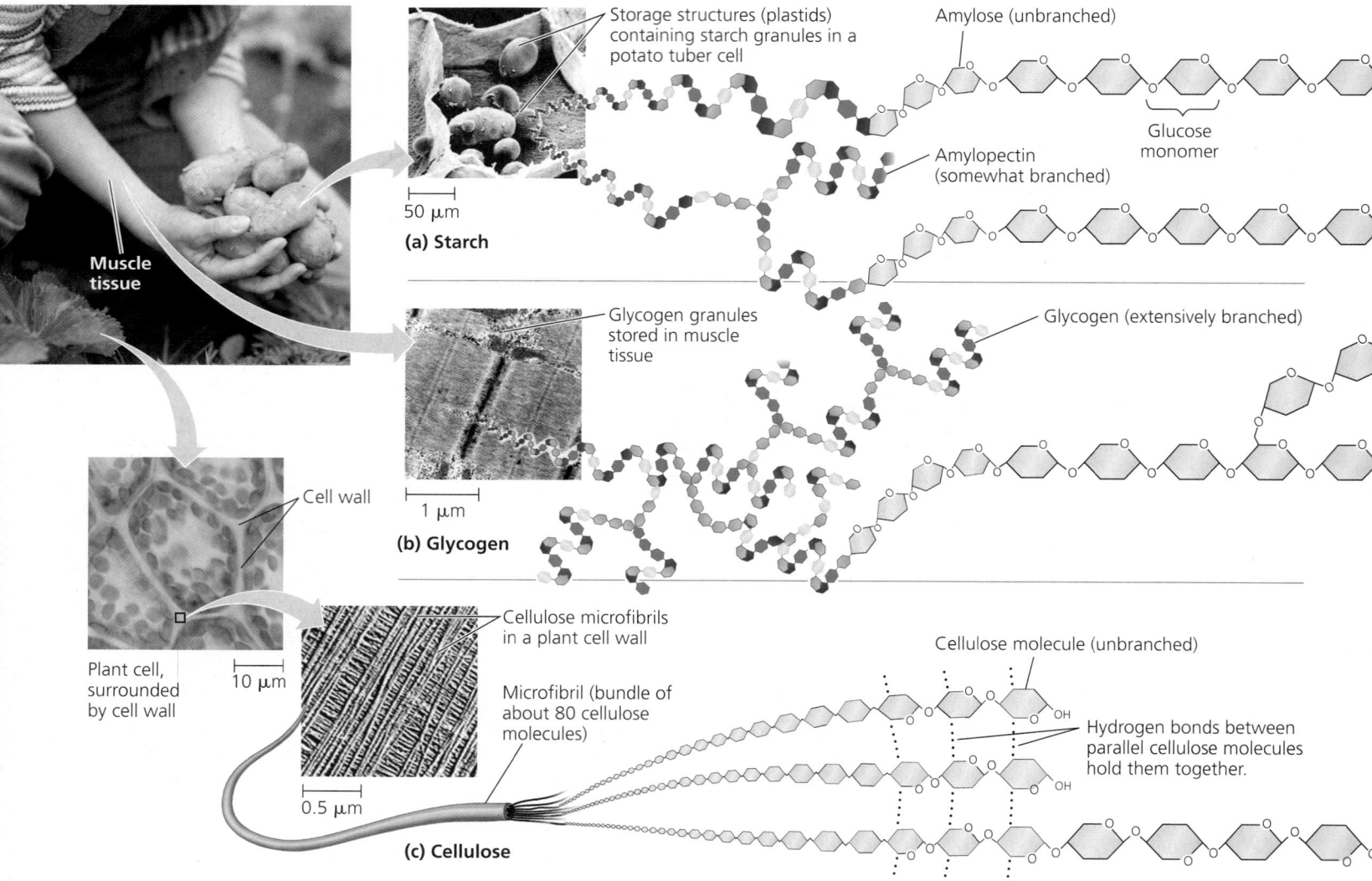

Storage structures (plastids) containing starch granules in a potato tuber cell

50 μm

(a) Starch

Amylose (unbranched)

Glucose monomer

Amylopectin (somewhat branched)

Muscle tissue

Glycogen granules stored in muscle tissue

1 μm

(b) Glycogen

Glycogen (extensively branched)

Cell wall

Plant cell, surrounded by cell wall

10 μm

Cellulose microfibrils in a plant cell wall

0.5 μm

Microfibril (bundle of about 80 cellulose molecules)

(c) Cellulose

Cellulose molecule (unbranched)

Hydrogen bonds between parallel cellulose molecules hold them together.

available as a nutrient for cells. Potato tubers and grains—the fruits of wheat, maize (corn), rice, and other grasses—are the major sources of starch in the human diet.

Most of the glucose monomers in starch are joined by 1–4 linkages (number 1 carbon to number 4 carbon), like the glucose units in maltose (see Figure 5.5a). The simplest form of starch, amylose, is unbranched. Amylopectin, a more complex starch, is a branched polymer with 1–6 linkages at the branch points. Both of these starches are shown in **Figure 5.6a**.

Animals store a polysaccharide called **glycogen**, a polymer of glucose that is like amylopectin but more extensively branched **(Figure 5.6b)**. Vertebrates store glycogen mainly in liver and muscle cells. Breakdown of glycogen in these cells releases glucose when the demand for energy increases. (The extensively branched structure of glycogen fits its function: More free ends are available for breakdown.) This stored fuel cannot sustain an animal for long, however. In humans, for example, glycogen stores are depleted in about a day unless they are replenished by eating. This is an issue of concern in low-carbohydrate diets, which can result in weakness and fatigue.

Structural Polysaccharides

Organisms build strong materials from structural polysaccharides. For example, the polysaccharide called **cellulose** is a major component of the tough walls that enclose plant cells **(Figure 5.6c)**. Globally, plants produce almost 10^{14} kg (100 billion tons) of cellulose per year; it is the most abundant organic compound on Earth.

Like starch, cellulose is a polymer of glucose with 1–4 glycosidic linkages, but the linkages in these two polymers differ. The

difference is based on the fact that there are actually two slightly different ring structures for glucose **(Figure 5.7a)**. When glucose forms a ring, the hydroxyl group attached to the number 1 carbon is positioned either below or above the plane of the ring. These two ring forms for glucose are called alpha (α) and beta (β), respectively. (Greek letters are often used as a "numbering" system for different versions of biological structures, much as we use the letters a, b, c, and so on for the parts of a question or a figure.) In starch, all the glucose monomers are in the α configuration **(Figure 5.7b)**, the arrangement we saw in Figures 5.4 and 5.5. In contrast, the glucose monomers of cellulose are all in the β configuration, making every glucose monomer "upside down" with respect to its neighbors **(Figure 5.7c; see also Figure 5.6c)**.

The differing glycosidic linkages in starch and cellulose give the two molecules distinct three-dimensional shapes. Certain starch molecules are largely helical, fitting their function of efficiently storing glucose units. Conversely, a cellulose molecule is straight. Cellulose is never branched, and some hydroxyl groups on its glucose monomers are free to hydrogen-bond with the hydroxyls of other cellulose molecules lying parallel to it. In plant cell walls, parallel cellulose molecules held together in this way are grouped into units called microfibrils (see Figure 5.6c). These cable-like microfibrils are a strong building material for plants and an important substance for humans because cellulose is the major constituent of paper and the only component of cotton. The unbranched structure of cellulose thus fits its function: imparting strength to parts of the plant.

Enzymes that digest starch by hydrolyzing its α linkages are unable to hydrolyze the β linkages of cellulose due to the different shapes of these two molecules. In fact,

▼ **Figure 5.7 Starch and cellulose structures.**

(a) α and β glucose ring structures. These two interconvertible forms of glucose differ in the placement of the hydroxyl group (highlighted in blue) attached to the number 1 carbon.

α Glucose β Glucose

"Insoluble Fiber" listed on food labels is mainly cellulose, shown below.

Nutrition Facts

Dietary Fiber 4g	16%
Soluble Fiber 2g	
Insoluble Fiber 2g	

(b) Starch: 1–4 linkage of α glucose monomers. All monomers are in the same orientation. Compare the positions of the —OH groups highlighted in yellow with those in cellulose (c).

(c) Cellulose: 1–4 linkage of β glucose monomers. In cellulose, every β glucose monomer is upside down with respect to its neighbors. (See the highlighted —OH groups.)

> **Mastering Biology** Animation: Starch, Cellulose, and Glycogen Structures

few organisms possess enzymes that can digest cellulose. Almost all animals, including humans, do not; the cellulose in our food passes through the digestive tract and is eliminated with the feces. Along the way, the cellulose abrades the wall of the digestive tract and stimulates the lining to secrete mucus, which aids in the smooth passage of food through the tract. Thus, although cellulose is not a nutrient for humans, it is an important part of a healthy diet. Most fruits, vegetables, and whole grains are rich in cellulose. On food packages, "insoluble fiber" refers mainly to cellulose (see Figure 5.7).

Some microorganisms can digest cellulose, breaking it down into glucose monomers. A cow harbors cellulose-digesting prokaryotes and protists in its gut. These microbes hydrolyze the cellulose of hay and grass and convert the glucose to other compounds that nourish the cow. Similarly, a termite, which is unable to digest cellulose by itself, has prokaryotes or protists living in its gut that can make a meal of wood. Some fungi can also digest cellulose in soil and elsewhere, thereby helping recycle chemical elements within Earth's ecosystems.

Another important structural polysaccharide is **chitin**, the carbohydrate used by arthropods (insects, spiders, crustaceans, and related animals) to build their exoskeletons—hard cases that surround the soft parts of an animal **(Figure 5.8)**. Made up of chitin embedded in a layer of proteins, the case is leathery and flexible at first, but becomes hardened when the proteins are chemically linked to each other (as in insects) or encrusted with calcium carbonate (as in crabs). Chitin is also found in fungi, which use this polysaccharide rather than cellulose as the building material for their cell walls. Chitin is similar to cellulose, with β linkages, except that the glucose monomer of chitin has a nitrogen-containing attachment (see Figure 5.8).

▼ **Figure 5.8 Chitin, a structural polysaccharide.**

◀ The structure of the chitin monomer

◀ Chitin, embedded in proteins, forms the exoskeleton of arthropods. This emperor dragonfly (*Anax imperator*) is molting—shedding its old exoskeleton (brown) and emerging upside down in adult form.

CONCEPT CHECK 5.2

1. Write the formula for a monosaccharide that has three carbons.
2. A dehydration reaction joins two glucose molecules to form maltose. The formula for glucose is $C_6H_{12}O_6$. What is the formula for maltose?
3. **WHAT IF?** After a cow is given antibiotics to treat an infection, a vet gives the animal a drink of "gut culture" containing various prokaryotes. Why is this necessary?

For suggested answers, see Appendix A.

CONCEPT 5.3

Lipids are a diverse group of hydrophobic molecules

Lipids are the one class of large biological molecules that does not include true polymers, and they are generally not big enough to be considered macromolecules. The compounds called **lipids** are grouped with each other because they share one important trait: They are hydrophobic: They mix poorly, if at all, with water. This behavior of lipids is based on their molecular structure. Although they may have some polar bonds associated with oxygen, lipids consist mostly of hydrocarbon regions with relatively non-polar C—H bonds. Lipids are varied in form and function. They include waxes and certain pigments, but we will focus on the types of lipids that are most important biologically: fats, phospholipids, and steroids.

⊝ **Mastering Biology** Animation: Lipids

Fats

Although fats are not polymers, they are large molecules assembled from smaller molecules by dehydration reactions, like the dehydration reaction described in Figure 5.2a. A **fat** consists of a glycerol molecule joined to three fatty acids **(Figure 5.9)**. Glycerol is an alcohol; each of its three carbons bears a hydroxyl group. A **fatty acid** has a long carbon skeleton, usually 16 or 18 carbon atoms in length. The carbon at one end of the skeleton is part of a carboxyl group, the functional group that gives these molecules the name fatty *acid*. The rest of the skeleton consists of a hydrocarbon chain. The relatively nonpolar C—H bonds in the hydrocarbon chains of fatty acids are the reason fats are hydrophobic. Fats separate from water because the water molecules hydrogen-bond to one another and exclude the fats. This is why vegetable oil (a liquid fat) separates from the aqueous vinegar solution in a bottle of salad dressing.

In making a fat, each fatty acid molecule is joined to glycerol by a dehydration reaction **(Figure 5.9a)**. This results in an ester linkage, a bond between a hydroxyl group and a carboxyl group. The completed fat consists of three fatty acids linked to one glycerol molecule. (Other names for a fat are *triacylglycerol* and *triglyceride*; levels of triglycerides are

▼ Figure 5.9 The synthesis and structure of a fat, or triacylglycerol.
The molecular building blocks of a fat are one molecule of glycerol and three molecules of fatty acids. The carbons of the fatty acids are arranged zigzag to suggest the actual orientations of the four single bonds extending from each carbon (see Figures 4.3a and 4.6b).

Glycerol

Fatty acid
(in this case, palmitic acid)

(a) One of three dehydration reactions in the synthesis of a fat.
One water molecule is removed for each fatty acid joined to the glycerol.

Ester linkage

(b) A fat molecule (triacylglycerol) with three fatty acid units.
In this example, two of the fatty acid units are identical.

▼ Figure 5.10 Saturated and unsaturated fats and fatty acids.

(a) Saturated fat

At room temperature, the molecules of a saturated fat, such as the fat in butter, are packed closely together, forming a solid.

Structural formula of a saturated fat molecule (Each hydrocarbon chain is represented as a zigzag line, where each bend represents a carbon atom; hydrogens are not shown.)

Space-filling model of stearic acid, a saturated fatty acid (red = oxygen, black = carbon, gray = hydrogen)

(b) Unsaturated fat

At room temperature, the molecules of an unsaturated fat such as olive oil cannot pack together closely enough to solidify because of the kinks in some of their fatty acid hydrocarbon chains.

Structural formula of an unsaturated fat molecule

Space-filling model of oleic acid, an unsaturated fatty acid

Cis double bond causes bending.

reported when blood is tested for lipids.) The fatty acids in a fat can all be the same, or they can be of two or three different kinds, as in **Figure 5.9b**.

The terms *saturated* fats and *unsaturated* fats are commonly used in the context of nutrition **(Figure 5.10)**. These terms refer to the structure of the hydrocarbon chains of the fatty acids. If there are no double bonds between carbon atoms composing a chain, then as many hydrogen atoms as possible are bonded to the carbon skeleton. Such a structure is said to be *saturated* with hydrogen, and the resulting fatty acid is therefore called a **saturated fatty acid (Figure 5.10a)**. An **unsaturated fatty acid** has one or more double bonds, with one fewer hydrogen atom on each double-bonded carbon. Nearly every double bond in naturally occurring fatty acids is a *cis* double bond, which creates a kink in the hydrocarbon chain wherever it occurs **(Figure 5.10b)**. (See Figure 4.7b to remind yourself about *cis* and *trans* double bonds.)

A fat made from saturated fatty acids is called a saturated fat. Most animal fats are saturated: The hydrocarbon chains of their fatty acids—the "tails" of the fat molecules—lack double bonds (see Figure 5.10a), and their flexibility allows the fat molecules to pack together tightly. Saturated animal

fats, such as lard and butter, are solid at room temperature. In contrast, the fats of plants and fishes are generally unsaturated, meaning that they are composed of one or more types of unsaturated fatty acids. Usually liquid at room temperature, plant and fish fats are referred to as oils—olive oil and cod liver oil are examples. The kinks where the *cis* double bonds are located (see Figure 5.10b) prevent the molecules from packing together closely enough to solidify at room temperature. The phrase *hydrogenated vegetable oils* on food labels means that unsaturated fats have been synthetically converted to saturated fats by adding hydrogen, allowing them to solidify. Peanut butter, margarine, and many other products are hydrogenated to prevent lipids from separating out in liquid (oil) form.

A diet rich in saturated fats is one of several factors that may contribute to the cardiovascular disease known as atherosclerosis. In this condition, deposits called plaques develop within the walls of blood vessels, causing inward bulges that impede blood flow and reduce the resilience of the vessels. The process of hydrogenating vegetable oils produces not only saturated fats but also unsaturated fats with *trans* double bonds. It appears that **trans fats** can contribute to coronary heart disease (see Concept 42.4). Because *trans* fats are especially common in baked goods and processed foods, the U.S. Food and Drug Administration (FDA) requires nutritional labels to include information on *trans* fat content. In addition, the FDA has ordered U.S. food manufacturers to stop producing *trans* fats in foods by 2021. Some countries, such as Denmark and Switzerland, have already banned artificially produced *trans* fats in foods.

The major function of fats is energy storage. The hydrocarbon chains of fats are similar to gasoline molecules and just as rich in energy. A gram of fat stores more than twice as much energy as a gram of a polysaccharide, such as starch. Because plants are relatively immobile, they can function with bulky energy storage in the form of starch. (Vegetable oils are generally obtained from seeds, where more compact storage is an asset to the plant.) Animals, however, must carry their energy stores with them, so there is an advantage to having a more compact reservoir of fuel—fat. Humans and other mammals stock their long-term food reserves in adipose cells (see Figure 4.6a), which swell and shrink as fat is deposited

and withdrawn from storage. In addition to storing energy, adipose tissue also cushions such vital organs as the kidneys, and a layer of fat beneath the skin insulates the body. This subcutaneous layer is especially thick in whales, seals, and most other marine mammals, insulating their bodies in cold ocean water.

Phospholipids

Cells as we know them could not exist without another type of lipid, called phospholipids. Phospholipids are essential for cells because they are major constituents of cell membranes. Their structure provides a classic example of how form fits function at the molecular level. As shown in **Figure 5.11**, a **phospholipid** is similar to a fat molecule but has only two fatty acids attached to glycerol rather than three. The third hydroxyl group of glycerol is joined to a phosphate group, which has a negative electrical charge in the cell. Typically, an additional small charged or polar molecule is also linked to the phosphate group. Choline is one such molecule (see Figure 5.11), but there are many others as well, allowing formation of a variety of phospholipids that differ from each other.

The two ends of phospholipids show different behaviors with respect to water. The hydrocarbon tails are hydrophobic and are excluded from water. However, the phosphate group and its attachments form a hydrophilic head that has an affinity for water. When phospholipids are added to water, they self-assemble into a double-layered sheet called a "bilayer" that shields their hydrophobic fatty acid tails from water (Figure 5.11d).

▼ **Figure 5.11 The structure of a phospholipid.** A phospholipid has a hydrophilic (polar) head and two hydrophobic (nonpolar) tails. This particular phospholipid, called a phosphatidylcholine, has a choline attached to a phosphate group. Shown here are **(a)** the structural formula, **(b)** the space-filling model (yellow = phosphorus, blue = nitrogen), **(c)** the symbol for a phospholipid that will appear throughout this book, and **(d)** the bilayer structure formed by self-assembly of phospholipids in an aqueous environment.

DRAW IT *Draw an oval around the hydrophilic head of the space-filling model.*

⊜ **Mastering Biology Figure Walkthrough**
Animation: Space-Filling Model of a Phospholipid

(a) Structural formula   (b) Space-filling model   (c) Phospholipid symbol   (d) Phospholipid bilayer

At the surface of a cell, phospholipids are arranged in a similar bilayer. The hydrophilic heads of the molecules are on the outside of the bilayer, in contact with the aqueous solutions inside and outside of the cell. The hydrophobic tails point toward the interior of the bilayer, away from the water. The phospholipid bilayer forms a boundary between the cell and its external environment and establishes separate compartments within eukaryotic cells; in fact, the existence of cells depends on the properties of phospholipids.

Steroids

Steroids are lipids characterized by a carbon skeleton consisting of four fused rings. Different steroids are distinguished by the particular chemical groups attached to this ensemble of rings. **Cholesterol**, a type of steroid, is a crucial molecule in animals **(Figure 5.12)**. It is a common component of animal cell membranes and is also the precursor from which other steroids, such as the vertebrate sex hormones, are synthesized. In vertebrates, cholesterol is synthesized in the liver and is also obtained from the diet. A high level of cholesterol in the blood may contribute to atherosclerosis, although some researchers are questioning the roles of cholesterol and saturated fats in the development of this condition.

▼ **Figure 5.12 Cholesterol, a steroid.** Cholesterol is the molecule from which other steroids, including the sex hormones, are synthesized. Steroids vary in the chemical groups attached to their four interconnected rings (shown in gold).

MAKE CONNECTIONS *Compare cholesterol with the sex hormones shown in the figure at the beginning of Concept 4.3. Circle the chemical groups that cholesterol has in common with estradiol; put a square around the chemical groups that cholesterol has in common with testosterone.*

> **Mastering Biology**
> **Interview with Lovell Jones: Investigating the effects of sex hormones on cancer**

CONCEPT CHECK 5.3

1. Compare the structure of a fat (triglyceride) with that of a phospholipid.
2. Why are human sex hormones considered lipids?
3. **WHAT IF?** Suppose a membrane surrounded an oil droplet, as it does in the cells of plant seeds and in some animal cells. Describe and explain the form it might take.

For suggested answers, see Appendix A.

Proteins include a diversity of structures, resulting in a wide range of functions

Nearly every dynamic function of a living being depends on proteins. In fact, the importance of proteins is underscored by their name, which comes from the Greek word *proteios*, meaning "first," or "primary." Proteins account for more than 50% of the dry mass of most cells, and they are instrumental in almost everything organisms do. Some proteins speed up chemical reactions, while others play a role in defense, storage, transport, cellular communication, movement, or structural support. **Figure 5.13** shows examples of proteins with these functions, which you'll learn more about in later chapters.

Life would not be possible without enzymes, most of which are proteins. Enzymatic proteins regulate metabolism by acting as **catalysts**, chemical agents that selectively speed up chemical reactions without being consumed in the reaction. Because an enzyme can perform its function over and over again, these molecules can be thought of as workhorses that keep cells running by carrying out the processes of life.

A human has tens of thousands of different proteins, each with a specific structure and function; proteins, in fact, are the most structurally sophisticated molecules known. Consistent with their diverse functions, they vary extensively in structure, each type of protein having a unique three-dimensional shape.

Proteins are all constructed from the same set of 20 amino acids, linked in unbranched polymers. The bond between amino acids is called a *peptide bond*, so a polymer of amino acids is called a **polypeptide**. A **protein** is a biologically functional molecule made up of one or more polypeptides, each folded and coiled into a specific three-dimensional structure.

Amino Acids (Monomers)

All amino acids share a common structure. An **amino acid** is an organic molecule with both an amino group and a carboxyl group (see Figure 4.9); the small figure shows the general formula for an amino acid. At the center of the amino acid is an asymmetric carbon atom called the *alpha (α) carbon*. Its four different partners are an amino group, a carboxyl group, a hydrogen atom, and a variable group symbolized by R. The R group, also called the side chain, differs with each amino acid. The R group may be as simple as a

Side chain (R group)

α carbon

Amino group

Carboxyl group

Enzymatic proteins

Function: Selective acceleration of chemical reactions
Example: Digestive enzymes catalyze the hydrolysis of bonds in food molecules.

Defensive proteins

Function: Protection against disease
Example: Antibodies inactivate and help destroy viruses and bacteria.

Storage proteins

Function: Storage of amino acids
Examples: Casein, the protein of milk, is the major source of amino acids for baby mammals. Plants have storage proteins in their seeds. Ovalbumin is the protein of egg white, used as an amino acid source for the developing embryo.

Ovalbumin   Amino acids for embryo

Transport proteins

Function: Transport of substances
Examples: Hemoglobin, the iron-containing protein of vertebrate blood, transports oxygen from the lungs to other parts of the body. Other proteins transport molecules across membranes, as shown here.

Transport protein

Cell membrane

Hormonal proteins

Function: Coordination of an organism's activities
Example: Insulin, a hormone secreted by the pancreas, causes other tissues to take up glucose, thus regulating blood sugar concentration.

High blood sugar   Insulin secreted   Normal blood sugar

Receptor proteins

Function: Response of cell to chemical stimuli
Example: Receptors built into the membrane of a nerve cell detect signaling molecules released by other nerve cells.

Receptor protein

Signaling molecules

Contractile and motor proteins

Function: Movement
Examples: Motor proteins are responsible for the undulations of cilia and flagella. Actin and myosin proteins are responsible for the contraction of muscles.

Actin   Myosin

Muscle tissue   30 μm

Structural proteins

Function: Support
Examples: Keratin is the protein of hair, horns, feathers, and other skin appendages. Insects and spiders use silk fibers to make their cocoons and webs, respectively. Collagen and elastin proteins provide a fibrous framework in animal connective tissues.

Collagen

Connective tissue   60 μm

hydrogen atom, or it may be a carbon skeleton with various functional groups attached. The physical and chemical properties of the side chain determine the unique characteristics of a particular amino acid, thus affecting its functional role in a polypeptide.

Figure 5.14 shows the 20 amino acids that cells use to build their thousands of proteins. Here the amino groups and carboxyl groups are all depicted in ionized form, the way they usually exist at the pH found in a cell. The amino acids are grouped according to the properties of their side chains. One group consists of amino acids with nonpolar side chains, which are hydrophobic. Another group consists of amino acids with polar side chains, which are hydrophilic. Acidic amino acids have side chains that are generally negative in charge due to the presence of a carboxyl group, which is usually dissociated (ionized) at cellular pH. Basic amino acids have amino groups in their side chains that are generally positive in charge. (The terms *acidic* and *basic* in this context refer only to groups in the side chains because *all* amino acids—as monomers—have carboxyl groups and amino groups.) Because they are charged, acidic and basic side chains are also hydrophilic.

▼ Figure 5.14 The 20 amino acids of proteins. At the top right are the non-ionized and ionized forms of a generic amino acid. The specific amino acids are shown below in their ionized form, prevalent at the pH within a cell (pH 7.2). The amino acids are grouped by properties of their side chains. The three-letter and one-letter abbreviations for the amino acids are in parentheses. All of the amino acids used in proteins are L enantiomers (see Figure 4.7c).

Nonpolar side chains; hydrophobic

Glycine (Gly or G)

Alanine (Ala or A)

Valine (Val or V)

Leucine (Leu or L)

Isoleucine (Ile or I)

Methionine (Met or M)

Phenylalanine (Phe or F)

Tryptophan (Trp or W)

Proline (Pro or P)

Polar side chains; hydrophilic

Since cysteine is only weakly polar, it is sometimes classified as a nonpolar amino acid.

Serine (Ser or S)

Threonine (Thr or T)

Cysteine (Cys or C)

Tyrosine (Tyr or Y)

Asparagine (Asn or N)

Glutamine (Gln or Q)

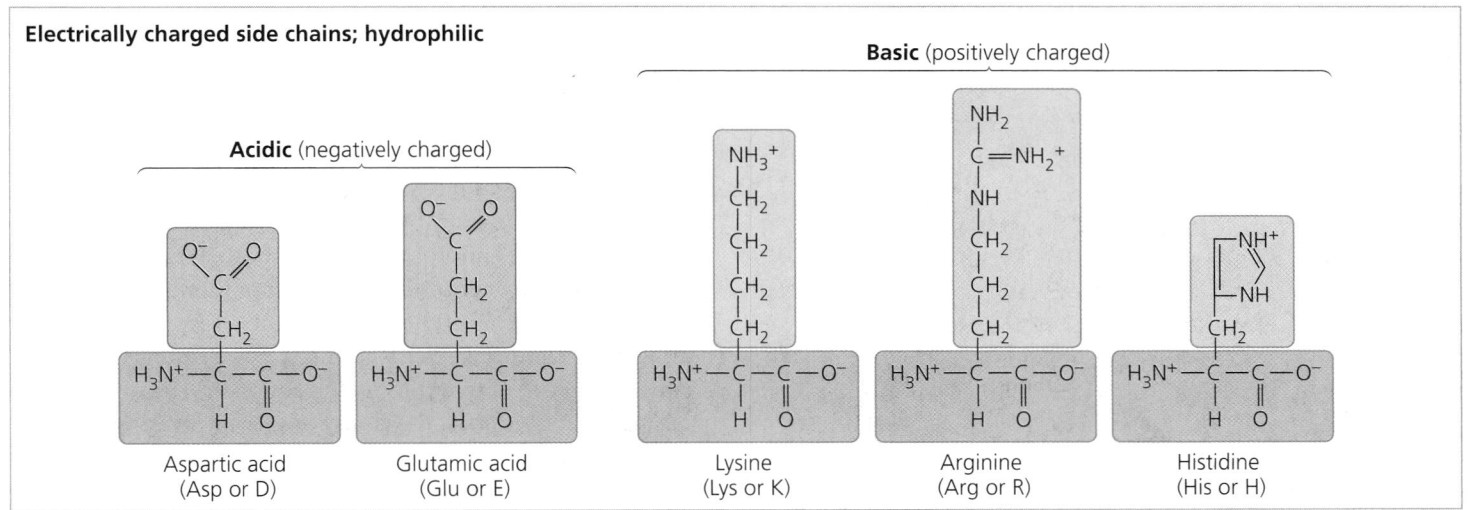

Electrically charged side chains; hydrophilic

Basic (positively charged)

Acidic (negatively charged)

Aspartic acid (Asp or D)

Glutamic acid (Glu or E)

Lysine (Lys or K)

Arginine (Arg or R)

Histidine (His or H)

Polypeptides (Amino Acid Polymers)

Now that we have examined amino acids, let's see how they are linked to form polymers **(Figure 5.15)**. When two amino acids are positioned so that the carboxyl group of one is adjacent to the amino group of the other, they can become joined by a dehydration reaction, with the removal of a water molecule. The resulting covalent bond is called a **peptide bond**. Repeated over and over, this process yields a polypeptide, a polymer of many amino acids linked by peptide bonds. You'll learn more about how cells synthesize polypeptides in Concept 17.4.

The repeating sequence of atoms highlighted in purple in Figure 5.15 is called the *polypeptide backbone*. Extending from this backbone are the different side chains (R groups) of the amino acids. Polypeptides range in length from a few amino acids to 1,000 or more. Each specific polypeptide has a unique linear sequence of amino acids. Note that one end of the polypeptide chain has a free amino group (the N-terminus of the polypeptide), while the opposite end has a free carboxyl

▼ **Figure 5.15 Making a polypeptide chain.** Peptide bonds are formed by dehydration reactions, which link the carboxyl group of one amino acid to the amino group of the next. The peptide bonds are formed one at a time, starting with the amino acid at the amino end (N-terminus). The polypeptide has a repetitive backbone (purple) to which the amino acid side chains (yellow and green) are attached.

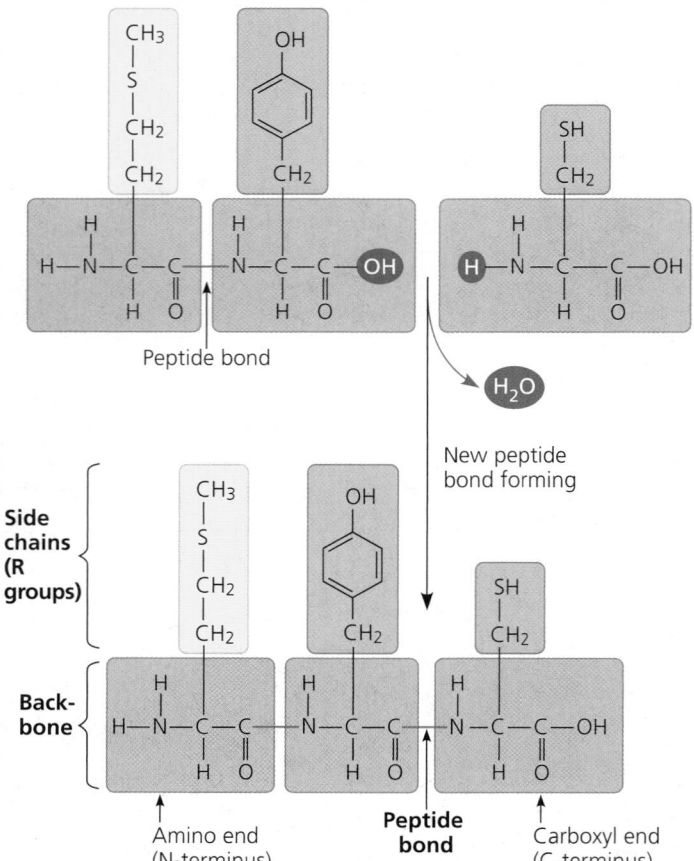

DRAW IT *Label the three amino acids in the upper part of the figure using three-letter and one-letter codes. Circle and label the carboxyl and amino groups that will form the new peptide bond.*

group (the C-terminus). The chemical nature of the molecule as a whole is determined by the kind and sequence of the side chains, which determine how a polypeptide folds and thus its final shape and chemical characteristics. The immense variety of polypeptides in nature illustrates an important concept introduced earlier—that cells can make many different polymers by linking a limited set of monomers into diverse sequences.

Protein Structure and Function

The specific activities of proteins result from their intricate three-dimensional architecture, the simplest level of which is the sequence of their amino acids. What can the amino acid sequence of a polypeptide tell us about the three-dimensional structure (commonly referred to simply as the "structure") of the protein and its function? The term *polypeptide* is not synonymous with the term *protein*. Even for a protein consisting of a single polypeptide, the relationship is somewhat analogous to that between a long strand of yarn and a sweater of particular size and shape that can be knitted from the yarn. A functional protein is not *just* a polypeptide chain, but one or more polypeptides precisely twisted, folded, and coiled into a molecule of unique shape, which can be shown in several different types of models **(Figure 5.16)**. And it is the amino acid sequence of each polypeptide that determines what three-dimensional structure the protein will have under normal cellular conditions.

When a cell synthesizes a polypeptide, the chain may fold spontaneously, assuming the functional structure for that protein. This folding is driven and reinforced by the formation of various bonds between parts of the chain, which in turn depends on the sequence of amino acids. Many proteins are roughly spherical (*globular proteins*), while others are shaped like long fibers (*fibrous proteins*). Even within these broad categories, countless variations exist.

A protein's specific structure determines how it works. In almost every case, the function of a protein depends on its ability to recognize and bind to some other molecule. In an especially striking example of the marriage of form and function, **Figure 5.17** shows the exact match of shape between an antibody (a protein in the body) and the particular foreign substance on a flu virus that the antibody binds to and marks for destruction. Also, you may recall another example of molecules with matching shapes from Concept 2.3: endorphin molecules (produced by the body) and morphine molecules (a manufactured drug), both of which fit into receptor proteins on the surface of brain cells in humans, producing euphoria and relieving pain. Morphine, heroin, and other opiate drugs are able to mimic endorphins because they all have a shape similar to that of endorphins and can thus fit into and bind to endorphin receptors in the brain. This fit is very specific, something like a handshake (see Figure 2.16). The endorphin receptor, like other receptor molecules, is a protein. The function of a protein—for instance, the ability of a receptor protein to bind to a particular pain-relieving signaling molecule—is an emergent property resulting from exquisite molecular order.

VISUALIZING PROTEINS

Proteins can be represented in different ways, depending on the goal of the illustration.

Structural Models

Using data from structural studies of proteins, computers can generate various types of models. Each model emphasizes a different aspect of the protein's structure, but no model can show what a protein actually looks like. These three models depict lysozyme, a protein in tears and saliva that helps prevent infection by binding to target molecules on bacteria.

1. *In which model is it easiest to follow the polypeptide backbone?*

➜ **Instructors:** The tutorial "Molecular Model: Lysozyme," in which students rotate 3-D models of lysozyme, can be assigned in **Mastering Biology**.

Target molecule

Space-filling model: Emphasizes the overall globular shape. Shows all the atoms of the protein (except hydrogen), which are color-coded: gray = carbon, red = oxygen, blue = nitrogen, and yellow = sulfur.

Ribbon model: Shows only the polypeptide backbone, emphasizing how it folds and coils to form a 3-D shape, in this case stabilized by disulfide bridges (yellow lines).

Wireframe model (blue): Shows the polypeptide backbone with side chains extending from it. A ribbon model (purple) is superimposed on the wireframe model. The bacterial target molecule (yellow) is bound.

Simplified Diagrams

It isn't always necessary to use a detailed computer model; simplified diagrams are useful when the focus of the figure is on the function of the protein, not the structure.

➜ **Instructors:** Additional questions related to this Visualizing Figure can be assigned in **Mastering Biology**.

A transparent shape is drawn around the contours of a ribbon model of the protein rhodopsin, showing the shape of the molecule as well as some internal details.

When structural details are not needed, a solid shape can be used.

Enzyme

A simple shape is used here to represent a generic enzyme because the diagram focuses on enzyme action in general.

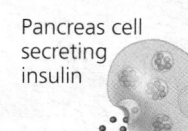

Pancreas cell secreting insulin

Sometimes a protein is represented simply as a dot, as shown here for insulin.

2. *Draw a simple version of lysozyme that shows its overall shape, based on the molecular models in the top section of the figure.*

3. *Why is it unnecessary to show the actual shape of insulin here?*

▶ **Figure 5.17 Complementarity of shape between two protein surfaces.** A technique called X-ray crystallography was used to generate a computer model of an antibody protein (blue and orange, left) bound to a flu virus protein (yellow and green, right). This is a wireframe model modified by adding an "electron density map" in the region where the two proteins meet. Computer software was then used to back the images away from each other slightly.

Four Levels of Protein Structure

In spite of their great diversity, proteins share three superimposed levels of structure, known as primary, secondary, and tertiary structure. A fourth level, quaternary structure, arises when a protein consists of two or more polypeptide chains. **Figure 5.18** describes these four levels of protein structure. Be sure to study this figure thoroughly before going on to the next section.

➜ **Mastering Biology** Animation: Protein Structure

Antibody protein   Protein from flu virus

VISUAL SKILLS *What do these computer models allow you to see about the two proteins?*

Primary Structure

Linear chain of amino acids

The **primary structure** of a protein is its sequence of amino acids. As an example, let's consider transthyretin, a globular blood protein that transports vitamin A and one of the thyroid hormones. Transthyretin is made up of four identical polypeptide chains, each composed of 127 amino acids. Shown here is one of these chains unraveled for a closer look at its primary structure. Each of the 127 positions along the chain is occupied by one of the 20 amino acids, indicated here by its three-letter abbreviation.

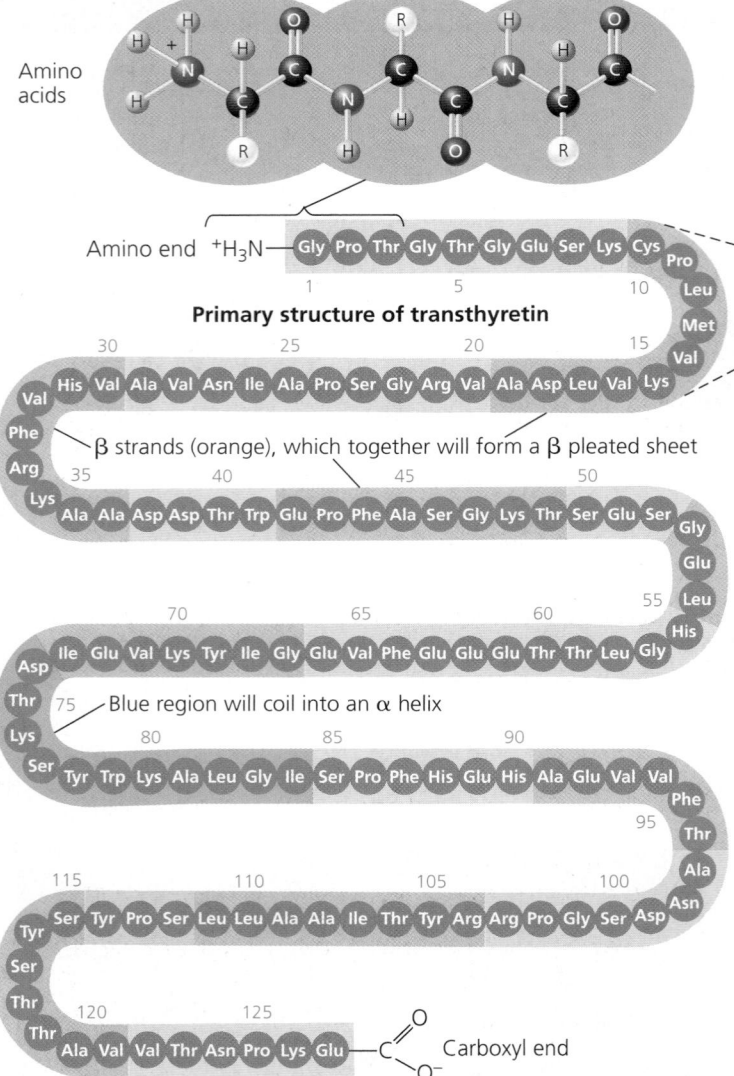

Amino acids

Amino end ^+H_3N — Gly Pro Thr Gly Thr Gly Glu Ser Lys Cys Pro Leu ...

Primary structure of transthyretin

β strands (orange), which together will form a β pleated sheet

Blue region will coil into an α helix

Carboxyl end

The primary structure is like the order of letters in a very long word. If left to chance, there would be 20^{127} different ways of making a polypeptide chain 127 amino acids long. However, the precise primary structure of a protein is determined not by the random linking of amino acids, but by inherited genetic information. The primary structure in turn dictates secondary structure (α helices and β pleated sheets) and tertiary structure, due to the chemical nature of the backbone and the side chains (R groups) of the amino acids along the polypeptide.

Secondary Structure

Regions stabilized by hydrogen bonds between atoms of the polypeptide backbone

A region of **α helix** in transthyretin

Hydrogen bond

A region of **β pleated sheet** (made up of adjacent β strands) in transthyretin

β strand, often shown as a folded or flat arrow pointing toward the carboxyl end

Hydrogen bond

Most proteins have segments of their polypeptide chains repeatedly coiled or folded in patterns that contribute to the protein's overall shape. These coils and folds, collectively referred to as **secondary structure**, are the result of hydrogen bonds between the repeating constituents of the polypeptide backbone (not the amino acid side chains). Within the backbone, the oxygen atoms have a partial negative charge, and the hydrogen atoms attached to the nitrogens have a partial positive charge (see Figure 2.14); therefore, hydrogen bonds can form between these atoms. Individually, these hydrogen bonds are weak, but because they are repeated many times over a relatively long region of the polypeptide chain, they can support a particular shape for that part of the protein.

One such secondary structure is the **α helix**, a delicate coil held together by hydrogen bonding between every fourth amino acid, as shown above. Although each transthyretin polypeptide has only one α helix region (see the Primary and Tertiary Structure sections), other globular proteins have multiple stretches of α helix separated by nonhelical regions (see hemoglobin in the Quaternary Structure section). Some fibrous proteins, such as α-keratin, the structural protein of hair, have the α helix formation over most of their length.

The other secondary structure is the **β pleated sheet**. As shown above, two or more segments of the polypeptide chain lying side by side (called β strands) are connected by hydrogen bonds between parts of the two parallel segments. β pleated sheets make up the core of many globular proteins, as is the case for transthyretin (see Tertiary Structure), and dominate some fibrous proteins, including the silk protein of a spider's web. The teamwork of so many hydrogen bonds makes each spider silk fiber stronger than a steel strand.

► Spiders secrete silk fibers made of a structural protein containing β pleated sheets, which allow the spiderweb to stretch and recoil.

Tertiary Structure

Three-dimensional shape stabilized by interactions between side chains

Quaternary Structure

Association of two or more polypeptides (some proteins only)

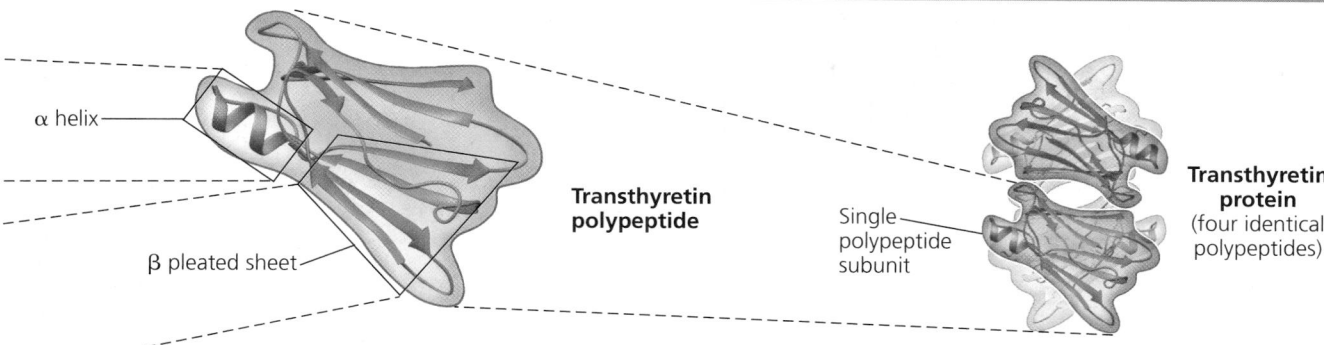

α helix

β pleated sheet

Transthyretin polypeptide

Single polypeptide subunit

Transthyretin protein (four identical polypeptides)

Superimposed on the patterns of secondary structure is a protein's tertiary structure, shown here in a ribbon model of the transthyretin polypeptide. While secondary structure involves interactions between backbone constituents, **tertiary structure** is the overall shape of a polypeptide resulting from interactions between the side chains (R groups) of the various amino acids. One type of interaction that contributes to tertiary structure is called—somewhat misleadingly—a **hydrophobic interaction**. As a polypeptide folds into its functional shape, amino acids with hydrophobic (nonpolar) side chains usually end up in clusters at the core of the protein, out of contact with water. Thus, a "hydrophobic interaction" is actually caused by the exclusion of nonpolar substances by water molecules. Once nonpolar amino acid side chains are close together, van der Waals interactions help hold them together. Meanwhile, hydrogen bonds between polar side chains and ionic bonds between positively and negatively charged side chains also help stabilize tertiary structure. These are all weak interactions in the aqueous cellular environment, but their cumulative effect helps give the protein a unique shape.

Covalent bonds called **disulfide bridges** may further reinforce the shape of a protein. Disulfide bridges form where two cysteine monomers, which have sulfhydryl groups (—SH) on their side chains (see Figure 4.9), are brought close together by the folding of the protein. The sulfur of one cysteine bonds to the sulfur of the second, and the disulfide bridge (—S—S—) rivets parts of the protein together (see yellow lines in Figure 5.16 ribbon model). All of these different kinds of interactions can contribute to the tertiary structure of a protein, as shown here in a small part of a hypothetical protein:

Some proteins consist of two or more polypeptide chains aggregated into one functional macromolecule. **Quaternary structure** is the overall protein structure that results from the aggregation of these polypeptide subunits. For example, shown above is the complete globular transthyretin protein, made up of its four polypeptides.

Another example is collagen, which is a fibrous protein that has three identical helical polypeptides intertwined into a larger triple helix, giving the long fibers great strength. This suits collagen fibers to their function as the girders of connective tissue in skin, bone, tendons, ligaments, and other body parts. (Collagen accounts for 40% of the protein in a human body.)

Collagen

Hemoglobin, the oxygen-binding protein of red blood cells, is another example of a globular protein with quaternary structure. It consists of four polypeptide subunits, two of one kind (α) and two of another kind (β). Both α and β subunits consist primarily of α-helical secondary structure. Each subunit has a nonpolypeptide component, called heme, with an iron atom that binds oxygen.

Heme

Iron

β subunit

α subunit

α subunit

β subunit

Hemoglobin

CH₂ / OH

NH₂ / C=O / CH₂

Hydrogen bond

CH / CH₃ CH₃

CH₃ CH₃ / CH

Hydrophobic interactions and van der Waals interactions

Disulfide bridge

CH₂ / S / S / CH₂

CH₂ / C=O / O⁻

NH₃⁺ / CH₂ / CH₂ / CH₂ / CH₂

Ionic bond

Polypeptide backbone of small part of a protein

R groups are colored as in Figure 5.14.

Mastering Biology Animation: Proteins
MP3 Tutor: Protein Structure and Function

▼ **Figure 5.19** A single amino acid substitution in a protein causes sickle-cell disease.

	Primary Structure	Secondary and Tertiary Structures	Quaternary Structure	Function	Red Blood Cell Shape
Normal hemoglobin	1 Val 2 His 3 Leu 4 Thr 5 Pro 6 Glu 7 Glu	Normal β subunit	Normal hemoglobin β α β α	Normal hemoglobin proteins do not associate with one another; each carries oxygen.	Normal red blood cells are full of individual hemoglobin proteins. 5 µm
Sickle-cell hemoglobin	1 Val 2 His 3 Leu 4 Thr 5 Pro 6 Val 7 Glu	Sickle-cell β subunit	Sickle-cell hemoglobin β α β α	Hydrophobic interactions between sickle-cell hemoglobin proteins lead to their aggregation into a fiber; capacity to carry oxygen is greatly reduced.	Fibers of abnormal hemoglobin deform red blood cell into sickle shape. 5 µm

MAKE CONNECTIONS *Considering the chemical characteristics of the amino acids valine and glutamic acid (see Figure 5.14), propose a possible explanation for the dramatic effect on protein function that occurs when valine is substituted for glutamic acid.*

→ **Mastering Biology** HHMI Animation: Sickle-Cell Disease

Sickle-Cell Disease: A Change in Primary Structure

Even a slight change in primary structure can affect a protein's shape and ability to function. For instance, **sickle-cell disease**, an inherited blood disorder, is caused by the substitution of one amino acid (valine) for the normal one (glutamic acid) at the position of the sixth amino acid in the primary structure of hemoglobin, the protein that carries oxygen in red blood cells. Normal red blood cells are disk-shaped, but in sickle-cell disease, the abnormal hemoglobin molecules tend to aggregate into chains, deforming some of the cells into a sickle shape **(Figure 5.19)**. A person with the disease has periodic "sickle-cell crises" when the angular cells clog tiny blood vessels, impeding blood flow. The toll taken on such patients is a dramatic example of how a simple change in protein structure can have devastating effects on protein function.

 → **Mastering Biology** Interview with Linus Pauling: Winner of the Nobel Prize in Chemistry and the Nobel Peace Prize

What Determines Protein Structure?

You've learned that a unique shape endows each protein with a specific function. But what are the key factors determining protein structure? You already know most of the answer: A polypeptide chain of a given amino acid sequence can be arranged into a three-dimensional shape determined by the interactions responsible for secondary and tertiary structure.

This folding normally occurs as the protein is being synthesized in the crowded environment within a cell, aided by other proteins. However, protein structure also depends on the physical and chemical conditions of the protein's environment. If the pH, salt concentration, temperature, or other aspects of its environment are altered, the weak chemical bonds and interactions within a protein may be destroyed, causing the protein to unravel and lose its native shape, a change called **denaturation (Figure 5.20)**. Because it is misshapen, the denatured protein is biologically inactive.

Most proteins become denatured if they are transferred from an aqueous environment to a nonpolar solvent, such as ether or chloroform; the polypeptide chain refolds so that its

▼ **Figure 5.20 Denaturation and renaturation of a protein.** High temperatures or various chemical treatments will denature a protein, causing it to lose its shape and hence its ability to function. If the denatured protein remains dissolved, it may renature when the chemical and physical aspects of its environment are restored to normal.

Denaturation
Renaturation
Normal protein
Denatured protein

hydrophobic regions face outward toward the solvent. Other denaturation agents include chemicals that disrupt the hydrogen bonds, ionic bonds, and disulfide bridges that maintain a protein's shape. Denaturation can also result from excessive heat, which agitates the polypeptide chain enough to overpower the weak interactions that stabilize the structure. The white of an egg becomes opaque during cooking because the denatured proteins are insoluble and solidify. This also explains why excessively high fevers can be fatal: Proteins in the blood tend to denature at very high body temperatures.

When a protein in a test-tube solution has been denatured by heat or chemicals, it can sometimes return to its functional shape when the denaturing agent is removed. (Sometimes this is not possible: For example, a fried egg will not become liquefied when placed back into the refrigerator!) We can conclude that the information for building specific shape is intrinsic to the protein's primary structure; this is often the case for small proteins. The sequence of amino acids determines the protein's shape—where an α helix can form, where β pleated sheets can exist, where disulfide bridges are located, where ionic bonds can form, and so on. But how does protein folding occur in the cell?

Protein Folding in the Cell

Biochemists now know the amino acid sequence for about 160 million proteins, with about 4.5–5 million added each month, and the three-dimensional shape for about 40,000. Researchers have tried to correlate the primary structure of many proteins with their three-dimensional structure to discover the rules of protein folding. Unfortunately, however, the protein-folding process is not that simple. Most proteins probably go through several intermediate structures on their way to a stable shape, and looking at the mature structure does not reveal the stages of folding required to achieve that form. However, biochemists have developed methods for tracking a protein through such stages and learning more about this important process.

Misfolding of polypeptides in cells is a serious problem that has come under increasing scrutiny by medical researchers. Many diseases—such as cystic fibrosis, Alzheimer's, Parkinson's, and mad cow disease—are associated with an accumulation of misfolded proteins. In fact, misfolded versions of the transthyretin protein featured in Figure 5.18 have been implicated in several diseases, including one form of senile dementia.

Even when scientists have a correctly folded protein in hand, determining its exact three-dimensional structure is not simple, for a single protein has thousands of atoms. The method most commonly used to determine the 3-D structure of a protein is **X-ray crystallography**, which depends on the diffraction of an X-ray beam by the atoms of a crystallized molecule. Using this technique, scientists can build a 3-D model that shows the exact position of every atom in a protein molecule **(Figure 5.21)**. Nuclear magnetic resonance (NMR) spectroscopy, cryo-electron microscopy (cryo-EM; see Concept 6.1)

▼ **Figure 5.21 Research Method**

X-Ray Crystallography

Application Scientists use X-ray crystallography to determine the three-dimensional (3-D) structure of macromolecules such as nucleic acids and proteins.

Technique Researchers aim an X-ray beam through a crystallized protein or nucleic acid. The atoms of the crystal diffract (bend) the X-rays into an orderly array that a digital detector records as a pattern of spots called an X-ray diffraction pattern, an example of which is shown here.

Diffracted X-rays

X-ray source X-ray beam Crystal Digital detector X-ray diffraction pattern

Results Using data from X-ray diffraction patterns and the sequence of monomers determined by chemical methods, researchers can build a 3-D computer model of the macromolecule being studied, such as the four-subunit protein transthyretin (see Figure 5.18) shown here.

and bioinformatics (see Concept 5.6) are complementary approaches to understanding protein structure and function.

The structure of some proteins is difficult to determine for a simple reason: A growing body of biochemical research has revealed that a significant number of proteins, or regions of proteins, do not have a distinct 3-D structure until they interact with a target protein or other molecule. Their flexibility and indefinite structure are important for their function, which may require binding with different targets at different times. These proteins, which may account for 20–30% of mammalian proteins, are called *intrinsically disordered proteins* and are the focus of current research.

CONCEPT CHECK 5.4

1. What parts of a polypeptide participate in the bonds that hold together secondary structure? Tertiary structure?
2. Thus far in the chapter, the Greek letters α and β have been used to specify at least three different pairs of structures. Name and briefly describe them.
3. Each amino acid has a carboxyl group and an amino group. Are these groups present in a polypeptide? Explain.
4. **WHAT IF?** Where would you expect a polypeptide region rich in the amino acids valine, leucine, and isoleucine to be located in a folded polypeptide? Explain.

For suggested answers, see Appendix A.

Nucleic acids store, transmit, and help express hereditary information

If the primary structure of polypeptides determines a protein's shape, what determines primary structure? The amino acid sequence of a polypeptide is programmed by a discrete unit of inheritance known as a **gene**. Genes consist of DNA, which belongs to the class of compounds called nucleic acids. **Nucleic acids** are polymers made of monomers called nucleotides.

The Roles of Nucleic Acids

The two types of nucleic acids, **deoxyribonucleic acid (DNA)** and **ribonucleic acid (RNA)**, enable living organisms to reproduce their complex components from one generation to the next. Unique among molecules, DNA provides directions for its own replication. DNA also directs RNA synthesis and, through RNA, controls protein synthesis; this entire process is called **gene expression (Figure 5.22)**.

DNA is the genetic material that organisms inherit from their parents. Each chromosome contains one long DNA molecule, usually carrying several hundred or more genes. When a cell reproduces itself by dividing, its DNA molecules are copied and passed along from one generation of cells to the next. The information that programs all the cell's activities is encoded in the structure of the DNA. The DNA, however, is not directly involved in running the operations of the cell, any more than computer software by itself can read the bar code on a box of cereal. Just as a scanner is needed to read a bar code, proteins are required to implement genetic programs. The molecular hardware of the cell—the tools that carry out biological functions—consists mostly of proteins. For example, the oxygen carrier in red blood cells is the protein hemoglobin that you saw earlier (see Figure 5.18), not the DNA that specifies its structure.

How does RNA, the other type of nucleic acid, fit into gene expression, the flow of genetic information from DNA to proteins? A given gene along a DNA molecule can direct synthesis of a type of RNA called *messenger RNA* (*mRNA*). The mRNA molecule interacts with the cell's protein-synthesizing machinery to direct production of a polypeptide, which folds into all or part of a protein. We can summarize the flow of genetic information as DNA → RNA → protein (see Figure 5.22). The sites of protein synthesis are cellular structures called ribosomes. In a eukaryotic cell, ribosomes are in the cytoplasm—the region between the nucleus and the cell's outer boundary, the plasma membrane—but DNA resides in the nucleus. Messenger RNA conveys genetic instructions for building proteins from the nucleus to the cytoplasm. Prokaryotic cells lack nuclei but still use mRNA to convey a message from the DNA to ribosomes and other cellular equipment that translate the coded information into amino acid sequences. Later, you'll read about other

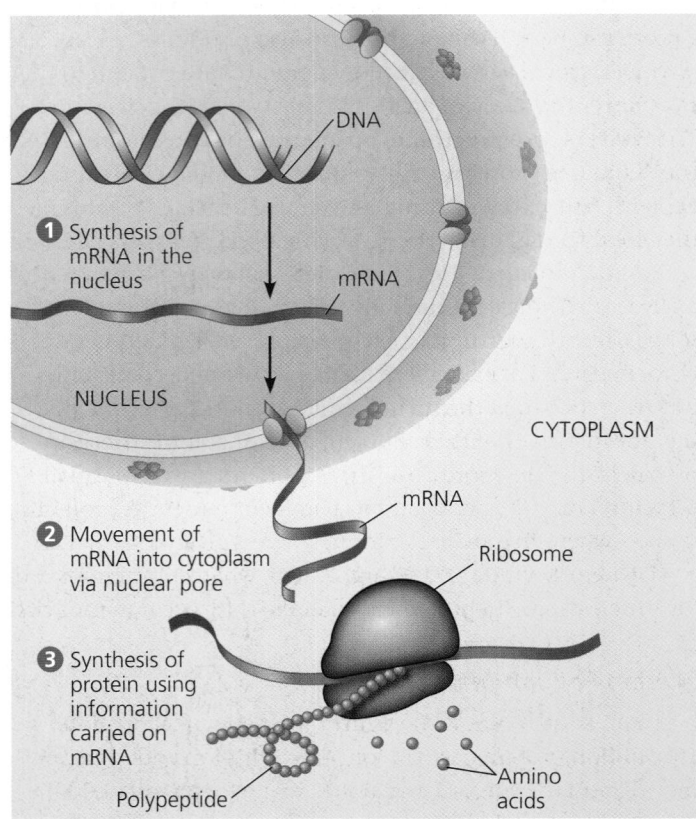

▼ **Figure 5.22 Gene expression: DNA → RNA → protein.** In a eukaryotic cell, DNA in the nucleus programs protein production in the cytoplasm by dictating synthesis of messenger RNA (mRNA).

① Synthesis of mRNA in the nucleus

DNA

mRNA

NUCLEUS

CYTOPLASM

② Movement of mRNA into cytoplasm via nuclear pore

mRNA

Ribosome

③ Synthesis of protein using information carried on mRNA

Polypeptide

Amino acids

→ **Mastering Biology** BioFlix® Animation: Gene Expression

functions of some recently discovered RNA molecules; the stretches of DNA that direct synthesis of these RNAs are also considered genes (see Concept 18.3).

The Components of Nucleic Acids

Nucleic acids are macromolecules that exist as polymers called **polynucleotides (Figure 5.23a)**. As indicated by the name, each polynucleotide consists of monomers called **nucleotides**. A nucleotide, in general, is composed of three parts: a five-carbon sugar (a pentose), a nitrogen-containing (nitrogenous) base, and one to three phosphate groups **(Figure 5.23b)**. The beginning monomer used to build a polynucleotide has three phosphate groups, but two are lost during the polymerization process. The portion of a nucleotide without any phosphate groups is called a *nucleoside*.

To understand the structure of a single nucleotide, let's first consider the nitrogenous bases **(Figure 5.23c)**. Each nitrogenous base has one or two rings that include nitrogen atoms. (They are called nitrogenous *bases* because the nitrogen atoms tend to take up H⁺ from solution, thus acting as bases.) There are two families of nitrogenous bases: pyrimidines and purines. A **pyrimidine** has one six-membered ring of carbon and nitrogen atoms. The members of the pyrimidine family are cytosine (C), thymine (T), and uracil (U).

Purines are larger, with a six-membered ring fused to a five-membered ring. The purines are adenine (A) and guanine (G). The specific pyrimidines and purines differ in the chemical groups attached to the rings. Adenine, guanine, and cytosine are found in both DNA and RNA; thymine is found only in DNA and uracil only in RNA.

Now let's add the sugar to which the nitrogenous base is attached. In DNA the sugar is **deoxyribose**; in RNA it is **ribose** (see Figure 5.23c). The only difference between these two sugars is that deoxyribose lacks an oxygen atom on the second carbon in the ring, hence the name *deoxy*ribose.

So far, we have built a nucleoside (base plus sugar). To complete the construction of a nucleotide, we attach one to three phosphate groups to the 5' carbon of the sugar (the carbon numbers in the sugar include ', the prime symbol; see Figure 5.23b). With one phosphate, this is a nucleoside monophosphate, more often called a nucleotide.

Nucleotide Polymers

The linkage of nucleotides into a polynucleotide involves a condensation reaction. (You will learn the details in Concept 16.2.) In the polynucleotide, adjacent nucleotides are joined by a phosphodiester linkage, which consists of a phosphate group that covalently links the sugars of two nucleotides. This bonding results in a repeating pattern of sugar-phosphate units called the *sugar-phosphate backbone* (see Figure 5.23a). (Note that the nitrogenous bases are not part of the backbone.) The two free ends of the polymer are distinctly different from each other. One end has a phosphate attached to a 5' carbon, and the other end has a hydroxyl group on a 3' carbon; we refer to these as the *5' end* and the *3' end*, respectively. We can say that a polynucleotide has a built-in directionality along its sugar-phosphate backbone, from 5' to 3', somewhat like a one-way street. The bases are attached all along the sugar-phosphate backbone.

The sequence of bases along a DNA (or mRNA) polymer is unique for each gene and provides very specific information to the cell. Because genes are hundreds to thousands of nucleotides long, the number of possible base sequences is effectively limitless. The information carried by the gene is encoded in its specific sequence of the four DNA bases. For example, the sequence 5'-AGGTAACTT-3' means one thing, whereas the sequence 5'-CGCTTTAAC-3' has a different meaning. (Entire genes, of course, are much longer.) The linear order of bases in a gene specifies the amino acid sequence—the primary structure—of a protein, which in turn specifies that protein's 3-D structure, thus enabling its function in the cell.

▼ **Figure 5.23 Components of nucleic acids. (a)** A polynucleotide has a sugar-phosphate backbone with variable appendages, the nitrogenous bases. **(b)** In a polynucleotide, each nucleotide monomer includes a nitrogenous base, a sugar, and a phosphate group. Note that carbon numbers in the sugar include primes ('). **(c)** A nucleoside includes a nitrogenous base (purine or pyrimidine) and a five-carbon sugar (deoxyribose or ribose).

(a) **Polynucleotide, or nucleic acid**

(b) **Nucleotide monomer in a polynucleotide**

(c) **Nucleoside components**

⊜ **Mastering Biology** HHMI Animation: The Chemical Structure of DNA **hhmi** BioInteractive

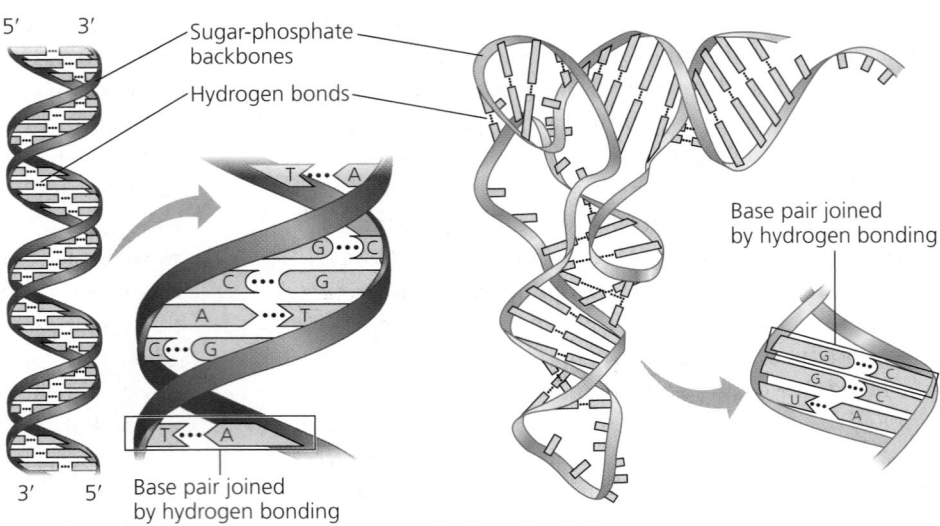

► **Figure 5.24 The structures of DNA and tRNA molecules. (a)** The DNA molecule is usually a double helix, with the sugar-phosphate backbones of the antiparallel polynucleotide strands (symbolized here by blue ribbons) on the outside of the helix. Hydrogen bonds between pairs of nitrogenous bases hold the two strands together. As illustrated here with symbolic shapes for the bases, adenine (A) can pair only with thymine (T), and guanine (G) can pair only with cytosine (C). Each DNA strand in this figure is the structural equivalent of the polynucleotide diagrammed in Figure 5.23a. **(b)** A tRNA molecule has a roughly L-shaped structure due to complementary base pairing of antiparallel stretches of RNA. In RNA, A pairs with U.

 Mastering Biology HHMI Animation: Paired DNA Strands

The Structures of DNA and RNA Molecules

A DNA molecule has two polynucleotides, or "strands," that wind around an imaginary axis, forming a **double helix (Figure 5.24a)**. The two sugar-phosphate backbones run in opposite 5′ → 3′ directions from each other; this arrangement is referred to as **antiparallel**, somewhat like a divided highway. The sugar-phosphate backbones are on the outside of the helix, and the nitrogenous bases are paired in the interior of the helix. The two strands are held together by hydrogen bonds between the paired bases (see Figure 5.24a). Most DNA molecules are very long, with thousands or even millions of base pairs. The one long DNA double helix in a eukaryotic chromosome includes many genes, each one a particular segment of the double-stranded molecule.

In base pairing, only certain bases in the double helix are compatible with each other. Adenine (A) in one strand always pairs with thymine (T) in the other, and guanine (G) always pairs with cytosine (C). Reading the sequence of bases along one strand of the double helix would tell us the sequence of bases along the other strand. If a stretch of one strand has the base sequence 5′-AGGTCCG-3′, then the base-pairing rules tell us that the same stretch of the other strand must have the sequence 3′-TCCAGGC-5′. The two strands of the double helix are *complementary*, each the predictable counterpart of the other. It is this feature of DNA that makes it possible to generate two identical copies of each DNA molecule in a cell that is preparing to divide. When the cell divides, the copies are distributed to the daughter cells, making them genetically identical to the parent cell. Thus, the structure of DNA accounts for its function of transmitting genetic information whenever a cell reproduces.

RNA molecules, by contrast, exist as single strands. Complementary base pairing can occur, however, between regions of two RNA molecules or even between two stretches of nucleotides in the *same* RNA molecule. In fact, base pairing within an RNA molecule allows it to take on the particular three-dimensional shape necessary for its function. Consider, for example, the type of RNA called *transfer RNA* (*tRNA*), which brings amino acids to the ribosome during the synthesis of a polypeptide. A tRNA molecule is about 80 nucleotides in length. Its functional shape results from base pairing between nucleotides where complementary stretches of the molecule can run antiparallel to each other **(Figure 5.24b)**.

Note that in RNA, adenine (A) pairs with uracil (U); thymine (T) is not present in RNA. Another difference between RNA and DNA is that DNA almost always exists as a double helix, whereas RNA molecules are more variable in shape. RNAs are versatile molecules, and many biologists believe RNA may have preceded DNA as the carrier of genetic information in early forms of life (see Concept 25.1).

⊙ **Mastering Biology Animation: Nucleic Acid Structure**

CONCEPT CHECK 5.5

1. **DRAW IT** In Figure 5.23a, number all the carbons of the top three nucleotides (use primes), circle the nitrogenous bases, and star the phosphates.

2. **DRAW IT** A region along one DNA strand has this sequence of nitrogenous bases: 5′-TAGGCCT-3′. Write down its complementary strand, labeling the 5′ and 3′ ends.

For suggested answers, see Appendix A.

CONCEPT 5.6

Genomics and proteomics have transformed biological inquiry and applications

Experimental work in the first half of the 20th century established the role of DNA as the carrier of genetic information, passed from generation to generation, that specified the functioning of living cells and organisms. Once the structure of the DNA molecule was described in 1953, and the linear sequence of nucleotide bases was understood to specify the amino acid sequence of proteins, biologists sought to "decode" genes by learning their nucleotide sequences (often called "base sequences").

The first chemical techniques for *DNA sequencing*, or determining the sequence of nucleotides along a DNA strand, one by one, were developed in the 1970s. Researchers began to study gene sequences, gene by gene, and the more they learned, the more questions they had: How was expression of genes regulated? Genes and their protein products clearly interacted with each other, but how? What was the function, if any, of the DNA that is not part of genes? To fully understand the genetic functioning of a living organism, the entire sequence of the full complement of DNA, the organism's *genome*, would be most enlightening. In spite of the apparent impracticality of this idea, in the late 1980s several prominent biologists put forth an audacious proposal to launch a project that would sequence the entire human genome—all 3 billion bases of it! This endeavor began in 1990 and was effectively completed in the early 2000s.

An unplanned but profound side benefit of this project—the Human Genome Project—was the rapid development of faster and less expensive methods of sequencing. This trend has continued: The cost for sequencing 1 million bases in 2001, well over $5,000, has decreased to less than $0.02 in 2017. And a human genome, the first of which took over 10 years to sequence, could be completed at today's pace in day or less **(Figure 5.25)**. The number of genomes that have been fully sequenced has exploded, generating reams of data and prompting development of **bioinformatics**, the use of computer software and other computational tools that can handle and analyze these large data sets.

The reverberations of these developments have transformed the study of biology and related fields. Biologists often look at problems by analyzing large sets of genes or even comparing whole genomes of different species, an approach called **genomics**. A similar analysis of large sets of proteins, including their sequences, is called **proteomics**. (Protein sequences can be determined either by using biochemical techniques or by translating the DNA sequences that code for them.) These approaches permeate all fields of biology, some examples of which are shown in **Figure 5.26**.

Perhaps the most significant impact of genomics and proteomics on the field of biology as a whole has been their contributions to our understanding of evolution. In addition to confirming evidence for evolution from the study of fossils and characteristics of currently existing species, genomics has helped us tease out relationships among different groups of organisms that had not been resolved by previous types of evidence, and thus infer evolutionary history.

DNA and Proteins as Tape Measures of Evolution

EVOLUTION We are accustomed to thinking of shared traits, such as hair and milk production in mammals, as evidence of shared ancestry. Because DNA carries heritable information in the form of genes, sequences of genes and their protein products document the hereditary background of an organism. The linear sequences of nucleotides in DNA molecules are passed from parents to offspring; these sequences determine the amino acid sequences of proteins. As a result, siblings have greater similarity in their DNA and proteins than do unrelated individuals of the same species.

Given our evolutionary view of life, we can extend this concept of "molecular genealogy" to relationships between species: We would expect two species that appear to be closely related based on anatomical evidence (and possibly fossil evidence) to also share a greater proportion of their DNA and protein sequences than do less closely related species. In fact, that is the case. An example is the comparison of the β polypeptide chain of human hemoglobin with the corresponding hemoglobin polypeptide in other vertebrates. In this chain of 146 amino acids, humans and gorillas differ in just 1 amino acid, while humans and frogs, more distantly related, differ in 67 amino acids. In the **Scientific Skills Exercise**, you can apply this sort of reasoning to additional species. And this conclusion holds true as well when comparing whole genomes: The human genome is 95–98% identical to that of the chimpanzee, but only roughly 85% identical to that of the mouse, a more distant evolutionary relative. Molecular biology has added a new tape measure to the toolkit biologists use to assess evolutionary kinship.

Comparing genomic sequences has practical applications as well. In the **Problem-Solving Exercise**, you can see how this type of genomic analysis can help you detect consumer fraud.

◀ **Figure 5.25 Automatic DNA-sequencing machines and abundant computing power enable rapid sequencing of genes and genomes.**

CONCEPT CHECK 5.6

1. How would sequencing the entire genome of an organism help scientists to understand how that organism functioned?

2. Given the function of DNA, why would you expect two species with very similar traits to also have very similar genomes?

For suggested answers, see Appendix A.

Contributions of **Genomics** and Proteomics **to Biology**

Nucleotide sequencing **and** the analysis of large **sets** of genes and proteins can be done rapidly **and** inexpensively due to advances in technology and information processing. Taken together, genomics and proteomics have advanced our understanding of biology across many different **fields.**

Paleontology

New DNA sequencing techniques have allowed decoding of minute quantities of DNA found in ancient tissues from **our** extinct relatives, the Neanderthals (*Homo neanderthalensis*). Sequencing the Neanderthal **genome** has informed our understanding of their physical appearance (as in this reconstruction), as **well** as their relationship with modern humans. (See Figures 34.51 and 34.52.)

Medical **Science**

Identifying the genetic basis for human diseases like cancer helps **researchers** focus their search for potential future treatments. Currently, sequencing the sets of genes expressed in an individual's tumor can allow a more targeted approach to treating the cancer, a type of "personalized medicine." (See Figures 12.20 and 18.27.)

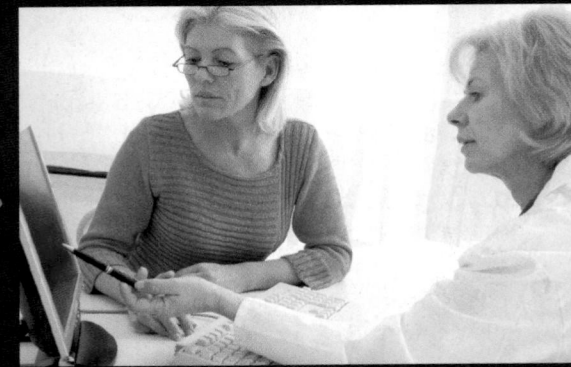

Evolution

A major aim of evolutionary biology is to understand the relationships among species, both living and extinct. For example, genome sequence comparisons have identified the hippopotamus as the land mammal sharing the most recent common ancestor with whales. (See Figure 22.20.)

Hippopotamus

Short-finned pilot whale

Conservation Biology

The tools of molecular genetics and genomics are increasingly used by forensic ecologists to identify which species of animals and plants are killed illegally. In one case, genomic sequences of DNA from illegal shipments of elephant tusks were used to track down poachers and pinpoint the territory where they were operating. (See Figure 56.8.)

Species Interactions

Most plant species exist in a mutually beneficial partnership with fungi (right) and bacteria associated with the plants' roots; these interactions improve plant growth. Genome sequencing and analysis of gene expression have allowed characterization of plant-associated communities. Such studies will help advance our understanding of such interactions and may improve agricultural practices. (See the Chapter 31 Scientific Skills Exercise and Figure 37.10.)

MAKE CONNECTIONS

Considering the examples provided here, describe how the approaches of genomics and proteomics help us to address a variety of biological questions.

→ **Mastering Biology HHMI Video: The Making of the Fittest: The Birth and Death of Genes (Icefish)**

► Human   ► Rhesus monkey   ► Gibbon

Analyzing Polypeptide Sequence Data

Are Rhesus Monkeys or Gibbons More Closely Related to Humans? In this exercise, you will look at amino acid sequence data for the β polypeptide chain of hemoglobin, often called β-globin. You will then interpret the data to hypothesize whether the monkey or the gibbon is more closely related to humans.

How Such Experiments Are Done Researchers can isolate the polypeptide of interest from an organism and then determine the amino acid sequence. More frequently, the DNA of the relevant gene is sequenced, and the amino acid sequence of the polypeptide is deduced from the DNA sequence of its gene.

Data from the Experiments In the data below, the letters give the sequence of the 146 amino acids in β-globin from humans,

rhesus monkeys, and gibbons. Because a complete sequence would not fit on one line here, the sequences are divided into three segments: amino acids 1–50, 51–100, and 101–146. The sequences for the three different species are aligned so that you can compare them easily. For example, you can see that for all three species, the first amino acid is V (valine) and the 146th amino acid is H (histidine).

INTERPRET THE DATA

1. Scan the monkey and gibbon sequences, letter by letter, circling any amino acids that do not match the human sequence. **(a)** How many amino acids differ between the monkey and the human sequences? **(b)** Between the gibbon and human?

2. For each nonhuman species, what percent of its amino acids are identical to the human sequence of β-globin?

3. Based on these data alone, state a hypothesis for which of these two species is more closely related to humans. What is your reasoning?

4. What other evidence could you use to support your hypothesis?

→ **Instructors:** A version of this Scientific Skills Exercise can be assigned in **Mastering Biology**.

Species	Alignment of Amino Acid Sequences of β-globin					
Human	1	VHLTPEEKSA	VTALWGKVNV	DEVGGEALGR	LLVVYPWTQR	FFESFGDLST
Monkey	1	VHLTPEEKNA	VTTLWGKVNV	DEVGGEALGR	LLLVYPWTQR	FFESFGDLSS
Gibbon	1	VHLTPEEKSA	VTALWGKVNV	DEVGGEALGR	LLVVYPWTQR	FFESFGDLST
Human	51	PDAVMGNPKV	KAHGKKVLGA	FSDGLAHLDN	LKGTFATLSE	LHCDKLHVDP
Monkey	51	PDAVMGNPKV	KAHGKKVLGA	FSDGLNHLDN	LKGTFAQLSE	LHCDKLHVDP
Gibbon	51	PDAVMGNPKV	KAHGKKVLGA	FSDGLAHLDN	LKGTFAQLSE	LHCDKLHVDP
Human	101	ENFRLLGNVL	VCVLAHHFGK	EFTPPVQAAY	QKVVAGVANA	LAHKYH
Monkey	101	ENFRLLGNVL	VCVLAHHFGK	EFTPPVQAAY	QKVVAGVANA	LAHKYH
Gibbon	101	ENFRLLGNVL	VCVLAHHFGK	EFTPPVQAAY	QKVVAGVANA	LAHKYH

Data from Human: http://www.ncbi.nlm.nih.gov/protein/AAA21113.1; rhesus monkey: http://www.ncbi.nlm.nih.gov/protein/122634; gibbon: http://www.ncbi.nlm.nih.gov/protein/122616

PROBLEM-SOLVING EXERCISE

Are you a victim of fish fraud?

WILD KING SALMON 17.99 lb $15.99

When buying salmon, perhaps you prefer the more expensive wild-caught Pacific salmon (*Oncorhynchus* species) over farmed Atlantic salmon (*Salmo salar*). But studies reveal that about 40% of the time, you aren't getting the fish you paid for!

→ **Instructors:** A version of this Problem-Solving Exercise can be assigned in Chapter 5 of **Mastering Biology**. A more extensive "Solve It" tutorial is in Chapter 26 of **Mastering Biology**.

In this exercise, you will investigate whether a piece of salmon has been fraudulently labeled.

Your Approach The principle guiding your investigation is that DNA sequences from within a species or from closely related species are more similar to each other than are sequences from more distantly related species.

Your Data You've been sold a piece of salmon labeled as coho salmon (*Oncorhynchus kisutch*). To see whether your fish was labeled correctly, you will compare a short DNA sequence from your sample with known sequences from the same gene for three salmon species. The sequences are as follows:

Sample labeled as *O. kisutch* (coho salmon)		5'-CGGCACCGCCCTAAGTCTCT-3'
Known sequences {	*O. kisutch* (coho salmon)	5'-AGGCACCGCCCTAAGTCTAC-3'
	O. keta (chum salmon)	5'-AGGCACCGCCCTGAGCCTAC-3'
	Salmo salar (Atlantic salmon)	5'-CGGCACCGCCCTAAGTCTCT-3'

Data from the International Barcode of Life.

Your Analysis

1. Circle any bases in the known sequences that do not match the sequence from your fish sample.

2. How many bases differ between **(a)** *O. kisutch* and your fish sample? **(b)** *O. keta* and the sample? **(c)** *S. salar* and the sample?

3. For each known sequence, what percentage of its bases are identical to your sample?

4. Based on these data alone, state a hypothesis for the species identity of your sample. What is your reasoning?

5 Chapter Review

→ Go to **Mastering Biology** for Assignments, the eText, the Study Area, and Dynamic Study Modules.

SUMMARY OF KEY CONCEPTS

→ To review key terms, go to the **Vocabulary Self-Quiz** in the **Mastering Biology** eText or Study Area, or go to goo.gl/zkjz9t.

CONCEPT 5.1

Macromolecules are polymers, built from monomers *(pp. 67–68)*

- Large carbohydrates (polysaccharides), proteins, and nucleic acids are **polymers,** chains of **monomers.** Components of lipids vary. Many monomers can form larger molecules by **dehydration reactions,** in which water molecules are released. Polymers can disassemble by the reverse process, **hydrolysis.** An immense variety of polymers can be built from a small set of monomers.

❓ *What is the fundamental basis for the differences between large carbohydrates, proteins, and nucleic acids?*

Large Biological Molecules	Components	Examples	Functions
CONCEPT 5.2 **Carbohydrates serve as fuel and building material** *(pp. 68–72)* ❓ *Compare starch and cellulose. What role does each play in the human body?*	Monosaccharide monomer	**Monosaccharides:** glucose, fructose	Fuel; carbon sources that can be converted to other molecules or combined into polymers
		Disaccharides: lactose, sucrose	
		Polysaccharides: • Cellulose (plants) • Starch (plants) • Glycogen (animals) • Chitin (animals and fungi)	• Cellulose strengthens plant cell walls • Starch stores glucose for energy in plants • Glycogen stores glucose for energy in animals • Chitin strengthens animal exoskeletons and fungal cell walls
CONCEPT 5.3 **Lipids are a diverse group of hydrophobic molecules** *(pp. 72–75)* ❓ *Why are lipids not considered to be polymers or macromolecules?*	Glycerol — 3 fatty acids	**Triacylglycerols** (fats or oils): glycerol + three fatty acids	Important energy source
	Head with ℗ — 2 fatty acids	**Phospholipids:** glycerol + phosphate group + two fatty acids	Lipid bilayers of membranes Hydrophilic heads / Hydrophobic tails
	Steroid backbone	**Steroids:** four fused rings with attached chemical groups	• Component of cell membranes (cholesterol) • Signaling molecules that travel through the body (hormones)
CONCEPT 5.4 **Proteins include a diversity of structures, resulting in a wide range of functions** *(pp. 75–83)* ❓ *Explain the basis for the great diversity of proteins.*	R Amino acid monomer (20 types)	• Enzymes • Defensive proteins • Storage proteins • Transport proteins • Hormones • Receptor proteins • Motor proteins • Structural proteins	• Catalyze chemical reactions • Protect against disease • Store amino acids • Transport substances • Coordinate organismal responses • Receive signals from outside cell • Function in cell movement • Provide structural support
CONCEPT 5.5 **Nucleic acids store, transmit, and help express hereditary information** *(pp. 84–86)* ❓ *What role does complementary base pairing play in nucleic acids?*	Nitrogenous base Phosphate group ℗—CH₂ O Sugar Nucleotide (monomer of a polynucleotide)	**DNA:** • Sugar = deoxyribose • Nitrogenous bases = C, G, A, T • Usually double-stranded	Stores hereditary information
		RNA: • Sugar = ribose • Nitrogenous bases = C, G, A, U • Usually single-stranded	Various functions in gene expression, including carrying instructions from DNA to ribosomes

CONCEPT 5.6

Genomics and proteomics have transformed biological inquiry and applications (pp. 86–89)

- Recent technological advances in DNA sequencing have given rise to **genomics**, an approach that analyzes large sets of genes or whole genomes, and **proteomics**, a similar approach for large sets of proteins. **Bioinformatics** is the use of computational tools and computer software to analyze these large data sets.
- The more closely two species are related evolutionarily, the more similar their DNA sequences are. DNA sequence data confirm models of evolution based on fossils and anatomical evidence.

? *Given the sequences of a particular gene in fruit flies, fish, mice, and humans, predict the relative similarity of the human sequence to that of each of the other species.*

TEST YOUR UNDERSTANDING

➡ For more multiple-choice questions, go to the **Practice Test** in the **Mastering Biology** eText or Study Area, or go to goo.gl/GruWRg.

Levels 1-2: Remembering/Understanding

1. Which of the following categories includes all others in the list?
   (A) disaccharide   (C) starch
   (B) polysaccharide   (D) carbohydrate

2. The enzyme amylase can break glycosidic linkages between glucose monomers only if the monomers are in the α form. Which of the following could amylase break down?
   (A) glycogen, starch, and amylopectin
   (B) glycogen and cellulose
   (C) cellulose and chitin
   (D) starch, chitin, and cellulose

3. Which statement about *unsaturated* fats is true?
   (A) They are more common in animals than in plants.
   (B) They have double bonds in their fatty acid chains.
   (C) They generally solidify at room temperature.
   (D) They contain more hydrogen than do saturated fats having the same number of carbon atoms.

4. The structural level of a protein *least* affected by a disruption in hydrogen bonding is the
   (A) primary level.   (C) tertiary level.
   (B) secondary level.   (D) quaternary level.

5. Enzymes that break down DNA catalyze the hydrolysis of the covalent bonds that join nucleotides together. What would happen to DNA molecules treated with these enzymes?
   (A) The two strands of the double helix would separate.
   (B) The phosphodiester linkages of the polynucleotide backbone would be broken.
   (C) The pyrimidines would be separated from the deoxyribose sugars.
   (D) All bases would be separated from the deoxyribose sugars.

Levels 3-4: Applying/Analyzing

6. The molecular formula for glucose is $C_6H_{12}O_6$. What would be the molecular formula for a polymer made by linking ten glucose molecules together by dehydration reactions?
   (A) $C_{60}H_{120}O_{60}$ (C) $C_{60}H_{100}O_{50}$
   (B) $C_{60}H_{102}O_{51}$ (D) $C_{60}H_{111}O_{51}$

7. Which of the following pairs of base sequences could form a short stretch of a normal double helix of DNA?
   (A) 5′-AGCT-3′ with 5′-TCGA-3′
   (B) 5′-GCGC-3′ with 5′-TATA-3′
   (C) 5′-ATGC-3′ with 5′-GCAT-3′
   (D) All of these pairs are correct.

8. Construct a table that organizes the following terms and label the columns and rows.

Monosaccharides	Polypeptides	Phosphodiester linkages
Fatty acids	Triacylglycerols	Peptide bonds
Amino acids	Polynucleotides	Glycosidic linkages
Nucleotides	Polysaccharides	Ester linkages

9. **DRAW IT** Copy the polynucleotide strand in Figure 5.23a and label the bases G, T, C, and T, starting from the 5′ end. Assuming this is a DNA polynucleotide, now draw the complementary strand, using the same symbols for phosphates (circles), sugars (pentagons), and bases. Label the bases. Draw arrows showing the 5′ → 3′ direction of each strand. Use the arrows to make sure the second strand is antiparallel to the first. Hint: After you draw the first strand vertically, turn the paper upside down; it is easier to draw the second strand from the 5′ toward the 3′ direction as you go from top to bottom.

Levels 5-6: Evaluating/Creating

10. **EVOLUTION CONNECTION** Comparisons of amino acid sequences can shed light on the evolutionary divergence of related species. If you were comparing two living species, would you expect all proteins to show the same degree of divergence? Why or why not? Justify your answer.

11. **SCIENTIFIC INQUIRY** Suppose you are a research assistant in a lab studying DNA-binding proteins. You have been given the amino acid sequences of all the proteins encoded by the genome of a certain species and have been asked to find candidate proteins that could bind DNA. What type of amino acids would you expect to see in the DNA-binding regions of such proteins? Explain your thinking.

12. **WRITE ABOUT A THEME: ORGANIZATION** Proteins, which have diverse functions in a cell, are all polymers of the same kinds of monomers—amino acids. Write a short essay (100–150 words) that discusses how the structure of amino acids allows this one type of polymer to perform so many functions.

13. **SYNTHESIZE YOUR KNOWLEDGE**

Given that the function of egg yolk is to nourish and support the developing chick, explain why egg yolks are so high in fat, protein, and cholesterol.

For selected answers, see Appendix A.

Dr. Diana Bautista is an Associate Professor of Molecular and Cell Biology at the University of California, Berkeley, and a Howard Hughes Medical Institute Faculty Scholar. Recipient of the Gill Transformative Investigator Award and the Society for Neuroscience Young Investigator Award, she received a B.S. in Biology from the University of Oregon and a Ph.D. in Neuroscience from Stanford University. Dr. Bautista grew up in Chicago, the first in her family to attend college and to pursue a career in science. She and her laboratory members investigate the cell-signaling pathways associated with pain and itch and how their misregulation can result in chronic pain and chronic itch. Her research is interdisciplinary: Dr. Bautista collaborates with computational biologists, immunologists, and physiologists. She believes that science benefits most from a diverse group of colleagues, each bringing a different perspective.

AN INTERVIEW WITH

Diana M. Bautista

What got you interested in biology, and in neuroscience in particular?

I always liked science as a kid, but it was never a big focus in my life. I went to college interested in art, then took some time off because I was a terrible artist. I ended up working with an environmental group helping low-income communities oppose a hazardous waste incinerator. That got me excited about science because it was at the interface of the environment and community empowerment, so I decided to return to college and major in biology. Once I returned to school, at the University of Oregon, I found out you could get a work-study job in a lab. I ended up working in Peter O'Day's lab studying vision in fruit flies by recording electrical signals in response to light. For

▼ Dr. Bautista studies cell responses to stimuli using equipment that measures electrical signals.

me, shining light on a fly eye and recording an electrical signal from their nervous system in real time was just the coolest thing I had ever seen. The lab was a really inclusive, warm environment, and Peter was an amazing mentor—if it weren't for him, I wouldn't be in science. I went to graduate school at Stanford in neuroscience, where I worked with Dr. Richard Lewis on a newly discovered calcium channel and how calcium signaling pathways drive cellular behaviors.

Why did you focus your research on pain?

After my Ph.D., I wanted to continue working on cell signaling but from a more organismal viewpoint, so I did my postdoctoral research with Dr. David Julius at the University of California, San Francisco. His lab had recently identified the cellular protein (TRPV1) that is responsive to capsaicin—the molecule that gives chili peppers their "hotness"—and showed that this protein is also activated by painful heat. That was really exciting to me because it involved thinking about proteins involved in cell signaling, but in the bigger context of pain. Proteins like TRPV1 serve important protective functions—for example, by stopping us from grabbing a hot frying pan that could cause a burn. One of my projects was to identify the wasabi receptor, a protein on the surface of nerve cells that plays a key role in inflammatory pain hypersensitivity. Once you activate this receptor protein—with irritants such as wasabi, mustard oil, or inflammatory mediators produced in the body by tissue injury such as a burn—it activates and makes pain cells more sensitive: Warmth triggers sensations of burning heat, and light touch triggers pain. Normally, pain hypersensitivity dies down after a few days

"It doesn't matter what you did before or where you come from—anybody can be a scientist."

as the tissue heals, but in some patients, it becomes a chronic debilitating disease.

What does your lab work on now?

We are interested in the sense of touch. Touch-sensitive neurons innervate the skin and mediate gentle touch sensations and texture preferences: Do you like wool sweaters, or do you find them itchy? Touch-sensitive neurons also innervate our musculature, allowing us to detect limb position and mediating coordinated movements, such as our ability to text or to walk in a straight line without staring at our feet. In someone suffering from chronic pain, a gentle touch feels incredibly painful, while a chronic itch sufferer experiences light touch as itchy and pain as pleasure. When you scratch an itch, you're scratching really hard—if you didn't have an itch it would hurt, but when you do have an itch it feels good. We have no idea how our body processes the same type of stimulus under different injury or disease conditions, and that's what really excited me when I started my own lab.

What is your advice to an undergraduate considering a career in biology?

I'm an advisor for undergraduates, so I talk to a lot of students. I like to tell them about my path and let them know that it doesn't matter what you did before or where you come from—anybody can be a scientist. I think it's sometimes tough early on in biology, because at first you're taught facts, but as you progress, biology is about figuring out the unknown. And if you can get into a lab early on, then you really see that side of science—that it's about learning what we know, and then going beyond that, doing experiments to identify new mechanisms.

6 A Tour of the Cell

KEY CONCEPTS

6.1 Biologists use microscopes and biochemistry to study cells *p. 94*

6.2 Eukaryotic cells have internal membranes that compartmentalize their functions *p. 97*

6.3 The eukaryotic cell's genetic instructions are housed in the nucleus and carried out by the ribosomes *p. 102*

6.4 The endomembrane system regulates protein traffic and performs metabolic functions *p. 104*

6.5 Mitochondria and chloroplasts change energy from one form to another *p. 109*

6.6 The cytoskeleton is a network of fibers that organizes structures and activities in the cell *p. 112*

6.7 Extracellular components and connections between cells help coordinate cellular activities *p. 118*

6.8 A cell is greater than the sum of its parts *p. 121*

Study Tip

Draw animal and plant cells: Draw an outline of an animal cell and add structures, labels, and functions. Draw a plant cell labeled with structures unique to plant cells.

Plasma membrane: a selective barrier

→ Go to Mastering Biology

For Students (in eText and Study Area)
• Get Ready for Chapter 6
• Figure 6.7 Walkthrough: Geometric Relationships between Surface Area and Volume
• BioFlix® Animations: Tour of an Animal Cell and Tour of a Plant Cell

For Instructors to Assign (in Item Library)
• Tutorial: Tour of an Animal Cell: The Endomembrane System
• Tutorial: Tour of a Plant Cell: Structures and Functions

Figure 6.1 The cell is an organism's basic unit of structure and function. Many forms of life exist as single-celled organisms, such as the *Paramecium* shown here. Larger, more complex organisms, including plants and animals, are multicellular. In this chapter, we focus mainly on eukaryotic cells—cells with a nucleus.

How does the internal organization of eukaryotic cells allow them to perform the functions of life?

Internal membranes divide a cell, such as this plant cell, into compartments where specific chemical reactions occur.

Energy and matter transformations

A system of internal membranes synthesizes and modifies proteins, lipids, and carbohydrates.

Chloroplasts convert light energy to chemical energy.

Mitochondria break down molecules, generating ATP.

Genetic information storage and transmission

DNA in the nucleus contains instructions for making proteins.

Ribosomes are the sites of protein synthesis.

Ribosome

Protein

Interactions with the environment

The plasma membrane controls what goes into and out of the cell.

Plant cells have a protective cell wall.

Biologists use microscopes and biochemistry to study cells

How can cell biologists investigate the inner workings of a cell, usually too small to be seen by the unaided eye? Before we tour the cell, it will be helpful to learn how cells are studied.

Microscopy

The development of instruments that extend the human senses allowed the discovery and early study of cells. Microscopes were invented in 1590 and further refined during the 1600s. Cell walls were first seen on dead cells of oak bark by Robert Hooke in 1665 and living cells by Antoni van Leeuwenhoek a few years later.

The microscopes first used by Renaissance scientists, as well as the microscopes you are likely to use in the laboratory, are all light microscopes. In a **light microscope (LM)**, visible light is passed through the specimen and then through glass lenses. The lenses refract (bend) the light in such a way that the image of the specimen is magnified as it is projected into the eye or into a camera (see Appendix C).

Three important parameters in microscopy are magnification, resolution, and contrast. *Magnification* is the ratio of an object's image size to its real size. Light microscopes can magnify effectively to about 1,000 times the actual size of the specimen; at greater magnifications, additional details cannot be seen clearly. *Resolution* is a measure of the clarity of the image; it is the minimum distance two points can be separated and still be distinguished as separate points. For example, what appears to the unaided eye as one star in the sky may be resolved as twin stars with a telescope, which has a higher resolving ability than the eye. Similarly, using standard techniques, the light microscope cannot resolve detail finer than about 0.2 micrometer (μm), or 200 nanometers (nm), regardless of the magnification **(Figure 6.2)**. The third parameter, *contrast*, is the difference in brightness between the light and dark areas of an image. Methods for enhancing contrast include staining or labeling cell components to stand out visually. **Figure 6.3** shows some different types of microscopy; study this figure as you read this section.

Until recently, the resolution barrier prevented cell biologists from using standard light microscopy when studying **organelles**, the membrane-enclosed structures within eukaryotic cells. To see these structures in any detail required the development of a new instrument. In the 1950s, the electron microscope was introduced to biology. Rather than focusing light, the **electron microscope (EM)** focuses a beam of electrons through the specimen or onto its surface (see Appendix C). Resolution is inversely related to the wavelength of the light (or electrons) a microscope uses for imaging, and electron beams

▼ **Figure 6.2 The size range of cells.** Most cells are between 1 and 100 µm in diameter (yellow region of chart), and their components are even smaller (see Figure 6.32), as are viruses. Notice that the scale along the left side is logarithmic, to accommodate the range of sizes shown. Starting at the top of the scale with 10 m and going down, each reference measurement marks a tenfold decrease in diameter or length. For a complete table of the metric system, see the back of the book.

1 centimeter (cm) = 10^{-2} meter (m) = 0.4 inch
1 millimeter (mm) = 10^{-3} m
1 micrometer (μm) = 10^{-3} mm = 10^{-6} m
1 nanometer (nm) = 10^{-3} μm = 10^{-9} m

⊙ **Mastering Biology** Animation: Metric System Review

have much shorter wavelengths than visible light. Modern electron microscopes can theoretically achieve a resolution of about 0.002 nm, though in practice they usually cannot resolve structures smaller than about 2 nm across. Still, this is a 100-fold improvement over the standard light microscope.

The **scanning electron microscope (SEM)** is especially useful for detailed study of the topography of a specimen (see Figure 6.3). The electron beam scans the surface of the sample, usually coated with a thin film of gold. The beam excites

Light Microscopy (LM)

Brightfield. Light passes directly through the specimen. Unstained (left), the image has little contrast. Staining with dyes (right) enhances contrast. Most stains require cells to be preserved, which kills them.

Phase-contrast. Variations in density within the specimen are amplified to enhance contrast in unstained cells; this is especially useful for examining living, unstained cells.

Differential interference contrast (Nomarski). As in phase-contrast microscopy, optical modifications are used to exaggerate differences in density; the image appears almost 3-D.

Fluorescence. Locations of specific molecules are revealed by labeling the molecules with fluorescent dyes or antibodies, which absorb ultraviolet radiation and emit visible light. In this fluorescently labeled uterine cell, DNA is blue, organelles called mitochondria are orange, and part of the cell's "skeleton" (called the *cytoskeleton*) is green.

Human cheek cells

50 μm

10 μm

Confocal. This image shows two types of fluorescence micrographs: confocal (top) and standard (bottom). (Nerve cells are green, support cells orange, areas of overlap yellow.) In confocal microscopy, a laser is used to create a single plane of fluorescence; out-of-focus light from other planes is eliminated. By capturing sharp images at many different planes, a 3-D reconstruction can be created. A standard fluorescence micrograph is blurry because out-of-focus light is not excluded.

50 μm

Deconvolution. The top of this image of a white blood cell was reconstructed from many blurry fluorescence images at different planes, each processed using deconvolution software. This process digitally removes out-of-focus light and reassigns it to its source, creating a much sharper 3-D image. The bottom is a compilation of standard fluorescent micrographs through the same cell.

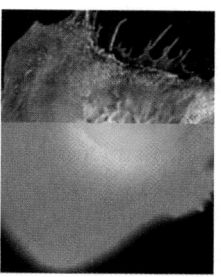

10 μm

Super-resolution. To make this super-resolution image of a cow aorta cell (top), individual fluorescent molecules were excited by UV light and their position recorded. (DNA is blue, mitochondria red, and part of the cytoskeleton green.) Combining information from many molecules in different places "breaks" the resolution limit, resulting in the sharp image on top. The size of each dot is well below the 200-nm resolution of a standard light microscope, as seen in the confocal image (bottom) of the same cell.

10 μm

Electron Microscopy (EM)

Scanning electron microscopy (SEM). Micrographs taken with a scanning electron microscope show a 3-D image of the surface of a specimen. This SEM shows the surface of a cell from a trachea (windpipe) covered with cell projections called cilia. Electron micrographs are black and white but are often artificially colorized to highlight particular structures, as has been done with all three electron micrographs shown here.

SEM   2 μm

Transmission electron microscopy (TEM). A transmission electron microscope profiles a thin section of a specimen. This TEM shows a section through a tracheal cell, revealing its internal structure. In preparing the specimen, some cilia were cut along their lengths, creating longitudinal sections, while other cilia were cut straight across, creating cross sections.

Cilia

Longitudinal section of cilium

Cross section of cilium

2 μm

TEM

Cryo-electron microscopy (cryo-EM). Specimens of tissue or aqueous solutions of proteins are frozen rapidly at temperatures less than −160°C, locking the molecules into a rigid state. A beam of electrons is passed through the sample to visualize the molecules by electron microscopy, and software is used to merge a series of such micrographs, creating a 3-D image like the one below.

Computer-generated image of the bacterial enzyme β-galactosidase, which breaks down lactose. This image was compiled from more than 90,000 cryo-EM images.

Abbreviations used in figure legends in this text:
LM = Light Micrograph
SEM = Scanning Electron Micrograph
TEM = Transmission Electron Micrograph

VISUAL SKILLS *When the tissue was sliced, what was the orientation of the cilia in the lower portion of the TEM? The upper portion? Explain how the orientation of the cilia determined the type of sections we see.*

electrons on the surface, and these secondary electrons are detected by a device that translates the pattern of electrons into an electronic signal sent to a video screen. The result is an image of the specimen's surface that appears three-dimensional.

The **transmission electron microscope (TEM)** is used to study the internal structure of cells (see Figure 6.3). The TEM aims an electron beam through a very thin section of the specimen, much as a light microscope aims light through a sample on a slide. For the TEM, the specimen has been stained with atoms of heavy metals, which attach to certain cellular structures, thus enhancing the electron density of some parts of the cell more than others. The electrons passing through the specimen are scattered more in the denser regions, so fewer are transmitted. The image displays the pattern of transmitted electrons. Instead of using glass lenses, both the SEM and TEM use electromagnets as lenses to bend the paths of the electrons, ultimately focusing the image onto a monitor for viewing.

Electron microscopes have revealed many subcellular structures that were impossible to resolve with the light microscope. But the light microscope offers advantages, especially in studying living cells. A disadvantage of electron microscopy is that the methods customarily used to prepare the specimen kill the cells and can introduce artifacts, structural features seen in micrographs that do not exist in the living cell.

In the past several decades, light microscopy has been revitalized by major technical advances (see Figure 6.3). Labeling individual cellular molecules or structures with fluorescent markers has made it possible to see such structures with increasing detail. In addition, both confocal and deconvolution microscopy have produced sharper images of three-dimensional tissues and cells. Finally, a group of new techniques and labeling molecules developed in recent years, called *super-resolution microscopy*, has allowed researchers to "break" the resolution barrier and distinguish subcellular structures as small as 10–20 nm across.

A recently developed new type of TEM called cryo-electron microscopy (cryo-EM) (see Figure 6.3) allows specimens to be preserved at extremely low temperatures. This avoids the use of preservatives, allowing visualization of structures in their cellular environment. This method is increasingly used to complement X-ray crystallography in revealing protein complexes and subcellular structures like ribosomes, described later. Cryo-EM has even been used to resolve some individual proteins. The Nobel Prize for Chemistry was awarded in 2017 to the developers of this valuable technique.

Microscopes are the most important tools of *cytology*, the study of cell structure. Understanding the function of each structure, however, required the integration of cytology and *biochemistry*, the study of the chemical processes (metabolism) of cells.

Cell Fractionation

A useful technique for studying cell structure and function is **cell fractionation (Figure 6.4)**, which takes cells apart

Cell Fractionation

Application　Cell fractionation is used to separate (fractionate) cell components based on size and density.

Technique　Cells are homogenized in a blender to break them up. The resulting mixture (*homogenate*) is centrifuged. The liquid above the pellet (*supernatant*) is poured into another tube and centrifuged at a higher speed for a longer period. This process is repeated several times. This process, called *differential centrifugation*, results in a series of pellets, each containing different cell components.

Tissue cells

Homogenization

Homogenate

Centrifugation

Centrifuged at 1,000 *g* (1,000 times the force of gravity) for 10 min

Supernatant poured into next tube

Differential centrifugation

20,000 *g* 20 min

Pellet rich in nuclei and cellular debris

80,000 *g* 60 min

Pellet rich in mitochondria (and chloroplasts if cells are from a plant)

150,000 *g* 3 hr

Pellet rich in "microsomes" (pieces of plasma membranes and cells' internal membranes)

Pellet rich in ribosomes

Results　In early experiments, researchers used microscopy to identify the organelles in each pellet and biochemical methods to determine their metabolic functions. These identifications established a baseline for this method, enabling today's researchers to know which cell fraction they should collect in order to isolate and study particular organelles.

MAKE CONNECTIONS　*If you wanted to study the process of translation of proteins from mRNA, which part of which fraction would you use? (See Figure 5.22.)*

and separates major organelles and other subcellular structures from one another. The piece of equipment that is used for this task is the centrifuge, which spins test tubes holding mixtures of disrupted cells at a series of increasing speeds, a process called *differential centrifugation*. At each speed, the resulting force causes a subset of the cell components to settle to the bottom of the tube, forming a pellet. At lower speeds, the pellet consists of larger components, and higher speeds result in a pellet with smaller components.

Cell fractionation enables researchers to prepare specific cell components in bulk and identify their functions, a task not usually possible with intact cells. For example, on one of the cell fractions, biochemical tests showed the presence of enzymes involved in cellular respiration, while electron microscopy revealed large numbers of the organelles called mitochondria. Together, these data helped biologists determine that mitochondria are the sites of cellular respiration. Biochemistry and cytology thus complement each other in correlating cell function with structure.

CONCEPT CHECK 6.1

1. How do stains used for light microscopy compare with those used for electron microscopy?
2. **WHAT IF?** Which type of microscope would you use to study (a) the changes in shape of a living white blood cell? (b) the details of surface texture of a hair?

For suggested answers, see Appendix A.

CONCEPT 6.2

Eukaryotic cells have internal membranes that compartmentalize their functions

Cells—the basic structural and functional units of every organism—are of two distinct types: prokaryotic and eukaryotic. Organisms of the domains Bacteria and Archaea consist of prokaryotic cells. Organisms of the domain Eukarya—protists, fungi, animals, and plants—all consist of eukaryotic cells. ("Protist" is an informal term referring to a diverse group of mostly unicellular eukaryotes.)

Comparing Prokaryotic and Eukaryotic Cells

All cells share certain basic features: They are all bounded by a selective barrier, called the *plasma membrane* (or the cell membrane). Inside all cells is a semifluid, jellylike substance called **cytosol**, in which subcellular components are suspended. All cells contain *chromosomes*, which carry genes in the form of DNA. And all cells have *ribosomes*, tiny complexes that make proteins according to instructions from the genes.

A major difference between prokaryotic and eukaryotic cells is the location of their DNA. In a **eukaryotic cell**, most of the DNA is in an organelle called the *nucleus*, which is bounded by a double membrane (see Figure 6.8). In a **prokaryotic cell**, the DNA is concentrated in a region that is not membrane-enclosed, called the **nucleoid (Figure 6.5)**.

▼ **Figure 6.5 A prokaryotic cell.** Lacking a true nucleus and the other membrane-enclosed organelles of the eukaryotic cell, the prokaryotic cell appears much simpler in internal structure. Prokaryotes include bacteria and archaea; the general cell structure of these two domains is quite similar.

Fimbriae: attachment structures on the surface of some prokaryotes (not visible on TEM)

Nucleoid: region where the cell's DNA is located (not enclosed by a membrane)

Ribosomes: complexes that synthesize proteins

Plasma membrane: membrane enclosing the cytoplasm

Cell wall: rigid structure outside the plasma membrane

Glycocalyx: outer coating of many prokaryotes, consisting of a capsule or a slime layer

Bacterial chromosome

Flagella: locomotion organelles of some prokaryotes

(a) A typical rod-shaped bacterium

0.5 μm

(b) A thin section through the bacterium *Corynebacterium diphtheriae* (colorized TEM)

Mastering Biology Animation: Prokaryotic Cell Structure and Function

Eukaryotic means "true nucleus" (from the Greek *eu*, true, and *karyon*, kernel, referring to the nucleus), and *prokaryotic* means "before nucleus" (from the Greek *pro*, before), reflecting the earlier evolution of prokaryotic cells.

The interior of either type of cell is called the **cytoplasm**; in eukaryotic cells, this term refers only to the region between the nucleus and the plasma membrane. Within the cytoplasm of a eukaryotic cell, suspended in cytosol, are a variety of organelles of specialized form and function. These membrane-bounded structures are absent in almost all prokaryotic cells, another distinction between prokaryotic and eukaryotic cells. In spite of the absence of organelles, though, the prokaryotic cytoplasm is not a formless soup. For example, some prokaryotes contain regions surrounded by proteins (not membranes), within which specific reactions take place.

Eukaryotic cells are generally much larger than prokaryotic cells (see Figure 6.2). Size is a general feature of cell structure that relates to function. The logistics of carrying out cellular metabolism sets limits on cell size. At the lower limit, the smallest cells known are bacteria called mycoplasmas, which have diameters between 0.1 and 1.0 μm. These are perhaps the smallest packages with enough DNA to program metabolism and enough enzymes and other cellular equipment to carry out the activities necessary for a cell to sustain itself and reproduce. Typical bacteria are 1–5 μm in diameter, about ten times the size of mycoplasmas. Eukaryotic cells are typically 10–100 μm in diameter.

Metabolic requirements also impose theoretical upper limits on the size that is practical for a single cell. At the boundary of every cell, the **plasma membrane** functions as a selective barrier that allows passage of enough oxygen, nutrients, and wastes to service the entire cell **(Figure 6.6)**. For each square micrometer of membrane, only a limited amount of a particular substance can cross per second, so the ratio of surface area to volume is critical. As a cell (or any other object) increases in size, its surface area grows proportionately less than its volume. (Area is proportional to a linear dimension squared, whereas volume is proportional to the linear dimension cubed.) Thus, a smaller cell has a greater ratio of surface area to volume: Compare the calculations for the first two "cells" in **Figure 6.7**. The **Scientific Skills Exercise** gives you a chance to calculate the volumes and surface areas of two actual cells—a mature yeast cell and a cell budding from it. To explore different ways that the surface area of cells is maximized in various organisms, see Make Connections Figure 33.8.

The need for a surface area large enough to accommodate the volume helps explain the microscopic size of most cells and the narrow, elongated shapes of some cells, such as nerve cells. Larger organisms do not generally have *larger* cells than smaller organisms—they simply have *more* cells (see the far right of Figure 6.7). A sufficiently high ratio of surface area to volume is especially important in cells that exchange a lot of material with their surroundings, such as intestinal cells. Such cells may have many long, thin projections from their

▼ **Figure 6.6 The plasma membrane.** The plasma membrane and the membranes of organelles consist of a double layer (bilayer) of phospholipids with various proteins attached to or embedded in it. The hydrophobic parts of phospholipids and membrane proteins are found in the interior of the membrane, while the hydrophilic parts are in contact with aqueous solutions on either side. Carbohydrate side chains may be attached to proteins or lipids on the outer surface of the plasma membrane.

(a) Colorized TEM of a plasma membrane. The plasma membrane appears as a pair of dark bands separated by a gold band.

(b) Structure of the plasma membrane

VISUAL SKILLS *What parts of the membrane diagram in (b) correspond to the dark bands in the TEM in (a)? What parts correspond to the gold band? (Review Figure 5.11.)*

➔ **Mastering Biology BioFlix® Animation: Membranes**

▼ **Figure 6.7 Geometric relationships between surface area and volume.** In this diagram, cells are represented as cubes. Using arbitrary units of length, we can calculate the cell's surface area (in square units, or units²), volume (in cubic units, or units³), and ratio of surface area to volume. A high surface area-to-volume ratio facilitates the exchange of materials between a cell and its environment.

Surface area increases while total volume remains constant

Total surface area [(height × width of 1 side) × 6 sides × number of cells]	6 units²	150 units²	750 units²
Total volume [(height × width × length of 1 cell) × number of cells]	1 unit³	125 units³	125 units³
Surface area-to- volume ratio [surface area ÷ volume]	6	1.2	6

➔ **Mastering Biology Figure Walkthrough**

Using a Scale Bar to Calculate Volume and Surface Area of a Cell

How Much New Cytoplasm and Plasma Membrane Are Made by a Growing Yeast Cell? The unicellular yeast *Saccharomyces cerevisiae* divides by budding off a small new cell that then grows to full size (see the yeast cells at the bottom of Figure 6.8). During its growth, the new cell synthesizes new cytoplasm, which increases its volume, and new plasma membrane, which increases its surface area. In this exercise, you will use a scale bar to determine the sizes of a mature parent yeast cell and a cell budding from it. You will then calculate the volume and surface area of each cell. You will use your calculations to determine how much cytoplasm and plasma membrane the new cell needs to synthesize to grow to full size.

How the Experiment Was Done Yeast cells were grown under conditions that promoted division by budding. The cells were then viewed with a differential interference contrast light microscope and photographed.

Data from the Experiment This light micrograph shows a budding yeast cell about to be released from the mature parent cell:

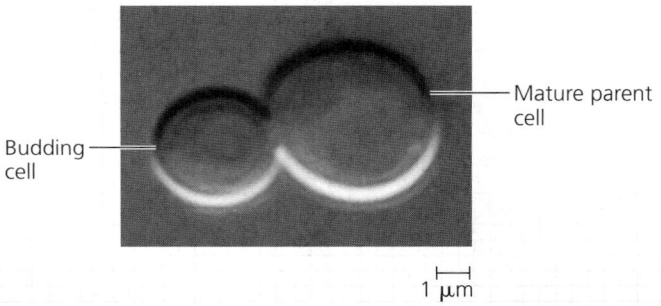

Budding cell

Mature parent cell

1 μm

Micrograph from K. Tatchell, using yeast cells grown for experiments described in L. Kozubowski et al., Role of the septin ring in the asymmetric localization of proteins at the mother-bud neck in *Saccharomyces cerevisiae, Molecular Biology of the Cell* 16:3455–3466 (2005).

INTERPRET THE DATA

1. Examine the micrograph of the yeast cells. The scale bar under the photo is labeled 1 μm. The scale bar works in the same way as a scale on a map, where, for example, 1 inch equals 1 mile. In this case the bar represents one thousandth of a millimeter. Using the scale bar as a basic unit, determine the diameter of the mature parent cell and the new cell. Start by measuring the scale bar and the diameter of each cell. The units you use are irrelevant, but working in millimeters is convenient. Divide each diameter by the length of the scale bar and then multiply by the scale bar's length value to give you the diameter in micrometers.

2. The shape of a yeast cell can be approximated by a sphere. **(a)** Calculate the volume of each cell using the formula for the volume of a sphere:

$$V = \frac{4}{3}\pi r^3$$

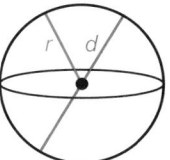

Note that π (the Greek letter pi) is a constant with an approximate value of 3.14, *d* stands for diameter, and *r* stands for radius, which is half the diameter. **(b)** What volume of new cytoplasm will the new cell have to synthesize as it matures? To determine this, calculate the difference between the volume of the full-sized cell and the volume of the new cell.

3. As the new cell grows, its plasma membrane needs to expand to contain the increased volume of the cell. **(a)** Calculate the surface area of each cell using the formula for the surface area of a sphere: $A = 4\pi r^2$. **(b)** How much area of new plasma membrane will the new cell have to synthesize as it matures?

4. When the new cell matures, it will be approximately how many times greater in volume and how many times greater in surface area than its current size?

➔ **Instructors:** A version of this Scientific Skills Exercise can be assigned in **Mastering Biology**.

---

surface called *microvilli*, which increase surface area without an appreciable increase in volume.

The evolutionary relationships between prokaryotic and eukaryotic cells will be discussed later in this chapter, and prokaryotic cells will be described in detail elsewhere (see Chapter 27). Most of the discussion of cell structure that follows in this chapter applies to eukaryotic cells.

A Panoramic View of the Eukaryotic Cell

In addition to the plasma membrane at its outer surface, a eukaryotic cell has extensive, elaborately arranged internal membranes that divide the cell into compartments—the organelles mentioned earlier. The cell's compartments provide different local environments that support specific metabolic functions, so incompatible processes can occur simultaneously in a single cell. The plasma membrane and organelle

membranes also participate directly in the cell's metabolism because many enzymes are built right into the membranes.

The basic fabric of most biological membranes is a double layer of phospholipids and other lipids. Embedded in this lipid bilayer or attached to its surfaces are diverse proteins (see Figure 6.6). However, each type of membrane has a unique composition of lipids and proteins suited to that membrane's specific functions. For example, enzymes embedded in the membranes of the organelles called mitochondria function in cellular respiration. Because membranes are so fundamental to the organization of the cell, Chapter 7 will discuss them in detail.

Before continuing with this chapter, examine the eukaryotic cells in **Figure 6.8**. The generalized diagrams of an animal cell and a plant cell introduce the various organelles and show the key differences between animal and plant cells. The micrographs at the bottom of the figure give you a glimpse of cells from different types of eukaryotic organisms.

Animal Cell (cutaway view of generalized cell)

ENDOPLASMIC RETICULUM (ER): network of membranous sacs and tubes; active in membrane synthesis and other synthetic and metabolic processes; has rough (ribosome-studded) and smooth regions

Rough ER Smooth ER

Flagellum: motility structure present in some animal cells, composed of a cluster of microtubules within an extension of the plasma membrane

Centrosome: region where the cell's microtubules are initiated; contains a pair of centrioles

CYTOSKELETON: reinforces cell's shape; functions in cell movement; components are made of protein. Includes:

Microfilaments
Intermediate filaments
Microtubules

Microvilli: membrane projections that increase the cell's surface area

Peroxisome: organelle with various specialized metabolic functions; produces hydrogen peroxide as a by-product and then converts it to water

Mitochondrion: organelle where cellular respiration occurs and most ATP is generated

Nuclear envelope: double membrane enclosing the nucleus; perforated by pores; continuous with ER

Nucleolus: nonmembranous structure involved in production of ribosomes; a nucleus has one or more nucleoli

Chromatin: material consisting of DNA and proteins; visible in a dividing cell as individual condensed chromosomes

NUCLEUS

Plasma membrane: membrane enclosing the cell

Ribosomes (small brown dots): complexes that make proteins; free in cytosol or bound to rough ER or nuclear envelope

Golgi apparatus: organelle active in synthesis, modification, sorting, and secretion of cell products

Lysosome: digestive organelle where macromolecules are hydrolyzed

➔ **Mastering Biology** BioFlix® Animation: Tour of an Animal Cell

Animal Cells

Cell

Nucleus

Nucleolus

Human cells from lining of uterus (colorized TEM)

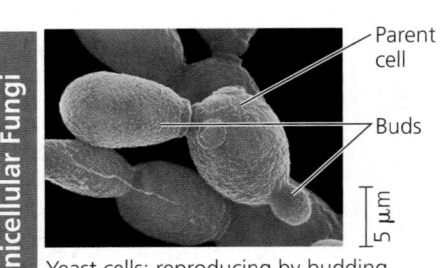

Unicellular Fungi

Parent cell

Buds

Yeast cells: reproducing by budding (above, colorized SEM) and a single cell (right, colorized TEM)

1 μm

Cell wall

Vacuole

Nucleus

Mitochondrion

Plant Cell (cutaway view of generalized cell)

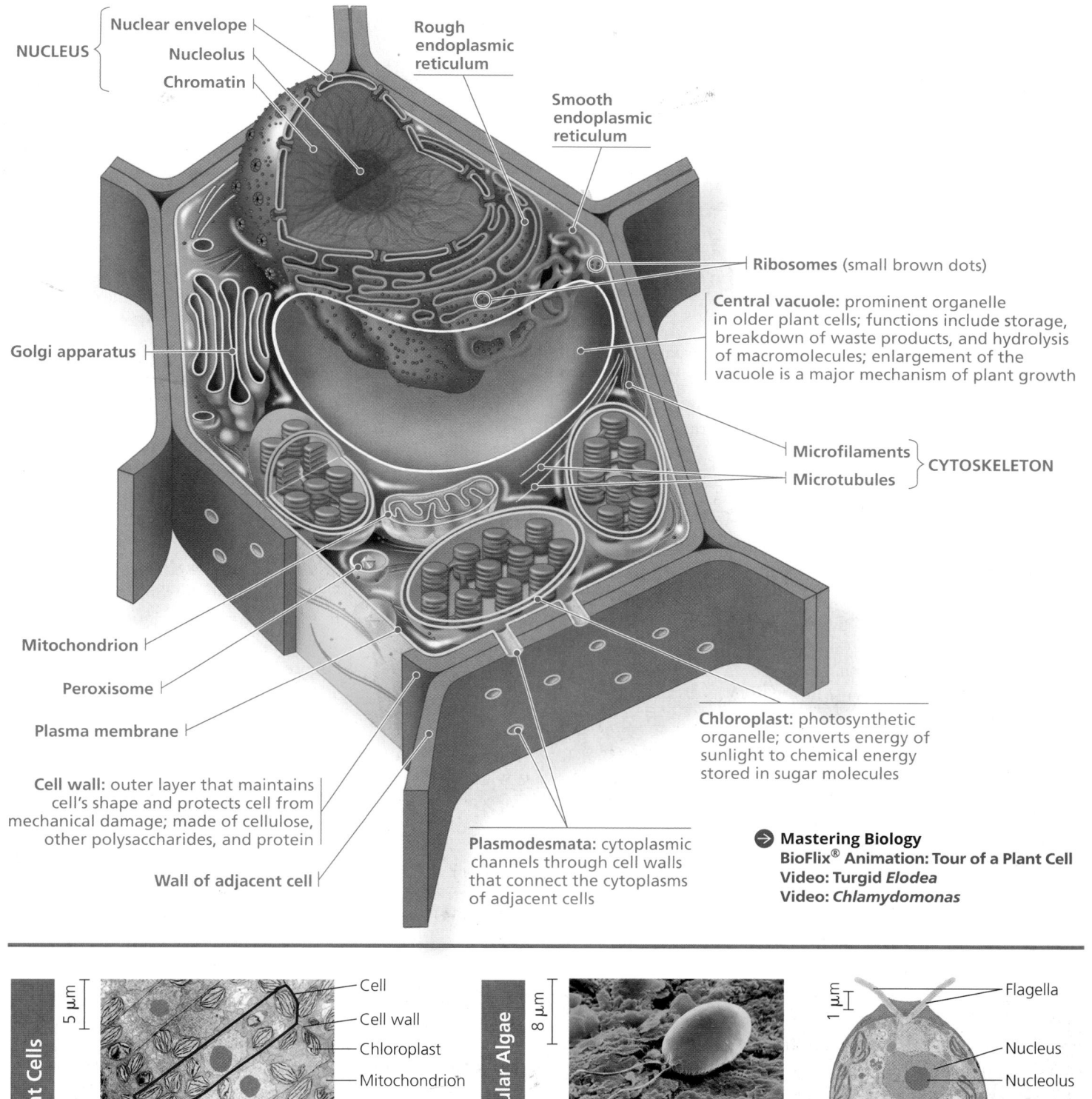

NUCLEUS
- Nuclear envelope
- Nucleolus
- Chromatin

Rough endoplasmic reticulum

Smooth endoplasmic reticulum

Ribosomes (small brown dots)

Central vacuole: prominent organelle in older plant cells; functions include storage, breakdown of waste products, and hydrolysis of macromolecules; enlargement of the vacuole is a major mechanism of plant growth

Golgi apparatus

Microfilaments
Microtubules
} CYTOSKELETON

Mitochondrion

Peroxisome

Plasma membrane

Cell wall: outer layer that maintains cell's shape and protects cell from mechanical damage; made of cellulose, other polysaccharides, and protein

Wall of adjacent cell

Plasmodesmata: cytoplasmic channels through cell walls that connect the cytoplasms of adjacent cells

Chloroplast: photosynthetic organelle; converts energy of sunlight to chemical energy stored in sugar molecules

→ Mastering Biology
BioFlix® Animation: Tour of a Plant Cell
Video: Turgid *Elodea*
Video: *Chlamydomonas*

Plant Cells

5 μm

- Cell
- Cell wall
- Chloroplast
- Mitochondrion
- Nucleus
- Nucleolus

Cells from duckweed (*Spirodela oligorrhiza*), a floating plant (colorized TEM)

Unicellular Algae

8 μm

Unicellular green alga *Chlamydomonas* (above, colorized SEM; right, colorized TEM)

1 μm

- Flagella
- Nucleus
- Nucleolus
- Vacuole
- Chloroplast
- Cell wall

1. Briefly describe the structure and function of the nucleus, the mitochondrion, the chloroplast, and the endoplasmic reticulum.

2. **DRAW IT** Draw a simplified elongated cell that measures $125 \times 1 \times 1$ arbitrary units. A nerve cell would be roughly this shape. Predict how its surface-to-volume ratio would compare with those in Figure 6.7. Then calculate the ratio and check your prediction.

For suggested answers, see Appendix A.

CONCEPT 6.3

The eukaryotic cell's genetic instructions are housed in the nucleus and carried out by the ribosomes

On the first stop of our detailed tour of the eukaryotic cell, let's look at two cellular components involved in the genetic control of the cell: the nucleus, which houses most of the cell's DNA, and the ribosomes, which use information from the DNA to make proteins.

The Nucleus: Information Central

The **nucleus** contains most of the genes in the eukaryotic cell. (Some genes are located in mitochondria and chloroplasts.) It is usually the most conspicuous organelle (see the purple structure in the fluorescence micrograph), averaging about 5 μm in diameter. The **nuclear envelope** encloses the nucleus **(Figure 6.9)**, separating its contents from the cytoplasm.

The nuclear envelope is a *double* membrane. The two membranes, each a lipid bilayer with associated proteins, are separated by a space of 20–40 nm. The envelope is perforated by pore structures that are about 100 nm in diameter. At the lip of each pore, the inner and outer membranes of the nuclear envelope are continuous. An intricate protein structure called a *pore complex* lines each pore and plays an important role in the cell by regulating the entry and exit of proteins and RNAs, as well as large complexes of macromolecules. Except at the pores, the nuclear side of the envelope is lined by the **nuclear lamina**, a netlike array of protein filaments (in animal cells, called *intermediate filaments*) that maintains the shape of the nucleus by mechanically supporting the nuclear envelope. There is also much evidence for a *nuclear matrix*, a framework of protein fibers extending throughout the nuclear interior. The nuclear lamina and matrix may help organize the genetic material so it functions efficiently.

Nucleus

5 μm

Within the nucleus, the DNA is organized into discrete units called **chromosomes**, structures that carry the genetic information. Each chromosome contains one long DNA molecule associated with many proteins, including small basic proteins called histones. Some of the proteins help coil the DNA molecule of each chromosome, reducing its length and allowing it to fit into the nucleus. The complex of DNA and proteins making up chromosomes is called **chromatin**. When a cell is not dividing, stained chromatin appears as a diffuse mass in micrographs, and the chromosomes cannot be distinguished from one another, even though discrete chromosomes are present. As a cell prepares to divide, however, the chromosomes form loops and coil, condensing and becoming thick enough to be distinguished under a microscope as separate structures (see Figure 16.23). Each eukaryotic species has a characteristic number of chromosomes. For example, a typical human cell has 46 chromosomes in its nucleus; the exceptions are human sex cells (eggs and sperm), which have only 23 chromosomes. A fruit fly cell has 8 chromosomes in most cells and 4 in the sex cells.

A prominent structure within the nondividing nucleus is the **nucleolus** (plural, *nucleoli*), which appears through the electron microscope as a mass of densely stained granules and fibers adjoining part of the chromatin. Here a type of RNA called *ribosomal RNA* (rRNA) is synthesized from genes in the DNA. Also in the nucleolus, proteins imported from the cytoplasm are assembled with rRNA into large and small subunits of ribosomes. These subunits then exit the nucleus through the nuclear pores to the cytoplasm, where a large and a small subunit can assemble into a ribosome. Sometimes there are two or more nucleoli; the number depends on the species and the stage in the cell's reproductive cycle. The nucleoli may also play a role in controlling cell division and the life span of a cell.

As we saw in Figure 5.22, the nucleus directs protein synthesis by synthesizing messenger RNA (mRNA) that carries information from the DNA. The mRNA is then transported to the cytoplasm via nuclear pores. Once an mRNA molecule reaches the cytoplasm, ribosomes translate the mRNA's genetic message into the primary structure of a specific polypeptide. (This process of transcribing and translating genetic information is described in detail in Chapter 17.)

→ **Mastering Biology** BioFlix® Animation: Nucleus and Ribosomes

Ribosomes: Protein Factories

Ribosomes, which are complexes made of ribosomal RNAs and proteins, are the cellular components that carry out protein synthesis **(Figure 6.10)**. (Note that ribosomes are not membrane bounded and thus are not considered organelles.)

▼ **Figure 6.9 The nucleus and its envelope.** Within the nucleus are the chromosomes, which appear as a mass of chromatin (DNA and associated proteins) and one or more nucleoli (singular, *nucleolus*), which function in ribosome synthesis. The nuclear envelope, which consists of two membranes separated by a narrow space, is perforated with pores and lined by the nuclear lamina.

1 μm

Nucleus

Nucleolus

Chromatin

Nuclear envelope:
Outer membrane
Inner membrane

Nuclear pores

Nuclear envelope:
Outer membrane
Inner membrane

Rough ER

▲ **Surface of nuclear envelope (TEM).** This specimen was prepared by a technique known as freeze-fracture, which cuts from the outer membrane to the inner membrane, revealing both.

Pore complex

Ribosome

◄ **Close-up of nuclear envelope**

0.25 μm

▲ **Pore complexes (TEM).** Each pore is ringed by protein particles.

► **Nuclear lamina (TEM).** The netlike lamina lines the inner surface of the nuclear envelope. (The light circular spots are nuclear pores.)

0.5 μm

DNA

Histone protein

▲ **Chromatin.** This segment of a chromosome shows two states of chromatin (DNA—blue—wrapped around histone proteins--purple) in a nondividing cell. In preparation for cell division, the chromatin will become more condensed.

MAKE CONNECTIONS *Chromosomes contain the genetic material and reside in the nucleus. How does the rest of the cell get access to the information they carry? (See Figure 5.22.)*

▼ **Figure 6.10 Ribosomes.** This electron micrograph of a pancreas cell shows both free and bound ribosomes. The simplified diagram and computer model show the two subunits of a ribosome.

⊘ **Mastering Biology**
Interview with Venki Ramakrishnan: Studying ribosome structure

0.25 μm

Ribosomes

ER

Free ribosomes in cytosol

Endoplasmic reticulum (ER)

Ribosomes bound to ER

Large subunit

Small subunit

TEM showing ER and ribosomes

Diagram of a ribosome

Computer model of a ribosome

DRAW IT *After you have read the section on ribosomes, circle a ribosome in the micrograph that might be making a protein that will be secreted.*

Cells with high rates of protein synthesis have particularly large numbers of ribosomes as well as prominent nucleoli, which makes sense, given the role of nucleoli in ribosome assembly. For example, a human pancreas cell, which makes many digestive enzymes, has a few million ribosomes.

Ribosomes build proteins in two cytoplasmic regions: At any given time, *free ribosomes* are suspended in the cytosol, while *bound ribosomes* are attached to the outside of the endoplasmic reticulum or nuclear envelope (see Figure 6.10). Bound and free ribosomes are structurally identical, and ribosomes can play either role at different times. Most of the proteins made on free ribosomes function within the cytosol; examples are enzymes that catalyze the first steps of sugar breakdown. Bound ribosomes generally make proteins that are destined for insertion into membranes, for packaging within certain organelles such as lysosomes (see Figure 6.8), or for export from the cell (secretion). Cells that specialize in protein secretion—for instance, the cells of the pancreas that secrete digestive enzymes—frequently have a high proportion of bound ribosomes. (You will learn more about ribosome structure and function in Concept 17.4.)

CONCEPT CHECK 6.3

1. What role do ribosomes play in carrying out genetic instructions?
2. Describe the molecular composition of nucleoli and explain their function.
3. **WHAT IF?** As a cell begins the process of dividing, its chromosomes become shorter, thicker, and individually visible in an LM (light micrograph). Explain what is happening at the molecular level.

For suggested answers, see Appendix A.

CONCEPT 6.4

The endomembrane system regulates protein traffic and performs metabolic functions

Many of the different membrane-bounded organelles of the eukaryotic cell are part of the **endomembrane system**, which includes the nuclear envelope, the endoplasmic reticulum, the Golgi apparatus, lysosomes, various kinds of vesicles and vacuoles, and the plasma membrane. This system carries out a variety of tasks in the cell, including synthesis of proteins, transport of proteins into membranes and organelles or out of the cell, metabolism and movement of lipids, and detoxification of poisons. The membranes of this system are related either through direct physical continuity or by the transfer of membrane segments as tiny **vesicles** (sacs made of membrane). Despite these relationships, the various membranes are not identical in structure and function. Moreover, the thickness, molecular composition, and types of chemical reactions carried out in a given membrane are not fixed, but may be modified several times during the membrane's life.

Having already looked at the nuclear envelope, we will now focus on the endoplasmic reticulum and the other endomembranes to which the endoplasmic reticulum gives rise.

The Endoplasmic Reticulum: Biosynthetic Factory

The **endoplasmic reticulum (ER)** is such an extensive network of membranes that it accounts for more than half the total membrane in many eukaryotic cells. (The word *endoplasmic* means "within the cytoplasm," and *reticulum* is Latin for "little net.") The ER consists of a network of membranous tubules and sacs called cisternae (from the Latin *cisterna*, a reservoir for a liquid). The ER membrane separates the internal compartment of the ER, called the *ER lumen* (cavity) or cisternal space, from the cytosol. And because the ER membrane is continuous with the nuclear envelope, the space between the two membranes of the envelope is continuous with the lumen of the ER **(Figure 6.11)**.

There are two distinct, though connected, regions of the ER that differ in structure and function: smooth ER and rough ER. **Smooth ER** is so named because its outer surface lacks ribosomes. **Rough ER** is studded with ribosomes on the outer surface of the membrane and thus appears rough through the electron microscope. As already mentioned, ribosomes are also attached to the cytoplasmic side of the nuclear envelope's outer membrane, which is continuous with rough ER.

Functions of Smooth ER

The smooth ER functions in diverse metabolic processes, which vary with cell type. These processes include synthesis of lipids, metabolism of carbohydrates, detoxification of drugs and poisons, and storage of calcium ions.

Enzymes of the smooth ER are important in the synthesis of lipids, including oils, steroids, and new membrane phospholipids. Among the steroids produced by the smooth ER in animal cells are the sex hormones of vertebrates and the various steroid hormones secreted by the adrenal glands. The cells that synthesize and secrete these hormones—in the testes and ovaries, for example—are rich in smooth ER, a structural feature that fits the function of these cells.

Other enzymes of the smooth ER help detoxify drugs and poisons, especially in liver cells. Detoxification usually involves adding hydroxyl groups to drug molecules, making them more water-soluble and easier to flush from the body. The sedative phenobarbital and other barbiturates are examples of drugs metabolized in this manner by smooth ER in liver cells. In fact, barbiturates, alcohol, and many other drugs induce the proliferation of smooth ER and its associated detoxification enzymes, thus increasing the rate of detoxification. This, in turn, increases tolerance to the drugs, meaning that higher doses are required to achieve a particular effect, such as sedation. Also, because some of the detoxification enzymes have relatively broad action, the proliferation of smooth ER in response to one drug can increase the need for higher dosages of other drugs as well. Barbiturate abuse, for instance, can decrease the effectiveness of certain antibiotics and other useful drugs.

▼ **Figure 6.11 Endoplasmic reticulum (ER).** A membranous system of interconnected tubules and flattened sacs called cisternae, the ER is also continuous with the nuclear envelope, as shown in the cutaway diagram at the top. The membrane of the ER encloses a continuous compartment called the ER lumen (or cisternal space). Rough ER, which is studded on its outer surface with ribosomes, can be distinguished from smooth ER in the electron micrograph (TEM). Transport vesicles bud off from a region of the rough ER called transitional ER and travel to the Golgi apparatus and other destinations.

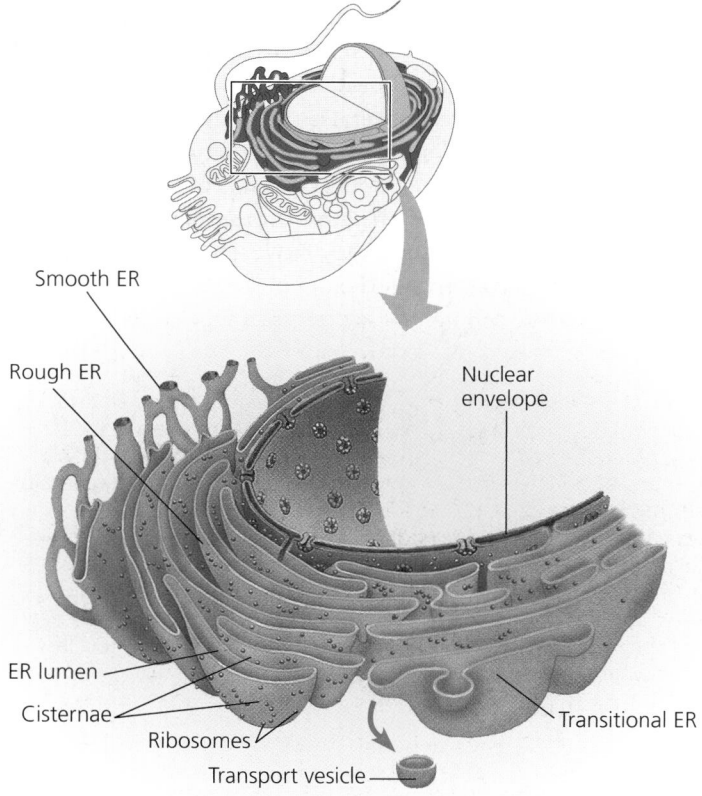

Smooth ER

Rough ER

Nuclear envelope

ER lumen

Cisternae

Ribosomes

Transport vesicle

Transitional ER

Smooth ER   Ribosomes   Rough ER   0.2 μm

The smooth ER also stores calcium ions. In muscle cells, for example, the smooth ER membrane pumps calcium ions from the cytosol into the ER lumen. When a muscle cell is stimulated by a nerve impulse, calcium ions rush back across the ER membrane into the cytosol and trigger contraction of the muscle cell. In other cell types, release of calcium ions from the smooth ER triggers different responses, such as secretion of vesicles carrying newly synthesized proteins.

Functions of Rough ER

Many cells secrete proteins that are produced by ribosomes attached to rough ER. For instance, certain pancreatic cells synthesize the protein insulin in the ER and secrete this hormone into the bloodstream. As a polypeptide chain grows from a bound ribosome, the chain is threaded into the ER lumen through a pore formed by a protein complex in the ER membrane. The new polypeptide folds into its functional shape as it enters the ER lumen. Most secretory proteins are **glycoproteins**, proteins with carbohydrates covalently bonded to them. The carbohydrates are attached to the proteins in the ER lumen by enzymes built into the ER membrane.

After secretory proteins are formed, the ER membrane keeps them separate from proteins in the cytosol, which are produced by free ribosomes. Secretory proteins depart from the ER wrapped in the membranes of vesicles that bud like bubbles from a specialized region called transitional ER (see Figure 6.11). Vesicles in transit from one part of the cell to another are called **transport vesicles**; we will examine their fate shortly.

In addition to making secretory proteins, rough ER is a membrane factory for the cell; it grows in place by adding membrane proteins and phospholipids to its own membrane. As polypeptides destined to be membrane proteins grow from the ribosomes, they are inserted into the ER membrane itself and anchored there by their hydrophobic portions. Like the smooth ER, the rough ER also makes membrane phospholipids; enzymes built into the ER membrane assemble phospholipids from precursors in the cytosol. The ER membrane expands, and portions of it are transferred in the form of transport vesicles to other components of the endomembrane system.

The Golgi Apparatus: Shipping and Receiving Center

After leaving the ER, many transport vesicles travel to the **Golgi apparatus**. We can think of the Golgi as a warehouse for receiving, sorting, shipping, and even some manufacturing. Here, products of the ER, such as proteins, are modified and stored and then sent to other destinations. Not surprisingly, the Golgi apparatus is especially extensive in cells specialized for secretion.

The Golgi apparatus consists of a group of associated, flattened membranous sacs—cisternae—looking like a stack of pita bread **(Figure 6.12)**. A cell may have many, even hundreds, of these stacks. The membrane of each cisterna in a stack separates its internal space from the cytosol. Vesicles concentrated in the vicinity of the Golgi apparatus are engaged in the transfer of material between parts of the Golgi and other structures.

A Golgi stack has a distinct structural directionality, with the membranes of cisternae on opposite sides of the stack differing in thickness and molecular composition. The two sides of a Golgi stack are referred to as the *cis* face and the *trans* face; these act, respectively, as the receiving and shipping departments of the Golgi apparatus. The term *cis* means "on the same side," and the *cis* face is usually located near the ER. Transport vesicles move material from the ER to the Golgi apparatus. A vesicle that buds from the ER can add its membrane and the contents of its lumen to the *cis* face by fusing with a Golgi membrane on that side. The *trans* face ("on the opposite side") gives rise to vesicles that pinch off and travel to other sites.

Products of the endoplasmic reticulum are usually modified during their transit from the *cis* region to the *trans* region of the Golgi apparatus. For example, glycoproteins formed in the ER have their carbohydrates modified, first in the ER itself, and then as they pass through the Golgi. The Golgi removes some sugar monomers and substitutes others,

producing a large variety of carbohydrates. Membrane phospholipids may also be altered in the Golgi.

In addition to its finishing work, the Golgi apparatus also manufactures some macromolecules. Many polysaccharides secreted by cells are Golgi products. For example, pectins and certain other noncellulose polysaccharides are made in the Golgi of plant cells and then incorporated along with cellulose into their cell walls. Like secretory proteins, nonprotein Golgi products that will be secreted depart from the *trans* face of the Golgi inside transport vesicles that eventually fuse with the plasma membrane. The contents are released and the vesicle membrane is incorporated into the plasma membrane, adding to the surface area.

The Golgi manufactures and refines its products in stages, with different cisternae containing unique teams of enzymes. Until recently, biologists viewed the Golgi as a static structure, with products in various stages of processing transferred from one cisterna to the next by vesicles. While this may occur, research from several labs has given rise to a new model of the Golgi as a more dynamic structure. According to the *cisternal maturation model*, the cisternae of the Golgi actually progress forward from the *cis* to the *trans* face, carrying and modifying their cargo as they move. Figure 6.12 shows the details of this model. The reality probably lies somewhere between the two models; recent research suggests the central

▼ **Figure 6.12 The Golgi apparatus.** The Golgi apparatus consists of stacks of associated, flattened sacs, or cisternae. Unlike the ER cisternae, these sacs are not physically connected, as you can see in the cutaway diagram. A Golgi stack receives and dispatches transport vesicles and the products they contain. A Golgi stack has a structural and functional directionality, with a *cis* face that receives vesicles containing ER products and a *trans* face that dispatches vesicles. The cisternal maturation model proposes that the Golgi cisternae themselves "mature," moving from the *cis* to the *trans* face while carrying some proteins along. In addition, some vesicles recycle enzymes that had been carried forward in moving cisternae, transporting them "backward" to a less mature region where their functions are needed.

Golgi apparatus

cis face
("receiving" side of Golgi apparatus)

❶ Vesicles move from ER to Golgi.

❷ Vesicles coalesce to form new *cis* Golgi cisternae.

❻ Vesicles also transport certain proteins back to ER, their site of function.

Cisternae

❸ Cisternal maturation: Golgi cisternae move in a *cis*-to-*trans* direction.

❹ Vesicles form and leave Golgi, carrying specific products to other locations or to the plasma membrane for secretion.

❺ Vesicles transport some proteins backward to less mature Golgi cisternae, where they function.

trans face
("shipping" side of Golgi apparatus)

0.1 μm

TEM of Golgi apparatus

regions of the cisternae may remain in place, while the outer ends are more dynamic.

Before a Golgi stack dispatches its products by budding vesicles from the *trans* face, it sorts these products and targets them for various parts of the cell. Molecular identification tags, such as phosphate groups added to the Golgi products, aid in sorting by acting like zip codes on mailing labels. Finally, transport vesicles budded from the Golgi may have external molecules on their membranes that recognize "docking sites" on the surface of specific organelles or on the plasma membrane, thus targeting the vesicles appropriately.

Lysosomes: Digestive Compartments

A **lysosome** is a membranous sac of hydrolytic enzymes that many eukaryotic cells use to digest (hydrolyze) macromolecules. Lysosomal enzymes work best in the acidic environment found in lysosomes. If a lysosome breaks open or leaks its contents, the released enzymes are not very active because the cytosol has a near-neutral pH. However, excessive leakage from a large number of lysosomes can destroy a cell by self-digestion.

Hydrolytic enzymes and lysosomal membrane are made by rough ER and then transferred to the Golgi apparatus for further processing. At least some lysosomes probably arise by budding from the *trans* face of the Golgi apparatus (see Figure 6.12). How are the proteins of the inner surface of the lysosomal membrane and the digestive enzymes themselves spared from destruction? Apparently, the three-dimensional shapes of these proteins protect vulnerable bonds from enzymatic attack.

Lysosomes carry out intracellular digestion in a variety of circumstances. Amoebas and many other unicellular protists eat by engulfing smaller organisms or food particles, a process called **phagocytosis** (from the Greek *phagein*, to eat, and *kytos*, vessel, referring here to the cell). The *food vacuole* formed in this way then fuses with a lysosome, whose enzymes digest the food (**Figure 6.13a**, top). Digestion products, including simple sugars, amino acids, and other monomers, pass into the cytosol and become nutrients for the cell. Some human cells also carry out phagocytosis. Among them are macrophages, a type of white blood cell that helps defend the body by engulfing and destroying bacteria and other invaders (see Figure 6.13a, bottom, and Figure 6.31).

▼ **Figure 6.13 Lysosomes.**

(a) Phagocytosis

❶ Lysosome contains active hydrolytic enzymes.

❷ Lysosome fuses with food vacuole.

❸ Hydrolytic enzymes digest food particles.

Digestive enzymes

Lysosome

Plasma membrane

Food vacuole

Digestion

A lysosome fuses with a food vacuole during the process of phagocytosis by a unicellular protist.

Nucleus

4 μm

Lysosome

In this colorized TEM of a macrophage (a type of white blood cell), lysosomes are purple. They contain enzymes that digest foreign particles such as bacteria and pollen.

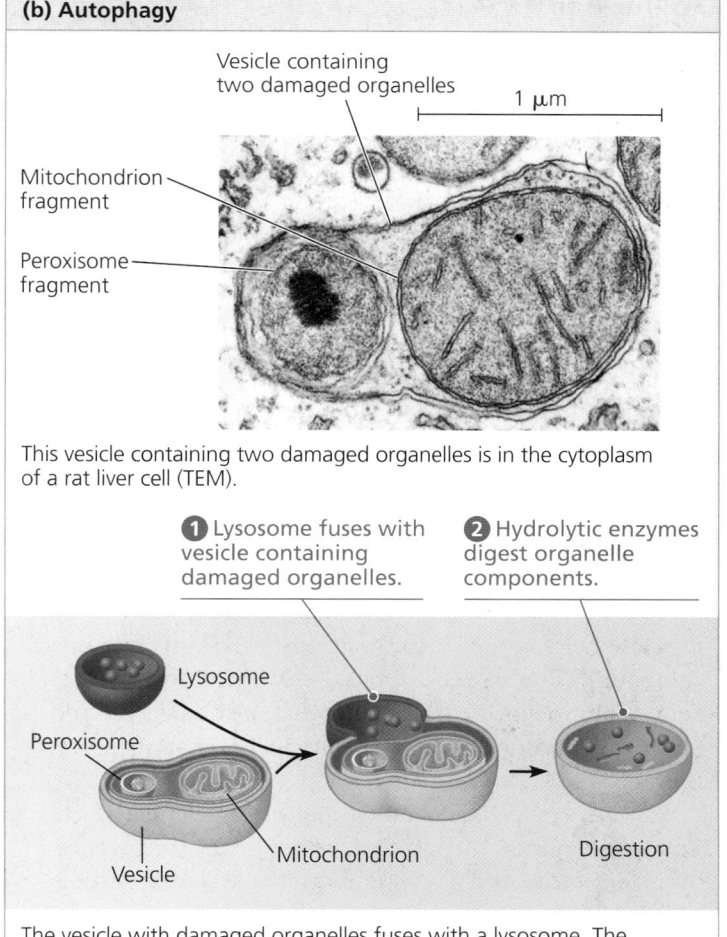

(b) Autophagy

Vesicle containing two damaged organelles

1 μm

Mitochondrion fragment

Peroxisome fragment

This vesicle containing two damaged organelles is in the cytoplasm of a rat liver cell (TEM).

❶ Lysosome fuses with vesicle containing damaged organelles.

❷ Hydrolytic enzymes digest organelle components.

Lysosome

Peroxisome

Vesicle

Mitochondrion

Digestion

The vesicle with damaged organelles fuses with a lysosome. The organelles are then digested and their components recycled.

Lysosomes also use their hydrolytic enzymes to recycle the cell's own organic material, a process called *autophagy*. During autophagy, a damaged organelle or small amount of cytosol becomes surrounded by a double membrane (of unknown origin), and a lysosome fuses with the outer membrane of this vesicle **(Figure 6.13b)**. The lysosomal enzymes dismantle the inner membrane and the enclosed material, and the resulting small organic compounds are released to the cytosol for reuse. With the help of lysosomes, the cell continually renews itself. A human liver cell, for example, recycles half of its macromolecules each week.

The cells of people with inherited lysosomal storage diseases lack a functioning hydrolytic enzyme normally present in lysosomes. The lysosomes become engorged with indigestible material, which begins to interfere with other cellular activities. In Tay-Sachs disease, for example, a lipid-digesting enzyme is missing or inactive, and the brain becomes impaired by an accumulation of lipids in the cells. Fortunately, lysosomal storage diseases are rare in the general population.

Vacuoles: Diverse Maintenance Compartments

Vacuoles are large vesicles derived from the endoplasmic reticulum and Golgi apparatus. Thus, vacuoles are an integral part of a cell's endomembrane system. Like all cellular membranes, the vacuolar membrane is selective in transporting solutes; as a result, the solution inside a vacuole differs in composition from the cytosol.

Vacuoles perform a variety of functions in different kinds of cells. **Food vacuoles**, formed by phagocytosis, have already been mentioned (see Figure 6.13a). Many unicellular protists living in fresh water have **contractile vacuoles** that pump excess water out of the cell, thereby maintaining a suitable concentration of ions and molecules inside the cell (see Figure 7.13). In plants and fungi, certain vacuoles carry out enzymatic hydrolysis, a function shared by lysosomes in animal cells. (In fact, some biologists consider these hydrolytic vacuoles to be a type of lysosome.) In plants, small vacuoles can hold reserves of important organic compounds, such as the proteins stockpiled in the storage cells in seeds. Vacuoles may also help protect the plant against herbivores by storing compounds that are poisonous or unpalatable to animals. Some plant vacuoles contain pigments, such as the red and blue pigments of petals that help attract pollinating insects to flowers.

Mature plant cells generally contain a large **central vacuole (Figure 6.14)**, which develops by the coalescence of smaller vacuoles. The solution inside the central vacuole, called cell sap, is the plant cell's main repository of inorganic ions, including potassium and chloride. The central vacuole plays a major role in the growth of plant cells, which enlarge

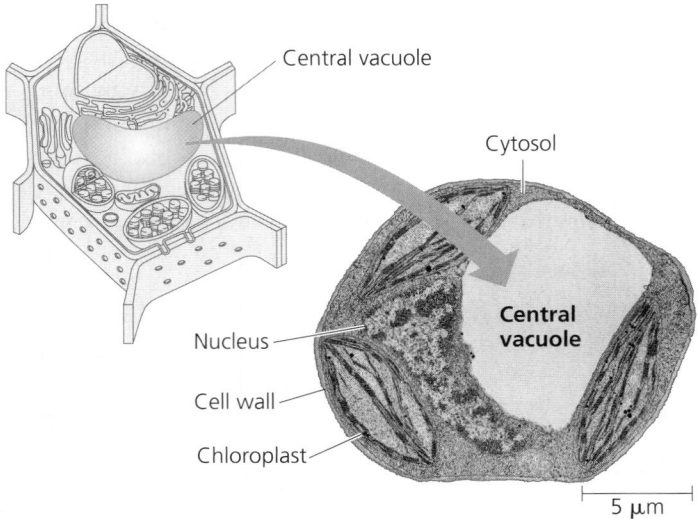

▼ **Figure 6.14 The plant cell vacuole.** The central vacuole is usually the largest compartment in a plant cell; the rest of the cytoplasm is often confined to a narrow zone between the vacuolar membrane and the plasma membrane (TEM).

Central vacuole

Cytosol

Central vacuole

Nucleus

Cell wall

Chloroplast

5 μm

→ **Mastering Biology BioFlix® Animation: Central Vacuole**

as the vacuole absorbs water, enabling the cell to become larger with a minimal investment in new cytoplasm. The cytosol often occupies only a thin layer between the central vacuole and the plasma membrane, so the ratio of plasma membrane surface to cytosolic volume is sufficient, even for a large plant cell.

The Endomembrane System: *A Review*

Figure 6.15 reviews the endomembrane system, showing the flow of membrane lipids and proteins through the various organelles. As the membrane moves from the ER to the Golgi and then elsewhere, its molecular composition and metabolic functions are modified, along with those of its contents. The endomembrane system is a complex and dynamic player in the cell's compartmental organization.

We'll continue our tour of the cell with some organelles that are not closely related to the endomembrane system but play crucial roles in the energy transformations carried out by cells.

CONCEPT CHECK 6.4

1. Describe the structural and functional distinctions between rough and smooth ER.

2. Describe how transport vesicles integrate the endomembrane system.

3. **WHAT IF?** Imagine a protein that functions in the ER but requires modification in the Golgi apparatus before it can achieve that function. Describe the protein's path through the cell, starting with the mRNA molecule that specifies the protein.

For suggested answers, see Appendix A.

1 The nuclear envelope is connected to the rough ER, which is also continuous with the smooth ER.

Nucleus

Nuclear envelope

Rough ER

Smooth ER

2 Membranes and proteins produced by the ER move via transport vesicles to the Golgi.

cis Golgi

3 The Golgi pinches off transport vesicles and other vesicles that give rise to lysosomes, other types of specialized vesicles, and vacuoles.

trans Golgi

Plasma membrane

4 The lysosome is available for fusion with another vesicle for digestion.

5 A transport vesicle carries proteins to the plasma membrane for secretion.

6 The plasma membrane expands by fusion of vesicles; proteins are secreted from the cell by exocytosis.

▲ **Figure 6.15 Review: Relationships among organelles of the endomembrane system.** The red arrows show some of the migration pathways for membranes and the materials they enclose.

⊃ **Mastering Biology**
BioFlix® Animation:
Endomembrane System

Mitochondria and chloroplasts change energy from one form to another

Organisms transform the energy they acquire from their surroundings. In eukaryotic cells, mitochondria and chloroplasts are the organelles that convert energy to forms that cells can use for work. **Mitochondria** (singular, *mitochondrion*) are the sites of cellular respiration, the metabolic process that uses oxygen to drive the generation of ATP by extracting energy from sugars, fats, and other fuels. **Chloroplasts**, found in plants and algae, are the sites of photosynthesis. This process in chloroplasts converts solar energy to chemical energy by absorbing sunlight and using it to drive the synthesis of organic compounds such as sugars from carbon dioxide and water.

In addition to having related functions, mitochondria and chloroplasts share similar evolutionary origins, which we'll look at briefly before examining their structures. In this section, we'll also consider the peroxisome, an oxidative organelle. The evolutionary origin of the peroxisome, as well as its relation to other organelles, is still a matter of some debate.

The Evolutionary Origins of Mitochondria and Chloroplasts

EVOLUTION Mitochondria and chloroplasts display similarities with bacteria that led to the **endosymbiont theory**, illustrated in **Figure 6.16**. This theory states that an early ancestor of eukaryotic cells (a *host* cell) engulfed an oxygen-using nonphotosynthetic prokaryotic cell. Eventually, the engulfed cell formed a relationship with the host cell in which it was enclosed, becoming an *endosymbiont* (a cell living within another cell). Indeed, over the course of evolution, the host cell and its endosymbiont merged into a single organism, a eukaryotic cell with the endosymbiont having become a mitochondrion. At least one of these cells may have then taken up a photosynthetic prokaryote, becoming the ancestor of eukaryotic cells that contain chloroplasts.

This is a widely accepted theory, which we will examine in more detail in Concept 25.3. This theory is consistent with many structural features of mitochondria and chloroplasts. First, rather than being bounded by a single membrane like organelles of the endomembrane system, mitochondria and typical chloroplasts have two membranes surrounding them. (Chloroplasts also have an internal system of membranous sacs.) There is evidence that the ancestral engulfed prokaryotes had two outer membranes, which became the double

▼ Figure 6.16 The endosymbiont theory of the origins of mitochondria and chloroplasts in eukaryotic cells. According to this theory, the proposed ancestors of mitochondria were oxygen-using nonphotosynthetic prokaryotes that were taken into host cells, while the proposed ancestors of chloroplasts were photosynthetic prokaryotes. The large arrows represent change over evolutionary time; the small arrows inside the cells show the process of the endosymbiont becoming an organelle, also over long periods of time.

Endoplasmic reticulum

Nucleus

Nuclear envelope

Ancestor of eukaryotic cells (host cell)

Engulfing of oxygen-using nonphotosynthetic prokaryote, which, over many generations of cells, becomes a mitochondrion

Mitochondrion

Engulfing of photosynthetic prokaryote

Chloroplast

At least one cell

Mitochondrion

Nonphotosynthetic eukaryote

Photosynthetic eukaryote

membranes of mitochondria and chloroplasts. Second, like prokaryotes, mitochondria and chloroplasts contain ribosomes, as well as circular DNA molecules—like bacterial chromosomes—associated with their inner membranes. The DNA in these organelles programs the synthesis of some organelle proteins on ribosomes that have been synthesized and assembled there as well. Third, also consistent with their probable evolutionary origins as cells, mitochondria and chloroplasts are autonomous (somewhat independent) organelles that grow and reproduce within the cell.

Next, we focus on the structures of mitochondria and chloroplasts, while providing an overview of their structures and functions. (In Chapters 9 and 10, we will examine their roles as energy transformers.)

Mitochondria: Chemical Energy Conversion

Mitochondria are found in nearly all eukaryotic cells, including those of plants, animals, fungi, and most protists. Some cells have a single large mitochondrion, but more often a cell has hundreds or even thousands of mitochondria; the

number correlates with the cell's level of metabolic activity. For example, cells that move or contract have proportionally more mitochondria per volume than less active cells.

Each of the two membranes enclosing the mitochondrion is a phospholipid bilayer with a unique collection of embedded proteins. The outer membrane is smooth, but the inner membrane is convoluted, with infoldings called **cristae (Figure 6.17a)**. The inner membrane divides the mitochondrion into two internal compartments. The first is the intermembrane space, the narrow region between the inner and outer membranes. The second compartment, the **mitochondrial matrix**, is enclosed by the inner membrane. The matrix contains many different enzymes as well as the mitochondrial DNA and ribosomes. Enzymes in the matrix catalyze some of the steps of cellular respiration. Other proteins that function in respiration, including the enzyme that makes ATP, are built into the inner membrane. As highly folded surfaces, the cristae give the inner mitochondrial membrane a large surface area, thus enhancing the productivity of cellular respiration. This is another example of structure fitting function. (Chapter 9 discusses cellular respiration in detail.)

Mitochondria are generally in the range of 1–10 μm long. Time-lapse films of living cells reveal mitochondria moving around, changing their shapes, and fusing or dividing into separate fragments, unlike the static structures seen in most diagrams and electron micrographs. These studies helped cell biologists understand that mitochondria in a living cell form a branched tubular network that is in a dynamic state of flux (see **Figure 6.17b** and **c**). In skeletal muscle, this network has been referred to by researchers as a "power grid."

Chloroplasts: Capture of Light Energy

Chloroplasts contain the green pigment chlorophyll, along with enzymes and other molecules that function in the photosynthetic production of sugar. These lens-shaped organelles, about 3–6 μm in length, are found in leaves and other green organs of plants and in algae (**Figure 6.18**; see also Figure 6.26c).

The contents of a chloroplast are partitioned from the cytosol by an envelope consisting of two membranes separated by a very narrow intermembrane space. Inside the chloroplast is another membranous system in the form of flattened, interconnected sacs called **thylakoids**. In some regions, thylakoids are stacked like poker chips; each stack is called a **granum** (plural, *grana*). The fluid outside the thylakoids is the **stroma**, which contains the chloroplast DNA and ribosomes as well as many enzymes. The membranes of the chloroplast divide the chloroplast space into three compartments: the intermembrane space, the stroma, and the thylakoid space. This compartmental organization enables the chloroplast to convert light energy to chemical energy

Mitochondrion

▼ **Figure 6.17 The mitochondrion, site of cellular respiration. (a)** The inner and outer membranes seen in the drawing and TEM establish two compartments: the intermembrane space and the mitochondrial matrix. The cristae increase the surface area of the inner membrane. Circular DNA molecules are associated with the inner mitochondrial membrane. **(b)** Mitochondria are dynamic, at times separating into fragments or fusing. **(c)** The LM shows one protist (*Euglena gracilis*) at a much lower magnification than the TEM. The mitochondria form a branched tubular network; the mitochondrial matrix is stained green. The nuclear DNA is stained red; molecules of mitochondrial DNA appear as bright yellow spots.

→ **Mastering Biology BioFlix® Animation: Mitochondria**

Intermembrane space
Outer membrane
Ribosomes
DNA
Inner membrane
Cristae
Matrix
0.1 μm

(a) Diagram and TEM of mitochondrion

Separating ⇅ Fusing

(b) Dynamic nature of mitochondrial networks

10 μm
Mitochondria
Mitochondrial DNA
Nuclear DNA

(c) Network of mitochondria in *Euglena* (LM)

during photosynthesis. (You will learn more about photosynthesis in Chapter 10.)

As with mitochondria, the static and rigid appearance of chloroplasts in micrographs or schematic diagrams cannot accurately depict their dynamic behavior in the living cell. Their shape is changeable, and they grow and occasionally pinch in two, reproducing themselves. They are mobile and, with mitochondria and other organelles, move around the cell along tracks of the cytoskeleton, a structural network we will consider in Concept 6.6.

The chloroplast is a specialized member of a family of closely related plant organelles called **plastids**. One type of plastid, the *amyloplast*, is a colorless organelle that stores starch (amylose), particularly in roots and tubers. Another is the *chromoplast*, which has pigments that give fruits and flowers their orange and yellow hues.

▼ **Figure 6.18 The chloroplast, site of photosynthesis. (a)** Many plants have lens-shaped chloroplasts, as shown here in a diagram and a TEM. A chloroplast has three compartments: the intermembrane space, the stroma, and the thylakoid space. Free ribosomes are present in the stroma, as are copies of chloroplast DNA molecules. **(b)** This light micrograph, at a much lower magnification than the TEM, shows a whole cell of the green alga *Spirogyra crassa*, which is named for the spiral arrangement of each cell's chloroplasts.

Chloroplast

Ribosomes
Stroma
Inner and outer membranes
Granum
DNA
Thylakoid
Intermembrane space
1 μm

(a) Diagram and TEM of chloroplast

100 μm
Chloroplasts

(b) Chloroplasts in an algal cell

→ **Mastering Biology BioFlix® Animation: Chloroplasts and Mitochondria**

▼ **Figure 6.19 A peroxisome.** Peroxisomes are roughly spherical and often have a granular or crystalline core that is thought to be a dense collection of enzyme molecules. Chloroplasts and mitochondria cooperate with peroxisomes in certain metabolic functions (TEM).

Peroxisome

Mitochon-drion

Chloroplasts

1 μm

Peroxisomes: Oxidation

The **peroxisome** is a specialized metabolic compartment bounded by a single membrane **(Figure 6.19)**. Peroxisomes contain enzymes that remove hydrogen atoms from various substrates and transfer them to oxygen (O_2), producing hydrogen peroxide (H_2O_2) as a by-product (from which the organelle derives its name). These reactions have many different functions. Some peroxisomes use oxygen to break fatty acids down into smaller molecules that are transported to mitochondria and used as fuel for cellular respiration. Peroxisomes in the liver detoxify alcohol and other harmful compounds by transferring hydrogen from the poisonous compounds to oxygen. The H_2O_2 formed by peroxisomes is itself toxic, but the organelle also contains an enzyme that converts H_2O_2 to water. This is an excellent example of how the cell's compartmental structure is crucial to its functions: The enzymes that produce H_2O_2 and those that dispose of this toxic compound are sequestered away from other cellular components that could be damaged.

Specialized peroxisomes called *glyoxysomes* are found in the fat-storing tissues of plant seeds. These organelles contain enzymes that initiate the conversion of fatty acids to sugar, which the emerging seedling uses as a source of energy and carbon until it can produce its own sugar by photosynthesis.

Peroxisomes grow larger by incorporating proteins made in the cytosol and ER, as well as lipids made in the ER and within the peroxisome itself. But how peroxisomes increase in number and how they arose in evolution—as well as what organelles they are related to—are still open questions.

CONCEPT CHECK 6.5

1. Describe two characteristics shared by chloroplasts and mitochondria. Consider both function and membrane structure.

2. Do plant cells have mitochondria? Explain.

3. **WHAT IF?** A classmate proposes that mitochondria and chloroplasts should be classified in the endomembrane system. Argue against the proposal.

For suggested answers, see Appendix A.

CONCEPT 6.6

The cytoskeleton is a network of fibers that organizes structures and activities in the cell

In the early days of electron microscopy, biologists thought that the organelles of a eukaryotic cell floated freely in the cytosol. But improvements in both light microscopy and electron microscopy have revealed the **cytoskeleton**, a network of fibers extending throughout the cytoplasm **(Figure 6.20)**. Bacterial cells also have fibers that form a type of cytoskeleton, constructed of proteins similar to eukaryotic ones, but here we will concentrate on eukaryotes. The eukaryotic cytoskeleton plays a major role in organizing the structures and activities of the cell.

Roles of the Cytoskeleton: Support and Motility

The most obvious function of the cytoskeleton is to give mechanical support to the cell and maintain its shape. This is especially important for animal cells, which lack walls. The remarkable strength and resilience of the cytoskeleton as a whole are based on its architecture. Like a dome tent, the cytoskeleton is stabilized by a balance between opposing forces exerted by its elements. And just as the skeleton of an animal helps fix the positions of other body parts, the cytoskeleton provides anchorage for many organelles and even cytosolic enzyme molecules. The cytoskeleton is more dynamic than an animal skeleton, however. It can be quickly dismantled in one part of the cell and reassembled in a new location, changing the shape of the cell.

Some types of cell motility (movement) also involve the cytoskeleton. The term *cell motility* includes both changes in

▼ **Figure 6.20 The cytoskeleton.** As shown in this fluorescence micrograph, the cytoskeleton extends throughout the cell. The cytoskeletal elements have been tagged with different fluorescent molecules: green for microtubules and reddish-orange for microfilaments. A third component of the cytoskeleton, intermediate filaments, is not evident here. (The blue color tags the DNA in the nucleus.)

10 μm

→ **Mastering Biology** BioFlix® Animation: Cytoskeleton

cell location and movements of cell parts. Cell motility generally requires interaction of the cytoskeleton with **motor proteins**. There are many such examples: Cytoskeletal elements and motor proteins work together with plasma membrane molecules to allow whole cells to move along fibers outside the cell. Inside the cell, vesicles and other organelles often use motor protein "feet" to "walk" to their destinations along a track provided by the cytoskeleton. For example, this is how vesicles containing neurotransmitter molecules migrate to the tips of axons, the long extensions of nerve cells that release these molecules as chemical signals to adjacent nerve cells **(Figure 6.21)**. The cytoskeleton also manipulates the plasma membrane, bending it inward to form food vacuoles or other phagocytic vesicles.

> ⟶ **Mastering Biology**
> **Interview with George Langford:**
> **Investigating how motor proteins move cargo**

Components of the Cytoskeleton

Now let's look more closely at the three main types of fibers that make up the cytoskeleton: *Microtubules* are the thickest of the three types; *microfilaments* (also called actin filaments) are the thinnest; and *intermediate filaments* are fibers with diameters in a middle range **(Table 6.1)**.

▼ **Figure 6.21 Motor proteins and the cytoskeleton.**

(a) A motor protein that attaches to a receptor on a vesicle can "walk" the vesicle along a microtubule or microfilament.

(b) Two vesicles containing neurotransmitters move along a microtubule toward the tip of a nerve cell extension called an axon (SEM).

Table 6.1 The Structure and Function of the Cytoskeleton			
Property	**Microtubules (Tubulin Polymers)**	**Microfilaments (Actin Filaments)**	**Intermediate Filaments**
Structure	Hollow tubes	Two intertwined strands of actin	Fibrous proteins coiled into cables
Diameter	25 nm with 15-nm lumen	7 nm	8–12 nm
Protein subunits	Tubulin, a dimer consisting of an α-tubulin and a β-tubulin	Actin	One of several different proteins (including keratins)
Main functions	Maintenance of cell shape; cell motility; chromosome movements in cell division; organelle movements	Maintenance of cell shape; changes in cell shape; muscle contraction; cytoplasmic streaming (plant cells); cell motility; cell division (animal cells)	Maintenance of cell shape; anchorage of nucleus and certain other organelles; formation of nuclear lamina
Fluorescence micrographs of fibroblasts. Connective tissue cells called fibroblasts are a favorite cell type for cell biology studies because they spread out flat and their internal structures are easy to see. In each fibroblast shown here, the structure of interest has been tagged with fluorescent molecules. In the third micrograph, the DNA in the nucleus has also been tagged (orange).	10 μm — Microtubule — Row of tubulin dimers — 25 nm — α β Tubulin dimer	10 μm — Microfilament — Actin subunit — 7 nm	5 μm — Intermediate filament — Keratin proteins — Fibrous subunit (keratins coiled together) — 8–12 nm

VISUAL SKILLS *How many tubulin dimers are in the boxed row?*

Microtubules

All eukaryotic cells have **microtubules**, hollow rods constructed from globular proteins called tubulins. Each tubulin protein is a *dimer*, a molecule made up of two components. A tubulin dimer consists of two slightly different polypeptides, α-tubulin and β-tubulin. Microtubules grow in length by adding tubulin dimers; they can also be disassembled and their tubulins used to build microtubules elsewhere in the cell. Because of the orientation of tubulin dimers, the two ends of a microtubule are slightly different. One end can accumulate or release tubulin dimers at a much higher rate than the other, thus growing and shrinking significantly during cellular activities. (This is called the "plus end," not because it can only add tubulin proteins but because it's the end where both "on" and "off" rates are much higher.)

Microtubules shape and support the cell and also serve as tracks along which organelles equipped with motor proteins can move. In addition to the example in Figure 6.21, microtubules guide vesicles from the ER to the Golgi apparatus and from the Golgi to the plasma membrane. Microtubules are also involved in the separation of chromosomes during cell division, as shown in Figure 12.7.

Centrosomes and Centrioles   In animal cells, microtubules grow out from a **centrosome**, a region that is often located near the nucleus. These microtubules function as compression-resisting girders of the cytoskeleton. Within the centrosome is a pair of **centrioles**, each composed of nine sets of triplet microtubules arranged in a ring **(Figure 6.22)**. Although centrosomes with centrioles may help organize microtubule assembly in animal cells, many other eukaryotic cells lack centrosomes with centrioles and instead organize microtubules by other means.

Cilia and Flagella   Some eukaryotic cells have **flagella** (singular, *flagellum*) and **cilia** (singular, *cilium*), cellular extensions that contain microtubules. A specialized arrangement of the microtubules is responsible for the beating of these structures. (The bacterial flagellum, shown in Figure 6.5, has a completely different structure.) Many unicellular protists are propelled through water by cilia or flagella that act as locomotor appendages, and the sperm of animals, algae, and some plants have flagella. When cilia or flagella extend from cells that are attached tightly together in a sheet that is part of a tissue layer, they can move fluid over the surface of the tissue. For example, the ciliated lining of the trachea (windpipe) sweeps mucus containing trapped debris out of the lungs (see the EMs in Figure 6.3). In a woman's reproductive tract, the cilia lining the oviducts help move an egg toward the uterus.

Motile cilia usually occur in large numbers on the cell surface. Flagella are usually limited to just one or a few per cell, and they are longer than cilia. Flagella and cilia differ in

▼ **Figure 6.22 Centrosome containing a pair of centrioles.** Most animal cells have a centrosome, a region near the nucleus where the cell's microtubules are initiated. Within the centrosome is a pair of centrioles, each about 250 nm (0.25 μm) in diameter. The two centrioles are at right angles to each other, and each is made up of nine sets of three microtubules. The blue portions of the drawing represent nontubulin proteins that connect the microtubule triplets.

VISUAL SKILLS   *How many microtubules are in a centrosome? In the drawing, circle and label one microtubule and describe its structure. Circle and label a triplet.*

their beating patterns. A flagellum has an undulating motion like the tail of a fish. In contrast, cilia have alternating power and recovery strokes, much like the oars of a racing crew boat **(Figure 6.23)**.

A cilium may also act as a signal-receiving "antenna" for the cell. Cilia that have this function are generally nonmotile, and there is only one per cell. (In fact, in vertebrate animals, it appears that almost all cells have such a cilium, which is called a *primary cilium.*) Membrane proteins on this kind of cilium transmit molecular signals from the cell's environment to its interior, triggering signaling pathways that may lead to changes in the cell's activities. Cilium-based signaling appears to be crucial to brain function and to embryonic development.

▼ **Figure 6.23 A comparison of the beating of flagella and motile cilia.**

(a) Motion of flagella. A flagellum usually undulates, its snakelike motion driving a cell in the same direction as the axis of the flagellum. Propulsion of a human sperm cell is an example of flagellate locomotion (LM).

Direction of swimming

5 μm

(b) Motion of cilia. Cilia have a back-and-forth motion. The rapid power stroke moves the cell in a direction perpendicular to the axis of the cilium. Then, during the slower recovery stroke, the cilium bends and sweeps sideways, closer to the cell surface. A dense nap of cilia, beating at a rate of about 40 to 60 strokes a second, covers this *Colpidium*, a freshwater protist (colorized SEM).

Direction of organism's movement

Power stroke   Recovery stroke

15 μm

→ **Mastering Biology**
Videos: Flagellum Movement in Swimming Sperm
Flagellum Beating with ATP
***Paramecium* Cilia**

Though different in length, number per cell, and beating pattern, motile cilia and flagella share a common structure. Each motile cilium or flagellum has a group of microtubules sheathed in an extension of the plasma membrane **(Figure 6.24a)**. Nine doublets of microtubules are arranged in a ring with two single microtubules in its center **(Figure 6.24b)**. This arrangement, referred to as the "9 + 2" pattern, is found in nearly all eukaryotic flagella and motile cilia. (Nonmotile primary cilia have a "9 + 0" pattern, lacking the central pair of microtubules.) The microtubule assembly of a cilium or flagellum is anchored in the cell by a **basal body**, which is structurally very similar to a centriole, with microtubule triplets in a "9 + 0" pattern **(Figure 6.24c)**. In fact, in many animals (including humans), the basal body of the fertilizing sperm's flagellum enters the egg and becomes a centriole.

How does the microtubule assembly produce the bending movements of flagella and motile cilia? Bending involves large motor proteins called **dyneins** (red in the diagram in Figure 6.24) that are attached along each outer microtubule doublet. A typical dynein protein has two "feet" that "walk" along the microtubule of the adjacent doublet, using ATP for energy. One foot maintains contact, while the other releases

and reattaches one step farther along the microtubule (see Figure 6.21). The outer doublets and two central microtubules are held together by flexible cross-linking proteins (blue in the diagram in Figure 6.24), and the walking movement is coordinated so that it happens on one side of the circle at a time. If the doublets were not held in place, the walking action would make them slide past each other. Instead, the movements of the dynein feet cause the microtubules—and the organelle as a whole—to bend.

Microfilaments (Actin Filaments)

Microfilaments are thin solid rods. They are also called actin filaments because they are built from molecules of **actin**, a globular protein. A microfilament is a twisted double chain of actin subunits (see Table 6.1). Besides occurring as linear filaments, microfilaments can form structural networks when certain proteins bind along the side of such a filament and allow a new filament to extend as a branch. Like microtubules, microfilaments seem to be present in all eukaryotic cells.

In contrast to the compression-resisting role of microtubules, the structural role of microfilaments in the cytoskeleton is to bear tension (pulling forces). A three-dimensional

▼ **Figure 6.24 Structure of a flagellum or motile cilium.**

0.1 μm

(a) A longitudinal section of a motile cilium shows microtubules running the length of the membrane-sheathed structure (TEM).

Microtubules

Plasma membrane

Basal body

0.5 μm

Outer microtubule doublet

Motor proteins (dyneins)

Central microtubule

Radial spoke

Cross-linking protein between outer doublets

Plasma membrane

(b) A cross section through a motile cilium shows the "9 + 2" arrangement of microtubules (TEM). The outer microtubule doublets are held together with the two central microtubules by flexible cross-linking proteins (blue in art), including the radial spokes. The doublets also have attached motor proteins called dyneins (red in art).

0.1 μm

Triplet

(c) Basal body: The nine outer doublets of a cilium or flagellum extend into the basal body, where each doublet joins another microtubule to form a ring of nine triplets. Each triplet is connected to the next by nontubulin proteins (thinner blue lines in diagram). This is a "9 + 0" arrangement: The two central microtubules are not present because they terminate above the basal body (TEM).

Cross section of basal body

DRAW IT *In (a) and (b), circle and label the central pair of microtubules. In (a), show where they terminate, and explain why they aren't seen in the cross section of the basal body in (c).*

→ **Mastering Biology Animation: Cilia and Flagella**

network formed by microfilaments just to the inside of the plasma membrane (*cortical microfilaments*) helps support the cell's shape (see Figure 6.8). This network gives the outer cytoplasmic layer of a cell, called the **cortex**, the semisolid consistency of a gel, in contrast with the more fluid state of the interior cytoplasm. In some kinds of animal cells, such as nutrient-absorbing intestinal cells, bundles of microfilaments make up the core of microvilli, delicate projections that increase the cell's surface area **(Figure 6.25)**.

Microfilaments are well known for their role in cell motility. Thousands of actin filaments and thicker filaments made

Microvillus

Plasma membrane

Microfilaments (actin filaments)

Intermediate filaments

0.25 μm

▶ **Figure 6.25 A structural role of microfilaments.** The surface area of this nutrient-absorbing intestinal cell is increased by its many microvilli (singular, *microvillus*), cellular extensions reinforced by bundles of microfilaments. These actin filaments are anchored to a network of intermediate filaments (TEM).

of a protein called **myosin** interact to cause contraction of muscle cells **(Figure 6.26a)**; muscle contraction is described in detail in Concept 50.5. In the unicellular protist *Amoeba* and some of our white blood cells, localized contractions brought about by actin and myosin are involved in the amoeboid (crawling) movement of the cells. The cell crawls along a surface by extending cellular extensions called **pseudopodia** (from the Greek *pseudes*, false, and *pod*, foot) and moving toward them **(Figure 6.26b)**. In plant cells, actin-protein interactions contribute to **cytoplasmic streaming**, a circular flow of cytoplasm within cells **(Figure 6.26c)**. This movement, which is especially common in large plant cells, speeds the movement of organelles and the distribution of materials within the cell.

Intermediate Filaments

Intermediate filaments are named for their diameter, which is larger than the diameter of microfilaments but smaller than that of microtubules (see Table 6.1). While microtubules and microfilaments are found in all eukaryotic cells, intermediate filaments are only found in the cells of some animals, including vertebrates. Specialized for bearing tension (like microfilaments), intermediate filaments are a diverse class of cytoskeletal elements. Each type is constructed from a particular molecular subunit belonging to a family of proteins whose members include the keratins. Microtubules and microfilaments, in contrast, are consistent in diameter and composition in all eukaryotic cells.

Intermediate filaments are more permanent fixtures of cells than are microfilaments and microtubules, which are often disassembled and reassembled in various parts of a cell. Even after cells die, intermediate filament networks often persist; for example, the outer layer of our skin consists of dead skin cells full of keratin filaments. Intermediate filaments are especially sturdy and play an important role in reinforcing the shape of a cell and fixing the position of certain organelles. For instance, the nucleus typically sits within a cage made of intermediate filaments. Other intermediate filaments make up the nuclear lamina, which lines the interior of the nuclear envelope (see Figure 6.9). In general, the various kinds of intermediate filaments seem to function together as the permanent framework of the entire cell.

CONCEPT CHECK 6.6

1. Describe how cilia and flagella bend.
2. **WHAT IF?** Males afflicted with Kartagener's syndrome are sterile because of immotile sperm, and they tend to suffer from lung infections. This disorder has a genetic basis. Suggest what the underlying defect might be.

For suggested answers, see Appendix A.

▼ **Figure 6.26 Microfilaments and motility.** In these three examples, interactions between actin filaments and motor proteins bring about cell movement.

(a) Myosin motors in muscle cell contraction. The "walking" of myosin projections (the so-called heads) drives the parallel myosin and actin filaments past each other so that the actin filaments approach each other in the middle (red arrows). This shortens the muscle cell. Muscle contraction involves the shortening of many muscle cells at the same time (TEM).

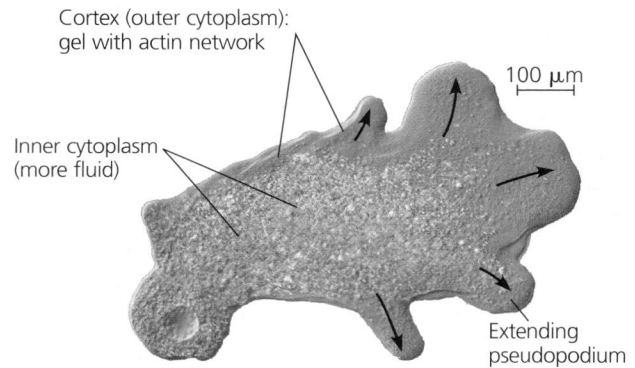

(b) Amoeboid movement. Interaction of actin filaments with myosin causes contraction of the cell, pulling the cell's trailing end (at left) forward (to the right) (LM).

(c) Cytoplasmic streaming in plant cells. A layer of cytoplasm cycles around the cell, moving over tracks of actin filaments. Myosin motors attached to some organelles drive the streaming by interacting with the actin (LM).

→ **Mastering Biology** BioFlix® Animation: Actin and Myosin in Muscle Contraction • Video: *Amoeba* Pseudopodia • Video: Cytoplasmic Streaming

Extracellular components and connections between cells help coordinate cellular activities

Having crisscrossed the cell to explore its inner components, we complete our tour of the cell by returning to the surface of this microscopic world, where there are additional structures with important functions. The plasma membrane is usually regarded as the boundary of the living cell, but most cells synthesize and secrete materials extracellularly (to the outside of the cell). Although these materials and the structures they form are outside the cell, their study is important to cell biology because they are involved in a great many essential cellular functions.

Cell Walls of Plants

The **cell wall** is an extracellular structure of plant cells **(Figure 6.27)**. This is one of the features that distinguishes plant cells from animal cells. The wall protects the plant cell, maintains its shape, and prevents excessive uptake of water. At the level of the whole plant, the strong walls of specialized cells hold the plant up against the force of gravity. Prokaryotes, some protists, and fungi also have cell walls, as you saw in Figures 6.5 and 6.8. Those organisms will be discussed in Chapters 27, 28, and 31.

Plant cell walls are much thicker than the plasma membrane, ranging from 0.1 μm to several micrometers. The exact chemical composition of the wall varies from species to species and even from one cell type to another in the same plant, but the basic design of the wall is consistent. Microfibrils made of the polysaccharide cellulose (see Figure 5.6) are synthesized by an enzyme called cellulose synthase and secreted to the extracellular space, where they become embedded in a matrix of other polysaccharides and proteins. This combination of materials, strong fibers in a "ground substance" (matrix), is the same basic architectural design found in steel-reinforced concrete and in fiberglass.

A young plant cell first secretes a relatively thin and flexible wall called the **primary cell wall** (see the micrograph in Figure 6.27). Between primary walls of adjacent cells is the **middle lamella**, a thin layer rich in sticky polysaccharides called pectins. The middle lamella glues adjacent cells together. (Pectin is used in cooking as a thickening agent in jams and jellies.) When the cell matures and stops growing, it strengthens its wall. Some plant cells do this simply by secreting hardening substances into the primary wall. Other cells add a **secondary cell wall** between the plasma membrane and the primary

▼ **Figure 6.27 Plant cell walls.** The drawing shows the relationship between primary and secondary cell walls in several mature plant cells. (Organelles aren't shown because many cells with secondary walls, such as the water-conducting cells, lack organelles.) The TEM shows the cell walls where two cells come together. The multilayered partition between plant cells consists of adjoining walls individually secreted by the cells. Adjacent cells are glued together by a very thin layer called the middle lamella.

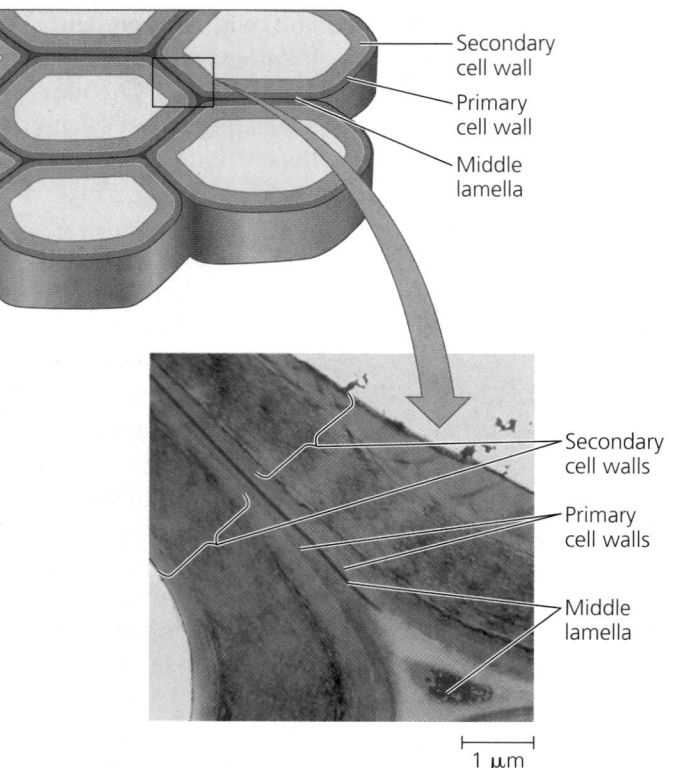

Secondary cell wall
Primary cell wall
Middle lamella

Secondary cell walls
Primary cell walls
Middle lamella

1 μm

wall. The secondary wall, often deposited in several laminated layers, has a strong and durable matrix that affords the cell protection and support. Wood, for example, consists mainly of secondary walls.

The Extracellular Matrix (ECM) of Animal Cells

Although animal cells lack walls akin to those of plant cells, they do have an elaborate **extracellular matrix (ECM)**. The main ingredients of the ECM are glycoproteins and other carbohydrate-containing molecules secreted by the cells. (Recall that glycoproteins are proteins with covalently bonded carbohydrates.) The most abundant glycoprotein in the ECM of most animal cells is **collagen**, which forms strong fibers outside the cells (see Figure 5.18). In fact, collagen accounts for about 40% of the total protein in the human body. The collagen fibers are embedded in a network woven out of **proteoglycans** secreted by

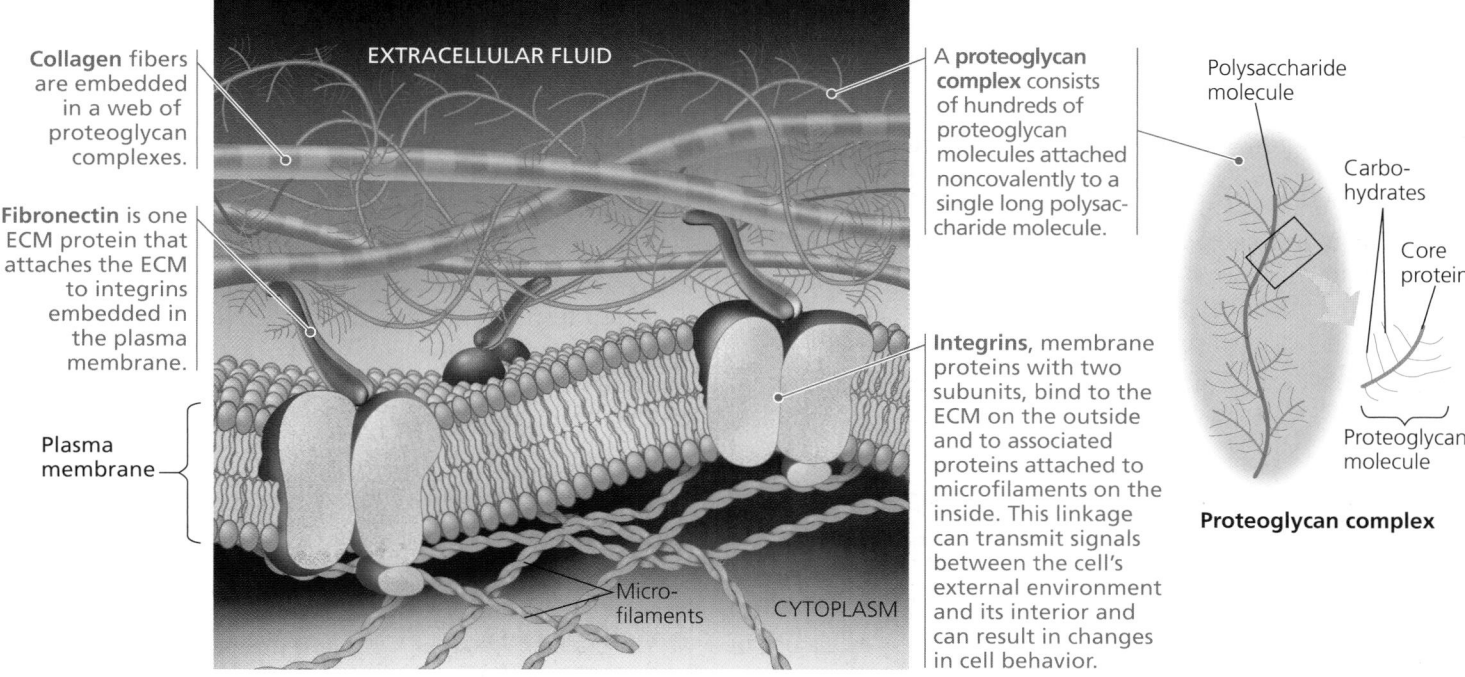

Collagen fibers are embedded in a web of proteoglycan complexes.

Fibronectin is one ECM protein that attaches the ECM to integrins embedded in the plasma membrane.

Plasma membrane

EXTRACELLULAR FLUID

A **proteoglycan complex** consists of hundreds of proteoglycan molecules attached noncovalently to a single long polysaccharide molecule.

Integrins, membrane proteins with two subunits, bind to the ECM on the outside and to associated proteins attached to microfilaments on the inside. This linkage can transmit signals between the cell's external environment and its interior and can result in changes in cell behavior.

Micro-filaments

CYTOPLASM

Polysaccharide molecule

Carbo-hydrates

Core protein

Proteoglycan molecule

Proteoglycan complex

▲ **Figure 6.28 Extracellular matrix (ECM) of an animal cell.** The molecular composition and structure of the ECM vary from one cell type to another. In this example, three different types of ECM molecules are present: collagen, fibronectin, and proteoglycans.

cells **(Figure 6.28)**. A proteoglycan molecule consists of a small core protein with many carbohydrate chains covalently attached, so that it may be up to 95% carbohydrate. Large proteoglycan complexes can form when hundreds of proteoglycan molecules become noncovalently attached to a single long polysaccharide molecule, as shown in Figure 6.28. Some cells are attached to the ECM by ECM glycoproteins such as **fibronectin**. Fibronectin and other ECM proteins bind to cell-surface receptor proteins called **integrins** that are built into the plasma membrane. Integrins span the membrane and bind on their cytoplasmic side to associated proteins attached to microfilaments of the cytoskeleton. The name *integrin* is based on the word *integrate*: Integrins are in a position to transmit signals between the ECM and the cytoskeleton and thus to integrate changes occurring outside and inside the cell.

Current research on fibronectin, other ECM molecules, and integrins reveals the influential role of the ECM in the lives of cells. By communicating with a cell through integrins, the ECM can regulate a cell's behavior. For example, some cells in a developing embryo migrate along specific pathways by matching the orientation of their microfilaments to the "grain" of fibers in the extracellular matrix. Researchers have also learned that the extracellular matrix around a cell can influence the activity of genes in the nucleus. Information about the ECM probably reaches the nucleus by a combination of mechanical and chemical signaling pathways. Mechanical signaling involves fibronectin, integrins, and microfilaments of the cytoskeleton. Changes in the cytoskeleton may in turn trigger signaling pathways inside the cell, leading to changes

in the set of proteins being made by the cell and therefore changes in the cell's function. In this way, the extracellular matrix of a particular tissue may help coordinate the behavior of all the cells of that tissue. Direct connections between cells also function in this coordination, as we'll explore next.

Cell Junctions

Cells in an animal or plant are organized into tissues, organs, and organ systems. Neighboring cells often adhere, interact, and communicate via sites of direct physical contact.

Plasmodesmata in Plant Cells

It might seem that the nonliving cell walls of plants would isolate plant cells from one another. But in fact, as shown in **Figure 6.29**, many plant cell walls are perforated with **plasmodesmata** (singular, *plasmodesma*; from the Greek *desma*, bond), channels that connect cells. The plasma

▼ **Figure 6.29 Plasmodesmata between plant cells.** The cytoplasm of one plant cell is continuous with the cytoplasm of its neighbors via plasmodesmata, cytoplasmic channels through the cell walls (TEM).

Cell walls

Interior of cell

Interior of adja-cent cell

0.5 μm

Plasmodesmata   Plasma membranes

membranes of adjacent cells line the channel of each plasmodesma and thus are continuous. Because the channels are filled with cytosol, the cells share the same internal chemical environment. By joining adjacent cells, plasmodesmata unify most of the plant into one living continuum. Water and small solutes can pass freely from cell to cell, and several experiments have shown that in some circumstances, certain proteins and RNA molecules can do this as well (see Concept 36.6). The macromolecules transported to neighboring cells appear to reach the plasmodesmata by moving along fibers of the cytoskeleton.

Tight Junctions, Desmosomes, and Gap Junctions in Animal Cells

In animals, there are three main types of cell junctions: *tight junctions, desmosomes,* and *gap junctions.* (Gap junctions are most like the plasmodesmata of plants, although gap junction pores are not lined with membrane—rather, they consist of proteins extending from each cell's membrane that form a connecting pore.) All three types of cell junctions are especially common in epithelial tissue, which lines the external and internal surfaces of the body. **Figure 6.30** uses epithelial cells of the intestinal lining to illustrate these junctions.

▼ Figure 6.30 Exploring Cell Junctions in Animal Tissues

Tight junctions prevent fluid from moving across a layer of cells.

Tight junction

Tight junction

Intermediate filaments

Desmosome

Gap junction

Ions or small molecules

Plasma membranes of adjacent cells

Space between cells

Extracellular matrix

Channels of a gap junction

TEM — 0.5 μm

TEM — 1 μm

TEM — 0.1 μm

Tight Junctions

At **tight junctions**, the plasma membranes of neighboring cells are very tightly pressed against each other, bound together by specific proteins. Forming continuous seals around the cells, tight junctions establish a barrier that prevents leakage of extracellular fluid across a layer of epithelial cells (see red dashed arrow). For example, tight junctions between skin cells make us watertight.

Desmosomes

Desmosomes (one type of *anchoring junction*) function like rivets, fastening cells together into strong sheets. Intermediate filaments made of sturdy keratin proteins anchor desmosomes in the cytoplasm. Desmosomes attach muscle cells to each other in a muscle. Some "muscle tears" involve the rupture of desmosomes.

Gap Junctions

Gap junctions (also called *communicating junctions*) provide cytoplasmic channels from one cell to an adjacent cell and in this way are similar in their function to the plasmodesmata in plants. Gap junctions consist of membrane proteins extending from the membranes of the two cells. These proteins create pores through which ions, sugars, amino acids, and other small molecules may pass. Gap junctions are necessary for communication between cells in many types of tissues, such as heart muscle, and in animal embryos.

→ **Mastering Biology** Animation: Cell Junctions

1. In what way are the cells of plants and animals structurally different from single-celled eukaryotes?

2. **WHAT IF?** If the plant cell wall or the animal extracellular matrix were impermeable, what effect would this have on cell function?

3. **MAKE CONNECTIONS** The polypeptide chain that makes up a tight junction weaves back and forth through the membrane four times, with two extracellular loops and one loop plus short C-terminal and N-terminal tails in the cytoplasm. Looking at Figure 5.14, what would you predict about the amino acid sequence of the tight junction protein?

For suggested answers, see Appendix A.

CONCEPT 6.8

A cell is greater than the sum of its parts

From our panoramic view of the cell's compartmental organization to our close-up inspection of each organelle's architecture, this tour of the cell has shown the correlation of structure with function. (See Figure 6.8 to review cell structure.)

→ **Mastering Biology** BioFlix® Animations: Tour of an Animal Cell • Tour of a Plant Cell

Remember that none of a cell's components work alone. For example, consider the microscopic scene in **Figure 6.31**. The large cell is a macrophage (see Figure 6.13a). It helps defend the mammalian body against infections by ingesting bacteria (the smaller cells) into phagocytic vesicles. The macrophage crawls along a surface and reaches out to the bacteria with thin pseudopodia (specifically, filopodia). Actin filaments interact with other elements of the cytoskeleton in these movements. After the macrophage engulfs the bacteria, they are destroyed by lysosomes produced by the elaborate endomembrane system. The digestive enzymes of the lysosomes and the proteins of the cytoskeleton are all made by ribosomes. And the synthesis of these proteins is programmed by genetic messages dispatched from the DNA in the nucleus. All these processes require energy, which mitochondria supply in the form of ATP.

Cellular functions arise from cellular order: The cell is a living unit greater than the sum of its parts. The cell in Figure 6.31 is a good example of integration of cellular processes, seen from the outside. But what about the internal organization of a cell? As you proceed in your study of biology to consider different cellular processes, it will be helpful to try to visualize the architecture and furnishings inside a cell. **Figure 6.32** is designed to introduce you to some important biological molecules and molecules and to help you get a sense of their relative sizes and organization in the context of cellular structures and organelles. As you study this figure, see if you can shrink yourself down to the size of a protein and contemplate your surroundings.

1. *Colpidium colpoda* is a unicellular protist that lives in freshwater, eats bacteria, and moves by cilia (see Figure 6.23b). Describe how the parts of this cell work together in the functioning of *C. colpoda*, including as many organelles and other cell structures as you can.

For suggested answers, see Appendix A.

▼ **Figure 6.31 Coordination of activities in a cell.** The ability of this macrophage to recognize, apprehend, and destroy bacteria involves coordination among components such as the cytoskeleton, lysosomes, and plasma membrane (colorized SEM).

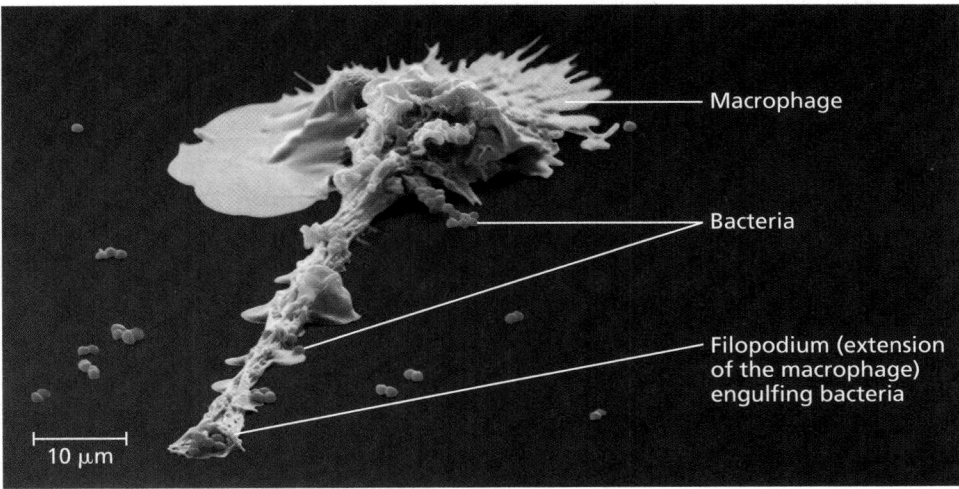

→ **Mastering Biology** Animation: Review of Animal Cell Structure and Function

A slice of a plant cell's interior is illustrated in the center panel, with all structures and molecules drawn to scale. Selected molecules and structures are shown above and below, all enlarged by the same factor so you can compare their sizes. All protein and nucleic acid structures are based on data from the Protein Data Bank; regions whose structure has not yet been determined are shown in gray.

This figure introduces a cast of characters that you will learn more about as you study biology. Refer back to this figure as you encounter these molecules in your studies.

Plant cell

Proton pump Calcium channel Aquaporin Receptor Outside cell

Plasma membrane

Enzyme in signaling pathway

Transport Signaling

(a) Membrane proteins (Chapter 7) Proteins embedded in cellular membranes help transport substances and conduct signals across membranes. They also participate in other crucial cellular functions. Many proteins are able to move within the membrane.

Plasma membrane

Cell wall

Mitochondrion

Chloroplast

Phosphofructokinase

Inner mitochondrial membrane Complex III Complex IV

Cyt c ATP synthase

Cyto-plasm

Q

Hexokinase

Complex II Matrix

Complex I Isocitrate dehydrogenase

(b) Cellular respiration (Chapter 9) Cellular respiration, a multi-step process, generates ATP from food molecules. The first two stages are carried out by enzymes in the cytoplasm and mitochondrial matrix; a few of these enzymes (pinkish-purple) are shown. The final stage is carried out by proteins (bluish-purple) that form a "chain" in the inner mitochondrial membrane.

Rubisco Stroma ATP synthase

NADP+ reductase

Cytochrome complex

Photosystem II Photosystem I Fd

Pq Pc Thylakoid membrane

Thylakoid

(c) Photosynthesis (Chapter 10) Photosynthesis produces sugars that provide food for all life on the planet. The process begins with large complexes of proteins and chlorophyll (green) embedded in the thylakoid membranes. These complexes trap light energy in molecules that are used by rubisco and other proteins in the stroma to make sugars.

(d) Transcription (Chapter 17) In the nucleus, the information contained in a DNA sequence is transferred to messenger RNA (mRNA) by an enzyme called RNA polymerase. After their synthesis, mRNA molecules leave the nucleus via nuclear pores.

mRNA

RNA polymerase

DNA

Nucleosome: DNA wrapped around 8 histone proteins

DNA

(e) Nuclear pore (Concept 6.3) The nuclear pore complex regulates molecular traffic in and out of the nucleus, which is bounded by a double membrane. Among the largest structures that pass through the pore are the ribosomal subunits, which are built in the nucleus.

(f) Translation (Chapter 17) In the cytoplasm, the information in mRNA is used to assemble a polypeptide with a specific sequence of amino acids. Both transfer RNA (tRNA) molecules and ribosomes play a role. The eukaryotic ribosome, which includes a large subunit and a small subunit, is a colossal complex composed of four large ribosomal RNA (rRNA) molecules and more than 80 proteins. Through transcription and translation, the nucleotide sequence of DNA in a gene determines the amino acid sequence of a polypeptide, via the intermediary mRNA.

Polypeptide made of amino acids

Ribosome

tRNA

Large subunit

Small subunit

mRNA

DNA

Nucleus

Nuclear pore

Cytoskeleton

Nuclear envelope

Motor protein

Ribosomes

Endoplasmic reticulum

Scale within cell

25 nm

Tubulin protein subunit (α + β dimer)

Microtubule

Microfilament

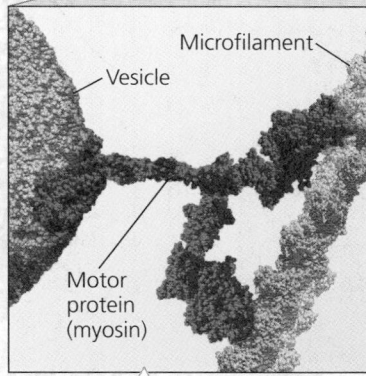

Microfilament

Vesicle

Motor protein (myosin)

25 nm
Scale of enlarged structures

(g) Cytoskeleton (Concept 6.6)
Cytoskeletal structures are polymers of protein subunits. Microtubules are hollow structural rods made of tubulin protein subunits, while microfilaments are cables that have two chains of actin proteins wound around each other.

(h) Motor proteins (Concept 6.6) Motor proteins, such as myosin, are responsible for transport of vesicles and movement of organelles within the cell.

1. List the following structures from largest to smallest: proton pump, nuclear pore, cyt c, ribosome.

2. Considering the structures of a nucleosome and of RNA polymerase, speculate about what must happen before RNA polymerase can transcribe the DNA that is wrapped around the histone proteins of a nucleosome.

3. Find another myosin motor protein walking on a microfilament in this figure. What organelle is being moved by that myosin protein?

Instructors: Additional questions related to this Visualizing Figure can be assigned in **Mastering Biology**.

6 Chapter Review

→ Go to **Mastering Biology** for Assignments, the eText, the Study Area, and Dynamic Study Modules.

SUMMARY OF KEY CONCEPTS

→ To review key terms, go to the **Vocabulary Self-Quiz** in the **Mastering Biology** eText or Study Area, or go to goo.gl/zkjz9t.

CONCEPT 6.1

Biologists use microscopes and biochemistry to study cells (pp. 94–97)

- Improvements in microscopy that affect the parameters of magnification, resolution, and contrast have catalyzed progress in the study of cell structure. **Light microscopy** (LM) and **electron microscopy** (EM), as well as other types, remain important tools.
- Cell biologists can obtain pellets enriched in particular cellular components by centrifuging disrupted cells at sequential speeds, a process known as **cell fractionation**.

? *How do microscopy and biochemistry complement each other to reveal cell structure and function?*

CONCEPT 6.2

Eukaryotic cells have internal membranes that compartmentalize their functions (pp. 97–102)

- All cells are bounded by a **plasma membrane**.
- **Prokaryotic cells** lack nuclei and other membrane-enclosed **organelles**, while **eukaryotic cells** have internal membranes that compartmentalize cellular functions.
- The surface-to-volume ratio is an important parameter affecting cell size and shape.
- Plant and animal cells have most of the same organelles: a nucleus, endoplasmic reticulum, Golgi apparatus, and mitochondria. Chloroplasts are present only in cells of photosynthetic eukaryotes.

? *Explain how the compartmental organization of a eukaryotic cell contributes to its biochemical functioning.*

	Cell Component	Structure	Function
CONCEPT 6.3 **The eukaryotic cell's genetic instructions are housed in the nucleus and carried out by the ribosomes** (pp. 102–104) ? *Describe the relationship between the nucleus and ribosomes.*	**Nucleus** / ER	Surrounded by nuclear envelope (double membrane) perforated by nuclear pores; nuclear envelope is continuous with endoplasmic reticulum (ER)	Houses chromosomes, which are made of chromatin (DNA and proteins); contains nucleoli, where ribosomal subunits are made; pores regulate entry and exit of materials
	Ribosome	Two subunits made of ribosomal RNAs and proteins; can be free in cytosol or bound to ER	Protein synthesis
CONCEPT 6.4 **The endomembrane system regulates protein traffic and performs metabolic functions** (pp. 104–109) ? *Describe the key role played by transport vesicles in the endomembrane system.*	**Endoplasmic reticulum (ER)** / Nuclear envelope	Extensive network of membrane-bounded tubules and sacs; ER membrane separates lumen from cytosol and is continuous with nuclear envelope	**Smooth ER:** synthesis of lipids, metabolism of carbohydrates, storage of calcium ions, detoxification of drugs and poisons **Rough ER:** aids in synthesis of secretory and other proteins on bound ribosomes; adds carbohydrates to proteins to make glycoproteins; produces new membrane
	Golgi apparatus	Stacks of flattened membranous sacs; has polarity (*cis* and *trans* faces)	Modification of proteins, carbohydrates on proteins, and phospholipids; synthesis of many polysaccharides; sorting of Golgi products, which are then released in vesicles
	Lysosome	Membranous sac of hydrolytic enzymes (in animal cells)	Breakdown of ingested substances, cell macromolecules, and damaged organelles for recycling
	Vacuole	Large membrane-bounded vesicle	Digestion, storage, waste disposal, water balance, cell growth, and protection

	Cell Component	Structure	Function
CONCEPT 6.5 **Mitochondria and chloroplasts change energy from one form to another** *(pp. 109–112)* ? *What does the endosymbiont theory propose as the origin for mitochondria and chloroplasts? Explain.*	Mitochondrion	Bounded by double membrane; inner membrane has infoldings	Cellular respiration
	Chloroplast	Typically two membranes around fluid stroma, which contains thylakoids stacked into grana	Photosynthesis (chloroplasts are present in cells of photosynthetic eukaryotes, including plants)
	Peroxisome	Specialized metabolic compartment bounded by a single membrane	Contains enzymes that transfer H atoms from substrates to oxygen, producing H_2O_2 (hydrogen peroxide), which is converted to H_2O

CONCEPT 6.6

The cytoskeleton is a network of fibers that organizes structures and activities in the cell *(pp. 112–117)*

- The **cytoskeleton** functions in structural support for the cell and in motility and signal transmission.
- **Microtubules** shape the cell, guide organelle movement, and separate chromosomes in dividing cells. **Cilia** and **flagella** are motile appendages containing microtubules. Primary cilia also play sensory and signaling roles. **Microfilaments** are thin rods that function in muscle contraction, amoeboid movement, **cytoplasmic streaming**, and support of microvilli. **Intermediate filaments** support cell shape and fix organelles in place.

? *Describe the role of motor proteins inside the eukaryotic cell and in whole-cell movement.*

CONCEPT 6.7

Extracellular components and connections between cells help coordinate cellular activities *(pp. 118–121)*

- Plant **cell walls** are made of cellulose fibers embedded in other polysaccharides and proteins.
- Animal cells secrete glycoproteins and **proteoglycans** that form the **extracellular matrix (ECM)**, which functions in support, adhesion, movement, and regulation.
- Cell junctions connect neighboring cells. Plants have **plasmodesmata** that pass through adjoining cell walls. Animal cells have **tight junctions, desmosomes**, and **gap junctions**.

? *Compare the structure and functions of a plant cell wall and the extracellular matrix of an animal cell.*

CONCEPT 6.8

A cell is greater than the sum of its parts *(pp. 121–123)*

Many components work together in a functioning cell.

? *When a cell ingests a bacterium, what role does the nucleus play?*

TEST YOUR UNDERSTANDING

→ For more multiple-choice questions, go to the **Practice Test** in the **Mastering Biology** eText or Study Area, or go to goo.gl/GruWRg.

Levels 1-2: Remembering/Understanding

1. Which structure is part of the endomembrane system?
(A) mitochondrion
(B) Golgi apparatus
(C) chloroplast
(D) centrosome

2. Which structure is common to plant *and* animal cells?
(A) chloroplast
(B) central vacuole
(C) mitochondrion
(D) centriole

3. Which of the following is present in a prokaryotic cell?
(A) mitochondrion
(B) ribosome
(C) nuclear envelope
(D) chloroplast

Levels 3-4: Applying/Analyzing

4. Cyanide binds to at least one molecule involved in producing ATP. In a cell exposed to cyanide, most of the cyanide will be in
(A) mitochondria.
(B) ribosomes.
(C) peroxisomes.
(D) lysosomes.

5. Which cell would be best for studying lysosomes?
(A) muscle cell
(B) nerve cell
(C) bacterial cell
(D) phagocytic white blood cell

6. **DRAW IT** Draw two eukaryotic cells. Label the structures listed here and show any physical connections between the structures of each cell: nucleus, rough ER, smooth ER, mitochondrion, centrosome, chloroplast, vacuole, lysosome, microtubule, cell wall, ECM, microfilament, Golgi apparatus, intermediate filament, plasma membrane, peroxisome, ribosome, nucleolus, nuclear pore, vesicle, flagellum, microvilli, plasmodesma.

Levels 5-6: Evaluating/Creating

7. **EVOLUTION CONNECTION** (a) What cell structures best reveal evolutionary unity? (b) Give an example of diversity related to specialized cellular modifications.

8. **SCIENTIFIC INQUIRY** Imagine protein X, destined to span the plasma membrane; assume the mRNA carrying the genetic message for protein X has been translated by ribosomes in a cell culture. If you fractionate the cells (see Figure 6.4), in which fraction would you find protein X? Explain by describing its transit.

9. **WRITE ABOUT A THEME: ORGANIZATION** Write a short essay (100–150 words) on this topic: Life is an emergent property that appears at the level of the cell. (See Concept 1.1.)

10. **SYNTHESIZE YOUR KNOWLEDGE**

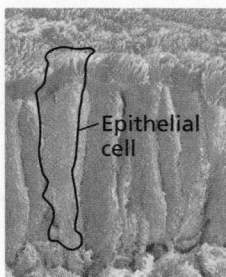

Epithelial cell

The cells in this SEM are epithelial cells from the small intestine. Discuss how their cellular structure contributes to their specialized functions of nutrient absorption and as a barrier between the intestinal contents and the blood supply on the other side of the sheet of epithelial cells.

For selected answers, see Appendix A.

Explore Scientific Papers with **Science in the Classroom** ⚛AAAS
How do actin networks affect myosin motors?
Go to "Fussy about Filaments" at www.scienceintheclassroom.org.
→ **Instructors:** Questions can be assigned in Mastering Biology.

7 Membrane Structure and Function

KEY CONCEPTS

7.1 Cellular membranes are fluid mosaics of lipids and proteins *p. 127*

7.2 Membrane structure results in selective permeability *p. 131*

7.3 Passive transport is diffusion of a substance across a membrane with no energy investment *p. 132*

7.4 Active transport uses energy to move solutes against their gradients *p. 136*

7.5 Bulk transport across the plasma membrane occurs by exocytosis and endocytosis *p. 139*

Study Tip

Make a visual study guide: Draw a plasma membrane (two lines) all the way down a piece of paper (or digitally). Label the cytoplasm and the extracellular fluid. At the top, draw the phospholipid bilayer in detail. As you read the chapter, draw and label membrane proteins that you encounter and diagram the different ways materials can enter or leave a cell.

Cytoplasm | Plasma membrane | Extracellular fluid

Phospholipid bilayer (draw in detail)

Glycoprotein (carbohydrate + protein)

→ Go to Mastering Biology

For Students (in eText and Study Area)
- Get Ready for Chapter 7
- Figure 7.12 Walkthrough: Osmosis
- BioFlix® Animation: Membrane Transport

For Instructors (in Item Library)
- Tutorial: Membrane Transport: Diffusion and Passive Transport
- Tutorial: Membrane Transport: Bulk Transport

Figure 7.1 Successful learning relies on communication between brain cells. Here, the vesicles fusing with the plasma membrane of the top cell release molecules (yellow) that bind to membrane proteins (light green) on the surface of the bottom cell, triggering the proteins to change shape. The plasma membrane that surrounds each cell regulates its exchanges with its environment and surrounding cells.

How does the plasma membrane regulate inbound and outbound traffic?

Plasma membrane

Vesicle

Passive transport of small molecules doesn't require energy; it may involve transport proteins.

Transport protein

Exocytosis: Large molecules are secreted when a vesicle fuses with the plasma membrane.

Bulk transport moves large molecules.

Endocytosis: Large molecules are taken in when the plasma membrane pinches inward, forming a vesicle.

ADP · ℗ · ATP

Active transport of small molecules requires energy and a transport protein.

Cellular membranes are fluid mosaics of lipids and proteins

Lipids and proteins are the staple ingredients of membranes, although carbohydrates are also important. The most abundant lipids in most membranes are phospholipids. Their ability to form membranes is inherent in their molecular structure. A phospholipid is an **amphipathic** molecule, meaning it has both a hydrophilic ("water-loving") region and a hydrophobic ("water-fearing") region (see Figure 5.11). A phospholipid bilayer can exist as a stable boundary between two aqueous compartments because the molecular arrangement shelters the hydrophobic tails of the phospholipids from water while exposing the hydrophilic heads to water **(Figure 7.2)**.

Like membrane lipids, most membrane proteins are amphipathic. Such proteins can reside in the phospholipid bilayer with their hydrophilic regions protruding. This molecular orientation maximizes contact of hydrophilic regions of a protein with water in the cytosol and extracellular fluid, while providing their hydrophobic parts with a nonaqueous environment.

▼ **Figure 7.2 Phospholipid bilayer (cross section).**

Two phospholipids

Hydrophilic head

Hydrophobic tail

WATER

WATER

VISUAL SKILLS *Consulting Figure 5.11, circle and label the hydrophilic and hydrophobic portions of the enlarged phospholipids on the right. Explain what each portion contacts when the phospholipids are in the plasma membrane.*

Figure 7.3 shows the currently accepted model of the arrangement of molecules in the plasma membrane. In this **fluid mosaic model**, the membrane is a mosaic of protein molecules bobbing in a fluid bilayer of phospholipids.

Fibers of extra-cellular matrix (ECM)

Glyco-protein

Carbohydrates

Glycolipid

EXTRACELLULAR SIDE OF MEMBRANE

Phospholipid   Cholesterol

Microfilaments of cytoskeleton

Peripheral proteins

Integral protein

CYTOPLASMIC SIDE OF MEMBRANE

▲ **Figure 7.3 Current model of an animal cell's plasma membrane (cutaway view). Lipids are colored gray and gold, proteins purple, and carbohydrates green.**

→ **Mastering Biology**
Animation: Structure of the Plasma Membrane

The proteins are not randomly distributed in the membrane, however. Groups of proteins are often associated in long-lasting, specialized patches, where they carry out common functions. Researchers have found specific lipids in these patches as well and have proposed naming them *lipid rafts*; however, the debate continues about whether such structures exist in living cells or are an artifact of biochemical techniques. In some regions, the membrane may be much more tightly packed with proteins than shown in Figure 7.3. Like all models, the fluid mosaic model is continually being refined as new research reveals more about membrane structure.

The Fluidity of Membranes

Membranes are not static sheets of molecules locked rigidly in place. A membrane is held together mainly by hydrophobic interactions, which are much weaker than covalent bonds (see Figure 5.18). Most of the lipids and some proteins can shift about sideways—that is, in the plane of the membrane, like partygoers elbowing their way through a crowded room. Very rarely, also, a lipid may flip-flop across the membrane, switching from one phospholipid layer to the other.

The sideways movement of phospholipids within the membrane is rapid. Adjacent phospholipids switch positions about 10^7 times per second, which means that a phospholipid can travel about 2 μm—the length of a typical bacterial cell—in 1 second. Proteins are much larger than lipids and move more slowly, when they do move. Many membrane proteins seem to be held immobile by their attachment to the cytoskeleton or to the extracellular matrix (see Figure 7.3).

Some membrane proteins seem to move in a highly directed manner, perhaps driven along cytoskeletal fibers by motor proteins. And other proteins simply drift in the membrane, as shown the classic experiment described in **Figure 7.4**.

A membrane remains fluid as temperature decreases until the phospholipids settle into a closely packed arrangement and the membrane solidifies, much as bacon grease forms lard when it cools. The temperature at which a membrane solidifies depends on the types of lipids it is made of. As the temperature decreases, the membrane remains fluid to a lower temperature if it is rich in phospholipids with unsaturated hydrocarbon tails (see Figures 5.10 and 5.11). Because of kinks in the tails where double bonds are located, unsaturated hydrocarbon tails cannot pack together as closely as saturated hydrocarbon tails, making the membrane more fluid **(Figure 7.5a)**.

The steroid cholesterol, which is wedged between phospholipid molecules in the plasma membranes of animal cells, has different effects on membrane fluidity at different temperatures **(Figure 7.5b)**. At relatively high temperatures—at 37°C, the body temperature of humans, for example—cholesterol makes the membrane less fluid by restraining phospholipid movement. However, because cholesterol also hinders the close packing of phospholipids, it lowers the temperature required for the membrane to solidify. Thus, cholesterol can be thought of as a "fluidity buffer" for the membrane, resisting changes in membrane fluidity that can be caused by changes in temperature. Compared to animals, plants have very low levels of cholesterol; rather, related steroid lipids buffer membrane fluidity in plant cells.

▼ **Figure 7.4** Inquiry

Do membrane proteins move?

Experiment Larry Frye and Michael Edidin, at Johns Hopkins University, labeled the plasma membrane proteins of a mouse cell and a human cell with two different markers and fused the cells. Using a microscope, they observed the markers on the hybrid cell.

Results

Membrane proteins

Mouse cell   +   Human cell   →   Hybrid cell   →   Mixed proteins after 1 hour

Conclusion The mixing of the mouse and human membrane proteins indicates that at least some membrane proteins move sideways within the plane of the plasma membrane.

Data from L. D. Frye and M. Edidin, The rapid intermixing of cell surface antigens after formation of mouse-human heterokaryons, *Journal of Cell Science* 7:319 (1970).

WHAT IF? *Suppose the proteins did not mix in the hybrid cell, even many hours after fusion. Could you conclude that proteins don't move within the membrane? What other explanation could there be?*

▼ **Figure 7.5 Factors that affect membrane fluidity.**

(a) Unsaturated versus saturated hydrocarbon tails.

Fluid | Viscous

Unsaturated hydrocarbon tails (kinked) prevent packing, enhancing membrane fluidity. | Saturated hydrocarbon tails pack together, increasing membrane viscosity.

(b) Cholesterol within the animal cell membrane.

Cholesterol

Cholesterol reduces membrane fluidity at moderate temperatures by reducing phospholipid movement, but at low temperatures it hinders solidification by disrupting the regular packing of phospholipids.

Membranes must be fluid to work properly; the fluidity of a membrane affects both its permeability and the ability of membrane proteins to move to where their function is needed. Usually, membranes are about as fluid as olive oil. When a membrane solidifies, its permeability changes, and enzymatic proteins in the membrane may become inactive if their activity requires movement within the membrane. However, membranes that are too fluid cannot support protein function either. Therefore, extreme environments (for example, those with extreme temperatures) pose a challenge for life, resulting in evolutionary adaptations that include differences in membrane lipid composition.

Evolution of Differences in Membrane Lipid Composition

EVOLUTION Variations in the cell membrane lipid compositions of many species appear to be evolutionary adaptations that maintain the appropriate membrane fluidity under specific environmental conditions. For instance, fishes that live in extreme cold have membranes with a high proportion of unsaturated hydrocarbon tails, enabling their membranes to remain fluid in spite of the low temperature (see Figure 7.5a). At the other extreme, some bacteria and archaea thrive at temperatures greater than 90°C (194°F) in thermal hot springs and geysers. Their membranes include unusual lipids that may prevent excessive fluidity at such high temperatures.

The ability to change the lipid composition of cell membranes in response to changing temperatures has evolved in organisms that live where temperatures vary. In many plants that tolerate extreme cold, such as winter wheat, the percentage of unsaturated phospholipids increases in autumn, an adjustment that keeps the membranes from solidifying during winter. Some bacteria and archaea also exhibit different proportions of unsaturated phospholipids in their cell membranes, depending on the temperature at which they are growing. Overall, natural selection has apparently favored organisms whose mix of membrane lipids ensures an appropriate level of membrane fluidity for their environment.

Membrane Proteins and Their Functions

Now we come to the *mosaic* aspect of the fluid mosaic model. Somewhat like a tile mosaic (shown here), a membrane is a collage of different proteins, often clustered together in groups, embedded in the fluid matrix of the lipid bilayer (see Figure 7.3). More than 50 kinds of proteins have been found in the plasma membrane of red blood cells, for example. Phospholipids form the main fabric of the membrane, but proteins determine most of the membrane's

◀ **Tile mosaic.**

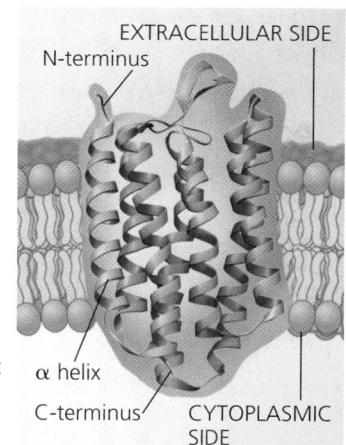

▶ **Figure 7.6 The structure of a transmembrane protein.** Bacteriorhodopsin (a bacterial transport protein) has a distinct orientation in the membrane, with its N-terminus outside the cell and its C-terminus inside. This ribbon model highlights the secondary structure of the hydrophobic parts, including seven transmembrane α helices, which lie mostly within the hydrophobic interior of the membrane. The nonhelical hydrophilic segments are in contact with the aqueous solutions on the extracellular and cytoplasmic sides of the membrane.

EXTRACELLULAR SIDE
N-terminus
α helix
C-terminus
CYTOPLASMIC SIDE

functions. Different types of cells contain different sets of membrane proteins, and the various membranes within a cell each have a unique collection of proteins.

Notice in Figure 7.3 that there are two major populations of membrane proteins: integral proteins and peripheral proteins. **Integral proteins** penetrate the hydrophobic interior of the lipid bilayer. The majority are **transmembrane proteins**, which span the membrane; other integral proteins extend only partway into the hydrophobic interior. The hydrophobic regions of an integral protein consist of one or more stretches of nonpolar amino acids (see Figure 5.14), typically 20–30 amino acids in length, usually coiled into α helices **(Figure 7.6)**. The hydrophilic parts of the molecule are exposed to the aqueous solutions on either side of the membrane. Some proteins also have one or more hydrophilic channels that allow passage through the membrane of hydrophilic substances (even of water itself). **Peripheral proteins** are not embedded in the lipid bilayer at all; they are loosely bound to the surface of the membrane, often to exposed parts of integral proteins (see Figure 7.3).

On the cytoplasmic side of the plasma membrane, some membrane proteins are held in place by attachment to the cytoskeleton. And on the extracellular side, certain membrane proteins may attach to materials outside the cell. For example, in animal cells, membrane proteins may be attached to fibers of the extracellular matrix (see Figure 6.28; *integrins* are one type of integral, transmembrane protein). These attachments combine to give animal cells a stronger framework than the plasma membrane alone could provide.

Figure 7.7 illustrates six major functions performed by proteins of the plasma membrane. A single cell may have different membrane proteins that carry out various functions, and one protein may itself carry out multiple functions. Thus, the membrane is a functional mosaic as well as a structural one.

⊖ **Mastering Biology** Animation: Functions of the Plasma Membrane

Proteins on a cell's surface are important in the medical field. For example, a protein called CD4 on the surface

▼ Figure 7.7 Some functions of membrane proteins. In many cases, a single protein performs multiple tasks.

(a) Transport. *Left:* A protein that spans the membrane may provide a hydrophilic channel across the membrane that is selective for a particular solute (see Figures 6.32a and 7.15a). *Right:* Other transport proteins shuttle a substance from one side to the other by changing shape (see Figure 7.15b). Some of these proteins hydrolyze ATP as an energy source to actively pump substances across the membrane.

(b) Enzymatic activity. A protein built into the membrane may be an enzyme with its active site (where the reactant binds) exposed to substances in the adjacent solution. In some cases, several enzymes in a membrane are organized as a team that carries out sequential steps of a metabolic pathway.

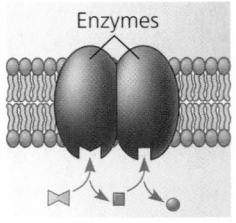

Enzymes

(c) Signal transduction. A membrane protein (receptor) may have a binding site with a specific shape that fits the shape of a chemical messenger, such as a hormone. The external messenger (signaling molecule) may cause the protein to change shape, allowing it to relay the message to the inside of the cell, usually by binding to a cytoplasmic protein (see Figures 6.32a and 11.6).

Signaling molecule

Receptor

Signal transduction

(d) Cell-cell recognition. Some glycoproteins serve as identification tags that are specifically recognized by membrane proteins of other cells. This type of cell-cell binding is usually short-lived compared with that shown in (e).

Glycoprotein

(e) Intercellular joining. Membrane proteins of adjacent cells may hook together in various kinds of junctions, such as gap junctions or tight junctions (see Figure 6.30). This type of binding is more long-lasting than that shown in (d).

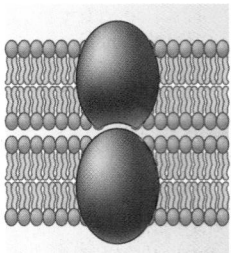

(f) Attachment to the cytoskeleton and extracellular matrix (ECM). Microfilaments or other elements of the cytoskeleton may be noncovalently bound to membrane proteins, a function that helps maintain cell shape and stabilizes the location of certain membrane proteins. Proteins that can bind to ECM molecules can coordinate extracellular and intracellular changes (see Figure 6.28).

VISUAL SKILLS *Some transmembrane proteins can bind to a particular ECM molecule and, when bound, transmit a signal into the cell. Use the proteins shown in (c) and (f) to explain how this might occur.*

of immune cells helps the human immunodeficiency virus (HIV) infect these cells, leading to acquired immune deficiency syndrome (AIDS). Despite multiple exposures to HIV, however, a small number of people do not develop AIDS and show no evidence of HIV-infected cells. Comparing their genes with the genes of infected individuals, researchers learned that resistant people have an unusual form of a gene that codes for an immune cell-surface protein called CCR5. Further work showed that although CD4 is the main HIV receptor, HIV must also bind to CCR5 as a "co-receptor" to infect most cells **(Figure 7.8a)**. An absence of CCR5 on the cells of resistant individuals, due to the gene alteration, prevents the virus from entering the cells **(Figure 7.8b)**.

This information has been key to developing a treatment for HIV infection. Interfering with CD4 causes dangerous side effects because of its many important functions in cells. Discovery of the CCR5 co-receptor provided a safer target for development of drugs that mask this protein and block HIV entry. One such drug, maraviroc (brand name Selzentry), was approved for treatment of HIV in 2007; ongoing trials to determine its ability to prevent HIV infection in uninfected, at-risk patients have been disappointing.

The Role of Membrane Carbohydrates in Cell-Cell Recognition

Cell-cell recognition, a cell's ability to distinguish one type of neighboring cell from another, is crucial to the functioning of an organism. It is important, for example, in the sorting of cells into tissues and organs in an animal embryo. It is also the basis for the rejection of foreign cells by the immune system, an important line of defense in vertebrate animals (see Concept 43.1). Cells recognize other cells by binding to molecules, often containing carbohydrates, on the extracellular surface of the plasma membrane (see Figure 7.7d).

▼ Figure 7.8 The genetic basis for HIV resistance.

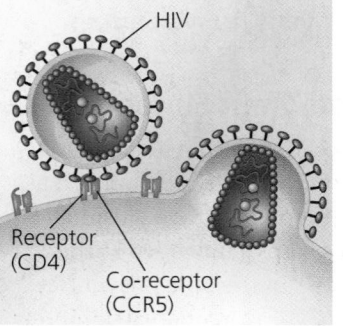

HIV

Receptor (CD4)

Co-receptor (CCR5)

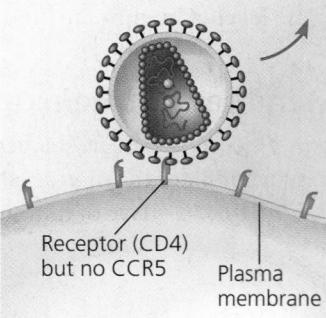

Receptor (CD4) but no CCR5

Plasma membrane

(a) HIV can infect a cell with CCR5 on its surface, as in most people.

(b) HIV cannot infect a cell lacking CCR5 on its surface, as in resistant individuals.

MAKE CONNECTIONS *Study Figures 2.16 and 5.17; each shows pairs of molecules binding to each other. What would you predict about CCR5 that would allow HIV to bind to it? How could a drug molecule interfere with this binding?*

Membrane carbohydrates are usually short, branched chains of fewer than 15 sugar units. Some are covalently bonded to lipids, forming molecules called **glycolipids**. (Recall that *glyco* refers to carbohydrate.) However, most are covalently bonded to proteins, which are thereby **glycoproteins** (see Figure 7.3).

The carbohydrates on the extracellular side of the plasma membrane vary from species to species, among individuals of the same species, and even from one cell type to another in a single individual. The diversity of the molecules and their location on the cell's surface enable membrane carbohydrates to function as markers that distinguish one cell from another. For example, the four human blood types designated A, B, AB, and O reflect variation in the carbohydrate part of glycoproteins on the surface of red blood cells.

Synthesis and Sidedness of Membranes

Membranes have distinct inside and outside faces. The two lipid layers may differ in lipid composition, and each protein has directional orientation in the membrane (see Figure 7.6). **Figure 7.9** shows how membrane sidedness arises: The asymmetrical arrangement of proteins, lipids, and their associated carbohydrates in the plasma membrane is determined as the membrane is being built.

CONCEPT CHECK 7.1

1. **VISUAL SKILLS** Carbohydrates are attached to plasma membrane proteins in the ER (see Figure 7.9). On which side of the vesicle membrane are the carbohydrates during transport to the cell surface?
2. **WHAT IF?** How might the membrane lipid composition of a native grass found in very warm soil around hot springs differ from that of a native grass found in cooler soil? Explain.

For suggested answers, see Appendix A.

CONCEPT 7.2

Membrane structure results in selective permeability

The biological membrane has emergent properties beyond those of the many individual molecules that make it up. The remainder of this chapter focuses on one of those properties: A membrane exhibits **selective permeability**; that is, it allows some substances to cross more easily than others. The ability to regulate transport across cellular boundaries is essential to the cell's existence. We will see once again that form fits function: The fluid mosaic model helps explain how membranes regulate the cell's molecular traffic.

A steady traffic of small molecules and ions moves across the plasma membrane in both directions. Consider the chemical

▼ **Figure 7.9 Synthesis of membrane components and their orientation in the membrane.**
The cytoplasmic (orange) face of the plasma membrane differs from the extracellular (aqua) face. The latter arises from the inside face of ER, Golgi, and vesicle membranes.

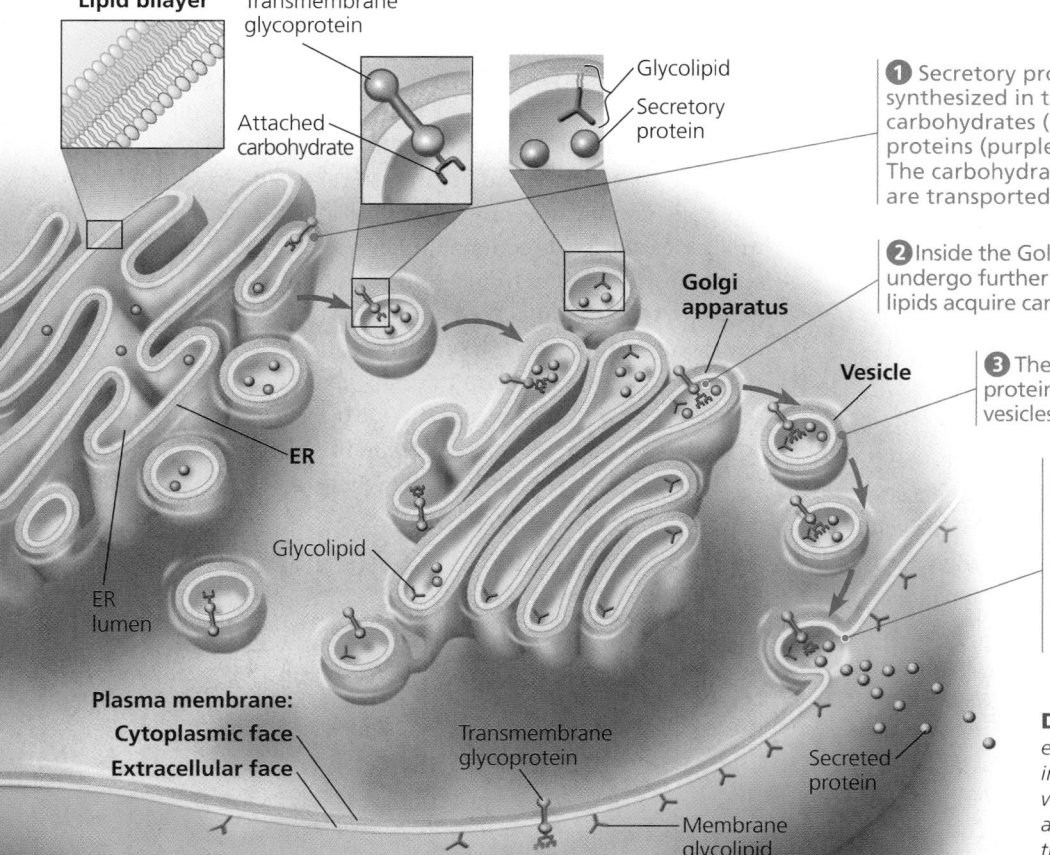

❶ Secretory proteins, membrane proteins, and lipids are synthesized in the endoplasmic reticulum (ER). In the ER, carbohydrates (green) are added to the transmembrane proteins (purple dumbbells), making them glycoproteins. The carbohydrate portions may then be modified. Materials are transported in vesicles to the Golgi apparatus.

❷ Inside the Golgi apparatus, the glycoproteins undergo further carbohydrate modification, and lipids acquire carbohydrates, becoming glycolipids.

❸ The glycoproteins, glycolipids, and secretory proteins (purple spheres) are transported in vesicles to the plasma membrane.

❹ As vesicles fuse with the plasma membrane, the outside face of the vesicle becomes continuous with the inside (cytoplasmic) face of the plasma membrane. This releases the secretory proteins from the cell, a process called *exocytosis*, and positions the carbohydrates of membrane glycoproteins and glycolipids on the outside (extracellular) face of the plasma membrane.

DRAW IT *Draw an integral membrane protein extending from partway through the ER membrane into the ER lumen. Next, draw the protein where it would be located in a series of numbered steps ending at the plasma membrane. Would the protein contact the cytoplasm or the extracellular fluid? Explain.*

exchanges between a muscle cell and the extracellular fluid that bathes it. Sugars, amino acids, and other nutrients enter the cell, and metabolic waste products leave it. The cell takes in O_2 for use in cellular respiration and expels CO_2. Also, the cell regulates its concentrations of inorganic ions, such as Na^+, K^+, Ca^{2+}, and Cl^-, by shuttling them one way or the other across the plasma membrane. In spite of heavy traffic through them, cell membranes are selective in their permeability: Substances do not cross the barrier indiscriminately. The cell is able to take up some small molecules and ions and exclude others.

The Permeability of the Lipid Bilayer

Nonpolar molecules, such as hydrocarbons, CO_2, and O_2, are hydrophobic, as are lipids. They can all therefore dissolve in the lipid bilayer of the membrane and cross it easily, without the aid of membrane proteins. However, the hydrophobic interior of the membrane impedes direct passage through the membrane of ions and polar molecules, which are hydrophilic. Polar molecules such as glucose and other sugars pass only slowly through a lipid bilayer, and even water, a very small polar molecule, does not cross rapidly relative to nonpolar molecules. A charged atom or molecule and its surrounding shell of water (see Figure 3.8) are even less likely to penetrate the hydrophobic interior of the membrane. Furthermore, the lipid bilayer is only one aspect of the gatekeeper system responsible for a cell's selective permeability. Proteins built into the membrane play key roles in regulating transport.

Transport Proteins

Specific ions and a variety of polar molecules can't move through cell membranes on their own. However, these hydrophilic substances can avoid contact with the lipid bilayer by passing through **transport proteins** that span the membrane.

Some transport proteins, called *channel proteins*, function by having a hydrophilic channel that certain molecules or ions use as a tunnel through the membrane (see Figure 7.7a, left). For example, the passage of water molecules through the membrane in certain cells is greatly facilitated by channel proteins known as **aquaporins (Figure 7.10)**. Most aquaporin proteins consist of four identical polypeptide subunits. Each polypeptide forms a channel that water molecules pass through, single-file, overall allowing entry of up to *3 billion* water molecules per second. Without aquaporins, only a tiny fraction of these water molecules would pass through the same area of the cell membrane in a second. Other transport proteins, called *carrier proteins*, hold on to their passengers and change shape in a way that shuttles them across the membrane (see Figure 7.7a, right).

A transport protein is specific for the substance it translocates (moves), allowing only a certain substance (or a small group of related substances) to cross the membrane. For example, a glucose carrier protein in the plasma membrane of red blood cells transports glucose across the membrane 50,000 times faster than glucose can pass through on its own.

This "glucose transporter" is so selective that it even rejects fructose, a structural isomer of glucose (see Figure 5.3). Thus, the selective permeability of a membrane depends on both the discriminating barrier of the lipid bilayer and the specific transport proteins built into the membrane.

> **Mastering Biology** Animation: Selective Permeability of Membranes

What establishes the *direction* of traffic across a membrane? And what mechanisms drive molecules across membranes? We will address these questions next as we explore two modes of membrane traffic: passive transport and active transport.

CONCEPT CHECK 7.2

1. What property allows O_2 and CO_2 to cross a lipid bilayer without the aid of membrane proteins?
2. **VISUAL SKILLS** Examine Figure 7.2. Why is a transport protein needed to move many water molecules rapidly across a membrane?
3. **MAKE CONNECTIONS** Aquaporins exclude passage of hydronium ions (H_3O^+), but some aquaporins allow passage of glycerol, a three-carbon alcohol (see Figure 5.9), as well as H_2O. Since H_3O^+ is closer in size to water than glycerol is, yet cannot pass through, what might be the basis of this selectivity?

For suggested answers, see Appendix A.

CONCEPT 7.3

Passive transport is diffusion of a substance across a membrane with no energy investment

Molecules have a type of energy called thermal energy, due to their constant motion (see Concept 3.2). One result of this motion is **diffusion**, the movement of particles of any substance so that they spread out into the available space. Each molecule moves randomly, yet diffusion of a *population* of molecules may be directional. Imagine a synthetic membrane separating pure water from a solution of a dye in water. Study **Figure 7.11a** carefully to see how diffusion would result in equal concentrations of dye molecules in both solutions. At that point, a dynamic equilibrium will exist, with as many dye molecules crossing per second in one direction as in the other.

Here is a simple rule of diffusion: In the absence of any other forces, a substance will diffuse from where it is more concentrated

▶ **Figure 7.10 An aquaporin.** This computer model shows water molecules (red and gray) passing through an aquaporin (blue ribbons), in a lipid bilayer (yellow, hydrophilic heads; green, hydrophobic tails).

> **Mastering Biology**
> Interview with Peter Agre: Discovering aquaporins
> Animation: Aquaporin

to where it is less concentrated. Put another way, a substance diffuses down its **concentration gradient**, the region along which the density of a chemical substance increases or decreases (in this case, decreases). Diffusion is a spontaneous process, needing no input of energy. Each substance diffuses down its *own* concentration gradient, unaffected by the concentration gradients of other substances **(Figure 7.11b)**.

Much of the traffic across cell membranes occurs by diffusion. When a substance is more concentrated on one side of a membrane than on the other, there is a tendency for it to diffuse across, down its concentration gradient (assuming that the membrane is permeable to that substance). One important example is the uptake of oxygen by a cell performing cellular respiration. Dissolved oxygen diffuses into the cell across the plasma membrane. As long as cellular respiration consumes the O_2 as it enters, diffusion into the cell will continue because the concentration gradient favors movement in that direction.

The diffusion of a substance across a biological membrane is called **passive transport** because it requires no energy. The concentration gradient itself represents potential energy

▼ **Figure 7.11 Diffusion of solutes across a synthetic membrane.** Each large arrow shows net diffusion of dye molecules of that color.

(a) Diffusion of one solute. Molecules of dye can pass through membrane pores. Random movement of dye molecules will cause some to pass through the pores; this happens more often on the side with more dye molecules. The dye diffuses from the more concentrated side to the less concentrated side (called diffusing down a concentration gradient). A dynamic equilibrium results: Solute molecules still cross, but at roughly equal rates in both directions.

(b) Diffusion of two solutes. Solutions of two different dyes are separated by a membrane that is permeable to both. Each dye diffuses down its own concentration gradient. There will be a net diffusion of the purple dye toward the left, even though the *total* solute concentration was initially greater on the left side.

→ **Mastering Biology Animation: Diffusion**

(see Concept 2.2 and Figure 8.5b) and drives diffusion. Remember, though, that membranes are selectively permeable and therefore have different effects on the rates of diffusion of various molecules. Water can diffuse very rapidly across the membranes of cells with aquaporins, compared with diffusion in the absence of aquaporins. The movement of water across the plasma membrane has important consequences for cells.

Effects of Osmosis on Water Balance

To see how two solutions with different solute concentrations interact, picture a U-shaped glass tube with a selectively permeable artificial membrane separating two sugar solutions **(Figure 7.12)**. Pores in this synthetic membrane are too small for sugar molecules to pass through but large enough for water molecules. However, tight clustering of water molecules around the hydrophilic solute molecules makes some

▼ **Figure 7.12 Osmosis.** Two sugar solutions of different concentrations are separated by a membrane that the solvent (water) can pass through but the solute (sugar) cannot. Water molecules move randomly and may cross in either direction, but overall, water diffuses from the solution with less concentrated solute to that with more concentrated solute. This passive transport of water, or osmosis, makes the sugar concentrations on both sides roughly equal.

Water moves from an area of higher to lower free water concentration (lower to higher solute concentration).

VISUAL SKILLS *If an orange dye capable of passing through the membrane was added to the left side of the tube above, how would it be distributed at the end of the experiment? (See Figure 7.11.) Would the final solution levels in the tube be affected? What cellular component does the membrane represent in this experiment?*

→ **Mastering Biology Figure Walkthrough**

of the water unavailable to cross the membrane. As a result, the solution with a higher solute concentration has a lower *free* water concentration. Water diffuses across the membrane from the region of higher free water concentration (lower solute concentration) to that of lower free water concentration (higher solute concentration) until the solute concentrations on both sides of the membrane are more nearly equal. The diffusion of free water across a selectively permeable membrane, whether artificial or cellular, is called **osmosis**. The movement of water across cell membranes and the balance of water between the cell and its environment are crucial to organisms. Let's now apply what we've learned about osmosis in this system to living cells.

Water Balance of Cells Without Cell Walls

To explain the behavior of a cell in a solution, we must consider both solute concentration and membrane permeability. Both factors are taken into account in the concept of **tonicity**, the ability of a surrounding solution to cause a cell to gain or lose water. The tonicity of a solution depends in part on its concentration of solutes that cannot cross the membrane (nonpenetrating solutes) relative to that inside the cell. If there is a higher concentration of nonpenetrating solutes in the surrounding solution, water will tend to leave the cell, and vice versa.

If a cell without a cell wall, such as an animal cell, is immersed in an environment that is **isotonic** to the cell (*iso* means "same"), there will be no *net* movement of water across the plasma membrane. Water diffuses across the membrane, but at the same rate in both directions. In an isotonic environment, the volume of an animal cell is stable **(Figure 7.13a)**.

Let's transfer the cell to a solution that is **hypertonic** to the cell (*hyper* means "more," in this case referring to nonpenetrating solutes). The cell will lose water, shrivel, and probably die. This is why an increase in the salinity (saltiness) of a lake can kill the animals there; if the lake water becomes hypertonic to the animals' cells, they might shrivel and die. However, taking up too much water can be just as hazardous to a cell as losing water. If we place the cell in a solution that is **hypotonic** to the cell (*hypo* means "less"), water will enter the cell faster than it leaves, and the cell will swell and lyse (burst) like an overfilled water balloon.

A cell without rigid cell walls can't tolerate either excessive uptake or excessive loss of water. This problem of water balance is automatically solved if such a cell lives in isotonic surroundings. Seawater is isotonic to many marine invertebrates. The cells of most terrestrial (land-dwelling) animals

are bathed in an extracellular fluid that is isotonic to the cells. In hypertonic or hypotonic environments, however, organisms that lack rigid cell walls must have other adaptations for **osmoregulation**, the control of solute concentrations and water balance. For example, the unicellular protist *Paramecium caudatum* lives in pond water, which is hypotonic to the cell. *Paramecium* has a plasma membrane that is much less permeable to water than the membranes of most other cells, but this only slows the uptake of water, which continually enters the cell. The reason the *Paramecium* cell doesn't burst is that it has a contractile vacuole, an organelle that functions as a pump to force water out of the cell as fast as it enters by osmosis **(Figure 7.14)**. In contrast, the bacteria and archaea that live in hypersaline (excessively salty) environments (see Figure 27.1) have cellular mechanisms that balance the internal and external solute concentrations to ensure that water does not move out of the cell. We'll examine other evolutionary adaptations for osmoregulation by animals in Concept 44.1.

Water Balance of Cells with Cell Walls

The cells of plants, prokaryotes, fungi, and some protists are surrounded by cell walls (see Figure 6.27). When such a cell is immersed in a hypotonic solution—bathed in rainwater, for example—the cell wall helps maintain the cell's water balance. Consider a plant cell. Like an animal cell, the plant cell swells as water enters by osmosis **(Figure 7.13b)**. However, the relatively inelastic cell wall will expand only so much before it exerts a back pressure on the cell, called *turgor pressure*, that

▼ **Figure 7.13 The water balance of living cells.** How living cells react to changes in the solute concentration of their environment depends on whether or not they have cell walls. **(a)** Animal cells, such as this red blood cell, do not have cell walls. **(b)** Plant cells do have cell walls. (Arrows indicate net water movement after the cells were first placed in these solutions.)

(a) Animal cell. An animal cell, such as this red blood cell, does not have a cell wall. Animal cells fare best in an isotonic environment unless they have special adaptations that offset the osmotic uptake or loss of water.

(b) Plant cell. Plant cells are turgid (firm) and generally healthiest in a hypotonic environment, where the uptake of water is eventually balanced by the wall pushing back on the cell.

Hypotonic solution	Isotonic solution	Hypertonic solution
Lysed	Normal	Shriveled
Turgid (normal)	Flaccid	Plasmolyzed

❓ *Why do limp celery stalks become crisp when placed in a glass of water?*

➔ **Mastering Biology** Animation: Osmosis and Water Balance in Cells
Video: Turgid *Elodea*
Video: Plasmolysis in *Elodea*

▼ **Figure 7.14 The contractile vacuole of *Paramecium*.** The vacuole collects fluid from canals in the cytoplasm. When full, the vacuole and canals contract, expelling fluid from the cell (LM).

Contractile vacuole
50 μm

⊕ **Mastering Biology** Video: *Paramecium* Vacuole

opposes further water uptake. At this point, the cell is **turgid** (very firm), which is the healthy state for most plant cells. Plants that are not woody, such as most houseplants, depend for mechanical support on cells kept turgid by a surrounding hypotonic solution. If a plant's cells and surroundings are isotonic, there is no net tendency for water to enter and the cells become **flaccid** (limp); the plant wilts.

However, a cell wall is of no advantage if the cell is immersed in a hypertonic environment. In this case, a plant cell, like an animal cell, will lose water to its surroundings and shrink. As the plant cell shrivels, its plasma membrane pulls away from the cell wall at multiple places. This phenomenon, called **plasmolysis**, causes the plant to wilt and can lead to plant death. The walled cells of bacteria and fungi also plasmolyze in hypertonic environments.

Facilitated Diffusion: Passive Transport Aided by Proteins

Let's look more closely at how water and certain hydrophilic solutes cross a membrane. As mentioned earlier, many polar molecules and ions blocked by the lipid bilayer of the membrane diffuse passively with the help of transport proteins that span the membrane. This phenomenon is called **facilitated diffusion**. Cell biologists are still trying to learn exactly how various transport proteins facilitate diffusion. Most transport proteins are very specific: They transport some substances but not others.

As mentioned earlier, the two types of transport proteins are channel proteins and carrier proteins. Channel proteins simply provide corridors that allow specific molecules or ions to cross the membrane **(Figure 7.15a)**. The hydrophilic passageways provided by these proteins can allow water molecules or small ions to diffuse very quickly from one side of the membrane to the other. Aquaporins, the water channel proteins, facilitate the massive levels of diffusion of water (osmosis) that occur in plant cells and in animal cells such as red blood cells (see Figure 7.13). Certain kidney cells also have a high number of aquaporins, allowing them to reclaim water from urine before it is excreted. If the kidneys didn't perform this function, you would excrete about 180 L of urine per day—and have to drink an equal volume of water!

▼ **Figure 7.15 Two types of transport proteins that carry out facilitated diffusion.** In both cases, the protein can transport the solute in either direction, but the net movement is down the concentration gradient of the solute.

(a) A channel protein has a channel through which water molecules or a specific solute can pass.

EXTRACELLULAR FLUID
Channel protein
CYTOPLASM
Solute

(b) A carrier protein alternates between two shapes, moving a solute across the membrane during the shape change.

Carrier protein
Solute

⊕ **Mastering Biology** Animation: Facilitated Diffusion

Channel proteins that transport ions are called **ion channels**. Many ion channels function as **gated channels**, which open or close in response to a stimulus (see Figure 11.8). For some gated channels, the stimulus is electrical. In a nerve cell, for example, a potassium ion channel protein (see computer model) opens in response to an electrical stimulus, allowing a stream of potassium ions to leave the cell. This restores the cell's ability to fire again. Other gated channels have a chemical stimulus: They open or close when a specific substance (not the one to be transported) binds to the channel. Ion channels are important in the functioning of the nervous system, as you'll learn in Chapter 48.

▼ **Potassium ion channel protein** (See also the side view of a calcium channel protein in Figure 6.32a.)

Potassium ion

⊕ **Mastering Biology**
Interview with Elba Serrano: Investigating how ion channels enable you to hear

Carrier proteins, such as the glucose transporter mentioned earlier, seem to undergo a subtle change in shape that somehow translocates the solute-binding site across the membrane **(Figure 7.15b)**. Such a change in shape may be triggered by the binding and release of the transported molecule. Like ion channels, carrier proteins involved in facilitated diffusion result in the net movement of a substance down

Interpreting a Scatter Plot with Two Sets of Data

Is Glucose Uptake into Cells Affected by Age? Glucose, an important energy source for animals, is transported into cells by facilitated diffusion using protein carriers. In this exercise, you will interpret a graph with two sets of data from an experiment that examined glucose uptake over time in red blood cells from guinea pigs of different ages. You will determine if the cells' rate of glucose uptake depended on the age of the guinea pig.

How the Experiment Was Done Researchers incubated guinea pig red blood cells in a 300 m*M* (millimolar) radioactive glucose solution at pH 7.4 at 25°C. Every 10 or 15 minutes, they removed a sample of cells and measured the concentration of radioactive glucose inside those cells. The cells came from either a 15-day-old or a 1-month-old guinea pig.

Data from the Experiment When you have multiple sets of data, it can be useful to plot them on the same graph for comparison. In the graph here, each set of dots (of the same color) forms a *scatter plot*, in which every data point represents two numerical values, one for each variable. For each data set, a curve that best fits the points has been drawn to make it easier to see the trends. (For additional information about graphs, see the Scientific Skills Review in Appendix D.)

INTERPRET THE DATA

1. First make sure you understand the parts of the graph. **(a)** Which variable is the independent variable—the variable controlled by the researchers? **(b)** Which variable is the dependent variable—the variable that depended on the treatment and was measured by the researchers? **(c)** What do the red dots represent? **(d)** The blue dots?

▼ **Glucose Uptake over Time in Guinea Pig Red Blood Cells.**

Data from T. Kondo and E. Beutler, Developmental changes in glucose transport of guinea pig erythrocytes, *Journal of Clinical Investigation* 65:1–4 (1980).

2. From the data points on the graph, construct a table of the data. Put "Incubation Time (min)" in the left column of the table.
3. What does the graph show? Compare and contrast glucose uptake in red blood cells from 15-day-old and 1-month-old guinea pigs.
4. Develop a hypothesis to explain the difference between glucose uptake in red blood cells from 15-day-old and 1-month-old guinea pigs. (Think about how glucose gets into cells.)
5. Design an experiment to test your hypothesis.

➔ **Instructors:** A version of this Scientific Skills Exercise can be assigned in **Mastering Biology**.

---

its concentration gradient. No energy input is thus required: This is passive transport. The **Scientific Skills Exercise** gives you an opportunity to work with data from an experiment related to glucose transport.

➔ **Mastering Biology** BioFlix® Animation: Passive Transport

CONCEPT CHECK 7.3

1. Speculate about how a cell performing cellular respiration might rid itself of the resulting CO_2.
2. **WHAT IF?** If a *Paramecium* swims from a hypotonic to an isotonic environment, will its contractile vacuole become more active or less? Why?

For suggested answers, see Appendix A.

CONCEPT 7.4

Active transport uses energy to move solutes against their gradients

Despite the help of transport proteins, facilitated diffusion is considered passive transport because the solute is moving down its concentration gradient, a process that requires no energy. Facilitated diffusion speeds transport of a solute by providing efficient passage through the membrane, but it does not alter the direction of transport. Some other transport proteins, however, can use energy to move solutes *against* their concentration gradients, across the plasma membrane from the side where they are less concentrated (whether inside or outside) to the side where they are more concentrated.

The Need for Energy in Active Transport

To pump a solute across a membrane against its gradient requires work; the cell must expend energy. Therefore, this type of membrane traffic is called **active transport**. The transport proteins that move solutes against their concentration gradients are all carrier proteins rather than channel proteins. This makes sense because when channel proteins are open, they merely allow solutes to diffuse down their concentration gradients rather than picking them up and transporting them against their gradients.

Active transport enables a cell to maintain internal concentrations of small solutes that differ from concentrations in its environment. For example, compared with its surroundings,

► **Figure 7.16 The sodium-potassium pump: a specific case of active transport.** This transport system pumps ions against steep concentration gradients. The pump oscillates between two shapes in a cycle that moves Na⁺ out of the cell (steps ❶ – ❸) and K⁺ into the cell (steps ❹ – ❻). The two shapes have different binding affinities for Na⁺ and K⁺. ATP hydrolysis powers the shape change by transferring a phosphate group to the transport protein (phosphorylating the protein).

VISUAL SKILLS *For each ion (Na⁺ and K⁺), describe its concentration inside the cell relative to outside. How many Na⁺ are moved out of the cell and how many K⁺ moved in per cycle?*

→ **Mastering Biology**
Animation: Active Transport

EXTRACELLULAR FLUID
[Na⁺] high
[K⁺] low

CYTOPLASM
[Na⁺] low
[K⁺] high

❶ Cytoplasmic Na⁺ binds to the sodium-potassium pump with high affinity when the protein has this shape.

❷ Binding of 3 Na⁺ ions stimulates phosphorylation by a kinase, using ATP.

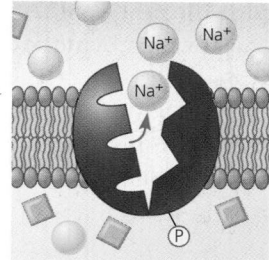

❸ Phosphorylation leads to a change in protein shape, reducing its affinity for Na⁺; 3 Na⁺ are released outside.

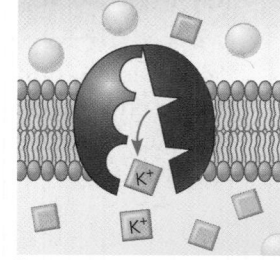

❻ 2 K⁺ are released; affinity for Na⁺ is high again, and the cycle repeats.

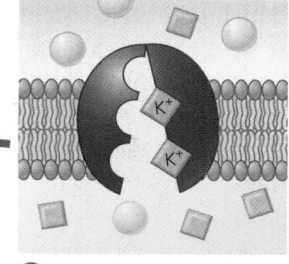

❺ Loss of the phosphate group restores the protein's original shape, which has a lower affinity for K⁺.

❹ The new shape has a high affinity for K⁺; 2 K⁺ bind on the extracellular side, triggering release of the phosphate group.

an animal cell has a much higher concentration of potassium ions (K⁺) and a much lower concentration of sodium ions (Na⁺). The plasma membrane helps maintain these steep gradients by pumping Na⁺ out of the cell and K⁺ into the cell.

As in other types of cellular work, ATP hydrolysis supplies the energy for most active transport. One way ATP can power active transport is when its terminal phosphate group is transferred directly to the transport protein. This can induce the protein to change its shape in a manner that translocates a solute bound to the protein across the membrane. One transport system that works this way is the **sodium-potassium pump**, which exchanges Na⁺ for K⁺ across the plasma membrane of animal cells **(Figure 7.16)**. The distinction between passive transport and active transport is reviewed in **Figure 7.17**.

How Ion Pumps Maintain Membrane Potential

All cells have voltages across their plasma membranes. Voltage is electrical potential energy (see Concept 2.2)—a separation of opposite charges. The cytoplasmic side of the membrane is negative in charge relative to the extracellular side because of an unequal distribution of anions and cations on the two sides. The voltage across a membrane, called a **membrane potential**, ranges from about −50 to −200 millivolts (mV). (The minus sign indicates that the inside of the cell is negative relative to the outside.)

▼ **Figure 7.17 Review: passive and active transport.**

Passive transport. Substances diffuse spontaneously down their concentration gradients, crossing a membrane with no expenditure of energy by the cell. The rate of diffusion can be greatly increased by transport proteins in the membrane.

Active transport. Some transport proteins expend energy and act as pumps, moving substances across a membrane against their concentration (or electrochemical) gradients. Energy is usually supplied by ATP hydrolysis.

Diffusion. Hydrophobic molecules and (at a slow rate) very small uncharged polar molecules can diffuse through the lipid bilayer.

Facilitated diffusion. Many hydrophilic substances diffuse through membranes with the assistance of transport proteins, either channel proteins (left) or carrier proteins (right).

VISUAL SKILLS *For each solute in the right panel, describe its direction of movement, and state whether it is moving with or against its concentration gradient.*

The membrane potential acts like a battery, an energy source that affects the traffic of all charged substances across the membrane. Because the inside of the cell is negative compared with the outside, the membrane potential favors the passive transport of cations into the cell and anions out of the cell. Thus, *two* forces drive the diffusion of ions across a membrane: a chemical force (the ion's concentration gradient, which has been our sole consideration thus far in the chapter) and an electrical force (the effect of the membrane potential on the ion's movement). This combination of forces acting on an ion is called the **electrochemical gradient**.

In the case of ions, then, we must refine our concept of passive transport: An ion diffuses not simply down its *concentration* gradient but, more exactly, down its *electrochemical* gradient. For example, the concentration of Na^+ inside a resting nerve cell is much lower than outside it. When the cell is stimulated, gated channels open that facilitate Na^+ diffusion. Sodium ions then "fall" down their electrochemical gradient, driven by the concentration gradient of Na^+ and by the attraction of these cations to the negative side (inside) of the membrane. In this example, both electrical and chemical contributions to the electrochemical gradient act in the same direction across the membrane, but this is not always so. In cases where electrical forces due to the membrane potential oppose the simple diffusion of an ion down its concentration gradient, active transport may be necessary. In Concepts 48.2 and 48.3, you'll learn about the importance of electrochemical gradients and membrane potentials in the transmission of nerve impulses.

Some membrane proteins that actively transport ions contribute to the membrane potential. Study Figure 7.16 to see if you can see why the sodium-potassium pump is a good example. Notice that the pump does not translocate Na^+ and K^+ one for one, but pumps *three* sodium ions out of the cell for every *two* potassium ions it pumps into the cell. With each "crank" of the pump, there is a net transfer of one positive charge from the cytoplasm to the extracellular fluid, a process that stores energy as voltage. A transport protein that generates voltage across a membrane is called an **electrogenic pump**.

The sodium-potassium pump appears to be the major electrogenic pump of animal cells. The main electrogenic pump of plants, fungi, and bacteria is a **proton pump**, which actively transports protons (hydrogen ions, H^+) out of the cell. The pumping of H^+ transfers positive charge from the cytoplasm to the extracellular solution **(Figure 7.18)**. By generating voltage across membranes, electrogenic pumps help store energy that can be tapped for cellular work. One important use of proton gradients in the cell is for ATP synthesis during cellular respiration, as you will see in Concept 9.4. Another is a type of membrane traffic called cotransport.

Cotransport: Coupled Transport by a Membrane Protein

A solute that exists in different concentrations across a membrane can do work as it moves across that membrane by diffusion down its concentration gradient. This is analogous to water that has been pumped uphill and performs work as it flows back down. In a mechanism called **cotransport**, a transport protein (a cotransporter) can couple the "downhill" diffusion of the solute to the "uphill" transport of a second substance against its own concentration gradient. For instance, a plant cell uses the gradient of H^+ generated by its ATP-powered proton pumps to drive the active transport of amino acids, sugars, and several other nutrients into the cell. In the example shown in **Figure 7.19**, a cotransporter couples the return of H^+ to the transport of sucrose into the cell. This protein can translocate sucrose into the cell against

▼ **Figure 7.19 Cotransport: active transport driven by a concentration gradient.** A carrier protein, such as this H^+/sucrose cotransporter in a plant cell (top), is able to use the diffusion of H^+ down its electrochemical gradient into the cell to drive the uptake of sucrose. (The cell wall is not shown.) Although not technically part of the cotransport process, an ATP-driven proton pump is shown here (bottom), which concentrates H^+ outside the cell. The resulting H^+ gradient represents potential energy that can be used for active transport—of sucrose, in this case. Thus, ATP hydrolysis indirectly provides the energy necessary for cotransport.

▼ **Figure 7.18 A proton pump.** Proton pumps are electrogenic pumps that store energy by generating voltage (charge separation) across membranes. A proton pump translocates positive charge in the form of hydrogen ions (H^+, or protons). The voltage and H^+ concentration gradient represent a dual energy source that can drive other processes, such as the uptake of nutrients. Most proton pumps are powered by ATP hydrolysis. (See Figure 6.32a.)

its concentration gradient, but only if the sucrose molecule travels in the company of an H^+. The H^+ uses the transport protein as an avenue to diffuse down its own electrochemical gradient, which is maintained by the proton pump. Plants use H^+/sucrose cotransport to load sucrose produced by photosynthesis into cells in the veins of leaves. The vascular tissue of the plant can then distribute the sugar to roots and other nonphotosynthetic organs that do not make their own sugars.

A similar cotransporter in animals transports Na^+ into intestinal cells together with glucose, which is moving down its concentration gradient into the cell. (The Na^+ is then pumped out of the cell into the blood on the other side by Na^+/K^+ pumps; see Figure 7.16.) Our understanding of Na^+/glucose cotransporters has helped us find more effective treatments for diarrhea, a serious problem in developing countries. Normally, sodium in waste is reabsorbed in the colon, maintaining constant levels in the body, but diarrhea expels waste so rapidly that reabsorption is not possible, and sodium levels fall precipitously. To treat this life-threatening condition, patients are given a solution to drink containing high concentrations of salt (NaCl) and glucose. The solutes are taken up by Na^+/glucose cotransporters on the surface of intestinal cells and passed through the cells into the blood. This simple treatment has lowered infant mortality worldwide.

→ **Mastering Biology** BioFlix® Animation: Active Transport

CONCEPT CHECK 7.4

1. Na^+/K^+ pumps help nerve cells establish a voltage across their plasma membranes. Do these pumps use ATP or produce ATP? Explain.

2. **VISUAL SKILLS** Compare the Na^+/K^+ pump in Figure 7.16 with the cotransporter in Figure 7.19. Explain why the Na^+/K^+ pump would not be considered a cotransporter.

3. **MAKE CONNECTIONS** Review the characteristics of the lysosome in Concept 6.4. Given the internal environment of a lysosome, what transport protein might you expect to see in its membrane?

For suggested answers, see Appendix A.

CONCEPT 7.5

Bulk transport across the plasma membrane occurs by exocytosis and endocytosis

Large molecules, such as proteins and polysaccharides, generally don't cross the membrane by diffusion or transport proteins. Instead, they usually enter and leave the cell in bulk, packaged in vesicles.

→ **Mastering Biology** BioFlix® Animation: Exocytosis and Endocytosis

Exocytosis

The cell secretes certain molecules by the fusion of vesicles with the plasma membrane; this process is called **exocytosis**

▼ **Figure 7.20 Exocytosis.**

❶ A vesicle containing secretory proteins buds off from the Golgi.

❷ The vesicle travels along a microtubule to the plasma membrane.

❸ The outside of the vesicle binds to the inside of the plasma membrane.

❹ The two membranes fuse, releasing (secreting) the proteins outside the cell.

❺ The vesicle membrane becomes part of the plasma membrane.

(Figure 7.20). A transport vesicle that has budded from the Golgi apparatus moves along a microtubule of the cytoskeleton to the plasma membrane. When the vesicle membrane and plasma membrane come into contact, specific proteins in both membranes rearrange the lipid molecules of the two bilayers so that the two membranes fuse. The contents of the vesicle spill out of the cell, and the vesicle membrane becomes part of the plasma membrane.

Many secretory cells use exocytosis to export products. For example, cells in the pancreas that make insulin secrete it into the extracellular fluid by exocytosis. In another example, nerve cells use exocytosis to release neurotransmitters that signal other neurons or muscle cells (see Figure 7.1). When plant cells are making cell walls, exocytosis delivers some of the necessary proteins and carbohydrates from Golgi vesicles to the outside of the cell.

Endocytosis

In **endocytosis**, the cell takes in molecules and particulate matter by forming new vesicles from the plasma membrane. Although the proteins involved in the processes are different, the events of endocytosis look like the reverse of exocytosis. First, a small area of the plasma membrane sinks inward to form a pocket. Then, as the pocket deepens, it pinches in, forming a vesicle containing material that had been outside the cell. Study **Figure 7.21** carefully to understand the three types of endocytosis: phagocytosis ("cellular eating"), pinocytosis ("cellular drinking"), and receptor-mediated endocytosis.

Human cells use receptor-mediated endocytosis to take in cholesterol for membrane synthesis and the synthesis of other steroids. Cholesterol travels in the blood in particles called low-density lipoproteins (LDLs), each a complex of lipids and a protein. LDLs bind to LDL receptors on plasma membranes and then enter the cells by endocytosis. In the inherited disease familial hypercholesterolemia, characterized by a very high level of cholesterol in the blood, LDLs cannot enter cells because the LDL receptor proteins are defective or missing. Cholesterol thus accumulates in the blood, contributing to

Phagocytosis

In **phagocytosis**, a cell engulfs a particle by extending pseudopodia (singular, *pseudopodium*) around it and packaging it within a membranous sac called a food vacuole. The particle will be digested after the food vacuole fuses with a lysosome containing hydrolytic enzymes (see Figure 6.13a).

An amoeba engulfing a green algal cell via phagocytosis (TEM).

Pinocytosis

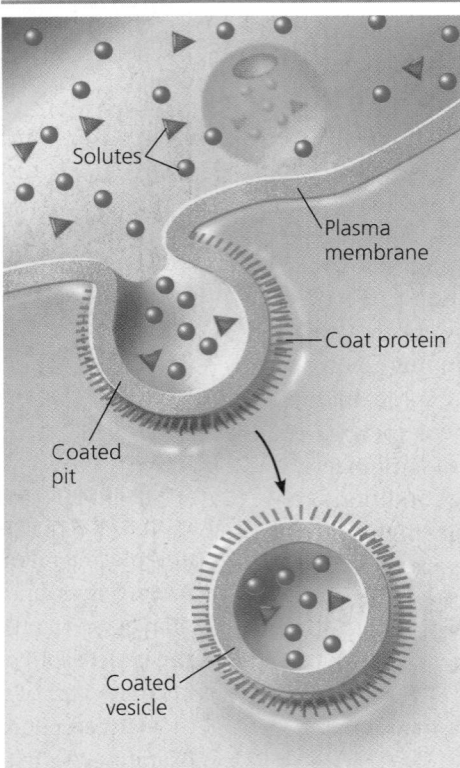

In **pinocytosis**, a cell continually "gulps" droplets of extracellular fluid into tiny vesicles, formed by infoldings of the plasma membrane. In this way, the cell obtains molecules dissolved in the droplets. Because any and all solutes are taken into the cell, pinocytosis as shown here is nonspecific for the substances it transports. In many cases, the parts of the plasma membrane that form vesicles are lined on their cytoplasmic side by a fuzzy layer of coat protein; the "pits" and resulting vesicles are called *coated pits*.

Pinocytotic vesicles forming (TEMs).

Receptor-Mediated Endocytosis

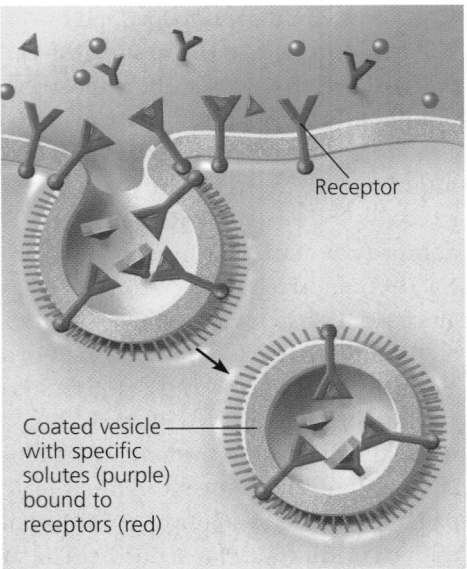

Receptor-mediated endocytosis is a specialized type of pinocytosis that enables the cell to acquire bulk quantities of specific substances, even though those substances may not be very concentrated in the extracellular fluid. Embedded in the plasma membrane are proteins with receptor sites exposed to the extracellular fluid. Specific solutes bind to the receptors. The receptor proteins then cluster in coated pits, and each coated pit forms a vesicle containing the bound molecules. The diagram shows only bound molecules (purple triangles) inside the vesicle, but other molecules from the extracellular fluid are also present. After the ingested material is liberated from the vesicle, the emptied receptors are recycled to the plasma membrane by the same vesicle (not shown).

Top: A coated pit. *Bottom*: A coated vesicle forming during receptor-mediated endocytosis (TEMs).

VISUAL SKILLS *Use the scale bars to estimate the diameters of (a) the food vacuole that will form around the algal cell (left micrograph; measure the length, not the width) and (b) the coated vesicle (lower right micrograph). (c) Which is larger, and by what factor?*

→ Mastering Biology Animation: Exocytosis and Endocytosis

early atherosclerosis, the buildup of lipids within blood vessel walls. This narrows the space in the vessels and impedes blood flow, potentially resulting in heart damage and stroke.

Endocytosis and exocytosis also provide mechanisms for rejuvenating or remodeling the plasma membrane. These processes occur continually in most eukaryotic cells, yet the amount of plasma membrane in a nongrowing cell remains fairly constant. The addition of membrane by one process appears to offset the loss of membrane by the other.

→ **Mastering Biology** BioFlix® Animation: Membrane Transport

CONCEPT CHECK 7.5

1. As a cell grows, its plasma membrane expands. Does this involve endocytosis or exocytosis? Explain.
2. **DRAW IT** Return to Figure 7.9, and circle a patch of plasma membrane that is coming from a vesicle involved in exocytosis.
3. **MAKE CONNECTIONS** In Concept 6.7, you learned that animal cells make an extracellular matrix (ECM). Describe the cellular pathway of synthesis and deposition of an ECM glycoprotein.

For suggested answers, see Appendix A.

7 Chapter Review

→ Go to **Mastering Biology** for Assignments, the eText, the Study Area, and Dynamic Study Modules.

SUMMARY OF KEY CONCEPTS

→ To review key terms, go to the **Vocabulary Self-Quiz** in the **Mastering Biology** eText or Study Area, or go to goo.gl/zkjz9t.

CONCEPT 7.1

Cellular membranes are fluid mosaics of lipids and proteins *(pp. 127–131)*

- In the **fluid mosaic model, amphipathic** proteins are embedded in the phospholipid bilayer.
- Phospholipids and some proteins move sideways within the membrane. The unsaturated hydrocarbon tails of some phospholipids keep membranes fluid at lower temperatures, while cholesterol helps membranes resist changes in fluidity caused by temperature changes.
- Membrane proteins function in transport, enzymatic activity, signal transduction, cell-cell recognition, intercellular joining, and attachment to the cytoskeleton and extracellular matrix. Short chains of sugars linked to proteins (in **glycoproteins**) and lipids (in **glycolipids**) on the exterior side of the plasma membrane interact with surface molecules of other cells.
- Membrane proteins and lipids are synthesized in the ER and modified in the ER and Golgi apparatus. The inside and outside faces of membranes differ in molecular composition.

? *In what ways are membranes crucial to life?*

CONCEPT 7.2

Membrane structure results in selective permeability *(pp. 131–132)*

- A cell must exchange molecules and ions with its surroundings, a process controlled by the **selective permeability** of the plasma membrane. Hydrophobic substances are soluble in lipids and pass through membranes rapidly, whereas polar molecules and ions generally require specific **transport proteins**.

? *How do aquaporins affect the permeability of a membrane?*

CONCEPT 7.3

Passive transport is diffusion of a substance across a membrane with no energy investment *(pp. 132–136)*

- **Diffusion** is the spontaneous movement of a substance down its **concentration gradient**. Water diffuses out of a cell (**osmosis**)

if the solution outside has a higher concentration of nonpenetrating solutes (**hypertonic**); water enters if the solution has a lower solute concentration (**hypotonic**). If the concentrations are equal (**isotonic**), no net osmosis occurs. Cell survival depends on balancing water uptake and loss.

- In **facilitated diffusion**, a transport protein speeds water or solute movement down its concentration gradient across a membrane. **Ion channels** facilitate the diffusion of ions across a membrane. Carrier proteins can undergo changes in shape that translocate bound solutes across the membrane.

Facilitated diffusion

Channel protein

Carrier protein

? *What happens to a cell placed in a hypertonic solution? Describe the free water concentration inside and out.*

CONCEPT 7.4

Active transport uses energy to move solutes against their gradients *(pp. 136–139)*

Active transport

- Specific membrane proteins use energy, usually in the form of ATP, to do the work of **active transport**.
- Ions can have both a concentration (chemical) gradient and an electrical gradient (voltage). These gradients combine in the **electrochemical gradient**, which determines the net direction of ionic diffusion.
- **Cotransport** of two solutes occurs when a membrane protein enables the "downhill" diffusion of one solute to drive the "uphill" transport of the other.

? *ATP is not directly involved in the functioning of a cotransporter. Why, then, is cotransport considered active transport?*

CONCEPT 7.5

Bulk transport across the plasma membrane occurs by exocytosis and endocytosis *(pp. 139–141)*

- In **exocytosis**, transport vesicles migrate to the plasma membrane, fuse with it, and release their contents. In **endocytosis**, molecules enter cells within vesicles that pinch inward from the plasma membrane. The three types of endocytosis are **phagocytosis, pinocytosis**, and **receptor-mediated endocytosis**.

? *Which type of endocytosis involves the binding of specific substances in the extracellular fluid to membrane proteins? What does this type of transport enable a cell to do?*

TEST YOUR UNDERSTANDING

→ For more multiple-choice questions, go to the **Practice Test** in the **Mastering Biology** eText or Study Area, or go to goo.gl/GruWRg.

Levels 1-2: Remembering/Understanding

1. In what way do the membranes of a eukaryotic cell vary?
   (A) Phospholipids are found only in certain membranes.
   (B) Certain proteins are unique to each membrane.
   (C) Only certain membranes of the cell are selectively permeable.
   (D) Only certain membranes are constructed from amphipathic molecules.

2. According to the fluid mosaic model of membrane structure, proteins of the membrane are mostly
   (A) spread in a continuous layer over the inner and outer surfaces of the membrane.
   (B) confined to the hydrophobic interior of the membrane.
   (C) embedded in a lipid bilayer.
   (D) randomly oriented in the membrane, with no fixed inside-outside polarity.

3. Which of the following factors would tend to increase membrane fluidity?
   (A) a greater proportion of unsaturated phospholipids
   (B) a greater proportion of saturated phospholipids
   (C) a lower temperature
   (D) a relatively high protein content in the membrane

Levels 3-4: Applying/Analyzing

4. Which of the following processes includes all the others?
   (A) osmosis
   (B) diffusion of a solute across a membrane
   (C) passive transport
   (D) transport of an ion down its electrochemical gradient

5. Based on Figure 7.19, which of these experimental treatments would increase the rate of sucrose transport into a plant cell?
   (A) decreasing extracellular sucrose concentration
   (B) decreasing extracellular pH
   (C) decreasing cytoplasmic pH
   (D) adding a substance that makes the membrane more permeable to hydrogen ions

6. **DRAW IT** An artificial "cell" consisting of an aqueous solution enclosed in a selectively permeable membrane is immersed in a beaker containing a different solution, the "environment," as shown in the accompanying diagram. The membrane is permeable to water and to the simple sugars glucose and fructose but impermeable to the disaccharide sucrose.
   (a) Draw solid arrows to indicate the net movement of solutes into and/or out of the cell.
   (b) Is the solution outside the cell isotonic, hypotonic, or hypertonic?

   (c) Draw a dashed arrow to show the net osmosis, if any.
   (d) Will the artificial cell become more flaccid, more turgid, or stay the same?
   (e) Eventually, will the two solutions have the same or different solute concentrations?

   "Cell"
   0.03 *M* sucrose
   0.02 *M* glucose

   "Environment"
   0.01 *M* sucrose
   0.01 *M* glucose
   0.01 *M* fructose

Levels 5-6: Evaluating/Creating

7. **EVOLUTION CONNECTION** *Paramecium* and other protists that live in hypotonic environments have cell membranes that limit water uptake, while those living in isotonic environments have membranes that are more permeable to water. Describe what water regulation adaptations might have evolved in protists in hypertonic habitats such as the Great Salt Lake and in habitats with changing salt concentration.

8. **SCIENTIFIC INQUIRY** An experiment is designed to study the mechanism of sucrose uptake by plant cells. Cells are immersed in a sucrose solution, and the pH of the solution is monitored. Samples of the cells are taken at intervals, and their sucrose concentration is measured. The pH is observed to decrease until it reaches a steady, slightly acidic level, and then sucrose uptake begins. (a) Evaluate these results and propose a hypothesis to explain them. (b) Predict what would happen if an inhibitor of ATP regeneration by the cell were added to the beaker once the pH was at a steady level. Explain.

9. **SCIENCE, TECHNOLOGY, AND SOCIETY** Extensive irrigation in arid regions causes salts to accumulate in the soil. (When water evaporates, salts that were dissolved in the water are left behind in the soil.) Based on what you learned about water balance in plant cells, explain why increased soil salinity (saltiness) might be harmful to crops.

10. **WRITE ABOUT A THEME: INTERACTIONS** A human pancreatic cell obtains O_2—and necessary molecules such as glucose, amino acids, and cholesterol—from its environment, and it releases CO_2 as a waste product. In response to hormonal signals, the cell secretes digestive enzymes. It also regulates its ion concentrations by exchange with its environment. Based on what you have just learned about the structure and function of cellular membranes, write a short essay (100–150 words) to describe how such a cell accomplishes these interactions with its environment.

11. **SYNTHESIZE YOUR KNOWLEDGE**

In the supermarket, lettuce and other produce are often sprayed with water. Explain why this makes vegetables crisp.

For selected answers, see Appendix A.

8 An Introduction to Metabolism

KEY CONCEPTS

8.1 An organism's metabolism transforms matter and energy *p. 144*

8.2 The free-energy change of a reaction tells us whether or not the reaction occurs spontaneously *p. 147*

8.3 ATP powers cellular work by coupling exergonic reactions to endergonic reactions *p. 150*

8.4 Enzymes speed up metabolic reactions by lowering energy barriers *p. 153*

8.5 Regulation of enzyme activity helps control metabolism *p. 159*

Study Tip

Make a table: Fill in the following table for each process you read about in this chapter, such as water spilling over a dam or a chemical reaction.

Process	Starting Materials; Relative Energy Level	Ending Materials; Relative Energy Level	Will this process occur spontaneously (without an input of energy)?
Water spilling over a dam	Water at the top of a dam; higher energy level	Water at the bottom of a dam; lower energy level	Yes
Hydrolysis of ATP (Figure 8.9)			

→ Go to Mastering Biology

For Students (in eText and Study Area)
- Get Ready for Chapter 8
- Animation: Exergonic and Endergonic Reactions
- Figure 8.10 Walkthrough: How ATP Drives Chemical Work

For Instructors to Assign (in Item Library)
- Activity: Energy Transformations
- Tutorial: How Enzymes Function

Figure 8.1 The green glowing spots on the outside of this Brazilian termite mound are larvae of the click beetle, *Pyrophorus nyctophanus*. These larvae convert the energy stored in organic molecules to light, a process called bioluminescence, which attracts termites that the larvae eat. Bioluminescence and other metabolic activities in a cell are energy transformations that are subject to physical laws.

How do the laws of thermodynamics relate to biological processes?

The first law of thermodynamics: *Energy can be transferred and transformed, but it cannot be created or destroyed.* For example:

Light energy from the sun

↓ *Transformed by plants to*

Chemical energy in organic molecules in plants

↓ *Transformed by termites to*

Chemical energy in organic molecules in termites

↓ *Transformed by click beetle larvae to*

Light energy produced by bioluminescence

The second law of thermodynamics: *Every energy transfer or transformation increases the entropy (disorder) of the universe.* For example, during every energy transformation, some energy is converted to thermal energy and released as heat:

Heat

Plant using light to make organic molecules

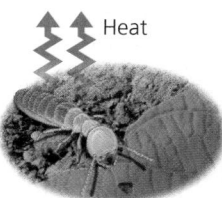

Heat

Termite digesting plant and making new molecules

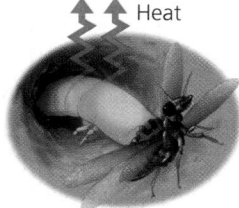

Heat

Click beetle larva digesting termite and emitting light

An organism's metabolism transforms matter and energy

The totality of an organism's chemical reactions is called **metabolism** (from the Greek *metabole*, change). Metabolism is an emergent property of life that arises from orderly inter-actions between molecules.

Metabolic Pathways

We can picture a cell's metabolism as an elaborate road map of many chemical reactions, arranged as intersecting metabolic pathways. In a **metabolic pathway**, a specific molecule is altered in a series of defined steps, resulting in a certain product. Each step is catalyzed by a specific *enzyme*, a macromolecule that speeds up a chemical reaction:

The mechanisms that regulate these enzymes balance meta-bolic supply and demand, just like traffic lights control the flow of traffic.

Metabolism as a whole manages the material and energy resources of the cell. Some metabolic pathways release energy by breaking down complex molecules to sim-pler compounds. These degradative processes are called **catabolic pathways**, or breakdown pathways. One major catabolic pathway is cellular respiration, which breaks down glucose and other organic fuels in the presence of oxygen to carbon dioxide and water. (Pathways can have more than one starting molecule and/or product.) Energy stored in the organic molecules becomes available to do cellular work, such as ciliary beating or membrane transport. **Anabolic pathways**, in contrast, consume energy to build compli-cated molecules from simpler ones; they are sometimes called biosynthetic pathways. Examples of anabolism are synthesis of an amino acid from simpler molecules and syn-thesis of a protein from amino acids. Catabolic and anabolic pathways are the "downhill" and "uphill" avenues of the metabolic landscape. Energy released from downhill reac-tions of catabolic pathways can be stored and then used to drive uphill reactions of anabolic pathways.

In this chapter, we will focus on mechanisms common to metabolic pathways. Because energy is fundamental to all metabolic processes, a basic knowledge of energy is neces-sary to understand how the living cell works. Although we'll look at some nonliving examples here to consider energetic principles, the concepts demonstrated by these examples also apply to **bioenergetics**, the study of how energy flows through living organisms.

Forms of Energy

Energy is the capacity to cause change. In everyday life, energy is important because some forms of energy can be used to do work—that is, to move matter against opposing forces, such as gravity and friction. Put another way, energy is the ability to rearrange a collection of matter. For example, you expend energy to turn the pages of this book, and your cells expend energy in transporting certain substances across membranes. Energy exists in various forms, and the work of life depends on the ability of cells to transform energy from one form to another.

Energy can be associated with the relative motion of objects; this energy is called **kinetic energy**. Moving objects can per-form work by imparting motion to other matter: A pool player uses the motion of the cue stick to push the cue ball, which in turn moves the other balls; water gushing through a dam turns turbines; and the contraction of leg muscles pushes bicycle pedals. **Thermal energy** is kinetic energy associated with the random movement of atoms or molecules; thermal energy in transfer from one object to another is called **heat**. Light is also a type of energy that can be harnessed to perform work, such as powering photosynthesis in green plants.

An object not presently moving may still possess energy. Energy that is not kinetic is called **potential energy**; it is energy that matter possesses because of its location or struc-ture. Water behind a dam, for instance, possesses energy because of its altitude above sea level. Molecules possess energy because of the arrangement of electrons in the bonds between their atoms. **Chemical energy** is a term used by biologists to refer to the potential energy available for release in a chemical reaction. Recall that catabolic pathways release energy by breaking down complex molecules. Biologists say that these complex molecules, such as glucose, are high in chemical energy. During a catabolic reaction, some bonds are broken and others are formed, releasing energy and resulting in lower-energy breakdown products. This transformation also occurs in the engine of a car when the hydrocarbons of gasoline react explosively with oxygen, releasing the energy that pushes the pistons and producing exhaust. Although less explosive, a similar reaction of food molecules with oxygen provides chemical energy in biological systems, producing carbon dioxide and water as waste products. Biochemical pathways, carried out in the context of cellular structures, enable cells to release chemical energy from food molecules and use the energy to power life processes.

How is energy converted from one form to another? Consider **Figure 8.2**. The woman climbing the ladder to the diving platform is releasing chemical energy from the food she ate for lunch and using some of that energy to perform

▼ Figure 8.2 Transformations between potential energy and kinetic energy.

A diver has more potential energy on the platform than in the water.

Diving converts potential energy to kinetic energy.

Climbing up converts the kinetic energy of muscle movement to potential energy.

A diver has less potential energy in the water than on the platform.

⊙ **Mastering Biology** Animation: Energy Transformations

the work of climbing. The kinetic energy of muscle movement is thus being transformed into potential energy due to her increasing height above the water. The man diving is converting his potential energy to kinetic energy, which is then transferred to the water as he enters it, resulting in splashing, noise, and increased movement of water molecules. A small amount of energy is lost as heat due to friction.

Now let's consider the original source of the organic food molecules that provided the necessary chemical energy for these divers to climb the steps. This chemical energy was itself derived from light energy absorbed by plants during photosynthesis. Organisms are energy transformers.

The Laws of Energy Transformation

The study of the energy transformations that occur in a collection of matter is called **thermodynamics**. Scientists use the word *system* to denote the matter under study; they refer to the rest of the universe—everything outside the system—as the *surroundings*. An *isolated system*, such as that approximated by liquid in a thermos bottle, is unable to exchange either energy or matter with its surroundings outside the thermos. In an *open system*, energy and matter can be transferred between the system and its surroundings. Organisms are open systems. They absorb energy—for instance, light energy or chemical energy in the form of organic molecules—and release heat and metabolic waste products, such as carbon dioxide, to the surroundings. Two laws of thermodynamics govern energy transformations in organisms and all other collections of matter.

The First Law of Thermodynamics

According to the **first law of thermodynamics**, the energy of the universe is constant: *Energy can be transferred and transformed, but it cannot be created or destroyed.* The first law is also known as the *principle of conservation of energy*. The electric company does not make energy, but merely converts it to a form that is convenient for us to use. By converting sunlight to chemical energy, a plant acts as an energy transformer, not an energy producer.

The brown bear in **Figure 8.3a** will convert the chemical energy of the organic molecules in its food to kinetic and other forms of energy as it carries out biological processes. What happens to this energy after it has performed work? The second law of thermodynamics helps to answer this question.

▼ Figure 8.3 The first two laws of thermodynamics.

Chemical energy in food

Kinetic energy

Heat

CO_2 + H_2O

(a) First law of thermodynamics: Energy can be transferred or transformed but neither created nor destroyed. For example, chemical reactions in this brown bear will convert the chemical (potential) energy in the fish into the kinetic energy of running.

(b) Second law of thermodynamics: Every energy transfer or transformation increases the disorder (entropy) of the universe. For example, as the bear runs, disorder is increased around its body by the release of heat and small molecules that are the by-products of metabolism. A brown bear can run at speeds up to 35 miles per hour (56 km/hr)—as fast as a racehorse.

⊙ **Mastering Biology** MP3 Tutor: Basic Energy Concepts

The Second Law of Thermodynamics

If energy cannot be destroyed, why can't organisms simply recycle their energy over and over again? It turns out that during every energy transfer or transformation, some energy is converted to thermal energy and released as heat, becoming unavailable to do work. Only a small fraction of the chemical energy from the food in Figure 8.3a is transformed into the motion of the brown bear shown in **Figure 8.3b**; most is lost as heat, which dissipates rapidly through the surroundings.

A system can put thermal energy to work only when there is a temperature difference that results in thermal energy flowing as heat from a warmer location to a cooler one. If temperature is uniform, as it is in a living cell, then the heat generated during a chemical reaction will simply warm a body of matter, such as the organism. (This can make a room crowded with people uncomfortably warm, as each person is carrying out a multitude of chemical reactions!)

A consequence of the loss of usable energy as heat to the surroundings is that each energy transfer or transformation makes the universe more disordered. We are all familiar with the word "disorder" in the sense of a messy room or a run-down building. The word "disorder" as used by scientists, however, has a specific molecular definition related to how dispersed the energy is in a system and how many different energy levels are present. For simplicity, we use "disorder" in the following discussion because our common understanding (the messy room) is a good analogy for molecular disorder.

Scientists use a quantity called **entropy** as a measure of molecular disorder, or randomness. The more randomly arranged a collection of matter is, the greater its entropy. We can now state the **second law of thermodynamics**: *Every energy transfer or transformation increases the entropy of the universe.* Although order can increase locally, there is an unstoppable trend toward randomization of the universe as a whole.

The physical disintegration of a system's organized structure is a good analogy for an increase in entropy. For example, you can observe increasing entropy in the gradual decay of an unmaintained building over time. Much of the increasing entropy of the universe is more abstract, however, because it takes the form of increasing amounts of heat and less ordered forms of matter. As the bear in Figure 8.3b converts chemical energy to kinetic energy, it is also increasing the disorder of its surroundings by producing heat and small molecules, such as the CO_2 it exhales, that are the breakdown products of food.

The concept of entropy helps us understand why certain processes are energetically favorable and occur on their own. It turns out that if a given process, by itself, leads to an increase in entropy, that process can proceed without requiring an input of energy. Such a process is called a **spontaneous process**. Note that as we're using it here, the word *spontaneous* does not imply that the process would occur quickly; rather, the word signifies that it is energetically favorable. (In fact, it may be helpful for you to think of the phrase "energetically favorable" when you read the formal term *spontaneous*, the word preferred by chemists.) Some spontaneous processes, such as an explosion, may be virtually instantaneous, while others, such as the rusting of an old car over time, are much slower.

A process that, on its own, leads to a decrease in entropy is said to be nonspontaneous: It will happen only if energy is supplied. We know from experience that certain events occur spontaneously and others do not. For instance, we know that water flows downhill spontaneously but moves uphill only with an input of energy, such as when a machine pumps the water against gravity. Some of that energy is inevitably lost as heat, increasing entropy in the surroundings, so usage of energy during a nonspontaneous process also leads to an increase in the entropy of the universe as a whole.

Biological Order and Disorder

Living systems increase the entropy of their surroundings, as predicted by thermodynamic law. You may wonder, then, how cells create ordered structures from less organized starting materials. For example, simpler molecules are ordered into the more complex structure of an amino acid, and amino acids are ordered into polypeptide chains. At the organismal level as well, complex and beautifully ordered structures result from biological processes that use simpler starting materials **(Figure 8.4)**. How can this happen, if entropy must constantly increase?

▼ **Figure 8.4 Order as a characteristic of life.** Order is evident in the detailed structures of **(a)** the Venus flower basket glass sponge, which inspired the Spanish architect Antoni Gaudí in his design of **(b)** the towers of La Sagrada Família church in Barcelona, Spain.

(a) Glass sponge   **(b)** La Sagrada Família towers

This increase in order is balanced by an organism's taking in organized forms of matter and energy from the surroundings and replacing them with less ordered forms. For example, an animal obtains starch, proteins, and other complex molecules from the food it eats. As catabolic pathways break these molecules down, the animal releases CO_2 and H_2O —small molecules that possess less chemical energy than the food did (see Figure 8.3b). The depletion of chemical energy is accounted for by heat generated during metabolism. On a larger scale, energy flows into most ecosystems in the form of light and exits in the form of heat (see Figure 1.9).

During the early history of life, complex organisms evolved from simpler ancestors. For instance, we can trace the ancestry of the plant kingdom from much simpler organisms called green algae to more complex flowering plants. However, this increase in organization over time in no way violates the second law. The entropy of a particular system, such as an organism, may actually decrease as long as the total entropy of the *universe*—the system plus its surroundings—increases. Thus, organisms are islands of low entropy in an increasingly random universe. The evolution of biological order is perfectly consistent with the laws of thermodynamics.

CONCEPT CHECK 8.1

1. **MAKE CONNECTIONS** How does the second law of thermodynamics help explain the diffusion of a substance across a membrane? (See Figure 7.11.)

2. Describe the forms of energy found in an apple as it grows on a tree, then falls, then is digested by someone who eats it.

3. **WHAT IF?** If you place a teaspoon of sugar in the bottom of a glass of water, it will dissolve completely over time. Left longer, eventually the water will disappear and the sugar crystals will reappear. Explain these observations in terms of entropy.

For suggested answers, see Appendix A.

CONCEPT 8.2

The free-energy change of a reaction tells us whether or not the reaction occurs spontaneously

The laws of thermodynamics that we've just explored apply to the universe as a whole. As biologists, we want to understand the chemical reactions of life—for example, which reactions occur spontaneously and which ones require some input of energy from outside. But how can we know this without assessing the energy and entropy changes in the entire universe for each separate reaction?

Free-Energy Change, ΔG

Recall that the universe is really equivalent to "the system" plus "the surroundings." In 1878, J. Willard Gibbs, a professor at Yale, defined a very useful function called the Gibbs free energy of a system (without considering its surroundings), symbolized by the letter G. We'll refer to the Gibbs free energy simply as free energy. **Free energy** is the portion of a system's energy that can perform work when temperature and pressure are uniform throughout the system, as in a living cell. Let's consider how we determine the free-energy change that occurs when a system changes—for example, during a chemical reaction.

The change in free energy, ΔG, can be calculated for a chemical reaction by applying the following equation:

$$\Delta G = \Delta H - T\Delta S$$

This equation uses only properties of the system (the reaction) itself: ΔH symbolizes the change in the system's *enthalpy* (in biological systems, equivalent to total energy); ΔS is the change in the system's entropy; and T is the absolute temperature in Kelvin (K) units (K = °C + 273; see the back of the book).

Using chemical methods, we can measure ΔG for any reaction. (The value will depend on conditions such as pH, temperature, and concentrations of reactants and products.) Once we know the value of ΔG for a process, we can use it to predict whether the process will be spontaneous (that is, whether it is energetically favorable and will occur without an input of energy). More than a century of experiments has shown that only processes with a negative ΔG are spontaneous. For ΔG to be negative, ΔH must be negative (the system gives up enthalpy and H decreases) or $T\Delta S$ must be positive (the system gives up order and S increases), or both: When ΔH and $T\Delta S$ are tallied, ΔG has a negative value ($\Delta G < 0$) for all spontaneous processes. In other words, every spontaneous process decreases the system's free energy, and processes that have a positive or zero ΔG are never spontaneous.

This information is immensely interesting to biologists, for it allows us to predict which kinds of change can happen without an input of energy. Such spontaneous changes can be harnessed by the cell to perform work. This principle is very important in the study of metabolism, where a major goal is to determine which reactions can supply energy for cellular work.

Free Energy, Stability, and Equilibrium

As we saw in the previous section, when a process occurs spontaneously in a system, we can be sure that ΔG is negative. Another way to think of ΔG is to realize that it represents the difference between the free energy of the final state and the free energy of the initial state:

$$\Delta G = G_{\text{final state}} - G_{\text{initial state}}$$

For a reaction to have a negative ΔG, the system must lose free energy during the change from initial state to final state. Because it has less free energy, the system in its final state is less likely to change and is therefore more stable than it was previously.

We can think of free energy as a measure of a system's instability—its tendency to change to a more stable state. Unstable systems (higher G) tend to change in such a way that they become more stable (lower G). For example, a diver on top of a platform is less stable (more likely to fall) than when floating in the water; a drop of concentrated dye is less stable (more likely to disperse) than when the dye is spread randomly through the liquid; and a glucose molecule is less stable (more likely to break down) than the simpler molecules into which it can be split **(Figure 8.5)**. Unless something prevents it, each of these systems will move toward greater stability: The diver falls, the solution becomes uniformly colored, and the glucose molecule is broken down into smaller molecules.

Another term that describes a state of maximum stability is *equilibrium*, which you learned about in Concept 2.4 in connection with chemical reactions. At equilibrium, the forward and reverse reactions occur at the same rate, and there is no further net change in the relative concentration of products and reactants. For a system at equilibrium, G is at its lowest possible value in that system. Free energy increases when a reaction is somehow pushed away from equilibrium, perhaps by removing some of the products (and thus changing their concentration relative to that of the reactants). We can think of the equilibrium state as a free-energy valley. Any change from the equilibrium position will have a positive ΔG and will not be spontaneous. For this reason, systems never spontaneously move away from equilibrium. Because a system at equilibrium cannot spontaneously change, it can do no work. *A process is spontaneous and can perform work only when it is moving toward equilibrium.*

Free Energy and Metabolism

We can now apply the free-energy concept more specifically to the chemistry of life's processes.

Exergonic and Endergonic Reactions in Metabolism

Based on their free-energy changes, chemical reactions can be classified as either exergonic ("energy outward") or endergonic ("energy inward"). An **exergonic reaction** proceeds with a net release of free energy **(Figure 8.6a)**. Because the chemical mixture loses free energy (G decreases), ΔG is negative for an exergonic reaction. Using ΔG as a standard for spontaneity, exergonic reactions are those that occur spontaneously. (Remember, the word *spontaneous* implies that it is energetically favorable, not that it will occur rapidly.) The magnitude of ΔG for an exergonic reaction

▼ **Figure 8.5 The relationship of free energy to stability, work capacity, and spontaneous change.** Unstable systems (top) are rich in free energy, G. They tend to change spontaneously to a more stable state (bottom). This "downhill" change can be harnessed to perform work.

- More free energy (higher G)
- Less stable
- Greater work capacity

In a **spontaneous change**
- The free energy of the system decreases ($\Delta G < 0$).
- The system becomes more stable.
- The released free energy can be harnessed to do work.

- Less free energy (lower G)
- More stable
- Less work capacity

(a) Gravitational motion. Objects move spontaneously from a higher altitude to a lower one.

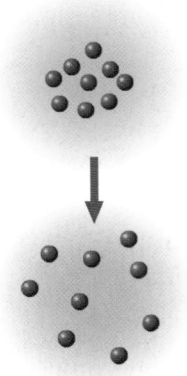

(b) Diffusion. Molecules in a drop of dye diffuse until they are randomly dispersed.

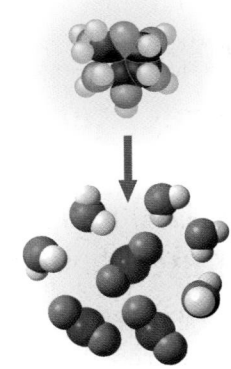

(c) Chemical reaction. In a cell, a glucose molecule is broken down into simpler molecules.

MAKE CONNECTIONS *Compare the redistribution of molecules shown in (b) to the transport of hydrogen ions (H^+) across a membrane by a proton pump, creating a concentration gradient (see Figure 7.18). Which process(es) result(s) in higher free energy? Which system(s) can do work?*

▼ Figure 8.6 Free energy changes (ΔG) in exergonic and endergonic reactions.

(a) Exergonic reaction: energy released, spontaneous

Free energy

Reactants

Energy

Products

Amount of energy released (ΔG < 0)

Progress of the reaction ⟶

(b) Endergonic reaction: energy required, nonspontaneous

Free energy

Products

Energy

Reactants

Amount of energy required (ΔG > 0)

Progress of the reaction ⟶

⊘ **Mastering Biology Animation: Exergonic and Endergonic Reactions**

represents the maximum amount of work the reaction can perform.* The greater the decrease in free energy, the greater the amount of work that can be done.

We can use the overall reaction for cellular respiration as an example:

$$C_6H_{12}O_6 + 6\,O_2 \rightarrow 6\,CO_2 + 6\,H_2O$$
$$\Delta G = -686 \text{ kcal/mol } (-2{,}870 \text{ kJ/mol})$$

For each mole (180 g) of glucose broken down by respiration under what are called "standard conditions" (1 *M* of each reactant and product, 25°C, pH 7), 686 kcal (2,870 kJ) of energy is made available for work. Because energy must be conserved, the chemical *products* of respiration store 686 kcal less free energy per mole than the *reactants*. The products are,

---

*The word *maximum* qualifies this statement because some of the free energy is released as heat and cannot do work. Therefore, ΔG represents a theoretical upper limit of available energy.

in a sense, the spent exhaust of a process that tapped the free energy stored in the bonds of the sugar molecules.

It is important to realize that the breaking of bonds does not release energy; on the contrary, as you will soon see, it requires energy. The phrase "energy stored in bonds" is short-hand for the potential energy that can be released when new bonds are formed after the original bonds break, as long as the products are of lower free energy than the reactants.

An **endergonic reaction** is one that absorbs free energy from its surroundings **(Figure 8.6b)**. Because this kind of reaction essentially *stores* free energy in molecules (*G* increases), ΔG is positive. Such reactions are non-spontaneous, and the magnitude of ΔG is the quantity of energy required to drive the reaction. If a chemical process is exergonic (downhill), releasing energy in one direction, then the reverse process must be endergonic (uphill), using energy. A reversible process cannot be downhill in both directions. If ΔG = −686 kcal/mol for respiration, which converts glucose and oxygen to CO_2 and H_2O, then the reverse process—the conversion of CO_2 and H_2O to glucose and oxygen (O_2)—must be strongly endergonic, with ΔG = +686 kcal/mol. Such a reaction would never happen by itself.

How, then, do plants make sugar? They get the required energy (686 kcal to make a mole of glucose) by capturing light from the sun and converting its energy to chemical energy. Next, in a long series of exergonic steps, they gradually spend that chemical energy to assemble glucose molecules.

Equilibrium and Metabolism

Reactions in an isolated system eventually reach equilibrium and can then do no work, as illustrated by the isolated hydroelectric system in **Figure 8.7**. The chemical reactions of metabolism are reversible, and they, too, would reach equilibrium if they occurred in the isolation of a test tube. Because systems at equilibrium are at a minimum of *G* and can do no work, a cell that has reached metabolic equilibrium is dead! *The fact that metabolism as a whole is never at equilibrium is one of the defining features of life.*

▼ Figure 8.7 Equilibrium and work in an isolated hydroelectric system. Water flowing downhill turns a turbine that drives a generator providing electricity to a lightbulb, but only until the system reaches equilibrium.

ΔG < 0   ΔG = 0

Like most systems, a living cell is not in equilibrium. Materials flow in and out, keeping metabolic pathways from ever reaching equilibrium, and the cell continues to do work throughout its life. This principle is illustrated by the open (and more realistic) hydroelectric system in **Figure 8.8a**. However, unlike this simple single-step system, a catabolic pathway in a cell releases free energy in a series of reactions. An example is cellular respiration, illustrated by analogy in **Figure 8.8b**. Some of the reversible reactions of respiration are constantly "pulled" in one direction—that is, they are kept out of equilibrium. The key to maintaining this lack of equilibrium is that the product of a reaction does not accumulate but instead becomes a reactant in the next step; finally, waste products are expelled from the cell. The overall sequence of reactions is kept going by the huge free-energy difference between glucose and O_2 at the top of the energy "hill" and CO_2 and H_2O at the "downhill" end. As long as our cells have a steady supply of glucose or other fuels and oxygen and are able to expel waste products to the surroundings, their metabolic pathways never reach equilibrium and can continue to do the work of life.

Organisms are open systems. Sunlight provides a daily source of free energy for an ecosystem's plants and other photosynthetic organisms. Animals and other nonphotosynthetic organisms in an ecosystem must have a source of free energy—the organic products of photosynthesis. Now we are ready to see how a cell actually performs the work of life.

CONCEPT 8.3

ATP powers cellular work by coupling exergonic reactions to endergonic reactions

A cell does three main kinds of work:

- *Chemical work*, the pushing of endergonic reactions that would not occur spontaneously, such as the synthesis of polymers from monomers (chemical work will be discussed further here; examples are shown in Chapters 9 and 10)
- *Transport work*, the pumping of substances across membranes against the direction of spontaneous movement (see Concept 7.4)
- *Mechanical work*, such as the beating of cilia (see Concept 6.6), the contraction of muscle cells, and the movement of chromosomes during cellular reproduction

A key feature in the way cells manage their energy resources to do this work is **energy coupling**, the use of an exergonic process to drive an endergonic one. ATP is responsible for mediating most energy coupling in cells, and in most cases it acts as the immediate source of energy that powers cellular work.

> **Mastering Biology** Animation: Energy Coupling

The Structure and Hydrolysis of ATP

ATP (adenosine triphosphate; see Concept 4.3) contains the sugar ribose, with the nitrogenous base adenine and a chain of three phosphate groups (the triphosphate group) bonded to it **(Figure 8.9a)**. In addition to its role in energy coupling, ATP is also one of the nucleoside triphosphates used to make RNA (see Figure 5.23).

▼ Figure 8.8 Equilibrium and work in open systems.

(a) An open hydroelectric system. Water flowing through a turbine keeps driving the generator because intake and outflow of water keep the system from reaching equilibrium.

$\Delta G < 0$

$\Delta G < 0$ $\Delta G < 0$ $\Delta G < 0$

(b) A multistep open hydroelectric system. Cellular respiration is analogous to this system: Glucose is broken down in a series of exergonic reactions that power the work of the cell. The product of each reaction is used as the reactant for the next, so no reaction reaches equilibrium.

The bonds between the phosphate groups of ATP can be broken by hydrolysis. When the terminal phosphate bond is broken by addition of a water molecule, a molecule of inorganic phosphate ($HOPO_3^{2-}$, abbreviated as $\textcircled{P}_i$ throughout this book) leaves the ATP, which becomes adenosine diphosphate, or ADP **(Figure 8.9b)**. The reaction is exergonic and releases 7.3 kcal of energy per mole of ATP hydrolyzed:

$$ATP + H_2O \rightarrow ADP + \textcircled{P}_i$$
$$\Delta G = -7.3 \text{ kcal/mol} (-30.5 \text{ kJ/mol})$$

This is the free-energy change measured under standard conditions. In the cell, conditions do not conform to standard conditions, primarily because reactant and product concentrations differ from 1 M. For example, when ATP

▼ **Figure 8.9 The structure and hydrolysis of adenosine triphosphate (ATP).** Throughout this book, the chemical structure of the triphosphate group seen in **(a)** will be represented by the three joined yellow circles shown in **(b)**.

(a) The structure of ATP. In the cell, most hydroxyl groups of phosphates are ionized (—O^-).

(b) The hydrolysis of ATP. The reaction of ATP and water yields inorganic phosphate ($\textcircled{P}_i$) and ADP and releases energy.

⮕ **Mastering Biology** Animation: The Structure of ATP
Animation: Space-Filling Model of ATP
Animation: Stick Model of ATP

hydrolysis occurs under typical cellular conditions, the actual ΔG is about −13 kcal/mol, 78% greater than the energy released by ATP hydrolysis under standard conditions.

Because their hydrolysis releases energy, the phosphate bonds of ATP are sometimes referred to as high-energy phosphate bonds, but the term is misleading. The phosphate bonds of ATP are not unusually strong bonds, as "high-energy" may imply; rather, the reactants (ATP and water) themselves have high energy relative to the energy of the products (ADP and $\textcircled{P}_i$). The release of energy during the hydrolysis of ATP comes from the chemical change of the system to a state of lower free energy, not from the phosphate bonds themselves.

ATP is useful to the cell because the energy it releases on losing a phosphate group is somewhat greater than the energy most other molecules could deliver. But why does this hydrolysis release so much energy? If we reexamine the ATP molecule in Figure 8.9a, we can see that all three phosphate groups are negatively charged. These like charges are crowded together, and their mutual repulsion contributes to the instability of this region of the ATP molecule. The triphosphate tail of ATP is the chemical equivalent of a compressed spring.

How ATP Provides Energy That Performs Work

When ATP is hydrolyzed in a test tube, the release of free energy merely heats the surrounding water. In an organism, this same generation of heat can sometimes be beneficial. For instance, the process of shivering uses ATP hydrolysis during muscle contraction to warm the body. In most cases in the cell, however, the generation of heat alone would be an inefficient (and potentially dangerous) use of a valuable energy resource. Instead, the cell's proteins harness the energy released during ATP hydrolysis in several ways to perform the three types of cellular work—chemical, transport, and mechanical.

For example, with the help of specific enzymes, the cell is able to use the high free energy of ATP to drive chemical reactions that, by themselves, are endergonic. If the ΔG of an endergonic reaction is less than the amount of energy released by ATP hydrolysis, then the two reactions can be coupled so that, overall, the coupled reactions are exergonic. This usually involves phosphorylation, the transfer of a phosphate group from ATP to some other molecule, such as the reactant. The recipient molecule with the phosphate group covalently bonded to it is then called a **phosphorylated intermediate**. The key to coupling exergonic and endergonic reactions is the formation of this phosphorylated intermediate, which is more

▼ Figure 8.10 How ATP drives chemical work: energy coupling using ATP hydrolysis.

In this example, the exergonic process of ATP hydrolysis drives an endergonic process—synthesis of the amino acid glutamine.

(a) Glutamic acid conversion to glutamine. Glutamine synthesis from glutamic acid (Glu) by itself is endergonic (ΔG is positive), so it is not spontaneous.

Glutamic acid   Ammonia   Glutamine $\Delta G_{Glu} = +3.4$ kcal/mol

(b) Conversion reaction coupled with ATP hydrolysis. In the cell, glutamine synthesis occurs in two steps, coupled by a phosphorylated intermediate (Glu-Ⓟ). ❶ ATP phosphorylates glutamic acid, making it less stable, with more free energy. ❷ Ammonia displaces the phosphate group, forming glutamine.

Glutamic acid   Phosphorylated intermediate   Glutamine

(c) Free-energy change for coupled reaction. ΔG for the glutamic acid conversion to glutamine (+3.4 kcal/mol) plus ΔG for ATP hydrolysis (−7.3 kcal/mol) gives the free-energy change for the overall reaction (−3.9 kcal/mol). Because the overall process is exergonic (net ΔG is negative), it occurs spontaneously.

$$\Delta G_{Glu} = +3.4 \text{ kcal/mol}$$
$$+ \Delta G_{ATP} = -7.3 \text{ kcal/mol}$$
$$\text{Net } \Delta G = -3.9 \text{ kcal/mol}$$

MAKE CONNECTIONS *Referring to Figure 5.14, explain why glutamine (Gln) is drawn in this figure as a glutamic acid (Glu) with an amino group attached.*

→ **Mastering Biology** Figure Walkthrough

---

reactive (less stable, with more free energy) than the original unphosphorylated molecule **(Figure 8.10)**.

Transport and mechanical work in the cell are also nearly always powered by the hydrolysis of ATP. In these cases, ATP hydrolysis leads to a change in a protein's shape and often its ability to bind another molecule. Sometimes this occurs via a phosphorylated intermediate, as seen for the transport protein in **Figure 8.11a**. In most instances of mechanical work involving motor proteins "walking" along cytoskeletal

elements **(Figure 8.11b)**, a cycle occurs in which ATP is first bound noncovalently to the motor protein. Next, ATP is hydrolyzed, releasing ADP and Ⓟ$_i$. Another ATP molecule can then bind. At each stage, the motor protein changes its shape and ability to bind the cytoskeleton, resulting in movement of the protein along the cytoskeletal track. Phosphorylation and dephosphorylation promote crucial protein shape changes during many other important cellular processes as well.

▼ Figure 8.11 How ATP drives transport and mechanical work.

ATP hydrolysis causes changes in the shapes and binding affinities of proteins. This can occur either **(a)** directly, by phosphorylation, as shown for a membrane protein carrying out active transport of a solute (see also Figure 7.16 and the proton pump in Figure 6.32, upper left), or **(b)** indirectly, via noncovalent binding of ATP and its hydrolytic products, as is the case for motor proteins that move vesicles (and other organelles) along cytoskeletal "tracks" in the cell (see also Figures 6.21 and 6.32, lower right).

(a) Transport work: ATP phosphorylates transport proteins, causing a shape change that allows transport of solutes.

(b) Mechanical work: ATP binds noncovalently to motor proteins and then is hydrolyzed, causing a shape change that walks the motor protein forward.

▼ Figure 8.12 The ATP cycle. Energy released by breakdown reactions (catabolism) in the cell is used to phosphorylate ADP, regenerating ATP. Chemical potential energy stored in ATP drives most cellular work.

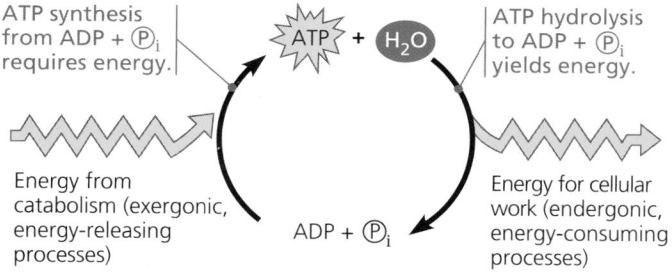

ATP synthesis from ADP + Ⓟᵢ requires energy.

ATP hydrolysis to ADP + Ⓟᵢ yields energy.

Energy from catabolism (exergonic, energy-releasing processes)

Energy for cellular work (endergonic, energy-consuming processes)

ADP + Ⓟᵢ

➡ **Mastering Biology Animation: Metabolism Overview**

The Regeneration of ATP

An organism at work uses ATP continuously, but ATP is a renewable resource that can be regenerated by the addition of phosphate to ADP **(Figure 8.12)**. The free energy required to phosphorylate ADP comes from exergonic breakdown reactions (catabolism) in the cell. This shuttling of inorganic phosphate and energy is called the ATP cycle, and it couples the cell's energy-yielding (exergonic) processes to the energy-consuming (endergonic) ones. The ATP cycle proceeds at an astonishing pace. For example, a working muscle cell recycles its entire pool of ATP in less than a minute. That turnover represents 10 million molecules of ATP consumed and regenerated per second per cell. If ATP could not be regenerated by the phosphorylation of ADP, humans would use up nearly their body weight in ATP each day.

Because both directions of a reversible process cannot be downhill, the regeneration of ATP from ADP and Ⓟᵢ is necessarily endergonic:

$$ADP + Ⓟᵢ \rightarrow ATP + H_2O$$
$$\Delta G = +7.3 \text{ kcal/mol} (+30.5 \text{ kJ/mol}) \text{ (standard conditions)}$$

Since ATP formation from ADP and Ⓟᵢ is not spontaneous, free energy must be spent to make it occur. Catabolic (exergonic) pathways, especially cellular respiration, provide the energy for the endergonic process of making ATP. Plants also use light energy to produce ATP. Thus, the ATP cycle is a key player in bioenergetics, functioning as a revolving door through which energy passes during its transfer from catabolic to anabolic pathways.

CONCEPT CHECK 8.3

1. How does ATP typically transfer energy from an exergonic to an endergonic reaction in the cell?
2. Which combination has more free energy: glutamic acid + ammonia + ATP or glutamine + ADP + Ⓟᵢ? Explain.
3. **MAKE CONNECTIONS** Does Figure 8.11a show passive or active transport? Explain. (See Concepts 7.3 and 7.4.)

For suggested answers, see Appendix A.

Enzymes speed up metabolic reactions by lowering energy barriers

The laws of thermodynamics tell us what will and will not happen under given conditions but say nothing about the rate of these processes. A spontaneous chemical reaction occurs without any requirement for outside energy, but it may occur so slowly that it is imperceptible. For example, even though the hydrolysis of sucrose (table sugar) to glucose and fructose is exergonic, occurring spontaneously with a release of free energy ($\Delta G = -7$ kcal/mol), a solution of sucrose dissolved in sterile water will sit for years at room temperature with no appreciable hydrolysis. However, if we add a small amount of the enzyme sucrase to the solution, then all the sucrose may be hydrolyzed within seconds, as shown here:

Sucrose ($C_{12}H_{22}O_{11}$)   Glucose ($C_6H_{12}O_6$)   Fructose ($C_6H_{12}O_6$)

How does the enzyme do this? An **enzyme** is a macromolecule that acts as a **catalyst**, a chemical agent that speeds up a reaction without being consumed by the reaction. In this chapter, we focus on enzymes that are proteins. (Some RNA molecules, called ribozymes, can function as enzymes; these will be discussed in Concepts 17.3 and 25.1.) Without regulation by enzymes, chemical traffic through the pathways of metabolism would become terribly congested because many chemical reactions would take such a long time. In the next two sections, we will see why spontaneous reactions can be slow and how an enzyme changes the situation.

The Activation Energy Barrier

Every chemical reaction between molecules involves both bond breaking and bond forming. For example, the hydrolysis of sucrose involves breaking the bond between glucose and fructose and one of the bonds of a water molecule and then forming two new bonds, as shown above. Changing one molecule into another generally involves contorting the starting molecule into a highly unstable state before the reaction can proceed. This contortion can be compared to the bending of a metal key ring when you pry it open to add a new key. The key ring is highly unstable in its opened form but returns to a stable state once the key is threaded all the way onto the ring. To reach the contorted state where bonds can change, reactant molecules must absorb energy from their surroundings. When the new bonds of the product molecules form, energy is released as heat, and the molecules return to stable shapes with lower energy than the contorted state.

The initial investment of energy for starting a reaction—the energy required to contort the reactant molecules so the bonds can break—is known as the *free energy of activation*, or **activation energy**, abbreviated E_A in this book. We can think of activation energy as the amount of energy needed to push the reactants to the top of an energy barrier, or "uphill," so that the "downhill" part of the reaction can begin. Activation energy is often supplied by heat in the form of thermal energy that the reactant molecules absorb from the surroundings. The absorption of thermal energy accelerates the reactant molecules, so they collide more often and more forcefully. It also agitates the atoms within the molecules, making the breakage of bonds more likely. When the molecules have absorbed enough energy for the bonds to break, the reactants are in an unstable condition known as the *transition state*.

Figure 8.13 graphs the energy changes for a hypothetical exergonic reaction that swaps portions of two reactant molecules:

$$AB + CD \rightarrow AC + B$$
$$\text{Reactants} \qquad \text{Products}$$

The activation of the reactants is represented by the uphill portion of the graph, in which the free-energy content of

▼ **Figure 8.13 Energy profile of an exergonic reaction.**
The "molecules" are hypothetical, with A, B, C, and D representing portions of the molecules. Thermodynamically, this is an exergonic reaction, with a negative ΔG, and the reaction occurs spontaneously. However, the activation energy (E_A) provides a barrier that determines the rate of the reaction.

The reactants AB and CD must absorb enough energy from the surroundings to reach the unstable transition state, where bonds can break.

After bonds have broken, new bonds form, releasing energy to the surroundings.

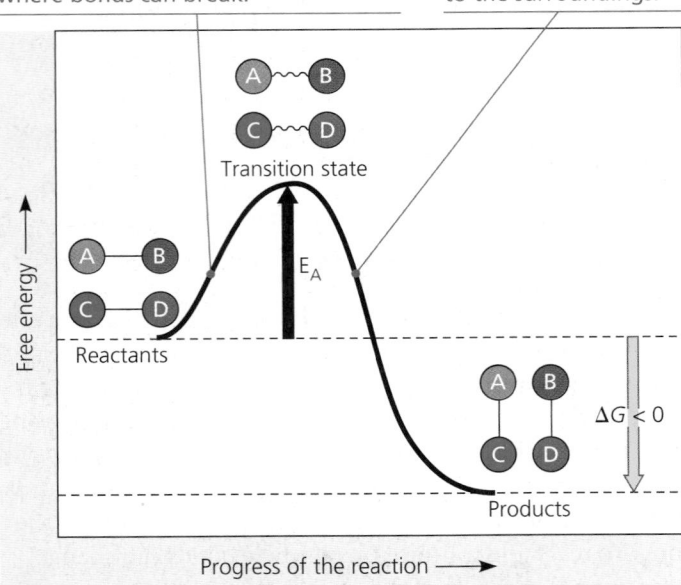

DRAW IT *Graph the progress of an endergonic reaction in which EF and GH form products EG and FH, assuming that the reactants must pass through a transition state.*

the reactant molecules is increasing. At the summit, when energy equivalent to E_A has been absorbed, the reactants are in the transition state: They are activated, and their bonds can be broken. As the atoms then settle into their new, more stable bonding arrangements, energy is released to the surroundings. This corresponds to the downhill part of the curve, which shows the loss of free energy by the molecules. The overall decrease in free energy means that E_A is repaid with dividends, as the formation of new bonds releases more energy than was invested in the breaking of old bonds.

The reaction shown in Figure 8.13 is exergonic and occurs spontaneously ($\Delta G < 0$). However, the activation energy provides a barrier that determines the rate of the reaction. The reactants must absorb enough energy to reach the top of the activation energy barrier before the reaction can occur. For some reactions, E_A is modest enough that even at room temperature there is sufficient thermal energy for many of the reactant molecules to reach the transition state in a short time. In most cases, however, E_A is so high and the transition state is reached so rarely that the reaction will hardly proceed at all. In these cases, the reaction will occur at a noticeable rate only if energy is provided, usually by heat. For example, the reaction of gasoline and oxygen is exergonic and will occur spontaneously, but energy is required for the molecules to reach the transition state and react. Only when the spark plugs fire in an automobile engine can there be the explosive release of energy that pushes the pistons. Without a spark, a mixture of gasoline hydrocarbons and oxygen will not react because the E_A barrier is too high.

How Enzymes Speed Up Reactions

Proteins, DNA, and other complex cellular molecules are rich in free energy and have the potential to decompose spontaneously; that is, the laws of thermodynamics favor their breakdown. These molecules only persist because at temperatures typical for cells, few molecules can make it over the hump of activation energy. The barriers for selected reactions must occasionally be surmounted, however, for cells to carry out the processes needed for life. Heat can increase the rate of a reaction by allowing reactants to attain the transition state more often, but this would not work well in biological systems. First, high temperature denatures proteins and kills cells. Second, heat would speed up *all* reactions, not just those that are needed. Instead of heat, organisms carry out **catalysis**, the process by which a catalyst selectively speeds up a reaction without itself being consumed. (You learned about catalysts earlier in this section.)

An enzyme catalyzes a reaction by lowering the E_A barrier **(Figure 8.14)**, enabling the reactant molecules to absorb enough energy to reach the transition state even at moderate temperatures, as we'll see shortly. It is crucial to note that *an enzyme cannot change the ΔG for a reaction; it cannot make an endergonic reaction exergonic.* Enzymes can only

Figure 8.14 The effect of an enzyme on activation energy.
Without affecting the free-energy change (ΔG) for a reaction, an enzyme speeds the reaction by reducing its activation energy (E_A).

→ **Mastering Biology** Animation: How Enzymes Work

hasten reactions that would eventually occur anyway, but this enables the cell to have a dynamic metabolism, routing chemicals smoothly through metabolic pathways. Also, enzymes are very specific for the reactions they catalyze, so they determine which chemical processes will be going on in the cell at any given time.

Substrate Specificity of Enzymes

The reactant an enzyme acts on is referred to as the enzyme's **substrate**. The enzyme binds to its substrate (or substrates, when there are two or more reactants), forming an **enzyme-substrate complex**. While enzyme and substrate are joined, the catalytic action of the enzyme converts the substrate to

the product (or products) of the reaction. The overall process can be summarized as follows:

$$\text{Enzyme} + \text{Substrate(s)} \rightleftharpoons \text{Enzyme-substrate complex} \rightleftharpoons \text{Enzyme} + \text{Product(s)}$$

Most enzyme names end in *-ase*. (See if you can find three examples of enzymes in the lower left of Figure 6.32.) For example, the enzyme sucrase catalyzes the hydrolysis of the disaccharide sucrose into its two monosaccharides, glucose and fructose (see the diagram at the beginning of Concept 8.4):

$$\text{Sucrase} + \text{Sucrose} + H_2O \rightleftharpoons \text{Sucrase-sucrose-}H_2O \text{ complex} \rightleftharpoons \text{Sucrase} + \text{Glucose} + \text{Fructose}$$

The reaction catalyzed by each enzyme is very specific; an enzyme can recognize its specific substrate even among closely related compounds. For instance, sucrase will act only on sucrose and will not bind to other disaccharides, such as maltose. What accounts for this molecular recognition? Recall that most enzymes are proteins, and that proteins are macromolecules with unique 3-D configurations. The specificity of an enzyme results from its shape, which is a consequence of its amino acid sequence.

Only a restricted region of the enzyme molecule actually binds to the substrate. This region, called the **active site**, is typically a pocket or groove on the surface of the enzyme where catalysis occurs (**Figure 8.15a**; see also Figure 5.16). Usually, the active site is formed by only a few of the enzyme's amino acids, with the rest of the protein molecule providing a framework that determines the shape of the active site. The specificity of an enzyme is attributed to a complementary fit between the shape of its active site and the shape of the substrate.

An enzyme is not a stiff structure locked into a given shape. In fact, recent work by biochemists has shown that enzymes (and other proteins) seem to "dance" between subtly different shapes in a dynamic equilibrium, with slight differences in free energy for each "pose." The shape that best fits the substrate isn't necessarily the one with the lowest energy, but during the very short time the enzyme takes on this shape, its active site can bind to the substrate. The active site itself is also not a rigid receptacle for the substrate. As shown in **Figure 8.15b**, when the substrate enters the active site, the enzyme changes shape slightly due to interactions between the substrate's chemical groups and chemical groups

Figure 8.15 Induced fit between an enzyme and its substrate.

(a) In this space-filling model of the enzyme hexokinase (blue), the active site forms a groove on the surface. The enzyme's substrate is glucose (red).

(b) When the substrate enters the active site, it forms weak bonds with the enzyme, inducing a change in the shape of the protein. This change allows additional weak bonds to form, causing the active site to enfold the substrate and hold it in place.

on the side chains of the amino acids that form the active site. This shape change makes the active site fit even more snugly around the substrate. The tightening of the binding after initial contact—called **induced fit**—is like a clasping handshake. Induced fit brings chemical groups of the active site into positions that enhance their ability to catalyze the chemical reaction.

Catalysis in the Enzyme's Active Site

In most enzymatic reactions, the substrate is held in the active site by so-called weak interactions, such as hydrogen bonds and ionic bonds. The R groups of a few of the amino acids that make up the active site catalyze the conversion of substrate to product, and the product departs from the active site. The enzyme is then free to take another substrate molecule into its active site. The entire cycle happens so fast that a single enzyme molecule typically acts on about 1,000 substrate molecules per second, and some enzymes are even faster. Enzymes, like other catalysts, emerge from the reaction in their original form. Therefore, very small amounts of enzyme can have a huge metabolic impact by functioning over and over again in catalytic cycles. **Figure 8.16** shows a catalytic cycle involving two substrates and two products.

Most metabolic reactions are reversible, and an enzyme can catalyze either the forward or the reverse reaction, depending on which direction has a negative ΔG. This in turn depends mainly on the relative concentrations of reactants and products. The net effect is always in the direction of equilibrium.

Enzymes use a variety of mechanisms that lower activation energy and speed up a reaction (see Figure 8.16, step ❸):

- In reactions involving two or more reactants, the active site provides a template on which the substrates can come together in the proper orientation for a reaction to occur between them.
- As the active site of an enzyme clutches the bound substrates, the enzyme may stretch the substrate molecules toward their transition state form, stressing and bending critical chemical bonds to be broken during the reaction. Because E_A is proportional to the difficulty of breaking the bonds, distorting the substrate helps it approach the transition state and thus reduces the amount of free energy that must be absorbed to achieve that state.
- The active site may also provide a microenvironment that is more conducive to a certain reaction than the solution itself would be without the enzyme. For example, if the active site has amino acids with acidic R groups, it may provide a pocket of low pH in an otherwise neutral cell. Here, an acidic amino acid may facilitate H^+ transfer to the substrate as a key step in catalyzing the reaction.
- Amino acids in the active site may directly participate in the chemical reaction. Sometimes this process involves

▼ **Figure 8.16 The active site and catalytic cycle of an enzyme.** An enzyme can convert one or more reactant molecules to one or more product molecules. The enzyme shown here converts two substrate molecules to two product molecules.

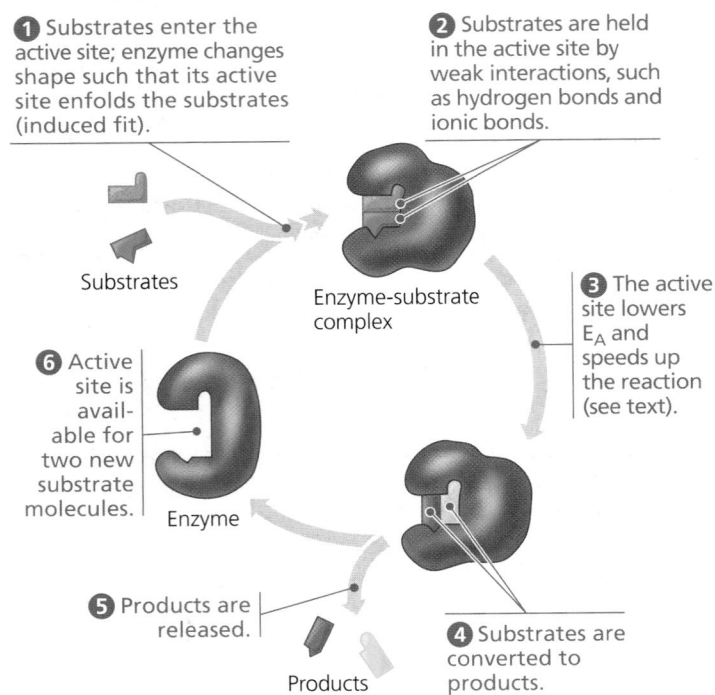

❶ Substrates enter the active site; enzyme changes shape such that its active site enfolds the substrates (induced fit).

❷ Substrates are held in the active site by weak interactions, such as hydrogen bonds and ionic bonds.

Substrates

Enzyme-substrate complex

❸ The active site lowers E_A and speeds up the reaction (see text).

❻ Active site is available for two new substrate molecules.

Enzyme

❺ Products are released.

❹ Substrates are converted to products.

Products

VISUAL SKILLS *The enzyme-substrate complex passes through a transition state (see Figure 8.13). Label the part of the cycle where the transition state occurs.*

→ **Mastering Biology** Animation: Enzymes: Steps in a Reaction

brief covalent bonding between the substrate and the side chain of an amino acid of the enzyme. Subsequent steps restore the side chains to their original states so that the active site is the same after the reaction as it was before.

The rate at which a particular amount of enzyme converts substrate to product is partly a function of the initial concentration of the substrate: The more substrate molecules that are available, the more frequently they access the active sites of the enzyme molecules. However, there is a limit to how fast the reaction can be pushed by adding more substrate to a fixed concentration of enzyme. At some point, the concentration of substrate will be high enough that all enzyme molecules will have their active sites engaged. As soon as the product exits an active site, another substrate molecule enters. At this substrate concentration, the enzyme is said to be *saturated*, and the rate of the reaction is determined by the speed at which the active site converts substrate to product. When an enzyme population is saturated, the only way to increase the rate of product formation is to add more enzyme. Cells often increase the rate of a reaction by producing more enzyme molecules. You can graph the overall progress of an enzymatic reaction in the **Scientific Skills Exercise.**

Making a Line Graph and Calculating a Slope

Does the Rate of Glucose 6-Phosphatase Activity Change over Time in Isolated Liver Cells? Glucose 6-phosphatase, which is found in mammalian liver cells, is a key enzyme in control of blood glucose level. The enzyme catalyzes the breakdown of glucose 6-phosphate into glucose and inorganic phosphate ($(P)_i$). These products are transported out of liver cells into the blood, increasing the blood glucose level. In this exercise, you will graph data from a time-course experiment that measured $(P)_i$ concentration in the buffer outside isolated liver cells, thus indirectly measuring glucose 6-phosphatase activity inside the cells.

How the Experiment Was Done Isolated rat liver cells were placed in a dish with buffer at physiological conditions (pH 7.4, 37°C). Glucose 6-phosphate (the substrate) was added so it could be taken up by cells and broken down by glucose 6-phosphatase. A sample of buffer was removed every 5 minutes and the concentration of $(P)_i$ that had been transported out of the cells was determined.

Data from the Experiment

Time (min)	Concentration of $(P)_i$ (μmol/mL)
0	0
5	10
10	90
15	180
20	270
25	330
30	355
35	355
40	355

Data from S. R. Commerford et al., Diets enriched in sucrose or fat increase gluconeogenesis and G-6-Pase but not basal glucose production in rats, *American Journal of Physiology-Endocrinology and Metabolism* 283:E545–E555 (2002).

INTERPRET THE DATA

1. To see patterns in the data from a time-course experiment like this, it is helpful to graph the data. First, determine which set of data goes on each axis. **(a)** What did the researchers intentionally vary in the experiment? This is the independent variable, which goes on the *x*-axis. **(b)** What are the units (abbreviated) for the independent variable? Explain in words what the abbreviation stands for. **(c)** What was measured by the researchers? This is the dependent variable, which goes on the *y*-axis. **(d)** What does the units abbreviation stand for? Label each axis, including the units.

2. Next, mark off the axes with just enough evenly spaced tick marks to accommodate the full set of data. Determine the span of data values for each axis. **(a)** What is the largest value to go on the *x*-axis? What is a reasonable spacing for the tick marks, and what should be the highest one? **(b)** What is the largest value to go on the *y*-axis? What is a reasonable spacing for the tick marks, and what should be the highest one?

3. Plot the data points on your graph. Match each *x*-value with its partner *y*-value and place a point on the graph at that coordinate. Draw a line that connects the points. (For additional information about graphs, see the Scientific Skills Review in Appendix D.)

4. Examine your graph and look for patterns in the data. **(a)** Does the concentration of $(P)_i$ increase evenly throughout the course of the experiment? To answer this question, describe the pattern you see in the graph. **(b)** What part of the graph shows the highest rate of enzyme activity? Consider that the rate of enzyme activity is related to the slope of the line, $\Delta y / \Delta x$ (the "rise" over the "run"), in μmol/(mL · min), with the steepest slope indicating the highest rate of enzyme activity. Calculate the rate of enzyme activity (slope) where the graph is steepest. **(c)** Can you think of a biological explanation for the pattern you see?

5. If your blood glucose level is low from skipping lunch, what reaction (discussed in this exercise) will occur in your liver cells? Write out the reaction and put the name of the enzyme over the reaction arrow. How will this reaction affect your blood glucose level?

→ **Instructors:** A version of this Scientific Skills Exercise can be assigned in **Mastering Biology**.

Effects of Local Conditions on Enzyme Activity

The activity of an enzyme—how efficiently the enzyme functions—is affected by general environmental factors, such as temperature and pH. It can also be affected by chemicals that specifically influence that enzyme. In fact, researchers have learned much about enzyme function by employing such chemicals.

Effects of Temperature and pH

The 3-D structures of proteins are sensitive to their environment (see Figure 5.20). As a consequence, each enzyme works better under some conditions than others, because these *optimal conditions* favor the most active shape for the enzyme.

Temperature and pH are environmental factors important in the activity of an enzyme. Up to a point, the rate of an enzymatic reaction increases with increasing temperature, partly because substrates collide with active sites more frequently when the molecules move rapidly. Above that temperature, however, the speed of the enzymatic reaction drops sharply. The thermal agitation of the enzyme molecule disrupts the hydrogen bonds, ionic bonds, and other weak interactions that stabilize the active shape of the enzyme, and the protein molecule eventually denatures. Each enzyme has an optimal temperature at which its reaction rate is greatest. Without denaturing the enzyme, this temperature allows the greatest number of molecular collisions and the fastest conversion of the reactants to product molecules. Most human enzymes have optimal temperatures of about 35–40°C (close to human body temperature). The thermophilic bacteria that live in hot springs contain enzymes with optimal temperatures of 70°C or higher **(Figure 8.17a)**.

▼ **Figure 8.17 Environmental factors affecting enzyme activity.** Each enzyme has an optimal **(a)** temperature and **(b)** pH that favor the most active shape of the protein molecule.

(a) The photo shows thermophilic cyanobacteria (green) thriving in the hot water of a Nevada geyser. The graph compares the optimal temperatures for an enzyme from the thermophilic bacterium *Thermus oshimai* (75°C) and human enzymes (37°C, body temperature).

(b) This graph shows the rate of reaction for two digestive enzymes over a range of pH values.

INTERPRET THE DATA *Looking at the graph in (b), what is the optimal pH for pepsin activity? Explain why natural selection might have resulted in the optimal pH for pepsin, a stomach enzyme (see Figure 3.11). What is the optimal pH for trypsin?*

Just as each enzyme has an optimal temperature, it also has a pH at which it is most active. The optimal pH values for most enzymes fall in the range of pH 6–8, but there are exceptions. For example, pepsin, a digestive enzyme in the human stomach, works best at a very low pH. Such an acidic environment denatures most enzymes, but pepsin is well-adapted evolutionarily to maintain its functional 3-D structure in the acidic environment of the stomach. In contrast, trypsin, a digestive enzyme residing in the more alkaline environment of the human intestine would be denatured in the stomach **(Figure 8.17b)**.

Cofactors

Many enzymes require nonprotein helpers for catalytic activity, often for chemical processes like electron transfers that cannot easily be carried out by the amino acids in proteins. These adjuncts, called **cofactors**, may be bound tightly to the enzyme as permanent residents, or they may bind loosely and reversibly along with the substrate. The cofactors of some enzymes are inorganic, such as the metal atoms zinc, iron, and copper in ionic form. If the cofactor is an organic molecule, it is referred to, more specifically, as a **coenzyme**. Most vitamins are important in nutrition because they act as coenzymes or raw materials from which coenzymes are made.

Enzyme Inhibitors

Certain chemicals selectively inhibit the action of specific enzymes. Sometimes the inhibitor attaches to the enzyme by covalent bonds, in which case the inhibition is usually irreversible. Many enzyme inhibitors, however, bind to the enzyme by weak interactions, and when this occurs the inhibition is reversible. Some reversible inhibitors resemble the normal substrate molecule and compete for admission into the active site **(Figure 8.18a and b)**. These mimics, called **competitive inhibitors**, reduce the productivity of enzymes by blocking substrates from entering active sites. This kind of inhibition can be overcome by increasing the concentration of substrate so that as active sites become available, more substrate molecules than inhibitor molecules are around to gain entry to the sites.

In contrast, **noncompetitive inhibitors** do not directly compete with the substrate to bind to the enzyme at the active site **(Figure 8.18c)**. Instead, they impede enzymatic reactions by binding to another part of the enzyme. This interaction causes the enzyme molecule to change its shape in such a way that the active site becomes much less effective at catalyzing the conversion of substrate to product.

Toxins and poisons are often irreversible enzyme inhibitors. An example is sarin, a nerve gas. In 2017, sarin was used in a chemical attack in Syria, killing about 100 people and injuring hundreds more. This small molecule binds covalently to the R group on the amino acid serine, which is found in the active site of acetylcholinesterase, an enzyme important in the nervous system. Other examples include the pesticides DDT and parathion, inhibitors of key enzymes in the nervous system. Finally, many antibiotics are inhibitors of specific

Figure 8.18 Inhibition of enzyme activity.

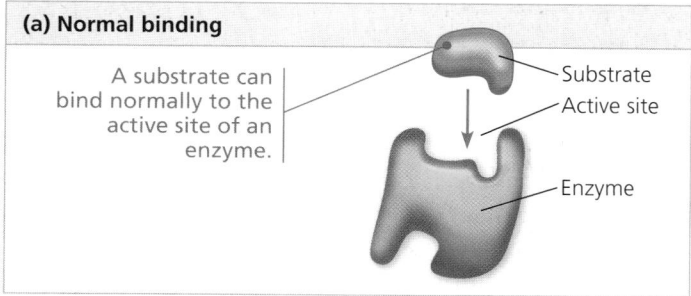

(a) Normal binding

A substrate can bind normally to the active site of an enzyme.

- Substrate
- Active site
- Enzyme

(b) Competitive inhibition

A competitive inhibitor mimics the substrate, competing for the active site.

- Competitive inhibitor

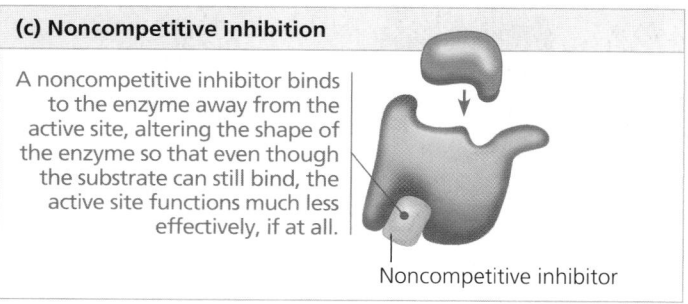

(c) Noncompetitive inhibition

A noncompetitive inhibitor binds to the enzyme away from the active site, altering the shape of the enzyme so that even though the substrate can still bind, the active site functions much less effectively, if at all.

- Noncompetitive inhibitor

→ **Mastering Biology** Animation: Enzymes: Competitive Inhibition • Animation: Enzymes: Noncompetitive Inhibition

enzymes in bacteria. For instance, penicillin blocks the active site of an enzyme that many bacteria use to make cell walls.

Citing enzyme inhibitors that are metabolic poisons may give the impression that enzyme inhibition is generally abnormal and harmful. In fact, molecules naturally present in the cell often regulate enzyme activity by acting as inhibitors. Such regulation—selective inhibition—is essential to the control of cellular metabolism, as you'll see in Concept 8.5.

The Evolution of Enzymes

EVOLUTION Thus far, biochemists have identified more than 4,000 different enzymes in various species, most likely a very small fraction of all enzymes. How did this grand profusion of enzymes arise? Recall that most enzymes are proteins, and proteins are encoded by genes. A permanent change in a gene, known as a *mutation*, can result in a protein with one or more changed amino acids. In the case of an enzyme, if the changed amino acids are in the active site or some other crucial region, the altered enzyme might have a novel activity or might bind to a different substrate. Under environmental conditions where the new function benefits the organism, natural selection would

Figure 8.19 Mimicking evolution of an enzyme with a new function.
Researchers tested whether the function of an enzyme called β-galactosidase, which breaks down the sugar lactose, could change over time in populations of the bacterium *Escherichia coli*. After seven rounds of mutation and selection in the lab, β-galactosidase evolved into an enzyme specialized for breaking down a sugar other than lactose. This ribbon model shows one subunit of the altered enzyme; six amino acids (blue dots) were different.

Two changed amino acids were found near the active site.

Active site

Two changed amino acids were found in the active site.

Two changed amino acids were found on the surface.

tend to favor the mutated form of the gene, causing it to persist in the population. This simplified model is generally accepted as the main way in which the multitude of different enzymes arose over the past few billion years of life's history. Data supporting this model have been collected by researchers using a lab procedure that mimics evolution in natural populations **(Figure 8.19)**.

CONCEPT CHECK 8.4

1. Many spontaneous reactions occur very slowly. Why don't all spontaneous reactions occur instantly?

2. Say you are using a Bunsen burner in the lab. Why doesn't the flame creep back into the gas line and set the gas supply on fire?

3. **WHAT IF?** Malonate is an inhibitor of the enzyme succinate dehydrogenase. How would you determine whether malonate is a competitive or noncompetitive inhibitor?

4. **DRAW IT** A mature lysosome has an internal pH of around 4.5. Using Figure 8.17b as a guide, draw a graph showing what you would predict for the rate of reaction for a lysosomal enzyme. Label its optimal pH, assuming its optimal pH matches its environment.

For suggested answers, see Appendix A.

CONCEPT 8.5

Regulation of enzyme activity helps control metabolism

Chemical chaos would result if all of a cell's metabolic pathways were operating simultaneously. Intrinsic to life's processes is a cell's ability to tightly regulate its metabolic pathways by controlling when and where its various enzymes are active. It does this either by switching on and off the genes that encode specific enzymes (which will be discussed in

Concepts 18.1 and 18.2) or, as discussed here, by regulating the activity of enzymes once they are made.

Allosteric Regulation of Enzymes

In many cases, the molecules that naturally regulate enzyme activity in a cell behave something like reversible noncompetitive inhibitors (see Figure 8.18c): These regulatory molecules change an enzyme's shape and the functioning of its active site by binding to a site elsewhere on the molecule, via noncovalent interactions. **Allosteric regulation** is the term used to describe any case in which a protein's function at one site is affected by the binding of a regulatory molecule to a separate site. It may result in either inhibition or stimulation of an enzyme's activity.

Allosteric Activation and Inhibition

Most enzymes known to be allosterically regulated are constructed from two or more subunits, each composed of a polypeptide chain with its own active site. The entire complex oscillates between two different shapes, one catalytically active and the other inactive (**Figure 8.20a**). In the simplest kind of allosteric regulation, an activating or inhibiting regulatory molecule binds to a regulatory site (sometimes called an allosteric site), often located where subunits join. The binding of an *activator* to a regulatory site stabilizes the shape that has functional active sites, whereas the binding of an *inhibitor* stabilizes the inactive form of the enzyme. The subunits of an allosteric enzyme fit together in such a way that a shape change in one subunit is transmitted to all others. Through this interaction of subunits, a single activator or inhibitor molecule that binds to one regulatory site will affect the active sites of all subunits.

Fluctuating concentrations of regulators can cause a sophisticated pattern of response in the activity of cellular enzymes. The products of ATP hydrolysis (ADP and P_i), for example, play a complex role in balancing the flow of traffic between anabolic and catabolic pathways by their effects on key enzymes. ATP binds to several catabolic enzymes allosterically, lowering their affinity for substrate and thus inhibiting their activity. ADP, however, functions as an activator of the same enzymes. This is logical because catabolism functions in regenerating ATP. If ATP production lags behind its use, ADP accumulates and activates the enzymes that speed up catabolism, producing more ATP. If the supply of ATP exceeds demand, then catabolism slows down as ATP molecules accumulate and bind to the same enzymes, inhibiting them. (You'll see specific examples of this type of regulation when you learn about cellular respiration in the next chapter; see, for example, Figure 9.19.) ATP, ADP, and other related molecules also affect key enzymes in anabolic pathways. In this way, allosteric enzymes control the rates of important reactions in both sorts of metabolic pathways.

In another kind of allosteric activation, a *substrate* molecule binding to one active site in a multisubunit enzyme triggers a shape change in all the subunits, thereby increasing catalytic activity at the other active sites (**Figure 8.20b**).

▼ **Figure 8.20 Allosteric regulation of enzyme activity.**

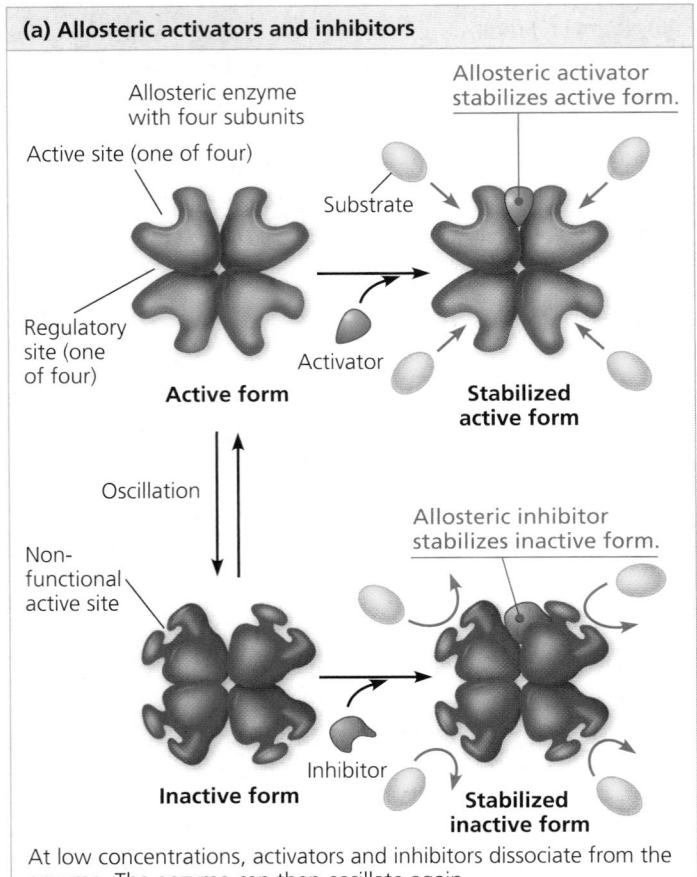

(a) Allosteric activators and inhibitors

Allosteric enzyme with four subunits
Active site (one of four)
Allosteric activator stabilizes active form.
Substrate
Regulatory site (one of four)
Activator
Active form
Stabilized active form
Oscillation
Non-functional active site
Allosteric inhibitor stabilizes inactive form.
Inhibitor
Inactive form
Stabilized inactive form

At low concentrations, activators and inhibitors dissociate from the enzyme. The enzyme can then oscillate again.

(b) Cooperativity: another type of allosteric activation

Binding of one substrate molecule to active site of one subunit locks all subunits in active conformation.
Substrate
Inactive form
Stabilized active form

The inactive form shown on the left oscillates with the active form when the active form is not stabilized by substrate.

Called **cooperativity**, this mechanism amplifies the response of enzymes to substrates: One substrate molecule primes an enzyme to act on additional substrate molecules more readily. Cooperativity is considered allosteric regulation because even though substrate is binding to an active site, its binding affects catalysis in another active site.

Although hemoglobin is not an enzyme (it carries O_2 rather than catalyzing a reaction), classic studies of hemoglobin have elucidated the principle of cooperativity. Hemoglobin is made up of four subunits, each with an O_2-binding site (see Figure 5.18). The binding of an O_2 to one binding site increases the affinity for O_2 of the remaining binding sites. Thus, where O_2 is at a high level, such as in the lungs or gills, hemoglobin's affinity for O_2 increases as more binding sites are filled. In O_2-deprived tissues, however, the release of each O_2 molecule decreases the O_2 affinity of the other binding sites, resulting in the release of O_2 where it is most needed. Cooperativity works similarly in multisubunit enzymes that have been studied.

Feedback Inhibition

Earlier, we looked at the allosteric inhibition of an enzyme in an ATP-generating pathway by ATP itself. This is a common mode of metabolic control, called **feedback inhibition**, in which a metabolic pathway is halted by the inhibitory binding of its end product to an enzyme that acts early in the pathway. **Figure 8.21** shows an example of feedback inhibition operating on an anabolic pathway. Some cells use this five-step pathway to synthesize the amino acid isoleucine from threonine, another amino acid. As isoleucine accumulates, it slows down its own synthesis by allosterically inhibiting the enzyme for the first step of the pathway. Feedback inhibition thereby prevents the cell from making more isoleucine than is necessary and thus wasting chemical resources.

▼ **Figure 8.21 Feedback inhibition in isoleucine synthesis.**

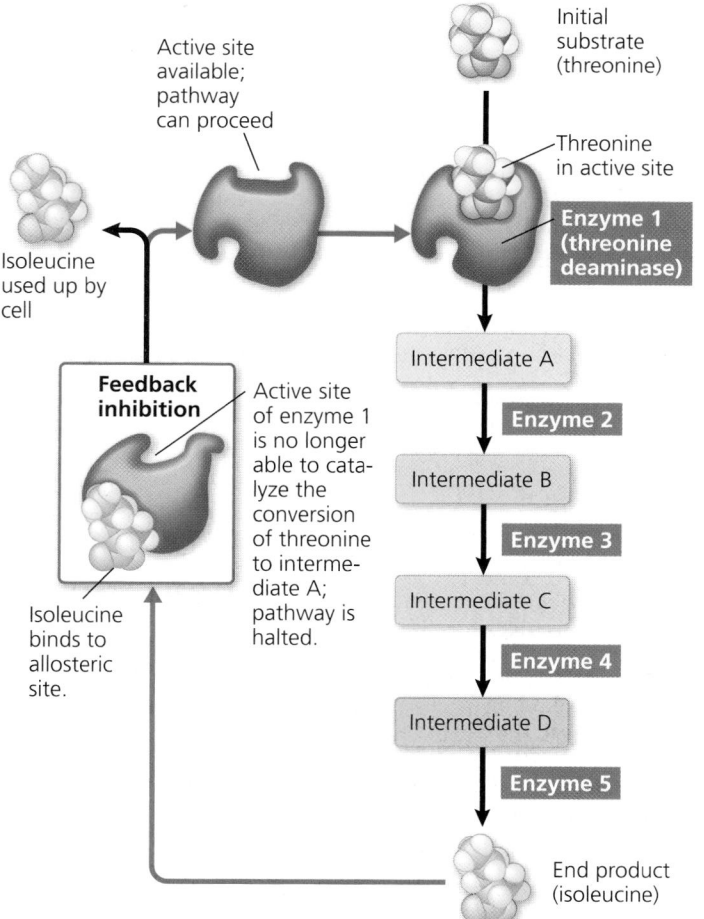

Localization of Enzymes Within the Cell

The cell is not just a bag of chemicals with thousands of different kinds of enzymes and substrates in a random mix. The cell is compartmentalized, and cellular structures help bring order to metabolic pathways. In some cases, a team of enzymes for several steps of a metabolic pathway are assembled into a multienzyme complex. The arrangement facilitates the sequence of reactions, with the product from the first enzyme becoming the substrate for an adjacent enzyme in the complex, and so on, until the end product is released. Some enzymes and enzyme complexes have fixed locations within the cell and act as structural components of particular membranes. Others are in solution within particular membrane-enclosed eukaryotic organelles, each with its own internal chemical environment. For example, in eukaryotic cells, the enzymes for the second and third stages of cellular respiration reside in specific locations within mitochondria (**Figure 8.22**; see also Figure 6.32b).

▼ **Figure 8.22 Organelles and structural order in metabolism.** Organelles such as the mitochondrion (TEM) contain enzymes that carry out specific functions, in this case the second and third stages of cellular respiration. (See also Figure 6.32b, lower left.)

Mitochondrion

| The matrix contains enzymes in solution that are involved in the second stage of cellular respiration. | Enzymes for the third stage of cellular respiration are embedded in the inner membrane. |

In this chapter, you have learned about the laws of thermodynamics that govern metabolism, the intersecting set of chemical pathways characteristic of life. We have explored the bioenergetics of breaking down and building up biological molecules. To continue the theme of energy flow, and specifically of bioenergetics, we will next examine cellular respiration. This is the major catabolic pathway that breaks down organic molecules and releases energy that can be used for the crucial processes of life.

CONCEPT CHECK 8.5

1. How do an activator and an inhibitor have different effects on an allosterically regulated enzyme?

2. **WHAT IF?** Regulation of isoleucine synthesis is an example of feedback inhibition of an anabolic pathway. With that in mind, explain how ATP might be involved in feedback inhibition of a catabolic pathway.

For suggested answers, see Appendix A.

8 Chapter Review

→ Go to **Mastering Biology** for Assignments, the eText, the Study Area, and Dynamic Study Modules.

SUMMARY OF KEY CONCEPTS

→ To review key terms, go to the **Vocabulary Self-Quiz** in the **Mastering Biology** eText or Study Area, or go to goo.gl/zkjz9t.

CONCEPT 8.1

An organism's metabolism transforms matter and energy (pp. 144–147)

- **Metabolism** is the collection of chemical reactions that occur in an organism. Enzymes catalyze reactions in intersecting **metabolic pathways**, which may be **catabolic** (breaking down molecules, releasing energy) or **anabolic** (building molecules, consuming energy). **Bioenergetics** is the study of the flow of energy through living organisms.
- **Energy** is the capacity to cause change; some forms of energy do work by moving matter. **Kinetic energy** is associated with motion and includes **thermal energy** associated with random motion of atoms or molecules. **Heat** is thermal energy in transfer from one object to another. **Potential energy** is related to the location or structure of matter and includes **chemical energy** possessed by a molecule due to its structure.
- The **first law of thermodynamics**, conservation of energy, states that energy cannot be created or destroyed, only transferred or transformed. The **second law of thermodynamics** states that **spontaneous processes**, those requiring no outside input of energy, increase the **entropy** (molecular disorder) of the universe.

? *Explain how the highly ordered structure of a cell does not conflict with the second law of thermodynamics.*

CONCEPT 8.2

The free-energy change of a reaction tells us whether or not the reaction occurs spontaneously (pp. 147–150)

- A living system's **free energy** is energy that can do work under cellular conditions. The change in free energy (ΔG) during a biological process is related directly to enthalpy change (ΔH) and to the change in entropy (ΔS): $\Delta G = \Delta H - T\Delta S$. Organisms live at the expense of free energy. A spontaneous process occurs with no energy input; during such a process, free energy decreases and the stability of a system increases. At maximum stability, the system is at equilibrium and can do no work.
- In an **exergonic** (spontaneous) chemical reaction, the products have less free energy than the reactants ($-\Delta G$). **Endergonic** (nonspontaneous) reactions require an input of energy ($+\Delta G$). The addition of starting materials and the removal of end products prevent metabolism from reaching equilibrium.

? *Explain the meaning of each component in the equation for the change in free energy of a spontaneous chemical reaction. Why are spontaneous reactions important in the metabolism of a cell?*

CONCEPT 8.3

ATP powers cellular work by coupling exergonic reactions to endergonic reactions (pp. 150–153)

- **ATP** is the cell's energy shuttle. Hydrolysis of its terminal phosphate yields ADP and ⓟ$_i$ and releases free energy.
- Through **energy coupling**, the exergonic process of ATP hydrolysis drives endergonic reactions by transfer of a phosphate group to specific reactants, forming a **phosphorylated intermediate**

that is more reactive. ATP hydrolysis (sometimes with protein phosphorylation) also causes changes in the shape and binding affinities of transport and motor proteins.
- Catabolic pathways drive regeneration of ATP from ADP + ⓟ$_i$.

? *Describe the ATP cycle: How is ATP used and regenerated in a cell?*

CONCEPT 8.4

Enzymes speed up metabolic reactions by lowering energy barriers (pp. 153–159)

- In a chemical reaction, the energy necessary to break the bonds of the reactants is the **activation energy**, E_A.
- **Enzymes** lower the E_A barrier:

- Each enzyme has a unique **active site** that binds one or more **substrate(s)**, the reactants on which it acts. It then changes shape, binding the substrate(s) more tightly (**induced fit**).
- The active site can lower an E_A barrier by orienting substrates correctly, straining their bonds, providing a favorable microenvironment, or even covalently bonding with the substrate.
- Each enzyme has an optimal temperature and pH.
- Inhibitors reduce enzyme function. A **competitive inhibitor** binds to the active site, whereas a **noncompetitive inhibitor** binds to a different site on the enzyme.
- Natural selection, acting on organisms with variant enzymes, is responsible for the diversity of enzymes found in organisms.

? *How do both activation energy barriers and enzymes help maintain the structural and metabolic order of life?*

CONCEPT 8.5

Regulation of enzyme activity helps control metabolism (pp. 159–161)

- Many enzymes are subject to **allosteric regulation**: Regulatory molecules, either activators or inhibitors, bind to specific regulatory sites, affecting the shape and function of the enzyme. In **cooperativity**, binding of one substrate molecule can stimulate binding or activity at other active sites. In **feedback inhibition**, the end product of a metabolic pathway allosterically inhibits the enzyme for a previous step in the pathway.
- Some enzymes are grouped into complexes, some are incorporated into membranes, and some are contained inside organelles, increasing the efficiency of metabolic processes.

? *What roles do allosteric regulation and feedback inhibition play in the metabolism of a cell?*

TEST YOUR UNDERSTANDING

➜ For more multiple-choice questions, go to the **Practice Test** in the **Mastering Biology** eText or Study Area, or go to goo.gl/GruWRg.

Levels 1-2: Remembering/Understanding

1. Choose the pair of terms that correctly completes this sentence: Catabolism is to anabolism as _____ is to _____.
(A) exergonic; spontaneous
(B) exergonic; endergonic
(C) free energy; entropy
(D) work; energy

2. Most cells cannot harness heat to perform work because
(A) heat does not involve a transfer of energy.
(B) cells do not have much thermal energy; they are relatively cool.
(C) temperature is usually uniform throughout a cell.
(D) heat can never be used to do work.

3. Which of the following metabolic processes can occur without a net influx of energy from some other process?
(A) $ADP + ⓅP_i + \rightarrow ATP + H_2O$
(B) $C_6H_{12}O_6 + 6\,O_2 \rightarrow 6\,CO_2 + 6\,H_2O$
(C) $6\,CO_2 + 6\,H_2O \rightarrow C_6H_{12}O_6 + 6\,O_2$
(D) Amino acids $\rightarrow$ Protein

4. If an enzyme in solution is saturated with substrate, the most effective way to obtain a faster yield of products is to
(A) add more of the enzyme.
(B) heat the solution to 90°C.
(C) add more substrate.
(D) add a noncompetitive inhibitor.

5. Some bacteria are metabolically active in hot springs because
(A) they are able to maintain a lower internal temperature.
(B) high temperatures make catalysis unnecessary.
(C) their enzymes have high optimal temperatures.
(D) their enzymes are completely insensitive to temperature.

Levels 3-4: Applying/Analyzing

6. If an enzyme is added to a solution where its substrate and product are in equilibrium, what will occur?
(A) Additional substrate will be formed.
(B) The reaction will change from endergonic to exergonic.
(C) The free energy of the system will change.
(D) Nothing; the reaction will stay at equilibrium.

Levels 5-6: Evaluating/Creating

7. DRAW IT Using a series of arrows, draw the branched metabolic reaction pathway described by the following statements, and then answer the question at the end. Use red arrows and minus signs to indicate inhibition.

L can form either M or N.

M can form O.

O can form either P or R.

P can form Q.

R can form S.

O inhibits the reaction of L to form M.

Q inhibits the reaction of O to form P.

S inhibits the reaction of O to form R.

Which reaction would prevail if both Q and S were present in the cell in high concentrations?
(A) L → M
(B) M → O
(C) L → N
(D) O → P

8. EVOLUTION CONNECTION Some people argue that biochemical pathways are too complex to have evolved because all intermediate steps in a given pathway must be present to produce the final product. Critique this argument. How could you use the diversity of metabolic pathways that produce the same or similar products to support your case?

9. SCIENTIFIC INQUIRY • DRAW IT A researcher has developed an assay to measure the activity of an important enzyme present in pancreatic cells growing in culture. She adds the enzyme's substrate to a dish of cells and then measures the appearance of reaction products. The results are graphed as the amount of product on the *y*-axis versus time on the *x*-axis. The researcher notes four sections of the graph. For a short period of time, no products appear (section A). Then (section B) the reaction rate is quite high (the slope of the line is steep). Next, the reaction gradually slows down (section C). Finally, the graph line becomes flat (section D). Draw and label the graph, and propose a model to explain the molecular events occurring at each stage of this reaction profile.

10. WRITE ABOUT A THEME: ENERGY AND MATTER Life requires energy. In a short essay (100–150 words), describe the basic principles of bioenergetics in an animal cell. How are the flow and transformation of energy different in a photosynthesizing cell? Include the role of ATP and enzymes in your discussion.

11. SYNTHESIZE YOUR KNOWLEDGE

Explain what is happening in this photo in terms of kinetic energy and potential energy. Include the energy conversions that occur when the penguins eat fish and climb back up on the glacier. Describe the role of ATP and enzymes in the underlying molecular processes, including what happens to the free energy of some of the molecules involved.

For selected answers, see Appendix A.

9 Cellular Respiration and Fermentation

KEY CONCEPTS

9.1 Catabolic pathways yield energy by oxidizing organic fuels *p. 165*

9.2 Glycolysis harvests chemical energy by oxidizing glucose to pyruvate *p. 170*

9.3 After pyruvate is oxidized, the citric acid cycle completes the energy-yielding oxidation of organic molecules *p. 171*

9.4 During oxidative phosphorylation, chemiosmosis couples electron transport to ATP synthesis *p. 174*

9.5 Fermentation and anaerobic respiration enable cells to produce ATP without the use of oxygen *p. 179*

9.6 Glycolysis and the citric acid cycle connect to many other metabolic pathways *p. 182*

Study Tip

Make a visual study guide: Draw a cell with a large mitochondrion, labeling the parts of the mitochondrion. As you go through the chapter, add key reactions for each stage of cellular respiration, linking the stages together. Label the carbon molecule(s) with the most energy and the carbon molecule(s) with the least energy. Your cell can be a simple sketch, as shown here.

Mitochondrion

→ Go to Mastering Biology

For Students (in eText and Study Area)
- Get Ready for Chapter 9
- BioFlix® Animation: Cellular Respiration
- Figure 9.12 Walkthrough: Free-Energy Change During Electron Transport

For Instructors to Assign (in Item Library)
- BioFlix® Tutorial: Glycolysis
- BioFlix® Tutorial: Cellular Respiration: Inputs and Outputs

Ready-to-Go Teaching Module
(in Instructor Resources)
- Oxidative Phosphorylation (Concept 9.4)

Figure 9.1 This hoary marmot (*Marmota caligata*) obtains energy for its cells by feeding on plants. In the process of cellular respiration, mitochondria in the cells of animals, plants, and other organisms break down organic molecules, generating ATP and waste products: carbon dioxide, water, and heat. Note that energy flows one way, but chemicals are recycled.

How is the chemical energy stored in food used to generate ATP, the molecule that drives most cellular work?

Light energy

used in

Photosynthesis

generates

$CO_2 + H_2O$

generates

Organic molecules $+ O_2$

used in

Cellular respiration in mitochondria

Plant cell

Animal cell

breaks down organic molecules, generating

ATP **powers most cellular work**

Heat

Catabolic pathways yield energy by oxidizing organic fuels

Living cells require transfusions of energy from outside sources to perform their many tasks—for example, assembling polymers, pumping substances across membranes, moving, and reproducing. The outside source of energy is food, and the energy stored in the organic molecules of food ultimately comes from the sun. As shown in **Figure 9.1**, energy flows into an ecosystem as sunlight and exits as heat; in contrast, the chemical elements essential to life are recycled. Photosynthesis generates oxygen, as well as organic molecules used by the mitochondria of eukaryotes as fuel for cellular respiration. Respiration breaks this fuel down, using oxygen (O_2) and generating ATP. The waste products of this type of respiration, carbon dioxide (CO_2) and water (H_2O), are the raw materials for photosynthesis.

> **Mastering Biology** Animation: Energy Flow and Chemical Recycling

Let's consider how cells harvest the chemical energy stored in organic molecules and use it to generate ATP, the molecule that drives most cellular work. Metabolic pathways that release stored energy by breaking down complex molecules are called catabolic pathways (see Concept 8.1). Transfer of electrons from food molecules (like glucose) to other molecules plays a major role in these pathways. In this section, we consider these processes, which are central to cellular respiration.

Catabolic Pathways and Production of ATP

Organic compounds possess potential energy as a result of the arrangement of electrons in the bonds between their atoms. Compounds that can participate in exergonic reactions can act as fuels. Through the activity of enzymes (see Concept 8.4), a cell systematically degrades complex organic molecules that are rich in potential energy to simpler waste products that have less energy. Some of the energy taken out of chemical storage can be used to do work; the rest is dissipated as heat.

One catabolic process, **fermentation**, is a partial degradation of sugars or other organic fuel that occurs without the use of oxygen. However, the most efficient catabolic pathway is **aerobic respiration**, in which oxygen is consumed as a reactant along with the organic fuel (*aerobic* is from the Greek *aer*, air, and *bios*, life). The cells of most eukaryotic and many prokaryotic organisms can carry out aerobic respiration. Some prokaryotes use substances other than oxygen as reactants in a similar process that harvests chemical energy without oxygen; this process is called *anaerobic respiration* (the prefix *an-* means "without"). Technically, the term **cellular respiration** includes both aerobic and anaerobic processes.

However, it originated as a synonym for aerobic respiration because of the relationship of that process to organismal respiration, in which an animal breathes in oxygen. Thus, *cellular respiration* is often used to refer to the aerobic process, a practice we follow in most of this chapter.

Although very different in mechanism, aerobic respiration is in principle similar to the combustion of gasoline in an automobile engine after oxygen is mixed with the fuel (hydrocarbons). Food provides the fuel for respiration, and the exhaust is carbon dioxide and water. The overall process can be summarized as follows:

$$\text{Organic compounds} + \text{Oxygen} \rightarrow \text{Carbon dioxide} + \text{Water} + \text{Energy}$$

Carbohydrates, fats, and proteins from food can all be processed and consumed as fuel. In animal diets, a major source of carbohydrates is starch, a storage polysaccharide that can be broken down into glucose ($C_6H_{12}O_6$) subunits. Here, we will learn the steps of cellular respiration by tracking the degradation of the sugar glucose:

$$C_6H_{12}O_6 + 6\,O_2 \rightarrow 6\,CO_2 + 6\,H_2O + \text{Energy (ATP + heat)}$$

> **Mastering Biology** BioFlix® Animation: Introduction to Cellular Respiration

This breakdown of glucose is exergonic, having a free-energy change of −686 kcal (−2,870 kJ) per mole of glucose decomposed ($\Delta G = -686$ kcal/mol). Recall that a negative $\Delta G (\Delta G < 0)$ indicates that the products of the chemical process store less energy than the reactants and that the reaction can happen spontaneously—in other words, without an input of energy (see Concept 8.2).

Catabolic pathways do not directly move flagella, pump solutes across membranes, polymerize monomers, or perform other cellular work. Catabolism is linked to work by a chemical drive shaft—ATP (see Concept 8.3). To keep working, the cell must regenerate its supply of ATP from ADP and P_i (see Figure 8.12). To understand how cellular respiration accomplishes this, let's examine the fundamental chemical processes known as oxidation and reduction.

Redox Reactions: Oxidation and Reduction

How do the catabolic pathways that decompose glucose and other organic fuels yield energy? The answer is based on the transfer of electrons during the chemical reactions. The relocation of electrons releases energy stored in organic molecules, and this energy ultimately is used to synthesize ATP.

The Principle of Redox

In many chemical reactions, there is a transfer of one or more electrons (e^-) from one reactant to another. These electron transfers are called oxidation-reduction reactions, or **redox reactions** for short. In a redox reaction, the loss of electrons from one substance is called **oxidation**, and the addition of electrons to another substance is known as **reduction**. (Note that *adding* electrons is called *reduction*; adding negatively charged electrons to an atom *reduces* the amount of positive charge of that atom.)

To take a simple, nonbiological example, consider the reaction between the elements sodium (Na) and chlorine (Cl) that forms table salt:

$$\underbrace{Na + Cl \longrightarrow Na^+ + Cl^-}$$

becomes oxidized (loses electron)
becomes reduced (gains electron)

We could generalize a redox reaction this way:

$$Xe^- + Y \longrightarrow X + Ye^-$$

becomes oxidized
becomes reduced

In the generalized reaction, substance Xe^-, the electron donor, is called the **reducing agent**; it reduces Y, which accepts the donated electron. Substance Y, the electron acceptor, is the **oxidizing agent**; it oxidizes Xe^- by removing its electron. Because an electron transfer requires both an electron donor and an acceptor, oxidation and reduction always go hand in hand.

Not all redox reactions involve the complete transfer of electrons from one substance to another; some change the degree of electron sharing in covalent bonds. Methane combustion, shown in **Figure 9.2**, is an example. The covalent electrons in methane are shared nearly equally between the bonded atoms because carbon and hydrogen have about the same affinity for valence electrons; they are about equally electronegative (see Concept 2.3). But when methane reacts with O_2, forming CO_2, electrons end up shared less equally between the carbon atom and its new covalent partners, the oxygen atoms, which are very electronegative. In effect, the carbon atom has partially "lost" its shared electrons; thus, methane has been oxidized.

Now let's examine the fate of the reactant O_2. The two atoms of O_2 share their electrons equally. But after the reaction with methane, when each O atom is bonded to two H atoms in H_2O, the electrons of those covalent bonds spend more time near the oxygen (see Figure 9.2). In effect, each O atom has partially "gained" electrons, so the oxygen molecule (O_2) has been reduced. Because the O atom is so electronegative, O_2 is one of the most powerful of all oxidizing agents.

Energy must be added to pull an electron away from an atom, just as energy is required to push a ball uphill. The more electronegative the atom (the stronger its pull on electrons), the more energy is required to take an electron away from it. An electron loses potential energy when it shifts from a less electronegative atom toward a more electronegative one, just as a ball loses potential energy when it rolls downhill. A redox reaction that moves electrons closer to an O atom, such as the

▼ **Figure 9.2 Methane combustion as an energy-yielding redox reaction.** The reaction releases energy to the surroundings because the electrons lose potential energy when they end up being shared unequally, spending more time near electronegative atoms such as oxygen.

VISUAL SKILLS *Is the carbon atom oxidized or reduced during this reaction? Explain.*

➔ **Mastering Biology** Animation: Redox Reactions

burning (oxidation) of methane, therefore releases chemical energy that can be put to work.

Oxidation of Organic Fuel Molecules During Cellular Respiration

The oxidation of methane by O_2 is the main combustion reaction that occurs at the burner of a gas stove. The combustion of gasoline in an automobile engine is also a redox reaction; the energy released pushes the pistons. But the energy-yielding redox process of greatest interest to biologists is respiration: the oxidation of glucose and other molecules in food. Examine again the summary equation for cellular respiration, but this time think of it as a redox process:

$$C_6H_{12}O_6 + 6 O_2 \longrightarrow 6 CO_2 + 6 H_2O + Energy$$

becomes oxidized
becomes reduced

As in the combustion of methane or gasoline, the fuel (glucose) is oxidized and O_2 is reduced. The electrons lose potential energy along the way, and energy is released.

In general, organic molecules that have an abundance of hydrogen are excellent fuels because their bonds are a source of "hilltop" electrons, whose energy may be released as these electrons "fall" down an energy gradient during their transfer to oxygen. The summary equation for respiration indicates that hydrogen is transferred from glucose to the O atoms in O_2. But the important point, not visible in the summary equation, is that the energy state of the electron changes as hydrogen (with its electron) is transferred to oxygen. In respiration, the oxidation of glucose transfers electrons to a lower energy state, liberating energy that becomes available for ATP synthesis. So, in general, we see fuels with multiple C—H bonds oxidized into products with multiple C—O bonds.

The main energy-yielding foods—carbohydrates and fats—are reservoirs of electrons associated with hydrogen, often in the form of C—H bonds. Only the barrier of activation energy holds back the flood of electrons to a lower energy state (see Figure 8.13). Without this barrier, a food substance like glucose would combine almost instantaneously with O_2. If we supply the activation energy by igniting glucose, it burns in air, releasing 686 kcal (2,870 kJ) of heat per mole of glucose (about 180 g). Body temperature is not high enough to initiate burning, of course. Instead, if you swallow some glucose, enzymes in your cells will lower the barrier of activation energy, allowing the sugar to be oxidized in a series of steps.

Stepwise Energy Harvest via NAD⁺ and the Electron Transport Chain

If energy is released from a fuel all at once, it cannot be harnessed efficiently for constructive work. For example, if a gasoline tank explodes, it cannot drive a car very far. Cellular respiration does not oxidize glucose (or any other organic fuel) in a single explosive step either. Rather, glucose is broken down in a series of steps, each one catalyzed by an enzyme. At key steps, electrons are stripped from the glucose. As is often the case in oxidation reactions, each electron travels with a proton—thus, as a hydrogen atom. The hydrogen atoms are not transferred directly to O_2, but instead are usually passed first to an electron carrier, a coenzyme called nicotinamide adenine dinucleotide, a derivative of the vitamin niacin. This coenzyme is well suited as an electron carrier because it can cycle easily between its oxidized form, **NAD⁺**, and its reduced form, **NADH**. As an electron acceptor, NAD⁺ functions as an oxidizing agent during respiration.

How does NAD⁺ trap electrons from glucose and the other organic molecules in food? Enzymes called dehydrogenases remove a pair of hydrogen atoms (2 electrons and 2 protons)

from the substrate (glucose, in the preceding example), thereby oxidizing it. The enzyme delivers the 2 electrons along with 1 proton to its coenzyme, NAD⁺, forming NADH **(Figure 9.3)**. The other proton is released as a hydrogen ion (H^+) into the surrounding solution:

$$\text{H}-\overset{|}{\underset{|}{\text{C}}}-\text{OH} + \text{NAD}^+ \xrightarrow{\text{Dehydrogenase}} \overset{|}{\underset{|}{\text{C}}}=\text{O} + \text{NADH} + \text{H}^+$$

By receiving 2 negatively charged electrons but only 1 positively charged proton, the nicotinamide portion of NAD⁺ has its charge neutralized when NAD⁺ is reduced to NADH. The name NADH shows the hydrogen that has been received in the reaction. NAD⁺ is the most versatile electron acceptor in cellular respiration and functions in several of the redox steps during the breakdown of glucose.

Electrons lose very little of their potential energy when they are transferred from glucose to NAD⁺. Each NADH molecule formed during respiration represents stored energy that can be tapped to make ATP when the electrons complete their "fall" down an energy gradient from NADH to O_2.

How do electrons that are extracted from glucose and stored as potential energy in NADH finally reach oxygen? It will help to compare the redox chemistry of cellular respiration to a much simpler reaction: the reaction between hydrogen and oxygen to form water **(Figure 9.4a)**. Mix H_2 and O_2, provide a spark for activation energy, and the gases combine explosively. In fact, combustion of liquid H_2 and O_2 is harnessed to help power the rocket engines that boost satellites into orbit and launch spacecraft. The explosion represents a release of energy as the electrons of hydrogen "fall" closer to the electronegative oxygen atoms. Cellular respiration also brings hydrogen and oxygen together to form water, but there are two important differences. First, in cellular respiration, the hydrogen that reacts with oxygen

▲ **Figure 9.3 NAD⁺ as an electron shuttle.** The full name for NAD⁺, nicotinamide adenine dinucleotide, describes its structure—the molecule consists of two nucleotides joined together at their phosphate groups (shown in yellow). (Nicotinamide is a nitrogenous base, although not one that is present in DNA or RNA; see Figure 5.23.) The enzymatic transfer of 2 electrons and 1 proton (H^+) from an organic molecule in food to NAD⁺ reduces the NAD⁺ to NADH: Most of the electrons removed from food are transferred initially to NAD⁺, forming NADH.

is derived from organic molecules rather than H_2. Second, instead of occurring in one explosive reaction, respiration uses an electron transport chain to break the fall of electrons to oxygen into several energy-releasing steps **(Figure 9.4b)**. An **electron transport chain** consists of a number of molecules, mostly proteins, built into the inner membrane of the mitochondria of eukaryotic cells (and the plasma membrane of respiring prokaryotes). Electrons removed from glucose are shuttled by NADH to the "top," higher-energy end of the chain. At the "bottom," lower-energy end, O_2 captures these electrons along with hydrogen nuclei (H^+), forming water. (Anaerobically respiring prokaryotes have an electron acceptor at the end of the chain that is different from O_2.)

Electron transfer from NADH to oxygen is an exergonic reaction with a free-energy change of -53 kcal/mol (-222 kJ/mol). Instead of this energy being released and wasted in a single explosive step, electrons cascade down the chain from one carrier molecule to the next in a series of redox reactions, losing a small amount of energy with each step until they finally reach oxygen, the terminal electron acceptor, which has a very great affinity for electrons. Each "downhill" carrier has a greater affinity for electrons than, and is thus capable of accepting electrons from (oxidizing), its "uphill" neighbor, with O_2 at the bottom of the chain. Therefore, the electrons transferred from glucose to NAD^+, reducing it to NADH, fall down an energy gradient in the electron transport chain to a far more stable location in an electronegative oxygen atom from O_2. Put another way, O_2 pulls electrons down the chain in an energy-yielding tumble analogous to gravity pulling objects downhill.

In summary, during cellular respiration, most electrons travel the following "downhill" route: glucose → NADH → electron transport chain → oxygen. Later in this chapter, you will learn more about how the cell uses the energy released from this exergonic electron fall to regenerate its supply of ATP. For now, having covered the basic redox mechanisms of cellular respiration, let's look at the entire process by which energy is harvested from organic fuels.

The Stages of Cellular Respiration: *A Preview*

The harvesting of energy from glucose by cellular respiration is a cumulative function of three metabolic stages. We list them here along with a color-coding scheme we will use throughout the chapter to help you keep track of the big picture:

1. GLYCOLYSIS (color-coded blue throughout the chapter)
2. PYRUVATE OXIDATION (light orange) and the CITRIC ACID CYCLE (dark orange)
3. OXIDATIVE PHOSPHORYLATION: Electron transport and chemiosmosis (purple)

Biochemists usually reserve the term *cellular respiration* for stages 2 and 3 together. In this text, however, we include glycolysis as a part of cellular respiration because most respiring cells deriving energy from glucose use glycolysis to produce the starting material for the citric acid cycle.

As diagrammed in **Figure 9.5**, glycolysis and then pyruvate oxidation and the citric acid cycle are the catabolic pathways that break down glucose and other organic fuels. **Glycolysis**, which occurs in the cytosol, begins the degradation process by breaking glucose into two molecules of a compound called pyruvate. In eukaryotes, pyruvate enters the mitochondrion and is oxidized to a compound called acetyl CoA, which enters the **citric acid cycle**. There, the breakdown of glucose to carbon dioxide is completed. (In prokaryotes, these processes take place in the cytosol.) Thus, the carbon dioxide produced by respiration represents fragments of oxidized organic molecules.

Some of the steps of glycolysis and the citric acid cycle are redox reactions in which dehydrogenases transfer electrons from substrates to NAD^+ or the related electron carrier FAD, forming NADH or $FADH_2$. (You'll learn more about FAD and $FADH_2$ later.) In the third stage of respiration, the electron transport chain accepts

▼ **Figure 9.4 An introduction to electron transport chains.**

(a) Uncontrolled reaction.
The one-step exergonic reaction of hydrogen with oxygen to form water releases a large amount of energy in the form of heat and light: an explosion.

(b) Cellular respiration. In cellular respiration, the same reaction occurs in stages: An electron transport chain breaks the "fall" of electrons in this reaction into a series of smaller steps and stores some of the released energy in a form that can be used to make ATP. (The rest of the energy is released as heat.)

▶ **Figure 9.5 An overview of cellular respiration.** During glycolysis, each glucose molecule is broken down into two molecules of pyruvate. In eukaryotic cells, as shown here, the pyruvate enters the mitochondrion. There it is oxidized to acetyl CoA, which will be further oxidized to CO_2 in the citric acid cycle. The electron carriers NADH and $FADH_2$ transfer electrons derived from glucose to electron transport chains. During oxidative phosphorylation, electron transport chains convert the chemical energy to a form used for ATP synthesis in the process called chemiosmosis. (During earlier steps of cellular respiration, a few molecules of ATP are synthesized in a process called substrate-level phosphorylation.) To visualize these processes in their cellular context, see Figure 6.32b.

⊙ **Mastering Biology** Animation: Overview of Cellular Respiration

electrons from NADH or $FADH_2$ generated during the first two stages and passes these electrons down the chain. At the end of the chain, the electrons are combined with molecular oxygen (O_2) and hydrogen ions (H^+), forming water (see Figure 9.4b). The energy released at each step of the chain is stored in a form the mitochondrion (or prokaryotic cell) can use to make ATP from ADP. This mode of ATP synthesis is called **oxidative phosphorylation** because it is powered by the redox reactions of the electron transport chain.

In eukaryotic cells, the inner membrane of the mitochondrion is the site of electron transport and another process called *chemiosmosis*, together making up oxidative phosphorylation. (In prokaryotes, these processes take place in the plasma membrane.) Oxidative phosphorylation accounts for almost 90% of the ATP generated by respiration. A smaller amount of ATP is formed directly in a few reactions of glycolysis and the citric acid cycle by a mechanism called **substrate-level phosphorylation (Figure 9.6)**. This mode of ATP synthesis

▼ **Figure 9.6 Substrate-level phosphorylation.** Some ATP is made by direct transfer of a phosphate group from an organic substrate to ADP by an enzyme. (For examples in glycolysis, see Figure 9.8, steps 7 and 10.)

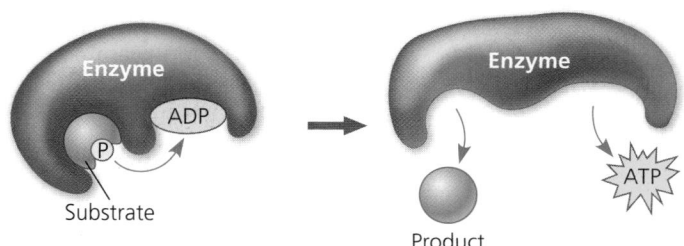

MAKE CONNECTIONS *Review Figure 8.9. In the reaction shown above, is the potential energy higher for the reactants or for the products? Explain.*

occurs when an enzyme transfers a phosphate group from a substrate molecule to ADP, rather than adding an inorganic phosphate to ADP as in oxidative phosphorylation. "Substrate molecule" here refers to an organic molecule generated as an intermediate during the catabolism of glucose. You'll see examples of substrate-level phosphorylation later in the chapter, in both glycolysis and the citric acid cycle.

You can think of the whole process this way: When you withdraw a relatively large sum of money from an ATM, it is not delivered to you in a single bill of a large denomination. Instead, the machine dispenses a number of smaller-denomination bills that you can spend more easily. This is analogous to ATP production during cellular respiration. For each molecule of glucose degraded to CO_2 and H_2O by respiration, the cell makes up to about 32 molecules of ATP, each with 7.3 kcal/mol of free energy. Respiration cashes in the large denomination of energy banked in a single molecule of glucose (686 kcal/mol under standard conditions) for the small change of many molecules of ATP, which is more practical for the cell to spend on its work.

This preview has introduced you to how glycolysis, the citric acid cycle, and oxidative phosphorylation fit into the process of cellular respiration so you can keep the big picture in mind as you take a closer look at each of these three stages of respiration. As you read about the chemical reactions, remember that each reaction is catalyzed by a specific enzyme, some of which are shown in Figure 6.32b.

CONCEPT CHECK 9.1

1. Compare and contrast aerobic and anaerobic respiration, including the processes involved.

2. **WHAT IF?** If the following redox reaction occurred, which compounds would be oxidized? Reduced?

$$C_4H_6O_5 + NAD^+ \rightarrow C_4H_4O_5 + NADH + H^+$$

For suggested answers, see Appendix A.

Glycolysis harvests chemical energy by oxidizing glucose to pyruvate

The word *glycolysis* means "sugar splitting," and that is exactly what happens during this pathway. Glucose, a six-carbon sugar, is split into two three-carbon sugars. These smaller sugars are then oxidized and their remaining atoms rearranged to form two molecules of pyruvate. (Pyruvate is the ionized form of pyruvic acid.)

As summarized in **Figure 9.7**, glycolysis can be divided into two phases: the energy investment phase and the energy payoff phase. During the energy investment phase, the cell actually spends ATP. This investment is repaid with interest during the energy payoff phase, when ATP is produced by substrate-level phosphorylation and NAD^+ is reduced to NADH by electrons released from the oxidation of glucose. The net energy yield from glycolysis, per glucose molecule, is 2 ATP plus 2 NADH. The ten steps of the glycolytic pathway are shown in **Figure 9.8**.

All of the carbon originally present in glucose is accounted for in the two molecules of pyruvate; no carbon is released as CO_2 during glycolysis. Glycolysis occurs whether or not O_2 is present. However, if O_2 *is* present, the chemical energy stored in pyruvate and NADH can be extracted by pyruvate oxidation, the citric acid cycle, and oxidative phosphorylation.

▼ **Figure 9.7** **The inputs and outputs of glycolysis.**

→ **Mastering Biology** Animation: Glycolysis

▼ **Figure 9.8** **The steps of glycolysis.** Glycolysis, a source of ATP and NADH, takes place in the cytosol. Two of the enzymes (in steps ❶ and ❸) are shown in Figure 6.32b.

GLYCOLYSIS: Energy Investment Phase

WHAT IF? *What would happen if you removed the dihydroxyacetone phosphate generated in step 4 as fast as it was produced?*

1. **VISUAL SKILLS** During the redox reaction in glycolysis (see step 6 in Figure 9.8), which one of the molecules acts as the oxidizing agent? The reducing agent?

For suggested answers, see Appendix A.

CONCEPT 9.3

After pyruvate is oxidized, the citric acid cycle completes the energy-yielding oxidation of organic molecules

Glycolysis releases less than a quarter of the chemical energy in glucose that can be harvested by cells; most of the energy remains stockpiled in the two molecules of pyruvate. When O_2 is present, the pyruvate in eukaryotic cells enters a mitochondrion, where the oxidation of glucose is completed. In aerobically respiring prokaryotic cells, this process occurs in the cytosol. (Later in the chapter, we'll examine other fates for pyruvate—for example, when O_2 is unavailable or in a prokaryote that is unable to use O_2.)

Oxidation of Pyruvate to Acetyl CoA

Upon entering the mitochondrion via active transport, pyruvate is first converted to a compound called acetyl coenzyme A, or **acetyl CoA (Figure 9.9)**. This step, linking glycolysis and the citric acid cycle, is carried out by a multienzyme complex that catalyzes three reactions: ❶ Pyruvate's carboxyl group

▼ **Figure 9.9 Oxidation of pyruvate to acetyl CoA, the step before the citric acid cycle.** Pyruvate enters the mitochondrion through a transport protein and is processed by a complex of several enzymes known as pyruvate dehydrogenase (shown in the computer-generated image based on cryo-EMs). This complex catalyzes the three numbered steps, which are described in the text. The CO_2 molecule will diffuse out of the cell. The NADH will be used in oxidative phosphorylation. The acetyl group of acetyl CoA will enter the citric acid cycle. (Coenzyme A is abbreviated S-CoA when it is attached to a molecule, emphasizing its sulfur atom, S.)

Pyruvate dehydrogenase

→ **Mastering Biology** BioFlix® Animation: Acetyl CoA

The energy payoff phase occurs after glucose is split into two three-carbon sugars. Thus, the coefficient 2 precedes all molecules in this phase.

GLYCOLYSIS: Energy Payoff Phase

6 Triose phosphate dehydrogenase
1,3-Bisphospho-glycerate
7 Phospho-glycerokinase
3-Phospho-glycerate
8 Phospho-glyceromutase
2-Phospho-glycerate
9 Enolase
Phosphoenol-pyruvate (PEP)
10 Pyruvate kinase
Pyruvate

6 Two sequential reactions: (1) G3P is oxidized by the transfer of electrons to NAD^+, forming NADH. (2) Using energy from this exergonic redox reaction, a phosphate group is attached to the oxidized substrate, making a high-energy product.

7 The phosphate group is transferred to ADP (substrate-level phosphorylation) in an exergonic reaction. The carbonyl group of G3P has been oxidized to the carboxyl group (—COO⁻) of an organic acid (3-phosphoglycerate).

8 This enzyme relocates the remaining phosphate group.

9 Enolase causes a double bond to form in the substrate by extracting a water molecule, yielding phosphoenolpyruvate (PEP), a compound with a very high potential energy.

10 The phosphate group is transferred from PEP to ADP (a second example of substrate-level phosphorylation), forming pyruvate.

→ **Mastering Biology** BioFlix® Animation: Glycolysis

(—COO⁻), already somewhat oxidized and thus carrying little chemical energy, is now fully oxidized and given off as a molecule of CO_2. This is the first step in which CO_2 is released during respiration. ❷ Next, the remaining two-carbon fragment is oxidized and the electrons transferred to NAD^+, storing energy in the form of NADH. ❸ Finally, coenzyme A (CoA), a sulfur-containing compound derived from a B vitamin, is attached via its sulfur atom to the two-carbon intermediate, forming acetyl CoA. Acetyl CoA has a high potential energy, which is used to transfer the acetyl group to a molecule in the citric acid cycle, a reaction that is therefore highly exergonic.

The Citric Acid Cycle

The citric acid cycle functions as a metabolic furnace that further oxidizes organic fuel derived from pyruvate. **Figure 9.10** summarizes the inputs and outputs as

▼ **Figure 9.10 An overview of pyruvate oxidation and the citric acid cycle.** The inputs and outputs per pyruvate molecule are shown with a focus on the carbon atoms involved. To calculate on a per-glucose basis, multiply by 2 because each glucose molecule is split during glycolysis into two pyruvate molecules.

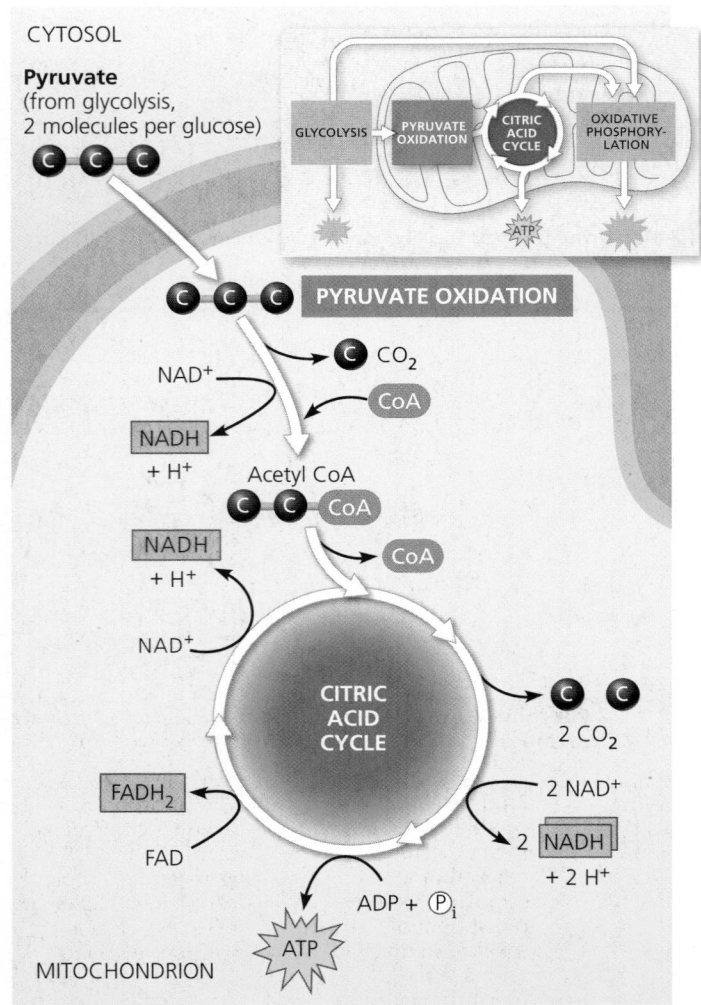

→ **Mastering Biology** Animation: The Citric Acid Cycle

pyruvate is broken down to three CO_2 molecules, including the molecule of CO_2 released during the conversion of pyruvate to acetyl CoA. The cycle generates 1 ATP per turn by substrate-level phosphorylation, but most of the chemical energy is transferred to NAD^+ and FAD during the redox reactions. The reduced coenzymes, NADH and $FADH_2$, shuttle their cargo of high-energy electrons into the electron transport chain. The citric acid cycle is also called the tricarboxylic acid cycle or the Krebs cycle, the latter honoring Hans Krebs. Krebs was the German-British scientist largely responsible for working out the pathway in the 1930s.

Now let's look at the citric acid cycle in more detail. The cycle has eight steps, each catalyzed by a specific enzyme. You can see in **Figure 9.11** that for each turn of the citric acid cycle, two carbons (red) enter in the relatively reduced form of an acetyl group (step ❶), and two different carbons (blue) leave in the completely oxidized form of CO_2 molecules (steps ❸ and ❹). The acetyl group of acetyl CoA joins the cycle by combining with the compound oxaloacetate, forming citrate (step ❶). Citrate is the ionized form of citric acid, for which the cycle is named. The next seven steps decompose the citrate back to oxaloacetate. It is this regeneration of oxaloacetate that makes the process a *cycle*.

Referring to Figure 9.11, we can tally the energy-rich molecules produced by the citric acid cycle. For each acetyl group entering the cycle, 3 NAD^+ are reduced to NADH (steps ❸, ❹, and ❽). In step ❻, electrons are transferred not to NAD^+, but to FAD, which accepts 2 electrons and 2 protons to become $FADH_2$. In many animal tissue cells, the reaction in step ❺ produces a guanosine triphosphate (GTP) molecule by substrate-level phosphorylation. GTP is a molecule similar to ATP in its structure and cellular function. This GTP may be used to make an ATP molecule (as shown) or directly power work in the cell. In the cells of plants, bacteria, and some animal tissues, step ❺ forms an ATP molecule directly by substrate-level phosphorylation. The output from step ❺ represents the only ATP generated during the citric acid cycle. Recall that each glucose gives rise to two molecules of acetyl CoA that enter the cycle. Because the numbers noted earlier are obtained from a single acetyl group entering the pathway, the total yield per glucose from the citric acid cycle turns out to be doubled, or 6 NADH, 2 $FADH_2$, and the equivalent of 2 ATP.

Most of the ATP produced by respiration is generated later, from oxidative phosphorylation, when the NADH and $FADH_2$ produced by the citric acid cycle and earlier steps relay the electrons extracted from food to the electron transport chain. In the process, they supply the necessary energy for the phosphorylation of ADP to ATP. We will explore this process in the next section.

▼ **Figure 9.11 A closer look at the citric acid cycle.** In the chemical structures, red type traces the fate of the two carbon atoms that enter the cycle via acetyl CoA (step **1**), and blue type indicates the two carbons that exit the cycle as CO_2 in steps **3** and **4**. (The red type goes only through step **5** because the succinate molecule is symmetrical; the two ends cannot be distinguished from each other.) Notice that the carbon atoms that enter the cycle from acetyl CoA do not leave the cycle in the same turn. They remain in the cycle, occupying a different location in the molecules on their next turn, after another acetyl group is added. Therefore, the oxaloacetate regenerated at step **8** is made up of different carbon atoms each time around. Carboxylic acids are represented in their ionized forms, as —COO⁻, because the ionized forms prevail at the pH within the mitochondrion. In eukaryotic cells, all the citric acid cycle enzymes are located in the mitochondrial matrix except for the enzyme that catalyzes step **6**, which resides in the inner mitochondrial membrane. (The enzyme that catalyzes step **3**, isocitrate dehydrogenase, is shown in Figure 6.32b.)

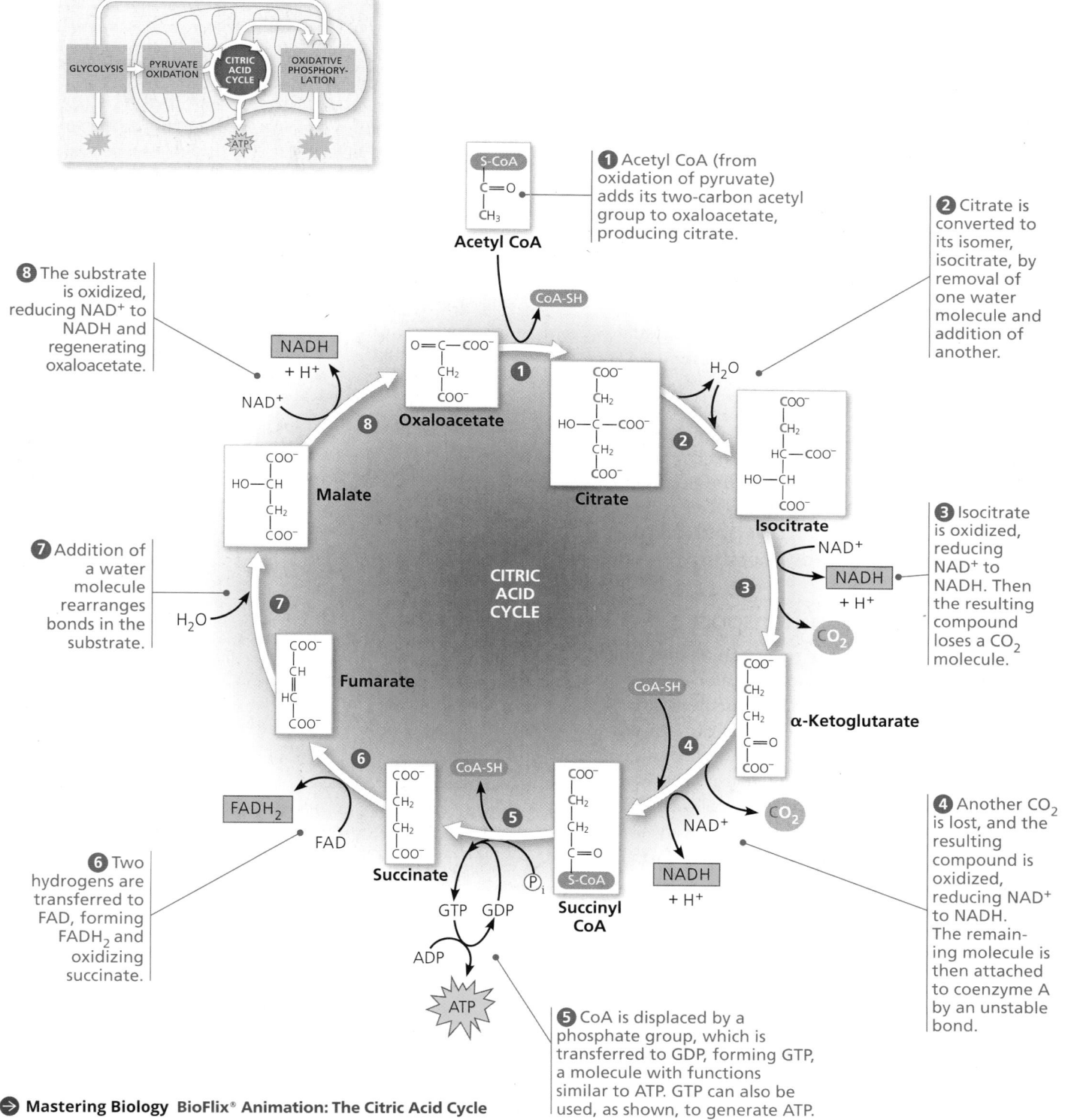

1 Acetyl CoA (from oxidation of pyruvate) adds its two-carbon acetyl group to oxaloacetate, producing citrate.

2 Citrate is converted to its isomer, isocitrate, by removal of one water molecule and addition of another.

3 Isocitrate is oxidized, reducing NAD^+ to NADH. Then the resulting compound loses a CO_2 molecule.

4 Another CO_2 is lost, and the resulting compound is oxidized, reducing NAD^+ to NADH. The remaining molecule is then attached to coenzyme A by an unstable bond.

5 CoA is displaced by a phosphate group, which is transferred to GDP, forming GTP, a molecule with functions similar to ATP. GTP can also be used, as shown, to generate ATP.

6 Two hydrogens are transferred to FAD, forming $FADH_2$ and oxidizing succinate.

7 Addition of a water molecule rearranges bonds in the substrate.

8 The substrate is oxidized, reducing NAD^+ to NADH and regenerating oxaloacetate.

→ **Mastering Biology** BioFlix® Animation: The Citric Acid Cycle

1. **VISUAL SKILLS** In the citric acid cycle shown in Figure 9.11, what molecules capture energy from the redox reactions? How is ATP produced?

2. What processes in your cells produce the CO_2 that you exhale?

3. **VISUAL SKILLS** The conversions shown in Figure 9.9 and step 4 of Figure 9.11 are each catalyzed by a large multienzyme complex. What similarities are there in the reactions that occur in these two cases?

For suggested answers, see Appendix A.

CONCEPT 9.4

During oxidative phosphorylation, chemiosmosis couples electron transport to ATP synthesis

Our main objective in this chapter is to learn how cells harvest the energy of glucose and other nutrients in food to make ATP. But the metabolic components of respiration we have examined so far, glycolysis and the citric acid cycle, produce only 4 ATP molecules per glucose molecule, all by substrate-level phosphorylation: 2 net ATP from glycolysis and 2 ATP from the citric acid cycle. At this point, molecules of NADH (and $FADH_2$) account for most of the energy extracted from each glucose molecule. These electron escorts link glycolysis and the citric acid cycle to the machinery of oxidative phosphorylation, which uses energy released by the electron transport chain to power ATP synthesis. In this section, you will learn first how the electron transport chain works and then how electron flow down the chain is coupled to ATP synthesis.

The Pathway of Electron Transport

The electron transport chain is a collection of molecules embedded in the inner membrane of the mitochondrion in eukaryotic cells. (In prokaryotes, these molecules reside in the plasma membrane.) The folding of the inner membrane to form cristae increases its surface area, providing space for thousands of copies of each component of the electron transport chain in a mitochondrion. Structure fits function: The infolded membrane with its concentration of electron carrier molecules is well-suited for the series of sequential redox reactions that take place along the electron transport chain. Most components of the chain are proteins, which exist in multiprotein complexes numbered I through IV. Tightly bound to these proteins are *prosthetic groups*, nonprotein components such as cofactors and coenzymes essential for the catalytic functions of certain enzymes.

Figure 9.12 shows the sequence of electron carriers in the electron transport chain and the drop in free energy as electrons travel down the chain. During this electron transport, electron carriers alternate between reduced and oxidized states as they accept and then donate electrons. Each component

▼ **Figure 9.12 Free-energy change during electron transport.** The overall energy drop (ΔG) for electrons traveling from NADH to oxygen is 53 kcal/mol, but this "fall" is broken up into a series of smaller steps by the electron transport chain. (An oxygen atom is represented here as $\frac{1}{2} O_2$ to show that O_2 is reduced, not individual oxygen atoms.) To view these proteins in their cellular context, see Figure 6.32b.

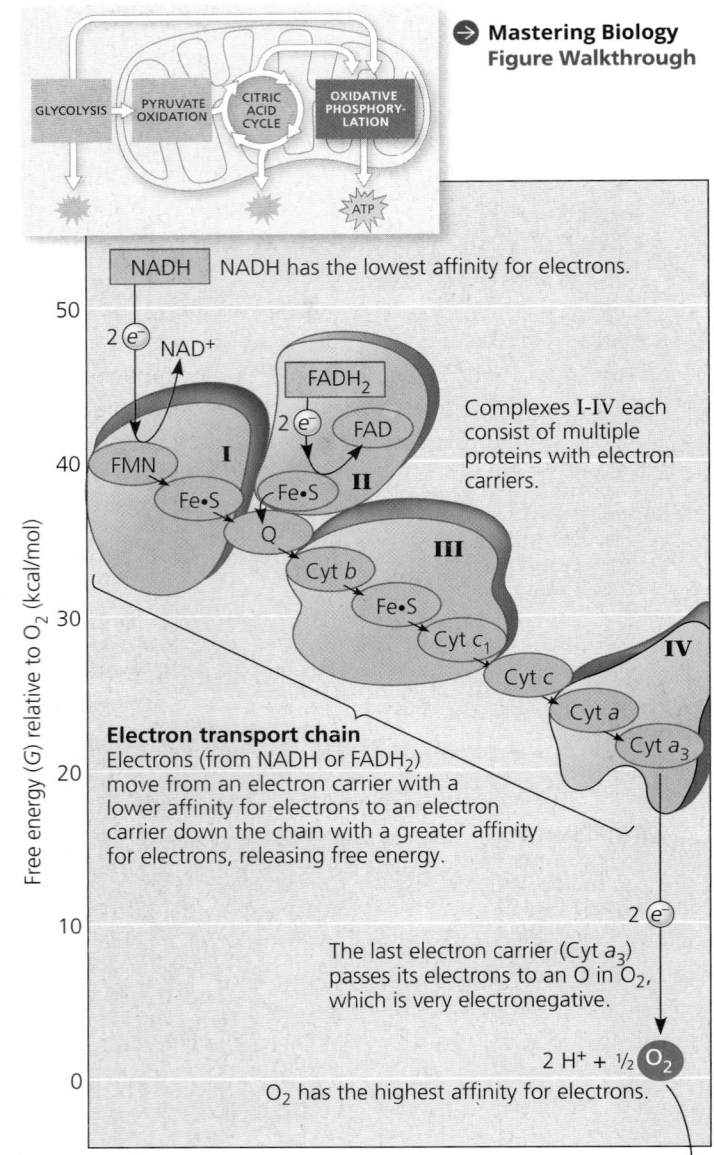

→ **Mastering Biology** Figure Walkthrough

VISUAL SKILLS *Compare the position of the electrons in NADH (see Figure 9.3) at the top of the chain with that in H_2O, at the bottom. Describe why the electrons in H_2O have less potential energy, using the term electronegativity.*

of the chain becomes reduced when it accepts electrons from its "uphill" neighbor, which has a lower affinity for electrons. It then returns to its oxidized form as it passes electrons to its "downhill" neighbor, which has a higher affinity for electrons.

Now let's take a closer look at the electrons as they drop in energy level, passing through the components of the electron transport chain in Figure 9.12. We'll first look at the passage of electrons through complex I in some detail as an illustration of the general principles involved in electron transport. Electrons acquired from glucose by NAD^+ during glycolysis and the citric acid cycle are transferred from NADH to the first molecule of the electron transport chain in complex I. This molecule is a flavoprotein, so named because it has a prosthetic group called flavin mononucleotide (FMN). In the next redox reaction, the flavoprotein returns to its oxidized form as it passes electrons to an iron-sulfur protein (Fe · S in complex I), one of a family of proteins with both iron and sulfur tightly bound. The iron-sulfur protein then passes the electrons to a compound called ubiquinone (Q in Figure 9.12). This electron carrier is a small hydrophobic molecule, the only member of the electron transport chain that is not a protein. Ubiquinone is individually mobile within the membrane rather than residing in a particular complex. (Another name for ubiquinone is coenzyme Q, or CoQ; you may have seen it sold as a nutritional supplement.)

Most of the remaining electron carriers between ubiquinone and oxygen are proteins called **cytochromes**. Their prosthetic group, called a heme group, has an iron atom that accepts and donates electrons. (The heme group in a cytochrome is similar to the heme group in hemoglobin, the protein of red blood cells, except that the iron in hemoglobin carries oxygen, not electrons.) The electron transport chain has several types of cytochromes, each named "cyt" with a letter and number to distinguish it as a different protein with a slightly different electron-carrying heme group. The last cytochrome of the chain, cyt a_3, passes its electrons to oxygen (in O_2), which is *very* electronegative. Each O also picks up a pair of hydrogen ions (protons) from the aqueous solution, neutralizing the −2 charge of the added electrons and forming water.

Another source of electrons for the electron transport chain is $FADH_2$, the other reduced product of the citric acid cycle. Notice in Figure 9.12 that $FADH_2$ adds its electrons from within complex II, at a lower energy level than NADH does. Consequently, although NADH and $FADH_2$ each donate an equivalent number of electrons (2) for oxygen reduction, the electron transport chain provides about one-third less energy for ATP synthesis when the electron donor is $FADH_2$ rather than NADH. We'll see why in the next section.

The electron transport chain makes no ATP directly. Instead, it eases the fall of electrons from food to oxygen, breaking a large free-energy drop into a series of smaller steps that release energy in manageable amounts, step by step. How does the mitochondrion (or the plasma membrane in prokaryotes) couple this electron transport and energy release to ATP synthesis? The answer is a mechanism called chemiosmosis.

Chemiosmosis: The Energy-Coupling Mechanism

Populating the inner membrane of the mitochondrion or the prokaryotic plasma membrane are many copies of a protein complex called **ATP synthase**, the enzyme that makes ATP from ADP and inorganic phosphate **(Figure 9.13)**. ATP synthase works like an ion pump running in reverse. Ion pumps usually use ATP as an energy source to transport ions against their gradients. Enzymes can catalyze a reaction in either direction, depending on the ΔG for the reaction, which is affected by the local concentrations of reactants and products (see Concepts 8.2 and 8.3). Under the conditions of cellular respiration, rather than hydrolyzing ATP to pump protons against their concentration gradient, ATP synthase uses the energy of an existing ion gradient to power ATP synthesis. The power source for ATP synthase is a difference in the concentration of H^+ (a pH difference) on opposite sides of the inner mitochondrial membrane. This process, in which energy stored in the form of a hydrogen ion gradient across a membrane is used to drive cellular work such as the synthesis of ATP, is

▼ **Figure 9.13 ATP synthase, a molecular mill.** Multiple ATP synthases reside in eukaryotic mitochondrial and chloroplast membranes and in prokaryotic plasma membranes. (See Figure 6.32b and c.)

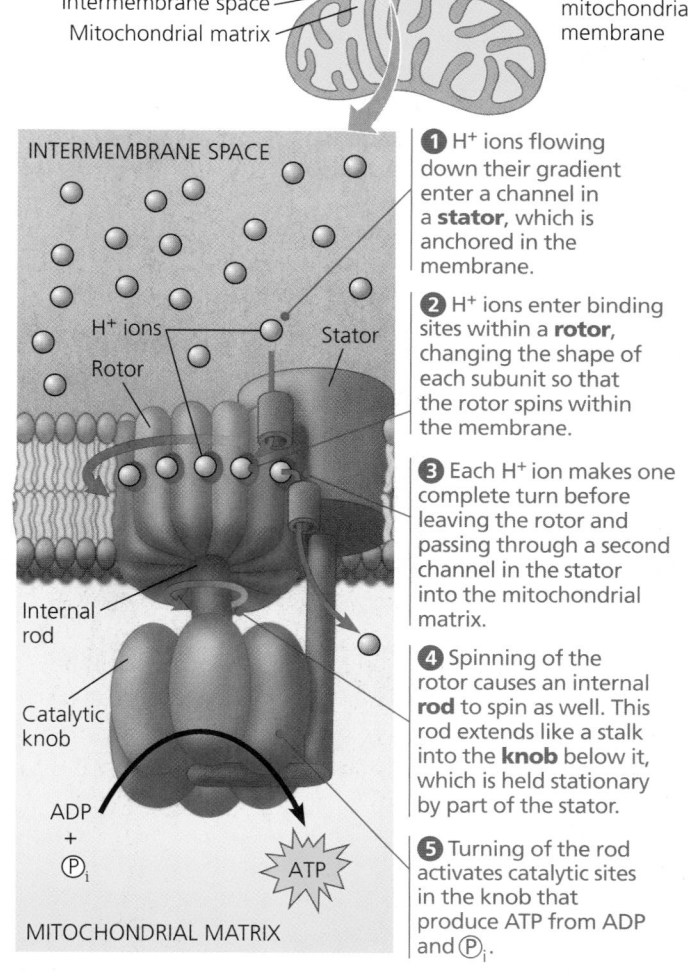

❶ H^+ ions flowing down their gradient enter a channel in a **stator**, which is anchored in the membrane.

❷ H^+ ions enter binding sites within a **rotor**, changing the shape of each subunit so that the rotor spins within the membrane.

❸ Each H^+ ion makes one complete turn before leaving the rotor and passing through a second channel in the stator into the mitochondrial matrix.

❹ Spinning of the rotor causes an internal **rod** to spin as well. This rod extends like a stalk into the **knob** below it, which is held stationary by part of the stator.

❺ Turning of the rod activates catalytic sites in the knob that produce ATP from ADP and P_i.

→ **Mastering Biology** BioFlix® Animation: ATP Synthase
Animation: Rotating ATP Synthase

called **chemiosmosis** (from the Greek *osmos*, push). The word *osmosis* was previously used in discussing water transport, but here it refers to the flow of H⁺ across a membrane.

From studying the structure of ATP synthase, scientists have learned how the flow of H⁺ through this enzyme powers ATP generation. ATP synthase is a multisubunit complex with four main parts, each made up of multiple polypeptides (see Figure 9.13). Protons move one by one into binding sites on the rotor, causing it to spin in a way that catalyzes ATP production from ADP and P_i. The flow of protons thus behaves somewhat like a rushing stream that turns a waterwheel. ATP synthase is the smallest molecular rotary motor known in nature.

How does the inner mitochondrial membrane (or the pro-karyotic plasma membrane) generate and maintain the H⁺ gradient that drives ATP synthesis by the ATP synthase protein complex? Establishing the H⁺ gradient is a major function of the electron transport chain, which is shown in its mitochondrial location in **Figure 9.14**. The chain is an energy converter that uses the exergonic flow of electrons from NADH and FADH_2 to pump H⁺ across the membrane, from the mitochondrial matrix into the intermembrane space. The H⁺ has a tendency to move back across the membrane, diffusing down its gradient. And the ATP synthases are the only sites that provide a route through the membrane for H⁺. As we described previously, the passage

▼ **Figure 9.14 Chemiosmosis couples the electron transport chain to ATP synthesis.** ❶ NADH and FADH_2 shuttle high-energy electrons extracted from food during glycolysis and the citric acid cycle into an electron transport chain built into the inner mitochondrial membrane. (See Figure 6.32b.) The gold arrows trace the transport of electrons, which are finally passed to a terminal acceptor (O₂, in the case of aerobic respiration) at the "downhill" end of the

chain, forming water. Most of the electron carriers of the chain are grouped into four complexes (I–IV). Two mobile carriers, ubiquinone (Q) and cytochrome c (Cyt c), move rapidly, ferrying electrons between the large complexes. As the complexes shuttle electrons, they pump protons from the mitochondrial matrix into the intermembrane space. FADH_2 deposits its electrons via complex II and so results in fewer protons being pumped into the intermembrane space

than occurs with NADH. Chemical energy originally harvested from food is transformed into a proton-motive force, a gradient of H⁺ across the membrane. ❷ During chemiosmosis, the protons flow back down their gradient via ATP synthase, which is built into the membrane nearby. The ATP synthase harnesses the proton-motive force to phosphorylate ADP, forming ATP. Together, electron transport and chemiosmosis make up oxidative phosphorylation.

WHAT IF? *If complex IV were nonfunctional, could chemiosmosis produce any ATP, and if so, how would the rate of synthesis differ?*

➡ **Mastering Biology** Animation: Electron Transport
BioFlix® Animation: Electron Transport

of H⁺ through ATP synthase uses the exergonic flow of H^+ to drive the phosphorylation of ADP. Thus, the energy stored in an H^+ gradient across a membrane couples the redox reactions of the electron transport chain to ATP synthesis.

At this point, you may be wondering how the electron transport chain pumps hydrogen ions into the intermembrane space. Researchers have found that certain members of the electron transport chain accept and release protons (H^+) along with electrons. (The aqueous solutions inside and surrounding the cell are a ready source of H^+.) At certain steps along the chain, electron transfers cause H^+ to be taken up and released into the surrounding solution. In eukaryotic cells, the electron carriers are spatially arranged in the inner mitochondrial membrane in such a way that H^+ is accepted from the mitochondrial matrix and deposited in the intermembrane space (see Figure 9.14). The H^+ gradient that results is referred to as a **proton-motive force**, emphasizing the capacity of the gradient to perform work. The force drives H^+ back across the membrane through the H^+ channels provided by ATP synthases.

In general terms, *chemiosmosis is an energy-coupling mechanism that uses energy stored in the form of an H^+ gradient across a membrane to drive cellular work*. In mitochondria, the energy for gradient formation comes from exergonic redox reactions along the electron transport chain, and ATP synthesis is the work performed. But chemiosmosis also occurs elsewhere and in other variations. Chloroplasts use chemiosmosis to generate ATP during photosynthesis; in these organelles, light (rather than chemical energy) drives both electron flow down an electron transport chain and the resulting H^+ gradient formation. Prokaryotes, as already mentioned, generate H^+ gradients across their plasma membranes. They then tap the proton-motive force not only to make ATP inside the cell but also to rotate their flagella and to pump nutrients and waste products across the membrane. Because of its central importance to energy conversions in prokaryotes and eukaryotes, chemiosmosis has helped unify the study of bioenergetics. Peter Mitchell was awarded the Nobel Prize in 1978 for originally proposing the chemiosmotic model.

An Accounting of ATP Production by Cellular Respiration

In the last few sections, we have looked rather closely at the key processes of cellular respiration. Now let's take a step back and remind ourselves of its overall function: harvesting the energy of glucose for ATP synthesis.

During respiration, most energy flows in this sequence: glucose → NADH → electron transport chain → proton-motive force → ATP. We can do some bookkeeping to calculate the ATP profit when cellular respiration oxidizes a molecule of glucose to six molecules of carbon dioxide. The three main departments of this metabolic enterprise are glycolysis, pyruvate oxidation and the citric acid cycle, and the electron transport chain, which drives oxidative phosphorylation. **Figure 9.15** gives a detailed accounting of the ATP yield for

▼ **Figure 9.15 ATP yield per molecule of glucose at each stage of cellular respiration.**

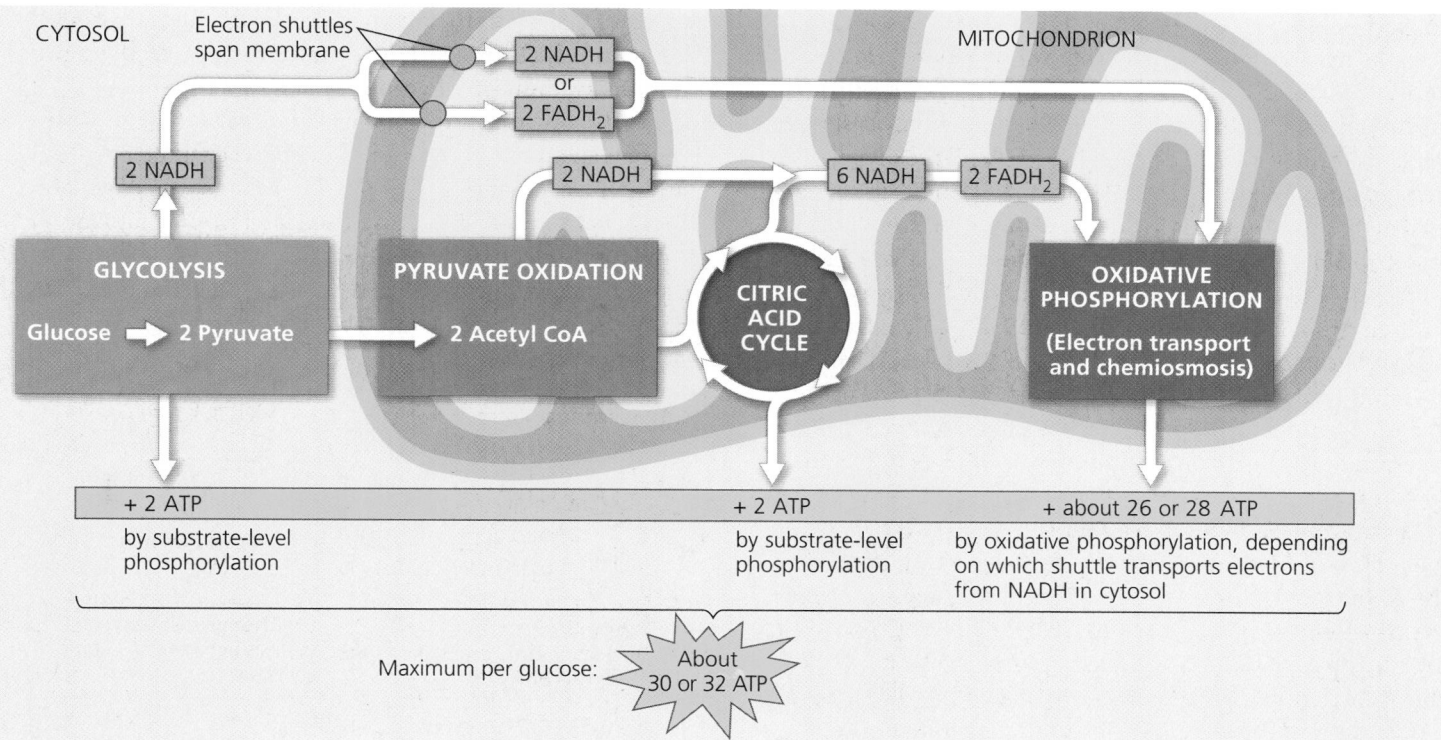

VISUAL SKILLS *After reading the discussion in the text, explain exactly how the total of 26 or 28 ATP from oxidative phosphorylation was calculated (see the yellow bar).*

⊘ **Mastering Biology** Animation: ATP Yield from Cellular Respiration each glucose molecule that is oxidized. The tally adds the 4 ATP produced directly by substrate-level phosphorylation during glycolysis and the citric acid cycle to the many more molecules of ATP generated by oxidative phosphorylation. Each NADH that transfers a pair of electrons from glucose to the electron transport chain contributes enough to the proton-motive force to generate a maximum of about 3 ATP.

Why are the numbers in Figure 9.15 inexact? There are three reasons we cannot state an exact number of ATP molecules generated by the breakdown of one molecule of glucose. First, phosphorylation and the redox reactions are not directly coupled to each other, so the ratio of the number of NADH molecules to the number of ATP molecules is not a whole number. We know that 1 NADH results in 10 H^+ being transported out across the inner mitochondrial membrane, but the exact number of H^+ that must reenter the mitochondrial matrix via ATP synthase to generate 1 ATP has long been debated. Based on experimental data, however, most biochemists now agree that the most accurate number is 4 H^+. Therefore, a single molecule of NADH generates enough proton-motive force for the synthesis of 2.5 ATP. The citric acid cycle also supplies electrons to the electron transport chain via $FADH_2$, but since its electrons enter later in the chain, each molecule of this electron carrier is responsible for transport of only enough H^+ for the synthesis of 1.5 ATP. These numbers also take into account the slight energetic cost of moving the ATP formed in the mitochondrion out into the cytosol, where it will be used.

Second, the ATP yield varies slightly depending on the type of shuttle used to transport electrons from the cytosol into the mitochondrion. The mitochondrial inner membrane is not permeable to NADH, so NADH in the cytosol is segregated from the machinery of oxidative phosphorylation. The 2 electrons of NADH captured in glycolysis must be conveyed into the mitochondrion by one of several electron shuttle systems. Depending on the kind of shuttle in a particular cell type, the electrons are passed either to NAD^+ or to FAD in the mitochondrial matrix (see Figures 9.14 and 9.15). If the electrons are passed to FAD, as in brain cells, only about 1.5 ATP can result from each NADH that was originally generated in the cytosol. If the electrons are passed to mitochondrial NAD^+, as in liver cells and heart cells, the yield is about 2.5 ATP per NADH.

A third variable that reduces the yield of ATP is the use of the proton-motive force generated by the redox reactions of respiration to drive other kinds of work. For example, the proton-motive force powers the mitochondrion's uptake of pyruvate from the cytosol (see Figure 9.9). However, if *all* the proton-motive force generated by the electron transport chain were used to drive ATP synthesis, one glucose molecule could generate a maximum of 28 ATP produced by oxidative phosphorylation plus 4 ATP (net) from substrate-level phosphorylation to give a total yield of about 32 ATP (or only about 30 ATP if the less efficient shuttle were functioning).

⊙ **Mastering Biology** Animation: Electron Transport Chain: Factors Affecting ATP Yield

We can now roughly estimate the efficiency of respiration—that is, the percentage of chemical energy in glucose that has been transferred to ATP. Recall that the complete oxidation of a mole of glucose releases 686 kcal of energy under standard conditions ($\Delta G = -686$ kcal/mol). Phosphorylation of ADP to form ATP stores at least 7.3 kcal per mole of ATP. Therefore, the efficiency of respiration is 7.3 kcal per mole of ATP times 32 moles of ATP per mole of glucose divided by 686 kcal per mole of glucose, which equals 0.34. Thus, about 34% of the potential chemical energy in glucose has been transferred to ATP; the actual percentage is bound to vary as ΔG varies under different cellular conditions. Cellular respiration is remarkably efficient in its energy conversion. By comparison, even the most efficient automobile converts only about 25% of the energy stored in gasoline to energy that moves the car.

The rest of the energy stored in glucose is lost as heat. We humans use some of this heat to maintain our relatively high body temperature (37°C), and we dissipate the rest through sweating and other cooling mechanisms.

Surprisingly, perhaps, it may be beneficial under certain conditions to reduce the efficiency of cellular respiration. A remarkable adaptation is shown by hibernating mammals, which overwinter in a state of inactivity and lowered metabolism. Although their internal body temperature is lower than normal, it still must be kept significantly higher than the external air temperature. One type of tissue, called brown fat, is made up of cells packed full of mitochondria. The inner mitochondrial membrane contains a channel protein called the uncoupling protein that allows protons to flow back down their concentration gradient without generating ATP. Activation of these proteins in hibernating mammals results in ongoing oxidation of stored fuel (fats), generating heat without any ATP production. In the absence of such an adaptation, the buildup of ATP would eventually cause cellular respiration to be shut down by regulatory mechanisms that will be discussed later. Brown fat is also used for heat generation in humans. In the **Scientific Skills Exercise**, you can work with data in a related but different case where a decrease in metabolic efficiency in cells is used to generate heat.

⊙ **Mastering Biology** BioFlix® Animation: Cellular Respiration

---

CONCEPT CHECK 9.4

1. **WHAT IF?** What effect would an absence of O_2 have on the process shown in Figure 9.14?

2. **WHAT IF?** In the absence of O_2, as in question 1, what do you think would happen if you decreased the pH of the intermembrane space of the mitochondrion? Explain your answer.

3. **MAKE CONNECTIONS** Membranes must be fluid to function properly (see Concept 7.1). How does the operation of the electron transport chain support that assertion?

For suggested answers, see Appendix A.

Making a Bar Graph and Evaluating a Hypothesis

Does Thyroid Hormone Level Affect O₂ Consumption in Cells? Some animals, such as mammals and birds, maintain a relatively constant body temperature, above that of their environment, by using heat produced as a by-product of metabolism. When the core temperature of these animals drops below an internal set point, their cells are triggered to reduce the efficiency of ATP production by the electron transport chains in mitochondria. At lower efficiency, extra fuel must be consumed to produce the same number of ATPs, generating additional heat. This response is moderated by the endocrine system, and researchers hypothesized that it might be triggered by thyroid hormone. In this exercise, you will use a bar graph to visualize data from an experiment that compared the metabolic rates (by measuring O₂ consumption) in mitochondria of cells from animals with different levels of thyroid hormone.

How the Experiment Was Done Liver cells were isolated from sibling rats that had low, normal, or elevated thyroid hormone levels. The oxygen consumption rate due to activity of the mitochondrial electron transport chains of each type of cell was measured under controlled conditions.

Data from the Experiment

Thyroid Hormone Level	Oxygen Consumption Rate [nmol O_2/(min · mg cells)]
Low	4.3
Normal	4.8
Elevated	8.7

Data from M. E. Harper and M. D. Brand, The quantitative contributions of mitochondrial proton leak and ATP turnover reactions to the changed respiration rates of hepatocytes from rats of different thyroid status, *Journal of Biological Chemistry* 268:14850–14860 (1993).

INTERPRET THE DATA

1. To visualize any differences in O₂ consumption between cell types, it will be useful to graph the data in a bar graph. First, set up the axes. **(a)** What is the independent variable (intentionally varied by the researchers), which goes on the x-axis? List the categories along the x-axis; because they are discrete rather than continuous, you can list them in any order. **(b)** What is the dependent variable (measured by the researchers), which goes on the y-axis? **(c)** What units (abbreviated) should go on the y-axis? Label the y-axis, including the units specified in the data table. Determine the range of values of the data that will need to go on the y-axis. What is the largest value? Draw evenly spaced tick marks and label them, starting with 0 at the bottom.

2. Graph the data for each sample. Match each x-value with its y-value and place a mark on the graph at that coordinate, then draw a bar from the x-axis up to the correct height for each sample. Why is a bar graph more appropriate than a scatter plot or line graph? (For additional information about graphs, see the Scientific Skills Review in Appendix D.)

3. Examine your graph and look for a pattern in the data. **(a)** Which cell type had the highest rate of O₂ consumption, and which had the lowest? **(b)** Does this support the researchers' hypothesis? Explain. **(c)** Based on what you know about mitochondrial electron transport and heat production, predict which rats had the highest, and which had the lowest, body temperature.

→ **Instructors:** A version of this Scientific Skills Exercise can be assigned in **Mastering Biology**.

Fermentation and anaerobic respiration enable cells to produce ATP without the use of oxygen

Because most of the ATP generated by cellular respiration is due to the work of oxidative phosphorylation, our estimate of ATP yield from aerobic respiration depends on an adequate supply of O₂ to the cell. Without the electronegative oxygen atoms in O₂ to pull electrons down the transport chain, oxidative phosphorylation eventually ceases. However, there are two general mechanisms by which certain cells can oxidize organic fuel and generate ATP *without* the use of O₂: anaerobic respiration and fermentation. The distinction between these

two is that an electron transport chain is used in anaerobic respiration but not in fermentation. (The electron transport chain is also called the respiratory chain because of its role in both types of cellular respiration.)

We have already mentioned anaerobic respiration, which takes place in certain prokaryotic organisms that live in environments without O₂. These organisms have an electron transport chain but do not use O₂ as a final electron acceptor at the end of the chain. O₂ performs this function very well because it consists of two extremely electronegative atoms, but other substances can also serve as final electron acceptors. Some "sulfate-reducing" marine bacteria, for instance, use the sulfate ion (SO_4^{2-}) at the end of their respiratory chain. Operation of the chain builds up a proton-motive force used to produce ATP, but H_2S (hydrogen sulfide) is made as a by-product rather than water. The rotten-egg odor you may have smelled while walking through a salt marsh or a mudflat signals the presence of sulfate-reducing bacteria.

Fermentation is a way of harvesting chemical energy without using either O_2 or any electron transport chain—in other words, without cellular respiration. How can food be oxidized without cellular respiration? Remember, oxidation simply refers to the loss of electrons to an electron acceptor, so it does not need to involve O_2. Glycolysis oxidizes glucose to two molecules of pyruvate. The oxidizing agent of glycolysis is NAD^+, and neither O_2 nor any electron transfer chain is involved. Overall, glycolysis is exergonic, and some of the energy made available is used to produce 2 ATP (net) by substrate-level phosphorylation. If O_2 *is* present, then additional ATP is made by oxidative phosphorylation when NADH passes electrons removed from glucose to the electron transport chain. But glycolysis generates 2 ATP whether oxygen is present or not—that is, whether conditions are aerobic or anaerobic.

As an alternative to respiratory oxidation of organic nutrients, fermentation is an extension of glycolysis that allows continuous generation of ATP by the substrate-level phosphorylation of glycolysis. For this to occur, there must be a sufficient supply of NAD^+ to accept electrons during the oxidation step of glycolysis. Without some mechanism to recycle NAD^+ from NADH, glycolysis would soon deplete the cell's pool of NAD^+ by reducing it all to NADH and would shut itself down for lack of an oxidizing agent. Under aerobic conditions, NAD^+ is recycled from NADH by the transfer of electrons to the electron transport chain. An anaerobic alternative is to transfer electrons from NADH to pyruvate, the end product of glycolysis.

Types of Fermentation

Fermentation consists of glycolysis plus reactions that regenerate NAD^+ by transferring electrons from NADH to pyruvate or derivatives of pyruvate. The NAD^+ can then be reused to oxidize sugar by glycolysis, which nets two molecules of ATP by substrate-level phosphorylation. There are many types of fermentation, differing in the end products formed from pyruvate. Two types are alcohol fermentation and lactic acid fermentation, and both are harnessed by humans for food and industrial production.

In **alcohol fermentation (Figure 9.16a)**, pyruvate is converted to ethanol (ethyl alcohol) in two steps. The first step releases CO_2 from the pyruvate, which is converted to the two-carbon compound acetaldehyde. In the second step, acetaldehyde is reduced by NADH to ethanol. This regenerates the supply of NAD^+ needed for the continuation of glycolysis. Many bacteria carry out alcohol fermentation under anaerobic conditions. Yeast (a fungus), in addition to aerobic respiration, also carries out alcohol fermentation. For thousands of years, humans have used yeast in brewing, winemaking, and baking. The CO_2 bubbles generated by baker's yeast during alcohol fermentation allow bread to rise.

▼ **Figure 9.16 Fermentation.** In the absence of oxygen, many cells use fermentation to produce ATP by substrate-level phosphorylation. NAD^+ is regenerated for use in glycolysis when pyruvate, the end product of glycolysis, serves as an electron acceptor for oxidizing NADH. Two of the common end products formed from fermentation are **(a)** ethanol and **(b)** lactate, the ionized form of lactic acid.

(a) Alcohol fermentation

(b) Lactic acid fermentation

➔ **Mastering Biology** Animation: Fermentation

During **lactic acid fermentation (Figure 9.16b)**, pyruvate is reduced directly by NADH to form lactate as an end product, regenerating NAD^+ with no release of CO_2. (Lactate is the ionized form of lactic acid.) Lactic acid fermentation by certain fungi and bacteria is used in the dairy industry to make cheese and yogurt. A complex series of fermentation and aerobic respiration pathways carried out by yeasts and bacteria on cacao beans is responsible for the production of chocolate.

▼ **Cacao beans and fruits**

What about lactate production in humans? Previously, we thought that human muscle cells only produced lactate when O_2 was in short supply, such as during intense exercise. Research done over the last few decades, though, indicates that the lactate story, in mammals at least, is more complicated. There are two types of skeletal muscle fibers. One (red muscle) preferentially oxidizes glucose completely to CO_2; the other (white muscle) produces significant amounts of lactate from the pyruvate made during glycolysis, even under aerobic conditions, offering fast but energetically inefficient ATP production. The lactate product is then mostly oxidized by red muscle cells in the vicinity, with the remainder exported to liver or kidney cells for glucose formation. Because this lactate production is not anaerobic, but the result of glycolysis in these cells, exercise physiologists prefer not to use the term *fermentation*.

During strenuous exercise, when carbohydrate catabolism outpaces the supply of O_2 from the blood to the muscle, lactate can't be oxidized to pyruvate. The lactate that accumulates was once thought to cause muscle fatigue during intense exercise and pain a day or so later. However, research suggests that, contrary to popular opinion, lactate production actually improves performance during exercise! Furthermore, within an hour, excess lactate is shuttled to other tissues for oxidation or to the liver and kidneys for production of glucose or its storage molecule, glycogen. (Nextday muscle soreness is more likely caused by trauma to cells in small muscle fibers, which leads to inflammation and pain.)

Comparing Fermentation with Anaerobic and Aerobic Respiration

Fermentation, anaerobic respiration, and aerobic respiration are three alternative cellular pathways for producing ATP by harvesting the chemical energy of food. All three use glycolysis to oxidize glucose and other organic fuels to pyruvate, with a net production of 2 ATP by substrate-level phosphorylation. And in all three pathways, NAD^+ is the oxidizing agent that accepts electrons from food during glycolysis.

A key difference is the contrasting mechanisms for oxidizing NADH back to NAD^+, which is required to sustain glycolysis. In fermentation, the final electron acceptor is an organic molecule such as pyruvate (lactic acid fermentation) or acetaldehyde (alcohol fermentation). In cellular respiration, by contrast, electrons carried by NADH are transferred to an electron transport chain, which regenerates the NAD^+ required for glycolysis.

Another major difference is the amount of ATP produced. Fermentation yields two molecules of ATP, produced by substrate-level phosphorylation. In the absence of an electron transport chain, the energy stored in pyruvate is unavailable. In cellular respiration, however, pyruvate is completely oxidized in the mitochondrion. Most of the chemical energy from this process is shuttled by NADH and $FADH_2$ in the form of electrons to the electron transport chain. There, the electrons move stepwise down a series of redox reactions to a final electron acceptor. (In aerobic respiration, the final electron acceptor is O_2; in anaerobic respiration, the final acceptor is another molecule with a high affinity for electrons, although less so than O_2.) Stepwise electron transport drives oxidative phosphorylation, yielding ATP. Thus, cellular respiration harvests much more energy from each sugar molecule than fermentation can. In fact, aerobic respiration yields up to 32 molecules of ATP per glucose molecule—up to 16 times as much as does fermentation.

Some organisms, called **obligate anaerobes**, carry out only fermentation or anaerobic respiration. In fact, these organisms cannot survive in the presence of oxygen, some forms of which can actually be toxic if protective systems are not present in the cell. A few cell types, such as cells of the vertebrate brain, can carry out only aerobic oxidation of pyruvate, and need O_2 to survive. Other organisms, including yeasts and many bacteria, can make enough ATP to survive using either fermentation or respiration. Such species are called **facultative anaerobes**. In yeast cells, for example, pyruvate is a fork in the metabolic road that leads to two alternative catabolic routes **(Figure 9.17)**. Under aerobic conditions, pyruvate can be converted to acetyl CoA, and oxidation continues in the citric acid cycle via aerobic respiration. Under anaerobic conditions, lactic acid fermentation occurs. Pyruvate is diverted from the citric acid cycle, serving instead

▼ **Figure 9.17 Pyruvate as a key juncture in catabolism.** Glycolysis is common to fermentation and cellular respiration. The end product of glycolysis, pyruvate, represents a fork in the catabolic pathways of glucose oxidation. In a facultative anaerobe, capable of both aerobic cellular respiration and fermentation, pyruvate is committed to one of those two pathways, usually depending on whether or not oxygen is present.

as an electron acceptor to recycle NAD^+. To make the same amount of ATP, a facultative anaerobe has to consume sugar at a much faster rate when fermenting than when respiring.

The Evolutionary Significance of Glycolysis

EVOLUTION The role of glycolysis in both fermentation and respiration has an evolutionary basis. Ancient prokaryotes are thought to have used glycolysis to make ATP long before oxygen was present in Earth's atmosphere. The oldest known fossils of bacteria date back 3.5 billion years, but appreciable quantities of oxygen probably did not begin to accumulate in the atmosphere until about 2.7 billion years ago. Cyanobacteria produced this O_2 as a by-product of photosynthesis. Therefore, early prokaryotes may have generated ATP exclusively from glycolysis. The fact that glycolysis is today the most widespread metabolic pathway among Earth's organisms suggests that it evolved very early in the history of life. The cytosolic location of glycolysis also implies great antiquity; the pathway does not require any of the membrane-enclosed organelles of the eukaryotic cell, which evolved approximately 1 billion years after the first prokaryotic cell. Glycolysis is a metabolic heirloom from early cells that continues to function in fermentation and as the first stage in the breakdown of organic molecules by respiration.

CONCEPT CHECK 9.5

1. Consider the NADH formed during glycolysis. What is the final acceptor for its electrons during fermentation? During aerobic respiration? During anaerobic respiration?

2. **WHAT IF?** A glucose-fed yeast cell is moved from an aerobic environment to an anaerobic one. How would its rate of glucose consumption change if ATP were to be generated at the same rate?

For suggested answers, see Appendix A.

CONCEPT 9.6

Glycolysis and the citric acid cycle connect to many other metabolic pathways

So far, we have looked at the oxidative breakdown of glucose in isolation from the cell's overall metabolic economy. In this section, you will learn that glycolysis and the citric acid cycle are major intersections of the cell's catabolic (breakdown) and anabolic (biosynthetic) pathways.

The Versatility of Catabolism

Throughout this chapter, glucose has been used as an example of a fuel for cellular respiration. But free glucose molecules are not common in the diets of humans and other

▼ **Figure 9.18 The catabolism of various molecules from food.** Carbohydrates, fats, and proteins can all be used as fuel for cellular respiration. Monomers of these molecules enter glycolysis or the citric acid cycle at various points. Glycolysis and the citric acid cycle are catabolic funnels through which electrons from all kinds of organic molecules flow on their exergonic fall to oxygen.

animals. We obtain most of our calories in the form of fats, proteins, and carbohydrates such as sucrose and other disaccharides, and starch, a polysaccharide. All these organic molecules in food can be used by cellular respiration to make ATP **(Figure 9.18)**.

Glycolysis can accept a wide range of carbohydrates for catabolism. In the digestive tract, starch is hydrolyzed to glucose, which is broken down in cells by glycolysis and the citric acid cycle. Glycogen, the polysaccharide that humans and many other animals store in their liver and muscle cells, can be hydrolyzed to glucose between meals as fuel for respiration. Digestion of disaccharides, including sucrose, provides glucose and other monosaccharides as fuel for respiration.

Proteins can also be used for fuel, but first they must be digested to their constituent amino acids. Many of the amino acids are used by the organism to build new proteins. Amino acids present in excess are converted by enzymes to intermediates of glycolysis and the citric acid cycle. Before amino acids can feed into glycolysis or the citric acid cycle, their amino groups must be removed, a process called *deamination*. The nitrogenous waste is excreted from the animal in the form of ammonia (NH_3), urea, or other waste products.

Catabolism can also harvest energy stored in fats obtained either from food or from fat cells in the body. After fats are digested to glycerol and fatty acids, the glycerol is converted to glyceraldehyde 3-phosphate, an intermediate of glycolysis. Most of the energy of a fat is stored in the fatty acids. A metabolic sequence called **beta oxidation** breaks the fatty acids down to two-carbon fragments, which enter the citric acid cycle as acetyl CoA. NADH and $FADH_2$ are also generated during beta oxidation; they can enter the electron transport chain, leading to further ATP production. Fats make excellent fuels, in large part due to their chemical structure and the high energy level of their electrons (present in many C—H bonds, equally shared between C and H) compared to those of carbohydrates. A gram of fat oxidized by respiration produces more than twice as much ATP as a gram of carbohydrate. Unfortunately, this also means that a person trying to lose weight must work hard to use up fat stored in the body because so many calories are stockpiled in each gram of fat.

Biosynthesis (Anabolic Pathways)

Cells need substance as well as energy. Not all the organic molecules of food are destined to be oxidized as fuel to make ATP. In addition to calories, food must also provide the carbon skeletons that cells require to make their own molecules. Some organic monomers obtained from digestion can be used directly. For example, as previously mentioned, amino acids from the hydrolysis of proteins in food can be incorporated into the organism's own proteins. Often, however, the body needs specific molecules that are not present as such in food. Compounds formed as intermediates of glycolysis and the citric acid cycle can be diverted into anabolic pathways as precursors from which the cell can synthesize the molecules it requires. For example, humans can make about half of the 20 amino acids in proteins by modifying compounds siphoned away from the citric acid cycle; the rest are "essential amino acids" that must be obtained in the diet. Also, glucose can be made from pyruvate, and fatty acids can be synthesized from acetyl CoA. Of course, these anabolic, or biosynthetic, pathways do not generate ATP, but instead consume it.

In addition, glycolysis and the citric acid cycle function as metabolic interchanges that enable our cells to convert some kinds of molecules to others as we need them. For example, an intermediate compound generated during glycolysis, dihydroxyacetone phosphate (see Figure 9.8, step 5), can be converted to one of the major precursors of fats. If we eat more food than we need, we store fat even if our diet is fat-free. Metabolism is remarkably versatile and adaptable.

Regulation of Cellular Respiration via Feedback Mechanisms

Basic principles of supply and demand regulate the metabolic economy. The cell does not waste energy making more of a particular substance than it needs. If there is a surplus of a certain amino acid, for example, the anabolic pathway that synthesizes that amino acid from an intermediate of the citric acid cycle is switched off. The most common mechanism for this control is feedback inhibition: The end product of the anabolic pathway inhibits the enzyme that catalyzes an early step of the pathway (see Figure 8.21). This prevents the needless diversion of key metabolic intermediates from uses that are more urgent.

The cell also controls its catabolism. If the cell is working hard and its ATP concentration begins to drop, cellular respiration speeds up. When there is plenty of ATP to meet demand, respiration slows down, sparing valuable organic molecules for other functions. Again, control is based mainly on regulating the activity of enzymes at strategic points in the catabolic pathway. As shown in **Figure 9.19**, one important switch is phosphofructokinase, the enzyme that catalyzes step 3 of glycolysis (see Figure 9.8). That is the first step that

▼ **Figure 9.19 The control of cellular respiration.** Allosteric enzymes at certain points in the respiratory pathway respond to inhibitors and activators that help set the pace of glycolysis and the citric acid cycle. Phosphofructokinase, which catalyzes an early step in glycolysis (see Figure 9.8, step ❸ and Figure 6.32b), is one such enzyme. It is stimulated by AMP (derived from ADP) but is inhibited by ATP and by citrate. This feedback regulation adjusts the rate of respiration as the cell's catabolic and anabolic demands change.

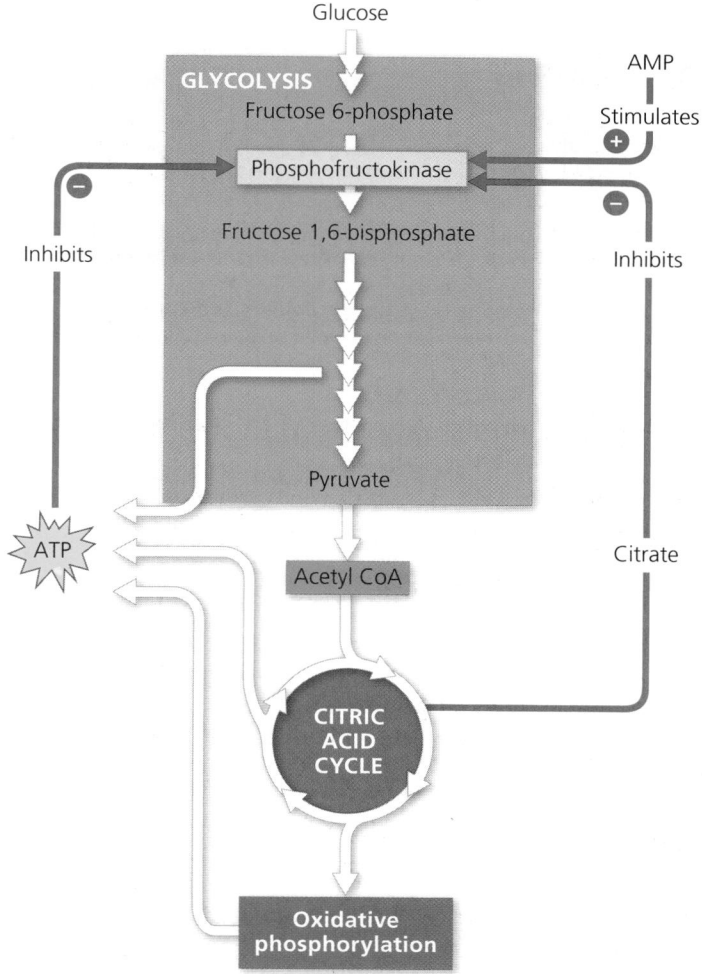

commits the substrate irreversibly to the glycolytic pathway. By controlling the rate of this step, the cell can speed up or slow down the entire catabolic process. Phosphofructokinase can thus be considered the pacemaker of cellular respiration.

Phosphofructokinase is an allosteric enzyme with receptor sites for specific inhibitors and activators. It is inhibited by ATP and stimulated by AMP (adenosine monophosphate), which the cell derives from ADP. As ATP accumulates, inhibition of the enzyme slows down glycolysis. The enzyme becomes active again as cellular work converts ATP to ADP (and AMP) faster than ATP is being regenerated. Phosphofructokinase is also sensitive to citrate, the first product of the citric acid cycle. If citrate accumulates in mitochondria, some of it passes into the cytosol and inhibits phosphofructokinase. This mechanism helps synchronize the rates of glycolysis and the citric acid cycle. As citrate accumulates, glycolysis slows down, and the supply of pyruvate and thus acetyl groups to the citric acid cycle decreases. If citrate consumption increases, either because of a demand for more ATP or because anabolic pathways are draining off intermediates of the citric acid cycle, glycolysis accelerates and meets the demand. Metabolic balance is augmented by the control of enzymes that catalyze other key steps of glycolysis and the citric acid cycle. Cells are thrifty, expedient, and responsive in their metabolism.

Review the first page of this chapter to put cellular respiration into the broader context of energy flow and chemical cycling in ecosystems. The energy that keeps us alive is *released*, not *produced*, by cellular respiration. We are tapping energy that was stored in food by photosynthesis, which captures light and converts it to chemical energy, a process you will learn about next, in Chapter 10.

CONCEPT CHECK 9.6

1. **MAKE CONNECTIONS** Compare the structures of a carbohydrate and a fat (see Figures 5.3 and 5.9). What features make fat a much better fuel?
2. Under what circumstances might your body synthesize fat molecules?
3. **VISUAL SKILLS** What will happen in a muscle cell that has used up its supply of O_2 and ATP? (Review Figures 9.17 and 9.19.)
4. **VISUAL SKILLS** During intense exercise, can a muscle cell use fat as a concentrated source of chemical energy? Explain. (Review Figures 9.17 and 9.18.)

For suggested answers, see Appendix A.

---

9 Chapter Review

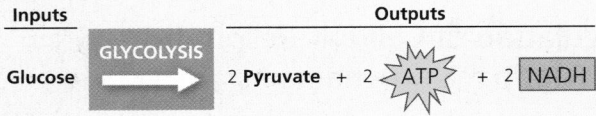

→ Go to **Mastering Biology** for Assignments, the eText, the Study Area, and Dynamic Study Modules.

SUMMARY OF KEY CONCEPTS

→ To review key terms, go to the **Vocabulary Self-Quiz** in the **Mastering Biology** eText or Study Area, or go to goo.gl/zkjz9t.

CONCEPT 9.1

Catabolic pathways yield energy by oxidizing organic fuels *(pp. 165–169)*

- Cells break down glucose and other organic fuels to yield chemical energy in the form of ATP. **Fermentation** is a process that results in the partial degradation of glucose without the use of oxygen (O_2). The process of **cellular respiration** is a more complete breakdown of glucose. In **aerobic respiration**, O_2 is used as a reactant; in *anaerobic respiration*, other substances are used in place of O_2.
- The cell taps the energy stored in food molecules through **redox reactions**, in which one substance partially or totally shifts electrons to another. **Oxidation** is the total or partial loss of electrons, while **reduction** is the total or partial addition of electrons. During aerobic respiration, glucose ($C_6H_{12}O_6$) is oxidized to CO_2, and O_2 is reduced to H_2O:

$$\overbrace{C_6H_{12}O_6}^{\text{becomes oxidized}} + \underbrace{6\,O_2}_{} \longrightarrow 6\,CO_2 + 6\,H_2O + \text{Energy}$$
becomes reduced

- Electrons lose potential energy during their transfer from glucose or other organic compounds to O_2. Electrons are usually passed first to **NAD$^+$**, reducing it to **NADH**, and are then passed from NADH to an **electron transport chain**, which conducts the electrons to O_2 in energy-releasing steps. The energy that is released is used to make ATP.
- Aerobic respiration occurs in three stages: (1) **glycolysis**, (2) pyruvate oxidation and the **citric acid cycle**, and (3) **oxidative phosphorylation** (electron transport and chemiosmosis).

? *Describe the difference between the two processes in cellular respiration that produce ATP: oxidative phosphorylation and substrate-level phosphorylation.*

CONCEPT 9.2

Glycolysis harvests chemical energy by oxidizing glucose to pyruvate *(pp. 170–171)*

- Glycolysis ("splitting of sugar") is a series of reactions that breaks down glucose into two pyruvate molecules, which may go on to enter the citric acid cycle, and nets 2 ATP and 2 NADH per glucose molecule.

Inputs		**Outputs**

Glucose → GLYCOLYSIS → 2 **Pyruvate** + 2 ATP + 2 NADH

? *Which reactions in glycolysis are the source of energy for the formation of ATP and NADH?*

CONCEPT 9.3

After pyruvate is oxidized, the citric acid cycle completes the energy-yielding oxidation of organic molecules (pp. 171–174)

- In eukaryotic cells, pyruvate enters the mitochondrion and is oxidized to **acetyl CoA**, which is further oxidized in the citric acid cycle.

Inputs		Outputs

2 Pyruvate ⟶ 2 Acetyl CoA

2 Oxaloacetate

CITRIC ACID CYCLE

2 ATP   8 NADH

6 CO_2 2 FADH₂

? *What molecular products indicate the complete oxidation of glucose during cellular respiration?*

CONCEPT 9.4

During oxidative phosphorylation, chemiosmosis couples electron transport to ATP synthesis (pp. 174–179)

- NADH and FADH₂ transfer electrons to the electron transport chain. Electrons move down the chain, losing energy in several energy-releasing steps. Finally, electrons are passed to O_2, reducing it to H_2O.

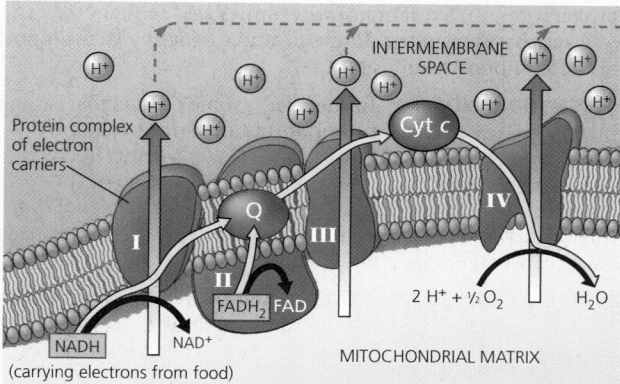

- Along the electron transport chain, electron transfer causes protein complexes to move H^+ from the mitochondrial matrix (in eukaryotes) to the intermembrane space, storing energy as a **proton-motive force** (H^+ gradient). As H^+ diffuses back into the matrix through **ATP synthase**, its passage drives the phosphorylation of ADP to form ATP, called **chemiosmosis**.
- About 34% of the energy stored in a glucose molecule is transferred to ATP during cellular respiration, producing a maximum of about 32 ATP.

? *Briefly explain the mechanism by which ATP synthase produces ATP. List three locations in which ATP synthases are found.*

CONCEPT 9.5

Fermentation and anaerobic respiration enable cells to produce ATP without the use of oxygen (pp. 179–182)

- Glycolysis nets 2 ATP by substrate-level phosphorylation, whether O_2 is present or not. Under anaerobic conditions, anaerobic respiration or fermentation can take place. In anaerobic respiration, an electron transport chain is present with a final electron acceptor other than oxygen. In fermentation, the electrons from NADH are passed to pyruvate or a derivative of pyruvate, regenerating the NAD^+ required to oxidize more glucose. Two common types of fermentation are **alcohol fermentation** and **lactic acid fermentation**.
- Fermentation, anaerobic respiration, and aerobic respiration all use glycolysis to oxidize glucose, but they differ in their final electron acceptor and whether an electron transport chain is used (respiration) or not (fermentation). Respiration yields more ATP; aerobic respiration, with O_2 as the final electron acceptor, yields about 16 times as much ATP as does fermentation.
- Glycolysis occurs in nearly all organisms and is thought to have evolved in ancient prokaryotes before there was O_2 in the atmosphere.

? *Which process yields more ATP, fermentation or anaerobic respiration? Explain.*

CONCEPT 9.6

Glycolysis and the citric acid cycle connect to many other metabolic pathways (pp. 182–184)

- Catabolic pathways funnel electrons from many kinds of organic molecules into cellular respiration. Many carbohydrates can enter glycolysis, most often after conversion to glucose. Amino acids of proteins must be deaminated before being oxidized. The fatty acids of fats undergo **beta oxidation** to two-carbon fragments and then enter the citric acid cycle as acetyl CoA. Anabolic pathways can use small molecules from food directly or build other substances using intermediates of glycolysis or the citric acid cycle.
- Cellular respiration is controlled by allosteric enzymes at key points in glycolysis and the citric acid cycle.

? *Describe how the catabolic pathways of glycolysis and the citric acid cycle intersect with anabolic pathways in the metabolism of a cell.*

TEST YOUR UNDERSTANDING

→ For more multiple-choice questions, go to the **Practice Test** in the **Mastering Biology** eText or Study Area, or go to goo.gl/GruWRg.

Levels 1-2: Remembering/Understanding

1. The *immediate* energy source that drives ATP synthesis by ATP synthase during oxidative phosphorylation is the
   - (A) oxidation of glucose and other organic compounds.
   - (B) flow of electrons down the electron transport chain.
   - (C) H^+ concentration gradient across the membrane holding ATP synthase.
   - (D) transfer of phosphate to ADP.

2. Which metabolic pathway is common to both fermentation and cellular respiration of a glucose molecule?
   - (A) the citric acid cycle
   - (B) the electron transport chain
   - (C) glycolysis
   - (D) reduction of pyruvate to lactate

3. The final electron acceptor of the electron transport chain that functions in aerobic oxidative phosphorylation is
   (A) O_2.
   (B) water.
   (C) NAD^+.
   (D) pyruvate.

4. In mitochondria, exergonic redox reactions
   (A) are the source of energy driving prokaryotic ATP synthesis.
   (B) provide the energy that establishes the proton gradient.
   (C) reduce carbon atoms to carbon dioxide.
   (D) are coupled via phosphorylated intermediates to endergonic processes.

Levels 3-4: Applying/Analyzing

5. What is the oxidizing agent in the following reaction?

   $$\text{Pyruvate} + \text{NADH} + H^+ \rightarrow \text{Lactate} + NAD^+$$

   (A) oxygen
   (B) NADH
   (C) lactate
   (D) pyruvate

6. When electrons flow along the electron transport chains of mitochondria, which of the following changes occurs?
   (A) The pH of the matrix increases.
   (B) ATP synthase pumps protons by active transport.
   (C) The electrons gain free energy.
   (D) NAD^+ is oxidized.

7. Most CO_2 from catabolism is released during
   (A) glycolysis.
   (B) the citric acid cycle.
   (C) lactate fermentation.
   (D) electron transport.

8. **MAKE CONNECTIONS** Step 3 in Figure 9.8 is a major point of regulation of glycolysis. The enzyme phosphofructokinase is allosterically regulated by ATP and related molecules (see Concept 8.5). Considering the overall result of glycolysis, would you expect ATP to inhibit or stimulate activity of this enzyme? Explain. (Hint: Make sure you consider the role of ATP as an allosteric regulator, not as a substrate of the enzyme.)

9. **MAKE CONNECTIONS** The proton pump shown in Figures 7.18 and 7.19 is a type of ATP synthase like that in Figure 9.13. Compare the processes shown in the three figures, and say whether they are involved in active or passive transport (see also Concepts 7.3 and 7.4).

10. **VISUAL SKILLS** This computer model shows the four parts of ATP synthase, each part consisting of a number of polypeptide subunits (the solid gray part is still an area of active research). Using Figure 9.13 as a guide, label the rotor, stator, internal rod, and catalytic knob of this molecular motor.

Levels 5-6: Evaluating/Creating

11. **INTERPRET THE DATA** Phosphofructokinase is an enzyme that acts on fructose 6-phosphate at an early step in glucose breakdown. Regulation of this enzyme controls whether the sugar will continue on in the glycolytic pathway. Considering this graph, under which condition is phosphofructokinase more active? Given what you know about glycolysis and regulation of metabolism by this enzyme, explain the mechanism by which phosphofructokinase activity differs depending on ATP concentration. Explain why it makes sense that regulation of this enzyme has evolved so that it works this way.

12. **DRAW IT** The graph here shows the pH difference across the inner mitochondrial membrane over time in an actively respiring cell. At the time indicated by the vertical arrow, a metabolic poison is added that specifically and completely inhibits all function of mitochondrial ATP synthase. Draw what you would expect to see for the rest of the graphed line over the next short period of time. Explain.

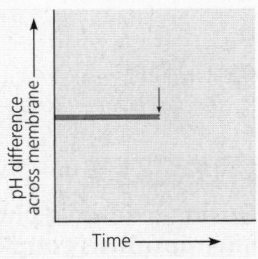

13. **EVOLUTION CONNECTION** ATP synthases are found in the prokaryotic plasma membrane and in mitochondria and chloroplasts. (a) Propose a hypothesis to account for an evolutionary relationship of these eukaryotic organelles and prokaryotes. (b) Explain how the amino acid sequences of the ATP synthases from the different sources might either support or fail to support your hypothesis.

14. **SCIENTIFIC INQUIRY** In the 1930s, some physicians prescribed low doses of a compound called dinitrophenol (DNP) to help patients lose weight. This unsafe method was abandoned after some patients died. DNP uncouples the chemiosmotic machinery by making the lipid bilayer of the inner mitochondrial membrane leaky to H^+. Explain how this could cause weight loss and death.

15. **WRITE ABOUT A THEME: ORGANIZATION** In a short essay (100–150 words), explain how oxidative phosphorylation—production of ATP using energy from the redox reactions of a spatially organized electron transport chain followed by chemiosmosis—is an example of how new properties emerge at each level of the biological hierarchy.

16. **SYNTHESIZE YOUR KNOWLEDGE**

Coenzyme Q (CoQ) is sold as a nutritional supplement. One company uses this marketing slogan for CoQ: "Give your heart the fuel it craves most." Considering the role of coenzyme Q, critique this claim. How do you think this product might function to benefit the heart? Is CoQ used as a "fuel" during cellular respiration?

For selected answers, see Appendix A.

10 Photosynthesis

KEY CONCEPTS

10.1 Photosynthesis feeds the biosphere *p. 188*

10.2 Photosynthesis converts light energy to the chemical energy of food *p. 189*

10.3 The light reactions convert solar energy to the chemical energy of ATP and NADPH *p. 192*

10.4 The Calvin cycle uses the chemical energy of ATP and NADPH to reduce CO_2 to sugar *p. 201*

10.5 Alternative mechanisms of carbon fixation have evolved in hot, arid climates *p. 203*

10.6 Photosynthesis is essential for life on Earth: a review *p. 206*

Study Tip

Make a visual study guide: Draw a cell with a large chloroplast, labeling the parts of the chloroplast. As you go through the chapter, add key reactions

Chloroplast

for each stage of photosynthesis, linking the stages together. Label the carbon molecule(s) with the most energy and the carbon molecule(s) with the least energy.

→ Go to Mastering Biology

For Students (in eText and Study Area)
• Get Ready for Chapter 10
• BioFlix® Animation: Photosynthesis
• Figure 10.13 Walkthrough: How Linear Electron Flow During the Light Reactions Generates ATP and NADPH

For Instructors to Assign (in Item Library)
• Activity: The Light Reactions
• BioFlix® Tutorial: Photosynthesis

Ready-to-Go Teaching Module (in Instructor Resources)
• The Light Reactions (Concept 10.3)

Figure 10.1 Each leaf of this tree is harvesting energy from sunlight and using it to convert CO_2 and H_2O into chemical energy stored in sugar and other organic molecules. The tree uses these sugars for energy and to build its own trunk, branches, and leaves. Remarkably, enough sugars are left over to feed other organisms (like moth larvae, shown below) that cannot carry out this extraordinary transformation.

How do photosynthetic cells use light to change carbon dioxide and water into organic molecules and oxygen?

Light energy

Photosynthesis in chloroplasts

used in

$CO_2 + H_2O$

Plant cell

Chloroplast

generates

Organic molecules $+ O_2$

used in

generates

Cellular respiration
generates

ATP

Heat

Photosynthesis feeds the biosphere

The conversion process that transforms the energy of sunlight into chemical energy stored in sugars and other organic molecules is called **photosynthesis**. Let's begin by placing photosynthesis in its ecological context.

Photosynthesis nourishes almost the entire living world directly or indirectly. An organism acquires the organic compounds it uses for energy and carbon skeletons by one of two major modes: autotrophic nutrition or heterotrophic nutrition. **Autotrophs** are "self-feeders" (*auto-* means "self," and *trophos* means "feeder"); they sustain themselves without eating anything derived from other living beings. Autotrophs produce their organic molecules from CO_2 and other inorganic raw materials obtained from the environment. They are the ultimate sources of organic compounds for all nonautotrophic organisms, and for this reason, biologists refer to autotrophs as the *producers* of the biosphere.

Almost all plants are autotrophs; the only nutrients they require are water and minerals from the soil and CO_2 from the air. Specifically, plants are *photo*autotrophs, organisms that use light as a source of energy to synthesize organic substances **(Figure 10.1)**. Photosynthesis also occurs in algae, certain other protists, and some prokaryotes **(Figure 10.2)**. In this chapter, we will touch on these other groups in passing, but our emphasis will be on plants. Variations in autotrophic nutrition that occur in prokaryotes and algae will be described in Concept 27.3.

Heterotrophs are unable to make their own food; they live on compounds produced by other organisms (*hetero-* means "other"). Heterotrophs are the biosphere's *consumers*. The most obvious "other-feeding" occurs when an animal eats plants or other organisms, but heterotrophic nutrition may be more subtle. Some heterotrophs consume the remains of other organisms and organic litter such as feces, and fallen leaves; these types of heterotrophs are known as *decomposers*. Most fungi and many types of prokaryotes get their nourishment this way. Earth's supply of fossil fuels was formed from remains of organisms that died hundreds of millions of years ago. In a sense, then, fossil fuels represent stores of the sun's energy from the distant past. Almost all heterotrophs, including humans, are completely dependent, either directly or indirectly, on photoautotrophs for food—and also for oxygen, a by-product of photosynthesis.

CONCEPT CHECK 10.1

1. Why are heterotrophs dependent on autotrophs?
2. **WHAT IF?** Fossil fuels are being depleted faster than they are replenished. Therefore, researchers are developing ways to produce "biodiesel" from the products of photosynthetic algae. They have proposed placing containers of these algae near industrial plants or highly congested city streets. Considering the process of photosynthesis, how does this arrangement make sense?

For suggested answers, see Appendix A.

▼ **Figure 10.2 Photoautotrophs.** These organisms use light energy to drive the synthesis of organic molecules from carbon dioxide and (in most cases) water. They feed themselves and the entire living world. On land, **(a)** plants are the predominant producers of food. In aquatic environments, photoautotrophs include unicellular and **(b)** multicellular algae, such as this kelp; **(c)** some non-algal unicellular protists, such as *Euglena*; **(d)** the prokaryotes called cyanobacteria; and **(e)** other photosynthetic prokaryotes, such as these purple sulfur bacteria, which produce sulfur (the yellow globules within the cells) (c–e, LMs).

(a) Plants

(b) Multicellular alga

10 μm

(c) Unicellular protists

(d) Cyanobacteria

20 μm

1 μm

(e) Purple sulfur bacteria

→ **Mastering Biology** BioFlix® **Animation: The Flow of Carbon Atoms in Producers, Consumers, and Decomposers**

Photosynthesis converts light energy to the chemical energy of food

The remarkable ability of an organism to harness light energy and use it to drive the synthesis of organic compounds emerges from structural organization in the cell: Photosynthetic enzymes and other molecules are grouped together as specialized molecular complexes in a biological membrane, enabling the necessary series of chemical reactions to be carried out efficiently. The process of photosynthesis most likely originated in a group of bacteria that had infolded regions of the plasma membrane containing clusters of such molecules. In existing photosynthetic bacteria, infolded photosynthetic membranes function similarly to the internal membranes of the **chloroplast**, the eukaryotic organelle that absorbs energy from sunlight and uses it to drive the synthesis of organic compounds from carbon dioxide (CO_2) and water (H_2O). According to what has come to be known as the *endosymbiont theory*, the original chloroplast was a photosynthetic prokaryote that lived inside an ancestor of eukaryotic cells. (You learned about this theory in Concept 6.5, and it will be described more fully in Concept 25.3.) Chloroplasts are present in a variety of photosynthesizing organisms (see some examples in Figure 10.2); in this chapter we focus on chloroplasts in plants, while mentioning prokaryotes from time to time.

Chloroplasts: The Sites of Photosynthesis in Plants

All green parts of a plant, including green stems and unripened fruit, have chloroplasts, but the leaves are the major sites of photosynthesis in most plants **(Figure 10.3)**. There are about half a million chloroplasts in a chunk of leaf with a top surface area of 1 mm^2. Chloroplasts are found mainly in the cells of the **mesophyll**, the tissue in the interior of the leaf. CO_2 enters the leaf, and O_2 exits, by way of microscopic pores called **stomata** (singular, *stoma*; from the Greek, meaning "mouth"). Water absorbed by the roots is delivered to the leaves in veins. Leaves also use veins to export sugar to roots and other nonphotosynthetic parts of the plant.

A typical mesophyll cell has about 30–40 chloroplasts, each measuring about 2–4 μm by 4–7 μm. A chloroplast has two membranes surrounding a dense fluid called the **stroma**. Suspended within the stroma is a third membrane system, made up of sacs called **thylakoids**, which segregates the stroma from the *thylakoid space* inside these sacs. In some places, thylakoid sacs are stacked in columns called *grana* (singular, *granum*). **Chlorophyll**, the green pigment that gives leaves their color, resides in the thylakoid membranes of the chloroplast. (The internal photosynthetic membranes of some prokaryotes are also called thylakoid membranes; see Figure 27.8b.) It is the light energy absorbed by chlorophyll that

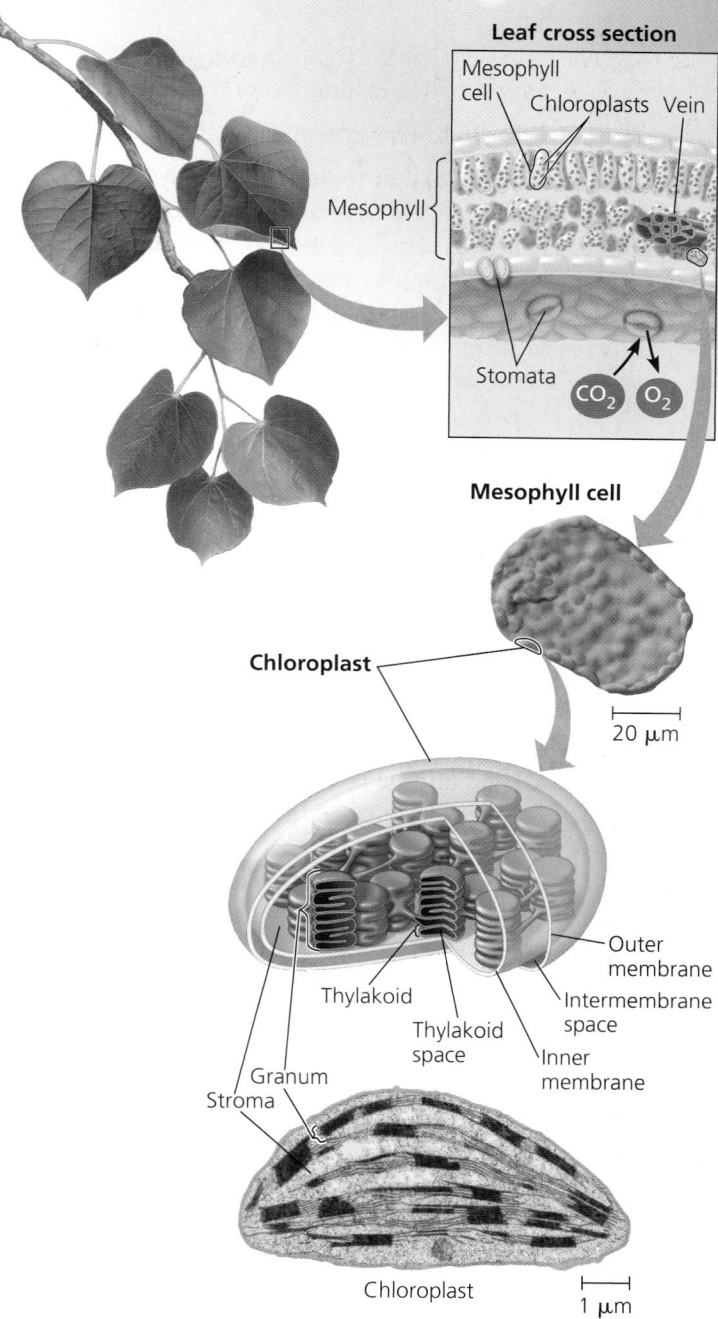

▲ **Figure 10.3 Zooming in on the location of photosynthesis in a plant.** Leaves are the major organs of photosynthesis in plants. These images take you into a leaf, then into a cell, and finally into a chloroplast, the organelle where photosynthesis occurs (middle, LM; bottom, TEM).

▶ **Mastering Biology** Video: Chloroplasts in Motion

drives the synthesis of organic molecules in the chloroplast. Now that we have looked at the sites of photosynthesis in plants, we are ready to look more closely at the process itself.

Tracking Atoms Through Photosynthesis

Scientists have tried for centuries to piece together the process by which plants make food. Although some of the steps are still not completely understood, the overall photosynthetic equation has been known since the 1800s: In the presence of light, the green parts of plants produce organic compounds and O_2 from CO_2

and H_2O. We can summarize the complex series of chemical reactions in photosynthesis with this chemical equation:

$$6\ CO_2 + 12\ H_2O + \text{Light energy} \rightarrow C_6H_{12}O_6 + 6\ O_2 + 6\ H_2O$$

We use glucose ($C_6H_{12}O_6$) here to simplify the relationship between photosynthesis and respiration, but the direct product of photosynthesis is actually a three-carbon sugar that can be used to make glucose. Water appears on both sides of the equation because 12 molecules are consumed and 6 molecules are newly formed during photosynthesis. We can simplify the equation by indicating only the net consumption of water:

$$6\ CO_2 + 6\ H_2O + \text{Light energy} \rightarrow C_6H_{12}O_6 + 6\ O_2$$

Writing the equation in this form, we can see that the overall chemical change during photosynthesis is the reverse of the one that occurs during cellular respiration (see Concept 9.1). It is important to realize that *both* of these metabolic processes occur in plant cells. However, as you will soon learn, chloroplasts do not synthesize sugars by simply reversing the steps of respiration.

Now let's divide the photosynthetic equation by 6 to put it in its simplest possible form:

$$CO_2 + H_2O \rightarrow [CH_2O] + O_2$$

Here, the brackets indicate that CH_2O is not an actual sugar but represents the general formula for a carbohydrate (see Concept 5.2). In other words, we are imagining the synthesis of a sugar molecule one carbon at a time (with six repetitions theoretically adding up to a glucose molecule: $C_6H_{12}O_6$). Let's now see how researchers tracked the elements C, H, and O from the reactants of photosynthesis to the products.

The Splitting of Water: Scientific Inquiry

One of the first clues to the mechanism of photosynthesis came from the discovery that the O_2 given off by plants is derived from H_2O and not from CO_2. The chloroplast splits water into hydrogen and oxygen atoms. Before this discovery, the prevailing hypothesis was that photosynthesis split carbon dioxide ($CO_2 \rightarrow C + O_2$) and then added water to the carbon ($C + H_2O \rightarrow [CH_2O]$). That hypothesis predicted that the O_2 released during photosynthesis came from CO_2, an idea challenged in the 1930s by C. B. van Niel, of Stanford University. Van Niel was investigating photosynthesis in bacteria that make their carbohydrate from CO_2 but do not release O_2. He concluded that, at least in these bacteria, CO_2 is not split into carbon and oxygen. One group of bacteria used hydrogen sulfide (H_2S) rather than water for photosynthesis, forming yellow globules of sulfur as a waste product (these globules are visible in **Figure 10.2e**). Here is the chemical equation for photosynthesis in these sulfur bacteria:

$$CO_2 + 2\ H_2S \rightarrow [CH_2O] + H_2O + 2\ S$$

Van Niel reasoned that the bacteria split H_2S and used the hydrogen atoms to make sugar. He then generalized that idea, proposing that all photosynthetic organisms require a hydrogen source but that the source varies:

$$\text{Sulfur bacteria: } CO_2 + 2\ H_2S \rightarrow [CH_2O] + H_2O + 2\ S$$

$$\text{Plants: } CO_2 + 2\ H_2O \rightarrow [CH_2O] + H_2O + O_2$$

$$\text{General: } CO_2 + 2\ H_2X \rightarrow [CH_2O] + H_2O + 2\ X$$

Thus, van Niel hypothesized that plants split H_2O as a source of electrons from hydrogen atoms, releasing O_2 as a by-product.

Nearly 20 years later, scientists confirmed van Niel's hypothesis by using oxygen-18 (^{18}O), a heavy isotope, as a tracer to follow the path of oxygen atoms during photosynthesis. The experiments showed that the O_2 produced by plants was labeled with ^{18}O *only* if water was the source of the tracer (experiment 1). If the ^{18}O was introduced to the plant in the form of CO_2, the label did not turn up in the released O_2 (experiment 2). In the following summary, green denotes labeled atoms of oxygen (^{18}O):

$$\text{Experiment 1: } CO_2 + 2\ H_2\mathbf{O} \rightarrow [CH_2O] + H_2O + \mathbf{O_2}$$

$$\text{Experiment 2: } \mathbf{C}O_2 + 2\ H_2O \rightarrow [CH_2\mathbf{O}] + H_2\mathbf{O} + O_2$$

A significant result of the shuffling of atoms during photosynthesis is the extraction of hydrogen from water and its incorporation into sugar. The waste product of photosynthesis, O_2, is released to the atmosphere. **Figure 10.4** shows the fates of all atoms in photosynthesis.

Photosynthesis as a Redox Process

Let's briefly compare photosynthesis with cellular respiration. Both processes involve redox reactions. During cellular respiration, energy is released from sugar when electrons associated with hydrogen are transported by carriers to oxygen, forming water as a by-product (see Concept 9.1). The electrons lose potential energy as they "fall" down the electron transport chain toward electronegative oxygen, and the mitochondrion harnesses that energy to synthesize ATP (see Figure 9.14). Photosynthesis reverses the direction of electron flow. Water is split, and its electrons are transferred along with hydrogen ions (H^+) from the water to carbon dioxide, reducing it to sugar.

$$\overbrace{\text{Energy} + 6\ CO_2 + \underbrace{6\ H_2O \longrightarrow C_6H_{12}O_6}_{\text{becomes oxidized}} + 6\ O_2}^{\text{becomes reduced}}$$

▼ **Figure 10.4 Tracking atoms through photosynthesis.** The atoms from CO_2 are shown in magenta, and the atoms from H_2O are shown in blue.

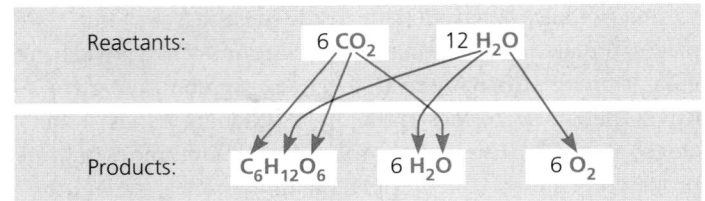

Because the electrons increase in potential energy as they move from water to sugar, this process requires energy—in other words, it is endergonic. This energy boost that occurs during photosynthesis is provided by light.

The Two Stages of Photosynthesis: *A Preview*

The equation for photosynthesis is a deceptively simple summary of a very complex process. Actually, photosynthesis is not a single process, but two processes, each of which has multiple steps. These two stages of photosynthesis are known as the **light reactions** (the *photo* part of photosynthesis) and the **Calvin cycle** (the *synthesis* part) **(Figure 10.5)**.

The light reactions are the steps of photosynthesis that convert solar energy to chemical energy. Water is split, providing a source of electrons and protons (hydrogen ions, H^+) and giving off O_2 as a by-product. Light absorbed by chlorophyll drives a transfer of the electrons and hydrogen ions from water to an acceptor called $NADP^+$ (nicotinamide adenine dinucleotide phosphate), where they are temporarily stored. (The electron acceptor $NADP^+$ is first cousin to NAD^+, which functions as an electron carrier in cellular respiration; the two molecules differ only by the presence of an extra phosphate group in the $NADP^+$ molecule.) The light reactions use solar energy to reduce $NADP^+$ to **NADPH** by adding a pair of electrons along with an H^+. The light reactions also generate ATP, using chemiosmosis to power the addition of a phosphate group to ADP, a process called **photophosphorylation**. Thus, light energy is initially converted to chemical energy in the form of

two compounds: NADPH and ATP. NADPH, a source of electrons, acts as "reducing power" that can be passed along to an electron acceptor, reducing it, while ATP is the versatile energy currency of cells. Notice that the light reactions produce no sugar; that happens in the second stage of photosynthesis, the Calvin cycle.

The Calvin cycle is named for Melvin Calvin, who, along with his colleagues James Bassham and Andrew Benson, began to elucidate its steps in the late 1940s. The cycle begins by incorporating CO_2 from the air into organic molecules already present in the chloroplast. This initial incorporation of carbon into organic compounds is known as **carbon fixation**. The Calvin cycle then reduces the fixed carbon to carbohydrate by the addition of electrons. The reducing power is provided by NADPH, which acquired its cargo of electrons in the light reactions. To convert CO_2 to carbohydrate, the Calvin cycle also requires chemical energy in the form of ATP, which is also generated by the light reactions. Thus, it is the Calvin cycle that makes sugar, but it can do so only with the help of the NADPH and ATP produced by the light reactions. The metabolic steps of the Calvin cycle are sometimes referred to as the dark reactions, or light-independent reactions, because none of the steps requires light *directly*. Nevertheless, the Calvin cycle in most plants occurs during daylight, for only then can the light reactions provide the NADPH and ATP that the Calvin cycle requires. In essence, the chloroplast uses light energy to make sugar by coordinating the two stages of photosynthesis.

▶ **Figure 10.5 An overview of photosynthesis: cooperation of the light reactions and the Calvin cycle.** In the chloroplast, the thylakoid membranes (green) are the sites of the light reactions, whereas the Calvin cycle occurs in the stroma (gray). The light reactions use solar energy to make ATP and NADPH, which supply chemical energy and reducing power, respectively, to the Calvin cycle. The Calvin cycle incorporates CO_2 into organic molecules, which are converted to sugar. (Recall that most simple sugars have formulas that are some multiple of CH_2O.) To visualize these processes in their cellular context, see Figure 6.32.

⊘ **Mastering Biology**
BioFlix® Animation: Photosynthesis

As Figure 10.5 indicates, the thylakoids of the chloroplast are the sites of the light reactions, while the Calvin cycle occurs in the stroma. On the outside of the thylakoids, during the light reactions, molecules of $NADP^+$ and ADP pick up electrons and phosphate, respectively, and NADPH and ATP are then released to the stroma, where they play crucial roles in the Calvin cycle. The two stages of photosynthesis are treated in this figure as metabolic modules that take in ingredients and crank out products. In the next two sections, we'll look more closely at how the two stages work, beginning with the light reactions.

CONCEPT CHECK 10.2

1. **MAKE CONNECTIONS** How do the CO_2 molecules used in photosynthesis reach and enter the chloroplasts inside leaf cells? (See Concept 7.2.)

2. Explain how the use of an oxygen isotope helped elucidate the chemistry of photosynthesis.

3. **WHAT IF?** The Calvin cycle requires ATP and NADPH, products of the light reactions. If a classmate asserted that the light reactions don't depend on the Calvin cycle and, with continual light, could just keep on producing ATP and NADPH, how would you respond?

For suggested answers, see Appendix A.

CONCEPT 10.3

The light reactions convert solar energy to the chemical energy of ATP and NADPH

Chloroplasts are chemical factories powered by the sun. Their thylakoids transform light energy into the chemical energy of ATP and NADPH. To better understand the conversion of light to chemical energy, we need to know about some important properties of light.

The Nature of Sunlight

Light is a form of energy known as electromagnetic energy, also called electromagnetic radiation. Electromagnetic energy travels in rhythmic waves analogous to those created by dropping a pebble into a pond. Electromagnetic waves, however, are disturbances of electric and magnetic fields rather than disturbances of a material medium such as water.

The distance between the crests of electromagnetic waves is called the **wavelength**. Wavelengths range from less than a nanometer (for gamma rays) to more than a kilometer (for radio waves). This entire range of radiation is known as the **electromagnetic spectrum (Figure 10.6)**. The segment most important to life is the narrow band from about 380 nm to 740 nm in wavelength. This radiation is known as

▼ **Figure 10.6 The electromagnetic spectrum.** White light is a mixture of all wavelengths of visible light. A prism can sort white light into its component colors by bending light of different wavelengths at different angles. (Droplets of water in the atmosphere can act as prisms, causing a rainbow to form.) Visible light drives photosynthesis.

visible light because it can be detected as various colors by the human eye.

The model of light as waves explains many of light's properties, but in certain respects light behaves as though it consists of discrete particles, called **photons**. Photons are not tangible objects, but they act like objects in that each of them has a fixed quantity of energy. The amount of energy is inversely related to the wavelength of the light: The shorter the wavelength, the greater the energy of each photon of that light. Thus, a photon of violet light packs nearly twice as much energy as a photon of red light (see Figure 10.6).

Although the sun radiates the full spectrum of electromagnetic energy, the atmosphere acts like a selective window, allowing visible light to pass through while screening out a substantial fraction of other radiation. The part of the spectrum we can see—visible light—is also the radiation that drives photosynthesis.

Photosynthetic Pigments: The Light Receptors

When light meets matter, it may be reflected, transmitted, or absorbed. Substances that absorb visible light are known as *pigments*. Different pigments absorb light of different wavelengths, and the wavelengths that are absorbed disappear. If a pigment is illuminated with white light, the color we see is the color most reflected or transmitted by the pigment. (If a pigment absorbs all wavelengths, it appears black.) We see green when we look at a leaf because chlorophyll absorbs violet-blue and red light while transmitting and reflecting green light

▼ **Figure 10.7 Why leaves are green: interaction of light with chloroplasts.** The chlorophyll molecules of chloroplasts absorb violet-blue and red light (the colors most effective in driving photosynthesis) and reflect or transmit green light. This is why leaves appear green.

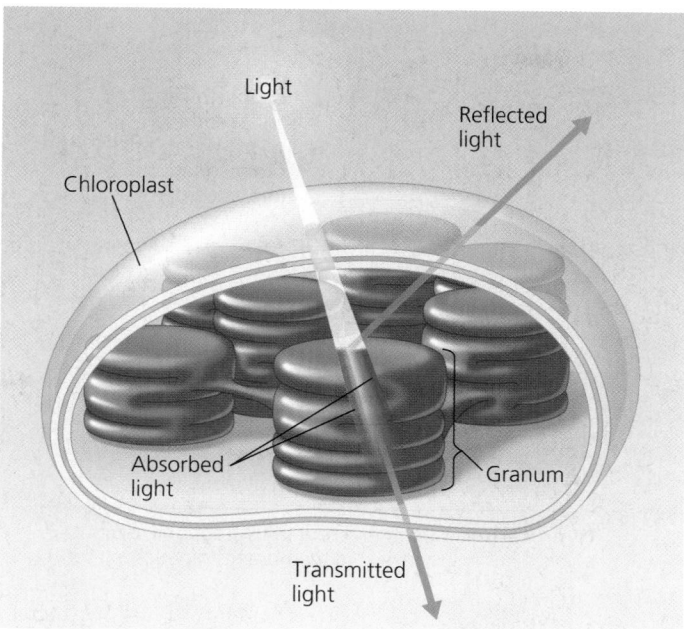

⊕ **Mastering Biology**
Animation: Light Energy and Pigments

(Figure 10.7). The ability of a pigment to absorb various wavelengths of light can be measured with an instrument called a **spectrophotometer**. This machine directs beams of light of different wavelengths through a solution of the pigment and measures the fraction of the light transmitted at each wavelength. A graph plotting a pigment's light absorption versus wavelength is called an **absorption spectrum (Figure 10.8)**.

The absorption spectra of chloroplast pigments provide clues to the relative effectiveness of different wavelengths for driving photosynthesis since light can perform work in chloroplasts only if it is absorbed. **Figure 10.9a** shows the absorption spectra of three types of pigments in chloroplasts: **chlorophyll *a***, the key light-capturing pigment that participates directly in the light reactions; the accessory pigment **chlorophyll *b***; and a separate group of accessory pigments called carotenoids. The spectrum of chlorophyll *a* suggests that violet-blue and red light work best for photosynthesis, since they are absorbed, while green is the least effective color. This is confirmed by an **action spectrum** for photosynthesis **(Figure 10.9b)**, which profiles the relative effectiveness of different wavelengths of radiation in driving the process. An action spectrum is prepared by illuminating chloroplasts with light of different colors and then plotting wavelength against some measure of photosynthetic rate, such as CO_2 consumption or O_2 release. The action spectrum for photosynthesis was first demonstrated by Theodor W. Engelmann, a German botanist, in 1883.

▼ **Figure 10.8** **Research Method**

Determining an Absorption Spectrum

Application An absorption spectrum is a visual representation of how well a particular pigment absorbs different wavelengths of visible light. Absorption spectra of various chloroplast pigments help scientists decipher the role of each pigment in a plant.

Technique A spectrophotometer measures the relative amounts of light of different wavelengths absorbed and transmitted by a pigment solution.

❶ White light is separated into colors (wavelengths) by a prism.

❷ One by one, the different colors of light are passed through the sample (chlorophyll in this example). Green light and blue light are shown here.

❸ The transmitted light strikes a photoelectric tube, which converts the light energy to electricity.

❹ The electric current is measured by a galvanometer. The meter indicates the fraction of light transmitted through the sample, from which we can determine the amount of light absorbed.

Results See Figure 10.9a for absorption spectra of three types of chloroplast pigments.

Before equipment for measuring O_2 levels had even been invented, Engelmann performed a clever experiment in which he used O_2-requiring bacteria to measure rates of photosynthesis in filamentous algae **(Figure 10.9c)**. His results are a striking match to the modern action spectrum shown in Figure 10.9b.

▼ Figure 10.9 Inquiry

Which wavelengths of light are most effective in driving photosynthesis?

Experiment Absorption and action spectra, along with a classic experiment by Theodor W. Engelmann, reveal which wavelengths of light are photosynthetically important.

Results

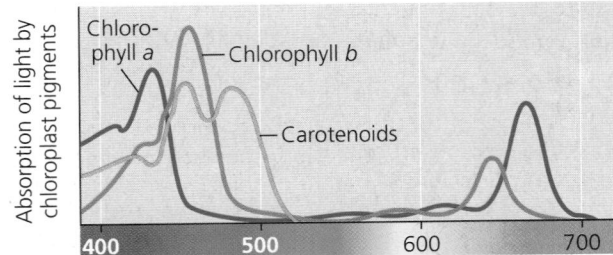

(a) Absorption spectra. The three curves show the wavelengths of light best absorbed by three types of chloroplast pigments.

(b) Action spectrum. This graph plots the rate of photosynthesis versus wavelength. The resulting action spectrum resembles the absorption spectrum for chlorophyll *a* but does not match exactly (see part a). This is partly due to the absorption of light by accessory pigments such as chlorophyll *b* and carotenoids.

(c) Engelmann's experiment. In 1883, Theodor W. Engelmann shined light through a prism onto an alga, exposing different segments of the alga to different wavelengths. He used aerobic bacteria, which congregate near an oxygen source, to determine which segments of the alga were releasing the most O_2 and thus photosynthesizing most. Bacteria congregated in greatest numbers around the parts of the alga illuminated with violet-blue or red light.

Conclusion The action spectra, confirmed by Engelmann's experiment, show which portions of the spectrum are most effective in driving photosynthesis.

Data from T. W. Engelmann, *Bacterium photometricum. Ein Beitrag zur vergleichen- den Physiologie des Licht-und Farbensinnes, Archiv. für Physiologie* 30:95–124 (1883).

→ **Instructors:** A related Experimental Inquiry Tutorial can be assigned in Mastering Biology.

INTERPRET THE DATA *According to the graph, which wave- lengths of light drive the highest rates of photosynthesis?*

▼ Figure 10.10 Structure of chlorophyll *a* and *b*.

Chlorophyll *a* and *b* differ in only one functional group:

CH₃ in chlorophyll *a*
CHO in chlorophyll *b*

Porphyrin ring: light-absorbing "head" of molecule; note magnesium atom at center

Hydrocarbon tail: interacts with hydrophobic regions of proteins inside thylakoid membranes of chloroplasts

→ **Mastering Biology**
Animation: Space-Filling Model of Chlorophyll

Notice by comparing Figures 10.9a and 10.9b that the action spectrum for photosynthesis is much broader than the absorption spectrum of chlorophyll *a*. The absorption spectrum of chlorophyll *a* alone underestimates the effectiveness of certain wavelengths in driving photosynthesis. This is partly because accessory pigments with different absorption spectra— including chlorophyll *b* and carotenoids—broaden the spectrum of colors that can be used for photosynthesis. **Figure 10.10** shows the structure of chlorophyll *a* compared with that of chlorophyll *b*. A slight structural difference between them is enough to cause the two pigments to absorb at slightly differ- ent wavelengths in the red and blue parts of the spectrum (see Figure 10.9a). As a result, chlorophyll *a* appears blue green and chlorophyll *b* olive green under visible light.

In the last decade, researchers have identified two other forms of chlorophyll—chlorophyll *d* and chlorophyll *f*—that absorb higher wavelengths of light. In 2018, researchers grew a species of cyanobacterium called *Chroococcidiopsis thermalis*

▶ **Colony of the cyanobacteria *Chroococcidiopsis thermalis*.** In this micrograph of cells becoming acclimated to higher wavelengths of light, cells that haven't yet acclimated and are still using chlorophyll *a* for photosynthesis appear purplish-pink, and cells that have acclimated and are using chlorophyll *f* appear yellow.

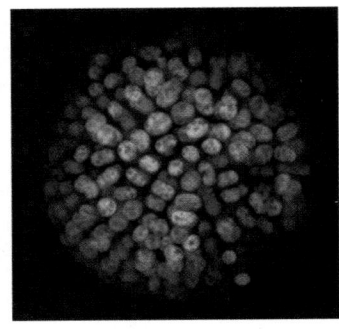

under only far-red light with a wavelength of 750 nm. They concluded that chlorophyll *f* was functioning in place of chlorophyll *a*, enabling this cyanobacterium to flourish in very shaded conditions. Higher-wavelength light has less energy, so this observation extends the lower limit of energy (the "red limit") needed for photosynthesis to occur.

Other accessory pigments include **carotenoids**, hydrocarbons that are various shades of yellow and orange because they absorb violet and blue-green light (see Figure 10.9a). Carotenoids may broaden the spectrum of colors that can drive photosynthesis. However, a more important function of at least some carotenoids seems to be *photoprotection*: These compounds absorb and dissipate excessive light energy that would otherwise damage chlorophyll or interact with oxygen, forming reactive oxidative molecules that are dangerous to the cell. Interestingly, carotenoids similar to the photoprotective ones in chloroplasts have a photoprotective role in the human eye. (Carrots, known for aiding night vision, are rich in carotenoids.) These and related molecules are, of course, found naturally in many vegetables and fruits. They are also often advertised in health food products as "phytochemicals" (from the Greek *phyton*, plant), some of which have antioxidant properties. Plants can synthesize all the antioxidants they require, but humans and other animals must obtain some of them from their diets.

Excitation of Chlorophyll by Light

What exactly happens when chlorophyll and other pigments absorb light? The colors corresponding to the absorbed wavelengths disappear from the spectrum of the transmitted and reflected light, but energy cannot disappear. When a molecule absorbs a photon of light, one of the molecule's electrons is elevated to an orbital where it has more potential energy (see Figure 2.6b). When the electron is in its normal orbital, the pigment molecule is said to be in its ground state. Absorption

of a photon boosts an electron to an orbital of higher energy, and the pigment molecule is then said to be in an excited state. The only photons absorbed are those whose energy is exactly equal to the energy difference between the ground state and an excited state, and this energy difference varies from one kind of molecule to another. Thus, a particular compound absorbs only photons corresponding to specific wavelengths, which is why each pigment has a unique absorption spectrum.

Once absorption of a photon raises an electron to an excited state, the electron cannot stay there long **(Figure 10.11a)**. The excited state, like all high-energy states, is unstable. Generally, when isolated pigment molecules absorb light, their excited electrons drop back down to the ground-state orbital in a billionth of a second, releasing their excess energy as heat. This conversion of light energy to heat is what makes the top of an automobile so hot on a sunny day. (White cars are coolest because their paint reflects all wavelengths of visible light.) In isolation, some pigments, including chlorophyll, emit light as well as heat after absorbing photons. As excited electrons fall back to the ground state, photons are given off, an afterglow called fluorescence. An illuminated solution of chlorophyll isolated from chloroplasts will fluoresce in the red part of the spectrum and also give off heat **(Figure 10.11b)**. This is best seen by illuminating with ultraviolet light, which chlorophyll can also absorb (see Figures 10.6 and 10.9a). Viewed under visible light, the fluorescence would be difficult to see against the green of the solution.

A Photosystem: A Reaction-Center Complex Associated with Light-Harvesting Complexes

Chlorophyll molecules excited by the absorption of light energy produce very different results in an intact chloroplast than they do in isolation (see Figure 10.11). In their native environment of the thylakoid membrane, chlorophyll

► **Figure 10.11 Excitation of isolated chlorophyll by light. (a)** Absorption of a photon causes a transition of the chlorophyll molecule from its ground state to its excited state. The photon boosts an electron to an orbital where it has more potential energy. If the illuminated molecule exists in isolation, its excited electron immediately drops back down to the ground-state orbital, and its excess energy is given off as heat and fluorescence (light). **(b)** A chlorophyll solution excited with ultraviolet light fluoresces with a red-orange glow.

WHAT IF? *If a leaf containing the same concentration of chlorophyll as in the solution was exposed to the same ultraviolet light, no fluorescence would be seen. Propose an explanation for the difference in fluorescence emission between the solution and the leaf.*

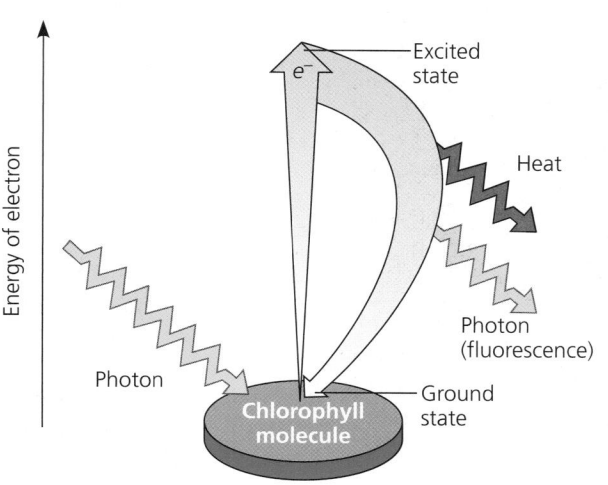

(a) Excitation of isolated chlorophyll molecule

(b) Fluorescence

molecules are organized along with other small organic molecules and proteins into complexes called photosystems.

A **photosystem** is composed of a reaction-center complex surrounded by several light-harvesting complexes **(Figure 10.12)**. The **reaction-center complex** is an

▼ **Figure 10.12 The structure and function of a photosystem.**

(a) How a photosystem harvests light. When a photon strikes a pigment molecule in a light-harvesting complex, the energy is passed from molecule to molecule until it reaches the reaction-center complex. Here, an excited electron from the special pair of chlorophyll *a* molecules is transferred to the primary electron acceptor.

(b) Structure of a photosystem. This computer model, based on X-ray crystallography, shows two photosystem complexes side by side. Chlorophyll molecules (bright green ball-and-stick models within the membrane; the tails are not shown) are interspersed with protein subunits (purple ribbons; notice the many α helices spanning the membrane). For simplicity, a photosystem will be shown as a single complex in the rest of the chapter.

organized association of proteins holding a special pair of chlorophyll *a* molecules and a primary electron acceptor. Each **light-harvesting complex** consists of various pigment molecules (which may include chlorophyll *a*, chlorophyll *b*, and multiple carotenoids) bound to proteins. The number and variety of pigment molecules enable a photosystem to harvest light over a larger surface area and a larger portion of the spectrum than could any single pigment molecule alone. Together, these light-harvesting complexes act as an antenna for the reaction-center complex. When a pigment molecule absorbs a photon, the energy is transferred from pigment molecule to pigment molecule within a light-harvesting complex, like a human "wave" at a sports arena, until it is passed to the pair of chlorophyll *a* molecules in the reaction-center complex. This pair of chlorophyll *a* molecules is special because their molecular environment—their location and the other molecules with which they are associated—enables them to use the energy from light not only to boost one of their electrons to a higher energy level, but also to transfer it to a different molecule—the **primary electron acceptor**, which is a molecule capable of accepting electrons and becoming reduced.

The solar-powered transfer of an electron from the reaction-center chlorophyll *a* pair to the primary electron acceptor is the first step of the light reactions. As soon as the chlorophyll electron is excited to a higher energy level, the primary electron acceptor captures it; this is a redox reaction. In the flask shown in Figure 10.11b, isolated chlorophyll fluoresces because there is no electron acceptor, so electrons of photoexcited chlorophyll drop right back to the ground state. In the structured environment of a chloroplast, however, an electron acceptor is readily available, and the potential energy represented by the excited electron is not dissipated as light and heat. Thus, each photosystem—a reaction-center complex surrounded by light-harvesting complexes—functions in the chloroplast as a unit. It converts light energy to chemical energy, which will ultimately be used for the synthesis of sugar.

The thylakoid membrane is populated by two types of photosystems that cooperate in the light reactions of photosynthesis: **photosystem II (PS II)** and **photosystem I (PS I)**. (They were named in order of their discovery, but photosystem II functions first in the light reactions.) Each has a characteristic reaction-center complex—a particular kind of primary electron acceptor next to a special pair of chlorophyll *a* molecules associated with specific proteins. The reaction-center chlorophyll *a* of photosystem II is known as P680 because this pigment is best at absorbing light having a wavelength of 680 nm (in the red part of the spectrum). The chlorophyll *a* at the reaction-center complex of photosystem I is called P700 because it most effectively absorbs light of wavelength 700 nm (in the far-red part of the spectrum). These two pigments, P680 and P700, are nearly identical chlorophyll *a* molecules. However, their association with different proteins in the thylakoid membrane affects the electron distribution in the two pigments and accounts for the slight differences in their light-absorbing properties. Now let's see how the two types of photosystems work together in using light energy to generate ATP and NADPH, the two main products of the light reactions.

Linear Electron Flow

Light drives the synthesis of ATP and NADPH by energizing the two types of photosystems embedded in the thylakoid membranes of chloroplasts. The key to this energy transformation is a flow of electrons through the photosystems and other molecular components built into the thylakoid membrane. This is called **linear electron flow**, and it occurs during the light reactions of photosynthesis, as shown in **Figure 10.13**. The numbered steps in the text correspond to the numbered steps in the figure.

1 A photon of light strikes one of the pigment molecules in a light-harvesting complex of PS II, boosting one of its electrons to a higher energy level. As this electron falls back to its ground state, an electron in a nearby pigment molecule is simultaneously raised to an excited state. The process continues, with the energy being relayed to other pigment molecules until it reaches the P680 pair of chlorophyll a molecules in the PS II reaction-center complex. It excites an electron in this pair of chlorophylls to a higher energy state.

2 This electron is transferred from the excited P680 to the primary electron acceptor. We can refer to the resulting form of P680, missing the negative charge of an electron, as P680$^+$.

3 An enzyme catalyzes the splitting of a water molecule into two electrons, two hydrogen ions (H$^+$), and an oxygen atom. The electrons are supplied one by one to the P680$^+$ pair, each electron replacing one transferred to the primary electron acceptor. (P680$^+$ is the strongest biological oxidizing agent known; its electron "hole" must be filled. This greatly facilitates the transfer of electrons from the split water molecule.) The H$^+$ are released into the thylakoid space (interior of the thylakoid). The oxygen atom immediately combines with an oxygen atom generated by the splitting of another water molecule, forming O$_2$.

▼ **Figure 10.13 How linear electron flow during the light reactions generates ATP and NADPH.** The gold arrows trace the flow of light-driven electrons from water to NADPH. The black arrows trace the transfer of energy from pigment molecule to pigment molecule. To see these proteins in their cellular context, see Figure 6.32b.

→ **Mastering Biology** Figure Walkthrough

Photosystem II (PS II)

Photosystem I (PS I)

4. Each photoexcited electron passes from the primary electron acceptor of PS II to PS I via an electron transport chain, the components of which are similar to those of the electron transport chain that functions in cellular respiration. The electron transport chain between PS II and PS I is made up of the electron carrier plastoquinone (Pq), a cytochrome complex, and a protein called plastocyanin (Pc). Each component carries out redox reactions as electrons flow down the electron transport chain, releasing free energy that is used to pump protons (H^+) into the thylakoid space, contributing to a proton gradient across the thylakoid membrane.

5. The potential energy stored in the proton gradient is used to make ATP in a process called chemiosmosis, to be discussed shortly.

6. Meanwhile, light energy has been transferred via light-harvesting complex pigments to the PS I reaction-center complex, exciting an electron of the P700 pair of chlorophyll *a* molecules located there. The photoexcited electron is then transferred to PS I's primary electron acceptor, creating an electron "hole" in the P700—which we now can call $P700^+$. In other words, $P700^+$ can now act as an electron acceptor, accepting an electron that reaches the bottom of the electron transport chain from PS II.

7. Photoexcited electrons are passed in a series of redox reactions from the primary electron acceptor of PS I down a second electron transport chain through the protein ferredoxin (Fd). (This chain does not create a proton gradient and thus does not produce ATP.)

8. The enzyme $NADP^+$ reductase catalyzes the transfer of electrons from Fd to $NADP^+$. Two electrons are required for its reduction to NADPH. Electrons in NADPH are at a higher energy level than they are in water (where they started), so they are more readily available for the reactions of the Calvin cycle. This process also removes an H^+ from the stroma.

▼ **Figure 10.14 A mechanical analogy for linear electron flow during the light reactions.**

The energy changes of electrons during their linear flow through the light reactions are shown in a mechanical analogy in **Figure 10.14**. Although the scheme shown in Figures 10.13 and 10.14 may seem complicated, do not lose track of the big picture: The light reactions use solar power to generate ATP and NADPH, which provide chemical energy and reducing power, respectively, to the carbohydrate-synthesizing reactions of the Calvin cycle.

Cyclic Electron Flow

In certain cases, photoexcited electrons can take an alternative path called **cyclic electron flow**, which uses photosystem I but not photosystem II. You can see in **Figure 10.15** that cyclic flow is a short circuit: The electrons cycle back from ferredoxin (Fd) to the cytochrome complex, then via a plastocyanin molecule (Pc) to a P700 chlorophyll in the PS I reaction-center complex. There is

◀ **Figure 10.15 Cyclic electron flow.** Photoexcited electrons from PS I are occasionally shunted back from ferredoxin (Fd) to chlorophyll via the cytochrome complex and plastocyanin (Pc). This electron shunt supplements the supply of ATP (via chemiosmosis) but produces no NADPH. The "shadow" of linear electron flow is included in the diagram for comparison with the cyclic route. The two Fd molecules in this diagram are actually one and the same—the final electron carrier in the electron transport chain of PS I—although it is depicted twice to clearly show its role in two parts of the process.

VISUAL SKILLS *Look at Figure 10.14 and explain how you would alter it to show a mechanical analogy for cyclic electron flow.*

no production of NADPH and no release of oxygen that results from this process. On the other hand, cyclic flow does generate ATP.

Rather than having both PS II and PS I, several of the currently existing groups of photosynthetic bacteria are known to have a single photosystem related to either PS II or PS I. For these species, which include the purple sulfur bacteria (see Figure 10.2e) and the green sulfur bacteria, cyclic electron flow is the one and only means of generating ATP during the process of photosynthesis. Evolutionary biologists hypothesize that these bacterial groups are descendants of ancestral bacteria in which photosynthesis first evolved, in a form similar to cyclic electron flow.

Cyclic electron flow can also occur in photosynthetic species that possess both photosystems; this includes some prokaryotes, such as the cyanobacteria shown in Figure 10.2d, as well as the eukaryotic photosynthetic species that have been tested thus far. Although the process is probably in part an "evolutionary leftover," research suggests it plays at least one beneficial role for these organisms. Plants with mutations that render them unable to carry out cyclic electron flow are capable of growing well in low light, but do not grow well where light is intense. This is evidence for the idea that cyclic electron flow may be photoprotective. Later you'll learn more about cyclic electron flow as it relates to a particular adaptation of photosynthesis (C$_4$ plants; see Concept 10.5).

Whether ATP synthesis is driven by linear or cyclic electron flow, the actual mechanism is the same. Before we move on to consider the Calvin cycle, let's review chemiosmosis, the process that uses membranes to couple redox reactions to ATP production.

A Comparison of Chemiosmosis in Chloroplasts and Mitochondria

Chloroplasts and mitochondria generate ATP by the same basic mechanism: chemiosmosis (see Figure 9.14). An electron transport chain pumps protons (H$^+$) across a membrane as electrons are passed through a series of carriers that have progressively more affinity for electrons. Thus, electron transport chains transform redox energy to a proton-motive force, potential energy stored in the form of an H$^+$ gradient across a membrane. An ATP synthase complex in the same membrane couples the diffusion of hydrogen ions down their gradient to the phosphorylation of ADP, forming ATP.

Some of the electron carriers, including the iron-containing proteins called cytochromes, are very similar in mitochondria and chloroplasts (see Figure 6.32b and c). The ATP synthase complexes of the two organelles are also quite similar. But there are noteworthy differences between photophosphorylation in chloroplasts and oxidative phosphorylation in mitochondria. Both work by way of chemiosmosis, but in chloroplasts, the high-energy electrons dropped down the transport chain come from water, while in mitochondria, they are extracted from organic molecules (which are thus oxidized). Chloroplasts do not need molecules from food to make ATP; their photosystems capture light energy and use it to drive the electrons from water to the top of the transport chain. In other words, mitochondria use chemiosmosis to transfer chemical energy from food molecules to ATP, whereas chloroplasts use it to transform light energy into chemical energy in ATP.

Although the spatial organization of chemiosmosis differs slightly between chloroplasts and mitochondria, it is easy to see similarities in the two **(Figure 10.16)**. Electron transport

▶ **Figure 10.16 Comparison of chemiosmosis in mitochondria and chloroplasts.** In both kinds of organelles, electron transport chains pump protons (H$^+$) across a membrane from a region of low H$^+$ concentration (light gray in this diagram) to one of high H$^+$ concentration (dark gray). The protons then diffuse back across the membrane through ATP synthase, driving the synthesis of ATP.

MAKE CONNECTIONS *Describe how you would change the pH in order to artificially cause ATP synthesis (a) outside an isolated mitochondrion (assume H$^+$ can freely cross the outer membrane; see Figure 9.14) and (b) in the stroma of a chloroplast. Explain.*

Mitochondrion

Chloroplast

MITOCHONDRION STRUCTURE

Intermembrane space

Inner membrane

Matrix

Key
- Higher [H$^+$]
- Lower [H$^+$]

Electron transport chain (ETC)

Pumping of H$^+$ by ETC

H$^+$

Diffusion of H$^+$ through ATP synthase

H$^+$

ATP synthase

ADP + ℗$_i$

ATP

H$^+$

Thylakoid space

Thylakoid membrane

Stroma

CHLOROPLAST STRUCTURE

chain proteins in the inner membrane of the mitochondrion pump protons from the mitochondrial matrix out to the intermembrane space, which then serves as a reservoir of hydrogen ions. Similarly, electron transport chain proteins in the thylakoid membrane of the chloroplast pump protons from the stroma into the thylakoid space, which functions as the H$^+$ reservoir. If you imagine the cristae of mitochondria pinching off from the inner membrane, this may help you see how the thylakoid space and the intermembrane space are comparable spaces in the two organelles, while the mitochondrial matrix is analogous to the stroma of the chloroplast.

In the mitochondrion, protons diffuse down their concentration gradient from the intermembrane space through ATP synthase to the matrix, driving ATP synthesis. In the chloroplast, ATP is synthesized as the hydrogen ions diffuse from the thylakoid space back to the stroma through ATP synthase complexes **(Figure 10.17)**, whose catalytic knobs are on the stroma side of the membrane. Thus, ATP forms in the stroma, where it is used to help drive sugar synthesis during the Calvin cycle.

The proton (H$^+$) gradient, or pH gradient, across the thylakoid membrane is substantial. When chloroplasts in an

▼ **Figure 10.17 The light reactions: organization of the thylakoid membrane.**

→ **Mastering Biology**
  Animation: The Light Reactions

❶ Energy from light excites an electron (e⁻) from chlorophyll, which is transferred to the primary electron acceptor in photosystem II.

❷ In the thylakoid space, water is split into:
  • 2 e⁻, which replace chlorophyll's electron twice,
  • 2 H⁺, which contribute to the high [H⁺] in the thylakoid space, and
  • an O atom, which joins another O, forming O₂ gas.

❸ Gold arrows track the flow of electrons. As electrons travel down the electron transport chain, 4 H⁺ are pumped into the thylakoid space, increasing [H⁺], decreasing the pH.

❹ In chemiosmosis, the diffusion of H⁺ from the thylakoid space back to the stroma down the [H⁺] gradient powers ATP synthase, producing ATP.

❺ Low-energy e⁻ from water end up in NADPH (high-energy e⁻). NADPH formation removes H⁺ from the stroma, decreasing its [H⁺], increasing the pH.

VISUAL SKILLS *Which three steps in the figure contribute to the [H⁺] gradient across the thylakoid membrane?*

experimental setting are illuminated, the pH in the thylakoid space drops to about 5 (the H^+ concentration increases), and the pH in the stroma increases to about 8 (the H^+ concentration decreases). This gradient of three pH units corresponds to a thousandfold difference in H^+ concentration. If the lights are then turned off, the pH gradient is abolished, but it can quickly be restored by turning the lights back on. Experiments such as this provided strong evidence in support of the chemiosmotic model.

The currently accepted model for the organization of the light-reaction "machinery" within the thylakoid membrane is based on several research studies. Each of the molecules and molecular complexes in Figure 10.17 is present in numerous copies in each thylakoid. Notice that NADPH, like ATP, is produced on the side of the membrane facing the stroma, where the Calvin cycle reactions take place.

Let's step back and see the big picture by summarizing the light reactions. Electron flow pushes electrons from water, where they are at a state of low potential energy, ultimately to NADPH, where they are stored at a state of high potential energy. The light-driven electron flow also generates ATP. Thus, the equipment of the thylakoid membrane converts light energy to chemical energy stored in ATP and NADPH. O_2 is produced as a by-product. Let's now see how the enzymes of the Calvin cycle use the products of the light reactions to synthesize sugar from CO_2.

→ **Mastering Biology** BioFlix® **Animation: The Light Reactions**

CONCEPT CHECK 10.3

1. What color of light is *least* effective in driving photosynthesis? Explain.
2. In the light reactions, what is the initial electron donor? Where do the electrons finally end up?
3. **WHAT IF?** In an experiment, isolated chloroplasts placed in an illuminated solution with the appropriate chemicals can carry out ATP synthesis. Predict what would happen to the rate of synthesis if a compound is added to the solution that makes membranes freely permeable to hydrogen ions.

For suggested answers, see Appendix A.

CONCEPT 10.4

The Calvin cycle uses the chemical energy of ATP and NADPH to reduce CO_2 to sugar

The Calvin cycle takes place in the stroma; it is similar to the citric acid cycle in that a starting material is regenerated after some molecules enter and others exit the cycle. However, the citric acid cycle is catabolic, oxidizing acetyl CoA and using the energy to synthesize ATP (see Figure 9.11), while the

Calvin cycle is anabolic, building carbohydrates from smaller molecules and consuming energy. Carbon enters the Calvin cycle in the form of CO_2 and leaves in the form of sugar. The cycle spends ATP as an energy source and consumes NADPH as reducing power for adding high-energy electrons to make the sugar.

As we mentioned in Concept 10.2, the carbohydrate produced directly from the Calvin cycle is not glucose. It is actually a three-carbon sugar called **glyceraldehyde 3-phosphate (G3P)**. For the net synthesis of *one* molecule of G3P, the cycle must take place three times, fixing *three* molecules of CO_2—one per turn of the cycle. (Recall that the term *carbon fixation* refers to the initial incorporation of CO_2 into organic material.) As we trace the steps of the Calvin cycle, keep in mind that we are following three molecules of CO_2 through the reactions. **Figure 10.18** divides the Calvin cycle into three phases: carbon fixation, reduction, and regeneration of the CO_2 acceptor.

Phase 1: Carbon fixation. The Calvin cycle incorporates each CO_2 molecule, one at a time, by attaching it to a five-carbon sugar named ribulose bisphosphate (which is abbreviated RuBP). The enzyme that catalyzes this first step is RuBP carboxylase-oxygenase, or **rubisco** (see Figure 6.32c). (This is the most abundant protein in chloroplasts and is also thought to be the most abundant protein on Earth.) The product of the reaction is a six-carbon intermediate that is short-lived because it is so energetically unstable that it immediately splits in half, forming two molecules of 3-phosphoglycerate (for each CO_2 fixed).

Phase 2: Reduction. Each molecule of 3-phosphoglycerate receives an additional phosphate group from ATP, becoming 1,3-bisphosphoglycerate. Next, a pair of electrons donated from NADPH reduces 1,3-bisphosphoglycerate, which also loses a phosphate group in the process, becoming glyceraldehyde 3-phosphate (G3P). Specifically, the electrons from NADPH reduce a carboxyl group on 1,3-bisphosphoglycerate to the aldehyde group of G3P, which stores more potential energy. G3P is a sugar—the same three-carbon sugar formed in glycolysis by the splitting of glucose (see Figure 9.8). Notice in Figure 10.18 that for every *three* molecules of CO_2 that enter the cycle, there are *six* molecules of G3P formed. But only one molecule of this three-carbon sugar can be counted as a net gain of carbohydrate because the rest are required to complete the cycle. The cycle began with 15 carbons' worth of carbohydrate in the form of three molecules of the five-carbon sugar RuBP. Now there are 18 carbons' worth of carbohydrate in the form of six molecules of G3P. One molecule exits the cycle to be used by the plant cell, but the other five molecules must be recycled to regenerate the three molecules of RuBP.

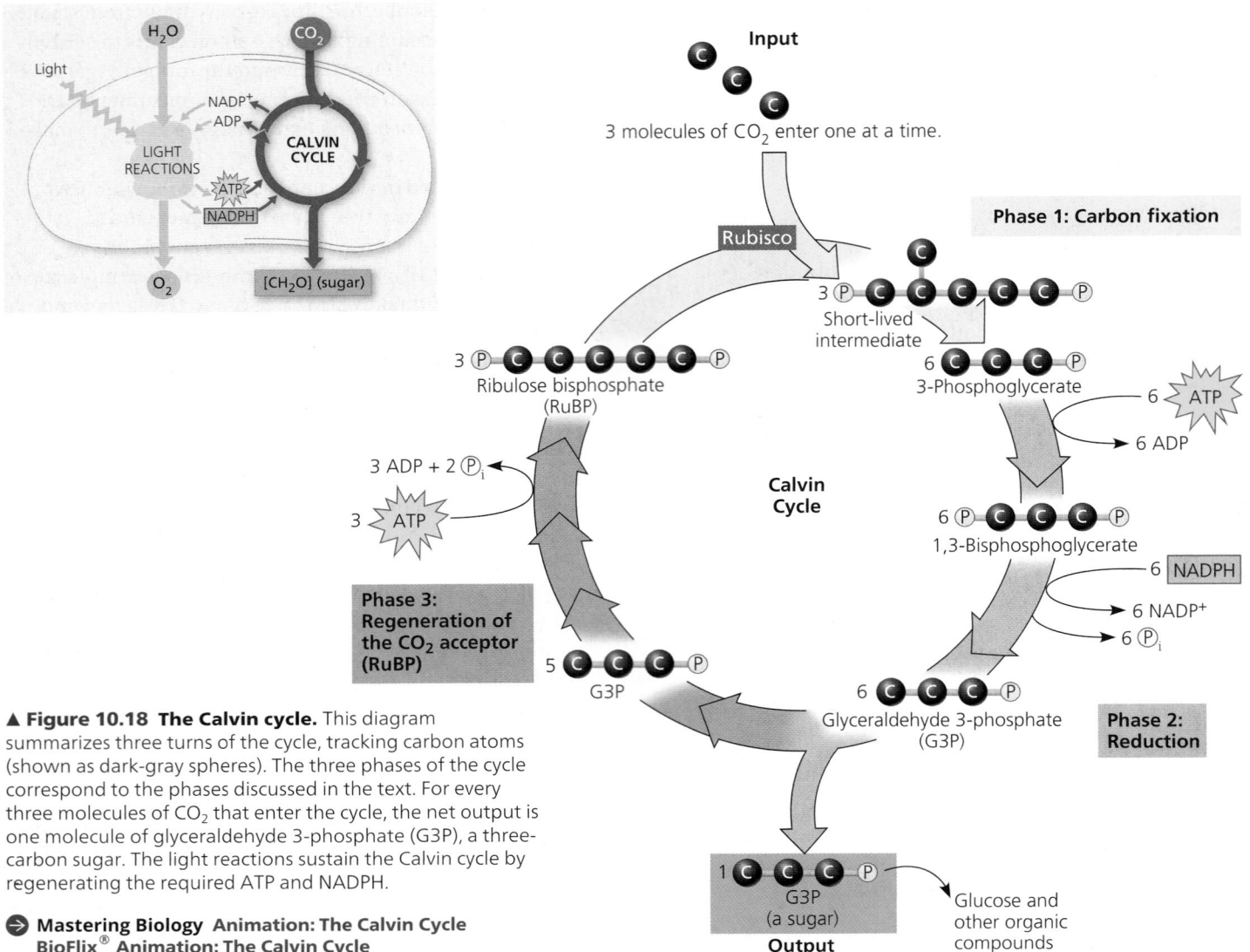

Input

3 molecules of CO_2 enter one at a time.

Phase 1: Carbon fixation

Rubisco

3 (P) C C C C C C (P)

Short-lived intermediate

3 (P) C C C C C (P)
Ribulose bisphosphate (RuBP)

6 C C C (P)
3-Phosphoglycerate

6 ATP
6 ADP

Calvin Cycle

3 ADP + 2 (P)i

3 ATP

6 (P) C C C (P)
1,3-Bisphosphoglycerate

6 NADPH
6 NADP+
6 (P)i

Phase 3: Regeneration of the CO_2 acceptor (RuBP)

5 C C C (P)
G3P

6 C C C (P)
Glyceraldehyde 3-phosphate (G3P)

Phase 2: Reduction

1 C C C (P)
G3P (a sugar)

Output

Glucose and other organic compounds

▲ **Figure 10.18 The Calvin cycle.** This diagram summarizes three turns of the cycle, tracking carbon atoms (shown as dark-gray spheres). The three phases of the cycle correspond to the phases discussed in the text. For every three molecules of CO_2 that enter the cycle, the net output is one molecule of glyceraldehyde 3-phosphate (G3P), a three-carbon sugar. The light reactions sustain the Calvin cycle by regenerating the required ATP and NADPH.

➔ **Mastering Biology** Animation: The Calvin Cycle
BioFlix® Animation: The Calvin Cycle

Phase 3: Regeneration of the CO_2 acceptor (RuBP).
In a complex series of reactions, the carbon skeletons of five molecules of G3P are rearranged by the last steps of the Calvin cycle into three molecules of RuBP. To accomplish this, the cycle spends three more molecules of ATP. The RuBP is now prepared to receive CO_2 again, and the cycle continues.

For the net synthesis of one G3P molecule, the Calvin cycle consumes a total of nine molecules of ATP and six molecules of NADPH. The light reactions regenerate the ATP and NADPH. The G3P spun off from the Calvin cycle becomes the starting material for metabolic pathways that synthesize other organic compounds, including glucose (from two molecules of G3P), sucrose, and other carbohydrates. Neither the light reactions nor the Calvin cycle alone can make sugar from CO_2. Photosynthesis is an emergent property of the intact chloroplast, which integrates the two stages of photosynthesis.

CONCEPT CHECK 10.4

1. To synthesize one glucose molecule, the Calvin cycle uses _____ molecules of CO_2, _____ molecules of ATP, and _____ molecules of NADPH.

2. How are the large numbers of ATP and NADPH molecules used during the Calvin cycle consistent with the high value of glucose as an energy source?

3. **WHAT IF?** Consider a poison that inhibits an enzyme of the Calvin cycle. Do you think such a poison will also inhibit the light reactions? Explain.

4. **DRAW IT** Draw a simple version of the Calvin cycle that shows only the number of carbon atoms at each step, using numerals rather than gray spheres (for example, "3 × 1C = 3C" at the beginning of the cycle). Explain how the total number of carbon atoms remains constant for every three turns of the cycle, noting the forms in which the carbon atoms enter and leave the cycle.

5. **MAKE CONNECTIONS** Review Figures 9.8 and 10.18, and then discuss the roles of intermediate and product played by glyceraldehyde 3-phosphate (G3P) in the two processes shown in these figures.

For suggested answers, see Appendix A.

Alternative mechanisms of carbon fixation have evolved in hot, arid climates

EVOLUTION Ever since plants first moved onto land about 475 million years ago, they have been adapting to the problems of terrestrial life, particularly the problem of dehydration. In Concept 36.4, we will consider anatomical adaptations that help plants conserve water, while in this chapter we are concerned with metabolic adaptations. The solutions often involve trade-offs. An important example is the balance between photosynthesis and the prevention of excessive water loss from the plant. The CO_2 required for photosynthesis enters a leaf (and the resulting O_2 exits) via stomata, the pores on the leaf surface (see Figure 10.3). However, stomata are also the main avenues of transpiration, the evaporative loss of water from leaves. On a hot, dry day, most plants close their stomata, a response that conserves water but also reduces CO_2 levels. With stomata even partially closed, CO_2 concentrations begin to decrease in the air spaces within the leaf, and the concentration of O_2 released from the light reactions begins to increase. These conditions within the leaf favor an apparently wasteful process called photorespiration.

Photorespiration: An Evolutionary Relic?

In most plants, initial fixation of carbon occurs via rubisco, the Calvin cycle enzyme that adds CO_2 to ribulose bisphosphate. Such plants are called **C_3 plants** because the first organic product of carbon fixation is a three-carbon compound, 3-phosphoglycerate (see Figure 10.18). C_3 plants include important agricultural plants such as rice, wheat, and soybeans. When their stomata close on hot, dry days, C_3 plants produce less sugar because the declining level of CO_2 in the leaf starves the Calvin cycle. In addition, rubisco is capable of binding O_2 in place of CO_2. As CO_2 becomes scarce within the air spaces of the leaf and O_2 builds up, rubisco adds O_2 to the Calvin cycle instead of CO_2. The product splits, and a two-carbon compound leaves the chloroplast. Peroxisomes and mitochondria within the plant cell rearrange and split this compound, releasing CO_2. The process is called **photorespiration** because it occurs in the light (*photo*) and consumes O_2 while producing CO_2 (*respiration*). However, unlike normal cellular respiration, photorespiration uses ATP rather than generating it. And unlike photosynthesis, photorespiration produces no sugar. In fact, photorespiration *decreases* photosynthetic output by siphoning organic material from the Calvin cycle and releasing CO_2 that would otherwise be fixed. This CO_2 can eventually be fixed if it is still in the leaf once the CO_2 concentration builds up to a high enough level. In the meantime, though, the process is energetically costly, much like a hamster running on its wheel.

How can we explain the existence of a metabolic process that seems to be counterproductive for the plant? According to one hypothesis, photorespiration is evolutionary baggage—a metabolic relic from a much earlier time when the atmosphere had less O_2 and more CO_2 than it does today. In the ancient atmosphere that prevailed when rubisco first evolved, the ability of the enzyme's active site to bind O_2 would have made little difference. The hypothesis suggests that modern rubisco retains some of its chance affinity for O_2, which is now so concentrated in the atmosphere that a certain amount of photorespiration is inevitable. There is also some evidence that photorespiration may provide protection against the damaging products of the light reactions, which build up when the Calvin cycle slows due to low CO_2.

In many types of plants—including a significant number of crop plants—photorespiration drains away as much as 50% of the carbon fixed by the Calvin cycle. Indeed, if photorespiration could be reduced in certain plant species without otherwise affecting photosynthetic productivity, crop yields and food supplies might increase.

In some plant species, alternate modes of carbon fixation have evolved that minimize photorespiration and optimize the Calvin cycle—even in hot, arid climates. The two most important of these photosynthetic adaptations are C_4 photosynthesis and crassulacean acid metabolism (CAM).

C_4 Plants

The **C_4 plants** are so named because they preface the Calvin cycle with an alternate mode of carbon fixation that forms a four-carbon compound as its first product. The C_4 pathway is believed to have evolved independently at least 45 separate times and is used by several thousand species in at least 19 plant families. Among the C_4 plants important to agriculture are sugarcane and corn, members of the grass family.

When the weather is hot and dry, a C_4 plant partially closes its stomata, conserving water but reducing the CO_2 concentration in the leaves. However, sugar is still made because C_4 plants use a multistep process that operates even under low CO_2 conditions. Photosynthesis begins in mesophyll cells but is completed in **bundle-sheath cells**, cells that are arranged into tightly packed sheaths around the veins of the leaf **(Figure 10.19)**. In C_4 leaves, the loosely arranged mesophyll cells are located between the bundle-sheath cells and the leaf surface, no more than two to three cells away from the bundle-sheath cells. In mesophyll cells, CO_2 is incorporated into organic compounds that then move into the bundle-sheath cells, where the Calvin cycle takes place. See the numbered steps in Figure 10.19, which are also described here:

1 The first step is carried out by an enzyme present only in mesophyll cells called **PEP carboxylase**, which has a much higher affinity for CO_2 than does rubisco, and no affinity for O_2. This enzyme adds CO_2 to

Photosynthetic cells of C₄ plant leaf {
- Mesophyll cell
- Bundle-sheath cell
}

Vein (vascular tissue)

C₄ leaf anatomy

Stoma

Plasmodesma

Mesophyll cell

PEP carboxylase

CO_2

Oxaloacetate (4C) PEP (3C)

Malate (4C) ADP ATP

Pyruvate (3C)

CO_2

Bundle-sheath cell

Calvin Cycle

Sugar

Vascular tissue

The C₄ pathway

1 In mesophyll cells, the enzyme PEP carboxylase adds CO_2 to PEP, forming a four-carbon compound.

2 The four-carbon compound (such as malate) moves into a bundle-sheath cell via plasmodesmata.

3 In bundle-sheath cells, an enzyme releases CO_2, which enters the Calvin cycle.

▲ **Figure 10.19 C₄ leaf anatomy and the C₄ pathway.** The structure and biochemical functions of the leaves of C₄ plants are an evolutionary adaptation to hot, dry climates. This adaptation maintains a CO_2 concentration in the bundle-sheath cells that favors photosynthesis over photorespiration.

phosphoenolpyruvate (PEP), forming the four-carbon product oxaloacetate; it does this even under conditions of lower CO_2 concentration and relatively higher O_2 concentration.

2 After the CO_2 is fixed in the mesophyll cells, the four-carbon products (malate in the example shown in Figure 10.19) are exported to bundle-sheath cells through plasmodesmata (see Figure 6.29).

3 Within the bundle-sheath cells, an enzyme releases CO_2 from the four-carbon compounds; the CO_2 is re-fixed into organic material by rubisco and the Calvin cycle. The same reaction regenerates pyruvate, which is transported to mesophyll cells. There, ATP is used to convert pyruvate to PEP, which can then accept addition of another CO_2, allowing the reaction cycle to continue. This ATP can be thought of, in a sense, as the "price" of concentrating CO_2 in the bundle-sheath cells. To generate this extra ATP, bundle-sheath cells carry out cyclic electron flow, the process described earlier in this chapter (see Figure 10.15). In fact, these cells contain PS I but no PS II, so cyclic electron flow is their only photosynthetic mode of generating ATP.

In effect, the mesophyll cells of a C₄ plant pump CO_2 into the bundle-sheath cells, keeping the CO_2 concentration in those cells high enough for rubisco to bind CO_2 rather than O_2. The cyclic series of reactions involving PEP carboxylase and the regeneration of PEP can be thought of as an ATP-powered pump that concentrates CO_2 where it can be fixed. In this way, C₄ photosynthesis spends ATP energy to minimize photorespiration and enhance sugar production. This adaptation is especially advantageous in hot regions with

intense sunlight, and it is in such environments that C₄ plants evolved and thrive today.

The concentration of CO_2 in the atmosphere has drastically increased since the Industrial Revolution began in the 1800s, and it continues to rise today due to human activities such as the burning of fossil fuels (see Concept 1.1). The resulting global climate change, including an increase in average temperatures around the planet, may have far-reaching effects on plant species. Scientists are concerned that increasing CO_2 concentration and temperature may affect C₃ and C₄ plants differently, thus changing the relative abundance of these species in a given plant community.

Which type of plant would stand to gain more from increasing CO_2 levels? Recall that in C₃ plants, the binding of O_2 rather than CO_2 by rubisco leads to photorespiration, lowering the efficiency of photosynthesis. C₄ plants overcome this problem by concentrating CO_2 in the bundle-sheath cells at the cost of ATP. Rising CO_2 levels should benefit C₃ plants by lowering the amount of photorespiration that occurs. At the same time, rising temperatures have the opposite effect, increasing photorespiration. (Other factors such as water availability may also come into play.) In contrast, many C₄ plants could be largely unaffected by increasing CO_2 levels or temperature. Researchers have investigated aspects of this question in several studies; you can work with data from one such experiment in the **Scientific Skills Exercise**. In different regions, the particular combination of CO_2 concentration and temperature is likely to alter the balance of C₃ and C₄ plants in varying ways. The effects of such a widespread and variable change in community structure are unpredictable and thus a cause of legitimate concern.

Making Scatter Plots with Regression Lines

Does Atmospheric CO_2 Concentration Affect the Productivity of Agricultural Crops? The atmospheric concentration of CO_2 has been rising globally, and scientists wondered whether this would affect C_3 and C_4 plants differently. In this exercise, you will make a scatter plot to examine the relationship between CO_2 concentration and growth of both corn (maize), a C_4 crop plant, and velvetleaf, a C_3 weed found in cornfields.

How the Experiment Was Done For 45 days, researchers grew corn and velvetleaf plants under controlled conditions, where all plants received the same amounts of water and light. The plants were divided into three groups, and each was exposed to a different concentration of CO_2 in the air: 350, 600, or 1,000 ppm (parts per million).

Data from the Experiment The table shows the dry mass (in grams) of corn and velvetleaf plants grown at the three concentrations of CO_2. The dry mass values are averages calculated from the leaves, stems, and roots of eight plants.

	350 ppm CO_2	600 ppm CO_2	1,000 ppm CO_2
Average dry mass of one corn plant (g)	91	89	80
Average dry mass of one velvetleaf plant (g)	35	48	54

Data from D. T. Patterson and E. P. Flint, Potential effects of global atmospheric CO_2 enrichment on the growth and competitiveness of C_3 and C_4 weed and crop plants, *Weed Science* 28(1):71–75 (1980).

INTERPRET THE DATA

1. To explore the relationship between the two variables, it is useful to graph the data in a scatter plot, and then draw a regression line. **(a)** First, place labels for the dependent and independent variables on the appropriate axes. Explain your choices. **(b)** Plot the data points for corn and velvetleaf using different symbols for each set of data, and add a key for the two symbols. (For additional information about graphs, see the Scientific Skills Review in Appendix D.)

2. Draw a "best-fit" line for each set of points. A best-fit line does not necessarily pass through all or even most points. Instead, it is a straight line that passes as close as possible to all data points from that set. Draw a best-fit line for each set of data. Because placement of the line is a matter of judgment, two

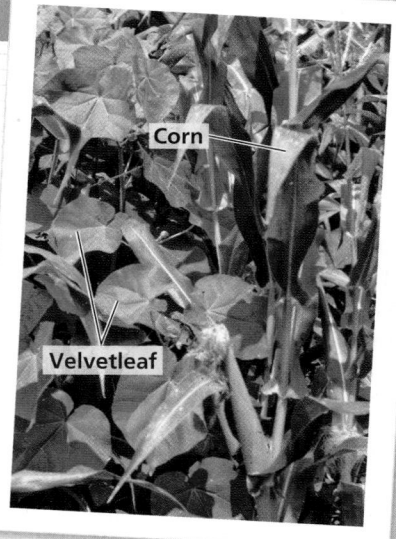

► Corn plant surrounded by invasive velvetleaf plants.

individuals may draw two slightly different lines for a given set of points. The line that actually fits best, a regression line, can be identified by squaring the distances of all points to any candidate line, then selecting the line that minimizes the sum of the squares. (See the graph in the Scientific Skills Exercise in Chapter 3 for an example of a linear regression line.) Excel or other software programs, including those on a graphing calculator, can plot a regression line once data points are entered. Using either Excel or a graphing calculator, enter the data points for each data set and have the program draw the two regression lines. Compare them to the lines you drew.

3. Describe the trends shown by the regression lines in your scatter plot. **(a)** Compare the relationship between increasing concentration of CO_2 and the dry mass of corn to that for velvetleaf. **(b)** Considering that velvetleaf is a weed invasive to cornfields, predict how increased CO_2 concentration may affect interactions between the two species.

4. Based on the data in the scatter plot, estimate the percentage change in dry mass of corn and velvetleaf plants if atmospheric CO_2 concentration increased from 390 ppm (current levels) to 800 ppm. **(a)** What is the estimated dry mass of corn and velvetleaf plants at 390 ppm? 800 ppm? **(b)** To calculate the percentage change in mass for each plant, subtract the mass at 390 ppm from the mass at 800 ppm (change in mass), divide by the mass at 390 ppm (initial mass), and multiply by 100. What is the estimated percentage change in dry mass for corn? For velvetleaf? **(c)** Do these results support the conclusion from other experiments that C_3 plants grow better than C_4 plants under increased CO_2 concentration? Why or why not?

→ **Instructors:** A version of this Scientific Skills Exercise can be assigned in **Mastering Biology.**

C_4 photosynthesis is considered more efficient than C_3 photosynthesis because it uses less water and resources. On our planet today, the world population and demand for food are rapidly increasing. At the same time, the amount of land suitable for growing crops is decreasing due to the effects of global climate change, which include an increase in sea level as well as a hotter, drier climate in many regions. To address issues of food supply, scientists in the Philippines have been working on genetically modifying rice—an important food staple that is a C_3 crop—so that it can instead carry out C_4 photosynthesis. Results so far seem promising, and these researchers estimate

that the yield of C_4 rice might be 30–50% higher than C_3 rice with the same input of water and resources.

CAM Plants

A second photosynthetic adaptation to arid conditions has evolved in pineapples, many cacti, and other succulent (water-storing) plants, such as aloe and jade plants. These plants open their stomata at night and close them during the day, just the reverse of how other plants behave. Closing stomata during the day helps desert plants conserve water, but it also prevents CO_2 from entering the leaves. During the night, when their stomata are open, these plants take up CO_2 and incorporate it into a variety of organic acids. This mode of carbon fixation is called **crassulacean acid metabolism**, or **CAM**, after the plant family Crassulaceae, the succulents in which the process was first discovered. The mesophyll cells of **CAM plants** store the organic acids they make during the night in their vacuoles until morning, when the stomata close. During the day, when the light reactions can supply ATP and NADPH for the Calvin cycle, CO_2 is released from the organic acids made the night before to become incorporated into sugar in the chloroplasts.

Notice in **Figure 10.20** that the CAM pathway is similar to the C_4 pathway in that CO_2 is first incorporated into organic intermediates before it enters the Calvin cycle. The difference is that in C_4 plants, the initial steps of carbon fixation are separated structurally from the Calvin cycle, whereas in CAM plants, the two steps occur within the same cell but at separate times. (Keep in mind that CAM, C_4, and C_3 plants all eventually use the Calvin cycle to make sugar from carbon dioxide.)

> **Mastering Biology** Animation: Photosynthesis in Dry Climates

▼ **Figure 10.20 C_4 and CAM photosynthesis compared.** The C_4 and CAM pathways are two evolutionary solutions to the problem of maintaining photosynthesis with stomata partially or completely closed on hot, dry days. Both adaptations are characterized by:
❶ preliminary incorporation of CO_2 into organic acids, followed by
❷ transfer of CO_2 to the Calvin cycle.

C_4 Sugarcane   **CAM**   Pineapple

(a) Spatial separation of steps. In C_4 plants, carbon fixation and the Calvin cycle occur in different types of cells.

(b) Temporal separation of steps. In CAM plants, carbon fixation and the Calvin cycle occur in the same cell at different times.

CONCEPT CHECK 10.5

1. Describe how photorespiration lowers photosynthetic output for plants.
2. The presence of only PS I, not PS II, in the bundle-sheath cells of C_4 plants has an effect on O_2 concentration. What is that effect, and how might that benefit the plant?
3. **MAKE CONNECTIONS** Refer to the discussion of ocean acidification in Concept 3.3. Ocean acidification and changes in the distribution of C_3 and C_4 plants may seem to be two very different problems, but what do they have in common? Explain.
4. **WHAT IF?** How would you expect the relative abundance of C_3 versus C_4 and CAM species to change in a geographic region whose climate becomes much hotter and drier, with no change in CO_2 concentration?

For suggested answers, see Appendix A.

CONCEPT 10.6

Photosynthesis is essential for life on Earth: a review

> **Mastering Biology** BioFlix® Animation: Photosynthesis

In this chapter, we have followed photosynthesis from photons to food. The light reactions capture solar energy and use it to make ATP and transfer electrons from H_2O to $NADP^+$, forming NADPH. The Calvin cycle uses the ATP and NADPH to produce sugar from CO_2. The energy that enters the chloroplasts as sunlight becomes stored as chemical energy in organic compounds. The entire process is reviewed visually in **Figure 10.21**, where photosynthesis is also put in its natural context.

As for the fates of photosynthetic products, enzymes in the chloroplast and cytosol convert the G3P made in the Calvin cycle to many other organic compounds. In fact, the sugar made in the chloroplasts supplies the entire plant with chemical energy and carbon skeletons for the synthesis of all the major organic molecules of plant cells. About 50% of the organic material made by photosynthesis is consumed as fuel for cellular respiration in plant cell mitochondria.

Technically, green cells are the only autotrophic parts of the plant. The rest of the plant depends on organic molecules exported from leaves through veins (see Figure 10.21, top). In most plants, carbohydrate is transported out of the leaves to the rest of the plant in the form of sucrose, a disaccharide. After arriving at nonphotosynthetic cells, the sucrose provides raw material for cellular respiration and a multitude of anabolic pathways that synthesize proteins, lipids, and other products. A considerable amount of sugar in the form of glucose is linked together to make the polysaccharide cellulose (see Figure 5.6c), especially in plant cells that are still growing and maturing. Cellulose, the main ingredient of cell walls, is the most abundant organic molecule in the plant—and probably on the surface of the planet.

Most plants and other photosynthesizers make more organic material per day than they need to use as respiratory fuel and precursors for biosynthesis. For example, flowering plants stockpile the extra sugar in the form of starch, storing some in the chloroplasts and some in cells of roots, tubers, seeds, and fruits. In accounting for the use of food molecules produced by photosynthesis, note that most flowering plants lose leaves, roots, stems, fruits, and often their entire bodies to heterotrophs, including humans.

On a global scale, photosynthesis is the process responsible for the presence of O_2 in our atmosphere. Furthermore, although each chloroplast is minuscule, their collective productivity in terms of food production is enormous: Photosynthesis makes an estimated 150 billion metric tons of carbohydrate per year (a metric ton is 1,000 kg, about 1.1 tons). That's organic matter equivalent in mass to a stack of about 60 trillion biology textbooks—and the stack would reach 17 times the distance from Earth to the sun! No chemical process is more important than photosynthesis to the welfare of life on Earth.

In Chapters 5 through 10, you have learned about many activities of cells. **Figure 10.22** integrates these cellular processes into the context of a working plant cell. As you study the figure, reflect on how each process fits into the big picture: As the most basic unit of living organisms, a cell performs all functions characteristic of life.

CONCEPT CHECK 10.6

1. **MAKE CONNECTIONS** How do plants use the sugar they produce during photosynthesis to directly power the work of the cell? Provide some examples of cellular work. (See Figures 8.10, 8.11, and 9.5.)

For suggested answers, see Appendix A.

▼ **Figure 10.21 A review of photosynthesis.** This diagram shows the main reactants and products of photosynthesis as they move through the tissues of a tree (left) and a chloroplast (right).

→ **Mastering Biology** HHMI Animation: Photosynthesis
MP3 Tutor: Photosynthesis

LIGHT REACTIONS	CALVIN CYCLE REACTIONS
• Are carried out by molecules in the thylakoid membranes • Convert light energy to the chemical energy of ATP and NADPH • Split H_2O and release O_2	• Take place in the stroma • Use ATP and NADPH to convert CO_2 to the sugar G3P • Return ADP, inorganic phosphate, and $NADP^+$ to the light reactions

MAKE CONNECTIONS

The Working Cell

This figure illustrates how a generalized plant cell functions, integrating the cellular activities you learned about in Chapters 5–10. To see some of the enzymes involved, look at Figure 6.32.

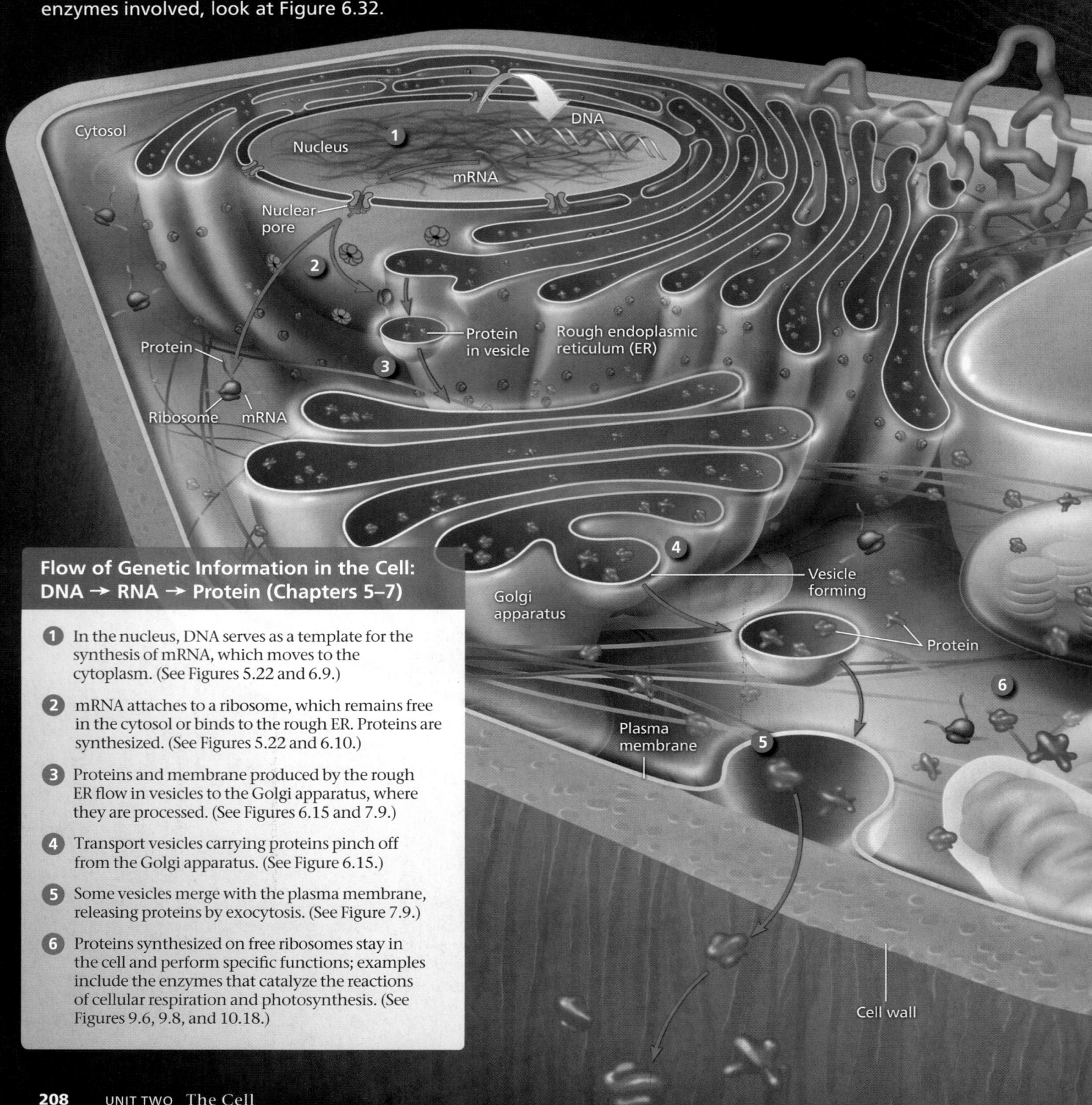

Cytosol

Nucleus

DNA

mRNA

Nuclear pore

Protein in vesicle

Rough endoplasmic reticulum (ER)

Protein

Ribosome   mRNA

Golgi apparatus

Vesicle forming

Protein

Plasma membrane

Cell wall

Flow of Genetic Information in the Cell: DNA → RNA → Protein (Chapters 5–7)

1. In the nucleus, DNA serves as a template for the synthesis of mRNA, which moves to the cytoplasm. (See Figures 5.22 and 6.9.)

2. mRNA attaches to a ribosome, which remains free in the cytosol or binds to the rough ER. Proteins are synthesized. (See Figures 5.22 and 6.10.)

3. Proteins and membrane produced by the rough ER flow in vesicles to the Golgi apparatus, where they are processed. (See Figures 6.15 and 7.9.)

4. Transport vesicles carrying proteins pinch off from the Golgi apparatus. (See Figure 6.15.)

5. Some vesicles merge with the plasma membrane, releasing proteins by exocytosis. (See Figure 7.9.)

6. Proteins synthesized on free ribosomes stay in the cell and perform specific functions; examples include the enzymes that catalyze the reactions of cellular respiration and photosynthesis. (See Figures 9.6, 9.8, and 10.18.)

Movement Across Cell Membranes (Chapter 7)

7 Water diffuses into and out of the cell directly through the plasma membrane and by facilitated diffusion through aquaporins. (See Figure 7.10.)

8 By passive transport, the CO_2 used in photosynthesis diffuses into the cell, and the O_2 formed as a by-product of photosynthesis diffuses out of the cell. Both solutes move down their concentration gradients. (See Figures 7.11 and 10.21.)

9 In active transport, energy (usually supplied by ATP) is used to transport a solute against its concentration gradient. (See Figure 7.17.)

Exocytosis (shown in step 5) and endocytosis move larger materials out of and into the cell. (See Figures 7.20 and 7.21.)

Energy Transformations in the Cell: Photosynthesis and Cellular Respiration (Chapters 8–10)

10 In chloroplasts, the process of photosynthesis uses the energy of light to convert CO_2 and H_2O to organic molecules, with O_2 as a by-product. (See Figures 10.1 and 10.21.)

11 In mitochondria, organic molecules are broken down by cellular respiration, capturing energy in molecules of ATP, which are used to power the work of the cell, such as protein synthesis and active transport. CO_2 and H_2O are by-products. (See Figures 8.9, 8.11, 9.1, and 9.15.)

Vacuole

10 Photosynthesis in chloroplast

CO_2

H_2O

Organic molecules

O_2

11 Cellular respiration in mitochondrion

ATP

ATP

ATP

ATP

Transport pump

9

8

7

Aquaporin

O_2

CO_2

H_2O

MAKE CONNECTIONS

The first enzyme that functions in glycolysis is hexokinase (see Figure 6.32a). How is hexokinase produced in this plant cell? Specify the locations of each step of the process. Where does hexokinase function? (See Figures 5.18, 5.22, and 9.8.)

→ **Mastering Biology**
BioFlix® Animation: Tour of an Animal Cell
BioFlix® Animation: Tour of a Plant Cell

10 Chapter Review

→ Go to **Mastering Biology** for Assignments, the eText, the Study Area, and Dynamic Study Modules.

SUMMARY OF KEY CONCEPTS

→ To review key terms, go to the **Vocabulary Self-Quiz** in the **Mastering Biology** eText or Study Area, or go to goo.gl/zkjz9t.

CONCEPT 10.1

Photosynthesis feeds the biosphere *(p. 188)*

- **Photosynthesis** converts the energy of sunlight into chemical energy stored in sugars.
- **Autotrophs** ("producers") produce their organic molecules from CO_2 and other inorganic raw materials; plants are *photoautotrophs* that use light energy to do this. **Heterotrophs** (consumers, including decomposers) are unable to make their own food and live on compounds produced by others.

❓ *Using the terms producers, consumers, and decomposers, explain how photosynthesis feeds organisms, either directly or indirectly.*

CONCEPT 10.2

Photosynthesis converts light energy to the chemical energy of food *(pp. 189–192)*

- In plants and other eukaryotic autotrophs, photosynthesis occurs in **chloroplasts**, organelles containing **thylakoids**. Stacks of thylakoids form grana. Photosynthesis is summarized as

$$6\ CO_2 + 12\ H_2O + \text{Light energy} \rightarrow C_6H_{12}O_6 + 6\ O_2 + 6\ H_2O.$$

Chloroplasts split water into hydrogen and oxygen, incorporating the electrons of hydrogen into sugar molecules. Photosynthesis is a redox process: H_2O is oxidized, and CO_2 is reduced. The **light reactions** in the thylakoid membranes split water, releasing O_2, producing ATP, and forming **NADPH**. The **Calvin cycle** in the **stroma** forms sugar from CO_2, using ATP for energy and NADPH for reducing power.

❓ *Compare the roles of CO_2 and H_2O in cellular respiration and photosynthesis.*

CONCEPT 10.3

The light reactions convert solar energy to the chemical energy of ATP and NADPH *(pp. 192–201)*

- Light is a form of electromagnetic energy. The colors we see as **visible light** include those **wavelengths** that drive photosynthesis. A pigment absorbs light of specific wavelengths; **chlorophyll *a*** is the main photosynthetic pigment in plants. Other accessory pigments absorb different wavelengths of light and pass the energy on to chlorophyll *a*.
- A pigment goes from a ground state to an excited state when a **photon** of light boosts one of the pigment's electrons to a higher-energy orbital. This excited state is unstable. Electrons from isolated pigments tend to fall back to the ground state, giving off heat and/or light.
- A **photosystem** is composed of a **reaction-center complex** surrounded by **light-harvesting complexes** that funnel the energy of photons to the reaction-center complex. When a special pair of reaction-center chlorophyll *a* molecules absorbs energy, one of its electrons is boosted to a higher energy level and transferred to the **primary electron acceptor**. **Photosystem II** contains P680 chlorophyll *a* molecules in the reaction-center complex; **photosystem I** contains P700 molecules.
- **Linear electron flow** during the light reactions uses both photosystems and produces NADPH, ATP, and oxygen:

- **Cyclic electron flow** employs only one photosystem, producing ATP but no NADPH or O_2.
- During chemiosmosis in both mitochondria and chloroplasts, electron transport chains generate an H^+ gradient across a membrane. ATP synthase uses this proton-motive force to make ATP.

❓ *The absorption spectrum of chlorophyll a differs from the action spectrum of photosynthesis. Explain this observation.*

CONCEPT 10.4

The Calvin cycle uses the chemical energy of ATP and NADPH to reduce CO_2 to sugar *(pp. 201–202)*

- The Calvin cycle occurs in the stroma, using electrons from NADPH and energy from ATP.

- One molecule of **G3P** exits the cycle per three CO_2 molecules fixed and is converted to glucose and other organic molecules.

[Diagram: Calvin Cycle — 3 CO_2 → Carbon fixation → 6 x 3C → Reduction → 1 G3P (3C); 5 x 3C → Regeneration of CO_2 acceptor → 3 x 5C]

DRAW IT *On the diagram above, draw where ATP and NADPH are used and where rubisco functions. Describe these steps.*

CONCEPT 10.5

Alternative mechanisms of carbon fixation have evolved in hot, arid climates *(pp. 203–206)*

- On dry, hot days, **C_3 plants** close their stomata, conserving water but keeping CO_2 out and O_2 in. Under these conditions, **photorespiration** can occur: **Rubisco** binds O_2 instead of CO_2, consuming ATP and releasing CO_2 without producing ATP or carbohydrate. Photorespiration may be an evolutionary relic, and it may play a photoprotective role.
- C_4 and CAM plants are adopted to hot, dry climates.

- **C₄ plants** minimize the cost of photorespiration by incorporating CO_2 into four-carbon compounds in mesophyll cells. These compounds are exported to **bundle-sheath cells**, where they release carbon dioxide for use in the Calvin cycle.
- **CAM plants** open their stomata at night, incorporating CO_2 into organic acids, which are stored in mesophyll cells. During the day, the stomata close, and the CO_2 is released from the organic acids for use in the Calvin cycle.
- Organic compounds produced by photosynthesis provide the energy and building material for Earth's ecosystems.

? *Why are C₄ and CAM photosynthesis more energetically expensive than C₃ photosynthesis? What climate conditions would favor C₄ and CAM plants?*

CONCEPT 10.6

Photosynthesis is essential for life on Earth: a review
(pp. 206–209)

- Organic compounds produced by photosynthesis provide the energy and building material for Earth's ecosystems.

? *How do plants use the products of photosynthesis?*

TEST YOUR UNDERSTANDING

➡ For more multiple-choice questions, go to the **Practice Test** in the **Mastering Biology** eText or Study Area, or go to goo.gl/GruWRg.

Levels 1-2: Remembering/Understanding

1. The light reactions supply the Calvin cycle with
   (A) light energy.
   (B) CO_2 and ATP.
   (C) H_2O and NADPH.
   (D) ATP and NADPH.

2. Which of the following sequences correctly represents the flow of electrons during photosynthesis?
   (A) NADPH $\rightarrow O_2 \rightarrow CO_2$
   (B) $H_2O \rightarrow$ NADPH $\rightarrow$ Calvin cycle
   (C) $H_2O \rightarrow$ photosystem I $\rightarrow$ photosystem II
   (D) NADPH $\rightarrow$ electron transport chain $\rightarrow O_2$

3. How is photosynthesis similar in C₄ plants and CAM plants?
   (A) In both cases, only photosystem I is used.
   (B) Both types of plants make sugar without the Calvin cycle.
   (C) In both cases, rubisco is not used to fix carbon initially.
   (D) Both types of plants make most of their sugar in the dark.

4. Which of the following statements is a correct distinction between autotrophs and heterotrophs?
   (A) Autotrophs, but not heterotrophs, can nourish themselves beginning with CO_2 and other nutrients that are inorganic.
   (B) Only heterotrophs require chemical compounds from the environment.
   (C) Cellular respiration is unique to heterotrophs.
   (D) Only heterotrophs have mitochondria.

5. Which of the following occurs during the Calvin cycle?
   (A) carbon fixation
   (B) reduction of NADP⁺
   (C) release of oxygen
   (D) generation of CO_2

Levels 3-4: Applying/Analyzing

6. In mechanism, photophosphorylation is most similar to
   (A) substrate-level phosphorylation in glycolysis.
   (B) oxidative phosphorylation in cellular respiration.
   (C) carbon fixation.
   (D) reduction of NADP⁺.

7. Which process is most directly driven by light energy?
   (A) creation of a pH gradient by pumping protons across the thylakoid membrane
   (B) reduction of NADP⁺ molecules
   (C) transfer of energy from pigment molecule to pigment molecule
   (D) ATP synthesis

Levels 5-6: Evaluating/Creating

8. **SCIENCE, TECHNOLOGY, AND SOCIETY** Scientific evidence indicates that the CO_2 added to the air by the burning of wood and fossil fuels is contributing to global warming, a rise in global temperature. Tropical rain forests are estimated to be responsible for approximately 20% of global photosynthesis, yet the consumption of large amounts of CO_2 by living trees is thought to make little or no *net* contribution to reduction of global warming. Explain why this might be the case. (*Hint*: What processes in both living and dead trees produce CO_2?)

9. **EVOLUTION CONNECTION** Photorespiration can decrease soybeans' photosynthetic output by about 50%. Would this figure be higher or lower in wild relatives of soybeans? Why?

10. **SCIENTIFIC INQUIRY • DRAW IT** The following diagram represents an experiment with isolated thylakoids. The thylakoids were first made acidic by soaking them in a solution at pH 4. After the thylakoid space reached pH 4, the thylakoids were transferred to a basic solution at pH 8. The thylakoids then made ATP in the dark. (See Concept 3.3 to review pH.)

Draw an enlargement of part of the thylakoid membrane in the beaker with the solution at pH 8. Draw ATP synthase. Label the areas of high H⁺ concentration and low H⁺ concentration. Show the direction protons flow through the enzyme, and show the reaction where ATP is synthesized. Would ATP end up in the thylakoid or outside of it? Explain why the thylakoids in the experiment were able to make ATP in the dark.

11. **WRITE ABOUT A THEME: ENERGY AND MATTER** In a short essay (100–150 words), describe how photosynthesis transforms solar energy into the chemical energy of sugar molecules.

12. **SYNTHESIZE YOUR KNOWLEDGE**

"Watermelon snow" in Antarctica is caused by a certain species of photosynthetic green algae that thrives in subzero temperatures (*Chlamydomonas nivalis*). These algae are also found in high-altitude, year-round snowfields. In both locations, UV light levels tend to be high. Propose an explanation for why this alga appears reddish-pink.

For selected answers, see Appendix A.

11 Cell Communication

11.1 External signals are converted to responses within the cell *p. 213*

11.2 Signal reception: A signaling molecule binds to a receptor, causing it to change shape *p. 217*

11.3 Signal transduction: Cascades of molecular interactions transmit signals from receptors to relay molecules in the cell *p. 221*

11.4 Cellular response: Cell signaling leads to regulation of transcription or cytoplasmic activities *p. 226*

11.5 Apoptosis requires integration of multiple cell-signaling pathways *p. 229*

Study Tip

Make a table: As you read about examples of cell signaling in this chapter, make a table and classify the events of each example into three stages: signal reception, signal transduction, and cellular response.

Example of Cell Signaling	Signal Reception	Signal Trans- duction	Cellular Response
Epinephrine	Epinephrine binds cell-surface receptor.	Relay molecules each activate the next molecule.	An enzyme is activated that breaks down glycogen into glucose for energy to fight or flee.

→ Go to Mastering Biology

For Students (in eText and Study Area)
• Get Ready for Chapter 11
• Animation: Overview of Cell Signaling
• Animation: Signal Transduction Pathways

For Instructors to Assign (in Item Library)
• Problem-Solving Exercise: Can a Skin Wound Turn Deadly?
• Tutorial: Cell-Signaling: Transduction and Response

Figure 11.1 This impala is fleeing for its life, racing to escape the predatory cheetah nipping at its heels. The impala is breathing rapidly, its heart pounding and its legs pumping furiously. These physiological functions are all part of the impala's "fight-or-flight" response, driven by hormones released from its adrenal glands at times of stress—in this case, upon sensing the cheetah.

How does cell signaling fuel the desperate flight of an impala?

The impala senses a cheetah.

↓

Its brain signals the adrenal glands to release epinephrine into the blood.

↓

Signal reception
An epinephrine molecule binds to a receptor on a muscle cell.

→

Signal transduction
Relay molecules transmit the signal, ultimately activating an enzyme.

→

Cellular response
The enzyme breaks down glycogen, releasing glucose that fuels the leg muscles.

Epinephrine

Receptor — *activates* → Pathway of relay molecules — *activates* → Enzyme → Glycogen → Glucose

Muscle cell in leg

External signals are converted to responses within the cell

Scientists think that signaling mechanisms first evolved hundreds of millions of years ago in ancient prokaryotes and single-celled eukaryotes and then were adapted for new uses in their multicellular descendants. So let's begin by considering signaling in some examples of single-celled organisms: bacteria and yeasts.

Evolution of Cell Signaling

EVOLUTION Research during the 1970s suggested that bacterial cells—somewhat surprisingly, since they are single-celled organisms—were capable of signaling to each other. Since then, we have come to understand that cell signaling is critical among prokaryotes. Bacterial cells secrete molecules that can be detected by other bacterial cells **(Figure 11.2)**. Sensing the concentration of such signaling molecules allows bacteria to monitor their own local cell density, a phenomenon called *quorum sensing*.

Quorum sensing allows bacterial populations to coordinate the behavior of all cells in a population in activities that require a given density of cells acting at the same time. One example is formation of a *biofilm*, an aggregation of bacterial cells attached to a surface by molecules secreted by the cells, but only after the cells have reached a certain density. The biofilm protects the cells in it, and they often derive nutrition from the surface they are on. Biofilms are believed to be involved in up to 80% of all human bacterial infections. You have probably encountered biofilms many times, perhaps without realizing it. The slimy coating on a fallen log or on leaves lying on a forest path, and even the film on your teeth each morning, are examples of bacterial biofilms. In fact, tooth-brushing and flossing disrupt biofilms that would otherwise cause cavities and gum disease.

Another example of bacterial behavior coordinated by quorum sensing is the secretion of toxins by infectious bacteria, which has serious medical implications. Sometimes treatment by antibiotics doesn't work with such infections because of antibiotic resistance that has evolved in a particular strain of bacteria. A promising alternative treatment would be to disrupt toxin production by interfering with the signaling pathways used in quorum sensing. In the **Problem-Solving Exercise**, you can participate in the process of scientific thinking involved in this novel approach.

Now let's look at an example of cell signaling in yeasts (single-celled fungi). Cells of the yeast *Saccharomyces cerevisiae*—which are used to make bread, wine, and beer—identify their sexual mates by chemical signaling when they reproduce sexually. There are two sexes, or mating types, called **a** and **α** (**Figure 11.3**). Each type secretes a specific

▼ **Figure 11.2 Communication among bacteria.** Soil-dwelling bacteria called myxobacteria ("slime bacteria") use chemical signals to share information about nutrient availability. When food is scarce, starving cells secrete a signaling molecule that stimulates neighboring cells to aggregate. The cells form a structure called a fruiting body that produces spores, thick-walled cells that can survive until the environment improves. The myxobacteria shown here are the species *Myxococcus xanthus* (steps 1–3, SEMs; lower photo, LM).

❶ Individual rod-shaped cells

0.5 mm

❷ Aggregation in progress

2.5 mm

❸ Spore-forming structure (fruiting body)

Fruiting bodies

▼ **Figure 11.3 Communication between mating yeast cells.** *Saccharomyces cerevisiae* cells use chemical signaling to identify cells of the opposite mating type and initiate the mating process. The two mating types and their corresponding chemical signaling molecules, or mating factors, are called **a** and **α**.

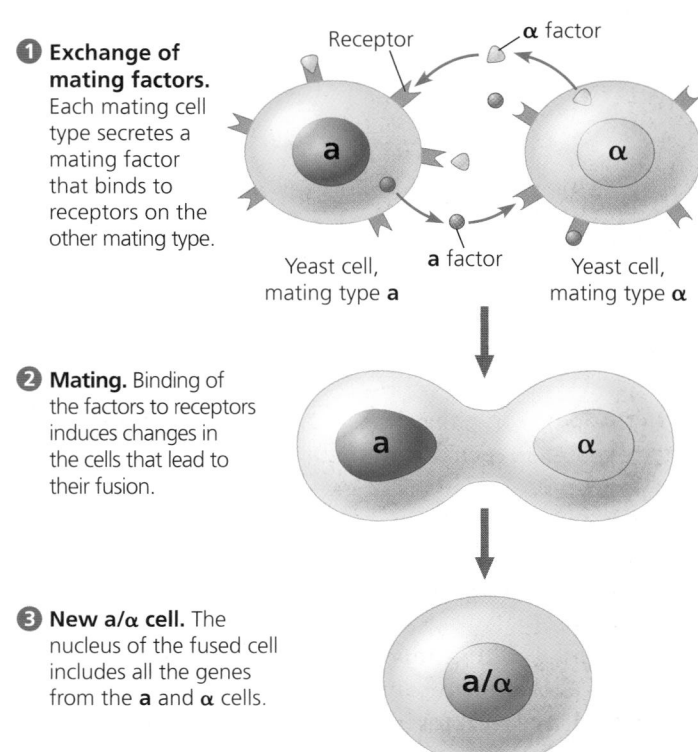

❶ **Exchange of mating factors.** Each mating cell type secretes a mating factor that binds to receptors on the other mating type.

Receptor

α factor

a

α

Yeast cell, mating type **a**

a factor

Yeast cell, mating type **α**

❷ **Mating.** Binding of the factors to receptors induces changes in the cells that lead to their fusion.

a

α

❸ **New a/α cell.** The nucleus of the fused cell includes all the genes from the **a** and **α** cells.

a/α

PROBLEM-SOLVING EXERCISE

Can a skin wound turn deadly?

Staphylococcus aureus (S. aureus) is a common bacterial species found on the surface of healthy skin that can turn into a serious pathogen if introduced into tissue through a cut or abrasion. Once *S. aureus* cells enter the body and reach a certain density, they secrete a toxin that kills body cells and contributes significantly to inflammation and damage. Because about one in 100 people carry a strain of *S. aureus* that is resistant to common antibiotics, a minor infection can turn permanently harmful or even deadly.

➡ **Instructors:** A version of this Problem-Solving Exercise can be assigned in **Mastering Biology**. Or a more extensive investigation called "Solve It: Is It Possible to Treat Bacterial Infections Without Traditional Antibiotics?" can be assigned.

Cells sense their own population density by *quorum sensing*, and at a certain density they start to secrete toxin. In this exercise, you will analyze whether blocking quorum sensing can stop *S. aureus* from producing toxin.

Your Approach In *S. aureus*, quorum sensing involves two separate signal transduction pathways. Two candidate synthetic peptides (short proteins), called peptides 1 and 2, have been proposed to interfere with *S. aureus* quorum-sensing pathways. Your job is to test each potential inhibitor of quorum sensing to see if it blocks either or both of the pathways that lead to toxin production.

For your experiment, you grow four cultures of *S. aureus* to a standardized high density and measure the concentration of toxin in the culture. The control culture contains no peptide. The other cultures have one or both candidate inhibitory peptides mixed into the growth medium before starting the cultures.

Your Data

Data from N. Balaban et al., Treatment of *Staphylococcus aureus* biofilm infection by the quorum-sensing inhibitor RIP, *Antimicrobial Agents and Chemotherapy* 51(6):2226–2229 (2007).

Your Analysis

1. Rank the cultures according to toxin production, from most to least.
2. Which, if any, of the cultures with peptide(s) resulted in a toxin concentration similar to the control culture? What is your evidence for this?
3. Was there an additive effect on toxin production when peptides 1 and 2 were both present in the growth medium? What is your evidence for this?
4. Based on these data, would you hypothesize that peptides 1 and 2 act on the same quorum-sensing pathway leading to toxin production or on two different pathways? What is your reasoning?
5. Do these data suggest a possible treatment for antibiotic-resistant *S. aureus* infections? What else would you want to know to investigate this further?

➡ **Mastering Biology**
HHMI Video: Interview with Bonnie Bassler
BBC Video: Brushing Your Teeth Can Save Your Life

➡ **Mastering Biology** Interview with Bonnie Bassler:
Exploring how bacteria communicate with each other

factor that binds only to receptors on the other type of cell. When exposed to each other's mating factors, a pair of cells of opposite type change shape, grow toward each other, and fuse (mate). The new **a/α** cell contains all the genes of both original cells, providing advantages to the cell's descendants, which arise by subsequent cell divisions.

The unique match between mating factor and receptor is key to ensuring mating only between cells of the same species of yeast. How does the binding of a mating factor by the yeast cell-surface receptor initiate a signal that brings about

the cellular response of mating? This occurs in a series of three major steps—signal reception, signal transduction, and cellular response—called a *signal transduction pathway*. Many such pathways exist in the cells of both unicellular and multicellular organisms. In fact, the molecular details of signal transduction in yeasts and mammals are strikingly similar, even though it's been over a billion years since they shared a common ancestor. This similarity suggests that early versions of cell-signaling mechanisms evolved well before the first multicellular organisms appeared on Earth.

Local and Long-Distance Signaling

Like bacteria or yeast cells, cells in a multicellular organism communicate by way of a signaling molecules targeted for cells that may or may not be immediately adjacent. As we saw in Concepts 6.7 and 7.1, eukaryotic cells may communicate by direct contact, which is one type of local signaling. Many animal and plant cells have cell junctions that directly connect the cytoplasms of adjacent cells **(Figure 11.4a)**. In these cases, signaling substances dissolved in the cytosol can pass between neighboring cells. Moreover, some animal cells may communicate by direct contact between cell-surface molecules, as shown in **Figure 11.4b**. This type of local signaling is especially important in embryonic development, the immune response, and in maintaining adult stem cell populations.

In many other cases of local signaling, signaling molecules are secreted by the signaling cell. Some molecules travel only short distances; such local regulators influence cells that are nearby. This type of local signaling in animals is called *paracrine signaling* **(Figure 11.5a)**. One class of local regulators in animals, *growth factors*, consists of compounds that stimulate nearby target cells to grow and divide. Numerous cells can simultaneously receive and respond to the growth factors produced by a single cell in their vicinity.

A highly specialized type of local signaling called *synaptic signaling* occurs in the animal nervous system **Figure 11.5b**; see Concept 48.4. An electrical signal along a nerve cell triggers the secretion of neurotransmitter molecules. These molecules act as chemical signals, diffusing across the synapse—the narrow space between the nerve cell and its target cell—triggering a response in the target cell. Drugs to treat

▼ Figure 11.4 **Communication requiring contact between cells.**

(a) Cell junctions. Both animals and plants have cell junctions that allow molecules, including signaling molecules, to pass readily between adjacent cells without crossing plasma membranes.

(b) Cell-surface molecules. In many animal cells, cell-surface molecules on adjacent cells interact with each other, resulting in a signal passing between the cells.

depression, anxiety, and post-traumatic stress disorder (PTSD) affect this signaling process.

Both animals and plants use molecules called **hormones** for long-distance signaling. In hormonal signaling in animals, also known as *endocrine signaling*, specialized cells release hormones, which travel through the circulatory system to other parts of the body, where they reach target cells that can recognize and respond to them **(Figure 11.5c)**. Many

▼ Figure 11.5 **Local and long-distance cell signaling by secreted molecules in animals.** In both local and long-distance signaling, only specific target cells that can recognize a given signaling molecule will respond to it.

(a) Paracrine signaling. A signaling cell acts on nearby target cells by secreting molecules of a local regulator (a growth factor, for example).

(b) Synaptic signaling. A nerve cell releases neurotransmitter molecules into a synapse, stimulating the target cell, such as a muscle or another nerve cell.

(c) Endocrine (hormonal) signaling. Specialized endocrine cells secrete hormones into body fluids, often blood. Hormones reach most body cells, but are bound by and affect only some cells.

plant hormones reach distant targets by traveling through cells (see Concept 39.2). Like local regulators, plant hormones vary widely in size and type. For instance, the plant hormone ethylene, which promotes fruit ripening and helps regulate growth, is a hydrocarbon of only six atoms (C_2H_4), small enough to diffuse through the air and pass through cell walls. (Small signaling molecules probably evolved early on among single-celled organisms.) In contrast, the mammalian hormone insulin, which regulates sugar levels in the blood, is a protein with thousands of atoms.

What happens when a potential target cell is exposed to a secreted signaling molecule? The ability of a cell to respond is determined by whether it has a specific receptor molecule that can bind to the signaling molecule. The information conveyed by this binding, the signal, must then be changed into another form—transduced—inside the cell before the cell can respond. The remainder of the chapter discusses this process, primarily as it occurs in animal cells. (Two signaling pathway proteins are shown in their cellular context in Figure 6.32a.)

The Three Stages of Cell Signaling: *A Preview*

Our current understanding of how signaling molecules act through signal transduction pathways had its origins in the pioneering work of Earl W. Sutherland, whose research led to a Nobel Prize in 1971. Sutherland and his colleagues at Vanderbilt University were investigating how the animal hormone epinephrine (also called adrenaline) triggers the "fight-or-flight" response in animals like the impala in Figure 11.1. One effect of epinephrine is to mobilize fuel reserves, which can be used by the animal to either defend itself (fight) or try to escape (flight). Epinephrine stimulates the breakdown of the storage polysaccharide glycogen within liver cells and skeletal muscle cells. The breakdown of glycogen releases the sugar glucose 1-phosphate, which the cell converts to glucose 6-phosphate. The liver or muscle cell can then use this compound, an early intermediate in glycolysis, for energy production (see Figure 9.8). Alternatively, the compound can be stripped of phosphate and released from the liver cell into the blood as glucose, which can fuel cells throughout the body.

But how, exactly, does epinephrine mobilize glucose for use? Sutherland's research team discovered that epinephrine outside the cell stimulates glycogen breakdown by somehow activating an enzyme, glycogen phosphorylase, inside the cell. However, when epinephrine was added to a solution containing the enzyme and its substrate, glycogen, no breakdown occurred. Glycogen phosphorylase could be activated by epinephrine only when the hormone was added to intact cells. This result told Sutherland two things. First, epinephrine does not interact directly with the enzyme responsible for glycogen breakdown; an intermediate step or series of steps must be occurring in the cell. Second, an intact, membrane-bound cell must be present for transmission of the signal to take place.

Sutherland's work suggested that the process going on at the receiving end of a cellular communication can be dissected into three stages: signal reception, signal transduction, and cellular response (**Figure 11.6**):

1. **Signal reception.** Reception is the target cell's detection of a signaling molecule coming from outside the cell. A chemical signal is "detected" when the signaling molecule binds to a receptor protein located at the cell's surface (or inside the cell, to be discussed later).

2. **Signal transduction.** The binding of the signaling molecule changes the receptor protein in some way, initiating the process of transduction. The transduction stage converts the signal to a form that can bring about a specific cellular response. In Sutherland's system, the binding of epinephrine to a receptor protein in a liver cell's plasma membrane leads to activation of glycogen phosphorylase in the cytosol. Transduction sometimes occurs in a single step but more often requires a sequence of changes in a series of different molecules—a **signal transduction pathway**. The molecules in the pathway are often called relay molecules; three are shown as an example.

► **Figure 11.6 Overview of cell signaling.** From the perspective of the cell receiving the message, cell signaling can be divided into three stages: signal reception, signal transduction, and cellular response. When reception occurs at the plasma membrane, as shown here, the transduction stage is usually a pathway of several steps (three are shown as an example), with each specific relay molecule in the pathway bringing about a change in the next molecule. The final molecule in the pathway triggers the cell's response.

VISUAL SKILLS *Where would the epinephrine in Sutherland's experiment fit into this diagram of cell signaling?*

EXTRACELLULAR FLUID   CYTOPLASM

Plasma membrane

1 **Signal reception**   2 **Signal transduction**   3 **Cellular response**

Receptor

Three relay molecules in a signal transduction pathway

Activation of cellular response, such as release of glucose from glycogen

Signaling molecule

→ **Mastering Biology Animation: Overview of Cell Signaling**

❸ Cellular response. The transduced signal finally triggers a specific cellular response. The response may be almost any imaginable cellular activity—such as catalysis by an enzyme (for example, glycogen phosphorylase), rearrangement of the cytoskeleton, or activation of specific genes in the nucleus. The cell-signaling process helps ensure that crucial activities like these occur in the right cells, at the right time, and in proper coordination with the activities of other cells of the organism. We'll now explore the mechanisms of cell signaling in more detail, including a discussion of regulation and termination of the process.

CONCEPT 11.2

Signal reception: A signaling molecule binds to a receptor, causing it to change shape

A wireless router may broadcast its network signal indiscriminately, but only computers with the correct password can connect to it: Reception of the signal depends on the receiver. Similarly, the signals emitted by an **a** mating type yeast cell are "heard" only by its prospective mates, **α** cells. In the case of the epinephrine circulating throughout the bloodstream of the impala in Figure 11.1, the hormone encounters many types of cells, but only certain target cells, those with the corresponding receptor protein, detect and respond to the epinephrine molecule. The signaling molecule is complementary in shape to a specific site on the receptor and attaches there, like a hand in a glove. The signaling molecule acts as a **ligand**, the term for a molecule that specifically binds to another (often larger) molecule. Ligand binding generally causes a receptor protein to undergo a change in shape. For many receptors, this shape change directly activates the receptor, enabling it to interact with other molecules in or on the cell. For other receptors, the immediate effect of ligand binding is to cause the aggregation of two or more receptor proteins, which leads to further molecular events inside the cell. Most signal receptors are plasma membrane proteins, but others are located inside the cell. We'll look at both of these types next.

→ Mastering Biology Animation: Reception

Receptors in the Plasma Membrane

Cell-surface transmembrane receptors play crucial roles in the biological systems of animals. The largest family of human cell-surface receptors is that of the G protein-coupled receptors (GPCRs). There are more than 800 GPCRs; an example is shown in **Figure 11.7**. Another example is the co-receptor hijacked by HIV to enter immune cells (see Figure 7.8); this GPCR is the target of the drug maraviroc, which has shown some success at treating AIDS.

Most water-soluble signaling molecules bind to specific sites on transmembrane receptor proteins that transmit information from the extracellular environment to the inside of the cell. We can see how cell-surface transmembrane receptors work by looking at three major types: G protein-coupled receptors (GPCRs), receptor tyrosine kinases, and ion channel receptors. These receptors are discussed and illustrated in **Figure 11.8**; study this figure before going on.

Given the many important functions of cell-surface receptors, it is not surprising that their malfunctions are associated with many human diseases, including cancer, heart disease, and asthma. To better understand and treat these conditions, a major focus of both university research teams and the pharmaceutical industry has been to analyze the structure of these receptors.

Although cell-surface receptors (half of which are GPCRs) represent 30% of all human proteins, determining their structures by X-ray crystallography (see Figure 5.21) has proved challenging. For one thing, cell-surface receptors tend to be flexible and inherently unstable, thus difficult to crystallize. It took years of persistent efforts for researchers to determine the first few of these structures, such as the GPCR

▼ **Figure 11.7 The structure of a G protein-coupled receptor (GPCR).** This is a model of the human β₂-adrenergic receptor, which binds epinephrine (adrenaline). The receptor was crystallized (discussed later in this section) in the presence of both a molecule that mimics epinephrine (green in the model) and cholesterol in the membrane (orange). Two receptors (blue) are shown as ribbon models in a side view. Caffeine can also bind to this receptor; see question 10 at the end of the chapter.

β₂-adrenergic receptors
Molecule that mimics ligand
Plasma membrane
Cholesterol

G Protein-Coupled Receptors

Signaling molecule binding site

Segment that interacts with G proteins inside the cell

G protein-coupled receptor

A **G protein-coupled receptor** (GPCR) is a cell-surface transmembrane receptor that works with the help of a **G protein**, a protein that binds the energy-rich molecule GTP. Many different signaling molecules—including yeast mating factors, neurotransmitters, and epinephrine (adrenaline) and many other hormones—use GPCRs.

G protein-coupled receptors vary in the binding sites for their ligands and also for different types of G proteins inside the cell. Nevertheless, GPCR proteins are all remarkably similar in structure. In fact, they make up a large family of eukaryotic receptor proteins with a secondary structure in which the single polypeptide, represented here in a ribbon model, has seven transmembrane α helices (outlined with cylinders and depicted in a row for clarity). Specific loops between the helices (here, the loops on the right) form binding sites for signaling molecules (outside the cell) and G proteins (on the cytoplasmic side).

GPCR-based signaling systems are extremely widespread and diverse in their functions, including roles in embryonic development and sensory reception. In humans, for example, vision, smell, and taste depend on GPCRs (see Concept 50.4). Similarities in the structures of G proteins and GPCRs in diverse organisms suggest that G proteins and their associated receptors evolved very early among eukaryotes.

Malfunctions of the associated G proteins themselves are involved in many human diseases, including bacterial infections. The bacteria that cause cholera, pertussis (whooping cough), and botulism, among others, make their victims ill by producing toxins that interfere with G protein function. Up to 60% of all medicines used today exert their effects by influencing G protein pathways.

❶ Attached but able to move along the cytoplasmic side of the membrane, a G protein functions as a molecular switch that is either on or off, depending on whether GDP or GTP is attached —hence the term *G protein*. (GTP, or guanosine triphosphate, is similar to ATP.) When GDP is bound to the G protein, as shown above, the G protein is inactive. The receptor and G protein work together with another protein, usually an enzyme.

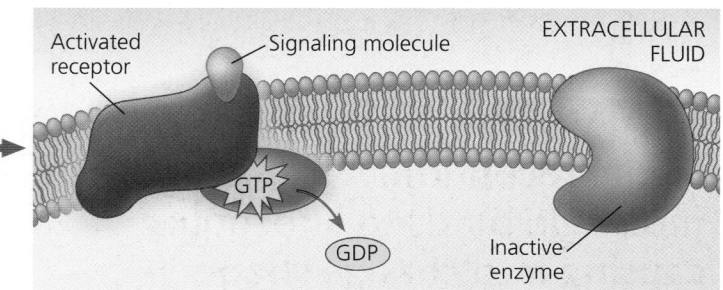

❷ When the appropriate signaling molecule binds to the extracellular side of the receptor, the receptor is activated and changes shape. Its cytoplasmic side then binds an inactive G protein, causing a GTP to displace the GDP. This activates the G protein.

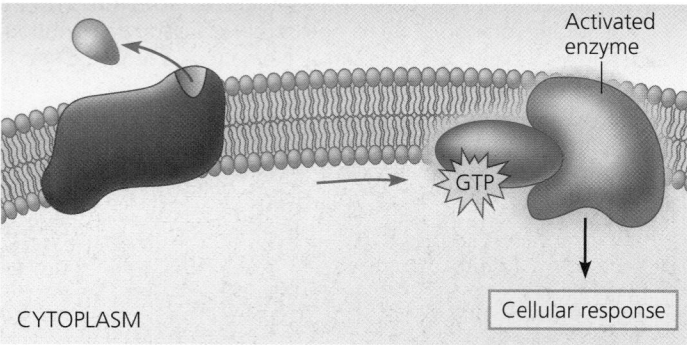

❸ The activated G protein dissociates from the receptor, diffuses along the membrane, and then binds to an enzyme, altering the enzyme's shape and activity. Once activated, the enzyme can trigger the next step leading to a cellular response. Binding of signaling molecules is reversible: Like other ligands, they bind and dissociate many times. The ligand concentration outside the cell determines how often a ligand is bound and initiates signaling.

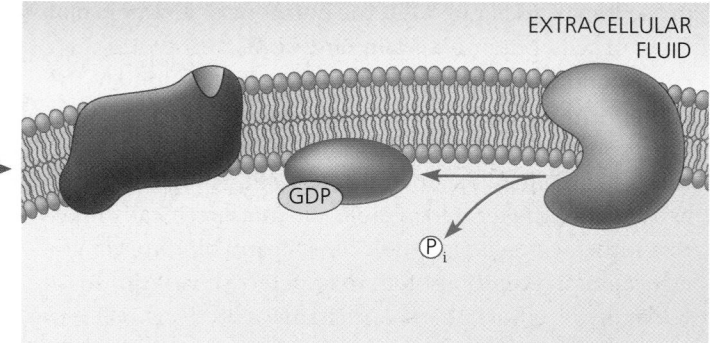

❹ The changes in the enzyme and G protein are only temporary: The G protein also acts as a GTPase, hydrolyzing its bound GTP to GDP and ⓟ$_i$; as a result, it can no longer activate the enzyme. The G protein leaves the enzyme, which returns to its inactive state. The G protein is now available for reuse. Its GTPase function allows the pathway shut down rapidly when the signaling molecule is no longer present.

Receptor Tyrosine Kinases

Receptor tyrosine kinases (RTKs) belong to a major class of plasma membrane receptors characterized by having enzymatic activity. An RTK is a *protein kinase*—an enzyme that catalyzes the transfer of phosphate groups from ATP to another protein. The part of the receptor protein extending into the cytoplasm functions more specifically as a tyrosine kinase, an enzyme that catalyzes the transfer of a phosphate group from ATP to the amino acid tyrosine of a substrate protein. Thus, RTKs are membrane receptors that attach phosphates to tyrosines.

Upon binding a ligand such as a growth factor, one RTK may activate ten or more different transduction pathways and cellular responses. Often, more than one signal transduction pathway can be triggered at once, helping the cell regulate and coordinate many aspects of cell growth and cell reproduction. The ability of a single ligand-binding event to trigger so many pathways is a key difference between RTKs and GPCRs; GPCRs generally activate a single transduction pathway. Abnormal RTKs that function even in the absence of signaling molecules are associated with many kinds of cancer.

1 Many receptor tyrosine kinases have the structure depicted schematically here. Before the signaling molecule binds, the receptors exist as individual units referred to as monomers. Notice that each monomer has an extracellular ligand-binding site, an α helix spanning the membrane, and an intracellular tail containing multiple tyrosines.

2 The binding of a signaling molecule (such as a growth factor) causes two receptor monomers to associate closely with each other, forming a complex known as a dimer, a process called dimerization. (In some cases, larger clusters form. The details of monomer association are a focus of current research.)

3 Dimerization activates the tyrosine kinase region of each monomer; each tyrosine kinase adds a phosphate from an ATP molecule to a tyrosine that is part of the tail of the other monomer.

4 Now that the receptor is fully activated, it is recognized by specific relay proteins inside the cell. Each such protein binds to a specific phosphorylated tyrosine, undergoing a resulting structural change that activates the bound relay protein. Each activated protein triggers a transduction pathway, leading to a cellular response.

Continued on next page

Ion Channel Receptors

A **ligand-gated ion channel** is a type of membrane channel receptor containing a region that can act as a "gate," opening or closing the channel when the receptor changes shape. When a signaling molecule binds as a ligand to the channel receptor, the channel opens or closes, allowing or blocking the flow of specific ions, such as Na^+ or Ca^{2+}. Like the other receptors we have discussed, these proteins bind the ligand at a specific site on their extracellular sides.

1 Here we show a ligand-gated ion channel receptor in which the channel remains closed until a ligand binds to the receptor.

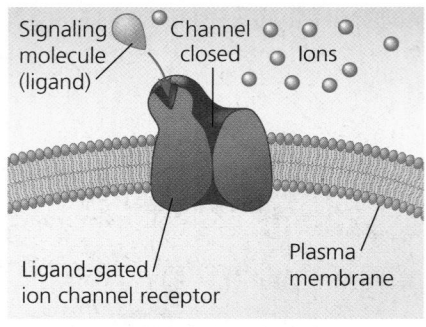

2 When the ligand binds to the receptor and the channel opens, specific ions can flow through the channel and rapidly change the local concentration of that ion inside the cell. This change may directly affect the activity of the cell in some way.

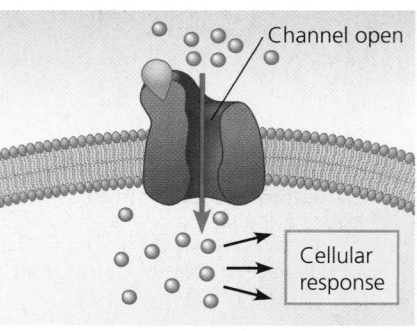

3 When the ligand dissociates from this receptor, the channel closes and ions no longer enter the cell.

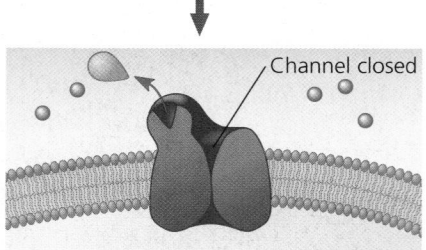

Ligand-gated ion channels are very important in the nervous system. For example, the neurotransmitter molecules released at a synapse between two nerve cells (see Figure 11.5b) bind as ligands to some ion channels on the receiving cell, causing the channels to open. Ions flow in (or, in some cases, out), triggering an electrical signal that propagates down the length of the receiving cell. Some of the gated ion channels are controlled by electrical signals instead of ligands; these voltage-gated ion channels are crucial to the functioning of the nervous system, as you'll see in Chapter 48. Some ion channels are present on membranes of organelles, such as the ER.

MAKE CONNECTIONS *Is the flow of ions through a ligand-gated channel an example of active or passive transport? (Review Concepts 7.3 and 7.4.)*

Mastering Biology Animation: Acetylcholine Receptor

shown in Figure 11.7. In that case, the β-adrenergic receptor was stable enough to be crystallized only while it was among membrane molecules and in the presence of a molecule mimicking its ligand. Newer techniques that do not require crystallization, like cryo-EM (see Figure 6.3), have shown promise in determining the structures of cell-surface receptors.

Abnormal functioning of receptor tyrosine kinases (RTKs) is associated with many types of cancers. For example, some breast cancer patients have tumor cells with excessive levels of a receptor tyrosine kinase called HER2 (see Concept 12.3 and Figure 18.27). Using molecular biological techniques, researchers have developed a protein called Herceptin that binds to HER2 on cells and inhibits cell division, thus thwarting further tumor development. In some clinical studies, treatment with Herceptin improved patient survival rates by more than one-third. One goal of ongoing research into these cell-surface receptors and other cell-signaling proteins is development of additional successful treatments.

Mastering Biology Interview with Diana Bautista: How cell signaling leads to itch and pain (see the interview before Chapter 6)

Intracellular Receptors

Intracellular receptor proteins are found in either the cytoplasm or nucleus of target cells. To reach such a receptor, a signaling molecule passes through the target cell's plasma membrane. A number of important signaling molecules can do this because they are either hydrophobic enough or small enough to cross the hydrophobic interior of the membrane (see Concept 7.1). Hydrophobic signaling molecules include steroid hormones and thyroid hormones of animals. Another chemical signaling molecule that possesses an intracellular receptor is nitric oxide (NO), a gas; this very small molecule readily passes between the membrane phospholipids. Once a hormone or other signaling molecule has entered a cell, its binding to an intracellular receptor changes the receptor into a hormone-receptor complex that is able to cause a response—in many cases, the turning on or off of particular genes.

The behavior of aldosterone is a representative example of how steroid hormones work. This hormone is secreted by cells of the adrenal gland (a gland that lies above the kidney), then travels through the blood and enters cells all over the body. However, a response occurs only in kidney cells because they alone contain receptors for aldosterone. In these cells, the hormone binds to and activates the receptor protein. With aldosterone attached, the active form of the receptor protein then enters the nucleus and turns on specific genes that control water and sodium flow in kidney cells, ultimately affecting blood volume **(Figure 11.9)**.

How does the activated hormone-receptor complex turn on genes? Recall that the genes in a cell's DNA

Figure 11.9 Steroid hormone interacting with an intracellular receptor.

Hormone (aldosterone)

EXTRACELLULAR FLUID

Receptor protein

Plasma membrane

Hormone-receptor complex

DNA

mRNA

NUCLEUS

New protein

CYTOPLASM

1. The steroid hormone aldosterone passes through the plasma membrane.

2. Aldosterone binds to a receptor protein in the cytoplasm, activating it.

3. The hormone-receptor complex enters the nucleus and binds to specific genes.

4. The bound protein acts as a transcription factor, stimulating the transcription of the gene into mRNA.

5. The mRNA is translated into a specific protein.

MAKE CONNECTIONS *Why is a cell-surface receptor protein not required for this steroid hormone to enter the cell? (See Concept 7.2.)*

➡️ **Mastering Biology** Animation: Steroid Hormone Pathway

function by being transcribed and processed into messenger RNA (mRNA), which leaves the nucleus and is translated into a specific protein by ribosomes in the cytoplasm (see Figure 5.22). Special proteins called *transcription factors* control which genes are turned on—that is, which genes are transcribed into mRNA—in a particular cell at a particular time. When the aldosterone receptor is activated, it acts as a transcription factor that turns on specific genes. (You'll learn more about transcription factors in Chapters 17 and 18.)

By acting as a transcription factor, the aldosterone receptor itself carries out two parts of the signaling pathway, as receptor and transducer. Most other intracellular receptors function in the same way, although many of them, such as the thyroid hormone receptor, are already in the nucleus before the signaling molecule reaches them. Interestingly, many of these intracellular receptor proteins are structurally similar, suggesting an evolutionary kinship.

CONCEPT CHECK 11.2

1. Nerve growth factor (NGF) is a water-soluble signaling molecule. Would you expect the receptor for NGF to be intracellular or in the plasma membrane? Explain.

2. **WHAT IF?** What would the effect be if a cell made defective receptor tyrosine kinase proteins that were unable to dimerize?

3. **MAKE CONNECTIONS** How is ligand binding similar to the process of allosteric regulation of enzymes? (See Figure 8.20.)

For suggested answers, see Appendix A.

CONCEPT 11.3

Signal transduction: Cascades of molecular interactions transmit signals from receptors to relay molecules in the cell

When receptors for signaling molecules are plasma membrane proteins, like most of those we have discussed, the transduction stage of cell signaling is usually a multistep pathway involving many molecules. Steps often include activation of proteins by addition or removal of phosphate groups or release of other small molecules or ions that act as signaling molecules. One benefit of using multiple steps is that a signal caused by a small number of signaling molecules can be greatly amplified. If each molecule transmits the signal to numerous molecules at the next step in the series, the result is a geometric increase in the number of activated molecules by the end (see Figure 11.16). A second benefit of using multistep pathways is that they provide more opportunities for coordination and control than do simpler systems. This allows regulation of the response, as we'll see later in the chapter.

Signal Transduction Pathways

The binding of a specific signaling molecule to a receptor in the plasma membrane triggers the first step in the signal transduction pathway—the chain of molecular interactions that leads to a particular response within the cell. Like falling dominoes, the signal-activated receptor activates another molecule, which activates yet another molecule, and so on, until the protein that produces the final cellular response is activated. The molecules that relay a signal from receptor to response, which we call relay molecules in this book, are often proteins. Protein-protein interactions are a major theme of cell signaling—indeed, a unifying theme of all cellular activities.

➡️ **Mastering Biology** Animation: Signal Transduction Pathways

Keep in mind that the original signaling molecule is not physically passed along a signaling pathway; in most cases, it never even enters the cell. When we say that the signal is relayed along a pathway, we mean that certain information is passed on. At each step, the signal is transduced into a different form, commonly a shape change in the next protein. Very often, the shape change is brought about by phosphorylation.

Protein Phosphorylation and Dephosphorylation

Previous chapters introduced the concept of activating a protein by adding one or more phosphate groups to it (see Figure 8.11a). In Figure 11.8, you have already seen how phosphorylation is involved in the activation of receptor tyrosine kinases. In fact, phosphorylation of proteins and its reverse, dephosphorylation, are commonly used in cells to regulate protein activity. An enzyme that transfers phosphate groups from ATP to a protein is generally known as a **protein kinase**. Recall that a receptor tyrosine kinase (RTK) is a specific kind of protein kinase that phosphorylates tyrosines on the other RTK in a dimer. Most cytoplasmic protein kinases, however, act on proteins different from themselves. Another distinction is that most cytoplasmic protein kinases phosphorylate either of two other amino acids, serine or threonine, rather than tyrosine. Serine/threonine kinases are widely involved in signaling pathways in animals, plants, and fungi.

Many signal transduction pathways use relay molecules that are protein kinases, and they often act on other protein kinases in the pathway. **Figure 11.10** depicts a hypothetical pathway containing two different protein kinases that create a **phosphorylation cascade**. The sequence of steps shown in the figure is similar to many known pathways, including those triggered in yeast by mating factors and in animal cells by many growth factors. The signal is transmitted by a cascade of protein phosphorylations, each causing a shape change in the phosphorylated protein. The shape change results from the interaction of the newly added phosphate groups with charged or polar amino acids on the protein being phosphorylated (see Figure 5.14). The shape change in turn alters the function of the protein, most often activating it. (In some cases, though, phosphorylation instead *decreases* the activity of the protein.)

A significant percentage of our own genes—about 2%—is thought to code for protein kinases. A single cell may have hundreds of different kinds, each specific for a different substrate protein. Together, protein kinases probably regulate the activity of a large proportion of the thousands of proteins in a cell. Among these are most of the proteins that, in turn, regulate cell division. Abnormal activity of such a

▶ **Figure 11.10 A phosphorylation cascade.** In a phosphorylation cascade, a series of different proteins in a pathway are phosphorylated in turn, each protein adding a phosphate group to the next one in line. Here, phosphorylation activates each protein, and dephosphorylation returns it to its inactive form. The active and inactive forms of each protein are represented by different shapes to remind you that activation is usually associated with a change in molecular shape.

WHAT IF? *What would happen if a mutation in protein kinase 2 made it incapable of being phosphorylated?*

 — Signaling molecule; Receptor; Activated relay molecule; Inactive protein kinase 1; ① A relay molecule activates protein kinase 1. Active protein kinase 1; ② Active protein kinase 1 activates protein kinase 2. Inactive protein kinase 2; ATP; ADP; Active protein kinase 2; ③ Active protein kinase 2 phosphorylates a protein (purple) that brings about the cell's response to the signal. ④ Protein phosphatases (PP) catalyze the removal of the phosphate groups from the proteins, making the proteins inactive again. Inactive protein; ATP; ADP; Active protein; Cellular response; Phosphorylation cascade kinase can cause abnormal cell division and contribute to the development of cancer.

Equally important in the phosphorylation cascade are the **protein phosphatases**, enzymes that can rapidly remove phosphate groups from proteins, a process called *dephosphorylation*. By dephosphorylating and thus inactivating protein kinases, phosphatases provide the mechanism for turning off the signal transduction pathway when the initial signal is no longer present. Phosphatases also make the protein kinases available for reuse, enabling the cell to respond again to an extracellular signal. The phosphorylation-dephosphorylation system acts as a molecular switch in the cell, turning activities on or off, or up or down, as required. At any given moment, the activity of a protein regulated by phosphorylation depends on the balance in the cell between active kinase molecules and active phosphatase molecules.

Small Molecules and Ions as Second Messengers

Not all components of signal transduction pathways are proteins. Many signaling pathways also involve small, nonprotein, water-soluble molecules or ions called **second messengers**. (The pathway's "first messenger" is considered to be the extracellular signaling molecule—the ligand—that binds to the membrane receptor.) Because they are small and water-soluble, they can readily spread through regions of the cell by diffusion. For example, as we'll see shortly, a second messenger called cyclic AMP carries the signal initiated by epinephrine from the plasma membrane of a liver or muscle cell into the cell's interior, where the signal eventually brings about glycogen breakdown. Second messengers participate in pathways that are initiated by both G protein-coupled receptors and receptor tyrosine kinases. The two most common second messengers are cyclic AMP and calcium ions, Ca^{2+}. A large variety of relay proteins respond to changes in the cytosolic concentration of one or the other of these second messengers.

Cyclic AMP

Having discovered that epinephrine somehow causes glycogen breakdown within cells, Earl Sutherland next looked for a second messenger that transmits the signal from the plasma membrane to the metabolic machinery in the cytoplasm. He found that binding of epinephrine to the G protein-coupled receptor (GPCR) in the plasma membrane results in a rise in the cytosolic concentration of **cyclic AMP (cAMP**; cyclic adenosine monophosphate), a small molecule produced from ATP. As shown in **Figure 11.11**, an enzyme embedded in the plasma membrane, **adenylyl cyclase** (also known as adenylate cyclase), converts ATP to cAMP in response to an extracellular signal—in this case, provided by epinephrine. When epinephrine outside the cell binds to a GPCR, it activates a G protein that in turn activates adenylyl cyclase. Adenylyl cyclase can then catalyze the synthesis of many molecules of cAMP. In this way, the normal cellular concentration of cAMP can be boosted 20-fold in a matter of seconds. The cAMP broadcasts the signal to the cytoplasm. It does not persist for long in the absence of the hormone because a different enzyme, called phosphodiesterase, converts cAMP to AMP. Another surge of epinephrine is needed to boost the cytosolic concentration of cAMP again.

Subsequent research has revealed that epinephrine is only one of many hormones and many other signaling molecules that lead to activation of adenylyl cyclase by G proteins and formation of cAMP **(Figure 11.12)**. The immediate effect of an elevation in cAMP level is usually the activation of a serine/threonine kinase called *protein kinase A*. The activated protein kinase A then phosphorylates various other proteins, depending on the cell type. (The complete pathway for epinephrine's stimulation of glycogen breakdown is shown later, in Figure 11.16.)

▼ **Figure 11.11 Cyclic AMP.** The second messenger cyclic AMP (cAMP) is made from ATP by adenylyl cyclase, an enzyme embedded in the plasma membrane. The phosphate group in cAMP is attached to both the 5′ and the 3′ carbons; cAMP's cyclic arrangement is the basis for its name. cAMP is inactivated by phosphodiesterase, an enzyme that converts it to AMP.

WHAT IF? *What would happen if a molecule that inactivated phosphodiesterase were introduced into the cell?*

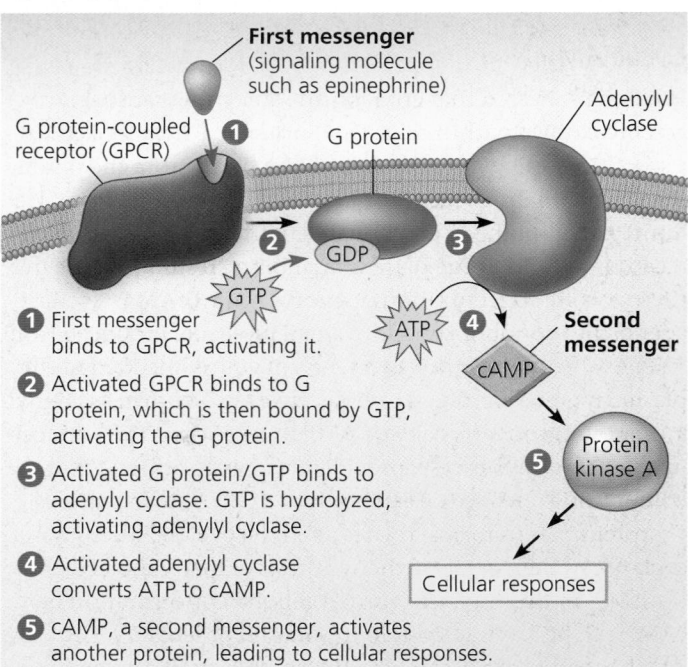

▼ Figure 11.12 cAMP as a second messenger in a G protein signaling pathway.

First messenger
(signaling molecule such as epinephrine)

G protein-coupled receptor (GPCR) ①

G protein

Adenylyl cyclase

②

GDP

GTP

③

ATP ④

Second messenger

cAMP

⑤ Protein kinase A

Cellular responses

① First messenger binds to GPCR, activating it.

② Activated GPCR binds to G protein, which is then bound by GTP, activating the G protein.

③ Activated G protein/GTP binds to adenylyl cyclase. GTP is hydrolyzed, activating adenylyl cyclase.

④ Activated adenylyl cyclase converts ATP to cAMP.

⑤ cAMP, a second messenger, activates another protein, leading to cellular responses.

DRAW IT *The bacterium that causes the disease cholera produces a toxin that locks the G protein in its activated state. Review Figure 11.8, then draw this figure as it would be if cholera toxin were present. (You do not need to draw the cholera toxin molecule.)*

Further regulation of cell metabolism is provided by other G protein systems that *inhibit* adenylyl cyclase. In these systems, a different signaling molecule activates a different receptor, which in turn activates an *inhibitory* G protein that blocks activation of adenylyl cyclase. Cell activities can be fine-tuned by the balance between these systems.

Now that we know about the role of cAMP in G protein signaling pathways, we can explain in molecular detail how certain microorganisms cause disease. Consider cholera, a disease that is frequently epidemic in places where the water supply is contaminated with human feces. People acquire the cholera bacterium, *Vibrio cholerae*, by drinking contaminated water. The bacteria form a biofilm on the lining of the small intestine and produce a toxin. The cholera toxin is an enzyme that chemically modifies a G protein involved in regulating salt and water secretion. Because the modified G protein is unable to hydrolyze GTP to GDP, it remains stuck in its active form, continuously stimulating adenylyl cyclase to make cAMP (see the question with Figure 11.12). The resulting high concentration of cAMP causes the intestinal cells to secrete large amounts of salts into the intestines, with water following by osmosis. An infected person quickly develops profuse diarrhea and if left untreated can soon die from the loss of water and salts.

Our understanding of signaling pathways involving cyclic AMP or related messengers has allowed us to develop treatments for certain conditions in humans such as erectile dysfunction. In one pathway, the gas nitric oxide (NO) is released by a cell and enters a neighboring muscle cell, where it causes production of a molecule similar to cAMP called *cyclic GMP* (*cGMP*). The cGMP then acts as a second messenger that causes relaxation of muscles, such as those in the walls of arteries. A compound that prolongs the signal (by inhibiting the hydrolysis of cGMP to GMP) was originally prescribed for chest pains because it relaxed blood vessels and increased blood flow to the heart muscle. Under the trade name Viagra, this compound is now widely used as a treatment for erectile dysfunction in human males. Because Viagra leads to dilation of blood vessels, it also allows increased blood flow to the penis, optimizing physiological conditions for penile erections.

Calcium Ions and Inositol Trisphosphate (IP₃)

Many of the signaling molecules that function in animals—including neurotransmitters, growth factors, and some hormones—induce responses in their target cells through signal transduction pathways that increase the cytosolic concentration of calcium ions (Ca^{2+}). Calcium is even more widely used than cAMP as a second messenger. Increasing the local cytosolic concentration of Ca^{2+} causes many responses in animal cells, including muscle cell contraction, exocytosis of molecules (secretion), and cell division. In plant cells, a wide range of hormonal and environmental stimuli can cause brief increases in cytosolic Ca^{2+} concentration, triggering various signaling pathways, such as the pathway for greening in response to light (see Figure 39.4). Cells use Ca^{2+} as a second messenger in pathways triggered by both G protein-coupled receptors and receptor tyrosine kinases.

Although cells always contain some Ca^{2+}, this ion can function as a second messenger because its concentration in the cytosol is normally much lower than the concentration outside the cell **(Figure 11.13)**. In fact, the level of Ca^{2+} in the blood and extracellular fluid of an animal is often more than 10,000 times higher than that in the cytosol. Calcium ions are actively transported out of the cell and are actively imported from the cytosol into the endoplasmic reticulum (and, under some conditions, into mitochondria and chloroplasts) by various protein pumps. As a result, the calcium concentration in the ER is usually much higher than that in the cytosol. Because the cytosolic calcium level is low, a small change in absolute numbers of ions represents a relatively large percentage change in local calcium concentration.

In response to a signal relayed by a signal transduction pathway, the cytosolic calcium level may rise, usually by a

▼ **Figure 11.13 The maintenance of calcium ion concentrations in an animal cell.** The Ca^{2+} concentration in the cytosol is usually much lower (light green) than in the extracellular fluid and ER (dark green). Protein pumps in the plasma membrane and the ER membrane, driven by ATP, move Ca^{2+} from the cytosol into the extracellular fluid and into the lumen of the ER. Mitochondrial pumps, driven by chemiosmosis (see Concept 9.4), move Ca^{2+} into mitochondria when the calcium level in the cytosol rises significantly.

Key   ▬ High $[Ca^{2+}]$ ▬ Low $[Ca^{2+}]$

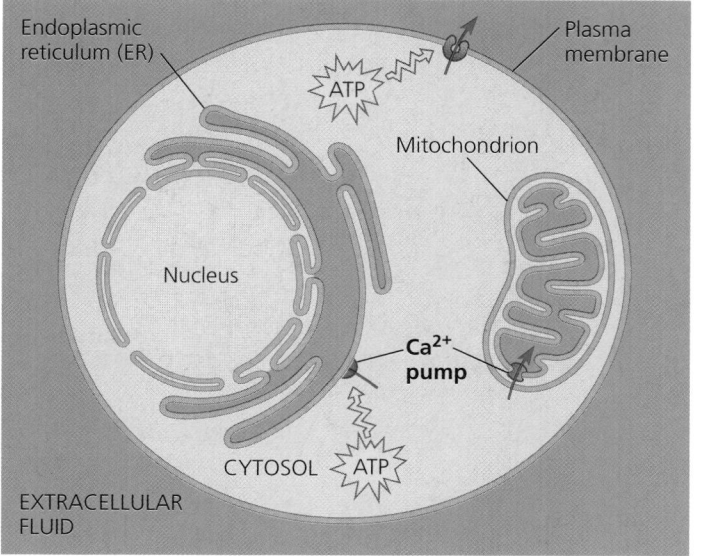

mechanism that releases Ca^{2+} from the cell's ER. The pathways leading to calcium release involve two other second messengers, **inositol trisphosphate (IP_3)** and **diacylglycerol (DAG)**. These two messengers are produced by cleavage of a certain kind of phospholipid in the plasma membrane. **Figure 11.14** shows the complete picture of how a signal causes IP_3 to stimulate the release of calcium from the ER. Because IP_3 acts before calcium in these pathways, calcium could be considered a "*third* messenger." However, scientists use the term *second messenger* for all small, nonprotein components of signal transduction pathways.

CONCEPT CHECK 11.3

1. What is a protein kinase, and what is its role in a signal transduction pathway?
2. When a signal transduction pathway involves a phosphorylation cascade, how does the cell's response get turned off?
3. What is the actual "signal" that is being transduced in any signal transduction pathway, such as those shown in Figures 11.6 and 11.10? In what way is this information being passed from the exterior to the interior of the cell?
4. **WHAT IF?** If you exposed a cell to a ligand that binds to a receptor and activates phospholipase C, predict the effect the IP_3-gated calcium channel would have on Ca^{2+} concentration in the cytosol.

For suggested answers, see Appendix A.

► **Figure 11.14 Calcium and IP_3 in signaling pathways.** Calcium ions (Ca^{2+}) and inositol trisphosphate (IP_3) function as second messengers in many signal transduction pathways. In this figure, the process is initiated by the binding of a signaling molecule to a G protein-coupled receptor. A receptor tyrosine kinase could also initiate this pathway by activating phospholipase C.

MAKE CONNECTIONS *Explain the difference in function (in terms of ion transport) between the Ca^{2+} pump in Figure 11.13 and the Ca^{2+} channel protein shown here. (See Figure 7.17.)*

❶ A signaling molecule binds to a receptor, leading to activation of phospholipase C.

❷ Phospholipase C cleaves a plasma membrane phospholipid called PIP_2 into DAG and IP_3.

❸ DAG functions as a second messenger in other pathways.

EXTRACELLULAR FLUID

Signaling molecule (first messenger)

G protein

CYTOSOL

G protein-coupled receptor

GDP

GTP

Phospholipase C

DAG

PIP_2

IP_3 (second messenger)

IP_3-gated calcium channel

Endoplasmic reticulum (ER) lumen

Ca^{2+}

Ca^{2+} (second messenger)

Various proteins activated

Cellular responses

❹ IP_3 quickly diffuses through the cytosol and binds to an IP_3-gated calcium channel in the ER membrane, causing it to open.

❺ Calcium ions flow out of the ER (down their concentration gradient), raising the Ca^{2+} level in the cytosol.

❻ The calcium ions activate the next protein in one or more signaling pathways.

Cellular response: Cell signaling leads to regulation of transcription or cytoplasmic activities

We now take a closer look at the cell's subsequent response to an extracellular signal—what some researchers call the "output response." What is the nature of the final step in a signaling pathway?

Nuclear and Cytoplasmic Responses

Ultimately, a signal transduction pathway leads to the regulation of one or more cellular activities. The response may occur in the nucleus or in the cytoplasm of the cell.

Many signaling pathways ultimately regulate protein synthesis, usually by turning specific genes on or off in the nucleus. Like an activated steroid receptor (see Figure 11.9), the final activated molecule in a signaling pathway may function as a transcription factor. **Figure 11.15** shows an example in which a signaling pathway activates a transcription factor that turns a gene on: The response to this growth factor signal is transcription, the synthesis of one or more specific mRNAs, which will be translated in the cytoplasm into specific proteins. In other cases, the transcription factor might regulate a gene by turning it off. Often, a transcription factor regulates several different genes.

Sometimes a signaling pathway may regulate the *activity* of proteins rather than causing their synthesis by activating gene expression. This directly affects proteins that function outside the nucleus. For example, a signal may cause the opening or closing of an ion channel in the plasma membrane or a change in the activity of a metabolic enzyme. As we have seen, the response of liver cells to the hormone epinephrine helps regulate cellular energy metabolism by affecting the activity of an enzyme. The final step in the epinephrine signaling pathway activates the enzyme that catalyzes the breakdown of glycogen. **Figure 11.16** shows the complete pathway leading to the release of glucose 1-phosphate molecules from glycogen. Notice that as each molecule is activated, the response is amplified, a subject we'll return to shortly.

Signal receptors, relay molecules, and second messengers participate in a variety of pathways, leading to both nuclear and cytoplasmic responses, including cell division. Malfunctioning of growth factor pathways like the one in Figure 11.15 can contribute to abnormal cell division and the development of cancer, as we'll see in Concept 18.5.

Regulation of the Response

Whether the response occurs in the nucleus or in the cytoplasm, it is not simply turned "on" or "off." Rather, the extent and specificity of the response are regulated in multiple ways.

▼ **Figure 11.15 Nuclear responses to a signal: the activation of a specific gene by a growth factor.** This diagram shows a typical signaling pathway that leads to regulation of gene activity in the cell nucleus. The initial signaling molecule, in this case a growth factor, triggers a phosphorylation cascade, as in Figure 11.10. (The ATP molecules and phosphate groups are not shown.) Once phosphorylated, the last kinase in the sequence enters the nucleus and activates a transcription factor, which stimulates transcription of a specific gene (or genes). The resulting mRNAs then direct the synthesis of a particular protein.

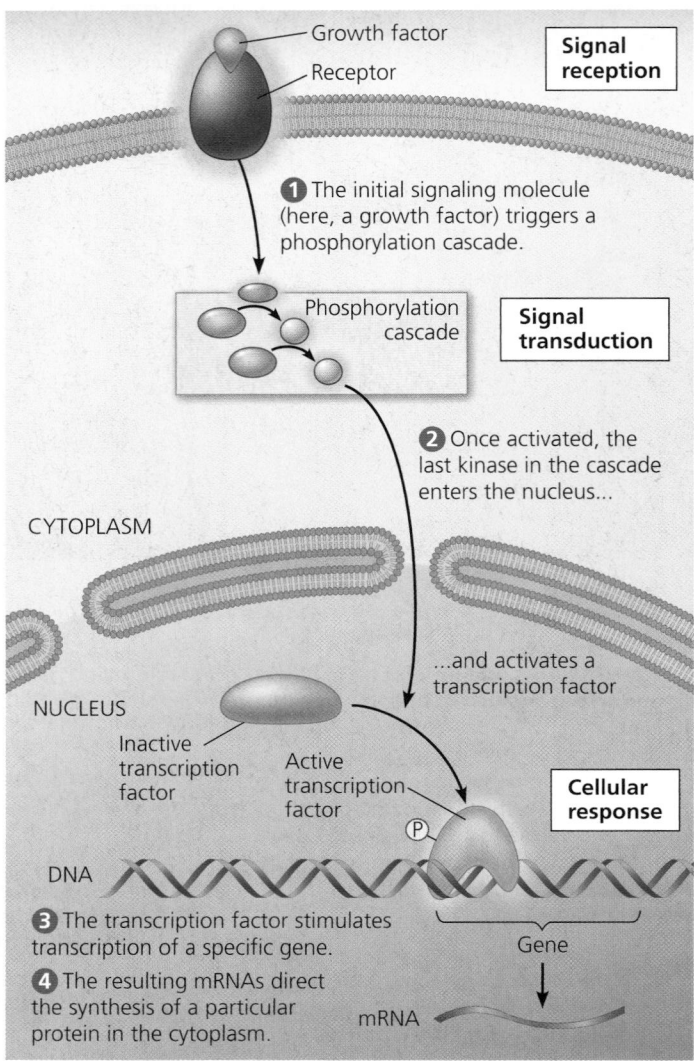

Growth factor

Receptor

Signal reception

❶ The initial signaling molecule (here, a growth factor) triggers a phosphorylation cascade.

Phosphorylation cascade

Signal transduction

❷ Once activated, the last kinase in the cascade enters the nucleus...

CYTOPLASM

...and activates a transcription factor

NUCLEUS

Inactive transcription factor

Active transcription factor

Cellular response

DNA

❸ The transcription factor stimulates transcription of a specific gene.

Gene

❹ The resulting mRNAs direct the synthesis of a particular protein in the cytoplasm.

mRNA

→ **Mastering Biology** Animation: Nuclear Response: Activating a Gene

Here we'll consider four aspects of this regulation. First, as mentioned earlier, signaling pathways generally amplify the cell's response to a single signaling event. The degree of amplification depends on the function of the specific molecules in the pathway. Second, the many steps in a multistep pathway provide control points at which the cell's response can be further regulated, contributing to the specificity of the response and allowing coordination with other signaling pathways. Third, the overall efficiency of the response is enhanced by the presence of proteins known as scaffolding

▼ **Figure 11.16 Cytoplasmic response to a signal: the stimulation of glycogen breakdown by epinephrine (adrenaline).** In this signaling system, the hormone epinephrine acts through a G protein-coupled receptor to activate a succession of relay molecules, including cAMP and two protein kinases (see also Figure 11.12). The final protein activated is the enzyme glycogen phosphorylase, which uses inorganic phosphate to release glucose monomers from glycogen in the form of glucose 1-phosphate molecules. This pathway amplifies the hormonal signal: One receptor protein can activate approximately 100 molecules of G protein, and each enzyme in the pathway, once activated, can act on many molecules of its substrate, the next molecule in the cascade. The number of activated molecules given for each step is approximate.

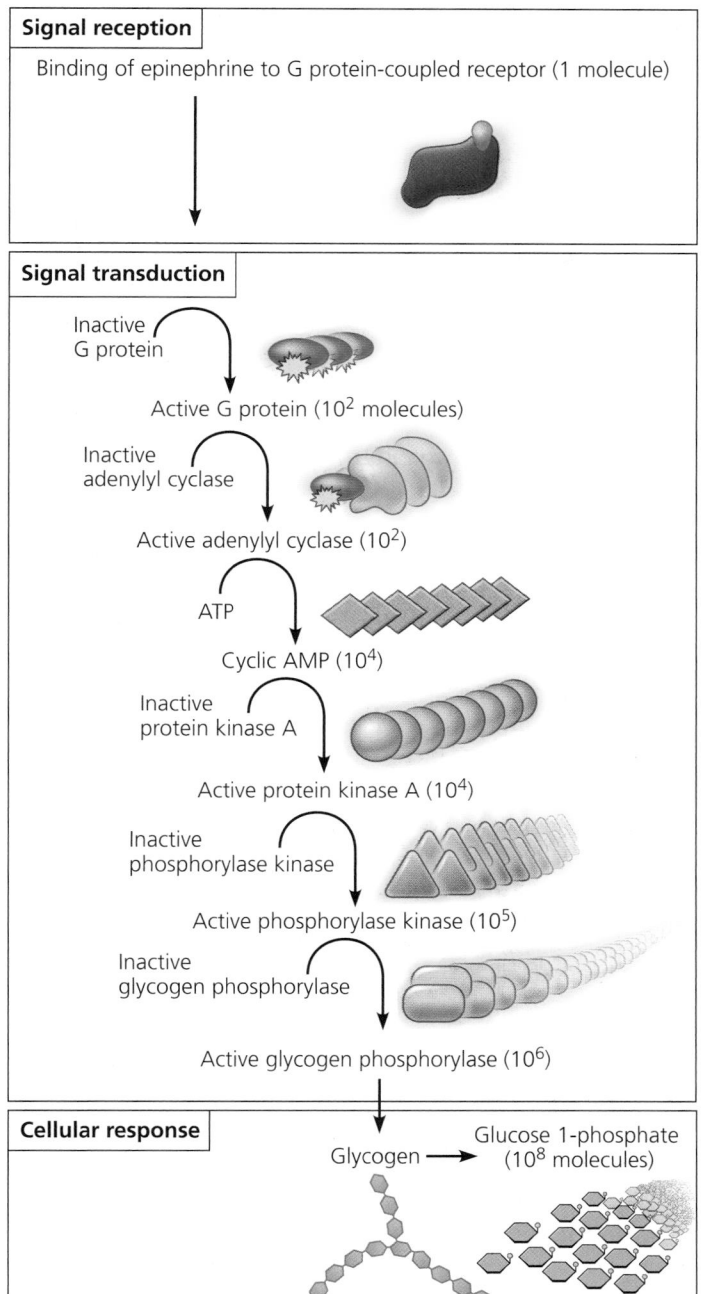

VISUAL SKILLS *In the figure, how many glucose 1-phosphate molecules are released in response to one signaling molecule? Calculate the factor by which the response is amplified in going from each step to the next.*

◉ Mastering Biology Animation: Cytoplasmic Response: Glycogen Breakdown

proteins. Finally, a crucial point in regulating the response is the termination of the signal.

Signal Amplification

Elaborate enzyme cascades amplify the cell's response to a signal. At each catalytic step in the cascade, the number of activated products can be much greater than in the preceding step. For example, in the epinephrine-triggered pathway in Figure 11.16, each adenylyl cyclase molecule catalyzes the formation of 100 or so cAMP molecules, each molecule of protein kinase A phosphorylates about 10 molecules of the next kinase in the pathway, and so on. The amplification effect stems from the fact that these proteins persist in their active form long enough to process multiple molecules of substrate before they become inactive again. As a result of the signal's amplification, a small number of epinephrine molecules binding to receptors on the surface of a liver cell or muscle cell can lead to the release of hundreds of millions of glucose molecules from glycogen.

The Specificity of Cell Signaling and Coordination of the Response

Consider two different cells in your body—a liver cell and a heart muscle cell, for example. Both are in contact with your bloodstream and are therefore constantly exposed to many different hormone molecules, as well as to local regulators secreted by nearby cells. Yet the liver cell responds to some signals but ignores others, and the same is true for the heart cell. And some kinds of signals trigger responses in both cells—but different responses. For instance, epinephrine stimulates the liver cell to break down glycogen, but the main response of the heart cell to epinephrine is contraction, leading to a more rapid heartbeat. How do we account for this difference?

The explanation for the specificity exhibited in cellular responses to signals is the same as the basic explanation for virtually all differences between cells: Because different kinds of cells turn on different sets of genes, *different kinds of cells have different collections of proteins*. The response of a cell to a signal depends on its particular collection of signal receptor proteins, relay proteins, and proteins needed to carry out the response. A liver cell, for example, is poised to respond appropriately to epinephrine by having the proteins listed in Figure 11.16 as well as those needed to manufacture glycogen.

Thus, two cells that respond differently to the same signal differ in one or more proteins that respond to the signal. Also, within some cells, there are more complex pathways that branch or converge, as shown in **Figure 11.17**. Notice, though, that these different pathways still have some molecules in common. For example, cells A, B, and C all use the same receptor protein for the red signaling molecule; differences in other proteins account for their differing responses. In cell D, a different receptor protein is used for the same signaling molecule, leading to yet another response. In cell B, a pathway triggered by one signal diverges to produce two responses; such branched pathways often involve receptor tyrosine kinases (which can activate multiple relay proteins) or second messengers (which can regulate numerous proteins). In cell C, two pathways triggered by separate signals converge to modulate a single response. Branching of pathways and "cross-talk" (interaction) between pathways are

▼ **Figure 11.17 The specificity of cell signaling.** The particular proteins a cell possesses determine what signaling molecules it responds to and the nature of the response. The four cells in these diagrams respond to the same signaling molecule (red) in different ways because each has a different set of proteins (purple and teal). Note, however, that the same kinds of molecules can participate in more than one pathway.

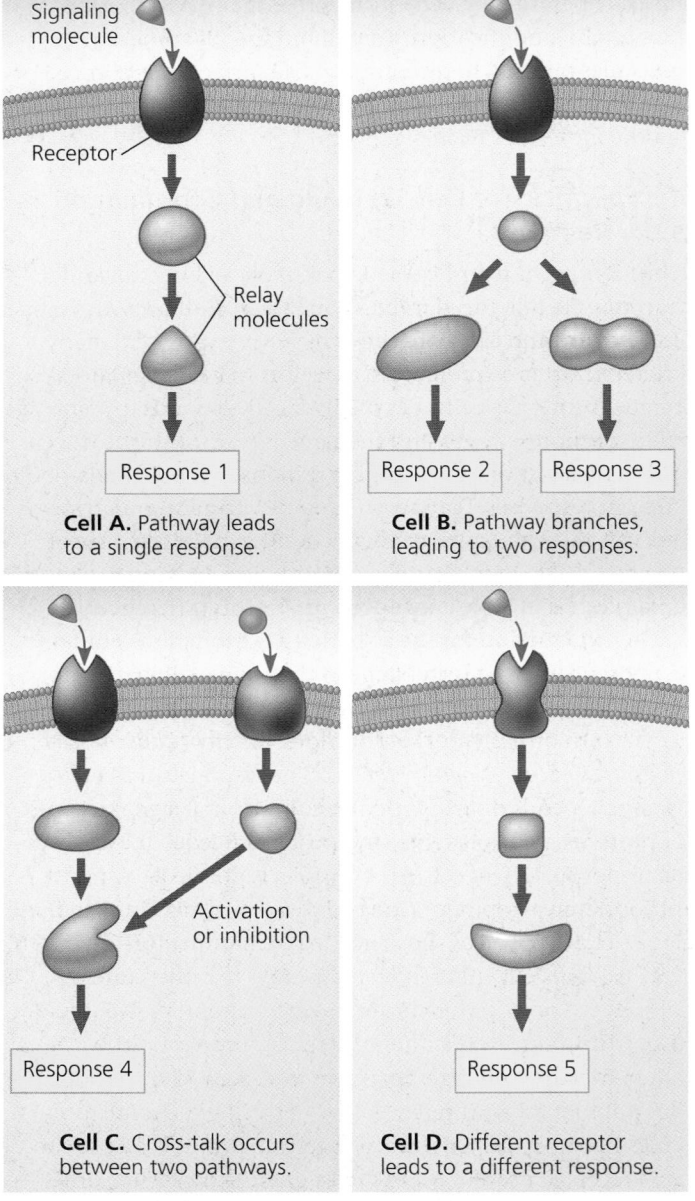

Cell A. Pathway leads to a single response.

Cell B. Pathway branches, leading to two responses.

Cell C. Cross-talk occurs between two pathways.

Cell D. Different receptor leads to a different response.

VISUAL SKILLS *Study the signaling pathway shown in Figure 11.14, and explain how the situation pictured for cell B in Figure 11.17 could apply to that pathway.*

important in regulating and coordinating a cell's responses to information coming in from different sources in the body. (You'll learn more about this coordination in Concept 11.5.) Moreover, the use of some of the same proteins in more than one pathway allows the cell to economize on the number of different proteins it must make.

⊙ **Instructors:** The Scientific Skills Exercise "Using Experiments to Test a Model" can be assigned in **Mastering Biology**. It explores the cellular response of a yeast cell to the signal initiated by a mating factor from a cell of the opposite mating type.

Signaling Efficiency: Scaffolding Proteins and Signaling Complexes

The illustrations of signaling pathways in Figure 11.17 (as well as diagrams of other pathways in this chapter) are greatly simplified. The diagrams show only a few relay molecules and, for clarity's sake, display these molecules spread out in the cytosol. If this were true in the cell, signaling pathways would operate very inefficiently because most relay molecules are proteins, and proteins are too large to diffuse quickly through the viscous cytosol. How does a given protein kinase, for instance, find its protein substrate?

In many cases, the efficiency of signal transduction is apparently increased by the presence of **scaffolding proteins**, large relay proteins to which several other relay proteins are simultaneously attached **(Figure 11.18)**. Researchers have found scaffolding proteins in brain cells that *permanently* hold together networks of signaling pathway proteins at synapses. This hardwiring enhances the speed and accuracy of signal transfer between cells because the rate of protein-protein interaction is not limited by diffusion. Furthermore, in some cases the scaffolding proteins themselves may directly activate relay proteins.

The importance of the relay proteins that serve as points of branching or intersection in signaling pathways is highlighted by the problems arising when these proteins are

▼ **Figure 11.18 A scaffolding protein.** The scaffolding protein shown here (orange) simultaneously binds to a specific activated membrane receptor and three different protein kinases. This physical arrangement facilitates signal transduction by these molecules.

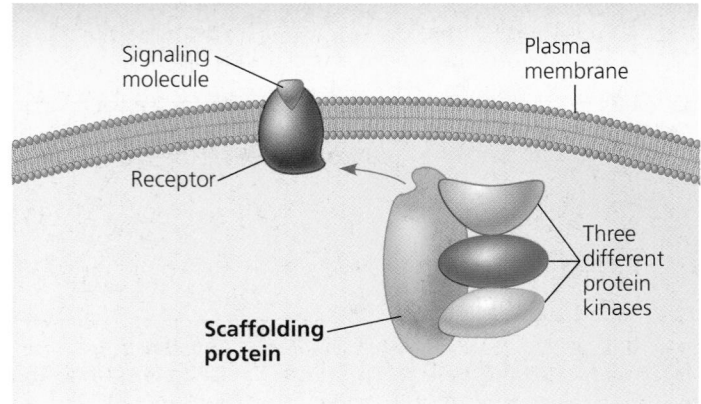

defective or missing. For instance, in an inherited disorder called Wiskott-Aldrich syndrome (WAS), the absence of a single relay protein leads to such diverse effects as abnormal bleeding, eczema, and a predisposition to infections and leukemia. These symptoms are thought to arise primarily from the absence of the protein in cells of the immune system. By studying normal cells, scientists found that the WAS protein is located just beneath the immune cell surface. The protein interacts both with microfilaments of the cytoskeleton and with several different components of signaling pathways that relay information from the cell surface, including pathways regulating immune cell proliferation. This multifunctional relay protein is thus both a branch point and an important intersection point in a complex signal transduction network that controls immune cell behavior. When the WAS protein is absent, the cytoskeleton is not properly organized and signaling pathways are disrupted, leading to the WAS symptoms.

Termination of the Signal

In the interest of keeping Figure 11.17 simple, we did not show the *inactivation* mechanisms that are an essential aspect of any cell-signaling pathway. For a cell of a multicellular organism to remain capable of responding to incoming signals, each molecular change in its signaling pathways must last only a short time. As we saw in the cholera example, if a signaling pathway component becomes locked into one state, whether active or inactive, consequences for the organism can be serious.

The ability of a cell to receive new signals depends on reversibility of the changes produced by prior signals. The binding of signaling molecules to receptors is reversible. As the external concentration of signaling molecules falls, fewer receptors are bound at any given moment, and the unbound receptors revert to their inactive form. The cellular response occurs only when the concentration of receptors with bound signaling molecules is above a certain threshold. When the number of active receptors falls below that threshold, the cellular response ceases. Then, by a variety of means, the relay molecules return to their inactive forms: The GTPase activity intrinsic to a G protein hydrolyzes its bound GTP; the enzyme phosphodiesterase converts cAMP to AMP; protein phosphatases inactivate phosphorylated kinases and other proteins; and so forth. As a result, the cell is soon ready to respond to a fresh signal.

> → **Mastering Biology** BioFlix® Animation: Mechanism of Hormone Action: Second Messenger cAMP

In this section, we explored the complexity of signaling initiation and termination in a single pathway, and we saw the potential for pathways to intersect with each other. In the next section, we'll consider one especially important network of interacting pathways in the cell.

CONCEPT 11.5

Apoptosis requires integration of multiple cell-signaling pathways

When signaling pathways were first discovered, they were thought to be linear, independent pathways. Our understanding of cellular communication has benefited from the realization that signaling pathway components interact with each other in various ways. For a cell to carry out the appropriate response, cellular proteins often must integrate multiple signals. Let's consider an important cellular process—cellular suicide—as an example.

Cells that are infected, are damaged, or have reached the end of their functional life span often undergo "programmed cell death" **(Figure 11.19)**. The best-understood type of this controlled cell suicide is **apoptosis** (from the Greek, meaning "falling off," and used in a classic Greek poem to refer to leaves falling from a tree). During this process, cellular agents chop up

▼ **Figure 11.19 Apoptosis of a human white blood cell.** On the left is a normal white blood cell, while on the right is a white blood cell undergoing apoptosis. The apoptotic cell is shrinking and forming lobes ("blebs"), which eventually are shed as membrane-bounded cell fragments (colorized SEMs).

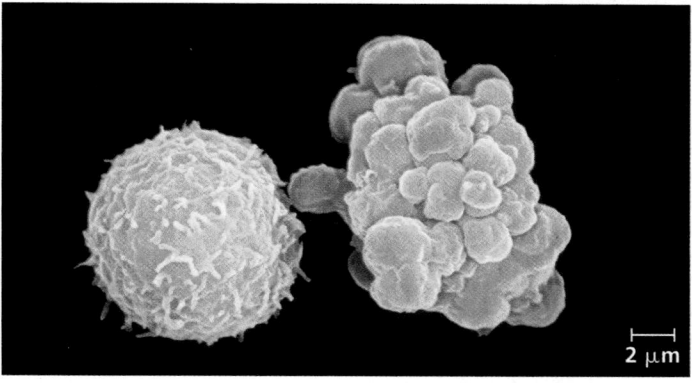

2 μm

> → **Mastering Biology** Video: Phosphate-Induced Apoptosis

the DNA and fragment the organelles and other cytoplasmic components. The cell shrinks and becomes lobed (a change called "blebbing"), and the cell's parts are packaged up in vesicles that are engulfed and digested by specialized scavenger cells, leaving no trace. Apoptosis protects neighboring cells from damage that they would otherwise suffer if a dying cell merely leaked out all its contents, including its many digestive enzymes.

The signal that triggers all of the complex events that occur during apoptosis can come from either outside or inside the cell. Outside the cell, signaling molecules released from other cells can initiate a signal transduction pathway that activates the genes and proteins responsible for carrying out cell death. Within a cell whose DNA has been irretrievably damaged, a series of protein-protein interactions can pass along a signal that similarly triggers cell death. Considering some examples of apoptosis can help us to see how signaling pathways are integrated in cells.

Apoptosis in the Soil Worm *Caenorhabditis elegans*

The molecular mechanisms of apoptosis were worked out by researchers studying embryonic development of a small soil worm, a nematode called *Caenorhabditis elegans*. Because the adult worm has only about 1,000 cells, the researchers were able to work out the entire ancestry of each cell. The timely suicide of cells occurs exactly 131 times during normal development of *C. elegans*, at precisely the same points in the cell lineage of each worm. In worms and other species, apoptosis is triggered by signals that activate a cascade of "suicide" proteins in the cells destined to die.

Genetic research on *C. elegans* initially revealed two key apoptosis genes, called *ced-3* and *ced-4* (*ced* stands for "cell death"), which encode proteins essential for apoptosis. The proteins are called Ced-3 and Ced-4, respectively. These and most other proteins involved in apoptosis are continually present in cells, but in inactive form; thus, regulation in this case occurs at the level of protein activity rather than through gene activity and protein synthesis. In *C. elegans*, a protein in the outer mitochondrial membrane, called Ced-9 (the product of the *ced-9* gene), serves as a master regulator of apoptosis, acting as a brake in the absence of a signal promoting apoptosis **(Figure 11.20a)**. When a death signal is received by the cell, signal transduction involves a change in Ced-9 that disables the brake, and the apoptotic pathway activates proteases and nucleases, enzymes that cut up the proteins and DNA of the cell. The main proteases of apoptosis are called *caspases*; in the nematode, the chief caspase is the Ced-3 protein.

Apoptotic Pathways and the Signals That Trigger Them

In humans and other mammals, several different pathways, involving about 15 different caspases, can carry out

▼ **Figure 11.20 Molecular basis of apoptosis in *C. elegans*.** Three proteins, Ced-3, Ced-4, and Ced-9, are critical to apoptosis and its regulation in the nematode. Apoptosis is more complicated in mammals but involves proteins similar to those in *C. elegans*.

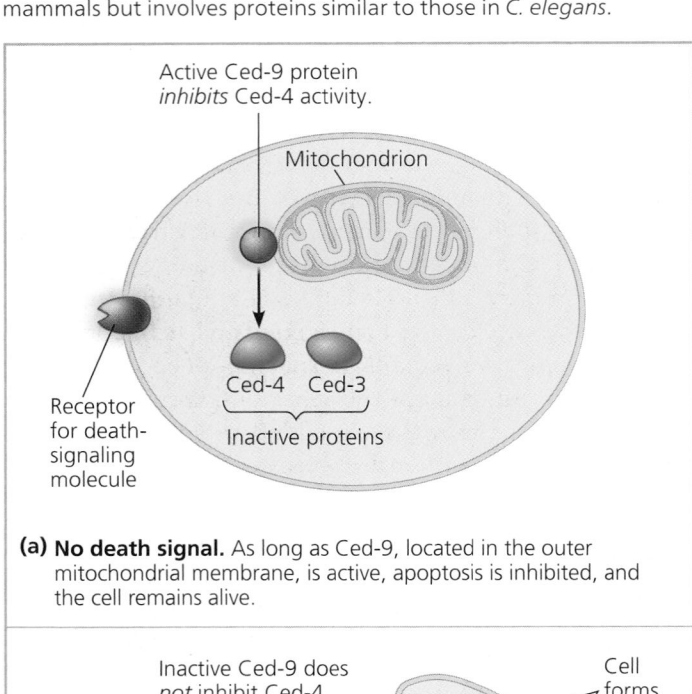

(a) No death signal. As long as Ced-9, located in the outer mitochondrial membrane, is active, apoptosis is inhibited, and the cell remains alive.

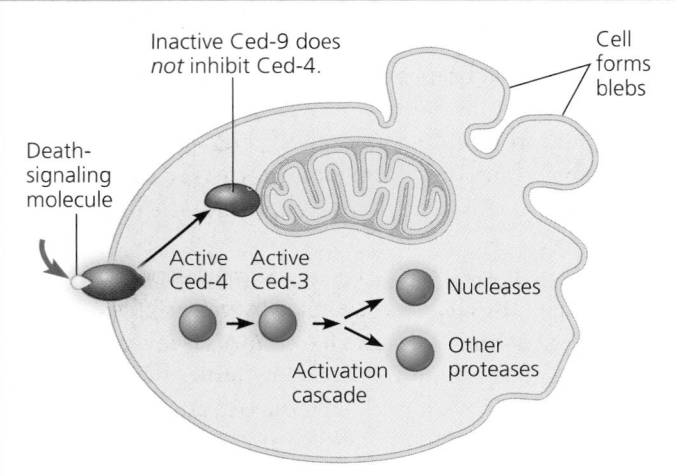

(b) Death signal. When a cell receives a death signal, Ced-9 is inactivated, relieving its inhibition of Ced-4. Active Ced-4 activates Ced-3, a protease, which triggers a cascade of reactions leading to the activation of nucleases and other proteases. The action of these enzymes causes the changes seen in apoptotic cells and eventually leads to cell death.

apoptosis. The pathway that is used depends on the type of cell and on the particular signal that initiates apoptosis. One major pathway involves certain mitochondrial proteins that are triggered to form molecular pores in the mitochondrial outer membrane, causing it to leak and release other proteins that promote apoptosis. Perhaps surprisingly, these latter include cytochrome *c*, which functions in mitochondrial electron transport in healthy cells (see Figure 9.14) but acts as a cell death factor when released from mitochondria. The process of mitochondrial apoptosis in mammals uses proteins similar to the nematode proteins Ced-3, Ced-4, and Ced-9. These can be thought of as relay proteins capable of transducing the apoptotic signal.

> **Mastering Biology** Animation: Apoptosis

At key gateways into the apoptotic program, relay proteins integrate signals from several different sources and can send a cell down an apoptotic pathway. Often, the signal originates outside the cell, like the death-signaling molecule depicted in **Figure 11.20b**, which presumably was released by a neighboring cell. When a death-signaling ligand occupies a cell-surface receptor, this binding leads to activation of caspases and other enzymes that carry out apoptosis, without involving the mitochondrial pathway. This process of signal reception, transduction, and response is similar to what we have discussed throughout this chapter. In a twist on the classic scenario, two other types of alarm signals that can lead to apoptosis originate from *inside* the cell rather than from a cell-surface receptor. One signal comes from the nucleus, generated when the DNA has suffered irreparable damage, and a second comes from the endoplasmic reticulum when excessive protein misfolding occurs. Mammalian cells make life-or-death "decisions" by somehow integrating the death signals and life signals they receive from these external and internal sources.

A built-in cell suicide mechanism is essential to development and maintenance in all animals. The similarities between apoptosis genes in nematodes and those in mammals, as well as the observation that apoptosis occurs in multicellular fungi and even in single-celled yeasts, indicate that the basic mechanism evolved early in the evolution of eukaryotes. In vertebrates, apoptosis is essential for normal development of the nervous system, for normal operation of the immune system, and for normal

morphogenesis of hands and feet in humans and paws in other mammals **(Figure 11.21)**. The level of apoptosis between the developing digits is lower in the webbed feet of ducks and other water birds than in the nonwebbed feet of land birds, such as chickens. In the case of humans, the failure of appropriate apoptosis can result in webbed fingers and toes.

Significant evidence points to the involvement of apoptosis in certain degenerative diseases of the nervous system, such as Parkinson's disease and Alzheimer's disease. In Alzheimer's disease, an accumulation of aggregated proteins in neuronal cells activates an enzyme that triggers apoptosis, resulting in the loss of brain function seen in these patients. Furthermore, cancer can result from a failure of cell suicide; some cases of human melanoma, for example, have been linked to faulty forms of the human version of the *C. elegans* Ced-4 protein. It is not surprising, therefore, that the signaling pathways feeding into apoptosis are quite elaborate. After all, the life-or-death question is the most fundamental one imaginable for a cell.

This chapter has introduced you to many of the general mechanisms of cell communication, such as ligand binding, protein-protein interactions and shape changes, cascades of interactions, and protein phosphorylation. Throughout your study of biology, you will encounter numerous examples of cell signaling.

CONCEPT CHECK 11.5

1. Give an example of apoptosis during embryonic development, and explain its function in the developing embryo.
2. **WHAT IF?** If apoptosis occurred when it should not, what types of protein defects might be the cause? What types could result in apoptosis not occurring when it should?

For suggested answers, see Appendix A.

▼ **Figure 11.21 Effect of apoptosis during paw development in the mouse.** In mice, humans, other mammals, and land birds, the embryonic region that develops into feet or hands initially has a solid, platelike structure. Apoptosis eliminates the cells in the interdigital regions, thus forming the digits. The embryonic mouse paws shown in these fluorescence light micrographs are stained so that cells undergoing apoptosis appear a bright yellowish green. Apoptosis of cells begins at the margin of each interdigital region (left), peaks as the tissue in these regions is reduced (middle), and is no longer visible when the interdigital tissue has been eliminated (right).

Interdigital tissue   Cells undergoing apoptosis   1 mm   Space between digits

11 Chapter Review

→ Go to **Mastering Biology** for Assignments, the eText, the Study Area, and Dynamic Study Modules.

SUMMARY OF KEY CONCEPTS

→ To review key terms, go to the **Vocabulary Self-Quiz** in the **Mastering Biology** eText or Study Area, or go to goo.gl/zkjz9t.

CONCEPT 11.1

External signals are converted to responses within the cell (pp. 213–217)

- **Signal transduction pathways** are crucial for many processes. Bacterial cells can sense the local density of bacterial cells (quorum sensing). Signaling during yeast cell mating has much in common with processes in multicellular organisms, suggesting an early evolutionary origin of signaling mechanisms.
- Local signaling by animal cells involves direct contact or the secretion of local regulators. For long-distance signaling, animal and plant cells use **hormones**; animals also pass signals electrically.
- Like epinephrine, other hormones that bind to membrane receptors trigger a three-stage cell-signaling pathway:

① Signal reception ② Signal transduction ③ Cellular response

Receptor

Signaling molecule

Relay molecules

Activation of cellular response

❓ *What determines whether a cell responds to a hormone such as epinephrine? What determines how a cell responds to such a hormone?*

CONCEPT 11.2

Signal reception: A signaling molecule binds to a receptor, causing it to change shape (pp. 217–221)

- The binding between signaling molecule (**ligand**) and receptor is highly specific. A specific shape change in a receptor is often the initial transduction of the signal.
- Three of the major types of cell-surface transmembrane receptors are the following: (1) **G protein-coupled receptors (GPCRs)** work with cytoplasmic **G proteins**. Ligand binding activates the receptor, which then activates a specific G protein, which activates yet another protein, thus propagating the signal. (2) **Receptor tyrosine kinases (RTKs)** react to the binding of signaling molecules by forming dimers and then adding phosphate groups to tyrosines on the cytoplasmic part of the other monomer making up the dimer. Relay proteins in the cell can then be activated by binding to different phosphorylated tyrosines, allowing this receptor to trigger several pathways at once. (3) **Ligand-gated ion channels** open or close in response to binding by specific signaling molecules, regulating the flow of specific ions across the membrane.
- The activity of all three types of receptors is crucial; abnormal GPCRs and RTKs are associated with many human diseases.
- Intracellular receptors are cytoplasmic or nuclear proteins. Signaling molecules that are hydrophobic or small enough to cross the plasma membrane bind to these receptors inside the cell.

❓ *How are the structures of a GPCR and an RTK similar? How does initiation of signal transduction differ for these two types of receptors?*

CONCEPT 11.3

Signal transduction: Cascades of molecular interactions transmit signals from receptors to relay molecules in the cell (pp. 221–225)

- At each step in a signal transduction pathway, the signal is transduced into a different form, which commonly involves a shape change in a protein. Many signal transduction pathways include **phosphorylation cascades**, in which a series of **protein kinases** each add a phosphate group to the next one in line, activating it. Enzymes called **protein phosphatases** remove the phosphate groups. The balance between phosphorylation and dephosphorylation regulates the activity of proteins involved in the sequential steps of a signal transduction pathway.
- **Second messengers**, such as the small molecule **cyclic AMP (cAMP)** and the ion Ca^{2+}, diffuse readily through the cytosol and thus help broadcast signals quickly. Many G proteins activate **adenylyl cyclase**, which makes cAMP from ATP. Cells use Ca^{2+} as a second messenger in GPCR and RTK pathways. The tyrosine kinase pathways can also involve two other second messengers, **diacylglycerol (DAG)** and **inositol trisphosphate** (IP_3). IP_3 can trigger a subsequent increase in Ca^{2+} level.

❓ *What is the difference between a protein kinase and a second messenger? Can both operate in the same signal transduction pathway?*

CONCEPT 11.4

Cellular response: Cell signaling leads to regulation of transcription or cytoplasmic activities (pp. 226–229)

- Some pathways lead to a nuclear response: Specific genes are turned on or off by activated transcription factors. In others, the response involves cytoplasmic regulation.
- Cellular responses are not simply on or off; they are regulated at many steps. Each protein in a signaling pathway amplifies the signal by activating multiple copies of the next component; for long pathways, the total amplification may be over a millionfold. The combination of proteins in a cell confers specificity in the signals it detects and the responses it carries out. **Scaffolding proteins** increase signaling efficiency. Pathway branching further helps the cell coordinate signals and responses. Signal response can be terminated quickly because ligand binding is reversible.

❓ *What mechanisms in the cell terminate its response to a signal and maintain its ability to respond to new signals?*

CONCEPT 11.5

Apoptosis requires integration of multiple cell-signaling pathways (pp. 229–231)

- **Apoptosis** is a type of programmed cell death in which cell components are disposed of in an orderly fashion. Studies of the soil worm *Caenorhabditis elegans* clarified molecular details of the relevant signaling pathways. A death signal leads to activation of caspases and nucleases, the main enzymes involved in apoptosis.
- Several apoptotic signaling pathways with related proteins exist in the cells of humans and other mammals, triggered in different ways. Signals eliciting apoptosis can originate from outside or inside the cell.

❓ *What is an explanation for the similarities between genes in yeasts, nematodes, and mammals that control apoptosis?*

TEST YOUR UNDERSTANDING

➡️ For more multiple-choice questions, go to the **Practice Test** in the **Mastering Biology** eText or Study Area, or go to goo.gl/GruWRg.

Levels 1-2: Remembering/Understanding

1. Binding of a signaling molecule to which type of receptor leads directly to a change in the distribution of substances on opposite sides of the membrane?
   (A) intracellular receptor
   (B) G protein-coupled receptor
   (C) phosphorylated receptor tyrosine kinase dimer
   (D) ligand-gated ion channel

2. The activation of receptor tyrosine kinases is characterized by
   (A) dimerization and phosphorylation.
   (B) dimerization and IP_3 binding.
   (C) a phosphorylation cascade.
   (D) GTP hydrolysis.

3. Lipid-soluble signaling molecules, such as aldosterone, cross the membranes of all cells but affect only target cells because
   (A) only target cells retain the appropriate DNA segments.
   (B) intracellular receptors are present only in target cells.
   (C) only target cells have enzymes that break down aldosterone.
   (D) only in target cells is aldosterone able to initiate the phosphorylation cascade that turns genes on.

4. Consider this pathway: epinephrine → G protein-coupled receptor → G protein → adenylyl cyclase → cAMP. Identify the second messenger.
   (A) cAMP   (C) GTP
   (B) G protein   (D) adenylyl cyclase

5. Which of the following occurs during apoptosis?
   (A) lysis of the cell
   (B) direct contact between signaling cells
   (C) fragmentation of the DNA
   (D) release of proteases outside the cell

Levels 3-4: Applying/Analyzing

6. Which observation suggested to Sutherland the involvement of a second messenger in epinephrine's effect on liver cells?
   (A) Enzymatic activity was proportional to the amount of calcium added to a cell-free extract.
   (B) Receptor studies indicated that epinephrine was a ligand.
   (C) Glycogen breakdown was observed only when epinephrine was administered to intact cells.
   (D) Glycogen breakdown was observed only when epinephrine and glycogen phosphorylase were mixed.

7. Protein phosphorylation is commonly involved with which of the following?
   (A) ligand binding by receptor tyrosine kinases.
   (B) activation of G protein-coupled receptors.
   (C) activation of protein kinase molecules.
   (D) release of Ca^{2+} from the ER lumen.

Levels 5-6: Evaluating/Creating

8. **DRAW IT** Draw the following apoptotic pathway, which operates in human immune cells. A death signal is received when a molecule called Fas binds its cell-surface receptor. The binding of many Fas molecules to receptors causes receptor clustering. The intracellular regions of the receptors, when together, bind proteins called adaptor proteins. These in turn bind to inactive molecules of caspase-8, which become activated and then activate caspase-3. Once activated, caspase-3 initiates apoptosis.

9. **EVOLUTION CONNECTION** Identify the evolutionary mechanisms that might account for the origin and persistence of cell-to-cell signaling systems in prokaryotes.

10. **SCIENTIFIC INQUIRY** Epinephrine initiates a signal transduction pathway that produces cyclic AMP (cAMP) and leads to the breakdown of glycogen to glucose, a major energy source for cells. But glycogen breakdown is only part of the fight-or-flight response that epinephrine brings about; the overall effect on the body includes an increase in heart rate and alertness, as well as a burst of energy. Given that caffeine blocks the activity of cAMP phosphodiesterase, propose a mechanism by which caffeine ingestion leads to heightened alertness and sleeplessness.

11. **SCIENCE, TECHNOLOGY, AND SOCIETY** The aging process is thought to be initiated at the cellular level. Among the changes that can occur after a certain number of cell divisions is the loss of a cell's ability to respond to growth factors and other signals. Much research into aging is aimed at understanding such losses, with the ultimate goal of extending the human life span. Not everyone, however, agrees that this is a desirable goal. If life expectancy were greatly increased, discuss what might be the social and ecological consequences.

12. **WRITE ABOUT A THEME: ORGANIZATION** The properties of life emerge at the biological level of the cell. The highly regulated process of apoptosis is not simply the destruction of a cell; it is also an emergent property. Write a short essay (about 100–150 words) that briefly explains the role of apoptosis in the development and proper functioning of an animal, and describe how this form of programmed cell death is a process that emerges from the orderly integration of signaling pathways.

13. **SYNTHESIZE YOUR KNOWLEDGE**

There are five basic tastes—sour, salty, sweet, bitter, and "umami." Salt is detected when the concentration of salt outside of a taste bud cell is higher than that inside of it, and ion channels allow the passive leakage of Na^+ into the cell. The resulting change in membrane potential (see Concept 7.4) sends the "salty" signal to the brain. Umami is a savory taste generated by glutamate (glutamic acid, found in monosodium glutamate, or MSG), which is used as a flavor enhancer in foods such as taco-flavored tortilla chips. The glutamate receptor is a GPCR, which, when bound, initiates a signaling pathway that ends with a cellular response, perceived by you as "taste." If you eat a regular potato chip and then rinse your mouth, you will no longer taste salt. But if you eat a flavored tortilla chip and then rinse, the taste persists. (Try it!) Propose a possible explanation for this difference.

For selected answers, see Appendix A.

12 The Cell Cycle

KEY CONCEPTS

12.1 Most cell division results in genetically identical daughter cells *p. 235*

12.2 The mitotic phase alternates with interphase in the cell cycle *p. 237*

12.3 The eukaryotic cell cycle is regulated by a molecular control system *p. 244*

Study Tip

Make a visual study guide: Figure 12.1 presents the events of the cell cycle as a simplified linear diagram. As you learn more about the cell cycle, draw a detailed linear diagram of the stages of interphase, mitosis, and cytokinesis. Include explanatory labels. Add the circular chart from Figure 12.6 and show how the two diagrams are related.

The Cell Cycle

Interphase
G₁ phase: Metabolic activity and growth

→ Go to Mastering Biology

For Students (in eText and Study Area)
- Get Ready for Chapter 12
- BioFlix® Animation: Mitosis
- Animation: Microtubule Depolymerization

For Instructors to Assign (in Item Library)
- BioFlix® Tutorial: Mitosis (2 of 3): Mechanisms of Mitosis
- Coaching Activity: Evaluating Science in the Media: Tanning and Skin Cancer

Ready-to-Go Teaching Module (in Instructor Resources)
- Mitosis (Concept 12.2)

Two cells   Four cells

Figure 12.1 A multicellular organism starts out as a single cell that divides into two. Those two cells then divide into four, as shown in these fluorescence micrographs of a marine worm embryo. Cell division continues throughout an organism's life, for growth or to replace worn-out or damaged cells. Each time a cell divides in this way, it is crucial that the daughter cells be genetically identical to the parent cell.

How does one parent cell give rise to two genetically identical daughter cells?

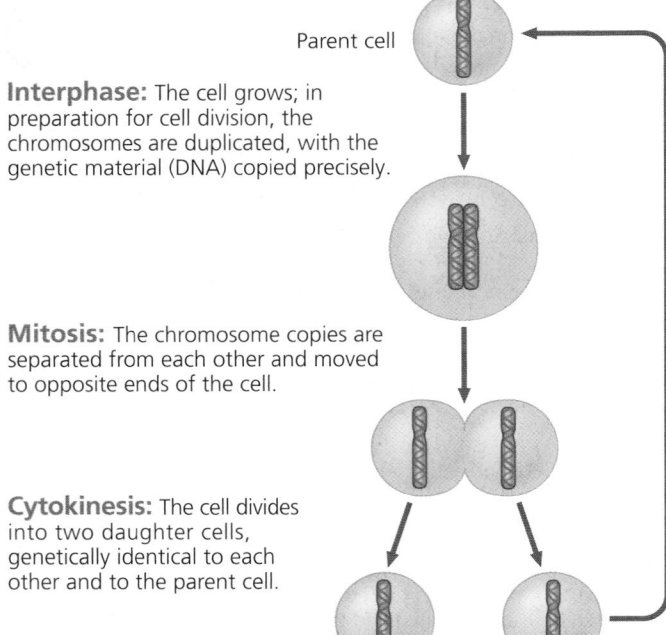

Parent cell

Interphase: The cell grows; in preparation for cell division, the chromosomes are duplicated, with the genetic material (DNA) copied precisely.

Mitosis: The chromosome copies are separated from each other and moved to opposite ends of the cell.

Cytokinesis: The cell divides into two daughter cells, genetically identical to each other and to the parent cell.

The daughter cells may go on to divide, repeating the cycle.

Daughter cells

Most cell division results in genetically identical daughter cells

The ability of organisms to produce more of their own kind is the one characteristic that best distinguishes living things from nonliving matter. This unique capacity to procreate, like all biological functions, has a cellular basis. The continuity of life is based on the reproduction of cells, or **cell division**.

Key Roles of Cell Division

Cell division plays several important roles in life. When a prokaryotic cell divides, it is actually reproducing because the process gives rise to a new organism (another cell). The same is true of any unicellular eukaryote, such as the amoeba shown in **Figure 12.2a**. As for multicellular eukaryotes, cell division enables each of these organisms to develop from a single cell—the fertilized egg. A two-celled embryo, the first stage in this process, is shown in **Figure 12.2b**. And cell division continues to function in renewal and repair in fully grown multicellular eukaryotes, replacing cells that die from accidents or normal wear and tear. For example, dividing cells in your bone marrow continuously make new blood cells **(Figure 12.2c)**.

The reproduction of a cell, with all of its complexity, cannot occur by a mere pinching in half; a cell is not like a soap bubble that simply enlarges and splits in two. In both prokaryotes and eukaryotes, a crucial function of most cell divisions is the distribution of identical genetic

▼ **Figure 12.2 The functions of cell division.**

◀ **(a) Asexual reproduction.** An amoeba, a single-celled eukaryote, is dividing into two cells. Each new cell will be an individual organism (LM).

100 μm

50 μm

▶ **(b) Growth and development.** This micrograph shows a sand dollar embryo shortly after the fertilized egg divided, forming two cells (LM).

◀ **(c) Tissue renewal.** These dividing bone marrow cells will give rise to new blood cells (LM).

20 μm

→ **Mastering Biology** Video: Cell Division in a Sea Urchin Embryo

material—DNA—to two daughter cells. (The exception is meiosis, the special type of eukaryotic cell division that can produce sperm and eggs.) What is most remarkable about cell division is the accuracy with which the DNA is passed from one generation of cells to the next. A dividing cell replicates its DNA, distributes the two copies to opposite ends of the cell, and then splits into daughter cells.

Cellular Organization of the Genetic Material

A cell's DNA, its genetic information, is called its **genome**. Although a prokaryotic genome is often a single DNA molecule, eukaryotic genomes usually consist of a number of DNA molecules. The overall length of DNA in a eukaryotic cell is enormous. A typical human cell, for example, has about 2 m of DNA—a length about 250,000 times greater than the cell's diameter. Before the cell can divide to form genetically identical daughter cells, all of this DNA must be copied, or replicated, and then the two copies must be separated so that each daughter cell ends up with a complete genome.

The replication and distribution of so much DNA are manageable because the DNA molecules are packaged into structures called **chromosomes** (from the Greek *chroma*, color, and *soma*, body), so named because they take up certain dyes used in microscopy **(Figure 12.3)**. Each eukaryotic chromosome consists of one very long, linear DNA molecule associated with many proteins (see Figure 6.9). The DNA molecule carries several hundred to a few thousand genes, the units of information that specify an organism's inherited traits. The associated proteins maintain the structure of the chromosome and help control the activity of the genes. Together, the entire complex of DNA and proteins that is the building material of chromosomes is referred to as **chromatin**. As you will soon see, the chromatin of a chromosome varies in its degree of condensation during the process of cell division.

▼ **Figure 12.3 Eukaryotic chromosomes.** Chromosomes (stained purple) are visible within the nucleus of this cell from an African blood lily. The thinner red threads in the surrounding cytoplasm are the cytoskeleton. The cell is preparing to divide (LM).

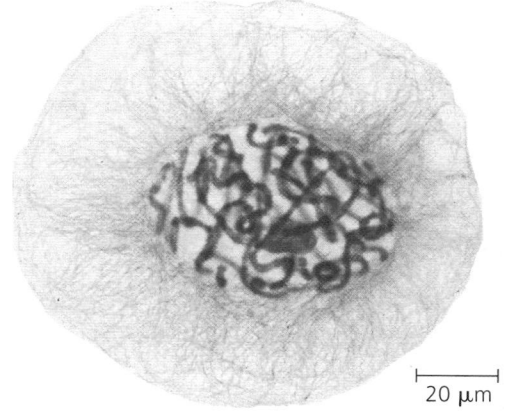

20 μm

Every eukaryotic species has a characteristic number of chromosomes in each cell's nucleus. For example, the nuclei of human **somatic cells** (all body cells except the reproductive cells) each contain 46 chromosomes, made up of two sets of 23, one set inherited from each parent. Reproductive cells, or **gametes**—such as sperm and eggs—have half as many chromosomes as somatic cells; in our example, human gametes have one set of 23 chromosomes. The number of chromosomes in somatic cells varies widely among species: 18 in cabbage plants, 48 in chimpanzees, 56 in elephants, 90 in hedgehogs, and 148 in one species of alga. We'll now consider how these chromosomes behave during cell division.

Distribution of Chromosomes During Eukaryotic Cell Division

When a cell is not dividing, and even as it replicates its DNA in preparation for cell division, each chromosome is in the form of a long, thin chromatin fiber. After DNA replication, however, the chromosomes condense as a part of cell division: Each chromatin fiber becomes densely coiled and folded, making the chromosomes much shorter and so thick that we can see them with a light microscope.

Each duplicated chromosome consists of two **sister chromatids**, which are joined copies of the original chromosome **(Figure 12.4)**. The two chromatids, each containing an identical DNA molecule, are often attached all along their lengths by protein complexes called *cohesins*; this attachment is known as *sister chromatid cohesion*. Each sister chromatid has a **centromere**, a region made up of repetitive sequences in the chromosomal DNA where the chromatid is attached most closely to its sister chromatid. This attachment is mediated by proteins that recognize and bind to the centromeric DNA; other bound proteins condense the DNA, giving the duplicated chromosome a narrow "waist." The portion of a chromatid to either side of the centromere is referred to as an *arm* of the chromatid. (An unduplicated chromosome has a single centromere, distinguished by the proteins that bind there, and two arms.)

Later in the cell division process, the two sister chromatids of each duplicated chromosome separate and move into two new nuclei, one forming at each end of

▼ **Figure 12.4 A highly condensed, duplicated human chromosome. (SEM)**

Sister chromatids

Centromeres, one on each sister chromatid

0.5 μm

DRAW IT *Circle one sister chromatid of this chromosome.*

the cell. Once the sister chromatids separate, they are no longer called sister chromatids but are considered individual chromosomes; this is the step that essentially doubles the number of chromosomes during cell division. Thus, each new nucleus receives a collection of chromosomes identical to that of the parent cell **(Figure 12.5)**. **Mitosis**, the division of the genetic material in the nucleus, is usually followed immediately by **cytokinesis**, the division of the cytoplasm. One cell has become two, each the genetic equivalent of the parent cell.

▼ **Figure 12.5 Chromosome duplication and distribution during cell division.**

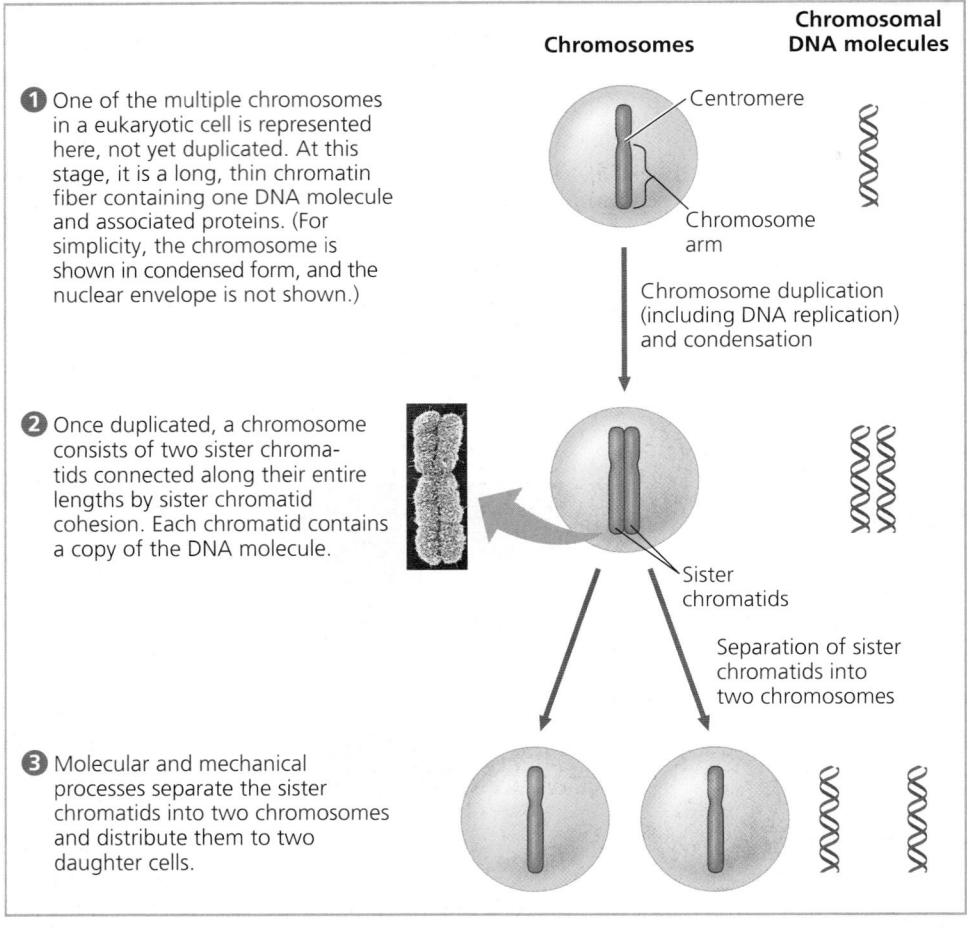

Chromosomes

Chromosomal DNA molecules

❶ One of the multiple chromosomes in a eukaryotic cell is represented here, not yet duplicated. At this stage, it is a long, thin chromatin fiber containing one DNA molecule and associated proteins. (For simplicity, the chromosome is shown in condensed form, and the nuclear envelope is not shown.)

Centromere

Chromosome arm

Chromosome duplication (including DNA replication) and condensation

❷ Once duplicated, a chromosome consists of two sister chromatids connected along their entire lengths by sister chromatid cohesion. Each chromatid contains a copy of the DNA molecule.

Sister chromatids

Separation of sister chromatids into two chromosomes

❸ Molecular and mechanical processes separate the sister chromatids into two chromosomes and distribute them to two daughter cells.

? *How many chromatid arms does the chromosome in* ❷ *have? Identify the point in the figure where one chromosome becomes two.*

→ **Mastering Biology** BioFlix® **Animation: Chromosome Duplication**

From a fertilized egg, mitosis and cytokinesis produced the 37 trillion somatic cells that now make up your body, and the same processes continue to generate new cells to replace dead and damaged ones. In contrast, you produce gametes—eggs or sperm—by a variation of cell division called *meiosis*, which yields daughter cells with only one set of chromosomes, half as many chromosomes as the parent cell. Meiosis in humans occurs only in special cells in the ovaries or testes (the gonads). Generating gametes, meiosis reduces the chromosome number from 46 (two sets) to 23 (one set). Fertilization fuses two gametes together and returns the chromosome number to 46 (two sets). Mitosis then conserves that number in every somatic cell nucleus of the new human individual. In Chapter 13, we will examine the role of meiosis in reproduction and inheritance in more detail. In the remainder of this chapter, we focus on mitosis and the rest of the cell cycle in eukaryotes.

CONCEPT CHECK 12.1

1. How many chromosomes are drawn in each part of Figure 12.5? (Ignore the micrograph in step 2.)

2. **WHAT IF?** A chicken has 78 chromosomes in its somatic cells. How many chromosomes did the chicken inherit from each parent? How many chromosomes are in each of the chicken's gametes? How many chromosomes will be in each somatic cell of the chicken's offspring?

For suggested answers, see Appendix A.

CONCEPT 12.2

The mitotic phase alternates with interphase in the cell cycle

In 1882, a German anatomist named Walther Flemming developed dyes that allowed him to observe, for the first time, the behavior of chromosomes during mitosis and cytokinesis. (In fact, Flemming coined the terms *mitosis* and *chromatin*.) During the period between one cell division and the next, it appeared to Flemming that the cell was simply growing larger. But we now know that many critical events occur during this stage in the life of a cell.

Phases of the Cell Cycle

Mitosis is just one part of the **cell cycle**, the life of a cell from the time it is first formed during division of a parent cell until its own division into two daughter cells (**Figure 12.6**). (Biologists use the words *daughter* or *sister* in relation to cells, but this is not meant to imply gender.) In fact, the **mitotic (M) phase**, which includes both mitosis and cytokinesis, is usually the shortest part of the cell cycle. The mitotic phase alternates with a much longer stage called **interphase**, which often accounts for about 90% of the cycle. Interphase can be divided into three phases: the **G₁ phase** ("first gap"), the **S phase** ("synthesis"), and the **G₂ phase** ("second gap").

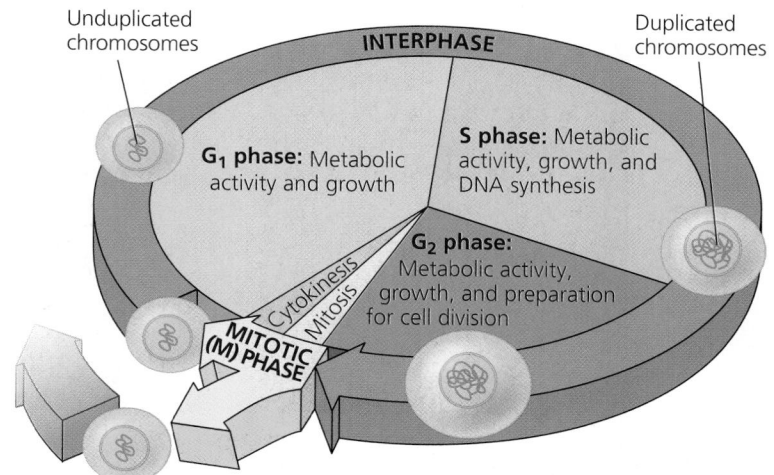

▼ **Figure 12.6 The cell cycle.** In a dividing cell, the mitotic (M) phase alternates with interphase, a growth period.

Unduplicated chromosomes
Duplicated chromosomes
INTERPHASE

G₁ phase: Metabolic activity and growth

S phase: Metabolic activity, growth, and DNA synthesis

G₂ phase: Metabolic activity, growth, and preparation for cell division

Cytokinesis
Mitosis
MITOTIC (M) PHASE

MITOTIC (M) PHASE:
Mitosis: Distribution of chromosomes into two daughter nuclei
Cytokinesis: Division of cytoplasm, producing two daughter cells. Each daughter cell can start a new cell cycle.

→ **Mastering Biology** Animation: The Cell Cycle

The G phases were misnamed as "gaps" when they were first observed because the cells appeared inactive, but we now know that intense metabolic activity and growth occur throughout interphase. During all three phases of interphase, actually, a cell grows by producing proteins and cytoplasmic organelles such as mitochondria and endoplasmic reticulum. Duplication of the chromosomes, crucial for eventual division of the cell, occurs entirely during the S phase. (We'll explore synthesis, or replication, of DNA in Concept 16.2.) Thus, a cell grows (G₁), continues to grow as it copies its chromosomes (S), grows more as it completes preparations for cell division (G₂), and divides (M). The daughter cells may then repeat the cycle.

A particular human cell might undergo one division in 24 hours. Of this time, the M phase would occupy less than 1 hour, while the S phase might occupy 10–12 hours, or about half the cycle. The rest of the time would be apportioned between the G₁ and G₂ phases. The G₂ phase usually takes 4–6 hours; in our example, G₁ would occupy about 5–6 hours. G₁ is the most variable in length in different types of cells. Some cells in a multicellular organism divide very infrequently or not at all. These cells spend their time in G₁ (or a related phase called G₀, to be discussed later in the chapter) doing their job in the organism—a cell of the pancreas secretes digestive enzymes, for example.

Mitosis is conventionally broken down into five stages: **prophase**, **prometaphase**, **metaphase**, **anaphase**, and **telophase**. Overlapping with the latter stages of mitosis, cytokinesis completes the mitotic phase. **Figure 12.7** describes these stages in an animal cell. Study this figure thoroughly before proceeding to the next two sections, which examine mitosis and cytokinesis more closely.

G₂ of Interphase

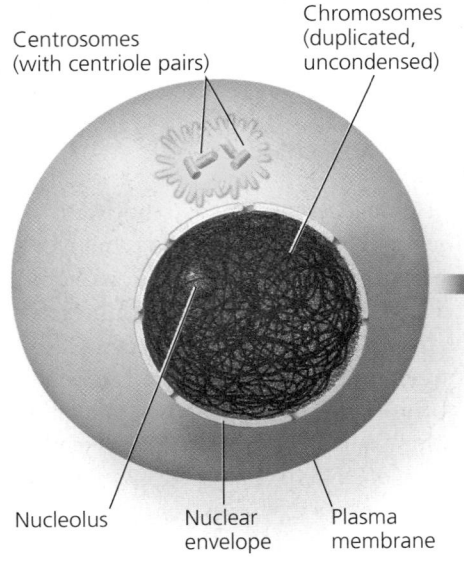

Centrosomes (with centriole pairs)

Chromosomes (duplicated, uncondensed)

Nucleolus

Nuclear envelope

Plasma membrane

Prophase

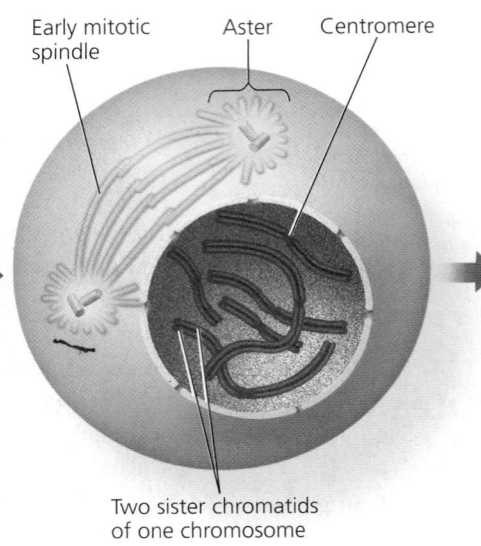

Early mitotic spindle

Aster

Centromere

Two sister chromatids of one chromosome

Prometaphase

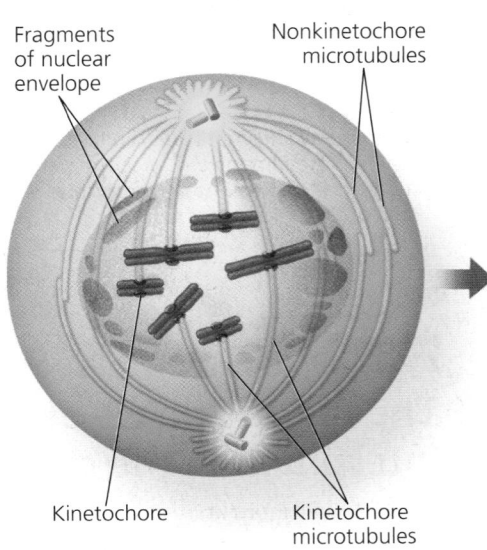

Fragments of nuclear envelope

Nonkinetochore microtubules

Kinetochore

Kinetochore microtubules

G₂ of Interphase

- A nuclear envelope encloses the nucleus.
- The nucleus contains one or more nucleoli (singular, *nucleolus*).
- Two centrosomes have formed by duplication of a single centrosome. Centrosomes are regions in animal cells that organize the microtubules of the spindle. Each centrosome contains two centrioles.
- Chromosomes, duplicated during S phase, cannot be seen individually because they have not yet condensed.

The fluorescence micrographs show dividing lung cells from a newt; this species has 22 chromosomes. Chromosomes appear blue, microtubules green, and intermediate filaments red. For simplicity, the drawings show only 6 chromosomes.

Prophase

- The chromatin fibers become more tightly coiled, condensing into discrete chromosomes observable with a light microscope.
- The nucleoli disappear.
- Each duplicated chromosome appears as two identical sister chromatids joined at their centromeres and, often, all along their arms by cohesins, resulting in sister chromatid cohesion.
- The mitotic spindle (named for its shape) begins to form. It is composed of the centrosomes and the microtubules that extend from them. The radial arrays of shorter microtubules that extend from the centrosomes are called asters ("stars").
- The centrosomes move away from each other, propelled partly by the lengthening microtubules between them.

Prometaphase

- The nuclear envelope fragments.
- The microtubules extending from each centrosome can now invade the nuclear area.
- The chromosomes have become even more condensed.
- A kinetochore, a specialized protein structure, has now formed at the centromere of each chromatid (thus, two per chromosome).
- Some of the microtubules attach to the kinetochores, becoming "kinetochore microtubules," which jerk the chromosomes back and forth.
- Nonkinetochore microtubules interact with those from the opposite pole of the spindle, lengthening the cell.

? *How many molecules of DNA are in the prometaphase drawing? How many molecules per chromosome? How many double helices are there per chromosome? Per chromatid?*

10 μm

| Metaphase | Anaphase | Telophase and Cytokinesis |

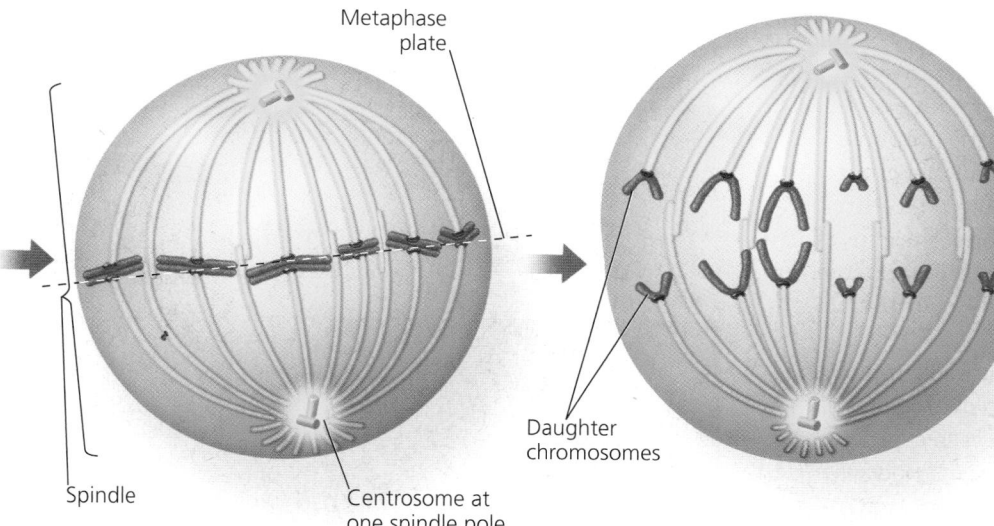

Metaphase plate

Spindle

Centrosome at one spindle pole

Daughter chromosomes

Cleavage furrow

Nucleolus forming

Nuclear envelope forming

Metaphase

- The centrosomes are now at opposite poles of the cell.

- The chromosomes have all arrived at the *metaphase plate*, a plane that is equidistant between the spindle's two poles. The chromosomes' centromeres lie at the metaphase plate.

- For each chromosome, the kinetochores of the sister chromatids are attached to kinetochore microtubules coming from opposite poles.

Anaphase

- Anaphase is the shortest stage of mitosis, often lasting only a few minutes.

- Anaphase begins when the cohesin proteins are cleaved. This allows the two sister chromatids of each pair to part suddenly. Each chromatid thus becomes an independent chromosome.

- The two new daughter chromosomes begin moving toward opposite ends of the cell as their kinetochore microtubules shorten. Because these microtubules are attached at the centromere region, the centromeres are pulled ahead of the arms, moving at a rate of about 1 μm/min.

- The cell elongates as the nonkinetochore microtubules lengthen.

- By the end of anaphase, the two ends of the cell have identical—and complete— collections of chromosomes.

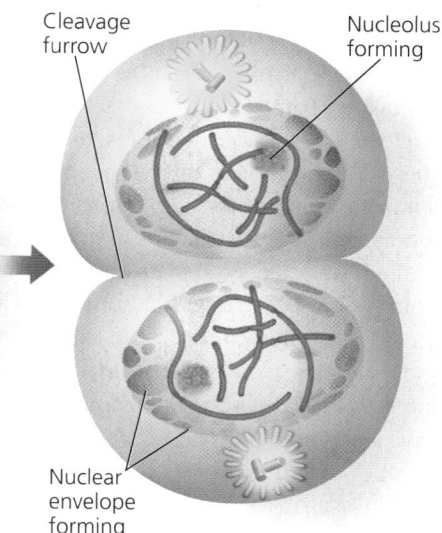

Telophase

- Two daughter nuclei form in the cell. Nuclear envelopes arise from the fragments of the parent cell's nuclear envelope and other portions of the endomembrane system.

- Nucleoli reappear.

- The chromosomes become less condensed.

- Any remaining spindle microtubules are depolymerized.

- Mitosis, the division of one nucleus into two genetically identical nuclei, is now complete.

Cytokinesis

- The division of the cytoplasm is usually well under way by late telophase, so the two daughter cells appear shortly after the end of mitosis.

- In animal cells, cytokinesis involves the formation of a cleavage furrow, which pinches the cell in two.

⊖ **Mastering Biology** BioFlix® Animation: Mitosis
Video: Animal Mitosis (time-lapse)

The Mitotic Spindle: *A Closer Look*

Many of the events of mitosis depend on the **mitotic spindle**, which begins to form in the cytoplasm during prophase. This structure consists of fibers made of microtubules and associated proteins. While the mitotic spindle assembles, the other microtubules of the cytoskeleton partially disassemble, providing the material used to construct the spindle. The spindle microtubules elongate (polymerize) by incorporating more subunits of the protein tubulin (see Table 6.1) and shorten (depolymerize) by losing subunits.

In animal cells, the assembly of spindle microtubules starts at the **centrosome**, a subcellular region containing material that functions throughout the cell cycle to organize the cell's microtubules. (It is also a type of *microtubule-organizing center*.) A pair of centrioles is located at the center of the centrosome, but they are not essential for cell division: If the centrioles are destroyed with a laser microbeam, a spindle nevertheless forms during mitosis. In fact, centrioles are not even present in plant cells, which do form mitotic spindles.

During interphase in animal cells, the single centrosome duplicates, forming two centrosomes, which remain near the nucleus. The two centrosomes move apart during prophase and prometaphase of mitosis as spindle microtubules grow out from them. By the end of prometaphase, the two centrosomes, one at each pole of the spindle, are at opposite ends of the cell. An *aster*, a radial array of short microtubules, extends from each centrosome. The spindle includes the centrosomes, the spindle microtubules, and the asters.

Each of the two sister chromatids of a duplicated chromosome has a **kinetochore**, a structure made up of proteins that have assembled on specific sections of DNA at each centromere. The chromosome's two kinetochores face in opposite directions. During prometaphase, some of the spindle microtubules attach to the kinetochores; these are called *kinetochore microtubules*. (The number of microtubules attached to a kinetochore varies among species, from one microtubule in yeast cells to 40 or so in some mammalian cells.) When one of a chromosome's kinetochores is "captured" by microtubules, the chromosome begins to move toward the pole from which those microtubules extend. However, this movement comes to a halt as soon as microtubules from the opposite pole attach to the kinetochore on the other chromatid. What happens next is like a tug-of-war that ends in a draw. The chromosome moves first in one direction and then in the other, back and forth, finally settling midway between the two ends of the cell. At metaphase, the centromeres of all the duplicated chromosomes are on a plane midway between the spindle's two poles. This plane is called the **metaphase plate**, which is an imaginary plate rather than an actual cellular structure **(Figure 12.8)**. Meanwhile, microtubules that do not attach to kinetochores have been elongating, and by metaphase they overlap and interact with other *nonkinetochore microtubules* from the opposite pole of the spindle. By metaphase, the microtubules of

▼ **Figure 12.8 The mitotic spindle at metaphase.** The kinetochores of each chromosome's two sister chromatids face in opposite directions. Here, each kinetochore is attached to a cluster of kinetochore microtubules (see TEM) extending from the nearest centrosome. Nonkinetochore microtubules overlap at the metaphase plate. The fluorescent micrograph is a rat kangaroo cell at metaphase.

Sister chromatids

Aster

Centrosome

Metaphase plate (imaginary)

Kinetochores

Overlapping nonkinetochore microtubules

Kinetochore microtubules

0.5 μm

Microtubules

Chromosomes

Kinetochores

5 μm

DRAW IT *On the lower micrograph, draw a line indicating the position of the metaphase plate. Draw arrows showing the directions of chromosome movement when anaphase begins.*

→ **Mastering Biology** Video: Spindle Formation During Mitosis
Animation: Mitosis

the asters have also grown and are in contact with the plasma membrane. The spindle is now complete.

The structure of the spindle correlates well with its function during anaphase. Anaphase begins suddenly when the cohesins holding together the sister chromatids of each chromosome are cleaved by an enzyme called *separase*. Once separated, the chromatids become individual chromosomes that move toward opposite ends of the cell.

How do the kinetochore microtubules function in this poleward movement of chromosomes? Apparently, two mechanisms are in play, both involving motor proteins. (To review how motor proteins move an object along a microtubule, see Figure 6.21.) Results of a cleverly designed experiment suggested that motor proteins on the kinetochores "walk" the chromosomes along the microtubules, which depolymerize at their kinetochore ends after the motor proteins have passed **(Figure 12.9)**. (This is referred to as the "Pac-man" mechanism because of its resemblance to the arcade game character that moves by eating all the dots in its path.) However, other researchers, working with different cell types or cells from other species, have shown that chromosomes are "reeled in" by motor proteins at the spindle poles and that the microtubules depolymerize after they pass by these motor proteins at the poles. The general consensus now is that both mechanisms are used and that their relative contributions vary among cell types.

In a dividing animal cell, the nonkinetochore microtubules are responsible for elongating the whole cell during anaphase. Nonkinetochore microtubules from opposite poles overlap each other extensively during metaphase (see Figure 12.8). During anaphase, the region of overlap is reduced as motor proteins attached to the microtubules walk them away from one another, using energy from ATP. As the microtubules push apart from each other, their spindle poles are pushed apart, elongating the cell. At the same time, the microtubules lengthen somewhat by the addition of tubulin subunits to their overlapping ends. As a result, the microtubules continue to overlap.

At the end of anaphase, duplicate groups of chromosomes have arrived at opposite ends of the elongated parent cell. Nuclei re-form during telophase. Cytokinesis generally begins during anaphase or telophase, and the spindle eventually disassembles by depolymerization of microtubules.

Cytokinesis: *A Closer Look*

In animal cells, cytokinesis occurs by a process known as **cleavage**. The first sign of cleavage is the appearance of a **cleavage furrow**, a shallow groove in the cell surface near the old metaphase plate **(Figure 12.10a)**. On the cytoplasmic side of the furrow is a contractile ring of actin microfilaments associated with molecules of the protein myosin. The actin microfilaments interact with the myosin molecules, causing the ring to contract. The contraction of the dividing cell's ring of microfilaments is like the pulling of a drawstring. The cleavage furrow deepens until the parent cell is pinched in two, producing two completely separated cells, each with its own nucleus and its own share of cytosol, organelles, and other subcellular structures.

Cytokinesis in plant cells, which have cell walls, is markedly different. There is no cleavage furrow. Instead, during telophase, vesicles derived from the Golgi apparatus move along microtubules to the middle of the cell, where they coalesce, producing a **cell plate**. Cell wall materials carried in the

---

▼ **Figure 12.9** Inquiry

At which end do kinetochore microtubules shorten during anaphase?

Experiment Gary Borisy and colleagues at the University of Wisconsin wanted to determine whether kinetochore microtubules depolymerize at the kinetochore end or the pole end as chromosomes move toward the poles during mitosis. First they labeled the microtubules of a pig kidney cell in early anaphase with a yellow fluorescent dye. (Nonkinetochore microtubules are not shown.)

Then they marked a region of the kinetochore microtubules between one spindle pole and the chromosomes by using a laser to eliminate the fluorescence from that region, while leaving the microtubules intact (see below). As anaphase proceeded, they monitored the changes in microtubule length on either side of the non-fluorescent mark.

Results As the chromosomes moved poleward, the microtubule segments on the kinetochore side of the mark shortened, while those on the spindle pole side stayed the same length.

Conclusion During anaphase in this cell type, chromosome movement is correlated with kinetochore microtubules shortening at their kinetochore ends and not at their spindle pole ends. This experiment supports the hypothesis that during anaphase, a chromosome is walked along a microtubule as the microtubule depolymerizes at its kinetochore end, releasing tubulin subunits.

Data from G. J. Gorbsky, P. J. Sammak, and G. G. Borisy, Chromosomes move poleward in anaphase along stationary microtubules that coordinately disassemble from their kinetochore ends, *Journal of Cell Biology* 104:9–18 (1987).

WHAT IF? *If this experiment had been done on a cell type in which "reeling in" at the poles was the main cause of chromosome movement, how would the mark have moved relative to the poles? How would the microtubule portions on either side of the mark have changed?*

→ **Mastering Biology** Animation: Microtubule Depolymerization

▼ **Figure 12.10 Cytokinesis in animal and plant cells.**

(a) Cleavage of an animal cell (SEM)

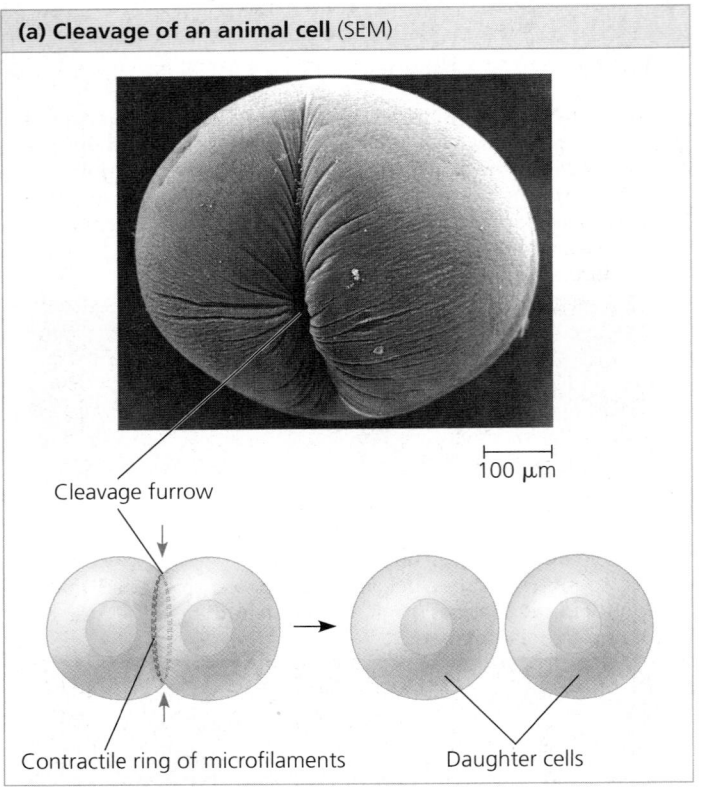

100 μm

Cleavage furrow

Contractile ring of microfilaments   Daughter cells

(b) Cell plate formation in a plant cell (TEM)

1 μm

Vesicles forming cell plate   Wall of parent cell   Cell plate   New cell wall

Daughter cells

➔ **Mastering Biology** Animation: Cytokinesis
Video: Cytokinesis in an Animal Cell

vesicles collect inside the cell plate as it grows **(Figure 12.10b)**. The cell plate enlarges until its surrounding membrane fuses with the plasma membrane along the perimeter of the cell. Two daughter cells result, each with its own plasma membrane. Meanwhile, a new cell wall arising from the contents of the cell plate forms between the daughter cells.

Figure 12.11 is a series of micrographs of a dividing plant cell. Examining this figure will help you review mitosis and cytokinesis.

Binary Fission in Bacteria

Prokaryotes (bacteria and archaea) can undergo a type of reproduction in which the cell grows to roughly double its size and then divides to form two cells. The term **binary fission**, meaning "division in half," refers to this process and to the asexual reproduction of single-celled eukaryotes, such as the amoeba in Figure 12.2a. However, the process in eukaryotes involves mitosis, while that in prokaryotes does not.

In bacteria, most genes are carried on a single bacterial chromosome that consists of a circular DNA molecule and associated proteins. Although bacteria are smaller and simpler than eukaryotic cells, the challenge of replicating their genomes in an orderly fashion and distributing the copies equally to two daughter cells is still formidable. For example, when it is fully stretched out, the chromosome of the bacterium *Escherichia coli* is about 500 times as long as the cell. For such a long chromosome to fit within the cell, it must be highly coiled and folded.

In some bacteria, the process of cell division is initiated when the DNA of the bacterial chromosome begins to replicate at a specific place on the chromosome called the **origin of replication**, producing two origins. As the chromosome continues to replicate, one origin moves rapidly toward the opposite end of the cell **(Figure 12.12)**. While the chromosome is replicating, the cell elongates. When replication is complete and the bacterium has reached about twice its initial size, proteins cause its plasma membrane to pinch inward, dividing the parent bacterial cell into two daughter cells. In this way, each cell inherits a complete genome.

Using the techniques of modern DNA technology to tag the origins of replication with molecules that glow green in fluorescence microscopy (see Figure 6.3), researchers have directly observed the movement of bacterial chromosomes. This movement is reminiscent of the poleward movements of the centromere regions of eukaryotic chromosomes during anaphase of mitosis, but bacteria don't have visible mitotic spindles or even microtubules. In most bacterial species studied, the two origins of replication end up at opposite ends of the cell or in some other very specific location, possibly anchored there by one or more proteins. How bacterial chromosomes move and how their specific location is established and maintained are active areas of research. Several proteins that play important roles have been identified. Polymerization of one protein resembling eukaryotic actin apparently functions

▼ **Figure 12.11 Mitosis in a plant cell.** These light micrographs show mitosis in cells of an onion root.

Nucleus
Nucleolus
Chromosomes condensing
Chromosomes
Cell plate
10 μm

❶ **Prophase.** The chromosomes are condensing and the nucleolus is beginning to disappear. Although not yet visible in the micrograph, the mitotic spindle is starting to form.

❷ **Prometaphase.** Discrete chromosomes are now visible; each consists of two aligned, identical sister chromatids. Later in prometaphase, the nuclear envelope will fragment.

❸ **Metaphase.** The spindle is complete, and the chromosomes, attached to microtubules at their kinetochores, are all at the metaphase plate.

❹ **Anaphase.** Chromatids of each chromosome have separated. The daughter chromosomes are moving to the ends of the cell as their kinetochore microtubules shorten.

❺ **Telophase.** Daughter nuclei are forming. Cytokinesis has started: The cell plate, which will divide the cytoplasm in two, grows toward the perimeter of the parent cell.

▼ **Figure 12.12 Bacterial cell division by binary fission.** The bacterium *E. coli*, shown here, has a single, circular chromosome.

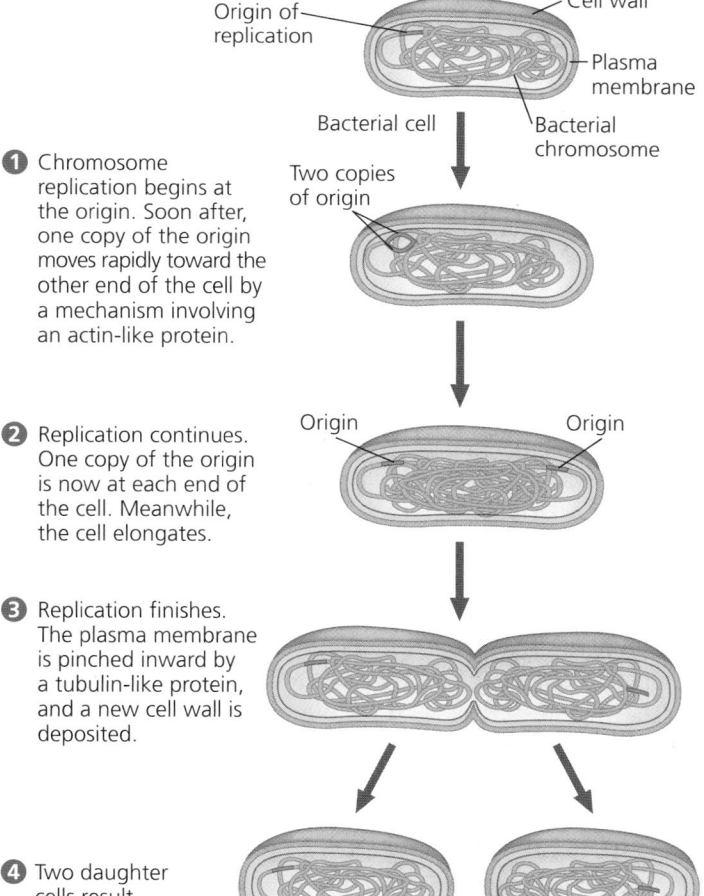

Origin of replication
Cell wall
Plasma membrane
Bacterial cell
Bacterial chromosome
Two copies of origin
Origin
Origin

❶ Chromosome replication begins at the origin. Soon after, one copy of the origin moves rapidly toward the other end of the cell by a mechanism involving an actin-like protein.

❷ Replication continues. One copy of the origin is now at each end of the cell. Meanwhile, the cell elongates.

❸ Replication finishes. The plasma membrane is pinched inward by a tubulin-like protein, and a new cell wall is deposited.

❹ Two daughter cells result.

➔ **Mastering Biology** Animation: Cell Division in Bacteria

in bacterial chromosome movement during cell division, and another protein that is related to tubulin helps pinch the plasma membrane inward, separating the two bacterial daughter cells.

The Evolution of Mitosis

EVOLUTION Given that prokaryotes preceded eukaryotes on Earth by more than a billion years, we might hypothesize that mitosis evolved from simpler prokaryotic mechanisms of cell reproduction. The fact that some of the proteins involved in bacterial binary fission are related to eukaryotic proteins that function in mitosis supports that hypothesis.

As eukaryotes with nuclear envelopes and larger genomes evolved, the ancestral process of binary fission, seen today in bacteria, somehow gave rise to mitosis. Variations on cell division exist in different groups of organisms. These variant processes may be similar to mechanisms used by ancestral species and thus may resemble steps in the evolution of mitosis from a binary fission-like process presumably carried out by very early bacteria. Possible intermediate stages are suggested by two unusual types of nuclear division found today in certain unicellular eukaryotes—dinoflagellates, diatoms, and some yeasts **(Figure 12.13).** These two modes of nuclear division are thought to be cases where ancestral mechanisms have remained relatively unchanged over evolutionary time. In both types, the nuclear envelope remains intact, in contrast to what happens in most eukaryotic cells. Keep in mind, however, that we can't observe cell division in cells of extinct species. This hypothesis uses only currently existing species as some possible examples of intermediate mechanisms. Other mechanisms may have existed in species that have gone extinct; we simply have no way of knowing.

▼ Figure 12.13 Mechanisms of cell division in several groups of organisms. Some unicellular eukaryotes existing today have mechanisms of cell division that may resemble intermediate steps in the evolution of mitosis. Except for **(a)**, cell walls are not shown.

Bacterial chromosome

(a) Bacteria. During binary fission in bacteria, the origins of the daughter chromosomes move to opposite ends of the cell. The mechanism involves polymerization of actin-like molecules and possibly proteins that may anchor the daughter chromosomes to specific sites on the plasma membrane.

Chromosomes

Microtubules

Intact nuclear envelope

(b) Dinoflagellates. In unicellular eukaryotes called dinoflagellates, the chromosomes attach to the nuclear envelope, which remains intact during cell division. Microtubules pass through the nucleus inside cytoplasmic tunnels, reinforcing the spatial orientation of the nucleus, which then divides in a process reminiscent of bacterial binary fission.

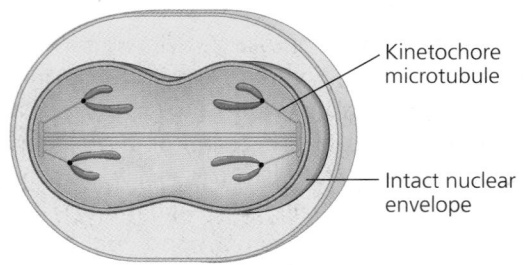

Kinetochore microtubule

Intact nuclear envelope

(c) Diatoms and some yeasts. In these two other groups of unicellular eukaryotes, the nuclear envelope also remains intact during cell division. In these organisms, the microtubules form a spindle *within* the nucleus. Microtubules separate the chromosomes, and the nucleus splits into two daughter nuclei.

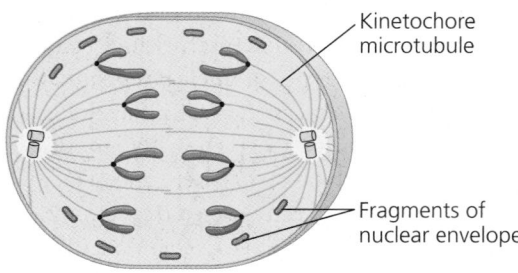

Kinetochore microtubule

Fragments of nuclear envelope

(d) Most eukaryotes. In most other eukaryotes, including plants and animals, the spindle forms outside the nucleus, and the nuclear envelope breaks down during mitosis. Microtubules separate the chromosomes, and two nuclear envelopes then form.

⊖ **Mastering Biology** Video: Nuclear Envelope Breakdown and Formation During Mitosis in *C. elegans*, a Eukaryote

1. How many chromosomes are shown in the illustration in Figure 12.8? Are they duplicated? How many chromatids are shown?
2. Compare cytokinesis in animal cells and plant cells.
3. During which stages of the cell cycle does a chromosome consist of two identical chromatids?
4. Compare the roles of tubulin and actin during eukaryotic cell division with the roles of tubulin-like and actin-like proteins during bacterial binary fission.
5. A kinetochore has been compared to a coupling device that connects a motor to the cargo that it moves. Explain.
6. **MAKE CONNECTIONS** What other functions do actin and tubulin carry out? Name the proteins they interact with to do so. (Review Figures 6.21a and 6.26a.)

For suggested answers, see Appendix A.

CONCEPT 12.3

The eukaryotic cell cycle is regulated by a molecular control system

The timing and rate of cell division in different parts of a plant or animal are crucial to normal growth, development, and maintenance. The frequency of cell division varies with the type of cell. For example, human skin cells divide frequently throughout life, whereas liver cells maintain the ability to divide but keep it in reserve until an appropriate need arises—say, to repair a wound. Some of the most specialized cells, such as fully formed nerve cells and muscle cells, do not divide at all in a mature human. These cell cycle differences result from regulation at the molecular level. The mechanisms of this regulation are of great interest, not only to understand the life cycles of normal cells but also to learn how cancer cells manage to escape the usual controls.

The Cell Cycle Control System

What controls the cell cycle? In the early 1970s, a variety of experiments led to the hypothesis that the cell cycle is driven by specific signaling molecules present in the cytoplasm. Some of the first strong evidence for this hypothesis came from experiments with mammalian cells grown in culture. In these experiments, two cells in different phases of the cell cycle were fused to form a single cell with two nuclei (**Figure 12.14**). If one of the original cells was in the S phase and the other was in G_1, the G_1 nucleus immediately entered the S phase, as though stimulated by signaling molecules present in the cytoplasm of the first cell. Similarly, if a cell undergoing mitosis (M phase) was fused with another cell in any stage of its cell cycle, even G_1, the second nucleus immediately entered mitosis, with condensation of the chromatin and formation of a mitotic spindle.

The experiment shown in Figure 12.14 and other experiments on animal cells and yeasts demonstrated that the

▼ Figure 12.14 Inquiry

Do molecular signals in the cytoplasm regulate the cell cycle?

Experiment Researchers at the University of Colorado wondered whether a cell's progression through the cell cycle is controlled by cytoplasmic molecules. They induced cultured mammalian cells at different phases of the cell cycle to fuse. Two experiments are shown.

Results

When a cell in S phase was fused with a cell in G_1, the G_1 nucleus immediately entered S phase—DNA was synthesized.

When a cell in M phase was fused with a cell in G_1, the G_1 nucleus immediately began mitosis—a spindle formed and the chromosomes condensed, even though the chromosomes had not been duplicated.

Conclusion The results of fusing a G_1 cell with a cell in the S or M phase of the cell cycle suggest that molecules present in the cytoplasm during the S or M phase control the progression to those phases.

Data from R. T. Johnson and P. N. Rao, Mammalian cell fusion: Induction of premature chromosome condensation in interphase nuclei, *Nature* 226:717–722 (1970).

WHAT IF? *If the progression of phases did not depend on cytoplasmic molecules and each phase began when the previous one was complete, how would the results have differed?*

sequential events of the cell cycle are directed by a distinct **cell cycle control system**, a cyclically operating set of molecules in the cell that both triggers and coordinates key events in the cell cycle **(Figure 12.15)**. The cell cycle control system has been compared to the control device of a washing machine. Like the washer's timing device, the cell cycle control system proceeds on its own, according to a built-in clock. However, just as a washer's cycle is subject to both internal control (such as the sensor that detects when the tub is filled with water) and external adjustment (such as starting or stopping the machine), the cell cycle is regulated at certain checkpoints by both internal and external signals. A **checkpoint** in the cell cycle is a control point where stop and go-ahead signals can regulate the cycle. Three important checkpoints are found in the G_1, G_2, and M phases (red gates in Figure 12.15), which will be discussed shortly.

To understand how cell cycle checkpoints work, we'll first identify some of the molecules that make up the cell cycle

▼ Figure 12.15 Mechanical analogy for the cell cycle control system.

In this diagram, the flat "stepping stones" around the perimeter represent sequential events. Like the control device of a washing machine, the cell cycle control system proceeds on its own, driven by a built-in clock. However, the system is subject to internal and external regulation at various checkpoints; three important checkpoints are shown (red).

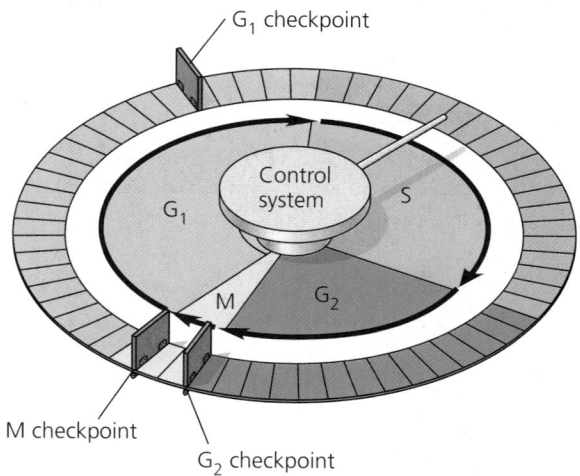

→ **Mastering Biology** Animation: Control of the Cell Cycle

control system (the molecular basis for the cell cycle clock) and describe how a cell progresses through the cycle. We'll then consider the internal and external checkpoint signals that can make the clock either pause or continue.

The Cell Cycle Clock: Cyclins and Cyclin-Dependent Kinases

Rhythmic fluctuations in the abundance and activity of cell cycle control molecules pace the sequential events of the cell cycle. These regulatory molecules are mainly proteins of two types: protein kinases and cyclins. As you learned in Concept 11.3, protein kinases are enzymes that activate or inactivate other proteins by phosphorylating them.

Many of the kinases that drive the cell cycle are actually present at a constant concentration in the growing cell, but much of the time they are in an inactive form. To be active, such a kinase must be attached to a **cyclin**, a protein that gets its name from its cyclically fluctuating concentration in the cell. Because of this requirement, these kinases are called **cyclin-dependent kinases**, or **Cdks**. The activity of a Cdk rises and falls with changes in the concentration of its cyclin partner. **Figure 12.16a** shows the fluctuating activity of **MPF**, the cyclin-Cdk complex that was discovered first (in frog eggs). Note that the peaks of MPF activity correspond to the peaks of cyclin concentration. The cyclin level rises during the S and G_2 phases and then falls abruptly during M phase.

The initials MPF stand for "maturation-promoting factor," but we can think of MPF as "M-phase-promoting factor" because it triggers the cell's passage into the M phase, past the G_2 checkpoint. When cyclins that accumulate during G_2 associate with Cdk molecules, the resulting MPF complex is active—it phosphorylates a variety of proteins, initiating mitosis (**Figure 12.16b**). MPF acts both directly as a kinase and indirectly by activating other kinases. For example, MPF causes phosphorylation of various proteins of the nuclear

▼ **Figure 12.16 Molecular control of the cell cycle at the G₂ checkpoint.** The steps of the cell cycle are timed by rhythmic fluctuations in the activity of cyclin-dependent kinases (Cdks). Here we focus on a cyclin-Cdk complex in animal cells called MPF, which acts at the G₂ checkpoint as a go-ahead signal, triggering the events of mitosis.

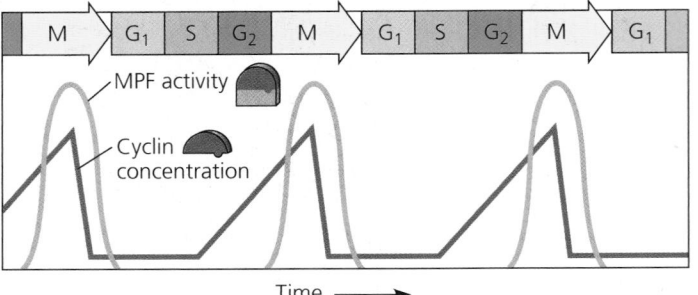

(a) Fluctuation of MPF activity and cyclin concentration during the cell cycle

1 Synthesis of cyclin begins in late S phase and continues through G₂. Because cyclin is protected from degradation during this stage, it accumulates.

5 During G₁, the degradation of cyclin continues, and the Cdk component of MPF is recycled.

4 During anaphase, the cyclin component of MPF is degraded, terminating the M phase. The cell enters the G₁ phase.

3 MPF promotes mitosis by phosphorylating various proteins. MPF's activity peaks during metaphase.

2 Cyclin combines with Cdk, producing MPF. When enough MPF molecules accumulate, the cell passes the G₂ checkpoint and begins mitosis.

(b) Molecular mechanisms that help regulate the cell cycle

VISUAL SKILLS *Explain how the events in the diagram in (b) are related to the "Time" axis of the graph in (a), beginning at the left.*

Mastering Biology
Interview with Paul Nurse: Discovering how protein kinases control the cell cycle

lamina (see Figure 6.9), which promotes fragmentation of the nuclear envelope during prometaphase of mitosis. There is also evidence that MPF contributes to molecular events required for chromosome condensation and spindle formation during prophase.

During anaphase, MPF helps switch itself off by initiating a process that leads to the destruction of its own cyclin. The noncyclin part of MPF, the Cdk, persists in the cell, inactive until it becomes part of MPF again by associating with new cyclin molecules synthesized during the S and G₂ phases of the next round of the cycle.

The fluctuating activities of different cyclin-Cdk complexes are of major importance in controlling all the stages of the cell cycle; they also give the go-ahead signals at some checkpoints. As mentioned above, MPF controls the cell's passage through the G₂ checkpoint. Cell behavior at the G₁ checkpoint is also regulated by the activity of cyclin-Cdk protein complexes. Animal cells appear to have at least three Cdk proteins and several different cyclins that operate at this checkpoint. Next, let's consider checkpoints in more detail.

Stop and Go Signs: Internal and External Signals at the Checkpoints

Animal cells generally have built-in stop signals that halt the cell cycle at checkpoints until overridden by go-ahead signals. Many signals registered at checkpoints come from cellular surveillance mechanisms inside the cell. These signals report whether crucial cellular processes that should have occurred by that point have in fact been completed correctly and thus whether or not the cell cycle should proceed. Checkpoints also register signals from outside the cell. The signals are transmitted within the cell by signal transduction pathways (see Figure 11.6). Three important checkpoints are those in the G₁, G₂, and M phases (see Figure 12.15).

For many cells, the G₁ checkpoint seems to be the most important. If a cell receives a go-ahead signal at the G₁ checkpoint, it will usually complete the G₁, S, G₂, and M phases and divide. If it does not receive a go-ahead signal at that point, it may exit the cycle, switching into a nondividing state called the **G₀ phase (Figure 12.17a)**. Most cells of the human body are actually in the G₀ phase. As mentioned earlier, mature nerve cells and muscle cells never divide. Other cells, such as liver cells, can be "called back" from the G₀ phase to the cell cycle by external cues, such as growth factors released during injury.

Biologists are currently working out the pathways that link signals originating inside and outside the cell with the responses by cyclin-dependent kinases and other proteins. An example of an internal signal occurs at the third important checkpoint, the M checkpoint **(Figure 12.17b)**. Anaphase, the separation of sister chromatids, does not begin until all the chromosomes are properly attached to the spindle at the metaphase plate. Researchers have learned that as long as some kinetochores are unattached to spindle microtubules, the sister chromatids remain together, delaying anaphase. Only when the kinetochores of all the chromosomes are properly attached to the spindle does the appropriate regulatory protein complex become activated. (In this case, the regulatory molecule is not a cyclin-Cdk complex but, instead, a different complex made up of several proteins.) Once activated, the complex sets off a chain of molecular events that activates the enzyme separase, which cleaves the cohesins, allowing the sister chromatids to separate. This mechanism ensures that daughter cells do not end up with missing or extra chromosomes.

There are checkpoints in addition to those in G_1, G_2, and M. For instance, a checkpoint in S phase stops cells with DNA damage from proceeding in the cell cycle. And, in 2014, researchers presented evidence for another checkpoint between anaphase and telophase that ensures anaphase is completed and the chromosomes are well separated before cytokinesis can begin, thus avoiding chromosomal damage.

What about the stop and go-ahead signals themselves—what are the signaling molecules? Studies using animal cells in culture have led to the identification of many external factors, both chemical and physical, that can influence cell division. For example, cells fail to divide if an essential nutrient is lacking in the culture medium. (This is analogous to trying to run a washing machine without the water supply hooked up; an internal sensor won't allow the machine to continue past the point where water is needed.) And even if all other conditions are favorable, most types of mammalian cells divide in culture only if the growth medium includes specific growth factors. As mentioned in Concept 11.1, a **growth factor** is a protein released by certain cells that stimulates other cells to divide. Different cell types respond specifically to different growth factors or combinations of growth factors.

Consider, for example, *platelet-derived growth factor* (*PDGF*), which is made by blood cell fragments called platelets. When an injury occurs, platelets release PDGF in the vicinity. The experiment illustrated in **Figure 12.18** demonstrates that PDGF is required for the division of cultured fibroblasts, a type of connective tissue cell. Fibroblasts have PDGF receptors on their plasma membranes. The binding of PDGF molecules to these receptors (which are receptor tyrosine kinases; see Figure 11.8) triggers a signal transduction pathway that allows the cells to pass the G_1 checkpoint and divide. PDGF stimulates fibroblast division not only in the artificial conditions of cell culture but also in an animal's body. Thus, injury results in a proliferation of fibroblasts that helps heal the wound.

The effect of an external physical factor on cell division is clearly seen in **density-dependent inhibition**, a phenomenon in which crowded cells stop dividing **(Figure 12.19a)**. Studies done many years ago showed that cultured cells normally divide until they form a single layer of cells on the inner surface

▼ **Figure 12.17 Two important checkpoints.** At certain checkpoints in the cell cycle (red gates), cells do different things depending on the signals they receive. Events of the **(a)** G_1 and **(b)** M checkpoints are shown. In **(b)**, the G_2 checkpoint has already been passed by the cell.

(a) G₁ checkpoint

In the absence of a go-ahead signal, a cell exits the cell cycle and enters G_0, a nondividing state.

If a cell receives a go-ahead signal, the cell continues on in the cell cycle.

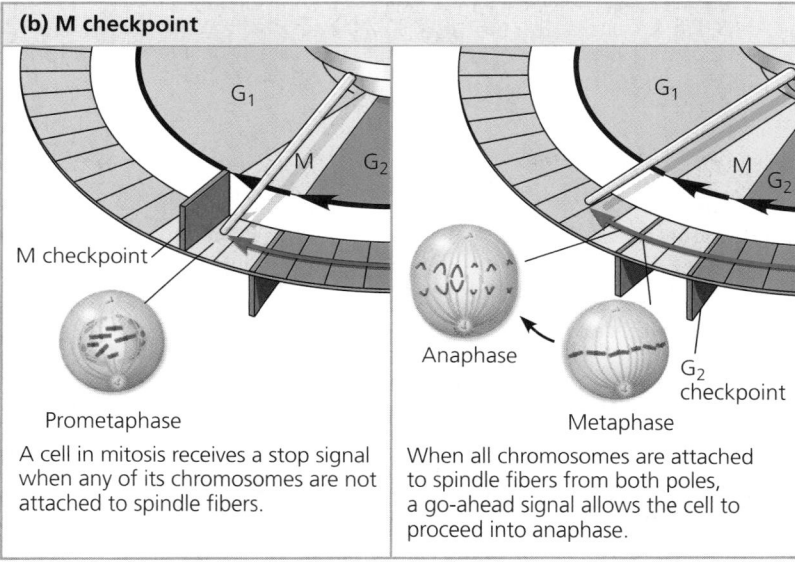

(b) M checkpoint

A cell in mitosis receives a stop signal when any of its chromosomes are not attached to spindle fibers.

When all chromosomes are attached to spindle fibers from both poles, a go-ahead signal allows the cell to proceed into anaphase.

WHAT IF? *What might be the result if the cell ignored either checkpoint and progressed through the cell cycle?*

of a culture flask, at which point the cells stop dividing. If some cells are removed, those bordering the open space begin dividing again and continue until the vacancy is filled. Follow-up studies revealed that the binding of a cell-surface protein to its counterpart on an adjoining cell sends a signal to both cells that inhibits cell division, preventing them from moving forward in the cell cycle, even in the presence of growth factors.

Most animal cells also exhibit **anchorage dependence** (see Figure 12.19a). To divide, they must be attached to something, such as the inside of a culture flask or the extracellular matrix of a tissue. Experiments suggest that, like cell density, anchorage is signaled to the cell cycle control system via pathways involving plasma membrane proteins and elements of the cytoskeleton linked to them.

Density-dependent inhibition and anchorage dependence appear to function not only in cell culture but also in the

▼ **Figure 12.18 The effect of platelet-derived growth factor (PDGF) on cell division.**

❶ A sample of human connective tissue is cut up into small pieces.

Scalpels

Petri dish

❷ Enzymes are used to digest the extracellular matrix in the tissue pieces, resulting in a suspension of free fibroblasts.

❸ Cells are transferred to culture vessels containing a basic growth medium consisting of glucose, amino acids, salts, and antibiotics (to prevent bacterial growth).

❹ PDGF is added to half the vessels. The culture vessels are incubated at 37°C for 24 hours.

Without PDGF

With PDGF

In the basic growth medium without PDGF (the control), the cells fail to divide.

In the basic growth medium plus PDGF, the cells proliferate. The SEM shows cultured fibroblasts.

MAKE CONNECTIONS
PDGF signals cells by binding to a cell-surface receptor tyrosine kinase. If you added a chemical that blocked phosphorylation, how would the results differ? (See Figure 11.8.)

10 μm

body's tissues, checking the growth of cells at some optimal density and location during embryonic development and throughout an organism's life. Cancer cells, which we examine next, exhibit neither density-dependent inhibition nor anchorage dependence **(Figure 12.19b)**.

Loss of Cell Cycle Controls in Cancer Cells

Cancer cells do not heed the normal signals that regulate the cell cycle. In culture, they do not stop dividing when growth factors are depleted. A logical hypothesis is that cancer cells do not need growth factors in their culture medium to grow and divide. They may make a required growth factor themselves, or they may have an abnormality in the signaling pathway that conveys the growth factor's signal to the cell cycle control system even in the absence of that factor. Another possibility is an abnormal cell cycle control system.

▼ **Figure 12.19 Density-dependent inhibition and anchorage dependence of cell division.** Individual cells are shown disproportionately large in the drawings.

Cells anchor to dish surface and divide (*anchorage dependence*).

When cells have formed a complete single layer, they stop dividing (*density-dependent inhibition*).

If some cells are scraped away, the remaining cells divide to fill the gap and then stop once they contact each other.

20 μm

(a) Normal mammalian cells. Cell density is limited to a single layer by contact with neighboring cells and the availability of nutrients, growth factors, and a substratum for attachment.

20 μm

(b) Cancer cells. Cancer cells usually continue to divide well beyond a single layer, forming a clump of overlapping cells. They do not exhibit anchorage dependence or density-dependent inhibition.

In these scenarios, the underlying basis of the abnormality is almost always a change in one or more genes (for example, a mutation) that alters the function of their protein products, resulting in faulty cell cycle control. We'll explore the molecular basis for such changes in Concept 18.5.

There are other important differences between normal cells and cancer cells that reflect malfunctions of the cell cycle. If and when they stop dividing, cancer cells do so at random points in the cycle, rather than at the normal checkpoints. Moreover, cancer cells can go on dividing indefinitely in culture if they are given a continual supply of nutrients; in essence, they are "immortal." A striking example is a cell line that has been reproducing in culture since 1951. Cells of this line are called HeLa cells because their original source was a tumor removed from a woman named Henrietta Lacks. (Neither Ms. Lacks nor her family gave permission or even knew about the propagation and use of her cells, which have helped biologists make countless significant discoveries over the years.) Cells in culture that acquire the ability to divide indefinitely are said to have undergone a process called **transformation**, causing them to behave (in cell division, at least) like cancer cells. By contrast, nearly all normal, non-transformed mammalian cells growing in culture divide only about 20 to 50 times before they stop dividing, age, and die. Finally, cancer cells evade the normal controls that trigger a cell to undergo apoptosis when something is wrong—for example, when an irreparable mistake has occurred during DNA replication preceding mitosis.

Abnormal cell behavior in the body can be catastrophic. The problem begins when a single cell in a tissue undergoes the first of many steps that convert a normal cell to a cancer cell. Such a cell often has altered proteins on its surface, and the body's immune system normally recognizes the cell as "nonself"—an insurgent—and destroys it. However, if the cell evades destruction, it may proliferate and form a tumor, a mass of abnormal cells within otherwise normal tissue. The abnormal cells may remain at the original site if their genetic and cellular changes don't allow them to move to or survive at another site. In that case, the tumor is called a **benign tumor**. Most benign tumors do not cause serious problems (depending on their location) and can be removed by surgery. In contrast, a **malignant tumor** includes cells whose genetic and cellular changes enable them to spread to new tissues and impair the functions of one or more organs; these cells are also sometimes called *transformed* cells (although usage of this term is generally restricted to cells in culture). An individual with a malignant tumor is said to have cancer **(Figure 12.20)**.

The changes that have occurred in cells of malignant tumors show up in many ways besides excessive proliferation. These cells may have unusual numbers of chromosomes, though whether this is a cause or an effect of tumor-related changes is an ongoing debate. Their metabolism may be altered, and they may cease to function in any constructive way. Abnormal changes on the cell surface cause cancer cells to lose attachments to neighboring cells and the extracellular matrix, allowing them to spread into nearby tissues. Cancer cells may also secrete signaling molecules that cause blood vessels to grow toward the tumor. A few tumor cells may separate from the original tumor, enter blood vessels and lymph vessels, and travel to other parts of the body. There, they may proliferate and form a new tumor. This spread of cancer cells to locations distant from their original site is called **metastasis** (see Figure 12.20).

A tumor that appears to be localized may be treated with high-energy radiation, which damages DNA in cancer cells much more than DNA in normal cells, apparently because the majority of cancer cells have lost the ability to repair such damage. To treat known or suspected metastatic tumors, chemotherapy is used, in which drugs that are toxic to actively dividing cells are administered through the circulatory system. As you might expect, chemotherapeutic drugs interfere with specific steps in the cell cycle. For example, the drug Taxol freezes the mitotic spindle by preventing microtubule depolymerization, which stops actively dividing cells from proceeding past metaphase and leads to their destruction. The side effects of chemotherapy are due to the effects of the drugs on normal cells that divide frequently, due to the function of that cell type in the organism. For example, nausea results from chemotherapy's effects on intestinal cells, hair loss from effects on hair follicle cells, and susceptibility to infection from effects on immune system cells. You can work with data from an experiment involving a potential chemotherapeutic agent in the **Scientific Skills Exercise**.

Over the past several decades, researchers have produced a flood of valuable information about cell-signaling pathways and how their malfunction contributes to the development of cancer through effects on the cell cycle. Coupled with new

▼ **Figure 12.20 The growth and metastasis of a malignant breast tumor.** A series of genetic and cellular changes contribute to a tumor becoming malignant (cancerous). The cells of malignant tumors grow in an uncontrolled way and can spread to neighboring tissues and, via lymph and blood vessels, to other parts of the body (metastasis).

→ **Mastering Biology** HHMI Animation: How Tumors Grow

5 μm

Breast cancer cell (colorized SEM)

Tumor

Glandular tissue

Lymph vessel

Blood vessel

Cancer cell

Metastatic tumor

1 A tumor grows from a single cancer cell.

2 Cancer cells invade neighboring tissue.

3 Cancer cells spread through lymph and blood vessels to other parts of the body.

4 A small percentage of cancer cells may metastasize to another part of the body.

Interpreting Histograms

▶ **Human brain cancer cell**

At What Phase Is the Cell Cycle Arrested by an Inhibitor? Many medical treatments are aimed at stopping cancer cell proliferation by blocking the cell cycle of cancerous tumor cells. One potential treatment is a cell cycle inhibitor derived from human umbilical cord stem cells. In this exercise, you will compare two histograms to determine where in the cell cycle the inhibitor blocks the division of cancer cells.

How the Experiment Was Done In the treated sample, human glioblastoma (brain cancer) cells were grown in tissue culture in the presence of the inhibitor, while a control sample of glioblastoma cells was grown in its absence. After 72 hours of growth, the two cell samples were harvested. To get a "snapshot" of the phase of the cell cycle each cell was in at that time, the samples were treated with a fluorescent chemical that binds to DNA and then run through a flow cytometer, an instrument that records the fluorescence level of each cell. Computer software then graphed the number of cells in each sample with a particular fluorescence level, as shown below.

Data from the Experiment

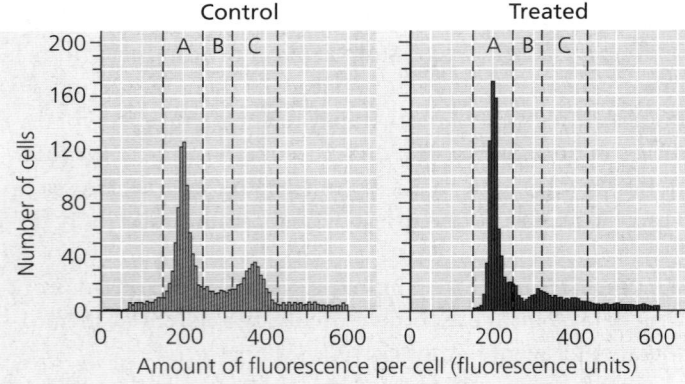

The data are plotted in a type of graph called a histogram (above), which groups the values for a numeric variable on the x-axis into intervals. A histogram allows you to see how all the experimental subjects (cells, in this case) are distributed along a continuous variable (amount of fluorescence). In these histograms, the bars are so narrow that the data appear to follow a curve for which you can detect peaks and dips. Each narrow bar represents the number of cells observed to have a level of fluorescence in the range of that interval. This in turn indicates the relative amount of DNA in those cells. Overall, comparing the two histograms allows you to see how the DNA content of this cell population is altered by the treatment.

INTERPRET THE DATA

1. Study the data in the histograms. **(a)** Which axis indirectly shows the relative amount of DNA per cell? Explain your answer. **(b)** In the control sample, compare the first peak in the histogram (in region A) to the second peak (in region C). Which peak shows the population of cells with the higher amount of DNA per cell? Explain. (For additional information about histograms, see the Scientific Skills Review in Appendix D.)

2. **(a)** In the control sample histogram, identify the phase of the cell cycle (G_1, S, or G_2) of the population of cells in each region delineated by vertical lines. Label the histogram with these phases and explain your answer. **(b)** Does the S phase population of cells show a distinct peak in the histogram? Why or why not?

3. The histogram representing the treated sample shows the effect of growing the cancer cells alongside human umbilical cord stem cells that produce the potential inhibitor. **(a)** Label the histogram with the cell cycle phases. Which phase of the cell cycle has the greatest number of cells in the treated sample? Explain. **(b)** Compare the distribution of cells among G_1, S, and G_2 phases in the control and treated samples. What does this tell you about the cells in the treated sample? **(c)** Based on what you learned in Concept 12.3, propose a mechanism by which the stem cell–derived inhibitor might arrest the cancer cell cycle at this stage. (More than one answer is possible.)

 Instructors: A version of this Scientific Skills Exercise can be assigned in **Mastering Biology**.

Data from K. K. Velpula et al., Regulation of glioblastoma progression by cord blood stem cells is mediated by downregulation of cyclin D1, *PLoS ONE* 6(3):e18017 (2011).

molecular techniques, such as the ability to rapidly sequence the DNA of cells in a particular tumor, medical treatments for cancer are beginning to become more "personalized" to a particular patient's tumor.

⊙ **Mastering Biology** BBC Video: Are Fruit Flies the Key in the Fight Against Cancer?

For example, the cells of roughly 20% of breast cancer tumors show abnormally high amounts of a cell-surface receptor tyrosine kinase called HER2, and many show an increase in the number of estrogen receptor (ER) molecules, intracellular receptors that can trigger cell division. Based on lab findings, a physician can prescribe chemotherapy with a molecule that blocks the function of the specific protein (Herceptin for HER2 and tamoxifen for ERs). Treatment using these agents, when appropriate, has led to increased survival rates and fewer cancer recurrences (see Make Connections Figure 18.27).

⊙ **Mastering Biology**
  Interview with Bruce Alberts:
  Cancer control and careers in science

CONCEPT CHECK 12.3

1. In Figure 12.14, why do the nuclei resulting from experiment 2 contain different amounts of DNA?

2. How does MPF allow a cell to pass the G_2 phase checkpoint and enter mitosis? (See Figure 12.16.)

3. **MAKE CONNECTIONS** Explain how receptor tyrosine kinases and intracellular receptors might function in triggering cell division. (Review Figures 11.8 and 11.9 and Concept 11.2.)

For suggested answers, see Appendix A.

12 Chapter Review

→ Go to **Mastering Biology** for Assignments, the eText, the Study Area, and Dynamic Study Modules.

SUMMARY OF KEY CONCEPTS

→ To review key terms, go to the **Vocabulary Self-Quiz** in the **Mastering Biology** eText or Study Area, or go to goo.gl/zkjz9t.

CONCEPT 12.1

Most cell division results in genetically identical daughter cells (pp. 235–237)

- Unicellular organisms reproduce by **cell division**; multicellular organisms depend on cell division for their development from a fertilized egg and for growth and repair.
- The genetic material (DNA) of a cell—its **genome**—is partitioned among **chromosomes**. Each eukaryotic chromosome consists of one DNA molecule associated with many proteins. Together, the complex of DNA and associated proteins is called **chromatin**. The chromatin of a chromosome exists in different states of condensation at different times. In animals, **gametes** have one set of chromosomes and **somatic cells** have two sets.
- Cells replicate their genetic material before they divide, each daughter cell receiving a copy of the DNA. Prior to cell division, chromosomes are duplicated. Each one then consists of two identical **sister chromatids** joined along their lengths by sister chromatid cohesion and held most tightly together at a constricted region at the **centromeres**. When this cohesion is broken, the chromatids separate during cell division, becoming the chromosomes of the daughter cells. Eukaryotic cell division consists of **mitosis** (division of the nucleus) and **cytokinesis** (division of the cytoplasm).

❓ *Differentiate between these terms: chromosome, chromatin, and chromatid.*

CONCEPT 12.2

The mitotic phase alternates with interphase in the cell cycle (pp. 237–244)

- Cell division is part of the **cell cycle**, an ordered sequence of events in the life of a cell.
- Between divisions, a cell is in **interphase**: the G_1, S, and G_2 phases. The cell grows throughout interphase, with DNA being replicated during the synthesis (S) phase. Mitosis and cytokinesis make up the **mitotic (M) phase** of the cell cycle.

- The **mitotic spindle**, made up of microtubules, controls chromosome movement during mitosis. In animal cells, it arises from the **centrosomes** and includes spindle microtubules and asters. Some spindle microtubules attach to the **kinetochores** of chromosomes and move the chromosomes to the **metaphase plate**. After sister chromatids separate, motor proteins move them along kinetochore microtubules toward opposite ends of the cell. The cell elongates when motor proteins push nonkinetochore microtubules from opposite poles away from each other.
- Mitosis is usually followed by cytokinesis. Animal cells carry out cytokinesis by **cleavage**, and plant cells form a **cell plate**.
- During **binary fission** in bacteria, the chromosome replicates and the daughter chromosomes actively move apart. Some of the proteins involved in bacterial binary fission are related to eukaryotic actin and tubulin.
- Since prokaryotes preceded eukaryotes by more than a billion years, it is likely that mitosis evolved from prokaryotic cell division. Certain unicellular eukaryotes exhibit mechanisms of cell division that may be similar to those of ancestors of existing eukaryotes. Such mechanisms might represent intermediate steps in the evolution of mitosis.

❓ *In which of the three phases of interphase and the stages of mitosis do chromosomes exist as single DNA molecules?*

CONCEPT 12.3

The eukaryotic cell cycle is regulated by a molecular control system (pp. 244–250)

- Signaling molecules present in the cytoplasm regulate progress through the cell cycle.
- In the **cell cycle control system**, cyclic changes in regulatory proteins—including **cyclins** and **cyclin-dependent kinases (Cdks)**—work as a cell cycle clock. The cell cycle stops at specific **checkpoints** until a go-ahead signal is received; important checkpoints occur in G_1, G_2, and M phases. Internal and external signals control the cell cycle checkpoints via signal transduction pathways. Most cells exhibit **density-dependent inhibition** of cell division as well as **anchorage dependence**.
- Cancer cells elude normal cell cycle regulation and divide unchecked, forming tumors. **Malignant tumors** invade nearby tissues and can undergo **metastasis**, exporting cancer cells to other sites, where they may form secondary tumors.

❓ *Explain the significance of the G_1, G_2, and M checkpoints and the go-ahead signals involved in the cell cycle control system.*

TEST YOUR UNDERSTANDING

→ For more multiple-choice questions, go to the **Practice Test** in the **Mastering Biology** eText or Study Area, or go to goo.gl/GruWRg.

Levels 1-2: Remembering/Understanding

1. Through a microscope, you can see a cell plate beginning to develop across the middle of a cell and nuclei forming on either side of the cell plate. This cell is most likely
   (A) an animal cell in the process of cytokinesis.
   (B) a plant cell in the process of cytokinesis.
   (C) a bacterial cell dividing.
   (D) a plant cell in metaphase.

2. Vinblastine is a standard chemotherapeutic drug used to treat cancer. Because it interferes with the assembly of microtubules, its effectiveness must be related to
   (A) disruption of mitotic spindle formation.
   (B) suppression of cyclin production.
   (C) myosin denaturation and inhibition of cleavage furrow formation.
   (D) inhibition of DNA synthesis.

3. One difference between cancer cells and normal cells is that cancer cells
   (A) are unable to synthesize DNA.
   (B) are arrested at the S phase of the cell cycle.
   (C) continue to divide even when they are tightly packed together.
   (D) cannot function properly because they are affected by density-dependent inhibition.

4. The decline of MPF activity at the end of mitosis is due to
   (A) the destruction of the protein kinase Cdk.
   (B) decreased synthesis of Cdk.
   (C) the degradation of cyclin.
   (D) the accumulation of cyclin.

5. In the cells of some organisms, mitosis occurs without cytokinesis. This will result in
   (A) cells with more than one nucleus.
   (B) cells that are unusually small.
   (C) cells lacking nuclei.
   (D) cell cycles lacking an S phase.

6. Which of the following occurs during S phase?
   (A) condensation of the chromosomes
   (B) replication of the DNA
   (C) separation of sister chromatids
   (D) spindle formation

Levels 3-4: Applying/Analyzing

7. Cell A has half as much DNA as cells B, C, and D in a mitotically active tissue. Cell A is most likely in
   (A) G_1.   (C) prophase.
   (B) G_2.   (D) metaphase.

8. The drug cytochalasin B blocks the function of actin. Which of the following aspects of the animal cell cycle would be most disrupted by cytochalasin B?
   (A) spindle formation
   (B) spindle attachment to kinetochores
   (C) cell elongation during anaphase
   (D) cleavage furrow formation and cytokinesis

9. **VISUAL SKILLS** The light micrograph shows dividing cells near the tip of an onion root. Identify a cell in each of the following stages: prophase, prometaphase, metaphase, anaphase, and telophase. Describe the major events occurring at each stage.

10. **DRAW IT** Draw one eukaryotic chromosome as it would appear during interphase, during each of the stages of mitosis, and during cytokinesis. Also draw and label the nuclear envelope and any microtubules attached to the chromosome(s).

Levels 5-6: Evaluating/Creating

11. **EVOLUTION CONNECTION** The result of mitosis is that the daughter cells end up with the same number of chromosomes that the parent cell had. Another potential way to maintain the number of chromosomes would be to carry out cell division first and then duplicate the chromosomes in each daughter cell. Assess whether this would be an equally good way of organizing the cell cycle. Explain why evolution has not led to this alternative.

12. **SCIENTIFIC INQUIRY** Although both ends of a microtubule can gain or lose subunits, one end (called the plus end) polymerizes and depolymerizes at a higher rate than the other end (the minus end). For spindle microtubules, the plus ends are in the center of the spindle, and the minus ends are at the poles. Motor proteins that move along microtubules specialize in walking either toward the plus end or toward the minus end; the two types are called plus end–directed and minus end–directed motor proteins, respectively. Given what you know about chromosome movement and spindle changes during anaphase, predict which type of motor proteins would be present on (a) kinetochore microtubules and (b) nonkinetochore microtubules.

13. **WRITE ABOUT A THEME: INFORMATION** The continuity of life is based on heritable information in the form of DNA. In a short essay (100–150 words), explain how the process of mitosis faithfully parcels out exact copies of this heritable information in the production of genetically identical daughter cells.

14. **SYNTHESIZE YOUR KNOWLEDGE**

Shown here are two HeLa cancer cells that are just completing cytokinesis. Explain how the cell division of cancer cells like these is misregulated. Identify genetic and other changes that might have caused these cells to escape normal cell cycle regulation.

For selected answers, see Appendix A.

Dr. Francisco Mojica is a Professor of Microbiology at the Universidad de Alicante in Spain, where he studies CRISPR sequences. His parents fostered in him a love of nature, and he realized in high school that he was most interested in biology. In college, when he first looked through a microscope and saw small single-celled "living beings" moving around, he was captivated, and he has spent his career focusing on prokaryotes.

While in graduate school in the 1990s, he found the prokaryotic DNA repeats that he later named CRISPR repeats, and he spent the next decade establishing their role as a prokaryotic immune system. Dr. Mojica's lab currently studies the mysteries that still surround CRISPR, trying to understand every single aspect of its mechanism of action and regulation.

AN INTERVIEW WITH

Francisco Mojica

When did you first decide to be a biologist?

In high school, my parents asked me, "What are you thinking you'll do with your life—become a doctor or a lawyer?" I said, "Not at all. I really like biology." In spite of his concern about my future career, my father told me I must study what I loved, so that I could be happy in my life. I really appreciated his words! So I went to university and studied biology. When you like biology, initially you get attracted because of the things you see—animals and plants. But when you see microscopic living beings for the first time, and they're moving around, that's really amazing! So, I decided to focus on microbiology.

What did you study as a graduate student?

I studied prokaryotes that live in salt ponds. These weird organisms can tolerate very high salt concentrations. I was interested in doing molecular biology, which had not been done in that university before—I knew that would be a challenge, and I love challenges. There weren't many genome sequences available at that time, in the early 1990s, but we had cloned a few sequences, and we had indirect evidence they might be related to adaptation to salinity. One of the DNA sequences had these regularly spaced repeats that were partially palindromic (reading the same forward and backward), so they could form secondary structures. We thought maybe these repeats were affected by salinity and would in turn affect gene expression, but we were absolutely wrong! At the end of my Ph.D. thesis, all I could say was, "We have found regularly interspaced repeats that are expressed by the cell, and they could be very important."

"Studying biology is basically just saying 'Wow!' all the time."

What made you continue to study these repeats?

When you find something in your thesis, you believe that it belongs to you—you have to try to answer that question. No one could figure out the role that these repeats played. However, others had found similar repeats in bacteria like *E. coli* and another very distantly-related bacterial species. That was an important finding that pushed me to keep working on the repeats, because it meant the repeats must be doing something

important for them to stay in the genomes of various prokaryotes over such long evolutionary time. Evolution is so fundamental to biology, to understanding everything.

How did you figure out what the repeats were doing?

I started wondering about the sequences *between* the repeats, the so-called "spacers." Where did they come from? I compared the spacer sequences to genome database sequences. By 2000 or 2001, many more genome sequences were available. Finally, in 2003, I found a spacer in *E. coli* whose sequence was identical to the sequence of a virus that infects *E. coli*. Even more interesting, the strain of *E. coli* with that spacer could not be infected by that virus. I checked more and more spacer sequences, and when a spacer sequence matched a sequence in the database, the match was always to the sequence of a virus that could no longer infect that species. So I realized these bacteria keep a "memory" of the viruses that infect them, and that the repeat regions were part of an immune system—an unexpected and very surprising finding. I proposed naming the repeats Clustered Regularly Interspaced Short Palindromic Repeats, or CRISPR, and to my surprise, the name took hold.

What is your advice to a student considering a career in biology?

Studying biology is basically just saying "Wow!" all the time. In science, the truth changes very fast. The most important thing is to be open. Be ready to hear something that destroys what you believe—because biology is a fast-moving field, a field that evolves a lot. Knowledge will change.

▲ Dr. Mojica cultured and analyzed prokaryotes

KEY CONCEPTS

13.1 Offspring acquire genes from parents by inheriting chromosomes *p. 255*

13.2 Fertilization and meiosis alternate in sexual life cycles *p. 256*

13.3 Meiosis reduces the number of chromosome sets from diploid to haploid *p. 259*

13.4 Genetic variation produced in sexual life cycles contributes to evolution *p. 265*

Study Tip

Make a visual study guide: Figure 13.1 presents the events of meiosis as a simplified diagram. As you learn more about meiosis, draw a detailed diagram of the stages of meiosis, starting with the sketch below. Include explanatory labels. Compare your visual study guides of mitosis (from Chapter 12) and meiosis, listing similarities and differences.

⊖ Go to Mastering Biology

For Students (in eText and Study Area)
- Figure 13.10 Walkthrough: A Comparison of Mitosis and Meiosis
- BioFlix® Animation: Meiosis
- Animation: Genetic Variation from Crossing Over

For Instructors to Assign (in Item Library)
- Activity: Meiosis Animation
- BioFlix® Tutorial: Meiosis (3 of 3): Determinants of Heredity and Genetic Variation

Ready-to-Go Teaching Module
(in Instructor Resources)
- Meiosis (Concept 13.3)

Figure 13.1 The family members shown in this photo have some similar features. Offspring resemble their parents more than they do unrelated individuals.

What biological mechanisms account for the resemblance between offspring and their parents?

Meiosis is a special type of cell division that produces cells with half the chromosomes of the parent cell. It occurs only in specialized cells, such as the cells of testes and ovaries in humans.

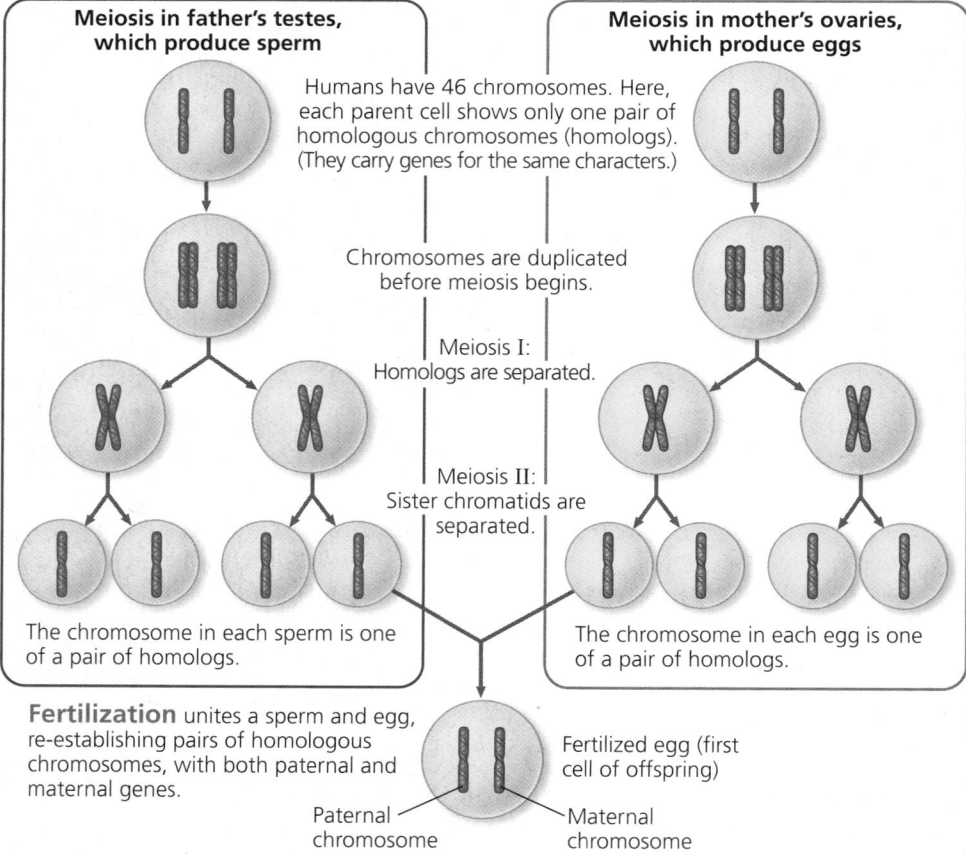

Meiosis in father's testes, which produce sperm

Meiosis in mother's ovaries, which produce eggs

Humans have 46 chromosomes. Here, each parent cell shows only one pair of homologous chromosomes (homologs). (They carry genes for the same characters.)

Chromosomes are duplicated before meiosis begins.

Meiosis I: Homologs are separated.

Meiosis II: Sister chromatids are separated.

The chromosome in each sperm is one of a pair of homologs.

The chromosome in each egg is one of a pair of homologs.

Fertilization unites a sperm and egg, re-establishing pairs of homologous chromosomes, with both paternal and maternal genes.

Fertilized egg (first cell of offspring)

Paternal chromosome

Maternal chromosome

Offspring acquire genes from parents by inheriting chromosomes

Friends may tell you that you have your mother's nose or your father's eyes—although they don't mean that literally, of course. The transmission of traits from one generation to the next is called inheritance, or **heredity** (from the Latin *heres*, heir). At the same time, sons and daughters are not identical copies of either parent or of their siblings. Along with inherited similarity, there is also **variation**. The study of both heredity and inherited variation is called **genetics**.

Inheritance of Genes

Parents endow their offspring with coded information in the form of hereditary units called **genes**. The genes we inherit from our mothers and fathers are our genetic link to our parents, and they account for family resemblances such as shared eye color or freckles. Our genes program specific traits that emerge as we develop from fertilized eggs into adults.

The genetic program is written in the language of DNA, the polymer of four different nucleotides you learned about in Concepts 1.1 and 5.5. Inherited information is passed on in the form of each gene's specific sequence of DNA nucleotides, much as printed information is communicated in the form of meaningful sequences of letters. In both cases, the language is symbolic. Just as your brain translates the word *apple* into a mental image of the fruit, cells translate genes into freckles and other features. Most genes program cells to synthesize specific enzymes and other proteins, whose cumulative action produces an organism's inherited traits. The programming of these traits in the form of DNA is one of the unifying themes of biology.

The transmission of hereditary traits has its molecular basis in the replication of DNA, which produces copies of genes that can be passed from parents to off-spring. In animals and plants, reproductive cells called **gametes** are the vehicles that transmit genes from one generation to the next. During fertilization, male and female gametes (sperm and eggs) unite, passing on genes of both parents to their offspring.

▲ **Sperm and egg**

Except for small amounts of DNA in mitochondria and chloroplasts, the DNA of a eukaryotic cell is packaged into chromosomes within the nucleus. Every species has a characteristic number of chromosomes. For example, humans have 46 chromosomes in their **somatic cells**—all cells of the body except the gametes and their precursors. Each chromosome consists of a single long DNA molecule, elaborately coiled in association with various proteins. One chromosome includes several hundred to a few thousand genes, each gene being a precise sequence of nucleotides along the DNA molecule. A gene's specific location along the length of a chromosome is called the gene's **locus** (plural, *loci*; from the Latin, meaning "place"). Our genetic endowment (our genome) consists of the genes and other DNA that make up the chromosomes we inherited from our parents.

Comparison of Asexual and Sexual Reproduction

Only organisms that reproduce asexually have offspring that are exact genetic copies of themselves. In **asexual reproduction**, a single individual (like a yeast cell or an amoeba; see Figure 12.2a) is the sole parent and passes copies of all its genes to its offspring without the fusion of gametes. For example, single-celled eukaryotic organisms can reproduce asexually by mitotic cell division, in which DNA is copied and allocated equally to two daughter cells. The genomes of the offspring are virtually exact copies of the parent's genome. Some multicellular organisms are also capable of reproducing asexually **(Figure 13.2)**. Because the cells of the offspring arise via mitosis in the parent, the offspring is usually genetically identical to its parent. An individual that reproduces asexually gives rise to a **clone**, an individual or group of individuals that are genetically identical to the parent. Genetic differences occasionally arise in asexually reproducing organisms as a result of changes in the DNA called mutations, which we will discuss in Concept 17.5.

In **sexual reproduction**, two parents give rise to offspring that have unique combinations of genes inherited from the two parents. In contrast to a clone, offspring of sexual reproduction vary genetically from their siblings and both parents: They are variations on a common theme of family resemblance, not exact replicas. Genetic variation like that shown in Figure 13.1 is an important consequence of sexual reproduction. What mechanisms generate this genetic variation? The key is the behavior of chromosomes during the sexual life cycle.

▼ **Figure 13.2 Asexual reproduction in two multicellular organisms. (a)** This relatively simple animal, a hydra, reproduces by budding. The bud, a localized mass of mitotically dividing cells, develops into a small hydra, which detaches from the parent (LM). **(b)** All the trees in this circle of redwoods arose asexually from a single parent tree, whose stump is in the center of the circle.

(a) Hydra   (b) Redwoods

→ **Mastering Biology** Video: Hydra Budding
Animation: Asexual Reproduction

1. **MAKE CONNECTIONS** Using your knowledge of gene expression in a cell, explain what causes the traits of parents (such as hair color) to show up in their offspring. (See Concept 5.5.)

2. Describe how an asexually reproducing eukaryotic organism produces offspring that are genetically identical to each other and to their parents.

3. **WHAT IF?** A horticulturalist breeds orchids, trying to obtain a plant with a unique combination of desirable traits. After many years, she finally succeeds, and wants to produce more plants like this one. Discuss whether she should crossbreed it with another plant or cause it to undergo asexual reproduction (forming a clone), and why.

For suggested answers, see Appendix A.

CONCEPT 13.2

Fertilization and meiosis alternate in sexual life cycles

A **life cycle** is the generation-to-generation sequence of stages in the reproductive history of an organism, from conception to production of its own offspring. In this section, we use humans as an example to track the behavior of chromosomes through the sexual life cycle. We begin by considering the chromosome count in human somatic cells and gametes. We will then explore how the behavior of chromosomes relates to the human life cycle and other types of sexual life cycles.

Sets of Chromosomes in Human Cells

In humans, each somatic cell has 46 chromosomes. Before mitosis begins, the chromosomes are duplicated. During mitosis, the chromosomes become condensed enough to be visible under a light microscope. At this point, they can be distinguished from one another by their size, the position of their centromeres, and the pattern of colored bands produced by certain chromatin-binding stains.

Careful examination of a micrograph of the 46 human chromosomes from a single cell in mitosis reveals that there are two chromosomes of each of 23 types. This becomes clear when images of the chromosomes are arranged in pairs, starting with the longest chromosomes. The resulting ordered display is called a **karyotype (Figure 13.3)**. The two chromosomes of a pair have the same length, centromere position, and staining pattern: These are called **homologous chromosomes** (or **homologs**). Both chromosomes of each pair carry genes controlling the same inherited characters. For example, if a gene for eye color is situated at a particular locus on a certain chromosome, then its homologous chromosome (its homolog) will also have a version of the eye-color gene at the equivalent locus.

The two chromosomes referred to as X and Y are an important exception to the general pattern of homologous chromosomes in human somatic cells. Typically, human females have a homologous pair of X chromosomes (XX), while males have one X and one Y chromosome (XY; see Figure 13.3). Only small

▼ **Figure 13.3** **Research Method**

Preparing a Karyotype

Application A karyotype is a display of condensed chromosomes arranged in pairs. Karyotyping can be used to screen for defective chromosomes or abnormal numbers of chromosomes associated with certain congenital disorders, such as Down syndrome.

Technique Karyotypes are prepared from isolated somatic cells, which are treated with a drug to stimulate mitosis and then grown in culture for several days. Cells arrested when the chromosomes are most highly condensed—at metaphase—are stained and then viewed with a microscope equipped with a digital camera. An image of the chromosomes is displayed on a computer monitor, and digital software is used to arrange them in pairs according to their appearance.

Results This karyotype shows the chromosomes from a human male (as seen by the presence of the XY chromosome pair), colored to emphasize their chromosome banding patterns. The size of the chromosome, position of the centromere, and pattern of stained bands help identify specific chromosomes. Although difficult to discern in the karyotype, each metaphase chromosome consists of two closely attached sister chromatids (see the diagram of the first pair of homologous duplicated chromosomes).

MAKE CONNECTIONS *Explain why each chromosome in the karyotype consists of two sister chromatids. (See Figures 12.6 and 12.7.)*

parts of the X and Y are homologous. Most of the genes carried on the X chromosome do not have counterparts on the tiny Y, and the Y chromosome has genes lacking on the X. Due to their role in sex determination, the X and Y chromosomes are called **sex chromosomes**. The other chromosomes are called **autosomes**.

The occurrence of pairs of homologous chromosomes in each human somatic cell is a consequence of our sexual origins. We inherit one chromosome of a pair from each parent. Thus, the 46 chromosomes in our somatic cells are actually two sets of 23 chromosomes—a maternal set (from our mother) and a paternal set (from our father). The number of chromosomes in a single set is represented by n. Any cell with two chromosome sets is called a **diploid cell** and has a diploid number of chromosomes, abbreviated $2n$. For humans, the diploid number is 46 ($2n = 46$), the number of chromosomes in our somatic cells. In a cell in which DNA synthesis has occurred, all the chromosomes are duplicated, and therefore each consists of two identical sister chromatids, associated closely at the centromere and along the arms. (Even though the chromosomes are duplicated, we still say the cell is diploid, or $2n$. This is because it has only two sets of information regardless of the number of chromatids, which are merely copies of the information in one set.) **Figure 13.4** helps clarify the various terms that we use to describe duplicated chromosomes in a diploid cell.

Unlike somatic cells, gametes contain a single set of chromosomes. Such cells are called **haploid cells**, and each has a haploid number of chromosomes (n). For humans, the haploid number is 23 ($n = 23$). The set of 23 consists of the 22 autosomes plus a single sex chromosome. An unfertilized egg contains an X chromosome; a sperm contains either an X or a Y chromosome.

Each sexually reproducing species has a characteristic diploid and haploid number. For example, the fruit fly *Drosophila melanogaster* has a diploid number ($2n$) of 8 and a haploid number (n) of 4, while for dogs, $2n$ is 78 and n is 39. The chromosome number generally does not correlate with the size or complexity of a species' genome; it simply reflects how many linear pieces of DNA make up the genome, which is a function of the evolutionary history of that species (see Concept 21.5). Now let's consider chromosome behavior during sexual life cycles. We'll use the human life cycle as an example.

Behavior of Chromosome Sets in the Human Life Cycle

The human life cycle begins when a haploid sperm from the father fuses with a haploid egg from the mother **(Figure 13.5)**. This union of gametes, culminating in fusion of their nuclei, is

▼ **Figure 13.5 The human life cycle.** In each generation, the number of chromosome sets per cell is halved during meiosis but doubles at fertilization. For humans, the number of chromosomes in a haploid cell is 23, consisting of one set ($n = 23$); the number of chromosomes in the diploid zygote and all somatic cells arising from it is 46, consisting of two sets ($2n = 46$).

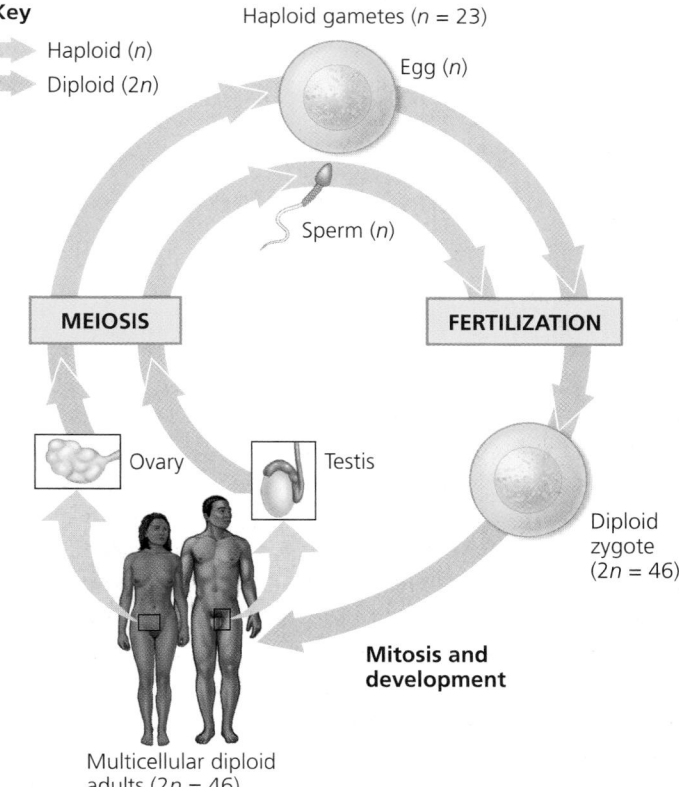

Key

➤ Haploid (n)
➤ Diploid ($2n$)

Haploid gametes ($n = 23$)

Egg (n)

Sperm (n)

MEIOSIS

FERTILIZATION

Ovary

Testis

Diploid zygote ($2n = 46$)

Mitosis and development

Multicellular diploid adults ($2n = 46$)

This figure introduces a color code that will be used for other life cycles later in this book. The aqua arrows identify haploid stages of a life cycle, and the tan arrows identify diploid stages.

➡ **Mastering Biology Animation: The Human Life Cycle**

▼ **Figure 13.4 Describing chromosomes.** A cell from an organism with a diploid number of 6 ($2n = 6$) is depicted here following chromosome duplication and condensation. Each of the six duplicated chromosomes consists of two sister chromatids associated closely along their lengths. Each homologous pair is composed of one chromosome from the maternal set (red) and one from the paternal set (blue). Each set is made up of three chromosomes in this example (long, medium, and short). Together, one maternal and one paternal chromatid in a pair of homologous chromosomes are called nonsister chromatids.

Key
▮ Maternal set of chromosomes ($n = 3$)
▮ Paternal set of chromosomes ($n = 3$) } $2n = 6$

Sister chromatids of one duplicated chromosome

Centromeres

Two nonsister chromatids in a homologous pair

Pair of homologous chromosomes (one from each set)

VISUAL SKILLS *How many sets of chromosomes are present in this diagram? How many pairs of homologous chromosomes are present? If the phases of mitosis in Figure 12.7 were redrawn using the color scheme shown in this figure, describe how the six chromosomes would be colored.*

➡ **Mastering Biology BioFlix® Animation: Chromosomes**

called **fertilization**. The resulting fertilized egg, or **zygote**, is diploid because it contains two haploid sets of chromosomes bearing genes representing the maternal and paternal family lines. As a human develops into a sexually mature adult, mitosis of the zygote and its descendant cells generates all the somatic cells of the body. Both chromosome sets in the zygote and all the genes they carry are passed with precision to the somatic cells.

The only cells of the human body not produced by mitosis are the gametes, which develop from specialized cells called *germ cells* in the gonads—ovaries in females and testes in males (see Figure 13.5). Imagine what would happen if human gametes were made by mitosis: They would be diploid like the somatic cells. At the next round of fertilization, when two gametes fused, the normal chromosome number of 46 would double to 92, and each subsequent generation would double the number of chromosomes yet again. This does not happen, however, because in sexually reproducing organisms, gamete formation involves a type of cell division called **meiosis**. This type of cell division reduces the number of sets of chromosomes from two in the parent cell to one in each gamete, counterbalancing the doubling that occurs at fertilization. As a result of meiosis, each human sperm and egg is haploid ($n = 23$). Fertilization restores the diploid condition by combining two sets of chromosomes, and the human life cycle is repeated, generation after generation (see Figure 13.5).

The steps of the human life cycle are typical of many sexually reproducing animals. Indeed, fertilization and meiosis are also the hallmarks of sexual reproduction in plants, fungi, and protists just as in animals. Fertilization and meiosis alternate in sexual life cycles, maintaining a constant number of chromosomes in a species from one generation to the next.

The Variety of Sexual Life Cycles

Although the alternation of meiosis and fertilization is common to all organisms that reproduce sexually, the timing of these two events in the life cycle varies, depending on the species. These variations can be grouped into three main types of life cycles. In the type that occurs in humans and most other animals, gametes are the only haploid cells **(Figure 13.6a)**. Meiosis occurs in germ cells during the production of gametes, which undergo no further cell division prior to fertilization. After fertilization, the diploid zygote divides by mitosis, producing a multicellular organism that is diploid.

Plants and some species of algae exhibit a second type of life cycle called **alternation of generations (Figure 13.6b)**. This type includes both diploid and haploid stages that are multicellular. The multicellular diploid stage is called the *sporophyte*. Meiosis in the sporophyte produces haploid cells called *spores*. Unlike a gamete, a haploid spore doesn't fuse with another cell but divides mitotically, generating a multicellular haploid stage called the *gametophyte*. Cells of the gametophyte give rise to gametes by mitosis. Fusion of two haploid gametes at fertilization results in a diploid zygote, which develops into the next sporophyte generation. Therefore, in this type of life cycle, the sporophyte generation produces a gametophyte as its offspring, and the gametophyte generation produces a sporophyte. The name *alternation of generations* fits well for this type of life cycle.

A third type of life cycle occurs in most fungi and some protists, including some algae **(Figure 13.6c)**. After gametes fuse and form a diploid zygote, meiosis occurs without a multicellular diploid offspring developing. Meiosis produces not gametes but

▼ **Figure 13.6 Three types of sexual life cycles.** The common feature of all three cycles is the alternation of meiosis and fertilization, key events that contribute to genetic variation among offspring. The cycles differ in the timing of these two key events. (Small circles are cells; large circles are organisms.)

(a) Animals

(b) Plants and some algae

(c) Most fungi and some protists

VISUAL SKILLS *For each type of life cycle, indicate whether haploid cells undergo mitosis, and if they do, describe the cells that are formed.*

haploid cells that then divide by mitosis and give rise to either unicellular descendants or a haploid multicellular adult organism. Subsequently, the haploid organism carries out further mitoses, producing the cells that develop into gametes. The only diploid stage found in these species is the single-celled zygote.

Note that *either* haploid or diploid cells can divide by mitosis, depending on the type of life cycle. Only diploid cells, however, can undergo meiosis: Haploid cells can't because they already have a single set of chromosomes that cannot be further reduced. Though the three types of sexual life cycles differ in the timing of meiosis and fertilization, they share a fundamental result: genetic variation among offspring.

CONCEPT CHECK 13.2

1. **MAKE CONNECTIONS** In Figure 13.4, how many DNA molecules (double helices) are present (see Figure 12.5)? What is the haploid number of this cell? Is a set of chromosomes haploid or diploid?

2. **VISUAL SKILLS** In the karyotype in Figure 13.3, how many pairs of chromosomes are present? How many sets?

3. Using shoes as an analogy for chromosomes, how would you describe the collection of "shoes" in human diploid and haploid cells?

4. **WHAT IF?** A certain eukaryote lives as a unicellular organism, but during environmental stress, it produces gametes. The gametes fuse, and the resulting zygote undergoes meiosis, generating new single cells. What type of organism could this be?

For suggested answers, see Appendix A.

CONCEPT 13.3

Meiosis reduces the number of chromosome sets from diploid to haploid

Several steps of meiosis closely resemble corresponding steps in mitosis. Meiosis, like mitosis, is preceded by interphase, which includes S phase (the duplication of chromosomes). However, this is followed by not one but two consecutive cell divisions, called **meiosis I** and **meiosis II**. These two divisions result in four daughter cells (rather than the two daughter cells of mitosis), each with only half as many chromosomes as the parent cell—one set, rather than two.

The Stages of Meiosis

The overview of meiosis in **Figure 13.7** shows, for a single pair of homologous chromosomes in a diploid cell, that both members of the pair are duplicated and the copies sorted into four haploid daughter cells. Recall that sister chromatids are two copies of *one* chromosome, closely associated all along their lengths; this association is called *sister chromatid cohesion*. Together, the sister chromatids make up one duplicated chromosome (see Figure 13.4). In contrast, the two chromosomes of a homologous pair are individual chromosomes that were inherited from each parent. Homologs appear alike in the microscope, but they

▼ **Figure 13.7 Overview of meiosis: how meiosis reduces chromosome number.** After the chromosomes duplicate in interphase, the diploid cell divides *twice*, yielding four haploid daughter cells. This overview tracks just one pair of homologous chromosomes, which for the sake of simplicity are drawn in the condensed state throughout. (They would not normally be condensed during interphase.)

Interphase

Pair of homologous chromosomes in diploid parent cell

Chromosomes are duplicated.

Pair of duplicated homologous chromosomes

Sister chromatids

Diploid cell with duplicated chromosomes

Meiosis I

① Homologous chromosomes are separated.

Haploid cells with duplicated chromosomes

Meiosis II

② Sister chromatids are separated.

Haploid cells with unduplicated chromosomes

DRAW IT *Redraw the cells in this figure using a simple double helix to represent each DNA molecule.*

➔ **Mastering Biology** Animation: Overview of Meiosis

may have different versions of genes at corresponding loci; each version is called an *allele* of that gene (see Figure 14.4). For example, one chromosome might have an allele for freckles, but the homologous chromosome may have an allele for the absence of freckles at the same locus. Homologs are not associated with each other in any obvious way except during meiosis.

Figure 13.8 describes in detail the stages of the two divisions of meiosis for an animal cell whose diploid number is 6. Study this figure thoroughly before going on.

▼ **Figure 13.8** **Exploring Meiosis in an Animal Cell**

MEIOSIS I: Separates homologous chromosomes

Prophase I	Metaphase I	Anaphase I	Telophase I and Cytokinesis

Centrosome (with centriole pair)

Sister chromatids

Chiasmata

Spindle microtubules

Fragments of nuclear envelope

Pair of homologous chromosomes

Centromere

Duplicated homologous chromosomes (red and blue) pair up and exchange segments; 2*n* = 6 in this example.

Kinetochore (at centromere)

Kinetochore microtubules

Metaphase plate

Chromosomes line up by homologous pairs.

Sister chromatids remain together

Homologous chromosomes separate

The two homologous chromosomes of each pair separate.

Cleavage furrow

Two haploid cells form; each chromosome still consists of two sister chromatids.

Prophase I

• Centrosome movement, spindle formation, and nuclear envelope breakdown occur as in mitosis. Chromosomes condense progressively throughout prophase I.

• During early prophase I, before the stage shown above, each chromosome pairs with its homolog, aligned gene by gene, and **crossing over** occurs: The DNA molecules of nonsister chromatids are broken (by proteins) and are rejoined to each other.

• At the stage shown above, each homologous pair has one or more X-shaped regions called **chiasmata** (singular, *chiasma*), where crossovers have occurred.

• Later in prophase I, microtubules from one pole or the other attach to the kinetochores, one at the centromere of each homolog. (The two kinetochores on the sister chromatids of a homolog are linked together by proteins and act as a single kinetochore.) Microtubules move the homologous pairs toward the metaphase plate (see the metaphase I diagram).

Metaphase I

• Pairs of homologous chromosomes are now arranged at the metaphase plate, with one chromosome of each pair facing each pole.

• Each pair has lined up independently of other pairs. (This arrangement is called *independent assortment*, to be discussed later.)

• Both chromatids of one homolog are attached to kinetochore microtubules from one pole; the chromatids of the other homolog are attached to microtubules from the opposite pole.

Anaphase I

• Breakdown of proteins that are responsible for sister chromatid cohesion along chromatid arms allows homologs to separate.

• The homologs move toward opposite poles, guided by the spindle apparatus.

• Sister chromatid cohesion persists at the centromere, causing the two chromatids of each chromosome to move as a unit toward the same pole.

Telophase I and Cytokinesis

• When telophase I begins, each half of the cell has a complete haploid set of duplicated chromosomes. Each chromosome is composed of two sister chromatids; one or both chromatids include regions of nonsister chromatid DNA.

• Cytokinesis (division of the cytoplasm) usually occurs simultaneously with telophase I, forming two haploid daughter cells.

• In animal cells like these, a cleavage furrow forms. (In plant cells, a cell plate forms.)

• In some species, chromosomes decondense and nuclear envelopes form.

• No chromosome duplication occurs between meiosis I and meiosis II.

➔ **Mastering Biology** Video: Meiosis I in Sperm Formation

MEIOSIS II: Separates sister chromatids

Prophase II	Metaphase II	Anaphase II	Telophase II and Cytokinesis

   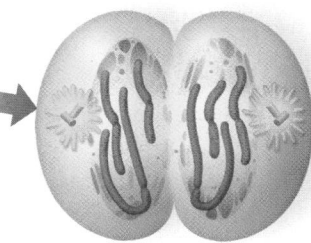

During another round of cell division, the sister chromatids finally separate; four haploid daughter cells result, containing unduplicated chromosomes.

Sister chromatids separate

Haploid daughter cells forming

Prophase II

- A spindle apparatus forms.

- In late prophase II (not shown here), chromosomes, each still composed of two chromatids associated at the centromere, are moved by microtubules toward the metaphase II plate.

Metaphase II

- The chromosomes are positioned at the metaphase plate as in mitosis.

- Because of crossing over in meiosis I, the two sister chromatids of each chromosome are not genetically identical.

- The kinetochores of sister chromatids are attached to microtubules extending from opposite poles.

Anaphase II

- Breakdown of proteins holding the sister chromatids together at the centromere allows the chromatids to separate and move toward opposite poles. Each chromatid has now become an individual chromosome.

Telophase II and Cytokinesis

- Nuclei form, the chromosomes begin decondensing, and cytokinesis occurs.

- The meiotic division of one parent cell produces four daughter cells, each with a haploid set of (unduplicated) chromosomes.

- The four daughter cells are genetically distinct from one another and from the parent cell.

MAKE CONNECTIONS *Look at Figure 12.7 and imagine the two daughter cells undergoing another round of mitosis, yielding four cells. Compare the number of chromosomes in each of those four cells, after mitosis, with the number in each cell in Figure 13.8, after meiosis. Explain how the process of meiosis results in this difference, even though meiosis also includes two cell divisions.*

➔ **Mastering Biology** BioFlix® Animation: Meiosis
Animation: Meiosis

Crossing Over and Synapsis During Prophase I

Prophase I of meiosis is a very busy time. The prophase I cell shown in Figure 13.8 is at a point fairly late in prophase I, when pairing of homologous chromosomes, crossing over, and chromosome condensation have already taken place. The sequence of events leading up to that point is shown in more detail in **Figure 13.9**.

After interphase, the chromosomes have been duplicated and the sister chromatids are held together by proteins called *cohesins*. ❶ Early in prophase I, the two members of a homologous pair associate loosely along their length. Each gene on one homolog is aligned precisely with the corresponding allele of that gene on the other homolog. The DNA of two nonsister chromatids—one maternal and one paternal—is broken by specific proteins at precisely matching points. ❷ Next, the formation of a zipper-like structure called the **synaptonemal complex** holds one homolog tightly to the other. ❸ During this association, called **synapsis**, the DNA breaks are closed up so that each broken end is joined to the corresponding segment of the *nonsister* chromatid. Thus, a paternal chromatid is joined to a piece of maternal chromatid beyond the crossover point, and vice versa.

❹ These points where crossing over has just occurred become visible as chiasmata (singular, *chiasma*) after the synaptonemal complex disassembles and the homologs move slightly apart from each other. The homologs remain attached because sister chromatids are still held together by sister chromatid cohesion, even though some of the DNA may no longer be attached to its original chromosomal DNA. At least one crossover per chromosome must occur in order for the homologous pair to stay together as it moves to the metaphase I plate, for reasons that will be explained shortly.

A Comparison of Mitosis and Meiosis

Figure 13.10 summarizes the key differences between meiosis and mitosis in diploid cells. Meiosis produces four cells and reduces the number of chromosome sets from two (diploid) to one (haploid), whereas mitosis produces two cells and conserves the number of chromosome sets. Meiosis produces cells that differ genetically from the parent cell and from each other, whereas mitosis produces daughter cells that are genetically identical to their parent cell and to each other.

Three events unique to meiosis occur during meiosis I:

1. **Synapsis and crossing over.** During prophase I, duplicated homologs pair up and crossing over occurs as described previously and in Figure 13.9. Synapsis and crossing over do not occur during prophase of mitosis.

2. **Alignment of homologous pairs at the metaphase plate.** At metaphase I of meiosis, pairs of homologs are positioned at the metaphase plate, rather than individual chromosomes, as in metaphase of mitosis.

▼ **Figure 13.9 Crossing over and synapsis in prophase I: a closer look.**

❶ After interphase, the chromosomes have been duplicated, and sister chromatids are held together by proteins called cohesins (purple). Each pair of homologs associate along their length. The DNA molecules of two nonsister chromatids are broken at precisely corresponding points. The chromatin of the chromosomes starts to condense.

❷ A zipper-like protein complex, the synaptonemal complex (green), begins to form, attaching one homolog to the other. The chromatin continues to condense.

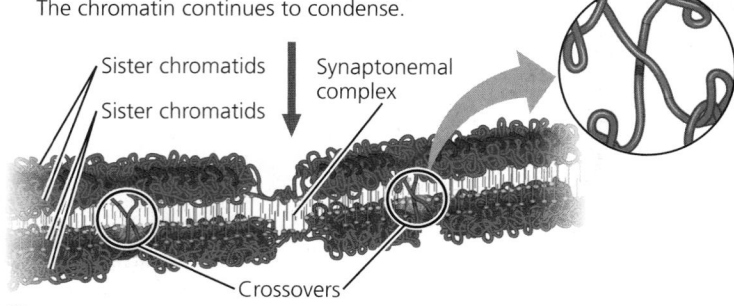

❸ The synaptonemal complex is fully formed; the two homologs are said to be in synapsis. During synapsis, the DNA breaks are closed up when each broken end is joined to the corresponding segment of the nonsister chromatid, producing crossovers.

❹ After the synaptonemal complex disassembles, the homologs move slightly apart from each other but remain attached because of sister chromatid cohesion, even though some of the DNA may no longer be attached to its original chromosome. The points of attachment where crossovers have occurred show up as chiasmata. The chromosomes continue to condense as they move toward the metaphase plate.

VISUAL SKILLS *Label the four DNA breaks in step 2.*

3. **Separation of homologs.** At anaphase I of meiosis, the duplicated chromosomes of each homologous pair move toward opposite poles, but the sister chromatids of each duplicated chromosome remain attached. In anaphase of mitosis, by contrast, sister chromatids separate.

▼ **Figure 13.10 A comparison of mitosis and meiosis.**

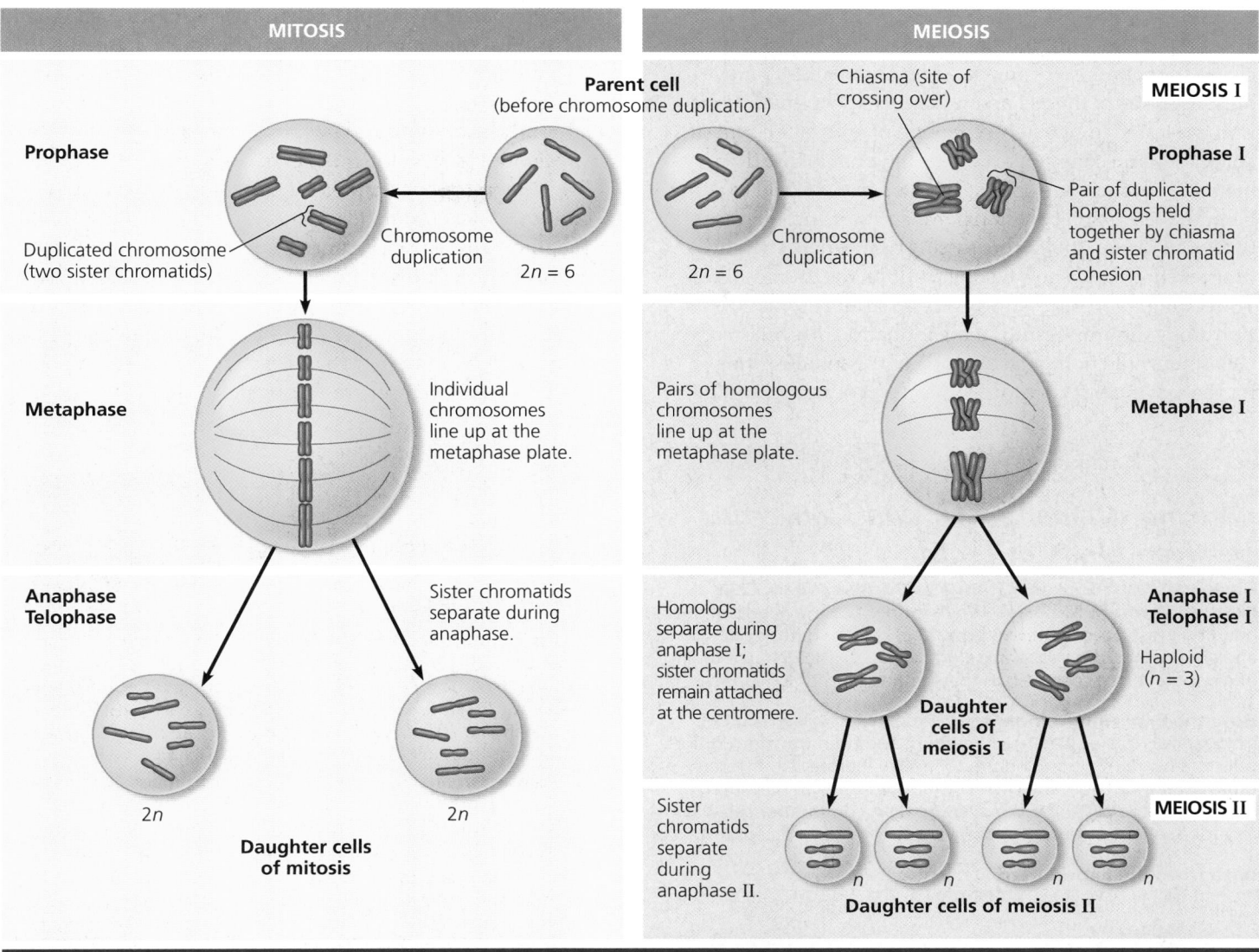

MITOSIS		MEIOSIS	

Parent cell (before chromosome duplication)

Prophase

Chiasma (site of crossing over)

MEIOSIS I

Prophase I

Pair of duplicated homologs held together by chiasma and sister chromatid cohesion

Duplicated chromosome (two sister chromatids)

Chromosome duplication

$2n = 6$

$2n = 6$

Chromosome duplication

Metaphase

Individual chromosomes line up at the metaphase plate.

Pairs of homologous chromosomes line up at the metaphase plate.

Metaphase I

Anaphase Telophase

Sister chromatids separate during anaphase.

Homologs separate during anaphase I; sister chromatids remain attached at the centromere.

Daughter cells of meiosis I

Anaphase I Telophase I

Haploid ($n = 3$)

$2n$

$2n$

Daughter cells of mitosis

Sister chromatids separate during anaphase II.

MEIOSIS II

n n n n

Daughter cells of meiosis II

SUMMARY		
Property	**Mitosis** (occurs in both diploid and haploid cells)	**Meiosis** (can only occur in diploid cells)
DNA replication	Occurs during interphase, before mitosis begins	Occurs during interphase before meiosis I but not before meiosis II
Number of divisions	One, including prophase, prometaphase, metaphase, anaphase, and telophase	Two, each including prophase, metaphase, anaphase, and telophase
Synapsis of homologous chromosomes	Does not occur	Occurs during prophase I along with crossing over between nonsister chromatids; resulting chiasmata hold pairs together due to sister chromatid cohesion
Number of daughter cells and genetic composition	Two, each genetically identical to the parent cell, with the same number of chromosomes	Four, each haploid (n); genetically different from the parent cell and from each other
Role in animals, fungi, and plants	Enables multicellular animal, fungus, or plant (gametophyte or sporophyte) to arise from a single cell; produces cells for growth, repair, and, in some species, asexual reproduction; produces gametes in the plant gametophyte	Produces gametes (in animals) or spores (in fungi and in plant sporophytes); reduces number of chromosome sets by half and introduces genetic variability among the gametes or spores

DRAW IT *Could any other combinations of chromosomes be generated during meiosis II from the specific cells shown in telophase I? Explain. (Hint: Draw the cells as they would appear in metaphase II.)*

→ **Mastering Biology Figure Walkthrough**

Sister chromatids stay together due to sister chromatid cohesion, mediated by cohesin proteins. In mitosis, this attachment lasts until the end of metaphase, when enzymes cleave the cohesins, freeing the sister chromatids to move to opposite poles of the cell. In meiosis, sister chromatid cohesion is released in two steps: cohesion of *arms* at the start of anaphase I and cohesion of *centromeres* at anaphase II. In metaphase I, the two homologs of each pair are held together because there is still cohesion between sister chromatid arms in regions beyond points of crossing over, where stretches of sister chromatids now belong to different chromosomes. The combination of crossing over and sister chromatid cohesion along the arms results in the formation of a chiasma. Chiasmata hold homologs together as the spindle forms for the first meiotic division. At the onset of anaphase I, the

release of cohesion along sister chromatid arms allows homologs to separate. At anaphase II, the release of sister chromatid cohesion at the centromeres allows the sister chromatids to separate. Thus, sister chromatid cohesion and crossing over, acting together, play an essential role in the lining up of chromosomes by homologous pairs at metaphase I.

Meiosis I reduces the number of chromosome sets: from two (diploid) to one (haploid). During the second meiotic division, sister chromatids separate, producing four haploid daughter cells. The mechanisms for separating sister chromatids in meiosis II and mitosis are virtually identical. The molecular basis of chromosome behavior during meiosis continues to be a focus of intense research. In the **Scientific Skills Exercise**, you can work with data that track the amount of DNA in cells as they progress through meiosis.

Scientific Skills Exercise

Making a Line Graph and Converting Between Units of Data

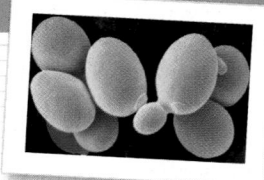

► **Budding yeast cells**

How Does DNA Content Change as Budding Yeast Cells Proceed Through Meiosis? When nutrients are low, cells of the budding yeast (*Saccharomyces cerevisiae*) exit the mitotic cell cycle and enter meiosis. In this exercise, you will track the DNA content of a population of yeast cells as they progress through meiosis.

How the Experiment Was Done Researchers grew a culture of yeast cells in a nutrient-rich medium and then transferred the cells to a nutrient-poor medium to induce meiosis. At different times after induction, the DNA content per cell was measured in a sample of the cells, and the average DNA content per cell was recorded in femtograms (fg; 1 femtogram = 1×10^{-15} gram).

Data from the Experiment

Time After Induction of Meiosis (hr)	Average Amount of DNA per Cell (fg)
0.0	24.0
1.0	24.0
2.0	40.0
3.0	47.0
4.0	47.5
5.0	48.0
6.0	48.0
7.0	47.5
7.5	25.0
8.0	24.0
9.0	23.5
9.5	14.0
10.0	13.0
11.0	12.5
12.0	12.0
13.0	12.5
14.0	12.0

INTERPRET THE DATA

1. First, set up your graph. **(a)** Place the labels for the independent and dependent variables on the appropriate axes, followed by units of measurement in parentheses. Explain your choices. **(b)** Add tick marks and values for each axis. Explain your choices. (For additional information about graphs, see the Scientific Skills Review in Appendix D.)

2. Because the variable on the *x*-axis varies continuously, it makes sense to plot the data on a line graph (if you are using Excel, choose "Scatter"). **(a)** Plot each data point from the table onto the graph. **(b)** Connect the data points with line segments.

3. Most of the yeast cells in the culture were in G_1 of the cell cycle before being moved to the nutrient-poor medium. **(a)** How many femtograms of DNA are there in each yeast cell in G_1? Estimate this value from the data in your graph. **(b)** How many femtograms of DNA should be present in each cell in G_2? (See Concept 12.2 and Figure 12.6.) At the end of meiosis I (MI)? At the end of meiosis II (MII)? (See Figure 13.7.) **(c)** Using these values as a guideline, distinguish the different phases by inserting vertical dashed lines in the graph between phases and label each phase (G_1, S, G_2, MI, MII). You can figure out where to put the dividing lines based on what you know about the DNA content of each phase (see Figure 13.7). **(d)** Think carefully about the point where the line at the highest value begins to slope downward. What specific point of meiosis does this "corner" represent? What stage(s) correspond(s) to the downward sloping line?

4. Given the fact that 1 fg of DNA $= 9.78 \times 10^5$ base pairs (on average), you can convert the amount of DNA per cell to the length of DNA in numbers of base pairs. **(a)** Calculate the number of base pairs of DNA in the haploid yeast genome. Express your answer in millions of base pairs (Mb), a standard unit for expressing genome size. Show your work. **(b)** How many base pairs per minute were synthesized during the S phase of these yeast cells?

➔ **Instructors**: A version of this Scientific Skills Exercise can be assigned in **Mastering Biology**.

Further Reading G. Simchen, Commitment to meiosis: What determines the mode of division in budding yeast? *BioEssays* 31:169–177 (2009).

1. **MAKE CONNECTIONS** Compare the chromosomes in a cell at metaphase of mitosis with those in a cell at metaphase II. (See Figures 12.7 and 13.8.)

2. **WHAT IF?** After the synaptonemal complex disappears, how would any pair of homologous chromosomes be associated if crossing over did not occur? What effect might this have on gamete formation?

For suggested answers, see Appendix A.

CONCEPT 13.4

Genetic variation produced in sexual life cycles contributes to evolution

How do we account for the genetic variation of the family members in Figure 13.1? As you will learn more about in later chapters, mutations are the original source of genetic diversity. These changes in an organism's DNA create the different versions of genes, known as alleles. Once these differences arise, reshuffling of the alleles during sexual reproduction produces the variation that results in each member of a sexually reproducing population having a unique combination of traits.

Origins of Genetic Variation Among Offspring

In species that reproduce sexually, the behavior of chromosomes during meiosis and fertilization is responsible for most of the variation that arises in each generation. Three mechanisms contribute to the genetic variation arising from sexual reproduction: independent assortment of chromosomes, crossing over, and random fertilization.

Independent Assortment of Chromosomes

One aspect of sexual reproduction that generates genetic variation is the random orientation of pairs of homologous chromosomes at metaphase of meiosis I. At metaphase I, the homologous pairs, each consisting of one maternal and one paternal chromosome, are situated at the metaphase plate. (Note that the terms *maternal* and *paternal* refer, respectively, to whether the chromosome in question was contributed by the mother or the father of the individual whose cells are undergoing meiosis.) Each pair may orient with either its maternal or paternal homolog closer to a given pole—its orientation is as random as the flip of a coin. Thus, there is a 50% chance that a particular daughter cell of meiosis I will get the maternal chromosome of a certain homologous pair and a 50% chance that it will get the paternal chromosome.

Because each pair of homologous chromosomes is positioned independently of the other pairs at metaphase I, the first meiotic division results in each pair sorting its maternal and paternal homologs into daughter cells independently of every other pair. This is called *independent assortment*. Each daughter cell represents one outcome of all possible combinations of maternal and paternal chromosomes. As shown in **Figure 13.11**, the number of combinations possible for daughter cells formed by meiosis of a diploid cell with two pairs of homologous chromosomes ($n = 2$) is four: two possible arrangements for the first pair times two possible arrangements for the second pair. Note that only two of the four combinations of daughter cells shown in the figure would result from meiosis of a *single* diploid cell because a single parent cell would have one or the other possible chromosomal arrangement at metaphase I, but not both. However, the population of daughter cells resulting from meiosis of a large number of diploid cells contains all four types in approximately equal numbers. In the case of $n = 3$, eight combinations ($2 \times 2 \times 2 = 2^3$) of chromosomes are possible for daughter cells. More generally, the number of possible combinations when chromosomes sort independently during meiosis is 2^n, where n is the haploid number of the species.

In the case of humans ($n = 23$), the number of possible combinations of maternal and paternal chromosomes in the resulting gametes is 2^{23}, or about 8.4 million. Each gamete that you produce in your lifetime contains one of roughly 8.4 million possible combinations of chromosomes. This is an underestimate, because it doesn't take into account crossing over, which we'll consider next.

Crossing Over

As a consequence of the independent assortment of chromosomes during meiosis, each of us produces a collection

▼ Figure 13.11 **The independent assortment of homologous chromosomes in meiosis.**

Possibility 1　　　　　　　　　　　Possibility 2

Two equally probable arrangements of chromosomes at metaphase I

Metaphase II

Daughter cells

Combination 1　　Combination 2　　　　Combination 3　　Combination 4

→ **Mastering Biology** Animation: Genetic Variation from Independent Assortment of Chromosomes of gametes differing greatly in their combinations of the chromosomes we inherited from our two parents. Figure 13.11 suggests that each chromosome in a gamete is exclusively maternal or paternal in origin. In fact, this is *not* the case, because crossing over produces **recombinant chromosomes**, individual chromosomes that carry genes (DNA) from two different parents **(Figure 13.12)**. In meiosis in humans, an average of one to three crossover events occurs per chromosome pair, depending on the size of the chromosomes and the position of their centromeres.

As you learned in Figure 13.9, crossing over produces chromosomes with new combinations of maternal and paternal alleles. At metaphase II, chromosomes that contain one or more recombinant chromatids can be oriented in two alternative, nonequivalent ways with respect to other chromosomes because their sister chromatids are no longer identical (see Figure 13.12). The different possible arrangements of nonidentical sister chromatids during meiosis II further increase the number of genetic types of daughter cells that can result from meiosis.

You'll learn more about crossing over in Concept 15.3. The important point for now is that crossing over, by combining DNA inherited from two parents into a single chromosome, is an important source of genetic variation in sexual life cycles.

Random Fertilization

The random nature of fertilization adds to the genetic variation arising from meiosis. In humans, each male and female gamete represents one of about 8.4 million (2^{23}) possible chromosome combinations due to independent assortment. The fusion of a male gamete with a female gamete during fertilization will produce a zygote with any of about 70 trillion ($2^{23} \times 2^{23}$) diploid combinations. If we factor in the variation brought about by crossing over, the number of possibilities is truly astronomical. You really *are* unique.

 Mastering Biology Animation: Genetic Variation from Random Fertilization

The Evolutionary Significance of Genetic Variation Within Populations

EVOLUTION Now that you've learned how new combinations of genes arise among offspring in a sexually reproducing population, how does the genetic variation in a population relate to evolution? Darwin recognized that a population evolves through the differential reproductive success of its variant members. On average, those individuals best suited to the local environment leave the most offspring, thereby transmitting their genes. Thus, natural selection results in the accumulation of genetic variations favored in a certain environment. As the environment changes, the population may survive if, in each generation, at least some of its members can cope effectively with the new conditions. Mutations are the original source of different alleles, which are then mixed and

▼ **Figure 13.12 The results of crossing over during meiosis.**

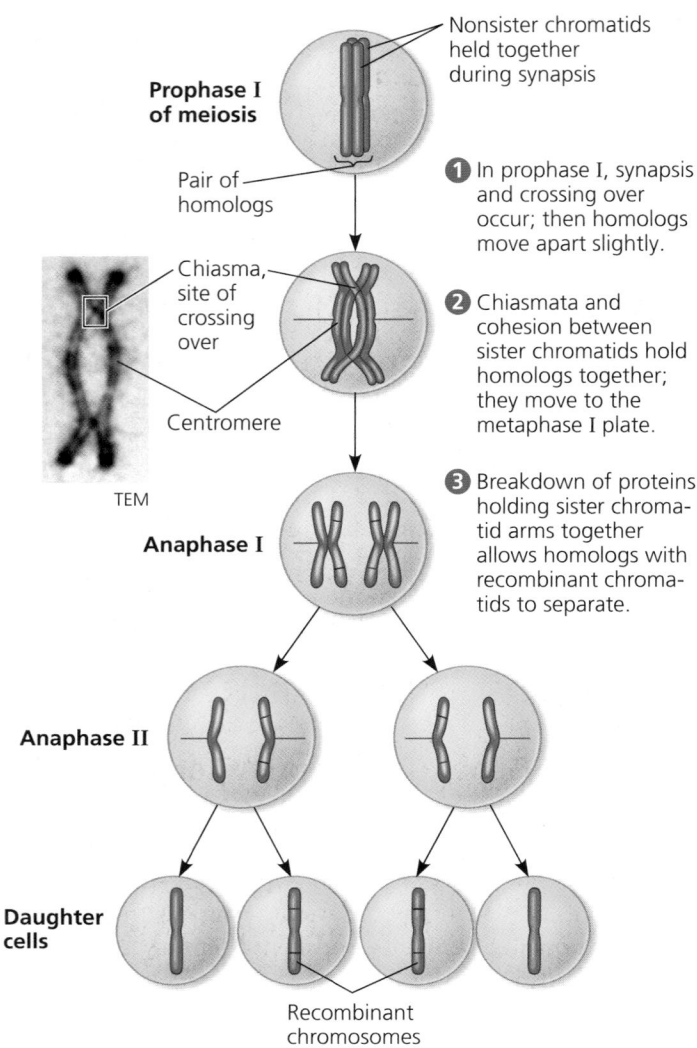

 Mastering Biology Animation: Genetic Variation from Crossing Over

matched during meiosis. New and different combinations of alleles may work better than those that previously prevailed.

In a stable environment, though, sexual reproduction seems as if it would be less advantageous than asexual reproduction, which ensures perpetuation of successful combinations of alleles. Furthermore, sexual reproduction is more expensive energetically than asexual reproduction. In spite of these apparent disadvantages, sexual reproduction is almost universal among animals. Why is this?

The ability of sexual reproduction to generate genetic diversity is the most commonly proposed explanation for the evolutionary persistence of this process. However, consider the unusual case of the bdelloid rotifer **(Figure 13.13)**. It appears that this group may not have reproduced sexually for more than 50 million years of their evolutionary history, a model that has been supported by recent analysis of the genetic sequences in its genome. Does this mean that genetic diversity is not advantageous in this species? It turns out that bdelloid rotifers are an exception to the "rule" that sex alone generates genetic diversity: Bdelloids have mechanisms other than sexual reproduction for generating genetic diversity. For example, they live in environments that can dry up for long periods of time, during which they can enter a state of suspended animation. In this state, their cell membranes may crack in places, allowing entry of DNA from other rotifer species and even from more distantly related species. Evidence suggests that this foreign DNA can become incorporated into the genome of the bdelloid, leading to increased genetic diversity. In fact, the genomic analysis shows that bdelloid rotifers pick up non-bdelloid DNA at a much higher rate than most other species pick up foreign DNA. The conclusion that bdelloid rotifers have developed other ways of achieving genetic diversity supports the idea that genetic diversity is advantageous, but points out as well that sexual reproduction is not the only way of generating such diversity.

▼ Figure 13.13 A bdelloid rotifer, an animal that reproduces only asexually.

200 μm

In this chapter, we have seen how sexual reproduction greatly increases the genetic variation present in a population. Although Darwin realized that heritable variation is what makes evolution possible, he could not explain why offspring resemble—but are not identical to—their parents. Ironically, Gregor Mendel, a contemporary of Darwin, published a theory of inheritance that helps explain genetic variation, but his discoveries had no impact on biologists until 1900, more than 15 years after Darwin (1809–1882) and Mendel (1822–1884) had died. In the next chapter, you'll learn how Mendel discovered the basic rules governing the inheritance of specific traits.

CONCEPT CHECK 13.4

1. What is the original source of variation among the different alleles of a gene?
2. The diploid number for fruit flies is 8, and the diploid number for grasshoppers is 46. If no crossing over took place, would the genetic variation among offspring from a given pair of parents be greater in fruit flies or grasshoppers? Explain.
3. **WHAT IF?** If maternal and paternal chromatids have the same two alleles for every gene, will crossing over lead to genetic variation?

For suggested answers, see Appendix A.

13 Chapter Review

→ Go to **Mastering Biology** for Assignments, the eText, the Study Area, and Dynamic Study Modules.

SUMMARY OF KEY CONCEPTS

→ To review key terms, go to the **Vocabulary Self-Quiz** in the **Mastering Biology** eText or Study Area, or go to goo.gl/zkjz9t.

CONCEPT 13.1

Offspring acquire genes from parents by inheriting chromosomes *(pp. 255–256)*

- Each **gene** in an organism's DNA exists at a specific **locus** on a certain chromosome.
- In **asexual reproduction**, a single parent produces genetically identical offspring by mitosis. **Sexual reproduction** combines genes from two parents, leading to genetically diverse offspring.

? *Explain why human offspring resemble their parents but are not identical to them.*

CONCEPT 13.2

Fertilization and meiosis alternate in sexual life cycles *(pp. 256–259)*

- Normal human **somatic cells** are **diploid**. They have 46 chromosomes made up of two sets of 23 chromosomes, one set from each parent. Human diploid cells have 22 pairs of **homologs** that are **autosomes**, and one pair of **sex chromosomes**; the latter typically determines whether the person is female (XX) or male (XY).
- In humans, ovaries and testes produce **haploid gametes** by **meiosis**, each gamete containing a single set of 23 chromosomes

(*n* = 23). During **fertilization**, an egg and sperm unite, forming a diploid (2*n* = 46) single-celled **zygote**, which develops into a multicellular organism by mitosis.
- Sexual life cycles differ in the timing of meiosis relative to fertilization and in the point(s) of the cycle at which a multicellular organism is produced by mitosis.

? *Compare and contrast the life cycles of animals and plants.*

CONCEPT 13.3

Meiosis reduces the number of chromosome sets from diploid to haploid *(pp. 259–265)*

- The two cell divisions of meiosis, **meiosis I** and **meiosis II**, produce four haploid daughter cells. The number of chromosome sets is reduced from two (diploid) to one (haploid) during meiosis I.
- Meiosis is distinguished from mitosis by three events of meiosis I:

Prophase I: Each pair of homologous chromosomes undergoes **synapsis** and **crossing over** between nonsister chromatids with the subsequent appearance of **chiasmata**.

Metaphase I: Chromosomes line up as homologous pairs on the metaphase plate.

Anaphase I: Homologs separate from each other; sister chromatids remain joined at the centromere.

- Meiosis II separates the sister chromatids.
- Sister chromatid cohesion and crossing over allow chiasmata to hold homologs together until anaphase I. Cohesins are cleaved along the arms at anaphase I, allowing homologs to separate, and at the centromeres in anaphase II, releasing sister chromatids from each other.

? *In prophase I, homologous chromosomes pair up and undergo synapsis and crossing over. Can this also occur during prophase II? Explain.*

CONCEPT 13.4

Genetic variation produced in sexual life cycles contributes to evolution *(pp. 265–267)*

- Three events in sexual reproduction contribute to genetic variation in a population: independent assortment of chromosomes during meiosis I, crossing over during meiosis I, and random fertilization of egg cells by sperm. During crossing over, DNA of nonsister chromatids in a homologous pair is broken and rejoined.
- Genetic variation is the raw material for evolution by natural selection. Mutations are the original source of this variation; recombination of variant genes generates additional genetic diversity.

? *Explain how three processes unique to sexual reproduction generate a great deal of genetic variation.*

TEST YOUR UNDERSTANDING

➡ For more multiple-choice questions, go to the **Practice Test** in the **Mastering Biology** eText or Study Area, or go to goo.gl/GruWRg.

Levels 1-2: Remembering/Understanding

1. A human cell containing 22 autosomes and a Y chromosome is
(A) a sperm.   (C) a zygote.
(B) an egg.   (D) a somatic cell of a male.

2. The two homologs of a pair move toward opposite poles of dividing cell during
(A) mitosis.   (C) meiosis II.
(B) meiosis I.   (D) fertilization.

Levels 3-4: Applying/Analyzing

3. Meiosis II is similar to mitosis in that
(A) sister chromatids separate during anaphase.
(B) DNA replicates before the division.
(C) the daughter cells are diploid.
(D) homologous chromosomes synapse.

4. If the DNA content of a diploid cell in the G_1 phase of the cell cycle is x, then the DNA content of the same cell at metaphase of meiosis I will be
(A) $0.25x$.   (B) $0.5x$.   (C) x.   (D) $2x$.

5. If we continue to follow the cell lineage from question 4, then the DNA content of a single cell at metaphase of meiosis II will be
(A) $0.25x$.   (B) $0.5x$.   (C) x.   (D) $2x$.

6. DRAW IT The diagram shows a cell in meiosis.
(a) Label the appropriate structures with these terms: chromosome (label as duplicated or unduplicated), centromere, kinetochore, sister chromatids, nonsister chromatids, homologous pair (use a bracket when labeling), homolog (label each one), chiasma, sister

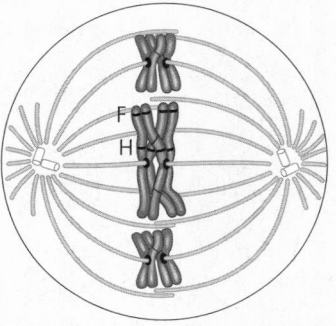

chromatid cohesion, and gene loci. Circle and label the alleles of the F gene.
(b) Precisely describe the makeup of a haploid set and a diploid set in this cell.
(c) Identify the stage of meiosis shown.

Levels 5-6: Evaluating/Creating

7. Explain how you can tell that the cell in question 6 is undergoing meiosis, not mitosis.

8. EVOLUTION CONNECTION Many species can reproduce either asexually or sexually. Explain what you think might be the evolutionary significance of the switch from asexual to sexual reproduction that occurs in some species when the environment becomes unfavorable.

9. SCIENTIFIC INQUIRY The diagram in question 6 represents just a few of the chromosomes of a meiotic cell in a certain person. Assume that the freckles gene is located at the locus marked F, and the hair-color gene is located at the locus marked H, both on the long chromosome. The individual from whom this cell was taken has inherited different alleles for each gene ("freckles" and "black hair" from one parent, and "no freckles" and "blond hair" from the other). Predict allele combinations in the gametes resulting from this meiotic event. (It will help if you draw out the rest of meiosis and label the alleles by name.) List other possible combinations of these alleles in this individual's gametes.

10. WRITE ABOUT A THEME: INFORMATION The continuity of life is based on heritable information in the form of DNA. In a short essay (100–150 words), explain how chromosome behavior during sexual reproduction in animals ensures perpetuation of parental traits in offspring and, at the same time, genetic variation among offspring.

11. SYNTHESIZE YOUR KNOWLEDGE

The Cavendish banana, the world's most popular fruit, is currently threatened by extinction due to a fungus. This banana variety is "triploid" ($3n$, with three sets of chromosomes) and can only reproduce through cloning by cultivators. Given what you know about meiosis, explain how the banana's triploid number accounts for its inability to form normal gametes. Considering genetic diversity, discuss how the absence of sexual reproduction might make this domesticated species vulnerable to infectious agents.

For selected answers, see Appendix A.

14 Mendel and the Gene Idea

KEY CONCEPTS

14.1 Mendel used the scientific approach to identify two laws of inheritance *p. 270*

14.2 Probability laws govern Mendelian inheritance *p. 276*

14.3 Inheritance patterns are often more complex than predicted by simple Mendelian genetics *p. 278*

14.4 Many human traits follow Mendelian patterns of inheritance *p. 284*

Study Tip

Solving genetics problems: In Figure 14.5, you'll learn how to predict the offspring of a genetic cross with a Punnett square, like the example shown below. For crosses shown in other figures, cover up the Punnett square and draw your own. Also, check out the Tips for Genetics Problems at the end of the chapter.

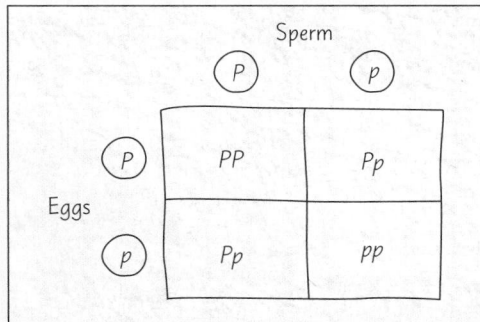

→ Go to Mastering Biology

For Students (in eText and Study Area)
• Get Ready for Chapter 14
• Figure 14.4 Walkthrough: Alleles, Alternative Versions of a Gene
• Animation: Cross of One Character in "MendAliens"

For Instructors to Assign (in Item Library)
• Tutorial: Mendel's Law of Independent Assortment
• Tutorial: Pedigree Analysis: Galactosemia

Figure 14.1 Flower color was one of the many characteristics of pea plants studied by Gregor Mendel. By breeding pea plants over many generations and carefully counting the types of offspring, Mendel developed a theory of inheritance that is the basis of modern genetics.

How are traits, such as the purple or white color of flowers, transmitted from parents to offspring?

Alternative versions of a gene (alleles) account for different traits.

Each parent cell has two alleles for each character, such as flower color. Here, the alleles are identical in each parent.

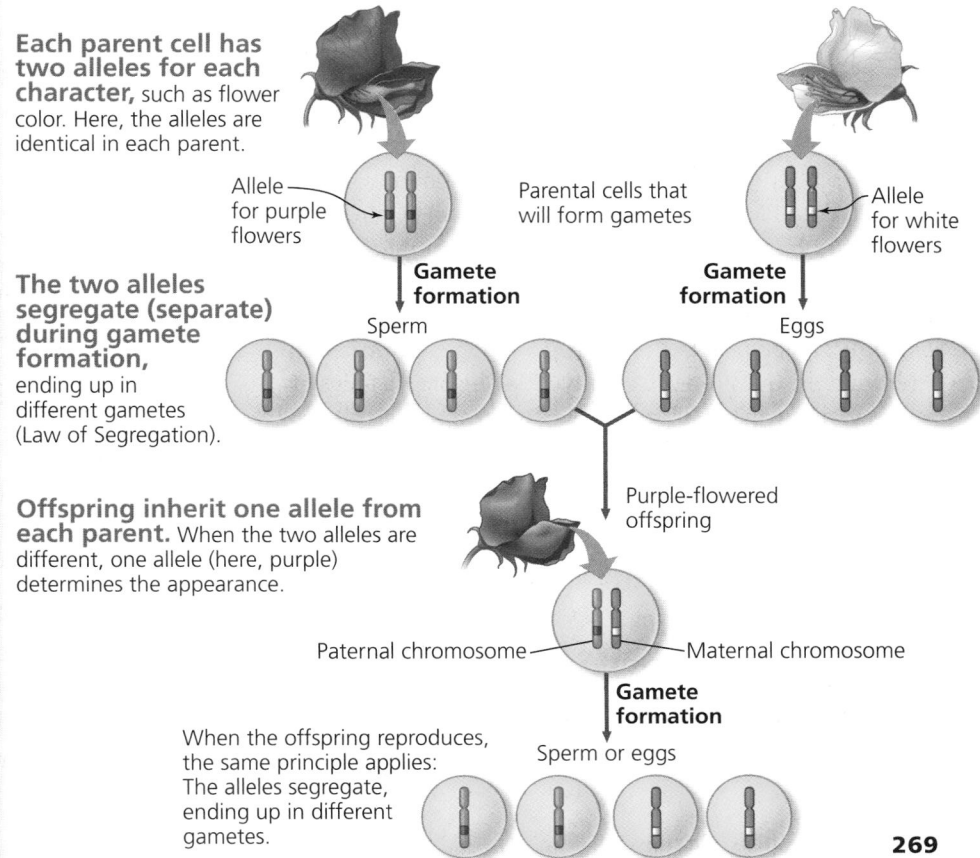

Allele for purple flowers

Parental cells that will form gametes

Allele for white flowers

Gamete formation

The two alleles segregate (separate) during gamete formation, ending up in different gametes (Law of Segregation).

Sperm

Eggs

Offspring inherit one allele from each parent. When the two alleles are different, one allele (here, purple) determines the appearance.

Purple-flowered offspring

Paternal chromosome — — Maternal chromosome

Gamete formation

When the offspring reproduces, the same principle applies: The alleles segregate, ending up in different gametes.

Sperm or eggs

Mendel used the scientific approach to identify two laws of inheritance

Modern genetics began during the mid-1800s with a monk named Gregor Mendel, who discovered the basic principles of heredity by breeding garden peas in carefully planned experiments. As we retrace his work, you will recognize the key elements of the scientific process that were introduced in Concept 1.3.

Mendel's Experimental, Quantitative Approach

Mendel grew up on his parents' small farm in a region of Austria that is now part of the Czech Republic. In this agricultural area, Mendel received agricultural training in school along with a basic education. As an adolescent, Mendel overcame financial hardship and illness to excel in high school and, later, at the Olmutz Philosophical Institute.

In 1843, at the age of 21, Mendel entered an Augustinian monastery, a reasonable choice at that time for someone who valued the life of the mind. He considered becoming a teacher but failed the necessary examination. In 1851, he left the monastery to pursue two years of study in physics and chemistry at the University of Vienna. These were very important years for Mendel's development as a scientist, in large part due to the strong influence of two professors. One was the physicist Christian Doppler, who encouraged his students to learn science through experimentation and trained Mendel to use mathematics to help explain natural phenomena. The other was a botanist named Franz Unger, who aroused Mendel's interest in the causes of variation in plants.

After attending university, Mendel returned to the monastery and was assigned to teach at a local school, where several other instructors were enthusiastic about scientific research. In addition, his fellow monks shared a long-standing fascination with the breeding of plants. Around 1857, Mendel began breeding garden peas in the abbey garden to study inheritance. Although the question of heredity had long been a focus of curiosity at the monastery, Mendel's fresh approach allowed him to deduce principles that had remained elusive to others.

One reason Mendel probably chose to work with peas is that there are many varieties. For example, one variety has purple flowers, while another variety has white flowers. A heritable feature that varies among individuals, such as flower color, is called a **character**. Each variant for a character, such as purple or white color for flowers, is called a **trait**.

Other advantages of using peas are their short generation time and the large number of offspring from each mating. Furthermore, Mendel could strictly control mating between

plants **(Figure 14.2)**. Each pea flower has both pollen-producing organs (stamens) and an egg-bearing organ (carpel). In nature, pea plants usually self-fertilize: Pollen grains from the stamens land on the carpel of the same flower, and sperm released from the pollen grains fertilize eggs present in the carpel.* To achieve

▼ Figure 14.2 Research Method

Crossing Pea Plants

Application By crossing, or mating, two true-breeding varieties of an organism, scientists can study patterns of inheritance. In this example, Mendel crossed pea plants that varied in flower color.

Technique

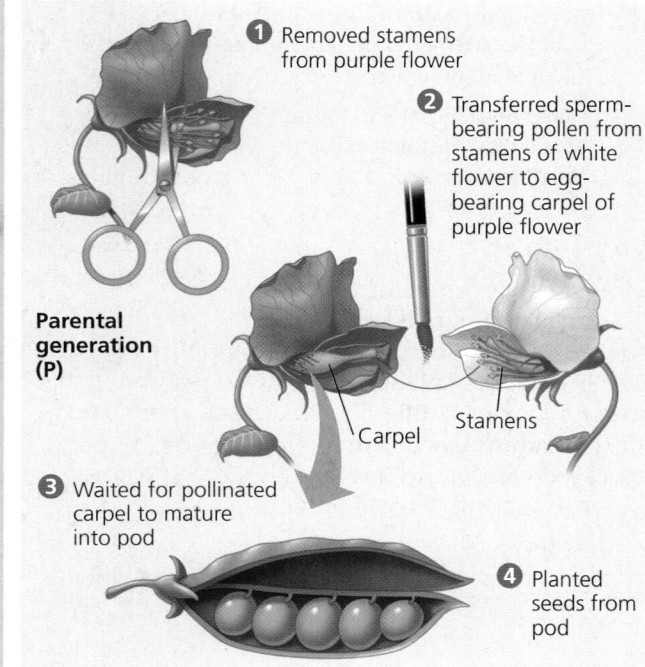

❶ Removed stamens from purple flower

❷ Transferred sperm-bearing pollen from stamens of white flower to egg-bearing carpel of purple flower

Parental generation (P)

Carpel

Stamens

❸ Waited for pollinated carpel to mature into pod

❹ Planted seeds from pod

Results When pollen from a white flower was transferred to a purple flower, the first-generation hybrids all had purple flowers. The result was the same for the reciprocal cross, which involved the transfer of pollen from purple flowers to white flowers.

First filial generation offspring (F₁)

❺ Examined offspring: all had purple flowers

---

*As you learned in Figure 13.6b, meiosis in plants produces spores, not gametes. In flowering plants like the pea, each spore develops into a microscopic haploid gametophyte that contains only a few cells and is located on the parent plant. The gametophyte produces sperm, in pollen grains, and eggs, in the carpel. For simplicity, we will not include the gametophyte stage in our discussion of fertilization in plants.

cross-pollination of two plants, Mendel removed the immature stamens of a plant before they produced pollen and then dusted pollen from another plant onto the altered flowers (see Figure 14.2). Each resulting zygote then developed into a plant embryo encased in a seed (a pea). His method allowed Mendel to always be sure of the parentage of new seeds.

Mendel chose to track only those characters that occurred in two distinct, alternative forms, such as purple or white flower color. He also made sure that he started his experiments with varieties that were **true-breeding**—that is, over many generations of self-pollination, these plants had produced only the same variety as the parent plant. For example, a plant with purple flowers is true-breeding if the seeds produced by self-pollination in successive generations all give rise to plants that also have purple flowers.

In a typical breeding experiment, Mendel cross-pollinated two contrasting, true-breeding pea varieties—for example, purple-flowered plants and white-flowered plants (see Figure 14.2). This mating, or *crossing*, of two true-breeding varieties is called **hybridization**. The true-breeding parents are referred to as the **P generation** (parental generation), and their hybrid offspring are the **F_1 generation** (first filial generation, the word *filial* from the Latin word for "son"). Allowing these F_1 hybrids to self-pollinate (or to cross-pollinate with other F_1 hybrids) produces an **F_2 generation** (second filial generation). Mendel usually followed traits for at least the P, F_1, and F_2 generations. Had he stopped his experiments with the F_1 generation, the basic patterns of inheritance would have eluded him. Mendel's quantitative analysis of the F_2 plants from thousands of genetic crosses like these allowed him to deduce two fundamental principles of heredity, now called the law of segregation and the law of independent assortment.

The Law of Segregation

The explanation of heredity most widely in favor during the 1800s was the "blending" hypothesis, the idea that genetic material contributed by the two parents mixes just as blue and yellow paints blend to make green. This hypothesis predicts that over many generations, a freely mating population will give rise to a uniform population of individuals, something we don't see.

If the blending model were correct, the F_1 hybrids from Mendel's cross between purple-flowered and white-flowered pea plants would have been pale purple flowers, a trait intermediate between those of the P generation. Notice in Figure 14.2 that the experiment produced a very different result: All the F_1 offspring had flowers of the same color as the purple-flowered parents. What happened to the white-flowered plants' genetic contribution to the hybrids? If it were lost, then the F_1 plants could produce only purple-flowered offspring in the F_2 generation. But when Mendel allowed the F_1 plants to self- or cross-pollinate and

planted their seeds, the white-flower trait reappeared in the F_2 generation. The blending hypothesis is also inconsistent with this reappearance of traits after they've skipped a generation.

Mendel used very large sample sizes and kept accurate records of his results: 705 of the F_2 plants had purple flowers, and 224 had white flowers. These data fit a ratio of approximately three purple to one white **(Figure 14.3)**. Mendel reasoned that the heritable factor for white flowers did not disappear in the F_1 plants but was somehow hidden, or masked,

▼ **Figure 14.3** Inquiry

When F_1 hybrid pea plants self- or cross-pollinate, which traits appear in the F_2 generation?

Experiment Mendel crossed true-breeding purple-flowered plants and white-flowered plants (crosses are symbolized by ×). The resulting F_1 hybrids were allowed to self-pollinate or were cross-pollinated with other F_1 hybrids. The F_2 generation plants were then observed for flower color.

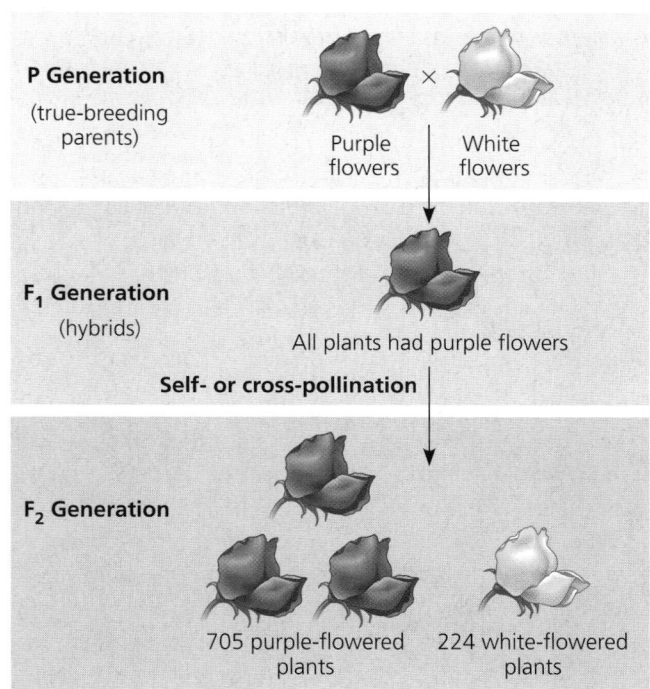

Results Both purple-flowered and white-flowered plants appeared in the F_2 generation, in a ratio of approximately 3:1.

Conclusion The "heritable factor" for the recessive trait (white flowers) had not been destroyed, deleted, or "blended" in the F_1 generation but was merely masked by the presence of the factor for purple flowers, which is the dominant trait.

Data from G. Mendel, Experiments in plant hybridization, *Proceedings of the Natural History Society of Brünn* 4:3–47 (1866).

WHAT IF? *If you mated two purple-flowered plants from the P generation, what ratio of traits would you expect to observe in the offspring? Explain. What might Mendel have concluded if he stopped his experiment after the F_1 generation?*

when the purple-flower factor was present. In Mendel's terminology, purple flower color is a *dominant* trait, and white flower color is a *recessive* trait. The reappearance of white-flowered plants in the F₂ generation was evidence that the heritable factor causing white flowers had not been diluted or destroyed by coexisting with the purple-flower factor in the F₁ hybrids. Mendel observed the same pattern of inheritance in six other characters, each represented by two distinctly different traits **(Table 14.1)**. For example, when Mendel crossed a true-breeding variety that produced smooth, round pea seeds with one that produced wrinkled seeds, all the F₁ hybrids produced round seeds; this is the dominant trait for seed shape. In the F₂ generation, approximately 75% of the seeds were round and 25% were wrinkled—a 3:1 ratio, as in Figure 14.3. Now let's see how Mendel deduced the law of segregation from his experimental results. In the discussion that follows, we will use modern terms instead of some of the terms used by Mendel. (For example, we'll use "gene" instead of Mendel's "heritable factor.")

Table 14.1 The Results of Mendel's F₁ Crosses for Seven Characters in Pea Plants

Character	Dominant Trait	×	Recessive Trait	F₂ Generation Dominant: Recessive	Ratio
Flower color	Purple	×	White	705:224	3.15:1
Seed color	Yellow	×	Green	6,022:2,001	3.01:1
Seed shape	Round	×	Wrinkled	5,474:1,850	2.96:1
Pod color	Green	×	Yellow	428:152	2.82:1
Pod shape	Inflated	×	Constricted	882:299	2.95:1
Flower position	Axial	×	Terminal	651:207	3.14:1
Stem length	Tall	×	Dwarf	787:277	2.84:1

Mendel's Model

Mendel developed a model to explain the 3:1 inheritance pattern that he consistently observed among the F₂ offspring in his pea experiments. We'll now explore four related concepts making up Mendel's model, the fourth of which is the law of segregation.

First, *alternative versions of genes account for variations in inherited characters.* The gene for flower color in pea plants, for example, exists in two versions, one for purple flowers and the other for white flowers. These alternative versions of a gene are called **alleles**. Today, we can relate this concept to chromosomes and DNA. As shown in **Figure 14.4**, each gene is a sequence of nucleotides at a specific place, or locus, along a particular chromosome. The DNA at that locus, however, can vary slightly in its nucleotide sequence. This variation in information content can affect the function of the encoded protein and thus an inherited character of the organism. The purple-flower allele and the white-flower allele are two DNA sequence variations possible at the flower-color locus on a pea plant's chromosomes. The purple-flower allele sequence allows synthesis of purple pigment, and the white-flower allele sequence does not.

Second, *for each character, an organism inherits two versions (that is, two alleles) of a gene, one from each parent.* Remarkably, Mendel made this deduction without knowing about the role, or even the existence, of chromosomes. Each somatic cell in a diploid organism has two sets of chromosomes, one set inherited from each parent (see Figure 13.4). Thus, a genetic locus is actually represented twice in a diploid cell, once on each homolog of a specific pair of chromosomes. The two alleles at a particular locus may be identical, as in the true-breeding plants of Mendel's P generation. Or the alleles may differ, as in the F₁ hybrids (see Figure 14.4).

Third, *if the two alleles at a locus differ, then one, the* **dominant allele**, *determines the organism's appearance; the other, the* **recessive allele**, *has no noticeable effect on the organism's appearance.* Accordingly, Mendel's F₁ plants had purple flowers because the allele for that trait is dominant and the allele for white flowers is recessive.

The fourth and final part of Mendel's model, the **law of segregation**, states that *the two alleles for a heritable character segregate (in other words, separate from each other) during gamete formation and end up in different gametes.* Thus, an egg or a sperm gets only one of the two alleles that are present in the diploid cells of the organism making the gamete. In terms of chromosomes, this segregation corresponds to the distribution of copies of the two members of a pair of homologous chromosomes to different gametes in meiosis (see Figure 13.7). Note that if an organism has identical alleles for a particular character, then that allele is present in all gametes. Because it is the only allele that can be passed on to offspring when the plant self-pollinates, the offspring always look the same as their parents in regard to that characteristic; this explains why these plants are true-breeding. But if different alleles are

► **Figure 14.4 Alleles, alternative versions of a gene.** This diagram shows a pair of homologous chromosomes in an F₁ hybrid pea plant, with the actual DNA sequence from the flower-color allele of each chromosome. The paternally inherited chromosome (blue) has an allele for purple flowers, which codes for a protein that indirectly controls synthesis of purple pigment. The maternally inherited chromosome (red) has an allele for white flowers, which results in no functional protein being made.

→ **Mastering Biology**
Figure Walkthrough

Allele for purple flowers

Locus for flower-color gene

Pair of homologous chromosomes

Allele for white flowers

DNA with nucleotide sequence CTAAATCGGT

Through a series of steps, this DNA sequence results in production of an enzyme that helps synthesize purple pigment.

Enzyme

This DNA sequence results in the absence of the enzyme.

DNA with nucleotide sequence ATAAATCGGT

One purple-flower allele results in sufficient pigment for purple flowers.

present, as in the F₁ hybrids, then 50% of the gametes receive the dominant allele and 50% receive the recessive allele.

Does Mendel's segregation model account for the 3:1 ratio he observed in the F₂ generation of his numerous crosses? For the flower-color character, the model predicts that the two different alleles present in an F₁ individual will segregate into gametes such that half the gametes will have the purple-flower allele and half will have the white-flower allele. During self-pollination, gametes of each class unite randomly. An egg with a purple-flower allele has an equal chance of being fertilized by a sperm with a purple-flower allele or by a sperm with a white-flower allele. Since the same is true for an egg with a white-flower allele, there are four equally likely combinations of sperm and egg. **Figure 14.5** illustrates these combinations using a **Punnett square**, a handy diagrammatic device for predicting the allele composition of offspring from a cross between individuals of known genetic makeup. Notice that we use a capital letter to symbolize a dominant allele and a lowercase letter for a recessive allele. In our example, *P* is the purple-flower allele, and *p* is the white-flower allele; it is often useful as well to be able to refer to the gene itself as the *P/p* gene.

In the F₂ offspring, what color will the flowers be? One-fourth of the

▼ **Figure 14.5 Mendel's law of segregation.** This diagram shows the genetic makeup of the generations in Figure 14.3. It illustrates Mendel's model for inheritance of the alleles of a single gene. Each plant has two alleles for the gene controlling flower color, one allele inherited from each of the plant's parents. To construct a Punnett square that predicts the F₂ generation offspring, we list all the possible gametes from one parent (here, the F₁ female) along the left side of the square and all the possible gametes from the other parent (here, the F₁ male) along the top. The boxes represent the offspring resulting from all the possible unions of male and female gametes.

P Generation

Appearance:
Genetic makeup:
Purple flowers
PP
White flowers
pp

Gametes:
P
p

Each true-breeding plant of the parental generation has two identical alleles, denoted as either *PP* or *pp*.

Gametes (circles) each contain only one allele for the flower-color gene. In this case, every gamete produced by a given parent has the same allele.

F₁ Generation

Appearance:
Genetic makeup:
Purple flowers
Pp

Gametes:
½ *P*
½ *p*

Union of parental gametes produces F₁ hybrids having a *Pp* combination. Because the purple-flower allele is dominant, all these hybrids have purple flowers.

When the hybrid plants produce gametes, the two alleles segregate. Half of the gametes receive the *P* allele and the other half the *p* allele.

F₂ Generation

Sperm from F₁ (*Pp*) plant

P   *p*

Eggs from F₁ (*Pp*) plant

P
PP   *Pp*

p
Pp   *pp*

3 : 1

This box, a Punnett square, shows all possible combinations of alleles in offspring that result from an F₁ × F₁ (*Pp* × *Pp*) cross. Each square represents an equally probable product of fertilization. For example, the bottom left box shows the genetic combination resulting from a *P* egg fertilized by a *P* sperm.

Random combination of the gametes results in the 3:1 ratio that Mendel observed in the F₂ generation.

→ **Mastering Biology** **Animation: Mendel's Cross of One Character: Flower Color**
Animation: Simplified Crosses of One Character in Humans
Animation: Cross of One Character in "MendAliens"

plants have inherited two purple-flower alleles; clearly, these plants will have purple flowers. One-half of the F_2 offspring have inherited one purple-flower allele and one white-flower allele; these plants will also have purple flowers, the dominant trait. Finally, one-fourth of the F_2 plants have inherited two white-flower alleles and will express the recessive trait. Thus, Mendel's model accounts for the 3:1 ratio of traits that he observed in the F_2 generation.

Useful Genetic Vocabulary

An organism that has a pair of identical alleles for a gene encoding a character is called a **homozygote** and is said to be **homozygous** for that gene. In the parental generation in Figure 14.5, the purple-flowered pea plant is homozygous for the dominant allele (*PP*), while the white plant is homozygous for the recessive allele (*pp*). Homozygous plants "breed true" because all of their gametes contain the same allele—either *P* or *p* in this example. If we cross dominant homozygotes with recessive homozygotes, every offspring will have two different alleles—*Pp* in the case of the F_1 hybrids of our flower-color experiment (see Figure 14.5). An organism that has two different alleles for a gene is called a **heterozygote** and is said to be **heterozygous** for that gene. Unlike homozygotes, heterozygotes produce gametes with different alleles, so they are not true-breeding. For example, *P*- and *p*-containing gametes are both produced by our F_1 hybrids. Self-pollination of the F_1 hybrids thus produces both purple-flowered and white-flowered offspring.

Because of the different effects of dominant and recessive alleles, an organism's traits do not always reveal its genetic composition. Therefore, we distinguish between an organism's appearance or observable traits, called its **phenotype**, and its genetic makeup, its **genotype**. As shown in Figure 14.5 for the case of flower color in pea plants, *PP* and *Pp* plants have the same phenotype (purple flowers) but different genotypes. **Figure 14.6** reviews these terms. Note that the term *phenotype* refers not only to traits that relate directly to physical appearance but also to physiological traits. For example, one pea variety lacks the normal ability to self-pollinate, which is a phenotypic trait (called non-self-pollination).

The Testcross

Given a purple-flowered pea plant, we cannot tell if it is homozygous (*PP*) or heterozygous (*Pp*) because both genotypes result in the same purple phenotype. To determine the genotype, we can cross this plant with a white-flowered plant (*pp*), which will make only gametes with the recessive allele (*p*). The allele in the gamete contributed by the purple-flowered plant of unknown genotype will therefore determine the appearance of the offspring **(Figure 14.7)**. If all the offspring of the cross have purple flowers, then the purple-flowered mystery plant must be homozygous for the dominant allele, because a *PP* × *pp* cross produces all *Pp* offspring. But if both the purple and the white phenotypes

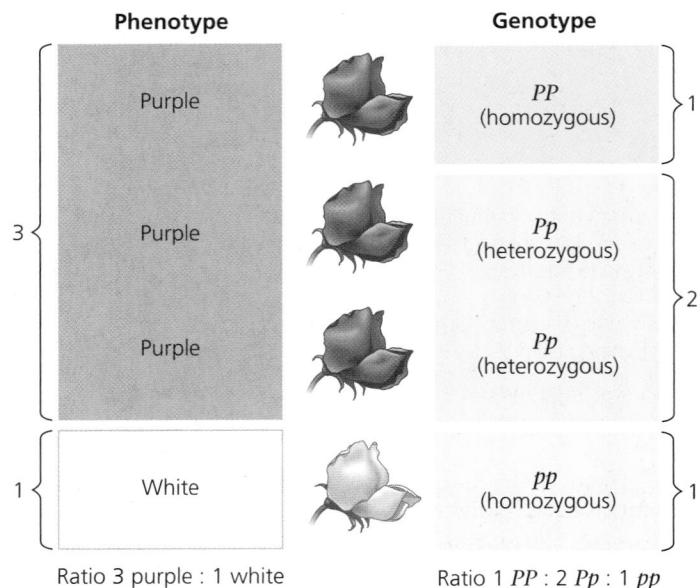

▼ **Figure 14.6 Phenotype versus genotype.** Grouping F_2 offspring from a cross for flower color according to phenotype results in the typical 3:1 phenotypic ratio. In terms of genotype, however, there are actually two categories of purple-flowered plants, *PP* (homozygous) and *Pp* (heterozygous), giving a 1:2:1 genotypic ratio.

Ratio 3 purple : 1 white

Ratio 1 *PP* : 2 *Pp* : 1 *pp*

appear among the offspring, then the purple-flowered parent must be heterozygous. The offspring of a *Pp* × *pp* cross will be expected to have a 1:1 phenotypic ratio. Breeding an organism of unknown genotype with a recessive homozygote is called a **testcross** because it can reveal the genotype of that organism. The testcross was devised by Mendel and continues to be used by geneticists.

The Law of Independent Assortment

Mendel derived the law of segregation from experiments in which he followed only a *single* character, such as flower color. All the F_1 progeny produced in his crosses of true-breeding parents were **monohybrids**, meaning that they were heterozygous for the one particular character being followed in the cross. We refer to a cross between such heterozygotes as a **monohybrid cross**.

Mendel worked out the second law of inheritance by following *two* characters at the same time, such as seed color and seed shape. Seeds (peas) may be either yellow or green. They also may be either round (smooth) or wrinkled. From single-character crosses, Mendel knew that the allele for yellow seeds is dominant (*Y*), and the allele for green seeds is recessive (*y*). For the seed-shape character, the allele for round is dominant (*R*), and the allele for wrinkled is recessive (*r*).

Imagine crossing two true-breeding pea varieties that differ in *both* of these characters—a cross between a plant with yellow round seeds (*YYRR*) and a plant with green wrinkled seeds (*yyrr*). The F_1 plants will be **dihybrids**, individuals heterozygous for the two characters being followed in the cross (*YyRr*). But are these two characters transmitted from parents to

▼ Figure 14.7 Research Method

The Testcross

Application An organism that shows a dominant trait in its phenotype, such as purple flowers in pea plants, can be either homozygous for the dominant allele or heterozygous. To determine the organism's genotype, geneticists can perform a testcross.

Technique In a testcross, the individual with the unknown genotype is crossed with a homozygous individual expressing the recessive trait (white flowers in this example), and Punnett squares are used to predict the possible outcomes.

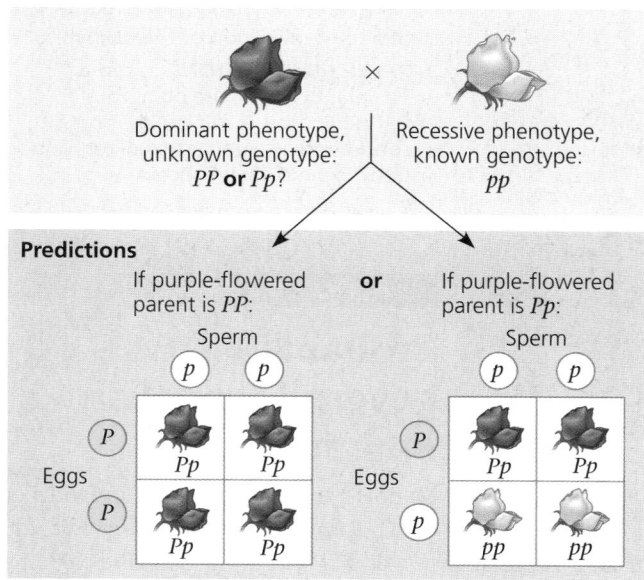

Results Matching the results to either prediction identifies the unknown parental genotype (either *PP* or *Pp* in this example). In this testcross, we transferred pollen from a white-flowered plant to the carpels of a purple-flowered plant; the opposite (reciprocal) cross would have led to the same results.

→ **Mastering Biology** Animation: Testcross in "MendAliens"

offspring as a package? That is, will the *Y* and *R* alleles always stay together, generation after generation? Or are seed color and seed shape inherited independently? **Figure 14.8** shows how a **dihybrid cross**, a cross between F₁ dihybrids, can determine which of these two hypotheses is correct.

The F₁ plants, of genotype *YyRr*, exhibit both dominant phenotypes, yellow seeds with round shapes, no matter which hypothesis is correct. The key step in the experiment is to see what happens when F₁ plants self-pollinate and produce F₂ offspring. If the hybrids must transmit their alleles in the same combinations in which the alleles were inherited

from the P generation, then the F₁ hybrids will produce only two classes of gametes: *YR* and *yr*. As shown on the left side of Figure 14.8, this "dependent assortment" hypothesis predicts that the phenotypic ratio of the F₂ generation will be 3:1, just as in a monohybrid cross:

The alternative hypothesis is that the two pairs of alleles segregate independently of each other. In other words, genes are packaged into gametes in all possible allelic combinations, as long as each gamete has one allele for each gene (see Figure 13.11). In our example, an F₁ plant will produce four classes of gametes in equal quantities: *YR, Yr, yR,* and *yr*. If sperm of the four classes fertilize eggs of the four classes, there will be 16 (4 × 4) equally probable ways in which the alleles can combine in the F₂ generation, as shown on the right side of Figure 14.8. These combinations result in four phenotypic categories with a ratio of 9:3:3:1 (9 yellow round to 3 green round to 3 yellow wrinkled to 1 green wrinkled):

When Mendel did the experiment and classified the F₂ offspring, his results were close to the predicted 9:3:3:1 phenotypic ratio, supporting the hypothesis that the alleles for one gene—the gene for seed color, for example—segregate into gametes independently of the alleles of any other gene, such as the gene for seed shape.

Mendel tested his seven pea characters in various dihybrid combinations and always observed a 9:3:3:1 phenotypic ratio in the F₂ generation. Is this consistent with the 3:1 phenotypic ratio seen for the monohybrid cross shown in Figure 14.5? To answer this question, count the number of yellow and green peas, ignoring shape, and calculate the ratio. The results of Mendel's dihybrid experiments are the basis for what we now call the **law of independent assortment**, which states that *two or more genes assort independently—that is, each pair of alleles segregates independently of any other pair of alleles—during gamete formation.*

This law applies only to genes (allele pairs) located on different chromosomes (that is, on chromosomes that are not homologous) or, alternatively, to genes that are very far apart on the same chromosome. (This will be explained in Concept 15.3, along with the more complex inheritance patterns of genes located near each other, alleles of which tend to be inherited together.) All the pea characters Mendel chose for analysis were controlled by genes on different chromosomes or were far apart on the same chromosome; this situation greatly simplified interpretation of his multicharacter pea crosses. All the examples we consider in the rest of this chapter involve genes located on different chromosomes.

→ **Mastering Biology** Animation: Independent Assortment

▼ Figure 14.8 Inquiry

Do the alleles for one character segregate into gametes dependently or independently of the alleles for a different character?

Experiment To follow the characters of seed color and seed shape through the F₂ generation, Mendel crossed a true-breeding plant with yellow round seeds with a true-breeding plant with green wrinkled seeds, producing dihybrid F₁ plants. Self-pollination of the F₁ dihybrids produced the F₂ generation. The two hypotheses (dependent and independent "assortment" of the two genes) predict different phenotypic ratios.

Results

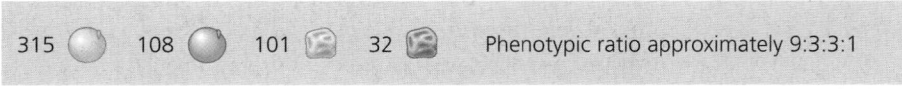

315 ◯   108 ◯   101 ▣   32 ▣   Phenotypic ratio approximately 9:3:3:1

Conclusion The results support the hypothesis of independent assortment, the only one that predicts two newly observed phenotypes: green round seeds and yellow wrinkled seeds (see the right-hand Punnett square). The alleles for each gene segregate independently of those of the other, and the two genes are said to assort independently.

Data from G. Mendel, Experiments in plant hybridization, *Proceedings of the Natural History Society of Brünn* 4:3–47 (1866).

WHAT IF? *Suppose Mendel had transferred pollen from an F₁ plant to the carpel of a plant that was homozygous recessive for both genes. Set up the cross and draw Punnett squares that predict the offspring for both hypotheses. Would this cross have supported the hypothesis of independent assortment equally well?*

⊛ **Mastering Biology**
   Animation: Mendel's Cross of Two Characters: Seed Shape and Seed Color
   Animation: A Simplified Cross of Two Characters in Humans
   Animation: Crosses of Two Characters in "MendAliens"

CONCEPT CHECK 14.1

1. **DRAW IT** Pea plants heterozygous for flower position and stem length (*AaTt*) are allowed to self-pollinate, and 400 of the resulting seeds are planted. Draw a Punnett square for this cross. How many offspring would be predicted to have terminal flowers and be dwarf? (See Table 14.1.)

2. **WHAT IF?** List all gametes that could be made by a pea plant heterozygous for seed color, seed shape, and pod shape (*YyRrIi*; see Table 14.1). How large a Punnett square would you need to draw to predict the offspring of a self-pollination of this "trihybrid"?

3. **MAKE CONNECTIONS** In some pea plant crosses, the plants are self-pollinated. Is self-pollination considered asexual or sexual reproduction? Explain. (See Concept 13.1.)

For suggested answers, see Appendix A.

CONCEPT 14.2

Probability laws govern Mendelian inheritance

Mendel's laws of segregation and independent assortment reflect the same rules of probability that apply to tossing coins, rolling dice, and drawing cards from a deck. The probability scale ranges from 0 to 1. An event that is certain to occur has a probability of 1, while an event that is certain *not* to occur has a probability of 0. With a coin that has heads on both sides, the probability of tossing heads is 1, and the probability of tossing tails is 0. With a normal coin, the chance of tossing heads is ½, and the chance of tossing tails is ½. The probability of drawing the ace of spades from a 52-card deck is ¹⁄₅₂. The probabilities of all possible outcomes for an event must add up to 1. With a deck of cards, the chance of picking a card other than the ace of spades is ⁵¹⁄₅₂.

Tossing a coin illustrates an important lesson about probability. For every toss, the probability of heads is ½. The outcome of any particular toss is unaffected by what has happened on previous trials. We refer to phenomena such as coin tosses as independent events. Each toss of a coin, whether done sequentially with one coin or simultaneously with many, is independent of every other toss. And like two separate coin tosses, the alleles of one gene segregate into gametes independently of another gene's alleles (the law of independent assortment). We'll now look at two basic rules of probability that help us predict the outcome of the fusion of such gametes in simple monohybrid crosses and more complicated crosses as well.

The Multiplication and Addition Rules Applied to Monohybrid Crosses

How do we determine the probability that two or more independent events will occur together in some specific combination? For example, what is the chance that two coins tossed simultaneously will both land heads up? The **multiplication rule** states that to determine the probability of one event *and* the other occurring, we multiply the probability of one event (one coin coming up heads) by the probability of the other event (the other coin coming up heads). By the multiplication rule, then, the probability that both coins will land heads up is $\frac{1}{2} \times \frac{1}{2} = \frac{1}{4}$.

We can apply the same reasoning to an F_1 monohybrid cross. With seed shape in pea plants as the heritable character, the genotype of F_1 plants is *Rr*. Segregation in a heterozygous plant is like flipping a coin in terms of calculating the probability of each outcome: Each egg produced has a $\frac{1}{2}$ chance of carrying the dominant allele (*R*) and a $\frac{1}{2}$ chance of carrying the recessive allele (*r*). The same odds apply to each sperm cell produced. For a particular F_2 plant to have wrinkled seeds, the recessive trait, both the egg and the sperm that come together must carry the *r* allele. The probability that an *r* allele will be present in the egg *and* in the sperm at fertilization is found by multiplying $\frac{1}{2}$ (the probability that the egg will have an *r*) $\times \frac{1}{2}$ (the probability that the sperm will have an *r*). Thus, the multiplication rule tells us that the probability of an F_2 plant having wrinkled seeds (*rr*) is $\frac{1}{4}$ **(Figure 14.9)**. Likewise, the probability of an F_2 plant carrying both dominant alleles for seed shape (*RR*) is $\frac{1}{4}$.

To figure out the probability that an F_2 plant from a monohybrid cross will be heterozygous rather than homozygous, we need to invoke a second rule. Notice in Figure 14.9 that the dominant allele can come from the egg and the recessive allele from the sperm, or vice versa. That is, F_1 gametes can combine to produce *Rr* offspring in two *mutually exclusive* ways: For any particular heterozygous F_2 plant, the dominant allele can come from the egg *or* the sperm, but not from both. According to the **addition rule**, the probability that any one of two or more mutually exclusive events (one event *or* the other) will occur is calculated by adding their individual probabilities. As we have just seen, the multiplication rule gives us the individual probabilities that we will now add together. The probability for one possible way of obtaining an F_2 heterozygote—the dominant allele from the egg and the recessive allele from the sperm—is $\frac{1}{4}$. The probability for the other possible way—the recessive allele from the egg and the dominant allele from the sperm—is also $\frac{1}{4}$ (see Figure 14.9). Using the rule of addition, then, we can calculate the probability of an F_2 heterozygote as $\frac{1}{4} + \frac{1}{4} = \frac{1}{2}$.

▼ **Figure 14.9 Segregation of alleles and fertilization as chance events.** When a heterozygote (*Rr*) forms gametes, whether a particular gamete ends up with an *R* or an *r* is like the toss of a coin. We can determine the probability for any genotype among the offspring of two heterozygotes by multiplying together the individual probabilities of an egg and sperm having a particular allele (*R* or *r* in this example).

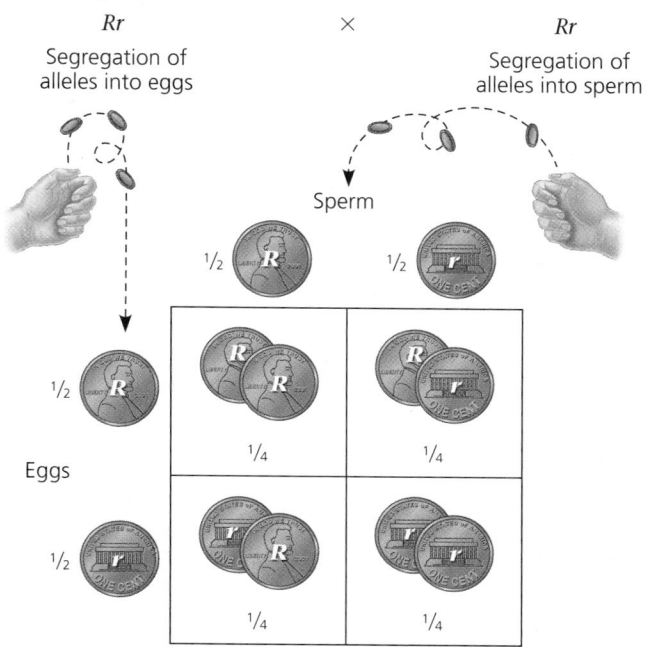

(Note that use of the word *and* in the statement is a clue that you should use the multiplication rule, while *or* suggests use of the addition rule.)

Solving Complex Genetics Problems with the Rules of Probability

We can also apply the rules of probability to predict the outcome of crosses involving multiple characters. Recall that each allelic pair (gene) segregates independently during gamete formation (the law of independent assortment). Thus, a dihybrid or other multicharacter cross is equivalent to two or more independent monohybrid crosses occurring simultaneously. By applying what we have learned about monohybrid crosses, we can determine the probability of specific genotypes occurring in the F_2 generation without having to construct unwieldy Punnett squares.

Consider the dihybrid cross between *YyRr* heterozygotes shown in Figure 14.8. We will focus first on the seed-color character. For a monohybrid cross of *Yy* plants, we can use a simple Punnett square to determine that the probabilities of the offspring genotypes are $\frac{1}{4}$ for *YY*, $\frac{1}{2}$ for *Yy*, and $\frac{1}{4}$ for *yy*. We can draw a second Punnett square to determine that the same probabilities apply to the offspring genotypes for seed shape: $\frac{1}{4}$ *RR*, $\frac{1}{2}$ *Rr*, and $\frac{1}{4}$ *rr*. Knowing these probabilities, we can simply use the multiplication rule to determine the probability of each of the genotypes in the F_2 generation. To give two examples, the calculations for finding the probabilities of two of the possible F_2 genotypes (*YYRR* and *YyRR*) are shown below:

Probability of *YYRR* = ¼ (probability of *YY*) × ¼ (*RR*) = ¹⁄₁₆

Probability of *YyRR* = ½ (*Yy*) × ¼ (*RR*) = ⅛

The *YYRR* genotype corresponds to the upper left box in the larger Punnett square in Figure 14.8 (one box = ¹⁄₁₆). Looking closely at the larger Punnett square in Figure 14.8, you will see that 2 of the 16 boxes (⅛) correspond to the *YyRR* genotype.

Now let's see how we can combine the multiplication and addition rules to solve even more complex problems in Mendelian genetics. Imagine a cross of two pea varieties in which we track the inheritance of three characters. Let's cross a trihybrid with purple flowers and yellow round seeds (heterozygous for all three genes) with a plant with purple flowers and green wrinkled seeds (heterozygous for flower color but homozygous recessive for the other two characters). Using Mendelian symbols, our cross is *PpYyRr* × *Ppyyrr*. What fraction of offspring from this cross are predicted to exhibit the recessive phenotypes for *at least two* of the three characters?

To answer this question, we can start by listing all genotypes we could get that fulfill this condition: *ppyyRr*, *ppYyrr*, *Ppyyrr*, *PPyyrr*, and *ppyyrr*. (Because the condition is *at least two* recessive traits, it includes the last genotype, which shows all three recessive traits.) Next, we calculate the probability for each of these genotypes resulting from our *PpYyRr* × *Ppyyrr* cross by multiplying together the individual probabilities for the allele pairs, just as we did in our dihybrid example. Note that in a cross involving heterozygous and homozygous allele pairs (for example, *Yy* × *yy*), the probability of heterozygous (*Yy*) offspring is ½ and the probability of homozygous (in this case, *yy*) offspring is ½. Finally, we use the addition rule to add the probabilities for all the different genotypes that fulfill the condition of at least two recessive traits resulting from our *PpYyRr* × *Ppyyrr* cross, as shown below:

ppyyRr	¼ (probability of *pp*) × ½ (*yy*) × ½ (*Rr*) = ¹⁄₁₆	
ppYyrr	¼ (*pp*) × ½ (*Yy*) × ½ (*rr*) = ¹⁄₁₆	
Ppyyrr	½ (*Pp*) × ½ (*yy*) × ½ (*rr*) = ²⁄₁₆	
PPyyrr	¼ (*PP*) × ½ (*yy*) × ½ (*rr*) = ¹⁄₁₆	
ppyyrr	¼ (*pp*) × ½ (*yy*) × ½ (*rr*) = ¹⁄₁₆	
Chance of *at least two* recessive traits		= ⁶⁄₁₆ = ⅜

In time, you'll be able to solve genetics problems faster by using the rules of probability than by filling in Punnett squares.

We cannot predict with certainty the exact numbers of progeny of different genotypes resulting from a genetic cross. But the rules of probability give us the *likelihood* of various outcomes. Usually, the larger the sample size, the closer the results will conform to our predictions. Mendel understood this statistical feature of inheritance and had a keen sense of the rules of chance. It was for this reason that he set up his experiments so as to generate, and then count, large numbers of offspring from his crosses.

CONCEPT CHECK 14.2

1. For any gene with a dominant allele *A* and recessive allele *a*, what proportions of the offspring from an *AA* × *Aa* cross are expected to be homozygous dominant, homozygous recessive, and heterozygous?

2. Two organisms, with genotypes *BbDD* and *BBDd*, are mated. Assuming independent assortment of the *B/b* and *D/d* genes, write the genotypes of all possible offspring from this cross and use the rules of probability to calculate the chance of each genotype occurring.

3. **WHAT IF?** Three characters (flower color, seed color, and pod shape) are considered in a cross between two pea plants: *PpYyIi* × *ppYyIi*. Using the rules of probability, determine the fraction of offspring predicted to be homozygous recessive for at least two of the three characters.

For suggested answers, see Appendix A.

CONCEPT 14.3

Inheritance patterns are often more complex than predicted by simple Mendelian genetics

In the 1900s, geneticists extended Mendelian principles not only to diverse organisms, but also to patterns of inheritance more complex than those described by Mendel. For the work that led to his two laws of inheritance, Mendel chose pea plant characters that turn out to have a relatively simple genetic basis: Each character is determined by one gene, for which there are only two alleles, one completely dominant and the other completely recessive. (There is one exception: Mendel's pod shape character is actually determined by two genes.) Not all heritable characters are determined so simply, and the relationship between genotype and phenotype is rarely so straightforward. Mendel himself realized that he could not explain the more complicated patterns he observed in crosses involving other pea characters or other plant species. This does not diminish the utility of Mendelian genetics, however, because the basic principles of segregation and independent assortment apply even to more complex patterns of inheritance. In this section, we will extend Mendelian genetics to hereditary patterns that were not reported by Mendel.

Extending Mendelian Genetics for a Single Gene

The inheritance of characters determined by a single gene deviates from simple Mendelian patterns when alleles are not completely dominant or recessive, when a particular gene has more than two alleles, or when a single gene produces multiple phenotypes. We will describe examples of each of these situations in this section.

Degrees of Dominance

Alleles can show different degrees of dominance and recessiveness in relation to each other. In Mendel's classic pea crosses, the F_1 offspring always looked like one of the two parental varieties because one allele in a pair showed **complete dominance** over the other. In such situations, the phenotypes of the heterozygote and the dominant homozygote are indistinguishable (see Figure 14.6).

For some genes, however, neither allele is completely dominant, and the F_1 hybrids have a phenotype somewhere between those of the two parental varieties. This phenomenon, called **incomplete dominance**, is seen when red snapdragons are crossed with white snapdragons: All the F_1 hybrids have pink flowers **(Figure 14.10)**. This third, intermediate phenotype results from flowers of the heterozygotes having less red pigment than the red homozygotes. (This is unlike the case of Mendel's pea plants, where the *Pp* heterozygotes make enough pigment for the flowers to be purple, indistinguishable from those of *PP* plants.)

At first glance, incomplete dominance of either allele seems to provide evidence for the blending hypothesis of inheritance, which would predict that the red or white trait could never reappear among offspring of the pink hybrids. In fact, interbreeding F_1 hybrids produces F_2 offspring with a phenotypic ratio of one red to two pink to one white. (Because heterozygotes have a separate phenotype from homozygotes, the genotypic and phenotypic ratios for the F_2 generation are the same, 1:2:1.) The segregation of the red-flower and white-flower alleles in the gametes produced by the pink-flowered plants confirms that the alleles for flower color are heritable factors that maintain their identity in the hybrids; that is, that the factors are discrete rather than "blendable."

Another variation on dominance relationships between alleles is called **codominance**; in this variation, the two alleles each affect the phenotype in separate, distinguishable ways. For example, the human MN blood group is determined by codominant alleles for two specific molecules located on the surface of red blood cells, the M and N molecules. A single gene (L), for which two allelic variations are possible (L^M or L^N), determines the phenotype of this blood group. Individuals homozygous for the L^M allele ($L^M L^M$) have red blood cells with only M molecules; individuals homozygous for the L^N allele ($L^N L^N$) have red blood cells with only N molecules. But *both* M and N molecules are present on the red blood cells of individuals heterozygous for the *M* and *N* alleles ($L^M L^N$). Note that the MN phenotype is *not* intermediate between the M and N phenotypes, which distinguishes codominance from incomplete dominance. Rather, *both* M and N phenotypes are exhibited by heterozygotes, since both molecules are present.

▼ **Figure 14.10 Incomplete dominance in snapdragon color.** When red snapdragons are crossed with white ones, the F_1 hybrids have pink flowers. Segregation of alleles into gametes of the F_1 plants results in an F_2 generation with a 1:2:1 ratio for both genotype and phenotype. Neither allele is dominant, so rather than using upper- and lowercase letters, we use the letter *C* with a superscript to indicate an allele for flower color: C^R for red and C^W for white.

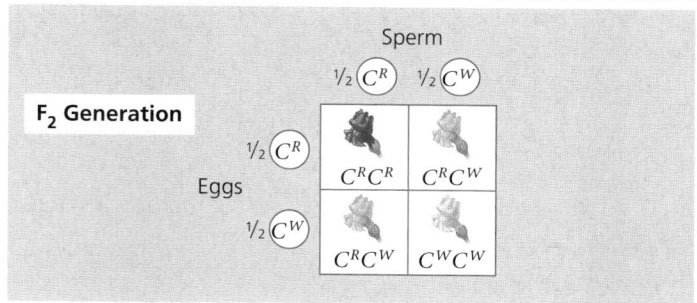

❓ *Suppose a classmate argues that this figure supports the blending hypothesis for inheritance. What might your classmate say, and how would you respond?*

⊛ **Mastering Biology** Animation: Incomplete Dominance in "MendAliens"

The Relationship Between Dominance and Phenotype

We've now seen that the relative effects of two alleles range from complete dominance of one allele to incomplete dominance of either allele to codominance of both alleles. It is important to understand that an allele is called *dominant* because it is seen in the phenotype, not because it somehow subdues a recessive allele. Alleles are simply variations in a gene's nucleotide sequence (see Figure 14.4). When a dominant allele coexists with a recessive allele in a heterozygote, they do not actually interact at all. It is in the pathway from genotype to phenotype that dominance and recessiveness come into play.

To illustrate the relationship between dominance and phenotype, we can use one of the characters Mendel studied—round versus wrinkled pea seed shape. The dominant allele (round) codes for an enzyme that helps convert an unbranched form of starch to a branched form in the seed. The recessive allele (wrinkled) codes for a defective form of this enzyme, leading to an accumulation of unbranched starch, which causes excess water to enter the seed by osmosis. Later, when the seed dries, it wrinkles. If a dominant allele is present, no excess water enters the seed and it does not wrinkle when it dries. One dominant allele results in enough of the enzyme to synthesize adequate amounts of branched starch, which means that dominant homozygotes and heterozygotes have the same phenotype: round seeds.

A closer look at the relationship between dominance and phenotype reveals an intriguing fact: For any character, the observed dominant/recessive relationship of alleles depends on the level at which we examine phenotype. **Tay-Sachs disease**, an inherited disorder in humans, is an example. The brain cells of a child with Tay-Sachs disease cannot metabolize certain lipids because a crucial enzyme does not work properly. As these lipids accumulate in brain cells, the child begins to suffer seizures, blindness, and degeneration of motor and mental performance and dies within a few years.

Only children who inherit two copies of the Tay-Sachs allele (homozygotes) have the disease. Thus, at the *organismal* level, the Tay-Sachs allele qualifies as recessive. However, the activity level of the lipid-metabolizing enzyme in heterozygotes is intermediate between the activity level in individuals homozygous for the normal allele and the activity level in individuals with Tay-Sachs disease. (The term *normal* is used in the genetic sense to refer to the allele coding for the enzyme that functions properly.) The intermediate phenotype observed at the *biochemical* level is characteristic of incomplete dominance of either allele. Fortunately, the heterozygote condition does not lead to disease symptoms, apparently because half the normal enzyme activity is sufficient to prevent lipid accumulation in the brain. Extending our analysis to yet another level, we find that heterozygous individuals produce equal numbers of normal and dysfunctional enzyme molecules. Thus, at the *molecular* level, the normal allele and the Tay-Sachs allele are codominant. As you can see, whether alleles appear to be completely dominant, incompletely dominant, or codominant depends on the level at which the phenotype is analyzed.

Frequency of Dominant Alleles While you might assume that the dominant allele for a particular character would be more common than the recessive allele, this is not always the case. For an example of a rare dominant allele, about one baby out of 400 in the United States is born with extra fingers or toes, a condition known as polydactyly. Some cases are caused by the presence of a dominant allele. The low frequency of these types of polydactyly indicates that the recessive allele, which results in five digits per appendage, is far more prevalent than the dominant allele in the population.

Multiple Alleles

Only two alleles exist for the pea characters that Mendel studied, but most genes have more than two alleles. The ABO blood groups in humans, for instance, are determined by the two alleles a person has of the blood group gene; the three possible alleles are I^A, I^B, and i. A person's blood group may be one of four types: A, B, AB, or O. These letters refer to two carbohydrates—A and B—that may be found attached to specific cell-surface molecules on red blood cells. An individual's blood cells may have carbohydrate A (type A blood), carbohydrate B (type B), both (type AB), or neither (type O), as shown in **Figure 14.11**, along with the relevant genotypes. Matching compatible blood groups is critical for safe blood transfusions (see Concept 43.3).

▼ **Figure 14.11 Multiple alleles for the ABO blood groups.** The four blood groups result from different combinations of three alleles.

(a) The three alleles for the ABO blood groups and their carbohydrates. Each allele codes for an enzyme that may add a specific carbohydrate (designated by the superscript on the allele and shown as a triangle or circle) to red blood cells.

Allele	I^A	I^B	i
Carbohydrate	A △	B ○	none

(b) Blood group genotypes and phenotypes. There are six possible genotypes, resulting in four different phenotypes.

Genotype	$I^A I^A$ or $I^A i$	$I^B I^B$ or $I^B i$	$I^A I^B$	ii
Red blood cell with surface carbohydrates				
Phenotype (blood group)	A	B	AB	O

VISUAL SKILLS *Based on the surface carbohydrate phenotype in part (b), what are the dominance relationships among the alleles?*

Pleiotropy

So far, we have treated Mendelian inheritance as though each gene affects only one phenotypic character. Most genes, however, have multiple phenotypic effects, a property called **pleiotropy** (from the Greek *pleion*, more). In humans, for example, pleiotropic alleles are responsible for the multiple symptoms associated with certain hereditary diseases, such as cystic fibrosis and sickle-cell disease, discussed later in this chapter. In the garden pea, the gene that determines flower color also affects the color of the coating on the outer surface of the seed, which can be gray or white. Given the intricate molecular and cellular interactions responsible for an organism's development and physiology, it isn't surprising that a single gene can affect a number of characters.

Extending Mendelian Genetics for Two or More Genes

Dominance relationships, multiple alleles, and pleiotropy all have to do with the effects of the alleles of a single gene. We now consider two situations in which two or more genes are involved in determining a particular phenotype. In the first case, called epistasis, one gene affects the phenotype of another because the two gene products interact; in the second case, called polygenic inheritance, multiple genes independently affect a single trait.

Epistasis

In **epistasis** (from the Greek for "standing upon"), the phenotypic expression of a gene at one locus alters that of a gene at a second locus. An example will help clarify this concept. In Labrador retrievers (commonly called "Labs"), black coat color is dominant to brown. Let's designate B and b as the two alleles for this character. For a Lab to have brown fur, its genotype must be bb; these dogs are called chocolate Labs. But there is more to the story. A second gene determines whether or not pigment will be deposited in the hair. The dominant allele, symbolized by E, results in the deposition of either black or brown pigment, depending on the genotype at the first locus. But if the Lab is homozygous recessive for the second locus (ee), then the coat is yellow, regardless of the genotype at the black/brown locus (yellow Labs). In this case, the gene for pigment deposition (E/e) is said to be *epistatic to* the gene that codes for black or brown pigment (B/b).

What happens if we mate black Labs that are heterozygous for both genes ($BbEe$)? Although the two genes affect the same phenotypic character (coat color), they follow the law of independent assortment. Thus, our breeding experiment represents an F_1 dihybrid cross, like those that produced a 9:3:3:1 ratio in Mendel's experiments. We can use a Punnett square to represent the genotypes of the F_2 offspring **(Figure 14.12)**. As a result of epistasis, the phenotypic ratio among the F_2 offspring is 9 black to 3 chocolate to 4 yellow Labs. Other types of epistatic interactions produce different ratios, but all are modified versions of 9:3:3:1.

Polygenic Inheritance

Mendel studied characters that could be classified on an either-or basis, such as purple versus white flower color. But many characters, such as human skin color and height, are not one of two discrete characters, but instead vary in the population in gradations along a continuum. These are called **quantitative characters**. Quantitative variation usually indicates **polygenic inheritance**, an additive effect of two or more genes on a single phenotypic character. (In a way, this is the converse of pleiotropy, where a single gene affects several phenotypic characters.) Height is a good example of polygenic inheritance: In 2014, a genomic study of over 250,000 people found almost 700 genetic variations in over 180 genes that affect height. Many variations were in or near

▼ **Figure 14.12 An example of epistasis.** This Punnett square illustrates the genotypes and phenotypes predicted for offspring of matings between two black Labrador retrievers of genotype *BbEe*. The *E/e* gene, which is epistatic to the *B/b* gene coding for hair pigment, controls whether or not pigment of any color will be deposited in the hair.

VISUAL SKILLS *Compare the four squares in the lower right of this Punnett square with those in Figure 14.8. Explain the genetic basis for the difference between the ratio (9:3:4) of phenotypes seen in this cross and the 9:3:3:1 ratio seen in Figure 14.8.*

genes involved in biochemical pathways affecting growth of the skeleton, but others were associated with genes not obviously related to growth. Another study in 2018, of 300,000 individuals, identified 124 genes that affect hair color. Although simplifying can sometimes be valuable in understanding genetic principles, as we'll see shortly, the classic idea of a single gene determining eye color or hair color or other such traits is an oversimplification. (Earlobe attachment, an example used until a decade ago in this textbook, was shown in 2017 to be affected by nearly 50 genes!)

Skin pigmentation in humans is also controlled by many separately inherited genes—378 at latest count, many of which are involved in the production of melanin skin pigments. Here, we'll simplify the story in order to understand the concept of polygenic inheritance. Let's consider three genes, with a dark-skin allele for each gene (*A, B,* or *C*) contributing one "unit" of darkness (also a simplification) to the phenotype and being incompletely dominant to the other, light-skin allele (*a, b,* or *c*). In our model, an *AABBCC* person would be very dark, whereas an *aabbcc* individual would be very light.

▼ **Figure 14.13 A simplified model for polygenic inheritance of skin color.** In this model, three separately inherited genes affect skin color. (The reported number is actually 378 genes.) The heterozygous individuals (*AaBbCc*) represented by the two rectangles at the top of this figure each carry three dark-skin alleles (black circles, which represent *A, B,* or *C*) and three light-skin alleles (white circles, which represent *a, b,* or *c*). The Punnett square shows all possible genetic combinations in gametes and offspring of hypothetical matings between these two. The results are summarized by phenotypic frequencies (fractions) under the Punnett square. (The phenotypic ratio is 1:6:15:20:15:6:1.)

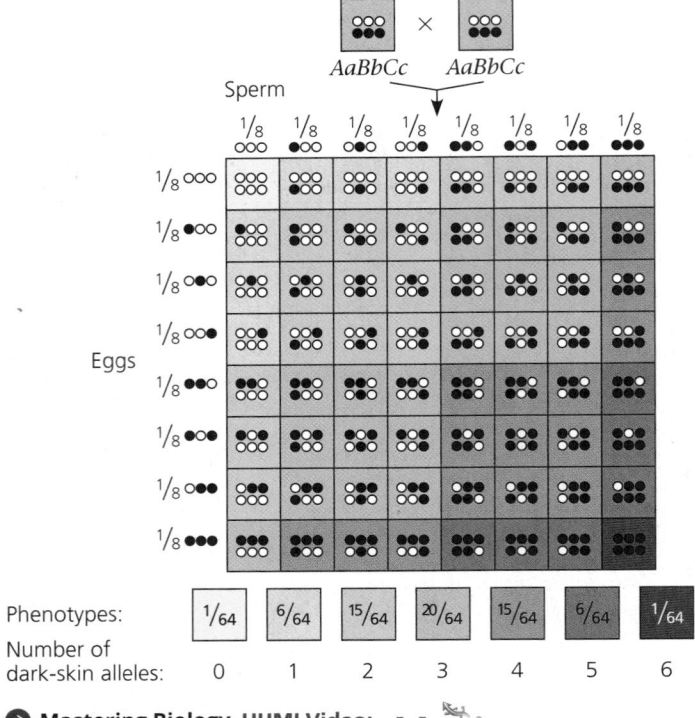

Phenotypes:

$^1/_{64}$	$^6/_{64}$	$^{15}/_{64}$	$^{20}/_{64}$	$^{15}/_{64}$	$^6/_{64}$	$^1/_{64}$

Number of dark-skin alleles: 0 1 2 3 4 5 6

➡ **Mastering Biology** HHMI Video: The Biology of Skin Color

An *AaBbCc* person would have skin of an intermediate shade. Because the alleles have a cumulative effect, the genotypes *AaBbCc* and *AABbcc* would make the same genetic contribution (three units) to skin darkness. There are seven skin color phenotypes that could result from a mating between *AaBbCc* heterozygotes, as shown in **Figure 14.13**. In a large number of such matings, the majority of offspring would be expected to have intermediate phenotypes (skin color in the middle range). You can graph the predictions from the Punnett square in the **Scientific Skills Exercise**. Environmental factors, such as exposure to the sun, also affect the skin color phenotype.

Nature and Nurture: The Environmental Impact on Phenotype

Another departure from simple Mendelian genetics arises when the phenotype for a character depends on environment as well as genotype. A single tree, locked into its inherited genotype, has leaves that vary in size, shape, and greenness, depending on their exposure to wind and sun. For humans, nutrition influences height, exercise alters build, sun-tanning darkens the skin, and experience improves performance on intelligence tests. Even identical twins, who are genetic equals, accumulate phenotypic differences as a result of their unique experiences.

Whether human characters are more influenced by genes or the environment—in everyday terms, nature versus nurture—is a debate that we will not attempt to settle here. We can say, however, that a genotype generally is not associated with a rigidly defined phenotype, but rather with a range of phenotypic possibilities due to environmental influences **(Figure 14.14)**. For some characters, such as the ABO blood group system, the phenotypic range is extremely narrow; that is, a given genotype mandates a very specific phenotype. Other characters, such as a person's blood count of red and white cells, vary quite a bit, depending on such factors as the altitude, the customary level of physical activity, and the presence of infectious agents.

▼ **Figure 14.14 The effect of environment on phenotype.** The outcome of a genotype lies within a phenotypic range that depends on the environment in which the genotype is expressed. For example, the acidity and free aluminum content of the soil affect the color of hydrangea flowers, which range from pink (basic soil) to blue-violet (acidic soil). Free aluminum is necessary for bluer colors.

(a) Hydrangeas grown in basic soil

(b) Hydrangeas of the same genetic variety grown in acidic soil with free aluminum

Generally, the phenotypic range is broadest for polygenic characters. Environment contributes to the quantitative nature of these characters, as we have seen in the continuous variation of skin color. Geneticists refer to such characters as **multifactorial**, meaning that many factors, both genetic and environmental, collectively influence phenotype.

➡ **Mastering Biology** BBC Video: Genetics vs. Environment

A Mendelian View of Heredity and Variation

We have now broadened our view of Mendelian inheritance by exploring degrees of dominance as well as multiple alleles, pleiotropy, epistasis, polygenic inheritance, and the phenotypic impact of the environment. How can we integrate these refinements into a comprehensive theory of Mendelian genetics? The key is to make the transition from the reductionist emphasis on single genes and phenotypic characters to the emergent properties of the organism as a whole, one of the themes of this book.

The term *phenotype* can refer not only to specific characters, such as flower color and blood group, but also to

Making a Histogram and Analyzing a Distribution Pattern

What Is the Distribution of Phenotypes Among Offspring of Two Parents Who Are Both Heterozygous for Three Additive Genes? Human skin color is a polygenic trait that is determined by the additive effects of many different genes. In this exercise, you will work with a simplified model of skin color genetics where only three genes are assumed to affect the darkness of skin color and where each gene has two alleles—dark or light (see Figure 14.13). In this model, each dark allele contributes equally to the darkness of skin color, and each pair of alleles segregates independently of any other pair. Using a type of graph called a histogram, you will determine the distribution of phenotypes of offspring with different numbers of dark-skin alleles. (For additional information about graphs, see the Scientific Skills Review in Appendix D.)

How This Model Is Analyzed To predict the phenotypes of the offspring of parents heterozygous for the three genes in our simplified model, we can use the Punnett square in Figure 14.13. The heterozygous individuals (*AaBbCc*) represented by the two rectangles at the top of that figure each carry three dark-skin alleles (black circles, which represent *A, B,* or *C*) and three light-skin alleles (white circles, which represent *a, b,* or *c*). The Punnett square shows all the possible genetic combinations in gametes and in offspring of a large number of hypothetical matings between these heterozygotes.

Predictions from the Punnett Square If we assume that each square in the Punnett square represents one offspring of the heterozygous *AaBbCc* parents, then the squares below show the possible skin color phenotypes and their predicted frequencies. Below the squares is the number of dark-skin alleles for each phenotype.

Phenotypes:	$1/64$	$6/64$	$15/64$	$20/64$	$15/64$	$6/64$	$1/64$
Number of dark-skin alleles:	0	1	2	3	4	5	6

1. A histogram is a bar graph that shows the distribution of numeric data (here, the number of dark-skin alleles). To make a histogram of the allele distribution, put skin color (as the number of dark-skin alleles) along the *x*-axis and predicted number of offspring (out of 64) with each phenotype on the *y*-axis. There are no gaps in these allele data, so draw the bars next to each other with no space in between.

2. You can see that the skin-color phenotypes are not distributed uniformly. **(a)** Which phenotype has the highest frequency? Draw a vertical dashed line through that bar. **(b)** Distributions of values like this one tend to show one of several common patterns. Sketch a rough curve that approximates the values and look at its shape. Is it symmetrically distributed around a central peak value (a "normal distribution," sometimes called a bell curve); is it skewed to one end of the *x*-axis or the other (a "skewed distribution"); or does it show two apparent groups of frequencies (a "bimodal distribution")? Explain the reason for the curve's shape. (It will help to read the text description that supports Figure 14.13.)

➔ **Instructors**: A version of this Scientific Skills Exercise can be assigned in **Mastering Biology**.

Further Reading R. A. Sturm, A golden age of human pigmentation genetics, *Trends in Genetics* 22:464–468 (2006).

---

an organism in its entirety—*all* aspects of its physical appearance, internal anatomy, physiology, and behavior. Similarly, the term *genotype* can refer to an organism's entire genetic makeup, not just its alleles for a single genetic locus. In most cases, a gene's impact on phenotype is affected by other genes and by the environment. In this integrated view of heredity and variation, an organism's phenotype reflects its overall genotype and unique environmental history.

Considering all that can occur in the pathway from genotype to phenotype, it is indeed impressive that Mendel could uncover the fundamental principles governing the transmission of individual genes from parents to offspring. Mendel's laws of segregation and of independent assortment explain heritable variations in terms of alternative forms of genes (hereditary "particles," now known as the alleles of genes) that are passed along, generation after generation, according to simple rules of probability. This theory of inheritance is equally valid for peas, flies, fishes, birds, and human beings—indeed, for any organism with a sexual life cycle. Furthermore, by extending the

principles of segregation and independent assortment to help explain such hereditary patterns as epistasis and quantitative characters, we begin to see how broadly Mendelian genetics applies. From Mendel's abbey garden came a theory of inheritance that anchors modern genetics. In the last section of this chapter, we will apply Mendelian genetics to human inheritance, with emphasis on the transmission of hereditary diseases.

CONCEPT CHECK 14.3

1. *Incomplete dominance* and *epistasis* are both terms that define genetic relationships. What is the most basic distinction between these terms?

2. If a man with type AB blood marries a woman with type O, what blood types would you expect in their children? What fraction would you expect of each type?

3. **WHAT IF?** A rooster with gray feathers and a hen of the same phenotype produce 15 gray, 6 black, and 8 white chicks. What is the simplest explanation for the inheritance of these colors in chickens? What phenotypes would you expect in the offspring of a cross between a gray rooster and a black hen?

For suggested answers, see Appendix A.

Many human traits follow Mendelian patterns of inheritance

The study of human genetics is fueled by our desire to understand our own inheritance and to develop treatments for genetically based diseases. While peas are convenient subjects for genetic research, humans are not. The human generation span is long (about 20 years), human parents produce many fewer offspring than peas, and breeding experiments would be unethical!

Pedigree Analysis

In place of breeding experiments, geneticists analyze the results of human matings that have already occurred. They collect information about a family's history for a particular trait and assemble this information into a family tree describing the trait across the generations—a family **pedigree**.

Figure 14.15a shows a three-generation pedigree that traces the occurrence of a pointed contour of the hairline on the forehead. We simplify the genetics of this trait, called a widow's peak, by assuming it is due to a dominant allele, W. (In reality, other genes are probably also involved.) In our model, because the widow's peak allele is dominant, all individuals who lack a widow's peak must be homozygous recessive (ww). The two grandparents with widow's peaks must have the Ww genotype since some of their offspring are homozygous recessive. The offspring in the second generation who *do* have widow's peaks must also be heterozygous because they are the products of $Ww \times ww$ matings. The third generation in this pedigree consists of two sisters. The one who has a widow's peak could be either homozygous (WW) or heterozygous (Ww), given what we know about the genotypes of her parents (both Ww).

Figure 14.15b is a pedigree of the same family, but this time we focus on a recessive trait, the inability of individuals to taste a chemical called PTC (phenylthiocarbamide). Compounds similar to PTC are found in broccoli, brussels sprouts, and related vegetables and account for the bitter taste some people report when eating these foods. We'll use t for the recessive allele and T for the dominant allele, which results in the ability to taste PTC. As you work your way through the pedigree, notice once again that you can apply what you have learned about Mendelian inheritance to understand the genotypes shown for the family members.

An important application of a pedigree is to help us calculate the probability that a future child will have a particular genotype and phenotype. Suppose that the couple represented in the second generation of Figure 14.15 decides to have one more child. What is the probability that the child will have a widow's peak? This is equivalent to a Mendelian F_1

▼ **Figure 14.15 Pedigree analysis.** Each pedigree traces a trait through three generations of the same family; the two traits have different inheritance patterns. For the sake of understanding genetic principles, each trait is shown as being determined by two alleles of a single gene—a simplification, since other genes likely also affect these two characters.

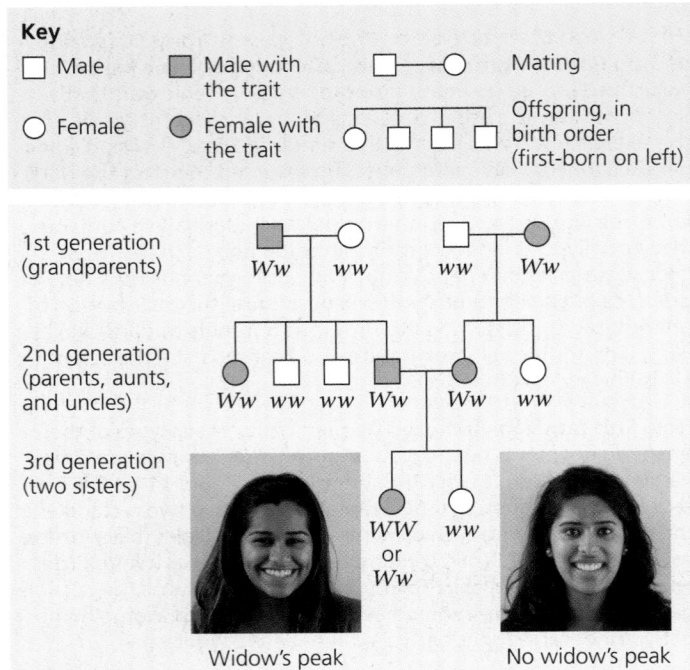

(a) Is a widow's peak a dominant or recessive trait?
Tips for pedigree analysis: Notice in the third generation that the second-born daughter lacks a widow's peak, although both of her parents had the trait. Such a pattern indicates that the trait is due to a dominant allele. If it were due to a *recessive* allele, and both parents had the recessive phenotype (straight hairline), *all* of their offspring would also have the recessive phenotype.

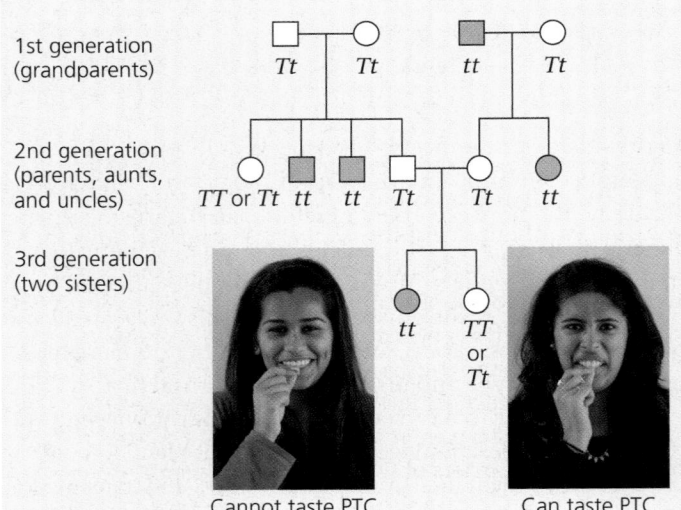

(b) Is the inability to taste a chemical called PTC a dominant or recessive trait?
Tips for pedigree analysis: Notice that the first-born daughter in the third generation has the trait (is unable to taste PTC), although both parents lack that trait (they *can* taste PTC). Such a pattern is explained if the non-taster phenotype is due to a recessive allele. If it were due to a *dominant* allele, then at least one parent would also have had the trait.

monohybrid cross ($Ww \times Ww$), and therefore the probability that a child will inherit a dominant allele and have a widow's peak is $\frac{3}{4}$ ($\frac{1}{4}$ WW + $\frac{1}{2}$ Ww). What is the probability that the child will be unable to taste PTC? We can also treat this as a monohybrid cross ($Tt \times Tt$), but this time we want to know the chance that the offspring will be homozygous recessive (tt). That probability is $\frac{1}{4}$. Finally, what is the chance that the child will have a widow's peak *and* be unable to taste PTC? Assuming that the genes for these two characters are on different chromosomes, the two pairs of alleles will assort independently in this dihybrid cross ($WwTt \times WwTt$). Therefore, we can use the multiplication rule: $\frac{3}{4}$ (chance of widow's peak) $\times$ $\frac{1}{4}$ (chance of inability to taste PTC) = $\frac{3}{16}$ (chance of widow's peak and inability to taste PTC).

Pedigrees are a more serious matter when the alleles in question cause disabling or deadly diseases instead of innocuous human variations such as hairline or inability to taste a harmless chemical. However, for disorders inherited as simple Mendelian traits, the same techniques of pedigree analysis apply.

Recessively Inherited Disorders

Thousands of genetic disorders are known to be inherited as simple recessive traits. These disorders range in severity from relatively mild, such as albinism (lack of pigmentation, which results in susceptibility to skin cancers and vision problems), to life-threatening, such as cystic fibrosis.

The Behavior of Recessive Alleles

How can we account for the behavior of alleles that cause recessively inherited disorders? Recall that genes code for proteins of specific function. An allele that causes a genetic disorder (let's call it allele a) codes for either a malfunctioning protein or no protein at all. In the case of disorders classified as recessive, heterozygotes (Aa) typically have the normal phenotype because one copy of the normal allele (A) produces a sufficient amount of the specific protein. Thus, a recessively inherited disorder shows up only in the homozygous individuals (aa) who inherit a recessive allele from each parent. Although phenotypically normal with regard to the disorder, heterozygotes may transmit the recessive allele to their offspring and thus are called **carriers**. **Figure 14.16** illustrates these ideas using albinism as an example.

Most people who have recessive disorders are born to parents who are carriers of the disorder but have a normal phenotype, as is the case shown in the Punnett square in Figure 14.16. A mating between two carriers corresponds to a Mendelian F_1 monohybrid cross, so the predicted genotypic ratio for the offspring is 1 AA : 2 Aa : 1 aa. Thus, each child has a $\frac{1}{4}$ chance of inheriting a double dose of the recessive allele; in the case of albinism, such a child will have albinism. From the genotypic ratio, we also can see that out of three offspring with the *normal* phenotype (one AA plus two Aa),

▼ **Figure 14.16 Albinism: a recessive trait.** One of the two sisters shown here does not have albinism; the other does. Most recessive homozygotes are born to parents who are carriers of the disorder but themselves have a normal phenotype, the case shown in the Punnett square.

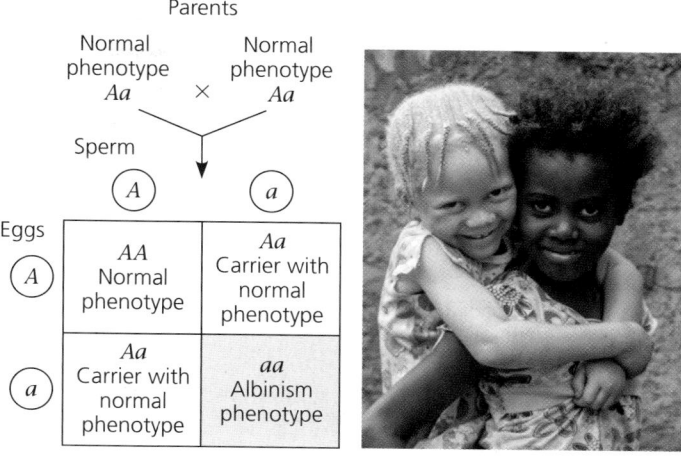

? *What is the probability that the sister without albinism is a carrier of the albinism allele?*

two are predicted to be heterozygous carriers, $\frac{2}{3}$ chance. Recessive homozygotes could also result from $Aa \times aa$ and $aa \times aa$ matings, but if the disorder is lethal before reproductive age or results in sterility (neither of which is true for albinism), no aa individuals will reproduce. Even if recessive homozygotes are able to reproduce, this will occur relatively rarely because such individuals are much less common in the population than heterozygous carriers (for reasons we'll examine in Concept 23.2).

In general, genetic disorders are not evenly distributed among all groups of people. For example, the incidence of Tay-Sachs disease, which we described earlier in this chapter, is disproportionately high among Ashkenazic Jews, Jewish people whose ancestors lived in central Europe. In that population, Tay-Sachs disease occurs in one out of 3,600 births, an incidence about 100 times greater than that among non-Jews or Mediterranean (Sephardic) Jews. This uneven distribution results from the different genetic histories of the world's peoples during less technological times, when populations were more geographically (and hence genetically) isolated.

When a disease-causing recessive allele is rare, it is relatively unlikely that two carriers of the same harmful allele will meet and mate. The probability of passing on recessive traits increases greatly, however, if the man and woman are close relatives (for example, siblings or first cousins). This is because people with recent common ancestors are more likely to carry the same recessive alleles than are unrelated people. Thus, these consanguineous ("same blood") matings, indicated in pedigrees by double lines, are more likely to produce offspring homozygous for recessive traits—including harmful ones. Such effects can be observed in many types of domesticated and zoo animals that have become inbred.

Although geneticists generally agree that inbreeding causes an increase in autosomal recessive conditions compared to those resulting from matings between unrelated parents, they debate exactly how much human consanguinity increases the risk of inherited diseases. For one thing, many harmful alleles have such severe effects that a homozygous embryo spontaneously aborts long before birth. Most societies and cultures have laws or taboos forbidding marriages between close relatives, some for social or economic reasons. These rules may have also resulted from the empirical observation that in most populations, stillbirths and physical or biochemical abnormalities at birth are more common when parents are closely related.

Cystic Fibrosis

The most common lethal genetic disease in the United States is **cystic fibrosis**, which strikes one out of every 2,500 people of European descent but is much rarer in other groups. Among people of European descent, one out of 25 (4%) are carriers of the cystic fibrosis allele. The normal allele for this gene codes for a membrane protein that functions in the transport of chloride ions between certain cells and the extracellular fluid. These chloride transport channels are defective or absent in the plasma membranes of children who inherit two recessive alleles for cystic fibrosis. The result is an abnormally high concentration of intracellular chloride, which causes an uptake of water due to osmosis. This in turn causes the mucus that coats certain cells to become thicker and stickier than normal and to build up in the pancreas, lungs, digestive tract, and other organs. Multiple (pleiotropic) effects result, including poor absorption of nutrients from the intestines, chronic bronchitis, and recurrent bacterial infections.

Untreated, cystic fibrosis can cause death by the age of 5. Daily doses of antibiotics to stop infection, gentle pounding on the chest to clear mucus from clogged airways, and other therapies can prolong life. In the United States, more than half of those with cystic fibrosis now survive into their 40s.

Sickle-Cell Disease: A Genetic Disorder with Evolutionary Implications

EVOLUTION The most common inherited disorder among people of African descent is **sickle-cell disease**, which affects one out of 400 African-Americans. Sickle-cell disease is caused by the substitution of a single amino acid in the hemoglobin protein of red blood cells; in homozygous individuals, all hemoglobin is of the sickle-cell (abnormal) variety. When the oxygen content of an affected individual's blood is low (at high altitudes or under physical stress, for instance), the sickle-cell hemoglobin proteins aggregate into long fibers that deform the red cells into a sickle shape (see Figure 5.19). Sickled cells may clump and clog small blood vessels, often leading to other symptoms throughout the body, including physical weakness, pain, organ damage, and even stroke and paralysis. Regular blood transfusions can ward off brain damage in children with sickle-cell disease, and new drugs can help prevent or treat other problems. There is currently no widely available cure, but the disease is the target of ongoing gene therapy research.

Although two sickle-cell alleles are necessary for an individual to have sickle-cell disease and thus the condition is considered recessive, the presence of one sickle-cell allele can affect the phenotype. Thus, at the organismal level, the normal allele is incompletely dominant to the sickle-cell allele **(Figure 14.17)**. At the molecular level, the two alleles are codominant; both normal and abnormal (sickle-cell) hemoglobins are made in heterozygotes (carriers), who are said to have *sickle-cell trait*. (Here the word "trait" is used to distinguish this condition from sickle-cell disease, not as it was defined earlier—as any variant of a character.) Heterozygotes are usually healthy but may suffer some symptoms during long periods of reduced blood oxygen.

About one out of ten African-Americans have sickle-cell trait, an unusually high frequency of heterozygotes for an allele with severe detrimental effects in homozygotes. Why haven't evolutionary processes resulted in the disappearance of the allele among this population? One explanation is that having a single copy of the sickle-cell allele reduces the frequency and severity of malaria attacks, especially among young children. The malaria parasite spends part of its life cycle in red blood cells (see Figure 28.18), and the presence of even heterozygous amounts of sickle-cell hemoglobin results in lower parasite densities and hence reduced malaria symptoms.

▼ **Figure 14.17 Sickle-cell disease and sickle-cell trait.**

Sickle-cell alleles

Low O₂

Sickle-cell hemoglobin proteins

Part of a fiber of sickle-cell hemoglobin proteins

Long fibers cause red blood cells to be sickle-shaped

Sickle-cell disease

(a) Homozygote with sickle-cell disease: Weakness, anemia, pain and fever, organ damage

Sickle-cell allele
Normal allele

Very low O₂

Sickle-cell and normal hemoglobin proteins

Part of a sickle-cell fiber and normal hemoglobin proteins

Sickled and normal red blood cells

Sickle-cell trait

(b) Heterozygote with sickle-cell trait: Some symptoms when blood oxygen is very low; reduction of malaria symptoms

Mastering Biology HHMI Animation: Sickle-Cell Disease

 hhmi BioInteractive

Thus, in tropical Africa, where infection with the malaria parasite is common, the sickle-cell allele confers an advantage to heterozygotes even though it is harmful in the homozygous state. (The balance between these two effects will be discussed in Concept 23.4; see Make Connections Figure 23.18.) The relatively high frequency of African-Americans with sickle-cell trait is a vestige of their tropical African ancestry.

 Mastering Biology HHMI Video: The Making of the Fittest: Natural Selection in Humans (Sickle-Cell Disease)

Dominantly Inherited Disorders

Although many harmful alleles are recessive, some disorders are due to dominant alleles. One example is *achondroplasia*, a form of dwarfism that occurs in one of every 25,000 people. Heterozygous individuals have the dwarf phenotype **(Figure 14.18)**. Therefore, all people who do not have achondroplasia—99.99% of the population—are homozygous recessive. Like the presence of extra fingers or toes mentioned earlier, achondroplasia is a trait for which the recessive allele is much more prevalent than the corresponding dominant allele.

Unlike achondroplasia, which is relatively harmless, some dominant alleles cause lethal diseases. Those that do are much less common than recessive alleles that have lethal effects. A lethal recessive allele is only lethal when homozygous; it can be passed from one generation to the next by heterozygous carriers because the carriers themselves have normal phenotypes. A lethal dominant allele, however, often causes the death of afflicted individuals before they can mature and reproduce, and in this case the allele is not passed on to future generations.

A lethal dominant allele may be passed on, though, if the lethal disease symptoms first appear after reproductive age.

▼ **Figure 14.18 Achondroplasia: a dominant trait.** Dr. Michael C. Ain has achondroplasia, a form of dwarfism caused by a dominant allele. This has inspired his work: He is a specialist in the repair of bone defects caused by achondroplasia and other disorders. The dominant allele (*D*) might have arisen as a mutation in the egg or sperm of a parent or could have been inherited from an affected parent, as shown for an affected father in the Punnett square.

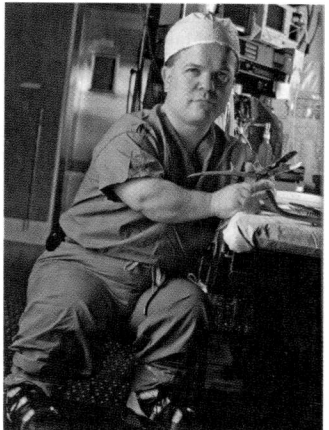

Parents

Dwarf phenotype *Dd* × Normal phenotype *dd*

Sperm

	D	*d*
Eggs *d*	*Dd* Dwarf phenotype	*dd* Normal phenotype
d	*Dd* Dwarf phenotype	*dd* Normal phenotype

In these cases, the individual may already have transmitted the allele to his or her children. For example, a degenerative disease of the nervous system, called **Huntington's disease**, is caused by a lethal dominant allele that has no obvious phenotypic effect until the individual is about 35 to 45 years old. Once the deterioration of the nervous system begins, it is irreversible and inevitably fatal. As with other dominant traits, a child born to a parent with the Huntington's disease allele has a 50% chance of inheriting the allele and the disorder (see the Punnett square in Figure 14.18). In the United States, this disease afflicts about one in 10,000 people.

At one time, the onset of symptoms was the only way to know if a person had inherited the Huntington's allele, but this is no longer the case. By analyzing DNA samples from a large family with a high incidence of the disorder, geneticists tracked the Huntington's allele to a locus near the tip of chromosome 4, and the gene was sequenced in 1993. This information led to the development of a test that could detect the presence of the Huntington's allele in an individual's genome. (The methods that make such tests possible are discussed in Concepts 20.1 and 20.4.) The availability of this test poses an agonizing dilemma for those with a family history of Huntington's disease. Some individuals may want to be tested for this disease, whereas others may decide it would be too stressful to find out.

 Mastering Biology Interview with Nancy Wexler: Mapping the gene that causes Huntington's disease

Multifactorial Disorders

The hereditary diseases we have discussed so far are sometimes described as simple Mendelian disorders because they result from abnormality of one or both alleles at a single genetic locus. Many more people are susceptible to diseases that have a multifactorial basis—a genetic component plus a significant environmental influence. Heart disease, diabetes, cancer, alcoholism, certain mental illnesses such as schizophrenia and bipolar disorder, and many other diseases are multifactorial. In these cases, the hereditary component is polygenic. For example, many genes affect cardiovascular health, making some of us more prone than others to heart attacks and strokes. No matter what our genotype, however, our lifestyle has a tremendous effect on phenotype for cardiovascular health and other multifactorial characters. Exercise, a healthful diet, abstinence from smoking, and an ability to handle stressful situations all reduce our risk of heart disease and some types of cancer.

Genetic Testing and Counseling

Avoiding simple Mendelian disorders is possible when the risk of a particular genetic disorder can be assessed before a child is conceived or during the early stages of the pregnancy. Many hospitals have genetic counselors who can provide information to prospective parents concerned about a family history for a specific disease. Fetal and newborn testing can also reveal genetic disorders.

Counseling Based on Mendelian Genetics and Probability Rules

Consider the case of a hypothetical couple, Tyler and Lily. Each had a brother who died from the same recessively inherited lethal disease. Before conceiving their first child, Tyler and Lily seek genetic counseling to determine the risk of having a child with the disease. From the information about their brothers, we know that both parents of Tyler and both parents of Lily must have been carriers of the recessive allele. Thus, Tyler and Lily are both products of $Aa \times Aa$ crosses, where a symbolizes the allele that causes this particular disease. We also know that Tyler and Lily are not homozygous recessive (aa) because they do not have the disease. Therefore, their genotypes are either AA or Aa.

Given a genotypic ratio of $1\,AA : 2\,Aa : 1\,aa$ for offspring of an $Aa \times Aa$ cross, Tyler and Lily each have a $2/3$ chance of being carriers (Aa). According to the rule of multiplication, the overall probability of their firstborn having the disorder is $2/3$ (the chance that Tyler is a carrier) times $2/3$ (the chance that Lily is a carrier) times $1/4$ (the chance of two carriers having a child with the disease), which equals $1/9$. Suppose that Lily and Tyler decide to have a child—after all, there is an $8/9$ chance that their baby will not have the disorder. If, despite these odds, their child is born with the disease, then we would know that *both* Tyler and Lily are, in fact, carriers (Aa genotype). If both Tyler and Lily are carriers, there is a $1/4$ chance that any subsequent child this couple has will have the disease. The probability is higher for subsequent children because the diagnosis of the disease in the first child established that both parents are carriers, not because the genotype of the first child affects in any way that of future children.

When we use Mendel's laws to predict possible outcomes of matings, it is important to remember that each child represents an independent event in the sense that its genotype is unaffected by the genotypes of older siblings. Suppose that Tyler and Lily have three more children, and *all three* have the hypothetical hereditary disease. There is only one chance in 64 ($1/4 \times 1/4 \times 1/4$) that such an outcome will occur. Despite these circumstances, the chance that a fourth child of this couple will have the disease remains $1/4$.

Tests for Identifying Carriers

Most children with recessive disorders are born to parents with normal phenotypes. The key to accurately assessing the genetic risk for a particular disease is therefore to find out whether the prospective parents are heterozygous carriers of the recessive allele. Tests are available that can distinguish individuals of normal phenotype who are dominant homozygotes from those who are heterozygous carriers, using blood or cells from the inside of the cheek. Hundreds of different recessive alleles can be detected, including those for Tay-Sachs disease, sickle-cell disease, and cystic fibrosis.

These tests for identifying carriers enable people with family histories of genetic disorders to make informed decisions about having children, including whether to do genetic testing of the fetus, should they decide to become pregnant. The results can also lead carriers to alert family members that they may also be carriers and might choose to be tested. The tests raise other issues as well: Could carriers be denied health or life insurance or lose the jobs providing those benefits, even though they themselves are healthy? The Genetic Information Nondiscrimination Act, in place since 2008, allays these concerns by prohibiting discrimination in employment or health insurance coverage based on genetic test results. This is an important act that protects patients' rights but is unknown to many, including physicians. Genetic counselors, trained in a growing number of professional programs, can help individuals understand their genetic test results. Once test results are clearly understood, those with certain results may face difficult decisions. Advances in biotechnology offer the potential to reduce human suffering, but along with them come ethical issues that require conscientious deliberation.

Fetal Testing

Two types of testing can be done to look for genetic disorders in the fetus: screening tests and diagnostic tests. Screening tests are generally noninvasive and can be followed by diagnostic tests if screening results suggest that a genetic disorder might be present or if there are other risks, as when both parents are carriers.

Screening tests include imaging and blood tests. Imaging techniques allow a physician to examine a fetus directly for major anatomical abnormalities that might not show up in genetic tests. In the *ultrasound* technique, for example, reflected sound waves are used to produce an image of the fetus by a simple noninvasive procedure. An ultrasound is often accompanied by a blood test that looks for fetal proteins that might signal a genetic disorder.

Medical scientists have also developed methods for isolating and analyzing fetal DNA that has escaped into the mother's blood (*cell-free fetal DNA*), using various techniques. Cell-free fetal DNA tests and other blood tests are increasingly being used as noninvasive prenatal screening tests for certain chromosomal defects and disorders; a positive result indicates to the parents that further diagnostic testing should be considered.

One diagnostic test that can determine whether the developing fetus has a serious recessive disease is **amniocentesis**, which can be performed starting at the 15th or 16th week of pregnancy **(Figure 14.19a)**. In this procedure, a physician inserts a needle into the uterus and extracts about 10 mL of amniotic fluid, the liquid that bathes the fetus. Some genetic disorders can be detected from the presence of certain proteins or hormones in the amniotic fluid itself. Tests

▼ **Figure 14.19 Diagnostic fetal testing for genetic disorders.** Most testing for genetic abnormalities is now DNA-based. In some cases biochemical tests are used to detect particular disorders, and karyotyping may be done to see whether the chromosomes of the fetus are normal in number and appearance.

for thousands of other disorders (such as cystic fibrosis) are performed on DNA from fetal cells sloughed off into the amniotic fluid using targeted DNA sequencing or a technique called a *DNA microarray* (see Figure 20.13) that tests for many alleles at the same time. A karyotype of these cultured cells can also confirm certain chromosomal defects that may not have been suggested by cell-free fetal DNA tests.

In an alternative diagnostic technique that is called **chorionic villus sampling (CVS)**, a physician inserts a narrow tube through the cervix into the uterus and suctions out a tiny sample of tissue from the placenta, the organ that transmits nutrients and fetal wastes between the fetus and the mother **(Figure 14.19b)**. The cells of the chorionic villi of the placenta—the portion sampled—are derived from the fetus and have the same genotype and DNA sequence as the new individual. DNA from these cells can be analyzed for the presence of many genetic disorders. The main advantage of CVS is that it can be performed as early as the 10th or 11th week of pregnancy.

Ultrasound and isolation of fetal cells or DNA from maternal blood pose no known risk to either mother or fetus, while amniocentesis and CVS can cause complications in a small percentage of cases (for both, between 1 in 1,000 and 1 in 500 cases). Amniocentesis or CVS for diagnostic testing is generally offered to women over age 35, due to their increased risk of bearing a child with Down syndrome or other chromosomal disorders, and

may also be offered to younger women if there are risk factors or conditions suggested by family history. If the fetal tests reveal a serious disorder, the parents face the difficult choice of either terminating the pregnancy or preparing to care for a child with a genetic disorder, one that might even be fatal. Parental and fetal screening for Tay-Sachs alleles done since 1980 has reduced the number of children born with this incurable disease by 90%.

Newborn Screening

Some genetic diseases and disorders can be detected at birth by simple biochemical tests that are routinely performed on blood from the baby's heel in most hospitals in the United States. One common screening program is for phenylketonuria (PKU), a recessively inherited disorder that occurs in about one out of every 10,000–15,000 births in the United States. Children with this disease cannot properly metabolize the amino acid phenylalanine. This compound and its byproduct, phenylpyruvate, can accumulate to toxic levels in the blood, causing severe intellectual disability. However, if PKU is detected in the newborn, a special diet low in phenylalanine will usually allow for typical development. (Among many other substances, this diet excludes the artificial sweetener aspartame, which contains phenylalanine.)

The number of newborn genetic tests, many for metabolic disorders or enzyme deficiencies, is over 100 and ever-increasing;

each state regulates which tests are used. If screening identifies a baby with a disease or disorder for which treatment can be started early on, that individual has improved chances for better life quality. Some diseases, like PKU, are treated by dietary changes and others by replacement of missing enzymes. Unfortunately, there are quite a few diseases for which no treatment is yet known.

Fetal and newborn screening for serious inherited diseases, tests for identifying carriers, and genetic counseling all rely on the Mendelian model of inheritance. We owe the "gene idea"—the concept of heritable factors transmitted according to simple rules of chance—to the elegant quantitative experiments of Gregor Mendel. The importance of his discoveries was overlooked by most biologists until the early 1900s, decades after he reported his findings. In the next chapter, you will learn how Mendel's laws have their physical basis in the behavior of chromosomes during sexual life cycles and how the synthesis of Mendelian genetics and a chromosome theory of inheritance catalyzed progress in genetics.

CONCEPT CHECK 14.4

1. Lucia and Jared each have a sibling with cystic fibrosis, but neither Lucia nor Jared nor any of their parents have the disease. Calculate the probability that if this couple has a child, the child will have cystic fibrosis. What would be the probability if a test revealed that Jared is a carrier but Lucia is not? Explain your answers.

2. **MAKE CONNECTIONS** Explain how the change of a single amino acid in hemoglobin leads to the aggregation of hemoglobin into long fibers. (Review Figures 5.14, 5.18, and 5.19.)

3. Juanita was born with six toes on each foot, a dominant trait called polydactyly. Two of her five siblings and her mother, but not her father, also have extra digits. What is Juanita's genotype for the number-of-digits character? Explain your answer. Use D and d to symbolize the alleles for this character.

4. **MAKE CONNECTIONS** In Table 14.1, note the phenotypic ratio of the dominant to recessive trait in the F$_2$ generation for the monohybrid cross involving flower color. Then determine the phenotypic ratio for the offspring of the second-generation couple in Figure 14.15b. What accounts for the difference in the two ratios?

For suggested answers, see Appendix A.

14 Chapter Review

→ Go to **Mastering Biology** for Assignments, the eText, the Study Area, and Dynamic Study Modules.

SUMMARY OF KEY CONCEPTS

→ To review key terms, go to the **Vocabulary Self-Quiz** in the **Mastering Biology** eText or Study Area, or go to goo.gl/zkjz9t.

CONCEPT 14.1

Mendel used the scientific approach to identify two laws of inheritance *(pp. 270–276)*

- Gregor Mendel formulated a theory of inheritance based on experiments with garden peas, proposing that parents pass on to their offspring discrete genes that retain their identity through generations. This theory includes two "laws."
- The **law of segregation** states that genes have alternative forms, or **alleles**. In a diploid organism, the two alleles of a gene segregate (separate) during meiosis and gamete formation; each sperm or egg carries only one allele of each pair. This law explains the 3:1 ratio of F$_2$ **phenotypes** observed when **monohybrids** cross-or self-pollinate. Each organism inherits one allele for each gene from each parent. In **heterozygotes**, the two alleles are different: Expression of the **dominant allele** masks the phenotypic effect of the **recessive allele. Homozygotes** have identical alleles of a given gene and are therefore **true-breeding**.
- The **law of independent assortment** states that the pair of alleles for a given gene segregates into gametes independently of the pair of alleles for any other gene. In a cross between **dihybrids** (individuals heterozygous for two genes), the offspring have four phenotypes in a 9:3:3:1 ratio.

? *When Mendel did crosses of true-breeding purple- and white-flowered pea plants, the white-flowered trait disappeared from the F$_1$ generation but reappeared in the F$_2$ generation. Use genetic terms to explain why that happened.*

CONCEPT 14.2

Probability laws govern Mendelian inheritance *(pp. 276–278)*

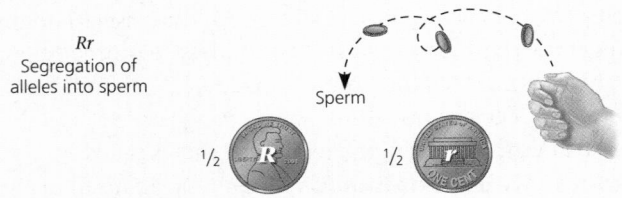

Rr
Segregation of alleles into sperm
Sperm
$^1/_2$ $^1/_2$

- The **multiplication rule** states that the probability of two or more events occurring together (a statement using *and*) is equal to the product of the individual probabilities of the independent single events. The **addition rule** states that the probability of an event that can occur in two or more independent, mutually exclusive ways (using *or*) is the sum of the individual probabilities.
- The rules of probability can be used to solve complex genetics problems. A dihybrid or other multicharacter cross is equivalent to two or more independent monohybrid crosses occurring simultaneously. In calculating the chances of the various offspring genotypes from such crosses, each character is first considered separately and then the individual probabilities are multiplied.

DRAW IT *Redraw the Punnett square on the right side of Figure 14.8 as two smaller monohybrid Punnett squares, one for each gene. Below each square, list the fractions of each phenotype produced. Use the rule of multiplication to compute the overall fraction of each of the possible dihybrid phenotypes. What is the phenotypic ratio?*

CONCEPT 14.3

Inheritance patterns are often more complex than predicted by simple Mendelian genetics (pp. 278–283)

• Extensions of Mendelian genetics for a single gene:

Relationship among alleles of a single gene	Description	Example
Complete dominance of one allele	Heterozygous phenotype same as that of homozygous dominant	PP 🌹 Pp 🌹
Incomplete dominance of either allele	Heterozygous phenotype intermediate between the two homozygous phenotypes	C^RC^R C^RC^W C^WC^W
Codominance	Both phenotypes expressed in heterozygotes	I^AI^B
Multiple alleles	In the population, some genes have more than two alleles	ABO blood group alleles I^A, I^B, i
Pleiotropy	One gene affects multiple phenotypic characters	Sickle-cell disease

• Extensions of Mendelian genetics for two or more genes:

Relationship among two or more genes	Description	Example
Epistasis	The phenotypic expression of one gene affects the expression of another gene	$BbEe \times BbEe$ 9 : 3 : 4
Polygenic inheritance	A single phenotypic character is affected by two or more genes	$AaBbCc \times AaBbCc$

• The expression of a genotype can be affected by environmental influences, resulting in a range of phenotypes. Polygenic characters that are also influenced by the environment are called **multifactorial** characters.

• An organism's overall phenotype reflects its overall genotype and unique environmental history. Even in more complex inheritance patterns, Mendel's fundamental laws still apply.

❓ *Which genetic relationships listed in the first column of the two tables above are demonstrated by the inheritance pattern of the ABO blood group alleles? For each genetic relationship, explain why this inheritance pattern is or is not an example.*

CONCEPT 14.4

Many human traits follow Mendelian patterns of inheritance (pp. 284–290)

• Analysis of family **pedigrees** can be used to deduce the possible genotypes of individuals and make predictions about future offspring. Such predictions are statistical probabilities rather than certainties.

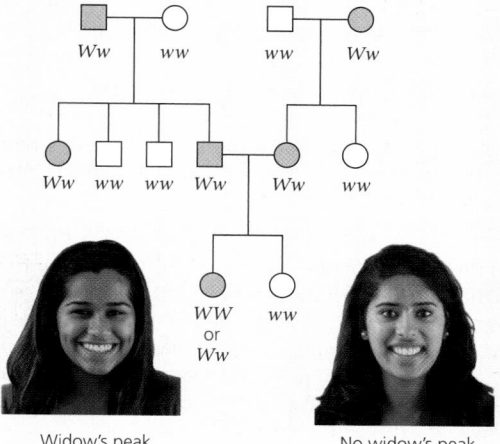

Widow's peak   No widow's peak

• Many genetic disorders are inherited as simple recessive traits. Most affected (homozygous recessive) individuals are children of phenotypically normal, heterozygous **carriers**.

• The sickle-cell allele has probably persisted for evolutionary reasons: Homozygotes have **sickle-cell disease**, but heterozygotes have an advantage because one copy of the sickle-cell allele reduces both the frequency and severity of malaria attacks.

Sickle-cell hemoglobin proteins — Part of a fiber of sickle-cell hemoglobin proteins — Long fibers cause red blood cells to be sickle-shaped — Sickle-cell disease

• Lethal dominant alleles are eliminated from the population if affected people die before reproducing. Nonlethal dominant alleles and lethal ones that strike relatively late in life can be inherited in a Mendelian way.

• Many human diseases are multifactorial—that is, they have both genetic and environmental components and do not follow simple Mendelian inheritance patterns.

• Using family histories, genetic counselors help couples determine the probability that their children will have genetic disorders. Genetic testing of prospective parents can reveal whether they are carriers of recessive alleles associated with specific disorders. Ultrasounds and blood tests can screen for certain disorders in a fetus. **Amniocentesis** and **chorionic villus sampling** can diagnose whether a suspected genetic disorder is present in a fetus. Other genetic tests can be performed after birth.

❓ *Both members of a couple know that they are carriers of the cystic fibrosis allele. None of their three children have cystic fibrosis, but any one of them might be a carrier. The couple would like to have a fourth child but are worried that he or she would very likely have the disease, since the first three do not. What would you tell the couple? Would it remove some uncertainty from their prediction if they could find out whether the three children are carriers?*

TIPS FOR GENETICS PROBLEMS

1. Write down symbols for the alleles. (These may be given in the problem.) When represented by single letters, the dominant allele is uppercase and the recessive is lowercase.

2. Write down the possible genotypes, as determined by the phenotype.
   a. If the phenotype is that of the dominant trait (for example, purple flowers), then the genotype is either homozygous dominant or heterozygous (*PP* or *Pp* in this example).
   b. If the phenotype is that of the recessive trait, the genotype must be homozygous recessive (for example, *pp*).
   c. If the problem says "true-breeding," the genotype is homozygous.

3. Determine what the problem is asking for. If asked to do a cross, write it out in the form [genotype] × [genotype], using the alleles you've decided on.

4. To figure out the outcome of a cross, set up a Punnett square.
   a. Put the gametes of one parent at the top and those of the other on the left. To determine the allele(s) in each gamete for a given genotype, set up a systematic way to list all the possibilities. (Remember, each gamete has one allele of each gene.) Note that there are 2^n possible types of gametes, where n is the number of gene loci that are heterozygous. For example, an individual with genotype *AaBbCc* would produce $2^3 = 8$ types of gametes. Write the genotypes of the gametes in circles above the columns and to the left of the rows.
   b. Fill in the Punnett square as if each possible sperm were fertilizing each possible egg, making all of the possible offspring. In a cross of *AaBbCc* × *AaBbCc*, for example, the Punnett square would have 8 columns and 8 rows, so there are 64 different offspring; you would know the genotype of each and thus the phenotype. Count genotypes and phenotypes to obtain the genotypic and phenotypic ratios. Because the Punnett square is so large, this method is not the most efficient. See tip 5.

5. You can use the rules of probability if a Punnett square would be too big. (For example, see the question at the end of the summary for Concept 14.2 and question 7 below.) You can consider each gene separately (see the section Solving Complex Genetics Problems with the Rules of Probability in Concept 14.2).

6. If the problem gives you the phenotypic ratios of offspring but not the genotypes of the parents in a given cross, the phenotypes can help you deduce the parents' unknown genotypes.
   a. For example, if $\frac{1}{2}$ of the offspring have the recessive phenotype and $\frac{1}{2}$ the dominant, you know that the cross was between a heterozygote and a homozygous recessive.
   b. If the ratio is 3:1, the cross was between two heterozygotes.
   c. If two genes are involved and you see a 9:3:3:1 ratio in the offspring, you know that each parent is heterozygous for both genes. Caution: Don't assume that the reported numbers will exactly equal the predicted ratios. For example, if there are 13 offspring with the dominant trait and 11 with the recessive, assume that the ratio is one dominant to one recessive.

7. For pedigree problems, use the tips in Figure 14.15 and below to determine what kind of trait is involved.
   a. If parents without the trait have offspring with the trait, the trait must be recessive and the parents both carriers.
   b. If the trait is seen in every generation, it is most likely dominant (see the next possibility, though).
   c. If both parents have the trait, then in order for it to be recessive, all offspring must show the trait.
   d. To determine the likely genotype of a certain individual in a pedigree, first label the genotypes of all the family members you can. Even if some of the genotypes are incomplete, label what you do know. For example, if an individual has the dominant phenotype, the genotype must be *AA* or *Aa*; you can write this as *A–*. Try different possibilities to see which fits the results. Use the rules of probability to calculate the probability of each possible genotype being the correct one.

TEST YOUR UNDERSTANDING

→ For multiple-choice questions, go to the **Practice Test** in the **Mastering Biology** eText or Study Area, or go to goo.gl/GruWRg.

Levels 1-2: Remembering/Understanding

1. **DRAW IT** A pea plant heterozygous for inflated pods (*Ii*) is crossed with a plant homozygous for constricted pods (*ii*). Draw a Punnett square for this cross to predict genotypic and phenotypic ratios. Assume that pollen comes from the *ii* plant.

2. A man with type A blood marries a woman with type B blood. Their child has type O blood. What are the genotypes of these three individuals? What genotypes, and in what frequencies, would you expect in future offspring from this marriage?

3. A man has six fingers on each hand and six toes on each foot. His wife and their daughter have the typical number of digits. Remember that extra digits is a dominant trait. What fraction of this couple's children would be expected to have extra digits?

4. **DRAW IT** Two pea plants heterozygous for the characters of pod color and pod shape are crossed. (See Table 14.1.) Draw a Punnett square to determine the phenotypic ratios of the offspring.

Levels 3-4: Applying/Analyzing

5. Flower position, stem length, and seed shape are three characters that Mendel studied. Each is controlled by an independently assorting gene and has dominant and recessive expression as indicated in Table 14.1. If a plant that is heterozygous for all three characters is allowed to self-fertilize, what proportion of the offspring would you expect to be each of the following? (Note: Use the rules of probability instead of a huge Punnett square.)
   (a) homozygous for the three dominant traits
   (b) homozygous for the three recessive traits
   (c) heterozygous for all three characters
   (d) homozygous for axial and tall, heterozygous for seed shape

6. Hemochromatosis is an inherited disease caused by a recessive allele. If a woman and her husband, who are both carriers, have three children, what is the probability of each of the following?
   (a) All three children are of normal phenotype.
   (b) One or more of the three children have the disease.
   (c) All three children have the disease.
   (d) At least one child is phenotypically normal.

   (*Note*: It will help to remember that the probabilities of all possible outcomes always add up to 1.)

7. The genotype of F$_1$ individuals in a tetrahybrid cross is *AaBbCcDd*. Assuming independent assortment of these four genes, what are the probabilities that F$_2$ offspring will have the following genotypes?
   (a) *aabbccdd*
   (b) *AaBbCcDd*
   (c) *AABBCCDD*
   (d) *AaBBccDd*
   (e) *AaBBCCdd*

8. What is the probability that each of the following pairs of parents will produce the indicated offspring? (Assume independent assortment of all gene pairs.)
   (a) $AABBCC \times aabbcc \rightarrow AaBbCc$
   (b) $AABbCc \times AaBbCc \rightarrow AAbbCC$
   (c) $AaBbCc \times AaBbCc \rightarrow AaBbCc$
   (d) $aaBbCC \times AABbcc \rightarrow AaBbCc$

9. Keisha and Jerome each have a sibling with sickle-cell disease. Neither Keisha nor Jerome nor any of their parents have the disease, and none of them have ever been tested to see if they carry the sickle-cell allele. Based on this incomplete information, calculate the probability that if this couple has a child, the child will have sickle-cell disease.

10. In 1981, a stray black cat with unusual rounded, curled-back ears was adopted by a family in California. Thousands of descendants of the cat have since been born, and cat fanciers have developed the curl cat into a show breed. Suppose you owned the first curl cat and wanted to develop a true-breeding variety. How would you determine whether the curl allele is dominant or recessive? How would you obtain true-breeding curl cats? How could you be sure they are true-breeding?

11. In tigers, a recessive allele of a particular gene causes both an absence of fur pigmentation (a white tiger) and a cross-eyed condition. If two phenotypically normal tigers that are heterozygous at this locus are mated, what percentage of their offspring will be cross-eyed? What percentage of tigers with crossed eyes will also be white?

12. In maize (corn) plants, a dominant allele I inhibits kernel color, while the recessive allele i permits color when homozygous. At a different locus, the dominant allele P causes purple kernel color, while the homozygous recessive genotype pp causes red kernels. If plants heterozygous at both loci are crossed, what will be the phenotypic ratio of the offspring?

13. The pedigree below traces the inheritance of alkaptonuria, a biochemical disorder. Affected individuals, indicated here by the colored circles and squares, are unable to metabolize a substance called alkapton, which colors the urine and stains body tissues. Does alkaptonuria appear to be caused by a dominant allele or by a recessive allele? Fill in the genotypes of the individuals whose genotypes can be deduced. What genotypes are possible for each of the other individuals?

14. Imagine that you are a genetic counselor, and a couple planning to start a family comes to you for information. Charles was married once before, and he and his first wife had a child with cystic fibrosis. The brother of his current wife, Elaine, died of cystic fibrosis. What is the probability that Charles and Elaine will have a baby with cystic fibrosis? (Neither Charles, Elaine, nor their parents have cystic fibrosis.)

Levels 5–6: Evaluating/Creating

15. **EVOLUTION CONNECTION** Over the past half century, there has been a trend in the United States and other developed countries for people to marry and start families later in life than did their parents and grandparents. What effects might this trend have on the incidence (frequency) of late-acting dominant lethal alleles in the population?

16. **SCIENTIFIC INQUIRY** You are handed a mystery pea plant with tall stems and axial flowers and asked to determine its genotype as quickly as possible. You know that the allele for tall stems (T) is dominant to that for dwarf stems (t) and that the allele for axial flowers (A) is dominant to that for terminal flowers (a).
   (a) Identify all the possible genotypes for your mystery plant.
   (b) Describe the one cross you would do in your garden to determine the exact genotype of your mystery plant.
   (c) While waiting for the results of your cross, you predict the results for each possible genotype listed in part a. Explain how you do this and why this is not called "performing a cross."
   (d) Explain how the results of your cross and your predictions will help you learn the genotype of your mystery plant.

17. **WRITE ABOUT A THEME: INFORMATION** The continuity of life is based on heritable information in the form of DNA. In a short essay (100–150 words), explain how the passage of genes from parents to offspring, in the form of particular alleles, ensures perpetuation of parental traits in offspring and, at the same time, genetic variation among offspring. Use genetic terms in your explanation.

18. **SYNTHESIZE YOUR KNOWLEDGE**

Just for fun, imagine that "shirt-striping" is a phenotypic character caused by a single gene. Construct a genetic explanation for the appearance of the family in the above photograph, consistent with their "shirt phenotypes." Include in your answer the presumed allele combinations for "shirt-striping" in each family member. Identify the inheritance pattern shown by the child.

For selected answers, see Appendix A.

15 The Chromosomal Basis of Inheritance

KEY CONCEPTS

15.1 Mendelian inheritance has its physical basis in the behavior of chromosomes *p. 295*

15.2 Sex-linked genes exhibit unique patterns of inheritance *p. 298*

15.3 Linked genes tend to be inherited together because they are located near each other on the same chromosome *p. 301*

15.4 Alterations of chromosome number or structure cause some genetic disorders *p. 306*

15.5 Some inheritance patterns are exceptions to standard Mendelian inheritance *p. 310*

Study Tip

Draw chromosomes: As you work on the figure legend questions in this chapter, draw the chromosomes for the crosses described in the questions. Below is a starting sketch for the What If? question that follows Figure 15.3. (Here you are including sex chromosomes to keep track of the offspring.)

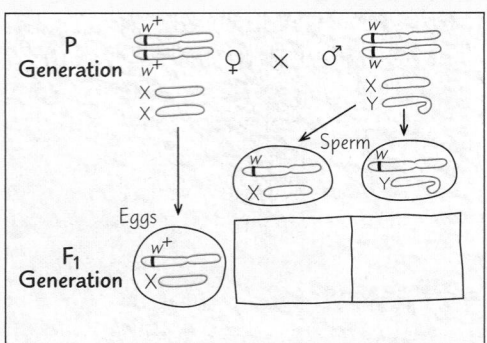

➔ Go to Mastering Biology

For Students (in eText and Study Area)
- Get Ready for Chapter 15
- Animation: The Chromosomal Basis of Independent Assortment
- Animation: Linked Genes and Crossing Over

For Instructors to Assign (in Item Library)
- Tutorial: Pedigrees and Sex-Linkage
- Tutorial: Recombination and Linkage Mapping

Figure 15.1 The four yellow dots mark the locations of a specific gene, tagged with a fluorescent yellow dye, on a pair of homologous chromosomes. The chromosomes have duplicated, so each chromosome has two sister chromatids, each with a copy of the gene. This provides a visual demonstration that genes—Mendel's "factors"—are segments of DNA located along chromosomes.

What is the relationship between genes and chromosomes?

Genes are located on chromosomes. In diploid cells, chromosomes and genes are present in pairs.

Chromosomes duplicate before cell division. Each duplicated chromosome has two copies of each allele, one on each sister chromatid.

During meiosis I, homologous chromosomes separate and alleles segregate. In meiosis II, sister chromatids separate.

Genes are passed on as discrete units. **Each chromosome has one version of a gene (one allele).**

Offspring inherit one allele from each parent.

Mendelian inheritance has its physical basis in the behavior of chromosomes

When Gregor Mendel proposed the existence of "hereditary factors" in 1860, no cellular structures had been identified that could house these imaginary units, and most biologists were skeptical. When the processes of mitosis and meiosis were worked out later that century, biologists saw parallels between the behavior of Mendel's proposed factors (genes) during sexual life cycles and the behavior of chromosomes: As shown in **Figure 15.1**, chromosomes and genes are both present in pairs in diploid cells, and homologous chromosomes separate and alleles segregate during the process of meiosis. After meiosis, fertilization restores the paired condition for both chromosomes and genes. Around 1902, Walter S. Sutton, Theodor Boveri, and others independently noted these parallels and began to develop the **chromosome theory of inheritance**. According to this theory, Mendelian genes have specific loci (sites) along chromosomes, and it is the chromosomes that undergo segregation and independent assortment.

The first solid evidence associating a specific gene with a specific chromosome came early in the 1900s from the work of Thomas Hunt Morgan, an experimental embryologist at Columbia University. Although Morgan was initially doubtful about both Mendelian genetics and the chromosome theory, his early experiments provided convincing evidence that chromosomes are indeed the location of Mendel's heritable factors.

Morgan's Choice of Experimental Organism

Many times in the history of biology, important discoveries have come to those insightful or lucky enough to choose an experimental organism suitable for the research problem being tackled. Mendel chose the garden pea because a number of distinct varieties were available. For his work, Morgan selected a species of fruit fly, *Drosophila melanogaster*, a common insect that feeds on the fungi growing on fruit. Fruit flies are prolific breeders; a single mating will produce hundreds of offspring, and a new generation can be bred every two weeks. Morgan's laboratory began using this convenient organism for genetic studies in 1907 and soon became known as "the fly room."

Another advantage of the fruit fly is that it has only four pairs of chromosomes, which are easily distinguishable with a light microscope. There are three pairs of autosomes and one pair of sex chromosomes. Female fruit flies have a pair of homologous X chromosomes, and males have one X chromosome and one Y chromosome.

While Mendel could readily obtain different pea varieties from seed suppliers, Morgan was probably the first person to want different varieties of the fruit fly. He faced the tedious task of carrying out many matings and then microscopically

▼ **Figure 15.2 Morgan's first mutant.** Wild-type *Drosophila* flies have red eyes (left). Among his flies, Morgan discovered a mutant male with white eyes (right). This variation made it possible for Morgan to trace a gene for eye color to a specific chromosome.

Wild type (red eyes)   **Mutant** (white eyes)

inspecting large numbers of offspring in search of naturally occurring variant individuals. After many months of this, he complained, "Two years' work wasted. I have been breeding those flies for all that time and I've got nothing out of it." Morgan persisted, however, and was finally rewarded with the discovery of a single male fly with white eyes instead of the usual red. The phenotype for a character most commonly observed in natural populations, such as red eyes in *Drosophila*, is called the **wild type (Figure 15.2)**. Traits that are alternatives to the wild type, such as white eyes in *Drosophila*, are called *mutant phenotypes* because they are due to alleles assumed to have originated as changes, or mutations, in the wild-type allele.

Morgan and his students invented a notation for symbolizing alleles in *Drosophila* that is still widely used for fruit flies. For a given character in flies, the gene takes its symbol from the first mutant (non–wild type) discovered. Thus, the allele for white eyes in *Drosophila* is symbolized by w. A superscript + identifies the allele for the wild-type trait: w^+ for the allele for red eyes, for example. Over the years, a variety of gene notation systems have been developed for different organisms. For example, human genes are usually written in all capitals, such as *HTT* for the gene involved in Huntington's disease. (More than one letter may be used for an allele, as in *HTT*.)

Correlating Behavior of a Gene's Alleles with Behavior of a Chromosome Pair: *Scientific Inquiry*

Morgan mated his white-eyed male fly with a red-eyed female. All the F_1 offspring had red eyes, suggesting that the wild-type allele is dominant. When Morgan bred the F_1 flies to each other, he observed the classical 3:1 phenotypic ratio among the F_2 offspring. However, there was a surprising additional result: The white-eye trait showed up only in males. All the F_2 females had red eyes, while half the males had red eyes and half had white eyes. Therefore, Morgan concluded that somehow a fly's eye color was linked to its sex. (If the eye color gene were unrelated to sex, half of the white-eyed flies would have been female.)

Recall that a female fly has two X chromosomes (XX), while a male fly has an X and a Y (XY). The correlation between the trait of white eye color and the male sex of the affected F_2 flies suggested to Morgan that the gene involved in his white-eyed mutant was located exclusively on the X chromosome, with no corresponding allele present on the Y chromosome. His reasoning can be followed in **Figure 15.3**. For a male, a single copy of the mutant allele would confer white eyes; since a male has only one X chromosome, there can be no wild-type allele (w^+) present to mask the recessive allele. However, a female could have white eyes only if both her X chromosomes carried the recessive mutant allele (w). This was impossible for the F_2 females in Morgan's experiment because all the F_1 fathers had red eyes, so each F_2 female received a w^+ allele on the X chromosome inherited from her father.

Morgan's finding of the correlation between a particular trait and an individual's sex provided support for the chromosome theory of inheritance: namely, that a specific gene is carried on a specific chromosome. In this case, an eye color gene is carried on the X chromosome.

Figure 15.4 illustrates the relationship between the chromosome theory of inheritance and Mendel's laws. The separation of homologs during anaphase I accounts for the segregation of the two alleles of a gene into separate gametes, and the random arrangement of chromosome pairs at metaphase I accounts for independent assortment of the alleles for two or more genes located on different homolog pairs. This figure traces the same dihybrid pea cross you learned about in Figure 14.8. By carefully studying Figure 15.4, you can see how the behavior of chromosomes during meiosis in the F_1 generation and subsequent random fertilization give rise to the F_2 phenotypic ratio observed by Mendel.

In addition to showing that a specific gene is carried on a specific chromosome, Morgan's work indicated that genes located on a sex chromosome exhibit unique inheritance patterns, which we will discuss in the next section. Recognizing the importance of Morgan's early work, many bright students were attracted to his fly room.

CONCEPT CHECK 15.1

1. **WHAT IF?** Propose a possible reason that the first naturally occurring mutant fruit fly Morgan saw involved a gene on a sex chromosome and was found in a male.

2. Which one of Mendel's laws describes the inheritance of alleles for a single character? Which law relates to the inheritance of alleles for two characters in a dihybrid cross?

3. **MAKE CONNECTIONS** Review the description of meiosis (see Figure 13.8) and Mendel's laws of segregation and independent assortment (see Concept 14.1). What is the physical basis for each of Mendel's laws?

For suggested answers, see Appendix A.

▼ Figure 15.3 **Inquiry**

In a cross between a wild-type female fruit fly and a mutant white-eyed male, what color eyes will the F_1 and F_2 offspring have?

Experiment Thomas Hunt Morgan wanted to analyze the behavior of two alleles of a fruit fly eye color gene. In crosses similar to those done by Mendel with pea plants, Morgan and his colleagues mated a wild-type (red-eyed) female with a mutant white-eyed male.

Morgan then bred an F_1 red-eyed female to an F_1 red-eyed male to produce the F_2 generation.

Results The F_2 generation showed a typical Mendelian ratio of 3 red-eyed flies: 1 white-eyed fly. However, all white-eyed flies were males; no females displayed the white-eye trait.

Conclusion All F_1 offspring had red eyes, so the mutant white-eye trait (w) must be recessive to the wild-type red-eye trait (w^+). Since the recessive trait—white eyes—was expressed only in males in the F_2 generation, Morgan deduced that this eye color gene is located on the X chromosome and that there is no corresponding locus on the Y chromosome.

Data from T. H. Morgan, Sex-limited inheritance in *Drosophila, Science* 32: 120–122 (1910).

→ **Instructors:** A related Experimental Inquiry Tutorial can be assigned in **Mastering Biology**.

WHAT IF? *Suppose this eye color gene were located on an autosome rather than a sex chromosome. Predict the phenotypes (including sex) of the F_2 flies in this hypothetical cross. (Hint: Draw Punnett squares, including chromosomes, for the F_1 and F_2 generations; see the Study Tip.)*

▼ **Figure 15.4 The chromosomal basis of Mendel's laws.** Here we correlate the results of one of Mendel's dihybrid crosses (see Figure 14.8) with the behavior of chromosomes during meiosis (see Figure 13.8). The arrangement of chromosomes at metaphase I of meiosis and their movement during anaphase I account for, respectively, the independent assortment and segregation of the alleles for seed color and shape. Each cell that undergoes meiosis in an F₁ plant produces two kinds of gametes. If we count the results for all cells, however, each F₁ plant produces equal numbers of all four kinds of gametes because the alternative chromosome arrangements at metaphase I are equally likely.

→ **Mastering Biology**
Animation: The Chromosomal Basis of Independent Assortment
MP3 Tutor: The Chromosomal Basis of Inheritance

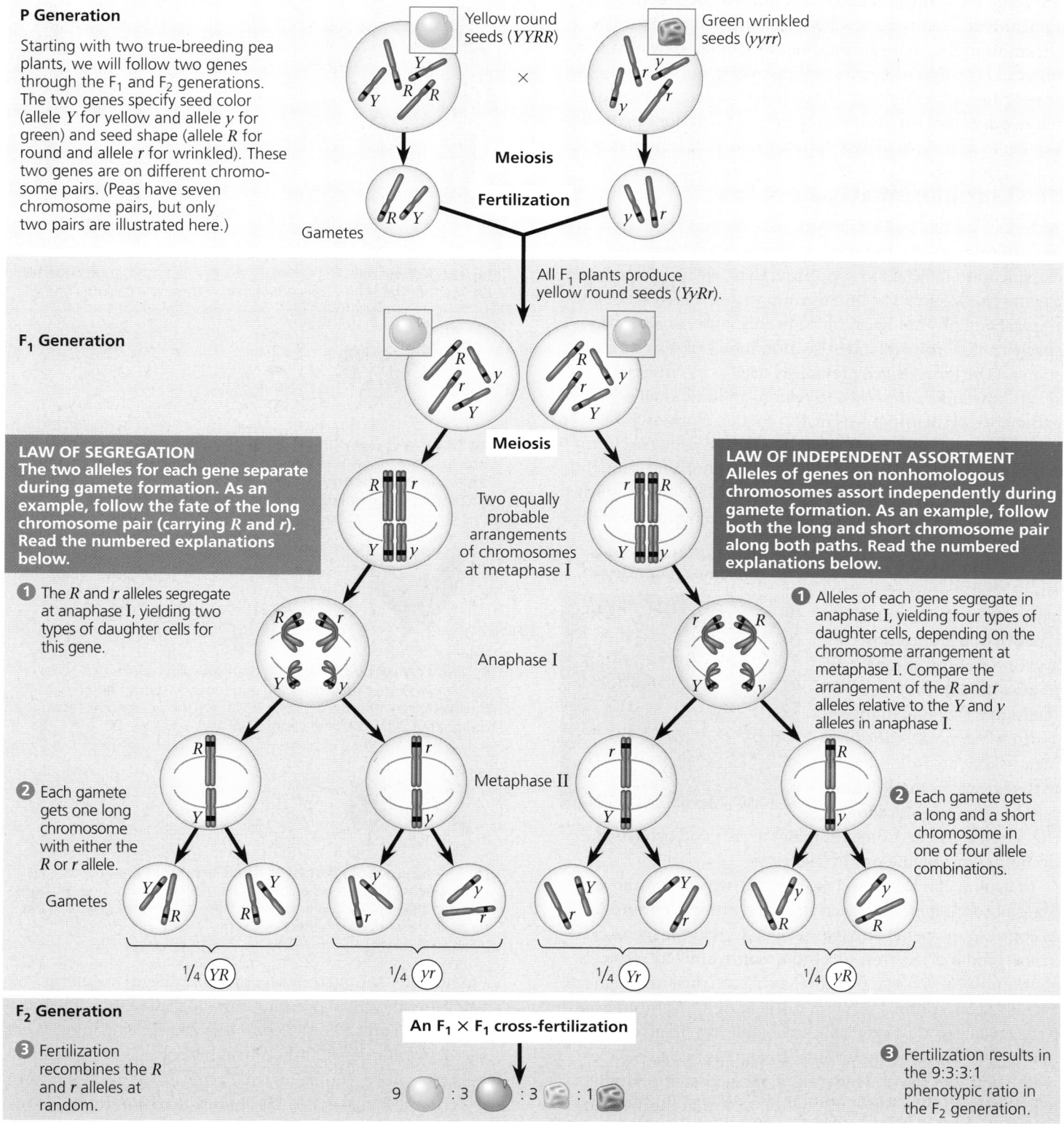

P Generation

Starting with two true-breeding pea plants, we will follow two genes through the F₁ and F₂ generations. The two genes specify seed color (allele *Y* for yellow and allele *y* for green) and seed shape (allele *R* for round and allele *r* for wrinkled). These two genes are on different chromosome pairs. (Peas have seven chromosome pairs, but only two pairs are illustrated here.)

Yellow round seeds (*YYRR*)

Green wrinkled seeds (*yyrr*)

Meiosis

Fertilization

Gametes

All F₁ plants produce yellow round seeds (*YyRr*).

F₁ Generation

LAW OF SEGREGATION
The two alleles for each gene separate during gamete formation. As an example, follow the fate of the long chromosome pair (carrying *R* and *r*). Read the numbered explanations below.

Meiosis

Two equally probable arrangements of chromosomes at metaphase I

LAW OF INDEPENDENT ASSORTMENT
Alleles of genes on nonhomologous chromosomes assort independently during gamete formation. As an example, follow both the long and short chromosome pair along both paths. Read the numbered explanations below.

❶ The *R* and *r* alleles segregate at anaphase I, yielding two types of daughter cells for this gene.

Anaphase I

❶ Alleles of each gene segregate in anaphase I, yielding four types of daughter cells, depending on the chromosome arrangement at metaphase I. Compare the arrangement of the *R* and *r* alleles relative to the *Y* and *y* alleles in anaphase I.

Metaphase II

❷ Each gamete gets one long chromosome with either the *R* or *r* allele.

❷ Each gamete gets a long and a short chromosome in one of four allele combinations.

Gametes

¼ (YR)   ¼ (yr)   ¼ (Yr)   ¼ (yR)

F₂ Generation

❸ Fertilization recombines the *R* and *r* alleles at random.

An F₁ × F₁ cross-fertilization

9 : 3 : 3 : 1

❸ Fertilization results in the 9:3:3:1 phenotypic ratio in the F₂ generation.

? If you crossed an F₁ plant above with a plant that was homozygous recessive for both genes (yyrr), how would the phenotypic ratio of the offspring compare with the 9:3:3:1 ratio seen here?

CHAPTER 15 The Chromosomal Basis of Inheritance **297**

CONCEPT 15.2

Sex-linked genes exhibit unique patterns of inheritance

Morgan's discovery of a trait (white eyes) that correlated with the sex of flies provided important support for the chromosome theory of inheritance. Because the identity of the sex chromosomes in an individual could be inferred by observing the sex of the fly, the behavior of the two members of the pair of sex chromosomes could be correlated with the behavior of the two alleles of the eye color gene. In this section, we'll take a closer look at the role of sex chromosomes in inheritance.

The Chromosomal Basis of Sex

Although sex has traditionally been described as binary—male or female—we are coming to understand that this classification may be too simplistic. Here, we use the term *sex* to refer to classification into a group with a shared set of anatomical and physiological traits. In this sense, sex in many species is determined largely by inheritance of sex chromosomes. (The term *gender*, previously used as a synonym of sex, is now more often used to refer to an individual's own experience of identifying as male, female, or otherwise.)

Humans and other mammals have two types of sex chromosomes, designated X and Y. The Y chromosome is much smaller than the X chromosome **(Figure 15.5)**. A person who inherits two X chromosomes, one from each parent, usually

develops anatomy we associate with the "female" sex, while "male" properties are associated with the inheritance of one X chromosome and one Y chromosome **(Figure 15.6a)**. Short segments at either end of the Y chromosome are the only regions that are homologous with regions on the X. These homologous regions allow the X and Y chromosomes in males to pair and behave like homologs during meiosis in the testes.

▼ **Figure 15.5 Human sex chromosomes (duplicated).**

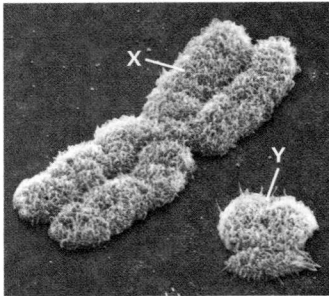

In mammalian testes and ovaries, the two sex chromosomes segregate during meiosis. Each egg receives one X chromosome. In contrast, sperm fall into two categories: Half the sperm cells a male produces receive an X chromosome, and half receive a Y chromosome. We can trace the sex of each offspring to the events of conception: If a sperm cell bearing an X chromosome fertilizes an egg, the zygote is XX, a female; if a sperm cell containing a Y chromosome fertilizes an egg, the zygote is XY, a male (see Figure 15.6a). Thus, in general, sex determination is a matter of chance—a fifty-fifty chance. Note that the mammalian X-Y system isn't the only chromosomal system for determining sex. **Figure 15.6b–d** illustrates three other systems.

▼ **Figure 15.6 Some chromosomal systems of sex determination.** Numerals indicate the number of autosomes in the species pictured. In *Drosophila*, males are XY, but sex depends on the ratio of the number of X chromosomes to the number of autosome sets, not simply on the presence of a Y chromosome. In some species (not shown here), sex is determined not by chromosomes but by environmental factors such as temperature.

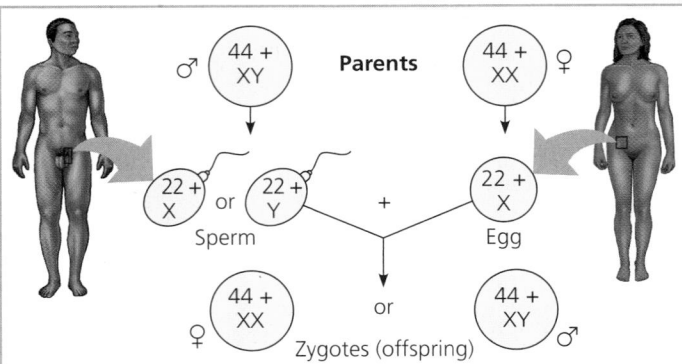

(a) The X-Y system. In mammals, the sex of an offspring depends on whether the sperm cell contains an X chromosome or a Y.

(b) The X-0 system. In grasshoppers, cockroaches, and some other insects, there is only one type of sex chromosome, the X. Females are XX; males have only one sex chromosome (X0). Sex of the offspring is determined by whether the sperm cell contains an X chromosome or no sex chromosome.

(c) The Z-W system. In birds, some fishes, and some insects, the sex chromosomes present in the egg (not the sperm) differ, and thus determine the sex of offspring. The sex chromosomes are designated Z and W. Females are ZW and males are ZZ.

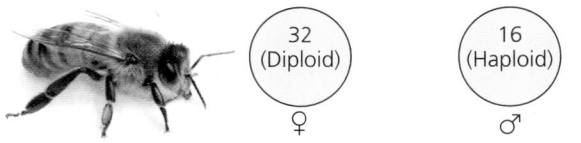

(d) The haplo-diploid system. There are no sex chromosomes in most species of bees and ants. Females develop from fertilized eggs and are thus diploid. Males develop from unfertilized eggs and are haploid; they have no fathers.

In humans, the anatomical signs of sex begin to emerge when the embryo is about 2 months old. Before then, the rudiments of the gonads are generic—they can develop into either testes or ovaries, depending on whether or not a Y chromosome is present, and depending on what genes are active. A gene on the Y chromosome—called *SRY*, for sex-determining region of Y—is required for the development of testes. In the absence of *SRY*, the gonads develop into ovaries, even in an XY embryo.

A gene located on either sex chromosome is called a **sex-linked gene**. The human X chromosome contains approximately 1,100 genes, which are called **X-linked genes**, while genes located on the Y chromosome are called *Y-linked genes*. On the human Y chromosome, researchers have identified 78 genes that code for about 25 proteins (some genes are duplicates). About half of these genes are expressed only in the testis, and some are required for normal testicular functioning and the production of normal sperm. The Y chromosome is passed along virtually intact from a father to all his sons. Because there are so few Y-linked genes, very few disorders are transferred from father to son on the Y chromosome.

The development of female gonads in humans requires a gene called *WNT4* (on chromosome 1, an autosome), which encodes a protein that promotes ovary development. An embryo that is XY but has extra copies of the *WNT4* gene can develop rudimentary female gonads. Overall, sex is determined by the interactions of a network of gene products like these.

The biochemical, physiological, and anatomical features associated with "males" and "females" are turning out to be more complicated than previously thought, with many genes involved in their development. Because of the complexity of this process, many variations exist: Some individuals vary in the number of sex chromosomes in their cells (see Concept 15.4), and others are born with intermediate sexual (*intersex*) characteristics, or with anatomical features that do not match an individual's sense of their own gender (*transgender* individuals). Sex determination is an active area of research that will likely yield a more sophisticated understanding in years to come.

Inheritance of X-Linked Genes

The fact that males and females inherit a different number of X chromosomes leads to a pattern of inheritance different from that produced by genes located on autosomes. While there are very few Y-linked genes, many of which help determine sex, the X chromosomes have numerous genes for characters unrelated to sex. X-linked genes in humans follow the same pattern of inheritance that Morgan observed for the eye color locus he studied in *Drosophila* (see Figure 15.3). Fathers pass X-linked alleles to all of their daughters but to none of their sons. In contrast, mothers can pass X-linked alleles to both sons and daughters, as shown in **Figure 15.7** for the inheritance of a mild X-linked disorder, red-green color blindness.

If an X-linked trait is due to a recessive allele, a female will express the phenotype only if she is homozygous for that allele. Because males have only one locus, the terms *homozygous* and *heterozygous* lack meaning for describing their X-linked genes; the term *hemizygous* is used in such cases. Any male receiving the recessive allele from his mother will express the trait. For this reason, far more males than females have X-linked recessive disorders. However, even though the chance of a female inheriting a double dose of the mutant allele is much less than the probability of a male inheriting a single dose, there *are* females with X-linked disorders. For instance, color blindness is almost always inherited as an X-linked trait. A color-blind daughter may be born to a color-blind father whose mate is a carrier (see Figure 15.7c). Because the X-linked allele for color blindness is relatively rare, though, the probability that such a man and woman will mate is low.

A number of human X-linked disorders are much more serious than color blindness, such as **Duchenne muscular dystrophy**, which affects about one out of 5,000 males born in the United States. The disease is characterized by a progressive weakening of the muscles and loss of coordination. Affected individuals' life expectancy is into the mid-20s. Researchers have traced the disorder to the absence of a key muscle protein called dystrophin and have mapped the gene for this protein to a specific locus on the X chromosome. Since the gene is known, gene therapy is being explored.

▼ **Figure 15.7 The transmission of X-linked recessive traits.** In this diagram, red-green color blindness is used as an example. The superscript N represents the dominant allele for normal color vision carried on the X chromosome, while n represents the recessive allele, which has a mutation causing color blindness. White boxes indicate unaffected individuals, light orange boxes indicate carriers, and dark orange boxes indicate color-blind individuals.

? *If a color-blind woman married a man who had normal color vision, what would be the probable phenotypes of their children?*

➜ Mastering Biology Animation: X-Linked Genes in "MendAliens"

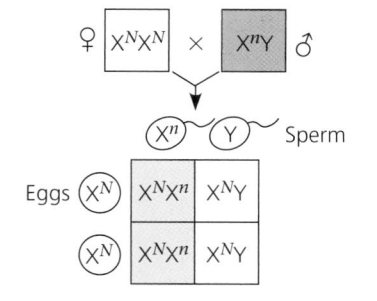

(a) A father with color-blindness will transmit the mutant allele to all daughters but to no sons. When the mother is a dominant homozygote, the daughters will have the normal phenotype but will be carriers of the mutation.

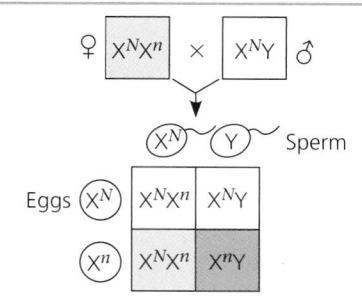

(b) If a carrier mates with a male who has normal color vision, there is a 50% chance that each daughter will be a carrier like her mother and a 50% chance that each son will have the disorder.

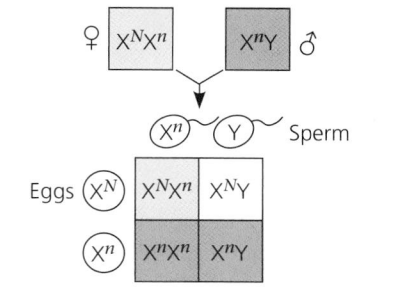

(c) If a carrier mates with a color-blind male, there is a 50% chance that each child born to them will have the disorder, regardless of sex. Daughters who have normal color vision will be carriers, whereas males who have normal color vision will be free of the recessive allele.

Hemophilia is an X-linked recessive disorder defined by the absence of one or more of the proteins required for blood clotting. When a person with hemophilia is injured, bleeding is prolonged because a firm clot is slow to form. Small cuts in the skin are usually not a problem, but bleeding in the muscles or joints can be painful and can lead to serious damage. In the 1800s, hemophilia was widespread among the royal families of Europe. Queen Victoria of England is known to have passed the allele to several of her descendants. Subsequent intermarriage with royal family members of other nations, such as Spain and Russia, further spread this X-linked trait, and its incidence is well documented in royal pedigrees. A few years ago, new genomic techniques allowed sequencing of DNA from tiny amounts isolated from the buried remains of royal family members. The genetic basis of the mutation, and how it resulted in a nonfunctional blood-clotting factor, is now understood. Today, people with hemophilia are treated as needed with intravenous injections of the protein that is missing.

X Inactivation in Female Mammals

Female mammals, including human females, inherit two X chromosomes—twice the number inherited by males—so you may wonder whether females make twice as many of the proteins encoded by X-linked genes as males. In fact, almost all of one X chromosome in each cell in female mammals becomes inactivated during early embryonic development. As a result, the cells of females and males have the same effective dose (one active copy) of most X-linked genes. The inactive X in each cell of a female condenses into a compact object called a **Barr body** (discovered by Canadian anatomist Murray Barr), which lies along the inside of the nuclear envelope. Most of the genes of the X chromosome that forms the Barr body are not expressed. In the ovaries, however, Barr body chromosomes are reactivated in the cells that give rise to eggs, resulting in every female gamete (egg) having an active X after meiosis.

British geneticist Mary Lyon demonstrated that the selection of which X chromosome will form the Barr body occurs randomly and independently in each embryonic cell present at the time of X inactivation. As a consequence, females consist of a *mosaic* of two types of cells: those with the active X derived from the father and those with the active X derived from the mother. After an X chromosome is inactivated in a particular cell, all mitotic descendants of that cell have the same inactive X. Thus, if a female is heterozygous for a sex-linked trait, about half of her cells will express one allele, while the others will express the alternate allele. **Figure 15.8** shows how this mosaicism results in the patchy coloration of a tortoiseshell cat. In humans, mosaicism can be observed in a recessive X-linked mutation that prevents the development of sweat glands. A woman who is heterozygous for this trait has patches of normal skin and patches of skin lacking sweat glands.

Inactivation of an X chromosome involves modification of the DNA and proteins bound to it called histones, including attachment of methyl groups ($-CH_3$) to DNA nucleotides.

▼ **Figure 15.8 X inactivation and the tortoiseshell cat.** The tortoiseshell gene is on the X chromosome, and the tortoiseshell phenotype requires the presence of two different alleles, one for orange fur and one for black fur. Normally, only females can have both alleles because only they have two X chromosomes. If a female cat is heterozygous for the tortoiseshell gene, she is tortoiseshell. Orange patches are formed by populations of cells in which the X chromosome with the orange allele is active; black patches have cells in which the X chromosome with the black allele is active. ("Calico" cats also have white areas, which are determined by another gene.)

⊙ **Mastering Biology HHMI Animation: X Inactivation**

(The regulatory role of DNA methylation is discussed in Concept 18.2.) A particular region of each X chromosome contains several genes involved in the inactivation process. The two regions, one on each X chromosome, associate briefly with each other in each cell at an early stage of embryonic development. Then one of the genes, called *XIST* (for X-inactive specific transcript), becomes active *only* on the chromosome that will become the Barr body (the inactive X). Multiple copies of the RNA product of this gene apparently attach to the X chromosome on which they are made, eventually almost covering it. Interaction of this RNA with the chromosome initiates X inactivation, and the RNA products of nearby genes help to regulate the process.

CONCEPT CHECK 15.2

1. A white-eyed female *Drosophila* is mated with a red-eyed (wild-type) male, the reciprocal cross of the one shown in Figure 15.3. What phenotypes and genotypes do you predict for the offspring from this cross?

2. Neither Tim nor Shonda has Duchenne muscular dystrophy, but their firstborn son does. What is the probability that a second child will have the disease? What is the probability if the second child is a boy? A girl?

3. **MAKE CONNECTIONS** Consider what you learned about dominant and recessive alleles in Concept 14.1. If a disorder were caused by a dominant X-linked allele, how would the inheritance pattern differ from what we see for recessive X-linked disorders?

For suggested answers, see Appendix A.

Linked genes tend to be inherited together because they are located near each other on the same chromosome

The number of genes in a cell is far greater than the number of chromosomes; in fact, each chromosome (except the Y) has hundreds or thousands of genes. Genes located near each other on the same chromosome tend to be inherited together in genetic crosses; such genes are said to be genetically linked and are called **linked genes**. When geneticists follow linked genes in breeding experiments, the results deviate from those expected from Mendel's law of independent assortment.

How Linkage Affects Inheritance

To see how linkage between genes affects the inheritance of two different characters, let's examine another of Morgan's *Drosophila* experiments. In this case, the characters are body color and wing size, each with two different phenotypes. Wild-type flies have gray bodies and normal-sized wings. In addition to these flies, Morgan had managed to obtain, through breeding, doubly mutant flies with black bodies and wings much smaller than normal, called vestigial wings. The mutant alleles are recessive to the wild-type alleles, and neither gene is on a sex chromosome. In his investigation of these two genes, Morgan carried out the crosses shown in **Figure 15.9**. The first was a P generation cross to generate F_1 dihybrid flies, and the second was essentially a testcross.

▼ Figure 15.9 Inquiry

How does linkage between two genes affect inheritance of characters?

Experiment Morgan wanted to know whether the genes for body color and wing size are genetically linked and, if so, how this affects their inheritance. The alleles for body color are b^+ (gray) and b (black), and those for wing size are vg^+ (normal) and vg (vestigial).

Morgan mated true-breeding P (parental) generation flies—wild-type flies with black, vestigial-winged flies—to produce heterozygous F_1 dihybrids ($b^+ b$ $vg^+ vg$), all of which are wild-type in appearance.

He then mated wild-type F_1 dihybrid females with homozygous recessive males. This testcross will reveal the genotype of the eggs made by the dihybrid female.

The male's sperm contributes only recessive alleles, so the phenotype of the offspring reflects the genotype of the female's eggs.

Note: Although only females (with pointed abdomens) are shown, half the offspring in each class would be males (with rounded abdomens).

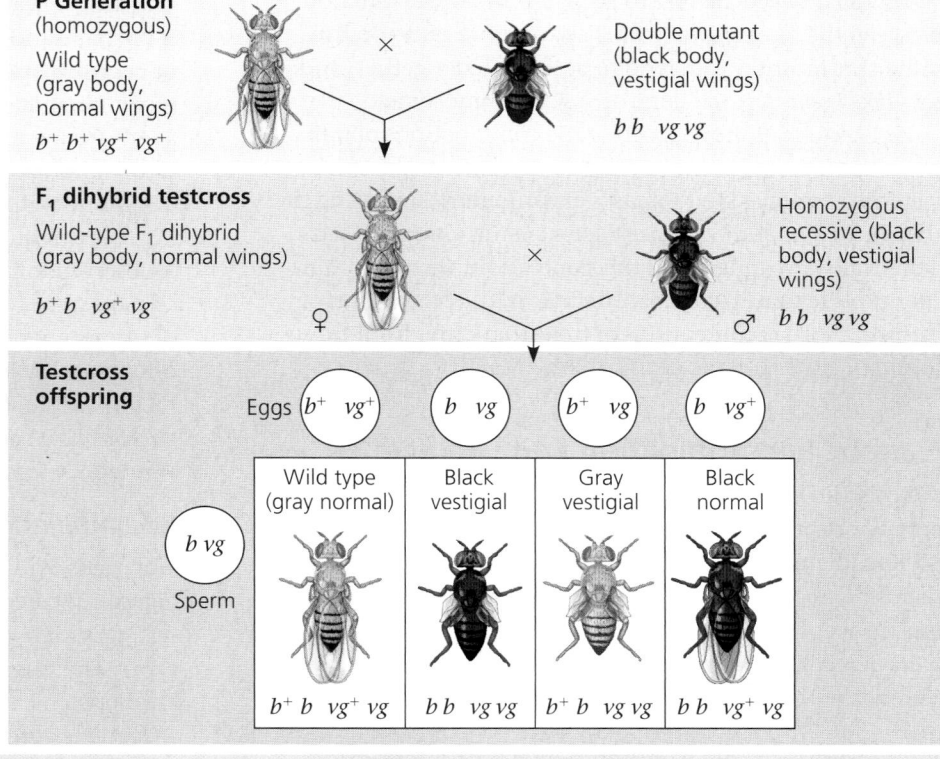

Predicted ratios of testcross offspring				
if genes are located on different chromosomes:	1 :	1 :	1 :	1
if genes are located on the same chromosome *and* parental alleles are always inherited together:	1 :	1 :	0 :	0

Results

Data from Morgan's experiment: 965 :	944 :	206 :	185

Conclusion Since most offspring had a parental (P generation) phenotype, Morgan concluded that the genes for body color and wing size are genetically linked on the same chromosome. However, the production of a relatively small number of offspring with nonparental phenotypes indicated that some mechanism occasionally breaks the linkage between specific alleles of genes on the same chromosome.

Data from T. H. Morgan and C. J. Lynch, The linkage of two factors in *Drosophila* that are not sex-linked, *Biological Bulletin* 23:174–182 (1912).

WHAT IF? *If the parental (P generation) flies had been true-breeding for gray body with vestigial wings and black body with normal wings, which phenotypic class(es) would be largest among the testcross offspring?*

The resulting flies had a much higher proportion of the combinations of traits seen in the P generation flies (called parental phenotypes) than would be expected if the two genes assorted independently. Morgan thus concluded that body color and wing size are usually inherited together in specific (parental) combinations because the genes are linked; they are near each other on the same chromosome:

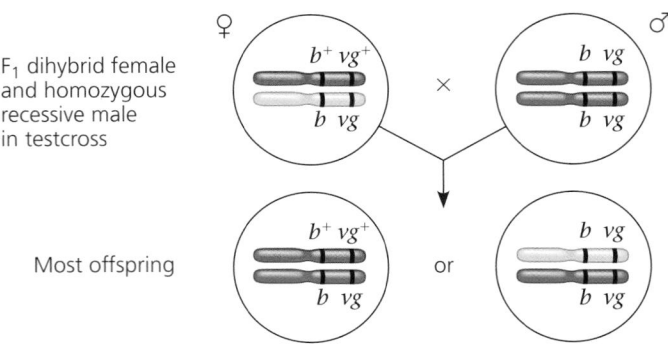

F₁ dihybrid female and homozygous recessive male in testcross

Most offspring

As you proceed, be sure to keep in mind the distinction between the terms *linked genes* (two or more genes on the same chromosome that tend to be inherited together) and *sex-linked gene* (a single gene on a sex chromosome).

As Figure 15.9 shows, both of the combinations of traits not seen in the P generation (called nonparental phenotypes) were also produced in Morgan's experiments, suggesting that the body color and wing size alleles are not always linked genetically. To understand this conclusion, we need to further explore **genetic recombination**, the production of offspring with combinations of traits that differ from those found in either P generation parent.

Genetic Recombination and Linkage

Meiosis and random fertilization generate genetic variation among offspring of sexually reproducing organisms due to independent assortment of chromosomes, crossing over in meiosis I, and the possibility of any sperm fertilizing any egg (see Concept 13.4). Here we'll examine the chromosomal basis of recombination of alleles in relation to the genetic findings of Mendel and Morgan.

Recombination of Unlinked Genes: Independent Assortment of Chromosomes

Mendel learned from crosses in which he followed two characters that some offspring have combinations of traits that do not match those of either parent. For example, consider a cross of a dihybrid pea plant with yellow round seeds, heterozygous for both seed color and seed shape (*YyRr*), with a plant homozygous for both recessive alleles (with green wrinkled seeds, *yyrr*). (This cross acts as a testcross because the results will reveal the genotype not only of the dihybrid *YyRr* plant,

which we know, but of the gametes made in that plant.) Let's represent the cross by the following Punnett square:

Notice in this Punnett square that one-half of the offspring are expected to inherit a phenotype that matches either of the phenotypes of the P (parental) generation originally crossed to produce the F₁ dihybrid (see Figure 15.2). These matching offspring are called **parental types** (short for phenotypes). But two nonparental phenotypes are also found among the offspring. Because these offspring have new combinations of seed shape and color, they are called **recombinant types**, or **recombinants** for short. When 50% of all offspring are recombinants, as in this example, geneticists say that there is a 50% frequency of recombination. The predicted phenotypic ratios among the offspring are similar to what Mendel actually found in his *YyRr × yyrr* crosses.

A 50% frequency of recombination in such testcrosses is observed for any two genes that are located on different chromosomes and thus cannot be linked. The physical basis of recombination between unlinked genes is the random orientation of homologous chromosomes at metaphase I of meiosis, which leads to the independent assortment of the two unlinked genes (see Figure 13.11 and the question in the Figure 15.4 legend).

Recombination of Linked Genes: Crossing Over

Now, let's explain the results of the *Drosophila* testcross in Figure 15.9. Recall that most (83%) of the offspring from the testcross for body color and wing size had parental phenotypes. That suggested that the two genes were on the same chromosome, since the occurrence of parental types with a frequency greater than 50% indicates that the genes are linked. About 17% of offspring, however, were recombinants.

Seeing these results, Morgan proposed that some process must occasionally break the physical connection between specific alleles of genes on the same chromosome. Later experiments showed that this process, now called **crossing over**, accounts for the recombination of linked genes. In crossing over, which occurs while replicated homologous chromosomes are paired during prophase of meiosis I, a set of proteins breaks the DNA molecules of one maternal and one paternal chromatid and rejoins each to the other

▶ **Figure 15.10 Chromosomal basis for recombination of linked genes.** In these diagrams re-creating the testcross in Figure 15.9, we track chromosomes as well as genes. The maternal chromosomes (those present in the wild-type F_1 dihybrid) are color-coded red and pink to distinguish one homolog from the other before any meiotic crossing over has occurred. Because crossing over between the b^+/b and vg^+/vg loci occurs in some, but not most, egg-producing cells, more eggs with parental-type chromosomes than with recombinant ones are produced in the mating females. Fertilization of the eggs by sperm of genotype $b\ vg$ gives rise to some recombinant offspring. The recombination frequency is the percentage of recombinant flies in the total pool of offspring.

DRAW IT *Suppose, as in the question at the bottom of Figure 15.9, the parental (P generation) flies were true-breeding for gray body with vestigial wings and black body with normal wings. Draw the chromosomes in each of the four possible kinds of eggs from an F_1 female, and label each chromosome as "parental" or "recombinant."*

⊙ **Mastering Biology** Animation: Linked Genes and Crossing Over

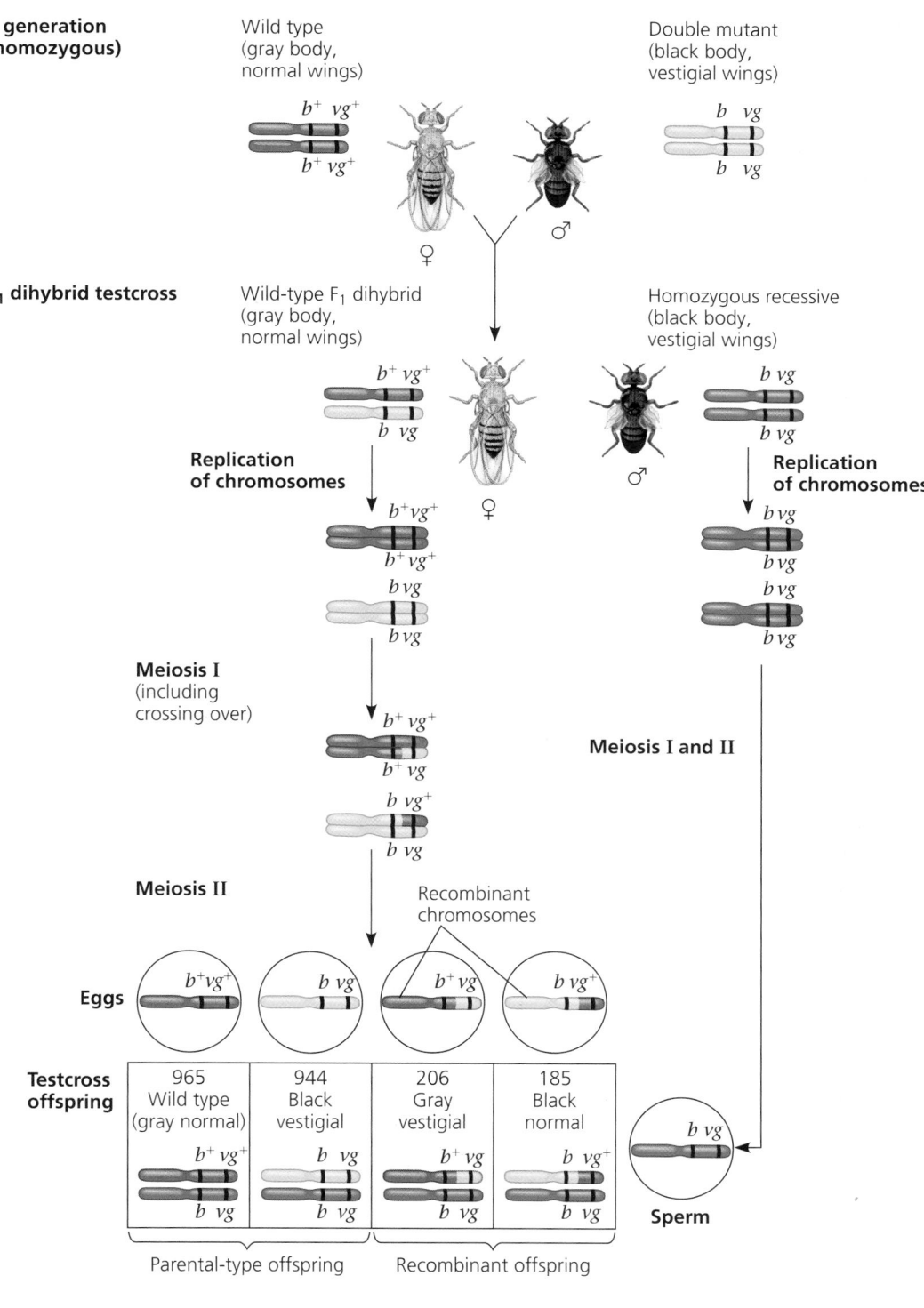

$$\text{Recombination frequency} = \frac{391 \text{ recombinants}}{2{,}300 \text{ total offspring}} \times 100 = 17\%$$

(see Figure 13.9). In effect, when a single crossover occurs, end portions of two nonsister chromatids trade places.

Figure 15.10 shows how crossing over in a dihybrid female fly resulted in recombinant eggs and ultimately recombinant offspring in Morgan's testcross. Most eggs had a chromosome with either the $b^+\ vg^+$ or $b\ vg$ parental genotype, but some eggs had a recombinant chromosome ($b^+\ vg$ or $b\ vg^+$). Fertilization of all classes of eggs by homozygous recessive sperm ($b\ vg$) produced an offspring population in which 17% exhibited a nonparental, recombinant phenotype, reflecting combinations of alleles not seen before in either P generation parent. In the **Scientific Skills Exercise**, you can use a statistical test to analyze the results from an F_1 dihybrid testcross and see whether the two genes assort independently or are linked.

Using the Chi-Square (χ^2) Test

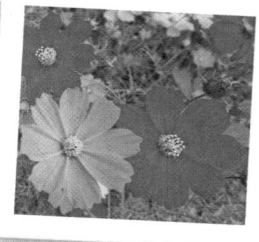

> **Cosmos plants**

Are Two Genes Linked or Unlinked? Genes that are in close proximity on the same chromosome will result in the linked alleles being inherited together more often than not. But how can you tell if certain alleles are inherited together due to linkage or whether they just happen to assort together randomly? In this exercise, you'll use a simple statistical test, the chi-square (χ^2) test, to analyze phenotypes of F_1 testcross progeny in order to see whether two genes are linked or unlinked.

How These Experiments Are Done If genes are unlinked and assorting independently, the phenotypic ratio of offspring from an F_1 testcross is expected to be 1:1:1:1 (see Figure 15.9). If the two genes are linked, however, the observed phenotypic ratio of the offspring will not match that ratio. Given that random fluctuations in the data do occur, how much must the observed numbers deviate from the expected numbers for us to conclude that the genes are not assorting independently but may instead be linked?

To answer this question, scientists use a statistical test. This test, called a chi-square (χ^2) test, compares an observed data set with an expected data set predicted by a hypothesis (here, that the genes are unlinked) and measures the discrepancy between the two, thus determining the "goodness of fit." If the discrepancy between the observed and expected data sets is so large that it is unlikely to have occurred by random fluctuation, we say there is statistically significant evidence against the hypothesis (or, more specifically, evidence for the genes being linked). If the discrepancy is small, then our observations are well explained by random variation alone. In this case, we say that the observed data are consistent with our hypothesis, or that the discrepancy is statistically insignificant. Note, however, that consistency with our hypothesis is not the same as proof of our hypothesis. Also, the size of the experimental data set is important: With small data sets like this one, even if the genes are linked, discrepancies might be small by chance alone if the linkage is weak. For simplicity, we overlook the effect of sample size here.

Data from the Simulated Experiment In cosmos plants, purple stem (*A*) is dominant to green stem (*a*), and short petals (*B*) is dominant to long petals (*b*). In a simulated cross, *AABB* plants were crossed with *aabb* plants to generate F_1 dihybrids (*AaBb*), which were then testcrossed (*AaBb* × *aabb*). A total of 900 offspring plants were scored for stem color and flower petal length.

Offspring from testcross of *AaBb* (F_1) × *aabb*	Purple stem/short petals (*A–B–*)*	Green stem/short petals (*aaB–*)	Purple stem/long petals (*A–bb*)	Green stem/long petals (*aabb*)
Expected ratio if the genes are unlinked	1	1	1	1
Expected number of offspring (of 900)				
Observed number of offspring (of 900)	220	210	231	239

*If the phenotype is dominant, a dash is used for the second allele; it could be either the dominant or recessive allele.

INTERPRET THE DATA

1. The results in the data table are from a simulated F_1 dihybrid testcross. The hypothesis that the two genes are unlinked predicts that the offspring phenotypic ratio will be 1:1:1:1. Using this ratio, calculate the expected number of each phenotype out of the 900 total offspring, and enter the values in that data table.

2. The goodness of fit is measured by χ^2. This statistic measures the amounts by which the observed values differ from their respective predictions to indicate how closely the two sets of values match. The formula for calculating this value is

$$\chi^2 = \Sigma \frac{(o - e)^2}{e}$$

where Σ = sum of, o = observed and e = expected. Calculate the χ^2 value for the data using the table below. Fill out that table, carrying out the operations indicated in the top row. Then add up the entries in the last column to find the χ^2 value.

Testcross Offspring	Expected (*e*)	Observed (*o*)	Deviation (*o* − *e*)	(*o* − *e*)²	(*o* − *e*)²/*e*
A–B–		220			
aaB–		210			
A–bb		231			
aabb		239			
				χ^2 = Sum	

3. The χ^2 value means nothing on its own—it is used to find the probability that, assuming the hypothesis is true, the observed data set could have resulted from random fluctuations. A low probability suggests that the observed data are not consistent with the hypothesis and thus the hypothesis should be rejected. A standard cutoff point used by biologists is a probability of 0.05 (5%). If the probability corresponding to the χ^2 value is 0.05 or less, the differences between observed and expected values are considered statistically significant and the hypothesis (that the genes are unlinked) should be rejected. If the probability is above 0.05, the results are not statistically significant; the observed data are consistent with the hypothesis that the genes are unlinked.

    To find the probability, locate your χ^2 value in the χ^2 distribution table in Appendix D. The "degrees of freedom" (df) of your data set is the number of categories (here, 4 phenotypes) minus 1, so df = 3. **(a)** Determine which values on the df = 3 line of the table your calculated χ^2 value lies between. **(b)** The column headings for these values show the probability range for your χ^2 number. Based on whether there are nonsignificant ($p > 0.05$) or significant ($p \leq 0.05$) differences between the observed and expected values, are the data consistent with the hypothesis that the two genes are unlinked and assorting independently, or is there enough evidence to reject this hypothesis?

→ **Instructors**: A version of this Scientific Skills Exercise can be assigned in **Mastering Biology**.

New Combinations of Alleles: Variation for Natural Selection

EVOLUTION The physical behavior of chromosomes during meiosis contributes to the generation of variation in offspring (see Concept 13.4). Each pair of homologous chromosomes lines up independently of other pairs during metaphase I, and crossing over prior to that, during prophase I, can mix and match parts of maternal and paternal homologs. Mendel's elegant experiments show that the behavior of the abstract entities known as genes—or, more concretely, alleles of genes—also leads to variation in offspring (see Concept 14.1). Now, putting these different ideas together, you can see that the recombinant chromosomes resulting from crossing over may bring alleles together in new combinations, and the subsequent events of meiosis distribute to gametes the recombinant chromosomes in a multitude of combinations, such as the new variants seen in Figures 15.9 and 15.10. Random fertilization then increases even further the number of variant allele combinations that can be created.

This abundance of genetic variation provides the raw material on which natural selection works. If the traits conferred by particular combinations of alleles are better suited for a given environment, organisms possessing those genotypes will be expected to thrive and leave more offspring, ensuring the continuation of their genetic complement. In the next generation, of course, the alleles will be shuffled anew. Ultimately, the interplay between environment and phenotype (and thus genotype) will determine which genetic combinations persist over time.

Mapping the Distance Between Genes Using Recombination Data: *Scientific Inquiry*

The discovery of linked genes and recombination due to crossing over motivated one of Morgan's students, Alfred H. Sturtevant, to work out a method for constructing a **genetic map**, an ordered list of the genetic loci along a particular chromosome.

Sturtevant hypothesized that the percentage of recombinant offspring, the *recombination frequency*, calculated from experiments like the one in Figures 15.9 and 15.10, depends on the distance between genes on a chromosome. He assumed that crossing over is a random event, with the chance of crossing over approximately equal at all points along a chromosome. Based on these assumptions, Sturtevant predicted that *the farther apart two genes are, the higher the probability that a crossover will occur between them and therefore the higher the recombination frequency*. His reasoning was simple: The greater the distance between two genes, the more points there are between them where crossing over can occur. Using recombination data from various fruit fly crosses, Sturtevant proceeded to assign relative positions to genes on the same chromosomes—that is, to *map* genes.

A genetic map based on recombination frequencies is called a **linkage map**. **Figure 15.11** shows Sturtevant's linkage map of three genes: the body color (*b*) and wing size (*vg*) genes depicted in Figure 15.10 and a third gene, called cinnabar (*cn*). Cinnabar is one of many *Drosophila* genes affecting eye color. Cinnabar eyes, a mutant phenotype, are a brighter red than the wild-type color. The recombination frequency between *cn* and *b* is 9%; that between *cn* and *vg*, 9.5%; and that between *b* and *vg*, 17%. In other words, crossovers between *cn* and *b* and between *cn* and *vg* are about half as frequent as crossovers between *b* and *vg*. Only a map that locates *cn* about midway between *b* and *vg* is consistent with these data, as you can prove to yourself by drawing alternative maps. Sturtevant expressed the distances between genes in **map units**, defining one map unit as equivalent to a 1% recombination frequency.

In practice, the interpretation of recombination data is more complicated than this example suggests. Some genes

---

▼ Figure 15.11 Research Method

Constructing a Linkage Map

Application A linkage map shows the relative locations of genes along a chromosome.

Technique A linkage map is based on the assumption that the probability of a crossover between two genetic loci is proportional to the distance separating the loci. The recombination frequencies used to construct a linkage map for a particular chromosome are obtained from experimental crosses, such as the cross depicted in Figures 15.9 and 15.10. The distances between genes are expressed as map units, with one map unit equivalent to a 1% recombination frequency. Genes are arranged on the chromosome in the order that best fits the data.

Results In this example, the observed recombination frequencies between three *Drosophila* gene pairs (*b* and *cn*, 9%; *cn* and *vg*, 9.5%; and *b* and *vg*, 17%) best fit a linear order in which *cn* is positioned about halfway between the other two genes:

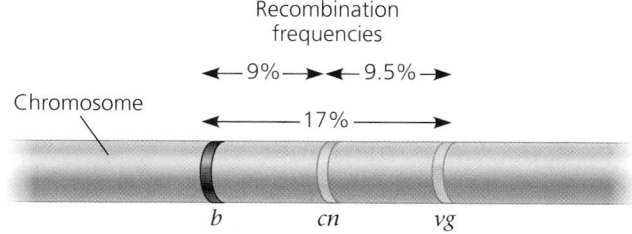

The *b-vg* recombination frequency (17%) is slightly less than the sum of the *b-cn* and *cn-vg* frequencies (9 + 9.5 = 18.5%) because of the few times that one crossover occurs between *b* and *cn* and another crossover occurs between *cn* and *vg*. The second crossover would "cancel out" the first, reducing the observed *b-vg* recombination frequency while contributing to the frequency between each of the closer pairs of genes. The value of 18.5% (18.5 map units) is closer to the actual distance between the genes. In practice, a geneticist would add the smaller distances in constructing a map.

on a chromosome are so far from each other that a crossover between them is virtually certain. The observed frequency of recombination in crosses involving two such genes can have a maximum value of 50%, a result indistinguishable from that for genes on different chromosomes. In this case, the physical connection between genes on the same chromosome is not reflected in the results of genetic crosses. Despite being on the same chromosome and thus being *physically connected*, the genes are *genetically unlinked*; alleles of such genes assort independently, as if they were on different chromosomes. In fact, at least two of the genes for pea characters that Mendel studied are now known to be on the same chromosome, but the distance between them is so great that linkage is not observed in genetic crosses. Consequently, the two genes behaved as if they were on different chromosomes in Mendel's experiments. Genes located far apart on a chromosome are mapped by adding the recombination frequencies from crosses involving closer pairs of genes lying between the two distant genes.

Using recombination data, Sturtevant and his colleagues were able to map numerous *Drosophila* genes in linear arrays. They found that the genes clustered into four groups of linked genes (*linkage groups*). Light microscopy had revealed four pairs of chromosomes in *Drosophila*, so the linkage map provided additional evidence that genes are located on chromosomes. Each chromosome has a linear array of specific genes, each gene with its own locus (**Figure 15.12**).

Because a linkage map is based strictly on recombination frequencies, it gives only an approximate picture of a chromosome. The frequency of crossing over is not actually uniform over the length of a chromosome, as Sturtevant assumed, and therefore map units do not correspond to actual physical distances (in nanometers, for instance). A linkage map does portray the order of genes along a chromosome, but it does not accurately portray the precise locations of those genes. Other methods enable geneticists to construct *cytogenetic maps* of chromosomes, which locate genes with respect to chromosomal features, such as stained bands, that can be seen in the microscope. Technical advances over recent decades have enormously increased the rate and affordability of DNA sequencing. Therefore, today, most researchers sequence whole genomes to map the locations of genes of a given species. The entire nucleotide sequence is the ultimate physical map of a chromosome, revealing the physical distances in DNA nucleotides between gene loci (see Concept 21.1). Comparing a linkage map with such a physical map or with a cytogenetic map of the same chromosome, we find that the linear order of genes is identical in all the maps, but the spacing between genes is not.

CONCEPT CHECK 15.3

1. When two genes are located on the same chromosome, what is the physical basis for the production of recombinant offspring in a testcross between a dihybrid parent and a double-mutant (recessive) parent?

2. **VISUAL SKILLS** For each type of offspring of the testcross in Figure 15.9, explain the relationship between its phenotype and the alleles contributed by the female parent. (It will be useful to draw out the chromosomes of each fly and follow the alleles throughout the cross.)

3. **WHAT IF?** Genes *A*, *B*, and *C* are located on the same chromosome. Testcrosses show that the recombination frequency between *A* and *B* is 28% and that between *A* and *C* is 12%. Can you determine the linear order of these genes? Explain.

For suggested answers, see Appendix A.

CONCEPT 15.4

Alterations of chromosome number or structure cause some genetic disorders

As you have learned so far in this chapter, the phenotype of an organism can be affected by small-scale changes involving individual genes. Random mutations are the source of all new alleles, which can lead to new phenotypic traits.

Large-scale chromosomal changes can also affect an organism's phenotype. Physical and chemical disturbances, as well as errors during meiosis, can damage chromosomes in major ways or alter their number in a cell. Large-scale chromosomal alterations in humans and other mammals often lead to spontaneous abortion (miscarriage) of a fetus, and individuals born with these types of genetic defects commonly exhibit various developmental disorders. Plants appear to tolerate such genetic defects better than animals do.

▼ **Figure 15.12 A partial genetic (linkage) map of a** ***Drosophila*** **chromosome.** This partial map shows just seven mapped genes on *Drosophila* chromosome II. (DNA sequencing has revealed over 9,000 genes on that chromosome.) The number at each gene locus is the number of map units from the arista length gene (far left).

Mastering Biology
Interview with Terry Orr-Weaver:
Why are humans so bad at meiosis?

Abnormal Chromosome Number

Ideally, the meiotic spindle distributes chromosomes to daughter cells without error. But there is an occasional mishap, called a **nondisjunction**, in which the members of a pair of homologous chromosomes do not move apart properly during meiosis I or sister chromatids fail to separate during meiosis II **(Figure 15.13)**. In nondisjunction, one gamete receives two of the same type of chromosome and another gamete receives no copy. The other chromosomes are usually distributed normally.

If either of the aberrant gametes unites with a normal one at fertilization, the zygote will also have an abnormal number of a particular chromosome, a condition known as **aneuploidy**. Fertilization involving a gamete that has no copy of a particular chromosome will lead to a missing chromosome in the zygote (so that the cell has $2n - 1$ chromosomes); the aneuploid zygote is said to be **monosomic** for that chromosome. If a chromosome is present in triplicate in the zygote (so that the cell has $2n + 1$ chromosomes), the aneuploid cell is **trisomic** for that chromosome. Mitosis will subsequently transmit the anomaly to all embryonic cells. Monosomy and trisomy are estimated to occur in 10–25% of human conceptions and are the main reason for pregnancy loss. If the organism survives, it usually has a set of traits caused by the abnormal dose of the genes associated with the extra or missing chromosome. Down syndrome is an example of trisomy in humans that will be discussed later.

▼ **Figure 15.13 Meiotic nondisjunction.** Gametes with an abnormal chromosome number can arise by nondisjunction in either meiosis I or meiosis II. For simplicity, the figure does not show the spores formed by meiosis in plants. Ultimately, spores form gametes that have the defects shown. (See Figure 13.6.)

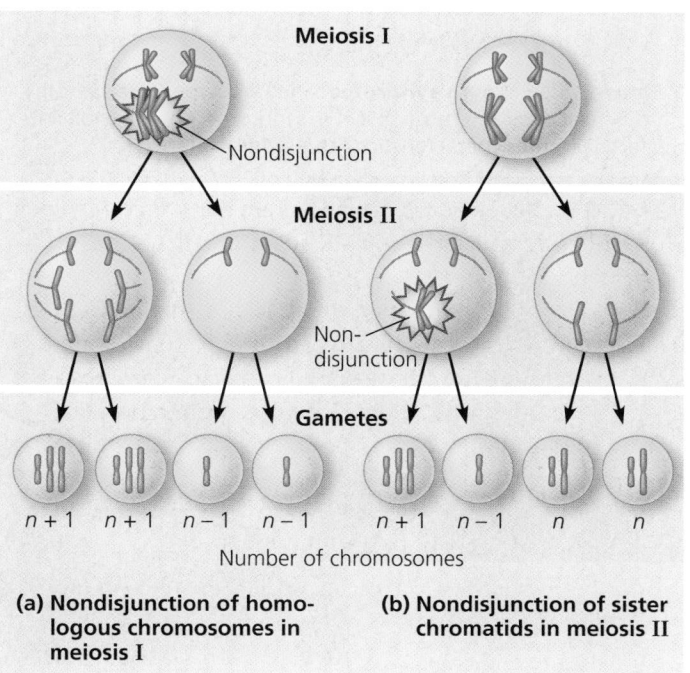

(a) Nondisjunction of homologous chromosomes in meiosis I

(b) Nondisjunction of sister chromatids in meiosis II

Nondisjunction can also occur during mitosis. If such an error takes place early in embryonic development, then the aneuploid condition is passed along by mitosis to a large number of cells and is likely to have a substantial effect on the organism.

Some organisms have more than two complete chromosome sets in all somatic cells. The general term for this chromosomal alteration is **polyploidy**; the specific terms *triploidy* ($3n$) and *tetraploidy* ($4n$) indicate three and four chromosomal sets, respectively. One way a triploid cell may arise is by the fertilization of an abnormal diploid egg produced by nondisjunction of all its chromosomes. Tetraploidy could result from the failure of a $2n$ zygote to divide after replicating its chromosomes. Subsequent normal mitotic divisions would then produce a $4n$ embryo.

Polyploidy is fairly common in the plant kingdom. The spontaneous origin of polyploid individuals plays an important role in plant evolution (see Concept 24.2). Many species we eat are polyploid: Bananas are triploid, wheat hexaploid ($6n$), and strawberries octoploid ($8n$). Polyploid animal species are much less common, but there are a few fishes and amphibians known to be polyploid. In general, polyploids are more nearly normal in appearance than aneuploids. One extra (or missing) chromosome apparently disrupts genetic balance more than does an entire extra set of chromosomes.

⊙ **Mastering Biology Animation: Polyploid Plants**

Alterations of Chromosome Structure

Errors in meiosis or damaging agents such as radiation can cause breakage of a chromosome, which can lead to four types of changes in chromosome structure **(Figure 15.14)**. A **deletion** occurs when a chromosomal fragment is lost. The affected chromosome is then missing certain genes. A broken fragment may become reattached as an extra segment to a sister or nonsister chromatid, producing a **duplication** of a portion of that chromosome. A chromosomal fragment may also reattach to the original chromosome but in the reverse orientation, producing an **inversion**. A fourth possible result of chromosomal breakage is for the fragment to join a nonhomologous chromosome, a rearrangement called a **translocation**.

Deletions and duplications are especially likely to occur during meiosis. In crossing over, nonsister chromatids sometimes exchange unequal-sized segments of DNA, so that one partner gives up more genes than it receives (see Figure 21.13). The products of such an unequal crossover are one chromosome with a deletion and one chromosome with a duplication.

A diploid embryo that is homozygous for a large deletion (both homologs have the deletion) or, if a male, has a single X chromosome with a large deletion is usually missing a number of essential genes. This condition is typically lethal.

Duplications and translocations also tend to be harmful. In reciprocal translocations, in which segments are exchanged between nonhomologous chromosomes, and in

▼ **Figure 15.14 Alterations of chromosome structure.** Red arrows indicate breakage points. Dark purple highlights the chromosomal parts affected by the rearrangements.

(a) Deletion

A deletion removes a chromosomal segment.

(b) Duplication

A duplication repeats a segment.

(c) Inversion

An inversion reverses a segment within a chromosome.

(d) Translocation

A translocation moves a segment from one chromosome to a nonhomologous chromosome. In a reciprocal translocation, the most common type, nonhomologous chromosomes exchange fragments.

Less often, a nonreciprocal translocation occurs: A chromosome transfers a fragment but receives none in return (not shown).

→ **Mastering Biology** Animation: Alterations of Chromosome Structure

inversions, the balance of genes is not abnormal—all genes are present in their normal doses. Nevertheless, translocations and inversions can alter phenotype because a gene's expression can be influenced by its location among neighboring genes, which can have devastating effects.

Human Conditions Due to Chromosomal Alterations

Alterations of chromosome number and structure are associated with a number of human conditions, which vary in severity. As described earlier, nondisjunction in meiosis results in

aneuploidy in gametes and any resulting zygotes. Although the frequency of aneuploid zygotes may be quite high in humans, most of these chromosomal alterations are so disastrous to development that the affected embryos are spontaneously aborted long before birth. However, some types of aneuploidy appear to upset the genetic balance less than others, so individuals with certain aneuploid conditions can survive to birth and beyond. These individuals have a set of traits—a *syndrome*—characteristic of the type of aneuploidy. Genetic disorders caused by aneuploidy can be diagnosed before birth by fetal testing (see Figure 14.19).

Down Syndrome (Trisomy 21)

One aneuploid condition, **Down syndrome**, affects approximately one out of every 830 children born in the United States **(Figure 15.15)**. Down syndrome is usually the result of an extra chromosome 21, so that each body cell has a total of 47 chromosomes. Because the cells are trisomic for chromosome 21, Down syndrome is often called *trisomy 21*. Down syndrome includes characteristic facial features, short stature, correctable heart defects, and developmental delays. Individuals with Down syndrome have an increased chance of developing leukemia and Alzheimer's disease but have a lower rate of high blood pressure, atherosclerosis (hardening of the arteries), stroke, and many types of solid tumors. People with Down syndrome, on average, have a life span shorter than is typical; however, most live to middle age and beyond with proper medical care. Many live independently or at home with their families, are employed, and are valuable contributors to their communities. Almost all males and about half of females with Down syndrome are sexually underdeveloped and sterile.

The frequency of Down syndrome increases with the age of the mother. While the syndrome occurs in just 0.1% (1 out of 900) births of children born to women at age 30, the risk

▼ **Figure 15.15 Down syndrome.** The karyotype shows trisomy 21, the most common cause of Down syndrome. The child exhibits the facial features characteristic of this syndrome.

climbs to 1% (1 out of 100) for mothers at age 40 and is even higher for older mothers. The correlation of Down syndrome with maternal age has not yet been explained. Most cases result from nondisjunction during meiosis I, and some research points to an age-dependent abnormality in meiosis. Trisomies of some other chromosomes (13 and 18) also increase in incidence with maternal age, although infants with other autosomal trisomies rarely survive for long. Medical experts recommend that prenatal screening for trisomies in the embryo be offered to all pregnant women, due to its low risk and useful results. In 2008, a law was passed stipulating that medical practitioners give accurate, up-to-date information about any prenatal or postnatal diagnosis received by parents and that they connect parents with appropriate support services.

Aneuploidy of Sex Chromosomes

Aneuploid conditions involving sex chromosomes appear to upset the genetic balance less than those involving autosomes. This may be because the Y chromosome carries relatively few genes. Also, extra copies of the X chromosome simply become inactivated as Barr bodies.

An extra X chromosome in a male, producing XXY, occurs approximately once in every 650 live male births. People with this condition, called *Klinefelter syndrome*, have male sex organs, but their testes are small and produce little or no sperm. Signs and symptoms vary widely, but may include taller than average height, less muscle mass, and enlarged breast tissue. Affected individuals may have learning disabilities. Testosterone replacement therapy helps stimulate changes that typically occur at puberty in males. About one of every 1,000 males is born with an extra Y chromosome (XYY). These males undergo typical sexual development and do not exhibit any well-defined syndrome, but tend to be taller than average.

Females with trisomy X (XXX), which occurs once in approximately 1,000 live female births, are generally healthy and have no unusual physical features other than being slightly taller than average. Females with trisomy X are at risk for learning disabilities. Monosomy X, which is called *Turner syndrome*, occurs about once in every 2,500 female births and is the only known viable monosomy in humans. Although these X0 individuals are phenotypically female, they are usually sterile because their sex organs do not mature. When provided with estrogen replacement therapy, girls with Turner syndrome do develop secondary sex characteristics. Most have typical intelligence.

Disorders Caused by Structurally Altered Chromosomes

Many deletions in human chromosomes, even in a heterozygous state, cause severe problems. One such syndrome, known as *cri du chat* ("cry of the cat"), results from a specific deletion in chromosome 5. A child born with this deletion has a small head with unusual facial features, severe intellectual disabilities, and a cry that sounds like the mewing of a distressed cat. Such individuals usually die in infancy or early childhood.

Chromosomal translocations can also occur during mitosis; some have been implicated in certain cancers, including *chronic myelogenous leukemia (CML)*. This disease occurs when a reciprocal translocation happens during mitosis of cells that are precursors of white blood cells. The exchange of a large portion of chromosome 22 with a small fragment from a tip of chromosome 9 produces a much shortened, easily recognized chromosome 22, called the *Philadelphia chromosome* (Figure 15.16). Such an exchange causes cancer by creating a new "fused" gene that leads to uncontrolled cell cycle progression. (The mechanism of gene activation will be discussed in Concept 18.5.)

▼ **Figure 15.16 Translocation associated with chronic myelogenous leukemia (CML).** The cancerous cells in nearly all CML patients contain an abnormally short chromosome 22, the so-called Philadelphia chromosome, and an abnormally long chromosome 9. These altered chromosomes result from the reciprocal translocation shown here, which presumably occurred in a single white blood cell precursor undergoing mitosis and was then passed along to all descendant cells.

Normal chromosome 9

Normal chromosome 22

Reciprocal translocation

Translocated chromosome 9

Translocated chromosome 22
(Philadelphia chromosome)

CONCEPT CHECK 15.4

1. About 5% of individuals with Down syndrome have a chromosomal translocation in which a third copy of chromosome 21 is attached to chromosome 14. If this translocation occurred in a parent's gonad, how could it lead to Down syndrome in a child?

2. **MAKE CONNECTIONS** The ABO blood type locus has been mapped on chromosome 9. A father who has type AB blood and a mother who has type O blood have a child with trisomy 9 and type A blood. Using this information, can you tell in which parent the nondisjunction occurred? Explain your answer. (See Figures 14.11 and 15.13.)

3. **MAKE CONNECTIONS** The gene that is activated on the Philadelphia chromosome codes for an intracellular tyrosine kinase. Review the discussion of cell cycle control in Concept 12.3, and explain how the activation of this gene could contribute to the development of cancer.

For suggested answers, see Appendix A.

Some inheritance patterns are exceptions to standard Mendelian inheritance

In the previous section, you learned about deviations from the usual patterns of chromosomal inheritance due to abnormal events in meiosis and mitosis. We conclude this chapter by describing two normally occurring exceptions to Mendelian genetics, one involving genes located in the nucleus and the other involving genes located outside the nucleus. In both cases, the sex of the parent contributing an allele is a factor in the pattern of inheritance.

Genomic Imprinting

Throughout our discussions of Mendelian genetics and the chromosomal basis of inheritance, we have assumed that a given allele will have the same effect whether it was inherited from the mother or the father. This is probably a safe assumption most of the time. For example, when Mendel crossed purple-flowered pea plants with white-flowered pea plants, he observed the same results regardless of whether the purple-flowered parent supplied the eggs or the sperm. In recent years, however, geneticists have identified a number of traits in placental mammals (and some flowering plants) that depend on which parent passed along the alleles for those traits. Such variation in phenotype depending on whether an allele is inherited from the male or female parent is called **genomic imprinting**. (Note that unlike sex-linked genes, most imprinted genes are on autosomes.) Using newer DNA sequence-based methods, about 100 imprinted genes have been identified in humans and 120–180 in mice.

Genomic imprinting occurs during gamete formation and results in the silencing of a particular allele of certain genes. Because these genes are imprinted differently in sperm and eggs, the offspring expresses only one allele of an imprinted gene, the one that has been inherited from a specific parent—either the female parent or the male parent, depending on the particular gene. The imprints are then transmitted to all body cells during early development. In each generation, the old imprints are "erased" in gamete-producing cells, and the chromosomes of the developing gametes are newly imprinted according to the sex of the individual forming the gametes. In a given species, the imprinted genes are always imprinted in the same way. For instance, a gene imprinted for maternal allele expression is always imprinted this way, generation after generation.

Consider, for example, the mouse gene for insulin-like growth factor 2 (*Igf2*), one of the first imprinted genes to be identified. Although this growth factor is required for normal prenatal growth, only the paternal allele is expressed **(Figure 15.17a)**. Evidence that the *Igf2* gene is imprinted came initially from crosses between normal-sized (wild-type) mice and dwarf

▼ **Figure 15.17 Genomic imprinting of the mouse *Igf2* gene.**

(a) Homozygote. A mouse homozygous for the wild-type *Igf2* allele is normal-sized. Only the paternal allele of this gene is expressed.

(b) Heterozygotes. Heterozygotes have different phenotypes depending on which allele is contributed by each parent. Matings between wild-type mice and those homozygous for the recessive mutant *Igf2* allele produce heterozygous offspring. Since the maternal *Igf2* allele is not expressed, the dwarf (mutant) phenotype is seen only when the father contributes the mutant allele.

(mutant) mice homozygous for a recessive mutation in the *Igf2* gene. The phenotypes of heterozygous offspring (with one normal allele and one mutant) differed depending on whether the mutant allele came from the mother or the father **(Figure 15.17b)**.

What exactly is a genomic imprint? It turns out that imprinting can involve either silencing an allele in one type of gamete (egg or sperm) or activating it in the other. In many cases, the imprint seems to consist of methyl ($-CH_3$) groups that are added to cytosine nucleotides of one of the alleles. Such methylation may silence the allele, an effect consistent with evidence that heavily methylated genes are usually inactive (see Concept 18.2). However, for a few genes, methylation has been shown to *activate* expression of the allele. This is the case for the *Igf2* gene: Methylation of certain cytosines on the paternal chromosome leads to expression of the paternal *Igf2* allele, by an indirect mechanism involving chromatin structure and protein-DNA interactions.

Genomic imprinting may affect only a small fraction of the genes in mammalian genomes, but most of the known imprinted genes are critical for embryonic development. Recent research has revealed that imprinted genes are involved in regulation of body temperature, sleep, and other metabolic functions. Some evolutionary biologists have proposed that genomic imprinting represents a kind of evolved "competition" between males (who benefit from larger offspring more likely to pass on the father's genes) and female mammals (who benefit from smaller offspring during birth). In experiments with mice, embryos engineered to inherit both copies of certain chromosomes from the same parent usually die before birth, whether that parent is male or female. Normal development seems to require that embryonic cells have exactly one active copy—not zero, not two—of certain genes. The association of improper imprinting with abnormal development and certain cancers has stimulated ongoing studies of how different genes are imprinted.

 Mastering Biology
Interview with Shirley Tilghman:
Discovering imprinted genes

Inheritance of Organelle Genes

Although our focus in this chapter has been on the chromosomal basis of inheritance, we end with an important amendment: Not all of a eukaryotic cell's genes are located on nuclear chromosomes, or even in the nucleus; some genes are located in organelles in the cytoplasm. Because they are outside the nucleus, these genes are sometimes called *extranuclear genes* or *cytoplasmic genes*. Mitochondria, as well as chloroplasts and other plastids in plants, contain small circular DNA molecules that carry a number of genes. These organelles reproduce themselves and transmit their genes to daughter organelles. Genes on organelle DNA are not distributed to offspring according to the same rules that direct the distribution of nuclear chromosomes during meiosis, so they do not display Mendelian inheritance.

The first hint that extranuclear genes exist came from studies by the German scientist Carl Correns on the inheritance of yellow or white patches on the leaves of an otherwise green plant. In 1909, he observed that the coloration of the offspring was determined only by the maternal parent (the source of eggs) and not by the paternal parent (the source of sperm). Subsequent research showed that such coloration patterns, or variegation **(Figure 15.18)**, are due to mutations in genes controlling pigmentation that are located in the DNA circles in the organelles called plastids (see Concept 6.5). In most plants, a zygote receives all its plastids from the cytoplasm of the egg and none from the sperm, which contributes little more than a haploid set of chromosomes. An egg may contain plastids with different alleles for a pigmentation gene. As the zygote develops, plastids containing wild-type or mutant pigmentation genes are distributed randomly to daughter cells. The pattern of leaf coloration exhibited by a plant depends on the ratio of wild-type to mutant plastids in its various tissues.

◀ **Figure 15.18 A painted nettle coleus plant.** The variegated (patterned) leaves on this coleus plant (*Plectranthus scutellarioides*) result from mutations that affect expression of pigment genes located in plastids, which generally are inherited from the maternal parent.

Similar maternal inheritance is also the rule for mitochondrial genes in most animals and plants, because the mitochondria passed on to a zygote come from the cytoplasm of the egg. (The few mitochondria contributed by the sperm appear to be destroyed in the egg.) The products of most mitochondrial genes help make up some of the protein complexes of the electron transport chain and ATP synthase (see Figure 9.14). Defects in one or more of these proteins, therefore, reduce the amount of ATP the cell can make and have been shown to cause a number of human disorders in as many as one out of every 5,000 births. Because the parts of the body most susceptible to energy deprivation are the nervous system and the muscles, most mitochondrial diseases primarily affect these systems. For example, *mitochondrial myopathy* causes weakness, intolerance of exercise, and muscle deterioration. Another mitochondrial disorder is *Leber's hereditary optic neuropathy*, which can produce sudden blindness in people as young as their 20s or 30s. The four mutations found thus far to cause this disorder affect oxidative phosphorylation during cellular respiration, a crucial function for the cell (see Concept 9.4).

The fact that mitochondrial disorders are inherited only from the mother suggests a way to avoid passing along these disorders. The chromosomes from the egg of an affected mother could be transferred to an egg of a healthy donor that has had its own chromosomes removed. This "two-mother" egg could then be fertilized by a sperm from the prospective father and transplanted into the womb of the prospective mother, becoming an embryo with three parents. After optimizing conditions for this approach in monkeys, researchers reported in 2013 that they have successfully carried out this procedure on human eggs. In 2016, clinics in Mexico and China reported the births of such "three-parent" babies. In the United Kingdom, this procedure was approved for use in 2017. In the United States, approval by the relevant federal agencies has not yet occurred.

In addition to the rarer diseases clearly caused by defects in mitochondrial DNA, mitochondrial mutations inherited from a person's mother may contribute to at least some types of diabetes and heart disease, as well as to other disorders that commonly debilitate the elderly, such as Alzheimer's disease. In the course of a lifetime, new mutations gradually accumulate in our mitochondrial DNA, and some researchers think that these mutations play a role in the normal aging process.

CONCEPT CHECK 15.5

1. Gene dosage—the number of copies of a gene that are actively being expressed—is important to proper development. Identify and describe two processes that establish the proper dosage of certain genes.

2. Reciprocal crosses between two primrose varieties, A and B, produced the following results: A female × B male → offspring with all green (nonvariegated) leaves; B female × A male → offspring with patterned (variegated) leaves. Explain these results.

3. **WHAT IF?** Mitochondrial genes are critical to the energy metabolism of cells, but mitochondrial disorders caused by mutations in these genes are generally not lethal. Why not?

For suggested answers, see Appendix A.

15 Chapter Review

→ Go to **Mastering Biology** for Assignments, the eText, the Study Area, and Dynamic Study Modules.

SUMMARY OF KEY CONCEPTS

→ To review key terms, go to the **Vocabulary Self-Quiz** in the **Mastering Biology** eText or Study Area, or go to goo.gl/zkjz9t.

CONCEPT 15.1

Morgan showed that Mendelian inheritance has its physical basis in the behavior of chromosomes (pp. 295–297)

- Morgan's work with an eye color gene in *Drosophila* led to the **chromosome theory of inheritance**, which states that genes are located on chromosomes and that the behavior of chromosomes during meiosis accounts for Mendel's laws.

? *What characteristic of the sex chromosomes allowed Morgan to correlate their behavior with that of the alleles of the eye color gene?*

CONCEPT 15.2

Sex-linked genes exhibit unique patterns of inheritance (pp. 298–300)

- Sex is often chromosomally based. Humans and other mammals have an X-Y system in which sex is largely determined by whether a Y chromosome is present. Other systems are found in birds, fishes, and insects.
- The sex chromosomes carry **sex-linked genes**, virtually all of which are on the X chromosome (**X-linked**). Any male who inherits a recessive X-linked allele (from his mother) will express the trait, such as color blindness.
- In mammalian females, one of the two X chromosomes in each cell is randomly inactivated during early embryonic development, becoming highly condensed into a **Barr body**.

? *Why are males affected by X-linked disorders much more often than females?*

CONCEPT 15.3

Linked genes tend to be inherited together because they are located near each other on the same chromosome (pp. 301–306)

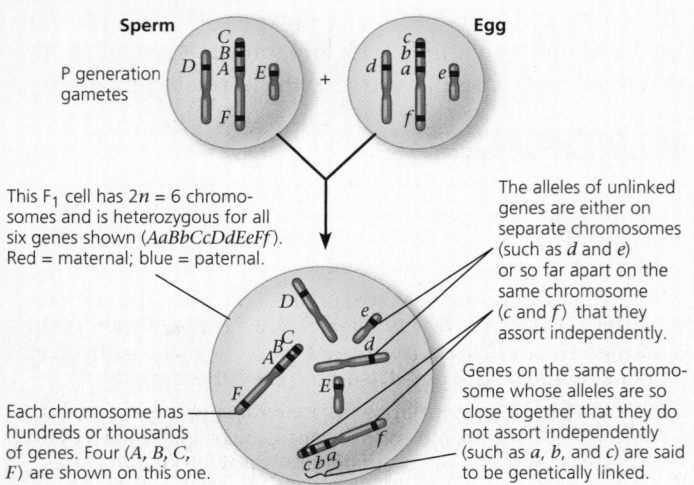

This F_1 cell has $2n = 6$ chromosomes and is heterozygous for all six genes shown (*AaBbCcDdEeFf*). Red = maternal; blue = paternal.

Each chromosome has hundreds or thousands of genes. Four (*A, B, C, F*) are shown on this one.

The alleles of unlinked genes are either on separate chromosomes (such as *d* and *e*) or so far apart on the same chromosome (*c* and *f*) that they assort independently.

Genes on the same chromosome whose alleles are so close together that they do not assort independently (such as *a*, *b*, and *c*) are said to be genetically linked.

- An F_1 dihybrid testcross yields **parental types** with the same combination of traits as those in the P generation parents and **recombinant types** (**recombinants**) with new combinations of traits not seen in either P generation parent. Because of the independent assortment of chromosomes, unlinked genes exhibit a 50% frequency of recombination in the gametes. For genetically **linked genes, crossing over** between nonsister chromatids during meiosis I accounts for the observed recombinants, always less than 50%.
- The order of genes on a chromosome and the relative distances between them can be deduced from recombination frequencies observed in genetic crosses. These data allow construction of a **linkage map** (a type of **genetic map**). The farther apart genes are, the more likely their allele combinations will be recombined during crossing over.

? *Why are specific alleles of two distant genes more likely to show recombination than those of two closer genes?*

CONCEPT 15.4

Alterations of chromosome number or structure cause some genetic disorders (pp. 306–309)

- **Aneuploidy**, an abnormal chromosome number, can result from **nondisjunction** during meiosis. When a normal gamete unites with one containing two copies or no copies of a particular chromosome, the resulting zygote and its descendant cells either have one extra copy of that chromosome (**trisomy**, $2n + 1$) or are missing a copy (**monosomy**, $2n - 1$). **Polyploidy** (extra sets of chromosomes) can result from nondisjunction of all chromosomes.
- Chromosome breakage can result in alterations of chromosome structure: **deletions, duplications, inversions**, and **translocations**. Translocations can be reciprocal or nonreciprocal.
- Changes in the number of chromosomes per cell or in the structure of individual chromosomes can affect the phenotype and, in some cases, lead to disorders. Such alterations cause **Down syndrome** (usually due to trisomy of chromosome 21), certain cancers associated with chromosomal translocations that occur during mitosis, and various other human disorders.

? *Why are inversions and reciprocal translocations less likely to be lethal than are aneuploidy, duplications, deletions, and nonreciprocal translocations?*

CONCEPT 15.5

Some inheritance patterns are exceptions to standard Mendelian inheritance (pp. 310–311)

- In mammals, the phenotypic effects of a small number of particular genes depend on which allele is inherited from each parent, a phenomenon called **genomic imprinting**. Imprints are formed during gamete production, with the result that one allele (either maternal or paternal) is not expressed in offspring.
- The inheritance of traits controlled by the genes present in mitochondria and plastids depends solely on the maternal parent because the zygote's cytoplasm containing these organelles comes from the egg. Some diseases affecting the nervous and muscular systems are caused by defects in mitochondrial genes that prevent cells from making enough ATP.

? *Explain how genomic imprinting and inheritance of mitochondrial and chloroplast DNA are exceptions to standard Mendelian inheritance.*

TEST YOUR UNDERSTANDING

➜ For multiple-choice questions, go to the **Practice Test** in the **Mastering Biology** eText or Study Area, or go to goo.gl/GruWRg.

Levels 1–2: Remembering/Understanding

1. A man with hemophilia (a recessive, sex-linked condition) has a daughter without hemophilia who marries a man without hemophilia. What is the probability of their daughter having hemophilia? Their son? If they have four sons, that all will be affected?

2. Pseudohypertrophic muscular dystrophy is an inherited disorder that causes gradual deterioration of the muscles. It is seen almost exclusively in boys born to unaffected parents and usually results in death in the early teens. Is this disorder caused by a dominant or a recessive allele? Is its inheritance sex-linked or autosomal? How do you know? Explain why this disorder is almost never seen in girls.

3. A wild-type fruit fly (heterozygous for gray body color and normal wings) is mated with a black fly with vestigial wings. The offspring have the following phenotypic distribution: wild-type, 778; black vestigial, 785; black normal, 158; gray vestigial, 162. What is the recombination frequency between these genes for body color and wing size? Is this consistent with the results of the experiment in Figure 15.9? Draw the chromosomes in the wild-type and black parents.

4. A planet is inhabited by creatures that reproduce with the same hereditary patterns seen in humans. Three phenotypic characters are height (T = tall, t = dwarf), head appendages (A = antennae, a = no antennae), and nose shape (S = upturned snout, s = downturned snout). Since the creatures are not "intelligent," Earth scientists are able to do some controlled breeding experiments using various heterozygotes in testcrosses. For tall heterozygotes with antennae, the offspring are tall/antennae, 46; dwarf/antennae, 7; dwarf/no antennae, 42; tall/no antennae, 5. For heterozygotes with antennae and an upturned snout, the offspring are antennae/upturned snout, 47; antennae/downturned snout, 2; no antennae/downturned snout, 48; no antennae/upturned snout, 3. Calculate the recombination frequencies for both experiments.

Levels 3–4: Applying/Analyzing

5. Scientists do a testcross of the creatures from problem 4 using heterozygotes for height and nose shape. The offspring are tall/upturned snout, 41; dwarf/upturned snout, 8; dwarf/downturned snout, 43; tall/downturned snout, 8. Calculate the recombination frequency from these data, and then use your answer from problem 4 to determine the correct order of the three linked genes.

6. A wild-type fruit fly (heterozygous for gray body color and red eyes) is mated with a black fruit fly with purple eyes. The offspring are wild-type, 721; black/purple, 751; gray/purple, 49; black/red, 45. What is the recombination frequency between these genes for body color and eye color? Using information from problem 3, what fruit flies (genotypes and phenotypes) would you mate to determine the order of the body color, wing size, and eye color genes on the chromosome?

7. Assume that genes A and B are 50 map units apart on the same chromosome. An animal heterozygous at both loci is crossed with one that is homozygous recessive at both loci. What percentage of the offspring will show recombinant phenotypes resulting from crossovers? Without knowing these genes are on the same chromosome, how would you interpret the results of this cross?

8. Two genes of a flower, one controlling blue (B) versus white (b) petals and the other controlling round (R) versus oval (r) stamens, are linked and are 10 map units apart. You cross a homozygous blue/oval plant with a homozygous white/round plant. The resulting F_1 progeny are crossed with homozygous white/oval plants, and 1,000 offspring plants are obtained. How many F_2 plants of each of the four phenotypes do you expect?

9. You design *Drosophila* crosses to provide recombination data for gene *a*, which is located on the chromosome shown in Figure 15.12. Gene *a* has recombination frequencies of 14% with the vestigial wing locus and 26% with the brown eye locus. Approximately where is *a* located along the chromosome relative to these two loci?

Levels 5–6: Evaluating/Creating

10. Banana plants, which are triploid, are seedless and therefore sterile. Thinking about meiosis, propose a possible explanation.

11. **EVOLUTION CONNECTION** Crossing over is thought to be evolutionarily advantageous because it continually shuffles genetic alleles into novel combinations. It was thought that Y-linked genes might degenerate because they lack homologous genes on the X chromosome with which to pair up prior to crossing over. However, when the Y chromosome was sequenced, scientists discovered eight large regions that were internally homologous to each other, and several genes are duplicates. (Y chromosome researcher David Page has called it a "hall of mirrors.") How might this be beneficial?

12. **SCIENTIFIC INQUIRY DRAW IT** Assume you are mapping genes *A, B, C,* and *D* in *Drosophila*. You know that these genes are linked on the same chromosome, and you determine the recombination frequencies between each pair of genes to be as follows: *A* and *B*, 8%; *A* and *C*, 28%; *A* and *D*, 25%; *B* and *C*, 20%; *B* and *D*, 33%.
    (a) Describe how you determined the recombination frequency for each pair of genes.
    (b) Draw a chromosome map based on your data.

13. **WRITE ABOUT A THEME: INFORMATION** The continuity of life is based on heritable information in the form of DNA. In a short essay (100–150 words), relate the structure and behavior of chromosomes to inheritance in both asexually and sexually reproducing species.

14. **SYNTHESIZE YOUR KNOWLEDGE**

Butterflies have an X-Y sex determination system that is different from that of flies or humans. Female butterflies may be either XY or X0, while butterflies with two or more X chromosomes are males. This photograph shows a tiger swallowtail *gynandromorph*, which is half male (left side) and half female (right side). Given that the first division of the zygote divides the embryo into the future right and left halves of the butterfly, propose a hypothesis that explains how nondisjunction during the first mitosis might have produced this unusual-looking butterfly.

For selected answers, see Appendix A.

16 The Molecular Basis of Inheritance

KEY CONCEPTS

16.1 DNA is the genetic material *p. 315*

16.2 Many proteins work together in DNA replication and repair *p. 320*

16.3 A chromosome consists of a DNA molecule packed together with proteins *p. 330*

Study Tip

Draw DNA replication: Make up a single-stranded DNA sequence that is 40 nucleotides in length. As you go through Concepts 16.1 and 16.2, use your sequence to draw a double-stranded DNA molecule, labeling the ends. Then diagram the process of replication of your DNA molecule.

DNA Structure and Replication

T C A G C G T A A G G T G A T G T A T A ...

→ Go to Mastering Biology

For Students (in eText and Study Area)
- Get Ready for Chapter 16
- Figure 16.15 Walkthrough: Addition of a Nucleotide to a DNA Strand
- BioFlix® Animation: DNA Replication

For Instructors to Assign (in Item Library)
- Activity: The Hershey-Chase Experiment
- Tutorial: Visualizing DNA

Ready-to-Go Teaching Module
(in Instructor Resources)
- DNA Replication (Concept 16.2)

Figure 16.1 The elegant double-helical structure of deoxyribonucleic acid (DNA) shook the scientific world when it was proposed in April 1953 by James Watson and Francis Crick. The DNA you inherited from your parents contains all your genes—your genetic information.

How does DNA replication transmit genetic information?

DNA replication allows genetic information to be inherited from a parent cell to daughter cells (by mitosis) and from generation to generation (starting with meiosis).

Unduplicated chromosome (one DNA molecule and proteins)

DNA segment from a chromosome

Each gene is a unit of hereditary information consisting of a specific DNA sequence.

DNA replication begins.

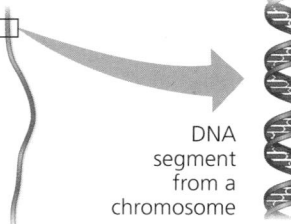

Replication begins at multiple sites (origins), each forming a replication bubble with a fork at each end.

New strands

DNA replication and chromosome condensation is completed.

Duplicated and condensed chromosome (two DNA molecules and proteins)

Two DNA molecules, which are distributed to daughter cells.

DNA is the genetic material

Today, even schoolchildren have heard of DNA, and scientists routinely manipulate DNA in the laboratory. Early in the 20th century, however, identifying the molecules of inheritance posed a major challenge to biologists.

The Search for the Genetic Material: *Scientific Inquiry*

Once T. H. Morgan's group showed that genes exist as parts of chromosomes (described in Concept 15.1), the two chemical components of chromosomes—DNA and protein—emerged as the leading candidates for the genetic material. Until the 1940s, the case for proteins seemed stronger: Biochemists had identified proteins as a class of macromolecules with great heterogeneity and specificity of function, essential requirements for the hereditary material. Moreover, little was known about nucleic acids, whose physical and chemical properties seemed far too uniform to account for the multitude of specific inherited traits exhibited by every organism. This view gradually changed as the role of DNA in heredity was worked out in studies of bacteria and the viruses that infect them, systems far simpler than fruit flies or humans. Let's trace the search for the genetic material as a case study in scientific inquiry.

Evidence That DNA Can Transform Bacteria

In 1928, a British medical officer named Frederick Griffith was trying to develop a vaccine against pneumonia. He was studying *Streptococcus pneumoniae*, a bacterium that causes pneumonia in mammals. Griffith had two strains (varieties) of the bacterium, one pathogenic (disease-causing) and one nonpathogenic (harmless). He was surprised to find that when he killed the pathogenic bacteria with heat and then mixed the cell remains with living bacteria of the nonpathogenic strain, some of the living cells became pathogenic **(Figure 16.2)**. Furthermore, this newly acquired trait of pathogenicity was inherited by all the descendants of the transformed bacteria. Apparently, some chemical component of the dead pathogenic cells caused this heritable change, although the identity of the substance was not known. Griffith called the phenomenon **transformation**, now defined as a change in genotype and phenotype due to the assimilation of external DNA by a cell. Later work by Oswald Avery, Maclyn McCarty, and Colin MacLeod identified the transforming substance as DNA.

Scientists remained skeptical, however, since many still viewed proteins as better candidates for the genetic material. Also, many biologists were not convinced that bacterial genes would be similar in composition and function to those of more complex organisms. But the major reason for the continued doubt was that so little was known about DNA.

Can a genetic trait be transferred between different bacterial strains?

Experiment Frederick Griffith studied two strains of the bacterium *Streptococcus pneumoniae*. The S (smooth) strain can cause pneumonia in mice; it is pathogenic because the cells have an outer capsule that protects them from an animal's immune system. Cells of the R (rough) strain lack a capsule and are nonpathogenic. To test for the trait of pathogenicity, Griffith injected mice with the two strains:

Living S cells (pathogenic control) | Living R cells (nonpathogenic control) | Heat-killed S cells (nonpathogenic control) | Mixture of heat-killed S cells and living R cells

Results

Mouse dies | Mouse healthy | Mouse healthy | Mouse dies

In blood sample, living S cells were found. They could reproduce, yielding more S cells.

Conclusion The living R bacteria had been transformed into pathogenic S bacteria by an unknown, heritable substance from the dead S cells that enabled the R cells to make capsules.

Data from F. Griffith, The significance of pneumococcal types, *Journal of Hygiene* 27:113–159 (1928).

WHAT IF? *How did this experiment rule out the possibility that the R cells simply used the dead S cells' capsules to become pathogenic?*

Evidence That Viral DNA Can Program Cells

Additional evidence that DNA was the genetic material came from studies of viruses that infect bacteria **(Figure 16.3)**. These viruses are called **bacteriophages** (meaning "bacteria-eaters"), or **phages** for short. Viruses are much simpler than cells. A **virus** is little more than DNA (or sometimes RNA) enclosed by a protective coat, which is often simply protein. To produce more viruses, a virus must infect a cell and take over the cell's metabolic machinery.

► **Figure 16.3 A virus infecting a bacterial cell.**
A phage called T2 attaches to a host cell and injects its genetic material through the plasma membrane, while the head and tail parts remain on the outer bacterial surface (colorized TEM).

Phage head

DNA

Tail sheath

Tail fiber

Genetic material

Bacterial cell

100 nm

Phages have been widely used as tools by researchers in molecular genetics. In 1952, Alfred Hershey and Martha Chase performed experiments showing that DNA is the genetic material of a phage known as T2. This is one of many phages that infect *Escherichia coli* (*E. coli*), a bacterium that normally lives in the intestines of mammals and is a model organism for molecular biologists. At that time, biologists already knew that T2, like many other phages, was composed almost entirely of DNA and protein. They also knew that the T2 phage could quickly turn an *E. coli* cell into a T2-producing factory that released many copies of new phages when the cell ruptured. Somehow, T2 could reprogram its host cell to produce viruses. But which viral component—protein or DNA—was responsible?

Hershey and Chase answered this question by devising an experiment showing that only one of the two components of T2 actually enters the *E. coli* cell during infection (**Figure 16.4**). In their experiment, they used a radioactive isotope of sulfur to tag protein in one batch of T2 and a radioactive isotope of phosphorus to tag DNA in a second batch. Because protein, but not DNA, contains sulfur, radioactive sulfur atoms were incorporated only into the protein of the phage. In a similar way, the atoms of radioactive phosphorus labeled only the DNA, not the protein, because nearly all the phage's phosphorus is in its DNA. In the experiment, separate samples of nonradioactive *E. coli* cells were infected separately with the protein-labeled and DNA-labeled batches of T2. The researchers then tested the two samples

▼ **Figure 16.4** Inquiry

Is protein or DNA the genetic material of phage T2?

Experiment Alfred Hershey and Martha Chase used radioactive sulfur and phosphorus to trace the fates of protein and DNA, respectively, of T2 phages that infected bacterial cells. They wanted to see which of these molecules entered the cells and could reprogram them to make more phages.

Results When proteins were labeled (batch 1), radioactivity remained outside the cells, but when DNA was labeled (batch 2), radioactivity was found inside the cells. Cells containing radioactive phage DNA released new phages with some radioactive phosphorus.

Conclusion Phage DNA entered bacterial cells, but phage proteins did not. Hershey and Chase concluded that DNA, not protein, functions as the genetic material of phage T2.

Data from A. D. Hershey and M. Chase, Independent functions of viral protein and nucleic acid in growth of bacteriophage, *Journal of General Physiology* 36:39–56 (1952).

WHAT IF? *How would the results have differed if proteins carried the genetic information?*

→ **Mastering Biology** Animation: The Hershey-Chase Experiment shortly after the onset of infection to see which type of radioactively labeled molecule—protein or DNA—had entered the bacterial cells and would therefore be capable of reprogramming them.

Hershey and Chase found that the phage DNA entered the host cells, but the phage protein did not. Moreover, when these bacteria were returned to a culture medium and the infection ran its course, the *E. coli* released phages that contained some radioactive phosphorus. This result further showed that the DNA inside the cell played an ongoing role during the infection process. They concluded that the DNA injected by the phage must be the molecule carrying the genetic information that makes the cells produce new viral DNA and proteins. The Hershey-Chase experiment was a landmark study because it provided powerful evidence that nucleic acids, rather than proteins, are the hereditary material, at least for certain viruses.

Additional Evidence That DNA Is the Genetic Material

Further evidence that DNA is the genetic material came from the laboratory of biochemist Erwin Chargaff. DNA was known to be a polymer of nucleotides, each having three components: a nitrogenous (nitrogen-containing) base, a pentose sugar called deoxyribose, and a phosphate group **(Figure 16.5)**. The base can be adenine (A), thymine (T), guanine (G), or cytosine (C). Chargaff analyzed the base composition of DNA from a number of different organisms. In 1950, he reported that the base composition of DNA varies from one species to another. For example, he found that 32.8% of sea urchin DNA nucleotides have the base A, whereas only 24.7% of those from the bacterium *E. coli* have an A. This evidence of molecular diversity among species, which most scientists had presumed to be absent from DNA, made DNA a more credible candidate for the genetic material.

Chargaff also noticed a peculiar regularity in the ratios of nucleotide bases. In the DNA of each species he studied, the number of adenines approximately equaled the number of thymines, and the number of guanines approximately equaled the number of cytosines. In sea urchin DNA, for example, the four bases are present in these percentages: A = 32.8% and T = 32.1%; G = 17.7% and C = 17.3%. (The percentages are not exactly the same because of limitations in Chargaff's techniques.)

These two findings became known as *Chargaff's rules*: (1) DNA base composition varies between species, and (2) for each species, the percentages of A and T bases are roughly equal, as are those of G and C bases. In the **Scientific Skills Exercise**, you can use Chargaff's rules to predict percentages of nucleotide bases. The basis for these rules remained unexplained until the discovery of the double helix.

 Mastering Biology HHMI Animation: Chargaff's Ratio **hhmi** BioInteractive

Building a Structural Model of DNA

Once most biologists were convinced that DNA was the genetic material, the challenge was to determine how the structure

▼ **Figure 16.5 The structure of a DNA strand.** Each DNA nucleotide monomer consists of the sugar deoxyribose (blue) attached to both a nitrogenous base (A, T, G, or C), and a phosphate group (yellow). The phosphate group of one nucleotide is attached to the sugar of the next by a covalent bond, forming a "backbone" of alternating phosphates and sugars from which the bases project. A polynucleotide strand has directionality, from the 5′ end (with the phosphate group) to the 3′ end (with the —OH group of the sugar). 5′ and 3′ refer to the numbers assigned to the carbons in the sugar ring (see magenta numbers).

 Mastering Biology Animation: DNA and RNA Structure

of DNA could account for its role in inheritance. By the early 1950s, the arrangement of covalent bonds in a nucleic acid polymer was well established (see Figure 16.5), and researchers focused on discovering the three-dimensional structure of DNA. Among the scientists working on the problem were Linus Pauling, at the California Institute of Technology, and Maurice Wilkins and Rosalind Franklin, at King's College in London. First to come up with the complete answer, however, were two scientists who were relatively unknown at the time—the American James Watson and the Englishman Francis Crick.

The collaborative work that solved the puzzle of DNA structure began soon after Watson went to Cambridge University, where Crick was studying protein structure with a technique called X-ray crystallography (see Figure 5.21).

Working with Data in a Table

Given the Percentage Composition of One Nucleotide in a Genome, Can We Predict the Percentages of the Other Three Nucleotides? Even before the structure of DNA was elucidated, Erwin Chargaff and his co-workers noticed a pattern in the base composition of nucleotides from different species: The percentage of adenine (A) bases roughly equaled that of thymine (T) bases, and the percentage of cytosine (C) bases roughly equaled that of guanine (G) bases. Further, the percentage of each pair (A–T or C–G) varied from species to species. We now know that the 1:1 A/T and C/G ratios are due to complementary base pairing between A and T and between C and G in the DNA double helix, and differences between species are due to the unique sequences of bases along a DNA strand. In this exercise, you will apply Chargaff's rules to predict the composition of bases in a genome.

How the Experiments Were Done In Chargaff's experiments, DNA was extracted from the given species, hydrolyzed to break apart the nucleotides, and then analyzed chemically. These studies provided approximate values for each type of nucleotide. (Today, whole-genome sequencing allows base composition analysis to be done more precisely directly from the sequence data.)

Data from the Experiments Tables are useful for organizing sets of data representing a common set of values (here, percentages of A, G, C, and T) for a number of different samples (in this case, from different species). You can apply the patterns that you see in the known data to predict unknown values. In the table, complete base distribution data are given for sea urchin DNA and salmon DNA; you will use Chargaff's rules to fill in the rest of the table with predicted values.

Source of DNA	Base Percentage			
	Adenine	Guanine	Cytosine	Thymine
Sea urchin	32.8	17.7	17.3	32.1
Salmon	29.7	20.8	20.4	29.1
Wheat	28.1	21.8	22.7	
E. coli	24.7	26.0		
Human	30.4			30.1
Ox	29.0			
Average %				

Data from several papers by Chargaff: for example, E. Chargaff et al., Composition of the desoxypentose nucleic acids of four genera of sea-urchin, *Journal of Biological Chemistry* 195: 155–160 (1952).

INTERPRET THE DATA

▶ **Sea urchin**

1. Explain how the sea urchin and salmon data demonstrate both of Chargaff's rules.

2. Using Chargaff's rules, fill in the table with your predictions of the missing percentages of bases, starting with the wheat genome and proceeding through *E. coli*, human, and ox. Show how you arrived at your answers.

3. If Chargaff's rule—that the amount of A equals the amount of T and the amount of C equals the amount of G—is valid, then hypothetically we could extrapolate this to the combined DNA of all species on Earth (like one huge Earth genome). To see whether the data in the table support this hypothesis, calculate the average percentage for each base in your completed table by averaging the values in each column. Does Chargaff's equivalence rule still hold true?

→ **Instructors:** A version of this Scientific Skills Exercise can be assigned in **Mastering Biology.**

While visiting the laboratory of Maurice Wilkins, Watson saw an X-ray diffraction image of DNA produced by Wilkins's accomplished colleague Rosalind Franklin **(Figure 16.6)**. Images produced by X-ray crystallography are not actually pictures of molecules. The spots in the image were produced by X-rays that were diffracted (deflected) as they passed through aligned fibers of purified DNA. Watson was familiar with the type of X-ray diffraction pattern of helical molecules, and seeing the

▼ **Figure 16.6 Rosalind Franklin and her X-ray diffraction photo of DNA.**

(a) Rosalind Franklin

(b) Franklin's X-ray diffraction photograph of DNA

photo that Wilkins showed him confirmed that DNA was helical in shape. The photo also added to earlier data obtained by Franklin and others suggesting the width of the helix and the spacing of the nitrogenous bases along it. The pattern in this photo implied that the helix was made up of two strands, contrary to a three-stranded model proposed by Linus Pauling a short time earlier. The presence of two strands accounts for the now-familiar term **double helix**. DNA is shown in some of its many different representations in **Figure 16.7**.

Watson and Crick began building models of a double helix that would conform to the X-ray measurements and what was then known about the chemistry of DNA, including Chargaff's rule of base equivalences. Having also read an unpublished annual report summarizing Franklin's work, they knew she had concluded that the sugar-phosphate backbones were on the outside of the DNA molecule, contrary to their working model. Franklin's arrangement was appealing because it put the negatively charged phosphate groups facing the aqueous surroundings, while the relatively hydrophobic nitrogenous bases were hidden in the interior. Watson constructed such a model, in which the two sugar-phosphate backbones are **antiparallel**—that is, their subunits run in opposite directions (see Figure 16.7). You can imagine the overall arrangement as a rope ladder with rigid rungs. The side ropes represent

DNA can be illustrated in many ways, but all diagrams represent the same basic structure. The level of detail shown depends on the process or the type of information being conveyed.

Here, the two DNA strands are shown untwisted so it's easier to see the chemical details. Note that the strands are **antiparallel**—they are oriented in opposite directions, like the lanes of a divided street.

Structural Images

The space-filling model on the left shows the 3-D shape of the DNA double helix; the diagram on the right shows chemical details of DNA's structure. In both images, phosphate groups are yellow, deoxyribose sugars are blue, and nitrogenous bases are shades of green and orange.

Bases 0.34 nm apart

One full turn every 10 base pairs (3.4 nm)

Diameter = 2 nm

The DNA double helix is "right-handed." Use your right hand as shown to follow the sugar-phosphate backbone up the helix and around to the back. (It won't work with your left hand.)

→ **Mastering Biology** Animation: **DNA Double Helix**

5' end   3' end

Phosphate group attached to 5' carbon

Nitrogenous base attached to 1' carbon

DNA nucleotide   Sugar-phosphate backbone

Sugar

Covalent sugar-phosphate bonds link nucleotides.

Hydrogen bonds between bases hold strands together.

Van der Waals interactions between stacked base pairs help hold the molecule together.

—OH attached to 3' carbon

3' end   5' end

1. *Describe the bonds that hold together the nucleotides in one DNA strand. Then compare them with the bonds that hold the two DNA strands together.*

2. *How do the two ends of a DNA strand differ in structure?*

Simplified Images

When molecular detail is not necessary, DNA is portrayed in a range of simplified diagrams, depending on the focus of the figure.

5' 3'

Nitrogenous bases

Sugar-phosphate backbone

3'   5'

The "ribbons" in these simplified double helix diagrams represent the sugar-phosphate backbones; these models emphasize the 3-D nature of DNA.

5'   3'

3'   5'

5'   3'

3'   5'

5' 3'

3' 5'

5' 3'

3' 5'

These flattened "ladder style" diagrams of DNA depict the sugar-phosphate backbones like the side rails of a ladder, with the base pairs as rungs. Light blue is used to indicate the more recently synthesized strand.

Sometimes the double-stranded DNA molecule is shown simply as two straight lines.

3. *Compare the information conveyed in the three ladder diagrams.*

DNA Sequences

Genetic information is carried in DNA as a linear sequence of nucleotides that may be transcribed into mRNA and translated into a polypeptide. When focusing on the DNA sequence, each nucleotide can be represented simply as the letter of its base: A, T, C, or G.

3'- A C G T A A G C G G T T A A T -5'
5'- T G C A T T C G C C A A T T A -3'

→ **Instructors**: Additional questions related to this Visualizing Figure can be assigned in **Mastering Biology**.

the sugar-phosphate backbones, and the rungs represent pairs of nitrogenous bases. Now imagine twisting the ladder to form a helix. Franklin's X-ray data indicated that the helix makes one full turn every 3.4 nm along its length. With the bases stacked just 0.34 nm apart, there are ten layers of base pairs, or rungs of the ladder, in each full turn of the helix.

The nitrogenous bases of the double helix are paired in specific combinations: adenine (A) with thymine (T), and guanine (G) with cytosine (C). It was mainly by trial and error that Watson and Crick arrived at this key feature of DNA. At first, Watson imagined that the bases paired like with like—for example, A with A and C with C. But this model did not fit the X-ray data, which suggested that the double helix had a uniform diameter. Why is this requirement inconsistent with like-with-like pairing of bases? Adenine and guanine are purines, nitrogenous bases with two organic rings, while cytosine and thymine are nitrogenous bases called pyrimidines, which have a single ring. Pairing a purine with a pyrimidine is the only combination that results in a uniform diameter for the double helix (**Figure 16.8**).

▼ **Figure 16.8 Possible base pairings in the DNA double helix.**

Purine + purine: too wide

Pyrimidine + pyrimidine: too narrow

Purine + pyrimidine: width consistent with X-ray data

Watson and Crick reasoned that there must be additional specificity of pairing dictated by the structure of the bases. Each base has chemical side groups that can form hydrogen bonds with its appropriate partner: Adenine can form two hydrogen bonds with thymine and only thymine; guanine forms three hydrogen bonds with cytosine and only cytosine. In shorthand, A pairs with T, and G pairs with C (**Figure 16.9**).

▼ **Figure 16.9 Base pairing in DNA.**

Nitrogenous base pairs are held together by hydrogen bonds.

| Adenine (A) | Thymine (T) | Guanine (G) | Cytosine (C) |

The Watson-Crick model took into account Chargaff's ratios and ultimately explained them. Wherever one strand of a DNA molecule has an A, the partner strand has a T. Similarly, a G in one strand is always paired with a C in the complementary strand. Therefore, in the DNA of any organism, the amount of adenine equals the amount of thymine, and the amount

of guanine equals the amount of cytosine. (Modern DNA-sequencing techniques have confirmed that the amounts are exactly equal.) Although the base-pairing rules dictate the combinations of nitrogenous bases that form the "rungs" of the double helix, they do not restrict the sequence of nucleotides *along* each DNA strand. The linear sequence of the four bases can be varied in countless ways, and each gene has a unique base sequence.

In April 1953, Watson and Crick surprised the scientific world with a succinct, one-page paper that reported their molecular model for DNA: the double helix, which has since become an icon of molecular biology. Watson and Crick, along with Maurice Wilkins, were awarded the Nobel Prize in 1962 for this work. (Sadly, Rosalind Franklin had died at the age of 37 in 1958 and was thus ineligible for the prize.) The beauty of the double helix model was that the structure of DNA suggested the basic mechanism of its replication.

→ **Mastering Biology** HHMI Video: Great Discoveries in Science: The Double Helix

CONCEPT 16.2

Many proteins work together in DNA replication and repair

Hereditary information in DNA directs the development of your biochemical, anatomical, physiological, and, to some extent, behavioral traits. Your resemblance to your parents has its basis in the accurate replication of DNA prior to meiosis and therefore its transmission from your parents' generation to yours. Replication prior to mitosis ensures faithful transmission of genetic information from a parent cell to two daughter cells.

Of all nature's molecules, nucleic acids are unique in their ability to dictate their own replication from monomers. The relationship between structure and function is evident in the double helix: The specific complementary pairing of nitrogenous bases in DNA has a functional significance. Watson and Crick ended their classic paper with this statement: "It has not escaped our notice that the specific pairing we have postulated immediately suggests a possible copying mechanism for the genetic material."[1] In this section, you will learn about the basic principle of **DNA replication**, the copying of DNA, as well as some important details of the process.

---

[1]J. D. Watson and F. H. C. Crick, Molecular structure of nucleic acids: a structure for deoxyribose nucleic acids, *Nature* 171:737–738 (1953).

▼ **Figure 16.10 A model for DNA replication: the basic concept.** In this simplified illustration, a short segment of DNA has been untwisted. Simple shapes symbolize the four kinds of bases. Dark blue represents DNA strands present in the parental molecule; light blue represents newly synthesized DNA.

→ **Mastering Biology** Animation: **DNA Replication: An Overview**

(a) The parental molecule has two complementary strands of DNA. Each base is paired by hydrogen bonding with its specific partner, A with T and G with C.

(b) First, the two DNA strands are separated. Each parental strand can now serve as a template for a new, complementary strand.

(c) Nucleotides complementary to the parental (dark blue) strand are connected to form the sugar-phosphate backbones of the new "daughter" (light blue) strands.

The Basic Principle: Base Pairing to a Template Strand

In a second paper, Watson and Crick stated their hypothesis for how DNA replicates: "Now our model for deoxyribonucleic acid is, in effect, a pair of templates, each of which is complementary to the other. We imagine that prior to duplication the hydrogen bonds are broken, and the two chains unwind and separate. Each chain then acts as a template for the formation on to itself of a new companion chain, so that eventually we shall have two pairs of chains, where we only had one before. Moreover, the sequence of the pairs of bases will have been duplicated exactly."[2]

Figure 16.10 illustrates the basic idea. If you cover a DNA strand in Figure 16.10a, its linear sequence of nucleotides is revealed by applying the base-pairing rules to the uncovered strand. The two strands are complementary; each stores the information necessary to reconstruct the other. When a cell copies a DNA molecule, each strand serves as a template for ordering nucleotides into a new, complementary strand. Nucleotides line up along the template strand according to the base-pairing rules and are linked to form the new strands. One double-stranded DNA molecule becomes two, each an exact replica of the "parental" molecule.

This model of DNA replication remained untested for several years following publication of the DNA structure. The necessary experiments were simple in concept but difficult to perform. Watson and Crick's model predicts that when a double helix replicates, each of the two daughter molecules will have one old strand, from the parental molecule, and one newly made strand. This **semiconservative model** can be distinguished from a *conservative model* of replication, in which the two parental strands somehow come back together after the process (that is, the parental molecule is conserved). In yet a third model, called the *dispersive model*, all four strands of DNA following replication have a mixture of old and new DNA **(Figure 16.11)**.

▼ **Figure 16.11 DNA replication: three alternative models.** Each short segment of double helix symbolizes the DNA within a cell. Beginning with a parent cell, we follow the DNA for two more generations of cells—two rounds of DNA replication. Parental DNA is dark blue; newly made DNA is light blue.

	Parent cell	First replication	Second replication

(a) Conservative model.
The two parental strands reassociate after acting as templates for new strands, thus restoring the parental double helix.

(b) Semiconservative model.
The two strands of the parental molecule separate, and each functions as a template for synthesis of a new, complementary strand.

(c) Dispersive model.
Each strand of *both* daughter molecules contains a mixture of old and newly synthesized DNA.

[2]J. D. Watson and F. H. C. Crick, Genetical implications of the structure of deoxyribonucleic acid, *Nature* 171:964–967 (1953).

Although mechanisms for conservative or dispersive DNA replication are not easy to devise, these models remained possibilities until they could be ruled out. After two years of preliminary work at the California Institute of Technology in the late 1950s, Matthew Meselson and Franklin Stahl devised a clever experiment that distinguished between the three models, described in **Figure 16.12**. Their results supported the semiconservative model of DNA replication, as predicted by Watson and Crick, and their experiment is widely recognized among biologists as a classic example of elegant design.

The basic principle of DNA replication is conceptually simple. However, the actual process involves some complicated biochemical gymnastics, as we will now see.

DNA Replication: *A Closer Look*

The bacterium *E. coli* has a single chromosome of about 4.6 million nucleotide pairs. In a favorable environment, an *E. coli* cell can copy all of this DNA and divide to form two genetically identical daughter cells in considerably less than an hour. Each of *your* somatic cells has 46 DNA molecules in its nucleus, one long double-helical molecule per chromosome. In all, that represents about 6 billion nucleotide pairs, or over 1,000 times more DNA than is found in most bacterial cells. If we were to print the one-letter symbols for these bases (A, G, C, and T) the size of the type you are now reading, the 6 billion nucleotide pairs of information in a diploid human cell would fill about 1,400 biology textbooks. Yet it takes one of your cells just a few hours to copy all of this DNA during S phase of interphase. This replication of an enormous amount of genetic information is achieved with very few errors—only about one per 10 billion nucleotides. The copying of DNA is remarkable in its speed and accuracy.

More than a dozen enzymes and other proteins participate in DNA replication. Much more is known about how this "replication machine" works in bacteria (such as *E. coli*) than in eukaryotes, and we will describe the basic steps of the process for *E. coli*, except where otherwise noted. What scientists have learned about eukaryotic DNA replication suggests, however, that most of the process is fundamentally similar for prokaryotes and eukaryotes.

Getting Started

The replication of chromosomal DNA begins at particular sites called **origins of replication**, short stretches of DNA that have a specific sequence of nucleotides. The *E. coli* chromosome, like many other bacterial chromosomes, is circular and has a single origin. Proteins that initiate DNA replication recognize this sequence and attach to the DNA, separating the two strands and opening up a replication "bubble" **(Figure 16.13a)**. Replication of DNA then proceeds in both directions until the entire molecule is copied. In contrast to a bacterial chromosome, a eukaryotic chromosome may have hundreds or even a few thousand replication origins. Multiple replication bubbles form and eventually fuse, thus speeding up the copying of the very

▼ **Figure 16.12** **Inquiry**

Does DNA replication follow the conservative, semiconservative, or dispersive model?

Experiment Matthew Meselson and Franklin Stahl cultured *E. coli* for several generations in a medium containing nucleotide precursors labeled with a heavy isotope of nitrogen, ^{15}N. They then transferred the bacteria to a medium with only ^{14}N, a lighter isotope. They took one sample after the first DNA replication and another after the second replication. They extracted DNA from the bacteria in the samples and then centrifuged each DNA sample to separate DNA of different densities.

❶ Bacteria cultured in medium with ^{15}N (heavy isotope)

❷ Bacteria transferred to medium with ^{14}N (lighter isotope)

Results

❸ DNA sample centrifuged after first replication

❹ DNA sample centrifuged after second replication

Less dense

More dense

Conclusion Meselson and Stahl compared their results to those predicted by each of the three models in Figure 16.11, as shown below. The first replication in the ^{14}N medium produced a band of many molecules of hybrid (^{15}N-^{14}N) DNA. This result eliminated the conservative model. The second replication produced both light and hybrid DNA, a result that refuted the dispersive model and supported the semiconservative model. They therefore concluded that DNA replication is semiconservative.

Predictions:	First replication	Second replication
Conservative model		
Semiconservative model		
Dispersive model		

Data from M. Meselson and F. W. Stahl, The replication of DNA in *Escherichia coli*, *Proceedings of the National Academy of Sciences USA* 44:671–682 (1958).

INQUIRY IN ACTION Read and analyze the original paper in *Inquiry in Action: Interpreting Scientific Papers.*

➔ **Instructors**: A related Experimental Inquiry Tutorial can be assigned in **Mastering Biology**.

WHAT IF? *If Meselson and Stahl had first grown the cells in ^{14}N-containing medium and then moved them into ^{15}N-containing medium before taking samples, what would have been the result after each of the two replications?*

▼ **Figure 16.13 Origins of replication in *E. coli* and eukaryotes.** The red arrows indicate the movement of the replication forks and thus the overall directions of DNA replication within each bubble.

(a) Origin of replication in an *E. coli* cell

The circular chromosome of *E. coli* and other bacteria has only one origin of replication. The parental strands separate there, forming a replication bubble with two forks (red arrows). Replication proceeds in both directions until the forks meet on the other side, resulting in two daughter DNA molecules. The TEM shows a bacterial chromosome with a replication bubble.

(b) Origins of replication in a eukaryotic cell

In a linear chromosome of a eukaryote, replication bubbles form at many sites along the giant DNA molecule during S phase of interphase. The bubbles expand as replication proceeds in both directions (red arrows). Eventually, the bubbles fuse and synthesis of the daughter strands is complete. The TEM shows three replication bubbles along the DNA of a cultured Chinese hamster cell.

DRAW IT *In the TEM, add arrows in the forks of the third bubble.*

→ **Mastering Biology** BioFlix® **Animation: The Replication Fork in *E. coli***

long DNA molecules **(Figure 16.13b)**. As in bacteria, eukaryotic DNA replication proceeds in both directions from each origin.

At each end of a replication bubble is a **replication fork**, a Y-shaped region where the parental strands of DNA are being unwound. Several kinds of proteins participate in the unwinding **(Figure 16.14)**. **Helicases** are enzymes that untwist the double helix at the replication forks, separating the two parental strands and making them available as template strands. After the parental strands separate, **single-strand binding proteins** bind to the unpaired DNA strands, keeping them from re-pairing. The untwisting of the double helix causes tighter twisting and strain ahead of the replication fork. **Topoisomerase** is an enzyme that helps relieve this strain by breaking, swiveling, and rejoining DNA strands.

Synthesizing a New DNA Strand

The unwound sections of parental DNA strands are now available to serve as templates for the synthesis of new complementary DNA strands. However, the enzymes that

▼ **Figure 16.14 Some of the proteins involved in the initiation of DNA replication.** The same proteins function at both replication forks in a replication bubble. For simplicity, only the left-hand fork is shown, and the DNA bases are drawn much larger in relation to the proteins than they are in reality.

Topoisomerase breaks, swivels, and rejoins the parental DNA ahead of the replication fork, relieving the strain caused by unwinding.

Primase synthesizes RNA primers, using the parental DNA as a template.

Helicase unwinds and separates the parental DNA strands.

Single-strand binding proteins stabilize the un-wound parental strands.

synthesize DNA cannot *initiate* the synthesis of a poly-nucleotide; they can only add DNA nucleotides to the end of an already existing chain that is base-paired with the template strand. An initial nucleotide chain that can be used as a pre-existing chain is produced during DNA synthesis; this is actually a short stretch of RNA, not DNA. The RNA chain is called a **primer** and is synthesized by the enzyme **primase** (see Figure 16.14). Primase starts a complementary RNA chain with a single RNA nucleotide and adds RNA nucleotides one at a time, using the parental DNA strand as a template. The completed primer, generally five to ten nucleotides long, is thus base-paired to the template strand. The new DNA strand will start from the 3′ end of the RNA primer.

Enzymes called **DNA polymerases** catalyze the synthesis of new DNA by adding nucleotides to the 3′ end of a pre-existing chain. In *E. coli*, there are several DNA polymerases, but two of them appear to play the major roles in DNA replication: DNA polymerase III and DNA polymerase I. The situation in eukaryotes is more complicated, with at least 11 different DNA polymerases discovered so far, although the general principles are the same.

Most DNA polymerases require a primer and a DNA template strand, along which complementary DNA nucleotides are lined up, one by one. In *E. coli*, DNA polymerase III (abbreviated DNA pol III) adds a DNA nucleotide to the RNA primer and then continues adding DNA nucleotides, which are complementary to the parental DNA template strand, to the growing end of the new DNA strand. The rate of elongation is about 500 nucleotides per second in bacteria and 50 per second in human cells.

Each nucleotide to be added to a growing DNA strand consists of a sugar attached to a base and to three phosphate groups. You have already encountered such a molecule—ATP (adenosine triphosphate; see Figure 8.9). The only difference between the ATP of energy metabolism and dATP, the adenine nucleotide used to make DNA, is the sugar component, which is deoxyribose in the building block of DNA but ribose in ATP. Like ATP, the nucleotides used for DNA synthesis are chemically reactive, partly because their triphosphate tails have an unstable cluster of negative charge. DNA polymerase catalyzes the addition of each monomer to the growing end of a DNA strand by a condensation reaction in which two phosphate groups are lost as a molecule of pyrophosphate (Ⓟ—Ⓟᵢ). Subsequent hydrolysis of the pyrophosphate to two molecules of inorganic phosphate (Ⓟᵢ) is a coupled exergonic reaction that helps drive the polymerization reaction **(Figure 16.15)**.

Antiparallel Elongation

As we have noted previously, the two ends of a DNA strand are different, giving each strand directionality, like a one-way street (see Figure 16.5). In addition, the two strands of DNA in a double helix are antiparallel, meaning that they are oriented in opposite directions to each other, like the two sides

▼ **Figure 16.15 Addition of a nucleotide to a DNA strand.** DNA polymerase catalyzes addition of a nucleotide to the 3′ end of a growing DNA strand, with the release of two phosphates.

DRAW IT *Circle the area where the new bond was made.*

→ **Mastering Biology Figure Walkthrough**

of a divided street (see Figure 16.15). Therefore, the two new strands formed during DNA replication must also be antiparallel to their template strands.

The antiparallel arrangement of the double helix, together with a constraint on the function of DNA polymerases, has an important effect on how replication occurs. Because of their structure, DNA polymerases can add nucleotides only to the free 3′ end of a primer or growing DNA strand, never to the 5′ end (see Figure 16.15). Thus, a new DNA strand can elongate only in the 5′ → 3′ direction. With this in mind, let's examine one of the two replication forks in a bubble **(Figure 16.16)**. Along one template strand, DNA polymerase III can synthesize a complementary strand continuously by elongating the new DNA in the mandatory 5′ → 3′ direction. DNA pol III remains in the replication fork on that template strand and continuously adds nucleotides to the new complementary strand as the fork progresses. The DNA strand made by this mechanism is called the **leading strand**. Only one primer is required for DNA pol III to synthesize the entire leading strand (see Figure 16.16).

To elongate the other new strand of DNA in the mandatory 5′ → 3′ direction, DNA pol III must work along the other template strand in the direction *away from* the replication fork. The DNA strand elongating in this direction is called the **lagging strand**. In contrast to the leading strand, which elongates continuously, the lagging strand is synthesized discontinuously, as a series of segments. These segments of the lagging strand are called **Okazaki fragments**, after Reiji Okazaki, the Japanese scientist who discovered them. The fragments are about 1,000–2,000 nucleotides long in *E. coli* and 100–200 nucleotides long in eukaryotes.

▼ Figure 16.16 Synthesis of the leading strand during DNA replication.

This diagram focuses on the left replication fork shown in the overview box. DNA polymerase III (DNA pol III), shaped like a cupped hand, is shown closely associated with a protein called the "sliding clamp" that encircles the newly synthesized double helix like a doughnut. The sliding clamp moves DNA pol III along the DNA template strand.

Overview

Template strand · Leading strand · Origin of replication · Lagging strand · Primer

5′ · 3′ · 3′ · 5′

Lagging strand · Leading strand

Overall directions of replication

1 After an RNA primer is made, DNA pol III starts to synthesize the leading strand.

Site where replication begins

Single-strand binding proteins

RNA primer

Sliding clamp

5′ · 3′ · 3′ · 5′

Parental DNA · Helicase

2 The leading strand is elongated continuously in the 5′ → 3′ direction as helicase opens up the fork further to the left.

5′ · 3′ · 3′ · 5′

→ **Mastering Biology** BioFlix® Animation: Synthesis of the Leading Strand

Figure 16.17 illustrates the steps in the synthesis of the lagging strand at one fork. Whereas only one primer is required on the leading strand, each Okazaki fragment on the lagging strand must be primed separately (steps **1** and **4**). After DNA pol III forms an Okazaki fragment (steps **2** to **4**), another DNA polymerase, DNA pol I, replaces the RNA nucleotides of the adjacent primer with DNA nucleotides one at a time (step **5**). But DNA pol I cannot join the final nucleotide of this replacement DNA segment to the first DNA nucleotide of the adjacent Okazaki fragment. Another enzyme, **DNA ligase**, accomplishes this task, joining the sugar-phosphate backbones of all the Okazaki fragments into a continuous DNA strand (step **6**).

→ **Mastering Biology** BioFlix® Animation: Synthesis of the Lagging Strand

▼ Figure 16.17 Synthesis of the lagging strand.

Overview

Origin of replication · Leading strand · Lagging strand

5′ · 3′ · 3′ · 5′

Template strand · Leading strand

Overall directions of replication

1 Primase joins RNA nucleotides into the first primer for the lagging strand.

Site where replication begins · Primer for leading strand

Template strand

3′ · 5′ · 3′ · 5′ · 3′ · 5′

2 DNA pol III adds DNA nucleotides to the primer, forming Okazaki fragment 1.

RNA primer for fragment 1

3′ · 5′ · 3′ · 5′ · 3′ · 5′

3 After reaching the next RNA primer to the right, DNA pol III detaches.

Okazaki fragment 1

3′ · 5′ · 3′ · 5′

4 Fragment 2 is primed. Then DNA pol III adds DNA nucleotides, detaching when it reaches the fragment 1 primer.

RNA primer for fragment 2 · Okazaki fragment 2

3′ · 5′ · 3′ · 5′

5 DNA pol I replaces the RNA with DNA, adding nucleotides to the 3′ end of fragment 1 (and, later, of fragment 2).

3′ · 5′ · 3′ · 5′

6 DNA ligase forms a bond between the newest DNA and the DNA of fragment 1.

7 The lagging strand in this region is now complete.

3′ · 5′ · 3′ · 5′

Overall direction of replication

Synthesis of the leading strand and synthesis of the lagging strand occur concurrently and at the same rate. The lagging strand is so named because its synthesis is delayed slightly relative to synthesis of the leading strand; each new fragment of the lagging strand cannot be started until enough template has been exposed at the replication fork.

Figure 16.18 and **Table 16.1** summarize DNA replication. Please study them carefully before proceeding.

The DNA Replication Complex

It is traditional—and convenient—to represent DNA polymerase molecules as locomotives moving along a DNA railroad track, but such a model is inaccurate in two important ways. First, the various proteins that participate in DNA replication actually form a single large complex, a "DNA replication machine." Many protein-protein interactions facilitate the efficiency of this complex. For example, by interacting with other proteins at the fork, primase apparently acts as a

molecular brake, slowing progress of the replication fork and coordinating the placement of primers and the rates of replication on the leading and lagging strands. Second, the DNA replication complex may not move along the DNA; rather, the DNA may move through the complex during the replication process. In eukaryotic cells, multiple copies of the complex, perhaps grouped into "factories," may be anchored to the nuclear matrix, a framework of fibers extending through the interior of the nucleus.

Some experimental evidence supports a model in which two DNA polymerase molecules, one on each template strand, "reel in" the parental DNA and extrude newly made daughter DNA molecules. In this so-called *trombone model*, the lagging strand is also looped back through the complex **(Figure 16.19)**. Whether the complex moves along the DNA or whether the DNA moves through the complex, either anchored or not, are still open, unresolved questions that are under active investigation.

▼ **Figure 16.18 A summary of bacterial DNA replication.** The detailed diagram shows the left-hand replication fork of the replication bubble shown in the overview (upper right). Viewing each daughter strand in its entirety in the overview, you can see that half of it is made continuously as the leading strand, while the other half (on the other side of the origin) is synthesized in fragments as the lagging strand.

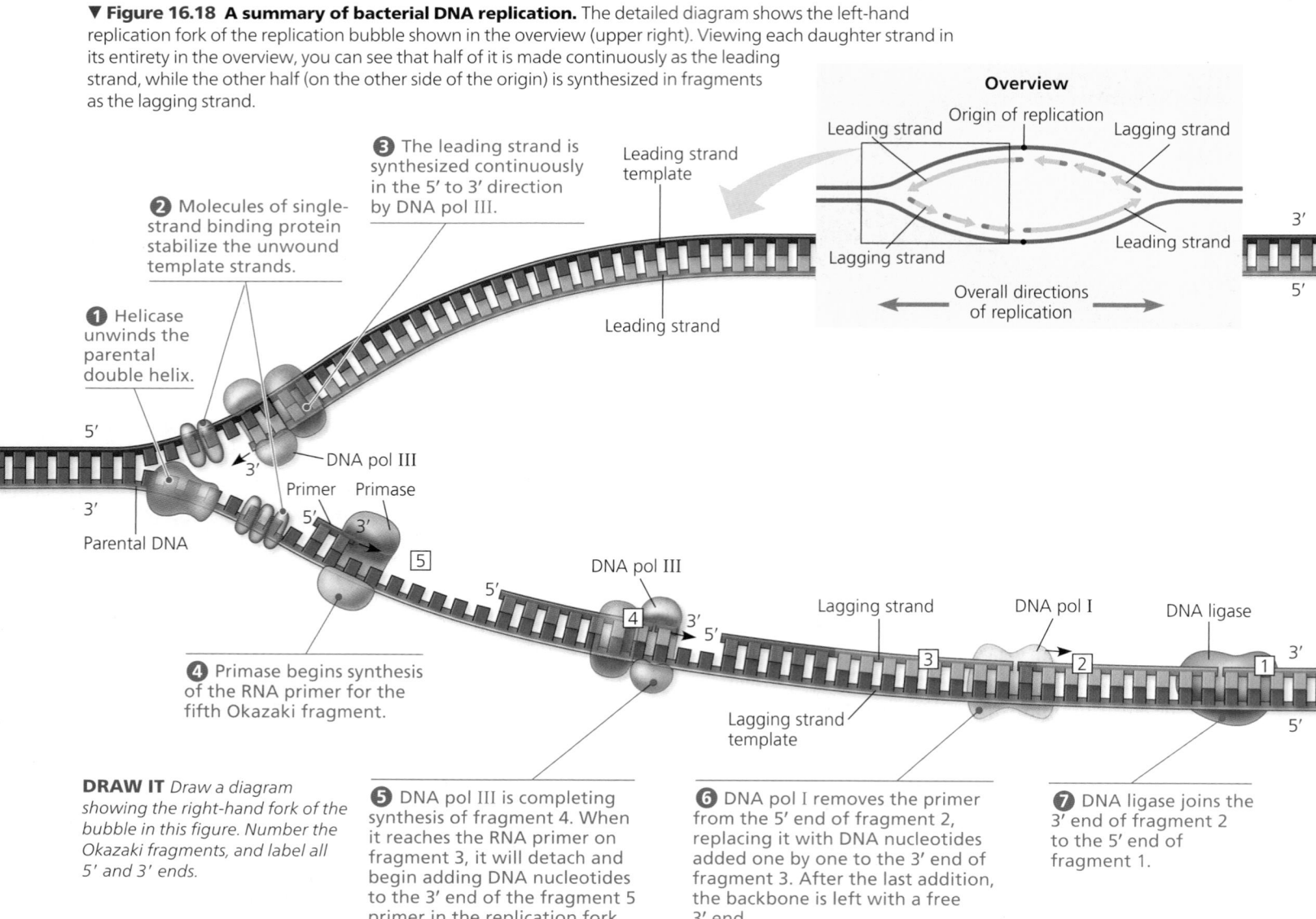

❸ The leading strand is synthesized continuously in the 5' to 3' direction by DNA pol III.

❷ Molecules of single-strand binding protein stabilize the unwound template strands.

❶ Helicase unwinds the parental double helix.

DNA pol III
Primer   Primase

Parental DNA

❹ Primase begins synthesis of the RNA primer for the fifth Okazaki fragment.

Overview

Leading strand   Origin of replication   Lagging strand

Lagging strand   Leading strand

Overall directions of replication

Leading strand template

Leading strand

DNA pol III

Lagging strand   DNA pol I   DNA ligase

Lagging strand template

DRAW IT *Draw a diagram showing the right-hand fork of the bubble in this figure. Number the Okazaki fragments, and label all 5' and 3' ends.*

❺ DNA pol III is completing synthesis of fragment 4. When it reaches the RNA primer on fragment 3, it will detach and begin adding DNA nucleotides to the 3' end of the fragment 5 primer in the replication fork.

❻ DNA pol I removes the primer from the 5' end of fragment 2, replacing it with DNA nucleotides added one by one to the 3' end of fragment 3. After the last addition, the backbone is left with a free 3' end.

❼ DNA ligase joins the 3' end of fragment 2 to the 5' end of fragment 1.

Mastering Biology Animation: DNA Replication: A Closer Look

Table 16.1 Bacterial DNA Replication Proteins and Their Functions

Protein	Function
Helicase	Unwinds parental double helix at replication forks
Single-strand binding protein	Binds to and stabilizes single-stranded DNA until it is used as a template
Topoisomerase	Relieves overwinding strain ahead of replication forks by breaking, swiveling, and rejoining DNA strands
Primase	Synthesizes an RNA primer at 5′ end of leading strand and at 5′ end of each Okazaki fragment of lagging strand
DNA pol III	Using parental DNA as a template, synthesizes new DNA strand by adding nucleotides to an RNA primer or a pre-existing DNA strand
DNA pol I	Removes RNA nucleotides of primer from 5′ end and replaces them with DNA nucleotides added to 3′ end of adjacent fragment
DNA ligase	Joins Okazaki fragments of lagging strand; on leading strand, joins 3′ end of DNA that replaces primer to rest of leading strand DNA

▼ **Figure 16.19 The "trombone" model of the DNA replication complex.** In this proposed model, two molecules of DNA polymerase III work together in a complex, one on each strand, with helicase and other proteins. The lagging strand template DNA loops through the complex, resembling the slide of a trombone.

DRAW IT *Draw a line tracing the lagging strand template in this figure.*

→ **Mastering Biology** BioFlix® Animation: DNA Replication

Proofreading and Repairing DNA

We cannot attribute the accuracy of DNA replication solely to the specificity of base pairing. Initial pairing errors between incoming nucleotides and those in the template strand occur at a rate of one in 10^5 nucleotides. However, errors in the completed DNA molecule amount to only one in 10^{10} (10 billion) nucleotides, an error rate that is 100,000 times lower. This is because during DNA replication, many DNA polymerases proofread each nucleotide against its template as soon as it is covalently bonded to the growing strand. Upon finding an incorrectly paired nucleotide, the polymerase removes the nucleotide and then resumes synthesis. (This action is similar to fixing a texting error by deleting the wrong letter and then entering the correct one.)

Mismatched nucleotides sometimes evade proofreading by a DNA polymerase. In **mismatch repair**, other enzymes remove and replace incorrectly paired nucleotides that have resulted from replication errors. Researchers highlighted the importance of such repair enzymes when they found that a hereditary defect in one of them is associated with a form of colon cancer. Apparently, this defect allows cancer-causing errors to accumulate in the DNA faster than normal.

Incorrectly paired or altered nucleotides can also arise after replication. In fact, maintenance of the genetic information encoded in DNA requires frequent repair of various kinds of damage to existing DNA. DNA molecules are constantly subjected to potentially harmful chemical and physical agents, such as X-rays, as we'll discuss in Concept 17.5. In addition, DNA bases may undergo spontaneous chemical changes under normal cellular conditions. However, these changes in DNA are usually corrected before they become permanent changes—*mutations*—perpetuated through successive replications. Each cell continuously monitors and repairs its genetic material. Because repair of damaged DNA is so important to the survival of an organism, it is no surprise that many different DNA repair enzymes have evolved. Almost 100 are known in *E. coli*, and about 170 have been identified so far in humans.

Most cellular systems for repairing incorrectly paired nucleotides, whether they are due to DNA damage or to replication errors, use a mechanism that takes advantage of the base-paired structure of DNA. In many cases, a segment of the strand containing the damage is cut out (excised) by a DNA-cutting enzyme—a **nuclease**—and the resulting gap is then filled in with nucleotides, using the undamaged strand as a template. The enzymes involved in filling the gap are a DNA polymerase and DNA ligase. There are several such DNA repair systems; one is called **nucleotide excision repair**.

❶ Teams of enzymes detect and repair damaged DNA, such as this thymine dimer (often caused by ultraviolet radiation), which distorts the DNA molecule.

Nuclease

❷ A nuclease enzyme cuts the damaged DNA strand at two points, and the damaged section is removed.

DNA polymerase

❸ Repair synthesis by a DNA polymerase fills in the missing nucleotides, using the undamaged strand as a template.

DNA ligase

❹ DNA ligase seals the free end of the new DNA to the old DNA, making the strand complete.

An example is shown in **Figure 16.20**. An important function of the DNA repair enzymes in our skin cells is to repair genetic damage caused by the UV rays of sunlight: For example, adjacent thymine bases on a DNA strand can become covalently linked into *thymine dimers*, causing the DNA to buckle and interfere with DNA replication. The importance of repairing this kind of damage is underscored by a disorder called xeroderma pigmentosum (XP), which in most cases is caused by an inherited defect in a nucleotide excision repair enzyme. Individuals with XP are hypersensitive to sunlight; mutations in their skin cells caused by ultraviolet light are left uncorrected, often resulting in skin cancer. The effects are extreme: Without sun protection, children who have XP can develop skin cancer by age 10.

Evolutionary Significance of Altered DNA Nucleotides

EVOLUTION Faithful replication of the genome and repair of DNA damage are important for the functioning of the organism and for passing on a complete, accurate genome to the next generation. The error rate after proofreading and repair is extremely low, but rare mistakes do slip through. Once a mismatched nucleotide pair is replicated, the sequence change is permanent in the daughter molecule that has the incorrect nucleotide as well as in any subsequent copies. As we mentioned earlier, a permanent change in the DNA sequence is called a mutation.

Mutations can change the phenotype of an organism (see Concept 17.5). And if they occur in germ cells, which give rise to gametes, mutations can be passed on from generation to generation. The vast majority of such changes either have no effect or are harmful, but a very small percentage can be beneficial. In either case, mutations are the original source of the variation on which natural selection operates during evolution and are ultimately responsible for the appearance of new species. (You'll learn more about this process in Unit Four.) The balance between complete fidelity of DNA replication or repair and a low mutation rate has resulted in new proteins that contribute to different phenotypes. Ultimately, over long periods of time, this process leads to new species and thus to the rich diversity of life on Earth today.

Replicating the Ends of DNA Molecules

For linear DNA, such as the DNA of eukaryotic chromosomes, the usual replication machinery cannot complete the 5′ ends of daughter DNA strands because there is no 3′ end of a preexisting polynucleotide for DNA polymerase to add onto. This is another consequence of the enzyme's requirements. Even if an Okazaki fragment can be started with an RNA primer hydrogen-bonded to the very end of the template strand, once that primer is removed, it cannot be replaced with DNA because there is no 3′ end available for nucleotide addition **(Figure 16.21)**. As a result, repeated rounds of replication produce shorter and shorter DNA molecules with uneven ("staggered") ends.

Most prokaryotes have a circular chromosome, with no ends, so the shortening of DNA does not occur. But what protects the genes of linear eukaryotic chromosomes from being eroded away during successive rounds of DNA replication? Eukaryotic chromosomal DNA molecules have special nucleotide sequences called **telomeres** at their ends **(Figure 16.22)**. Telomeres do not contain genes; instead, the DNA typically consists of multiple repetitions of one short nucleotide sequence. In each human telomere, for example, the six-nucleotide sequence TTAGGG is repeated between 100 and 1,000 times.

Telomeres have two protective functions. First, specific proteins associated with telomeric DNA prevent the staggered ends of the daughter molecule from activating the cell's systems for monitoring DNA damage. (Staggered ends of a DNA molecule, which often result from double-strand breaks, can trigger signal transduction pathways leading to cell cycle arrest or cell death.) Second, telomeric DNA acts as a kind of buffer zone that provides some protection against the organism's genes shortening, somewhat like how the plastic-wrapped ends of a shoelace slow down its unraveling. Telomeres do not prevent the erosion of genes near the ends of chromosomes; they merely postpone it.

▼ **Figure 16.21 Shortening of the ends of linear DNA molecules.** Here we follow the left end of one DNA molecule through two rounds of replication. After the first round, the new lagging strand is shorter than its template. After a second round, both the leading and lagging strands have become shorter than the original parental DNA. Although not shown here, the other ends of these chromosomal DNA molecules (not shown) also become shorter.

▼ **Figure 16.22 Telomeres.** Eukaryotes have repetitive, noncoding sequences called telomeres at the ends of their DNA. Telomeres are stained orange in these mouse chromosomes (LM).

their original length and compensating for the shortening that occurs during DNA replication. This enzyme contains its own RNA molecule that it uses as a template to artificially "extend" the leading strand, allowing the lagging strand to maintain a given length. Telomerase is not active in most human somatic cells, but its activity varies from tissue to tissue. The activity of telomerase in germ cells results in telomeres of maximum length in the zygote.

⊘ **Mastering Biology** BBC Video: Can Humans Live Forever?

Normal shortening of telomeres may protect organisms from cancer by limiting the number of divisions that somatic cells can undergo. Cells from large tumors often have unusually short telomeres, as we would expect for cells that have undergone many cell divisions. Further shortening would presumably lead to self-destruction of the tumor cells. Telomerase activity is abnormally high in cancerous somatic cells, suggesting that its ability to stabilize telomere length may allow these cancer cells to persist. Many cancer cells do seem capable of unlimited cell division, as do immortal strains of cultured cells (see Concept 12.3). For several years, researchers have studied inhibition of telomerase as a possible cancer therapy. While studies that inhibited telomerase in mice with tumors have led to the death of cancer cells, eventually the cells have restored the length of their telomeres by an alternative pathway. This is an area of ongoing research that may eventually yield useful cancer treatments.

As shown in Figure 16.21, telomeres become shorter during every round of replication. Thus, as expected, telomeric DNA tends to be shorter in dividing somatic cells of older individuals and in cultured cells that have divided many times. It has been proposed that shortening of telomeres is somehow connected to the aging process of certain tissues and even to aging of the organism as a whole.

But what about germ cells, whose genome must persist virtually unchanged from an organism to its offspring over many generations? If the chromosomes of germ cells became shorter in every cell cycle, essential genes would eventually be missing from the gametes they produce. However, this does not occur: An enzyme called *telomerase* catalyzes the lengthening of telomeres in eukaryotic germ cells, thus restoring

CONCEPT CHECK 16.2

1. What role does complementary base pairing play in the replication of DNA?
2. Identify two major functions of DNA pol III in DNA replication.
3. **MAKE CONNECTIONS** What is the relationship between DNA replication and the S phase of the cell cycle? See Figure 12.6.
4. **VISUAL SKILLS** If the DNA pol I in a given cell were nonfunctional, how would that affect the synthesis of a *leading* strand? In the overview box in Figure 16.18, point out where DNA pol I would normally function on the top leading strand.

For suggested answers, see Appendix A.

A chromosome consists of a DNA molecule packed together with proteins

Now let's examine how DNA is packaged into chromosomes, the structures that carry genetic information. The main component of the genome in most bacteria is a double-stranded, circular DNA molecule that is associated with specific proteins. A bacterial chromosome differs from a eukaryotic chromosome in that a eukaryotic chromosome consists of a single linear DNA molecule associated with a large number of proteins. In *E. coli*, the chromosomal DNA consists of about 4.6 million nucleotide pairs, representing about 4,400 genes. This is 100 times more DNA than is found in a typical virus, but only about one-thousandth as much DNA as in a human somatic cell. Even so, that is a tremendous amount of DNA to be packaged in such a small container.

Stretched out, the DNA of an *E. coli* cell would measure about a millimeter in length, 1,000 times longer than the region it occupies in the cell. Within a bacterium, however, certain proteins cause the chromosome to coil and "supercoil," densely packing it so that it fills only part of the cell.

▼ Figure 16.23 Exploring Chromatin Packing in a Eukaryotic Chromosome

DNA, the Double Helix

Shown below is a ribbon model of DNA, with each ribbon representing one of the polynucleotide strands. The TEM shows a molecule of naked (protein-free) DNA; the double helix is 2 nm across.

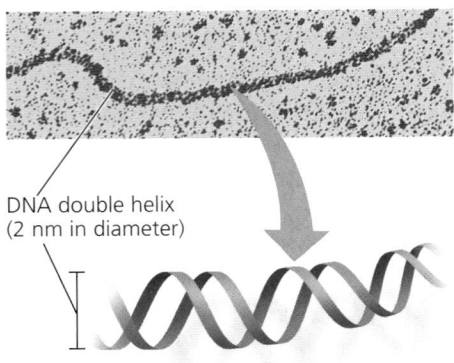

DNA double helix
(2 nm in diameter)

Histones

Proteins called **histones** are responsible for the main level of DNA packing in interphase chromatin. More than a fifth of a histone's amino acids are positively charged (Lys or Arg) and therefore bind tightly to the negatively charged DNA.

Four types of histones are most common in chromatin. The histones are very similar among eukaryotes, probably reflecting their important role in chromatin structure.

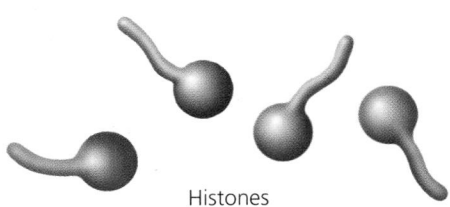

Histones

Nucleosomes in a 10-nm Fiber

In electron micrographs, unfolded chromatin is roughly 10 nm in diameter (the *10-nm fiber*). Such chromatin resembles beads on a string (see the TEM). Each "bead" is a **nucleosome**, the basic unit of DNA packing; the "string" between beads is called *linker DNA*.

A nucleosome consists of DNA wound twice around a protein core of eight histones, two each of the four main histone types. The amino end of each histone (the *histone tail*) extends outward from the nucleosome and is involved in regulation of gene expression.

In the cell cycle, the histones leave the DNA only briefly during DNA replication. Generally, they do the same during the process of transcription, which requires access to the DNA by transcription proteins.

Nucleosome
(10 nm in diameter)

Histone tail

DNA

Euchromatin (loosely arranged 10-nm fiber)

Euchromatin/Heterochromatin

A recent technique called ChromEMT allows scientists to visualize chromatin in intact cells. ChromEMT and other new techniques have shown that the 10-nm fiber is the basic constituent of interphase chromatin.

During interphase, different regions of a chromosome may exist as euchromatin (below left) or heterochromatin (below). In euchromatin, the 10-nm fiber is loosely arranged in a more open configuration than heterochromatin; higher degrees of organization, including the 30-nm fiber in one older model, may exist in specific cells or at certain times. (This is an area of very active research.) The DNA in euchromatin is accessible to the proteins that carry out transcription, and its genes can be expressed. In heterochromatin, the 10-nm fiber is more densely arranged and less accessible to these proteins; genes in heterochromatin are generally not expressed.

Euchromatin and heterochromatin are organized into regions by other proteins not shown here; this organization is dynamic but disappears once mitosis begins.

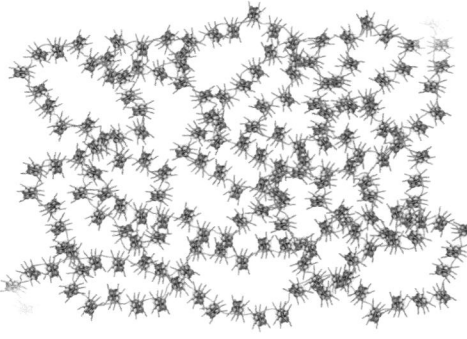

Heterochromatin (densely arranged 10-nm fiber)

Unlike the nucleus of a eukaryotic cell, this dense region of DNA in a bacterium, called the nucleoid, is not bounded by membrane (see Figure 6.5).

Each eukaryotic chromosome contains a single linear DNA double helix that, in humans, averages about 1.5×10^8 nucleotide pairs. This is an enormous amount of DNA relative to a chromosome's condensed length. If completely stretched out, such a DNA molecule would be about 4 cm long, thousands of times the diameter of a cell nucleus—and that's not even considering the DNA of the other 45 human chromosomes!

In the cell, eukaryotic DNA is precisely combined with a large amount of protein. The exact way in which this complex of DNA and protein, called **chromatin**, fits into the nucleus has long been debated. **Figure 16.23** outlines the current view of the organization of chromatin during interphase and how it is condensed during mitosis into the metaphase chromosome. Study this figure carefully before reading further.

Chromatin undergoes striking changes in how densely it is packed during the course of the cell cycle (see Figure 12.7). In interphase cells stained for light microscopy, the

Prophase

When mitosis begins, DNA replication has already occurred, so each chromosome consists of two sister chromatids. During prophase, the chromatin of each sister chromatid begins to condense. Two related proteins called condensin II and condensin I play important roles. First, condensin II proteins (red) bind to the 10-nm fiber of DNA and form DNA loops that get larger and larger. The condensin II proteins form a central scaffold from which the loops extend. As the loops grow, the chromosome gets wider and shorter. By the end of prophase, the chromosome is half as long.

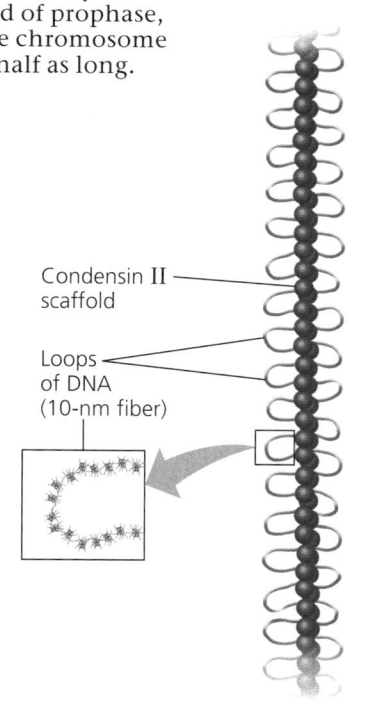

Condensin II scaffold

Loops of DNA (10-nm fiber)

Prometaphase

When prometaphase begins, condensin I proteins (green) bind to DNA outside the central scaffold, making smaller loops (not shown) out of the larger loops generated by condensin II. The process continues, with more and more loops extending outward, and the chromosome getting denser, shorter, and wider. The scaffold itself also begins to twist into a helix (suggested by the curved gray arrows), allowing even more loops per given length of chromosome.

Condensin I

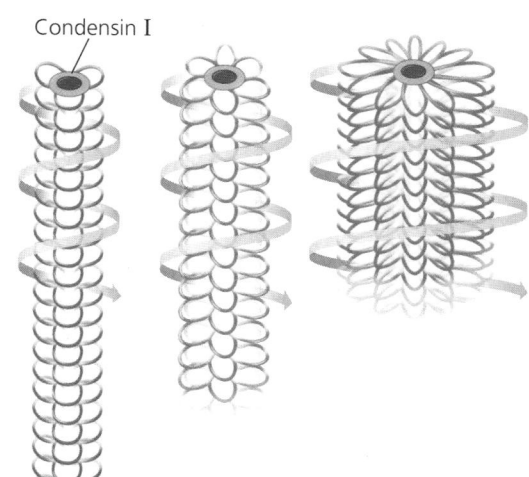

Metaphase

At metaphase, the chromosome is at its most dense, with the most loops per turn, and therefore is at its shortest length. The two sister chromatids are fully condensed.

Sister chromatid (700 nm)

Replicated chromosome (1,400 nm)

→ **Mastering Biology Animation: DNA Packing**

chromatin usually appears as a diffuse mass within the nucleus, with some denser clumps, including in regions of centromeres and telomeres. The less compacted, more dispersed interphase chromatin is called **euchromatin** ("true chromatin") to distinguish it from the more compacted, denser-appearing **heterochromatin** (see Figure 16.23). Recent research has clarified our understanding of chromatin structure: For both types of chromatin, the basic organizing unit is the 10-nm fiber—nucleosomes joined by linker DNA. In heterochromatin, this 10-nm fiber is folded and bent back on itself to a much greater degree than in euchromatin, accounting for its denser appearance. Because heterochromatin is so compacted, it is largely inaccessible to the proteins responsible for transcribing the genetic information, a crucial early step in gene expression. In contrast, the looser packing of euchromatin makes its DNA accessible to those proteins, and the genes present in euchromatin are available for transcription.

Early on, biologists assumed that interphase chromatin was a tangled mass in the nucleus, like a bowl of spaghetti, but this is far from the case. Although an interphase chromosome lacks an obvious scaffold, there are proteins that further organize the 10-nm fiber into larger compartments and smaller looped domains. Some of the looped domains appear to be attached to the nuclear lamina, on the inside of the nuclear envelope, and perhaps also to fibers of the nuclear matrix. These attachments may help organize regions of chromatin where genes are active. The chromatin of each chromosome occupies a specific restricted area within the interphase nucleus, and the chromatin fibers of different chromosomes do not appear to be entangled **(Figure 16.24a)**.

As a cell prepares for mitosis, its chromatin becomes organized into loops and coils, eventually condensing into a characteristic number of short, thick metaphase chromosomes that are distinguishable from each other with the light microscope **(Figure 16.24b)**.

The chromosome is a dynamic structure that is condensed, loosened, modified, and remodeled as necessary for various cell processes, including DNA replication, mitosis, meiosis, and gene expression. Certain chemical modifications of histones affect the state of chromatin condensation and also have multiple effects on gene expression, as you'll see in Concept 18.2.

In this chapter, you have learned how DNA molecules are arranged in chromosomes and how DNA replication provides the copies of genes that parents pass to offspring. However, it is not enough that genes be copied and transmitted; the information they carry must be used by the cell. In other words, genes must also be expressed. In the next chapter, we will examine how the cell expresses the genetic information encoded in DNA.

▼ **Figure 16.24 "Painting" chromosomes.** Researchers can treat ("paint") human chromosomes with molecular tags that cause each chromosome pair to appear a different color.

(a) The ability to visually distinguish among chromosomes makes it possible to see how the chromosomes are arranged in the interphase nucleus. Each chromosome appears to occupy a specific territory during interphase. In general, the two homologs of a pair are not located together.

(b) These metaphase chromosomes have been "painted" so that the two homologs of a pair are the same color. On the left is a spread of treated chromosomes; on the right, they have been organized into a karyotype.

MAKE CONNECTIONS *If you arrested a human cell in metaphase I of meiosis and applied this technique, what would you observe? How would this differ from what you would see in metaphase of mitosis? Review Figure 13.8 and Figure 12.7.*

CONCEPT CHECK 16.3

1. Describe the structure of a nucleosome, the basic unit of DNA packing in eukaryotic cells.

2. How does euchromatin differ from heterochromatin in structure and function?

3. **MAKE CONNECTIONS** Interphase chromosomes appear to be attached to the nuclear lamina and perhaps also the nuclear matrix. Describe these two structures. See Figure 6.9 and the associated text.

For suggested answers, see Appendix A.

16 Chapter Review

→ Go to **Mastering Biology** for Assignments, the eText, the Study Area, and Dynamic Study Modules.

SUMMARY OF KEY CONCEPTS

→ To review key terms, go to the **Vocabulary Self-Quiz** in the **Mastering Biology** eText or Study Area, or go to goo.gl/zkjz9t.

CONCEPT 16.1

DNA is the genetic material (pp. 315–320)

- Experiments with bacteria and with **phages** provided the first strong evidence that the genetic material is DNA.
- Watson and Crick deduced that DNA is a **double helix** and built a structural model. Two **antiparallel** sugar-phosphate chains wind around the outside of the molecule; the nitrogenous bases project into the interior, where they hydrogen-bond in specific pairs, A with T, G with C.

Nitrogenous bases

Sugar-phosphate backbone

Hydrogen bond

? *What does it mean when we say that the two DNA strands in the double helix are antiparallel? What would an end of the double helix look like if the strands were parallel?*

CONCEPT 16.2

Many proteins work together in DNA replication and repair (pp. 320–329)

- The Meselson-Stahl experiment showed that **DNA replication** is **semiconservative**: The parental molecule unwinds, and each strand then serves as a template for the synthesis of a new strand according to base-pairing rules.
- DNA replication at one **replication fork** is summarized here:

DNA pol III synthesizes **leading strand** continuously

Parental DNA

5′
3′

Helicase

Primase synthesizes a short RNA **primer**

DNA pol III starts DNA synthesis at 3′ end of primer, continues in 5′ → 3′ direction

Lagging strand is synthesized in short **Okazaki fragments**, later joined by **DNA ligase**

DNA pol I replaces the RNA primer with DNA nucleotides

Origin of replication

3′
5′

3′
5′

- **DNA polymerases** proofread new DNA, replacing incorrect nucleotides. In **mismatch repair**, enzymes correct errors that persist. **Nucleotide excision repair** is a process by which **nucleases** cut out and other enzymes replace damaged stretches of DNA.
- The ends of eukaryotic chromosomal DNA get shorter with each round of replication. The presence of **telomeres**, repetitive sequences at the ends of linear DNA molecules, postpones the erosion of genes. Telomerase catalyzes the lengthening of telomeres in germ cells.

? *Compare DNA replication on the leading and lagging strands, including similarities and differences.*

CONCEPT 16.3

A chromosome consists of a DNA molecule packed together with proteins (pp. 330–332)

- The chromosome of most bacterial species is a circular DNA molecule with some associated proteins, making up the nucleoid.
- The **chromatin** making up a eukaryotic chromosome is composed of DNA, **histones**, and other proteins. The histones bind to each other and to the DNA to form **nucleosomes**, the most basic units of DNA packing, which exist as part of the 10-nm fiber. Histone tails extend outward from each bead-like nucleosome core. Additional coiling and folding lead ultimately to the highly condensed chromatin of the metaphase chromosome.
- Chromosomes occupy restricted areas in the interphase nucleus. In interphase cells, most 10-nm fiber chromatin is loosely arranged (**euchromatin**), but some is more densely arranged (**heterochromatin**). Euchromatin, but not heterochromatin, is generally accessible for gene transcription.

? *Describe the levels of chromatin packing you'd expect to see in an interphase nucleus.*

TEST YOUR UNDERSTANDING

→ For more multiple-choice questions, go to the **Practice Test** in the **Mastering Biology** eText or Study Area, or go to goo.gl/GruWRg.

Levels 1-2: Remembering/Understanding

1. In his work with pneumonia-causing bacteria and mice, Griffith found that
   - (A) the protein coat from pathogenic cells was able to transform nonpathogenic cells.
   - (B) heat-killed pathogenic cells caused pneumonia.
   - (C) some substance from pathogenic cells was transferred to nonpathogenic cells, making them pathogenic.
   - (D) the polysaccharide coat of bacteria caused pneumonia.

2. What is the basis for the difference in how the leading and lagging strands of DNA molecules are synthesized?
   - (A) The origins of replication occur only at the 5′ end.
   - (B) Helicases and single-strand binding proteins work at the 5′ end.
   - (C) DNA polymerase can join new nucleotides only to the 3′ end of a pre-existing strand, and the strands are antiparallel.
   - (D) DNA ligase works only in the 3′ → 5′ direction.

3. In analyzing the number of different bases in a DNA sample, which result would be consistent with the base-pairing rules?
   - (A) A = G
   - (B) A + G = C + T
   - (C) A + T = G + C
   - (D) A = C

4. The elongation of the leading strand during DNA synthesis
   - (A) progresses away from the replication fork.
   - (B) occurs in the 3′ → 5′ direction.
   - (C) produces Okazaki fragments.
   - (D) depends on the action of DNA polymerase.

5. In a nucleosome, the DNA is wrapped around
   - (A) histones.
   - (B) ribosomes.
   - (C) polymerase molecules.
   - (D) a thymine dimer.

Levels 3-4: Applying/Analyzing

6. *E. coli* cells grown on ^{15}N medium are transferred to ^{14}N medium and allowed to grow for two more generations (two rounds of DNA replication). DNA extracted from these cells is centrifuged. What density distribution of DNA would you expect in this experiment?

(A) one high-density and one low-density band
(B) one intermediate-density band
(C) one high-density and one intermediate-density band
(D) one low-density and one intermediate-density band

7. A student isolates, purifies, and combines in a test tube a variety of molecules needed for DNA replication. When she adds some DNA to the mixture, replication occurs, but each DNA molecule consists of a normal strand paired with numerous segments of DNA a few hundred nucleotides long. What has she probably left out of the mixture?

(A) DNA polymerase
(B) DNA ligase
(C) Okazaki fragments
(D) primase

8. The spontaneous loss of amino groups from adenine in DNA results in hypoxanthine, an uncommon base, opposite thymine. What combination of proteins could repair such damage?

(A) nuclease, DNA polymerase, DNA ligase
(B) telomerase, primase, DNA polymerase
(C) telomerase, helicase, single-strand binding protein
(D) DNA ligase, replication fork proteins, adenylyl cyclase

9. MAKE CONNECTIONS Although the proteins that cause the *E. coli* chromosome to coil are not histones, what property would you expect them to share with histones, given their ability to bind to DNA (see Figure 5.14)?

Levels 5-6: Evaluating/Creating

10. EVOLUTION CONNECTION Some bacteria may be able to respond to environmental stress by increasing the rate at which mutations occur during cell division. How might this be accomplished? Might there be an evolutionary advantage to this ability? Explain.

11. SCIENTIFIC INQUIRY

DRAW IT Model building can be an important part of the scientific process. The illustration shown above is a computer-generated model of a DNA replication complex. The parental and newly synthesized DNA strands are color-coded differently, as are each of the following three proteins: DNA pol III, the sliding clamp, and single-strand binding protein.

(a) Using what you've learned in this chapter to clarify this model, label each DNA strand and each protein.
(b) Draw an arrow to indicate the overall direction of DNA replication.

12. WRITE ABOUT A THEME: ORGANIZATION The continuity of life is based on heritable information in the form of DNA, and structure and function are correlated at all levels of biological organization. In a short essay (100–150 words), describe how the structure of DNA is correlated with its role as the molecular basis of inheritance.

13. SYNTHESIZE YOUR KNOWLEDGE

This image shows DNA (gray) interacting with a computer-generated model of a TAL protein (multicolored), one of a family of proteins found only in a species of the bacterium *Xanthomonas*. The bacterium uses proteins like this one to find specific gene sequences in cells of the organisms it infects, such as tomatoes, rice, and citrus fruits. Given what you know about DNA structure and considering the image above, discuss how the TAL protein's structure suggests that it functions.

For selected answers, see Appendix A.

17 Gene Expression: From Gene to Protein

KEY CONCEPTS

17.1 Genes specify proteins via transcription and translation *p. 336*

17.2 Transcription is the DNA-directed synthesis of RNA: *A Closer Look* *p. 342*

17.3 Eukaryotic cells modify RNA after transcription *p. 345*

17.4 Translation is the RNA-directed synthesis of a polypeptide: *A Closer Look* *p. 347*

17.5 Mutations of one or a few nucleotides can affect protein structure and function *p. 357*

Study Tip

Make a visual study guide: Sketch the process shown below, and add labels and details as you read the chapter. (In this exercise, assume all processes take place in a eukaryotic cell.)

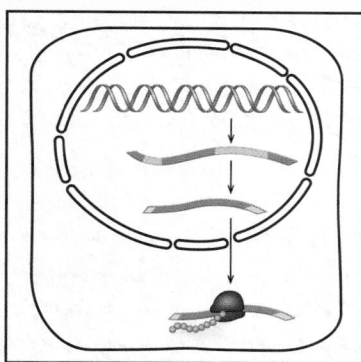

→ Go to Mastering Biology

For Students (in eText and Study Area)
- Get Ready for Chapter 17
- BioFlix® Animation: Protein Synthesis
- Figure 17.27 Walkthrough: Types of Small-Scale Mutations that Affect mRNA Sequence

For Instructors to Assign (in Item Library)
- BioFlix® Tutorial: Protein Synthesis (1 of 3): Overview
- Tutorial: CRISPR: A Revolution in Genome Editing

Ready-to-Go Teaching Module (in Instructor Resources)
- Gene Expression: Mutations (Concept 17.5)

Figure 17.1 A population of albino donkeys grazes on vegetation on the hillsides of Asinara, an Italian island. Several centuries ago, a recessive mutation that disables pigment synthesis arose in the DNA of one donkey and was passed down through the generations. Inbreeding has resulted in a large number of homozygous albino donkeys living on the island today.

How can one change in DNA result in such a dramatic change in appearance?

Proteins are the link between genotype and phenotype. Gene expression is the process by which DNA directs the synthesis of proteins:

Genotype: **DNA**

Gene for pigmentation

TRANSCRIPTION

mRNA

Protein: Enzyme required for pigment synthesis

TRANSLATION

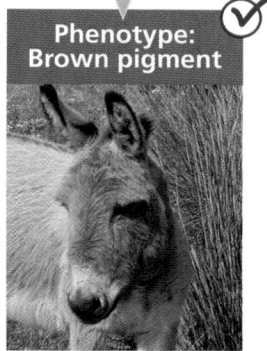

Phenotype: Brown pigment ✓

Genotype: **DNA**

Gene for pigmentation with mutation

mRNA

Protein: Enzyme required for pigment synthesis does not work

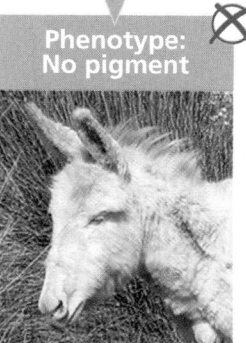

Phenotype: No pigment ✗

CONCEPT 17.1

Genes specify proteins via transcription and translation

Inherited traits like albinism are determined by genes, which have information content in the form of specific sequences of nucleotides along stretches of DNA, the genetic material. The DNA inherited by an organism leads to specific traits by dictating the synthesis of proteins and of RNA molecules involved in protein synthesis. In the case of coat and skin color, the gene containing the information to synthesize an enzyme that makes pigment allows normal coloring, and a faulty version of that gene leads to lack of color, or the albino phenotype. Proteins are the link between genotype and phenotype. **Gene expression** is the process by which DNA directs the synthesis of proteins (or, in some cases, just RNAs). The expression of genes that code for proteins includes two stages: transcription and translation.

Before going into the details of how genes direct protein synthesis, let's step back and examine how the fundamental relationship between genes and proteins was discovered.

Evidence from Studying Metabolic Defects

In 1902, British physician Archibald Garrod was the first to suggest that genes dictate phenotypes through enzymes, proteins that catalyze specific chemical reactions in the cell. He postulated that the symptoms of an inherited disease reflect an inability to make a particular enzyme. He later referred to such diseases as "inborn errors of metabolism." For example, people with a disease called alkaptonuria have black urine because it contains a chemical called alkapton, which darkens upon exposure to air. Garrod reasoned that these people cannot make an enzyme that breaks down alkapton, so alkapton is expelled in their urine.

Several decades later, research supported Garrod's hypothesis that a gene dictates the production of a specific enzyme, later named the *one gene–one enzyme hypothesis*. Biochemists learned that cells synthesize and degrade most organic molecules via metabolic pathways, in which each chemical reaction in a sequence is catalyzed by a specific enzyme (see Concept 8.1). Such metabolic pathways lead, for instance, to the synthesis of the pigments that give the brown donkey in **Figure 17.1** its fur color or fruit flies (*Drosophila*) their eye color (see Figure 15.3). In the 1930s, the American biochemist and geneticist George Beadle and his French colleague Boris Ephrussi speculated that in *Drosophila*, each mutation affecting eye color blocks pigment synthesis at a specific step by preventing production of the enzyme that catalyzes that step. But neither the chemical reactions nor the enzymes that catalyze them were known at the time.

Nutritional Mutants in Neurospora: Scientific Inquiry

A breakthrough in demonstrating the relationship between genes and enzymes came a few years later at Stanford University, where Beadle and Edward Tatum began working with a bread mold, *Neurospora crassa*. Unlike the diploid

organisms studied by Mendel (peas) and T. H. Morgan (fruit flies), *Neurospora* is a haploid species. Because its genome contains only one copy of each gene, that single copy determines the individual's expressed phenotype. Therefore, when Beadle and Tatum wanted to discover the function of any gene, they only needed to mutate and disable that one allele. They chose to study the protein-coding genes required for a specific nutritional activity. They caused mutations in genes by bombarding *Neurospora* with X-rays and looked among the survivors for mutants that differed in their nutritional needs from the wild-type bread mold.

Wild-type *Neurospora* has modest food requirements. It can grow in the laboratory on *minimal medium*—a simple solution containing minimal nutrients for growth—inorganic salts, glucose, and the vitamin biotin—incorporated into agar, a support medium. From these basics, wild-type mold cells use their metabolic pathways to produce all the other nutrient molecules (like amino acids) they need for growth, dividing repeatedly and forming visible colonies of genetically identical cells. As shown in **Figure 17.2**, Beadle and Tatum generated different "nutritional mutants" of *Neurospora* cells, each of which had a mutation in one gene and was unable to synthesize a particular

▼ **Figure 17.2 Beadle and Tatum's experimental approach.** To obtain nutritional mutants, Beadle and Tatum exposed *Neurospora* cells to X-rays, inducing mutations, then screened for (tested for) mutants that had new nutritional requirements, such as arginine, as shown here.

❶ Individual *Neurospora* cells were placed on complete medium.

❷ The cells were subjected to X-rays to induce a mutation in one gene per cell.

❸ Each surviving cell formed a colony of genetically identical cells.

❹ Cells from each colony were placed in a vial with only minimal medium. Cells that did not grow were identified as nutritional mutants.

No growth

❺ Cells from one nutritional mutant colony were placed in a series of vials, each containing minimal medium plus one additional nutrient.

Growth

| Minimal medium + valine | Minimal medium + lysine | Minimal medium + arginine | **Control:** Wild-type cells growing on minimal medium |

❻ The vials were observed for growth. In this example, the mutant cells grew only on minimal medium + arginine, indicating that this mutant was missing the enzyme for the synthesis of arginine.

essential nutrient. Such cells could not grow on minimal medium but could grow on *complete medium*, which contains all nutrients needed for growth. For *Neurospora*, the complete medium consists of the minimal medium supplemented with all 20 amino acids and a few other nutrients. Beadle and Tatum hypothesized that in each nutritional mutant, the gene for the enzyme that synthesizes a particular nutrient had been disabled by the mutation. They were able to determine experimentally which nutrient each mutant strain (original cell and its descendants) was unable to synthesize.

This approach resulted in a valuable collection of mutant strains of *Neurospora*, each catalogued by its defect in a particular pathway. Two colleagues, Adrian Srb and Norman Horowitz, used a collection of arginine-requiring mutants to investigate the biochemical pathway for arginine synthesis in *Neurospora* (**Figure 17.3**). Srb and Horowitz pinned down each mutant's defect more specifically, using additional tests to distinguish among three classes of arginine-requiring mutants. Mutants in each class required a different set of compounds along the arginine-synthesizing pathway, which has three steps. These results, and those of many similar experiments done by Beadle and Tatum, suggested that each class was blocked at a different step in this pathway because mutants in that class lacked the enzyme that catalyzes the blocked step.

Because Beadle and Tatum set up their experimental conditions so that each mutant was defective in a single gene, the collected results, taken together, provided strong support for the *one gene–one enzyme hypothesis*, as they dubbed it: that the function of a gene is to dictate the production of a specific enzyme. Further support for this hypothesis came from experiments that identified the specific enzymes lacking in the mutants. Beadle and Tatum shared a Nobel Prize in 1958 for "their discovery that genes act by regulating definite chemical events" (in the words of the Nobel committee).

Today, we know of countless examples in which a mutation in a gene causes a faulty enzyme that in turn leads to an identifiable condition. The albino donkeys in Figure 17.1 lack a key enzyme called tyrosinase in the metabolic pathway that produces melanin, a dark pigment. The absence of melanin causes white fur and other effects throughout the donkey's body. Its nose, ears, and hooves, as well as its eyes, are pink because no melanin is present to mask the reddish color of the blood vessels that run through those structures.

The Products of Gene Expression: A Developing Story

As researchers learned more about proteins, they made revisions to the one gene–one enzyme hypothesis. First of all, not all proteins are enzymes. Keratin, the structural protein of animal hair, and the hormone insulin are two examples of nonenzyme proteins. Because proteins that are not enzymes are nevertheless gene products, molecular biologists began to think in terms of one gene–one protein. However, many proteins are constructed from two or more different polypeptide chains, and each polypeptide is specified by its own gene. For example, hemoglobin—the oxygen-transporting protein of vertebrate red blood cells—contains two kinds of polypeptides, and thus two genes code for this protein, one for each type of polypeptide (see Figure 5.18). Beadle and Tatum's idea was therefore restated as the *one gene–one polypeptide hypothesis*. Even this description is not entirely accurate, though. First, in many cases, a eukaryotic gene can code for a set of closely related polypeptides via a process called alternative splicing, which you will learn about later in this chapter. Second, quite a few genes code for RNA molecules that have important functions in cells even though they are never translated into protein. For now, we will focus on genes that do code for polypeptides. (Note that it is common to refer to these gene products as proteins—a practice you will encounter in this book—rather than more precisely as polypeptides.)

Basic Principles of Transcription and Translation

Genes provide the instructions for making specific proteins, but a gene does not build a protein directly. The bridge between DNA and protein synthesis is the nucleic acid RNA. RNA is chemically similar to DNA except that it contains ribose instead of deoxyribose as its sugar and has the nitrogenous base uracil rather than thymine (see Figure 5.23). Thus, each nucleotide along a DNA strand has A, G, C, or T as its base, and each nucleotide along an RNA strand has A, G, C, or U as its base. An RNA molecule usually consists of a single strand.

We often describe the flow of information from gene to protein in terms of languages. Just as specific sequences of letters communicate information in a language such as English, specific sequences of monomers convey information in polymers like nucleic acids and proteins. In DNA or RNA, the monomers are the four types of nucleotides listed previously, which differ in their nitrogenous bases. Genes are typically hundreds or thousands of nucleotides long, each gene having a specific sequence of nucleotides. Each polypeptide of a protein also has monomers arranged in a particular linear order (the protein's primary structure; see Figure 5.18), but its monomers are amino acids. Thus, nucleic acids and proteins contain information written in two different chemical languages. Getting from DNA to protein involves two major stages: transcription and translation.

Transcription is the synthesis (production) of RNA using information in the DNA. The two nucleic acids are written in different forms of the same language, and the information is simply transcribed, or "rewritten," from DNA to RNA. Just as a DNA strand provides a template for making a new complementary strand during DNA replication (see Concept 16.2), it also can serve as a template for assembling a complementary sequence of RNA nucleotides. For a protein-coding gene, the resulting RNA molecule is a faithful transcript of the gene's

▼ Figure 17.3 Inquiry

Do individual genes specify the enzymes that function in a biochemical pathway?

Experiment Working with the mold *Neurospora crassa*, Adrian Srb and Norman Horowitz, then at Stanford University, used Beadle and Tatum's experimental approach (see Figure 17.2) to isolate mutants that required arginine in their growth medium. The researchers showed that these mutants fell into three classes, each defective in a different gene. From studies by others on mammalian liver cells, they suspected that the metabolic pathway of arginine biosynthesis involved a precursor nutrient and the intermediate molecules ornithine and citrulline, as shown in the diagram on the right.

Their most famous experiment, shown here, tested both the *one gene–one enzyme hypothesis* and their postulated arginine-synthesizing pathway. In this experiment, they grew their three classes of mutants under the four different conditions shown in the results table below. They included minimal medium (MM) as a control, knowing that wild-type cells could grow on MM but mutant cells could not. (See test tubes below.)

Growth:
Wild-type cells can grow and divide.

No growth: Mutant cells cannot grow and divide.

Control: Minimal medium

Results As shown in the table on the right, the wild-type strain was capable of growth under all experimental conditions, requiring only the minimal medium. The three classes of mutants each had a specific set of growth requirements. For example, class II mutants could not grow when ornithine alone was added but could grow when either citrulline or arginine was added.

Results Table	Classes of *Neurospora crassa*			
Condition	**Wild type**	**Class I mutants**	**Class II mutants**	**Class III mutants**
Minimal medium (MM) (control)				
MM + ornithine				
MM + citrulline				
MM + arginine (control)				
Summary of results	Can grow with or without any supplements	Can grow on ornithine, citrulline, or arginine	Can grow only on citrulline or arginine	Require arginine to grow

Conclusion From the growth requirements of the mutants, Srb and Horowitz deduced that each class of mutant was unable to carry out one step in the pathway for synthesizing arginine, presumably because it lacked the necessary enzyme, as shown in the table on the right. Because each of their mutants was mutated in a single gene, they concluded that each mutated gene must normally dictate the production of one enzyme. Their results supported the one gene–one enzyme hypothesis, proposed by Beadle and Tatum, and also confirmed that the arginine pathway described in the mammalian liver also operates in *Neurospora*. (Notice in the results table that a mutant can grow only if supplied with a compound made *after* the defective step because this bypasses the defect.)

Data from A. M. Srb and N. H. Horowitz, The ornithine cycle in *Neurospora* and its genetic control, *Journal of Biological Chemistry* 154:129–139 (1944).

WHAT IF? *Suppose the experiment had shown that class I mutants could grow only in MM supplemented by ornithine or arginine and that class II mutants could grow in MM supplemented by citrulline, ornithine, or arginine. What conclusions would the researchers have drawn from those results regarding the biochemical pathway and the defect in class I and class II mutants?*

protein-building instructions. This type of RNA molecule is called **messenger RNA (mRNA)** because it carries a genetic message from the DNA to the protein-synthesizing machinery of the cell. (Transcription is the general term for the synthesis of *any* kind of RNA on a DNA template. Later, you will learn about some other types of RNA produced by transcription.)

Translation is the synthesis of a polypeptide using the information in the mRNA. During this stage, there is a change in language: The cell must translate the nucleotide sequence of an mRNA molecule into the amino acid sequence of a polypeptide. The sites of translation are **ribosomes**, molecular complexes that facilitate the orderly linking of amino acids into polypeptide chains.

Transcription and translation occur in all organisms. Because most studies have involved bacteria and eukaryotic cells, they are our main focus in this chapter. Our understanding of transcription and translation in archaea lags behind, but we do know that archaeal cells share some features of gene expression with bacteria and others with eukaryotes.

The basic mechanics of transcription and translation are similar for bacteria and eukaryotes, but there is an important difference in the flow of genetic information within the cells. Bacteria do not have nuclei. Therefore, nuclear membranes do not separate bacterial DNA and mRNA from ribosomes and the other protein-synthesizing equipment **(Figure 17.4a)**. This lack of compartmentalization allows translation of an mRNA to begin while its transcription is still in progress, as you'll see later. By contrast, eukaryotic cells have nuclei. The presence of a nuclear envelope separates transcription from translation in space and time **(Figure 17.4b)**. Transcription occurs in the nucleus, but the mRNA must be transported to the cytoplasm for translation. In eukaryotes, before RNA transcripts from protein-coding genes can leave the nucleus, they are modified in various ways to produce the final, functional mRNA. The transcription of a protein-coding eukaryotic gene results in *pre-mRNA*, and further RNA processing yields the finished mRNA. The initial RNA transcript from any gene, including those specifying RNA that is not translated into protein (such as tRNA and rRNA, which will be described in Concept 17.4), is more generally called a **primary transcript**.

To summarize: Genes program protein synthesis via genetic messages in the form of messenger RNA. Put another way, cells are governed by a molecular chain of command with a directional flow of genetic information:

This idea, that the flow of information went only one way, was named the *central dogma* by Francis Crick in 1956. In the 1970s, however, scientists were surprised to discover some enzymes that use RNA molecules as templates for DNA synthesis, an example of information flow from RNA to DNA

▼ **Figure 17.4 Overview: the roles of transcription and translation in the flow of genetic information.** In a cell, inherited information flows from DNA to RNA to protein. The two main stages of information flow are transcription and translation. A miniature version of part (a) or (b) accompanies several figures later in the chapter as an orientation diagram to help you see where a particular figure fits into the overall scheme of gene expression.

(a) Bacterial cell. In a bacterial cell, which lacks a nucleus, mRNA produced by transcription is immediately translated without additional processing.

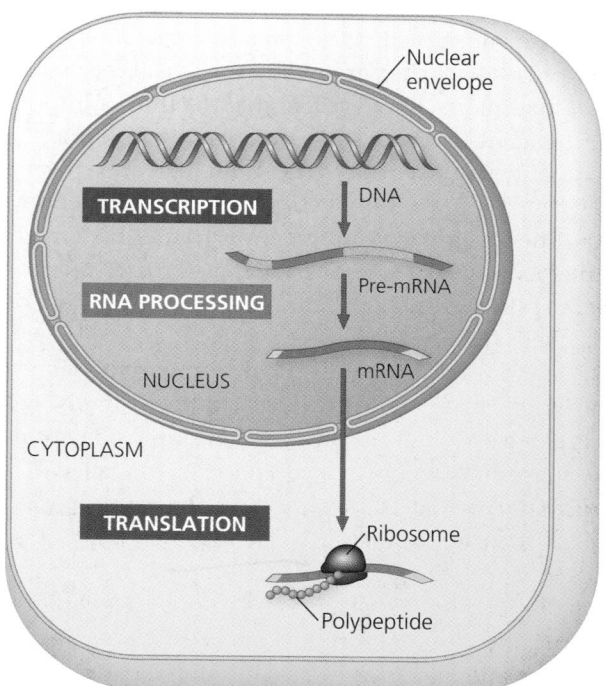

(b) Eukaryotic cell. The nucleus provides a separate compartment for transcription. The original RNA transcript, called pre-mRNA, is processed in various ways before leaving the nucleus as mRNA.

→ **Mastering Biology**
Animation: Overview of Protein Synthesis in Bacteria
Animation: Overview of Protein Synthesis in Eukaryotes

(see Concept 19.2). Still, these exceptions do not discredit the idea that, in general, genetic information flows from DNA to RNA to protein. In the next section, we discuss how the instructions for assembling amino acids into a specific order are encoded in nucleic acids.

The Genetic Code

When biologists began to suspect that the instructions for protein synthesis were encoded in DNA, they recognized a problem: There are only four nucleotide bases to specify 20 amino acids. Thus, the genetic code cannot be a language like Chinese, where each written symbol corresponds to a word. How many nucleotides, then, would turn out to correspond to an amino acid?

Codons: Triplets of Nucleotides

If each kind of nucleotide base were translated into an amino acid, only four amino acids could be specified, one per nucleotide base. Would a language of two-letter code words suffice? The two-nucleotide sequence AG, for example, could specify one amino acid, and GT could specify another. Since there are four possible nucleotide bases in each position, this would give us 16 (that is, 4×4, or 4^2) possible arrangements—still not enough to code for all 20 amino acids.

Triplets of nucleotide bases are the smallest units of uniform length that can code for all the amino acids. If each arrangement of three consecutive nucleotide bases specifies an amino acid, there can be 64 (that is, 4^3) possible code words—more than enough to specify all the amino acids. Experiments have verified that the flow of information from gene to protein is based on a **triplet code**: The genetic instructions for a polypeptide chain are written in the DNA as a series of nonoverlapping, three-nucleotide words. The series of words in a gene is transcribed into a complementary series of nonoverlapping, three-nucleotide words in mRNA, which is then translated into a chain of amino acids **(Figure 17.5)**.

During transcription, the gene determines the sequence of nucleotide bases along the length of the RNA molecule that is being synthesized. For each gene, only one of the two DNA strands is transcribed. This strand is called the **template strand** because it provides the pattern, or template, for the sequence of nucleotides in an RNA transcript. For any given gene, the same strand is used as the template every time that gene is transcribed. However, farther along on the same chromosomal DNA molecule, the opposite strand may function as the template for a different gene. Particular DNA sequences associated with a gene determine how the enzyme that transcribes genes is oriented when it binds, and this establishes which strand will be used as the template.

An mRNA molecule is complementary rather than identical to its DNA template because RNA nucleotides are assembled on the template according to base-pairing rules (see Figure 17.5). The pairs are similar to those that form during DNA replication, except that U (the RNA substitute for T) pairs with A and the mRNA nucleotides contain ribose instead of deoxyribose. Like a new strand of DNA, the RNA molecule is synthesized in an antiparallel direction to

▼ **Figure 17.5 The triplet code.** For each gene, one DNA strand functions as a template for transcription of RNAs, such as mRNA. The base-pairing rules for DNA synthesis also guide transcription, except that RNA is made with uracil (U) instead of thymine (T). During translation, the mRNA is read as a sequence of nucleotide triplets, called codons. Each codon specifies an amino acid to be added to the growing polypeptide chain. The mRNA is read in the 5′ → 3′ direction.

VISUAL SKILLS *By convention, the nontemplate strand, also called the coding strand, is used to represent a DNA sequence. Write the sequence of the mRNA strand and the nontemplate strand—in both cases reading from 5′ to 3′—and compare them. Why do you think this convention was adopted? (Hint: Why is this called the coding strand?)*

→ **Mastering Biology** MP3 Tutor: DNA to RNA to Protein

the template strand of DNA. (To review what is meant by "antiparallel" and the 5′ and 3′ ends of a nucleic acid chain, see Figure 16.7.) In the example in Figure 17.5, the nucleotide triplet ACC along the DNA template strand (written as 3′-ACC-5′) provides a template for 5′-UGG-3′ in the mRNA molecule. The mRNA nucleotide triplets are called **codons**, and they are customarily written in the 5′ → 3′ direction. In our example, UGG is the codon for the amino acid tryptophan (abbreviated Trp, or W). The term *codon* is also used for the DNA nucleotide triplets along the *nontemplate* strand. These codons are complementary to the template strand and thus identical in sequence to the mRNA (except for T's instead of U's). For this reason, the nontemplate DNA strand is often called the **coding strand**; by convention, the sequence of the coding strand is used when a gene's sequence is reported.

During translation, the sequence of codons along an mRNA molecule is decoded, or translated, into a sequence of amino acids making up a polypeptide chain. The codons are read by the translation machinery in the 5′ → 3′

direction along the mRNA. Each codon specifies which one of the 20 amino acids will be incorporated at the corresponding position along a polypeptide. Because codons are nucleotide triplets, the number of nucleotides making up a genetic message must be three times the number of amino acids in the protein product. For example, 300 nucleotides along an mRNA strand code for 100 amino acids in the polypeptide it encodes.

Cracking the Code

Molecular biologists cracked the genetic code of life in the early 1960s when a series of elegant experiments disclosed the amino acid translations of each of the mRNA codons. The first codon was deciphered in 1961 by Marshall Nirenberg, of the National Institutes of Health, and colleagues. They synthesized an artificial mRNA made up of only uracil-bearing nucleotides, thus containing only one codon (UUU) over and over. When added to a test-tube mixture that contained all 20 amino acids, ribosomes, and the other components required for protein synthesis, the "poly-U" mRNA was translated into a polypeptide containing many units of the phenylalanine (Phe, or F) amino acids, strung together as a polyphenylalanine chain. Thus, Nirenberg determined that the mRNA codon UUU specifies the amino acid phenylalanine. Soon, the amino acids specified by the codons AAA, GGG, and CCC were similarly identified.

Although more elaborate techniques were required to decode mixed triplets such as AUA and CGA, all 64 codons were deciphered by the mid-1960s. As **Figure 17.6** shows, 61 of the 64 triplets code for amino acids. The other three codons act as "stop" signals, or termination codons, marking the end of translation. Notice that the codon AUG has a dual function: It codes for the amino acid methionine (Met, or M) and also functions as a "start" signal, or initiation codon. Genetic messages usually begin with the mRNA codon AUG, which signals the protein-synthesizing machinery to begin translating the mRNA at that location. (Because AUG also stands for methionine, polypeptide chains all begin with methionine when they are synthesized. However, an enzyme may subsequently remove this starter amino acid from the chain.)

Notice in Figure 17.6 that there is redundancy in the genetic code, but no ambiguity. For example, although codons GAA and GAG both specify glutamic acid (redundancy), neither of them ever specifies any other amino acid (no ambiguity). The redundancy in the code is not altogether random. In many cases, codons that are synonyms for a particular amino acid differ only in the third nucleotide base of the triplet. We will see why this is later in the chapter.

Our ability to extract the intended message from a written language depends on reading the symbols in the correct groupings—that is, in the correct **reading frame**. Consider this statement: "The red dog ate the bug." Group the letters incorrectly by starting at the wrong point, and the result

▼ **Figure 17.6 The codon table for mRNA.** The three nucleotide bases of an mRNA codon are designated here as the first, second, and third bases, reading in the 5' → 3' direction along the mRNA. The codon AUG not only stands for the amino acid methionine (Met, or M) but also functions as a "start" signal for ribosomes to begin translating the mRNA at that point. Three of the 64 codons function as "stop" signals, marking where ribosomes end translation. Both one- and three-letter codes are shown for the amino acids; see Figure 5.14 for their full names.

VISUAL SKILLS *A segment in the middle of an mRNA has the sequence 5'-AGAGAACCGCGA-3'. Using the codon table, translate this sequence, assuming the first three nucleotides are a codon.*

→ **Mastering Biology** Animation: Translation: The Genetic Code

will probably be gibberish: for example, "her edd oga tet heb ug." The reading frame is also important in the molecular language of cells. The short stretch of polypeptide shown in Figure 17.5, for instance, will be made correctly only if the mRNA nucleotides are read from left to right (5' → 3') in the groups of three shown in the figure: UGG UUU GGC UCA. Although a genetic message is written with no spaces between the codons, the cell's protein-synthesizing machinery reads the message as a series of nonoverlapping three-letter words. The message is *not* read as a series of overlapping words—UGGUUU, and so on—which would convey a very different message.

Evolution of the Genetic Code

EVOLUTION The genetic code is nearly universal, shared by organisms from the simplest bacteria to the most complex plants and animals. The mRNA codon CCG, for instance, is translated as the amino acid proline in all organisms whose

▼ **Figure 17.7 Evidence for evolution: expression of genes from different species.** Because diverse forms of life share a common genetic code due to their shared ancestry, one species can be programmed to produce proteins characteristic of a second species by introducing DNA from the second species into the first.

 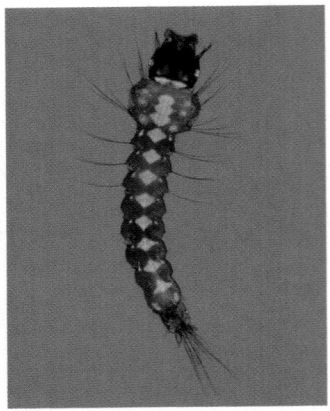

(a) Tobacco plant expressing a firefly gene. The yellow glow is produced by a chemical reaction catalyzed by the protein product of the firefly gene.

(b) Mosquito larva expressing a jellyfish gene. The gene codes for green fluorescent protein (GFP) and is inserted into organisms as a *reporter gene* so researchers can tell if a gene of interest is being expressed.

⊙ **Mastering Biology** Video: GFP Transgenic Mice

genetic code has been examined. In laboratory experiments, genes can be transcribed and translated after being transplanted from one species to another, sometimes with quite striking results, as shown in **Figure 17.7**. Bacteria can be programmed by the insertion of human genes to synthesize certain human proteins for medical use, such as insulin. Such applications have produced many exciting developments in the area of biotechnology (see Concept 20.4).

The evolutionary significance of the code's near universality is clear. A language shared by all living things must have been operating very early in the history of life—early enough to be present in the common ancestor of all present-day organisms. A shared genetic vocabulary is a reminder of the kinship of all life.

CONCEPT CHECK 17.1

1. **MAKE CONNECTIONS** In a research article about alkaptonuria published in 1902, Garrod suggested that humans inherit two "characters" (alleles) for a particular enzyme and that both parents must contribute a faulty version for the offspring to have alkaptonuria. Today, would this disorder be called dominant or recessive? (See Concept 14.4.)

2. Describe the polypeptide product you would expect from a poly-G mRNA that is 30 nucleotides long.

3. **DRAW IT** The template strand of a gene contains the sequence 3'-TTCAGTCGT-5'. Suppose that the nontemplate sequence was transcribed instead of the template sequence. Draw the nontemplate sequence in 3' to 5' order. Then draw the mRNA sequence and translate it using Figure 17.6. Predict how well the protein synthesized from the nontemplate strand would function, if at all.

For suggested answers, see Appendix A.

Transcription is the DNA-directed synthesis of RNA: *A Closer Look*

Now that we have considered the linguistic logic and evolutionary significance of the genetic code, we are ready to reexamine transcription, the first stage of gene expression, in greater detail.

Molecular Components of Transcription

Messenger RNA, the carrier of information from DNA to the cell's protein-synthesizing machinery, is transcribed from the template strand of a gene. An enzyme called an **RNA polymerase** pries the two strands of DNA apart and joins together RNA nucleotides complementary to the DNA template strand, thus elongating the RNA polynucleotide **(Figure 17.8)**. Like the DNA polymerases that function in DNA replication, RNA polymerases can assemble a polynucleotide only in its 5' → 3' direction, adding onto its 3' end. Unlike DNA polymerases, however, RNA polymerases are able to start a chain from scratch; they don't need to add the first nucleotide onto a pre-existing primer.

Specific sequences of nucleotides along the DNA mark where transcription of a gene begins and ends. The DNA sequence where RNA polymerase attaches and initiates transcription is known as the **promoter**; in bacteria, the sequence that signals the end of transcription is called the **terminator**. (The termination mechanism in eukaryotes will be described later.) Molecular biologists refer to the direction of transcription as "downstream" and the other direction as "upstream." These terms are also used to describe the positions of nucleotide sequences within the DNA or RNA. Thus, the promoter sequence in DNA is said to be upstream from the terminator. The stretch of DNA downstream from the promoter that is transcribed into an RNA molecule is called a **transcription unit**.

Bacteria have a single type of RNA polymerase that synthesizes not only mRNA but also other types of RNA that function in gene expression, such as ribosomal RNA. In contrast, eukaryotes have at least three types of RNA polymerase in their nuclei; the one used for pre-mRNA synthesis is called RNA polymerase II. The other RNA polymerases transcribe RNA molecules that are not translated into protein. In the discussion that follows, we start with the features of mRNA synthesis common to both bacteria and eukaryotes and then describe some key differences.

Synthesis of an RNA Transcript

The three stages of transcription, as shown in Figure 17.8 and described next, are initiation, elongation, and termination of the RNA chain. Study Figure 17.8 to familiarize yourself with the stages and the terms used to describe them.

RNA Polymerase Binding and Initiation of Transcription

The promoter sequence of a gene includes within it the transcription **start point**—the nucleotide where RNA polymerase actually begins synthesizing the mRNA—and typically extends several dozen or so nucleotide pairs upstream from the start point (**Figure 17.9**). Based on interactions with proteins (transcription factors), RNA polymerase binds in a precise location and orientation on the promoter. This binding determines where transcription starts and the direction it will travel, thus which strand of DNA is used as the template.

▼ **Figure 17.8 The stages of transcription: initiation, elongation, and termination.** This general depiction of transcription applies to both bacteria and eukaryotes, but the details of termination differ, as described in the text. Also, in a bacterium, the RNA transcript is immediately usable as mRNA; in a eukaryote, the RNA transcript must first undergo processing.

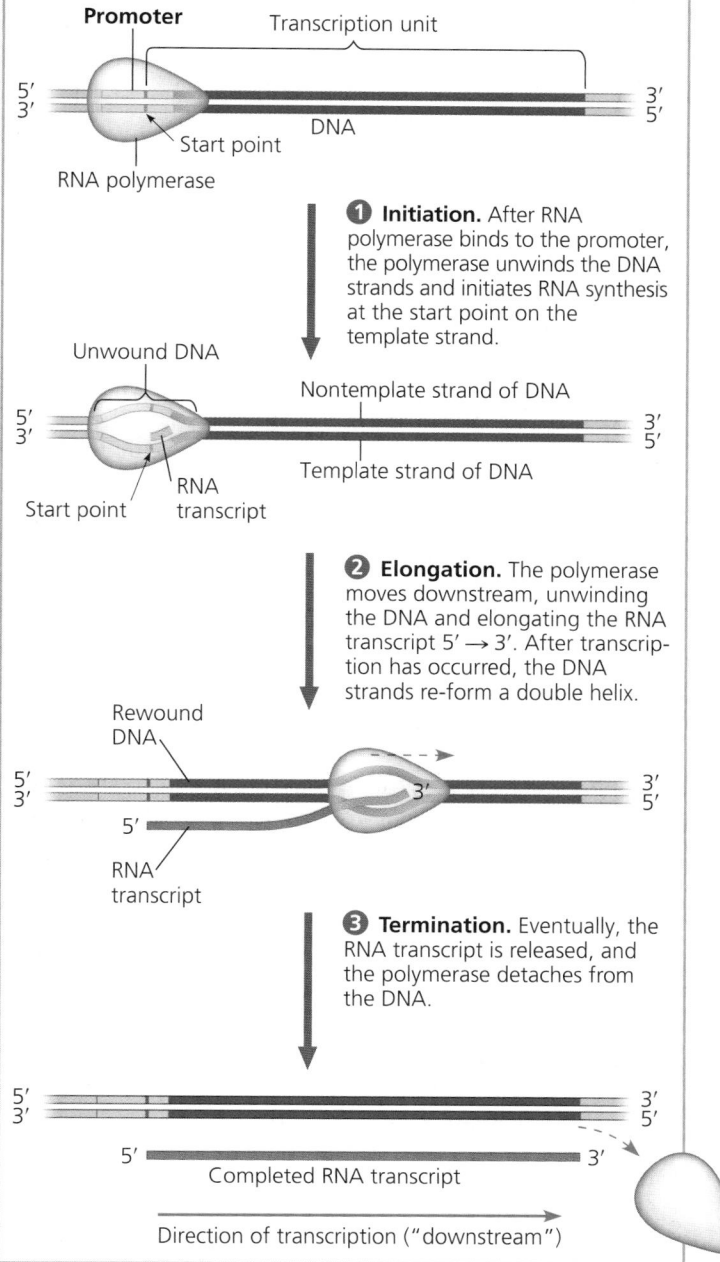

MAKE CONNECTIONS *Compare the use of a template strand during transcription and replication. See Figure 16.18.*

→ **Mastering Biology** **Animation: Overview of Transcription**
Animation: Overview of Transcription in Bacteria

▼ **Figure 17.9 The initiation of transcription at a eukaryotic promoter.** In eukaryotic cells, proteins called transcription factors mediate the initiation of transcription by RNA polymerase II.

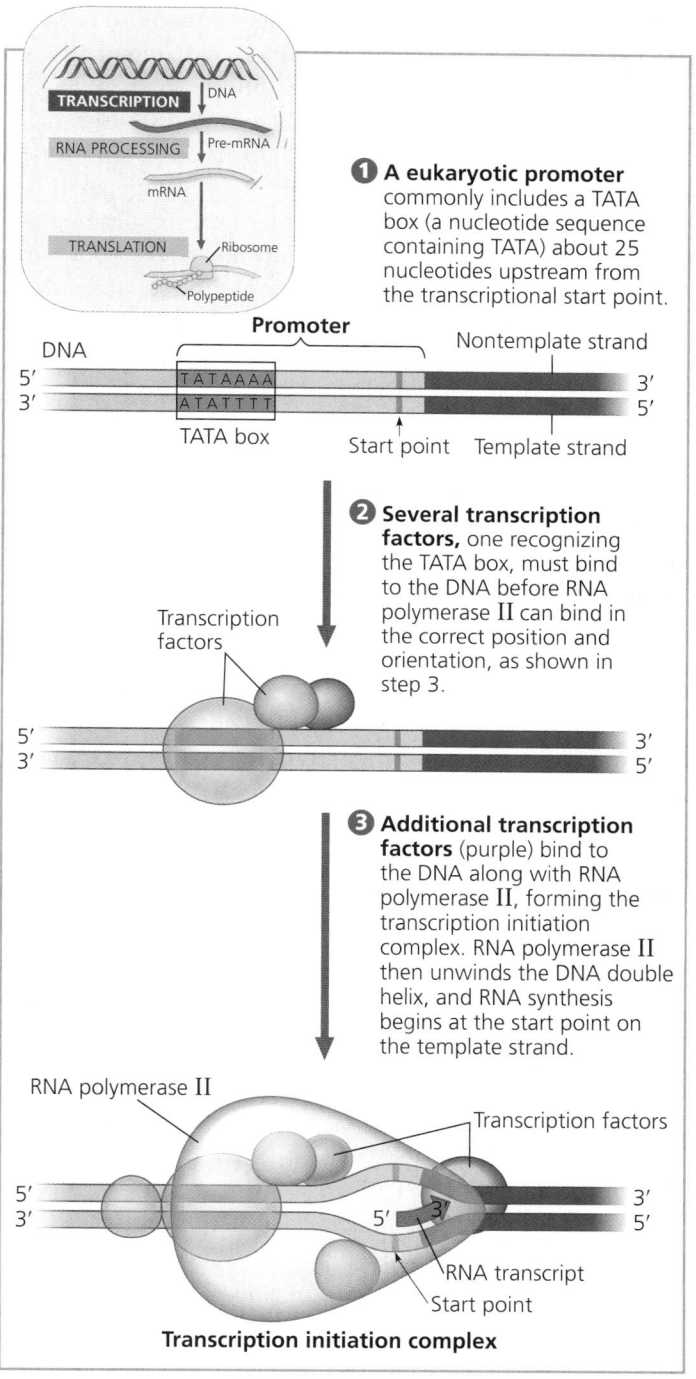

Transcription initiation complex

❓ *Explain how the interaction of RNA polymerase with the promoter would differ if the figure showed transcription initiation for bacteria.*

Certain sections of a promoter sequence are especially important for binding RNA polymerase in a way that ensures that transcription will begin at the correct start point. In bacteria, part of the RNA polymerase itself specifically recognizes and binds to the promoter. In eukaryotes, a collection of proteins called **transcription factors** (purple proteins in Figure 17.9), help guide the binding of RNA polymerase and the initiation of transcription. Only after transcription factors are attached to the promoter does RNA polymerase II bind to it. The whole complex of transcription factors and RNA polymerase II bound to the promoter is called a **transcription initiation complex**. Figure 17.9 shows the role of transcription factors and a crucial promoter DNA sequence called the **TATA box** in forming the initiation complex at a eukaryotic promoter.

The interaction between eukaryotic RNA polymerase II and transcription factors is an example of the importance of protein-protein interactions in controlling eukaryotic transcription. Once the appropriate transcription factors are firmly attached to the promoter DNA and the polymerase is bound to them in the correct orientation on the DNA, the enzyme unwinds the two DNA strands and begins transcribing the template strand at the start point.

Elongation of the RNA Strand

As RNA polymerase moves along the DNA, it untwists the double helix, exposing about 10–20 DNA nucleotides at a time for pairing with RNA nucleotides **(Figure 17.10)**. The enzyme adds nucleotides to the 3′ end of the growing RNA molecule as it moves along the double helix. As transcription proceeds forward, the newly synthesized RNA molecule behind the RNA polymerase peels away from its DNA template, and the DNA double helix re-forms. Transcription progresses at a rate of about 40 nucleotides per second in eukaryotes.

A single gene can be transcribed simultaneously by several molecules of RNA polymerase following each other like trucks in a convoy. A growing strand of RNA trails off from each polymerase, with the length of each new strand reflecting how far along the template the enzyme has traveled from the start point (see the mRNA molecules in Figure 17.23). The congregation of many polymerase molecules simultaneously transcribing a single gene increases the amount of mRNA transcribed from it, which helps the cell make the encoded protein in large amounts.

Termination of Transcription

Bacteria and eukaryotes differ in the way they terminate transcription. In bacteria, transcription proceeds through a terminator sequence in the DNA. The transcribed terminator (an RNA sequence) functions as the termination signal, causing the polymerase to detach from the DNA and release the transcript, which requires no further modification before translation. In eukaryotes, RNA polymerase II

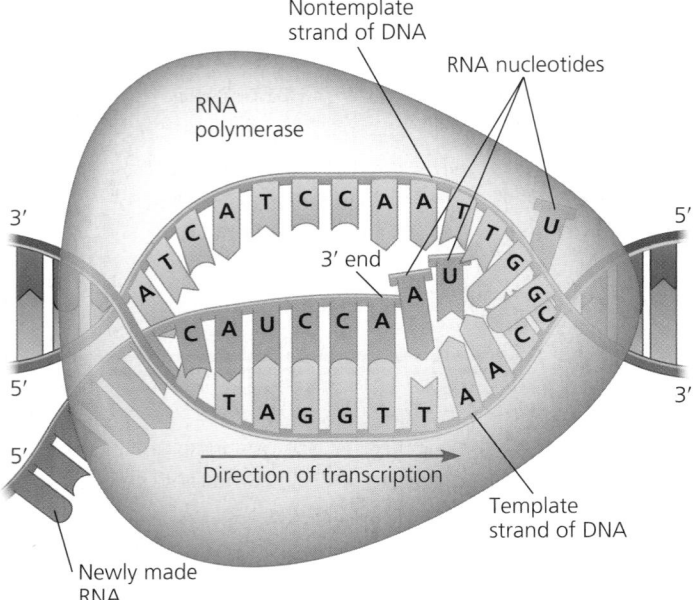

▼ **Figure 17.10 Transcription elongation.** RNA polymerase moves along the DNA template strand, joining complementary RNA nucleotides to the 3′ end of the growing RNA transcript. Behind the polymerase, the new RNA peels away from the template strand, which re-forms a double helix with the nontemplate strand.

→ **Mastering Biology BioFlix® Animation: Transcription**
Animation: Elongation of the RNA Strand

transcribes a sequence on the DNA called the polyadenylation signal sequence, which specifies a polyadenylation signal (AAUAAA) in the pre-mRNA. This is called a "signal" because once this stretch of six RNA nucleotides appears, it is immediately bound by certain proteins in the nucleus. Then, at a point about 10–35 nucleotides downstream from the AAUAAA, these proteins cut the RNA transcript free from the polymerase, releasing the pre-mRNA. The pre-mRNA then undergoes processing, the topic of the next section. Although that cleavage marks the end of the mRNA, the RNA polymerase II continues to transcribe. Enzymes begin to degrade the RNA made after cleavage, starting at its newly exposed 5′ end. The polymerase continues transcribing, pursued by the enzymes, until they catch up to the polymerase and it dissociates from the DNA.

CONCEPT CHECK 17.2

1. What is a promoter? Is it located at the upstream or downstream end of a transcription unit?

2. What enables RNA polymerase to start transcribing a gene at the right place on the DNA in a bacterial cell? In a eukaryotic cell?

3. **WHAT IF?** Suppose X-rays caused a sequence change in the TATA box of a particular gene's promoter. How would that affect transcription of the gene? (See Figure 17.9.)

For suggested answers, see Appendix A.

Eukaryotic cells modify RNA after transcription

Enzymes in the eukaryotic nucleus modify pre-mRNA in specific ways before the genetic message is dispatched to the cytoplasm. During this **RNA processing**, both ends of the primary transcript are altered. Also, in most cases, certain interior sections of the RNA molecule are cut out and the remaining parts spliced together. These modifications produce an mRNA molecule ready for translation.

Alteration of mRNA Ends

Each end of a pre-mRNA molecule is modified in a particular way **(Figure 17.11)**. The 3′ end, which is synthesized first, receives a **5′ cap**, a modified form of a guanine (G) nucleotide added onto the 5′ end after transcription of the first 20–40 nucleotides have been transcribed. The 3′ end of the pre-mRNA molecule is also modified before the mRNA exits the

nucleus. Recall that the pre-mRNA is cut and released soon after the polyadenylation signal, AAUAAA, is transcribed. At the 3′ end, an enzyme then adds 50–250 more adenine (A) nucleotides, forming a **poly-A tail**. The 5′ cap and poly-A tail share several important functions. First, they seem to facilitate the export of the mature mRNA from the nucleus. Second, they help protect the mRNA from degradation by hydrolytic enzymes. And third, they help ribosomes attach to the 5′ end of the mRNA once it reaches the cytoplasm. In addition to the cap and tail on a eukaryotic mRNA molecule, Figure 17.11 shows the untranslated regions (UTRs) at the 5′ and 3′ ends of the mRNA (referred to as the 5′ UTR and 3′ UTR). The UTRs are parts of the mRNA that will not be translated into protein, but they have other functions, such as ribosome binding.

Split Genes and RNA Splicing

A remarkable stage of RNA processing in the eukaryotic nucleus is called **RNA splicing (Figure 17.12)**, in which large portions of the RNA primary transcript molecules are removed and the remaining portions are reconnected. This

▼ **Figure 17.11 RNA processing: addition of the 5′ cap and poly-A tail.** Enzymes modify the two ends of a eukaryotic pre-mRNA molecule. The modified ends may promote the export of mRNA from the

nucleus, and they help protect the mRNA from degradation. When the mRNA reaches the cytoplasm, the modified ends, in conjunction with certain cytoplasmic proteins, help with ribosome attachment. The 5′ cap and

poly-A tail are not translated into protein, nor are the regions called the 5′ untranslated region (5′ UTR) and 3′ untranslated region (3′ UTR). The pink segments are introns, which will be described shortly (see Figure 17.12).

→ **Mastering Biology Animation: Adding the 5′ Cap and Poly-A Tail**

▼ **Figure 17.12 RNA processing: RNA splicing.** The RNA molecule shown here codes for β-globin, one of the polypeptides of hemoglobin. The numbers under the RNA refer to codons; β-globin is 146 amino acids

long. The β-globin gene and its pre-mRNA transcript have three exons, corresponding to sequences that will exit the nucleus as mRNA. (The 5′ UTR and 3′ UTR do not code for proteins, but they are parts of

exons because they remain in the mRNA.) During RNA processing, the introns are cut out and the exons spliced together. In many genes, the introns are much longer than the exons.

DRAW IT *On the mRNA, indicate the sites of the start and stop codons.*

→ **Mastering Biology BioFlix® Animation: RNA Processing**

cut-and-paste job is similar to editing a movie. The average length of a transcription unit along a human DNA molecule is about 27,000 nucleotide pairs, so the primary RNA transcript is also that long. However, the average-sized protein of 400 amino acids requires only 1,200 nucleotides in RNA to code for it. (Remember, each amino acid is encoded by a *triplet* of nucleotides.) This is because most eukaryotic genes and their RNA transcripts have long noncoding stretches of nucleotides, regions that are not translated. Even more surprising is that most of these noncoding sequences are interspersed between coding segments of the gene and thus between coding segments of the pre-mRNA. In other words, the sequence of DNA nucleotides that codes for a eukaryotic polypeptide is usually not continuous; it is split into segments. The noncoding segments of nucleic acid that lie between coding regions are called <u>int</u>ervening sequences, or **introns**. The other regions are called **exons**, because they are eventually <u>ex</u>pressed, usually by being translated into amino acid sequences. (Exceptions include the UTRs of the exons at the ends of the RNA, which make up part of the mRNA but are not translated into protein. Because of these exceptions, you may prefer to think of exons as sequences of RNA that *ex*it the nucleus.) The terms *intron* and *exon* are used for both RNA sequences and the DNA sequences that specify them.

In making a primary transcript from a gene, RNA polymerase II transcribes both introns and exons from the DNA, but the mRNA molecule that enters the cytoplasm is an abridged version. In the process of RNA splicing, the introns are cut out from the molecule and the exons joined together, forming an mRNA molecule with a continuous coding sequence.

How is pre-mRNA splicing carried out? The removal of introns is accomplished by a large complex made of proteins and small RNAs called a **spliceosome**. This complex binds to several short nucleotide sequences along an intron, including key sequences at each end **(Figure 17.13)**. The intron is then released (and rapidly degraded), and the spliceosome joins together the two exons that flanked the intron. The small RNAs in the spliceosome not only participate in spliceosome assembly and splice site recognition, but also catalyze the splicing reaction; like proteins, RNAs can act as catalysts.

> **Mastering Biology**
> **Interview with Joan Steitz:**
> **Studying RNA, her "favorite molecule"**

Ribozymes

The idea of a catalytic role for the RNAs in the spliceosome arose from the discovery of **ribozymes**, RNA molecules that function as enzymes. In some organisms, RNA splicing can occur without proteins or even additional RNA molecules: The intron RNA functions as a ribozyme and catalyzes its own removal. For example, in the ciliate protist *Tetrahymena*, self-splicing occurs in the production of ribosomal RNA (rRNA),

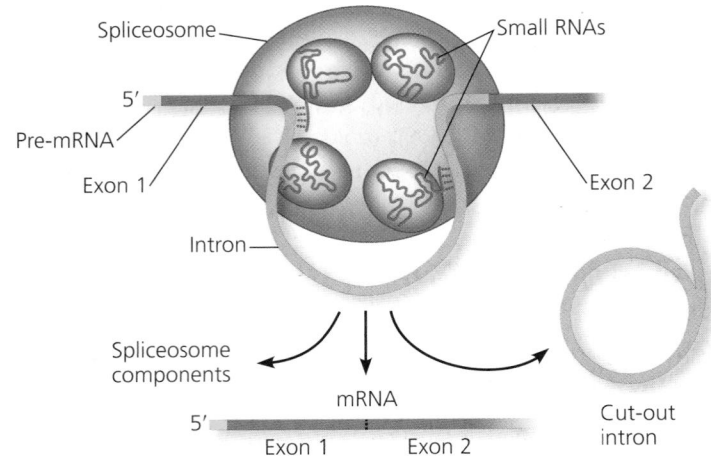

▼ **Figure 17.13 A spliceosome splicing a pre-mRNA.** The diagram shows a portion of a pre-mRNA transcript, with an intron (pink) flanked by two exons (red). Small RNAs within the spliceosome base-pair with nucleotides at specific sites along the intron. Next, small spliceosome RNAs catalyze cutting of the pre-mRNA and the splicing together of the exons, releasing the intron for rapid degradation.

➔ **Mastering Biology Animation: A Spliceosome**

a component of the organism's ribosomes. The pre-rRNA actually removes its own introns! The discovery of ribozymes invalidated the idea that all biological catalysts are proteins.

Three properties of RNA enable some RNA molecules to function as enzymes. First, because RNA is single-stranded, a region of an RNA molecule may base-pair, in an antiparallel arrangement, with a complementary region elsewhere in the same molecule; this gives the molecule a particular three-dimensional structure. A specific structure is essential to the catalytic function of ribozymes, just as it is for enzymatic proteins. Second, like certain amino acids in an enzymatic protein, some of the bases in RNA contain functional groups that can participate in catalysis. Third, the ability of RNA to hydrogen-bond with other nucleic acid molecules (either RNA or DNA) adds specificity to its catalytic activity. For example, complementary base pairing between the RNA of the spliceosome and the RNA of a primary RNA transcript precisely locates the region where the ribozyme catalyzes splicing. Later in this chapter, you will see how these properties of RNA also allow it to perform important noncatalytic roles in the cell, such as recognition of the three-nucleotide codons on mRNA.

The Functional and Evolutionary Importance of Introns

EVOLUTION Whether or not RNA splicing and the presence of introns have provided selective advantages during evolutionary history is a matter of some debate. In any case, it is informative to consider their possible adaptive benefits. Specific functions have not been identified for most introns, but at least some contain sequences that regulate gene expression, and many affect gene products.

One important consequence of the presence of introns in genes is that a single gene can encode more than one kind of polypeptide. Many genes are known to give rise to two or more different polypeptides, depending on which segments are treated as exons during RNA processing; this is called **alternative RNA splicing** (see Figure 18.14). Results from the Human Genome Project (discussed in Concept 21.1) suggest that alternative RNA splicing is one reason humans can get along with about the same number of genes as a nematode (roundworm). Because of alternative splicing, the number of different protein products an organism produces can be much greater than its number of genes.

Proteins often have a modular architecture consisting of discrete structural and functional regions called **domains**. One domain of an enzyme, for example, might include the active site, while another might allow the enzyme to bind to a cellular membrane. In quite a few cases, different exons code for the different domains of a protein **(Figure 17.14)**.

The presence of introns in a gene may facilitate the evolution of new and potentially beneficial proteins as a result of a process known as *exon shuffling* (see Figure 21.16). Introns increase the probability of crossing over between the exons of alleles of a gene—simply by providing more terrain for crossovers without interrupting coding sequences. This might result in new combinations of exons and proteins with altered structure and function. We can also imagine the occasional mixing and matching of exons between completely different (nonallelic) genes. Exon shuffling of either sort could lead to new proteins with novel combinations of functions. While most of the shuffling would result in nonbeneficial changes, occasionally a beneficial variant might arise.

▼ **Figure 17.14 Correspondence between exons and protein domains.**

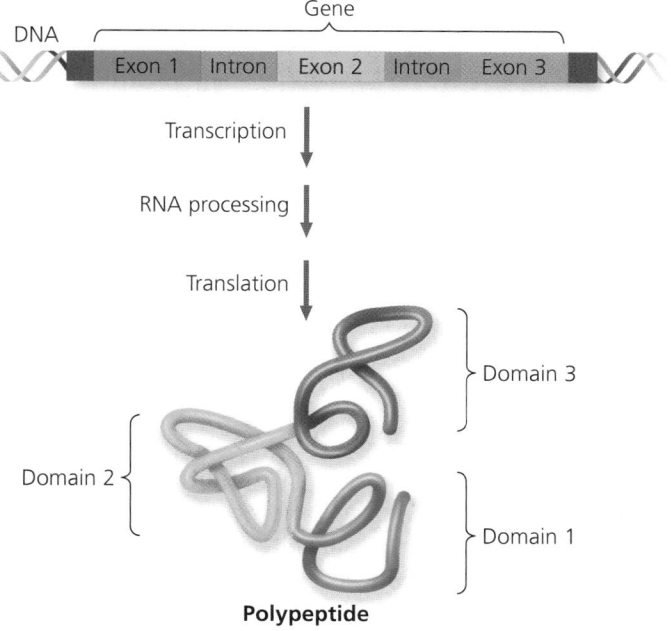

Polypeptide

CONCEPT CHECK 17.3

1. Given that there are about 20,000 human protein-coding genes, how can human cells make 75,000–100,000 different proteins?

2. Compare RNA splicing to how you would watch a pre-recorded television show. What would introns correspond to in this analogy?

3. **WHAT IF?** What would be the effect of treating cells with an agent that removed the 5′ cap from mRNAs?

For suggested answers, see Appendix A.

CONCEPT 17.4

Translation is the RNA-directed synthesis of a polypeptide: *A Closer Look*

We will now examine how genetic information flows from mRNA to protein—the process of translation **(Figure 17.15)**. We'll focus on the basic steps of translation that occur in both bacteria and eukaryotes, while pointing out key differences.

▼ **Figure 17.15 Translation: the basic concept.** As a molecule of mRNA is moved through a ribosome, codons are translated into amino acids, one by one. The translators, or interpreters, are tRNA molecules, each type with a specific anticodon at one end and a corresponding amino acid at the other end. A tRNA adds its amino acid cargo to a growing polypeptide chain when the anticodon hydrogen-bonds to the complementary codon on the mRNA.

→ **Mastering Biology** BioFlix® Animation: Translation
Animation: Overview of Translation

Molecular Components of Translation

In the process of translation, a cell "reads" a genetic message and builds a polypeptide accordingly. The message is a series of codons along an mRNA molecule, and the translator is called a **transfer RNA (tRNA)**. The function of a tRNA is to transfer an amino acid from the cytoplasmic pool of amino acids to a growing polypeptide in a ribosome. A cell keeps its cytoplasm stocked with all 20 amino acids, either by synthesizing them from other compounds or by taking them up from the surrounding solution. The ribosome, a structure made of proteins and RNAs, adds each amino acid brought to it by a tRNA to the growing end of a polypeptide chain (see Figure 17.15).

Translation is simple in principle but complex in its biochemistry and mechanics, especially in the eukaryotic cell. In dissecting translation, we'll focus on the slightly less complicated version of the process that occurs in bacteria. We'll first look at the major players in this process.

The Structure and Function of Transfer RNA

The key to translating a genetic message into a specific amino acid sequence is the fact that each tRNA molecule enables translation of a given mRNA codon into a certain amino acid. This is possible because a tRNA bears a specific amino acid at one end of its three-dimensional structure, while at the other end is a nucleotide triplet that can base-pair with the complementary codon on mRNA.

A tRNA molecule consists of a single RNA strand that is only about 80 nucleotides long (compared to hundreds of nucleotides for most mRNA molecules). The presence of complementary stretches of nucleotide bases that can hydrogen-bond to each other allows this single strand to fold back on itself and form a molecule with a particular 3-D structure. Flattened into one plane to clarify this base pairing, a tRNA molecule looks like a cloverleaf **(Figure 17.16a)**. The tRNA actually twists and folds into a compact 3-D structure that is roughly L-shaped **(Figure 17.16b)**, with the 5′ and 3′ ends of the linear tRNA both located near one end of the structure. The protruding 3′ end acts as the attachment site for an amino acid. The loop extending from the other end of the L includes the **anticodon**, the particular nucleotide triplet that base-pairs to a specific mRNA codon. Thus, the structure of a tRNA molecule fits its function.

Anticodons are conventionally written 3′ → 5′ to align properly with codons written 5′ → 3′ (see Figure 17.15). (For base pairing, RNA strands must be antiparallel, like DNA.) As an example of how tRNAs work, consider the mRNA codon 5′-GGC-3′, which is translated as the amino acid glycine. The tRNA that base-pairs with this codon by hydrogen bonding has 3′-CCG-5′ as its anticodon and carries glycine at its other end (see the incoming tRNA approaching the ribosome in Figure 17.15). As an mRNA molecule is moved through

▼ **Figure 17.16 The structure of transfer RNA (tRNA).**

(a) **Two-dimensional structure.** The four base-paired regions and three loops are characteristic of all tRNAs, as is the base sequence of the amino acid attachment site at the 3′ end. The anticodon triplet is unique to each tRNA type, as are some sequences in the other two loops. (The asterisks mark bases that have been chemically modified, a characteristic of tRNA. The modified bases contribute to tRNA function in a way that is not yet understood.)

(b) **Three-dimensional structure**

(c) **Symbol** used in this book

VISUAL SKILLS *Look at the tRNA shown in this figure. Based on its anticodon, identify the codon it would bind to, as well as the amino acid that it would carry.*

→ **Mastering Biology** HHMI Video: RNA Folding **hhmi** BioInteractive

a ribosome, glycine will be added to the polypeptide chain whenever the codon 5′-GGC-3′ is presented for translation. Codon by codon, the genetic message is translated as tRNAs position each amino acid in the order prescribed, and the ribosome adds that amino acid onto the growing polypeptide chain. The tRNA molecule is a translator in the sense that, in the context of the ribosome, it can read a nucleic acid word (the mRNA codon) and interpret it as a protein word (the amino acid).

Like mRNA and other types of cellular RNA, transfer RNA molecules are transcribed from DNA templates. In a eukaryotic cell, tRNA, like mRNA, is made in the nucleus and then travels to the cytoplasm, where it will participate in the process of translation. In both bacterial and eukaryotic cells, each tRNA molecule is used repeatedly, picking up its designated amino acid in the cytosol, depositing this cargo onto a polypeptide chain at the ribosome, and then leaving the ribosome, ready to pick up another of the same amino acid.

The accurate translation of a genetic message requires two instances of molecular recognition. First, a tRNA that binds to an mRNA codon specifying a particular amino acid must carry that amino acid, and no other, to the ribosome. The correct matching up of tRNA and amino acid is carried out by a family of related enzymes that are aptly named **aminoacyl-tRNA synthetases (Figure 17.17)**. The active site of each type of aminoacyl-tRNA synthetase fits only a specific combination of amino acid and tRNA. (Regions of both the amino acid attachment end and the anticodon end of the tRNA ensure the specific fit.) There are 20 different synthetases, one for each amino acid. A synthetase joins a given amino acid to an appropriate tRNA; one synthetase is able to bind to all the different tRNAs for its particular amino acid. The synthetase catalyzes the covalent attachment of the amino acid to its tRNA in a process driven by the hydrolysis of ATP. The resulting aminoacyl tRNA, also called a *charged tRNA*, is released from the enzyme and is then available to deliver its amino acid to a growing polypeptide chain on a ribosome.

The second instance of molecular recognition is the pairing of the tRNA anticodon with the appropriate mRNA codon. If one tRNA type existed for each mRNA codon specifying an amino acid, there would be 61 tRNAs (see Figure 17.6). In bacteria, however, there are only about 45 tRNAs, signifying that some tRNAs must be able to bind to more than one codon. This is possible because the rules for base pairing between the third nucleotide base of a codon and the corresponding base of a tRNA anticodon are relaxed compared to those at other codon positions. For example, the nucleotide base U at the 5′ end of a tRNA anticodon can pair with either A or G in the third position (at the 3′ end) of an mRNA codon. The flexible base pairing at this codon position is called **wobble**. Wobble explains why the synonymous codons for a given amino acid most often differ in their third nucleotide base. Accordingly, a tRNA with the anticodon 3′-UCU-5′ can base-pair with either the mRNA codon 5′-AGA-3′ or 5′-AGG-3′, both of which code for arginine (see Figure 17.6).

▼ **Figure 17.17 Aminoacyl-tRNA synthetases provide specificity in joining amino acids to their tRNAs.** Linkage of a tRNA to its amino acid is an endergonic process that occurs at the expense of ATP, which loses two phosphate groups, becoming AMP.

1 The amino acid and the appropriate tRNA enter the active site of the specific synthetase.

Tyrosine (Tyr) (amino acid)

Tyrosyl-tRNA synthetase (enzyme), which can only bind tyrosine and Tyr-tRNA

Tyr-tRNA

A U A

Anticodon on tRNA complementary to the Tyr codon on mRNA

ATP

AMP + 2 (P)ᵢ

Aminoacyl-tRNA synthetase

2 Using ATP, the synthetase catalyzes the covalent bonding of the amino acid to its specific tRNA.

tRNA

Amino acid

3 The tRNA, charged with its amino acid, is released by the synthetase.

Computer model

⊙ **Mastering Biology Animation: Transfer RNA**

The Structure and Function of Ribosomes

Although the ribosomes of bacteria and eukaryotes are very similar in structure and function, eukaryotic ribosomes are slightly larger and differ somewhat from bacterial ribosomes in their molecular composition. The differences are medically significant. Certain antibiotic drugs can inactivate bacterial ribosomes without affecting the ability of eukaryotic ribosomes to make proteins. These drugs, including tetracycline and streptomycin, are used to combat bacterial infections.

Ribosomes facilitate the specific coupling of tRNA anticodons with mRNA codons during protein synthesis. A ribosome consists of a large subunit and a small subunit, each made up of proteins and one or more **ribosomal RNAs**, or **rRNAs (Figure 17.18)**. In eukaryotes, the subunits are made in the nucleolus. Ribosomal RNA genes are transcribed, and the RNA is processed and assembled with proteins imported from the

▼ Figure 17.18 The anatomy of a functioning ribosome.

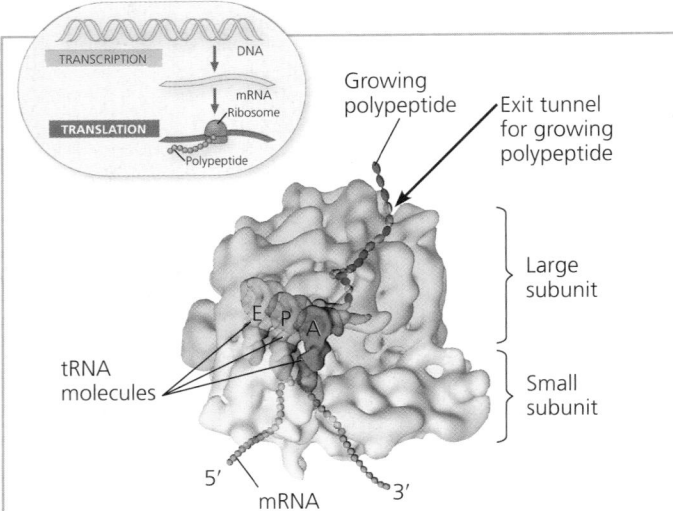

(a) Computer model of functioning ribosome. This is a model of a bacterial ribosome, showing its overall shape. The eukaryotic ribosome is roughly similar. A ribosomal subunit is a complex of ribosomal RNA molecules and proteins.

(b) Schematic model showing binding sites. A ribosome has an mRNA binding site and three tRNA binding sites, known as the A, P, and E sites. This schematic ribosome will appear in later diagrams.

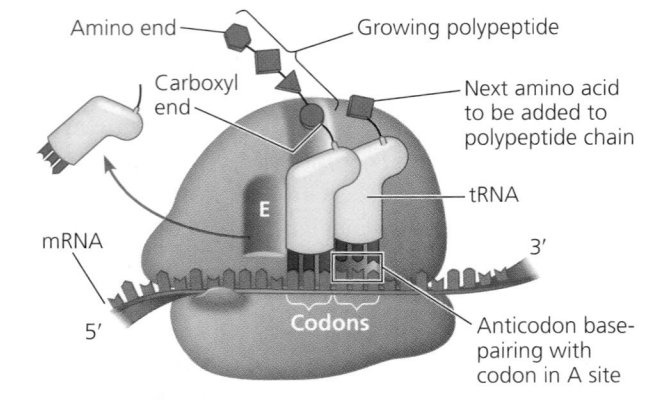

(c) Schematic model with mRNA and tRNA. A tRNA fits into the A site when its anticodon base-pairs with an mRNA codon. The P site holds the tRNA attached to the growing polypeptide. The A site holds the tRNA carrying the next amino acid to be added to the polypeptide chain. Discharged tRNAs leave from the E site. The polypeptide grows at its carboxyl end.

→ **Mastering Biology Animation: Ribosomes**

cytoplasm. Completed ribosomal subunits are then exported via nuclear pores to the cytoplasm. In both bacteria and eukaryotes, a large and a small subunit join to form a functional ribosome only when attached to an mRNA molecule. About one-third of the mass of a ribosome is made up of proteins; the rest consists of three rRNA molecules (in bacteria) or four (in eukaryotes). Because most cells contain thousands of ribosomes, rRNA is the most abundant type of cellular RNA.

The structure of a ribosome reflects its function of bringing an mRNA molecule together with tRNAs carrying amino acids. The mRNA itself has a binding site for the ribosome. (In the **Scientific Skills Exercise**, you can work with DNA sequences representing this binding site in a group of *Escherichia coli* genes.) The ribosome, in turn, has a binding site for mRNA, as well as three binding sites for tRNA (see Figure 17.18). The **P site** (peptidyl-tRNA binding site) holds the tRNA carrying the growing polypeptide chain, while the **A site** (aminoacyl-tRNA binding site) holds the tRNA carrying the next amino acid to be added to the chain. Discharged tRNAs leave the ribosome from the **E site** (exit site). The ribosome holds the tRNA and mRNA in close proximity and positions the new amino acid so that it can be added to the carboxyl end of the growing polypeptide. It then catalyzes the formation of the peptide bond. As the polypeptide becomes longer, it passes through an *exit tunnel* in the ribosome's large subunit. When the polypeptide is complete, it is released through the exit tunnel.

The widely accepted model is that rRNAs, rather than ribosomal proteins, are primarily responsible for both the structure and the function of the ribosome. The proteins, which are largely on the exterior, support the shape changes of the rRNA molecules as they carry out catalysis during translation. Ribosomal RNA is the main constituent of the A and P sites and of the interface between the two subunits; it also acts as the catalyst of peptide bond formation. Thus, a ribosome could actually be considered one colossal ribozyme!

Building a Polypeptide

We can divide translation, the synthesis of a polypeptide, into three stages: initiation, elongation, and termination. All three require protein "factors" that aid in the translation process. Some steps of initiation and elongation also require energy, provided by the hydrolysis of guanosine triphosphate (GTP).

Ribosome Association and Initiation of Translation

The initiation stage of translation brings together an mRNA, a tRNA bearing the first amino acid of the polypeptide, and the two subunits of a ribosome. First, a small ribosomal subunit binds to both the mRNA and a specific initiator tRNA, which carries the amino acid methionine. In bacteria, the small subunit can bind the two in either order; it binds the mRNA at a specific RNA sequence, just upstream of the AUG start codon. In eukaryotes, the small subunit, with the initiator tRNA already bound, binds to the 5′ cap of the mRNA and then moves, or *scans*,

Interpreting a Sequence Logo

How Can a Sequence Logo Be Used to Identify Ribosome Binding Sites on Bacterial mRNAs? When initiating translation, ribosomes bind to an mRNA at a ribosome binding site upstream of the AUG start codon. Because mRNAs from different genes all bind to a ribosome, the genes encoding these mRNAs are likely to have a similar base sequence where the ribosomes bind. Therefore, candidate ribosome binding sites on mRNA can be identified by comparing DNA sequences (and thus the mRNA sequences) of multiple genes in a species, searching the region upstream of the start codon for shared (conserved) stretches of bases. In this exercise, you will analyze DNA sequences from multiple such genes, represented by a visual graphic called a sequence logo.

How the Experiment Was Done The DNA sequences of 149 genes from the *E. coli* genome were aligned using computer software. The aim was to identify similar base sequences—at the appropriate location in each gene—as potential ribosome binding sites. Rather than presenting the data as a series of 149 sequences aligned in a column (a sequence alignment), the researchers used a sequence logo.

Data from the Experiment To show how sequence logos are made, the potential ribosome binding regions from ten *E. coli* genes are shown below in a sequence alignment, followed by the sequence logo derived from the aligned sequences. Note that the DNA shown is the nontemplate (coding) strand, which is how DNA sequences are typically presented. (All data from Thomas D. Schneider.)

▲ Sequence alignment

▼ Sequence logo

Ribosome binding site on mRNA

INTERPRET THE DATA

1. In the sequence logo, the horizontal axis shows the primary sequence of the DNA by nucleotide position. Letters for each base are stacked on top of each other according to their relative frequency at that position among the aligned sequences, with the most common base as the largest letter at the top of the stack. The height of each letter represents the relative frequency of that base *at that position*. **(a)** In the sequence alignment, count the number of each base at position –9 and order them from most to least frequent. Compare this to the size and placement of each base at –9 in the logo. **(b)** Do the same for positions 0 and 1.

2. The height of a stack of letters in a logo indicates the predictive power of that stack (determined statistically). If the stack is tall, we can be more confident in predicting what base will be in that position if a new sequence is added to the logo. For example, at position 2 in the sequence alignment, all ten sequences have a G; the probability of finding a G there in a new sequence is very high, as is the stack in the sequence logo. For short stacks, the bases all have about the same frequency, so it's hard to predict a base at those positions. **(a)** Looking at the sequence logo, which two positions have the most predictable bases? What bases do you predict would be at those positions in a newly sequenced gene? **(b)** Which 12 positions have the least predictable bases? How do you know? How does this reflect the relative frequencies of the bases shown at these positions in the sequence alignment? Use the two leftmost positions of the 12 as examples in your answer.

3. In the actual experiment, the researchers used 149 sequences to build their sequence logo, which is shown below. There is a stack at each position, even if short, because the sequence logo includes more data. **(a)** Which three positions in this sequence logo have the most predictable bases? Name the most frequent base at each. **(b)** Which four positions have the least predictable bases? How can you tell?

4. A consensus sequence identifies the base occurring most often at each position in the set of sequences. **(a)** Write out the consensus sequence of this (the nontemplate) strand. In any position where the base can't be determined, put a dash. **(b)** Which provides more information—the consensus sequence or the sequence logo? What is lost in the less informative method?

5. **(a)** Based on the logo, what five adjacent base positions in the 5′ UTR region are most likely to be involved in ribosome binding? Explain. **(b)** What is represented by the bases in positions 0–2?

Further Reading T. D. Schneider and R. M. Stephens, Sequence logos: A new way to display consensus sequences, *Nucleic Acids Research* 18:6097–6100 (1990).

➔ **Instructors:** A version of this Scientific Skills Exercise can be assigned in **Mastering Biology**.

downstream along the mRNA until it reaches the AUG start codon, where the initiator tRNA then hydrogen-bonds, as shown in step 1 of **Figure 17.19**. In either bacteria or eukaryotes, the AUG signals the start of translation; this is important because it establishes the codon reading frame for the mRNA.

The union of mRNA, initiator tRNA, and the small ribosomal subunit is followed by the attachment of a large ribosomal subunit, completing the *translation initiation complex* (see Figure 17.19). Proteins called *initiation factors* are required to bring all these components together. The cell also expends energy obtained by hydrolysis of a GTP molecule to form the initiation complex. At the completion of the initiation process, the initiator tRNA sits in the P site of the ribosome, and the vacant A site is ready for the next aminoacyl tRNA. Note that a polypeptide is always synthesized in one direction, from the initial methionine at the amino end, also called the N-terminus, toward the final amino acid at the carboxyl end, also called the C-terminus (see Figure 5.15).

▼ **Figure 17.19 The initiation of translation.**

1 A small ribosomal subunit binds to an mRNA. In a bacterial cell, the mRNA binding site on this subunit recognizes a specific nucleotide sequence on the mRNA just upstream of the start codon (see the Scientific Skills Exercise). An initiator tRNA, with the anticodon UAC, base-pairs with the start codon, AUG. This tRNA carries the amino acid methionine (Met).

2 The arrival of a large ribosomal subunit completes the initiation complex. Proteins called initiation factors (not shown) are required to bring all the translation components together. Hydrolysis of GTP provides the energy for the assembly. The initiator tRNA is in the P site; the A site is available to the tRNA bearing the next amino acid.

➔ **Mastering Biology Animation: Initiation of Translation**

Elongation of the Polypeptide Chain

In the elongation stage of translation, amino acids are added one by one to the previous amino acid at the C-terminus of the growing chain. Each addition involves several proteins called *elongation factors* and occurs in a three-step cycle described in **Figure 17.20**. Energy expenditure occurs in the first and third steps. Codon recognition requires hydrolysis of one molecule of GTP, which increases the accuracy and efficiency of this step. One more GTP is hydrolyzed (broken up) to provide energy for the translocation step.

The mRNA is moved through the ribosome in one direction only, 5' end first; this is equivalent to the ribosome moving 5' → 3' on the mRNA. The main point is that the ribosome and the mRNA move relative to each other, unidirectionally, codon by codon. The elongation cycle takes less than a tenth of a second in bacteria and is repeated as each amino acid is added until the polypeptide is completed. The empty tRNAs that are released from the E site return to the cytoplasm, where they will be reloaded with the appropriate amino acid (see Figure 17.17).

Termination of Translation

The final stage of translation is termination **(Figure 17.21)**. Elongation continues until a stop codon in the mRNA reaches the A site. The nucleotide base triplets UAG, UAA, and UGA (all written 5' → 3') do not code for amino acids but instead act as signals to stop translation. A *release factor*, a protein shaped like an aminoacyl tRNA, binds directly to the stop codon in the A site. The release factor causes the addition of a water molecule instead of an amino acid to the polypeptide chain. (Water molecules are abundant in the cytosol.) This reaction hydrolyzes (breaks) the bond between the completed polypeptide and the tRNA in the P site, releasing the polypeptide through the exit tunnel of the ribosome's large subunit. The remainder of the translation assembly then comes apart in a multistep process, aided by other protein factors. Breakdown of the translation assembly requires the hydrolysis of two more GTP molecules.

Completing and Targeting the Functional Protein

The process of translation is often not sufficient to make a functional protein. In this section, you will learn about modifications that polypeptide chains undergo after the translation process as well as some of the mechanisms used to target completed proteins to specific sites in the cell.

Protein Folding and Post-translational Modifications

During its synthesis, a polypeptide chain begins to coil and fold spontaneously as a consequence of its amino acid sequence (primary structure), forming a protein with a specific shape: a three-dimensional molecule with secondary and tertiary structure (see Figure 5.18). Thus, a gene determines primary structure, which in turn determines shape.

▼ **Figure 17.20 The elongation cycle of translation.** The hydrolysis of GTP plays an important role in the elongation process; elongation factors are not shown.

→ **Mastering Biology** Animation:
Elongation Cycle of Translation

❶ Codon recognition. The anticodon of an incoming aminoacyl tRNA base-pairs with the complementary mRNA codon in the A site. Hydrolysis of GTP increases the accuracy and efficiency of this step. Although not shown, many different aminoacyl tRNAs are present, but only the one with the appropriate anticodon will bind and allow the cycle to progress.

Amino end of polypeptide

Carboxyl end of polypeptide

Ribosome ready for next aminoacyl tRNA

mRNA

5′

3′

GTP

GDP + Ⓟ_i

E

P A
site site

❸ Translocation. The ribosome translocates the tRNA in the A site to the P site. At the same time, the empty tRNA in the P site is moved to the E site, where it is released. The mRNA moves along with its bound tRNAs, bringing the next codon to be translated into the A site.

GDP + Ⓟ_i

GTP

❷ Peptide bond formation. An rRNA molecule of the large ribosomal subunit catalyzes the formation of a peptide bond between the carboxyl end of the growing polypeptide in the P site and the amino group of the new amino acid in the A site. As shown in the next diagram, this step removes the polypeptide from the tRNA in the P site and attaches it to the amino acid on the tRNA in the A site.

▼ **Figure 17.21 The termination of translation.** Like elongation, termination requires GTP hydrolysis as well as additional protein factors, which are not shown here.

→ **Mastering Biology** Animation:
Termination of Translation

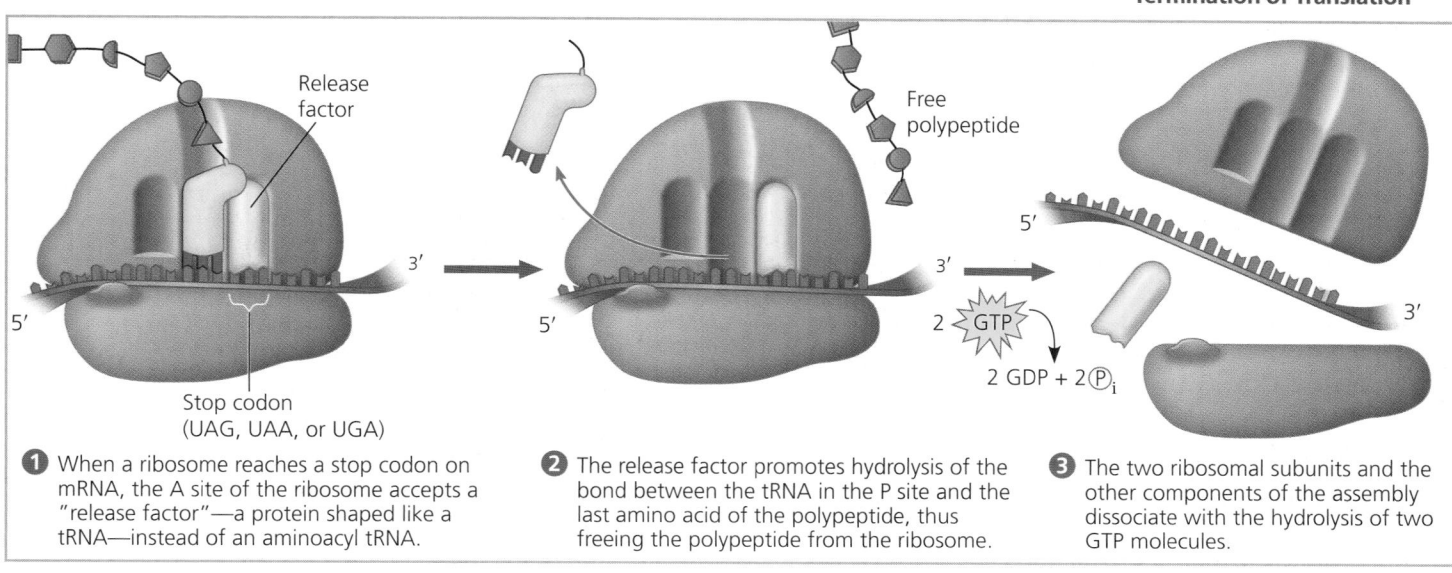

Release factor

Free polypeptide

5′

3′

3′

5′

Stop codon
(UAG, UAA, or UGA)

5′

2 GTP

2 GDP + 2Ⓟ_i

3′

3′

❶ When a ribosome reaches a stop codon on mRNA, the A site of the ribosome accepts a "release factor"—a protein shaped like a tRNA—instead of an aminoacyl tRNA.

❷ The release factor promotes hydrolysis of the bond between the tRNA in the P site and the last amino acid of the polypeptide, thus freeing the polypeptide from the ribosome.

❸ The two ribosomal subunits and the other components of the assembly dissociate with the hydrolysis of two GTP molecules.

Additional steps—*post-translational modifications*—may be required before the protein can begin doing its particular job in the cell. Certain amino acids may be chemically modified by the attachment of sugars, lipids, phosphate groups, or other additions. Enzymes may remove one or more amino acids from the leading (amino) end of the polypeptide chain. In some cases, a polypeptide chain may be enzymatically cleaved into two or more pieces. In other cases, two or more polypeptides that are synthesized separately may come together, if the protein has quaternary structure; an example is hemoglobin (see Figure 5.18).

→ **Mastering Biology BioFlix® Animation: Protein Processing**

Targeting Polypeptides to Specific Locations

In electron micrographs of eukaryotic cells active in protein synthesis, two populations of ribosomes are evident: free and bound (see Figure 6.10). Free ribosomes are suspended in the cytosol and mostly synthesize proteins that stay in the cytosol and function there. In contrast, bound ribosomes are attached to the cytosolic side of the endoplasmic reticulum (ER) or to the nuclear envelope. Bound ribosomes make proteins of the endomembrane system (see Figure 6.15) as well as proteins secreted from the cell, such as insulin. It is important to note that the ribosomes themselves are identical and can alternate between being free ribosomes one time they are used and being bound ribosomes the next.

What determines whether a ribosome is free in the cytosol or bound to rough ER? Polypeptide synthesis always begins in the cytosol as a free ribosome starts to translate an mRNA molecule. There, the process continues to completion—*unless* the growing polypeptide itself cues the ribosome to attach to the ER. The polypeptides of proteins destined for the endomembrane system or for secretion are marked by a **signal peptide**, which targets the protein to the ER **(Figure 17.22)**. The signal peptide, a sequence of about 20 amino acids at or near the leading end (N-terminus) of the polypeptide, is recognized as it emerges from the ribosome by a protein-RNA complex called a **signal-recognition particle (SRP)**. This particle escorts the ribosome to a receptor protein built into the ER membrane. The receptor is part of a multiprotein translocation complex. Polypeptide synthesis continues there, and the growing polypeptide snakes across the membrane into the ER lumen via a protein pore. The signal peptide is removed by an enzyme. The rest of the completed polypeptide, if it is to be secreted from the cell, is released into solution within the ER lumen (see Figure 17.22). Or, if the polypeptide is to be a membrane protein, amino acid sequences further in the chain cause that part to remain embedded in the ER membrane. In either case, it travels in a transport vesicle to its destination (see Figure 7.9).

Other kinds of signal peptides are used to target polypeptides to mitochondria, chloroplasts, the interior of the nucleus, and other organelles that are not part of the endomembrane system. The critical difference in these cases is that translation

▼ **Figure 17.22 The signal mechanism for targeting proteins to the ER.**

1 Polypeptide synthesis begins on a free ribosome in the cytosol.

2 An SRP binds to the signal peptide, (if present), halting synthesis momentarily.

3 The SRP binds to a receptor protein in the ER membrane, part of a protein complex that forms a pore.

4 The SRP leaves, and polypeptide synthesis resumes, with simultaneous translocation across the membrane.

5 The signal peptide is cleaved by an enzyme in the receptor protein complex.

6 The rest of the completed polypeptide leaves the ribosome and folds into its final conformation.

MAKE CONNECTIONS *If this protein were destined for secretion, what would happen to it after its synthesis was completed? See Figure 7.9.*

→ **Mastering Biology Animation: Cotranslational Translocation** is completed in the cytosol before the polypeptide is imported into the organelle. Translocation mechanisms also vary, but in all cases studied to date, the "postal zip codes" that address proteins for secretion or to cellular locations are signal peptides of some sort. Bacteria also employ signal peptides to target proteins to the plasma membrane or for secretion.

Making Multiple Polypeptides in Bacteria and Eukaryotes

In previous sections, you learned how a single polypeptide is synthesized using the information encoded in an mRNA molecule. When a polypeptide is required in a cell, though, the need is for many copies, not just one.

In both bacteria and eukaryotes, multiple ribosomes translate an mRNA at the same time (**Figure 17.23**); that is, a single mRNA is used to make many copies of a polypeptide simultaneously. Once a ribosome is far enough past the start codon, a second ribosome can attach to the mRNA, eventually resulting in a number of ribosomes trailing along the mRNA. Such strings of ribosomes, called **polyribosomes** (or **polysomes**), can be seen with an electron microscope; they can be either free or bound. They enable a cell to rapidly make many copies of a polypeptide.

Another way both bacteria and eukaryotes increase the number of copies of a polypeptide is by transcribing multiple mRNAs from the same gene. However, the coordination of the two processes—transcription and translation—differs in the two groups. The most important differences between bacteria

▼ **Figure 17.23 Polyribosomes.**

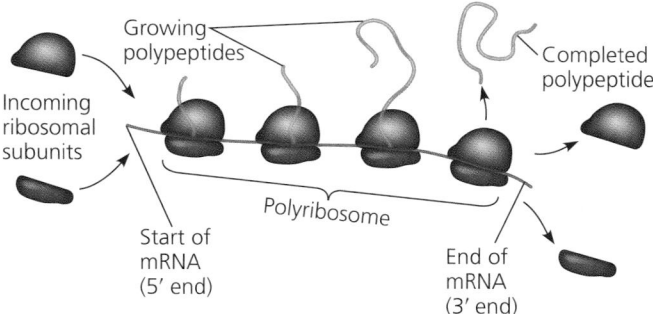

(a) An mRNA molecule is generally translated simultaneously by several ribosomes in clusters called polyribosomes.

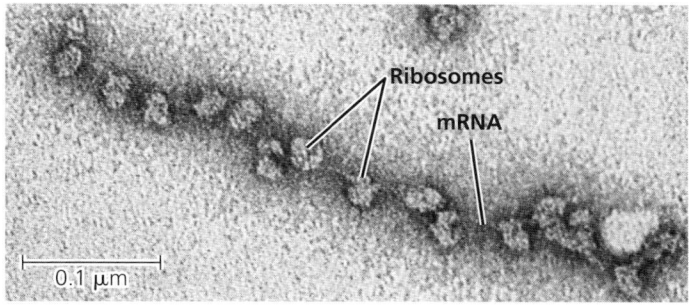

(b) This micrograph shows a large polyribosome in a bacterial cell. Growing polypeptides are not visible here (TEM).

▼ **Figure 17.24 Coupled transcription and translation in bacteria.** In bacterial cells, the translation of mRNA can begin as soon as the leading (5′) end of the mRNA molecule peels away from the DNA template. The micrograph (TEM) shows a strand of *E. coli* DNA being transcribed by RNA polymerase molecules. Attached to each RNA polymerase molecule is a growing strand of mRNA, which is already being translated by ribosomes. The newly synthesized polypeptides are not visible in the micrograph but are shown in the diagram.

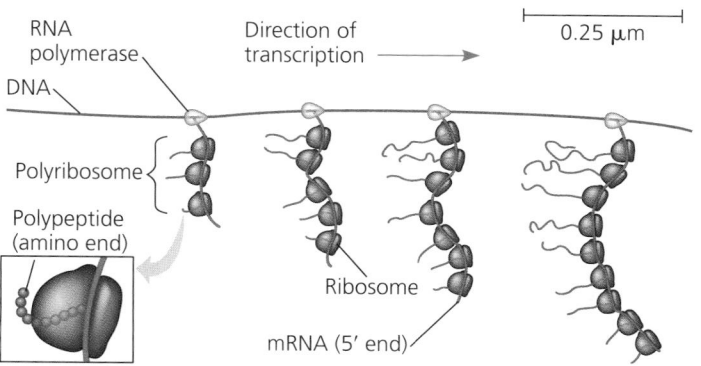

VISUAL SKILLS *Which one of the mRNA molecules started being transcribed first? On that mRNA, which ribosome started translating the mRNA first?*

and eukaryotes arise from the bacterial cell's lack of compartmental organization. Like a one-room workshop, a bacterial cell ensures a streamlined operation by coupling the two processes. With no nuclear envelope, it can simultaneously transcribe and translate the same gene (**Figure 17.24**), and the newly made protein can quickly diffuse to its site of function.

In contrast, the eukaryotic cell's nuclear envelope segregates transcription from translation and provides a compartment for extensive RNA processing. This processing stage includes additional steps, discussed earlier, the regulation of which can help coordinate the eukaryotic cell's elaborate activities. **Figure 17.25** summarizes the path from gene to polypeptide in a eukaryotic cell.

CONCEPT CHECK 17.4

1. What two processes ensure that the correct amino acid is added to a growing polypeptide chain?
2. Describe how a polypeptide to be secreted reaches the endomembrane system.
3. **DRAW IT** Draw a tRNA with the anticodon 3′-CGU-5′. Given wobble, what two different codons could it bind to? Draw each codon on an mRNA, labeling all 5′ and 3′ ends, the tRNA, and the amino acid it carries.
4. **WHAT IF?** In eukaryotic cells, mRNAs have been found to have a circular arrangement in which proteins hold the poly-A tail near the 5′ cap. How might this increase translation efficiency?

For suggested answers, see Appendix A.

▼ **Figure 17.25 A summary of transcription and translation in a eukaryotic cell.** This diagram shows the path from one gene to one polypeptide. Each gene in the DNA can be transcribed repeatedly into many identical RNA molecules, and each mRNA can be translated repeatedly to yield many identical polypeptide molecules. (Remember that the final products of some genes are not polypeptides but RNA molecules that don't get translated, including tRNAs and rRNAs.) In general, the steps of transcription and translation are similar in bacterial, archaeal, and eukaryotic cells. The major difference is the occurrence of RNA processing in the eukaryotic nucleus. Other significant differences are found in the initiation stages of both transcription and translation and in the termination of transcription. To visualize these processes in their cellular context, see Figure 6.32d-f.

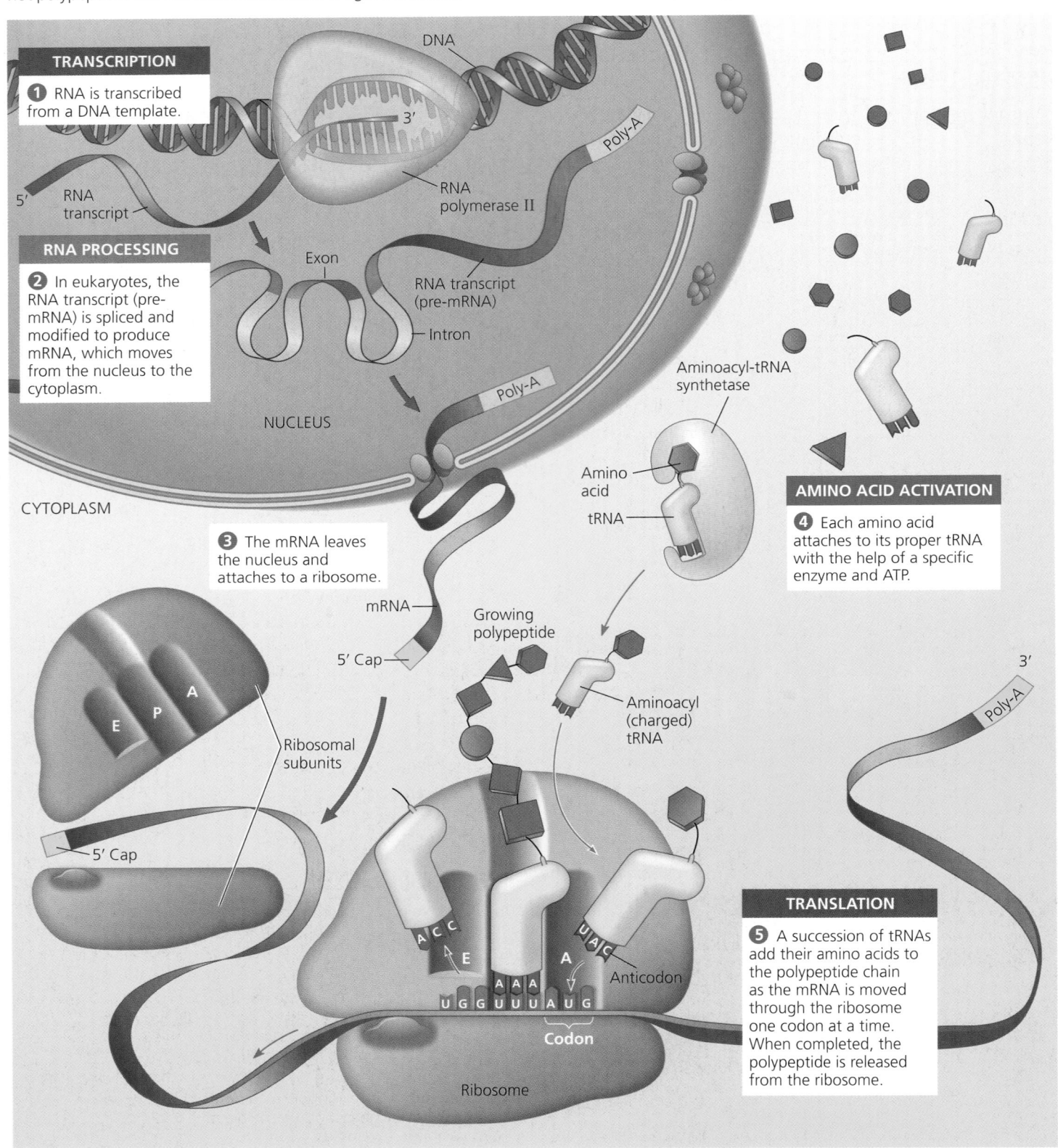

TRANSCRIPTION

1 RNA is transcribed from a DNA template.

DNA

3′

RNA transcript

5′

RNA polymerase II

Poly-A

RNA PROCESSING

2 In eukaryotes, the RNA transcript (pre-mRNA) is spliced and modified to produce mRNA, which moves from the nucleus to the cytoplasm.

Exon

RNA transcript (pre-mRNA)

Intron

Poly-A

NUCLEUS

CYTOPLASM

3 The mRNA leaves the nucleus and attaches to a ribosome.

mRNA

5′ Cap

Growing polypeptide

Aminoacyl-tRNA synthetase

Amino acid

tRNA

AMINO ACID ACTIVATION

4 Each amino acid attaches to its proper tRNA with the help of a specific enzyme and ATP.

Aminoacyl (charged) tRNA

3′

Poly-A

E P A

Ribosomal subunits

5′ Cap

A C C

E

A A A

U G G U U U A U G

Codon

U A C

A

Anticodon

TRANSLATION

5 A succession of tRNAs add their amino acids to the polypeptide chain as the mRNA is moved through the ribosome one codon at a time. When completed, the polypeptide is released from the ribosome.

Ribosome

➲ **Mastering Biology BioFlix® Animation: Protein Synthesis**

Mutations of one or a few nucleotides can affect protein structure and function

Now that you have explored the process of gene expression, you can consider the effects of changes to the genetic information of a cell. Because these changes, called **mutations**, are the ultimate source of new genes, they are responsible for the huge diversity of genes found among organisms. Earlier, we considered chromosomal rearrangements that affect long segments of DNA (see Figure 15.14); these are considered large-scale mutations. Here we examine small-scale mutations of one or a few nucleotide pairs, including **point mutations**, changes in a single nucleotide pair of a gene.

If a point mutation occurs in a gamete or in a cell that gives rise to gametes, it may be transmitted to offspring and to future generations. If the mutation has an adverse effect on the phenotype of a person, the mutant condition is referred to as a genetic disorder or hereditary disease. For example, we can trace the genetic basis of sickle-cell disease to the mutation of a single nucleotide pair in the gene that encodes the β-globin polypeptide of hemoglobin. The change of a single nucleotide in the DNA's template strand leads to an altered mRNA and the production of an abnormal protein (**Figure 17.26**; also see Figure 5.19). In individuals who are homozygous for the mutant allele, the sickling of red blood cells caused by the altered hemoglobin produces the multiple symptoms associated with sickle-cell disease (see Concept 14.4

and Figure 23.18). Another disorder caused by a point mutation is a heart condition called familial cardiomyopathy, which is responsible for some of the tragic incidents of sudden death in young athletes. Point mutations in several genes encoding muscle proteins have been identified, any of which can lead to this disorder.

Types of Small-Scale Mutations

Many mutations occur in regions outside of protein-coding genes, and any potential effect they have on the phenotype of the organism may be subtle and hard to detect. For this reason, here we'll concentrate on mutations within protein-coding genes. Small-scale mutations within a gene can be divided into two general categories: (1) single nucleotide-pair substitutions and (2) nucleotide-pair insertions or deletions. Insertions and deletions can involve one or more nucleotide pairs.

Substitutions

A **nucleotide-pair substitution** is the replacement of one nucleotide and its partner with another pair of nucleotides (**Figure 17.27a**). Some substitutions have no effect on the encoded protein, owing to the redundancy of the genetic code. For example, if 3′-CCG-5′ on the template strand mutated to 3′-CCA-5′, the mRNA codon that used to be GGC would become GGU, but a glycine would still be inserted at the proper location in the protein (see Figure 17.6). In other words, a change in a nucleotide pair may transform one codon into another that is translated into the same amino acid. Such a change is an example of a **silent mutation**, which has no observable effect on the phenotype. (Silent mutations can occur outside genes as well.) Interestingly, there is evidence that some silent mutations may indirectly affect where or at what level the gene gets expressed, even though the actual protein is the same.

Substitutions that change one amino acid to another one are called **missense mutations**. Such a mutation may have little effect on the protein: The new amino acid may have properties similar to those of the amino acid it replaces, or it may be in a region of the protein where the exact sequence of amino acids is not essential to the protein's function.

However, the nucleotide-pair substitutions of greatest interest are those that cause a major change in a protein. The alteration of a single amino acid in a crucial area of a protein—such as in the part of the β-globin subunit of

▼ **Figure 17.26 The molecular basis of sickle-cell disease: a point mutation.** The allele that causes sickle-cell disease differs from the wild-type (normal) allele by a single DNA nucleotide pair. The micrographs are SEMs of a normal red blood cell (on the left) and a sickled red blood cell (right) from individuals homozygous for wild-type and mutant alleles, respectively.

Wild-type β-globin	Sickle-cell β-globin	
Wild-type β-globin DNA	**Mutant β-globin DNA**	In the DNA, the mutant (sickle-cell) template strand (top) has an A where the wild-type template has a T.
3′ C T C 5′ 5′ G A G 3′	3′ C A C 5′ 5′ G T G 3′	
mRNA	**mRNA**	The mutant mRNA has a U instead of an A in one codon.
5′ G A G 3′	5′ G U G 3′	
Normal hemoglobin	**Sickle-cell hemoglobin**	The mutant β-globin has a valine (Val) instead of a glutamic acid (Glu).
Glu	Val	

 Mastering Biology HHMI Animation: Sickle-Cell Disease

hhmi BioInteractive

▼ **Figure 17.27 Types of small-scale mutations that affect mRNA sequence.** All but one of the types shown here also affect the amino acid sequence of the encoded polypeptide.

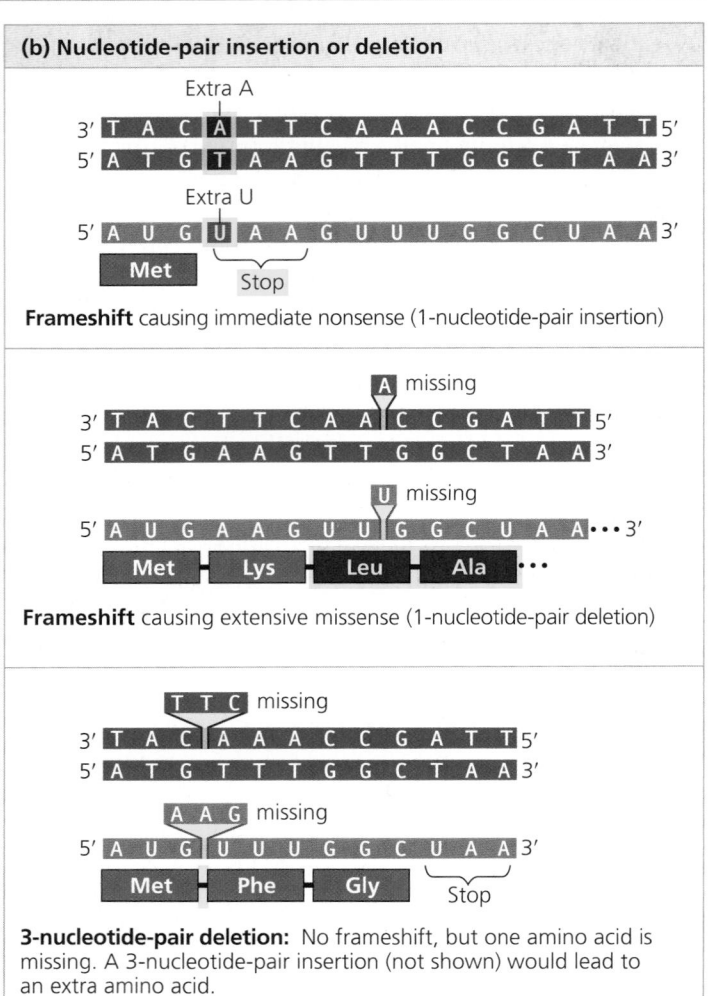

→ **Mastering Biology Figure Walkthrough • Animation: Mutation Types**

hemoglobin shown in Figure 17.26 or in the active site of an enzyme as shown in Figure 8.19—can significantly alter protein activity. Occasionally, such a mutation leads to an improved protein or one with novel capabilities, but much more often such mutations are neutral or detrimental, leading to a useless or less active protein that impairs cellular function. A second example of a missense mutation is a mutation in the tyrosinase gene that causes the albino phenotype in the Asinara donkey. Recently, Italian researchers sequenced the tyrosinase gene in donkeys from this wild population and showed that a recessive mutation is responsible for the lack of pigment. A C is changed to a G in this gene,

resulting in a histidine being added to the protein instead of an aspartic acid. The changed amino acid is in a site in tyrosinase that normally binds a copper atom, and it makes the enzyme incapable of binding copper, rendering it nonfunctional. As in the case of sickle-cell disease, just one change in a nucleotide base results in a dramatic phenotypic change.

Substitution mutations are usually missense mutations; that is, the altered codon still codes for an amino acid and thus makes sense, although not necessarily the *right* sense. But a point mutation can also change a codon for an amino acid into a stop codon. This is called a **nonsense mutation**, and it causes translation to be terminated prematurely; the

Are insulin mutations the cause of three infants' neonatal diabetes?

Insulin is a hormone that acts as a key regulator of blood glucose level, and insulin deficiency can lead to diabetes. In some cases of neonatal (newborn) diabetes, the gene coding for the insulin protein has a nucleotide-pair substitution mutation that alters the protein structure enough to cause it to malfunction. Doctors can use DNA sequence information to diagnose diabetes and other diseases. For example, the insulin gene sequence of a patient with neonatal diabetes can be analyzed to determine if it has a mutation and, if so, its effect.

→ **Instructors**: A version of this Problem-Solving Exercise can be assigned in **Mastering Biology**. Or a more extensive investigation called "Solve It: Which Insulin Mutations May Result in Disease?" can be assigned.

In this exercise, you will determine the effect of mutations within insulin gene sequences in infant diabetes patients.

Your Approach Suppose you are a medical geneticist presented with three infant patients, all of whom have a nucleotide-pair substitution in their insulin gene. It is your job to analyze each mutation to figure out its effect on the amino acid sequence of the insulin protein. To identify the mutation in each patient, you will compare his or her individual insulin complementary DNA (cDNA) sequence to that of the wild-type cDNA. (cDNA is a double-stranded DNA molecule that is based on the mRNA sequence and thus contains only the portion of a gene that is translated—introns are not included. cDNA sequences are commonly used to compare the coding regions of genes.) Identifying the codons that have been changed will tell you which, if any, amino acids are altered in the patient's insulin protein.

Your Data You will analyze the cDNA codons for amino acids 35–54 (of the 110 amino acids) of each patient's insulin protein, so the start codon (AUG) is not present. The sequences of the wild-type cDNA and the patients' cDNA are shown below, arranged in codons.

Wild-type cDNA 5'– CTG GTG GAA GCT CTC TAC CTA GTG TGC GGG GAA CGA GGC TTC TTC TAC ACA CCC AAG ACC –3'

Patient 1 cDNA 5'– CTG GTG GAA GCT CTC TAC CTA GTG TGC GGG GAA CGA GGC TGC TTC TAC ACA CCC AAG ACC –3'

Patient 2 cDNA 5'– CTG GTG GAA GCT CTC TAC CTA GTG TGC GGG GAA CGA GGC TCC TTC TAC ACA CCC AAG ACC –3'

Patient 3 cDNA 5'– CTG GTG GAA GCT CTC TAC CTA GTG TGC GGG GAA CGA GGC TTC TTG TAC ACA CCC AAG ACC –3'

Data from N. Nishi and K. Nanjo, Insulin gene mutations and diabetes, *Journal of Diabetes Investigation* 2:92–100 (2011).

Your Analysis

1. Compare each patient's cDNA sequence to the wild-type cDNA sequence. Circle the codons where a nucleotide-pair substitution mutation has occurred.
2. Use a codon table (see Figure 17.6) to identify the amino acid encoded by the codon with the mutation in each patient's insulin sequence, and compare it to the amino acid encoded by the codon in the corresponding wild-type sequence. As is standard practice with DNA sequences, the *coding* (nontemplate) strand of the cDNA has been provided, so to convert it to mRNA for use with the codon table, you only need to change T to U. Classify each patient's nucleotide-pair substitution mutation: Is it a silent, missense, or nonsense mutation? Explain, for each answer.
3. Compare the structure of the amino acid you identified in each patient's insulin sequence to that of the corresponding amino acid in the wild-type sequence (see Figure 5.14). Given that each patient has neonatal diabetes, discuss how the change of amino acid in each might have affected the insulin protein and thus resulted in the disease. (Consider the chemical nature of the side chains.)

---

resulting polypeptide will be shorter than the polypeptide encoded by the normal gene. Most nonsense mutations lead to nonfunctional proteins.

In the **Problem-Solving Exercise**, you'll work with a few common single nucleotide-pair substitution mutations in the gene encoding insulin, some or all of which may lead to diabetes. You will classify these mutations into one of the types we just described and characterize the change in the amino acid sequence.

Insertions and Deletions

Insertions and **deletions** are additions or losses of nucleotide pairs in a gene **(Figure 17.27b)**. These mutations have a disastrous effect on the resulting protein more often than substitutions do. Insertion or deletion of nucleotides may alter the reading frame of the genetic message, the triplet grouping of nucleotides on the mRNA that is read during translation. Such a mutation, called a **frameshift mutation**, occurs whenever the number of nucleotides inserted or deleted is not a multiple of three. All nucleotides downstream of the deletion or insertion will be improperly grouped into codons; the result will be extensive missense mutations, usually ending sooner or later in a nonsense mutation that leads to premature termination. Unless the frameshift is very near the end of the gene, the protein is almost certain to be nonfunctional. Insertions and deletions also occur outside of coding regions; these are not called frameshift mutations, but can have effects on the phenotype—for instance, they can affect how a gene is expressed.

New Mutations and Mutagens

Mutations can arise in a number of ways. Errors during DNA replication or recombination can lead to nucleotide-pair substitutions, insertions, or deletions, as well as to mutations affecting longer stretches of DNA. If an incorrect nucleotide is added to a growing chain during replication, for example, the base on that nucleotide will then be mismatched with the nucleotide base on the other strand. In many cases, the error will be corrected by DNA proofreading and repair systems (see Concept 16.2). Otherwise, the incorrect base will be used as a template in the next round of replication, resulting in a mutation. Such mutations are called *spontaneous mutations*. It is difficult to calculate the rate at which such mutations occur. Rough estimates have been made of the rate of mutation during DNA replication for both *E. coli* and eukaryotes, and the numbers are similar: About one nucleotide in every 10^{10} is altered, and the change is passed on to the next generation of cells.

A number of physical and chemical agents, called **mutagens**, interact with DNA in ways that cause mutations. In the 1920s, Hermann Muller discovered that X-rays caused genetic changes in fruit flies, and he used X-rays to make *Drosophila* mutants for his genetic studies (just as Beadle and Tatum did with their *Neurospora* cells). But Muller also recognized an alarming implication of his discovery: X-rays and other forms of high-energy radiation pose hazards to the genetic material of people as well as laboratory organisms. Mutagenic radiation, a physical mutagen, includes ultraviolet (UV) light, which can cause disruptive thymine dimers in DNA (see Figure 16.20).

Chemical mutagens fall into several categories. Nucleotide analogs are chemicals similar to normal DNA nucleotides but that pair incorrectly during DNA replication. Other chemical mutagens interfere with correct DNA replication by inserting themselves into the DNA and distorting the double helix. Still other mutagens cause chemical changes in bases that change their pairing properties.

Researchers have developed a variety of methods to test the mutagenic activity of chemicals. A major application of these tests is the preliminary screening of chemicals to identify those that may cause cancer. This approach makes sense because most carcinogens (cancer-causing chemicals) are mutagenic, and conversely, most mutagens are carcinogenic.

Using CRISPR to Edit Genes and Correct Disease-Causing Mutations

Ever since biologists have understood how disease-causing proteins were the result of mutations in genes, they have sought techniques for **gene editing**—altering genes in a specific, predictable way. Their aim has been to use such a technique to change specific genes in living cells, in part to try to correct genes that cause disease. Over the past 15 years, biologists have developed a powerful new technique for gene editing, called the **CRISPR-Cas9 system**, that is transforming the field of *genetic engineering*, the direct manipulation of genes for practical purposes. Cas9 is a bacterial protein that helps defend bacteria against the viruses that infect them (bacteriophages). In bacterial cells, Cas9 acts together with "guide RNA" made from the CRISPR region of the bacterial genome. (Figure 19.7 explains how this defense system works in bacteria.)

➡ **Mastering Biology** HHMI Video: Jennifer Doudna: CRISPR Provides a Tool for Changing Genes

➡ **Mastering Biology** Interview with Francisco Mojica: Discovering CRISPR in bacteria (see the interview before Chapter 13)

Similar to other DNA-cutting enzymes involved in DNA repair (described in Concept 16.2), Cas9 is a nuclease that cuts double-stranded DNA molecules. The power of Cas9 for gene editing is that the Cas9 protein will cut any sequence to which it is directed. Cas9 is directed to its target by a guide RNA molecule that it binds and uses as a homing device, cutting both strands of any DNA sequence that is complementary to the guide RNA. Scientists have been able to exploit the function of Cas9 by introducing a Cas9–guide RNA complex into a cell they wish to alter **(Figure 17.28)**. The guide RNA in the complex is engineered to be complementary to the "target" gene. Cas9 cuts both strands of the target DNA, and the resulting broken ends of DNA trigger a DNA repair system (similar to that shown in Figure 16.20). When there is no undamaged DNA for the enzymes of the repair system to use as a template, as shown at the bottom left of Figure 17.28, the repair enzymes introduce or remove random nucleotides while rejoining the ends. Generally, this process alters the DNA sequence so that the gene no longer works properly. This technique is a highly successful way for researchers to "knock out" (disable) a given gene to study what that gene does in an organism.

But what about using this system to help treat genetic diseases? Researchers have modified the technique so that the CRISPR-Cas9 system can be used to repair a gene that has a harmful mutation. They introduce a segment from the normal (functional) gene along with the CRISPR-Cas9 system. After Cas9 cuts the target DNA, repair enzymes can use the normal DNA as a template to repair the target DNA at the break point. In this way, the CRISPR-Cas9 system edits the defective gene so that it is corrected (see the bottom right of Figure 17.28).

In 2018, researchers reported promising results using the CRISPR-Cas 9 system in an attempt to correct the genetic

▼ Figure 17.28 Gene editing using the CRISPR-Cas9 system.

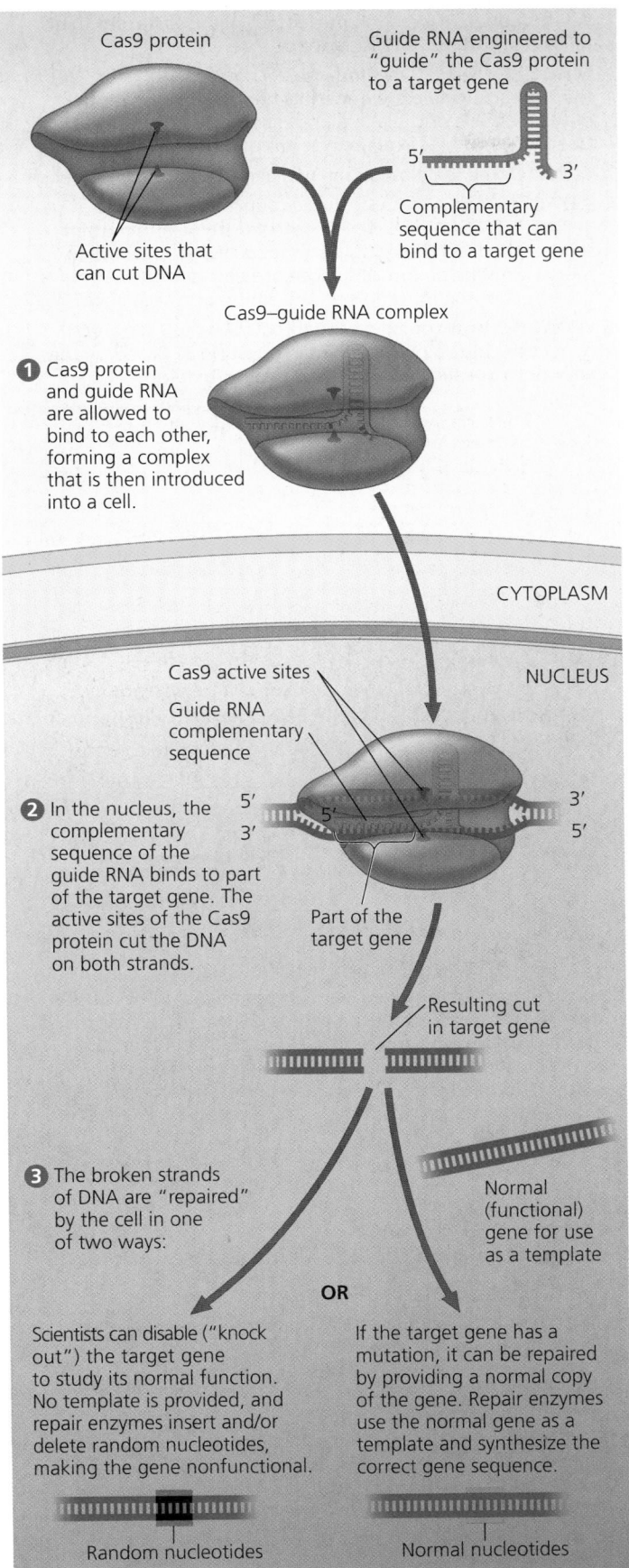

Cas9 protein

Guide RNA engineered to "guide" the Cas9 protein to a target gene

5′ 3′

Active sites that can cut DNA

Complementary sequence that can bind to a target gene

Cas9–guide RNA complex

1 Cas9 protein and guide RNA are allowed to bind to each other, forming a complex that is then introduced into a cell.

CYTOPLASM

NUCLEUS

Cas9 active sites

Guide RNA complementary sequence

2 In the nucleus, the complementary sequence of the guide RNA binds to part of the target gene. The active sites of the Cas9 protein cut the DNA on both strands.

5′   5′   3′
3′   5′

Part of the target gene

Resulting cut in target gene

Normal (functional) gene for use as a template

3 The broken strands of DNA are "repaired" by the cell in one of two ways:

OR

Scientists can disable ("knock out") the target gene to study its normal function. No template is provided, and repair enzymes insert and/or delete random nucleotides, making the gene nonfunctional.

If the target gene has a mutation, it can be repaired by providing a normal copy of the gene. Repair enzymes use the normal gene as a template and synthesize the correct gene sequence.

Random nucleotides

Normal nucleotides

defect that causes sickle-cell disease. They edited human cells from patients with sickle-cell disease and injected them into bone marrow in mice. After 19 weeks, the gene remained corrected in 20–40% of the injected cells. Researchers, physicians, and patients alike are excited about the potential of CRISPR technology to treat or even cure human diseases that have a genetic basis, such as sickle-cell, Alzheimer's, and Parkinson's diseases, as well as some types of cancer. However, there are still concerns about using CRISPR-treated cells in humans because of possible effects on genes that are not being targeted; this is an active area of research.

Jennifer Doudna, a co-discoverer of CRISPR-Cas9, recognized not only its incredible potential but also the danger of its misapplication. After having a dream in which Adolf Hitler asked her what could be done with this gene-editing tool, Doudna realized it was imperative to spend some time reflecting on ethical considerations. In 2015, she convened a conference of biologists who agreed at the end of the meeting to urge extreme caution as the field moves forward—and the discussion continues.

What Is a Gene? *Revisiting the Question*

Our definition of a gene has evolved over the past few chapters, as it has through the history of genetics. We began with the Mendelian concept of a gene as a discrete unit of inheritance that affects a phenotypic character (Chapter 14). We saw that Morgan and his colleagues assigned such genes to specific loci on chromosomes (Chapter 15). We went on to view a gene as a region of specific nucleotide sequence along the length of the DNA molecule of a chromosome (Chapter 16). Finally, in this chapter, we have considered a functional definition of a gene as a DNA sequence that codes for a specific polypeptide chain or a functional RNA molecule, such as a tRNA. All these definitions are useful, depending on the context in which genes are being studied.

We have noted that merely saying a gene codes for a polypeptide is an oversimplification. Most eukaryotic genes contain noncoding segments (such as introns), so large portions of these genes have no corresponding segments in polypeptides. Molecular biologists also often include promoters and certain other regulatory regions of DNA within the boundaries of a gene. These DNA sequences are not transcribed, but they can be considered part of the functional gene because they must be present for transcription to occur. Our definition of a gene must also be broad enough to include the DNA that is transcribed into rRNA, tRNA, and other RNAs that are not translated. These genes have no polypeptide products but play crucial roles in the cell. Thus, we arrive at the following definition: *A gene is a region of DNA that can be expressed to produce a final functional product that is either a polypeptide or an RNA molecule.*

When considering phenotypes, however, it is often useful to start by focusing on genes that code for polypeptides. In this chapter, you have learned in molecular terms how a typical gene is expressed—by transcription into RNA and then translation into a polypeptide that forms a protein of specific structure and function. Proteins, in turn, bring about an organism's observable phenotype.

A given type of cell expresses only a subset of its genes. This is an essential feature in multicellular organisms: You'd be in trouble if the lens cells in your eyes started expressing the genes for hair proteins, which are normally expressed only in hair follicle cells! Gene expression is precisely regulated, which we'll explore in the next chapter, beginning with the simpler case of bacteria and continuing with eukaryotes.

CONCEPT CHECK 17.5

1. What happens when one nucleotide pair is lost from the middle of the coding sequence of a gene?

2. **MAKE CONNECTIONS** Individuals heterozygous for the sickle-cell allele are generally healthy but show phenotypic effects of the allele under some circumstances (see Figure 14.17). Explain in terms of gene expression.

3. **DRAW IT** The template strand of a gene includes this sequence:
3′-TACTTGTCCGATATC-5′. The sequence is mutated to 3′-TACTTGTCCAATATC-5′. For both wild-type and mutant sequences, draw the double-stranded DNA, the resulting mRNA, and the amino acid sequence each encodes. What is the effect of the mutation on the amino acid sequence?

4. **WHAT IF?** How could you use the CRISPR-Cas9 system to correct the mutation described in question 3? Based on the answer to question 3, would this be worth doing?

For suggested answers, see Appendix A.

17 Chapter Review

→ Go to **Mastering Biology** for Assignments, the eText, the Study Area, and Dynamic Study Modules.

SUMMARY OF KEY CONCEPTS

→ To review key terms, go to the **Vocabulary Self-Quiz** in the **Mastering Biology** eText or Study Area, or go to goo.gl/zkjz9t.

CONCEPT 17.1

Genes specify proteins via transcription and translation (pp. 336–342)

- Beadle and Tatum's studies of mutant strains of *Neurospora* led to the one gene–one polypeptide hypothesis. During **gene expression**, the information encoded in genes is used to make specific polypeptide chains (enzymes and other proteins) or RNA molecules.
- **Transcription** is the synthesis of RNA complementary to a **template strand** of DNA. **Translation** is the synthesis of a polypeptide whose amino acid sequence is specified by the nucleotide sequence in **messenger RNA (mRNA)**.
- Genetic information is encoded as a sequence of nonoverlapping nucleotide triplets, or **codons**. A codon in mRNA either is translated into an amino acid (61 of the 64 codons) or serves as a stop signal (3 codons). Codons must be read in the correct **reading frame**.

❓ *Describe the process of gene expression, by which a gene affects the phenotype of an organism.*

CONCEPT 17.2

Transcription is the DNA-directed synthesis of RNA: A Closer Look (pp. 342–344)

- RNA synthesis is catalyzed by **RNA polymerase**, which links together RNA nucleotides complementary to a DNA template strand. Transcription follows the same base-pairing rules as

DNA replication, except that in RNA, uracil (U) substitutes for thymine (T).

- The three stages of transcription are initiation, elongation, and termination. A **promoter** sequence, often including a **TATA box** in eukaryotes, establishes where RNA synthesis is initiated. **Transcription factors** help eukaryotic RNA polymerase recognize promoter sequences, forming a **transcription initiation complex**. Termination differs in bacteria and eukaryotes.

❓ *Compare the initiation of transcription in bacteria and eukaryotes.*

CONCEPT 17.3

Eukaryotic cells modify RNA after transcription (pp. 345–347)

- Eukaryotic mRNAs undergo **RNA processing**, which includes RNA splicing, the addition of a modified nucleotide **5′ cap** to the 5′ end, and the addition of a **poly-A tail** to the 3′ end. The processed mRNA includes an untranslated region (5′ UTR or 3′ UTR) at each end of the coding segment.
- Most eukaryotic genes are split into segments: They have **introns** interspersed among the **exons** (the regions included in the mRNA). In **RNA splicing**, introns are removed and exons joined. RNA splicing is typically carried out by **spliceosomes**, but in

some cases, RNA alone catalyzes its own splicing. The properties of RNA allow some RNAs (called **ribozymes**) to act as catalysts. The presence of introns allows for **alternative RNA splicing**.

? *What function do the 5' cap and the poly-A tail serve on a eukaryotic mRNA?*

CONCEPT 17.4

Translation is the RNA-directed synthesis of a polypeptide: *A Closer Look* (pp. 347–356)

- A cell translates an mRNA message into protein using **transfer RNAs (tRNAs)**. After being bound to a specific amino acid by an **aminoacyl-tRNA synthetase**, a tRNA lines up via its **anticodon** at the complementary codon on mRNA. A **ribosome**, made up of **ribosomal RNAs (rRNAs)** and proteins, facilitates this coupling with binding sites for mRNA and tRNA.
- Ribosomes coordinate the three stages of translation: initiation, elongation, and termination. The formation of peptide bonds between amino acids is catalyzed by rRNAs as tRNAs move through the **A** and **P sites** and exit through the **E site**.

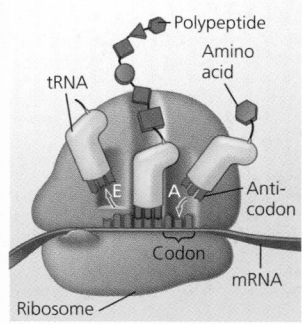

- After translation, during protein processing, proteins may be modified by cleavage or by attachment of sugars, lipids, phosphates, or other chemical groups.
- Free ribosomes in the cytosol initiate synthesis of all proteins, but proteins with a **signal peptide** are synthesized on the ER.
- A gene can be transcribed by multiple RNA polymerases simultaneously. Also, a single mRNA molecule can be translated simultaneously by a number of ribosomes, forming a **polyribosome**. In bacteria, these processes are coupled, but in eukaryotes they are separated in space and time by the nuclear membrane.

? *How do tRNAs function in the context of the ribosome during translation?*

CONCEPT 17.5

Mutations of one or a few nucleotides can affect protein structure and function (pp. 357–362)

- Small-scale **mutations** include **point mutations**, changes in one DNA nucleotide pair, which may lead to production of

nonfunctional proteins. **Nucleotide-pair substitutions** can cause **missense** or **nonsense mutations**. Nucleotide-pair **insertions** or **deletions** may produce **frameshift mutations**.

- Spontaneous mutations can occur during DNA replication, recombination, or repair. Chemical and physical **mutagens** cause DNA damage that can alter genes.
- The **CRISPR-Cas9 system** is a powerful new **gene-editing** technique with the potential to correct genetic mutations that cause disease. Significant technical and ethical questions have been raised about how this could or should be used.

? *What will be the results of chemically modifying one nucleotide base of a gene? What role is played by DNA repair systems in the cell?*

TEST YOUR UNDERSTANDING

→ For more multiple-choice questions, go to the **Practice Test** in the **Mastering Biology** eText or Study Area, or go to goo.gl/GruWRg.

Levels 1-2: Remembering/Understanding

1. In eukaryotic cells, transcription cannot begin until
(A) the two DNA strands have completely separated and exposed the promoter.
(B) several transcription factors have bound to the promoter.
(C) the 5' caps are removed from the mRNA.
(D) the DNA introns are removed from the template.

2. Which of the following is true of a codon?
(A) It never codes for the same amino acid as another codon.
(B) It can code for more than one amino acid.
(C) It can be either in DNA or in RNA
(D) It is the basic unit of protein structure.

3. The anticodon of a particular tRNA molecule is
(A) complementary to the corresponding mRNA codon.
(B) complementary to the corresponding triplet in rRNA.
(C) the part of tRNA that bonds to a specific amino acid.
(D) catalytic, making the tRNA a ribozyme.

4. Which of the following is true of RNA processing?
(A) Exons are cut out before mRNA leaves the nucleus.
(B) Nucleotides are added at both ends of the RNA.
(C) Ribozymes may function in the addition of a 5' cap.
(D) RNA splicing adds a poly-A tail to the mRNA.

5. Which component is directly involved in translation?
(A) RNA polymerase   (C) spliceosome
(B) ribosome   (D) DNA

Levels 3-4: Applying/Analyzing

6. Using Figure 17.6, identify a 5' → 3' sequence of nucleotides in the DNA template strand for an mRNA coding for the polypeptide sequence Phe-Pro-Lys.
(A) 5'-UUUCCCAAA-3'
(B) 5'-GAACCCCTT-3'
(C) 5'-CTTCGGGAA-3'
(D) 5'-AAACCCUUU-3'

7. Which of the following mutations would be *most* likely to have a harmful effect on an organism?
(A) a deletion of three nucleotides near the middle of a gene
(B) a single nucleotide deletion in the middle of an intron
(C) a single nucleotide deletion near the end of the coding sequence
(D) a single nucleotide insertion downstream of, and close to, the start of the coding sequence

8. Would the coupling of the processes shown in Figure 17.24 be found in a eukaryotic cell? Explain why or why not.

9. Complete the following table:

Type of RNA	Functions
Messenger RNA (mRNA)	
Transfer RNA (tRNA)	
	In a ribosome, plays a structural role; as a ribozyme, plays a catalytic role (catalyzes peptide bond formation)
Primary transcript	
Small RNAs in the spliceosome	

Levels 5-6: Evaluating/Creating

10. **EVOLUTION CONNECTION** Most amino acids are coded for by a set of similar codons (see Figure 17.6). Propose at least one evolutionary explanation to account for this pattern.

11. **SCIENTIFIC INQUIRY** Knowing that the genetic code is almost universal, a scientist uses molecular biological methods to insert the human β-globin gene (shown in Figure 17.12) into bacterial cells, hoping the cells will express it and synthesize functional β-globin protein. Instead, the protein produced is nonfunctional and is found to contain many fewer amino acids than does β-globin made by a eukaryotic cell. Explain why.

12. **WRITE ABOUT A THEME: INFORMATION** Evolution accounts for the unity and diversity of life, and the continuity of life is based on heritable information in the form of DNA. In a short essay (100–150 words), discuss how the fidelity with which DNA is inherited is related to the processes of evolution. (Review the discussion of proofreading and DNA repair in Concept 16.2.)

13. **SYNTHESIZE YOUR KNOWLEDGE**

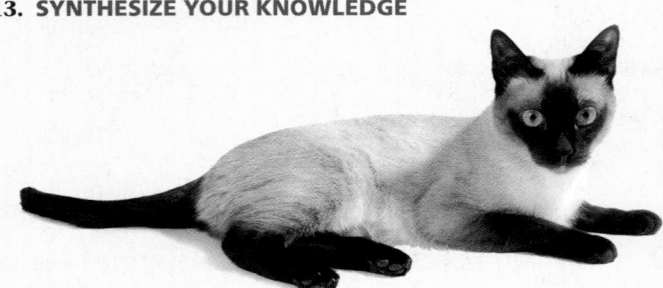

Some mutations result in proteins that function well at one temperature but are nonfunctional at a different (usually higher) temperature. Siamese cats have such a "temperature-sensitive" mutation in a gene encoding an enzyme that makes dark pigment in the fur. The mutation results in the breed's distinctive point markings and lighter body color (see the photo). Using this information and what you learned in the chapter, explain the pattern of the cat's fur pigmentation.

For selected answers, see Appendix A.

18 Regulation of Gene Expression

KEY CONCEPTS

18.1 Bacteria often respond to environmental change by regulating transcription *p. 366*

18.2 Eukaryotic gene expression is regulated at many stages *p. 370*

18.3 Noncoding RNAs play multiple roles in controlling gene expression *p. 379*

18.4 A program of differential gene expression leads to the different cell types in a multicellular organism *p. 381*

18.5 Cancer results from genetic changes that affect cell cycle control *p. 388*

Study Tip

Make a visual study guide: Draw a region of DNA representing a gene. For each level of gene regulation that you read about, draw a sketch of your gene being regulated at that level. A drawing of chromatin modification regulation is shown as an example.

DNA — Gene

Chromatin modification regulation

Gene in region of tightly condensed chromatin (heterochromatin); gene not expressed

Nucleosome (DNA + histones)

Chromatin modification and loosening

Gene in region of loosely condensed chromatin (euchromatin); gene may be expressed

➔ Go to Mastering Biology

For Students (in eText and Study Area)
- Get Ready for Chapter 18
- Animation: Control of Gene Expression
- Animation: Causes of Cancer

For Instructors to Assign (in Item Library)
- Tutorial: Regulation of Gene Expression in Eukaryotes
- Scientific Skills Exercise: Analyzing DNA Deletion Experiments

Ready-to-Go Teaching Module
(in Instructor Resources)
- The *trp* and *lac* Operons (Concept 18.1)

Figure 18.1 *Anableps anableps*, or "cuatro ojos" ("four eyes"), glides through lakes and ponds in South America, the upper half of each eye protruding from the water. The eye's upper half is well-suited for aerial vision and the lower half for aquatic vision. All eye cells, however, contain the same genes.

How can two cells with the same set of genes function differently?

To be expressed, each gene requires a particular set of transcription factors. Different cells have different sets of specific transcription factors.

Eye cell specialized for aerial vision

Specific transcription factors

RNA polymerase II

Gene 1

Gene 2

Gene 3

Expressed — Gene 1 product (optimizes eyesight in air)

No transcription factors for Gene 2 in this cell, so Gene 2 is not expressed.

Expressed — Gene 3 product (required in all eye cells)

Eye cell specialized for aquatic vision

Gene 1

Gene 2

Gene 3

No transcription factors for Gene 1 in this cell, so Gene 1 is not expressed.

Expressed — Gene 2 product (optimizes eyesight in water)

Expressed — Gene 3 product (required in all eye cells)

Differential gene expression is the expression of different genes by cells with the same genome. The uniquely expressed genes allow cells to carry out their specific function.

Bacteria often respond to environmental change by regulating transcription

Bacterial cells that can conserve resources and energy have a selective advantage over cells that are unable to do so. Thus, natural selection has favored bacteria that express only the genes whose products are needed by the cell.

Consider, for instance, an individual *Escherichia coli* (*E. coli*) cell living in a human colon, dependent for its nutrients on the whimsical eating habits of its host. If the environment is lacking in the amino acid tryptophan, which the bacterium needs to survive, the cell responds by activating a metabolic pathway that makes tryptophan from another compound. If the human host later eats a tryptophan-rich meal, the bacterial cell stops producing tryptophan, thus avoiding wasting resources to produce a substance that is readily available from the surrounding solution.

A metabolic pathway can be controlled on two levels, as shown for the synthesis of tryptophan in **Figure 18.2**. First, cells can adjust the activity of enzymes already present. This is a fairly rapid physiological response, which relies on the sensitivity of many enzymes to chemical cues that increase or decrease their catalytic activity (see Concept 8.5). The activity of the first enzyme in the pathway is inhibited by the pathway's end product—tryptophan, in this case (Figure 18.2a). Thus, if tryptophan accumulates in a cell, it shuts down the synthesis of more tryptophan by inhibiting enzyme activity. Such *feedback inhibition*, typical of anabolic (biosynthetic) pathways, allows a cell to adapt to short-term fluctuations in the supply of a substance it needs (see Figure 8.21).

Second, cells can adjust the production level of certain enzymes via a genetic mechanism; that is, they can regulate the expression of the genes encoding the enzymes. If, in our example, the environment provides all the tryptophan the cell needs, the cell stops making the enzymes that catalyze the synthesis of tryptophan (Figure 18.2b). In this case, the control of enzyme production occurs at the level of transcription, the synthesis of messenger RNA from the genes that code for these enzymes.

Regulation of the tryptophan synthesis pathway is just one example of how bacteria tune their metabolism to changing environments. Many genes of the bacterial genome are switched on or off by changes in the metabolic status of the cell; some genes are regulated singly and others as groups of related genes. One basic mechanism for this type of regulation of groups of genes in bacteria, described as the *operon model*, was discovered in 1961 by François Jacob and Jacques Monod at the Pasteur Institute in Paris. Let's see what an operon is and how it works.

▼ **Figure 18.2 Regulation of a metabolic pathway.** In the pathway for tryptophan synthesis, an abundance of tryptophan can both **(a)** inhibit the activity of the first enzyme in the pathway (feedback inhibition), a rapid response, and **(b)** repress expression of the genes encoding all subunits of the enzymes in the pathway, a longer-term response. The symbol ⊖ stands for inhibition.

(a) Regulation of enzyme activity

(b) Regulation of enzyme production

Operons: The Basic Concept

E. coli synthesizes the amino acid tryptophan from a precursor molecule in the three-step pathway shown in Figure 18.2. Each reaction in the pathway is catalyzed by a specific enzyme, and the five genes that code for the subunits of these enzymes are clustered together on the bacterial chromosome. A single promoter serves all five genes, which together constitute a transcription unit. (Recall that a promoter is a site where RNA polymerase can bind to DNA and begin transcription; see Figure 17.8.) Thus, transcription gives rise to one long mRNA molecule that codes for the five polypeptides making up the enzymes in the tryptophan pathway **(Figure 18.3a)**. The cell can translate this one mRNA into five separate polypeptides because the mRNA is punctuated with start and stop codons that signal where the coding sequence for each polypeptide begins and ends.

A key advantage of grouping genes of related function into one transcription unit is that a single "on-off switch" can control the whole cluster of functionally related genes; in other words, these genes are *coordinately controlled*. When an *E. coli* cell must make tryptophan for itself because its surrounding environment lacks this amino acid, all the enzymes for the metabolic pathway are synthesized at the same time. The on-off switch is a segment of DNA called an **operator**. Both its location and name suit its function: Positioned within the promoter or, in some cases, between the promoter and the enzyme-coding genes, the operator controls the access of

(a) Tryptophan absent, repressor inactive, operon on. RNA polymerase attaches to the DNA at the operon's promoter and transcribes the operon's genes. Enzymes for tryptophan synthesis are made.

(b) Tryptophan present, repressor active, operon off. As tryptophan accumulates, it inhibits its own production by activating the repressor protein, which binds to the operator, blocking transcription. Enzymes for tryptophan synthesis are not made.

▲ **Figure 18.3 The *trp* operon in *E. coli*: regulated synthesis of repressible enzymes.** Tryptophan is an amino acid produced by an anabolic pathway catalyzed by three enzymes (see Figure 18.2). **(a)** The five genes encoding the polypeptide subunits of the enzymes in this pathway are grouped, along with a promoter, into the *trp* operon. The *trp* operator (the repressor binding site) is located within the *trp* promoter (the RNA polymerase binding site). **(b)** Accumulation of tryptophan, the end product of the pathway, represses transcription of the *trp* operon, thus blocking synthesis of all the enzymes in the pathway and shutting down tryptophan production.

VISUAL SKILLS *Describe what happens to the* trp *operon as the cell uses up its store of tryptophan.*

RNA polymerase to the genes. Together, the operator, the promoter, and the genes they control—the entire stretch of DNA required for enzyme production for the tryptophan pathway—constitute an **operon**. The *trp* operon (*trp* for tryptophan) is one of many operons in the *E. coli* genome (see Figure 18.3a).

If the operator is the operon's switch for controlling transcription, how does this switch work? By itself, the *trp* operon is turned on; that is, RNA polymerase can bind to the promoter and transcribe the genes of the operon. The *trp* operon can be switched off by a protein that is called the *trp* repressor. A **repressor** binds to the operator, preventing RNA polymerase from transcribing the genes, often by preventing RNA polymerase from binding **(Figure 18.3b)**. A repressor protein is specific for the operator of a particular operon. For example, the *trp* repressor, which switches off the *trp* operon by binding to the *trp* operator, has no effect on other operons in the *E. coli* genome.

A repressor protein is encoded by a **regulatory gene**—in this case, a gene called *trpR; trpR* is located some distance from the *trp* operon and has its own promoter. Regulatory genes are among the bacterial genes that are expressed continuously,

although at a low rate, and a few *trp* repressor molecules are always present in *E. coli* cells. Why, then, is the *trp* operon not switched off permanently? First, the binding of repressors to operators is reversible. An operator alternates between two states: one with the repressor bound and one without the repressor bound. The relative duration of the repressor-bound state is higher when more active repressor molecules are present. Second, the *trp* repressor, like most regulatory proteins, is an allosteric protein, with two alternative shapes: active and inactive (see Figure 8.20). The *trp* repressor is synthesized in the inactive form, which has little affinity for the *trp* operator. Only when a tryptophan molecule binds to the *trp* repressor at an allosteric site does the repressor protein change to the active form that can attach to the operator, turning the operon off.

Tryptophan functions in this system as a **corepressor**, a small molecule that cooperates with a repressor protein to switch an operon off. As tryptophan accumulates, more tryptophan molecules associate with *trp* repressor molecules, which can then bind to the *trp* operator and shut down production of the tryptophan pathway enzymes. If the cell's tryptophan level drops, many fewer *trp* repressor proteins would have tryptophan bound, rendering them inactive; they would dissociate from the operator, allowing transcription of the operon's genes to resume. The *trp* operon is one example of how gene expression can respond to changes in the cell's internal and external environment.

Repressible and Inducible Operons: Two Types of Negative Gene Regulation

The *trp* operon is said to be a *repressible operon* because its transcription is usually on but can be inhibited (repressed) when a specific small molecule (in this case, tryptophan) binds allosterically to a regulatory protein. In contrast, an *inducible operon* is usually off but can be stimulated (induced) to be on when a specific small molecule interacts with a different regulatory protein. The classic example of an inducible operon is the *lac* operon (*lac* stands for "lactose").

The disaccharide lactose (milk sugar) is available to *E. coli* when the bacterium is in contact with any dairy product. Lactose metabolism by *E. coli* begins with hydrolysis of the disaccharide into its component monosaccharides (glucose and galactose), a reaction catalyzed by the enzyme β-galactosidase. Only a few molecules of this enzyme are present in an *E. coli* cell growing in the absence of lactose. If lactose is added to the bacterium's environment, however,

the number of β-galactosidase molecules in the cell increases 1,000-fold within about 15 minutes. How can a cell ramp up enzyme production this quickly?

The gene for β-galactosidase (*lacZ*) is part of the *lac* operon, which includes two other genes coding for enzymes that function in the use of lactose **(Figure 18.4)**. The entire transcription unit is under the command of one main operator and promoter. The regulatory gene, *lacI*, located outside the *lac* operon, codes for an allosteric repressor protein that can switch off the *lac* operon by binding to the *lac* operator. So far, this sounds just like regulation of the *trp* operon, but there is one important difference. Recall that the *trp* repressor protein is inactive by itself and requires tryptophan as a corepressor in order to bind to the operator. The *lac* repressor, in contrast, is active by itself, binding to the operator and switching the *lac* operon off. In this case, a specific small molecule, called an **inducer**, *inactivates* the repressor.

For the *lac* operon, the inducer is allolactose, an isomer of lactose formed in small amounts from lactose that enters the cell. In the absence of lactose (and therefore allolactose), the *lac* repressor is in its active shape and binds to the operator; thus, the genes of the *lac* operon are silenced (Figure 18.4a). If lactose is added to the cell's surroundings, allolactose binds to the *lac* repressor and alters its shape so the repressor can no longer bind to the operator. Without the *lac* repressor bound, the *lac* operon is transcribed into mRNA, and the enzymes for using lactose are made (Figure 18.4b).

In the context of gene regulation, the enzymes of the lactose pathway are referred to as *inducible enzymes* because their synthesis is induced by a chemical signal (allolactose, in this case). Analogously, the enzymes for tryptophan synthesis are said to be repressible. *Repressible enzymes* generally function in anabolic pathways, which synthesize essential end

(a) Lactose absent, repressor active, operon off. The *lac* repressor is innately active, and in the absence of lactose it switches off the operon by binding to the operator. The enzymes for using lactose are not made.

(b) Lactose present, repressor inactive, operon on. Allolactose, an isomer of lactose, binds to the repressor, inactivating it and "derepressing" the operon. The inactive repressor cannot bind to the operator, and so the genes of the *lac* operon are transcribed, and the enzymes for using lactose are made.

→ **Mastering Biology**
Animation: The *lac* Operon in *E. coli*

◄ **Figure 18.4 The *lac* operon in *E. coli*: regulated synthesis of inducible enzymes.** *E. coli* uses three enzymes to take up and metabolize lactose, the genes for which are clustered in the *lac* operon. The first gene, *lacZ*, codes for β-galactosidase, which hydrolyzes lactose to glucose and galactose. The second, *lacY*, codes for a permease, the membrane protein that transports lactose into the cell. The third, *lacA*, codes for transacetylase, an enzyme that detoxifies other molecules entering the cell via the permease. Unusually, the gene for the *lac* repressor, *lacI*, is adjacent to the *lac* operon. The function of the teal region within the promoter will be revealed in Figure 18.5.

products from raw materials (precursors). By suspending production of an end product when it is already present in sufficient quantity, the cell can allocate its organic precursors and energy for other uses. In contrast, inducible enzymes usually function in catabolic pathways, which break down a nutrient to simpler molecules. By producing the appropriate enzymes only when the nutrient is available, the cell avoids wasting energy and precursors making proteins that are not needed.

Regulation of both the *trp* and *lac* operons involves the *negative* control of genes because the operons are switched off by the active form of their respective repressor protein. It may be easier to see this for the *trp* operon, but it is also true for the *lac* operon. In the case of the *lac* operon, allolactose induces enzyme synthesis not by directly activating the *lac* operon, but by freeing it from the negative effect of the repressor (see Figure 18.4b). Gene regulation is said to be *positive* only when a regulatory protein interacts directly with the genome to increase transcription.

Positive Gene Regulation

When glucose and lactose are both present in its environment, *E. coli* preferentially uses glucose. The enzymes for glucose breakdown in glycolysis (see Figure 9.8) are continually present. Only when lactose is present *and* glucose is in short supply does *E. coli* use lactose as an energy source, and only then does it synthesize appreciable quantities of the enzymes for lactose breakdown.

How does the *E. coli* cell sense the glucose concentration and relay this information to the *lac* operon? Again, the mechanism depends on the interaction of an allosteric regulatory protein with a small organic molecule, **cyclic AMP (cAMP)** in this case, which accumulates when glucose is scarce (see Figure 11.11 for the structure of cAMP). The regulatory protein, called *cAMP receptor protein* (*CRP*), is an **activator**, a protein that binds to DNA and stimulates transcription of a gene. (CRP is also called *catabolite activator protein*, or CAP.) When cAMP binds to this regulatory protein, CRP assumes its active shape and can attach to a specific site at the upstream end of the *lac* promoter **(Figure 18.5a)**. This attachment increases the affinity of RNA polymerase for the *lac* promoter, which is actually rather low even when no *lac* repressor is bound to the operator. By facilitating the binding of RNA polymerase to the promoter and thereby increasing the rate of transcription of the *lac* operon, the attachment of CRP to the promoter directly stimulates gene expression. Therefore, this mechanism qualifies as positive regulation.

If the amount of glucose in the cell increases, the cAMP concentration falls, and without cAMP, CRP detaches from the *lac* operon. Because CRP is inactive, RNA polymerase binds less efficiently to the promoter, and transcription of the *lac* operon proceeds only at a low level, even when lactose is present **(Figure 18.5b)**. Thus, the *lac* operon is under dual control: negative control by the *lac* repressor and positive control by CRP. Whether or not transcription occurs is controlled by allolactose:

Without allolactose, the *lac* repressor is active and the operon is off (transcription does not occur; see Figure 18.4a); with allolactose, the *lac* repressor is inactive and the operon is on (transcription occurs; see Figure 18.4b). The *rate* of transcription is controlled by whether CRP has cAMP bound to it: With bound cAMP, the rate is high; without it, the rate is low. It is as though the operon has both an on-off switch and a volume control.

In addition to regulating the *lac* operon, CRP helps regulate other operons that encode enzymes used in catabolic pathways. All told, it may affect the expression of more than 100 genes in *E. coli*. When glucose is plentiful and CRP is inactive, the synthesis of enzymes that catabolize compounds other than glucose generally slows down. The ability to catabolize other compounds, such as lactose, enables a cell deprived of

▼ **Figure 18.5 Positive control of the lac operon by cAMP receptor protein (CRP).** RNA polymerase has high affinity for the *lac* promoter only when CRP is bound to a DNA site at the upstream end of the promoter. CRP, in turn, attaches to its DNA site only when associated with cyclic AMP (cAMP), whose concentration in the cell rises when the glucose concentration falls. Thus, when glucose is present, even if lactose is also available, the cell preferentially catabolizes glucose and makes very low levels of the enzymes for using lactose.

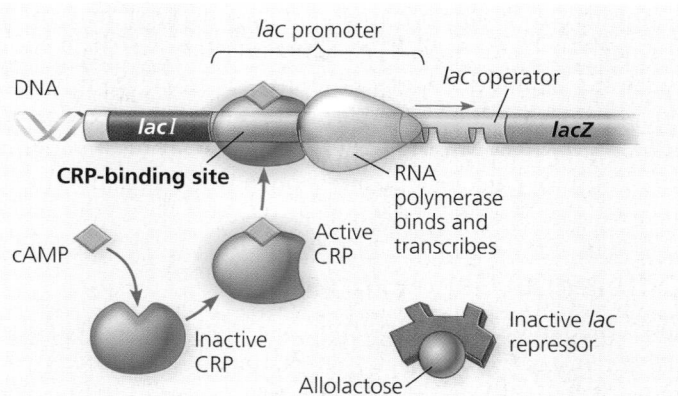

(a) Lactose present, glucose scarce (cAMP level high): abundant *lac* mRNA synthesized. If glucose is scarce, the high level of cAMP activates CRP, which binds to the promoter and increases RNA polymerase binding there. The *lac* operon produces large amounts of mRNA coding for the enzymes that the cell needs for use of lactose.

(b) Lactose present, glucose present (cAMP level low): little *lac* mRNA synthesized. When glucose is present, cAMP is scarce, and CRP is unable to stimulate transcription at a significant rate, even though no repressor is bound.

glucose to survive. The compounds present in any given cell at a certain moment determine which operons are switched on—the result of simple interactions of activator and repressor proteins with the promoters of the genes in question.

CONCEPT CHECK 18.1

1. How does binding of the *trp* corepressor to the *trp* repressor alter repressor function and transcription? How does binding of the *lac* inducer alter the function of the *lac* repressor?

2. Describe the binding of repressors and activators to the *lac* operon when both lactose and glucose are scarce. What is the effect of these scarcities on transcription of the *lac* operon?

3. **WHAT IF?** A certain mutation in *E. coli* changes the *lac* operator so that the active repressor cannot bind. How would this affect the cell's production of β-galactosidase?

For suggested answers, see Appendix A.

CONCEPT 18.2

Eukaryotic gene expression is regulated at many stages

All organisms, whether prokaryotes or eukaryotes, must regulate which genes are expressed at any given time. Both unicellular organisms and the cells of multicellular organisms continually turn genes on and off in response to signals from their external and internal environments. Regulation of gene expression is also essential for cell specialization in multicellular organisms, which are made up of different types of cells. To perform its own distinct role, each cell type must maintain a specific program of gene expression in which certain genes are expressed and others are not (see Figure 18.1).

Differential Gene Expression

A typical human cell might express about a third to a half of its protein-coding genes at any given time. Highly differentiated cells, such as muscle or nerve cells, express a smaller fraction of their genes. Almost all the cells in a multicellular organism contain an identical genome. (Cells of the immune system are one exception, as you will see in Figure 43.13.) A subset of genes is expressed in each cell type; some of these—about 35%—are "housekeeping" genes, expressed by many cell types, while others are unique to that cell type. The uniquely expressed genes allow these cells to carry out their specific function. The differences between cell types, therefore, are due not to different genes being present, but to **differential gene expression**, the expression of different genes by cells with the same genome.

The function of any cell depends on its expressing the appropriate set of genes. The transcription factors of a cell must locate the right genes at the right time, a task like finding a needle in a haystack. Abnormal gene expression can cause serious imbalances and diseases, including cancer.

Figure 18.6 summarizes the process of gene expression in a eukaryotic cell, highlighting key stages in the expression of a protein-coding gene. Each stage depicted in Figure 18.6

▼ **Figure 18.6 Stages in gene expression that can be regulated in eukaryotic cells.** In this diagram, the colored boxes indicate the processes most often regulated; each color indicates the type of molecule that is affected (blue = DNA, red/orange = RNA, purple = protein). The nuclear envelope separating transcription from translation in eukaryotic cells allows for RNA processing, a form of post-transcriptional control that is not possible in prokaryotes. Eukaryotes also have a greater variety of control mechanisms operating before transcription and after translation. Some genes are regulated at multiple stages.

→ **Mastering Biology**
Animation: Control of Gene Expression

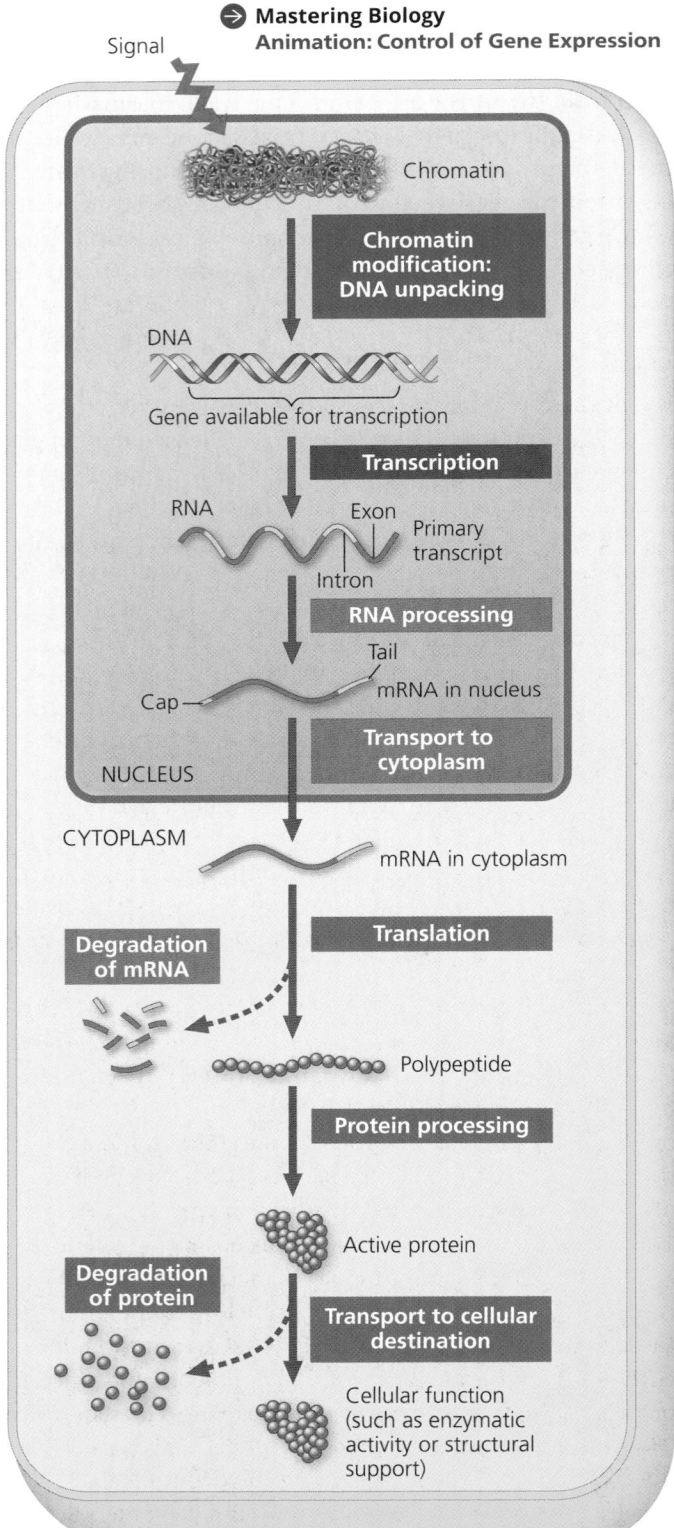

is a potential control point at which gene expression can be turned on or off, accelerated, or slowed down. Many genes have more than one control point.

When the structure of DNA was determined in 1953, an understanding of the mechanisms that control gene expression in eukaryotes seemed almost hopelessly out of reach. Since then, advances in DNA technology (see Chapter 20) have enabled molecular biologists to uncover many details of eukaryotic gene regulation. In all organisms, gene expression is commonly controlled at transcription; regulation at this stage often occurs in response to signals coming from outside the cell, such as hormones or other signaling molecules. For this reason, the term *gene expression* is often equated with transcription for both bacteria and eukaryotes. While this may be the case for bacteria, the greater complexity of eukaryotic cell structure and function provides opportunities for regulating gene expression at many stages besides transcription (see Figure 18.6). Let's examine some of the important control points of eukaryotic gene expression more closely.

Regulation of Chromatin Structure

Recall that the DNA of eukaryotic cells is packaged with proteins in an elaborate complex known as chromatin, the basic unit of which is the nucleosome (see Figure 16.23). The structural organization of chromatin not only packs a cell's DNA into a compact form that fits inside the nucleus, but also helps regulate gene expression in several ways. Genes within heterochromatin, which is more densely arranged than euchromatin, are usually not expressed. In euchromatin, whether or not a gene is transcribed is affected by the location of nucleosomes along a gene's promoter and also the sites where the DNA attaches to the protein scaffolding of the chromosome.

Chromatin structure and gene expression can be influenced by chemical modifications of both the histone proteins of the nucleosomes around which DNA is wrapped and the nucleotides that make up that DNA. Here we examine the effects of these modifications, which are catalyzed by specific enzymes.

Histone Modifications and DNA Methylation

Chemical modifications to histones, found in all eukaryotic organisms, play a direct role in the regulation of gene transcription. The N-terminus of each histone protein in a nucleosome protrudes outward from the nucleosome **(Figure 18.7a)**. These so-called *histone tails* are accessible to various modifying enzymes that catalyze the addition or removal of specific chemical groups, such as acetyl ($-COCH_3$), methyl, and phosphate groups (see Figure 4.9). Generally, **histone acetylation**—the addition of an acetyl group to an amino acid in a histone tail—appears to promote transcription by opening up chromatin structure **(Figure 18.7b)**, while the addition of methyl groups to histones can lead to the condensation of chromatin and reduced transcription. Often, the addition of a particular chemical group may create a new binding site for enzymes that further modify chromatin structure.

Rather than modifying histone proteins, a different set of enzymes can methylate the DNA itself on certain bases, usually cytosine. Such **DNA methylation** occurs in most plants, animals, and fungi. Long stretches of inactive DNA, such as that of inactivated mammalian X chromosomes (see Figure 15.8), are generally more methylated than regions of actively transcribed DNA. On a smaller scale, the DNA of individual genes is usually more heavily methylated in cells in which those genes are not expressed. Removal of the extra methyl groups can turn on some of these genes.

▼ **Figure 18.7 A simple model of histone tails and the effect of histone acetylation.** Amino acids in histone tails may be chemically modified by addition of acetyl groups (green balls) or other groups (such as methyl or phosphate groups). Such modifications affect chromatin structure in a region, sometimes by providing binding sites for other chromatin-modifying enzymes.

 Mastering Biology HHMI **Animation: Role of MECP2 in Gene Silencing**

(a) **Histone tails protrude outward from a nucleosome.** The amino acids in the histone tails are accessible for chemical modification.

(b) **Acetylation of histone tails promotes loose chromatin structure that permits transcription.** A region of chromatin in which nucleosomes are unacetylated forms a more condensed arrangement of chromatin (left) in which the DNA is not transcribed. When nucleosomes become highly acetylated (right), the chromatin becomes less condensed, and the DNA is accessible for transcription.

Once methylated, genes usually stay that way through successive cell divisions in a given individual. At DNA sites where one strand is already methylated, enzymes methylate the correct daughter strand after each round of DNA replication. Methylation patterns are thus passed on to daughter cells, and cells forming specialized tissues keep a chemical record of what occurred during embryonic development. A methylation pattern maintained in this way also accounts for *genomic imprinting* in mammals, where methylation permanently regulates expression of either the maternal or paternal allele of particular genes at the start of development (see Figure 15.17). DNA methylation and histone modification are believed to be coordinated in their regulation.

Epigenetic Inheritance

The chromatin modifications that we just discussed do not change the DNA sequence, yet they still may be passed along to future generations of cells. Inheritance of traits transmitted by mechanisms not involving the nucleotide sequence itself is called *epigenetic inheritance*, the study of which is called **epigenetics**. Whereas mutations in the DNA are permanent changes, modifications to the chromatin can be reversed. For example, DNA methylation patterns are largely erased during gamete formation and reestablished during embryonic development. Furthermore, they are changeable, thus responding more rapidly to environmental conditions.

Research on epigenetics has skyrocketed over the past 20 years. The importance of epigenetic information in regulating gene expression is now widely accepted. One key study by Robert Waterland and Randy Jirtle at Duke University, published in 2003, used a mouse mutant whose genome had been altered so that a gene called *agouti* that determines coat color, normally expressed only briefly during fur formation, was instead expressed throughout development. This overexpression resulted in yellow mice (**Figure 18.8a**, left mouse) rather than the usual brownish ("agouti") color. Previous work had shown that merely supplementing the diet of pregnant mothers with methyl group–containing compounds (like folic acid) could shift the range of coat colors of the offspring back to normal (see Figure 18.8a, right mouse). Waterland and Jirtle reproduced this result, analyzing the state of methylation of the DNA. They showed that the extent of the color shift correlated with the level of DNA methylation. In other words, feeding methyl groups to the mothers at a key time during gestation led to a change in gene expression in the offspring's phenotype. Further studies showed that the effects were even observed in the next generation—the "grandpups" of the original female mouse.

A similar epigenetic effect due to changes in methylation occurs in humans as well. Near the end of World War II, during the winter of 1944–45, Dutch railway workers went on strike to try to prevent the Nazis from bringing in more troops. In retaliation, the Nazis blocked all deliveries of food. Over 20,000 Dutch died in the "Dutch Hunger Winter"

(a) Effects of maternal diet on genetically identical mice. These two mice are genetically identical (*agouti* mutants), but their mothers were fed different diets. A maternal diet without compounds that donate methyl groups resulted in the obese, yellow offspring on the left. A diet supplemented with methyl-donating compounds, such as folic acid, led to the speckled brown fur (normal *agouti* phenotype) and normal weight of the mouse on the right.

(b) The Dutch Hunger Winter. During the Nazi blockade of food transport into the Netherlands near the end of World War II, the population was starving, including this mother and her two children. Children of mothers who were pregnant during that time have experienced long-term effects.

(**Figure 18.8b**). Over time, doctors found that the offspring of women in early pregnancy at that time experienced adverse health effects as adults: higher rates of obesity, high triglyceride and cholesterol levels, type 2 diabetes, and schizophrenia. Furthermore, the affected individuals had a 10% higher mortality rate after the age of 68 than their siblings who were *in utero* when food was readily available. A collaboration between labs in the Netherlands and the United States compared those adults with their siblings and published their findings in 2018. Statistical analysis enabled the researchers to conclude that differences between the siblings in DNA methylation of certain genes caused these long-term adverse medical conditions—examples of epigenetic inheritance.

Epigenetic variations might help explain cases where one identical twin acquires a genetically based disease, such as schizophrenia, but the other does not, despite their identical genomes. Alterations in normal patterns of DNA methylation are seen in some cancers, where they are associated with inappropriate gene expression. Evidently, enzymes that modify chromatin structure are integral parts of the eukaryotic cell's machinery for regulating transcription.

Regulation of Transcription Initiation

Chromatin-modifying enzymes provide initial control of gene expression by making a region of DNA either more or less able to bind the transcription machinery. Once the chromatin of a gene is optimally modified for expression, the initiation of transcription is the next major step at which gene expression is regulated. As in bacteria, the regulation of transcription initiation in eukaryotes involves proteins that bind to DNA and either facilitate or inhibit binding of RNA polymerase. The process is more complicated in eukaryotes, however. Before looking at how eukaryotic cells control their transcription, let's review the structure of a eukaryotic gene.

Organization of a Typical Eukaryotic Gene and Its Transcript

A eukaryotic gene and the DNA elements (segments) that control it are typically organized as shown in **Figure 18.9**, which extends what you learned about eukaryotic genes in Chapter 17. Recall that a cluster of proteins called a *transcription initiation complex* assembles on the promoter sequence at the "upstream" end of the gene (see Figure 17.9).

One of these proteins, RNA polymerase II, then proceeds to transcribe the gene, synthesizing a primary RNA transcript (pre-mRNA). RNA processing includes enzymatic addition of a 5' cap and a poly-A tail, as well as splicing out of introns, to yield a mature mRNA. Associated with most eukaryotic genes are multiple **control elements**, segments of noncoding DNA that serve as binding sites for the proteins called transcription factors, which bind to the control elements and regulate transcription. Control elements on the DNA and the transcription factors that bind to them are critical to the precise regulation of gene expression seen in different cell types.

The Roles of General and Specific Transcription Factors

There are two types of transcription factors: General transcription factors act at the promoter of all genes, while some genes require specific transcription factors that bind to control elements that may be close to or farther away from the promoter.

General Transcription Factors at the Promoter To initiate transcription, eukaryotic RNA polymerase II requires the assistance of transcription factors. Some transcription factors (such as those illustrated in Figure 17.9) are essential for the transcription of *all* protein-coding genes; therefore, they are often called *general transcription factors*. A few general transcription factors bind to a DNA sequence, such as the TATA box in most promoters, but many bind to proteins, including

▼ **Figure 18.9 A eukaryotic gene and its transcript.** Each eukaryotic gene has a promoter—a DNA sequence where RNA polymerase II binds and starts transcription, proceeding "downstream." A number of control elements (gold) are involved in regulating the initiation of transcription; these are DNA sequences located near (proximal to) or far from

(distal to) the promoter. Distal control elements can be grouped together as enhancers, one of which is shown for this gene. At the other end of the gene, a polyadenylation (poly-A) signal sequence in the last exon of the gene is transcribed into an RNA sequence that signals where the transcript is cleaved and the poly-A tail added. Transcription may continue for

hundreds of nucleotides beyond the poly-A signal before terminating. RNA processing of the primary transcript into a functional mRNA involves three steps: addition of the 5' cap, addition of the poly-A tail, and splicing. In the cell, the 5' cap is added soon after transcription is initiated, and splicing occurs while transcription is still under way (see Figures 17.11 and 17.12).

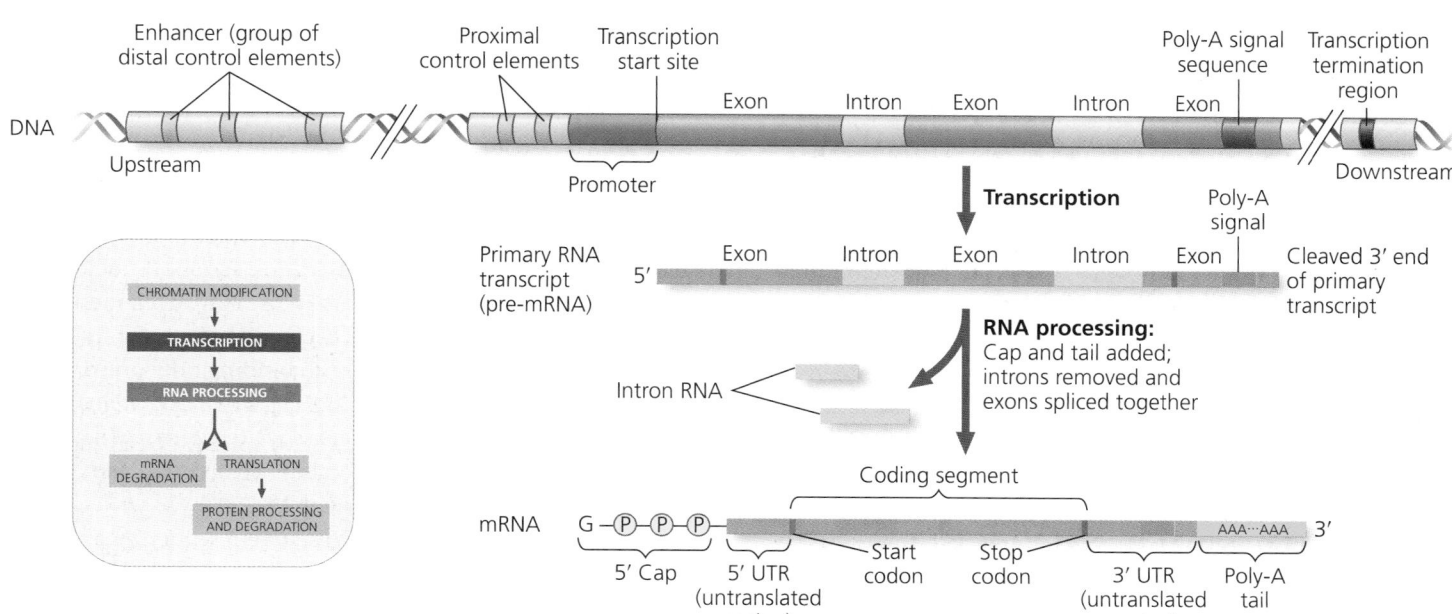

other transcription factors as well as RNA polymerase II. Protein-protein interactions are crucial to the initiation of eukaryotic transcription. Only when the complete initiation complex has assembled can the polymerase begin to move along the DNA template strand and transcribe.

Some genes are expressed all the time, but others are not; instead, they are regulated. For these genes, the interaction of general transcription factors and RNA polymerase II with a promoter usually leads to a low rate of initiation and production of few RNA transcripts from genes that are not expressed at significant levels all the time or in all cells. In eukaryotes, high levels of transcription of these particular genes at the appropriate time and place depend on the interaction of control elements with another set of proteins, which can be thought of as *specific transcription factors*.

Enhancers and Specific Transcription Factors
As you can see in Figure 18.9, some control elements, named *proximal control elements*, are close to the promoter. The more distant *distal control elements*, groupings of which are called **enhancers**, may be thousands of nucleotides upstream or downstream of a gene or even within an intron. A given gene may have multiple enhancers, each active at a different time, cell type, or location in the organism. Each enhancer, however, is generally associated with only that gene and no other.

In eukaryotes, the rate of gene expression can be strongly increased or decreased by the binding of specific transcription factors, either activators or repressors, to the control elements of enhancers. Hundreds of transcription activators have been discovered in eukaryotes; the structure of one is shown in **Figure 18.10**. Researchers have identified two types of structural domains that are commonly found in a large number of transcription activators: a *DNA-binding domain*—a part of the protein's three-dimensional structure that binds to DNA— and one or more *activation domains*. Activation domains bind other regulatory proteins or components of the transcription machinery, facilitating a series of protein-protein interactions that result in enhanced transcription of a given gene.

Figure 18.11 shows the currently accepted model for how binding of activators to an enhancer located far from the promoter influences transcription. Protein-mediated bending of the DNA brings the bound activators into contact with a group of *mediator proteins*, which in turn interact with general transcription factors at the promoter. These protein-protein interactions help assemble and position the initiation complex on the promoter. One of the studies supporting this model shows that proteins regulating one of the mouse globin genes contact both the gene's promoter and an enhancer located about 50,000 nucleotides upstream. Protein interactions allow these two DNA regions to come together in a very specific fashion, in spite of the many nucleotide pairs between them.

Specific transcription factors that function as repressors can inhibit gene expression in several different ways. Some repressors bind directly to control element DNA (in enhancers

▼ **Figure 18.10 The structure of MyoD, a transcriptional activator.** The MyoD protein is made up of two polypeptide subunits (purple and salmon) with extensive regions of α helix. Each subunit has one DNA-binding domain (lower half) and one activation domain (upper half). The latter includes binding sites for the other subunit and for other proteins. MyoD is involved in muscle development in vertebrate embryos (see Concept 18.4).

VISUAL SKILLS *Describe how the two functional domains of the MyoD protein relate to the two polypeptide subunits.*

or elsewhere), blocking activator binding. Other repressors interfere with the activator itself so it can't bind the DNA.

In addition to influencing transcription directly, some activators and repressors indirectly affect chromatin structure. Studies using yeast and mammalian cells show that some activators recruit proteins that acetylate histones near the promoters of specific genes, thus promoting transcription (see Figure 18.7). Similarly, some repressors recruit proteins that remove acetyl groups from histones, leading to reduced transcription, a phenomenon referred to as *silencing*. Indeed, recruitment of chromatin-modifying proteins seems to be the most common mechanism of repression in eukaryotic cells.

⊙ **Mastering Biology** HHMI Animation: Regulation of Eukaryotic DNA Transcription

Combinatorial Control of Gene Activation
In eukaryotes, the precise control of transcription depends largely on the binding of activators to DNA control elements. Considering that many genes must be regulated in a typical animal or plant cell, the number of different nucleotide sequences in control elements is surprisingly small. A dozen or so short nucleotide sequences appear again and again in the control elements for different genes. On average, each enhancer is composed of about ten control elements, each binding only one or two specific transcription factors. It is the particular *combination* of control elements in an enhancer associated with a gene, rather than a single unique control element, that is important in regulating transcription of the gene.

Even with only a dozen control element sequences available, many combinations are possible. Each combination of control elements can activate transcription only when the appropriate

▼ Figure 18.11 A model for the action of enhancers and transcription activators.

Bending of the DNA by a protein enables enhancers to influence a promoter hundreds or even thousands of nucleotides away. Specific transcription factors (activators) bind to the enhancer DNA sequences and then to a group of mediator proteins. These in turn bind to general transcription factors and then RNA polymerase II, thus assembling the transcription initiation complex. These protein-protein interactions lead to correct positioning of the complex on the promoter and the initiation of RNA synthesis. Only one enhancer (with three gold control elements) is shown here, but a gene may have several enhancers that act at different times or in different cell types.

1 Activator proteins bind to distal control elements grouped as an enhancer in the DNA. This enhancer has three binding sites, each called a distal control element.

2 A DNA-bending protein brings the bound activators closer to the promoter. General transcription factors, mediator proteins, and RNA polymerase II are nearby.

3 The activators bind to certain mediator proteins and general transcription factors, helping them form an active transcription initiation complex on the promoter.

CHROMATIN MODIFICATION → TRANSCRIPTION → RNA PROCESSING → mRNA DEGRADATION / TRANSLATION → PROTEIN PROCESSING AND DEGRADATION

DNA — Activators — Enhancer — Distal control element — Promoter — Gene — TATA box — General transcription factors — DNA-bending protein — Group of mediator proteins — RNA polymerase II — RNA synthesis

Transcription initiation complex

◉ **Mastering Biology**
Animation: Regulation of Transcription by Transcription Activators and Enhancers

transcription activators are present, which may occur at a precise time during development or in a particular cell type. In the **Scientific Skills Exercise**, you can work with data from an experiment that identified the control elements in an enhancer of a particular human gene. **Figure 18.12** illustrates how the use of different combinations of just a few control elements can allow differential regulation of transcription in two representative cell types—liver cells and lens cells. This can occur because each cell type contains a different group of transcription activators. How cell types come to differ during this process, even though they all arise from one cell (the fertilized egg), will be explored in Concept 18.4.

Coordinately Controlled Genes in Eukaryotes

How does the eukaryotic cell deal with a group of genes of related function that need to be turned on or off at the same time? Earlier in this chapter, you learned that in bacteria, such *coordinately controlled* genes are often clustered into an operon, which is regulated by a single promoter and transcribed into a single mRNA molecule. Thus, the genes are expressed together, and the encoded proteins are produced at the same time. With a few exceptions, operons that work in this way have *not* been found in eukaryotic cells.

Eukaryotic genes that are co-expressed, such as genes coding for the enzymes of a metabolic pathway, are typically scattered over different chromosomes. Here, coordinate gene expression depends on every gene of a dispersed group having a specific combination of control elements. Transcription activators in the nucleus that recognize the control elements bind to them, promoting simultaneous transcription of the genes, no matter where they are in the genome.

Coordinate control of dispersed genes in a eukaryotic cell often occurs in response to chemical signals from outside the cell. A steroid hormone, for example, enters a cell and binds to a specific intracellular receptor protein, forming a hormone-receptor complex that serves as a transcription activator (see Figure 11.9). Every gene whose transcription is stimulated by a given steroid hormone, regardless of its chromosomal location, has a control element recognized by that hormone-receptor complex. This is how estrogen activates a group of genes that stimulate cell division in uterine cells, preparing the uterus for pregnancy.

Many signaling molecules, such as nonsteroid hormones and growth factors, bind to receptors on a cell's surface and never actually enter the cell. Such molecules can control gene expression indirectly by triggering signal transduction pathways that activate particular transcription factors (see Figure 11.15). Coordinate regulation in such pathways is the same as for steroid hormones: Genes with the same sets of control elements are activated by the same chemical signals. Because this system for coordinating gene regulation is so widespread, biologists think that it probably arose early in evolutionary history.

Scientific Skills Exercise

Analyzing DNA Deletion Experiments

▲ Tissue inflammation

What Control Elements Regulate Expression of the *mPGES-1* Gene? The promoter of a gene includes the DNA immediately upstream of the transcription start site, but the control elements (grouped in an enhancer) that regulate the level of transcription of the gene may be thousands of base pairs upstream of the promoter. Because the distance and spacing of control elements make them difficult to identify, scientists begin by deleting possible control elements and measuring the effect on gene expression. In this exercise, you will analyze data obtained from DNA deletion experiments that tested possible control elements for the human gene *mPGES-1*. This gene codes for an enzyme that synthesizes a type of prostaglandin, a chemical made during tissue inflammation.

How the Experiment Was Done The researchers hypothesized that there were three possible control elements in an enhancer region located 8–9 kilobases (kb) upstream of the *mPGES-1* gene. Control elements regulate whatever gene is in the appropriate downstream location. Thus, to test the activity of the possible elements, researchers first synthesized molecules of DNA *(DNA constructs)* that had the intact enhancer region upstream of a *reporter gene*, a gene whose mRNA product could be easily measured experimentally. Next, they made three more DNA constructs, with one of the three proposed control elements deleted in each (see the left side of the figure). The researchers then introduced each DNA construct into a separate human cell culture, where the cells took up the DNA. After 48 hours, the amount of reporter gene mRNA made by the cells was measured. Comparing these amounts allowed researchers to determine if any of the deletions had an effect on expression of the reporter gene, mimicking the effect of deletions on *mPGES-1* gene expression. (The *mPGES-1* gene itself couldn't be used to measure expression levels because the cells express their own *mPGES-1* gene, and the mRNA from that gene would confuse the results.)

Data from the Experiment The diagrams on the left side of the figure show the intact DNA sequence (top) and the three experimental DNA constructs. A red X is located on the possible control element (1, 2, or 3) that was deleted in each experimental DNA construct. The area between the slashes represents the approximately 8 kb of DNA located between the promoter and the enhancer region. The horizontal bar graph on the right shows

Data from J. N. Walters et al., Regulation of human microsomal prostaglandin E synthase-1 by IL-1b requires a distal enhancer element with a unique role for C/EBPb, *Biochemical Journal* 443:561–571 (2012).

the amount of reporter gene mRNA that was present in each cell culture after 48 hours relative to the amount that was in the culture containing the intact enhancer region (top bar = 100%).

INTERPRET THE DATA

1. **(a)** What is the independent variable in the graph? **(b)** What is the dependent variable? **(c)** What was the control treatment in this experiment? Label it on the diagram.

2. Do the data suggest that any of these possible control elements are actual control elements? Explain.

3. **(a)** Did deletion of any of the possible control elements cause a *reduction* in reporter gene expression? If so, which one(s), and how can you tell? **(b)** If loss of a control element causes a reduction in gene expression, what must be the normal role of that control element? Provide a biological explanation for how the loss of such a control element could lead to a reduction in gene expression.

4. **(a)** Did deletion of any of the possible control elements cause an *increase* in reporter gene expression relative to the control? If so, which one(s), and how can you tell? **(b)** If loss of a control element causes an increase in gene expression, what must be the normal role of that control element? Propose a biological explanation for how the loss of such a control element could lead to an increase in gene expression.

➔ **Instructors**: A version of this Scientific Skills Exercise can be assigned in **Mastering Biology**.

► **Figure 18.12 Cell type–specific transcription.** Both liver cells and lens cells have the genes for making the proteins albumin and crystallin, but only liver cells make albumin (a blood protein) and only lens cells make crystallin (the main protein of the lens of the eye). The specific transcription factors made in a cell determine which genes are expressed. In this example, the genes for albumin and crystallin are shown at the top, each with an enhancer made up of three different control elements. Although the enhancers for the two genes both have a gray control element, each enhancer has a unique *combination* of elements. All the activator proteins required for high-level expression of the albumin gene are present in liver cells only (left), whereas the activators needed for expression of the crystallin gene are present in lens cells only (right). For simplicity, we consider only the role of specific transcription factors that are activators here, although repressors may also influence transcription in certain cell types.

VISUAL SKILLS *Describe the enhancer for the albumin gene in each type of cell. How would the nucleotide sequence of this enhancer in the liver cell compare with that in the lens cell?*

Nuclear Architecture and Gene Expression

Each chromosome in the interphase nucleus of animal cells occupies a distinct territory (see Figure 16.24a). Recently, *chromosome conformation capture* techniques (*3C* and others) have been developed that allow researchers to cross-link and identify regions of chromosomes associating with each other during interphase. These studies reveal two organizational details: First, the territory of each chromosome is divided into regions of chromatin loops (*topologically associated domains,* or *TADs*), within which chromatin sites associate mainly with each other. Second, loops of chromatin, each likely a TAD, extend from individual chromosomal territories into specific sites in the nucleus **(Figure 18.13)**. Different loops from the same chromosome and loops from other chromosomes may congregate in such sites, some of which are rich in RNA polymerases and other transcription-associated proteins. Like a recreation center that draws members from many different neighborhoods, these so-called *transcription factories* are thought to be areas specialized for a common function.

The old view that the nuclear contents are like a bowl of amorphous chromosomal spaghetti has given way to a new model of a nucleus with a defined architecture and regulated

movements of chromatin. Several lines of evidence suggest that unexpressed genes are located in the outer edges of the nucleus, while those that are being expressed are found in its interior region. Relocation of particular genes from their chromosomal territories to transcription factories in the interior may be part of the process of readying genes for transcription. In 2017, a consortium of researchers funded by the National Institutes of Health began investigating genome organization over time and its relationship to genome function.

Mechanisms of Post-transcriptional Regulation

Transcription alone does not constitute gene expression. The expression of a protein-coding gene is ultimately measured by the amount of functional protein a cell makes, and much happens between the synthesis of the RNA transcript and the activity of the protein in the cell. Many regulatory mechanisms operate at the various stages after transcription (see Figure 18.6). These mechanisms allow a cell to rapidly fine-tune gene expression in response to environmental changes without altering its transcription patterns. Here we explore how cells regulate gene expression after transcription.

▼ Figure 18.13 Chromosomal interactions in the interphase nucleus.
Although each chromosome has its own territory (see Figure 16.24a), loops of chromatin may extend into other sites in the nucleus. (Each loop may be a topologically associated domain, or TAD.) Some of these sites are transcription factories that are occupied by multiple chromatin loops from the same chromosome (blue loops) or other chromosomes (red and green loops).

Chromosomes in the interphase nucleus (fluorescence micrograph)

5 μm

Chromosome territory

Chromatin loop

Transcription factory

RNA Processing

RNA processing in the nucleus and the export of mature RNA to the cytoplasm provide opportunities for regulating gene expression not available in prokaryotes. One example of regulation at the RNA-processing level is **alternative RNA splicing**, in which different mRNA molecules are produced from the same primary transcript, depending on which RNA segments are treated as exons and which as introns. Regulatory proteins specific to a cell type control intron/exon choices by binding to regulatory sequences within the primary transcript.

A simple example of alternative RNA splicing is shown in **Figure 18.14** for the troponin T gene, which encodes multiple closely related proteins with slightly different effects on muscle contraction. Other genes code for many more products. For instance, researchers have found a *Drosophila* gene with enough alternatively spliced exons to generate about 19,000 membrane proteins that have different extracellular domains. At least 17,500 (94%) of the alternative mRNAs are actually synthesized. Each developing nerve cell in the fly appears to synthesize a different form of the protein, which acts as a unique identifier on the cell surface and helps prevent excessive overlap of nerve cells during development of the nervous system.

Alternative RNA splicing can significantly expand the repertoire of a eukaryotic genome. In fact, alternative splicing was proposed as one explanation for the surprisingly low number of human genes counted when the human genome was sequenced. The number of human genes was found to be similar to that of a soil worm (nematode), a mustard plant, or a sea anemone. This discovery prompted questions about what, if not the number of genes, accounts for the more complex form and structure of humans. More than 90% of human protein-coding genes likely undergo alternative splicing. Thus, the extent of alternative splicing greatly multiplies the number of possible human proteins, which may be better correlated with complexity of form than the number of genes.

Initiation of Translation and mRNA Degradation

Translation is another stage where gene expression is regulated, most commonly at the initiation stage (see Figure 17.19). For some mRNAs, the initiation of translation can be blocked by regulatory proteins that bind to specific sequences or structures within the untranslated region (UTR) at the 5′ or 3′ end, preventing the attachment of ribosomes. (Recall from Concept 17.3 that both the 5′ cap and the poly-A tail of an mRNA molecule are important for ribosome binding.)

Alternatively, translation of *all* the mRNAs in a cell may be regulated simultaneously. In a eukaryotic cell, such "global" control usually involves the activation or inactivation of one or more protein factors required to initiate translation. This mechanism plays a role in starting translation of mRNAs that are stored in eggs. Just after

▼ Figure 18.14 Alternative RNA splicing of the troponin T gene.
The primary transcript of this gene can be spliced in more than one way, generating different mRNA molecules. Notice that one mRNA molecule has ended up with exon 3 (green) and the other with exon 4 (purple). These two mRNAs are translated into different but related muscle proteins.

CHROMATIN MODIFICATION
↓
TRANSCRIPTION
↓
RNA PROCESSING
↙
mRNA DEGRADATION   TRANSLATION
↓
PROTEIN PROCESSING AND DEGRADATION

Exons

DNA   1 2 3 4 5

Troponin T gene

Primary RNA transcript   1 2 3 4 5

RNA splicing

mRNA   1 2 3 5   OR   1 2 4 5

fertilization, translation is triggered by the sudden activation of translation initiation factors. The response is a burst of synthesis of the proteins encoded by the stored mRNAs. Some plants and algae store mRNAs during periods of darkness; light then triggers the reactivation of the translational apparatus.

The life span of mRNA molecules in the cytoplasm is important in determining the pattern of protein synthesis in a cell. Bacterial mRNA molecules typically are degraded by enzymes within a few minutes. This short life span of mRNAs is one reason bacteria can change their patterns of protein synthesis so quickly in response to environmental changes. In contrast, some mRNAs in multicellular eukaryotes typically survive for hours, days, or even weeks. For instance, the mRNAs for the hemoglobin polypeptides (α-globin and β-globin) in developing red blood cells are unusually stable, and these long-lived mRNAs are translated repeatedly in red blood cells.

Nucleotide sequences that affect how long an mRNA remains intact are often found in the untranslated region at the 3' end of the molecule (see Figure 18.9). In one experiment, researchers transferred such a sequence from the short-lived mRNA for a growth factor to the 3' end of a normally stable globin mRNA. The globin mRNA was quickly degraded.

Other mechanisms that degrade or block expression of mRNA molecules have come to light. They involve a group of recently discovered RNA molecules that regulate gene expression at several levels, as you'll see shortly.

Protein Processing and Degradation

The final opportunities for controlling gene expression occur after translation. Often, eukaryotic polypeptides must be processed to yield functional protein molecules. For instance, cleavage of the initial insulin polypeptide (pro-insulin) forms the active hormone. In addition, many proteins undergo chemical modifications that make them functional. Regulatory proteins are commonly activated or inactivated by the reversible addition of phosphate groups (see Figure 11.10), and proteins destined for the surface of animal cells acquire sugars (see Figure 6.12). Cell-surface proteins and many others must also be transported to target destinations in the cell in order to function (see Figure 17.22). Regulation might occur at any of the steps involved in modifying or transporting a protein.

Finally, the length of time each protein functions in the cell is strictly regulated by selective degradation. Many proteins, such as the cyclins involved in regulating the cell cycle, must be relatively short-lived if the cell is to function appropriately (see Figure 12.16). To mark a protein for destruction, the cell commonly attaches molecules of a small protein called ubiquitin to the protein. Giant protein complexes called proteasomes then recognize the ubiquitin tagged proteins and degrade them.

⊙ **Mastering Biology**
Animation: Post-Transcriptional Control Mechanisms

CONCEPT 18.3

Noncoding RNAs play multiple roles in controlling gene expression

Genome sequencing has revealed that protein-coding DNA accounts for only 1.5% of the human genome and a similarly small percentage of the genomes of many other multicellular eukaryotes. A very small fraction of the non-protein-coding DNA consists of genes for RNAs such as ribosomal RNA and transfer RNA. Until recently, scientists assumed that most of the remaining DNA was not transcribed, thinking that since it didn't specify proteins or the few known types of RNA, such DNA didn't contain meaningful genetic information—in fact, it was called "junk DNA." However, some genomic studies have cast doubt on this description. For example, a massive study showed that roughly 75% of the human genome is transcribed at some point in any given cell. Introns account for only a fraction of this transcribed RNA, most of which is untranslated. The verdict is still out on how much of the transcribed RNA is functional, but at least some of the genome is transcribed into non-protein-coding RNAs (also called *noncoding RNAs*, or *ncRNAs*), including a variety of small RNAs. Researchers are uncovering more evidence of the biological roles of these ncRNAs every day.

These discoveries have revealed a large and diverse population of RNA molecules in the cell that play crucial roles in regulating gene expression and have gone largely unnoticed until recently. The longstanding view that mRNAs are the most important RNAs because they code for proteins needs revision. This represents a major shift in thinking by biologists, one that you are witnessing as students entering this field of study.

Effects on mRNAs by MicroRNAs and Small Interfering RNAs

Regulation by both small and large ncRNAs occurs at several points in the pathway of gene expression, including mRNA translation and chromatin modification. We'll examine two types of small ncRNAs, the importance of which was acknowledged when their discovery was the focus of the 2006 Nobel Prize in Physiology or Medicine, which was awarded for work completed only eight years earlier.

Since 1993, a number of research studies have uncovered **microRNAs (miRNAs)**—small, single-stranded RNA molecules capable of binding to complementary sequences in mRNA molecules. A longer RNA precursor is processed by cellular enzymes into an miRNA, a single-stranded RNA of about 22 nucleotides that forms a complex with one or more proteins **(Figure 18.15)**. The miRNA allows the complex to bind to any mRNA molecule with at least seven or eight nucleotides of complementary sequence. The miRNA-protein complex then degrades the target mRNA or, less often, simply blocks its translation. There are approximately 1,500 genes for miRNAs in the human genome, and biologists estimate that expression of at least one-half of all human genes may be regulated by miRNAs, a remarkable figure given that the existence of miRNAs was unknown until the early 1990s.

Another class of small noncoding RNAs, similar in size and function to miRNAs, is called **small interfering RNAs**

(siRNAs). Both miRNAs and siRNAs can associate with the same proteins, producing similar results. In fact, if siRNA precursor RNA molecules are injected into a cell, the cell's machinery can process them into siRNAs that turn off expression of genes with related sequences, similarly to how miRNAs function. The distinction between miRNAs and siRNAs is based on subtle differences in the structure of their precursors, which in both cases are RNA molecules that are mostly double-stranded. The blocking of gene expression by siRNAs, referred to as **RNA interference (RNAi)**, is used in the laboratory as a means of disabling specific genes to investigate their function.

EVOLUTION How did the RNAi pathway evolve? As you will learn in Concept 19.2, some viruses have double-stranded RNA genomes. Given that the cellular RNAi pathway can process double-stranded RNAs into homing devices that lead to destruction of related RNAs, some scientists think that this pathway may have evolved as a natural defense against infection by such viruses. However, the fact that RNAi can also affect the expression of nonviral cellular genes may reflect a different evolutionary origin for the RNAi pathway.

While this section has focused on ncRNAs in eukaryotes, small ncRNAs are also used by bacteria as a defense system, called the CRISPR-Cas9 system, against viruses that infect them (see Concept 17.5). The use of ncRNAs thus evolved long ago, but we don't yet know how bacterial ncRNAs are related to those of eukaryotes.

Chromatin Remodeling and Effects on Transcription by ncRNAs

In addition to regulating mRNAs, small noncoding RNAs can cause remodeling of chromatin structure. In the S phase of the cell cycle, for example, the centromeric regions of DNA must be loosened for chromosomal replication and then re-condensed into heterochromatin in preparation for mitosis. In some yeasts, siRNAs produced by the yeast cells from the centromeric DNA are required to re-form the heterochromatin at the centromeres. Exactly how the process starts is still debated, but biologists agree on the general idea: The siRNA system in yeast interacts with other, larger noncoding RNAs and with chromatin-modifying enzymes to condense the centromere chromatin into heterochromatin. In most mammalian cells, siRNAs have not been found, and the mechanism for centromere DNA condensation is not yet understood. However, it may also turn out to involve small noncoding RNAs.

A recently discovered class of small ncRNAs called *piwi-interacting RNAs*, or *piRNAs*, also induces formation of heterochromatin, blocking expression of some parasitic DNA elements in the genome known as transposons. (Transposons are discussed in Concept 21.4.) Usually 24–31 nucleotides in length, piRNAs are processed from a longer, single-stranded RNA precursor. They play an indispensable role in the germ

▼ **Figure 18.15 Regulation of gene expression by microRNAs (miRNAs).** An miRNA of about 22 nucleotides, formed by enzymatic processing of an RNA precursor, associates with one or more proteins in a complex that can degrade or block translation of target mRNAs.

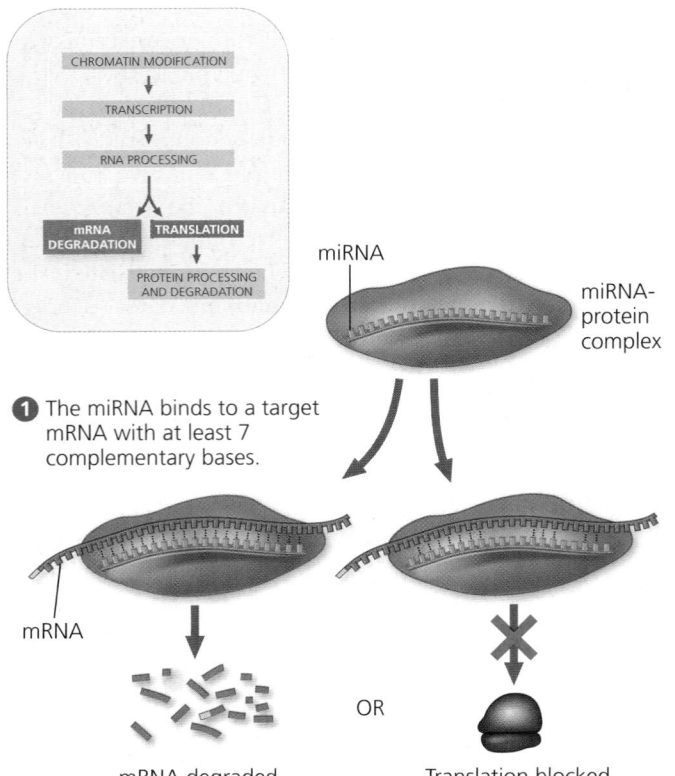

1 The miRNA binds to a target mRNA with at least 7 complementary bases.

mRNA degraded OR Translation blocked

2 If miRNA and mRNA bases are complementary all along their length, the mRNA is degraded (left); if the match is less complete, translation is blocked (right).

cells of many animal species, where they appear to help reestablish appropriate methylation patterns in the genome during gamete formation.

 Mastering Biology Interview with Haifan Lin: Discovering piwi-interacting RNAs and connections to testicular cancer

Researchers have also found a relatively large number of **long noncoding RNAs (lncRNAs)**, ranging from 200 to hundreds of thousands of nucleotides in length, that are expressed at significant levels in specific cell types at particular times. The functional significance of these lncRNAs has been debated, but in 2017 a large international research consortium published an atlas of almost 28,000 such RNAs; their analysis supported the idea that almost 20,000 were functional and that some were associated with specific diseases. One lncRNA, long known to be functional, is responsible for X chromosome inactivation, which prevents expression of genes located on one of the X chromosomes in most female mammals (see Figure 15.8). In this case, lncRNAs—transcripts of the *XIST* gene located on the chromosome to be inactivated—bind back to and coat that chromosome. This binding leads to condensation of the entire chromosome into heterochromatin.

The examples just described involve chromatin remodeling in large regions of the chromosome. Because chromatin structure affects transcription and thus gene expression, RNA-based regulation of chromatin structure is sure to play an important role in gene regulation. Additionally, some experimental evidence supports the idea of an alternate role for lncRNAs in which they can act as a scaffold, bringing DNA, proteins, and other RNAs together into complexes. These associations may act either to condense chromatin or, in some cases, to help bring the enhancer of a gene together with mediator proteins and the gene's promoter, activating gene expression in a more direct fashion.

Given the extensive functions of noncoding RNAs, it is not surprising that many of the ncRNAs characterized thus far play important roles in embryonic development—the topic we turn to in the next section. Embryonic development is perhaps the ultimate example of precisely regulated gene expression.

CONCEPT CHECK 18.3

1. Compare miRNAs and siRNAs, including their functions.
2. **WHAT IF?** Suppose the mRNA being degraded in Figure 18.14 coded for a protein that promotes cell division in a multicellular organism. What would happen if a mutation disabled the gene for the miRNA that triggers this degradation?
3. **MAKE CONNECTIONS** Inactivation of one of the X chromosomes in female mammals involves a noncoding RNA (*XIST*). Suggest a model for how *XIST* RNA initiates Barr body formation (see Concept 15.2).

For suggested answers, see Appendix A.

A program of differential gene expression leads to the different cell types in a multicellular organism

In the embryonic development of multicellular organisms, a fertilized egg (a zygote) gives rise to cells of many different types, each with a different structure and corresponding function. Typically, cells are organized into tissues, tissues into organs, organs into organ systems, and organ systems into the whole organism. Thus, any developmental program must produce cells of different types that form higher-level structures arranged in a particular way in three dimensions. The processes that occur during animal development are detailed in Chapter 47; in this chapter, we focus on the program of regulation of gene expression that orchestrates development, using a few animal species as examples.

A Genetic Program for Embryonic Development

The photos in **Figure 18.16** illustrate the dramatic difference between a frog zygote (fertilized egg) and the tadpole it becomes. This remarkable transformation results from three interrelated processes: cell division, cell differentiation, and morphogenesis. Through a succession of mitotic cell divisions, the zygote gives rise to a large number of cells. Cell division alone, however, would merely produce a great ball of identical cells, nothing like a tadpole. During embryonic development, cells not only increase in number, but also undergo cell **differentiation**, the process by which cells become specialized in structure and function. Moreover, the different kinds of cells are not

▼ **Figure 18.16 From fertilized egg to animal: What a difference four days makes.** It takes just four days for cell division, differentiation, and morphogenesis to transform each of the fertilized frog eggs shown in **(a)** into a tadpole like the one in **(b)**.

(a) Fertilized eggs of a frog   **(b) Newly hatched tadpole**

randomly distributed but are organized into tissues and organs in a particular three-dimensional arrangement. The physical processes that give an organism its shape constitute **morphogenesis**, the development of the form of an organism and its structures.

All three processes are rooted in cellular behavior. Even morphogenesis, the shaping of the organism, can be traced back to changes in the shape, motility, and other characteristics of the cells that make up various regions of the embryo. As you have seen, the activities of a cell depend on the genes it expresses and the proteins it produces. Almost all cells in an organism have the same genome; therefore, differential gene expression results from the genes being regulated differently in each cell type.

In Figure 18.12, you saw a simplified view of how differential gene expression occurs in two cell types, a liver cell and a lens cell. Each of these fully differentiated cells has a particular mix of specific transcription factor activators that turn on the collection of genes whose products are required in the cell. The fact that both cells arose through a series of mitoses from a common fertilized egg inevitably leads to a question: How do different sets of activators come to be present in the two cells?

It turns out that materials placed into the egg by maternal cells set up a sequential program of gene regulation that is carried out as embryonic cells divide, and this program coordinates cell differentiation during embryonic development. To understand how this works, we will consider two basic developmental processes. First, we'll explore how cells that arise from early embryonic mitoses develop the differences that start each cell along its own differentiation pathway. Second, we'll see how cellular differentiation leads to one particular cell type, using muscle development as an example.

Cytoplasmic Determinants and Inductive Signals

What generates the first differences among cells in an early embryo? And what controls the differentiation of all the various cell types as development proceeds? You can probably deduce the answer: The specific genes expressed in any particular cell of a developing organism determine its path. Two sources of information, used to varying extents in different species, "tell" a cell which genes to express at any given time during embryonic development.

One important source of information early in development is the egg's cytoplasm, which contains both RNA and proteins encoded by the mother's DNA. The cytoplasm of an unfertilized egg is not homogeneous. Messenger RNAs, proteins, other substances, and organelles are distributed unevenly in the unfertilized egg, and this unevenness has a profound impact on the development of the future embryo in

many species. Maternal substances in the egg that influence the course of early development are called **cytoplasmic determinants (Figure 18.17a)**. After fertilization, early mitotic divisions distribute the zygote's cytoplasm into separate cells. The nuclei of these cells may thus be exposed to different cytoplasmic determinants, depending on which portions of the zygotic cytoplasm a cell received. The combination of cytoplasmic determinants in a cell helps determine its developmental fate by regulating expression of the cell's genes during the course of cell differentiation.

▼ Figure 18.17 **Sources of developmental information for the early embryo.**

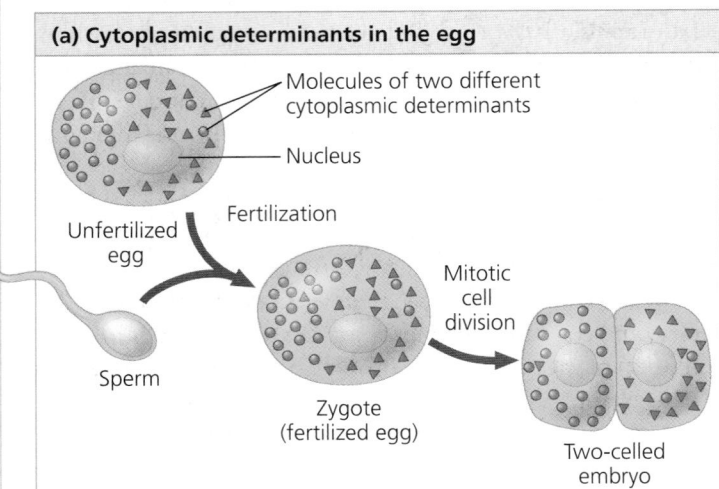

(a) Cytoplasmic determinants in the egg

- Molecules of two different cytoplasmic determinants
- Nucleus

Unfertilized egg

Fertilization

Sperm

Zygote (fertilized egg)

Mitotic cell division

Two-celled embryo

The unfertilized egg has molecules in its cytoplasm, encoded by the mother's genes, that influence development. Many of these cytoplasmic determinants, like the two shown here, are unevenly distributed in the egg. After fertilization and mitotic division, the cell nuclei of the embryo are exposed to different combinations of cytoplasmic determinants and, as a result, express different genes.

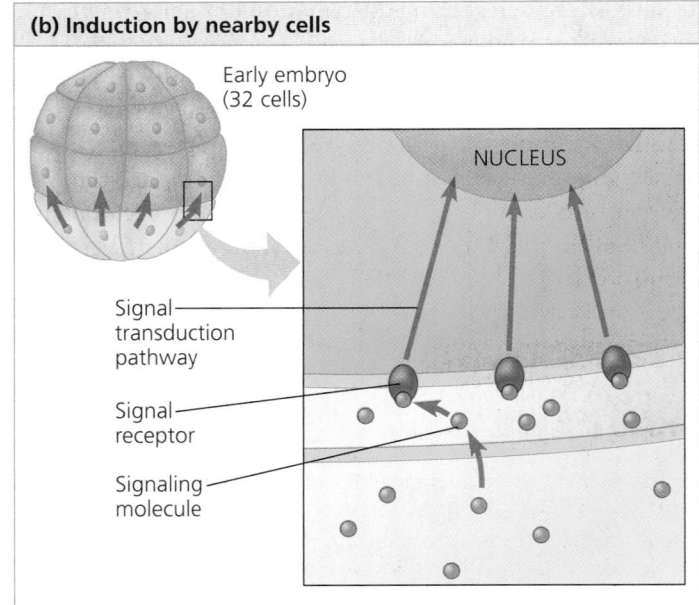

(b) Induction by nearby cells

Early embryo (32 cells)

NUCLEUS

Signal transduction pathway

Signal receptor

Signaling molecule

Cells in the lowest tier of cells are releasing molecules that signal (induce) nearby cells to change their gene expression.

The other major source of developmental information, which becomes increasingly important as the number of embryonic cells increases, is the environment around a particular cell. Most influential are the signals conveyed to an embryonic cell from other embryonic cells in the vicinity, including contact with cell-surface molecules on neighboring cells and the binding of growth factors secreted by neighboring cells (see Concept 11.1). Such signals cause changes in the target cells, a process called **induction (Figure 18.17b)**. The molecules that transmit these signals within the target cell are cell-surface receptors and other signaling pathway proteins. In general, the signal sends a cell down a specific developmental path by causing changes in its gene expression that lead to observable cellular changes. Thus, interactions between embryonic cells help induce differentiation into the many specialized cell types making up a new organism.

Sequential Regulation of Gene Expression During Cellular Differentiation

The earliest changes that set a cell on its path to specialization are subtle ones, showing up only at the molecular level. Before biologists knew much about the molecular changes occurring in embryos, they coined the term **determination** to refer to the point at which an embryonic cell is irreversibly committed to becoming a particular cell type. Once it has undergone determination, an embryonic cell can be experimentally placed in another location in the embryo and it will still differentiate into the cell type that is its normal fate. Differentiation, then, is the process by which a cell attains its determined fate. As the tissues and organs of an embryo develop and their cells differentiate, the cells become more noticeably different in structure and function.

Today we understand determination in terms of molecular changes that result in observable cell differentiation, marked by the expression of genes for *tissue-specific proteins*. Such proteins are found only in a specific cell type and give the cell its characteristic structure and function. The first sign of differentiation is the appearance of mRNAs for tissue-specific proteins. Later, differentiation is observable with a microscope as changes in cellular structure. On the molecular level, different sets of genes are sequentially expressed in a regulated manner as new cells arise when their precursors divide. Multiple steps in gene expression may be regulated during differentiation, transcription being the most common. In the fully differentiated cell, transcription remains the principal regulatory point for maintaining appropriate gene expression.

Differentiated cells are specialists at making tissue-specific proteins. For example, as a result of transcriptional regulation, liver cells specialize in making albumin, and lens cells specialize in making crystallin (see Figure 18.12). Skeletal muscle cells in vertebrates are another instructive example. Each of these cells is a long fiber containing many nuclei within a single plasma membrane. Skeletal muscle cells have high concentrations of muscle-specific versions of the contractile proteins myosin and actin, as well as membrane receptor proteins that detect signals from nerve cells.

Muscle cells develop from embryonic precursor cells that have the potential to develop into a number of cell types, including cartilage cells and fat cells, but particular conditions commit them to becoming muscle cells. Although the committed cells appear unchanged under the microscope, determination has occurred, and they are now a cell type called *myoblasts*. Eventually, myoblasts start to churn out large amounts of muscle-specific proteins and fuse to form mature, elongated, multinucleate skeletal muscle cells.

Researchers have worked out what happens at the molecular level during muscle cell determination. To do so, they grew embryonic precursor cells in culture and analyzed them using molecular techniques you will learn about in Concepts 20.1 and 20.2. In a series of experiments, they isolated different genes, caused each to be expressed in a separate embryonic precursor cell, and then looked for differentiation into myoblasts and muscle cells. In this way, they identified several so-called "master regulatory genes" whose protein products commit the cells to becoming skeletal muscle cells. Thus, in the case of muscle cells, the molecular basis of determination is the expression of one or more of these master regulatory genes.

To understand more about how determination occurs in muscle cell differentiation, let's focus on the master regulatory gene called *myoD*. The *myoD* gene deserves its designation as a master regulatory gene. Researchers have shown that the MyoD protein it encodes is capable of changing some kinds of fully differentiated nonmuscle cells, such as fat cells and liver cells, into muscle cells. Why doesn't MyoD work on *all* kinds of cells? One likely explanation is that activation of muscle-specific genes is not solely dependent on MyoD but requires a particular *combination* of regulatory proteins, some of which are lacking in cells that do not respond to MyoD.

What is the molecular basis for muscle cell differentiation? The MyoD protein is a transcription factor (see Figure 18.10) that binds to specific control elements in the enhancers of various target genes and stimulates their expression **(Figure 18.18)**. Some target genes for MyoD encode still other muscle-specific transcription factors. MyoD also stimulates expression of the *myoD* gene itself, an example of positive feedback that perpetuates MyoD's effect in maintaining the cell's differentiated state. Presumably, all the genes activated by MyoD have enhancer control elements recognized by MyoD and are thus coordinately controlled. Finally, the secondary transcription factors activate

▼ **Figure 18.18 Determination and differentiation of muscle cells.** Skeletal muscle cells arise from embryonic cells as a result of changes in gene expression. (In this depiction, the process of gene activation is greatly simplified.)

❶ **Determination.** Signals from other cells lead to activation of a master regulatory gene called *myoD*, and the cell makes MyoD protein, a specific transcription factor activator. The cell, now called a myoblast, is irreversibly committed to becoming a skeletal muscle cell.

❷ **Differentiation.** MyoD protein stimulates the *myoD* gene further and activates genes encoding other muscle-specific transcription factors, which in turn activate genes for muscle proteins. MyoD also turns on genes that block the cell cycle, thus stopping cell division. The nondividing myoblasts fuse to become mature multinucleate muscle cells, also called muscle fibers.

Nucleus
Embryonic precursor cell
Myoblast (determined)
Part of a muscle fiber (fully differentiated cell)

Master regulatory gene *myoD*   Other muscle-specific genes
DNA   OFF   OFF
mRNA
MyoD protein (transcription factor)
OFF
mRNA
MyoD
mRNA
A different transcription factor
mRNA   mRNA
Myosin, other muscle proteins, and proteins that block the cell cycle

WHAT IF? *What would happen if a mutation in the* myoD *gene resulted in the production of an altered MyoD protein that could not activate the* myoD *gene?*

the genes for proteins such as myosin and actin that confer the unique properties of skeletal muscle cells.

The determination and differentiation of other kinds of tissues may play out in a similar fashion. Experimental results support the idea that master regulatory proteins like MyoD might function by opening the chromatin in certain regions. This allows access to transcription machinery for activation of the next set of cell-type-specific genes.

Pattern Formation: Setting Up the Body Plan

We have now seen how different programs of gene expression that are activated in the fertilized egg can result in differentiated cells and tissues. But for the tissues to function effectively in the organism as a whole, the organism's *body plan*—its overall three-dimensional arrangement—must be established and superimposed on the differentiation process. Let's look at the molecular basis for the establishment of the body plan, using the well-studied fruit fly *Drosophila melanogaster* as an example.

Cytoplasmic determinants and inductive signals both contribute to spatially organizing the tissues and organs of an organism in their characteristic places. This developmental process is referred to as **pattern formation**.

Just as the locations of the front, back, and sides of a new building are determined before construction begins, pattern formation in animals begins in the early embryo, when the major axes of an animal are established. In a bilaterally symmetrical animal, the relative positions of head and tail, right and left sides, and back and front—the three major body axes—are set up before the organs appear. The molecular cues that control pattern formation, collectively called **positional information**, are provided by cytoplasmic determinants and inductive signals (see Figure 18.17). These cues tell a cell its location relative to the body axes and to neighboring cells, and determine how the cell and its descendants will respond to future molecular signals.

During the early 20th century, classical embryologists made detailed anatomical observations of embryonic development in a number of species and performed experiments in which they manipulated embryonic tissues. Although this research laid the groundwork for understanding the mechanisms of development, it did not reveal the specific molecules that guide development or determine how patterns are established.

In the 1940s, scientists began using the genetic approach—the study of mutants—to investigate *Drosophila* development. That approach has had spectacular success. These studies have established that genes control development and have led to an understanding of the key roles that specific molecules play in defining position and directing differentiation. By combining anatomical, genetic, and biochemical approaches to the study of *Drosophila* development, researchers have discovered developmental principles common to many other species, including humans.

The Life Cycle of Drosophila

Fruit flies and other arthropods have a modular construction, an ordered series of segments. These segments make up the body's three major parts: the head, the thorax (the mid-body, from which the wings and legs extend), and the abdomen (**Figure 18.19a**). Like other bilaterally symmetrical animals, *Drosophila* has an anterior-posterior (head-to-tail) axis, a dorsal-ventral (back-to-belly) axis, and a right-left axis. In *Drosophila*, cytoplasmic determinants that are localized in the unfertilized egg provide positional information for the placement of anterior-posterior and dorsal-ventral axes even before fertilization. We'll focus here on the molecules involved in establishing the anterior-posterior axis as a case in point.

The *Drosophila* egg develops in one of the female's ovaries, next to the nurse cells, which supply the egg with nutrients, mRNAs, and other substances needed for development. The egg and nurse cells are surrounded by follicle cells, which make the eggshell (**Figure 18.19b**, top). After fertilization and laying of the egg, embryonic development results in the formation of a segmented larva, which goes through three larval stages. Then, in a process much like that by which a caterpillar becomes a butterfly, the fly larva forms a pupa in which it metamorphoses into the adult fly pictured in Figure 18.19a.

Genetic Analysis of Early Development: Scientific Inquiry

Edward B. Lewis was a visionary American biologist who, in the 1940s, first showed the value of the genetic approach to studying embryonic development in *Drosophila*. Lewis studied bizarre mutant flies with developmental defects that led to extra wings or legs in the wrong place (**Figure 18.20**). He located the mutations on the fly's genetic map, thus connecting the developmental abnormalities to specific genes. This research supplied the first concrete evidence that genes somehow direct the developmental processes studied by embryologists. The genes Lewis discovered, called **homeotic genes**, are regulatory genes that control pattern formation in the fly.

Further insight into pattern formation during early embryonic development did not come for another 30 years,

▼ **Figure 18.19 Key events in *Drosophila* development.**

(a) Adult. The adult fly is segmented, and multiple segments make up each of the three main body parts—head, thorax, and abdomen. The body axes are shown by arrows.

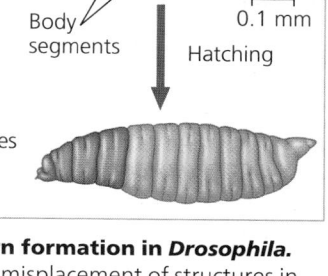

BODY AXES

0.5 mm

(b) Development from egg to larva.

❶ **Developing egg within one ovarian follicle** (among many in an ovary). The egg (yellow) and nurse cells are surrounded by follicle cells.

Follicle cell
Nucleus
Egg
Nurse cell

❷ **Mature, unfertilized egg.** The developing egg enlarges as nutrients and mRNAs are supplied to it by the nurse cells, which shrink. Eventually, the mature egg fills the eggshell that has been secreted by the follicle cells.

Egg-shell
Depleted nurse cells

Fertilization
Laying of egg

❸ **Fertilized egg.** The egg is fertilized within the mother and then laid.

Embryonic development

❹ **Segmented embryo.** The egg develops into a segmented embryo.

Body segments
0.1 mm

❺ **Larva.** The embryo develops into a larva, which has three stages. The third stage forms a pupa (not shown), within which the larva metamorphoses into the adult shown in (a).

Hatching

▼ **Figure 18.20 Abnormal pattern formation in *Drosophila*.** Mutations in homeotic genes cause misplacement of structures in an animal, such as the legs extending from the mutant fly's head in place of antenna (colorized SEMs).

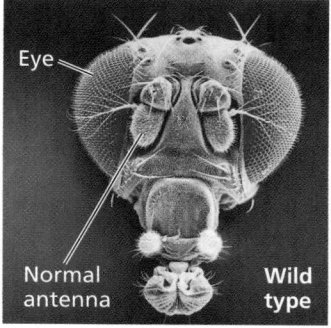

Eye
Normal antenna
Wild type

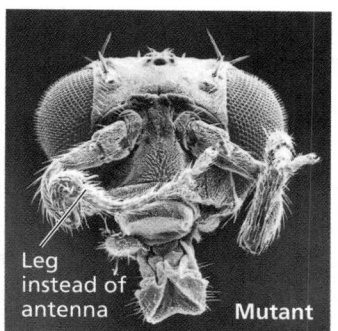

Leg instead of antenna
Mutant

when two researchers in Germany, Christiane Nüsslein-Volhard and Eric Wieschaus, set out to identify *all* the genes that affect segment formation in *Drosophila*. The project was daunting for three reasons. The first was the sheer number of *Drosophila* genes, now known to total about 14,000. The genes affecting segmentation might be just a few needles in a haystack or might be so numerous and varied that the scientists would be unable to make sense of them. Second, mutations affecting a process as fundamental as segmentation would surely be **embryonic lethals**, mutations with phenotypes causing death at the embryonic or larval stage. Because organisms with embryonic lethal mutations never reproduce, they cannot be bred for study. The researchers dealt with this problem by looking for recessive mutations, which can be propagated in heterozygous flies that act as genetic carriers. Third, cytoplasmic determinants in the egg were known to play a role in axis formation, so the researchers knew they would have to study the mother's genes as well as those of the embryo. It is the mother's genes that we will discuss further as we focus on how the anterior-posterior body axis is set up in the developing egg.

Nüsslein-Volhard and Wieschaus began their search for segmentation genes by exposing flies to mutagenic agents and scanning their descendants for dead embryos or larvae with abnormal segmentation or other defects. For example, to find genes that might set up the anterior-posterior axis, they looked for embryos or larvae with abnormal ends, such as two heads or two tails, predicting that such abnormalities would arise from mutations in maternal genes required for correctly setting up the offspring's head or tail end.

Using this approach, Nüsslein-Volhard and Wieschaus eventually identified about 1,200 genes essential for pattern formation during embryonic development. Of these, about 120 were essential for normal segmentation patterns. Next, the researchers were able to group these segmentation genes by general function and to isolate many of them for further study. The result was a detailed molecular understanding of the early steps in pattern formation in *Drosophila*.

When the results of Nüsslein-Volhard and Wieschaus were combined with Lewis's earlier work, a coherent picture of *Drosophila* development emerged. In recognition of their discoveries, the three researchers were awarded a Nobel Prize in 1995. Let's consider a specific example of the genes that Nüsslein-Volhard, Wieschaus, and co-workers found.

Axis Establishment

As we mentioned earlier, cytoplasmic determinants in the egg are the substances that initially establish the axes of the *Drosophila* body. These substances are encoded by genes of the mother, fittingly called maternal effect genes. A gene classified as a **maternal effect gene** is one that, when mutant in the mother, results in a mutant phenotype in the offspring, regardless of the offspring's own genotype. In fruit fly

development, the mRNA or protein products of maternal effect genes are placed in the egg while it is still in the mother's ovary. When the mother has a mutation in such a gene, she makes a defective gene product (or none at all), and her eggs are abnormal; when these eggs are fertilized, they fail to develop properly.

Because maternal effect genes control the orientation (polarity) of the egg and consequently that of the fly, they are also called *egg-polarity genes*. Two groups of these genes set up the anterior-posterior and dorsal-ventral axes of the embryo. Like mutations in segmentation genes, mutations in maternal effect genes are generally embryonic lethals.

Bicoid: A Morphogen That Determines Head Structures

To see how maternal effect genes determine the body axes of the offspring, we will focus on one such gene, called *bicoid*, a term meaning "two-tailed." An embryo or larva whose mother has two mutant *bicoid* alleles lacks the front half of its body and has posterior structures at both ends **(Figure 18.21)**. This phenotype suggested to Nüsslein-Volhard and her colleagues that the product of the mother's *bicoid* gene is essential for setting up the anterior end of the fly and might be concentrated at the future anterior end of the embryo. This hypothesis is an example of the *morphogen gradient hypothesis* first proposed by embryologists a century ago, in which gradients of substances called **morphogens** establish an embryo's axes and other features of its form.

DNA technology and other modern biochemical methods enabled the researchers to test whether the *bicoid* product, a protein called Bicoid, is in fact a morphogen that determines the anterior end of the fly. First, they asked whether the location of the mRNA and protein products of this gene in the egg was consistent with the hypothesis. They found that

▼ **Figure 18.21 Effect of the *bicoid* gene on *Drosophila* development.** A wild-type fruit fly larva has a head, three thoracic (T) segments, eight abdominal (A) segments, and a tail. A larva whose mother has two mutant alleles of the *bicoid* gene has two tails and lacks all anterior structures (LMs).

Wild-type larva
250 μm

Mutant larva (*bicoid*)

bicoid mRNA is highly concentrated at the extreme anterior end of the mature egg **(Figure 18.22)**. After the egg is fertilized, the mRNA is translated into protein. The Bicoid protein then diffuses from the anterior end toward the posterior, resulting in a gradient of protein within the early embryo, most highly concentrated at the anterior end. These results are consistent with the hypothesis that Bicoid protein specifies the fly's anterior end. To test this more specifically, scientists injected pure *bicoid* mRNA into various regions of early embryos. The protein that resulted from its translation caused anterior structures to form at the injection sites.

The *bicoid* research was groundbreaking for several reasons. First, it led to the identification of a specific protein required for some of the earliest steps in pattern formation. It thus helped us understand how different regions of the egg can give rise to cells that go down different developmental pathways. Second, it increased our understanding of the mother's critical role in the initial phases of embryonic development. Third, the principle that a gradient of morphogens can determine polarity and position has proved to be a key developmental concept for a number of species, just as early embryologists had hypothesized.

Maternal mRNAs are crucial during development of many species. In *Drosophila*, gradients of specific proteins encoded by maternal mRNAs not only determine the posterior and anterior ends but also establish the dorsal-ventral axis. As the fly embryo grows, it reaches a point when the embryonic program of gene expression takes over, and the maternal mRNAs must be destroyed. Later, positional information encoded by the embryo's genes, operating on an ever finer scale, establishes a specific number of correctly oriented segments and triggers the formation of each segment's characteristic structures. Mutant genes affecting this final step can lead to an abnormal adult pattern, as you saw in Figure 18.20.

> **Mastering Biology** Interview with Nancy Hopkins: Studying the genetic basis of development in zebrafish

Evolutionary Developmental Biology ("Evo-Devo")

EVOLUTION The fly with legs emerging from its head in Figure 18.20 is the result of a single mutation in one gene, a homeotic gene. The gene does not encode any antenna protein, however. Instead, it encodes a transcription factor that regulates other genes, and its malfunction leads to misplaced structures, such as legs instead of antennae. The observation that a change in gene regulation during development could lead to such a fantastic change in body form prompted some scientists to consider whether these types of mutations could contribute to evolution by generating novel body shapes. Ultimately, this line of inquiry gave rise to the field of evolutionary developmental biology, so-called "evo-devo," which will be further discussed in Concept 21.6.

▼ **Figure 18.22** Inquiry

Could Bicoid be a morphogen that determines the anterior end of a fruit fly?

Experiment Using a genetic approach to study *Drosophila* development, Christiane Nüsslein-Volhard and colleagues at two research institutions in Germany analyzed expression of the *bicoid* gene. The researchers hypothesized that *bicoid* normally codes for a morphogen that specifies the head (anterior) end of the embryo. To begin to test this hypothesis, they used molecular techniques to determine whether the mRNA and protein encoded by this gene were found in the anterior end of the fertilized egg and early embryo of wild-type flies.

Results *Bicoid* mRNA (dark blue in the light micrographs and drawings) was confined to the anterior end of the unfertilized egg. Later in development, Bicoid protein (dark orange) was seen to be concentrated in cells at the anterior end of the embryo.

Bicoid mRNA in mature unfertilized egg

Bicoid protein in early embryo

Conclusion The location of *bicoid* mRNA and the diffuse gradient of Bicoid protein seen later are consistent with the hypothesis that Bicoid protein is a morphogen specifying formation of head-specific structures.

Further Reading C. Nüsslein-Volhard et al., Determination of anteroposterior polarity in *Drosophila, Science* 238:1675–1681 (1987); W. Driever and C. Nüsslein-Volhard, A gradient of Bicoid protein in *Drosophila* embryos, *Cell* 54:83–93 (1988); T. Berleth et al., The role of localization of *bicoid* RNA in organizing the anterior pattern of the *Drosophila* embryo, *EMBO Journal* 7:1749–1756 (1988).

WHAT IF? *The researchers needed further evidence, so they injected bicoid mRNA into the anterior end of an egg from a female with a mutation disabling the bicoid gene. Given that the hypothesis was supported, what must their results have been?*

> **Mastering Biology** Animation: Role of *bicoid* Gene in *Drosophila* Development

In this section, we have seen how a carefully orchestrated program of sequential gene regulation controls the transformation of a fertilized egg into a multicellular organism. The program is carefully balanced between turning on the genes for differentiation in the right place and turning off other genes. Even when an organism is fully developed, gene expression is regulated in a similarly fine-tuned manner. In the final section of the chapter, we'll consider how fine this tuning is by looking at how specific changes in expression of just a few genes can lead to the development of cancer.

1. **MAKE CONNECTIONS** As you learned in Chapter 12, mitosis gives rise to two daughter cells that are genetically identical to the parent cell. Yet you, the product of many mitotic divisions, are not composed of identical, zygote-like cells. Why?

2. **MAKE CONNECTIONS** Explain how the signaling molecules released by an embryonic cell can induce changes in a neighboring cell without entering the cell. (See Figures 11.15 and 11.16.)

3. How do fruit fly maternal effect genes determine the polarity of the egg and the embryo?

4. **WHAT IF?** In Figure 18.17b, the lower cell is synthesizing signaling molecules, whereas the upper cell is expressing receptors for these molecules. In terms of gene regulation and cytoplasmic determinants, explain how these cells came to synthesize different molecules.

For suggested answers, see Appendix A.

CONCEPT 18.5

Cancer results from genetic changes that affect cell cycle control

In Concept 12.3, we considered cancer as a disease in which cells escape from their normal cell division control mechanisms. Now that we have discussed the molecular basis of gene expression and its regulation, we can look at cancer more closely. The gene regulation systems that go wrong during cancer turn out to be the very same systems that play important roles in embryonic development, the immune response, and many other biological processes. Thus, research into the molecular basis of cancer has both benefited from and informed many other fields of biology.

Types of Genes Associated with Cancer

The genes that normally regulate cell growth and division during the cell cycle include genes for growth factors, their receptors, and the intracellular molecules of signaling pathways. (To review cell signaling, see Concept 11.2; for regulation of the cell cycle, see Concept 12.3.) Mutations that alter any of these genes in somatic cells can lead to cancer. The agent of such change can be random spontaneous mutation. Many cancer-causing mutations likely also result from environmental influences: chemical carcinogens like tobacco, X-rays and other high-energy radiation, and some viruses.

Cancer research uncovered cancer-causing genes called **oncogenes** (from the Greek *onco*, tumor) in certain types of viruses. Later, related versions of viral oncogenes were found in the genomes of humans and other animals. The normal versions of the cellular genes, called **proto-oncogenes**, code for proteins that stimulate normal cell growth and division.

How might a proto-oncogene—a gene that has an essential function in normal cells—become an oncogene, a cancer-causing gene? In general, an oncogene arises from a genetic change that leads to an increase either in the amount of the proto-oncogene's protein product or in the intrinsic activity of each protein molecule. The genetic changes that convert proto-oncogenes to oncogenes fall into four main categories **(Figure 18.23)**: epigenetic changes, translocations, gene amplification, and point mutations.

First, alterations in epigenetic modifications that can lead to abnormal chromatin condensation in a cell are often found in tumor cells. If a mutation in a gene for a chromatin-modifying enzyme leads to loosening of chromatin in a region that is normally not being expressed, a proto-oncogene in that region could be expressed at abnormally high levels (see Figure 18.23a). For example, a gene for one such enzyme has been shown to be mutated in 20% of tumor cells analyzed.

Second, cancer cells are frequently found to contain chromosomes that have broken and rejoined incorrectly, translocating fragments from one chromosome to another (see Figure 15.14). If a translocated proto-oncogene ends up near an especially active promoter (or other control element), its transcription may increase, making it an oncogene (see Figure 18.23b).

The third main type of genetic change, amplification, increases the number of copies of the proto-oncogene in the cell through repeated gene duplication (discussed in Concept 21.5; see Figure 18.23c).

Fourth, a point mutation either in the promoter or an enhancer that controls a proto-oncogene, could cause an increase in its expression. A point mutation in the coding sequence of the proto-oncogene could change the gene's product to a protein that is more active or more resistant to degradation than the normal protein (see Figure 18.23d).

Any of these four mechanisms can lead to abnormal stimulation of the cell cycle and put the cell on the path to becoming a cancer cell.

In addition to genes whose products normally promote cell division, cells contain genes whose normal products *inhibit* cell division. Such genes are called **tumor-suppressor genes** since the proteins they encode help prevent uncontrolled cell growth. Any mutation that decreases the normal activity of a tumor-suppressor protein may contribute to the onset of cancer, in effect stimulating growth through the absence of suppression.

The protein products of tumor-suppressor genes have various functions. Some repair damaged DNA, which prevents the cell from accumulating cancer-causing mutations. Other tumor-suppressor proteins control adhesion of cells to each other or to the extracellular matrix; proper cell anchorage is crucial in normal tissues and is often absent in cancers. Still other tumor-suppressor proteins are components of cell-signaling pathways that inhibit the cell cycle.

(a) Epigenetic changes: Mutation in a gene for a chromatin-modifying enzyme can lead to loosening of the chromatin and inappropriate expression of a proto-oncogene.

Proto-oncogene

Normal growth-stimulating protein in excess

(b) Translocation or transposition: gene moved to new locus, under new controls

Proto-oncogene

New promoter   Oncogene

Normal growth-stimulating protein in excess

(c) Gene amplification: multiple copies of the gene

Proto-oncogene

Normal growth-stimulating protein in excess

(d) Point mutation within a control element   **Point mutation within the gene**

Proto-oncogene

Oncogene   Oncogene

Normal growth-stimulating protein in excess   Hyperactive or degradation-resistant protein

Interference with Normal Cell-Signaling Pathways

Let's consider how protein components of cell-signaling pathways function in normal cells and what goes wrong with their function in cancer cells. We will focus on the products of two key genes, the *ras* proto-oncogene and the *p53* tumor-suppressor gene. Mutations in *ras* occur in about 30% of human cancers and mutations in *p53* in more than 50%.

The Ras protein, encoded by the **_ras_ gene** (named for r̲a̲t s̲arcoma, a connective tissue cancer), is a G protein that relays a signal from a growth factor receptor on the plasma membrane to a cascade of protein kinases (see Figures 11.8 and 11.10). The cellular response at the end of the pathway is the synthesis of a protein that stimulates the cell cycle **(Figure 18.24a)**. Normally, such a pathway will not operate unless triggered by the appropriate growth factor. But certain mutations in the *ras* gene can lead to production of a hyperactive Ras protein that triggers the kinase cascade even in

the absence of growth factor, resulting in increased cell division **(Figure 18.24b)**. In fact, hyperactive versions or excess amounts of any of the pathway's components can have the same outcome: excessive cell division.

Figure 18.25a shows a pathway in which an intracellular signal leads to the synthesis of a protein that suppresses the cell cycle. In this case, the signal is damage to the cell's DNA, perhaps as the result of exposure to ultraviolet light. Operation of this signaling pathway blocks the cell cycle until the damage has been repaired. Otherwise, the damage might contribute to tumor formation by causing mutations or chromosomal abnormalities. Thus, the genes for the components of the pathway act as tumor-suppressor genes. The **_p53_ gene**, named for the apparent molecular weight of its protein product, is a tumor-suppressor gene. The protein it encodes is a specific transcription factor that promotes the synthesis of cell cycle–inhibiting proteins. That is why a mutation that knocks out the *p53* gene or in a gene required to activate the p53 protein (for example, a gene

▼ Figure 18.24 Normal and mutant cell cycle–stimulating pathway.

(a) Normal cell cycle–stimulating pathway.
The normal pathway is triggered by ❶ a growth factor that binds to ❷ its receptor in the plasma membrane. The signal is relayed to ❸ a G protein called Ras. Like all G proteins, Ras is active when GTP is bound to it. Ras passes the signal to ❹ a series of protein kinases. The last kinase activates ❺ a transcription factor (activator) that turns on one or more genes for ❻ a protein that stimulates the cell cycle.

(b) Mutant cell cycle–stimulating pathway.
If a mutation makes Ras or any other pathway component abnormally active, excessive cell division and cancer may result.

▼ Figure 18.25 Normal and mutant cell cycle–inhibiting pathway.

(a) Normal cell cycle–inhibiting pathway. In the normal pathway, ❶ DNA damage is an intracellular signal that is passed via ❷ protein kinases, leading to activation of ❸ p53. ❹ Activated p53 promotes transcription of the gene for ❺ a protein that inhibits the cell cycle. The resulting suppression of cell division ensures that the damaged DNA is not replicated. If the DNA damage is irreparable, then the p53 signal leads to apoptosis.

(b) Mutant cell cycle–inhibiting pathway. Mutations causing deficiencies in any pathway component can contribute to the development of cancer.

❓ *Explain whether a cancer-causing mutation in a tumor-suppressor gene, such as p53, is more likely to be a recessive or a dominant mutation.*

called *ATM*) can lead to excessive cell growth and cancer **(Figure 18.25b).**

The *p53* gene has been called the "guardian angel of the genome." Once the p53 protein is activated—say, by the ATM protein, a protein kinase, after DNA damage—p53 activates several other genes, such as *p21*. The p21 protein halts the cell cycle by binding to cyclin-dependent kinases, allowing time for the cell to repair the DNA. Researchers recently showed that p53 also activates expression of a group of miRNAs that inhibit the cell cycle. The p53 protein can also turn on genes directly involved in DNA repair. If DNA damage is irreparable, p53 activates "suicide" genes, whose protein products bring about programmed cell death (apoptosis; see Figure 11.20). Thus, p53 acts in several ways to prevent a cell from passing on mutations due to DNA damage. If mutations do accumulate and the cell survives through many divisions—as is more likely if the *p53* tumor-suppressor gene is defective or missing—cancer may ensue. The many functions of p53 suggest a complex picture of regulation in normal cells, one that we do not yet fully understand.

A recent study of elephants may underscore the protective role of *p53*. The incidence of cancer among elephants in zoo-based studies has been estimated at about 3%, compared with closer to 30% for humans. Genome sequencing revealed that elephants have 20 copies of the *p53* gene, compared to one copy in humans, other mammals, and even manatees, elephants' closest living relatives. There are undoubtedly other underlying reasons, but the correlation between low cancer rate and extra copies of the *p53* gene bears further investigation.

 Mastering Biology HHMI
Animation: Using *p53* to Fight Cancer   **hhmi** BioInteractive

For the present, the diagrams in Figure 18.24 and Figure 18.25 are an accurate view of how mutations can contribute to cancer, but we still don't know exactly how a particular cell becomes a cancer cell. Recent studies have shown, for instance, that DNA methylation and histone modification patterns in normal cells differ from those in cancer cells and that miRNAs probably participate in cancer development. There is still a lot to learn, and you and your classmates may be the ones to make important discoveries about cancer biology.

The Multistep Model of Cancer Development

More than one somatic mutation or epigenetic change is generally needed to produce all the changes characteristic of a full-fledged cancer cell. This may help explain why the incidence of cancer increases greatly with age. If cancer results from an accumulation of mutations that occur throughout life, then the longer we live, the more likely we are to develop cancer.

The model of a multistep path to cancer is well supported by studies of one of the best-understood types of human cancer: colorectal cancer, which affects the colon and/or rectum. About 140,000 new cases of colorectal cancer are diagnosed each year in the United States, and the disease causes 50,000 deaths per year. Like most cancers, colorectal cancer develops gradually **(Figure 18.26)**. The first sign is often a polyp, a small, benign growth in the colon lining. The cells of the polyp look normal, although they divide unusually frequently. The tumor grows and may eventually become malignant, invading other tissues. The development of a malignant tumor is paralleled by a gradual accumulation of mutations that convert proto-oncogenes to oncogenes and knock out tumor-suppressor genes. A *ras* oncogene and a mutated *p53* tumor-suppressor gene are often involved.

About half a dozen changes must occur at the DNA level for a cell to become fully cancerous. These changes usually include the appearance of at least one active oncogene and the mutation or loss of several tumor-suppressor genes. Furthermore, since mutant tumor-suppressor alleles are usually recessive, in most cases mutations must knock out *both* alleles in a cell's genome to block tumor suppression. (Most oncogenes, on the other hand, behave as dominant alleles.)

Since we understand the progression of this type of cancer, routine screenings (colonoscopies, for example) are recommended to identify and remove any suspicious polyps. The colorectal cancer mortality rate has been declining for the past 20 years due to increased screening and improved treatments. Treatments for other cancers have improved as well. Advances in the sequencing of DNA and mRNA allow medical researchers to compare the genes expressed by different types of tumors and by the same type in different people. These comparisons have led to personalized treatments based on the molecular characteristics of a person's tumor; **Figure 18.27** shows how this approach has been applied to breast cancer.

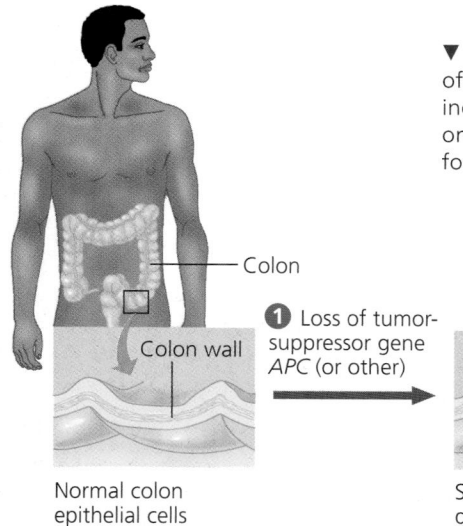

▼ **Figure 18.26 A multistep model for the development of colorectal cancer.** This type of cancer is one of the best understood. Changes in a tumor parallel a series of genetic changes, including mutations affecting several tumor-suppressor genes (such as *p53*) and the *ras* proto-oncogene. Mutations of tumor-suppressor genes often entail loss (deletion) of the gene. *APC* stands for adenomatous polyposis coli, and *SMAD4* is a gene involved in signaling that results in apoptosis.

Colon

Colon wall

1 Loss of tumor-suppressor gene *APC* (or other)

Normal colon epithelial cells

Small benign growth (polyp)

2 Activation of *ras* oncogene

3 Loss of tumor-suppressor gene *SMAD4*

Larger benign growth (adenoma)

4 Loss of tumor-suppressor gene *p53*

5 Additional mutations

Malignant tumor (carcinoma)

MAKE CONNECTIONS

Genomics, Cell Signaling, and Cancer

Modern medicine that melds genome-wide molecular studies with cell-signaling research is transforming the treatment of many diseases, such as breast cancer. Using microarray analysis (see Figure 20.13) and other techniques, researchers measured the relative levels of mRNA transcripts for every gene in hundreds of breast cancer tumor samples. They identified four major subtypes of breast cancer that differ in their expression of three signal receptors involved in regulating cell growth and division:

- Estrogen receptor alpha (ERα)
- Progesterone receptor (PR)
- HER2, a type of receptor called a receptor tyrosine kinase (RTK); other RTKs are also present, such as HER3

(ERα and PR are steroid receptors; see Figure 11.9.) The absence or excess expression of these receptors can cause aberrant cell signaling, leading in some cases to inappropriate cell division, which can contribute to cancer (see Figure 18.24).

▲ **A research scientist examines DNA-sequencing data from breast cancer samples.**

Milk duct

Mammary gland lobule

Epithelial milk-secreting cell

Support cell

Normal Breast Cells in a Milk Duct

In a normal breast cell, the three signal receptors are at normal levels (indicated by +):
- ERα+
- PR+
- HER2+

Duct interior

Estrogen receptor alpha (ERα)

Progesterone receptor (PR)

HER2

HER3

Extracellular matrix

Breast Cancer Subtypes

Each breast cancer subtype is characterized by the overexpression (indicated by ++ or +++) or absence (–) of three signal receptors: ERα, PR, and HER2. Breast cancer treatments are becoming more effective because they can be tailored to the specific cancer subtype.

Luminal A

ERα
PR

- ERα+++
- PR++
- HER2–
- 40% of breast cancers
- Best prognosis

Luminal B

- ERα++
- PR++
- HER2– (shown here); some HER2++
- Divide rapidly
- 15–20% of breast cancers
- Poorer prognosis than luminal A

Both luminal subtypes overexpress ERα (luminal A more than luminal B) and PR, and usually lack expression of HER2. Both can be treated with drugs that target ERα and inactivate it, the most well-known drug being tamoxifen. These subtypes can also be treated with drugs that inhibit estrogen synthesis.

Basal-like

- ERα–
- PR–
- HER2–
- 15–20% of breast cancers
- More aggressive; poorer prognosis than other subtypes

The basal-like subtype is "triple negative"—it usually does not express ERα, PR, or HER2. It often has a mutation in the tumor-suppressor gene *BRCA1* (see Concept 18.5). Treatments that target ERα, PR, or HER2 are not effective, but new treatments are being developed. Currently, patients are treated with cytotoxic chemotherapy, which selectively kills fast-growing cells.

HER2

HER3
HER2

- ERα– or ERα+
- PR–
- HER2++
- 10–15% of breast cancers
- Poorer prognosis than luminal A subtype

The HER2 subtype overexpresses HER2. Because it does not express PR (and, in many cases, ERα) at normal levels, the cells are unresponsive to therapies that target PR (and, for ERα– cells, ERα). However, patients with the HER2 subtype can be treated with Herceptin, an antibody protein that inactivates HER2 (see Concept 12.3).

1 A signaling molecule (such as a growth factor) binds as a ligand to a single HER3 (a monomer), allowing it to associate with another HER.

2 Even without a ligand bound, HER2 can associate with another HER. Here, HER3 and HER2 associate closely together, forming a dimer.

3 Formation of a dimer activates each monomer.

4 Each monomer adds phosphate from ATP to the other monomer, triggering a signal transduction pathway.

5 The signal is transduced through the cell, which leads to a cellular response—in this case, turning on genes that trigger cell division. HER2 cells have up to 100 times as many HER2 receptors as normal cells, so they undergo uncontrolled cell division.

Treatment with Herceptin for the HER2 subtype

1 Patients with the HER2 subtype are treated with the drug Herceptin, which can bind to HER2.

2 When bound to HER2, Herceptin blocks signaling and excessive cell division in certain patients.

MAKE CONNECTIONS

When researchers compared gene expression in normal breast cells and cells from breast cancers, they found that the genes showing the most significant differences in expression encoded signal receptors, as shown here. Given what you learned in Chapters 11, 12, and this chapter, explain why this result is not surprising.

Breast cancer is the second most common form of cancer in the United States, and the first among women. Each year, this cancer strikes over 230,000 women (and some men) in the United States and kills 40,000 (450,000 worldwide). A major problem with understanding breast cancer is its heterogeneity: Tumors differ in significant ways. Identifying differences between types of breast cancer is expected to improve treatment and decrease the mortality rate. In 2012, the Cancer Genome Atlas Network, sponsored by the National Institutes of Health, published the results of a multi-team effort that used a genomics approach to profile subtypes of breast cancer based on their molecular signatures. Four major types of breast cancer were identified (Figure 18.27). It is now routine to screen for the presence of particular signaling receptors in any breast cancer tumors, and individuals with breast cancer, along with their physicians, can now make more informed decisions about their treatments.

Inherited Predisposition and Environmental Factors Contributing to Cancer

The fact that multiple genetic changes are required to produce a cancer cell helps explain the observation that cancers can run in families. An individual inheriting an oncogene or a mutant allele of a tumor-suppressor gene is one step closer to accumulating the necessary mutations for cancer to develop than is an individual without any such mutations.

Geneticists are working to identify inherited cancer alleles so that predisposition to certain cancers can be detected early in life. About 15% of colorectal cancers, for example, involve inherited mutations. One syndrome, called hereditary nonpolyposis colon cancer (HNPCC), increases an individual's lifetime risk of colon cancer to 50–70%. HNPCC, also known as Lynch syndrome, is caused by an autosomal dominant allele of any one of a group of DNA repair genes, underscoring the importance of DNA repair systems. This syndrome is responsible for 2–5% of colon cancers. Other inherited mutations that cause colon cancer affect the tumor-suppressor gene called *adenomatous polyposis coli*, or *APC* (see Figure 18.26). This gene has multiple functions in the cell, including regulation of cell migration and adhesion. Even in patients with no family history of the disease, the *APC* gene is mutated in 60% of colorectal cancers. In these individuals, new mutations must have occurred in both *APC* alleles before the gene's function is lost. Currently, only 15% of colorectal cancers are associated with known inherited mutations, so researchers continue to try to identify "markers" that could predict the risk of developing this type of cancer.

Given the prevalence and significance of breast cancer, it is not surprising that it was one of the first cancers for which the role of inheritance was investigated. It turns out that for 5–10% of patients with breast cancer, there is evidence of a strong inherited predisposition. Geneticist Mary-Claire King began

→ **Mastering Biology**
Interview with Mary-Claire King: Discovering breast cancer genes

working on this problem in the mid-1970s. After 16 years of research, she convincingly demonstrated that mutations in one gene—*BRCA1*—were associated with increased susceptibility to breast cancer, a finding that flew in the face of medical opinion at the time. (*BRCA* stands for <u>br</u>east <u>ca</u>ncer.) Mutations in that gene or a gene called *BRCA2* are found in at least half of inherited breast cancers, and tests using DNA sequencing can detect these mutations. A woman who inherits one mutant *BRCA1* allele has a 60% probability of developing breast cancer before the age of 50, compared with only a 2% probability for an individual homozygous for the normal allele.

BRCA1 and *BRCA2* are considered tumor-suppressor genes because their wild-type alleles protect against breast cancer and their mutant alleles are recessive. (Mutations in *BRCA1* are commonly found in the genomes of cells from basal-like breast cancers; see Figure 18.27.) The BRCA1 and BRCA2 proteins both appear to function in the cell's DNA damage repair pathway. More is known about BRCA2: Along with another protein, it helps repair breaks that occur in both strands of DNA, a function crucial for maintaining undamaged DNA.

Because DNA breakage can contribute to cancer, it makes sense that the risk of cancer can be lowered by minimizing exposure to DNA-damaging agents, such as the ultraviolet radiation in sunlight and chemicals found in cigarette smoke. Novel genomics-based analyses of specific cancers, such as the approach described in Figure 18.27, are contributing to both early diagnosis and development of treatments that interfere with expression of key genes in tumors. Ultimately, such approaches are expected to lower the death rate from cancer.

The Role of Viruses in Cancer

The study of genes associated with cancer, inherited or not, increases our basic understanding of how disruption of normal gene regulation can result in this disease. In addition to the mutations and other genetic alterations described in this section, a number of *tumor viruses* can cause cancer in various animals, including humans. In fact, one of the earliest breakthroughs in understanding cancer came in 1911, when Peyton Rous, an American pathologist, discovered a virus that causes cancer in chickens. Also, the Epstein-Barr virus, which causes infectious mononucleosis, has been linked to several types of cancer in humans, notably Burkitt's lymphoma. Papillomaviruses cause cancer of the cervix, and a virus called HTLV-1 causes a type of adult leukemia. Viruses play a role in about 15% of the cases of human cancer.

Viruses may at first seem very different from mutations as a cause of cancer. However, we now know that viruses can interfere with gene regulation in several ways if they integrate their genetic material into the DNA of a cell. Viral integration may donate an oncogene to the cell, disrupt a tumor-suppressor gene, or convert a proto-oncogene to an oncogene. Some viruses produce proteins that inactivate p53 and other tumor-suppressor proteins, making the cell more prone to becoming cancerous. Viruses are powerful biological agents; you'll learn more about them in Chapter 19.

→ **Mastering Biology** Animation: Causes of Cancer

CONCEPT CHECK 18.5

1. Cancer-promoting mutations are likely to have different effects on the activity of proteins encoded by proto-oncogenes than they do on proteins encoded by tumor-suppressor genes. Explain.
2. Under what circumstances is cancer considered to have a hereditary component?
3. **MAKE CONNECTIONS** The p53 protein can activate genes involved in apoptosis. Review Concept 11.5, and discuss how mutations in genes coding for proteins that function in apoptosis could contribute to cancer.

For suggested answers, see Appendix A.

18 Chapter Review

→ Go to **Mastering Biology** for Assignments, the eText, the Study Area, and Dynamic Study Modules.

SUMMARY OF KEY CONCEPTS

→ To review key terms, go to the **Vocabulary Self-Quiz** in the **Mastering Biology** eText or Study Area, or go to goo.gl/zkjz9t.

CONCEPT 18.1

Bacteria often respond to environmental change by regulating transcription *(pp. 366–370)*

- Cells control metabolism by regulating enzyme activity or the expression of genes coding for enzymes. In bacteria, genes are often clustered into **operons**, with one promoter serving several adjacent genes. An **operator** site on the DNA switches the operon on or off, resulting in coordinate regulation of the genes.

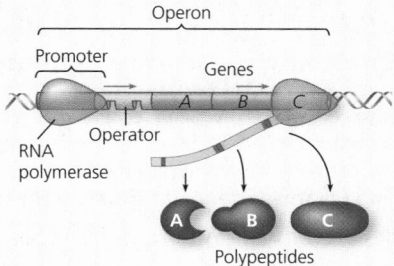

- Both repressible and inducible operons are examples of negative gene regulation. In either type of operon, binding of a specific **repressor** protein to the operator shuts off transcription. (The repressor is encoded by a separate **regulatory gene**.) In a repressible operon (usually encoding anabolic enzymes), the repressor is active when bound to a **corepressor**, often the end product of the pathway.

Repressible operon:

- In an inducible operon (usually encoding catabolic enzymes), binding of an **inducer** to an innately active repressor inactivates the repressor and turns on transcription.

Inducible operon:

- Some operons have positive gene regulation via a stimulatory **activator** protein (such as CRP, when activated by **cyclic AMP**), binds to a site within the promoter and stimulates transcription.

? *Compare and contrast the roles of a corepressor and an inducer in negative regulation of an operon.*

CONCEPT 18.2

Eukaryotic gene expression is regulated at many stages *(pp. 370–379)*

Chromatin modification
- Genes in highly compacted chromatin are generally not transcribed.
- **Histone acetylation** loosens chromatin structure, enhancing transcription.
- **DNA methylation** generally reduces transcription.

Transcription
- Regulation of transcription initiation: DNA **control elements** in **enhancers** bind specific transcription factors. Bending of the DNA enables **activators** to contact proteins at the promoter, initiating transcription.
- Coordinate regulation in each cell type:
  Enhancer for liver-specific genes   Enhancer for lens-specific genes

RNA processing
- **Alternative RNA splicing:**
  Primary RNA transcript
  mRNA   OR

mRNA degradation
- Each mRNA has a characteristic life span, determined in part by sequences in the 5′ and 3′ UTRs.

Translation
- Initiation of translation can be controlled via regulation of initiation factors.

Protein processing and degradation
- Protein processing and degradation are subject to regulation.

? *Describe what must happen in a cell for a gene specific to that cell type to be transcribed.*

CONCEPT 18.3

Noncoding RNAs play multiple roles in controlling gene expression (pp. 379–381)

Chromatin modification
• Small and/or large noncoding RNAs can promote heterochromatin formation in certain regions, which can block transcription.

CHROMATIN MODIFICATION
↓
TRANSCRIPTION
↓
RNA PROCESSING
↓

Translation
• miRNA or siRNA can block the translation of specific mRNAs.

mRNA DEGRADATION   TRANSLATION
↓
PROTEIN PROCESSING AND DEGRADATION

mRNA degradation
• miRNA or siRNA can target specific mRNAs for destruction.

❓ *Why are miRNAs called noncoding RNAs? Explain how they participate in gene regulation.*

CONCEPT 18.4

A program of differential gene expression leads to the different cell types in a multicellular organism (pp. 381–388)

• Embryonic cells become committed to a certain fate (**determination**), and undergo **differentiation**, becoming specialized in structure and function for their determined fate. Cells differ in structure and function not because they contain different genomes but because they express different genes. **Morphogenesis** encompasses the processes that give shape to the organism and its various structures.
• Localized **cytoplasmic determinants** in the unfertilized egg are distributed differentially to daughter cells, where they regulate the expression of those cells' developmental fates. In the process called **induction**, signaling molecules from embryonic cells cause transcriptional changes in nearby target cells.
• Differentiation is marked by the appearance of tissue-specific proteins, which enable differentiated cells to carry out their specialized roles.
• In animals, **pattern formation**, the development of a spatial organization of tissues and organs, begins in the early embryo. **Positional information**, the molecular cues that control pattern formation, tells a cell its location relative to the body's axes and to other cells. In *Drosophila*, gradients of **morphogens** encoded by **maternal effect genes** determine the body axes. For example, the gradient of Bicoid protein determines the anterior-posterior axis.

❓ *Describe the two main processes that cause embryonic cells to head down different pathways to their final fates.*

CONCEPT 18.5

Cancer results from genetic changes that affect cell cycle control (pp. 388–395)

• The products of **proto-oncogenes** and **tumor-suppressor genes** control cell division. A DNA change that makes a proto-oncogene overly active converts it to an **oncogene**, which may promote extra cell division and cancer. A tumor-suppressor gene encodes a protein that inhibits abnormal cell division. A mutation that reduces the activity of its protein product may lead to excessive cell division and cancer.

• Many proto-oncogenes and tumor-suppressor genes encode components of growth-stimulating and growth-inhibiting signaling pathways, respectively, and mutations in them can interfere with normal cell-signaling pathways. A hyperactive version of a protein in a stimulatory pathway, such as **Ras** (a G protein), functions as an oncogene protein. A defective version of a protein in an inhibitory pathway, such as **p53** (a transcription activator), fails to act as a tumor suppressor.

EFFECTS OF MUTATIONS

Protein (such as Ras) overexpressed

Protein (such as p53) nonfunctional

Cell cycle overstimulated   Increased cell division   Cell cycle not inhibited

• In the multistep model of cancer development, normal cells are converted to cancer cells by the accumulation of mutations affecting proto-oncogenes and tumor-suppressor genes. Technical advances in DNA and mRNA sequencing are enabling cancer treatments that are more individually based.
• Genomics-based studies have resulted in researchers proposing four subtypes of breast cancer, based on expression of genes by tumor cells.
• Individuals who inherit a mutant allele of a proto-oncogene or tumor-suppressor gene have a predisposition to develop a particular cancer. Certain viruses promote cancer by integration of viral DNA into a cell's genome.

❓ *Compare the usual functions of proteins encoded by proto-oncogenes with those of proteins encoded by tumor-suppressor genes.*

TEST YOUR UNDERSTANDING

➔ For more multiple-choice questions, go to the **Practice Test** in the **Mastering Biology** eText or Study Area, or go to goo.gl/GruWRg.

Levels 1-2: Remembering/Understanding

1. If a particular operon encodes enzymes for making an essential amino acid and is regulated like the *trp* operon, then
   (A) the amino acid inactivates the repressor.
   (B) the repressor is active in the absence of the amino acid.
   (C) the amino acid acts as a corepressor.
   (D) the amino acid turns on transcription of the operon.

2. Muscle cells differ from nerve cells mainly because they
   (A) express different genes.
   (B) contain different genes.
   (C) use different genetic codes.
   (D) have unique ribosomes.

3. The functioning of enhancers is an example of
   (A) a eukaryotic equivalent of prokaryotic promoter functioning.
   (B) transcriptional control of gene expression.
   (C) the stimulation of translation by initiation factors.
   (D) post-translational control that activates certain proteins.

4. Cell differentiation always involves
   (A) transcription of the *myoD* gene.
   (B) the movement of cells.
   (C) the production of tissue-specific proteins.
   (D) the selective loss of certain genes from the genome.

5. Which of the following is an example of post-transcriptional control of gene expression?
   (A) the addition of methyl groups to cytosine bases of DNA
   (B) the binding of transcription factors to a promoter
   (C) the removal of introns and alternative splicing of exons
   (D) gene amplification contributing to cancer

Levels 3-4: Applying/Analyzing

6. What would occur if the repressor of an inducible operon were mutated so it could not bind the operator?
(A) irreversible binding of the repressor to the promoter
(B) reduced transcription of the operon's genes
(C) buildup of a substrate for the pathway controlled by the operon
(D) continuous transcription of the operon's genes

7. Absence of *bicoid* mRNA from a *Drosophila* egg leads to the absence of anterior larval body parts and mirror-image duplication of posterior parts. This is evidence that the product of the *bicoid* gene
(A) normally leads to formation of head structures.
(B) normally leads to formation of tail structures.
(C) is transcribed in the early embryo.
(D) is a protein present in all head structures.

8. Which of the following statements about the DNA in one of your brain cells is true?
(A) Most of the DNA codes for protein.
(B) The majority of genes are likely to be transcribed.
(C) It is the same as the DNA in one of your liver cells.
(D) Each gene lies immediately adjacent to an enhancer.

9. Within a cell, the amount of protein made using a given mRNA molecule in that cell depends partly on
(A) the degree of DNA methylation.
(B) the rate at which the mRNA is degraded.
(C) the number of introns present in the mRNA.
(D) the types of ribosomes present in the cytoplasm.

10. Proto-oncogenes can change into oncogenes that cause cancer. Which of the following best explains the presence of these potential time bombs in eukaryotic cells?
(A) Proto-oncogenes first arose from viral infections.
(B) Proto-oncogenes are mutant versions of normal genes.
(C) Proto-oncogenes are genetic "junk."
(D) Proto-oncogenes normally help regulate cell division.

Levels 5-6: Evaluating/Creating

11. DRAW IT The diagram below shows five genes, including their enhancers, from the genome of a certain species. Imagine that yellow, blue, green, black, red, and purple activator proteins exist that can bind to the appropriately color-coded control elements in the enhancers of these genes.

(a) Draw an X above enhancer elements (of all the genes) that would have activators bound in a cell in which only gene 5 is transcribed. Identify which colored activators would be present.
(b) Draw a dot above all enhancer elements that would have activators bound in a cell in which the green, blue, and yellow activators are present. Identify which gene(s) would be transcribed.

(c) Imagine that genes 1, 2, and 4 code for nerve-specific proteins, and genes 3 and 5 are skin-specific. Identify which activators would be present in each cell type to ensure transcription of the appropriate genes.

12. EVOLUTION CONNECTION DNA sequences can act as "tape measures of evolution" (see Concept 5.6). Scientists analyzing the human genome sequence were surprised to find that some of the regions of the human genome that are most highly conserved (similar to comparable regions in other species) don't code for proteins. Propose a possible explanation for this observation.

13. SCIENTIFIC INQUIRY Prostate cells usually require testosterone and other androgens to survive. But some prostate cancer cells thrive despite treatments that eliminate androgens. One hypothesis is that estrogen, often considered a female hormone, may be activating genes normally controlled by an androgen in these cancer cells. Describe one or more experiments to test this hypothesis. (See Figure 11.9 to review the action of these steroid hormones.)

14. WRITE ABOUT A THEME: INTERACTIONS In a short essay (100–150 words), discuss how the processes shown in Figure 18.2 are examples of feedback mechanisms regulating biological systems in bacterial cells.

15. SYNTHESIZE YOUR KNOWLEDGE

The flashlight fish has an organ under its eye that emits light, which serves to startle predators and attract prey, and allows the fish to communicate with other fish. Some species can rotate the organ inside and then out, so the light appears to flash on and off. The light is actually emitted by bacteria (of the genus *Vibrio*) that live in the organ in a mutualistic relationship with the fish. (The bacteria receive nutrients from the fish.) The bacteria must multiply until they reach a certain density in the organ (a "quorum"; see Concept 11.1), at which point they all begin emitting light at the same time. There is a group of six or so genes, called *lux* genes, whose gene products are necessary for light formation. Given that these bacterial genes are regulated together, propose a hypothesis for how the genes are organized and regulated.

For selected answers, see Appendix A.

19 Viruses

KEY CONCEPTS

19.1 A virus consists of a nucleic acid surrounded by a protein coat *p. 399*

19.2 Viruses replicate only in host cells *p. 401*

19.3 Viruses and prions are formidable pathogens in animals and plants *p. 408*

Study Tip

Make flowcharts: Make a simple flowchart for each viral replicative cycle in this chapter. Below is the beginning of an example for the lysogenic cycle. Note the similarities and differences between the replicative cycles.

> **Lysogenic cycle**
>
> Virus attaches to bacterial cell and injects *genome*.
>
> ↓
>
> Viral genome is integrated into the bacterial chromosome.
>
> ↓

→ Go to Mastering Biology

For Students (in eText and Study Area)
- Get Ready for Chapter 19
- HHMI Animation: HIV Replicative Cycle
- HHMI Video: Interview with Katie Walter

For Instructors to Assign (in Item Library)
- Tutorial: Viral Replication
- Activity: Phage Lysogenic and Lytic Cycles
- Tutorial: CRISPR: The Discovery of Bacterial Adaptive Immunity

Figure 19.1 A human immune cell (blue) infected by human immunodeficiency virus (HIV) is releasing more HIV viruses (gold dots), which will go on to infect other cells. Left untreated, HIV destroys vital immune system cells, causing AIDS. This disease has killed about 35 million people worldwide.

How does a virus make more viruses?

A virus consists only of nucleic acid, proteins, and sometimes a membranous envelope. After infecting a host cell, it uses the host cell's molecules to make new viruses, as shown in this simplified diagram.

With some viruses, the host cell bursts when the new viruses are released.

OR

With some viruses, the host cell remains alive after the new viruses are released.

A virus consists of a nucleic acid surrounded by a protein coat

Compared with eukaryotic and even prokaryotic cells, most viruses are much smaller and simpler in structure. Lacking the structures and metabolic machinery found in a cell, a **virus** is an infectious particle consisting of little more than genes packaged in a protein coat.

Are viruses living or nonliving? Early on, they were considered biological chemicals; the Latin root for *virus* means "poison." Viruses can cause a wide variety of diseases, so researchers in the late 1800s saw a parallel with bacteria and proposed that viruses were the simplest of living forms. However, viruses cannot reproduce or carry out metabolic activities outside of a host cell. Most biologists would probably agree that viruses are not alive, but instead exist in a shady area between life-forms and chemicals. The simple phrase used by two researchers describes them aptly: Viruses lead "a kind of borrowed life."

The Discovery of Viruses: *Scientific Inquiry*

Scientists detected viruses indirectly long before they were able to see them. The story of how viruses were discovered begins in 1883. A German scientist named Adolf Mayer was studying tobacco mosaic disease, which stunts the growth of tobacco plants and gives their leaves a mottled, or mosaic, coloration. Mayer discovered that he could transmit the disease from plant to plant by rubbing sap extracted from diseased leaves onto healthy plants. After an unsuccessful search for an infectious microorganism in the sap, he suggested that the disease was caused by unusually small bacteria that were invisible under a microscope. This hypothesis was tested a decade later by Dmitri Ivanowsky, a Russian biologist who passed sap from infected tobacco leaves through a filter designed to remove bacteria. After filtration, the sap still produced mosaic disease.

But Ivanowsky reasoned that perhaps the bacteria were small enough to pass through the filter or made a toxin that could do so. The second possibility was ruled out when the Dutch botanist Martinus Beijerinck carried out a classic series of experiments that showed that the infectious agent in the filtered sap could replicate **(Figure 19.2)**.

In fact, the pathogen replicated only within the host it infected. In further experiments, Beijerinck showed that unlike bacteria used in the lab at that time, the mysterious agent of mosaic disease could not be cultivated on nutrient media in test tubes or petri dishes. Beijerinck imagined a replicating particle much smaller and simpler than a bacterium, and he is generally credited with being the first scientist to voice the concept of a virus. His suspicions were confirmed in 1935 when the American scientist Wendell Stanley crystallized the infectious particle, now known as tobacco mosaic virus (TMV). Subsequently, TMV and many other viruses were actually seen with the help of the electron microscope.

▼ Figure 19.2 Inquiry

What causes tobacco mosaic disease?

Experiment In the late 1800s, Martinus Beijerinck, of the Technical School in Delft, the Netherlands, investigated the properties of the agent that causes tobacco mosaic disease (then called spot disease).

❶ Extracted sap from tobacco plant with tobacco mosaic disease

❷ Passed sap through a porcelain filter known to trap bacteria

❸ Rubbed filtered sap on healthy tobacco plants

❹ Healthy plants became infected

Results When the filtered sap was rubbed on healthy plants, they became infected. Their sap, extracted and filtered, could then act as a source of infection for another group of plants. Each successive group of plants developed the disease to the same extent as earlier groups.

Conclusion The infectious agent was apparently not a bacterium because it could pass through a bacterium-trapping filter. The pathogen must have been replicating in the plants because its ability to cause disease was undiluted after several transfers from plant to plant.

Data from M. J. Beijerinck, Concerning a *contagium vivum fluidum* as cause of the spot disease of tobacco leaves, *Verhandelingen der Koninkyke akademie Wettenschappen te Amsterdam* 65:3–21 (1898). Translation published in English as Phytopathological Classics Number 7 (1942), American Phytopathological Society Press, St. Paul, MN.

WHAT IF? *If Beijerinck had observed that the infection of each group was weaker than that of the previous group and that ultimately the sap could no longer cause disease, what might he have concluded?*

Structure of Viruses

The tiniest viruses are only 20 nm in diameter—smaller than a ribosome. Millions could easily fit on a pinhead. Even the largest known virus, which has a diameter of 1,500 nanometers (1.5 μm), is barely visible under the light microscope. Stanley's discovery that some viruses could be crystallized was exciting and puzzling news. Not even the simplest of cells can aggregate into regular crystals. But if viruses are not cells, then what are they? Examining the structure of a virus more closely reveals that it is an infectious particle consisting of one or more molecules of a nucleic acid enclosed in a protein coat and, for some viruses, surrounded by a membranous envelope.

The simple structure of viruses make them a useful biological system: To a large extent, molecular biology was born in the laboratories of biologists studying viruses that infect bacteria. Experiments using these viruses provided evidence that genes are made of nucleic acids, and they were critical in working out the molecular mechanisms of the fundamental processes of DNA replication, transcription, and translation.

Viral Genomes

We usually think of genes as being made of double-stranded DNA, but many viruses defy this convention. Their genomes may consist of double-stranded DNA, single-stranded DNA, double-stranded RNA, or single-stranded RNA, depending on the type of virus. A virus is called a DNA virus or an RNA virus based on the kind of nucleic acid that makes up its genome. In either case, the genome is usually organized as a single linear or circular molecule of nucleic acid, although the genomes of some viruses consist of multiple molecules of nucleic acid. The smallest viruses known have only three genes in their genome, while the largest have several hundred to 2,000. For comparison, bacterial genomes contain about 200 to a few thousand genes.

Capsids and Envelopes

The protein shell enclosing the viral genome is called a **capsid**. Depending on the type of virus, the capsid may be rod-shaped, polyhedral, or more complex in shape. Capsids are built from a large number of protein subunits called *capsomeres*, but the number of different *kinds* of proteins in a capsid is usually small. Tobacco mosaic virus has a rigid, rod-shaped capsid made from over 1,000 molecules of a single type of protein arranged in a helix; rod-shaped viruses are commonly called *helical viruses* for this reason **(Figure 19.3a)**. Adenoviruses, which infect the respiratory tracts of animals, have 252 identical protein molecules arranged in a polyhedral capsid with 20 triangular facets—an icosahedron; thus, these and other similarly shaped viruses are referred to as *icosahedral viruses* **(Figure 19.3b)**.

Some viruses have accessory structures that help them infect their hosts. For instance, a membranous envelope surrounds the capsids of influenza viruses and many other viruses found in animals **(Figure 19.3c)**. These **viral envelopes**, which are derived from the membranes of the host cell, contain host cell phospholipids and membrane proteins. They also contain proteins and glycoproteins of viral origin. (Glycoproteins are proteins with

▼ **Figure 19.3 Viral structure.** Viruses are made up of nucleic acid (DNA or RNA) enclosed in a protein coat (the capsid) and sometimes further wrapped in a membranous envelope. The individual protein subunits making up the capsid are called capsomeres. Although diverse in size and shape, viruses have many common structural features. (All micrographs are colorized TEMs.)

RNA

Capsomeres making up capsid

Capsid

DNA

Protein spike

RNA-protein complex

Viral polymerase

Capsid

Membranous envelope

Glycoprotein spikes

Capsid:
Head

DNA

Tail sheath

Tail fiber

18 × 250 nm

70–90 nm (diameter)

80–200 nm (diameter)

80 × 225 nm

40 nm

40 nm

40 nm

40 nm

(a) Tobacco mosaic virus has a helical capsid, made up of capsomeres, surrounding a single helical RNA molecule, with the overall shape of a rigid rod.

(b) Adenoviruses have an icosahedral capsid with a protein spike at each vertex (corner). The capsid contains double-stranded DNA.

(c) Influenza viruses have an outer envelope, with glycoprotein spikes, around a capsid that contains eight double-helical RNA-protein complexes, each associated with a viral polymerase.

(d) Bacteriophage T4, like other "T-even" phages, has a complex capsid consisting of an icosahedral head and a tail apparatus. T4's double-stranded DNA is enclosed in the head.

carbohydrates covalently attached.) Some viruses carry a few viral enzymes, such as viral polymerase, within their capsids.

Many of the most complex capsids are found among the viruses that infect bacteria, called **bacteriophages**, or simply **phages**. The first phages studied included seven that infect *Escherichia coli* (*E. coli*). These seven phages were named type 1 (T1), type 2 (T2), and so forth, in the order of their discovery. (The T2 phage was used in the experiment that established DNA as the genetic material; see Figure 16.4.) The three "T-even" phages (T2, T4, and T6) turned out to be very similar in structure. Their capsids have elongated icosahedral heads enclosing their DNA. Attached to the head is a protein tail piece with fibers by which the phages attach to a bacterial cell **(Figure 19.3d)**. In the next section, we'll examine how these few viral parts function together with cellular components to produce large numbers of viral progeny.

CONCEPT CHECK 19.1

1. **VISUAL SKILLS** Describe one similarity and two differences between the structures of tobacco mosaic virus (TMV) and influenza virus (see Figure 19.3).
2. **MAKE CONNECTIONS** Bacteriophages were used to provide evidence that DNA carries genetic information (see Figure 16.4). Briefly describe the experiment carried out by Hershey and Chase, including in your description why the researchers chose to use phages.

For suggested answers, see Appendix A.

CONCEPT 19.2

Viruses replicate only in host cells

Viruses lack metabolic enzymes and equipment for making proteins, such as ribosomes. They are obligate intracellular parasites; in other words, they can replicate only within a host cell. It is fair to say that viruses in isolation are merely packaged sets of genes in transit from one host cell to another.

Each particular virus can infect cells of only a limited number of host species, called the **host range** of the virus. This host specificity results from the evolution of recognition systems by the virus. Viruses usually identify host cells by a "handshake" fit between viral surface proteins and specific receptor molecules on the outside of cells. Such receptor molecules are most often proteins that carry out necessary functions for the host cell and have been adopted by viruses as portals of entry. Some viruses have broad host ranges. For example, West Nile virus and equine encephalitis virus are distinctly different viruses that can each infect mosquitoes, birds, horses, and humans. Other viruses have host ranges so narrow that they infect only a single species. Measles virus, for instance, can infect only humans. Furthermore, viral infection of multicellular eukaryotes is usually limited to particular tissues. Human cold viruses infect only the cells lining the upper respiratory tract, and the HIV seen in Figure 19.1 binds to receptors present only on certain types of immune cells.

General Features of Viral Replicative Cycles

A viral infection begins when a virus binds to a host cell and the viral genome makes its way inside **(Figure 19.4)**. The mechanism of genome entry depends on the type of virus and the type of host cell. For example, T-even phages use their elaborate tail apparatus to inject DNA into a bacterium (see Figure 19.3d). Other viruses are taken up by endocytosis or, in the case of enveloped viruses, by fusion of the viral envelope with the host's plasma membrane. Once the viral genome is inside, the proteins it encodes can commandeer the host, reprogramming the cell to copy the viral genome and manufacture viral proteins. The host provides the nucleotides for making viral nucleic acids, as well as enzymes, ribosomes,

▼ **Figure 19.4 A simplified viral replicative cycle.** A virus is an intracellular parasite that uses the equipment and small molecules of its host cell to replicate. In this simplest of viral cycles, the parasite is a DNA virus with a capsid consisting of a single type of protein.

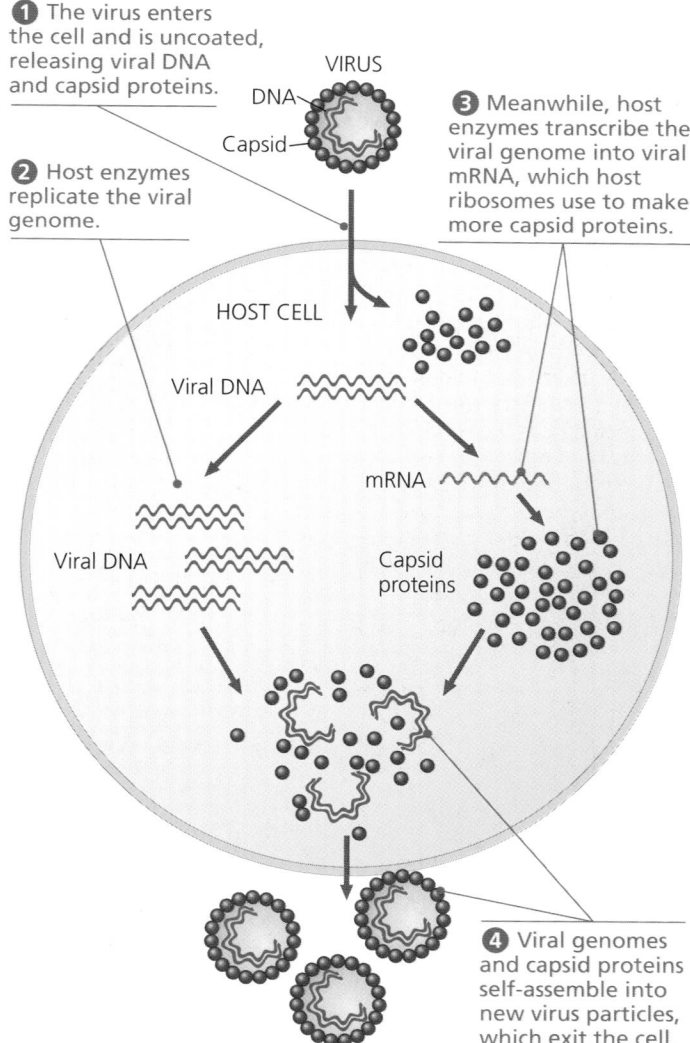

1 The virus enters the cell and is uncoated, releasing viral DNA and capsid proteins.

VIRUS

DNA

Capsid

2 Host enzymes replicate the viral genome.

3 Meanwhile, host enzymes transcribe the viral genome into viral mRNA, which host ribosomes use to make more capsid proteins.

HOST CELL

Viral DNA

mRNA

Viral DNA

Capsid proteins

4 Viral genomes and capsid proteins self-assemble into new virus particles, which exit the cell.

MAKE CONNECTIONS *Label each of the straight gray arrows to identify the process that is occurring. Review Figure 17.25.*

➡ **Mastering Biology Animation: Simplified Viral Replicative Cycle** tRNAs, amino acids, ATP, and other components needed for making the viral proteins. Many DNA viruses use the DNA polymerases of the host cell to synthesize new genomes along the templates provided by the viral DNA. In contrast, to replicate their genomes, RNA viruses use virally encoded RNA polymerases that can use RNA as a template. (Uninfected cells generally make no enzymes for carrying out this process.)

After the viral nucleic acid molecules and capsomeres are produced, they spontaneously self-assemble into new viruses. In fact, researchers can separate the RNA and capsomeres of TMV and then reassemble complete viruses simply by mixing the components together under the right conditions. The simplest type of viral replicative cycle ends with the exit of hundreds or thousands of viruses from the infected host cell, a process that often damages or destroys the cell. Such cellular damage and death, as well as the body's responses to this destruction, cause many of the symptoms associated with viral infections. The viral progeny that exit a cell have the potential to infect additional cells, spreading the viral infection.

There are many variations on the simplified viral replicative cycle we have just described. We will now take a look at some of these variations in bacterial viruses (phages) and animal viruses; later in the chapter, we will consider plant viruses.

Replicative Cycles of Phages

Phages are the best understood of all viruses, although some of them are also among the most complex. Research on phages led to the discovery that some double-stranded DNA viruses can replicate by two alternative mechanisms: the lytic cycle and the lysogenic cycle.

The Lytic Cycle

A phage replicative cycle that culminates in death of the host cell is known as a **lytic cycle**. The term *lytic* refers to the last stage of infection, during which the bacterium lyses (breaks open) and releases the phages that were produced within the cell. Each of these phages can then infect a healthy cell, and a few successive lytic cycles can destroy an entire bacterial population in just a few hours. A phage that replicates only by a lytic cycle is a **virulent phage**. **Figure 19.5** illustrates the major steps in the lytic cycle of T4, a typical virulent phage.

The Lysogenic Cycle

Instead of lysing their host cells, many phages coexist with them in a state called lysogeny. In contrast to the lytic cycle, which kills the host cell, the **lysogenic cycle** allows

▶ **Figure 19.5 The lytic cycle of phage T4, a virulent phage.** Phage T4 has almost 300 genes, which are transcribed and translated using the host cell's machinery. One of the first phage genes translated after the viral DNA enters the host cell codes for an enzyme that degrades the host cell's DNA (step ❷); the phage DNA is protected from breakdown because it contains a modified form of cytosine that is not recognized by the phage enzyme. The entire lytic cycle, from the phage's first contact with the cell surface to cell lysis, takes only 20–30 minutes at 37°C.

Phage assembly

Head   Tail   Tail fibers

➔ **Mastering Biology Animation: Phage Lytic Cycle**

❶ **Attachment.** The T4 phage uses its tail fibers to bind to specific surface proteins on an *E. coli* cell that act as receptors.

❺ **Release.** The phage directs production of an enzyme that damages the bacterial cell wall, allowing fluid to enter. The cell swells and finally bursts, releasing 100 to 200 phage particles.

❷ **Entry of phage DNA and degradation of host DNA.** The sheath of the tail contracts, injecting the phage DNA into the cell and leaving an empty capsid outside. The cell's DNA is hydrolyzed.

❹ **Self-assembly.** Three separate sets of proteins self-assemble to form phage heads, tails, and tail fibers. The phage genome is packaged inside the capsid as the head forms.

❸ **Synthesis of viral genomes and proteins.** The phage DNA directs production of phage proteins and copies of the phage genome by host and viral enzymes, using components within the cell.

replication of the phage genome without destroying the host. Phages capable of using both modes of replicating within a bacterium are called **temperate phages**. A temperate phage called lambda, written with the Greek letter λ, has been widely used in biological research. Phage λ resembles T4, but its tail has only one short tail fiber.

Infection of an *E. coli* cell by phage λ begins when the phage binds to the surface of the cell and injects its linear DNA genome **(Figure 19.6)**. Within the host, the λ DNA molecule forms a circle. What happens next depends on the replicative mode: lytic cycle or lysogenic cycle. During a lytic cycle, the viral genes immediately turn the host cell into a λ-producing factory, and the cell soon lyses and releases its virus progeny. During a lysogenic cycle, however, the λ DNA molecule is incorporated into a specific site on the *E. coli* chromosome by viral proteins that break both circular DNA molecules and join them to each other. When integrated into the bacterial chromosome in this way, the viral DNA is known as a **prophage**. One prophage gene codes for a protein that prevents transcription of most of the other prophage genes. Thus, the phage genome is mostly silent within the bacterium. Every time the *E. coli* cell prepares to divide, it replicates the phage DNA along with its own chromosome such that each daughter cell

inherits a prophage. A single infected cell can quickly give rise to a large population of bacteria carrying the virus in prophage form. This mechanism enables viruses to propagate without killing the host cells on which they depend.

The term *lysogenic* signifies that prophages are capable of generating active phages that lyse their host cells. This occurs when the λ genome (or that of another temperate phage) is induced to exit the bacterial chromosome and initiate a lytic cycle. An environmental signal, such as a certain chemical or high-energy radiation, usually triggers the switchover from the lysogenic to the lytic mode.

In addition to the gene for the viral protein that prevents transcription, a few other prophage genes may be expressed during lysogeny. Expression of these genes may alter the host's phenotype, a phenomenon that can have important medical significance. For example, the three species of bacteria that cause the human diseases diphtheria, botulism, and scarlet fever would not be so harmful to humans without certain prophage genes that cause the host bacteria to make toxins. And the difference between the *E. coli* strain in our intestines and the O157:H7 strain that has caused several deaths by food poisoning appears to be the presence of toxin genes of prophages in the O157:H7 strain.

▼ **Figure 19.6 The lytic and lysogenic cycles of phage λ, a temperate phage.** After entering the bacterial cell and circularizing, the λ DNA can immediately initiate the production of a large number of progeny phages (lytic cycle) or integrate into the bacterial chromosome (lysogenic cycle). In most cases, phage λ follows the lytic pathway, which is similar to that detailed in Figure 19.5. However, once a lysogenic cycle begins, the prophage may be carried in the host cell's chromosome for many generations. Phage λ has one main tail fiber, which is short.

⊘ **Mastering Biology**
Animation: Phage Lysogenic and Lytic Cycles

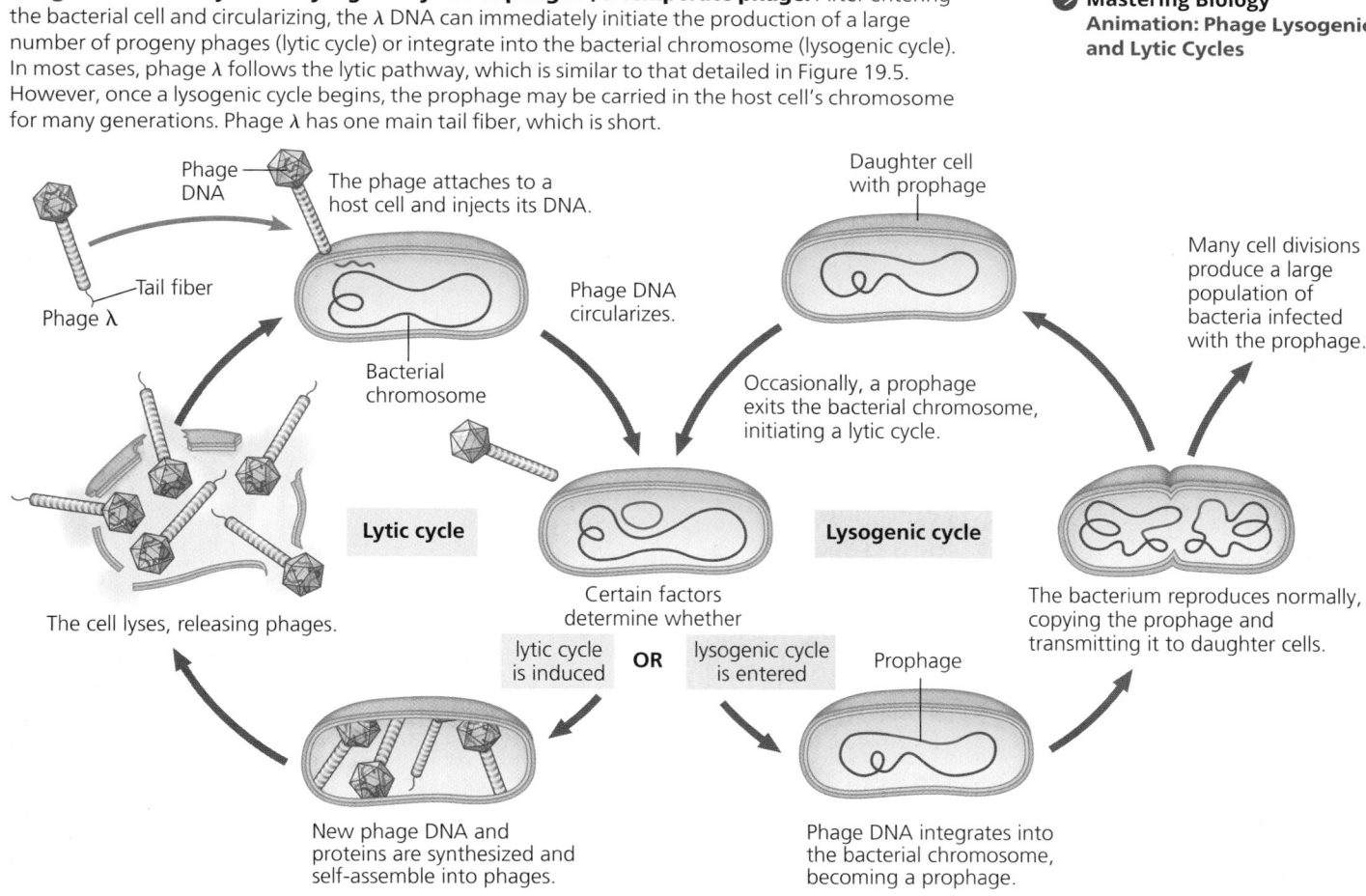

Phage DNA

The phage attaches to a host cell and injects its DNA.

Daughter cell with prophage

Tail fiber

Phage λ

Phage DNA circularizes.

Bacterial chromosome

Many cell divisions produce a large population of bacteria infected with the prophage.

Occasionally, a prophage exits the bacterial chromosome, initiating a lytic cycle.

Lytic cycle

Lysogenic cycle

The cell lyses, releasing phages.

Certain factors determine whether

lytic cycle is induced **OR** lysogenic cycle is entered

Prophage

The bacterium reproduces normally, copying the prophage and transmitting it to daughter cells.

New phage DNA and proteins are synthesized and self-assemble into phages.

Phage DNA integrates into the bacterial chromosome, becoming a prophage.

Bacterial Defenses Against Phages

After reading about the lytic cycle, you may have wondered why phages haven't exterminated all bacteria. Lysogeny is one major reason why bacteria have been spared from extinction caused by phages. Bacteria also have their own defenses against phages. First, natural selection favors bacterial mutants with surface proteins that are no longer recognized as receptors by a particular type of phage. Second, when phage DNA does enter a bacterium, the DNA often is identified as foreign and cut up by cellular enzymes called **restriction enzymes**, which are so named because they *restrict* a phage's ability to replicate within the bacterium. (Restriction enzymes are used in molecular biology and DNA cloning techniques; see Concept 20.1.) The bacterium's own DNA is methylated in a way that prevents attack by its own restriction enzymes. A third defense is a system present in both bacteria and archaea called the *CRISPR-Cas system*, which you learned about in Concept 17.5.

→ **Mastering Biology** HHMI Videos: Jennifer Doudna: CRISPR-Cas9 was First Discovered in Bacteria • Jennifer Doudna: Science is a Serendipitous Endeavor

The CRISPR-Cas system was discovered during a study of repetitive DNA sequences present in the genomes of many prokaryotes. These sequences, which puzzled scientists, were named clustered regularly interspaced short palindromic repeats (CRISPRs) because each sequence reads the same forward and backward (a palindrome), with different stretches of "spacer DNA" in between the repeats. At first, scientists assumed the spacer DNA sequences were random and meaningless, but analysis by several research groups showed that each spacer sequence corresponded to DNA from a particular phage that had infected the cell. Further studies revealed that particular nuclease proteins interact with the CRISPR region. These nucleases, called Cas (CRISPR-associated) proteins, can identify and cut phage DNA, thereby defending the bacterium against phage infection.

When a phage infects a bacterial cell that has the CRISPR-Cas system, the DNA of the invading phage is stored, integrated into the genome between two repeat sequences **(Figure 19.7)**. If the cell survives the infection, any further attempt by the same type of phage to infect this cell (or its offspring) triggers transcription of the CRISPR region into RNA molecules. These RNAs are cut into pieces and then bound by Cas proteins, such as the Cas9 protein (see Figure 17.28). The Cas protein uses a portion of the phage-related RNA as a homing device to identify the invading phage DNA and cut it, leading to its destruction.

Just as natural selection favors bacteria that have receptors altered by mutation or that have enzymes that cut phage DNA, it also favors phage mutants that can bind to altered receptors or that are resistant to enzymes. Thus, the bacterium-phage relationship is in constant evolutionary flux.

Replicative Cycles of Animal Viruses

Everyone has suffered from viral infections, whether cold sores, influenza, or the common cold. Like all viruses, those

▼ **Figure 19.7 The CRISPR-Cas system: a type of bacterial immune system.**

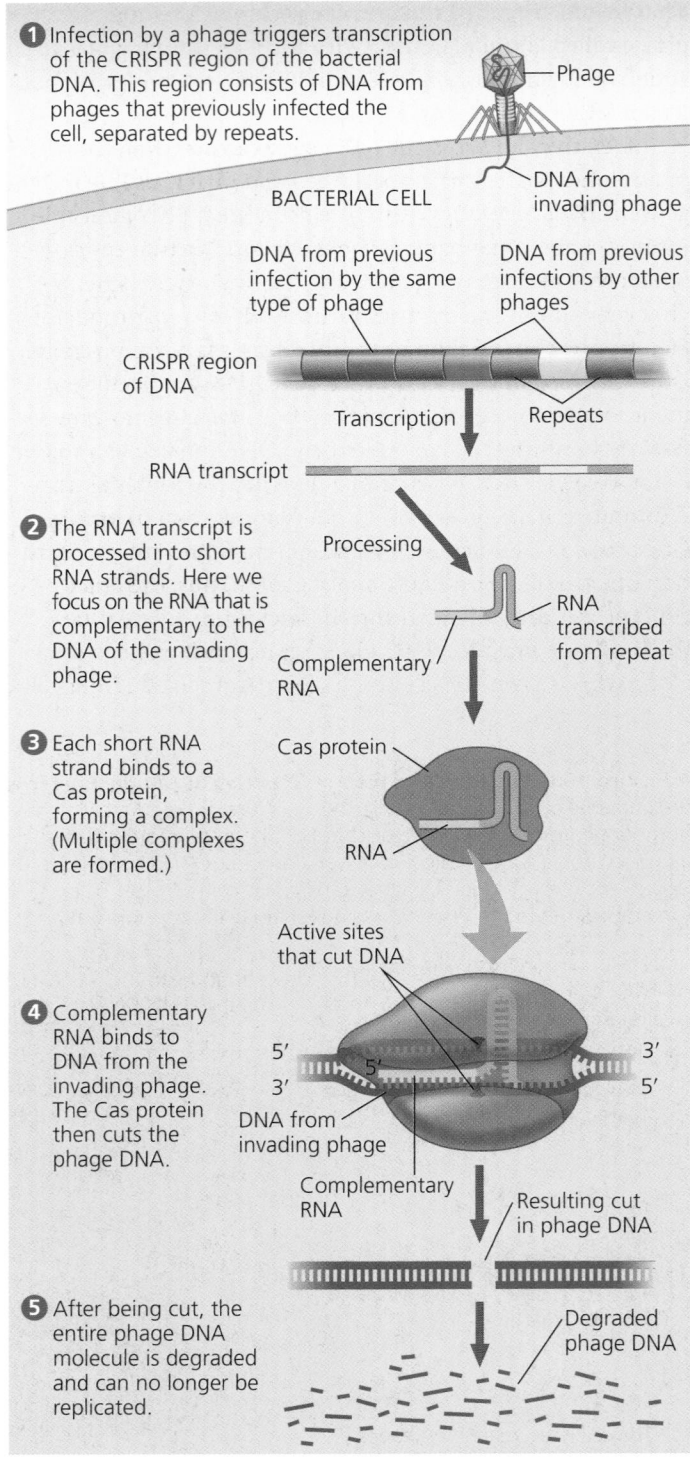

❶ Infection by a phage triggers transcription of the CRISPR region of the bacterial DNA. This region consists of DNA from phages that previously infected the cell, separated by repeats.

Phage

DNA from invading phage

BACTERIAL CELL

DNA from previous infection by the same type of phage

DNA from previous infections by other phages

CRISPR region of DNA

Transcription

Repeats

RNA transcript

❷ The RNA transcript is processed into short RNA strands. Here we focus on the RNA that is complementary to the DNA of the invading phage.

Processing

RNA transcribed from repeat

Complementary RNA

❸ Each short RNA strand binds to a Cas protein, forming a complex. (Multiple complexes are formed.)

Cas protein

RNA

Active sites that cut DNA

❹ Complementary RNA binds to DNA from the invading phage. The Cas protein then cuts the phage DNA.

5′   5′   3′
3′   5′

DNA from invading phage

Complementary RNA

Resulting cut in phage DNA

❺ After being cut, the entire phage DNA molecule is degraded and can no longer be replicated.

Degraded phage DNA

▶ Computer model of CRISPR-Cas9 gene editing complex from *Streptococcus pyogenes*

→ **Mastering Biology Interview with Francisco Mojica: Discovering CRISPR in bacteria (see the interview before Chapter 13)**

that cause illness in humans and other animals can replicate only inside host cells. Many variations on the basic scheme of viral infection and replication are represented among the animal viruses. Key variables are the nature of the viral genome (double- or single-stranded DNA or RNA) and the presence or absence of an envelope.

Whereas there are only a few bacteriophages that have an envelope or RNA genome, many animal viruses have both. In fact, nearly all animal viruses with RNA genomes have an envelope, as do some with DNA genomes. Rather than consider all the mechanisms of viral infection and replication, we will focus first on the roles of viral envelopes and then on the functioning of RNA as the genetic material of many animal viruses.

Viral Envelopes

An animal virus equipped with an envelope—that is, a membranous outer layer—uses it to enter the host cell. Protruding from the outer surface of this envelope are viral glycoproteins that bind to specific receptor molecules on the surface of a host cell. **Figure 19.8** outlines the events in the replicative cycle of an enveloped virus with an RNA genome. Ribosomes bound to the endoplasmic reticulum (ER) of the host cell make the protein parts of the envelope glycoproteins; cellular enzymes in the ER and Golgi apparatus then add the sugars. The resulting viral

glycoproteins, embedded in membrane derived from the host cell, are transported to the cell surface. In a process much like exocytosis (see Figure 7.20), new viral capsids are wrapped in membrane as they bud from the cell. In other words, the viral envelope is usually derived from the host cell's plasma membrane, although all or most of the molecules of this membrane are specified by viral genes. The enveloped viruses are now free to infect other cells. This replicative cycle does not necessarily kill the host cell, in contrast to the lytic cycles of phages.

Some viruses have envelopes that are not derived from plasma membrane. Herpesviruses, for example, are temporarily cloaked in membrane derived from the nuclear envelope of the host; they then shed this membrane in the cytoplasm and acquire a new envelope made from membrane of the Golgi apparatus. These viruses have a double-stranded DNA genome and replicate within the host cell nucleus, using a combination of viral and cellular enzymes to replicate and transcribe their DNA. In the case of herpesviruses, copies of the viral DNA can remain behind as mini-chromosomes in the nuclei of certain nerve cells. There they remain latent until some sort of physical or emotional stress triggers a new round of active virus production. The infection of other cells by these new viruses causes the blisters characteristic of herpes, such as cold sores or genital sores. Once someone acquires a herpesvirus infection, flare-ups may recur throughout the person's life.

▼ **Figure 19.8 The replicative cycle of an enveloped RNA virus.** Shown here is a virus with a single-stranded RNA genome that functions as a template for synthesis of mRNA.

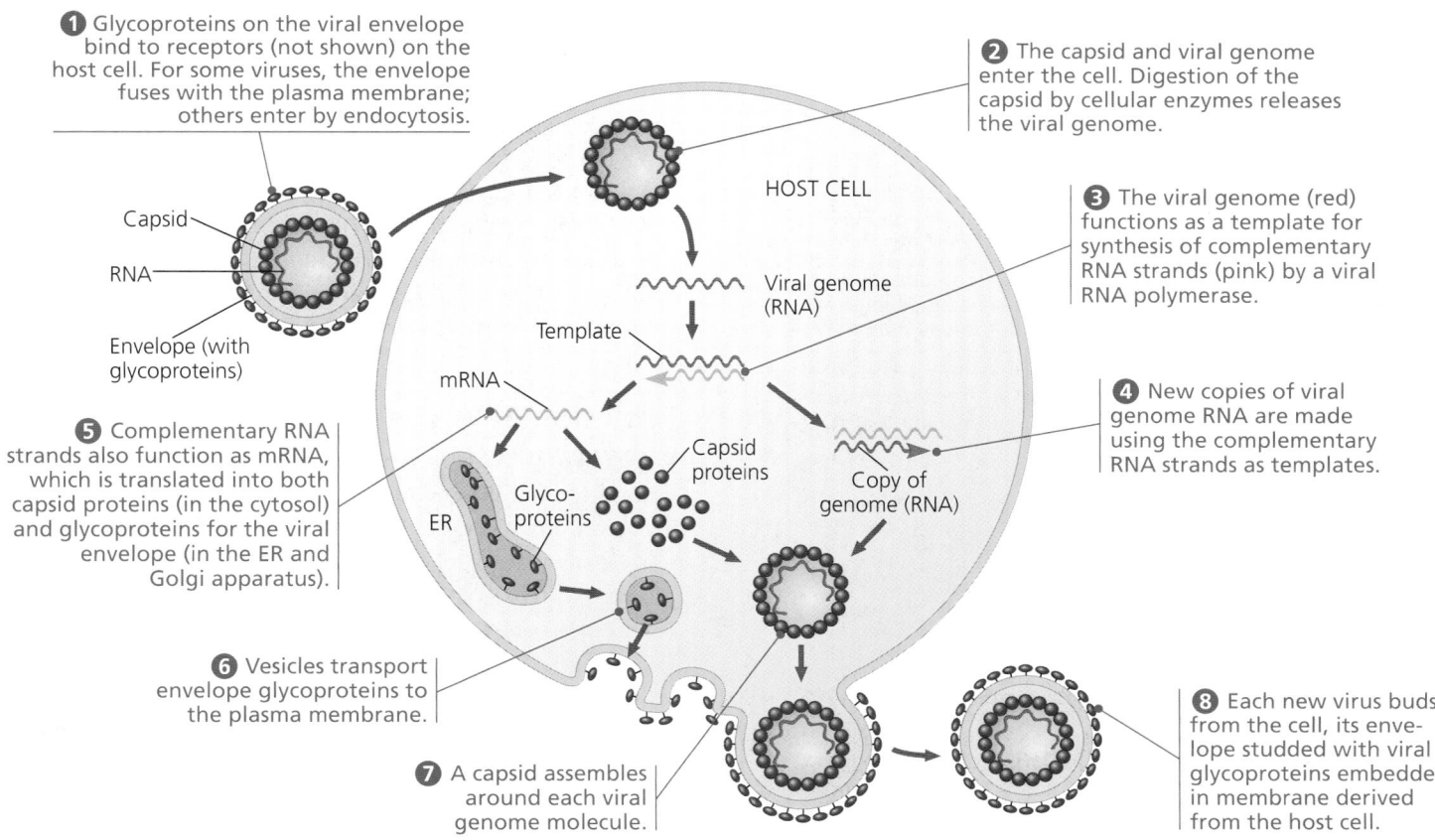

① Glycoproteins on the viral envelope bind to receptors (not shown) on the host cell. For some viruses, the envelope fuses with the plasma membrane; others enter by endocytosis.

② The capsid and viral genome enter the cell. Digestion of the capsid by cellular enzymes releases the viral genome.

③ The viral genome (red) functions as a template for synthesis of complementary RNA strands (pink) by a viral RNA polymerase.

④ New copies of viral genome RNA are made using the complementary RNA strands as templates.

⑤ Complementary RNA strands also function as mRNA, which is translated into both capsid proteins (in the cytosol) and glycoproteins for the viral envelope (in the ER and Golgi apparatus).

⑥ Vesicles transport envelope glycoproteins to the plasma membrane.

⑦ A capsid assembles around each viral genome molecule.

⑧ Each new virus buds from the cell, its envelope studded with viral glycoproteins embedded in membrane derived from the host cell.

Capsid

RNA

Envelope (with glycoproteins)

HOST CELL

Viral genome (RNA)

Template

mRNA

ER

Glyco-proteins

Capsid proteins

Copy of genome (RNA)

Viral Genetic Material

Table 19.1 shows the common classification system for animal viruses, which is based on their genetic material: double- or single-stranded DNA, or double- or single-stranded RNA, with further classification of RNA viruses. Although some phages and most plant viruses are RNA viruses, the broadest variety of RNA genomes is found among the viruses that infect animals. There are three types of single-stranded RNA genomes found in

Class/Family	Envelope?	Examples That Cause Human Diseases
Table 19.1 Classes of Animal Viruses		
I. Double-Stranded DNA (dsDNA)		
Adenovirus (see Figure 19.3b)	No	Respiratory viruses
Papillomavirus	No	Warts, cervical cancer
Polyomavirus	No	Tumors
Herpesvirus	Yes	Herpes simplex I and II (cold sores, genital sores); varicella zoster (shingles, chicken pox); Epstein-Barr virus (mononucleosis, Burkitt's lymphoma)
Poxvirus	Yes	Smallpox virus; cowpox virus
II. Single-Stranded DNA (ssDNA)		
Parvovirus	No	B19 parvovirus (mild rash)
III. Double-Stranded RNA (dsRNA)		
Reovirus	No	Rotavirus (diarrhea); Colorado tick fever virus
IV. Single-Stranded RNA (ssRNA); Serves as mRNA		
Picornavirus	No	Rhinovirus (common cold); poliovirus; hepatitis A virus; other intestinal viruses
Coronavirus	Yes	Severe acute respiratory syndrome (SARS); Middle East respiratory syndrome (MERS)
Flavivirus	Yes	Zika virus (see Figure 19.10c); yellow fever virus; dengue virus; West Nile virus; hepatitis C virus
Togavirus	Yes	Chikungunya virus (see Figure 19.10b); rubella virus; equine encephalitis viruses
V. ssRNA; Serves as Template for mRNA Synthesis		
Filovirus	Yes	Ebola virus (hemorrhagic fever; see Figure 19.10a)
Orthomyxovirus	Yes	Influenza virus (see Figure 19.3c)
Paramyxovirus	Yes	Measles virus; mumps virus
Rhabdovirus	Yes	Rabies virus
VI. ssRNA; Serves as Template for DNA Synthesis		
Retrovirus	Yes	Human immunodeficiency virus (HIV/AIDS; see Figure 19.9); human T-lymphotropic virus type 1 (HTLV-1) (leukemia)

animal viruses (classes IV–VI in Table 19.1). The genome of class IV viruses can directly serve as mRNA and thus can be translated into viral protein immediately after infection. Figure 19.8 shows a virus of class V, in which the RNA genome serves instead as a *template* for mRNA synthesis. The RNA genome is transcribed into complementary RNA strands, which function both as mRNA and as templates for the synthesis of additional copies of genomic RNA. All viruses that use an RNA genome as a template for mRNA transcription require RNA → RNA synthesis. These viruses use a viral enzyme capable of carrying out this process; there are no such enzymes in most cells. The enzyme used in this process is encoded by the viral genome. After the protein is synthesized, it is packaged during viral self-assembly with the genome inside the viral capsid.

The RNA animal viruses with the most complicated replicative cycles are the **retroviruses** (class VI). These viruses have an enzyme called **reverse transcriptase** that transcribes an RNA template into a DNA copy, an RNA → DNA information flow that is the opposite of the usual direction. This unusual phenomenon is the source of the name retroviruses (*retro* means "backward"). Of particular medical importance is **HIV (human immunodeficiency virus)**, the retrovirus shown in Figure 19.1 that causes **AIDS (acquired immunodeficiency syndrome)**. HIV and other retroviruses are enveloped viruses that contain two identical molecules of single-stranded RNA and two molecules of reverse transcriptase.

The HIV replicative cycle (illustrated in **Figure 19.9**) is typical of a retrovirus. After HIV enters a host cell, its reverse transcriptase molecules are released into the cytoplasm, where they catalyze synthesis of viral DNA. The newly made viral DNA then enters the cell's nucleus and integrates into the DNA of a chromosome. The integrated viral DNA, called a **provirus**, never leaves the host's genome, remaining a permanent resident of the cell. (Recall that a prophage, in contrast, leaves the host's genome at the start of a lytic cycle.) The RNA polymerase of the host transcribes the proviral DNA into RNA molecules, which can function both as mRNA for the synthesis of viral proteins and as genomes for the new viruses that will be assembled and released from the cell. In Concept 43.4, we describe how HIV causes the deterioration of the immune system that occurs in AIDS.

Evolution of Viruses

EVOLUTION We began this chapter by asking whether or not viruses are alive. Viruses do not really fit our definition of living organisms. An isolated virus is biologically inert, unable to replicate its genes or regenerate its own ATP. Yet it has a genetic program written in the universal language of life. Do we think of viruses as nature's most complex associations of molecules or as the simplest forms of life? Either way, we must bend our usual definitions. Although viruses cannot replicate or carry out metabolic activities independently, their use of the genetic code makes it hard to deny their evolutionary connection to the living world.

How did viruses originate? Viruses have been found that infect every form of life—not only bacteria, animals, and plants,

① The envelope glycoproteins enable the virus to bind to specific receptors (not shown) on certain white blood cells.

Glycoprotein · Viral envelope · Capsid · RNA (two identical strands) · Reverse transcriptase

HIV

▼ **Figure 19.9 The replicative cycle of HIV, the retrovirus that causes AIDS.** The photos on the left (artificially colored TEMs) show HIV entering and leaving a human white blood cell. See Figure 7.8 for the cell-surface proteins that act as receptors for HIV. Note in step ⑥ that DNA synthesized from the viral RNA genome is integrated as a provirus into the host cell chromosomal DNA, a characteristic unique to retroviruses.

HIV · Membrane of white blood cell

0.25 μm

HIV entering a cell

② The virus fuses with the host cell's plasma membrane.

③ The capsid proteins are removed, releasing the viral proteins, RNA, and reverse transcriptase.

④ Reverse transcriptase catalyzes the synthesis of a DNA strand, using the viral RNA as a template and then hydrolyzing it.

HOST CELL

Viral capsid proteins

Reverse transcriptase

Viral RNA

⑤ Reverse transcriptase catalyzes the synthesis of a second DNA strand complementary to the first.

Single strand of DNA

Double-stranded DNA

⑥ The double-stranded DNA is incorporated as a provirus into the cell's DNA.

NUCLEUS

Chromosomal DNA · Provirus

⑦ Proviral genes are transcribed into RNA molecules, which serve as genomes for progeny viruses and as mRNAs for translation into viral protein.

RNA genome for the progeny viruses

mRNA

ER

⑧ Viral envelope glycoproteins are made in the ER.

⑨ Viral capsid proteins and reverse transcriptase are made in the cytosol.

⑩ Vesicles transport the glycoproteins to the cell's plasma membrane.

⑪ Capsids are assembled around viral genomes and reverse transcriptase molecules.

⑫ New viruses, with viral envelope glycoproteins, bud from the host cell.

New HIV leaving a cell

MAKE CONECTIONS *Describe what is known about binding of HIV to immune system cells. (See Figure 7.8.) How was this discovered?*

➔ **Mastering Biology** HHMI Animation: HIV Replicative Cycle
Animation: Retrovirus (HIV) Replicative Cycle

hhmi BioInteractive

but also archaea, fungi, and algae and other protists. Researchers estimate that each milliliter (one-fifth of a teaspoon) of seawater contains between one and 100 million viruses, ten times the number of microorganisms! Because viruses depend on cells for their own propagation, it seems likely that viruses are not the descendants of precellular forms of life but evolved—possibly multiple times—*after* the first cells appeared. Most molecular biologists favor the hypothesis that viruses originated from naked bits of cellular nucleic acids that moved from one cell to another, perhaps via injured cell surfaces. The evolution of genes coding for capsid proteins may have allowed viruses to bind cell membranes, thus facilitating the infection of uninjured cells.

Candidates for the original sources of viral genomes include plasmids and transposons. *Plasmids* are small, circular DNA molecules found in bacteria and in the unicellular fungi called yeasts. Plasmids exist apart from the genome, can replicate independently of the genome, and are occasionally transferred between cells. *Transposons* are DNA segments that can move from one location to another within a cell's genome. Thus, plasmids, transposons, and viruses all share an important feature: They are *mobile genetic elements*. (We'll discuss plasmids in more detail in Concepts 20.1 and 27.2 and transposons in Concept 21.4.)

Consistent with this notion of pieces of DNA shuttling from cell to cell is the observation that a viral genome can have more in common with the genome of its host than with the genomes of viruses that infect other hosts. Indeed, some viral genes are essentially identical to genes of the host.

Debate about the origin of viruses was reinvigorated about 20 years ago when an extremely large virus was discovered: Mimivirus is a double-stranded DNA (dsDNA) virus with an icosahedral capsid that is 400 nm in diameter, the size of a small bacterium. Its genome contains 1.2 million bases (Mb)—about 100 times as many as the influenza virus genome—and an estimated 1,000 genes. Perhaps the most surprising aspect of mimivirus, however, was that its genome included genes previously found only in cellular genomes. Some of these genes code for proteins involved in translation, DNA repair, protein folding, and polysaccharide synthesis. Whether mimivirus evolved *before* the first cells and then developed an exploitative relationship with them or evolved more recently and simply scavenged genes from its hosts is not yet settled.

In the past decade, several even larger viruses have been discovered that cannot be classified with any existing known virus. One such virus is 1 μm (1,000 nm) in diameter, with a dsDNA genome of around 2–2.5 Mb, larger than that of some small eukaryotes. What's more, over 90% of its 2,000 or so genes are unrelated to cellular genes, inspiring the name it was given, pandoravirus. The number of genes in pandoraviruses varies from 1,500 to 2,500 genes. A second virus, called *Pithovirus sibericum*, with a diameter of 1.5 μm and 500 genes, was discovered in permanently frozen soil in Siberia. This virus, once thawed, was able to infect an amoeba after being frozen for 30,000 years! How these and all other viruses fit in the tree of life is an intriguing, unresolved question.

The ongoing evolutionary relationship between viruses and the genomes of their host cells is an association that continues to make viruses very useful experimental systems in molecular biology. Knowledge about viruses also allows many practical applications, since viruses have a tremendous impact on all organisms through their ability to cause disease.

CONCEPT CHECK 19.2

1. Compare the effect on the host cell of a lytic (virulent) phage and a lysogenic (temperate) phage.
2. **MAKE CONNECTIONS** Compare the CRISPR-Cas system with the miRNA system discussed in Concept 18.3, including their mechanisms and their functions.
3. **MAKE CONNECTIONS** The RNA virus in Figure 19.8 has a viral RNA polymerase that functions in step 3 of the virus's replicative cycle. Compare this with a cellular RNA polymerase in terms of template and overall function (see Figure 17.10).
4. Why is HIV called a retrovirus?
5. **VISUAL SKILLS** Looking at Figure 19.9, imagine you are a researcher trying to combat HIV infection. What molecular processes could you attempt to block?

For suggested answers, see Appendix A.

CONCEPT 19.3

Viruses and prions are formidable pathogens in animals and plants

Diseases caused by viral infections afflict humans, agricultural crops, and livestock worldwide. Other smaller, less complex entities known as prions also cause disease in animals. We'll first consider animal viruses.

Viral Diseases in Animals

A viral infection can produce symptoms by a number of different mechanisms. Viruses may damage or kill cells by causing the release of hydrolytic enzymes from lysosomes. Some viruses cause infected cells to produce toxins that lead to disease symptoms, and some have molecular components that are toxic, such as envelope proteins. How much damage a virus causes depends partly on the ability of the infected tissue to regenerate by cell division. People usually recover completely from colds because the epithelium of the respiratory tract, which the viruses infect, can efficiently repair itself. In contrast, damage inflicted by poliovirus to mature nerve cells is permanent because these cells do not divide and usually cannot be replaced. Many of the temporary symptoms associated with viral infections, such as fever and body aches, actually result from the body's own efforts to defend itself against infection rather than from cell death caused by the virus.

The immune system is a critical part of the body's natural defenses (see Chapter 43). It is also the basis for the major medical tool used to prevent viral infections—vaccines. A **vaccine** is a harmless derivative of a pathogen that stimulates the immune system to mount defenses against the harmful pathogen. Smallpox, a viral disease that was once a devastating scourge in many parts of the world, was eradicated by 1980 due to a vaccination program carried out by the World Health Organization (WHO). The very narrow host range of the smallpox virus—it infects only humans—was a critical factor in the success of this program. Similar worldwide vaccination campaigns are under way to eradicate polio, the incidence of which has dropped by 99%, and measles. Although an effective vaccine exists for measles, a large outbreak occurred in the Pacific Northwest in 2019, correlated with lower vaccination rates in that region. Effective vaccines are also available to protect against rubella, mumps, hepatitis B, and a number of other viral diseases.

Although vaccines can prevent some viral illnesses, medical care can do little, at present, to cure most viral infections once they occur. The antibiotics that help us recover from bacterial infections are powerless against viruses. Antibiotics kill bacteria by inhibiting enzymes specific to bacteria but have little or no effect on eukaryotic or virally encoded enzymes. However, the few enzymes that are encoded only by viruses have provided targets for other drugs. Most antiviral drugs resemble nucleosides and thus interfere with viral nucleic acid synthesis. One such drug is acyclovir, which impedes herpesvirus replication by inhibiting the viral polymerase that synthesizes viral DNA but not the eukaryotic one. Similarly, azidothymidine (AZT) curbs HIV replication by interfering with the synthesis of DNA by reverse transcriptase. In the past 30 years, much effort has gone into developing drugs to treat HIV. Currently, multidrug treatments, sometimes called "cocktails," are considered to be most effective. Such treatments commonly include a combination of two nucleoside mimics and a protease inhibitor, which interferes with an enzyme required for assembly of the viruses. Multidrug treatments originally involved taking up to 20 pills multiple times per day but now usually consist of a single daily tablet. Another effective treatment involves a drug called maraviroc, which blocks a protein on the surface of human immune cells that helps bind the HIV virus (see Figure 7.8). This drug has also been used successfully to prevent infection in individuals who either have been exposed to, or are at risk of exposure to, HIV.

 Mastering Biology BBC Video: Know Your Enemy: Bacteria vs. Viruses • HHMI Animation: Treating HIV Infection with a Reverse Transcriptase Inhibitor **hhmi** BioInteractive

in 1999 and has spread to 49 U.S. states, by 2019 resulting in over 50,000 cases and about 2,000 deaths.

The deadly Ebola virus **(Figure 19.10a)**, recognized initially in 1976 in central Africa, is one of several emerging viruses that cause *hemorrhagic fever*, an often fatal illness characterized by fever, vomiting, massive bleeding, and circulatory system collapse. In 2014, a widespread outbreak, or **epidemic**, of Ebola virus occurred. By 2016, it had resulted in over 11,000 deaths. In 2017, 2018, and 2019, smaller outbreaks occurred in the Democratic Republic of the Congo.

The mosquito-borne chikungunya virus **(Figure 19.10b)** causes an acute illness with fever, rashes, and persistent joint pain. Chikungunya has long been considered a tropical virus, but it has now appeared in Italy, France, and Spain. The Zika virus **(Figure 19.10c)** was first observed in Uganda in 1947, but for decades only a few cases occurred per year. In the spring of 2015, however, it became an emerging virus when it caused a large outbreak in Brazil. Although symptoms of Zika are often mild, the infection of pregnant women was correlated with a striking increase in the number of babies born with abnormally small brains, a condition called microcephaly. Zika is a mosquito-borne flavivirus (like West Nile virus) that infects neural cells, posing a particular danger to fetal brain development.

 Mastering Biology HHMI Videos: Interview with Katie Walter • Genetically Modified Mosquitoes **hhmi** BioInteractive

Where do these new strains of virus come from? One cause of rapidly emerging viral diseases in humans is mutation of existing viruses into new viruses that can spread more easily. RNA viruses have a high rate of mutation because viral RNA polymerases do not proofread and correct errors in replicating their RNA genomes. Some mutations change existing viruses into new viral strains that can cause disease, even in people immune to the original virus. A well-known related example is how three or four mutations causing changes in a surface protein of a cat virus (feline panleukopenia virus) resulted in the emergence in 1978 of canine parvovirus, a very contagious deadly virus infecting dogs.

Emerging Viral Diseases

Viruses that suddenly become apparent are often referred to as *emerging viruses*. HIV, the AIDS virus, is a classic example: It appeared in San Francisco in the early 1980s, seemingly out of nowhere, although later studies uncovered a case in the Belgian Congo in 1959. Some other dangerous emerging viruses cause encephalitis, inflammation of the brain. One example is the West Nile virus, which appeared in North America

▼ **Figure 19.10 Emerging viruses.**

(a) Ebola viruses budding from a monkey cell (colorized SEM).

(b) Chikungunya viruses emerging from a cell in the upper left and packing together (colorized TEM).

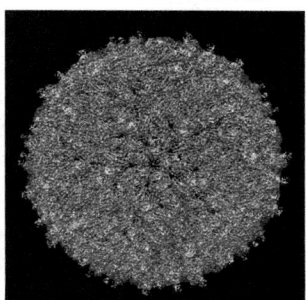

(c) Computer-generated image of a **Zika virus**, based on a technique called cryo-electron microscopy.

A second cause of the emergence of viral diseases is the spread of a viral disease from a small, isolated human population. HIV, the virus that causes AIDS, went unnamed and virtually unnoticed for decades before spreading around the world. In this case, technological and social factors, including affordable international travel, blood transfusions, unprotected sexual intercourse, and reuse of needles to inject intravenous drugs, allowed a previously rare human disease to become a global scourge.

→ Mastering Biology
Interview with David Satcher: The role of the CDC in recognizing AIDS and in public health

A third cause of new viral diseases in humans is the spread of existing viruses from other animals. Scientists estimate that about three-quarters of new human diseases originate in this way. Animals infected with a virus that can be transmitted to humans are said to be a natural reservoir for that virus. HIV is an example, as scientists believe that it originated from a version of the virus found in chimpanzees in central Africa, after people ate chimpanzee meat for food and were infected by exposure to chimpanzee blood.

In general, flu epidemics provide an instructive example of these three causes of emerging viruses. There are three types of influenza virus: types B and C, which infect only humans and have never caused an epidemic, and type A, which infects a wide range of animals, including birds, pigs, horses, and humans. The influenza type A viruses present in pigs and wild and domestic birds are potential emerging viruses that represent a long-term threat to human health.

A case in point is the H5N1 strain of avian influenza virus, which is highly contagious and deadly in birds. The first transmission from birds to humans was documented in Hong Kong in 1997. Since then, about 700 people have been infected, with an alarming mortality rate of greater than 50%. The high mortality rate is partly because H5N1 is very different from strains of influenza that have circulated among humans for a long time. Individuals are therefore not able to mount a strong immune response against viruses like H5N1.

Different strains of influenza A are given standardized names; for example, the name H5N1 identifies which forms of two viral surface proteins are present—hemagglutinin and neuraminidase (HA and NA, respectively; see the glycoprotein spikes in Figure 19.3c). These two proteins together help determine the host range and severity of disease caused by each virus. As of 2017, 18 types of hemagglutinin, a protein that helps the flu virus attach to host cells, and 11 types of neuraminidase, an enzyme that helps release new virus particles from infected cells, have been identified. All possible combinations of HA and NA have been found in some waterbirds. In 2018, researchers using advanced microscopy techniques reported that the HA protein is quite dynamic, stretching toward the target host cell, retracting, then refolding to a new shape and re-approaching the target cell.

Although deadly, the H5N1 strain has not yet caused an epidemic because nearly all cases have been the result of

transmission from birds to people, rather than person-to-person. Epidemics occur when genetic changes allow a new viral strain to be easily transmitted between humans. An event like this occurred in 2009, when a strain of influenza virus (H1N1) appeared that was very different from the virus that causes the seasonal flu. The H1N1 influenza virus spread rapidly, prompting WHO to declare a global epidemic, or **pandemic**. Within half a year, the disease had reached 207 countries, infecting over 600,000 people and killing almost 8,000. In addition to the 2009 H1N1 pandemic, influenza A strains have caused three other major flu epidemics among humans in the last 100 years. The most notable of these was the "Spanish flu" pandemic of 1918–1919, which killed 40–50 million people worldwide. In the **Scientific Skills Exercise**, you'll analyze genetic changes in variants of the 2009 H1N1 influenza virus and correlate them with spread of the disease.

The disease caused by H1N1 was originally called "swine flu" because parts of the viral genome were very similar to strains of influenza in pigs. However, studies revealed that the virus was not transmitted from pigs to humans. Instead, the story was more complex: H1N1 was a unique combination of swine, avian, and human influenza genes that allowed it to spread among humans.

Influenza viruses can change quickly because they have a genome made up of nine segments of RNA rather than a single RNA molecule. When an animal like a pig or a bird is infected with multiple strains of influenza virus, the RNA molecules making up the viral genomes can mix and match (*reassort*) during viral assembly, resulting in new genetic combinations. If a flu virus from pigs recombines with viruses that circulate widely among humans, it may acquire the ability to spread easily from person to person, dramatically increasing the potential for a major human outbreak. Pigs are believed to have been the main hosts for recombination that led to the 2009 H1N1 flu virus.

Influenza viruses also have a high rate of mutation, for reasons mentioned earlier. Coupled with reassortments, these mutations can lead to the emergence of a viral strain from animals that can infect human cells. Scientists are worried that the H5N1 strain will evolve in a way that enables it to spread as easily as the H1N1 strain. This would represent a major global health threat like that of the 1918 pandemic.

How easily could this capability evolve in the H5N1 strain? In 2011, scientists working with ferrets, small mammals used as model organisms for the human flu, found that only a few mutations of the avian flu virus were sufficient to allow infection of cells in the human nasal cavity and windpipe. Furthermore, when the scientists transferred nasal swabs serially from ferret to ferret, the virus became transmissible through the air. Reports of this startling discovery ignited a firestorm of debate about whether to publish the results. Ultimately, the scientific community decided the benefits of preventing pandemics outweighed the risks of the information being used for harmful purposes, and the work was published in 2012.

Normal seasonal flu viruses (including influenza types A and B) are not considered emerging viruses because variations

Analyzing a Sequence-Based Phylogenetic Tree to Understand Viral Evolution

► H1N1 flu vaccination

How Can Sequence Data Be Used to Track Flu Virus Evolution? In 2009, an influenza A H1N1 virus caused a pandemic, and the virus has continued to resurface in outbreaks across the world. Researchers in Taiwan were curious about why the virus kept appearing despite widespread flu vaccine initiatives. They hypothesized that newly evolved variant strains of the H1N1 virus were able to evade human immune system defenses. To test this hypothesis, they needed to determine if each wave of flu infection was caused by a different H1N1 variant strain.

How the Experiment Was Done Scientists obtained the genome sequences for 4,703 virus isolates collected from patients with H1N1 flu in Taiwan, each named by type/location/identifying number/year. They compared the sequences in different strains for the viral hemagglutinin (HA) gene, and based on mutations that had occurred, arranged the isolates into a phylogenetic tree (see Figure 26.5 for information on how to read phylogenetic trees).

Data from the Experiment In the phylogenetic tree, each branch tip is one variant strain of the H1N1 virus with a unique HA gene sequence. The tree is a way to visualize a working hypothesis about the evolutionary relationships between H1N1 variants.

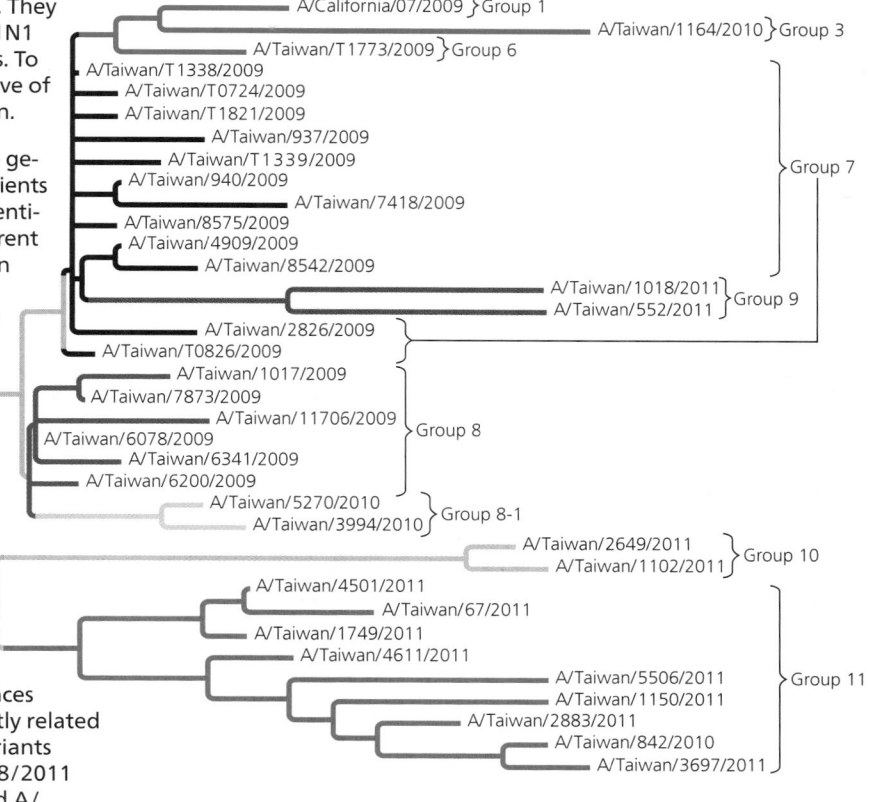

INTERPRET THE DATA

1. The more closely connected two variants are in the tree, the more alike they are in terms of HA gene sequence. Each fork in a branch, called a node, shows where two lineages separate due to different accumulated mutations. The length of the branches is a measure of how many sequence differences there are between the variants, indicating how distantly related they are. Referring to the phylogenetic tree, which variants are more closely related to each other: A/Taiwan/1018/2011 and A/Taiwan/552/2011 or A/Taiwan/1018/2011 and A/Taiwan/8542/2009? Explain your answer.

2. The scientists arranged the branches into groups made up of one ancestral variant and all of its descendant, mutated variants. They are color-coded in the figure. Using group 11 as an example, trace the lineage of its variants. **(a)** Do all of the nodes have the same number of branches or branch tips? **(b)** Are all of the branches in the group the same length? **(c)** What do these results indicate?

3. The graph shows the number of isolates collected (each from an ill patient) on the *y*-axis and the month and year that the isolates were collected on the *x*-axis. Each group of variants is plotted separately with a line color that matches the tree diagram. **(a)** Which group of variants was the earliest to cause the first wave of H1N1 flu in over 100 patients in Taiwan? **(b)** After a group of variants had a peak number of infections, did members of that same group cause another (later) wave of infection? **(c)** One variant in group 1 (green, uppermost branch) was used to make a vaccine that was distributed very early in the pandemic. Based on the graphed data, does it look like the vaccine was effective?

4. Groups 9, 10, and 11 all had H1N1 variants that caused a large number of infections at the same time in Taiwan. Does this mean that the scientists' hypothesis, that new variants cause new waves of infection, was incorrect? Explain your answer.

▲ Scientists graphed the number of isolates by the month and year of isolate collection to show the period in which each viral variant was actively causing illness in people.

Data from J.-R. Yang et al., New variants and age shift to high fatality groups contribute to severe successive waves in the 2009 influenza pandemic in Taiwan, *PLoS ONE* 6(11): e28288 (2011).

⊙ **Instructors:** A version of this Scientific Skills Exercise can be assigned in **Mastering Biology**.

of seasonal flu viruses have been circulating among humans for long enough that most components are recognized by the immune system. However, these viruses still undergo mutation and reassortment of genome segments, and variations of the HA protein are used each year to generate vaccines against the strains predicted most likely to occur the following year.

As we have seen, emerging viruses are usually existing viruses that mutate, spread more widely in the current host species, or spread to new host species. Changes in host behavior or environmental changes can increase the viral traffic responsible for emerging diseases. For instance, new roads built through remote areas can allow viruses to spread between previously isolated human populations. Also, the destruction of forests to expand cropland can bring humans into contact with animals that host infectious viruses. Finally, genetic mutations and changes in host ranges can allow viruses to jump between species. Many viruses are transmitted by mosquitoes. A dramatic expansion of the disease caused by the chikungunya virus occurred in the mid-2000s when a mutation allowed it to infect not only the mosquito species *Aedes aegypti* but also the related *Aedes albopictus*. Insecticides and mosquito netting over beds are crucial tools in public health attempts to prevent diseases carried by mosquitoes **(Figure 19.11)**.

▼ **Figure 19.11 Netting as protection against virus-carrying mosquitoes.**

Recently, scientists have become concerned about the possible effects of climate change on worldwide viral transmission. Dengue fever, also mosquito-borne, has appeared in Florida and Portugal, regions where it had not been seen before. The possibility that global climate change has allowed mosquito species carrying these viruses to expand their ranges and interact more is troubling because of the increased chance of a mutation allowing a virus to jump to a new host. This is an area of active research by scientists applying climate change models to what is known about the habitat requirements of mosquito species.

Viral Diseases in Plants

More than 2,000 types of viral diseases of plants are known, accounting for an annual loss of over $30 billion worldwide

due to destruction of crops. Common signs of viral infection include bleached or brown spots on leaves and fruits **(Figure 19.12)**, stunted growth, and damaged flowers or roots, all of which can diminish the yield and quality of crops.

▶ **Figure 19.12 Immature tomato infected by a virus.**

Plant viruses have the same basic structure and mode of replication as animal viruses. Most known plant viruses, including tobacco mosaic virus (TMV), have an RNA genome. Many have a helical capsid, like TMV, while others have an icosahedral capsid (see Figure 19.3b).

Viral diseases of plants spread by two major routes. In the first route, *horizontal transmission*, an external source infects the plant. Because the invading virus must get past the plant's outer protective layer of cells (the epidermis), a plant becomes more susceptible to viral infections if it has been damaged by wind, injury, or herbivores. Herbivores, especially insects, pose a double threat because they can also carry viruses, transmitting disease from plant to plant. Moreover, gardeners may transmit plant viruses inadvertently on pruning shears and other tools. The other route of viral infection is *vertical transmission*, in which a plant inherits a viral infection from a parent. Vertical transmission can occur in asexual propagation (for example, through cuttings) or in sexual reproduction via infected seeds.

Once a virus enters a plant cell and begins replicating, viral genomes and associated proteins can spread throughout the plant through plasmodesmata, the cytoplasmic connections that penetrate the walls between adjacent plant cells (see Figure 36.19). The passage of viral macromolecules from cell to cell is facilitated by virally encoded proteins that cause enlargement of plasmodesmata. Scientists have not yet devised cures for most viral plant diseases, so research efforts are focused largely on reducing disease transmission and on breeding resistant varieties of crop plants.

Prions: Proteins as Infectious Agents

The viruses discussed in this chapter are infectious agents that spread diseases, and their genetic material is composed of nucleic acids, whose ability to be replicated is well known. Surprisingly, there are also *proteins* that are infectious. Proteins called **prions** appear to cause degenerative brain diseases in various animal species. These diseases include scrapie in sheep; mad cow disease, which plagued the European beef industry about 20 years ago; and Creutzfeldt-Jakob disease in humans, which has caused the death of some 175 people in the United Kingdom since 1996. Prions can be transmitted in food, as may occur when people eat beef from cattle with mad cow disease. Kuru, another human disease caused by prions, was identified in the early 1900s among the South Fore indigenous people of New Guinea. When a kuru epidemic peaked there in the 1960s, scientists at first thought the disease had a genetic basis because family members also often contracted the disease. Eventually, however, investigations revealed a

► **Figure 19.13 Model for how prions propagate.** Prions are misfolded versions of normal brain proteins. When a prion contacts a normally folded version of the same protein, it may induce the normal protein to assume the abnormal shape. The resulting chain reaction may continue until high levels of prion aggregation cause cellular malfunction and eventual degeneration of the brain.

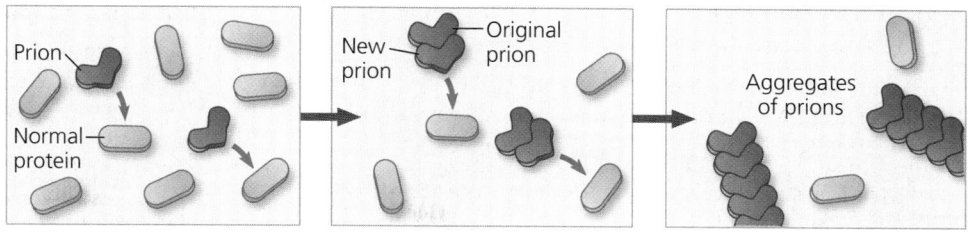

Prion

Normal protein

New prion — Original prion

Aggregates of prions

→ **Mastering Biology** Animation: Prions: Characteristics • Animation: Prions: Diseases

different story: After a death, family members practiced ritual cannibalism, eating organs of the deceased, and prions were transmitted primarily in brain tissue. Women got kuru more often than men because men ate the more "prestigious" organs, like the heart, while women and children ate the brains.

Two characteristics of prions are especially alarming. First, prions act very slowly, with an incubation period of at least ten years before symptoms develop. The lengthy incubation period prevents sources of infection from being identified until long after the first cases appear, allowing many more infections to occur. Second, prions are not destroyed or deactivated by heating to normal cooking temperatures. To date, there is no known cure for prion diseases, and the only hope for developing effective treatments lies in understanding the process of infection.

How can a protein, which cannot replicate itself, be a transmissible pathogen? According to the leading model, a prion is a misfolded form of a protein normally present in brain cells. When the prion gets into a cell containing the normal form of the protein, the prion somehow converts

normal protein molecules to the misfolded prion versions. Several prions then aggregate into a complex that can convert other normal proteins to prions, which join the chain **(Figure 19.13)**. Prion aggregation interferes with normal cellular functions and causes disease symptoms. This model was greeted with much skepticism when it was first proposed by Stanley Prusiner in the early 1980s, but it is now widely accepted. Prusiner was awarded the Nobel Prize in 1997 for his work on prions. He has also proposed that prions are involved in neurodegenerative diseases such as Alzheimer's and Parkinson's disease. There are many outstanding questions about these small infectious agents.

CONCEPT CHECK 19.3

1. Describe two ways in which a preexisting virus can become an emerging virus.
2. Contrast horizontal and vertical transmission of viruses in plants.
3. **WHAT IF?** TMV has been isolated from virtually all commercial tobacco products. Why, then, is TMV infection not an additional hazard for smokers?

For suggested answers, see Appendix A.

19 Chapter Review

→ Go to **Mastering Biology** for Assignments, the eText, the Study Area, and Dynamic Study Modules.

SUMMARY OF KEY CONCEPTS

→ To review key terms, go to the **Vocabulary Self-Quiz** in the **Mastering Biology** eText or Study Area, or go to goo.gl/zkjz9t.

CONCEPT 19.1

A virus consists of a nucleic acid surrounded by a protein coat (pp. 399–401)

- Researchers discovered viruses in the late 1800s by studying a plant disease, tobacco mosaic disease.
- A **virus** is a small nucleic acid genome enclosed in a protein **capsid** and sometimes a membranous **viral envelope**. The genome may be single- or double-stranded DNA or RNA.

? *Are viruses generally considered living or nonliving? Explain.*

CONCEPT 19.2

Viruses replicate only in host cells (pp. 401–408)

- Viruses use enzymes, ribosomes, and small molecules of host cells to synthesize progeny viruses during replication.

- Each type of virus has a characteristic **host range**, affected by whether cell-surface proteins are present that viral surface proteins can bind to.
- **Phages** (viruses that infect bacteria) can replicate by two alternative mechanisms: the **lytic cycle** and the **lysogenic cycle**.

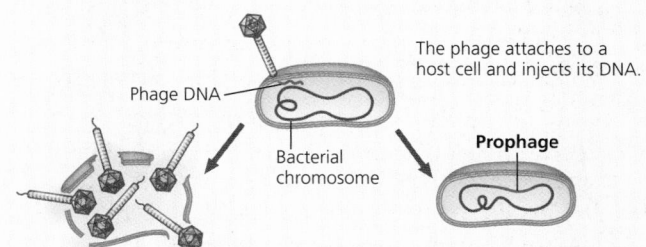

The phage attaches to a host cell and injects its DNA.

Phage DNA

Bacterial chromosome

Prophage

Lytic cycle
- **Virulent** or **temperate phage**
- Destruction of host DNA
- Production of new phages
- Lysis of host cell causes release of progeny phages

Lysogenic cycle
- **Temperate phage** only
- Genome integrates into bacterial chromosome as **prophage**, which
  (1) is replicated and passed on to daughter cells and
  (2) can be induced to leave the chromosome and initiate a lytic cycle

- Bacteria have various ways of defending themselves against phage infections, including the CRISPR-Cas system.
- Many animal viruses have an envelope. **Retroviruses** (such as **HIV**) use the enzyme **reverse transcriptase** to copy their RNA genome into DNA, which can be integrated into the host genome as a **provirus**.
- Since viruses can replicate only within cells, they probably evolved after the first cells appeared, perhaps as packaged fragments of cellular nucleic acid.

? *Describe enzymes that are not found in most cells but are necessary for the replication of certain types of viruses.*

<hr>

CONCEPT 19.3

Viruses and prions are formidable pathogens in animals and plants *(pp. 408–413)*

- Symptoms of viral diseases may be caused by direct viral harm to cells or by the body's immune response. **Vaccines** stimulate the immune system to defend the host against specific viruses.
- An **epidemic**, a widespread outbreak of a disease, can become a **pandemic**, a global epidemic.
- Outbreaks of emerging viral diseases in humans are usually not new, but rather are caused by existing viruses that expand their host territory. The H1N1 2009 flu virus was a new combination of pig, human, and avian viral genes that caused a pandemic. The H5N1 avian flu virus has the potential to cause a high-mortality flu pandemic.
- Viruses enter plant cells through damaged cell walls (horizontal transmission) or are inherited from a parent (vertical transmission).
- **Prions** are slow-acting, virtually indestructible infectious proteins that cause brain diseases in mammals.

? *What aspect of an RNA virus makes it more likely than a DNA virus to become an emerging virus?*

TEST YOUR UNDERSTANDING

→ For more multiple-choice questions, go to the **Practice Test** in the **Mastering Biology** eText or Study Area, or go to goo.gl/GruWRg.

Levels 1–2: Remembering/Understanding

1. Which of the following characteristics, structures, or processes is common to both bacteria and viruses?
   (A) metabolism
   (B) ribosomes
   (C) genetic material composed of nucleic acid
   (D) cell division

2. Emerging viruses arise by
   (A) mutation of existing viruses.
   (B) the spread of existing viruses to new host species.
   (C) the spread of existing viruses more widely within their host species.
   (D) all of the above.

3. To cause a human pandemic, the H5N1 avian flu virus would have to
   (A) spread to primates such as chimpanzees.
   (B) develop into a virus with a different host range.
   (C) become capable of human-to-human transmission.
   (D) become much more pathogenic.

Levels 3–4: Applying/Analyzing

4. A bacterium is infected with an experimentally constructed bacteriophage composed of the T2 phage protein coat and T4 phage DNA. The new phages produced would have
   (A) T2 protein and T4 DNA.   (C) T2 protein and T2 DNA.
   (B) T4 protein and T2 DNA.   (D) T4 protein and T4 DNA.

5. RNA viruses require their own supply of certain enzymes because
   (A) host cells rapidly destroy the viruses.
   (B) host cells lack enzymes that can replicate the viral genome.
   (C) these enzymes translate viral mRNA into proteins.
   (D) these enzymes penetrate host cell membranes.

6. **DRAW IT** Redraw Figure 19.8 to show the replicative cycle of a virus with a single-stranded genome that can function as mRNA (a class IV virus).

Levels 5–6: Evaluating/Creating

7. **EVOLUTION CONNECTION** The success of some viruses lies in their ability to evolve rapidly within the host. Such viruses evade the host's defenses by mutating and producing many altered progeny viruses before the body can mount an attack. Thus, the viruses present late in infection differ from those that initially infected the body. Discuss this as an example of evolution in microcosm. Which viral lineages tend to predominate?

8. **SCIENTIFIC INQUIRY** When bacteria infect an animal, the number of bacteria in the body increases in an exponential fashion (graph A). After infection by a virulent animal virus with a lytic replicative cycle, there is no evidence of infection for a while. Later, however, the number of viruses rises suddenly and subsequently increases in a series of steps (graph B). Explain the difference in the curves.

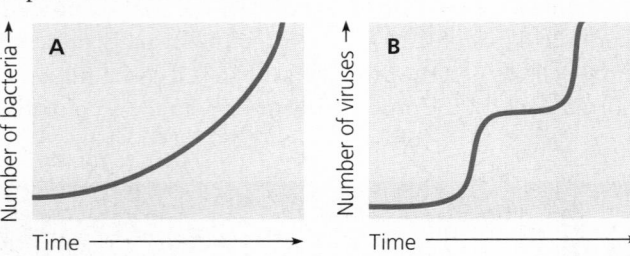

9. **WRITE ABOUT A THEME: ORGANIZATION** While viruses are considered by most scientists to be nonliving, they do show some characteristics of life, including the correlation of structure and function. In a short essay (100–150 words), discuss how the structure of a virus correlates with its function.

10. **SYNTHESIZE YOUR KNOWLEDGE**

Oseltamivir (Tamiflu), an antiviral drug prescribed for the flu, inhibits the enzyme neuraminidase. Explain how this drug could prevent infection in someone exposed to the flu or could shorten the course of flu in an infected patient (the reasons for which it is prescribed).

For selected answers, see Appendix A.

20 DNA Tools and Biotechnology

KEY CONCEPTS

20.1 DNA sequencing and DNA cloning are valuable tools for genetic engineering and biological inquiry *p. 416*

20.2 Biologists use DNA technology to study gene expression and function *p. 423*

20.3 Cloned organisms and stem cells are useful for basic research and other applications *p. 428*

20.4 The practical applications of DNA-based biotechnology affect our lives in many ways *p. 433*

Study Tip

Apply what you've learned: Write down a gene-related question you've wondered about. Then make a table listing techniques you could use to investigate your question and how you would apply them. Below is an example.

Is autism caused by genes, the environment, or both?	
Technique	Application to question
DNA sequencing	–Compare DNA sequences from people with autism with sequences from nonaffected people.

→ Go to Mastering Biology

For Students (in eText and Study Area)
• Get Ready for Chapter 20
• HHMI Video: Genetic Engineering
• Animation: Creating Recombinant DNA
• HHMI Video: Polymerase Chain Reaction (PCR)

For Instructors to Assign (in Item Library)
• Tutorial: Restriction Enzymes, Recombinant DNA, and Gene Cloning
• Activity: Cloning a Gene in Bacteria

Figure 20.1 This model shows a technique in which a DNA strand is passed through a small pore in a membrane. The resulting changes in an electrical current are used to determine the nucleotide sequence. The first human genome sequence, completed in 2003, took 13 years and cost $1 billion; the time and cost of genome sequencing have been greatly reduced by improved methods like the one shown here.

What are the main techniques and applications of biotechnology?

TECHNIQUES

DNA sequencing

Gene cloning

PCR amplification

Expressing genes
Ribosome
Protein

Analyzing gene expression
RNA sequencing

Gene editing
CRISPR-Cas9

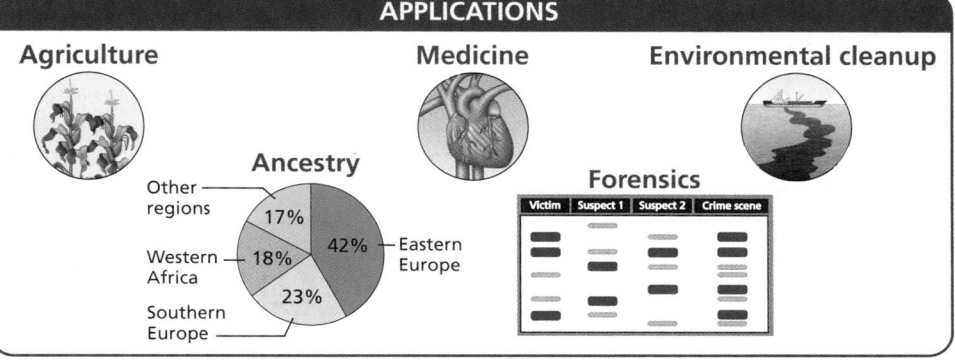

APPLICATIONS

Agriculture

Medicine

Environmental cleanup

Ancestry
Other regions 17%
Eastern Europe 42%
Western Africa 18%
Southern Europe 23%

Forensics

| Victim | Suspect 1 | Suspect 2 | Crime scene |

DNA sequencing and DNA cloning are valuable tools for genetic engineering and biological inquiry

The discovery of the structure of the DNA molecule, and specifically the recognition that its two strands are complementary to each other, opened the door for the development of DNA sequencing and other techniques for manipulating DNA—known as **DNA technology**—used in biological research today. Key to these techniques is **nucleic acid hybridization**, the base pairing of one strand of a nucleic acid to a complementary sequence from another nucleic acid strand, either DNA or RNA. Nucleic acid hybridization forms the foundation of virtually every technique used in **genetic engineering**, the direct manipulation of genes for practical purposes. Genetic engineering has launched a revolution in fields as varied as criminal law, medicine, and basic biological research. In this section, we'll explore several important techniques and their uses.

DNA Sequencing

Researchers can exploit the principle of complementary base pairing to determine the complete nucleotide sequence of a DNA molecule, a process called **DNA sequencing**. The first automated procedure, called dideoxy sequencing, was developed in the 1970s by biochemist Frederick Sanger, who received the Nobel Prize in 1980 for this accomplishment. Dideoxy sequencing is still used for routine small-scale sequencing jobs.

 Mastering Biology HHMI Video: Sanger Method of DNA Sequencing

During the first decade of this century, "next-generation sequencing" techniques were developed that are rapid and inexpensive **(Figure 20.2)**. DNA fragments are amplified (copied) to yield an enormous number of identical fragments **(Figure 20.3)**. A single template strand of each fragment is immobilized, and the complementary strand is synthesized, one nucleotide at a time. A chemical technique enables electronic monitors to identify in real time which of the four nucleotides is added; this method is thus called *sequencing by synthesis*. Thousands or hundreds of thousands of fragments, each about 300 nucleotides long, are sequenced in parallel in machines like those shown in Figure 20.2, accounting for the high rate of nucleotides sequenced per hour. This is an example of "high-throughput" DNA technology and is currently the method of choice for studies where massive numbers of

▼ **Figure 20.2 Next-generation DNA sequencing machines.**

DNA samples—even a set of numerous fragments representing an entire genome—are being sequenced.

More and more often, next-generation sequencing is complemented (or in some cases replaced) by "third-generation sequencing," with each new technique being faster and less expensive than the previous one. In some new methods, the DNA is neither cut into fragments nor amplified. Instead, a single, very long DNA molecule is sequenced on its own. Several groups have developed techniques that move a single strand of a DNA molecule through a very small pore (a *nanopore*) in a membrane, identifying the bases one by one by the distinct way each interrupts an electrical current. One model of this concept is shown in Figure 20.1, in which the pore is a protein channel embedded in a lipid membrane. (Other researchers are using artificial membranes and nanopores.) Each type of base interrupts the electrical current for a slightly different length of time. In 2015, the first nanopore sequencer went on the market; this device is the size of a small candy bar and connects to a computer via a USB port. Associated software allows immediate identification and analysis of the sequence. This is only one of many approaches to further increase the rate and cut the cost of sequencing, while also allowing the methodology to move out of the laboratory and into the field.

Improved DNA-sequencing techniques have transformed the way in which we can explore fundamental biological questions about evolution and how life works (see Make Connections Figure 5.26). Little more than 15 years after the human genome sequence was announced, researchers had completed the sequencing of thousands of genomes, with tens of thousands in progress. Complete genome sequences have been determined for cells from several cancers, for ancient humans, and for the many bacteria that live in the human intestine. In Chapter 21, you'll learn more about how this rapid acceleration of sequencing technology has revolutionized our study of the evolution of species and the genome itself. Now, let's consider how individual genes are studied.

▼ **Figure 20.3** **Research Method**

Sequencing by Synthesis: Next-Generation Sequencing

1 Genomic DNA is fragmented, and fragments of 300 base pairs are selected.

2 Each fragment is placed in a droplet with a bead.

Bead

3 Many copies of each fragment are made using a technique called PCR (see Figure 20.7). One million identical copies are made and attached by their 5′ end to the bead.

4 The bead is placed into a small well along with DNA polymerases and primers that can hybridize to the 3′ end of the single (template) strand.

DNA polymerase

Template strand of DNA

5′   3′   3′ 5′
Primer

A T G C

Application In current next-generation sequencing techniques, each fragment is about 300 nucleotides long; by sequencing the fragments in parallel, about 2 billion nucleotides can be sequenced in 24 hours.

Technique See numbered steps and diagrams.

Results Each of the 2,000,000 wells in the multiwell plate, which holds a different fragment, yields a different sequence. The results for one fragment are shown below as a "flow-gram." The sequences of the entire set of fragments are analyzed using computer software, which "stitches" them together into a whole sequence—here, an entire genome.

No. of nucleotides in sequence

■ A
□ T
▦ G
▨ C

TTCTGCGAA

INTERPRET THE DATA *If the template strand has two or more identical nucleotides in a row, their complementary nucleotides will be added one after the other in the same flow step. How are two or more of the same nucleotide (in a row) detected in the flow-gram? (See sample on the right.) Write out the sequence of the first 25 nucleotides in the flow-gram above, starting from the left. (Ignore the very short lines.)*

5 The well is one of 2 million on a multiwell plate, each containing a different DNA fragment to be sequenced. A solution of one of the four nucleotides required for DNA synthesis (deoxynucleoside triphosphates, or dNTPs) is added to all wells and then washed off. This is done sequentially for all four nucleotides: dATP, dTTP, dGTP, and then dCTP. The entire process is then repeated over and over again.

A T G C

DNA polymerase

Template strand of DNA

C C A A T G T A G C G C A G T A
dATP → PP$_i$
Primer

6 In each well, if the next base on the template strand (T in this example) is complementary to the added nucleotide (A, here), the nucleotide is joined to the growing strand, releasing PP$_i$, which causes a flash of light that is recorded.

A T G C

C C A A T G T A G C G C A G T A
dTTP → dTTP

7 The nucleotide is washed off and a different nucleotide (dTTP, here) is added. If the nucleotide is not complementary to the next template base (G, here), it is not joined to the strand, no reaction occurs, and there is no flash.

A T G C

C C A A T G T A G C G C A G T A
dGTP → dGTP

A T G C

C C A A T G C T A G C G C A G T A
dCTP → PP$_i$

8 The process of adding and washing off the four nucleotides is repeated until every fragment has a complete complementary strand. The pattern of flashes reveals the sequence of the original fragment in each well.

Making Multiple Copies of a Gene or Other DNA Segment

A molecular biologist studying a particular gene or group of genes faces a challenge. Naturally occurring DNA molecules are very long, and a single molecule usually carries hundreds or even thousands of genes. Moreover, in many eukaryotic genomes, protein-coding genes occupy only a small proportion of the chromosomal DNA, the rest being noncoding nucleotide sequences. A single human gene, for example, might constitute only 1/100,000 of a chromosomal DNA molecule. As a further complication, it's not easy to distinguish a gene from the surrounding DNA because they differ only in nucleotide sequence. To study a specific gene, scientists have developed methods to isolate a segment of DNA carrying that gene and make multiple identical copies of it—a process called **DNA cloning**.

Most methods for cloning pieces of DNA in the laboratory share certain general features. One common approach uses bacteria, most often *Escherichia coli*. Recall from Figure 16.13 that the *E. coli* chromosome is a large, circular molecule of DNA. In addition, *E. coli* and many other bacteria also have **plasmids**, small, circular DNA molecules that are replicated separately. A plasmid has only a small number of genes; these genes may be useful when the bacterium is in a particular environment but may not be required for survival or reproduction under most conditions.

To clone pieces of DNA using bacteria, scientists have isolated plasmids from bacterial cells and altered them by genetic engineering. Researchers insert DNA they want to study ("foreign" DNA) into the plasmid **(Figure 20.4)**. The resulting plasmid is now a **recombinant DNA molecule**, a molecule containing DNA from two different sources, very often different species. The plasmid is then returned to a bacterial cell, producing a *recombinant bacterium*. This single cell reproduces through repeated cell divisions to form a clone of cells, a population of genetically identical cells. Because the dividing bacteria replicate the recombinant plasmid and pass it on to their descendants, the foreign DNA and any genes it carries are cloned at the same time. The production of multiple copies of a single gene is a type of DNA cloning called **gene cloning**.

In Figure 20.4, the plasmid acts as a **cloning vector**, a DNA molecule that can carry foreign DNA into a host cell and be replicated there. Bacterial plasmids are widely used as cloning vectors for several reasons: They can be readily obtained from commercial suppliers, manipulated to form recombinant plasmids by insertion of foreign DNA in a test tube (referred to as *in vitro*, from the Latin meaning "in glass"), and then easily introduced into bacterial cells. The foreign DNA in Figure 20.4 is a gene from a eukaryotic cell; we will describe in more detail how the foreign DNA segment was obtained later in this section.

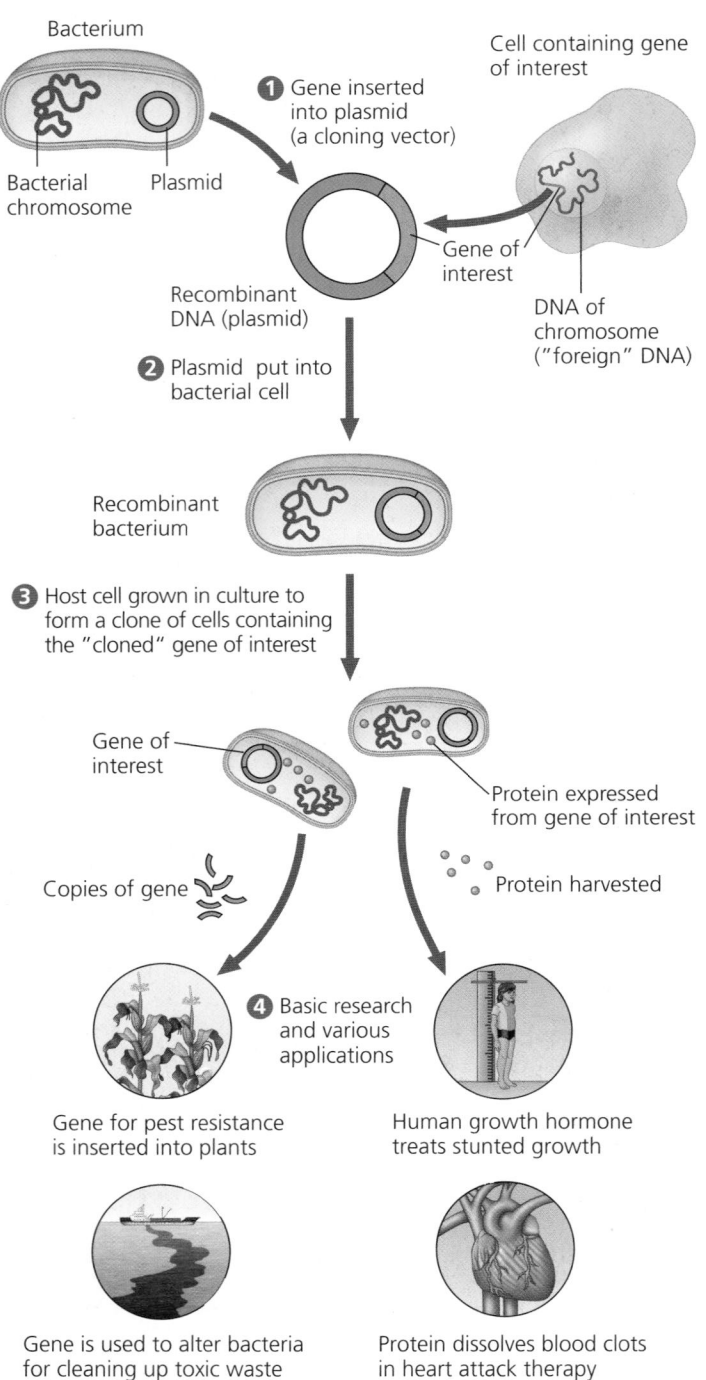

▼ **Figure 20.4 Gene cloning and some uses of cloned genes.** In this simplified diagram of gene cloning, we start with a plasmid (originally isolated from a bacterial cell) and a gene of interest from another organism. Only one plasmid and one copy of the gene of interest are shown at the top of the figure, but the starting materials would include many of each.

Bacterium

Cell containing gene of interest

❶ Gene inserted into plasmid (a cloning vector)

Bacterial chromosome   Plasmid

Gene of interest

DNA of chromosome ("foreign" DNA)

Recombinant DNA (plasmid)

❷ Plasmid put into bacterial cell

Recombinant bacterium

❸ Host cell grown in culture to form a clone of cells containing the "cloned" gene of interest

Gene of interest

Protein expressed from gene of interest

Copies of gene

Protein harvested

❹ Basic research and various applications

Gene for pest resistance is inserted into plants

Human growth hormone treats stunted growth

Gene is used to alter bacteria for cleaning up toxic waste

Protein dissolves blood clots in heart attack therapy

→ **Mastering Biology** HHMI Video: Genetic Engineering **hhmi** BioInteractive

Gene cloning is useful for two basic purposes: to make many copies of, or *amplify*, a particular gene and to produce a protein product from it (see Figure 20.4). Researchers can isolate copies of a cloned gene from bacteria for use in basic research or to endow another organism with a new metabolic capability, such as pest resistance. For example, a resistance gene present in one crop species might be cloned and transferred into plants of another species. (Such organisms are called *genetically modified organisms*, or *GMOs* for short; they will be discussed later in the chapter.) Alternatively, a protein with medical uses, such as human growth hormone, can be harvested in large quantities from cultures of bacteria carrying a cloned gene for the protein. (We'll look at the techniques for expressing cloned genes later.) Since one gene is only a very small part of the total DNA in a cell, the ability to amplify such a DNA fragment is crucial for any application involving a single gene.

Using Restriction Enzymes to Make a Recombinant DNA Plasmid

Gene cloning and genetic engineering generally rely on the use of enzymes that cut DNA molecules at a limited number of specific locations. These enzymes, called restriction endonucleases, or **restriction enzymes**, were discovered in the late 1960s by biologists doing basic research on bacteria. Restriction enzymes protect the bacterial cell by cutting up foreign DNA from other organisms or phages (see Concept 19.2).

Hundreds of different restriction enzymes have been identified and isolated. Each restriction enzyme is very specific, recognizing a particular short DNA sequence, or **restriction site**, and cutting both DNA strands at precise points within this restriction site. The DNA of a bacterial cell is protected from the cell's own restriction enzymes by the addition of methyl groups ($-CH_3$) to adenines or cytosines within the sequences recognized by the enzymes.

⊙ **Mastering Biology** Animation: Restriction Enzymes

Figure 20.5 shows how restriction enzymes are used to clone a foreign DNA fragment into a bacterial plasmid. At the top of the figure is a bacterial plasmid (like the one shown in Figure 20.4) that has a single restriction site recognized by a particular restriction enzyme. As shown in this example, most restriction sites are symmetrical. In other words, the sequence of nucleotides is the same on both strands when read in the 5' → 3' direction. The most commonly used restriction enzymes recognize sequences containing four to eight nucleotide pairs. Because any sequence that is this short usually occurs (by chance) many times in a long DNA molecule, a restriction enzyme will make many cuts in such a DNA molecule, yielding a set of **restriction fragments**. Since restriction enzymes always cut at the same exact DNA sequence, copies of any given DNA molecule exposed to the same restriction enzyme always yield the same set of restriction fragments.

▼ **Figure 20.5 Using a restriction enzyme and DNA ligase to make a recombinant DNA plasmid.** The restriction enzyme in this example (called *Eco*RI) recognizes a single six-base-pair restriction site present in this plasmid. It makes staggered cuts in the sugar-phosphate backbones, producing fragments with "sticky ends." Foreign DNA fragments with complementary sticky ends can base-pair with the plasmid ends; the ligated product is a recombinant plasmid. (If the two plasmid sticky ends base-pair, the original nonrecombinant plasmid is reformed.)

Bacterial plasmid

Restriction site

DNA

1 A restriction enzyme cuts the sugar-phosphate backbones at each arrow.

Sticky end

2 A DNA fragment from another source is added. Base pairing of sticky ends produces various combinations.

Fragment from different DNA molecule cut by the same restriction enzyme

One possible combination

3 DNA ligase seals the strands.

Recombinant DNA molecule

Recombinant plasmid

DRAW IT *The restriction enzyme* Hind*III recognizes the sequence 5'-AAGCTT-3', cutting between the two A's. Draw the double-stranded sequence before and after the enzyme cuts it.*

⊙ **Mastering Biology** Animation: Recombinant DNA

The most useful restriction enzymes cleave the sugar-phosphate backbones in the two DNA strands in a staggered manner, as shown in step ❶ of Figure 20.5. The resulting double-stranded restriction fragments have at least one single-stranded end, called a **sticky end**. These short extensions can form hydrogen-bonded base pairs with complementary sticky ends on any other DNA molecules cut with the same restriction enzyme, such as the inserted DNA shown in step ❷ of Figure 20.5. The associations formed in this way are only temporary but can be made permanent by DNA ligase, an enzyme that catalyzes the formation of covalent bonds that close up the sugar-phosphate backbones of DNA strands (see step ❸ of Figure 20.5). At the bottom of Figure 20.5, you can see the stable recombinant DNA molecule that was produced by the ligase-catalyzed joining of DNA from two different sources. The end result, in this example, is the formation of a stable recombinant plasmid containing foreign DNA.

→ **Mastering Biology Animation: Creating Recombinant DNA**

To check the recombinant plasmids after they have been copied many times in host cells to make sure the fragment has been inserted (see Figure 20.4), a researcher might cut the products again using the same restriction enzyme. If the insert is there, there would be two DNA fragments, one the size of the plasmid and one the size of the inserted DNA. To separate and visualize the fragments, researchers carry out a technique called **gel electrophoresis**, which uses a gel made of a polymer that has microscopic holes of different sizes, through which shorter fragments can travel faster. The gel works as a molecular sieve to separate out a mixture of nucleic acid fragments by length **(Figure 20.6)**. Gel electrophoresis is used in conjunction with many different techniques in molecular biology.

Amplifying DNA: The Polymerase Chain Reaction (PCR) and Its Use in DNA Cloning

Now that we have examined the cloning vector in some detail, let's consider how biologists obtain the foreign DNA to be inserted. Most researchers have some sequence information about the DNA fragment they want to clone. Using this information, they can start with genomic DNA from the particular species of interest and obtain many copies of the desired gene by using a technique called the **polymerase chain reaction**, or **PCR**. **Figure 20.7** illustrates the steps in PCR. Within a few hours, this technique can make billions of copies of a specific target DNA segment in a sample, even if that segment makes up less than 0.001% of the total DNA in the sample.

→ **Mastering Biology HHMI Video: Polymerase Chain Reaction (PCR)**

▼ **Figure 20.6 Gel electrophoresis.** A gel made of a polymer acts as a molecular sieve to separate nucleic acids or proteins differing in size, electrical charge, or other physical properties as they move in an electric field. In the example shown here, DNA molecules are separated by length in a gel made of a polysaccharide called agarose.

(a) Each sample, a mixture of different DNA molecules, is placed in a separate well near one end of a thin slab of agarose gel. The gel is set into a small plastic support and immersed in an aqueous, buffered solution in a tray with electrodes at each end. The current is then turned on, causing the negatively charged DNA molecules to move toward the positive electrode.

(b) Shorter molecules are slowed down less than longer molecules, so shorter molecules move faster through the gel. After the current is turned off, a DNA-binding dye is added that fluoresces pink in ultraviolet (UV) light. Each pink band corresponds to many thousands of DNA molecules of the same length. The bands at the upper and lower edges of the gel are restriction fragments of standard lengths for comparison with samples of unknown length.

→ **Mastering Biology Animation: Gel Electrophoresis of DNA**

In the PCR procedure, a three-step cycle causes a chain reaction that produces an exponentially growing population of identical DNA molecules. During each cycle, the reaction mixture is ❶ heated to high temperatures to denature (separate) the strands of the double-stranded DNA and then ❷ cooled to allow annealing (hydrogen bonding) of short, single-stranded DNA primers complementary to sequences on opposite strands at each end of the target sequence; finally, ❸ a special DNA polymerase extends the primers in the 5' → 3' direction. This cycle is then repeated 30–40 times. If a standard DNA polymerase were used, this enzyme

The Polymerase Chain Reaction (PCR)

Application With PCR, any specific segment—the so-called target sequence—in a DNA sample can be copied many times (amplified) within a test tube.

Technique PCR requires double-stranded DNA containing the target sequence, a heat-resistant DNA polymerase, all four nucleotides, and two 15- to 20-nucleotide single DNA strands that serve as primers. One primer is complementary to one end of the target sequence on one strand; the second primer is complementary to the other end of the sequence on the other strand.

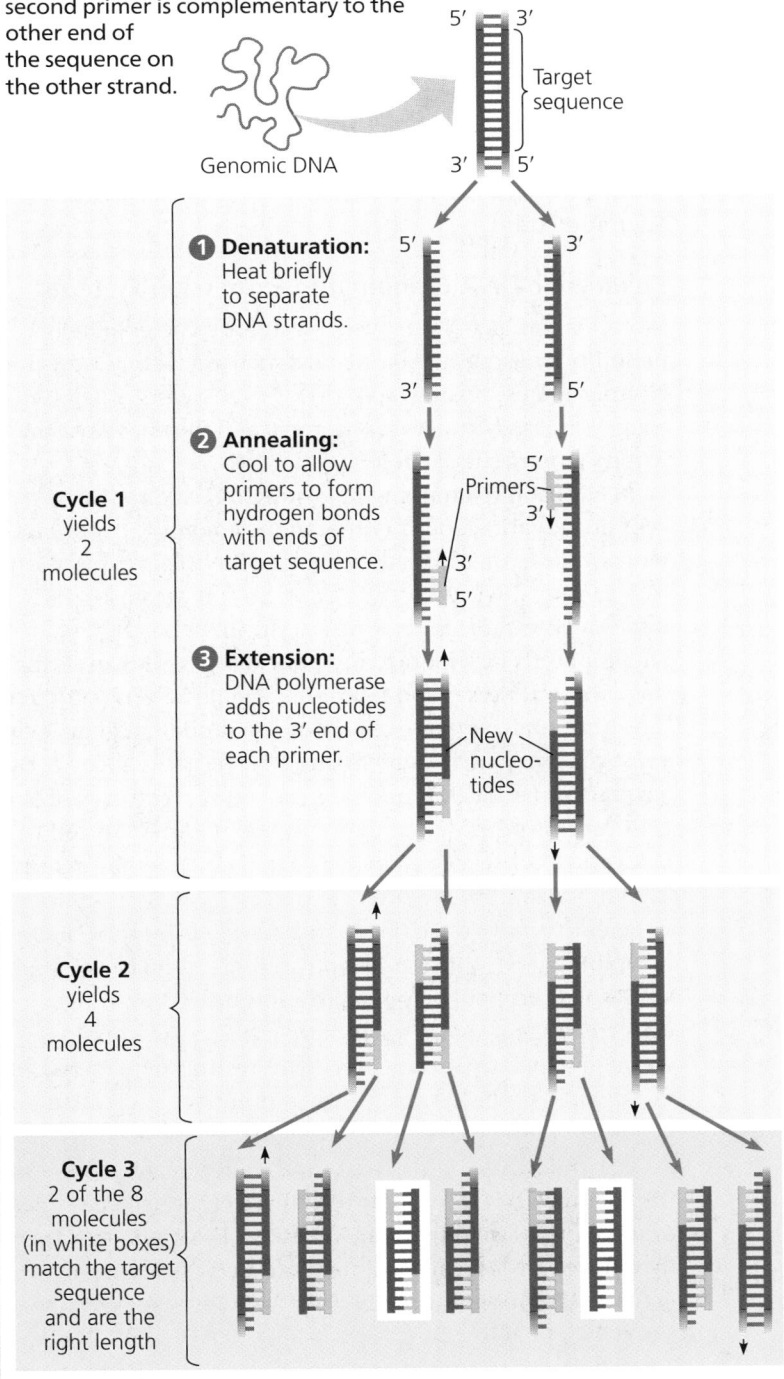

Genomic DNA

Target sequence

Cycle 1 yields 2 molecules

1 **Denaturation:** Heat briefly to separate DNA strands.

2 **Annealing:** Cool to allow primers to form hydrogen bonds with ends of target sequence.

Primers

3 **Extension:** DNA polymerase adds nucleotides to the 3′ end of each primer.

New nucleotides

Cycle 2 yields 4 molecules

Cycle 3 2 of the 8 molecules (in white boxes) match the target sequence and are the right length

Results After three cycles, two molecules match the target sequence exactly. After 30 more cycles, over 1 billion (10^9) molecules match.

→ **Mastering Biology** Animation: Copying DNA Through PCR

would be denatured along with the DNA during the first heating step and would have to be replaced after each cycle. The key to automating PCR was the discovery of an unusual heat-stable DNA polymerase enzyme called *Taq* polymerase, named after the bacterial species from which it was first isolated. This bacterial species, *Thermus aquaticus*, lives in hot springs, and the stability of its DNA polymerase at high temperatures is an evolutionary adaptation that enables the enzyme to function at temperatures up to 95°C. Today, researchers also use a DNA polymerase from the archaean species *Pyrococcus furiosus*. This enzyme, called *Pfu* polymerase, is more accurate and stable but more expensive than *Taq* polymerase.

PCR is speedy and very specific. Only a minuscule amount of DNA need be present in the starting material, and this DNA can be partially degraded, as long as there are a few copies of the complete target sequence. The key to the high specificity is the pair of primers used for each PCR amplification. The primer sequences are chosen so that they hybridize *only* to sequences at opposite ends of the target segment, one on the 3′ end of each strand. (For high specificity, the primers must be at least 15 nucleotides long.) With each successive cycle, the number of target segment molecules of the correct length doubles, so the number of molecules equals 2^n, where *n* is the number of cycles. After 30 or so cycles, about a billion copies of the target sequence are present!

Despite its speed and specificity, PCR amplification cannot substitute for gene cloning in cells to make large amounts of a gene. This is because the polymerases that are used have no proofreading function, and occasional errors during PCR replication limit the number of good copies and the length of DNA fragments that can be copied. Instead, PCR is used to provide the specific DNA fragment for cloning. PCR primers are synthesized to include a restriction site at each end of the DNA fragment that matches the site in the cloning vector, and the fragment and vector are cut and ligated together **(Figure 20.8)**. The resulting plasmids are sequenced so that those with error-free inserts can be selected.

Devised in 1985, PCR has had a major impact on biological research and genetic engineering. PCR has been used to amplify DNA from a wide variety of sources: a 40,000-year-old frozen woolly mammoth; fingerprints or tiny amounts of blood, tissue, or semen found at crime scenes; single embryonic cells for rapid prenatal diagnosis of genetic disorders (see Figure 14.19); and cells infected with viruses that are

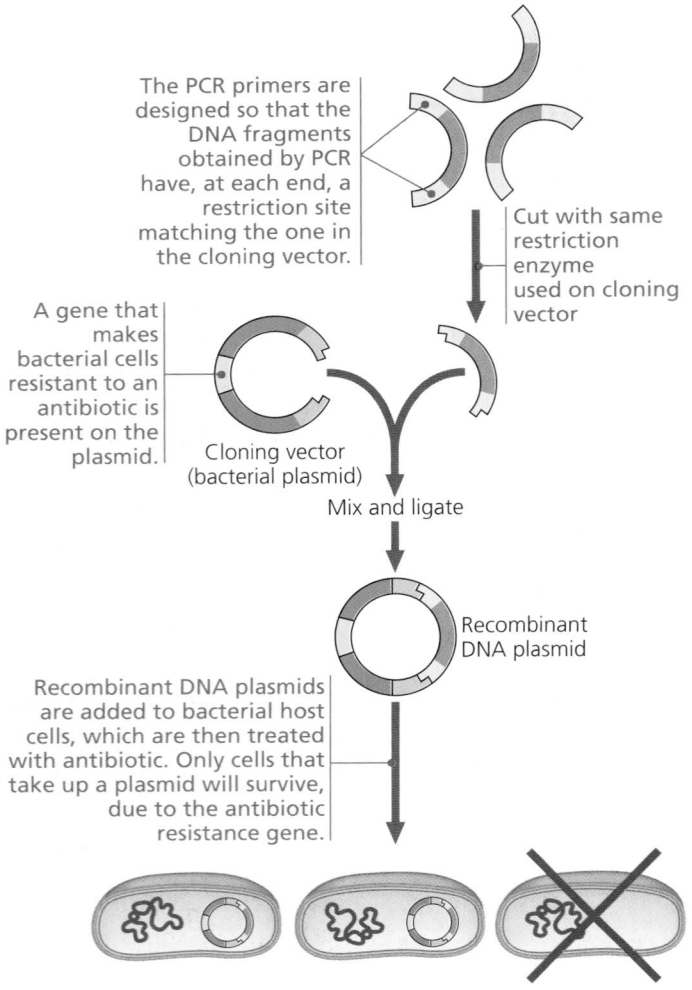

The PCR primers are designed so that the DNA fragments obtained by PCR have, at each end, a restriction site matching the one in the cloning vector.

Cut with same restriction enzyme used on cloning vector

A gene that makes bacterial cells resistant to an antibiotic is present on the plasmid.

Cloning vector (bacterial plasmid)

Mix and ligate

Recombinant DNA plasmid

Recombinant DNA plasmids are added to bacterial host cells, which are then treated with antibiotic. Only cells that take up a plasmid will survive, due to the antibiotic resistance gene.

difficult to detect, such as HIV. (To test for HIV, viral genes are amplified.) We'll return to applications of PCR later.

Expressing Cloned Eukaryotic Genes

Once a gene has been cloned in host cells, its protein product can be expressed in large amounts for research or for practical applications, which we'll explore in Concept 20.4. Cloned genes can be expressed in either bacterial or eukaryotic cells; each option has advantages and disadvantages.

Bacterial Expression Systems

Getting a cloned eukaryotic gene to function in bacterial host cells can be difficult because certain aspects of gene expression are different in eukaryotes and bacteria. To overcome differences in promoters and other DNA control sequences

(see Concept 17.2), scientists usually employ an **expression vector**, a cloning vector that contains a highly active bacterial promoter just upstream of a restriction site where the eukaryotic gene can be inserted in the correct reading frame. The bacterial host cell will recognize the promoter and proceed to express the foreign gene now linked to that promoter. Such expression vectors allow the synthesis of many eukaryotic proteins in bacterial cells.

Another problem with expressing cloned eukaryotic genes in bacteria is the presence of noncoding regions (introns) in most eukaryotic genes (see Concept 17.3). Introns can make a eukaryotic gene very long and unwieldy, and they prevent correct expression of the gene by bacterial cells, which do not have RNA-splicing machinery. This problem can be surmounted by using a form of the gene that includes only the exons. (This is called *complementary DNA*, or *cDNA*; see Figure 20.10.)

Eukaryotic DNA Cloning and Expression Systems

Molecular biologists can avoid eukaryotic-bacterial incompatibility by using eukaryotic cells such as yeasts as hosts for cloning and expressing eukaryotic genes. Yeasts, single-celled fungi, are as easy to grow as bacteria, and they have plasmids, a rarity among eukaryotes.

In addition to enabling RNA splicing, eukaryotic host cells are advantageous because many eukaryotic proteins will not function unless they are modified after translation—for example, by the addition of carbohydrate groups (glycosylation) or lipid groups in the ER and Golgi. Bacterial cells lack membrane-bound organelles and cannot carry out these modifications, and if the gene product requiring such processing is from a mammalian gene, even yeast cells may not be able to modify the protein correctly. Several cultured cell types have proved successful as host cells for this purpose, including some mammalian cell lines and an insect cell line that can be infected by a virus carrying recombinant DNA.

Scientists have developed other methods for introducing recombinant DNA into eukaryotic cells. In **electroporation**, a brief electrical pulse applied to a solution containing cells creates temporary holes in their plasma membranes, through which DNA can enter. (This technique is now commonly used for bacteria as well.) Alternatively, scientists can inject DNA directly into single eukaryotic cells using microscopically thin needles. Another way to get DNA into plant cells is by using the soil bacterium *Agrobacterium tumefaciens*, as we'll see later. Whatever the method, if the introduced DNA is incorporated into a cell's genome by genetic recombination, then it can be stably expressed by the cell. Expressing different versions of genes in cells allows researchers to study protein function, a topic we'll return to in Concept 20.2.

Cross-Species Gene Expression and Evolutionary Ancestry

EVOLUTION The ability to express eukaryotic proteins in bacteria (even if the proteins can't be modified properly) is quite remarkable when we consider how different eukaryotic and bacterial cells are. In fact, examples abound of genes that are taken from one species and function perfectly well when transferred into another very different species, such as a firefly gene in a tobacco plant and a jellyfish gene in a pig (see Figure 17.7). These observations underscore the shared evolutionary ancestry of species living today.

One example involves a gene called *Pax-6*, which has been found in animals as diverse as vertebrates and fruit flies. The vertebrate *Pax-6* gene product (the PAX-6 protein) triggers a complex program of gene expression resulting in formation of the vertebrate eye, which has a single lens. Expression of the fly *Pax-6* gene leads to formation of the compound fly eye, which is quite different from the vertebrate eye. When the mouse *Pax-6* gene was cloned and introduced into a fly embryo so that it replaced the fly's own *Pax-6* gene, researchers were surprised to see that the mouse version of the gene led to formation of a compound fly eye (see Figure 50.16). Conversely, when the fly *Pax-6* gene was transferred into a vertebrate embryo—a frog, in this case—a frog eye formed. Although the genetic programs triggered in vertebrates and flies generate very different eyes, the two versions of the *Pax-6* gene can substitute for each other to trigger lens development, evidence of their evolution from a gene in a very ancient common ancestor. Because of their ancient evolutionary roots, all living organisms share the same basic mechanisms of gene expression. This commonality is the basis of many recombinant DNA techniques described in this chapter.

CONCEPT CHECK 20.1

1. **MAKE CONNECTIONS** The restriction site for an enzyme called *Pvu*I is the following sequence:

    5'-CGATCG-3'

    3'-GCTAGC-5'

    Staggered cuts are made between the T and C on each strand. What type of bonds are being cleaved? (See Concept 5.5.)

2. **DRAW IT** One strand of a DNA molecule has the following sequence:

    5'-CTTGACGATCGTTACCG-3'

    Draw the other strand. Will *Pvu*I (see question 1) cut this molecule? If so, draw the products.

3. What are some potential difficulties in using plasmid vectors and bacterial host cells to produce large quantities of proteins from cloned eukaryotic genes?

4. **VISUAL SKILLS** Compare Figure 20.7 with Figure 16.21. How does replication of DNA ends during PCR proceed without shortening the fragments each time?

For suggested answers, see Appendix A.

Biologists use DNA technology to study gene expression and function

To see how a biological system works, scientists seek to understand the functioning of the system's component parts. Analysis of when and where a gene or group of genes is expressed can provide important clues about their function and how they contribute to the organism as a whole.

Analyzing Gene Expression

Biologists driven to understand the assorted cell types of a multicellular organism, cancer cells, or the developing tissues of an embryo first try to discover which genes are expressed by the cells of interest. The most straightforward way to do this is usually to identify the mRNAs being made. We'll first examine techniques that look for patterns of expression of specific individual genes. Next, we'll explore ways to characterize groups of genes being expressed by cells or tissues of interest. As you will see, all of these procedures depend in some way on base pairing between complementary nucleotide sequences.

Studying the Expression of Single Genes

Suppose we have cloned a gene that we suspect plays an important role in the embryonic development of *Drosophila melanogaster* (the fruit fly). The first thing we might wonder is which embryonic cells express the gene—in other words, where in the embryo is the corresponding mRNA found? We can detect the mRNA in an embryonic sample using nucleic acid hybridization with molecules of complementary sequence to the mRNA we want to follow. Using our cloned gene as a template, we can synthesize a short, single-stranded nucleic acid (either RNA or DNA) complementary to the mRNA of interest; this is called a **nucleic acid probe**. For example, if part of the sequence on the mRNA were

5' ···C U C A U C A C C G G C ··· 3'

then we would synthesize this single-stranded DNA probe:

3' G A G T A G T G G C C G 5'

Each probe molecule is labeled during synthesis with a fluorescent tag so we can follow it. A solution containing probe molecules is applied to *Drosophila* embryos, allowing the probe to hybridize specifically with any complementary sequences on the many mRNAs in embryonic cells in which the gene is being transcribed. Because this technique allows us to see the mRNA in place (or *in situ*, in Latin) in the intact organism, this technique is called ***in situ* hybridization**.

▼ **Figure 20.9 Determining where single genes are expressed by *in situ* hybridization analysis.** This *Drosophila* embryo was incubated in a solution containing DNA probes for five different mRNAs, each probe labeled with a different fluorescently colored tag. The embryo was then viewed using fluorescence microscopy. Each color marks where a specific gene is expressed as mRNA. The arrows from the groups of yellow and blue cells above the micrograph show a magnified view of nucleic acid hybridization of the appropriately colored probe to the mRNA. The thorax (trunk) and abdomen are made up of repeating segments. Yellow cells (expressing the *wg* gene) interact with blue cells (expressing the *en* gene); their interaction helps establish the pattern in a body segment.

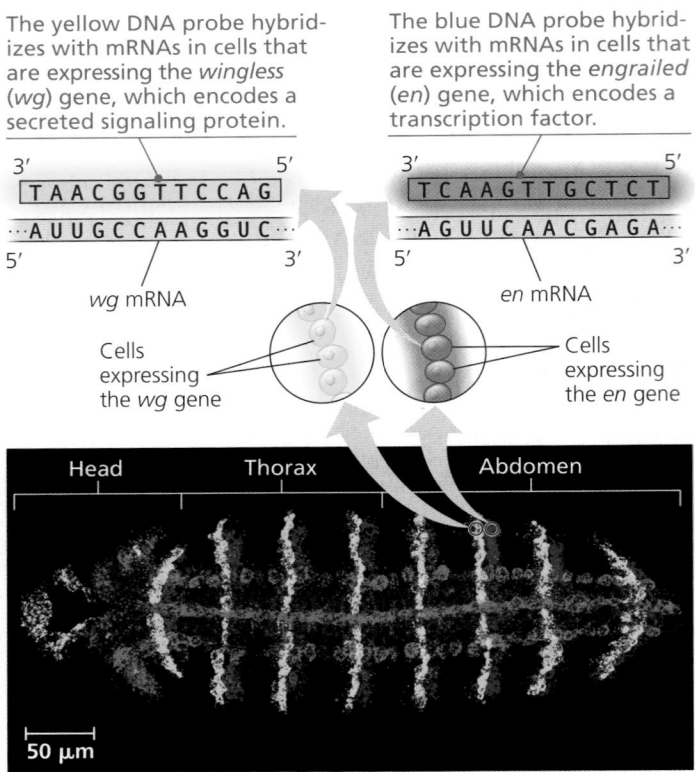

The yellow DNA probe hybridizes with mRNAs in cells that are expressing the *wingless* (*wg*) gene, which encodes a secreted signaling protein.

The blue DNA probe hybridizes with mRNAs in cells that are expressing the *engrailed* (*en*) gene, which encodes a transcription factor.

3′ TAACGGTTCCAG 5′
AUUGCCAAGGUC 5′ ··· 3′

wg mRNA

Cells expressing the *wg* gene

3′ TCAAGTTGCTCT 5′
AGUUCAACGAGA 5′ ··· 3′

en mRNA

Cells expressing the *en* gene

Head | Thorax | Abdomen

50 μm

▼ **Figure 20.10 Making complementary DNA (cDNA) from eukaryotic genes.** Complementary DNA is made in a test tube using mRNA as a template for the first strand. Only one mRNA is shown after step 1, but the final collection of cDNAs would reflect all the mRNAs present in the cell.

❶ Reverse transcriptase is added to a test tube containing mRNAs, isolated from a sample of cells.

DNA in nucleus

mRNAs in cytoplasm

❷ Reverse transcriptase makes the first DNA strand using the mRNA as a template and a short poly-dT as a DNA primer.

Reverse transcriptase

Poly-A tail

mRNA 5′ ────── 3′ AAAAAA 3′
                    TTTTT 5′

DNA strand

Primer (poly-dT)

❸ mRNA is degraded by another enzyme.

5′ ─ ─ ─ ─ AAA AAA 3′
3′ ────────── TTTTT 5′

❹ DNA polymerase synthesizes the second DNA strand, using a primer in the reaction mixture. (Several options exist for primers.)

5′ ────── 3′
3′ ────── 5′

DNA polymerase

❺ The result is cDNA, which carries the complete coding sequence of the gene but no introns.

5′ ────── 3′
3′ ────── 5′

cDNA

Different probes can be labeled with different fluorescent dyes, sometimes with strikingly beautiful results **(Figure 20.9)**.

Other mRNA detection techniques may be preferable for comparing the amounts of a specific mRNA in several samples at the same time—for example, in different cell types or in embryos of different stages. One method that is widely used is called the **reverse transcriptase polymerase chain reaction**, or **RT-PCR**.

RT-PCR begins by turning sample sets of mRNAs into double-stranded DNAs with the corresponding sequences. First, the enzyme reverse transcriptase (from a retrovirus; see Figure 19.9) is used to synthesize a complementary DNA copy (a *reverse transcript*) of each mRNA in the sample **(Figure 20.10)**. The mRNA is then degraded by addition of a specific enzyme, and a second DNA strand, complementary to the first, is synthesized by DNA polymerase. The resulting double-stranded DNA is called **complementary DNA**

(cDNA). (Made from mRNA, cDNA lacks introns and can be used for protein expression in bacteria, as mentioned earlier.) To analyze the timing of expression of the *Drosophila* gene of interest, for example, we would first isolate all the mRNAs from different stages of *Drosophila* embryos and make cDNA from each stage **(Figure 20.11)**.

Next in RT-PCR is the PCR step (see Figure 20.7). As you will recall, PCR is a way of rapidly making many copies of one specific stretch of double-stranded DNA, using primers that hybridize to the opposite ends of the segment that we are interested in studying. In our case, we would add primers corresponding to a segment of our *Drosophila* gene, using the cDNA from each embryonic stage as a template for PCR amplification in separate samples. When the products are analyzed on a gel, copies of the amplified region will be observed as bands only in samples that originally contained mRNA from the gene being studied. This method can tell researchers whether an mRNA is present. However, accurate measurement of mRNA levels requires newer, more quantitative PCR machines that use a dye that fluoresces only when bound to a double-stranded PCR product. This technique, called *quantitative RT-PCR* (*qRT-PCR*), can detect the light and measure the PCR product, thus avoiding the need for electrophoresis while providing quantitative data, a distinct advantage. RT-PCR or qRT-PCR can also be carried out with mRNAs collected from different tissues at one time to discover which tissue is producing a specific mRNA.

→ **Instructors**: The Scientific Skills Exercise "Analyzing Quantitative and Spatial Gene Expression Data" can be assigned in **Mastering Biology**. Students can work with data from an experiment that investigated mRNA expression using both *in situ* hybridization and quantitative RT-PCR.

Studying the Expression of Interacting Groups of Genes

A major goal of biologists is to learn how genes act together to produce and maintain a functioning organism. Now that the genomes of a number of species have been sequenced, it is possible to study the expression of large groups of genes—an approach called the *systems approach*. Researchers use what is known about the whole genome to investigate which genes are transcribed in different tissues or at different stages of development. One aim is to identify networks of gene expression across an entire genome.

To accomplish such genome-wide expression studies, today's rapid, inexpensive DNA-sequencing methods allow researchers to discover which genes are expressed by simply sequencing the cDNA samples from different tissues or different embryonic stages. This straightforward method is called **RNA sequencing**, or **RNA-seq** (pronounced "RNA-seek"), even though it is the cDNA that is actually sequenced. In RNA-seq, the mRNA (or other RNA) samples are isolated, cut into shorter, similar-sized fragments, and converted into cDNAs **(Figure 20.12)**. These short cDNA stretches are sequenced, and a computer program reassembles them, either mapping them onto the genome of the species in question (when available) or simply putting the fragments in order from scratch based on overlapping sequences of multiple RNAs.

▼ **Figure 20.11 Research Method**

RT-PCR Analysis of the Expression of Single Genes

Application RT-PCR uses the enzyme reverse transcriptase (RT) in combination with PCR and gel electrophoresis. RT-PCR can be used to compare gene expression between samples—for instance, in different embryonic stages, in different tissues, or in the same type of cell under different conditions.

Technique In this example, samples containing mRNAs from six embryonic stages of *Drosophila* were analyzed for a specific mRNA as shown below. (In steps 1 and 2, the mRNA from only one stage is shown.)

1 **cDNA synthesis** is carried out by incubating the mRNAs from each stage with reverse transcriptase and other necessary components.

2 **PCR amplification** is then performed using primers specific to the *Drosophila* gene of interest to see whether its mRNA was present in each sample.

3 **Gel electrophoresis** will reveal amplified DNA products only in samples that contained mRNA transcribed from the specific *Drosophila* gene.

Results The mRNA for this gene is expressed from stage 2 through stage 6. The size of the amplified fragment (shown by its position on the gel) depends on the distance between the primers that were used (not on the size of the mRNA).

▼ **Figure 20.12 Use of RNA sequencing (RNA-seq) to analyze expression of many genes.** RNA-seq yields a wide range of information about expression of genes, including their level of expression.

1 mRNAs are isolated from the tissue being studied.

2 mRNAs are cut into similar-sized, small fragments.

3 mRNAs are reverse-transcribed into cDNAs of the same size.

4 cDNAs are sequenced.

5 The short sequences are mapped by computer onto the genome sequence. The resulting data, including the number of times a sequence is present, indicate which genes are expressed in a given tissue and at what level.

Several characteristics make RNA-seq a powerful technique. First, it is not based on hybridization with a labeled probe, so it doesn't depend on knowing genomic sequences. Second, it can measure levels of expression over a very wide range. Third, a careful analysis provides a wealth of information about expression of a particular gene, such as relative levels of alternatively spliced mRNAs. In most cases, however, expression of individual genes still needs to be confirmed by RT-PCR.

An older method of genome-wide expression studies, called **DNA microarray assays**, is less powerful than RNA-seq but is still used for some applications, such as fetal testing (see Figure 14.19). A DNA microarray consists of tiny amounts of a large number of single-stranded DNA fragments representing different genes fixed to a glass slide in a tightly spaced array, or grid, of dots. (The microarray is also called a *DNA chip* by analogy to a computer chip.) Ideally, these fragments represent all the genes of an organism. The mRNAs from the cells being studied are reverse-transcribed into cDNAs (see Figure 20.10), and a fluorescent label is added so the cDNAs can be used as probes on the microarray. Different fluorescent labels are used for different cell samples so that multiple samples can be tested in the same experiment. The resulting pattern of colored dots, shown in an actual-size microarray in **Figure 20.13**, reveals the dots to which each probe was bound and thus the genes that are expressed in the cell samples being tested.

▼ **Figure 20.13 Use of microarrays to analyze expression of many genes.** Researchers extracted mRNAs from two different human tissues and synthesized two sets of cDNAs, fluorescently labeled red (tissue 1) or green (tissue 2). Labeled cDNAs were hybridized with a microarray containing 5,760 human genes (about 25% of human genes), part of which is shown in the enlargement. Red indicates that the gene in that well was expressed in tissue 1, green in tissue 2, yellow in both, and black in neither. The fluorescence intensity at each spot indicates the relative expression of the gene.

▼ DNA microarray (actual size). Each dot is a well containing identical copies of DNA fragments that carry a specific gene.

cDNA

Genes that were expressed in tissue 1 bind to red cDNAs made from mRNAs in that tissue.

Genes that were expressed in tissue 2 bind to green cDNAs.

Genes that were expressed in both tissues bind both red and green cDNAs; these wells appear yellow.

Genes that were not expressed in either tissue do not bind either cDNA; these wells appear black.

Scientists can now measure the expression of thousands of genes at one time. DNA technology makes such studies possible; with automation, they are easily performed on a large scale. By uncovering gene interactions and providing clues to gene function, DNA microarray assays and RNA-seq may contribute to a better understanding of diseases and suggest new diagnostic techniques or therapies. For instance, comparing patterns of gene expression in breast cancer tumors and noncancerous breast tissue has already resulted in more informed and effective treatment protocols (see Figure 18.27). Ultimately, information from these methods should provide a grander view of how ensembles of genes interact to form an organism and maintain its vital systems.

Determining Gene Function

Once they identify a gene of interest, how do scientists determine its function? A gene's sequence can be compared with sequences in other species. If the function of a similar gene in another species is known, one might suspect that the gene product in question performs a comparable task. Data about the location and timing of gene expression may reinforce the suggested function. To obtain stronger evidence, one approach is to disable the gene and then observe the consequences in the cell or organism.

Editing Genes and Genomes

Molecular biologists have long sought techniques for altering, or editing, the genetic material of cells or organisms in a predictable way. In one such technique, called *in vitro mutagenesis*, specific mutations are introduced into a cloned gene, and the mutated gene is returned to a cell in such a way that it disables ("knocks out") the normal cellular copies of the same gene. If the introduced mutations alter or destroy the function of the gene product, the phenotype of the mutant cell may help reveal the function of the missing normal protein.

In Concept 17.5, you read about the CRISPR-Cas9 system, the powerful new technique for gene editing in living cells and organisms that is taking the field of genetic engineering by storm (see Figure 17.28). This system, worked out by Jennifer Doudna **(Figure 20.14)** and Emmanuelle Charpentier, is a highly effective way for researchers to knock out a given gene in order to study what that gene does. It has already been used in many organisms, including bacteria, fish, mice, insects, human cells, and various crop plants. Modifications of the technique allow researchers to repair a gene that has a mutation. This approach is used for gene therapy, which will be discussed later in the chapter.

In another application of the CRISPR-Cas9 system, scientists are attempting to address the global problem of insect-borne diseases by altering genes in insects so that,

▼ **Figure 20.14 Jennifer Doudna holding a model of CRISPR-Cas9.**

 Mastering Biology HHMI Video: Jennifer Doudna: CRISPR-Cas Proteins are the Swiss Army Knife of the Genome

hhmi BioInteractive

for example, they cannot transmit disease. An extra twist to this approach is to engineer the new allele so that it is much more highly favored for inheritance than the wild-type allele. This technology is called a **gene drive** because the biased inheritance of the engineered gene during reproduction rapidly "drives" the new allele through the population.

Other Methods for Studying Gene Function

Another method for silencing expression of selected genes doesn't alter the genome; instead, it exploits the phenomenon of **RNA interference (RNAi)**, described in Concept 18.3. This experimental approach uses synthetic double-stranded RNA molecules matching the sequence of a particular gene to trigger breakdown of the gene's messenger RNA or to block its translation. In organisms such as the nematode and the fruit fly, RNAi has already proved valuable for analyzing the functions of genes on a large scale. This method is quicker than using the CRISPR-Cas9 system, but it leads to only a temporary reduction of gene expression rather than a permanent gene knockout or alteration.

In humans, ethical considerations prohibit knocking out genes to determine their functions. An alternative approach is to analyze the genomes of large numbers of people with a certain phenotypic condition or disease, such as heart disease or diabetes, to try to find differences they all share compared with people without that condition. The logic is that these differences may be associated with one or more malfunctioning genes and thus in a sense are naturally occurring gene knockouts. In these large-scale analyses, called **genome-wide association studies**, researchers look for *genetic markers*, DNA sequences that vary in the population. In a gene, such sequence variation is the basis of different alleles,

as we have seen for sickle-cell disease (see Figure 17.26). And just like the coding sequences of genes, noncoding DNA at a specific locus on a chromosome may exhibit small nucleotide differences among individuals. Variations in coding or noncoding DNA sequences among a population are called polymorphisms (from the Greek for "many forms").

 Mastering Biology Interview with Charles Rotimi: Using genomics to study health-related conditions in African-Americans

Among the most useful genetic markers in tracking down genes that contribute to diseases and disorders are single base-pair variations in the genomes of the human population. For more than 99% of the nucleotides in the human genome, virtually all people have the same nucleotide in each position. However, once in every 100–300 base pairs of both coding and noncoding DNA sequences are positions where the sequence varies between individuals. A single base-pair site where variation is found in at least 1% of the population is called a **single nucleotide polymorphism** (**SNP**, pronounced "snip"); a few million SNPs occur in the human genome. To find SNPs in large numbers of people, it isn't necessary to sequence their DNA; SNPs can be detected by very sensitive microarray assays, RNA-seq, or PCR.

Once a SNP is identified that is found in all people affected by the disease being studied, researchers focus on that region and sequence it. In nearly all cases, the SNP itself does not contribute directly to the disease in question by altering the encoded protein; in fact, most SNPs are in noncoding regions. Instead, having a particular SNP associated with a disease suggests that the gene whose mutation causes the disease is located very close to that SNP on that chromosome. This closeness (*genetic linkage*; see Concept 15.3) means that crossing over between the SNP and the gene is very unlikely during gamete formation. Therefore, the SNP and gene are almost always inherited together, so the SNP acts as a genetic marker for the disease-causing allele **(Figure 20.15)**. SNPs have been

▼ **Figure 20.15 Single nucleotide polymorphisms (SNPs) as genetic markers for disease-associated alleles.** This diagram depicts the same region of the genome from two groups of individuals, one group having a particular disease or condition with a genetic basis. Unaffected people have an AT pair at a given SNP locus, while affected people have a CG pair there. Once the allele is confirmed as being associated with the disease in question, the SNP that varies in this way can be used as a marker for the disease-associated allele.

found that correlate with diabetes, heart disease, and several types of cancer, and the search is on for genes that might be involved in these and other conditions that are based on genetics.

The experimental approaches you have learned about thus far focus on working with molecules, mainly DNA and proteins. In a parallel line of inquiry, biologists have been developing powerful techniques for cloning whole multicellular organisms. One aim of this work is to obtain special types of cells, called stem cells, that can give rise to all types of tissues. Being able to manipulate stem cells would allow scientists to use the DNA-based methods previously discussed to alter stem cells for the treatment of diseases. Methods involving the cloning of organisms and production of stem cells are the subject of the next section.

CONCEPT CHECK 20.2

1. Describe the role of complementary base pairing during RT-PCR, RNA sequencing, and DNA microarray analysis.
2. **VISUAL SKILLS** Consider the microarray in Figure 20.13. If a sample from normal tissue is labeled with a green fluorescent dye and a sample from cancerous tissue is labeled red, what color spots would represent genes you would be interested in if you were studying cancer? Explain.

For suggested answers, see Appendix A.

CONCEPT 20.3

Cloned organisms and stem cells are useful for basic research and other applications

Along with advances in DNA technology, scientists have been developing and refining methods for cloning whole multicellular organisms from single cells. In this context, cloning produces one or more organisms that are genetically identical to the "parent" that donated the single cell. This is often called *organismal cloning* to differentiate it from gene cloning and, more significantly, from cell cloning—the division of an asexually reproducing cell such as a bacterium into a group of genetically identical cells. (The common theme is that the product is genetically identical to the parent.) The current interest in organismal cloning is primarily due to its ability to generate stem cells. A **stem cell** is a relatively unspecialized cell that can both reproduce itself indefinitely and, under appropriate conditions, differentiate into specialized cells of one or more types. Stem cells have great potential for regenerating damaged tissues.

The cloning of plants and animals was first attempted over 50 years ago in experiments designed to answer basic questions about the genetic potential of single cells. For example, researchers wondered if all the cells of an organism have the same genes or whether cells lose genes during the process of differentiation (see Concept 18.4). One way to answer this question is to see whether a differentiated cell can generate a whole organism—in other words, whether cloning an organism is possible. Let's discuss these early experiments before we consider more recent progress in organismal cloning and procedures for producing stem cells.

Cloning Plants: Single-Cell Cultures

The successful cloning of whole plants from single differentiated cells was accomplished at Cornell University during the 1950s by F. C. Steward and his students, who worked with carrot plants. They found that differentiated cells taken from the root (the carrot) and incubated in culture medium could grow into normal adult plants, each genetically identical to the parent plant. These results showed that differentiation does not necessarily involve irreversible changes in the DNA. In plants, mature cells can "dedifferentiate" and then give rise to all the specialized cell types of the organism; any cell with this potential is said to be **totipotent**.

Plant cloning is used extensively in agriculture. For plants such as orchids, cloning is the only commercially practical means of producing new plants. In other cases, cloning has been used to reproduce a plant with valuable characteristics, such as resistance to plant pathogens. In fact, you yourself may be a plant cloner: If you have ever grown a new plant from a cutting, you have practiced cloning!

Cloning Animals: Nuclear Transplantation

Differentiated cells from animals generally do not divide in culture, much less develop into the multiple cell types of a new organism. Therefore, early researchers had to use a different approach to answer the question of whether differentiated animal cells are totipotent. Their approach was to remove the nucleus of an egg (creating an *enucleated* egg) and replace it with the nucleus of a differentiated cell, a procedure called *nuclear transplantation*, now more commonly called *somatic cell nuclear transfer*. If the nucleus from the differentiated donor cell retains its full genetic potential, then it should be able to direct development of the recipient cell into all the tissues and organs of an organism. Such experiments were conducted on one species of frog (*Rana pipiens*) by Robert Briggs and Thomas King in the 1950s and on another frog species (*Xenopus laevis*) by John Gurdon in the 1970s **(Figure 20.16)**. These researchers transplanted a nucleus from an embryonic or tadpole cell into an enucleated egg of the same species. In Gurdon's experiments, the transplanted nucleus was often able to support normal development of the egg into a tadpole. However, he found that the potential of a transplanted nucleus to direct normal development was inversely related to the age of the donor: The older the donor nucleus, the lower the percentage of normal tadpoles.

From these results, Gurdon concluded that something in the nucleus *does* change as animal cells differentiate. In frogs and most other animals, nuclear potential tends to be restricted more and more as embryonic development and cell differentiation progress. These were foundational experiments that ultimately led to stem cell technology, and Gurdon received the 2012 Nobel Prize in Medicine for this work.

Reproductive Cloning of Mammals

In addition to cloning frogs, researchers were able to clone mammals using early embryonic cells as a source of donor nuclei. Until about 25 years ago, though, it was not known whether a nucleus from a fully differentiated cell could be reprogrammed successfully to act as a donor nucleus. In 1997, researchers in Scotland announced the birth of Dolly, a lamb cloned from an adult sheep by nuclear transfer from a differentiated mammary gland cell **(Figure 20.17)**. Using a technique related to that in Figure 20.16, the researchers implanted early embryos into surrogate mothers. Out of several hundred embryos, one successfully completed normal development, and Dolly was born, a genetic clone of the nucleus donor. By the age of 6, Dolly had developed premature arthritis, and complications from a lung infection led to her euthanization. This led to speculation that this sheep's cells were in some way not quite as healthy as those of a normal sheep, possibly reflecting incomplete reprogramming of the original transplanted nucleus. Reprogramming involves epigenetic changes that lead to changes in chromatin structure (see Concept 18.2), to be discussed shortly.

▼ **Figure 20.16** Inquiry

Can the nucleus from a differentiated animal cell direct development of an organism?

Experiment John Gurdon and colleagues at Oxford University, in England, destroyed the nuclei of frog (*Xenopus laevis*) eggs by exposing the eggs to ultraviolet light. They then transplanted nuclei from cells of frog embryos and tadpoles into the enucleated eggs.

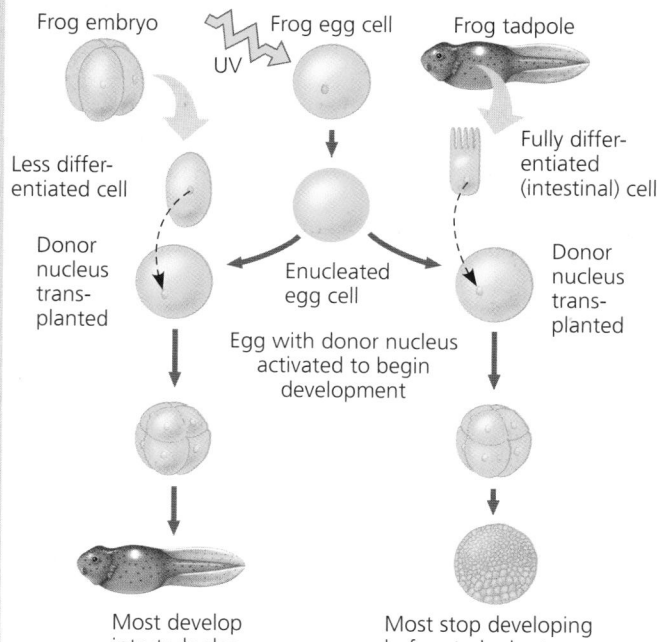

Results When the transplanted nuclei came from an early embryo, the cells of which are relatively undifferentiated, most of the recipient eggs developed into tadpoles. But when the nuclei came from the fully differentiated intestinal cells of a tadpole, fewer than 2% of the eggs developed into normal tadpoles, and most of the embryos stopped developing at a much earlier stage.

Conclusion The nucleus from a differentiated frog cell can direct development of a tadpole. However, its ability to do so decreases as the donor cell becomes more differentiated, presumably because of changes in the nucleus.

Data from J. B. Gurdon et al., The developmental capacity of nuclei transplanted from keratinized cells of adult frogs, *Journal of Embryology and Experimental Morphology* 34:93–112 (1975).

WHAT IF? *If each cell in a four-cell embryo was already so specialized that it was not totipotent, what results would you predict for the experiment on the left side of the figure?*

 Mastering Biology HHMI Video: Somatic Cell Nuclear Transfer

▶ **Figure 20.17 Reproductive cloning of a mammal by nuclear transfer.** Dolly, shown here as a lamb, has a very different appearance from her surrogate mother, standing beside her.

Since that time, researchers have cloned many other mammals, including mice, cats, cows, horses, pigs, and dogs. In 2018, Chinese biologists reported the first cloning of a primate, the long-tailed macaque. In most cases, the research goal has been the production of new individuals, known as *reproductive cloning*. We have learned from such experiments that cloned animals of the same species are not always identical. For example, in 2016, scientists examined four 7- to 9-year-old clones genetically identical to Dolly and concluded that, unlike Dolly, they were healthy and aging normally. Another example of nonidentity in clones is the

▼ **Figure 20.18 CC ("Carbon Copy"), the first cloned cat (right), and her single parent.** Rainbow (left) donated the nucleus in a cloning procedure that resulted in CC. However, the two cats are not identical: Rainbow is a classic calico cat with orange patches on her fur and has a "reserved personality," while CC has a gray and white coat and is more playful.

first cloned cat, named CC for Carbon Copy **(Figure 20.18)**. The color and pattern of her calico coat differed from that of her single female parent because of random X chromosome inactivation, which is a normal occurrence during embryonic development (see Figure 15.8). And identical human twins, which are naturally occurring "clones," are always slightly different. Clearly, environmental influences and random phenomena play a significant role during development.

Epigenetic Differences in Cloned Animals

In most nuclear transplantation studies thus far, only a small percentage of cloned embryos develop normally to birth. And like Dolly, many cloned animals exhibit defects. Cloned mice, for instance, are prone to obesity, pneumonia, liver failure, and premature death. Scientists assert that even cloned animals that appear normal are likely to have subtle defects.

Researchers have uncovered some reasons for the low efficiency of cloning and the high incidence of abnormalities. In the nuclei of fully differentiated cells, a small subset of genes is turned on and expression of the rest of the genes is repressed. This regulation often is the result of epigenetic changes in chromatin, such as acetylation of histones or methylation of DNA (see Figure 18.7). During the nuclear transfer procedure, many of these changes must be reversed in the later-stage nucleus from a donor animal for genes to be expressed or repressed appropriately in earlier stages of development. Researchers have found that the DNA in cells from cloned embryos, like that of differentiated cells, often has abnormally high numbers of methyl groups. This finding suggests that the reprogramming of donor nuclei requires

more accurate and complete chromatin restructuring than occurs during cloning procedures. Because DNA methylation helps regulate gene expression, misplaced or extra methyl groups in the DNA of donor nuclei may interfere with the pattern of gene expression necessary for normal embryonic development. In fact, the success of a cloning attempt may depend in large part on whether or not the chromatin in the donor nucleus can be artificially modified to resemble that of a newly fertilized egg.

Stem Cells of Animals

Progress in cloning mammalian embryos, including primates, has heightened speculation about the cloning of humans, which has not yet been achieved past very early embryonic stages. The main reason researchers have been trying to clone human embryos is not for reproduction, but for the production of stem cells to treat human diseases. Recall that a stem cell is a relatively unspecialized cell that can both reproduce itself indefinitely and, under appropriate conditions, differentiate into specialized cells of one or more types **(Figure 20.19)**.

Mastering Biology HHMI Animation: Somatic Cell Nuclear Transfer

Embryonic and Adult Stem Cells

Many early animal embryos contain stem cells capable of giving rise to differentiated cells of any type. Stem cells can be isolated from early embryos at a stage called the blastula stage or its human equivalent, the blastocyst stage. In culture, these

▼ **Figure 20.19 How stem cells maintain their own population and generate differentiated cells.**

❶ A stem cell can divide into another stem cell and a progenitor cell (or into two stem cells or two progenitor cells).

Stem cell

Cell division

Stem cell   and   Progenitor cell

❷ A progenitor cell can differentiate into one of several cell types, depending on external factors. This example is from a stem cell derived from bone marrow.

Fat cells   or   Bone cells   or White blood cells

Mastering Biology HHMI Video: Cultured Human Embryonic Stem Cells *embryonic stem (ES) cells* reproduce indefinitely, and depending on culture conditions, they can be made to differentiate into a wide variety of specialized cells **(Figure 20.20)**, including even eggs and sperm.

The adult body also has stem cells, which serve to replace nonreproducing specialized cells as needed. In contrast to ES cells, *adult stem cells* are not able to give rise to all cell types in the organism, though they can generate several defined types. For example, one of the several types of stem cells in bone marrow can generate all the different kinds of blood cells (see Figure 20.20), and another type of bone marrow stem cell can differentiate into bone, cartilage, fat, muscle, and the linings of blood vessels. To the surprise of many, the adult brain has been found to contain stem cells that continue to produce certain kinds of nerve cells there. Researchers have also reported finding stem cells in skin, hair, eyes, and dental pulp. Although adult animals have only tiny

▼ **Figure 20.20 Working with stem cells.** Animal stem cells, which can be isolated from early embryos or adult tissues and grown in culture, are self-perpetuating, relatively undifferentiated cells. Embryonic stem cells are easier to grow than adult stem cells and can theoretically give rise to *all* types of cells in an organism. The range of cell types that can arise from adult stem cells is not yet fully understood.

Mastering Biology HHMI Animation: **Creating Embryonic Stem Cell Lines** hhmi BioInteractive

numbers of stem cells, scientists are learning to identify and isolate these cells from various tissues and, in some cases, to grow them in culture. With the right culture conditions (for instance, the addition of specific growth factors), cultured stem cells from adult animals have been made to differentiate into various defined types of specialized cells, although none are as versatile as ES cells.

Research with embryonic or adult stem cells is a source of valuable data about differentiation and has enormous potential for medical applications. The ultimate aim is to supply cells for the repair of damaged or diseased organs: for example, insulin-producing pancreatic cells for people with type 1 diabetes or certain kinds of brain cells for people with Parkinson's disease or Huntington's disease. Adult stem cells from bone marrow have long been used in bone marrow transplants as a source of immune system cells in patients whose own immune systems are nonfunctional because of genetic disorders or radiation treatments for cancer.

The developmental potential of adult stem cells is limited to certain tissues. ES cells hold more promise than adult stem cells for most medical applications because ES cells are **pluripotent**, capable of differentiating into many different cell types. One source of ES cells, first reported in 2013, is cloned human blastocysts produced by transferring a nucleus from a differentiated cell into an enucleated egg (similar to cloning Dolly the sheep). Prior to that report, cells were obtained only from embryos donated by patients undergoing infertility treatments or from long-term cell cultures originally established with cells isolated from donated embryos, an issue that prompts ethical and political discussions. Although the techniques for cloning early human embryos are still being optimized and the ethics discussed by researchers, they represent a potential new source for ES cells that may be less controversial. Furthermore, with a donor nucleus from a person with a particular disease, researchers should be able to produce ES cells that match the patient and are thus not rejected by his or her immune system when used for treatment. When the main aim of cloning is to produce ES cells to treat disease, the process is called *therapeutic cloning*. Although most people believe that reproductive cloning of humans is unethical, opinions vary about the morality of therapeutic cloning.

Induced Pluripotent Stem (iPS) Cells

In another approach to make stem cells for research and therapy, researchers succeeded in 2007 in learning how to turn back the clock in fully differentiated cells, reprogramming them to act like ES cells. Differentiated cells can be transformed into a type of ES cell by using a modified retrovirus to introduce extra, cloned copies of four "stem cell" master regulatory genes. The "deprogrammed" cells are known as *induced pluripotent stem* (iPS) cells because, in using this fairly simple laboratory technique to return them to their undifferentiated state, pluripotency has been restored. The experiments that first transformed human differentiated cells into iPS cells are described in **Figure 20.21**. Shinya Yamanaka

▼ Figure 20.21 Inquiry

Can a fully differentiated human cell be "deprogrammed" to become a stem cell?

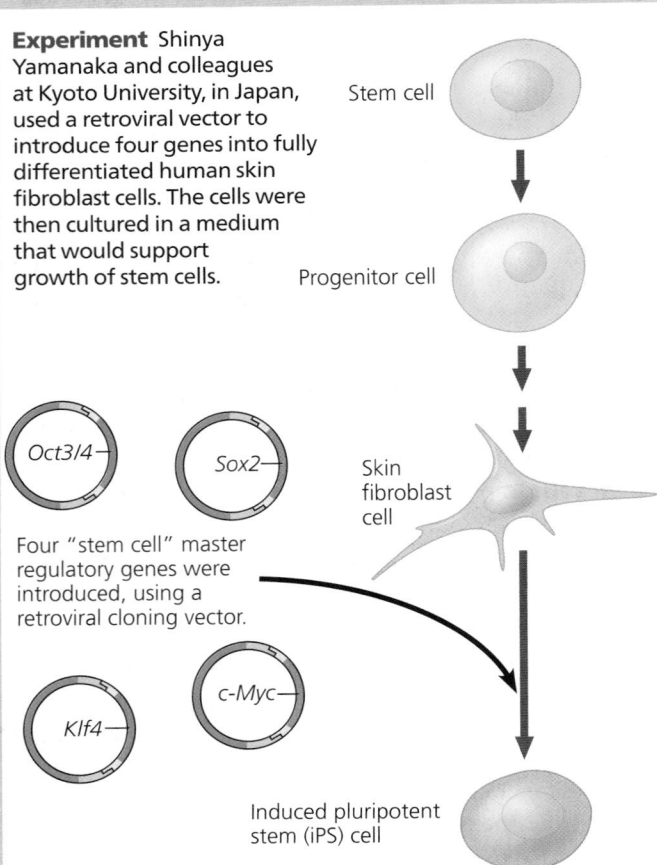

Experiment Shinya Yamanaka and colleagues at Kyoto University, in Japan, used a retroviral vector to introduce four genes into fully differentiated human skin fibroblast cells. The cells were then cultured in a medium that would support growth of stem cells.

Stem cell

Progenitor cell

Oct3/4   Sox2

Skin fibroblast cell

Four "stem cell" master regulatory genes were introduced, using a retroviral cloning vector.

Klf4   c-Myc

Induced pluripotent stem (iPS) cell

Results Two weeks later, the cells resembled embryonic stem cells in appearance and were actively dividing. Their gene expression patterns, gene methylation patterns, and other characteristics were also consistent with those of embryonic stem cells. The iPS cells were able to differentiate into heart muscle cells, as well as other cell types.

Conclusion The four genes induced differentiated skin cells to become pluripotent stem cells, with characteristics of embryonic stem cells.

Data from K. Takahashi et al., Induction of pluripotent stem cells from adult human fibroblasts by defined factors, *Cell* 131:861–872 (2007).

WHAT IF? *Patients with diseases such as heart disease or Alzheimer's could have their own skin cells reprogrammed to become iPS cells. Once affordable procedures have been developed for this and for converting iPS cells into heart or nervous system cells, the patients' own iPS cells might be used to treat their disease. When organs are transplanted from a donor to a recipient, the recipient's immune system may reject the transplant, a dangerous condition. Would using iPS cells be expected to carry the same risk? Why or why not? Given that these cells are actively dividing, undifferentiated cells, what risks might this procedure carry?*

received the 2012 Nobel Prize in Medicine for this work, shared with John Gurdon, whose work you read about in Figure 20.16.

By many criteria, iPS cells can perform most of the functions of ES cells, but there are some differences in gene expression and other cellular functions, such as cell division. At least until these differences are fully understood, the study of ES cells will continue to make important contributions to the development of stem cell therapies. (In fact, it is likely that ES cells will always be a focus of basic research as well.) In the meantime, work is proceeding using the iPS cells that have been experimentally produced.

There are two major potential uses for human iPS cells. First, cells from patients with diseases have been reprogrammed to become iPS cells, which act as model cells for studying the disease and potential treatments. Human iPS cell lines have already been developed from individuals with type 1 diabetes, Parkinson's disease, Huntington's disease, Down syndrome, and many other diseases. Second, in the field of regenerative medicine, a patient's own cells could be reprogrammed into iPS cells and then used to replace nonfunctional tissues, such as cells of the retina of the eye that have been damaged by a condition called age-related macular degeneration (AMD). In fact, in 2014, Japanese researchers made iPS cells out of an AMD patient's skin cells, caused them to differentiate into retinal cells, and implanted them into her retina. Although her vision was not markedly improved, further deterioration was halted. Unfortunately, the cost of reprogramming cells into iPS cells is so high that it can't be used as a standard treatment. Over time, the procedure may become less expensive, and some researchers have suggested creating a bank for storage of donor iPS cells that could be matched to patients' tissues. Development of techniques that direct iPS cells to become specific cell types for regenerative medicine is an area of intense research. If the challenges are met, iPS cells created in this way could eventually provide tailor-made "replacement" cells for patients without using any human eggs or embryos, thus circumventing most ethical objections.

CONCEPT CHECK 20.3

1. Based on current knowledge, how would you explain the difference in the percentage of tadpoles that developed from the two kinds of donor nuclei in Figure 20.16?

2. A few companies in China and South Korea provide the service of cloning dogs, using cells from their clients' pets to provide nuclei in procedures like that in Figure 20.17. Should their clients expect the clone to look identical to their original pet? Why or why not? What ethical questions does this bring up?

3. **MAKE CONNECTIONS** Based on what you know about muscle differentiation (see Figure 18.18) and genetic engineering, propose the first experiment you might try if you wanted to direct an embryonic stem cell or iPS cell to develop into a muscle cell.

For suggested answers, see Appendix A.

The practical applications of DNA-based biotechnology affect our lives in many ways

In the last section, we'll survey the practical applications of DNA-based **biotechnology**, the manipulation of organisms or their components to make useful products. Today, major applications of DNA technology and genetic engineering include medicine, forensic evidence and genetic profiles, environmental cleanup, and agriculture.

Medical Applications

One important use of DNA technology is the identification of human genes whose mutation plays a role in genetic diseases. These discoveries may lead to ways of diagnosing, treating, and even preventing such conditions. DNA technology has also identified genes that play a role in a number of "nongenetic" diseases, from arthritis to AIDS, by influencing susceptibility to these diseases. Furthermore, a wide variety of diseases involve changes in gene expression within the affected cells and often within the patient's immune system. By using RNA-seq and DNA microarray assays (see Figures 20.12 and 20.13) or other techniques to compare gene expression in healthy and diseased tissues, researchers are finding genes that are turned on or off in particular diseases. These genes and their products are potential targets for prevention or therapy.

Diagnosis and Treatment of Diseases

A new chapter in the diagnosis of infectious diseases has been opened by DNA technology, in particular the use of PCR and labeled nucleic acid probes to track down pathogens. For example, because the sequence of the RNA genome of HIV is known, RT-PCR can be used to amplify and therefore detect and quantify even a small amount of HIV RNA in blood or tissue samples (see Figure 20.11). RT-PCR is often the best way to detect an otherwise elusive infectious agent.

Medical scientists can now diagnose hundreds of human genetic disorders by using PCR with primers that target the genes associated with these disorders. The amplified DNA product is then sequenced to reveal the presence or absence of the disease-causing mutation. Among the genes for human diseases targeted in this way are those for sickle-cell disease, hemophilia, cystic fibrosis, Huntington's disease, and Duchenne muscular dystrophy. Individuals with such diseases can often be identified before the onset of symptoms, even before birth (see Figure 14.19). PCR can also be used to identify symptomless carriers of potentially harmful recessive alleles as part of genetic counseling (see Concept 14.4).

Personal Genome Analysis

As you learned earlier, genome-wide association studies have pinpointed SNPs (single nucleotide polymorphisms) that are linked to disease-associated alleles (see Figure 20.15). Individuals can be tested by PCR and sequencing for a SNP that is correlated with the abnormal allele. The presence of particular SNPs is correlated with increased risk for particular adverse health conditions such as heart disease, Alzheimer's disease, and some types of cancer.

Direct-to-consumer genome analysis companies offer kits allowing individuals to send in a swab containing cheek cells that the company will analyze genetically. Genetic testing for risk factors like heart disease, Alzheimer's disease, and some types of cancer is carried out by looking for linked SNPs that have been previously identified (see Figure 20.15). It may be helpful for individuals to learn about their health risks, with the understanding that such genetic tests merely reflect correlations and do not make predictions.

In addition to health-related genetic information, these companies compare a person's DNA segments with those from reference populations around the world, established from thousands of individuals of known ancestry. Based on how closely the segments match up, the report can tell an individual about their likely ancestral breakdown. Females can also learn about their maternal lines of descent, based on comparisons of mitochondrial DNA, which is contributed to the egg only by the mother. For males, an analysis of the sequence of the Y chromosome can trace their paternal lines of descent. As the size of the database used by these companies increases, the results become more refined and more accurate.

Personalized Medicine

The techniques described in this chapter have also prompted improvements in disease treatments—for example, by knowing which specific cancer-related genes have been mutated in a particular patient's tumor. By analyzing the expression of many genes in large numbers of breast cancer patients, researchers can refine their understanding of the different subtypes of breast cancer (see Figure 18.27). Knowing the expression levels of particular genes in a given individual can help physicians determine the likelihood that the cancer will recur, thus helping them design an appropriate treatment. Given that some low-risk patients have a 96% survival rate over a ten-year period with no treatment, gene expression analysis allows doctors and patients access to valuable information when they are considering treatment options.

Many researchers envision a future of **personalized medicine**, a type of medical care in which each person's specific genetic profile can provide information about diseases or conditions for which the person is especially at risk and help make health-care decisions. As we will see later in the chapter, a *genetic profile* is currently taken to mean a set of genetic markers such as SNPs. Ultimately, however, it will likely mean the complete DNA sequence of an individual—after sequencing becomes inexpensive enough. In fact, a 2016 study suggested that for difficult-to-diagnose patients it might be faster and more economical to sequence just the expressed regions (the exons, together called the *exome*) or even all of the genome right away, rather than running through a standard series of diagnostic tests. The ability to sequence a person's genome quickly and inexpensively is advancing rapidly, perhaps faster than development of appropriate treatments for the conditions. Still, the identification of genes involved in these conditions provides targets for therapeutic interventions.

An individual's genomic information can be used to predict the benefits and risks of particular medications, an approach called *pharmacogenetics*. There are over 300 medications for which the Food and Drug Administration recommends genetic testing for patients prior to their use.

Human Gene Therapy and Gene Editing

Gene therapy—the introduction of genes into an afflicted individual for therapeutic purposes—holds great potential for treating the relatively small number of disorders traceable to a single defective gene. The aim of this approach is to insert a normal allele of the defective gene into the somatic cells of the tissue affected by the disorder.

> **Mastering Biology** BBC Video: Cystic Fibrosis and the Promise of Gene Therapy

For gene therapy of somatic cells to be permanent, the cells that receive the normal allele must be cells that multiply throughout the patient's life. Bone marrow cells, which include the stem cells that give rise to all the cells of the blood and immune system, are prime candidates. **Figure 20.22** outlines one procedure for gene therapy in an individual whose bone marrow cells do not produce a vital enzyme because of a single defective gene. One type of severe combined immunodeficiency (SCID) is caused by this kind of defect. If the treatment is successful, the patient's bone marrow cells will begin producing the missing protein, and the patient may be cured.

The procedure shown in Figure 20.22 was used in gene therapy trials for SCID in France in 2000, resulting in the first indisputable success of gene therapy. However, three patients subsequently developed leukemia, a type of blood cell cancer, and one of them died. Researchers have concluded it is likely that the insertion of the retroviral vector occurred near a gene that triggers the proliferation of blood cells. Using a viral vector that does not come from a retrovirus, clinical researchers have treated several other genetic diseases somewhat successfully with gene therapy, including

▼ **Figure 20.22 Gene therapy using a retroviral vector.**
A retrovirus that has been rendered harmless is used as a vector in this procedure, which exploits the ability of a retrovirus to insert a DNA transcript of its RNA genome into the chromosomal DNA of its host cell (see Figure 19.9). If the foreign gene carried by the retroviral vector is expressed, the cell and its descendants will possess the gene product. Cells that reproduce throughout life, such as bone marrow cells, are ideal candidates for gene therapy.

Cloned gene (normal allele, absent from patient's cells)

① Insert RNA version of normal allele into retrovirus or other viral vector.

Viral RNA

② Let virus infect bone marrow cells that have been removed from the patient and cultured.

Viral capsid

③ Viral DNA carrying the normal allele inserts into chromosome.

Bone marrow cell from patient

Bone marrow

④ Inject engineered cells into patient.

> **Mastering Biology** HHMI Video: Gene Therapy
> HHMI Video: The Search for a Mutated Gene
> HHMI Video: Genes as Medicine

hhmi BioInteractive

a type of progressive blindness (see Concept 50.3), a degenerative disease of the nervous system, and a blood disorder involving the β-globin gene.

Gene therapy raises many technical issues. For example, how can the activity of the transferred gene be controlled so that cells make appropriate amounts of the gene product at the right time and in the right place? How can we be sure that the insertion of the therapeutic gene does not harm some other necessary cell function? As more is learned about DNA control elements and gene interactions, researchers may be able to answer such questions.

A more direct approach that avoids the complications of using a viral vector in gene therapy is made possible by gene editing, especially using the CRISPR-Cas9 system (see Figure 17.28). In this approach, the existing defective gene is edited to correct the mutation.

 Mastering Biology HHMI Video: Jennifer Doudna: CRISPR Provides a Tool for Changing Genes

In 2018, researchers reported promising results using the CRISPR-Cas9 system in an attempt to correct the genetic defect that causes sickle-cell disease. They edited cells from patients with sickle-cell disease and injected them into bone marrow in mice. After 19 weeks, the gene remained corrected in 20–40% of the injected cells. Researchers, physicians, and patients alike are excited about the potential of CRISPR technology to treat or even cure human diseases that have a genetic basis, such as sickle-cell, Alzheimer's, and Parkinson's diseases, as well as some types of cancer. However, there are still concerns about using CRISPR-treated cells in humans because of possible effects on genes that are not being targeted; this is an active area of research.

In addition to technical challenges, gene therapy and gene editing provoke ethical questions. Some critics believe that tampering with human genes in any way is unethical. Other observers see no fundamental difference between transplanting genes into somatic cells and transplanting organs from one person to another.

An even more pressing question is the issue of engineering human germ-line cells to try to correct a defect in future generations. Such genetic engineering is now routinely done in laboratory mice, and some methods for genetic engineering of human embryos have been developed.

The development of the CRISPR-Cas9 system has engendered much debate about the ethics of gene editing. Jennifer Doudna (Figure 20.14), a co-discoverer of CRISPR-Cas9, and other biologists have agreed to exercise extreme caution as the field moves forward. Together, they called for the research community to "strongly discourage" any experimental work on human eggs or embryos.

In spite of this general consensus among biologists, a Chinese researcher reported in 2018 that he had used the CRISPR-Cas9 system to edit genes in embryos that completed fetal development and were born as twins and a third individual. He claimed to have edited the *CCR5* gene, which codes for a co-receptor for HIV (see Figure 7.8), so that HIV would be unable to bind and infect the cells. The twins' father was HIV-positive, which was the rationale the scientist used for carrying out this genetic engineering. However, this act was roundly condemned as highly unethical by the biological community; the researcher lost his job and may face a criminal investigation. In 2019, an advisory panel to the World Health Organization proposed establishing a registry to monitor any gene-editing research in humans.

There are many potential technical problems with the CRISPR-Cas9 technique, like the off-target effects mentioned previously. Beyond those technical concerns, however, are underlying ethical considerations: Under what circumstances, if any, should genomes of human germ lines be altered? Would alterations lead to the practice of eugenics, a deliberate effort to control the genetic makeup of human populations? It is imperative to consider these questions.

Pharmaceutical Products

The pharmaceutical industry derives significant benefit from advances in DNA technology and genetic research, applying them to the development of useful drugs to treat diseases. Pharmaceutical products are synthesized using methods of either organic chemistry or biotechnology, depending on the nature of the product.

Synthesis of Small Molecules for Use as Drugs

Determining the sequence and structure of proteins crucial for tumor cell survival has led to the identification of small molecules that combat certain cancers by blocking the function of these proteins. One drug, imatinib (trade name Gleevec), is a small molecule that inhibits one tyrosine kinase (see Figure 11.8). The overexpression of this kinase, resulting from a chromosomal translocation, is instrumental in causing chronic myelogenous leukemia (CML; see Figure 15.16). Patients in the early stages of CML who are treated with imatinib have exhibited nearly complete, sustained remission from the cancer. Drugs that work in a similar way have also been used with success to treat a few types of lung and breast cancers. This approach is feasible only for cancers whose molecular basis is fairly well understood.

 Mastering Biology HHMI Video: Gleevec

In many cases of such drug-treated tumors, though, cells later arise that are resistant to the new drug. In one study, the whole genome of the tumor cells was sequenced both before and after the appearance of drug resistance. Comparison of the sequences showed genetic changes that allowed the tumor cells to "get around" the drug-inhibited protein. Here, we can see that cancer cells demonstrate the principles of evolution: Certain tumor cells have a random mutation that allows them to survive in the presence of a particular drug, and as a consequence of natural selection in the presence of the drug, these are the cells that survive and reproduce.

Protein Production in Cell Cultures
Pharmaceutical products that are proteins are commonly synthesized on a large scale using cell cultures. You learned earlier in the chapter about DNA cloning and gene expression systems for producing large quantities of a chosen protein that is present naturally in only minute amounts. The host cells used in such expression systems can even be engineered to secrete a protein as it is made, thereby simplifying the task of purifying it by traditional biochemical methods.

Among the first pharmaceutical products manufactured in this way were human insulin and human growth hormone (HGH). Some 2 million people with diabetes in the United States depend on insulin treatment to control their disease. Human growth hormone has been a boon to children born with a form of dwarfism caused by inadequate amounts of HGH, as well as helping patients with AIDS gain weight. Another important pharmaceutical product produced by genetic engineering is tissue plasminogen activator (TPA). If administered shortly after a heart attack, TPA helps dissolve blood clots and reduces the risk of subsequent heart attacks.

Protein Production by "Pharm" Animals   In some cases, instead of using cell systems to produce large quantities of protein products, pharmaceutical scientists can use whole organisms. To do this, they use a **transgene**, a gene that has been transferred into one organism from another. They first remove eggs from a female of the recipient species and fertilize them *in vitro*. Meanwhile, they clone the desired gene from a donor organism. They then inject the cloned DNA directly into the nuclei of the fertilized eggs. Some of the cells integrate the foreign DNA, the transgene, into their genome and are able to express the foreign gene. The engineered embryos that arise from these zygotes are then surgically implanted in a surrogate female. If an embryo develops successfully, the result is a **transgenic** organism that expresses its new, "foreign" gene.

Assuming that the introduced gene encodes a protein desired in large quantities, transgenic animals can act as pharmaceutical "factories." For example, a transgene for a human blood protein such as antithrombin, which prevents blood clots, can be inserted into the genome of a goat in such a way that the transgene's product is secreted in the animal's milk **(Figure 20.23)**. The protein is then purified from the milk (which is easier than purification from a cell culture). Such

▼ **Figure 20.23 Goats as "pharm" animals.** This transgenic goat carries a gene for a human blood protein, antithrombin, which she secretes in her milk. Patients with a rare hereditary disorder in which this protein is lacking suffer from formation of blood clots in their blood vessels. Easily purified from the goat's milk, the protein is used to prevent blood clots in these patients during surgery or childbirth.

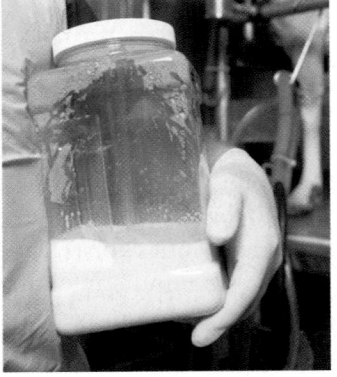

proteins must be tested to ensure that they (or contaminants from the farm animals) will not cause allergic reactions or other adverse effects in patients who receive them.

Forensic Evidence and Genetic Profiles

In violent crimes, body fluids or small pieces of tissue may be left at the scene or on the clothes or other possessions of the victim or assailant. If enough blood, semen, or tissue is available, forensic laboratories can determine the blood type or tissue type by using antibodies to detect specific cell-surface proteins. However, such tests require fairly fresh samples in relatively large amounts. Also, because many people have the same blood or tissue type, this approach can only exclude a suspect; it cannot provide strong evidence of guilt.

DNA testing, on the other hand, can identify the guilty individual with a high degree of certainty because the DNA sequence of every person is unique (except for identical twins). Genetic markers that vary in the population can be analyzed for a given person to determine that individual's unique set of genetic markers, or **genetic profile**. (This term is preferred over "DNA fingerprint" by forensic scientists, who want to emphasize the heritable aspect of these markers rather than the fact that they produce a pattern on a gel that, like a fingerprint, is visually recognizable.) The FBI started applying DNA technology to forensics in 1988, using a method involving gel electrophoresis and nucleic acid hybridization to detect similarities and differences in DNA samples. This method required much smaller samples of blood or tissue than earlier methods—only about 1,000 cells.

> **Mastering Biology Animation: Genetic Profiles**

Today, forensic scientists use an even more sensitive method that takes advantage of variations in length of genetic markers called **short tandem repeats (STRs)**. These are tandemly repeated units of two- to five-nucleotide sequences in specific regions of the genome. The number of repeats present in these regions is highly variable from person to person (polymorphic); even for a single individual, the two alleles of an STR may differ from each other. For example, one individual may have the sequence ACAT repeated 30 times at one genome locus and 15 times at the same locus on the other homolog, whereas another individual may have 18 repeats at this locus on each homolog. (These two genotypes can be expressed by the two repeat numbers: 30,15 and 18,18.) PCR is used to amplify particular STRs, using sets of primers that are labeled with colored fluorescent tags; the length of the region, and thus the number of repeats, can then be determined by electrophoresis. The PCR step allows use of this method even when the DNA is in poor condition or available only in minute quantities: A tissue sample containing as few as 20 cells can be sufficient.

In a murder case, for example, this method can be used to compare DNA samples from the suspect, the victim, and a small amount of blood found at the crime scene. The forensic scientist tests only a few selected portions of the DNA—usually 13 STR markers. However, even this small set of markers can provide a forensically useful genetic profile because the probability that two people (who are not identical twins) would have exactly the same set of STR markers is vanishingly small. The Innocence Project, a nonprofit organization dedicated to overturning wrongful convictions, uses STR analysis of archived samples from crime scenes to revisit old cases. As of 2019, according to that group, 362 innocent people have been released from prison as a result of DNA-based forensic and legal work **(Figure 20.24)**.

Genetic profiles can also be useful for other purposes. A comparison of the DNA of a mother, her child, and the purported father can conclusively settle a question of paternity. Sometimes paternity is of historical interest: Genetic profiles provided strong evidence that Thomas Jefferson or one of

his close male relatives fathered at least one of the children of Sally Hemings, one of his slaves. Genetic profiles can also identify victims of mass casualties. The largest such effort occurred after the attack on the World Trade Center in 2001; more than 10,000 samples of victims' remains were compared with DNA samples from personal items, such as toothbrushes, provided by families. Ultimately, forensic scientists succeeded in identifying almost 3,000 victims using these methods.

Just how reliable is a genetic profile? The greater the number of markers examined in a DNA sample, the more likely it is that the profile is unique to one individual. In forensic cases using STR analysis with 13 markers, the probability of two people having identical DNA profiles is somewhere between one in 10 billion and one in several trillion. (For comparison, the world's population is between 7 and 8 billion.) The exact probability depends on the frequency of those markers in the general population. Information on how common various markers are in different ethnic groups is critical because these marker frequencies may vary considerably among ethnic groups and between a particular ethnic group and the population as a whole. With the increasing availability of frequency data, forensic scientists can make extremely accurate statistical calculations. Thus, despite problems that can arise from insufficient data, human error, or flawed evidence, genetic profiles are now accepted as compelling evidence by legal experts and scientists alike.

Environmental Cleanup

Increasingly, the diverse abilities of certain microorganisms to transform chemicals are being exploited for environmental cleanup. If the growth needs of such microorganisms make them unsuitable for direct use, scientists can now transfer the genes for their valuable metabolic capabilities into other microorganisms, which can then be used to treat environmental problems. For example, many bacteria can extract heavy metals, such as copper, lead, and nickel, from their environments and incorporate the metals into compounds such as copper sulfate or lead sulfate, which are readily recoverable. Genetically engineered microorganisms may become important in both mining (especially as ore reserves are depleted) and cleaning up highly toxic mining wastes. Biotechnologists are also trying to engineer microorganisms that can degrade chlorinated hydrocarbons and other harmful compounds. These microorganisms could be used in wastewater treatment plants or by manufacturers before the compounds are ever released into the environment.

Agricultural Applications

Scientists are working to learn more about the genomes of agriculturally important plants and animals. For a number of years, they have been using DNA technology in an effort to improve agricultural productivity. The selective breeding of

▼ **Figure 20.24 STR analysis used to release an innocent man from prison.**

(a) In 1984, Earl Washington was convicted and sentenced to death for the 1982 rape and murder of Rebecca Williams. His sentence was commuted to life in prison in 1993 due to new doubts about the evidence. In 2000, STR analysis by forensic scientists associated with the Innocence Project showed conclusively that he was innocent. This photo shows Washington just before his release in 2001, after 17 years in prison.

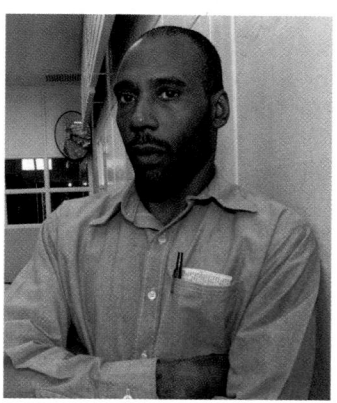

Source of sample	STR marker 1	STR marker 2	STR marker 3
Semen on victim	17,19	13,16	12,12
Earl Washington	16,18	14,15	11,12
Kenneth Tinsley	17,19	13,16	12,12

(b) In STR analysis, selected STR markers in a DNA sample are amplified by PCR, and the PCR products are separated by electrophoresis. The procedure reveals how many repeats are present for each STR locus in the sample. An individual has two alleles per STR locus, each with a certain number of repeats. This table shows the number of repeats for three STR markers in three samples: from semen found on the victim, from Washington, and from another man (Kenneth Tinsley), who was in prison because of an unrelated conviction. These and other STR data (not shown) exonerated Washington and led Tinsley to plead guilty to the murder.

both livestock (animal husbandry) and crops has exploited naturally occurring mutations and genetic recombination for thousands of years.

As we described earlier, DNA technology enables scientists to produce transgenic animals, which speeds up the selective breeding process. The goals of creating a transgenic animal are often the same as the goals of traditional breeding—for instance, to make a sheep with better quality wool, a pig with leaner meat, or a cow that will mature in a shorter time. Scientists might, for example, identify and clone a gene that causes the development of larger muscles (muscles make up most of the meat we eat) in one breed of cattle and transfer it to other cattle or even to sheep. However, health problems are not uncommon among farm animals carrying genes from other species, and modification of the animal's own genes using the CRISPR-Cas9 system will likely emerge as a more useful technique. Animal health and welfare are important issues to consider when genetically altering animals.

Agricultural scientists have already endowed a number of crop plants with genes for desirable traits, such as delayed ripening and resistance to spoilage, disease, and drought. Modifications can also add value to food crops, giving them a longer shelf life or improved flavor or nutritional value. For many plant species, a single tissue cell grown in culture can give rise to an adult plant. Thus, genetic manipulations can be performed on an ordinary somatic cell and the cell then can be used to generate a plant with new traits.

Genetic engineering is rapidly replacing traditional plant-breeding programs, especially for useful traits, such as herbicide or pest resistance, determined by one or a few genes. Crops engineered with a bacterial gene making the plants resistant to an herbicide can grow while weeds are destroyed, and genetically engineered crops that can resist destructive insects reduce the need for chemical insecticides.

The Food and Agriculture Organization of the United Nations has predicted that we will need 70% more food by the year 2050 than the planet is currently producing. One novel approach is being taken by researchers working on the international C_4 Rice Project. The common form of rice, a global food staple, uses the C_3 form of photosynthesis (see Concept 10.5). The aim of researchers engaged in this project is to genetically engineer a strain of rice that can use C_4 photosynthesis, which is more efficient. In 2017, the first step was accomplished when they were able to genetically engineer a corn gene into rice plants.

Safety and Ethical Questions Raised by DNA Technology

Early concerns about potential dangers associated with recombinant DNA technology and genetic engineering focused on the possibility that hazardous new pathogens might be created. What might happen, for instance, if in a research study cancer cell genes were transferred into bacteria or viruses? To guard against such rogue microorganisms, scientists developed a set of guidelines that were adopted as formal government regulations in the United States and some other countries. One safety measure is a set of strict laboratory procedures designed to prevent engineered microorganisms from either infecting researchers or accidentally leaving the laboratory. In addition, strains of microorganisms to be used in recombinant DNA experiments are genetically crippled to ensure that they cannot survive outside the laboratory. Finally, certain obviously dangerous experiments have been banned.

Today, most public concern about possible hazards centers not on recombinant microorganisms but on **genetically modified organisms (GMOs)** used as food. A GMO is a transgenic organism that has acquired one or more genes from another species or from another variety of the same species. Some salmon, for example, have been genetically modified by addition of a more active salmon growth hormone gene. However, the majority of the GMOs that contribute to our food supply are not animals, but crop plants.

GM crops are widespread in the United States, Argentina, and Brazil; together, these countries account for over 80% of the world's acreage devoted to such crops. In the United States, most corn, soybean, and canola crops are genetically modified, and a recent law requires labeling of GM products. The same foods are an ongoing subject of controversy in Europe, where the GM revolution has been met with strong opposition. Many Europeans are concerned about the safety of GM foods and the possible environmental consequences of growing GM plants. Although a small number of GM crops have been grown on European soil, the European Union established a comprehensive legal framework regarding GMOs in 2015. Among other regulations, individual member states may ban either the growing or importing of GM crops, which must be clearly labeled. The high degree of consumer distrust in Europe makes the future of GM crops there uncertain.

Advocates of a cautious approach toward GM crops fear that transgenic plants might pass their new genes to close relatives in nearby wild areas. We know that lawn and crop grasses, for example, commonly exchange genes with wild relatives via pollen transfer. If crop plants carrying genes for resistance to herbicides, diseases, or insect pests pollinated wild ones, the offspring might become "super weeds" that are very difficult to control. Another worry involves possible risks to human health from GM foods. Some people fear that the protein products of transgenes might lead to allergic reactions. Although there is some evidence that this could happen, advocates claim that these proteins could be tested in advance to avoid producing ones that cause allergic reactions. (For further discussion of plant biotechnology and GM crops, see Concept 38.3.)

Today, governments and regulatory agencies throughout the world are grappling with how to facilitate the use of biotechnology in agriculture, industry, and medicine while ensuring that new products and procedures are safe. In the United States, such applications of biotechnology must be evaluated for potential risks by various regulatory agencies, including the Food and Drug Administration, the Environmental Protection Agency, the National Institutes of Health, and the Department of Agriculture. Meanwhile, these same agencies and the public must consider the ethical implications of biotechnology.

Advances in biotechnology have allowed us to obtain complete genome sequences for humans and many other species, providing a vast treasure trove of information about genes. We can ask how certain genes differ from species to species, as well as how genes and, ultimately, entire genomes have evolved. (These are the subjects of Chapter 21.) At the same time, the increasing speed and falling cost of sequencing the genomes of individuals are raising significant ethical questions. Who should have the right to examine someone else's genetic information? How should that information be used? Should a person's genome be a factor in determining eligibility for a job or insurance? Ethical considerations, as well as concerns about potential environmental and health hazards, will likely slow some applications of biotechnology. There is always a danger that too much regulation will stifle basic research and its potential benefits. On the other hand, genetic engineering—especially gene editing with the CRISPR-Cas9 system—enables us to profoundly and rapidly alter species that have been evolving for millennia. A good example is the potential use of a gene drive that would eliminate the ability of mosquito species to carry diseases or even eradicate certain mosquito species. There would probably be health benefits to this approach, at least initially, but unforeseen problems could easily arise. Given the tremendous power of DNA technology, we must proceed with humility and caution.

> → **Mastering Biology** Interview with David Suzuki: Exploring ethical issues related to DNA technology

CONCEPT CHECK 20.4

1. What is the advantage of using stem cells for gene therapy or gene editing?
2. List at least three different properties that have been acquired by crop plants via genetic engineering.
3. **WHAT IF?** As a physician, you have a patient with symptoms that suggest a hepatitis A infection, but you have not been able to detect viral proteins in the blood. Knowing that hepatitis A is an RNA virus, what lab tests could you perform to support your diagnosis? Explain the results that would support your hypothesis.

For suggested answers, see Appendix A.

20 Chapter Review

> → Go to **Mastering Biology** for Assignments, the eText, the Study Area, and Dynamic Study Modules.

SUMMARY OF KEY CONCEPTS

> → To review key terms, go to the **Vocabulary Self-Quiz** in the **Mastering Biology** eText or Study Area, or go to goo.gl/zkjz9t.

CONCEPT 20.1

DNA sequencing and DNA cloning are valuable tools for genetic engineering and biological inquiry (pp. 416–423)

- **Nucleic acid hybridization**, the base pairing of one strand of a nucleic acid to the complementary sequence on a strand from another nucleic acid molecule, is widely used in **DNA technology**.
- **DNA sequencing** can be carried out using the dideoxy sequencing method in automated sequencing machines.
- Fast and inexpensive next-generation (high-throughput) techniques for sequencing DNA are based on sequencing by synthesis: DNA polymerase is used to synthesize a stretch of DNA using a single-stranded template, and the order in which nucleotides are added reveals the sequence. Third-generation sequencing methods, including nanopore technology, sequence long DNA molecules one at a time.

- **Gene cloning** (or **DNA cloning**) produces multiple copies of a gene (or DNA segment) that can be used to manipulate and analyze DNA and to produce useful new products or organisms with beneficial traits.
- In **genetic engineering**, bacterial **restriction enzymes** are used to cut DNA molecules within short, specific nucleotide sequences (**restriction sites**), yielding a set of double-stranded **restriction fragments** with single-stranded **sticky ends**:

  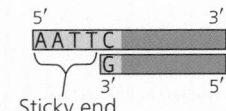
Sticky end

- The sticky ends on restriction fragments from one DNA source can base-pair with complementary sticky ends on fragments from other DNA molecules. Sealing the base-paired fragments with DNA ligase produces **recombinant DNA molecules**.
- DNA restriction fragments of different lengths can be separated by **gel electrophoresis**.
- The **polymerase chain reaction (PCR)** can amplify (produce many copies of) a specific target segment of DNA, using primers that bracket the desired sequence and a heat-resistant DNA polymerase.

- To clone a eukaryotic gene:

Cloning vector (often a bacterial plasmid)

DNA fragments obtained by PCR or from another source (cut by same restriction enzyme used on cloning vector)

Mix and ligate

Recombinant DNA plasmids

Recombinant **plasmids** are returned to host cells, each of which divides to form a clone of cells.

- Expressing cloned eukaryotic genes in bacterial host cells poses several technical difficulties. The use of cultured eukaryotic cells as host cells, coupled with appropriate **expression vectors**, helps avoid these problems.

? *Describe how the process of gene cloning results in a cell clone containing a recombinant plasmid.*

CONCEPT 20.2

Biologists use DNA technology to study gene expression and function *(pp. 423–428)*

- Several techniques use hybridization of a **nucleic acid probe** to detect the presence of specific mRNAs.
- *In situ* **hybridization** and **RT-PCR** can detect the presence of a given mRNA in a tissue or an RNA sample, respectively.
- Sets of genes co-expressed by a group of cells can be detected by **RNA sequencing (RNA-seq)**—sequencing the **cDNAs** corresponding to mRNAs from the cells. **DNA microarrays** are also used for this purpose.
- For a gene of unknown function, experimental inactivation of the gene (a gene knockout) and observation of the resulting phenotypic effects can provide clues to its function. The CRISPR-Cas9 system allows researchers to edit genes in living cells in a specific, desired way. The new alleles can be altered so that they are inherited in a biased way through a population (**gene drive**).
- In humans, **genome-wide association studies** identify and use **single nucleotide polymorphisms (SNPs)** as genetic markers for alleles that are associated with particular conditions.

? *What useful information is obtained by detecting expression of specific genes?*

CONCEPT 20.3

Cloned organisms and stem cells are useful for basic research and other applications *(pp. 428–432)*

- The question of whether all the cells in an organism have the same genome prompted the first attempts at organismal cloning.
- Single differentiated cells from plants are often **totipotent**: capable of generating all the tissues of a complete new plant.
- Transplantation of the nucleus from a differentiated animal cell into an enucleated egg can sometimes give rise to a new animal.
- Certain embryonic **stem cells** (ES cells) from animal embryos and particular adult stem cells from adult tissues can reproduce and differentiate both in the lab and in the organism, offering the

potential for medical use. ES cells are **pluripotent** but difficult to acquire. Induced pluripotent stem (iPS) cells resemble ES cells in their capacity to differentiate; they can be generated by reprogramming differentiated cells. iPS cells hold promise for medical research and regenerative medicine.

? *Describe how, using mice, a researcher could carry out (1) organismal cloning, (2) production of ES cells, and (3) generation of iPS cells, focusing on how the cells are reprogrammed. (The procedures are basically the same in humans and mice.)*

CONCEPT 20.4

The practical applications of DNA-based biotechnology affect our lives in many ways *(pp. 433–439)*

- DNA technology, including the analysis of genetic markers such as SNPs, is increasingly being used in the diagnosis of genetic and other diseases and in personal genome analysis. Personalized medicine offers potential for an individual minimizing their known risk for a disease, as well as better treatment of genetic disorders or cancers. **Gene therapy** or gene editing with the CRISPR-Cas9 system may also lead to permanent cures. DNA technology is used with cell cultures in the large-scale production of protein hormones and other proteins with therapeutic uses. Some therapeutic proteins are being produced in **transgenic** "pharm" animals.
- Analysis of genetic markers such as **short tandem repeats (STRs)** in DNA isolated from tissue or body fluids found at crime scenes leads to a **genetic profile**. Use of genetic profiles can provide definitive evidence that a suspect is innocent or strong evidence of guilt. Such analysis is also useful in parenthood disputes and in identifying the remains of crime victims.
- Genetically engineered microorganisms can be used to extract minerals from the environment or degrade various types of toxic waste materials.
- The aims of developing transgenic plants and animals are to improve agricultural productivity and food quality.
- The potential benefits of genetic engineering must be carefully weighed against the potential for harm to humans or the environment.

? *What factors affect whether a given genetic disease would be a good candidate for successful gene therapy?*

TEST YOUR UNDERSTANDING

→ For more multiple-choice questions, go to the **Practice Test** in the **Mastering Biology** eText or Study Area, or go to goo.gl/GruWRg.

Levels 1-2: Remembering/Understanding

1. In DNA technology, the term *vector* can refer to
   (A) the enzyme that cuts DNA into restriction fragments.
   (B) the sticky end of a DNA fragment.
   (C) a SNP marker.
   (D) a plasmid used to transfer DNA into a living cell.

2. Which of the following tools of DNA technology is incorrectly paired with its use?
   (A) electrophoresis—separation of DNA fragments
   (B) DNA ligase—cutting DNA, creating sticky ends of restriction fragments
   (C) DNA polymerase—polymerase chain reaction to amplify sections of DNA
   (D) reverse transcriptase—production of cDNA from mRNA

3. Plants are more readily manipulated by genetic engineering than are animals because
   (A) plant genes do not contain introns.
   (B) more vectors are available for transferring recombinant DNA into plant cells.
   (C) a somatic plant cell can often give rise to a complete plant.
   (D) plant cells have larger nuclei.

4. A paleontologist has recovered a bit of tissue from the 400-year-old preserved skin of an extinct dodo (a bird). To compare a specific region of the DNA from a sample with DNA from living birds, which of the following would be most useful for increasing the amount of dodo DNA available for testing?
   (A) SNP analysis
   (B) polymerase chain reaction (PCR)
   (C) electroporation
   (D) gel electrophoresis

Levels 3-4: Applying/Analyzing

5. Which of the following is true of cDNA produced using human brain tissue as the starting material?
   (A) The procedure to make it requires amplification by the polymerase chain reaction.
   (B) It is produced from pre-mRNA using reverse transcriptase.
   (C) It can be labeled and used as a probe to detect genes expressed in the brain.
   (D) It includes the introns of the pre-mRNA.

6. Expression of a cloned eukaryotic gene in a bacterial cell involves many challenges. The use of mRNA and reverse transcriptase is part of a strategy to solve the problem of
   (A) post-transcriptional processing.
   (B) post-translational processing.
   (C) nucleic acid hybridization.
   (D) restriction fragment ligation.

7. Which of the following sequences in double-stranded DNA is most likely to be recognized as a cutting site for a restriction enzyme?
   (A) AAGG
       TTCC
   (B) GGCC
       CCGG
   (C) ACCA
       TGGT
   (D) AAAA
       TTTT

Levels 5-6: Evaluating/Creating

8. **MAKE CONNECTIONS** Imagine you want to study one of the human crystallins, proteins present in the lens of the eye (see Figure 1.8). To obtain a sufficient amount of the protein of interest, you decide to clone the gene that codes for it. Assume you know the sequence of this gene. Explain how you would go about this.

9. **MAKE CONNECTIONS** Looking at Figure 20.15, what does it mean for a SNP to be "linked" to a disease-associated allele? How does this allow the SNP to be used as a genetic marker? (See Concept 15.3.)

10. **DRAW IT** You are cloning an aardvark gene, using a bacterial plasmid as a vector. The green diagram shows the plasmid, which contains the restriction site for the enzyme used in Figure 20.5. Above the plasmid is a segment of linear aardvark DNA that was synthesized using PCR. Diagram your cloning procedure, and show what would happen to these two molecules during each step. Use one color for the aardvark

DNA and its bases and another color for those of the plasmid. Label each step and all 5′ and 3′ ends.

5′ GAATTCTAAAGCGCTTATGAATTC 3′
3′ CTTAAGATTTCGCGAATACTTAAG 5′
Aardvark DNA

Plasmid

11. **EVOLUTION CONNECTION** Ethical considerations aside, if DNA-based technologies became widely used, discuss how they might change the way evolution proceeds, as compared with the natural evolutionary mechanisms that have operated for the past 4 billion years.

12. **SCIENTIFIC INQUIRY** You hope to study a gene that codes for a neurotransmitter protein produced in human brain cells. You know the amino acid sequence of the protein. Explain how you might (a) identify what genes are expressed in a specific type of brain cell, (b) identify (and isolate) the neurotransmitter gene, (c) produce multiple copies of the gene for study, and (d) produce large quantities of the neurotransmitter for evaluation as a potential medication.

13. **WRITE ABOUT A THEME: INFORMATION** In a short essay (100–150 words), discuss how the genetic basis of life plays a central role in biotechnology.

14. **SYNTHESIZE YOUR KNOWLEDGE**

The water in the Yellowstone National Park hot springs shown here is around 160°F (70°C). Biologists assumed that no species of organisms could live in water above about 130°F (55°C), so they were surprised to find several species of bacteria there, now called *thermophiles* ("heat-lovers"). You've learned in this chapter how an enzyme from one species, *Thermus aquaticus*, made feasible one of the most important DNA-based techniques used in labs today. Identify the enzyme, and indicate the value of its being isolated from a thermophile. Suggest other reasons why enzymes from this bacterium (or other thermophiles) might also be valuable.

For selected answers, see Appendix A.

Explore Scientific Papers with Science in the Classroom AAAS
How can CRISPR-Cas9 be used to create a gene drive?
Go to "Can We Handle the Power of CRISPR?"
at www.scienceintheclassroom.org.
→ **Instructors:** Questions can be assigned in Mastering Biology.

21 Genomes and Their Evolution

KEY CONCEPTS

21.1 The Human Genome Project fostered development of faster, less expensive sequencing techniques *p. 443*

21.2 Scientists use bioinformatics to analyze genomes and their functions *p. 444*

21.3 Genomes vary in size, number of genes, and gene density *p. 448*

21.4 Multicellular eukaryotes have a lot of noncoding DNA and many multigene families *p. 450*

21.5 Duplication, rearrangement, and mutation of DNA contribute to genome evolution *p. 454*

21.6 Comparing genome sequences provides clues to evolution and development *p. 459*

Study Tip

Write genomic analysis questions: As you go through the chapter, make a table of genomic analyses that are discussed and write down some questions they can address. You can include questions not covered in the text.

Genomic projects and websites	Questions that can be addressed
Human Genome Project	How much DNA is in the human genome?
	How much of the human genome codes for proteins?
	What functions are carried out by other regions of the genome?
ENCODE project	

→ Go to Mastering Biology

For Students (in eText and Study Area)
- Get Ready for Chapter 21
- HHMI Animation: Shotgun Sequencing
- Animation: The Human Genome Project: The Genes on Human Chromosome 17

For Instructors to Assign (in Item Library)
- Tutorial: Scientific Skills Exercise: Reading an Amino Acid Sequence Identity Table
- Tutorial: Using BLAST: What Can a Protein Sequence Reveal About Cancer?

Figure 21.1 The elephant shark (*Callorhinchus milii*) looks vaguely prehistoric and has been called a "living fossil." In fact, it has the slowest evolving genome of any vertebrate sequenced so far. Comparing rates of genome change in different species provides insights into the evolutionary past.

What are some questions that can be explored by sequencing and comparing genomes?

What are the functions of the human genome?

Genes for proteins and RNA (1.5%)

Noncoding DNA (98.5%)

How do genomes differ in **number of genes?**

Escherichia coli 4,400 genes

Homo sapiens 21,300 genes

Zea mays (corn) 32,000 genes

What do gene sequences tell us about **evolutionary relationships** between species?

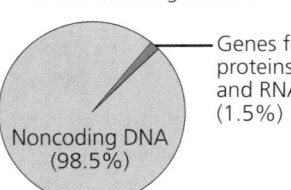

Chimpanzee

Human

Mouse

70 60 50 40 30 20 10 0
Millions of years ago

How do genomes **evolve** over time?

Elephant shark: slower-evolving genome

Tiger tail sea horse: faster-evolving genome

The Human Genome Project fostered development of faster, less expensive sequencing techniques

Upon sequencing the genome of any species, scientists can study whole sets of genes and their interactions, an approach called **genomics**. The sequencing efforts that feed this approach generate enormous volumes of data. The necessity of dealing with this ever-increasing flood of information has energized the field of **bioinformatics**, the application of computational methods to store and analyze biological data.

The sequencing of the human genome, an ambitious undertaking, officially began as the **Human Genome Project** in 1990. Organized by an international, publicly funded consortium of scientists at universities and research institutes, the project involved 20 large sequencing centers in six countries plus a host of other labs working on smaller parts of the project.

After the human genome sequence was largely completed in 2003, the sequence of each chromosome was analyzed and described in a series of papers, the last of which covered chromosome 1 and was published in 2006. The sequence is over 99% complete—the figure is not 100% because numerous gaps remain, usually in regions with much repetitive DNA that are difficult to sequence. The sequenced DNA was pooled from a few individuals; scientists reviewed the results and agreed on a **reference genome**, a full sequence that best represents the genome of a species. The human reference genome is continuously being revised and re-released.

The ultimate goal in mapping any genome is to determine the complete nucleotide sequence of each chromosome. For the human genome, this was accomplished by scientists using sequencing machines and the dideoxy chain termination method mentioned in Concept 20.1. Two approaches complemented each other in this endeavor. The initial approach was a methodical one that ordered each fragment based on earlier genetic mapping of the human genome. In 1998, however, molecular biologist J. Craig Venter set up a company (Celera Genomics) and declared his intention to sequence the entire human genome using an alternative strategy. The **whole-genome shotgun approach** starts with the cloning and sequencing of DNA fragments from randomly cut DNA. Powerful computer programs then assemble the resulting very large number of overlapping short sequences into a single continuous sequence **(Figure 21.2)**. The whole-genome shotgun approach is widely used today.

A major thrust of the Human Genome Project was the development of technology for faster sequencing (see Concept 20.1). Improvements over the years chipped away at each time-consuming step, enabling the rate of sequencing to accelerate impressively: Whereas a productive lab could typically sequence 1,000 base pairs a day in the 1980s, by the year 2000 each research center working on the Human Genome Project was sequencing 1,000 base pairs *per second*. Currently, the most widely used automated "next-generation" sequencing machines can sequence nearly 35 million base pairs per second (see Figure 20.3). Also, because of the sensitivity of these techniques, the DNA can be sequenced directly, either in fragments or as a single whole molecule; the cloning step (❷ in Figure 21.2) is unnecessary. Methods that can analyze biological materials very rapidly and produce enormous volumes of data are said to be "high-throughput." Sequencing machines using rapid sequencing techniques are an example of high-throughput devices.

Along with massive increases in sequencing speed, the cost of sequencing entire genomes has plummeted. Whereas sequencing the first human genome took 13 years and cost between $500 million and $1 billion, by 2007 the same task took four months at a cost of $1 million. In 2019, the fastest machines are able to sequence the complete genomes of 48 individuals in 44 hours for under $1,000 per genome.

▼ **Figure 21.2 Whole-genome shotgun approach to sequencing.** In this approach, random DNA fragments are cloned (see Figure 20.4), sequenced, and then ordered relative to each other.

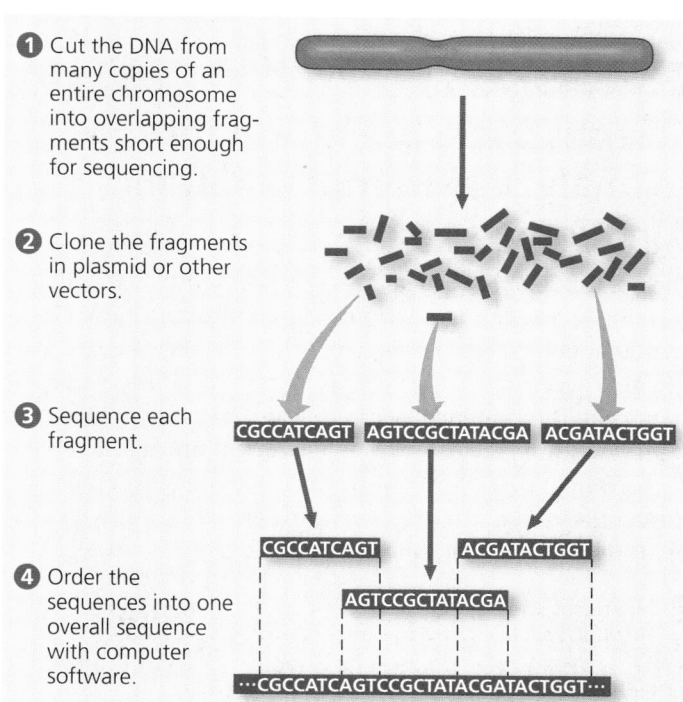

❶ Cut the DNA from many copies of an entire chromosome into overlapping fragments short enough for sequencing.

❷ Clone the fragments in plasmid or other vectors.

❸ Sequence each fragment.

CGCCATCAGT   AGTCCGCTATACGA   ACGATACTGGT

❹ Order the sequences into one overall sequence with computer software.

CGCCATCAGT

ACGATACTGGT

AGTCCGCTATACGA

···CGCCATCAGTCCGCTATACGATACTGGT···

VISUAL SKILLS *The fragments in step 2 of this figure are depicted as scattered, rather than arranged in an ordered array. How does this depiction reflect the approach?*

➡ **Mastering Biology** HHMI Animation: Shotgun Sequencing

These technological advances have also facilitated an approach called **metagenomics** (from the Greek *meta*, beyond), in which DNA from an entire community of species (a *metagenome*) is collected from an environmental sample and sequenced. Again, computer software sorts out the partial sequences and assembles them into the individual species' partial of complete genomes. An advantage of this technique is the ability to sequence the DNA of mixed microbial populations, which eliminates the need to culture each species separately in the lab, a challenge that has limited the study of microorganisms. So far, this approach has been applied to communities found in environments as diverse as the human intestine and extreme habitats like thermal springs where the temperature exceeds 80°C.

→ **Mastering Biology**
   BBC Video: Collecting New Genes to Create New Life

At first glance, genome sequences of humans and other organisms are simply dry lists of nucleotide bases—millions of A's, T's, C's, and G's in mind-numbing succession. Making sense of this massive amount of data has called for new analytical approaches, which we discuss next.

CONCEPT CHECK 21.1

1. Describe the whole-genome shotgun approach.

For suggested answers, see Appendix A.

CONCEPT 21.2

Scientists use bioinformatics to analyze genomes and their functions

Each of the 20 or so sequencing centers around the world working on the Human Genome Project churned out voluminous amounts of DNA sequence day after day. As the data began to accumulate, the need to coordinate efforts to keep track of all the sequences became clear. Thanks to the foresight of research scientists and government officials involved in the Human Genome Project, its goals included establishing centralized databases and refining analytical software, all made readily accessible on the Internet.

→ **Mastering Biology**
   HHMI Video: Leading Edge Bioinformatics

Centralized Resources for Analyzing Genome Sequences

Making bioinformatics resources available to researchers worldwide and speeding up the dissemination of information served to accelerate progress in DNA sequence analysis.

The National Library of Medicine (NLM) and the National Institutes of Health (NIH) maintain the National Center for Biotechnology Information (NCBI), which today maintains a website (www.ncbi.nlm.nih.gov), which has extensive bioinformatics resources. On this site are links to databases, software, and a wealth of information about genomics and related topics. Similar websites have also been established by three genome centers with which the NCBI collaborates: the European Molecular Biology Laboratory, the DNA Data Bank of Japan, and BGI (formerly known as the Beijing Genomics Institute) in Shenzhen, China. These large, comprehensive websites are complemented by others maintained by individual or small groups of laboratories. Smaller websites often provide databases and software designed for a narrower purpose, such as studying genetic and genomic changes in one particular type of cancer.

The NCBI database of sequences is called GenBank. As of August 2019, it included the sequences of 214 million fragments of genomic DNA, totaling 366 billion base pairs! GenBank is constantly updated, and the amount of data it contains increases rapidly. Any sequence in the database can be retrieved and analyzed using software from the NCBI website or elsewhere.

One very widely used software program available on the NCBI website, called BLAST (Basic Local Alignment Search Tool), allows the user to compare a DNA sequence with every sequence in GenBank, base by base. A researcher might search for similar regions in other genes of the same species or among the genes of other species. Another program allows comparison of protein sequences. A third program can search any protein sequence for *conserved* (common) stretches of amino acids (domains) for which a function is known or suspected, and it can show a three-dimensional model of the domain alongside other relevant information (**Figure 21.3**). There are even various software programs that can align and compare a collection of sequences, either nucleic acids or polypeptides, and diagram them in the form of an evolutionary tree based on the sequence relationships. (One such diagram is shown in Figure 21.17.)

Two research institutions, Rutgers University and the University of California, San Diego, also maintain a worldwide database of all three-dimensional protein structures that have been experimentally determined, called the Protein Data Bank (www.wwpdb.org). These structures can be rotated by the viewer to show all sides of the protein. Throughout this book, you'll find images of protein structures that have been obtained from the Protein Data Bank.

There is a vast array of resources available for researchers anywhere in the world to use free of charge. Let us now consider the types of questions scientists can address using these resources.

→ **Instructors:** BLAST Data Analysis Tutorials, which teach students how to work with real data from the BLAST database, can be assigned in **Mastering Biology**.

❶ In this window, a partial amino acid sequence from an unknown muskmelon protein ("Query") is aligned with sequences from other proteins that the program found to be similar. Each sequence represents a domain called WD40.

❷ Four hallmarks of the WD40 domain are highlighted in yellow. (Sequence similarity is based on chemical aspects of the amino acids, so the amino acids in each hallmark region are not always identical.)

❸ The Cn3D ("See in 3D") program displays 3D models of domains, such as this ribbon model of cow transducin (the protein highlighted in purple in the Sequence Alignment Viewer). This protein is the only one of those shown for which a structure has been determined. The sequence similarity of the other proteins to cow transducin suggests that their structures are likely to be similar.

❻ This window displays information about the WD40 domain from the Conserved Domain Database (CDD), which can find and describe similar domains in related proteins.

WD40 - Sequence Alignment Viewer

```
                            Query   ~~~ktGGIRL~RHfksVSAVEWHRk~~gDYLSTlvLreSRAVLIHQlsk
             Cow [transducin]       ~nvrvSRELA~GHtgyLSCCRFLDd~~nQIVTs~~Sg~DTTCALWDie~
      Mustard weed [transducin]     gtvpvSRMLT~GHrgyVSCCQYVPnedaHLITs~~Sg~DQTCILWDvtt
             Corn [GNB protein]     gnmpvSRILT~GHkgyVSCCQYVPdgetRLITS~~Sg~DQTCVLWDvt~
           Human [PAFA protein]     ~~~ecIRTMH~GHdhnVSSVAIMPng~dHIVSA~~Sr~DKTIKMWEvg~
  Nematode [unknown protein #1]     ~~~rcVKTLK~GHtnyVFCCCFNPs~~gTLIAS~~GsfDETIRIWCar~
  Nematode [unknown protein #2]     ~~~rmTKTLK~GHnnyVFCCNFNPq~~sSLVVS~~GsfDESVRIWDvk~
        Fission yeast [FWDR protein]  ~~~seCISILhGHtdsVLCLTFDS~~~~TLLVS~~GsaDCTVKLWHfs~
```

WD40 - Cn3D 4.1

CDD Descriptive Items

Name: WD40

WD40 domain, found in a number of eukaryotic proteins that cover a wide variety of functions including adaptor/regulatory modules in signal transduction, pre-mRNA processing and cytoskeleton assembly; typically contains a GH dipeptide 11-24 residues from its N-terminus and the WD dipeptide at its C-terminus and is 40 residues long, hence the name WD40;

❹ Cow transducin contains seven WD40 domains, one of which is highlighted here in gray.

❺ The yellow segments correspond to the WD40 hallmarks highlighted in yellow in the window above.

Identifying Protein-Coding Genes and Understanding Their Functions

Using available DNA sequences, geneticists can study genes directly, rather than taking the classical genetic approach, which requires determining the function of an unknown gene from the phenotype. But this more recent approach poses a new challenge: What does the gene actually do? Given a long DNA sequence from a database like GenBank, scientists aim to identify all protein-coding genes in the sequence and ultimately their functions. This process, called **gene annotation**, uses three lines of evidence to identify a gene.

First, computers are utilized in a search for patterns that indicate the presence of genes. The usual approach is to use software to scan the stored sequences for those that represent transcriptional and translational start and stop signals, RNA-splicing sites, and other telltale signs of protein-coding genes, such as promoter sequences. The software also looks for certain short sequences that specify known mRNAs. Thousands of such sequences, called *expressed sequence tags*, or *ESTs*, have been collected from cDNA sequences and are cataloged in computer databases. This type of analysis identifies sequences that may turn out to be previously unknown protein-coding genes.

Although the identities of about half of the human genes were known before the Human Genome Project began, the other genes, previously unknown, were revealed by DNA sequence analysis. Once such suspected genes are identified, the second step is to obtain clues about their identities and functions. Software is used to compare their sequences with those of known genes from other organisms. Due to redundancy in the genetic code, the DNA sequence itself may vary more among species than the protein sequence does. Thus, scientists interested in proteins often compare the predicted amino acid sequence of a protein to that of other proteins. The final step is to confirm the identities of these genes using RNA-seq (see Figure 20.12) or some other method to show that the relevant RNA is actually expressed from the proposed gene.

→ **Mastering Biology** Animation: The Human Genome Project: Genes on Human Chromosome 17

Sometimes a newly identified sequence will match, at least partially, the sequence of a gene or protein in another species whose function is well known. For example, a plant researcher working on signaling pathways in the muskmelon would be excited to see that a partial amino acid sequence from a gene she had identified matched sequences in other species encoding a functional part of a protein called a WD40 domain (see Figure 21.3). WD40 domains are present in many eukaryotic proteins and are known to function in signal transduction pathways. Alternatively, a new gene sequence might be similar to a previously encountered sequence whose function is still unknown. Another possibility is that the sequence is entirely unlike anything ever seen before. This was true for about a third of the genes of *Escherichia coli* when its genome was sequenced. In such cases, protein function is usually deduced through a combination of biochemical and functional studies. The biochemical approach aims to determine the three-dimensional structure of the protein as well as other attributes, such as potential binding sites for other molecules. Functional studies usually involve *knocking out* (blocking or disabling) the gene in an organism to see how the phenotype is affected. The CRISPR-Cas 9 system, described in Figure 17.28, is an example of an experimental technique used to block gene function.

Understanding Genes and Gene Expression at the Systems Level

The impressive computational power provided by the tools of bioinformatics allows the study of whole sets of genes and their interactions, as well as the comparison of genomes from different species. Genomics is a rich source of new insights into fundamental questions about genome organization, regulation of gene expression, embryonic development, and evolution.

One informative approach was taken by a long-term research project called ENCODE (Encyclopedia of DNA Elements), which began in 2003. The aim of the project is to learn everything possible about the functionally important elements in the human genome, at first using multiple experimental techniques on different types of cultured cells. Investigators sought to identify protein-coding genes and genes for noncoding RNAs, along with sequences that regulate gene expression, such as enhancers and promoters. In addition, they extensively characterized DNA and histone modifications and chromatin structure—features termed "epigenetic" since they affect gene expression without changing the sequence of nucleotide bases (see Concept 18.3). The second phase of the project, involving more than 440 scientists in 32 research groups, culminated in 2012 with the simultaneous publication of 30 papers describing over 1,600 large data sets. Now in its fourth phase, the ENCODE project is expanding its analysis of the human genome and that of the mouse (a mammalian model organism), seeking to identify regulatory and other important sequences. The considerable power of ENCODE has been that it provides the opportunity to compare results from specific projects with each other, yielding a much richer picture of the human and mouse genomes.

Perhaps the most striking finding was that about 75% of the human genome is transcribed at some point in at least one of the cell types studied, even though less than 2% codes for proteins. Furthermore, biochemical functions have been assigned to DNA elements making up at least 80% of the genome. To learn more about the different types of functional elements, parallel projects analyzed in a similar way the genomes of two model organisms, the soil nematode *Caenorhabditis elegans* and the fruit fly *Drosophila melanogaster*. Because genetic and biochemical experiments using DNA technology can be performed on these species, testing the activities of potentially functional DNA elements in their genomes is expected to illuminate the workings of the human genome.

Because the ENCODE project analyzed cells in culture, its potential for clinical applications was limited. A related project called the Roadmap Epigenomics Project set out to characterize the *epigenome*—the epigenetic features of the genome of hundreds of human cell types and tissues. The aim was to focus on the epigenomes of stem cells, normal tissues from mature adults, and relevant tissues from individuals with diseases such as cancer and neurodegenerative and autoimmune disorders. In 2015, a series of papers reported on the results from 111 tissues. One of the most useful findings was that the original tissue in which a cancer arose can be identified in cells of a secondary tumor based on characterization of their epigenomes.

Systems Biology

The scientific progress resulting from sequencing genomes and studying large sets of genes has encouraged scientists to attempt similar systematic studies of sets of proteins and their properties (such as their abundance, chemical modifications, and interactions), an approach called **proteomics**. (A **proteome** is the entire set of proteins expressed by a cell or group of cells.) Proteins, not the genes that encode them, carry out most of the activities of the cell. Therefore, if we are to understand the functioning of cells and organisms, we must study when and where proteins are produced in an organism, as well as how they interact in networks.

Genomics and proteomics enable molecular biologists to approach the study of life from an increasingly integrated perspective. Using the tools we have described, biologists have begun to compile catalogs of genes and proteins, listing all the "parts" that contribute to the operation of cells, tissues, and organisms. With such catalogs in hand, researchers have shifted their attention from the individual parts—genes and proteins—to their functional integration in biological systems. As you may recall, Concept 1.1 discussed this approach, called **systems biology**, which aims to model the dynamic behavior of whole biological systems based on the study of the interactions among the system's parts. Because of the vast amounts of data generated in these types of studies, advances in computer technology and bioinformatics are crucial to studying systems biology.

One important use of the systems biology approach is to define gene and protein interaction networks. To map the protein interaction network in the yeast *Saccharomyces cerevisiae*, for instance, researchers used sophisticated techniques to knock out pairs of genes, one pair at a time, creating doubly mutant cells. They then compared the fitness of each double mutant (based in part on the size of the cell colony it formed) to that predicted from the fitness of each of the two single mutants. The researchers reasoned that if the observed fitness matched the prediction, then the products of the two genes didn't interact with each other, but if the observed fitness was greater or less than predicted, then the gene products interacted in the cell. They then used computer software to build a graphic model by "mapping" the gene products to certain locations in the model, based on the similarity of their interactions. This resulted in the network-like "functional map" of protein interactions shown in **Figure 21.4**. Processing the vast number of protein-protein interactions generated by this experiment and integrating them into the completed map required powerful computers, mathematical tools, and newly developed software.

Application of Systems Biology to Medicine

The Cancer Genome Atlas began in 2007 and culminated in 2018 with publications called the Pan-Cancer Atlas. This project is another example of systems biology in which many interacting genes and gene products are analyzed together as a group. Under the joint leadership of the National Cancer Institute and the NIH, the project aimed to determine how changes in biological systems lead to cancer. A pilot project set out to find all the common mutations in three types of cancer—lung cancer, ovarian cancer, and glioblastoma of the brain—by comparing gene sequences and patterns of gene expression in cancer cells with those in normal cells. Work on glioblastoma confirmed the role of several suspected genes and identified a few previously unknown ones, suggesting possible new targets for therapies.

As high-throughput techniques become more rapid and less expensive, they are increasingly being applied to the problem of cancer. The approach described previously proved so fruitful for those three types of cancer that it was extended to ten other types, chosen because they are common and often lethal in humans, as well as to *metastatic* tumors (see Figure 12.20), those that have dispersed from primary tumors and invaded organs far away in the body. Ninety percent of cancer deaths are primarily caused by metastasis. Results from the study of metastatic tumors, published in 2017, highlighted several key genes whose mutations were frequently found in

▼ **Figure 21.4 The systems biology approach to protein interactions.** This global protein interaction map shows the likely interactions (lines) among about 4,500 gene products (dots) in *Saccharomyces cerevisiae,* the budding yeast. Dots of the same color represent gene products involved in one of the 13 similarly colored cellular functions listed around the map. The white dots represent proteins that haven't been assigned to any color-coded function. The expanded area shows additional details of one map region where the gene products (blue dots) carry out amino acid biosynthesis, uptake, and related functions.

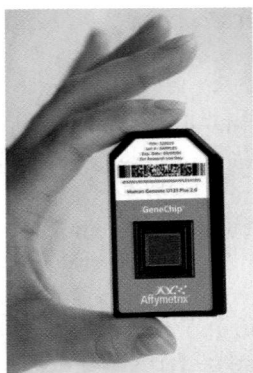

◀ **Figure 21.5 A human gene microarray chip.** Tiny spots of DNA arranged in a grid on this silicon wafer represent almost all of the genes in the human genome. Using this chip, researchers can analyze expression patterns for all these genes at the same time (see Figure 20.13).

→ **Mastering Biology**
HHMI Animation: Gene Chip Manufacturing

hhmi
BioInteractive

metastases and could be targets for chemotherapy. Overall, the Pan-Cancer Atlas contributed significantly to understanding how, where, and why tumors arise, underscoring the value of an integrative systems biology approach to treating cancer.

In addition to whole-genome sequencing, DNA microarrays on glass or silicon chips **(Figure 21.5)** and, increasingly, RNA-seq (see Figure 20.12) are used to analyze gene expression patterns in patients who have various cancers and other diseases. Analyzing which genes are overexpressed or underexpressed in a particular cancer allows physicians to tailor patients' treatment to their unique genetic makeup and the specifics of their cancers. This approach has been used to characterize subsets of particular cancers, enabling more refined treatments. Breast cancer is one example (see Figure 18.27).

Eventually, medical records may include an individual's DNA sequence, a sort of genetic bar code, with regions highlighted that predispose the person to specific diseases. The use of such sequences for personalized medicine—disease prevention and treatment—has great potential.

Systems biology is a very efficient way to study emergent properties at the molecular level. Novel properties arise at each successive level of biological complexity as a result of the arrangement of building blocks at the underlying level (see Concept 1.1). The more we can learn about the arrangement and interactions of the components of genetic systems, the deeper will be our understanding of whole organisms. The rest of this chapter surveys what we've learned from genomic studies.

CONCEPT CHECK 21.2

1. What role does the Internet play in current genomics and proteomics research?
2. Explain the advantage of the systems biology approach to studying cancer versus the approach of studying a single gene at a time.
3. **MAKE CONNECTIONS** The ENCODE pilot project found that at least 75% of the genome is transcribed into RNAs, far more than could be accounted for by protein-coding genes. Review Concepts 17.3 and 18.3 and suggest some roles that these RNAs might play.
4. **MAKE CONNECTIONS** In Concept 20.2, you learned about genome-wide association studies. Explain how these studies use the systems biology approach.

For suggested answers, see Appendix A.

CONCEPT 21.3

Genomes vary in size, number of genes, and gene density

The sequences of thousands of genomes have been completed, with tens of thousands of genomes either in progress or considered permanent drafts (because they require more work than it would be worth to complete them). Among the sequences in progress are roughly 22,000 metagenomes. In the completely sequenced group, about 137,500 are genomes of bacteria, and 1,200 are archaeal genomes. There are 430 completed eukaryotic species, along with 4,500 permanent drafts. Among these are vertebrates, invertebrates, protists, fungi, and plants. Next, we'll discuss what we've learned about genome size, number of genes, and gene density, focusing on general trends.

Genome Size

Comparing the three domains (Bacteria, Archaea, and Eukarya), we find a general difference in genome size between prokaryotes and eukaryotes **(Table 21.1)**. While

Table 21.1 Genome Sizes and Estimated Numbers of Genes*

Organism	Haploid Genome Size (Mb)	Number of Genes	Genes per Mb
Bacteria			
Haemophilus influenzae	1.8	1,700	940
Escherichia coli	4.6	4,400	950
Archaea			
Archaeoglobus fulgidus	2.2	2,500	1,130
Methanosarcina barkeri	4.8	3,600	750
Eukaryotes			
Saccharomyces cerevisiae (yeast, a fungus)	12	6,300	525
Utricularia gibba (floating bladderwort)	82	28,500	348
Caenorhabditis elegans (nematode)	100	20,100	200
Arabidopsis thaliana (mustard family plant)	120	27,000	225
Drosophila melanogaster (fruit fly)	165	14,000	85
Daphnia pulex (water flea)	200	31,000	155
Zea mays (corn)	2,300	32,000	14
Ailuropoda melanoleuca (giant panda)	2,400	21,000	9
Homo sapiens (human)	3,000	21,300	7
Paris japonica (Japanese canopy plant)	149,000	ND	ND

*Some values given here are likely to be revised as genome analysis continues. Mb = million base pairs; the haploid number is used because it represents a complete set of genetic information. ND = not determined.

there are some exceptions, most bacterial genomes have between 1 and 6 million base pairs (Mb); for example, the genome of *E. coli* has 4.6 Mb. Genomes of archaea are mostly within the size range of bacterial genomes. (Keep in mind, however, that many fewer archaeal genomes have been completely sequenced, so this picture may change.) Eukaryotic genomes tend to be larger: The genome of the single-celled yeast *Saccharomyces cerevisiae* (a fungus) has about 12 Mb, while most animals and plants, which are multicellular, have genomes of at least 100 Mb. There are 165 Mb in the fruit fly genome, while humans have 3,000 Mb, about 500 to 3,000 times as many as a typical bacterium.

Aside from this general difference between prokaryotes and eukaryotes, a comparison of genome sizes among eukaryotes fails to reveal any systematic relationship between genome size and the organism's phenotype. For instance, the range among plants is enormous: The genome of *Paris japonica*, the Japanese canopy plant, contains 149 billion base pairs (149,000 Mb), while that of another plant, *Utricularia gibba*, a bladderwort, contains only 82 Mb. Even more striking, there is a single-celled amoeba, *Polychaos dubium*, whose genome size has been estimated at 670 billion base pairs (670,000 Mb). (This genome has not yet been sequenced.) On a finer scale, comparing two insect species, the cricket (*Anabrus simplex*) genome turns out to have 11 times as many base pairs as the fruit fly (*Drosophila melanogaster*) genome. There is a wide range of genome sizes within the groups of protists, insects, amphibians, and plants and less of a range within mammals and reptiles.

Number of Genes

The number of genes also varies between prokaryotes and eukaryotes: Bacteria and archaea, in general, have fewer genes than eukaryotes. Free-living bacteria and archaea have from 1,500 to 7,500 genes, while the number of genes in eukaryotes ranges from about 5,000 for unicellular fungi (yeasts) to at least 40,000 for some multicellular eukaryotes.

Within the eukaryotes, the number of genes in a species is often lower than expected from considering simply the size of its genome. Looking at Table 21.1, you can see that the genome of the nematode *C. elegans* is 100 Mb in size and contains roughly 20,100 genes. In comparison, the genome of *Drosophila melanogaster* is much bigger (165 Mb) but has only about two-thirds the number of genes—14,000 genes.

Considering an example closer to home, we noted that the human genome contains 3,000 Mb, well over ten times the size of either the *D. melanogaster* or *C. elegans* genome. At the outset of the Human Genome Project, biologists expected somewhere between 50,000 and 100,000 genes to be identified in the completed sequence, based on the number of known human proteins. As the project progressed, the estimate was revised downward several times. There is still debate about the number, but it is much lower, probably around 21,300. This estimate, similar to the number of genes in the nematode *C. elegans*, surprised biologists, who had been expecting many more human genes.

What genetic attributes allow humans (and other vertebrates) to get by with no more genes than nematodes? An important factor is that vertebrate genomes "get more bang for the buck" from their coding sequences because of extensive alternative splicing of RNA transcripts. Recall that this process generates more than one polypeptide from a single gene (see Figure 18.14). A typical human gene contains about ten exons, and an estimated 90% or more of these multi-exon genes are spliced in at least two different ways. Some genes are expressed in hundreds of alternatively spliced forms, others in just two. Scientists have not yet catalogued all of the different forms, but it is clear that the number of different proteins encoded in the human genome far exceeds the proposed number of genes.

Additional polypeptide diversity could result from post-translational modifications such as cleavage or the addition of carbohydrate groups in different cell types or at different developmental stages. Finally, the discovery of miRNAs and other RNAs that play regulatory roles has added a new variable to the mix (see Concept 18.3). Some scientists think that this added level of regulation of some genes, when present, may contribute to greater organismal complexity.

Gene Density and Noncoding DNA

We can take both genome size and number of genes into account by comparing gene density in different species. In other words, we can ask how many genes are in a given length of DNA. When we compare the genomes of bacteria, archaea, and eukaryotes, we see that eukaryotes generally have larger genomes but fewer genes in a given number of base pairs. Humans have hundreds or thousands of times as many base pairs in their genome as most bacteria, as we already noted, but only 5 to 15 times as many genes; thus, gene density is lower in humans (see Table 21.1). Even unicellular eukaryotes, such as yeasts, have fewer genes per million base pairs than bacteria and archaea. Among the genomes that have been sequenced completely, humans and other mammals have the lowest gene density.

In all bacterial genomes studied so far, most of the DNA consists of genes for protein, tRNA, or rRNA; the small amount remaining consists mainly of nontranscribed regulatory sequences, such as promoters. The sequence of nucleotides along a bacterial protein-coding gene is not interrupted by noncoding sequences (introns). In eukaryotic genomes, by contrast, most of the DNA neither encodes protein nor is transcribed into RNA molecules of known function, and the DNA includes more complex regulatory sequences. In fact, humans have 10,000 times as much noncoding DNA as bacteria. Some of this DNA in multicellular eukaryotes is present as introns within genes. Indeed, introns account for most of the difference in average length between human genes (27,000 base pairs) and bacterial genes (1,000 base pairs).

In addition to introns, multicellular eukaryotes have a vast amount of non-protein-coding DNA between genes. In the next section, we will describe the composition and arrangement of these great stretches of DNA in the human genome.

CONCEPT CHECK 21.3

1. The current best estimate is that the human genome contains around 21,300 genes. However, there is evidence that human cells produce many more than 21,300 different polypeptides. What processes might account for this discrepancy?

2. The Genomes Online Database (GOLD) website of the Joint Genome Institute has information about genome sequencing projects. Scroll through the page at https://gold.jgi.doe.gov/statistics and describe the information you find. What percent of bacterial genome projects have medical relevance?

3. **WHAT IF?** What evolutionary processes might account for prokaryotes having smaller genomes than eukaryotes?

For suggested answers, see Appendix A.

CONCEPT 21.4

Multicellular eukaryotes have a lot of noncoding DNA and many multigene families

We have spent most of our time focusing on genes that code for proteins. Yet the coding regions of these genes and the genes for small RNAs like tRNAs make up a small portion of most multicellular eukaryotic genomes. For example, only a tiny part—about 1.5%—codes for proteins or is transcribed into rRNAs or tRNAs. **Figure 21.6** shows what is known about the makeup of the remaining 98.5% of the genome.

Gene-related regulatory sequences and introns account, respectively, for 5% and about 20%, respectively, of the human genome. The rest, located between functional genes, includes some unique (single-copy) noncoding DNA, such as gene fragments and **pseudogenes**, former genes that have accumulated mutations over a long time and no longer produce functional proteins. (The genes that produce small noncoding RNAs are a tiny percentage of the genome, distributed between the 20% introns and the 15% unique noncoding DNA.) Most of the DNA between functional genes, however, is **repetitive DNA**, which consists of sequences that are present in multiple copies in the genome.

▼ **Figure 21.6 Types of DNA sequences in the human genome.** The gene sequences that code for proteins or are transcribed into rRNA or tRNA molecules make up only about 1.5% of the human genome (dark purple in the pie chart), while introns and regulatory sequences associated with genes (lighter purple) make up about a quarter. The vast majority of the human genome does not code for proteins (although much of it gives rise to RNAs), and a large amount is repetitive DNA (dark and light green and teal).

> **Mastering Biology** HHMI Animation: **Coding Sequences in DNA**

The bulk of many eukaryotic genomes consists of DNA sequences that neither code for proteins nor are transcribed to produce RNAs with known functions; this noncoding DNA was often described in the past as "junk DNA." However, genome comparisons over the past ten years have revealed the persistence of this DNA in diverse genomes over many hundreds of generations. For example, the genomes of humans, rats, and mice contain almost 500 regions of noncoding DNA that are *identical* in sequence in all three species. This is a higher level of sequence conservation than is seen for protein-coding regions in these species, strongly suggesting that the noncoding regions have important functions. The results of the ENCODE project discussed earlier have underscored the key roles played by much of this noncoding DNA. Next, we'll examine how genes and noncoding DNA sequences are organized within genomes of multicellular eukaryotes, using the human genome as our main example. Genome organization tells us a lot about how genomes have evolved and continue to evolve, as we'll see in Concept 21.5.

Transposable Elements and Related Sequences

Both prokaryotes and eukaryotes have stretches of DNA that can move from one location to another within the genome. These stretches are known as *transposable genetic elements*, or simply **transposable elements**. During the process called *transposition*, a transposable element moves from one site in a cell's DNA to a different target site by a type of recombination process. Transposable elements are sometimes called "jumping genes," but actually they never entirely detach from the cell's DNA. Instead, the original and new DNA sites are brought very close together by enzymes and other proteins that bend the DNA. Surprisingly, about 75% of human repetitive DNA (44% of the entire human genome) is made up of transposable elements and sequences related to them.

The first evidence for wandering DNA segments came from American geneticist Barbara McClintock's breeding experiments with calico corn (maize) in the 1940s and 1950s **(Figure 21.7)**. Tracking corn plants through many generations, McClintock analyzed changes in the color of corn kernels. The patterns she saw led her to propose that there were genetic elements capable of moving from other locations in the genome into the genes for kernel color, disrupting the genes and changing the kernel color. McClintock's hypothesis provoked great interest among her maize colleagues, but most other scientists thought the phenomenon she had observed might occur only in maize. Her careful work and insightful ideas were finally validated many years later when transposable elements were found in bacteria. In 1983, at the age of 81, McClintock received the Nobel Prize for her pioneering research.

Movement of Transposons and Retrotransposons

Eukaryotic transposable elements are of two types. The first type, **transposons**, move within a genome by means of a DNA intermediate. Transposons can move by a "cut-and-paste"

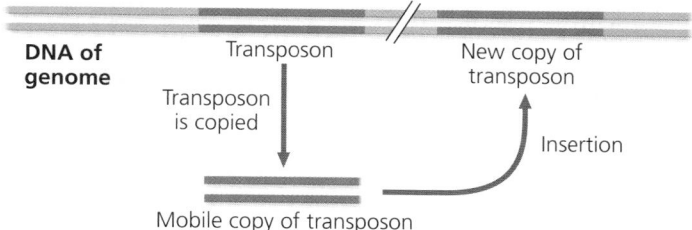

▼ Figure 21.8 Transposon movement. Movement of transposons by either the copy-and-paste mechanism (shown here) or the cut-and-paste mechanism involves a double-stranded DNA intermediate that is inserted into the genome.

VISUAL SKILLS *How would this figure differ if it showed the cut-and-paste mechanism?*

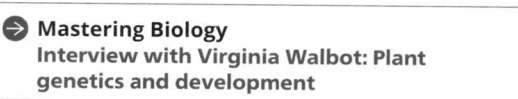

⊘ Mastering Biology
Interview with Virginia Walbot: Plant genetics and development

mechanism, which removes the element from the original site, or by a "copy-and-paste" mechanism, which leaves a copy behind **(Figure 21.8)**. Both mechanisms require an enzyme called *transposase*, which is generally encoded by the transposon.

Most transposable elements in eukaryotic genomes are of the second type, **retrotransposons**, which move by means of an RNA intermediate that is a transcript of the retrotransposon DNA. Thus, retrotransposons always leave a copy at the original site during transposition **(Figure 21.9)**. To insert at another site, the RNA intermediate is first converted back to DNA by reverse transcriptase, an enzyme encoded by the retrotransposon. (Reverse transcriptase is also encoded by retroviruses, as you learned in Concept 19.2. In fact, retroviruses may have evolved from retrotransposons, or vice versa.) Another cellular enzyme catalyzes insertion of the reverse-transcribed DNA at a new site.

▼ Figure 21.9 Retrotransposon movement. Movement begins with synthesis of a single-stranded RNA intermediate. The remaining steps are essentially identical to part of the retrovirus replicative cycle (see Figure 19.8).

◀ **Figure 21.7 The effect of transposable elements on corn kernel color.** Barbara McClintock first proposed the idea of mobile genetic elements after observing variegations in the color of the kernels on a corn cob (top right).

Sequences Related to Transposable Elements

Multiple copies of transposable elements and sequences related to them are scattered throughout eukaryotic genomes. A single unit is usually hundreds to thousands of base pairs long, and the dispersed copies are similar but usually not identical to each other. Some of these are transposable elements that can move; the enzymes required for this movement may be encoded by any transposable element, including the one that is moving. Others are related sequences that have lost the ability to move altogether. Transposable elements and related sequences make up 25–50% of most mammalian genomes (see Figure 21.6) and even higher percentages in amphibians and many plants. In fact, the very large size of some plant genomes is accounted for by extra transposable elements rather than by extra genes. For example, transposable elements make up 85% of the corn genome!

In humans and other primates, a large portion of transposable element–related DNA consists of a family of similar sequences called *Alu elements*. These sequences alone account for approximately 10% of the human genome. *Alu* elements are about 300 nucleotides long, much shorter than most functional transposable elements, and they do not code for any protein. However, many *Alu* elements are transcribed into RNA, and at least some of these RNAs are thought to help regulate gene expression.

An even larger percentage (17%) of the human genome is made up of a type of retrotransposon called *LINE-1*, or *L1*. These sequences are much longer than *Alu* elements—about 6,500 base pairs—and typically have a very low rate of transposition. However, researchers working with mice have discovered that transcription of L1 retrotransposons is crucial for the development of early (one- and two-cell stage) embryos. They have proposed that transcription of L1 retrotransposons may affect the chromatin structure in ways important for embryonic development.

Although some transposable elements encode proteins, these proteins do not carry out normal cellular functions. Therefore, transposable elements are usually included in the "noncoding" DNA category, along with other repetitive sequences.

Other Repetitive DNA, Including Simple Sequence DNA

Repetitive DNA that is not related to transposable elements has probably arisen from mistakes during DNA replication or recombination. Such DNA accounts for about 14% of the human genome (see Figure 21.6). About a third of this (5–6% of the human genome) consists of duplications of long stretches of DNA, with each unit ranging from 10,000 to 300,000 base pairs. These long segments seem to have been copied from one chromosomal location to another site on the same or a different chromosome and probably include some functional genes.

In contrast to scattered copies of long sequences, stretches of DNA known as **simple sequence DNA** contain many copies of tandemly repeated short sequences, as in the following example (showing one DNA strand only):

… GTTACGTTACGTTACGTTACGTTACGTTAC…

In this case, the repeated unit (GTTAC) consists of 5 nucleotides, but the number can range from 2 to 500. When the unit contains 2–5 nucleotides, the series of repeats is called a **short tandem repeat**, or **STR**; we discussed the use of STR analysis in preparing genetic profiles by use of PCR in Concept 20.4 (see Figure 20.24). The number of copies of the repeated unit can vary from site to site within a given genome. There could be as many as several hundred thousand repetitions of the GTTAC unit at one site, but only half that number at another. STR analysis is performed on sites selected because they have relatively few repeats. The repeat number varies from person to person, and since humans are diploid, each person has two alleles per repeat site; these can differ in repeat number. This diversity produces the variation represented in the genetic profiles that result from STR analysis.

Simple sequence DNA makes up 3% of the human genome, much of it located at chromosomal telomeres and centromeres, where it may play a structural role. The DNA at centromeres is essential for the separation of chromatids in cell division (see Concept 12.2) and, along with simple sequence DNA located elsewhere, may also help organize the chromatin within the interphase nucleus. The simple sequence DNA located at telomeres binds proteins that protect chromosomal ends from degradation and from joining to other chromosomes.

Short repetitive sequences like those described here provide a challenge for whole-genome shotgun sequencing because the presence of many short repeats hinders accurate reassembly of fragment sequences by computers. Regions of simple sequence DNA account for much of the uncertainty present in estimates of whole-genome sizes and are the reason some sequences are considered "permanent drafts."

Genes and Multigene Families

Now, let's take a look at genes. Recall that DNA sequences that code for proteins or give rise to tRNA or rRNA make up only 1.5% of the human genome (see Figure 21.6). If we include introns and regulatory sequences, the total amount of DNA that is gene-related—coding and noncoding—constitutes about 25% of the human genome. Put another way, only about 6% (1.5% out of 25%) of the length of the average gene is represented in the final gene product.

Many eukaryotic genes are present as unique sequences, with only one copy per haploid set of chromosomes. But in the human genome and the genomes of many other animals and plants, these unique genes make up less than half of the total gene-related DNA. The rest occur in **multigene families**, collections of two or more identical or very similar genes.

In multigene families that consist of *identical* DNA sequences, those sequences are usually clustered tandemly and, with the notable exception of the genes for histone proteins, have RNAs as their final products. An example is the family of identical DNA sequences that each include the genes for the three largest rRNA molecules **(Figure 21.10a)**. These rRNA molecules are transcribed from a single transcription unit that is repeated tandemly hundreds to thousands of times in one or several clusters in the genome of a multicellular eukaryote. The many copies of this rRNA transcription unit help cells to quickly make the millions of ribosomes needed for active protein synthesis. The primary transcript is cleaved to yield three rRNA molecules, which combine with proteins and one other kind of rRNA (5S rRNA) to form ribosomal subunits.

The classic examples of multigene families of *nonidentical* genes are two related families of genes that encode globins, a group of proteins that include the α and β polypeptide subunits of hemoglobin. One family, located on chromosome 16 in humans, encodes various forms of α-globin; the other, on chromosome 11, encodes forms of β-globin **(Figure 21.10b)**. The different forms of each globin subunit are expressed at different times in development, allowing hemoglobin to function effectively in the changing environment of the developing animal. In humans, for example, the embryonic and fetal forms of hemoglobin have a higher affinity for oxygen than the adult forms, ensuring the efficient transfer of oxygen from mother to fetus. Also found in the globin gene family clusters are several pseudogenes.

In Concept 21.5, we'll consider the evolution of these two globin gene families as we explore how arrangements of genes provide insight into the evolution of genomes. We'll also examine some processes that have shaped the genomes of different species over evolutionary time.

CONCEPT CHECK 21.4

1. Discuss the characteristics of mammalian genomes that make them larger than prokaryotic genomes.

2. **VISUAL SKILLS** Which of the three mechanisms described in Figures 21.8 and 21.9 result(s) in a copy remaining at the original site as well as a copy appearing in a new location?

3. Contrast the organizations of the rRNA gene family and the globin gene families. For each, explain how the existence of a family of genes benefits the organism.

4. **MAKE CONNECTIONS** Assign each DNA segment at the top of Figure 18.9 to a sector in the pie chart in Figure 21.6.

For suggested answers, see Appendix A.

▼ **Figure 21.10 Gene families.**

(a) Part of the ribosomal RNA gene family. The TEM at the top shows three of the hundreds of copies of rRNA transcription units in the rRNA gene family of a salamander genome. Each "feather" corresponds to a single unit being transcribed by about 100 molecules of RNA polymerase (dark dots along the DNA), moving left to right (red arrow). The growing RNA transcripts extend from the DNA, accounting for the feather-like appearance. In the diagram of a transcription unit below the TEM, the genes for three types of rRNA (darker blue) are adjacent to regions that are transcribed but later removed (medium blue). A single transcript is processed to yield one of each of the three rRNAs (red), key components of the ribosome.

(b) The human α-globin and β-globin gene families. Adult hemoglobin is composed of two α-globin and two β-globin polypeptide subunits, as shown in the molecular model. The genes (darker blue) encoding α- and β-globins are found in two families, organized as shown here. The noncoding DNA (light blue) separating the functional genes within each family includes pseudogenes (ψ; gold), versions of the functional genes that no longer encode functional polypeptides. Genes and pseudogenes are named with Greek letters, as you have seen previously for the α- and β-globins. Some genes are expressed only in the embryo or fetus.

VISUAL SKILLS *In the TEM at the top of part (a), how could you determine the direction of transcription if it weren't indicated by the red arrow?*

Duplication, rearrangement, and mutation of DNA contribute to genome evolution

EVOLUTION Now that we have explored the makeup of the human genome, let's see what its composition reveals about how the genome evolved. The basis of change at the genomic level is mutation, which underlies much of genome evolution. It seems likely that the earliest forms of life had a minimal number of genes—those necessary for survival and reproduction. If this were indeed the case, one aspect of evolution must have been an increase in the size of the genome, with the extra genetic material providing the raw material for gene diversification. In this section, we'll first look at how extra copies of all or part of a genome can arise and then consider subsequent processes that can lead to the evolution of proteins (or RNA products) with slightly different or entirely new functions.

Duplication of Entire Chromosome Sets

An accident in meiosis, such as failure to separate homologs during meiosis I, can result in one or more extra sets of chromosomes, a condition known as polyploidy (see Concept 15.4). Although such accidents would most often be lethal, in rare cases they could facilitate the evolution of genes. In a polyploid organism, one set of genes can provide essential functions for the organism. The genes in the one or more extra sets can diverge by accumulating mutations; these variations may persist if the organism carrying them survives and reproduces. In this way, genes with novel functions can evolve. As long as one copy of an essential gene is expressed, the divergence of another copy can lead to its encoded protein acting in a novel way, thereby changing the organism's phenotype.

The outcome of this accumulation of mutations may eventually be the branching off of a new species. While polyploidy is rare among animals, it is relatively common among plants, especially flowering plants. Some botanists estimate that as many as 80% of the plant species that are alive today show evidence of polyploidy having occurred among their ancestral species. You'll learn more about the details of how polyploidy leads to plant speciation in Concept 24.2.

Alterations of Chromosome Structure

With the recent explosion in genomic sequence information, we can now compare the chromosomal organizations of many different species in detail. This information allows us to make inferences about the evolutionary processes that shape chromosomes and may drive speciation. For example, scientists have long known that sometime in the last 6 million years, when the ancestors of humans and chimpanzees diverged as species, the fusion of two ancestral chromosomes

▼ **Figure 21.11 Human and chimpanzee chromosomes.** The positions of telomere-like and centromere-like sequences on human chromosome 2 (left) match those of telomeres on chimpanzee chromosomes 12 and 13 and the centromere on chimpanzee chromosome 13 (right). This suggests that chromosomes 12 and 13 in a human ancestor fused end to end to form human chromosome 2. The centromere from ancestral chromosome 12 remained functional on human chromosome 2, while the one from ancestral chromosome 13 did not.

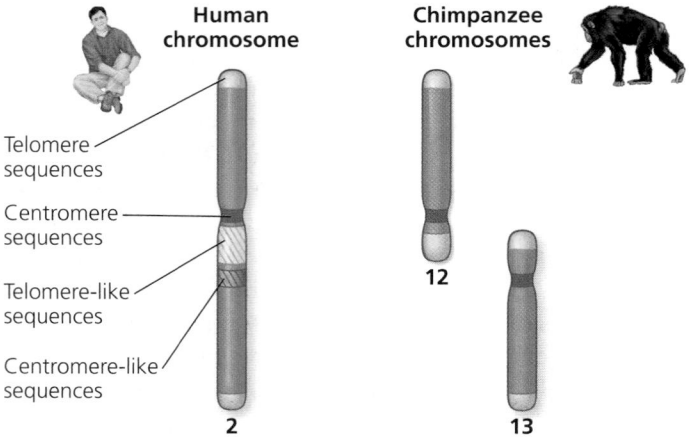

in the human line led to different haploid numbers for humans ($n = 23$) and chimpanzees ($n = 24$). The banding patterns in stained chromosomes suggested that the ancestral versions of current chimpanzee chromosomes 12 and 13 fused end to end, forming chromosome 2 in an ancestor of the human lineage **(Figure 21.11)**.

How do we know a chromosome didn't just split into two in an ancestor of chimps? A comparison of chromosomes in other great apes—gorillas, chimps, bonobos, and orangutans, our closest relatives—shows that those species have 24 chromosomes, so the simplest conclusion is that the last common ancestor of all great apes had 24 chromosomes. (You'll learn more about this type of reasoning in Concept 26.3.) Sequencing and analysis of human chromosome 2 during the Human Genome Project revealed sequences for telomeres and an extra, unused centromere in the middle of it, among other very strong supporting evidence for the model described previously (see Figure 21.11).

In another study of broader scope, researchers compared the DNA sequence of each human chromosome with the whole-genome sequence of the mouse **(Figure 21.12)**. One part of their study showed that large blocks of genes on human chromosome 16 are found on four mouse chromosomes, indicating that the genes in each block stayed together in both the mouse and the human lineages during their divergent evolution from a common ancestor.

Performing the same comparison of chromosomes of humans and six other mammalian species allowed the researchers to reconstruct the evolutionary history of chromosomal rearrangements in these eight species. They found many duplications and inversions of large portions of chromosomes, the result of errors during meiotic recombination in which the DNA was broken and rejoined incorrectly. The rate of

▼ **Figure 21.12 Human and mouse chromosomes.** Here, we can see that DNA sequences very similar to large blocks of human chromosome 16 (colored areas in this diagram) are found on mouse chromosomes 7, 8, 16, and 17. This finding suggests that the DNA sequence in each block has stayed together in the mouse and human lineages since the time they diverged from a common ancestor.

Human chromosome   **Mouse chromosomes**

16   7   8   16   17

→ **Mastering Biology**
Interview with Eric Lander:
Exploring the human genome

these events seems to have begun accelerating about 100 million years ago, around 35 million years before large dinosaurs became extinct and the number of mammalian species began rapidly increasing. The apparent coincidence is interesting because chromosomal rearrangements are thought to contribute to the generation of new species. Although two individuals with different arrangements could still mate and produce offspring, the offspring would have two nonequivalent sets of chromosomes, making meiosis inefficient or even impossible. Thus, chromosomal rearrangements would lead to two populations that could not successfully mate with each other, a step on the way to their becoming two separate species. (You'll learn more about this in Concept 24.2.)

The same study also unearthed a pattern with medical relevance. Analysis of the chromosomal breakage points associated with the rearrangements showed that specific sites were used over and over again. A number of these recombination "hot spots" correspond to locations of chromosomal rearrangements within the human genome that are associated with congenital diseases (see Concept 15.4).

Duplication and Divergence of Gene-Sized Regions of DNA

Errors during meiosis can also lead to the duplication of chromosomal regions that are smaller than the ones we've just discussed, including segments the length of individual genes. Unequal crossing over during prophase I of meiosis, for instance, can result in one chromosome with a deletion and another with a duplication of a particular gene or genes. Transposable elements can provide homologous sites where nonsister chromatids can cross over, even when other chromatid regions are not correctly aligned **(Figure 21.13)**.

Also, slippage can occur during DNA replication, such that the template shifts with respect to the new complementary strand, and a part of the template strand is either skipped by the replication machinery or used twice as a template. As a

▼ **Figure 21.13 Gene duplication due to unequal crossing over.** One mechanism by which a gene (or other DNA segment) can be duplicated is recombination during meiosis between copies of a transposable element (yellow) flanking the gene (blue). Such recombination between misaligned nonsister chromatids of homologous chromosomes produces one chromatid with two copies of the gene and one chromatid with no copy. (Genes and transposable elements are shown only in the region of interest.)

MAKE CONNECTIONS *Examine how crossing over occurs in Figure 13.9. In the middle panel above, draw a line through the portions that result in the upper chromatid in the bottom panel. Use a different color to do the same for the other chromatid.*

result, a segment of DNA is deleted or duplicated. It is easy to imagine how such errors could occur in regions of repeats. (See the question in Figure 21.13.) The variable number of repeated units of simple sequence DNA at a given site, used for STR analysis, is probably due to errors like these. Evidence that unequal crossing over and template slippage during DNA replication lead to duplication of genes is found in the existence of multigene families, such as the globin family.

Evolution of Genes with Related Functions: The Human Globin Genes

In Figure 21.10b, you saw the organization of the α-globin and β-globin gene families as they exist in the human genome today. Now, let's consider how events such as duplications can lead to the evolution of genes with related functions like the globin genes. A comparison of gene sequences within a multigene family can suggest the order in which the genes arose. Re-creating the evolutionary history of the globin genes using this approach indicates that they all evolved from one common ancestral globin gene that underwent duplication and divergence into the α-globin and β-globin ancestral genes about 450–500 million years ago

▼ Figure 21.14 A proposed model for the sequence of events in the evolution of the human α-globin and β-globin gene families from a single ancestral globin gene.

Ancestral globin gene

Duplication of ancestral gene

Mutation in both copies

Transposition to different chromosomes

Further duplications and mutations

Evolutionary time

α-Globin gene family on chromosome 16

β-Globin gene family on chromosome 11

? *The gold elements are pseudogenes. Explain how they could have arisen after gene duplication.*

(Figure 21.14). Each of these genes was later duplicated several times, and the copies then diverged from each other in sequence, yielding the current family members. In fact, the common ancestral globin gene also gave rise to the oxygen-binding muscle protein myoglobin and to the plant protein leghemoglobin. The latter two proteins function as monomers, and their genes are included in a "globin superfamily."

After the duplication events, the differences between the genes in the globin families undoubtedly arose from mutations that accumulated in the gene copies over many generations. The current model is that the necessary function provided by an α-globin protein, for example, was fulfilled by one gene, while other copies of the α-globin gene accumulated random mutations. Many mutations may have had an adverse effect on the organism, and others may have had no effect. However, a few mutations must have altered the function of the protein product in a way that benefitted the organism at a particular life stage without substantially changing the protein's oxygen-carrying function. Presumably, natural selection acted on these altered genes, maintaining them in the population.

In the **Scientific Skills Exercise**, you can compare amino acid sequences of the globin family proteins and see how such comparisons were used to generate the model for globin gene evolution shown in Figure 21.14. The existence of several pseudogenes among the functional globin genes provides additional evidence for this model: Random mutations in these "genes" over evolutionary time have destroyed their function.

Evolution of Genes with Novel Functions

In the evolution of the globin gene families, gene duplication and subsequent divergence produced family members whose protein products performed similar functions

(oxygen transport). Alternatively, one copy of a duplicated gene can undergo alterations that lead to a completely new function for the protein product. The genes for lysozyme and α-lactalbumin are a good example.

Lysozyme is an enzyme that helps protect animals against bacterial infection by hydrolyzing bacterial cell walls (see Visualizing Figure 5.16); α-lactalbumin is a nonenzymatic protein that plays a role in milk production in mammals. The two proteins are quite similar in their amino acid sequences and three-dimensional structures **(Figure 21.15)**. Both genes are found in mammals, but only the lysozyme gene is present in birds. These findings suggest that at some time after the lineages leading to mammals and birds had separated, the lysozyme gene was duplicated in the mammalian lineage but not in the avian lineage. Subsequently, one copy of the duplicated lysozyme gene evolved into a gene encoding α-lactalbumin, a protein with a completely new function associated with a key characteristic of mammals—milk production. In one study, evolutionary biologists searched vertebrate genomes for genes with similar sequences. There appear to be at least eight members of the lysozyme family, distributed widely among mammalian species. The functions of all the encoded gene products are not yet known, but it will be exciting to discover whether they are as different as the functions of lysozyme and α-lactalbumin.

Besides the duplication and divergence of whole genes, rearrangement of existing DNA sequences within genes has also contributed to genome evolution. The presence of introns may have promoted the evolution of new proteins by facilitating the duplication or shuffling of exons, as we'll see next.

Rearrangements of Parts of Genes: Exon Duplication and Exon Shuffling

Recall from Concept 17.3 that an exon often codes for a protein domain, a distinct structural and functional region of a protein molecule, such as the WD40 domain in Figure 21.3. We've already seen that unequal crossing over during meiosis can lead to duplication of a gene on one chromosome and its loss from the homologous chromosome (see Figure 21.13). By a similar process, a particular exon within a gene could be duplicated on one chromosome and deleted from the other. The gene with the duplicated exon would code for a protein containing a second copy of the encoded domain. This change in the protein's structure might

▼ **Figure 21.15 Comparison of lysozyme and α-lactalbumin proteins.** Computer-generated ribbon models of the similar structures of **(a)** lysozyme and **(b)** α-lactalbumin are shown, along with **(c)** a comparison of the amino acid sequences of the two proteins. Single-letter amino acid codes are used (see Figure 5.14). Identical amino acids are highlighted in yellow, and dashes indicate gaps in one sequence that have been introduced by the software to optimize the alignment.

MAKE CONNECTIONS *Even though two amino acids are not identical, they may be structurally and chemically similar and therefore behave similarly. Using Figure 5.14 as a reference, examine the nonidentical amino acids in positions 1–30 and note cases where the amino acids in the two sequences are similarly acidic or basic.*

(a) Lysozyme (b) α-lactalbumin

Lysozyme	1	KVFERCELARTLKRLGMDGYRGISLANWMCLAKWESGYNTRATN
α-lactalbumin	1	KQFTKCELSQLLK--DIDGYGGIALPELICTMFHTSGYDTQAIV

Lysozyme	51	STDYGIFQINSRYWCNDGKTPGAVNACHLSCSALLQDNIADAVA
α-lactalbumin	51	STEYGLFQISNKLWCKSSQVPQSRNICDISCDKFLDDDITDDIM

Lysozyme	101	RDPQGIRAWVAWRNRCQ-NRDVRQYVQGCGV
α-lactalbumin	101	D-IKGIDYWLAHKALCT--EKLEQWLCEKL-

(c) Amino acid sequence alignments of lysozyme and α-lactalbumin

augment its function by increasing its stability, enhancing its ability to bind a particular ligand, or altering some other property. Quite a few protein-coding genes have multiple copies of related exons, which presumably arose by duplication and then diverged. The gene encoding the extracellular matrix protein collagen is a good example. Collagen is a structural protein (see Figure 5.18) with a highly repetitive amino acid sequence, which reflects the repetitive pattern of exons in the collagen gene.

As an alternative possibility, we can imagine the occasional mixing and matching of different exons either within a gene or between two different (nonallelic) genes owing to errors in meiotic recombination. This process, termed *exon shuffling*, could lead to new proteins with novel combinations of functions. As an example, let's consider the gene for tissue plasminogen activator (TPA). The TPA protein is an extracellular protein that helps control blood clotting. It has four domains of three types, each encoded by an exon, and one of those exons is present in two copies. Because each type of exon is also found in other proteins, the current version of the gene for TPA is thought to have arisen by several instances of exon shuffling during errors in meiotic recombination and subsequent duplication **(Figure 21.16)**.

▼ **Figure 21.16 Evolution of a new gene by exon shuffling.** Meiotic errors could have moved exons, each encoding a particular domain, from ancestral forms of the genes for epidermal growth factor, fibronectin, and plasminogen (left) into the evolving gene for tissue plasminogen activator, TPA (right). Subsequent duplication of the "kringle" exon (K) from the plasminogen gene after its movement into the TPA gene could account for the two copies of this exon in the TPA gene existing today.

Epidermal growth factor gene with multiple EGF exons

Fibronectin gene with multiple "finger" exons

Plasminogen gene with a "kringle" exon

Exon shuffling

Exon duplication

Exon shuffling

Portions of ancestral genes   **TPA gene as it exists today**

VISUAL SKILLS *Looking at Figure 21.13, describe the steps by which transposable elements within introns might have facilitated the exon shuffling shown here.*

Reading an Amino Acid Sequence Identity Table

How Have Amino Acid Sequences of Human Globin Genes Diverged During Their Evolution? To build a model of the evolutionary history of the globin genes (see Figure 21.14), researchers compared the amino acid sequences of the polypeptides they encode. In this exercise, you will analyze comparisons of the amino acid sequences of the globin polypeptides to shed light on their evolutionary relationships.

How the Experiment Was Done Scientists obtained the DNA sequences for each of the eight globin genes and "translated" them into amino acid sequences. They then used a computer program to align the sequences (with dashes indicating gaps in one sequence) and calculate a percent identity value for each pair of globins. The percent identity reflects the number of positions with identical amino acids relative to the total number of amino acids in a globin polypeptide. The data were displayed in a table to show the pairwise comparisons.

Data from the Experiment The following table shows an example of a pairwise alignment—that of the α_1-globin (alpha-1 globin) and ζ-globin (zeta globin) amino acid sequences—using the standard single-letter symbols for amino acids. To the left of each line of amino acid sequence is the number of the first amino acid in that line. The percent identity value for the α_1- and ζ-globin amino acid sequences was calculated by counting the number of

matching amino acids (86, highlighted in yellow), dividing by the total number of amino acid positions (143), and then multiplying by 100. This resulted in a 60% identity value for the α_1-ζ pair, as shown in the amino acid identity table at the bottom of the page. The values for other globin pairs were calculated in the same way.

▲ Hemoglobin

INTERPRET THE DATA

1. Note that in the amino acid identity table, the data are arranged so each globin pair can be compared. **(a)** Some cells in the table have dashed lines. Given the pairs that are being compared for these cells, what percent identity value is implied by the dashed lines? **(b)** Notice that the cells in the lower left half of the table are blank. Using the information already provided in the table, fill in the missing values. Why does it make sense that these cells were left blank in the table?

2. The earlier that two genes arose from a duplicated gene, the more their nucleotide sequences can have diverged, which may result in amino acid differences in the protein products. **(a)** Based on that premise, identify which two genes are most divergent from each other. What is the percent amino acid identity between their polypeptides? **(b)** Using the same approach, identify which two globin genes are the most recently duplicated. What is the percent identity between them?

3. The model of globin gene evolution shown in Figure 21.14 suggests that an ancestral gene duplicated and mutated to become α- and β-globin genes, and then each one was further duplicated and mutated. What features of the data set support the model?

4. Make an ordered list of all the percent identity values from the table, starting with 100% at the top. Next to each number write the globin pair(s) with that percent identity value. Use one color for the globins from the α family and a different color for the globins from the β family. **(a)** Compare the order of pairs on your list with their positions in the model shown in Figure 21.14. Does the order of pairs describe the same relative "closeness" of globin family members seen in the model? **(b)** Compare the percent identity values for pairs within the α or β family to the values for between-family pairs.

→ **Instructors:** A version of this Scientific Skills Exercise can be assigned in **Mastering Biology**.

Further Reading R. C. Hardison, Globin genes on the move, *Journal of Biology* 7:35.1–35.5 (2008).

Globin	Alignment of Globin Amino Acid Sequences
α_1	1 MVLSPADKTNVKAAWGKVGAHAGEYGAEAL
ζ	1 MSLTKTERTIIVSMWAKISTQADTIGTETL
α_1	31 ERMFLSFPTTKTYFPHFDLSH-GSAQVKGH
ζ	31 ERLFLSHPQTKTYFPHFDL-HPGSAQLRAH
α_1	61 GKKVADALTNAVAHVDDMPNALSALSDLHA
ζ	61 GSKVVAAVGDAVKSIDDIGGALSKLSELHA
α_1	91 HKLRVDPVNFKLLSHCLLVTLAAHLPAEFT
ζ	91 YILRVDPVNFKLLSHCLLVTLAARFPADFT
α_1	121 PAVHASLDKFLASVSTVLTSKYR
ζ	121 AEAHAAWDKFLSVVSSVLTEKYR

Amino Acid Identity Table								
		α Family			**β Family**			
		α_1 (alpha 1)	α_2 (alpha 2)	ζ (zeta)	β (beta)	δ (delta)	ϵ (epsilon)	A_γ (gamma A)	G_γ (gamma G)
α Family	α_1	-----	100	60	45	44	39	42	42
	α_2		-----	60	45	44	39	42	42
	ζ			-----	38	40	41	41	41
β Family	β				-----	93	76	73	73
	δ					-----	73	71	72
	ϵ						-----	80	80
	A_γ							-----	99
	G_γ								-----

Compiled using data from the National Center for Biotechnology Information (NCBI).

How Transposable Elements Contribute to Genome Evolution

The persistence of transposable elements as a large fraction of some eukaryotic genomes is consistent with the idea that they play an important role in shaping a genome over evolutionary time. These elements can contribute to the evolution of the genome in several ways. They can promote recombination, disrupt cellular genes or control elements, and carry entire genes or individual exons to new locations.

Transposable elements of similar sequence scattered throughout the genome facilitate recombination between different (nonhomologous) chromosomes by providing homologous regions for crossing over (see Figure 21.13). Most such recombination events are probably detrimental, causing chromosomal translocations and other changes in the genome that may be lethal to the organism. But over the course of evolutionary time, an occasional recombination event of this sort may be advantageous to the organism. (For the change to be heritable, of course, it must happen in a cell that will give rise to a gamete.)

The movement of a transposable element can have a variety of consequences. For instance, a transposable element that "jumps" into a protein-coding sequence will prevent the production of a normal transcript of the gene. (Introns provide a sort of "safety zone" that does not affect the transcript because the transposable element will be spliced out—unless it affects the splicing process.) If a transposable element inserts within a regulatory sequence, the transposition may lead to increased or decreased production of one or more proteins. Transposition caused both types of effects on the genes coding for pigment-synthesizing enzymes in McClintock's corn kernels. Again, while such changes are usually harmful, in the long run some may provide a survival advantage. A possible example was mentioned earlier: At least some of the *Alu* transposable elements in the human genome are known to produce RNAs that regulate expression of human genes.

During transposition, a transposable element may carry along a gene or even a group of genes to a new position in the genome. This occurrence probably accounts for the location of the α-globin and β-globin gene families on different human chromosomes, as well as the dispersion of the genes of certain other gene families. By a similar tag-along process, an exon from one gene may be inserted into another gene in a mechanism similar to that of exon shuffling during recombination. For example, an exon may be inserted by transposition into the intron of a protein-coding gene. If the inserted exon is retained in the RNA transcript during RNA splicing, the protein that is synthesized will have an additional domain, which may confer a new function on the protein.

Most often, the processes discussed in this section produce harmful effects, which may be lethal, or have no effect at all. In a few cases, however, small heritable changes that are beneficial may occur. Over many generations, the resulting genetic diversity provides valuable raw material for natural selection. Diversification of genes and their products is an important factor in the evolution of new species. Thus, the accumulation of changes in the genome of each species provides a record of its evolutionary history. To read this record, we must be able to identify genomic changes. Comparing the genomes of different species allows us to do that, increasing our understanding of how genomes evolve. You will learn more about these topics next.

CONCEPT CHECK 21.5

1. Describe three examples of errors in cellular processes that lead to DNA duplications.
2. Explain how multiple exons might have arisen in the ancestral EGF and fibronectin genes shown in Figure 21.16 (left).
3. What are three ways that transposable elements are thought to contribute to genome evolution?
4. **WHAT IF?** In 2005, Icelandic scientists reported finding a large chromosomal inversion present in 20% of northern Europeans, and they noted that Icelandic women with this inversion had significantly more children than women without it. What would you expect to happen to the frequency of this inversion in the Icelandic population in future generations?

For suggested answers, see Appendix A.

CONCEPT 21.6

Comparing genome sequences provides clues to evolution and development

EVOLUTION In the last 30 years, we have seen rapid advances in genome sequencing and data collection, new techniques for assessing gene activity across the whole genome and for editing a gene sequence in a specific way in living cells, and refined approaches for understanding how genes and their products work together in complex systems. In the field of biology, we are truly poised on the brink of a new world.

Comparing genome sequences from different species reveals a lot about the evolutionary history of life, from very ancient to more recent. Similarly, comparative studies of the genetic programs that direct embryonic development in different species are uncovering the mechanisms that generated the great diversity of life-forms present today. We'll now look at what has been learned from these two approaches.

Comparing Genomes

The more similar in sequence the genes and genomes of two species are, the less time has passed for mutations and other changes to accumulate, and therefore the more closely related those species are in their evolutionary history. Comparing genomes of closely related species sheds light on more recent evolutionary events, whereas comparing genomes of very distantly related species helps us understand ancient evolutionary history. In either case, learning about characteristics that are shared or divergent between groups enhances our picture of the evolution of organisms and biological processes. Evolutionary relationships between species can be represented by a diagram in the form of a tree (often turned sideways), where each branch point marks the divergence of two lineages (see Figure 1.20). **Figure 21.17** shows the evolutionary relationships of some groups and species we'll now examine.

Comparing Distantly Related Species

Determining which genes have remained similar—that is, are *highly conserved*—in distantly related species can help clarify evolutionary relationships among species that diverged from each other long ago. Indeed, comparisons of the specific gene sequences of bacteria, archaea, and eukaryotes indicate that these three groups diverged between 2 and 4 billion years ago and strongly support the theory that they are the fundamental domains of life (see Figure 21.17).

In addition to their value in evolutionary biology, comparative genomic studies confirm the relevance of research on model organisms to our understanding of biology in general and human biology in particular. Very ancient genes can still be surprisingly similar in disparate species. One experimental study tested the ability of the human version of each of 414 important yeast genes to function equivalently in yeast cells. Remarkably, the researchers concluded that 47% of these

▼ **Figure 21.17 Evolutionary relationships of the three domains of life.** The tree diagram at the top shows the ancient divergence of bacteria, archaea, and eukaryotes. A portion of the eukaryote lineage is expanded in the inset to show the more recent divergence of three mammalian species discussed in this chapter.

yeast genes could be replaced by the corresponding human gene. This striking result underscores the common origin of yeasts and humans—two distantly related species.

Comparing Closely Related Species

The genomes of two closely related species are likely to be organized similarly because of their relatively recent divergence. Their long shared history also means that only a small number of gene differences are found when their genomes are compared. These genetic differences can thus be more easily correlated with phenotypic differences between the two species. An exciting application of this type of analysis is seen as researchers compare the human genome with the genomes of the chimpanzee, mouse, rat, and other mammals. Identifying the genes shared by all of these species but not by nonmammals gives us clues about what it takes to make a mammal, while finding the genes shared by chimpanzees and humans but not by rodents tells us something about primates. And, of course, comparing the human genome with that of the chimpanzee helps us answer a tantalizing question: What genomic information defines a human or a chimpanzee?

An analysis of the overall composition of the human and chimpanzee genomes, which are thought to have diverged only about 6 million years ago (see Figure 21.17), reveals some general differences. Considering single nucleotide substitutions, the two genomes differ by only 1.2%. When researchers looked at longer stretches of DNA, however, they were surprised to find a further 2.7% difference due to insertions or deletions of larger regions in the genome of one or the other species; many of the insertions were duplications or other repetitive DNA. In fact, a third of the human duplications are not present in the chimpanzee genome, and some of these duplications contain regions associated with human diseases. More *Alu* elements have undergone transposition in the human genome, leading to 7,000 human-specific elements compared with 2,300 chimpanzee-specific elements. The chimpanzee genome, in turn, contains many copies of a retroviral provirus not present in humans. All of these observations provide clues to the forces that might have swept the two genomes along different paths, but we don't have a complete picture yet.

Along with chimpanzees, bonobos are the other African ape species that are the closest living relatives to humans. The sequencing of the bonobo genome revealed that in some regions, human sequences were more closely related to either chimpanzee or bonobo sequences than chimpanzee or bonobo sequences were to each other. Such a fine-grained comparison of three closely related species allows even more detail to be worked out in reconstructing their related evolutionary history.

We also don't know how the genetic differences revealed by genome sequencing might account for the distinct characteristics of each species. To discover the basis for the phenotypic differences between chimpanzees and humans, biologists are studying specific genes and types of genes that differ between the two species and comparing them with their counterparts in other mammals. This approach has revealed a number of genes that are apparently changing (evolving) faster in the human than in either the chimpanzee or the mouse. Among them are genes involved in defense against malaria and tuberculosis as well as at least one gene that regulates brain size. When genes are classified by function, the genes that seem to be evolving the fastest are those that code for transcription factors. This discovery makes sense because transcription factors regulate gene expression and thus play a key role in orchestrating the overall genetic program.

One transcription factor of interest, whose gene is called *FOXP2*, might be involved with the acquisition of speech in humans. Attention was first drawn to *FOXP2* in 2002 when a research publication reported that this gene was evolving faster in humans (*Homo sapiens*) than in other primates, based in part on two amino acids found only in the human protein sequence. The researchers proposed that this was a human-specific version of the gene that resulted from strong natural selection and might have conferred on humans the ability to speak.

In 2014, a high-quality sequence of the Neanderthal (*Homo neanderthalensis*) genome was achieved from a small amount of preserved genomic DNA, and shortly thereafter the DNA of another member of the species *Homo*, the Denisovan, was sequenced as well (see Concept 34.7). It turns out that some DNA sequences from both species encode the same two amino acids, so the two amino acids are not human-specific. In a paper published in 2018, researchers analyzed genomic sequences from a larger and more diverse population of modern humans. In this more robust study, the researchers found no evidence for strong natural selection for those two amino acids in the human lineage during the time frame relevant to language acquisition.

However, the *FOXP2* gene is still interesting in its own right. Several lines of evidence suggest that the *FOXP2* gene product regulates genes that function in vocalization in vertebrates. First, mutations in this gene can produce severe speech and language impairment in humans. Second, the *FOXP2* gene is expressed in the brains of zebra finches and canaries at the time when these songbirds are learning their songs. And third, perhaps the strongest evidence comes from a "knockout" experiment in which researchers disrupted the *FOXP2* gene in mice and analyzed the resulting phenotype **(Figure 21.18)**. The homozygous mutant mice had malformed brains and failed to emit normal ultrasonic vocalizations; mice with one faulty copy of the gene also showed significant problems with vocalization. These results support the idea that the *FOXP2* gene product turns on genes involved in vocalization.

Expanding on this analysis, another research group replaced the *FOXP2* gene in mice with a "humanized" copy coding for the human versions of the FOXP2 protein, with the two amino acids that differ between human and chimpanzee sequences. The researchers reported that although the mice with the human *FOXP2* gene were generally healthy,

they had subtly different vocalizations and showed changes in brain cells in circuits associated with speech in human brains. While the *FOXP2* gene clearly affects vocalization and speech, the precise mechanism by which it does this is still to be determined.

The *FOXP2* story is an excellent example of how different approaches can complement each other in uncovering biological phenomena of widespread importance. The *FOXP2* experiments used mice as a model for humans because it would be unethical (as well as impractical) to carry out such experiments in humans. Mice and humans, which diverged about 65.5 million years ago (see Figure 21.17), share about 85% of their genes. This genetic similarity can be exploited in studying human genetic disorders. If researchers know the organ or tissue that is affected by a particular genetic disorder, they can look for genes that are expressed in these locations in mice.

Even though more distantly related to humans, fruit flies have also been a useful model species for study of such human disorders as Parkinson's disease and alcoholism, while nematodes (soil worms) have yielded a wealth of information about aging. Further research efforts are under way to extend genomic studies to many more species, including neglected species from diverse branches of the tree of life. These studies will advance our understanding of evolution, of course, as well as all aspects of biology, from human health to ecology.

Comparing Genomes Within a Species

Another exciting consequence of our ability to analyze genomes is our growing understanding of the spectrum of genetic variation in humans. Because the history of the human species is so short—probably about 200,000 years—the amount of DNA variation among humans is small compared to that of many other species. Much of our diversity seems to be in the form of single nucleotide polymorphisms (SNPs). SNPs are single base-pair sites where genetic variation is found in at least 1% of the population (see Concept 20.2); they are usually detected by DNA sequencing. In the human genome, SNPs occur on average about once in 100–300 base pairs. Scientists have already identified the locations of several million SNP sites in the human genome and continue to find additional SNPs. These are stored in databases around the world, one of which is run by the National Center for Biotechnology Information (NCBI) and can be accessed at http://www.ncbi.nlm.nih.gov/SNP/.

In the course of this search, they have also found other variations—including chromosomal regions with inversions, deletions, and duplications. The most surprising discovery has been the widespread occurrence of *copy-number variants* (*CNVs*), loci where some individuals have one or multiple copies of a particular gene or genetic region rather than the standard two copies (one on each homolog). CNVs result from regions of the genome being duplicated or deleted inconsistently within the population. One study of 40 people found more than 8,000 CNVs involving 13% of the genes in

What is the function of a gene (*FOXP2*) that may be involved in language acquisition?

Experiment Several lines of evidence support a role for the *FOXP2* gene in the development of speech and language in humans and of vocalization in other vertebrates. In 2005, Joseph Buxbaum and collaborators at the Mount Sinai School of Medicine and several other institutions tested the function of *FOXP2*. They used the mouse, a model organism in which genes can be easily knocked out, as a representative vertebrate that vocalizes: Mice produce ultrasonic squeaks (whistles) to communicate stress. The researchers used genetic engineering to produce mice in which one or both copies of *FOXP2* were disrupted.

| Wild type: two normal copies of *FOXP2* | Heterozygote: one copy of *FOXP2* disrupted | Homozygote: both copies of *FOXP2* disrupted |

They then compared the phenotypes of these mice. Two of the characters they examined are included here: brain anatomy and vocalization.

Experiment 1: Researchers cut thin sections of brain and stained them with reagents that allow visualization of brain anatomy in a UV fluorescence microscope.

Experiment 2: To induce stress, researchers separated each newborn mouse pup from its mother and recorded the number of ultrasonic whistles produced by the pup.

Results

Experiment 1 Results: Disruption of both copies of *FOXP2* led to brain abnormalities in which the cells were disorganized. Phenotypic effects on the brain of heterozygotes, with one disrupted copy, were less severe. (Each color in the micrographs below reveals a different cell or tissue type.)

Experiment 2 Results: Disruption of both copies of *FOXP2* led to an absence of ultrasonic vocalization in response to stress. The effect on vocalization in the heterozygote was also extreme.

Wild type

Heterozygote

Homozygote

Conclusion *FOXP2* plays a significant role in the development of functional communication systems in mice. The results augment evidence from studies of birds and humans, supporting the hypothesis that *FOXP2* may act similarly in diverse organisms.

Data from W. Shu et al., Altered ultrasonic vocalization in mice with a disruption in the *Foxp2* gene, *Proceedings of the National Academy of Sciences USA* 102:9643–9648 (2005).

WHAT IF? *Since the results support a role for mouse FOXP2 in vocalization, you might wonder whether the human FOXP2 protein is a key regulator of speech. If you were given the amino acid sequences of wild-type and mutant human FOXP2 proteins and the wild-type chimpanzee FOXP2 protein, how would you investigate this question? What further clues could you obtain by comparing these sequences to that of the mouse FOXP2 protein?*

the genome, and these CNVs probably represent just a small subset of the total. Since these variants encompass much longer stretches of DNA than the single nucleotides of SNPs, CNVs are more likely to have phenotypic consequences and to play a role in complex diseases and disorders. At the very least, the high incidence of copy-number variation blurs the meaning of the phrase "a normal human genome."

Copy-number variants, SNPs, and variations in repetitive DNA such as short tandem repeats (STRs) are useful genetic markers for studying human evolution. In one study, the genomes of two Africans from different communities were

sequenced: Archbishop Desmond Tutu, the South African civil rights advocate of Xhosa and Tswana tribal heritage and a speaker of a Bantu language, and !Gubi, a hunter-gatherer from a Khoisan language-speaking community in the Kalahari Desert of Namibia. This community is believed to be the human group with the oldest known lineage. The comparison revealed many differences, as you might expect. The analysis was then broadened to compare the protein-coding regions of !Gubi's genome with those of three other Khoisan-speaking men also living in the Kalahari. Remarkably, the four genomes of these men living in the same area differed more from each other than a European would from an Asian. These data highlight the extensive diversity among African genomes, consistent with our understanding that humans and their ancestors arose in Africa and that native Africans have therefore been evolving for longer than any migrant (non-African) population. Extending this approach will help answer questions about the differences between human populations and the migratory routes of human populations throughout history.

> **Mastering Biology**
> Interview with Charles Rotimi: Using genomics to study health-related conditions in African-Americans

Widespread Conservation of Developmental Genes Among Animals

Biologists in the field of evolutionary developmental biology, or **evo-devo**, as it is often called, compare developmental processes of different multicellular organisms. Their aim is to understand how these processes have evolved and how changes in them can modify existing organismal features or lead to new ones. With the advent of molecular techniques and the recent flood of genomic information, we are beginning to realize that the genomes of related species with strikingly different forms may have only minor differences in gene sequence or, perhaps more importantly, in gene regulation. Discovering the molecular basis of these differences in turn helps us understand the origins of the myriad diverse forms that cohabit this planet, thus informing our study of the evolution of life.

In Concept 18.4, you learned about the homeotic genes in *Drosophila melanogaster*, which encode transcription factors that regulate gene expression and specify the identity of body segments in the fruit fly (see Figure 18.20). Molecular analysis of the homeotic genes in *Drosophila* has shown that they all include a 180-nucleotide sequence called a **homeobox**, which codes for a 60-amino-acid *homeodomain* in the encoded proteins. An identical or very similar nucleotide sequence has been discovered in the homeotic genes of many invertebrates and vertebrates. The resemblance even extends to the organization of these genes: The vertebrate genes homologous to the homeotic genes of fruit flies have kept the same chromosomal arrangement **(Figure 21.19)**. The similarities in sequence and organization are so striking that one researcher has whimsically referred to flies as "little people with wings." Homeobox-containing sequences have also been found in regulatory genes of much more distantly related eukaryotes, including plants and yeasts. From these similarities, we can deduce that the homeobox DNA sequence evolved very early and was sufficiently valuable to organisms to have been conserved in animals and plants virtually unchanged for hundreds of millions of years.

Homeotic genes in animals were named *Hox* genes, short for homeobox-containing genes, because homeotic genes

▼ **Figure 21.19 Conservation of homeotic genes in a fruit fly and a mouse.** Homeotic genes that control the form of anterior and posterior structures of the body occur in the same linear sequence on chromosomes in Drosophila and mice. Each colored band on the chromosomes shown here represents a homeotic gene. In fruit flies, all homeotic genes are found on one chromosome. The mouse and other mammals have the same or similar sets of genes on four chromosomes. The color code indicates the parts of the embryos in which these genes are expressed and the adult body regions that result. All of these genes are essentially identical in flies and mice, except for those represented by black bands, which are less similar in the two animals.

were the first genes found to have this sequence. Other homeobox-containing genes were later found that do not act as homeotic genes; that is, they do not directly control the identity of body parts. However, most of these genes, in animals at least, are associated with development, suggesting their ancient and fundamental importance in that process. In *Drosophila*, for example, homeoboxes are present not only in the homeotic genes but also in the egg-polarity gene *bicoid* (see Figures 18.21 and 18.22), in several of the segmentation genes, and in a master regulatory gene for eye development.

Researchers have discovered that the homeobox-encoded homeodomain is the part of the protein that binds to DNA when the protein functions as a transcription factor. Elsewhere in the protein, domains that are more variable interact with other transcription factors, allowing the homeodomain-containing protein to recognize specific enhancers and regulate the associated genes. Proteins with homeodomains probably regulate development by coordinating the transcription of batteries of developmental genes, switching them on or off. In embryos of *Drosophila* and other animal species, different combinations of homeobox genes are active in different parts of the embryo. This selective expression of regulatory genes, varying over time and space, is central to pattern formation.

Developmental biologists have found that in addition to homeotic genes, many other genes involved in development are highly conserved from species to species. These include numerous genes encoding components of signaling pathways. The extraordinary similarity among some developmental genes in different animal species raises a question: How can the same genes be involved in the development of animals whose forms are so very different from each other?

In some cases, small changes in regulatory sequences of particular genes cause changes in gene expression patterns that can lead to major changes in body form. For example, the differing patterns of expression of the *Hox* genes along the body axis in a crustacean and an insect can explain the variation in number of leg-bearing segments among these closely related animals **(Figure 21.20)**. In other cases, similar genes direct different developmental processes in various organisms, resulting in diverse body shapes. Several *Hox* genes, for instance, are expressed in the embryonic and larval stages of the sea urchin, a nonsegmented animal that has a body plan quite different from those of insects and mice. Sea urchin adults make the pincushion-shaped shells you may have seen on the beach; two species of live sea urchins are shown in the photo. Sea urchins are among the organisms long used in classical embryological studies (see Concept 47.2).

▼ **Two species of sea urchins**

▼ **Figure 21.20 Effect of differences in Hox gene expression in a crustacean and an insect.** Changes in the expression patterns of Hox genes have occurred over evolutionary time since insects diverged from a crustacean ancestor. These changes account in part for the different body plans of **(a)** the brine shrimp Artemia, a crustacean, and **(b)** the grasshopper, an insect. Shown here are regions of the adult body color-coded for expression of four Hox genes that determine the formation of particular body parts during embryonic development. Each color represents a specific Hox gene.

Swimmerets

(a) Expression of four *Hox* genes in the brine shrimp *Artemia*. Three of the *Hox* genes are expressed together in one region (indicated by stripes), specifying the identity of the segments that have swimmerets. The fourth (teal) specifies the identity of the genital segments.

(b) Expression of the grasshopper versions of the same four *Hox* genes. In the grasshopper, each *Hox* gene is expressed in a discrete region and specifies the identity of that region.

In this final chapter of the genetics unit, you have learned how studying genomic composition and comparing the genomes of different species can illuminate the process by which genomes evolve. Furthermore, comparing developmental programs, we can see that the unity of life is reflected in the similarity of molecular and cellular mechanisms used to establish body pattern, although the genes directing development may differ among organisms. The similarities between genomes reflect the common ancestry of life on Earth. But the differences are also crucial, for they have created the huge diversity of organisms that have evolved. In the remaining chapters, we expand our perspective beyond the level of molecules, cells, and genes to explore this diversity on the organismal level.

CONCEPT CHECK 21.6

1. Would you expect the genome of the macaque (a monkey) to be more like that of a mouse or that of a human? Explain.

2. DNA sequences called homeoboxes help homeotic genes in animals direct development. Given that they are common to flies and mice, explain why these animals are so different.

3. **WHAT IF?** There are three times as many human-specific insertions of *Alu* elements in the human genome as in the chimpanzee genome. How did these extra *Alu* elements arise in the human genome? Propose a role they might have played in the divergence of these two species.

For suggested answers, see Appendix A.

SUMMARY OF KEY CONCEPTS

→ To review key terms, go to the **Vocabulary Self-Quiz** in the **Mastering Biology** eText or Study Area, or go to goo.gl/zkjz9t.

CONCEPT 21.1

The Human Genome Project fostered development of faster, less expensive sequencing techniques (pp. 443–444)

- The **Human Genome Project** was largely completed in 2003, aided by major advances in sequencing technology.
- In the **whole-genome shotgun** approach, the whole genome is cut into many small, overlapping fragments that are sequenced; computer software then assembles the genome sequence.

? *How did the Human Genome Project result in more rapid, less expensive DNA-sequencing technology?*

CONCEPT 21.2

Scientists use bioinformatics to analyze genomes and their functions (pp. 444–448)

- Computer analysis of genome sequences aids **gene annotation**, the identification of protein-coding sequences. Methods to determine gene function include comparing sequences of newly discovered genes with those of known genes in other species and observing the effects of experimentally inactivating the genes.
- In **systems biology**, scientists use the computer-based tools of **bioinformatics** to compare genomes and study sets of genes and proteins as whole systems (**genomics** and **proteomics**). Studies include large-scale analyses of protein interactions, functional DNA elements, and genes contributing to medical conditions.

? *What has been the most significant finding of the ENCODE project? Why was the project expanded to include nonhuman species?*

CONCEPT 21.3

Genomes vary in size, number of genes, and gene density (pp. 448–450)

	Bacteria	Archaea	Eukarya
Genome size	Most are 1–6 Mb		Most are 10–4,000 Mb, but a few are much larger
Number of genes	1,500–7,500		Most are 5,000–45,000
Gene density	Higher than in eukaryotes		Lower than in prokaryotes (Within eukaryotes, lower density is correlated with larger genomes)
Introns	None in protein-coding genes	Present in some genes	Present in most genes of multicellular eukaryotes, but only in some genes of unicellular eukaryotes
Other noncoding DNA	Very little		Can exist in large amounts; generally more repetitive noncoding DNA in multicellular eukaryotes

? *Compare genome size, gene number, and gene density (a) in the three domains and (b) among eukaryotes.*

CONCEPT 21.4

Multicellular eukaryotes have a lot of noncoding DNA and many multigene families (pp. 450–453)

- Only 1.5% of the human genome codes for proteins or gives rise to rRNAs or tRNAs; the rest is noncoding DNA, including **pseudogenes** and **repetitive DNA** of unknown function.
- The most abundant type of repetitive DNA in multicellular eukaryotes consists of **transposable elements** and related sequences. In eukaryotes, there are two types of transposable elements: **transposons**, which move via a DNA intermediate, and **retrotransposons**, which are more prevalent and move via an RNA intermediate.
- Other repetitive DNA includes short, noncoding sequences that are tandemly repeated thousands of times (**simple sequence DNA**, which includes **STRs**); these sequences are especially prominent in centromeres and telomeres, where they probably play structural roles in the chromosome.
- Though many eukaryotic genes are present in one copy per haploid chromosome set, others (most, in some species) are members of a gene family, such as the human globin gene families:

? *Explain how the function of transposable elements might account for their prevalence in human noncoding DNA.*

CONCEPT 21.5

Duplication, rearrangement, and mutation of DNA contribute to genome evolution (pp. 454–459)

- Errors in cell division can lead to extra copies of all or part of entire chromosome sets, which may then diverge if one set accumulates sequence changes. Polyploidy occurs more often among plants than animals and contributes to speciation.
- The chromosomal organization of genomes can be compared among species, providing information about evolutionary relationships. Within a given species, rearrangements of chromosomes are thought to contribute to the emergence of new species.
- The genes encoding the various related but different globin proteins evolved from one common ancestral globin gene, which duplicated and diverged into the α-globin and β-globin ancestral genes. Subsequent duplication and random mutation gave rise to the present globin genes, all of which code for oxygen-binding proteins. The copies of some duplicated genes have diverged so much that the functions of their encoded proteins (such as lysozyme and α-lactalbumin) are now substantially different.
- Rearrangement of exons within and between genes during evolution has led to genes containing multiple copies of similar exons and/or several different exons derived from other genes.
- Movement of transposable elements or recombination between copies of the same element can generate new sequence combinations that are beneficial to the organism. These may alter the functions of genes or their patterns of expression and regulation.

? *How could chromosomal rearrangements lead to the emergence of new species?*

Comparing genome sequences provides clues to evolution and development *(pp. 459–464)*

- Comparisons of genomes from widely divergent and closely related species provide valuable information about ancient and more recent evolutionary history, respectively. Analysis of single nucleotide polymorphisms (SNPs) and copy-number variants (CNVs) among individuals in a species can also shed light on the evolution of that species.
- Evolutionary developmental (**evo-devo**) biologists have shown that homeotic genes and some other genes associated with animal development contain a **homeobox** region whose sequence is highly conserved among diverse species.

? *What type of information can be obtained by comparing the genomes of closely related species? Of very distantly related species?*

TEST YOUR UNDERSTANDING

➜ For more multiple-choice questions, go to the **Practice Test** in the **Mastering Biology** eText or Study Area, or go to goo.gl/GruWRg.

Levels 1-2: Remembering/Understanding

1. Bioinformatics includes
   (A) using DNA technology to clone genes.
   (B) using computer programs to align DNA sequences.
   (C) using a person's genomic sequence to inform decisions about medical treatment.
   (D) amplifying DNA segments from a species' genome.

2. Homeotic genes
   (A) encode transcription factors that control the expression of genes responsible for specific anatomical structures.
   (B) are found only in *Drosophila* and other arthropods.
   (C) are the only genes that contain the homeobox domain.
   (D) encode proteins that form anatomical structures in the fly.

Levels 3-4: Applying/Analyzing

3. Two eukaryotic proteins have one domain in common but are otherwise very different. Which of the following processes is most likely to have contributed to this similarity?
   (A) gene duplication   (C) exon shuffling
   (B) alternative splicing   (D) random point mutations

4. **DRAW IT** Below are the amino acid sequences (using single letters; see Figure 5.14) of three short segments of the FOXP2 protein from five species. These segments contain all amino acid differences between the FOXP2 proteins of these species. Compare the amino acid sequences by answering parts (a)–(d).

Chimpanzee	PKSSD	... TSSTT ...	NARRD
Mouse	PKSSE	... TSSTT ...	NARRD
Gorilla	PKSSD	... TSSTT ...	NARRD
Human	PKSSD	... TSSNT ...	SARRD
Rhesus monkey	PKSSD	... TSSTT ...	NARRD

   (a) Circle the names of any species that have identical amino acid sequences for the FOXP2 protein.
   (b) In the sequence for the mouse, circle any amino acid that differs from the sequence for the chimpanzee, gorilla, and rhesus monkey. Then draw a box around any amino acid that differs from the human sequence.
   (c) In the human sequence, underline any amino acid that differs from the sequence for the chimpanzee, gorilla, and rhesus monkey.

   (d) Primates and rodents diverged about 65 million years ago, and chimpanzees and humans diverged about 6 million years ago (see Figure 21.17). How many amino acid differences are there between the sequence for the mouse and the sequence for the chimpanzee, gorilla, and rhesus monkey? How many amino acid differences are there between the human sequence and the sequence for the chimpanzee, gorilla, and rhesus monkey? Based solely on the numbers of amino acid differences occurring over these time periods, what might you hypothesize about the rate of evolution of the *FOXP2* gene? Based on the information in the chapter regarding the *FOXP2* gene, is your hypothesis correct?

Levels 5-6: Evaluating/Creating

5. **EVOLUTION CONNECTION** Genes important in the embryonic development of animals, such as homeobox-containing genes, have been relatively well conserved during evolution; that is, they are more similar among different species than are many other genes. Explain why this is.

6. **SCIENTIFIC INQUIRY** The scientists mapping the SNPs in the human genome noticed that groups of SNPs tended to be inherited together, in blocks known as haplotypes, ranging in length from 5,000 to 200,000 base pairs. There are as few as four or five commonly occurring combinations of SNPs per haplotype. Integrating what you've learned throughout this chapter and this unit, propose an explanation for this observation.

7. **WRITE ABOUT A THEME: INFORMATION** The continuity of life is based on heritable information in the form of DNA. In a short essay (100–150 words), explain how mutations in protein-coding genes and regulatory DNA contribute to evolution.

8. **SYNTHESIZE YOUR KNOWLEDGE**

Insects have three thoracic (trunk) segments. While researchers have found insect fossils with wings on all segments, modern insects have wings or related structures on only the second and third segment. In modern insects, *Hox* gene products act to inhibit wing formation on the first segment. The treehopper insect (above) is an exception. In addition to the pair of prominent wings on its second segment, its first segment has an ornate helmet resembling a set of thorns, which is a modified, fused pair of "wings." (This provides camouflage in tree branches, reducing the risk of predation.) Explain how changes in gene regulation could have led to the evolution of such a structure.

For selected answers, see Appendix A.

Cassandra Extavour is Professor in the Department of Organismic and Evolutionary Biology at Harvard University, where she and her students are studying genes that control early embryonic development, as well as the evolutionary origins of these genes and how their functions have changed over time. Born in Toronto, Canada, Dr. Extavour received a B.Sc. (honors) in Molecular Biology and Genetics from the University of Toronto, followed by a Ph.D. in the same subject area from Universidad Autónoma de Madrid, Spain. After completing two postdoctoral fellowships (one in Greece, the other in the United Kingdom), she worked for several years as a Research Associate at Cambridge University, UK, before joining the faculty at Harvard. Dr. Extavour is a dedicated advocate for women and underrepresented groups in science, and she also has performed worldwide as a professional soprano.

AN INTERVIEW WITH

Cassandra Extavour

How did you first become interested in science?

Although my first love was music, and music remains an essential part of my life, toward the end of high school I also became quite interested in biology. My conversations with a friend who was fascinated by human behavior led me to wonder how the brain causes people to exhibit different behaviors. But how could I learn about that? My parents did not go to college. I didn't even know that there were careers in science that were not in medicine. I thought the only way to learn about the brain was to be a doctor. So I entered university as a pre-med. But by the time I was finishing my undergraduate studies, I had learned about research labs. I got a summer internship in a developmental genetics lab, and I loved it—it was like entering a new world where I could figure out the answers to questions that interested me.

What are the main questions you are asking in your research?

I'm interested in understanding the genes that control embryonic development and how the function of those genes has changed in different organisms due to evolution. Take a human, for example. Our bodies have many different types of cells. The instructions for everything that each of our cells needs to do are contained in the cell's DNA. All of our cells have exactly same DNA. What's mysterious is, since every cell in the body has the same set of instructions, why aren't they all doing the same thing? I work mostly with insects, but the same question applies to them (and to all multicellular organisms). We're trying to figure out how the same set of genes is used in different ways by different cells. We're also interested in where these genes came from in the first place and how they have evolved over time.

Much of your research has been on "evo-devo." Say more about what that means.

Evo-devo stands for evolutionary developmental biology. Scientists in this field try to understand the difference in development between, let's say, a coconut and a rose. They're both plants, they both make seeds, they both make pollen, but they don't look the same at all. Why is that? Or the fact that, when you compare the development of a monkey and a human, you can't detect many differences in their embryos or their genes. How can they be similar in so many ways and yet end up as such different organisms? In my lab, we are particularly interested in egg and sperm cells—known as germ cells. We're trying to figure out how an embryo determines which cells become germ cells and how it tells those cells, "Okay, listen: When the animal you are part of becomes an adult, you will need to make eggs and sperm." We've found that in some cases, the same genes are used by a variety of different species to make germ cells. But in other cases, we've been surprised to find that a gene used to produce germ cells in one organism is used to make a completely different type of cell in another closely related organism.

What do you find most rewarding about your job?

When I was in high school, I never would have thought that I could get paid to think of interesting questions and then try to answer them. I love doing that, and I love the amazing freedom that comes with my job—I get to decide what I'm going to do each day. But the best part is working with students as they gain the independence, knowledge, and confidence to listen to my suggestions and then say, "I'm not sure that's the best experiment. Here's an experiment that I think would be better, and here's why." That is a great moment. What they are really saying is, "I'm not learning to be a scientist anymore—I am a scientist." They've come to understand what scientists do: We make the best proposal we can, we test it as best we can, and we say, "This is what I think the results from my experiments mean."

> *"What's mysterious is, since every cell in the body has the same set of instructions, why aren't they all doing the same thing?"*

▲ Embryo of the crustacean *Parhyale hawaiensis* showing cells (in red) that will develop into germ cells.

22 Descent with Modification: A Darwinian View of Life

KEY CONCEPTS

22.1 The Darwinian revolution challenged traditional views of a young Earth inhabited by unchanging species *p. 469*

22.2 Descent with modification by natural selection explains the adaptations of organisms and the unity and diversity of life *p. 471*

22.3 Evolution is supported by an overwhelming amount of scientific evidence *p. 476*

Study Tip

Make a word cloud: This chapter covers key topics related to evolution, such as descent with modification, heritable characteristics, natural selection, adaptation, convergent evolution, homology, and unity of life. Draw a "word cloud" of these terms: As you read the chapter, draw a line between terms that affect or relate to one another. Next to that line, briefly explain how those terms are connected.

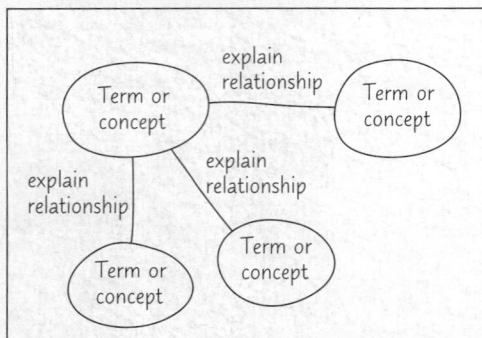

→ Go to Mastering Biology

For Students (in eText and Study Area)
• Get Ready for Chapter 22
• Figure 22.17 Walkthrough: Tree Thinking
• Figure 22.20 Walkthrough: Transition to Life in the Sea
• BBC Video: A Future Without Antibiotics?

For Instructors to Assign (in Item Library)
• HHMI Video: The Making of the Fittest: Natural Selection and Adaptation (Pocket Mouse)

Figure 22.1 This Malaysian orchid mantis (*Hymenopus coronatus*) closely resembles the flower on which it rests, waiting for unwary prey to come within its reach. Other mantises have diverse shapes and colors that have evolved in different environments—yet all mantises also share certain features, such as grasping forelimbs, large eyes, and six legs.

What causes the similarities and differences among Earth's many different species?

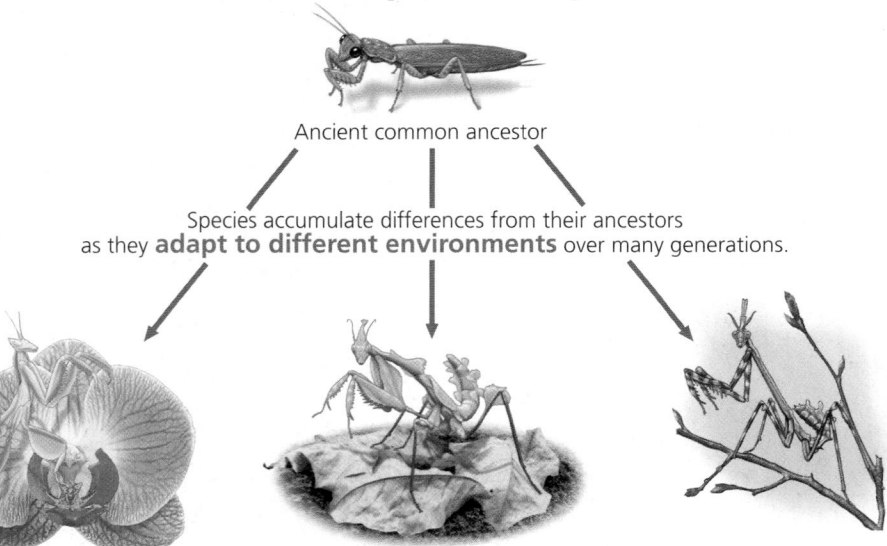

Ancient common ancestor

Species accumulate differences from their ancestors as they **adapt to different environments** over many generations.

While different in some ways, these species share many similar features because they descended from a common ancestor. This process of

descent with modification

shared ancestry, resulting in shared characteristics

accumulation of differences

has given rise to the **diversity of life**.

The Darwinian revolution challenged traditional views of a young Earth inhabited by unchanging species

More than a century and a half ago, Charles Darwin was inspired to develop a scientific explanation for the *diversity of life*, the great number and remarkable variety of species on Earth. When he published his hypothesis in his book *The Origin of Species*, Darwin ushered in a scientific revolution—the era of evolutionary biology. As we'll see, Darwin developed his revolutionary ideas over time, influenced by the work of others and his travels **(Figure 22.2)**.

Endless Forms Most Beautiful

To set the stage for our study of Darwin and evolutionary biology, let's return to the well-hidden orchid mantis in Figure 22.1. This species is a member of a diverse group of insects, the Mantodea, which includes 2,300 species in 430 genera. All these mantises share certain features, such as three pairs of legs, triangular heads with bulging eyes, and a flexible "neck." Such shared features illustrate the unity of life, a phrase that highlights the fact that all organisms share characteristics.

But Earth's many different species also differ from one another. The mantises, for example, differ in features such as their size, shape, and color. Overall, the orchid mantis and its many close relatives illustrate three key observations about life:

- organisms are well suited (adapted) for life in their environments (Here and throughout this text, the term *environment* refers to other organisms as well as to the physical aspects of an organism's surroundings.)
- the many shared characteristics (unity) of life
- the rich diversity of life

Darwin, a keen observer of nature since childhood, set out to explain these three broad observations—an effort that eventually led him to conclude that life evolves over time.

For now, we will define **evolution** as *descent with modification*, a phrase Darwin used to summarize the process

1809
Lamarck publishes his hypothesis of evolution.

1798
Malthus publishes "Essay on the Principle of Population."

1795
Hutton proposes his principle of gradualism.

1812
Cuvier publishes his extensive studies of vertebrate fossils.

1830
Lyell publishes *Principles of Geology*.

Sketch of a flying frog by Wallace

1858
While studying species in the Malay Archipelago, Wallace (shown above in 1848) sends Darwin his hypothesis of natural selection.

1790

1809
Charles Darwin is born.

1831–1836
Darwin travels around the world on HMS *Beagle*.

1844
Darwin writes his essay on descent with modification.

Darwin saw marine iguanas in the Galápagos Islands.

1859
The Origin of Species is published.

1870

▲ **Figure 22.2 The intellectual context of Darwin's ideas.**

THE ORIGIN OF SPECIES

by which species accumulate differences from their ancestors as they adapt to different environments over time. Evolution can also be defined as a change in the genetic composition of a population from generation to generation (see Concept 23.3).

However it is defined, we can view evolution in two related but different ways: as a pattern and as a process. The *pattern* of evolutionary change is revealed by data from many scientific disciplines, including biology, geology, physics, and chemistry. These data are facts—they are observations about the natural world—and these observations show that life has evolved over time. The *process* of evolution consists of the mechanisms that cause the observed pattern of change. These mechanisms represent natural causes of the natural phenomena we observe. Indeed, the power of evolution as a unifying theory is its ability to explain and connect a vast array of observations about the living world.

As with all general theories in science, we continue to test our understanding of evolution by examining whether it can account for new observations and experimental results. In this and the following chapters, we'll examine how ongoing discoveries shape what we know about the pattern and process of evolution.

We'll begin our exploration of these discoveries by retracing Darwin's quest to explain the adaptations, unity, and diversity of what he called life's "endless forms most beautiful."

Scala Naturae and Classification of Species

Long before Darwin was born, several Greek philosophers suggested that life might have changed gradually over time. But one philosopher who greatly influenced early Western science, Aristotle (384–322 BCE), viewed species as fixed (unchanging). Through his observations of nature, Aristotle recognized certain "affinities" among organisms. He concluded that life-forms could be arranged on a ladder, or scale, of increasing complexity, later called the *scala naturae* ("scale of nature"). Each form of life, perfect and permanent, had its allotted rung on this ladder.

These ideas were generally consistent with the Old Testament account of creation, which holds that species were individually designed by God and therefore perfect. In the 1700s, many scientists interpreted the often remarkable ways in which organisms are well suited for life in their environment as evidence that the Creator had designed each species for a particular purpose.

One such scientist was Carolus Linnaeus (1707–1778), a Swedish physician and botanist who sought to classify life's diversity, in his words, "for the greater glory of God." In the 1750s, Linnaeus developed the two-part, or *binomial*, format for naming species (such as *Homo sapiens* for humans) that is still used today. In contrast to the linear hierarchy of the *scala naturae*, Linnaeus used a nested classification system, grouping similar species into increasingly inclusive categories. For example, similar species are grouped in the same genus,

similar genera (plural of genus) are grouped in the same family, and so on.

Linnaeus did not ascribe the resemblances among species to evolutionary kinship, but rather to the pattern of their creation. A century later, however, Darwin argued that classification should be based on evolutionary relationships. He also noted that scientists using the Linnaean system often grouped organisms in ways that reflected those relationships.

Ideas About Change over Time

Among other sources of information, Darwin drew from the work of scientists studying **fossils**, the remains or traces of organisms from the past. Many fossils are found in sedimentary rocks formed from the sand and mud that settle to the bottom of seas, lakes, and swamps **(Figure 22.3)**. New layers of sediment cover older ones and compress them into superimposed layers of rock called **strata** (singular, *stratum*). The fossils in a particular stratum provide a glimpse of some of the organisms that populated Earth at the time that layer formed. Later, erosion may carve through upper (younger) strata, revealing deeper (older) strata that had been buried.

Paleontology, the study of fossils, was developed in large part by French scientist Georges Cuvier (1769–1832). In examining strata near Paris, Cuvier noted that the older the stratum, the more dissimilar its fossils were to current life-forms. He also observed that from one layer to the next, some new species appeared while others disappeared. He inferred that extinctions must have been a common occurrence, but he staunchly opposed the idea of evolution. Cuvier speculated that each boundary between strata represented a sudden catastrophic

▼ **Figure 22.3 Formation of sedimentary strata with fossils.**

1 Rivers carry sediment into aquatic habitats such as seas and swamps. Over time, sedimentary rock layers (strata) form under water. Some strata contain fossils.

2 As water levels change and geological activity pushes layers of rock upward, the strata and their fossils are exposed.

Younger stratum with more recent fossils

Older stratum with older fossils

→ **Mastering Biology** Video: Grand Canyon

event, such as a flood, that had destroyed many of the species living in that area. Such regions, he reasoned, were later repopulated by different species immigrating from other areas.

In contrast to Cuvier's emphasis on sudden events, other scientists suggested that profound change could take place through the cumulative effect of slow but continuous processes. In 1795, Scottish geologist James Hutton (1726–1797) proposed that Earth's geologic features could be explained by gradual mechanisms, such as valleys being formed by rivers. The leading geologist of Darwin's time, Charles Lyell (1797–1875), incorporated Hutton's thinking into his proposal that the same geologic processes are operating today as in the past, and at the same rate.

Hutton's and Lyell's ideas strongly influenced Darwin's thinking. Darwin agreed that if geologic change results from slow, continuous actions rather than from sudden events, then Earth must be much older than the widely accepted age of a few thousand years. It would, for example, take a very long time for a river to carve a canyon by erosion. He later reasoned that perhaps similarly slow and subtle processes could produce substantial biological change. However, Darwin was not the first to apply the idea of gradual change to biological evolution.

Lamarck's Hypothesis of Evolution

Although some 18th-century naturalists suggested that life evolves as environments change, only one proposed a mechanism for *how* life changes over time: French biologist Jean-Baptiste de Lamarck (1744–1829). Alas, Lamarck is primarily remembered today *not* for his visionary recognition that evolutionary change explains patterns in fossils and how organisms are well suited for their environments, but for the incorrect mechanism he proposed.

Lamarck published his hypothesis in 1809, the year Darwin was born. By comparing living species with fossil forms, Lamarck had found what appeared to be several lines of descent, each a chronological series of older to younger fossils leading to a living species. He explained his findings using two principles that were widely accepted at the time. The first was *use and disuse*, the idea that parts of the body that are used extensively become larger and stronger, while those that are not used deteriorate. Among many examples, he cited a giraffe stretching its neck to reach leaves on high branches. The second principle, *inheritance of acquired characteristics*, stated that an organism could pass these modifications to its offspring. Lamarck reasoned that the long, muscular neck of the living giraffe had evolved over many generations as giraffes stretched their necks ever higher.

Lamarck also thought that evolution happens because organisms have an innate drive to become more complex. Darwin rejected this idea, but he, too, thought that variation was introduced into the evolutionary process in part by inheritance of acquired characteristics. Today, however, our understanding of genetics refutes this mechanism: Experiments

▶ **Figure 22.4**
Acquired traits cannot be inherited. This bonsai tree was "trained" to grow as a dwarf by pruning and shaping. However, seeds from this tree would produce offspring of normal size.

show that traits acquired by use during an individual's life are not inherited in the way proposed by Lamarck (**Figure 22.4**).

Lamarck was vilified in his own time, especially by Cuvier, who denied that species ever evolve. In retrospect, however, Lamarck did recognize that the fact that organisms are well-suited for life in their environments can be explained by gradual evolutionary change, and he did propose a testable explanation for how this change occurs.

CONCEPT CHECK 22.1

1. How did Hutton's and Lyell's ideas influence Darwin's thinking about evolution?
2. **MAKE CONNECTIONS** Scientific hypotheses must be testable (see Concept 1.3). Applying this criterion, are Cuvier's explanation of the fossil record and Lamarck's hypothesis of evolution scientific? Explain your answer in each case.

For suggested answers, see Appendix A.

CONCEPT 22.2

Descent with modification by natural selection explains the adaptations of organisms and the unity and diversity of life

As the 19th century dawned, it was generally thought that species had remained unchanged since their creation. A few clouds of doubt about the permanence of species were beginning to gather, but no one could have forecast the thundering storm just beyond the horizon. How did Charles Darwin become the lightning rod for a revolutionary view of life?

Darwin's Research

Charles Darwin (1809–1882) was born in Shrewsbury, in western England. Even as a boy, he had a consuming interest in nature. When he was not reading nature books, he was fishing, hunting, riding, and collecting insects. However,

Darwin's father, a physician, could see no future for his son as a naturalist and sent him to medical school in Edinburgh. But Charles found medicine boring and surgery before the days of anesthesia horrifying. He quit medical school and enrolled at Cambridge University, intending to become a clergyman. (At that time, many scholars of science belonged to the clergy.)

At Cambridge, Darwin became the protégé of John Henslow, a botany professor. Soon after Darwin graduated, Henslow recommended him to Captain Robert FitzRoy, who was readying the survey ship HMS *Beagle* for a voyage around the world. Darwin would pay his own way and serve as a conversation partner to the young captain. FitzRoy, who was himself an adept scientist, accepted Darwin because he was a skilled naturalist and they were of similar age and social class.

The Voyage of the Beagle

Darwin embarked from England on the *Beagle* in December 1831. The primary mission of the voyage was to chart stretches of the South American coast that were poorly known to Europeans. Darwin, however, spent most of his time on shore, observing and collecting thousands of plants and animals. He described features of organisms that made them well suited to such diverse environments as the humid jungles of Brazil, the expansive grasslands of Argentina, and the towering peaks of the Andes. He also noted that the plants and animals in temperate regions of South America more closely resembled species living in the South American tropics than species living in temperate regions of Europe. Furthermore, the fossils he found, though clearly different from living species, distinctly resembled the living organisms of South America.

Darwin also spent much time thinking about geology. Despite repeated bouts of seasickness, he read Lyell's *Principles*

of Geology during the voyage. He experienced geologic change firsthand when a violent earthquake shook the coast of Chile, and he observed afterward that rocks along the coast had been thrust upward by several meters. Finding fossils of ocean organisms high in the Andes, Darwin inferred that the rocks containing the fossils must have been raised there by many similar earthquakes. These observations reinforced what he had learned from Lyell: Physical evidence did not support the traditional view that Earth was only a few thousand years old.

Darwin's interest in the species (or fossils) found in an area was further stimulated by the *Beagle*'s stop at the Galápagos, a group of volcanic islands located near the equator about 900 km west of South America **(Figure 22.5)**. Darwin was fascinated by the unusual organisms there. The birds he collected included several kinds of mockingbirds. These mockingbirds, though similar to each other, seemed to be different species. Some were unique to individual islands, while others lived on two or more adjacent islands. Furthermore, although the animals on the Galápagos resembled species living on the South American mainland, most of the Galápagos species were not known from anywhere else in the world. Darwin hypothesized that the Galápagos had been colonized by organisms that had strayed from South America and then diversified, giving rise to new species on the various islands.

Darwin's Focus on Adaptation

During the voyage of the *Beagle*, Darwin observed many examples of **adaptations**, inherited characteristics of organisms that enhance their survival and reproduction in specific environments. Later, as he reassessed his observations, he began to perceive adaptation to the environment and the origin of new species as closely related processes. Could a new species arise

▼ **Figure 22.5 The voyage of HMS *Beagle* (December 1831–October 1836).**

Darwin in 1840, after his return from the voyage

HMS *Beagle* in port

▼ Figure 22.6 Three examples of beak variation in Galápagos finches. The Galápagos Islands are home to more than a dozen species of closely related finches, some found only on a single island. A striking difference among them is their beaks, which are adapted for specific diets.

(a) Cactus-eater. The long, sharp beak of the common cactus finch (*Geospiza scandens*) helps it tear and eat cactus flowers and pulp.

(b) Insect-eater. The green warbler finch (*Certhidea olivacea*) uses its narrow, pointed beak to grasp insects.

(c) Seed-eater. The large ground finch (*Geospiza magnirostris*) has a large beak adapted for cracking seeds found on the ground.

MAKE CONNECTIONS *Review Figure 1.20. Circle the most recent common ancestor shared by the three species that eat insects. Are all of the descendants of that ancestor insect-eaters?*

from an ancestral form by the gradual accumulation of adaptations to a different environment? From studies made years after Darwin's voyage, biologists have concluded that this is indeed what happened to a diverse group of finches found on the Galápagos Islands (see Figure 1.20). The finches' various beaks and behaviors are adapted to the specific foods available on their home islands **(Figure 22.6)**. Darwin realized that explaining such adaptations was essential to understanding evolution. His explanation of how adaptations arise centered on **natural selection**, a process in which individuals that have certain inherited traits tend to survive and reproduce at higher rates than do other individuals *because of* those traits. By the early 1840s, Darwin had worked out the major features of his hypothesis. He set these ideas on paper in 1844, when he wrote a long essay on descent with modification and its underlying mechanism, natural selection. Yet he was still reluctant to publish his ideas, in part because he anticipated the uproar they would cause. During this time, Darwin continued to compile evidence in support of his hypothesis. By the mid-1850s, he had described his ideas to Lyell and a few others. Lyell, who was not yet convinced of evolution, nevertheless urged Darwin to publish on the subject before someone else came to the same conclusions and published first.

In June 1858, Lyell's prediction came true. Darwin received a manuscript from Alfred Russel Wallace (1823–1913), a British naturalist working in the South Pacific islands of the Malay Archipelago (see Figure 22.2). Wallace had developed a hypothesis of natural selection nearly identical to Darwin's. He asked Darwin to evaluate his paper and forward it to Lyell if it merited publication. Darwin complied, writing to Lyell: "Your words have come true with a vengeance.... I never saw a more striking coincidence... so all my originality, whatever it may amount to, will be smashed." On July 1, 1858, Lyell and a colleague presented Wallace's paper, along with extracts from Darwin's unpublished 1844 essay, to the Linnean Society of London. Darwin quickly finished his

book, titled *On the Origin of Species by Means of Natural Selection* (commonly referred to as *The Origin of Species*), and published it the next year. Although Wallace had submitted his ideas for publication first, he admired Darwin and thought that Darwin had developed and tested the idea of natural selection so extensively that he should be known as its main architect.

Within a decade, Darwin's book and its proponents had convinced most scientists that life's diversity is the product of evolution. Darwin succeeded where previous evolutionists had failed by presenting a plausible scientific mechanism with immaculate logic and an avalanche of supporting evidence.

> **Mastering Biology** HHMI Video: The Origin of Species: The Making of a Theory

Ideas from *The Origin of Species*

In his book, Darwin amassed evidence that descent with modification by natural selection explains three broad observations about nature—the unity of life, the diversity of life, and the striking ways in which organisms are suited for life in their environments.

Descent with Modification

In the first edition of *The Origin of Species*, Darwin never used the word *evolution* (although the final word of the book is "evolved"). Rather, he discussed *descent with modification*, a phrase that summarized his view of life. Organisms share many characteristics, leading Darwin to perceive unity in life. He attributed the unity of life to the descent of all organisms from an ancestor that lived in the remote past. He also thought that as the descendants of that ancestral organism lived in various habitats, they gradually accumulated diverse modifications, or adaptations, that fit them to specific ways of life. Thus, Darwin thought of evolution as a process in which both *descent* (shared ancestry, resulting in shared characteristics) and *modification* (the accumulation of differences) can be observed.

▶ **Figure 22.7 "I think . . . ."**
In this 1837 sketch, Darwin envisioned the branching pattern of evolution. Branches that end in twigs labeled A–D represent particular groups of living organisms; all other branches represent extinct groups.

Darwin reasoned that over a long period of time, descent with modification eventually led to the rich diversity of life we see today. He viewed the history of life as a tree, with multiple branchings from a common trunk out to the tips of the youngest twigs **(Figure 22.7)**. In his diagram, the tips of the twigs that are labeled A–D represent several groups of organisms living in the present day, while the unlabeled branches represent groups that are extinct. Each fork of the tree represents the most recent common ancestor of all the lines of evolution that subsequently branch from that point.

Darwin thought that such a branching process, along with past extinction events, could explain the large morphological gaps that sometimes exist between related groups of organisms. As an example, let's consider the three living species of elephants: the Asian elephant (*Elephas maximus*) and two species of African elephants (*Loxodonta africana* and *L. cyclotis*). These closely related species are very similar because they shared the same line of descent until a relatively recent split from their common ancestor, as shown in the tree diagram in **Figure 22.8**. Note that seven lineages related to elephants have become extinct over the past 32 million years. As a result, there are no living species that fill the morphological gap between the elephants and their nearest relatives today, the hyraxes and manatees.

Extinctions like those in Figure 22.8 are common. In fact, many evolutionary branches, even some major ones, are dead ends: Scientists estimate that over 99% of all species that have ever lived are now extinct. As in Figure 22.8, fossils of extinct species can document the divergence of present-day groups by "filling in" gaps between them.

Artificial Selection, Natural Selection, and Adaptation

Darwin proposed the mechanism of natural selection to explain the observable patterns of evolution. He crafted his argument carefully, hoping to persuade even the most skeptical readers. First he discussed familiar examples of selective breeding of domesticated plants and animals. Humans have modified other species over many generations by selecting and breeding individuals that possess desired traits, a process

▼ **Figure 22.8 Descent with modification.** This evolutionary tree of elephants and their relatives is based mainly on fossils—their anatomy, order of appearance in strata, and geographic distribution. Note that most branches of descent ended in extinction (denoted by the dagger symbol, †). (Time line not to scale.)

Hyracoidea (Hyraxes)

Sirenia (Manatees and relatives)

†*Moeritherium*

†*Barytherium*

†*Deinotherium*

†*Mammut*

†*Platybelodon*

†*Stegodon*

†*Mammuthus*

Elephas maximus (Asia)

Common ancestor of living elephant species

Loxodonta africana (Africa)

Loxodonta cyclotis (Africa)

60   34   24   5.5 2 10⁴ 0

Millions of years ago   Years ago

VISUAL SKILLS *Based on this tree, approximately when did the most recent ancestor shared by* Mammuthus *(woolly mammoths), Asian elephants, and African elephants live?*

► **Figure 22.9 Artificial selection.** These different vegetables have all been selected from one species of wild mustard (*Brassica oleracea*). By selecting variations in different parts of the plant, breeders have obtained these divergent results.

Brussels sprouts

Selection for axillary (side) buds

Cabbage

Selection for apical (tip) bud

Broccoli

Selection for flowers and stems

Kale

Selection for leaves

Wild mustard

Selection for stems

Kohlrabi

called **artificial selection (Figure 22.9)**. As a result of artificial selection, crops, livestock animals, and pets often bear little resemblance to their wild ancestors.

Darwin then argued that a similar process occurs in nature. He based his argument on two observations, from which he drew two inferences:

Observation #1: Members of a population often vary in their inherited traits **(Figure 22.10)**.

Observation #2: All species can produce more offspring than their environment can support **(Figure 22.11)**, and many of these offspring fail to survive and reproduce.

Inference #1: Individuals whose inherited traits give them a higher probability of surviving and reproducing in a given environment tend to leave more offspring than do other individuals.

Inference #2: This unequal ability of individuals to survive and reproduce will lead to the accumulation of favorable traits in the population over generations.

As these two inferences suggest, Darwin saw a connection between natural selection and the capacity of organisms to "overreproduce." He began to make this connection after reading an essay by economist Thomas Malthus, who contended that much of human suffering—disease, famine, and war—resulted from the human population's potential to increase faster than food supplies and other resources. Darwin similarly realized that the capacity to overreproduce was typical of all species. Of the many eggs laid, young born, and seeds spread, only a fraction complete their development and leave offspring of their own. The rest are starved, eaten, diseased, unmated, or unable to tolerate physical conditions of the environment such as salinity or temperature.

An organism's heritable traits can influence not only its own performance, but also how well its offspring cope with

▼ **Figure 22.10 Variation in a population.** Individuals in this population of Asian ladybird beetles (*Harmonia axyridis*) vary in color and spot pattern. Natural selection may act on these variations only if (1) they are heritable and (2) they affect the beetles' ability to survive and reproduce.

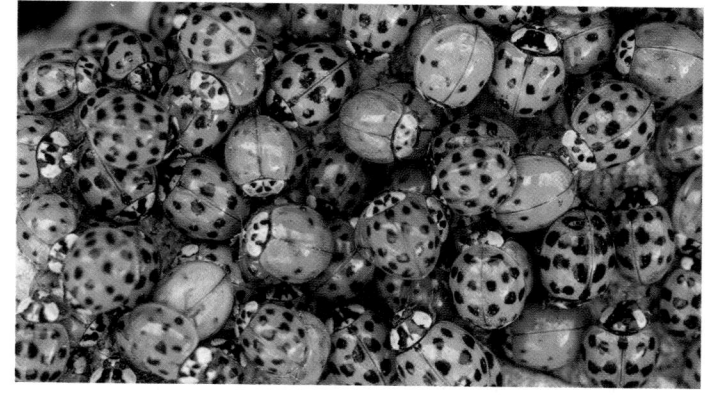

► **Figure 22.11 Overproduction of offspring.** A single puffball fungus can produce billions of spores that give rise to offspring. If all of these offspring and their descendants survived to maturity, they would carpet the surrounding land surface.

Spore cloud

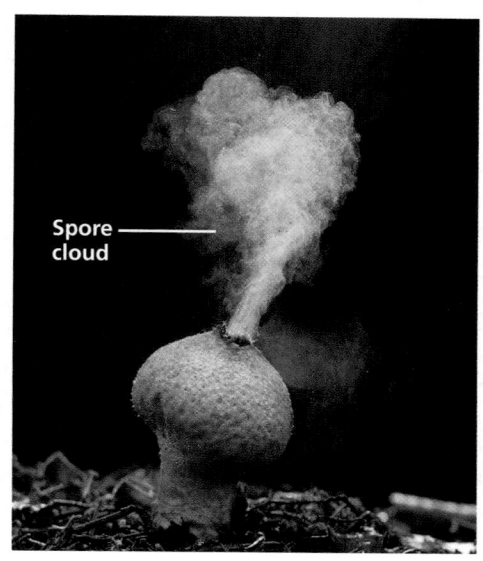

environmental challenges. For example, an organism might have a trait that gives its offspring an advantage in escaping predators, obtaining food, or tolerating physical conditions. When such advantages increase the number of offspring that survive and reproduce, the traits that are favored will likely appear at a greater frequency in the next generation. Thus, over time, natural selection resulting from factors such as predators, lack of food, or adverse physical conditions can lead to an increase in the proportion of favorable traits in a population.

How rapidly do such changes occur? Darwin reasoned that if artificial selection can bring about dramatic change in a relatively short period of time, then natural selection should be capable of substantial modification of species over many hundreds of generations. Even if the advantages of some heritable traits over others are slight, the advantageous variations will gradually accumulate in the population, and less favorable variations will diminish. Over time, this process will increase the frequency of individuals with favorable adaptations, hence increasing the degree to which organisms are well suited for life in their environment.

Key Features of Natural Selection

Let's now summarize some of the main ideas of natural selection:

- Natural selection is a process in which individuals that have certain heritable traits survive and reproduce at a higher rate than do other individuals because of those traits.
- Over time, natural selection can increase the frequency of adaptations that are favorable in a given environment **(Figure 22.12)**.
- If an environment changes, or if individuals move to a new environment, natural selection may result in adaptation to these new conditions, sometimes giving rise to new species.

One subtle but important point is that although natural selection occurs through interactions between individual organisms and their environment, *individuals do not evolve.* Rather, it is the population that evolves over time.

A second key point is that natural selection can amplify or diminish only those heritable traits that differ among the individuals in a population. Thus, even if a trait is heritable, if all the individuals in a population are genetically identical for that trait, evolution by natural selection cannot occur.

Third, environmental factors vary from place to place and over time. A trait that is favorable in one place or time may be useless—or even detrimental—in other places or times. Natural selection is always operating, but which traits are favored depends on the context in which a species lives and mates.

Next, we'll survey the wide range of observations that support a Darwinian view of evolution by natural selection.

▲ **Figure 22.12 Camouflage as an example of evolutionary adaptation.** Related species of moths have diverse shapes and colors that evolved in different environments, as seen in this dead-leaf moth (*Oxytenis modestia*) in Peru **(a)** and buff-end moth (*Phalera bucephala*) in Scotland **(b)**.

CONCEPT CHECK 22.2

1. How does the concept of descent with modification explain both the unity and diversity of life?
2. **WHAT IF?** If you discovered a fossil of an extinct mammal that lived high in the Andes, would you predict that it would more closely resemble present-day mammals from South American jungles or present-day mammals that live high in Asian mountains? Explain.
3. **MAKE CONNECTIONS** Review the relationship between genotype and phenotype (see Figures 14.5 and 14.6). Suppose that in a particular pea population, flowers with the white phenotype are favored by natural selection. Predict what would happen over time to the frequency of the white-flower allele (the *p* allele) in the population.

For suggested answers, see Appendix A.

CONCEPT 22.3

Evolution is supported by an overwhelming amount of scientific evidence

In *The Origin of Species*, Darwin marshaled a broad range of evidence to support the concept of descent with modification. Still—as he readily acknowledged—there were instances in which key evidence was lacking. For example, Darwin referred to the origin of flowering plants as an "abominable mystery," and he lamented the lack of fossils showing how earlier groups of organisms gave rise to new groups.

In the last 150 years, new discoveries have filled many of the gaps that Darwin identified. The origin of flowering plants, for example, is much better understood (see Concept 30.3), and many fossils have been discovered that signify the origin of new groups of organisms (see Concept 25.2). In this section, we'll consider four types of data that document the pattern of evolution and illuminate how it occurs: direct observations, homology, the fossil record, and biogeography.

Direct Observations of Evolutionary Change

Biologists have documented evolutionary change in thousands of scientific studies. We'll examine many such studies throughout this unit, but let's look at two examples here.

Natural Selection in Response to Introduced Species

Herbivores, animals that eat plants, often have adaptations that help them feed efficiently on their primary food sources. What happens when herbivores switch to a new food source with different characteristics?

An opportunity to study this question in nature is provided by soapberry bugs, which use their "beak"—a hollow, needlelike mouthpart—to feed on seeds located within the fruits of various plants. In southern Florida, the soapberry bug (*Jadera haematoloma*) feeds on the seeds of a native plant, the balloon vine (*Cardiospermum corindum*). In central Florida, however, where balloon vines have become rare, soapberry bugs feed on the seeds of the golden rain tree (*Koelreuteria elegans*), a species recently introduced from Asia.

Soapberry bugs feed most effectively when the length of their beak is similar to the depth at which seeds are found within the fruit. Golden rain tree fruit consists of three flat lobes, and its seeds are much closer to the fruit surface than are the seeds of the plump, round fruit of the native balloon vine. These differences led researchers to predict that in populations that feed on golden rain tree, natural selection would result in beaks that are *shorter* than those in populations that feed on balloon vine (**Figure 22.13**). Indeed, beak lengths are shorter in the populations that feed on golden rain tree.

Researchers have also studied beak length evolution in soapberry bug populations that feed on plants introduced to Louisiana, Oklahoma, and Australia. In each of these locations, the fruit of the introduced plants is larger than the fruit of the native plant. Thus, in populations feeding on introduced species in these regions, researchers predicted that natural selection would result in the evolution of *longer* beaks. Again, data collected in field studies upheld this prediction.

The observed changes in beak lengths had important consequences. In Australia, for example, the increase in beak

▼ **Figure 22.13** Inquiry

Can a change in a population's food source result in evolution by natural selection?

Field Study Soapberry bugs feed most effectively when the length of their "beak" is similar to the depth of the seeds within the fruit. Researchers measured beak lengths in soapberry bug populations feeding on the native balloon vine. They also measured beak lengths in populations feeding on the introduced golden rain tree. The researchers then compared the measurements with those of museum specimens collected in the two areas before the golden rain tree was introduced.

Soapberry bug with beak inserted in balloon vine fruit

Results Beak lengths were shorter in populations feeding on the introduced species than in populations feeding on the native species, in which the seeds are buried more deeply. The average beak length in museum specimens from each population (indicated by red arrows) was similar to beak lengths in populations feeding on native species.

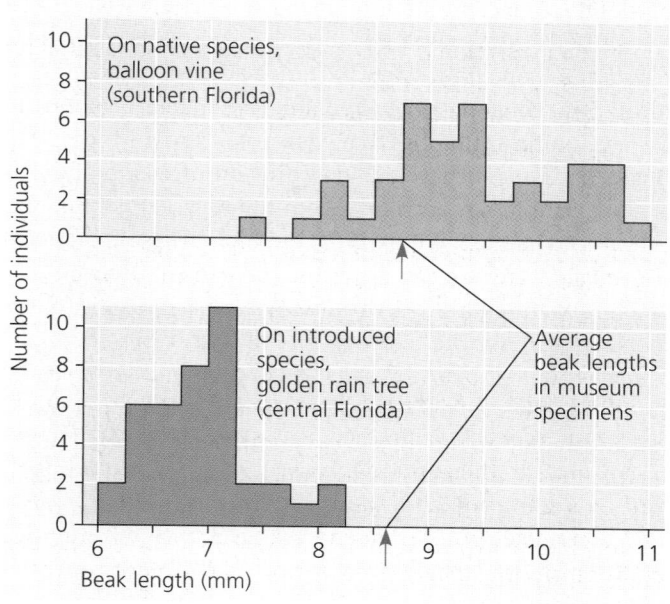

Conclusion Museum specimens and contemporary data suggest that a change in the seed depth of the soapberry bug's food source can result in evolution by natural selection for a corresponding change in beak size.

Data from S. P. Carroll and C. Boyd, Host race radiation in the soapberry bug: natural history with the history, *Evolution* 46:1052–1069 (1992).

WHAT IF? *Data from additional studies showed that when soapberry bug eggs from a population that fed on balloon vine fruits were then reared on golden rain tree fruits (or vice versa), the beak lengths of the adult insects were most similar to those in the population from which the eggs were initially obtained. Interpret these results.*

length nearly doubled the success with which soapberry bugs could eat the seeds of the introduced species. Furthermore, since historical data show that the golden rain tree reached central Florida just 35 years before the scientific studies were initiated, the results demonstrate that natural selection can cause rapid evolution in a wild population.

The Evolution of Drug-Resistant Bacteria

An example of ongoing natural selection that dramatically affects humans is the evolution of drug-resistant pathogens (disease-causing organisms and viruses). This is a particular problem with bacteria and viruses because they can produce new generations in a short period of time; as a result, resistant strains of these pathogens can proliferate very quickly.

Consider the evolution of drug resistance in the bacterium *Staphylococcus aureus*. About one in three people harbor this species on their skin or in their nasal passages with no negative effects. However, certain genetic varieties (strains) of this species, known as methicillin-resistant *S. aureus* (MRSA), are formidable pathogens. Most MRSA infections are caused by recently appearing strains such as clone USA300, which can cause "flesh-eating disease" and potentially fatal infections **(Figure 22.14)**. How did clone USA300 and other strains of MRSA become so dangerous?

The story begins in 1943, when penicillin became the first widely used antibiotic. Since then, penicillin and other antibiotics have saved millions of lives. However, by 1945, more than 20% of the *S. aureus* strains seen in hospitals were already resistant to penicillin. These bacteria had an enzyme, penicillinase, that could destroy penicillin. Researchers responded by developing antibiotics that were not destroyed by penicillinase, but resistance to each new drug occurred rapidly.

For example, a promising new antibiotic, methicillin, was introduced in 1959. But by 1961, methicillin-resistant strains of *S. aureus* were observed. How did these resistant strains emerge? Methicillin works by deactivating an enzyme that bacteria use to synthesize their cell walls. However, some *S. aureus* populations included individuals that were able to synthesize their cell walls using a different enzyme that was not affected by methicillin. These individuals survived the methicillin treatments and reproduced at higher rates than did other individuals. Over time, these resistant individuals became increasingly common, leading to the spread of MRSA.

Initially, MRSA could be controlled by antibiotics that work differently from the way methicillin works. But this has become less effective because some MRSA strains are resistant to multiple antibiotics—probably because bacteria can exchange genes with members of their own and other species. Thus, the multidrug-resistant strains of today may have emerged over time as MRSA strains that were resistant to different antibiotics exchanged genes.

Finally, it is important to note that *S. aureus* is not the only pathogenic bacterium that has evolved resistance to multiple antibiotics. Furthermore, in recent decades, antibiotic resistance has spread much faster than new antibiotics have been

▼ **Figure 22.14 Clone USA300: a virulent strain of methicillin-resistant *Staphylococcus aureus* (MRSA).** Resistant to multiple antibiotics and highly contagious, this strain and its close relatives can cause lethal infections of the skin, lungs, and blood. As shown here, researchers have identified key areas of the USA300 genome that code for adaptations that cause its virulent properties.

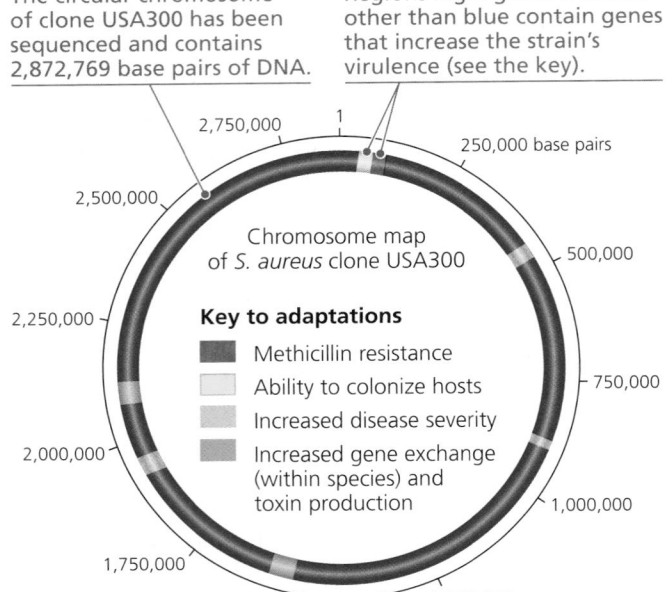

WHAT IF? *Some drugs being developed specifically target and kill only S. aureus; others slow the growth of MRSA but do not kill it. Based on how natural selection works and on the fact that bacterial species can exchange genes, explain why each of these strategies might be effective.*

discovered—a problem of great public health concern. Hope may loom on the horizon, however. For example, scientists recently reported the discovery of "teixobactin," a new antibiotic that shows promise for treating MRSA and other pathogens. In addition, as we'll describe in the Scientific Skills Exercise in Chapter 27, the methods used in the discovery of teixobactin may lead to the discovery of other new antibiotics as well.

The *S. aureus* and soapberry bug examples highlight three key points about natural selection. First, natural selection is a process of editing, not a creative mechanism. A drug does not *create* resistant pathogens; it *selects for* resistant individuals that are already present in the population. Second, in species that produce new generations in short periods of time, evolution by natural selection can occur rapidly—in just a few years (*S. aureus*) or decades (soapberry bugs). Third, natural selection depends on time and place. It favors those characteristics in a genetically variable population that provide an advantage in the current, local environment. What is beneficial in one situation may be useless or even harmful in another. Beak lengths suitable for the size of the typical fruit eaten by members of a particular soapberry bug population are favored by natural selection. However, a beak length suitable for fruit of one size can be disadvantageous when the bug is feeding on fruit of another size.

⊙ **Mastering Biology BBC Video: A Future Without Antibiotics?**

Homology

A second type of evidence for evolution comes from analyzing similarities among different organisms. As we've discussed, evolution is a process of descent with modification: Characteristics present in an ancestral organism are altered (by natural selection) in its descendants over time as they face different environmental conditions. As a result, related species can have characteristics that have an underlying similarity yet function differently. Similarity resulting from common ancestry is known as **homology**. As we'll describe, an understanding of homology can be used to make testable predictions and explain observations that are otherwise puzzling.

▼ **Figure 22.15 Mammalian forelimbs: homologous structures.** Even though they have become adapted for different functions, the forelimbs of all mammals are constructed from the same basic skeletal elements: one large bone (purple), attached to two smaller bones (orange and tan), attached to several small bones (gold), attached to several metacarpals (green), attached to approximately five digits, each of which is composed of multiple phalanges (blue).

Humerus

Radius

Ulna

Carpals

Metacarpals

Phalanges

Human   Cat   Whale   Bat

Anatomical and Molecular Homologies

The view of evolution as a remodeling process leads to the prediction that closely related species should share similar features—and they do. Of course, closely related species share the features used to determine their relationship, but they also share many other features. Some of these shared features make little sense except in the context of evolution. For example, the forelimbs of all mammals—including humans, cats, whales, and bats—show the same arrangement of bones from the shoulder to the tips of the digits, even though the appendages have very different functions: lifting, walking, swimming, and flying **(Figure 22.15)**. Such striking anatomical resemblances would be highly unlikely if these structures had arisen anew in each species. Rather, the underlying skeletons of the arms, forelegs, flippers, and wings of different mammals are **homologous structures** that represent variations on a structural theme that was present in their common ancestor.

Comparing early stages of development in different animal species reveals additional anatomical homologies not visible in adult organisms. For example, at some point in their development, all vertebrate embryos have a tail located posterior to (behind) the anus, as well as structures called pharyngeal (throat) arches **(Figure 22.16)**. These homologous arches ultimately develop into structures with very different functions, such as gills in fishes and parts of the ears and throat in humans and other mammals.

Some of the most intriguing homologies concern "leftover" structures of marginal, if any, importance to the organism. These **vestigial structures** are remnants of features that served a function in the organism's ancestors. For instance, snakes arose from ancestors with legs, and the skeletons of some snakes retain vestiges of the pelvis and leg bones of their ancestors. Likewise, blind species of cave fishes descended from ancestors with eyes—which explains why these blind fishes have eye remnants buried under their scales. We would not expect to see these vestigial structures if snakes and blind cave fishes had origins separate from those of other vertebrate animals.

Biologists also observe similarities among organisms at the molecular level. All forms of life use essentially the same genetic code, suggesting that all species descended from common ancestors that used this code. But molecular homologies go beyond a shared code. For example, organisms as dissimilar as humans and bacteria share genes inherited from a very distant common ancestor. Some of these homologous genes

▼ **Figure 22.16 Anatomical similarities in vertebrate embryos.** At some stage in their embryonic development, all vertebrates have a tail located posterior to the anus (referred to as a post-anal tail), as well as pharyngeal (throat) arches. Descent from a common ancestor can explain such similarities.

Pharyngeal arches

Post-anal tail

Chick embryo (LM)   Human embryo

have acquired new functions, while others, such as those coding for the ribosomal subunits used in protein synthesis (see Figure 17.18), have retained their original functions. It is also common for organisms to have genes that have lost their function, even though the homologous genes in related species may be fully functional. Like vestigial structures, it appears that such inactive "pseudogenes" may be present simply because a common ancestor had them.

Homologies and "Tree Thinking"

Some homologous characteristics, such as the genetic code, are shared by all species because they date to the deep ancestral past. In contrast, homologous characteristics that evolved more recently are shared only within smaller groups of organisms.

Consider the *tetrapods* (from the Greek *tetra*, four, and *pod*, foot), the vertebrate group that consists of amphibians, mammals, and reptiles. Like all vertebrates, tetrapods have a backbone. But unlike other vertebrates, tetrapods also have limbs with digits (see Figure 22.15). As suggested by this example, homologous characteristics form a nested pattern: All life shares the deepest layer (in this case, all vertebrates have a backbone), and each successive smaller group adds its own homologies to those it shares with larger groups (in this case, all tetrapods have a backbone *and* limbs with digits). This nested pattern is exactly what we would expect to result from descent with modification from a common ancestor.

Biologists often represent the pattern of descent from common ancestors with an **evolutionary tree**, a diagram that reflects evolutionary relationships among groups of organisms. We will explore evolutionary trees in more detail in Chapter 26, but for now, let's consider how we can interpret and use such trees.

Figure 22.17 is an evolutionary tree of tetrapods and their closest living relatives, the lungfishes. In this diagram, each branch point represents the most recent common ancestor of the two lineages diverging from that point. For example, lungfishes and all tetrapods

descended from ancestor ❶, whereas mammals, lizards and snakes, crocodiles, and birds all descended from ancestor ❸. As expected, the three homologies shown on the tree—limbs with digits, the amnion (a protective embryonic membrane), and feathers—form a nested pattern. Limbs with digits were present in common ancestor ❷ and hence are found in all of the descendants of that ancestor (the tetrapods). The amnion was present only in ancestor ❸ and hence is shared only by some tetrapods (mammals and reptiles). Feathers were present only in ancestor ❻ and hence are found only in birds.

To explore "tree thinking" further, note that in Figure 22.17 mammals are positioned closer to amphibians than to birds. As a result, you might conclude that mammals are more closely related to amphibians than they are to birds. However, mammals are actually more closely related to birds than to amphibians because mammals and birds share a more recent common ancestor (ancestor ❸) than do mammals and amphibians (ancestor ❷). Ancestor ❷ is also the most recent common ancestor of birds and amphibians, making mammals and birds equally related to amphibians. Finally, note that the tree in Figure 22.17 shows the relative timing of events but not their actual dates. Thus, we can conclude that ancestor ❷ lived before ancestor ❸, but we do not know when that was.

Evolutionary trees are hypotheses that summarize our current understanding of patterns of descent. Our confidence in these relationships, as with any hypothesis, depends on the strength of the supporting data. In the case of Figure 22.17, the tree is supported by many different data sets, including both anatomical and DNA sequence data. As a result, biologists are confident that it accurately reflects evolutionary history. Scientists can use such well-supported evolutionary trees to make specific and sometimes surprising predictions about organisms (see Figure 26.17).

▶ **Figure 22.17 Tree thinking: information provided in an evolutionary tree.** This evolutionary tree for tetrapods and their closest living relatives, the lungfishes, is based on anatomical and DNA sequence data. The purple hatch marks indicate the origin of three important homologies, each of which evolved only once. Birds are nested within and evolved from reptiles; hence, the group of organisms called "reptiles" technically includes birds.

VISUAL SKILLS *Based on this evolutionary tree, are crocodiles more closely related to lizards or birds? Explain.*

 Mastering Biology Figure Walkthrough

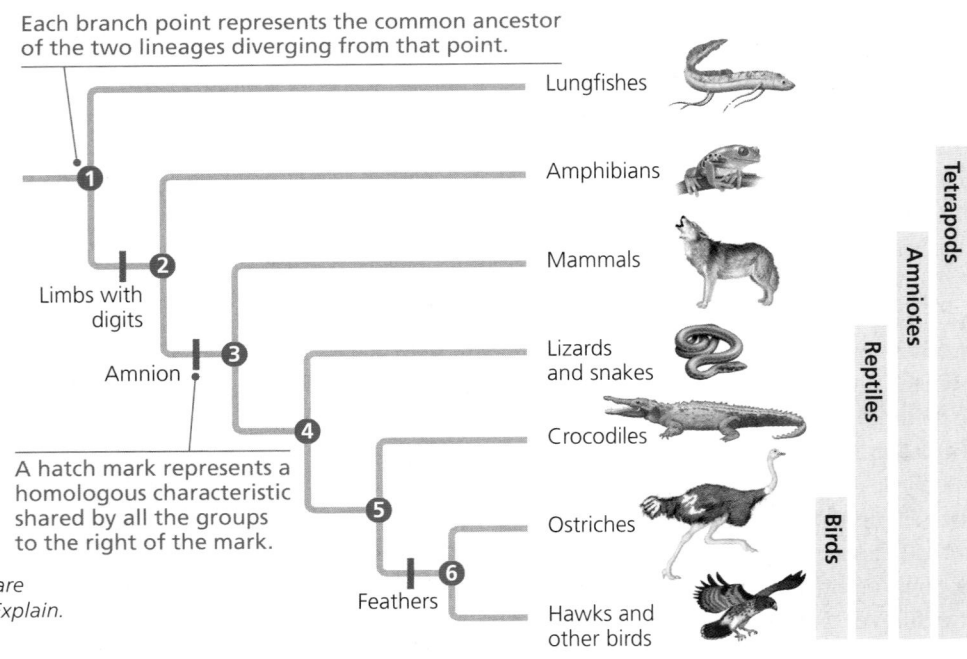

Each branch point represents the common ancestor of the two lineages diverging from that point.

Limbs with digits

Amnion

A hatch mark represents a homologous characteristic shared by all the groups to the right of the mark.

Feathers

Lungfishes
Amphibians
Mammals
Lizards and snakes
Crocodiles
Ostriches
Hawks and other birds

Tetrapods
Amniotes
Reptiles
Birds

A Different Cause of Resemblance: Convergent Evolution

Although organisms that are closely related share characteristics because of common descent, distantly related organisms can resemble one another for a different reason: **convergent evolution**, the independent evolution of similar features in different lineages. Consider marsupial mammals, many of which live in Australia. Marsupials are distinct from another group of mammals—the eutherians, or placental mammals—few of which live in Australia. (Eutherians complete their embryonic development in the uterus, whereas marsupials are born as embryos and complete their development while nursing, often in an external pouch.) Some Australian marsupials have eutherian look-alikes with superficially similar adaptations. For instance, a forest-dwelling Australian marsupial called the sugar glider is superficially very similar to flying squirrels, gliding eutherians that live in North American forests **(Figure 22.18)**. But the sugar glider has many other characteristics that make it a marsupial, much more closely related to kangaroos and other Australian marsupials than to flying squirrels or other eutherians. Once again, our understanding of evolution can explain these observations. Although they evolved independently from different ancestors, these two mammals have adapted to similar environments in similar ways. In such examples in which species share features because of convergent evolution, the resemblance is said to be **analogous**, not homologous. Analogous features share similar function, but not common ancestry, while homologous features share common ancestry, but not necessarily similar function.

▲ **Figure 22.18 Convergent evolution.** The ability to glide through the air evolved independently in these two distantly related mammals.

The Fossil Record

A third type of evidence for evolution comes from fossils. The fossil record documents the pattern of evolution, showing that past organisms differed from present-day organisms and that many species have become extinct. Fossils also show the evolutionary changes that have occurred in various groups of organisms. To give one of hundreds of possible examples, researchers found that over several thousand years, the pelvic bone in fossil stickleback fish became greatly reduced in size. The consistent nature of this change over time suggests that the reduction in the size of the pelvic bone may have been driven by natural selection.

Fossils can also shed light on the origins of new groups of organisms. An example is the fossil record of cetaceans, the mammalian order that includes whales, dolphins, and porpoises. Some of these fossils **(Figure 22.19)** provided strong support for a hypothesis based on DNA sequence data: that cetaceans are closely related to even-toed ungulates, a group that includes hippopotamuses, pigs, deer, and cows.

What else can fossils tell us about cetacean origins? The earliest cetaceans lived 50–60 million years ago. The fossil record indicates that prior to that time, most mammals were terrestrial. Although scientists had long realized that whales and other cetaceans originated from land mammals, few fossils had been found that revealed how cetacean limb structure had changed over time, leading eventually to the development of flippers as well as to the loss of hind limbs and the development of flukes (the lobes on a whale's tail). In the past few decades, however, a series of remarkable fossils have been discovered in Pakistan, Egypt, and North America. These fossils document steps in the transition from life on land to life in the sea, filling in some of the gaps between ancestral and living cetaceans **(Figure 22.20)**.

Finally, although living cetaceans look very different from living even-toed ungulates, the fossil discoveries show that extinct early cetaceans such as *Pakicetus* looked fairly similar

▼ **Figure 22.19 Ankle bones: one piece of the puzzle.** Comparing fossils and present-day examples of the astragalus (a type of ankle bone) indicates that cetaceans are closely related to even-toed ungulates. **(a)** In most mammals, the astragalus is shaped like that of a dog, with a double hump on one end (red arrows) but not at the opposite end (blue arrow). **(b)** Fossils show that the early cetacean *Pakicetus* had an astragalus with double humps at both ends, a shape otherwise found only in pigs **(c)**, deer **(d)**, and all other even-toed ungulates.

Most mammals   Cetaceans and even-toed ungulates

(a) *Canis* (dog)   (b) *Pakicetus*   (c) *Sus* (pig)   (d) *Odocoileus* (deer)

▼ **Figure 22.20 The transition to life in the sea.** Multiple lines of evidence support the hypothesis that cetaceans (yellow labels) evolved from terrestrial mammals. Fossils document the reduction over time in the pelvis and hind limb bones of extinct (†) cetacean ancestors, including *Pakicetus, Rodhocetus,* and *Dorudon*. DNA sequence data support the hypothesis that cetaceans are most closely related to hippopotamuses.

Other even-toed ungulates

Hippopotamuses

Pelvis and hind limb bones

†*Pakicetus*

†*Rodhocetus*

†*Dorudon*

Living cetaceans

Common ancestor of cetaceans

70   60   50   40   30   20   10   0
Millions of years ago

Key to cetacean pelvis and hind limb bones   ■ Pelvis   ■ Tibia   ■ Femur   ■ Foot

VISUAL SKILLS *Use the diagram to determine which happened first during the evolution of cetaceans: changes in hind limb structure or the origin of tail flukes (the lobes on a whale's tail). Explain.*

➔ **Mastering Biology Figure Walkthrough**

to deer and other living even-toed ungulates. The extinct early cetaceans also looked very similar to extinct early even-toed ungulates, such as *Diacodexis* **(Figure 22.19)**. Similar patterns are seen in fossils documenting the origins of other groups of organisms, including mammals (see Figure 25.7), flowering plants (see Concept 30.3), and tetrapods (see Figure 34.21). In each of these cases, the fossil record shows that over time, descent with modification produced increasingly large differences among related groups of organisms, ultimately resulting in the diversity of life today.

▼ **Figure 22.21 *Diacodexis,* an early even-toed ungulate.**

20 cm

Biogeography

A fourth type of evidence comes from **biogeography**, the scientific study of the geographic distributions of species. These distributions are influenced by many factors, including *continental drift*, the slow movement of Earth's continents over time. About 250 million years ago, these movements united all of Earth's landmasses into a single large continent called **Pangaea**

▼ **Figure 22.22 Closely related freshwater fish separated by 9,000 km of ocean.** There are three species in these two closely related genera of freshwater fish within the family Galaxiidae.

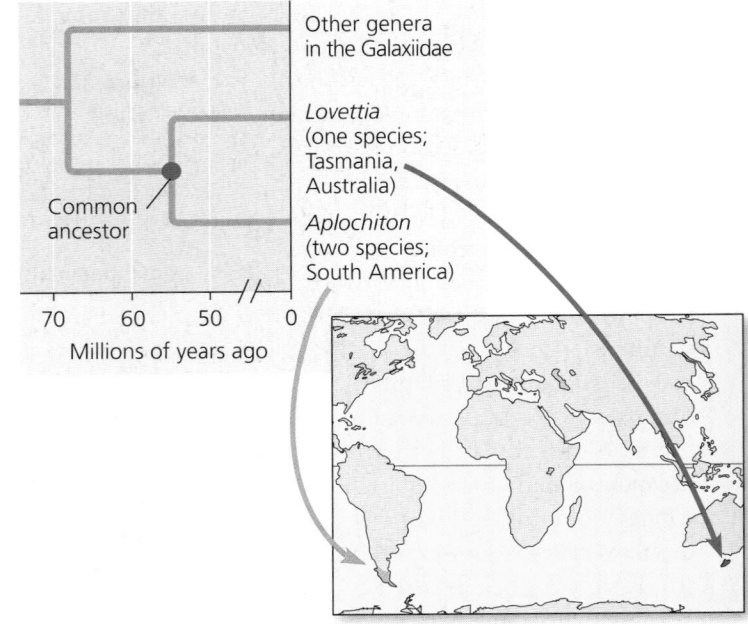

Other genera in the Galaxiidae

Lovettia (one species; Tasmania, Australia)

Aplochiton (two species; South America)

Common ancestor

70   60   50   0
Millions of years ago

(see Figure 25.16). Roughly 200 million years ago, Pangaea began to break apart; by 20 million years ago, the continents were within a few hundred kilometers of their present locations.

We can use our understanding of evolution and continental drift to predict where fossils of different groups of organisms might be found. For example, scientists have constructed evolutionary trees for horses based on anatomical data. These trees and the ages of fossils of horse ancestors suggest that the genus that includes present-day horses (*Equus*) originated 5 million years ago in North America. Geologic evidence indicates that at that time, North and South America were not yet connected, making it difficult for horses to travel between them. Thus, we would predict that the oldest *Equus* fossils should be found only on the continent on which the group originated—North America. This prediction and others like it for different groups of organisms have been upheld, providing more evidence for evolution.

Evolution also explains biogeographic data. For example, a group of three species of freshwater fish in the family Galaxiidae are found in regions separated by wide stretches of open ocean (**Figure 22.22**). None of these three species can survive in salt water, so how can we explain their geographic distributions? These three species represent all known living descendants of a common ancestor. Since all of the living species cannot survive in salt water, it is likely that their common ancestor also was restricted to freshwater habitats. At first glance, this only deepens the mystery—how could that ancestor or its descendants have swum across vast stretches of ocean? A clue comes from genetic analyses indicating that the evolutionary lineages leading to these three species began to diverge about 55 million years ago. By that time, the southernmost portion of Pangaea was breaking apart into what would eventually become South America, Australia, and Antarctica. This suggests that their ancestors had spread across southern Pangaea before it broke apart. At later times, these three freshwater species originated on the continental regions where they currently live, resulting in their present-day geographic distribution.

What Is Theoretical About Darwin's View of Life?

Some people dismiss Darwin's ideas as "just a theory." But the *pattern* of evolution—the observation that life has evolved

---

Scientific Skills Exercise

Making and Testing Predictions

Can Predation Result in Natural Selection for Color Patterns in Guppies? Our understanding of evolution changes constantly as new observations lead to new hypotheses—and hence to new ways to test our understanding of evolutionary theory. Consider the wild guppies (*Poecilia reticulata*) that live in pools connected by streams on the Caribbean island of Trinidad. Male guppies have highly varied color patterns that are controlled by genes that are only expressed in adult males. Female guppies choose males with bright color patterns as mates more often than they choose males with drab coloring. But the bright colors that attract females also can make the males more conspicuous to predators. Researchers observed that in pools with few predator species, the benefits of bright colors appear to "win out," and males are more brightly colored than in pools where predation is more intense.

One guppy predator, the killifish, preys on juvenile guppies that have not yet displayed their adult coloration. Researchers predicted that if adult guppies with drab colors were transferred to a pool with only killifish, eventually the descendants of these guppies would be more brightly colored (because of the female preference for brightly colored males).

Guppies transplanted

Pools with pike-cichlids and guppies

Pools with killifish, but no guppies prior to transplant

How the Experiment Was Done Researchers transplanted 200 adult guppies from pools containing pike-cichlid fish, predators of adult guppies, to pools with only killifish, predators of juvenile guppies. They tracked the number of bright-colored spots and the total area of those spots on adult male guppies in each generation.

Data from the Experiment After 22 months (15 generations), researchers compared the color pattern data for guppies from the source and transplanted populations.

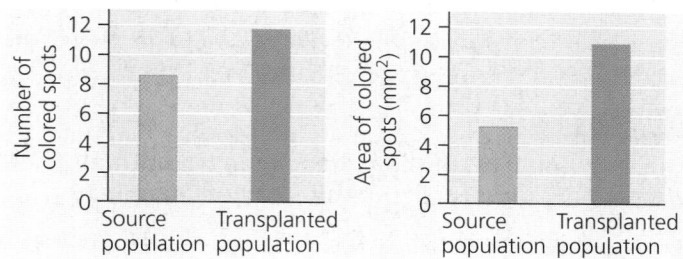

Data from J. A. Endler, Natural selection on color patterns in *Poecilia reticulata*, *Evolution* 34:76–91 (1980).

INTERPRET THE DATA

1. Identify the following elements of hypothesis-based science in this example: **(a)** question, **(b)** hypothesis, **(c)** prediction, **(d)** control group, and **(e)** experimental group. (For additional information about hypothesis-based science, see Concept 1.3 and the Scientific Skills Review in Appendix D.)
2. Explain how the types of data the researchers chose to collect enabled them to test their prediction.
3. What conclusion do you draw from the data presented above?
4. **(a)** Predict what would happen if, after 22 months, guppies from the transplanted population were returned to the source pool. **(b)** Describe an experiment to test your prediction.

➡ **Instructors:** A version of this Scientific Skills Exercise can be assigned in **Mastering Biology**.

over time—is supported by a great deal of evidence. In addition, Darwin's explanation of the *process* of evolution—that natural selection is the primary cause of the observed pattern of evolutionary change—makes sense of massive amounts of data. As the **Scientific Skills Exercise** describes, the effects of natural selection also can be observed and tested in nature.

What, then, is theoretical about evolution? Keep in mind that the scientific meaning of the term *theory* is very different from its meaning in everyday use. The colloquial use of the word *theory* comes close to what scientists mean by a hypothesis. In science, a theory is much more comprehensive than a hypothesis. A theory, such as the theory of evolution by natural selection, accounts for many observations and explains and integrates a great variety of phenomena. Such a unifying theory does not become widely accepted unless its predictions stand up to thorough and continual testing by experiment and additional observation (see Concept 1.3). As the rest of this unit demonstrates, this has certainly been the case with the theory of evolution by natural selection.

The skepticism of scientists as they continue to test theories prevents these ideas from becoming dogma. For example, although Darwin thought that evolution was a very slow process, we now know that this isn't always true. Populations can evolve rapidly, and new species can form in relatively short periods of time: a few thousand years or less. Furthermore, evolutionary biologists now recognize that natural selection is not the only mechanism responsible for evolution. Indeed, the study of evolution today is livelier than ever as scientists use a wide range of experimental approaches and genetic analyses to test predictions based on natural selection and other evolutionary mechanisms.

Although Darwin's theory attributes life's diversity to natural processes, the diverse products of evolution are nevertheless elegant and inspiring. As Darwin wrote in the final sentence of *The Origin of Species*, "There is grandeur in this view of life . . . [in which] endless forms most beautiful and most wonderful have been, and are being, evolved."

CONCEPT CHECK 22.3

1. Explain how the following statement is inaccurate: "Antibiotics have created drug resistance in MRSA."
2. How does evolution account for (a) the similar mammalian forelimbs with different functions shown in Figure 22.15 and (b) the similar forms of the two distantly related mammals shown in Figure 22.18?
3. **WHAT IF?** Fossils show that dinosaurs originated 200–250 million years ago. Would you expect the geographic distribution of early dinosaur fossils to be broad (on many continents) or narrow (on one or a few continents only)? Explain.

For suggested answers, see Appendix A.

22 Chapter Review

→ Go to **Mastering Biology** for Assignments, the eText, the Study Area, and Dynamic Study Modules.

SUMMARY OF KEY CONCEPTS

→ To review key terms, go to the **Vocabulary Self-Quiz** in the **Mastering Biology** eText or Study Area, or go to goo.gl/zkjz9t.

CONCEPT 22.1

The Darwinian revolution challenged traditional views of a young Earth inhabited by unchanging species (pp. 469–471)

- Darwin proposed that life's diversity arose over long periods of time from ancestral species through natural selection, a departure from prevailing views. For example, Cuvier studied **fossils** but denied that **evolution** occurs; he proposed that sudden catastrophic events in the past caused species to disappear from an area. Lamarck hypothesized that species evolve, but the underlying mechanisms he proposed are not supported by evidence.

? *Why was the age of Earth important for Darwin's ideas about evolution?*

CONCEPT 22.2

Descent with modification by natural selection explains the adaptations of organisms and the unity and diversity of life (pp. 471–476)

- Darwin's experiences during the voyage of the *Beagle* gave rise to his idea that new species originate from ancestral forms through the accumulation of **adaptations**. He refined his theory for many years and finally published it in 1859 after learning that Wallace had come to the same idea.
- In *The Origin of Species*, Darwin proposed that over long periods of time, descent with modification produced the rich diversity of life through the mechanism of **natural selection**.

Observations

| Individuals in a population vary in their heritable characteristics. | Organisms produce more offspring than the environment can support. |

Inferences

Individuals that are well suited to their environment tend to leave more offspring than other individuals.

and

Over time, favorable traits accumulate in the population.

? *Describe how overreproduction and heritable variation relate to evolution by natural selection.*

CONCEPT 22.3

Evolution is supported by an overwhelming amount of scientific evidence (pp. 476–484)

- Researchers have directly observed natural selection leading to evolution in many studies in diverse organisms.
- Organisms share characteristics because of common descent (**homology**) or because natural selection affects independently evolving species in similar environments in similar ways (**convergent evolution**).
- Fossils show that past organisms differed from living organisms, that many species have become extinct, and that species have evolved over long periods of time; fossils also document the evolutionary origin of new groups of organisms.
- Evolutionary theory can explain some **biogeographic** patterns.

? *Summarize the different lines of evidence supporting the hypothesis that cetaceans descended from land mammals and are closely related to even-toed ungulates.*

TEST YOUR UNDERSTANDING

→ For more multiple-choice questions, go to the **Practice Test** in the Mastering Biology eText or Study Area, or go to goo.gl/GruWRg.

Levels 1-2: Remembering/Understanding

1. Which of the following is an observation or inference on which natural selection is based?
   (A) Individuals do not vary in their heritable characteristics.
   (B) Only well-adapted individuals produce offspring.
   (C) Species produce more offspring than the environment can support.
   (D) Nearly all of each individual's offspring will survive and reproduce.

2. Which of the following observations helped Darwin shape his concept of descent with modification?
   (A) Species diversity declines farther from the equator.
   (B) Fewer species live on islands than on the nearest continents.
   (C) Birds live on islands located farther from the mainland than the birds' maximum nonstop flight distance.
   (D) South American temperate plants are more similar to South American tropical plants than to European temperate ones.

Levels 3-4: Applying/Analyzing

3. Within six months of effectively using methicillin to treat *S. aureus* infections in a community, all new *S. aureus* infections were caused by a resistant strain (MRSA). How can this best be explained?
   (A) A patient must have become infected with MRSA from another community.
   (B) In response to the drug, *S. aureus* began making drug-resistant versions of the protein targeted by the drug.
   (C) Some drug-resistant bacteria were present at the start of treatment, and natural selection increased their frequency.
   (D) *S. aureus* evolved to resist vaccines.

4. DNA sequences in many human genes are very similar to the sequences of corresponding genes in chimpanzees. The most likely explanation for this result is that
   (A) humans and chimpanzees share a relatively recent common ancestor.
   (B) humans evolved from chimpanzees.
   (C) chimpanzees evolved from humans.
   (D) convergent evolution led to the DNA similarities.

5. The upper forelimbs of humans and bats have fairly similar skeletal structures, whereas the corresponding bones in whales have very different shapes and proportions. However, genetic data suggest that all three kinds of organisms diverged from a common ancestor at about the same time. Which of the following is the most likely explanation for these data?
   (A) Forelimb evolution was adaptive in people and bats, but not in whales.
   (B) Natural selection in an aquatic environment resulted in significant changes to whale forelimb anatomy.
   (C) Genes mutate faster in bats than in humans or whales.
   (D) Whales are not properly classified as mammals.

Levels 5-6: Evaluating/Creating

6. **EVOLUTION CONNECTION** Explain why anatomical and molecular features often fit a similar nested pattern. In addition, describe a process that can cause this not to be the case.

7. **SCIENTIFIC INQUIRY • DRAW IT** Mosquitoes resistant to the pesticide DDT first appeared in India in 1959, but now are found throughout the world. (a) Graph the data in the table below. (b) After examining the graph, hypothesize why the percentage of mosquitoes resistant to DDT rose rapidly. (c) Suggest an explanation for the global spread of DDT resistance.

Month	0	8	12
Mosquitoes Resistant* to DDT	4%	45%	77%

*Mosquitoes were considered resistant if they were not killed within 1 hour of receiving a dose of 4% DDT.

Data from C. F. Curtis et al., Selection for and against insecticide resistance and possible methods of inhibiting the evolution of resistance in mosquitoes, *Ecological Entomology* 3:273–287 (1978).

8. **WRITE ABOUT A THEME: INTERACTIONS** Write a short essay (about 100–150 words) evaluating whether changes to an organism's physical environment are likely to result in evolutionary change. Use an example to support your reasoning.

9. **SYNTHESIZE YOUR KNOWLEDGE**

This honeypot ant (genus *Myrmecocystus*) can store liquid food inside its expandable abdomen. Consider other ants you are familiar with, and explain how a honeypot ant exemplifies three key features of life: adaptation, unity, and diversity.

For selected answers, see Appendix A.

23 The Evolution of Populations

KEY CONCEPTS

23.1 Genetic variation makes evolution possible *p. 487*

23.2 The Hardy-Weinberg equation can be used to test whether a population is evolving *p. 489*

23.3 Natural selection, genetic drift, and gene flow can alter allele frequencies in a population *p. 493*

23.4 Natural selection is the only mechanism that consistently causes adaptive evolution *p. 497*

Study Tip

Make a table: This chapter discusses different types of selection—disruptive, stabilizing, sexual, balancing, frequency-dependent, and heterozygote advantage. These types of selection are not all mutually exclusive, and some evolutionary changes are examples of more than one. To help you study, make a table like the one shown listing each type of selection, its distinguishing features, and example(s) from the chapter.

Type of selection	Distinguishing features	Example
Balancing selection	Maintains two or more phenotypes in a population; includes frequency-dependent selection and heterozygote advantage	Figures 23.17 and 23.18
Directional selection		

Figure 23.1 The medium ground finch (*Geospiza fortis*) is a seed-eating bird that inhabits the Galápagos Islands. In 1977, the *G. fortis* population on the island of Daphne Major was decimated by a long period of drought: Of some 1,200 birds, only 180 survived. The surviving birds had larger, deeper beaks, indicating that this population of finches had evolved.

What mechanisms can cause the evolution of populations?

Evolution occurs when natural selection, genetic drift, or gene flow alter the allele frequencies of a population over time.

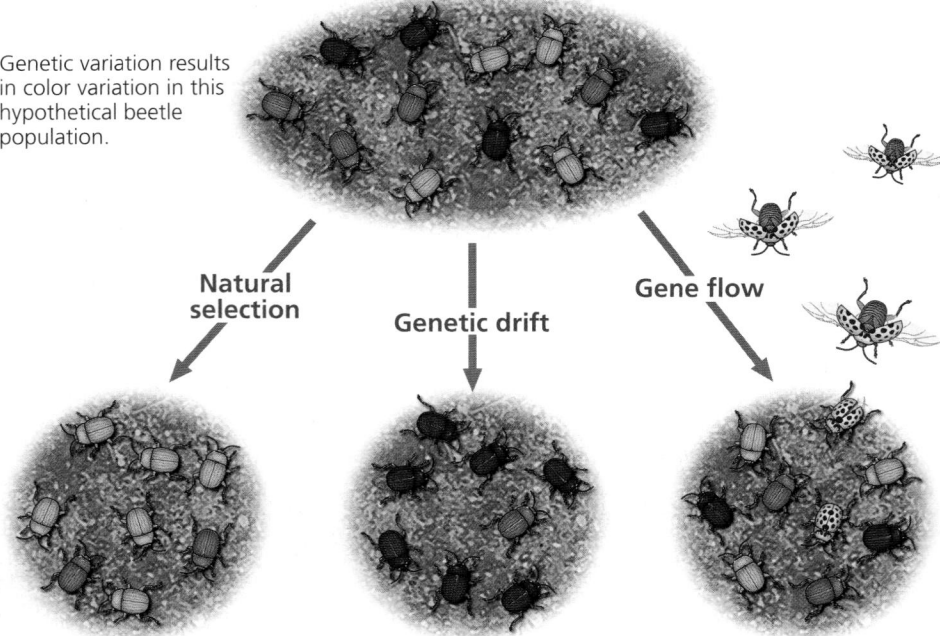

Genetic variation results in color variation in this hypothetical beetle population.

Natural selection

Genetic drift

Gene flow

The more visible darker beetles have been preyed on at a higher rate, causing lighter beetles (and the alleles they carry) to become more common over time.

By chance alone, alleles encoding darker color have become more common over several generations.

The arrival of beetles with a spotted color pattern has brought new alleles into the population.

A common misconception about evolution is that individual organisms evolve. It is true that natural selection acts on individuals: Each organism's traits affect its survival and reproductive success compared with other individuals. But the impact of natural selection only becomes apparent in how a *population* of organisms changes over time.

To see why, consider the population of medium ground finches decimated by a drought, resulting in only 180 survivors out of some 1,200 birds. Researchers Peter and Rosemary Grant observed that during the drought, small, soft seeds were scarce, while large, hard seeds were more plentiful. Birds with larger, deeper beaks were better able to crack and eat the larger seeds and survived at a higher rate than finches with smaller beaks.

Since beak depth is an inherited trait in these birds, the offspring of surviving birds also tended to have deep beaks. As a result, the average beak depth in the next generation of *G. fortis* was greater than it had been in the pre-drought population **(Figure 23.2)**. The finch population had evolved by natural selection. However, the *individual* finches did not evolve. Each bird had a beak of a particular size, which did not grow larger during the drought. Rather, the proportion of large beaks in the population increased from generation to generation: The population evolved, not its individual members.

Focusing on evolutionary change in populations, we can define evolution on its smallest scale, called **microevolution**, as a change in allele frequencies in a population over generations. In the finch population, for example, birds that had alleles encoding large beaks survived at higher rates than did other birds—causing those alleles to be more common after the drought than they had been before it. There are three main mechanisms that can cause such changes in allele frequencies: natural selection, genetic drift (chance events that alter allele frequencies), and gene flow (the transfer of alleles between populations). Each of these mechanisms affects the genetic composition of populations, but only natural selection consistently improves the degree to which organisms are well suited for life in their environment (adaptation). Before we examine this more closely, let's revisit a prerequisite for these processes: genetic variation.

 Mastering Biology
Interview with Rosemary and Peter Grant: Studying evolution in the Galápagos finches

▶ **Figure 23.2 Evidence of selection by food source.**
The data represent adult beak depths of finches hatched before and after the 1977 drought. In one generation, natural selection resulted in a larger average beak size in the population.

 Instructors: A related Experimental Inquiry Tutorial can be assigned in **Mastering Biology**.

Genetic variation makes evolution possible

In *The Origin of Species*, Darwin provided abundant evidence that life on Earth has evolved over time, and he proposed natural selection as the primary mechanism for that change. He observed that individuals differ in their inherited traits and that selection acts on such differences, leading to evolutionary change. Although Darwin realized that variation in heritable traits is a prerequisite for evolution, he did not know precisely how organisms pass heritable traits to their offspring.

Just a few years after Darwin published *The Origin of Species*, Gregor Mendel wrote a groundbreaking paper on inheritance in pea plants proposing a model of inheritance in which organisms transmit discrete heritable units (now called genes) to their offspring. Although Darwin did not know about genes, Mendel's paper set the stage for understanding the genetic differences on which evolution is based.

Genetic Variation

Individuals within all species vary in their phenotypic traits. Humans, for example, vary noticeably in facial features, height, and voice. And though you cannot identify blood group (A, B, AB, or O) from a person's appearance, this and many other molecular traits also vary extensively among individuals.

Such phenotypic variations often reflect **genetic variation**, differences among individuals in the composition of their genes or other DNA sequences. Some heritable phenotypic differences occur on an "either-or" basis, such as the flower colors of Mendel's pea plants: Each plant had flowers that were either purple or white (see Figure 14.3). Characters that vary in this way are typically determined by a single gene locus, with different alleles producing distinct phenotypes. In contrast, other phenotypic differences vary in gradations along a continuum. Such variation usually results from the influence of two or more genes on a single phenotypic character. In fact, many phenotypic characters are influenced by multiple genes, including coat color in horses **(Figure 23.3)**, seed number in maize (corn), and height in humans.

How much do genes and other DNA sequences vary from one individual to another? Genetic variation at the

▼ **Figure 23.3 Phenotypic variation in horses.**

▼ Figure 23.4 Extensive genetic variation at the molecular level.

This diagram summarizes data from a study comparing the DNA sequence of the alcohol dehydrogenase (*Adh*) gene in several fruit flies (*Drosophila melanogaster*). The *Adh* gene has four exons (dark blue) separated by introns (light blue); the exons include the coding regions that are ultimately translated into the amino acids of the Adh enzyme (see Figure 5.1). Only one substitution has a phenotypic effect, producing a different form of the Adh enzyme.

MAKE CONNECTIONS *Review Figures 17.6 and 17.11. Explain how a base-pair substitution that alters a coding region of the Adh locus could have no effect on amino acid sequence. Then explain how an insertion in an exon could have no effect on the protein produced.*

Base-pair substitutions are shown in orange.

A red arrow indicates an insertion site.

The substitution at this site results in the translation of a different amino acid.

A deletion of 26 base pairs occurred here.

1   500   1,000   1,500   2,000   2,500

Exon   Intron

whole-gene level (*gene variability*) can be quantified as the average percentage of loci that are heterozygous. (Recall that a heterozygous individual has two different alleles for a given locus, whereas a homozygous individual has two identical alleles for that locus.) As an example, on average the fruit fly *Drosophila melanogaster* is heterozygous for about 1,920 of its 13,700 loci (14%) and homozygous for all the rest.

Considerable genetic variation can also be measured at the molecular level of DNA (*nucleotide variability*). But little of this variation results in phenotypic variation. Why? Many nucleotide variations occur within *introns*, noncoding segments of DNA lying between *exons*, the regions retained in mRNA after RNA processing (see Figure 17.12). And of the variations that occur within exons, most do not cause a change in the amino acid sequence of the protein encoded by the gene. For example, in the sequence comparison shown in **Figure 23.4**, there are 43 nucleotide sites with variable base pairs (where substitutions have occurred), as well as several sites where insertions or deletions have occurred. Although 18 variable sites occur within the four exons of the *Adh* gene, only one of these variations (at site 1,490) results in an amino acid change. Note, however, that this single variable site is enough to cause genetic variation at the level of the gene, and hence two different forms of the Adh enzyme are produced.

It is important to bear in mind that some phenotypic variation does not result from genetic differences among individuals (**Figure 23.5** shows a striking example in a caterpillar of the southwestern United States). Phenotype is the product of an inherited genotype and many environmental influences (see Concept 14.3). In a human example, bodybuilders alter their phenotypes dramatically but do not pass their huge muscles on to the next generation. In general, only the genetically determined part of phenotypic variation can have evolutionary consequences. As such, genetic variation provides the raw material for evolutionary change: *Without genetic variation, evolution cannot occur.*

Sources of Genetic Variation

Genetic variation originates when mutation, gene duplication, or other processes produce new alleles and new genes. Genetic variants can be produced rapidly in organisms with short generation times. Sexual reproduction can also result in genetic variation as existing genes are arranged in new ways.

Formation of New Alleles

New alleles can arise by *mutation*, a change in the nucleotide sequence of an organism's DNA. Mutations can be caused by factors such as errors in DNA replication, exposure to UV light and other high-energy forms of radiation, and exposure to certain chemicals (see Concept 17.5). A change of as little as one base in a gene—a "point mutation"—can have a significant impact on phenotype, as in sickle-cell disease (see Figure 23.18). Furthermore, organisms reflect many generations of past selection, and hence their phenotypes tend to be well suited for life in their environments. As a result, most new mutations that alter a phenotype are at least slightly harmful.

In some cases, natural selection quickly removes such harmful alleles. In diploid organisms, however, harmful alleles that are recessive can be hidden from selection. Indeed, a harmful recessive allele can persist for generations by propagation in heterozygous individuals (where its harmful effects can be masked by the more favorable dominant allele). Such "heterozygote protection" maintains a huge pool of alleles

▼ **Figure 23.5 Nonheritable variation.** These caterpillars of the moth *Nemoria arizonaria* owe their different appearances to chemicals in their diets, not to differences in their genotypes. **(a)** Caterpillars raised on a diet of oak flowers resemble the flowers, whereas **(b)** their siblings raised on oak leaves resemble oak twigs.

(a)

(b)

that may be harmful under present conditions, but that could be beneficial if the environment changes.

While many mutations are harmful, many others are not. Recall that much of the DNA in eukaryotic genomes does not encode proteins (see Figure 21.6). Point mutations in these noncoding regions generally result in **neutral variation**, differences in DNA sequence that do not confer a selective advantage or disadvantage. The redundancy in the genetic code is another source of neutral variation: Even a point mutation in a gene that encodes a protein will have no effect on the protein's function if the amino acid doesn't change. And even a change in the amino acid may not affect the protein's shape and function. Moreover, a mutant allele may on rare occasions actually make its bearer better suited to the environment, enhancing reproductive success.

Finally, note that in multicellular organisms, only mutations in cell lines that produce gametes can be passed to offspring. In plants and fungi, this is not as limiting as it may sound, since many different cell lines can produce gametes. But in most animals, the majority of mutations occur in somatic cells and are not passed to offspring.

Altering Gene Number or Position

Chromosomal changes that delete, disrupt, or rearrange many loci are usually harmful. However, when such large-scale changes leave genes intact, they may not affect the organisms' phenotype. In rare cases, chromosomal rearrangements may even be beneficial. For example, the translocation of part of one chromosome to a different chromosome could link genes in a way that produces a positive effect.

A key potential source of variation is the duplication of genes due to errors in meiosis (such as unequal crossing over), slippage during DNA replication, or the activities of transposable elements (see Concept 21.5). Duplications of large chromosome segments, like other chromosomal aberrations, are often harmful, but the duplication of smaller pieces of DNA may not be. Gene duplications that do not have severe effects can persist over generations, allowing mutations to accumulate. The result is an expanded genome with new genes that may take on new functions.

Such increases in gene number appear to have played a major role in evolution. For example, the remote ancestors of mammals had a single gene for detecting odors that has since been duplicated many times. As a result, humans today have about 380 functional olfactory receptor genes, and mice have about 1,200. This dramatic proliferation of olfactory genes probably helped early mammals, enabling them to detect faint odors and to distinguish among many different smells.

Rapid Reproduction

Mutation rates tend to be low in plants and animals, averaging about one mutation in every 100,000 genes per generation, and are often even lower in prokaryotes. But prokaryotes have many more generations per unit of time, so mutations can quickly generate genetic variation in their populations. The same is true of viruses. For instance, HIV has a generation time of about two days (that is, it takes two days for a newly formed virus to produce the next generation). HIV also has an RNA genome, which has a much higher mutation rate than a typical DNA genome due to the lack of RNA repair mechanisms in host cells (see Concept 19.2). Thus, single-drug treatments are less effective against HIV: Mutant forms of the virus, resistant to a particular drug, tend to proliferate quickly. The most effective AIDS treatments to date have been drug "cocktails" that combine several medications. This approach has worked well as it is less likely that a set of mutations that together confer resistance to *all* the drugs will occur in a short time period.

Sexual Reproduction

In organisms that reproduce sexually, most of the genetic variation in a population results from the unique combination of alleles that each individual receives from its parents. Of course, at the nucleotide level, all the differences among these alleles have originated from past mutations. Sexual reproduction then shuffles existing alleles and deals them at random to produce individual genotypes.

Three mechanisms contribute to this shuffling: crossing over, independent assortment of chromosomes, and fertilization (see Concept 13.4). During meiosis, homologous chromosomes, one inherited from each parent, trade some of their alleles by crossing over. The resulting recombinant chromosomes are then distributed at random into gametes. Then, because myriad possible mating combinations exist in a population, fertilization typically brings together gametes with different genetic backgrounds. Together, these three mechanisms ensure that sexual reproduction rearranges existing alleles into fresh combinations each generation, providing much of the genetic variation that makes evolution possible.

→ **Mastering Biology** Animation: Origins of Genetic Variation

---

CONCEPT CHECK 23.1

1. Explain why genetic variation within a population is a prerequisite for evolution.
2. Of all the mutations that occur in a population, why do only a small fraction become widespread?
3. **MAKE CONNECTIONS** If a population stopped reproducing sexually but still reproduced asexually, how would its genetic variation be affected over time? (See Concept 13.4.)

For suggested answers, see Appendix A.

CONCEPT 23.2

The Hardy-Weinberg equation can be used to test whether a population is evolving

The presence of genetic variation in a population does not guarantee that the population will evolve. For evolution to occur,

▼ **Figure 23.6 One species, two populations.** These two caribou populations in the Yukon are not totally isolated; they sometimes share the same area (shaded purple). Still, members of each population are most likely to breed within their own population.

MAP AREA
ALASKA CANADA

Porcupine herd range

Fortymile herd range

Porcupine herd

Fortymile herd

one or more factors must be at work. In this section, we'll explore one way to test whether evolution is occurring in a population.

Gene Pools and Allele Frequencies

A **population** is a group of individuals of the same species that live in the same area and interbreed, producing fertile offspring. Different populations of a species may be isolated geographically from one another, exchanging genetic material only rarely. This is common for species that live on widely separated islands or in different lakes, but not all populations are isolated **(Figure 23.6)**. Still, members of a population typically breed with one another and thus are more closely related to each other than to members of other populations.

We can characterize a population's genetic makeup by describing its **gene pool**, which consists of all copies of every type of allele at every locus in all members of the population. If only one allele exists for a particular locus in a population, that allele is said to be *fixed* in the gene pool, and all individuals are homozygous for that allele. But if there are two or more alleles for a particular locus in a population, individuals may be either homozygous or heterozygous.

For example, imagine a population of 500 wildflower plants with two alleles, C^R and C^W, for a locus that codes for flower pigment. These alleles show incomplete dominance; thus, each genotype has a distinct phenotype. Plants homozygous for the C^R allele ($C^R C^R$) produce red pigment and have red flowers; plants homozygous for the C^W allele ($C^W C^W$) produce no red pigment and have white flowers; and heterozygotes ($C^R C^W$) produce some red pigment and have pink flowers.

$C^R C^R$

$C^W C^W$

$C^R C^W$

Each genotype and each allele has a frequency (proportion) in the population. To see how these frequencies are calculated, suppose our population has 320 plants with red flowers (genotype $C^R C^R$), 160 with pink flowers (genotype $C^R C^W$), and 20 with white flowers (genotype $C^W C^W$). Since 320 of the 500 plants have red flowers, the frequency of the $C^R C^R$ genotype is 0.64 (320/500). Likewise, the frequency of the $C^R C^W$ genotype is 0.32 (160/500) and the frequency of the $C^W C^W$ genotype is 0.04 (20/500).

With respect to allele frequencies, wildflowers are diploid organisms, so each individual has two alleles for each of its genes. Hence, the 500 individuals in our population have a total of 1,000 copies of the gene for flower color. The C^R allele accounts for 800 of these copies (320 × 2 = 640 for $C^R C^R$ plants, plus 160 × 1 = 160 for $C^R C^W$ plants). Thus, the frequency of the C^R allele is 800/1,000 = 0.8 (80%).

When studying a locus with two alleles, the convention is to use p to represent the frequency of one allele and q to represent the frequency of the other allele. Thus, p, the frequency of the C^R allele in the gene pool of this population, is $p = 0.8$ (80%). And because there are only two alleles for this gene, the frequency of the C^W allele, represented by q, must be $q = 1 - p = 0.2$ (20%). For loci that have more than two alleles, the sum of all allele frequencies must still equal 1 (100%).

The Hardy-Weinberg Equation

One way to assess whether natural selection or other factors are causing evolution at a particular locus is to determine what the genetic makeup of a population would be if it were *not* evolving at that locus. We can then compare that with the data we actually observed for the population. If there are no differences, this suggests that the population is not evolving. If there are differences, this suggests that the population is evolving—and then we can investigate why.

Hardy-Weinberg Equilibrium

In a population that is not evolving, allele and genotype frequencies will remain constant from generation to generation, provided that only Mendelian segregation and recombination of alleles are at work. Such a population is said to be in **Hardy-Weinberg equilibrium**, named for the British mathematician and German physician, respectively, who independently developed this idea in 1908.

To determine whether a population is in Hardy-Weinberg equilibrium, it is helpful to think about genetic crosses in a new way. Previously, we used Punnett squares to determine the genotypes of offspring in a genetic cross (see Figure 14.5). Here, instead of considering the possible allele combinations from one cross, we'll consider the combination of alleles in *all* of the crosses in a population. Imagine that all the alleles for a given locus from all the individuals in a population are placed

▼ **Figure 23.7 Selecting alleles at random from a gene pool.**

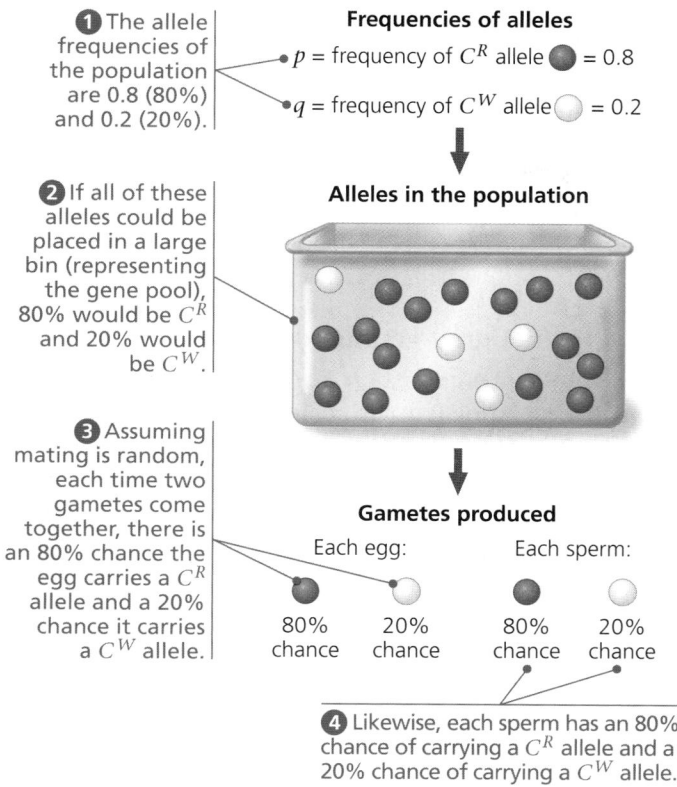

❶ The allele frequencies of the population are 0.8 (80%) and 0.2 (20%).

Frequencies of alleles

p = frequency of C^R allele ● = 0.8
q = frequency of C^W allele ○ = 0.2

❷ If all of these alleles could be placed in a large bin (representing the gene pool), 80% would be C^R and 20% would be C^W.

Alleles in the population

❸ Assuming mating is random, each time two gametes come together, there is an 80% chance the egg carries a C^R allele and a 20% chance it carries a C^W allele.

Gametes produced

Each egg: Each sperm:

80% chance   20% chance   80% chance   20% chance

❹ Likewise, each sperm has an 80% chance of carrying a C^R allele and a 20% chance of carrying a C^W allele.

DRAW IT *Draw a bin with six white balls instead of four. For the frequency of C^R to remain equal to 0.8, how many red balls should the bin contain?*

→ **Mastering Biology** BioFlix® **Animation: Allele Frequencies**

in a large bin **(Figure 23.7)**. We can think of this bin as holding the population's gene pool for that locus. "Reproduction" occurs by selecting alleles at random from the bin; somewhat similar events occur in nature when fishes release sperm and eggs into the water or when plant sperm (in pollen) is blown about by the wind. By viewing reproduction as a process of randomly selecting and combining alleles from the bin (the gene pool), we are in effect assuming that mating occurs at random: that all male-female matings are equally likely.

Let's apply the bin analogy to the hypothetical wildflower population discussed earlier. In that population of 500 flowers, the frequency of the allele for red flowers (C^R) is $p = 0.8$, and the frequency of the allele for white flowers (C^W) is $q = 0.2$. This implies that a bin holding all 1,000 copies of the flower-color gene in the population would contain 800 C^R alleles and 200 C^W alleles. Assuming that gametes are formed by selecting alleles at random from the bin, the probability that an egg or sperm contains a C^R or C^W allele is equal to the frequency of these alleles in the bin. Thus, as shown in Figure 23.7, each egg (and each sperm) has an 80% chance of containing a C^R allele and a 20% chance of containing a C^W allele.

Using the rule of multiplication (see Figure 14.9), we can now calculate the frequencies of the three possible genotypes, assuming random unions of sperm and eggs.

The probability that two C^R alleles will come together is $p \times p = p^2 = 0.8 \times 0.8 = 0.64$. Thus, about 64% of the plants in the next generation will have the genotype $C^R C^R$. The frequency of $C^W C^W$ individuals is expected to be about $q \times q = q^2 = 0.2 \times 0.2 = 0.04$, or 4%. $C^R C^W$ heterozygotes can arise in two different ways. If the sperm provides the C^R allele and the egg provides the C^W allele, the resulting heterozygotes will be $p \times q = 0.8 \times 0.2 = 0.16$, or 16% of the total. If the sperm provides the C^W allele and the egg the C^R allele, the heterozygous offspring will make up $q \times p = 0.2 \times 0.8 = 0.16$, or 16%. The frequency of heterozygotes is thus the sum of these possibilities: $pq + qp = 2pq = 0.16 + 0.16 = 0.32$, or 32%.

As shown in **Figure 23.8**, the genotype frequencies in the next generation must add up to 1 (100%). Thus, the equation

▼ **Figure 23.8 Hardy-Weinberg equilibrium.** In our wildflower population, the gene pool remains constant from one generation to the next. Mendelian processes alone do not alter frequencies of alleles or genotypes.

Gene pool of the initial (parental) generation:

80% C^R ($p = 0.8$)   20% C^W ($q = 0.2$)

Sperm

C^R $p = 0.8$ C^W $q = 0.2$

C^R
$p = 0.8$

Eggs

C^W
$q = 0.2$

0.64 (p^2)
$C^R C^R$ 0.16 (pq)
$C^R C^W$

0.16 (qp)
$C^R C^W$ 0.04 (q^2)
$C^W C^W$

Genotypes in this offspring generation:

64% $C^R C^R$, 32% $C^R C^W$, and 4% $C^W C^W$

Alleles in this offspring generation:

64% C^R (from $C^R C^R$ plants) + 16% C^R (from $C^R C^W$ plants) = 80% C^R = 0.8 = p

4% C^W (from $C^W C^W$ plants) + 16% C^W (from $C^R C^W$ plants) = 20% C^W = 0.2 = q

With random mating, allele frequencies will result in the same mix of genotypes in the next generation:

64% $C^R C^R$, 32% $C^R C^W$, and 4% $C^W C^W$ plants

WHAT IF? *If the frequency of the C^R allele were 0.6, predict the frequencies of the $C^R C^R$, $C^R C^W$, and $C^W C^W$ genotypes.*

→ **Mastering Biology** Figure Walkthrough

for Hardy-Weinberg equilibrium states that if a population is not evolving at a locus with two alleles, the three genotypes will appear in the following proportions:

$$p^2 \quad + \quad 2pq \quad + \quad q^2 \quad = \quad 1$$

p^2	$2pq$	q^2
Expected frequency of genotype $C^R C^R$	Expected frequency of genotype $C^R C^W$	Expected frequency of genotype $C^W C^W$

Note that for a locus with two alleles, only three genotypes are possible (in this case, $C^R C^R$, $C^R C^W$, and $C^W C^W$). As a result, the sum of the frequencies of the three genotypes must equal 1 (100%) in *any* population regardless of whether the population is in Hardy-Weinberg equilibrium. The key point is that a population is in Hardy-Weinberg equilibrium only if the observed genotype frequency of one homozygote is p^2, the observed frequency of the other homozygote is q^2, and the observed frequency of heterozygotes is $2pq$. Finally, as suggested by Figure 23.8, if a population such as our wildflowers is in Hardy-Weinberg equilibrium and its members continue to mate randomly generation after generation, allele and genotype frequencies will remain constant. The system operates somewhat like a deck of cards: No matter how many times the deck is reshuffled to deal out new hands, the deck itself remains the same. Aces do not grow more numerous than jacks. And the repeated shuffling of a population's gene pool over the generations cannot, in itself, change the frequency of one allele relative to another.

Conditions for Hardy-Weinberg Equilibrium

The Hardy-Weinberg approach describes a population that is not evolving. This can occur if a population meets all five of the conditions for Hardy-Weinberg equilibrium listed in **Table 23.1**. But in nature, the allele and genotype frequencies of a population often *do* change over time. Such changes

can occur when at least one of the conditions for Hardy-Weinberg equilibrium is not met.

Although departure from the conditions in Table 23.1 is common—resulting in evolutionary change—it is also common for natural populations to be in Hardy-Weinberg equilibrium for specific genes. One way this can happen is if selection alters allele frequencies at some loci but not others. In addition, some populations evolve so slowly that the changes in their allele and genotype frequencies are difficult to distinguish from those predicted for a non-evolving population.

Applying the Hardy-Weinberg Equation

The Hardy-Weinberg equation is often used as an initial test of whether evolution is occurring in a population (Concept Check 23.2, question 3 is an example). The equation also has medical applications, such as estimating the percentage of a population carrying the allele for an inherited disease. For example, consider phenylketonuria (PKU), a metabolic disorder that results from homozygosity for a recessive allele. This disorder occurs in about one out of every 10,000 babies born in the United States. Left untreated, PKU results in mental disability and other problems. (As described in Concept 14.4, newborns are now routinely tested for PKU, and symptoms can be largely avoided with a diet very low in phenylalanine.)

To apply the Hardy-Weinberg equation, we must assume that no new PKU mutations are being introduced into the population (condition 1) and that people neither choose their mates on the basis of whether or not they carry this gene nor generally mate with close relatives (condition 2). We must also ignore any effects of differential survival and reproductive success among PKU genotypes (condition 3) and assume that there are no effects of genetic drift (condition 4) or of gene flow from other populations into the United States (condition 5). These assumptions are reasonable: The mutation rate for the PKU gene is low, inbreeding and other forms of nonrandom mating are not common in the United States, selection occurs only against the rare homozygotes (and then only if dietary restrictions are not followed), the U.S. population is very large, and populations outside the country have PKU allele frequencies similar to those seen in the United States.

If all these assumptions hold, then the frequency of individuals in the population born with PKU will correspond to q^2 in the Hardy-Weinberg equation (q^2 = frequency of homozygotes). Because the allele is recessive, we must estimate the number of heterozygotes rather than counting them directly as we did with the pink flowers. Recall that there is one PKU occurrence per 10,000 births, which indicates that $q^2 = 0.0001$. Thus, the frequency (q) of the recessive allele for PKU is

$$q = \sqrt{0.0001} = 0.01$$

and the frequency of the dominant allele is

$$p = 1 - q = 1 - 0.01 = 0.99$$

Table 23.1 Conditions for Hardy-Weinberg Equilibrium	
Condition	**Consequence if Condition Does Not Hold**
1. No mutations	The gene pool is modified if mutations occur or if entire genes are deleted or duplicated.
2. Random mating	If individuals mate within a subset of the population, such as near neighbors or close relatives (inbreeding), random mixing of gametes does not occur and genotype frequencies change.
3. No natural selection	Allele frequencies change when individuals with different genotypes show consistent differences in their survival or reproductive success.
4. Extremely large population size	In small populations, allele frequencies fluctuate by chance over time (genetic drift).
5. No gene flow	By moving alleles into or out of populations, gene flow can alter allele frequencies.

▶ **Mastering Biology** Animation: Causes of Evolutionary Change

Using the Hardy-Weinberg Equation to Interpret Data and Make Predictions

Is Evolution Occurring in a Soybean Population? One way to test whether evolution is occurring in a population is to compare the observed genotype frequencies at a locus with those expected for a non-evolving population based on the Hardy-Weinberg equation. In this exercise, you will test whether a soybean population is evolving at a locus with two alleles, C^G and C^Y, that affect chlorophyll production and hence leaf color.

How the Experiment Was Done Students planted soybean seeds and then counted the number of seedlings of each genotype at day 7 and again at day 21. Seedlings of each genotype could be distinguished visually because the C^G and C^Y alleles show incomplete dominance: $C^G C^G$ seedlings have green leaves, $C^G C^Y$ seedlings have green-yellow leaves, and $C^Y C^Y$ seedlings have yellow leaves.

Data from the Experiment

Time (days)	Number of Seedlings			
	Green ($C^G C^G$)	Green-yellow ($C^G C^Y$)	Yellow ($C^Y C^Y$)	Total
7	49	111	56	216
21	47	106	20	173

INTERPRET THE DATA

1. Use the observed genotype frequencies from the day 7 data to calculate the frequencies of the C^G allele (p) and the C^Y allele (q).

2. Next, use the Hardy-Weinberg equation ($p^2 + 2pq + q^2 = 1$) to calculate the day 7 expected frequencies of genotypes $C^G C^G$, $C^G C^Y$, and $C^Y C^Y$ for a population in Hardy-Weinberg equilibrium.

3. (a) Calculate the observed frequencies of genotypes $C^G C^G$, $C^G C^Y$, and $C^Y C^Y$ at day 7. (b) Compare these frequencies to the expected frequencies calculated in question 2. Is the seedling population in Hardy-Weinberg equilibrium at day 7, or is evolution occurring? Explain your reasoning and identify which genotypes, if any, appear to be selected for or against.

4. (a) Calculate the observed frequencies of genotypes $C^G C^G$, $C^G C^Y$, and $C^Y C^Y$ at day 21. (b) Compare these frequencies to the expected frequencies calculated in question 2 and to the observed frequencies at day 7. Is the seedling population in Hardy-Weinberg equilibrium at day 21, or is evolution occurring? Explain your reasoning and identify which genotypes, if any, appear to be selected for or against.

5. Homozygous $C^Y C^Y$ individuals cannot produce chlorophyll. The ability to photosynthesize becomes more critical as seedlings age and begin to exhaust the supply of food that was stored in the seed from which they emerged. Develop a hypothesis that explains the data for days 7 and 21. Based on this hypothesis, predict how the frequencies of the C^G and C^Y alleles will change beyond day 21.

➔ **Instructors:** A version of this Scientific Skills Exercise can be assigned in **Mastering Biology**.

The frequency of carriers, heterozygous people who do not have PKU but may pass the PKU allele to offspring, is

$$2pq = 2 \times 0.99 \times 0.01 = 0.0198$$
(approximately 2% of the U.S. population)

Remember, the assumption of Hardy-Weinberg equilibrium yields an approximation; the real number of carriers may differ. Still, our calculations suggest that harmful recessive alleles at this and other loci can be concealed in a population because they are carried by healthy heterozygotes. The **Scientific Skills Exercise** provides another opportunity for you to apply the Hardy-Weinberg equation to allele data.

CONCEPT CHECK 23.2

1. A population has 85 individuals of genotype AA, 320 of genotype Aa, and 295 of genotype aa. Calculate the genotype frequencies and the frequencies of alleles A and a.

2. The frequency of allele a is 0.45 for a population in Hardy-Weinberg equilibrium. What are the expected frequencies of genotypes AA, Aa, and aa?

3. **WHAT IF?** A locus that affects susceptibility to a degenerative brain disease has two alleles, V and v. In a population, 16 people have genotype VV, 92 have genotype Vv, and 12 have genotype vv. Is this population evolving? Explain.

For suggested answers, see Appendix A.

CONCEPT 23.3

Natural selection, genetic drift, and gene flow can alter allele frequencies in a population

Note again the five conditions required for a population to be in Hardy-Weinberg equilibrium (see Table 23.1). A deviation from any of these conditions is a potential cause of evolution. New mutations (violation of condition 1) can alter allele frequencies, but because mutations are rare, the change from one generation to the next is likely to be very small. Nonrandom mating (violation of condition 2) can affect the frequencies of homozygous and heterozygous genotypes but by itself has no effect on allele frequencies in the gene pool. (Allele frequencies can change if individuals with certain inherited traits are more likely than other individuals to obtain mates. However, such a situation not only causes a deviation from random mating, but also violates condition 3, no natural selection.)

For the rest of this section we will focus on the three mechanisms that alter allele frequencies directly and cause most evolutionary change: natural selection, genetic drift, and gene flow (violations of conditions 3–5).

Natural Selection

The concept of natural selection is based on differential success in survival and reproduction: Individuals in a population exhibit variations in their heritable traits, and those with traits that are better suited to their environment tend to produce more offspring than those with traits that are not as well suited.

In genetic terms, selection results in alleles being passed to the next generation in proportions that differ from the present generation. For example, the fruit fly *D. melanogaster* has an allele that confers resistance to the insecticide DDT. This allele has a frequency of 0% in laboratory strains of *D. melanogaster* from flies collected in the wild in the early 1930s, prior to DDT use. However, in strains established from flies collected after 1960 (following 20 or more years of DDT use), the allele frequency is 37%. Thus, this allele either arose by mutation between 1930 and 1960 or was present in 1930, but very rare. In any case, the rise in frequency of this allele most likely occurred because DDT is an effective poison that is a strong selective force in exposed fly populations.

As the *D. melanogaster* example suggests, an allele that confers resistance to an insecticide will increase in frequency in a population exposed to that insecticide. Such changes are not coincidental. By consistently favoring some alleles over others, natural selection can cause **adaptive evolution**, in which traits that enhance survival or reproduction tend to increase in frequency over time. We'll explore this process in Concept 23.4.

Genetic Drift

If you flip a coin 1,000 times, a result of 700 heads and 300 tails might make you suspicious about that coin. But if you flip a coin only 10 times, an outcome of 7 heads and 3 tails would not be surprising. The smaller the number of coin flips, the more likely it is that chance alone will cause a deviation from the predicted result. (In this case, the prediction is an equal number of heads and tails.) Chance events can also cause allele frequencies to fluctuate unpredictably from one generation to the next, especially in small populations—a process called **genetic drift**.

Figure 23.9 models how genetic drift might affect a small population of our wildflowers. In this example, drift leads to the loss of an allele from the gene pool, but it is a matter of chance that the C^W allele is lost and not the C^R allele. Such unpredictable changes in allele frequencies can be caused by chance events associated with survival and reproduction. Perhaps a large animal such as a moose stepped on the three $C^W C^W$ individuals in generation 2, killing them and increasing the chance that only the C^R allele would be passed to the next generation. Allele frequencies can also be affected by chance events that occur during fertilization. For example, suppose two individuals of genotype $C^R C^W$ had a small number of offspring. By chance alone, every egg and sperm pair that generated offspring could happen to have carried the C^R allele and not the C^W allele.

Certain circumstances can result in genetic drift having a significant impact on a population. Two examples are the founder effect and the bottleneck effect.

The Founder Effect

When a few individuals become isolated from a larger population, this smaller group may establish a new population whose gene pool differs from the source population; this is called the **founder effect**. The founder effect might occur, for example, when a few members of a population are blown by a storm to a new island. Genetic drift, in which chance events alter allele frequencies, can occur in such a case because the

▼ **Figure 23.9 Genetic drift.** This small wildflower population has a stable size of ten plants. Suppose that by chance only five plants of generation 1 (those highlighted in yellow) produce fertile offspring. Again by chance, only two plants of generation 2 leave fertile offspring.

VISUAL SKILLS *Based on this diagram, summarize how the frequency of the C^W allele changes over time.*

Generation 1	Generation 2	Generation 3
p (frequency of C^R) = 0.7	p = 0.5	p = 1.0
q (frequency of C^W) = 0.3	q = 0.5	q = 0.0

storm indiscriminately transports some individuals (and their alleles), but not others, from the source population.

The founder effect probably accounts for the relatively high frequency of certain inherited disorders among isolated human populations. For example, in 1814, 15 British colonists founded a settlement on Tristan da Cunha, a group of small islands in the Atlantic Ocean midway between Africa and South America. Apparently, one of the colonists carried a recessive allele for retinitis pigmentosa, a progressive form of blindness that affects homozygous individuals. Of the founding colonists' 240 descendants on the island in the late 1960s, four had retinitis pigmentosa. The frequency of the allele that causes this disease is ten times higher on Tristan da Cunha than in the populations from which the founders came.

The Bottleneck Effect

A sudden change in the environment, such as a fire or flood, may drastically reduce the size of a population. A severe drop in population size can cause the **bottleneck effect**, so named because the population has passed through a "bottleneck" that greatly reduces its size **(Figure 23.10)**. By chance alone, certain alleles may be overrepresented among the survivors, others may be underrepresented, and some may be absent altogether. Ongoing genetic drift is likely to have substantial effects on the gene pool until the population becomes large enough that chance events have less impact. But even if a population that has passed through a bottleneck ultimately recovers in size, it may have low levels of genetic variation for a long period of time—a legacy of the genetic drift that occurred when the population was small.

Case Study: *Impact of Genetic Drift on the Greater Prairie Chicken*

Human actions sometimes create severe bottlenecks. Millions of greater prairie chickens (*Tympanuchus cupido*) once lived on the prairies of Illinois. As these prairies were converted to farmland during the 1800s and 1900s, the number of greater

▼ **Figure 23.10 The bottleneck effect.** Shaking a few marbles through the neck of a bottle is analogous to a drastic reduction in the size of a population. By chance, blue marbles are overrepresented in the surviving population and gold marbles are absent.

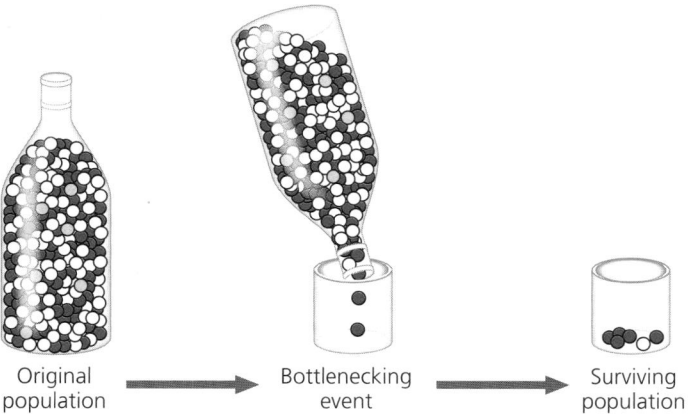

Original population   Bottlenecking event   Surviving population

prairie chickens plummeted **(Figure 23.11a)**. By 1993 fewer than 50 birds remained. These few surviving birds had low levels of genetic variation, and less than 50% of their eggs hatched, compared with much higher hatching rates of the larger populations in Kansas and Nebraska **(Figure 23.11b)**.

These data suggest that genetic drift during the bottleneck may have led to a loss of genetic variation and an increase in the frequency of harmful alleles. To investigate this hypothesis, researchers extracted DNA from 15 museum specimens of Illinois greater prairie chickens. Of the 15 birds, 10 had been collected in the 1930s, when there were 25,000 greater prairie chickens in Illinois, and 5 had been collected in the 1960s, when there were 1,000 greater prairie chickens in Illinois. By studying the DNA of these specimens, the researchers were able to obtain a minimum, baseline estimate of how much genetic variation was present in the Illinois population *before* the population shrank to extremely low numbers. This baseline estimate is a key piece of information that is not usually available in cases of population bottlenecks.

▼ **Figure 23.11 Genetic drift and loss of genetic variation.**

Greater prairie chicken

Pre-bottleneck (Illinois, 1820)   Post-bottleneck (Illinois, 1993)

☐ Range of greater prairie chicken

Grasslands in which the prairie chickens live once covered most of the state.

In 1993, with less than 1% of the grasslands remaining, the prairie chickens were found in just two locations.

(a) The Illinois population of greater prairie chickens dropped from millions of birds in the 1800s to fewer than 50 birds in 1993.

Location	Population size	Number of alleles per locus	Percentage of eggs hatched
Illinois			
1930–1960s	1,000–25,000	5.2	93
1993	<50	3.7	<50
Kansas, 1998 (no bottleneck)	750,000	5.8	99
Nebraska, 1998 (no bottleneck)	75,000–200,000	5.8	96

(b) In the small Illinois population, genetic drift led to decreases in the number of alleles per locus and the percentage of eggs hatched.

The researchers surveyed six loci and found that the 1993 population had fewer alleles per locus than the pre-bottleneck Illinois or the current Kansas and Nebraska populations (see Figure 23.11b). Thus, as predicted, drift had reduced the genetic variation of the small 1993 population. Drift may also have increased the frequency of harmful alleles, leading to the low egg-hatching rate. To counteract these negative effects, 271 birds from neighboring states were added to the Illinois population over four years. This strategy succeeded: New alleles entered the population, and the egg-hatching rate improved to over 90%. Overall, studies on the Illinois greater prairie chicken illustrate the powerful effects of genetic drift in small populations and provide hope that in at least some populations, these effects can be reversed.

Effects of Genetic Drift: A Summary

The examples we've described highlight four key points:

1. **Genetic drift is significant in small populations.** Chance events can cause an allele to be disproportionately over- or underrepresented in the next generation. Although chance events occur in populations of all sizes, they tend to alter allele frequencies substantially only in small populations.

2. **Genetic drift can cause allele frequencies to change at random.** Because of genetic drift, an allele may increase in frequency one year, then decrease the next; the change from year to year is not predictable. Thus, unlike natural selection, which in a given environment consistently favors some alleles over others, genetic drift causes allele frequencies to change at random over time.

3. **Genetic drift can lead to a loss of genetic variation within populations.** By causing allele frequencies to fluctuate randomly over time, genetic drift can eliminate alleles from a population. Because evolution depends on genetic variation, such losses can influence how effectively a population can adapt to a change in the environment.

4. **Genetic drift can cause harmful alleles to become fixed.** Alleles that are neither harmful nor beneficial can be lost or become fixed (reach a frequency of 100%) by chance through genetic drift. In very small populations, genetic drift can also cause alleles that are slightly harmful to become fixed. When this occurs, the population's survival can be threatened (as in greater prairie chickens).

Gene Flow

Natural selection and genetic drift are not the only phenomena affecting allele frequencies. Allele frequencies can also change by **gene flow**, the transfer of alleles into or out of a population due to the movement of fertile individuals or their gametes. For example, suppose that near our original hypothetical wildflower population there is another population consisting primarily of white-flowered individuals ($C^W C^W$). Insects carrying pollen from these plants may fly to and pollinate

plants in our original population. The introduced C^W alleles would modify our original population's allele frequencies in the next generation. Because alleles are transferred between populations, gene flow tends to reduce the genetic differences between populations. In fact, if it is extensive enough, gene flow can result in two populations combining into a single population with a common gene pool.

Alleles transferred by gene flow can also affect how well populations are adapted to local environmental conditions. For instance, mainland and island populations of the Lake Erie water snake (*Nerodia sipedon*) differ in their color patterns **(Figure 23.12)**: Nearly all snakes from the Ohio or Ontario mainlands are strongly banded, whereas the majority of snakes

▼ **Figure 23.12 Gene flow and local adaptation in the Lake Erie water snake (*Nerodia sipedon*).** Researchers assigned letters to variations in coloration in *N. sipedon* populations. Color pattern A is strong banding, patterns B and C are intermediate banding, and pattern D is no banding. Banding is advantageous for camouflage in mainland environments, whereas having no bands is advantageous in island environments. However, gene flow from the mainland causes banding to persist in island populations.

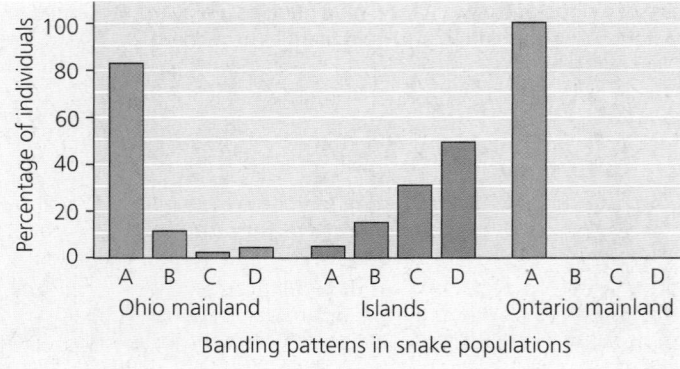

WHAT IF? *Suppose a severe weather event caused island populations to decrease in size but did not affect the size of mainland populations. Predict how gene flow from the mainland would affect color patterns in island populations in the year immediately after the storm. Explain.*

from islands are unbanded or intermediate. Banding coloration is an inherited trait, determined by a few loci (with alleles that encode bands being dominant to alleles that encode the absence of bands). On islands, water snakes live along rocky shorelines, while on the mainland, they live in marshes. Snakes without bands are more well camouflaged in island habitats than are snakes with bands. Hence, on islands, snakes without bands survive at higher rates than do snakes with bands.

These data indicate that snakes without bands are favored by natural selection in island populations. Thus, we might expect that *all* snakes on islands would lack bands. Why is this not the case? The answer lies in gene flow from the mainland. In any given year, three to ten snakes from the mainland swim to the islands and join the populations there. As a result, each year such migrants transfer alleles for banded coloration from the mainland (where nearly all snakes have bands) to the islands. This ongoing gene flow has prevented selection from removing all of the alleles for banded coloration from island populations—thereby preventing island populations from adapting fully to local conditions.

Gene flow can also transfer alleles that improve the ability of populations to adapt to local conditions. For example, gene flow has resulted in the worldwide spread of several insecticide resistance alleles in the mosquito *Culex pipiens*, a vector of West Nile virus and other diseases. Each of these alleles has a unique genetic signature that allowed researchers to document that it arose by mutation in only one or a few geographic locations. In their population of origin, these alleles increased because they provided insecticide resistance. These alleles were then transferred by gene flow to other populations exposed to insecticides, where again, their frequencies increased as a result of natural selection.

Finally, gene flow has become an increasingly important agent of evolutionary change in human populations. Humans today move much more freely about the world than in the past. As a result, mating is more common between members of populations that previously had very little contact, leading to an exchange of alleles and fewer genetic differences between those populations.

→ **Mastering Biology** BioFlix® Animation: Natural Selection, Genetic Drift, and Gene Flow

CONCEPT CHECK 23.3

1. In what sense is natural selection more "predictable" than genetic drift?

2. Distinguish genetic drift from gene flow in terms of (a) how they occur and (b) their implications for future genetic variation in a population.

3. **WHAT IF?** Suppose two plant populations exchange pollen and seeds. In one population, individuals of genotype *AA* are most common (9,000 *AA*, 900 *Aa*, 100 *aa*), while the opposite is true in the other population (100 *AA*, 900 *Aa*, 9,000 *aa*). If neither allele has a selective advantage, what will happen over time to the allele and genotype frequencies of these populations?

For suggested answers, see Appendix A.

Natural selection is the only mechanism that consistently causes adaptive evolution

Evolution by natural selection is a blend of chance and "sorting": chance in the creation of new genetic variations (as in mutation) and sorting as natural selection favors some alleles over others. Because of this favoring process, the outcome of natural selection is *not* random. Instead, natural selection consistently increases the frequencies of alleles that provide reproductive advantage, thus leading to adaptive evolution.

Natural Selection: *A Closer Look*

To see how natural selection can cause adaptive evolution, we'll begin with the concept of relative fitness and the different ways that selection acts on an organism's phenotype.

Relative Fitness

The phrases "struggle for existence" and "survival of the fittest" are commonly used to describe natural selection, but these expressions are misleading if always taken to mean direct competitive contests among individuals. There *are* animal species in which individuals, usually the males, lock horns or otherwise do combat to determine mating privilege. But reproductive success is generally more subtle and depends on many factors besides outright battle. For example, a barnacle that is more efficient at collecting food than its neighbors may have greater stores of energy and hence be able to produce a larger number of eggs. A moth may have more offspring than other moths in the same population because its body colors more effectively conceal it from predators, improving its chance of surviving long enough to produce more offspring. These examples illustrate how in a given environment, certain traits can lead to greater **relative fitness**: the contribution an individual makes to the gene pool of the next generation *relative to* the contributions of other individuals.

Although we often refer to the relative fitness of a genotype, remember that the entity that is subjected to natural selection is the whole organism, not the underlying genotype. Thus, selection acts more directly on the phenotype than on the genotype; it acts on the genotype indirectly, via how the genotype affects the phenotype.

Directional, Disruptive, and Stabilizing Selection

Natural selection can alter the frequency distribution of heritable traits in three ways, depending on which phenotypes in a population are favored: through directional selection, disruptive selection, and stabilizing selection.

Directional selection occurs when conditions favor individuals exhibiting one extreme of a phenotypic range, thereby shifting a population's frequency curve for the phenotypic character in one direction or the other **(Figure 23.13a)**. Directional selection is common when a population's environment changes or when members of a population migrate to a new (and different) habitat. For instance, an increase in the relative abundance of large seeds over small seeds led to an increase in beak depth in a population of Galápagos finches (see Figure 23.2).

Disruptive selection (Figure 23.13b) occurs when conditions favor individuals at both extremes of a phenotypic range over individuals with intermediate phenotypes. One example is a population of black-bellied seedcracker finches in Cameroon whose members display two distinctly different beak sizes. Small-billed birds feed mainly on soft seeds, whereas large-billed birds specialize in cracking hard seeds. It appears that birds with intermediate-sized bills are relatively inefficient at cracking both types of seeds and thus have lower relative fitness.

Stabilizing selection (Figure 23.13c) acts against both extreme phenotypes and favors intermediate variants. This mode of selection reduces variation and tends to maintain the status quo for a particular phenotypic character. For example, the birth weights of most human babies lie in the range of 3–4 kg (6.6–8.8 pounds). Babies who are much smaller suffer considerably higher rates of mortality; babies who are much larger show a slight increase in mortality.

Regardless of which type of selection occurs, however, the basic mechanism remains the same. Selection favors individuals whose heritable phenotypic traits provide higher reproductive success than do the traits of other individuals.

⊖ **Mastering Biology** HHMI Video: Selection for Tuskless Elephants

The Key Role of Natural Selection in Adaptive Evolution

The adaptations of organisms include many striking examples. Certain octopuses, for example, have the ability to change color rapidly, enabling them to blend into different backgrounds. Another example is the remarkable jaws of snakes **(Figure 23.14)**, which allow them to swallow prey much larger than their own head (a feat analogous to a person swallowing a whole watermelon). Other adaptations, such as a version of an enzyme that shows improved function in cold environments, may be less visually dramatic but just as important for survival and reproduction.

Such adaptations can arise gradually over time as natural selection increases the frequencies of alleles that enhance survival or reproduction. As the proportion of individuals that have favorable traits increases, the degree to which a

▼ **Figure 23.13 Modes of selection.** These cases describe three ways in which a hypothetical deer mouse population with heritable variation in fur coloration might evolve. The graphs show how the frequencies of individuals with different fur colors change over time. The large white arrows symbolize selective pressures against certain phenotypes.

MAKE CONNECTIONS *Review Figure 22.13. Which mode of selection has occurred in soapberry bug populations that feed on the introduced golden rain tree? Explain.*

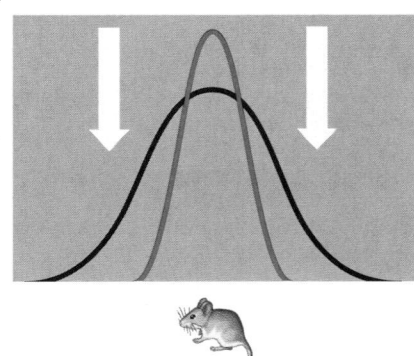

(a) Directional selection shifts the overall makeup of the population by favoring variants that are at one extreme of the distribution. In this case, lighter mice are selected against because they live among dark rocks, making it harder for them to hide from predators.

(b) Disruptive selection favors variants at both ends of the distribution. These mice have colonized a patchy habitat made up of light and dark rocks, with the result that mice of an intermediate color are selected against.

(c) Stabilizing selection removes extreme variants from the population and preserves intermediate types. If the environment consists of rocks of an intermediate color, both light and dark mice will be selected against.

▼ Figure 23.14 Movable jaw bones in snakes.

The bones of the upper jaw that are shown in green are movable.

Ligament

The skull bones of most terrestrial vertebrates are relatively rigidly attached to one another, limiting jaw movement. In contrast, most snakes have movable bones in their upper jaw, allowing them to swallow food much larger than their head.

species is well suited for life in its environment improves; that is, adaptive evolution occurs. However, the physical and biological components of an organism's environment may change over time. As a result, what constitutes a "good match" between an organism and its environment can be a moving target, making adaptive evolution a continuous, dynamic process. Environmental conditions can also differ from place to place, causing different alleles to be favored in different locations. When this occurs, natural selection can cause the populations of a species to differ genetically from one another.

And what about genetic drift and gene flow? Both can, in fact, increase the frequencies of alleles that enhance survival or reproduction, but neither does so consistently. Genetic drift can cause the frequency of a slightly beneficial allele to increase, or, alternately, to decrease. Similarly, gene flow may introduce alleles that are advantageous or ones that are disadvantageous. Natural selection is the only evolutionary mechanism that consistently leads to adaptive evolution.

 Mastering Biology HHMI Video: Got Lactase? The Co-Evolution of Genes and Culture **hhmi** BioInteractive

Sexual Selection

Charles Darwin was the first to explore the implications of **sexual selection**, a process in which individuals with certain inherited characteristics are more likely than other individuals of the same sex to obtain mates. Sexual selection can result in **sexual dimorphism**, a difference in secondary

▲ Figure 23.15 Sexual dimorphism and sexual selection. A peacock (left) and a peahen (right) show extreme sexual dimorphism. There is intrasexual selection between competing males, followed by intersexual selection when the females choose among the showiest males.

sexual characteristics between males and females of the same species **(Figure 23.15)**. These distinctions include differences in size, color, ornamentation, and behavior.

How does sexual selection operate? There are several ways. In **intrasexual selection**, meaning selection within the same sex, individuals of one sex compete directly for mates of the opposite sex. In many species, intrasexual selection occurs among males. For example, a single male may patrol a group of females and prevent other males from mating with them. The patrolling male may defend his status by defeating smaller, weaker, or less fierce males in combat. More often, this male is the psychological victor in ritualized displays that discourage would-be competitors but do not risk injury that would reduce his own fitness (see Figure 51.16). Intrasexual selection also occurs among females in a variety of species, including ring-tailed lemurs and broadnosed pipefish.

In **intersexual selection**, also called *mate choice*, individuals of one sex (usually the females) are choosy in selecting their mates from the other sex. In many cases, the female's choice depends on the showiness of the male's appearance or behavior (see Figure 23.15). What intrigued Darwin about mate choice is that male showiness may not seem adaptive in any other way and may in fact pose some risk. For example, bright plumage may make male birds more visible to predators. But if such characteristics help a male gain a mate, and if this benefit outweighs the risk from predation, then both the bright plumage and the female preference for it will be reinforced because they enhance overall reproductive success.

How do female preferences for certain male characteristics evolve in the first place? One hypothesis is that females prefer male traits that are correlated with "good genes." If the trait preferred by females is indicative of a male's overall genetic quality, both the male trait and female preference

Do females select mates based on traits indicative of "good genes"?

Experiment Female gray tree frogs (*Hyla versicolor*) prefer to mate with males that give long mating calls. To test whether the genetic makeup of long-calling (LC) males is superior to that of short-calling (SC) males, researchers fertilized half the eggs of each female with sperm from an LC male and the other half with sperm from an SC male. In two separate experiments (one in 1995, one in 1996), the resulting half-sibling offspring were raised in a common environment, and their survival and growth were monitored.

Recording of SC male's call

Recording of LC male's call

Female gray tree frog

SC male gray tree frog

LC male gray tree frog

SC sperm × Eggs × LC sperm

Offspring of SC father   Offspring of LC father

Survival and growth of these half-sibling offspring compared

Results

Offspring Performance	1995	1996
Larval survival	LC better	NSD
Larval growth	NSD	LC better
Time to metamorphosis	LC better (shorter)	LC better (shorter)

NSD = no significant difference; LC better = offspring of LC males superior to offspring of SC males.

Conclusion Because offspring fathered by an LC male outperformed their half-siblings fathered by an SC male, the team concluded that the duration of a male's mating call is indicative of the male's overall genetic quality. This result supports the hypothesis that female mate choice can be based on a trait that indicates whether the male has "good genes."

Data from A. M. Welch et al., Call duration as an indicator of genetic quality in male gray tree frogs, *Science* 280:1928–1930 (1998).

INQUIRY IN ACTION Read and analyze the original paper in *Inquiry in Action: Interpreting Scientific Papers.*

WHAT IF? *Why did the researchers split each female frog's eggs into two batches for fertilization by different males? Why didn't they mate each female with a single male frog?*

for it should increase in frequency. **Figure 23.16** describes one experiment testing this hypothesis in gray tree frogs.

In several bird species, the traits preferred by females are related to overall male health. Here, too, female preference appears to be based on traits that reflect "good genes"—in this case, alleles indicative of a robust immune system.

Balancing Selection

As we've seen, genetic variation is often found at loci affected by selection. What prevents natural selection from reducing the variation at those loci by culling all unfavorable alleles? As mentioned earlier, in diploid organisms, many unfavorable recessive alleles persist because they are hidden from selection when in heterozygous individuals. In addition, selection itself may preserve variation at some loci, thus maintaining two or more phenotypic forms in a population. Known as **balancing selection**, this type of selection includes frequency-dependent selection and heterozygote advantage.

Frequency-Dependent Selection

In **frequency-dependent selection**, the fitness of a phenotype depends on how common it is in the population. Consider the scale-eating fish (*Perissodus microlepis*) of Lake Tanganyika, in Africa. These fish attack other fish from behind, darting in to remove a few scales from the flank of their prey. Of interest here is a peculiar feature of the scale-eating fish: Some are "left-mouthed" and some are "right-mouthed."

Because their mouth twists to the left, left-mouthed fish typically attack their prey's right flank **(Figure 23.17)**. (To see why, twist your lower jaw and lips to the left and imagine trying to take a bite from the left side of a fish, approaching it from behind.) Similarly, right-mouthed fish attack from the left. Prey species guard against attack from whatever phenotype of scale-eating fish is most common. Thus, from year to year, selection favors whichever mouth phenotype is least common. As a result, the frequency of left- and right-mouthed fish oscillates over time, and this balancing selection keeps the frequency of each phenotype close to 50%.

Heterozygote Advantage

If individuals who are heterozygous at a particular locus have greater fitness than do both kinds of homozygotes, they exhibit **heterozygote advantage**. In such a case, natural selection tends to maintain two or more alleles at that locus. Note that heterozygote advantage is defined in terms of *genotype*, not phenotype. Thus, whether heterozygote advantage represents stabilizing or directional selection depends on the relationship between the genotype and the phenotype. For example, if the phenotype of a heterozygote is intermediate to the phenotypes of both homozygotes, then heterozygote advantage is a form of stabilizing selection.

An example of heterozygote advantage occurs at the locus in humans that codes for the β polypeptide subunit of hemoglobin, the oxygen-carrying protein of red blood cells. In homozygous individuals, a recessive allele at that locus causes sickle-cell disease. The red blood cells of people with sickle-cell disease become distorted in shape, or *sickled*, under low-oxygen conditions (see Figure 5.19). These sickled cells can clump together and block the flow of blood in the capillaries, damaging organs such as the kidney, heart, and brain. Although some red blood cells become sickled in

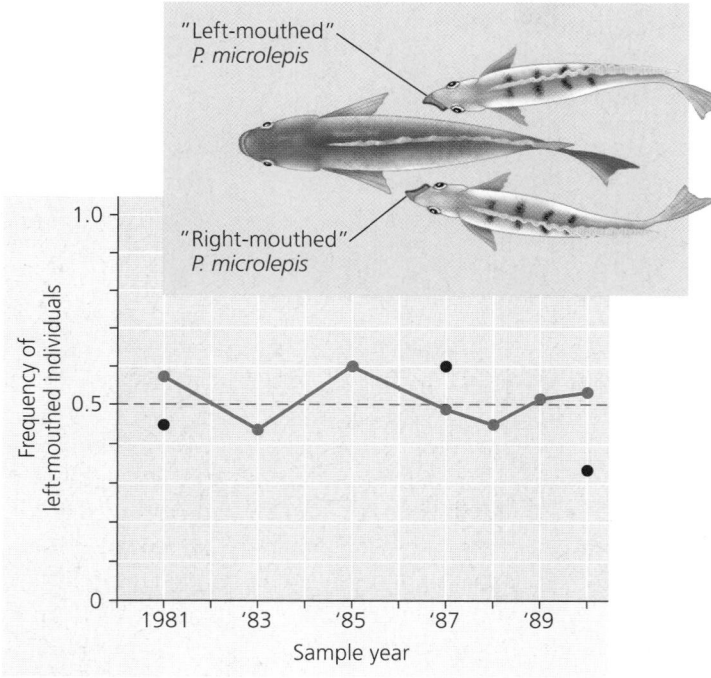

▼ **Figure 23.17 Frequency-dependent selection.** In a population of the scale-eating fish *Perissodus microlepis*, the frequency of left-mouthed individuals (red data points) rises and falls in a regular manner. The black data points indicate the frequency of left-mouthed individuals among adults that reproduced in three sample years.

"Left-mouthed"
P. microlepis

"Right-mouthed"
P. microlepis

INTERPRET THE DATA *For 1981, 1987, and 1990, compare the frequency of left-mouthed individuals among breeding adults to the frequency of left-mouthed individuals in the entire population. What do the data indicate about when natural selection favors left-mouthed individuals over right-mouthed individuals (or vice versa)? Explain.*

heterozygotes, not enough become sickled to cause sickle-cell disease.

As described in **Figure 23.18**, heterozygotes for the sickle-cell allele are protected against the most severe effects of malaria, a disease caused by a parasite that infects red blood cells. One reason for this partial protection is that the body destroys sickled red blood cells rapidly, killing the parasites they harbor. Malaria is a major killer in some tropical regions. In such regions, selection favors heterozygotes over homozygous dominant individuals, who are more vulnerable to the effects of malaria, and also over homozygous recessive individuals, who develop sickle-cell disease. These selective pressures have caused the frequency of the sickle-cell allele to reach relatively high levels in areas where the malaria parasite, which is carried by mosquitoes, is common.

Why Natural Selection Cannot Fashion Perfect Organisms

Though natural selection leads to adaptation, nature abounds with examples of organisms that are less than ideally suited for their lifestyles. There are several reasons why.

1. **Selection can act only on existing variations.**
   Natural selection favors only the fittest phenotypes among those currently in the population, which may not be the ideal traits. New advantageous alleles do not arise on demand. For example, historically, snowshoe hares (*Lepus americanus*) have molted, changing their coats from brown to white, at a time of year that matched the onset of snowfall, providing camouflage all winter. But due to climate change, the first snowfall now occurs later in the year. In some populations, the date at which the hares change their coat color has remained the same, causing the hares to be poorly camouflaged early in winter and thus easier for predators to spot and kill **(Figure 23.19)**. Because their gene pools lack alleles that could delay when molting occurs, these populations have been unable to adapt to changing conditions.

2. **Evolution is limited by historical constraints.**
   Each species has a legacy of descent with modification from ancestral forms. Evolution does not scrap the ancestral anatomy and build each new complex structure from scratch; rather, evolution co-opts existing structures and adapts them to new situations. We could imagine that if a terrestrial animal were to adapt to an environment in which flight would be advantageous, it might be best just to grow an extra pair of limbs that would serve as wings. However, evolution does not work this way; instead, it operates on the traits an organism already has. Thus, in birds and bats, an existing pair of limbs took on new functions for flight as these organisms evolved from nonflying ancestors.

3. **Adaptations are often compromises.** Each organism must do many different things. A seal spends part of its time on rocks; it could probably walk better if it had legs instead of flippers, but then it would not swim nearly as well. We humans owe much of our versatility and athleticism to our prehensile hands and flexible limbs, but these also make us prone to sprains, torn ligaments, and dislocations: Structural reinforcement has been compromised for agility. Organisms face many such *trade-offs* in which the ability to perform one function may reduce the ability to perform another—and as with seals and humans, those trade-offs can restrict adaptive evolution.

▼ **Figure 23.19 Lack of variation in a population can limit adaptation.** For the snowshoe hare, changing its coat color from brown **(a)** to white **(b)** too early is disadvantageous, but the population lacks alleles that would encode a delay in molting.

MAKE CONNECTIONS

The Sickle-Cell Allele

This child has sickle-cell disease, a genetic disorder that strikes individuals who have two copies of the sickle-cell allele. This allele causes an abnormality in the structure and function of hemoglobin, the oxygen-carrying protein in red blood cells. Although sickle-cell disease is lethal if not treated, in some regions the sickle-cell allele can reach frequencies as high as 15–20%. How can such a harmful allele be so common?

Events at the Molecular Level

- Due to a point mutation, the sickle-cell allele differs from the wild-type allele by a single nucleotide. (See Figure 17.26.)
- The resulting change in one amino acid leads to hydrophobic interactions between the sickle-cell hemoglobin proteins under low-oxygen conditions.
- As a result, the sickle-cell proteins bind to each other in chains that together form a fiber.

Consequences for Cells

- The abnormal hemoglobin fibers distort the red blood cell into a sickle shape under low-oxygen conditions, such as those found in small blood vessels returning to the heart.

Template strand

Sickle-cell allele on chromosome

C C T C T T T C
G G A G A A A G

An adenine has replaced a thymine in the template strand of the sickle-cell allele, changing one codon in the mRNA produced during transcription. This change causes an amino acid change in sickle-cell hemoglobin: A valine replaces a glutamic acid at one position. (See Figure 5.19.)

Sickle-cell hemoglobin

Low-oxygen conditions

Fiber

Sickled red blood cell

Wild-type allele

C C T C T T T C
G G A G A A A G

Normal hemoglobin (does not aggregate into fibers)

Normal red blood cell

> Mastering Biology HHMI Video:
> **The Making of the Fittest:**
> **Natural Selection in**
> **Humans (Sickle-Cell Disease)**

Infected mosquitoes spread malaria when they bite people. (See Figure 28.16.)

Evolution in Populations

- Homozygotes with two sickle-cell alleles are strongly selected against because of mortality caused by sickle-cell disease. In contrast, heterozygotes experience few harmful effects from sickling yet are more likely to survive malaria than are homozygotes.
- In regions where malaria is common, the net effect of these opposing selective forces is heterozygote advantage. This has caused evolutionary change in populations—the products of which are the areas of relatively high frequencies of the sickle-cell allele shown in the map.

Effects on Individual Organisms

- The formation of sickled red blood cells causes homozygotes with two copies of the sickle-cell allele to have sickle-cell disease.
- Some sickling also occurs in heterozygotes, but not enough to cause the disease; they have sickle-cell trait. (See Figure 14.17.)

The sickled blood cells of a homozygote block small blood vessels, causing great pain and damage to organs such as the heart, kidney, and brain.

Normal red blood cells are flexible and are able to flow freely through small blood vessels.

Key
Frequencies of the sickle-cell allele

- 3.0–6.0%
- 6.0–9.0%
- 9.0–12.0%
- 12.0–15.0%
- >15.0%

Distribution of malaria caused by *Plasmodium falciparum* (a parasitic unicellular eukaryote carried by mosquitoes)

MAKE CONNECTIONS

In a region free of malaria, would individuals who are heterozygous for the sickle-cell allele be selected for or selected against? Explain.

4. Chance, natural selection, and the environment interact. Chance events can affect the subsequent evolutionary history of populations. For instance, when a storm blows insects or birds hundreds of kilometers over an ocean to an island, the wind does not necessarily transport those individuals that are best suited to the new environment. Thus, not all alleles present in the founding population's gene pool are better suited to the new environment than the alleles that are "left behind." In addition, the environment at a particular location may change unpredictably from year to year, again limiting the extent to which adaptive evolution results in organisms being well suited for current environmental conditions.

With these four constraints, evolution does not result in perfect organisms. Natural selection operates on a "better than" basis. We can, in fact, see evidence for evolution in the many imperfections of the organisms it produces.

CONCEPT CHECK 23.4

1. What is the relative fitness of a sterile mule? Explain.
2. Explain why natural selection is the only mechanism that consistently leads to adaptive evolution in a population.
3. **VISUAL SKILLS** Consider a population in which heterozygotes at a certain locus have an extreme phenotype (such as being larger than homozygotes) that confers a selective advantage. Compare this description to the three types of selection shown in Figure 23.13. Does this situation represent directional, disruptive, or stabilizing selection? Explain.

For suggested answers, see Appendix A.

23 Chapter Review

→ Go to **Mastering Biology** for Assignments, the eText, the Study Area, and Dynamic Study Modules.

SUMMARY OF KEY CONCEPTS

→ To review key terms, go to the **Vocabulary Self-Quiz** in the **Mastering Biology** eText or Study Area, or go to goo.gl/zkjz9t.

CONCEPT 23.1

Genetic variation makes evolution possible (pp. 487–489)

- **Genetic variation** refers to genetic differences among individuals within a population.
- The nucleotide differences that provide the basis of genetic variation originate when mutation and gene duplication produce new alleles and new genes. New genetic variants are produced rapidly in organisms with short generation times. In sexually reproducing organisms, most of the genetic differences among individuals result from crossing over, the independent assortment of chromosomes, and fertilization.

? *Typically, most of the nucleotide variability that occurs within a genetic locus does not affect the phenotype. Explain why.*

CONCEPT 23.2

The Hardy-Weinberg equation can be used to test whether a population is evolving (pp. 489–493)

- A **population**, a localized group of organisms belonging to one species, is united by its **gene pool**, the aggregate of all the alleles in the population.
- For a population in **Hardy-Weinberg equilibrium**, the allele and genotype frequencies will remain constant if the population is large, mating is random, mutation is negligible, there is no gene flow, and there is no natural selection. For such a population, if p and q represent the frequencies of the only two possible alleles at a particular locus, then p^2 is the frequency of one kind of homozygote, q^2 is the frequency of the other kind of homozygote, and $2pq$ is the frequency of the heterozygous genotype.

? *Is it circular reasoning to calculate p and q from observed genotype frequencies and then use those values of p and q to test if the population is in Hardy-Weinberg equilibrium? Explain your answer.*

CONCEPT 23.3

Natural selection, genetic drift, and gene flow can alter allele frequencies in a population (pp. 493–497)

- In natural selection, individuals that have certain inherited traits tend to survive and reproduce at higher rates than other individuals *because of* those traits.
- In **genetic drift**, chance fluctuations in allele frequencies over generations tend to reduce genetic variation.
- **Gene flow**, the transfer of alleles between populations, tends to reduce genetic differences between populations over time.

? *Would two small, geographically isolated populations in very different environments be likely to evolve in similar ways? Explain.*

CONCEPT 23.4

Natural selection is the only mechanism that consistently causes adaptive evolution (pp. 497–504)

- One organism has greater **relative fitness** than another organism if it leaves more fertile descendants. The modes of natural selection differ in their effect on phenotype:

Original population   Evolved population

Directional selection

Disruptive selection

Stabilizing selection

- Unlike genetic drift and gene flow, natural selection consistently increases the frequencies of alleles that enhance survival and reproduction, thus improving the degree to which organisms are well-suited for life in their environment.
- **Sexual selection** can result in secondary sex characteristics that can give individuals advantages in mating.
- **Balancing selection** occurs when natural selection maintains two or more forms in a population.

- There are constraints to evolution: Natural selection can act only on available variation; structures result from modified ancestral anatomy; adaptations are often compromises; and chance, natural selection, and the environment interact.

? *How might secondary sex characteristics in males differ from those in females in a species in which females compete for mates?*

TEST YOUR UNDERSTANDING

➔ For more multiple-choice questions, go to the **Practice Test** in the **Mastering Biology** eText or Study Area, or go to goo.gl/GruWRg.

Levels 1-2: Remembering/Understanding

1. Natural selection changes allele frequencies because some _____ survive and reproduce better than others.
   - (A) alleles
   - (B) loci
   - (C) species
   - (D) individuals

2. No two people are genetically identical, except for identical twins. The main source of genetic variation among humans is
   - (A) new mutations that occurred in the preceding generation.
   - (B) genetic drift.
   - (C) the reshuffling of alleles in sexual reproduction.
   - (D) environmental effects.

Levels 3-4: Applying/Analyzing

3. If the nucleotide variability of a locus equals 0%, what is the gene variability and number of alleles at that locus?
   - (A) gene variability = 0%; number of alleles = 0
   - (B) gene variability = 0%; number of alleles = 1
   - (C) gene variability = 0%; number of alleles = 2
   - (D) gene variability > 0%; number of alleles = 2

4. One population of a species has 25 individuals, all with genotype *AA*; a second population of this species has 40 individuals, all with genotype *aa*. Assume that these populations live far apart but in similar environmental conditions. Based on this information, the observed genetic variation most likely resulted from
   - (A) genetic drift.
   - (B) gene flow.
   - (C) nonrandom mating.
   - (D) directional selection.

5. A fruit fly population has a gene with two alleles, *A1* and *A2*. Tests show that 70% of the gametes produced in the population contain the *A1* allele. If the population is in Hardy-Weinberg equilibrium, what proportion of the flies carry both *A1* and *A2*?
   - (A) 0.7
   - (B) 0.49
   - (C) 0.42
   - (D) 0.21

Levels 5-6: Evaluating/Creating

6. **EVOLUTION CONNECTION** Using at least two examples, explain how the process of evolution is revealed by the imperfections of living organisms.

7. **SCIENTIFIC INQUIRY • INTERPRET THE DATA** For populations of the marine mussel *Mytilus edulis* around Long Island, New York, researchers measured the frequency of a particular allele (*lap*^{94}) for an enzyme involved in regulating the mussel's saltwater balance. They presented their data as a series of pie charts linked to sampling sites within Long Island Sound, where the salinity is highly variable, and along the coast of the open ocean, where salinity is constant. (a) Create a data table for the 11 sampling sites by estimating the frequency of *lap*^{94} from the pie charts. (b) Graph the frequencies for sites 1–8 to show how the frequency of this allele changes with increasing salinity in Long Island Sound (from southwest to northeast). Evaluate how the data from sites 9–11 compare with the data from the sites within the Sound. (c) Considering the various mechanisms that can alter allele frequency, construct a hypothesis that explains the patterns

in the data and that accounts for the following observations: (1) The *lap*^{94} allele helps mussels maintain osmotic balance in water with a high salt concentration but is costly to use in less salty water; and (2) mussels produce larvae that can disperse long distances before they settle on rocks and grow into adults.

Data from R. K. Koehn and T. J. Hilbish, The adaptive importance of genetic variation, *American Scientist* 75:134–141 (1987).

8. **WRITE ABOUT A THEME: ORGANIZATION** Heterozygotes at the sickle-cell locus produce both normal and abnormal (sickle-cell) hemoglobin (see Concept 14.4). When hemoglobin molecules are packed into a heterozygote's red blood cells, some cells receive relatively large quantities of abnormal hemoglobin, making these cells prone to sickling. In a short essay (approximately 100–150 words), explain how these molecular and cellular events lead to emergent properties at the individual and population levels of biological organization.

9. **SYNTHESIZE YOUR KNOWLEDGE**

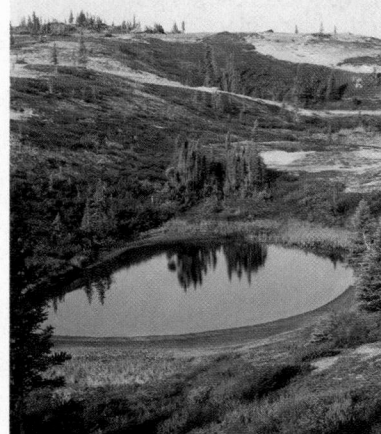

The lake shown in this photograph, called a kettle lake, formed 14,000 years ago when a glacier that covered the surrounding area melted. Initially devoid of animal life, over time the lake was colonized by invertebrates and other animals. Hypothesize how mutation, natural selection, genetic drift, and gene flow may have affected populations that colonized the lake.

For selected answers, see Appendix A.

24 The Origin of Species

KEY CONCEPTS

24.1 The biological species concept emphasizes reproductive isolation *p. 507*

24.2 Speciation can take place with or without geographic separation *p. 511*

24.3 Hybrid zones reveal factors that cause reproductive isolation *p. 516*

24.4 Speciation can occur rapidly or slowly and can result from changes in few or many genes *p. 520*

Study Tip

Make a table: Some processes that can lead to speciation only occur in allopatric (geographically separated) populations, while others also occur in sympatric (geographically overlapping) populations. To help you keep track of these processes and the geographic conditions under which they can occur, fill in the table as you read the chapter.

Process (include page # or figure #)	Can occur in allopatric populations (yes/no)?	Can occur in sympatric populations (yes/no)?
Sexual selection (Figure 24.12)	Yes	Yes

→ Go to Mastering Biology

For Students (in eText and Study Area)
- Get Ready for Chapter 24
- Figure 24.10 Walkthrough: One Mechanism for Allopolyploid Speciation in Plants

For Instructors to Assign (in Item Library)
- Interpreting Data: Evolution of Reproductive Barriers in Lab Populations of Fruit Flies
- Problem-Solving Exercise: Is Hybridization Promoting Insecticide Resistance?

Figure 24.1 This flightless cormorant is one of many species in the Galápagos Islands that are found nowhere else in the world. How did a bird that cannot fly come to live in this isolated place? When Darwin visited the Galápagos, he also was intrigued by its unique species, and later concluded that they must have originated in the islands from ancestors that had traveled from South America.

How do new species originate from existing species?

Over time, populations of a single species connected by gene flow can diverge genetically, giving rise to a new species:

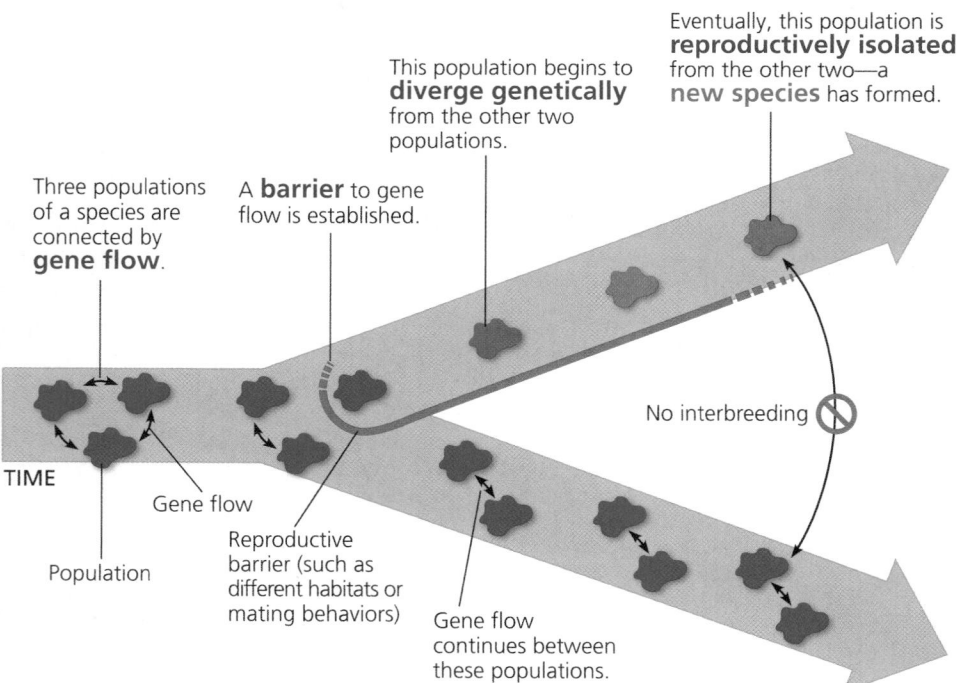

Three populations of a species are connected by **gene flow**.

A **barrier** to gene flow is established.

This population begins to **diverge genetically** from the other two populations.

Eventually, this population is **reproductively isolated** from the other two—a **new species** has formed.

No interbreeding

TIME

Gene flow

Population

Reproductive barrier (such as different habitats or mating behaviors)

Gene flow continues between these populations.

When Darwin came to the Galápagos Islands, he noted that they were teeming with unique species, like the flightless bird in Figure 24.1. Later, he realized that these species had formed relatively recently, writing in his diary, "Both in space and time, we seem to be brought somewhat near to that great fact—that mystery of mysteries—the first appearance of new beings on this Earth."

The "mystery of mysteries" that captivated Darwin is **speciation**, the process by which one species splits into two species. Speciation fascinated Darwin (and many biologists since) because it has produced the tremendous diversity of life, repeatedly yielding new species that differ from existing ones. Speciation also helps to explain the many features that organisms share (the unity of life): When one species splits into two, the species that result share many characteristics because they are descended from this common ancestor. For example, at the DNA sequence level, such similarities indicate that the flightless cormorant in Figure 24.1 is closely related to flying cormorants found in the Americas. This suggests that the flightless cormorant originated from an ancestral cormorant species that flew from the mainland to the Galápagos.

Speciation forms a conceptual bridge between **microevolution**, changes over time in allele frequencies in a population, and **macroevolution**, the broad pattern of evolution above the species level. An example of macroevolutionary change is the origin of new groups of organisms, such as mammals or flowering plants, through a series of speciation events. We examined microevolutionary mechanisms in Chapter 23, and we'll turn to macroevolution in Chapter 25. In this chapter, we'll explore the mechanisms by which new species originate from existing ones. First, let's establish what we actually mean by a "species."

CONCEPT 24.1

The biological species concept emphasizes reproductive isolation

The word *species* is Latin for "kind" or "appearance." In daily life, we commonly distinguish between various "kinds" of organisms—dogs and cats, for instance—based on differences in their appearance. But are organisms truly divided into the discrete units we call species? To answer this question, biologists compare not only the morphology (body form) of different groups of organisms but also their physiology, biochemistry, and DNA sequences. The results generally confirm that morphologically distinct species are indeed discrete groups, differing in many ways besides their body forms.

The Biological Species Concept

The primary definition of species used in this textbook is the **biological species concept**. According to this concept, a

▼ **Figure 24.2 The biological species concept is based on the potential to interbreed, not on physical similarity.**

(a) Similarity between different species. The eastern meadowlark (*Sturnella magna*, left) and the western meadowlark (*Sturnella neglecta*, right) have similar body shapes and colorations. Nevertheless, they are distinct biological species because their songs and other behaviors are different enough to prevent interbreeding should they meet in the wild.

(b) Diversity within a species. As diverse as we may be in appearance, all humans belong to a single biological species (*Homo sapiens*), defined by our capacity to interbreed successfully.

species is a group of populations whose members have the potential to interbreed in nature and produce viable, fertile offspring—but do not produce viable, fertile offspring with members of other such groups **(Figure 24.2)**. Thus, the members of a biological species are united by being reproductively compatible, at least potentially. All human beings, for example, belong to the same species. A businesswoman in Manhattan may be unlikely to meet a dairy farmer in Mongolia, but if the two should happen to meet and mate, they could have viable babies who develop into fertile adults. In contrast, humans and chimpanzees remain distinct biological species, even where they live in the same region, because many factors keep them from interbreeding and producing fertile offspring.

What holds the gene pool of a species together, causing its members to resemble each other more than they resemble members of other species? Recall the evolutionary mechanism of *gene flow*, the transfer of alleles between populations (see Concept 23.3). Typically, gene flow occurs between the different populations of a species. This ongoing exchange of alleles tends to hold the populations together genetically. But as we'll explore in this chapter, a reduction or lack of gene flow can play a key role in the formation of new species.

Reproductive Isolation

Because biological species are defined in terms of reproductive compatibility, the formation of a new species hinges on **reproductive isolation**—the existence of biological

▼ Figure 24.3 Exploring Reproductive Barriers

Prezygotic barriers impede mating or hinder fertilization if mating does occur

| Habitat Isolation | Temporal Isolation | Behavioral Isolation | Mechanical Isolation |

Individuals of different species

MATING ATTEMPT

Habitat Isolation

Two species that occupy different habitats within the same area may encounter each other rarely, if at all.

Example: These two fly species in the genus *Rhagoletis* occur in the same geographic areas, but the apple maggot fly (*Rhagoletis pomonella*) feeds and mates on hawthorns and apples **(a)** while its close relative, the blueberry maggot fly (*R. mendax*), mates and lays its eggs only on blueberries **(b)**.

Temporal Isolation

Species that breed during different times of the day, different seasons, or different years cannot mix their gametes.

Example: In North America, the geographic ranges of the western spotted skunk (*Spilogale gracilis*) **(c)** and the eastern spotted skunk (*Spilogale putorius*) **(d)** overlap, but *S. gracilis* mates in late summer and *S. putorius* mates in late winter.

Behavioral Isolation

Courtship rituals that attract mates and other behaviors unique to a species are effective reproductive barriers, even between closely related species.

Example: Blue-footed boobies, inhabitants of the Galápagos, mate only after a courtship display unique to their species. Part of the "script" calls for the male to high-step **(e)**, a behavior that calls the female's attention to his bright blue feet.

→ **Mastering Biology Video: Blue-Footed Boobies Courtship Ritual**

Mechanical Isolation

Mating is attempted, but morphological differences prevent its successful completion.

Example: Snails in the genus *Bradybaena* approach each other headfirst when they attempt to mate. Once their heads have moved slightly past each other, the snails' genitals emerge, and if their shells spiral in the same direction, mating can occur **(f)**. But if a snail attempts to mate with a snail whose shell spirals in the opposite direction **(g)**, the two snails' genital openings (indicated by arrows) will not be aligned, and mating cannot be completed.

factors (barriers) that impede members of two species from interbreeding and producing viable, fertile offspring **(Figure 24.3)**. Such barriers block gene flow between the species and limit the formation of **hybrids**, offspring that result from an interspecific mating. Although a single barrier may not prevent all gene flow, a combination of several barriers can effectively isolate a species' gene pool.

Clearly, a fly cannot mate with a frog or a fern, but the reproductive barriers between more closely related species are not so obvious. As described in Figure 24.3, these barriers can be

Gametic Isolation	Postzygotic barriers prevent a hybrid zygote from developing into a viable, fertile adult		
	Reduced Hybrid Viability	Reduced Hybrid Fertility	Hybrid Breakdown

FERTILIZATION → VIABLE, FERTILE OFFSPRING

Sperm of one species may not be able to fertilize the eggs of another species. For instance, sperm may not be able to survive in the reproductive tract of females of the other species, or biochemical mechanisms may prevent the sperm from penetrating the membrane surrounding the other species' eggs.

Example: Gametic isolation separates certain closely related species of aquatic animals, such as sea urchins **(h)**. Sea urchins release their sperm and eggs into the surrounding water, where they fuse and form zygotes. It is difficult for gametes of different species, such as the red and purple urchins shown here (*Strongylocentrotus franciscanus* and *S. purpuratus*, respectively), to fuse because proteins on the surfaces of the eggs and sperm bind very poorly to each other.

The genes of different parent species may interact in ways that impair the hybrid's development or survival in its environment.

Example: Some salamander subspecies of the genus *Ensatina* live in the same regions and habitats, where they may occasionally hybridize. But most of the hybrids do not complete development, and those that do are frail **(i)**.

Even if hybrids are vigorous, they may be sterile. If the chromosomes of the two parent species differ in number or structure, meiosis in the hybrids may fail to produce normal gametes. Since the infertile hybrids cannot produce offspring when they mate with either parent species, genes cannot flow freely between the species.

Example: The hybrid offspring of a male donkey **(j)** and a female horse **(k)** is a mule **(l)**, which is robust but sterile. A "hinny" (not shown), the offspring of a female donkey and a male horse, is also sterile.

Some first-generation hybrids are viable and fertile, but when they mate with one another or with either parent species, offspring of the next generation are feeble or sterile.

Example: Strains of cultivated rice have accumulated different mutant recessive alleles at two loci in the course of their divergence from a common ancestor. Hybrids between them are vigorous and fertile (**m**, left and right), but plants in the next generation that carry too many of these recessive alleles are small and sterile (**m**, center). Although these rice strains are not yet considered different species, they have begun to be separated by postzygotic barriers.

classified according to whether they contribute to reproductive isolation before or after fertilization. **Prezygotic barriers** ("before the zygote") block fertilization from occurring. Such barriers typically act in one of three ways: by impeding members of different species from attempting to mate, by preventing an attempted mating from being completed successfully, or by hindering fertilization if mating is completed successfully. If a sperm cell from one species overcomes prezygotic barriers and fertilizes an ovum from another species, a variety of **postzygotic barriers** ("after the zygote") may contribute to reproductive isolation after the hybrid zygote is formed. Developmental errors may reduce survival among hybrid embryos. Or problems after birth may cause hybrids to be infertile or decrease their chance of surviving long enough to reproduce.

→ **Mastering Biology** Video: Prezygotic Barriers to Mating in Galápagos Finches by Peter and Rosemary Grant

Limitations of the Biological Species Concept

One strength of the biological species concept is that it directs our attention to a way by which speciation can occur: by the evolution of reproductive isolation. However, the number of species to which this concept can be usefully applied is limited. There is, for example, no way to evaluate the reproductive isolation of fossils. The biological species concept also does not apply to organisms that reproduce asexually all or most of the time, such as prokaryotes. (Many prokaryotes do transfer genes among themselves, as we will discuss in Concept 27.2, but this is not part of their reproductive process.) Furthermore, in the biological species concept, species are designated by the *absence* of gene flow. But there are many pairs of species that are morphologically and ecologically distinct, and yet gene flow occurs between them. An example is the grizzly bear (*Ursus arctos*) and polar bear (*Ursus maritimus*), whose hybrid offspring have been dubbed "grolar bears" **(Figure 24.4)**. As we'll discuss, natural selection can cause such species to remain distinct even though some gene flow occurs between them. Because of the limitations to the biological species concept, alternative species concepts are useful in certain situations.

Other Definitions of Species

While the biological species concept emphasizes the *separateness* of different species due to reproductive barriers, several other definitions emphasize the *unity within* a species. For example, the **morphological species concept** distinguishes a species by body shape and other structural features. The morphological species concept can be applied to asexual and sexual organisms, and it can be useful even without information on the extent of gene flow. In practice, scientists often distinguish species using morphological criteria. A disadvantage of this approach, however, is that it relies on subjective criteria; researchers may disagree on which structural features distinguish a species.

◀ Grizzly bear (*U. arctos*)

▼ Polar bear (*U. maritimus*)

◀ Hybrid "grolar bear"

▲ **Figure 24.4 Hybridization between two species of bears in the genus *Ursus*.**

The **ecological species concept** defines a species in terms of its ecological niche, the sum of how members of the species interact with the nonliving and living parts of their environment (see Concept 54.1). For example, two species of oak trees might differ in their size or in their ability to tolerate dry conditions, yet still occasionally interbreed. Because they occupy different ecological niches, these oaks would be considered separate species even though they are connected by some gene flow. Unlike the biological species concept, the ecological species concept can accommodate asexual as well as sexual species. It also emphasizes the role of disruptive natural selection as organisms adapt to different environments.

In addition to those discussed here, more than 20 other species definitions have been proposed. The usefulness of each definition depends on the situation and the research questions being asked. For our purposes of studying how species originate, the biological species concept, with its focus on reproductive barriers, is particularly helpful.

→ **Mastering Biology** HHMI Video: The Origin of Species: The Beak of the Finch

CONCEPT CHECK 24.1

1. (a) Which species concept(s) could you apply to both asexual and sexual species? (b) Which would be most useful for identifying species in the field? Explain.
2. **WHAT IF?** Suppose two bird species live in a forest and are not known to interbreed. One species feeds and mates in the treetops and the other on the ground. But in captivity, the birds can interbreed and produce viable, fertile offspring. What type of reproductive barrier most likely keeps these species separate in nature? Explain.

For suggested answers, see Appendix A.

Speciation can take place with or without geographic separation

Having discussed what constitutes a unique species, let's return to the process by which such species arise from existing species. We'll describe this process by focusing on the geographic setting in which gene flow is interrupted between populations of the existing species—in allopatric speciation the populations are geographically isolated, while in sympatric speciation they are not **(Figure 24.5)**.

Allopatric ("Other Country") Speciation

In **allopatric speciation** (from the Greek *allos*, other, and *patra*, homeland), gene flow is interrupted when a population is divided into geographically isolated subpopulations. For example, the water level in a lake may subside, resulting in two or more smaller lakes that are now home to separated populations (see Figure 24.5a). Or a river may change course and divide a population of animals that cannot cross it. Allopatric speciation can also occur without geologic change, such as when individuals colonize a remote area and their descendants become geographically isolated from the parent population. The flightless cormorant shown in Figure 24.1 probably originated in this way from an ancestral flying species that reached the Galápagos Islands.

The Process of Allopatric Speciation

How formidable must a geographic barrier be to promote allopatric speciation? The answer depends on the ability of the organisms to move about. Birds, mountain lions, and coyotes can cross rivers and canyons—as can windblown pollen

▼ **Figure 24.5 The geography of speciation.**

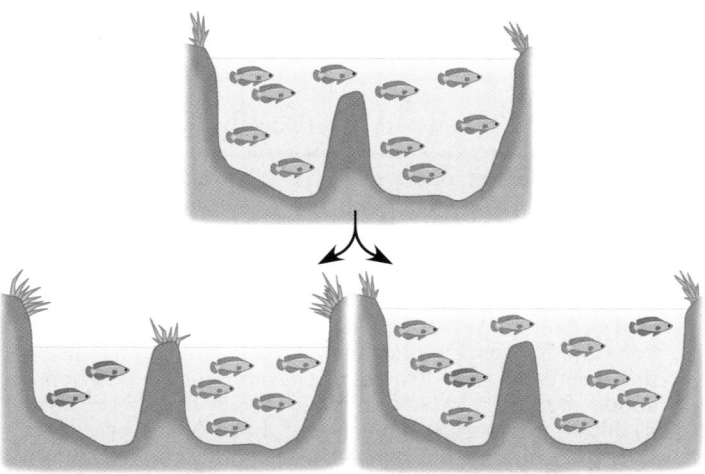

(a) Allopatric speciation. A population forms a new species while geographically isolated from its parent population.

(b) Sympatric speciation. A subset of a population forms a new species without geographic separation.

and the seeds of some flowering plants. In contrast, small rodents may find a river or canyon a formidable barrier.

Once geographic isolation has occurred, the separated gene pools may diverge. Different mutations arise, and natural selection and genetic drift may alter allele frequencies in different ways in the separated populations. Reproductive isolation may then evolve as a by-product of the genetic divergence that results from selection or drift.

Figure 24.6 describes an example. On Andros Island, in the Bahamas, populations of the mosquitofish *Gambusia hubbsi* colonized a series of ponds that later became isolated from one another. Genetic analyses indicate that little or no gene flow currently occurs between the ponds. The environments of these ponds are very similar except that some contain predatory fishes, while others do not. In ponds with predatory fishes, selection has favored the evolution of a mosquitofish body shape that enables rapid bursts of speed (see Figure 24.6a). In ponds without predatory fishes, selection has favored a different body shape, one that improves the ability to swim for long periods of time (see Figure 24.6b).

How have these different selective pressures affected the evolution of reproductive barriers? Researchers studied this question by bringing together mosquitofish from the two types of ponds. They found that female mosquitofish prefer to mate with males whose body shape is similar to their own. This preference establishes a behavioral barrier

▼ **Figure 24.6 Evolution in mosquitofish populations.**
Different body shapes have evolved in mosquitofish populations from ponds with and without predators. These differences affect how quickly the fishes can accelerate to escape and their survival rate when exposed to predators.

In ponds with predatory fishes, the mosquitofish's head is streamlined and the tail is powerful, enabling rapid bursts of speed.

In ponds without predatory fishes, mosquitofish have a different body shape that favors long, steady swimming.

(a) Differences in body shape

 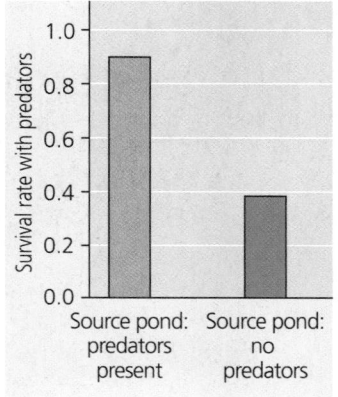

(b) Differences in escape acceleration and survival

to reproduction between mosquitofish from ponds with predators and those from ponds without predators. Thus, as a by-product of selection for avoiding predators, reproductive barriers have formed in these allopatric populations.

Evidence of Allopatric Speciation

Many studies provide evidence that speciation can occur in allopatric populations. For example, laboratory studies show that reproductive barriers can develop when populations are isolated experimentally and subjected to different environmental conditions **(Figure 24.7)**.

Field studies indicate that allopatric speciation also can occur in nature. Consider the 30 species of snapping shrimp in the genus *Alpheus* that live off the Isthmus of Panama, the land bridge that connects South and North America **(Figure 24.8)**. Fifteen of these species live on the Atlantic side of the isthmus, while the other 15 live on the Pacific side. Before the isthmus formed, gene flow could occur between the Atlantic and Pacific populations of snapping shrimp. Did the species on different sides of the isthmus originate by allopatric speciation? Morphological and genetic data group these shrimp into 15 pairs of *sister species*, pairs whose member species are each other's closest relative. In each of these

▼ **Figure 24.8 Allopatric speciation in snapping shrimp (*Alpheus*).** The shrimp pictured are just 2 of the 15 pairs of sister species that arose as populations were divided by the formation of the Isthmus of Panama. The color-coded type indicates the sister species.

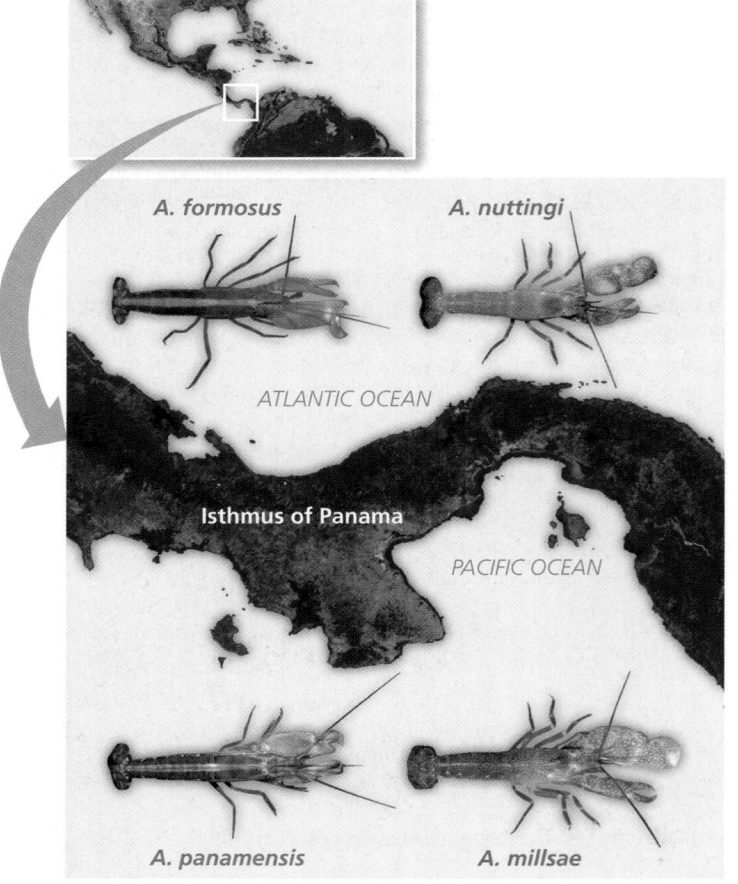

A. formosus   A. nuttingi

ATLANTIC OCEAN

Isthmus of Panama

PACIFIC OCEAN

A. panamensis   A. millsae

▼ **Figure 24.7** **Inquiry**

Can divergence of allopatric populations lead to reproductive isolation?

Experiment A researcher divided a laboratory population of the fruit fly *Drosophila pseudoobscura*, raising some flies on a starch medium and others on a maltose medium. After one year (about 40 generations), natural selection resulted in divergent evolution: Populations raised on starch digested starch more efficiently, while those raised on maltose digested maltose more efficiently. The researcher then put flies from the same or different populations in mating cages and measured mating frequencies. All flies used in the mating preference tests were reared for one generation on a standard cornmeal medium.

Initial population of fruit flies (*Drosophila pseudoobscura*)

Some flies raised on starch medium

Some flies raised on maltose medium

Mating experiments after 40 generations

Results Mating patterns among populations of flies raised on different media are shown below. When flies from "starch populations" were mixed with flies from "maltose populations," the flies tended to mate with like partners (see the lefthand table). But in a control group consisting of flies from two different populations adapted to starch, the flies were about as likely to mate with flies of the other population as with flies of their own population (see the righthand table); similar results were obtained for control groups adapted to maltose.

		Female	
		Starch	Maltose
Male	Starch	22	9
	Maltose	8	20

Number of matings in experimental group

		Female	
		Starch population 1	Starch population 2
Male	Starch population 1	18	15
	Starch population 2	12	15

Number of matings in control group

Conclusion In the experimental group, the strong preference of "starch flies" and "maltose flies" to mate with like-adapted flies indicates that a reproductive barrier was forming between these fly populations. Although this barrier was not absolute (some mating between starch flies and maltose flies did occur), after 40 generations reproductive isolation appeared to be increasing. This barrier may have been caused by differences in courtship behavior that arose as an incidental by-product of differing selective pressures as these allopatric populations adapted to different sources of food.

Data from D. M. B. Dodd, Reproductive isolation as a consequence of adaptive divergence in *Drosophila pseudoobscura, Evolution* 43:1308–1311 (1989).

WHAT IF? *Why were all flies used in the mating preference tests reared on a standard medium (rather than on starch or maltose)?*

15 pairs, one of the sister species lives on the Atlantic side of the isthmus, while the other lives on the Pacific side. This fact strongly suggests that the two species arose as a consequence of geographic separation. Furthermore, genetic analyses indicate that the *Alpheus* species originated from 9 to 3 million years ago, with the sister species that live in the deepest water diverging first. These divergence times are consistent with geologic evidence that the isthmus formed gradually, starting 10 million years ago, and closing completely about 3 million years ago.

The importance of allopatric speciation is also suggested by the fact that regions that are isolated or highly subdivided by barriers typically have more species than do otherwise similar regions that lack such features. For example, many unique plants and animals are found on the geographically isolated Hawaiian Islands (we'll return to the origin of Hawaiian species in Concept 25.4). Field studies also show that reproductive isolation between two populations generally increases as the geographic distance between them increases, a finding consistent with allopatric speciation. In the **Scientific Skills Exercise**, you can analyze data from one such study that examined reproductive isolation in geographically separated salamander populations.

Note that while geographic isolation prevents interbreeding between members of allopatric populations, physical separation is not a biological barrier to reproduction. Biological reproductive barriers such as those described in Figure 24.3 are intrinsic to the organisms themselves. Hence, it is biological barriers that can prevent interbreeding when members of different populations come into contact with one another.

Sympatric ("Same Country") Speciation

In **sympatric speciation** (from the Greek *syn*, together), speciation occurs in populations that live in the same geographic area (see Figure 24.5b). How can reproductive barriers form between sympatric populations while their members remain in contact with each other? Although such contact (and the ongoing gene flow that results) makes sympatric speciation less common than allopatric speciation, sympatric speciation can occur if gene flow is reduced by such factors as polyploidy, sexual selection, and habitat differentiation. (Note that sexual selection and habitat differentiation also can promote allopatric speciation.)

Scientific Skills Exercise

Identifying Independent and Dependent Variables, Making a Scatter Plot, and Interpreting Data

Does Distance Between Salamander Populations Increase Their Reproductive Isolation? Allopatric speciation begins when populations become geographically isolated, reducing mating between individuals in different populations and thus reducing gene flow. It is logical that as distance between populations increases, so will their degree of reproductive isolation. To test this hypothesis, researchers studied populations of the dusky salamander (*Desmognathus ochrophaeus*) living on different mountain ranges in the southern Appalachians.

How the Experiment Was Done The researchers tested the reproductive isolation of pairs of salamander populations by leaving one male and one female together and later checking the females for the presence of sperm. Four mating combinations were tested for each pair of populations (A and B)—two *within* the same population (female A with male A and female B with male B) and two *between* populations (female A with male B and female B with male A).

Data from the Experiment The researchers used an index of reproductive isolation that ranged from a value of 0 (no isolation) to a value of 2 (full isolation). The proportion of successful matings for each mating combination was measured, with 100% success = 1 and no success = 0. The reproductive isolation value for two populations is the sum of the proportion of successful matings of each type within populations (AA + BB) minus the sum of the proportion

of successful matings of each type between populations (AB + BA). The table provides distance and reproductive isolation data for 27 pairs of dusky salamander populations.

INTERPRET THE DATA

1. State the researchers' hypothesis, and identify the independent and dependent variables in this study. Explain why the researchers used four mating combinations for each pair of populations.

2. Calculate the value of the reproductive isolation index **(a)** if *all* of the matings within a population were successful, but *none* of the matings between populations were successful; **(b)** if salamanders are equally successful in mating with members of their own population and members of another population.

3. Make a scatter plot to help you visualize any patterns that might indicate a relationship between the variables. Plot the independent variable on the *x*-axis and the dependent variable on the *y*-axis. (For additional information about graphs, see the Scientific Skills Review in Appendix D.)

4. Interpret your graph by **(a)** explaining in words any pattern indicating a possible relationship between the variables and **(b)** hypothesizing the possible cause of such a relationship.

➡ **Instructors:** A version of this Scientific Skills Exercise can be assigned in **Mastering Biology**.

Data from S. G. Tilley, A. Verrell, and S. J. Arnold, Correspondence between sexual isolation and allozyme differentiation: a test in the salamander *Desmognathus ochrophaeus*, *Proceedings of the National Academy of Sciences USA* 87:2715–2719 (1990).

Geographic Distance (km)	15	32	40	47	42	62	63	81	86	107	107	115	137	147
Reproductive Isolation Value	0.32	0.54	0.50	0.50	0.82	0.37	0.67	0.53	1.15	0.73	0.82	0.81	0.87	0.87
Distance (continued)	137	150	165	189	219	239	247	53	55	62	105	179	169	
Isolation (continued)	0.50	0.57	0.91	0.93	1.5	1.22	0.82	0.99	0.21	0.56	0.41	0.72	1.15	

Polyploidy

A species may originate from an accident during cell division that results in extra sets of chromosomes, a condition called **polyploidy**. Polyploid speciation occasionally occurs in animals; for example, the gray tree frog *Hyla versicolor* (see Figure 23.16) is thought to have originated in this way. However, polyploidy is far more common in plants. In fact, botanists estimate that more than 80% of the plant species alive today are descended from ancestors that formed by polyploid speciation.

Two distinct forms of polyploidy have been observed in plant (and a few animal) populations. An **autopolyploid** (from the Greek *autos*, self) is an individual that has more than two chromosome sets that are all derived from a single species. In plants, for example, a failure of cell division could double a cell's chromosome number from the original number ($2n$) to a tetraploid number ($4n$) **(Figure 24.9)**.

▼ **Figure 24.9 Sympatric speciation by autopolyploidy.**

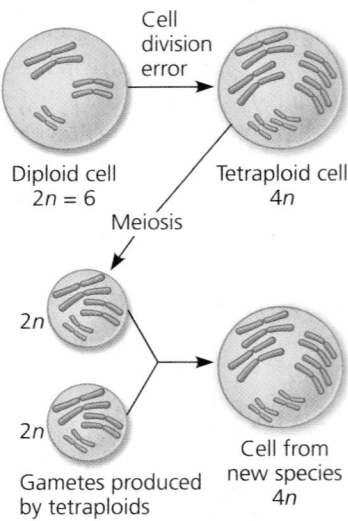

Cell division error

Diploid cell $2n = 6$

Tetraploid cell $4n$

Meiosis

$2n$

$2n$

Gametes produced by tetraploids

Cell from new species $4n$

A tetraploid can produce fertile tetraploid offspring by self-pollinating or by mating with other tetraploids. In addition, the tetraploids are reproductively isolated from $2n$ plants of the original population because the triploid ($3n$) offspring of such unions have reduced fertility. Thus, in just one generation, autopolyploidy can generate reproductive isolation without any geographic separation.

A second form of polyploidy can occur when two different species interbreed and produce hybrid offspring. Most such hybrids are sterile because the set of chromosomes from one species cannot pair during meiosis with the set of chromosomes from the other species. However, an infertile hybrid may be able to propagate itself asexually (as many plants can do). In subsequent generations, various mechanisms can change a sterile hybrid into a fertile polyploid called an **allopolyploid** **(Figure 24.10)**. The allopolyploids are fertile when mating with each other but cannot interbreed with either parent species; thus, they represent a new biological species. The new species has a diploid chromosome number equal to the sum of the diploid chromosome numbers of the two parent species.

Although it can be challenging to study speciation in the field, scientists have documented at least five new plant species that have originated by polyploid speciation since 1850. One of these examples involves the origin of a new species of goatsbeard plant (genus *Tragopogon*) in the Pacific Northwest. *Tragopogon* first arrived in the region when humans introduced

▼ **Figure 24.10 One mechanism for allopolyploid speciation in plants.** Most hybrids are sterile because their chromosomes are not homologous and cannot pair during meiosis. However, such a hybrid may be able to reproduce asexually. This diagram traces one mechanism by which sterile hybrids can give rise to fertile allopolyploids that are members of a new species.

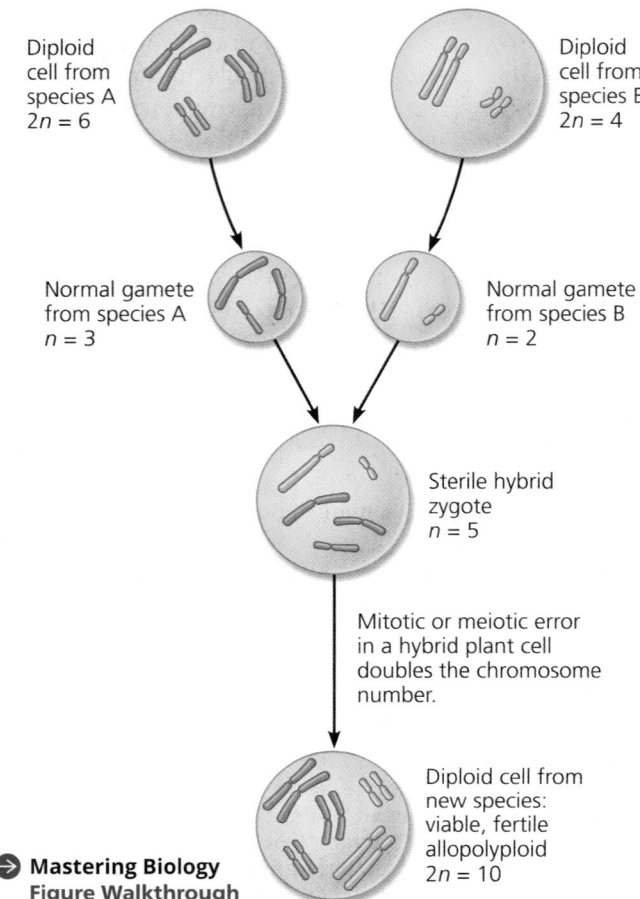

Diploid cell from species A $2n = 6$

Diploid cell from species B $2n = 4$

Normal gamete from species A $n = 3$

Normal gamete from species B $n = 2$

Sterile hybrid zygote $n = 5$

Mitotic or meiotic error in a hybrid plant cell doubles the chromosome number.

Diploid cell from new species: viable, fertile allopolyploid $2n = 10$

→ **Mastering Biology** Figure Walkthrough

three European species in the early 1900s: *T. pratensis, T. dubius,* and *T. porrifolius*. These three species are now common in the Pacific Northwest and many other parts of North America. In 1950, a new *Tragopogon* species was discovered near the Idaho-Washington border, a region where all three European species also were found. Genetic analyses revealed that this new species, *Tragopogon miscellus*, is an allopolyploid that originated from a hybrid of two of the European species **(Figure 24.11)**. Although the *T. miscellus* population grows mainly by reproduction of its own members, hybridization between the parent species (followed by a mitotic or meiotic error that doubles the chromosome number) continues to add new members to the *T. miscellus* population. Later, scientists discovered another new *Tragopogon* species, *T. mirus*, which also arose by polyploid speciation.

Many important agricultural crops—such as oats, cotton, potatoes, tobacco, and wheat—are polyploids. For example, the wheat used for bread, *Triticum aestivum*, is an allohexaploid (six sets of chromosomes, two sets from each of three different species). The events that eventually led to modern wheat probably began about 8,000 years ago in the Middle

▼ **Figure 24.11 Allopolyploid speciation in *Tragopogon*.** The gray boxes indicate the parent species of the two new polyploid species. The diploid chromosome number of each species is shown in parentheses.

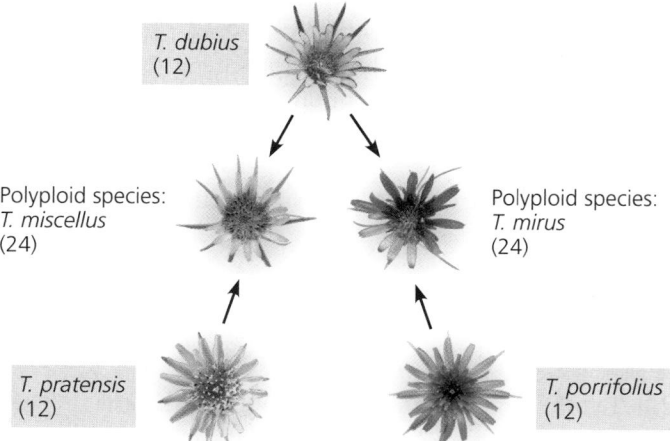

T. dubius
(12)

Polyploid species:
T. miscellus
(24)

Polyploid species:
T. mirus
(24)

T. pratensis
(12)

T. porrifolius
(12)

VISUAL SKILLS *Based on this diagram, identify the two parent species of each polyploid species.*

East when an early cultivated wheat species hybridized with a wild grass. Today, plant geneticists generate new polyploids in the laboratory by using chemicals that induce meiotic and mitotic errors. By harnessing the evolutionary process, researchers can produce new polyploid species with desired qualities, such as a polyploid that combines the high yield of wheat with the hardiness of rye.

⊖ **Mastering Biology** Animation: Speciation by Changes in Ploidy

Sexual Selection

There is evidence that sympatric speciation can also be driven by sexual selection. Clues to how this can occur have been found in cichlid fish from one of Earth's hot spots of animal speciation, East Africa's Lake Victoria. This lake was once home to as many as 600 species of cichlids. Genetic data indicate that these species originated within the last 100,000 years from a small number of colonizing species that arrived from other lakes and rivers. How did so many species—more than double the number of freshwater fish species known in all of Europe—originate within a single lake?

One hypothesis is that subgroups of the original cichlid populations adapted to different food sources and the resulting genetic divergence contributed to speciation in Lake Victoria. But sexual selection, in which (typically) females select males based on their appearance (see Concept 23.4), may also have been a factor. Researchers have studied two closely related sympatric species of cichlids that differ mainly in the coloration of breeding males: Breeding *Pundamilia pundamilia* males have a blue-tinged back, whereas breeding *Pundamilia nyererei* males have a red-tinged back **(Figure 24.12)**. Their results suggest that mate choice based on male breeding coloration can act as a reproductive barrier that keeps the gene pools of these two species separate.

▼ **Figure 24.12** Inquiry

Does sexual selection in cichlids result in reproductive isolation?

Experiment Researchers placed males and females of *Pundamilia pundamilia* and *P. nyererei* together in two aquarium tanks, one with natural light and one with a monochromatic orange lamp. Under normal light, the two species are noticeably different in male breeding coloration; under monochromatic orange light, the two species are very similar in color. The researchers then observed the mate choices of the females in each tank.

Normal light

Monochromatic orange light

P. pundamilia

P. nyererei

Results Under normal light, females of each species strongly preferred males of their own species. But under orange light, females of each species responded indiscriminately to males of both species. The resulting hybrids were viable and fertile.

Conclusion The researchers concluded that mate choice by females based on male breeding coloration can act as a reproductive barrier that keeps the gene pools of these two species separate. Since the species can still interbreed when this prezygotic barrier is breached in the laboratory, the genetic divergence between the species is likely to be small. This suggests that speciation in nature has occurred relatively recently.

Data from O. Seehausen and J. J. M. van Alphen, The effect of male coloration on female mate choice in closely related Lake Victoria cichlids (*Haplochromis nyererei* complex), *Behavioral Ecology and Sociobiology* 42:1–8 (1998).

WHAT IF? *Suppose that female cichlids living in the murky waters of a polluted lake could not distinguish colors well. In such waters, how might the gene pools of these species change over time?*

Habitat Differentiation

Sympatric speciation can also occur when a subpopulation exploits a habitat or resource not used by the parent population. Consider the North American apple maggot fly (*Rhagoletis pomonella*), a pest of apples. The fly's original habitat was the native hawthorn tree (see Figure 24.3a), but about 200 years ago, some populations colonized apple trees that had been introduced by European settlers. Apple maggot flies usually mate on or near their host plant. This results in a prezygotic barrier (habitat isolation) between populations that feed on apples and populations that feed on hawthorns. Furthermore, because apples mature more quickly than hawthorn fruit, natural selection has favored apple-feeding flies with rapid development. These apple-feeding populations now show temporal isolation from the hawthorn-feeding

R. pomonella, providing a second prezygotic barrier to gene flow between the two populations. Researchers also have identified alleles that benefit the flies that use one host plant but harm the flies that use the other host plant. Natural selection operating on these alleles has provided a postzygotic barrier to reproduction, further limiting gene flow. Altogether, although the two populations are still classified as subspecies rather than separate species, sympatric speciation appears to be well under way.

Allopatric and Sympatric Speciation: *A Review*

Now let's recap the processes by which new species form. In allopatric speciation, a new species forms in geographic isolation from its parent population. Geographic isolation severely restricts gene flow. Intrinsic barriers to reproduction with members of the parent population may then arise as a by-product of genetic changes that occur within the isolated population. Many different processes can produce such genetic changes, including natural selection under different environmental conditions, genetic drift, and sexual selection. Once formed, reproductive barriers that arise in allopatric populations can prevent interbreeding with the parent population even if the populations come back into contact.

Sympatric speciation, in contrast, requires the emergence of a reproductive barrier that isolates a subset of a population from the remainder of the population in the same area. Though rarer than allopatric speciation, sympatric speciation can occur when gene flow to and from the isolated subpopulation is blocked. This can occur as a result of polyploidy, a condition in which an organism has extra sets of chromosomes. Sympatric speciation also can result from sexual selection. Finally, sympatric speciation can occur when a subset of a population becomes reproductively isolated because of natural selection that results from a switch to a habitat or food source not used by the parent population.

Having reviewed the geographic context in which species originate, we'll next explore in more detail what can happen when new or partially formed species come into contact.

 Mastering Biology HHMI Video: Anole Lizards: An Example of Speciation

CONCEPT CHECK 24.2

1. Summarize key differences between allopatric and sympatric speciation. Which type of speciation is more common, and why?
2. Describe two mechanisms that can decrease gene flow in sympatric populations, thereby making sympatric speciation more likely to occur.
3. **WHAT IF?** Is allopatric speciation more likely to occur on an island close to a mainland or on a more isolated island of the same size? Explain your prediction.
4. **MAKE CONNECTIONS** Review the process of meiosis in Figure 13.8. Describe how an error during meiosis could lead to polyploidy.

For suggested answers, see Appendix A.

Hybrid zones reveal factors that cause reproductive isolation

What happens if species with incomplete reproductive barriers come into contact with one another? One possible outcome is the formation of a **hybrid zone**, a region in which members of different species meet and mate, producing at least some offspring of mixed ancestry. In this section, we'll explore hybrid zones and what they reveal about factors that cause the evolution of reproductive isolation.

Patterns Within Hybrid Zones

Some hybrid zones form as narrow bands, such as the one depicted in **Figure 24.13** for the yellow-bellied toad (*Bombina variegata*) and its close relative, the fire-bellied toad (*B. bombina*). This hybrid zone, represented by the red line on the map, extends for 4,000 km but is less than 10 km wide in most places. The hybrid zone occurs where the higher-altitude habitat of the yellow-bellied toad meets the lowland habitat of the fire-bellied toad. Across a given "slice" of the zone, the frequency of alleles specific to yellow-bellied toads typically decreases from close to 100% at the edge where only yellow-bellied toads are found to around 50% in the central portion of the zone and to close to 0% at the edge where only fire-bellied toads are found.

What causes such a pattern of allele frequencies across a hybrid zone? We can infer that there is an obstacle to gene flow—otherwise, alleles from one parent species would also be common in the gene pool of the other parent species. Are geographic barriers reducing gene flow? Not in this case, since the toads can move throughout the hybrid zone. A more important factor is that hybrid toads have increased rates of embryonic mortality and a variety of morphological abnormalities, including ribs that are fused to the spine and malformed tadpole mouthparts. Because the hybrids have poor survival and reproduction, they produce few viable offspring with members of the parent species. As a result, hybrid individuals rarely serve as a stepping-stone from which alleles are passed from one species to the other. Outside the hybrid zone, additional obstacles to gene flow may be provided by natural selection in the different environments in which the parent species live.

Hybrid zones typically are located wherever the habitats of the interbreeding species meet. Those regions often resemble a group of isolated patches scattered across the landscape—more like the pattern of spots on a Dalmatian than the continuous band shown in Figure 24.13. But regardless of whether they have complex or simple spatial patterns, hybrid zones form when two species lacking complete barriers to reproduction come into contact. What happens when the habitats of the interbreeding species change over time?

Hybrid zone (red line) occurs where the habitats of the two species meet.

Fire-bellied toad, *Bombina bombina*: lives at lower altitudes

Fire-bellied toad range

Hybrid zone

Yellow-bellied toad range

▲ **Figure 24.13 A narrow hybrid zone for *Bombina* toads in Europe.** The graph shows species-specific allele frequencies across the width of the zone near Krakow, Poland, averaged over six genetic loci. A value of 1.0 indicates that all individuals were yellow-bellied toads, 0 indicates that all individuals were fire-bellied toads, and intermediate frequencies indicate that some individuals were of mixed ancestry.

❓ *Does the graph indicate that gene flow is spreading fire-bellied toad alleles into the range of the yellow-bellied toad? Explain.*

Yellow-bellied toad, *Bombina variegata*: lives at higher altitudes

Hybrid zone

Frequency of *B. variegata*-specific alleles

1.0
0.75
0.5
0.25
0

Yellow-bellied toad range

Fire-bellied toad range

40   30   20   10   0   10   20
Distance from hybrid zone center (km)

Hybrid Zones and Environmental Change

A change in environmental conditions can alter where the habitats of interbreeding species meet. When this happens, an existing hybrid zone can move to a new location, or a novel hybrid zone may form.

For example, black-capped chickadees (*Poecile atricapillus*) and Carolina chickadees (*P. carolinensis*) interbreed in a narrow hybrid zone that runs from New Jersey to Kansas. Recent studies have shown that the location of this hybrid zone has shifted northward as the climate has warmed **(Figure 24.14)**. In another example, a series of warm winters prior to 2003 enabled the southern flying squirrel (*Glaucomys volans*) to expand northward into the range of the northern flying squirrel, *G. sabrinus*. Previously, the ranges of these two species had not overlapped. Genetic analyses showed that these flying squirrels began to hybridize where their ranges came into contact, thereby forming a novel hybrid zone induced by climate change.

Finally, note that a hybrid zone can be a source of novel genetic variation that improves the ability of one or both parent species to cope with changing environmental conditions. This can occur when an allele found only in one parent species is transferred first to hybrid individuals and then to the other parent species when hybrids breed with the second parent species. Recent genetic analyses have shown that hybridization has been a source for such novel genetic variation in some insect, mammal, bird, and plant species. In the **Problem-Solving Exercise**, you can examine one such example: a case in which hybridization may have led to the transfer of insecticide-resistance alleles between species of mosquitoes that transmit malaria.

▼ **Figure 24.14 A shift in a hybrid zone resulting from climate change.** Black-capped (a) and Carolina chickadees (b) interbreed in a hybrid zone that runs from Kansas to New Jersey. In Pennsylvania, the center of this hybrid zone moved 12 km to the north from 2002 to 2012. This shift is consistent with predictions based on the warmer winter temperatures that have resulted from climate change.

(a) **Black-capped chickadee** (*Poecile atricapillus*)

(b) **Carolina chickadee** (*Poecile carolinensis*)

Is hybridization promoting insecticide resistance in mosquitoes that transmit malaria?

Malaria is a leading cause of human illness and mortality worldwide, with 200 million people infected and 600,000 deaths each year. In the 1960s, the incidence of malaria was reduced, owing to the use of insecticides that killed mosquitoes in the genus *Anopheles*, which transmit the disease from person to person. But today, mosquitoes are becoming resistant to insecticides—causing a resurgence in malaria.

Insecticide-treated bed nets have helped reduce cases of malaria in many countries, but resistance to insecticides is rising in mosquito populations.

→ **Instructors:** A version of this Problem-Solving Exercise can be assigned in **Mastering Biology.**

In this exercise, you will investigate whether alleles encoding resistance to insecticides have been transferred between closely related species of *Anopheles*.

Your Approach The principle guiding your investigation is that DNA analyses can detect the transfer of resistance alleles between closely related mosquito species. To find out whether such transfers have occurred, you will analyze DNA results from two species of mosquitoes that transmit malaria (*Anopheles gambiae* and *A. coluzzii*) and from *A. gambiae* × *A. coluzzii* hybrids.

Your Data Resistance to DDT and other insecticides in *Anopheles* is affected by a sodium channel gene, *kdr*. The *r* allele of this gene confers resistance, while the wild-type (+/+) genotype is not resistant. Researchers sequenced the *kdr* gene from mosquitoes collected in Mali during three time periods: pre-2006 (2002 and 2004), 2006, and post-2006 (2009–2012). *A. gambiae* and *A. coluzzii* were collected during all three time periods, but their hybrids only occurred in 2006, the first year that insecticide-treated bed nets were used to reduce the spread of malaria. A likely explanation is that the introduction of the treated bed nets may have briefly favored hybrid individuals, which are usually at a selective disadvantage.

Observed numbers of mosquitoes by *kdr* genotype			
	+/+	+/r	r/r
A. gambiae:			
Pre-2006	3	5	2
2006	8	8	7
Post-2006	3	3	57
Hybrids: 2006	10	7	0
A. coluzzii:			
Pre-2006	226	0	0
2006	70	7	0
Post-2006	79	127	94

Your Analysis

1. **(a)** Calculate the *kdr* genotype frequencies in *A. gambiae* for each time period. To do this, divide the number of individuals that have a given genotype by the total number of individuals observed for that time period. **(b)** How did the *kdr* genotype frequencies change over time? Describe a hypothesis that accounts for these observations.

2. How did the frequencies of *kdr* genotypes change over time in *A. coluzzii*? Describe a hypothesis that accounts for these observations.

3. Do these results indicate that hybridization can lead to the transfer of adaptive alleles? Explain.

4. Predict how the transfer of the *r* allele to *A. coluzzii* populations could affect the number of malaria cases in the years immediately following the transfer.

Hybrid Zones over Time

Studying a hybrid zone is like observing a naturally occurring experiment on speciation. Will the hybrids become reproductively isolated from their parents and form a new species? This has occurred by polyploid speciation in some cases, as we saw for the formation of two new species of goatsbeard plants (see Figure 24.11). Hybridization also has led to what may turn out to be a new species of finch in the Galápagos archipelago: A 2018 study found that descendants of hybrids between the large cactus finch (*Geospiza conirostris*) and the medium ground finch (*G. fortis*) are becoming reproductively isolated from both parent species. As of 2019, this emerging species is known informally as "Big Bird."

In other cases, however, interspecific hybrids do not become reproductively isolated from their parent species.

In such situations, there are three common outcomes for a hybrid zone over time: reinforcement of barriers, fusion of species, or stability **(Figure 24.15)**. Let's examine what studies suggest about these possibilities.

Reinforcement: Strengthening Reproductive Barriers

Hybrids often are less fit than members of their parent species. In such cases, natural selection should strengthen prezygotic barriers to reproduction, reducing the formation of unfit hybrids. Because this process involves *reinforcing* reproductive barriers, it is called **reinforcement**. If reinforcement is occurring, a logical prediction is that barriers to reproduction between species should be stronger for sympatric populations than for allopatric populations.

▼ **Figure 24.15 Formation of a hybrid zone and common outcomes for hybrids over time.**
The thick gray and purple arrows represent the passage of time.

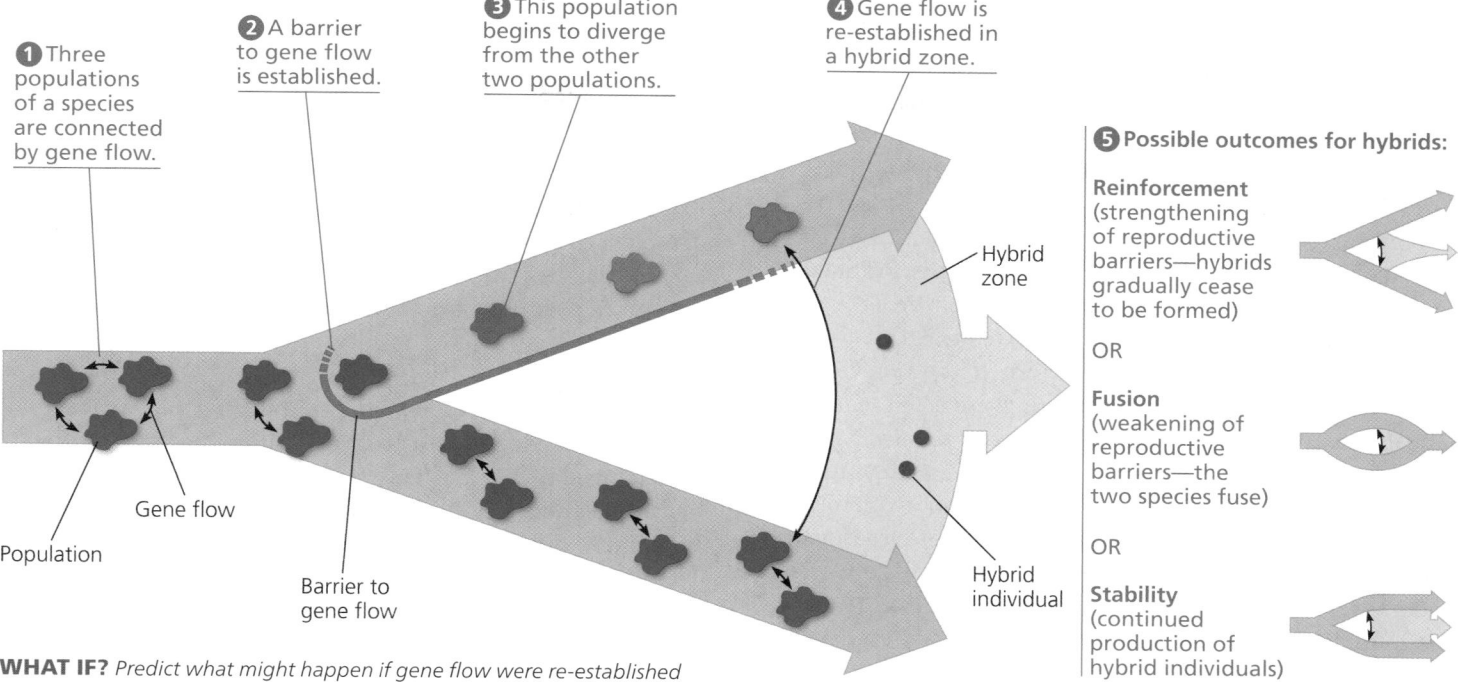

1 Three populations of a species are connected by gene flow.

2 A barrier to gene flow is established.

3 This population begins to diverge from the other two populations.

4 Gene flow is re-established in a hybrid zone.

Hybrid zone

Gene flow

Population

Barrier to gene flow

Hybrid individual

5 Possible outcomes for hybrids:

Reinforcement (strengthening of reproductive barriers—hybrids gradually cease to be formed)

OR

Fusion (weakening of reproductive barriers—the two species fuse)

OR

Stability (continued production of hybrid individuals)

WHAT IF? *Predict what might happen if gene flow were re-established at step 3 in this process.*

As an example, let's consider two species of European flycatcher, the pied flycatcher (*Ficedula hypoleuca*) and the collared flycatcher (*Ficedula albicollis*). In allopatric populations of these birds, males of the two species closely resemble one another, while in sympatric populations, the males look very different. Female flycatchers do not select males of the other species when given a choice between males from sympatric populations, but they frequently do make mistakes when selecting between males from allopatric populations. Thus, barriers to reproduction are stronger in birds from sympatric populations than in birds from allopatric populations, as you would predict if reinforcement were occurring. Similar results have been observed in a number of organisms, including some fishes, insects, plants, and other birds.

Fusion: Weakening Reproductive Barriers

Barriers to reproduction may be weak when two species meet in a hybrid zone. Indeed, so much gene flow may occur that reproductive barriers weaken further and the gene pools of the two species become increasingly alike. In effect, the speciation process reverses, eventually causing the two hybridizing species to fuse into a single species.

For example, genetic and morphological evidence indicates that the recent loss of the large tree finch from the Galápagos island of Floreana resulted from extensive hybridization with another finch species on that island. Such a situation also may be occurring among Lake Victoria cichlids (**Figure 24.16**). Many pairs of ecologically similar cichlid

▼ **Figure 24.16 Fusion: the breakdown of reproductive barriers.**
Increasingly cloudy water in Lake Victoria over the past several decades may have weakened reproductive barriers between *P. nyererei* and *P. pundamilia*. In areas of cloudy water, the two species have hybridized extensively, causing their gene pools to fuse.

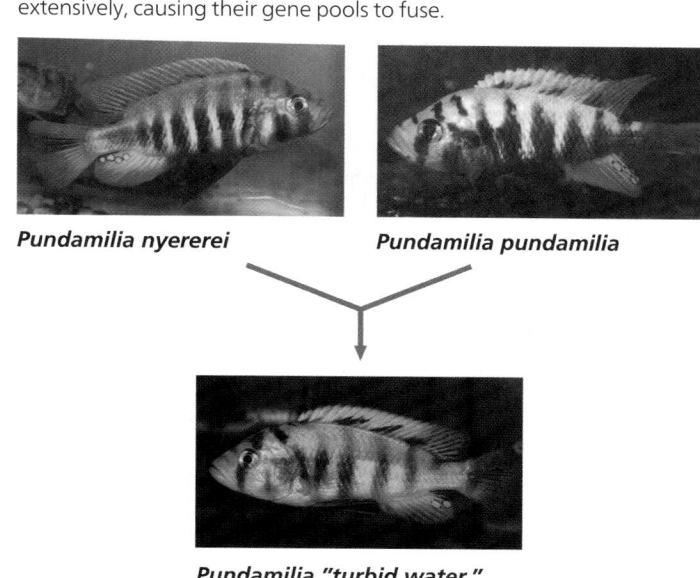

Pundamilia nyererei

Pundamilia pundamilia

Pundamilia "turbid water," hybrid offspring from a location with turbid water

species are reproductively isolated because the females of one species prefer to mate with males of one color, while females of the other species prefer to mate with males of a different color (see Figure 24.12). Results from field and laboratory studies indicate that murky waters caused by pollution have reduced the ability of females to use color to distinguish males of their own species from males of closely related species. In some polluted waters, many hybrids have been produced, leading to fusion of the parent species' gene pools and a loss of species.

Stability: Continued Formation of Hybrid Individuals

Many hybrid zones are stable in the sense that hybrids continue to be produced. In some cases, this occurs because the hybrids survive or reproduce better than members of either parent species, at least in certain habitats or years. But stable hybrid zones have also been observed in cases where the hybrids are selected *against*—an unexpected result.

For example, hybrids continue to form in the *Bombina* hybrid zone even though they are strongly selected against. One explanation relates to the narrowness of the *Bombina* hybrid zone (see Figure 24.13). Evidence suggests that members of both parent species migrate into the zone from the parent populations located outside the zone, thus leading to the continued production of hybrids. If the hybrid zone were wider, this would be less likely to occur, since the center of the zone would receive little gene flow from distant parent populations located outside the hybrid zone.

Sometimes the outcomes in hybrid zones match our predictions (European flycatchers and cichlid fishes), and sometimes they don't (*Bombina*). But whether our predictions are upheld or not, events in hybrid zones can shed light on how barriers to reproduction between closely related species change over time. In the next section, we'll examine how interactions between hybridizing species can also provide a glimpse into the speed and genetic control of speciation.

CONCEPT CHECK 24.3

1. What are hybrid zones, and why can they be viewed as "natural laboratories" in which to study speciation?

2. **WHAT IF?** Consider two species that diverged while geographically separated but resumed contact before reproductive isolation was complete. Predict the outcome over time if the two species mated indiscriminately and (a) hybrid offspring survived and reproduced more poorly than offspring from intraspecific matings or (b) hybrid offspring survived and reproduced as well as offspring from intraspecific matings.

For suggested answers, see Appendix A.

CONCEPT 24.4

Speciation can occur rapidly or slowly and can result from changes in few or many genes

Darwin faced many questions when he began to ponder that "mystery of mysteries"—speciation. He found answers to some of those questions when he realized that evolution by natural selection helps explain both the diversity of life and the adaptations of organisms (see Concept 22.2). But biologists since Darwin have continued to ask fundamental questions about speciation. How long does it take for new species to form? And how many genes change when one species splits into two? Answers to these questions are also emerging.

The Time Course of Speciation

We can gather information about how long it takes new species to form from broad patterns in the fossil record and from studies that use morphological data (including fossils) or molecular data to assess the time interval between speciation events in particular groups of organisms.

Patterns in the Fossil Record

The fossil record includes many episodes in which new species appear suddenly in a geologic stratum, persist essentially unchanged through several strata, and then disappear. For example, there are dozens of species of marine invertebrates that make their debut in the fossil record with novel morphologies, but then change little for millions of years before becoming extinct. The term **punctuated equilibria** is used to describe these periods of apparent stasis punctuated by sudden change **(Figure 24.17a)**. Other species do not show a punctuated pattern; instead, they appear to have changed more gradually over long periods of time **(Figure 24.17b)**.

What might punctuated and gradual patterns tell us about how long it takes new species to form? Suppose that a species survived for 5 million years, but most of the morphological changes that caused it to be designated a new species occurred during its first 50,000 years—just 1% of the total existence of

▼ **Figure 24.17 Two models for the tempo of speciation.**

(a) Punctuated model. New species change most as they branch from a parent species and then change little for the rest of their existence.

Time

(b) Gradual model. Species diverge from one another more slowly and steadily over time.

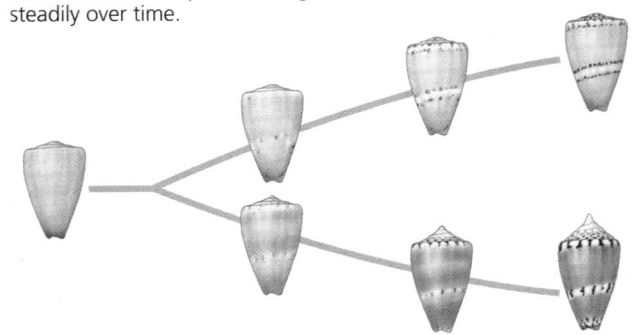

the species. Time periods this short (in geologic terms) often cannot be distinguished in fossil strata, in part because the rate of sediment accumulation may be too slow to separate layers this close in time. Thus, based on its fossils, the species would seem to have appeared suddenly and then lingered with little or no change before becoming extinct. Even though such a species may have originated more slowly than its fossils suggest (in this case taking up to 50,000 years), a punctuated pattern indicates that speciation occurred relatively rapidly. For species whose fossils changed much more gradually, we also cannot tell exactly when a new biological species formed, since information about reproductive isolation does not fossilize. However, it is likely that speciation in such groups occurred relatively slowly, perhaps taking millions of years.

→ **Mastering Biology**
Interview with Stephen Jay Gould: An "architect" of the concept of punctuated equilibria

Speciation Rates

The existence of fossils that display a punctuated pattern suggests that once the process of speciation begins, it can be completed relatively rapidly—a suggestion supported by many studies. For example, rapid speciation appears to have produced the wild sunflower *Helianthus anomalus*. Genetic evidence indicates that this species originated by the hybridization of two other sunflower species, *H. annuus* and *H. petiolaris*. The hybrid species *H. anomalus* is ecologically distinct and reproductively isolated from both parent species **(Figure 24.18)**. Unlike the outcome of allopolyploid speciation, in which there is a change in chromosome number after hybridization, in these sunflowers the two parent species and the hybrid all have the same number of chromosomes ($2n = 34$). How, then, did speciation occur? To study this question, researchers performed an experiment designed to mimic events in nature **(Figure 24.19)**. Their results indicated that natural selection could produce extensive genetic changes in hybrid populations over short periods of time. These changes appear to have caused the hybrids to diverge reproductively from their parents and form a new species, *H. anomalus*.

▼ **Figure 24.18 A hybrid sunflower species and its dry sand dune habitat.** The wild sunflower *Helianthus anomalus* shown here originated via the hybridization of two other sunflowers, *H. annuus* and *H. petiolaris*, which live in nearby but moister environments.

▼ **Figure 24.19** **Inquiry**

How does hybridization lead to speciation in sunflowers?

Experiment Loren Rieseberg and his colleagues crossed the two parent sunflower species, *H. annuus* and *H. petiolaris*, to produce experimental hybrids in the laboratory (for each gamete, only two of the $n = 17$ chromosomes are shown).

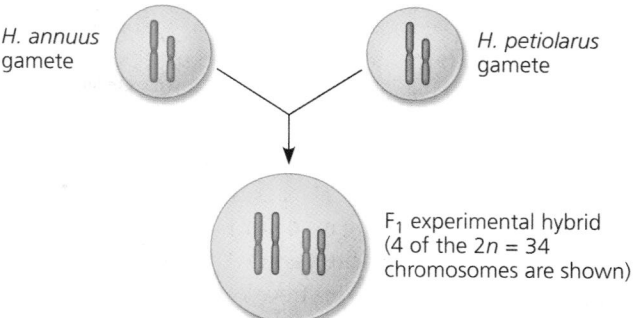

Note that in the first (F_1) generation, each chromosome of the experimental hybrids consisted entirely of DNA from one or the other parent species. The researchers then tested whether the F_1 and subsequent generations of experimental hybrids were fertile. They also used species-specific genetic markers to compare the chromosomes in the experimental hybrids with the chromosomes in the naturally occurring hybrid *H. anomalus*.

Results Although only 5% of the F_1 experimental hybrids were fertile, after just four more generations the hybrid fertility rose to more than 90%. The chromosomes of individuals from this fifth hybrid generation differed from those in the F_1 generation (see above) but were similar to those in *H. anomalus* individuals from natural populations:

▪ Comparison region containing *H. annuus*–specific marker
▪ Comparison region containing *H. petiolarus*–specific marker

Conclusion Over time, the chromosomes in the population of experimental hybrids became similar to the chromosomes of *H. anomalus* individuals from natural populations. This suggests that the observed rise in the fertility of the experimental hybrids may have occurred as selection eliminated regions of DNA from the parent species that were not compatible with one another. Overall, it appeared that the initial steps of the speciation process occurred rapidly and could be mimicked in a laboratory experiment.

Data from L. H. Rieseberg et al., Role of gene interactions in hybrid speciation: evidence from ancient and experimental hybrids, *Science* 272:741–745 (1996).

WHAT IF? *The increased fertility of the experimental hybrids could have resulted from natural selection for thriving under laboratory conditions. Evaluate this alternative explanation for the result.*

The sunflower example, along with the apple maggot fly, Lake Victoria cichlid, and fruit fly examples discussed earlier, suggests that new species can arise rapidly *once divergence begins.* But what is the total length of time between speciation events? This interval consists of the time that elapses before populations of a newly formed species start to diverge from one another plus the time it takes for speciation to be complete once divergence begins. It turns out that the total time between speciation events varies considerably. In a survey of data from 84 groups of plants and animals, speciation intervals ranged from 4,000 years (in cichlids of Lake Nabugabo, Uganda) to 40 million years (in some beetles). Overall, the time between speciation events averaged 6.5 million years and was rarely less than 500,000 years.

These data suggest that on average, millions of years may pass before a newly formed plant or animal species will itself give rise to another new species. As you'll read in Concept 25.4, this finding has implications for how long it takes life on Earth to recover from mass extinction events. Moreover, the extreme variability in the time it takes new species to form indicates that organisms do not have an internal "speciation clock" that causes them to produce new species at regular intervals. Instead, speciation begins only after gene flow between populations is interrupted, perhaps by changing environmental conditions or by unpredictable events, such as a storm that transports a few individuals to a new area. Furthermore, once gene flow is interrupted, the populations must diverge genetically to such an extent that they become reproductively isolated—all before other events cause gene flow to resume, possibly reversing the speciation process (see Figure 24.16).

Studying the Genetics of Speciation

Studies of ongoing speciation (as in hybrid zones) can reveal traits that cause reproductive isolation. By identifying the genes that control those traits, scientists can explore a fundamental question of evolutionary biology: How many genes influence the formation of new species?

In some cases, the evolution of reproductive isolation results from the effects of a single gene. For example, in Japanese snails of the genus *Euhadra*, a change in a single gene results in a mechanical barrier to reproduction. This gene controls the direction in which the shells spiral. When their shells spiral in different directions, the snails' genitalia are oriented in a manner that prevents mating (Figure 24.3f and g show a similar example). Recent genetic analyses have uncovered other single genes that cause reproductive isolation in fruit flies or mice.

A major barrier to reproduction between two closely related species of monkey flower, *Mimulus cardinalis* and *M. lewisii*, also appears to be influenced by a relatively small number of genes. These two species are isolated by several prezygotic and postzygotic barriers. Of these, one prezygotic barrier, pollinator choice, accounts for most of the isolation: In a hybrid zone between *M. cardinalis* and *M. lewisii*, nearly 98% of pollinator visits were restricted to one species or the other.

▼ **Figure 24.20 A locus that influences pollinator choice.** Pollinator preferences provide a strong barrier to reproduction between *Mimulus lewisii* and *M. cardinalis*. After transferring the *M. lewisii* allele for a flower-color locus into *M. cardinalis* and vice versa, researchers observed a shift in some pollinators' preferences.

(a) Typical *Mimulus lewisii*

(b) *M. lewisii* with an *M. cardinalis* flower-color allele

(c) Typical *Mimulus cardinalis*

(d) *M. cardinalis* with an *M. lewisii* flower-color allele

WHAT IF? *If* M. cardinalis *individuals that had the* M. lewisii yup *allele were planted in an area that housed both monkey flower species, how might the production of hybrid offspring be affected?*

The two monkey flower species are visited by different pollinators: Hummingbirds prefer the red-flowered *M. cardinalis*, and bumblebees prefer the pink-flowered *M. lewisii*. Pollinator choice is affected by at least two loci in the monkey flowers, one of which, the "yellow upper," or *yup*, locus, influences flower color **(Figure 24.20)**. By crossing the two parent species to produce F_1 hybrids and then performing repeated backcrosses of these F_1 hybrids to each parent species, researchers succeeded in transferring the *M. cardinalis* allele at this locus into *M. lewisii*, and vice versa. In a field experiment, *M. lewisii* plants with the *M. cardinalis yup* allele received 68-fold more visits from hummingbirds than did wild-type *M. lewisii*. Similarly, *M. cardinalis* plants with the *M. lewisii yup* allele received 74-fold more visits from bumblebees than did wild-type *M. cardinalis*. Thus, a mutation at a single locus can influence pollinator preference and hence contribute to reproductive isolation in monkey flowers.

In other organisms, the speciation process is influenced by larger numbers of genes and gene interactions. For example, hybrid sterility between two subspecies of the fruit fly *Drosophila pseudoobscura* results from gene interactions among at least four loci, and postzygotic isolation in the sunflower hybrid zone discussed earlier is influenced by at least 26 chromosome segments (and an unknown number of genes). Overall, studies suggest that few or many genes can influence the evolution of reproductive isolation and hence the emergence of a new species.

From Speciation to Macroevolution

As you've seen, speciation may begin with differences as small as the color on a cichlid's back. However, as speciation occurs again and again, such differences can accumulate and become more pronounced, eventually leading to the formation of new groups of organisms that differ greatly from their ancestors (as in the origin of whales from terrestrial mammals; see Figure 22.20). Moreover, as one group of organisms increases in size by producing many new species, another group of organisms may shrink, losing species to extinction. The cumulative effects of

many such speciation and extinction events have helped shape the sweeping evolutionary changes that are documented in the fossil record. In the next chapter, we turn to such large-scale evolutionary changes as we begin our study of macroevolution.

24 Chapter Review

→ Go to **Mastering Biology** for Assignments, the eText, the Study Area, and Dynamic Study Modules.

SUMMARY OF KEY CONCEPTS

→ To review key terms, go to the **Vocabulary Self-Quiz** in the **Mastering Biology** eText or Study Area, or go to goo.gl/zkjz9t.

CONCEPT 24.1

The biological species concept emphasizes reproductive isolation (pp. 507–510)

- A biological **species** is a group of populations whose individuals may interbreed and produce viable, fertile offspring with each other but not with members of other species.
- New species form when **reproductive isolation** between populations develops through the establishment of **prezygotic** or **postzygotic barriers** that separate gene pools.

? *Explain the role of gene flow in the biological species concept.*

CONCEPT 24.2

Speciation can take place with or without geographic separation (pp. 511–516)

- In **allopatric speciation**, gene flow is reduced when two populations of one species become geographically separated from each other. One or both populations may undergo evolutionary change during the period of separation, resulting in the establishment of barriers to reproduction.

Original population

Allopatric speciation

Sympatric speciation

- In **sympatric speciation**, a new species originates while remaining in the same geographic area as the parent species. Plant species (and, more rarely, animal species) have evolved sympatrically through **polyploidy**. Sympatric speciation can also result from sexual selection and habitat shifts.

? *Can factors that cause sympatric speciation also cause allopatric speciation? Explain.*

CONCEPT 24.3

Hybrid zones reveal factors that cause reproductive isolation (pp. 516–520)

- Many groups of organisms form **hybrid zones** in which members of different species meet and mate, producing at least some offspring of mixed ancestry.
- Many hybrid zones are *stable*, in that hybrid offspring continue to be produced over time. In others, **reinforcement** strengthens prezygotic barriers to reproduction, thus decreasing the formation of unfit hybrids. In still other hybrid zones, barriers to reproduction may weaken over time, resulting in the *fusion* of the species' gene pools (reversing the speciation process).

? *What factors can support the long-term stability of a hybrid zone if the parent species live in different environments?*

CONCEPT 24.4

Speciation can occur rapidly or slowly and can result from changes in few or many genes (pp. 520–523)

- New species can form rapidly once divergence begins, but it can take millions of years for that to happen. The time interval between speciation events varies considerably, from a few thousand years to tens of millions of years.
- Researchers have identified particular genes involved in some cases of speciation. Speciation can be driven by a few or many genes.

? *Is speciation something that happened only in the distant past, or are new species continuing to arise today? Explain.*

➡ For more multiple-choice questions, go to the **Practice Test** in the **Mastering Biology** eText or Study Area, or go to goo.gl/GruWRg.

Levels 1–2: Remembering/Understanding

1. The *largest* unit within which gene flow can readily occur is a
   (A) population.
   (B) species.
   (C) genus.
   (D) hybrid.

2. Males of different species of the fruit fly *Drosophila* that live in the same parts of the Hawaiian Islands have different elaborate courtship rituals. These rituals involve fighting other males and making stylized movements that attract females. What type of reproductive isolation does this represent?
   (A) habitat isolation
   (B) temporal isolation
   (C) behavioral isolation
   (D) gametic isolation

3. According to the punctuated equilibria model,
   (A) given enough time, most existing species will branch gradually into new species.
   (B) most new species accumulate their unique features relatively rapidly as they come into existence, then change little for the rest of their duration as a species.
   (C) most evolution occurs in sympatric populations.
   (D) speciation is usually due to a single mutation.

Levels 3–4: Applying/Analyzing

4. Bird guides once listed the myrtle warbler and Audubon's warbler as distinct species. Recently, these birds have been classified as eastern and western forms of a single species, the yellow-rumped warbler. Which of the following pieces of evidence, if true, would be cause for this reclassification?
   (A) The two forms interbreed often in nature, and their offspring survive and reproduce well.
   (B) The two forms live in similar habitats and have similar food requirements.
   (C) The two forms have many genes in common.
   (D) The two forms are very similar in appearance.

5. Which of the following factors would be the most likely to contribute to allopatric speciation?
   (A) The separated population is large, and genetic drift occurs.
   (B) Selection pressures in the isolated population are similar to those in the ancestral population.
   (C) Gene flow between the two populations is extensive.
   (D) Different mutations begin to distinguish the gene pools of the separated populations.

6. Plant species A has a diploid chromosome number of 12. Plant species B has a diploid number of 16. A new species, C, arises as an allopolyploid from A and B. The diploid number for species C would probably be
   (A) 14.
   (B) 16.
   (C) 28.
   (D) 56.

Levels 5–6: Evaluating/Creating

7. **EVOLUTION CONNECTION** Explain the biological basis for assigning all human populations to a single species. Can you think of a scenario by which a second human species could originate in the future?

8. **SCIENTIFIC INQUIRY • DRAW IT** In this chapter, you read that bread wheat (*Triticum aestivum*) is an allohexaploid, containing two sets of chromosomes from each of three different parent species. Genetic analysis suggests that each of the three species pictured following this question contributed chromosome sets to *T. aestivum*. (The capital letters represent sets of chromosomes, each of which can be traced to a particular species, *not* individual genes. The numbers in parentheses are the species' chromosome numbers.) Evidence also indicates that the first polyploidy event began with the spontaneous hybridization of the early cultivated wheat species *T. monococcum* and a wild *Triticum* grass species. Draw a diagram of one possible chain of events that could have produced the allohexaploid *T. aestivum*.

Ancestral species:

AA	BB	DD
Triticum monococcum (14)	Wild *Triticum* (14)	Wild *T. tauschii* (14)

Product:

AA BB DD

T. aestivum (bread wheat) (42)

9. **WRITE ABOUT A THEME: INFORMATION** In sexually reproducing species, each individual inherits DNA from both parent organisms. In a short essay (100–150 words), apply this idea to what occurs when organisms of two species that have homologous chromosomes mate and produce (F_1) hybrid offspring. What percentage of the DNA in the F_1 hybrids' chromosomes comes from each parent species? As the hybrids mate and produce F_2 and later-generation hybrid offspring, describe how recombination and natural selection may affect whether the DNA in hybrid chromosomes is derived from one parent species or the other.

10. **SYNTHESIZE YOUR KNOWLEDGE**

Suppose that females of one population of strawberry poison dart frogs (*Dendrobates pumilio*) prefer to mate with males that are orange-red in color. In a different population, females prefer males with yellow skin. Explain how such differences could arise and how they could affect the evolution of reproductive isolation in allopatric versus sympatric populations.

For selected answers, see Appendix A.

25 The History of Life on Earth

KEY CONCEPTS

25.1 Conditions on early Earth made the origin of life possible *p. 526*

25.2 The fossil record documents the history of life *p. 528*

25.3 Key events in life's history include the origins of unicellular and multicellular organisms and the colonization of land *p. 532*

25.4 The rise and fall of groups of organisms reflect differences in speciation and extinction rates *p. 537*

25.5 Major changes in body form can result from changes in the sequences and regulation of developmental genes *p. 544*

25.6 Evolution is not goal oriented *p. 547*

Study Tip

Draw a time line: Draw a time line like the example shown here, and mark when the following key evolutionary events occurred: "oxygen revolution," origin of eukaryotes, first multicellular organisms, colonization of land by large eukaryotes, origin of tetrapods, and the Permian mass extinction. Briefly describe the significance of each event. You can also add more events to your time line as you study.

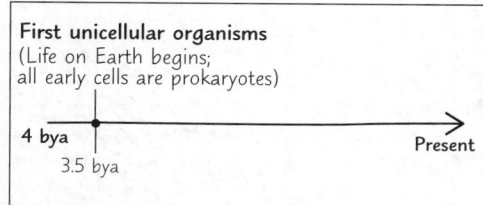

First unicellular organisms
(Life on Earth begins;
all early cells are prokaryotes)

4 bya
3.5 bya
Present

Go to Mastering Biology

For Students (in eText and Study Area)
- Get Ready for Chapter 25
- Figure 25.10 Walkthrough: Serial Endosymbiosis
- HHMI Video: The Making of the Fittest

For Instructors to Assign (in Item Library)
- Tutorial: Visualizing the Scale of Geologic Time
- HHMI Video: The Day the Mesozoic Died

Figure 25.1 In the 1870s, researchers working in the searing heat of the Sahara Desert were startled to discover fossils of ancient whales. These fossils were spectacular not only for where they were found, but also for documenting the transition from life on land to life in the sea—just one example of the sweeping changes seen in the history of life on Earth.

How has life on Earth changed over time?

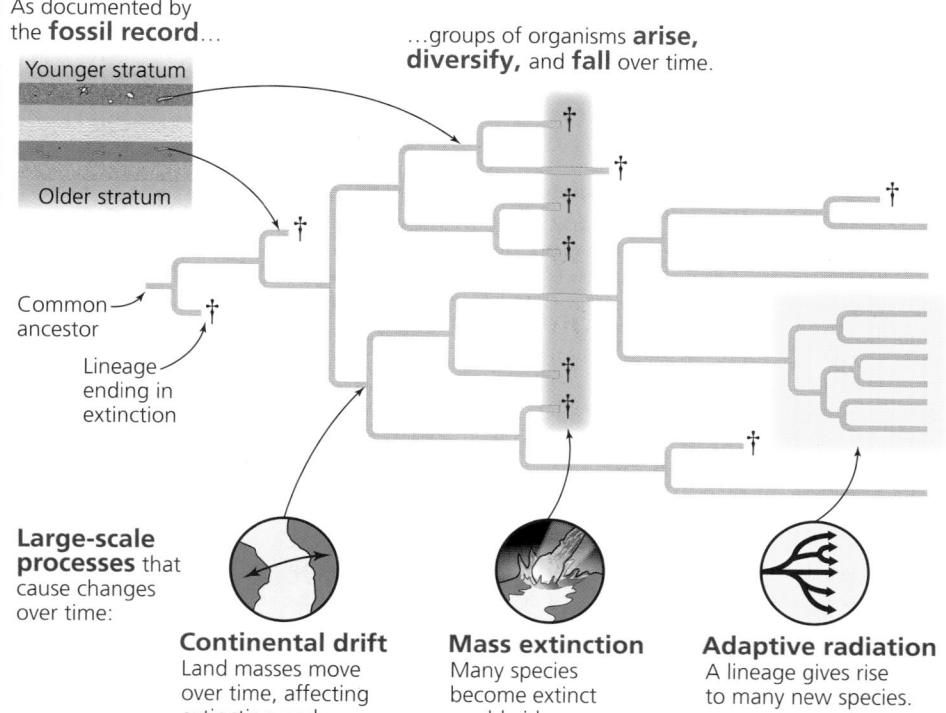

As documented by the **fossil record**...

Younger stratum

Older stratum

...groups of organisms **arise, diversify,** and **fall** over time.

Common ancestor

Lineage ending in extinction

Large-scale processes that cause changes over time:

Continental drift
Land masses move over time, affecting extinction and speciation rates.

Mass extinction
Many species become extinct worldwide.

Adaptive radiation
A lineage gives rise to many new species.

The whale fossils discovered in the Sahara Desert (see Figure 25.1) are one example of evidence that past organisms were very different from those presently living. The sweeping changes in life on Earth as revealed by such fossils illustrate **macroevolution**, the broad pattern of evolution above the species level. Examples of macroevolutionary change include the emergence of terrestrial vertebrates through a series of speciation events, the impact of mass extinctions on biodiversity, and the origin of key adaptations such as flight.

Taken together, such changes provide a grand view of the evolutionary history of life. We'll examine that history in this chapter, beginning with a discussion of hypotheses regarding the origin of life. This is the most speculative topic of the entire unit, because no fossil evidence of that seminal episode exists.

CONCEPT 25.1

Conditions on early Earth made the origin of life possible

Direct evidence of life on early Earth comes from fossils of microorganisms that lived 3.5 billion years ago. But how did the first living cells appear? Observations and experiments in chemistry, geology, and physics have led scientists to propose one scenario that we'll examine here. They hypothesize that chemical and physical processes could have produced simple cells through a sequence of four main stages:

1. The abiotic (nonliving) synthesis of small organic molecules, such as amino acids and nitrogenous bases

2. The joining of these small molecules into macromolecules, such as proteins and nucleic acids

3. The packaging of these molecules into **protocells**, droplets with membranes that maintained an internal chemistry different from that of their surroundings

4. The origin of self-replicating molecules that eventually made inheritance possible

Though speculative, this scenario leads to predictions that can be tested in the laboratory. In this section, we'll examine some of the evidence for each stage.

Synthesis of Organic Compounds on Early Earth

Our planet formed 4.6 billion years ago, condensing from a vast cloud of dust and rocks that surrounded the young sun. For its first few hundred million years, Earth was bombarded by huge chunks of rock and ice left over from the formation of the solar system. The collisions generated so much heat that all of the available water was vaporized, preventing the formation of seas and lakes.

This massive bombardment ended 4 billion years ago, setting the stage for the origin of life. The first atmosphere had little oxygen and was likely thick with water vapor, along with compounds released by volcanic eruptions, such as nitrogen and its oxides, carbon dioxide, methane, ammonia, and hydrogen. As Earth cooled, the water vapor condensed into oceans, and much of the hydrogen escaped into space.

During the 1920s, Russian chemist A. I. Oparin and British scientist J. B. S. Haldane independently hypothesized that Earth's early atmosphere was a reducing (electron-adding) environment, in which organic compounds could have formed from simpler molecules. The energy for this synthesis could have come from lightning and UV radiation. Haldane suggested that the early oceans were a solution of organic molecules, a "primitive soup" from which life arose.

In 1953, Stanley Miller, working with Harold Urey at the University of Chicago, tested the Oparin-Haldane hypothesis by creating laboratory conditions comparable to those that scientists at the time thought existed on early Earth (see Figure 4.2). His apparatus yielded a variety of amino acids found in organisms today, along with other organic compounds. Many laboratories have since repeated Miller's classic experiment using different recipes for the atmosphere, some of which also produced organic compounds.

However, some evidence suggests that the early atmosphere was made up primarily of nitrogen and carbon dioxide and was neither reducing nor oxidizing (electron removing). Recent Miller/Urey-type experiments using such "neutral" atmospheres have also produced organic molecules. In addition, small pockets of the early atmosphere, such as those near the openings of volcanoes, may have been reducing. Perhaps the first organic compounds formed near volcanoes. In a test of this hypothesis **(Figure 25.2)**, researchers used modern equipment

▼ **Figure 25.2 Amino acid synthesis in a simulated volcanic eruption.** In addition to his classic 1953 study, Miller also conducted an experiment simulating a volcanic eruption. In a 2008 reanalysis of those results, researchers found that far more amino acids were produced under simulated volcanic conditions than were produced in the conditions of the classic 1953 study.

MAKE CONNECTIONS *Explain how more than 20 amino acids could have been obtained in the 2008 study. (See Concept 5.4.)*

◄ **Figure 25.3 Did life originate in deep-sea alkaline vents?** The first organic compounds may have arisen in warm alkaline vents similar to this one from the 40,000-year-old "Lost City" vent field in the mid-Atlantic Ocean. These vents contain hydrocarbons and are full of tiny pores (inset) lined with iron and other catalytic minerals. Early oceans were acidic, so a pH gradient would have formed between the interior of the vents and the surrounding ocean water. Energy for the synthesis of organic compounds could have been harnessed from this pH gradient.

to reanalyze molecules that Miller had saved from one of his experiments. They found that numerous amino acids had formed under conditions that simulated a volcanic eruption.

Another hypothesis is that organic compounds were first produced in deep-sea **hydrothermal vents**, areas on the seafloor where heated water and minerals gush from Earth's interior into the ocean. Some of these vents, known as "black smokers," release water so hot (300–400°C) that organic compounds formed there may have been unstable. But other deep-sea vents, called **alkaline vents**, release water that has a high pH (9–11) and is warm (40–90°C) rather than hot, an environment that may have been more suitable for the origin of life **(Figure 25.3)**.

Studies related to the volcanic atmosphere and alkaline vent hypotheses show that the abiotic synthesis of organic molecules is possible under various conditions. Another source of organic molecules may have been meteorites. For example, fragments of the Murchison meteorite, a 4.5-billion-year-old rock that landed in Australia in 1969, contain more than 80 amino acids, some in large amounts. These amino acids cannot be contaminants from Earth because they consist of an equal mix of D and L isomers (see Figure 4.7). Organisms make and use only L isomers, with a few rare exceptions. Studies have shown that the Murchison meteorite also contained other key organic molecules, including lipids, simple sugars, and nitrogenous bases such as uracil.

Abiotic Synthesis of Macromolecules

The presence of small organic molecules, such as amino acids and nitrogenous bases, is not sufficient for the emergence of life as we know it. Every cell has many types of macromolecules, including enzymes and other proteins and the nucleic acids needed for self-replication. Could such macromolecules have formed on early Earth? A 2016 study demonstrated that one key step, the abiotic synthesis of RNA's two purine bases, adenine (A) and guanine (G), can occur spontaneously from simple precursor molecules; the abiotic synthesis of the smaller cytosine (C) and uracil (U) bases was accomplished in 2009. In addition, by dripping solutions of amino acids or RNA nucleotides onto hot sand, clay, or rock, researchers have produced polymers of these molecules. The polymers formed spontaneously, without the help of enzymes or ribosomes. Unlike proteins, the amino acid polymers are a complex mix of linked and cross-linked amino acids. Still, it is possible that such polymers acted as weak catalysts for a variety of chemical reactions on early Earth.

Protocells

All organisms must be able to carry out both reproduction and energy processing (metabolism). DNA molecules carry genetic information, including the instructions needed to replicate themselves accurately during reproduction. But DNA replication requires elaborate enzymatic machinery, along with an abundant supply of nucleotide building blocks provided by the cell's metabolism. This suggests that self-replicating molecules and a metabolic source of building blocks may have appeared together in early protocells. The necessary conditions may have been met in *vesicles*, fluid-filled compartments enclosed by a membrane-like structure. Recent experiments show that abiotically produced vesicles can exhibit certain properties of life, including simple reproduction and metabolism, as well as the maintenance of an internal chemical environment different from that of their surroundings **(Figure 25.4)**.

For example, vesicles can form spontaneously when lipids or other organic molecules are added to water. When this occurs, molecules that have both a hydrophobic region and a hydrophilic region can organize into a bilayer similar to the lipid bilayer of a plasma membrane. Adding substances such as *montmorillonite*, a soft mineral clay produced by the weathering of volcanic ash, greatly increases the rate of vesicle self-assembly (see Figure 25.4a). This clay, which is thought to have been common on early Earth, provides surfaces on which organic molecules become concentrated, increasing the likelihood that the molecules will react with each other and form vesicles. Abiotically produced vesicles can "reproduce" on their own (see Figure 25.4b), and they can increase in size ("grow") without dilution of their contents. Vesicles also can absorb montmorillonite particles, including those on which RNA and other organic molecules have become attached (see Figure 25.4c). Finally, experiments have shown that some vesicles have a selectively permeable bilayer and can perform metabolic reactions using an external source of reagents—another important prerequisite for life.

▼ Figure 25.4 Features of abiotically produced vesicles.

(a) Self-assembly. The presence of montmorillonite clay greatly increases the rate of vesicle self-assembly.

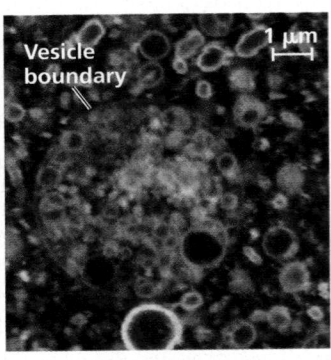

(b) Reproduction. Vesicles can divide on their own, as in this vesicle "giving birth" to smaller vesicles (LM).

(c) Absorption of RNA. This vesicle has incorporated montmorillonite clay particles coated with RNA (orange).

MAKE CONNECTIONS *Explain how molecules with both a hydrophobic region and a hydrophilic region can self-assemble into a bilayer when in water. (See Concept 5.3.)*

Self-Replicating RNA

The first genetic material was most likely RNA, not DNA. RNA plays a central role in protein synthesis, but it can also function as an enzyme-like catalyst (see Concept 17.3). Such RNA catalysts are called **ribozymes**. Some ribozymes can make complementary copies of short pieces of RNA, provided that they are supplied with nucleotide building blocks.

Natural selection on the molecular level has produced ribozymes capable of self-replication in the laboratory. How does this occur? Unlike double-stranded DNA, which takes the form of a uniform helix, single-stranded RNA molecules assume a variety of specific three-dimensional shapes mandated by their nucleotide sequences. In a given environment, RNA molecules with certain nucleotide sequences may have shapes that enable them to replicate faster and with fewer errors than other sequences. The RNA molecule with the greatest ability to replicate itself will leave the most descendant molecules. Occasionally, a copying error will result in a molecule with a shape that is even more adept at self-replication. Similar selection events may have occurred on early Earth. Thus, life as we

know it may have been preceded by an "RNA world," in which small RNA molecules were able to replicate and to store genetic information about the vesicles that carried them.

Recently, Dr. Jack Szostak and colleagues at Harvard University succeeded in building a vesicle in which copying of a template strand of RNA could occur—a key step toward constructing a vesicle with self-replicating RNA. On early Earth, a vesicle with such self-replicating, catalytic RNA would differ from its many neighbors that lacked such molecules. If that vesicle could grow, split, and pass its RNA molecules to its "daughters," the daughters would be protocells. Although the first such protocells likely carried only limited amounts of genetic information, specifying only a few properties, their inherited characteristics could have been acted on by natural selection. The most successful of the early protocells would have increased in number because they could exploit their resources effectively and pass their abilities on to subsequent generations.

Once RNA sequences that carried genetic information appeared in protocells, many additional changes would have been possible. For example, RNA could have provided the template on which DNA nucleotides were assembled. Double-stranded DNA is a more chemically stable repository for genetic information than is the more fragile RNA. DNA also can be replicated more accurately. Accurate replication was advantageous as genomes grew larger through gene duplication and other processes and as more properties of the protocells became coded in genetic information. Once DNA appeared, the stage was set for a blossoming of new forms of life—a change we see documented in the fossil record.

➔ **Mastering Biology**
Interview with Jack Szostak: Studying the origin of life

CONCEPT CHECK 25.1

1. What hypothesis did Miller test in his classic experiment?
2. How would the appearance of protocells have represented a key step in the origin of life?
3. **MAKE CONNECTIONS** In changing from an "RNA world" to today's "DNA world," genetic information must have flowed from RNA to DNA. After reviewing Figures 17.4 and 19.9, suggest how this could have occurred. Does such a flow occur today?

For suggested answers, see Appendix A.

CONCEPT 25.2

The fossil record documents the history of life

Starting with the earliest traces of life, the fossil record opens a window into the world of long ago and provides glimpses of the evolution of life over billions of years. In this section, we'll examine fossils as a form of scientific evidence: how fossils

▶ **Figure 25.5 A gallery of fossil types. (a) Sedimentary rock.** Most fossils are found in sedimentary rocks, such as this plant fossil. **(b) Mineralized organic matter.** Some fossils, such as this petrified tree trunk, form as minerals seep into and replace organic matter. **(c) Trace fossils.** Footprints, burrows, or other traces of an organism's activities can be preserved in the fossil record. **(d) Amber.** Entire organisms are sometimes found preserved in hardened resin from a tree. **(e) Frozen soil, ice, and acid bogs.** Rarely, frozen soil, ice, or an acid bog can preserve the body of larger organisms, such as this wolf pup.

(a) Plant fossil, about 300 million years old

(b) Petrified tree in Arizona, about 215 million years old

(c) 150-million-year-old dinosaur tracks in Colorado

(e) A 50,000-year-old wolf pup found in frozen soil in Yukon, Canada

(d) Insect in amber, about 40 million years old

form, how scientists date and interpret them, and what they can and cannot tell us about changes in the history of life.

The Fossil Record

Sedimentary rocks are the richest source of fossils. As a result, the fossil record is based primarily on the sequence in which fossils have accumulated in sedimentary rock layers, called *strata* (see Figure 22.3). Useful information is also provided by other types of fossils, such as insects preserved in amber (fossilized tree sap) and mammals frozen in soil **(Figure 25.5)**.

The fossil record shows that there have been great changes in the kinds of organisms on Earth at different points in time. Many past organisms were unlike organisms living today, and many organisms that once were common are now extinct. As we'll see later in this section, fossils also document how new groups of organisms arose from previously existing ones.

As substantial and significant as the fossil record is, keep in mind that it is an incomplete chronicle of evolution. Many of Earth's organisms did not die in the right place and time to be preserved as fossils. Of the fossils that were formed, many were destroyed by later geologic processes, and only a fraction of the others have been discovered. As a result, the known fossil record is biased in favor of species that existed for a long time,

were abundant and widespread in certain kinds of environments, and had hard shells, skeletons, or other parts that facilitated their fossilization. Even with its limitations, however, the fossil record is a remarkably detailed account of biological change over the vast scale of geologic time. Furthermore, as shown by recently unearthed fossils of whale ancestors with hind limbs (see Figure 22.19 and Figure 25.1), gaps in the fossil record continue to be filled by new discoveries.

How Rocks and Fossils Are Dated

Fossils provide valuable data for reconstructing the history of life, but only if we can determine where they fit in that story. While the order of fossils in rock strata tells us the sequence in which the fossils were laid down—their relative ages—it does not tell us their actual ages. Examining the relative positions of fossils is like peeling off layers of wallpaper in an old house. You can infer the sequence in which the layers were applied, but not the year each layer was added.

How can we determine the age of a fossil? One of the most common techniques is **radiometric dating**, which is based on the decay of radioactive isotopes (see Concept 2.2). In this process, a radioactive "parent" isotope decays to a "daughter" isotope at a characteristic rate. The rate of decay is expressed by the **half-life**, the time required for 50% of the parent

▼ **Figure 25.6 Radiometric dating.** In this diagram, each unit of time represents one half-life of a radioactive isotope.

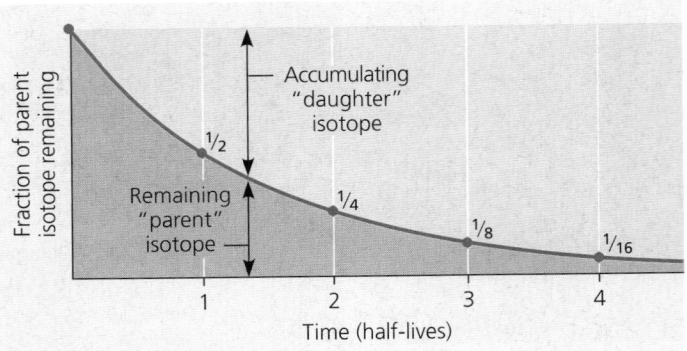

DRAW IT *Relabel the x-axis of this graph in years to illustrate the radioactive decay of uranium-238 (half-life = 4.5 billion years).*

isotope to decay **(Figure 25.6)**. Each type of radioactive isotope has a characteristic half-life, which is not affected by temperature, pressure, or other environmental variables. For example, carbon-14 decays relatively quickly; its half-life is 5,730 years. Uranium-238 decays slowly; its half-life is 4.5 billion years.

Fossils contain isotopes of elements that accumulated in the organisms when they were alive. For example, a living organism contains the most common carbon isotope, carbon-12, as well as a radioactive isotope, carbon-14. When the organism dies, it stops accumulating carbon, and the amount of carbon-12 in its tissues does not change over time. However, the carbon-14 that it contains at the time of death slowly decays into another element, nitrogen-14. Thus, by measuring the ratio of carbon-14 to carbon-12 in a fossil, we can determine the fossil's age. This method works for fossils up to about 75,000 years old; fossils older than that contain too little carbon-14 to be detected with current techniques. Radioactive isotopes with longer half-lives are used to date older fossils.

Determining the age of these older fossils in sedimentary rocks can be challenging. Organisms do not use radioisotopes with long half-lives, such as uranium-238, to build their bones or shells. In addition, sedimentary rocks are often composed of sediments of differing ages. Although we cannot date these older fossils directly, an indirect method can be used to infer the age of fossils that are sandwiched between two layers of volcanic rock. As lava cools into volcanic rock, radioisotopes from the surrounding environment become trapped in the newly formed rock. Some of the trapped radioisotopes have long half-lives, allowing geologists to estimate the ages of ancient volcanic rocks. If two volcanic layers surrounding fossils are found to be 525 million and 535 million years old, for example, then the fossils are roughly 530 million years old.

The study of fossils has helped geologists establish a geologic record: a standard time scale that divides Earth's history into four eons and further subdivisions **(Table 25.1)**. We'll describe key events that occurred during these different time periods in Concept 25.3.

Table 25.1 The Geologic Record

Eon	Era	Period	Epoch	Age (Millions of Years Ago)
Phanerozoic	Cenozoic	Quaternary	Holocene	0.01
			Pleistocene	2.6
		Neogene	Pliocene	5.3
			Miocene	23
		Paleogene	Oligocene	33.9
			Eocene	56
			Paleocene	66
	Mesozoic	Cretaceous		145
		Jurassic		201
		Triassic		252
	Paleozoic	Permian		299
		Carboniferous		359
		Devonian		419
		Silurian		444
		Ordovician		485
		Cambrian		541
Proterozoic	Neoproterozoic	Ediacaran		635
				1,000
				2,500
Archaean				4,000
Hadean			Approx. 4,600	

The Origin of New Groups of Organisms

Some fossils document the origin of new groups of organisms. Such fossils are central to our understanding of evolution; they illustrate how new features arise and how long it takes for such changes to occur. We'll examine one such case here and in **Figure 25.7**: the origin of mammals.

Along with amphibians and reptiles, mammals belong to the group of animals called *tetrapods* (from the Greek *tetra*, four, and *pod*, foot), named for having four limbs. Mammals have a number of unique anatomical features that fossilize readily, allowing scientists to trace their origin. For example, the lower jaw is composed of one bone (the dentary) in mammals but several bones in other tetrapods. In addition, the lower and upper jaws in mammals hinge between a different set of bones than in other tetrapods. Mammals also have a unique set of three bones that transmit sound in the middle ear, the hammer, anvil, and stirrup, whereas other tetrapods have only one such bone, the stirrup (see Concept 34.6). Finally, the teeth of mammals are differentiated into incisors (for tearing), canines (for piercing), and the multi-pointed premolars and molars

Over the course of 120 million years, mammals originated gradually from a group of tetrapods called synapsids. Shown here are a few of the many fossil organisms whose morphological features represent intermediate steps between living mammals and their early synapsid ancestors. The evolutionary context of the origin of mammals is shown in the tree diagram at right (the dagger symbol † indicates extinct lineages).

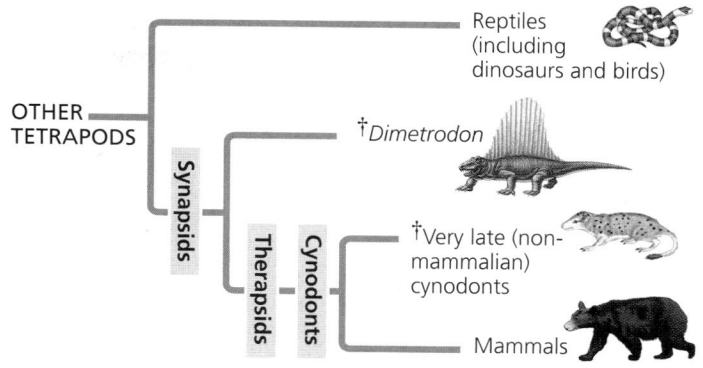

Key to skull bones

█ Articular   ▒ Dentary
█ Quadrate   ▒ Squamosal

Synapsid (300 mya)

Early synapsids had multiple bones in the lower jaw and single-pointed teeth. The jaw hinge was formed by the articular and quadrate bones. Early synapsids also had an opening called the *temporal fenestra* behind the eye socket. Powerful cheek muscles for closing the jaws probably passed through the temporal fenestra. Over time, this opening enlarged and moved in front of the hinge between the lower and upper jaws, thereby increasing the power and precision with which the jaws could be closed (much as moving a doorknob away from the hinge makes a door easier to close).

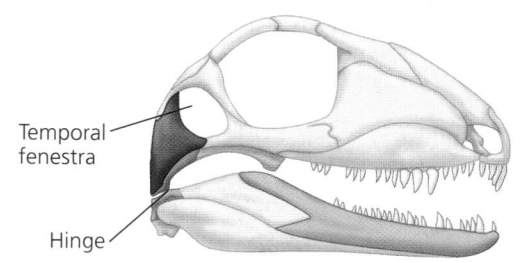

Therapsid (280 mya)

Later, a group of synapsids called therapsids appeared. Therapsids had large dentary bones, long faces, and the first examples of specialized teeth, large canines. These trends continued in a group of therapsids called cynodonts.

Early cynodont (260 mya)

In early cynodont therapsids, the dentary was the largest bone in the lower jaw, the temporal fenestra was large and positioned forward of the jaw hinge, and teeth with several cusps first appeared (not visible in the diagram). As in earlier synapsids, the jaw had an articular-quadrate hinge.

Later cynodont (220 mya)

Later cynodonts had teeth with complex cusp patterns, and their lower and upper jaws hinged in two locations: They retained the original articular-quadrate hinge and formed a new, second hinge between the dentary and squamosal bones. (The temporal fenestra is not visible in this or the below cynodont skull at the angles shown.)

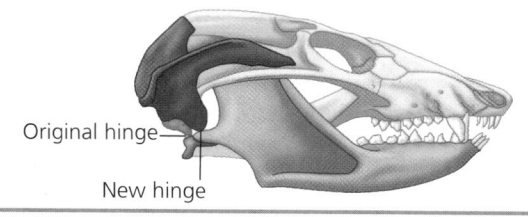

Very late cynodont (195 mya)

In some very late (nonmammalian) cynodonts and early mammals, the original articular-quadrate hinge was lost, leaving the dentary-squamosal hinge as the only hinge between the lower and upper jaws, as in living mammals. The articular and quadrate bones migrated into the ear region (not shown), where they functioned in transmitting sound. In the mammal lineage, these two bones later evolved into the familiar hammer (malleus) and anvil (incus) bones of the ear.

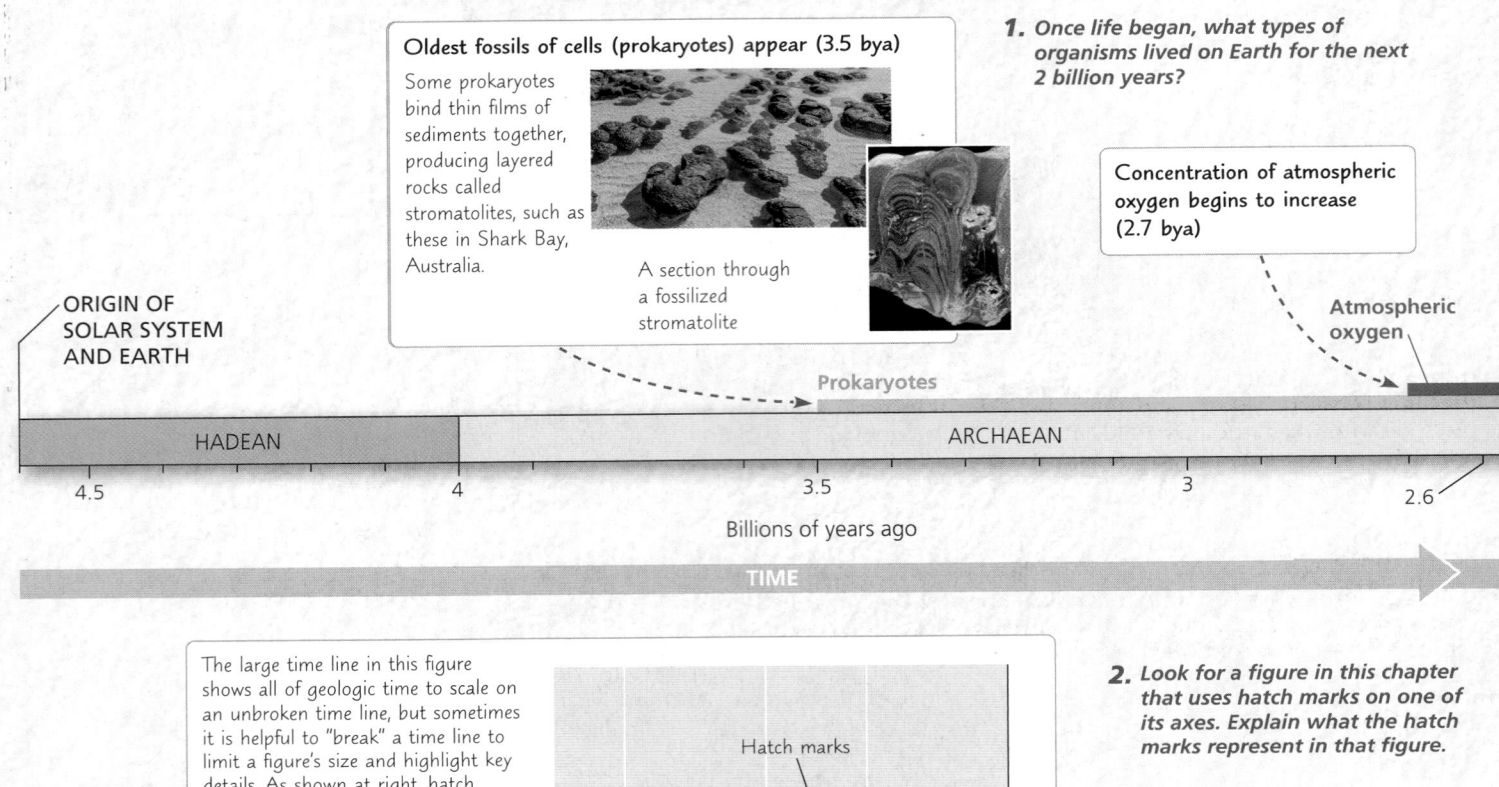

▼ Figure 25.8 VISUALIZING THE SCALE OF GEOLOGIC TIME

Geologic time is so vast that it can be difficult to visualize when key events in the history of life on Earth occurred. In this time line, time runs from left to right, from 4.6 billion years ago to the present. The fossils along the top illustrate representative organisms from different points in time.

Oldest fossils of cells (prokaryotes) appear (3.5 bya)

Some prokaryotes bind thin films of sediments together, producing layered rocks called stromatolites, such as these in Shark Bay, Australia.

A section through a fossilized stromatolite

Concentration of atmospheric oxygen begins to increase (2.7 bya)

1. *Once life began, what types of organisms lived on Earth for the next 2 billion years?*

ORIGIN OF SOLAR SYSTEM AND EARTH

Atmospheric oxygen

Prokaryotes

HADEAN   ARCHAEAN

4.5   4   3.5   3   2.6

Billions of years ago

TIME

The large time line in this figure shows all of geologic time to scale on an unbroken time line, but sometimes it is helpful to "break" a time line to limit a figure's size and highlight key details. As shown at right, hatch marks are often used to represent such an interruption.

Hatch marks

125   100   25   0

Millions of years ago

2. *Look for a figure in this chapter that uses hatch marks on one of its axes. Explain what the hatch marks represent in that figure.*

(for crushing and grinding). In contrast, the teeth of other tetrapods usually consist of a row of undifferentiated, single-pointed teeth.

As detailed in Figure 25.7, the fossil record shows that the unique features of mammalian jaws and teeth evolved gradually, in a series of steps. Keep in mind that the figure includes just a few examples of the fossil skulls that document the origin of mammals. If all the known fossils in the sequence were arranged by shape and placed side by side, their features would blend smoothly from one group to the next. Some of these fossils would reflect how the features of a group that dominates life today, the mammals, gradually arose in a previously existing group, the cynodonts. Others would reveal side branches on the tree of life—groups of organisms that thrived for millions of years but ultimately left no descendants that survive today.

CONCEPT CHECK 25.2

1. Describe an example from the fossil record that shows how life has changed over time.

2. **WHAT IF?** A fossilized skull you unearthed has a carbon-14/carbon-12 ratio about $1/16$ that of the skulls of present-day animals. Approximately how many years old is the fossilized skull?

For suggested answers, see Appendix A.

CONCEPT 25.3

Key events in life's history include the origins of unicellular and multicellular organisms and the colonization of land

The first three eons in the geologic record—the Hadean, the Archaean, and the Proterozoic—together lasted about

Oldest fossils of eukaryotic cells appear (1.8 bya)

Early eukaryote fossil with a cell wall structure like that of some living algae

Origin of jawed vertebrates (440 mya)

Coccosteus cuspidatus, a placoderm (fishlike vertebrate) that had a bony shield covering its head and front end

Cambrian explosion (535–525 mya) Sudden increase in diversity of many animal phyla

Steps in the origin of tetrapods (375 mya)

Tiktaalik, an extinct aquatic organism; the closest known relative of the tetrapods (four-legged vertebrates) that went on to colonize land

Flowering plants (angiosperms) appear (140 mya)

Archaefructus sinensis, an early flowering plant

PRESENT

Single-celled eukaryotes   Multicellular eukaryotes   Animals   Colonization of land   Humans

PROTEROZOIC   Paleozoic   Mesozoic   Ceno-zoic

2   1.5   1   0.5

Billions of years ago

PHANEROZOIC

TIME

If we redraw the timeline as a circle, we can apply the visual analogy of a timer that begins with the origin of Earth and counts down for 1 hour. In this way, we can relate the relative timing and duration of events that occurred billions of years ago to a familiar time scale. On a 1-hour time scale, animals originated about 9 minutes ago, while humans appeared less than 0.2 second ago.

PRESENT
Humans
Colonization of land
Animals
Multicellular eukaryotes
Single-celled eukaryotes

ORIGIN OF THE SOLAR SYSTEM AND EARTH

60
Minutes
15   45
30

Prokaryotes

Atmospheric oxygen

3. *Using the analogy of a 1-hour countdown timer, when did prokaryotes originate? When did the colonization of land occur?*

➤ **Instructors:** Additional questions related to this Visualizing Figure can be assigned in **Mastering Biology**.

4 billion years. The Phanerozoic eon, roughly the last half billion years, encompasses most of the time that animals have existed on Earth. It is divided into three eras: the Paleozoic, Mesozoic, and Cenozoic. Each era represents a distinct age in the history of Earth and its life. For example, the Mesozoic era is sometimes called the "age of reptiles" because of its abundance of reptilian fossils, including those of dinosaurs. The boundaries between the eras correspond to major extinction events, when many forms of life disappeared and were replaced by forms that evolved from the survivors.

As we've seen, the fossil record provides a sweeping overview of the history of life over geologic time. Here we will focus on a few major events in that history, returning to study the details in Unit Five. **Figure 25.8** will help you visualize how long ago

these key events occurred against the vast backdrop of geologic time.

The First Single-Celled Organisms

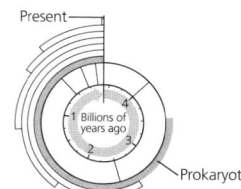

Present
Billions of years ago
4
2
1
Prokaryotes

Earth's first organisms were single-celled prokaryotes that lived in the ocean. The earliest direct evidence of these organisms, dating from 3.5 billion years ago, comes from fossilized stromatolites. **Stromatolites** are layered rocks that form when certain prokaryotes bind thin films of sediment together. Stromatolites and other early prokaryotes were Earth's sole inhabitants for about 1.5 billion years. As we will see, these prokaryotes transformed life on our planet.

Photosynthesis and the Oxygen Revolution

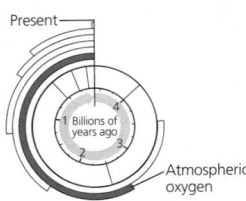

Present —
1 Billions of years ago
Atmospheric oxygen

Most atmospheric oxygen gas (O_2) is of biological origin, produced during the water-splitting step of photosynthesis. When oxygenic photosynthesis first evolved—in photosynthetic prokaryotes similar to today's cyanobacteria—the free O_2 it produced probably dissolved in the surrounding water until it reached a high enough concentration to react with elements dissolved in water, including iron. This would have caused the iron to precipitate as iron oxide, which accumulated as sediments. These sediments were compressed into banded iron formations, red layers of rock containing iron oxide that are a source of iron ore today. Once all of the dissolved iron had precipitated, additional O_2 dissolved in the water until the seas and lakes became saturated with O_2. After this occurred, the O_2 finally began to "gas out" of the water and enter the atmosphere. This change left its mark in the rusting of iron-rich terrestrial rocks, a process that began about 2.7 billion years ago.

As shown in **Figure 25.9**, the amount of atmospheric O_2 increased gradually from about 2.7 to 2.4 billion years ago, but then shot up relatively rapidly to between 1% and 10% of its present level. This "oxygen revolution" had an enormous impact on life. In some of its chemical forms, oxygen attacks chemical bonds and can inhibit enzymes and damage cells. As a result, the rising concentration of atmospheric O_2 probably doomed many prokaryotic groups. Some species survived in habitats that remained anaerobic, where we find their descendants living today (see Concept 27.4). Among other survivors, diverse adaptations to the changing atmosphere evolved, including cellular respiration, which uses O_2 in the process of harvesting the energy stored in organic molecules.

▼ **Figure 25.9 The rise of atmospheric oxygen.** Chemical analyses of ancient rocks have enabled this reconstruction of atmospheric oxygen levels during Earth's history.

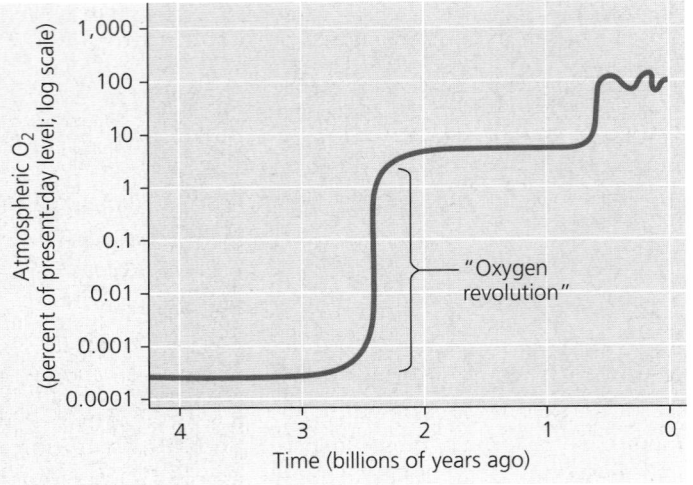

"Oxygen revolution"

The rise in atmospheric O_2 levels left a huge imprint on the history of life. A few hundred million years later, another fundamental change occurred: the origin of the eukaryotic cell.

The First Eukaryotes

Present —
1 Billions of years ago
Single-celled eukaryotes

The oldest widely accepted fossils of eukaryotes are of single-celled organisms that lived 1.8 billion years ago. Recall that eukaryotic cells have more complex organization than prokaryotic cells: Eukaryotic cells have a nuclear envelope, mitochondria, endoplasmic reticulum, and other internal structures that prokaryotes lack. Also, unlike prokaryotic cells, eukaryotic cells have a well-developed cytoskeleton, a feature that enables eukaryotic cells to change their shape and thereby surround and engulf other cells.

How did the eukaryotes evolve from their prokaryotic ancestors? Current evidence indicates that the eukaryotes originated by **endosymbiosis** when a prokaryotic cell engulfed a small cell that would evolve into an organelle found in all eukaryotes, the mitochondrion. The small, engulfed cell is an example of an *endosymbiont*, a cell that lives within another cell, called the *host cell*. The prokaryotic ancestor of the mitochondrion probably entered the host cell as undigested prey or an internal parasite. Though such a process may seem unlikely, scientists have directly observed cases in which endosymbionts that began as prey or parasites developed a mutually beneficial relationship with a host in as little as five years.

By whatever means the relationship began, we can hypothesize how the symbiosis could have become beneficial. For example, in a world that was becoming increasingly aerobic, a host that was itself an anaerobe would have benefited from endosymbionts that could make use of the oxygen. Over time, the host and endosymbionts would have become a single organism, its parts inseparable. Although all eukaryotes have mitochondria or remnants of these organelles, they do not all have plastids (a general term for chloroplasts and related organelles). Thus, the **serial endosymbiosis** hypothesis supposes that mitochondria evolved before plastids through a sequence of endosymbiotic events. As shown in **Figure 25.10**, both mitochondria and plastids are thought to have descended from bacterial cells. The original host—the cell that engulfed the bacterium whose descendants gave rise to the mitochondrion—is thought to have been an archaean or a close relative of the archaea.

A great deal of evidence supports the endosymbiotic origin of mitochondria and plastids:

- The inner membranes of both organelles have enzymes and transport systems that are homologous to those found in the plasma membranes of living bacteria.

▼ **Figure 25.10 A hypothesis for the origin of mitochondria and plastids through serial endosymbiosis.** The proposed host cell was an archaean or a close relative of the archaea. The proposed ancestors of mitochondria were aerobic, heterotrophic bacteria, while the proposed ancestors of plastids were photosynthetic bacteria. In this figure, the arrows represent change over evolutionary time.

Mastering Biology Figure Walkthrough

- Mitochondria and plastids replicate by a splitting process that is similar to that of certain bacteria. In addition, each of these organelles contains circular DNA molecules

that, like the chromosomes of bacteria, are not associated with histones or large amounts of other proteins.
- As might be expected of organelles descended from free-living organisms, mitochondria and plastids also have the cellular machinery (including ribosomes) needed to transcribe and translate their DNA into proteins.
- Finally, in terms of size, RNA sequences, and sensitivity to certain antibiotics, the ribosomes of mitochondria and plastids are more similar to bacterial ribosomes than they are to the cytoplasmic ribosomes of eukaryotic cells.

In Concept 28.1, we'll return to the origin of eukaryotes, focusing on what genomic data have revealed about the prokaryotic lineages that gave rise to the host and endosymbiont cells.

The Origin of Multicellularity

An orchestra can play a greater variety of musical compositions than a violin soloist can; the increased complexity of the orchestra makes more variations possible. Likewise, the origin of structurally complex eukaryotic cells sparked the evolution of greater morphological diversity than was possible for the simpler prokaryotic cells. After the first eukaryotes appeared, a great range of unicellular forms evolved, giving rise to the diversity of single-celled eukaryotes that continue to flourish today. Another wave of diversification also occurred: Some single-celled eukaryotes gave rise to multicellular forms, whose descendants include a variety of algae, plants, fungi, and animals.

Early Multicellular Eukaryotes

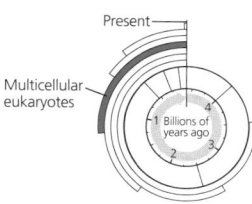

Fossil evidence and DNA sequence data suggest that multicellular eukaryotes emerged about 1.3 billion years ago. The oldest known fossils of multicellular eukaryotes that can be resolved taxonomically are of relatively small red algae that lived 1.2 billion years ago. Larger and more diverse multicellular eukaryotes do not appear in the fossil record until about 600 million years ago (**Figure 25.11**). These fossils, referred to as the

▼ **Figure 25.11 Life becomes large.** Fossils from the Ediacaran period document the oldest known macroscopic eukaryotes, such as **(a)** *Doushantuophyton*, an alga that lived 600 mya, and **(b)** *Kimberella*, a mollusc (or close relative) that lived 560 mya.

Ediacaran biota, were of mostly soft-bodied organisms—some over 1 m long—that lived from 635 to 541 million years ago (mya). The Ediacaran biota included both algae and animals, along with various organisms of unknown taxonomic affinity.

The rise of large eukaryotes in the Ediacaran period represents an enormous change in the history of life. Before that time, Earth was a microbial world: Its only inhabitants were single-celled prokaryotes and eukaryotes, along with an assortment of microscopic, multicellular eukaryotes. As the diversification of the Ediacaran biota came to a close about 541 million years ago, the stage was set for another, even more spectacular burst of evolutionary change: the "Cambrian explosion."

The Cambrian Explosion

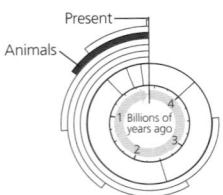

Many present-day animal phyla appear suddenly in fossils formed 535–525 million years ago, early in the Cambrian period. This phenomenon is referred to as the **Cambrian explosion**. Fossils of several animal groups—sponges, cnidarians (sea anemones and their relatives), and molluscs (snails, clams, and their relatives)—appear in even older rocks dating from the late Proterozoic **(Figure 25.12)**.

Prior to the Cambrian explosion, all large animals were soft-bodied. The fossils of large pre-Cambrian animals reveal little

▼ **Figure 25.12 Appearance of selected animal groups.** The white bars indicate the earliest fossil record of these animal groups.

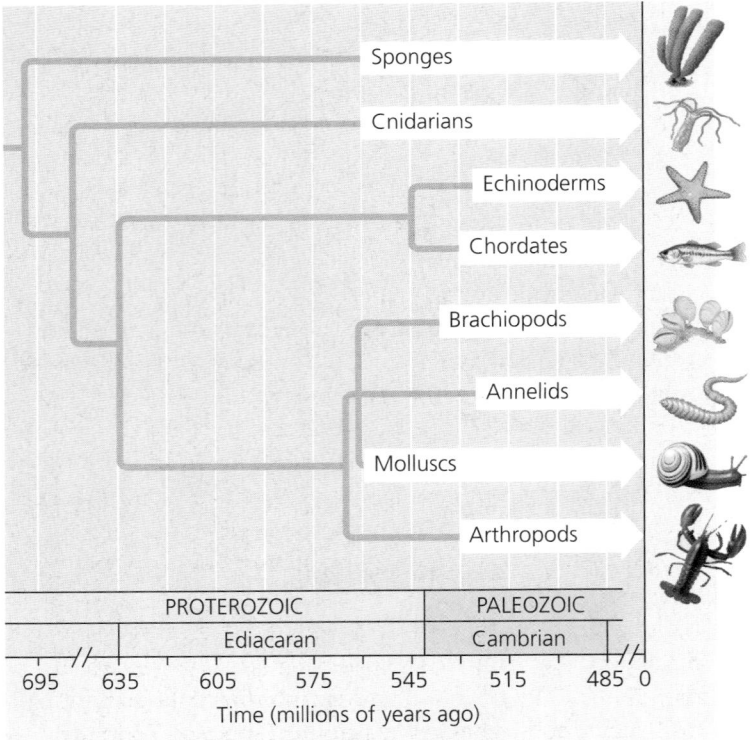

VISUAL SKILLS *Circle the branch point that represents the most recent common ancestor of chordates and annelids. What is a minimum estimate of that ancestor's age?*

evidence of predation. Instead, these animals appear to have been grazers (feeding on algae), filter feeders, or scavengers, not hunters. The Cambrian explosion changed all of that. In a relatively short period of time (10 million years), predators over 1 m in length emerged that had claws and other features for capturing prey; simultaneously, new defensive adaptations, such as sharp spines and heavy body armor, appeared in their prey.

Although the Cambrian explosion had an enormous impact on life on Earth, it appears that many animal phyla originated long before that time. Recent DNA analyses suggest that sponges had evolved by 700 million years ago; such analyses also indicate that the common ancestor of arthropods, chordates, and other animal phyla that radiated during the Cambrian explosion lived 670 million years ago. Researchers have unearthed 710-million-year-old sediments containing steroids indicative of a particular group of sponges—a finding that supports the molecular data. In contrast, the oldest macroscopic fossils of animals date from 560 million years ago and include *Kimberella*, shown in Figure 25.11. Overall, fossils and DNA analyses suggest that animals originated about 700 million years ago and then remained small for over 100 million years—until they diversified explosively during the Cambrian and beyond.

The Colonization of Land

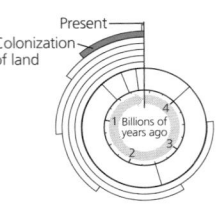

The colonization of land was another milestone in the history of life. There is fossil evidence that some prokaryotes lived on terrestrial surfaces as early as 3.2 billion years ago. However, larger forms of life, such as fungi, plants, and animals, did not begin to colonize land until about 500 million years ago. This gradual evolutionary venture out of aquatic environments was associated with adaptations that made it possible to reproduce on land and that helped prevent dehydration. For example, many plants today have a vascular system for transporting materials internally and a waterproof coating of wax on their leaves that slows the loss of water to the air. Early signs of these adaptations were present 420 million years ago, at which time small plants (about 10 cm high) existed that had a vascular system but lacked true roots or leaves. By 40 million years later, plants had diversified greatly and included trees and many other plants with true roots and leaves.

Plants appear to have colonized land in the company of fungi. Even today, the roots of most plants are associated with fungi that aid in the absorption of water and minerals from the soil (see Concept 31.1). These root fungi (or *mycorrhizae*), in turn, obtain their organic nutrients from the plants. Such mutually beneficial associations of plants and fungi are evident in some of the oldest fossilized plants, dating this relationship back to the early spread of life onto land **(Figure 25.13)**.

Although many animal groups are now represented in terrestrial environments, the most widespread and diverse land

▼ **Figure 25.13 An ancient symbiosis.** This 405-million-year-old fossil stem (cross section) documents mycorrhizae in the early plant *Aglaophyton major*. The inset shows an enlarged view of a cell containing a branched fungal structure called an arbuscule; the fossil arbuscule resembles those seen in plant cells today.

Branched hyphae (arbuscule)

Zone of arbuscule-containing cells

animals are arthropods (particularly insects and spiders) and tetrapods. Arthropods were among the first animals to colonize land, roughly 450 million years ago. The earliest tetrapods found in the fossil record lived about 365 million years ago and appear to have evolved from a group of lobe-finned fishes (see Concept 34.3). Tetrapods include humans, although we are late arrivals on the scene. The human lineage diverged from other primates around 6–7 million years ago, and our species originated only about 195,000 years ago. If the clock of Earth's history were rescaled to represent an hour, humans appeared less than 0.2 second ago.

→ **Mastering Biology**
Interview with Geerat Vermeij: What fossils reveal about the history of life

CONCEPT CHECK 25.3

1. The first appearance of free oxygen in the atmosphere likely triggered a massive wave of extinctions among the prokaryotes of the time. Why?
2. What evidence supports the hypothesis that mitochondria preceded plastids in the evolution of eukaryotic cells?
3. **WHAT IF?** What would a fossil record of life today look like?

For suggested answers, see Appendix A.

CONCEPT 25.4

The rise and fall of groups of organisms reflect differences in speciation and extinction rates

From its beginnings, life on Earth has been marked by the rise and fall of groups of organisms. Anaerobic prokaryotes originated, flourished, and then declined as the oxygen content of the atmosphere rose. Billions of years later, the first tetrapods emerged from the sea, giving rise to several major new groups of organisms. One of these, the amphibians, went

▼ **Figure 25.14 How speciation and extinction affect diversity.** The species diversity of an evolutionary lineage will increase when more new member species originate than are lost to extinction. In this hypothetical phylogenetic tree, each horizontal branch represents one species. By 2 million years ago both lineage A and lineage B have given rise to four species, and no species have become extinct (denoted by a dagger symbol, †). By time 0, however, lineage A contains only one species, while lineage B contains eight species.

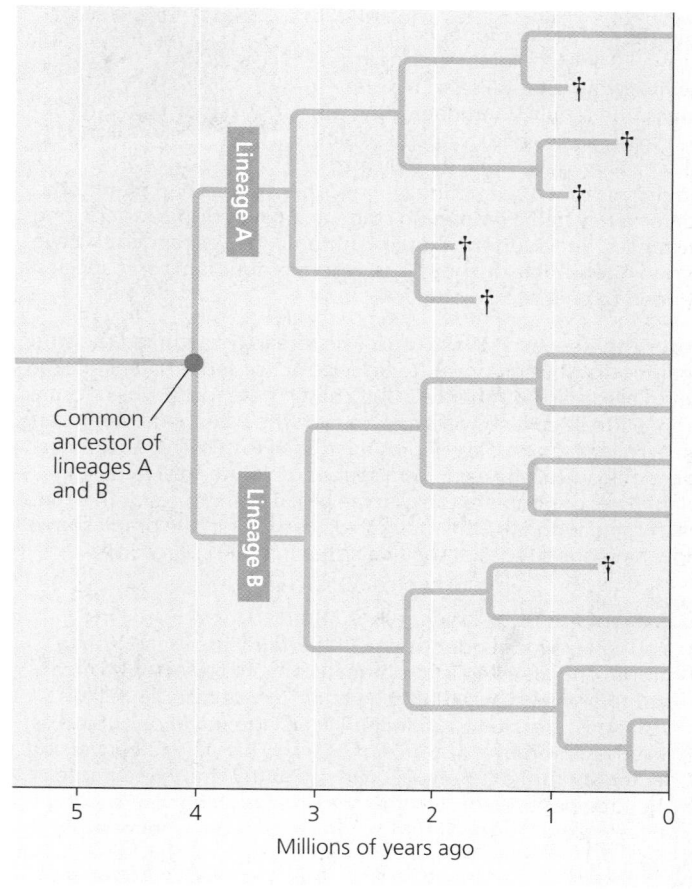

Common ancestor of lineages A and B

Lineage A

Lineage B

Millions of years ago

VISUAL SKILLS *Consider the period from 2 million years ago to time 0. For each lineage, determine how many speciation and extinction events occurred during that time.*

on to dominate life on land for 100 million years, until other tetrapods (including dinosaurs and, later, mammals) replaced them as the dominant terrestrial vertebrates.

The rise and fall of these and other major groups of organisms have shaped the history of life. Narrowing our focus, we can also see that the rise or fall of any particular group is related to the speciation and extinction rates of its member species **(Figure 25.14)**. Just as a population increases in size when there are more births than deaths, the rise of a group of organisms occurs when more new species are produced than are lost to extinction. The reverse occurs when a group is in decline. In the **Scientific Skills Exercise**, you can interpret data from the fossil record about changes in a group of snail species in the early Paleogene period. Such changes in the fates of groups of organisms have been influenced by large-scale processes such as plate tectonics, mass extinctions, and adaptive radiations.

Estimating Quantitative Data from a Graph and Developing Hypotheses

Do Ecological Factors Affect Evolutionary Rates? Researchers studied the fossil record to investigate whether differing modes of larval dispersal might explain species longevity within one taxon of marine snails, the family Volutidae. Some of the snail species had nonplanktonic larvae: They developed directly into adults without a swimming stage. Other species had planktonic larvae: They had a swimming stage and could disperse very long distances. The adults of these planktonic species tended to have broad geographic distributions, whereas nonplanktonic species tended to be more isolated.

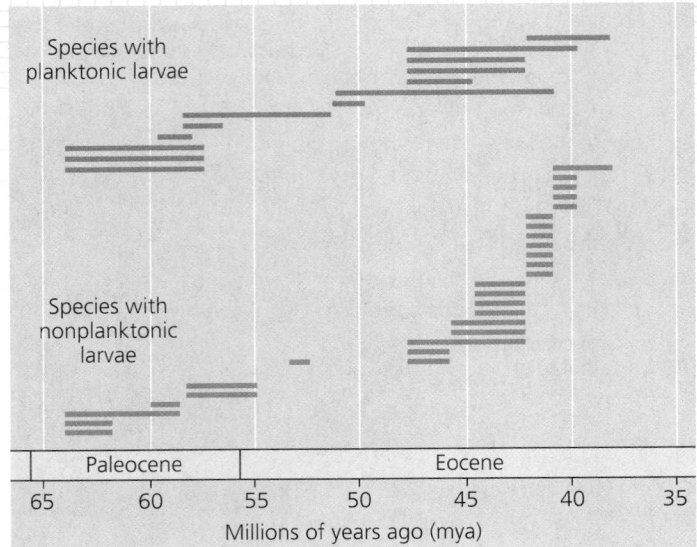

How the Research Was Done The researchers studied the stratigraphic distribution of volutes in outcrops of sedimentary rocks located along North America's Gulf coast. These rocks, which formed from 66 to 37 million years ago, early in the Paleogene period, are an excellent source of well-preserved snail fossils. The researchers were able to classify each fossil species of volute snail as having planktonic or nonplanktonic larvae based on features of the earliest formed whorls of the snail's shell. Each bar in the graph shows how long one species of snail persisted in the fossil record.

INTERPRET THE DATA

1. You can estimate quantitative data (fairly precisely) from a graph. The first step is to obtain a conversion factor by measuring along an axis that has a scale. In this case, 25 million years (my; from 60 to 35 million years ago [mya] on the x-axis) is represented by a distance of 7.0 cm. This yields a conversion factor (a ratio) of 25 my/7.0 cm = 3.6 my/cm. To estimate the time period represented by a horizontal bar on this graph, measure the length of that bar in centimeters and multiply

that measurement by the conversion factor, 3.6 my/cm. For example, a bar that measures 1.1 cm on the graph represents a persistence time of 1.1 cm × 3.6 my/cm = 4 million years.

2. Calculate the mean (average) persistence times for species with planktonic larvae and species with nonplanktonic larvae.

3. Count the number of new species that form in each group beginning at 60 mya (the first three species in each group were present around 64 mya, the first time period sampled, so we don't know when those species first appear in the fossil record).

4. Propose a hypothesis to explain the differences in longevity of snail species with planktonic and nonplanktonic larvae.

Data from T. A. Hansen, Larval dispersal and species longevity in Lower Tertiary gastropods, *Science* 199:885–887 (1978). Reprinted with permission from AAAS.

➔ **Instructors:** A version of this Scientific Skills Exercise can be assigned in **Mastering Biology**.

Plate Tectonics

If photographs of Earth were taken from space every 10,000 years and spliced together to make a movie, it would show something many of us find hard to imagine: The seemingly "rock solid" continents we live on move over time. Over the past billion years, there have been three occasions (1 billion, 600 million, and 250 million years ago) when most of the landmasses of Earth came together to form a supercontinent, then later broke apart. Each time, this breakup yielded a different configuration of continents. Based on the directions in which the continents are moving today, some geologists have estimated that a new supercontinent will form roughly 250 million years from now.

According to the theory of **plate tectonics**, the continents are part of great plates of Earth's crust that essentially float on the hot, underlying portion of the mantle (**Figure 25.15**). Movements in the mantle cause the plates to move over time in a process called *continental drift*. Geologists can measure the rate at which the plates are moving now, usually only a few

centimeters per year. They can also infer the past locations of the continents using the magnetic signal recorded in rocks at the time of their formation. This method works because as a continent shifts its position over time, the direction of magnetic north recorded in its newly formed rocks also changes.

Earth's major tectonic plates are shown

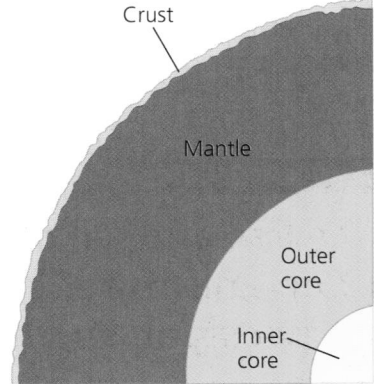

▼ **Figure 25.15 Cutaway view of Earth.** The thickness of the crust is exaggerated here.

Crust

Mantle

Outer core

Inner core

in **Figure 25.16**. Many important geologic processes, including the formation of mountains and islands, occur at plate boundaries. In some cases, two plates are moving away from each other, as are the North American and Eurasian plates, which are currently drifting apart at a rate of about 2 cm per year. In other cases, two plates are sliding past each other, forming regions where earthquakes are common. California's infamous San Andreas Fault is part of a border where two plates slide past each other. In still other cases, two plates collide, producing violent upheavals and forming new mountains along the plate boundaries. One spectacular example of this occurred 45 million years ago, when the Indian plate crashed into the Eurasian plate, starting the formation of the Himalayan mountains.

Consequences of Continental Drift

Plate movements rearrange geography slowly, but their cumulative effects are dramatic. In addition to reshaping the physical features of our planet, continental drift also has a major impact on life on Earth.

One reason for this is that continental drift alters the habitats in which organisms live. Consider the changes shown in **Figure 25.17**. About 250 million years ago, plate movements brought previously separated landmasses together into a supercontinent named **Pangaea**. Ocean basins became deeper, which drained shallow coastal seas. At that time, as now, most marine species inhabited shallow waters, and the formation of Pangaea destroyed much of that habitat. Pangaea's interior was cold and dry, probably an even more severe environment than that of central Asia today. Overall, the formation of Pangaea greatly altered the physical environment and climate, which drove some species to extinction and provided new opportunities for groups of organisms that survived the crisis.

Organisms are also affected by the climate change that results when a continent shifts its location. The southern tip of Labrador, Canada, for example, once was located in the tropics but has moved 40° to the north over the last 200 million years. When faced with the changes in climate that such shifts in position entail, organisms adapt, move to a new location, or become extinct (this last outcome occurred for many organisms stranded on Antarctica, which separated from Australia 40 million years ago).

Continental drift also promotes allopatric speciation on a grand scale. When supercontinents break apart, regions that once were connected become isolated. As the continents drifted apart over the last 200 million years, each became a separate evolutionary arena, with lineages of plants and animals that diverged from those on other continents. For example, marsupial mammals fill ecological roles in Australia analogous to those filled by eutherians (placental mammals) on other continents (see Figure 22.18). Fossil evidence shows

▼ **Figure 25.16 Earth's major tectonic plates.** The blue arrows indicate direction of movement. The reddish orange dots represent zones of violent tectonic activity.

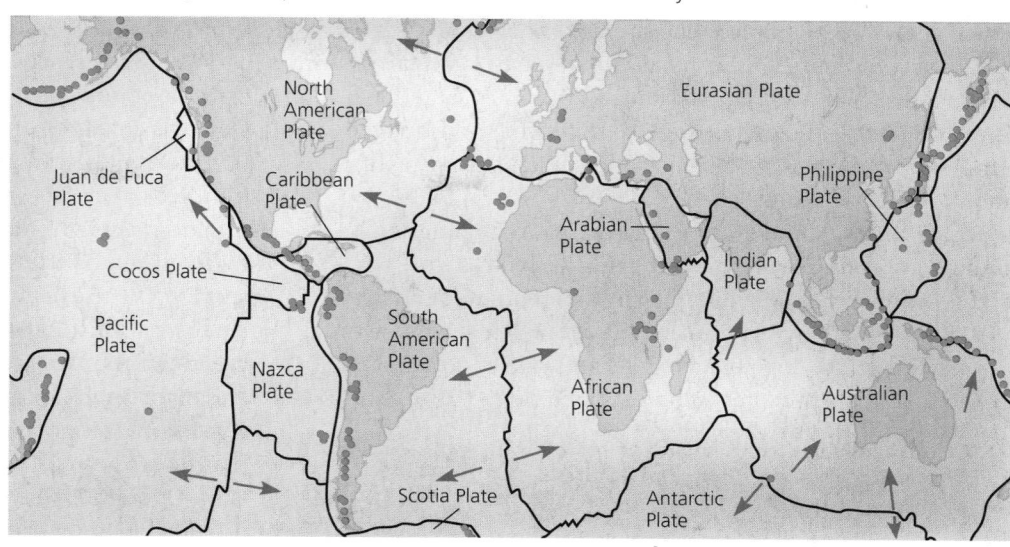

→ **Mastering Biology** HHMI Animation: Plate Tectonics **hhmi** BioInteractive

▼ **Figure 25.17 The history of continental drift during the Phanerozoic eon.**

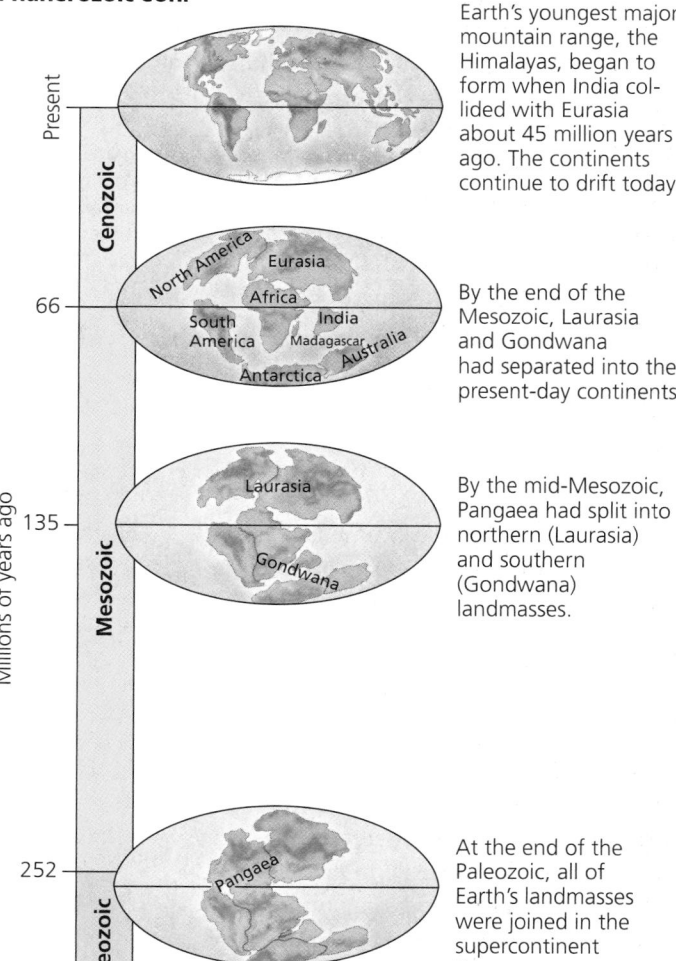

Earth's youngest major mountain range, the Himalayas, began to form when India collided with Eurasia about 45 million years ago. The continents continue to drift today.

By the end of the Mesozoic, Laurasia and Gondwana had separated into the present-day continents.

By the mid-Mesozoic, Pangaea had split into northern (Laurasia) and southern (Gondwana) landmasses.

At the end of the Paleozoic, all of Earth's landmasses were joined in the supercontinent Pangaea.

VISUAL SKILLS *Is the Australian plate's current direction of movement (see Figure 25.16) similar to the direction it traveled over the past 66 million years?*

CHAPTER 25 The History of Life on Earth **539**

that marsupial mammals originated in what is now Asia and reached Australia via South America and Antarctica while the continents were still joined. The subsequent breakup of the southern continents set Australia "afloat," like a giant raft of marsupials. In Australia, marsupials diversified, and the few eutherians (placental mammals) that lived there became extinct; on other continents, most marsupials became extinct and the eutherians diversified.

Finally, continental drift can help explain puzzling geographic distributions of extinct organisms, such as fossils of the same species of Permian freshwater reptiles found in both Brazil and the West African nation of Ghana. These two parts of the world, now separated by 3,000 km of ocean, were joined together when these reptiles were living.

➡️ **Mastering Biology** HHMI Video: Animated Life: Pangea, Wegener, and Continental Drift

Mass Extinctions

The fossil record shows that the overwhelming majority of species that ever lived are now extinct. A species may become extinct for many reasons. Its habitat may have been destroyed, or its environment may have changed in a manner unfavorable to the species. For example, if ocean temperatures fall by even a few degrees, species that are otherwise well adapted may perish. Even if physical factors in the environment remain stable, biological factors may change—the origin of one species can spell doom for another.

Although extinction occurs regularly, at certain times disruptive changes to the global environment have caused the rate of extinction to increase dramatically. The result is a **mass extinction**, in which large numbers of species become extinct worldwide.

The "Big Five" Mass Extinction Events

Five mass extinctions are documented in the fossil record over the past 500 million years **(Figure 25.18)**. These events are particularly well documented for the decimation of hard-bodied animals that lived in shallow seas, the organisms for which the fossil record is most complete. In each mass extinction, 50% or more of marine species became extinct.

Two mass extinctions—the Permian and the Cretaceous—have received the most attention. The Permian mass extinction, which defines the boundary between the Paleozoic and Mesozoic eras (252 million years ago), claimed about 96% of marine animal species and drastically altered life in the ocean. Terrestrial life was also affected. For example, 8 out of

27 known orders of insects were wiped out. This mass extinction occurred in less than 500,000 years, possibly in just a few thousand years—an instant in the context of geologic time.

The Permian mass extinction occurred during the most extreme episode of volcanism in the past 500 million years. Geologic data indicate that 1.6 million km² (roughly half the size of Western Europe) in Siberia was covered with lava hundreds of meters thick. The eruptions are thought to have produced enough carbon dioxide to warm the global climate by an estimated 6°C, harming many temperature-sensitive species. The rise in atmospheric CO_2 levels would also have led to ocean acidification, thereby reducing the availability of calcium carbonate, which is required by reef-building corals and many shell-building species (see Figure 3.12). The eruptions would also have added nutrients such as phosphorus to marine ecosystems, stimulating the growth of microorganisms. Upon their deaths, these microorganisms would have provided food for bacterial decomposers. Bacteria use oxygen as they decompose the bodies of dead organisms, thus causing oxygen concentrations to drop. This would have harmed oxygen-breathers and promoted the growth of anaerobic bacteria that emit a poisonous metabolic by-product, hydrogen sulfide (H_2S) gas. Overall, the volcanic eruptions may have triggered a series of catastrophic events that together resulted in the Permian mass extinction.

▼ **Figure 25.18 Mass extinction and the diversity of life.** The five generally recognized mass extinction events, indicated by red arrows, represent peaks in the extinction rate of marine animal families (red line and left vertical axis). These mass extinctions interrupted the overall increase, over time, in the number of extant families of marine animals (blue line and right vertical axis).

INTERPRET THE DATA *As mentioned in the text, 96% of marine animal species became extinct in the Permian mass extinction (252 million years ago). Explain why the blue curve shows only a 50% drop at that time.*

The Cretaceous mass extinction occurred 66 million years ago. This event extinguished more than half of all marine species and eliminated many families of terrestrial plants and animals, including all dinosaurs (except birds, which are members of the same group; see Figure 34.25). One clue to a possible cause of the Cretaceous mass extinction is a thin layer of clay enriched in iridium that dates to the time of the mass extinction. Iridium is an element that is very rare on Earth but common in many of the meteorites and other extraterrestrial objects that occasionally fall to Earth. As a result, researchers proposed that this clay is fallout from a huge cloud of debris that billowed into the atmosphere when an asteroid collided with Earth. This cloud would have blocked sunlight and caused a sudden drop in global temperatures lasting for several months to years.

Is there evidence of such an asteroid? Research has focused on the Chicxulub crater, a 66-million-year-old scar beneath sediments off the coast of Mexico (**Figure 25.19**). The crater is the right size to have been caused by an object with a diameter of 10 km. When the impact occurred, many species may have been particularly vulnerable to extinction because recent large-scale volcanic eruptions had already caused

▼ **Figure 25.19 A trauma for Cretaceous life.** Beneath the Caribbean Sea, the 66-million-year-old Chicxulub crater measures 180 km across. The horseshoe shape of the crater and the pattern of debris in sedimentary rocks indicate that an asteroid struck at a low angle. This drawing represents the impact and its immediate effect: a cloud of hot vapor and debris that could have killed many of the plants and animals in North America within hours and ignited massive wildfires.

 Mastering Biology HHMI Video: The Day the Mesozoic Died

their populations to decline. Moreover, a 2018 study estimated that wildfires resulting from the impact contributed to a sharp rise in atmospheric CO_2 levels, causing a period of global warming that lasted 100,000 years—yet another stress for populations already in decline. Critical evaluation of these and other hypotheses for mass extinctions continues.

Is a Sixth Mass Extinction Under Way?

As you will read further in Concept 56.1, human actions, such as habitat destruction, are modifying the global environment to such an extent that many species are threatened with extinction. More than 1,000 species have become extinct in the last 400 years. Scientists estimate that this rate is 100 to 1,000 times the typical background rate seen in the fossil record. Is a sixth mass extinction now in progress?

This question is difficult to answer, in part because it is hard to document the total number of extinctions occurring today. Tropical rain forests, for example, harbor many undiscovered species. As a result, destroying tropical forest may drive species to extinction before we even learn of them.

Such uncertainties make it hard to assess the full extent of the current extinction crisis. Even so, it is clear that losses to date have not reached those of the "big five" mass extinctions, in which large percentages of Earth's species became extinct. This does not in any way discount the seriousness of today's situation. Monitoring programs show that many species are declining at an alarming rate due to habitat loss, introduced species, overharvesting, and other factors. Recent studies on a variety of organisms, including lizards, pine trees, and polar bears, suggest that climate change may hasten some of these declines. The fossil record also highlights the potential importance of climate change: Over the last 500 million years, extinction rates have tended to increase when global temperatures were high (**Figure 25.20**).

▼ **Figure 25.20 Fossil extinctions and temperature.** Extinction rates increased when global temperatures were high. Temperatures were estimated using ratios of oxygen isotopes and converted to an index in which 0 is the overall average temperature.

Overall, the evidence suggests that unless dramatic actions are taken, a sixth, human-caused mass extinction is likely to occur within the next few centuries.

Consequences of Mass Extinctions

Mass extinctions have significant and long-term effects. By eliminating large numbers of species, a mass extinction can reduce a thriving and complex ecological community to a pale shadow of its former self. And once an evolutionary lineage disappears, it cannot reappear. The course of evolution is changed forever. Consider what would have happened if the early primates living 66 million years ago had died out in the Cretaceous mass extinction. Humans would not exist, and life on Earth would differ greatly from what it is today.

The fossil record shows that it typically takes 5–10 million years for the diversity of life to recover to previous levels after a mass extinction. In some cases, it has taken much longer than that: It took about 100 million years for the number of marine families to recover after the Permian mass extinction (see Figure 25.18). These data have sobering implications. If current trends continue and a sixth mass extinction occurs, it will take millions of years for life on Earth to recover.

Mass extinctions can also alter ecological communities by changing the types of organisms residing there. For example, after the Permian and Cretaceous mass extinctions, the percentage of marine organisms that were predators increased substantially **(Figure 25.21)**. A rise in the number of predators can increase both the risks faced by prey and the competition among predators for food. In addition, mass extinctions can curtail lineages with novel and advantageous features. For example, in the late Triassic period, a group of gastropods (snails and their relatives) arose that could drill through the shells of bivalves (such as clams) and feed on the animals inside. Although shell drilling provided access to a new and abundant source of food, this newly formed group was wiped out during the mass extinction at the end of the Triassic (about 200 million years ago). Another 120 million years passed before another group of gastropods (the oyster drills) exhibited the ability to drill through shells. As their predecessors might have done if they had not originated at an unfortunate time, oyster drills have since diversified into many new species. Finally, by eliminating so many species, mass extinctions can pave the way for adaptive radiations, in which new groups of organisms proliferate.

Adaptive Radiations

The fossil record shows that the diversity of life has increased over the past 250 million years (see blue line in Figure 25.18). This increase has been fueled by **adaptive radiations**, periods of evolutionary change in which groups of organisms form many new species whose adaptations allow them to fill different ecological roles, or niches, in their communities. Large-scale adaptive radiations occurred after each of the big five mass extinctions, when survivors became adapted to the many vacant ecological niches. Adaptive radiations have also occurred in groups of organisms that possessed major evolutionary innovations, such as seeds or armored body coverings, or that colonized regions in which they faced little competition from other species.

Worldwide Adaptive Radiations

Fossil evidence indicates that mammals underwent a dramatic adaptive radiation after the extinction of terrestrial dinosaurs 66 million years ago **(Figure 25.22)**. Although mammals originated about 180 million years ago, the mammal fossils older than 66 million years are mostly small and show less morphological diversity than found today. Many species appear to have been nocturnal based on their large eye sockets, similar to those in living nocturnal mammals. A few early mammals were intermediate in size, such as *Repenomamus giganticus*, a 1-m-long predator that lived 130 million years ago—but none approached the size of many dinosaurs. Early mammals may have been restricted in size and diversity because they were eaten or outcompeted by the larger and more diverse dinosaurs. With the disappearance of the dinosaurs (except for birds), mammals expanded greatly in both diversity and size, filling the ecological roles once occupied by terrestrial dinosaurs.

The history of life has also been greatly altered by radiations in which groups of organisms increased in diversity as they came to play entirely new ecological roles in their communities. As we'll explore in later chapters, examples include the rise of photosynthetic prokaryotes, the evolution

▼ **Figure 25.21 Mass extinctions and ecology.** The Permian and Cretaceous mass extinctions (indicated by red arrows) altered the ecology of the oceans by increasing the percentage of marine genera that were predators.

▼ **Figure 25.22 Adaptive radiation of mammals.**

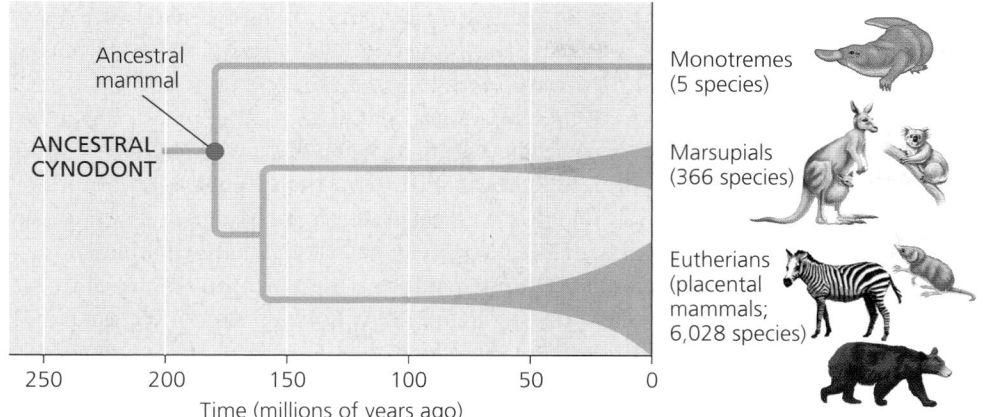

Monotremes (5 species)

Marsupials (366 species)

Eutherians (placental mammals; 6,028 species)

ANCESTRAL CYNODONT

Ancestral mammal

Time (millions of years ago)

250 200 150 100 50 0

of large predators in the Cambrian explosion, and the radiations following the colonization of land by plants, insects, and tetrapods. Each of these last three radiations was associated with major evolutionary innovations that facilitated life on land. The radiation of plants, for example, was associated with key adaptations, such as stems that support plants against gravity and a waxy coat that protects leaves from water loss. Finally, organisms that arise in an adaptive radiation can serve as a new source of food for still other organisms. In fact, the diversification of plants stimulated a series of adaptive radiations in insects that ate or pollinated plants, one reason that insects are the most diverse group of animals on Earth today.

Regional Adaptive Radiations

Striking adaptive radiations have also occurred over more limited geographic areas. Such radiations can be initiated when a few organisms make their way to a new, often distant location in which they face relatively little competition from other organisms. The Hawaiian archipelago is one of the world's great showcases of this type of adaptive radiation **(Figure 25.23)**.

▼ **Figure 25.23 Adaptive radiation on the Hawaiian Islands.** Molecular analysis indicates that these remarkably varied Hawaiian plants, known collectively as the "silversword alliance," are all descended from an ancestral tarweed that arrived on the islands from North America about 5 million years ago. Silverswords have since spread into different habitats and formed new species with strikingly different adaptations.

Dubautia laxa

Close North American relative, the tarweed *Carlquistia muirii*

KAUAI 5.1 million years

OAHU 3.7 million years

MOLOKAI

LANAI

MAUI 1.3 million years

HAWAII 0.4 million years

N

Argyroxiphium sandwicense

Dubautia waialealae

→ **Mastering Biology**
Animation: Overview of Macroevolution

Dubautia scabra

Dubautia linearis

Located about 3,500 km from the nearest continent, the volcanic islands are progressively older as one follows the chain toward the northwest; the youngest island, Hawaii, is less than a million years old and still has active volcanoes. Each island was born "naked" and was gradually populated by stray organisms that rode the ocean currents and winds either from far-distant land areas or from older islands of the archipelago itself. The physical diversity of each island, including immense variation in soil conditions, elevation, and rainfall, provides many opportunities for evolutionary divergence by natural selection. Multiple invasions followed by speciation events have ignited an explosion of adaptive radiation in Hawaii. As a result, thousands of species that inhabit the islands are found nowhere else on Earth. Among plants, for example, about 1,100 species are unique to the Hawaiian Islands. Unfortunately, many of these species are now facing an elevated risk of extinction due to human actions such as habitat destruction and the introduction of non-native plant species.

CONCEPT CHECK 25.4

1. Explain the consequences of plate tectonics for life on Earth.
2. What factors promote adaptive radiations?
3. **WHAT IF?** Based on evidence from previous mass extinctions, describe ecological and evolutionary consequences that could result if a sixth human-caused mass extinction occurs.

For suggested answers, see Appendix A.

CONCEPT 25.5

Major changes in body form can result from changes in the sequences and regulation of developmental genes

The fossil record tells us what the great changes in the history of life have been and when they occurred. Moreover, an understanding of plate tectonics, mass extinction, and adaptive radiation provides a picture of how those changes came about. But we can also seek to understand the intrinsic biological mechanisms that underlie changes seen in the fossil record. For this, we turn to genetic mechanisms of change, paying particular attention to genes that influence development.

Effects of Developmental Genes

As you read in Concept 21.6, "evo-devo"—research at the interface between evolutionary biology and developmental biology—is illuminating how slight genetic differences can produce major morphological differences between species. In particular, large morphological differences can result from genes that alter the rate, timing, and spatial pattern of change in an organism's form as it develops from a zygote into an adult.

Changes in Rate and Timing

Many striking evolutionary transformations are the result of **heterochrony** (from the Greek *hetero*, different, and *chronos*, time), an evolutionary change in the rate or timing of developmental events. For example, an organism's shape depends in part on the relative growth rates of different body parts during development. Changes to these rates can alter the adult form substantially, as seen in the contrasting shapes of human and chimpanzee skulls **(Figure 25.24)**. Other examples of the dramatic evolutionary effects of heterochrony include how increased growth rates of finger bones yielded the skeletal structure of wings in bats (see Figure 22.15) and how slowed growth of leg and pelvic bones led to the reduction and eventual loss of hind limbs in whales (see Figure 22.20).

Heterochrony can also alter the timing of reproductive development relative to the development of nonreproductive organs. If the development of reproductive organs accelerates compared to that of other organs, the sexually mature stage of a species may retain body features that were juvenile structures in an ancestral species, a condition called

▼ **Figure 25.24 Relative skull growth rates.** In the human evolutionary lineage, mutations slowed the growth of the jaw relative to other parts of the skull. As a result, in humans the skull of an adult is more similar to the skull of an infant than is the case for chimpanzees.

Chimpanzee infant   Chimpanzee adult

Chimpanzee fetus   Chimpanzee adult

Human fetus   Human adult

▲ **Figure 25.25 Paedomorphosis.** The adults of some species retain features that were juvenile in ancestors. This salamander is an axolotl, an aquatic species becomes a sexually mature adult while retaining certain larval (tadpole) characteristics, including gills.

paedomorphosis (from the Greek *paedos*, of a child, and *morphosis*, formation). For example, most salamander species have aquatic larvae that undergo metamorphosis in becoming adults. But some species grow to adult size and become sexually mature while retaining gills and other larval features **(Figure 25.25)**. Such an evolutionary alteration of developmental timing can produce animals that appear very different from their ancestors, even though the overall genetic change may be small. Indeed, recent evidence indicates that a change at a single locus was probably sufficient to bring about paedomorphosis in the axolotl salamander, although other genes may have contributed as well.

Changes in Spatial Pattern

Substantial evolutionary changes can also result from alterations in genes that control the spatial organization of body parts. For example, master regulatory genes called **homeotic genes** (see Concept 18.4) determine such basic features as where a pair of wings and legs will develop on a bird or how a plant's flower parts are arranged.

The products of one class of homeotic genes, the *Hox* genes, provide positional information in an animal embryo. This information prompts cells to develop into structures appropriate for a particular location. Changes in *Hox* genes or in how they are expressed can have a profound impact on morphology. For example, among crustaceans, a change in the location where two *Hox* genes (*Ubx* and *Scr*) are expressed correlates with the conversion of a swimming appendage to a feeding appendage. Similarly, when comparing plant species, changes to the expression of homeotic genes known as *MADS-box* genes can produce flowers that differ dramatically in form (see Concept 35.5).

The Evolution of Development

The 560-million-year-old fossils of Ediacaran animals (see Figure 25.11) suggest that a set of genes sufficient to produce complex animals existed at least 25 million years *before* the Cambrian explosion. If such genes have existed for so long, how can we explain the astonishing increases in diversity seen during and since the Cambrian explosion?

Adaptive evolution by natural selection provides one answer to this question. As we've seen throughout this unit, by sorting among differences in the sequences of protein-encoding genes, selection can improve adaptations rapidly. In addition, new genes (created by gene duplication events) can take on new metabolic and structural functions, as can existing genes that are regulated in new ways.

Examples in the previous section suggest that developmental genes may have been particularly important. Thus, we'll turn next to how new morphological forms can arise from changes in the nucleotide sequences or regulation of developmental genes.

Changes in Gene Sequence

New developmental genes arising after gene duplication events probably facilitated the origin of novel morphological forms. But since other genetic changes also may have occurred at such times, it can be difficult to establish causal links between genetic and morphological changes that occurred in the past.

This difficulty was sidestepped in a study of developmental changes associated with the divergence of six-legged insects from crustacean ancestors that had more than six legs. (As discussed in Concept 33.4, insects arose from within a subgroup of the crustaceans, the traditional name for organisms such as shrimp, crabs, and lobsters.) Crustaceans and insects differ in the pattern of expression and the effects of the *Hox* gene *Ubx*: In particular, in insects, *Ubx* suppresses leg formation where it is expressed **(Figure 25.26)**.

To examine the workings of this gene, researchers cloned the *Ubx* gene from an insect, the fruit fly *Drosophila*, and from

▼ **Figure 25.26 Effects of the *Hox* gene *Ubx* on the insect body plan.** In crustaceans, the *Hox* gene *Ubx* is expressed in the region shaded green, the body segments between the head and genital segments. In insects, *Ubx* is expressed in only a subset (shaded pink) of the homologous body segments, where it suppresses leg formation.

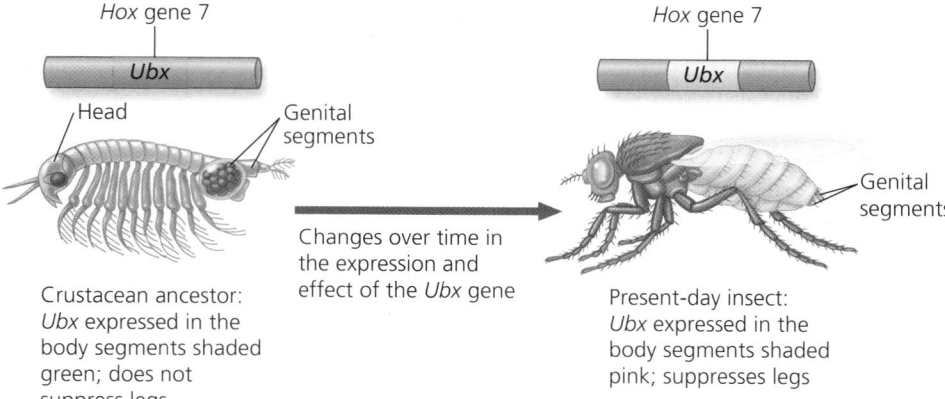

Crustacean ancestor: *Ubx* expressed in the body segments shaded green; does not suppress legs

Changes over time in the expression and effect of the *Ubx* gene

Present-day insect: *Ubx* expressed in the body segments shaded pink; suppresses legs a crustacean, the brine shrimp *Artemia*. Next, they genetically engineered fruit fly embryos to express either the *Drosophila Ubx* gene or the *Artemia Ubx* gene throughout their bodies. The *Drosophila* gene suppressed 100% of the limbs in the embryos, as expected, whereas the *Artemia* gene suppressed only 15%.

The researchers then sought to uncover key steps involved in the evolutionary transition from an ancestral *Ubx* gene to an insect *Ubx* gene. Their approach was to identify mutations that would cause the *Artemia Ubx* gene to suppress leg formation, thus making its gene act more like an insect *Ubx* gene. To do this, they constructed a series of "hybrid" *Ubx* genes, each of which contained known segments of the *Drosophila Ubx* gene and known segments of the *Artemia Ubx* gene. By inserting these hybrid genes into fruit fly embryos (one hybrid gene per embryo) and observing their effects on leg development, the researchers were able to pinpoint the exact amino acid changes responsible for the suppression of additional limbs in insects. In so doing, this study provided evidence that particular changes in the nucleotide sequence of a developmental gene contributed to a major evolutionary change: the origin of the six-legged insect body plan.

Changes in Gene Regulation

A change in the nucleotide sequence of a gene may affect its function wherever the gene is expressed, while changes in the regulation of gene expression can be limited to one cell type (see Concept 18.4). Thus, a change in the regulation of a developmental gene may have fewer harmful side effects than a change to the sequence of the gene. This reasoning has prompted researchers to suggest that changes in the form of organisms may often be caused by mutations that affect the regulation of developmental genes—not their sequences.

This idea is supported by studies of a variety of species, including threespine stickleback fish. These fish live in the open ocean and in shallow, coastal waters. In western Canada, they also live in lakes formed when the coastline receded during the past 12,000 years.

▼ Figure 25.27 Inquiry

What causes the loss of spines in lake stickleback fish?

Experiment Marine populations of the threespine stickleback fish (*Gasterosteus aculeatus*) have a set of protective spines on their lower (ventral) surface; however, these spines have been lost or reduced in some lake populations of this fish. Researchers performed genetic crosses and found that most of the reduction in spine size resulted from the effects of a single developmental gene, *Pitx1*. The researchers then tested two hypotheses about how *Pitx1* causes this morphological change.

▲ **Threespine stickleback (*Gasterosteus aculeatus*)**

Hypothesis A: A change in the DNA sequence of *Pitx1* had caused spine reduction in lake populations. To test this idea, the team used DNA sequencing to compare the coding sequence of the *Pitx1* gene between marine and lake stickleback populations.

Hypothesis B: A change in the regulation of the expression of *Pitx1* had caused spine reduction. To test this idea, the researchers monitored where in the developing embryo the *Pitx1* gene was expressed. They conducted whole-body *in situ* hybridization experiments (see Concept 20.2) using *Pitx1* DNA as a probe to detect *Pitx1* mRNA in the fish.

Results

Test of Hypothesis A:	Are there differences in the coding sequence of the *Pitx1* gene in marine and lake stickleback fish?	**Result: No** →	The 283 amino acids of the *Pitx1* protein are identical in marine and lake stickleback populations.
Test of Hypothesis B:	Are there any differences in the regulation of expression of *Pitx1*?	**Result: Yes** →	Red arrows (→) indicate regions of *Pitx1* gene expression in the photographs below. *Pitx1* is expressed in the ventral spine and mouth regions of developing marine stickleback fish but only in the mouth region of developing lake stickleback fish.

Marine stickleback embryo   **Lake stickleback embryo**

Close-up of mouth

Close-up of ventral surface

Conclusion The loss or reduction of ventral spines in lake populations of threespine stickleback fish appears to have resulted primarily from a change in the regulation of *Pitx1* gene expression, not from a change in the gene's sequence.

Data from M. D. Shapiro et al., Genetic and developmental basis of evolutionary pelvic reduction in threespine sticklebacks, *Nature* 428:717–723 (2004).

WHAT IF? *Describe the set of results that would have led researchers to the conclusion that a change in the coding sequence of the* Pitx1 *gene was more important than a change in regulation of gene expression.*

Marine stickleback fish have a pair of spines on their ventral (lower) surface, which deter some predators. These spines are often reduced or absent in stickleback fish living in lakes that lack predatory fishes and that are also low in calcium. Spines may have been lost in such lakes because they are not advantageous in the absence of predators, and the limited calcium is needed for purposes other than constructing spines.

At the genetic level, the developmental gene *Pitx1* was known to influence whether stickleback fish have ventral spines. Was the reduction of spines in some lake populations due to changes in the *Pitx1* gene or to changes in how the gene is expressed **(Figure 25.27)**? The researchers' results indicate that the regulation of gene expression has changed, not the DNA sequence. Moreover, lake stickleback fish do express the *Pitx1* gene in tissues not related to the production of spines (for example, the mouth), illustrating how morphological change can be caused by altering the expression of a developmental gene in some parts of the body but not others. In a follow-up study, researchers showed that changes to the *Pel* enhancer, a noncoding DNA region that affects expression of the *Pitx1* gene, resulted in the reduction of ventral spines in lake sticklebacks. Overall, results from studies on stickleback fish provide a clear and detailed example of how changes in gene regulation can alter the form of individual organisms and ultimately lead to evolutionary change in populations.

 Mastering Biology HHMI Video: The Making of the Fittest: Evolving Switches, Evolving Bodies (Stickleback)

 Mastering Biology Interview with Cassandra Extavour: Studying the evolution of genes that control development (see the interview before Chapter 22)

CONCEPT CHECK 25.5

1. Explain how new body forms can originate by heterochrony.

2. Why is it likely that *Hox* genes have played a major role in the evolution of novel morphological forms?

3. **MAKE CONNECTIONS** Given that changes in morphology are often caused by changes in the regulation of gene expression, predict whether noncoding DNA is likely to be affected by natural selection. (Review Concept 18.3.)

For suggested answers, see Appendix A.

CONCEPT 25.6

Evolution is not goal oriented

What does our study of macroevolution tell us about how evolution works? One lesson is that throughout the history of life, the origin of new species has been affected by both the small-scale factors described in Concept 23.3 (such as natural selection operating in populations) and the large-scale factors described in this chapter (such as continental drift promoting bursts of speciation throughout the globe).

Moreover, to paraphrase the Nobel Prize–winning geneticist François Jacob, evolution is like tinkering—a process in which new forms arise by the modification of existing structures or existing developmental genes. Over time, such tinkering has led to the three key features of the natural world described on the opening page of Chapter 22: the striking ways in which organisms are suited for life in their environments, the many shared characteristics of life, and the rich diversity of life.

Evolutionary Novelties

François Jacob's view of evolution harkens back to Darwin's concept of descent with modification. As new species form, novel and complex structures can arise as gradual modifications of ancestral structures. In many cases, complex structures have evolved in increments from simpler versions that performed the same basic function. For example, consider the human eye, an intricate organ constructed from numerous parts that work together in forming an image and transmitting it to the brain. How could the human eye have evolved in gradual increments? Some argue that if the eye needs all of its components to function, a partial eye could not have been of use to our ancestors.

The flaw in this argument, as Darwin himself noted, lies in the assumption that only complicated eyes are useful. In fact, many animals depend on eyes that are far less complex than our own. The simplest eyes that we know of are patches of light-sensitive photoreceptor cells. These simple eyes appear to have had a single evolutionary origin and are now found in a variety of animals, including small molluscs called limpets. Such eyes have no equipment for focusing images, but they do enable the animal to distinguish light from dark. Limpets cling more tightly to their rock when a shadow falls on them, a behavioral adaptation that reduces the risk of being eaten **(Figure 25.28)**. Limpets have had a long evolutionary history, demonstrating that their "simple" eyes are quite adequate to support their survival and reproduction.

In the animal kingdom, complex eyes have evolved independently from such basic structures many times. Some

▼ **Figure 25.28** Limpets (*Patella vulgata*), molluscs that can sense light and dark with a simple patch of photoreceptor cells.

▼ Figure 25.29 A range of eye complexity among molluscs.

(a) Patch of pigmented cells

Pigmented cells (photoreceptors)

Epithelium

Nerve fibers

The limpet *Patella* has a simple patch of photoreceptors.

(b) Eyecup

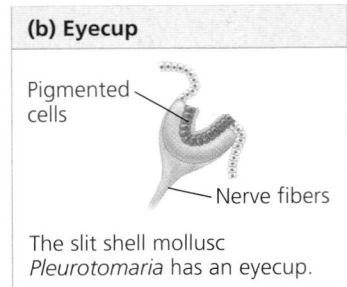

Pigmented cells

Nerve fibers

The slit shell mollusc *Pleurotomaria* has an eyecup.

(c) Pinhole camera-type eye

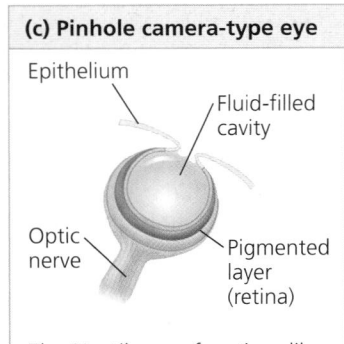

Epithelium

Fluid-filled cavity

Optic nerve

Pigmented layer (retina)

The *Nautilus* eye functions like a pinhole camera (an early type of camera lacking a lens).

(d) Eye with primitive lens

Cellular mass (lens)

Cornea

Optic nerve

The marine snail *Murex* has a primitive lens consisting of a mass of crystal-like cells. The cornea is a transparent region of tissue that protects the eye and helps focus light.

(e) Complex camera lens-type eye

Cornea

Lens

Retina

Optic nerve

The squid *Loligo* has a complex eye with features (cornea, lens, and retina) similar to those of vertebrate eyes. However, the squid eye evolved independently from vertebrate eyes.

→ **Mastering Biology** BBC Video: How Did Eyes Evolve?

molluscs, such as squids and octopuses, have eyes as complex as those of humans and other vertebrates **(Figure 25.29)**. Although complex mollusc eyes evolved independently of vertebrate eyes, both evolved from a simple cluster of photoreceptor cells present in a common ancestor. In each case, the complex eye evolved through a series of steps that benefited the eyes' owners at every stage. Evidence of their independent evolution can be found in their structure: Vertebrate eyes detect light at the back layer of the retina and conduct nerve impulses toward the front, while complex mollusc eyes do the reverse.

Throughout their evolutionary history, eyes retained their basic function of vision. But evolutionary novelties can also arise when structures that originally played one role gradually acquire a different one. For example, as cynodonts gave rise to early mammals, bones that formerly comprised the jaw hinge (the articular and quadrate; see Figure 25.7) were incorporated into the ear region of mammals, where they eventually took on a new function: the transmission of sound (see Concept 34.6).

Structures that evolve in one context but become co-opted for another function are sometimes called *exaptations* to distinguish them from the adaptive origin of the original structure. Note that the concept of exaptation does not imply that a structure somehow evolves in anticipation of future use. Natural selection cannot predict the future; it can only improve a structure in the context of its *current* utility. Novel features, such as the new jaw hinge and ear bones of early mammals, can arise gradually via a series of intermediate stages, each of which has some function in the organism's current context.

→ **Mastering Biology** HHMI Video: The Origin of Flight: What Use Is Half a Wing?

Evolutionary Trends

What else can we learn from patterns of macroevolution? Consider evolutionary "trends" observed in the fossil record. For instance, some evolutionary lineages exhibit a trend toward larger or smaller body size. An example is the evolution of the present-day horse (genus *Equus*), a descendant of the 55-million-year-old *Hyracotherium* **(Figure 25.30)**. About the size of a large dog, *Hyracotherium* had four toes on its front feet, three toes on its hind feet, and teeth adapted for browsing on bushes and trees. In comparison, present-day horses are larger, have only one toe on each foot, and possess teeth modified for grazing on grasses.

Extracting a single evolutionary progression from the fossil record can be misleading, however; it is like describing a bush as growing toward a single point by tracing only the branches that lead to that twig. For example, by selecting certain species from the available fossils, it is possible to arrange a succession of animals intermediate between *Hyracotherium* and living horses that shows a trend toward large, single-toed species (follow the yellow highlighting in Figure 25.30). However, if we consider *all* fossil horses known today, this apparent trend vanishes. The genus *Equus* did not evolve in a straight line; it is the only surviving twig of an evolutionary tree that is so branched that it is more like a bush. *Equus* actually descended through a series of speciation episodes that included several adaptive radiations, not all of which led to large, one-toed, grazing horses. In fact, phylogenetic analyses suggest that all lineages that include grazers are closely related to *Parahippus*; the many other horse lineages, all of which are now extinct, remained multi-toed browsers for 35 million years.

Branching evolution *can* result in a real evolutionary trend even if some species counter the trend. One model of long-term trends views species as analogous to individuals: Speciation is their birth, extinction is their death, and new species that diverge from them are their offspring. In this model, just as populations of individual organisms undergo natural selection, species undergo *species selection*. The species that endure the longest and generate the most new offspring species determine the direction of major evolutionary trends. The species selection model suggests that "differential speciation success" plays a role in macroevolution similar to the role of differential reproductive success

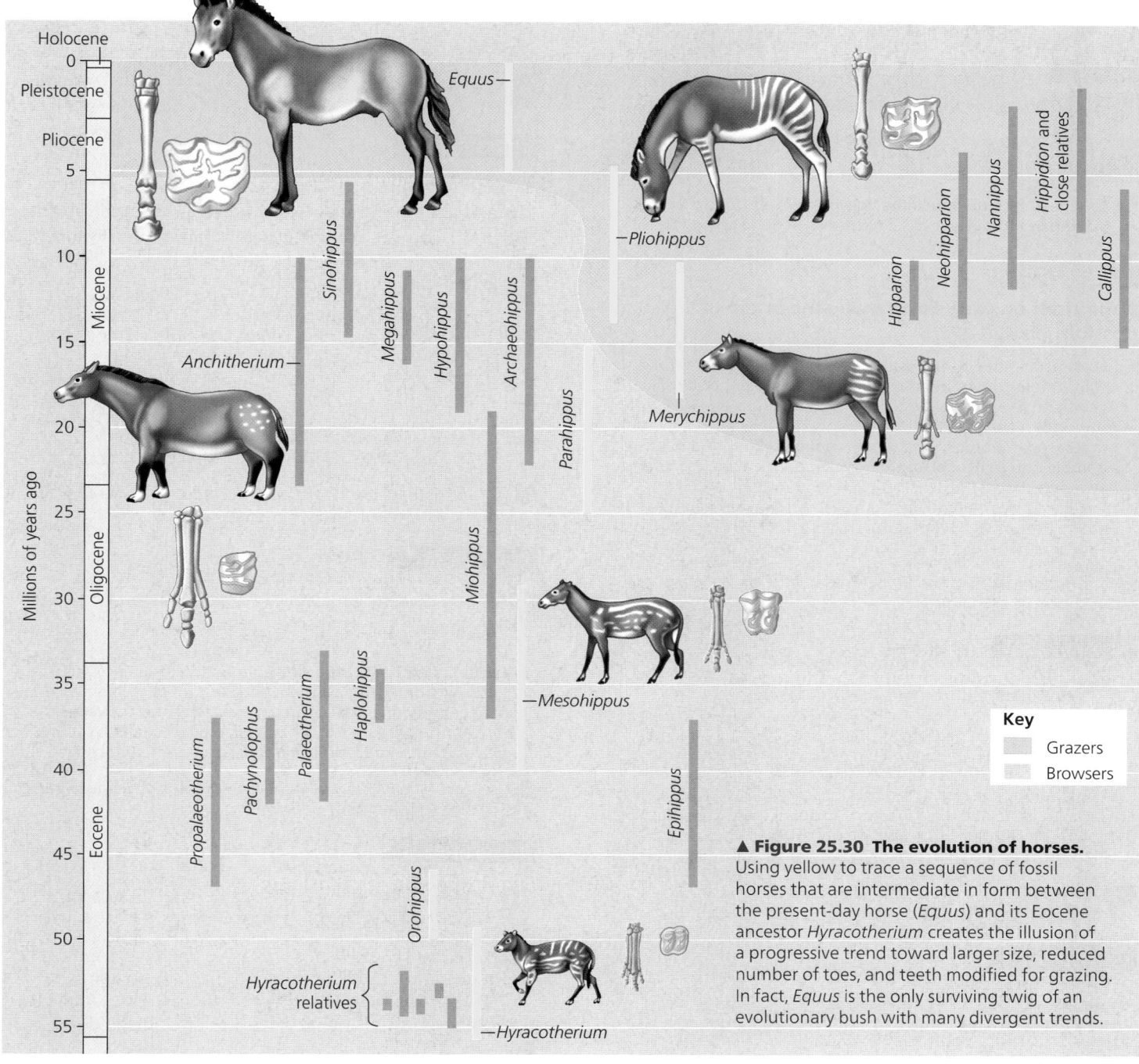

Figure 25.30 The evolution of horses. Using yellow to trace a sequence of fossil horses that are intermediate in form between the present-day horse (*Equus*) and its Eocene ancestor *Hyracotherium* creates the illusion of a progressive trend toward larger size, reduced number of toes, and teeth modified for grazing. In fact, *Equus* is the only surviving twig of an evolutionary bush with many divergent trends.

Key
- Grazers
- Browsers

in microevolution. Evolutionary trends can also result directly from natural selection. For example, when horse ancestors invaded the grasslands that spread during the mid-Cenozoic, there was strong selection for grazers that could escape predators by running faster. This trend would not have occurred without open grasslands.

Whatever its cause, an evolutionary trend does not imply that there is some intrinsic drive toward a particular phenotype. Evolution is the result of the interactions between organisms and their current environments; if environmental conditions change, an evolutionary trend may cease or even reverse itself. The cumulative effect of these ongoing interactions between organisms and their environments is enormous: It is through them that the staggering diversity of life—Darwin's "endless forms most beautiful"—has arisen.

CONCEPT CHECK 25.6

1. How can the Darwinian concept of descent with modification explain the evolution of such complex structures as the vertebrate eye?

2. **WHAT IF?** The myxoma virus kills up to 99.8% of infected European rabbits in populations with no previous exposure to the virus. The virus is transmitted between living rabbits by mosquitoes. Describe an evolutionary trend (in either the rabbit or virus) that might occur after a rabbit population first encounters the virus.

For suggested answers, see Appendix A.

→ Go to **Mastering Biology** for Assignments, the eText, the Study Area, and Dynamic Study Modules.

SUMMARY OF KEY CONCEPTS

→ To review key terms, go to the **Vocabulary Self-Quiz** in the **Mastering Biology** eText or Study Area, or go to goo.gl/zkjz9t.

CONCEPT 25.1

Conditions on early Earth made the origin of life possible (pp. 526–528)

- Experiments simulating possible early atmospheres have produced organic molecules from inorganic precursors. Amino acids, lipids, sugars, and nitrogenous bases have also been found in meteorites.
- Amino acids and RNA nucleotides polymerize when dripped onto hot sand, clay, or rock. Organic compounds can spontaneously assemble into **protocells**, membrane-bounded droplets that have some properties of cells.
- The first genetic material may have self-replicating, catalytic RNA. Early protocells containing such RNA would have increased through natural selection.

? *Describe the roles that montmorillonite clay and vesicles may have played in the origin of life.*

CONCEPT 25.2

The fossil record documents the history of life (pp. 528–532)

- The fossil record, based largely on fossils found in sedimentary rocks, documents the rise and fall of different groups of organisms over time.
- Sedimentary strata reveal the relative ages of fossils. The ages of fossils can be estimated by **radiometric dating** and other methods.
- The fossil record shows how new groups of organisms can arise via the gradual modification of preexisting organisms.

? *What are the challenges of estimating the ages of old fossils? Explain how these challenges may be overcome in some circumstances.*

CONCEPT 25.3

Key events in life's history include the origins of unicellular and multicellular organisms and the colonization of land (pp. 532–537)

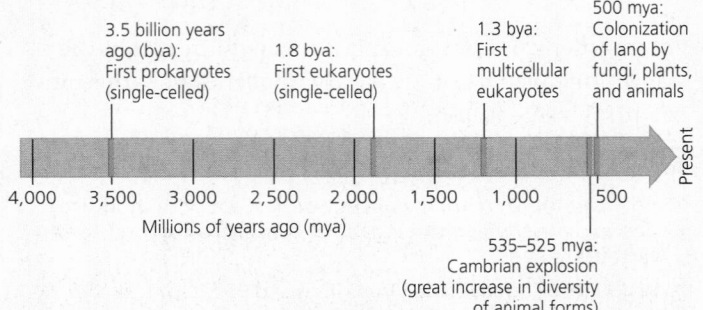

3.5 billion years ago (bya): First prokaryotes (single-celled)

1.8 bya: First eukaryotes (single-celled)

1.3 bya: First multicellular eukaryotes

500 mya: Colonization of land by fungi, plants, and animals

Millions of years ago (mya)

535–525 mya: Cambrian explosion (great increase in diversity of animal forms)

? *What is the "Cambrian explosion," and why is it significant?*

CONCEPT 25.4

The rise and fall of groups of organisms reflect differences in speciation and extinction rates (pp. 537–544)

- In **plate tectonics**, continental plates move gradually over time, altering the physical geography and climate of Earth, leading to extinctions in some groups and speciation in others.
- Evolutionary history has been punctuated by five **mass extinctions** that radically altered life's history. Possible causes for these extinctions include continental drift, volcanic activity, and impacts from asteroids.
- Large increases in the diversity of life have resulted from **adaptive radiations** that followed mass extinctions. Adaptive radiations have also occurred in groups of organisms that possessed major evolutionary innovations or that colonized new regions in which there was little competition from other organisms.

? *Explain how the broad evolutionary changes seen in the fossil record are the cumulative result of speciation and extinction events.*

CONCEPT 25.5

Major changes in body form can result from changes in the sequences and regulation of developmental genes (pp. 544–547)

- Developmental genes affect morphological differences between species by influencing the rate, timing, and spatial patterns of change in an organism's form as it develops into an adult.
- The evolution of new forms can be caused by changes in the nucleotide sequences or regulation of developmental genes.

? *How could changes in a single gene or DNA region ultimately lead to the origin of a new group of organisms?*

CONCEPT 25.6

Evolution is not goal oriented (pp. 547–549)

- Novel and complex biological structures can evolve through a series of incremental modifications, each of which benefits the organism that possesses it.
- Evolutionary trends can be caused by natural selection in a changing environment or species selection, resulting from interactions between organisms and their current environments.

? *Explain the reasoning behind the statement "Evolution is not goal oriented."*

TEST YOUR UNDERSTANDING

→ For more multiple-choice questions, go to the **Practice Test** in the **Mastering Biology** eText or Study Area, or go to goo.gl/GruWRg.

Levels 1-2: Remembering/Understanding

1. Fossilized stromatolites
   (A) formed around deep-sea vents.
   (B) resemble structures formed by bacterial communities that are found today in some shallow marine bays.
   (C) provide evidence that plants moved onto land in the company of fungi around 500 million years ago.
   (D) contain the first undisputed fossils of eukaryotes.

2. The oxygen revolution changed Earth's environment dramatically. Which of the following took advantage of the presence of free oxygen in the oceans and atmosphere?
   (A) the evolution of cellular respiration, which used oxygen to help harvest energy from organic molecules
   (B) the persistence of some animal groups in anaerobic habitats
   (C) the evolution of photosynthetic pigments that protected early algae from the corrosive effects of oxygen
   (D) the evolution of chloroplasts after early protists incorporated photosynthetic cyanobacteria

3. Which factor most likely caused animals and plants in India to differ greatly from species in nearby southeast Asia?
   (A) The species became separated by convergent evolution.
   (B) The climates of the two regions are similar.
   (C) India is in the process of separating from the rest of Asia.
   (D) India was a separate continent until 45 million years ago.

4. Large-scale, worldwide adaptive radiations have occurred in which of the following situations?
   (A) when there are no available ecological niches
   (B) after each of the big five mass extinctions
   (C) after colonization of an isolated island that contains suitable habitat and few competitor species
   (D) whenever an evolutionary innovation was needed for organisms to thrive

5. Scientists studying the origin of life have accomplished which of the following steps?
   (A) abiotic synthesis of protocells with self-replicating, catalytic RNA
   (B) formation of vesicles that use RNA as a template for DNA synthesis
   (C) formation of protocells that use DNA to direct the polymerization of amino acids
   (D) abiotic synthesis of RNA's bases (A, C, G, U)

Levels 3-4: Applying/Analyzing

6. A genetic change that caused a certain *Hox* gene to be expressed along the tip of a vertebrate limb bud instead of farther back helped make possible the evolution of the tetrapod limb. This type of change is illustrative of
   (A) the influence of environment on development.
   (B) paedomorphosis.
   (C) a change in a developmental gene or in its regulation that altered the spatial organization of body parts.
   (D) heterochrony.

7. A swim bladder is a gas-filled sac that helps fish maintain buoyancy. The evolution of the swim bladder from the air-breathing organ (a simple lung) of an ancestral fish is an example of
   (A) exaptation.
   (B) changes in *Hox* gene expression.
   (C) paedomorphosis.
   (D) adaptive radiation.

Levels 5-6: Evaluating/Creating

8. **EVOLUTION CONNECTION** Describe how gene flow, genetic drift, and natural selection all can influence macroevolution.

9. **SCIENTIFIC INQUIRY** Herbivory (plant eating) has evolved repeatedly in insects, typically from meat-eating or detritus-feeding ancestors (detritus is dead organic matter). Moths and butterflies, for example, eat plants, whereas their "sister group" (the insect group to which they are most closely related), the caddisflies, feed on animals, fungi, or detritus. As illustrated in the following phylogenetic tree, the combined moth/butterfly and caddisfly group shares a common ancestor with flies and fleas. Like caddisflies, flies and fleas are thought to have evolved from ancestors that did not eat plants.

There are 140,000 species of moths and butterflies and 7,000 species of caddisflies. State a hypothesis about the impact of herbivory on adaptive radiations in insects. How could this hypothesis be tested?

10. **WRITE ABOUT A THEME: ORGANIZATION** You have seen many examples of how form fits function at all levels of the biological hierarchy. However, we can imagine forms that would function better than some forms actually found in nature. For example, if the wings of a bird were not formed from its forelimbs, such a hypothetical bird could fly yet also hold objects with its forelimbs. In a short essay (100–150 words), use the concept of "evolution as tinkering" to explain why there are limits to the functionality of forms in nature.

11. **SYNTHESIZE YOUR KNOWLEDGE**

In 2010, the Soufriere Hills volcano on the Caribbean island of Montserrat erupted violently, spewing huge clouds of ash and gases into the sky. Explain how the volcanic eruptions at the end of the Permian period and the formation of Pangaea, both of which occurred about 252 million years ago, set in motion events that altered evolutionary history.

For selected answers, see Appendix A.

Sallie (Penny) Chisholm, Institute Professor in Civil and Environmental Engineering at the Massachusetts Institute of Technology, has received the 2011 National Medal of Science, the 2019 Crafoord Prize in Biosciences, and many other prestigious awards. Dr. Chisholm received a B.A. in Biology and Chemistry from Skidmore College and a Ph.D. in Biology from the State University of New York, Albany. Elected to the U.S. National Academy of Sciences in 2003, Dr. Chisholm has a long-standing interest in the essential roles that photosynthetic microbes play in marine ecosystems. Currently, she and her students are studying these organisms from the genome to global scale.

AN INTERVIEW WITH

Penny Chisholm

How did you first become interested in science?

I grew up in Michigan's Upper Peninsula, in the sticks. Growing up, my family was all about business; I wasn't from an academic background. So, when I started studying biology in college, I was kind of surprised to find out that I liked science as a way of knowing. I had never been exposed to doing experiments and discovering new things on your own. I had the chance to do an independent research project on a local lake to measure its manganese budget. We measured the movement of manganese into and out of the lake over the course of a year, drilling holes in the ice to sample in winter. All of this was really cool and got me excited about the dynamics of aquatic ecosystems. Once I looked under the microscope and saw a world that I didn't even know existed—the microbes living in lakes—there was no turning back.

Much of your research is on the bacterium *Prochlorococcus*. Tell us about this organism.

Prochlorococcus is the smallest photosynthetic cell on the planet and the most abundant—it is less than 1/100th the width of a human hair, and there are over a billion billion billion of these tiny cells in the ocean! *Prochlorococcus* feeds small protists, which in turn are eaten by larger marine organisms, and thus they play a key role in ocean ecosystems. And when *Prochlorococcus* cells are infected by viruses, they break apart and their organic compounds serve as a food source for other bacteria in the ocean. Despite these organisms' importance, we didn't even know they existed when I was a student. And then, in the mid-1980s, I was fortunate enough to be involved in their discovery. I was fascinated by them—they were almost everywhere in the oceans. Eventually we learned to grow them in the lab. I was hooked!

What has *Prochlorococcus* taught you?

We've known for a long time about diatoms and other relatively large photosynthetic organisms—phytoplankton—in the ocean. Those organisms were like the trees and the bushes. Suddenly, with the discovery of small photosynthetic bacteria such as *Prochlorococcus*, we learned, "Oh! There's grass, too!" Once we knew *Prochlorococcus* was there, we developed methods to monitor it in the field and sort out individual cells. We wondered how a single species could be so abundant in so many different areas of the ocean. We found out that this organism has extremely high levels of genetic diversity. This enables it to thrive in many different ocean environments, despite ever-changing conditions. Overall, studying *Prochlorococcus* has taught me to be humbled by the mind-blowing complexity of the natural world.

To date, what is the most surprising discovery from your lab?

The discovery of "grass" in the ocean was certainly a surprise! More recently, we've learned that the ocean is full of vesicles, very small lipid-bound particles that are budded off a cell. They look like little round peas on top of the cell. For years I had a picture of them that one of my students had taken, and I would ask people, "What do you think this is?" No one knew, until one of my post-docs speculated that they were vesicles, and demonstrated this to be true. He discovered that they contain DNA, as well as RNA and a variety of proteins and lipids. We can determine which species produced a vesicle by sequencing the DNA it contains. We found out that many different species are producing vesicles. This is a beautiful example of how a casual observation in the lab—a photograph that I kept in a file cabinet for years—eventually led to the discovery of a new component of ocean ecosystems. Now the challenge is figuring out what they are doing!

What do you find most rewarding about your job?

Working with *Prochlorococcus* is such a gift! Over the years, it's been an almost endless process of discovery—the organism is such a fundamental part of life in the ocean, and there are so many different questions you can ask. It has also been fruitful and exciting to work on *Prochlorococcus* with students and colleagues from many different backgrounds. It is wonderful to see my students working together to answer really hard questions. And I've learned so much from them—it has been a privilege and a joy to help them discover what they're best at and what they really enjoy doing. I love what I do. What could be more important than studying life?

> *"Studying* Prochlorococcus *has taught me to be humbled by the mind-blowing complexity of the natural world."*

▼ *Prochlorococcus* cells and vesicles.

Cell

Vesicle

1 μm

26 Phylogeny and the Tree of Life

KEY CONCEPTS

26.1 Phylogenies show evolutionary relationships *p. 554*

26.2 Phylogenies are inferred from morphological and molecular data *p. 558*

26.3 Shared characters are used to construct phylogenetic trees *p. 559*

26.4 An organism's evolutionary history is documented in its genome *p. 565*

26.5 Molecular clocks help track evolutionary time *p. 566*

26.6 Our understanding of the tree of life continues to change based on new data *p. 568*

Study Tip

Label a drawing: The members of a *sister group* are each other's closest relatives, making sister groups a useful way to describe evolutionary relationships. As in this example, identify the sister groups associated with each branch point in Figures 26.11, 26.12, 26.14, and 26.16.

Frog

Frogs form a sister group with the combined chicken/human/mouse lineages (the "frog–chicken/human/mouse" sister group)

Chicken

Chicken–human/mouse

Human

Human–mouse

Mouse

→ Go to Mastering Biology

For Students (in eText and Study Area)
- Get Ready for Chapter 26
- Figure 26.12 Walkthrough: Using Derived Characters to Infer Phylogeny

For Instructors to Assign (in Item Library)
- Misconception Questions 126, 128, 129
- Tutorial: Visualizing Phylogenetic Relationships

Ready-to-Go Teaching Module (in Instructor Resources)
- Phylogenetic Trees (Concept 26.3)

Figure 26.1 Although this animal resembles a snake, it is actually a legless lizard: the glass lizard *Ophisaurus apodus*. Why isn't the glass lizard considered a snake? Biologists based this decision in part on a comparison of its traits with those of snakes—and the glass lizard does not have a highly mobile jaw or many vertebrae, two traits shared by all snakes.

How do biologists distinguish and categorize the millions of species on Earth?

Traits shared due to **common ancestry** are used to classify organisms into groups that reflect their **evolutionary history:**

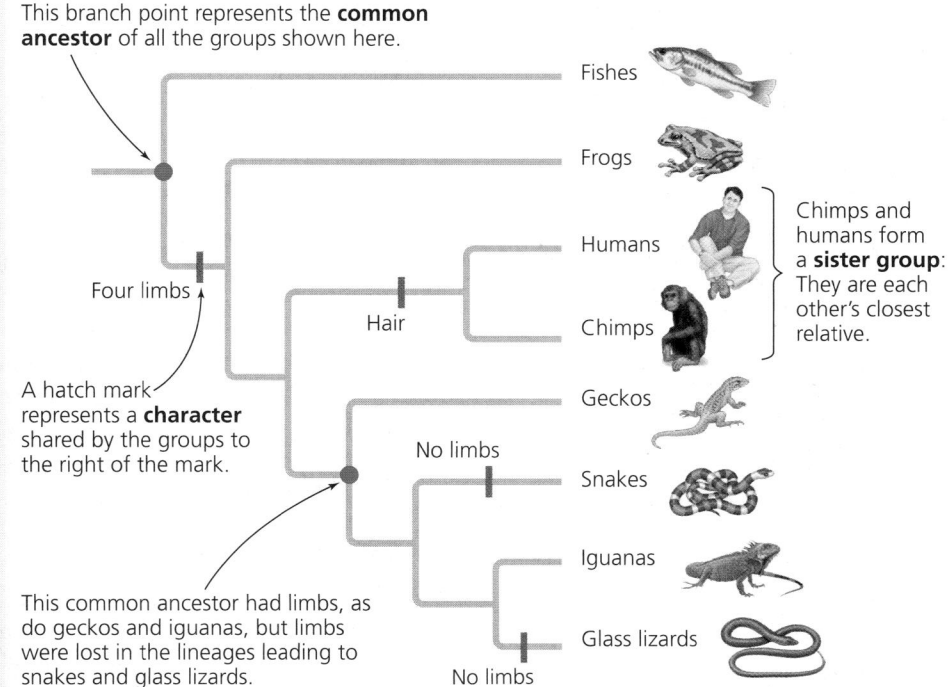

This branch point represents the **common ancestor** of all the groups shown here.

Fishes

Frogs

Humans

Chimps and humans form a **sister group**: They are each other's closest relative.

Chimps

Hair

Four limbs

A hatch mark represents a **character** shared by the groups to the right of the mark.

Geckos

No limbs

Snakes

This common ancestor had limbs, as do geckos and iguanas, but limbs were lost in the lineages leading to snakes and glass lizards.

No limbs

Iguanas

Glass lizards

553

Phylogenies show evolutionary relationships

The glass lizard in Figure 26.1 is part of the continuum of life extending from the earliest organisms to the great variety of species alive today. In this unit, we will survey this diversity and describe hypotheses regarding how it evolved.

To set the stage for surveying life's diversity, in this chapter we consider how biologists trace **phylogeny**, the evolutionary history of a species or group of species. A phylogeny of lizards and snakes, for example, indicates that both glass lizards and snakes evolved from lizards with legs—but they evolved from different lineages of legged lizards **(Figure 26.2)**. Thus, it appears that their legless conditions evolved independently. As we'll see, biologists reconstruct phylogenies like that in Figure 26.2 using **systematics**, a discipline focused on classifying organisms and determining their evolutionary relationships. We'll begin by describing how organisms are named.

Binomial Nomenclature

Common names for organisms—such as monkey, finch, and lilac—convey meaning in casual usage, but they can also cause confusion. Each of these names, for example, refers to more than one species. Moreover, some common names do not accurately reflect the kind of organism they signify. Consider these three "fishes": jellyfish (a cnidarian), crayfish (a small, lobster-like crustacean), and silverfish (an insect). And, of course, a given organism has different names in different languages.

To avoid ambiguity when communicating about their research, biologists refer to organisms by Latin scientific names. The two-part format of the scientific name, commonly called a **binomial**, was instituted in the 18th century by Carolus Linnaeus (see Concept 22.1). The first part of a binomial is the name of the **genus** (plural, *genera*) to which the species belongs. The second part, called the specific epithet, is unique for each species within the genus. An example of a binomial is *Panthera pardus*, the scientific name for the leopard. Notice that the first letter of the genus is capitalized and the entire binomial is italicized. (Newly created scientific names are also "latinized": You can name an insect you discover after a friend, but you must add a Latin ending.) Many of the more than 11,000 binomials assigned by Linnaeus are still used today, including the optimistic name he gave our own species—*Homo sapiens*, meaning "wise man."

Hierarchical Classification

In addition to naming species, Linnaeus also grouped them into a hierarchy of increasingly inclusive categories. The first grouping is built into the binomial: Species that appear to be closely related are grouped into the same genus. For example, the leopard (*Panthera pardus*) belongs to a genus that also includes the African lion (*Panthera leo*), the tiger (*Panthera tigris*), and the jaguar (*Panthera onca*). Beyond genera, biologists employ progressively more comprehensive categories of classification. The classification system named after Linnaeus, the Linnaean system, places related genera in the same **family**, families into **orders**, orders into **classes**, classes into **phyla** (singular, *phylum*), phyla into **kingdoms**, and, more recently, kingdoms into **domains (Figure 26.3)**. The resulting biological classification of a particular organism is somewhat like a postal address identifying a person in a particular apartment, in a building with many apartments, on a street with many apartment buildings, in a city with many streets, and so on.

The named group at any level of the hierarchy is called a **taxon** (plural, *taxa*). In the leopard example, *Panthera* is a taxon at the genus level, and Mammalia is a taxon at the class level that includes all the many orders of mammals. Note that in the Linnaean system, taxa broader than the genus are not italicized, though they are capitalized.

Classifying species is a way to structure our human view of the world. We lump together various species of trees to which we give the common name of pines and distinguish them from other trees that we call firs. Systematists have decided that pines and firs are different enough to be placed in separate genera, yet similar enough to be grouped into the same family, Pinaceae. As with pines and firs, higher levels of classification are usually defined by particular characters chosen by systematists. However, characters that are useful for classifying one group of organisms may not be appropriate for other organisms. For this reason, the larger categories often are not comparable between lineages; that is, an order of snails does not exhibit the same degree of morphological or genetic diversity as an order of mammals. As we'll see, the placement of species into orders, classes, and so on also does not necessarily reflect evolutionary history.

▼ **Figure 26.2 Convergent evolution of limbless bodies.** A phylogeny based on DNA sequence data reveals that a legless body form evolved independently from legged ancestors in the lineages leading to glass lizards and to snakes.

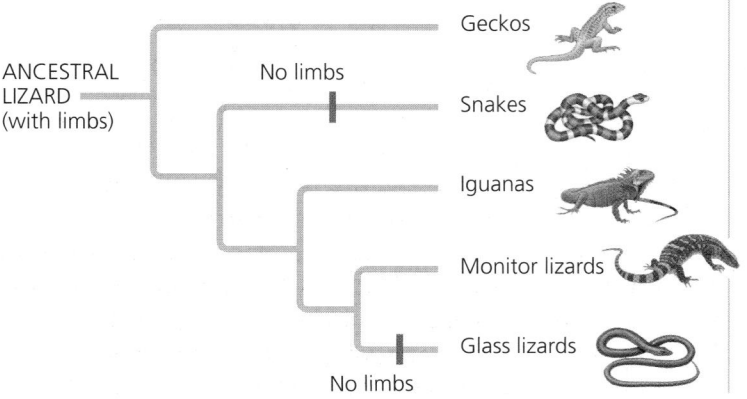

▼ **Figure 26.3 Linnaean classification.** At each level, or "rank," species are placed in groups within more inclusive groups.

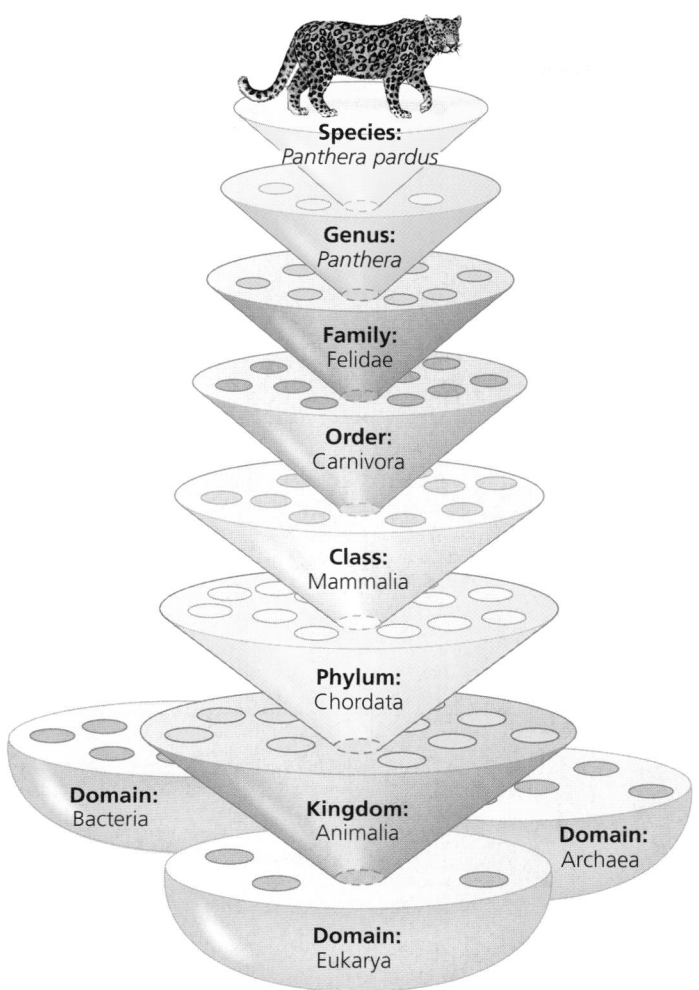

Species: *Panthera pardus*

Genus: *Panthera*

Family: Felidae

Order: Carnivora

Class: Mammalia

Phylum: Chordata

Domain: Bacteria

Kingdom: Animalia

Domain: Archaea

Domain: Eukarya

▼ **Figure 26.4 The connection between classification and phylogeny.** Hierarchical classification can reflect the branching patterns of phylogenetic trees. This tree shows evolutionary relationships between some of the taxa within order Carnivora, itself a branch of class Mammalia.

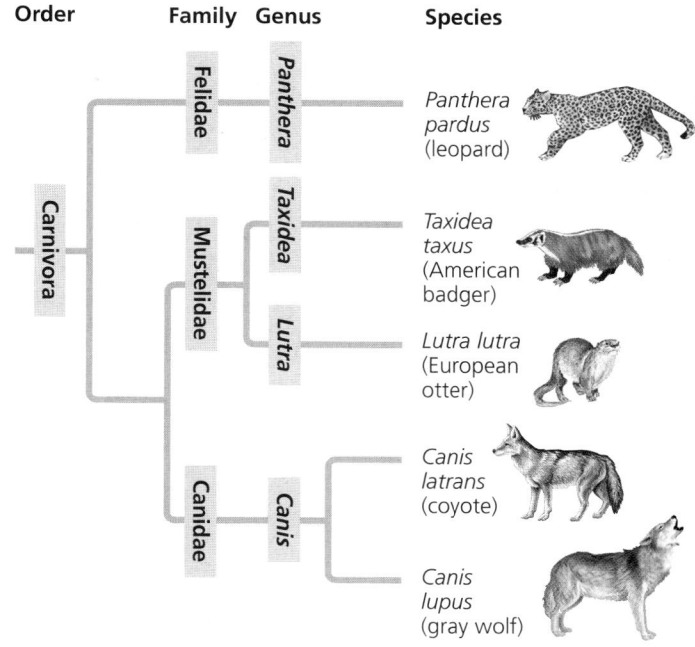

Order | Family | Genus | Species

Carnivora

Felidae — Panthera — *Panthera pardus* (leopard)

Mustelidae — Taxidea — *Taxidea taxus* (American badger)

Lutra — *Lutra lutra* (European otter)

Canidae — Canis — *Canis latrans* (coyote)

Canis lupus (gray wolf)

Linking Classification and Phylogeny

The evolutionary history of a group of organisms can be represented in a branching diagram called a **phylogenetic tree**. As in **Figure 26.4**, the branching pattern often matches how systematists have classified groups of organisms nested within more inclusive groups. Sometimes, however, systematists have placed a species within a genus (or other group) to which it is *not* most closely related. One reason for such a mistake might be that over the course of evolution, a species has lost a key feature shared by its close relatives. If DNA or other new evidence indicates that an organism has been misclassified, the organism may be reclassified to accurately reflect its evolutionary history. Another issue is that while the Linnaean system may distinguish groups, such as amphibians, mammals, reptiles, and other classes of vertebrates, it tells us nothing about these groups' evolutionary relationships to one another.

Such difficulties in aligning Linnaean classification with phylogeny have led many systematists to propose that classification be based entirely on evolutionary relationships. In such systems, names are assigned only to groups that include a common

ancestor and all of its descendants. Using this approach, some commonly recognized groups become part of other groups previously at the same level of the Linnaean system. For example, because birds evolved from a group of reptiles, Aves (the Linnaean class to which birds are assigned) is considered a subgroup of Reptilia (also a class in the Linnaean system).

What We Can and Cannot Learn from Phylogenetic Trees

Regardless of how groups are named, a phylogenetic tree represents a hypothesis about evolutionary relationships. These relationships often are depicted as a series of dichotomies, or two-way branch points. Each **branch point** represents the common ancestor of the two evolutionary lineages diverging from it. An **evolutionary lineage** is a sequence of ancestral organisms leading to a particular descendant taxon. **Figure 26.5** illustrates these elements of a phylogenetic tree and offers more tips on interpreting these diagrams.

Each tree in Figure 26.5 has a branch point that represents the common ancestor of chimpanzees and humans. Chimps and humans are an example of **sister taxa**, groups of organisms that share an immediate common ancestor that is not shared by any other group. Note that there is a sister group associated with each branch point in a tree (since each branch point represents the common ancestor of the lineages diverging from it).

The members of a sister group are each other's closest relatives, making sister groups a useful way to describe the

VISUALIZING PHYLOGENETIC RELATIONSHIPS

A phylogenetic tree visually represents a hypothesis of how a group of organisms are related. This figure explores how the way a tree is drawn conveys information.

→ **Instructors:** Additional questions related to this Visualizing Figure can be assigned in **Mastering Biology**.

Parts of a Phylogenetic Tree

This tree shows how the five groups of organisms at the tips of the branches, called taxa, are related. Each **branch point** represents the common ancestor of the evolutionary lineages diverging from it.

This branch point represents the common ancestor of all the animal groups shown in this tree.

Each horizontal branch represents an **evolutionary lineage**. The length of the branch is arbitrary unless the diagram specifies that branch lengths represent information such as time or amount of genetic change (see Figure 26.13).

Each position along a branch represents an ancestor in the lineage leading to the taxon named at the tip.

Chimps and humans are an example of **sister taxa**, groups of organisms that share a common ancestor that is not shared by any other group. The members of a sister group are each other's closest relatives.

1. According to this tree, which group or groups of organisms are most closely related to frogs?

2. Label the part of the diagram that represents the most recent common ancestor of frogs and humans.

Alternative Forms of Tree Diagrams

These diagrams are referred to as "trees" because they use the visual analogy of branches to represent evolutionary lineages diverging over time. In this text, trees are usually drawn horizontally, as shown above, but the same tree can be drawn vertically or diagonally without changing the relationships it conveys.

3. How many sister taxa are shown in these two trees? Identify them.

4. Redraw the horizontal tree in Figure 26.2 as a vertical tree and a diagonal tree.

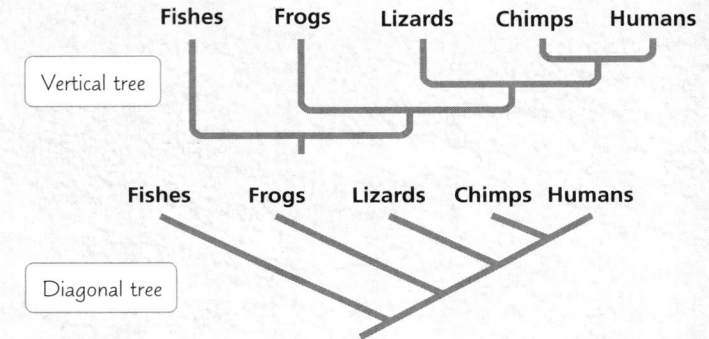

Vertical tree

Diagonal tree

Rotating Around Branch Points

Rotating the branches of a tree around a branch point does not change what they convey about evolutionary relationships. As a result, the order in which taxa appear at the branch tips is not significant. What matters is the branching pattern, which signifies the order in which the lineages have diverged from common ancestors.

Rotation

If you rotate the branches of the tree at left around the three blue points, the result is the tree at right.

Note: The order of the taxa does NOT represent a sequence of evolution "leading to" the last taxon shown (in this tree, humans).

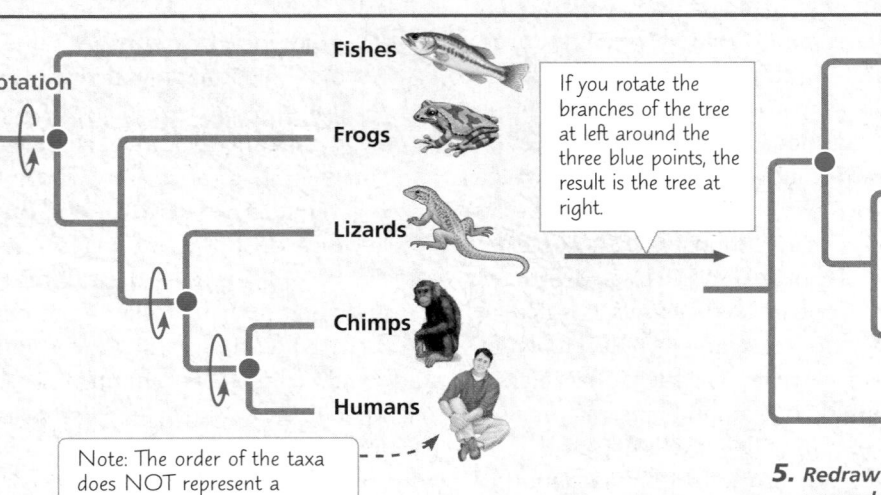

5. Redraw the tree above, rotating around the green branch point. Identify the two closest relatives of humans as shown in each of the three trees. Explain your answer.

evolutionary relationships shown in a tree. For example, in Figure 26.5, the evolutionary lineage leading to lizards shares an immediate common ancestor with the lineage leading to chimpanzees and humans. Thus, we can describe this portion of the tree by saying that lizards are the sister taxon to a group consisting of chimpanzees and humans.

As also discussed in Figure 26.5, the branches of a tree can be rotated around branch points without changing the relationships shown in the tree. That is, the order in which the taxa appear at the right side of the tree does not represent a *sequence* of evolution—in this case, it does not imply a sequence leading from fishes to humans.

This tree, like all of the phylogenetic trees in this book, is **rooted**, which means that a branch point within the tree (often drawn farthest to the left) represents the most recent common ancestor of all taxa in the tree. A lineage that diverges from all other members of its group early in the history of the group is called a **basal taxon**. Hence, like the fishes in Figure 26.5, a basal taxon lies on one of the two branches diverging from the branch point that represents the common ancestor of the group. In general, the lineages represented by these two branches originated at the same point in time and hence have been evolving for the same length of time.

What other key points do we need to keep in mind when interpreting phylogenetic trees? First, they are intended to show patterns of descent, not phenotypic similarity. Although closely related organisms often resemble one another due to their common ancestry, they may not if their lineages have evolved at different rates or faced very different environmental conditions. For example, even though crocodiles are more closely related to birds than to lizards (see Figure 22.17), they look more like lizards because morphology has changed dramatically in the bird lineage.

Second, we cannot necessarily infer the ages of the taxa or branch points shown in a tree. For example, the tree in Figure 26.5 does not indicate that chimpanzees evolved before humans. Rather, the tree shows only that chimpanzees and humans share a recent common ancestor, but we cannot tell when that ancestor lived or when the first chimpanzees or humans arose. Generally, unless we are given specific information about what the branch lengths in a tree mean—for example, that they are proportional to time—we should interpret the diagram solely in terms of patterns of descent. No assumptions should be made about when particular species evolved or how much change occurred in each lineage.

Third, we should not assume that a taxon on a phylogenetic tree evolved from the taxon next to it. Figure 26.5 does not indicate that humans evolved from chimpanzees or vice versa. We can infer only that the lineage leading to humans and the lineage leading to chimpanzees both evolved from a recent common ancestor. That ancestor, which is now extinct, was neither a human nor a chimpanzee.

Applying Phylogenies

Understanding phylogeny can have practical applications. Consider maize (corn), which originated in the Americas and is now an important food crop worldwide. From a phylogeny of maize based on DNA data, researchers have identified two species of wild grasses that may be maize's closest living relatives. These two close relatives may be useful as "reservoirs" of beneficial alleles that can be transferred to cultivated maize by cross-breeding or genetic engineering.

A different use of phylogenetic trees is to infer species identities by analyzing the relatedness of DNA sequences from different organisms. Researchers have used this approach to investigate whether "whale meat" had been harvested illegally from whale species protected under international law rather than from species that can be harvested legally **(Figure 26.6)**.

▼ **Figure 26.6** Inquiry

What is the species identity of food being sold as whale meat?

Experiment C. S. Baker and S. R. Palumbi purchased 13 samples of "whale meat" from Japanese fish markets. They sequenced part of the mitochondrial DNA (mtDNA) from each sample and compared their results with the comparable mtDNA sequence from known whale species. To infer the species identity of each sample, the team constructed a *gene tree*, a phylogenetic tree that shows patterns of relatedness among DNA sequences rather than among taxa.

Results Of the species in the resulting gene tree, only Minke whales caught in the Southern Hemisphere can be sold legally in Japan.

Minke (Southern Hemisphere) mtDNA
Unknown mtDNA #1a, 2, 3, 4, 5, 6, 7, 8

Minke (North Atlantic) mtDNA
Unknown mtDNA #9

Humpback mtDNA

Unknown mtDNA #1b

Blue mtDNA

Unknown mtDNA #10, 11, 12, 13

Fin mtDNA

Conclusion This analysis indicated that mtDNA sequences of six of the unknown samples (in red) were most closely related to mtDNA sequences of whales that are not legal to harvest.

Data from C. S. Baker and S. R. Palumbi, Which whales are hunted? A molecular genetic approach to monitoring whaling, *Science* 265:1538–1539 (1994). Reprinted with permission from AAAS.

WHAT IF? *What different results would have indicated that all of the whale meat had been harvested legally?*

How do researchers construct phylogenetic trees like those we've considered here? In the next section, we'll begin to answer that question by examining the data used to determine phylogenies.

CONCEPT CHECK 26.1

1. **VISUAL SKILLS** Based on the tree in Figure 26.4, are leopards more closely related to badgers or wolves, or are they equally related to these two species? Explain.

2. **VISUAL SKILLS** Which of the trees shown here depicts an evolutionary history different from the other two? Explain.

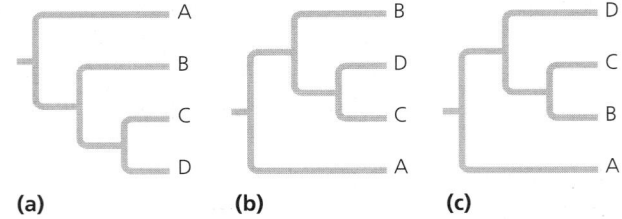

(a)   (b)   (c)

3. **DRAW IT** The bear family (Ursidae) is more closely related to the badger/otter family (Mustelidae) than to the dog family (Canidae). Use this information to redraw Figure 26.4.

For suggested answers, see Appendix A.

CONCEPT 26.2

Phylogenies are inferred from morphological and molecular data

To infer phylogeny, systematists must gather as much information as possible about the morphology, genes, and biochemistry of the relevant organisms. It is important to focus on features that result from common ancestry because only those features reflect evolutionary relationships.

Morphological and Molecular Homologies

Recall that phenotypic and genetic similarities due to shared ancestry are called **homologies**. For example, the similarity in the number and arrangement of bones in the forelimbs of mammals is due to their descent from a common ancestor with the same bone structure; this is an example of a morphological homology (see Figure 22.15). In the same way, genes or other DNA sequences are homologous if they are descended from sequences carried by a common ancestor.

In general, organisms that share very similar morphologies or similar DNA sequences are likely to be more closely related than organisms with vastly different structures or sequences. In some cases, however, the morphological divergence between related species can be great and their genetic divergence small (or vice versa). Consider the Hawaiian silversword plants: some of these species are tall, twiggy trees, while others are dense, ground-hugging shrubs (see Figure 25.23). But despite these striking phenotypic differences, the silverswords' genes are very similar. Based on these small molecular divergences, scientists estimate that the silversword group began to diverge 5 million years ago. We'll discuss how scientists use molecular data to estimate such divergence times later in this chapter.

Sorting Homology from Analogy

A potential source of confusion in constructing a phylogeny is similarity between organisms that is due to convergent evolution—called **analogy**—rather than to shared ancestry (homology). Convergent evolution occurs when similar environmental pressures and natural selection produce similar (analogous) adaptations in organisms from different evolutionary lineages. For example, the two mole-like animals shown in **Figure 26.7** look very similar. However, their internal anatomy, physiology, and reproductive systems are very dissimilar. Indeed, genetic and fossil evidence indicate that the common ancestor of these animals lived 160 million years ago. This common ancestor and most of its descendants were not mole-like. It appears that analogous characteristics evolved independently in these two lineages as they became adapted to similar lifestyles—hence, the similar features of these animals should not be considered when reconstructing their phylogeny.

Another clue to distinguishing between homology and analogy is the complexity of the characters being compared. The more elements that are similar in two complex structures, the more likely it is that the structures evolved from a common ancestor. For instance, the skulls of an adult human and an adult chimpanzee both consist of many bones fused together. The compositions of the skulls match almost perfectly, bone for bone. It is highly improbable that such complex structures, matching in so many details, have separate origins. More likely, the genes involved in the development of both skulls were inherited from a common ancestor.

The same argument applies to comparisons at the gene level. Genes are sequences of thousands of nucleotides, each of which represents an inherited character in the form of one of the four DNA bases: A (adenine), G (guanine), C (cytosine), or T (thymine). If genes in two organisms share many portions of their nucleotide sequences, it is likely that the genes are homologous.

▶ **Figure 26.7 Convergent evolution in burrowers.** A long body, large front paws, small eyes, and a pad of thick skin that protects the nose all evolved independently in these species.

Australian "mole"

African golden mole

Evaluating Molecular Homologies

Comparing DNA molecules often poses technical challenges for researchers. The first step after sequencing the DNA is to align comparable sequences from the species being studied. If the species are very closely related, the sequences probably differ at only one or a few sites. In contrast, comparable nucleic acid sequences in distantly related species usually have different bases at many sites and may have different lengths. This is because insertions and deletions accumulate over long periods of time.

Suppose, for example, that certain noncoding DNA sequences near a particular gene are very similar in two species, except that the first base of the sequence has been deleted in one of the species. The effect is that the remaining sequence shifts back one notch. A comparison of the two sequences that does not take this deletion into account would overlook what in fact is a very good match. Computer programs can be used to identify such matches by testing possible alignments for comparable DNA segments of differing lengths (**Figure 26.8**).

Such molecular comparisons reveal that many base substitutions and other differences have accumulated in the comparable genes of an Australian "mole" and a golden mole. The many differences indicate that their lineages have diverged greatly since their common ancestor; thus, we say that the living species are not closely related. In contrast, the high degree of gene sequence similarity among the silversword

▼ **Figure 26.8 Aligning segments of DNA.** Systematists search for similar sequences along DNA segments from two species (only one DNA strand is shown for each species). In this example, 11 of the original 12 bases have not changed since the species diverged. Hence, those portions of the sequences still align once the length is adjusted.

1 These homologous DNA sequences are identical as species 1 and species 2 begin to diverge from their common ancestor.

1 C C A T C A G A G T C C
2 C C A T C A G A G T C C

2 Deletion and insertion mutations shift what had been matching sequences in the two species.

Deletion

1 C C A T C A (G) A G T C C
2 C C A T C A G A G T C C

(G T A) Insertion

3 Of the regions of the species 2 sequence that match the species 1 sequence, those shaded orange no longer align because of these mutations.

1 C C A T C A A G T C C
2 C C A T G T A C A G A G T C C

4 The matching regions realign after a computer program adds gaps in sequence 1.

1 C C A T _ _ _ C A _ A G T C C
2 C C A T G T A C A G A G T C C

▼ **Figure 26.9 A coincidental match.** These two DNA sequences from distantly related species coincidentally share 23% of their bases (those highlighted in yellow).

A C G G A T A G T C C A C T A G G C A C T A
T C A C C G A C A G G T C T T T G A C T A G

WHAT IF? *Why might you expect organisms that are not closely related to nevertheless share roughly 25% of their bases?*

plants indicates that they are all very closely related, in spite of their considerable morphological differences.

Just as with morphological characters, it is necessary to distinguish homology from analogy in evaluating molecular similarities for evolutionary studies. Two sequences that resemble each other at many points along their length most likely are homologous (see Figure 26.8). But in organisms that do not appear to be closely related, the bases that their otherwise very different sequences happen to share may simply be coincidental matches (**Figure 26.9**). Scientists have developed statistical tools that can help distinguish "distant" homologies from such coincidental matches in extremely divergent sequences.

CONCEPT CHECK 26.2

1. Decide whether each of the following pairs of structures more likely represents analogy or homology, and explain your reasoning: (a) a porcupine's quills and a cactus's spines; (b) a cat's paw and a human's hand; (c) an owl's wing and a hornet's wing.
2. **WHAT IF?** Suppose that two species, A and B, have similar appearances but very divergent gene sequences, while species B and C have very different appearances but similar gene sequences. Which pair of species is more likely to be closely related: A and B or B and C? Explain.

For suggested answers, see Appendix A.

CONCEPT 26.3

Shared characters are used to construct phylogenetic trees

As we've discussed, a key step in reconstructing phylogenies is to distinguish homologous features from analogous ones (since only homology reflects evolutionary history). Next, we'll describe cladistics, a widely used set of methods for inferring phylogeny.

Cladistics

In the approach to systematics called **cladistics**, common ancestry is the primary criterion used to classify organisms. Using this methodology, biologists attempt to place species into groups called **clades**, each of which includes an ancestral species and all of its descendants. Clades, like the categories of the Linnaean system, are nested within larger clades. In

▼ Figure 26.10 Monophyletic, paraphyletic, and polyphyletic groups.

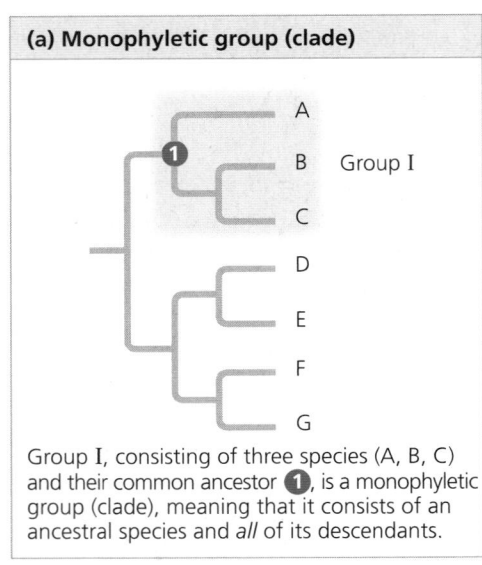

(a) Monophyletic group (clade)

Group I, consisting of three species (A, B, C) and their common ancestor **1**, is a monophyletic group (clade), meaning that it consists of an ancestral species and *all* of its descendants.

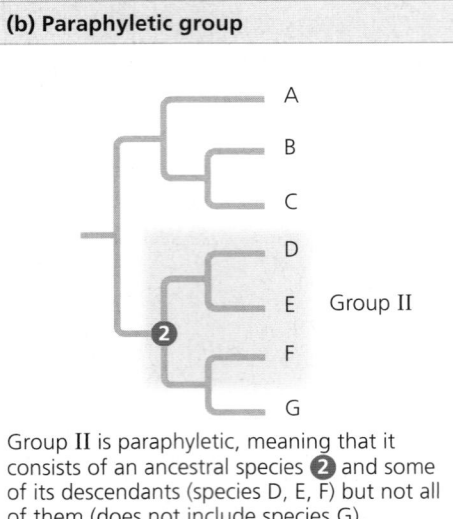

(b) Paraphyletic group

Group II is paraphyletic, meaning that it consists of an ancestral species **2** and some of its descendants (species D, E, F) but not all of them (does not include species G).

(c) Polyphyletic group

Group III, consisting of four species (A, B, C, D), is polyphyletic, meaning that the most recent common ancestor **3** of its members is *not* part of the group.

Figure 26.4, for example, the cat group (Felidae) represents a clade within a larger clade (Carnivora) that also includes the dog group (Canidae).

A taxon is equivalent to a clade only if it is **monophyletic** (from the Greek, meaning "single tribe"), signifying that it consists of an ancestral species and all of its descendants **(Figure 26.10a)**. Contrast this with a **paraphyletic** ("beside the tribe") group, which consists of an ancestral species and some, but not all, of its descendants **(Figure 26.10b)**, or a **polyphyletic** ("many tribes") group, which includes distantly related species but does not include their most recent common ancestor **(Figure 26.10c)**.

Note that in a paraphyletic group, the most recent common ancestor of all members of the group *is* part of the group, whereas in a polyphyletic group the most recent common ancestor *is not* part of the group. For example, a group consisting of even-toed ungulates (hippopotamuses, deer, and their relatives) and their common ancestor is paraphyletic because it does not include cetaceans (whales, dolphins, and porpoises), which descended from that ancestor **(Figure 26.11)**. In contrast, a group consisting of seals and cetaceans (based on their similar body forms) is polyphyletic because it does not include the common ancestor of seals and cetaceans. Biologists avoid defining such polyphyletic groups; if new evidence indicates that an existing group is polyphyletic, its members are reclassified.

Shared Ancestral and Shared Derived Characters

As a result of descent with modification, organisms have characters they share with their ancestors, and they also have characters that differ from those of their ancestors. For example, all mammals have backbones, but a backbone does not distinguish mammals from other vertebrates because *all* vertebrates have backbones. The backbone predates the branching of mammals from other vertebrates. Thus for mammals, the backbone is a **shared ancestral character**, a character that originated in

an ancestor of the taxon. In contrast, hair is a character shared by all mammals but *not* found in their ancestors. Thus, in mammals, hair is considered a **shared derived character**, an evolutionary novelty unique to a clade.

▼ Figure 26.11 Paraphyletic versus polyphyletic groups.

This group is paraphyletic because it does not include all the descendants of the common ancestor (it excludes cetaceans).

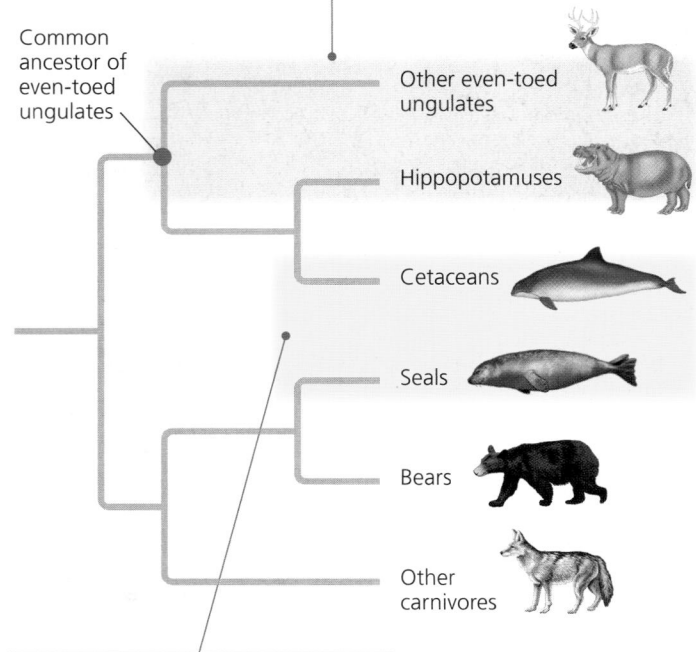

Common ancestor of even-toed ungulates

Other even-toed ungulates

Hippopotamuses

Cetaceans

Seals

Bears

Other carnivores

This group is polyphyletic because it does not include the most recent common ancestor of its members.

DRAW IT *Circle the branch point that represents the most recent common ancestor of cetaceans and seals. Explain why that ancestor would not be part of a cetacean–seal group defined by similar body form.*

Note that a shared derived character can refer to the loss of a feature, such as the loss of limbs in snakes or whales. Also, keep in mind that it is a relative matter whether a character is considered ancestral or derived. A backbone can also qualify as a shared derived character, but only at a deeper branch point that distinguishes all vertebrates from other animals.

Inferring Phylogenies Using Derived Characters

Shared derived characters are unique to particular clades. Because all features of organisms arose at some point in the history of life, it should be possible to determine the clade in which each shared derived character first appeared and to use that information to infer evolutionary relationships.

To give an example of this approach, consider the set of characters shown in **Figure 26.12a** for each of five vertebrates—a leopard, turtle, frog, bass, and lamprey (a jawless aquatic vertebrate). As a basis of comparison, we need to select an outgroup. An **outgroup** is a species or group of species from an evolutionary lineage that is closely related to but not part of the group of species that we are studying (the **ingroup**). A suitable outgroup can be determined based on evidence from morphology, paleontology, embryonic development, and gene sequences. An appropriate outgroup for our example is the lancelet, a small animal that lives in mudflats and (like vertebrates) is a member of the more inclusive group called the chordates. Unlike the vertebrates, however, the lancelet does not have a backbone.

In our analysis, a character found in both the outgroup and the ingroup is assumed to be ancestral. We'll also assume that each derived character in Figure 26.12a arose only once

in the ingroup. Thus, for a character that only occurs in a subset of the ingroup, we'll assume that the character arose in the lineage leading to those members of the ingroup.

By comparing members of the ingroup with each other and with the outgroup, we can determine which characters were derived at the various branch points of vertebrate evolution. In our example, *all* of the vertebrates in the ingroup have backbones: This character was present in the ancestral vertebrate, but not in the outgroup. Now note that hinged jaws are absent in the outgroup and in lampreys, but present in all other members of the ingroup. This indicates that hinged jaws arose in a lineage leading to all members of the ingroup *except* lampreys. Hence, we can conclude that lampreys are the sister taxon to the other vertebrates in the ingroup. Proceeding in this way, we can translate the data in our table of characters into a phylogenetic tree that places all the ingroup taxa into a hierarchy based on their derived characters **(Figure 26.12b)**.

Phylogenetic Trees with Proportional Branch Lengths

In the phylogenetic trees we have presented so far, the lengths of the tree's branches do not indicate the degree of evolutionary change in each lineage. Furthermore, the chronology represented by the branching pattern of the tree is relative (earlier versus later) rather than absolute (how many millions of years ago). But in some tree diagrams, branch lengths are proportional to the amount of evolutionary change or to the times at which particular events occurred.

▼ **Figure 26.12 Using derived characters to infer phylogeny.** The derived characters used here include the amnion, a membrane that encloses the embryo inside a fluid-filled sac (see Figure 34.26). Note that a different set of characters could lead us to infer a different phylogenetic tree.

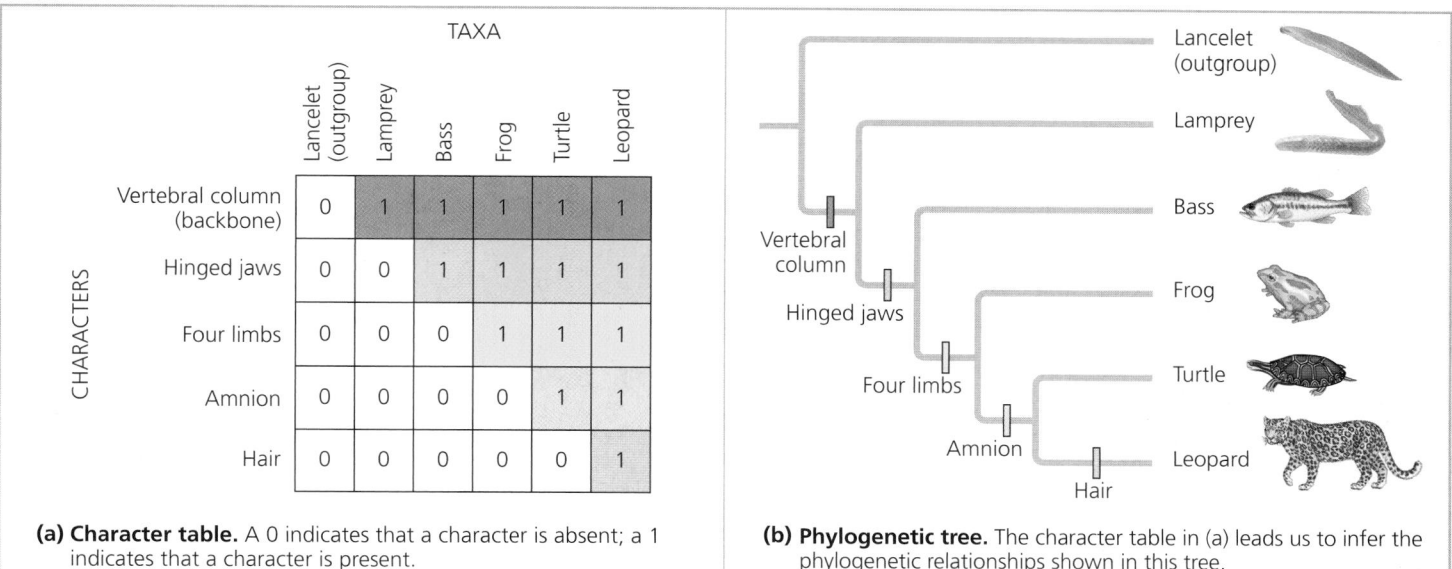

(a) **Character table.** A 0 indicates that a character is absent; a 1 indicates that a character is present.

	Lancelet (outgroup)	Lamprey	Bass	Frog	Turtle	Leopard
Vertebral column (backbone)	0	1	1	1	1	1
Hinged jaws	0	0	1	1	1	1
Four limbs	0	0	0	1	1	1
Amnion	0	0	0	0	1	1
Hair	0	0	0	0	0	1

(b) **Phylogenetic tree.** The character table in (a) leads us to infer the phylogenetic relationships shown in this tree.

DRAW IT *In (b), circle the most inclusive clade for which a hinged jaw is a shared ancestral character.*

→ **Mastering Biology** Figure Walkthrough

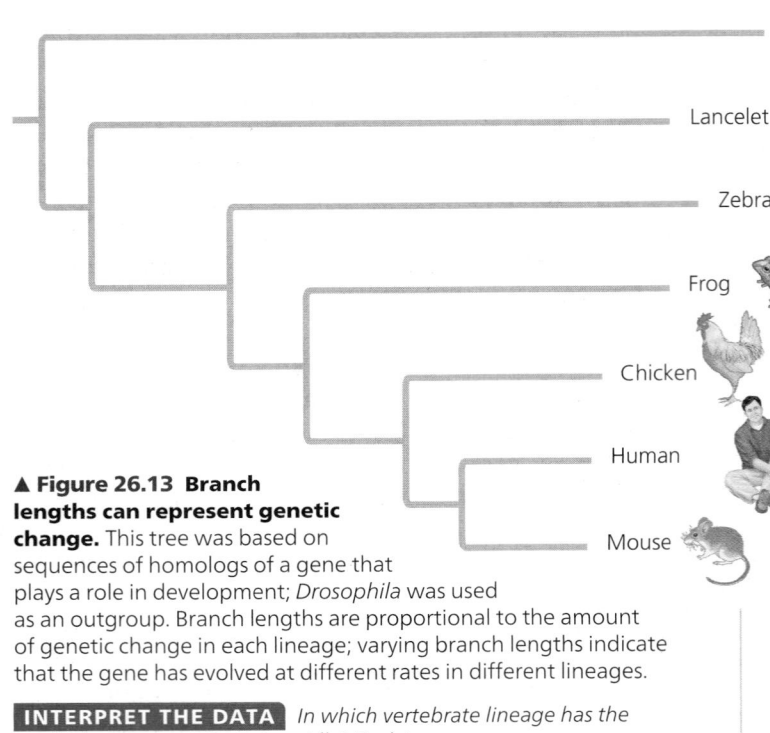

▲ **Figure 26.13 Branch lengths can represent genetic change.** This tree was based on sequences of homologs of a gene that plays a role in development; *Drosophila* was used as an outgroup. Branch lengths are proportional to the amount of genetic change in each lineage; varying branch lengths indicate that the gene has evolved at different rates in different lineages.

INTERPRET THE DATA *In which vertebrate lineage has the studied gene evolved most rapidly? Explain.*

In **Figure 26.13**, for example, each branch length of the phylogenetic tree reflects the number of changes that have taken place in a particular DNA sequence in that lineage. Note that the total length of the horizontal lines from the base of the tree to the mouse is less than that of the line leading to the fruit fly, *Drosophila*. This indicates that fewer genetic changes have occurred in the mouse lineage than in the *Drosophila* lineage. In addition, because the mouse and fly lineages have evolved for an equal amount of time after they diverged from a common ancestor, the *rate* of change was lower in mouse lineage than in the *Drosophila* lineage.

In general, although the branches of a phylogenetic tree may have different lengths, among organisms alive today, all the different lineages that descend from a common ancestor have survived for the same number of years. To take an extreme example, humans and bacteria had a common ancestor that lived over 3 billion years ago. Fossils and genetic evidence indicate that this ancestor was a single-celled prokaryote. Even though bacteria have apparently changed little in their morphology since that common ancestor, there have nonetheless been 3 billion years of evolution in the bacterial lineage, just as there have been 3 billion years of evolution in the lineage that ultimately gave rise to humans.

These equal spans of chronological time can be represented in a phylogenetic tree whose branch lengths are proportional to time **(Figure 26.14)**. Such a tree draws on fossil data to place branch points in the context of geologic time. Additionally, it is possible to combine these two types of trees by labeling branch points with information about rates of genetic change or dates of divergence.

Maximum Parsimony and Maximum Likelihood

As the database of DNA sequences that enables us to study more species grows, the difficulty of building the phylogenetic tree that best describes their evolutionary history also grows. What if you are analyzing data for 50 species? There are 3×10^{76} different ways to arrange 50 species into a tree! And which tree in this huge forest reflects the true phylogeny? Systematists can never be sure of finding the most accurate tree in such a large data set, but they can narrow the possibilities by applying the principles of maximum parsimony and maximum likelihood.

According to the principle of **maximum parsimony**, we should first investigate the simplest explanation that is consistent with the facts. (The parsimony principle is also called "Occam's razor" after William of Occam, a 14th-century English philosopher who advocated solving problems by "shaving away" unnecessary complications.) In the case of trees based on morphology, the most parsimonious tree requires the fewest evolutionary events, as measured by the origin of shared derived morphological

▼ **Figure 26.14 Branch lengths can indicate time.** This tree is based on the same DNA data as that in Figure 26.13, but here the branch points are dated based on fossil evidence. Thus, the branch lengths are proportional to time. Each lineage has the same total length from the base of the tree to the branch tip, indicating that all the lineages have diverged from the common ancestor for equal amounts of time.

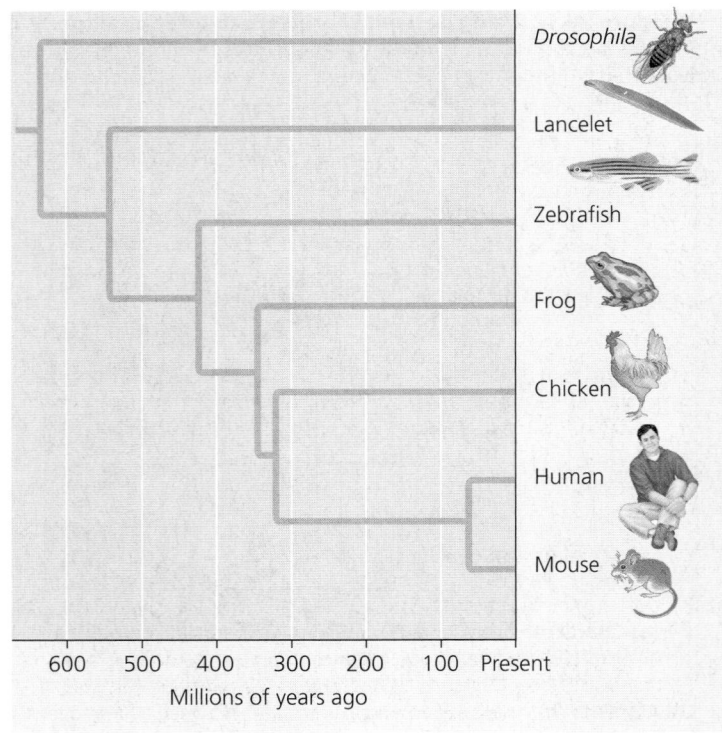

Millions of years ago

characters. For phylogenies based on DNA, the most parsimonious tree requires the fewest base changes.

A **maximum likelihood** approach identifies the tree most likely to have produced a given set of DNA data based on certain probability rules about how DNA sequences change over time. For example, the underlying probability rules could be based on the assumption that all nucleotide substitutions are equally likely. If evidence suggests that this assumption is not correct, more complex rules could

be devised to account for different rates of change among different nucleotides or at different positions in a gene.

Scientists have developed many computer programs to search for trees that are parsimonious and likely. When a large amount of accurate data is available, the methods used in these programs usually yield similar trees. As an example of one method, **Figure 26.15** walks you through the process of identifying the most parsimonious molecular tree for a three-species problem. Computer programs use the

---

▼ **Figure 26.15** **Research Method**

Applying Parsimony to a Problem in Molecular Systematics

Application In considering possible phylogenies for a group of species, systematists compare molecular data for the species. An efficient way to begin is by identifying the most parsimonious hypothesis—the one that requires the fewest evolutionary events (molecular changes) to have occurred.

Technique Follow the numbered steps as we apply the principle of parsimony to a hypothetical phylogenetic problem involving three closely related beetle species.

① First, draw the three possible trees for the species. (Although only three trees are possible when ordering three species, the number of possible trees increases rapidly with the number of species: There are 15 trees for four species and 34,459,425 trees for 10 species.)

Three phylogenetic hypotheses:

② Tabulate the molecular data for the species. In this simplified example, the data represent a DNA sequence consisting of just four nucleotide bases. Data from several outgroup species (not shown) were used to infer the ancestral DNA sequence.

	Site 1	2	3	4
Species I	C	T	A	T
Species II	C	T	T	C
Species III	A	G	A	C
Ancestral sequence	A	G	T	T

③ Now focus on site 1 in the DNA sequence. In the tree on the left, a single base-change event, represented by the purple hatch mark on the branch leading to species I and II (and labeled 1/C, indicating a change at site 1 to nucleotide C), is sufficient to account for the site 1 data. In the other two trees, two base-change events are necessary.

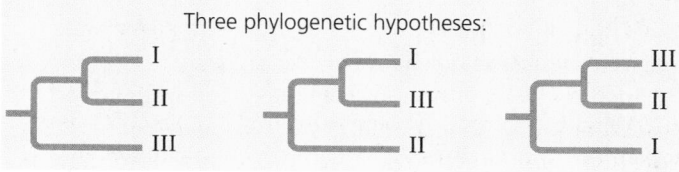

④ Continuing the comparison of bases at sites 2, 3, and 4 reveals that each of the three trees requires a total of five additional base-change events (purple hatch marks).

Results To identify the most parsimonious tree, we total all of the base-change events noted in steps 3 and 4. We conclude that the first tree is the most parsimonious of the three possible phylogenies. (In a real example, many more sites would be analyzed. Hence, the trees would often differ by more than one base-change event.)

6 events   7 events   7 events

principle of parsimony to estimate phylogenies in a similar way: They examine large numbers of possible trees and identify those that require the fewest evolutionary changes.

Phylogenetic Trees as Hypotheses

This is a good place to reiterate that any phylogenetic tree represents a hypothesis about how the organisms in the tree are related to one another. The best hypothesis is the one that best fits all the available data. A phylogenetic hypothesis may be modified when new evidence compels systematists to revise their trees. Indeed, while many older phylogenetic hypotheses have been supported by new morphological and molecular data, others have been changed or rejected.

Thinking of phylogenies as hypotheses also allows us to use them in a powerful way: We can make and test predictions based on the assumption that a particular phylogeny—our hypothesis—is correct. For example, in an approach known as *phylogenetic bracketing*, we can predict (by parsimony) that features shared by two groups of closely related organisms are present in their common ancestor and all of its descendants unless independent data indicate otherwise. (Note that "prediction" can refer to unknown past events as well as to evolutionary changes yet to occur.)

This approach has been used to make novel predictions about dinosaurs. For example, there is evidence that birds descended from the theropods, a group of bipedal dinosaurs. As seen in **Figure 26.16**, the closest living relatives of birds are crocodiles. Birds and crocodiles share numerous features: They have four-chambered hearts, they "sing" to defend territories and attract mates (although a crocodile's "song" is more like a bellow), and they build nests. Both birds and crocodiles also care for their eggs by warming them. Birds warm their eggs by sitting on them, whereas crocodiles cover

their eggs with their neck. Reasoning that any feature shared by birds and crocodiles is likely to have been present in their common ancestor (denoted by the blue dot in Figure 26.16) and *all* of its descendants, biologists predicted that dinosaurs had four-chambered hearts, sang, built nests, and cared for their eggs.

Internal organs, such as the heart, rarely fossilize, and it is, of course, difficult to test whether dinosaurs sang to defend territories and attract mates. However, fossil discoveries have supported the prediction that dinosaurs built nests and cared for their eggs. First, a fossil embryo of an *Oviraptor* dinosaur was found, still inside its egg. This egg was identical to those found in another fossil, one that showed an *Oviraptor* crouching over a group of eggs in a posture similar to how birds sit on their nests today **(Figure 26.17)**. Researchers suggested that the *Oviraptor* dinosaur preserved in this second fossil died while incubating or protecting its eggs. The broader conclusion that emerged from this work—that dinosaurs built nests and cared for their eggs—has since been strengthened by additional fossil discoveries that show that other species

▼ **Figure 26.17 Fossil support for a phylogenetic prediction: Dinosaurs built nests and cared for their eggs.**

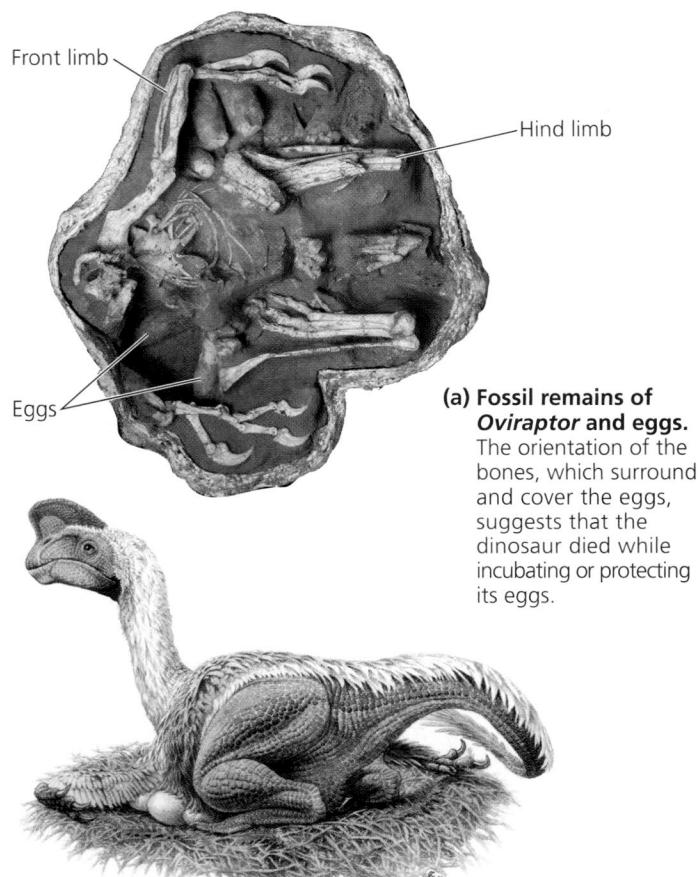

(a) Fossil remains of *Oviraptor* and eggs. The orientation of the bones, which surround and cover the eggs, suggests that the dinosaur died while incubating or protecting its eggs.

Front limb

Hind limb

Eggs

▼ **Figure 26.16 A phylogenetic tree of birds and their close relatives.** (A dagger, †, indicates extinct lineages.)

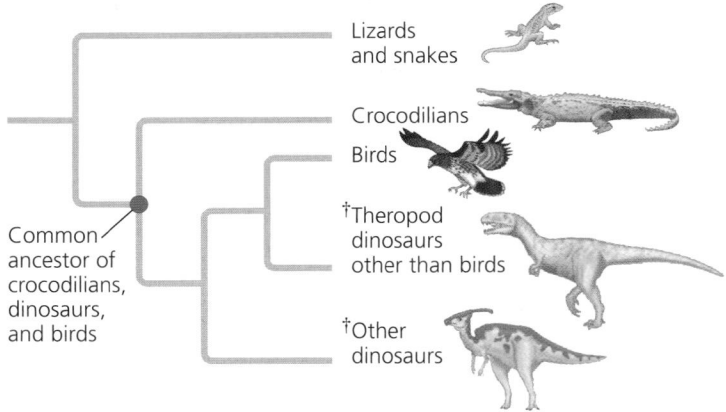

Lizards and snakes

Crocodilians

Birds

†Theropod dinosaurs other than birds

†Other dinosaurs

Common ancestor of crocodilians, dinosaurs, and birds

VISUAL SKILLS *In this tree diagram, what is the sister taxon of the clade that includes dinosaurs and their most recent common ancestor? Explain.*

(b) Artist's reconstruction of the dinosaur's posture on its nest based on the fossil findings.

of dinosaurs built nests and sat on their eggs. Finally, by supporting predictions based on the phylogenetic hypothesis shown in Figure 26.16, fossil discoveries of nests and parental care in dinosaurs provide independent data that suggest that the hypothesis is correct.

CONCEPT CHECK 26.3

1. To distinguish a particular clade of mammals within the larger clade that corresponds to class Mammalia, would hair be a useful character? Why or why not?
2. The most parsimonious tree of evolutionary relationships can be inaccurate. How can this occur?
3. **WHAT IF?** Draw a phylogenetic tree that includes the relationships from Figure 25.7 and Figure 26.16. Traditionally, all the taxa shown besides birds and mammals were classified as reptiles. Would a cladistic approach support that classification? Explain.

For suggested answers, see Appendix A.

CONCEPT 26.4

An organism's evolutionary history is documented in its genome

As you have seen in this chapter, comparisons of nucleic acids or other molecules can be used to deduce relatedness. In some cases, such comparisons can reveal phylogenetic relationships that cannot be determined by nonmolecular methods such as comparative anatomy. For example, the analysis of molecular data helps us uncover evolutionary relationships between groups that have little common ground for morphological comparison, such as animals and fungi. And molecular methods allow us to reconstruct phylogenies among groups of present-day organisms for which the fossil record is poor or lacking entirely.

Different genes can evolve at different rates, even in the same evolutionary lineage. As a result, molecular trees can represent short or long periods of time, depending on which genes are used. For example, the DNA that codes for ribosomal RNA (rRNA) changes relatively slowly. Therefore, comparisons of DNA sequences in these genes are useful for investigating relationships between taxa that diverged hundreds of millions of years ago. Studies of rRNA sequences indicate, for instance, that fungi are more closely related to animals than to plants. In contrast, mitochondrial DNA (mtDNA) evolves relatively rapidly and can be used to explore recent evolutionary events. One research team has traced the relationships among Native American groups through their mtDNA sequences. The molecular findings corroborate other evidence that the Pima of Arizona, the Maya of Mexico, and the Yanomami of Venezuela are closely related, probably descending from the first of three waves of immigrants that crossed the Bering Land Bridge from Asia to the Americas about 15,000 years ago.

Gene Duplications and Gene Families

What do molecular data reveal about the evolutionary history of genome change? Consider gene duplication, which can play an important role in evolution because it increases the number of genes in the genome, providing more opportunities for further evolutionary changes. Molecular techniques now allow us to trace the phylogenies of gene duplications. These molecular phylogenies must account for repeated duplications that have resulted in *gene families*, groups of related genes within an organism's genome (see Figure 21.10).

Accounting for such duplications leads us to distinguish two types of homologous genes **(Figure 26.18)**: orthologous genes and paralogous genes. In **orthologous genes** from the Greek *orthos*, exact), the homology is the result of a speciation event

▼ **Figure 26.18 Two types of homologous genes.** Colored bands mark regions of the genes where differences in base sequences have accumulated.

(a) Formation of orthologous genes: a product of speciation

Ancestral gene

Ancestral species

Speciation with divergence of gene

Orthologous genes

Species A   Species B

(b) Formation of paralogous genes: within a species

Ancestral gene

Species C

Gene duplication and divergence

Paralogous genes

Species C after many generations

and hence occurs between genes found in different species (see Figure 26.18a). For example, the genes that code for cytochrome *c* (a protein that functions in electron transport chains) in humans and dogs are orthologous. In **paralogous genes** (from the Greek *para*, in parallel), the homology results from gene duplication; hence, multiple copies of these genes have diverged from one another within a species (see Figure 26.18b). In Concept 23.1, you encountered the example of olfactory receptor genes, which have undergone many gene duplications in vertebrates; humans have 380 functional copies of these paralogous genes, while mice have 1,200.

Note that orthologous genes can only diverge after speciation has taken place, that is, after the genes are found in separate gene pools. Thus, differences among orthologous genes reflect the history of speciation events, making them well-suited for inferring phylogeny. For example, although the cytochrome *c* genes in humans and dogs serve the same function, the gene's sequence in humans has diverged from that in dogs in the time since these species last shared a common ancestor. Paralogous genes, on the other hand, can diverge within a species because they are present in more than one copy in the genome. The paralogous genes that make up the olfactory receptor gene family in humans have diverged from each other during our long evolutionary history. They now specify proteins that confer sensitivity to a wide variety of molecules, ranging from food odors to sex pheromones.

Genome Evolution

Now that we can compare the entire genomes of different organisms, including our own, two patterns have emerged. First, lineages that diverged long ago often share many orthologous genes. For example, though the human and mouse lineages diverged about 65 million years ago, 99% of the genes of humans and mice are orthologous. And 50% of human genes are orthologous with those of yeast, despite 1 billion years of divergent evolution. Such commonalities explain why disparate organisms nevertheless share many biochemical and developmental pathways. As a result of these shared pathways, the functioning of genes linked to diseases in humans can often be investigated by studying yeast and other organisms distantly related to humans.

Second, the number of genes a species has doesn't seem to increase through duplication at the same rate as perceived phenotypic complexity. Humans have only about four times as many genes as yeast, a single-celled eukaryote, even though—unlike yeast—we have a large, complex brain and a body with more than 200 different types of tissues. Evidence is emerging that many human genes are more versatile than those of yeast: A single human gene can encode multiple proteins that perform different tasks in various body tissues. Unraveling the mechanisms that cause this genomic versatility and phenotypic variation is an exciting challenge.

CONCEPT CHECK 26.4

1. Explain how comparing proteins of two species can yield data about the species' evolutionary relationship.
2. Distinguish orthologous from paralogous genes. Which of these two types of genes should be used to infer phylogeny? Explain your answer.
3. **MAKE CONNECTIONS** Review Figure 18.12, then suggest how a particular gene could have different functions in different tissues within an organism.

For suggested answers, see Appendix A.

CONCEPT 26.5

Molecular clocks help track evolutionary time

One goal of evolutionary biology is to understand the relationships among all organisms. It is also helpful to know when lineages diverged from one another, including those for which there is no fossil record. But how can we determine the timing of phylogenies that extend beyond the fossil record?

Molecular Clocks

We stated earlier that researchers have estimated that the common ancestor of Hawaiian silversword plants lived about 5 million years ago. How did they make this estimate? They relied on the concept of a **molecular clock**, an approach for measuring the absolute time of evolutionary change based on the observation that some genes and other regions of genomes appear to evolve at constant rates. An assumption underlying the molecular clock is that the number of nucleotide substitutions in orthologous genes is proportional to the time that has elapsed since the genes branched from their common ancestor. In the case of paralogous genes, the number of substitutions is proportional to the time since the ancestral gene was duplicated.

We can calibrate the molecular clock of a gene that has a reliable average rate of evolution by graphing the number of genetic differences—for example, nucleotide, codon, or amino acid differences—against the dates of evolutionary branch points that are known from the fossil record **(Figure 26.19)**. The average rates of genetic change inferred from such graphs can then be used to estimate the dates of events that cannot be discerned from the fossil record, such as the origin of the silverswords discussed earlier.

Of course, no gene marks time with complete precision. In fact, some portions of the genome appear to have evolved in irregular bursts that are not at all clocklike. And even those genes that seem to act as reliable molecular clocks are accurate only in the statistical sense of showing a fairly smooth *average* rate of change. Over time, there may still be deviations from that average rate. Furthermore, the same gene may evolve at

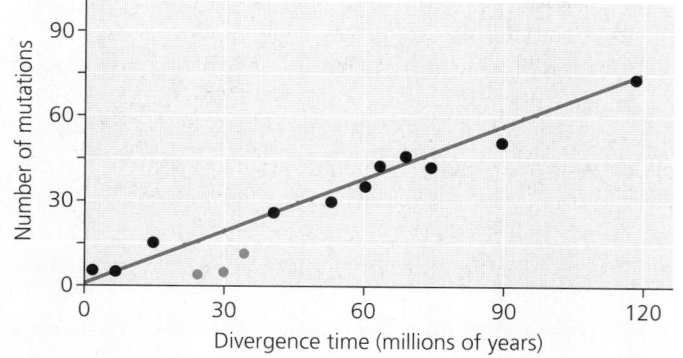

▼ **Figure 26.19 A molecular clock for mammals.** The number of accumulated mutations in seven well-studied proteins has increased over time in a consistent manner for most mammal species. The three green data points represent primate species, whose proteins appear to have evolved more slowly than those of other mammals. The divergence time for each data point was based on fossil evidence.

INTERPRET THE DATA *Use the graph to estimate the divergence time for a mammal with a total of 30 mutations in the seven proteins.*

different rates in different groups of organisms. Finally, when comparing genes that are clocklike, the rate of the clock may vary greatly from one gene to another; some genes evolve a million times faster than others.

Differences in Clock Speed

What causes such differences in the speed at which clocklike genes evolve? The answer stems from the fact that some mutations are selectively neutral—neither beneficial nor detrimental. Of course, many new mutations are harmful and are removed quickly by selection. But if most of the rest are neutral and have little or no effect on fitness, then the rate of evolution of those neutral mutations should indeed be regular, like a clock. Differences in the clock rate for different genes are related to how important a gene is. If the exact sequence of amino acids that a gene specifies is essential to survival, most of the mutational changes will be harmful and only a few will be neutral. As a result, such genes change only slowly. But if the exact sequence of amino acids is less critical, fewer of the new mutations will be harmful and more will be neutral. Such genes change more quickly.

Potential Problems with Molecular Clocks

As we've seen, molecular clocks do not run as smoothly as would be expected if the underlying mutations were selectively neutral. Many irregularities are likely to be the result of natural selection in which certain DNA changes are favored over others. Indeed, evidence suggests that almost half the amino acid differences in proteins of two *Drosophila* species, *D. simulans* and *D. yakuba*, are not neutral but have resulted from natural selection. But because the direction of natural

selection may change repeatedly over long periods of time (and hence may average out), some genes experiencing selection can nevertheless serve as approximate markers of elapsed time.

Another question arises when researchers attempt to extend molecular clocks beyond the time span documented by the fossil record. Although an abundant fossil record extends back only about 550 million years, molecular clocks have been used to date evolutionary divergences that occurred a billion or more years ago. These estimates assume that the clocks have been constant for all that time. Such estimates are highly uncertain.

In some cases, problems may be avoided by calibrating molecular clocks with data on the rates at which genes have evolved in different taxa. In other cases, problems may be avoided by using many genes rather than just using one or a few genes. By using many genes, fluctuations in evolutionary rate due to natural selection or other factors that vary over time may average out. For example, one group of researchers constructed molecular clocks of vertebrate evolution from published sequence data for 658 nuclear genes. Despite the broad period of time covered (nearly 600 million years) and the fact that natural selection probably affected some of these genes, their estimates of divergence times agreed closely with fossil-based estimates. As this example suggests, if used with care, molecular clocks can aid our understanding of evolutionary relationships.

Applying a Molecular Clock: Dating the Origin of HIV

Researchers have used a molecular clock to date the origin of HIV infection in humans. Phylogenetic analysis shows that HIV, the virus that causes AIDS, is descended from viruses that infect chimpanzees and other primates. (Most of these viruses do not cause AIDS-like diseases in their native hosts.) When did HIV jump to humans? There is no simple answer because the virus has spread to humans more than once. The multiple origins of HIV are reflected in the variety of strains (genetic types) of the virus. HIV's genetic material is made of RNA, and like other RNA viruses, it evolves quickly.

The most widespread strain in humans is HIV-1 M. To pinpoint the earliest HIV-1 M infection, researchers compared samples of the virus from various times during the epidemic, including a sample from 1959. Comparisons of gene sequences showed that the virus has evolved in a clocklike fashion. Extrapolating backward in time using the molecular clock indicates that the HIV-1 M strain first spread to humans around 1930 **(Figure 26.20)**. A later study, which dated the origin of HIV using a more advanced molecular clock approach than that covered in this book, estimated that the HIV-1 M strain first spread to humans around 1910.

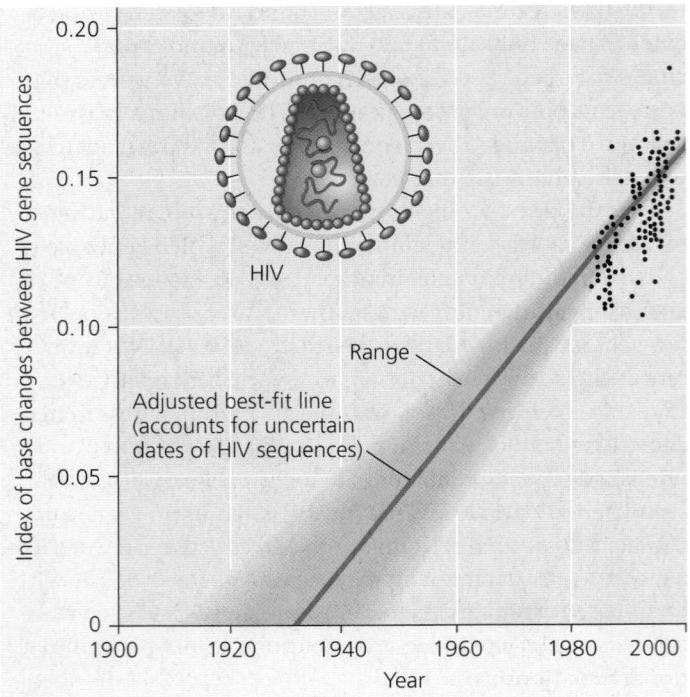

▼ **Figure 26.20 Dating the origin of HIV-1 M.** The black data points are based on DNA sequences of an HIV gene in patients' blood samples. (The dates when these individual HIV gene sequences arose are not certain because a person can harbor the virus for years before symptoms occur.) Projecting the gene's rate of change backward in time by this method suggests that the virus originated in the 1930s.

HIV

Range

Adjusted best-fit line (accounts for uncertain dates of HIV sequences)

Index of base changes between HIV gene sequences

Year

CONCEPT CHECK 26.5

1. What is a molecular clock? What assumption underlies the use of a molecular clock?
2. **MAKE CONNECTIONS** Review Concept 17.5. Then explain how numerous base changes could occur in an organism's DNA yet have no effect on its fitness.

For suggested answers, see Appendix A.

CONCEPT 26.6

Our understanding of the tree of life continues to change based on new data

The discovery that the glass lizard in Figure 26.1 evolved from a different lineage of legless lizards than did snakes is one example of how our understanding of life's diversity is informed by systematics. Indeed, in recent decades, systematists have gained insight into even the very deepest branches of the tree of life by analyzing DNA sequence data.

From Two Kingdoms to Three Domains

Biologists once classified all known species into two kingdoms: plants and animals. Classification schemes with more than two kingdoms gained broad acceptance in the late 1960s, when many biologists recognized five kingdoms: Monera (prokaryotes), Protista (a diverse kingdom consisting mostly of unicellular organisms), Plantae, Fungi, and Animalia. This system highlighted the two fundamentally different types of cells, prokaryotic and eukaryotic, and set the prokaryotes apart from all eukaryotes by placing them in their own kingdom, Monera.

However, phylogenies based on genetic data soon revealed a problem with this system: Some prokaryotes differ as much from each other as they do from eukaryotes. Such difficulties led biologists to adopt a three-domain system. The three domains—Bacteria, Archaea, and Eukarya—are a classification level higher than the kingdom level. The validity of these domains has been supported by many studies, including a recent study that analyzed nearly 100 completely sequenced genomes.

The domain Bacteria contains most of the currently known prokaryotes, while the domain Archaea consists of a diverse group of prokaryotic organisms that inhabit a wide variety of environments. The domain Eukarya consists of all the organisms that have cells containing true nuclei. This domain includes many groups of single-celled organisms as well as multicellular plants, fungi, and animals. **Figure 26.21** represents one possible phylogenetic tree for the three domains and some of the many lineages they encompass.

The three-domain system highlights the fact that much of the history of life has been about single-celled organisms. The two prokaryotic domains consist entirely of single-celled organisms, and even in Eukarya, only the branches labeled in red type (plants, fungi, and animals) are dominated by multicellular organisms. Of the five kingdoms previously recognized by systematists, most biologists continue to recognize Plantae, Fungi, and Animalia, but not Monera and Protista. The kingdom Monera is obsolete because it would have members in two different domains. The kingdom Protista has also crumbled because it includes members that are more closely related to plants, fungi, or animals than to other protists (see Figure 28.5).

New research continues to change our understanding of the tree of life. For example, recent metagenomic studies have uncovered the genomes of many new species of archaea, leading to discovery of the Thaumarchaeota and other previously unknown phyla of archaea. One of these newly discovered groups, the lokiarcheaota, may represent the long-sought-after sister group of the eukaryotes. If confirmed, this discovery could help shed light on how eukaryotes arose from their prokaryotic ancestors (see Concepts 27.4 and 28.1).

The Important Role of Horizontal Gene Transfer

In the phylogeny shown in Figure 26.21, the first major split in the history of life occurred when bacteria diverged from other organisms. If this tree is correct, eukaryotes and archaea are more closely related to each other than either is to bacteria.

▼ **Figure 26.21 The three domains of life.** This phylogenetic tree is based on sequence data for rRNA and other genes. For simplicity, only some of the major branches in each domain are shown. Lineages within Eukarya that are dominated by multicellular organisms (plants, fungi, and animals) are in red type, while the two lineages denoted by an asterisk are based on DNA from cellular organelles. All other lineages consist solely or mainly of single-celled organisms.

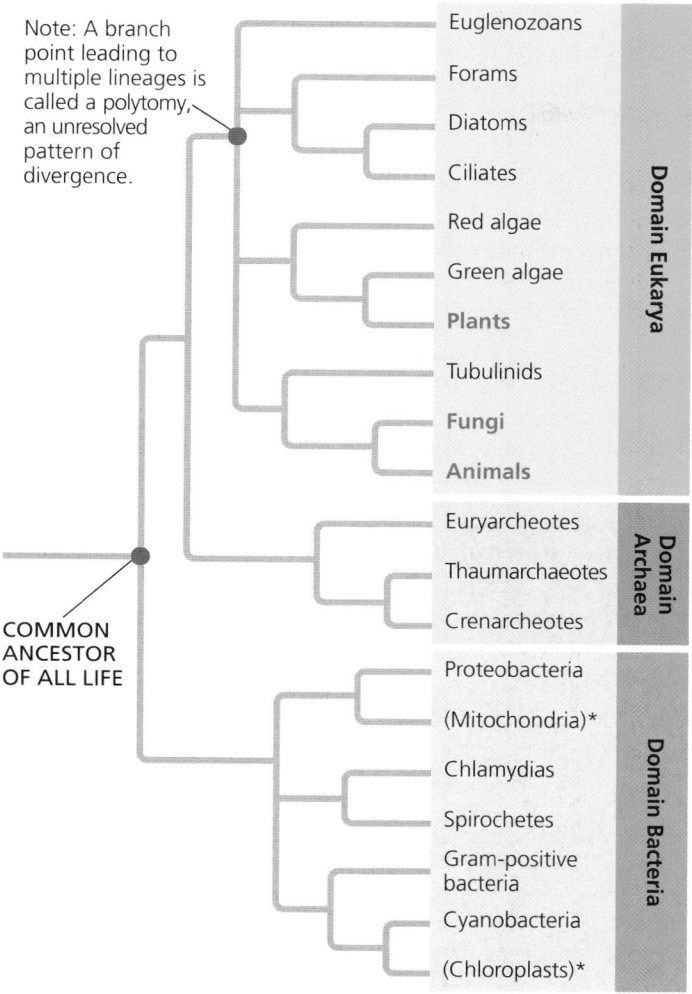

Note: A branch point leading to multiple lineages is called a polytomy, an unresolved pattern of divergence.

COMMON ANCESTOR OF ALL LIFE

Euglenozoans
Forams
Diatoms
Ciliates
Red algae
Green algae
Plants
Tubulinids
Fungi
Animals

Domain Eukarya

Euryarcheotes
Thaumarchaeotes
Crenarcheotes

Domain Archaea

Proteobacteria
(Mitochondria)*
Chlamydias
Spirochetes
Gram-positive bacteria
Cyanobacteria
(Chloroplasts)*

Domain Bacteria

MAKE CONNECTIONS *After reviewing endosymbiont theory (see Figure 6.16), explain the specific positions of the mitochondrion and chloroplast lineages on this tree.*

This reconstruction of the tree of life is based in part on sequence comparisons of rRNA genes, which code for the RNA components of ribosomes. However, some other genes reveal a different set of relationships. For example, researchers have found that many of the genes that influence metabolism in yeast (a unicellular eukaryote) are more similar to genes in the domain Bacteria than they are to genes in the domain Archaea—a finding that suggests that the eukaryotes may share a more recent common ancestor with bacteria than with archaea.

What causes trees based on data from different genes to yield such different results? Comparisons of complete genomes from the three domains show that there have been substantial movements of genes between organisms in the different domains. These took place through **horizontal gene transfer,**

a process in which genes are transferred from one genome to another through mechanisms such as exchange of transposable elements and plasmids, viral infection (see Concept 19.2), and fusions of organisms (as when a host and its endosymbiont become a single organism). Recent research reinforces the view that horizontal gene transfer is important. For example, one study found that on average, 80% of the genes in 181 prokaryotic genomes had moved between species at some point during the course of evolution. Because phylogenetic trees are based on the assumption that genes are passed vertically from one generation to the next, the occurrence of such horizontal transfer events helps to explain why trees built using different genes can give inconsistent results.

Horizontal gene transfer can also occur between eukaryotes. For example, over 200 cases of the horizontal transfer of transposons have been reported in eukaryotes, including humans and other primates, plants, birds, and lizards. Nuclear genes have also been transferred horizontally from one eukaryote to another. The **Scientific Skills Exercise** describes one such example, giving you the opportunity to interpret data on the transfer of a pigment gene to an aphid from another species.

Studies show that eukaryotes can even acquire genes by horizontal gene transfer from prokaryotes. For example, a recent genomic analysis showed that the alga *Galdieria sulphuraria* **(Figure 26.22)** acquired about 5% of its nuclear genes from various bacterial and archaeal species. Unlike most eukaryotes, this alga can survive in environments that are highly acidic or extremely hot, as well as those with high concentrations of heavy metals. The researchers identified specific genes transferred from prokaryotes that have enabled *G. sulphuraria* to thrive in such extreme habitats.

Overall, horizontal gene transfer has played a key role throughout the evolutionary history of life, and it continues to occur today. Some biologists have argued that horizontal gene transfer was so common that the early history of life should be represented not as a dichotomously branching tree like that in Figure 26.21, but rather as a tangled network of

▼ **Figure 26.22 A recipient of transferred genes: the alga *Galdieria sulphuraria*.** Genes received from prokaryotes enable *G. sulphuraria* (inset) to grow in extreme environments, including on sulfur-encrusted rocks around volcanic hot springs similar to this one in Yellowstone National Park.

5 μm

Using Protein Sequence Data to Test an Evolutionary Hypothesis

Did Aphids Acquire Their Ability to Make Carotenoids Through Horizontal Gene Transfer? Carotenoids are colored molecules that have diverse functions in many organisms, such as photosynthesis in plants and light detection in animals. Plants and many microorganisms can synthesize carotenoids from scratch, but animals generally cannot (they must obtain carotenoids from their diet). One exception is the pea aphid *Acyrthosiphon pisum*, a small, plant-dwelling insect whose genome includes a full set of genes for the enzymes needed to make carotenoids. Because other animals lack these genes, it is unlikely that aphids inherited them from a single-celled common ancestor shared with microorganisms and plants. So where did they come from? Evolutionary biologists hypothesized that an aphid ancestor acquired these genes by horizontal gene transfer from distantly related organisms.

How the Experiment Was Done Scientists obtained the DNA sequences for the carotenoid-biosynthesis genes from several species, including aphids, fungi, bacteria, and plants. A computer "translated" these sequences into amino acid sequences of the encoded polypeptides and aligned the amino acid sequences. This allowed the team to compare the corresponding polypeptides in the different organisms.

Data from the Experiment The sequences below show the first 60 amino acids of one polypeptide of the carotenoid-biosynthesis enzymes in the plant *Arabidopsis thaliana* (bottom) and the corresponding amino acids in five nonplant species,

using the one-letter abbreviations for the amino acids (see Figure 5.14). A dash (–) indicates a gap inserted in a sequence to optimize its alignment with the corresponding sequence in *Arabidopsis*.

INTERPRET THE DATA

1. In the rows of data for the organisms being compared with the aphid, highlight the amino acids that are identical to the corresponding amino acids in the aphid.

2. Which organism has the most amino acids in common with the aphid? Rank the partial polypeptides from the other four organisms in degree of similarity to that of the aphid.

3. **(a)** Do these data support the hypothesis that aphids acquired the gene for this polypeptide by horizontal gene transfer? Why or why not? **(b)** If horizontal gene transfer did occur, what type of organism is likely to have been the source?

4. What additional sequence data would support your hypothesis?

5. How would you account for the similarities between the aphid sequence and the sequences for the bacteria and plant?

→ **Instructors:** A version of this Scientific Skills Exercise can be assigned in **Mastering Biology**.

Data from Nancy A. Moran, Yale University. See N. A. Moran and T. Jarvik, Lateral transfer of genes from fungi underlies carotenoid production in aphids, *Science* 328:624–627 (2010).

Organism	Alignment of Amino Acid Sequences					
Acyrthosiphon (aphid)	IKIIIIGSGV	GGTAAAARLS	KKGFQVEVYE	KNSYNGGRCS	IIR-HNGHRF	DQGPSL--YL
Ustilago (fungus)	KKVVIIGAGA	GGTALAARLG	RRGYSVTVLE	KNSFGGGRCS	LIH-HDGHRW	DQGPSL--YL
Gibberella (fungus)	KSVIVIGAGV	GGVSTAARLA	KAGFKVTILE	KNDFTGGRCS	LIH-NDGHRF	DQGPSL--LL
Staphylococcus (bacterium)	MKIAVIGAGV	TGLAAAARIA	SQGHEVTIFE	KNNNVGGRMN	QLK-KDGFTF	DMGPTI--VM
Pantoea (bacterium)	KRTFVIGAGF	GGLALAIRLQ	AAGIATTVLE	QHDKPGGRAY	VWQ-DQGFTF	DAGPTV--IT
Arabidopsis (plant)	WDAVVIGGGH	NGLTAAAYLA	RGGLSVAVLE	RRHVIGGAAV	TEEIVPGFKF	SRCSYLQGLL

▶ **Figure 26.23 A tangled web of life.** Horizontal gene transfer may have been so common in the early history of life that the base of a "tree of life" might be more accurately portrayed as a tangled web.

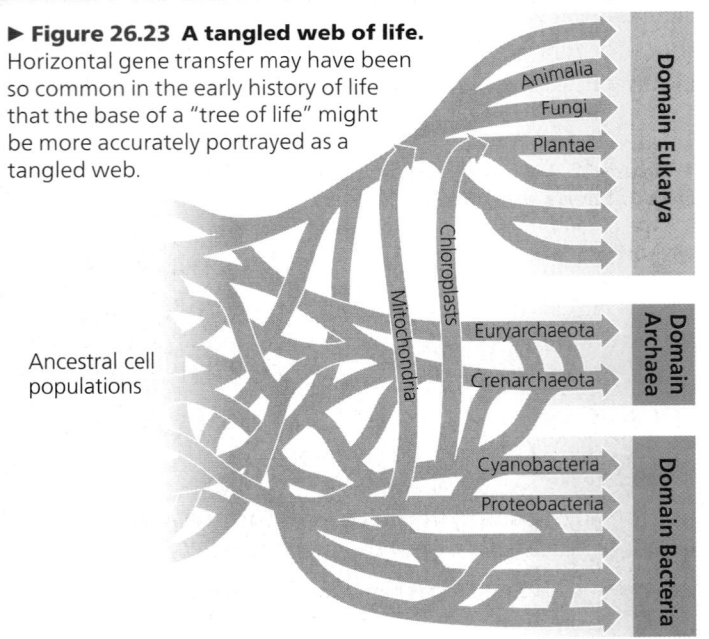

connected branches **(Figure 26.23)**. Although scientists continue to debate how best to portray the earliest steps in the history of life, in recent decades there have been many exciting discoveries about evolutionary events that occurred over time. We'll explore such discoveries in the rest of this unit, beginning with Earth's earliest inhabitants, the prokaryotes.

→ **Mastering Biology**
Interview with Nancy Moran: Studying symbioses and horizontal gene transfer

CONCEPT CHECK 26.6

1. Why is the kingdom Monera no longer considered a valid taxon?

2. Explain why phylogenies based on different genes can yield different branching patterns for the tree of all life.

3. **MAKE CONNECTIONS** Explain how the origin of eukaryotes is thought to have represented a fusion of organisms, leading to extensive horizontal gene transfer. (See Figure 25.10.)

For suggested answers, see Appendix A.

26 Chapter Review

→ Go to **Mastering Biology** for Assignments, the eText, the Study Area, and Dynamic Study Modules.

SUMMARY OF KEY CONCEPTS

→ To review key terms, go to the **Vocabulary Self-Quiz** in the **Mastering Biology** eText or Study Area, or go to goo.gl/zkjz9t.

CONCEPT 26.1

Phylogenies show evolutionary relationships (pp. 554–558)

- Linnaeus's **binomial** classification system gives organisms two-part names: a **genus** plus a specific epithet.
- In the Linnaean system, species are grouped in increasingly broad **taxa**: Related genera are placed in the same **family**, families in **orders**, orders in **classes**, classes in **phyla**, phyla in **kingdoms**, and (more recently) kingdoms in **domains**.
- Systematists depict evolutionary relationships as branching **phylogenetic trees**. Many systematists propose that classification be based entirely on evolutionary relationships.

- Unless branch lengths are proportional to time or genetic change, a phylogenetic tree indicates only patterns of descent.
- Much information can be learned about a species from its evolutionary history; hence, phylogenies are useful in a wide range of applications.

? *Humans and chimpanzees are sister species. Explain what this statement means.*

CONCEPT 26.2

Phylogenies are inferred from morphological and molecular data (pp. 558–559)

- Organisms with similar morphologies or DNA sequences are likely to be more closely related than organisms with very different structures and genetic sequences.
- To infer phylogeny, **homology** (similarity due to shared ancestry) must be distinguished from **analogy** (similarity due to convergent evolution).
- Computer programs are used to align comparable DNA sequences and to distinguish molecular homologies from coincidental matches between taxa that diverged long ago.

? *Why is it necessary to distinguish homology from analogy to infer phylogeny?*

CONCEPT 26.3

Shared characters are used to construct phylogenetic trees (pp. 559–565)

- A **clade** is a monophyletic group that includes an ancestral species and all of its descendants.

- Clades can be distinguished by their **shared derived characters**.
- Among phylogenies, the most parsimonious tree is the one that requires the fewest evolutionary changes. The most likely tree is the one based on the most likely pattern of changes.
- Well-supported phylogenetic hypotheses are consistent with a wide range of data.

? *Explain the logic of using shared derived characters to infer phylogeny.*

CONCEPT 26.4

An organism's evolutionary history is documented in its genome (pp. 565–566)

- **Orthologous genes** are homologous genes found in different species as a result of speciation. **Paralogous genes** are homologous genes within a species that result from gene duplication; such genes can diverge and potentially take on new functions.
- Distantly related species often have many orthologous genes. The small variation in gene number in organisms of varying complexity suggests that genes are versatile and may have multiple functions.

? *When reconstructing phylogenies, is it more useful to compare orthologous or paralogous genes? Explain.*

CONCEPT 26.5

Molecular clocks help track evolutionary time (pp. 566–568)

- Some regions of DNA change at a rate consistent enough to serve as a **molecular clock**, a method of estimating the date of past evolutionary events based on the amount of genetic change. Other DNA regions change in a less predictable way.
- Molecular clock analyses suggest that the most common strain of HIV jumped from primates to humans in the early 1900s.

? *Describe some assumptions and limitations of molecular clocks.*

CONCEPT 26.6

Our understanding of the tree of life continues to change based on new data (pp. 568–570)

- Past classification systems have given way to the current view of the tree of life, which consists of three great domains: Bacteria, Archaea, and Eukarya.

- Phylogenies based in part on rRNA genes suggest that eukaryotes are most closely related to archaea, while data from some other genes suggest a closer relationship to bacteria.
- Genetic analyses indicate that extensive **horizontal gene transfer** has occurred throughout the evolutionary history of life.

? *Why was the five-kingdom system abandoned for a three-domain system?*

TEST YOUR UNDERSTANDING

➔ For more multiple-choice questions, go to the **Practice Test** in the **Mastering Biology** eText or Study Area, or go to goo.gl/GruWRg.

Levels 1-2: Remembering/Understanding

1. In a comparison of birds and mammals, the condition of having four limbs is
   (A) a shared ancestral character.
   (B) a shared derived character.
   (C) a character useful for distinguishing birds from mammals.
   (D) an example of analogy rather than homology.

2. To apply parsimony to constructing a phylogenetic tree,
   (A) choose the tree that assumes all evolutionary changes are equally probable.
   (B) choose the tree in which the branch points are based on as many shared derived characters as possible.
   (C) choose the tree that represents the fewest evolutionary changes, in either DNA sequences or morphology.
   (D) choose the tree with the fewest branch points.

Levels 3-4: Applying/Analyzing

3. **VISUAL SKILLS** In Figure 26.4, which similarly inclusive taxon is represented as descending from the same common ancestor as Canidae?
   (A) Felidae   (C) Carnivora
   (B) Mustelidae   (D) *Lutra*

4. Three living species X, Y, and Z share a common ancestor T, as do extinct species U and V. A grouping that consists of species T, X, Y, and Z (but not U or V) makes up
   (A) a monophyletic taxon.
   (B) an ingroup, with species U as the outgroup.
   (C) a paraphyletic group.
   (D) a polyphyletic group.

5. **VISUAL SKILLS** Based on the tree below, which statement is correct?

   (A) Lizards and goats form a sister group.
   (B) Salamanders are a sister group to the group containing lizards, goats, and humans.
   (C) Salamanders are more closely related to lizards than to humans.
   (D) Goats and humans are the only sister group shown in this tree.

6. If you were using cladistics to build a phylogenetic tree of cats, which of the following would be the best outgroup?
   (A) wolf   (C) lion
   (B) domestic cat   (D) leopard

7. **VISUAL SKILLS** The relative lengths of the frog and mouse branches in the phylogenetic tree in Figure 26.13 indicate that
   (A) frogs evolved before mice.
   (B) mice evolved before frogs.
   (C) the homolog evolved more rapidly in the mouse lineage.
   (D) the homolog evolved more slowly in the mouse lineage.

Levels 5-6: Evaluating/Creating

8. **EVOLUTION CONNECTION** Darwin suggested looking at a species' close relatives to learn what its ancestors may have been like. Explain how his suggestion anticipates recent methods, such as phylogenetic bracketing and the use of outgroups in cladistic analysis.

9. **SCIENTIFIC INQUIRY • DRAW IT** (a) Draw a phylogenetic tree based on the characters in the table below. Draw hatch marks on the tree to indicate the origin of characters 1–6. (b) Assume that tuna and dolphins are sister species and redraw the phylogenetic tree accordingly. Use hatch marks to indicate the origin of characters 1–6. (c) Determine how many evolutionary changes are required in each tree. Identify the most parsimonious tree.

Character	Lancelet (outgroup)	Lamprey	Tuna	Salamander	Turtle	Leopard	Dolphin
(1) Backbone	0	1	1	1	1	1	1
(2) Hinged jaw	0	0	1	1	1	1	1
(3) Four limbs	0	0	0	1	1	1	1*
(4) Amnion	0	0	0	0	1	1	1
(5) Milk	0	0	0	0	0	1	1
(6) Dorsal fin	0	0	1	0	0	0	1

*Although adult dolphins have only two obvious limbs (their flippers), as embryos they have two hind-limb buds, for a total of four limbs.

10. **WRITE ABOUT A THEME: INFORMATION** In a short essay (100–150 words), explain how genetic information—along with an understanding of the process of descent with modification—enables scientists to reconstruct phylogenies that extend hundreds of millions of years back in time.

11. **SYNTHESIZE YOUR KNOWLEDGE**

This West Indian manatee (*Trichechus manatus*) is an aquatic mammal. Like amphibians and reptiles, mammals are tetrapods (vertebrates with four limbs). Explain why manatees are considered tetrapods even though they lack hind limbs, and suggest traits that manatees likely share with leopards and other mammals (see Figure 26.12b). Discuss how early members of the manatee lineage might have differed from today's manatees.

For selected answers, see Appendix A.

27 Bacteria and Archaea

KEY CONCEPTS

27.1 Structural and functional adaptations contribute to prokaryotic success *p. 574*

27.2 Rapid reproduction, mutation, and genetic recombination promote genetic diversity in prokaryotes *p. 578*

27.3 Diverse nutritional and metabolic adaptations have evolved in prokaryotes *p. 581*

27.4 Prokaryotes have radiated into a diverse set of lineages *p. 583*

27.5 Prokaryotes play crucial roles in the biosphere *p. 586*

27.6 Prokaryotes have both beneficial and harmful impacts on humans *p. 587*

Study Tip

Make a table: As you read the chapter, identify examples of challenging environments in which prokaryotes thrive. List the structural, metabolic, or other features of prokaryotes that contribute to their success in each of these environments.

Environmental challenge	Prokaryotic feature enabling success
Lack of water	Capsule or slime layer
Exposure to antibiotics	

⊖ Go to Mastering Biology

For Students (in eText and Study Area)
- Get Ready for Chapter 27
- Figure 27.13 Walkthrough: Conjugation and Recombination in *E. coli*
- Animation: Structure and Reproduction of Bacteria

For Instructors to Assign (in Item Library)
- Scientific Skills Exercise: Calculating and Interpreting Means and Standard Errors
- Tutorial: Genetic Transfer in Bacteria

Figure 27.1 The pink color of this lake in Spain is a sign that its waters are much saltier than seawater. Its color is caused not by minerals or other nonliving sources, but by trillions of prokaryotes that thrive in salinities that kill other cells. Other prokaryotes live in environments that are far too cold or hot or acidic for most other organisms.

What characteristics enable prokaryotes to reach huge population sizes and thrive in diverse environments?

Due to the organisms' **small size** and **rapid reproduction**, prokaryotic populations can reach enormous numbers.

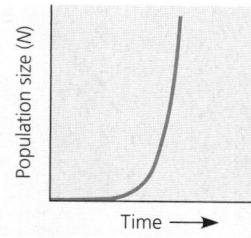

Because prokaryotic populations are so large, **mutations** produce high genetic diversity, enabling rapid evolution.

Their **diverse adaptations** enable prokaryotes to live in a wide range of environments.

Rapid evolution in prokaryotic populations results in diverse metabolic and structural adaptations.

Protective coat

Endospore (resistant cell that can enable bacterium to survive harsh conditions)

Prokaryotes, the single-celled organisms that make up domains Bacteria and Archaea, can thrive in a wide range of extreme environments. Prokaryotic species are also very well adapted to more "normal" habitats—the lands and waters in which most other species are found. Their ability to adapt to a broad range of habitats helps explain why prokaryotes are the most abundant organisms on Earth. Indeed, the number of prokaryotes in a handful of fertile soil is greater than the number of people who have ever lived. In this chapter, we'll examine the adaptations, diversity, and enormous ecological impact of these remarkable organisms.

CONCEPT 27.1

Structural and functional adaptations contribute to prokaryotic success

The first organisms to inhabit Earth were prokaryotes that lived 3.5 billion years ago (see Concept 25.3). Throughout their long evolutionary history, prokaryotic populations have been (and continue to be) subjected to natural selection in all kinds of environments, resulting in their enormous diversity today.

We'll begin by describing prokaryotes. Most prokaryotes are unicellular, although the cells of some species remain attached to each other after cell division. Prokaryotic cells typically have diameters of 0.5–5 µm, much smaller than the 10- to 100-µm diameter of many eukaryotic cells. (One notable exception, *Thiomargarita namibiensis*, can be as large as 750 µm in diameter—bigger than a poppy seed.) Prokaryotic cells have a variety of shapes **(Figure 27.2)**.

▼ **Figure 27.2 The most common shapes of prokaryotes.**
(a) Cocci (singular, *coccus*) are spherical prokaryotes. They can occur singly, in chains of two or more cells, and in clusters resembling bunches of grapes. **(b)** Bacilli (singular, *bacillus*) are rod-shaped prokaryotes. They are usually solitary, but in some forms the rods are arranged in chains. **(c)** Spiral prokaryotes include spirochetes (shown here), which are corkscrew-shaped; other spiral prokaryotes resemble commas or loose coils (colorized SEMs).

(a) Spherical   **(b) Rod-shaped**   **(c) Spiral**

Finally, although they are unicellular and small, prokaryotes are well organized, achieving all of an organism's life functions within a single cell.

Cell-Surface Structures

A key feature of nearly all prokaryotic cells is the cell wall, which maintains cell shape, protects the cell, and prevents it from bursting in a hypotonic environment (see Figure 7.13). In a hypertonic environment, most prokaryotes lose water and shrink away from their wall. Such water losses can inhibit cell reproduction. Thus, salt can be used to preserve foods because it causes food-spoiling prokaryotes to lose water, preventing them from reproducing rapidly.

The cell walls of prokaryotes differ in structure from those of eukaryotes. In eukaryotes that have cell walls, such as plants and fungi, the walls are usually made of cellulose or chitin (see Concept 5.2). In contrast, most bacterial cell walls contain **peptidoglycan**, a polymer composed of modified sugars cross-linked by short polypeptides. This molecular fabric encloses the entire bacterium and anchors other molecules that extend from its surface. Archaeal cell walls contain a variety of polysaccharides and proteins but lack peptidoglycan.

Using a technique called the **Gram stain**, developed by the 19th-century Danish physician Hans Christian Gram, scientists can categorize many bacterial species according to differences in cell wall composition. To do this, samples are first stained with crystal violet dye and iodine, then rinsed in alcohol, and finally stained with the red dye safranin, which enters the cell and binds to its DNA. The structure of a bacterium's cell wall determines the staining response **(Figure 27.3)**. **Gram-positive** bacteria have relatively simple walls composed of a thick layer of peptidoglycan. The walls of **gram-negative** bacteria have less peptidoglycan and are structurally more complex, with an outer membrane that contains lipopolysaccharides (carbohydrates bonded to lipids).

Gram staining is a valuable tool in medicine for quickly determining if a patient's infection is due to gram-negative or to gram-positive bacteria. This information has treatment implications. The lipid portions of the lipopolysaccharides in the walls of many gram-negative bacteria are toxic, causing fever or shock. Furthermore, the outer membrane of a gram-negative bacterium helps protect it from the body's defenses. Gram-negative bacteria also tend to be more resistant than gram-positive species to antibiotics because the outer membrane impedes the entry of some drugs. However, some gram-positive species have virulent strains that are resistant to one or more antibiotics. (Figure 22.14 discusses one example: methicillin-resistant *Staphylococcus aureus*, or MRSA, which can cause lethal skin infections.)

The effectiveness of certain antibiotics, such as penicillin, derives from their inhibition of peptidoglycan cross-linking. The resulting cell wall may not be functional, particularly in

(a) Gram-positive bacteria

Gram-positive bacteria stain a darker color and have a thick cell wall made of peptidoglycan.

(b) Gram-negative bacteria

Gram-negative bacteria have a thinner layer of peptidoglycan located between the plasma membrane and an outer membrane.

gram-positive bacteria. Such drugs destroy many species of pathogenic bacteria without adversely affecting human cells, which do not have peptidoglycan.

The cell wall of many prokaryotes is surrounded by a sticky layer of polysaccharide or protein. This layer is called a **capsule** if it is dense and well-defined **(Figure 27.4)** or a *slime layer* if it is not as well organized. Both kinds of sticky outer layers enable prokaryotes to adhere to their substrate or to other individuals in a colony. Capsules and slime layers also protect against dehydration, and some capsules shield pathogenic prokaryotes from attacks by their host's immune system.

In another way of withstanding harsh conditions, certain bacteria develop resistant cells called **endospores** when they

lack water or essential nutrients **(Figure 27.5)**. The original cell produces a copy of its chromosome and surrounds that copy with a multilayered structure, forming the endospore. Water is removed from the endospore, and its metabolism halts. The original cell then lyses, releasing the endospore. Most endospores are so durable that they can survive in boiling water; killing them requires heating lab equipment to 121°C under high pressure. In less hostile environments, endospores can remain dormant but viable for centuries, able to rehydrate and resume metabolism when their environment improves.

Finally, some prokaryotes stick to their substrate or to one another by means of hairlike appendages called **fimbriae**

▼ **Figure 27.4 Capsule.** The polysaccharide capsule around this *Streptococcus* bacterium enables the prokaryote to attach to cells in the respiratory tract—in this colorized TEM, a tonsil cell.

▼ **Figure 27.5 An endospore.** *Bacillus anthracis*, the bacterium that causes the disease anthrax, produces endospores (TEM). An endospore's protective, multilayered coat helps it survive in the soil for years.

▼ **Figure 27.6 Fimbriae.** These numerous protein-containing appendages enable some prokaryotes to attach to surfaces or to other cells (colorized TEM).

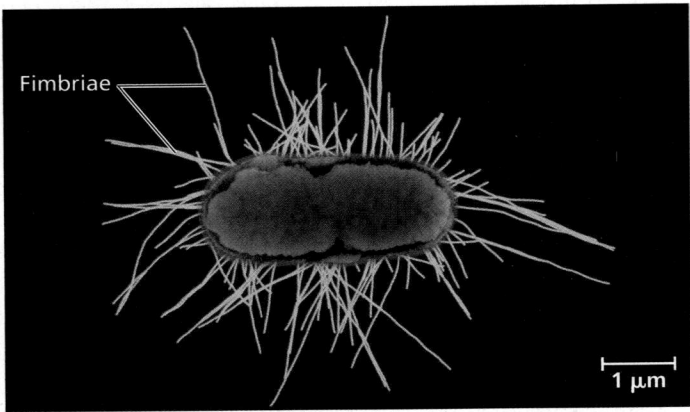

Fimbriae

1 μm

(singular, *fimbria*) **(Figure 27.6)**. For example, the bacterium that causes gonorrhea, *Neisseria gonorrhoeae*, uses fimbriae to fasten itself to the mucous membranes of its host. Fimbriae are usually shorter and more numerous than **pili** (singular, *pilus*), appendages that pull two cells together prior to DNA transfer from one cell to the other (see Figure 27.12); pili are sometimes referred to as *sex pili*.

Motility

About half of all prokaryotes are capable of **taxis**, a directed movement toward or away from a stimulus (from the Greek *taxis*, to arrange). For example, prokaryotes that exhibit *chemotaxis* change their movement pattern in response to chemicals. They may move *toward* nutrients or oxygen (positive chemotaxis) or *away from* a toxic substance (negative chemotaxis). Some species can move at velocities exceeding 50 μm/sec—up to 50 times their body length per second. For perspective, consider that a person 1.7 m tall moving that fast would be running 306 km (190 miles) per hour!

Of the various structures that enable prokaryotes to move, the most common are flagella **(Figure 27.7)**. Flagella (singular, *flagellum*) may be scattered over the entire surface of the cell or concentrated at one or both ends. Prokaryotic flagella differ greatly from eukaryotic flagella: They are one-tenth the width and typically are not covered by an extension of the plasma membrane (see Figure 6.24). The flagella of prokaryotes and eukaryotes also

differ in their molecular composition and their mechanism of propulsion. Among prokaryotes, bacterial and archaeal flagella are similar in size and rotational mechanism, but they are composed of entirely different and unrelated proteins. Overall, these structural and molecular comparisons indicate that the flagella of bacteria, archaea, and eukaryotes arose independently. Since current evidence shows that the flagella of organisms in the three domains perform similar functions but are not related by common descent, they are described as analogous, not homologous, structures (see Concept 22.3).

Evolutionary Origins of Bacterial Flagella

The bacterial flagellum shown in Figure 27.7 has three main parts (the motor, hook, and filament) that are themselves composed of 42 different kinds of proteins. How could such a complex structure evolve? In fact, much evidence indicates that bacterial flagella originated as simpler structures that were modified in a stepwise fashion over time. As in the case of the human eye (see Concept 25.6), biologists asked whether a less complex version of the flagellum could still benefit its owner. Analyses of hundreds of bacterial genomes indicate that only half of the flagellum's protein components appear to be necessary for it to function; the others are inessential or not encoded in the genomes of some species. Of the 21 proteins required by

▼ **Figure 27.7 A prokaryotic flagellum.** The motor of a prokaryotic flagellum consists of a system of protein rings embedded in the cell wall and plasma membrane (TEM). The electron transport chain pumps protons out of the cell. The diffusion of protons back into the cell provides the force that turns a curved hook and thereby causes the attached filament to rotate and propel the cell. (This diagram shows flagellar structures characteristic of gram-negative bacteria.)

Flagellum

Filament

20 nm

Hook

Motor

Cell wall

Plasma membrane

Rod

Peptidoglycan layer

VISUAL SKILLS *Predict which of the four protein rings shown in this diagram are likely hydrophobic. Explain your answer.*

→ **Mastering Biology** Video: Prokaryotic Flagella

all species studied to date, 19 are modified versions of proteins that perform other tasks in bacteria. For example, a set of 10 proteins in the motor is homologous to 10 similar proteins in a secretory system found in bacteria. (A secretory system is a protein complex that enables a cell to produce and release certain macromolecules.) Two other proteins in the motor are homologous to proteins that function in ion transport. The proteins that comprise the rod, hook, and filament are all related to each other and are descended from an ancestral protein that formed a pilus-like tube. These findings suggest that the bacterial flagellum evolved as other proteins were added to an ancestral secretory system. This is an example of *exaptation*, the process in which structures originally adapted for one function take on new functions through descent with modification.

Internal Organization and DNA

The cells of prokaryotes are simpler than those of eukaryotes in both their internal structure and the physical arrangement of their DNA (see Figure 6.5). Prokaryotic cells lack the complex compartmentalization associated with the membrane-enclosed organelles found in eukaryotic cells. However, some prokaryotic cells do have specialized membranes that perform metabolic functions **(Figure 27.8)**. These membranes are usually infoldings of the plasma membrane. Recent discoveries also indicate that some prokaryotes can store metabolic by-products in simple compartments that are made out of proteins; these compartments do not have a membrane.

The genome of a prokaryote is structurally different from a eukaryotic genome and in most cases has considerably less DNA. Prokaryotes typically have one circular chromosome **(Figure 27.9)**, whereas eukaryotes usually have several to many linear chromosomes. In addition, in prokaryotes the chromosome is associated with many fewer proteins than are the chromosomes of eukaryotes. Also unlike eukaryotes,

▼ **Figure 27.8 Specialized membranes of prokaryotes.**
(a) Infoldings of the plasma membrane, reminiscent of the cristae of mitochondria, function in cellular respiration in some aerobic prokaryotes (TEM). **(b)** Photosynthetic prokaryotes called cyanobacteria have thylakoid membranes, much like those in chloroplasts (TEM).

0.2 μm

Respiratory membrane

(a) Aerobic prokaryote

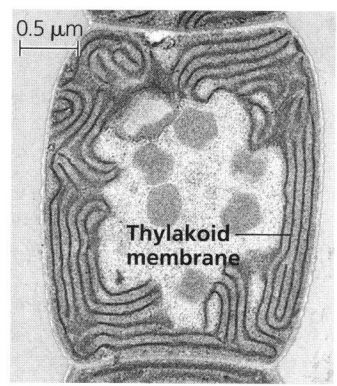

0.5 μm

Thylakoid membrane

(b) Photosynthetic prokaryote

▼ **Figure 27.9 A prokaryotic chromosome and plasmids.** The thin, tangled loops surrounding this ruptured *Escherichia coli* cell are parts of the cell's large, circular chromosome (colorized TEM). Three of the cell's plasmids, the much smaller rings of DNA, are also shown.

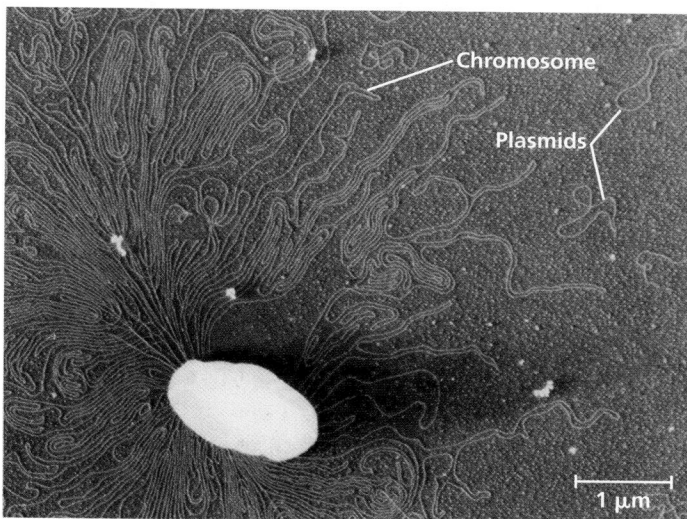

Chromosome

Plasmids

1 μm

VISUAL SKILLS *Identify and label the third plasmid visible in this micrograph.*

prokaryotes lack a nucleus; their chromosome is located in the **nucleoid**, a region of cytoplasm that is not enclosed by a membrane. In addition to its single chromosome, a typical prokaryotic cell may also have much smaller rings of independently replicating DNA molecules called **plasmids** (see Figure 27.9), most carrying only a few genes.

Although DNA replication, transcription, and translation are fundamentally similar processes in prokaryotes and eukaryotes, some of the details differ (see Chapter 17). For example, prokaryotic ribosomes are slightly smaller than eukaryotic ribosomes and differ in their protein and RNA content. These differences allow certain antibiotics, such as erythromycin and tetracycline, to bind to ribosomes and block protein synthesis in prokaryotes but not in eukaryotes. As a result, people can use these antibiotics to kill or inhibit the growth of bacteria without harming themselves.

Reproduction

Many prokaryotes can reproduce quickly in favorable environments. By *binary fission* (see Figure 12.12), a single prokaryotic cell divides into 2 cells, which then divide into 4, 8, 16, and so on. Under optimal conditions, many prokaryotes can divide every 1–3 hours; some species can produce a new generation in only 20 minutes. At this rate, a single prokaryotic cell could give rise to a colony outweighing Earth in only two days!

In reality, of course, this does not occur. The cells eventually exhaust their nutrient supply, poison themselves with metabolic wastes, face competition from other microorganisms, or are consumed by other organisms. Still, the potential of many prokaryotic species for rapid population growth emphasizes three key features of their biology: *They are small, they reproduce by binary fission, and they often have short generation times*. As a result, prokaryotic populations can consist of many trillions of individuals—far more than populations of multicellular eukaryotes, such as plants or animals.

→ **Mastering Biology** Animation: Structure and Reproduction of Bacteria

CONCEPT CHECK 27.1

1. Describe two adaptations that enable prokaryotes to survive in environments too harsh for other organisms.

2. Contrast the cellular and DNA structures of prokaryotes and eukaryotes.

3. **MAKE CONNECTIONS** Suggest a hypothesis that explains why the thylakoid membranes of chloroplasts resemble those of cyanobacteria. Refer to Figures 6.18 and 26.21.

For suggested answers, see Appendix A.

CONCEPT 27.2

Rapid reproduction, mutation, and genetic recombination promote genetic diversity in prokaryotes

As we discussed in Unit Four, evolution cannot occur without genetic variation. The diverse adaptations exhibited by prokaryotes suggest that their populations must have considerable genetic variation—and they do. In this section, we'll examine three factors that give rise to high levels of genetic diversity in prokaryotes: rapid reproduction, mutation, and genetic recombination.

Rapid Reproduction and Mutation

Most of the genetic variation in sexually reproducing species results from the way existing alleles are arranged in new combinations during meiosis and fertilization (see Concept 13.4). Prokaryotes do not reproduce sexually, so at first glance their extensive genetic variation may seem puzzling. But in many species, this variation can result from a combination of rapid reproduction and mutation.

Consider the bacterium *Escherichia coli* as it reproduces by binary fission in a human intestine, one of its natural environments. After repeated rounds of division, most of the offspring cells are genetically identical to the original parent cell. However, if errors occur during DNA replication, some of the offspring cells may differ genetically. The probability of such a mutation occurring in a given *E. coli* gene is about one in 10 million (1×10^{-7}) per cell division. But among the 2×10^{10} new *E. coli* cells that arise each day in a person's intestine, there will be approximately $(2 \times 10^{10}) \times (1 \times 10^{-7}) = 2,000$ bacteria that have a mutation in that gene. The total number of new

mutations when all 4,300 *E. coli* genes are considered is about $4,300 \times 2,000$—more than 8 million per day per human host.

The key point is that new mutations, though rare on a per gene basis, can increase genetic diversity quickly in species with short generation times and large populations. This diversity, in turn, can lead to rapid evolution **(Figure 27.10)**:

▼ **Figure 27.10** **Inquiry**

Can prokaryotes evolve rapidly in response to environmental change?

Experiment Researchers tested the ability of *E. coli* populations to adapt to a new environment. They established 12 populations, each founded by a single cell from an *E. coli* strain, and followed these populations for 20,000 generations (3,000 days). To maintain a continual supply of resources, each day the researchers performed a *serial transfer*: They transferred 0.1 mL of each population to a new tube containing 9.9 mL of fresh growth medium. The growth medium used throughout the experiment provided a challenging environment that contained only low levels of glucose and other resources needed for growth.

Samples were periodically removed from the 12 populations and grown in competition with the common ancestral strain in the experimental (low-glucose) environment.

Daily serial transfer

0.1 mL (population sample)

Old tube (discarded after transfer)

New tube (9.9 mL growth medium)

Results The fitness of the experimental populations, as measured by the growth rate of each population, increased rapidly for the first 5,000 generations (2 years) and more slowly for the next 15,000 generations. The graph shows the averages for the 12 populations.

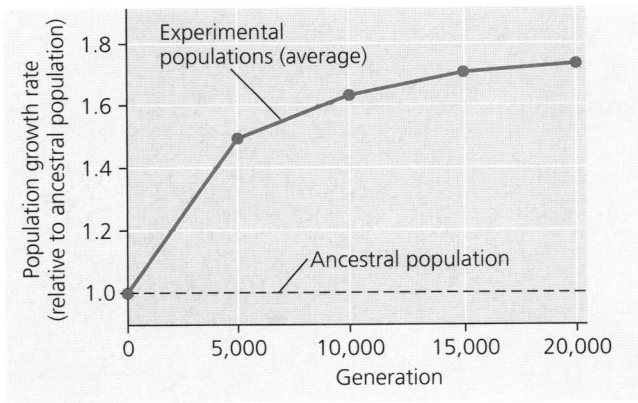

Conclusion Populations of *E. coli* continued to accumulate beneficial mutations for 20,000 generations, allowing rapid evolution of increased population growth rates in their new environment.

Data from V. S. Cooper and R. E. Lenski, The population genetics of ecological specialization in evolving *Escherichia coli* populations, *Nature* 407:736–739 (2000).

WHAT IF? *Suggest possible functions of the genes whose sequence or expression was altered as the experimental populations evolved in the low-glucose environment.*

Individuals that are genetically better equipped for their environment tend to survive and reproduce at higher rates than other individuals. The ability of prokaryotes to adapt rapidly to new conditions highlights the point that although the structure of their cells is simpler than that of eukaryotic cells, prokaryotes are not "primitive" or "inferior" in an evolutionary sense. They are, in fact, highly evolved: For 3.5 billion years, prokaryotic populations have responded successfully to many types of environmental challenges.

Genetic Recombination

Although new mutations are a major source of variation in prokaryotic populations, additional diversity arises from *genetic recombination*, the combining of DNA from two sources. In eukaryotes, the sexual processes of meiosis and fertilization combine DNA from two individuals in a single zygote. But meiosis and fertilization do not occur in prokaryotes. Instead, three other mechanisms—transformation, transduction, and conjugation—can bring together prokaryotic DNA from different individuals (that is, different cells). When the individuals are members of different species, this movement of genes from one organism to another is called *horizontal gene transfer*. Scientists have found evidence that each of these mechanisms can transfer DNA within and between species in both domain Bacteria and domain Archaea. To date, however, most of our knowledge comes from research on bacteria.

Transformation and Transduction

In **transformation**, the genotype and possibly phenotype of a prokaryotic cell are altered by the uptake of foreign DNA from its surroundings. For example, a harmless strain of *Streptococcus pneumoniae* can be transformed into pneumonia-causing cells if the cells are exposed to DNA from a pathogenic strain (see Figure 16.2). This transformation occurs when a nonpathogenic cell takes up a piece of DNA carrying the allele for pathogenicity and replaces its own allele with the foreign allele, an exchange of homologous DNA segments. The cell is now a recombinant: Its chromosome contains DNA derived from two different cells.

For many years after transformation was discovered in laboratory cultures, most biologists thought it was too rare and haphazard to play an important role in natural bacterial populations. But researchers have since learned that many bacteria have cell-surface proteins that recognize DNA from closely related species and transport it into the cell. Once inside the cell, the foreign DNA can be incorporated into the genome by homologous DNA exchange.

In **transduction**, phages (short for "bacteriophages," the viruses that infect bacteria) carry prokaryotic genes from one host cell to another. In most cases, transduction results from accidents that occur during the phage

▼ **Figure 27.11 Transduction.** Phages may carry pieces of a bacterial chromosome from one cell (the donor) to another (the recipient). If crossing over occurs after the transfer, genes from the donor may be incorporated into the recipient's genome.

1 A phage infects a bacterial cell that carries the A^+ and B^+ alleles on its chromosome (brown). This bacterium will be the "donor" cell.

2 The phage DNA is replicated, and the cell makes many copies of phage proteins (represented as purple dots). Certain phage proteins halt the synthesis of proteins encoded by the host cell's DNA, and the host cell's DNA may be fragmented, as shown here.

3 As new phage particles assemble, a fragment of bacterial DNA carrying the A^+ allele happens to be packaged in a phage capsid.

4 The phage carrying the A^+ allele from the donor cell infects a recipient cell with alleles A^- and B^-. Crossing over at two sites (dotted lines) allows donor DNA (brown) to be incorporated into recipient DNA (green).

5 The genotype of the resulting recombinant cell (A^+B^-) differs from the genotypes of both the donor (A^+B^+) and the recipient (A^-B^-).

Phage DNA
Donor cell
A^+
Crossing over
Recipient cell
Recombinant cell

VISUAL SKILLS *Based on this diagram, describe the circumstances in which a transduction event would result in horizontal gene transfer.*

replicative cycle **(Figure 27.11).** A virus that carries prokaryotic DNA may not be able to replicate because it lacks some or all of its own genetic material. However, the virus can attach to another prokaryotic cell (a recipient) and inject prokaryotic DNA acquired from the first cell (the donor). If some of this DNA is then incorporated into the recipient cell's chromosome by crossing over, a recombinant cell is formed.

Conjugation and Plasmids

In a process called **conjugation**, DNA is transferred between two prokaryotic cells (usually of the same species) that are temporarily joined. In bacteria, the DNA transfer is always one-way: One cell donates the DNA, and the other receives it. We'll focus here on the mechanism used by *E. coli*.

▼ Figure 27.12 Bacterial conjugation. The *E. coli* donor cell (left) extends a pilus that attaches to a recipient cell, a key first step in the transfer of DNA. The pilus is a flexible tube of protein subunits (TEM).

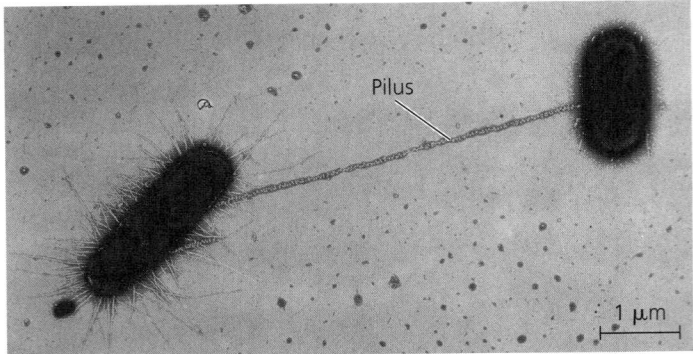

Pilus

1 μm

First, a pilus of the donor cell attaches to the recipient **(Figure 27.12)**. The pilus then retracts, pulling the two cells together, like a grappling hook. The next step is thought to be the formation of a temporary structure between the two cells, a "mating bridge" through which the donor may transfer DNA to the recipient. However, the mechanism by which DNA transfer occurs is unclear; indeed, recent evidence indicates that DNA may pass directly through the hollow pilus.

The ability to form pili and donate DNA during conjugation results from the presence of a particular piece of DNA called the **F factor** (F for fertility). The F factor of *E. coli* consists of about 25 genes, most required for the production of pili. As shown in **Figure 27.13**, the F factor can

▼ Figure 27.13 Conjugation and recombination in *E. coli*. The DNA replication that accompanies transfer of an F plasmid or part of an Hfr bacterial chromosome is called *rolling circle replication*. In effect, the intact circular parental DNA strand "rolls" as its other strand peels off and a new complementary strand is synthesized.

⬡ **Mastering Biology**
Figure Walkthrough

① A cell carrying an F plasmid (an F⁺ cell) forms a mating bridge with an F⁻ cell. One strand of the plasmid's DNA breaks at the point marked by the arrowhead.

② Using the unbroken strand as a template, the cell synthesizes a new strand (light blue). Meanwhile, the broken strand peels off (red arrow), and one end enters the F⁻ cell. There, synthesis of its complementary strand begins.

③ DNA replication continues in both the donor and recipient cells, as the transferred plasmid strand moves farther into the recipient cell.

④ Once DNA transfer and synthesis are completed, the plasmid in the recipient cell circularizes. The recipient cell is now a recombinant F⁺ cell.

(a) Conjugation and transfer of an F plasmid

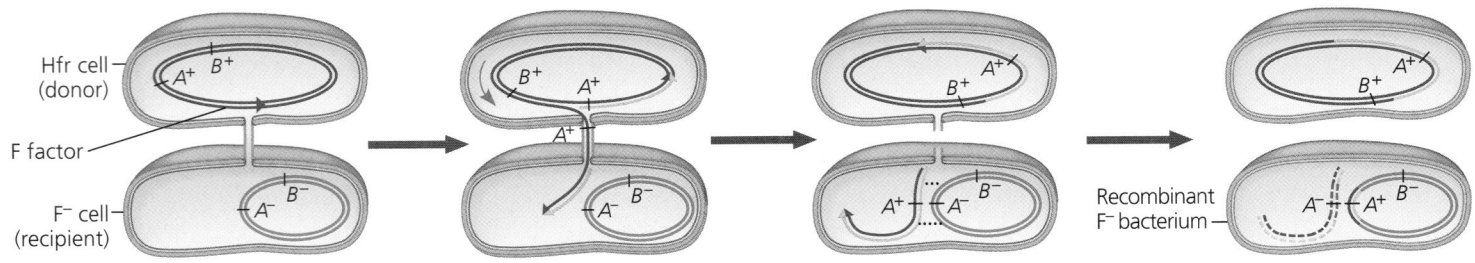

① In an Hfr cell, the F factor (dark blue) is integrated into the bacterial chromosome. Since an Hfr cell has all of the F factor genes, it can form a mating bridge with an F⁻ cell and transfer DNA.

② A single strand of the F factor breaks and begins to move through the bridge. DNA replication occurs in both donor and recipient cells, resulting in double-stranded DNA (daughter strands shown in lighter color).

③ The mating bridge usually breaks before the entire chromosome is transferred. Crossing over at two sites (dotted lines) can result in the exchange of homologous genes (here, A^+ and A^-) between the transferred DNA (brown) and the recipient's chromosome (green).

④ Cellular enzymes degrade any linear DNA not incorporated into the chromosome. The recipient cell, with a new combination of genes but no F factor, is now a recombinant F⁻ cell.

(b) Conjugation and transfer of part of an Hfr bacterial chromosome, resulting in recombination. A^+/A^- and B^+/B^- indicate alleles for gene A and gene B, respectively.

exist either as a plasmid or as a segment of DNA within the bacterial chromosome.

The F Factor as a Plasmid   The F factor in its plasmid form is called the **F plasmid**. Cells containing the F plasmid, designated F⁺ cells, function as DNA donors during conjugation (see Figure 27.13a). Cells lacking the F factor, designated F⁻, function as DNA recipients during conjugation. The F⁺ condition is transferable in the sense that an F⁺ cell converts an F⁻ cell to F⁺ if a copy of the entire F plasmid is transferred. Even if this does not occur, as long as some of the F plasmid's DNA is transferred successfully to the recipient cell, that cell is now a recombinant cell.

The F Factor in the Chromosome   Chromosomal genes can be transferred during conjugation when the donor cell's F factor is integrated into the chromosome. A cell with the F factor built into its chromosome is called an *Hfr cell* (for high frequency of recombination). Like an F⁺ cell, an Hfr cell functions as a donor during conjugation with an F⁻ cell (see Figure 27.13b). When chromosomal DNA from an Hfr cell enters an F⁻ cell, homologous regions of the Hfr and F⁻ chromosomes may align, allowing segments of their DNA to be exchanged. As a result, the recipient cell becomes a recombinant bacterium that has genes derived from the chromosomes of two different cells—a new genetic variant on which evolution can act.

R Plasmids and Antibiotic Resistance   Many species of bacteria have genes that code for enzymes that specifically destroy or otherwise hinder the effectiveness of certain antibiotics, such as tetracycline or ampicillin. Such "resistance genes" are often carried by plasmids known as **R plasmids** (R for resistance). Many R plasmids, like F plasmids, have genes that encode pili and enable DNA transfer from one bacterial cell to another by conjugation. As a result, R plasmids and the resistance genes they carry can spread rapidly through a bacterial population—a problem of great public health concern (see Concept 27.6). Making the problem still worse, some R plasmids carry genes for resistance to as many as ten antibiotics.

⊙ **Mastering Biology** BBC Video: A Future Without Antibiotics?

CONCEPT CHECK 27.2

1. Although rare on a per gene basis, new mutations can add considerable genetic variation to prokaryotic populations in each generation. Explain how this occurs.

2. Distinguish between the three mechanisms by which bacteria can transfer DNA from one bacterial cell to another.

3. **WHAT IF?** If a nonpathogenic bacterium were to acquire resistance to antibiotics, could this strain pose a health risk to people? In general, how does DNA transfer among bacteria affect the spread of resistance genes?

For suggested answers, see Appendix A.

Diverse nutritional and metabolic adaptations have evolved in prokaryotes

The extensive genetic variation found in prokaryotes is reflected in their diverse nutritional adaptations. Like all organisms, prokaryotes can be categorized by how they obtain energy and the carbon used in building the organic molecules that make up cells. Every type of nutrition observed in eukaryotes is represented among prokaryotes, along with some nutritional modes unique to prokaryotes. In fact, prokaryotes have an astounding range of metabolic adaptations, much broader than that found in eukaryotes.

Organisms that obtain energy from light are called *phototrophs*, and those that obtain energy from chemicals are called *chemotrophs*. Organisms that need only CO_2 or related compounds as a carbon source are called *autotrophs*. In contrast, *heterotrophs* require at least one organic nutrient, such as glucose, to make other organic compounds. Combining the two energy sources with the two carbon sources results in four major modes of nutrition, summarized in **Table 27.1**.

Table 27.1 Major Nutritional Modes

Mode	Energy Source	Carbon Source	Types of Organisms
AUTOTROPH			
Photoautotroph	Light	CO_2, HCO_3^-, or related compound	Photosynthetic prokaryotes (for example, cyanobacteria); plants; certain protists (for example, algae)
Chemoautotroph	Inorganic chemicals (such as H_2S, NH_3, or Fe^{2+})	CO_2, HCO_3^-, or related compound	Unique to certain prokaryotes (for example, *Sulfolobus*)
HETEROTROPH			
Photoheterotroph	Light	Organic compounds	Unique to certain aquatic and salt-loving prokaryotes (for example, *Rhodobacter*, *Chloroflexus*)
Chemoheterotroph	Organic compounds	Organic compounds	Many prokaryotes (for example, *Clostridium*) and protists; fungi; animals; some plants

The Role of Oxygen in Metabolism

Prokaryotic metabolism also varies with respect to oxygen (O_2). **Obligate aerobes** must use O_2 for cellular respiration and cannot grow without it. **Obligate anaerobes**, on the other hand, are poisoned by O_2. Some obligate anaerobes live exclusively by fermentation; others extract chemical energy by **anaerobic respiration**, in which substances other than O_2, such as nitrate ions (NO_3^-) or sulfate ions (SO_4^{2-}), accept electrons at the "downhill" end of electron transport chains. **Facultative anaerobes** use O_2 if it is present but can also carry out fermentation or anaerobic respiration in an anaerobic environment.

Nitrogen Metabolism

Nitrogen is essential for the production of amino acids and nucleic acids in all organisms. Whereas eukaryotes can obtain nitrogen only from a limited group of nitrogen compounds, prokaryotes can metabolize nitrogen in many forms. For example, some cyanobacteria and some methanogens (various archaea) convert atmospheric nitrogen (N_2) to ammonia (NH_3), a process called **nitrogen fixation**. The cells can then incorporate this "fixed" nitrogen into amino acids and other organic molecules. In terms of nutrition, nitrogen-fixing cyanobacteria are some of the most self-sufficient organisms since they need only light, CO_2, N_2, water, and some minerals to grow.

Nitrogen fixation has a large impact on other organisms. For example, nitrogen-fixing prokaryotes can increase the nitrogen available to plants, which cannot use atmospheric nitrogen but can use the nitrogen compounds that the prokaryotes produce from ammonia. Concept 55.4 discusses this and other essential roles that prokaryotes play in the nitrogen cycles of ecosystems.

Metabolic Cooperation

Cooperation between prokaryotic cells allows them to use environmental resources they could not use as individual cells. In some cases, this cooperation takes place between specialized cells of a filament. For instance, the cyanobacterium *Anabaena* has genes that encode proteins for photosynthesis and for nitrogen fixation. However, a single cell cannot carry out both processes at the same time because photosynthesis produces O_2, which inactivates the enzymes involved in nitrogen fixation. Instead of living as isolated cells, *Anabaena* forms filamentous chains **(Figure 27.14)**. Most cells in a filament carry out only photosynthesis, while a few specialized cells called **heterocysts** (sometimes called *heterocytes*) carry out only nitrogen fixation. Each heterocyst is surrounded by a thickened cell wall that restricts entry of O_2 produced by neighboring photosynthetic cells. Intercellular connections allow heterocysts to transport fixed nitrogen to neighboring cells and to receive carbohydrates.

Metabolic cooperation among the cells of one or more prokaryotic species often occurs in surface-coating colonies known as **biofilms**. Cells in a biofilm secrete signaling

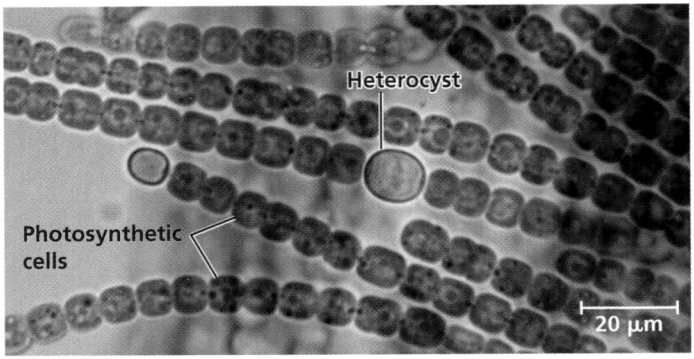

▲ **Figure 27.14 Metabolic cooperation in a prokaryote.** In the filamentous freshwater cyanobacterium *Anabaena*, heterocysts fix nitrogen, while the other cells carry out photosynthesis (LM).

molecules that recruit nearby cells, causing the colonies to grow. The cells also produce polysaccharides and proteins that stick the cells to the substrate and to one another; these polysaccharides and proteins form the capsule, or slime layer, mentioned earlier in the chapter. Channels in the biofilm allow nutrients to reach cells in the interior and wastes to be expelled.

Biofilms are common in nature, but they can cause a wide range of problems for humans. For example, biofilms growing in pipes slow the flow of liquids such as water or oil and degrade the pipes themselves. Biofilms also corrode boat hulls, oil platforms, and many other structures and industrial products. In medical settings, biofilms can contaminate contact lenses, catheters, pacemakers, heart valves, artificial joints, and many other devices. They also contribute to tooth decay and a wide range of chronic infections **(Figure 27.15)**, some of which can be lethal. Many biofilms can evade host immune responses and are extremely resistant to antibiotics, making them hard to treat. Altogether, each year biofilms cause billions of dollars in damage and inflict tens of millions of people with chronic infections.

▼ **Figure 27.15 *Pseudomonas aeruginosa* forming a biofilm (colorized SEM).** This widespread bacterial species can be found in soil and water and as part of the normal community in the human intestines. *P. aeruginosa* can cause serious infections of other human organs, including the skin, urinary tract, and lungs. These infections can be difficult to treat, in part because the bacteria can grow as biofilms that have increased resistance to antibiotics.

1. Distinguish between the four major modes of nutrition, noting which are unique to prokaryotes.
2. A bacterium requires only the amino acid methionine as an organic nutrient and lives in lightless caves. What mode of nutrition does it employ? Explain.
3. **WHAT IF?** Describe what you might eat for a typical meal if humans, like cyanobacteria, could fix nitrogen.

For suggested answers, see Appendix A.

CONCEPT 27.4

Prokaryotes have radiated into a diverse set of lineages

Since their origin 3.5 billion years ago, prokaryotic populations have radiated extensively as a wide range of structural and metabolic adaptations have evolved in them. Collectively, these adaptations have enabled prokaryotes to inhabit every environment known to support life—if there are organisms in a particular place, some of those organisms are prokaryotes. Yet despite their obvious success, it is only in recent decades that advances in genomics have begun to reveal the full extent of prokaryotic diversity.

An Overview of Prokaryotic Diversity

In the 1970s, microbiologists began using small-subunit ribosomal RNA as a marker for evolutionary relationships. Their results indicated that many prokaryotes once classified as bacteria are actually more closely related to eukaryotes and belong in a domain of their own: Archaea. Microbiologists have since analyzed larger amounts of genetic data—including more than 1,700 entire genomes—and have concluded that a few traditional groups, such as cyanobacteria, are monophyletic. However, other traditional groups, such as gram-negative bacteria, are scattered throughout several lineages. **Figure 27.16** shows one phylogenetic hypothesis for some of the major taxa of prokaryotes based on molecular systematics.

One lesson from studying prokaryotic phylogeny is that the genetic diversity of prokaryotes is immense. When researchers began to sequence the genes of prokaryotes, they could investigate only the small fraction of species that could be cultured in the laboratory. In the 1980s, researchers began using the polymerase chain reaction (PCR; see Figure 20.8) to analyze the genes of prokaryotes collected from the environment (such as from soil or water samples). Such "genetic prospecting" is now widely used; in fact, today entire prokaryotic genomes can be obtained from environmental samples using *metagenomics* (see Concept 21.1). Each year these techniques add new branches to the tree of life. While only about 16,000 prokaryotic species worldwide have been assigned scientific names, a single handful of soil could contain 10,000 prokaryotic species

▼ **Figure 27.16 A simplified phylogeny of prokaryotes.** This tree shows relationships among major prokaryotic groups; some of these relationships are shown as polytomies to reflect their uncertain order of divergence. Recent studies indicate that within Archaea, the thaumarchaeotes, aigarchaeotes, crenarchaeotes, and korarchaeotes are closely related; systematists have placed them in a supergroup called "TACK" in reference to the first letters of their names.

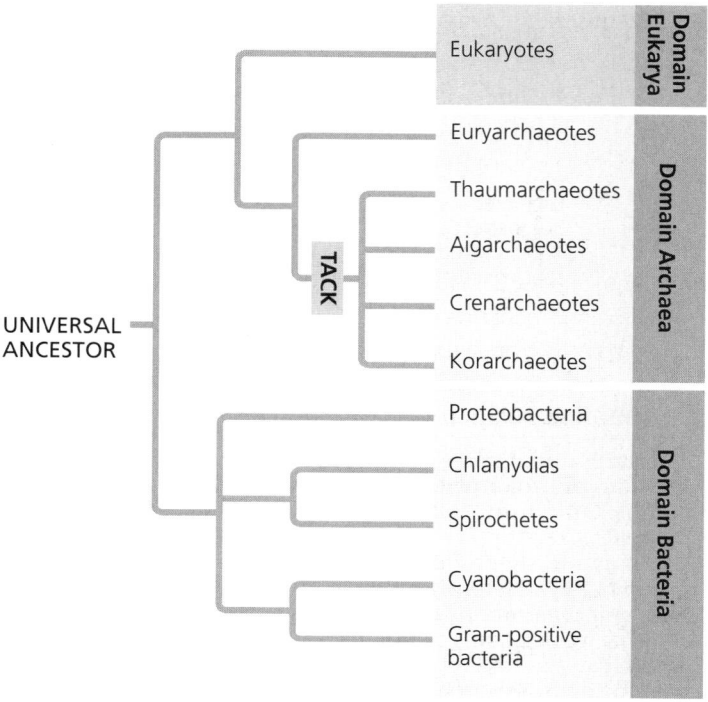

VISUAL SKILLS *Based on this phylogenetic tree diagram, which domain is the sister group of Archaea?*

by some estimates. Taking full stock of this diversity will require many years of research.

Another important lesson from molecular systematics is that horizontal gene transfer has played a key role in the evolution of prokaryotes. Over hundreds of millions of years, prokaryotes have acquired genes from even distantly related species, and they continue to do so today. As a result, significant portions of the genomes of many prokaryotes are actually mosaics of genes imported from other species. For example, a study of 329 sequenced bacterial genomes found that an average of 75% of the genes in each genome had been transferred horizontally at some point in their evolutionary history. As we saw in Concept 26.6, such gene transfers can make it difficult to determine phylogenetic relationships. Still, it is clear that for billions of years, the prokaryotes have evolved in two separate lineages, the bacteria and the archaea (see Figure 27.16).

Bacteria

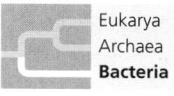
Eukarya
Archaea
Bacteria

As surveyed in **Figure 27.17**, bacteria include the vast majority of prokaryotic species familiar to most people, from the pathogenic species that cause strep throat and tuberculosis to the beneficial species used to make Swiss cheese and yogurt.

This figure highlights a few key groups of bacteria, but their actual diversity is far greater than shown here. Likewise, gene sequence data indicate that the 16,000 known species of bacteria represent a small fraction of the actual number, estimated at 700,000–1.4 million species.

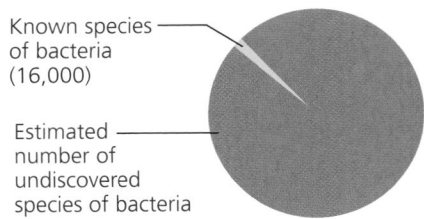

Known species of bacteria (16,000)

Estimated number of undiscovered species of bacteria

Proteobacteria

This large and diverse clade of gram-negative bacteria includes photoautotrophs, chemoautotrophs, and heterotrophs. One autotroph, the sulfur bacterium *Thiomargarita namibiensis*, obtains energy by oxidizing H_2S, producing sulfur as a waste product (the small globules in the photograph below). Heterotrophs include pathogens such as *Neisseria gonorrhoeae*, which causes gonorrhea; *Vibrio cholerae*, which causes cholera; and *Helicobacter pylori*, which causes stomach ulcers. Current evidence indicates that mitochondria evolved by endosymbiosis from a heterotroph in the subgroup alpha proteobacteria.

Thiomargarita namibiensis containing sulfur wastes (LM)

200 μm

Chlamydias

These parasites can survive only within animal cells, depending on their hosts for resources as basic as ATP. The gram-negative walls of chlamydias are unusual in that they lack peptidoglycan. One species, *Chlamydia trachomatis*, is the most common cause of blindness in the world and also causes nongonococcal urethritis, the most common sexually transmitted disease in the United States.

Chlamydia (arrows inside an animal cell (colorized TEM)

2.5 μm

Spirochetes

These helical gram-negative heterotrophs spiral through their environment by means of rotating, internal, flagellum-like filaments. Many spirochetes are free-living, but others are notorious pathogenic parasites: *Treponema pallidum* causes syphilis, and *Borrelia burgdorferi* causes Lyme disease.

Leptospira, a spirochete (colorized TEM)

5 μm

Cyanobacteria

These gram-negative photoautotrophs are the only prokaryotes with plantlike, oxygen-generating photosynthesis. (In fact, chloroplasts are thought to have evolved from an endosymbiotic cyanobacterium.) Both solitary and filamentous cyanobacteria are abundant components of freshwater and marine *phytoplankton*, the collection of small photosynthetic organisms that drift near the water's surface.

Cylindrospermum, a filamentous cyanobacterium

40 μm

Gram-Positive Bacteria

Gram-positive bacteria rival the proteobacteria in diversity. Species in one subgroup, the actinomycetes, form colonies containing branched chains of cells; two of these species cause tuberculosis and leprosy, but most are decomposers living in soil. Soil-dwelling species in the genus *Streptomyces* are cultured as a source of antibiotics, including tetracycline and erythromycin. Other subgroups of gram-positive bacteria include pathogens such as *Staphylococcus aureus* (see Figure 22.14), *Bacillus anthracis*, which causes anthrax, and *Clostridium botulinum*, which causes botulism.

Streptomyces, the source of many antibiotics (SEM)

5 μm

Every major mode of nutrition and metabolism is represented among the 80 or so phyla of bacteria, and even a small group of bacteria may contain species exhibiting many different nutritional modes. As we'll see, the diverse nutritional and metabolic capabilities of bacteria—and archaea—are behind the great impact these organisms have on Earth and its life.

Archaea

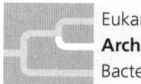

Archaea share certain traits with bacteria and other traits with eukaryotes **(Table 27.2)**. However, archaea also have many unique characteristics, as we would expect in a taxon that has followed a separate evolutionary path for so long.

The first prokaryotes assigned to domain Archaea live in environments so extreme that few other organisms can survive there. Such organisms are called **extremophiles**, meaning "lovers" of extreme conditions (from the Greek *philos*, lover), and include extreme halophiles and extreme thermophiles.

Extreme halophiles (from the Greek *halo*, salt) live in highly saline environments, such as the Great Salt Lake in Utah, the Dead Sea in Israel, and the Spanish lake shown in Figure 27.1. Some species merely tolerate salinity, while others require

▲ **Figure 27.18 Extreme thermophiles.** Orange and yellow colonies of thermophilic prokaryotes grow in the hot water of Yellowstone National Park's Grand Prismatic Spring.

MAKE CONNECTIONS *How might the enzymes of thermophiles differ from those of other organisms? (Review enzymes in Concept 8.4.)*

an environment that is several times saltier than seawater (which has a salinity of 3.5%). For example, the proteins and cell wall of *Halobacterium* have unusual features that improve function in extremely salty environments but render these organisms incapable of survival if the salinity drops below 9%.

Extreme thermophiles (from the Greek *thermos*, hot) thrive in very hot environments. For example, the archaean *Pyrococcus furiosus* lives in geothermally heated marine sediments that reach temperatures of 100°C. At temperatures this high, the cells of most organisms die because their DNA does not remain in a double helix and many of their proteins denature. *P. furiosus* and other extreme thermophiles avoid this fate because they have structural and biochemical adaptations that make their DNA and proteins stable at high temperatures—a feature that has enabled *P. furiosus* to be used in biotechnology as a source of DNA polymerase for the PCR technique (see Figure 20.8). Other examples of extreme thermophiles include archaea in the genus *Sulfolobus*, which live in sulfur-rich volcanic springs **(Figure 27.18)**. One extreme thermophile that lives near deep-sea hot springs called *hydrothermal vents* is informally known as "strain 121" since it can reproduce even at 121°C.

Many other archaea live in more moderate environments. Consider **methanogens**, archaea that release methane as a by-product of how they obtain energy. Many methanogens use CO_2 to oxidize H_2, a process that produces both energy and methane waste. Among the strictest of anaerobes, methanogens are poisoned by O_2. Although some methanogens live in extreme environments, such as under kilometers of ice in Greenland, others live in swamps and marshes where other microorganisms have consumed all the O_2. The "marsh gas" found in such environments is the methane released by these archaea. Other species inhabit the anaerobic guts of cattle, termites, and other herbivores, playing an essential role in the nutrition of these animals. Methanogens

Table 27.2 A Comparison of the Three Domains of Life

CHARACTERISTIC	DOMAIN		
	Bacteria	Archaea	Eukarya
Nuclear envelope	Absent	Absent	Present
Membrane-enclosed organelles	Absent	Absent	Present
Peptidoglycan in cell wall	Present	Absent	Absent
Membrane lipids	Unbranched hydrocarbons	Some branched hydrocarbons	Unbranched hydrocarbons
RNA polymerase	One kind	Several kinds	Several kinds
Initiator amino acid for protein synthesis	Formyl-methionine	Methionine	Methionine
Introns in genes	Very rare	Present in some genes	Present in many genes
Response to the antibiotics streptomycin and chloramphenicol	Growth usually inhibited	Growth not inhibited	Growth not inhibited
Histones associated with DNA	Absent	Present in some species	Present
Circular chromosome	Present	Present	Absent
Growth at temperatures > 100°C	No	Some species	No

are also useful to humans as decomposers in sewage treatment facilities.

Many extreme halophiles and most known methanogens are archaea in the clade Euryarchaeota (from the Greek *eurys*, broad, a reference to their wide habitat range). The euryarchaeotes also include some extreme thermophiles, though most thermophilic species belong to a second clade, Crenarchaeota (*cren* means "spring," such as a hydrothermal spring). Metagenomic studies have identified many species of euryarchaeotes and crenarchaeotes that are not extremophiles. These archaea exist in habitats ranging from farm soils to lake sediments to the surface of the open ocean.

New findings continue to inform our understanding of archaeal phylogeny. For example, recent metagenomic studies have uncovered the genomes of many species that are not members of Euryarchaeota or Crenarchaeota. Moreover, phylogenomic analyses show that three of these newly discovered groups—the Thaumarchaeota, Aigarchaeota, and Korarchaeota—are more closely related to the Crenarchaeota than they are to the Euryarchaeota. These findings have led to the identification of a "supergroup" that contains the Thaumarchaeota, Aigarchaeota, Crenarchaeota, and Korarchaeota (see Figure 27.16). This supergroup is referred to as "TACK" based on the names of the groups it includes.

The importance of the TACK supergroup was highlighted by the recent discovery of the lokiarchaeotes, a group that is closely related to TACK archaea and that may represent the long-sought-after sister group of the eukaryotes. As such, the characteristics of lokiarchaeotes may shed light on one of the major puzzles of biology today—how eukaryotes arose from their prokaryotic ancestors. The pace of these and other recent discoveries suggests that as metagenomic prospecting continues, the tree in Figure 27.16 will likely undergo further changes.

CONCEPT CHECK 27.4

1. Explain how molecular systematics and metagenomics have contributed to our understanding of the phylogeny and evolution of prokaryotes.
2. **DRAW IT** Redraw Figure 27.16 assuming that eukaryotes are more closely related to TACK archaea than to euryarchaeotes. Under this assumption, would the three domains shown in Figure 27.16 be valid? Explain.

For suggested answers, see Appendix A.

CONCEPT 27.5

Prokaryotes play crucial roles in the biosphere

If people were to disappear from the planet tomorrow, life on Earth would change for many species, but few would be driven to extinction. In contrast, prokaryotes are so important to the biosphere that if they were to disappear, the prospects of survival for many other species would be dim.

Chemical Recycling

The atoms that make up the organic molecules in all living things were at one time part of inorganic substances in the soil, air, and water. Sooner or later, those atoms will return to the nonliving environment. Ecosystems depend on the continual recycling of chemical elements between the living and nonliving components of the environment, and prokaryotes play a major role in this process. For example, some chemoheterotrophic prokaryotes function as **decomposers**, breaking down dead organisms as well as waste products and thereby unlocking supplies of carbon, nitrogen, and other elements. Without the actions of prokaryotes and other decomposers such as fungi, life as we know it would cease. (See Concept 55.4 for a detailed discussion of chemical cycles.)

Prokaryotes also convert some molecules to forms that can be taken up by other organisms. Cyanobacteria and other autotrophic prokaryotes use CO_2 to make organic compounds such as sugars, which are then passed up through food chains. Cyanobacteria also produce atmospheric O_2, and a variety of prokaryotes fix atmospheric nitrogen (N_2) into forms that other organisms can use to make the building blocks of proteins and nucleic acids. Under some conditions, prokaryotes can increase the availability of nutrients that plants require for growth, such as nitrogen, phosphorus, and potassium **(Figure 27.19)**. Prokaryotes can also *decrease* the availability of key plant nutrients; this occurs when prokaryotes "immobilize" nutrients by using them to synthesize molecules

▼ **Figure 27.19 Impact of bacteria on soil nutrient availability.** Pine seedlings grown in sterile soils to which one of three strains of the bacterium *Burkholderia glathei* had been added absorbed more potassium (K^+) than did seedlings grown in soil without any bacteria. Other results (not shown) demonstrated that strain 3 increased the amount of K^+ released from mineral crystals to the soil.

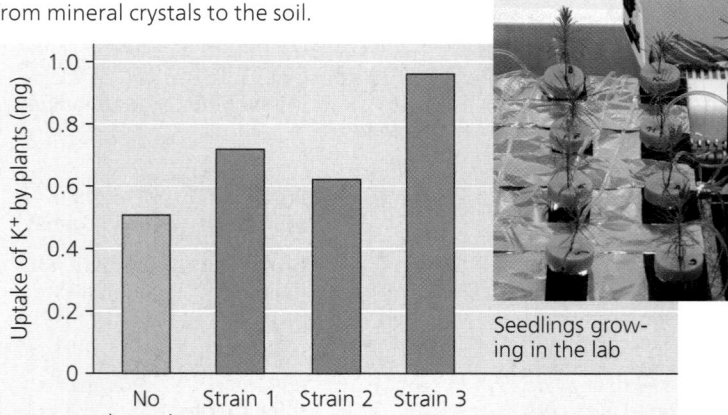

Seedlings growing in the lab

WHAT IF? *Estimate the average uptake of K^+ for seedlings in soils with bacteria. What would you expect this average to be if bacteria had no effect on nutrient availability?*

that remain within their cells. Thus, prokaryotes can have complex effects on soil nutrient concentrations. In marine environments, an archaean from the clade Crenarchaeota can perform nitrification, a key step in the nitrogen cycle (see Figure 55.14). Crenarchaeotes dominate the oceans by numbers, comprising an estimated 10^{28} cells. The sheer abundance of these organisms suggests that they may have a large impact on the global nitrogen cycle.

Ecological Interactions

Prokaryotes play a central role in many ecological interactions. Consider **symbiosis** (from a Greek word meaning "living together"), an ecological relationship in which two species live in close contact with each other. Prokaryotes often form symbiotic associations with much larger organisms. In general, the larger organism in a symbiotic relationship is known as the **host**, and the smaller is known as the **symbiont**. There are many cases in which a prokaryote and its host participate in **mutualism**, an ecological interaction between two species in which both benefit **(Figure 27.20)**. Other interactions take the form of **commensalism**, an ecological relationship in which one species benefits while the other is not harmed or helped in any significant way. For example, more than 150 bacterial species live on the outer surface of your body, covering portions of your skin with up to 10 million cells per square centimeter. Some of these species are commensalists: You provide them with food, such as the oils that exude from your pores, and a place to live, while they neither harm nor benefit you. Finally, some symbiotic prokaryotes engage in **parasitism**, an ecological relationship in which a **parasite** feeds on the cell contents, tissues, or body fluids of its host. As a group, parasites harm but usually do not kill their host, at least not immediately (unlike a predator). Parasites that cause disease are known as **pathogens**, many of which are prokaryotic. (We'll discuss

mutualism, commensalism, and parasitism in greater detail in Concept 54.1.)

The very existence of an ecosystem can depend on prokaryotes. For example, consider the diverse ecological communities found at hydrothermal vents. These communities are densely populated by many different kinds of animals, including worms, clams, crabs, and fishes. But since sunlight does not penetrate to the deep ocean floor, the community does not include photosynthetic organisms. Instead, the energy that supports the community is derived from the metabolic activities of chemoautotrophic bacteria. These bacteria harvest chemical energy from compounds such as hydrogen sulfide (H_2S) that are released from the vent. An active hydrothermal vent may support hundreds of eukaryotic species, but when the vent stops releasing chemicals, the chemoautotrophic bacteria cannot survive. As a result, the entire vent community collapses.

Mastering Biology Interview with Penny Chisholm: Studying the role of cyanobacteria in marine ecosystems (see the interview before Chapter 26)

CONCEPT CHECK 27.5

1. Explain how prokaryotes, though small, can be considered giants in their collective impact on Earth and its life.
2. **MAKE CONNECTIONS** Review Figure 10.5. Then summarize the main steps by which cyanobacteria produce O_2 and use CO_2 to make organic compounds.

For suggested answers, see Appendix A.

CONCEPT 27.6

Prokaryotes have both beneficial and harmful impacts on humans

Although the best-known prokaryotes tend to be the bacteria that cause human illness, these pathogens represent only a small fraction of prokaryotic species. Many other prokaryotes have positive interactions with people, and some play essential roles in agriculture and industry.

Mutualistic Bacteria

As is true for many other eukaryotes, human well-being can depend on mutualistic prokaryotes. For example, our intestines are home to an estimated 500–1,000 species of bacteria; collectively, their cells outnumber all human cells in the body by a factor of ten. Different species live in different portions of the intestines, and they vary in their ability to process different foods. Many of these species are mutualists, digesting food that our own intestines cannot break down. The genome of one of these gut mutualists, *Bacteroides thetaiotaomicron*, includes a large array of genes involved in synthesizing carbohydrates, vitamins, and other nutrients needed by

▼ **Figure 27.20 Mutualism: bacterial "headlights."** The glowing oval below the eye of the flashlight fish (*Photoblepharon palpebratus*) is an organ harboring bioluminescent bacteria. The fish uses the light to attract prey and to signal potential mates. The bacteria receive nutrients from the fish.

▲ **Figure 27.21 Lyme disease.** Ticks in the genus *Ixodes* spread the disease by transmitting the spirochete *Borrelia burgdorferi* (colorized SEM). A rash may develop at the site of the tick's bite; the rash may be large and ring-shaped (as shown) or much less distinctive.

humans. Signals from the bacterium activate human genes that build the network of intestinal blood vessels necessary to absorb nutrient molecules. Other signals induce human cells to produce antimicrobial compounds to which *B. thetaiotaomicron* is not susceptible. This action may reduce the population sizes of other, competing species, thus potentially benefiting both *B. thetaiotaomicron* and its human host.

Pathogenic Bacteria

All the pathogenic prokaryotes known to date are bacteria, and they deserve their negative reputation. Bacteria cause about half of all human diseases. For example, more than 1.5 million people die each year of the lung disease tuberculosis, caused by *Mycobacterium tuberculosis*. And another 2 million people die each year from diarrheal diseases caused by various bacteria.

Some bacterial diseases are transmitted by other species, such as fleas or ticks. In the United States, the most widespread pest-carried disease is Lyme disease, which infects about 300,000 people each year according to a recent CDC estimate **(Figure 27.21)**. Caused by a bacterium carried by ticks, Lyme disease can result in debilitating arthritis, heart disease, nervous disorders, and death if untreated.

Pathogenic prokaryotes usually cause illness by producing poisons, which are classified as exotoxins or endotoxins. **Exotoxins** are proteins secreted by certain bacteria and other organisms. Cholera, a dangerous diarrheal disease, is caused by an exotoxin secreted by the proteobacterium *Vibrio cholerae*. The exotoxin stimulates intestinal cells to release chloride ions into the gut, and water follows by osmosis. In another example, the potentially fatal disease botulism is caused by botulinum toxin, an exotoxin secreted by the gram-positive bacterium *Clostridium botulinum* as it ferments various foods, including improperly canned meat,

seafood, and vegetables. Like other exotoxins, the botulinum toxin can produce disease even if the bacteria that manufacture it are no longer present when the food is eaten.

Endotoxins are lipopolysaccharide components of the outer membrane of gram-negative bacteria. In contrast to exotoxins, endotoxins are released only when the bacteria die and their cell walls break down. Endotoxin-producing bacteria include species in the genus *Salmonella*, such as *Salmonella typhi*, which causes typhoid fever. You might have heard of food poisoning caused by other *Salmonella* species that can be found in poultry and some fruits and vegetables.

Finally, horizontal gene transfer can spread genes associated with virulence, turning normally harmless bacteria into potent pathogens. *E. coli*, for instance, is ordinarily a harmless symbiont in the human intestines, but pathogenic strains that cause bloody diarrhea have emerged. One of the most dangerous strains, O157:H7, is a global threat; in the United States alone, there are around 75,000 cases of O157:H7 infection per year, often from contaminated beef or produce. Scientists have discovered over 1,000 genes in O157:H7 that have no counterpart in harmless *E. coli* strains. Some of the genes found only in O157:H7 are associated with virulence and were likely transferred by phage-mediated horizontal gene transfer (transduction) from pathogenic bacterial species.

Antibiotic Resistance

Since their initial use in the 1940s, antibiotics have saved many lives and reduced the incidence of disease caused by pathogenic bacteria. Unfortunately, as **Figure 27.22** shows,

▼ **Figure 27.22 The rise of antibiotic resistance.** As illustrated by the examples shown here, bacteria have developed resistance to every antibiotic currently in clinical use, often within a few years. It took more than 50 years for resistance to develop to colistin, most likely because this antibiotic has toxic side effects in humans and hence was used rarely (as a treatment of last resort).

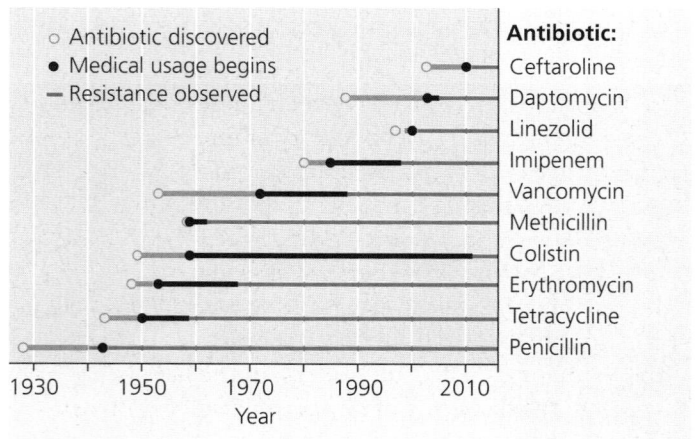

VISUAL SKILLS *Based on this diagram, identify the antibiotic for which resistance appeared before the antibiotic began to be used medicinally.*

resistance to antibiotics has evolved in many bacteria, often rapidly. Compounding this problem, the discovery of new antibiotics has not kept pace with the rate at which resistance has evolved in bacteria. We now face a major public health concern: For every antibiotic now in use, at least one species of bacteria has developed resistance to it.

The rise of antibiotic resistance is driven by several factors. The rapid reproduction of bacteria enables cells carrying resistance genes to quickly produce large numbers of resistant offspring. Thus, when humans use antibiotics in medical or agricultural settings, natural selection can cause the fraction of a bacterial population carrying genes for antibiotic resistance to increase rapidly. Resistance genes also can spread within and among bacterial species by horizontal gene transfer. As a result, bacterial strains that are resistant to multiple antibiotics are becoming more common, making the treatment of certain bacterial infections very difficult.

Consider tuberculosis, the leading cause of death by infectious disease worldwide. The recent rise of drug-resistant strains of the tuberculosis bacterium (*Mycobacterium tuberculosis*) threatens decades of progress in combating tuberculosis. "Superbug" tuberculosis strains came to public notice in 2006, when an outbreak in a rural area of South Africa killed 52 of the 53 patients infected with what is now called extensively drug-resistant tuberculosis (XDR-TB). As of 2017, more than 100 countries had confirmed cases of XDR-TB. There are few treatments for XDR-TB, and those we have are difficult to administer and more toxic than are treatments for standard strains of the tuberculosis bacterium.

M. tuberculosis is not the only bacterium in which resistance to multiple antibiotics has evolved. Highly resistant strains have also been found in species that cause potentially lethal infections of the skin, lungs, digestive tract, and other human organs. Even so, recent discoveries give cause for hope. In 2018, for example, researchers used a metagenomic approach to sequence genes from soil bacteria that could not be grown in the laboratory. They found that some of these genes encoded a new class of antibiotics, which have been named *malacidins*. Malacidins have been effective against multidrug-resistant gram-positive pathogens, including clone USA300 **(Figure 27.23)**, a highly virulent strain of methicillin-resistant *Staphylococcus aureus* (MRSA; see Figure 22.14). Also in 2018, other researchers developed virus-like particles capable of targeting (and killing) particular species of multidrug-resistant bacteria. In recent years, novel approaches for growing soil bacteria in the laboratory also have led to the discovery of new antibiotics, as you can explore in the **Scientific Skills Exercise**.

▼ **Figure 27.23 Effectiveness of malacidin against MRSA clone USA300.** Treatment with malacidin rapidly cleared a skin infection in rats caused by clone USA300, a highly virulent strain of methicillin-resistant *Staphylococcus aureus* (MRSA). Malacidin was as effective as daptomycin, an antibiotic used to treat life-threatening infections.

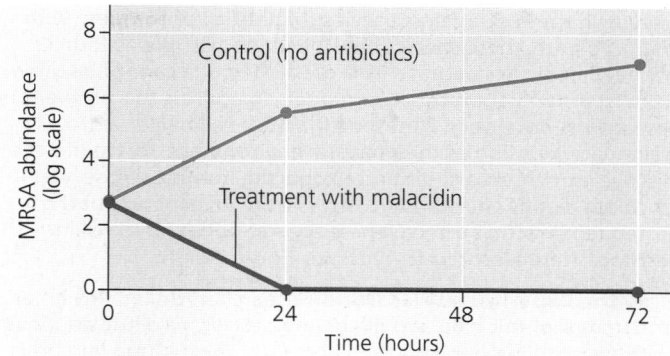

Prokaryotes in Research and Technology

On a positive note, we reap many benefits from the metabolic capabilities of both bacteria and archaea. For example, people have long used bacteria to convert milk to cheese and yogurt. Bacteria are also used in the production of beer and wine, pepperoni, fermented cabbage (sauerkraut), and soy sauce. In recent decades, our greater understanding of prokaryotes has led to an explosion of new applications in biotechnology. Examples include the use of *E. coli* in gene cloning (see Figure 20.4) and the use of DNA polymerase from *Pyrococcus furiosus* in the PCR technique (see Figure 20.8). Through genetic engineering, we can modify bacteria to produce vitamins, antibiotics, hormones, and other products (see Concept 20.1).

Recently, the prokaryotic CRISPR-Cas system, which helps bacteria and archaea defend against attack by viruses (see Figure 19.7), has been developed into a powerful new tool for altering genes in virtually any organism. The genomes of many prokaryotes contain short DNA repeats, called CRISPRs, that interact with proteins known as the Cas (CRISPR-associated) proteins. Cas proteins, acting together with "guide RNA" made from the CRISPR region, can cut any DNA sequence to which they are directed. Scientists have been able to exploit this system by introducing a Cas protein (Cas9) to guide RNA into cells whose DNA they want to alter (see Figure 17.28). Among other applications, this **CRISPR-Cas9 system** has already opened new lines of research on HIV, the virus that causes AIDS **(Figure 27.24)**. While the CRISPR-Cas9 system can potentially be used in many different ways, care must be taken to guard against the unintended consequences that could arise when applying such a new and powerful technology.

Another valuable application of bacteria is to reduce our use of petroleum. Consider the plastics industry. Globally,

Scientific Skills Exercise

Calculating and Interpreting Means and Standard Errors

Can Antibiotics Obtained from Soil Bacteria Help Fight Drug-Resistant Bacteria? Soil bacteria synthesize antibiotics, which they use against species that attack or compete with them. To date, these species have been inaccessible as sources for new medicines because 99% of soil bacteria cannot be grown using standard laboratory techniques. To address this problem, researchers developed a method in which soil bacteria grow in a simulated version of their natural environment; this led to the discovery of a new antibiotic, teixobactin. In this exercise, you'll calculate means and standard errors from an experiment that tested teixobactin's effectiveness against MRSA (methicillin-resistant *Staphylococcus aureus*; see Figure 22.14).

How the Experiment Was Done Researchers drilled tiny holes into a small plastic chip and filled the holes with a dilute aqueous solution containing soil bacteria and agar. The dilution had been calibrated so that only one bacterium was likely to grow in each hole. After the agar solidified, the chip was placed in a container containing the original soil; nutrients and other essential materials from the soil diffused into the agar, allowing the bacteria to grow.

After isolating teixobactin from a soil bacterium, researchers performed the following experiment: Mice infected with MRSA were given low (1 mg/kg) or high (5 mg/kg) doses of teixobactin or vancomycin, an existing antibiotic; in the control, mice infected with MRSA were not given an antibiotic. After 26 hours, researchers sampled infected mice and estimated the number of *S. aureus* colonies in each sample. Results were reported on a log scale; note that a decrease of 1.0 on this scale reflects a tenfold decrease in MRSA abundance.

Data from the Experiment

Treatment	Dose (mg/kg)	Log of Number of Colonies	Mean ($\bar{x}$)
Control	—	9.0, 9.5, 9.0, 8.9	
Vancomycin	1.0	8.5, 8.4, 8.2	
	5.0	5.3, 5.9, 4.7	
Teixobactin	1.0	8.5, 6.0, 8.4, 6.0	
	5.0	3.8, 4.9, 5.2, 4.9	

► **Plastic chip used to grow soil bacteria**

Data from L. Ling et al. A new antibiotic kills pathogens without detectable resistance, *Nature* 517:455–459 (2015).

INTERPRET THE DATA

1. The mean ($\bar{x}$) of a variable is the sum of the data values divided by the number of observations (n):

$$\bar{x} = \frac{\sum x_i}{n}$$

   In this formula, x_i is the value of the ith observation of the variable (observation number "i" out of a total of n such observations); the $\sum$ symbol indicates that the n values of x are to be added together. Calculate the mean for each treatment.

2. Use your results from question 1 to evaluate the effectiveness of vancomycin and teixobactin.

3. The variation found in a set of data can be estimated by the standard deviation, s:

$$s = \sqrt{\frac{\sum (x_i - \bar{x})^2}{n - 1}}$$

   Calculate the standard deviation for each treatment.

4. The standard error (SE), which indicates how greatly the mean would likely vary if the experiment was repeated, is calculated as:

$$SE = \frac{s}{\sqrt{n}}$$

   As a rough rule of thumb, if an experiment were to be repeated, the new mean typically would lie within two standard errors of the original mean (that is, within the range $\bar{x} \pm 2SE$). Calculate $\bar{x} \pm 2SE$ for each treatment, determine whether these ranges overlap, and interpret your results.

→ **Instructors:** A version of this Scientific Skills Exercise can be assigned in **Mastering Biology**.

▼ **Figure 27.24 CRISPR: opening new avenues of research for treating HIV infection. (a)** In laboratory experiments, untreated (control) human cells were susceptible to infection by HIV, the virus that causes AIDS. **(b)** In contrast, cells treated with a CRISPR-Cas9 system that targets HIV were resistant to viral infection. The CRISPR-Cas9 system was also able to remove HIV proviruses (see Figure 19.8) that had become incorporated into the DNA of human cells.

(a) Control cells. The green color indicates infection by HIV.

(b) Experimental cells. These cells were treated with a CRISPR-Cas9 system that targets HIV.

each year more than 750 billion pounds of plastic are produced from petroleum and used to make toys, storage containers, soft drink bottles, and many other items. Many of these products end up as plastic waste that contaminates natural habitats and degrades slowly, creating environmental problems (see Concept 56.4). One promising avenue for addressing these problems is to use bacteria that produce natural plastics **(Figure 27.25)**. For example, some bacteria synthesize a type of polymer known as PHA (polyhydroxyalkanoate), which they use to store chemical energy. The PHA can be extracted, formed into pellets, and used to make durable, yet biodegradable, plastics. Researchers are also seeking to reduce the use of petroleum and other fossil fuels by engineering bacteria that can produce ethanol from various forms of biomass, including agricultural waste, switchgrass, and corn.

Another way to harness prokaryotes is **bioremediation**, the use of organisms to remove pollutants from soil, air, or

◄ Figure 27.25
Bacteria synthesizing and storing PHA, a component of biodegradable plastics.

► Figure 27.26
Bioremediation of an oil spill.
Spraying fertilizer stimulates the growth of native bacteria that metabolize oil, increasing the speed of the breakdown process up to fivefold.

water. For example, anaerobic bacteria and archaea decompose the organic matter in sewage, converting it to material that can be used as landfill or fertilizer after chemical sterilization. Other bioremediation applications include cleaning up oil spills **(Figure 27.26)** and precipitating radioactive material (such as uranium) out of groundwater.

The usefulness of prokaryotes largely derives from their diverse forms of nutrition and metabolism. All this metabolic versatility evolved prior to the appearance of the structural novelties that heralded the evolution of eukaryotic organisms, discussed in the rest of this unit.

CONCEPT CHECK 27.6

1. Identify at least two ways that prokaryotes have affected you positively today.
2. A pathogenic bacterium's toxin causes symptoms that increase the bacterium's chance of spreading from host to host. Does this information indicate whether the poison is an exotoxin or endotoxin? Explain.
3. **WHAT IF?** How might a sudden and dramatic change in your diet affect the diversity of prokaryotic species that live in your digestive tract?

For suggested answers, see Appendix A.

27 Chapter Review

→ Go to **Mastering Biology** for Assignments, the eText, the Study Area, and Dynamic Study Modules.

SUMMARY OF KEY CONCEPTS

→ To review key terms, go to the **Vocabulary Self-Quiz** in the **Mastering Biology** eText or Study Area, or go to goo.gl/zkjz9t.

CONCEPT 27.1

Structural and functional adaptations contribute to prokaryotic success (pp. 574–578)

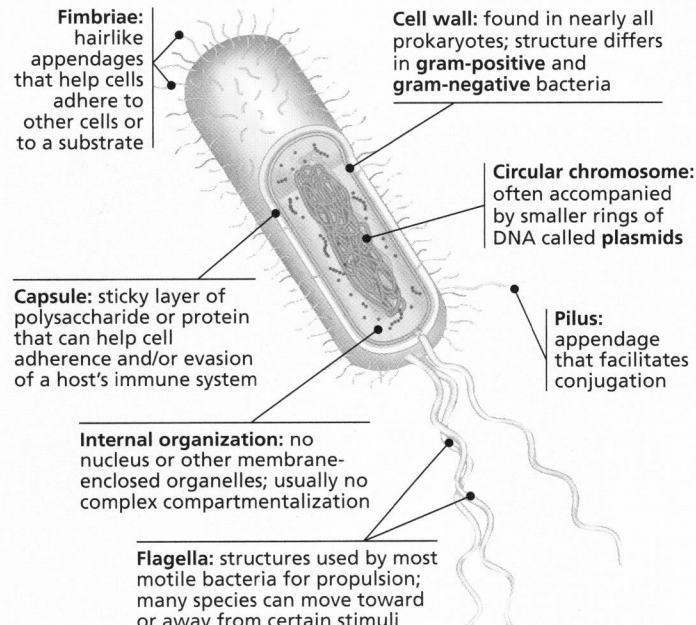

Fimbriae: hairlike appendages that help cells adhere to other cells or to a substrate

Cell wall: found in nearly all prokaryotes; structure differs in **gram-positive** and **gram-negative** bacteria

Circular chromosome: often accompanied by smaller rings of DNA called **plasmids**

Capsule: sticky layer of polysaccharide or protein that can help cell adherence and/or evasion of a host's immune system

Pilus: appendage that facilitates conjugation

Internal organization: no nucleus or other membrane-enclosed organelles; usually no complex compartmentalization

Flagella: structures used by most motile bacteria for propulsion; many species can move toward or away from certain stimuli

• Many **prokaryotic** species can reproduce quickly by binary fission, leading to the formation of extremely large populations.

❓ *Describe features of prokaryotes that enable them to thrive in a wide range of different environments.*

CONCEPT 27.2

Rapid reproduction, mutation, and genetic recombination promote genetic diversity in prokaryotes (pp. 578–581)

• Because prokaryotes can often proliferate rapidly, mutations can quickly increase a population's genetic variation. As a result, prokaryotic populations often can evolve in short periods of time in response to changing conditions.
• Genetic diversity in prokaryotes also can arise by recombination of the DNA from two different cells (via **transformation**, **transduction**, or **conjugation**). By transferring advantageous alleles, such as ones for antibiotic resistance, recombination can promote adaptive evolution in prokaryotic populations.

❓ *Mutations are rare and prokaryotes reproduce asexually, yet their populations can have high genetic diversity. Explain how this can occur.*

CONCEPT 27.3

Diverse nutritional and metabolic adaptations have evolved in prokaryotes (pp. 581–583)

• Nutritional diversity is much greater in prokaryotes than in eukaryotes and includes all four modes of nutrition: photoautotrophy, chemoautotrophy, photoheterotrophy, and chemoheterotrophy.
• Among prokaryotes, **obligate aerobes** require O_2, **obligate anaerobes** are poisoned by O_2, and **facultative anaerobes** can survive with or without O_2.

- Unlike eukaryotes, prokaryotes can metabolize nitrogen in many different forms. Some can convert atmospheric nitrogen to ammonia, a process called **nitrogen fixation**.
- Prokaryotic cells and even species may cooperate metabolically, including in surface-coating **biofilms**.

? *Describe the range of prokaryotic metabolic adaptations.*

CONCEPT 27.4

Prokaryotes have radiated into a diverse set of lineages *(pp. 583–586)*

- Molecular systematics is helping biologists classify prokaryotes and identify new clades.
- Diverse nutritional types are scattered among the major groups of bacteria. The two largest groups are proteobacteria and gram-positive bacteria.
- Some archaea, such as **extreme thermophiles** and **extreme halophiles**, live in extreme environments. Other archaea live in moderate environments such as soils and lakes.

? *How have molecular data informed prokaryotic phylogeny?*

CONCEPT 27.5

Prokaryotes play crucial roles in the biosphere *(pp. 586–587)*

- Decomposition by heterotrophic prokaryotes and the synthetic activities of autotrophic and nitrogen-fixing prokaryotes contribute to the recycling of elements in ecosystems.
- Many prokaryotes have a **symbiotic** relationship with a **host**; the relationships between prokaryotes and their hosts range from **mutualism** to **commensalism** to **parasitism**.

? *In what ways are prokaryotes key to the survival of many species?*

CONCEPT 27.6

Prokaryotes have both beneficial and harmful impacts on humans *(pp. 587–591)*

- People depend on mutualistic prokaryotes, including hundreds of species that live in our intestines and help digest food.
- Pathogenic bacteria typically cause disease by releasing **exotoxins** or **endotoxins**. Horizontal gene transfer can spread genes associated with virulence to harmless species or strains.
- Resistance to multiple antibiotics has evolved in many pathogenic bacteria, a problem of great concern for public health.
- Prokaryotes can be used in **bioremediation** and production of plastics, vitamins, antibiotics, and other products.

? *Describe beneficial and harmful impacts of prokaryotes on humans.*

TEST YOUR UNDERSTANDING

→ For more multiple-choice questions, go to the **Practice Test** in the **Mastering Biology** eText or Study Area, or go to goo.gl/GruWRg.

Levels 1-2: Remembering/Understanding

1. A process that cannot produce genetic variation in bacterial populations is
   (A) transduction.   (C) mutation.
   (B) conjugation.   (D) meiosis.

2. Photoautotrophs use
   (A) light as an energy source and CO_2 as a carbon source.
   (B) light as an energy source and methane as a carbon source.
   (C) N_2 as an energy source and CO_2 as a carbon source.
   (D) CO_2 as both an energy source and a carbon source.

3. Which of the following statements is true?
   (A) Archaea and bacteria have identical membrane lipids.
   (B) The cell walls of archaea lack peptidoglycan.
   (C) Prokaryotes have low levels of genetic diversity.
   (D) No archaea are capable of using CO_2 to oxidize H_2, releasing methane.

4. Which of the following involves metabolic cooperation among prokaryotic cells?
   (A) binary fission   (C) biofilms
   (B) endospore formation   (D) photoautotrophy

5. Which of the following describes a bacterium that lives in the human intestine and causes disease?
   (A) commensalist   (C) gut mutualist
   (B) decomposer   (D) symbiotic pathogen

6. Photosynthesis that releases O_2 occurs in
   (A) cyanobacteria.   (C) gram-positive bacteria.
   (B) archaea.   (D) chemoautotrophic bacteria.

Levels 3-4: Applying/Analyzing

7. **EVOLUTION CONNECTION** In patients with nonresistant strains of the tuberculosis bacterium, antibiotics can relieve symptoms in a few weeks. However, it takes much longer to halt the infection, and patients may discontinue treatment while bacteria are still present. Explain how this could result in the evolution of drug-resistant pathogens.

Levels 5-6: Evaluating/Creating

8. **SCIENTIFIC INQUIRY • INTERPRET THE DATA** The nitrogen-fixing bacterium *Rhizobium* infects the roots of some plant species, forming a mutualism in which the bacterium provides nitrogen, and the plant provides carbohydrates. Scientists measured the 12-week growth of one such plant species (*Acacia irrorata*) when infected by six different *Rhizobium* strains. (a) Graph the data. (b) Interpret the graph.

Rhizobium strain	1	2	3	4	5	6
Plant mass (g)	0.91	0.06	1.56	1.72	0.14	1.03

Data from J. J. Burdon et al., Variation in the effectiveness of symbiotic associations between native rhizobia and temperate Australian *Acacia*: within species interactions, *Journal of Applied Ecology* 36:398–408 (1999).

Note: Without *Rhizobium*, after 12 weeks, *Acacia* plants have a mass of about 0.1 g.

9. **WRITE ABOUT A THEME: ENERGY** In a short essay (about 100–150 words), discuss how prokaryotes and other members of hydrothermal vent communities transfer and transform energy.

10. **SYNTHESIZE YOUR KNOWLEDGE**

Explain how the small size and rapid reproduction rate of bacteria (such as the population shown here on the tip of a pin) contribute to their large population sizes and high genetic variation.

For selected answers, see Appendix A.

28 Protists

KEY CONCEPTS

28.1 Most eukaryotes are single-celled organisms *p. 594*

28.2 Excavates include protists with modified mitochondria and protists with unique flagella *p. 597*

28.3 SAR is a highly diverse group of protists defined by DNA similarities *p. 601*

28.4 Red algae and green algae are the closest relatives of plants *p. 609*

28.5 Unikonts include protists that are closely related to fungi and animals *p. 611*

28.6 Protists play key roles in ecological communities *p. 614*

Study Tip

Label a drawing: Protist life cycles vary greatly. To make sense of this variation, it can be helpful to identify when mitosis occurs in different life cycles, as shown here for a portion of Figure 28.14. Identify the stages at which mitosis occurs in Figures 28.14, 28.18, 28.25, and 28.28.

→ Go to Mastering Biology

For Students (in eText and Study Area)
- Get Ready for Chapter 28
- Animation: Life Cycle of a Malaria Parasite

For Instructors to Assign (in Item Library)
- Scientific Skills Exercise: Interpreting Comparisons of Genetic Sequences
- Misconception Questions 136–140

Figure 28.1 Knowing that prokaryotes are extremely small, you might assume that this image shows six prokaryotes and one much larger eukaryote. But in fact, the only prokaryote is the organism immediately above the scale bar—the rest are members of diverse, mostly unicellular groups of eukaryotes informally known as **protists**.

What gave rise to the great diversity of protists, and how have their lineages diverged over time?

Much of protistan diversity can be traced to endosymbiosis:

Over time, early heterotrophic and photosynthetic eukaryotes have given rise to four highly diverse **supergroups** of eukaryotes:

- Excavata
- SAR
- Archaeplastida
- Unikonta

Most eukaryotes are single-celled organisms

Protists—along with plants, animals, and fungi—are eukaryotes; they are in domain Eukarya, one of the three domains of life. Genetic and morphological studies have shown that some protists are more closely related to plants, fungi, or animals than they are to other protists. As a result, the kingdom in which all protists once were classified, Protista, has been abandoned, and various protist lineages are now recognized as major groups in their own right. Most biologists still use the term *protist*, but only as a convenient way to refer to eukaryotes that are not plants, animals, or fungi.

Unlike the cells of prokaryotes, the cells of protists and other eukaryotes have a nucleus and other membrane-enclosed organelles, such as mitochondria and the Golgi apparatus. Such organelles provide specific locations where particular cellular functions are accomplished, making the structure and organization of eukaryotic cells more complex than those of prokaryotic cells. Eukaryotic cells also have a well-developed cytoskeleton that extends throughout the cell (see Figure 6.20). The cytoskeleton provides the structural support that enables eukaryotic cells to have asymmetric (irregular) forms, as well as to change in shape as they feed, move, or grow. In contrast, prokaryotic cells lack a well-developed cytoskeleton, thus limiting the extent to which they can maintain asymmetric forms or change shape over time.

We'll survey the diversity of eukaryotes throughout the rest of this unit, beginning in this chapter with the protists. As you explore this material, bear in mind that

- the organisms in most eukaryotic lineages are protists, and
- most protists are unicellular.

Thus, life differs greatly from how most of us commonly think of it. The large, multicellular organisms that we know best (plants, animals, and fungi) are the tips of just a few branches on the great tree of life (see Figure 26.21).

 Mastering Biology HHMI Video: Seeing the Invisible: Van Leeuwenhoek's First Glimpses of the Microbial World   **hhmi** BioInteractive

Structural and Functional Diversity in Protists

Given that they are classified in a number of different groups, it isn't surprising that few general characteristics of protists can be cited without exceptions. In fact, protists exhibit more structural and functional diversity than the eukaryotes with which we are most familiar—plants, animals, and fungi.

For example, most protists are unicellular, although there are some colonial and multicellular species. Single-celled protists are justifiably considered the simplest eukaryotes, but at the cellular level, many protists are very complex—the most elaborate of all cells. In multicellular organisms, essential biological functions are carried out by organs. Unicellular protists carry out the same essential functions, but they do so using subcellular organelles, not multicellular organs. The organelles that protists use are mostly those discussed in Figure 6.8, including the nucleus, endoplasmic reticulum, Golgi apparatus, and lysosomes. Certain protists also rely on organelles not found in most other eukaryotic cells, such as contractile vacuoles that pump excess water from the protistan cell (see Figure 7.13). One group of dinoflagellate protists has the most complex subcellular structure known: a unique, eye-like organelle called an *ocelloid* with components resembling the lens and retina in the eyes of vertebrates and some other animals **(Figure 28.2)**.

Protists are also very diverse in their nutrition. Some protists are photoautotrophs and contain chloroplasts. Some are heterotrophs, absorbing organic molecules or ingesting larger food particles. Still other protists, called **mixotrophs**, combine photosynthesis and heterotrophic nutrition. Photoautotrophy, heterotrophy, and mixotrophy have all arisen independently in many different protist lineages.

Reproduction and life cycles also are highly varied among protists. Some protists are only known to reproduce asexually; others can also reproduce sexually or at least employ the sexual processes of meiosis and fertilization. All three basic types of sexual life cycles (see Figure 13.6) are represented among protists, along with some variations that do not quite fit any of these types. We will examine the life cycles of several protist groups later in this chapter.

Endosymbiosis in Eukaryotic Evolution

What gave rise to the enormous diversity of protists that exist today? There is abundant evidence that much of protistan diversity has its origins in **endosymbiosis**, a relationship

▶ **Figure 28.2 The eye-like ocelloid of *Erythropsidinium*, a dinoflagellate.** This complex, subcellular structure resembles the camera-like eye found in vertebrates and some molluscs. In some dinoflagellates, the ocelloid may function in prey detection.

"Retina"

"Lens"

between two species in which one organism lives inside the cell or cells of another organism (the host). In particular, as we discussed in Concept 25.3, structural, biochemical, and DNA sequence data indicate that mitochondria and plastids are derived from bacteria that were engulfed by the ancestors of early eukaryotic cells. The evidence also suggests that mitochondria evolved before plastids. Thus, a defining moment in the origin of eukaryotes occurred when a host cell engulfed a bacterium that would later become an organelle found in all eukaryotes—the mitochondrion.

To determine which bacterial lineage gave rise to mitochondria, researchers have compared the DNA sequences of mitochondrial genes (mtDNA) to those found in major clades of bacteria. In the **Scientific Skills Exercise**, you will interpret one such set of DNA sequence comparisons. Collectively, such studies indicate that mitochondria arose from an alpha proteobacterium (see Figure 27.17). Results from mtDNA sequence analyses also indicate that the mitochondria of protists, animals, fungi, and plants descended from a single common ancestor, thus suggesting that mitochondria arose only once over the course of evolution. Similar analyses provide evidence that plastids descended from a single common ancestor—a cyanobacterium that was engulfed by a eukaryotic host cell.

Scientific Skills Exercise

Interpreting Comparisons of Genetic Sequences

Which Bacteria Are Most Closely Related to Mitochondria?
Early eukaryotes acquired mitochondria by endosymbiosis: A host cell engulfed an aerobic bacterium that persisted within the cytoplasm to the mutual benefit of both cells. In studying which living bacteria might be most closely related to mitochondria, researchers compared DNA sequences encoding ribosomal RNA (rRNA), a component of ribosomes. Ribosomes perform critical cell functions. Hence, genes encoding rRNA are under strong selection and change slowly over time, making them suitable for comparing even distantly related species. In this exercise, you will interpret some of the research data to draw conclusions about the phylogeny of mitochondria.

How the Research Was Done Researchers isolated and cloned DNA sequences from the gene that codes for the small-subunit rRNA molecule for the wheat (a eukaryote) mitochondrion and for five bacterial species:

- Wheat, used as the source of genes encoding mitochondrial rRNA
- *Agrobacterium tumefaciens*, an alpha proteobacterium that lives within plant tissue and produces tumors in the host
- *Comamonas testosteroni*, a beta proteobacterium
- *Escherichia coli*, a well-studied gamma proteobacterium that inhabits human intestines
- *Mycoplasma capricolum*, a gram-positive mycoplasma, which is the only group of bacteria lacking cell walls
- *Anacystis nidulans,* a cyanobacterium

Data from the Research Gene sequences encoding rRNA for the wheat mitochondrion and the five bacteria were aligned and compared. The data table below, called a *comparison matrix*,

 ▶ Wheat, used as the source of mitochondrial RNA

summarizes the comparison of 617 nucleotide positions from the gene sequences. Each value in the table is the percentage of the 617 nucleotide positions for which the pair of organisms have the same base. Any positions that were identical across all six RNA-encoding gene sequences were omitted from this comparison matrix.

INTERPRET THE DATA

1. First, make sure you understand how to read the comparison matrix. **(a)** Find the cell that represents the comparison of *C. testosteroni* and *E. coli*. What value is given in this cell? **(b)** What does that value signify about the comparable gene sequences in those two organisms? **(c)** Explain why some cells have a dash rather than a value. **(d)** Why are some cells shaded gray, with no value?

2. Why did the researchers choose one plant mitochondrion and five bacterial species to include in the comparison matrix?

3. **(a)** Which bacterium has an rRNA-encoding gene that is most similar to that of the wheat mitochondrion? **(b)** What is the significance of this similarity?

⊘ **Instructors:** A version of this Scientific Skills Exercise can be assigned in **Mastering Biology**.

	Wheat mitochondrion	A. tumefaciens	C. testosteroni	E. coli	M. capricolum	A. nidulans
Wheat mitochondrion	–	48	38	35	34	34
A. tumefaciens		–	55	57	52	53
C. testosteroni			–	61	52	52
E. coli				–	48	52
M. capricolum					–	50
A. nidulans						–

Data from D. Yang et al., Mitochondrial origins, *Proceedings of the National Academy of Sciences USA* 82:4443–4447 (1985).

Progress has also been made toward identifying the host cell that engulfed an alpha proteobacterium, thereby setting the stage for the origin of eukaryotes. For example, researchers recently reported the discovery of a new group of archaea, the lokiarchaeotes. In phylogenomic analyses, this group was identified as the sister group of the eukaryotes and its genome was found to encode many eukaryote-specific features. Was the host cell that engulfed an alpha proteobacterium a lokiarchaeote? While this could be correct, it is also possible that the host belonged to a different lineage of archaea or was closely related to the archaea (but was not itself an archaean). In any case, current evidence indicates that the host was a relatively complex cell in which certain features of eukaryotic cells had evolved, such as a cytoskeleton that enabled it to change shape (and thereby engulf the alpha proteobacterium).

Plastid Evolution: A Closer Look

As you've seen, current evidence indicates that mitochondria are descended from a bacterium that was engulfed by a host cell that was an archaean (or a close relative of the archaea). This event gave rise to the eukaryotes. There is also much evidence that later in eukaryotic history, a lineage of heterotrophic eukaryotes acquired an additional endosymbiont—a photosynthetic cyanobacterium—that then evolved into plastids. According to the hypothesis illustrated in **Figure 28.3**, this plastid-bearing lineage gave rise to two lineages of photosynthetic protists, or **algae**: red algae and green algae.

Let's examine some of the steps in Figure 28.3 more closely. First, recall that cyanobacteria are gram-negative and that gram-negative bacteria have two cell membranes, an inner plasma membrane and an outer membrane that is part of the cell wall (see Figure 27.3). Plastids in red algae and green algae are also surrounded by two membranes. Transport proteins in these membranes are homologous to proteins in the inner and outer membranes of cyanobacteria, providing further support for the hypothesis that plastids originated from a cyanobacterial endosymbiont.

On several occasions during eukaryotic evolution, red algae and green algae underwent **secondary endosymbiosis**, meaning they were ingested in the food vacuoles of heterotrophic eukaryotes and became endosymbionts themselves. For example, protists known as chlorarachniophytes likely evolved when a heterotrophic eukaryote engulfed a green alga. Evidence for this process can be found within the

▼ **Figure 28.3 Diversity of plastids produced by endosymbiosis.** Studies of plastid-bearing eukaryotes suggest that plastids evolved from a cyanobacterium that was engulfed by an ancestral heterotrophic eukaryote (primary endosymbiosis). That ancestor then diversified into red algae and green algae, some of which were subsequently engulfed by other eukaryotes (secondary endosymbiosis).

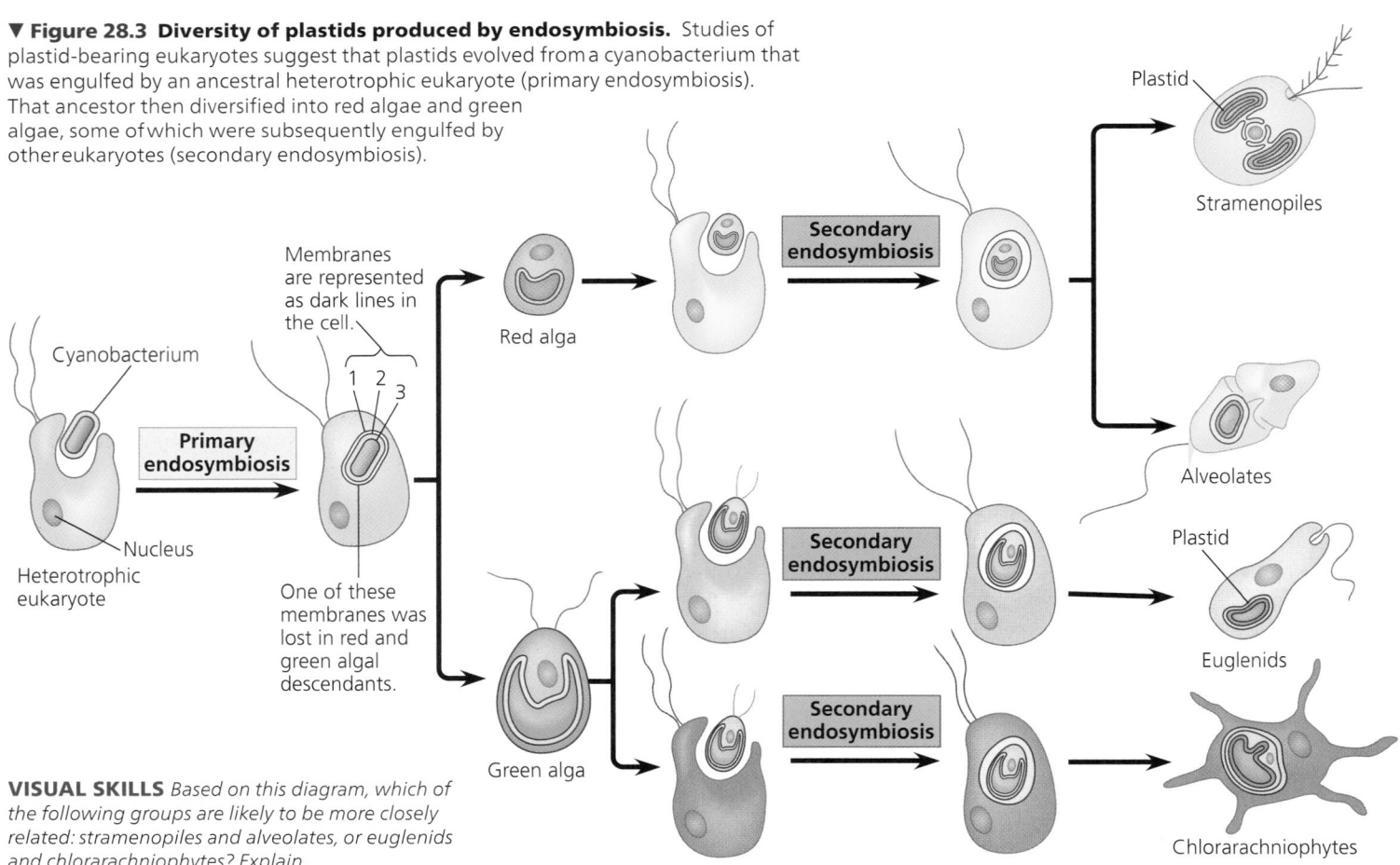

VISUAL SKILLS *Based on this diagram, which of the following groups are likely to be more closely related: stramenopiles and alveolates, or euglenids and chlorarachniophytes? Explain.*

► Figure 28.4
Nucleomorph within a plastid of a chlorarachniophyte.

Inner plastid membrane

Outer plastid membrane

Nuclear pore-like gap

Nucleomorph

Thylakoid membranes

▼ Figure 28.6 **A haptophyte and a cryptophyte.** Although they are major groups of protists, haptophytes and cryptophytes are not shown in Figure 28.5 because their phylogenetic relationships have yet to be resolved. **(a)** The haptophyte *Emiliania huxleyi* is a key producer in the ocean. **(b)** The outer surface of this photosynthetic cryptophyte is covered with protective plates. (SEM)

2 μm   8 μm

engulfed cell, which contains a tiny vestigial nucleus, called a *nucleomorph* **(Figure 28.4)**. Genes from the nucleomorph are still transcribed, and their DNA sequences indicate that the engulfed cell was a green alga.

Four Supergroups of Eukaryotes

Our understanding of the evolutionary history of eukaryotic diversity has been in flux in recent years. Not only has kingdom Protista been abandoned, but other hypotheses have been discarded as well. For example, many biologists once thought that the first lineage to have diverged from all other eukaryotes was the *amitochondriate protists*, organisms without conventional mitochondria and with fewer membrane-enclosed organelles than other protist groups. But recent structural and DNA data have undermined this hypothesis. Many of the so-called amitochondriate protists have been shown to have mitochondria—though reduced ones—and some of these organisms are now classified in distantly related groups.

The ongoing changes in our understanding of the phylogeny of protists pose challenges to students and instructors alike. Hypotheses about these relationships are a focus of scientific activity, changing rapidly as new data cause previous ideas to be modified or discarded. We'll focus here on one current hypothesis: the four supergroups of eukaryotes shown in **Figure 28.5**. Because the root of the eukaryotic tree is not known, all four supergroups are shown as diverging simultaneously from a common ancestor. We know that this is not correct, but we do not know which supergroup was the first to diverge from the other three. In addition, major new groups of protists have been discovered in recent years, but their relationship to the supergroups shown in Figure 28.5 has yet to be resolved. Examples of such unresolved groups include haptophytes and cryptophytes **(Figure 28.6)**, as well as the hemimastigophores, a novel group described in 2018. Overall, as

you read this chapter, it may be helpful to focus less on the specific names of groups of organisms and more on why the organisms are important and how ongoing research is elucidating their evolutionary relationships.

CONCEPT CHECK 28.1

1. Cite at least four examples of structural and functional diversity among protists.
2. Summarize the role of endosymbiosis in eukaryotic evolution.
3. **MAKE CONNECTIONS** After studying Figure 28.3, predict how many distinct genomes are contained within the cell of a chlorarachniophyte. Explain. (See Figures 6.17 and 6.18.)

For suggested answers, see Appendix A.

CONCEPT 28.2

Excavates include protists with modified mitochondria and protists with unique flagella

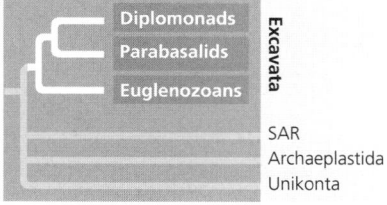

Diplomonads
Parabasalids
Euglenozoans

Excavata

SAR
Archaeplastida
Unikonta

Now that we have examined some of the broad patterns in eukaryotic evolution, we will look more closely at the four main groups of protists shown in Figure 28.5.

We begin with **Excavata** (the excavates), a clade that was originally proposed based on morphological studies of the

The tree below represents a phylogenetic hypothesis for the relationships among eukaryotes on Earth today. The eukaryotic groups at the branch tips are related in larger "supergroups," labeled vertically at the far right of the tree. Groups that were formerly classified in the kingdom Protista are highlighted in yellow. Dotted lines indicate evolutionary relationships that are uncertain. For clarity, this tree only includes representative clades from each supergroup. In addition, the recent discoveries of many new groups of eukaryotes indicate that eukaryotic diversity is actually much greater than shown here.

Diplomonads — Excavata
Parabasalids
Euglenozoans

Diatoms — Stramenopiles
Brown algae
Oomycetes
Dinoflagellates — Alveolates — SAR
Apicomplexans
Ciliates
Radiolarians — Rhizarians
Forams
Cercozoans

Red algae — Archaeplastida
Chlorophytes — Green algae
Charophytes
Plants

Tubulinids — Amoebozoans
Slime molds
Entamoebas
Nucleariids — Opisthokonts — Unikonta
Fungi
Choanoflagellates
Animals

■ Excavata

Some members of this supergroup (whose monophyly has been called into question by some recent genomic studies) have an "excavated" groove on one side of the cell body. Two major clades (the parabasalids and diplomonads) have highly reduced mitochondria; members of a third clade (the euglenozoans) have flagella that differ in structure from those of other organisms. Excavates include parasites such as *Giardia*, as well as many predatory and photosynthetic species.

5 μm

***Giardia intestinalis*, a diplomonad parasite.** This diplomonad (colorized SEM), which lacks the characteristic surface groove of the Excavata, inhabits the intestines of mammals. It can infect people when they drink water contaminated with feces containing *Giardia* cysts. Drinking such water—even from a seemingly pristine stream—can cause severe diarrhea. Boiling the water kills the parasite.

DRAW IT *Draw a simplified version of this phylogenetic tree that shows only the four supergroups of eukaryotes. Now sketch how the tree would look if the unikonts were the sister group to all other eukaryotes.*

■ SAR

This supergroup contains (and is named after) three large and very diverse clades: Stramenopila, Alveolata, and Rhizaria. Stramenopiles include some of the most important photosynthetic organisms on Earth, such as the diatoms shown here. Alveolates also include photosynthetic species, as well as important pathogens, such as *Plasmodium*, which causes malaria. According to one current hypothesis, stramenopiles and alveolates originated by secondary endosymbiosis when a heterotrophic protist engulfed a red alga.

50 μm

Diatom diversity. These beautiful single-celled protists are important photosynthetic organisms in aquatic communities (LM).

The rhizarian subgroup of SAR includes many species of amoebas, most of which have pseudopodia that are threadlike in shape. Pseudopodia are extensions that can bulge from any portion of the cell; they are used in movement and in the capture of prey.

100 μm

***Globigerina*, a rhizarian in SAR.** This species is a foram, a group whose members have threadlike pseudopodia that extend through pores in the shell, or test (LM). The inset shows a foram test, which is hardened by calcium carbonate.

■ Archaeplastida

This supergroup of eukaryotes includes red algae and green algae, along with plants. Red algae and green algae include unicellular species, colonial species, and multicellular species (including the green alga *Volvox*). Many of the large algae known informally as "seaweeds" are multicellular red or green algae. Protists in Archaeplastida include key photosynthetic species that form the base of the food web in many aquatic communities.

20 μm   25 μm

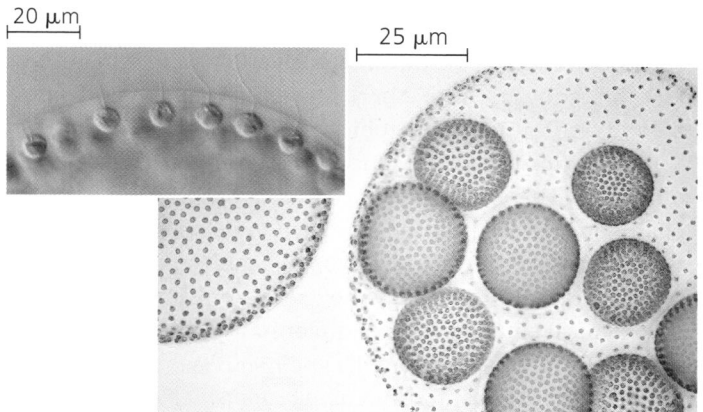

***Volvox*, a multicellular freshwater green alga.** This alga has two types of differentiated cells, and so it is considered multicellular rather than colonial. It resembles a hollow ball whose wall is composed of hundreds of biflagellated cells (see inset LM) embedded in a gelatinous extracellular matrix; if isolated, these cells cannot reproduce. However, the alga also contains cells that are specialized for either sexual or asexual reproduction. The large algae shown here will eventually release the small "daughter" algae that can be seen within them (LM).

→ **Mastering Biology** Video: *Volvox*

■ Unikonta

This supergroup of eukaryotes, also called Amorphea, includes amoebas that have lobe- or tube-shaped pseudopodia, as well as animals, fungi, and non-amoeba protists that are closely related to animals or fungi. According to one current hypothesis, the unikonts were the first eukaryotic supergroup to diverge from all other eukaryotes; however, this hypothesis has yet to be widely accepted.

A unikont amoeba. This amoeba, the tubulinid *Amoeba proteus*, is using its pseudopodia to move.

→ **Mastering Biology** Video: *Amoeba* Pseudopodium

cytoskeleton. Some members of this diverse group also have an "excavated" feeding groove on one side of the cell body. The excavates include the diplomonads, parabasalids, and euglenozoans. Molecular data indicate that each of these three groups is monophyletic. In contrast, some recent genomic studies indicate that the excavates are not monophyletic, making Excavata the most controversial of our four supergroups.

Diplomonads and Parabasalids

The protists in these two groups lack plastids and have highly reduced mitochondria (until recently, they were thought to lack mitochondria altogether). Most diplomonads and parabasalids are found in anaerobic environments.

Diplomonads have reduced mitochondria called *mitosomes*. These organelles lack functional electron transport chains and hence cannot use oxygen to help extract energy from carbohydrates and other organic molecules. Instead, diplomonads get the energy they need from anaerobic biochemical pathways. Many diplomonads are parasites, including the infamous *Giardia intestinalis* (see Figure 28.5), which inhabits the intestines of mammals.

Structurally, diplomonads have two equal-sized nuclei and multiple flagella. Recall that eukaryotic flagella are extensions of the cytoplasm, consisting of bundles of microtubules covered by the cell's plasma membrane (see Figure 6.24). They are quite different from prokaryotic flagella, which are filaments composed of globular proteins attached to the cell surface (see Figure 27.7).

Parabasalids also have reduced mitochondria; called *hydrogenosomes*, these organelles generate some energy anaerobically, releasing hydrogen gas as a by-product. The best-known parabasalid is *Trichomonas vaginalis*, a sexually transmitted parasite that infects about 140 million people each year worldwide. *T. vaginalis* travels along the mucus-coated lining of the human reproductive and urinary tracts by moving its flagella and by undulating part of its plasma membrane **(Figure 28.7)**. In females, if the vagina's normal acidity is disturbed, *T. vaginalis* can outcompete beneficial microorganisms there and infect the vagina. One of *T. vaginalis*'s genes encodes a product that allows it to feed

on the vaginal lining, promoting infection. Studies suggest that *T. vaginalis* acquired this gene by horizontal gene transfer from bacterial parasites in the vagina. (*Trichomonas* infections also can occur in the urethra of males, though often without symptoms.)

Euglenozoans

Protists called **euglenozoans** belong to a diverse clade that includes predatory heterotrophs, photosynthetic autotrophs, mixotrophs, and parasites. The main morphological feature that distinguishes protists in this clade is the presence of a rod with either a spiral or a crystalline structure inside each of their flagella **(Figure 28.8)**. The two best-studied groups of euglenozoans are the kinetoplastids and the euglenids.

Kinetoplastids

Protists called **kinetoplastids** have a single, large mitochondrion that contains an organized mass of DNA called a *kinetoplast*. These protists include species that feed on prokaryotes in freshwater, marine, and moist terrestrial ecosystems, as well as species that parasitize animals, plants, and other protists. For example, kinetoplastids in the genus *Trypanosoma* infect humans and cause sleeping sickness **(Figure 28.9)**, a neurological disease that is invariably fatal if not treated. This disease currently afflicts about 10,000 people each year, mostly in rural areas of Africa. The infection occurs via the bite of a vector (carrier) organism, the African tsetse fly. Trypanosomes also cause Chagas' disease, which is transmitted by bloodsucking insects and can lead to congestive heart failure.

Trypanosomes evade immune responses with an effective "bait-and-switch" defense. The surface of a trypanosome is coated with millions of copies of a single protein. However, before the host's immune system can recognize the protein and mount an attack, new generations of the

▼ **Figure 28.7 The parabasalid parasite *Trichomonas vaginalis.*** (colorized SEM)

▼ **Figure 28.8 Euglenozoan flagellum.** Most euglenozoans have a crystalline rod inside one of their flagella (the TEM is a flagellum shown in cross section). The rod lies alongside the 9 + 2 ring of microtubules found in all eukaryotic flagella (compare with Figure 6.24).

▼ **Figure 28.9** *Trypanosoma*, the kinetoplastid that causes sleeping sickness. (colorized SEM)

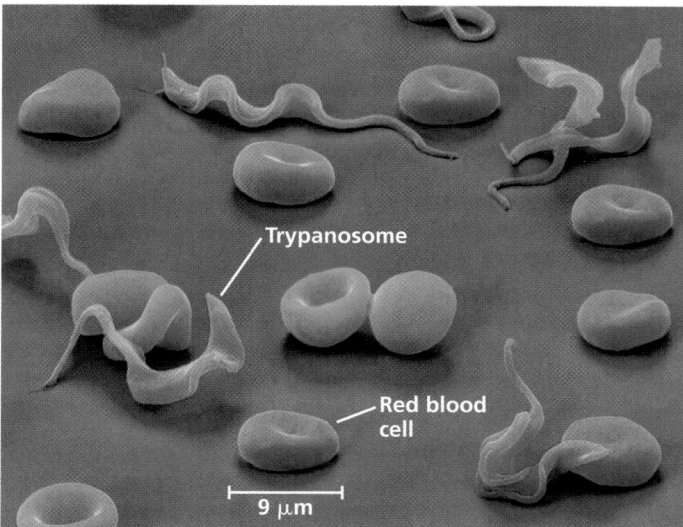

Trypanosome

Red blood cell

9 μm

parasite switch to another surface protein with a different molecular structure. Frequent changes in the surface protein prevent the host from developing immunity. (The Scientific Skills Exercise in Chapter 43 explores this topic further.) About a third of *Trypanosoma*'s genome is dedicated to producing these surface proteins.

Euglenids

A **euglenid** has a pocket at one end of the cell from which one or two flagella emerge **(Figure 28.10)**. Some euglenids are mixotrophs: They perform photosynthesis when sunlight is available, but when it is not, they can become heterotrophic, absorbing organic nutrients from their environment. Many other euglenids engulf prey by phagocytosis.

CONCEPT CHECK 28.2

1. Why do some biologists describe the mitochondria of diplomonads and parabasalids as "highly reduced"?
2. **WHAT IF?** DNA sequence data for a diplomonad, a euglenid, a plant, and an unidentified protist suggest that the unidentified species is most closely related to the diplomonad. Further studies reveal that the unknown species has fully functional mitochondria. Based on these data, at what point on the phylogenetic tree in Figure 28.5 did the mystery protist's lineage probably diverge from other eukaryote lineages? Explain.

For suggested answers, see Appendix A.

CONCEPT 28.3

SAR is a highly diverse group of protists defined by DNA similarities

Excavata

Diatoms
Brown algae — **Stramenopiles**
Oomycetes

Dinoflagellates
Apicomplexans — **Alveolates**
Ciliates

SAR

Radiolarians
Forams — **Rhizarians**
Cercozoans

Archaeplastida
Unikonta

Our second supergroup, referred to as **SAR**, was proposed recently based on whole-genome DNA sequence analyses. These studies found that three major clades of protists—the stramenopiles, alveolates, and rhizarians—form a monophyletic supergroup. This supergroup contains a large, extremely diverse collection of protists. To date, this supergroup has not

▶ **Figure 28.10** *Euglena*, a euglenid commonly found in pond water.

→ **Mastering Biology** Video: *Euglena*

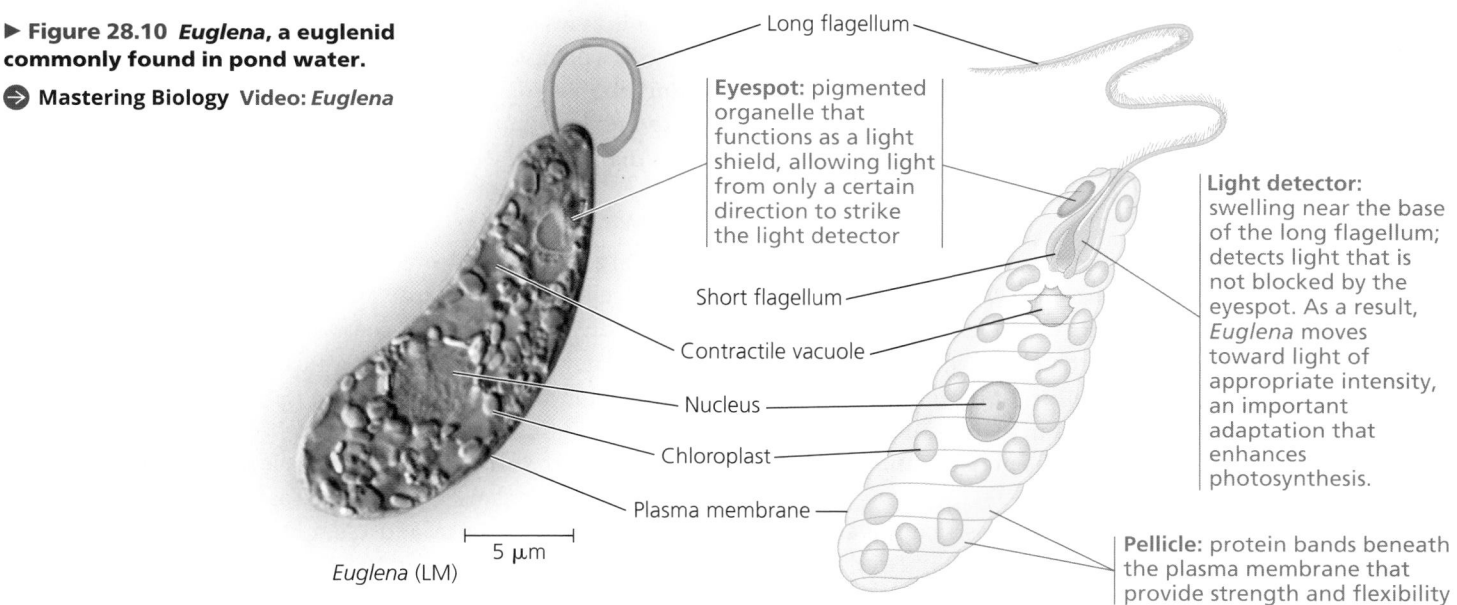

Long flagellum

Eyespot: pigmented organelle that functions as a light shield, allowing light from only a certain direction to strike the light detector

Light detector: swelling near the base of the long flagellum; detects light that is not blocked by the eyespot. As a result, *Euglena* moves toward light of appropriate intensity, an important adaptation that enhances photosynthesis.

Short flagellum

Contractile vacuole

Nucleus

Chloroplast

Plasma membrane

Pellicle: protein bands beneath the plasma membrane that provide strength and flexibility

5 μm

Euglena (LM)

received a formal name but is instead known by the first letters of its major clades: SAR.

Some morphological and DNA sequence data suggest that two of these groups, the stramenopiles and alveolates, originated more than a billion years ago, when a common ancestor of these two clades engulfed a single-celled, photosynthetic red alga. Because red algae are thought to have originated by primary endosymbiosis (see Figure 28.3), such an origin for the stramenopiles and alveolates is referred to as secondary endosymbiosis. Others question this idea, noting that some species in these groups lack plastids or their remnants (including any trace of plastid genes in their nuclear DNA).

Despite its lack of a formal name, the SAR supergroup represents the best current hypothesis for the phylogeny of the three large protist clades to which we now turn.

Stramenopiles

One major subgroup of SAR, the **stramenopiles**, includes some of the most important photosynthetic organisms on the planet. Their name (from the Latin *stramen*, straw, and *pilos*, hair) refers to their characteristic flagellum, which has numerous fine, hairlike projections. In most stramenopiles, this "hairy" flagellum is paired with a shorter "smooth" (nonhairy) flagellum **(Figure 28.11)**. Here we'll focus on three groups of stramenopiles: diatoms, oomycetes, and brown algae.

▼ **Figure 28.11 Stramenopile flagella.** Most stramenopiles, such as *Synura petersenii*, have two flagella: one covered with fine, stiff hairs and a shorter one that is smooth.

Hairy flagellum

Smooth flagellum

2.5 μm

Diatoms

A key group of photosynthetic protists, **diatoms** are unicellular algae that have a unique glass-like wall made of silicon dioxide embedded in an organic matrix **(Figure 28.12)**. The wall consists of two parts that overlap like a shoe box and its lid. These walls provide effective protection from the crushing jaws of predators: Live diatoms can withstand pressures as great as 1.4 million kg/m^2, equal to the pressure under each leg of a table supporting an elephant!

With an estimated 100,000 living species, diatoms are a highly diverse group of protists (see Figure 28.5). They are among the most abundant photosynthetic organisms both in the ocean and in lakes: One bucket of water scooped from

▶ **Figure 28.12 The diatom *Triceratium morlandii.*** (colorized SEM)

40 μm

the surface of the sea may contain millions of these microscopic algae. The abundance of diatoms in the past is also evident in the fossil record, where massive accumulations of fossilized diatom walls are major constituents of sediments known as *diatomaceous earth*. These sediments are mined for their quality as a filtering medium and for many other uses.

Diatoms are so widespread and abundant that their photosynthetic activity affects global carbon dioxide (CO_2) levels. Diatoms have this effect in part because of events that occur during episodes of rapid population growth, or *blooms*, when ample nutrients are available. Typically, diatoms are eaten by a variety of protists and invertebrates, but during a bloom, many escape this fate. When these uneaten diatoms die, their bodies sink to the ocean floor. It takes decades, or even centuries, for dead diatoms that sink to the ocean floor to be broken down by bacteria and other decomposers. As a result, the carbon in their bodies remains there for some time, rather than being released immediately as CO_2 as the decomposers respire. The overall effect of these events is that CO_2 absorbed by diatoms during photosynthesis is transported, or "pumped," to the ocean floor.

With an eye toward reducing global warming by lowering atmospheric CO_2 levels, some scientists advocate promoting diatom blooms by fertilizing the ocean with essential nutrients such as iron. In one recent study, researchers found that CO_2 was indeed pumped to the ocean floor after iron was added to a small region of the ocean. Further tests are planned to examine whether iron fertilization has undesirable side effects (such as oxygen depletion or the production of nitrous oxide, a more potent greenhouse gas than CO_2).

Brown Algae

The largest and most complex algae are **brown algae**. All are multicellular, and most are marine. Brown algae are especially common along temperate coasts that have cold-water currents. They owe their characteristic brown or olive color to the carotenoids in their plastids.

Many of the species commonly called "seaweeds" are brown algae. Some brown algal seaweeds have specialized structures that resemble organs in plants, such as a rootlike **holdfast**, which anchors the alga, and a stemlike **stipe**, which supports the leaflike **blades (Figure 28.13)**. Unlike plants, however, brown algae lack true tissues and organs. Moreover, morphological and DNA data show that these similarities evolved

Figure 28.13 Seaweeds: adapted to life at the ocean's margins.

▼ Figure 28.13 Seaweeds: adapted to life at the ocean's margins. The sea palm (*Postelsia*) lives on rocks along the coast of the northwestern United States and western Canada. The body of this brown alga is well adapted to maintaining a firm foothold despite the crashing surf.

Blade

Stipe

Holdfast

independently in the algal and plant lineages and are thus analogous, not homologous. In addition, while plants have adaptations (such as rigid stems) that provide support against gravity, brown algae have adaptations that enable their main photosynthetic surfaces (the leaflike blades) to be near the water surface. Some brown algae accomplish this task with gas-filled, bubble-shaped floats. Giant brown algae known as kelps that live in deep waters have such floats in their blades, which are attached to stipes that can rise as much as 60 m from the seafloor—more than half the length of a football field.

Brown algae are important commodities for humans. Some species are eaten, such as *Laminaria* (Japanese "kombu"), which is used in soups. In addition, the cell walls of brown algae contain a gel-forming substance, called algin, which is used to thicken many processed foods, including pudding and salad dressing.

Alternation of Generations

A variety of life cycles have evolved among the multicellular algae. The most complex life cycles include an **alternation of generations**, the alternation of multicellular haploid and diploid forms, as shown in **Figure 28.14** for the brown alga

▼ Figure 28.14 The life cycle of the brown alga *Laminaria*: an example of alternation of generations.

1 The sporophytes are usually found in water just below the line of the lowest tides, attached to rocks by branching holdfasts.

Key
- Haploid (*n*)
- Diploid (*2n*)

Sporangia

2 Cells on the surface of the blade develop into sporangia.

Sporophyte (*2n*)

MEIOSIS

3 Sporangia produce zoospores by meiosis.

Zoospore

4 The zoospores are all structurally alike, but about half of them divide asexually (by mitosis) and develop into male gametophytes; the other half divide asexually and develop into female gametophytes.

Female

Gametophytes (*n*)

7 The zygotes grow into new sporophytes while attached to the remains of the female gametophyte.

Developing sporophyte

Zygote (*2n*)

Male

Mature female gametophyte (*n*)

FERTILIZATION

Egg

Sperm

5 Male gametophytes release sperm, and female gametophytes produce eggs, which remain attached to the female gametophyte. Eggs secrete a chemical signal that attracts sperm of the same species, thereby increasing the probability of fertilization in the ocean.

6 Sperm fertilize the eggs.

VISUAL SKILLS *Based on this diagram, are the sperm shown in* **5** *genetically identical to one another? Explain.*

Laminaria. Although haploid and diploid conditions alternate in *all* sexual life cycles—human gametes, for example, are haploid—the term *alternation of generations* applies only to life cycles in which both haploid and diploid stages are multicellular. As you will read in Concept 29.1, alternation of generations also evolved in plants.

As detailed in Figure 28.14, the diploid individual in this complex algal life cycle is called the *sporophyte* because it produces spores. The spores are haploid and move by means of flagella; they are called zoospores. The zoospores divide by mitosis and develop into haploid, multicellular male and female *gametophytes*, which produce gametes. The union of two gametes (fertilization) results in a diploid zygote, which matures and gives rise to a new multicellular sporophyte.

In *Laminaria*, the two generations are **heteromorphic**, meaning that the sporophytes and gametophytes are structurally different. Other algal life cycles have an alternation of **isomorphic** generations, in which the sporophytes and gametophytes look similar to each other, although they differ in chromosome number.

Oomycetes (Water Molds and Their Relatives)

Oomycetes include the water molds, the white rusts, and the downy mildews. Based on their morphology, these organisms were previously classified as fungi (in fact, oomycete means "egg fungus"). For example, many oomycetes have multinucleate filaments (hyphae) that resemble fungal hyphae **(Figure 28.15a)**. However, there are key differences between oomycetes and fungi. Among them, oomycetes typically have cell walls made of cellulose, whereas the walls of fungi consist mainly of another polysaccharide, chitin. Data from molecular systematics have confirmed that oomycetes are not closely related to fungi. In both oomycetes and fungi, the high surface-to-volume ratio of filamentous structures enhances the uptake of nutrients from the environment.

Although oomycetes descended from plastid-bearing ancestors, they no longer have plastids and do not perform photosynthesis. Instead, they typically acquire nutrients as decomposers or parasites. Most water molds are decomposers that grow as cottony masses on dead algae and animals, mainly in freshwater habitats **(Figure 28.15b)**. White rusts and downy mildews generally live on land as plant parasites.

The ecological impact of oomycetes can be significant. For example, the oomycete *Phytophthora infestans* causes potato late blight, which turns the stalks and stems of potato (and tomato) plants into black slime. Late blight contributed to the devastating Irish famine of the 19th century, in which a million people died and at least that many were forced to leave Ireland. The disease remains a major problem today, causing crop losses as high as 70% in some regions. Globally, about $1 billion is spent each year on pesticides to control the parasite.

Alveolates

Members of the next subgroup of SAR, the **alveolates**, have membrane-enclosed sacs (alveoli) just under the plasma membrane **(Figure 28.16)**. Alveolates are abundant in many habitats and include a wide range of photosynthetic and heterotrophic protists. We'll discuss three alveolate clades here: a group of flagellates (the dinoflagellates), a group of parasites (the apicomplexans), and a group of protists that move using cilia (the ciliates).

▼ **Figure 28.15 Oomycetes.**

(a) Closeup of oomycete hyphae (LM)

100 μm

(b) Water mold hyphae growing on a goldfish

▼ **Figure 28.16 Alveoli.** These sacs under the plasma membrane are a characteristic that distinguishes alveolates from other eukaryotes (TEM).

Flagellum   Alveoli

Alveolate

0.2 μm

▼ **Figure 28.17 Dinoflagellates.**

(a) Dinoflagellate flagella. Beating of the spiral flagellum, which lies in a groove that encircles the cell, makes this specimen of *Pfiesteria shumwayae* spin (colorized SEM).

Flagella

4 μm

(b) Red tide in the Gulf of Carpentaria in northern Australia. The red color is due to high concentrations of a carotenoid-containing dinoflagellate.

→ **Mastering Biology** Video: Dinoflagellate

Dinoflagellates

The cells of many **dinoflagellates** are reinforced by cellulose plates. Grooves in this "armor" house two flagella, one of which causes dinoflagellates (from the Greek *dinos*, whirling) to spin as they move through the waters of their marine and freshwater communities **(Figure 28.17a)**.

Although their ancestors may have originated by secondary endosymbiosis (see Figure 28.3), roughly half of all dinoflagellate species are now purely heterotrophic. Others are important species of *phytoplankton* (photosynthetic plankton, which include photosynthetic bacteria as well as algae); many photosynthetic dinoflagellates are mixotrophic.

Periods of explosive population growth (blooms) in dinoflagellates sometimes cause a phenomenon called "red tide" **(Figure 28.17b)**. The blooms make coastal waters appear brownish red or pink because of the presence of carotenoids, the most common pigments in dinoflagellate plastids. When blooms occur, toxins produced by certain dinoflagellates have caused massive kills of invertebrates and fishes. Humans who eat molluscs that have accumulated the toxins are affected as well, sometimes fatally. A 2017 study found that ocean warming (caused by ongoing climate change) has facilitated more frequent blooms of toxic dinoflagellates, making them a growing threat to human health.

Apicomplexans

Nearly all **apicomplexans** are parasites of animals—and virtually all animal species examined so far are attacked by these parasites. The parasites spread through their host as tiny infectious cells called *sporozoites*. Apicomplexans are so named because one end (the *apex*) of the sporozoite cell contains a *complex* of organelles specialized for penetrating host cells and tissues. Although apicomplexans are not photosynthetic, recent data show that they retain a modified plastid (apicoplast), most likely of red algal origin.

Most apicomplexans have intricate life cycles with both sexual and asexual stages. Those life cycles often require two or more host species for completion. For example, *Plasmodium*, the parasite that causes malaria, lives in both mosquitoes and humans **(Figure 28.18)**.

Historically, malaria has rivaled tuberculosis as the leading cause of human death by infectious disease. The incidence of malaria was diminished in the 1960s by insecticides that reduced carrier populations of *Anopheles* mosquitoes and by drugs that killed *Plasmodium* in humans. But the emergence of resistant varieties of both *Anopheles* and *Plasmodium* has led to a resurgence of malaria. About 220 million people in the tropics are currently infected, and 450,000 die each year. In regions where malaria is common, the lethal effects of this disease have resulted in the evolution of high frequencies of the sickle-cell allele; for an explanation of this connection, see Figure 23.18.

The search for malarial vaccines has been hampered by the fact that *Plasmodium* lives mainly inside cells, hidden from the host's immune system. And, like trypanosomes, *Plasmodium* continually changes its surface proteins. Even so, significant progress was made when European regulators recently approved the world's first licensed malarial vaccine. In 2019, routine use of this vaccine began in selected regions of Africa. However, this vaccine, which targets a protein on the surface of sporozoites, provides only partial protection against malaria. As a result, researchers continue to study other potential vaccine targets, including the apicoplast. This approach may be effective because the apicoplast is a modified plastid; as such, it descended from a cyanobacterium and hence has different metabolic pathways from those in human patients.

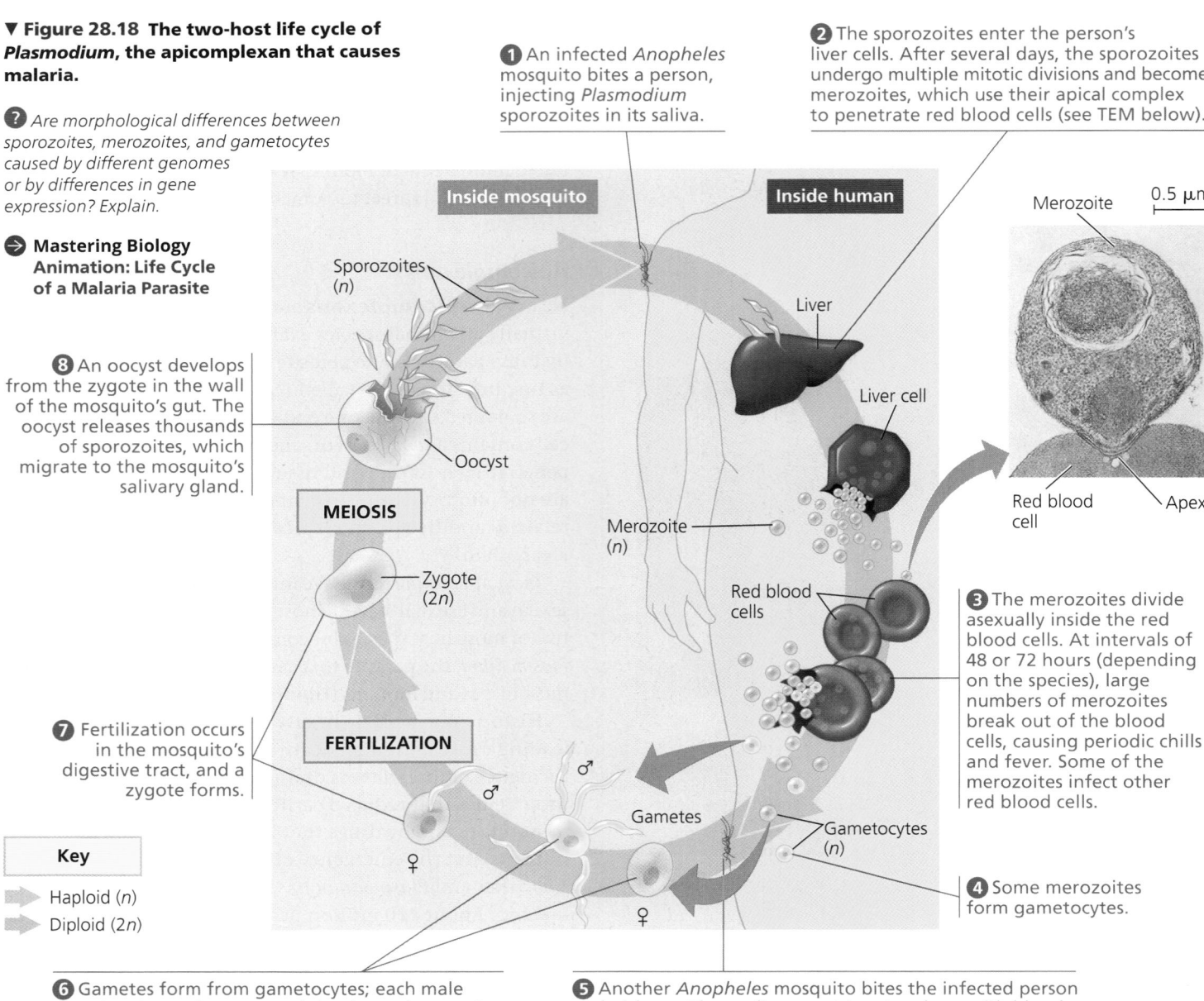

▼ **Figure 28.18 The two-host life cycle of _Plasmodium_, the apicomplexan that causes malaria.**

? _Are morphological differences between sporozoites, merozoites, and gametocytes caused by different genomes or by differences in gene expression? Explain._

→ **Mastering Biology Animation: Life Cycle of a Malaria Parasite**

Inside mosquito

Inside human

❶ An infected _Anopheles_ mosquito bites a person, injecting _Plasmodium_ sporozoites in its saliva.

❷ The sporozoites enter the person's liver cells. After several days, the sporozoites undergo multiple mitotic divisions and become merozoites, which use their apical complex to penetrate red blood cells (see TEM below).

0.5 μm

Merozoite

Sporozoites (_n_)

Liver

Oocyst

Liver cell

Red blood cell

Apex

❽ An oocyst develops from the zygote in the wall of the mosquito's gut. The oocyst releases thousands of sporozoites, which migrate to the mosquito's salivary gland.

MEIOSIS

Merozoite (_n_)

Zygote (2_n_)

Red blood cells

❸ The merozoites divide asexually inside the red blood cells. At intervals of 48 or 72 hours (depending on the species), large numbers of merozoites break out of the blood cells, causing periodic chills and fever. Some of the merozoites infect other red blood cells.

❼ Fertilization occurs in the mosquito's digestive tract, and a zygote forms.

FERTILIZATION

♂

♂

Gametes

Gametocytes (_n_)

♀

♀

❹ Some merozoites form gametocytes.

Key

▶ Haploid (_n_)

▶ Diploid (2_n_)

❻ Gametes form from gametocytes; each male gametocyte produces several slender male gametes.

❺ Another _Anopheles_ mosquito bites the infected person and picks up _Plasmodium_ gametocytes along with blood.

Ciliates

The **ciliates** are a large and varied group of protists named for their use of cilia to move and feed **(Figure 28.19a)**. Most ciliates are predators, typically of bacteria or of other protists. Their cilia may completely cover the cell surface or may be clustered in a few rows or tufts. In certain species, rows of tightly packed cilia function collectively in locomotion. Other ciliates scurry about on leg-like structures constructed from many cilia bonded together.

A distinctive feature of ciliates is the presence of two types of nuclei: tiny micronuclei and large macronuclei. A cell has one or more nuclei of each type. Genetic variation results from **conjugation**, a sexual process in which two individuals exchange haploid micronuclei but do not reproduce

(Figure 28.19b). Ciliates generally reproduce asexually by binary fission, during which the existing macronucleus disintegrates and a new one is formed from the cell's micronuclei. Each macronucleus typically contains multiple copies of the ciliate's genome. Genes in the macronucleus control the everyday functions of the cell, such as feeding, waste removal, and maintaining water balance.

→ **Mastering Biology Video: Ciliate Movement in _Stentor_**

Rhizarians

Our next subgroup of SAR is the **rhizarians**. Many species in this group are **amoebas**, protists that move and feed by

▼ **Figure 28.19** Structure and function in the ciliate *Paramecium caudatum*.

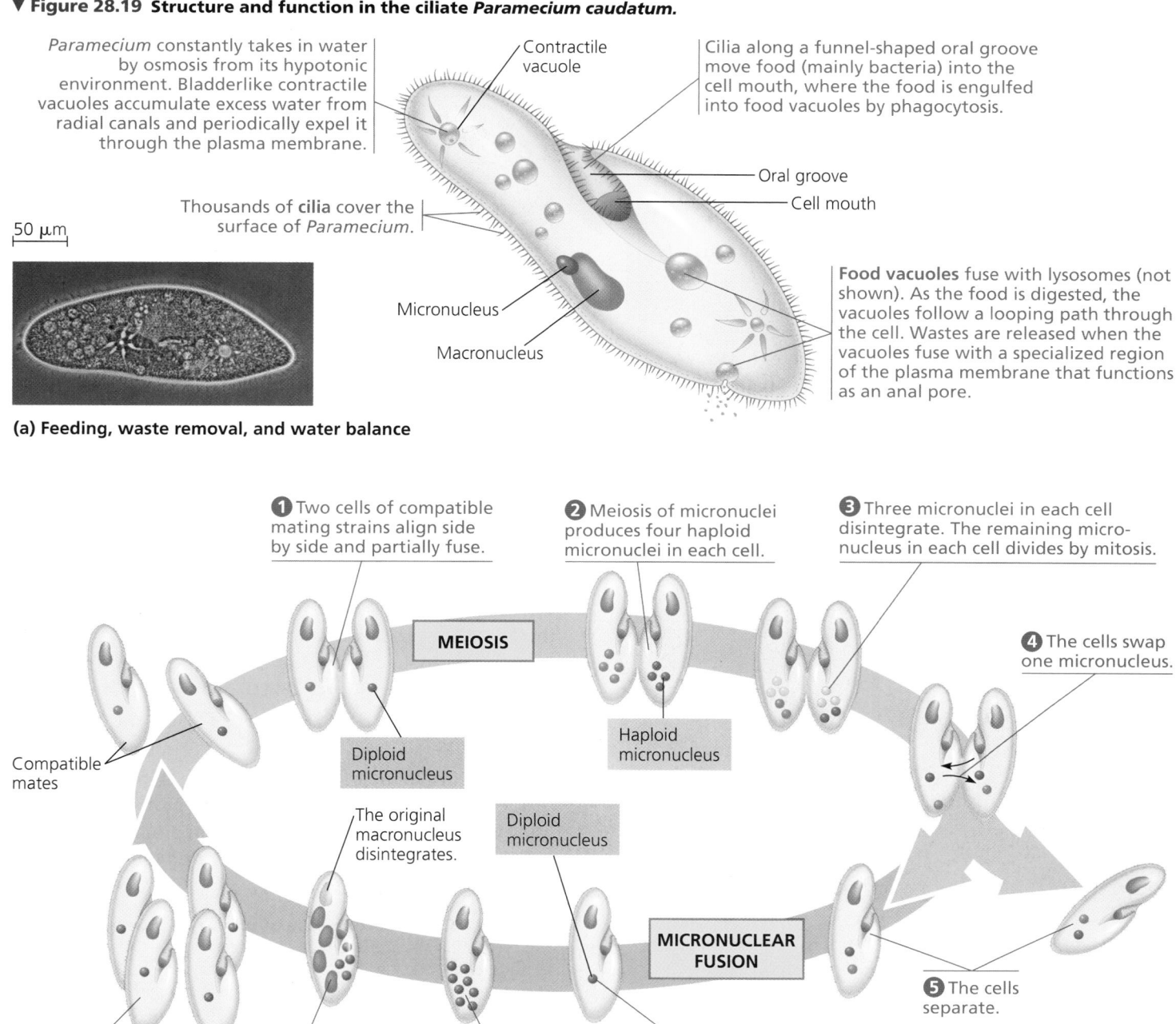

Paramecium constantly takes in water by osmosis from its hypotonic environment. Bladderlike contractile vacuoles accumulate excess water from radial canals and periodically expel it through the plasma membrane.

Contractile vacuole

Cilia along a funnel-shaped oral groove move food (mainly bacteria) into the cell mouth, where the food is engulfed into food vacuoles by phagocytosis.

Oral groove

Cell mouth

Thousands of **cilia** cover the surface of *Paramecium*.

50 μm

Micronucleus

Macronucleus

Food vacuoles fuse with lysosomes (not shown). As the food is digested, the vacuoles follow a looping path through the cell. Wastes are released when the vacuoles fuse with a specialized region of the plasma membrane that functions as an anal pore.

(a) Feeding, waste removal, and water balance

❶ Two cells of compatible mating strains align side by side and partially fuse.

❷ Meiosis of micronuclei produces four haploid micronuclei in each cell.

❸ Three micronuclei in each cell disintegrate. The remaining micronucleus in each cell divides by mitosis.

MEIOSIS

❹ The cells swap one micronucleus.

Compatible mates

Diploid micronucleus

Haploid micronucleus

The original macronucleus disintegrates.

Diploid micronucleus

Diploid micronucleus

MICRONUCLEAR FUSION

❺ The cells separate.

❾ Two rounds of binary fission yield four daughter cells.

❽ Four micronuclei become macronuclei.

❼ Three rounds of mitosis produce eight micronuclei.

❻ The two micronuclei fuse.

Key

Conjugation

Asexual reproduction

(b) Conjugation and reproduction

MAKE CONNECTIONS *The events shown in steps* ❹ *through* ❻ *of this diagram have a similar overall effect to what event in the human life cycle (see Figure 13.5)? Explain.*

⊙ **Mastering Biology** Video: *Paramecium*

means of **pseudopodia**, extensions that may bulge from almost anywhere on the cell surface. An amoeba moves by extending a pseudopodium and anchoring its tip; more cytoplasm then streams into the pseudopodium. Amoebas do not constitute a monophyletic group; instead, they are dispersed across many distantly related eukaryotic taxa. Most amoebas that are rhizarians differ morphologically from other amoebas by having threadlike pseudopodia. Rhizarians also include flagellated (non-amoeboid) protists that feed using threadlike pseudopodia.

We'll examine three groups of rhizarians here: radiolarians, forams, and cercozoans.

Radiolarians

The protists called **radiolarians** have delicate, intricately symmetrical internal skeletons that are generally made of silica. The pseudopodia of these mostly marine protists radiate from the central body **(Figure 28.20)** and are reinforced by bundles of microtubules. The microtubules are covered by a thin layer of cytoplasm, which engulfs smaller microorganisms that become attached to the pseudopodia. Cytoplasmic streaming then carries the captured prey into the main part of the cell. After radiolarians die, their skeletons settle to the seafloor, where they have accumulated as an ooze that is hundreds of meters thick in some locations.

▼ **Figure 28.20 A radiolarian.** Numerous threadlike pseudopodia radiate from the central body of this radiolarian (LM).

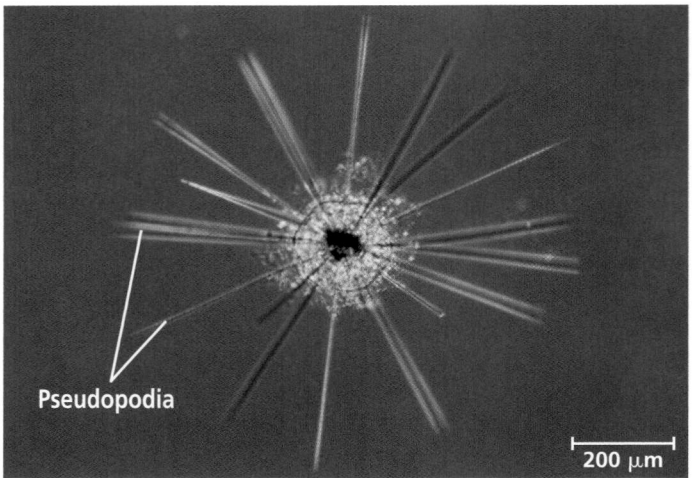

Pseudopodia

200 μm

Forams

The protists called **foraminiferans** (from the Latin *foramen*, little hole, and *ferre*, to bear), or **forams**, are named for their porous shells, called **tests** (see Figure 28.5). Foram tests consist of a single piece of organic material that typically is hardened with calcium carbonate. The pseudopodia that extend through the pores function in swimming, test formation, and feeding. Many forams also derive nourishment from the photosynthesis of symbiotic algae that live within the tests.

Forams are found in both the ocean and fresh water. Most species live in sand or attach themselves to rocks or algae, but some live as plankton. The largest forams, though single-celled, have tests several centimeters in diameter.

Ninety percent of all identified species of forams are known from fossils. Along with the calcium-containing

▼ **Figure 28.21 Fossil forams.** By measuring the magnesium content in fossilized forams like these, researchers seek to learn how ocean temperatures have changed over time. Forams take up more magnesium in warmer water than in colder water.

remains of other protists, the fossilized tests of forams are part of marine sediments, including sedimentary rocks that are now land formations. Foram fossils are excellent markers for correlating the ages of sedimentary rocks in different parts of the world. Researchers are also studying these fossils to obtain information about climate change and its effects on the oceans and their life **(Figure 28.21)**.

Cercozoans

First identified in molecular phylogenies, the **cercozoans** are a large group of amoeboid and flagellated protists that feed using threadlike pseudopodia. Cercozoan protists are common inhabitants of marine, freshwater, and soil ecosystems.

▼ **Figure 28.22 A second case of primary endosymbiosis?** The cercozoan *Paulinella* conducts photosynthesis in a unique sausage-shaped structure called a chromatophore (LM). Chromatophore membranes include a peptidoglycan layer, indicating that they are derived from a bacterium. DNA evidence shows that chromatophores are derived from a different cyanobacterium than that from which plastids are derived.

Chromatophore

5 μm

Most cercozoans are heterotrophs. Many are parasites of plants, animals, or other protists; many others are predators. The predators include the most important consumers of bacteria in aquatic and soil ecosystems, along with species that eat other protists, fungi, and even small animals. One small group of cercozoans, the chlorarachniophytes (mentioned earlier in the discussion of secondary endosymbiosis), are mixotrophic: These organisms ingest smaller protists and bacteria as well as perform photosynthesis. At least one other cercozoan, *Paulinella chromatophora*, is an autotroph, deriving its energy from light and its carbon from CO_2. *Paulinella* **(Figure 28.22)** appears to represent an intriguing additional evolutionary example of a eukaryotic lineage that obtained its photosynthetic apparatus directly from a cyanobacterium.

CONCEPT CHECK 28.3

1. Explain why forams have such a well-preserved fossil record.
2. **WHAT IF?** Would you expect the plastid DNA of photosynthetic dinoflagellates and diatoms to be more similar to the nuclear DNA of plants (domain Eukarya) or to the chromosomal DNA of cyanobacteria (domain Bacteria)? Explain.
3. **MAKE CONNECTIONS** Which of the three life cycles in Figure 13.6 exhibits alternation of generations? How does it differ from the other two?
4. **MAKE CONNECTIONS** Review Figures 9.2 and 10.5, and then summarize how CO_2 and O_2 are both used and produced by chlorarachniophytes and other aerobic algae.

For suggested answers, see Appendix A.

CONCEPT 28.4

Red algae and green algae are the closest relatives of plants

As described earlier, morphological and molecular evidence indicates that plastids arose when a heterotrophic protist acquired a cyanobacterial endosymbiont. Later, photosynthetic descendants of this ancient protist evolved into red algae and green algae (see Figure 28.3), and the lineage that produced green algae then gave rise to plants. Together, red algae, green algae, and plants make up our third eukaryotic supergroup, which is called **Archaeplastida**. Archaeplastida is a monophyletic group that descended from the ancient protist that engulfed a cyanobacterium. We will examine

plants in Chapters 29 and 30; here we will look at the diversity of their closest algal relatives, red algae and green algae.

Red Algae

Many of the 6,000 known species of **red algae** (rhodophytes, from the Greek *rhodos*, red) are reddish, owing to the photosynthetic pigment phycoerythrin, which masks the green of chlorophyll **(Figure 28.23)**. However, other species (those adapted to shallow water) have less phycoerythrin. As a result, red algal species may be greenish red in very shallow water, bright red at moderate depths, and almost black in deep water. Some species lack pigmentation altogether and live as heteroptrophic parasites on other red algae.

▼ **Figure 28.23 Red algae.**

▶ ***Bonnemaisonia hamifera.*** This red alga has a filamentous form.

⊢——⊣
8 mm

◀ **Dulse (*Palmaria palmata*).** This edible species has a "leafy" form.

▼ **Nori.** The red alga *Porphyra* is the source of a traditional Japanese food.

The seaweed is grown on nets in shallow coastal waters.

Paper-thin, glossy sheets of dried nori make a mineral-rich wrap for rice, seafood, and vegetables in sushi.

Red algae are abundant in the warm coastal waters of tropical oceans. Some of their photosynthetic pigments, including phycoerythrin, allow them to absorb blue and green light, which penetrate relatively far into the water. A species of red alga has been discovered near the Bahamas at a depth of more than 260 m. There are also a small number of freshwater and terrestrial species.

Most red algae are multicellular. Although none are as big as the giant brown kelps, the largest multicellular red algae are included in the informal designation "seaweeds." You may have eaten one of these multicellular red algae, *Porphyra* (Japanese "nori"), as crispy sheets or as a wrap for sushi (see Figure 28.23). Red algae reproduce sexually and have diverse life cycles in which alternation of generations is common. However, unlike other algae, red algae do not have flagellated gametes, so they depend on water currents to bring gametes together for fertilization.

Green Algae

The grass-green chloroplasts of **green algae** have a structure and pigment composition much like the chloroplasts of plants. Molecular systematics and cellular morphology leave little doubt that green algae and plants are closely related. In fact, some systematists now advocate including green algae in an expanded "plant" kingdom, Viridiplantae (from the Latin *viridis*, green). Phylogenetically, this change makes sense, since otherwise the green algae are a paraphyletic group.

Green algae can be divided into two main groups, the charophytes and the chlorophytes. The charophytes include the algae most closely related to plants, and we will discuss them along with plants in Chapter 29.

The second group, the chlorophytes (from the Greek *chloros*, green), includes more than 7,000 species. Most live in fresh water, but there are also many marine and some terrestrial species. The simplest chlorophytes are unicellular organisms such as *Chlamydomonas*, which resemble gametes of more complex chlorophytes. Various species of unicellular chlorophytes live independently in aquatic habitats as phytoplankton or inhabit damp soil. Some live symbiotically within other eukaryotes, contributing part of their photosynthetic output to the food supply of their hosts. Still other chlorophytes live in environments exposed to intense visible and ultraviolet radiation; these species are protected by radiation-blocking compounds in their cytoplasm, cell wall, or zygote coat.

Larger size and greater complexity evolved in green algae by three different mechanisms:

1. The formation of colonies of individual cells, as seen in *Pediastrum* (**Figure 28.24a**) and other species that contribute to the stringy masses known as pond scum.

▼ **Figure 28.24 Examples of large chlorophytes.**

10 μm

(a) *Pediastrum,* a pond alga. Chlorophytes in this genus form daughter colonies with the same number and arrangement of cells as the parent colony.

2 cm

(b) *Ulva,* or sea lettuce. This multicellular, edible chlorophyte has differentiated structures, such as its leaflike blades and a rootlike holdfast that anchors the alga.

(c) *Caulerpa,* an intertidal chlorophyte. The branched filaments lack cross-walls and thus are multinucleate. In effect, the body of this alga is one huge "supercell."

2. The formation of true multicellular bodies by cell division and differentiation, as in *Volvox* (see Figure 28.5) and *Ulva* (**Figure 28.24b**).

3. The repeated division of nuclei with no cytoplasmic division, as in *Caulerpa* (**Figure 28.24c**).

Most chlorophytes have complex life cycles, with both sexual and asexual reproductive stages. Nearly all species of chlorophytes reproduce sexually by means of biflagellated gametes that have cup-shaped chloroplasts (**Figure 28.25**). Alternation of generations has evolved in some chlorophytes, including *Ulva*.

> **Mastering Biology** Animation: Alternation of Generations in a Protist

CONCEPT CHECK 28.4

1. Contrast red algae and brown algae.
2. Why is it accurate to say that *Ulva* is truly multicellular but *Caulerpa* is not?
3. **WHAT IF?** Suggest a possible reason why species in the green algal lineage may have been more likely to colonize land than species in the red algal lineage.

For suggested answers, see Appendix A.

▼ **Figure 28.25** The life cycle of *Chlamydomonas*, a unicellular chlorophyte.

Flagella

Cell wall

Nucleus

Cross section of cup-shaped chloroplast

1 μm

(TEM)

❶ In *Chlamydomonas*, mature cells are haploid and contain a single cup-shaped chloroplast.

❷ In response to a nutrient shortage, drying of the enviroment, or other stress, cells develop into gametes.

❸ Gametes of different mating types (designated + and –) fuse (fertilization), forming a diploid zygote.

Gamete (*n*)

❼ These daughter cells develop flagella and cell walls and then emerge as swimming zoospores from the parent cell. The zoospores develop into mature haploid cells.

Zoospore

Mature cell (*n*)

FERTILIZATION

ASEXUAL REPRODUCTION

SEXUAL REPRODUCTION

Zygote (2*n*)

Key

Haploid (*n*)

Diploid (2*n*)

❻ When a mature cell reproduces asexually, it resorbs its flagella and then undergoes two rounds of mitosis, forming four cells (more in some species).

MEIOSIS

❹ The zygote secretes a durable coat that protects the cell from harsh conditions.

VISUAL SKILLS *Circle the stage(s) in the diagram in which clones are formed, producing additional new daughter cells that are genetically identical to the parent cell(s).*

❺ After a dormant period, meiosis produces four haploid individuals (two of each mating type) that emerge and mature.

CONCEPT 28.5

Unikonts include protists that are closely related to fungi and animals

Excavata
SAR
Archaeplastida
Tubulinids
Slime molds
Entamoebas
Nucleariids
Fungi
Choanoflagellates
Animals
Unikonta

Unikonta (also called Amorphea) is an extremely diverse supergroup of eukaryotes that includes animals, fungi, and some protists. There are two major clades of unikonts, the amoebozoans (tubulinids and close protist relatives) and the opisthokonts (animals, fungi, and closely related protist

groups). Each of these two major clades is strongly supported by molecular systematics. Support for the close relationship between amoebozoans and opisthokonts is provided by comparisons of myosin proteins and by some (but not all) studies based on multiple genes or whole genomes.

An ongoing debate involving the unikonts concerns the root of the eukaryotic tree. Recall that the root of a phylogenetic tree anchors the tree in time: Branch points close to the root are the oldest. At present, the root of the eukaryotic tree is uncertain; hence, we do not know which supergroup of eukaryotes was the first to diverge from all other eukaryotes. If the root of the eukaryotic tree were known, it would help scientists infer characteristics of the common ancestor of all eukaryotes.

In trying to determine the root of the eukaryotic tree, researchers have based their phylogenies on different sets of genes, with conflicting results. Researchers have also tried a different approach, based on tracing the evolutionary history of genes that were transferred to eukaryotes from bacterial

▼ Figure 28.26 Inquiry

What is the root of the eukaryotic tree?

Experiment Responding to the difficulty in determining the root of the eukaryotic phylogenetic tree, Romain Derelle and colleagues used a new approach. They studied 39 nuclear genes that eukaryotes acquired by horizontal gene transfer from a single bacterial source: the alpha proteobacterium endosymbiont that gave rise to the mitochondrion. These 39 genes were known to be orthologous (see Figure 26.18). As a result, differences among these genes reflect the history of speciation events that have occurred since the eukaryotes began to diverge from their last common ancestor—making these genes well suited for inferring the root of the eukaryotic tree. Derelle and colleagues obtained the sequences of these genes for 67 species of eukaryotes representing a wide range of eukaryotic taxa. Those sequences were then used to infer evolutionary relationships among the four eukaryotic supergroups shown in Figure 28.5.

Results

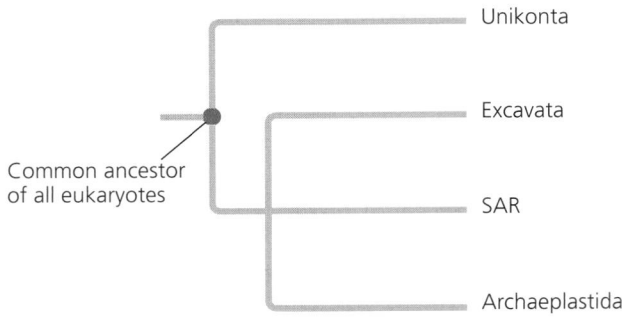

Conclusion The results show that Excavata, SAR, and Archaeplastida form a clade, which supports the hypothesis that the root of the tree is located between the unikonts and all other eukaryotes.

Data from R. Derelle et al., Bacterial proteins pinpoint a single eukaryotic root, *Proceedings of the National Academy of Sciences USA* 112:E693–699 (2015).

WHAT IF? *Other recent evidence indicates that the root of the eukaryotic tree may lie between a clade that includes unikonts and excavates and all other eukaryotes. Draw the tree suggested by this result.*

donors **(Figure 28.26)**. Results from this approach indicate that Excavata, SAR, and Archaeplastida share a more recent common ancestor than any of them does with Unikonta. This suggests that the root of the tree is located between the unikonts and all other eukaryotes, which implies that the unikonts were the first eukaryotic supergroup to diverge from all other eukaryotes. This idea remains controversial and will require more supporting evidence to be widely accepted.

Amoebozoans

The **amoebozoan** clade includes many species of amoebas that have lobe- or tube-shaped pseudopodia rather than the threadlike pseudopodia found in rhizarians. Amoebozoans include tubulinids, slime molds, and entamoebas.

Tubulinids

Tubulinids constitute a large and varied group of amoebozoans that have lobe- or tube-shaped pseudopodia. These unicellular protists are ubiquitous in soil as well as freshwater and marine environments. Most are heterotrophs that actively seek and consume bacteria and other protists; one such tubulinid species, *Amoeba proteus*, is shown in Figure 28.5. Some tubulinids also feed on detritus (nonliving organic matter).

Slime Molds

Slime molds, or mycetozoans (from the Latin, meaning "fungus animals"), once were thought to be fungi because, like fungi, they produce fruiting bodies that aid in spore dispersal. However, DNA sequence analyses indicate that the resemblance between slime molds and fungi is a case of evolutionary convergence. Slime molds have diverged into two main branches, plasmodial slime molds and cellular slime molds. We'll compare their characteristics and life cycles.

Plasmodial Slime Molds Many plasmodial slime molds are brightly colored, often yellow or orange **(Figure 28.27)**. As they grow, they form a mass called a plasmodium, which can be many centimeters in diameter. (Don't confuse a slime mold's plasmodium with the genus *Plasmodium*, which includes the parasitic apicomplexan that causes malaria.) Despite its size, the plasmodium is not multicellular; it is a single mass of cytoplasm that is undivided by plasma membranes and that contains many nuclei. This "supercell" is the product of mitotic nuclear divisions that are not followed by cytokinesis. The plasmodium extends pseudopodia through moist soil, leaf mulch, or rotting logs, engulfing food particles by phagocytosis as it grows. If the habitat begins to dry up or there is no food left, the plasmodium stops growing and differentiates into fruiting bodies that function in sexual reproduction.

▶ **Figure 28.27 A mature plasmodium, the feeding stage in the life cycle of a plasmodial slime mold.**

4 cm

Cellular Slime Molds The life cycle of the protists called cellular slime molds can prompt us to question what it means to be an individual organism. The feeding stage of these organisms consists of solitary cells that function individually, but when food is depleted, the cells form a slug-like aggregate that functions as a unit **(Figure 28.28)**. Unlike the feeding stage (plasmodium) of a plasmodial slime mold, these aggregated cells remain separated by their individual plasma membranes. Ultimately, the aggregated cells form an asexual fruiting body.

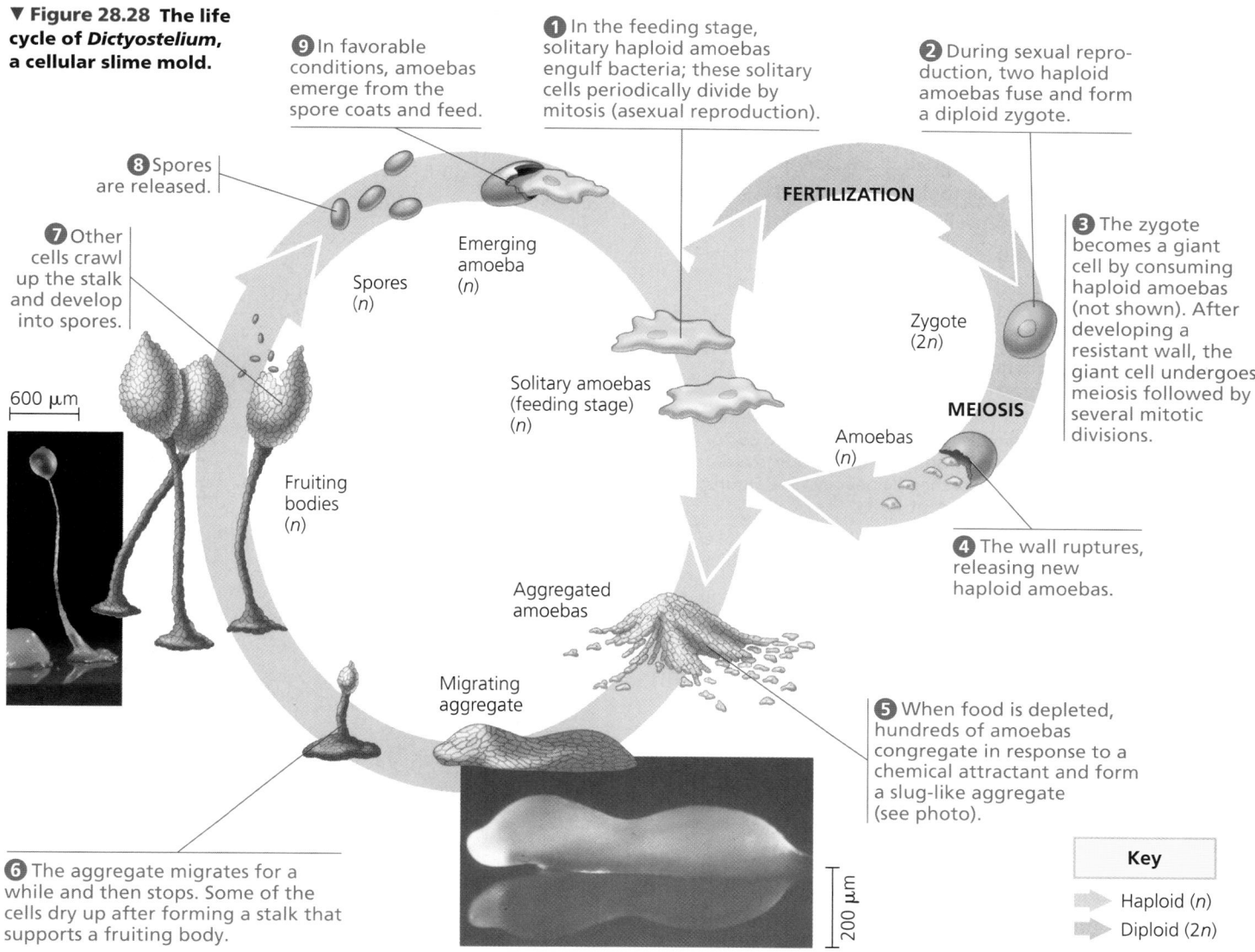

▼ **Figure 28.28 The life cycle of *Dictyostelium*, a cellular slime mold.**

❶ In the feeding stage, solitary haploid amoebas engulf bacteria; these solitary cells periodically divide by mitosis (asexual reproduction).

❷ During sexual reproduction, two haploid amoebas fuse and form a diploid zygote.

❾ In favorable conditions, amoebas emerge from the spore coats and feed.

❽ Spores are released.

❼ Other cells crawl up the stalk and develop into spores.

❸ The zygote becomes a giant cell by consuming haploid amoebas (not shown). After developing a resistant wall, the giant cell undergoes meiosis followed by several mitotic divisions.

FERTILIZATION

Emerging amoeba (*n*)

Spores (*n*)

Zygote (*2n*)

Solitary amoebas (feeding stage) (*n*)

MEIOSIS

Amoebas (*n*)

600 μm

Fruiting bodies (*n*)

❹ The wall ruptures, releasing new haploid amoebas.

Aggregated amoebas

Migrating aggregate

❺ When food is depleted, hundreds of amoebas congregate in response to a chemical attractant and form a slug-like aggregate (see photo).

200 μm

❻ The aggregate migrates for a while and then stops. Some of the cells dry up after forming a stalk that supports a fruiting body.

Key

Haploid (*n*)

Diploid (*2n*)

VISUAL SKILLS *Suppose cells were removed from the slug-like aggregate shown in the photo. Use the information in the life cycle to infer whether these cells would be haploid or diploid. Explain.*

Dictyostelium discoideum, a cellular slime mold commonly found on forest floors, has become a model organism for studying the evolution of multicellularity. One line of research has focused on the slime mold's fruiting body stage. During this stage, the cells that form the stalk die as they dry out, while the spore cells at the top survive and have the potential to reproduce (see Figure 28.28). Scientists have found that mutations in a single gene can turn individual *Dictyostelium* cells into "cheaters" that never become part of the stalk. Because these mutants gain a strong reproductive advantage over noncheaters, why don't all *Dictyostelium* cells cheat?

Recent discoveries suggest an answer to this question. Cheating cells lack a specific surface protein, and noncheating cells can recognize this difference. Noncheaters preferentially aggregate with other noncheaters, thus depriving cheaters of the chance to exploit them. Such a recognition system may have been important in the evolution of other multicellular eukaryotes, such as animals and plants.

Entamoebas

Whereas most amoebozoans are free-living, those that belong to the genus *Entamoeba* are symbiotic parasites. They infect all classes of vertebrate animals as well as some invertebrates. Humans are host to at least six species of *Entamoeba*, but only one, *E. histolytica*, is known to be pathogenic. *E. histolytica* causes amoebic dysentery and is spread via contaminated drinking water, food, or eating utensils. Responsible for up to 110,000 deaths worldwide every year, the disease is the third-leading cause of death due to eukaryotic parasites, after malaria (see Figure 28.18) and schistosomiasis (see Figure 33.10).

Opisthokonts

Opisthokonts are an extremely diverse group of eukaryotes that includes animals, fungi, and several groups of protists. We will discuss the evolutionary history of fungi and animals in Chapters 31–34. Of the opisthokont protists, we will discuss the nucleariids in Chapter 31 because they are more closely related to fungi than they are to other protists. Similarly, we will discuss choanoflagellates in Chapter 32, since they are more closely related to animals than they are to other protists. The nucleariids and choanoflagellates illustrate why scientists have abandoned the former kingdom Protista: A monophyletic group that includes these single-celled eukaryotes would also have to include the multicellular animals and fungi that are closely related to them.

CONCEPT 28.6

Protists play key roles in ecological communities

Most protists are aquatic, and they are found almost anywhere there is water, including moist terrestrial habitats such as damp soil and leaf litter. In oceans, ponds, and lakes, many protists are bottom-dwellers that attach to rocks and other substrates or creep through the sand and silt. As we've seen, other protists are important constituents of plankton. We'll focus here on two key roles that protists play in the varied habitats in which they live: that of symbiont and that of producer.

Symbiotic Protists

Many protists form symbiotic associations with other species. For example, photosynthetic dinoflagellates are food-providing symbiotic partners of the animals (coral polyps) that build coral reefs. Coral reefs are highly diverse ecological communities. That diversity ultimately depends on corals—and on the mutualistic protists that nourish them. Corals support reef diversity by providing food to some species and habitat to many others.

Another example is the wood-digesting protists that inhabit the gut of many termite species (**Figure 28.29**). Unaided, termites cannot digest wood, and they rely on protistan or prokaryotic symbionts to do so. Termites cause over $3.5 billion in damage annually to wooden homes in the United States.

Symbiotic protists also include parasites that have compromised the economies of entire countries. Consider the malaria-causing protist *Plasmodium*: Income levels in countries hard hit by malaria are 33% lower than in similar countries free of the disease. Protists can have devastating effects on other

▶ **Figure 28.29 A symbiotic protist.** This organism is a hypermastigote, a member of a group of parabasalids that live in the gut of termites and certain cockroaches and enable the hosts to digest wood (SEM).

10 μm

▼ **Figure 28.30 Sudden oak death.** Many dead oak trees are visible in this Monterey County, California landscape. Infected trees lose their ability to adjust to cycles of wet and dry weather.

species too. Massive fish kills have been attributed to *Pfiesteria shumwayae* (see Figure 28.17), a dinoflagellate parasite that attaches to its victims and eats their skin. Among species that parasitize plants, the stramenopile *Phytophthora ramorum* (an oomycete) has emerged as a major new forest pathogen. This species causes sudden oak death (SOD), a disease that has killed millions of oaks and other trees in the United States and Great Britain (**Figure 28.30**; also see Concept 54.5).

Photosynthetic Protists

Many protists are important **producers**, organisms that use energy from light (or in some prokaryotes, inorganic chemicals) to convert CO_2 to organic compounds. Producers form the base of ecological food webs. In aquatic communities, the main producers are photosynthetic protists and prokaryotes (**Figure 28.31**). All other organisms in the community are **consumers** that depend on producers for food, either directly (by eating them) or indirectly (by eating an organism that ate a producer). Scientists estimate that roughly 30% of the world's photosynthesis is performed by diatoms, dinoflagellates, multicellular algae, and other aquatic protists. Photosynthetic prokaryotes contribute another 20%, and plants are responsible for the remaining 50%.

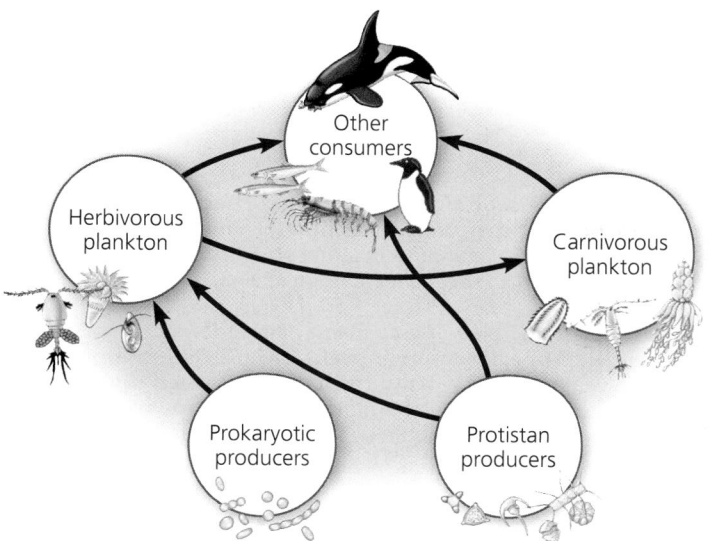

▲ **Figure 28.31 Protists: key producers in aquatic communities.** Arrows in this simplified food web lead from food sources to the organisms that eat them.

Because producers form the foundation of food webs, factors that affect producers can dramatically affect their entire community. In aquatic environments, photosynthetic protists are often held in check by low concentrations of nitrogen, phosphorus, or iron. Various human actions can increase the concentrations of these elements in aquatic communities. For example, when fertilizer is applied to a field, some of the fertilizer may be washed by rainfall into a river that drains into a lake or ocean. When people add nutrients to aquatic communities in this or other ways, the abundance of photosynthetic protists can increase spectacularly. Such increases can have major ecological consequences, including the formation of large "dead zones" in marine ecosystems (see Figure 56.23).

A pressing question is how global warming will affect photosynthetic protists and other producers. As shown in **Figure 28.32**, the growth and biomass of photosynthetic protists and prokaryotes have declined in many ocean regions as sea surface temperatures have increased. By what mechanism do rising sea surface temperatures reduce the growth of marine producers? One hypothesis relates to the rise or upwelling of cold, nutrient-rich waters from below. Many marine producers rely on nutrients brought to the surface in this way. However, rising sea surface temperatures can cause the formation of a layer of light, warm water that acts as a barrier to nutrient upwelling—thus reducing the growth of marine producers. If sustained, the changes shown in Figure 28.32 would likely have far-reaching effects on marine ecosystems, fishery yields, and the global carbon cycle (see Figure 55.14). Global warming can also affect producers on land, but there the base of food webs is occupied not by protists but by plants, which we will discuss in Chapters 29 and 30.

CONCEPT CHECK 28.6

1. Justify the claim that photosynthetic protists are among the biosphere's most important organisms.
2. Describe three symbioses that include protists.
3. **WHAT IF?** High water temperatures and pollution can cause corals to expel their dinoflagellate symbionts. How might such "coral bleaching" affect corals and other species?
4. **MAKE CONNECTIONS** The bacterium *Wolbachia* is a symbiont that lives in mosquito cells and spreads rapidly through mosquito populations. *Wolbachia* can make mosquitoes resistant to infection by *Plasmodium*; researchers are seeking a strain that confers resistance and does not harm mosquitoes. Compare evolutionary changes that could occur if malaria control is attempted using such a *Wolbachia* strain versus using insecticides to kill mosquitoes. (Review Figure 28.18 and Concept 23.4.)

For suggested answers, see Appendix A.

▼ **Figure 28.32 Effects of climate change on marine producers.**

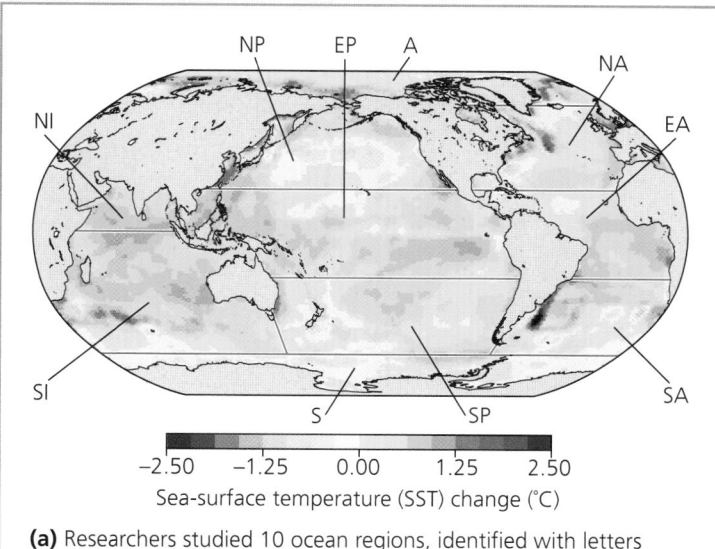

(a) Researchers studied 10 ocean regions, identified with letters on the map [see (b) for the corresponding names]. SSTs have increased since 1950 in most areas of these regions.

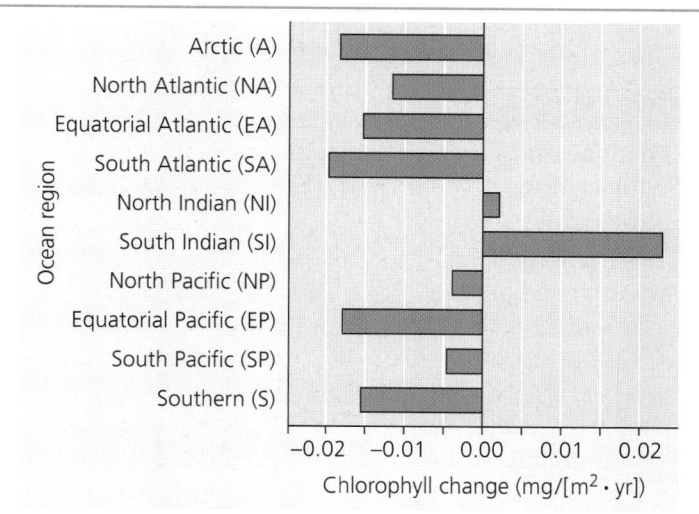

(b) The concentration of chlorophyll, an index for the biomass and growth of marine producers, has decreased over the same time period in most ocean regions.

28 Chapter Review

→ Go to **Mastering Biology** for Assignments, the eText, the Study Area, and Dynamic Study Modules.

SUMMARY OF KEY CONCEPTS

→ To review key terms, go to the **Vocabulary Self-Quiz** in the **Mastering Biology** eText or Study Area, or go to goo.gl/zkjz9t.

CONCEPT 28.1

Most eukaryotes are single-celled organisms *(pp. 594–597)*

- Domain Eukarya includes many groups of **protists**, along with plants, animals, and fungi. Unlike prokaryotes, protists and other eukaryotes have a nucleus and other membrane-enclosed organelles, as well as a cytoskeleton that enables them to have asymmetric forms and to change shape as they feed, move, or grow.
- Protists are structurally and functionally diverse and have a wide variety of life cycles. Most are unicellular. Protists include photoautotrophs, heterotrophs, and **mixotrophs**.

- Current evidence indicates that eukaryotes originated by **endosymbiosis** when an archaeal host (or a host closely related to the archaea) engulfed an alpha proteobacterium that would evolve into an organelle found in all eukaryotes, the mitochondrion.
- Plastids are thought to be descendants of cyanobacteria that were engulfed by early eukaryotic cells. The plastid-bearing lineage eventually evolved into **red algae** and **green algae**. Other protist groups evolved from **secondary endosymbiotic** events in which red algae or green algae were themselves engulfed.
- In one hypothesis, eukaryotes are grouped into four supergroups: **Excavata, SAR, Archaeplastida**, and **Unikonta**.

? *Describe similarities and differences between protists and other eukaryotes.*

Key Concept/Eukaryote Supergroup	Major Groups	Key Morphological Characteristics	Specific Examples

CONCEPT 28.2
 	
Excavates include protists with modified mitochondria and protists with unique flagella *(pp. 597–601)*
 	
? *What evidence indicates that the excavates form a clade?*
	Diplomonads and parabasalids
	
	Euglenozoans
	  Kinetoplastids
	  Euglenids
		Modified mitochondria
		
		Spiral or crystalline rod inside flagella
			Giardia, Trichomonas
			
			Trypanosoma, Euglena

CONCEPT 28.3
 	
SAR is a highly diverse group of protists defined by DNA similarities *(pp. 601–609)*
 	
? *Although they are not photosynthetic, apicomplexan parasites such as* Plasmodium *have modified plastids. Describe a current hypothesis that explains this observation.*
	Stramenopiles
	  Diatoms
	  Oomycetes
	  Brown algae
	
	Alveolates
	  Dinoflagellates
	  Apicomplexans
	  Ciliates
	
	Rhizarians
	  Radiolarians
	  Forams
	  Cercozoans
		Hairy and smooth flagella
		
		Membrane-enclosed sacs (alveoli) beneath plasma membrane
		
		Amoebas with threadlike pseudopodia
			Phytophthora, Laminaria
			
			Pfiesteria, Plasmodium, Paramecium
			
			Globigerina

CONCEPT 28.4 **Red algae and green algae are the closest relatives of plants** *(pp. 609–611)* ? *On what basis do systematists place plants in the same supergroup (Archaeplastida) as red and green algae?*	**Red algae** **Green algae** **Plants**	Phycoerythrin (photosynthetic pigment) Plant-type chloroplasts (See Chapters 29 and 30.)	*Porphyridium, Palmaria* *Chlamydomonas, Ulva* Mosses, ferns, conifers, flowering plants

CONCEPT 28.5
 	
Unikonts include protists that are closely related to fungi and animals *(pp. 611–614)*
 	
? *Describe a key feature for each of the main protist subgroups of Unikonta.*
	Amoebozoans
	  Tubulinids
	  Slime molds
	  Entamoebas
	
	Opisthokonts
		Amoebas with lobe-shaped or tube-shaped pseudopodia
		
		(Highly variable; see Chapters 31–34.)
			Amoeba, Dictyostelium
			
			Choanoflagellates, nucleariids, animals, fungi

Protists play key roles in ecological communities (pp. 614–615)

- Protists form mutualistic and parasitic relationships that affect their symbiotic partners and many other members of the community.
- Photosynthetic protists are among the most important **producers** in aquatic communities. Because they are at the base of the food web, factors that affect photosynthetic protists affect many other species in the community.

? *Describe several protists that are ecologically important.*

TEST YOUR UNDERSTANDING

→ For more multiple-choice questions, go to the **Practice Test** in the **Mastering Biology** eText or Study Area, or go to goo.gl/GruWRg.

Levels 1-2: Remembering/Understanding

1. Plastids that are surrounded by more than two membranes are evidence of
   (A) evolution from mitochondria.
   (B) fusion of plastids.
   (C) origin of the plastids from archaea.
   (D) secondary endosymbiosis.

2. Biologists think that endosymbiosis gave rise to mitochondria before plastids partly because
   (A) the products of photosynthesis could not be metabolized without mitochondrial enzymes.
   (B) all eukaryotes have mitochondria (or their remnants), whereas many eukaryotes do not have plastids.
   (C) mitochondrial DNA is less similar to prokaryotic DNA than is plastid DNA.
   (D) without mitochondrial CO_2 production, photosynthesis could not occur.

3. Which group is correctly paired with its description?
   (A) diatoms—important consumers in aquatic communities
   (B) diplomonads—protists with modified mitochondria
   (C) apicomplexans—producers with intricate life cycles
   (D) red algae—acquired plastids by secondary endosymbiosis

4. According to the phylogeny presented in this chapter, which protists are in the same eukaryotic supergroup as plants?
   (A) green algae
   (B) dinoflagellates
   (C) red algae
   (D) both A and C

5. In a life cycle with alternation of generations, multicellular haploid forms alternate with
   (A) unicellular haploid forms.
   (B) unicellular diploid forms.
   (C) multicellular haploid forms.
   (D) multicellular diploid forms.

Levels 3-4: Applying/Analyzing

6. Based on the phylogenetic tree in Figure 28.5, which of the following statements is correct?
   (A) Excavata and SAR form a sister group.
   (B) The most recent common ancestor of SAR is older than that of Unikonta.
   (C) The most basal (first to diverge) eukaryotic supergroup cannot be determined.
   (D) Excavata is the most basal eukaryotic supergroup.

7. **EVOLUTION CONNECTION • DRAW IT** Medical researchers seek to develop drugs that can kill or restrict the growth of human pathogens yet have few harmful effects on patients. These drugs often work by disrupting the metabolism of the pathogen or by targeting its structural features.
   Draw and label a phylogenetic tree that includes an ancestral prokaryote and the following groups of organisms: Excavata, SAR, Archaeplastida, Unikonta, and, within Unikonta, amoebozoans, animals, choanoflagellates, fungi, and nucleariids. Based on this tree, hypothesize whether it would be most difficult to develop drugs to combat human pathogens that are prokaryotes, protists, animals, or fungi. (You do not need to consider the evolution of drug resistance by the pathogen.)

Levels 5-6: Evaluating/Creating

8. **SCIENTIFIC INQUIRY** Applying the "If … then" logic of deductive reasoning (see Concept 1.3), what are a few of the predictions that arise from the hypothesis that plants evolved from green algae? Put another way, how could you test this hypothesis?

9. **WRITE ABOUT A THEME: INTERACTIONS** Organisms interact with each other and the physical environment. In a short essay (100–150 words), explain how the response of diatom populations to a drop in nutrient availability can affect both other organisms and aspects of the physical environment (such as carbon dioxide concentrations).

10. **SYNTHESIZE YOUR KNOWLEDGE**

50 μm

This micrograph shows a single-celled eukaryote, the ciliate *Didinium* (left), about to engulf its *Paramecium* prey, which is also a ciliate. Identify the eukaryotic supergroup to which ciliates belong and describe the role of endosymbiosis in the evolutionary history of that supergroup. Are these ciliates more closely related to all other protists than they are to plants, fungi, or animals? Explain.

For selected answers, see Appendix A.

29 Plant Diversity I: How Plants Colonized Land

KEY CONCEPTS

29.1 Plants evolved from green algae *p. 619*

29.2 Mosses and other nonvascular plants have life cycles dominated by gametophytes *p. 623*

29.3 Ferns and other seedless vascular plants were the first plants to grow tall *p. 629*

Figure 29.1 For much of Earth's history, the land surface was largely lifeless. Some prokaryotes lived on land 3.2 billion years ago, but it was only within the last 500 million years that small plants, fungi, and animals joined them ashore. Finally, by 385 million years ago, the first forests appeared (but with different species than those of today).

Study Tip

Make a table: As you read the chapter, make a table of adaptations for life on land that evolved in plants after they diverged from their close aquatic relatives, a group of charophyte algae.

Adaptation	Description	How it facilitates life on land
Cuticle	Waxy covering on outer body surface	Reduces dessication

→ Go to Mastering Biology

For Students (in eText and Study Area)
- Get Ready for Chapter 29
- Animation: Moss Life Cycle
- Animation: Fern Life Cycle

For Instructors to Assign (in Item Library)
- Visualizing the Concept: Haploid and Diploid Generations Alternate in Plant Life Cycles
- Scientific Skills Exercise: Making Bar Graphs and Interpreting Data

What were the major developments in the evolution of plants?

Plants originated from green algae about 470 million years ago.

By 425 million years ago, some early plants had **traits facilitating life on land.**

Reproductive structures

Photosynthetic branches

Structures that anchor the plant to the soil

Over time, early plants gave rise to a **rich diversity of plants.**

Nonvascular plants
(such as mosses)

Seedless vascular plants
(such as ferns)

Seed plants
(cone-bearing plants and flowering plants)

CONCEPT 29.1

Plants evolved from green algae

Today, there are more than 325,000 known plant species. Although a few plant species returned to aquatic habitats during their evolution, most present-day plants live on land. In this text, we distinguish plants from algae, which are photosynthetic protists.

Plants enabled other life-forms to survive on land. For example, plants supply oxygen and are a key source of food for terrestrial animals. Also, by their very presence, plants such as the trees of a forest physically create the habitats required by animals and many other organisms.

As you read in Concept 28.4, green algae called charophytes are the closest relatives of plants. We'll begin with a closer look at the evidence for this relationship.

Evidence of Algal Ancestry

Many key traits of plants are found in some algae. For example, plants are multicellular, eukaryotic, photosynthetic autotrophs, as are brown, red, and certain green algae. Plants have cell walls made of cellulose, and so do green algae, dinoflagellates, and brown algae. And chloroplasts with chlorophylls *a* and *b* are present in green algae, euglenids, and a few dinoflagellates, as well as in plants.

However, the charophytes are the only present-day algae that share certain distinctive traits with plants, suggesting that they are the closest living relatives of plants. For example, the cells of both plants and charophytes have distinctive circular rings of proteins embedded in the plasma membrane **(Figure 29.2)**; these protein rings synthesize the cellulose found in the cell wall. In contrast, noncharophyte algae have linear sets of proteins that synthesize cellulose. Likewise, in species of plants that have flagellated sperm, the structure of the sperm closely resembles that of charophyte sperm. Analyses of nuclear, chloroplast, and mitochondrial DNA also support the close relationship between plants and charophytes.

More specifically, a recent analysis of nearly 900 protein-coding nuclear genes found that charophytes in the clade Zygnematophyceae—such as *Zygnema*—are the closest living relatives of plants **(Figure 29.3)**. Although this evidence shows that plants arose from within a group of charophyte algae, it does not mean that plants are descended from these living algae. Even so, present-day charophytes may tell us something about the algal ancestors of plants.

▼ **Figure 29.2 Rings of cellulose-synthesizing proteins.** (inverted-contrast TEM)

20 nm

▶ **Figure 29.3 *Zygnema*, an alga closely related to plants.** *Zygnema* is a filamentous pond alga in the Zygnematophyceae, the group of charophytes that is most closely related to plants.

20 µm

Adaptations Enabling the Move to Land

Many species of charophyte algae inhabit shallow waters around the edges of ponds and lakes, where they are subject to occasional drying. In such environments, natural selection favors individual algae that can survive periods when they are not submerged. In charophytes, a layer of a durable polymer called **sporopollenin** prevents exposed zygotes from drying out. A similar chemical adaptation is found in the tough sporopollenin walls that encase plant spores.

The accumulation of such traits by at least one population of charophyte algae (now extinct) enabled their descendants—the first plants—to live permanently above the waterline. This ability opened a new frontier: a terrestrial habitat that offered enormous benefits. The bright sunlight was unfiltered by water and plankton; the atmosphere offered more plentiful carbon dioxide than did water; and the soil by the water's edge was rich in some mineral nutrients. But these benefits were accompanied by challenges: a relative scarcity of water and a lack of structural support against gravity. (To appreciate why such support is important, picture how the soft body of a jellyfish (jelly) collapses when waves strand it on a beach.) Plants diversified as new adaptations arose that enabled them to thrive despite these challenges.

Today, what adaptations are unique to plants? The answer depends on where you draw the boundary dividing plants from algae **(Figure 29.4)**. Since the placement of this

Red algae

ANCESTRAL ALGA

Chlorophytes

Other charophytes

Closest charophyte relative

Embryophytes

Viridiplantae

Streptophyta

Plantae

▶ **Figure 29.4 Three possible "plant" kingdoms, the clades Viridiplantae, Streptophyta, and Plantae.**

Charophyte algae lack the key traits of plants described in this figure: alternation of generations and the associated trait of multicellular, dependent embryos. As described in the main text, charophyte algae also lack walled spores produced in sporangia and apical meristems. This suggests that these four traits were absent in the ancestor common to plants and charophytes but instead evolved as derived traits of plants. Not every plant exhibits all of these traits: Certain lineages of plants have lost some traits over time.

Alternation of Generations

The life cycles of all plants alternate between two generations of distinct multicellular organisms: gametophytes and sporophytes. As shown in the diagram below (using a fern as an example), each generation gives rise to the other, a process that is called **alternation of generations**. This type of reproductive cycle evolved in various groups of algae but does not occur in the charophytes, the algae most closely related to plants. Take care not to confuse the alternation of generations in plants with the haploid and diploid stages in the life cycles of other sexually reproducing organisms. Alternation of generations is distinguished by the fact that the life cycle

includes both multicellular haploid organisms and multicellular diploid organisms. The multicellular haploid **gametophyte** ("gamete-producing plant") is named for its production by mitosis of haploid gametes—eggs and sperm—that fuse during fertilization, forming diploid zygotes. Mitotic division of the zygote produces a multicellular diploid **sporophyte** ("spore-producing plant"). Meiosis in a mature sporophyte produces haploid **spores**, reproductive cells that can develop into a new haploid organism without fusing with another cell. Mitotic division of the spore cell produces a new multicellular gametophyte, and the cycle begins again.

Alternation of generations: five generalized steps

❶ The gametophyte produces haploid gametes by mitosis.

Gametophyte (*n*)

Gamete from another plant

Mitosis

n

n — Spore

Gamete

n

Mitosis

n

❷ Two gametes unite (fertilization) and form a diploid zygote.

MEIOSIS

FERTILIZATION

❺ The spores develop into multicellular haploid gametophytes.

❹ The sporophyte produces unicellar haploid spores by meiosis.

Sporophyte (*2n*)

Zygote

2n

Mitosis

❸ The zygote develops into a multicellular diploid sporophyte.

Key	
	Haploid (*n*)
	Diploid (*2n*)

Multicellular, Dependent Embryos

As part of a life cycle with alternation of generations, multicellular plant embryos develop from zygotes that are retained within the tissues of the female parent (a gametophyte). The parental tissues protect the developing embryo from harsh environmental conditions and provide nutrients such as sugars and amino acids. The embryo has specialized *placental transfer cells* that enhance the transfer of nutrients to the embryo through elaborate ingrowths of the wall surface (plasma membrane and cell wall). The multicellular, dependent embryo of plants is such a significant derived trait that plants are also known as **embryophytes**.

Embryo (LM) and placental transfer cell (TEM) of *Marchantia* (a liverwort)

Embryo

Maternal tissue

10 µm

Wall ingrowths

Placental transfer cell (blue outline)

2 µm

MAKE CONNECTIONS *Review sexual life cycles in Figure 13.6. Identify which type of sexual life cycle has alternation of generations, and summarize how it differs from other life cycles.*

boundary is the subject of ongoing debate, this text uses a traditional definition that equates the kingdom Plantae with embryophytes (plants with embryos). In this context, let's now examine the derived traits that separate plants from their closest algal relatives.

Derived Traits of Plants

Several adaptations that facilitate survival and reproduction on dry land emerged after plants diverged from their algal relatives. Examples of such traits that are found in plants but not in charophyte algae include the following:

- **Alternation of generations.** This type of life cycle, consisting of multicellular forms that give rise to each other in turn, is described in **Figure 29.5**.
- **Walled spores produced in sporangia.** The sporophyte stage of the plant life cycle has multicellular organs called **sporangia** (singular, *sporangium*) that produce spores **(Figure 29.6)**. The polymer sporopollenin makes the walls of these spores resistant to harsh environments, enabling plant spores to be dispersed through dry air without harm.
- **Apical meristems.** Plants also differ from their algal relatives in having **apical meristems**, localized regions of cell division at the tips of roots and shoots **(Figure 29.7)**. Apical meristem cells can divide throughout the plant's life, enabling its roots and shoots to elongate, thus increasing the plant's exposure to environmental resources.

Additional derived traits that relate to terrestrial life have evolved in many plant species. For example, the epidermis in many species has a covering, the **cuticle**, which consists of wax and other polymers. Permanently exposed to the air, plants run a far greater risk of desiccation (drying out) than do their algal relatives. The cuticle acts as waterproofing, helping prevent excessive water loss from the aboveground plant

▼ **Figure 29.6 Sporophytes and sporangia of a moss in the genus *Mnium*.** Each of the many spores produced by a sporangium is encased by a durable, sporopollenin-enriched wall.

Sporangium

Spores

Longitudinal section of sporangium (LM)

Sporophyte

Gametophyte

Apical meristem of shoot

Developing leaves

◀ **Figure 29.7 Apical meristem of a plant shoot.** Cells produced by apical meristems differentiate into the outer epidermis, which protects the body, and various types of internal tissues, such as those in leaves (LM).

100 μm

organs, while also providing some protection from microbial attack. Most plants also have specialized pores called **stomata** (singular, *stoma*), which support photosynthesis by allowing the exchange of CO_2 and O_2 between the outside air and the plant (see Figure 10.3). Stomata are also the main avenues by which water evaporates from the plant; in hot, dry conditions, the stomata close, minimizing water loss.

The earliest plants lacked true roots and leaves. Without roots, how did these plants absorb nutrients from the soil? Fossils dating from 420 million years ago reveal an adaptation that may have aided early plants in nutrient uptake: They formed symbiotic associations with fungi. We'll explore these associations, called *mycorrhizae*, and their benefits to both plants and fungi in more detail in Concept 31.1. For now, the main point is that mycorrhizal fungi form extensive networks of filaments through the soil and transfer nutrients to their symbiotic plant partner. This benefit may have helped plants without roots to colonize land.

The Origin and Diversification of Plants

The algae most closely related to plants include many unicellular species and small colonial species. Since it is likely that the first plants were similarly small, the search for the earliest fossils of plants has focused on the microscopic world. As mentioned earlier, microorganisms colonized land as early as 3.2 billion years ago. But the microscopic fossils that document life on land changed dramatically 470 million years ago with the appearance of spores from early plants.

What distinguishes these spores from those of algae or fungi? One clue comes from their chemical composition, which matches the composition of plant spores today but differs from that of the spores of other organisms. In addition, the walls of these ancient spores have structural features that today are found only in the spores of certain plants (liverworts). And in rocks dating to 450 million years ago,

Figure 29.8 Ancient plant spores and tissue. (colorized SEMs)

(a) Fossilized spores.
The chemical composition and the physical structure of the walls of these 450-million-year-old spores match those found in certain plants today.

(b) Fossilized sporophyte tissue.
The spores in (a) were embedded in tissue that appears to be from plants.

50 μm

Table 29.1 Ten Phyla of Extant Plants		
	Common Name	Number of Known Species
Nonvascular Plants (Bryophytes)		
Phylum Hepatophyta	Liverworts	9,000
Phylum Bryophyta	Mosses	13,000
Phylum Anthocerophyta	Hornworts	225
Vascular Plants		
Seedless Vascular Plants		
Phylum Lycophyta	Lycophytes	1,200
Phylum Monilophyta	Monilophytes	12,000
Seed Plants		
Gymnosperms		
Phylum Ginkgophyta	Ginkgo	1
Phylum Cycadophyta	Cycads	350
Phylum Gnetophyta	Gnetophytes	75
Phylum Coniferophyta	Conifers	600
Angiosperms		
Phylum Anthophyta	Flowering plants	290,000

researchers have discovered similar spores embedded in plant cuticle material that resembles spore-bearing tissue in living plants **(Figure 29.8)**.

Fossils of larger plant structures, such as the *Cooksonia* sporangium in **Figure 29.9**, date to 425 million years ago, which is 45 million years after the appearance of plant spores in the fossil record. While the precise age (and form) of the first plants has yet to be discovered, those ancestral species gave rise to the vast diversity of living plants. **Table 29.1** provides basic information about the ten extant phyla. (Extant lineages are those that have surviving members.) As you read the rest of this section, look at Table 29.1 together with **Figure 29.10**, which reflects a view of plant phylogeny that is based on plant anatomy, biochemistry, and genetics.

One way to distinguish groups of plants is whether or not they have an extensive system of **vascular tissue**, cells joined into tubes that transport water and nutrients throughout the plant body. Most present-day plants have a complex vascular tissue system and are therefore called **vascular plants**. Plants that do not have an extensive transport system—liverworts, mosses, and hornworts—are described as "nonvascular" plants, even though some mosses do have simple vascular tissue. Nonvascular plants are often

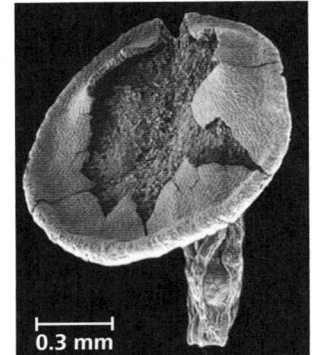

▶ **Figure 29.9 *Cooksonia* sporangium fossil.** With the exception of the shape of its sporangia, *Cooksonia*'s overall form was very similar to the early plant shown in Figures 29.1 and 29.16.

0.3 mm

informally called **bryophytes** (from the Greek *bryon*, moss, and *phyton*, plant). Although the term *bryophyte* is commonly used to refer to all nonvascular plants, molecular studies and morphological analyses of sperm structure have concluded that bryophytes do not form a monophyletic group (a clade).

Vascular plants, which form a clade that comprises about 93% of all extant plant species, can be categorized further into smaller clades. Two of these clades are the **lycophytes** (the club mosses and their relatives) and the **monilophytes** (ferns and their relatives). The plants in each of these clades lack seeds, which is why collectively the two clades are informally called **seedless vascular plants**. However, notice in Figure 29.10 that, like bryophytes, seedless vascular plants do not form a clade.

A group such as the bryophytes or the seedless vascular plants represents a collection of organisms that share key biological features. It can be informative to group organisms according to their features, such as having a vascular system but lacking seeds. But members of such a group, unlike members of a clade, do not necessarily share the same ancestry. For example, even though monilophytes and lycophytes are all seedless vascular plants, monilophytes share a more recent common ancestor with seed plants. As a result, we would expect monilophytes and seed plants to share key traits not found in lycophytes—and they do, as you'll read in Concept 29.3.

A third clade of vascular plants consists of seed plants, which represent the vast majority of living plant species.

▼ **Figure 29.10 Highlights of plant evolution.** The phylogeny shown here illustrates a leading hypothesis about the relationships between plant groups.

MAKE CONNECTIONS *The figure identifies which lineages are plants, nonvascular plants, vascular plants, seedless vascular plants, and seed plants. Which of these categories are monophyletic, and which are paraphyletic? Explain. (See Figure 26.10 to review these terms.)*

A **seed** is an embryo packaged with a supply of nutrients inside a protective coat. Seed plants can be divided into two groups, gymnosperms and angiosperms, based on the absence or presence of enclosed chambers in which seeds mature. **Gymnosperms** (from the Greek *gymnos*, naked, and *sperm*, seed) are known as "naked seed" plants because their seeds are not enclosed in chambers. Living gymnosperm species, the most familiar of which are the conifers, form a clade. **Angiosperms** (from the Greek *angion*, container) are a huge clade consisting of all flowering plants; their seeds develop inside chambers that originate within flowers. Nearly 90% of living plant species are angiosperms.

Note that the phylogeny depicted in Figure 29.10 focuses only on the relationships between extant plant lineages. Paleobotanists have also discovered fossils belonging to extinct plant lineages. As you'll read later in the chapter, these fossils can reveal intermediate steps in the emergence of plant groups found on Earth today.

CONCEPT CHECK 29.1

1. Why do researchers identify the charophytes rather than another group of algae as the closest living relatives of plants?
2. Identify four derived traits that distinguish plants from charophyte green algae and facilitate life on land. Explain.
3. **WHAT IF?** What would the human life cycle be like if we had alternation of generations? Assume that the multicellular diploid stage would be similar in form to an adult human.

For suggested answers, see Appendix A.

CONCEPT 29.2

Mosses and other nonvascular plants have life cycles dominated by gametophytes

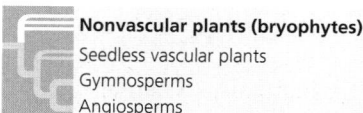

Nonvascular plants (bryophytes)
Seedless vascular plants
Gymnosperms
Angiosperms

The nonvascular plants (bryophytes) are represented today by three phyla of small, herbaceous (nonwoody) plants: **liverworts** (phylum Hepatophyta), **mosses** (phylum Bryophyta), and **hornworts** (phylum Anthocerophyta). Liverworts and hornworts are named for their shapes, plus the suffix *wort* (from the Anglo-Saxon for "herb"). Mosses are familiar to many people, although some plants commonly called "mosses" are not really mosses at all. These include Irish moss (a red seaweed), reindeer moss (a lichen), club mosses (seedless vascular plants), and Spanish mosses (lichens in some regions and flowering plants in others).

Phylogenetic analyses indicate that liverworts, mosses, and hornworts diverged from other plant lineages early in the history of plant evolution (see Figure 29.10). Fossil evidence provides some support for this idea: The earliest spores of plants (dating from 450 to 470 million years ago) have structural features found only in the spores of liverworts, and by 430 million years ago spores similar to those of mosses and hornworts also occur in the fossil record. The earliest fossils of vascular plants date to about 425 million years ago.

Over the long course of their evolution, liverworts, mosses, and hornworts have acquired many unique adaptations. Next, we'll examine some of those features.

Bryophyte Gametophytes

Unlike vascular plants, in all three bryophyte phyla the haploid gametophytes are the dominant stage of the life cycle: They are usually larger and longer-living than the sporophytes, as shown in the moss life cycle in **Figure 29.11**. The sporophytes are typically present only part of the time.

▼ **Figure 29.11 The life cycle of a moss.**

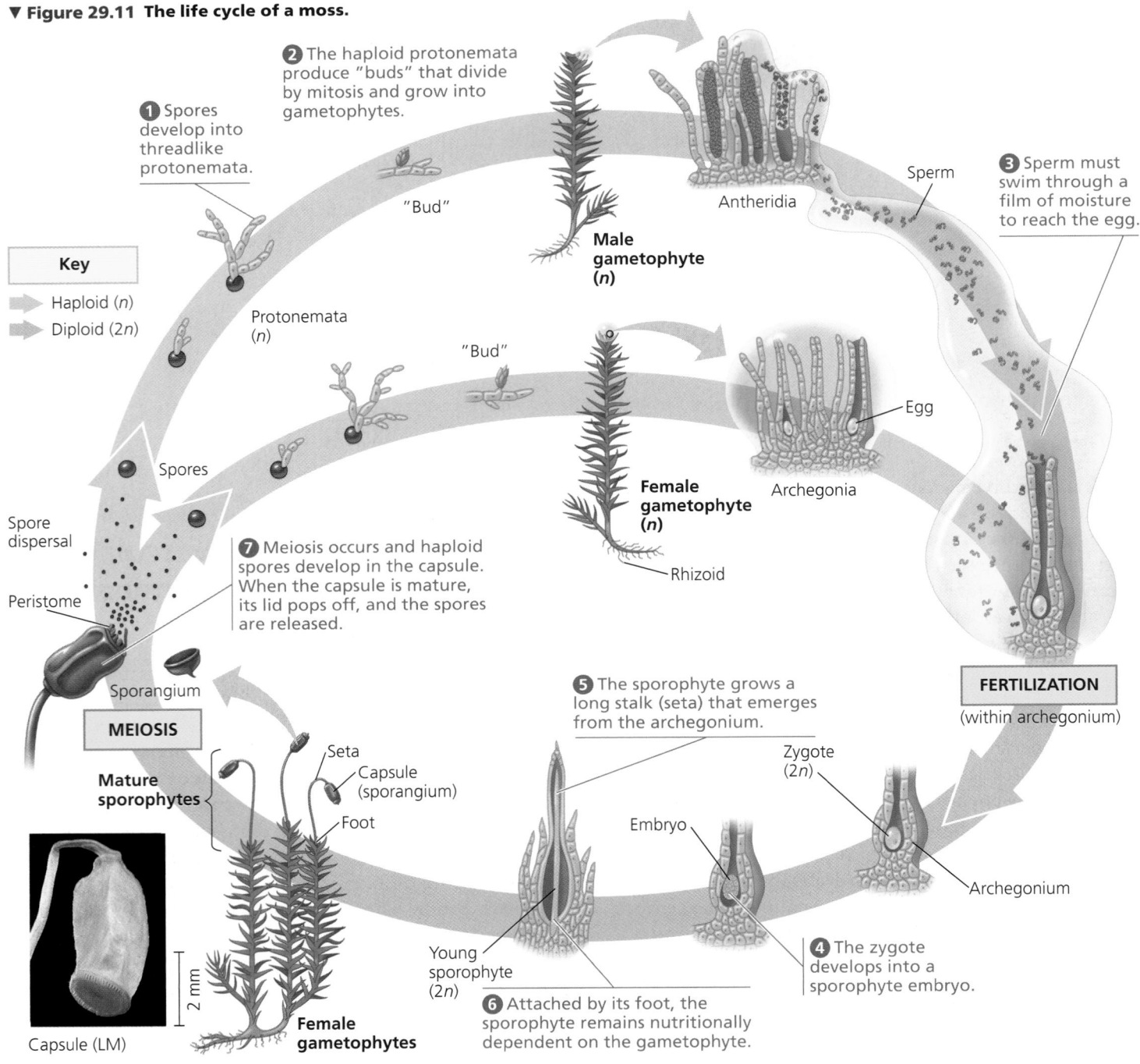

1 Spores develop into threadlike protonemata.

2 The haploid protonemata produce "buds" that divide by mitosis and grow into gametophytes.

3 Sperm must swim through a film of moisture to reach the egg.

Sperm

Antheridia

Male gametophyte (*n*)

"Bud"

Key

Haploid (*n*)
Diploid (2*n*)

Protonemata (*n*)

"Bud"

Egg

Archegonia

Female gametophyte (*n*)

Rhizoid

Spores

Spore dispersal

Peristome

7 Meiosis occurs and haploid spores develop in the capsule. When the capsule is mature, its lid pops off, and the spores are released.

FERTILIZATION

(within archegonium)

Sporangium

MEIOSIS

Seta
Capsule (sporangium)
Foot

Mature sporophytes

5 The sporophyte grows a long stalk (seta) that emerges from the archegonium.

Zygote (2*n*)

Embryo

Archegonium

Young sporophyte (2*n*)

6 Attached by its foot, the sporophyte remains nutritionally dependent on the gametophyte.

4 The zygote develops into a sporophyte embryo.

2 mm

Capsule (LM)

Female gametophytes

VISUAL SKILLS *In this diagram, does the sperm cell that fertilizes the egg cell differ genetically from the egg? Explain.*

→ **Mastering Biology** Animation: Moss Life Cycle

When bryophyte spores are dispersed to a favorable habitat, such as moist soil or tree bark, they may germinate and grow into gametophytes. Germinating moss spores, for example, characteristically produce a mass of green, branched, one-cell-thick filaments known as a **protonema** (plural, *protonemata*). A protonema has a large surface area that enhances absorption of water and minerals. In favorable conditions, a protonema produces one or more "buds," each of which develops into a moss gametophyte. (Note that when referring to nonvascular plants, we often use quotation marks for structures similar to the buds, stems, and leaves of vascular plants because the definitions of these terms are based on vascular plant organs.)

Bryophyte gametophytes generally form ground-hugging carpets, partly because their body parts are too thin to support a tall plant. A second constraint on the height of many bryophytes is the absence of vascular tissue, which would enable long-distance transport of water and nutrients. (The thin structure of bryophyte organs makes it possible to distribute materials for short distances without specialized vascular tissue.) However, some mosses have conducting tissues in the center of their "stems." A few of these mosses can grow as tall as 60 cm (2 feet) as a result. Phylogenetic analyses suggest that conducting tissues similar to those of vascular plants arose independently in these mosses by convergent evolution.

The gametophytes are anchored to the ground by delicate **rhizoids**, which are long, tubular single cells (in liverworts and hornworts) or filaments of cells (in mosses). Unlike roots, which are found in vascular plant sporophytes, rhizoids lack specialized conducting cells and do not play a primary role in water and mineral absorption.

Gametophytes can form multiple **gametangia**, multicellular structures that produce gametes and are covered by protective tissue. Female gametangia are called **archegonia**, and male gametangia are called **antheridia**. Each archegonium produces one egg, whereas each antheridium produces many sperm. Some bryophyte gametophytes are bisexual, but in mosses the archegonia and antheridia are typically carried on separate female and male gametophytes. Flagellated sperm swim through a film of water toward eggs, entering the archegonia in response to chemical attractants. Eggs are not released but instead remain within the bases of archegonia. After fertilization, embryos are retained within the archegonia. Layers of placental transfer cells help transport nutrients to the embryos as they develop into sporophytes.

Bryophyte sperm typically require a film of water to reach the eggs. Given this requirement, it is not surprising

▶ Figure 29.12 **A moss brood body.**

that many bryophyte species are found in moist habitats. The fact that sperm swim through water to reach the egg also means that in species with separate male and female gametophytes (most species of mosses), sexual reproduction is likely to be more successful when individuals are located close to one another.

Many bryophyte species can increase the number of individuals in a local area through various methods of asexual reproduction. For example, some mosses reproduce asexually by forming *brood bodies*, small plantlets that detach from the parent plant and grow into new, genetically identical copies of their parent **(Figure 29.12)**.

Bryophyte Sporophytes

The cells of bryophyte sporophytes contain plastids that are usually green and photosynthetic when the sporophytes are young. Even so, bryophyte sporophytes cannot live independently. A bryophyte sporophyte remains attached to its parental gametophyte throughout the sporophyte's lifetime, dependent on the gametophyte for supplies of sugars, amino acids, minerals, and water.

Bryophytes have the smallest sporophytes of all extant plant groups. A typical bryophyte sporophyte consists of a foot, a seta, and a sporangium. Embedded in the archegonium, the **foot** absorbs nutrients from the gametophyte. The **seta** (plural, *setae*), or stalk, conducts these materials to the sporangium, also called a **capsule**, which produces spores by meiosis.

Bryophyte sporophytes can produce enormous numbers of spores. A single moss capsule, for example, can generate over 5 million spores. In most mosses, the seta becomes elongated, enhancing spore dispersal by elevating the capsule. Typically, the upper part of the capsule features a ring of interlocking, tooth-like structures known as the **peristome** (see Figure 29.11). These "teeth" open under dry conditions and close again when it is moist. This allows moss spores to be discharged gradually, via periodic gusts of wind that can carry them long distances.

Moss and hornwort sporophytes are often larger and more complex than those of liverworts. For example, hornwort sporophytes, which superficially resemble grass blades, have a cuticle. Moss and hornwort sporophytes also have stomata, as do all vascular plants (but not liverworts).

Figure 29.13 shows some examples of gametophytes and sporophytes in the bryophyte phyla.

Liverworts (Phylum Hepatophyta)

This phylum's common and scientific names (from the Latin *hepaticus*, liver) refer to the liver-shaped gametophytes of its members, such as *Marchantia*, shown below. In medieval times, their shape was thought to be a sign that the plants could help treat liver diseases. Some liverworts, including *Marchantia*, are described as "thalloid" because of the flattened shape of their gametophytes. *Marchantia* gametangia are elevated on stalks that look like miniature trees. You would need a magnifying glass to see the sporophytes, which have a short seta (stalk) with an oval or round capsule. Other liverworts, such as *Plagiochila*, below, are called "leafy" because their stemlike gametophytes have many leaflike appendages. There are many more species of leafy liverworts than thalloid liverworts.

Thallus

Female gametangia (elevated on stalks)

Sporophyte

Foot

Seta

Capsule (sporangium)

500 μm

Gametophytes of *Marchantia polymorpha*, a "thalloid" liverwort

***Marchantia* sporophyte (LM)**

***Plagiochila deltoidea*, a "leafy" liverwort**

Hornworts (Phylum Anthocerophyta)

This phylum's common and scientific names (from the Greek *keras*, horn) refer to the long, tapered shape of the sporophyte. A typical sporophyte can grow to about 5 cm high. Unlike a liverwort or moss sporophyte, a hornwort sporophyte lacks a seta and consists only of a sporangium. The sporangium releases mature spores by splitting open, starting at the tip of the horn. The gametophytes, which are usually 1–2 cm in diameter, grow mostly horizontally and often have multiple sporophytes attached. Hornworts are frequently among the first species to colonize open areas with moist soils; a symbiotic relationship with nitrogen-fixing cyanobacteria contributes to their ability to do this (nitrogen is often in short supply in such areas).

Mosses (Phylum Bryophyta)

Moss gametophytes, which range in height from less than 1 mm to 60 cm, are less than 15 cm tall in most species. The familiar carpet of moss you observe consists mainly of gametophytes. The blades of their "leaves" are usually only one cell thick, but more complex "leaves" that have ridges coated with cuticle can be found on the common hairy-cap moss (*Polytrichum*, below) and its close relatives. Moss sporophytes are typically elongated and visible to the naked eye, with heights ranging up to about 20 cm. Though green and photosynthetic when young, they turn tan or brownish red when ready to release spores.

An *Anthoceros* hornwort species

Sporophyte

Gametophyte

***Polytrichum commune*, hairy-cap moss**

Capsule

Seta

Sporophyte

Gametophyte

The Ecological and Economic Importance of Mosses

Wind dispersal of lightweight spores has distributed mosses throughout the world. These plants are particularly common and diverse in moist forests and wetlands. Some mosses colonize bare, sandy soil, where, researchers have found, they help retain nitrogen in the soil **(Figure 29.14)**. In northern coniferous forests, species such as the feather moss *Pleurozium* harbor nitrogen-fixing cyanobacteria that increase the availability of nitrogen in the ecosystem. Other mosses inhabit such extreme environments as mountaintops, tundra, and deserts. Many mosses are able to live in very cold or dry habitats because they can survive the loss of most of their body water, then rehydrate when moisture is available. Few vascular plants can survive the same degree of desiccation. Moreover, phenolic compounds in moss cell walls absorb damaging levels of UV radiation present in deserts or at high altitudes.

One wetland moss genus, *Sphagnum*, or peat moss, is often a major component of deposits of partially decayed organic material known as **peat (Figure 29.15a)**. Boggy regions with thick layers of peat are called peatlands. *Sphagnum* does not decay readily, in part because of phenolic compounds embedded in its cell walls. The low temperature, pH, and oxygen level of peatlands also inhibit decay of moss and other organisms in these boggy wetlands. As a result, some peatlands have preserved corpses for thousands of years **(Figure 29.15b)**.

Peat has long been a fuel source in Europe and Asia, and it is still harvested for fuel today, notably in Ireland and

---

▼ **Figure 29.14** **Inquiry**

Can bryophytes reduce the rate at which key nutrients are lost from soils?

Experiment Soils in terrestrial ecosystems are often low in nitrogen, a nutrient required for normal plant growth. Richard Bowden, of Allegheny College, measured annual inputs (gains) and outputs (losses) of nitrogen in a sandy soil ecosystem dominated by the moss *Polytrichum*. Nitrogen inputs were measured from rainfall (dissolved ions, such as nitrate, NO_3^-), biological N_2 fixation, and wind deposition. Nitrogen losses were measured in leached water (dissolved ions, such as NO_3^-) and gaseous emissions (such as N_2O emitted by bacteria). Bowden measured losses for soils with *Polytrichum* and for soils where the moss was removed two months before the experiment began.

Results A total of 10.5 kg of nitrogen per hectare (kg/ha) entered the ecosystem each year. Little nitrogen was lost by gaseous emissions [0.10 kg/(ha · yr)]. The results of comparing nitrogen losses by leaching are shown below.

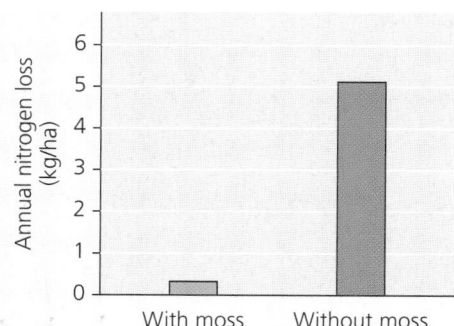

Conclusion The moss *Polytrichum* greatly reduced the loss of nitrogen by leaching in this ecosystem. Each year, the moss ecosystem retained over 95% of the 10.5 kg/ha of total nitrogen inputs (only 0.1 kg/ha and 0.3 kg/ha were lost to gaseous emissions and leaching, respectively).

Data from R. D. Bowden, Inputs, outputs, and accumulation of nitrogen in an early successional moss (*Polytrichum*) ecosystem, *Ecological Monographs* 61:207–223 (1991).

WHAT IF? *How might the presence of* Polytrichum *affect plant species that typically colonize the sandy soils after the moss?*

▼ **Figure 29.15** *Sphagnum,* **or peat moss: a bryophyte with economic, ecological, and archaeological significance.**

(a) Peat being harvested from a peatland

(b) "Tollund Man," a bog mummy dating from 405–100 BCE.
The acidic, oxygen-poor conditions produced by *Sphagnum* can preserve human or other animal bodies for thousands of years.

Making Bar Graphs and Interpreting Data

Could Nonvascular Plants Have Caused Weathering of Rocks and Contributed to Climate Change During the Ordovician Period? The oldest traces of terrestrial plants are fossilized spores formed 470 million years ago. Between that time and the end of the Ordovician period 444 million years ago, the atmospheric CO_2 level dropped by half, and the climate cooled dramatically.

One possible cause of the drop in CO_2 during the Ordovician period is the breakdown, or weathering, of rock. As rock weathers, calcium silicate (Ca_2SiCO_3) is released and combines with CO_2 from the air, producing calcium carbonate ($CaCO_3$). Today, the roots of vascular plants increase rock weathering and mineral release by producing acids that break down rock and soil. Although nonvascular plants lack roots, they require the same mineral nutrients as vascular plants. Could nonvascular plants also increase the chemical weathering of rock? If so, they could have contributed to the decline in atmospheric CO_2 during the Ordovician. In this exercise, you will interpret data from a study of the effects of moss on releasing minerals from two types of rock.

How the Experiment Was Done The researchers set up experimental and control microcosms, or small artificial ecosystems, to measure mineral release from rocks. First, they placed rock fragments of volcanic origin, either granite or andesite, into small glass containers. Then they mixed water and macerated (chopped and crushed) moss of the species *Physcomitrella patens*. They added this mixture to the experimental microcosms (72 granite and 41 andesite). For the control microcosms (77 granite and 37 andesite), they filtered out the moss and just added the water. After 130 days, they measured the amounts of various minerals found in the water in the control microcosms and in the water and moss in the experimental microcosms.

Data from the Experiment The moss grew (increased its biomass) in the experimental microcosms. The table shows the mean amounts in micromoles (μmol) of several minerals measured in the water and the moss in the microcosms.

INTERPRET THE DATA

1. Why did the researchers add filtrate from which macerated moss had been removed to the control microcosms?

2. Make two bar graphs (for granite and andesite) comparing the mean amounts of each element weathered from rocks in the control and experimental microcosms. (Hint: For an experimental microcosm, what sum represents the total amount weathered from rocks?) (For additional information on bar graphs, see the Scientific Skills Review in Appendix D.)

3. Overall, what is the effect of moss on chemical weathering of rock? Are the results similar or different for granite and andesite?

4. Based on their experimental results, the researchers added weathering of rock by nonvascular plants to simulation models of the Ordovician climate. The new models predicted decreased CO_2 levels and global cooling sufficient to produce the glaciations in the late Ordovician period. What assumptions did the researchers make in using results from their experiments in climate simulation models?

5. "Life has profoundly changed the Earth." Explain whether or not these experimental results support this statement.

➔ **Instructors:** A version of this Scientific Skills Exercise can be assigned in **Mastering Biology**.

	Ca^{2+} (μmol)		Mg^{2+} (μmol)		K$^+$ (μmol)	
	Granite	Andesite	Granite	Andesite	Granite	Andesite
Mean weathered amount released in water in the control microcosms	1.68	1.54	0.42	0.13	0.68	0.60
Mean weathered amount released in water in the experimental microcosms	1.27	1.84	0.34	0.13	0.65	0.64
Mean weathered amount taken up by moss in the experimental microcosms	1.09	3.62	0.31	0.56	1.07	0.28

Data from T. M. Lenton et al., First plants cooled the Ordovician, *Nature Geoscience* 5:86–89 (2012).

Canada. Peat moss is also useful as a soil conditioner and for packing plant roots during shipment because it has large dead cells that can absorb roughly 20 times the moss's weight in water.

Peatlands cover just 3% of Earth's land surface but contain one-third of the world's soil carbon: Globally, an estimated 500 billion tons of organic carbon is stored as peat. Current overharvesting of *Sphagnum*—primarily for use in peat-fired power stations—contributes to global warming by releasing stored CO_2. In addition, if global temperatures continue to rise, the water levels of some peatlands are expected to drop. Such a change would expose peat to air and cause it to decompose, thereby releasing additional stored CO_2 and contributing further to global warming. The historical and expected future effects of *Sphagnum* on the global climate underscore the importance of preserving and managing peatlands.

Mosses may have a long history of affecting climate change. In the **Scientific Skills Exercise**, you will explore the question of whether they did so during the Ordovician period by contributing to the weathering of rocks.

1. How do bryophytes differ from other plants?

2. Give three examples of how structure fits function in bryophytes.

3. **MAKE CONNECTIONS** Review the discussion of feedback regulation in Concept 1.1. Could effects of global warming on peatlands alter CO_2 concentrations in ways that result in negative or positive feedback? Explain.

For suggested answers, see Appendix A.

CONCEPT 29.3

Ferns and other seedless vascular plants were the first plants to grow tall

Nonvascular plants (bryophytes)
Seedless vascular plants
Gymnosperms
Angiosperms

During the first 100 million years of plant evolution, bryophytes were prominent types of vegetation. But it is vascular plants that dominate most landscapes today. The earliest fossils of vascular plants date to 425 million years ago. These plants lacked seeds but had well-developed vascular systems, an evolutionary novelty that set the stage for vascular plants to grow taller than their bryophyte counterparts. As in bryophytes, however, the sperm of ferns and all other seedless vascular plants are flagellated and swim through a film of water to reach eggs. In part because of these swimming sperm, seedless vascular plants today are most common in damp environments.

Origins and Traits of Vascular Plants

Unlike the nonvascular plants, ancient relatives of vascular plants had branched sporophytes that were not dependent on gametophytes for nutrition **(Figure 29.16)**. Although these early plants were less than 20 cm tall, their branching enabled their bodies to become more complex and to have multiple sporangia. As plant bodies became more complex over time, competition for space and sunlight probably increased. As we'll see, that competition may have stimulated still more evolution in vascular plants, eventually leading to the formation of the first forests.

Early vascular plants had some derived traits of today's vascular plants, but they lacked roots and some other adaptations that evolved later. The main traits that characterize living vascular plants are life cycles with dominant sporophytes, transport in vascular tissues called xylem and phloem, and well-developed roots and leaves, including spore-bearing leaves called sporophylls.

▼ **Figure 29.16 Sporophytes of *Aglaophyton major*, an ancient relative of living vascular plants.** This reconstruction from 405-million-year-old fossils exhibits dichotomous (Y-shaped) branching with sporangia at the ends of branches. Sporophyte branching characterizes living vascular plants but is lacking in living nonvascular plants (bryophytes). *Aglaophyton* had structures called rhizoids that anchored it to the ground. The inset shows a fossilized stoma of *A. major* (colorized LM).

Sporangia

25 μm

Fossilized stoma

2 cm

Rhizoids

Life Cycles with Dominant Sporophytes

As mentioned earlier, mosses and other bryophytes have life cycles dominated by gametophytes (see Figure 29.11). Fossil evidence suggests that a change began to develop in some of the earliest vascular plants, whose gametophytes and sporophytes were about equal in size. Further reductions in gametophyte size occurred among extant vascular plants; in these groups, the sporophyte generation is the larger and more complex form in the alternation of generations **(Figure 29.17)**. In ferns, for example, the familiar leafy plants are the sporophytes. You would have to get down on your hands and knees and search the ground carefully to find fern gametophytes, which are tiny structures that often grow on or just below the soil surface.

Transport in Xylem and Phloem

Vascular plants have two types of vascular tissue: xylem and phloem (see Figure 35.10). **Xylem** conducts most of the water and minerals. The xylem of all vascular plants includes **tracheids**, tube-shaped cells that carry water and minerals up from the roots. The water-conducting cells of the xylem are dead at functional maturity and are *lignified*; that is, their cell walls are strengthened by the polymer **lignin**. The tissue called **phloem** has cells arranged into tubes that distribute sugars, amino acids, and other organic products; these cells are alive at functional maturity.

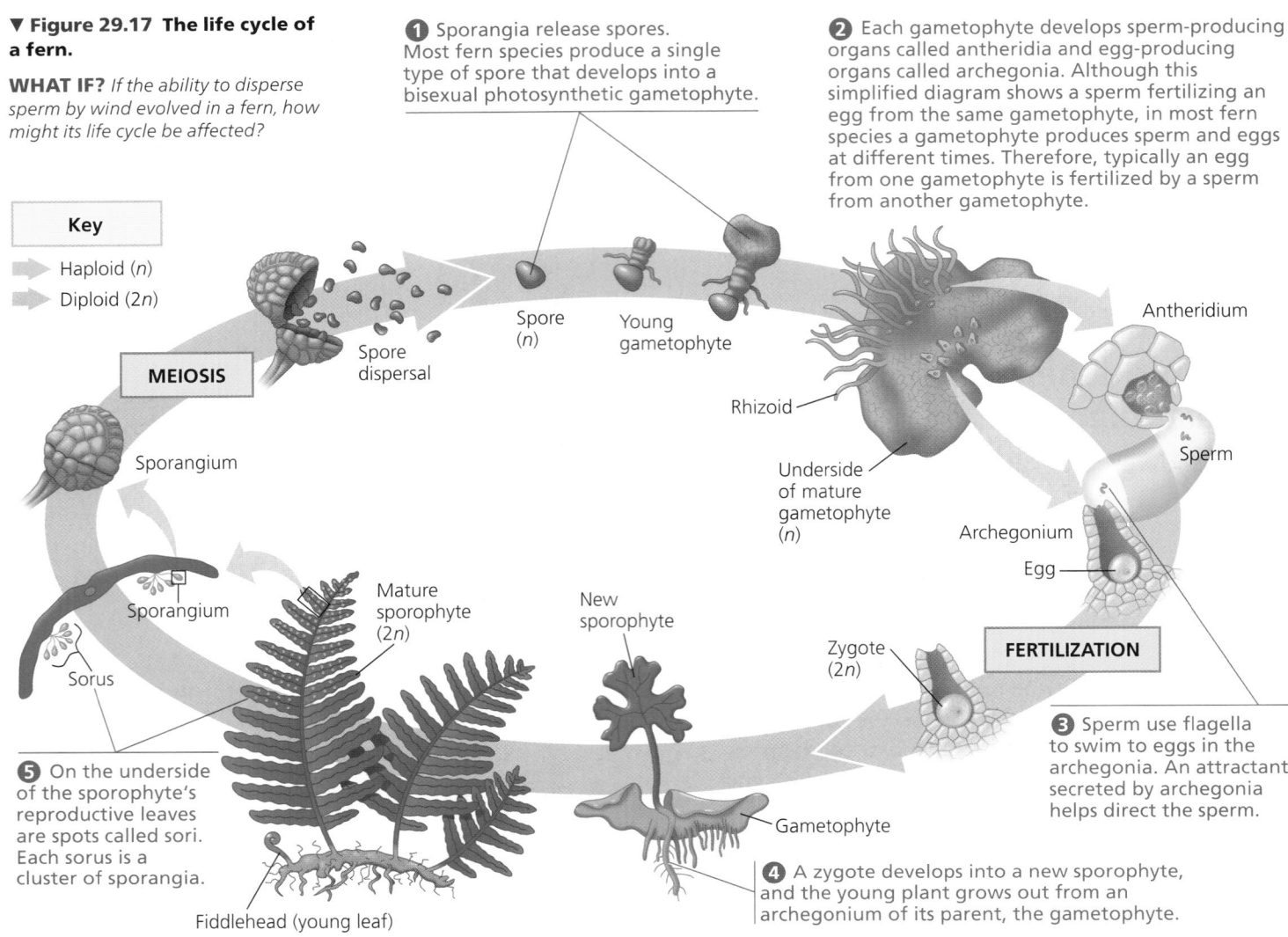

▼ **Figure 29.17 The life cycle of a fern.**

WHAT IF? *If the ability to disperse sperm by wind evolved in a fern, how might its life cycle be affected?*

Key
→ Haploid (*n*)
→ Diploid (*2n*)

1 Sporangia release spores. Most fern species produce a single type of spore that develops into a bisexual photosynthetic gametophyte.

2 Each gametophyte develops sperm-producing organs called antheridia and egg-producing organs called archegonia. Although this simplified diagram shows a sperm fertilizing an egg from the same gametophyte, in most fern species a gametophyte produces sperm and eggs at different times. Therefore, typically an egg from one gametophyte is fertilized by a sperm from another gametophyte.

MEIOSIS

Spore dispersal

Spore (*n*)

Young gametophyte

Rhizoid

Antheridium

Sperm

Underside of mature gametophyte (*n*)

Sporangium

Archegonium

Egg

Sporangium

Sorus

Mature sporophyte (*2n*)

New sporophyte

Zygote (*2n*)

FERTILIZATION

5 On the underside of the sporophyte's reproductive leaves are spots called sori. Each sorus is a cluster of sporangia.

3 Sperm use flagella to swim to eggs in the archegonia. An attractant secreted by archegonia helps direct the sperm.

Gametophyte

Fiddlehead (young leaf)

4 A zygote develops into a new sporophyte, and the young plant grows out from an archegonium of its parent, the gametophyte.

→ **Mastering Biology** Animation: Fern Life Cycle

Lignified vascular tissue helped enable vascular plants to grow tall. Their stems became strong enough to provide support against gravity, and they could transport water and mineral nutrients high above the ground. Tall plants could also outcompete short plants for access to the sunlight needed for photosynthesis. In addition, the spores of tall plants could disperse farther than those of short plants, enabling tall species to colonize new environments more rapidly. Overall, the ability to grow tall gave vascular plants a competitive edge over nonvascular plants, which typically are less than 5 cm in height. Over time, competition among vascular plants also would have increased, leading to selection for taller growth forms—a process that eventually gave rise to the trees that formed the first forests 385 million years ago.

Evolution of Roots

Vascular tissue also provides benefits below ground. Instead of the rhizoids seen in bryophytes, roots evolved in the sporophytes of almost all vascular plants. **Roots** are organs that absorb water

and nutrients from the soil. Roots also anchor vascular plants to the ground, hence allowing the shoot system to grow taller.

Root tissues of living plants closely resemble stem tissues of early vascular plants preserved in fossils. This suggests that roots may have evolved from the lowest belowground portions of stems in ancient vascular plants. It is unclear whether roots evolved only once in the common ancestor of all vascular plants or independently in different lineages. Although the roots of living members of these lineages of vascular plants share many similarities, fossil evidence hints at convergent evolution. The oldest fossils of lycophytes, for example, already displayed simple roots 400 million years ago, when the ancestors of ferns and seed plants still had none. Studying genes that control root development in different vascular plant species may help resolve this question.

Evolution of Leaves

Leaves are structures that serve as the primary photosynthetic organ of vascular plants. In terms of size and

▼ **Figure 29.18 Microphyll and megaphyll leaves.**

Microphyll leaves

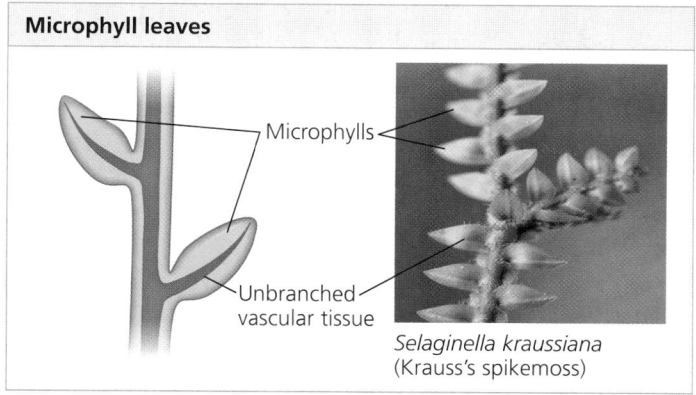

Microphylls

Unbranched vascular tissue

Selaginella kraussiana
(Krauss's spikemoss)

Megaphyll leaves

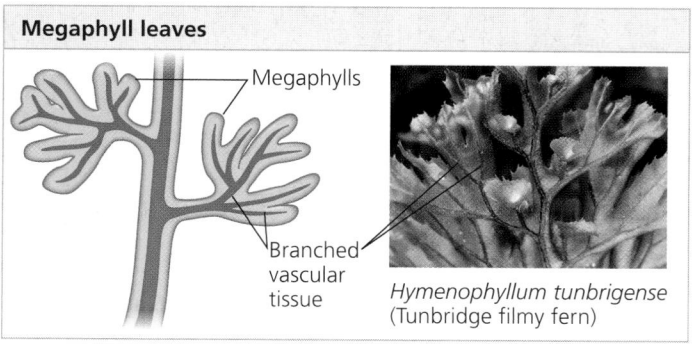

Megaphylls

Branched vascular tissue

Hymenophyllum tunbrigense
(Tunbridge filmy fern)

complexity, leaves can be classified as either microphylls or megaphylls **(Figure 29.18)**. All of the lycophytes—and only the lycophytes—have **microphylls**, which are small, often spine-shaped leaves supported by a single strand of vascular tissue. Almost all other vascular plants have **megaphylls**, leaves with a highly branched vascular system; a few species have reduced leaves that appear to have evolved from megaphylls. Megaphylls are typically larger than microphylls and therefore support greater photosynthetic productivity than microphylls. Microphylls first appear in the fossil record 410 million years ago, but megaphylls do not emerge until 370 million years ago, toward the end of the Devonian period.

Sporophylls and Spore Variations

One milestone in the evolution of plants was the emergence of **sporophylls**, modified leaves that bear sporangia. Sporophylls vary greatly in structure. For example, fern sporophylls produce clusters of sporangia known as **sori** (singular, *sorus*), usually on the undersides of the sporophylls (see Figure 29.17). In many lycophytes and in most gymnosperms, groups of sporophylls form cone-like structures called **strobili** (singular, *strobilus*). The sporophylls of angiosperms are called *carpels* and *stamens* (see Figure 30.8).

Most seedless vascular plant species are **homosporous**: They have one type of sporophyll bearing one type of sporangium that produces one type of spore, which typically develops into a bisexual gametophyte, as in most ferns. In contrast, a **heterosporous** species has two types of sporophylls, called

megasporophylls and microsporophylls. Megasporophylls have megasporangia, which produce **megaspores**, spores that develop into female gametophytes. Microsporophylls have microsporangia, which produce **microspores**, smaller spores that develop into male gametophytes. All seed plants and a few seedless vascular plants are heterosporous. The following diagram compares the two conditions:

Homosporous spore production
(most seedless vascular plants)

Sporangium on sporophyll → Single type of spore → Typically a bisexual gametophyte ⟨ Eggs / Sperm ⟩

Heterosporous spore production
(all seed plants)

Megasporangium on megasporophyll → Megaspore → Female gametophyte → Eggs

Microsporangium on microsporophyll → Microspore → Male gametophyte → Sperm

Classification of Seedless Vascular Plants

As we noted earlier, biologists recognize two clades of living seedless vascular plants: the lycophytes (phylum Lycophyta) and the monilophytes (phylum Monilophyta). The lycophytes include the club mosses, the spikemosses, and the quillworts. The monilophytes include the ferns, the horsetails, and the whisk ferns and their relatives. Although ferns, horsetails, and whisk ferns differ greatly in appearance, recent anatomical and molecular comparisons provide convincing evidence that these three groups make up a clade. Accordingly, many systematists now classify them together as the phylum Monilophyta, as we do in this chapter. Others refer to these groups as three separate phyla within a clade. **Figure 29.19** describes the two main groups of seedless vascular plants.

Phylum Lycophyta: Club Mosses, Spikemosses, and Quillworts

Present-day species of lycophytes are relics of a far more impressive past. By the Carboniferous period (359–299 million years ago), the lycophyte evolutionary lineage included small herbaceous plants and giant trees with diameters of more than 2 m and heights of more than 40 m. The giant lycophyte trees thrived for millions of years in moist swamps, but their diversity declined when Earth's climate became drier during the Permian period (299–252 million years ago). The small lycophytes survived, represented today by about 1,200 species. Though some are commonly called club mosses and spikemosses, they are not true mosses (which, as discussed earlier, are nonvascular plants).

Lycophytes (Phylum Lycophyta)

Many lycophytes grow on tropical trees as epiphytes, plants that use other plants as a substrate but are not parasites. Other species grow on temperate forest floors. In some species, the tiny gametophytes live above ground and are photosynthetic. Others live below ground, nurtured by symbiotic fungi.

Sporophytes have upright stems with many small leaves, as well as ground-hugging stems that produce dichotomously branching roots. Spikemosses are usually relatively small and often grow horizontally. In many club mosses and spikemosses, sporophylls are clustered into club-shaped cones (strobili). Quillworts, named for their leaf shape, form a single genus whose members live in marshy areas or as submerged aquatic plants. Club mosses are all homosporous, whereas spikemosses and quillworts are all heterosporous. The spores of club mosses are released in clouds and are so rich in oil that magicians and photographers once ignited them to create smoke or flashes of light.

Selaginella moellendorffii, a spikemoss

Isoetes gunnii, a quillwort

Strobili (clusters of sporophylls)

2.5 cm

1 cm

Diphasiastrum tristachyum, a club moss

Monilophytes (Phylum Monilophyta)

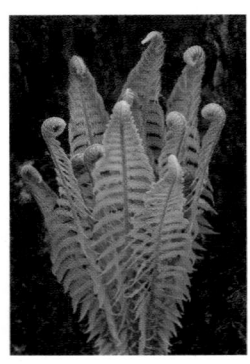

Matteuccia struthiopteris (ostrich fern)

2.5 cm

Ferns

Unlike the lycophytes, ferns have megaphylls (see Figure 29.18). The sporophytes typically have horizontal stems that give rise to large leaves called fronds, often divided into leaflets. A frond grows as its coiled tip, the fiddlehead, unfurls.

Almost all species are homosporous. The gametophyte in some species shrivels and dies after the young sporophyte detaches itself. In most species, sporophytes have stalked sporangia with springlike devices that catapult spores several meters. Airborne spores can be carried far from their origin. Some species produce more than a trillion spores in a plant's lifetime.

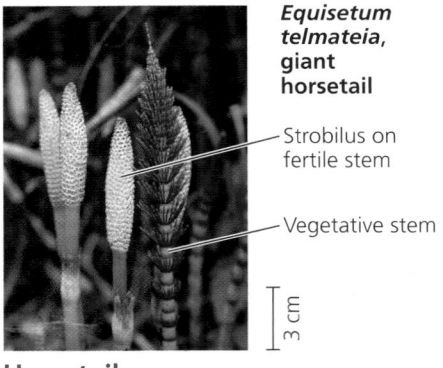

Equisetum telmateia, giant horsetail

Strobilus on fertile stem

Vegetative stem

3 cm

Horsetails

The group's name refers to the brushy appearance of the stems, which have a gritty texture that made them historically useful as "scouring rushes" for pots and pans. Some species have separate fertile (cone-bearing) and vegetative stems. Horsetails are homosporous, with cones releasing spores that typically give rise to bisexual gametophytes.

Horsetails are also called arthrophytes ("jointed plants") because their stems have joints. Rings of small leaves or branches emerge from each joint, but the stem is the main photosynthetic organ. Large air canals carry oxygen to the roots, which often grow in waterlogged soil.

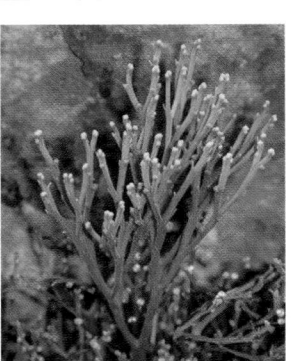

Psilotum nudum, a whisk fern

4 cm

Whisk Ferns and Relatives

Like primitive vascular plant fossils, the sporophytes of whisk ferns (genus *Psilotum*) have dichotomously branching stems but no roots. Stems have scalelike outgrowths that lack vascular tissue and may have resulted from the evolutionary reduction of leaves. Each yellow knob on a stem consists of three fused sporangia. Species of the genus *Tmesipteris* closely related to whisk ferns and found only in the South Pacific, also lack roots but have small, leaflike outgrowths in their stems, giving them a vine-like appearance. Both genera are homosporous, with spores giving rise to bisexual gametophytes that grow underground and are only about a centimeter long.

Phylum Monilophyta: Ferns, Horsetails, and Whisk Ferns and Relatives

Ferns radiated extensively from their Devonian origins and grew alongside lycophyte trees and horsetails in the great Carboniferous swamp forests. Today, ferns are by far the most widespread seedless vascular plants, numbering more than 12,000 species. Though most diverse in the tropics, many ferns thrive in temperate forests, and some species are even adapted to arid habitats.

As mentioned earlier, ferns and other monilophytes are more closely related to seed plants than to lycophytes. As a result, monilophytes and seed plants share traits that are not found in lycophytes, including megaphyll leaves and roots that can branch at various points along the length of an existing root. In lycophytes, by contrast, roots branch only at the growing tip of the root, forming a Y-shaped structure.

The monilophytes called horsetails were very diverse during the Carboniferous period, some growing as tall as 15 m. Today, only 15 species survive as a single, widely distributed genus, *Equisetum*, often found in marshy places and along streams.

Psilotum (whisk ferns) and a closely related genus, *Tmesipteris*, form a clade consisting mainly of tropical epiphytes. Plants in these two genera, the only vascular plants lacking true roots, once were called "living fossils" because of their resemblance to fossils of ancient relatives of living vascular plants (see Figures 29.16 and 29.19). However, much evidence, including analyses of DNA sequences and sperm structure, indicates that the genera *Psilotum* and *Tmesipteris* are closely related to ferns. This hypothesis suggests that their ancestor's true roots were lost during evolution. Today, plants in these two genera absorb water and nutrients through numerous absorptive rhizoids.

The Significance of Seedless Vascular Plants

The ancestors of living lycophytes, horsetails, and ferns, along with their extinct seedless vascular relatives, grew to great heights during the Devonian and early Carboniferous, forming the first forests **(Figure 29.20)**. How did their dramatic growth affect Earth and its other life?

One major effect was that early forests contributed to a large drop in CO_2 levels during the Carboniferous period, causing global cooling that resulted in widespread glacier formation. The trees of early forests contributed to this drop in CO_2 levels in part by the actions of their roots. The roots of vascular plants secrete acids that break down rocks, thereby increasing the rate at which calcium and magnesium are released from rocks into the soil. These chemicals react with carbon dioxide dissolved in rain water, forming compounds that ultimately wash into the oceans, where they are incorporated into rocks (calcium or magnesium carbonates). The net effect of these processes—which were accelerated by plants—is that CO_2 removed from the air is stored in marine rocks. Although carbon stored in these rocks can be returned to the atmosphere, it typically takes millions of years for this to occur (as when geological uplift brings the rocks to the surface, exposing them to erosion).

In addition, the seedless vascular plants that formed the first forests eventually became coal, again removing CO_2 from the atmosphere for long periods of time. In the stagnant waters of Carboniferous swamps, the dead bodies of early trees did not completely decay. This organic material turned to thick layers of peat, later covered by the sea. Marine sediments piled on top, and over millions of years, heat and pressure converted the peat to coal. In fact, Carboniferous coal deposits are the most extensive ever formed. Coal was crucial to the Industrial Revolution, and people worldwide still burn 6 billion tons a year. It is ironic that coal, formed from plants that contributed to

▶ **Figure 29.20 Artist's conception of a Carboniferous forest based on fossil evidence.** Lycophyte trees, with trunks covered with small leaves, thrived in the "coal forests" of the Carboniferous, along with giant ferns and horsetails.

Lycophyte trees   Horsetail   Fern

a global cooling, now contributes to global warming by returning carbon to the atmosphere (see Figure 56.29).

Growing along with the seedless plants in Carboniferous swamps were primitive seed plants. Though seed plants were not dominant at that time, they rose to prominence after the swamps began to dry up at the end of the Carboniferous period. The next chapter traces the origin and diversification of seed plants, continuing our story of adaptation to life on land.

CONCEPT CHECK 29.3

1. List the key derived traits found in monilophytes and seed plants, but not in lycophytes.

2. How do the main similarities and differences between seedless vascular plants and nonvascular plants affect function in these plants?

3. **MAKE CONNECTIONS** In Figure 29.17, if fertilization occurred between gametes from one gametophyte, how would this affect the production of genetic variation from sexual reproduction? See Concept 13.4.

For suggested answers, see Appendix A.

29 Chapter Review

→ Go to **Mastering Biology** for Assignments, the eText, the Study Area, and Dynamic Study Modules.

SUMMARY OF KEY CONCEPTS

→ To review key terms, go to the **Vocabulary Self-Quiz** in the **Mastering Biology** eText or Study Area, or go to goo.gl/zkjz9t.

CONCEPT 29.1

Plants evolved from green algae *(pp. 619–623)*

- Morphological and biochemical traits, as well as similarities in nuclear and chloroplast genes, indicate that certain charophyte algae are the closest living relatives of plants.
- A protective layer of **sporopollenin** and other traits allow charophytes to tolerate occasional drying along the edges of ponds and lakes. Such traits may have enabled the algal ancestors of plants to survive in terrestrial conditions, opening the way to the colonization of dry land.
- Derived traits that distinguish plants from charophytes, their closest algal relatives, include **cuticles, stomata,** multicellular dependent embryos, walled spores produced in sporangia, and the two shown here:

1 Alternation of generations   **2** Apical meristems

- Fossils show that plants arose more than 470 million years ago. Subsequently, plants diverged into several major groups, including nonvascular plants (**bryophytes**); **seedless vascular plants,** such as **lycophytes** and ferns; and the two groups of seed plants: **gymnosperms** and **angiosperms**.

? *Draw a phylogenetic tree illustrating our current understanding of plant phylogeny; label the common ancestor of plants and the origins of vascular tissue, apical meristems, and seeds.*

CONCEPT 29.2

Mosses and other nonvascular plants have life cycles dominated by gametophytes *(pp. 623–629)*

- Lineages leading to the three extant clades of nonvascular plants, or **bryophytes—liverworts, mosses,** and **hornworts**—diverged from other plants early in plant evolution.
- In bryophytes, the dominant generation consists of haploid **gametophytes,** such as those that make up a carpet of moss. **Rhizoids** anchor gametophytes to the substrate on which they grow. The flagellated sperm produced by **antheridia** require a film of water to travel to the eggs in the **archegonia**.
- The diploid stage of the life cycle—the **sporophytes**—grow out of archegonia and are attached to the gametophytes and dependent on them for nourishment. Smaller and simpler than vascular plant sporophytes, they typically consist of a **foot, seta** (stalk), and **sporangium**.
- *Sphagnum,* or peat moss, is common in large regions known as peatlands and has many practical uses, including as a fuel.

? *Summarize the ecological importance of mosses.*

CONCEPT 29.3

Ferns and other seedless vascular plants were the first plants to grow tall *(pp. 629–634)*

- Fossils of the forerunners of today's vascular plants date back about 425 million years and show that these small plants had independent, branching sporophytes and a vascular system.
- Over time, other derived traits of living vascular plants arose, such as a life cycle with dominant sporophytes, lignified vascular tissue, well-developed **roots** and **leaves,** and **sporophylls**.
- Seedless vascular plants include the **lycophytes** (phylum Lycophyta: club mosses, spikemosses, and quillworts) and the **monilophytes** (phylum Monilophyta: ferns, horsetails, and whisk ferns and relatives). Current evidence indicates that seedless vascular plants, like bryophytes, do not form a clade.
- Ancient lineages of lycophytes included both small herbaceous plants and large trees. Present-day lycophytes are small herbaceous plants.
- Seedless vascular plants formed the earliest forests 385 million years ago. Their growth may have contributed to a major global cooling that took place during the Carboniferous period. The decaying remnants of the first forests eventually became coal.

? *What trait(s) allowed vascular plants to grow tall, and why might increased height have been advantageous?*

➡ For more multiple-choice questions, go to the **Practice Test** in the **Mastering Biology** eText or Study Area, or go to goo.gl/GruWRg.

Levels 1-2: Remembering/Understanding

1. Three of the following are evidence that charophytes are the closest algal relatives of plants. Select the exception.
   (A) similar sperm structure
   (B) the presence of chloroplasts
   (C) similarities in cell wall formation during cell division
   (D) genetic similarities in chloroplasts

2. Which of the following characteristics of plants is absent in their closest relatives, the charophyte algae?
   (A) chlorophyll *b*
   (B) cellulose in cell walls
   (C) sexual reproduction
   (D) alternation of multicellular generations

3. In plants, which of the following are produced by meiosis?
   (A) haploid gametes
   (B) diploid gametes
   (C) haploid spores
   (D) diploid spores

4. Microphylls are found in which plant group?
   (A) lycophytes
   (B) liverworts
   (C) ferns
   (D) hornworts

Levels 3-4: Applying/Analyzing

5. Suppose an efficient conducting system evolved in a moss that could transport water and other materials as high as a tall tree. Which of the following statements about "trees" of such a species would be true?
   (A) Spore dispersal distances would probably decrease.
   (B) Females could produce only one archegonium.
   (C) Unless its body parts were strengthened, such a "tree" would probably flop over.
   (D) Individuals would probably compete less effectively for access to light.

6. Identify each of the following structures as haploid or diploid.
   a. sporophyte
   b. spore
   c. gametophyte
   d. zygote

7. **EVOLUTION CONNECTION • DRAW IT** Draw a phylogenetic tree that represents our current understanding of evolutionary relationships between a moss, a gymnosperm, a lycophyte, and a fern. Use a charophyte alga as the outgroup. (See Figure 26.5 to review phylogenetic trees.) Label each branch point of the phylogeny with at least one derived character unique to the clade descended from the common ancestor represented by the branch point.

Levels 5-6: Evaluating/Creating

8. **SCIENTIFIC INQUIRY • INTERPRET THE DATA** The feather moss *Pleurozium schreberi* harbors species of symbiotic nitrogen-fixing bacteria. Scientists studying this moss in northern forests found that the percentage of the ground surface "covered" by the moss increased from about 5% in forests that burned 35 to 41 years ago to about 70% in forests that burned 170 or more years ago. From mosses growing in these forests, they also obtained the following data on nitrogen fixation:

Age (years after fire)	N fixation rate [kg N/(ha · yr)]
35	0.001
41	0.005
78	0.08
101	0.3
124	0.9
170	2.0
220	1.3
244	2.1
270	1.6
300	3.0
355	2.3

Data from O. Zackrisson et al., Nitrogen fixation increases with successional age in boreal forests, *Ecology* 85:3327–3334 (2006).

   (a) Use the data to draw a line graph, with age on the *x*-axis and the nitrogen fixation rate on the *y*-axis.
   (b) Along with the nitrogen added by nitrogen fixation, about 1 kg of nitrogen per hectare per year is deposited into northern forests from the atmosphere as rain and small particles. Evaluate the extent to which *Pleurozium* affects nitrogen availability in northern forests of different ages.

9. **WRITE ABOUT A THEME: INTERACTIONS** Giant lycophyte trees had microphylls, whereas ferns and seed plants have megaphylls. Write a short essay (100–150 words) describing how a forest of lycophyte trees may have differed from a forest of large ferns or seed plants. In your answer, consider how the type of forest may have affected interactions among small plants growing beneath the tall ones.

10. **SYNTHESIZE YOUR KNOWLEDGE**

These stomata are from the leaf of a common horsetail. Describe how stomata and other adaptations facilitated life on land and ultimately led to the formation of the first forests.

For selected answers, see Appendix A.

30 Plant Diversity II: The Evolution of Seed Plants

KEY CONCEPTS

30.1 Seeds and pollen grains are key adaptations for life on land *p. 637*

30.2 Gymnosperms bear "naked" seeds, typically on cones *p. 640*

30.3 The reproductive adaptations of angiosperms include flowers and fruits *p. 644*

30.4 Human welfare depends on seed plants *p. 651*

Study Tip

Make flowcharts: Make simple flowcharts showing the main events in the life cycles of a pine and flowering plant. The example shown is the beginning of a flowchart summarizing the pine life cycle.

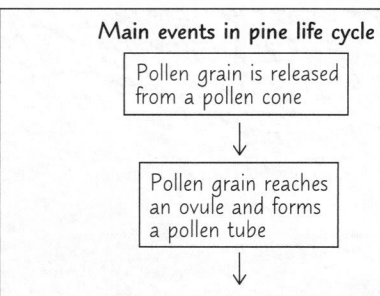

Main events in pine life cycle

Pollen grain is released from a pollen cone

↓

Pollen grain reaches an ovule and forms a pollen tube

↓

→ Go to Mastering Biology

For Students (in eText and Study Area)
- Get Ready for Chapter 30
- Figure 30.3 Walkthrough: From Ovule to Seed in a Gymnosperm
- Figure 30.12 Walkthrough: The Life Cycle of an Angiosperm
- Animation: Angiosperm Life Cycle

For Instructors to Assign (in Item Library)
- Scientific Skills Exercise: Using Natural Logarithms to Interpret Data
- Tutorial: Gymnosperms

Figure 30.1 In 1980, Mount St. Helens in Washington state erupted, leaving the region covered in ash and devoid of visible life. Just a few years later, seed plants such as fireweed (*Chamerion angustifolium*) had already colonized the barren landscape—an example of seed plants' versatility and success as the dominant producers on land.

Fireweed seed

What adaptations have enabled seed plants to make up the vast majority of plant biodiversity?

Reduced male and female gametophytes develop within parental sporophytes, protected from environmental stresses.

In **gymnosperms,** the gametophytes develop within cones.

In **angiosperms,** the gametophytes develop within flowers.

Pollen protects male gametophytes (which produce sperm) and can be transported by wind or animals.

Pollen-producing structures

Pollen-producing structures

Ovules protect female gametophytes (which produce eggs).

Ovules

Ovules

An ovule fertilized by pollen develops into a **seed**.

Embryo

Seed coat (protects embryo)

Food supply (nourishes embryo)

Fireweed and other early arrivals reached the blast zone in Figure 30.1 as seeds. A **seed** consists of an embryo and its food supply, surrounded by a protective coat. When mature, seeds are dispersed from their parent by wind or other means, enabling them to colonize distant locations.

Plants not only have affected the recovery of regions such as Mount St. Helens but also have transformed Earth. Continuing the saga of how this occurred, this chapter follows the emergence and diversification of the group to which fireweed belongs, the seed plants.

CONCEPT 30.1

Seeds and pollen grains are key adaptations for life on land

We begin with an overview of terrestrial adaptations that seed plants added to those already present in nonvascular plants (bryophytes) and seedless vascular plants (see Concept 29.1). In addition to seeds, all seed plants have reduced gametophytes, heterospory, ovules, and pollen. As we'll see, these adaptations helped seed plants cope with conditions such as drought and exposure to ultraviolet (UV) radiation in sunlight. They also freed seed plants from requiring water for fertilization, enabling reproduction under a broader range of conditions than in seedless plants.

Advantages of Reduced Gametophytes

Mosses and other bryophytes have life cycles dominated by gametophytes, whereas ferns and other seedless vascular plants have sporophyte-dominated life cycles. The evolutionary trend of gametophyte reduction continued further in the vascular plant lineage that led to seed plants **(Figure 30.2)**. While the gametophytes of seedless vascular plants are visible to the naked eye, the gametophytes of most seed plants are microscopic.

This miniaturization allowed for an important evolutionary innovation in seed plants: Their tiny gametophytes can develop from spores retained within the sporangia of the parental sporophyte. This arrangement can protect the gametophytes from environmental stresses. For example, the moist reproductive tissues of the sporophyte shield the gametophytes from UV radiation and protect them from drying out. This relationship

▼ Figure 30.2 **Gametophyte-sporophyte relationships in different plant groups.**

	PLANT GROUP		
	Mosses and other nonvascular plants	**Ferns and other seedless vascular plants**	**Seed plants (gymnosperms and angiosperms)**
Gametophyte	Dominant	Reduced, independent (photosynthetic and free-living)	Reduced (usually microscopic), dependent on surrounding sporophyte tissue for nutrition
Sporophyte	Reduced, dependent on gametophyte for nutrition	Dominant	Dominant
Example			

MAKE CONNECTIONS *In seed plants, how might retaining the gametophyte within the sporophyte affect embryo fitness? (See Concepts 17.5, 23.1, and 23.4 to review mutagens, mutations, and fitness.)*

also enables the developing gametophytes to obtain nutrients from the parental sporophyte. In contrast, the free-living gametophytes of seedless vascular plants must fend for themselves.

Heterospory: The Rule Among Seed Plants

You read in Concept 29.3 that most seedless plants are *homosporous*—they produce one kind of spore, which usually gives rise to a bisexual gametophyte. Ferns and other close relatives of seed plants are homosporous, suggesting that seed plants had homosporous ancestors. At some point, seed plants or their ancestors became *heterosporous*, producing two kinds of spores: Megasporangia on modified leaves called megasporophylls produce *megaspores* that give rise to female gametophytes, and microsporangia on modified leaves called microsporophylls produce *microspores* that give rise to male gametophytes. Each megasporangium has one megaspore, whereas each microsporangium has many microspores.

As noted previously, the miniaturization of seed plant gametophytes probably contributed to the great success of this clade. Next, we'll look at the development of the female gametophyte within an ovule and the development of the male gametophyte in a pollen grain. Then we'll follow the transformation of a fertilized ovule into a seed.

Ovules and Production of Eggs

Although a few species of seedless plants are heterosporous, seed plants are unique in retaining the megasporangium within the parent sporophyte. A layer of sporophyte tissue called **integument** envelops and protects the megasporangium. Gymnosperm megasporangia are surrounded by one integument, whereas those in angiosperms usually have two integuments. The whole structure—megasporangium, megaspore, and their integument(s)—is called an **ovule (Figure 30.3a)**. Inside each ovule (from the Latin *ovulum*, little egg), a female gametophyte develops from a megaspore and produces one or more eggs.

Pollen and Production of Sperm

A microspore develops into a **pollen grain** that consists of a male gametophyte enclosed within the pollen wall. (The wall's outer layer is made of molecules secreted by sporophyte cells, so we refer to the male gametophyte as being *in* the pollen grain, not *equivalent to* the pollen grain.) Sporopollenin in the pollen wall protects the pollen grain as it is transported by wind or by hitchhiking on an animal. The transfer of pollen to the part of a seed plant that contains the ovules is called **pollination**. If a pollen grain germinates (begins growing), it gives rise to a pollen tube that discharges sperm into the female gametophyte within the ovule, as shown in **Figure 30.3b**.

In nonvascular plants and seedless vascular plants such as ferns, free-living gametophytes release flagellated sperm that swim through a film of water to reach eggs. Given this requirement, it is not surprising that many of these species live in moist habitats. But a pollen grain can be carried by wind or animals, eliminating the dependence on water for sperm transport. The ability of seed plants to transfer sperm without water likely contributed to their colonization of dry habitats. The sperm of seed plants also do not require motility because they are carried to the eggs by pollen tubes. The sperm of some gymnosperm species (such as cycads and ginkgos, shown in Figure 30.7) retain the ancient flagellated condition, but flagella have been lost in the sperm of most gymnosperms and all angiosperms.

▼ **Figure 30.3 From ovule to seed in a gymnosperm.**

Immature ovulate cone — Integument (2n) — Spore wall — Megasporangium (2n) — Megaspore (n) — Micropyle — Pollen grain (n) — Male gametophyte (within a germinated pollen grain) (n) — Female gametophyte (n) — Egg nucleus (n) — Discharged sperm nucleus (n) — Pollen tube — Seed coat (derived from integument) — Spore wall (surrounded by megasporangium remnant) — Food supply (female gametophyte tissue) (n) — Embryo (2n) (new sporophyte)

(a) Unfertilized ovule. In this longitudinal section through the ovule of a pine (a gymnosperm), a fleshy megasporangium is surrounded by a protective layer of tissue called an integument. The micropyle, the only opening through the integument, allows entry of a pollen grain.

(b) Fertilized ovule. A megaspore develops into a female gametophyte, which produces an egg. The pollen grain, which had entered through the micropyle, contains a male gametophyte. The male gametophyte develops a pollen tube that discharges sperm, thereby fertilizing the egg.

(c) Gymnosperm seed. Fertilization initiates the transformation of the ovule into a seed, which consists of a sporophyte embryo, a food supply, and a protective seed coat derived from the integument. The megasporangium dries out and collapses.

VISUAL SKILLS *Based on this figure, a gymnosperm seed contains cells from how many different plant generations? Identify the cells and whether each is haploid or diploid.*

▶ **Mastering Biology** Figure Walkthrough

Using Natural Logarithms to Interpret Data

How Long Can Seeds Remain Viable in Dormancy?
Environmental conditions can vary greatly over time, and they may not be favorable for germination when seeds are produced. One way that plants cope with such variation is through seed dormancy. Under favorable conditions, seeds of some species can germinate after many years of dormancy.

One unusual opportunity to test how long seeds can remain viable occurred when seeds from date palm trees (*Phoenix dactylifera*) were discovered under the rubble of a 2,000-year-old fortress near the Dead Sea. As you saw in the Chapter 2 Scientific Skills Exercise and Concept 25.2, scientists use radiometric dating to estimate the ages of fossils and other old objects. In this exercise, you will estimate the ages of three of these ancient seeds by using natural logarithms.

How the Experiment Was Done
Scientists measured the fraction of carbon-14 that remained in three ancient date palm seeds: two that were not planted and one that was planted and germinated. For the germinated seed, the scientists used a seed coat fragment found clinging to a root of the seedling. (The seedling grew into the plant in the photo.)

Data from the Experiment
This table shows the fraction of carbon-14 remaining from the three ancient date palm seeds.

	Fraction of Carbon-14 Remaining
Seed 1 (not planted)	0.7656
Seed 2 (not planted)	0.7752
Seed 3 (germinated)	0.7977

INTERPRET THE DATA
A logarithm is the power to which a base is raised to produce a given number x. For example, if the base is 10 and $x = 100$, the logarithm of 100 equals 2 (because $10^2 = 100$). A natural logarithm (ln) is the logarithm of a number x to the base e, where e is about 2.718. Natural logarithms are useful in calculating rates of some natural processes, such as radioactive decay.

The equation $F = e^{-kt}$ describes the fraction F of an original isotope remaining after a period of t years; the exponent is negative because it refers to a *decrease* over time. The constant k provides a measure of how rapidly the original isotope decays. For the decay of carbon-14 to nitrogen-14, $k = 0.00012097$.

To estimate t, the age of the three seeds, we rearrange the equation $F = e^{-kt}$ to find an equation for t:

$$t = -\left(\frac{\ln F}{k}\right)$$

1. Using the equation for t, the data from the table, and a calculator, estimate the ages of seed 1, seed 2, and seed 3.
2. Why do you think there was more carbon-14 in the germinated seed?

➡️ **Instructors:** A version of this Scientific Skills Exercise can be assigned in **Mastering Biology**.

Data from S. Sallon et al., Germination, genetics, and growth of an ancient date seed, *Science* 320:1464 (2008).

The Evolutionary Advantage of Seeds

If a sperm fertilizes an egg of a seed plant, the zygote grows into a sporophyte embryo. As shown in **Figure 30.3c**, the ovule develops into a seed: the embryo, with a food supply, packaged in a protective coat derived from the integument(s).

Seeds provide protection from harsh conditions and facilitate dispersal to new habitats—as do the spores of seedless plants. Moss spores, for example, may survive even if the local environment becomes too cold, too hot, or too dry for the mosses themselves to live. Their tiny size enables the spores to be dispersed in a dormant state to a new area, where they can germinate and give rise to new moss gametophytes if and when conditions are favorable enough for them to break dormancy. Spores were the main way that mosses, ferns, and other seedless plants spread over Earth for the first 100 million years of plant life on land.

Although mosses and other seedless plants continue to be very successful today, seeds represent a major evolutionary innovation that contributed to the opening of new ways of life for seed plants. What advantages do seeds provide over spores? Spores are usually single-celled, whereas seeds are multicellular, consisting of an embryo protected by a layer of tissue, the seed coat. A seed can remain dormant for days, months, or even years after being released from the parent plant, whereas most spores have shorter lifetimes. Also, unlike spores, seeds have a supply of stored food. Most seeds land close to their parent sporophyte plant, but some are carried long distances (up to hundreds of kilometers) by wind or animals. If conditions are favorable where it lands, the seed can emerge from dormancy and germinate, with its stored food providing critical support for growth as the sporophyte embryo emerges as a seedling. As we explore in the **Scientific Skills Exercise**, some seeds have germinated after more than 1,000 years.

CONCEPT CHECK 30.1

1. Contrast how sperm reach the eggs of seedless plants with how sperm reach the eggs of seed plants.
2. What features not present in seedless plants have contributed to the success of seed plants on land?
3. **WHAT IF?** If a seed could not enter dormancy, how might that affect the embryo's transport or survival?

For suggested answers, see Appendix A.

Gymnosperms bear "naked" seeds, typically on cones

Nonvascular plants (bryophytes)
Seedless vascular plants
Gymnosperms
Angiosperms

Extant seed plants form two sister clades: gymnosperms and angiosperms. Recall that gymnosperms have "naked" seeds exposed on sporophylls that usually form cones. (Angiosperm seeds are enclosed in chambers that mature into fruits.) Most gymnosperms are cone-bearing plants called **conifers**, such as pines, firs, and redwoods.

The Life Cycle of a Pine

As you read earlier, seed plant evolution has included three key reproductive adaptations: the miniaturization of their gametophytes; the advent of the seed as a resistant, dispersible stage in the life cycle; and the appearance of pollen as an airborne agent that brings gametes together. **Figure 30.4** shows how these adaptations come into play during the life cycle of a pine, a familiar conifer.

▼ **Figure 30.4 The life cycle of a pine.**

❶ In most conifer species, each tree has both ovulate and pollen cones.

❸ An ovulate cone scale has two ovules, each containing a megasporangium. Only one ovule is shown.

Ovule

Ovulate cone

Pollen cones

Longitudinal section of ovulate cone

Megasporocyte (2n)

Integument

❹ Pollination occurs when a pollen grain reaches the ovule. The pollen grain then germinates, forming a pollen tube that slowly digests its way through the megasporangium.

Microsporangia

Microsporocytes (2n)

Pollen grains (n)

Germinating pollen grain

Megasporangium (2n)

MEIOSIS

MEIOSIS

Mature sporophyte (2n)

Microsporangium (2n)

Longitudinal section of pollen cone

Surviving megaspore (n)

❺ While the pollen tube develops, the megasporocyte undergoes meiosis, producing four haploid cells. One survives as a megaspore.

Seedling

❷ Microsporocytes divide by meiosis, producing haploid microspores. A microspore develops into a pollen grain (a male gametophyte enclosed within the pollen wall).

Archegonium

Female gametophyte

❻ The megaspore develops into a female gametophyte that contains two or three archegonia, each of which will form an egg.

Seeds on surface of ovulate scale

Food reserves (gametophyte tissue) (n)

Seed coat (2n)

Discharged sperm nucleus (n)

Pollen tube

Embryo (new sporophyte) (2n)

FERTILIZATION

Egg nucleus (n)

❼ By the time the eggs are mature, sperm cells have developed in the pollen tube, which extends to the female gametophyte. Fertilization occurs when sperm and egg nuclei unite.

Key

➡ Haploid (n)
➡ Diploid (2n)

❽ Fertilization usually occurs more than a year after pollination. All eggs may be fertilized, but usually only one zygote develops into an embryo. The ovule becomes a seed, consisting of an embryo, food supply, and seed coat.

MAKE CONNECTIONS *What type of cell division occurs as a megaspore develops into a female gametophyte? Explain. (See Figure 13.10.)*

➔ **Mastering Biology Animation: Pine Life Cycle**

The pine tree is the sporophyte; its sporangia are located on scalelike structures packed densely in cones. Like all seed plants, conifers are heterosporous. As such, they have two types of sporangia that produce two types of spores: microsporangia that produce microspores and megasporangia that produce megaspores. In conifers, the two types of spores are produced by separate cones: small pollen cones and large ovulate cones.

Pollen cones have a relatively simple structure: Their scales are modified leaves (microsporophylls) that bear microsporangia. Within each microsporangium, cells called microsporocytes undergo meiosis, producing haploid microspores. Each microspore develops into a pollen grain containing a male gametophyte. In conifers, the yellow pollen is released in large amounts and carried by the wind, dusting everything in its path.

Ovulate cones are more complex: Their scales are compound structures composed of both modified leaves (megasporophylls bearing megasporangia) and modified stem tissue. Within each megasporangium, megasporocytes undergo meiosis and produce haploid megaspores inside the ovule. Surviving megaspores develop into female gametophytes, which are retained within the sporangia.

In most pine species, each tree has both types of cones. From the time pollen and ovulate cones appear on the tree, it takes nearly three years for the male and female gametophytes to be produced and brought together and for mature seeds to form from fertilized ovules. The scales of each ovulate cone then separate, and seeds are dispersed by the wind. A seed that lands in a suitable environment germinates, its embryo emerging as a pine seedling.

Early Seed Plants and the Rise of Gymnosperms

The origins of characteristics found in pines and other living seed plants date back to the late Devonian period (380 million years ago). Fossils from that time reveal that some plants had acquired features that are also present in seed plants, such as megaspores and microspores. For example, *Archaeopteris* was a heterosporous tree with a woody stem. It grew up to 20 m tall and had fernlike leaves. But it did not bear seeds and therefore is not classified as a seed plant.

The earliest evidence of seed plants comes from 360-million-year-old fossils of plants in the genus *Elkinsia* **(Figure 30.5)**. These and other early seed plants lived 55 million years before the first fossils classified as gymnosperms and more than 200 million years before the first fossils of angiosperms. These early seed plants became extinct, and we don't know which extinct lineage gave rise to the gymnosperms.

The oldest fossils of species from an extant lineage of gymnosperms are 305 million years old. These early gymnosperms lived in moist Carboniferous ecosystems that were

▼ **Figure 30.5 A fossil of the early seed plant *Elkinsia*.**

Ovule

▼ **Figure 30.6 An ancient pollinator.** This 110-million-year-old fossil shows pollen on an insect, the thrip *Gymnopollisthrips minor*. Structural features of the pollen suggest that it was produced by gymnosperms (most likely by species related to extant ginkgos or cycads). Although most gymnosperms today are wind-pollinated, many cycads are insect-pollinated.

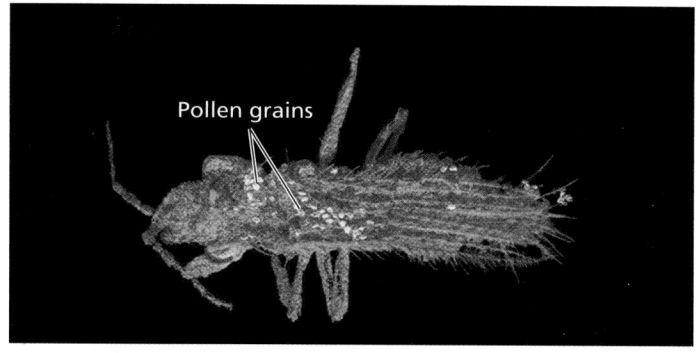

Pollen grains

dominated by lycophytes, horsetails, ferns, and other seedless vascular plants. As the Carboniferous period gave way to the Permian (299 to 252 million years ago), the climate became much drier. As a result, the lycophytes, horsetails, and ferns that dominated Carboniferous swamps were largely replaced by gymnosperms, which were better suited to the drier climate.

Gymnosperms thrived as the climate dried, in part because they have the key terrestrial adaptations found in all seed plants, such as seeds and pollen. In addition, some gymnosperms were particularly well suited to arid conditions because of the thick cuticles and relatively small surface areas of their needle-shaped leaves.

Gymnosperms dominated terrestrial ecosystems throughout much of the Mesozoic era, which lasted from 252 to 66 million years ago. In addition to serving as the food supply for giant herbivorous dinosaurs, these gymnosperms were involved in many other interactions with animals. Recent fossil discoveries, for example, show that some gymnosperms were pollinated by insects more than 100 million years ago—the earliest evidence of insect pollination in any plant group **(Figure 30.6)**. Late in the Mesozoic, angiosperms began to replace gymnosperms in some ecosystems.

Gymnosperm Diversity

Although angiosperms now dominate most terrestrial ecosystems, gymnosperms remain an important part of Earth's flora. For example, vast regions in northern latitudes are covered by forests of conifers (see Figure 52.13).

Of the ten plant phyla (see Table 29.1), four are gymnosperms: Cycadophyta, Ginkgophyta, Gnetophyta, and Coniferophyta. It is uncertain how the four phyla of gymnosperms are related to each other. **Figure 30.7** surveys the diversity of extant gymnosperms.

Phylum Cycadophyta

The 350 species of living cycads have large cones and palmlike leaves (true palm species are angiosperms). Unlike most seed plants, cycads have flagellated sperm, indicating their descent from seedless vascular plants that had motile sperm. Cycads thrived during the Mesozoic era, known as the age of cycads as well as the age of dinosaurs. Today, however, cycads are the most endangered of all plant groups: 75% of their species are threatened by habitat destruction and other human actions.

Cycas revoluta

Phylum Ginkgophyta

Ginkgo biloba is the only surviving species of this phylum; like cycads, ginkgos have flagellated sperm. Also known as the maidenhair tree, *Ginkgo biloba* has deciduous fanlike leaves that turn gold in autumn. It is a popular ornamental tree in cities because it tolerates air pollution well. Landscapers often plant only pollen-producing trees because the fleshy seeds smell rancid as they decay.

Phylum Gnetophyta

Phylum Gnetophyta includes plants in three genera: *Gnetum*, *Ephedra*, and *Welwitschia*. Some species are tropical, whereas others live in deserts. Although very different in appearance, the genera are grouped together based on molecular data.

▶ **Welwitschia** (also below). This genus consists of one species, *Welwitschia mirabilis*, a plant that can live for thousands of years and is found only in the deserts of southwestern Africa. Its straplike leaves are among the largest leaves known.

Ovulate cones

◀ **Gnetum.** This genus includes about 35 species of tropical trees, shrubs, and vines, mainly native to Africa and Asia. Their leaves look similar to those of flowering plants, and their seeds look somewhat like fruits.

▶ **Ephedra.** This genus includes about 40 species that inhabit arid regions worldwide. These desert shrubs, commonly called "Mormon tea," produce the compound ephedrine, which is used medicinally as a decongestant.

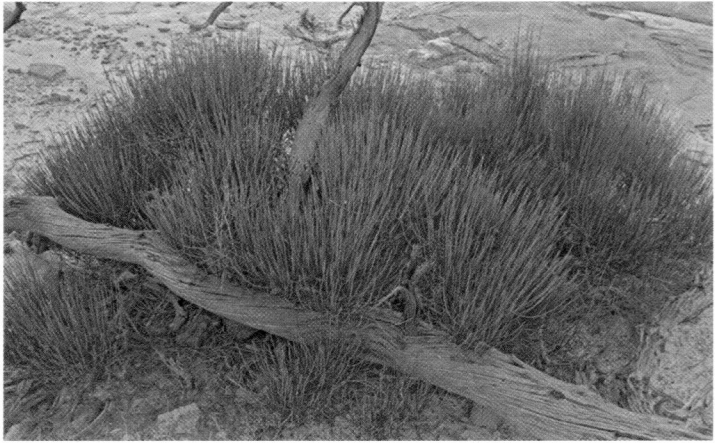

Phylum Coniferophyta

Phylum Coniferophyta, the largest gymnosperm phyla, consists of about 600 species of conifers (from the Latin *conus*, cone, and *ferre*, to carry), including many large trees. Most species have woody cones, but a few have fleshy cones. Some, such as pines, have needlelike leaves. Others, such as redwoods, have scalelike leaves. Some species dominate vast northern forests, whereas others are native to the Southern Hemisphere.

Most conifers are evergreens; they retain their leaves throughout the year. Even during winter, a limited amount of photosynthesis occurs on sunny days. When spring comes, conifers already have fully developed leaves that can take advantage of the sunnier, warmer days. Some conifers, such as the dawn redwood, tamarack, and larch, are deciduous trees that lose leaves each autumn.

▶ **Douglas fir.** This evergreen tree (*Pseudotsuga menziesii*) provides more timber than any other North American tree species. Some uses include house framing, plywood, pulpwood for paper, railroad ties, and boxes and crates.

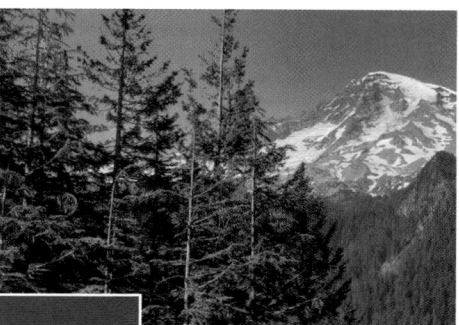

▶ **Common juniper.** The "berries" of the common juniper (*Juniperus communis*) are actually ovule-producing cones consisting of fleshy sporophylls.

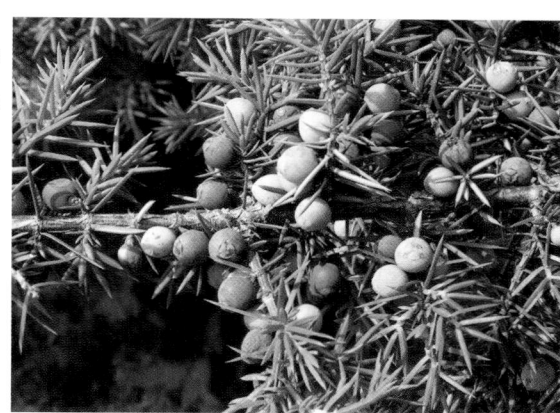

◀ **European larch.** The needlelike leaves of this deciduous conifer (*Larix decidua*) turn yellow before they are shed in autumn. Native to the mountains of central Europe, including Switzerland's Matterhorn, depicted here, this species is extremely cold-tolerant, able to survive winter temperatures that plunge to –50°C.

◀ **Wollemi pine.** Survivors of a conifer group once known only from fossils, living *Wollemi* pines (*Wollemia nobilis*) were discovered in 1994 in a national park near Sydney, Australia. At that time, the species consisted of 40 known trees. As a result of conservation efforts, it is now widely propagated. The inset photo compares the leaves of this "living fossil" with actual fossils.

▶ **Sequoia.** This giant sequoia (*Sequoiadendron giganteum*) in California's Sequoia National Park weighs about 2,500 metric tons, equivalent to about 24 blue whales (the largest animals) or 40,000 people. The giant sequoia is one of the largest living organisms and also among the most ancient, with some individuals estimated to be between 1,800 and 2,700 years old. Their cousins, the coast redwoods (*Sequoia sempervirens*), grow to heights of more than 110 m (taller than the Statue of Liberty) and are found only in a narrow coastal strip of northern California and southern Oregon.

▶ **Bristlecone pine.** This species (*Pinus longaeva*), which is found in the White Mountains of California, includes some of the world's oldest living organisms. One bristlecone pine was recently found to be more than 5,000 years old, and some may be even older.

1. Explain how the pine life cycle in Figure 30.4 reflects the five adaptations common to all seed plants.
2. **VISUAL SKILLS** Based on Figure 30.4, compare and contrast the function of pollination and fertilization in sexual reproduction in pines.
3. **MAKE CONNECTIONS** Early seed plants in genus *Elkinsia* are a sister group to a clade consisting of gymnosperms and angiosperms. Draw a phylogenetic tree of seed plants that shows *Elkinsia*, gymnosperms, and angiosperms; date the branch points on this tree using fossil evidence. (See Figure 26.5.)

For suggested answers, see Appendix A.

The reproductive adaptations of angiosperms include flowers and fruits

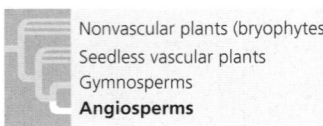

Nonvascular plants (bryophytes)
Seedless vascular plants
Gymnosperms
Angiosperms

Commonly known as flowering plants, angiosperms are seed plants with the reproductive structures called flowers and fruits. The name *angiosperm* (from the Greek *angion*, container) refers to seeds contained in fruits. Angiosperms are the most diverse and widespread of all plants, with more than 290,000 species (about 90% of all plant species).

Characteristics of Angiosperms

All angiosperms are classified in a single phylum, Anthophyta. Before considering the evolution of angiosperms, we will examine two of their key adaptations—flowers and fruits— and the roles of these structures in the angiosperm life cycle.

Flowers

The **flower** is a unique angiosperm structure that is specialized for sexual reproduction. In many angiosperm species, insects or other animals transfer pollen from one flower to the sex organs on another flower, which makes pollination more directed than the wind-dependent pollination of most species of gymnosperms. However, some angiosperms *are* wind-pollinated, particularly those species that occur in dense populations, such as grasses and tree species in temperate forests.

A flower is a specialized shoot that can have up to four types of modified leaves called floral organs: sepals, petals, stamens, and carpels **(Figure 30.8)**. Starting at the base of the flower are the **sepals**, which are usually green and enclose the flower before it opens (think of a rosebud). Interior to the sepals are the **petals**, which are brightly colored in most flowers and can aid in attracting pollinators. Flowers that are wind-pollinated, such as grasses, generally lack brightly colored parts. In all angiosperms, the sepals and petals are sterile floral organs, meaning that they do not produce sperm or eggs.

▼ **Figure 30.8 The structure of an idealized flower.**

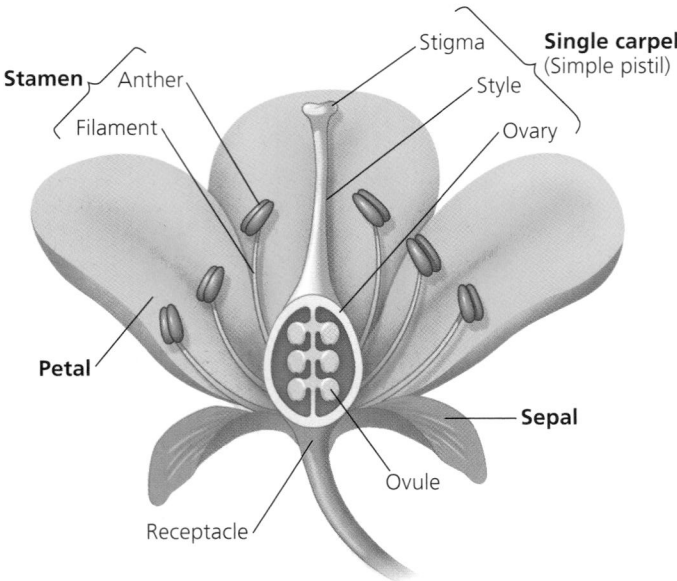

Within the petals are two types of fertile floral organs that produce spores, the stamens and carpels. Stamens and carpels are sporophylls, modified leaves that are specialized for reproduction. **Stamens** are microsporophylls: They produce microspores that develop into pollen grains containing male gametophytes. A stamen consists of a stalk called the **filament** and a terminal sac, the **anther**, where pollen is produced. **Carpels** are megasporophylls: They produce megaspores that give rise to female gametophytes. The carpel is the "container" mentioned earlier in which seeds are enclosed; as such, it is a key structure that distinguishes angiosperms from gymnosperms. At the tip of the carpel is a sticky **stigma** that receives pollen. A **style** leads from the stigma to a structure at the base of the carpel, the **ovary**; the ovary contains one or more ovules. As in gymnosperms, each angiosperm ovule contains a female gametophyte. If fertilized, an ovule develops into a seed.

A flower may have one or more carpels. In many species, multiple carpels are fused into one structure. The term **pistil** is sometimes used to refer to a single carpel (a simple pistil) or two or more fused carpels (a compound pistil). Flowers also vary in symmetry **(Figure 30.9)** and other aspects of shape, as

▼ **Figure 30.9 Flower symmetry.**

In radial symmetry, the sepals, petals, stamens, and carpels radiate out from a center. Any line through the central axis divides the flower into two equal parts.

Sepal

Radial symmetry (daffodil)

Fused petals

In bilateral symmetry, the flower can only be divided into two equal parts by a single line.

Bilateral symmetry (orchid)

DRAW IT *Draw the single line that can divide the bilaterally symmetrical flower into two equal parts.*

well as size, color, and odor. Much of this diversity results from adaptation to specific pollinators (see Figures 38.4 and 38.5).

Fruits

As seeds develop from ovules after fertilization, the ovary wall thickens and the ovary matures into a **fruit**. A pea pod is an example of a fruit, with seeds (mature ovules, the peas) encased in the ripened ovary (the pod).

Fruits protect seeds and aid in their dispersal. Mature fruits can be either fleshy or dry (**Figure 30.10**). Tomatoes, plums, and grapes are examples of fleshy fruits, in which the wall (pericarp) of the ovary becomes soft during ripening. Dry fruits include beans, nuts, and grains. Some dry fruits split open at maturity to release seeds, whereas others remain closed. The dry, wind-dispersed fruits of grasses, harvested while on the plant, are major staple foods for humans. The cereal grains of maize, rice, wheat, and other grasses, though easily mistaken for seeds, are each actually a fruit with a dry outer covering (the former wall of the ovary) that adheres to the seed coat of the seed within.

▼ **Figure 30.10 Some variations in fruit structure.**

▼ Tomato, a fleshy fruit with soft outer and inner layers of pericarp (fruit wall)

▼ Ruby grapefruit, a fleshy fruit with a firm outer layer and soft inner layer of pericarp

▼ Nectarine, a fleshy fruit with a soft outer layer and hard inner layer (pit) of pericarp

▼ Hazelnut, a dry fruit that remains closed at maturity

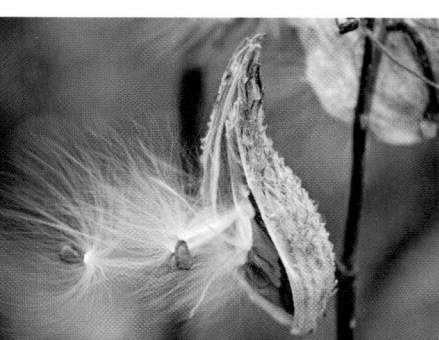

◄ Milkweed, a dry fruit that splits open at maturity

As shown in **Figure 30.11**, various adaptations of fruits and seeds help to disperse seeds (see also Figure 38.12). The seeds of some flowering plants, such as dandelions and maples, are contained within fruits that function like parachutes or propellers, adaptations that enhance dispersal by wind. Some fruits, such as coconuts, are adapted to dispersal by water. And the seeds of many angiosperms are carried by animals. Some angiosperms have fruits modified as burrs that cling to animal fur (or the clothes of humans). Others produce edible fruits, which are usually nutritious, sweet tasting, and vividly colored, advertising their ripeness. When an animal eats the fruit, it digests the fruit's fleshy part, but the tough seeds usually pass unharmed through the animal's digestive tract. When the animal defecates, it may deposit the seeds, along with a supply of natural fertilizer, many kilometers from where the fruit was eaten.

▼ **Figure 30.11 Fruit adaptations that enhance seed dispersal.**

◄ Some plants have mechanisms that disperse seeds by explosive action.

► Wings enable maple fruits to be carried by the wind.

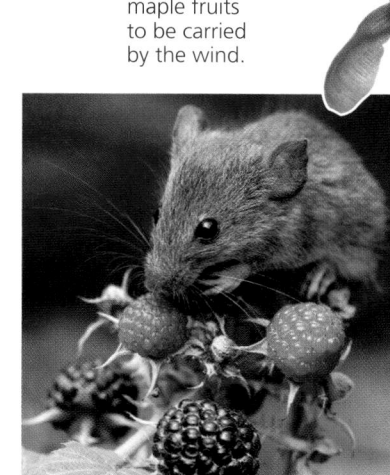

◄ Seeds within berries and other edible fruits are often dispersed in animal feces.

► The barbs of cockleburs facilitate seed dispersal by allowing the fruits to "hitchhike" on animals.

→ **Mastering Biology** Animation: Fruit Structure and Seed Dispersal

The Angiosperm Life Cycle

You can follow a typical angiosperm life cycle in **Figure 30.12**. The flower of the sporophyte produces microspores that form male gametophytes and megaspores that form female gametophytes. The male gametophytes are in the pollen grains, which develop within microsporangia in the anthers. Each male gametophyte has two haploid cells: a *generative cell* that divides, forming two sperm, and a *tube cell* that produces a pollen tube. Each ovule, which develops in the ovary, contains a female gametophyte, also known as an **embryo sac**. The embryo sac consists of only a few cells, one of which is the egg.

After its release from the anther, the pollen is carried to the sticky stigma at the tip of a carpel. Although some flowers self-pollinate, most have mechanisms that ensure **cross-pollination**, which in angiosperms is the transfer of pollen from an anther of a flower on one plant to the stigma of a

▼ **Figure 30.12 The life cycle of an angiosperm.**

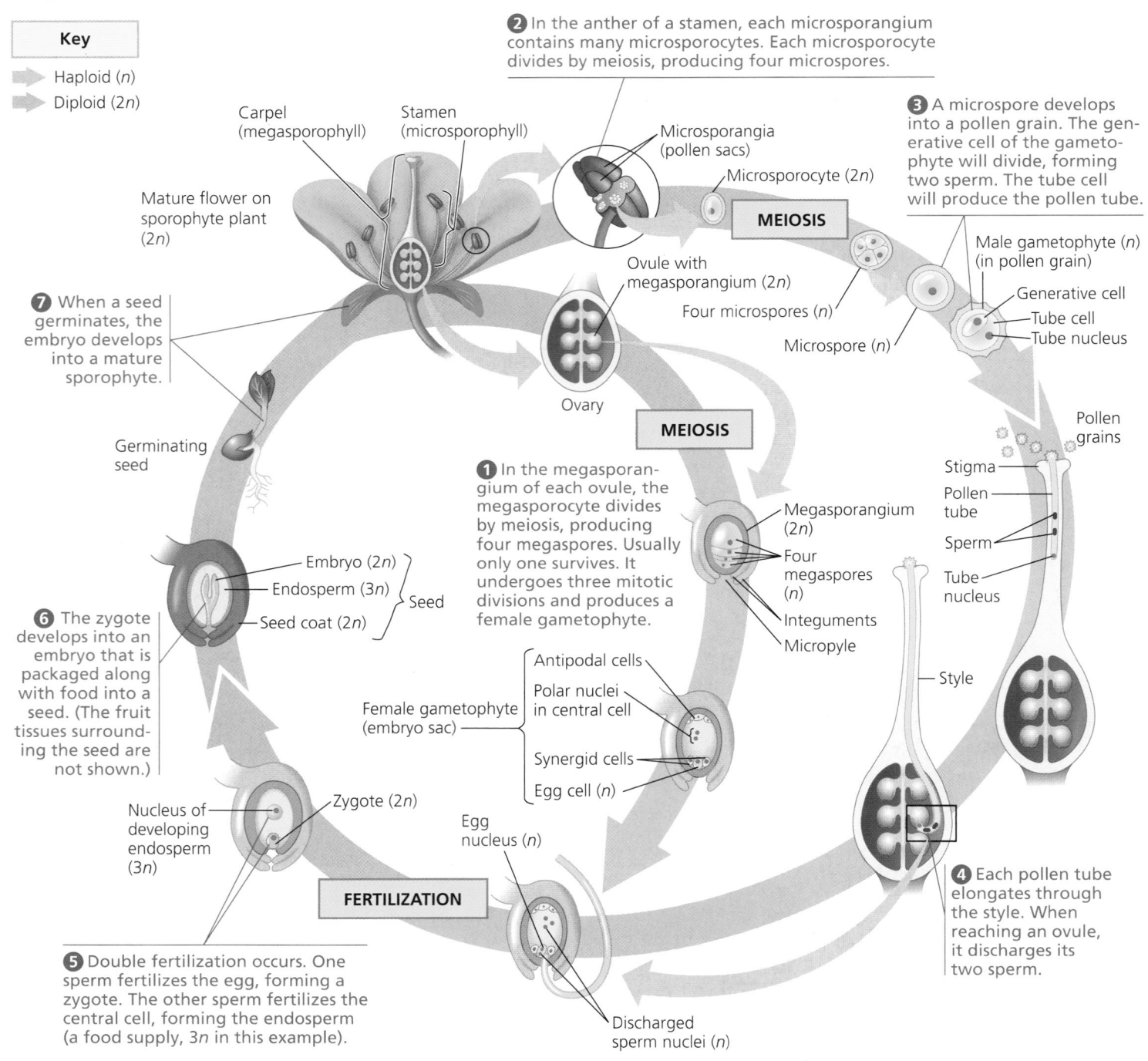

Key

Haploid (n)
Diploid (2n)

❷ In the anther of a stamen, each microsporangium contains many microsporocytes. Each microsporocyte divides by meiosis, producing four microspores.

❸ A microspore develops into a pollen grain. The generative cell of the gametophyte will divide, forming two sperm. The tube cell will produce the pollen tube.

Carpel (megasporophyll)

Stamen (microsporophyll)

Microsporangia (pollen sacs)

Microsporocyte (2n)

MEIOSIS

Mature flower on sporophyte plant (2n)

Ovule with megasporangium (2n)

Four microspores (n)

Male gametophyte (n) (in pollen grain)

Generative cell
Tube cell
Tube nucleus

Microspore (n)

❼ When a seed germinates, the embryo develops into a mature sporophyte.

Ovary

MEIOSIS

Pollen grains

Germinating seed

❶ In the megasporangium of each ovule, the megasporocyte divides by meiosis, producing four megaspores. Usually only one survives. It undergoes three mitotic divisions and produces a female gametophyte.

Megasporangium (2n)

Stigma

Pollen tube

Four megaspores (n)

Sperm

Tube nucleus

Embryo (2n)
Endosperm (3n)   ⎱ Seed
Seed coat (2n)

Integuments
Micropyle

Style

❻ The zygote develops into an embryo that is packaged along with food into a seed. (The fruit tissues surrounding the seed are not shown.)

Antipodal cells

Polar nuclei in central cell

Female gametophyte (embryo sac)

Synergid cells

Egg cell (n)

Nucleus of developing endosperm (3n)

Zygote (2n)

Egg nucleus (n)

FERTILIZATION

❺ Double fertilization occurs. One sperm fertilizes the egg, forming a zygote. The other sperm fertilizes the central cell, forming the endosperm (a food supply, 3n in this example).

❹ Each pollen tube elongates through the style. When reaching an ovule, it discharges its two sperm.

Discharged sperm nuclei (n)

VISUAL SKILLS *Based on this figure, what is the maximum number of seeds this flower could produce? To produce that number of seeds, at least how many pollen grains would have to germinate?*

→ **Mastering Biology** Figure Walkthrough
Animation: Angiosperm Life Cycle

flower on another plant of the same species. Cross-pollination enhances genetic variability. In some species, stamens and carpels of a single flower may mature at different times, or they may be so arranged that self-pollination is unlikely.

The pollen grain absorbs water and germinates after it adheres to the stigma of a carpel. The tube cell produces a pollen tube that grows down within the style of the carpel. After reaching the ovary, the pollen tube penetrates through the **micropyle**, a pore in the integuments of the ovule, and discharges two sperm cells into the female gametophyte (embryo sac). One sperm fertilizes the egg, forming a diploid zygote. The other sperm fuses with the two nuclei in the large central cell of the female gametophyte, producing a triploid cell ($3n$). This type of **double fertilization**, in which one fertilization event produces a zygote and the other produces a triploid cell, is unique to angiosperms.

After double fertilization, the ovule matures into a seed. The zygote develops into a sporophyte embryo with a rudimentary root and one or two seed leaves called **cotyledons**. The triploid central cell of the female gametophyte develops into **endosperm**, tissue rich in starch and other food reserves that nourish the developing embryo.

What is the function of double fertilization in angiosperms? One hypothesis is that double fertilization synchronizes the development of food storage in the seed with the development of the embryo. If a particular flower is not pollinated or sperm cells are not discharged into the embryo sac, fertilization does not occur, and neither endosperm nor embryo forms. So perhaps double fertilization is an adaptation that prevents flowering plants from squandering nutrients on infertile ovules.

Another type of double fertilization occurs in some gymnosperm species belonging to the phylum Gnetophyta. However, double fertilization in these species gives rise to two embryos rather than to an embryo and endosperm.

As you read earlier, the seed consists of the embryo, the endosperm, and a seed coat derived from the integuments. An ovary develops into a fruit as its ovules become seeds. After being dispersed, a seed may germinate if environmental conditions are favorable. The coat ruptures and the embryo emerges as a seedling, using food stored in the endosperm and cotyledons until it can produce its own food by photosynthesis.

Angiosperm Evolution

Charles Darwin once referred to the origin of angiosperms as an "abominable mystery." He was particularly troubled by the relatively sudden and geographically widespread appearance of angiosperms in the fossil record (about 100 million years ago, based on fossils known to Darwin). Recent fossil evidence and phylogenetic analyses have led to progress in solving Darwin's mystery, but we still do not fully understand how angiosperms arose from earlier seed plants.

Fossil Angiosperms

Angiosperms are now thought to have originated in the early Cretaceous period, about 140 million years ago. By the mid-Cretaceous (100 million years ago), angiosperms began to dominate some terrestrial ecosystems. Landscapes changed dramatically as conifers and other gymnosperms gave way to flowering plants in many parts of the world. The Cretaceous ended 66 million years ago with mass extinctions of dinosaurs and many other animal groups and with further increases in the diversity and importance of angiosperms.

What evidence suggests that angiosperms arose 140 million years ago? First, although pollen grains are common in rocks from the Jurassic period (201 to 145 million years ago), none of these pollen fossils have features characteristic of angiosperms, suggesting that angiosperms may have originated after the Jurassic. Indeed, the earliest fossils with distinctive angiosperm features are of 130-million-year-old pollen grains discovered in China, Israel, and England. Early fossils of larger flowering plant structures include those of *Archaefructus* **(Figure 30.13)** and *Leefructus*, both of which were discovered in China in rocks that are 125 million years old.

▼ **Figure 30.13 An early flowering plant.**

(a) *Archaefructus sinensis*, a 125-million-year-old fossil. This herbaceous species had simple flowers and bulbous structures that may have served as floats, suggesting it was aquatic. Recent phylogenetic analyses indicate that *Archaefructus* may belong to the water lily group.

Carpel

Stamen

Possible floats

(b) Artist's reconstruction of *Archaefructus sinensis*

Overall, early angiosperm fossils indicate that the group arose and began to diversify over a 20- to 30-million-year period—a less sudden event than was suggested by the fossils known during Darwin's lifetime.

Can we infer traits of the angiosperm common ancestor from traits found in early fossil angiosperms? *Archaefructus*, for example, was herbaceous and had bulbous structures that may have served as floats, suggesting it was aquatic. But investigating whether the angiosperm common ancestor was herbaceous and aquatic also requires examining fossils of other seed plants thought to have been closely related to angiosperms. All of those plants were woody, indicating that the common ancestor was probably woody and probably not aquatic. As we'll see, this conclusion has been supported by recent phylogenetic analyses.

Angiosperm Phylogeny

Molecular and morphological evidence suggests that extant gymnosperm lineages had diverged from the lineage leading to angiosperms by 305 million years ago. Note that this does not imply that angiosperms originated 305 million years ago, but that the most recent common ancestor of extant gymnosperms and angiosperms lived at that time. Indeed, angiosperms may be more closely related to several extinct lineages of woody seed plants than they are to gymnosperms. One such lineage is the Bennettitales, an extinct group with flowerlike structures that may have been pollinated by insects **(Figure 30.14a)**. However, the Bennettitales and other similar lineages of extinct woody seed plants did not have carpels or flowers and hence are not classified as angiosperms.

Making sense of the origin of angiosperms also depends on working out the order in which angiosperm clades diverged from one another. Here, dramatic progress has been made in recent years. Molecular and morphological evidence suggests that the shrub *Amborella trichopoda*, water lilies, and star anise are living representatives of lineages that diverged from other angiosperms early in the history of the group **(Figure 30.14b)**. *Amborella* is woody, supporting the conclusion mentioned earlier that the angiosperm common ancestor was probably woody. Like the Bennettitales, *Amborella*, water lilies, and star anise lack *vessel elements*, efficient water-conducting cells that are found in most present-day angiosperms. Overall, based on the features of ancestral species and angiosperms like *Amborella*, researchers have hypothesized that early angiosperms were woody shrubs that had small flowers and relatively simple water-conducting cells.

Evolutionary Links with Animals

Plants and animals have interacted for hundreds of millions of years, and those interactions have led to evolutionary change. For example, herbivores can reduce a plant's reproductive success by eating its roots, leaves, or seeds. As a result,

▼ **Figure 30.14 Angiosperm evolutionary history.**

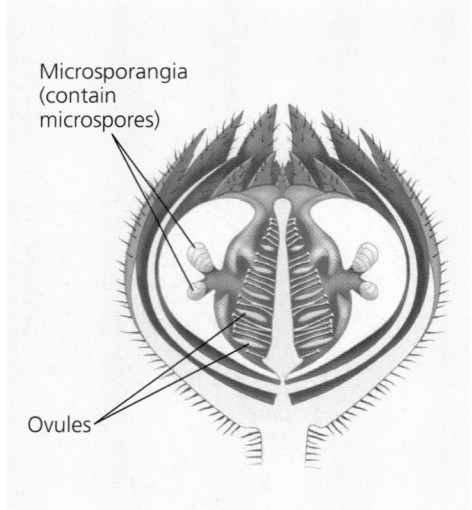

(a) A close relative of the angiosperms? This reconstruction shows a longitudinal section through the flowerlike structures found in the Bennettitales, an extinct group of woody seed plants hypothesized to be more closely related to extant angiosperms than to extant gymnosperms.

Microsporangia (contain microspores)

Ovules

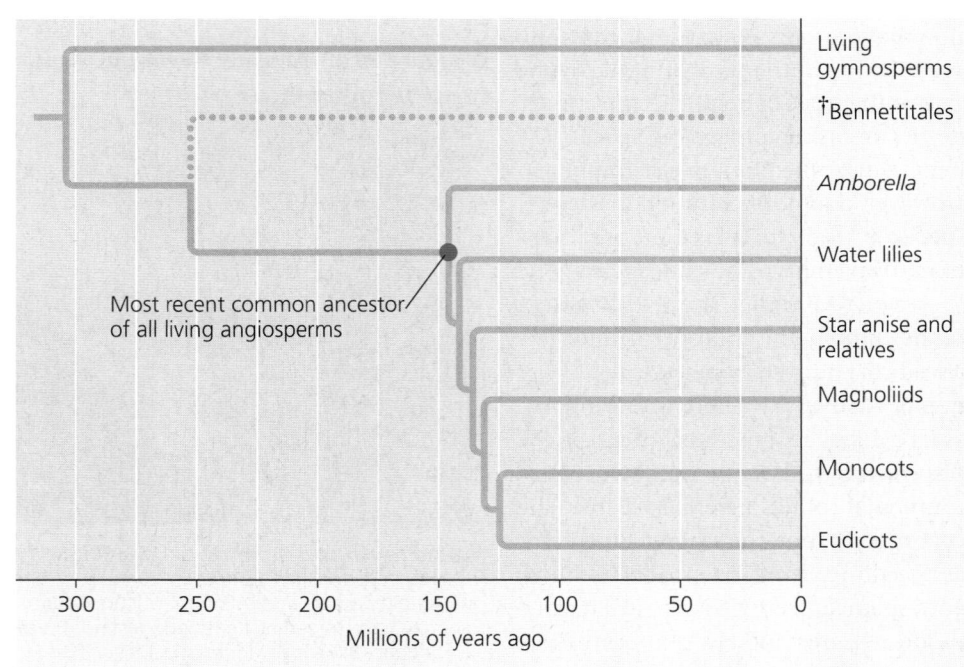

Living gymnosperms

†Bennettitales

Amborella

Water lilies

Star anise and relatives

Magnoliids

Monocots

Eudicots

Most recent common ancestor of all living angiosperms

300 250 200 150 100 50 0
Millions of years ago

(b) Angiosperm phylogeny. This tree represents a current hypothesis of angiosperm evolutionary relationships, based on morphological and molecular evidence. Angiosperms originated about 140 million years ago. The dotted line indicates the uncertain position of the Bennettitales, which may be the sister taxon to the angiosperms.

VISUAL SKILLS *Would the branching order of the phylogeny in (b) necessarily have to be redrawn if a 150-million-year-old fossil monocot were discovered? Explain.*

◀ **Figure 30.15 A bee pollinating a bilaterally symmetrical flower.** To harvest nectar from this Scottish broom flower, a honeybee must land as shown. This releases a tripping mechanism that arches the flower's stamens over the bee and dusts it with pollen. Later, some of this pollen may rub off onto the stigma of the next flower the bee visits.

→ **Mastering Biology Video: Bee Pollinating**

if an effective defense against herbivores originates in a group of plants, those plants may be favored by natural selection—as will herbivores that overcome this new defense. Plant-pollinator and other mutually beneficial interactions also can have such reciprocal evolutionary effects.

Plant-pollinator interactions also may have affected the rates at which new species form. Consider the impact of a flower's symmetry (see Figure 30.9). On a flower with bilateral symmetry, an insect pollinator can obtain nectar (a sugary solution secreted by flower glands) only when landing in a certain position **(Figure 30.15)**. This constraint makes it more likely that pollen is placed on a part of the insect's body that will come into contact with the stigma of a flower of the same species. Such specificity of pollen transfer reduces gene flow between diverging populations and could lead to increased rates of speciation in plants with bilateral symmetry. This hypothesis can be tested using the approach illustrated in this diagram:

A key step in this approach is to identify cases in which a clade with bilaterally symmetric flowers shares an immediate common ancestor with a clade whose members have radially symmetric flowers. One recent study identified 19 pairs of closely related "bilateral" and "radial" clades. On average, the clade with bilaterally symmetric flowers had nearly 2,400 more species than did the related clade with radial symmetry. This result suggests that flower shape can affect the rate at which new species form, perhaps by affecting the behavior of insect pollinators. Overall, plant-pollinator interactions may have contributed to the increasing dominance of flowering plants in the Cretaceous period, making angiosperms centrally important to ecological communities.

Angiosperm Diversity

From their humble beginnings in the Cretaceous period, angiosperms have diversified into more than 290,000 living species. Until the late 1990s, most systematists divided flowering plants into two groups, based partly on the number of cotyledons, or seed leaves, in the embryo. Species with one cotyledon were called **monocots**, and those with two were called **dicots**. Other features, such as flower and leaf structure, were also used to define the two groups. Recent DNA studies, however, indicate that the species traditionally called dicots are paraphyletic. The vast majority of species once categorized as dicots form a large clade, now known as **eudicots** ("true" dicots). **Figure 30.16** compares the main characteristics of monocots and eudicots. The rest of the former dicots are now grouped into four small lineages. Three of these lineages—*Amborella*, water lilies, and star anise and relatives—are informally called **basal angiosperms** because they diverged from other angiosperms early in the history of the group (see Figure 30.14b). A fourth lineage, the **magnoliids**, evolved later. **Figure 30.17** provides an overview of angiosperm diversity.

▼ **Figure 30.16 Characteristics of monocots and eudicots.**

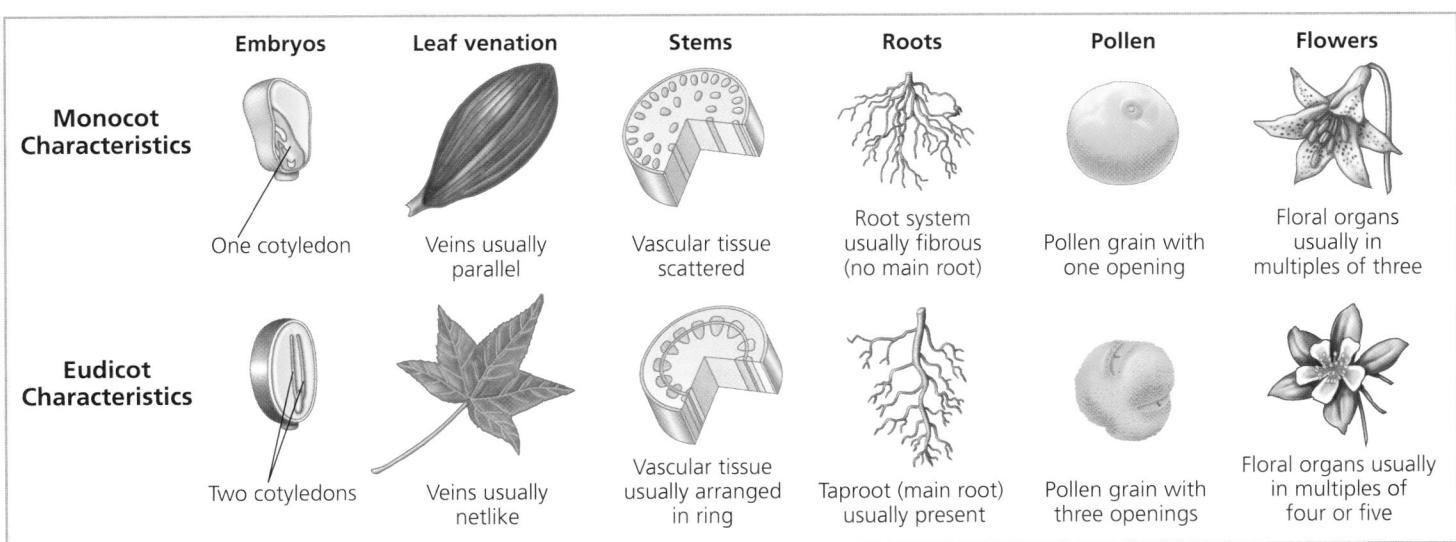

	Embryos	Leaf venation	Stems	Roots	Pollen	Flowers
Monocot Characteristics	One cotyledon	Veins usually parallel	Vascular tissue scattered	Root system usually fibrous (no main root)	Pollen grain with one opening	Floral organs usually in multiples of three
Eudicot Characteristics	Two cotyledons	Veins usually netlike	Vascular tissue usually arranged in ring	Taproot (main root) usually present	Pollen grain with three openings	Floral organs usually in multiples of four or five

Basal Angiosperms

Surviving basal angiosperms consist of three lineages comprising only about 100 species. The first lineage to have diverged from other angiosperms is represented today by a single species, *Amborella trichopoda* (right). The other surviving lineages diverged later: a clade that includes water lilies and a clade consisting of the star anise and its relatives.

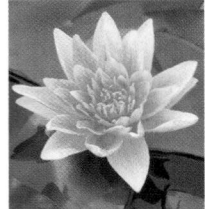

◀ **Water lily (*Nymphaea "Rene Gerard"*).** Species of water lilies are found in aquatic habitats throughout the world. Water lilies belong to a clade that diverged from other angiosperms early in the group's history.

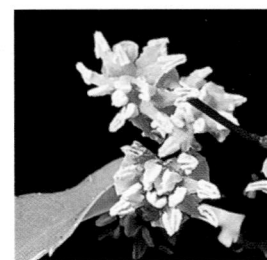

Amborella trichopoda. ▶ This small shrub, found only on the South Pacific island of New Caledonia, may be the sole survivor of a branch at the base of the angiosperm tree.

◀ **Star anise (*Illicium*).** This genus belongs to a third surviving lineage of basal angiosperms.

Magnoliids

Magnoliids consist of about 8,500 species, most notably magnolias, laurels, and black pepper plants. They include both woody and herbaceous species. Although they share some traits with basal angiosperms, such as a typically spiral rather than whorled arrangement of floral organs, magnoliids are more closely related to eudicots and monocots.

◀ **Southern magnolia (*Magnolia grandiflora*).** This member of the magnolia family is a large tree. The variety of southern magnolia shown here, called "Goliath," has flowers that measure up to about a foot across.

Monocots

About one-quarter of angiosperm species are monocots —about 72,000 species. Some of the largest groups are the orchids, grasses, and palms. Grasses include some of the most agriculturally important crops, such as maize, rice, and wheat.

◀ **Orchid (*Paphiopedilum callosum*)**

Barley (*Hordeum vulgare*), a grass ▶

▲ **Pygmy date palm (*Phoenix roebelenii*)**

Eudicots

More than two-thirds of angiosperm species are eudicots—roughly 210,000 species. The largest group is the legume family, which includes such crops as peas and beans. Also important economically is the rose family, which includes many plants with ornamental flowers as well as some species with edible fruits, such as strawberry plants and apple and pear trees. Most of the familiar flowering trees are eudicots, such as oak, walnut, maple, willow, and birch.

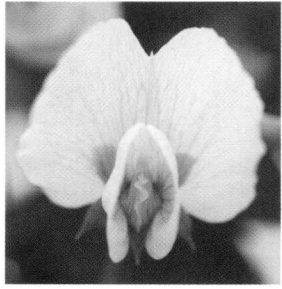

◀ **Snow pea (*Pisum sativum*), a legume**

Dog rose ▶ **(*Rosa canina*), a wild rose**

▲ **Armenian oak (*Quercus pontica*)**

1. It is said that an oak is an acorn's way of making more acorns. Write an explanation that includes these terms: sporophyte, gametophyte, ovule, seed, ovary, and fruit.

2. Compare and contrast a pine cone and a flower in terms of structure and function.

3. **WHAT IF?** Do speciation rates in closely related clades of flowering plants show that flower shape is *correlated with* the rate at which new species form or that flower shape is *responsible for* this rate? Explain.

For suggested answers, see Appendix A.

CONCEPT 30.4

Human welfare depends on seed plants

In forests and on farms, seed plants are key sources of food, fuel, wood products, and medicine. Our reliance on them makes the preservation of plant diversity critical.

Products from Seed Plants

Most of our food comes from angiosperms. Just six crops—maize, rice, wheat, potatoes, cassava, and sweet potatoes—yield 80% of all the calories consumed by humans. We also depend on angiosperms to feed livestock: It takes 5–7 kg of grain to produce 1 kg of grain-fed beef.

Today's crops are the products of artificial selection—the result of plant domestication that began about 12,000 years ago. To appreciate the scale of this transformation, note how the number and size of seeds in domesticated plants are greater than those of their wild relatives, as in the case of maize and the grass teosinte (see Figure 38.16). Scientists can glean information about domestication by comparing the genes of crops with those of wild relatives. With maize, dramatic changes such as increased cob size and loss of the hard coating around teosinte kernels may have been initiated by as few as five mutations.

Flowering plants also provide other edible products. Two popular beverages come from tea leaves and coffee beans, and you can thank the cacao tree for cocoa and chocolate. Spices are derived from various plant parts, such as flowers (cloves, saffron), fruits and seeds (vanilla, black pepper, mustard), leaves (basil, mint, sage), and even bark (cinnamon).

Many seed plants are sources of wood, which is absent in all living seedless plants. Wood consists of tough-walled xylem cells (see Figure 35.22). It is the primary source of fuel for much of the world, and wood pulp, typically derived from conifers such as fir and pine, is used to make paper. Wood remains the most widely used construction material.

For centuries, humans have also depended on seed plants for medicines. Many cultures use herbal remedies, and scientists have extracted and identified medicinally active compounds from many of these plants and later synthesized

Table 30.1 Examples of Plant-Derived Medicines		
Compound	**Source**	**Use**
Atropine	Belladonna plant	Eye pupil dilator
Digitalin	Foxglove	Heart medication
Menthol	Eucalyptus tree	Throat soother
Quinine	Cinchona tree	Malaria preventive
Taxol	Pacific yew	Ovarian cancer drug
Tubocurarine	Curare tree	Muscle relaxant
Vinblastine	Periwinkle	Leukemia drug

them. Willow leaves and bark have long been used in pain-relieving remedies, including prescriptions by the Greek physician Hippocrates. In the 1800s, scientists traced the willow's medicinal property to the chemical salicin. A synthesized derivative, acetylsalicylic acid, is what we call aspirin. Plants are also a direct source of medicinal compounds **(Table 30.1)**. In the United States, about 25% of prescription drugs contain an active ingredient from plants, usually seed plants.

Threats to Plant Diversity

Although plants may be a renewable resource, plant diversity is not. The exploding human population and its demand for space and resources are threatening plant species across the globe. The problem is especially severe in the tropics, where more than two-thirds of the human population live and where population growth is fastest. About 63,000 km^2 (15 million acres) of tropical rain forest are cleared each year **(Figure 30.18)**, a rate that would completely eliminate the remaining 11 million km^2 of tropical forests in 175 years. The loss of forests reduces the absorption of atmospheric carbon dioxide (CO_2) that occurs during photosynthesis, potentially contributing to global warming. Also, as forests disappear, so do large numbers of plant species. Of course, once a species becomes extinct, it can never return.

The loss of plant species is often accompanied by the loss of insects and other rain forest animals. Scientists estimate that if current rates of loss in the tropics and elsewhere

▼ **Figure 30.18 Clear-cutting of tropical forests.** Over the past several hundred years, nearly half of Earth's tropical forests have been cut down and converted to farmland and other uses. A satellite image from 1975 (left) shows a dense forest in Brazil. By 2012, much of this forest had been cut down. Deforested and urban areas are shown as light purple.

continue, 50% or more of Earth's species will become extinct within the next few centuries. Such losses would constitute a global mass extinction, rivaling the Permian and Cretaceous mass extinctions and forever changing the evolutionary history of plants (and many other organisms).

Many people have ethical concerns about contributing to the extinction of species. In addition, there are practical reasons to be concerned about the loss of plant diversity. So far, we have explored the potential uses of only a tiny fraction of the more than 325,000 known plant species. For example, almost all our food is based on the cultivation of only about two dozen species of seed plants. And fewer than 5,000 plant species have been studied as potential sources of medicines.

The tropical rain forest may be a medicine chest of healing plants that could be extinct before we even know they exist. If we begin to view rain forests and other ecosystems as living treasures that can regenerate only slowly, we may learn to harvest their products at sustainable rates.

CONCEPT CHECK 30.4

1. Explain why plant diversity can be considered a nonrenewable resource.
2. **WHAT IF?** How could phylogenies be used to help researchers search more efficiently for novel medicines derived from seed plants?

For suggested answers, see Appendix A.

30 Chapter Review

→ Go to **Mastering Biology** for Assignments, the eText, the Study Area, and Dynamic Study Modules.

SUMMARY OF KEY CONCEPTS

→ To review key terms, go to the **Vocabulary Self-Quiz** in the **Mastering Biology** eText or Study Area, or go to goo.gl/zkjz9t.

CONCEPT 30.1

Seeds and pollen grains are key adaptations for life on land (pp. 637–639)

Five Derived Traits of Seed Plants		
Reduced gametophytes	Microscopic male and female gametophytes (*n*) are nourished and protected by the sporophyte (2*n*)	Male gametophyte / Female gametophyte
Heterospory	Microspore (gives rise to a male gametophyte) / Megaspore (gives rise to a female gametophyte)	
Ovules	Ovule (gymnosperm) { Integument (2*n*) / Megaspore (*n*) / Megasporangium (2*n*) }	
Pollen	Pollen grains make water unnecessary for fertilization	
Seeds	Seeds: survive better than unprotected spores, can be transported long distances	Seed coat / Food supply / Embryo

❓ *Describe how the parts of an ovule (integument, megaspore, megasporangium) correspond to the parts of a seed.*

CONCEPT 30.2

Gymnosperms bear "naked" seeds, typically on cones (pp. 640–644)

- Dominance of the sporophyte generation, the development of seeds from fertilized ovules, and the role of pollen in transferring sperm to ovules are key features of a typical gymnosperm life cycle.
- Gymnosperms appear early in the plant fossil record and dominated many Mesozoic terrestrial ecosystems. Living seed plants can be divided into two monophyletic groups: gymnosperms and angiosperms. Extant gymnosperms include cycads, *Ginkgo biloba*, gnetophytes, and **conifers**.

❓ *Although there are just over 1,000 species of gymnosperms, the group is still very successful in terms of its evolutionary longevity, adaptations, and geographic distribution. Explain.*

CONCEPT 30.3

The reproductive adaptations of angiosperms include flowers and fruits (pp. 644–651)

- **Flowers** generally consist of four types of modified leaves: **sepals, petals, stamens** (which produce pollen), and **carpels** (which produce ovules). **Ovaries** ripen into **fruits**, which often carry seeds by wind, water, or animals to new locations.
- Flowering plants originated about 140 million years ago, and by the mid-Cretaceous (100 mya) had begun to dominate some terrestrial ecosystems. Fossils and phylogenetic analyses offer insights into the origin of flowers.
- Several groups of **basal angiosperms** have been identified. Other major clades of angiosperms include **magnoliids, monocots,** and **eudicots**.
- Pollination and other interactions between angiosperms and animals may have contributed to the success of flowering plants during the last 100 million years.

❓ *Explain why Darwin called the origin of angiosperms an "abominable mystery," and describe what has been learned from fossil evidence and phylogenetic analyses.*

CONCEPT 30.4

Human welfare depends on seed plants *(pp. 651–652)*

- Humans depend on seed plants for products such as food, wood, and many medicines.
- Destruction of habitat threatens the extinction of many plant species and the animal species they support.

❓ *Explain why destroying the remaining tropical forests might harm humans and lead to a mass extinction.*

TEST YOUR UNDERSTANDING

➔ For more multiple-choice questions, go to the **Practice Test** in the **Mastering Biology** eText or Study Area, or go to goo.gl/GruWRg.

Levels 1-2: Remembering/Understanding

1. Where in an angiosperm would you find a megasporangium?
(A) in the style of a flower
(B) enclosed in the stigma of a flower
(C) within an ovule contained within an ovary of a flower
(D) packed into pollen sacs within the anthers found on a stamen

2. Which of the following is a key feature of seed plants facilitating life on land?
(A) homospory
(B) pollen
(C) reduced sporophytes
(D) spores

3. In angiosperms, which of the following is correctly paired with its chromosome count?
(A) microspore—n
(C) egg—$2n$
(B) zygote—n
(D) megaspore—$2n$

4. Which of the following is a characteristic that distinguishes gymnosperms and angiosperms from other plants?
(A) alternation of generations
(B) independent gametophytes
(C) vascular tissue
(D) ovules

5. Which of the following are present in angiosperms but not in gymnosperms?
(A) seeds
(C) ovaries
(B) pollen
(D) ovules

Levels 3-4: Applying/Analyzing

6. DRAW IT Use the letters a–d to label where on the phylogenetic tree each of the following derived characters appears.
a. flowers
c. seeds
b. embryos
d. vascular tissue

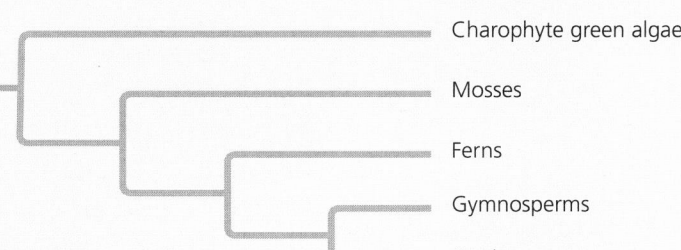

Charophyte green algae
Mosses
Ferns
Gymnosperms
Angiosperms

7. EVOLUTION CONNECTION The history of life has been punctuated by several mass extinctions. For example, the impact of a meteorite may have wiped out most of the dinosaurs and many forms of marine life at the end of the Cretaceous period (see Concept 25.4). Fossils indicate that plants were less severely affected by this mass extinction. What adaptations may have enabled plants to withstand this disaster better than animals?

Levels 5-6: Evaluating/Creating

8. SCIENTIFIC INQUIRY • DRAW IT As will be described in detail in Concept 38.1, the female gametophyte of angiosperms typically has seven cells, one of which, the central cell, contains two haploid nuclei. After double fertilization, the central cell develops into endosperm, which is triploid. Because magnoliids, monocots, and eudicots typically have female gametophytes with seven cells and triploid endosperm, scientists assumed that this was the ancestral state for angiosperms. Consider, however, the following recent discoveries:
- Our understanding of angiosperm phylogeny has changed to that shown in Figure 30.14b.
- *Amborella trichopoda* has eight-celled female gametophytes and triploid endosperm.
- Water lilies and star anise have four-celled female gametophytes and diploid endosperm.
(a) Draw a phylogeny of the angiosperms (see Figure 30.14b), incorporating the data given above about the number of cells in female gametophytes and the ploidy of the endosperm. Assume that all of the star anise relatives have four-celled female gametophytes and diploid endosperm.
(b) What does your labeled phylogeny suggest about the evolution of the female gametophyte and endosperm in angiosperms?

9. WRITE ABOUT A THEME: ORGANIZATION Cells are the basic units of structure and function in all organisms. A key feature in the life cycle of plants is the alternation of multicellular haploid and diploid generations. Imagine a lineage of flowering plants in which mitotic cell division did not occur between the events of meiosis and fertilization (see Figure 30.12). In a short essay (100–150 words), describe how this change in the timing of cell division would affect the structure and life cycle of plants in this lineage.

10. SYNTHESIZE YOUR KNOWLEDGE

This photograph shows a milkweed seed in flight. Describe how seeds and other adaptations in seed plants contributed to the rise of seed plants and their dominant role in plant communities today.

For selected answers, see Appendix A.

31 Fungi

KEY CONCEPTS

31.1 Fungi are heterotrophs that feed by absorption *p. 655*

31.2 Fungi produce spores through sexual or asexual life cycles *p. 657*

31.3 The ancestor of fungi was an aquatic, single-celled, flagellated protist *p. 659*

31.4 Fungi have radiated into a diverse set of lineages *p. 660*

31.5 Fungi play key roles in nutrient cycling, ecological interactions, and human welfare *p. 667*

Study Tip

Draw a diagram: To help you recognize key differences between the life cycles of fungi and humans, draw and label simple diagrams like the partial example shown here depicting the life stages at which meiosis, mitosis, gamete formation, a multicellular organism, and fertilization occur in humans and in fungi.

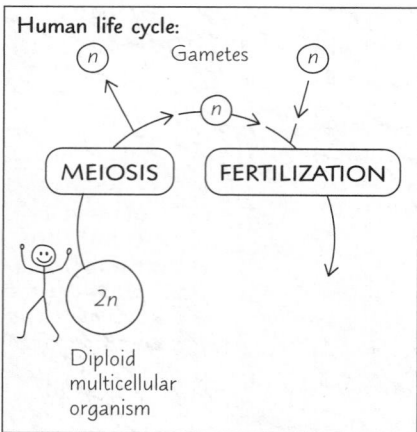

→ Go to Mastering Biology

For Students (in eText and Study Area)
- Get Ready for Chapter 31
- Animation: Life Cycle of a Mushroom

For Instructors to Assign (in Item Library)
- Activity: Fungal Reproduction and Nutrition
- Scientific Skills Exercise: Synthesizing Information from Multiple Data Sets

Figure 31.1 These little mushrooms are just the small aboveground extension of a vast network of filaments located below the forest floor. Such underground fungal networks, called mycelia, in some cases link mushrooms that are hundreds of meters apart. In fact, the largest known mycelium spreads beneath 965 hectares of forest—more than the area of 1,800 football fields.

How do structure and function in fungi relate to their role in ecosystems?

As they grow, multicellular fungi extend filaments called **hyphae** into their surroundings.

Spores enable fungi to colonize new environments. The spores germinate and grow when conditions are favorable.

Hyphae **secrete enzymes** that break down organic matter, releasing nutrients.

Hyphae **absorb** the released nutrients.

Almost any organic molecule can be digested by at least some fungi, making them highly effective **decomposers in ecosystems.**

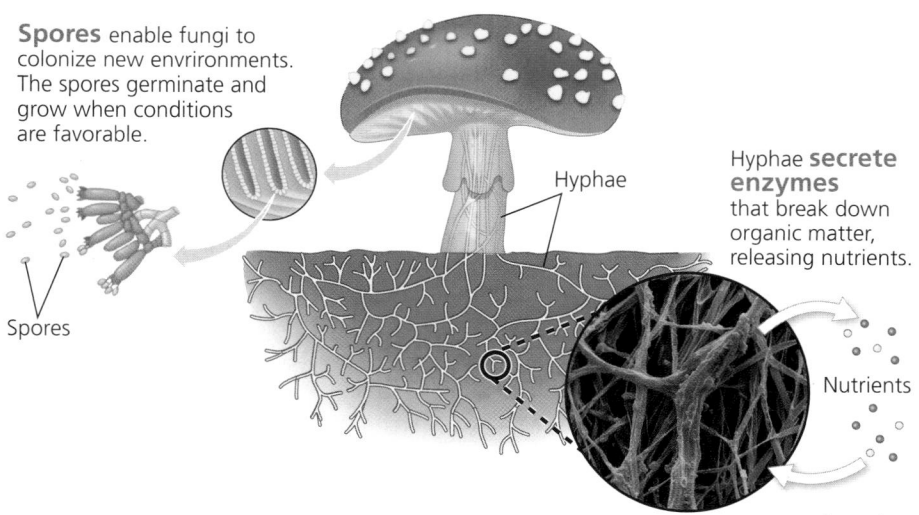

The hidden network of fungal filaments in Figure 31.1 is a fitting symbol of the neglected grandeur of the kingdom Fungi. Most of us are barely aware of these eukaryotes beyond the mushrooms we eat or the occasional brush with athlete's foot. Yet fungi are a huge and important component of the biosphere. Some fungi are exclusively single-celled, though most have complex multicellular bodies. These diverse organisms are found in just about every imaginable terrestrial and aquatic habitat.

Fungi are not only diverse and widespread but also essential for the well-being of most ecosystems. They break down organic material and recycle nutrients, allowing other organisms to assimilate essential chemical elements. In this chapter, we will investigate the structure and evolutionary history of fungi, survey the major groups of fungi, and discuss their ecological and commercial significance.

CONCEPT 31.1

Fungi are heterotrophs that feed by absorption

Despite their vast diversity, all fungi share some key traits—most importantly, the way they derive nutrition. Another key characteristic of many fungi is that they grow by forming multicellular filaments, a body structure that plays an important role in how they obtain food.

Nutrition and Ecology

Like animals, fungi are heterotrophs: They cannot make their own food as plants and algae can. But unlike animals, fungi do not ingest (eat) their food. Instead, a fungus absorbs nutrients from the environment outside of its body. Many fungi do this by secreting hydrolytic enzymes into their surroundings. These enzymes break down complex molecules to smaller organic compounds that the fungi can absorb into their cells and use. Other fungi use enzymes to penetrate the walls of cells, enabling the fungi to absorb nutrients from the cells. Collectively, the different enzymes found in various fungal species can digest compounds from a wide range of sources, living or dead.

This diversity of food sources corresponds to the varied roles of fungi in ecological communities: Different species live as decomposers, parasites, or mutualists. Fungi that are decomposers break down and absorb nutrients from nonliving organic material, such

as fallen logs, animal corpses, and the wastes of organisms. Parasitic fungi absorb nutrients from the cells of living hosts. Some parasitic fungi are pathogenic, including many species that cause diseases in plants and others that cause diseases in animals. Mutualistic fungi also absorb nutrients from a host, but they reciprocate with actions that benefit the host. For example, mutualistic fungi that live within the digestive tracts of certain termite species use their enzymes to break down wood, as do mutualistic protists in other termites (see Figure 28.29).

The versatile enzymes that enable fungi to digest a wide range of food sources are not the only reason for their ecological success. Another important factor is how their body structure increases the efficiency of nutrient absorption.

Body Structure

The most common fungal body structures are multicellular filaments and single cells (**yeasts**). Many fungal species can grow as both filaments and yeasts, but even more grow only as filaments; relatively few species grow only as single-celled yeasts. Yeasts often inhabit moist environments, including plant sap and animal tissues, where there is a ready supply of soluble nutrients, such as sugars and amino acids.

The morphology of multicellular fungi enhances their ability to grow into and absorb nutrients from their surroundings **(Figure 31.2)**. The bodies of these fungi typically form a

▼ **Figure 31.2 Structure of a multicellular fungus.** The top photograph shows the sexual structures, in this case called mushrooms, of the penny bun fungus (*Boletus edulis*). The bottom photograph shows a mycelium growing on fallen conifer needles. The inset SEM shows hyphae.

Reproductive structure. Tiny haploid cells called spores are produced inside the mushroom.

Hyphae. The mushroom and its subterranean mycelium are a continuous network of hyphae.

Spore-producing structures

Mycelium

60 μm

❓ *Although the mushrooms in the top photograph appear to be different individuals, could their DNA be identical? Explain.*

→ **Mastering Biology Animation: Fungal Growth and Nutrition**

▼ Figure 31.3 Two forms of hyphae.

(a) Septate hypha   **(b) Coenocytic hypha**

network of tiny filaments called **hyphae** (singular, *hypha*). Hyphae consist of tubular cell walls surrounding the plasma membrane and cytoplasm of the cells. The cell walls are strengthened by **chitin**, a strong but flexible polysaccharide. Chitin-rich walls can enhance feeding by absorption. As a fungus absorbs nutrients from its environment, the concentrations of those nutrients in its cells increases, causing water to move into the cells by osmosis. The movement of water into fungal cells creates pressure that could cause their cells to burst if they were not surrounded by a chitin-strengthened, rigid cell wall.

Another important structural feature of most fungi is that their hyphae are divided into cells by cross-walls, or **septa** (singular, *septum*) **(Figure 31.3a)**. Septa generally have pores large enough to allow ribosomes, mitochondria, and even nuclei to flow from cell to cell. Some fungi lack septa **(Figure 31.3b)**. Known as **coenocytic fungi**, these organisms consist of a continuous cytoplasmic mass having hundreds or thousands of nuclei. The coenocytic condition results from the repeated division of nuclei without cytokinesis.

Fungal hyphae form an interwoven mass called a **mycelium** (plural, *mycelia*) that infiltrates the material on which the fungus feeds (see Figure 31.2). The structure of a mycelium maximizes its surface-to-volume ratio, making feeding very efficient. Just 1 cm³ of rich soil may contain as much as 1 km of hyphae with a total surface area of 300 cm² in contact with the soil. A fungal mycelium grows rapidly, as proteins and other materials synthesized by the fungus move through cytoplasmic streaming to the tips of the extending hyphae. The fungus concentrates its energy and resources on adding hyphal length and thus overall absorptive surface area, rather than on increasing hyphal girth. Multicellular fungi are not motile in the typical sense—they cannot run, swim, or fly in search of food or mates. However, as they grow, such fungi can move into new territory, swiftly extending the tips of their hyphae.

Specialized Hyphae in Mycorrhizal Fungi

Some fungi have specialized hyphae that allow them to feed on living animals **(Figure 31.4a)**, while others have modified hyphae called *haustoria* that enable them to extract nutrients from plants. Our focus here, however, will be on fungi that have specialized branching hyphae such as **arbuscules (Figure 31.4b)** through which fungi exchange nutrients with their plant hosts. Such mutually beneficial relationships between fungi and plant roots are called **mycorrhizae** (the term means "fungus roots").

▼ Figure 31.4 Specialized hyphae.

(a) Hyphae adapted for trapping and killing prey. In *Arthrobotrys*, a soil fungus, portions of the hyphae are modified as hoops that can constrict around a nematode (roundworm) in less than a second. The growing hyphae then penetrate the worm's body, and the fungus digests its prey's inner tissues (SEM).

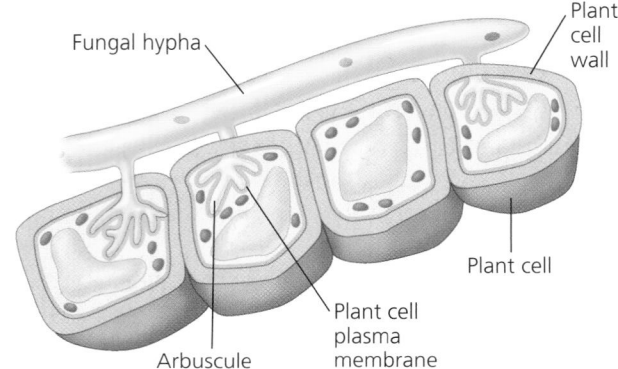

(b) Arbuscules. Some mutualistic fungi have specialized hyphae called arbuscules that can exchange nutrients with living plant cells. Arbuscules remain separated from a plant cell's cytoplasm by the plasma membrane of the plant cell (orange).

Mycorrhizal fungi (fungi that form mycorrhizae) can improve delivery of phosphate ions and other minerals to plants because the vast mycelial networks of the fungi are more efficient than the plants' roots at acquiring these minerals from the soil. In exchange, the plants supply the fungi with organic nutrients such as carbohydrates.

There are two main types of mycorrhizal fungi (see Figure 37.14). **Ectomycorrhizal fungi** (from the Greek *ektos*, out) form sheaths of hyphae over the surface of a root and typically grow into the extracellular spaces of the root cortex. **Arbuscular mycorrhizal fungi** extend arbuscules through the root cell wall and into tubes formed by invagination (pushing inward, as in Figure 31.4b) of the root cell plasma membrane. In the **Scientific Skills Exercise**, you'll compare genomic data from fungi that form mycorrhizae and fungi that do not.

Mycorrhizae are enormously important both in natural ecosystems and in agriculture. Almost all vascular plants have mycorrhizae and rely on their fungal partners for essential nutrients. Foresters commonly inoculate pine seedlings

Interpreting Genomic Data and Generating Hypotheses

What Can Genomic Analysis of a Mycorrhizal Fungus Reveal About Mycorrhizal Interactions? The first genome of a mycorrhizal fungus to be sequenced was that of the basidio-mycete *Laccaria bicolor* (see photo). In nature, *L. bicolor* is a common ectomy-corrhizal fungus of trees such as poplar and fir, as well as a free-living soil organism. In forest nurseries, it is often added to soil to enhance seedling growth. The fungus can easily be grown alone in culture and can establish mycorrhizae with tree roots in the laboratory. Researchers hope that studying the genome of *Laccaria* will yield clues to the processes by which it interacts with its mycorrhizal partners— and by extension, to mycorrhizal interactions involving other fungi.

How the Study Was Done Using the whole-genome shot-gun method (see Figure 21.2) and bioinformatics, researchers sequenced the genome of *L. bicolor* and compared it with the genomes of some nonmycorrhizal basidiomycete fungi. The team used microarrays to compare gene expression levels for different protein-coding genes and for the same genes in a mycorrhizal mycelium and a free-living mycelium. They could thus identify the genes for fungal proteins that are made specifically in mycorrhizae.

Data from the Study

Table 1 Numbers of Genes in *L. bicolor* and Four Nonmycorrhizal Fungal Species

	L. bicolor	1	2	3	4
Protein-coding genes	20,614	13,544	10,048	7,302	6,522
Genes for membrane transporters	505	412	471	457	386
Genes for small secreted proteins (SSPs)	2,191	838	163	313	58

Table 2 *L. bicolor* Genes Most Highly Upregulated in Ectomycorrhizal Mycelium (ECM) of Douglas Fir or Poplar Versus Free-Living Mycelium (FLM)

Protein ID	Protein Feature or Function	Douglas Fir ECM/FLM Ratio	Poplar ECM/FLM Ratio
298599	SSP	22,877	12,913
293826	Enzyme inhibitor	14,750	17,069
333839	SSP	7,844	1,931
316764	Enzyme	2,760	1,478

Data from F. Martin et al., The genome of *Laccaria bicolor* provides insights into mycorrhizal symbiosis, *Nature* 452:88–93 (2008).

INTERPRET THE DATA

1. **(a)** In Table 1, which fungal species has the most genes encoding membrane transporters (membrane transport proteins; see Concept 7.2)? **(b)** Why might these genes be of particular importance to *L. bicolor*?

2. The phrase "small secreted proteins" (SSPs) refers to proteins less than 100 amino acids in length that the fungi secrete; their function is not yet known. **(a)** Describe the Table 1 data on SSPs. **(b)** The researchers found that the SSP genes shared a common feature that indicated the encoded proteins were destined for secretion. Based on Figure 17.22 and the text discussion of that figure, predict what this common feature of the SSP genes was. **(c)** Suggest a hypothesis for the roles of SSPs in mycorrhizae.

3. Table 2 shows data from gene expression studies for the four *L. bicolor* genes whose transcription was most increased ("upregulated") in mycorrhizae. **(a)** For the gene encoding the first protein listed, what does the number 22,877 indicate? **(b)** Do the data in Table 2 support your hypothesis in question 2(c)? Explain. **(c)** Compare the data for poplar mycorrhizae with those for Douglas fir and hypothesize what might account for any differences.

→ **Instructors:** A version of this Scientific Skills Exercise can be assigned in **Mastering Biology**.

---

with mycorrhizal fungi to promote growth. In the absence of human intervention, mycorrhizal fungi colonize soils by dispersing haploid cells called **spores** that form new mycelia after germinating. Spore dispersal is a key component of how fungi reproduce and spread to new areas, as we discuss next.

1. Compare and contrast the nutritional mode of a fungus with your own nutritional mode.

2. **WHAT IF?** Suppose a certain fungus is a mutualist that lives within an insect host, yet its ancestors were parasites that grew in and on the insect's body. What derived traits might you find in this mutualistic fungus?

3. **MAKE CONNECTIONS** Review Figure 10.3 and Figure 10.5. If a plant has mycorrhizae, where might carbon that enters the plant's stomata as CO_2 eventually be deposited: in the plant, in the fungus, or both? Explain.

For suggested answers, see Appendix A.

CONCEPT 31.2

Fungi produce spores through sexual or asexual life cycles

Most fungi propagate themselves by producing vast numbers of spores, either sexually or asexually. For example, puffballs, the reproductive structures of certain fungal species, may release trillions of spores (see Figure 31.17). Spores can be carried long distances by wind or water. If they land in a moist place where there is food, they germinate, producing a new mycelium. To appreciate how effective spores are at dispersing, leave a slice of melon exposed to the air. Even without a visible source of spores nearby, within a week, you will likely observe fuzzy mycelia growing from microscopic spores that have fallen onto the melon.

Figure 31.5 Generalized life cycle of fungi. Many fungi reproduce both sexually and asexually, as shown here; others, however, reproduce only sexually or asexually.

Key

Haploid (*n*)

Heterokaryotic (unfused nuclei from different parents)

Diploid (2*n*)

PLASMOGAMY (fusion of cytoplasm)

Heterokaryotic stage

Spore-producing structures

Spores (*n*)

ASEXUAL REPRODUCTION

Mycelium

KARYOGAMY (fusion of nuclei)

SEXUAL REPRODUCTION

Zygote (2*n*)

GERMINATION

GERMINATION

MEIOSIS

Spores (*n*)

? *Compare the genetic variation found in spores produced in the sexual and asexual portions of the life cycle. Why do these differences occur?*

Hours, days, or (in some fungi) even centuries may pass between plasmogamy and the next stage in the sexual cycle, **karyogamy**. During karyogamy, the haploid nuclei contributed by the two parents fuse, producing diploid cells. Zygotes and other transient structures form during karyogamy, the only diploid stage in most fungi. Meiosis then restores the haploid condition, ultimately leading to the formation of genetically diverse spores. Meiosis is a key step in sexual reproduction, so spores produced in this way are sometimes referred to as "sexual spores."

The sexual processes of karyogamy and meiosis generate extensive genetic variation, a prerequisite for natural selection. (See Concepts 13.2 and 23.1 to review how sex can increase genetic diversity.) The heterokaryotic condition also offers some of the advantages of diploidy in that one haploid genome may compensate for harmful mutations in the other.

Figure 31.5 generalizes the many different life cycles that can produce fungal spores. In this section, we will survey the main aspects of sexual and asexual reproduction in fungi.

Sexual Reproduction

The nuclei of fungal hyphae and the spores of most fungi are haploid, although many species have transient diploid stages that form during sexual life cycles. Sexual reproduction often begins when hyphae from two mycelia release signaling molecules called **pheromones**. If the mycelia are of different mating types, the pheromones from each partner bind to receptors on the other, and the hyphae extend toward the source of the pheromones. When the hyphae meet, they fuse. In species with such a "compatibility test," this process contributes to genetic variation by preventing hyphae from fusing with other hyphae from the same mycelium or another genetically identical mycelium.

The union of the cytoplasms of two parent mycelia is known as **plasmogamy** (see Figure 31.5). In most fungi, the haploid nuclei contributed by each parent do not fuse right away. Instead, parts of the fused mycelium contain coexisting, genetically different nuclei. Such a mycelium is said to be a **heterokaryon** (meaning "different nuclei"). In some species, the haploid nuclei pair off two to a cell, one from each parent. Such a mycelium is **dikaryotic** (meaning "two nuclei"). As a dikaryotic mycelium grows, the two nuclei in each cell divide in tandem without fusing. Because these cells retain two separate haploid nuclei, they differ from diploid cells, which have pairs of homologous chromosomes within a single nucleus.

Asexual Reproduction

Many fungi reproduce both sexually and asexually, as shown in Figure 31.5; others, however, reproduce only sexually or only asexually. As with sexual reproduction, the processes of asexual reproduction vary widely among fungi.

Many fungi reproduce asexually by growing as filamentous fungi that produce (haploid) spores by mitosis; such species are informally referred to as **molds** if they form visible mycelia. Depending on your housekeeping habits, you may have observed molds in your kitchen, forming furry carpets on bread or fruit **(Figure 31.6)**. Molds typically grow rapidly and produce many spores asexually, enabling the fungi to colonize new sources of food. Many species that

▼ Figure 31.6 *Penicillium*, a mold commonly encountered as a decomposer of food. The bead-like clusters in the colorized SEM are conidia, structures involved in asexual reproduction.

1.5 μm

produce such spores can also reproduce sexually if they happen to contact a member of their species of a different mating type.

Other fungi reproduce asexually by growing as single-celled yeasts. Instead of producing spores, asexual reproduction in yeasts occurs by ordinary cell division or by the pinching of small "bud cells" off a parent cell **(Figure 31.7)**. As already mentioned, some fungi that grow as yeasts can also grow as filamentous mycelia.

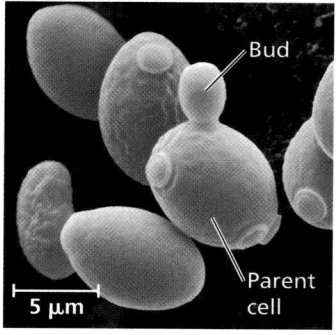

▼ Figure 31.7 **The yeast *Saccharomyces cerevisiae* in stages of budding.** (SEM)

Bud

Parent cell

5 μm

Many yeasts and filamentous fungi have no known sexual stage in their life cycle. Since early mycologists (biologists who study fungi) classified fungi based mainly on their type of sexual structure, this posed a problem. Mycologists have traditionally lumped all fungi lacking sexual reproduction into a group called **deuteromycetes** (from the Greek *deutero*, second, and *mycete*, fungus). Whenever a sexual stage is discovered for a so-called deuteromycete, the species is reclassified in a particular phylum, depending on the type of sexual structures it forms. In addition to searching for sexual stages of such unassigned fungi, mycologists can now use genomic techniques to classify them.

CONCEPT CHECK **31.2**

1. **MAKE CONNECTIONS** Compare Figure 31.5 with Figure 13.6. In terms of haploidy versus diploidy, how do the life cycles of fungi and humans differ?

2. **WHAT IF?** Suppose that you sample the DNA of two mushrooms on opposite sides of your yard and find that they are identical. Propose two hypotheses that could reasonably account for this result.

For suggested answers, see Appendix A.

CONCEPT **31.3**

The ancestor of fungi was an aquatic, single-celled, flagellated protist

Data from molecular systematics offer insights into the early evolution of fungi. As a result, systematists now recognize that fungi and animals are more closely related to each other than either group is to plants or to most other eukaryotes.

The Origin of Fungi

Phylogenetic analyses suggest that fungi evolved from a flagellated ancestor. While the majority of fungi lack flagella, two basal lineages of fungi (the cryptomycetes and the chytrids, as we'll discuss shortly) do have flagella. Moreover, most of

▼ Figure 31.8 **Fungi and their close relatives.** Molecular evidence indicates that the nucleariids, a group of single-celled protists, are the closest living relatives of fungi.

UNICELLULAR, FLAGELLATED ANCESTOR

Animals (and their close protistan relatives)

Nucleariids

Fungi

Opisthokonts

the protists that share a close common ancestor with animals and fungi also have flagella. DNA sequence data indicate that these three groups of eukaryotes—the fungi, the animals, and their protistan relatives—form a monophyletic group, or clade **(Figure 31.8)**. As discussed in Concept 28.5, members of this clade are called **opisthokonts**, a name that refers to the posterior (*opistho-*) location of the flagellum in these organisms.

Within the opisthokont clade, fungi are more closely related to several groups of single-celled protists than they are to animals, suggesting that the ancestor of fungi was unicellular. One such group of unicellular protists, the **nucleariids**, consists of amoebas that feed on algae and bacteria. DNA evidence further indicates that animals are more closely related to a *different* group of protists (the choanoflagellates) than they are to either fungi or nucleariids. Together, these results suggest that multicellularity evolved in animals and fungi independently, from different single-celled ancestors.

Using molecular clock analyses, scientists have estimated that the ancestors of animals and fungi diverged into separate lineages more than a billion years ago. Fossils of certain unicellular, marine eukaryotes that lived as early as 1.5 billion years ago have been interpreted as fungi, but those claims remain controversial. Furthermore, although fungi probably originated in aquatic environments, the oldest fossils that are widely accepted as fungi are of terrestrial species that lived 440 million years ago **(Figure 31.9)**. Fungi may have colonized land as early as 505 million years ago: Soils of that age

Septa

Central filament

▶ Figure 31.9 **Fossil hyphae from the fungus *Tortotubus* (440 million years ago).** The central filament is surrounded by two partially overlapping filaments (LM).

have a chemical "signature" similar to that found in soils where fungi are active today. Overall, more fossils are needed to help clarify when fungi originated and what features were present in their earliest lineages.

The Move to Land

Plants colonized land about 470 million years ago (see Concept 29.1), and fungi may well have colonized land before plants. Indeed, some researchers have described life on land before the arrival of plants as a "green slime" that consisted of cyanobacteria, algae, and a variety of small, heterotrophic species, including fungi. With their capacity for extracellular digestion, fungi would have been well suited for feeding on other early terrestrial organisms (or their remains).

Once on land, some fungi formed symbiotic associations with early plants. For example, 405-million-year-old fossils of the early plant *Aglaophyton* contain evidence of mycorrhizal relationships between plants and fungi (see Figure 25.13). This evidence includes fossils of hyphae that have penetrated within plant cells and formed structures that resemble the arbuscules formed today by arbuscular mycorrhizae. Similar structures have been found in a variety of other early plants, suggesting that plants probably existed in beneficial relationships with fungi from the earliest periods of colonization of land. The earliest plants lacked roots, limiting their ability to extract nutrients from the soil. As occurs in mycorrhizal associations today, it is likely that soil nutrients were transferred to early plants via the extensive mycelia formed by their symbiotic fungal partners.

Support for the antiquity of mycorrhizal associations has also come from molecular studies. For a mycorrhizal fungus and its plant partner to establish a symbiotic relationship, certain genes must be expressed by the fungus and other genes must be expressed by the plant. Researchers focused on three plant genes (called *sym* genes) whose expression is required for the formation of mycorrhizae in flowering plants. They found that these genes were present in all major plant lineages, including basal lineages such as liverworts (see Figure 29.13). Furthermore, after they transferred a liverwort *sym* gene to a flowering plant mutant that could not form mycorrhizae, the mutant recovered its ability to form mycorrhizae. These results suggest that mycorrhizal *sym* genes were present in early plants—and that the function of these genes has been conserved for hundreds of millions of years as plants continued to adapt to life on land.

CONCEPT CHECK 31.3

1. Why are fungi classified as opisthokonts despite the fact that most fungi lack flagella?

2. Describe the importance of mycorrhizae, both today and in the colonization of land. What evidence supports the antiquity of mycorrhizal associations?

3. **WHAT IF?** If fungi colonized land before plants, where might the fungi have lived? How would their food sources have differed from what they feed on today?

For suggested answers, see Appendix A.

CONCEPT 31.4

Fungi have radiated into a diverse set of lineages

In the past decade, molecular analyses have reshaped our understanding of the evolutionary relationships between fungal groups. In addition, metagenomic studies have led to the discovery of entirely new groups of fungi. As a result, the phylogeny of fungi is undergoing dramatic change. For example, one traditional group, the Zygomycota, has been abandoned because it was paraphyletic, and its members have been reassigned to other groups. Recent studies also indicate that the *microsporidians*, an enigmatic group of unicellular parasites, should be classified as fungi and may belong to a basal fungal lineage (one that diverged from other fungi early in the history of the group).

Figure 31.10 presents a current hypothesis of the relationships among fungal groups. In this section, we'll survey the groups identified in this phylogenetic tree. However, the groups shown in Figure 31.10 may represent only a small fraction of the diversity of extant fungal groups. (Extant lineages are those that have surviving members.) While there are about 145,000 known species of fungi, in recent years more than 2,000 new species have been discovered annually. By some estimates the actual number of fungal species lies between 2.2 and 3.8 million—more than all of the 1.9 million species of organisms (of every type) that biologists have currently identified and named.

▼ **Figure 31.10 Phylogeny of fungi.** This phylogenetic hypothesis shows major clades of extant fungi; dotted lines indicate evolutionary relationships that are uncertain.

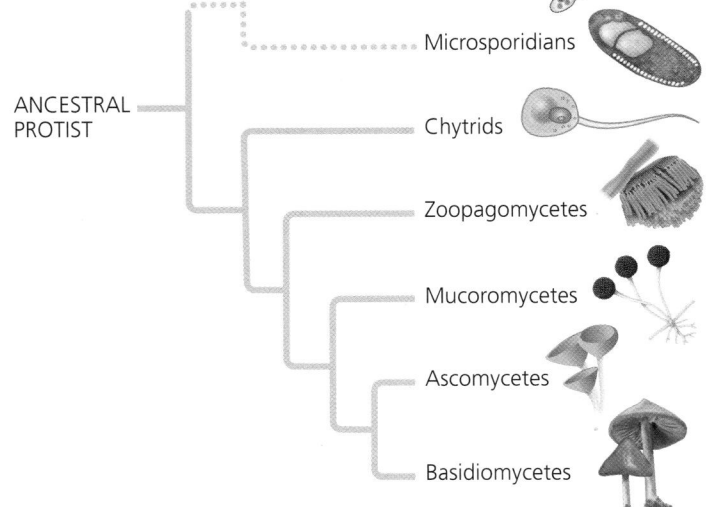

Cryptomycetes

Microsporidians

ANCESTRAL PROTIST

Chytrids

Zoopagomycetes

Mucoromycetes

Ascomycetes

Basidiomycetes

Cryptomycetes and Microsporidians

Cryptomycetes
Microsporidians
Chytrids
Zoopagomycetes
Mucoromycetes
Ascomycetes
Basidiomycetes

Genomic studies indicate that **cryptomycetes** (fungi in the phylum Cryptomycota) and **microsporidians** (fungi in the phylum Microsporidia) form a sister group and are a basal fungal lineage (see Figure 31.10). While most molecular comparisons support the placement of cryptomycetes and microsporidians at the base of the fungal tree, more data are needed to help resolve this phylogeny.

Cryptomycetes

Although only 30 species have been identified to date, genetic data suggest that the cryptomycetes are a large and diverse group. DNA sequences from members of this group have been found in marine and freshwater communities, as well as soils. Cryptomycetes also have been found in aerobic and anaerobic environments, and in geographical locations across the globe. Like the species shown in **Figure 31.11**, *Rozella allomycis*, many of the cryptomycetes identified to date are parasites of protists and other fungi.

▼ **Figure 31.11** The cryptomycete *Rozella allomycis* parasitizing another fungus.

Hypha of host fungus

Rozella spores

Cryptomycetes are unicellular and have flagellated spores. Cryptomycetes also can synthesize a chitin-rich cell wall, a key structural feature of the fungi (see Concept 31.1).

Microsporidians

The 1,300 species of microsporidians are unicellular parasites of protists and animals, including humans **(Figure 31.12)**. Infections in humans can cause reduced longevity and weight loss. The microsporidian *Nosema ceranae* is a parasite of honeybees and may contribute to *Colony Collapse Disorder*, a devastating outbreak that has led to the loss of honeybee colonies throughout the world.

Like all fungi, microsporidians can synthesize a chitin-rich cell wall. Other aspects of their biology are unusual. For example, microsporidians have highly reduced mitochondria and

▼ **Figure 31.12** A spore of the microsporidian *Fibrillanosema crangonycis*. (TEM)

1 μm

small genomes, with only 2,000 genes in some species. The genome of one microsporidian, *Encephalitozoon intestinalis*, has just 2.3 Mb of DNA—the smallest genome of any eukaryote sequenced to date. Unlike other basal fungi, microsporidians lack flagellated spores; instead, they produce unique spores that infect host cells via a harpoon-like organelle.

Chytrids

Cryptomycetes
Microsporidians
Chytrids
Zoopagomycetes
Mucoromycetes
Ascomycetes
Basidiomycetes

The fungi classified in phylum Chytridiomycota, called **chytrids**, are ubiquitous in lakes and soil; recent metagenomic studies have uncovered new clades of chytrids in hydrothermal vent and other marine communities. Some of the approximately 1,000 chytrid species are decomposers, while others are parasites of protists, other fungi, plants, or animals; as we'll see later in the chapter, two chytrid parasites have contributed to the global decline of amphibian populations. Still other chytrids are important mutualists. For example, anaerobic chytrids that live in the digestive tracts of sheep and cattle help to break down plant matter, thereby contributing significantly to the animal's growth.

Nearly all chytrids have flagellated spores, called **zoospores** **(Figure 31.13)**. Like other fungi, chytrids have cell walls made of chitin, and they also share certain key enzymes and metabolic pathways with other fungal groups. Some chytrids form colonies with hyphae, while others exist as single spherical cells.

⊝ **Mastering Biology** Video: *Phlyctochytrium* Zoospore Release

▶ **Figure 31.13 Flagellated chytrid zoospore.** (TEM)

Flagellum

4 μm

Zoopagomycetes

Cryptomycetes
Microsporidians
Chytrids
Zoopagomycetes
Mucoromycetes
Ascomycetes
Basidiomycetes

Most of the 900 species of **zoopagomycetes**, fungi in the phylum Zoopagomycota, live as parasites or as commensal (neutral) symbionts of animals; some are parasites of other fungi or protists. Zoopagomycetes form filamentous hyphae and reproduce asexually by producing nonflagellated spores. Some zoopagomycetes induce insects that they parasitize to perch near the top of plants; the insects subsequently die and fungal spores are released to infect new victims **(Figure 31.14)**. Sexual reproduction, where known, involves the formation of a durable structure called a zygosporangium, which houses and protects the zygote.

The loss of flagellated spores in the zoopagomycetes and other fungal lineages may have been associated with a

◀ **Figure 31.14
A fly covered with fungal hyphae.**
This fly was killed by the zoopagomycete *Entomophthora muscae*, also known as the "fly death fungus," for obvious reasons.

transition to life on land. Basal fungal lineages had flagellated spores, enabling dispersal through water. In contrast, zoopagomycetes and all of their closest fungal relatives (the clade consisting of the mucoromycetes, ascomycetes, and

▼ **Figure 31.15 The life cycle of the mucoromycete *Rhizopus stolonifer* (black bread mold).**

1 Mycelia have various mating types (here designated (–), with red nuclei, and (+), with blue nuclei).

2 Neighboring mycelia of different mating types form hyphal extensions (gametangia), each of which encloses several haploid nuclei.

Key

Haploid (*n*)
Heterokaryotic (*n* + *n*)
Diploid (2*n*)

Rhizopus growing on bread

Mating type (–)

Mating type (+)

Gametangia with haploid nuclei

PLASMOGAMY

3 A zygosporangium forms, containing multiple haploid nuclei from the two parents.

Young zygosporangium (heterokaryotic)

100 µm

Zygosporangium

8 The spores germinate and grow into new mycelia.

SEXUAL REPRODUCTION

KARYOGAMY

4 The zygosporangium develops a rough, thick-walled coating that can resist harsh conditions for months.

9 Mycelia can also reproduce asexually by forming sporangia that produce genetically identical haploid spores.

Dispersal and germination

7 The sporangium disperses genetically diverse haploid spores.

Diploid nuclei

Sporangia

ASEXUAL REPRODUCTION

Sporangium

MEIOSIS

5 When conditions are favorable, karyogamy occurs, then meiosis.

50 µm

Dispersal and germination

Mycelium

6 The zygosporangium germinates into a sporangium on a short stalk.

basidiomycetes; see Figure 31.10) have nonflagellated spores, which are dispersed by wind in terrestrial fungi.

Mucoromycetes

Cryptomycetes
Microsporidians
Chytrids
Zoopagomycetes
Mucoromycetes
Ascomycetes
Basidiomycetes

There are approximately 750 known species of **mucoromycetes**, fungi in the phylum Mucoromycota. This phylum includes species of fast-growing molds responsible for causing foods such as bread, peaches, strawberries, and sweet potatoes to rot during storage. Although some mucoromycetes are decomposers, most are associated with plants. Many mucoromycetes live as parasites or pathogens of plants, while others live as mutualists (including some mycorrhizae).

The life cycle of *Rhizopus stolonifer* (black bread mold) is fairly typical of mucoromycete species **(Figure 31.15)**. Its hyphae spread out over the food surface, penetrate it, and absorb nutrients. The hyphae are coenocytic, with septa found only where reproductive cells are formed. In the asexual phase, bulbous black sporangia develop at the tips of upright hyphae. Within each sporangium, hundreds of genetically identical haploid spores develop and are dispersed through the air. Spores that happen to land on moist food germinate, growing into new mycelia.

If environmental conditions deteriorate—for instance, if the mold consumes all its food—*Rhizopus* may reproduce sexually. The parents in a sexual union are mycelia of different mating types, which possess different chemical markers but may appear identical. Plasmogamy produces a sturdy structure called a **zygosporangium** (plural, *zygosporangia*), in which karyogamy and then meiosis occur. Note that while a zygosporangium represents the zygote (*2n*) stage in the life cycle, it is not a zygote in the usual sense (that is, a cell with one diploid nucleus). Rather, a zygosporangium is a multinucleate structure, first heterokaryotic with many haploid nuclei from the two parents, then with many diploid nuclei after karyogamy.

Zygosporangia are resistant to freezing and drying and are metabolically inactive. When conditions improve, the nuclei of the zygosporangium undergo meiosis, the zygosporangium germinates into a sporangium, and the sporangium releases genetically diverse haploid spores that may colonize a new substrate. Some mucoromycetes can actually "aim" and then shoot their sporangia toward bright light. **Figure 31.16** shows one example, *Pilobolus*, which decomposes animal dung. Its sporebearing hyphae bend toward light, where there are likely to be openings in the vegetation through which spores may reach fresh grass. The fungus then launches its sporangia in a jet of water that can travel up to 2.5 m. Grazing animals ingest the fungi with the grass and then scatter the spores in feces, thereby enabling the next generation of fungi to grow.

Finally, the phylum Mucoromycota also includes the glomeromycetes, a clade of fungi that form arbuscular

▶ **Figure 31.16** *Pilobolus* aiming its sporangia.

1 mm

mycorrhizae (see Figure 31.4b and Figure 37.14). The tips of the hyphae that push into plant root cells branch into tiny treelike arbuscules. About 85% of all plant species have mutualistic partnerships with arbuscular mycorrhizae.

Ascomycetes

Cryptomycetes
Microsporidians
Chytrids
Zoopagomycetes
Mucoromycetes
Ascomycetes
Basidiomycetes

Mycologists have described 90,000 species of **ascomycetes**, fungi in the phylum Ascomycota, from a wide variety of marine, freshwater, and terrestrial habitats. The defining feature of ascomycetes is the production of spores (called ascospores) in saclike **asci** (singular, *ascus*); thus, they are commonly called *sac fungi*. During their sexual stage, most ascomycetes develop fruiting bodies, called **ascocarps**, which range in size from microscopic to macroscopic **(Figure 31.17)**. The ascocarps contain the spore-forming asci.

Ascomycetes vary in size and complexity from unicellular yeasts to elaborate cup fungi and morels (see Figure 31.17). They include some of the most devastating plant pathogens, which we will discuss later. However, many ascomycetes are important decomposers, particularly of plant material. More than 25% of all ascomycete species live with green algae or cyanobacteria in beneficial symbiotic associations called lichens. Some ascomycetes form mycorrhizae with plants.

▼ **Figure 31.17 Ascomycetes (sac fungi).**

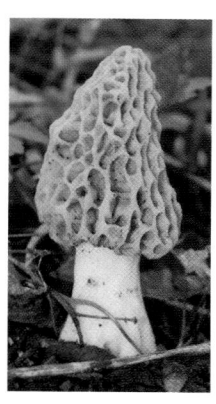

▶ *Tuber melanosporum* is a truffle species that forms ectomycorrhizae with trees. The ascocarp grows underground and emits a strong odor. These ascocarps have been dug up and the middle one sliced open.

◀ The edible ascocarp of *Morchella esculenta*, the tasty morel, is often found under trees in orchards.

❓ *Ascomycetes vary greatly in morphology (see also Figure 31.10). How could you confirm that a fungus is an ascomycete?*

Many others live between mesophyll cells in leaves; some of these species release toxic compounds that help protect the plant from insects.

Although the life cycles of various ascomycete groups differ in the details of their reproductive structures and processes, we'll illustrate some common elements using the bread mold *Neurospora crassa* **(Figure 31.18)**. Ascomycetes reproduce asexually by producing enormous numbers of asexual spores called **conidia** (singular, *conidium*). Unlike the asexual spores of most mucoromycetes, in most ascomycetes, conidia are not formed inside sporangia. Rather, they are produced externally at the tips of specialized hyphae called

conidiophores, often in clusters or long chains, from which they may be dispersed by the wind.

Conidia may also be involved in sexual reproduction, fusing with hyphae from a mycelium of a different mating type, as occurs in *Neurospora*. Fusion of two different mating types is followed by plasmogamy, resulting in the formation of dikaryotic cells, each with two haploid nuclei representing the two parents. The cells at the tips of these dikaryotic hyphae develop into many asci. Within each ascus, karyogamy combines the two parental genomes, and then meiosis forms four genetically different nuclei. This is usually followed by a mitotic division, forming eight ascospores. The

▼ **Figure 31.18 The life cycle of *Neurospora crassa*, an ascomycete.** *Neurospora* is a bread mold and research organism that also grows in the wild on burned vegetation.

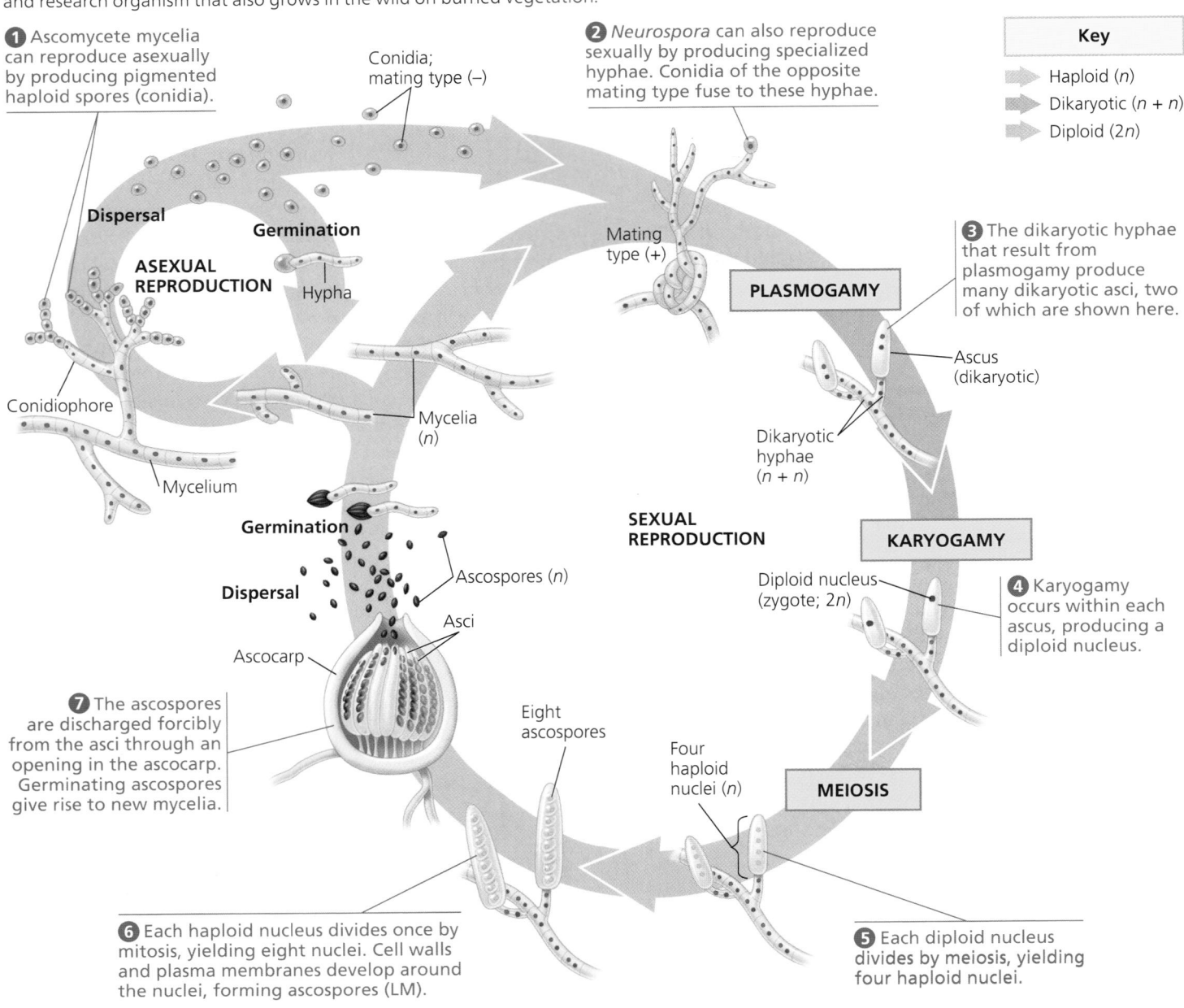

Key
Haploid (*n*)
Dikaryotic (*n* + *n*)
Diploid (2*n*)

❶ Ascomycete mycelia can reproduce asexually by producing pigmented haploid spores (conidia).

Conidia; mating type (–)

❷ *Neurospora* can also reproduce sexually by producing specialized hyphae. Conidia of the opposite mating type fuse to these hyphae.

Dispersal

Germination

ASEXUAL REPRODUCTION

Hypha

Conidiophore

Mycelium

Mycelia (*n*)

Mating type (+)

PLASMOGAMY

❸ The dikaryotic hyphae that result from plasmogamy produce many dikaryotic asci, two of which are shown here.

Ascus (dikaryotic)

Dikaryotic hyphae (*n* + *n*)

SEXUAL REPRODUCTION

KARYOGAMY

Diploid nucleus (zygote; 2*n*)

❹ Karyogamy occurs within each ascus, producing a diploid nucleus.

Germination

Dispersal

Ascospores (*n*)

Asci

Ascocarp

Eight ascospores

Four haploid nuclei (*n*)

MEIOSIS

❼ The ascospores are discharged forcibly from the asci through an opening in the ascocarp. Germinating ascospores give rise to new mycelia.

❻ Each haploid nucleus divides once by mitosis, yielding eight nuclei. Cell walls and plasma membranes develop around the nuclei, forming ascospores (LM).

❺ Each diploid nucleus divides by meiosis, yielding four haploid nuclei.

VISUAL SKILLS *What is the ploidy of a cell in the specialized hypha shown in* ❷*?*

ascospores develop in and are eventually discharged from the ascocarp.

Compared to the life cycle of mucoromycetes, the extended dikaryotic stage of ascomycetes (and also basidiomycetes) provides additional opportunities for genetic recombination. In *Neurospora*, for example, many dikaryotic cells can develop into asci. The haploid nuclei in these asci fuse, and their genomes then recombine during meiosis, resulting in a multitude of genetically different offspring from one mating event (see steps 3–5 in Figure 31.18).

As described in Figure 17.2, biologists in the 1930s used *Neurospora* in research that led to the one gene–one enzyme hypothesis. Today, this ascomycete continues to serve as a model research organism. In 2003, its entire genome was published. This tiny fungus has about three-fourths as many genes as the fruit fly *Drosophila* and about half as many as a human **(Table 31.1)**. The *Neurospora* genome is relatively compact, having few of the stretches of noncoding DNA that occupy so much space in the genomes of humans and many other eukaryotes. In fact, there is evidence that *Neurospora* has a genomic defense system that prevents noncoding DNA such as transposons from accumulating.

Table 31.1 Comparison of Gene Density in *Neurospora*, *Drosophila*, **and** *Homo sapiens*

	Genome Size (million base pairs)	Number of Genes	Gene Density (genes per million base pairs)
Neurospora crassa (ascomycete fungus)	41	9,700	236
Drosophila melanogaster (fruit fly)	165	14,000	85
Homo sapiens (human)	3,000	<21,000	7

Basidiomycetes

Cryptomycetes
Microsporidians
Chytrids
Zoopagomycetes
Mucoromycetes
Ascomycetes
Basidiomycetes

About 50,000 species, including mushrooms, puffballs, and shelf fungi, are called **basidiomycetes** and are classified in the phylum Basidiomycota **(Figure 31.19)**. This phylum also includes mutualists that form mycorrhizae and two groups of destructive plant parasites: rusts and smuts. The name of the phylum derives from the **basidium** (plural, *basidia*; Latin for "little pedestals"), a cell in which karyogamy occurs, followed immediately by meiosis. The club-like shape of the basidium also gives rise to the common name *club fungus*.

▼ **Figure 31.19 Basidiomycetes (club fungi).**

► Shelf fungi, important decomposers of wood

◄ Puffballs emitting spores

► Maiden veil fungus (*Dictyphora*), a fungus with an odor like rotting meat

Basidiomycetes are important decomposers of wood and other plant material. Of all the fungi, certain basidiomycetes are the best at decomposing the complex polymer lignin, an abundant component of wood. Many shelf fungi break down the wood of weak or damaged trees and continue to decompose the wood after the tree dies.

The life cycle of a basidiomycete usually includes a long-lived dikaryotic mycelium. As in ascomycetes, this extended dikaryotic stage provides many opportunities for genetic recombination events, in effect multiplying the result of a single mating. Periodically, in response to environmental stimuli, the mycelium reproduces sexually by producing

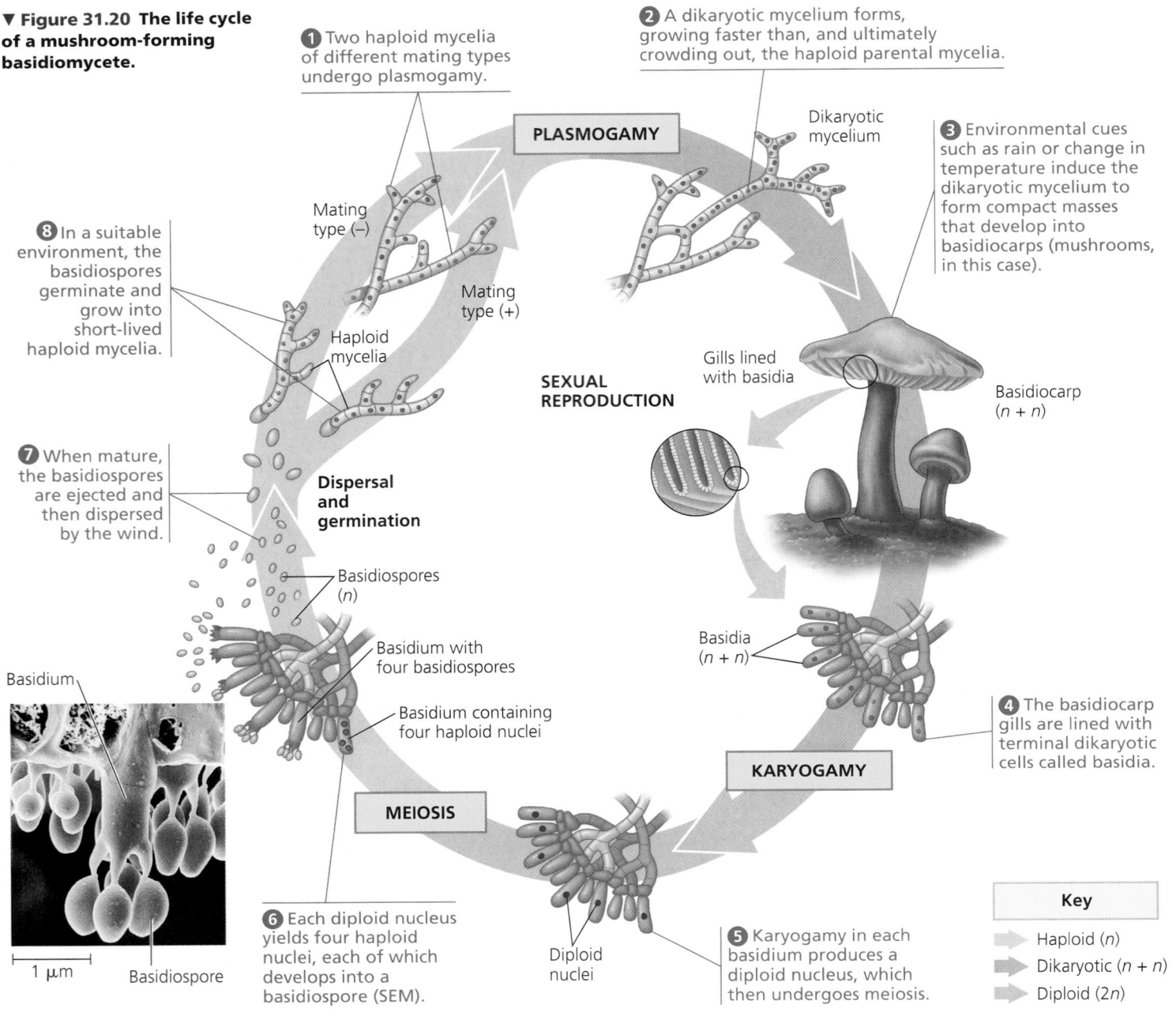

▼ **Figure 31.20 The life cycle of a mushroom-forming basidiomycete.**

PLASMOGAMY

❶ Two haploid mycelia of different mating types undergo plasmogamy.

❷ A dikaryotic mycelium forms, growing faster than, and ultimately crowding out, the haploid parental mycelia.

Dikaryotic mycelium

❸ Environmental cues such as rain or change in temperature induce the dikaryotic mycelium to form compact masses that develop into basidiocarps (mushrooms, in this case).

Mating type (−)

Mating type (+)

❽ In a suitable environment, the basidiospores germinate and grow into short-lived haploid mycelia.

Haploid mycelia

SEXUAL REPRODUCTION

Gills lined with basidia

Basidiocarp $(n + n)$

❼ When mature, the basidiospores are ejected and then dispersed by the wind.

Dispersal and germination

Basidiospores (n)

Basidium with four basidiospores

Basidium containing four haploid nuclei

Basidia $(n + n)$

❹ The basidiocarp gills are lined with terminal dikaryotic cells called basidia.

Basidium

MEIOSIS

KARYOGAMY

❻ Each diploid nucleus yields four haploid nuclei, each of which develops into a basidiospore (SEM).

Diploid nuclei

❺ Karyogamy in each basidium produces a diploid nucleus, which then undergoes meiosis.

1 μm   Basidiospore

Key	
➤	Haploid (n)
➤	Dikaryotic $(n + n)$
➤	Diploid $(2n)$

VISUAL SKILLS *Use the diagram to determine the ploidy of a cell in the aboveground stalk of a mushroom.*

⊙ **Mastering Biology** Animation: Life Cycle of a Mushroom

elaborate fruiting bodies called **basidiocarps (Figure 31.20)**. The common white mushrooms in the supermarket are familiar examples of a basidiocarp.

By concentrating growth in the hyphae of mushrooms, a basidiomycete mycelium can erect its fruiting structures in just a few hours; a mushroom pops up as it absorbs water and as cytoplasm streams in from the dikaryotic mycelium. By this process, in some species a ring of mushrooms, popularly called a "fairy ring," may appear literally overnight **(Figure 31.21)**. The mycelium below the fairy ring expands outward at a rate of about 30 cm per year, decomposing

organic matter in the soil as it grows. Some giant fairy rings are produced by mycelia that are centuries old.

After a mushroom forms, its cap supports and protects a large surface area of dikaryotic basidia on gills. During karyogamy, the two nuclei in each basidium fuse, producing a diploid nucleus (see Figure 31.20). This nucleus then undergoes meiosis, yielding four haploid nuclei, each of which ultimately develops into a basidiospore. Large numbers of basidiospores are produced: The gills of a common white mushroom have a surface area of about 200 cm² and may drop a billion basidiospores, which blow away.

▼ **Figure 31.21 A fairy ring.** According to legend, mushroom rings spring up where fairies have danced on a moonlit night. The text provides a biological explanation of how these rings form.

CONCEPT CHECK 31.4

1. What feature of chytrids supports the hypothesis that they include members of basal fungal lineages?

2. Give examples of how form fits function in mucoromycetes, ascomycetes, and basidiomycetes.

3. **WHAT IF?** Suppose that the mutation of an ascomycete changed its life cycle so that plasmogamy, karyogamy, and meiosis occurred in quick succession. How might this affect the ascospores and ascocarps?

For suggested answers, see Appendix A.

CONCEPT 31.5

Fungi play key roles in nutrient cycling, ecological interactions, and human welfare

In our survey of fungal classification, we've touched on some of the ways fungi influence other organisms. We will now look more closely at these impacts, focusing on how fungi act as decomposers, mutualists, and pathogens.

Fungi as Decomposers

Fungi are well adapted as decomposers of organic material, including the cellulose and lignin of plant cell walls. In fact, almost any carbon-containing substrate—even jet fuel and house paint—can be consumed by at least some fungi. The same is true of bacteria. As a result, fungi and bacteria are primarily responsible for keeping ecosystems stocked with the inorganic nutrients essential for plant growth. Without these decomposers, carbon, nitrogen, and other elements would remain tied up in organic matter. If that were to happen, plants and the animals that eat them could not exist because elements taken from the soil would not be returned. Without decomposers, life as we know it would cease.

Fungi as Mutualists

Fungi may form mutualistic relationships with plants, algae, cyanobacteria, and animals. Mutualistic fungi absorb

nutrients from a host organism, but they reciprocate with actions that benefit the host—as we already saw for the key mycorrhizal associations that fungi form with most vascular plants. We turn now to other examples of mutualistic fungi.

Fungus-Plant Mutualisms

All plant species studied to date appear to harbor symbiotic **endophytes**, fungi (or bacteria) that live inside leaves or other plant parts without causing harm. Most fungal endophytes identified to date are ascomycetes but some are mucoromycetes. Fungal endophytes benefit certain grasses and other nonwoody plants by making toxins that deter herbivores or by increasing host plant tolerance of heat, drought, or heavy metals. As described in **Figure 31.22**, researchers

---

▼ **Figure 31.22 Inquiry**

Do fungal endophytes benefit a woody plant?

Experiment Fungal endophytes are symbiotic fungi found within the bodies of all plants examined to date. Researchers tested whether fungal endophytes benefit the cacao tree (*Theobroma cacao*). This tree, whose name means "food of the gods" in Greek, is the source of the beans used to make chocolate, and it is cultivated throughout the tropics. A particular mixture of fungal endophytes was added to the leaves of some cacao seedlings, but not others. (In cacao, fungal endophytes colonize leaves after the seedling germinates.) The seedlings were then inoculated with a virulent pathogen, the protist *Phytophthora*.

Results Fewer leaves were killed by the pathogen in seedlings with fungal endophytes than in seedlings without endophytes. Among leaves that survived, pathogens damaged less of the leaf surface area in seedlings with endophytes than in seedlings without endophytes.

▢ Endophyte not present; pathogen present (E–P+)
▢ Both endophyte and pathogen present (E+P+)

Conclusion The presence of endophytes appears to benefit cacao trees by reducing the leaf mortality and damage caused by *Phytophthora*.

Data from A. E. Arnold et al., Fungal endophytes limit pathogen damage in a tropical tree, *Proceedings of the National Academy of Sciences USA* 100:15649–15654 (2003).

WHAT IF? *The researchers also performed control treatments. Suggest two controls they might have used, and explain how each would be helpful in interpreting the results described here.*

▼ **Figure 31.23 Fungus-gardening insects.** These leaf-cutting ants depend on fungi to convert plant material to a form the insects can digest. The fungi, in turn, depend on the nutrients from the leaves the ants feed them.

▼ **Figure 31.24 Variation in lichen growth forms.**

◄ A fruticose (shrublike) lichen

► A foliose (leaflike) lichen

◄ Crustose (encrusting) lichens

studying how fungal endophytes affect a woody plant tested whether leaf endophytes benefit seedlings of the cacao tree, *Theobroma cacao*. Their findings show that the fungal endophytes of woody flowering plants can play an important role in defending against pathogens.

Fungus-Animal Mutualisms

As mentioned earlier, some fungi share their digestive services with animals, helping break down plant material in the guts of cattle and other grazing mammals. Many species of ants take advantage of the digestive power of fungi by raising them in "farms." Leaf-cutter ants, for example, scour tropical forests in search of leaves, which they cannot digest on their own but carry back to their nests and feed to the fungi **(Figure 31.23)**. As the fungi grow, their hyphae develop specialized swollen tips that are rich in proteins and carbohydrates. The ants feed primarily on these nutrient-rich tips. Not only do the fungi break down plant leaves into substances the insects can digest, but they also detoxify plant defensive compounds that would otherwise kill or harm the ants. In some tropical forests, the fungi have helped these insects become the major consumers of leaves.

The evolution of such farmer ants and that of their fungal "crops" have been tightly linked for over 50 million years. The fungi have become so dependent on their caretakers that in many cases they can no longer survive without the ants, and vice versa.

Lichens

A **lichen** is a symbiotic association between a photosynthetic microorganism and a fungus in which millions of photosynthetic cells are held in a mass of fungal hyphae. Lichens grow on the surfaces of rocks, rotting logs, trees, and roofs in various forms **(Figure 31.24)**. The photosynthetic partners are unicellular or filamentous green algae or cyanobacteria. The fungal component is most often an ascomycete, but some glomeromycete and basidiomycete lichens are known. Recent studies have found that many lichens also have a basidiomycete yeast as a second fungal component. As the role of these yeasts remains unknown, our discussion will focus on the primary fungal partner.

The fungus usually gives a lichen its overall shape and structure, and tissues formed by hyphae account for most of the lichen's mass. The cells of the alga or cyanobacterium generally occupy an inner layer below the lichen surface **(Figure 31.25)**. The merger of fungus and alga or cyanobacterium is so complete that lichens are given scientific names as though they were single organisms. As might be expected of such "dual organisms," asexual reproduction as a symbiotic unit is common. This can occur either by fragmentation of the parental lichen or by the formation of **soredia** (singular, *soredium*), small clusters of hyphae with embedded algae (see Figure 31.25). The fungi of many lichens also reproduce sexually.

In most lichens, each partner provides something the other could not obtain on its own. The alga or cyanobacterium provides carbon compounds; a cyanobacterium also fixes nitrogen (see Concept 27.3) and provides organic

▶ **Figure 31.25 Anatomy of an ascomycete lichen.** (colorized SEM)

Ascocarp of fungus

Fungal hyphae

Algal layer

Soredia

50 μm

Fungal hyphae

Algal cell

nitrogen compounds. The fungus provides its photosynthetic partner with a suitable environment for growth. The physical arrangement of hyphae allows for gas exchange, protects the photosynthetic partner, and retains water and minerals, most of which are absorbed from airborne dust or from rain. The fungus also secretes acids, which aid in the uptake of minerals.

Lichens are important pioneers on cleared rock and soil surfaces, such as volcanic flows and burned forests. They break down the surface by physically penetrating and chemically attacking it, and they trap windblown soil. Nitrogen-fixing lichens also add organic nitrogen to some ecosystems. These processes make it possible for a succession of plants to grow. Fossils show that lichens were on land 420 million years ago. These early lichens may have modified rocks and soil much as they do today, helping pave the way for plants.

Fungi as Parasites

Like mutualistic fungi, parasitic fungi absorb nutrients from the cells of living hosts, but they provide no benefits in return. About 30% of the 145,000 known species of fungi make a living as parasites or pathogens, mostly of plants **(Figure 31.26)**. An example of a plant pathogen is *Cryphonectria parasitica*, the ascomycete fungus that causes chestnut blight, which dramatically changed the landscape of the northeastern United States. Accidentally introduced via trees imported from Asia in the early 1900s, spores of the fungus entered cracks in the bark of American chestnut trees and produced hyphae, killing many trees. The once-common chestnuts now survive mainly as sprouts from the stumps of former trees. Another ascomycete, *Fusarium circinatum*, causes pine pitch canker, a disease that threatens pines throughout the world. In addition, between 10% and 50% of the world's

fruit harvest is lost annually due to fungi, and grain crops also suffer major losses each year.

Some fungi that attack food crops produce compounds that are toxic to humans. One example is the ascomycete *Claviceps purpurea*, which grows on rye plants, forming purple structures called ergots (see Figure 31.26c). If infected rye is milled into flour, toxins from the ergots can cause ergotism, characterized by gangrene, nervous spasms, burning sensations, hallucinations, and temporary insanity. An epidemic of ergotism around 944 CE killed up to 40,000 people in France. One compound that has been isolated from ergots is lysergic acid, the raw material from which the hallucinogen LSD is made.

Although animals are less susceptible to parasitic fungi than are plants, about 1,000 fungi are known to parasitize animals. Two such parasites, the chytrids *Batrachochytrium dendrobatidis* (discovered in 1998) and *B. salamandrivorans* (discovered in 2013; this species primarily attacks salamanders), have been implicated in the recent decline or extinction of 500 species of frogs and other amphibians. These chytrids can cause severe skin infections, leading

▼ **Figure 31.26 Examples of fungal diseases of plants.**

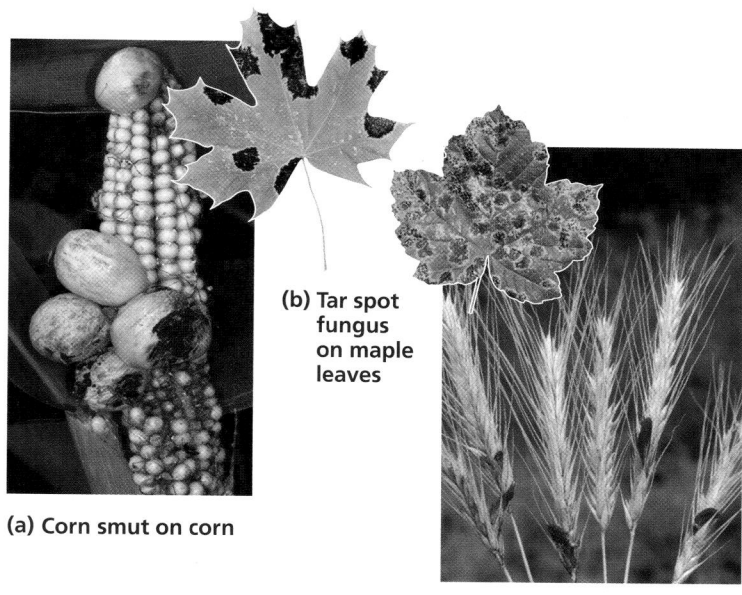

(a) Corn smut on corn

(b) Tar spot fungus on maple leaves

(c) Ergots on rye

▼ **Figure 31.27 Amphibians under attack.** The number of yellow-legged frogs (*Rana muscosa*) plummeted after the chytrid *Batrachochytrium dendrobatidis* reached the Sixty Lake Basin area of California. In the years leading up to the chytrid's 2004 arrival, there had been more than 2,300 frogs in these lakes. By 2009, only 38 frogs remained; all the survivors were in two lakes (yellow) where frogs had been treated with a fungicide to reduce the chytrid's impact.

California

Sixty Lake Basin

2004
2005
2006
2007 2008

N

▲ Yellow-legged frogs killed by *B. dendrobatidis* infection

Key

- - - Boundary of chytrid spread

Lake status in 2009:

■ Frog population extinct

□ Treatment lake: frogs treated with fungicides and released

INTERPRET THE DATA *Do the data depicted indicate that the chytrid caused or is correlated to the drop in frog numbers? Explain.*

to massive die-offs **(Figure 31.27)**. Field observations and studies of museum specimens show that *B. dendrobatidis* and *B. salamandrivorans* first appeared in amphibian populations shortly before their declines in Australia, Costa Rica, Germany, the United States, and other countries. Genetic analyses indicate that both *B. dendrobatidis* and *B. salamandrivorans* originated in Asia and spread from there via the commercial trade of frogs and salamanders.

The general term for an infection in an animal by a fungal parasite is **mycosis**. In humans, skin mycoses include the disease ringworm, so named because it appears as circular red areas on the skin. Most commonly, the ascomycetes that cause ringworm grow on the feet, causing the intense itching and blisters known as athlete's foot. Though highly contagious, athlete's foot and other ringworm infections can be treated with fungicidal lotions and powders.

Systemic mycoses, by contrast, spread through the body and usually cause very serious illnesses. They are typically caused by inhaled spores. For example, coccidioidomycosis is a systemic mycosis that produces tuberculosis-like symptoms

in the lungs. Each year, hundreds of cases in North America require treatment with antifungal drugs, without which the disease could be fatal.

Some mycoses are opportunistic, occurring only when a change in the body's microorganisms, chemical environment, or immune system allows fungi to grow unchecked. *Candida albicans*, for example, is one of the normal inhabitants of moist epithelia, such as the vaginal lining. Under certain circumstances, *C. albicans* can grow too rapidly and become pathogenic, leading to so-called "yeast infections." A related species, *C. auris*, has emerged as a global threat, often in healthcare facilities. Resistant to multiple antifungal drugs, this species can infect the bloodstream and cause life-threatening infections.

Practical Uses of Fungi

The dangers posed by fungi should not overshadow their immense benefits. We depend on their ecological services as decomposers and recyclers of organic matter. In addition, mushrooms are not the only fungi of interest for human consumption. Fungi are used to ripen Roquefort and other blue cheeses. Morels and truffles, the edible fruiting bodies of various ascomycetes, are highly prized for their complex flavors (see Figure 31.17). These fungi can sell for hundreds to thousands of dollars a pound. Truffles release strong odors that attract mammals and insects, which in nature feed on them and disperse their spores. In some cases, the odors mimic the pheromones (sex attractants) of certain mammals. For example, the odors of several European truffles mimic the pheromones released by male pigs, which explains why truffle hunters sometimes use female pigs to help find these delicacies.

Humans have used yeasts to produce alcoholic beverages and bread for thousands of years. Under anaerobic conditions, yeasts ferment sugars to alcohol and CO_2, which causes dough to rise. Only relatively recently have the yeasts involved been separated into pure cultures for more controlled use. The yeast *Saccharomyces cerevisiae* is the most important of all cultured fungi (see Figure 31.7). It is available as many strains of baker's yeast and brewer's yeast.

Many fungi have great medical value as well. For example, a compound extracted from ergots is used to reduce high blood pressure and to stop maternal bleeding after childbirth. Some fungi produce antibiotics that are effective in treating bacterial infections. In fact, the first antibiotic discovered was penicillin, made by the ascomycete mold *Penicillium*. Other examples of pharmaceuticals derived from fungi include cholesterol-lowering drugs and cyclosporine, a drug used to suppress the immune system after organ transplants.

Fungi also figure prominently in basic research. For example, the yeast *Saccharomyces cerevisiae* is used to study the molecular genetics of eukaryotes because its cells are easy to culture and manipulate. Scientists are gaining insight into the genes involved in Parkinson's disease by examining the functions of homologous genes in *S. cerevisiae*.

Genetically modified fungi also hold much promise. For example, scientists have succeeded in engineering a strain of *S. cerevisiae* that produces human glycoproteins, including insulin-like growth factor. Such fungus-produced glycoproteins have the potential to treat people with medical conditions that prevent them from producing these compounds. Meanwhile, other researchers are sequencing the genome of *Gliocladium roseum*, an ascomycete that can grow on wood or agricultural waste and that naturally produces hydrocarbons similar to those in diesel fuel **(Figure 31.28)**. They hope to decipher the metabolic pathways by which *G. roseum* synthesizes hydrocarbons, with the goal of harnessing those pathways to produce biofuels without reducing land area for growing food crops (as occurs when ethanol is produced from corn).

Having now completed our survey of the kingdom Fungi, we will turn in the rest of this unit to the closely related kingdom Animalia, to which we humans belong.

▶ **Figure 31.28 Can this fungus be used to produce biofuels?** The ascomycete *Gliocladium roseum* can produce hydrocarbons similar to those in diesel fuel (colorized SEM).

CONCEPT CHECK 31.5

1. What are some of the benefits that lichen algae can derive from their relationship with fungi?
2. What characteristics of pathogenic fungi result in their being efficiently transmitted?
3. **WHAT IF?** How might life on Earth differ from what we know today if no mutualistic relationships between fungi and other organisms had ever evolved?

For suggested answers, see Appendix A.

31 Chapter Review

➔ Go to **Mastering Biology** for Assignments, the eText, the Study Area, and Dynamic Study Modules.

SUMMARY OF KEY CONCEPTS

➔ To review key terms, go to the **Vocabulary Self-Quiz** in the **Mastering Biology** eText or Study Area, or go to goo.gl/zkjz9t.

CONCEPT 31.1

Fungi are heterotrophs that feed by absorption *(pp. 655–657)*

- All **fungi** (including decomposers and symbionts) are heterotrophs that acquire nutrients by absorption. Many fungi secrete enzymes that break down complex molecules.
- Most fungi grow as thin, multicellular filaments called **hyphae**; relatively few species grow only as single-celled **yeasts**. In their multicellular form, fungi consist of **mycelia**, networks of branched hyphae adapted for absorption. Mycorrhizal fungi have specialized hyphae that enable them to form a mutually beneficial relationship with plants.

? *How does the morphology of multicellular fungi affect the efficiency of nutrient absorption?*

CONCEPT 31.2

Fungi produce spores through sexual or asexual life cycles *(pp. 657–659)*

- In fungi, the sexual life cycle involves cytoplasmic fusion (**plasmogamy**) and nuclear fusion (**karyogamy**), with an

intervening **heterokaryotic** stage in which cells have haploid nuclei from two parents. The diploid cells resulting from karyogamy are short-lived and undergo meiosis, producing genetically diverse haploid **spores**.
- Many fungi can reproduce asexually as filamentous fungi or yeasts.

DRAW IT *Draw a generalized fungal life cycle, labeling asexual and sexual reproduction, meiosis, plasmogamy, karyogamy, and the points in the cycle when spores and the zygote are produced.*

CONCEPT 31.3

The ancestor of fungi was an aquatic, single-celled, flagellated protist *(pp. 659–660)*

- Molecular evidence indicates that fungi and animals diverged over a billion years ago from a common unicellular ancestor that had a flagellum. However, the oldest fossils that are widely accepted as fungi are 440 million years old.
- Chytrids and other basal fungal lineages have flagellated spores.
- Fungi were among the earliest colonizers of land; fossil evidence indicates that these colonizers included species that were symbionts with early plants.

? *Did multicellularity originate independently in fungi and animals? Explain.*

CONCEPT 31.4

Fungi have radiated into a diverse set of lineages
(pp. 660–667)

Fungal Phylum	Distinguishing Features	
Cryptomycota (**cryptomycetes**)	Parasites with flagellated spores	
Microsporidia (**microsporidians**)	Parasitic cells that form resistant spores	
Chytridiomycota (**chytrids**)	Flagellated spores	
Zoopagomycota (**zoopagomycetes**)	Resistant zygosporangium as sexual stage	
Mucuromycota (**mucuromycetes**)	Include fungi that form arbuscular mycorrhizae with plants	
Ascomycota (**ascomycetes**)	Sexual spores (ascospores) borne internally in sacs called asci; vast numbers of asexual spores (conidia) produced	
Basidiomycota (**basidiomycetes**)	Elaborate fruiting body (basidiocarp) containing many basidia that produce sexual spores (basidiospores)	

DRAW IT *Draw a phylogenetic tree of the major groups of fungi.*

CONCEPT 31.5

Fungi play key roles in nutrient cycling, ecological interactions, and human welfare *(pp. 667–671)*

- Fungi perform essential recycling of chemical elements between the living and nonliving world.
- **Lichens** are highly integrated symbiotic associations of fungi and algae or cyanobacteria.
- Many fungi are parasites, mostly of plants.
- Humans use fungi for food and to make antibiotics.

? *How are fungi important as decomposers, mutualists, and pathogens?*

TEST YOUR UNDERSTANDING

⊘ For more multiple-choice questions, go to the **Practice Test** in the **Mastering Biology** eText or Study Area, or go to goo.gl/GruWRg.

Levels 1-2: Remembering/Understanding

1. *All* fungi are
   (A) symbiotic.
   (B) heterotrophic.
   (C) flagellated.
   (D) decomposers.

2. Which of the following cells or structures are associated with *asexual* reproduction in fungi?
   (A) ascospores
   (B) basidiospores
   (C) zygosporangia
   (D) conidiophores

3. The closest relatives of fungi are thought to be the
   (A) animals.
   (B) vascular plants.
   (C) mosses.
   (D) slime molds.

Levels 3-4: Applying/Analyzing

4. The most important adaptive advantage associated with the filamentous nature of fungal mycelia is
   (A) the ability to form haustoria and parasitize other organisms.
   (B) the potential to inhabit almost all terrestrial habitats.
   (C) the increased chance of contact between mating types.
   (D) an extensive surface area well suited for invasive growth and absorptive nutrition.

5. **SCIENTIFIC INQUIRY • INTERPRET THE DATA** The grass *Dichanthelium lanuginosum* lives in hot soils and houses fungi of the genus *Curvularia* as endophytes. Researchers tested the impact of *Curvularia* on the heat tolerance of this grass. They grew plants without (E−) and with (E+) *Curvularia* endophytes at different temperatures and measured plant mass and the number of new shoots the plants produced. Draw a bar graph for plant mass versus temperature and interpret it.

Soil Temp.	*Curvularia* + or −	Plant Mass (g)	No. of New Shoots
30°C	E−	16.2	32
	E+	22.8	60
35°C	E−	21.7	43
	E+	28.4	60
40°C	E−	8.8	10
	E+	22.2	37
45°C	E−	0	0
	E+	15.1	24

Data from R. S. Redman et al., Thermotolerance generated by plant/fungal symbiosis, *Science* 298:1581 (2002).

Levels 5-6: Evaluating/Creating

6. **EVOLUTION CONNECTION** The fungus-alga symbiosis that makes up a lichen is thought to have evolved multiple times independently in different fungal groups. However, lichens fall into three well-defined growth forms (see Figure 31.24). How could you test the following hypotheses? Hypothesis 1: Crustose, foliose, and fruticose lichens each represent a monophyletic group. Hypothesis 2: Each lichen growth form represents convergent evolution by diverse fungal groups.

7. **WRITE ABOUT A THEME: ORGANIZATION** As you read in this chapter, fungi have long formed symbiotic associations with plants and with algae. In a short essay (100–150 words), describe how these two types of associations may lead to emergent properties in biological communities.

8. **SYNTHESIZE YOUR KNOWLEDGE**

This wasp is the unfortunate victim of an entomopathogenic fungus (a parasitic fungus of insects). Write a paragraph describing what this image illustrates about the nutritional mode, body structure, and ecological role of the fungus.

For selected answers, see Appendix A.

32 An Overview of Animal Diversity

KEY CONCEPTS

32.1 Animals are multicellular, heterotrophic eukaryotes with tissues that develop from embryonic layers *p. 674*

32.2 The history of animals spans more than half a billion years *p. 675*

32.3 Animals can be characterized by body plans *p. 679*

32.4 Views of animal phylogeny continue to be shaped by new molecular and morphological data *p. 682*

Study Tip

Draw a time line: Create a time line like the one shown here, and mark when the following occurred: first biochemical evidence of animals, Edicaran biota, Cambrian explosion, colonization of land by vertebrates, and dinosaur extinction. Briefly describe the significance of each event.

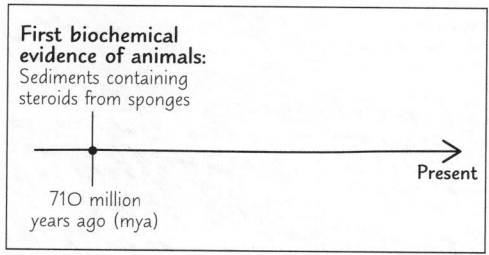

First biochemical evidence of animals: Sediments containing steroids from sponges

710 million years ago (mya)

Present

⮕ Go to Mastering Biology

For Students (in eText and Study Area)
- Get Ready for Chapter 32
- Interview with Nicole King: Investigating the ancestry of choanoflagellates

For Instructors to Assign (in Item Library)
- Tutorial: Visualizing Animal Body Symmetry and Axes
- Tutorial: Animal Body Plans
- Scientific Skills Exercise: Calculating and Interpreting Correlation Coefficients

Figure 32.1 This chameleon can wield its long, sticky tongue with blinding speed to capture its prey. Other predatory animals use strength, concealed traps, or toxins to capture prey. Likewise, herbivorous animals can strip plants bare of leaves or seeds, while parasitic animals weaken their hosts by consuming their tissues or body fluids. Overall, we can think of animals as a kingdom of consumers.

What key characteristics of animals make them such efficient consumers?

Animals are **heterotrophs**: They obtain energy and nutrients by eating other organisms.

Animals have **tissues**, collections of specialized cells that function as a unit. Tissues form from layers of embryonic cells.

Animal embryo (cross section)

Animals process their food inside their bodies. Many animals have an efficient **digestive system** that has a mouth at one end and an anus at the other.

Mouth

Anus

Nerve cell

The tissues of animals include those composed of unique **nerve and muscle cells**.

Muscle cells

Their nerve and muscle cells enable animals to **move** as well as to **detect and capture** potential prey.

673

Animals are multicellular, heterotrophic eukaryotes with tissues that develop from embryonic layers

Listing features shared by all animals is challenging, as there are exceptions to nearly every criterion we might select. When taken together, however, several characteristics of animals sufficiently describe the group for our discussion.

Nutritional Mode

Animals differ from both plants and fungi in their mode of nutrition. Plants are autotrophic eukaryotes capable of generating organic molecules through photosynthesis. Fungi are heterotrophs that grow on or near their food and that feed by absorption (often after they have released enzymes that digest the food outside their bodies). Unlike plants, animals cannot construct all of their own organic molecules, and so, in most cases, they ingest them—either by eating other living organisms or by eating nonliving organic material. But unlike fungi, most animals feed by ingesting their food and then using enzymes to digest it within their bodies.

Cell Structure and Specialization

Animals are eukaryotes, and like plants and most fungi, animals are multicellular. In contrast to plants and fungi, however, animals lack the structural support of cell walls. Instead, proteins external to the cell membrane provide structural support to animal cells and connect them to one another (see Figure 6.28). The most abundant of these proteins is collagen, which is not found in plants or fungi.

The cells of most animals are organized into **tissues**, groups of similar cells that act as a functional unit. For example, muscle tissue and nervous tissue are responsible for moving the body and conducting nerve impulses, respectively. The ability to move and conduct nerve impulses underlies many of the adaptations that differentiate animals from plants and fungi (which lack muscle and nerve cells). For this reason, muscle and nerve cells are central to the animal lifestyle.

Reproduction and Development

Most animals reproduce sexually, and the diploid stage usually dominates the life cycle. In the haploid stage, sperm and egg cells are produced directly by meiotic division, unlike what occurs in plants and fungi (see Figure 13.6). In most animal species, a small, flagellated sperm fertilizes a larger, nonmotile egg, forming a diploid zygote. The zygote then undergoes **cleavage**, a succession of mitotic cell divisions without cell growth between the divisions. During the development of

most animals, cleavage leads to the formation of a multicellular embryonic stage called a **blastula**, which in many animals takes the form of a hollow ball **(Figure 32.2)**. Following this stage is the process of **gastrulation**, during which the layers of embryonic tissues that will develop into adult body parts are produced (see also Figure 47.8). The resulting developmental stage is called a **gastrula**.

Although some animals, including humans, develop directly into adults, the life cycles of most animals include at least one larval stage. A **larva** is a sexually immature form of an animal that is morphologically distinct from the adult, usually eats different food, and may even have a different habitat than the adult, as in the case of the aquatic larva of a mosquito or dragonfly. Animal larvae eventually undergo **metamorphosis**, a developmental transformation that

▼ **Figure 32.2 Early embryonic development in animals.**

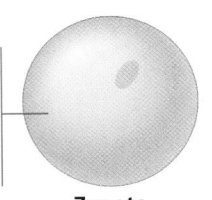

❶ The zygote of an animal undergoes a series of mitotic cell divisions called cleavage.

Zygote

Cleavage

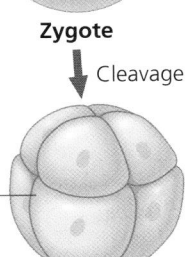

❷ An eight-cell embryo is formed by three rounds of cell division.

Eight-cell stage

Cleavage

❸ In most animals, cleavage produces a multicellular stage called a blastula. The blastula is typically a hollow ball of cells that surround a cavity called the blastocoel.

Blastula

Blastocoel

Cross section of blastula

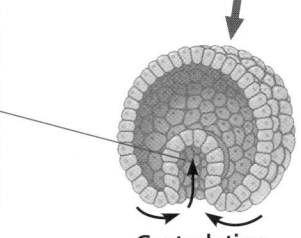

❹ Most animals also undergo gastrulation, a process in which one end of the embryo folds inward, expands, and eventually fills the blastocoel, producing layers of embryonic tissues: the ectoderm (outer layer) and the endoderm (inner layer).

Gastrulation

❺ The pouch formed by gastrulation, called the archenteron, opens to the outside via the blastopore.

❻ The endoderm of the archenteron develops into the tissue lining the animal's digestive tract.

Blastocoel

Endoderm

Ectoderm

Archenteron

Blastopore

Cross section of gastrula

turns the animal into a juvenile that resembles an adult but is not yet sexually mature.

Though adult animals vary widely in morphology, the genes that control animal development are similar across a broad range of taxa. All animals have developmental genes that regulate the expression of other genes, and many of these regulatory genes contain sets of DNA sequences called *homeoboxes* (see Concept 21.6). In particular, most animals share a unique homeobox-containing family of genes, known as *Hox* genes. *Hox* genes play important roles in the development of animal embryos, controlling the expression of many other genes that influence morphology.

CONCEPT CHECK 32.1

1. Summarize the main stages of animal development. What family of control genes plays a major role?
2. **WHAT IF?** What animal characteristics would be needed by an imaginary plant that could chase, capture, and digest its prey—yet could also extract nutrients from soil and conduct photosynthesis?

For suggested answers, see Appendix A.

CONCEPT 32.2

The history of animals spans more than half a billion years

To date, biologists have identified 1.3 million extant species of animals, and estimates of the actual number run far higher. This vast diversity encompasses a spectacular range of morphological variation, from corals to cockroaches to crocodiles. Various studies suggest that this great diversity originated during the last billion years. For example, researchers have unearthed 710-million-year-old sediments containing chemical evidence of steroids that today are primarily produced by a particular group of sponges. Since sponges are animals, these "fossil steroids" suggest that animals had arisen by 710 million years ago.

DNA analyses generally agree with this fossil biochemical evidence; for example, one recent molecular clock study estimated that sponges originated about 700 million years ago. These findings are also consistent with molecular analyses suggesting that the common ancestor of all extant animal species lived about 770 million years ago. What was this common ancestor like, and how did animals arise from their single-celled ancestors?

Steps in the Origin of Multicellular Animals

One way to gather information about the origin of animals is to identify protist groups that are closely related to animals. As shown in **Figure 32.3**, morphological and molecular evidence point to choanoflagellates as the closest living relatives of animals. Based on such evidence, researchers have hypothesized that the common ancestor of choanoflagellates and living animals may have been a suspension feeder similar to present-day choanoflagellates.

Scientists exploring *how* animals may have arisen from their single-celled ancestors have noted that the origin of multicellularity requires the evolution of new ways for cells to adhere (attach) and signal (communicate) to each other. To learn more about such mechanisms, researchers compared

▼ **Figure 32.3 Three lines of evidence that choanoflagellates are closely related to animals.**

? *Are the data described in* **3** *consistent with predictions that could be made from the evidence in* **1** *and* **2***? Explain.*

1 Morphologically, choanoflagellate cells and the collar cells (or *choanocytes*) of sponges, a basal group of animals, are almost indistinguishable.

Individual choanoflagellate

Collar cell (choanocyte)

Choanoflagellates

OTHER EUKARYOTES

Animals

Sponges

Other animals

3 DNA sequence data indicate that choanoflagellates and animals are sister groups. In addition, genes for signaling and adhesion proteins previously known only from animals have been discovered in choanoflagellates.

2 Similar collar cells have been identified in other animals, including cnidarians, flatworms, and echinoderms—but they have never been observed in non-choanoflagellate protists or in plants or fungi.

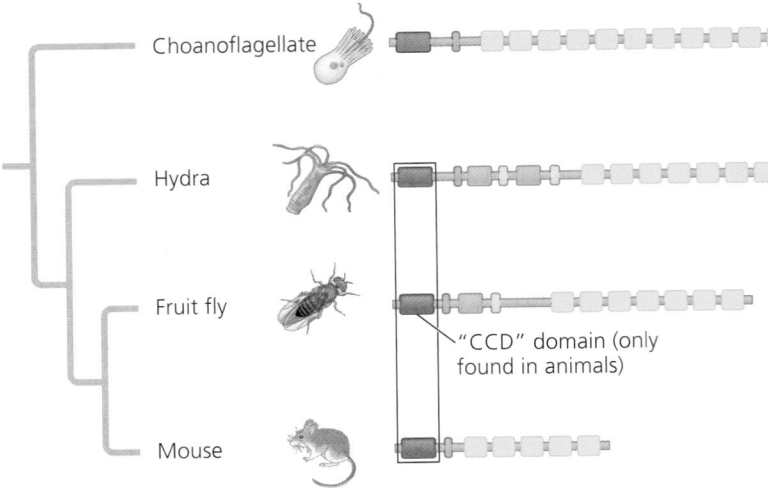

Choanoflagellate

Hydra

Fruit fly

"CCD" domain (only found in animals)

Mouse

◀ **Figure 32.4 Cadherin proteins in choanoflagellates and animals.** The ancestral cadherin-like protein of choano-flagellates has seven kinds of domains (regions), each represented here by a particular symbol. With the exception of the "CCD" domain, which is found only in animals, the domains of animal cadherin proteins are present in the choanoflagellate cadherin-like protein. The cadherin protein domains shown here were identified from whole-genome sequence data; evolutionary relationships are based on morphological and DNA sequence data.

the genome of the unicellular choanoflagellate *Monosiga brevicollis* with those of representative animals. This analysis uncovered 78 protein domains in *M. brevicollis* that were otherwise only known to occur in animals. (A *domain* is a key structural or functional region of a protein.) For example, *M. brevicollis* has genes that encode domains of certain proteins (known as cadherins) that play key roles in how animal cells attach to one another, as well as genes that encode protein domains that animals (and only animals) use in cell-signaling pathways.

Let's take a closer look at the cadherin attachment proteins we just mentioned. DNA sequence analyses show that animal cadherin proteins are composed primarily of domains that are also found in a cadherin-like protein of choanoflagellates **(Figure 32.4)**. However, animal cadherin proteins also contain a highly conserved region not found in the choanoflagellate protein (the "CCD" domain labeled in Figure 32.4). These data suggest that the cadherin attachment protein originated by the rearrangement of protein domains found in choanoflagellates plus the incorporation of a novel domain, the conserved CCD region. Overall, comparisons of choano-flagellate and animal genomes suggest that key steps in the transition to multicellularity in animals involved new ways of using proteins or parts of proteins that were encoded by genes found in choanoflagellates.

Next, we'll survey the fossil evidence for how animals evolved from their distant common ancestor over four geologic eras (see Table 25.1 to review the geologic time scale).

⊙ Mastering Biology
Interview with Nicole King: Investigating the ancestry of choanoflagellates

Neoproterozoic Era (1 Billion–541 Million Years Ago)

Although data from fossil steroids and molecular clocks indicate an earlier origin, the first generally accepted

macroscopic fossils of animals date from about 560 million years ago. These fossils are members of an early group of mostly soft-bodied multicellular eukaryotes, known collectively as the **Ediacaran biota**. The name comes from the Ediacara Hills of Australia, where fossils of these organisms were first discovered **(Figure 32.5)**. Similar fossils have since been found on other continents. Among the oldest Ediacaran fossils that resemble animals, some have been classified as molluscs (snails and their relatives) or close relatives of the molluscs, while others are thought to be sponges or cnidarians (sea anemones and their relatives). Still others have proved difficult to classify, as they do not seem to be closely related to any living animal or algal groups. In addition to these macroscopic fossils, Neoproterozoic rocks have also yielded what may be microscopic fossils of early animal embryos. Although these microfossils appear to exhibit the basic structural organization of present-day animal

(a) *Dickinsonia costata* (animal phylum unknown)

2.5 cm

0.5 cm

(b) *Kimberella*, a mollusc (or close relative)

▲ **Figure 32.5 Ediacaran fossil animals.** Fossils dating to 560 million years ago document the earliest known macroscopic animals, including these two species.

▼ **Figure 32.6 Early evidence of predation.** This 550-million-year-old fossil of the animal *Cloudina* shows evidence of having been attacked by a predator that bored through its shell.

Bore hole

0.1 mm

embryos, debate continues about whether these fossils are indeed of animals.

The fossil record from the Ediacaran period (635–541 million years ago) also provides early evidence of predation. Consider *Cloudina*, a small animal whose body was protected by a shell resembling a series of nested cones **(Figure 32.6)**. Some *Cloudina* fossils show signs of attack: round "bore holes" that resemble those formed today by predators that drill through the shells of their prey to gain access to the soft-bodied organisms lying within. Like *Cloudina*, some other small Ediacaran animals had shells or other defensive structures that may have been selected for by predators. Overall, the fossil evidence indicates that the Ediacaran was a time of increasing animal diversity—a trend that continued in the Paleozoic.

Paleozoic Era (541–252 Million Years Ago)

Another wave of animal diversification occurred 535–525 million years ago, early in the Cambrian period of the Paleozoic era—a phenomenon referred to as the **Cambrian explosion** (see Concept 25.3). In strata formed before the Cambrian explosion, only a few animal phyla have been observed. But in strata that are 535–525 million years old, paleontologists have found the oldest fossils of about half of all extant animal phyla, including the first arthropods, chordates, and echinoderms. Many of these fossils, which include the first large animals with hard, mineralized skeletons, look very different from most living animals **(Figure 32.7)**. Even so, paleontologists have established that these Cambrian fossils are members of extant animal phyla, or at least are close relatives. In particular, most of the fossils from the Cambrian explosion are of **bilaterians**, an enormous clade whose members (unlike sponges and cnidarians) typically have a two-sided or bilaterally symmetric form and a

508-million-year-old fossil of *Hallucigenia*

complete digestive tract, an efficient digestive system that has a mouth at one end and an anus at the other. As we'll discuss later in the chapter, bilaterians include molluscs, arthropods, chordates, and most other living animal phyla.

As the diversity of animal phyla increased during the Cambrian, the diversity of Ediacaran life-forms declined. What caused these trends? Fossil evidence suggests that during the early Cambrian period, as predators acquired novel adaptations for catching prey, such as claws, prey species acquired new defenses, such as protective shells. As new predator-prey relationships emerged, natural selection may have led to the decline of the soft-bodied Ediacaran species and the rise of various bilaterian phyla. Another hypothesis focuses on an increase in atmospheric oxygen that preceded the Cambrian explosion. More plentiful oxygen would have enabled animals with higher metabolic rates and larger body sizes to thrive, while potentially harming other species. A third hypothesis proposes that genetic changes affecting development, such as the origin of *Hox* genes and the addition of new microRNAs (small RNAs involved in gene regulation), facilitated the evolution of new body forms. In the **Scientific Skills Exercise**, you can investigate whether there is a correlation between microRNAs (miRNAs; see Figure 18.15) and body complexity in various animal phyla. These various hypotheses are not mutually exclusive; predator-prey relationships, atmospheric changes, and changes in development may each have played a role.

The Cambrian period was followed by the Ordovician, Silurian, and Devonian periods, when animal diversity continued to increase, although punctuated by episodes of mass extinction (see Figure 25.18). Vertebrates (fishes) emerged

1 cm

▲ **Figure 32.7 A Cambrian seascape.** This artist's reconstruction depicts a diverse array of organisms found in fossils from the Burgess Shale site in British Columbia, Canada. The animals include *Pikaia* (eel-like chordate at top left), *Marella* (small arthropod swimming at left), *Anomalocaris* (large animal with grasping limbs near a circular mouth), and *Hallucigenia* (animals with toothpick-like spikes on the seafloor and in inset).

Calculating and Interpreting Correlation Coefficients

Is Animal Complexity Correlated with miRNA Diversity? Animal phyla vary greatly in morphology, from simple sponges that lack tissues and symmetry to complex vertebrates. Members of different animal phyla have similar developmental genes, but the number of miRNAs varies considerably. In this exercise, you will explore whether miRNA diversity is correlated to morphological complexity.

How the Study Was Done In the analysis, miRNA diversity is represented by the average number of miRNAs in a phylum (x), while morphological complexity is represented by the average number of cell types (y). Researchers examined the relationship between these two variables by calculating the correlation coefficient (r). The correlation coefficient indicates the extent and direction of a linear relationship between two variables (x and y) and ranges in value between -1 and 1. When $r < 0$, x and y are negatively correlated, meaning that in a plot of the data points, values of y become smaller as values of x become larger. When $r > 0$, x and y are positively correlated (y becomes larger as x becomes larger). When $r = 0$, the variables are not correlated.

The formula for the correlation coefficient r is the following:

$$r = \frac{\sum (x_i - \bar{x})(y_i - \bar{y})}{(n - 1)(s_x s_y)}$$

In this formula, n is the number of observations, x_i is the value of the ith observation of variable x, and y_i is the value of the ith observation of variable y. $\bar{x}$ and $\bar{y}$ are the means of variables x and y, and s_x and s_y are the standard deviations of variables x and y. The $\sum$ symbol indicates that the n values of the product $(x_i - \bar{x})(y_i - \bar{y})$ are to be added together.

Data from the Study

Animal Phylum	i	No. of miRNAs (x_i)	($x_i - \bar{x}$)	($x_i - \bar{x}$)^2	No. of Cell Types (y_i)	($y_i - \bar{y}$)	($y_i - \bar{y}$)^2	($x_i - \bar{x}$)($y_i - \bar{y}$)
Porifera	1	5.8			25			
Platyhelminthes	2	35			30			
Cnidaria	3	2.5			34			
Nematoda	4	26			38			
Echinodermata	5	38.6			45			
Cephalochordata	6	33			68			
Arthropoda	7	59.1			73			
Urochordata	8	25			77			
Mollusca	9	50.8			83			
Annelida	10	58			94			
Vertebrata	11	147.5			172.5			
		$\bar{x} =$ $s_x =$		$\sum =$	$\bar{y} =$ $s_y =$		$\sum =$	$\sum =$

Data from B. Deline et al., Evolution of metazoan morphological disparity, *Proceedings of the National Academy of Sciences USA* 115:E8909–E8918 (2018).

INTERPRET THE DATA

1. First, practice reading the data table. For the eighth observation ($i = 8$), what are x_i and y_i? For which phylum are these data?

2. Next, we'll calculate the mean and standard deviation for each variable (to review these topics, see the Scientific Skills Review in Appendix D). **(a)** The **mean** ($\bar{x}$) is the sum of the data values divided by n, the number of observations: $\bar{x} = \dfrac{\sum x_i}{n}$ Calculate the mean number of miRNAs ($\bar{x}$) and the mean number of cell types ($\bar{y}$) and enter them in the data table (to calculate $\bar{y}$, replace each x in the formula with a y). **(b)** Next, calculate ($x_i - \bar{x}$) and ($y_i - \bar{y}$) for each observation, recording your results in the appropriate column. Square each of those results to complete the ($x_i - \bar{x}$)^2 and ($y_i - \bar{y}$)^2 columns; sum the results for those columns. **(c)** Next, we'll calculate the **standard deviation**, s, which describes the variation found in a data set. Calculate s_x, the standard deviation of the miRNA data, using the following formula, where n = the number of observations. [Use the sum of ($x_i - \bar{x}$)^2 you found in part (b).]

$$s_x = \sqrt{\frac{\sum (x_i - \bar{x})^2}{n - 1}}$$

Repeat to find s_y by substituting y for x in this equation.

3. Next, we'll calculate the correlation coefficient r for the variables x and y. **(a)** First, use the results from question 2(b) to complete the ($x_i - \bar{x}$)($y_i - \bar{y}$) column; sum the results in that column. **(b)** Now find the product of $s_x s_y$ using your results from 2(c) above. Plug these values into the formula for r.

4. Do these data indicate that miRNA diversity and animal complexity are negatively correlated, positively correlated, or uncorrelated? Explain.

5. What does your analysis suggest about the role of miRNA diversity in the evolution of animal complexity?

➜ **Instructors:** A version of this Scientific Skills Exercise can be assigned in **Mastering Biology**.

---

as the top predators of the marine food web. By 450 million years ago, groups that diversified during the Cambrian period began to make an impact on land. Arthropods were the first animals to adapt to terrestrial habitats, as indicated by fragments of arthropod remains and by well-preserved fossils from several continents of millipedes, centipedes, and spiders. Another clue is seen in fossilized fern galls—enlarged cavities that fern plants form in response to stimulation by

resident insects, which then use the galls for protection. Fossils indicate that fern galls date back at least 302 million years, suggesting that insects and plants were influencing each other's evolution by that time.

Vertebrates colonized land around 365 million years ago and diversified into numerous terrestrial groups. Two of these survive today: the amphibians (such as frogs and salamanders) and the amniotes (reptiles, including birds, and mammals). We will explore these groups, known collectively as the tetrapods, in more detail in Chapter 34.

Mesozoic Era (252–66 Million Years Ago)

The animal phyla that had evolved during the Paleozoic began to spread into new habitats. In the oceans, the first coral reefs formed, providing other marine animals with new places to live. Some reptiles returned to the water, leaving plesiosaurs (see Figure 34.25) and other large aquatic predators as their descendants. On land, descent with modification in some tetrapods led to the origin of wings and other flight equipment in pterosaurs and birds. Large and small dinosaurs emerged, as both predators and herbivores. At the same time, the first mammals—tiny nocturnal insect-eaters—appeared on the scene. In addition, as you read in Concept 30.3, flowering plants (angiosperms) and insects both underwent dramatic diversifications during the late Mesozoic.

Cenozoic Era (66 Million Years Ago to the Present)

Mass extinctions of both terrestrial and marine animals ushered in a new era, the Cenozoic. Among the groups of species that disappeared were the large, nonflying dinosaurs and the marine reptiles. The fossil record of the early Cenozoic documents the rise of large mammalian herbivores and predators as mammals began to exploit the vacated ecological niches. The global climate gradually cooled throughout the Cenozoic, triggering significant shifts in many animal lineages. Among primates, for example, some species in Africa adapted to the open woodlands and savannas that replaced many of the former dense forests. The ancestors of our own species were among those grassland apes.

CONCEPT CHECK 32.2

1. Put the following milestones in animal evolution in order from oldest to most recent: (a) origin of mammals, (b) earliest evidence of terrestrial arthropods, (c) Ediacaran fauna, (d) extinction of large, nonflying dinosaurs.
2. **VISUAL SKILLS** Explain what is represented by the red-colored portion of the branch leading to animals. (See Figure 26.5, "Visualizing Phylogenetic Relationships," to review phylogenetic tree diagrams.)

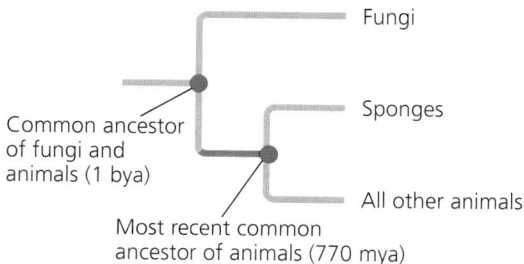

Fungi

Common ancestor of fungi and animals (1 bya)

Sponges

Most recent common ancestor of animals (770 mya)

All other animals

3. **MAKE CONNECTIONS** Evaluate whether the origin of cell-to-cell attachment proteins in animals illustrates descent with modification. (See Concept 22.2.)

For suggested answers, see Appendix A.

Animals can be characterized by body plans

Animal species vary tremendously in morphology, but their great diversity in form can be described by a relatively small number of major "body plans." A **body plan** is a particular set of morphological and developmental traits that are integrated into a functional whole—the living animal. The term *plan* here does not imply that animal forms are the result of conscious planning or invention. But body plans do provide a succinct way to compare and contrast key animal features. They also are of interest in the study of *evo-devo*, the interface between evolution and development.

Like all features of organisms, animal body plans have evolved over time. In some cases, including key stages in gastrulation, novel body plans emerged early in the history of animal life and have not changed since. As we'll discuss, however, other aspects of animal body plans have changed multiple times over the course of evolution. As we explore the major features of animal body plans, bear in mind that similar body forms may have evolved independently in different lineages. In addition, body features can be lost over the course of evolution, causing some closely related species to look very different from one another.

Symmetry

A basic feature of animal bodies is their type of symmetry—radial or bilateral—or absence of symmetry. (Many sponges, for example, lack symmetry altogether.) Use **Figure 32.8** to help you visualize radial and bilateral symmetry and how axes of orientation are used to describe the position of different parts of an animal's body.

The symmetry of an animal generally fits its lifestyle. Many radial animals are sessile (living attached to a substrate) or planktonic (drifting or weakly swimming, such as jellies, commonly called jellyfishes). Their symmetry equips them to meet the environment equally well from all sides. In contrast, bilateral animals typically move actively from place to place. Nearly all animals with a bilaterally symmetric body plan (such as arthropods and mammals) have sensory equipment concentrated at the head end of their body, including a central nervous system ("brain"). This central nervous system enables them to coordinate the complex movements involved in crawling, burrowing, flying, or swimming. Fossil evidence indicates that these two fundamentally different kinds of symmetry have existed for at least 550 million years.

Tissues

Animal body plans also vary with regard to tissue organization. Recall that tissues are collections of specialized cells that act as a functional unit. Sponges and a few other groups lack tissues. In all other animals, the embryo becomes layered during gastrulation (see Figure 47.8, "Visualizing Gastrulation,"

A basic feature of animal bodies is their type of symmetry. We'll focus on two common types of symmetry: radial symmetry and bilateral symmetry.

Radial Symmetry

The body parts of a radially symmetrical animal are arranged around a single main axis that passes through the center of the animal. Any imaginary slice through the central axis, as shown here in blue, divides the animal into mirror images.

1. Does a radial animal have a front side and a back side? A left side and a right side? Explain.

Central axis

The symmetrical central body cavity of this hydra has a single opening (the "mouth/anus") surrounded by a circular rim of tentacles. Prey captured by the tentacles are pushed into the body cavity and digested.

Many sponges have an irregular shape, lacking symmetry altogether. But like this barrel sponge, some sponges exhibit radial symmetry, with their cells arranged in a regular pattern around a central axis.

Bilateral Symmetry and Body Axes

The body parts of a bilaterally symmetrical animal are arranged around two axes of orientation, the head-tail axis and the dorsal-ventral axis. Only one imaginary slice, again shown in blue, divides the animal into mirror-image halves, referred to as a right side and a left side.

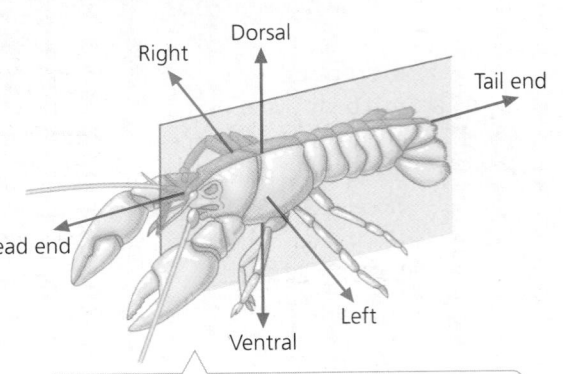

Dorsal
Right
Tail end
Head end
Ventral
Left

Lobsters and other bilateral animals have a head end and a tail end, a dorsal (top) side and a ventral (bottom) side, as well as a right side and a left side. In nearly all bilateral animals, eyes and other sensory organs are concentrated in the head end of the body.

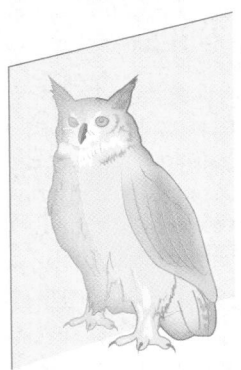

An upright posture evolved in some groups of bilateral animals, including humans and birds, such as the owl shown here.

2. Draw the two axes of orientation on this owl and label its head end, tail end, and dorsal, ventral, right, and left sides.

➔ **Instructors:** Additional questions related to this Visualizing Figure can be assigned in **Mastering Biology**.

---

which will help you understand this three-dimensional folding process). As development progresses, these layers, called *germ layers*, form the various tissues and organs of the body. **Ectoderm**, the germ layer covering the surface of the embryo, gives rise to the outer covering of the animal and, in some phyla, to the central nervous system. **Endoderm**, the innermost germ layer, lines the pouch that forms during gastrulation (the archenteron) and gives rise to the lining of the digestive tract (or cavity) and to the lining of organs such as the liver and lungs of vertebrates.

Cnidarians and a few other animal groups that have only these two germ layers are said to be **diploblastic**. All bilaterally symmetrical animals have a third germ layer, called the **mesoderm**, which fills much of the space between the ectoderm and endoderm. Thus, animals with bilateral symmetry are also said to be **triploblastic** (having three germ layers). In triploblasts, the mesoderm forms the muscles and most

other organs between the digestive tract and the outer covering of the animal. Triploblasts include a broad range of animals, from flatworms to arthropods to vertebrates. (Although some diploblasts actually do have a third germ layer, it is not nearly as well developed as the mesoderm of animals considered to be triploblastic.)

Body Cavities

Nearly all animals have a **body cavity**, a fluid- or air-filled space located between the digestive tract (endoderm) and the outer body wall (ectoderm). Body cavities have diverse functions, such as to provide structural support and to facilitate the internal transport of nutrients, gases, and wastes.

Many triploblastic animals have a **coelom** (from the Greek *koilos*, hollow), a body cavity that forms from tissue derived from mesoderm. The inner and outer layers of mesoderm that surround the cavity connect and form structures that suspend the internal organs **(Figure 32.9a)**. A coelom's fluid cushions the suspended organs, helping to prevent internal injury. In soft-bodied animals, such as earthworms, the fluid in the coelom acts like a skeleton against which muscles can work. A coelom also enables the internal organs to grow and move independently of the outer body wall. If it were not for your coelom, for example, every beat of your heart or ripple of your intestine would warp your body's surface. Animals possessing coeloms are sometimes called *coelomates*.

Other triploblastic animals have a **hemocoel**, a body cavity that forms between the mesoderm and endoderm **(Figure 32.9b)**. A hemocoel contains *hemolymph*, a fluid that functions in the internal transport of nutrients and waste.

▼ **Figure 32.9 Body cavities of triploblastic animals.** The organ systems develop from the three embryonic germ layers.

(a) Coelom

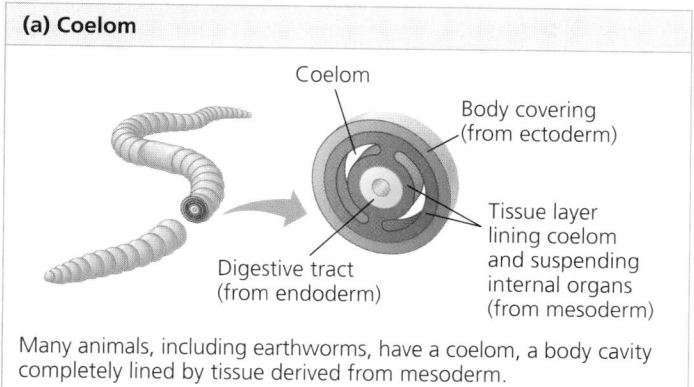

Many animals, including earthworms, have a coelom, a body cavity completely lined by tissue derived from mesoderm.

(b) Hemocoel

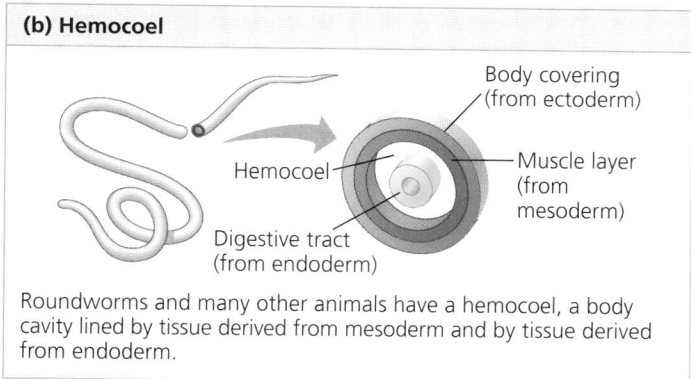

Roundworms and many other animals have a hemocoel, a body cavity lined by tissue derived from mesoderm and by tissue derived from endoderm.

(c) Compact (no body cavity)

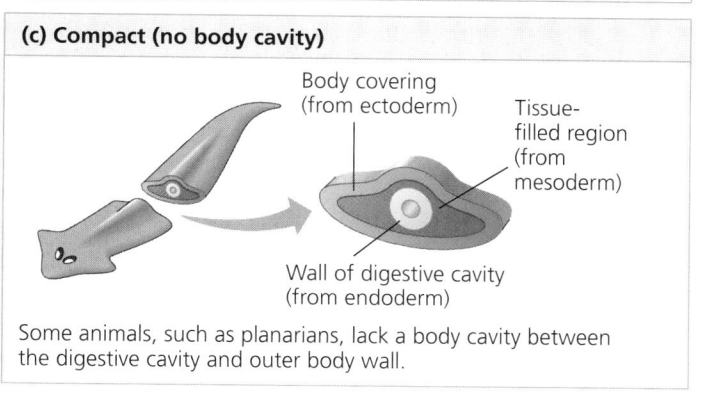

Some animals, such as planarians, lack a body cavity between the digestive cavity and outer body wall.

Key

■ Ectoderm ■ Mesoderm ▫ Endoderm

Hemolymph is analogous to your blood and is circulated through the body cavity in an open system by the heart.

Many animals have a hemocoel and a coelom. Molluscs, for instance, have a hemocoel as the primary body cavity, but they also have a reduced coelom that surrounds the heart and reproductive structures (see Figure 33.16). Other animals, such as rotifers and nematodes (see Figures 33.12 and 33.26), have only a hemocoel. Animals with only a hemocoel once were called *pseudocoelomates* (from the Greek *pseudo*, false) and the cavity, a pseudocoelom. Hemocoels and coeloms have been independently gained or lost multiple times in the course of animal evolution; hence, their presence or absence is not a good indicator of phylogenetic relationships.

Finally, some triploblastic animals lack a body cavity altogether **(Figure 32.9c)**. These compact animals tend to have thin, flat bodies. Such animals don't require an internal transport system: With bodies that are only a few cells thick, the exchange of nutrients, gases, and wastes can occur across the entire body surface. These animals are sometimes called *acoelomates* (from the Greek *a-*, without).

Protostome and Deuterostome Development

Many animals can be described as having one of two developmental modes: **protostome development** or **deuterostome development**. These modes can generally be distinguished by differences in cleavage, coelom formation, and fate of the blastopore.

Cleavage

Many animals with protostome development undergo **spiral cleavage**, in which the planes of cell division are diagonal to the vertical axis of the embryo; as seen in the eight-cell stage of the embryo, smaller cells are centered over the grooves between larger, underlying cells **(Figure 32.10a**, left). Furthermore, the so-called **determinate cleavage** of some animals with protostome development rigidly casts ("determines") the developmental fate of each embryonic cell very early. A cell isolated from a snail at the four-cell stage, for example, cannot develop into a whole animal. Instead, after repeated divisions, such a cell will form an inviable embryo that lacks many parts.

In contrast to the spiral cleavage pattern, deuterostome development is predominantly characterized by **radial cleavage**. The cleavage planes are either parallel or perpendicular to the vertical axis of the embryo; as seen at the eight-cell stage, the tiers of cells are aligned, one directly above the other (see Figure 32.10a, right). Most animals with deuterostome development also have **indeterminate cleavage**, meaning that each cell produced by early cleavage divisions retains the capacity to develop into a complete embryo. For example, if the cells of a sea urchin embryo are separated at the four-cell stage, each can form a complete larva. Similarly, it is the indeterminate cleavage of the human zygote that makes identical twins possible.

Coelom Formation

During gastrulation, an embryo's developing digestive tube initially forms as a blind pouch, the **archenteron**, which becomes the gut **(Figure 32.10b)**. As the archenteron forms in protostome development, initially solid masses of mesoderm split and form the coelom. In contrast, in deuterostome development, the mesoderm buds from the wall of the archenteron, and its cavity becomes the coelom.

Fate of the Blastopore

Protostome and deuterostome development often differ in the fate of the **blastopore**, the indentation that during gastrulation leads to the formation of the archenteron **(Figure 32.10c)**. After the archenteron develops, in most animals a second opening forms at the opposite end of the gastrula. In many species,

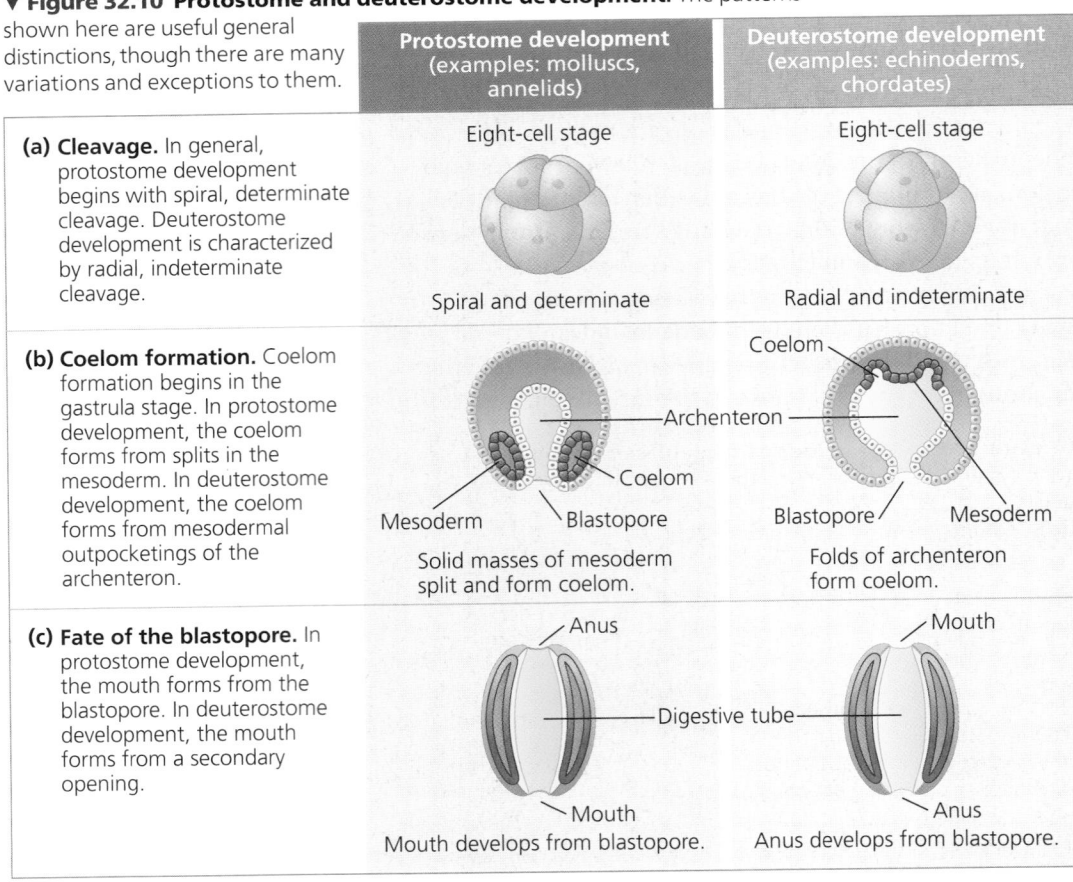

▼ **Figure 32.10 Protostome and deuterostome development.** The patterns shown here are useful general distinctions, though there are many variations and exceptions to them.

	Protostome development (examples: molluscs, annelids)	Deuterostome development (examples: echinoderms, chordates)
(a) Cleavage. In general, protostome development begins with spiral, determinate cleavage. Deuterostome development is characterized by radial, indeterminate cleavage.	Eight-cell stage — Spiral and determinate	Eight-cell stage — Radial and indeterminate
(b) Coelom formation. Coelom formation begins in the gastrula stage. In protostome development, the coelom forms from splits in the mesoderm. In deuterostome development, the coelom forms from mesodermal outpocketings of the archenteron.	Mesoderm, Archenteron, Coelom, Blastopore — Solid masses of mesoderm split and form coelom.	Coelom, Archenteron, Blastopore, Mesoderm — Folds of archenteron form coelom.
(c) Fate of the blastopore. In protostome development, the mouth forms from the blastopore. In deuterostome development, the mouth forms from a secondary opening.	Anus, Digestive tube, Mouth — Mouth develops from blastopore.	Mouth, Digestive tube, Anus — Anus develops from blastopore.

Key ▪ Ectoderm ▪ Mesoderm ▫ Endoderm

the blastopore and this second opening become the two openings of the digestive tube: the mouth and the anus. In protostome development, the mouth generally develops from the first opening, the blastopore, and it is for this characteristic that the term *protostome* derives (from the Greek *protos*, first, and *stoma*, mouth). In deuterostome development (from the Greek *deuteros*, second), the mouth is derived from the secondary opening, and the blastopore usually forms the anus.

CONCEPT CHECK 32.3

1. Compare three aspects of the early development of a snail (a mollusc) and a human (a chordate).

2. Describe how animals that lack a body cavity exchange materials without an internal transport system.

3. **WHAT IF?** Evaluate this claim: Ignoring the details of their specific anatomy, worms, humans, and most other triploblasts have a shape analogous to that of a doughnut.

For suggested answers, see Appendix A.

CONCEPT 32.4

Views of animal phylogeny continue to be shaped by new molecular and morphological data

As animals with diverse body plans radiated during the early Cambrian, some lineages arose, thrived for a period of time,

and then became extinct, leaving no descendants. However, by 500 million years ago, most animal phyla with members alive today were established. Next, we'll examine relationships among these taxa along with some remaining questions that are currently being addressed using genomic data.

The Diversification of Animals

Zoologists currently recognize about three dozen phyla of extant animals, 15 of which are shown in **Figure 32.11**. Researchers infer evolutionary relationships among these phyla by analyzing whole genomes, as well as morphological traits, ribosomal RNA (rRNA) genes, *Hox* genes, protein-coding nuclear genes, and mitochondrial genes. Notice how the following points are reflected in Figure 32.11.

1. **All animals share a common ancestor.** Current evidence indicates that animals are monophyletic, forming a clade called Metazoa. All extant and extinct animal lineages have descended from a common ancestor.

2. **Sponges are the sister group to all other animals.** Sponges (phylum Porifera) are basal animals, having diverged from all other animals early in the history of the group. Recent morphological and phylogenomic analyses indicate that sponges are monophyletic, as shown here.

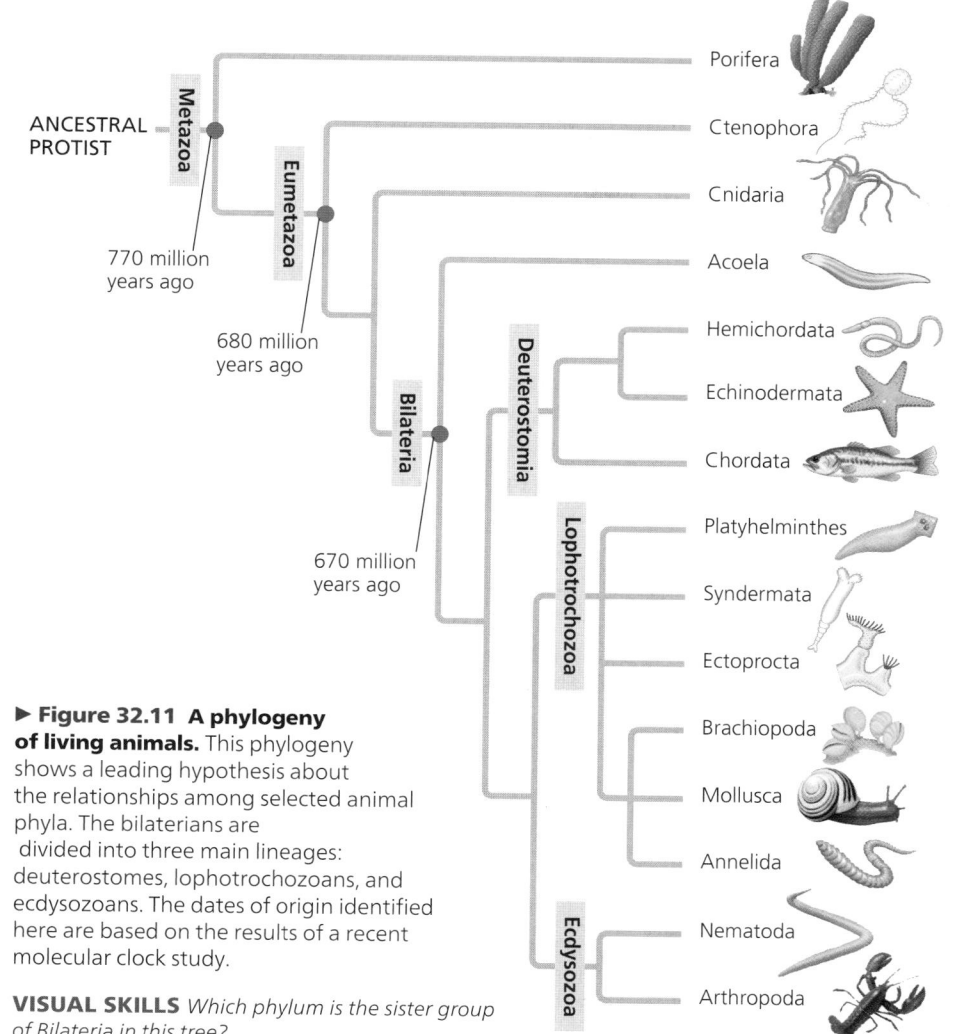

Porifera

Ctenophora

Cnidaria

Acoela

Hemichordata

Echinodermata

Chordata

Platyhelminthes

Syndermata

Ectoprocta

Brachiopoda

Mollusca

Annelida

Nematoda

Arthropoda

ANCESTRAL PROTIST

Metazoa

770 million years ago

Eumetazoa

680 million years ago

Bilateria

670 million years ago

Deuterostomia

Lophotrochozoa

Ecdysozoa

▶ **Figure 32.11 A phylogeny of living animals.** This phylogeny shows a leading hypothesis about the relationships among selected animal phyla. The bilaterians are divided into three main lineages: deuterostomes, lophotrochozoans, and ecdysozoans. The dates of origin identified here are based on the results of a recent molecular clock study.

VISUAL SKILLS *Which phylum is the sister group of Bilateria in this tree?*

Lophophore

(a) Lophophore feeding structures of an ectoproct

Apical tuft of cilia

Mouth

Anus

(b) Structure of a trochophore larva

▲ **Figure 32.12 Morphological characteristics of lophotrochozoans.**

3. **Eumetazoa is a clade of animals with tissues.** All animals except for sponges and a few others belong to a clade of **eumetazoans** ("true animals"). Members of this group have tissues, such as muscle tissue and nervous tissue. Basal eumetazoans, which include the phyla Ctenophora (comb jellies) and Cnidaria, are diploblastic and generally have radial symmetry.

4. **Most animal phyla belong to the clade Bilateria.** Bilateral symmetry and the presence of three prominent germ layers are shared derived characters that help define the clade Bilateria. This clade contains the majority of animal phyla, and its members are known as *bilaterians*. The Cambrian explosion was primarily a rapid diversification of bilaterians.

5. **There are three major clades of bilaterian animals.** Bilaterians have diversified into three main lineages, Deuterostomia, Lophotrochozoa, and Ecdysozoa. With one exception, the phyla in these clades consist entirely of **invertebrates**, animals that lack a backbone; Chordata is the only phylum that also includes **vertebrates**, animals with a backbone.

As depicted in Figure 32.11, hemichordates (acorn worms), echinoderms (sea stars and relatives), and chordates are members of the bilaterian clade **Deuterostomia**; thus, the term *deuterostome* refers not only to a mode of animal development, but also to the members of this clade. (The dual meaning of this term can be confusing since some organisms with a deuterostome developmental pattern are *not* members of clade Deuterostomia.) Hemichordates share some characteristics with chordates, such as gill slits and a dorsal nerve cord; echinoderms lack these characteristics. These shared traits may have been present in the common ancestor of the deuterostome clade (and lost in the echinoderm lineage). As mentioned above, phylum Chordata, the only phylum with vertebrate members, also includes invertebrates.

Bilaterians also diversified in two major clades that are composed entirely of invertebrates: the *lophotrochozoans* and the *ecdysozoans*. The name **Lophotrochozoa** refers to two different features observed in some animals belonging to this clade **(Figure 32.12).** Some lophotrochozoans, such as ectoprocts, develop a unique structure called a **lophophore** (from the Greek *lophos,* crest, and *pherein,* to carry), a crown of ciliated tentacles that function in feeding (see Figure 32.12a). Individuals in other phyla, including molluscs and annelids, go through a distinctive developmental stage called the **trochophore larva** (see Figure 32.12b)—hence the name lophotrochozoan.

The clade name **Ecdysozoa** refers to a characteristic shared by nematodes, arthropods, and some of the other ecdysozoan phyla that are not included in our survey. These animals secrete external skeletons (exoskeletons); the stiff covering of a cricket and the flexible cuticle of a nematode are examples. As the animal grows, it molts, squirming out of its old exoskeleton and secreting a larger one. The process of shedding the old exoskeleton is called *ecdysis*. Though named for this characteristic, the clade was proposed mainly on the basis of molecular data that support the common ancestry of its members. Furthermore, some taxa excluded from this clade by their molecular data, such as certain species of leeches, do in fact molt.

Future Directions in Animal Systematics

While many scientists think that current evidence supports the phylogeny in Figure 32.11, aspects continue to be debated. Although it can be frustrating that the phylogenies in textbooks are not set-in-stone truths, the uncertainty inherent in these diagrams is a healthy reminder that science is an ongoing, dynamic process of inquiry. We'll conclude with two questions that are the focus of ongoing research.

1. **Are ctenophores basal metazoans?** Many researchers have concluded that sponges are basal metazoans (see Figure 32.11). This conclusion was supported in a 2017 phylogenomic analysis, but several other studies have placed the comb jellies (phylum Ctenophora) at the base of the animal tree. In addition to the most recent phylogenomic results, data consistent with placing sponges at the base of the animal tree include fossil steroid

evidence, molecular clock analyses, the morphological similarity of sponge collar cells to the cells of choanoflagellates (see Figure 32.3), and the fact that sponges are one of the few animal groups that lack tissues (as might be expected for basal animals). Ctenophores, on the other hand, have tissues, and their cells do not resemble the cells of choanoflagellates. At present, the idea that ctenophores are basal metazoans remains an intriguing but controversial hypothesis.

2. **Are the Acoela basal bilaterians?** A series of recent papers—including a 2016 phylogenomic study—have indicated that flatworms in phylum Acoela are basal bilaterians, as shown in Figure 32.11. A different conclusion was supported by one recent analysis, which placed members of the Acoela within Deuterostomia. Researchers are sequencing the genomes of additional species in the phylum and in closely related groups to provide a more definitive test of the hypothesis that the Acoela are basal bilaterians. If further evidence supports this hypothesis, bilaterians may have descended from a common ancestor that resembled this group of living flatworms—that is, an ancestor that had a simple nervous system, a saclike gut with a single opening (the "mouth"), and no excretory system.

CONCEPT CHECK 32.4

1. Describe the evidence that cnidarians share a more recent common ancestor with other animals than with sponges.

2. **WHAT IF?** Suppose ctenophores are basal metazoans and sponges are the sister group of all remaining animals. Under this hypothesis, redraw Figure 32.11 and discuss whether animals with tissues would form a clade.

3. **MAKE CONNECTIONS** Based on the phylogeny in Figure 32.11 and the information in Figure 25.11, evaluate this statement: "The Cambrian explosion actually consists of three explosions, not one."

For suggested answers, see Appendix A.

---

32 Chapter Review

→ Go to **Mastering Biology** for Assignments, the eText, the Study Area, and Dynamic Study Modules.

SUMMARY OF KEY CONCEPTS

→ To review key terms, go to the **Vocabulary Self-Quiz** in the **Mastering Biology** eText or Study Area, or go to goo.gl/zkjz9t.

CONCEPT 32.1

Animals are multicellular, heterotrophic eukaryotes with tissues that develop from embryonic layers
(pp. 674–675)

- Animals are heterotrophs that ingest their food.

- Animals are multicellular eukaryotes. Their cells are supported and connected to one another by collagen and other structural proteins located outside the cell membrane. Nervous tissue and muscle tissue are key animal features.

- In most animals, **gastrulation** follows the formation of the **blastula** and leads to the formation of embryonic tissue layers. Most animals have *Hox* genes that regulate the development of body form. Although *Hox* genes have been highly conserved over the course of evolution, they can produce a wide diversity of animal morphology.

? *Describe key ways that animals differ from plants and fungi.*

CONCEPT 32.2

The history of animals spans more than half a billion years *(pp. 675–679)*

- Fossil biochemical evidence and molecular clock analyses indicate that animals arose over 700 million years ago.
- Genomic analyses suggest that key steps in the origin of animals involved new ways of using proteins that were encoded by genes found in choanoflagellates.

? *What caused the Cambrian explosion? Describe current hypotheses.*

CONCEPT 32.3

Animals can be characterized by body plans *(pp. 679–682)*

- Animals may lack symmetry or may have **radial** or **bilateral symmetry**. Bilaterally symmetrical animals have **dorsal** and **ventral** sides, as well as **head** and **tail** ends.
- Eumetazoan embryos may be **diploblastic** (two germ layers) or **triploblastic** (three germ layers). Triploblastic animals with a **body cavity** may have a **coelom** or a **hemocoel** (or both).
- **Protostome** and **deuterostome development** often differ in patterns of **cleavage**, coelom formation, and **blastopore** fate.

? *Describe how body plans provide useful information yet should be interpreted cautiously as evidence of evolutionary relationships.*

CONCEPT 32.4

Views of animal phylogeny continue to be shaped by new molecular and morphological data *(pp. 682–684)*

- This phylogenetic tree shows key steps in animal evolution:

? *Consider clades Bilateria, Lophotrochozoa, Metazoa, Chordata, Ecdysozoa, Eumetazoa, and Deuterostomia. List the clades to which humans belong in order from the most to the least inclusive clade.*

TEST YOUR UNDERSTANDING

➔ For more multiple-choice questions, go to the **Practice Test** in the **Mastering Biology** eText or Study Area, or go to goo.gl/GruWRg.

Levels 1-2: Remembering/Understanding

1. One of the characteristics unique to animals is
   (A) gastrulation.
   (B) multicellularity.
   (C) sexual reproduction.
   (D) flagellated sperm.

2. The distinction between sponges and other animal phyla is based mainly on the absence versus the presence of
   (A) a body cavity.
   (B) a complete digestive tract.
   (C) mesoderm.
   (D) tissues.

3. Which of the following was probably an important factor in bringing about the Cambrian explosion?
   (A) the movement of animals onto land
   (B) an increase in the concentration of atmospheric nitrogen
   (C) the emergence of predator-prey relationships
   (D) the origin of bilaterian animals

Levels 3-4: Applying/Analyzing

4. Based on the tree in Figure 32.11, which statement is true?
   (A) The animal kingdom is not monophyletic.
   (B) The Acoela are more closely related to echinoderms than to annelids.
   (C) Sponges are basal animals.
   (D) Bilaterians do not form a clade.

Levels 5-6: Evaluating/Creating

5. **EVOLUTION CONNECTION** A professor begins a lecture on animal phylogeny (as shown in Figure 32.11) by saying, "We are all worms." In this context, what did she mean?

6. **WRITE ABOUT A THEME: INTERACTIONS** Animal life changed greatly during the Cambrian explosion, with some groups expanding in diversity and others declining. Write a short essay (100–150 words) interpreting these events as feedback regulation at the level of the biological community.

7. **SYNTHESIZE YOUR KNOWLEDGE**

This organism is an animal. What can you infer about its body structure and lifestyle (that might not be obvious from its appearance)? This animal has a protostome developmental pattern and a trochophore larva. Identify the major clades that this animal belongs to. Explain your selection, and describe when these clades originated and how they are related to one another.

For selected answers, see Appendix A.

33 An Introduction to Invertebrates

KEY CONCEPTS

33.1 Sponges are basal animals that lack tissues *p. 690*

33.2 Cnidarians are an ancient phylum of eumetazoans *p. 691*

33.3 Lophotrochozoans, a clade identified by molecular data, have the widest range of animal body forms *p. 694*

33.4 Ecdysozoans are the most species-rich animal group *p. 705*

33.5 Echinoderms and chordates are deuterostomes *p. 713*

Study Tip

Make a table: Invertebrates differ in how they obtain food, remove wastes, reproduce, and move throughout their habitat. As you read, summarize how members of the 11 phyla of invertebrates described in detail in this chapter solve these shared challenges of life.

Phylum	Feeding method	Waste removal	Repro-duction	Movement
Porifera	Filter feeders	Diffusion	Herma-phrodites	Sessile

→ Go to Mastering Biology

For Students (in eText and Study Area)
• Get Ready for Chapter 33
• Video: Echinoderm Tube Feet
• HHMI Animation: Coral Bleaching

For Instructors to Assign (in Item Library)
• Tutorial: Ecdysozoans
• Scientific Skills Exercise: Understanding Experimental Design and Interpreting Data

Figure 33.1 The blue dragon (*Glaucus atlanticus*) has thin, finger-like structures that help it to float (upside down) on the sea's surface. It eats Portuguese men-of-war and absorbs their stinging cells, which it then uses to deliver a deadly sting of its own. The blue dragon is just one of more than a million species of invertebrates (animals without a backbone), which account for over 95% of animal species.

How can we make sense of the great number and morphological diversity of invertebrates?

Classifying invertebrate species into groups based on **evolutionary relationships**—as in this simplified phylogeny of all animals—helps us to understand the great diversity of invertebrate life.

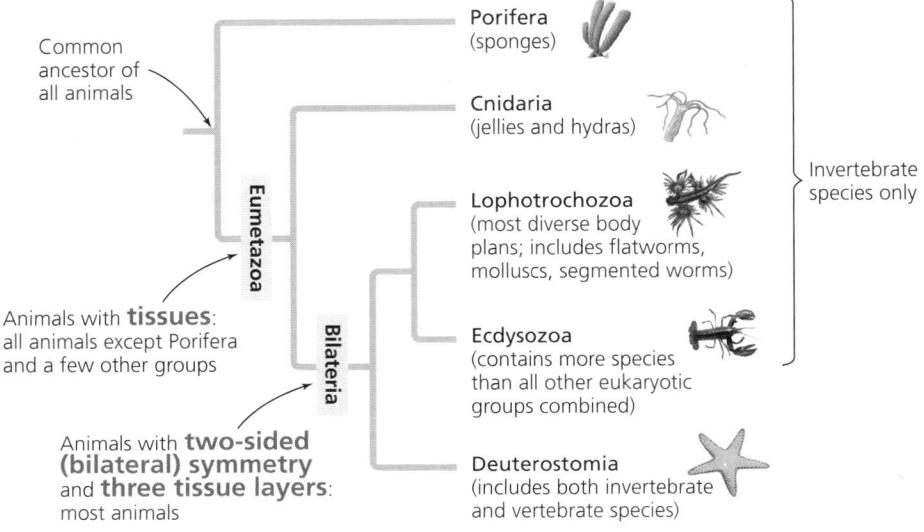

Common ancestor of all animals

Eumetazoa

Bilateria

Animals with **tissues**: all animals except Porifera and a few other groups

Animals with **two-sided (bilateral) symmetry** and **three tissue layers**: most animals

Porifera (sponges)

Cnidaria (jellies and hydras)

Lophotrochozoa (most diverse body plans; includes flatworms, molluscs, segmented worms)

Ecdysozoa (contains more species than all other eukaryotic groups combined)

Deuterostomia (includes both invertebrate and vertebrate species)

Invertebrate species only

Invertebrates are animals that lack a backbone. They occupy almost every habitat on Earth, from the scalding water released by deep-sea hydrothermal vents to the frozen ground of Antarctica. Evolution in these varied environments has produced an immense diversity of forms, including species that are microscopic and species whose members can grow to 18 m long (1.5 times the length of a school bus). In this chapter, we'll take a tour of the invertebrates, using the phylogenetic tree in Figure 33.1 as a guide. **Figure 33.2** surveys 23 phyla as representatives of invertebrate diversity.

▼ Figure 33.2 Exploring Invertebrate Diversity

Kingdom Animalia encompasses 1.3 million known species, and estimates of the total number of species range as high as 10–20 million. Of the 23 phyla surveyed here, 13 are discussed more fully in this chapter, Chapter 32, or Chapter 34; cross-references are given at the end of their descriptions.

Porifera (5,500 species)

Animals in this phylum are informally called sponges. Sponges are sessile animals that lack tissues. They live as filter feeders, trapping particles that pass through the internal channels of their body (see Concept 33.1).

A sponge

Cnidaria (10,000 species)

Cnidarians include corals, jellies, and hydras. These animals have a diploblastic, radially symmetrical body plan that includes a gastrovascular cavity with a single opening that serves as both mouth and anus (see Concept 33.2).

A jelly

Acoela (400 species)

Flatworms in this phylum have a simple nervous system and a saclike gut, and thus were once placed in phylum Platyhelminthes. Recent molecular analyses, however, indicate that Acoela is a separate lineage that diverged before the three main bilaterian clades (see Concept 32.4).

Acoela

Placozoa (1 species)

The single known species in this phylum, *Trichoplax adhaerens*, doesn't look like an animal. It consists of a simple bilayer of a few thousand cells and reproduces by dividing into two individuals or by budding off many multicellular individuals. Placozoans are thought to be basal animals, but it is not yet known how they are related to Porifera, Cnidaria, and other early-diverging phyla.

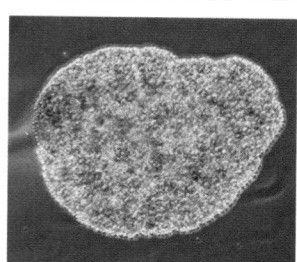

A placozoan (LM)

0.5 mm

Ctenophora (100 species)

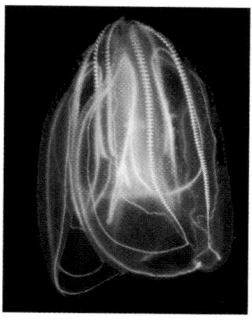

Ctenophores (comb jellies) are diploblastic and radially symmetrical like cnidarians, suggesting that both phyla diverged early from other animals. Comb jellies make up much of the ocean's plankton. They have eight "combs" of cilia that propel the animals through the water. When a small animal contacts the tentacles of some comb jellies, specialized cells burst open, covering the prey with sticky threads.

A ctenophore, or comb jelly

Lophotrochozoa

Platyhelminthes (20,000 species)

Flatworms in this phylum (including tapeworms, planarians, and flukes) have bilateral symmetry and a central nervous system that processes information from sensory structures. They have no body cavity or specialized organs for circulation (see Concept 33.3).

A marine flatworm

Syndermata (2,900 species)

This recently established phylum includes two groups formerly classified as separate phyla: the rotifers, microscopic animals with complex organ systems, and the acanthocephalans, highly modified parasites of vertebrates (see Concept 33.3).

A rotifer (LM)

Continued on next page

Ectoprocta (4,500 species)

Ectoprocts (also known as bryozoans) live as sessile colonies and are covered by a tough exoskeleton (see Concept 33.3).

Ectoprocts

Gastrotricha (800 species)

Gastrotrichs are tiny worms whose ventral surface is covered with cilia, leading them to be called hairy bellies. Most species live on the bottoms of lakes or oceans, where they feed on small organisms and partially decayed organic matter. This individual has consumed algae, visible as the greenish material inside its gut.

A gastrotrich (differential interference contrast LM)

Brachiopoda (335 species)

Brachiopods, or lamp shells, may be easily mistaken for clams or other molluscs. However, most brachiopods have a unique stalk that anchors them to their substrate, as well as a crown of cilia called a lophophore (see Concept 33.3).

A brachiopod

Cycliophora (1 species)

A cycliophoran (colorized SEM)

The only known cycliophoran species, *Symbion pandora*, was discovered in 1995 on the mouthparts of a lobster. This tiny creature has a unique body plan and a bizarre life cycle: Males impregnate females that are still developing in their mothers' bodies. The fertilized females escape, settle elsewhere on the lobster, and release their offspring.

Nemertea (900 species)

Also called ribbon worms, nemerteans swim through water or burrow in sand, extending a unique proboscis to capture prey. Nemerteans have a reduced coelom, so their bodies are relatively solid. They also have an alimentary canal and a closed circulatory system in which the blood is contained in vessels and hence is distinct from fluid in the body cavity.

A ribbon worm

Annelida (16,500 species)

Annelids, or segmented worms, are distinguished by their body segmentation. Earthworms are the most familiar annelids, but most annelids live in marine or freshwater habitats (see Concept 33.3).

A marine annelid

Mollusca (100,000 species)

Molluscs (including snails, clams, squids, and octopuses) have a soft body that in many species is protected by a hard shell (see Concept 33.3).

An octopus

Ecdysozoa

Loricifera (10 species)

Loriciferans (from the Latin *lorica*, corset, and *ferre*, to bear) are tiny animals that inhabit sediments on the seafloor. A loriciferan can telescope its head, neck, and thorax in and out of the lorica, a pocket formed by six plates surrounding the abdomen. Though the natural history of loriciferans is mostly a mystery, at least some species likely eat bacteria.

A loriciferan (LM)

Priapula (16 species)

A priapulan

Priapulans are worms with a large, rounded proboscis at the anterior end. (They are named after Priapos, the Greek god of fertility, who was symbolized by a giant penis.) Ranging from 0.5 mm to 20 cm in length, most species burrow through seafloor sediments. Fossil evidence suggests that priapulans were among the major predators during the Cambrian period.

Onychophora (110 species)

An onychophoran

Onychophorans, also called velvet worms, originated during the Cambrian explosion (see Concept 25.3). Originally, they thrived in the ocean, but at some point they succeeded in colonizing land. Today they live only in humid forests. Onychophorans have fleshy antennae and several dozen pairs of saclike legs.

Nematoda (25,000 species)

A roundworm

Also called roundworms, nematodes are enormously abundant and diverse in the soil and in aquatic habitats; many species parasitize plants and animals. Their most distinctive feature is a tough cuticle that coats their body (see Concept 33.4).

Tardigrada (800 species)

Tardigrades (from the Latin *tardus*, slow, and *gradus*, step) are sometimes called water bears for their overall shape and lumbering, bearlike gait. Most tardigrades are less than 0.5 mm in length. Some live in oceans or fresh water, while others live on plants or animals. Harsh conditions may cause tardigrades to enter a state of dormancy; while dormant, they can survive for days at temperatures as low as −200°C! A recent phylogenomic study found that over 15% of tardigrade genes entered their genome by horizontal gene transfer, the largest fraction known for any animal.

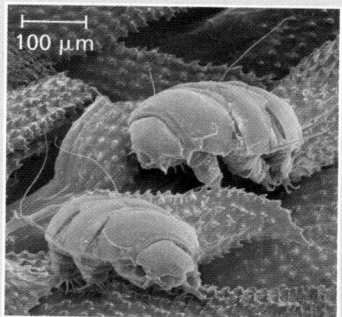

Tardigrades (colorized SEM)

Arthropoda (1,000,000 species)

The vast majority of known animal species, including insects, crustaceans, and arachnids, are arthropods. All arthropods have a segmented exoskeleton and jointed appendages (see Concept 33.4).

A spider
(an arachnid)

Deuterostomia

Hemichordata (85 species)

An acorn worm

Like echinoderms and chordates, hemichordates are members of the deuterostome clade (see Figure 32.11). Hemichordates share some traits with chordates, such as gill slits and a dorsal nerve cord. The largest group of hemichordates is the enteropneusts, or acorn worms. Acorn worms are marine and generally live buried in mud or under rocks; they may grow to more than 2 m in length.

Chordata (60,000 species)

More than 90% of all known chordate species have backbones (and thus are vertebrates). However, the phylum Chordata also includes two groups of invertebrates: lancelets and tunicates. See Chapter 34 for a full discussion of this phylum.

A tunicate

Echinodermata (7,000 species)

A sea urchin

Echinoderms, such as sand dollars, sea stars, and sea urchins, are marine animals in the deuterostome clade that are bilaterally symmetrical as larvae but not as adults. They move and feed by using a network of internal canals to pump water to different parts of their body (see Concept 33.5).

Sponges are basal animals that lack tissues

Porifera
Cnidaria
Lophotrochozoa
Ecdysozoa
Deuterostomia

Animals in the phylum Porifera are known informally as sponges. Among the simplest of animals, sponges are sedentary and were mistaken for plants by the ancient Greeks. Most species are marine, and they range in size from a few millimeters to a few meters. Sponges are **filter feeders**: They filter out food particles suspended in the surrounding water as they draw it through their body, which in some species resembles a sac perforated with pores. Water is drawn through the pores into a central cavity, the **spongocoel**, and then flows out of the sponge through a larger opening called the **osculum (Figure 33.3)**. More complex sponges have folded body walls, and many contain branched water canals and several oscula.

Sponges represent a lineage that diverged from other animals early in the history of the group; thus, they are said to be *basal animals*. Unlike nearly all other animals, sponges lack tissues, groups of similar cells that act as a functional unit as in muscle tissue and nervous tissue. However, the sponge body does contain several different cell types. For example, lining the interior of the spongocoel are flagellated **choanocytes**, or collar cells (named for the finger-like projections that form a "collar" around the flagellum). These cells engulf bacteria and other food particles by phagocytosis. The similarity between choanocytes and the cells of choanoflagellates supports molecular evidence indicating that animals and choanoflagellates are sister groups (see Figure 32.3).

The body of a sponge consists of two layers of cells separated by a gelatinous region called the **mesohyl**. Because both cell layers are in contact with water, processes such as gas exchange and waste removal can occur by diffusion across the membranes of these cells. Other tasks are performed by cells called **amoebocytes**, named for their use of pseudopodia. These cells move through the mesohyl and have many functions. For example, they take up food from the surrounding water and from choanocytes, digest it, and carry nutrients to other cells. Amoebocytes also manufacture tough skeletal fibers within the mesohyl. In some sponges, these fibers are sharp spicules made from calcium carbonate or silica. Other sponges produce more flexible fibers composed of a protein called spongin; you may have seen these pliant skeletons being sold as brown bath sponges. Finally, and perhaps most

▼ **Figure 33.3 Anatomy of a sponge.** In the main diagram, portions of the front and back wall are cut away to show the sponge's internal structure.

Azure vase sponge (*Callyspongia plicifera*)

❺ **Choanocytes.** The spongocoel is lined with flagellated cells called choanocytes. By beating flagella, the choanocytes create a current that draws water in through the pores and out through the osculum.

Food particles in mucus

Collar

Choanocyte

Flagellum

Phagocytosis of food particles

Amoebocyte

Osculum

❹ **Spongocoel.** Water passing through pores enters a cavity called the spongocoel.

❸ **Pores.** Water enters the sponge through pores formed by doughnut-shaped cells that span the body wall.

❷ **Epidermis.** The outer layer consists of tightly packed epidermal cells.

❶ **Mesohyl.** The wall of this sponge consists of two layers of cells separated by a gelatinous matrix, the mesohyl ("middle matter").

Water flow

Spicules

❻ The movement of a choanocyte's flagellum also draws water through its collar of finger-like projections. Food particles are trapped in the mucus that coats the projections, engulfed by phagocytosis, and either digested or transferred to amoebocytes.

❼ **Amoebocytes.** These cells can transport nutrients to other cells of the sponge body, produce materials for skeletal fibers (spicules), or become any type of sponge cell as needed.

importantly, amoebocytes are *totipotent* (capable of becoming other types of sponge cells). This gives the sponge body remarkable flexibility, enabling it to adjust its shape in response to changes in its physical environment (such as the direction of water currents).

Most sponges are **hermaphrodites**, meaning that each individual functions as both male and female in sexual reproduction by producing sperm *and* eggs. Almost all sponges exhibit sequential hermaphroditism: They function first as one sex and then as the other. Cross-fertilization can result when sperm released into the water current by an individual functioning as a male is drawn into a neighboring individual that is functioning as a female. The resulting zygotes develop into flagellated, swimming larvae that disperse from the parent sponge. After settling on a suitable substrate, a larva develops into a sessile adult.

Sponges produce a variety of antibiotics and other defensive compounds, which hold promise for fighting human diseases. For example, a compound called cribrostatin isolated from marine sponges can kill both cancer cells and penicillin-resistant strains of the bacterium *Streptococcus*. Other sponge-derived compounds are also being tested as possible anticancer agents.

CONCEPT CHECK 33.1

1. Describe how sponges feed.
2. **WHAT IF?** Some molecular evidence suggests that the sister group of animals is not the choanoflagellates, but rather a group of parasitic protists, Mesomycetozoa. Given that these parasites lack collar cells, can this hypothesis be correct? Explain.

For suggested answers, see Appendix A.

CONCEPT 33.2

Cnidarians are an ancient phylum of eumetazoans

All animals except sponges and a few other groups are *eumetazoans* ("true animals"), members of a clade of animals with tissues. One of the first lineages to have diverged from others in this clade is the phylum Cnidaria, which originated about 680 million years ago according to DNA analyses. Cnidarians have diversified into a wide range of sessile and motile forms, including hydras, corals, and jellies (commonly called "jellyfish"). Yet most cnidarians still exhibit the relatively simple, diploblastic, radial body plan that existed in early members of the group some 560 million years ago.

The basic body plan of a cnidarian is a sac with a central digestive compartment, the **gastrovascular cavity**. A single opening to this cavity functions as both mouth and anus. There are two variations on this body plan: the largely sessile

▼ **Figure 33.4 Polyp and medusa forms of cnidarians.** The body wall of a cnidarian has two layers of cells: an outer layer of epidermis (darker blue; derived from ectoderm) and an inner layer of gastrodermis (yellow; derived from endoderm). Digestion begins in the gastrovascular cavity and is completed inside food vacuoles in the gastrodermal cells. Sandwiched between the epidermis and gastrodermis is a gelatinous layer, the mesoglea.

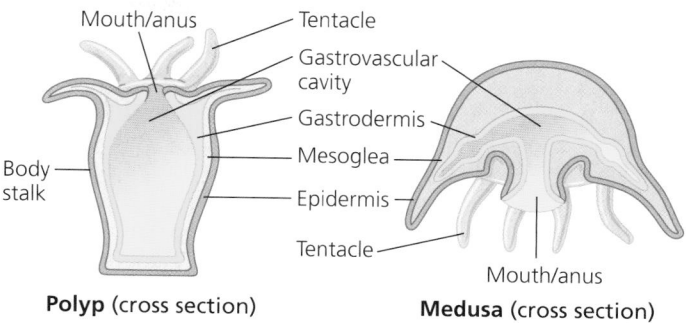

Polyp (cross section)   **Medusa** (cross section)

polyp and the more motile medusa **(Figure 33.4)**. **Polyps** are cylindrical forms that adhere to the substrate by the aboral end of their body (the end opposite the mouth) and extend their tentacles, waiting for prey. Examples of the polyp form include hydras and sea anemones. Although they are primarily sedentary, many polyps can move slowly across their substrate using muscles at the aboral end of their body. When threatened by a predator, some sea anemones can detach from the substrate and "swim" by bending their body column back and forth, or thrashing their tentacles. A **medusa** (plural, *medusae*) resembles a flattened, mouth-down version of the polyp. It moves freely in the water by a combination of passive drifting and contractions of its bell-shaped body. Medusae include free-swimming jellies. The tentacles of a jelly dangle from the oral surface, which points downward. Some cnidarians exist only as polyps or only as medusae; others have both a polyp stage and a medusa stage in their life cycle.

Cnidarians are predators that often use tentacles arranged in a ring around their mouth to capture prey and push the food into their gastrovascular cavity, where digestion begins. Enzymes are secreted into the cavity, thus breaking down the prey into a nutrient-rich broth. Cells lining the cavity then absorb these nutrients and complete the digestive process; any undigested remains are expelled through the cnidarian's mouth/anus. The tentacles are armed with batteries of **cnidocytes**, cells unique to cnidarians that function in defense and prey capture. Cnidocytes contain *cnidae* (from the Greek *cnide*, nettle), capsule-like organelles that are capable of exploding outward and that give phylum Cnidaria its name **(Figure 33.5)**. Specialized cnidae called **nematocysts** contain a stinging thread that can penetrate the body surface of the cnidarian's prey. Other kinds of cnidae have long threads that stick to or entangle small prey that bump into the cnidarian's tentacles.

Contractile tissues and nerves occur in their simplest forms in cnidarians. Cells of the epidermis (outer layer) and

▼ **Figure 33.5 A cnidocyte of a hydra.** This type of cnidocyte contains a stinging capsule, the nematocyst, which contains a coiled thread. When a "trigger" is stimulated by touch or by certain chemicals, the thread shoots out, puncturing and injecting poison into prey.

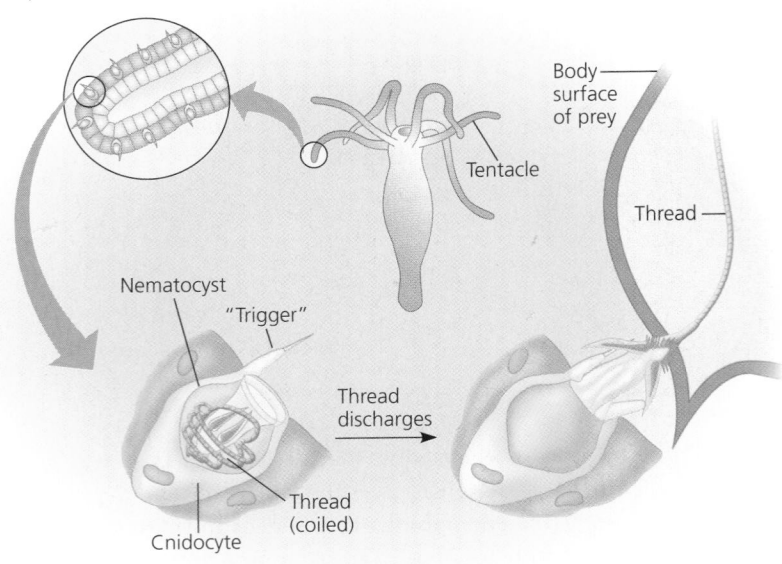

gastrodermis (inner layer) have bundles of microfilaments arranged into contractile fibers. The gastrovascular cavity acts as a hydrostatic skeleton (see Concept 50.6) against which the contractile cells can work. When a cnidarian closes its mouth, the volume of the cavity is fixed, and contraction of selected cells causes the animal to change shape. Cnidarians have no brain. Instead, movements are coordinated by a noncentralized *nerve net* that is associated with sensory structures distributed around the body. Thus, even though it lacks a brain, the animal can detect and respond to stimuli from all directions.

Fossil and molecular evidence suggests that early in its evolutionary history, the phylum Cnidaria diverged into two major clades, Medusozoa and Anthozoa **(Figure 33.6)**.

Medusozoans

All cnidarians that produce a medusa are members of clade Medusozoa, a group that includes the *scyphozoans* (jellies) and *cubozoans* (box jellies) shown in Figure 33.6a, along with the *hydrozoans*. Most hydrozoans alternate between the polyp and medusa forms, as seen in the life cycle of *Obelia* **(Figure 33.7)**. The polyp stage, a colony of interconnected polyps in the case of *Obelia*, is more conspicuous than the medusa. Hydras, among the few cnidarians found in fresh water, are also unusual hydrozoans in that they exist only in polyp form.

Unlike hydrozoans, most scyphozoans and cubozoans spend the majority of their life cycles in the medusa stage. Coastal scyphozoans, for example, often have a brief polyp stage during their life cycle, whereas those that live in the open ocean generally lack the polyp stage altogether. As their name (which means "cube animals") suggests, cubozoans have a box-shaped medusa stage. Most cubozoans live in tropical oceans and are equipped with highly toxic cnidocytes. For example, the sea wasp (*Chironex fleckeri*), a cubozoan that lives off the coast of northern Australia, is one of the deadliest organisms known: Its sting causes intense pain and can lead to respiratory failure, cardiac arrest, and death within minutes.

▼ **Figure 33.6 Cnidarians.**

(a) Medusozoans

Many scyphozoans (jellies) are bioluminescent. Food captured by nematocyst-bearing tentacles is transferred to specialized oral arms for transport to the mouth.

This cubozoan, a sea wasp, produces a poison that can subdue fish, crustaceans (as seen here), and other large prey. The poison is more potent than cobra venom.

(b) Anthozoans

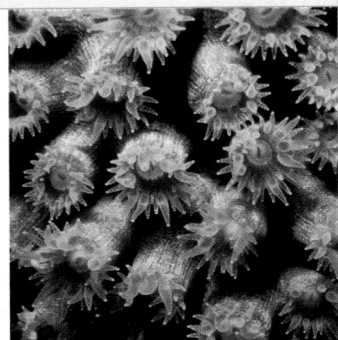

Sea anemones and other anthozoans exist only as polyps. Many anthozoans form symbiotic relationships with photosynthetic algae.

These star corals live as colonies of polyps. Their soft bodies are enclosed at the base by a hard exoskeleton.

→ **Mastering Biology**
Video: Jelly Swimming • Video: Hydra Eating *Daphnia*

▼ **Figure 33.7** **The life cycle of the hydrozoan *Obelia*.** The polyp is asexual, and the medusa is sexual, releasing eggs and sperm. These two stages alternate, one producing the other.

❸ Other polyps, specialized for reproduction, lack tentacles and produce tiny medusae by asexual budding.

❷ Some of the colony's polyps, equipped with tentacles, are specialized for feeding.

❹ Medusae swim off, grow, and reproduce sexually.

Reproductive polyp

Feeding polyp

❶ A colony of interconnected polyps (inset, LM) results from asexual reproduction by budding.

Medusa bud

MEIOSIS

Gonad

Medusa

SEXUAL REPRODUCTION

Egg

Sperm

ASEXUAL REPRODUCTION (BUDDING)

Portion of a colony of polyps

FERTILIZATION

Zygote

Developing polyp

Larva

Mature polyp

❻ The larva eventually settles and develops into a new polyp.

❺ The zygote develops into a solid ciliated larva.

1 mm

Key

Haploid (*n*)

Diploid (2*n*)

MAKE CONNECTIONS *Compare and contrast the* Obelia *life cycle to the life cycles in Figure 13.6. Which life cycle in that figure is most similar to that of* Obelia? *Explain. (See also Figure 29.5.)*

Anthozoans

Sea anemones and corals belong to the clade Anthozoa (see Figure 33.6b). These cnidarians occur only as polyps. Corals live as solitary or colonial forms, often forming symbioses with algae. Many species secrete a hard **exoskeleton** (external skeleton) of calcium carbonate. Each polyp generation builds on the skeletal remains of earlier generations, constructing rocklike reefs with shapes characteristic of their species. These skeletons are what we usually think of as coral.

Coral reefs are to tropical seas what rain forests are to tropical land areas: They provide habitat for many other species. Unfortunately, these reefs are being destroyed at an alarming rate. Pollution, overharvesting, and ocean acidification (see Figure 3.12) are major threats. Climate change is also contributing to their demise by raising seawater temperatures above the range in which corals thrive (see Concept 56.4).

→ **Mastering Biology** HHMI Animation: Coral Bleaching

CONCEPT CHECK 33.2

1. Compare and contrast the polyp and medusa forms of cnidarians.

2. **VISUAL SKILLS** Use the cnidarian life cycle diagram in Figure 33.7 to determine the ploidy of a feeding polyp and of a medusa.

3. **MAKE CONNECTIONS** Many new animal body plans emerged during and after the Cambrian explosion. In contrast, cnidarians today retain the same diploblastic, radial body plan found in cnidarians 560 million years ago. Are cnidarians therefore less successful or less "highly evolved" than other animal groups? Explain. (See Concepts 25.3 and 25.6.)

For suggested answers, see Appendix A.

Lophotrochozoans, a clade identified by molecular data, have the widest range of animal body forms

Porifera
Cnidaria
Lophotrochozoa
Ecdysozoa
Deuterostomia

Bilateria

The vast majority of animal species belong to the clade Bilateria, whose members exhibit bilateral symmetry and are triploblastic (see Concept 32.3). Most bilaterians also have a digestive tract with two openings (a mouth and an anus) and a body cavity (a coelom or hemocoel; see Figure 32.9). Recent DNA analyses suggest that the common ancestor of living bilaterians lived about 670 million years ago. To date, however, the oldest fossil that is widely accepted as a bilaterian is of *Kimberella*, a mollusc (or close relative) that lived 560 million years ago (see Figure 32.5b). Many other bilaterian groups first appeared in the fossil record during the Cambrian explosion (535 to 525 million years ago).

Molecular evidence suggests that today there are three major clades of bilaterally symmetrical animals: Lophotrochozoa, Ecdysozoa, and Deuterostomia. This section will focus on the first of these clades, the lophotrochozoans. Concepts 33.4 and 33.5 will explore the other two clades.

Although the clade Lophotrochozoa was identified by molecular data, its name comes from features found in some of its members. Some lophotrochozoans develop a structure called a *lophophore*, a crown of ciliated tentacles that functions in feeding, while others go through a distinctive stage called the *trochophore larva* (see Figure 32.12). Other members of the group have neither of these features. Few other unique morphological features are widely shared within the group—in fact, the lophotrochozoans are the most diverse bilaterian clade in terms of body plan. This diversity in form is reflected in the number of phyla classified in the group: Lophotrochozoa includes 18 phyla, more than twice the number in any other clade of bilaterians.

We'll now introduce six of the diverse lophotrochozoan phyla: the flatworms, rotifers and acanthocephalans, ectoprocts, brachiopods, molluscs, and annelids.

Flatworms

Flatworms (phylum Platyhelminthes) live in marine, freshwater, and damp terrestrial habitats. In addition to free-living species, flatworms include many parasitic species, such as flukes and tapeworms. Flatworms are so named because they have thin bodies that are flattened dorsoventrally (between the dorsal and ventral surfaces); the word *platyhelminth*

means "flat worm." (Note that *worm* is not a formal scientific name but rather refers to animals with long, thin bodies.) The smallest flatworms are nearly microscopic free-living species, while some tapeworms are more than 20 m long.

Although flatworms are triploblastic, they lack a body cavity. Their flat shape increases their surface area, placing all their cells close to water in the surrounding environment or in their gut. Because of this proximity to water, gas exchange and the elimination of nitrogenous waste (ammonia) can occur by diffusion across the body surface. As shown in **Figure 33.8**, a flat shape is one of several structural features that maximize surface area and have arisen (by convergent evolution) in different groups of animals and other organisms.

As you might expect since all their cells are close to water, flatworms have no organs specialized for gas exchange, and their relatively simple excretory apparatus functions mainly to maintain osmotic balance with their surroundings. This apparatus consists of **protonephridia**, networks of tubules with ciliated structures called *flame bulbs* that pull fluid through branched ducts opening to the outside (see Figure 44.8). Most flatworms have a gastrovascular cavity with only one opening. Though flatworms lack a circulatory system, the fine branches of the gastrovascular cavity distribute food directly to the animal's cells.

Early in their evolutionary history, flatworms separated into two lineages, Catenulida and Rhabditophora. Catenulida is a small clade of about 100 flatworm species, most of which live in freshwater habitats. Catenulids typically reproduce asexually by budding at their posterior end. The offspring often produce their own buds before detaching from the parent, thereby forming a chain of two to four genetically identical individuals—hence their informal name, "chain worms."

The other ancient flatworm lineage, Rhabditophora, is a diverse clade of about 20,000 freshwater and marine species, one example of which is shown in Figure 33.8. We'll explore the rhabditophorans in more detail, focusing on free-living and parasitic members of this clade.

Free-Living Species

Free-living rhabditophorans are important as predators and scavengers in a wide range of freshwater and marine habitats. The best-known members of this group are freshwater species in the genus *Dugesia*, commonly called **planarians**. Abundant in unpolluted ponds and streams, planarians prey on smaller animals or feed on dead animals. They move by using cilia on their ventral surface, gliding along a film of mucus they secrete. Some other rhabditophorans also use their muscles to swim through water with an undulating motion.

A planarian's head features a pair of light-sensitive eyespots as well as lateral flaps that function mainly to detect specific chemicals. The planarian nervous system is more complex and centralized than the nerve nets of cnidarians

Maximizing Surface Area

In general, the amount of metabolic or chemical activity a cell or organism can carry out is proportional to its mass or volume. Maximizing metabolic rate, however, requires the efficient uptake of energy and raw materials, such as nutrients and oxygen, as well as the effective disposal of waste products. For large cells, plants, and animals, these exchange processes have the potential to be limiting due to simple geometry. When a cell or organism grows without changing shape, its volume increases more rapidly than its surface area (see Figure 6.7). As a result, there is proportionately less surface area over which exchange processes can occur. The challenge posed by the relationship between surface area and volume occurs in diverse contexts and organisms, but the evolutionary adaptations that meet this challenge are similar. Structures that maximize surface area through branching, flattening, folding, and projections have an essential role in biological systems.

These diagrams compare surface area (SA) for two different shapes with the same volume (V). Note which shape has the greater surface area.

SA: $6\ (3\ \text{cm} \times 3\ \text{cm}) = 54\ \text{cm}^2$

V: $3\ \text{cm} \times 3\ \text{cm} \times 3\ \text{cm} = 27\ \text{cm}^3$

SA: $2\ (3\ \text{cm} \times 1\ \text{cm}) + 2\ (9\ \text{cm} \times 1\ \text{cm}) + 2\ (3\ \text{cm} \times 9\ \text{cm}) = 78\ \text{cm}^2$

V: $1\ \text{cm} \times 3\ \text{cm} \times 9\ \text{cm} = 27\ \text{cm}^3$

Branching

Water uptake relies on passive diffusion. The highly branched filaments of a fungal mycelium increase the surface area across which water and minerals can be absorbed from the environment. (See Figure 31.2.)

Folding

This TEM shows portions of two chloroplasts in a plant leaf. Photosynthesis occurs in chloroplasts, which have a flattened and interconnected set of internal membranes called thylakoid membranes. The foldings of the thylakoid membranes increase their surface area, enhancing the exposure to light and thus increasing the rate of photosynthesis. (See Figure 10.4.)

1 μm

Thylakoid

Flattening

By having a body that is only a few cells thick, an organism such as this flatworm can use its entire body surface for exchange. (See Figure 40.3)

Projections

In vertebrates, the small intestine is lined with finger-like projections called villi that absorb nutrients released by the digestion of food. Each of the villi shown here is covered with large numbers of microscopic projections called microvilli, resulting in a total surface area of about 300 m² in a human, as large as a tennis court. (See Figure 41.12.)

MAKE CONNECTIONS

Find other examples of flattening, folding, branching, and projections (see Chapters 6, 9, 35, and 42). How is maximizing surface area important to the structure's function in each example?

▼ **Figure 33.9** Anatomy of a planarian.

Digestion is completed within the cells lining the gastrovascular cavity, which has many fine subbranches that provide an extensive surface area.

Undigested wastes are eliminated through an opening at the tip of the pharynx.

Pharynx. A muscular pharynx can be extended through the mouth. Digestive juices are spilled onto prey, and the pharynx sucks small pieces of food into the gastrovascular cavity, where digestion continues.

Gastrovascular cavity

Mouth

Eyespots

Ventral nerve cords. From the ganglia, a pair of ventral nerve cords runs the length of the body.

Ganglia. At the anterior end of the worm, near the main sources of sensory input, is a pair of ganglia, dense clusters of nerve cells.

(Figure 33.9). Experiments have shown that planarians can learn to modify their responses to stimuli.

Some planarians can reproduce asexually through fission. The parent constricts roughly in the middle of its body, separating into a head end and a tail end; each end then regenerates the missing parts. Sexual reproduction also occurs. Planarians are hermaphrodites, and copulating mates typically cross-fertilize each other.

Parasitic Species

More than half of the known species of rhabditophorans live as parasites in or on other animals. Many have suckers that attach to the internal organs or outer surfaces of the host animal. In most species, a tough covering helps protect the parasites within their hosts. We'll discuss two ecologically and economically important subgroups of parasitic rhabditophorans, the trematodes and the tapeworms.

Trematodes As a group, trematodes parasitize a wide range of hosts, and most species have complex life cycles with alternating sexual and asexual stages. Many trematodes require an intermediate host in which larvae develop before infecting the final host (usually a vertebrate), where the adult worms live. For example, various trematodes that parasitize humans spend part of their lives in snail hosts **(Figure 33.10)**. Around the world, about 200 million people are infected with trematodes called blood flukes (*Schistosoma*) and suffer from schistosomiasis, a disease whose symptoms include pain, anemia, and diarrhea.

Living within more than one kind of host puts demands on trematodes that free-living animals don't face. A blood fluke, for instance, must evade the immune systems of both

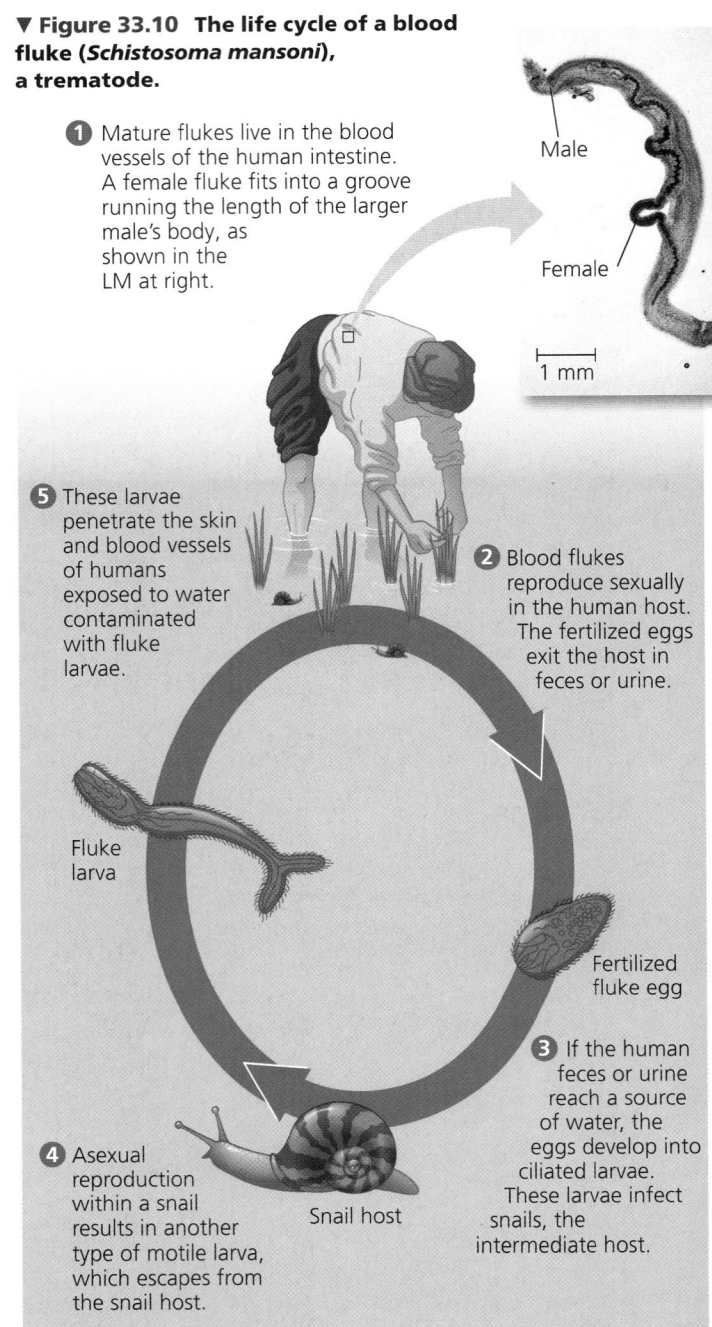

▼ **Figure 33.10** The life cycle of a blood fluke (*Schistosoma mansoni*), a trematode.

❶ Mature flukes live in the blood vessels of the human intestine. A female fluke fits into a groove running the length of the larger male's body, as shown in the LM at right.

Male

Female

1 mm

❺ These larvae penetrate the skin and blood vessels of humans exposed to water contaminated with fluke larvae.

❷ Blood flukes reproduce sexually in the human host. The fertilized eggs exit the host in feces or urine.

Fluke larva

Fertilized fluke egg

❸ If the human feces or urine reach a source of water, the eggs develop into ciliated larvae. These larvae infect snails, the intermediate host.

❹ Asexual reproduction within a snail results in another type of motile larva, which escapes from the snail host.

Snail host

WHAT IF? *Snails eat algae, whose growth is stimulated by nutrients found in fertilizer. How would the contamination of irrigation water with fertilizer likely affect the occurrence of schistosomiasis? Explain.*

snails and humans. By mimicking the surface proteins of its hosts, the blood fluke creates a partial immunological camouflage for itself. It also releases molecules that manipulate the hosts' immune systems into tolerating the parasite's existence. These defenses are so effective that individual blood flukes can survive in humans for more than 40 years.

Tapeworms The tapeworms are a second large and diverse group of parasitic rhabditophorans **(Figure 33.11)**. The adults live mostly inside vertebrates, including humans. In many tapeworms, the anterior end, or *scolex*, is armed with suckers and often hooks that the worm uses to attach itself to the intestinal lining of its host. Tapeworms lack a mouth and gastrovascular cavity; they simply absorb nutrients released by digestion in the host's intestine. Absorption occurs across the tapeworm's body surface.

Posterior to the scolex is a long ribbon of units called *proglottids*, which are little more than sacs of sex organs. After sexual reproduction, proglottids loaded with thousands of fertilized eggs are released from the posterior end of a tapeworm and leave the host's body in feces. In one type of life cycle, feces carrying the eggs contaminate the food or water of intermediate hosts, such as pigs or cattle, and the tapeworm eggs develop into larvae that encyst in muscles of these animals. A human acquires the larvae by eating undercooked meat containing the cysts, and the worms develop into mature adults within the human. Large tapeworms can block the intestines and rob enough nutrients from the human host to cause nutritional deficiencies. Several different oral medications can kill the adult worms.

Proglottids with reproductive structures

100 μm

Hooks

Scolex

Sucker

▲ **Figure 33.11 Anatomy of a tapeworm.** The inset shows a close-up of the scolex (colorized SEM).

Rotifers and Acanthocephalans

Recent phylogenetic analyses have shown that two traditional animal phyla, the rotifers (former phylum Rotifera) and the acanthocephalans (former phylum Acanthocephala), should be combined into a single phylum, Syndermata. Each of the two groups has distinctive characteristics.

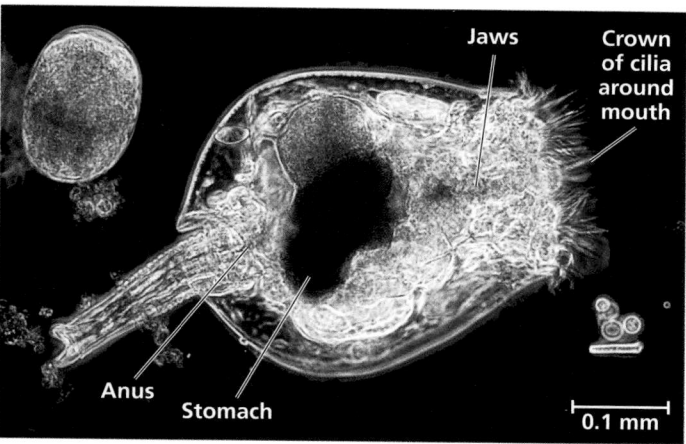

▼ **Figure 33.12 A rotifer.** Although smaller than many unicellular protists, rotifers are multicellular and have specialized organ systems (LM).

Jaws

Crown of cilia around mouth

Anus

Stomach

0.1 mm

Rotifers

There are roughly 1,800 species of rotifers, tiny animals that inhabit freshwater, marine, and damp soil habitats. Ranging in size from about 50 μm to 2 mm, rotifers are smaller than many protists but nevertheless are multicellular and have specialized organ systems **(Figure 33.12)**. In contrast to cnidarians and flatworms, which have a gastrovascular cavity, rotifers have an **alimentary canal**, a digestive tube with two openings, a mouth and an anus. Internal organs lie within the *hemocoel* (see Figure 32.9b). Fluid in the hemocoel serves as a hydrostatic skeleton. Movement of a rotifer's body distributes the fluid throughout the body, circulating nutrients.

The word *rotifer* is derived from the Latin meaning "wheel-bearer," a reference to the crown of cilia that draws a vortex of water into the mouth. Posterior to the mouth, rotifers have jaws called trophi that grind up food, mostly microorganisms suspended in the water. Digestion is then completed farther along the alimentary canal. Most other bilaterians also have an alimentary canal, which enables the stepwise digestion of a wide range of food particles.

Rotifers exhibit some unusual forms of reproduction. Some species consist only of females that produce more females from unfertilized eggs, a type of asexual reproduction called **parthenogenesis**. Some other invertebrates (for example, aphids and some bees) and even some vertebrates (for example, some lizards and some fishes) can also reproduce in this way. In addition to being able to produce females by parthenogenesis, some rotifers can also reproduce sexually under certain conditions, such as high levels of crowding. The resulting embryos can remain dormant for years. Once they break dormancy, the embryos develop into another generation of females that reproduce asexually.

It is puzzling that many rotifer species persist without males. The vast majority of animals and plants reproduce sexually at least some of the time, and sexual reproduction has certain advantages over asexual reproduction (see Concept 46.1). For example, species that reproduce asexually tend to accumulate harmful mutations in their genomes faster than sexually reproducing species. As a result, asexual species should experience higher rates of extinction.

Seeking to understand how they persist without males, researchers have been studying a clade of asexual rotifers named Bdelloidea. Some 360 species of bdelloid rotifers are known, and all of them reproduce by parthenogenesis without any males. Paleontologists have discovered bdelloid rotifers preserved in 35-million-year-old amber, and the morphology of these fossils resembles only the female form, with no evidence of males. Molecular clock analyses indicate that bdelloids have been asexual for over 50 million years. While it appears that they do not reproduce sexually, bdelloid rotifers may be able to generate genetic diversity in other ways. For example, bdelloids can tolerate very high levels of desiccation. When conditions improve and their cells rehydrate, DNA from other species enters their cells through cracks in the plasma membrane. Recent evidence suggests that this foreign DNA can be incorporated into the bdelloids' genome, thereby leading to increased genetic diversity.

Acanthocephalans

Acanthocephalans (1,100 species) are sexually reproducing parasites of vertebrates that lack a complete digestive tract and usually are less than 20 cm long. They are commonly called spiny-headed worms because of the curved hooks on the proboscis at the anterior end of their body **(Figure 33.13)**. Acanthocephalans are triploblastic and, like rotifers, they have a hemocoel. Recent studies have shown that acanthocephalans originated from within the group traditionally known as Rotifera. In particular, rotifers in the genus *Seison* share a more recent common ancestor with acanthocephalans than they do with other rotifers, making the acanthocephalans a group of highly modified "rotifers."

All acanthocephalans are parasites that have complex life cycles with two or more hosts. Some species manipulate the behavior of their intermediate hosts (generally arthropods)

in ways that increase their chances of reaching their final hosts (generally vertebrates). For example, acanthocephalans that infect New Zealand mud crabs cause their hosts to move to more visible areas on the beach, where the crabs are more likely to be eaten by birds, the worms' final hosts.

Ectoprocts and Brachiopods

Bilaterians in the phyla Ectoprocta and Brachiopoda have a *lophophore*, a crown of ciliated tentacles around their mouth (see Figure 32.12a). As the cilia draw water toward the mouth, the tentacles trap suspended food particles. Other similarities, such as a U-shaped alimentary canal and the absence of a distinct head, reflect these organisms' sessile existence. In contrast to flatworms, which lack a body cavity, and rotifers, which have a hemocoel, ectoprocts and brachiopods have a coelom (see Figure 32.9a).

Ectoprocts (from the Greek *ecto*, outside, and *procta*, anus) are colonial animals that superficially resemble clumps of moss. (In fact, their common name, bryozoans, means "moss animals.") In most species, the colony is encased in a hard exoskeleton studded with pores through which the lophophores extend **(Figure 33.14a)**. Most ectoproct species live in the sea, where they are among the most widespread and numerous sessile animals. Several species are important reef builders. Ectoprocts also live in lakes and rivers. Colonies of the freshwater ectoproct *Pectinatella magnifica* grow on submerged sticks or rocks and can grow into a gelatinous, ball-shaped mass more than 10 cm across.

Brachiopods, or lamp shells, superficially resemble clams and other hinge-shelled molluscs, but the two halves of the brachiopod shell are dorsal and ventral rather than lateral, as in clams **(Figure 33.14b)**. All brachiopods are marine. Most live attached to the seafloor by a stalk, opening their shell slightly to allow water to flow through the lophophore. The living brachiopods are remnants of a much richer past that included

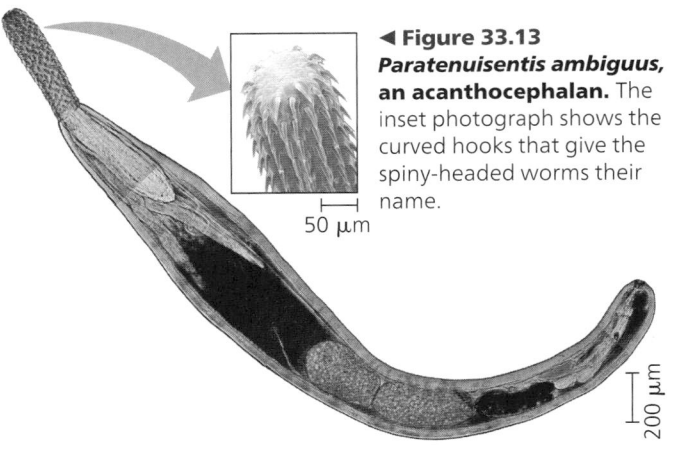

◀ **Figure 33.13**
Paratenuisentis ambiguus,
an acanthocephalan. The inset photograph shows the curved hooks that give the spiny-headed worms their name.

50 μm

200 μm

▼ **Figure 33.14 Ectoprocts and brachiopods.**

Lophophore

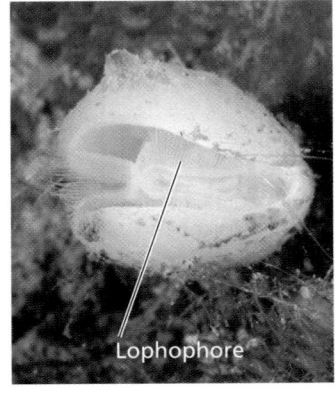

Lophophore

(a) Ectoprocts, such as this creeping bryozoan (*Plumatella repens*), are colonial.

(b) Brachiopods, such as this lampshell (*Terebratulina retusa*), have a hinged shell. The two parts of the shell are dorsal and ventral.

30,000 species in the Paleozoic and Mesozoic eras. Some living brachiopods, such as those in the genus *Lingula*, appear nearly identical to fossils of species that lived 400 million years ago.

Molluscs

Snails and slugs, oysters and clams, and octopuses and squids are all molluscs (phylum Mollusca). There are over 100,000 known species, making them the second most diverse phylum of animals (after the arthropods, discussed later). Although the majority of molluscs are marine, roughly 8,000 species inhabit fresh water, and 28,000 species of snails and slugs live on land. All molluscs are soft-bodied, and most secrete a hard protective shell made of calcium carbonate. Slugs, squids, and octopuses have a reduced internal shell or have lost their shell completely during their evolution.

Despite their apparent differences, all molluscs have a similar body plan **(Figure 33.15)**. The primary body cavity is a hemocoel, but they also have a reduced coelom. The body of a mollusc has three main parts: a muscular **foot**, usually used for movement; a **visceral mass** containing most of the internal organs; and a **mantle**, a fold of tissue that drapes over the visceral mass and secretes a shell (if one is present). In many molluscs, the mantle extends beyond the visceral mass, producing a water-filled chamber, the **mantle cavity**, which houses the gills, anus, and excretory pores. Many molluscs feed by using a straplike organ called a **radula** to scrape up food.

Most molluscs have separate sexes, and their gonads (ovaries or testes) are located in the visceral mass. Many snails, however, are hermaphrodites. The life cycle of many marine molluscs includes a ciliated larval stage, the trochophore (see Figure 32.12b), which is also characteristic of marine annelids (segmented worms) and some other lophotrochozoans.

The basic body plan of molluscs has evolved in various ways in the phylum's eight major clades. We'll examine four of those clades here: Polyplacophora (chitons), Gastropoda (snails and slugs), Bivalvia (clams, oysters, and other bivalves), and Cephalopoda (squids, octopuses, cuttlefishes, and chambered nautiluses). We will then focus on threats facing some groups of molluscs.

Chitons

Chitons have an oval-shaped body and a shell composed of eight dorsal plates **(Figure 33.16)**. The chiton's body itself, however, is unsegmented. You can find these marine animals clinging to rocks along the shore during low tide. If you try to dislodge a chiton by hand, you will be surprised at how well its foot, acting as a suction cup, grips the rock. A chiton can also use its foot to creep slowly over the rock surface. Chitons use their radula to scrape algae off the rock surface.

▼ **Figure 33.16   A chiton.** Note the eight-plate shell characteristic of molluscs in the clade Polyplacophora.

▼ **Figure 33.15   The basic body plan of a mollusc.**

Metanephridium. Excretory organs called metanephridia remove metabolic wastes from the hemolymph.

Visceral mass

Heart. Most molluscs have an open circulatory system. The dorsally located heart pumps circulatory fluid called hemolymph through arteries into sinuses (body spaces). The organs of the mollusc are thus continually bathed in hemolymph.

The long digestive tract is coiled in the visceral mass.

Mantle

Coelom

Intestine

Gonads

Stomach

Shell

Radula

Mantle cavity

Anus

Gill

The nervous system consists of a nerve ring around the esophagus, from which nerve cords extend.

Foot

Nerve cords

Hemocoel

Esophagus

Mouth

Mouth

Radula. The mouth region in many mollusc species contains a rasp-like feeding organ called a radula. This belt of backward-curved teeth repeatedly thrusts outward and then retracts into the mouth, scraping and scooping like a backhoe.

Gastropods

About three-quarters of all living species of molluscs are gastropods **(Figure 33.17)**. Most gastropods are marine, but there are also freshwater species. Still other gastropods have adapted to life on land, where snails and slugs thrive in habitats ranging from deserts to rain forests.

Gastropods move literally at a snail's pace by a rippling motion of their foot or by means of cilia—a slow process that can leave them vulnerable to attack. Most gastropods have a single, spiraled shell into which the animal can retreat when threatened. The shell, which is secreted by glands at the edge of the mantle, has several functions, including protecting the animal's soft body from injury and dehydration. One of its most important roles is as a defense against predators, as is demonstrated by comparing populations with different histories of predation (see the **Scientific Skills Exercise**). As they move slowly about, most gastropods use their radula to graze on algae

(a) A land snail

▲ **Figure 33.17 Gastropods.** The many species of gastropods have colonized terrestrial as well as aquatic environments.

(b) A sea slug. Nudibranchs, or sea slugs, lost their shell during their evolution.

or plants. Several groups, however, are predators, and their radula has become modified for boring holes in the shells of other molluscs or for tearing apart prey. In the cone snails, the teeth of the radula act as poison darts that are used to subdue prey.

Scientific Skills Exercise

Understanding Experimental Design and Interpreting Data

Is There Evidence of Selection for Defensive Adaptations in Mollusc Populations Exposed to Predators? The fossil record shows that historically, increased risk to prey species from predators is often accompanied by increased incidence and expression of prey defenses. Researchers tested whether populations of the predatory European green crab (*Carcinus maenas*) have exerted similar selective pressures on its gastropod prey, the flat periwinkle (*Littorina obtusata*). Periwinkles from southern sites in the Gulf of Maine have experienced predation by European green crabs for over 100 generations, at about one generation per year. Periwinkles from northern sites in the Gulf have been interacting with the invasive green crabs for relatively few generations, as the invasive crabs spread to the northern Gulf comparatively recently.

Previous research shows that (1) periwinkle shells recently collected from the Gulf are thicker than those collected in the late 1800s, and (2) periwinkle populations from southern sites have thicker shells than periwinkle populations from northern sites. In this exercise, you'll interpret the design and results of the researchers' experiment studying the rates of predation by crabs on periwinkles from northern and southern populations.

▼ **A periwinkle**

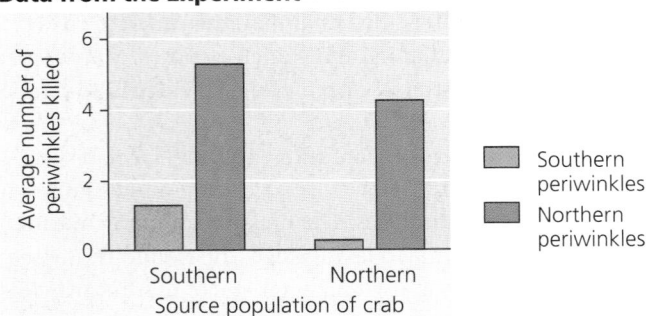

How the Experiment Was Done The researchers collected periwinkles and crabs from sites in the northern and southern Gulf of Maine, separated by 450 km of coastline. A single crab was placed in a cage with eight periwinkles of different sizes. After three days, researchers assessed the fate of the eight periwinkles. Four different treatments were set up, with crabs from northern or southern populations offered periwinkles from northern and southern populations. All crabs were of similar size and included equal numbers of males and females. Each experimental treatment was tested 12 to 14 times.

In a second part of the experiment, the bodies of periwinkles from northern and southern populations were removed from their shells and presented to crabs from northern and southern populations.

Data from the Experiment

Graph: Average number of periwinkles killed (y-axis, 0 to 6) vs. Source population of crab (x-axis: Southern, Northern). Legend: Southern periwinkles (light gray), Northern periwinkles (dark gray).

Data from R. Rochette et al., Interaction between an invasive decapod and a native gastropod: Predator foraging tactics and prey architectural defenses, *Marine Ecology Progress Series* 330:179–188 (2007).

When the researchers presented the crabs with unshelled periwinkles, all the unshelled periwinkles were consumed in less than an hour.

INTERPRET THE DATA

1. What hypotheses were the researchers testing in this study? What are the independent variables? What are the dependent variables?

2. Why did the research team set up four different treatments?

3. Why did researchers present unshelled periwinkles to the crabs? What do the results of this part of the experiment indicate?

4. Summarize the results of the experiment. Do these results support the hypothesis you identified in question 1? Explain.

5. Suggest how natural selection may have affected populations of flat periwinkles in the southern Gulf of Maine over the last 100 years.

➔ **Instructors:** A version of this Scientific Skills Exercise can be assigned in **Mastering Biology**.

Many gastropods have a head with eyes at the tips of tentacles. Terrestrial snails lack the gills typical of most aquatic gastropods. Instead, the lining of their mantle cavity functions as a lung, exchanging respiratory gases with the air.

Bivalves

The molluscs of the clade Bivalvia are all aquatic and include many species of clams, oysters, mussels, and scallops. Bivalves have a shell divided into two halves **(Figure 33.18)**. The halves are hinged, and powerful adductor muscles draw them tightly together to protect the animal's soft body. Bivalves have no distinct head, and the radula has been lost. Some bivalves have eyes and sensory tentacles along the outer edge of their mantle.

The gills of bivalves are used for feeding as well as gas exchange in most species **(Figure 33.19)**. Most bivalves are

▼ **Figure 33.18 A bivalve.** This scallop has many eyes (dark blue spots) lining each half of its hinged shell.

▼ **Figure 33.19 Anatomy of a clam.** Food particles suspended in water that enters through the incurrent siphon are collected by the gills and passed via cilia and the palps to the mouth.

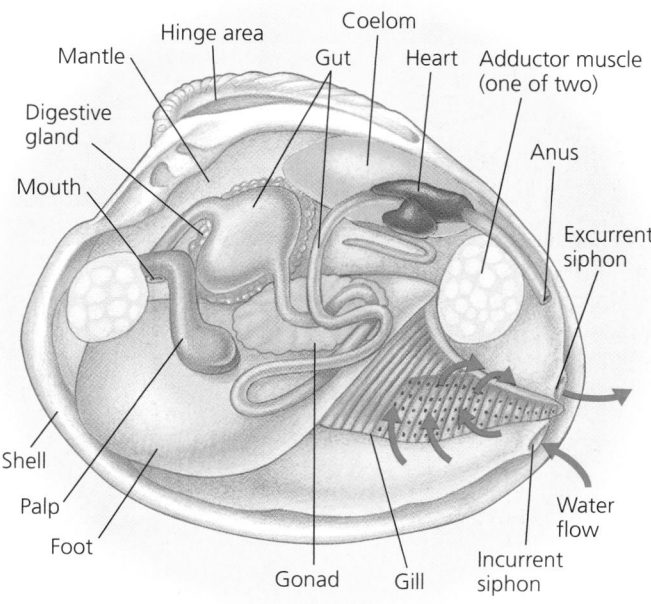

suspension feeders. They trap small food particles in mucus that coats their gills, and cilia then convey those particles to the mouth. Water enters the mantle cavity through an incurrent siphon, passes over the gills, and then exits the mantle cavity through an excurrent siphon.

Most bivalves lead sedentary lives, a characteristic suited to suspension feeding. Mussels secrete strong threads that tether them to rocks, docks, boats, and the shells of other animals. However, clams can pull themselves into the sand or mud, using their muscular foot for an anchor, and scallops can skitter along the seafloor by flapping their shells, rather like the mechanical false teeth sold in novelty shops.

Cephalopods

Cephalopods are active marine predators **(Figure 33.20)**. They use their tentacles to grasp prey, which they then bite with beak-like jaws and immobilize with a poison present in their saliva. The foot of a cephalopod has become modified into a muscular excurrent siphon and part of the tentacles. Squids dart about by drawing water into their mantle cavity and then firing a jet of water through the excurrent siphon; they steer by pointing the siphon in different directions. Octopuses use a similar mechanism to escape predators.

The mantle covers the visceral mass of cephalopods, but the shell is generally reduced and internal (in most species) or missing altogether (in some cuttlefishes and some octopuses).

▼ **Figure 33.20 Cephalopods.**

▶ Squids are speedy carnivores with beak-like jaws and well-developed eyes.

◀ Octopuses are considered among the most intelligent invertebrates.

▶ Chambered nautiluses are the only living cephalopods with an external shell.

One small group of cephalopods with external shells, the chambered nautiluses, survives today.

Cephalopods are the only molluscs with a *closed circulatory system*, in which the blood remains separate from fluid in the body cavity. They also have well-developed sense organs and a complex brain. The ability to learn and behave in a complex manner is probably more critical to fast-moving predators than to sedentary animals such as clams.

The ancestors of octopuses and squids were probably shelled molluscs that took up a predatory lifestyle; the shell was lost in later evolution. Shelled cephalopods called **ammonites**, some of them as large as truck tires, were the dominant invertebrate predators of the seas for hundreds of millions of years until their disappearance during the mass extinction at the end of the Cretaceous period, 66 million years ago.

Most species of squid are less than 75 cm long, but some are much larger. The giant squid (*Architeuthis dux*), for example, has an estimated maximum length of 13 m for females and 10 m for males. The colossal squid (*Mesonychoteuthis hamiltoni*), is even larger, with an estimated maximum length of 14 m. Unlike *A. dux*, which has large suckers and small teeth on its tentacles, *M. hamiltoni* has two rows of sharp hooks at the ends of its tentacles that can inflict deadly lacerations.

It is likely that *A. dux* and *M. hamiltoni* spend most of their time in the deep ocean, where they may feed on large fishes. Remains of both giant squid species have been found in the stomachs of sperm whales, which are probably their only natural predator. Scientists first photographed *A. dux* in the wild in 2005 while it was attacking baited hooks at a depth of 900 m. *M. hamiltoni* has yet to be observed in nature. Overall, much remains to be learned about these marine giants.

Protecting Freshwater and Terrestrial Molluscs

Species extinction rates have increased dramatically in the last 400 years, raising concern that a sixth, human-caused mass extinction may be under way (see Concept 25.4). Among the many taxa under threat, molluscs have the dubious distinction of being the animal group with the largest number of documented extinctions **(Figure 33.21)**.

Threats to molluscs are especially severe in two groups: freshwater bivalves and terrestrial gastropods. For example, the pearl mussels, a group of freshwater bivalves that can make natural pearls (gems that form when a mussel or oyster secretes layers of a lustrous coating around a grain of sand or other small irritant), are among the world's most endangered animals. Roughly 10% of the 300 pearl mussel species that once lived in North America have become extinct in the last 100 years, and over two-thirds of those that remain are threatened by extinction. Terrestrial gastropods, such as the snail in Figure 33.21, are faring no better. Hundreds of Pacific island land snails have disappeared since 1800. Overall, more than 50% of the Pacific island land snails are extinct or under imminent threat of extinction.

Threats faced by freshwater and terrestrial molluscs include habitat loss, pollution, competition or predation by non-native species, and overharvesting by humans. Is it too late to protect these molluscs? In some locations, reducing water pollution and changing how water is released from dams have led to dramatic rebounds in pearl mussel populations. Such results provide hope that with corrective measures, other endangered mollusc species can be revived.

▼ **Figure 33.21 The silent extinction.** Molluscs account for a largely unheralded but sobering 40% of all documented extinctions of animal species. These extinctions have resulted from human actions such as habitat loss and overharvesting. Many pearl mussel populations, for example, were driven to extinction by overharvesting for their shells, which were used to make buttons and other goods. Land snails are highly vulnerable to the same threats; like pearl mussels, they are among the world's most imperiled animal groups.

MAKE CONNECTIONS *Freshwater bivalves feed on and can reduce the abundance of photosynthetic protists and bacteria. As such, what types of effects would the extinction of freshwater bivalves likely have on aquatic communities (see Concept 28.6)?*

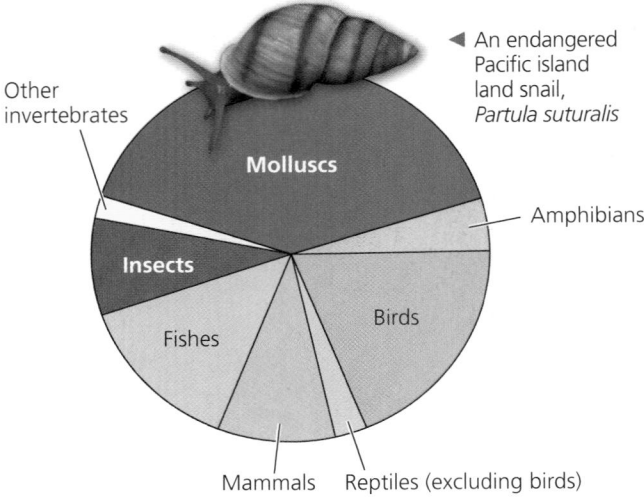

◀ An endangered Pacific island land snail, *Partula suturalis*

Other invertebrates

Molluscs

Insects

Fishes

Mammals   Reptiles (excluding birds)

Birds

Amphibians

▲ Recorded extinctions of animal species

▲ Workers on a mound of pearl mussels killed to make buttons (ca. 1919)

Annelids

Annelida means "little rings," referring to the annelid body's resemblance to a series of fused rings. Annelids are segmented worms that live in the sea, in most freshwater habitats, and in damp soil. Annelids, which have a coelom (and no hemocoel), range in length from less than 1 mm to more than 3 m.

Traditionally, the phylum Annelida was divided into three main groups, Polychaeta (the polychaetes), Oligochaeta (the oligochaetes), and Hirudinea (the leeches). The names of the first two of these groups reflected the relative number of chaetae, bristles made of chitin, on their bodies: polychaetes (from the Greek *poly*, many, and *chaitē*, long hair) have many more chaetae per segment than do oligochaetes.

However, a 2011 phylogenomic study and other recent molecular analyses have indicated that the oligochaetes are a subgroup of the polychaetes, making the polychaetes (as defined morphologically) a paraphyletic group. Likewise, the leeches have been shown to be a subgroup of the oligochaetes. As a result, these traditional names are no longer used to describe the evolutionary history of the annelids. Instead, current evidence indicates that the annelids can be divided into two major clades, Errantia and Sedentaria—a grouping that reflects broad differences in lifestyle.

Errantians

Clade Errantia (from the Old French *errant*, traveling) is a large and diverse group, most of whose members are marine **(Figure 33.22)**. As their name suggests, many errantians are mobile; some swim among the plankton (small, drifting organisms), while many others crawl on or burrow in the seafloor. Many are predators, while others are grazers that feed on large, multicellular algae. The group also includes some relatively immobile species, such as the tube-dwelling *Platynereis*, a marine species that recently has become a model organism for studying neurobiology and development.

In many errantians, each body segment has a pair of prominent paddle-like or ridge-like structures called parapodia ("beside feet") that function in locomotion (see Figure 33.22).

▼ **Figure 33.22 An errantian, the predator *Nereimyra punctata*.** This marine annelid ambushes prey from burrows it has constructed on the seafloor. *N. punctata* hunts by touch, detecting its prey with long sensory organs called cirri that extend from the burrow.

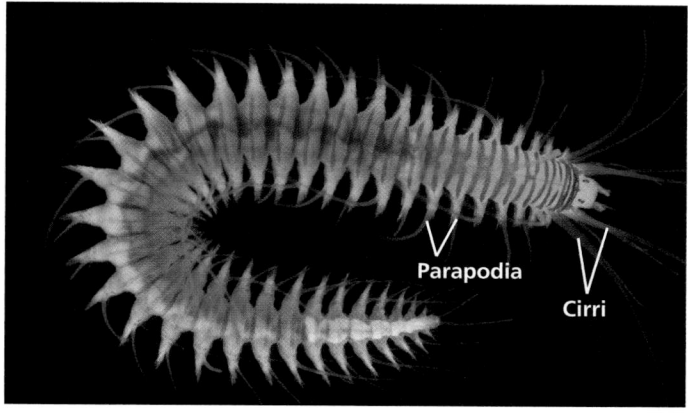

Each parapodium has numerous chaetae. (Possession of parapodia with numerous chaetae is not unique to Errantia, however, as some members of the other major clade of annelids, Sedentaria, also have these features.) In many species, the parapodia are richly supplied with blood vessels and also function as gills. Errantians also tend to have well-developed jaws and sensory organs, as might be expected of predators or grazers that move about in search of food.

Sedentarians

Species in the other major clade of annelids, Sedentaria (from the Latin *sedere*, sit), tend to be less mobile than those in Errantia. Some species burrow slowly through marine sediments or soil, while others live within tubes that protect and support their soft bodies. Tube-dwelling sedentarians often have elaborate gills or tentacles used for filter feeding **(Figure 33.23)**.

Although the Christmas tree worm shown in Figure 33.23 once was classified as a "polychaete," current evidence indicates it is a sedentarian. The clade Sedentaria also contains former "oligochaetes," including the two groups we turn to next, the leeches and the earthworms.

Leeches   Some leeches are parasites that suck blood by attaching temporarily to other animals, including humans **(Figure 33.24)**, but most are predators that feed on other

▼ **Figure 33.23 The Christmas tree worm, *Spirobranchus giganteus*.** This sedentarian's two tree-shaped whorls are tentacles, which it uses in gas exchange and to collect food particles from the water. The tentacles emerge from a calcium carbonate tube secreted by the worm that protects and supports its soft body.

► **Figure 33.24 A leech.** A nurse applied this medicinal leech (*Hirudo medicinalis*) to a patient's sore thumb to drain blood from a hematoma (an abnormal accumulation of blood around an internal injury).

invertebrates. Leeches range in length from 1 to 30 cm. Most leeches inhabit fresh water, but there are also marine species and terrestrial leeches, which live in moist vegetation. Some parasitic species use bladelike jaws to slit the skin of their host. The host is usually oblivious to this attack because the leech secretes an anesthetic. After making the incision, the leech secretes a chemical, hirudin, which keeps the blood of the host from coagulating near the incision. The parasite then sucks as much blood as it can hold, often more than ten times its own weight. After this gorging, a leech can last for months without another meal.

Until the 20th century, leeches were frequently used for bloodletting. Today they are used to drain blood that accumulates in tissues following certain injuries or surgeries. In addition, forms of hirudin produced with recombinant DNA techniques can be used to dissolve unwanted blood clots that form during surgery or as a result of heart disease.

Earthworms Earthworms eat their way through the soil, extracting nutrients as the soil passes through the alimentary canal. Undigested material, mixed with mucus secreted into the canal, is eliminated as fecal castings through the anus. Farmers value earthworms because the animals till and aerate the earth, and their castings improve the texture of the soil. (Charles Darwin estimated that one acre of farmland contains about 50,000 earthworms, producing 18 tons of castings per year.)

A guided tour of the anatomy of an earthworm, which is representative of annelids, is shown in **Figure 33.25**.

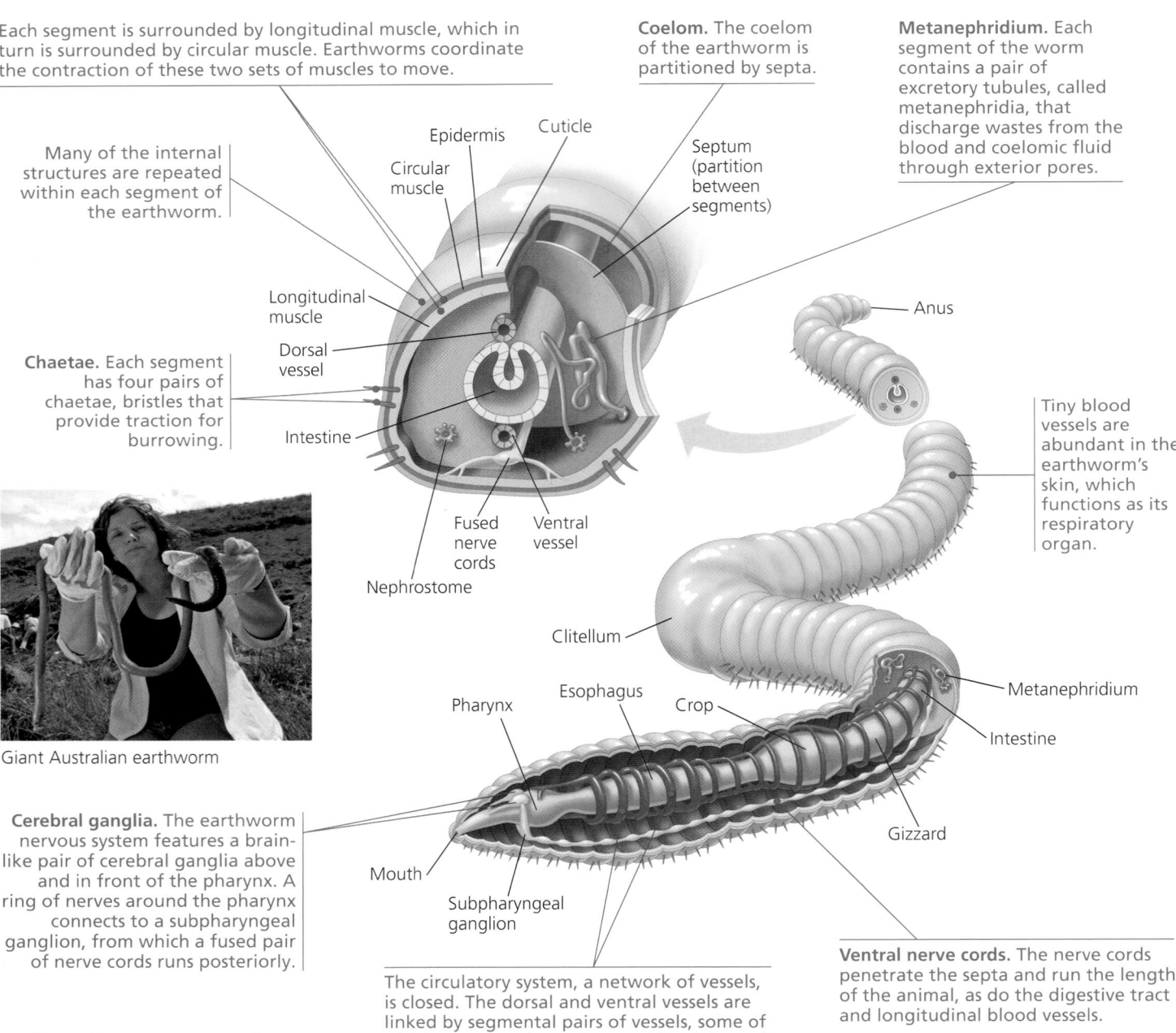

Each segment is surrounded by longitudinal muscle, which in turn is surrounded by circular muscle. Earthworms coordinate the contraction of these two sets of muscles to move.

Many of the internal structures are repeated within each segment of the earthworm.

Coelom. The coelom of the earthworm is partitioned by septa.

Metanephridium. Each segment of the worm contains a pair of excretory tubules, called metanephridia, that discharge wastes from the blood and coelomic fluid through exterior pores.

Epidermis

Cuticle

Circular muscle

Septum (partition between segments)

Longitudinal muscle

Dorsal vessel

Chaetae. Each segment has four pairs of chaetae, bristles that provide traction for burrowing.

Intestine

Fused nerve cords

Ventral vessel

Nephrostome

Anus

Tiny blood vessels are abundant in the earthworm's skin, which functions as its respiratory organ.

Clitellum

Metanephridium

Intestine

Pharynx

Esophagus

Crop

Mouth

Subpharyngeal ganglion

Gizzard

Giant Australian earthworm

Cerebral ganglia. The earthworm nervous system features a brain-like pair of cerebral ganglia above and in front of the pharynx. A ring of nerves around the pharynx connects to a subpharyngeal ganglion, from which a fused pair of nerve cords runs posteriorly.

The circulatory system, a network of vessels, is closed. The dorsal and ventral vessels are linked by segmental pairs of vessels, some of which are muscular and pump blood through the circulatory system.

Ventral nerve cords. The nerve cords penetrate the septa and run the length of the animal, as do the digestive tract and longitudinal blood vessels.

▲ **Figure 33.25 Anatomy of an earthworm, a sedentarian.**

Earthworms are hermaphrodites, but they do cross-fertilize. Two earthworms mate by aligning themselves in opposite directions in such a way that they exchange sperm, and then they separate. Some earthworms can also reproduce asexually by fragmentation followed by regeneration.

As a group, Lophotrochozoa encompasses a remarkable range of body plans, as illustrated by members of such phyla as Syndermata, Ectoprocta, Mollusca, and Annelida. Next we'll explore the diversity of Ecdysozoa, a dominant presence on Earth in terms of sheer number of species.

> ## CONCEPT CHECK 33.3
>
> 1. Explain how tapeworms can survive without a body cavity, a mouth, a digestive system, or an excretory system.
> 2. Annelid anatomy can be described as "a tube within a tube." Explain.
> 3. **MAKE CONNECTIONS** Explain how the molluscan foot in gastropods and the excurrent siphon in cephalopods represent examples of descent with modification (see Concept 22.2).
>
> *For suggested answers, see Appendix A.*

CONCEPT 33.4

Ecdysozoans are the most species-rich animal group

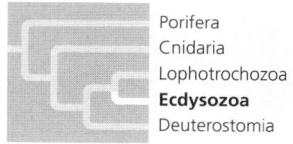

Porifera
Cnidaria
Lophotrochozoa
Ecdysozoa
Deuterostomia

Although defined primarily by molecular evidence, the clade Ecdysozoa includes animals that shed a **cuticle** (a tough external coat) as they grow. In fact, the group derives its name from this process, which is called *ecdysis*, or **molting**. Ecdysozoa includes about eight animal phyla and contains more known species than all other animal, protist, fungus, and plant groups combined. Here we'll focus on the two largest ecdysozoan phyla, the nematodes and arthropods, which are among the most successful and abundant of all animal groups.

Nematodes

Among the most ubiquitous of animals, nematodes (phylum Nematoda), or roundworms, are found in most aquatic habitats, in the soil, in the moist tissues of plants, and in the body fluids and tissues of animals. The cylindrical bodies of nematodes range from less than 1 mm to more than 1 m long, often tapering to a fine tip at the posterior end and to a blunter tip at the anterior end **(Figure 33.26)**. A nematode's body is covered by a tough cuticle (a type of exoskeleton); as the worm grows, it periodically sheds its old cuticle and secretes a new, larger one. Nematodes have an alimentary canal, though they lack a circulatory system. Nutrients are transported throughout the body via fluid in the hemocoel. The body wall muscles are all longitudinal, and their contraction produces a thrashing motion.

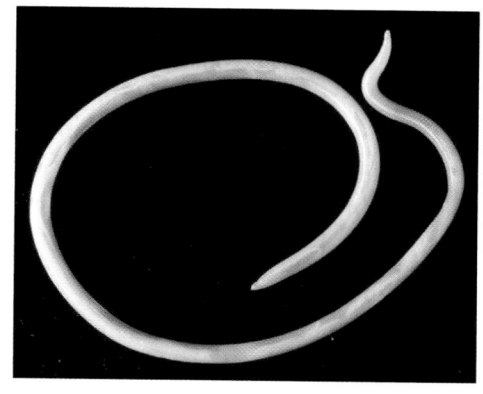

► **Figure 33.26 A free-living nematode.** (colorized SEM)

Multitudes of nematodes live in moist soil and in decomposing organic matter on the bottoms of lakes and oceans. While 25,000 species are known, perhaps 20 times that number actually exist. These free-living worms play an important role in decomposition and nutrient cycling, but little is known about most species. One species of soil nematode, *Caenorhabditis elegans*, however, is very well studied and has become a model research organism in biology (see Concept 47.3). Ongoing studies of *C. elegans* are providing insight into mechanisms involved in aging in humans, as well as many other topics.

Phylum Nematoda includes many species that parasitize plants, and some are major agricultural pests that attack the roots of crops. Other nematodes parasitize animals. Some of these species benefit humans by attacking insects such as cutworms that feed on the roots of crop plants. On the other hand, humans are hosts to at least 50 nematode species, including various pinworms and hookworms. One notorious nematode is *Trichinella spiralis*, the worm that causes trichinosis **(Figure 33.27)**. Humans acquire this nematode by eating raw or undercooked pork or other meat (including wild game such as bear or walrus) that has juvenile worms encysted in the muscle tissue. Within the human intestines, the juveniles develop into sexually mature adults. Females

▼ **Figure 33.27 Juveniles of the parasitic nematode** *Trichinella spiralis* **encysted in human muscle tissue.** (LM)

Encysted juveniles   Muscle tissue   50 µm

burrow into the intestinal muscles and produce more juveniles, which bore through the body or travel in lymphatic vessels to other organs, including skeletal muscles, where they encyst.

Parasitic nematodes have an extraordinary molecular toolkit that enables them to redirect some of the cellular functions of their hosts. Some species inject their plant hosts with molecules that induce the development of root cells, which then supply nutrients to the parasites. When *Trichinella* parasitizes animals, it regulates the expression of specific muscle cell genes encoding proteins that make the cell elastic enough to house the nematode. Additionally, the infected muscle cell releases signals that promote the growth of new blood vessels, which then supply the nematode with nutrients.

Arthropods

Zoologists estimate that there are about a billion billion (10^{18}) arthropods living on Earth. More than 1 million arthropod species have been described, most of which are insects. In fact, two out of every three known species are arthropods, and members of the phylum Arthropoda can be found in nearly all habitats of the biosphere. By the criteria of species diversity, distribution, and sheer numbers, arthropods must be regarded as the most successful of all animal phyla.

Arthropod Origins

Biologists hypothesize that the diversity and success of **arthropods** are related to their body plan—their segmented body, hard exoskeleton, and jointed appendages. How did this body plan arise and what advantages did it provide?

The earliest fossils of arthropods are from the Cambrian explosion (535–525 million years ago), indicating that the arthropods are at least that old. The fossil record of the Cambrian explosion also contains many species of *lobopods*, a group from which arthropods may have evolved. Lobopods such as *Hallucigenia* (see Figure 32.7) had segmented bodies, but most of their body segments were identical to one another. Early arthropods, such as the trilobites, also showed little variation from segment to segment **(Figure 33.28)**. As arthropods

▶ **Figure 33.28 A trilobite fossil.** Trilobites were common denizens of the shallow seas throughout the Paleozoic era but disappeared with the great Permian extinctions about 250 million years ago. Paleontologists have described about 4,000 trilobite species.

continued to evolve, groups of segments tended to become functionally united into "body regions" specialized for tasks such as feeding, walking, or swimming. These evolutionary changes resulted not only in great diversification but also in efficient body plans that permit the division of labor among different body regions.

▼ **Figure 33.29 Inquiry**

Did the arthropod body plan result from new *Hox* genes?

Experiment One hypothesis suggests that the arthropod body plan resulted from the origin (by gene duplication and subsequent mutations) of two unusual *Hox* genes found in arthropods: *Ultrabithorax* (*Ubx*) and *abdominal-A* (*abd-A*). Researchers tested this hypothesis using onychophorans, a group of invertebrates closely related to arthropods. Unlike many living arthropods, onychophorans have a body plan in which most body segments are identical to one another. If the origin of the *Ubx* and *abd-A Hox* genes drove the evolution of body segment diversity in arthropods, these genes probably arose on the arthropod branch of the evolutionary tree:

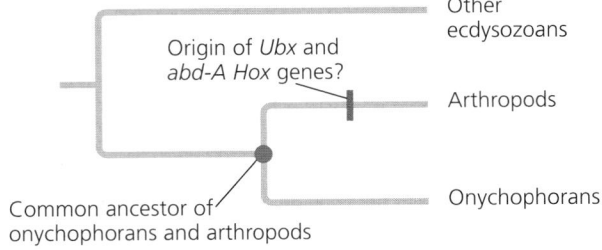

According to this hypothesis, *Ubx* and *abd-A* would not have been present in the common ancestor of arthropods and onychophorans; hence, onychophorans should not have these genes. The researchers examined the *Hox* genes of the onychophoran *Acanthokara kaputensis*.

Results The onychophoran *A. kaputensis* has all arthropod *Hox* genes, including *Ubx* and *abd-A*.

Red indicates the body regions of this onychophoran embryo in which *Ubx* or *abd-A* genes were expressed. (The inset shows this area enlarged.)

Ant = antenna
J = jaws
L1–L15 = body segments

Conclusion The evolution of increased body segment diversity in arthropods was not related to the origin of new *Hox* genes.

Data from J. K. Grenier et al., Evolution of the entire arthropod *Hox* gene set predated the origin and radiation of the onychophoran/arthropod clade, *Current Biology* 7:547–553 (1997).

WHAT IF? *Suppose A. kaputensis did not have the Ubx and abd-A Hox genes. How would the conclusions of this study have been affected? Explain.*

What genetic changes led to the increasing complexity of the arthropod body plan? Arthropods today have two unusual *Hox* genes, both of which influence segmentation. To test whether these genes could have driven the evolution of increased body segment diversity in arthropods, researchers studied *Hox* genes in onychophorans (see Figure 33.2), close relatives of arthropods **(Figure 33.29)**. Their results indicate that the diversity of arthropod body plans did *not* arise from the acquisition of new *Hox* genes. Instead, the evolution of body segment diversity in arthropods was probably driven by changes in the sequence or regulation of existing *Hox* genes (see Concept 25.5).

General Characteristics of Arthropods

Over the course of evolution, the appendages of some arthropods have become modified, specializing in functions such as walking, feeding, sensory reception, reproduction, and defense. Like the appendages from which they were derived, these modified structures are jointed and come in pairs. **Figure 33.30** illustrates the diverse appendages and other arthropod characteristics of a lobster.

The body of an arthropod is completely covered by the cuticle, an exoskeleton constructed from layers of protein and the polysaccharide chitin. As you know if you've ever eaten a crab or lobster, the cuticle can be thick and hard over some parts of the body and thin and flexible over others, such as the joints. The rigid exoskeleton protects the animal and provides points of attachment for the muscles that move the appendages. But it also prevents the arthropod from growing, unless it occasionally sheds its exoskeleton and produces a larger one. This molting process is energetically expensive, and it leaves the arthropod vulnerable to predation and other dangers until its new, soft exoskeleton hardens.

When the arthropod exoskeleton first evolved in the sea, its main functions were probably protection and anchorage for muscles, but it later enabled certain arthropods to live on land. The exoskeleton's relative impermeability to water helped prevent desiccation, and its strength provided support when arthropods left the buoyancy of water. Fossil evidence suggests that arthropods were among the first animals to colonize land, roughly 450 million years ago. These fossils include fragments of arthropod remains, as well as possible millipede burrows. Arthropod fossils from several continents indicate that by 410 million years ago,

millipedes, centipedes, spiders, and a variety of wingless insects all had colonized land.

Arthropods have well-developed sensory organs, including eyes, olfactory (smell) receptors, and antennae that function in both touch and smell. Most sensory organs are concentrated at the anterior end of the animal, although there are interesting exceptions. Female butterflies, for example, "taste" plants using sensory organs on their feet.

Like many molluscs, arthropods have an **open circulatory system**, in which fluid called *hemolymph* is propelled by a heart through short arteries and then into the hemocoel—the body cavity surrounding the tissues and organs. (The term *blood* is generally reserved for fluid in a closed circulatory system.) Hemolymph reenters the arthropod heart through pores that are usually equipped with valves. In most arthropods, the coelom that forms in the embryo becomes much reduced as development progresses, and the hemocoel becomes the main body cavity in adults.

A variety of specialized gas exchange organs have evolved in arthropods. These organs allow the diffusion of respiratory gases in spite of the exoskeleton. Most aquatic species have gills with thin, feathery extensions that expose a large surface area to the surrounding water. Terrestrial arthropods generally have internal surfaces specialized for gas exchange. For example, most insects have tracheal systems, branched air ducts leading into the interior of the body from pores in the cuticle.

Morphological and molecular data suggest that living arthropods consist of three major lineages that diverged early in the phylum's evolution: **chelicerates** (sea spiders, horseshoe crabs, scorpions, ticks, mites, and spiders); **myriapods**

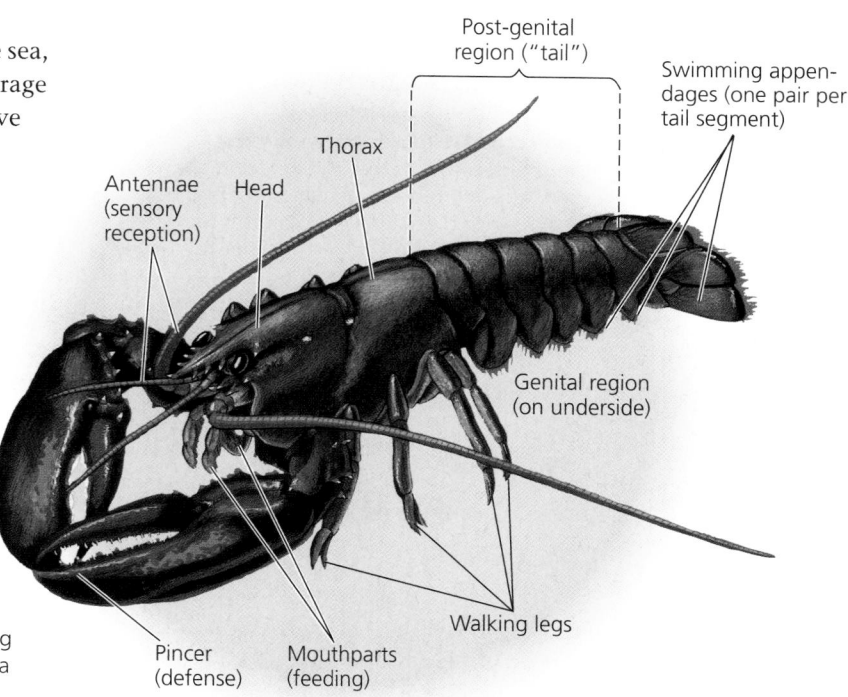

▶ **Figure 33.30 External anatomy of an arthropod.** Many of the distinctive features of arthropods are apparent in this dorsal view of a lobster. The body is segmented, but this character is obvious only in the post-genital region or "tail," located behind the genitals. The appendages (including antennae, pincers, mouthparts, walking legs, and swimming appendages) are jointed. The head bears a pair of compound (multilens) eyes. The whole body, including appendages, is covered by an exoskeleton.

(centipedes and millipedes); and **pancrustaceans** (a recently defined, diverse group that includes insects as well as lobsters, shrimp, barnacles, and other crustaceans).

Chelicerates

Chelicerates (clade Chelicerata) are named for clawlike feeding appendages called **chelicerae**, which serve as pincers or fangs. Chelicerates lack antennae, and most have simple eyes (eyes with a single lens).

The earliest chelicerates were **eurypterids**, or water scorpions. These marine and freshwater predators grew up to 3 m long; it is thought that some species could have walked on land, much as land crabs do today. Most of the marine chelicerates, including all of the eurypterids, are extinct. Among the marine chelicerates that survive today are the sea spiders (pycnogonids) and horseshoe crabs **(Figure 33.31)**.

The bulk of modern chelicerates are **arachnids**, a group that includes scorpions, spiders, ticks, and mites **(Figure 33.32)**. Nearly all ticks are bloodsucking parasites that live on the body surfaces of reptiles or mammals. Parasitic mites live on or in a wide variety of vertebrates, invertebrates, and plants.

Arachnids have six pairs of appendages: the chelicerae; a pair of appendages called *pedipalps* that function in sensing, feeding, defense, or reproduction; and four pairs of walking legs. Spiders use their fang-like chelicerae, which are equipped with poison glands, to attack prey. As the chelicerae pierce the prey, the spider secretes digestive juices onto the prey's torn tissues. The food softens, and the spider sucks up the liquid meal. In most spiders, gas exchange is carried out by **book lungs**, stacked platelike structures contained in an internal chamber **(Figure 33.33)**. The extensive surface area of these respiratory organs enhances the exchange of O_2 and CO_2 between the hemolymph and air.

A unique adaptation of many spiders is the ability to catch insects by constructing webs of silk, a liquid protein produced

▼ **Figure 33.31 Horseshoe crabs (*Limulus polyphemus*).** Common on the Atlantic and Gulf coasts of the United States, these "living fossils" have changed little in hundreds of millions of years. They are surviving members of a rich diversity of chelicerates that once filled the seas.

▲ Scorpions have pedipalps that are pincers specialized for defense and the capture of food. The tip of the tail bears a poisonous stinger.

▲ Dust mites are ubiquitous scavengers in human dwellings but are harmless except to those people who are allergic to them (colorized SEM).

◄ Web-building spiders are generally most active during the daytime.

▲ **Figure 33.32 Arachnids.**

► **Figure 33.33 Book lungs.**

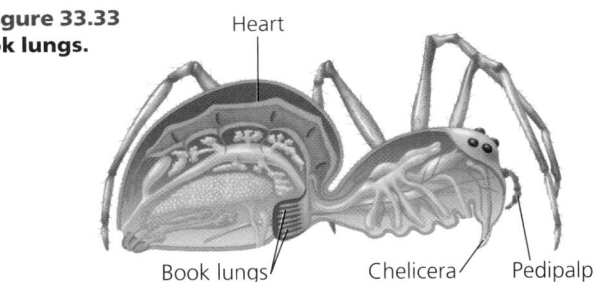

Heart

Book lungs   Chelicera   Pedipalp

by specialized abdominal glands. The silk is spun by organs called spinnerets into fibers that then solidify. Each spider engineers a web characteristic of its species and builds it perfectly on the first try, indicating that this complex behavior is inherited. Various spiders also use silk in other ways: as droplines for rapid escape, as a cover for eggs, and even as "gift wrap" for food that males offer females during courtship. Many small spiders also extrude silk into the air and let themselves be transported by wind, a behavior known as "ballooning."

Myriapods

Millipedes and centipedes belong to the clade Myriapoda **(Figure 33.34)**. All living myriapods are terrestrial. The myriapod head has a pair of antennae and three pairs of

(a) Millipede

(b) Centipede

▲ **Figure 33.34** **Myriapods.**

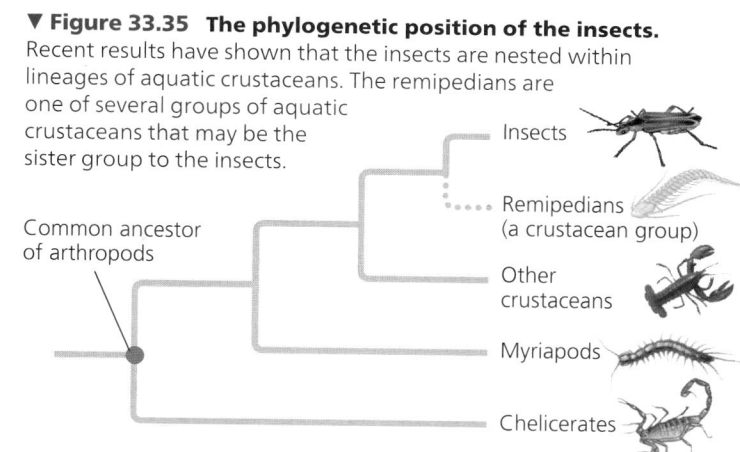

▼ **Figure 33.35** **The phylogenetic position of the insects.** Recent results have shown that the insects are nested within lineages of aquatic crustaceans. The remipedians are one of several groups of aquatic crustaceans that may be the sister group to the insects.

Common ancestor of arthropods

Insects

Remipedians (a crustacean group)

Other crustaceans

Myriapods

Chelicerates

DRAW IT *Circle the portions of this tree that comprise the clade Pancrustacea.*

appendages modified as mouthparts, including the jaw-like mandibles.

Millipedes have a large number of legs, though fewer than the thousand their name implies. Each trunk segment is formed from two fused segments and bears two pairs of legs (see Figure 33.34a). Millipedes eat decaying leaves and other plant matter. They may have been among the earliest animals on land, living on mosses and early vascular plants.

Unlike millipedes, centipedes are carnivores. Each segment of a centipede's trunk region has one pair of legs (see Figure 33.34b). Centipedes have poison claws on their foremost trunk segment that paralyze prey and aid in defense.

Pancrustaceans

A series of recent papers present evidence that terrestrial insects are more closely related to lobsters and other crustaceans than they are to the terrestrial group we just discussed, the myriapods (millipedes and centipedes). These studies also suggest that the diverse group of organisms referred to as crustaceans are paraphyletic: Some lineages of crustaceans are more closely related to insects than they are to other crustaceans **(Figure 33.35)**. However, together the insects and crustaceans form a clade, which systematists have named Pancrustacea (from the Greek *pan*, all). We turn next to a description of the members of Pancrustacea, focusing first on crustaceans and then on the insects.

Crustaceans Crustaceans (crabs, lobsters, shrimps, barnacles, and many others) thrive in a broad range of marine, freshwater, and terrestrial environments. Many crustaceans have highly specialized appendages. Lobsters and crayfishes, for instance, have a toolkit of 19 pairs of appendages (see Figure 33.30). The anterior-most appendages form two pairs of antennae; crustaceans are the only arthropods with two pairs. Three or more pairs of appendages are modified as mouthparts, including the hard mandibles. Walking legs are present on the thorax, and, unlike their terrestrial relatives, the insects, crustaceans also have appendages on their post-genital region, or "tail."

Small crustaceans exchange gases across thin areas of the cuticle; larger species have gills. Nitrogenous wastes also diffuse through thin areas of the cuticle, but a pair of glands regulates the salt balance of the hemolymph.

Sexes are separate in most crustaceans. In the case of lobsters and crayfishes, the male uses a specialized pair of abdominal appendages to transfer sperm to the reproductive pore of the female during copulation. Most aquatic crustaceans go through one or more swimming larval stages.

One of the largest groups of crustaceans (numbering over 11,000 species) is the *isopods*, which include terrestrial, freshwater, and marine species. Some isopod species are abundant in habitats at the bottom of the deep ocean. Among the terrestrial isopods are the pill bugs, or wood lice, common on the undersides of moist logs and leaves.

Lobsters, crayfishes, crabs, and shrimps are all relatively large crustaceans called *decapods* **(Figure 33.36)**. The cuticle of decapods is hardened by calcium carbonate. Most decapod species are marine. Crayfishes, however, live in fresh water, and some tropical crabs live on land.

▼ **Figure 33.36** **A ghost crab, an example of a decapod.** Ghost crabs live on sandy ocean beaches worldwide. Primarily nocturnal, they take shelter in burrows during the day.

▲ **Figure 33.37 Krill.** These planktonic crustaceans are consumed in vast quantities by some whales.

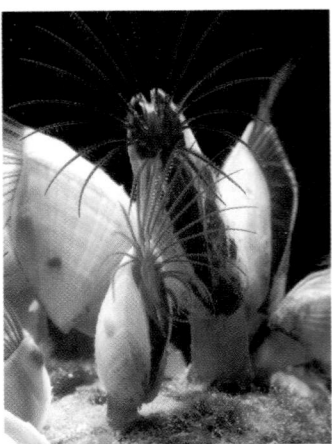

▲ **Figure 33.38 Barnacles.** The jointed appendages projecting from the barnacles' shells capture organisms and organic particles suspended in the water.

Many small crustaceans are important members of marine and freshwater plankton communities. Planktonic crustaceans include many species of *copepods*, which are among the most numerous of all animals. Some copepods are grazers that feed upon algae, while others are predators that eat small animals (including smaller copepods!). Copepods are rivaled in abundance by the shrimplike krill, which grow to about 5 cm long **(Figure 33.37)**. A major food source for baleen whales (including blue whales, humpbacks, and right whales), krill are now being harvested in great numbers by humans for food and agricultural fertilizer. The larvae of many larger-bodied crustaceans are also planktonic.

With the exception of a few parasitic species, barnacles are a group of sessile crustaceans whose cuticle is hardened into a shell containing calcium carbonate **(Figure 33.38)**. Most barnacles anchor themselves to rocks, boat hulls, pilings, and other submerged surfaces. Their natural adhesive is as strong as synthetic glues. These barnacles feed by extending appendages from their shell to strain food from the water. Barnacles were not recognized as crustaceans until the 1800s, when naturalists discovered that barnacle larvae resemble the larvae of other crustaceans. The remarkable mix of unique traits and crustacean homologies found in barnacles was a major inspiration to Charles Darwin as he developed his theory of evolution.

We turn now to a group nested within the paraphyletic crustaceans, the insects.

Insects Insects and their six-legged terrestrial relatives form an enormous clade, Hexapoda; we'll focus here on the insects, since as a group they have more described species than all other eukaryotic groups combined. Insects live in almost every terrestrial habitat and in fresh water, and flying insects fill the air. Insects are rare, though not absent, in marine habitats. The internal anatomy of an insect includes several complex organ systems, which are highlighted in **Figure 33.39**.

The oldest insect fossils date to about 415 million years ago. Later, an explosion in insect diversity took place when insect flight evolved during the Carboniferous and Permian periods (359–252 million years ago). An animal that can fly can escape predators, find food and mates, and disperse to new habitats more effectively than an animal that must crawl about on the

▼ **Figure 33.39 Anatomy of a grasshopper, an insect.** The insect body has three regions: head, thorax, and post-genital region. The segmentation of the thorax and post-genital region is obvious, but the segments that form the head are fused.

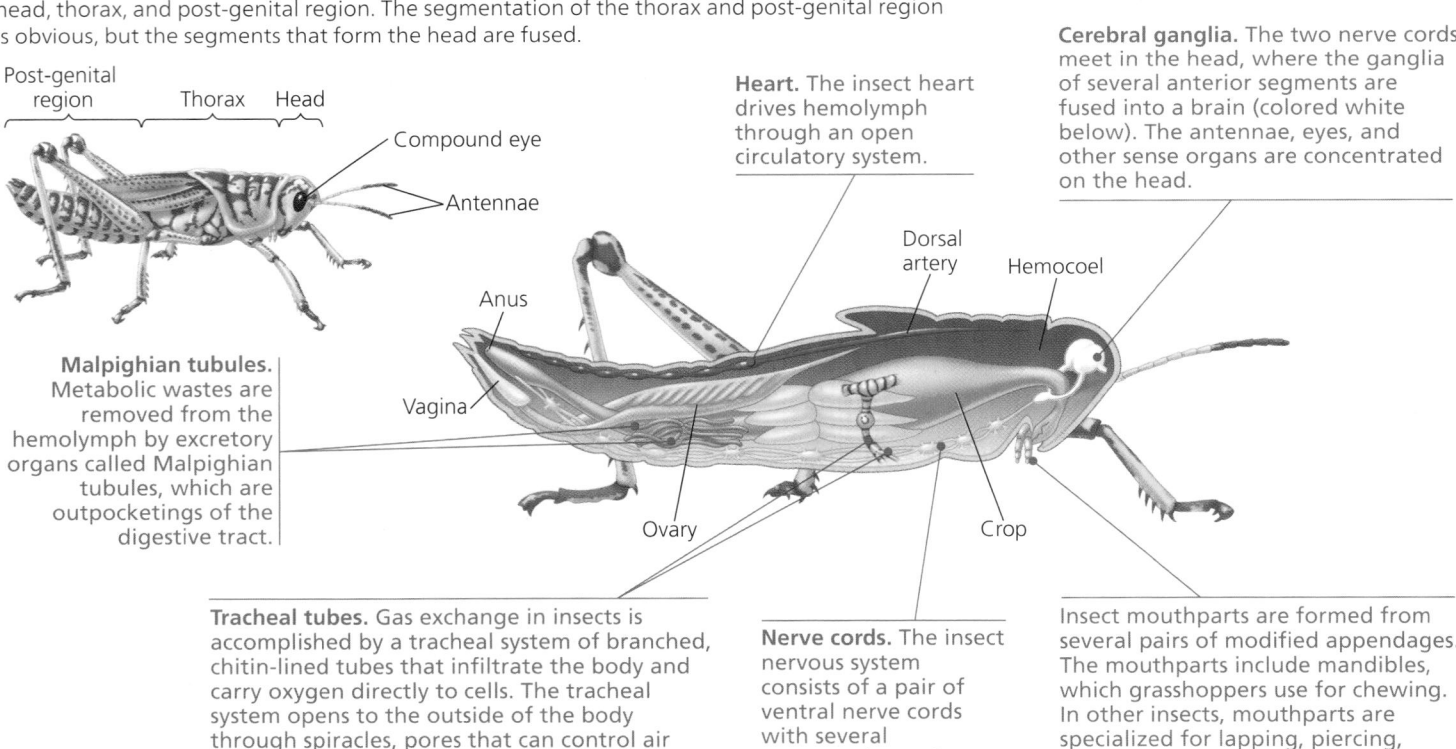

Cerebral ganglia. The two nerve cords meet in the head, where the ganglia of several anterior segments are fused into a brain (colored white below). The antennae, eyes, and other sense organs are concentrated on the head.

Heart. The insect heart drives hemolymph through an open circulatory system.

Post-genital region | Thorax | Head

Compound eye

Antennae

Dorsal artery

Hemocoel

Anus

Malpighian tubules. Metabolic wastes are removed from the hemolymph by excretory organs called Malpighian tubules, which are outpocketings of the digestive tract.

Vagina

Ovary

Crop

Tracheal tubes. Gas exchange in insects is accomplished by a tracheal system of branched, chitin-lined tubes that infiltrate the body and carry oxygen directly to cells. The tracheal system opens to the outside of the body through spiracles, pores that can control air flow and water loss by opening or closing.

Nerve cords. The insect nervous system consists of a pair of ventral nerve cords with several segmental ganglia.

Insect mouthparts are formed from several pairs of modified appendages. The mouthparts include mandibles, which grasshoppers use for chewing. In other insects, mouthparts are specialized for lapping, piercing, or sucking.

ground. Many insects have one or two pairs of wings that emerge from the dorsal side of the thorax. Because the wings are extensions of the cuticle, insects can fly without sacrificing any walking legs **(Figure 33.40)**. By contrast, the flying vertebrates—birds and bats—have one of their two pairs of walking legs modified into wings, making some of these species clumsy on the ground.

▲ **Figure 33.40 Ladybird beetle in flight.**

Insects also radiated in response to the origin of new plant species, which provided new sources of food. By the speciation mechanisms described in Concept 24.2, an insect population feeding on a new plant species can diverge from other populations, eventually forming a new species of insect. A fossil record of diverse insect mouthparts, for example, suggests that specialized modes of feeding on gymnosperms and other Carboniferous plants contributed to early adaptive radiations of insects. Later, a major increase in insect diversity appears to have been stimulated by the evolutionary expansion of flowering plants during the mid-Cretaceous period (about 100 million years ago). Although insect and plant diversity decreased during the Cretaceous mass extinction, both groups have rebounded over the past 66 million years. Increases in the diversity of particular insect groups have often been associated with radiations of the flowering plants on which they fed.

Many insects undergo metamorphosis during their development. In the **incomplete metamorphosis** of grasshoppers and some other insect groups, the young (called nymphs) resemble adults but are smaller, have different body proportions, and lack wings. The nymph undergoes a series of molts, each time looking more like an adult. With the final molt, the insect reaches full size, acquires wings, and becomes sexually mature. Insects with **complete metamorphosis** have larval stages specialized for eating and growing that are known by such names as caterpillar, maggot, or grub. The larval stage looks entirely different from the adult stage, which is specialized for dispersal and reproduction. Metamorphosis from the larval stage to the adult occurs during a pupal stage **(Figure 33.41)**.

Reproduction in insects is usually sexual, with separate male and female individuals. Adults come together and recognize each other as members of the same species by advertising with bright colors (as in butterflies), sounds (as in crickets), or odors (as in moths). Fertilization is generally internal. In most species, sperm are deposited directly into the female's vagina at the time of copulation, though in some species the male deposits a sperm packet outside the female, and the female picks it up. An internal structure in the female called the spermatheca stores the sperm, usually enough to fertilize more than one batch of eggs. Many insects mate only once in a lifetime. After mating, a female often lays her eggs on an appropriate food source where the next generation can begin eating as soon as it hatches.

Insects are classified in more than 30 orders, 8 of which are introduced in **Figure 33.42**.

(a) Larva (caterpillar)

(b) Pupa

(c) Later-stage pupa

(d) Emerging adult

(e) Adult

▲ **Figure 33.41 Complete metamorphosis of a butterfly. (a)** The larva (caterpillar) spends its time eating and growing, molting as it grows. **(b)** After several molts, the larva develops into a pupa. **(c)** Within the pupa, the larval tissues are broken down, and the adult is built by the division and differentiation of cells that were quiescent in the larva. **(d)** Eventually, the adult begins to emerge from the pupal cuticle. **(e)** Hemolymph is pumped into veins of the wings and then withdrawn, leaving the hardened veins as struts supporting the wings. The insect will fly off and reproduce, deriving much of its nourishment from the food reserves stored by the feeding larva.

⊙ **Mastering Biology** Video: Butterfly Emerging

▼ **Figure 33.42** **Exploring Insect Diversity**

Although there are more than 30 orders of insects, we'll focus on just 8 here. Two orders of wingless insects, the bristletails (Archaeognatha) and silverfish (Zygentoma), diverged from other insects early in insect evolution. Evolutionary relationships among the other groups discussed here are under debate and so are not depicted on the tree.

Archaeognatha (bristletails; 350 species)

These wingless insects are found under bark and in other moist, dark habitats such as leaf litter, compost piles, and rock crevices. They feed on algae, plant debris, and lichens.

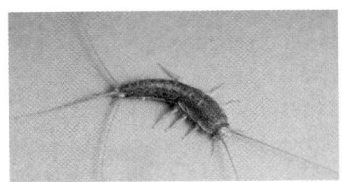

Zygentoma (silverfish; 450 species)

These small, wingless insects have a flattened body and reduced eyes. They live in leaf litter or under bark. They can also infest buildings, where they can become pests.

Winged insects (many orders; six are shown below)

Complete metamorphosis

Coleoptera (beetles; 350,000 species)

Beetles, such as this weevil in the genus *Rhinastus*, constitute the most species-rich order of insects. They have two pairs of wings, one of which is thick and stiff, the other membranous. They have an armored exoskeleton and mouthparts adapted for biting and chewing.

Diptera (151,000 species)

Dipterans have one pair of wings; the second pair has become modified into balancing organs called halteres. Their mouthparts are adapted for sucking, piercing, or lapping. Flies and mosquitoes are among the best-known dipterans, which live as scavengers, predators, and parasites. Like many other insects, flies such as this red tachinid (*Adejeania vexatrix*) have well-developed compound eyes that provide a wide-angle view and excel at detecting fast movements.

Hymenoptera (125,000 species)

Most hymenopterans, which include ants, bees, and wasps, are highly social insects. They have two pairs of membranous wings, a mobile head, and chewing or sucking mouthparts. The females of many species have a posterior stinging organ. Many species, such as this European paper wasp (*Polistes dominulus*), build elaborate nests.

Lepidoptera (120,000 species)

Proboscis

Butterflies and moths have two pairs of wings covered with tiny scales. To feed, they uncoil a long proboscis, visible in this photograph of a hummingbird hawkmoth (*Macroglossum stellatarum*). This moth's name refers to its ability to hover in the air while feeding from a flower. Most lepidopterans feed on nectar, but some species feed on other substances, including animal blood or tears.

Incomplete metamorphosis

Hemiptera (85,000 species)

Hemipterans include so-called "true bugs," such as stink bugs, bed bugs, and assassin bugs. (Insects in other orders are sometimes erroneously called bugs.) Hemipterans have two pairs of wings, one pair partly leathery, the other pair membranous. They have piercing or sucking mouthparts and undergo incomplete metamorphosis, as shown in this image of an adult stink bug guarding its offspring (nymphs).

Orthoptera (13,000 species)

Grasshoppers, crickets, and their relatives are mostly herbivorous. They have large hind legs adapted for jumping, two pairs of wings (one leathery, one membranous), and biting or chewing mouthparts. This aptly named spiny devil katydid (*Panacanthus cuspidatus*) has a face and legs specialized for making a threatening display. Male orthopterans commonly make courtship sounds by rubbing together body parts, such as ridges on their hind legs.

Animals as numerous, diverse, and widespread as insects are bound to affect the lives of most other terrestrial organisms, including humans. Insects consume enormous quantities of plant matter; play key roles as predators, parasites, and decomposers; and are an essential source of food for larger animals such as lizards, rodents, and birds. Humans depend on bees, flies, and many other insects to pollinate crops and orchards. In addition, people in many parts of the world eat insects as an important source of protein. On the other hand, insects are carriers for many diseases, including African sleeping sickness (spread by tsetse flies that carry the protist *Trypanosoma*; see Figure 28.9) and malaria (spread by mosquitoes that carry the protist *Plasmodium*; see Figure 23.18 and Figure 28.18).

Insects also compete with humans for food. In parts of Africa, for instance, insects claim about 75% of the crops. In the United States, billions of dollars are spent each year on pesticides. But insect populations often evolve pesticide resistance after short periods of time. Try as we may, not even humans have challenged the preeminence of insects and their arthropod kin. As one prominent entomologist put it: "Bugs are not going to inherit the Earth. They own it now. So we might as well make peace with the landlord."

CONCEPT CHECK 33.4

1. How do nematode and annelid body plans differ?
2. Describe two adaptations that have enabled insects to thrive on land.
3. **MAKE CONNECTIONS** Historically, annelids and arthropods were viewed as closely related because both have body segmentation. Yet DNA sequence data indicate that annelids belong to one clade (Lophotrochozoa) and arthropods to another (Ecdysozoa). Could traditional and molecular hypotheses be tested by studying the *Hox* genes that control body segmentation (see Concept 21.6)? Explain.

For suggested answers, see Appendix A.

CONCEPT 33.5

Echinoderms and chordates are deuterostomes

Porifera
Cnidaria
Lophotrochozoa
Ecdysozoa
Deuterostomia

Sea stars, sea urchins, and other echinoderms (phylum Echinodermata) may seem to have little in common with vertebrates (animals that have a backbone) and other members of phylum Chordata. Nevertheless, DNA evidence indicates that echinoderms and chordates are closely related, with both phyla belonging to the Deuterostomia clade of bilaterian animals. Echinoderms and chordates also share features characteristic of a deuterostome mode of development, such as radial cleavage and formation of the anus from the blastopore (see Figure 32.10). As discussed in Concept 32.4, however, some animal phyla with members that have deuterostome developmental features, including ectoprocts and brachiopods, are not in the deuterostome clade. Hence, despite its name, the clade Deuterostomia is defined primarily by DNA similarities, not developmental similarities.

Echinoderms

Sea stars (commonly called starfish) and most other groups of **echinoderms** (from the Greek *echin*, spiny, and *derma*, skin) are slow-moving or sessile marine animals. Echinoderms have a coelom. A thin epidermis covers an endoskeleton of hard calcareous plates, and most species are prickly from skeletal bumps and spines. Unique to echinoderms is the **water vascular system**, a network of hydraulic canals branching into extensions called **tube feet** that function in locomotion and feeding **(Figure 33.43)**. Sexual reproduction of echinoderms usually involves separate male and female individuals that release their gametes into the water.

Echinoderms descended from bilaterally symmetrical ancestors, yet on first inspection most species seem to have a radially symmetrical form. The internal and external parts of most adult echinoderms radiate from the center, often as five spokes. However, echinoderm larvae have bilateral symmetry. Furthermore, the symmetry of adult echinoderms is not truly radial. For example, the opening (madreporite) of a sea star's water vascular system is not central but shifted to one side.

Living echinoderms are divided into five clades.

Asteroidea: Sea Stars and Sea Daisies

Sea stars have arms radiating from a central disk; the undersurfaces of the arms bear tube feet. By a combination of muscular and chemical actions, the tube feet can attach to or detach from a substrate. The sea star adheres firmly to rocks or creeps along slowly as its tube feet extend, grip, release, extend, and grip again. Although the base of the tube foot has a flattened disk that resembles a suction cup, the gripping action results from adhesive chemicals, not suction (see Figure 33.43).

Sea stars also use their tube feet to grasp prey, such as clams and oysters. The arms of the sea star embrace the closed bivalve, clinging tightly with their tube feet. The sea star then turns part of its stomach inside out, everting it through its mouth and into the narrow opening between the halves of the bivalve's shell. Next, the digestive system of the sea star secretes juices that begin digesting the mollusc within its own shell. The sea star then brings its stomach back inside its body, where digestion of the mollusc's (now liquefied) body is completed. The ability to begin the digestive process outside of its body allows a sea star to consume bivalves and other prey species that are much larger than its mouth.

Sea stars and some other echinoderms have considerable powers of regeneration. Sea stars can regrow lost arms, and

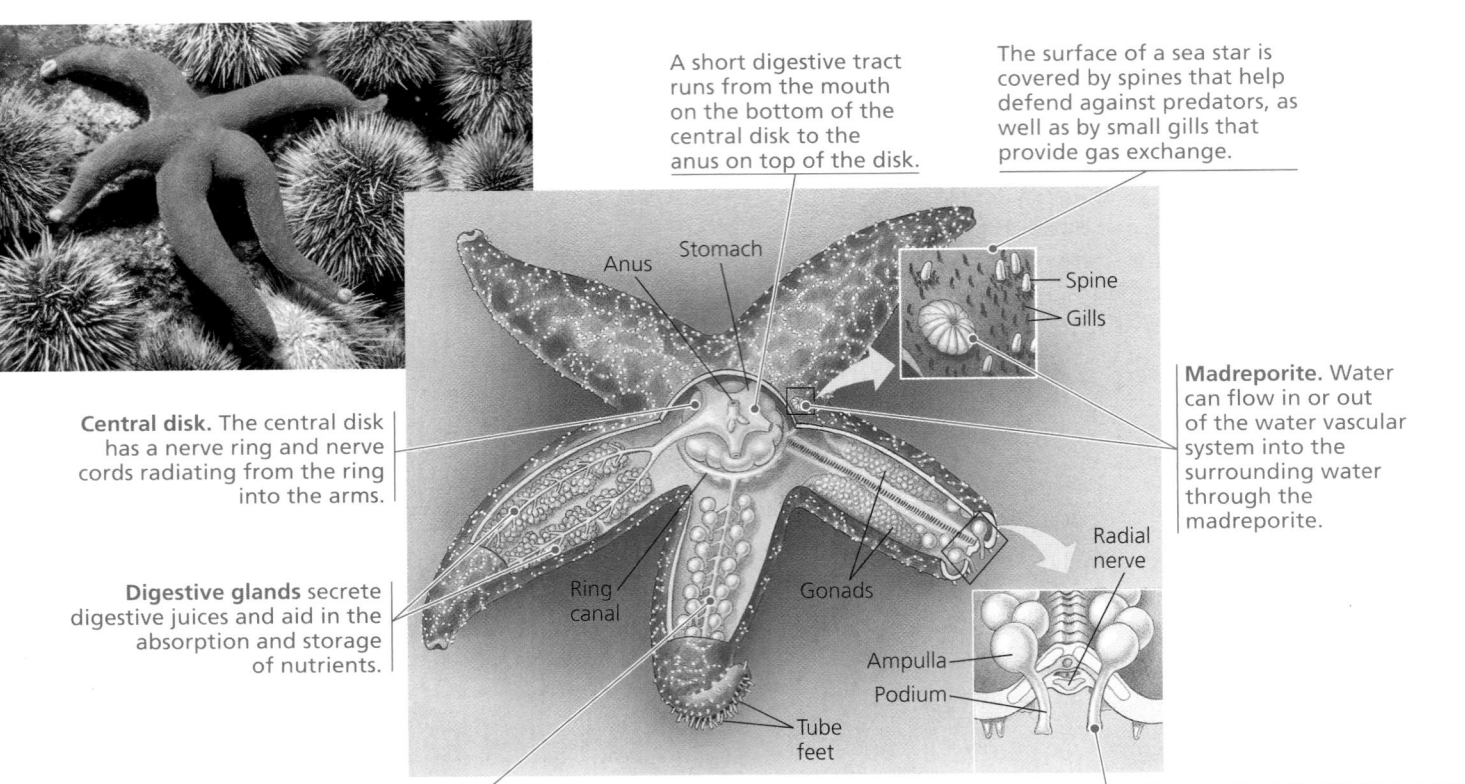

A short digestive tract runs from the mouth on the bottom of the central disk to the anus on top of the disk.

The surface of a sea star is covered by spines that help defend against predators, as well as by small gills that provide gas exchange.

Central disk. The central disk has a nerve ring and nerve cords radiating from the ring into the arms.

Anus   Stomach

Spine

Gills

Madreporite. Water can flow in or out of the water vascular system into the surrounding water through the madreporite.

Radial nerve

Digestive glands secrete digestive juices and aid in the absorption and storage of nutrients.

Ring canal

Gonads

Ampulla

Podium

Tube feet

Radial canal. The water vascular system consists of a ring canal in the central disk and five radial canals, each running down in a groove down the entire length of an arm. Branching from each radial canal are hundreds of hollow, muscular tube feet filled with fluid.

Each tube foot consists of a bulb-like ampulla and a podium (foot portion). When the ampulla squeezes, water is forced into the podium, which expands and contacts the substrate. Adhesive chemicals are then secreted from the base of the podium, attaching it to the substrate. To detach the tube foot, de-adhesive chemicals are secreted and muscles in the podium contract, forcing water back into the ampulla and shortening the podium. As it moves, a sea star leaves an observable "footprint" of adhesive material on the substrate.

▲ **Figure 33.43 Anatomy of a sea star, an echinoderm (top view).** In the diagram, the digestive glands, radial canal, and gonads are illustrated in separate arms for clarity, but each is present in every arm. The photograph shows a sea star surrounded by sea urchins, which are also echinoderms.

⊙ **Mastering Biology Video: Echinoderm Tube Feet**

members of one genus can even regrow an entire body from a single arm if part of the central disk remains attached.

The clade Asteroidea, to which sea stars belong, also includes a small group of armless species, the *sea daisies*. Only three species of sea daisies are known, all of which live on submerged wood. A sea daisy's body is typically disk-shaped; it has a five-sided organization and measures less than a centimeter in diameter **(Figure 33.44)**. The edge of the body is ringed with small spines. Sea daisies absorb nutrients through a membrane that surrounds their body.

▶ **Figure 33.44 A sea daisy (clade Asteroidea).**

Ophiuroidea: Brittle Stars

Brittle stars have a distinct central disk and long, flexible arms **(Figure 33.45)**. They move primarily by lashing their arms in serpentine movements. The base of a brittle star tube foot lacks the flattened disk found in sea stars but does secrete adhesive chemicals. Hence, like sea stars and other

▼ **Figure 33.45 A brittle star (clade Ophiuroidea).**

▼ Figure 33.46 A sea urchin (clade Echinoidea).

echinoderms, brittle stars can use their tube feet to grip substrates. Some species are suspension feeders; others are predators or scavengers.

Echinoidea: Sea Urchins and Sand Dollars

Sea urchins and sand dollars have no arms, but they do have five radially arranged groups of tube feet that function in slow movement. Sea urchins also have muscles that pivot their long spines, which aid in locomotion as well as protection **(Figure 33.46)**. A sea urchin's mouth, located on its underside, is ringed by highly complex, jaw-like structures that are well adapted to eating seaweed. Sea urchins are roughly spherical, whereas sand dollars are flat disks.

Crinoidea: Sea Lilies and Feather Stars

Sea lilies live attached to the substrate by a stalk; feather stars crawl about by using their long, flexible arms. Both use their arms in suspension feeding. The arms encircle the mouth, which is directed upward, away from the substrate **(Figure 33.47)**. Crinoidea is an ancient group whose

▼ Figure 33.47 A feather star (clade Crinoidea).

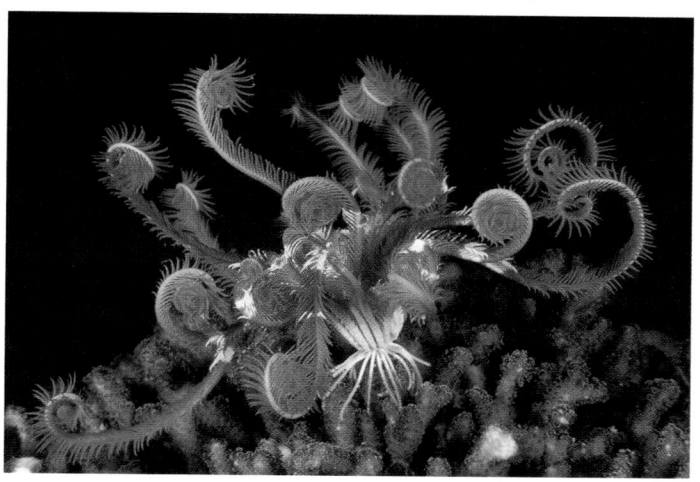

▼ Figure 33.48 A sea cucumber (clade Holothuroidea).

morphology has changed little over the course of evolution; fossilized sea lilies some 500 million years old are extremely similar to present-day members of the clade.

Holothuroidea: Sea Cucumbers

On casual inspection, sea cucumbers do not look much like other echinoderms. They lack spines, and their endoskeleton is much reduced. They are also elongated in their oral-aboral axis, giving them the shape for which they are named and further disguising their relationship to sea stars and sea urchins **(Figure 33.48)**. Closer examination, however, reveals that sea cucumbers have five radially arranged sections of tube feet, as in other echinoderms. Some of the tube feet around the mouth are developed as feeding tentacles.

Chordates

Phylum Chordata consists of two basal groups of invertebrates, the lancelets and the tunicates, as well as the vertebrates. Chordates are bilaterally symmetrical animals with a coelom and segmented bodies. The close relationship between echinoderms and chordates does not mean that one phylum evolved from the other. In fact, echinoderms and chordates have evolved independently of one another for over 500 million years. We will trace the phylogeny of chordates in Chapter 34, focusing on the history of vertebrates.

CONCEPT CHECK 33.5

1. How do sea star tube feet attach to substrates?

2. **WHAT IF?** The insect *Drosophila melanogaster* and the nematode *Caenorhabditis elegans* are prominent model organisms. Are these species the most appropriate invertebrates for making inferences about humans and other vertebrates? Explain.

3. **MAKE CONNECTIONS** Describe how the features and diversity of echinoderms illustrate the unity of life, the diversity of life, and the match between organisms and their environments (see Concept 22.2).

For suggested answers, see Appendix A.

33 Chapter Review

→ Go to **Mastering Biology** for Assignments, the eText, the Study Area, and Dynamic Study Modules.

SUMMARY OF KEY CONCEPTS

→ To review key terms, go to the **Vocabulary Self-Quiz** in the **Mastering Biology** eText or Study Area, or go to goo.gl/zkjz9t.

This table recaps the animal groups surveyed in this chapter.

Key Concept						Phylum		Description
CONCEPT 33.1 **Sponges are basal animals that lack tissues** (pp. 690–691) ❓ *Lacking tissues and organs, how do sponges accomplish tasks such as gas exchange, nutrient transport, and waste disposal?*						Porifera (sponges)		Lack tissues; have choanocytes (collar cells—flagellated cells that ingest bacteria and tiny food particles)
CONCEPT 33.2 **Cnidarians are an ancient phylum of eumetazoans** (pp. 691–693) ❓ *Describe the cnidarian body plan and its two major variations.*						Cnidaria (hydras, jellies, sea anemones, corals)		Unique stinging structures (nematocysts) housed in specialized cells (cnidocytes); diploblastic; radially symmetrical; gastrovascular cavity (digestive compartment with a single opening)
CONCEPT 33.3 **Lophotrochozoans, a clade identified by molecular data, have the widest range of animal body forms** (pp. 694–705) ❓ *Is the lophotrochozoan clade united by unique morphological features shared by all of its members? Explain.*	Metazoa	Eumetazoa	Bilateria	Lophotrochozoa		Platyhelminthes (flatworms)		No body cavity; dorsoventrally flattened; gastrovascular cavity or no digestive tract
						Syndermata (rotifers and acanthocephalans)		Hemocoel; rotifers have alimentary canal (digestive tube with mouth and anus) and jaws (trophi); acanthocephalans are parasites of vertebrates
						Ectoprocta and Brachiopoda		Coelom; have lophophores (feeding structures bearing ciliated tentacles)
						Mollusca (clams, snails, squids)		Hemocoel; reduced coelom; three main body parts (muscular foot, visceral mass, mantle); most have hard shell made of calcium carbonate
						Annelida (segmented worms)		Coelom; body wall and internal organs are segmented (except digestive tract, which is unsegmented)
CONCEPT 33.4 **Ecdysozoans are the most species-rich animal group** (pp. 705–713) ❓ *Describe some ecological roles of nematodes and arthropods.*				Ecdysozoa		Nematoda (roundworms)		Hemocoel; cylindrical body with tapered ends; no circulatory system; undergo ecdysis
						Arthropoda (spiders, centipedes, crustaceans, and insects)		Hemocoel; reduced coelom. Have segmented body, jointed appendages, and exoskeleton made of protein and chitin
CONCEPT 33.5 **Echinoderms and chordates are deuterostomes** (pp. 713–715) ❓ *You've read that echinoderms and chordates are closely related and have evolved independently for over 500 million years. Explain how both of these statements can be correct.*				Deuterostomia		Echinodermata (sea stars, sea urchins)		Coelom; bilaterally symmetrical larvae and five-part body organization as adults; unique water vascular system; endoskeleton
						Chordata (lancelets, tunicates, vertebrates)		Coelom; have notochord; dorsal, hollow nerve cord; pharyngeal slits; post-anal tail (see Figure 34.3)

For more multiple-choice questions, go to the **Practice Test** in the **Mastering Biology** eText or Study Area, or go to goo.gl/GruWRg.

Levels 1-2: Remembering/Understanding

1. A land snail, a clam, and an octopus all share
   (A) a mantle.
   (B) a radula.
   (C) gills.
   (D) distinct cephalization.

2. Which phylum is characterized by animals that have a segmented body?
   (A) Cnidaria
   (B) Platyhelminthes
   (C) Arthropoda
   (D) Mollusca

3. The water vascular system of echinoderms
   (A) functions as a circulatory system that distributes nutrients to body cells.
   (B) functions in locomotion and feeding.
   (C) is bilateral in organization, even though the adult animal is not bilaterally symmetrical.
   (D) moves water through the animal's body during filter feeding.

4. Which of the following combinations of phylum and description is correct?
   (A) Echinodermata—radial symmetry as a larva, coelom
   (B) Nematoda—roundworms, internal skeleton
   (C) Platyhelminthes—flatworms, gastrovascular cavity, no body cavity
   (D) Porifera—gastrovascular cavity, coelom

Levels 3-4: Applying/Analyzing

5. In Figure 33.1, which two main clades branch from the most recent common ancestor of the eumetazoans?
   (A) Porifera and Cnidaria
   (B) Lophotrochozoa and Ecdysozoa
   (C) Cnidaria and Bilateria
   (D) Deuterostomia and Bilateria

6. **MAKE CONNECTIONS** In Figure 33.7, assume that the two medusae shown at step 4 were produced by one polyp colony. Review Concept 12.1 and Concept 13.3, and then use your understanding of mitosis and meiosis to select which of the following is true.
   (A) Both the medusae and the gametes are genetically identical.
   (B) Neither the medusae nor the gametes are genetically identical.
   (C) The medusae are not genetically identical but the gametes are genetically identical.
   (D) The medusae are genetically identical but the gametes differ genetically.

Levels 5-6: Evaluating/Creating

7. **EVOLUTION CONNECTION • INTERPRET THE DATA** Based on Figure 32.11, draw a phylogenetic tree of Bilateria that includes the ten phyla of bilaterians discussed in detail in this chapter. Label each branch that leads to a phylum with a C, H, HC, or N, depending on whether members of the phylum have a coelom only (C), hemocoel only (H), hemocoel and reduced coelom (HC), or no body cavity (N). Use your labeled tree to answer the following questions. (a) For each of the three major clades of bilaterians, what (if anything) can be inferred about whether the common ancestor of the clade had a coelom? (b) To what extent has the presence of a coelom in animals changed over the course of evolution?

8. **SCIENTIFIC INQUIRY** Bats emit ultrasonic sounds and then use the returning echoes of those sounds to locate and capture flying insects, such as moths, in the dark. In response to bat attacks, some tiger moths make ultrasonic clicks of their own. Researchers hypothesize that tiger moth clicks likely either (1) jam the bat's sonar or (2) warn the bat about the moth's toxic chemical defenses. The graph below shows two patterns observed in studies of moth capture rates over time.

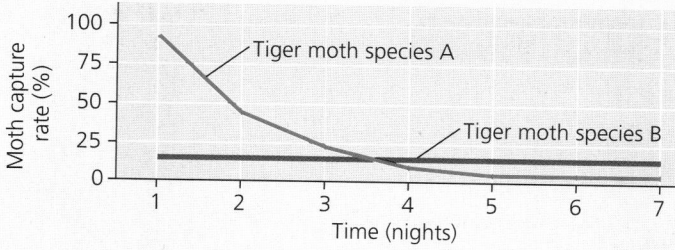

Bats in these experiments were "naive," meaning that prior to the study the bats had not previously hunted tiger moths. Indicate whether the results support hypothesis (1), hypothesis (2), or both. Explain why the researchers used naive bats in this study.

9. **WRITE ABOUT A THEME: ORGANIZATION** Write a short essay (100–150 words) that explains how the structure of the digestive tract in different invertebrate groups affects the size of the organisms that they can eat.

10. **SYNTHESIZE YOUR KNOWLEDGE**

Collectively, do these beetles and all other invertebrate species combined form a monophyletic group? Explain your answer and provide an overview of the evolutionary history of invertebrate life.

For selected answers, see Appendix A.

34

The Origin and Evolution of Vertebrates

KEY CONCEPTS

34.1 Chordates have a notochord and a dorsal, hollow nerve cord *p. 719*

34.2 Vertebrates are chordates that have a backbone *p. 722*

34.3 Gnathostomes are vertebrates that have jaws *p. 725*

34.4 Tetrapods are gnathostomes that have limbs *p. 730*

34.5 Amniotes are tetrapods that have a terrestrially adapted egg *p. 734*

34.6 Mammals are amniotes that have hair and produce milk *p. 741*

34.7 Humans are mammals that have a large brain and bipedal locomotion *p. 748*

Study Tip

Make a table: Humans are mammals—and we are also amniotes, tetrapods, lobe-fins, osteichthyans, gnathostomes, vertebrates, and chordates (see Figure 34.2). As you read the chapter, list derived characters for these clades and describe whether these traits have been retained, modified, or lost in the human lineage.

Clade	Derived characters	Found in humans?
Chordates	Notochord	Modified (gelatinous disks between vertebrae)
	Post-anal tail	Lost

→ Go to Mastering Biology

For Students (in eText and Study Area)
- Get Ready for Chapter 34
- Video: Sea Horse Camouflage
- HHMI Video: Great Transitions: The Origins of Humans

For Instructors to Assign (in Item Library)
- Tutorial: Chordates
- Problem-Solving Exercise: Declining Amphibian Populations

Figure 34.1 Small and slender, this inconspicuous species (*Myllokunmingia fengjiaoa*) lived in the ocean 530 million years ago, along with an immense variety of dangerous and well-defended invertebrate animals. Although lacking claws or armor, this ancient species was closely related to one of the most successful groups of animals on Earth: the vertebrates, animals with a backbone.

What are some key characteristics that have appeared during vertebrate evolution?

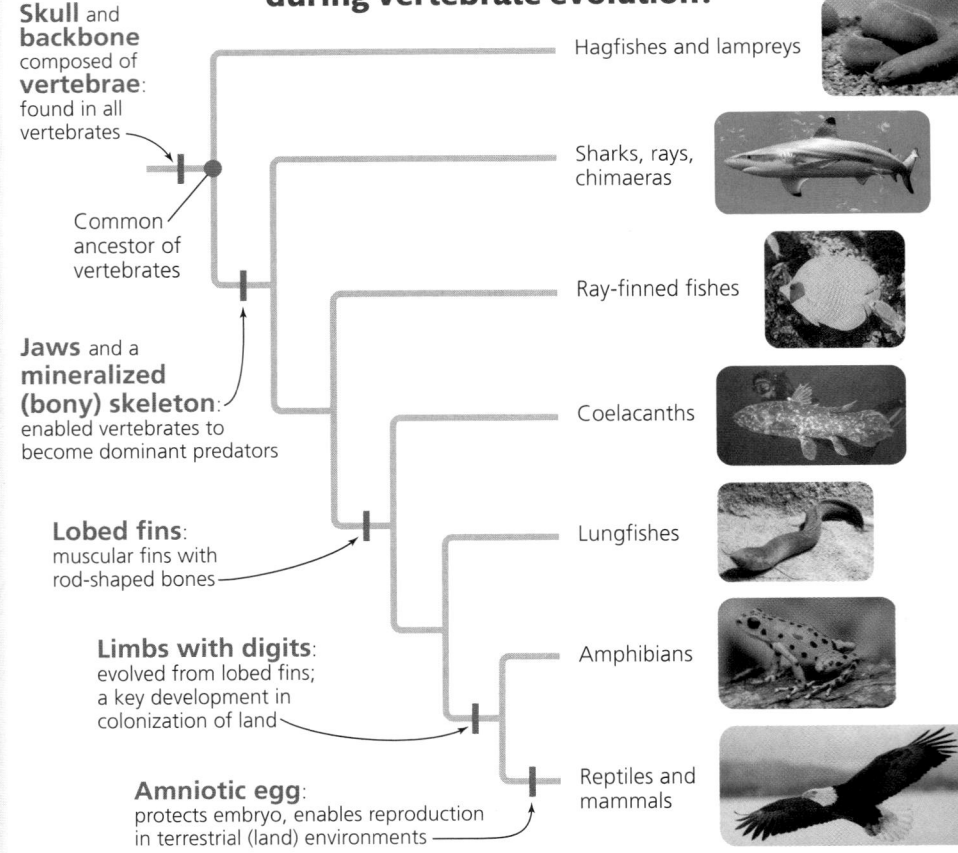

Skull and **backbone** composed of **vertebrae**: found in all vertebrates

Common ancestor of vertebrates

Jaws and a **mineralized (bony) skeleton**: enabled vertebrates to become dominant predators

Lobed fins: muscular fins with rod-shaped bones

Limbs with digits: evolved from lobed fins; a key development in colonization of land

Amniotic egg: protects embryo, enables reproduction in terrestrial (land) environments

Hagfishes and lampreys

Sharks, rays, chimaeras

Ray-finned fishes

Coelacanths

Lungfishes

Amphibians

Reptiles and mammals

There are more than 60,000 species of **vertebrates**, animals with a backbone. This is a relatively small number compared to, say, the 1 million insect species on Earth. But what vertebrates lack in number of species, they make up for in *disparity*, varying enormously in characteristics such as body mass. Vertebrates include the heaviest animals ever to walk on land, plant-eating dinosaurs that were as massive as 40,000 kg (heavier than 13 pickup trucks). The biggest animal ever to exist on Earth is also a vertebrate—the blue whale, which can exceed 100,000 kg. On the other end of the spectrum, the fish *Schindleria brevipinguis* is just 8.4 mm long and has a mass roughly 100 billion times smaller than that of a blue whale.

In this chapter, we will track the origin and evolution of the vertebrate body plan, from a notochord to a head to a mineralized skeleton to limbs with digits (such as toes). We'll also explore the major groups of vertebrates (both living and extinct), as well as the evolutionary history of our own species—*Homo sapiens*.

CONCEPT 34.1

Chordates have a notochord and a dorsal, hollow nerve cord

Vertebrates are members of the phylum Chordata, the chordates. **Chordates** are bilaterian (bilaterally symmetrical) animals, and within Bilateria, they belong to the clade of animals known as Deuterostomia (see Figure 32.11). As shown in **Figure 34.2**, there are two groups of invertebrate deuterostomes that are more closely related to vertebrates than they are to other invertebrates: the cephalochordates and the urochordates. Thus, along with the vertebrates, these two invertebrate groups are classified within the chordates.

Derived Characters of Chordates

All chordates share a set of derived characters, though many species possess some of these traits only during embryonic

► **Figure 34.2 Phylogeny of living chordates.** This phylogenetic hypothesis shows the major clades of chordates in relation to the other main deuterostome clade, Echinodermata (see Concept 33.5). Derived characters are listed for selected clades; for example, only gnathostomes have a jaw. In some lineages, derived traits have been lost over time or occur in reduced form; hagfishes and lampreys, for example, are vertebrates with reduced vertebrae.

DRAW IT *Label the branch point representing the common ancestor of vertebrates "1" and the branch point representing the common ancestor of tetrapods "2." Redraw the vertebrate portion of this tree by rotating its branches around these two branch points. Use the redrawn tree to explain why it is not correct to represent evolution as a sequence of events "leading to" humans and other mammals.*

▼ **Figure 34.3 Chordate characteristics.** All chordates possess the four highlighted structural trademarks at some point during their development.

development. **Figure 34.3** illustrates four key characters of chordates: a notochord; a dorsal, hollow nerve cord; pharyngeal slits or clefts; and a muscular, post-anal tail.

Notochord

Chordates are named for a skeletal structure, the notochord, present in all chordate embryos as well as in some adult chordates. The **notochord** is a longitudinal, flexible rod located between the digestive tube and the nerve cord. It is composed of large, fluid-filled cells encased in fairly stiff, fibrous tissue. The notochord provides skeletal support throughout most of the length of a chordate, and in larvae or adults that retain it, it also provides a firm but flexible structure against which muscles can work during swimming. In most vertebrates, a more complex, jointed skeleton develops around the ancestral notochord, and the adult retains only remnants of the embryonic notochord. In humans, for example, the notochord is reduced and forms part of the gelatinous disks sandwiched between the vertebrae.

Dorsal, Hollow Nerve Cord

The nerve cord of a chordate embryo develops from a plate of ectoderm that rolls into a neural tube located dorsal to the notochord. The resulting dorsal, hollow nerve cord is unique to chordates. Other animal phyla have solid nerve cords, and in most cases they are ventrally located. The nerve cord of a chordate embryo develops into the central nervous system: the brain and spinal cord.

Pharyngeal Slits or Clefts

The digestive tube of chordates extends from the mouth to the anus. The region just posterior to the mouth is the pharynx. In all chordate embryos, a series of arches separated by grooves forms along the outer surface of the pharynx. In most chordates, these grooves (known as **pharyngeal clefts**) develop into slits that open into the pharynx. These **pharyngeal slits** allow water entering the mouth to exit the body without passing through the entire digestive tract.

Pharyngeal slits function as suspension-feeding devices in many invertebrate chordates. In vertebrates (with the exception of vertebrates with limbs, the *tetrapods*), these slits and the pharyngeal arches that support them have been modified for gas exchange and are called gills. In tetrapods, the pharyngeal clefts do not develop into slits. Instead, the pharyngeal arches that surround the clefts develop into parts of the ear and other structures in the head and neck.

Muscular, Post-Anal Tail

Chordates have a tail that extends posterior to the anus, although in many species it is greatly reduced during embryonic development. In contrast, most nonchordates have a digestive tract that extends nearly the whole length of the body. The chordate tail contains skeletal elements and muscles, and it helps propel many aquatic species in the water.

Lancelets

The sister group to all other living chordates is the group of animals called **lancelets** (Cephalochordata), which get their name from their bladelike shape **(Figure 34.4)**. As larvae, lancelets develop a notochord; a dorsal, hollow nerve cord; numerous pharyngeal slits; and a post-anal tail. The larvae feed on plankton in the water column, alternating between upward swimming and passive sinking. As the larvae sink, they trap plankton and other suspended particles in their pharynx.

Adult lancelets can reach 6 cm in length. They retain key chordate traits, closely resembling the generalized chordate shown in Figure 34.3. Following metamorphosis, an adult lancelet swims down to the seafloor and wriggles backward into the sand, leaving only its anterior end exposed. Cilia draw seawater into the lancelet's mouth. A net of mucus secreted across the pharyngeal slits removes tiny food particles as the water passes through the slits, and the trapped food enters the intestine. The pharynx and pharyngeal slits play a minor role in gas exchange, which occurs mainly across the external body surface.

A lancelet frequently leaves its burrow to swim to a new location. Though feeble swimmers, these invertebrate chordates display, in a simple form, the swimming mechanism of fishes. Coordinated contraction of muscles arranged like rows of chevrons (>>>>) along the sides of the notochord flexes the notochord, producing side-to-side undulations that thrust the body forward. This serial arrangement of muscles is evidence of the lancelet's segmentation. The muscle segments develop from blocks of mesoderm called *somites*, which are found along each side of the notochord in all chordate embryos.

▼ **Figure 34.4 The lancelet *Branchiostoma*, a cephalochordate.**
This small invertebrate displays all four main chordate characters.
Water enters the mouth and passes through the pharyngeal
slits into the atrium, a chamber that vents to the outside via the
atriopore; large particles are blocked from entering the mouth by
tentacle-like cirri. The serially arranged segmental muscles produce
the lancelet's wavelike swimming movements.

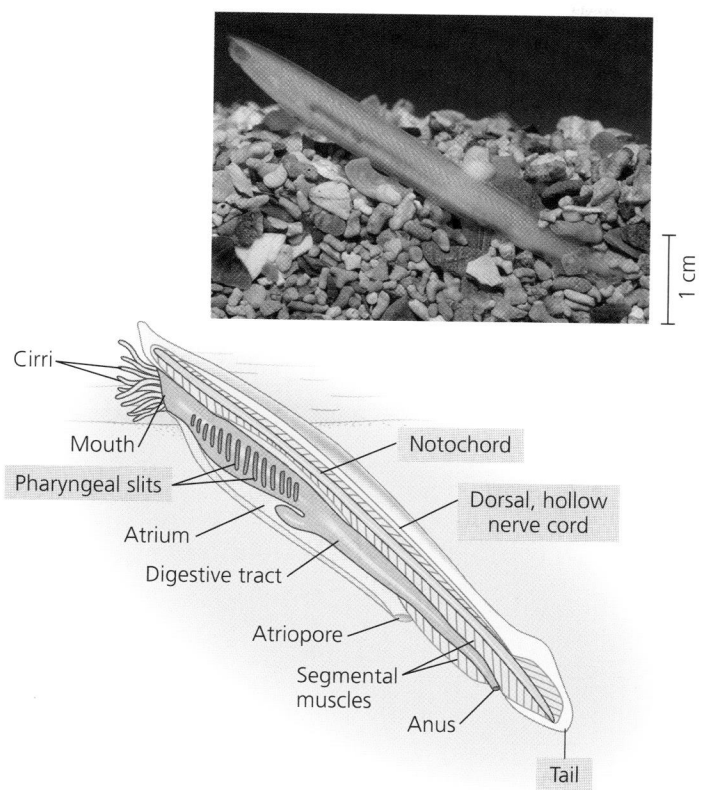

Globally, lancelets are rare, but in a few areas (such as
Tampa Bay, on the Florida coast), they may reach densities of
more than 5,000 individuals per square meter.

Tunicates

Recent molecular stud-
ies indicate that the
tunicates (Urochordata)
are more closely related to
other chordates than are
lancelets. The chordate
characters of tunicates are
most apparent during their
larval stage, which may be
as brief as a few minutes
(Figure 34.5a). In many species, the larva uses its tail muscles
and notochord to swim through water in search of a suitable
substrate on which it can settle, guided by cues it receives from
light- and gravity-sensitive cells.

Once a tunicate has settled on a substrate, it undergoes a
radical metamorphosis in which many of its chordate characters disappear. Its tail and notochord are resorbed; its nervous
system degenerates; and its remaining organs rotate 90°. As an
adult, a tunicate draws in water through an incurrent siphon;
the water then passes through the pharyngeal slits into a chamber called the atrium and exits through an excurrent siphon
(Figure 34.5b and c). Food particles are filtered from the water
by a mucous net and transported by cilia to the esophagus. The
anus empties into the excurrent siphon. Some tunicate species shoot a jet of water through their excurrent siphon when
attacked, earning them the informal name of "sea squirts."

▼ **Figure 34.5 A tunicate, a urochordate.**

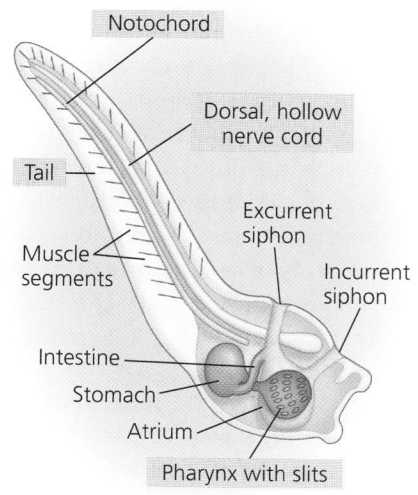

(a) A tunicate larva is a free-swimming but
nonfeeding "tadpole" in which all four
main characters of chordates are evident.

(b) In the adult, prominent pharyngeal slits
function in suspension feeding, but other
chordate characters are not obvious.

(c) An adult tunicate, or sea squirt, is a
sessile animal (photo is approximately
life-sized).

The loss of chordate characters in the adult stage of tunicates appears to have occurred after the tunicate lineage branched off from other chordates. Even the tunicate larva appears to be highly derived. For example, tunicates have nine *Hox* genes, whereas all other chordates studied to date—including the early-diverging lancelets—share a set of 13 *Hox* genes. The apparent loss of four *Hox* genes indicates that the chordate body plan of a tunicate larva is built using a different set of genetic controls than other chordates.

Early Chordate Evolution

Although lancelets and tunicates are relatively obscure animals, they occupy key positions in the history of life and can provide clues about the evolutionary origin of vertebrates. For example, as you have read, lancelets display key chordate characters as adults, and their lineage branches from the base of the chordate phylogenetic tree. These findings suggest that the ancestral chordate may have looked something like a lancelet—that is, it had an anterior end with a mouth; a notochord; a dorsal, hollow nerve cord; pharyngeal slits; and a post-anal tail.

Research on lancelets has also revealed important clues about the evolution of the chordate brain. Rather than a full-fledged brain, lancelets have only a slightly swollen tip on the anterior end of their dorsal nerve cord **(Figure 34.6)**. But the same *Hox* genes that organize major regions of the forebrain, midbrain, and hindbrain of vertebrates express themselves in a corresponding pattern in this small cluster of cells in the lancelet's nerve cord. This suggests that the vertebrate brain is an elaboration of an ancestral structure similar to the lancelet's simple nerve cord tip.

As for tunicates, several of their genomes have been completely sequenced and can be used to identify genes likely to have been present in early chordates. Researchers have

▼ **Figure 34.6 Expression of developmental genes in lancelets and vertebrates.** *Hox* genes (including *BF1, Otx,* and *Hox3*) control the development of major regions of the vertebrate brain. These genes are expressed in the same anterior-to-posterior order in lancelets and vertebrates. Each colored bar is positioned above the portion of the brain whose development that gene controls.

Nerve cord of lancelet embryo

Brain of vertebrate embryo (shown straightened)

Forebrain Midbrain Hindbrain

MAKE CONNECTIONS *What do these expression patterns and those in Figure 21.19 indicate about* Hox *genes and their evolution?*

suggested that ancestral chordates had genes associated with vertebrate organs such as the heart and thyroid gland. These genes are found in tunicates and vertebrates but are absent from nonchordate invertebrates. Another recent study found that tunicates (but not lancelets) have embryonic cells that have some of the characteristics of the *neural crest*, a derived trait found in all vertebrates (see Figure 34.7). This suggests that embryonic cells similar to those in tunicates may represent an intermediate cell population from which the vertebrate neural crest evolved.

CONCEPT CHECK 34.1

1. Identify the four derived characters that all chordates have at some point during their life.
2. You are a chordate, yet you lack most of the main derived characters of chordates. Explain.
3. **VISUAL SKILLS** Based on the phylogenetic tree in Figure 34.2, predict which vertebrate groups should have lungs or lung derivatives. Explain.

For suggested answers, see Appendix A.

CONCEPT 34.2

Vertebrates are chordates that have a backbone

During the Cambrian period, half a billion years ago, a lineage of chordates gave rise to vertebrates. With a skeletal system and a more complex nervous system than that of their ancestors, vertebrates became more efficient at two essential tasks: capturing food and avoiding being eaten.

Derived Characters of Vertebrates

Living vertebrates share a set of derived characters that distinguish them from other chordates. For example, as a result of gene duplication, vertebrates possess two or more sets of *Hox* genes (lancelets and tunicates have only one). Other important families of genes that produce transcription factors and signaling molecules are also duplicated in vertebrates. The resulting additional genetic complexity may be associated with innovations in the vertebrate nervous system and skeleton, including the development of a skull and a backbone composed of vertebrae. In some vertebrates, the vertebrae are little more than small prongs of cartilage arrayed dorsally along the notochord. In the majority of vertebrates, however, the vertebrae enclose the spinal cord and have taken over the mechanical roles of the notochord.

Another feature unique to vertebrates is the **neural crest**, a collection of cells that appears along the edges of the closing neural tube of an embryo **(Figure 34.7)**. Neural crest cells disperse throughout the embryo, where they give rise to a variety of structures, including teeth, some of the bones and cartilage of the skull, several types of neurons, and the sensory capsules in which eyes and other sense organs develop.

▼ **Figure 34.7 The neural crest, embryonic source of many unique vertebrate traits.**

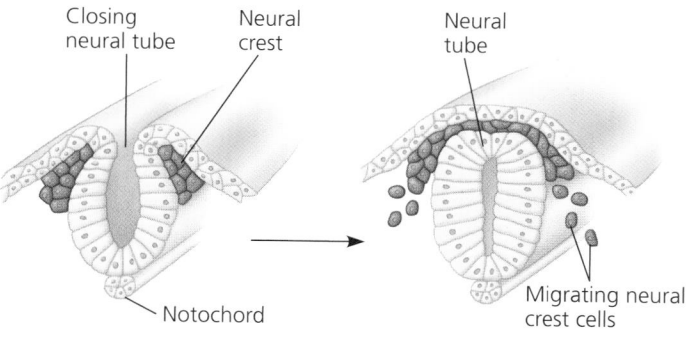

Closing neural tube

Neural crest

Neural tube

Notochord

Migrating neural crest cells

(a) The neural crest consists of bilateral bands of cells near the margins of the embryonic folds that form the neural tube.

(b) Neural crest cells migrate to distant sites in the embryo.

(c) The migrating neural crest cells give rise to some of the anatomical structures unique to vertebrates, including some of the bones and cartilage of the skull. (A fetal human skull is depicted here.)

Hagfishes and Lampreys

Cephalochordata
Urochordata
Myxini
Petromyzontida
Chondrichthyes
Actinopterygii
Actinistia
Dipnoi
Amphibia
Reptilia
Mammalia

The **hagfishes** (Myxini) and the **lampreys** (Petromyzontida) are the only lineages of living vertebrates whose members lack jaws. Unlike most vertebrates, lampreys and hagfishes also do not have a backbone. Even so, lampreys are classified as vertebrates because they have rudimentary vertebrae (composed of cartilage, not bone). The hagfishes, in contrast, once were thought to lack vertebrae altogether; hence, they were classified as invertebrate chordates closely related to vertebrates.

In the past few years, however, this interpretation has changed. Recent research has shown that hagfishes, like lampreys, have rudimentary vertebrae. In addition, a series of molecular phylogenetic studies have supported the hypothesis that hagfishes are vertebrates. Molecular analyses also indicate that hagfishes and lampreys are sister groups, as shown in Figure 34.2. Together, the hagfishes and lampreys form a clade of living jawless vertebrates, the **cyclostomes**. (All other vertebrates have jaws and make up a much larger clade, the gnathostomes, which we will discuss in Concept 34.3.)

Hagfishes

The hagfishes are jawless vertebrates that have highly reduced vertebrae and a skull that is made of cartilage. They swim in a snakelike fashion by using their segmental muscles to exert

▼ **Figure 34.8 A hagfish.**

Slime glands

force against their notochord, which they retain in adulthood as a strong, flexible rod of cartilage. Hagfishes have a small brain, eyes, ears, and a nasal opening that connects with the pharynx. Their mouths contain tooth-like formations made of the protein keratin.

All of the 30 living species of hagfishes are marine. Measuring up to 60 cm in length, most are bottom-dwelling scavengers **(Figure 34.8)** that feed on worms and sick or dead fish. Rows of slime glands on a hagfish's flanks secrete a substance that absorbs water, forming a slime that may repel other scavengers when a hagfish is feeding. When attacked by a predator, a hagfish can produce several liters of slime in less than a minute. The slime coats the gills of the attacking fish, sending it into retreat or even suffocating it. Biologists and engineers are investigating the properties of hagfish slime as a model for developing a space-filling gel that could be used, for instance, to stop bleeding during surgery.

Lampreys

The second group of living jawless vertebrates, the lampreys, consists of about 38 species inhabiting various marine and freshwater environments **(Figure 34.9)**. Some are parasites

▼ **Figure 34.9 A sea lamprey.** Parasitic lampreys use their mouth (inset) and tongue to bore a hole in the side of a fish. The lamprey then ingests the blood and other tissues of its host.

that feed by clamping their round, jawless mouth onto the flank of a live fish, their "host." Parasitic lampreys use their rasping mouth and tongue to penetrate the skin of the fish and ingest the fish's blood and other tissues.

As larvae, lampreys live in freshwater streams. The larva is a suspension feeder that resembles a lancelet and spends much of its time partially buried in sediment. About 20 species of lampreys are not parasitic. These species feed only as larvae; following several years in streams, they mature sexually, reproduce, and die within a few days. In contrast, parasitic species of lampreys migrate to the sea or lakes as they mature into adults. One such parasite, the sea lamprey (*Petromyzon marinus*), has invaded the Great Lakes over the past 170 years and devastated a number of fisheries there.

The skeleton of lampreys is made of cartilage. Unlike the cartilage found in most vertebrates, lamprey cartilage contains no collagen. Instead, it is a stiff matrix of other proteins. The notochord of lampreys persists as the main axial skeleton in the adult, as it does in hagfishes. However, lampreys also have a flexible sheath around their rodlike notochord. Along the length of this sheath, pairs of cartilaginous projections related to vertebrae extend dorsally, partially enclosing the nerve cord.

Early Vertebrate Evolution

In the late 1990s, paleontologists working in China discovered a vast collection of fossils of early chordates that appear to straddle the transition to vertebrates. The fossils were formed 530 million years ago during the Cambrian explosion, when many animal groups were undergoing rapid diversification (see Concept 32.2).

The most primitive of the fossils are the 3-cm-long *Haikouella* (**Figure 34.10**). In many ways, *Haikouella* resembled a lancelet. Its mouth structure indicates that, like lancelets, it probably was a suspension feeder. However, *Haikouella* also had some of the characters of vertebrates. For example, it had a well-formed brain, small eyes, and muscle segments along the body, as do the vertebrate fishes. Unlike the vertebrates, however, *Haikouella* did not have a skull or ear organs, suggesting that these characters emerged with further innovations to the chordate nervous system. (The earliest "ears" were organs for maintaining balance, a function still performed by the ears of humans and other living vertebrates.)

Early signs of a skull can be seen in *Myllokunmingia* (see Figure 34.1). About the same size as *Haikouella*, *Myllokunmingia* had ear capsules and eye capsules, parts of the skull that surround these organs. Based on these and other characters, *Myllokunmingia* is considered the first chordate to have a head. The origin of a head—consisting of a brain at the anterior end of the dorsal nerve cord, eyes and other sensory organs, and a skull—enabled chordates to coordinate more complex movement and feeding behaviors. Although it had a head, *Myllokunmingia* lacked vertebrae and hence is not classified as a vertebrate.

▼ **Figure 34.10 Fossil of an early chordate.** Discovered in 1999 in southern China, *Haikouella* had eyes and a brain but lacked a skull, a trait found in vertebrates. The organism's color in the drawing is fanciful.

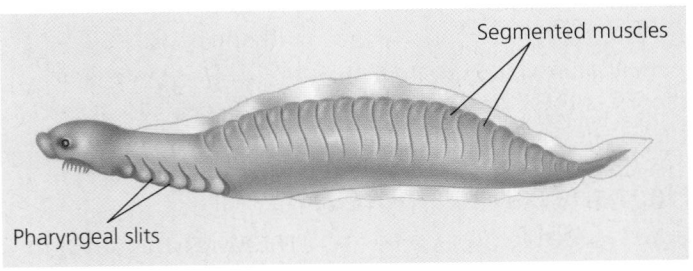

Segmented muscles

Pharyngeal slits

▼ **Figure 34.11 A conodont.** Conodonts were early jawless vertebrates that lived from 500 million to 200 million years ago. Unlike hagfishes and lampreys, conodonts had mineralized mouthparts, which they used for either predation or scavenging.

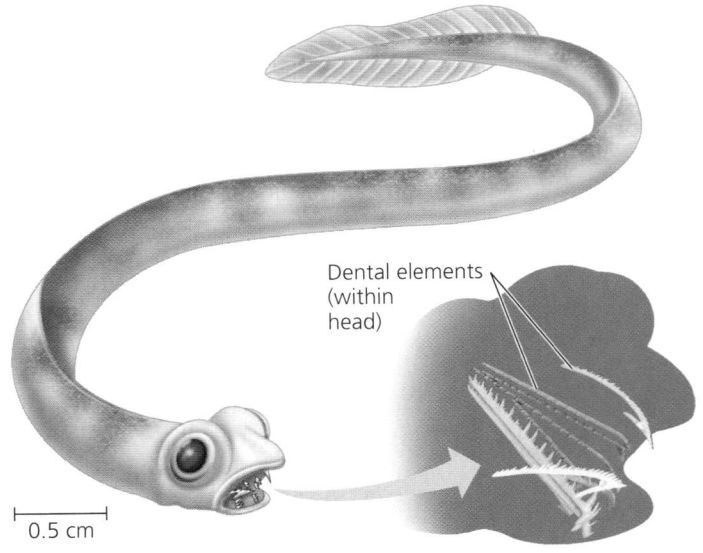

Dental elements (within head)

0.5 cm

The earliest fossils of vertebrates date to 500 million years ago and include those of **conodonts (Figure 34.11)**, a group of slender, soft-bodied vertebrates that lacked jaws and whose internal skeleton was composed of cartilage. Conodonts had large eyes, which they may have used in locating prey that were then impaled on a set of barbed hooks at the anterior end of their mouth (see Figure 34.11). These hooks were made of dental tissues that were *mineralized*—hardened by the incorporation of minerals such as calcium. The food was then passed back to the pharynx, where a different set of dental elements sliced and crushed the food.

Conodonts were extremely abundant for 300 million years. Their fossilized dental elements are so plentiful that they have been used for decades by petroleum geologists as guides to the age of rock layers in which they search for oil.

Vertebrates with additional innovations emerged during the Ordovician, Silurian, and Devonian periods (485–359 million years ago). These vertebrates had paired fins and, as in lampreys, an inner ear with two semicircular canals that provided a sense of balance. Like conodonts, these vertebrates lacked jaws, but they had a muscular pharynx, which they may have used to suck in bottom-dwelling organisms or detritus. They were also armored with mineralized bone, which covered varying amounts of their body and may have offered protection from predators **(Figure 34.12)**. There were many species of these jawless, armored swimming vertebrates, but they all became extinct by the end of the Devonian.

Finally, note that the human skeleton is heavily mineralized bone, whereas cartilage plays a fairly minor role. But a bony internal skeleton was a relatively late development in the history of vertebrates. Instead, the vertebrate skeleton evolved initially as a structure made of unmineralized cartilage. Steps toward a bony skeleton began 470 million years ago, with the appearance of mineralized bone on the outer surface of the skull in some jawless vertebrates. Shortly after that time, the internal skeleton began to mineralize, first as calcified cartilage. By 430 million years ago, some vertebrates had a thin layer of bone lining the cartilage of their internal skeleton. The bones of vertebrates underwent even more mineralization in the group we turn to next, the jawed vertebrates.

▼ **Figure 34.12 Jawless armored vertebrates.** *Pteraspis* and *Pharyngolepis* were two of many genera of jawless vertebrates that emerged during the Ordovician, Silurian, and Devonian periods.

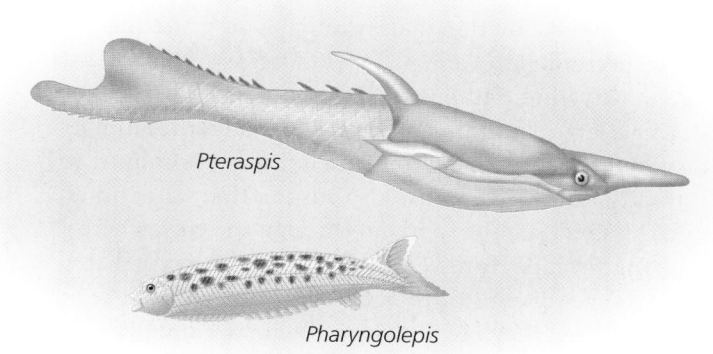

Pteraspis

Pharyngolepis

CONCEPT CHECK 34.2

1. How are differences in the anatomy of lampreys and conodonts reflected in each animal's feeding method?
2. **WHAT IF?** In several different animal lineages, organisms with a head first appeared around 530 million years ago. Does this finding constitute proof that having a head is favored by natural selection? Explain.
3. **WHAT IF?** Suggest key roles that mineralized bone might have played in early vertebrates.

For suggested answers, see Appendix A.

CONCEPT 34.3

Gnathostomes are vertebrates that have jaws

Hagfishes and lampreys are survivors from the early Paleozoic era, when jawless vertebrates were common. Since then, jawless vertebrates have been far outnumbered by the jawed vertebrates, the **gnathostomes**. Living gnathostomes are a diverse group that includes sharks and their relatives, ray-finned fishes, lobe-finned fishes, amphibians, reptiles (including birds), and mammals.

Derived Characters of Gnathostomes

Gnathostomes ("jaw mouth") are named for their jaws, hinged structures that, especially with the help of teeth, enable gnathostomes to grip food items firmly and slice them. According to one hypothesis, gnathostome jaws evolved by modification of the skeletal rods that had previously supported the anterior pharyngeal (gill) slits. **Figure 34.13** shows a stage in this evolutionary process in which several of these skeletal rods have been modified into precursors of jaws (green) and their structural supports (red). The remaining gill slits, no longer required for suspension feeding, remained as the major sites of respiratory gas exchange with the external environment.

Gnathostomes share other derived characters besides jaws. The common ancestors of all gnathostomes underwent an additional duplication of *Hox* genes, such that the single set present in early chordates became four. In fact, the entire genome appears to have duplicated, and together these genetic changes likely enabled the origin of jaws and other novel features in gnathostomes. The gnathostome forebrain is enlarged compared to that of other vertebrates, and it is associated with enhanced senses of smell and vision. Another characteristic of aquatic gnathostomes is the

▼ **Figure 34.13 Possible step in the evolution of jawbones.**

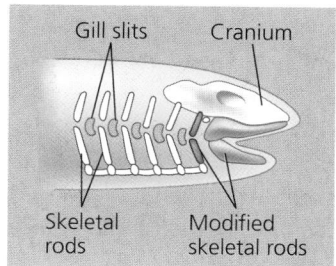

Gill slits

Cranium

Skeletal rods

Modified skeletal rods

lateral line system, organs that form a row along each side of the body and are sensitive to vibrations in the surrounding water. Precursors of these organs were present in the head shields of some jawless vertebrates.

Fossil Gnathostomes

Gnathostomes appeared in the fossil record about 440 million years ago and steadily became more diverse. Their success probably resulted from a combination of anatomical features: Their paired fins and tail (which were also found in jawless vertebrates) allowed them to swim efficiently after prey, and their jaws enabled them to grab prey or simply bite off chunks of flesh. Over time, dorsal, ventral, and anal fins stiffened by bony structures called fin rays also evolved in some early gnathostomes. Fin rays provide thrust and steering control when aquatic vertebrates swim after prey or away from predators. Faster swimming was supported by other adaptations, including a more efficient gas exchange system in the gills.

The earliest gnathostomes include extinct lineages of armored vertebrates known collectively as **placoderms**, which means "plate-skinned." Most placoderms were less than a meter long, though some giants measured more than 10 m **(Figure 34.14)**. Other jawed vertebrates, called **acanthodians**, emerged at roughly the same time and radiated during the Silurian and Devonian periods (444–359 million years ago). Placoderms had disappeared by 359 million years ago, and acanthodians became extinct about 70 million years later.

Overall, a series of recent fossil discoveries have revealed that 440–420 million years ago was a period of tumultuous evolutionary change. Gnathostomes that lived during this period had highly variable forms, and by 420 million years ago, they had

▼ **Figure 34.14 Fossil of an early gnathostome.** A formidable predator, the placoderm *Dunkleosteus* grew up to 10 m in length. Its jaw structure indicates that *Dunkleosteus* could exert a force of 560 kg/cm² (8,000 pounds per square inch) at the tip of its jaws.

0.5 m

diverged into the three lineages of jawed vertebrates that survive today: chondrichthyans, ray-finned fishes, and lobe-fins.

Chondrichthyans (Sharks, Rays, and Their Relatives)

Cephalochordata
Urochordata
Myxini
Petromyzontida
Chondrichthyes
Actinopterygii
Actinistia
Dipnoi
Amphibia
Reptilia
Mammalia

Sharks, rays, and their relatives include some of the biggest and most successful vertebrate predators in the oceans. They belong to the clade Chondrichthyes, which means "cartilage fish." As their name indicates, the **chondrichthyans** have a skeleton composed predominantly of cartilage, though often impregnated with calcium.

When the name Chondrichthyes was first coined, in the 1800s, scientists thought that chondrichthyans represented an early stage in the evolution of the vertebrate skeleton and that mineralization had evolved only in more derived lineages (such as "bony fishes"). However, as armored jawless vertebrates demonstrate, the mineralization of the vertebrate skeleton had already begun before the chondrichthyan lineage branched off from other vertebrates. Moreover, bone-like tissues have been found in early chondrichthyans, such as the fin skeleton of a shark that lived in the Carboniferous period. Traces of bone can also be found in living chondrichthyans—in their scales, at the base of their teeth, and, in some sharks, in a thin layer on the surface of their vertebrae. Such findings indicate that the restricted distribution of bone in the chondrichthyan body is a derived condition, emerging after chondrichthyans diverged from other gnathostomes.

There are about 1,000 species of living chondrichthyans. The largest and most diverse group consists of the sharks, rays, and skates **(Figure 34.15a and b)**. A second group is composed of a few dozen species of ratfishes, also called chimaeras **(Figure 34.15c)**.

Most sharks have a streamlined body and are swift swimmers, but they do not maneuver very well. Powerful movements of the trunk and the tail fin propel them forward. The dorsal fins function mainly as stabilizers, and the paired pectoral (fore) and pelvic (hind) fins are important for maneuvering. Although a shark gains buoyancy by storing a large amount of oil in its huge liver, the animal is still more dense than water, and if it stops swimming it sinks. Continual swimming also ensures that water flows into the shark's mouth and out through the gills, where gas exchange occurs. However, some sharks and many skates and rays spend a good deal of time resting on the seafloor. When resting, they use muscles of their jaws and pharynx to pump water over the gills.

▼ Figure 34.15 Chondrichthyans.

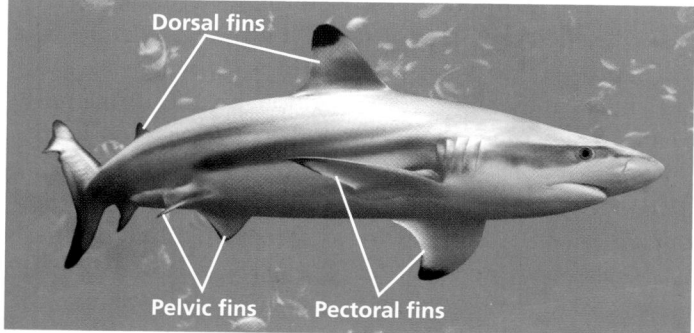

(a) Blacktip reef shark (*Carcharhinus melanopterus*). Sharks are fast swimmers with acute senses. Like all gnathostomes, they have paired pectoral and pelvic fins.

(b) Southern stingray (*Dasyatis americana*). Most rays are bottom-dwellers that feed on molluscs and crustaceans. Some rays cruise in open water and scoop food into their gaping mouths.

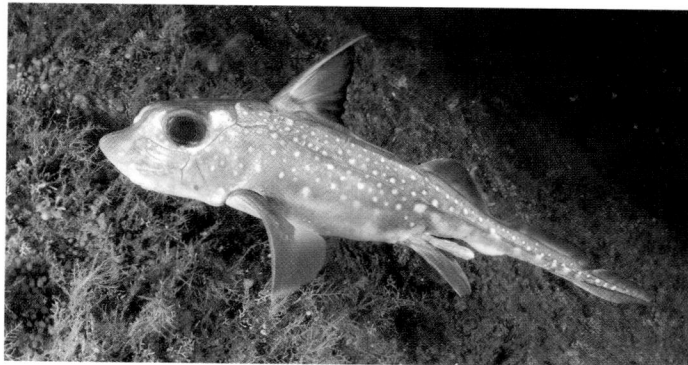

(c) Spotted ratfish (*Hydrolagus colliei*). Ratfishes, or chimaeras, typically live at depths greater than 80 m and feed on shrimp, molluscs, and sea urchins. Some species have a venomous spine at the front of their first dorsal fin.

The largest sharks and rays are suspension feeders that consume plankton. Most sharks, however, are carnivores that swallow their prey whole or use their powerful jaws and sharp teeth to tear flesh from animals too large to swallow in one piece. Sharks have several rows of teeth that gradually move to the front of the mouth as old teeth are lost. The digestive tract of many sharks is proportionately shorter than that of many other vertebrates. Within the shark intestine is a *spiral valve*, a corkscrew-shaped ridge that increases surface area and prolongs the passage of food through the digestive tract.

Acute senses are adaptations that go along with the active, carnivorous lifestyle of sharks. Sharks have sharp vision but cannot distinguish colors. The nostrils of sharks, like those of most aquatic vertebrates, open into dead-end cups. They function only for olfaction (smelling), not for breathing. Like some other vertebrates, sharks have a pair of regions in the skin of their head that can detect electric fields generated by the muscle contractions of nearby animals. Like all nonmammalian aquatic vertebrates, sharks have no eardrums, structures that terrestrial vertebrates use to transmit sound waves in air to the auditory organs. Sound reaches a shark through water, and the animal's entire body transmits the sound to the hearing organs of the inner ear.

Shark eggs are fertilized internally. The male has a pair of claspers on its pelvic fins that transfer sperm into the female's reproductive tract. Some species of sharks are **oviparous**; they lay eggs that hatch outside the mother's body. These sharks release their fertilized eggs after encasing them in protective coats. Other species are **ovoviviparous**; they retain the fertilized eggs in the oviduct. Nourished by the egg yolk, the embryos develop into young that are born after hatching within the uterus. A few species are **viviparous**; the young develop within the uterus and obtain nourishment prior to birth by receiving nutrients from the mother's blood through a yolk sac placenta, by absorbing a nutritious fluid produced by the uterus, or by eating other eggs. The reproductive tract of the shark empties along with the excretory system and digestive tract into the **cloaca**, a common chamber that has a single opening to the outside.

Although rays are closely related to sharks, they have adopted a very different lifestyle. Most rays are bottom-dwellers that feed by using their jaws to crush molluscs and crustaceans. They have a flattened shape and use their greatly enlarged pectoral fins like water wings to propel themselves through the water. The tail of many rays is whiplike and, in some species, bears venomous barbs that function in defense.

Chondrichthyans have thrived for over 400 million years. Today, however, they are severely threatened by overfishing. For example, a recent report indicated that shark populations in the Pacific have plummeted by up to 95%, and shark populations that live closest to people have declined the most.

Ray-Finned Fishes and Lobe-Fins

Cephalochordata
Urochordata
Myxini
Petromyzontida
Chondrichthyes
Actinopterygii
Actinistia
Dipnoi
Amphibia
Reptilia
Mammalia

The vast majority of vertebrates belong to the clade of gnathostomes called Osteichthyes. Unlike chondrichthyans, nearly all living **osteichthyans** have an ossified (bony) endoskeleton with a hard matrix of calcium phosphate. The name Osteichthyes ("bony fish") was coined long before the advent of phylogenetic systematics. When it was originally defined, the group excluded tetrapods, but we now know that such a taxon would be paraphyletic (see Figure 34.2). Therefore, systematists today include tetrapods along with bony fishes in the clade Osteichthyes. Clearly, the name of the group does not accurately describe all of its members.

This section discusses the aquatic osteichthyans known informally as fishes. Most fishes breathe by drawing water over four or five pairs of gills located in chambers covered by a protective bony flap called the **operculum (Figure 34.16)**. Water is drawn into the mouth, through the pharynx, and out between the gills by movement of the operculum and contraction of muscles surrounding the gill chambers.

Most fishes can maintain a buoyancy equal to the surrounding water by filling an air sac known as a **swim bladder**. (If a fish swims to greater depths or toward the surface, where water pressure differs, the fish shuttles gas between its blood and swim bladder, keeping the volume of gas in the bladder constant.) Charles Darwin proposed that the lungs of tetrapods evolved from swim bladders, but strange as it may sound, the opposite seems to be true: Swim bladders arose from lungs. Osteichthyans in many early-branching lineages have lungs, which they use to breathe air as a supplement to gas exchange in their gills. This suggests that lungs arose in early osteichthyans; later, swim bladders evolved from lungs in some lineages.

In nearly all fishes, the skin is covered by flattened, bony scales that differ in structure from the tooth-like scales of sharks. Glands in the skin secrete a slimy mucus over the skin, an adaptation that reduces drag during swimming. Like the ancient aquatic gnathostomes mentioned earlier, fishes have a lateral line system, which is evident as a row of tiny pits in the skin on either side of the body.

The details of fish reproduction vary extensively. Most species are oviparous, reproducing by external fertilization after the female sheds large numbers of small eggs. However, internal fertilization and birthing characterize other species.

Ray-Finned Fishes

Nearly all the aquatic osteichthyans familiar to us are among the over 27,000 species of **ray-finned fishes** (Actinopterygii) **(Figure 34.17)**. Named for the bony rays that support their fins, the ray-finned fishes originated during the Silurian period (444–419 million years ago). The group has diversified greatly since that time, resulting in numerous species and many modifications in body form and fin structure that affect maneuvering, defense, and other functions.

Ray-finned fishes serve as a major source of protein for humans, who have harvested them for thousands of years. However, industrial-scale fishing operations have driven some of the world's biggest fisheries to collapse. For example, after decades of abundant harvests, in the 1990s the catch of cod (*Gadus morhua*) in the northwest Atlantic plummeted to just 5% of its historic maximum, bringing cod fishing there to a near halt. Despite ongoing restrictions on the fishery, cod populations have yet to recover to sustainable levels. Ray-finned fishes also face other pressures from humans, such as the diversion of rivers by dams. Changing water flow patterns can hamper the fishes' ability to obtain food and interferes with migratory pathways and spawning grounds.

Lobe-Fins

Like the ray-finned fishes, the other major lineage of osteichthyans, the **lobe-fins** (Sarcopterygii), also originated during

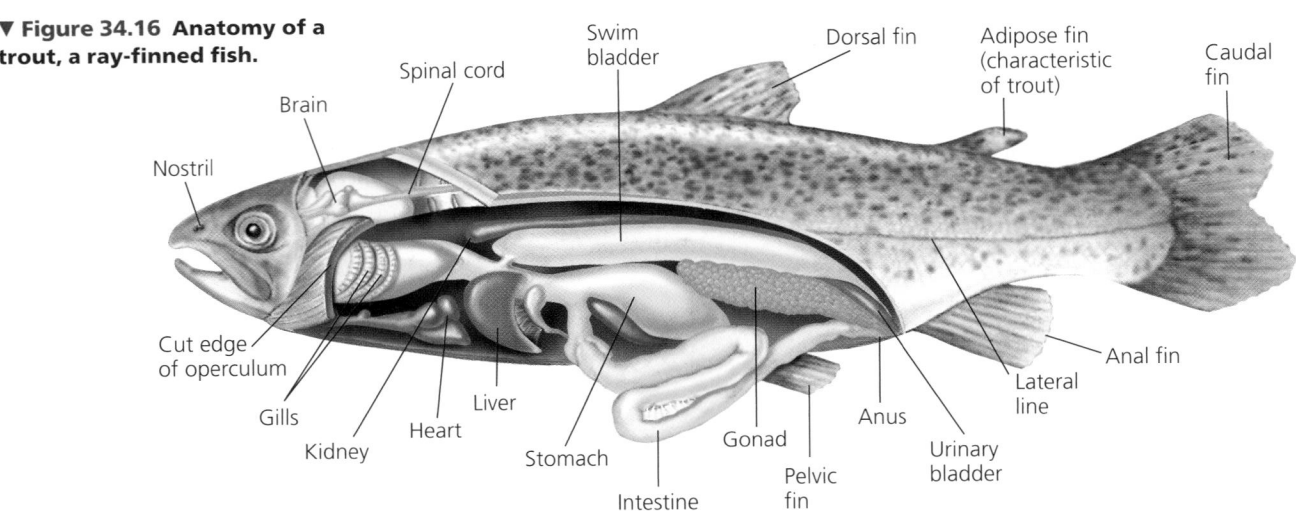

▼ **Figure 34.16 Anatomy of a trout, a ray-finned fish.**

Brain
Spinal cord
Swim bladder
Dorsal fin
Adipose fin (characteristic of trout)
Caudal fin
Nostril
Cut edge of operculum
Gills
Kidney
Heart
Liver
Stomach
Intestine
Pelvic fin
Gonad
Anus
Urinary bladder
Lateral line
Anal fin

▲ Yellowfin tuna (*Thunnus albacares*) is a fast-swimming schooling fish that is commercially important worldwide.

► Native to coral reefs of the Pacific Ocean, the brightly colored red lionfish (*Pterois volitans*) can inject venom through its spines, causing a severe and painful reaction in humans.

▲ The sea horse has a highly modified body form, as exemplified by *Hippocampus ramulosus,* shown above. Sea horses are unusual among animals in that the male carries the young during their embryonic development.

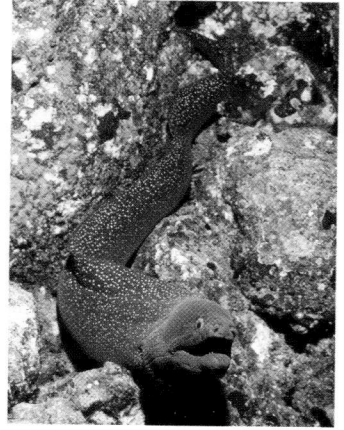

▲ The fine-spotted moray eel (*Gymnothorax dovii*) is a predator that ambushes prey from crevices in its coral reef habitat.

▲ **Figure 34.17 Ray-finned fishes (Actinopterygii).**

⊘ **Mastering Biology** Video: Sea Horse Camouflage

▲ **Figure 34.18 A reconstruction of an ancient lobe-fin.** Discovered in 2009, *Guiyu oneiros* is the earliest known lobe-fin, dating to 420 million years ago. The fossil of this species was nearly complete, allowing for an accurate reconstruction; regions shown in gray were missing from the fossil.

the Silurian period **(Figure 34.18)**. The key derived character of lobe-fins is the presence of rod-shaped bones surrounded by a thick layer of muscle in their pectoral and pelvic fins. During the Devonian (419–359 million years ago), many lobe-fins lived in brackish waters, such as in coastal wetlands. There they may have used their lobed fins to help them move across logs or the muddy bottom (as do some living lobe-fins). Some Devonian lobe-fins were gigantic predators. It is not uncommon to find spike-shaped fossils of Devonian lobe-fin teeth as big as your thumb.

By the end of the Devonian period, lobe-fin diversity was dwindling, and today only three lineages survive. One lineage,

the coelacanths (Actinistia), was thought to have become extinct 75 million years ago. However, in 1938, fishermen caught a living coelacanth off the east coast of South Africa **(Figure 34.19)**. Until the 1990s, all subsequent discoveries were near the Comoros Islands in the western Indian Ocean. Since 1999, coelacanths have also been found at various places along the eastern coast of Africa and in the eastern Indian Ocean, near Indonesia. The Indonesian population may represent a second species.

The second lineage of living lobe-fins, the lungfishes (Dipnoi), is represented today by six species in three genera, all of which are found in the Southern Hemisphere. Lungfishes arose in the ocean but today are found only in fresh water, generally in stagnant ponds and swamps. They surface to gulp air into lungs connected to their pharynx. Lungfishes also have gills, which are the main organs for gas exchange in Australian lungfishes. When ponds shrink during the dry season, some lungfishes can burrow into the mud and estivate (wait in a state of torpor; see Concept 40.4).

▼ **Figure 34.19 A coelacanth (*Latimeria*).** These lobe-fins were found living off the coasts of southern Africa and Indonesia.

The third lineage of lobe-fins that survives today is far more diverse than the coelacanths or the lungfishes. During the mid-Devonian, these organisms adapted to life on land and gave rise to vertebrates with limbs and feet, called tetrapods—a lineage that includes humans.

CONCEPT CHECK 34.3

1. What derived characters do sharks and tuna share? What features distinguish tuna from sharks?

2. Describe key adaptations of aquatic gnathostomes.

3. **DRAW IT** Redraw Figure 34.2 to show four lineages: cyclostomes, lancelets, gnathostomes, and tunicates. Label the vertebrate common ancestor and circle the lineage that includes humans.

4. **WHAT IF?** Imagine that we could replay the history of life. Is it possible that a group of vertebrates that colonized land could have arisen from aquatic gnathostomes other than the lobe-fins? Explain.

For suggested answers, see Appendix A.

CONCEPT 34.4

Tetrapods are gnathostomes that have limbs

One of the most significant events in vertebrate history took place by 365 million years ago, as the fins of a lineage of lobe-fins gradually evolved into the limbs and feet of tetrapods. Until then, all vertebrates had shared the same basic fishlike anatomy. After the colonization of land, tetrapods diversified greatly and gave rise to many new forms, from leaping frogs to flying eagles to bipedal humans.

Derived Characters of Tetrapods

The most significant character of **tetrapods** gives the group its name, which means "four feet" in Greek. In place of pectoral and pelvic fins, tetrapods have limbs with digits. Limbs support a tetrapod's weight on land, while feet with digits efficiently transmit muscle-generated forces to the ground when it walks.

Life on land selected for numerous other changes to the tetrapod body plan. In tetrapods, the head is separated from the body by a neck that originally had one vertebra on which the skull could move up and down. Later, with the origin of a second vertebra in the neck, the head could also swing from side to side. The bones of the pelvic girdle, to which the hind legs are attached, are fused to the backbone, permitting forces generated by the hind legs against the ground to be transferred to the rest of the body. Except for some fully aquatic species (such as the axolotl; see Figure 42.1), the adults of living tetrapods do not have gills; during embryonic development, the pharyngeal clefts instead give rise to parts of the ears, certain glands, and other structures.

We'll discuss later how some of these characters were dramatically altered or lost in various lineages of tetrapods. In birds, for example, the pectoral limbs became wings, and in whales, the entire body converged toward a fishlike shape.

Shoulder bones

Ribs

Scales

Neck

Head

Eyes on top of skull

Flat skull

Humerus

Ulna

"Wrist"

Elbow

Radius

Fin skeleton

Fin

Fish Characters	Tetrapod Characters
Scales	Neck
Fins	Ribs
Gills and lungs	Fin skeleton
	Flat skull
	Eyes on top of skull

▲ **Figure 34.20 Discovery of a "fishapod": *Tiktaalik*.** Paleontologists were on the hunt for fossils that could shed light on the evolutionary origin of tetrapods. Based on the ages of previously discovered fossils, researchers were looking for a dig site with rocks about 365–385 million years old. Ellesmere Island, in the Canadian Arctic, was one of the few such sites that was also likely to contain fossils, because it was once a river. The search at this site was rewarded by the discovery of fossils of a 375-million-year-old lobe-fin, named *Tiktaalik*. As shown in the chart and photographs, *Tiktaalik* exhibits both fish and tetrapod characters.

MAKE CONNECTIONS *Describe how* Tiktaalik's *features illustrate Darwin's concept of descent with modification (see Concept 22.2).*

→ **Mastering Biology HHMI Video: Great Transitions: The Origin of Tetrapods**

hhmi BioInteractive

The Origin of Tetrapods

As you have read, the Devonian coastal wetlands were home to a wide range of lobe-fins. Those that entered shallow, oxygen-poor water could have used their lungs to breathe air. Some species probably used their stout fins to swim and "walk" underwater across the bottom (moving their fins in an alternating gait, as do some living lobe-fins). This suggests that the tetrapod body plan did not evolve "out of nowhere" but was simply a modification of a pre-existing body plan.

The recent discovery of a fossil called *Tiktaalik* provided new details on how this process of modification occurred **(Figure 34.20)**. Like a fish, this species had fins, gills, and lungs, and its body was covered in scales. But unlike a fish, *Tiktaalik* had a full set of ribs that would have helped it breathe air and support its body. Also unlike a fish, *Tiktaalik* had a neck, allowing it to move its head about. In addition, the bones of *Tiktaalik*'s front fin have the same basic pattern found in all limbed animals: one bone (the humerus), followed by two bones (the radius and ulna), followed by a group of small bones that comprise the wrist. Finally, *Tiktaalik*'s pelvis and rear fin were larger and more robust than those of a fish; the pelvis is the bony structure to which hind limbs are attached in tetrapods. Although it is unlikely that *Tiktaalik* could walk on land, the skeletal structure of its fins and pelvis suggests that it could prop itself up and walk in water on its fins. Since *Tiktaalik* predates the oldest known tetrapod, its features suggest that key "tetrapod" traits, such as a wrist, ribs, and a neck, were in fact ancestral to the tetrapod lineage.

Tiktaalik and other extraordinary fossil discoveries have allowed paleontologists to reconstruct how fins became progressively more limb-like over time, culminating in the appearance in the fossil record of the first tetrapods 365 million years ago **(Figure 34.21)**. Over the next 60 million years, a great diversity of tetrapods arose. Some of these species retained functional gills and had weak limbs, while others lost their gills and had stronger limbs that facilitated walking on land. Overall, judging from the morphology and locations of their fossils, most of these early tetrapods probably remained tied to water, a characteristic they share with some members of the most basal group of living tetrapods, the amphibians.

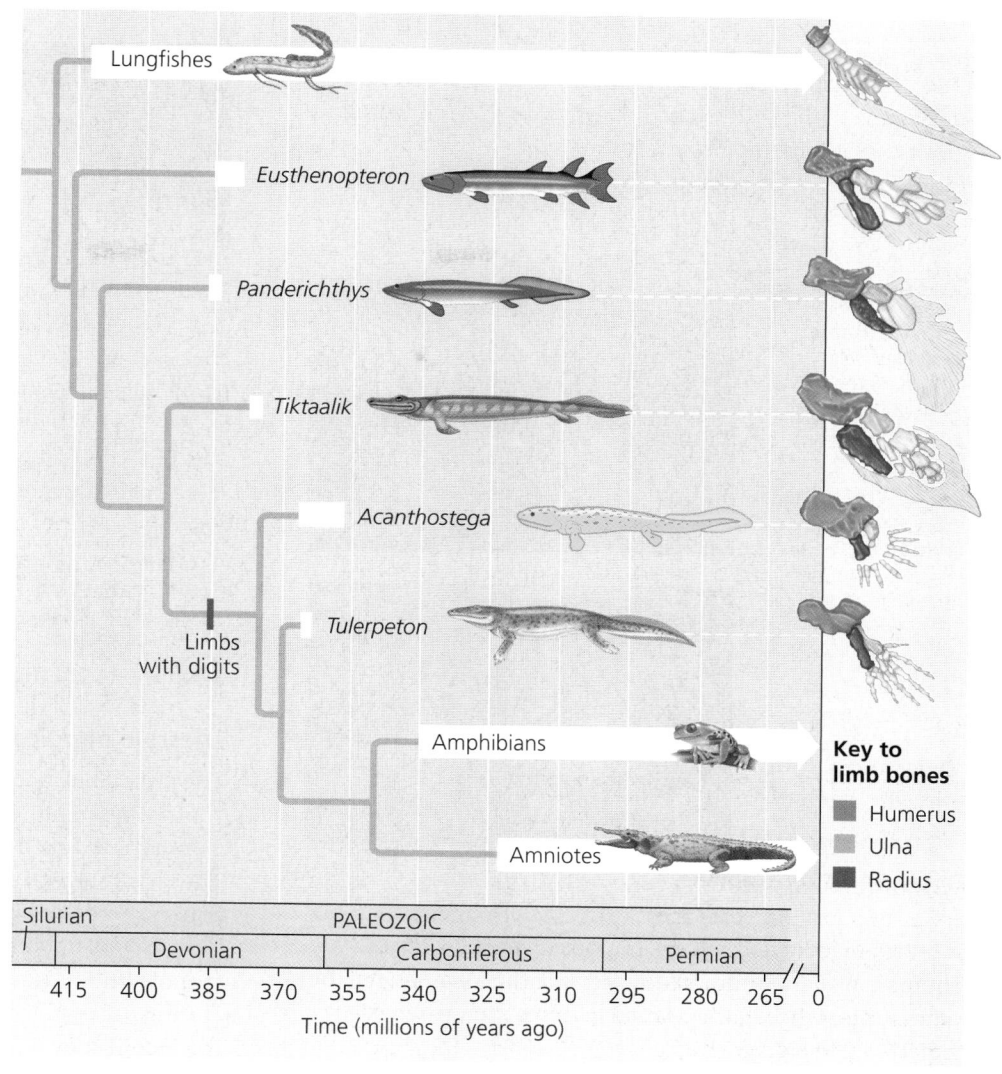

▲ **Figure 34.21 Steps in the origin of limbs with digits.** The white bars on the branches of this diagram place known fossils in time; arrowheads indicate lineages that extend to today. The drawings of extinct organisms are based on fossilized skeletons, but the colors are fanciful.

VISUAL SKILLS *Assuming the branch point dates shown in this diagram are accurate, what range of dates would include the origin of amphibians?*

Amphibians

The **amphibians** are represented today by about 6,150 species in three clades: salamanders (clade Urodela, "tailed ones"), frogs (clade Anura, "tailless ones"), and caecilians (clade Apoda, "legless ones").

Salamanders

There are about 550 known species of urodeles, or salamanders. Some are entirely aquatic, but others live on land as adults or throughout life. Most salamanders that live on land walk with a side-to-side bending of the body, a trait

(a) Order Urodela. Urodeles (salamanders) retain their tail as adults.

(b) Order Anura. Anurans (toads and frogs) lack a tail as adults.

(c) Order Apoda. Apodans, or caecilians, are legless, mainly burrowing amphibians.

▲ **Figure 34.22 Amphibians.**

also found in early terrestrial tetrapods **(Figure 34.22a).** Paedomorphosis is common among aquatic salamanders; the axolotl, for instance, retains larval features even when sexually mature (see Figure 25.25).

Frogs

Numbering about 5,420 species, anurans, or frogs, are better suited than salamanders to locomotion on land **(Figure 34.22b).** Adult frogs use their powerful hind legs to hop along the terrain. Although often distinctive in appearance, the animals known as "toads" are simply frogs that have leathery skin or other adaptations for life on land. A frog nabs insects and other prey by flicking out its long, sticky tongue, which is attached to the front of the mouth. Frogs display a great variety of adaptations that help them avoid being eaten by larger predators. Their skin glands secrete distasteful or even poisonous mucus. Many poisonous species have color patterns that camouflage them or have bright coloration, which predators appear to associate with danger (see Figure 54.6).

Caecilians

The approximately 170 species of apodans, or caecilians, are legless and nearly blind, and superficially they resemble earthworms **(Figure 34.22c).** Their absence of legs is a secondary adaptation, as they evolved from a legged ancestor. Caecilians inhabit tropical areas, where most species burrow in moist forest soil.

(a) The tadpole is an aquatic herbivore with a fishlike tail and internal gills.

(b) During metamorphosis, the gills and tail are resorbed, and walking legs develop. The adult frog will live on land.

(c) The adults return to water to mate. The male grasps the female, stimulating her to release eggs. The eggs are laid and fertilized in water. They have a jelly coat but lack a shell and would desiccate in air.

▲ **Figure 34.23 The "dual life" of a frog (*Rana temporaria*).**

Lifestyle and Ecology of Amphibians

The term *amphibian* (derived from *amphibious*, meaning "both ways of life") refers to the life stages of many frog species that live first in water and then on land **(Figure 34.23).** The larval stage of a frog, called a tadpole, is usually an aquatic herbivore with gills, a lateral line system resembling that of aquatic vertebrates, and a long, finned tail. The tadpole initially lacks legs; it swims by undulating its tail. During the metamorphosis that leads to the "second life," the tadpole develops legs, lungs, a pair of external eardrums, and a digestive system adapted to a carnivorous diet. At the same time, the gills disappear; the lateral line system also disappears in most species. The young frog crawls onto shore and becomes a terrestrial hunter. In spite of their name, however, many amphibians do not live a dual—aquatic and terrestrial—life. There are some strictly aquatic or strictly terrestrial frogs, salamanders, and caecilians. Moreover, salamander and caecilian larvae look much like the adults, and typically both the larvae and the adults are carnivorous.

Most amphibians are found in damp habitats such as swamps and rain forests. Even those adapted to drier habitats spend much of their time in burrows or under moist leaves, where humidity is high. One reason amphibians require relatively wet habitats is that they rely heavily on their moist skin for gas exchange—if their skin dries out, they cannot get enough oxygen. In addition, amphibians typically lay their eggs in water or in moist environments on land; their eggs lack a shell and dehydrate quickly in dry air.

Fertilization is external in most amphibians; the male grasps the female and spills his sperm over the eggs as the female sheds them (see Figure 34.23c). Some amphibian species lay vast numbers of eggs in temporary pools, and egg mortality is high. In contrast, other species lay relatively few

◄ **Figure 34.24 A mobile nursery.** A female marsupial frog (*Flectonotus fitzgeraldi*) incubates her eggs in pouches of skin on her back.

eggs and display various types of parental care. Depending on the species, either males or females may house eggs on their back (**Figure 34.24**), in their mouth, or even in their stomach. Certain tropical tree frogs stir their egg masses into moist, foamy nests that resist drying.

Many amphibians exhibit complex and diverse social behaviors, especially during breeding seasons. Frogs are usually quiet, but the males of many species vocalize to defend their breeding territory or to attract females. In some species, migrations to specific breeding sites may involve vocal communication, celestial navigation, or chemical signaling.

Over the past 30 years, zoologists have documented a rapid and alarming decline in amphibian populations in locations throughout the world. There appear to be several causes, including the spread of a disease-causing chytrid fungus (see Figure 31.25), habitat loss, climate change, and pollution. In some cases, declines have become extinctions. Recent studies indicate that at least nine amphibian species have become extinct within the last four decades; more than 100 other species have not been observed in that time and are considered possibly extinct. In the **Problem-Solving Exercise**, you can explore one possible strategy to prevent amphibian deaths from fungal infections.

PROBLEM-SOLVING EXERCISE

Can declining amphibian populations be saved by a vaccine?

Amphibian populations are declining rapidly worldwide. The fungus *Batrachochytrium dendrobatidis* (*Bd*) has contributed to this decline: This pathogen causes severe skin infections in many amphibian species, leading to massive die-offs. Efforts to save amphibians from *Bd* have had limited success, and there is little evidence that frogs and other amphibians have acquired resistance to *Bd* on their own.

Yellow-legged frogs (*Rana muscosa*) in California killed by *Bd* infection

In this exercise, you will investigate whether amphibians can acquire resistance to the fungal pathogen *Bd*.

Your Approach The principle guiding your investigation is that prior exposure to a pathogen can enable amphibians to acquire immunological resistance to that pathogen. To see whether this occurs after exposure to *Bd*, you will analyze data on acquired resistance in Cuban tree frogs (*Osteopilus septentrionalis*).

Your Data To create variation in number of prior exposures to *Bd*, Cuban tree frogs were exposed to *Bd* and cleared of their infection (using heat treatments) from zero to three times; frogs with no prior exposures are referred to as "naive." Researchers then exposed frogs to *Bd* and measured mean abundance of *Bd* on the frog's skin, frog survival, and abundance of lymphocytes (a type of white blood cell involved in the vertebrate immune response).

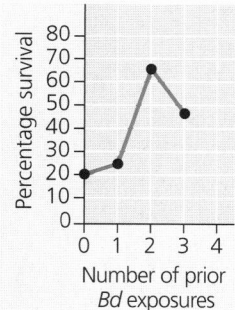

Number of prior *Bd* exposures	Thousands of lymphocytes per g of frog
0	134
1	240
2	244
3	227

Your Analysis

1. Describe and interpret the results shown in the figure.
2. **(a)** Graph the data in the table. **(b)** Based on these data, develop a hypothesis that explains the results discussed in question 1.
3. Breeding populations of amphibian species threatened by *Bd* have been established in captivity. In addition, evidence suggests that Cuban tree frogs can acquire resistance after exposure to dead *Bd*. Based on this information and your answers to questions 1 and 2, suggest a strategy for repopulating regions decimated by *Bd*.

→ **Instructors:** A version of this Problem-Solving Exercise can be assigned in **Mastering Biology**.

1. Describe the origin of tetrapods and identify some of their key derived traits.

2. Some amphibians never leave the water, whereas others can survive in relatively dry terrestrial environments. Contrast the adaptations that facilitate these two lifestyles.

3. **WHAT IF?** Scientists think that amphibian populations may provide an early warning system of environmental problems. What features of amphibians might make them particularly sensitive to environmental problems?

For suggested answers, see Appendix A.

CONCEPT 34.5

Amniotes are tetrapods that have a terrestrially adapted egg

The **amniotes** are a group of tetrapods whose extant members are the reptiles (including birds, as we'll discuss in this section) and mammals **(Figure 34.25)**. During their evolution, amniotes acquired a number of new adaptations to life on land.

Derived Characters of Amniotes

Amniotes are named for the major derived character of the clade, the **amniotic egg**, which contains four specialized membranes: the amnion, the chorion, the yolk sac, and the allantois **(Figure 34.26)**. Called *extraembryonic membranes* because they are not part of the body of the embryo itself, these membranes develop from tissue layers that grow out from the embryo. The amniotic egg is named for the amnion, which encloses a compartment of fluid that bathes the embryo and acts as a hydraulic shock absorber. The other membranes in the egg function in gas exchange, the transfer of stored nutrients to the embryo, and waste storage. The amniotic egg was a key evolutionary innovation for terrestrial life: It allowed the embryo to develop on land in its own private "pond," hence reducing the dependence of tetrapods on an aqueous environment for reproduction.

In contrast to the shell-less eggs of amphibians, the amniotic eggs of most reptiles and some mammals have a shell. A shell slows dehydration of the egg in air, an adaptation that helped amniotes to occupy a wider range of terrestrial habitats than amphibians, their closest living relatives. (Seeds played a similar role in the evolution of plants, as discussed in Concept 30.1.)

▶ **Figure 34.25 A phylogeny of amniotes.** Extant groups are named at the tips of the branches in boldface type. The dagger symbols (†) indicate extinct groups.

VISUAL SKILLS *Based on this phylogeny, are pterosaurs dinosaurs? Are birds? Explain.*

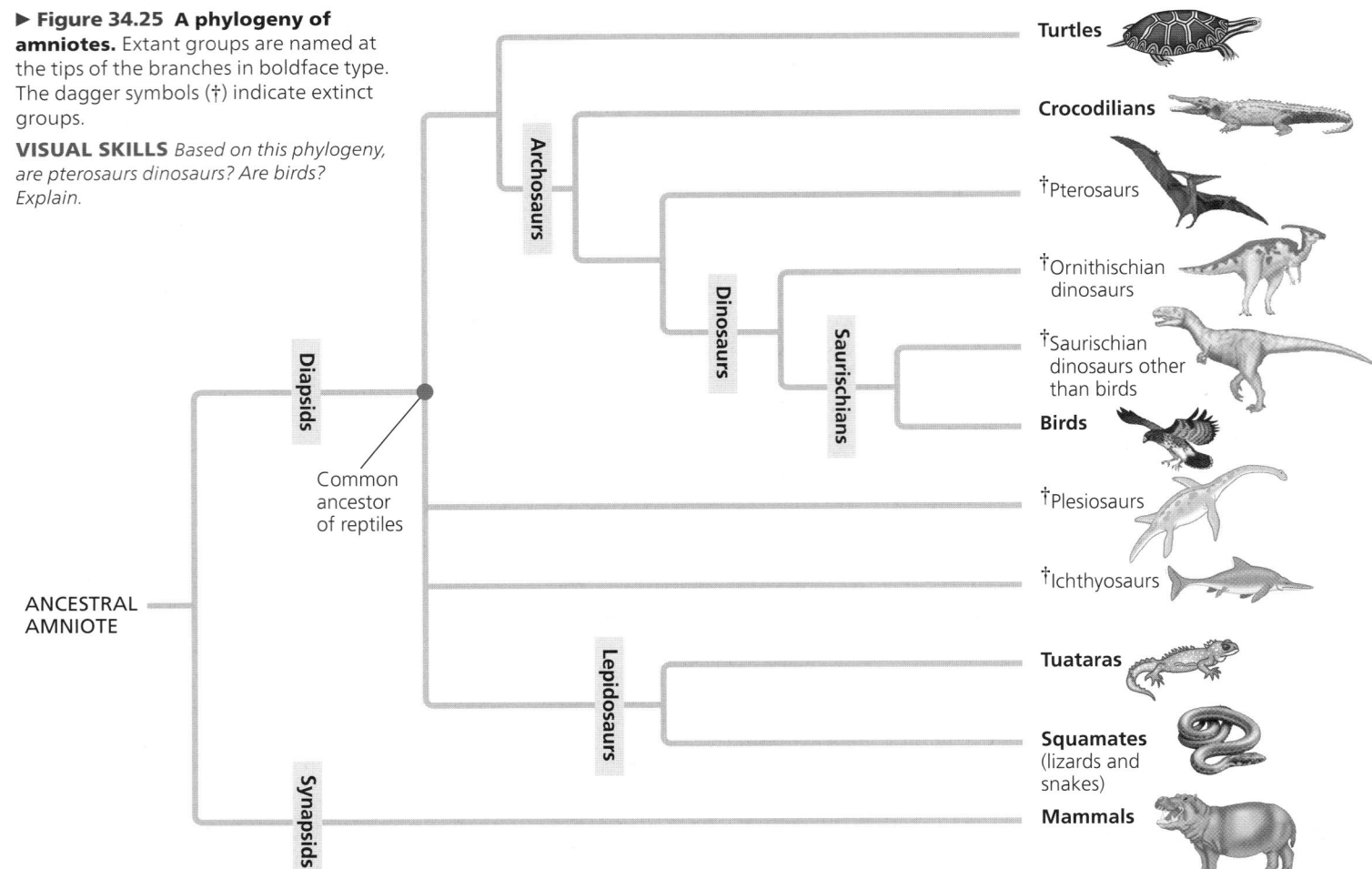

▼ Figure 34.26 The amniotic egg. The embryos of reptiles and mammals form four extraembryonic membranes: the allantois, chorion, amnion, and yolk sac. This diagram shows these membranes in the shelled egg of a reptile.

Extraembryonic membranes

Allantois. The allantois is a disposal sac for certain metabolic wastes produced by the embryo.

Chorion. The chorion and the membrane of the allantois exchange gases between the embryo and the air.

Embryo

Amniotic cavity with amniotic fluid

Yolk (nutrients)

Shell

Albumen

Amnion. The amnion protects the embryo in a fluid-filled cavity that cushions against mechanical shock.

Yolk sac. The yolk sac contains the yolk, a stockpile of nutrients. Other nutrients are stored in the albumen ("egg white").

Extraembryonic membranes

Most mammals have lost the eggshell over the course of their evolution, and the embryo avoids desiccation by developing within the amnion inside the mother's body.

Amniotes have acquired other key adaptations to life on land. For example, amniotes use their rib cage to ventilate their lungs. This method is more efficient than throat-based ventilation, which amphibians use as a supplement to breathing through their skin. The increased efficiency of rib cage ventilation may have allowed amniotes to abandon breathing through their skin and develop less permeable skin, thereby conserving water.

Early Amniotes

The most recent common ancestor of living amphibians and amniotes lived 350 million years ago. Based on where their fossils have been found, the earliest amniotes lived in warm, moist areas, as did the first tetrapods. Over time, early amniotes expanded into a wide range of new environments, including dry and high-latitude regions. Fossil evidence shows that the earliest amniotes resembled small lizards with sharp teeth, a sign that they were predators **(Figure 34.27)**. Later groups of amniotes also included herbivores, as evidenced by their grinding teeth and other features.

▼ Figure 34.27 Artist's reconstruction of *Hylonomus*, an early amniote. About 25 cm long, this species lived 310 million years ago and probably ate insects and other small invertebrates.

Present-day amniotes include two large clades of terrestrial vertebrates: reptiles and mammals.

Reptiles

Cephalochordata
Urochordata
Myxini
Petromyzontida
Chondrichthyes
Actinopterygii
Actinistia
Dipnoi
Amphibia
Reptilia
Mammalia

Living members of the **reptile** clade include turtles, tuataras, lizards and snakes, crocodilians, and birds (see Figure 34.25). There are about 20,800 species of reptiles, the majority of which are squamates (lizards and snakes; 10,425 species) or birds (10,000 species).

As a group, the reptiles share several derived characters that distinguish them from other tetrapods. For example, unlike amphibians, reptiles have scales that contain the protein keratin (as does a human nail). Scales help protect the animal's skin from desiccation and abrasion. In addition, most reptiles lay their shelled eggs on land; the shell protects the egg from drying out **(Figure 34.28)**. Fertilization occurs internally, before the eggshell is secreted.

Reptiles such as lizards and snakes are sometimes described as "cold-blooded" because they do not use their metabolism

▼ Figure 34.28 Hatching reptiles. These baby panther chameleons (*Furcifer pardalis*) are breaking out of their parchment-like shells, a common type of shell among living reptiles other than birds.

extensively to control their body temperature. However, they do regulate their body temperature by using behavioral adaptations. For example, many lizards bask in the sun when the air is cool and seek shade when the air is too warm. A more accurate description of these reptiles is to say that they are **ectothermic**, which means that they absorb external heat as their main source of body heat. By warming themselves directly with solar energy rather than through the metabolic breakdown of food, an ectothermic reptile can survive on less than 10% of the food energy required by a mammal of the same size. But the reptile clade is not entirely ectothermic; birds are **endothermic**, capable of maintaining body temperature through metabolic activity.

The Origin and Evolutionary Radiation of Reptiles

Fossil evidence indicates that the earliest reptiles lived about 310 million years ago and resembled lizards. Like all reptiles today, these early reptiles were **diapsids**. A key derived character of diapsids is a pair of holes on each side of the skull, behind the eye sockets; muscles pass through these holes and attach to the jaw, controlling jaw movement.

The diapsids are composed of three main lineages. The first lineage includes the turtles. A second lineage gave rise to the **lepidosaurs**, which include tuataras, lizards, and snakes. This lineage also produced some marine reptiles, including the giant mososaurs. Some of these marine species rivaled today's whales in length; all of them are extinct. The third diapsid lineage, the **archosaurs**, produced the crocodilians, pterosaurs, and dinosaurs. Our focus here will be on extinct lineages of archosaurs; we'll discuss living reptiles shortly.

Pterosaurs, which originated in the late Triassic, were the first tetrapods to exhibit flapping flight. The pterosaur wing was completely different from the wings of birds and bats. It consisted of a collagen-strengthened membrane that stretched between the trunk or hind leg and a very long digit on the foreleg. The smallest pterosaurs were no bigger than a sparrow, and the largest had a wingspan of nearly 11 m. They appear to have converged on many of the ecological roles later played by birds; some were insect-eaters, others grabbed fish out of the ocean, and still others filtered small animals through thousands of fine needlelike teeth. But by 66 million years ago, pterosaurs had become extinct.

On land, the **dinosaurs** diversified into a vast range of shapes and sizes, from bipeds the size of a pigeon to 45-m-long quadrupeds with necks long enough to let them browse the tops of trees. One lineage of dinosaurs, the ornithischians, were herbivores; they included many species with elaborate defenses against predators, such as tail clubs and horned crests. The other main lineage of dinosaurs, the saurischians, included the long-necked giants and a group called the **theropods**, which were bipedal carnivores. Theropods included the famous *Tyrannosaurus rex* as well as the ancestors of birds.

Dinosaurs once were considered slow, sluggish creatures. Since the 1970s, however, fossil discoveries and research have led to the conclusion that many dinosaurs were agile and fast moving. Dinosaurs had a limb structure that enabled them to walk and run more efficiently than could earlier tetrapods, which had a sprawling gait. Fossilized footprints and other evidence suggest that some species were social—they lived and traveled in groups, much as many mammals do today. Paleontologists have also discovered evidence that some dinosaurs built nests and protected their eggs, as birds do today (see Figure 26.17). Finally, some anatomical evidence supports the hypothesis that at least some dinosaurs were endotherms.

All dinosaurs except birds became extinct by the end of the Cretaceous period (66 million years ago). Their extinction may have been caused at least in part by the asteroid or comet impact described in Concept 25.4. Some analyses of the fossil record are consistent with this idea in that they show a sudden decline in dinosaur diversity at the end of the Cretaceous. However, other analyses indicate that the number of dinosaur species had begun to decline several million years before the Cretaceous ended. Further fossil discoveries and new analyses will be needed to resolve this debate.

Next, we'll discuss the three extant lineages of reptiles, the turtles, lepidosaurs (tuataras, lizards, and snakes), and archosaurs (crocodilians and birds).

 Mastering Biology
Interview with Paul Sereno: Adventures hunting dinosaurs

 Mastering Biology HHMI Video: How to Find a Dinosaur

Turtles

Turtles are one of the most distinctive groups of reptiles alive today. For example, turtles do not have any holes in their skull behind the eye sockets, whereas other reptiles have two holes behind each eye socket. Recall that such skull holes are a key derived trait of the diapsids. Thus, until recently it was not clear whether turtles—like all other living reptiles—should be classified within the diapsid clade. However, in 2015, new fossil discoveries showed that early turtles had the skull openings found in other diapsids. This suggests that turtles are diapsids that have lost the holes in their skull over

▼ **Figure 34.29 Extant reptiles (other than birds).**

(a) Black-breasted hill turtle (*Geoemyda spengleri*)

the course of evolution. Moreover, recent phylogenomic studies indicate that turtles are nested within the diapsid clade and are more closely related to crocodilians and birds than to other reptiles (see Figure 34.25).

All turtles have a boxlike shell made of upper and lower shields that are fused to the vertebrae, clavicles (collarbones), and ribs **(Figure 34.29a)**. Most of the 351 known species of turtles have a hard shell, providing excellent defense against predators. Fossil evidence shows that *Pappochelys*, a turtle that lived 240 million years ago, had a series of hard, shell-like bones along its belly. By 220 million years ago, another early turtle had a complete lower shell but an incomplete upper shell, suggesting that turtles acquired full shells in stages.

The earliest turtles could not retract their head into their shell, but mechanisms for doing so evolved independently in two separate branches of turtles. The side-necked turtles fold their neck horizontally, while the vertical-necked turtles fold their neck vertically.

Some turtles have adapted to deserts, and others live almost entirely in ponds and rivers. Still others live in the sea. Sea turtles have a reduced shell and enlarged forelimbs that function as flippers. They include the largest living turtles, the deep-diving leatherbacks, which can exceed a mass of 1,500 kg and feed on jellies. Leatherbacks and other sea turtles are endangered by being caught in fishing nets, as well as by the residential and commercial development of the beaches where the turtles lay their eggs.

Lepidosaurs

One surviving lineage of lepidosaurs is represented by a single species, a lizard-like reptile called the tuatara **(Figure 34.29b)**. Fossil evidence indicates that tuatara ancestors lived at least 220 million years ago. These organisms thrived on many continents well into the Cretaceous period and reached up to a meter in length. Today, however, tuataras are found only on 30 islands off the coast of New Zealand. When humans arrived in New Zealand 750 years ago, the rats that accompanied them devoured tuatara eggs, eventually eliminating the reptiles on the main islands. The tuataras that remain on the outlying islands are about 50 cm long and feed on insects, small lizards, and bird eggs and chicks. They can live to be over 100 years old. Their future survival depends on whether their remaining habitats are kept rat-free.

The other major living lineage of lepidosaurs consists of the lizards and snakes, or squamates, which number 10,425 species **(Figure 34.29c and d)**. Many squamates are small; the Jaragua lizard, discovered recently in the Dominican Republic, is only 16 mm long—small enough to fit comfortably on a dime. In contrast, the Komodo dragon of Indonesia is a lizard

(b) Tuatara (*Sphenodon punctatus*)

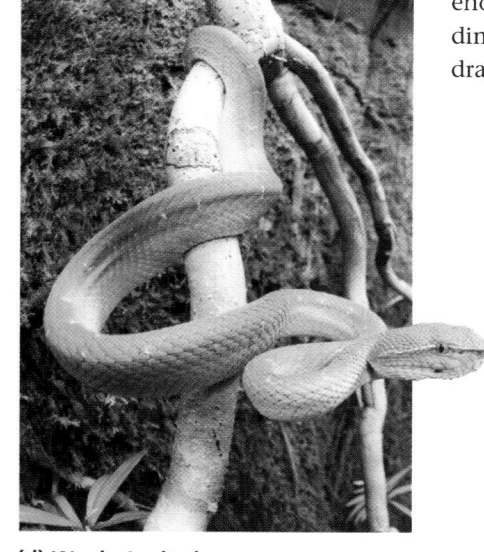

(d) Wagler's pit viper (*Tropidolaemus wagleri*)

(c) Australian thorny devil lizard (*Moloch horridus*)

(e) American alligator (*Alligator mississippiensis*)

that can reach a length of 3 m. It hunts deer and other large prey, delivering venom with its bite.

Snakes descended from lizards with legs (see Figure 26.1); thus they are classified as legless lizards. Today, some species of snakes retain vestigial pelvic and limb bones, providing evidence of their ancestry. Despite their lack of legs, snakes are quite proficient at moving on land, most often by producing waves of lateral bending that pass from head to tail. Force exerted by the bends against solid objects pushes the snake forward. Snakes can also move by gripping the ground with their belly scales at several points along the body while the scales at intervening points are lifted slightly off the ground and pulled forward.

Snakes are carnivorous, and a number of adaptations aid them in hunting and eating prey. They have acute chemical sensors, and though they lack eardrums, they are sensitive to ground vibrations, which helps them detect the movements of prey. Heat-detecting organs between the eyes and nostrils of pit vipers, including rattlesnakes, are sensitive to minute temperature changes, enabling these night hunters to locate warm animals. Venomous snakes inject their toxin through a pair of sharp teeth that may be hollow or grooved. The flicking tongue is not venomous but helps fan odors toward olfactory (smell) organs on the roof of the mouth. Loosely articulated jawbones and elastic skin enable most snakes to swallow prey larger than the diameter of the snake's head (see Figure 23.14).

Next, we'll turn to the third (and final) clade of reptiles with living members, the archosaurs. There are two extant groups of archosaurs: the crocodilians and the birds.

Crocodilians

Alligators and crocodiles (collectively called crocodilians) belong to a lineage that reaches back to the late Triassic. The earliest members of this lineage were small terrestrial quadrupeds with long, slender legs. Later species became larger and adapted to aquatic habitats, breathing air through their upturned nostrils. Some Mesozoic crocodilians grew as long as 12 m and may have attacked dinosaurs and other prey at the water's edge.

The 24 known species of living crocodilians are confined to warm regions of the globe. In the southeastern United States, the American alligator **(Figure 34.29e)** has made a comeback after spending years on the endangered species list.

Birds

There are about 10,000 species of birds in the world. Like crocodilians, birds are archosaurs, but almost every feature of their anatomy has been modified in their adaptation to flight **(Figure 34.30)**.

Derived Characters of Birds Many of the characters of birds are adaptations that facilitate flight, including

(a) Wing

(b) Bone structure

(c) Feather structure

◀ **Figure 34.30 Form fits function: the avian wing and feather. (a)** A wing is a remodeled version of the tetrapod forelimb. **(b)** The bones of many birds have a honeycombed internal structure and are filled with air. **(c)** A feather consists of a central air-filled shaft, from which radiate the vanes. The vanes are made up of barbs, which bear small branches called barbules. Birds have contour feathers and downy feathers. Contour feathers are stiff and contribute to the aerodynamic shapes of the wings and body. Their barbules have hooks that cling to barbules on neighboring barbs. When a bird preens, it runs the length of each contour feather through its beak, engaging the hooks and uniting the barbs into a precise shape. Downy feathers lack hooks, and the free-form arrangement of their barbs produces a fluffiness that provides insulation by trapping air.

⊘ **Mastering Biology Video: Soaring Hawk**

weight-saving modifications that make flying more efficient. For example, birds lack a urinary bladder, and the females of most species have only one ovary. The gonads of both females and males are usually small, except during the breeding season, when they increase in size. Living birds are also toothless, an adaptation that trims the weight of the head.

A bird's most obvious adaptations for flight are its wings and feathers (see Figure 34.30). Feathers are made of the protein β-keratin, which is also found in the scales of other reptiles. The shape and arrangement of the feathers form the wings into airfoils, and they illustrate some of the same principles of aerodynamics as the wings of an airplane. Power for flapping the wings comes from contractions of large pectoral (breast) muscles anchored to a keel on the sternum (breastbone). Some birds, such as eagles and hawks, have wings adapted for soaring on air currents and flap their wings only occasionally; other birds, including hummingbirds, must flap continuously to stay aloft (see Figure 34.34). Among the fastest birds are the appropriately named swifts, which can fly at speeds up to 170 km/hr.

Flight provides numerous benefits. It enhances scavenging and hunting, including enabling many birds to feed on flying insects, an abundant, nutritious food resource. Flight also provides ready escape from earthbound predators and enables some birds to migrate great distances to exploit different food resources and seasonal breeding areas.

Flying requires a great expenditure of energy from an active metabolism. Birds are endothermic; they use their own metabolic heat to maintain a high, constant body temperature. Feathers and in some species a layer of fat provide insulation that enables birds to retain body heat. The lungs have tiny tubes leading to and from elastic air sacs that improve airflow and oxygen uptake. This efficient respiratory system and a circulatory system with a four-chambered heart keep tissues well supplied with oxygen and nutrients, supporting a high rate of metabolism.

Flight also requires both acute vision and fine muscle control. Birds have color vision and excellent eyesight. The visual and motor areas of the brain are well developed, and the brain is proportionately larger than those of amphibians and nonbird reptiles.

Birds generally display very complex behaviors, particularly during breeding season, when they engage in elaborate courtship rituals. Because eggs have shells by the time they are laid, fertilization must be internal. Copulation usually involves contact between the openings to the birds' cloacas. After eggs are laid, the avian embryo must be kept warm through brooding by the mother, the father, or both, depending on the species.

The Origin of Birds Cladistic analyses of birds and reptilian fossils indicate that birds belong to the group of bipedal

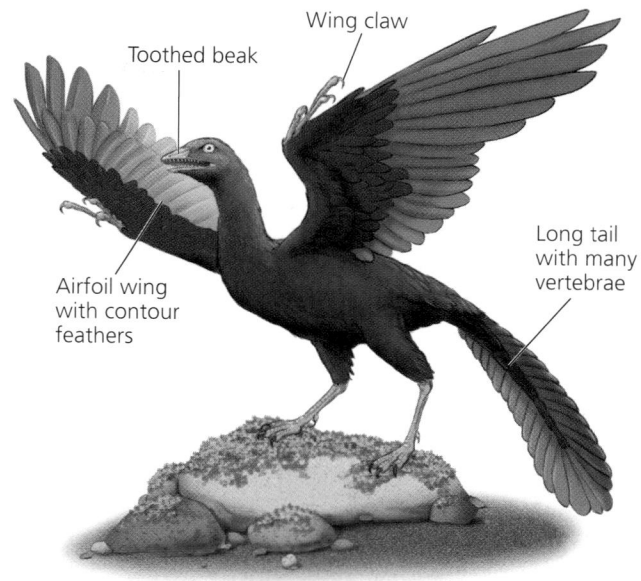

▼ **Figure 34.31 Artist's reconstruction of the early bird** *Archaeopteryx.* Fossil evidence indicates that *Archaeopteryx* was capable of powered flight but retained many characters of nonbird dinosaurs. Colors in this drawing were based on pigments extracted from fossilized *Archaeopteryx* feathers.

Wing claw

Toothed beak

Airfoil wing with contour feathers

Long tail with many vertebrae

Mastering Biology HHMI Video: Great Transitions: The Origin of Birds

saurischian dinosaurs called theropods. Since the late 1990s, Chinese paleontologists have unearthed a spectacular trove of feathered theropod fossils that are shedding light on the origin of birds. Several species of dinosaurs closely related to birds had feathers with vanes, and a wider range of species had filamentous feathers. Such findings imply that feathers evolved long before powered flight. Among the possible functions of these early feathers were insulation, camouflage, and courtship display.

By about 160 million years ago, a lineage of feathered theropods had evolved into birds. Many researchers consider *Archaeopteryx*, which was discovered in a German limestone quarry in 1861, to be the earliest known bird **(Figure 34.31)**. It had feathered wings but retained ancestral characters such as teeth, clawed digits in its wings, and a long tail. *Archaeopteryx* flew well at high speeds, but unlike a present-day bird, it could not take off from a standing position. Fossils of later birds from the Cretaceous show a gradual loss of certain ancestral dinosaur features, such as teeth and clawed forelimbs, as well as the acquisition of innovations found in extant birds, including a short tail covered by a fan of feathers.

Living Birds Clear evidence of Neornithes, the clade that includes the 28 orders of living birds, can be found before the Cretaceous-Paleogene boundary 66 million years ago.

▼ **Figure 34.32** An emu (*Dromaius novaehollandiae*), a flightless bird native to Australia.

▼ **Figure 34.34 Hummingbird feeding while hovering.** A hummingbird can rotate its wings in all directions, enabling it to hover and fly backward.

Several groups of living and extinct birds include one or more flightless species. The **ratites**, an order of birds that includes the ostrich, rhea, kiwi, cassowary, and emu, are all flightless **(Figure 34.32)**. In ratites, the sternal keel is absent, and the pectoral muscles are small relative to those of birds that can fly.

Penguins make up another flightless order of birds, but, like flying birds, they have powerful pectoral muscles. They use these muscles to "fly" in the water: As they swim, they flap their flipper-like wings in a manner that resembles the flight stroke of a more typical bird **(Figure 34.33)**. Certain species of rails, ducks, and pigeons are also flightless.

Although the demands of flight have rendered the general body forms of many flying birds similar to one another, experienced bird-watchers can distinguish species by their profile, colors, flying style, behavior, and beak shape. The skeleton of a hummingbird's wing is

▼ **Figure 34.35 A specialized beak.** This American flamingo (*Phoenicopterus ruber*) dips its beak into the water and strains out the food.

▼ **Figure 34.33 A king penguin (*Aptenodytes patagonicus*) "flying" underwater.** With their streamlined shape and powerful pectoral muscles, penguins are fast and agile swimmers.

unique, making it the only bird that can hover and fly backward **(Figure 34.34)**. Adult birds lack teeth, but during the course of avian evolution their beaks have taken on a variety of shapes suited to different diets. Some birds, such as parrots, have crushing beaks with which they can crack open hard nuts and seeds. Other birds, such as flamingoes, are filter feeders. Their beaks have "strainers" that enable them to capture food particles from the water **(Figure 34.35)**. Foot structure, too, shows considerable variation. Various birds use their feet for perching on branches **(Figure 34.36)**, grasping food, defense, swimming or walking, and even courtship (see Figure 24.3e).

▼ **Figure 34.36 Feet adapted to perching.** This great tit (*Parus major*) is a member of the Passeriformes, the perching birds. The toes of these birds can lock around a branch or wire, enabling the bird to rest for long periods.

CONCEPT CHECK 34.5

1. Describe three key amniote adaptations for life on land.
2. Are snakes tetrapods? Explain.
3. Identify four avian adaptations for flight.
4. **VISUAL SKILLS** Based on the phylogeny shown in Figure 34.25, identify the sister group for (a) reptiles, (b) squamates, and (c) the clade that includes crocodilians and birds.

For suggested answers, see Appendix A.

CONCEPT 34.6

Mammals are amniotes that have hair and produce milk

Cephalochordata
Urochordata
Myxini
Petromyzontida
Chondrichthyes
Actinopterygii
Actinistia
Dipnoi
Amphibia
Reptilia
Mammalia

The reptiles we have been discussing represent one of the two living lineages of amniotes. The other amniote lineage is our own, the **mammals**. Today, there are about 6,400 known species of mammals on Earth.

Derived Characters of Mammals

Mammals are named for their distinctive mammary glands, which produce milk for offspring. All mammalian mothers nourish their young with milk, a balanced diet rich in fats, sugars, proteins, minerals, and vitamins. Hair, another mammalian character, and a fat layer under the skin provide insulation that can conserve water and protect the body against extremes of heat or cold. Another mammalian adaptation for life on land is the kidney (see Figure 44.13), which is efficient at conserving water when removing wastes from the body. Some mammals, such as kangaroo rats, are so adept at conserving water that they can survive

▼ **Figure 34.37 Adaptations of the kangaroo rat to its extremely dry habitat.**

❷ The animal stays in its cool, relatively humid burrow during the heat of the day, emerging at night to forage.

❶ The kangaroo rat's thick, oily skin limits evaporative water loss.

❸ The large surface area of its nasal passages increases the efficiency of water reabsorption when the animal exhales.

❹ A kangaroo rat rarely drinks: It obtains nearly all of the water it needs from catabolic pathways and from moisture in food.

❺ The large intestine and kidney absorb water so effectively that little water is lost in feces and urine.

MAKE CONNECTIONS *Explain how the catabolic pathways mentioned in ❹ could provide a kangaroo rat with water. (See Concept 9.1.)*

in arid environments while drinking little or no water at all **(Figure 34.37)**.

Like birds, mammals are endothermic, and most have a high metabolic rate. Efficient respiratory and circulatory systems (including a four-chambered heart) support a mammal's metabolism. Also as in birds, mammals generally have a larger brain than other vertebrates of equivalent size, and many species are capable learners. The relatively long duration of parental care extends the time for offspring to learn important survival skills by observing their parents. In addition, whereas the teeth of reptiles are generally uniform in size and shape, the jaws of mammals typically bear a variety of teeth with sizes and shapes adapted for chewing many kinds of foods. Humans, like most mammals, have teeth modified for shearing (incisors and canine teeth) and for crushing and grinding (premolars and molars).

Early Evolution of Mammals

Mammals belong to a group of amniotes known as **synapsids**. Early nonmammalian synapsids lacked hair, had a sprawling gait, and laid eggs. A distinctive characteristic of synapsids is the single temporal fenestra, a hole behind the eye socket on each side of the skull. Humans retain this feature; your jaw muscles pass through the temporal fenestra and anchor on your temple. Fossil evidence shows that the jaw was remodeled as mammalian features arose gradually in successive lineages of earlier synapsids (see Figure 25.7); in all, these changes took more than 100 million years. In addition, two of the bones that formerly made up the jaw joint (the quadrate and the articular) were incorporated into the mammalian middle ear **(Figure 34.38)**. This evolutionary change is reflected in changes that occur during development. For example, as a mammalian embryo grows, the posterior region of its jaw—which in a reptile forms the articular bone—can be observed to detach from the jaw and migrate to the ear, where it forms the malleus.

Synapsids evolved into large herbivores and carnivores during the Permian period (299–252 million years ago), and for a time they were the dominant tetrapods. However, the Permian-Triassic extinctions took a heavy toll on them, and their diversity fell during the Triassic (252–201 million years ago). Increasingly mammal-like synapsids emerged by the end of the Triassic. While not true mammals, these synapsids had acquired a number of the derived characters that distinguish mammals from other amniotes. They were small and probably hairy, and they likely fed on insects at night. Their bones show that they grew faster than other synapsids, suggesting that they probably had a relatively high metabolic rate; however, they still laid eggs.

During the Jurassic (201–145 million years ago), the first true mammals arose and diversified into many short-lived lineages. A diverse set of mammal species coexisted with dinosaurs in the Jurassic and Cretaceous periods, but most of these species were small, measuring less than 1 m in length. One factor that may have contributed to their small size is that dinosaurs already occupied many ecological niches of large-bodied animals.

Fossil and molecular evidence indicate that by 160 million years ago, the three major lineages of mammals had emerged: those leading to monotremes (egg-laying mammals), marsupials (mammals with a pouch), and eutherians (placental mammals). After the extinction of large dinosaurs, pterosaurs, and marine reptiles during the late Cretaceous period, mammals underwent an adaptive radiation, giving rise to large predators and herbivores as well as flying and aquatic species.

Monotremes

Monotremes are found only in Australia and New Guinea and are represented by one species of platypus and four species of echidnas (spiny anteaters; **Figure 34.39**). Monotremes lay eggs, a character that is ancestral for amniotes and retained in most reptiles. Like all mammals, monotremes have hair and produce milk, but they lack nipples. Milk is secreted by glands on the belly of the mother. After hatching, the baby sucks the milk from the mother's fur.

▶ **Figure 34.38 The evolution of the mammalian ear bones.** *Biarmosuchus* was a synapsid, a lineage that eventually gave rise to the mammals. Bones that transmit sound in the ear of mammals arose from the modification of bones in the jaw of nonmammalian synapsids.

MAKE CONNECTIONS *Review the definition of exaptation in Concept 25.6. Summarize the process by which exaptation occurs and explain how the incorporation of the articular and quadrate bones into the mammalian inner ear is an example.*

(a) In *Biarmosuchus*, the meeting of the articular and quadrate bones formed the jaw joint.

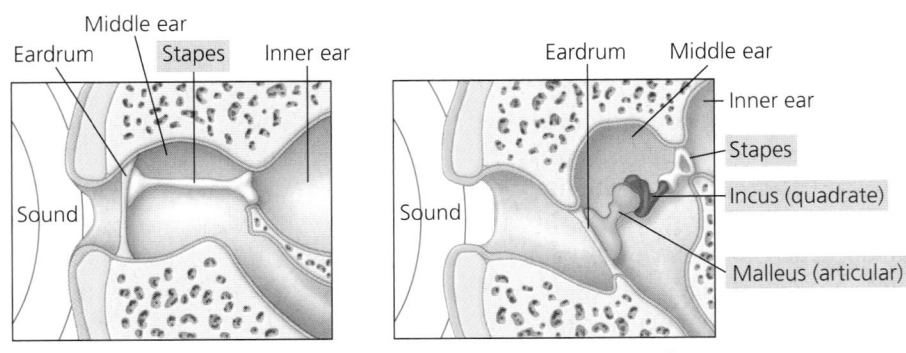

(b) During the evolutionary remodeling of the mammalian skull, a new jaw joint formed between the dentary and squamosal bones (see Figure 25.7). No longer used in the jaw, the quadrate and articular bones became incorporated into the middle ear as two of the three bones that transmit sound from the eardrum to the inner ear.

Figure 34.39 Short-beaked echidna (*Tachyglossus aculeatus*), an Australian monotreme. Monotremes have hair and produce milk, but they lack nipples. Monotremes are the only mammals that lay eggs (inset).

(a) A young brushtail possum (Australia). The offspring of marsupials are born very early in their development. They finish their growth while nursing from a nipple (in their mother's pouch in most species).

Marsupials

Opossums, kangaroos, and koalas are examples of the group called **marsupials**. Both marsupials and eutherians share derived characters not found among monotremes. They have higher metabolic rates and nipples that provide milk, and they give birth to live young. The embryo begins development inside the uterus of the female's reproductive tract. The lining of the uterus and the extraembryonic membranes that arise from the embryo form a **placenta**, a structure in which nutrients diffuse into the embryo from the mother's blood.

A marsupial is born very early in its development and completes its embryonic development while nursing **(Figure 34.40a)**. In most species, the nursing young are held within a maternal pouch called a *marsupium*. A red kangaroo, for instance, is about the size of a honeybee at its birth, just 33 days after fertilization. Its back legs are merely buds, but its front legs are strong enough for it to crawl from the exit of its mother's reproductive tract to a pouch that opens to the front of her body, a journey that lasts a few minutes. In other species, the marsupium opens to the rear of the mother's body; in greater bilbies, this protects the young as their mother burrows in the dirt **(Figure 34.40b)**.

Marsupials existed worldwide during the Mesozoic era, but today they are found only in the Australian region and in the Americas **(Figure 34.40c)**. The biogeography of marsupials illustrates the interplay between biological and geologic evolution (see Concept 25.4). After the breakup of the supercontinent Pangaea, South America and Australia became island continents, and their marsupials diversified in isolation from the eutherians that began an adaptive radiation on the northern continents. Australia has not been in contact with another continent since early in the Cenozoic era, 66 million years ago. In Australia, convergent evolution has resulted in a diversity of marsupials that resemble eutherians in similar ecological roles in other

(b) A greater bilby (Australia). The greater bilby is a digger and burrower that eats termites and other insects, along with the seeds, roots, and bulbs of various plants. The female's rear-opening pouch helps protect the young from dirt as the mother digs. Other marsupials, such as kangaroos, have a pouch that opens to the front.

(c) Virginia opossum (Central and North America). This species is the only marsupial found north of Mexico. After the young complete development in the mother's pouch, they spend another four to five months riding on her back before living independently.

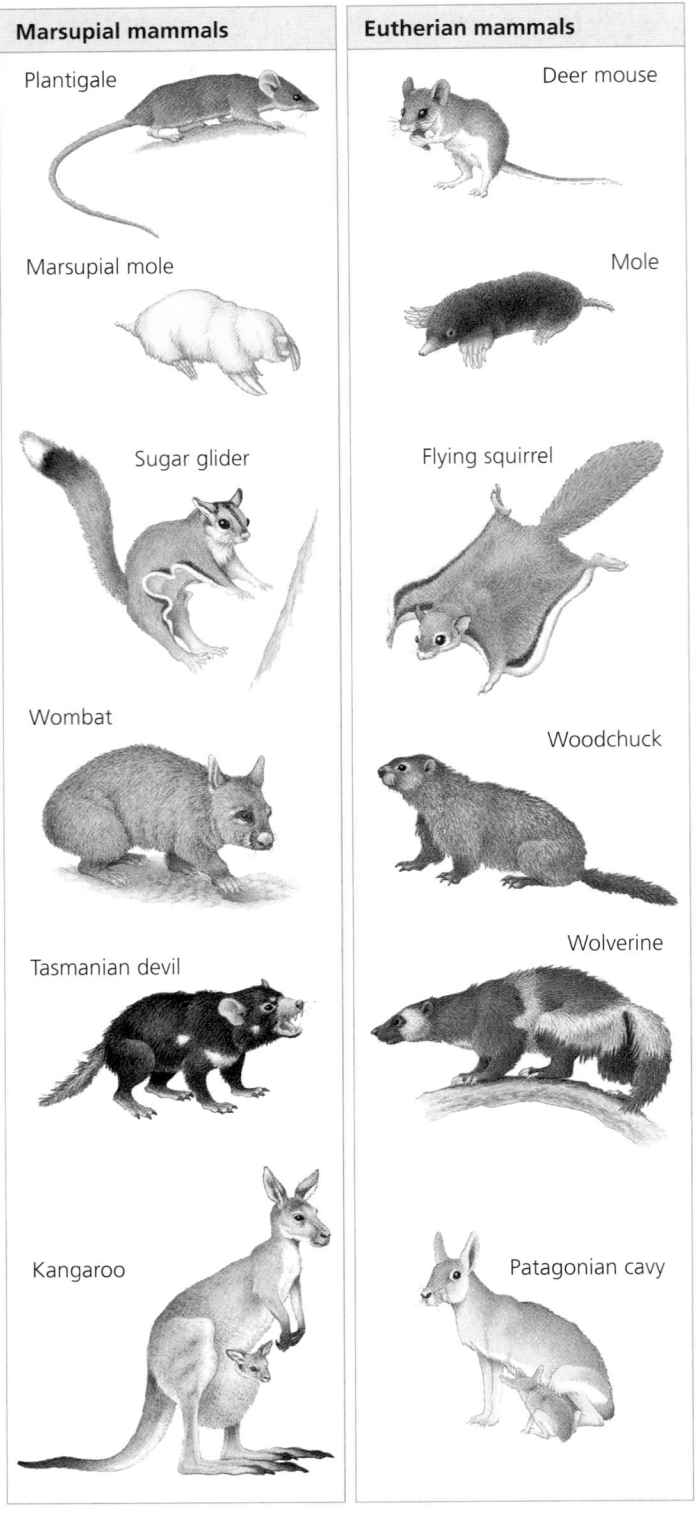

▼ **Figure 34.41 Convergent evolution of marsupials and eutherians (placental mammals).** (Drawings not to scale.)

Marsupial mammals	Eutherian mammals
Plantigale	Deer mouse
Marsupial mole	Mole
Sugar glider	Flying squirrel
Wombat	Woodchuck
Tasmanian devil	Wolverine
Kangaroo	Patagonian cavy

at the Panamanian isthmus and extensive two-way traffic of animals took place over the land bridge. Today, only three families of marsupials live outside the Australian region, and the only marsupials found in the wild in North America are a few species of opossum.

Eutherians (Placental Mammals)

Eutherians are commonly called placental mammals because their placentas are more complex than those of marsupials. Eutherians have a longer pregnancy than marsupials. Young eutherians complete their embryonic development within the uterus, joined to their mother by the placenta. The eutherian placenta provides an intimate and long-lasting association between the mother and her developing young.

The major groups of living eutherians are thought to have diverged from one another in a burst of evolutionary change. The timing of this burst is uncertain: Molecular data suggest it occurred about 100 million years ago, while morphological data suggest it was about 60 million years ago. **Figure 34.42** explores several major eutherian orders and their phylogenetic relationships with each other as well as with the monotremes and marsupials.

Primates

The mammalian order Primates includes the lemurs, tarsiers, monkeys, and apes. Humans are members of the ape group.

Derived Characters of Primates   Most primates have hands and feet adapted for grasping, and their digits have flat nails instead of the narrow claws of other mammals. There are other characteristic features of the hands and feet, too, such as skin ridges on the fingers (which account for human fingerprints). Relative to other mammals, primates have a large brain and short jaws, giving them a flat face. Their forward-looking eyes are close together on the front of the face. Primates also exhibit relatively well-developed parental care and complex social behavior.

The earliest known primates were tree-dwellers, and many of the characteristics of primates are adaptations to the demands of living in the trees. Grasping hands and feet allow primates to hang onto tree branches. All living primates except humans have a big toe that is widely separated from the other toes, enabling them to grasp branches with their feet. All primates also have a thumb that is relatively movable and separate from the fingers, but monkeys and apes have a fully **opposable thumb**; that is, they can touch the ventral surface (fingerprint side) of the tip of all four fingers with the ventral surface of the thumb of the same hand. In monkeys and apes other than humans, the opposable thumb functions in a grasping "power grip." In humans, a distinctive bone structure at the base of the thumb allows it to be used for more precise manipulation. The unique dexterity of humans represents descent with modification from our tree-dwelling

parts of the world **(Figure 34.41)**. In contrast, although South America had a diverse marsupial fauna throughout the Paleogene, it has experienced several immigrations of eutherians. One of the most important occurred about 3 million years ago, when North and South America joined

One hypothesis for the evolutionary relationships of mammals, represented by the tree shown here, clusters the eutherian orders into four main clades. All 20 extant orders of mammals are listed; the orders in bold type are surveyed in more detail around the tree.

ANCESTRAL SYNAPSID

180 million years ago (mya)

160 mya

Monotremes (5 species)

Monotremata

Marsupials (366 species)

Marsupialia

Eutherians (6,028 species)

Proboscidea
Sirenia
Hyracoidea
Tubulidentata
Afrosoricida (golden moles and tenrecs)
Macroscelidea (elephant shrews)

Xenarthra

Rodentia
Lagomorpha
Primates
Dermoptera (flying lemurs)
Scandentia (tree shrews)

Carnivora
Perissodactyla
Cetartiodactyla
Chiroptera
Eulipotyphla
Pholidota (pangolins)

Monotremata
(platypuses, echidnas)
Lay eggs; no nipples; young suck milk from fur of mother

Echidna

Marsupialia
(kangaroos, opossums, koalas)
Complete embryonic development in pouch on mother's body

Koala

Proboscidea (elephants)
Long, muscular trunk; thick, loose skin; upper incisors elongated as tusks

African elephant

Sirenia (manatees, dugongs)
Aquatic; finlike forelimbs and no hind limbs; herbivorous

Manatee

Hyracoidea (hyraxes)
Short legs; stumpy tail; herbivorous; complex, multi-chambered stomach

Rock hyrax

Xenarthra (sloths, anteaters, armadillos)
Reduced teeth or no teeth; herbivorous (sloths) or carnivorous (anteaters, armadillos)

Anteater

Rodentia (squirrels, beavers, rats, porcupines, mice)
Chisel-like, continuously growing incisors worn down by gnawing; herbivorous

Red squirrel

Lagomorpha (rabbits, hares, picas)
Chisel-like incisors; hind legs longer than forelegs and adapted for running and jumping; herbivorous

Jackrabbit

Primates (lemurs, monkeys, chimpanzees, gorillas, humans)
Opposable thumbs; forward-facing eyes; well-developed cerebral cortex; omnivorous

Golden lion tamarin

Carnivora (dogs, wolves, bears, cats, weasels, otters, seals, walruses)
Sharp, pointed canine teeth and molars for grasping and cutting; carnivorous

Coyote

Perissodactyla (horses, zebras, tapirs, rhinoceroses)
Hooves with an odd number of toes on each foot; herbivorous

Indian rhinoceros

Cetartiodactyla
• Artiodactyls
(sheep, pigs, cattle, deer, giraffes)
Hooves with an even number of toes on each foot; herbivorous

Bighorn sheep

• Cetaceans
(whales, dolphins, porpoises)
Aquatic; streamlined body; paddle-like forelimbs and no hind limbs; thick layer of insulating blubber; carnivorous

Pacific white-sided porpoise

Chiroptera (bats)
Adapted for flight; broad skinfold that extends from elongated fingers to body and legs; carnivorous or herbivorous

Frog-eating bat

◀ **Figure 34.43 Verreaux's sifakas (Propithecus verreauxi), a type of lemur.**

reached South America from Africa by rafting on logs or other debris. What is certain is that New World monkeys and Old World monkeys underwent separate adaptive radiations during their many millions of years of separation **(Figure 34.45)**. All species of New World monkeys are arboreal, whereas Old World monkeys include ground-dwelling as well as arboreal species. Most monkeys in both groups are diurnal (active during the day) and usually live in bands held together by social behavior.

The other group of anthropoids consists of primates informally called apes **(Figure 34.46)**. The ape group includes four genera of gibbons, along with the so-called "great apes," the genera *Pongo* (orangutans), *Gorilla* (gorillas), *Pan* (chimpanzees and bonobos), and *Homo* (humans). The apes diverged from Old World monkeys

ancestors. Arboreal maneuvering also requires excellent eye-hand coordination. The overlapping visual fields of the two forward-facing eyes enhance depth perception, an obvious advantage when brachiating (traveling by swinging from branch to branch in trees).

Living Primates There are three main groups of living primates: (1) the lemurs of Madagascar **(Figure 34.43)** and the lorises and bush babies of tropical Africa and southern Asia; (2) the tarsiers, which live in southeastern Asia; and (3) the **anthropoids**, which include monkeys and apes and are found worldwide. The first group—lemurs, lorises, and bush babies—probably resemble early arboreal primates. The oldest known tarsier fossils date to 55 million years ago; along with DNA evidence, these fossils indicate that tarsiers are more closely related to anthropoids than to the lemur group **(Figure 34.44)**.

You can see in Figure 34.44 that monkeys do not form a clade but rather consist of two groups, the New and Old World monkeys. Both of these groups are thought to have originated in Africa or Asia. The fossil record indicates that New World monkeys first colonized South America roughly 25 million years ago. By that time, South America and Africa had drifted apart, and monkeys may have

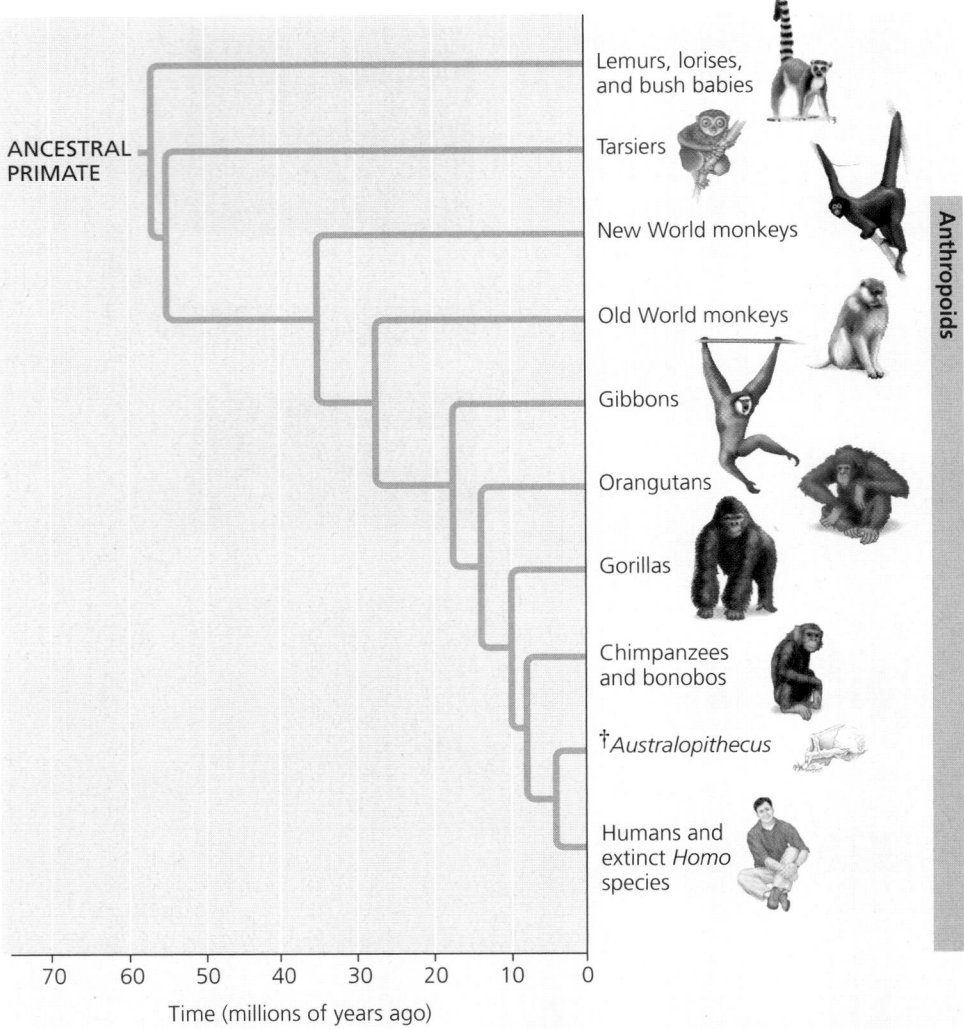

▲ **Figure 34.44 A phylogenetic tree of primates.** The fossil record indicates that the lineage leading to monkeys and apes diverged from other primates 55 million years ago. The lineage leading to humans and *Australopithecus* branched off from other apes about 8 million years ago; the dagger symbol (†) denotes an extinct group.

VISUAL SKILLS *Is the phylogeny shown here consistent with the idea that humans evolved from chimpanzees? Explain.*

(a) New World monkeys, such as spider monkeys (shown here), squirrel monkeys, and capuchins, have a prehensile tail (one adapted for grasping) and nostrils that open to the sides.

(b) Old World monkeys lack a prehensile tail, and their nostrils open downward. This group includes macaques (shown here), mandrils, baboons, and rhesus monkeys.

▲ **Figure 34.45 New World monkeys and Old World monkeys.**

about 25–30 million years ago. Today, nonhuman apes are found exclusively in tropical regions of the Old World. With the exception of gibbons, living apes are larger than either New or Old World monkeys. All living apes have relatively long arms, short legs, and no tail. Although all nonhuman apes spend time in trees, only gibbons and orangutans are primarily arboreal. Social organization varies among the apes; gorillas and chimpanzees are highly social. Finally, compared to other primates, apes have a larger brain in proportion to their body size, and their behavior is more flexible. These two characteristics are especially prominent in our next group, the hominins.

CONCEPT CHECK 34.6

1. Contrast monotremes, marsupials, and eutherians in terms of how they bear young.
2. Identify at least five derived traits of primates.
3. **MAKE CONNECTIONS** Develop a hypothesis to explain why the diversity of mammals increased in the Cenozoic. Your explanation should consider mammalian adaptations as well as factors such as mass extinctions and continental drift (review Concept 25.4).

For suggested answers, see Appendix A.

▼ **Figure 34.46 Nonhuman apes.**

(a) Gibbons (18 species), such as this Muller's gibbon, are found only in south-eastern Asia. Their very long arms and fingers are adaptations for brachiating (swinging by the arms from branch to branch).

(b) Orangutans (3 species) are shy apes that live in the rain forests of Sumatra and Borneo. Named in 2017, this species (*Pongo tapanuliensis*) is the first great ape discovered in nearly 100 years.

(c) Gorillas (2 species) are the largest apes; some males are almost 2 m tall and weigh about 200 kg. Found only in Africa, these herbivores usually live in groups of up to about 20 individuals.

(d) Chimpanzees live in tropical Africa. They feed and sleep in trees but also spend a great deal of time on the ground. Chimpanzees are intelligent, communicative, and social.

(e) Bonobos are in the same genus (*Pan*) as chimpanzees but are smaller. They survive today only in the African nation of Congo.

⊘ **Mastering Biology Videos:**
Chimp Cracking Nut •
Gibbons Brachiating

Humans are mammals that have a large brain and bipedal locomotion

In our tour of Earth's biodiversity, we turn now to our own species, *Homo sapiens*, which is about 200,000 years old. When you consider that life has existed on Earth for at least 3.5 billion years, we are clearly evolutionary newcomers.

Derived Characters of Humans

Many characters distinguish humans from other apes. Most obviously, humans stand upright and are bipedal (walk on two legs). Humans have a much larger brain and are capable of language, symbolic thought, artistic expression, and the manufacture and use of complex tools. Humans also have reduced jawbones and jaw muscles, along with a shorter digestive tract.

At the molecular level, the list of derived characters of humans is growing as scientists compare the human and chimpanzee genomes. Although the two genomes are 99%

identical, a difference of 1% can represent many changes in a genome with 3 billion base pairs. Furthermore, changes in a small number of genes can have large effects. This point was highlighted by recent results showing that humans and chimpanzees differ in the expression of 19 regulatory genes. These genes turn other genes on and off and hence may account for many differences between humans and chimpanzees.

Bear in mind that such genomic differences—and whatever derived phenotypic traits they encode—separate humans from other *living* apes. But many of these new characters first emerged in our ancestors, long before our own species appeared. We will consider some of these ancestors to see how these characters originated.

→ **Mastering Biology** HHMI Video: **Great Transitions: The Origin of Humans**

The Earliest Hominins

The study of human origins is known as **paleoanthropology**. Paleoanthropologists have unearthed fossils of approximately 25 extinct species that are more closely related to humans than to chimpanzees. These species are known as **hominins** **(Figure 34.47)**. Since 1994, fossils of four hominin species

Paranthropus robustus

Homo ergaster

Homo neanderthalensis

Homo sapiens

Paranthropus boisei

Australopithecus africanus

Kenyanthropus platyops

Australopithecus anamensis

Australopithecus garhi

Homo erectus

Homo rudolfensis

Homo habilis

Australopithecus afarensis

Ardipithecus ramidus

Orrorin tugenensis

Sahelanthropus tchadensis

Millions of years ago (y-axis: 0, 0.5, 1.0, 1.5, 2.0, 2.5, 3.0, 3.5, 4.0, 4.5, 5.0, 5.5, 6.0, 6.5, 7.0)

▲ **Figure 34.47 A timeline for selected hominin species.** Most of the fossils illustrated here come from sites in eastern and southern Africa. Note that at most times in hominin history, two or more hominin species were contemporaries.

dating to more than 4 million years ago have been discovered. The oldest of these hominins, *Sahelanthropus tchadensis*, lived about 6.5 million years ago.

Sahelanthropus and other early hominins shared some of the derived characters of humans. For example, they had reduced canine teeth, and some fossils suggest that they had relatively flat faces. They also show signs of having been more upright and bipedal than other apes. One clue to their upright stance can be found in the foramen magnum, the hole at the base of the skull through which the spinal cord passes. In chimpanzees, the foramen magnum is relatively far back on the skull, while in early hominins (and in humans), it is located underneath the skull. This position allows us to hold our head directly over our body, as early hominins apparently did as well. The pelvis, leg bones, and feet of the 4.4-million-year-old *Ardipithecus ramidus* also suggest that early hominins were increasingly bipedal **(Figure 34.48)**. (We will return to the subject of bipedalism later in the chapter.)

Note that the characters that distinguish humans from other living apes did not all evolve in tight unison. While early hominins were showing signs of bipedalism, their brains remained small—about 300–450 cm^3 in volume, compared with an average of 1,300 cm^3 for *Homo sapiens*. The earliest hominins were also small overall. *A. ramidus*, for example, is estimated to have been about 1.2 m tall, with relatively large teeth and a jaw that projected beyond the upper part of the face. Humans, in contrast, average about 1.7 m in height and have a relatively flat face; compare your own face with that of the chimpanzees in Figure 34.46d.

It's important to avoid two common misconceptions about early hominins. One is to think of them either as chimpanzees or as having evolved from chimpanzees. Chimpanzees represent the tip of a separate branch of evolution, and they acquired derived characters of their own after they diverged from their common ancestor with humans.

Another misconception is to think of human evolution as a ladder leading directly from an ancestral ape to *Homo sapiens*. This

▼ **Figure 34.48 "Ardi," a 4.4-million-year-old fossil of *Ardipithecus ramidus*.**

error is often illustrated as a parade of fossil species that become progressively more like ourselves as they march across the page. If human evolution is a parade, it is a very disorderly one, with many groups breaking away to wander other evolutionary paths. At times, several hominin species coexisted. These species often differed in skull shape, body size, and diet (as inferred from their teeth). Ultimately, all but one lineage—the one that gave rise to *Homo sapiens*—ended in extinction. Overall, considering the characteristics of all hominins that lived over the past 6.5 million years, *H. sapiens* appears not as the end result of a straight evolutionary path, but rather as the only surviving member of a highly branched evolutionary tree.

Australopiths

The fossil record indicates that hominin diversity increased dramatically between 4 and 2 million years ago. Many of the hominins from this period are collectively called australopiths. Their phylogeny remains unresolved on many points, but as a group, they are almost certainly paraphyletic. The earliest member of the group, *Australopithecus anamensis*, lived 4.2–3.9 million years ago, close in time to older hominins such as *Ardipithecus ramidus*.

Australopiths got their name from the 1924 discovery in South Africa of *Australopithecus africanus* ("southern ape of Africa"), which lived between 3 and 2.4 million years ago. With the discovery of more fossils, it became clear that *A. africanus* walked fully erect (was bipedal) and had human-like hands and teeth. However, its brain was only about one-third the size of the brain of a present-day human.

In 1974, in the Afar region of Ethiopia, scientists discovered a 3.2-million-year-old *Australopithecus* skeleton that was 40% complete. "Lucy," as the fossil was named, was short—only about 1 m tall. Lucy and similar fossils have been given the species name *Australopithecus afarensis* (for the Afar region). Fossil evidence shows that *A. afarensis* existed as a species for at least 1 million years.

At the risk of oversimplifying, we could say that *A. afarensis* had fewer of the derived characters of humans above the neck than below. Lucy's brain was the size of a softball, a size similar to that expected for a chimpanzee of Lucy's body size. *A. afarensis* skulls also have a long lower jaw. Skeletons of *A. afarensis* suggest that these hominins were capable of arboreal locomotion, with arms that were relatively long in proportion to body size (compared to the proportions in humans). However, fragments of pelvic and skull bones indicate that *A. afarensis* walked on two legs. Fossilized footprints in Laetoli, Tanzania, corroborate the skeletal evidence that hominins living at the time of *A. afarensis* were bipedal **(Figure 34.49)**.

Another lineage of australopiths consisted of the "robust" australopiths. These hominins, which included species such as *Paranthropus boisei*, had sturdy skulls with powerful jaws

▶ Figure 34.49 Evidence that hominins walked upright 3.5 million years ago.

(a) The Laetoli footprints, more than 3.5 million years old, confirm that upright posture evolved quite early in hominin history.

(b) An artist's reconstruction of *A. afarensis*, a hominin alive at the time of the Laetoli footprints.

and large teeth, adapted for grinding and chewing hard, tough foods. They contrast with the "gracile" (slender) australopiths, including *A. afarensis* and *A. africanus*, which had lighter feeding equipment adapted for softer foods.

Combining evidence from the earliest hominins with the much richer fossil record of later australopiths makes it possible to formulate hypotheses about significant trends in hominin evolution. In the **Scientific Skills Exercise**, you'll examine one such trend: how hominin brain volume has changed over time. Here we'll consider two other trends: the emergence of bipedalism and tool use.

Bipedalism

Our anthropoid ancestors of 30–35 million years ago were still tree-dwellers. By about 10 million years ago, the Himalayan mountain range had formed, thrust up in the aftermath of the Indian plate's collision with the Eurasian plate (see Figure 25.16). The climate became drier, and the forests of what are now Africa and Asia contracted. The result was an increased area of savanna (grassland) habitat, with fewer trees. Researchers have hypothesized that as the habitat

changed, natural selection may have favored adaptations that made moving over open ground more efficient. Underlying this idea is the fact that while nonhuman apes are superbly adapted for climbing trees, they are less well suited for ground travel. For example, as a chimpanzee walks, it uses four times the amount of energy used by a human.

Although elements of this hypothesis survive, the picture now appears more complex. Even though all fossils of early hominins show indications of bipedalism, none of these hominins lived in savannas. Instead, they lived in mixed habitats ranging from forests to open woodlands. Furthermore, whatever the selective pressure that led to bipedalism, hominins did not become more bipedal in a simple, linear fashion. *Ardipithecus* had skeletal elements indicating that it could switch to upright walking but also was well suited for climbing trees. Australopiths seem to have had various locomotor styles, and some species spent more time on the ground than others. Only about 1.9 million years ago did hominins begin to walk long distances on two legs. These hominins lived in more arid environments, where bipedal walking requires less energy than walking on all fours.

Tool Use

As you read earlier, the manufacture and use of complex tools are derived behavioral characters of humans. Determining the origin of tool use in hominin evolution is very difficult. Other apes are capable of surprisingly sophisticated tool use. Orangutans, for example, can fashion sticks into probes for retrieving insects from their nests. Chimpanzees are even more adept, using rocks to smash open food and putting leaves on their feet to walk over thorns. It's likely that early hominins were capable of this sort of simple tool use, but finding fossils of modified sticks or leaves that were used as shoes is practically impossible.

The oldest generally accepted evidence of tool use by hominins is 2.5-million-year-old cut marks on animal bones found in Ethiopia. These marks suggest that hominins cut flesh from the bones of animals using stone tools. Interestingly, the hominins whose fossils were found near the site where the bones were discovered had a relatively small brain. If these hominins, which have been named *Australopithecus garhi*, were in fact the creators of the stone tools used on the bones, that would suggest that stone tool use originated before the evolution of large brains in hominins.

⊘ **Mastering Biology** HHMI Video: The Tool-Making Animal

Early *Homo*

The earliest fossils that paleoanthropologists place in our genus, *Homo*, include those of the species *Homo habilis*. These fossils, ranging in age from about 2.4 to 1.6 million years, show clear signs of certain derived hominin characters above

Determining the Equation of a Regression Line

How Has Brain Volume Changed Over Time in the Hominin Lineage? The hominin taxon includes *Homo sapiens* and about 20 extinct species that are thought to represent early relatives of humans. Researchers have found that the brain volume of the earliest hominins ranged between 300 and 450 cm³, similar to the brain volume of chimpanzees. The brain volumes of modern humans range between 1,200 and 1,800 cm³. In this exercise, you'll examine how mean brain volume changed over time and across various hominin species.

How the Study Was Done In this table, x_i is the mean age of each hominin species, and y_i is the mean brain volume (cm³). Ages with negative values represent millions of years before the present (which has an age of 0.0).

Hominin Species	Mean age (millions of years; x_i)	$x_i - \bar{x}$	Mean Brain Volume (cm³; y_i)	$y_i - \bar{y}$	$(x_i - \bar{x})$ × $(y_i - \bar{y})$
Ardipithecus ramidus	−4.4		325		
Australopithecus afarensis	−3.4		375		
Homo habilis	−1.9		550		
Homo ergaster	−1.6		850		
Homo erectus	−1.2		1,000		
Homo heidelbergensis	−0.5		1,200		
Homo neanderthalensis	−0.1		1,400		
Homo sapiens	0.0		1,350		

Data from Dean Falk, Florida State University, 2013.

INTERPRET THE DATA

How did the brain volume of hominin species change over time? In particular, is there a linear (straight-line) relationship between brain volume and time?

To find out, we'll perform a linear regression, a technique for determining the equation for the straight line that provides a "best fit" to a set of data. Recall that the equation for a straight line between two variables, x and y, is

$$y = mx + b$$

In this equation, m represents the slope of the line, while b represents the y-intercept (the point at which the straight line crosses the y-axis). When $m < 0$, the line has a negative slope, indicating that the values of y become *smaller* as values of x become *larger*. When $m > 0$, the line has a positive slope, meaning that the

values of y become larger as values of x become larger. When $m = 0$, y has a constant value (b).

The correlation coefficient, r, can be used to calculate the values of m and b in a linear regression:

$$m = r\frac{s_y}{s_x} \quad \text{and} \quad b = \bar{y} - m\bar{x}.$$

In these equations, s_x and s_y are the standard deviations of variables x and y, respectively, while $\bar{x}$ and $\bar{y}$ are the means of those two variables. (See the Scientific Skills Exercise for Chapter 32 for more information about the correlation coefficient, mean, and standard deviation.)

1. Calculate the means ($\bar{x}$ and $\bar{y}$) from the $n = 8$ data points in the table. Next, fill in the ($x_i - \bar{x}$) and ($y_i - \bar{y}$) columns in the data table, and use those results to calculate the standard deviations s_x and s_y.

2. As described in the Scientific Skills Exercise for Chapter 32, the formula for a correlation coefficient is

$$r = \frac{\sum (x_i - \bar{x})(y_i - \bar{y})}{(n - 1)(s_x s_y)}$$

Fill in the column in the data table for the product $(x_i - \bar{x}) \times (y_i - \bar{y})$. Use these values and the standard deviations calculated in question 1 to calculate the correlation coefficient r between the brain volume of hominin species (y) and the ages of those species (x).

3. Based on the value of r that you calculated in question 2, describe in words the correlation between mean brain volume of hominin species and the mean age of the species.

4. **(a)** Use your calculated value of r to calculate the slope (m) and the y-intercept (b) of a regression line for this data set. **(b)** Graph the regression line for the mean brain volume of hominin species versus the mean age of the species. Be careful to select and label your axes correctly. **(c)** Plot the data from the table on the same graph that shows the regression line. Does the regression line appear to provide a reasonable fit to the data?

5. The equation for a regression line can be used to calculate the value of y expected for any particular value of x. For example, suppose that a linear regression indicated that $m = 2$ and $b = 4$. In this case, when $x = 5$, we expect that $y = 2x + 4 = (2 \times 5) + 4 = 14$. Based on the values of m and b that you determined in question 4, use this approach to determine the expected mean brain volume for a hominin that lived 4 million years ago (that is, $x = -4$).

6. The slope of a line can be defined as $m = \frac{y_2 - y_1}{x_2 - x_1}$, where (x_1, y_1) and (x_2, y_2) are the coordinates of two points on the line. As such, the slope represents the ratio of the rise of a line (how much the line rises vertically) to the run of the line (how much the line changes horizontally). Use the definition of the slope to estimate how long it took for mean brain volume to increase by 100 cm³ over the course of hominin evolution.

➔ **Instructors:** A version of this Scientific Skills Exercise can be assigned in **Mastering Biology**.

the neck. Compared to the australopiths, *H. habilis* had a shorter jaw and a larger brain volume, about 550–750 cm³. Sharp stone tools have also been found with some fossils of *H. habilis* (the name means "handy man").

Fossils from 1.9 to 1.5 million years ago mark a new stage in hominin evolution. A number of paleoanthropologists recognize these fossils as those of a distinct species, *Homo ergaster*. *Homo ergaster* had a substantially larger brain than

H. habilis (over 900 cm³), as well as long, slender legs with hip joints well adapted for long-distance walking **(Figure 34.50)**. The fingers were relatively short and straight, suggesting that *H. ergaster* did not climb trees like earlier hominins. *Homo ergaster* fossils have been discovered in far more arid environments than earlier hominins and have been associated with more sophisticated stone tools. Its smaller teeth also suggest that *H. ergaster* either ate different foods than australopiths (more meat and less plant material) or prepared some of its food before chewing, perhaps by cooking or mashing the food.

Homo ergaster marks an important shift in the relative sizes of the sexes. In primates, a size difference between males and females is a major component of sexual dimorphism (see Concept 23.4). On average, male gorillas and orangutans weigh about twice as much as females of their species. In *Australopithecus afarensis*, males were 1.5 times as heavy as females. The extent of sexual dimorphism decreased further in early *Homo*, a trend that continues through our own species: Human males weigh only about 1.2 times as much as females.

The reduced sexual dimorphism may offer some clues to the social systems of extinct hominins. In extant primates, extreme sexual dimorphism is associated with intense male-male competition for multiple females. In species that undergo more pair-bonding (including our own), sexual dimorphism is less dramatic. In *H. ergaster*, therefore, males and females may have engaged in more pair-bonding than earlier hominins did.

Fossils now generally recognized as *H. ergaster* once were considered early members of another species, *Homo erectus*, and some paleoanthropologists still hold this position. *Homo erectus* originated in Africa and was the first hominin

▲ **Figure 34.50 Fossil of *Homo ergaster*.** This 1.7-million-year-old fossil from Kenya belongs to a young *Homo ergaster* male. This individual was tall, slender, and fully bipedal, and he had a relatively large brain.

to migrate out of Africa. The oldest fossils of hominins outside Africa, dating back 1.8 million years, were discovered in the present-day country of Georgia. *Homo erectus* eventually migrated as far as the Indonesian archipelago. Fossil evidence indicates that *H. erectus* became extinct between 200,000 and 70,000 years ago.

Neanderthals

In 1856, miners discovered some mysterious human fossils in a cave in the Neander Valley in Germany. The 40,000-year-old fossils belonged to a thick-boned hominin with a prominent brow. The hominin was named *Homo neanderthalensis* and is commonly called a Neanderthal. Neanderthals were living in Europe by 350,000 years ago and later spread to the Near East, central Asia, and southern Siberia. They had a brain larger than that of present-day humans, buried their dead, and made hunting tools from stone and wood. But despite their adaptations and culture, Neanderthals became extinct at some point between 28,000 and 40,000 years ago.

What is the evolutionary relationship of Neanderthals to *Homo sapiens*? Genetic data indicate that the lineages leading to *H. sapiens* and to Neanderthals diverged about 600,000 years ago **(Figure 34.51)**. This indicates that while Neanderthals and humans share a relatively recent common ancestor, humans did not descend directly from Neanderthals (as was once thought). Neanderthals were, in fact, more closely related to the "Denisovans," a hominin whose fossils have been discovered in Siberia and Tibet, than to humans.

Another long-standing question is whether mating occurred between Neanderthals and humans, leading to interspecific gene flow. Some researchers have argued that evidence of gene flow can be found in fossils that show

▼ **Figure 34.51 Evolutionary relationships and interbreeding among humans, Neanderthals, and Denisovans.** Genomic data and fossil evidence (see Figure 34.52) document multiple examples of interbreeding among humans, Neanderthals, and Denisovans.

a mixture of human and Neanderthal characteristics. Recent analyses of the DNA sequence of the Neanderthal genome indicated that limited gene flow did occur between the two species. In 2015, the most extensive evidence yet of such gene flow was reported: DNA extracted from the fossil of a human jawbone was found to contain long stretches of Neanderthal DNA **(Figure 34.52a)**. In fact, the amount of Neanderthal DNA in this fossil indicated that this individual's great-great-great-grandparent was a Neanderthal.

Other recent genomic studies have shown that gene flow also occurred between Neanderthals and the Denisovans, whose DNA was isolated from bone fragments discovered in a Siberian cave. In 2018, for example, DNA analyses revealed that the 90,000-year-old bone fragment shown in **Figure 34.52b** came from a woman who had a Neanderthal mother and a Denisovan father—a remarkable discovery documenting the direct offspring of two different species of hominins. Genomic analyses also indicate that gene flow occurred between the Denisovans and our own species, *H. sapiens*.

Homo sapiens

Evidence from fossils, archaeology, and DNA studies has improved our understanding about how our own species, *Homo sapiens*, emerged and spread around the world.

Fossil evidence indicates that the ancestors of humans originated in Africa. Older species gave rise to later species, ultimately including *H. sapiens*. A 2017 study indicates that by 315,000 years ago, some hominin populations in Morocco had facial features like those of our species, while the back of the skull remained elongated, as in earlier species. The oldest known fossils of our species include

specimens from Ethiopia that are 195,000 and 160,000 years old **(Figure 34.53)**. These early humans had less pronounced browridges than those found in *H. erectus* and Neanderthals, and they were more slender than other recent hominins. The Ethiopian fossils support inferences about the origin of humans from molecular evidence. DNA analyses show that Europeans and Asians share a relatively recent common ancestor and that many African lineages branched off more basal positions on the human family tree. These findings suggest that all living humans have ancestors that originated as *H. sapiens* in Africa.

The oldest fossils of *H. sapiens* outside Africa are from the Middle East and date back about 180,000 years. Fossil evidence and genetic analyses suggest that humans spread beyond Africa in one or more waves, first into Asia and then to Europe and Australia. The date of the first arrival of humans in the New World is uncertain, although the oldest generally accepted evidence puts that date at about 15,000 years ago.

New findings continually update our understanding of the human evolutionary lineage. For example, in 2015, the human family gained a new member, *Homo naledi*. The structure of its foot indicates that *H. naledi* was fully bipedal, and the shape of its hand suggests that *H. naledi* had fine motor skills **(Figure 34.54)**, as in *H. sapiens*, Neanderthals, and other species that used tools extensively. But *H. naledi* also had a small brain, a broadly flared upper pelvis, and other features that have led researchers to conclude that it was an early member of our genus.

As an early member of our genus, it is likely that *H. naledi* originated more than 1.5 million years ago—yet initial estimates of the age of the *H. naledi* fossils ranged from 3 million

▼ **Figure 34.53 A 160,000-year-old-fossil of *Homo sapiens*.**

▼ **Figure 34.54 Fossils of hand bones and foot bones (top and side views) of *Homo naledi*.**

years old to just 100,000 years old. Scientists didn't know how old the fossils are because they were found on the floor of a deep cave, not encased in rocks that could be dated using radioactive isotopes. In 2017, however, teeth from a newly discovered *H. naledi* skull were dated using radioactive isotopes and found to be 300,000 years old. Overall, fossil evidence suggests that *H. naledi* arose 1–2 million years ago (like other early members of our genus) and then persisted almost until the present.

About ten years before the discovery of *H. naledi*, researchers reported another astonishing find: skeletal remains of adult hominins dating from 100,000 to 60,000 years ago and representing a previously unknown species, *Homo floresiensis*. First discovered in a limestone cave on the Indonesian island of Flores, the individuals were much shorter and had a much smaller brain volume than *H. sapiens*—more similar, in fact, to an australopith. The researchers who discovered these fossils argue that certain features of the skeletons, such as the shape of the teeth and the thickness and proportions of the skull, suggest that *H. floresiensis* descended from the larger *H. erectus*. Not convinced, some researchers have argued that the fossils represent small *H. sapiens* individuals with a disorder such as Down syndrome or microcephaly (a condition in which a person has a deformed, miniature brain).

Over the past few years, most studies have supported the designation of *H. floresiensis* as a new hominin. One such study found that the wrist bones of the Flores fossils are similar in shape to those of nonhuman apes and early hominins, but different from those of Neanderthals and *H. sapiens*. These researchers concluded that the Flores fossils represent a species whose lineage branched off before the origin of the clade that includes Neanderthals and humans. A different study comparing the foot bones of the Flores fossils with those of other hominins also concluded that *H. floresiensis* arose before *H. sapiens*. Finally, in 2016, *H. floresiensis*-like fossils dating to 700,000 years ago were discovered at a second location on Flores— again indicating that the *H. floresiensis* lineage branched off long before the origin of *H. sapiens*. Open questions that may yet be answered by new discoveries include how *H. floresiensis* originated and whether it survived long enough to encounter *H. sapiens*, which was living in Indonesia by 50,000 years ago.

The rapid expansion of our species may have been spurred by changes in human cognition as *H. sapiens* evolved in Africa. Evidence of sophisticated thought in *H. sapiens* includes the recent discovery in South Africa of 77,000-year-old art—geometric markings made on pieces of ochre **(Figure 34.55)**. Similarly, archaeologists working in southern and eastern Africa recently found 75,000-year-old ostrich eggs and snail shells with holes neatly drilled through them. By 30,000 years ago, humans were producing spectacular cave paintings.

While these developments can help us understand the spread of *H. sapiens*, it is not clear whether they played a

▼ **Figure 34.55 Art, a human hallmark.** The engravings on this 77,000-year-old piece of ochre, discovered in South Africa's Blombos Cave, are among the earliest signs of symbolic thought in humans.

role in the extinction of other hominins. Neanderthals, for example, also made complex tools and showed a capacity for symbolic thought. As a result, while some scientists have suggested that Neanderthals were driven to extinction by competition with *H. sapiens*, others question that idea.

Our discussion of humans brings this unit on biological diversity to an end. But keep in mind that our sequence of topics isn't meant to imply that life consists of a ladder leading from lowly microorganisms to lofty humanity. Biological diversity is the product of branching phylogeny, not ladderlike "progress." The fact that there are almost as many species of ray-finned fishes alive today as in all other vertebrate groups combined shows that our finned relatives are not outmoded underachievers that failed to leave the water. The tetrapods—amphibians, reptiles, and mammals—are derived from one lineage of lobe-finned vertebrates. As tetrapods diversified on land, fishes continued their branching evolution in the greatest portion of the biosphere's volume. Similarly, the ubiquity of diverse prokaryotes throughout the biosphere today is a reminder of the enduring ability of these relatively simple organisms to keep up with the times through adaptive evolution. Biology exalts life's diversity, past and present.

CONCEPT CHECK 34.7

1. Identify some characters that distinguish hominins from other apes.

2. Provide an example in which different features of organisms in the hominin evolutionary lineage evolved at different rates.

3. **WHAT IF?** Some genetic studies suggest that the most recent common ancestor of *Homo sapiens* that lived outside of Africa left Africa about 50,000 years ago. Compare this date with the dates of fossils given in the text. Can both the genetic results and the dates ascribed to the fossils be correct? Explain.

For suggested answers, see Appendix A.

34 Chapter Review

SUMMARY OF KEY CONCEPTS

→ To review key terms, go to the **Vocabulary Self-Quiz** in the **Mastering Biology** eText or Study Area, or go to goo.gl/zkjz9t.

Key Concept							Clade	Description
CONCEPT 34.1 **Chordates have a notochord and a dorsal, hollow nerve cord** (pp. 719–722) ? *Describe likely features of the common ancestor of chordates. Explain your answer.*	**Chordates:** *Hox* genes; notochord; dorsal, hollow nerve cord; pharyngeal slits; post-anal tail						Cephalochordata (lancelets)	Basal chordates; marine suspension feeders that exhibit four key derived characters of chordates
							Urochordata (tunicates)	Marine suspension feeders; larvae display the derived traits of chordates
CONCEPT 34.2 **Vertebrates are chordates that have a backbone** (pp. 722–725) ? *Identify the shared features of early fossil vertebrates.*		**Vertebrates:** *Hox* genes duplication, backbone of vertebrae	Cyclostomes: jawless vertebrates				Myxini (hagfishes) and Petromyzontida (lampreys)	Jawless aquatic vertebrates with reduced vertebrae; hagfishes have a head with a skull and brain, eyes, and other sensory organs; some lampreys feed by attaching to a live fish and ingesting its blood
CONCEPT 34.3 **Gnathostomes are vertebrates that have jaws** (pp. 725–730) ? *How would the appearance of organisms with jaws have altered ecological interactions? Provide supporting evidence.*				**Gnathostomes:** hinged jaws, four sets of *Hox* genes	Osteichthyans: bony skeleton	Lobe-fins: muscular fins or limbs	Chondrichthyes (sharks, rays, skates, ratfishes)	Aquatic gnathostomes; have a cartilaginous skeleton, a derived trait formed by the reduction of an ancestral mineralized skeleton
							Actinopterygii (ray-finned fishes)	Aquatic gnathostomes; have a bony skeleton and maneuverable fins supported by rays
							Actinistia (coelacanths)	Ancient lineage of aquatic lobe-fins still surviving in Indian Ocean
							Dipnoi (lungfishes)	Freshwater lobe-fins with both lungs and gills; sister group of tetrapods
CONCEPT 34.4 **Tetrapods are gnathostomes that have limbs** (pp. 730–734) ? *Which features of amphibians restrict most species to living in aquatic or moist terrestrial habitats?*						**Tetrapods:** four limbs, neck, fused pelvic girdle	Amphibia (salamanders, frogs, caecilians)	Have four limbs descended from modified fins; most have moist skin that functions in gas exchange; many live both in water (as larvae) and on land (as adults)
CONCEPT 34.5 **Amniotes are tetrapods that have a terrestrially adapted egg** (pp. 734–741) ? *Explain why birds are considered reptiles.*						**Amniotes:** amniotic egg, rib cage ventilation	Reptilia (tuataras, lizards and snakes, turtles, crocodilians, birds)	One of two groups of living amniotes; have amniotic eggs and rib cage ventilation, key adaptations for life on land
CONCEPT 34.6 **Mammals are amniotes that have hair and produce milk** (pp. 741–747) ? *Describe the origin and early evolution of mammals.*							Mammalia (monotremes, marsupials, eutherians)	Evolved from synapsid ancestors; include egg-laying monotremes (echidnas, platypus); pouched marsupials (such as kangaroos, opossums); and eutherians (placental mammals, such as rodents, primates)

CONCEPT 34.7

Humans are mammals that have a large brain and bipedal locomotion (pp. 748–754)

- Derived characters of humans include bipedalism and a larger brain and reduced jaw compared with other apes.
- **Hominins**—humans and species that are more closely related to humans than to chimpanzees—originated in Africa about 8 million years ago. Early hominins had a small brain but probably walked upright.
- The oldest evidence of tool use is 2.5 million years old.
- *Homo ergaster* was the first fully bipedal, large-brained hominin. *Homo erectus* was the first hominin to leave Africa.
- Neanderthals lived in Europe and the Near East from about 350,000 to 28,000 years ago.
- *Homo sapiens* originated in Africa about 195,000 years ago and began to spread to other continents about 180,000 years ago.

? *Based on fossil evidence, summarize how key hominin features have changed over time.*

TEST YOUR UNDERSTANDING

→ For more multiple-choice questions, go to the **Practice Test** in the **Mastering Biology** eText or Study Area, or go to goo.gl/GruWRg.

Levels 1-2: Remembering/Understanding

1. Vertebrates and tunicates share
   - (A) jaws adapted for feeding.
   - (B) a high degree of cephalization.
   - (C) an endoskeleton that includes a skull.
   - (D) a notochord and a dorsal, hollow nerve cord.

2. Living vertebrates can be divided into two major clades. Select the appropriate pair.
   - (A) the chordates and the tetrapods
   - (B) the urochordates and the cephalochordates
   - (C) the cyclostomes and the gnathostomes
   - (D) the marsupials and the eutherians

3. Unlike eutherians, *both* monotremes and marsupials
   - (A) lack nipples.
   - (B) have some embryonic development outside the uterus.
   - (C) lay eggs.
   - (D) are found in Australia and Africa.

4. In Figure 34.25, which of the following is the sister taxon to the archosaurs?
   - (A) synapsids
   - (B) crocodilians
   - (C) turtles
   - (D) lepidosaurs

5. As hominins diverged from other primates, which of the following appeared first?
   - (A) reduced jawbones
   - (B) an enlarged brain
   - (C) the making of stone tools
   - (D) bipedal locomotion

Levels 3-4: Applying/Analyzing

6. Which of the following could be considered the most recent common ancestor of living tetrapods?
   - (A) a sturdy-finned, shallow-water lobe-fin whose appendages had skeletal supports similar to those of terrestrial vertebrates
   - (B) an armored, jawed placoderm with two pairs of appendages
   - (C) an early ray-finned fish that developed bony skeletal supports in its paired fins
   - (D) a salamander that had legs supported by a bony skeleton but moved with the side-to-side bending typical of fishes

7. **EVOLUTION CONNECTION** Living members of a vertebrate lineage can be very different from early members of the lineage, and evolutionary reversals (character losses) are common. Give examples that illustrate these observations, and explain their evolutionary causes.

Levels 5-6: Evaluating/Creating

8. **SCIENTIFIC INQUIRY • DRAW IT** As a consequence of size alone, larger organisms tend to have larger brains than smaller organisms. However, some organisms have brains that are considerably larger than expected for their size. There are high energetic costs associated with the development and maintenance of brains that are large relative to body size.
   - (a) The fossil record documents trends in which brains that are large relative to body size evolved in certain lineages, including hominins. In such lineages, what can you infer about the costs and benefits of large brains?
   - (b) Hypothesize how natural selection might favor the evolution of large brains despite their high maintenance costs.
   - (c) Data for 14 bird species are listed below. Graph the data, placing deviation from expected brain size on the *x*-axis and mortality rate on the *y*-axis. What can you conclude about the relationship between brain size and mortality?

Deviation from Expected Brain Size*	−2.4	−2.1	−2.0	−1.8	−1.0	0.0	0.3	0.7	1.2	1.3	2.0	2.3	3.0	3.2
Mortality Rate	0.9	0.7	0.5	0.9	0.4	0.7	0.8	0.4	0.8	0.3	0.6	0.6	0.3	0.6

Data from D. Sol et al., Big-brained birds survive better in nature, *Proceedings of the Royal Society B* 274:763–769 (2007).

*Values < 0 indicate brain sizes smaller than expected; values > 0 indicate sizes larger than expected.

9. **WRITE ABOUT A THEME: ORGANIZATION** Early tetrapods had a sprawling gait (like that of a lizard): As the right front foot moved forward, the body twisted to the left and the left rib cage and lung were compressed; the reverse occurred with the next step. Normal breathing, in which both lungs expand equally with each breath, was hindered during walking and prevented during running. In a short essay (100–150 words), explain how the origin of organisms such as dinosaurs, whose gait allowed them to move without compressing their lungs, could have led to emergent properties.

10. **SYNTHESIZE YOUR KNOWLEDGE**

This animal is a vertebrate with hair. What can you infer about its phylogeny? Identify as many key derived characters as you can that distinguish this animal from invertebrate chordates.

For selected answers, see Appendix A.

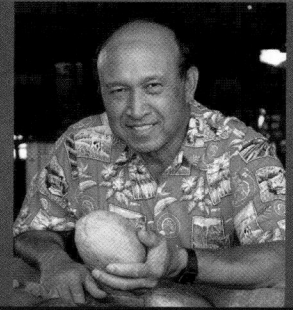

Meet the man who saved the papaya industry in Hawaii, Dennis Gonsalves. Dr. Gonsalves received a B.S. in Horticulture and a M.S. in Plant Pathology from the University of Hawaii, Manoa, and a Ph.D. in Plant Pathology from the University of California, Davis. He became an Associate Professor of plant pathology at the University of Florida and later a Liberty Hyde Bailey Professor at Cornell University. He and his colleagues created virus-resistant papaya plants that rescued the papaya-growing industry in Hawaii from the devastating effects of the ringspot virus. He then served as the Director of the Agricultural Research Center in Hilo, Hawaii, until he retired in 2012. He received the Humboldt Prize for renowned scientific research in 1992 and the Lee Hutchison Award for accomplishments in research, mentoring, and outreach to developing countries.

AN INTERVIEW WITH

Dennis Gonsalves

How did you get interested in plant biology?

I was born on a sugar plantation in Kohala, Hawaii, so I grew up in an agricultural setting. My dad was a laborer on the Kohala Sugar Plantation. I attended the University of Hawaii to study agriculture engineering because my goal was to get a degree and then work on the sugar plantations as an agriculture engineer. But when I was a sophomore, the sugar plantations were closing in Hawaii, and the program was disbanded. I took up horticulture as an alternative. After graduation, I moved to Kauai, where there was a University of Hawaii experiment station where I worked as a technician in a plant pathology lab. There, I was a given a small project of my own to work on, and I fell in love with research.

◀ Top: A papaya tree infected with ringspot virus. Inset: An infected papaya fruit. Bottom:

Tell us about your experiences in graduate school.

I went to UC Davis for my Ph.D. I understood the material. I could do the research. But when it came to giving lectures, I was hampered by a stuttering problem. A few months before I got my Ph.D., I applied for a job at the University of Florida and was invited for an interview. And then I thought, "I have to interview. I am a poor speaker. I don't stand a chance if I do a poor job." I was sick with stress. I practiced and practiced, and then it dawned on me, "I know more about this topic than my professors do. I'm the expert." So, I flew to Florida, and I will never forget that day: I got up and gave my talk, and I have not stuttered since!

What is the ringspot virus and how did you become interested in it?

The virus produces a ring spot on the fruit, but more importantly, the leaves turn yellow, the growth gets stunted, and often the plant won't produce fruit. It is rapidly transmitted by aphids, small sap-sucking insects. An aphid can land on an infected papaya, feed for less than a minute, then jump to a healthy papaya plant and infect it. In 1978, I returned to Hawaii for a short vacation and I met a colleague who told me, "Dennis, there is this virus disease on the Big Island where they grow 95% of the papaya, and it's close to the papaya-growing areas. People are trying to prevent it from spreading." So, when I returned to Cornell, I tried to figure out how to control the papaya ringspot virus.

What was your strategy in fighting the disease?

In our initial attempt, we used artificial selection to isolate a mild strain of papaya ringspot

virus to protect plants against infection, but it didn't grow well under field conditions. I have learned, however, as a scientist, sometimes you have to change directions. I was a classical virologist, but I decided to retrain myself as a molecular biologist. Research by others with tobacco had established that certain viral genes, if introduced into the host plant genome, induce resistance. Other colleagues at Cornell had invented a "gene gun" that uses fine, DNA-coated tungsten beads to literally shoot DNA into cells. We shot tungsten beads coated with DNA of the coat protein gene of the ringspot virus into papaya embryos growing in culture. Some of these genetically modified embryos could be grown into plants. When I inoculated them with the virus, one line was completely resistant; it became a parent of the resistant "Rainbow" variety. "Rainbow" was developed just in time because the virus had spread into the main papaya-growing areas and was decimating the papaya crop.

"As a scientist, sometimes you have to change directions."

Have foes of genetically modified foods opposed your research?

In 1994, we teamed up with Thai researchers to repeat our success using a large-fruited papaya variety popular in Thailand. Ten years later, we were successful, but opponents of genetic engineering stormed the fences surrounding our field trials and destroyed many of the fruit. That transgenic papaya has never seen the light of day. Although I've had a pretty good career, that was a heartbreaking thing for me.

Vascular Plant Structure, Growth, and Development

35.1 Plants have a hierarchical organization consisting of organs, tissues, and cells *p. 759*

35.2 Different meristems generate new cells for primary and secondary growth *p. 766*

35.3 Primary growth lengthens roots and shoots *p. 768*

35.4 Secondary growth increases the diameter of stems and roots in woody plants *p. 772*

35.5 Growth, morphogenesis, and cell differentiation produce the plant body *p. 775*

Study Tip

Make a table: To help keep track of what different plant cells do, make the following table:

Type of plant cell	What it does	How structure fits function

➔ Go to Mastering Biology

For Students (in eText and Study Area)
- Get Ready for Chapter 35
- BioFlix® Animation: Tour of a Plant Cell
- Figure 35.19 Walkthrough: Secondary Growth of a Woody Stem

For Instructors to Assign (in Item Library)
- Tutorial: Visualizing Primary and Secondary Growth
- Activity: Primary and Secondary Growth

Figure 35.1 There is beauty to behold at every level of plant organization: Every cell, every tissue, and every organ has a function, and the structure of each has been molded by natural selection.

How does structure fit function in vascular plants?

At the organ level

Leaves provide surface area for absorbing sunlight and exchanging gases.

Stems support and elevate leaves, maximizing photosynthesis.

Roots anchor the plant and absorb water and minerals.

At the tissue level

Dermal tissue protects organs.

Vascular tissue provides support and transports resources.

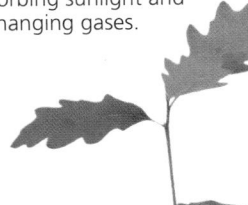

Leaf cross section

Ground tissue includes cells that carry out photosynthesis and store sugars.

At the cellular level

Photosynthetic cells are packed with chloroplasts that convert sunlight into chemical energy.

Chloroplasts

Tube-shaped cells transport resources. The cell shown here carries water and minerals. Others conduct sugars.

Cells with root hairs near the tips of roots increase the surface area for absorbing water and minerals.

Chapters 29 and 30 provided an overview of plant diversity, including both nonvascular and vascular plants. In this chapter and throughout Unit Six, we'll focus on vascular plants, especially angiosperms because flowering plants are the primary producers in many terrestrial ecosystems and are of great agricultural importance. This chapter mainly explores nonreproductive growth—roots, stems, and leaves—and focuses primarily on the two main groups of angiosperms: eudicots and monocots (see Figure 30.16). Later, in Chapter 38, we'll examine angiosperm reproductive growth: flowers, seeds, and fruits.

CONCEPT 35.1

Plants have a hierarchical organization consisting of organs, tissues, and cells

Plants, like most animals, are composed of cells, tissues, and organs. A **cell** is the fundamental unit of life. A **tissue** is a group of cells consisting of one or more cell types that together perform a specialized function. An **organ** consists of several types of tissues that together carry out particular functions. As you learn about plant structure, keep in mind how natural selection has produced plant forms that fit plant function at all levels of structure. We begin by discussing plant organs because their structures are most familiar.

Vascular Plant Organs: Roots, Stems, and Leaves

EVOLUTION The basic morphology, or shape, of vascular plants reflects their evolutionary history as terrestrial organisms that inhabit and draw resources from two very different environments—below the ground and above the ground. They must absorb water and minerals from below the ground surface and CO_2 and light from above the ground surface. The ability to acquire these resources efficiently is traceable to the evolution of roots, stems, and leaves as the three basic organs. These organs form a **root system** and a **shoot system**, the latter consisting of stems and leaves **(Figure 35.2)**. Vascular plants, with few exceptions, rely on both systems for survival. Roots are almost never photosynthetic; they starve unless *photosynthates*, the sugars and the other carbohydrates produced during photosynthesis, are imported from the shoot system. Conversely, the shoot system depends on the water and minerals that roots absorb from the soil.

Roots

A **root** is an organ that anchors a vascular plant in the soil, absorbs minerals and water, and often stores carbohydrates and other reserves. The *primary root*, originating in the seed

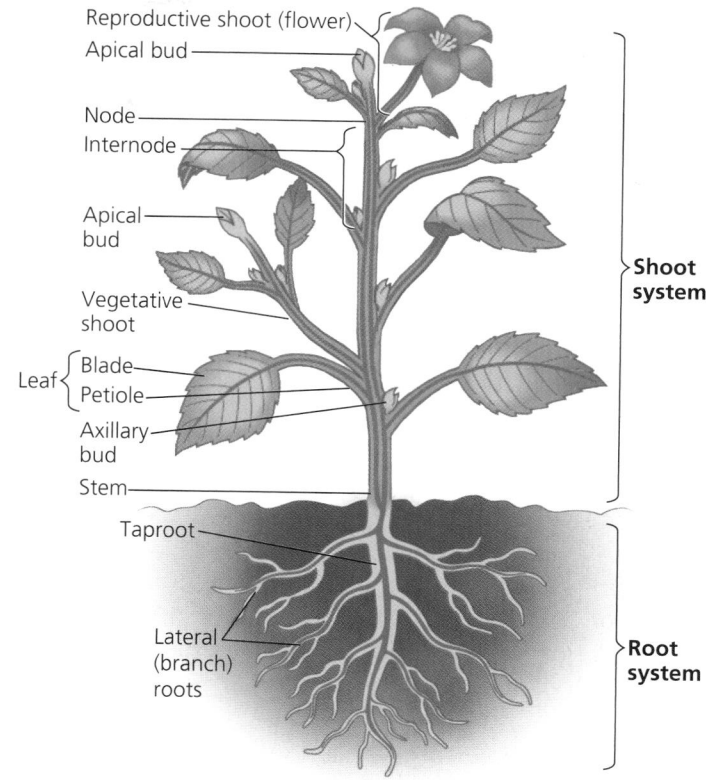

▼ **Figure 35.2 An overview of a flowering plant.** The plant body is divided into a root system and a shoot system, connected by vascular tissue (purple strands in this diagram) that is continuous throughout the plant. The plant shown is an idealized eudicot.

Reproductive shoot (flower)
Apical bud
Node
Internode
Apical bud
Vegetative shoot
Leaf { Blade / Petiole
Axillary bud
Stem
Taproot
Lateral (branch) roots
Shoot system
Root system

embryo, is the first root (and the first organ) to emerge from a germinating seed. It soon branches to form **lateral roots** (see Figure 35.2) that can also branch, greatly enhancing the ability of the root system to anchor the plant and to acquire resources such as water and minerals from the soil.

Tall, erect plants with large shoot masses generally have a *taproot system*, consisting of one main vertical root, the **taproot**, which usually develops from the primary root. In taproot systems, the role of absorption is restricted largely to the tips of lateral roots. A taproot, although energetically expensive to make, facilitates the anchorage of the plant in the soil. By preventing toppling, the taproot enables the plant to grow taller, thereby giving it access to more favorable light conditions and, in some cases, providing an advantage for pollen and seed dispersal. Taproots can also be specialized for food storage.

Small vascular plants or those that have a trailing growth habit are particularly susceptible to grazing animals that can uproot the plant and kill it. Such plants are most efficiently anchored by a *fibrous root system*, a thick mat of slender roots spreading out below the soil surface (see Figure 30.16). In plants that have fibrous root systems, including most monocots, the primary root dies early on and does not

▶ **Figure 35.3 Root hairs of a radish seedling.** Root hairs grow by the thousands just behind the tip of each root. By increasing the root's surface area, they greatly enhance the absorption of water and minerals from the soil.

⊙ **Mastering Biology** Video: Root Growth in a Radish Seedling

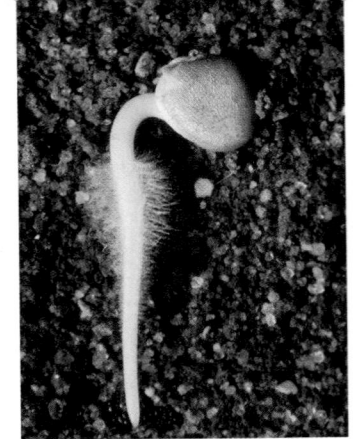

form a taproot. Instead, many small roots emerge from the stem. Such roots are said to be *adventitious*, a term describing a plant organ that grows in an unusual location, such as roots arising from stems or leaves. Each root forms its own lateral roots, which in turn form their own lateral roots. Because this mat of roots holds the topsoil in place, plants such as grasses that have dense fibrous root systems are especially good at preventing soil erosion.

In most plants, the absorption of water and minerals occurs primarily near the tips of elongating roots, where vast numbers of **root hairs**, thin, finger-like extensions of root epidermal cells, emerge and increase the surface area of the root enormously **(Figure 35.3)**. Most root systems also form *mycorrhizal associations*, symbiotic interactions with soil fungi that increase a plant's ability to absorb minerals (see Figure 37.14). The roots of many plants are adapted for specialized functions **(Figure 35.4)**.

▼ **Figure 35.4 Evolutionary adaptations of roots.**

▲ **Prop roots.** The aerial, adventitious roots of maize (corn) are prop roots, so named because they support tall, top-heavy plants. All roots of a mature maize plant are adventitious whether they emerge above or below ground.

▲ **Storage roots.** Many plants, such as the common beet, store food and water in their roots.

▲ **Pneumatophores.** Also known as air roots, pneumatophores are produced by trees such as mangroves that inhabit tidal swamps. By projecting above the water's surface at low tide, they enable the root system to obtain oxygen, which is lacking in the thick, waterlogged mud.

◀ **Buttress roots.** Because of moist conditions in the tropics, root systems of many of the tallest trees are surprisingly shallow. Aerial roots that look like buttresses, such as seen in *Gyranthera caribensis* in Venezuela, give architectural support to the trunks of trees.

▶ **"Strangling" aerial roots.** Strangler fig seeds germinate in the crevices of tall trees. Aerial roots grow to the ground, wrapping around the host tree and objects such as this Cambodian temple. Shoots grow upward and shade out the host tree, killing it.

Stems

A **stem** is a plant organ bearing leaves and buds. Its chief function is to elongate and orient the shoot in a way that maximizes photosynthesis by the leaves. Another function of stems is to elevate reproductive structures, thereby facilitating the dispersal of pollen and fruit. Green stems may also perform a limited amount of photosynthesis. Each stem consists of an alternating system of **nodes**, the points at which leaves are attached, and **internodes**, the stem segments between nodes (see Figure 35.2). Most of the growth of a young shoot is concentrated near the growing shoot tip, or **apical bud**. Apical buds are not the only types of buds found in shoots. In the upper angle (axil) formed by each leaf and the stem is an **axillary bud**, which can potentially form a lateral branch or, in some cases, a thorn or flower.

Some plants have stems with alternative functions, such as food storage or asexual reproduction. Many of these modified stems, including rhizomes, stolons, and tubers, are often mistaken for roots **(Figure 35.5)**.

▼ **Figure 35.5 Evolutionary adaptations of stems.**

◄ **Rhizomes.** The base of this iris plant is an example of a rhizome, a horizontal shoot that grows just below the surface. Vertical shoots emerge from axillary buds on the rhizome.

► **Stolons.** Shown here on a strawberry plant, stolons are horizontal shoots that grow along the surface. These "runners" enable a plant to reproduce asexually, as plantlets grow from axillary buds along each runner.

◄ **Tubers.** Tubers, such as these potatoes, are enlarged ends of rhizomes or stolons specialized for storing food. The "eyes" of a potato are clusters of axillary buds.

❓ *Which of these three examples has nodes?*

Leaves

In most vascular plants, the **leaf** is the main photosynthetic organ. In addition to intercepting light, leaves exchange gases with the atmosphere, dissipate heat, and defend themselves from herbivores and pathogens. These functions may have conflicting anatomical and physiological requirements. For example, a dense covering of hairs may help repel herbivorous insects but may also trap air near the leaf surface, thereby reducing gas exchange and, consequently, photosynthesis. Because of these conflicting demands and trade-offs, leaves vary extensively in form. In general, however, a leaf consists of a flattened **blade** and a stalk, the **petiole**, which joins the leaf to the stem at a node (see Figure 35.2). Grasses and many other monocots lack petioles; instead, the base of the leaf forms a sheath that envelops the stem.

Monocots and eudicots differ in the arrangement of **veins**, the vascular tissue of leaves. Most monocots have parallel major veins of equal diameter that run the length of the blade. Eudicots generally have a branched network of veins arising from a major vein (the *midrib*) that runs down the center of the blade (see Figure 30.16).

In identifying angiosperms according to structure, taxonomists rely mainly on floral morphology, but they also use variations in leaf morphology, such as leaf shape, the branching pattern of veins, and the spatial arrangement of leaves. **Figure 35.6** illustrates a difference in leaf shape: simple versus compound. Unlike leaves, the leaflets of compound leaves are not associated with axillary buds. Compound leaves may help confine invading pathogens to a single leaflet, rather than allowing them to spread to the entire leaf.

▼ **Figure 35.6 Simple versus compound leaves.**

Simple leaf

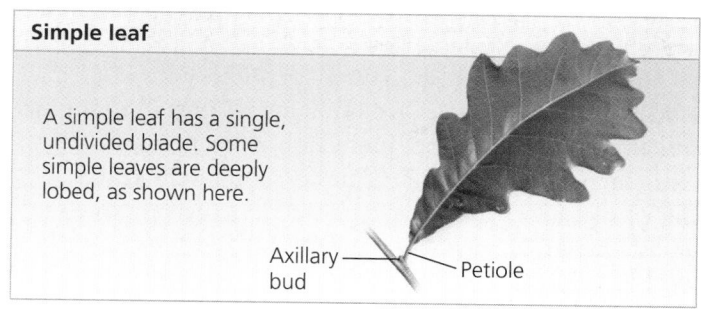

A simple leaf has a single, undivided blade. Some simple leaves are deeply lobed, as shown here.

Axillary bud — Petiole

Compound leaf

In a compound leaf, the blade consists of multiple leaflets. A leaflet has no axillary bud at its base. In some plants, each leaflet is further divided into smaller leaflets.

Leaflet

Axillary bud — Petiole

▼ Figure 35.7 Evolutionary adaptations of leaves.

▶ **Tendrils.** The tendrils by which this pea plant clings to a support are modified leaves. After it has "lassoed" a support, a tendril forms a coil that brings the plant closer to the support. Tendrils are typically modified leaves, but some tendrils are modified stems, as in grapevines.

◀ **Spines.** The spines of cacti, such as this prickly pear, are actually leaves; photosynthesis is carried out by the fleshy green stems.

◀ **Storage leaves.** Bulbs, such as this cut onion, have a short underground stem and modified leaves that store food.

Plantlet

Storage leaves
Stem

◀ **Reproductive leaves.** The leaves of some succulents, such as *Kalanchoë daigremontiana*, produce adventitious plantlets, which fall off the leaf and take root in the soil.

The shapes of leaves are often products of genetic programs that are tweaked by environmental influences. Interpret the data in the **Scientific Skills Exercise** to explore the roles of genetics and the environment in determining leaf morphology in red maple trees.

Almost all leaves are specialized for photosynthesis. However, in some species evolution has resulted in additional functions, such as support, protection, storage, or asexual reproduction **(Figure 35.7).** Some are sporophylls, leaves highly specialized for sexual reproduction, such as carpels and stamens in flowers (see Figure 30.12).

Dermal, Vascular, and Ground Tissues

All three basic vascular plant organs—roots, stems, and leaves—are composed of three fundamental tissue types: dermal, vascular, and ground tissues. Each of these general types forms a **tissue system** that is continuous throughout the plant, connecting all the organs. However, specific characteristics of the tissues and the spatial relationships of tissues to one another vary in different organs **(Figure 35.8).**

Dermal tissue serves as the outer protective covering of the plant. Like our skin, it forms the first line of defense against physical damage and pathogens. In nonwoody plants, it is usually

Using Bar Graphs to Interpret Data

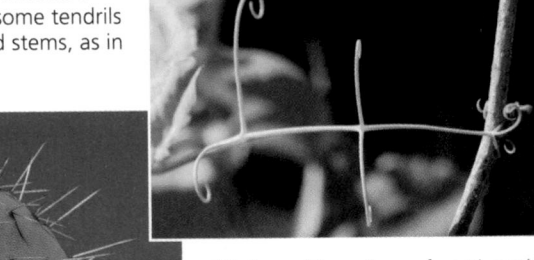

Nature Versus Nurture: Why Are Leaves from Northern Red Maples "Toothier" Than Leaves from Southern Red Maples? Not all leaves of the red maple (*Acer rubrum*) are the same. The "teeth" along the margins of leaves growing in northern locations differ in size and number from those of their southern counterparts. (The leaf seen here has an intermediate appearance.) Are these differences due to genetic differences between northern and southern *Acer rubrum* populations, or do they arise from environmental differences between northern and southern locations, such as average temperature, that affect gene expression?

How the Experiment Was Done Seeds of *Acer rubrum* were collected from four latitudinally distinct sites: Ontario (Canada), Pennsylvania, South Carolina, and Florida. The seeds from the four sites were then grown in a northern location (Rhode Island) and a southern location (Florida). After a few years of growth, leaves were harvested from the four sets of plants growing in the two locations. The average area of single teeth and the average number of teeth per leaf area were determined.

Data from the Experiment

Seed Collection Site	Average Area of a Single Tooth (cm²)		Number of Teeth per cm² of Leaf Area	
	Grown in Rhode Island	Grown in Florida	Grown in Rhode Island	Grown in Florida
Ontario (43.32°N)	0.017	0.017	3.9	3.2
Pennsylvania (42.12°N)	0.020	0.014	3.0	3.5
South Carolina (33.45°N)	0.024	0.028	2.3	1.9
Florida (30.65°N)	0.027	0.047	2.1	0.9

Data from D. L. Royer et al., Phenotypic plasticity of leaf shape along a temperature gradient in *Acer rubrum*, *PLoS ONE* 4(10):e7653 (2009).

INTERPRET THE DATA

1. Make a bar graph for tooth size and a bar graph for number of teeth. (For information on bar graphs, see the Scientific Skills Review in Appendix D.) From north to south, what is the general trend in tooth size and number of teeth in leaves of *Acer rubrum*?

2. Based on the data, would you conclude that leaf tooth traits in the red maple are largely determined by genetic heritage (genotype), by the capacity for responding to environmental change within a single genotype (phenotypic plasticity), or by both? Make specific reference to the data in answering the question.

3. The "toothiness" of leaf fossils of known age has been used by paleoclimatologists to estimate past temperatures in a region. If a 10,000-year-old fossilized red maple leaf from South Carolina had an average of 4.2 teeth per square centimeter of leaf area, what could you infer about the temperature of South Carolina 10,000 years ago compared with the temperature today? Explain your reasoning.

➲ **Instructors:** A version of this Scientific Skills Exercise can be assigned in **Mastering Biology**.

▼ **Figure 35.8 The three tissue systems.** The dermal tissue system (blue) provides a protective cover for the entire body of a plant. The vascular tissue system (purple), which transports materials between the root and shoot systems, is also continuous throughout the plant but is arranged differently in each organ. The ground tissue system (yellow), which is responsible for most of the metabolic functions, is located between the dermal tissue and the vascular tissue in each organ.

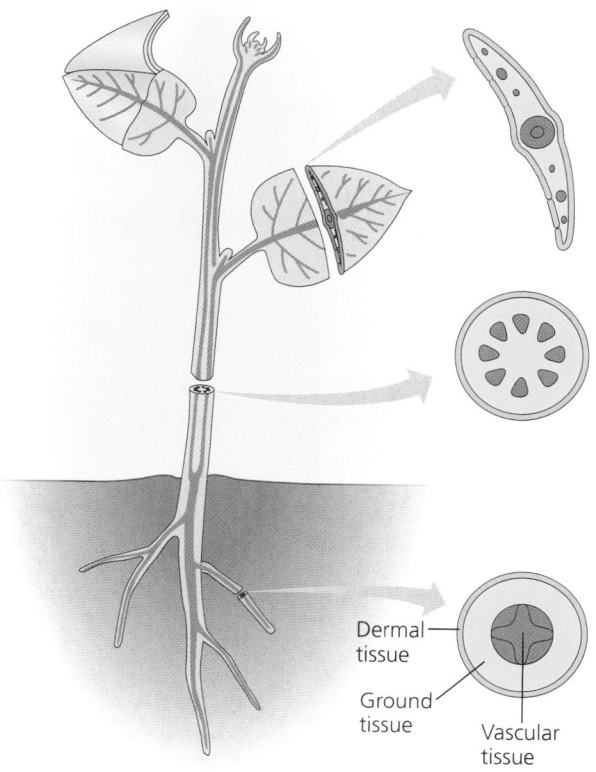

Dermal tissue

Ground tissue

Vascular tissue

a single tissue called the **epidermis**, a layer of tightly packed cells. In leaves and most stems, the **cuticle**, a waxy epidermal coating, helps prevent water loss. In woody plants, protective tissues called **periderm** replace the epidermis in older regions of stems and roots. In addition to protecting the plant from water loss and disease, the epidermis has specialized characteristics in each organ. In roots, water and minerals absorbed from the soil enter through the epidermis, especially in root hairs. In shoots, specialized epidermal cells called **guard cells** are involved in gaseous exchange. Another class of highly specialized epidermal cells found in shoots consists of outgrowths called **trichomes**. In some desert species, hairlike trichomes reduce water loss and reflect excess light. Some trichomes defend against insects through shapes that hinder movement or glands that secrete sticky fluids or toxic compounds **(Figure 35.9)**.

The two major functions of **vascular tissue** are to facilitate the transport of materials through the plant and to provide mechanical support. Vascular tissues are of two types: xylem and phloem. **Xylem** conducts water and dissolved minerals upward from roots into the shoots. **Phloem** transports sugars, the products of photosynthesis, from where they are

▼ **Figure 35.9 Trichome diversity on the surface of a leaf.** Three types of trichomes are found on the surface of marjoram (*Origanum majorana*). Spear-like trichomes help hinder the movement of crawling insects, while the other two types of trichomes secrete oils and other chemicals involved in defense (colorized SEM).

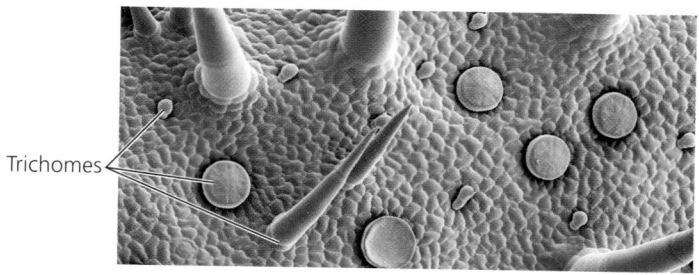

Trichomes

300 μm

made (usually the leaves) to where they are needed or stored— usually roots and sites of growth, such as developing leaves and fruits. The vascular tissue of a root or stem is collectively called the **stele** (the Greek word for "pillar"). The arrangement of the stele varies, depending on the species and organ. In angiosperms, for example, the root stele is a solid central *vascular cylinder* of xylem and phloem, whereas the stele of stems and leaves consists of *vascular bundles*, separate strands containing xylem and phloem (see Figure 35.8). Both xylem and phloem are composed of a variety of cell types, including cells that are highly specialized for transport or support.

Tissue that is neither dermal nor vascular is **ground tissue**. Ground tissue that is internal to the vascular tissue is known as **pith**, and ground tissue that is external to the vascular tissue is called **cortex**. Ground tissue is not just filler: It includes cells specialized for functions such as storage, photosynthesis, support, and short-distance transport.

Common Types of Plant Cells

In a plant, as in any multicellular organism, cells undergo cell *differentiation*; that is, they become specialized in structure and function during the course of development. Cell differentiation may involve changes both in the cytoplasm and its organelles and in the cell wall. **Figure 35.10**, on the next two pages, focuses on the major types of plant cells. Notice the structural adaptations that make specific functions possible. You may also wish to review basic plant cell structure (see Figures 6.8 and 6.27).

→ **Mastering Biology** BioFlix® Animation: **Tour of a Plant Cell**

CONCEPT CHECK 35.1

1. How does the vascular tissue system enable leaves and roots to function together in supporting growth and development of the whole plant?

2. **WHAT IF?** If humans were photoautotrophs, making food by capturing light energy for photosynthesis, how might our anatomy be different?

3. **MAKE CONNECTIONS** Explain how central vacuoles and cellulose cell walls contribute to plant growth (see Concepts 6.4 and 6.7).

For suggested answers, see Appendix A.

Exploring Examples of Differentiated Plant Cells

Parenchyma Cells

Mature **parenchyma cells** have primary walls that are relatively thin and flexible, and most lack secondary walls. (See Figure 6.27 to review primary and secondary cell walls.) When mature, parenchyma cells generally have a large central vacuole. Parenchyma cells perform most of the metabolic functions of the plant, synthesizing and storing various organic products. For example, photosynthesis occurs within the chloroplasts of parenchyma cells in the leaf. Some parenchyma cells in stems and roots have colorless plastids called amyloplasts that store starch. The fleshy tissue of many fruits is composed mainly of parenchyma cells. Most parenchyma cells retain the ability to divide and differentiate into other types of plant cells under particular conditions—during wound repair, for example. It is even possible to grow an entire plant from a single parenchyma cell.

Parenchyma cells in a privet (*Ligustrum*) leaf (LM)   25 μm

Collenchyma Cells

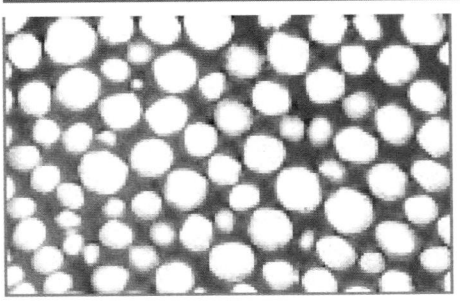

Grouped in strands, **collenchyma cells** (seen here in cross section) help support young parts of the plant shoot. Collenchyma cells are generally elongated cells that have thicker primary walls than parenchyma cells, though the walls are unevenly thickened. Young stems and petioles often have strands of collenchyma cells just below their epidermis. Collenchyma cells provide flexible support without restraining growth. At maturity, these cells are living and flexible, elongating with the stems and leaves they support—unlike sclerenchyma cells, which we discuss next.

Collenchyma cells in a common nettle (*Urtica dioica*) stem (LM)   5 μm

Sclerenchyma Cells

5 μm

Sclereid cells (in pear) (LM)

25 μm

Cell wall

Fiber cells (cross section from ash tree) (LM)

Sclerenchyma cells also function as supporting elements in the plant but are much more rigid than collenchyma cells. In sclerenchyma cells, the secondary cell wall, produced after cell elongation has ceased, is thick and contains large amounts of **lignin**, a relatively indigestible strengthening polymer that accounts for more than a quarter of the dry mass of wood. Lignin is present in all vascular plants but not in bryophytes. Mature sclerenchyma cells cannot elongate, and they occur in regions of the plant that have stopped growing in length. Sclerenchyma cells are so specialized for support that many are dead at functional maturity, but they produce secondary walls before the protoplast (the living part of the cell) dies. The rigid walls remain as a "skeleton" that supports the plant, in some cases for hundreds of years.

Two types of sclerenchyma cells, known as *sclereids* and *fibers*, are specialized entirely for support and strengthening. Sclereids, which are boxier than fibers and irregular in shape, have very thick, lignified secondary walls. Sclereids impart the hardness to nutshells and seed coats and the gritty texture to pear fruits. Fibers, which are usually grouped in strands, are long, slender, and tapered. Some are used commercially, such as hemp fibers for making rope and flax fibers for weaving into linen.

Water-Conducting Cells of the Xylem

The two types of water-conducting cells, **tracheids** and **vessel elements**, are tubular, elongated cells that are dead and lignified at functional maturity. Tracheids occur in the xylem of all vascular plants. In addition to tracheids, most angiosperms, as well as a few gymnosperms and a few seedless vascular plants, have vessel elements. When the living cellular contents of a tracheid or vessel element disintegrate, the cell's thickened walls remain behind, forming a nonliving conduit through which water can flow. The secondary walls of tracheids and vessel elements are often interrupted by pits, thinner regions where only primary walls are present (see Figure 6.27 to review primary and secondary walls). Water can migrate laterally between neighboring cells through pits.

Tracheids are long, thin cells with tapered ends. Water moves from cell to cell mainly through the pits, where it does not have to cross thick secondary walls.

Vessel elements are generally wider, shorter, thinner walled, and less tapered than the tracheids. They are aligned end to end, forming long pipes known as **vessels** that in some cases are visible with the naked eye. The end walls of vessel elements have perforation plates that enable water to flow freely through the vessels.

The secondary walls of tracheids and vessel elements are hardened with lignin. This hardening provides support and prevents collapse under the tension of water transport.

Tracheids and vessels (colorized SEM)

Vessel elements, with perforated end walls

Tracheids

Sugar-Conducting Cells of the Phloem

Unlike the water-conducting cells of the xylem, the sugar-conducting cells of the phloem are alive at functional maturity. In seedless vascular plants and gymnosperms, sugars and other organic nutrients are transported through long, narrow cells called sieve cells. In the phloem of angiosperms, these nutrients are transported through sieve tubes, which consist of chains of cells that are called **sieve-tube elements**, or sieve-tube members.

Though alive, sieve-tube elements lack a nucleus, ribosomes, a distinct vacuole, and cytoskeletal elements. This reduction in cell contents enables nutrients to pass more easily through the cell. The end walls between sieve-tube elements, called **sieve plates**, have pores that facilitate the flow of fluid from cell to cell along the sieve tube. Alongside each sieve-tube element is a nonconducting cell called a **companion cell**, which is connected to the sieve-tube element by numerous plasmodesmata (see Figure 6.27). The nucleus and ribosomes of the companion cell serve not only that cell itself but also the adjacent sieve-tube element. In some plants, the companion cells in leaves also help load sugars into the sieve-tube elements, which then transport the sugars to other parts of the plant.

Sieve-tube element (left) and companion cell: cross section (TEM)

Sieve-tube elements: longitudinal view

Sieve-tube elements: longitudinal view (LM)

Sieve plate with pores (LM)

Different meristems generate new cells for primary and secondary growth

A major difference between plants and most animals is that plant growth is not limited to an embryonic or juvenile period. Instead, growth occurs throughout the plant's life, a process called **indeterminate growth**. Plants can keep growing because they have undifferentiated tissues called **meristems** containing cells that can divide, leading to new cells that elongate and become differentiated **(Figure 35.11)**. Except for dormant periods, most plants grow continuously. In contrast, most animals and some plant organs—such as leaves, thorns, and flowers—undergo **determinate growth**; they stop growing after reaching a certain size.

There are two main types of meristems: apical meristems and lateral meristems. **Apical meristems**, located at root and shoot tips, provide cells that enable **primary growth**, growth in length. Primary growth allows roots to extend throughout the soil and shoots to increase exposure to light. In herbaceous (non-woody) plants, it produces all, or almost all, of the plant body. Woody plants, however, also grow in circumference in the parts of stems and roots that no longer grow in length. This growth in thickness, known as **secondary growth**, is made possible by **lateral meristems**: the vascular cambium and cork cambium. These cylinders of dividing cells extend along the length of roots and stems. The **vascular cambium** adds vascular tissue called secondary xylem (wood) and secondary phloem. Most of the thickening is from secondary xylem. The **cork cambium** replaces the epidermis with the thicker, tougher periderm.

Cells in apical and lateral meristems divide frequently during the growing season, generating additional cells. Some new cells remain in the meristem and produce more cells, while others differentiate and are incorporated into tissues and organs. Cells that remain as sources of new cells have traditionally been called *initials* but are increasingly being called *stem cells* to correspond to animal stem cells that also divide and remain functionally undifferentiated.

Cells displaced from the meristem may divide several more times as they differentiate into mature cells. During primary growth, these cells give rise to three tissues called **primary meristems**—the *protoderm, ground meristem,* and *procambium*—that will produce, respectively, the three mature tissues of a root or shoot: the dermal, ground, and vascular tissues. The lateral meristems in woody plants also have stem cells, which give rise to all secondary growth.

The relationship between primary and secondary growth is seen in the winter twig of a deciduous tree. At the shoot tip is the dormant apical bud, enclosed by scales that protect its apical meristem **(Figure 35.12)**. In spring, the bud sheds its scales and begins a new spurt of primary growth, producing a series of nodes

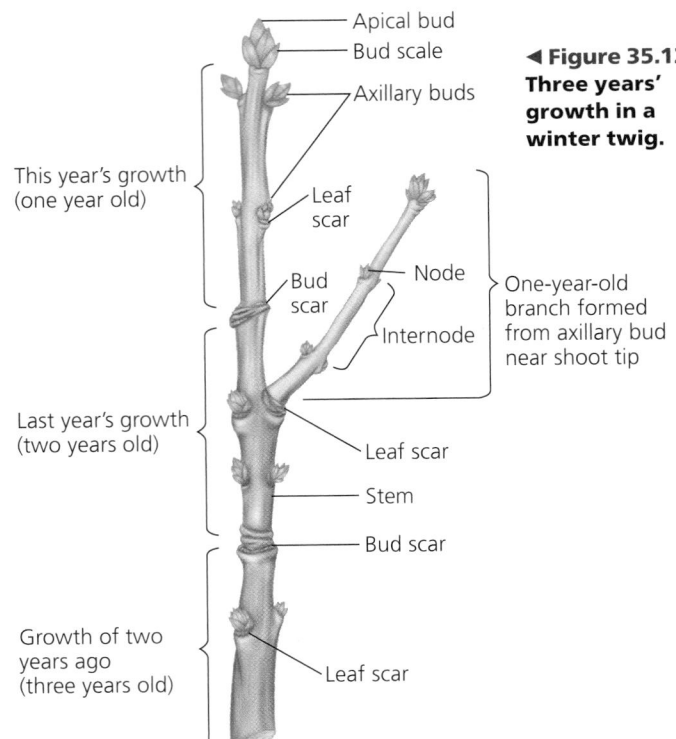

▲ **Figure 35.12 Three years' growth in a winter twig.**

Labels: Apical bud; Bud scale; Axillary buds; This year's growth (one year old); Leaf scar; Node; Bud scar; Internode; One-year-old branch formed from axillary bud near shoot tip; Last year's growth (two years old); Leaf scar; Stem; Bud scar; Growth of two years ago (three years old); Leaf scar

and internodes. On each growth segment, nodes are marked by scars left when leaves fell. Leaf scars are prominent in many twigs. Above each scar is an axillary bud or a branch formed by an axillary bud. Farther down are bud scars from whorls of scales that enclosed the apical bud during the previous winter. In each growing season, primary growth extends shoots, and secondary growth increases the diameter of parts formed in previous years.

Although meristems enable plants to grow throughout their lives, plants do die, of course. Based on the length of their life cycle, flowering plants can be categorized as annuals, biennials, or perennials. *Annuals* complete their life cycle—from germination to flowering to seed production to death—in a single year or less. Many wildflowers are annuals, as are most staple food crops, including legumes and cereal grains such as wheat and rice. Dying after producing seeds and fruits enables plants to transfer the maximum amount of energy to reproduction. *Biennials*, such as turnips, generally require two growing seasons to complete their life cycle, flowering and fruiting only in their second year. *Perennials* live many years and include trees, shrubs, and some grasses. Some buffalo grass of the North American plains is thought to have been growing for 10,000 years from seeds that sprouted at the close of the last ice age.

CONCEPT CHECK 35.2

1. Would primary and secondary growth ever occur simultaneously in the same plant?
2. Roots and stems grow indeterminately, but leaves do not. How might this benefit the plant?
3. **WHAT IF?** After growing carrots for one season, a gardener decides that the carrots are too small. Since carrots are biennials, the gardener leaves the crop in the ground for a second year, thinking the carrot roots will grow larger. Is this a good idea? Explain.

For suggested answers, see Appendix A.

VISUALIZING PRIMARY AND SECONDARY GROWTH

All vascular plants have primary growth: growth in length. Woody plants also have secondary growth: growth in thickness. As you study the diagrams, visualize how shoots and roots grow longer and thicker.

→ **Mastering Biology Animation:**
Primary and Secondary Growth

Primary Growth (growth in length)

Primary growth is made possible by apical meristems at the tips of shoots and roots.

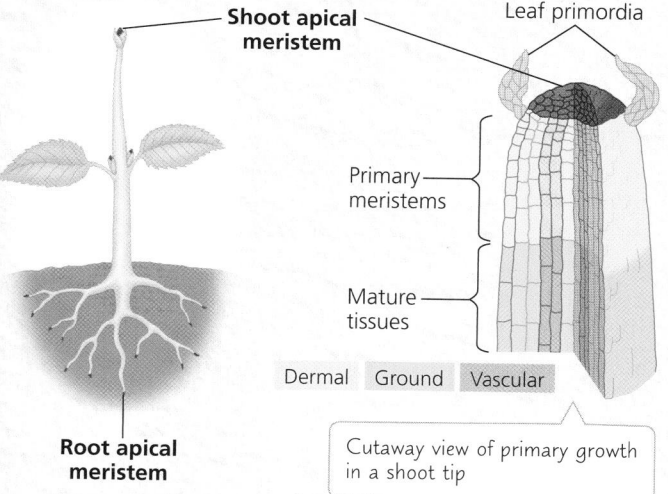

Shoot apical meristem

Leaf primordia

Primary meristems

Mature tissues

Root apical meristem

Dermal   Ground   Vascular

Cutaway view of primary growth in a shoot tip

Apical meristem cells are undifferentiated. When they divide, some daughter cells remain in the apical meristem, ensuring a continuing population of undifferentiated cells. Other daughter cells become partly differentiated as primary meristem cells. After dividing and growing in length, they become fully differentiated cells in the mature tissues.

Cell division in apical meristem

Daughter cell in primary meristem

Cell division in primary meristem

Growing cells in primary meristem

Differentiated cells (for example, vessel elements)

Time   Growth

Youngest differentiated cells

Older differentiated cells

The addition of elongated, differentiated cells lengthens a stem (as shown here) or root.

1. *A thimble-like root cap protects each root apical meristem. Draw and label a simple outline of a root divided into four sections: root cap (bottom), root apical meristem, primary meristems, and mature tissues.*

Secondary Growth (growth in thickness)

Secondary growth is made possible by two lateral meristems extending along the length of a shoot or root where primary growth has ceased.

The **lateral meristems**, called the vascular cambium and cork cambium, are cylinders of dividing cells that are one cell thick.

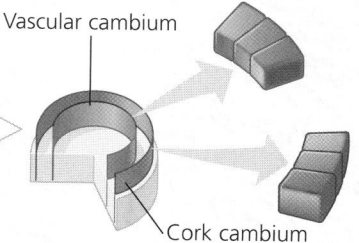

Vascular cambium

Cork cambium

Increased circumference: When a cambium cell divides, sometimes both daughter cells remain in the cambium and grow, increasing the cambium circumference.

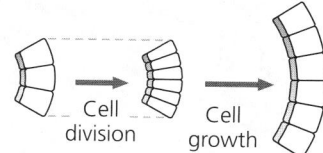

Cell division   Cell growth

Addition of secondary xylem and phloem cells: When a vascular cambium cell divides, sometimes one daughter cell becomes a secondary xylem cell (X) to the inside of the cambium or a secondary phloem cell (P) to the outside. Although xylem and phloem cells are shown being added equally here, usually many more xylem cells are produced.

Direction of secondary growth

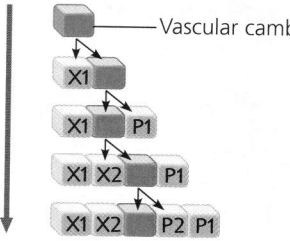

Vascular cambium cell

Time

X1

X1   P1

X1 X2   P1

X1 X2   P2 P1

Addition of cork cells: When a cork cambium cell divides, sometimes one daughter cell becomes a cork cell (C) to the outside of the cambium.

Direction of secondary growth

Cork cambium cell

Time

C1

C2 C1

Lateral meristems

Completed primary growth

Vascular cambium cell

Cork cambium cell

When the vascular cambium and cork cambium become active in a stem (or root), primary growth has ceased in that area.

Direction of secondary growth

Youngest xylem cell   Youngest phloem cell   Cork cells

A stem (or root) thickens as secondary xylem, secondary phloem, and cork cells are added. Most of the cells are secondary xylem (wood).

Oldest xylem cell   Oldest phloem cell

2. *Draw the row of cells from the boxed area below and label the vascular cambium cell (V), 5 xylem cells from oldest (X1) to youngest (X5), and 3 phloem cells (P1 to P3). Show what happens after growth continues by drawing and labeling a row with twice as many xylem and phloem cells. How does the vascular cambium's location change?*

→ **Instructors:** Additional questions related to this Visualizing Figure can be assigned in **Mastering Biology**.

Primary growth lengthens roots and shoots

Primary growth arises directly from cells produced by apical meristems. In herbaceous plants, almost the entire plant is created through primary growth, whereas in woody plants only the nonwoody, more recently formed parts of the plant represent primary growth. Although both roots and shoots lengthen as a result of cells derived from apical meristems, the details of their primary growth differ in many ways.

Primary Growth of Roots

The entire biomass of a primary root is derived from the root apical meristem. The root apical meristem also makes a thimble-like **root cap**, which protects the delicate apical meristem as the root pushes through the abrasive soil. The root cap secretes a polysaccharide slime that lubricates the soil around the tip of the root. Growth occurs just behind the tip in three overlapping zones of cells at successive stages of primary growth. These are the zones of cell division, elongation, and differentiation **(Figure 35.13)**.

The *zone of cell division* includes the stem cells of the root apical meristem and their immediate products. New root cells are produced in this region, including cells of the root cap.

▼ **Figure 35.13 Primary growth of a eudicot root.** In the micrograph, mitotic cells in the apical meristem are revealed by staining for cyclin, a protein involved in cell division (LM).

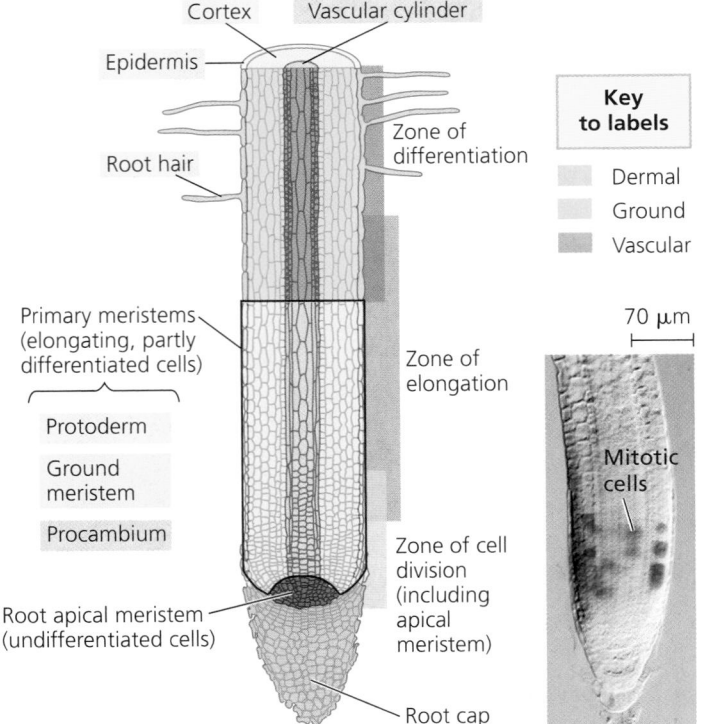

Typically, a few millimeters behind the tip of the root is the *zone of elongation*, where most of the growth occurs as root cells elongate—sometimes to more than ten times their original length. Cell elongation in this zone pushes the tip farther into the soil. Meanwhile, the root apical meristem keeps adding cells to the younger end of the zone of elongation. Even before the root cells finish lengthening, many begin specializing in structure and function. As this occurs, the three primary meristems—the protoderm, ground meristem, and procambium—become evident. In the *zone of differentiation*, or zone of maturation, cells complete their differentiation and become distinct cell types.

The protoderm, the outermost primary meristem, gives rise to the epidermis, a single layer of cuticle-free cells covering the root. Root hairs are the most prominent feature of the root epidermis. These modified epidermal cells function in the absorption of water and minerals. Root hairs typically only live a few weeks but together make up 70–90% of the total root surface area. It has been estimated that a four-month-old rye plant has about 14 billion root hairs. Laid end to end, the root hairs of a single rye plant would cover 10,000 km, one-quarter the length of the equator.

Sandwiched between the protoderm and the procambium is the ground meristem, which gives rise to mature ground tissue. The ground tissue of roots, consisting mostly of parenchyma cells, is found in the cortex, the region between the vascular tissue and epidermis. In addition to storing carbohydrates, cells in the cortex transport water and salts from the root hairs to the center of the root. The cortex also allows for *extracellular* diffusion of water, minerals, and oxygen from the root hairs inward because there are large spaces between cells. The innermost layer of the cortex is called the **endodermis**, a cylinder one cell thick that forms the boundary with the vascular cylinder. The endodermis is a selective barrier that regulates passage of substances from the soil into the vascular cylinder (see Figure 36.9).

The procambium gives rise to the vascular cylinder, which consists of a solid core of xylem and phloem tissues surrounded by a cell layer called the **pericycle**. In most eudicot roots, the xylem has a starlike appearance in cross section, and the phloem occupies the indentations between the arms of the xylem "star" **(Figure 35.14a)**. In many monocot roots, the vascular tissue consists of a core of undifferentiated parenchyma cells surrounded by a ring of alternating xylem and phloem tissues **(Figure 35.14b)**.

By increasing the length of roots, primary growth facilitates their penetration and exploration of the soil. If a resource-rich pocket is located in the soil, the branching of roots may be stimulated. Branching, too, is a form of primary growth. Lateral (branch) roots arise from meristematically active regions of the pericycle, the outermost cell layer in the vascular cylinder, which is adjacent to and just inside the endodermis (see Figure 35.14). The emerging lateral roots

▼ **Figure 35.14 Organization of primary tissues in young roots.** Parts (a) and (b) show cross sections of the roots of a *Ranunculus* (buttercup) species and *Zea mays* (maize), respectively. These represent two basic patterns of root organization, of which there are many variations, depending on the plant species (all LMs).

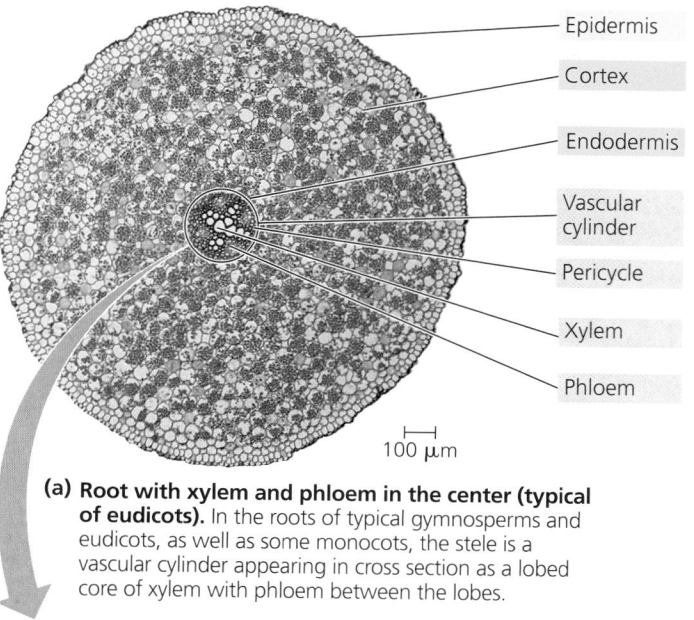

Epidermis
Cortex
Endodermis
Vascular cylinder
Pericycle
Xylem
Phloem

100 μm

(a) Root with xylem and phloem in the center (typical of eudicots). In the roots of typical gymnosperms and eudicots, as well as some monocots, the stele is a vascular cylinder appearing in cross section as a lobed core of xylem with phloem between the lobes.

Endodermis
Pericycle
Xylem
Phloem

70 μm

Key to labels

Dermal
Ground
Vascular

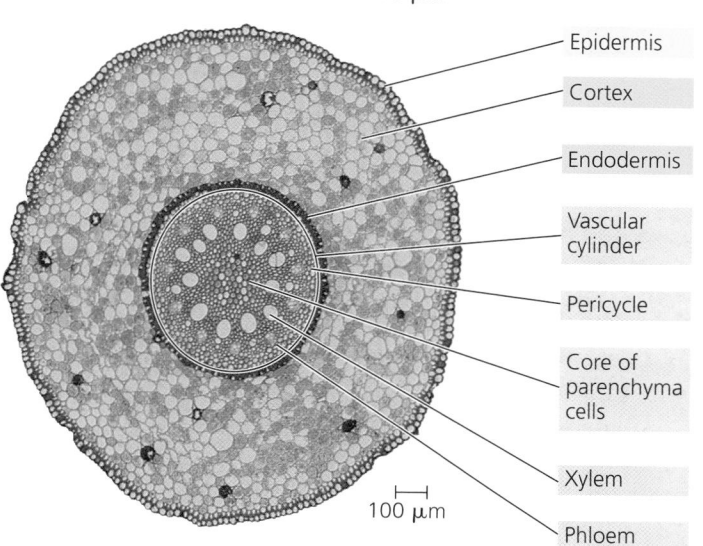

Epidermis
Cortex
Endodermis
Vascular cylinder
Pericycle
Core of parenchyma cells
Xylem
Phloem

100 μm

(b) Root with parenchyma in the center (typical of monocots). The stele of many monocot roots is a vascular cylinder with a core of parenchyma surrounded by a ring of xylem and a ring of phloem.

⊙ **Mastering Biology** Animation: Root Cross Sections

▼ **Figure 35.15 The formation of a lateral root.** A lateral root originates in the pericycle, the outermost layer of the vascular cylinder of a root, and destructively pushes through the outer tissues before emerging. In this light micrograph, the view of the original root is a cross section, but the view of the lateral root is a longitudinal section (a view along the length of the lateral root).

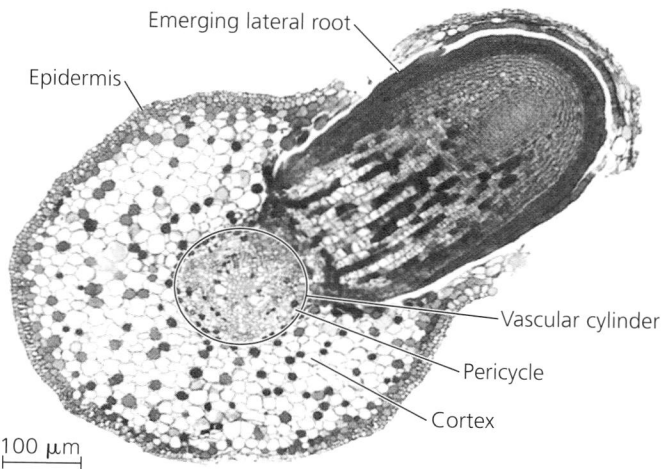

Emerging lateral root
Epidermis
Vascular cylinder
Pericycle
Cortex

100 μm

DRAW IT *Draw what the original root and lateral root would look like when viewed from the side, labeling both roots.*

disruptively push through the outer tissues until they emerge from the established root **(Figure 35.15)**.

Primary Growth of Shoots

The entire biomass of a primary shoot—all its leaves and stems—derives from its shoot apical meristem, a dome-shaped mass of dividing cells at the shoot tip **(Figure 35.16)**. The shoot apical meristem is a delicate structure protected by the leaves of the apical bud. These young leaves are spaced close together because the

▼ **Figure 35.16 The shoot tip.** Leaf primordia arise from the flanks of the dome of the apical meristem. This is a longitudinal section of the shoot tip of *Coleus* (LM).

Leaf primordia
Young leaf
Shoot apical meristem
Protoderm
Procambium
Ground meristem
Axillary bud meristems

0.25 mm

internodes are very short. Shoot elongation is due to the lengthening of internode cells below the shoot tip. As with the root apical meristem, the shoot apical meristem gives rise to three types of primary meristems in the shoot—the protoderm, ground meristem, and procambium. These three primary meristems in turn give rise to the mature primary tissues of the shoot.

The branching of shoots, which is also part of primary growth, arises from the activation of axillary buds, each of which has its own shoot apical meristem. Because of chemical communication by plant hormones, the closer an axillary bud is to an active apical bud, the more inhibited it is, a phenomenon called **apical dominance**. (The specific hormonal changes underlying apical dominance are discussed in Concept 39.2.) If an animal eats the end of the shoot or if shading results in the light being more intense on the side of the shoot, the chemical communication underlying apical dominance is disrupted. As a result, the axillary buds break dormancy and start to grow. Released from dormancy, an axillary bud eventually gives rise to a lateral shoot, complete with its own apical bud, leaves, and axillary buds. When gardeners prune shrubs and pinch back houseplants, they are reducing the number of apical buds a plant has, thereby allowing branches to develop and giving the plants a fuller, bushier appearance.

Stem Growth and Anatomy

The stem is covered by an epidermis that is usually one cell thick and covered with a waxy cuticle that prevents water loss. Some examples of specialized epidermal cells in the stem include guard cells and trichomes.

The ground tissue of stems consists mostly of parenchyma cells. However, collenchyma cells just beneath the epidermis strengthen many stems during primary growth. Sclerenchyma cells, especially fiber cells, also provide support in those parts of the stems that are no longer elongating.

Vascular tissue runs the length of a stem in vascular bundles. Unlike lateral roots, which arise from vascular tissue deep within a root and disrupt the vascular cylinder, cortex, and epidermis as they emerge (see Figure 35.15), lateral shoots develop from axillary bud meristems on the stem's surface and do not disrupt other tissues (see Figure 35.16). Near the soil surface, in the transition zone between shoot and root, the bundled vascular arrangement of the stem converges with the solid vascular cylinder of the root.

The vascular tissue of stems in most eudicot species consists of vascular bundles arranged in a ring **(Figure 35.17a)**. The xylem in each bundle faces the pith, and the phloem in each bundle faces the cortex. In most monocot stems, the vascular bundles do not form a ring but have a more scattered arrangement in the ground tissue **(Figure 35.17b)**.

Leaf Growth and Anatomy

Figure 35.18 provides an overview of leaf anatomy. Leaves develop from **leaf primordia** (singular, *primordium*), projections shaped like a cat's ear that emerge along the sides of the

▼ **Figure 35.17 Organization of primary tissues in young stems.**

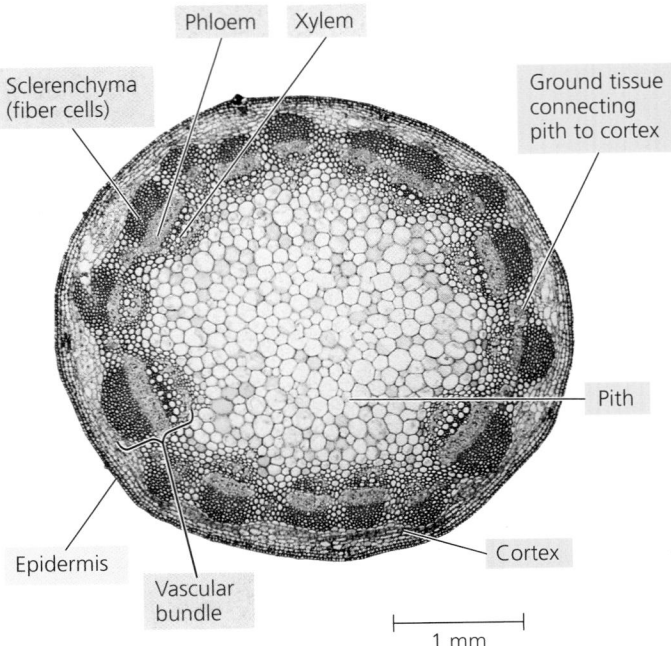

(a) Cross section of stem with vascular bundles forming a ring (typical of eudicots). Ground tissue toward the inside is called pith, and ground tissue toward the outside is called cortex (LM).

Key to labels	
	Dermal
	Ground
	Vascular

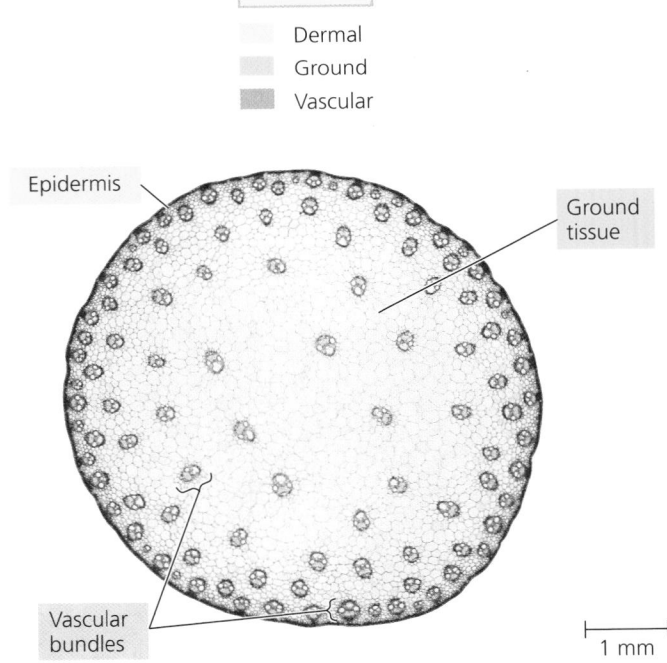

(b) Cross section of stem with scattered vascular bundles (typical of monocots). In such an arrangement, ground tissue is not partitioned into pith and cortex (LM).

VISUAL SKILLS *Compare the locations of the vascular bundles in eudicot and monocot stems. Then explain why the terms pith and cortex are not used in describing the ground tissue of monocot stems.*

➔ **Mastering Biology** Animation: Stem Cross Sections

▼ **Figure 35.18 Leaf anatomy.**

Key to labels
- Dermal
- Ground
- Vascular

Cuticle
Sclerenchyma fibers
Stoma
Bundle-sheath cell
Xylem
Phloem
Vein
Guard cells

(a) Cutaway drawing of leaf tissues

⬤ **Mastering Biology** Animation: Leaf Anatomy

Guard cells
Stomatal pore
Epidermal cell
50 μm

(b) Surface view of a spiderwort (*Tradescantia*) leaf (LM)

Upper epidermis
Palisade mesophyll
Spongy mesophyll
Lower epidermis
Cuticle

Vein   Air spaces   Guard cells
100 μm

(c) Cross section of a lilac (*Syringa*) leaf (LM)

shoot apical meristem (see Figure 35.16). Unlike roots and stems, secondary growth in leaves is minor or nonexistent. As with roots and stems, the three primary meristems give rise to the tissues of the mature organ.

The leaf epidermis is covered by a waxy cuticle that greatly reduces water loss except where it is interrupted by **stomata** (singular, *stoma*), which allow exchange of CO_2 and O_2 between the surrounding air and the photosynthetic cells inside the leaf. In addition to regulating CO_2 uptake for photosynthesis, stomata are major avenues for the evaporative loss of water. The term *stoma* can refer to the stomatal pore or to the entire stomatal complex consisting of a pore flanked by the two specialized epidermal cells known as guard cells, which regulate the opening and closing of the pore. (We will discuss stomata in detail in Concept 36.4.)

The leaf's ground tissue, called the **mesophyll** (from the Greek *mesos*, middle, and *phyll*, leaf), is sandwiched between the upper and lower epidermal layers. Mesophyll consists mainly of parenchyma cells specialized for photosynthesis. The mesophyll in many eudicot leaves has two distinct layers: palisade and spongy. *Palisade mesophyll*, located beneath the upper epidermis, consists of one or more layers of elongated, chloroplast-rich cells that are specialized for light capture. *Spongy mesophyll*, located inward from the lower epidermis, consists of irregularly shaped cells that have fewer chloroplasts. These cells form a labyrinth of air spaces through which

CO_2 and O_2 circulate to and from the palisade layer. The air spaces are particularly large in the vicinity of stomata, where CO_2 is taken up from the outside air and O_2 is released.

The vascular tissue of each leaf is continuous with the vascular tissue of the stem. Veins subdivide repeatedly and branch throughout the mesophyll. This network brings xylem and phloem into close contact with the photosynthetic tissue, which obtains water and minerals from the xylem and loads its sugars and other organic products into the phloem for transport to other parts of the plant. The vascular structure also functions as a framework that reinforces the shape of the leaf. Each vein is enclosed by a protective *bundle sheath*, a layer of cells that regulates the movement of substances between the vascular tissue and the mesophyll. Bundle-sheath cells are very prominent in leaves of species that carry out C_4 photosynthesis (see Concept 10.5).

CONCEPT CHECK 35.3

1. Contrast primary growth in roots and shoots.

2. **WHAT IF?** A fossil leaf from a region that in the geological past was intermittently very dry and very swampy has stomata only on its upper epidermis. Was the leaf from a desert plant or from a floating aquatic plant? Explain.

3. **MAKE CONNECTIONS** How are root hairs and microvilli analogous structures? (See Figure 6.8 and the discussion of analogy in Concept 26.2.)

For suggested answers, see Appendix A.

Secondary growth increases the diameter of stems and roots in woody plants

Many land plants display secondary growth, the growth in thickness produced by lateral meristems. The advent of secondary growth during plant evolution allowed the production of novel plant forms ranging from massive forest trees to woody vines. All gymnosperm species and many eudicot species undergo secondary growth, but it is unusual in monocots. It occurs in stems and roots of woody plants, but rarely in leaves.

Secondary growth consists of the tissues produced by the vascular cambium and cork cambium. The vascular cambium adds secondary xylem (wood) and secondary phloem, thereby increasing vascular flow and support for the shoots. The cork cambium produces a tough, thick covering of waxy cells that protect the stem from water loss and from invasion by insects, bacteria, and fungi.

In woody plants, primary growth and secondary growth occur simultaneously. As primary growth adds leaves and lengthens stems and roots in the younger regions of a plant, secondary growth increases the diameter of stems and roots in older regions where primary growth has ceased. The process is similar in shoots and roots. **Figure 35.19** provides an overview of growth in a woody stem.

▼ **Figure 35.19 Secondary growth of a woody stem.**

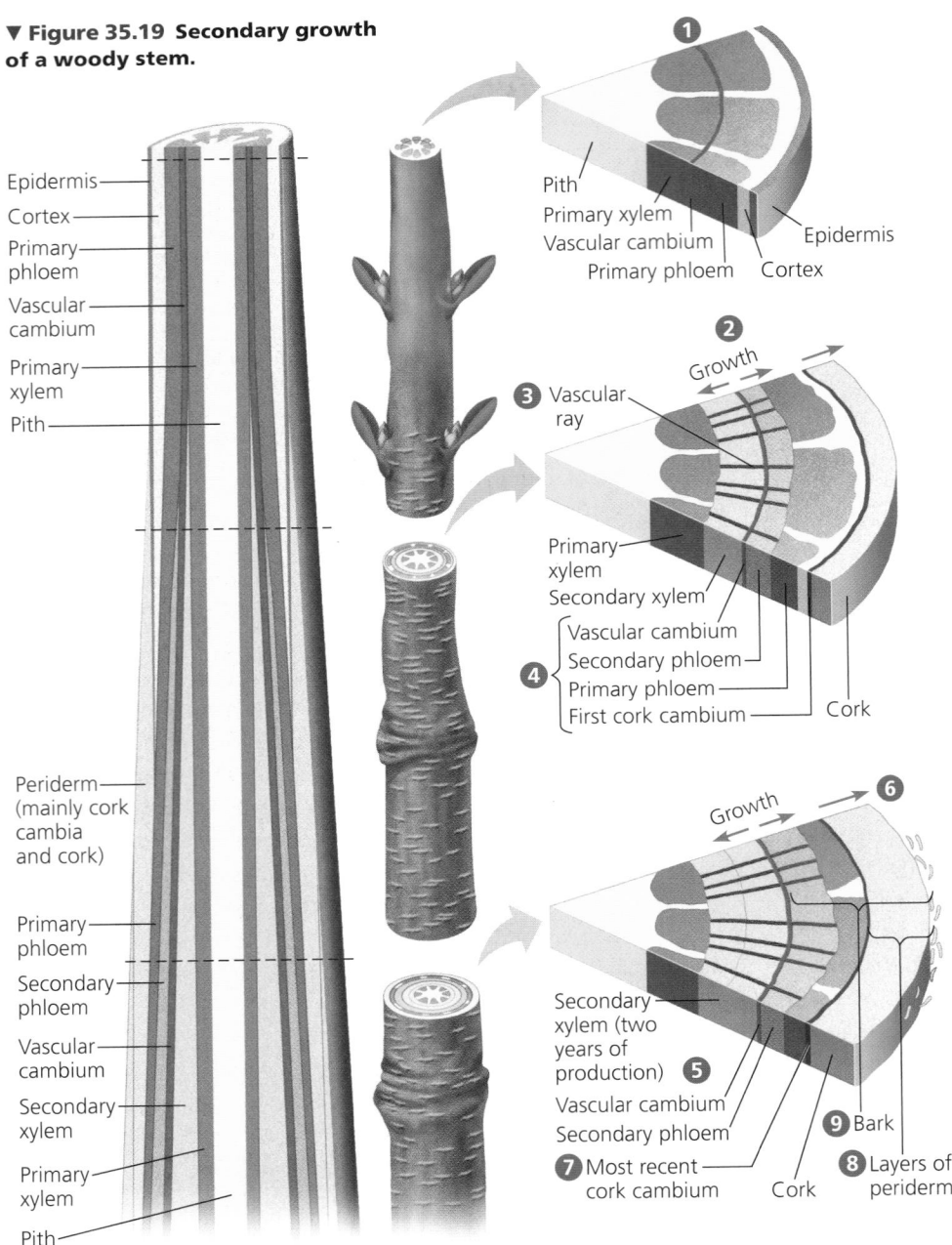

❶ Primary growth from the activity of the apical meristem is complete here. The vascular cambium has formed, and its cell divisions will give rise to the bulk of secondary growth.

❷ Although primary growth continues in the apical bud, only secondary growth occurs in this region. The stem thickens as the vascular cambium forms secondary xylem to the inside and secondary phloem to the outside.

❸ Some stem cells of the vascular cambium give rise to vascular rays.

❹ As the vascular cambium's diameter increases, the secondary phloem and other tissues external to the cambium can't keep pace because their cells no longer divide. As a result, these tissues, including the epidermis, will eventually rupture. A second lateral meristem, the cork cambium, develops from parenchyma cells in the cortex. The cork cambium produces cork cells, which replace the epidermis.

❺ In year 2 of secondary growth, the vascular cambium produces more secondary xylem and phloem. Most of the thickening is from secondary xylem. Meanwhile, the cork cambium produces more cork.

❻ As the stem's diameter increases, the outermost tissues exterior to the cork cambium rupture and are sloughed off.

❼ In many cases, the cork cambium re-forms deeper in the cortex. When none of the cortex is left, the cambium develops from phloem parenchyma cells.

❽ Each cork cambium and the tissues it produces form a layer of periderm.

❾ Bark consists of all tissues exterior to the vascular cambium.

→ **Mastering Biology Figure Walkthrough Animation: Secondary Growth**

VISUAL SKILLS *Based on the diagram, explain how the vascular cambium causes some tissues to rupture.*

The Vascular Cambium and Secondary Vascular Tissue

The vascular cambium, a cylinder of meristematic cells only one cell thick, is wholly responsible for the production of secondary vascular tissue. In a typical woody stem, the vascular cambium is located outside the pith and primary xylem and to the inside of the primary phloem and the cortex. In a typical woody root, the vascular cambium forms exterior to the primary xylem and interior to the primary phloem and pericycle.

In cross section, the vascular cambium appears as a ring of meristematic cells (see step 4 of Figure 35.19). As these cells divide, they increase the cambium's circumference and add secondary xylem to the inside and secondary phloem to the outside. Each ring is larger than the previous ring, increasing the diameter of roots and stems.

Some of the stem cells in the vascular cambium are elongated and oriented with their long axis parallel to the axis of the stem or root. The cells they produce give rise to mature cells such as the tracheids, vessel elements, and fibers of the xylem, as well as the sieve-tube elements, companion cells, axially oriented parenchyma, and fibers of the phloem. Other stem cells in the vascular cambium are shorter and are oriented perpendicular to the axis of the stem or root: they give rise to *vascular rays*—radial files of mostly parenchyma cells that connect the secondary xylem and phloem (see step 3 of Figure 35.19). These cells move water and nutrients between the secondary xylem and phloem, store carbohydrates and other reserves, and aid in wound repair.

As secondary growth continues, layers of secondary xylem (wood) accumulate, consisting mainly of tracheids and vessel elements (see Figure 35.10), as well as fibers. In most species of gymnosperms, tracheids are the only water-conducting cells. Most angiosperms also have vessel elements. The walls of secondary xylem cells are heavily lignified, giving wood its hardness and strength.

In temperate regions, wood that develops early in the spring, known as early (or spring) wood, usually has secondary xylem cells with large diameters and thin cell walls (Figure 35.20). This structure maximizes delivery of water to leaves. Wood produced later in the growing season is called late (or summer) wood. It has thick-walled cells that do not transport as much

water but provide more support. Because there is a marked contrast between the large cells of the new early wood and the smaller cells of the late wood of the previous growing season, a year's growth appears as a distinct *growth ring* in cross sections of most tree trunks and roots. Therefore, researchers can estimate a tree's age by counting growth rings. *Dendrochronology* is the science of analyzing tree growth ring patterns. Growth rings vary in thickness, depending on seasonal growth. Trees grow well in wet and warm years but may grow hardly at all in cold or dry years. Since a thick ring indicates a warm year and a thin ring indicates a cold or dry one, scientists use ring patterns to study climate changes (Figure 35.21).

As a tree or woody shrub ages, older layers of secondary xylem no longer transport water and minerals (a solution

▼ **Figure 35.20** Cross section of a three-year-old *Tilia* (linden) stem. (LM)

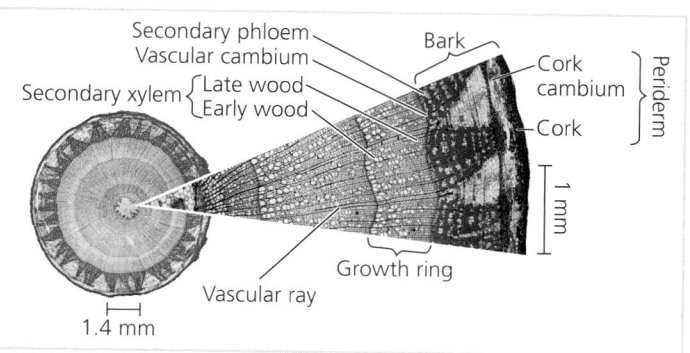

▼ **Figure 35.21** Research Method

Using Dendrochronology to Study Climate

Application Dendrochronology, the science of analyzing growth rings, is useful in studying climate change. Most scientists attribute recent global warming to the burning of fossil fuels and release of CO_2 and other greenhouse gases, whereas a small minority think it is a natural variation. Studying climate patterns requires comparing past and present temperatures, but instrumental climate records span only the last two centuries and apply only to some regions. By examining growth rings of Mongolian conifers dating back to the mid-1500s, Gordon C. Jacoby and Rosanne D'Arrigo, of the Lamont-Doherty Earth Observatory, and colleagues sought to learn whether Mongolia has experienced similar warm periods in the past.

Technique Researchers can analyze patterns of rings in living and dead trees. They can even study wood used for building long ago by matching samples with those from naturally situated specimens of overlapping age. Core samples, each about the diameter of a pencil, are taken from the bark to the center of the trunk. Each sample is dried and sanded to reveal the rings. By comparing, aligning, and averaging many samples from the conifers, the researchers compiled a chronology. The trees became a chronicle of environmental change.

Results This graph summarizes a composite record of the ring-width indexes for the Mongolian conifers from 1550 to 1993. The higher indexes indicate wider rings and higher temperatures.

Data from G. C. Jacoby et al., Mongolian tree rings and 20th-century warming, *Science* 273:771–773 (1996).

INTERPRET THE DATA *What does the graph indicate about environmental change during the period 1550–1993?*

called xylem sap). These layers are called *heartwood* because they are closer to the center of a stem or root **(Figure 35.22)**. The newest, outer layers of secondary xylem still transport xylem sap and are therefore known as *sapwood*. Sapwood allows a large tree to survive even if the center of its trunk is hollow **(Figure 35.23)**. Because each new layer of secondary xylem has a larger circumference, secondary growth enables the xylem to transport more sap each year, supplying an increasing number of leaves. Heartwood is generally darker than sapwood because of resins and other compounds that permeate the cell cavities and help protect the core of the tree from fungi and wood-boring insects.

▼ **Figure 35.22 Anatomy of a tree trunk.**

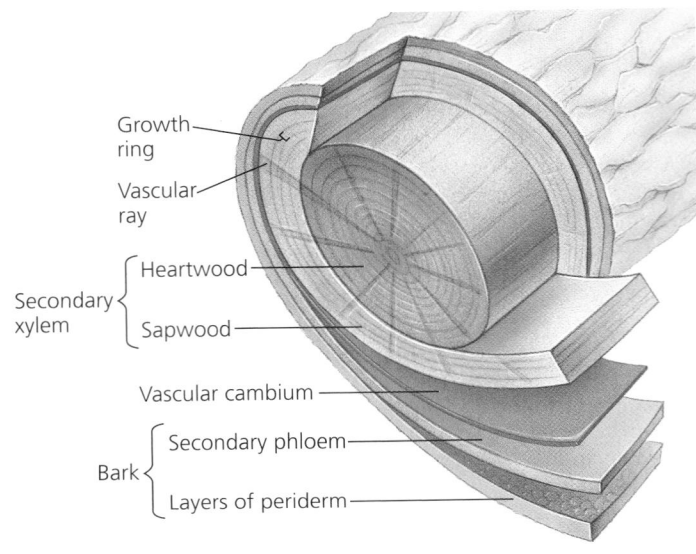

Growth ring
Vascular ray
Heartwood
Secondary xylem
Sapwood
Vascular cambium
Secondary phloem
Bark
Layers of periderm

◄ **Figure 35.23 Is this tree living or dead?** The Wawona Sequoia tunnel in Yosemite National Park in California was cut in 1881 as a tourist attraction. This giant sequoia (*Sequoiadendron giganteum*) lived for another 88 years before falling during a severe winter. It was 71.3 m tall and estimated to be 2,100 years old. Though conservation policies today would forbid the mutilation of such an important specimen, the Wawona Sequoia did teach a valuable botanical lesson: Trees can endure the excision of large portions of their heartwood for decades.

VISUAL SKILLS *Name in sequence the tissues that were destroyed as the lumberjacks excavated through the base of the tree to its center. Refer also to Figure 35.19.*

Only the youngest secondary phloem, closest to the vascular cambium, functions in sugar transport. As a stem or root increases in circumference, the older secondary phloem is sloughed off, which is one reason secondary phloem does not accumulate as extensively as secondary xylem.

The Cork Cambium and the Production of Periderm

During the early stages of secondary growth, the epidermis is pushed outward, causing it to split, dry, and fall off the stem or root. It is replaced by tissues produced by the first cork cambium, a cylinder of dividing cells that arises in the outer cortex of stems (see Figure 35.19) and in the pericycle in roots. The cork cambium gives rise to cork cells that accumulate to the outside of the cork cambium. As cork cells mature, they deposit a waxy, hydrophobic material called suberin in their walls before dying. Because cork cells have suberin and are usually compacted together, most of the periderm is impermeable to water and gases, unlike the epidermis. Cork thus functions as a barrier that helps protect the stem or root from water loss, physical damage, and pathogens. It should be noted that "cork" is commonly and incorrectly referred to as "bark." In plant biology, **bark** includes all tissues external to the vascular cambium. Its main components are the secondary phloem (produced by the vascular cambium) and, external to that, the most recent periderm and all the older layers of periderm (see Figure 35.22). As this process continues, older layers of periderm are sloughed off, as evident in the cracked, peeling exteriors of many tree trunks.

How can living cells in the interior tissues of woody organs absorb oxygen and respire if they are surrounded by a waxy periderm? Dotting the periderm are small, raised areas called **lenticels**, in which there is more space between cork cells, enabling living cells within a woody stem or root to exchange gases with the outside air. Lenticels often appear as horizontal slits, as shown on the stem in Figure 35.19.

Figure 35.24 summarizes the relationships between the primary and secondary tissues of a woody shoot.

Evolution of Secondary Growth

EVOLUTION Surprisingly, some insights into the evolution of secondary growth have been achieved by studying the herbaceous plant *Arabidopsis thaliana*. Researchers have found that they can stimulate some secondary growth in *Arabidopsis* stems by adding weights to the plant. These findings suggest that weight carried by the stem activates a developmental program leading to wood formation. Moreover, several developmental genes that regulate shoot apical meristems in *Arabidopsis* have been found to regulate vascular cambium activity in poplar (*Populus*) trees. This suggests that the processes of primary and secondary growth are evolutionarily more closely related than was previously thought.

▼ **Figure 35.24 A summary of primary and secondary growth in a woody shoot.** Woody roots have the same meristems and tissues. However, the ground tissue of a root is not divided into pith and cortex, and the cork cambium arises instead from the pericycle, the outermost layer of the vascular cylinder.

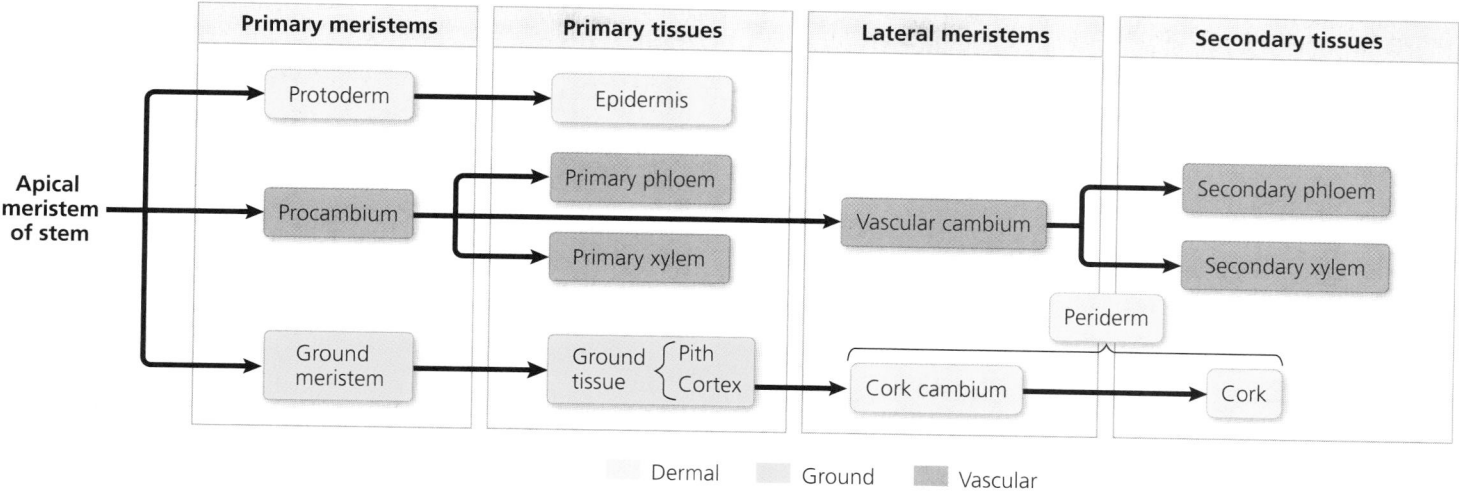

Dermal   Ground   Vascular

CONCEPT CHECK 35.4

1. A sign is hammered into a tree 2 m from the tree's base. If the tree is 10 m tall and elongates 1 m each year, how high will the sign be after ten years?

2. Stomata and lenticels are both involved in exchange of CO_2 and O_2. Why do stomata need to be able to close, but lenticels do not?

3. Would you expect a tropical tree to have distinct growth rings? Why or why not?

4. **WHAT IF?** If a complete ring of bark is removed from around a tree trunk (a technique called girdling), would the tree die slowly (in weeks) or quickly (in days)? Explain why.

For suggested answers, see Appendix A.

CONCEPT 35.5

Growth, morphogenesis, and cell differentiation produce the plant body

The specific series of changes by which cells form tissues, organs, and organisms is called **development**. Development unfolds according to the genetic information that an organism inherits from its parents but is also influenced by the external environment. A single genotype can produce different phenotypes in different environments. For example, the aquatic plant *Cabomba aquatica* forms two very different types of leaves, depending on whether the shoot apical meristem is submerged **(Figure 35.25)**. This ability to alter form in response to local environmental conditions is called *developmental plasticity*. Dramatic examples of plasticity, as in *Cabomba*, are much more common in plants than in animals and may help compensate for plants' inability to escape adverse conditions by moving.

▼ **Figure 35.25 Developmental plasticity in the aquatic plant *Cabomba aquatica*.** The underwater leaves of *Cabomba* are feathery, an adaptation that protects them from damage by lessening their resistance to moving water. In contrast, the surface leaves are pads that aid in flotation. The two leaf types have genetically identical cells, but their different environments result in the turning on or off of different genes during leaf development.

Surface leaf

Underwater leaves

The three overlapping processes involved in the development of a multicellular organism are growth, morphogenesis, and cell differentiation. *Growth* is an irreversible increase in size. *Morphogenesis* (from the Greek *morphê*, shape, and *genesis*, creation) is the process that gives a tissue, organ, or organism its shape and determines the positions of cell types. *Cell differentiation* is the process by which cells with the same genes become different from one another. We'll examine these three processes in turn, but first we'll discuss how applying techniques of modern molecular biology to model organisms, particularly *Arabidopsis thaliana*, has revolutionized the study of plant development.

Model Organisms: Revolutionizing the Study of Plants

As in other branches of biology, techniques of molecular biology and a focus on model organisms such as *Arabidopsis thaliana* have catalyzed a research explosion in the last few decades. *Arabidopsis*, a tiny weed in the mustard family, has no inherent agricultural value but is a favored model organism of plant geneticists and molecular biologists for many reasons. It is so small that thousands of plants can be cultivated in a few square meters of lab space. It also has a short generation time, taking about six weeks for a seed to grow into a mature plant that produces more seeds. This rapid maturation enables biologists to conduct genetic cross experiments in a relatively short time. One plant can produce over 5,000 seeds, another property that makes *Arabidopsis* useful for genetic analysis.

Beyond these basic traits, the plant's genome makes it particularly well suited for analysis by molecular genetic methods. The *Arabidopsis* genome, which includes about 27,000 protein-encoding genes, is among the smallest known in plants. Furthermore, the plant has only five pairs of chromosomes, making it easier for geneticists to locate specific genes. Because *Arabidopsis* has such a small genome, it was the first plant to have its entire genome sequenced.

The natural range of *Arabidopsis* includes varied climates and elevations, from the high mountains of Central Asia to the European Atlantic coast, and from North Africa to the Arctic Circle. These local varieties can differ markedly in outward appearance **(Figure 35.26)**. Genome-sequencing efforts are being expanded to include hundreds of populations of Arabidopsis from throughout its natural range in Eurasia. Contained in the genomes of these populations is information about evolutionary adaptations that enabled Arabidopsis to expand its range into new environments following the retreat of the last ice age. This information may provide plant breeders with new insights and strategies for crop improvement.

Another property that makes *Arabidopsis* attractive to molecular biologists is that its cells can be easily transformed with *transgenes*, genes from a different organism that are stably introduced into the genome of another. CRISPR technology (see Figure 20.14), which is rapidly becoming the technique of choice for creating plants with specific mutations,

▼ **Figure 35.26 Variations in leaf arrangement, leaf shape, and shoot growth between different populations of *Arabidopsis thaliana*.** Information in the genomes of these populations may provide insights into strategies for expanding crop production into new environments.

has been used successfully in *Arabidopsis*. By disrupting or "knocking out" a specific gene, scientists can garner important information about the gene's normal function.

Large-scale projects are under way to determine the function of every gene in *Arabidopsis*. By identifying each gene's function and tracking every biochemical pathway, researchers aim to determine the blueprints for plant development, a major goal of systems biology. It may one day be possible to make a computer-generated "virtual plant" that enables researchers to visualize which genes are activated in different parts of the plant as the plant develops.

Basic research on model organisms such as *Arabidopsis* has accelerated the pace of discovery in the plant sciences, including the identification of the complex genetic pathways underlying plant structure. As you read more about this, you'll be able to appreciate not only the power of studying model organisms but also the history of investigation that underpins all modern plant research.

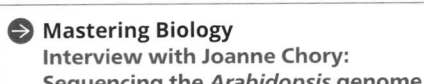 **Mastering Biology**
Interview with Joanne Chory: Sequencing the *Arabidopsis* genome

Growth: Cell Division and Cell Expansion

Cell division enhances the potential for growth by increasing the number of cells, but plant growth itself is brought about by cell enlargement. The process of plant cell division is described more fully in Chapter 12 (see Figure 12.10), and Chapter 39 discusses the process of cell elongation (see Figure 39.7). Here we are concerned with how cell division and enlargement contribute to plant form.

Cell Division

The new cell walls that bisect plant cells during cytokinesis develop from the cell plate (see Figure 12.10). The precise plane of cell division, determined during late interphase, usually corresponds to the shortest path that halves the cytoplasm of the parent cell. However, during certain points in development the cytoplasm may not be divided equally, resulting in one daughter cell being larger than the other, even though they have the same number of chromosomes. Such cases of *asymmetrical cell division* usually signal a key event in development. For example, the formation of guard cells involves an asymmetrical cell division. An epidermal cell divides asymmetrically, forming a large cell that remains an undifferentiated epidermal cell and a small cell that becomes the guard cell "mother cell." Guard cells form when this small mother cell divides in a plane perpendicular to the first cell division **(Figure 35.27)**. Thus, asymmetrical cell division generates cells with different fates—that is, cells that mature into different types.

Cell Expansion

Cell divisions do not constitute growth because there is no increase of mass involved. Rather, it is cell expansion that is responsible for plant growth. Before discussing how cell expansion contributes to plant growth and form, it is useful to consider the difference in cell expansion between plants and animals. Animal cells grow mainly by synthesizing protein-rich cytoplasm, a metabolically expensive process. Growing plant cells also produce additional protein-rich material in their cytoplasm, but water uptake typically accounts for about 90% of expansion. Most of this water is stored in the large central vacuole. The vacuolar solution, or *vacuolar sap*, is very dilute and nearly devoid of the energetically expensive macromolecules that are found in great abundance in the rest of the cytoplasm. Large vacuoles are therefore a "cheap" way of filling space, enabling a plant to grow rapidly and economically. Bamboo shoots, for instance, can elongate more than 2 m per week. Rapid and efficient extensibility of shoots and roots was an important evolutionary adaptation that increased their exposure to light and soil.

Plant cells rarely expand equally in all directions. Their greatest expansion is usually oriented along the plant's main axis. For example, cells near the tip of the root may elongate 20 times or more their original length, with relatively little increase in width. The orientation of cellulose microfibrils in the innermost layers of the cell wall causes this differential growth. The microfibrils do not stretch, so the cell expands mainly perpendicular to the main orientation of the microfibrils, as shown in **Figure 35.28**. A leading hypothesis proposes that microtubules positioned just beneath the plasma membrane organize the cellulose-synthesizing enzyme complexes and guide their movement through the plasma membrane as they create the microfibrils that form much of the cell wall.

▼ **Figure 35.27 Asymmetrical cell division and stomatal development.** An asymmetrical cell division precedes the development of epidermal guard cells, the cells that border stomata (see Figure 35.18).

▼ **Figure 35.28 The orientation of plant cell expansion.** Growing plant cells expand mainly through water uptake. In a growing cell, enzymes weaken cross-links in the cell wall, allowing it to expand as water diffuses into the vacuole by osmosis; at the same time, more microfibrils are made. The orientation of the cell expansion is mainly perpendicular to the orientation of cellulose microfibrils in the wall. The orientation of microtubules in the cell's outermost cytoplasm determines the orientation of cellulose microfibrils (fluorescent LM). The microfibrils are embedded in a matrix of other (noncellulose) polysaccharides, some of which form the cross-links visible in the TEM.

Morphogenesis and Pattern Formation

A plant's body is more than a collection of dividing and expanding cells. During morphogenesis, cells acquire different identities in an ordered spatial arrangement. For example, dermal tissue forms on the exterior and vascular tissue in the interior—never the other way around. The development of specific structures in specific locations is called **pattern formation**.

Two types of hypotheses have been put forward to explain how the fate of plant cells is determined during pattern formation. Hypotheses based on *lineage-based mechanisms* propose that cell fate is determined early in development and that cells pass on this destiny to their progeny. In this view, the basic pattern of cell differentiation is mapped out according to the directions in which meristematic cells divide and expand. On the other hand, hypotheses based on *position-based mechanisms* propose that the cell's final position in an emerging organ determines what kind of cell it will become. In support of this view, experiments in which neighboring cells have been destroyed with lasers have demonstrated that a plant cell's fate is established late in the cell's development and largely depends on signaling from its neighbors.

In contrast, cell fate in animals is largely determined by lineage-dependent mechanisms involving transcription factors. The homeotic (*Hox*) genes that encode such transcription factors are critical for the proper number and placement of embryonic structures, such as legs and antennae, in the fruit fly *Drosophila* (see Figure 18.19). Interestingly, maize has a homolog of *Hox* genes called *KNOTTED-1*, but unlike its counterparts in the animal world, *KNOTTED-1* does not affect the number or placement of plant organs. As you will see, an unrelated class of transcription factors called *MADS-box* proteins plays that role in plants. *KNOTTED-1* is, however, important in the development of leaf shape, including the production of compound leaves. If the *KNOTTED-1* gene is expressed in greater quantity than normal in the genome of tomato plants, the normally compound leaves will then become "super-compound" **(Figure 35.29)**.

▼ **Figure 35.29 Overexpression of a *Hox*-like gene in leaf formation.** *KNOTTED-1* is a gene that is involved in leaf and leaflet formation. An increase in its expression in tomato plants results in leaves that are "super-compound" (right) compared with normal leaves (left).

Gene Expression and the Control of Cell Differentiation

The cells of a developing organism can synthesize different proteins and diverge in structure and function even though they share a common genome. If a mature cell removed from a root or leaf can dedifferentiate in tissue culture and give rise to the diverse cell types of a plant, then it must possess all the genes necessary to make any kind of cell in the plant. Therefore, cell differentiation depends, to a large degree, on the control of gene expression—the regulation of

transcription and translation, resulting in the production of specific proteins.

Evidence suggests that the activation or inactivation of specific genes involved in cell differentiation results largely from cell-to-cell communication. Cells receive information about how they should specialize from neighboring cells. For example, two cell types arise in the root epidermis of *Arabidopsis*: root hair cells and hairless epidermal cells. Cell fate is associated with the position of the epidermal cells relative to other plant cells. The immature epidermal cells that are in contact with two underlying cells of the root cortex differentiate into root hair cells, whereas the immature epidermal cells in contact with only one cell in the cortex differentiate into mature hairless cells. The differential expression of a homeotic gene called *GLABRA-2* (from the Latin *glaber*, bald) is needed for proper distribution of root hairs **(Figure 35.30)**. Researchers have demonstrated this requirement by coupling the *GLABRA-2* gene to a "reporter gene" that causes every cell expressing *GLABRA-2* in the root to turn pale blue following a certain protocol. The *GLABRA-2* gene is normally expressed only in epidermal cells that will not develop root hairs.

▼ **Figure 35.30 Control of root hair differentiation by a homeotic gene. (LM)**

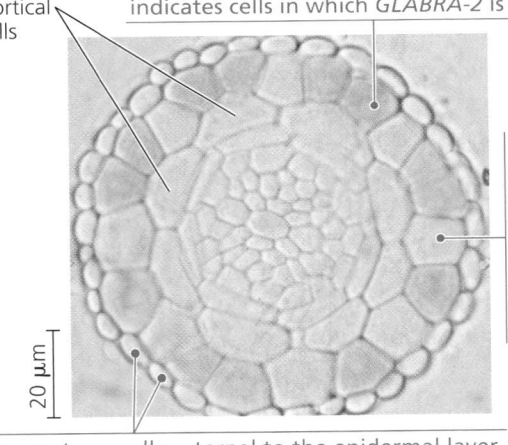

When an epidermal cell borders a single cortical cell, the homeotic gene *GLABRA-2* is expressed, and the cell remains hairless. (The blue color indicates cells in which *GLABRA-2* is expressed.)

Cortical cells

20 μm

Here an epidermal cell borders two cortical cells. *GLABRA-2* is not expressed, and the cell will develop a root hair.

The root cap cells external to the epidermal layer will be sloughed off before root hairs emerge.

WHAT IF? *What would the roots look like if GLABRA-2 were rendered dysfunctional by a mutation?*

→ **Mastering Biology**
Interview with Philip Benfey: Studying root development

Shifts in Development: Phase Changes

Multicellular organisms generally pass through developmental stages. In humans, these are infancy, childhood, adolescence, and adulthood, with puberty as the dividing line between the nonreproductive and reproductive stages. Plants also pass through stages, developing from a juvenile stage to an adult

▼ **Figure 35.31 Phase change in *Muehlenbeckia australis*.** This native climbing vine of New Zealand has two types of leaves: fiddle-shaped juvenile leaves (top) and oval, smooth-margined mature leaves. This dual foliage reflects a phase change in the development of the apical meristem of each shoot. Once a node forms, the developmental phase—juvenile or adult—is fixed; fiddle-shaped leaves do not mature into oval-shaped leaves.

Juvenile leaf

Mature leaf

vegetative stage to an adult reproductive stage. In animals, the developmental changes take place throughout the entire organism, such as when a larva develops into an adult animal. In contrast, plant developmental stages, called *phases*, occur within a single region, the shoot apical meristem. The morphological changes that arise from these transitions in shoot apical meristem activity are called **phase changes**. In the transition from a juvenile phase to an adult phase, some species exhibit conspicuous changes in leaf shape **(Figure 35.31)**. Juvenile nodes and internodes retain their juvenile status even after the shoot apical meristem of the main shoot has changed to the adult phase. Therefore, any *new* leaves that develop on branches that emerge from axillary buds at juvenile nodes will also be juvenile, even though the apical meristem of the stem's main axis may have been producing mature nodes for years.

If environmental conditions permit, an adult plant is induced to flower. Biologists have made great progress in explaining the genetic control of floral development—the topic of the next section.

Genetic Control of Flowering

Flower formation involves a phase change from vegetative growth to reproductive growth. This transition is triggered by a combination of environmental cues, such as day length, and internal signals, such as hormones. (You will learn more about

the roles of these signals in flowering in Concept 39.3.) Unlike vegetative growth, which is indeterminate, floral growth is usually determinate: The production of a flower by a shoot apical meristem generally stops the primary growth of that shoot. The transition from vegetative growth to flowering is associated with the switching on of flower-inducing genes. The protein products of these genes are transcription factors that regulate the genes required for the conversion of the indeterminate vegetative meristems to determinate floral meristems.

When a shoot apical meristem is induced to flower, the order of each primordium's emergence determines its development into a specific type of floral organ—a sepal, petal, stamen, or carpel (see Figure 30.8 to review basic flower structure). These floral organs form four whorls that can be described roughly as concentric "circles" when viewed from above. Sepals form the first (outermost) whorl; petals form the second; stamens form the third; and carpels form the fourth (innermost) whorl. Plant biologists have identified several genes that encode transcription factors that regulate the development of this characteristic floral pattern. Positional information determines which genes are expressed in a particular floral organ primordium. The result is the development of an emerging floral primordium into a specific floral organ. A mutation in a flower-inducing gene can cause abnormal floral development, such as petals growing in place of stamens **(Figure 35.32)**. Some homeotic mutants with increased petal numbers produce showier flowers that are prized by gardeners.

▼ **Figure 35.32 Genes and pattern formation in flower development.**

▲ **Normal *Arabidopsis* flower.** *Arabidopsis* normally has four whorls of flower parts: sepals (Se), petals (Pe), stamens (St), and carpels (Ca).

▶ **Abnormal *Arabidopsis* flower.** Researchers have identified several mutations that cause abnormal flowers to develop. This flower has an extra set of petals in place of stamens and an internal flower where normal plants have carpels.

MAKE CONNECTIONS *Provide another example of a homeotic gene mutation that leads to organs being produced in the wrong place (see Concept 18.4).*

By studying mutants with abnormal flowers, researchers have identified and cloned three classes of genes that determine floral organ identity. **Figure 35.33a** shows a simplified version of the **ABC hypothesis** of flower formation, which proposes that three classes of genes direct the formation of the four types of floral organs. According to the ABC hypothesis, each class of genes is switched on in two specific whorls of the floral meristem. Normally, *A* genes are switched on in the two outer whorls (sepals and petals); *B* genes are switched on in the two middle whorls (petals and stamens); and *C* genes are switched on in the two inner whorls (stamens and carpels). Sepals arise from those parts of floral meristems in which only *A* genes are active; petals arise where *A* and *B* genes are active; stamens where *B* and *C* genes are active; and carpels where only *C* genes are active. The ABC hypothesis can account for the phenotypes of mutants lacking *A*, *B*, or *C* gene activity, with one addition: Where *A* gene activity is present, it inhibits *C*, and vice versa. If either the *A* gene or *C* gene is suppressed, the other gene is expressed. **Figure 35.33b** shows the floral patterns of mutants lacking each of the three classes of genes and depicts how the hypothesis accounts for the

floral phenotypes. By constructing such hypotheses and designing experiments to test them, researchers are tracing the genetic basis of plant development.

In dissecting the plant to examine its parts, as we have done in this chapter, we must remember that the whole plant functions as an integrated organism. Plant structures largely reflect evolutionary adaptations to the challenges of a photo-autotrophic existence on land.

> **Mastering Biology**
> **Interview with Virginia Walbot:**
> **Researching plant genetics and development**

CONCEPT CHECK 35.5

1. How can two cells in a plant have vastly different structures even though they have the same genome?

2. What are three differences between animal development and plant development?

3. **WHAT IF?** In some species, sepals look like petals, and both are collectively called "tepals." Suggest an extension to the ABC hypothesis that could account for tepals.

For suggested answers, see Appendix A.

▼ **Figure 35.33 The ABC hypothesis for the functioning of genes in flower development.**

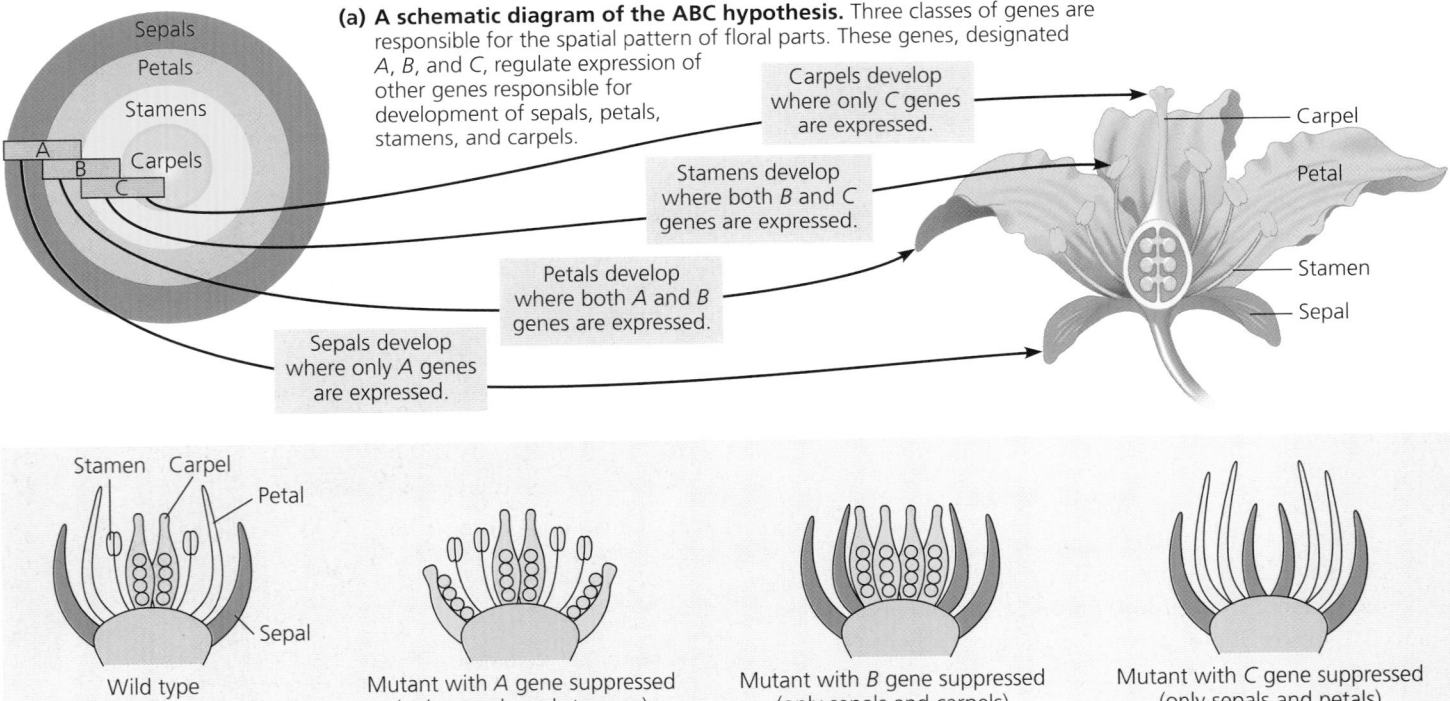

(a) A schematic diagram of the ABC hypothesis. Three classes of genes are responsible for the spatial pattern of floral parts. These genes, designated *A*, *B*, and *C*, regulate expression of other genes responsible for development of sepals, petals, stamens, and carpels.

Carpels develop where only *C* genes are expressed.

Stamens develop where both *B* and *C* genes are expressed.

Petals develop where both *A* and *B* genes are expressed.

Sepals develop where only *A* genes are expressed.

Sepals / Petals / Stamens / Carpels

A / B / C

Carpel / Petal / Stamen / Sepal

Stamen Carpel Petal Sepal

Wild type

Mutant with *A* gene suppressed (only carpels and stamens)

Mutant with *B* gene suppressed (only sepals and carpels)

Mutant with *C* gene suppressed (only sepals and petals)

(b) Side view of wild-type flower and flowers with organ identity mutations. The phenotype of mutants lacking a functional *A*, *B*, or *C* gene can be explained by the model in part (a) and the observation that if the *A* gene or *C* gene is suppressed, then the other gene is expressed in that whorl. For example, if the *A* gene is suppressed in a mutant, then the *C* gene is expressed where the *A* gene would normally be expressed. Therefore, carpels (*C* gene expressed) develop in the outermost whorl, and stamens (*B* and *C* genes expressed) develop in the next whorl.

DRAW IT *(a) For each mutant, draw a "bull's-eye" diagram like the one in part (a), labeling the type of organ and gene(s) expressed in each whorl. (b) Draw and label a "bull's-eye" diagram for a mutant flower in which the A and B genes were suppressed.*

SUMMARY OF KEY CONCEPTS

→ To review key terms, go to the **Vocabulary Self-Quiz** in the **Mastering Biology** eText or Study Area, or go to goo.gl/zkjz9t.

CONCEPT 35.1

Plants have a hierarchical organization consisting of organs, tissues, and cells *(pp. 759–765)*

- Vascular plants have shoots consisting of **stems**, **leaves**, and, in angiosperms, flowers. **Roots** anchor the plant, absorb and conduct water and minerals, and store food. Leaves are attached to stem **nodes** and are the main **organs** of photosynthesis. The **axillary buds**, in axils of leaves and stems, give rise to branches. Plant organs may be adapted for specialized functions.

- Vascular plants have three **tissue systems**—dermal, vascular, and ground—which are continuous throughout the plant. The **dermal tissue** is a continuous layer of cells that covers the plant exterior. **Vascular tissues (xylem** and **phloem)** facilitate the long-distance transport of substances. **Ground tissues** function in storage, metabolism, and regeneration.

- **Parenchyma cells** are relatively undifferentiated and thin-walled cells that retain the ability to divide; they perform most of the metabolic functions of synthesis and storage. **Collenchyma cells** have unevenly thickened walls; they support young, growing parts of the plant. **Sclerenchyma cells**—sclereids and fibers—have thick, lignified walls that help support mature, nongrowing parts of the plant. **Tracheids** and **vessel elements**, the water-conducting cells of xylem, have thick walls and are dead at functional maturity. **Sieve-tube elements** are living but highly modified cells that are largely devoid of internal organelles; they function in the transport of sugars through the phloem of angiosperms.

? *Describe at least three specializations in plant organs and plant cells that are adaptations to life on land.*

CONCEPT 35.2

Different meristems generate new cells for primary and secondary growth *(pp. 766–767)*

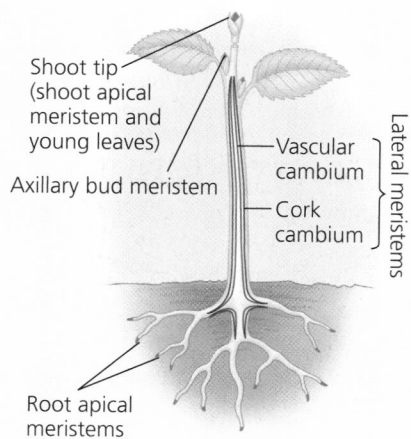

? *What is the difference between primary and secondary growth?*

CONCEPT 35.3

Primary growth lengthens roots and shoots *(pp. 768–771)*

- The root **apical meristem** is located near the tip of the root, where it generates cells for the growing root axis and the **root cap**.
- The apical meristem of a shoot is located in the **apical bud**, where it gives rise to alternating **internodes** and leaf-bearing nodes.
- Eudicot stems have vascular bundles in a ring, whereas monocot stems have scattered vascular bundles.
- **Mesophyll** cells are adapted for photosynthesis. **Stomata**, epidermal pores formed by pairs of **guard cells**, allow for gaseous exchange and are major avenues for water loss.

? *How does branching in roots differ from that in stems?*

CONCEPT 35.4

Secondary growth increases the diameter of stems and roots in woody plants *(pp. 772–775)*

- The **vascular cambium** is a meristematic cylinder that produces secondary xylem and secondary phloem during **secondary growth**. Older layers of secondary xylem (heartwood) become inactive, whereas younger layers (sapwood) still conduct water.

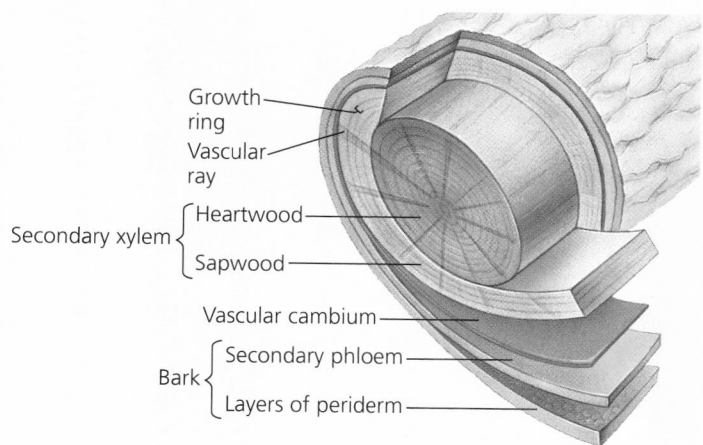

- The **cork cambium** gives rise to a thick protective covering called the periderm, which consists of the cork cambium plus the layers of cork cells it produces.

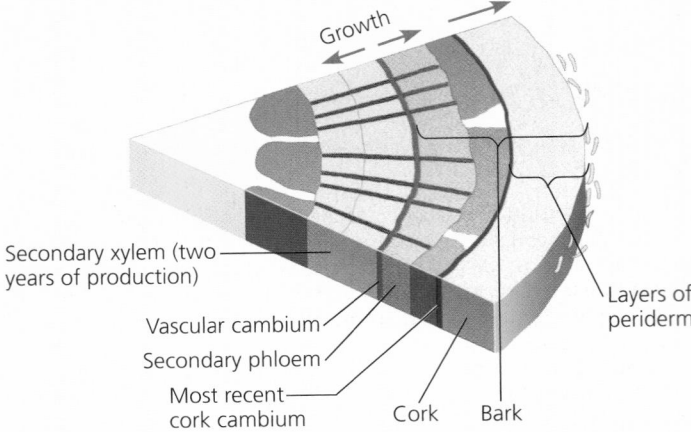

Secondary xylem (two years of production)

Vascular cambium

Secondary phloem

Most recent cork cambium

Cork

Bark

Layers of periderm

Growth

? *What advantages did plants gain from the evolution of secondary growth?*

CONCEPT 35.5

Growth, morphogenesis, and cell differentiation produce the plant body (pp. 775–780)

- Cell division and cell expansion are the primary determinants of growth.
- Morphogenesis, the development of body shape and organization, depends on cells responding to positional information from their neighbors.
- Cell differentiation, arising from differential gene activation, enables cells within the plant to assume different functions despite having identical genomes. The way in which a plant cell differentiates is determined largely by the cell's position in the developing plant.
- Internal or environmental cues may cause a plant to switch from one developmental stage to another—for example, from developing juvenile leaves to developing mature leaves. Such changes in shape are called **phase changes**.
- Studies of floral development have provided a model system for studying **pattern formation**. The **ABC hypothesis** identifies how three classes of genes control formation of sepals, petals, stamens, and carpels.

? *By what mechanism do plant cells tend to elongate along one axis instead of expanding in all directions?*

TEST YOUR UNDERSTANDING

→ For more multiple-choice questions, go to the **Practice Test** in the **Mastering Biology** eText or Study Area, or go to goo.gl/GruWRg.

Levels 1-2: Remembering/Understanding

1. Most of the growth of a plant body is the result of
   (A) cell differentiation.
   (B) morphogenesis.
   (C) cell division.
   (D) cell elongation.

2. The innermost layer of the root cortex is the
   (A) core.   (C) endodermis.
   (B) pericycle.   (D) pith.

3. Heartwood and sapwood consist of
   (A) bark.   (C) secondary xylem.
   (B) periderm.   (D) secondary phloem.

4. The phase change of an apical meristem from the juvenile to the mature vegetative phase is often revealed by
   (A) a change in the shape of the leaves produced.
   (B) the initiation of secondary growth.
   (C) the formation of lateral roots.
   (D) the activation of flower-inducing genes.

5. The vascular cambium gives rise to
   (A) all xylem.
   (B) all phloem.
   (C) primary xylem and phloem.
   (D) secondary xylem and phloem.

6. The root pericycle is the site where
   (A) secondary growth originates.
   (B) root hairs originate.
   (C) lateral roots originate.
   (D) the endodermis originates.

7. Root apical meristems are found
   (A) only in taproots.
   (B) only in lateral roots.
   (C) only in adventitious roots.
   (D) in all roots.

Levels 3-4: Applying/Analyzing

8. Suppose a flower had normal expression of genes A and C and expression of gene B in all four whorls. Based on the ABC hypothesis, what would be the structure of that flower, starting at the outermost whorl?
   (A) carpel-petal-petal-carpel
   (B) petal-petal-stamen-stamen
   (C) sepal-carpel-carpel-sepal
   (D) sepal-sepal-carpel-carpel

9. Which of the following arise(s), directly or indirectly, from meristematic activity?
   (A) secondary xylem
   (B) leaves
   (C) dermal tissue
   (D) all of the above

10. A strawberry plant mutant that fails to make stolons would suffer from
    (A) too little mineral absorption.
    (B) a tendency to topple over.
    (C) too little water absorption.
    (D) a reduction in asexual reproduction.

11. **DRAW IT** On this cross section from a woody eudicot, label a growth ring, late wood, early wood, and a vessel element. Then draw an arrow in the pith-to-cork direction.

12. VISUAL SKILLS How does the internal anatomy of a tea leaf differ from that of an iris leaf? (Review Figure 35.18.) How might this difference relate to the orientations of the leaves? Explain how the difference shows that structure fits function.

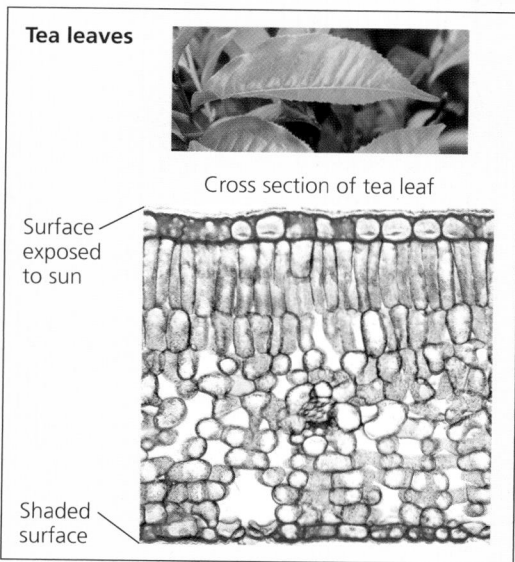

Tea leaves

Cross section of tea leaf

Surface exposed to sun

Shaded surface

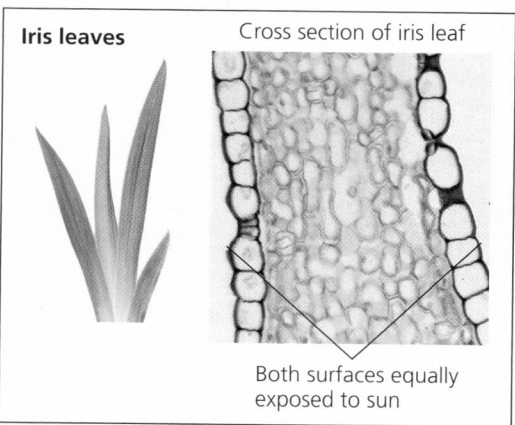

Iris leaves

Cross section of iris leaf

Both surfaces equally exposed to sun

13. VISUAL SKILLS Visitors to the state of Washington can marvel at an unusual sight: a bicycle embedded 4 m above the ground in a living tree. Use your knowledge of primary and secondary growth to visualize and explain how this happened.

Levels 5-6: Evaluating/Creating

14. EVOLUTION CONNECTION Evolutionary biologists have coined the term *exaptation* to describe a common occurrence in the evolution of life: A limb or organ evolves in a particular context but over time takes on a new function (see Concept 25.6). What are some examples of exaptations in plant organs?

15. SCIENTIFIC INQUIRY Grasslands typically do not flourish when large herbivores are removed. Instead, grasslands are replaced by broad-leaved herbaceous eudicots, shrubs, and trees. Based on your knowledge of the structure and growth habits of monocots versus eudicots, suggest a reason why.

16. SCIENCE, TECHNOLOGY, AND SOCIETY Hunger and malnutrition are urgent problems for many poor countries, yet plant biologists in wealthy nations have focused most of their research efforts on *Arabidopsis thaliana*. Some people have argued that if plant biologists are truly concerned about fighting world hunger, they should study cassava and plantain because these two crops are staples for many of the world's poor. If you were an Arabidopsis researcher, how might you respond to this argument?

17. WRITE ABOUT A THEME: ORGANIZATION In a short essay (100–150 words), explain how the evolution of lignin affected vascular plant structure and function.

18. SYNTHESIZE YOUR KNOWLEDGE

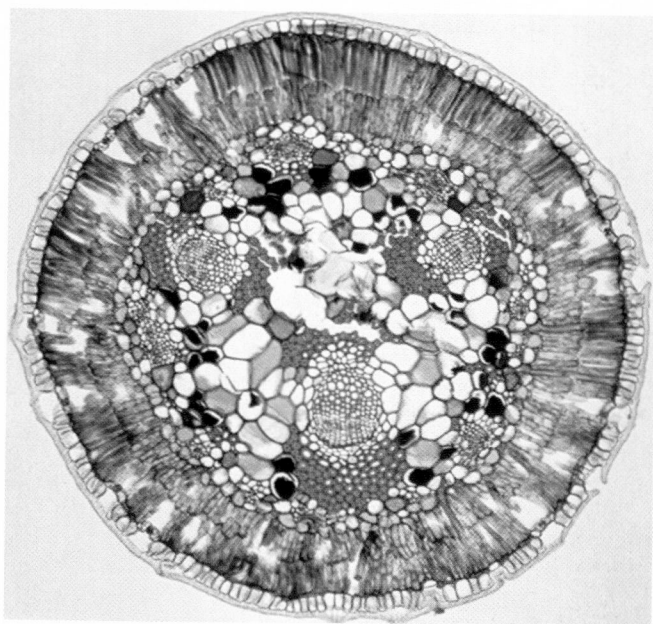

This stained light micrograph shows a cross section through a plant organ from *Hakea purpurea*, a shrub native to some arid regions of Australia. (a) Review Figures 35.14, 35.17, and 35.18 to identify whether this is a root, stem, or leaf. Explain your reasoning. (b) How might this organ be an adaptation for dry conditions?

For selected answers, see Appendix A.

Explore Scientific Papers with Science in the Classroom △AAAS
How does one gene affect leaf shape?
Go to "Shape Shifters of the Plant World"
at www.scienceintheclassroom.org.
→ **Instructors:** Questions can be assigned in Mastering Biology.

36 Resource Acquisition and Transport in Vascular Plants

KEY CONCEPTS

36.1 Adaptations for acquiring resources were key steps in the evolution of vascular plants *p. 785*

36.2 Different mechanisms transport substances over short or long distances *p. 787*

36.3 Transpiration drives the transport of water and minerals from roots to shoots via the xylem *p. 792*

36.4 The rate of transpiration is regulated by stomata *p. 796*

36.5 Sugars are transported from sources to sinks via the phloem *p. 799*

36.6 The symplast is highly dynamic *p. 801*

Study Tip

Make a flowchart: Make a simple flowchart of the movement of water molecules from the soil through the tissues of a plant to the outside air.

→ Go to Mastering Biology

For Students (in eText and Study Area)
• Get Ready for Chapter 36
• Figure 36.9 Walkthrough: Transport of Water and Minerals from Root Hairs to the Xylem
• BioFlix® Animation: Water Transport in Plants

For Instructors to Assign (in Item Library)
• BioFlix® Tutorial: Water Transport in Plants: The Transpiration-Cohesion-Tension Mechanism
• BioFlix® Tutorial: Transpiration

Ready-to-Go Teaching Module
(in Instructor Resources)
• Transport in Plants (Concept 36.2)

Figure 36.1 This English ivy has covered every square centimeter of a wall with foliage, an effective way to acquire light energy for photosynthesis. The sugars produced by the leaves and the water and minerals absorbed by the roots have to be transported throughout the plant.

What causes the movement of water, minerals, and sugars in most vascular plants?

Water and minerals are pulled up from the roots by **negative pressure** (tension) generated by evaporation from the leaves.

Water molecules are joined in chains as they move upward.

Sugars are pushed by **positive pressure** from where they are produced or stored to where they are needed. They can move both ways between leaves and roots.

Transport of sugars from mature leaves to roots

Transport of sugars from a storage root to new leaves

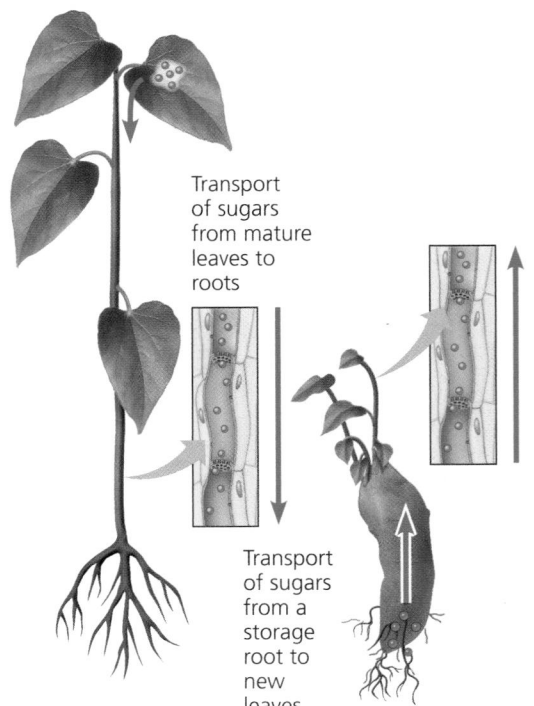

Adaptations for acquiring resources were key steps in the evolution of vascular plants

EVOLUTION Most plants grow in soil and therefore inhabit two worlds—above ground, where shoots acquire sunlight and CO_2, and below ground, where roots acquire water and minerals. The successful colonization of the land by plants depended on adaptations that allowed early plants to acquire resources from these two different settings.

The algal ancestors of plants absorbed water, minerals, and CO_2 directly from the water in which they lived. Transport in these algae was relatively simple because every cell was close to the source of these substances. The earliest plants were nonvascular and produced photosynthetic shoots above the shallow fresh water in which they lived. These leafless shoots typically had waxy cuticles and few stomata, which allowed them to avoid excessive water loss while still permitting some exchange of CO_2 and O_2 for photosynthesis. The anchoring and absorbing functions of early plants were assumed by the base of the stem or by threadlike rhizoids (see Figure 29.16).

As plants evolved and increased in number, competition for light, water, and nutrients intensified. Taller plants with broad, flat appendages had an advantage in absorbing light. This increase in surface area, however, resulted in more evaporation and therefore a greater need for water. Larger shoots also required stronger anchorage. These needs favored the production of multicellular, branching roots. Meanwhile, as greater shoot heights further separated the top of the photosynthetic shoot from the nonphotosynthetic parts below ground, natural selection favored plants capable of efficient long-distance transport of water, minerals, and products of photosynthesis.

The evolution of vascular tissue consisting of xylem and phloem made possible the development of extensive root and shoot systems that carry out long-distance transport (see Figure 35.10). The **xylem** transports water and minerals from roots to shoots. The **phloem** transports products of photosynthesis from where they are made or stored to where they are needed. **Figure 36.2** provides an overview of resource acquisition and transport in an actively photosynthesizing plant.

Shoot Architecture and Light Capture

Because most plants are photoautotrophs, their success depends ultimately on their ability to photosynthesize. Over the course of evolution, plants have developed a wide variety

▼ **Figure 36.2 An overview of resource acquisition and transport in a vascular plant during the day.**

In leaves and green stems, gas exchange occurs through stomata. During photosynthesis, CO_2 is taken up and O_2 is produced. Cells use some of that O_2 for cellular respiration, and the rest is released into the air.

Sugars are produced by photosynthesis in the leaves.

Transpiration, the loss of water from leaves (mostly through stomata), creates a force within leaves that pulls xylem sap upward.

Water and minerals are transported upward from roots to shoots as xylem sap.

Phloem sap can flow both ways between shoots and roots. It moves from sites of sugar production (usually leaves) or storage (usually roots) to sites of sugar use or storage.

Water and minerals in the soil are absorbed by roots.

H_2O and minerals

As root cells carry out their cellular respiration, they exchange gases with the air spaces of the soil, taking in O_2 and discharging CO_2.

MAKE CONNECTIONS *How would you change the figure in order to show gas exchanges at night? (See Figure 10.22.)*

of shoot architectures that enable each species to compete successfully for light absorption in the ecological niche it occupies. For example, the lengths and widths of stems, as well as the branching pattern of shoots, are all architectural features affecting light capture.

Stems serve as supporting structures for leaves and as conduits for the transport of water and nutrients. Plants that grow tall avoid shading from neighboring plants. Most tall plants require thick stems, which enable greater vascular flow to and from the leaves and stronger mechanical support for them. Vines are an exception, relying on other objects (usually other plants) to support their stems. In woody plants, stems become thicker through secondary growth (see Figure 35.11). Branching generally enables plants to harvest sunlight for photosynthesis more effectively. However, some species, such as the coconut palm, do not branch at all. Why is there so much variation in branching patterns? Plants have only a finite amount of energy to devote to shoot growth. If most of that energy goes into branching, there is less available for growing tall, and the risk of being shaded by taller plants increases. Conversely, if most of the energy goes into growing tall, the plants are not optimally harvesting sunlight.

Leaf size and structure account for much of the outward diversity in plant form. Leaves range in length from 1.3 mm in the pygmyweed (*Crassula connata*), a native of dry, sandy regions in the western United States, to 20 m in the palm *Raphia regalis*, a native of African rain forests. These species represent extreme examples of a general correlation observed between water availability and leaf size. The largest leaves are typically found in species from tropical rain forests, whereas the smallest are usually found in species from dry or very cold environments, where liquid water is scarce and evaporative loss is more problematic.

The arrangement of leaves on a stem, known as *phyllotaxy*, is an architectural feature important in light capture. Phyllotaxy is determined by the shoot apical meristem (see Figure 35.16) and is specific to each species **(Figure 36.3)**. A species may have one leaf per node (alternate, or spiral, phyllotaxy), two leaves per node (opposite phyllotaxy), or more (whorled phyllotaxy). Most angiosperms have alternate phyllotaxy, with leaves arranged in an ascending spiral around the stem, each successive leaf emerging about 137.5° from the site of the previous one. Why 137.5°? One hypothesis is that this angle minimizes shading of the lower leaves by those above. In environments where intense sunlight can harm leaves, the greater shading provided by oppositely arranged leaves may be advantageous.

The total area of the leafy portions of all the plants in a community, from the top layer of vegetation to the bottom layer, affects the productivity of each plant. When there are many layers of vegetation, the shading of the lower leaves is so great that they photosynthesize less than they respire. When this happens, the nonproductive leaves or branches

▼ **Figure 36.3 Emerging phyllotaxy of Norway spruce.** This SEM, taken from above a shoot tip, shows the pattern of emergence of leaves. The leaves are numbered, with 1 being the youngest. (Some numbered leaves are not visible in the close-up.)

VISUAL SKILLS *With your finger, trace the progression of leaf emergence, moving from leaf number 29 to 28 and so on. What is the pattern? Based on this pattern of phyllotaxy, predict between which two developing leaf primordia the next primordium will emerge.*

undergo programmed cell death and are eventually shed, a process called *self-pruning*.

Plant features that reduce self-shading increase light capture. A useful measurement in this regard is the *leaf area index*, the ratio of the total upper leaf surface of a single plant or an entire crop divided by the surface area of the land on which the plant or crop grows **(Figure 36.4)**. Leaf area index values of up to 7 are common for many mature crops, and

▼ **Figure 36.4 Leaf area index.** The leaf area index of a single plant is the ratio of the total area of the top surfaces of the leaves to the area of ground covered by the plant, as shown in this illustration of two plants viewed from the top. With many layers of leaves, a leaf area index value can easily exceed 1.

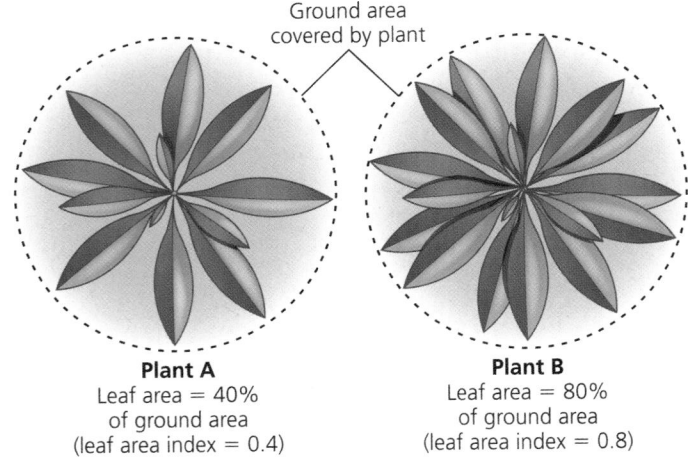

Ground area covered by plant

Plant A
Leaf area = 40% of ground area
(leaf area index = 0.4)

Plant B
Leaf area = 80% of ground area
(leaf area index = 0.8)

❓ *Would a higher leaf area index always increase the amount of photosynthesis? Explain.*

there is little agricultural benefit to leaf area indexes higher than this value. Adding more leaves increases shading of lower leaves to the point that self-pruning occurs.

Another factor affecting light capture is leaf orientation. Some plants have horizontally oriented leaves; others, such as grasses, have leaves that are vertically oriented. In low-light conditions, horizontal leaves capture sunlight much more effectively than vertical leaves. In grasslands or other sunny regions, however, horizontal orientation may expose upper leaves to overly intense light, injuring leaves and reducing photosynthesis. But if a plant's leaves are nearly vertical, light rays are essentially parallel to the leaf surfaces, so no leaf receives too much light, and light penetrates more deeply to the lower leaves.

The Photosynthesis–Water Loss Compromise

The broad surface of most leaves favors light capture, while open stomatal pores allow for the diffusion of CO_2 into the photosynthetic tissues. Open stomatal pores, however, also promote evaporation of water from the plant. Over 90% of the water lost by plants is by evaporation from stomatal pores. Consequently, shoot adaptations represent compromises between enhancing photosynthesis and minimizing water loss, particularly in environments where water is scarce. Later in the chapter, we'll discuss the mechanisms by which plants enhance CO_2 uptake and minimize water loss by regulating the opening of stomatal pores.

Root Architecture and Acquisition of Water and Minerals

Just as carbon dioxide and sunlight are resources exploited by the shoot system, soil contains resources mined by the root system. Plants rapidly adjust the architecture and physiology of their roots to exploit patches of available nutrients in the soil. The roots of many plants, for example, respond to pockets of low nitrate availability in soils by extending straight through the pockets instead of branching within them. Conversely, when encountering a pocket rich in nitrate, a root will often branch extensively there. Root cells also respond to high soil nitrate levels by synthesizing more proteins involved in nitrate transport and assimilation. Thus, not only does the plant devote more of its mass to exploiting a nitrate-rich patch, but the cells also absorb nitrate more efficiently.

The efficient absorption of limited nutrients is also enhanced by reduced competition within the root system. For example, cuttings taken from stolons of buffalo grass (*Bouteloua dactyloides*) develop fewer and shorter roots in the presence of cuttings from the same plant than they do in the presence of cuttings from another buffalo grass plant. Researchers are trying to uncover how the plant distinguishes self from nonself.

Plant roots also form mutually beneficial relationships with microorganisms that enable the plant to exploit soil resources more efficiently. For example, the evolution of mutualistic associations between roots and fungi called **mycorrhizae** was a critical step in the successful colonization of land by plants. Mycorrhizal hyphae indirectly endow the root systems of many plants with an enormous surface area for absorbing water and minerals, particularly phosphate. The role of mycorrhizae in plant nutrition will be examined in Concept 37.3.

Once acquired, resources must be transported to other parts of the plant that need them. In the next section, we examine the processes and pathways that enable resources such as water, minerals, and sugars to be transported throughout the plant.

CONCEPT CHECK 36.1

1. Why is long-distance transport important for vascular plants?

2. Some plants can detect increased levels of light reflected from leaves of encroaching neighbors. This detection elicits stem elongation, production of erect leaves, and reduced lateral branching. How do these responses help the plant compete?

3. **WHAT IF?** If you prune a plant's shoot tips, what will be the short-term effect on the plant's branching and leaf area index?

For suggested answers, see Appendix A.

CONCEPT 36.2

Different mechanisms transport substances over short or long distances

Given the diversity of substances that move through plants and the great range of distances and barriers over which such substances must be transported, it is not surprising that plants employ a variety of transport processes. Before examining these processes, however, we'll look at the two major pathways of transport: the apoplast and the symplast.

The Apoplast and Symplast: Transport Continuums

Plant tissues have two major compartments—the apoplast and the symplast. The **apoplast** consists of everything external to the plasma membranes of living cells and includes cell walls, extracellular spaces, and the interior of dead cells such as vessel elements and tracheids (see Figure 35.10). The **symplast** consists of the entire mass of cytosol of all the living cells in a plant, as well as the plasmodesmata, the cytoplasmic channels that interconnect them.

Figure 36.5 Cell compartments and routes for short-distance transport.

Some substances may use more than one transport route.

The apoplast is the continuum of cell walls and extracellular spaces.

The symplast is the continuum of cytosol connected by plasmodesmata.

Cell wall

Apoplastic route

Cytosol

Symplastic route

Transmembrane route

Plasmodesma

Plasma membrane

Key

Apoplast

Symplast

The compartmental structure of plants provides three routes for transport within a plant tissue or organ: the apoplastic, symplastic, and transmembrane routes **(Figure 36.5)**. In the *apoplastic route*, water and solutes (dissolved chemicals) move along the continuum of cell walls and extracellular spaces. In the *symplastic route*, water and solutes move along the continuum of cytosol. This route requires substances to cross a plasma membrane once, when they first enter the plant. After entering one cell, substances can move from cell to cell via plasmodesmata. In the *transmembrane route*, water and solutes move out of one cell, across the cell wall, and into the neighboring cell, which may pass them to the next cell in the same way. The transmembrane route requires repeated crossings of plasma membranes as substances exit one cell and enter the next. These three routes are not mutually exclusive, and some substances may use more than one route to varying degrees.

Short-Distance Transport of Solutes Across Plasma Membranes

In plants, as in any organism, the selective permeability of the plasma membrane controls the short-distance movement of substances into and out of cells (see Concept 7.2). Both active and passive transport mechanisms occur in plants, and plant cell membranes are equipped with the same *general* types of pumps and transport proteins (channel proteins, carrier proteins, and cotransporters) that function in other cells. There are, however, *specific* differences between the membrane transport processes of plant and animal cells. In this section, we'll focus on some of those differences.

Unlike in animal cells, hydrogen ions (H^+) rather than sodium ions (Na^+) play the primary role in basic transport processes in plant cells. For example, in plant cells the membrane potential (the voltage across the membrane) is established mainly through the pumping of H^+ by proton pumps **(Figure 36.6a)** rather than the pumping of Na^+ by sodium-potassium pumps. Also, H^+ is most often

cotransported in plants, whereas Na^+ is typically cotransported in animals. During cotransport, plant cells use the energy in the H^+ gradient and membrane potential to drive the active transport of many different solutes. For instance, cotransport with H^+ is responsible for absorption of neutral solutes, such as the sugar sucrose, by phloem cells and other plant cells. An H^+/sucrose cotransporter couples movement of sucrose against its concentration gradient with movement of H^+ down its electrochemical gradient **(Figure 36.6b)**. Cotransport with H^+ also facilitates movement of ions, as in the uptake of nitrate (NO_3^-) by root cells **(Figure 36.6c)**.

The membranes of plant cells also have ion channels that allow only certain ions to pass **(Figure 36.6d)**. As in animal cells, most channels are gated, opening or closing in response to stimuli such as chemicals, pressure, or voltage. Later in this chapter, we'll discuss how potassium ion channels in guard cells function in opening and closing stomata. Ion channels are also involved in producing electrical signals analogous to the action potentials of animals (see Concept 48.2). However, these signals are 1,000 times slower and employ Ca^{2+}-activated anion channels rather than the sodium ion channels used by animal cells.

Short-Distance Transport of Water Across Plasma Membranes

The absorption or loss of water by a cell occurs by **osmosis**, the diffusion of free water—water that is not bound to solutes or surfaces—across a membrane (see Figure 7.12). The physical property that predicts the direction in which water will flow is called **water potential**, a quantity that includes the effects of solute concentration and physical pressure. Free water moves from regions of higher water potential to regions of lower water potential if there is no barrier to its flow. The word *potential* in the term *water potential* refers to water's potential energy—water's capacity to perform work when it moves from a region of higher water potential to a region of lower water potential. For example, if a plant cell or seed is immersed in a solution that has a higher water potential, water will move into the cell or seed, causing it to expand. The expansion of plant cells and seeds can be a powerful force: The expansion of cells in tree roots can break concrete sidewalks, and the swelling of wet grain seeds within the holds of damaged ships can produce catastrophic hull failure and sink the ships. Given the strong forces generated by swelling seeds, it is interesting to consider whether water uptake by seeds is an active process. This question is

▼ Figure 36.6 Solute transport across plant cell plasma membranes.

(a) H⁺ and membrane potential. The plasma membranes of plant cells use ATP-dependent proton pumps to pump H⁺ out of the cell. These pumps contribute to the membrane potential and the establishment of a pH gradient across the membrane. These two forms of potential energy can drive the transport of solutes.

(b) H⁺ and cotransport of neutral solutes. Neutral solutes such as sugars can be loaded into plant cells by cotransport with H⁺ ions. H⁺/sucrose cotransporters, for example, play a key role in loading sugar into the phloem prior to sugar transport throughout the plant.

(c) H⁺ and cotransport of ions. Cotransport mechanisms involving H⁺ also participate in regulating ion fluxes into and out of cells. For example, H⁺/NO₃⁻ cotransporters in the plasma membranes of root cells are important for the uptake of NO₃⁻ by plant roots.

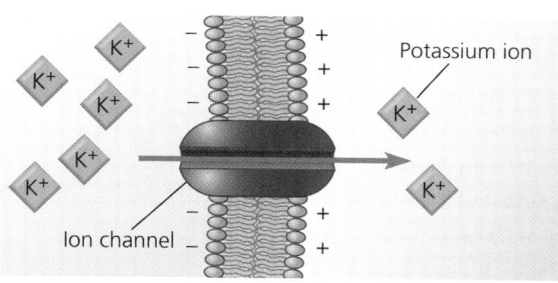

(d) Ion channels. Plant ion channels open and close in response to voltage, stretching of the membrane, and chemical factors. When open, ion channels allow specific ions to diffuse across membranes. For example, a K⁺ ion channel is involved in the release of K⁺ from guard cells when stomata close.

? *Assume that a plant cell has all four of the plasma membrane transport proteins shown above and that you have a specific inhibitor for each protein. Predict the effect of each inhibitor on the cell's membrane potential.*

examined in the **Scientific Skills Exercise**, which explores the effect of temperature on this process.

Water potential is abbreviated by the Greek letter Ψ (psi, pronounced "sigh"). Plant biologists measure Ψ in a unit of pressure called a **megapascal** (abbreviated MPa). By definition, the Ψ of pure water in a container open to the atmosphere under standard conditions (at sea level and at room temperature) is 0 MPa. One MPa is equal to about 10 times atmospheric pressure at sea level. The internal pressure of a living plant cell due to the osmotic uptake of water is approximately 0.5 MPa, about twice the air pressure inside an inflated car tire.

How Solutes and Pressure Affect Water Potential

Solute concentration and physical pressure are the major determinants of water potential in hydrated plants, as expressed in the water potential equation:

$$\Psi = \Psi_S + \Psi_P$$

where Ψ is the water potential, Ψ_S is the solute potential (osmotic potential), and Ψ_P is the pressure potential. The **solute potential** (Ψ_S) of a solution is directly proportional to its molarity. Solute potential is also called *osmotic potential* because solutes affect the direction of osmosis. The solutes in plants are typically mineral ions and sugars. By definition, the Ψ_S of pure water is 0. When solutes are added, they bind water molecules. As a result, there are fewer free water molecules, reducing the capacity of the water to move and do work. In this way, an increase in solute concentration has a negative effect on water potential, which is why the Ψ_S of a solution is always expressed as a negative number. For example, a 0.1 M solution of a sugar has a Ψ_S of −0.23 MPa. As the solute concentration increases, Ψ_S will become more negative.

Pressure potential (Ψ_P) is the physical pressure on a solution. Unlike Ψ_S, Ψ_P can be positive or negative relative to atmospheric pressure. For example, when a solution is being withdrawn by a syringe, it is under negative pressure; when it is being expelled from a syringe, it is under positive pressure. The water in living cells is usually under positive pressure due to the osmotic uptake of water. Specifically, the **protoplast** (the living part of the cell, which also includes the plasma membrane) presses against the cell wall, creating what is known as **turgor pressure**. This pushing effect of internal pressure, much like the air in an inflated tire, is critical for plant function because it helps maintain the stiffness of plant tissues and also serves as the driving force for cell elongation. Conversely, the water in the hollow nonliving xylem cells (tracheids and vessel elements) of a plant is often under a negative pressure potential (tension) of less than −2 MPa.

Calculating and Interpreting Temperature Coefficients

Does the Initial Uptake of Water by Seeds Depend on Temperature? One way to answer this question is to soak seeds in water at different temperatures and measure the rate of water uptake at each temperature. The data can be used to calculate the temperature coefficient, Q_{10}, the factor by which a physiological reaction (or process) rate increases when the temperature is raised by 10°C:

$$Q_{10} = \left(\frac{k_2}{k_1}\right)^{\frac{10}{t_2-t_1}}$$

where t_2 is the higher temperature (°C), t_1 is the lower temperature, k_2 is the reaction (or process) rate at t_2, and k_1 is the reaction (or process) rate at t_1. (If $t_2 - t_1 = 10$, as here, the math is simplified.)

Q_{10} values may be used to make inferences about the physiological process under investigation. Chemical (metabolic) processes involving large-scale protein shape changes are highly dependent on temperature and have higher Q_{10} values, closer to 2 or 3. In contrast, many, but not all, physical parameters are relatively independent of temperature and have Q_{10} values closer to 1. For example, the Q_{10} of the change in the viscosity of water is 1.2–1.3. In this exercise, you will calculate Q_{10} using data from radish seeds (*Raphanus sativum*) to assess whether the initial uptake of water by seeds is more likely to be a physical or a chemical process.

How the Experiment Was Done Samples of radish seeds were weighed and placed in water at four different temperatures. After 30 minutes, the seeds were removed, blotted dry, and reweighed. The researchers then calculated the percent increase in mass due to water uptake for each sample.

Data from the Experiment

Temperature	% Increase in Mass Due to Water Uptake After 30 Minutes
5°C	18.5
15°C	26.0
25°C	31.0
35°C	36.2

Data from J. D. Murphy and D. L. Noland, Temperature effects on seed imbibition and leakage mediated by viscosity and membranes, *Plant Physiology* 69:428–431 (1982).

INTERPRET THE DATA

1. Based on the data, does the initial uptake of water by radish seeds vary with temperature? What is the relationship between temperature and water uptake?

2. **(a)** Using the data for 35°C and 25°C, calculate Q_{10} for water uptake by radish seeds. Repeat the calculation using the data for 25°C and 15°C and the data for 15°C and 5°C. **(b)** What is the average Q_{10}? **(c)** Do your results imply that the uptake of water by radish seeds is mainly a physical process or a chemical (metabolic) process? **(d)** Given that the Q_{10} for the change in the viscosity of water is 1.2–1.3, could the slight temperature dependence of water uptake by seeds be a reflection of the slight temperature dependence of the viscosity of water?

3. Besides temperature, what other independent variables could you alter to test whether radish seed swelling is essentially a physical process or a chemical process?

4. Would you expect plant growth to have a Q_{10} closer to 1 or 3? Why?

→ **Instructors:** A version of this Scientific Skills Exercise can be assigned in **Mastering Biology**.

As you learn to apply the water potential equation, keep in mind the key point: *Water moves from regions of higher water potential to regions of lower water potential.*

Water Movement Across Plant Cell Membranes

Now let's consider how water potential affects absorption and loss of water by a living plant cell. First, imagine a cell that is **flaccid** (limp) as a result of losing water. The cell has a Ψ_P of 0 MPa. Suppose this flaccid cell is bathed in a solution of higher solute concentration (more negative solute potential) than the cell itself **(Figure 36.7a)**. Since the external solution has the lower (more negative) water potential, water diffuses out of the cell. The cell's protoplast undergoes **plasmolysis**—that is, it shrinks and pulls away from the cell wall. If we place the same flaccid cell in pure water ($\Psi = 0$ MPa) **(Figure 36.7b)**, the cell, because it contains solutes, has a lower water potential than the water, and water enters the cell by osmosis. The contents of the cell begin to swell and press the plasma membrane against the cell wall. The partially elastic wall, exerting turgor pressure, confines the pressurized protoplast. When this pressure is enough to offset the tendency for water to enter because of the solutes in the cell, then Ψ_P and Ψ_S are equal, and $\Psi = 0$. This matches the water potential of the extracellular environment— in this example, 0 MPa. A dynamic equilibrium has been reached, and there is no further *net* movement of water.

In contrast to a flaccid cell, a walled cell with a greater solute concentration than its surroundings is **turgid**, or very firm. When turgid cells in a nonwoody tissue push against each other, the tissue is stiffened. The effects of turgor loss are seen during **wilting**, when leaves and stems droop as a result of cells losing water.

Turgid

Wilted

Aquaporins: Facilitating Diffusion of Water

A difference in water potential determines the *direction* of water movement across membranes, but how do water molecules actually cross the membranes? Water molecules are small enough to diffuse across the phospholipid bilayer, even though the bilayer's interior is hydrophobic. However, their movement across biological membranes is too rapid to be explained by unaided diffusion. Transport proteins called **aquaporins** (see Figure 7.10 and Concept 7.2) facilitate the

▼ **Figure 36.7 Water relations in plant cells.** In these experiments, flaccid cells (cells in which the protoplast contacts the cell wall but lacks turgor pressure) are placed in two environments. The blue arrows indicate initial net water movement.

Initial flaccid cell:
$\Psi_P = 0$
$\Psi_S = -0.7$
$\overline{\Psi = -0.7 \text{ MPa}}$

Environment
0.4 M sucrose solution:
$\Psi_P = 0$
$\Psi_S = -0.9$
$\overline{\Psi = -0.9 \text{ MPa}}$

Final plasmolyzed cell at osmotic equilibrium with its surroundings:
$\Psi_P = 0$
$\Psi_S = -0.9$
$\overline{\Psi = -0.9 \text{ MPa}}$

(a) Initial conditions: cellular Ψ > environmental Ψ. The protoplast loses water, and the cell plasmolyzes. After plasmolysis is complete, the water potentials of the cell and its surroundings are the same.

→ **Mastering Biology** Video: Plasmolysis

Initial flaccid cell:
$\Psi_P = 0$
$\Psi_S = -0.7$
$\overline{\Psi = -0.7 \text{ MPa}}$

Environment
Pure water:
$\Psi_P = 0$
$\Psi_S = 0$
$\overline{\Psi = 0 \text{ MPa}}$

Final turgid cell at osmotic equilibrium with its surroundings:
$\Psi_P = 0.7$
$\Psi_S = -0.7$
$\overline{\Psi = 0 \text{ MPa}}$

(b) Initial conditions: cellular Ψ < environmental Ψ. Net uptake of water by osmosis makes the cell turgid. When this tendency for water to enter is offset by the back pressure of the elastic wall, water potentials are equal for the cell and its surroundings.

transport of water molecules across plant cell plasma membranes. Aquaporin channels, which can open and close, affect the *rate* at which water moves osmotically across the membrane. Their permeability is decreased by increases in cytosolic Ca^{2+} or decreases in cytosolic pH.

Long-Distance Transport: The Role of Bulk Flow

Diffusion is an effective transport mechanism over the spatial scales typically found at the cellular level. However, diffusion is much too slow to function in long-distance transport within a plant. Although diffusion from one end of a cell to the other takes just seconds, diffusion from the roots to the top of a giant redwood would take several centuries. Instead, long-distance transport occurs through **bulk flow**, the movement of liquid in response to a pressure gradient. The bulk flow of material always occurs from higher to lower pressure. Unlike osmosis, bulk flow is independent of solute concentration.

Long-distance bulk flow occurs within specialized cells in the vascular tissue, namely, the tracheids and vessel elements of the xylem and the sieve-tube elements of the phloem. In leaves, the branching of veins ensures that no cell is more than a few cells away from the vascular tissue **(Figure 36.8)**.

The structures of the conducting cells of the xylem and phloem facilitate bulk flow. Mature tracheids and vessel elements are dead cells and therefore have no cytoplasm, and the cytoplasm of sieve-tube elements (also called sieve-tube members) is almost devoid of organelles (see Figure 35.10). If you have dealt with a partially clogged drain, you know that the volume of flow depends on the pipe's diameter. Clogs reduce the effective diameter of the drainpipe. Such

experiences help us understand how the structures of plant cells specialized for bulk flow fit their function. Like unclogging a drain, the absence or reduction of cytoplasm in a plant's "plumbing" facilitates bulk flow through the xylem and phloem. Bulk flow is also enhanced by the perforation plates at the ends of vessel elements and the porous sieve plates connecting sieve-tube elements.

Diffusion, active transport, and bulk flow act in concert to transport resources throughout the whole plant. For example, bulk flow due to a pressure difference is the mechanism of long-distance transport of sugars in the phloem, but active transport of sugar at the cellular level maintains this pressure difference. In the next three sections, we'll examine in more detail the transport of water and minerals from roots to shoots, the control of evaporation, and the transport of sugars.

▼ **Figure 36.8 Venation in an aspen leaf.** The finer and finer branching of leaf veins in eudicot leaves ensures that no leaf cell is far removed from the vascular system.

Cell

VISUAL SKILLS *In this leaf, what is the maximum number of cells any mesophyll cell is from a vein?*

1. If a plant cell immersed in distilled water has a Ψ_S of -0.7 MPa and a Ψ of 0 MPa, what is the cell's Ψ_P? If you put it in an open beaker of solution that has a Ψ of -0.4 MPa, what would be its Ψ_P at equilibrium?

2. How would a reduction in the number of aquaporin channels affect a plant cell's ability to adjust to new osmotic conditions?

3. How would the long-distance transport of water be affected if tracheids and vessel elements were alive at maturity? Explain.

4. **WHAT IF?** What would happen if you put plant protoplasts in pure water? Explain.

For suggested answers, see Appendix A.

CONCEPT 36.3

Transpiration drives the transport of water and minerals from roots to shoots via the xylem

Picture yourself struggling to carry a 19-liter (5-gallon) container of water weighing 19 kilograms (42 pounds) up several flights of stairs. Imagine doing this 40 times a day. Then consider the fact that an average-sized tree, despite having neither heart nor muscle, transports a similar volume of water effortlessly on a daily basis. How do trees accomplish this feat? To answer this question, we'll follow each step in the journey of water and minerals from roots to leaves.

Absorption of Water and Minerals by Root Cells

Although all living plant cells absorb nutrients across their plasma membranes, the cells near the tips of roots are particularly important because most of the absorption of water and minerals occurs there. In this region, the epidermal cells are permeable to water, and many are differentiated into root hairs, modified cells that account for much of the absorption of water by roots (see Figure 35.3). The root hairs absorb the soil solution, which consists of water molecules and dissolved mineral ions that are not bound tightly to soil particles. The soil solution is drawn into the hydrophilic walls of epidermal cells and passes freely along the cell walls and the extracellular spaces into the root cortex. This flow enhances the exposure of the cells of the cortex to the soil solution, providing a much greater membrane surface area for absorption than the surface area of the epidermis alone. Although the soil solution usually has a low mineral concentration, active transport enables roots to accumulate essential minerals, such as K^+, to concentrations hundreds of times greater than in the soil.

Transport of Water and Minerals into the Xylem

Water and minerals that pass from the soil into the root cortex cannot be transported to the rest of the plant until they enter the xylem of the vascular cylinder, or stele. The **endodermis**, the innermost layer of cells in the root cortex, functions as a last checkpoint for the selective passage of minerals from the cortex into the vascular cylinder **(Figure 36.9)**. Minerals already in the symplast when they reach the endodermis continue through the plasmodesmata of endodermal cells and pass into the vascular cylinder. These minerals were already screened by the plasma membrane they had to cross to enter the symplast in the epidermis or cortex.

Minerals that reach the endodermis via the apoplast encounter a dead end that blocks their passage into the vascular cylinder. This barrier, located in the transverse and radial walls of each endodermal cell, is the **Casparian strip**, a belt made of suberin, a waxy material impervious to water and dissolved minerals (see Figure 36.9). Because of the Casparian strip, water and minerals cannot cross the endodermis and enter the vascular cylinder via the apoplast. Instead, water and minerals that are passively moving through the apoplast must cross the *selectively permeable* plasma membrane of an endodermal cell before they can enter the vascular cylinder. In this way, the endodermis transports needed minerals from the soil into the xylem and keeps many unneeded or toxic substances out. The endodermis also prevents solutes that have accumulated in the xylem from leaking back into the soil solution.

The last segment in the soil-to-xylem pathway is the passage of water and minerals into the tracheids and vessel elements of the xylem. These water-conducting cells lack protoplasts when mature and are therefore parts of the apoplast. Endodermal cells, as well as living cells within the vascular cylinder, discharge minerals from their protoplasts into their own cell walls. Both diffusion and active transport are involved in this transfer of solutes from the symplast to the apoplast, and the water and minerals can now enter the tracheids and vessel elements, where they are transported to the shoot system by bulk flow.

Bulk Flow Transport via the Xylem

Water and minerals from the soil enter the plant through the epidermis of roots, cross the root cortex, and pass into the vascular cylinder. From there the **xylem sap**, the water and dissolved minerals in the xylem, is transported long distances by bulk flow to the veins that branch throughout each leaf. As noted earlier, bulk flow is much faster than diffusion or active transport. Peak velocities in the transport of xylem sap can range from 15 to 45 m/hr for trees with wide vessel elements. The stems and leaves depend on this rapid delivery system for their supply of water and minerals.

The process of transporting xylem sap involves the loss of an astonishing amount of water by **transpiration**, the loss of water vapor from leaves and other aerial parts of the plant. A single maize plant, for example, transpires 60 L of water (the equivalent of 170 12-ounce bottles) during a growing season.

▼ Figure 36.9 Transport of water and minerals from root hairs to the xylem.

VISUAL SKILLS *After studying the figure, explain how the Casparian strip forces water and minerals to pass through the plasma membranes of endodermal cells.*

→ **Mastering Biology Figure Walkthrough**
BioFlix® Animation: Water Transport from Soil into Roots

Casparian strip
Pathway along apoplast
Endodermal cell
Pathway through symplast
Plasmodesmata
Water moves upward in vascular cylinder

Casparian strip
Plasma membrane
Apoplastic route
Symplastic route
Root hair
Vessels (xylem)
Epidermis
Endodermis
Vascular cylinder (stele)
Cortex

❶ **Apoplastic route.** Uptake of soil solution by the hydrophilic walls of root hairs provides access to the apoplast. Water and minerals can then diffuse into the cortex along this matrix of walls and extracellular spaces.

❷ **Symplastic route.** Minerals and water that cross the plasma membranes of root hairs can enter the symplast.

❸ **Transmembrane route.** As soil solution moves along the apoplast, some water and minerals are transported into the protoplasts of cells of the epidermis and cortex and then move inward via the symplast.

❹ **The endodermis: controlled entry to the vascular cylinder (stele).** Within the transverse and radial walls of each endodermal cell is the Casparian strip, a belt of waxy material (purple band) that blocks the passage of water and dissolved minerals. Only minerals already in the symplast or entering that pathway by crossing the plasma membrane of an endodermal cell can detour around the Casparian strip and pass into the vascular cylinder (stele).

❺ **Transport in the xylem.** Endodermal cells and also living cells within the vascular cylinder discharge water and minerals into their walls (apoplast). The xylem vessels then transport the water and minerals by bulk flow upward into the shoot system.

A maize crop growing at a typical density of 60,000 plants per hectare transpires almost 4 million L of water per hectare (about 400,000 gallons of water per acre) every growing season. If the transpired water is not replaced by water transported up from the roots, the leaves will wilt, and the plants will eventually die.

Xylem sap rises to heights of more than 120 m in the tallest trees. Is the sap mainly *pushed* upward from the roots, or is it mainly *pulled* upward? Let's evaluate the relative contributions of these two mechanisms.

Pushing Xylem Sap: Root Pressure

At night, when there is almost no transpiration, root cells continue actively pumping mineral ions into the xylem of the vascular cylinder. Meanwhile, the Casparian strip of the endodermis prevents the ions from leaking back out into the cortex and soil. The resulting accumulation of minerals lowers the water potential within the vascular cylinder.

Water flows in from the root cortex, generating **root pressure**, a push of xylem sap. The root pressure sometimes causes more water to enter the leaves than is transpired, resulting in **guttation**, the exudation of water droplets that can be seen in the morning on the tips or edges of some plant leaves **(Figure 36.10)**. Guttation fluid should not be confused with dew, which is condensed atmospheric moisture.

▶ **Figure 36.10 Guttation.** Root pressure is forcing excess water from this strawberry leaf.

In most plants, root pressure is a minor mechanism driving the ascent of xylem sap, pushing water only a few meters at most. The positive pressures produced are simply too weak to overcome the gravitational force of the water column in the xylem, particularly in tall plants. Many plants do not generate any root pressure or do so only during part of the growing season. Even in plants that display guttation, root pressure cannot keep pace with transpiration after sunrise. For the most part, xylem sap is not pushed from below by root pressure but is pulled up.

Pulling Xylem Sap: The Cohesion-Tension Hypothesis

As we have seen, root pressure, which depends on the active transport of solutes by plants, is only a minor force in the ascent of xylem sap. Far from depending on the metabolic activity of cells, most of the xylem sap that rises through a tree does not even require living cells to do so. As demonstrated by Eduard Strasburger in 1891, leafy stems with their lower end immersed in toxic solutions of copper sulfate or acid will readily draw these poisons up if the stem is cut below the surface of the liquid. As the toxic solutions ascend, they kill all living cells in their path, eventually arriving in the transpiring leaves and killing the leaf cells as well. Nevertheless, as Strasburger noted, the uptake of the toxic solutions and the loss of water from the dead leaves can continue for weeks.

In 1894, a few years after Strasburger's findings, two Irish scientists, John Joly and Henry Dixon, put forward a hypothesis that remains the leading explanation of the ascent of xylem sap. According to their **cohesion-tension hypothesis**, transpiration provides the pull for the ascent of xylem sap, and the cohesion of water molecules transmits this pull along the entire length of the xylem from shoots to roots. Hence, xylem sap is normally under negative pressure, or tension. Since transpiration is a "pulling" process, our exploration of the rise of xylem sap by the cohesion-tension mechanism begins not with the roots but with the leaves, where the driving force for transpirational pull begins.

Transpirational Pull Stomata on a leaf's surface lead to a maze of internal air spaces that expose the mesophyll cells to the CO_2 they need for photosynthesis. The air in these spaces is saturated with water vapor because it is in contact with the moist walls of the cells. On most days, the air outside the leaf is drier; that is, it has lower water potential than the air inside the leaf. Therefore, water vapor in the air spaces of a leaf diffuses down its water potential gradient and exits the leaf via the stomata. It is this loss of water vapor by diffusion and evaporation that we call transpiration.

But how does loss of water vapor from the leaf translate into a pulling force for upward movement of water through a plant? The negative pressure potential that causes water to move up through the xylem develops at the surface of mesophyll cell walls in the leaf **(Figure 36.11)**. The cell wall acts like a very thin capillary network. Water adheres to the cellulose microfibrils and other hydrophilic components

▼ **Figure 36.11 Generation of transpirational pull.** Negative pressure (tension) at the air-water interface in the leaf is the basis of transpirational pull, which draws water out of the xylem.

5 Water from the xylem is pulled into the surrounding cells and air spaces to replace the water that was lost.

4 The increased surface tension shown in step **3** pulls water from surrounding cells and air spaces.

3 The evaporation of the water film causes the air-water interface to retreat farther into the cell wall and to become more curved. This curvature increases the surface tension and the rate of transpiration.

2 At first, the water vapor lost by transpiration is replaced by evaporation from the water film that coats mesophyll cells.

1 In transpiration, water vapor (shown as blue dots) diffuses from the moist air spaces of the leaf to the drier air outside via stomata.

→ **Mastering Biology BioFlix® Animation: Transpiration**

of the cell wall. As water evaporates from the water film that covers the cell walls of mesophyll cells, the air-water interface retreats farther into the cell wall. Because of the high surface tension of water, the curvature of the interface induces a tension, or negative pressure potential, in the water. As more water evaporates from the cell wall, the curvature of the air-water interface increases and the pressure of the water becomes more negative. Water molecules from the more hydrated parts of the leaf are then pulled toward this area, reducing the tension. These pulling forces are transferred to the xylem because each water molecule is cohesively bound to the next by hydrogen bonds. Thus, transpirational pull depends on several of the properties of water discussed in Concept 3.2: adhesion, cohesion, and surface tension.

The role of negative pressure potential in transpiration is consistent with the water potential equation because negative pressure potential (tension) *lowers* water potential. Because water moves from areas of higher water potential to areas of lower water potential, the more negative pressure potential at the air-water interface causes water in xylem cells to be "pulled" into mesophyll cells, which lose water to the air spaces, the water diffusing out through stomata. In this way, the negative water potential of leaves provides the "pull" in transpirational pull. The transpirational pull on xylem sap is transmitted all the way from the leaves to the young roots and even into the soil solution **(Figure 36.12)**.

Cohesion and Adhesion in the Ascent of Xylem Sap

Cohesion and adhesion facilitate the transport of water by bulk flow. Cohesion is the attractive force between molecules of the same substance. Water has an unusually high cohesive force due to the hydrogen bonds each water molecule can potentially make with other water molecules. Water's cohesive force within the xylem gives it a tensile strength equivalent to that of a steel wire of similar diameter. The cohesion of water makes it possible to pull a column of xylem sap from above

▼ **Figure 36.12 Ascent of xylem sap.** Hydrogen bonding forms an unbroken chain of water molecules extending from leaves to the soil. The force driving the ascent of xylem sap is a gradient of water potential (Ψ). For bulk flow over long distance, the Ψ gradient is due mainly to a gradient of the pressure potential (Ψ_P). Transpiration results in the Ψ_P at the leaf end of the xylem being lower than the Ψ_P at the root end. The Ψ values shown at the left are a "snapshot." They may vary during daylight, but the direction of the Ψ gradient remains the same.

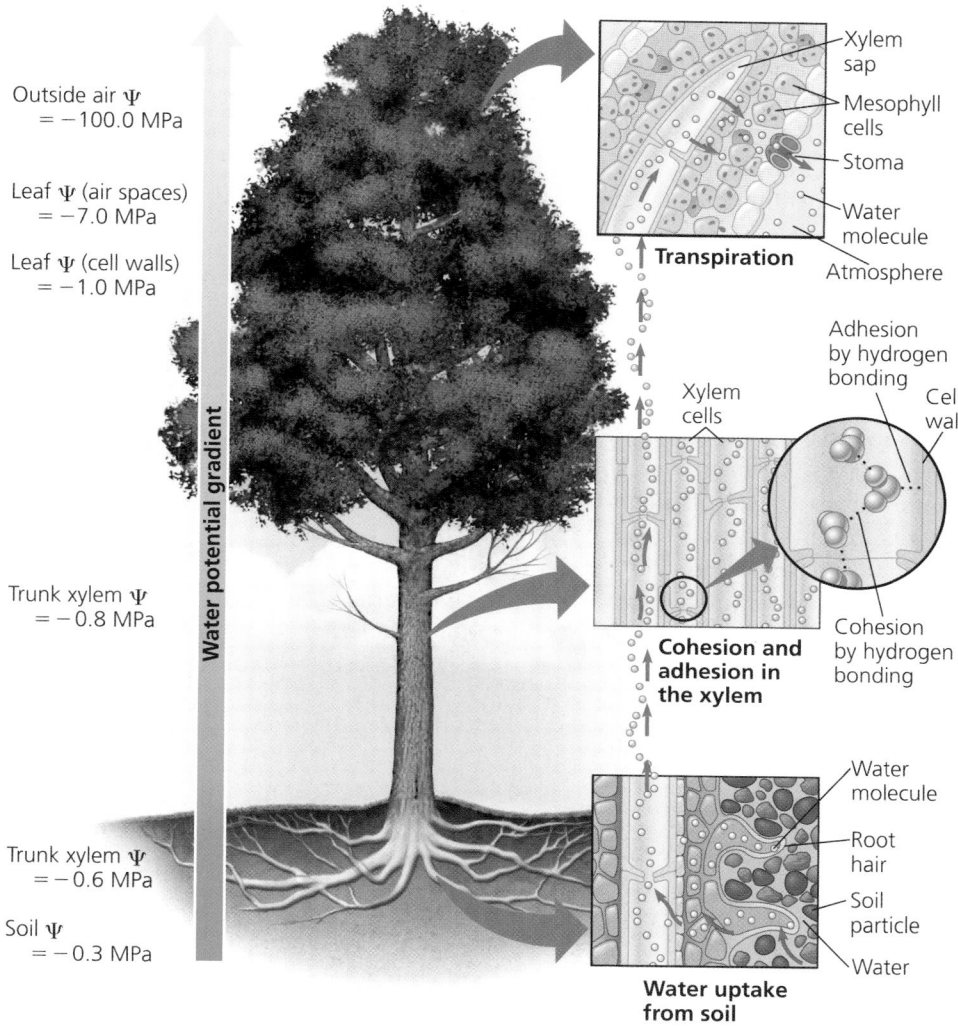

Outside air Ψ = −100.0 MPa

Leaf Ψ (air spaces) = −7.0 MPa

Leaf Ψ (cell walls) = −1.0 MPa

Trunk xylem Ψ = −0.8 MPa

Trunk xylem Ψ = −0.6 MPa

Soil Ψ = −0.3 MPa

Water potential gradient

Xylem sap
Mesophyll cells
Stoma
Water molecule
Transpiration
Atmosphere

Adhesion by hydrogen bonding
Cell wall
Xylem cells
Cohesion by hydrogen bonding
Cohesion and adhesion in the xylem

Water molecule
Root hair
Soil particle
Water
Water uptake from soil

> **Mastering Biology** BioFlix® Animation: Transpiration-Cohesion-Tension Mechanism
> Animation: Transport of Xylem Sap
> BioFlix® Animation: Water Transport from Roots to Leaves

without the water molecules separating. Water molecules exiting the xylem in the leaf tug on adjacent water molecules, and this pull is relayed, molecule by molecule, down the entire column of water in the xylem. Meanwhile, the strong adhesion of water molecules (again by hydrogen bonds) to the hydrophilic walls of xylem cells helps offset the downward force of gravity.

The upward pull on the sap creates tension within the vessel elements and tracheids, which are like elastic pipes. Positive pressure causes an elastic pipe to swell, whereas tension pulls the walls of the pipe inward. On a warm day, a decrease in the diameter of a tree trunk can even be measured.

As transpirational pull puts the vessel elements and tracheids under tension, their thick secondary walls prevent them from collapsing, much as wire rings maintain the shape of a vacuum-cleaner hose. The tension produced by transpirational pull lowers water potential in the root xylem to such an extent that water flows passively from the soil, across the root cortex, and into the vascular cylinder.

Transpirational pull can extend down to the roots only through an unbroken chain of water molecules. Cavitation, the formation of a water vapor pocket, breaks the chain. It is more common in wide vessel elements than in tracheids and can occur during drought stress or when xylem sap freezes in winter. The air bubbles resulting from cavitation expand and block water channels of the xylem. The rapid expansion of air bubbles produces clicking noises that can be heard by placing sensitive microphones at the surface of the stem.

The interruption of xylem sap transport by cavitation is not always permanent. The chain of water molecules can detour around the air bubbles through pits between adjacent tracheids or vessel elements (see Figure 35.10). Moreover, root pressure enables small plants to refill blocked vessel elements. Recent evidence suggests that cavitation may even be repaired when the xylem sap is under negative pressure, although the mechanism by which this occurs is uncertain. In addition, secondary growth adds a layer of new xylem each year. Only the youngest, outermost secondary xylem layers transport water. Although the older secondary xylem no longer transports water, it does provide support for the tree (see Figure 35.22).

Xylem Sap Ascent by Bulk Flow: *A Review*

The cohesion-tension mechanism that transports xylem sap against gravity is an excellent example of how physical principles apply to biological processes. In the long-distance transport of water from roots to leaves by bulk flow, the movement of fluid is driven by a water potential difference at opposite ends of xylem tissue. The water potential difference is created at the leaf end of the xylem by the evaporation of water from leaf cells. Evaporation lowers the water potential at the air-water interface, thereby generating the negative pressure (tension) that pulls water through the xylem.

Bulk flow in the xylem differs from diffusion in some key ways. First, it is driven by differences in pressure potential (Ψ_P); solute potential (Ψ_S) is not a factor. Therefore, the water potential gradient within the xylem is essentially a pressure gradient. Also, the flow does not occur across plasma membranes of living cells, but instead within hollow, dead cells. Furthermore, it moves the entire solution together—not just water or solutes—and at much greater speed than diffusion.

The plant expends no energy to lift xylem sap by bulk flow. Instead, the absorption of sunlight drives most of transpiration by causing water to evaporate from the moist walls of

mesophyll cells and by lowering the water potential in the air spaces within a leaf. Thus, the ascent of xylem sap, like the process of photosynthesis, is ultimately solar powered.

> **Mastering Biology BioFlix® Animation: Water Transport in Plants**

CONCEPT CHECK 36.3

1. A horticulturalist notices that when *Zinnia* flowers are cut at dawn, a small drop of water collects at the surface of the rooted stump. However, when the flowers are cut at noon, no drop is observed. Suggest an explanation.
2. **WHAT IF?** Suppose an *Arabidopsis* mutant lacking functional aquaporin proteins has a root mass three times greater than that of wild-type plants. Suggest an explanation.
3. **MAKE CONNECTIONS** How are the Casparian strip and tight junctions similar (see Figure 6.30)?

For suggested answers, see Appendix A.

CONCEPT 36.4

The rate of transpiration is regulated by stomata

Leaves generally have large surface areas and high surface-to-volume ratios. The large surface area enhances light absorption for photosynthesis. The high surface-to-volume ratio aids in CO_2 absorption during photosynthesis as well as in the release of O_2, a by-product of photosynthesis. Upon diffusing through the stomata, CO_2 enters a honeycomb of air spaces formed by the spongy mesophyll cells (see Figure 35.18). Because of the irregular shapes of these cells, the leaf's internal surface area may be 10 to 30 times greater than the external surface area.

Although large surface areas and high surface-to-volume ratios increase the rate of photosynthesis, they also increase water loss by way of the stomata. Thus, a plant's tremendous requirement for water is largely a consequence of the shoot system's need for ample exchange of CO_2 and O_2 for photosynthesis. By opening and closing the stomata, guard cells help balance the plant's requirement to conserve water with its requirement for photosynthesis.

Stomata: Major Pathways for Water Loss

About 95% of the water a plant loses escapes through stomata, although these pores account for only 1–2% of the external leaf surface. The waxy cuticle limits water loss through the remaining surface of the leaf. Each stoma is flanked by a pair of guard cells. Guard cells control the diameter of the stoma by changing shape, thereby widening or narrowing the gap between the guard cell pair. Under the same environmental conditions, the amount of water lost per leaf depends largely on stomatal density **(Figure 36.13)** and the average size of the stomatal pores.

▼ Figure 36.13 Partially open stomata in the epidermis of a bean (*Phaseolus*) leaf.

VISUAL SKILLS *If the micrograph above is of a section of epidermis 500 μm by 300 μm, what is the stomatal density of this bean leaf? Give your answer in number of stomata per square centimeter.*

Stomatal density can be under both genetic and environmental control. As a result of evolution by natural selection, shade-tolerant species tend to have lower stomatal densities than shade-intolerant species because CO_2 uptake doesn't limit photosynthesis under shady conditions. Stomatal density, however, can also be regulated by environmental changes. Low CO_2 levels during leaf development, for example, lead to increased stomatal densities in many species, an adaptation that facilitates CO_2 uptake under these conditions. By measuring the stomatal density of leaf fossils, scientists have gained insight into the levels of atmospheric CO_2 in past climates. A recent British survey found that stomatal density of many woodland species has decreased since 1927, when a similar survey was made. This observation is consistent with other findings that atmospheric CO_2 levels increased dramatically during the late 1900s.

Mechanisms of Stomatal Opening and Closing

When guard cells take in water from neighboring cells by osmosis, they become more turgid. In most angiosperm guard cells, the cell walls facing the stomatal pore are thicker, and the cellulose microfibrils are oriented in a direction that causes the guard cells to bow outward when turgid **(Figure 36.14a)**. This bowing outward increases the size of the pore between the guard cells. When the cells lose water and become flaccid, they become less bowed, and the pore closes.

The changes in turgor pressure in guard cells result primarily from the reversible absorption and loss of K^+. Stomata open when guard cells actively accumulate K^+ from neighboring epidermal cells **(Figure 36.14b)**. The flow of K^+ across the plasma membrane of the guard cell is coupled to the generation of a membrane potential by proton pumps (see Figure 36.6a). Stomatal opening correlates with active transport of H^+ out of the guard cell. The resulting voltage

(membrane potential) drives K^+ into the cell through specific membrane channels. The absorption of K^+ causes the water potential to become more negative within the guard cells, and the cells become more turgid as water enters by osmosis. Because most of the K^+ and water are stored in the vacuole, the vacuolar membrane also plays a role in regulating guard

▼ Figure 36.14 Mechanisms of stomatal opening and closing.

 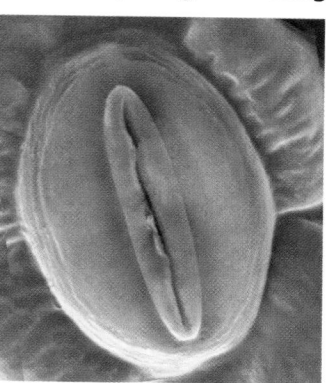

Guard cells turgid/Stoma open   **Guard cells flaccid/Stoma closed**

 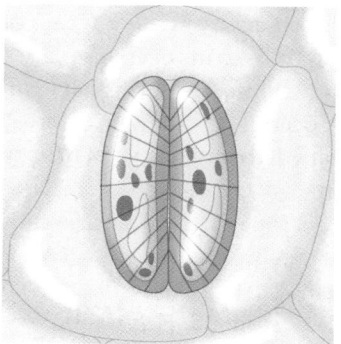

Radially oriented cellulose microfibrils

Cell wall

Nucleus

Vacuole

Chloroplast   Guard cell

(a) Changes in guard cell shape and stomatal opening and closing (surface view). Guard cells of a typical angiosperm are illustrated in their turgid (stoma open) and flaccid (stoma closed) states. The radial orientation of cellulose microfibrils in the cell walls causes the guard cells to increase more in length than width when turgor increases. Since the two guard cells are tightly joined at their tips, they bow outward when turgid, causing the stomatal pore to open.

 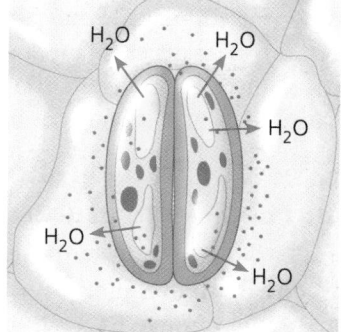

H_2O H_2O H_2O H_2O

K^+ H_2O H_2O

H_2O H_2O H_2O H_2O

(b) Role of potassium ions (K^+) in stomatal opening and closing. The transport of K^+ (symbolized here as red dots) across the plasma membrane and vacuolar membrane causes the turgor changes of guard cells. The uptake of anions, such as malate and chloride ions (not shown), also contributes to guard cell swelling.

CHAPTER 36   Resource Acquisition and Transport in Vascular Plants   **797** cell dynamics. Stomatal closing results from a loss of K^+ from guard cells to neighboring cells, which leads to an osmotic loss of water. Aquaporins also help regulate the osmotic swelling and shrinking of guard cells.

Stimuli for Stomatal Opening and Closing

In general, stomata are open during the day and mostly closed at night, preventing the plant from losing water under conditions when photosynthesis cannot occur. At least three cues contribute to stomatal opening at dawn: light, CO_2 depletion, and an internal "clock" in guard cells.

Light stimulates guard cells to accumulate K^+ and become turgid. This response is triggered by illumination of blue-light receptors in the plasma membrane of guard cells. Activation of these receptors stimulates the activity of proton pumps in the plasma membrane of the guard cells, in turn promoting absorption of K^+.

Stomata also open in response to depletion of CO_2 within the leaf's air spaces as a result of photosynthesis. As CO_2 concentrations decrease during the day, the stomata progressively open if sufficient water is supplied to the leaf.

An internal "clock" in the guard cells ensures that stomata continue their daily rhythm of opening and closing. This rhythm occurs even if a plant is kept in a dark location. All eukaryotic organisms have internal clocks that regulate cyclic processes. Cycles with intervals of approximately 24 hours are called **circadian rhythms** (which you'll learn more about in Concept 39.3).

Drought stress can also cause stomata to close. A hormone called **abscisic acid (ABA)**, which is produced in roots and leaves in response to water deficiency, signals guard cells to close stomata. This response reduces wilting but also restricts CO_2 absorption, thereby slowing photosynthesis. ABA also directly inhibits photosynthesis. Water availability is closely tied to plant productivity not because water is needed as a substrate in photosynthesis, but because freely available water allows plants to keep stomata open and take up more CO_2.

Guard cells control the photosynthesis-transpiration compromise on a moment-to-moment basis by integrating a variety of internal and external stimuli. Even the passage of a cloud or a transient shaft of sunlight through a forest can affect the rate of transpiration.

Effects of Transpiration on Wilting and Leaf Temperature

As long as most stomata remain open, transpiration is greatest on a day that is sunny, warm, dry, and windy because these environmental factors increase evaporation. If transpiration cannot pull sufficient water to the leaves, the shoot becomes slightly wilted as cells lose turgor pressure. Although plants respond to such mild drought stress by rapidly closing stomata, some evaporative water loss still occurs through the cuticle. Under prolonged drought conditions, leaves can become severely wilted and irreversibly injured.

Transpiration also results in evaporative cooling, which can lower a leaf's temperature by as much as 10°C compared with the surrounding air. This cooling prevents the leaf from reaching temperatures that could denature enzymes involved in photosynthesis and other metabolic processes.

Adaptations That Reduce Evaporative Water Loss

Water availability is a major determinant of plant productivity. The main reason water availability is tied to plant productivity is not related to photosynthesis's direct need for water as a substrate but rather because freely available water allows plants to keep stomata open and take up more CO_2. The problem of reducing water loss is especially acute for desert plants. Plants adapted to arid environments are called **xerophytes** (from the Greek *xero*, dry).

Many species of desert plants avoid drying out by completing their short life cycles during the brief rainy seasons. Rain comes infrequently in deserts, but when it arrives, the vegetation is transformed as dormant seeds of annual species quickly germinate and bloom, completing their life cycle before dry conditions return.

Other xerophytes have unusual physiological or morphological adaptations that enable them to withstand harsh desert conditions. The stems of many xerophytes are fleshy because they store water for use during long dry periods. Cacti have highly reduced leaves that resist excessive water loss; photosynthesis is carried out mainly in their stems. Another adaptation common in arid habitats is crassulacean acid metabolism (CAM), a specialized form of photosynthesis found in succulents of the family Crassulaceae and several other families (see Figure 10.20). Because the leaves of CAM plants take in CO_2 at night, the stomata can remain closed during the day, when evaporative stresses are greatest. Other examples of xerophytic adaptations are discussed in **Figure 36.15**.

→ **Mastering Biology MP3 Tutor: Transpiration**

CONCEPT CHECK 36.4

1. What are the stimuli that control the opening and closing of stomata?
2. The pathogenic fungus *Fusicoccum amygdali* secretes a toxin called fusicoccin that activates the plasma membrane proton pumps of plant cells and leads to uncontrolled water loss. Suggest a mechanism by which the activation of proton pumps could lead to severe wilting.
3. **WHAT IF?** If you buy cut flowers, why might the florist recommend cutting the stems underwater and then transferring the flowers to a vase while the cut ends are still wet?
4. **MAKE CONNECTIONS** Explain why the evaporation of water from leaves lowers their temperature (see Concept 3.2).

For suggested answers, see Appendix A.

▼ Figure 36.15 Some xerophytic adaptations.

▶ Ocotillo (*Fouquieria splendens*) is common in the southwestern region of the United States and northern Mexico. It is leafless during most of the year, thereby avoiding excessive water loss (right). Immediately after a heavy rainfall, it produces small leaves (below and inset). As the soil dries, the leaves quickly shrivel and die.

▼ Oleander (*Nerium oleander*), shown in the inset, is commonly found in arid climates. Its leaves have a thick cuticle and multiple-layered epidermal tissue that reduce water loss. Stomata are recessed in cavities called "crypts," an adaptation that reduces the rate of transpiration by protecting the stomata from hot, dry wind. Trichomes help minimize transpiration by breaking up the flow of air, allowing the chamber of the crypt to have a higher humidity than the surrounding atmosphere (LM).

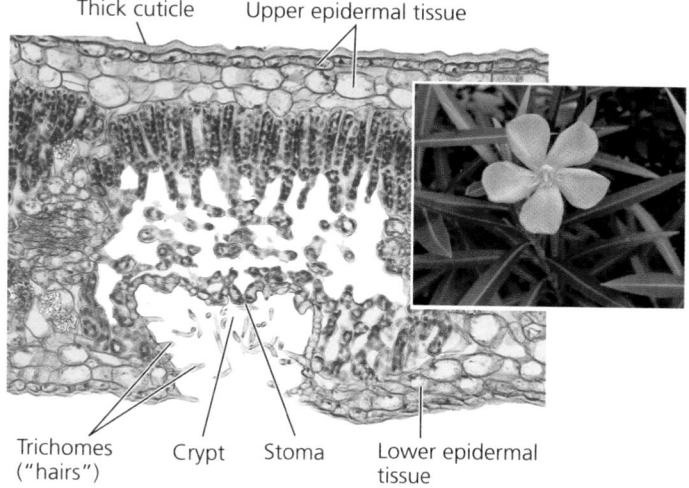

Thick cuticle   Upper epidermal tissue

100 μm

Trichomes ("hairs")   Crypt   Stoma   Lower epidermal tissue

▶ The long, white hairlike bristles along the stem of the old man cactus (*Cephalocereus senilis*) help reflect the intense sunlight of the Mexican desert.

CONCEPT 36.5

Sugars are transported from sources to sinks via the phloem

The unidirectional flow of water and minerals from soil to roots to leaves through the xylem is largely in an upward direction. In contrast, the movement of photosynthates often runs in the opposite direction, transporting sugars from mature leaves to lower parts of the plant, such as root tips that require large amounts of sugars for energy and growth. The transport of the products of photosynthesis, known as **translocation**, is carried out by another tissue, the phloem.

Movement from Sugar Sources to Sugar Sinks

Sieve-tube elements are specialized cells in angiosperms that serve as conduits for translocation. Arranged end to end, they form long sieve tubes (see Figure 35.10). Between these cells are sieve plates, structures that allow the flow of sap along the sieve tube. **Phloem sap**, the aqueous solution that flows through sieve tubes, differs markedly from the xylem sap that is transported by tracheids and vessel elements. By far the most prevalent solute in phloem sap is sugar, typically sucrose in most species. The sucrose concentration may be as high as 30% by weight, giving the sap a syrupy thickness. Phloem sap may also contain amino acids, hormones, and minerals.

In contrast to the unidirectional transport of xylem sap from roots to leaves, phloem sap moves from sites of sugar production to sites of sugar use or storage (see Figure 36.2). A **sugar source** is a plant organ that is a net producer of sugar, by photosynthesis or by breakdown of starch. In contrast, a **sugar sink** is an organ that is a net consumer or depository of sugar. Growing roots, buds, stems, and fruits are sugar sinks. Although expanding leaves are sugar sinks, mature leaves, if well illuminated, are sugar sources. A storage organ, such as a tuber or a bulb, may be a source or a sink, depending on the season. When stockpiling carbohydrates in the summer, it is a sugar sink. After breaking dormancy in the spring, it is a sugar source because its starch is broken down to sugar, which is carried to the growing shoot tips.

Sinks usually receive sugar from the nearest sugar sources. The upper leaves on a branch, for example, may export sugar to the growing shoot tip, whereas the lower leaves may export sugar to the roots. A growing fruit may monopolize the sugar sources that surround it. For each sieve tube, the direction of transport depends on the locations of the sugar source and sugar sink that are connected by that tube. Therefore, neighboring sieve tubes may carry sap in opposite directions if they originate and end in different locations.

Sugar must be transported, or loaded, into sieve-tube elements before being exported to sugar sinks. In some species, it moves from mesophyll cells to sieve-tube elements via the symplast, passing through plasmodesmata. In other species, it moves by symplastic and apoplastic pathways. In maize leaves, for example, sucrose diffuses through the symplast from photosynthetic mesophyll cells into small veins. Much of it then moves into the apoplast and is accumulated by nearby sieve-tube elements, either directly or, as shown in **Figure 36.16a**, through companion cells. In some plants, the walls of the companion cells feature many ingrowths, enhancing solute transfer between apoplast and symplast.

In many plants, sugar movement into the phloem requires active transport because sucrose is more concentrated in sieve-tube elements and companion cells than in mesophyll. Proton pumping and H^+/sucrose cotransport enable sucrose to move from mesophyll cells to sieve-tube elements or companion cells **(Figure 36.16b)**.

Sucrose is unloaded at the sink end of a sieve tube. The process varies by species and organ. However, the concentration of free sugar in the sink is always lower than in the sieve tube because the unloaded sugar is consumed during growth and metabolism of the cells of the sink or converted to insoluble polymers such as starch. As a result of this sugar concentration gradient, sugar molecules diffuse from the phloem into the sink tissues, and water follows by osmosis.

Bulk Flow by Positive Pressure: The Mechanism of Translocation in Angiosperms

Phloem sap flows from source to sink at rates that are as great as 1 m/hr, which is much faster than diffusion or cytoplasmic streaming. Researchers concluded that it moves through the sieve tubes of angiosperms by bulk flow driven by positive pressure, known as *pressure flow* **(Figure 36.17)**. The building of pressure at the source and reduction of that pressure at the sink cause sap to flow from source to sink.

The pressure-flow hypothesis explains why phloem sap flows from source to sink, and experiments build a strong case for pressure flow as the mechanism of translocation in

▼ **Figure 36.16 Loading of sucrose into phloem.**

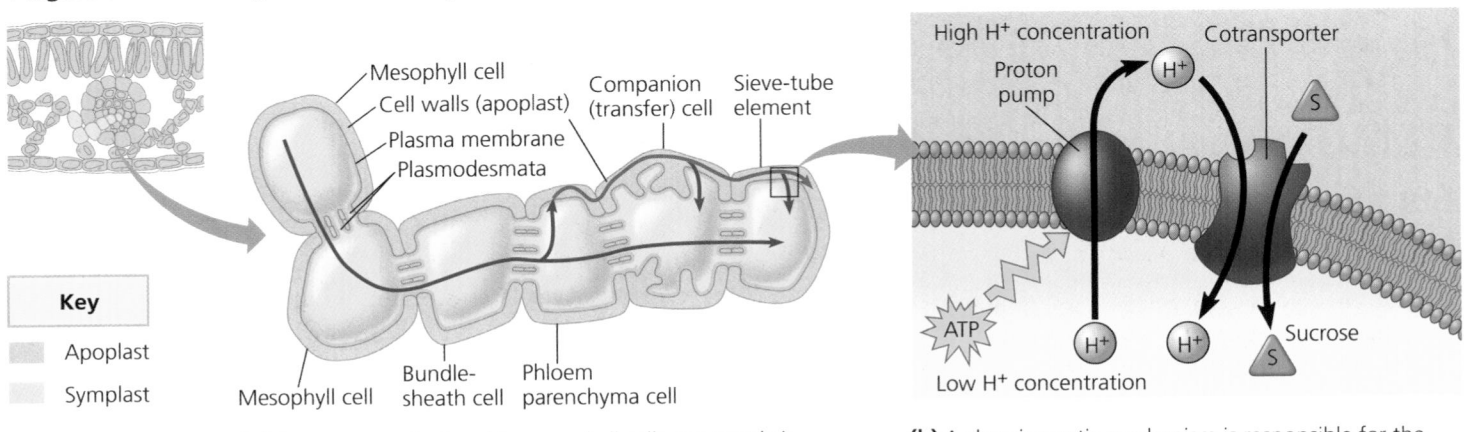

Key
Apoplast
Symplast

(a) Sucrose manufactured in mesophyll cells can travel via the symplast (blue arrows) to sieve-tube elements. In some species, sucrose exits the symplast near sieve tubes and travels through the apoplast (red arrow). It is then actively accumulated from the apoplast by sieve-tube elements and their companion cells.

(b) A chemiosmotic mechanism is responsible for the active transport of sucrose into companion cells and sieve-tube elements. Proton pumps generate an H^+ gradient, which drives sucrose accumulation with the help of a cotransport protein that couples sucrose transport to the diffusion of H^+ back into the cell.

▼ Figure 36.17 Bulk flow by positive pressure (pressure flow) in a sieve tube.

① Loading of sugar (green dots) into the sieve tube at the source (in this example, a mesophyll cell in a leaf) reduces water potential inside the sieve-tube elements. This causes the tube to take up water by osmosis.

② This uptake of water generates a positive pressure that forces the sap to flow along the tube.

③ The pressure is relieved by the unloading of sugar and the consequent loss of water at the sink.

④ In leaf-to-root translocation, xylem recycles water from sink to source.

→ **Mastering Biology** Animation: Phloem Transport

angiosperms **(Figure 36.18)**. However, studies using electron microscopes suggest that in nonflowering vascular plants, the pores between phloem cells may be too small or obstructed to permit pressure flow.

Sinks vary in energy demands and capacity to unload sugars. Sometimes there are more sinks than can be supported by sources. In such cases, a plant might abort some flowers, seeds, or fruits—a phenomenon called *self-thinning*. Removing sinks can also be a horticulturally useful practice. For example, since large apples command a much better price than small ones, growers sometimes remove flowers or young fruits so that their trees produce fewer but larger apples.

CONCEPT CHECK 36.5

1. Compare and contrast the forces that move phloem sap and xylem sap over long distances.
2. Identify plant organs that are sugar sources, organs that are sugar sinks, and organs that might be either. Explain.
3. Why can xylem transport water and minerals using dead cells, whereas phloem requires living cells?
4. **WHAT IF?** Apple growers in Japan sometimes make a non-lethal spiral slash around the bark of trees that are destined for removal after the growing season. This practice makes the apples sweeter. Why?

For suggested answers, see Appendix A.

▼ Figure 36.18 Inquiry

Does phloem sap contain more sugar near sources than near sinks?

Experiment The pressure-flow hypothesis predicts that phloem sap near sources should have a higher sugar content than phloem sap near sinks. To test this idea, researchers used aphids that feed on phloem sap. An aphid probes with a hypodermic-like mouthpart called a stylet that penetrates a sieve-tube element. As sieve-tube pressure forced out phloem sap into the stylets, the researchers separated the aphids from the stylets, which then acted as taps exuding sap for hours. Researchers measured the sugar concentration of sap from stylets at different points between a source and sink.

Aphid feeding

Stylet in sieve-tube element

Separated stylet exuding sap

Results The closer the stylet was to a sugar source, the higher its sugar concentration was.

Conclusion The results of such experiments support the pressure-flow hypothesis, which predicts that sugar concentrations should be higher in sieve tubes closer to sugar sources.

Data from S. Rogers and A. J. Peel, Some evidence for the existence of turgor pressure in the sieve tubes of willow (*Salix*), *Planta* 126:259–267 (1975).

WHAT IF? Spittlebugs (Clasirptora sp.) *are xylem sap feeders that use strong muscles to pump xylem sap through their guts. Could you isolate xylem sap from the excised stylets of spittlebugs?*

CONCEPT 36.6

The symplast is highly dynamic

Although we have been discussing transport in mostly physical terms, almost like the flow of solutions through pipes, plant transport is a dynamic and finely tuned process that changes during development. A leaf, for example, may begin as a sugar sink but spend most of its life as a sugar source. Also, environmental changes may trigger responses in plant transport processes. Water stress may activate signal transduction pathways that greatly alter the membrane transport proteins governing the overall transport of water and minerals. Because the symplast is living tissue, it is largely responsible for the dynamic changes in plant transport processes.

We'll look now at some other examples: changes in plasmodesmata, chemical signaling, and electrical signaling.

Changes in Plasmodesmatal Number and Pore Size

Based mostly on the static images provided by electron microscopy, biologists formerly considered plasmodesmata to be unchanging, pore-like structures. More recent studies, however, have revealed that plasmodesmata are highly dynamic. They can open or close rapidly in response to changes in turgor pressure, cytosolic Ca^{2+} level, or cytosolic pH. Although some plasmodesmata form during cytokinesis, they can also form much later. Moreover, loss of function is common during differentiation. For example, as a leaf matures from a sink to a source, its plasmodesmata either close or are eliminated, causing phloem unloading to cease.

Early studies by plant physiologists and pathologists came to differing conclusions regarding pore sizes of plasmodesmata. Physiologists injected fluorescent probes of different molecular sizes into cells and recorded whether the molecules passed into adjacent cells. Based on these observations, they concluded that the pore sizes were approximately 2.5 nm—too small for macromolecules such as proteins to pass. In contrast, pathologists provided electron micrographs showing evidence of the passage of virus particles with diameters of 10 nm or greater **(Figure 36.19)**.

Subsequently, it was learned that plant viruses produce *viral movement proteins* that cause the plasmodesmata to dilate, enabling the viral RNA to pass between cells. More recent evidence shows that plant cells themselves regulate plasmodesmata as part of a communication network. The viruses can subvert this network by mimicking the cell's regulators of plasmodesmata.

A high degree of cytosolic interconnectedness exists only within certain groups of cells and tissues, which are known as *symplastic domains*. Informational molecules, such as proteins and RNAs, coordinate development between cells within each symplastic domain. If symplastic communication is disrupted, development can be grossly affected.

Phloem: An Information Superhighway

In addition to transporting sugars, the phloem is a "superhighway" for the transport of macromolecules and viruses. This transport is systemic (throughout the body), affecting many or all of the plant's systems or organs. Macromolecules translocated through the phloem include proteins and various types of RNA that enter the sieve tubes through plasmodesmata. Although they are often likened to the gap junctions between animal cells, plasmodesmata are unique in their ability to traffic proteins and RNA.

Systemic communication through the phloem helps integrate the functions of the whole plant. One classic example is the delivery of a flower-inducing chemical signal from leaves to vegetative meristems. Another is a defensive response to localized infection, in which chemical signals traveling through the phloem activate defense genes in noninfected tissues.

Electrical Signaling in the Phloem

Rapid, long-distance electrical signaling through the phloem is another dynamic feature of the symplast. Electrical signaling has been studied extensively in plants that have rapid leaf movements, such as the sensitive plant (*Mimosa pudica*) and Venus flytrap (*Dionaea muscipula*). However, its role in other species is less clear. Some studies have revealed that a stimulus in one part of a plant can trigger an electrical signal in the phloem that affects another part, where it may elicit a change in gene transcription, respiration, photosynthesis, phloem unloading, or hormonal levels. Thus, the phloem can serve a nerve-like function, allowing for swift electrical communication between widely separated organs.

The coordinated transport of materials and information is central to plant survival. Plants can acquire only so many resources in the course of their lifetimes. Ultimately, the successful acquisition of these resources and their optimal distribution are the most critical determinants of whether the plant will compete successfully.

▼ **Figure 36.19 Virus particles moving cell to cell through plasmodesma connecting turnip leaf cells. (TEM)**

Plasmodesma

Cytoplasm of cell 2

Virus particles

Cytoplasm of cell 1

Cell walls

100 nm

➔ **Mastering Biology**
Interview with Patricia Zambryski:
Exploring the dynamics of plasmodesmata

CONCEPT CHECK 36.6

1. How do plasmodesmata differ from gap junctions?
2. Nerve-like signals in animals are thousands of times faster than their plant counterparts. Suggest a behavioral reason for the difference.
3. **WHAT IF?** Suppose plants were genetically modified to be unresponsive to viral movement proteins. Would this be a good way to prevent the spread of infection? Explain.

For suggested answers, see Appendix A.

36 Chapter Review

Go to **Mastering Biology** for Assignments, the eText, the Study Area, and Dynamic Study Modules.

SUMMARY OF KEY CONCEPTS

To review key terms, go to the **Vocabulary Self-Quiz** in the **Mastering Biology** eText or Study Area, or go to goo.gl/zkjz9t.

CONCEPT 36.1

Adaptations for acquiring resources were key steps in the evolution of vascular plants (pp. 785–787)

- Leaves typically function in gathering sunlight and CO_2. Stems serve as supporting structures for leaves and as conduits for the long-distance transport of water and nutrients. Roots mine the soil for water and minerals and anchor the whole plant.
- Natural selection has produced plant architectures that optimize resource acquisition in the ecological niche in which the plant species naturally exists.

? *How did the evolution of xylem and phloem contribute to the successful colonization of land by vascular plants?*

CONCEPT 36.2

Different mechanisms transport substances over short or long distances (pp. 787–792)

- The selective permeability of the plasma membrane controls the movement of substances into and out of cells. Both active and passive transport mechanisms occur in plants.
- Plant tissues have two major compartments: the **apoplast** (everything outside the cells' plasma membranes) and the **symplast** (the cytosol and connecting plasmodesmata).
- Direction of water movement depends on the **water potential**, a quantity that incorporates solute concentration and physical pressure. The **osmotic** uptake of water by plant cells and the resulting internal pressure that builds up make plant cells **turgid**.
- Long-distance transport occurs through **bulk flow**, the movement of liquid in response to a pressure gradient. Bulk flow occurs within the tracheids and vessel elements of the **xylem** and within the sieve-tube elements of the **phloem**.

? *Is xylem sap usually pulled or pushed up the plant?*

CONCEPT 36.3

Transpiration drives the transport of water and minerals from roots to shoots via the xylem (pp. 792–796)

- Water and minerals from the soil enter the plant through the epidermis of roots, cross the root cortex, and then pass into the vascular cylinder by way of the selectively permeable cells of the **endodermis**. From the vascular cylinder, the **xylem sap** is transported long distances by bulk flow to the veins that branch throughout each leaf.

- The **cohesion-tension hypothesis** proposes that the movement of xylem sap is driven by a water potential difference created at the leaf end of the xylem by the evaporation of water from leaf cells. Evaporation lowers the water potential at the air-water interface, thereby generating the negative pressure that pulls water through the xylem.

? *Why is the ability of water molecules to form hydrogen bonds important for the movement of xylem sap?*

CONCEPT 36.4

The rate of transpiration is regulated by stomata (pp. 796–799)

- **Transpiration** is the loss of water vapor from plants. **Wilting** occurs when the water lost by transpiration is not replaced by absorption from roots. Plants respond to water deficits by closing their stomata. Under prolonged drought conditions, plants can become irreversibly injured.
- Stomata are the major pathway for water loss from plants. A stoma opens when guard cells bordering the stomatal pore take up K^+. The opening and closing of stomata are controlled by light, CO_2, the drought hormone **abscisic acid**, and a **circadian rhythm**.
- **Xerophytes** are plants that are adapted to arid environments. Reduced leaves and CAM photosynthesis are examples of adaptations to arid environments.

? *Why are stomata necessary?*

CONCEPT 36.5

Sugars are transported from sources to sinks via the phloem (pp. 799–801)

- Mature leaves are the main **sugar sources**, although storage organs can be seasonal sources. Growing organs such as roots, stems, and fruits are the main **sugar sinks**. The direction of phloem transport is always from sugar source to sugar sink.
- Phloem loading depends on the active transport of sucrose. Sucrose is cotransported with H^+, which diffuses down a gradient generated by proton pumps. Loading of sugar at the source and unloading at the sink maintain a pressure difference that keeps **phloem sap** flowing through a sieve tube.

? *Why is phloem transport considered an active process?*

CONCEPT 36.6

The symplast is highly dynamic (pp. 801–802)

- Plasmodesmata can change in permeability and number. When dilated, they provide a passageway for the symplastic transport of proteins, RNAs, and other macromolecules over long distances. The phloem also conducts nerve-like electrical signals that help integrate whole-plant function.

? *By what mechanisms is symplastic communication regulated?*

CHAPTER 36   Resource Acquisition and Transport in Vascular Plants   **803**

⊘ For more multiple-choice questions, go to the **Practice Test** in the **Mastering Biology** eText or Study Area, or go to goo.gl/GruWRg.

Levels 1-2: Remembering/Understanding

1. Which of the following is an adaptation that enhances the uptake of water and minerals by roots?
   (A) mycorrhizae
   (B) pumping through plasmodesmata
   (C) active uptake by vessel elements
   (D) rhythmic contractions by cells in the root cortex

2. Which structure or compartment is part of the symplast?
   (A) the interior of a vessel element
   (B) the interior of a sieve tube
   (C) the cell wall of a mesophyll cell
   (D) an extracellular air space

3. Movement of phloem sap from a source to a sink
   (A) occurs through the apoplast of sieve-tube elements.
   (B) depends ultimately on the activity of proton pumps.
   (C) depends on tension, or negative pressure potential.
   (D) results mainly from diffusion.

Levels 3-4: Applying/Analyzing

4. Photosynthesis ceases when leaves wilt, mainly because
   (A) the chlorophyll in wilting leaves is degraded.
   (B) accumulation of CO_2 in the leaf inhibits enzymes.
   (C) stomata close, preventing CO_2 from entering the leaf.
   (D) photolysis, the water-splitting step of photosynthesis, cannot occur when there is a water deficiency.

5. What would enhance water uptake by a plant cell?
   (A) decreasing the Ψ of the surrounding solution
   (B) positive pressure on the surrounding solution
   (C) the loss of solutes from the cell
   (D) increasing the Ψ of the cytoplasm

6. A plant cell with a Ψ_S of -0.65 MPa maintains a constant volume when bathed in a solution that has a Ψ_S of -0.30 MPa and is in an open container. The cell has a
   (A) Ψ_P of $+0.65$ MPa.
   (B) Ψ of -0.65 MPa.
   (C) Ψ_P of $+0.35$ MPa.
   (D) Ψ_P of 0 MPa.

7. Compared with a cell with few aquaporin proteins in its membrane, a cell containing many aquaporin proteins will
   (A) have a faster rate of osmosis.
   (B) have a lower water potential.
   (C) have a higher water potential.
   (D) accumulate water by active transport.

8. Which of the following would tend to increase transpiration?
   (A) spiny leaves
   (B) sunken stomata
   (C) a thicker cuticle
   (D) higher stomatal density

Levels 5-6: Evaluating/Creating

9. **EVOLUTION CONNECTION** Large brown algae called kelps can grow as tall as 25 m. Kelps consist of a holdfast anchored to the ocean floor, blades that float at the surface and collect light, and a long stalk connecting the blades to the holdfast (see Figure 28.13). Specialized cells in the stalk, although nonvascular, can transport sugar. Suggest a reason why these structures analogous to sieve-tube elements might have evolved in kelps.

10. **SCIENTIFIC INQUIRY • INTERPRET THE DATA** A Minnesota gardener notes that the plants immediately bordering a walkway are stunted compared with those farther away. Suspecting that the soil near the walkway may be contaminated from salt added to the walkway in winter, the gardener tests the soil. The composition of the soil near the walkway is identical to that farther away except that it contains an additional 50 mM NaCl. Assuming that the NaCl is completely ionized, calculate how much it will lower the solute potential of the soil at 20°C using the *solute potential equation*:

$$\Psi_S = -iCRT$$

where *i* is the ionization constant (2 for NaCl), C is the molar concentration (in mol/L), R is the pressure constant $[R = 0.00831\ (\text{L} \cdot \text{MPa})/(\text{mol} \cdot \text{K})]$, and T is the temperature in Kelvin $(273 + °C)$.
   How would this change in the solute potential of the soil affect the water potential of the soil? In what way would the change in the water potential of the soil affect the movement of water in or out of the roots?

11. **SCIENTIFIC INQUIRY** Cotton plants wilt within a few hours of flooding of their roots. The flooding leads to low-oxygen conditions, increases in cytosolic Ca^{2+} concentration, and decreases in cytosolic pH. Suggest a hypothesis to explain how flooding leads to wilting.

12. **WRITE ABOUT A THEME: ORGANIZATION** Natural selection has led to changes in the architecture of plants that enable them to photosynthesize more efficiently in the ecological niches they occupy. In a short essay (100–150 words), explain how shoot architecture enhances photosynthesis.

13. **SYNTHESIZE YOUR KNOWLEDGE**

Imagine yourself as a water molecule in the soil solution of a forest. In a short essay (100–150 words), explain what pathways and what forces would be necessary to carry you to the leaves of these trees.

For selected answers, see Appendix A.

37 Soil and Plant Nutrition

KEY CONCEPTS

37.1 Soil contains a living, complex ecosystem *p. 806*

37.2 Plant roots absorb many types of essential elements from the soil *p. 809*

37.3 Plant nutrition often involves relationships with other organisms *p. 812*

Study Tip

Use mnemonics: Mnemonics are memory tools to help remember facts. For example, "See Hopkins, California—Mighty good" (CHOPKNSCaMg) is a mnemonic for remembering the nine plant macronutrients, as shown below. Make your own mnemonics for remembering some functions of plant macronutrients (see Table 37.1).

> See Hopkins, California—Mighty good
>
> See = C = carbon
> H = H = hydrogen
> o = O = oxygen

Go to Mastering Biology

For Students (in eText and Study Area)
- Get Ready for Chapter 37
- Animation: How Plants Obtain Minerals from Soil
- Figure 37.11 Walkthrough: The Roles of Soil Bacteria in the Nitrogen Nutrition of Plants

For Instructors to Assign (in Item Library)
- Scientific Skills Exercise: Making Observations
- Tutorial: Nitrogen Nutrition in Plants

Figure 37.1 This farmer in India is applying chemical fertilizers to a field of rice. Nitrogen (N), phosphorous (P), and potassium (K) are commonly depleted from soils to such an extent as to limit crop productivity. That is why they are the major ingredients of most chemical fertilizers.

Why do plants need minerals from the soil?

Nitrogen, phosphorus, potassium, and other minerals are essential for plant growth because of their roles in the structure and function of plant cells.

Nitrogen is a component of
- DNA and RNA (throughout the cell)
- proteins (throughout the cell)
- chlorophyll

Phosphorus is a component of
- DNA and RNA (throughout the cell)
- ATP produced by mitochondria
- phospholipids in cell membranes

Potassium is a cofactor for enzymes used throughout the cell; plays a major role in maintaining turgor

Soil contains a living, complex ecosystem

The upper layers of the soil, from which plants absorb nearly all of the water and minerals they require, contain a wide range of living organisms that interact with each other and with the physical environment. This complex ecosystem may take centuries to form but can be destroyed by human mismanagement in just a few years. To understand why soil must be conserved and why particular plants grow where they do, we must first consider the basic physical properties of soil: its texture and composition.

Soil Texture

The texture of soil depends on the sizes of its particles. Soil particles can range from coarse sand (0.02–2 mm in diameter) to silt (0.002–0.02 mm) to microscopic clay particles (less than 0.002 mm). These different-sized particles arise ultimately from the weathering of rock. Water freezing in crevices of rocks causes mechanical fracturing, and weak acids in the soil break rocks down chemically. When organisms penetrate the rock, they accelerate breakdown by chemical and mechanical means. Roots, for example, secrete acids that dissolve the rock, and their growth in fissures leads to mechanical fracturing. Mineral particles released by weathering become mixed with living organisms and **humus**, the remains of dead organisms and other organic matter, forming **topsoil**. The topsoil and other soil layers are called **soil horizons (Figure 37.2)**. The topsoil, or A horizon, can range in depth from millimeters to meters. We focus mostly on properties of topsoil because it is generally the most important soil layer for plant growth.

The topsoils that are the most fertile—supporting the most abundant growth—are **loams**, which are composed of roughly equal amounts of sand, silt, and clay. Loamy soils have enough small silt and clay particles to provide ample surface area for the adhesion and retention of minerals and water.

Plants are actually nourished by the soil solution, which consists of the water and dissolved minerals in the pores between soil particles. After a heavy rain, water drains from the larger spaces in the soil, but smaller spaces retain water because water molecules are attracted to the negatively charged surfaces of clay and other particles. The large spaces between soil particles in sandy soils generally don't retain enough water to support vigorous plant growth, but they do enable efficient diffusion of oxygen to the roots. Soils with large amounts of clay tend to retain too much water, and when soil does not drain adequately, the air is replaced by water, and the roots suffocate from lack of oxygen. Typically, the most fertile topsoils have pores

▼ **Figure 37.2 Soil horizons.**

The A horizon is the topsoil, a mixture of broken-down rock of various textures, living organisms, and decaying organic matter.

The B horizon contains much less organic matter than the A horizon and is less weathered.

The C horizon is composed mainly of partially broken-down rock. Some of the rock served as "parent" material for minerals that later helped form the upper horizons.

containing about half water and half air, providing a good balance between aeration, drainage, and water storage capacity. The physical properties of soils can be adjusted by adding soil amendments, such as peat moss, compost, manure, or sand.

Topsoil Composition

In addition to water and air, topsoil contains inorganic (mineral) and organic chemical components. The organic components include the many life-forms that inhabit the soil.

Inorganic Components

The surface charges of soil particles determine their ability to bind many nutrients. Most of the soil particles in productive soils are negatively charged and therefore do not bind negatively charged ions (anions), such as the plant nutrients nitrate (NO_3^-), phosphate ($H_2PO_4^-$), and sulfate (SO_4^{2-}). As a result, these nutrients are easily lost by *leaching*, percolation of water through the soil. Positively charged ions (cations)—such as potassium (K^+), calcium (Ca^{2+}), and magnesium (Mg^{2+})—adhere to negatively charged soil particles, so are less easily lost by leaching.

Roots, however, do not absorb mineral cations directly from soil particles; they absorb them from the soil solution. Mineral cations enter the soil solution by **cation exchange**, a process in which cations are displaced from soil particles by other cations, particularly H$^+$ **(Figure 37.3)**. Therefore, a soil's capacity to exchange cations is determined by the number of cation adhesion sites and by the soil's pH. In general, the more clay and organic matter in the soil, the higher the

▼ Figure 37.3 Cation exchange in soil.

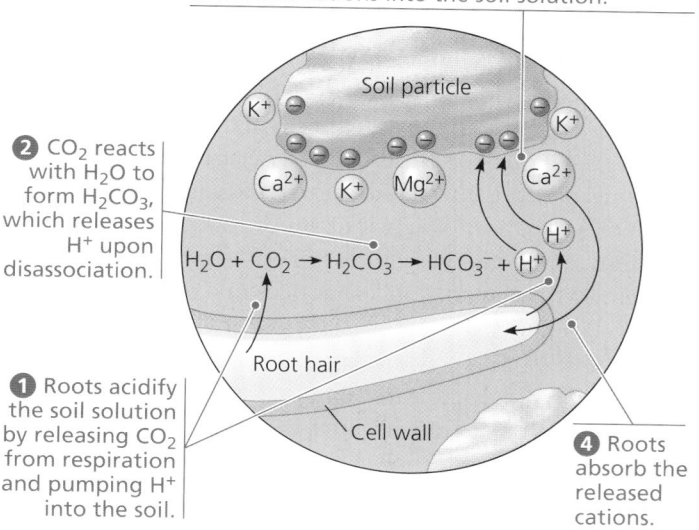

❸ H^+ ions in the soil solution neutralize the negative charge of soil particles, causing release of mineral cations into the soil solution.

❷ CO_2 reacts with H_2O to form H_2CO_3, which releases H^+ upon disassociation.

Soil particle

K^+ Ca^{2+} K^+ Mg^{2+} Ca^{2+}

H^+

$H_2O + CO_2 \rightarrow H_2CO_3 \rightarrow HCO_3^- + H^+$

Root hair

Cell wall

❶ Roots acidify the soil solution by releasing CO_2 from respiration and pumping H^+ into the soil.

❹ Roots absorb the released cations.

VISUAL SKILLS *Which are more likely to be leached from the soil by decreasing pH—cations or anions? Explain.*

→ Mastering Biology Animation: How Plants Obtain Minerals from Soil

cation exchange capacity. The clay content is important because these small particles have a high ratio of surface area to volume, allowing for the ample binding of cations.

Organic Components

The major organic component of topsoil is humus, which consists of organic material produced by the decomposition of fallen leaves, dead organisms, feces, and other organic matter by bacteria and fungi. Humus prevents clay particles from packing together and forms a crumbly soil that retains water but is still porous enough to aerate roots. Humus also increases the soil's capacity to exchange cations and is a reservoir of mineral nutrients that return gradually to the soil as microorganisms decompose the organic matter.

Topsoil is home to an astonishing number and variety of organisms. A teaspoon of topsoil has about 5 billion bacteria, which cohabit with fungi, algae and other protists, insects, earthworms, nematodes, and plant roots. The activities of all these organisms affect the soil's physical and chemical properties. Earthworms, for example, consume organic matter and derive their nutrition from the bacteria and fungi growing on this material. They excrete wastes and move large amounts of material to the soil surface. In addition, they move organic matter into deeper layers. Earthworms mix and clump the soil particles, allowing for better gaseous diffusion and water retention. Roots also affect soil texture and composition. For example, they reduce erosion by binding the soil, and they lower soil pH by excreting acids.

Soil Conservation and Sustainable Agriculture

Ancient farmers recognized that crop yields on a particular plot of land decreased over the years. Moving to uncultivated areas, they observed the same pattern of reduced yields over time. Eventually, they realized that **fertilization**, the addition of mineral nutrients to the soil, could make soil a renewable resource that enabled crops to be cultivated season after season at a fixed location. This sedentary agriculture facilitated a new way of life. Humans began to build permanent dwellings—the first villages. They also stored food for use between harvests, and food surpluses enabled some people to specialize in nonfarming occupations. In short, soil management, by fertilization and other practices, helped prepare the way for modern societies.

Unfortunately, soil *mis*management has been a recurrent problem throughout human history. A dramatic example is the American Dust Bowl, an ecological and human disaster that ravaged the southwestern Great Plains of the United States in the 1930s. This region suffered through devastating dust storms that resulted from a prolonged drought and decades of inappropriate farming techniques. Before farmers arrived, the Great Plains had been covered by hardy grasses that held the soil in place in spite of recurring droughts and torrential rains. But in the late 1800s and early 1900s, many settlers planted wheat and raised cattle. These land uses left the soil exposed to erosion by winds. A few years of drought made the problem worse. During the 1930s, huge quantities of fertile soil were blown away in "black blizzards," rendering millions of hectares of farmland useless **(Figure 37.4)**. In one of the worst dust storms, clouds of dust blew eastward to Chicago, where soil fell like snow, and even reached the Atlantic coast. Hundreds of thousands of people in the Dust Bowl region were forced to abandon their homes and land, a plight immortalized in John Steinbeck's novel *The Grapes of Wrath*.

▼ Figure 37.4 A massive dust storm in the American Dust Bowl during the 1930s.

❓ *Which soil horizon contributed to these dust clouds?*

Soil mismanagement continues to be a major problem to this day. More than 30% of the world's farmland has reduced productivity caused by poor soil conditions, such as chemical contamination, mineral deficiencies, acidity, salinity, and poor drainage. As the world's population grows, the demand for food increases. Because soil quality greatly affects crop yield, soil resources must be managed prudently. Today, the most productive lands are already being used for agriculture, so there are no more frontiers for farmers to clear. Thus, it is critical that farmers embrace **sustainable agriculture**, a commitment to farming practices that are conservation minded, environmentally safe, and profitable. Sustainable agriculture includes the prudent use of irrigation and soil amendments, the protection of topsoil from salinization and erosion, and the restoration of degraded lands.

Irrigation

Because water shortage often limits plant growth, perhaps no technology has increased crop yield as much as irrigation. However, irrigation drains freshwater resources. Globally, about 75% of all freshwater use is devoted to agriculture. Many rivers in arid regions have been reduced to trickles by diverting water for irrigation. The primary source of irrigation water, however, is not surface waters, such as rivers and lakes, but underground water reserves called *aquifers*. In some regions, the rate of water removal is exceeding the natural refilling of the aquifers, making this practice unsustainable. A further consequence of aquifer depletion is *land subsidence*, a gradual settling or sudden sinking of the ground surface directly above the aquifer **(Figure 37.5)**. Land subsidence can destroy property and negatively impact agriculture.

Irrigation, particularly from groundwater, can also lead to soil *salinization*—the addition of salts to the soil that make it too salty for cultivating plants. Salts dissolved in irrigation

▼ **Figure 37.5 Sudden land subsidence.** Overuse of groundwater for irrigation triggered formation of this sinkhole in Florida.

water accumulate in the soil as the water evaporates, making the water potential of the soil solution more negative. The water potential gradient from soil to roots is reduced, diminishing water uptake (see Figure 36.12).

Many forms of irrigation, such as the flooding of fields, are wasteful because much of the water evaporates. To use water efficiently, farmers must understand the water-holding capacity of their soil, the water needs of their crops, and the appropriate irrigation technology. One popular technology is *drip irrigation*, the slow release of water to soil and plants from perforated plastic tubing placed directly at the root zone. Because drip irrigation requires less water and reduces salinization, it is used mainly in arid agricultural regions.

Fertilization

In natural ecosystems, mineral nutrients are usually recycled by the excretion of animal wastes and the decomposition of humus. Agriculture, however, can interfere with this recycling. The lettuce you eat, for example, contains minerals extracted from a farmer's field. As you excrete wastes, these minerals are deposited far from their original source. Over many harvests, the farmer's field eventually becomes depleted of nutrients. Nutrient depletion is a major cause of global soil degradation. Farmers must reverse nutrient depletion by means of fertilization.

Today, most farmers in industrialized nations use fertilizers containing minerals that are either mined or prepared by energy-intensive processes. These fertilizers are usually enriched in nitrogen (N), phosphorus (P), and potassium (K)—the nutrients most commonly deficient in depleted soils. You may have seen fertilizers labeled with a three-number code, called the N–P–K ratio. A fertilizer marked "15–10–5," for instance, is 15% N (as ammonium or nitrate), 10% P (as phosphate), and 5% K (as the mineral potash).

Manure, fishmeal, and compost are called "organic" fertilizers because they contain organic material. Before plants can use organic material, however, it must be decomposed into the inorganic nutrients that roots can absorb. Whether from organic fertilizer or a chemical factory, the minerals a plant extracts are in the same form. However, organic fertilizers release them gradually, whereas minerals in commercial fertilizers are immediately available but may not be retained by the soil for long. Minerals not absorbed by roots are often leached from the soil by rainwater or irrigation. Mineral runoff into lakes may lead to explosions in algal populations that can deplete oxygen levels and reduce fish populations.

Adjusting Soil pH

Soil pH is an important factor that influences mineral availability by its effect on cation exchange and the chemical form of minerals. Depending on the soil pH, a particular mineral may be bound too tightly to clay particles or may be in a chemical form that the plant cannot absorb. Most plants prefer slightly acidic soil because the high H⁺ concentrations can displace positively charged minerals from soil particles, making them more available for absorption. Adjusting soil pH is tricky because a change in H⁺ concentration may make one mineral more available but another less available. At pH 8, for instance, plants can absorb calcium, but iron is almost unavailable. The soil pH should be matched to a crop's mineral needs. If the soil is too alkaline, adding sulfate will lower the pH. Soil that is too acidic can be adjusted by adding lime (calcium carbonate or calcium hydroxide).

When the soil pH dips to 5 or lower, toxic aluminum ions (Al^{3+}) become more soluble and are absorbed by roots, stunting root growth and preventing the uptake of calcium, a needed plant nutrient. Low soil pH and Al^{3+} toxicity are especially serious problems in tropical regions, where the pressure of producing food for a growing population is often most acute. Some plants can cope with high Al^{3+} levels by secreting organic anions that bind Al^{3+} and render it harmless. Scientists have altered tobacco and papaya plants by introducing a citrate synthase gene from a bacterium into the plants' genomes. The resulting overproduction of citric acid increased aluminum resistance.

Controlling Erosion

As happened most dramatically in the Dust Bowl, water and wind erosion can remove large amounts of topsoil. Erosion is a major cause of soil degradation because nutrients are carried away by wind and streams. To limit erosion, farmers can plant rows of trees as windbreaks, terrace hillside crops, and cultivate crops in a contour pattern **(Figure 37.6)**. Crops such as alfalfa and wheat provide good ground cover and protect the soil better than maize and other crops that are usually planted in more widely spaced rows.

Erosion can also be reduced by a plowing technique called **no-till agriculture**. In traditional plowing, the entire field is tilled, or turned over. This practice helps control weeds but disrupts the meshwork of roots that holds the soil in place, leading to increased surface runoff and erosion. In no-till agriculture, a special plow creates narrow furrows for seeds and fertilizer. In this way, the field is seeded with minimal disturbance to the soil, while also using less fertilizer.

Phytoremediation

Some land areas are unfit for cultivation because toxic metals or organic pollutants have contaminated the soil or groundwater. Traditionally, soil remediation, the detoxification of contaminated soils, has focused on nonbiological technologies, such as removing and storing contaminated soil in landfills, but these techniques are costly and often disrupt the landscape. **Phytoremediation** is a nondestructive biotechnology that harnesses the ability of some plants to extract soil pollutants and concentrate them in portions of the plant that can be easily removed for safe disposal. For

▼ **Figure 37.6 Contour tillage.** These crops are planted in rows that go around, rather than up and down, the hills. Contour tillage helps slow water runoff and topsoil erosion after heavy rains.

example, alpine pennycress (*Thlaspi caerulescens*) can accumulate zinc in its shoots at concentrations 300 times higher than most plants can tolerate. The shoots can be harvested and the zinc removed. Such plants show promise for cleaning up areas contaminated by smelters, mines, or nuclear tests. Phytoremediation is a type of bioremediation, which also uses prokaryotes and protists to detoxify polluted sites (see Concepts 27.6 and 55.5).

We have examined the importance of soil conservation for sustainable agriculture. Mineral nutrients contribute greatly to soil fertility, but which minerals are most important, and why do plants need them? These are the topics of the next section.

CONCEPT CHECK 37.1

1. Explain how the phrase "too much of a good thing" can apply to watering and fertilizing plants.
2. Some lawn mowers collect clippings. What is a drawback of this practice with respect to plant nutrition?
3. **WHAT IF?** How would adding clay to loamy soil affect the soil's capacity to exchange cations and retain water? Explain.
4. **MAKE CONNECTIONS** Note three ways the properties of water contribute to soil formation. See Concept 3.2.

For suggested answers, see Appendix A.

CONCEPT 37.2

Plant roots absorb many types of essential elements from the soil

Water, air, and soil minerals all contribute to plant growth. A plant's water content can be measured by comparing the mass before and after drying. Typically, 80–90% of a plant's fresh mass is water. About 96% of the remaining dry mass consists of carbohydrates such as cellulose and starch that are produced by photosynthesis. Thus, the components of carbohydrates—carbon, oxygen, and hydrogen—are the most abundant elements in dried plant residue. Inorganic substances from the soil, although essential for plant survival, account for only about 4% of a plant's dry mass.

Essential Elements

The inorganic substances in plants contain more than 50 chemical elements. In studying the chemical composition of plants, we must distinguish elements that are essential from those that are merely present in the plant. A chemical element is considered an **essential element** only if it is required for a plant to complete its life cycle and reproduce.

To determine which chemical elements are essential, researchers use **hydroponic culture**, in which plants are grown in mineral solutions instead of soil **(Figure 37.7)**. Such studies have helped identify 17 essential elements needed by all plants **(Table 37.1)**. Hydroponic culture is also used on a small scale to grow some greenhouse crops.

Nine of the essential elements are called **macronutrients** because plants require them in relatively large amounts. Six of these are the major components of organic compounds forming a plant's structure: carbon, oxygen, hydrogen, nitrogen, phosphorus, and sulfur. The other three macronutrients are potassium, calcium, and magnesium. Of all the mineral nutrients, nitrogen contributes the most to plant growth and crop yields. Plants require nitrogen as a component of proteins, nucleic acids, chlorophyll, and other important organic molecules.

The other essential elements are called **micronutrients** because plants need them in only tiny quantities. They are chlorine, iron, manganese, boron, zinc, copper, nickel, and molybdenum. Sodium is a ninth essential micronutrient for plants that use the C_4 or CAM pathways of photosynthesis (see Concept 10.5) because it is needed for the regeneration of phosphoenolpyruvate, the CO_2 acceptor used in these two types of carbon fixation.

Micronutrients function in plants mainly as enzyme cofactors (see Concept 8.4). Iron, for example, is a metallic component of cytochromes, the proteins in the electron transport chains of chloroplasts and mitochondria. It is because micronutrients generally just aid enzymes that plants need only tiny quantities. The requirement for molybdenum, for instance, is so modest that there is only one atom of this rare element for every 60 million atoms of hydrogen in dried plant material. Yet a deficiency of molybdenum or any other micronutrient can weaken or kill a plant.

Many animals, including humans, get much of their essential mineral nutrients from plants. Low dietary concentrations of essential micronutrients, such as iron (Fe), iodine (I), and zinc (Zn), contribute to the problem of micronutrient malnutrition ("hidden hunger"). Human malnourishment, as opposed to undernourishment, results when people receive enough calories from their ingestion of carbohydrates but lack other dietary factors (such as essential minerals, vitamins, or amino acids) required for good health.

Symptoms of Mineral Deficiency

The symptoms of a deficiency depend partly on the mineral's function as a nutrient. For example, a deficiency of magnesium, a component of chlorophyll, causes *chlorosis*, yellowing of leaves. In some cases, the relationship between a deficiency and its symptoms is less direct. For instance, iron deficiency can cause chlorosis even though chlorophyll contains no iron because iron ions are required as a cofactor in an enzymatic step of chlorophyll synthesis.

Mineral deficiency symptoms depend not only on the role of the nutrient but also on its mobility within the plant. If a nutrient moves about freely, symptoms appear first in older organs because young, growing tissues are a greater sink for nutrients in short supply. For example, a plant deficient in magnesium, a relatively mobile ion, first shows signs of chlorosis in its older leaves. In contrast, a deficiency of a relatively immobile mineral, such as iron, affects young parts of the plant first. The older tissues have adequate amounts of iron that they retain during periods of short supply. The mineral requirements of a plant also change with the age of the plant. Young seedlings, for example, rarely show mineral deficiency symptoms because their mineral needs are met largely by the mineral reserves stored in the seeds themselves.

The symptoms of a mineral deficiency may vary between species but in a given plant are often distinctive enough to aid in diagnosis. Deficiencies of phosphorus, potassium, and nitrogen are most common, as in the example of maize leaves in **Figure 37.8**.

▼ **Figure 37.7**   **Research Method**

Hydroponic Culture

Application In hydroponic culture, plants are grown in mineral solutions without soil. One use of hydroponic culture is to identify essential elements in plants.

Technique Plant roots are bathed in aerated solutions of known mineral composition. Aerating the water provides the roots with oxygen for cellular respiration. (Note: The flasks would normally be opaque to prevent algal growth.) A mineral, such as iron, can be omitted to test whether it is essential.

Control: Solution containing all minerals

Experimental: Solution without iron

Results If the omitted mineral is essential, mineral deficiency symptoms occur, such as stunted growth and discolored leaves. By definition, the plant would not be able to complete its life cycle. Deficiencies of different elements may have different symptoms, which can aid in diagnosing mineral deficiencies in soil.

Table 37.1 Essential Elements in Plants

Element (Form Primarily Absorbed by Plants)	% Mass in Dry Tissue	Major Function(s)	Early Visual Symptom(s) of Nutrient Deficiencies
Macronutrients			
Carbon (CO_2)	45%	Major component of organic compounds	Poor growth
Oxygen (CO_2)	45%	Major component of organic compounds	Poor growth
Hydrogen (H_2O)	6%	Major component of organic compounds	Wilting, poor growth
Nitrogen (NO_3^-, NH_4^+)	1.5%	Component of nucleic acids, proteins, and chlorophyll	Chlorosis at tips of older leaves (common in heavily cultivated soils or soils low in organic material)
Potassium (K^+)	1.0%	Enzyme cofactor; major solute functioning in water balance; operation of stomata	Mottling of older leaves, drying of leaf edges; weak stems; roots poorly developed (common in acidic or sandy soils)
Calcium (Ca^{2+})	0.5%	Important component of middle lamella and cell walls; maintains membrane function; signal transduction	Crinkling of young leaves; death of terminal buds (common in acidic or sandy soils)
Magnesium (Mg^{2+})	0.2%	Component of chlorophyll; cofactor of many enzymes	Chlorosis between veins, found in older leaves (common in acidic or sandy soils)
Phosphorus ($H_2PO_4^-$, HPO_4^{2-})	0.2%	Component of nucleic acids, phospholipids, ATP	Healthy appearance but very slow development; thin stems; purpling of veins; poor flowering and fruiting (common in acidic, wet, or cold soils)
Sulfur (SO_4^{2-})	0.1%	Component of proteins	General chlorosis in young leaves (common in sandy or very wet soils)
Micronutrients			
Chlorine (Cl^-)	0.01%	Photosynthesis (water-splitting); functions in water balance	Wilting; stubby roots; leaf mottling (uncommon)
Iron (Fe^{3+}, Fe^{2+})	0.01%	Respiration; photosynthesis: chlorophyll synthesis; N_2 fixation	Chlorosis between veins, found in young leaves (common in basic soils)
Manganese (Mn^{2+})	0.005%	Active in formation of amino acids; activates some enzymes; required for water-splitting step of photosynthesis	Chlorosis between veins, found in young leaves (common in basic soils rich in humus)
Boron ($H_2BO_3^-$)	0.002%	Cofactor in chlorophyll synthesis; role in cell wall function; pollen tube growth	Death of meristems; thick, leathery, and discolored leaves (occurs in any soil; most common micronutrient deficiency)
Zinc (Zn^{2+})	0.002%	Active in formation of chlorophyll; cofactor of some enzymes; needed for DNA transcription	Reduced internode length; crinkled leaves (common in some geographic regions)
Copper (Cu^+, Cu^{2+})	0.001%	Component of many redox and lignin-biosynthetic enzymes	Light green color throughout young leaves, with drying of leaf tips; roots stunted and excessively branched (common in some geographic regions)
Nickel (Ni^{2+})	0.001%	Nitrogen metabolism	General chlorosis in all leaves; death of leaf tips (common in acidic or sandy soils)
Molybdenum (MoO_4^{2-})	0.0001%	Nitrogen metabolism	Death of root and shoot tips; chlorosis in older leaves (common in acidic soils in some geographic areas)

MAKE CONNECTIONS *Explain why CO_2, rather than O_2, is the source of much of the dry mass oxygen in plants. See Concept 10.1.*

> **Mastering Biology**
> **Interview with Gloria Coruzzi:**
> **Assimilating more nitrogen into plants**

▶ **Figure 37.8 The most common mineral deficiencies, as seen in maize leaves.** Mineral deficiency symptoms may vary in different species. In maize, nitrogen deficiency is evident in a yellowing that starts at the tip and moves along the center (midrib) of older leaves. Phosphorus-deficient maize plants have reddish purple margins, particularly in young leaves. Potassium-deficient maize plants exhibit "firing," or drying, along tips and margins of older leaves.

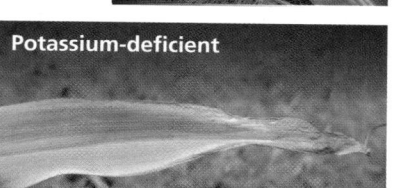

Healthy

Nitrogen-deficient

Phosphorus-deficient

Potassium-deficient

In the **Scientific Skills Exercise**, you can diagnose a mineral deficiency in leaves of an orange tree. Micronutrient shortages are less common than macronutrient shortages and tend to occur in certain geographic regions because of differences in soil composition. One way to confirm a diagnosis is to analyze the mineral content of the plant or soil. The amount of a micronutrient needed to correct a deficiency is usually small. For example, a zinc deficiency in fruit trees can usually be cured by hammering a few zinc nails into each tree trunk. Moderation is important because overdoses of a micronutrient or macronutrient can be detrimental or toxic. Too much nitrogen, for example, can lead to excessive vine growth in tomato plants at the expense of good fruit production.

Scientific Skills Exercise

Making Observations

What Mineral Deficiency Is This Plant Exhibiting? Plant growers often diagnose deficiencies in their crops by examining changes to the foliage, such as chlorosis (yellowing), death of some leaves, discoloring, mottling, scorching, or changes in size or texture. In this exercise, you will diagnose a mineral deficiency by observing a plant's leaves and applying what you have learned about symptoms from the text and Table 37.1.

Data The data for this exercise come from the photograph below of leaves on an orange tree exhibiting a mineral deficiency.

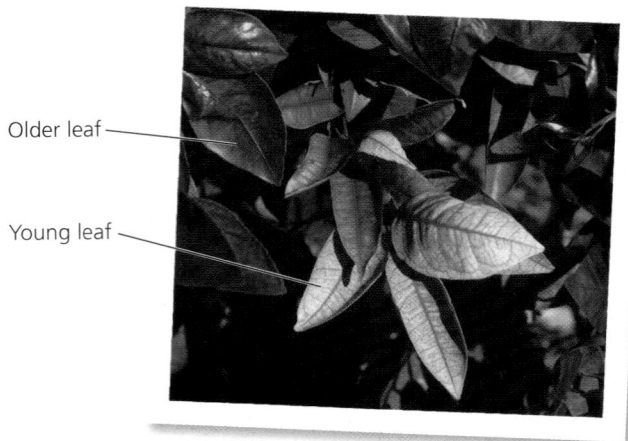

Older leaf

Young leaf

INTERPRET THE DATA

1. How do the young leaves differ in appearance from the older leaves?
2. What change do you observe in the leaves, and where in the leaves does it appear? List the three nutrients whose deficiencies give rise to this symptom. Based on the symptom's location, which one of these three nutrients can be ruled out, and why? What does the location suggest about the other two nutrients?
3. How would your hypothesis about the cause of this deficiency be influenced if tests showed that the soil was low in humus?

➔ **Instructors:** A version of this Scientific Skills Exercise can be assigned in **Mastering Biology**.

Global Climate Change and Food Quality

Since plant growth is generally enhanced by increases in atmospheric CO_2 and by warming temperatures, global food production, as measured by plant biomass, is expected to increase in response to global climate change in certain parts of the world. But food production must also be judged by its quality. Unfortunately, there are indications that plants today, compared with plants in preindustrial times, are not taking up nutrients sufficiently to keep pace with their increased fixation of CO_2 into carbohydrates. For example, a study of 43 crops found significant declines in protein, Ca, P, Fe, riboflavin, and ascorbic acid from 1950 to 1999. Although this decline in food quality might be caused by a shift toward higher-yielding crops, studies of wild plants suggest that global climate change is the culprit. For example, a study revealed that the protein content of goldenrod (*Solidago canadensis*) pollen has declined by one third since the Industrial Revolution, and the change closely correlates with the rise in CO_2 that has occurred. Declines in the qualities of pollen as a food source might play a role in the widespread decline of honeybees, important pollinators whose decreasing population threatens crop production.

CONCEPT CHECK 37.2

1. Are some essential elements more important than others? Explain.
2. **WHAT IF?** If an element increases the growth rate of a plant, can it be defined as an essential element?
3. **MAKE CONNECTIONS** Based on Figure 9.17, explain why hydroponically grown plants would grow much more slowly if they were not sufficiently aerated.

For suggested answers, see Appendix A.

CONCEPT 37.3

Plant nutrition often involves relationships with other organisms

To this point, we have looked at plants as exploiters of soil resources, but plants and soil have a two-way relationship. Dead plants provide much of the energy needed by soil-dwelling bacteria and fungi. Many of these organisms also benefit from sugar-rich secretions produced by living roots. Meanwhile, plants derive benefits from their associations with soil bacteria and fungi. As shown in **Figure 37.9**, mutually beneficial relationships across kingdoms and domains are not rare in nature. However, they are of particular importance to plants. We'll explore some important *mutualisms* between plants and soil bacteria and fungi, as well as some unusual, nonmutualistic forms of plant nutrition.

MAKE CONNECTIONS

Mutualism Across Kingdoms and Domains

Some toxic species of fish don't make their own poison. How is that possible? Some species of ants chew leaves but don't eat them. Why? The answers lie in some amazing mutualisms, relationships between different species in which each species provides a substance or service that benefits the other (see Concept 54.1). Sometimes mutualisms occur within the same kingdom, such as between two species of animals. Many mutualisms, however, involve species from different kingdoms or domains, as in these examples.

Fungus–Bacterium

A lichen is a mutualistic association between a fungus and a photosynthetic partner. In the lichen *Peltigera*, the photosynthetic partner is a species of cyanobacterium. The cyanobacterium supplies carbohydrates, while the fungus provides anchorage, protection, minerals, and water. (See Figure 31.25.)

The lichen *Peltigera*

A longitudinal section of the lichen *Peltigera* showing green photosynthetic bacteria sandwiched between layers of fungus

Animal–Bacterium

Fugu is the Japanese name for puffer fish and the delicacy made from it, which can be deadly. Most species of puffer fish contain lethal amounts of the nerve toxin tetrodotoxin in their organs, especially the liver, ovaries, and intestines. Therefore, a specially trained chef must remove the poisonous parts. The tetrodotoxin is synthesized by mutualistic bacteria (various *Vibrio* species) associated with the fish. The fish gains a potent chemical defense, while the bacteria live in a high-nutrient, low-competition environment.

Puffer fish (fugu)

Plant–Bacterium

The floating fern *Azolla* provides carbohydrates for a nitrogen-fixing cyanobacterium that resides in the air spaces of the leaves. In return, the fern receives nitrogen from the cyanobacterium. (See Concept 27.5.)

The floating fern *Azolla*

Animal–Fungus

Leaf-cutter ants harvest leaves that they carry back to their nest, but the ants do not eat the leaves. Instead, a fungus grows by absorbing nutrients from the leaves, and the ants eat part of the fungus that they have cultivated.

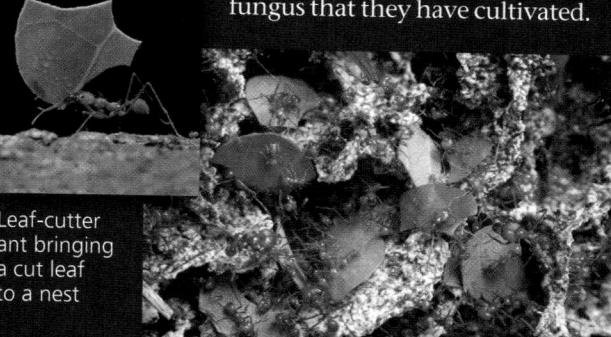

Leaf-cutter ant bringing a cut leaf to a nest

Ants tending a fungal garden in a nest

Plant–Fungus

Most plant species have mycorrhizae, mutualistic associations between roots and fungi. The fungus absorbs carbohydrates from the roots. In return, the fungus's mycelium, a dense network of filaments called hyphae, increases the surface area for the uptake of water and minerals by the roots. (See Figure 31.4.)

A fungus growing on the root of a sorghum plant (SEM)

Plant–Animal

Some species of plants are aggressively defended from predators and competitors by ants. The plant provides nourishment for the ants in the form of carbohydrate-rich nectar from glands called nectaries. (See Figure 54.9 for another example of a mutualism between a plant and ants.)

Protective ant harvesting nectar

MAKE CONNECTIONS

Describe three more examples of mutualisms. (See Figure 27.20, Figure 38.4, and Concept 41.4.)

Bacteria and Plant Nutrition

A variety of soil bacteria play roles in plant nutrition. Some engage in mutually beneficial chemical exchanges with plant roots. Others enhance the decomposition of organic materials and increase nutrient availability.

Rhizobacteria

Rhizobacteria are bacteria that live either in close association with plant roots or in the **rhizosphere**, the soil closely surrounding plant roots. Many rhizobacteria form mutually beneficial associations with plant roots. Rhizobacteria depend on nutrients such as sugars, amino acids, and organic acids that are secreted by plant cells. Up to 20% of a plant's photosynthetic production may be used to fuel these complex bacterial communities. In return, plants reap many benefits from these mutualistic associations. Some rhizobacteria produce antibiotics that protect roots from disease. Others absorb toxic metals or make nutrients more available to roots. Still others convert gaseous nitrogen into forms usable by the plant or produce chemicals that stimulate plant growth. Inoculation of seeds with plant-growth-promoting rhizobacteria can increase crop yield and reduce the need for fertilizers and pesticides.

Some rhizobacteria are free-living in the rhizosphere, whereas others are **endophytes** that live between cells within the plant. Both the intercellular spaces occupied by endophytic bacteria and the rhizosphere associated with each plant root system contain a unique and complex cocktail of root secretions and microbial products that differ from those of the surrounding soil. A recent study revealed that the compositions of bacterial communities living endophytically and in the rhizosphere are not identical **(Figure 37.10)**. A better

---

▼ Figure 37.10 Inquiry

How variable are the compositions of bacterial communities inside and outside of roots?

Experiment The bacterial communities found within and immediately outside of root systems are known to improve plant growth. In order to devise agricultural strategies to increase the benefits of these bacterial communities, it is necessary to determine how complex they are and what factors affect their composition. A problem with studying these bacterial communities is that a handful of soil contains as many as 10,000 types of bacteria, more than all the bacterial species that have been described. Researchers cannot simply culture each species and use a taxonomic key to identify them; a molecular approach is needed.

▼ Bacteria (green) on surface of root (fluorescent LM)

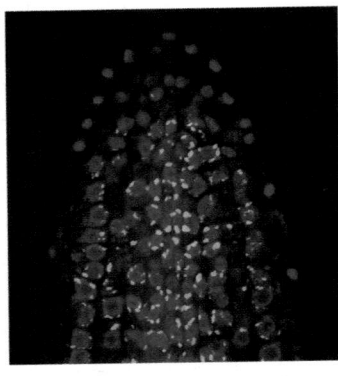

Jeffery Dangl and his colleagues estimated the number of bacterial "species" in samples using a technique called *metagenomics* (see Concept 21.1). The bacterial community samples differed in location (endophytic, inside the rhizosphere, or outside the rhizosphere), soil type (having large amounts of clay or porous), and the developmental stage of the root system with which they were associated (old or young). The DNA from each sample was purified, and the polymerase chain reaction (PCR) was used to amplify the DNA that codes for the 16S ribosomal RNA subunits. Many thousands of DNA sequence variations were found in each sample. The researchers lumped the sequences that were more than 97% identical into "taxonomic units" or "species." (The word *species* is in quotation marks because "two organisms having a single gene that is more than 97% identical" is not explicit in any definition of species.) Having established the types of "species" in each community, the researchers made a tree diagram showing the percent of bacterial "species" found in common in each community.

Results This tree diagram breaks down the relatedness of bacterial communities into finer and finer levels of detail.

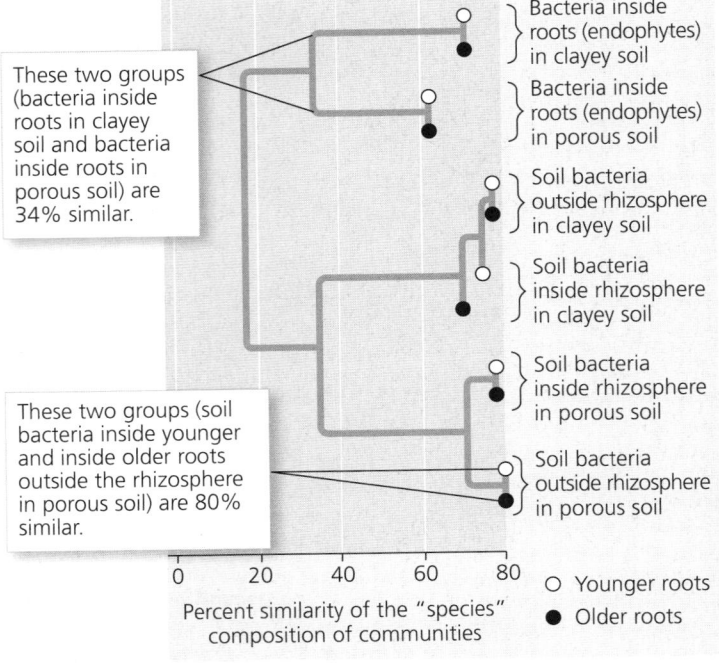

These two groups (bacteria inside roots in clayey soil and bacteria inside roots in porous soil) are 34% similar.

These two groups (soil bacteria inside younger and inside older roots outside the rhizosphere in porous soil) are 80% similar.

Bacteria inside roots (endophytes) in clayey soil

Bacteria inside roots (endophytes) in porous soil

Soil bacteria outside rhizosphere in clayey soil

Soil bacteria inside rhizosphere in clayey soil

Soil bacteria inside rhizosphere in porous soil

Soil bacteria outside rhizosphere in porous soil

Percent similarity of the "species" composition of communities

○ Younger roots
● Older roots

Data from D. S. Lundberg et al., Defining the core *Arabidopsis thaliana* root microbiome, *Nature* 488:86–94 (2012).

The two explanatory labels give examples of how to interpret the diagram.

Conclusion The "species" composition of the bacterial communities varied markedly according to the location inside the root versus outside the root and according to soil type.

INTERPRET THE DATA **(a)** Which of the three community locations was least like the other two? **(b)** Rank the three variables (community location, developmental stage of roots, and soil type) in terms of how strongly they affect the "species" composition of the bacterial communities.

---

understanding of the types of bacteria within and around roots could potentially have profound agricultural benefits.

Bacteria in the Nitrogen Cycle

Because nitrogen is required in large amounts for synthesizing proteins and nucleic acids, no mineral deficiency is more limiting to plant growth than a lack of nitrogen. The forms of nitrogen that plants can use include NO_3^- and NH_4^+. Some soil nitrogen derives from the weathering of rocks, and lightning produces small amounts of NO_3^- that get carried to the soil in rain. However, most of the nitrogen available to plants comes from the activity of bacteria **(Figure 37.11)**. This activity is part of the **nitrogen cycle**, a series of natural processes by which certain nitrogen-containing substances from the air and soil are made available to living things, are used by them, and are returned to the air and soil (see Figure 55.14).

Plants commonly acquire nitrogen in the form of NO_3^- (nitrate). Soil NO_3^- is largely formed by a two-step process called *nitrification*, which consists of the oxidation of ammonia (NH_3) to nitrite (NO_2^-), followed by oxidation of NO_2^- to NO_3^-. Different types of *nitrifying bacteria* mediate each step, as shown at the bottom of Figure 37.11. After the roots absorb NO_3^-, a plant enzyme reduces it back to NH_4^+, which other enzymes incorporate into amino acids and other organic compounds. Most plant species transport nitrogen from roots to shoots via the xylem as NO_3^- or as organic compounds synthesized in the roots. Some soil nitrogen is lost, particularly in anaerobic soils, when denitrifying bacteria convert NO_3^- to N_2, which diffuses into the atmosphere.

In addition to NO_3^-, plants can acquire nitrogen in the form of NH_4^+ (ammonium) through two processes, as shown on the left in Figure 37.11. In one process, *nitrogen-fixing bacteria* convert gaseous nitrogen (N_2) to NH_3, which then picks up another H^+ in the soil solution, forming NH_4^+. In the other process, called ammonification, decomposers convert the organic nitrogen from dead organic material into NH_4^+.

Bacteria and Nitrogen Fixation

Although Earth's atmosphere is 79% nitrogen, plants cannot use free gaseous nitrogen (N_2) because there is a triple bond between the two nitrogen atoms, making the molecule almost inert. For atmospheric N_2 to be of use to plants, it must be reduced to NH_3 by a process known as **nitrogen fixation**. All nitrogen-fixing organisms are bacteria. Some nitrogen-fixing bacteria are free-living in the soil (see Figure 37.11), whereas others live in the rhizosphere. Among this latter group, members of the genus *Rhizobium* form efficient and intimate associations with the roots of legumes (such as beans, alfalfa, and peanuts), altering the structure of the hosts' roots markedly, as will be discussed shortly.

The multistep conversion of N_2 to NH_3 by nitrogen fixation can be summarized as follows:

$$N_2 + 8e^- + 8 H^+ + 16 ATP \longrightarrow 2 NH_3 + H_2 + 16 ADP + 16 \text{\textcircled{P}}_i$$

▼ **Figure 37.11 The roles of soil bacteria in the nitrogen nutrition of plants.** Ammonium (NH_4^+) is made available to plants by two types of soil bacteria: those that fix atmospheric N_2 (nitrogen-fixing bacteria) and those that decompose organic material (ammonifying bacteria). Although plants absorb some ammonium from the soil, they absorb mainly nitrate, which is produced from ammonium by nitrifying bacteria. Plants reduce nitrate back to ammonium before incorporating the nitrogen into organic compounds.

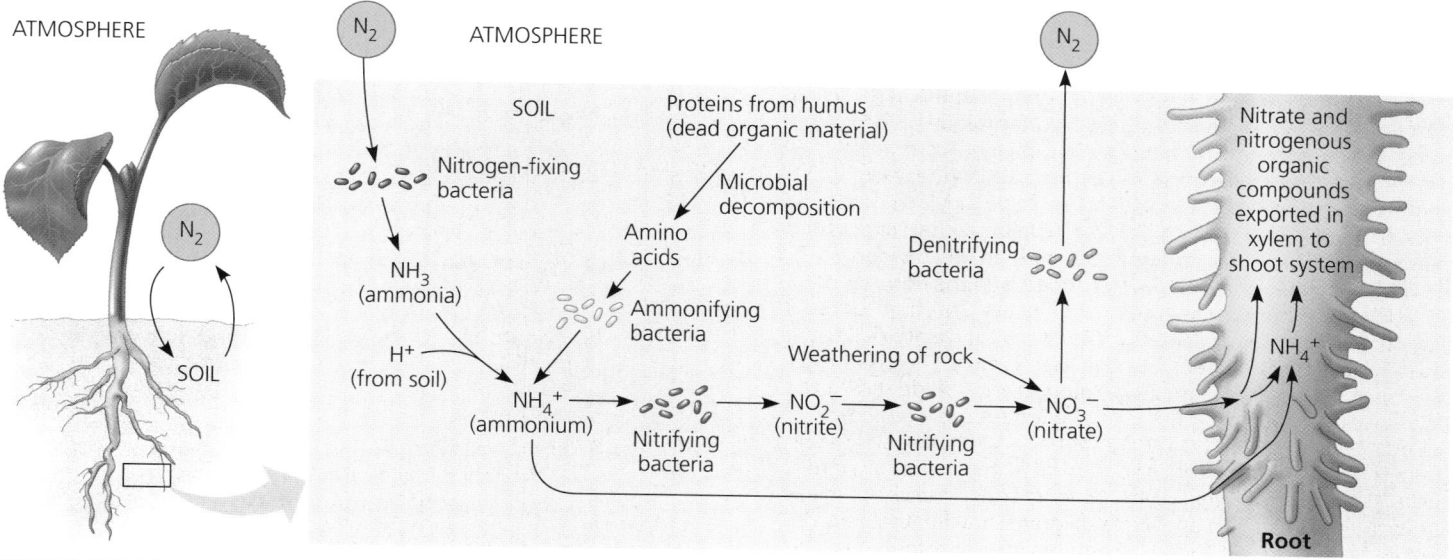

VISUAL SKILLS *If an animal died near a root, would the plant have greater access to ammonium, nitrate, or both?*

→ **Mastering Biology Figure Walkthrough**

The reaction is driven by the enzyme complex *nitrogenase*. Because the process of nitrogen fixation requires 16 ATP molecules for every 2 NH₃ molecules synthesized, nitrogen-fixing bacteria require a rich supply of carbohydrates from decaying material, root secretions, or (in the case of the *Rhizobium* bacteria) the vascular tissue of roots.

The mutualism between *Rhizobium* ("root-living") bacteria and legume roots involves dramatic changes in root structure. Along a legume's roots are swellings called **nodules**, composed of plant cells "infected" by *Rhizobium* bacteria (**Figure 37.12**).

▶ **Figure 37.12 Root nodules on a legume.** The spherical structures along this soybean root system are nodules containing *Rhizobium* bacteria. The bacteria fix nitrogen and obtain photosynthetic products supplied by the plant.

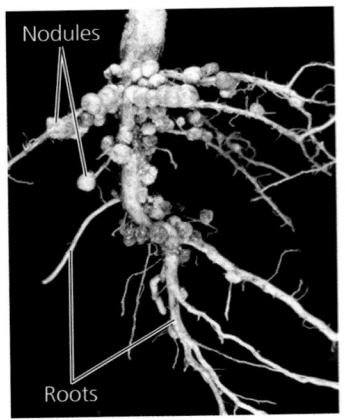

Nodules

Roots

? *How is the relationship between legume plants and* Rhizobium *bacteria mutualistic?*

Inside each nodule, *Rhizobium* bacteria assume a form called **bacteroids**, which are contained within vesicles formed in the root cells. Legume-*Rhizobium* relationships generate more usable nitrogen for plants than all industrial fertilizers used today—and at virtually no cost to the farmer.

Nitrogen fixation by *Rhizobium* requires an anaerobic environment, a condition facilitated by the location of the bacteroids inside living cells in the root cortex. The woody external layers of root nodules also help to limit gas exchange. Some root nodules appear reddish because of a molecule called leghemoglobin (*leg-* for "legume"), an iron-containing protein that binds reversibly to oxygen (similar to the hemoglobin in human red blood cells). This protein is an oxygen "buffer," reducing the concentration of free oxygen and thereby providing an anaerobic environment for nitrogen fixation while regulating the oxygen supply for the intense cellular respiration required to produce ATP for nitrogen fixation.

Each legume species is associated with a strain of *Rhizobium* bacteria. **Figure 37.13** describes how a root nodule develops after bacteria enter through an "infection thread" in a root hair. The symbiotic relationship between a legume and nitrogen-fixing bacteria is mutualistic. The bacteria supply the host plant with fixed nitrogen while the plant

▼ **Figure 37.13 Development of a soybean root nodule.**

❶ Roots emit chemical signals that attract *Rhizobium* bacteria. The bacteria then emit signals that stimulate root hairs to elongate and to form an infection thread by an invagination of the plasma membrane.

Infection thread

Rhizobium bacteria

Infected root hair

Dividing cells in root cortex

Bacteroid

❷ The infection thread containing the bacteria penetrates the root cortex. Cells of the cortex and pericycle begin dividing, and vesicles containing the bacteria bud into cortical cells from the branching infection thread. Bacteria within the vesicles develop into nitrogen-fixing bacteroids.

Nodule vascular tissue

Dividing cells in pericycle

Bacteroid

Bacteroids

Root hair sloughed off

Developing root nodule

❸ Growth continues in the affected regions of the cortex and pericycle, and these two masses of dividing cells fuse, forming the nodule.

Sclerenchyma cells

❺ The mature nodule grows to be many times the diameter of the root. A layer of lignin-rich sclerenchyma cells forms, reducing absorption of oxygen and thereby helping maintain the anaerobic environment needed for nitrogen fixation.

Nodule vascular tissue

Bacteroid

❹ The nodule develops vascular tissue (individual cells not shown) that supplies nutrients to the nodule and carries nitrogenous compounds into the vascular cylinder for distribution throughout the plant.

VISUAL SKILLS *What plant tissue systems are modified by root nodule formation?*

provides the bacteria with carbohydrates and other organic compounds. The root nodules use most of the ammonium produced to make amino acids, which are then transported up to the shoot through the xylem.

How does a legume species recognize a certain strain of *Rhizobium*? And how does an encounter with that specific *Rhizobium* strain lead to development of a nodule? Each partner responds to chemical signals from the other by expressing certain genes whose products contribute to nodule formation. By understanding the formation of root nodules, researchers hope to learn how to induce *Rhizobium* uptake and nodule formation in crop plants that do not normally form such nitrogen-fixing mutualistic relationships.

Nitrogen Fixation and Agriculture

The benefits of nitrogen fixation underlie most types of **crop rotation**. In this practice, a nonlegume such as maize is planted one year, and the following year alfalfa or some other legume is planted to restore the concentration of fixed nitrogen in the soil. To ensure that the legume encounters its specific *Rhizobium* strain, the seeds are exposed to bacteria before sowing. Instead of being harvested, the legume crop is often plowed under so that it will decompose as "green manure," reducing the need for manufactured fertilizers.

Many plant families besides legumes include species that benefit from mutualistic nitrogen fixation. For example, red alder (*Alnus rubra*) trees host nitrogen-fixing actinomycete bacteria. Rice, a crop of great commercial importance, benefits indirectly from mutualistic nitrogen fixation. Rice farmers culture a free-floating aquatic fern, *Azolla*, which has mutualistic cyanobacteria that fix nitrogen (see the bottom of Figure 37.9). The growing rice eventually shades and kills the *Azolla*, and decomposition of this nitrogen-rich organic material increases the rice paddy's fertility. Ducks also eat the *Azolla*, providing the rice paddy with an extra source of manure and providing the rice farmers with a source of meat.

Fungi and Plant Nutrition

Certain species of soil fungi also form mutualistic relationships with roots and play a major role in plant nutrition. Some of these fungi are endophytic, but the most important relationships are **mycorrhizae** ("fungus roots"), the intimate mutualistic associations of roots and fungi (see Figure 31.4b). The host plant provides the fungus with a steady supply of sugar. Meanwhile, the fungus increases the surface area for water uptake and also supplies the plant with phosphorus and other minerals absorbed from the soil. The fungi of mycorrhizae also secrete growth factors that stimulate roots to grow and branch, as well as antibiotics that help protect the plant from soil pathogens.

Mycorrhizae and Plant Evolution

EVOLUTION Mycorrhizae are not oddities; they are formed by most plant species. In fact, this plant-fungus mutualism might have been one of the evolutionary adaptations that helped plants initially colonize land (see Concept 29.1). When the earliest plants, which evolved from green algae, began to invade the land 400 to 500 million years ago, they encountered a harsh environment. Although the soil contained mineral nutrients, it lacked organic matter. Therefore, rain probably quickly leached away many of the soluble mineral nutrients. The barren land, however, was also a place of opportunities because light and carbon dioxide were abundant, and there was little competition or herbivory.

Neither the early land plants nor early land fungi were fully equipped to exploit the terrestrial environment. The early plants lacked the ability to extract essential nutrients from the soil, while the fungi were unable to manufacture carbohydrates. Instead of the fungi becoming parasitic on the rhizoids of the evolving plants (roots or root hairs had not yet evolved), the two organisms formed mycorrhizal associations, a mutualistic symbiosis that allowed both of them to exploit the terrestrial environment. Fossil evidence supports the idea that mycorrhizal associations occurred in the earliest land plants. The small minority of extant angiosperms that are nonmycorrhizal probably lost this ability through gene loss.

Types of Mycorrhizae

Mycorrhizae come in two forms called ectomycorrhizae and arbuscular mycorrhizae. **Ectomycorrhizae** form a dense sheath, or mantle, of mycelia (mass of branching hyphae) over the *surface* of the root. Fungal hyphae extend from the mantle into the soil, greatly increasing the surface area for water and mineral absorption. Hyphae also grow into the root cortex. These hyphae do not penetrate the root cells but form a network in the apoplast, or extracellular space, which facilitates nutrient exchange between the fungus and the plant. Compared with "uninfected" roots, the roots of ectomycorrhizae are generally thicker, shorter, and more branched. They typically do not form root hairs, which would be superfluous given the extensive surface area of the fungal mycelium. Only about 10% of plant families have species that form ectomycorrhizae. The vast majority of these species are woody, including members of the pine, oak, birch, and eucalyptus families.

Unlike ectomycorrhizae, **arbuscular mycorrhizae** (also called endomycorrhizae) do not ensheath the root but are embedded within it. They start when microscopic soil hyphae respond to the presence of a root by growing toward it, establishing contact, and growing along its surface. The hyphae penetrate between epidermal cells and then enter the root cortex, where they digest small patches of the cell walls

▼ Figure 37.14 Mycorrhizae.

Ectomycorrhizae. The mantle of the fungal mycelium ensheathes the root. Fungal hyphae extend from the mantle into the soil, absorbing water and minerals, especially phosphorus. Hyphae also extend into the extracellular spaces of the root cortex, providing extensive surface area for nutrient exchange between the fungus and its host plant.

Arbuscular mycorrhizae (endomycorrhizae). No mantle forms around the root, but microscopic fungal hyphae extend into the root. Within the root cortex, the fungus makes extensive contact with the plant through branching of hyphae that form arbuscules, providing an enormous surface area for nutrient swapping. The hyphae penetrate the cell walls, but not the plasma membranes, of cells within the cortex.

but don't pierce the plasma membrane. Instead of entering the cytoplasm, a hypha grows into a tube formed by invagination of the root cell's membrane. This invagination is like poking a finger gently into a balloon without popping it; your finger is like the fungal hypha, and the balloon skin is like the root cell's membrane. After the hyphae have penetrated in this way, some of them branch densely, forming structures called arbuscules ("little trees"), which are important sites of nutrient transfer between the fungus and the plant. Within the hyphae themselves, oval vesicles may form, possibly serving as food storage sites for the fungus. Arbuscular mycorrhizae are far more common than ectomycorrhizae, being found in over 85% of plant species, including most crops. About 5% of plant species don't form mycorrhizal associations. **Figure 37.14** provides an overview of mycorrhizae.

Agricultural and Ecological Importance of Mycorrhizae

Good crop yields often depend on the formation of mycorrhizae. Roots can form mycorrhizal symbioses only if exposed to the appropriate species of fungus. In most ecosystems, these fungi are present in the soil, and seedlings develop mycorrhizae. But if crop seeds are collected in one environment and planted in foreign soil, the plants may show signs of malnutrition (particularly phosphorus deficiency) resulting from

the absence of fungal partners. Treating seeds with spores of mycorrhizal fungi can help seedlings form mycorrhizae, facilitating recovery of damaged natural ecosystems (see Concept 55.5) and improving crop yield.

Mycorrhizal associations are also important in understanding ecological relationships. Arbuscular mycorrhizae fungi exhibit little host specificity; a single fungus may form a shared mycorrhizal network with several plants, even plants of different species. Mycorrhizal networks in a plant community may benefit one plant species more than another. Other examples of how mycorrhizae may affect the structures of plant communities come from studies of introduced plant species. For instance, garlic mustard (*Alliaria petiolata*), a European species that has been introduced into the forests of the eastern half of the United States, does not form mycorrhizae but hinders the growth of other plant species by preventing the growth of arbuscular mycorrhizal fungi.

Epiphytes, Parasitic Plants, and Carnivorous Plants

Almost all plant species have mutualistic relationships with soil fungi, bacteria, or both. Some plant species, including epiphytes, parasites, and carnivores, have unusual adaptations that facilitate exploiting other organisms **(Figure 37.15)**.

▼ Figure 37.15 Exploring Unusual Nutritional Adaptations in Plants

Epiphytes

An **epiphyte** (from the Greek *epi*, upon, and *phyton*, plant) is a plant that grows on another plant. Epiphytes produce and gather their own nutrients; they do not tap into their hosts for sustenance. Usually anchored to the branches or trunks of living trees, epiphytes absorb water and minerals from rain, mostly through leaves rather than roots. Some examples are staghorn ferns, bromeliads, and many orchids, including the vanilla plant.

▶ **Staghorn fern**, an epiphyte

Parasitic Plants

Unlike epiphytes, parasitic plants absorb water, minerals, and sometimes products of photosynthesis from their living hosts. Many species have roots that function as haustoria, nutrient-absorbing projections that tap into the host plant. Some parasitic species, such as orange-colored, spaghetti-like dodder (genus *Cuscuta*), lack chlorophyll entirely, whereas others, such as mistletoe (genus *Phoradendron*), are photosynthetic. Still others, such as Indian pipe (*Monotropa uniflora*), absorb nutrients from the hyphae of mycorrhizae associated with other plants.

◀ **Mistletoe**, a photosynthetic parasite

▲ **Dodder**, a nonphotosynthetic parasite (orange)

▲ **Indian pipe**, a nonphotosynthetic parasite of mycorrhizae

Carnivorous Plants

Carnivorous plants are photosynthetic but supplement their mineral diet by capturing insects and other small animals. They live in acid bogs and other habitats where soils are poor in nitrogen and other minerals. Pitcher plants such as *Nepenthes* and *Sarracenia* have modified leaves that form water-filled funnels into which prey slip and drown, eventually to be digested by enzymes. Sundews (genus *Drosera*) exude a sticky fluid from tentacle-like glands on highly modified leaves. Stalked glands secrete sweet mucilage that attracts and ensnares insects, and they also release digestive enzymes. Other glands then absorb the nutrient "soup." The highly modified leaves of Venus flytrap (*Dionaea muscipula*) close quickly but partially when a prey hits two trigger hairs in rapid enough succession. Smaller insects can escape, but larger ones are trapped by the teeth lining the margins of the lobes. Excitation by the prey causes the trap to narrow more and digestive enzymes to be released.

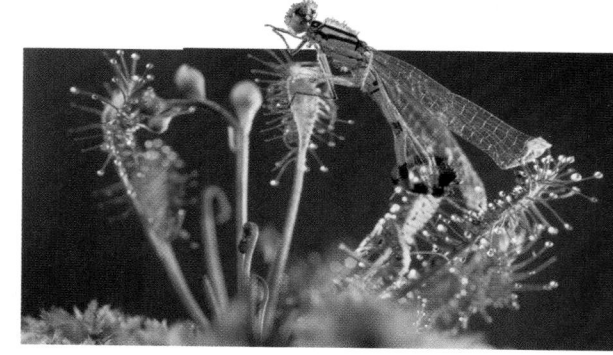

▲ **Sundew**

◀ **Pitcher plants**

▼ **Venus flytraps**

A recent study suggests that exploiting other organisms may be the norm. Chanyarat Paungfoo-Lonhienne and her colleagues at the University of Queensland in Australia have provided evidence that *Arabidopsis* and tomato can take up bacteria and yeast into their roots and digest them. Due to the small pore size of the cell wall (less than 10 nm) relative to the size of bacterial cells (about 1,000 nm), taking in microorganisms may depend on digestion of the cell wall. A study with wheat suggests that microorganisms provide only a tiny fraction of the plant's nitrogen needs, but this may not be true for all plants. These findings suggest that many plant species might engage in a limited amount of carnivory.

CONCEPT CHECK 37.3

1. Why is the study of the rhizosphere critical to understanding plant nutrition?
2. How do soil bacteria and mycorrhizae contribute to plant nutrition?
3. **MAKE CONNECTIONS** What is a general term that is used to describe the strategy of using photosynthesis *and* heterotrophy for nutrition (see Concept 28.1)? What is a well-known class of protists that uses this strategy?
4. **WHAT IF?** A peanut farmer finds that the older leaves of his plants are turning yellow following a long period of wet weather. Suggest a reason why.

For suggested answers, see Appendix A.

37 Chapter Review

→ Go to **Mastering Biology** for Assignments, the eText, the Study Area, and Dynamic Study Modules.

SUMMARY OF KEY CONCEPTS

→ To review key terms, go to the **Vocabulary Self-Quiz** in the **Mastering Biology** eText or Study Area, or go to goo.gl/zkjz9t.

CONCEPT 37.1

Soil contains a living, complex ecosystem (pp. 806–809)

- Soil particles of various sizes derived from the breakdown of rock are found in soil. Soil particle size affects the availability of water, oxygen, and minerals in the soil.
- A soil's composition encompasses its inorganic and organic components. **Topsoil** is a complex ecosystem teeming with bacteria, fungi, protists, animals, and the roots of plants.
- Some agricultural practices can deplete the mineral content of soil, tax water reserves, and promote erosion. The goal of soil conservation is to minimize this damage.

? How is soil a complex ecosystem?

CONCEPT 37.2

Plant roots absorb many types of essential elements from the soil (pp. 809–812)

- **Macronutrients**, elements required in relatively large amounts, include carbon, oxygen, hydrogen, nitrogen, and other major ingredients of organic compounds. **Micronutrients**, elements required in very small amounts, typically have catalytic functions as cofactors of enzymes.
- Deficiency of a mobile nutrient usually affects older organs more than younger ones; the reverse is true for nutrients that are less mobile within a plant. Macronutrient deficiencies are most common, particularly deficiencies of nitrogen, phosphorus, and potassium.
- Rather than tailoring the soil to match the plant, genetic engineers are tailoring the plant to match the soil.

? Do plants need soil to grow? Explain.

CONCEPT 37.3

Plant nutrition often involves relationships with other organisms (pp. 812–820)

- **Rhizobacteria** derive their energy from the **rhizosphere**, a microorganism-enriched ecosystem intimately associated with roots. Plant secretions support the energy needs of the rhizosphere. Some rhizobacteria produce antibiotics, whereas others make nutrients more available for plants. Most are free-living, but some live inside plants. Plants satisfy most of their huge need for nitrogen from the bacterial decomposition of **humus** and the fixation of gaseous nitrogen.

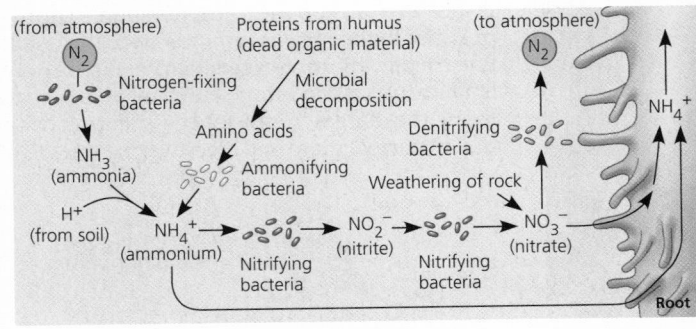

- Nitrogen-fixing bacteria convert atmospheric N_2 to nitrogenous minerals that plants can absorb as a nitrogen source for organic synthesis. The most efficient mutualism between plants and nitrogen-fixing bacteria occurs in the **nodules** that are formed by *Rhizobium* bacteria growing in the roots of legumes. These bacteria obtain sugar from the plant and supply the plant with fixed nitrogen. In agriculture, legume crops are rotated with other crops to restore nitrogen to the soil.
- **Mycorrhizae** are mutualistic associations of fungi and roots. The fungal hyphae of mycorrhizae absorb water and minerals, which they supply to their plant hosts.

- **Epiphytes** grow on the surfaces of other plants but acquire water and minerals from rain. Parasitic plants absorb nutrients from host plants. Carnivorous plants supplement their mineral nutrition by digesting animals.

? *Do all plants gain energy directly from photosynthesis? Explain.*

TEST YOUR UNDERSTANDING

→ For more multiple-choice questions, go to the **Practice Test** in the **Mastering Biology** eText or Study Area, or go to goo.gl/GruWRg.

Levels 1-2: Remembering/Understanding

1. The inorganic nutrient most often lacking in crops is
   (A) carbon.
   (C) phosphorus.
   (B) nitrogen.
   (D) potassium.

2. Micronutrients are needed in very small amounts because
   (A) most of them are mobile in the plant.
   (B) most serve mainly as cofactors of enzymes.
   (C) most are supplied in large enough quantities in seeds.
   (D) they play only a minor role in the growth and health of the plant.

3. Mycorrhizae enhance plant nutrition mainly by
   (A) absorbing water and minerals through the fungal hyphae.
   (B) providing sugar to root cells, which have no chloroplasts.
   (C) converting atmospheric nitrogen to ammonia.
   (D) enabling the roots to parasitize neighboring plants.

4. Epiphytes are
   (A) fungi that attack plants.
   (B) fungi that form mutualistic associations with roots.
   (C) nonphotosynthetic parasitic plants.
   (D) plants that grow on other plants.

5. A problem with intensive irrigation is
   (A) overfertilization.
   (B) aquifer depletion.
   (C) the long-term depletion of soil oxygen.
   (D) the clogging of waterways by vegetation debris.

Levels 3-4: Applying/Analyzing

6. A mineral deficiency is likely to affect older leaves more than younger leaves if
   (A) the mineral is a micronutrient.
   (B) the mineral is very mobile within the plant.
   (C) the mineral is required for chlorophyll synthesis.
   (D) the mineral is a macronutrient.

7. The greatest difference in health between two groups of plants of the same species, one group with mycorrhizae and one group without mycorrhizae, would be in an environment
   (A) where nitrogen-fixing bacteria are abundant.
   (B) that has soil with poor drainage.
   (C) that has hot summers and cold winters.
   (D) in which the soil is relatively deficient in mineral nutrients.

8. Two groups of tomatoes were grown under laboratory conditions, one with humus added to the soil and one a control without humus. The leaves of the plants grown without humus were yellowish (less green) compared with those of the plants grown in humus-enriched soil. The best explanation is that
   (A) the healthy plants used the food in the decomposing leaves of the humus for energy to make chlorophyll.
   (B) the humus made the soil more loosely packed, so water penetrated more easily to the roots.
   (C) the humus contained minerals such as magnesium and iron needed for the synthesis of chlorophyll.
   (D) the heat released by the decomposing leaves of the humus caused more rapid growth and chlorophyll synthesis.

9. The specific relationship between a legume and its mutualistic *Rhizobium* strain probably depends on
   (A) each legume having a chemical dialogue with a fungus.
   (B) each *Rhizobium* strain having a form of nitrogenase that works only in the appropriate legume host.
   (C) each legume being found where the soil has only the *Rhizobium* specific to that legume.
   (D) specific recognition between chemical signals and signal receptors of the *Rhizobium* strain and legume species.

10. **DRAW IT** Draw a simple sketch of cation exchange, showing a root hair, a soil particle with anions, and a hydrogen ion displacing a mineral cation.

Levels 5-6: Evaluating/Creating

11. **EVOLUTION CONNECTION** Imagine taking the plant out of the picture in Figure 37.11. Write a paragraph explaining how soil bacteria could sustain the recycling of nitrogen *before* land plants evolved.

12. **SCIENTIFIC INQUIRY** Acid precipitation has an abnormally high concentration of hydrogen ions (H^+). One effect of acid precipitation is to deplete the soil of nutrients such as calcium (Ca^{2+}), potassium (K^+), and magnesium (Mg^{2+}). Suggest a hypothesis to explain how acid precipitation washes these nutrients from the soil. How might you test your hypothesis?

13. **SCIENCE, TECHNOLOGY, AND SOCIETY** In many countries, irrigation is depleting aquifers to such an extent that land is subsiding, harvests are decreasing, and it is becoming necessary to drill wells deeper. In many cases, the withdrawal of groundwater has now greatly surpassed the aquifers' rates of natural recharge. Discuss the possible consequences of this trend. What can society and science do to help alleviate this growing problem?

14. **WRITE ABOUT A THEME: INTERACTIONS** The soil in which plants grow teems with organisms from every taxonomic kingdom. In a short essay (100–150 words), discuss examples of how the mutualistic interactions of plants with bacteria, fungi, and animals improve plant nutrition.

15. **SYNTHESIZE YOUR KNOWLEDGE**

Making a footprint in the soil seems like an insignificant event. In a short essay (100–150 words), explain how a footprint would affect the properties of the soil and how these changes would affect soil organisms and the emergence of seedlings.

For selected answers, see Appendix A.

38 Angiosperm Reproduction and Biotechnology

KEY CONCEPTS

38.1 Flowers, double fertilization, and fruits are key features of the angiosperm life cycle *p. 823*

38.2 Flowering plants reproduce sexually, asexually, or both *p. 833*

38.3 People modify crops by breeding and genetic engineering *p. 836*

Study Tip

Make a flowchart: Complete the steps in the following flowchart to compare angiosperm sexual reproduction with human reproduction.

Sexual reproduction in angiosperms	Human reproduction
Meiosis occurs in	Meiosis occurs in
↓	↓
Meiosis produces	Meiosis produces
↓	↓
↓	↓
Diploid embryo	Diploid embryo

→ Go to Mastering Biology

For Students (in eText and Study Area)
- Get Ready for Chapter 38
- Figure 38.6 Walkthrough: The Life Cycle of Angiosperms
- Animation: Angiosperm Life Cycle

For Instructors to Assign (in Item Library)
- Tutorial: Angiosperm Life Cycle
- Scientific Skills Exercise: Using Positive and Negative Numbers to Interpret Data

Figure 38.1 The orchid *Ophrys speculum*, shown above, is pollinated only by male wasps of the species *Dasyscolia ciliata*. The flowers resemble the female wasps visually and tactilely, and even release a similar scent. Male wasps who attempt to mate with the flowers get dusted with pollen (see inset). After the males fly off, they often transfer pollen to other *Ophrys* flowers. As shown in the simplified diagram, pollination is critical to sexual reproduction in angiosperms.

How do angiosperms reproduce sexually?

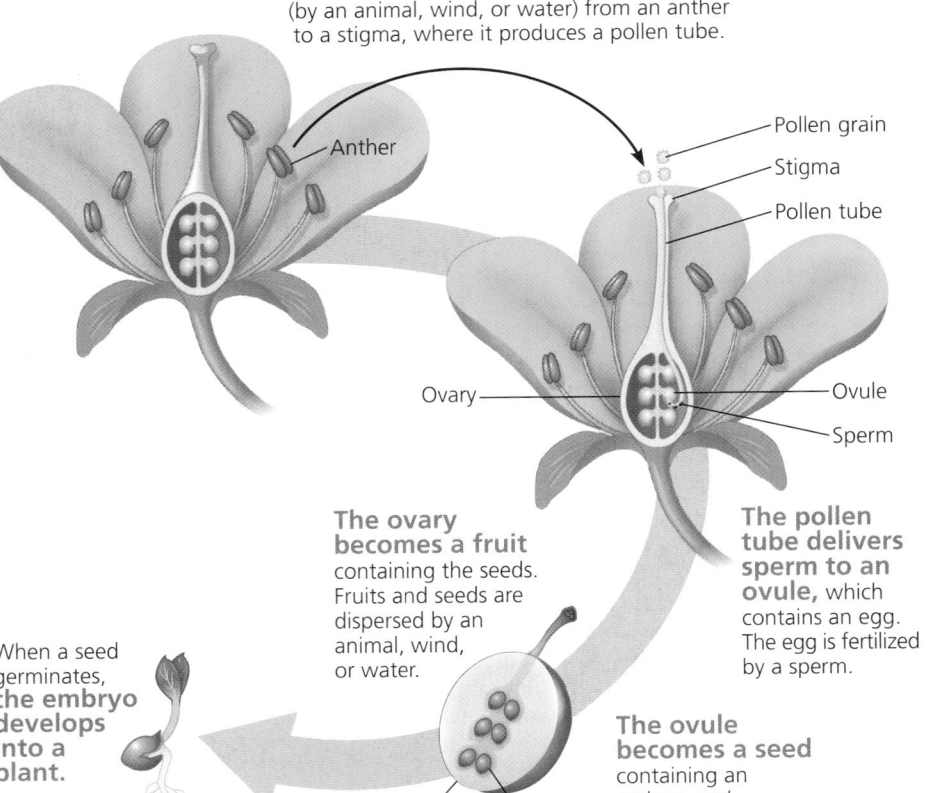

A pollen grain is transferred (by an animal, wind, or water) from an anther to a stigma, where it produces a pollen tube.

Anther

Pollen grain

Stigma

Pollen tube

Ovary

Ovule

Sperm

The ovary becomes a fruit containing the seeds. Fruits and seeds are dispersed by an animal, wind, or water.

The pollen tube delivers sperm to an ovule, which contains an egg. The egg is fertilized by a sperm.

When a seed germinates, **the embryo develops into a plant.**

The ovule becomes a seed containing an embryo and food supply.

Fruit

Seed

In Chapters 29 and 30, we approached plant reproduction from an evolutionary perspective, tracing the descent of plants from algal ancestors. Because angiosperms are the most important group of plants in most terrestrial ecosystems, we'll examine their reproductive biology, both sexual and asexual, in detail in this chapter. The importance of angiosperm reproduction, however, is not limited to natural ecosystems: the cultivation of angiosperms forms the basis for much of agriculture. For over 10,000 years, plant breeders have genetically manipulated the traits of a few hundred wild angiosperm species by artificial selection, transforming them into the crops we grow today. In recent decades, genetic engineering has dramatically increased the variety of ways and the speed with which we can modify plants, but many controversies have followed in the wake of this new biotechnology.

CONCEPT 38.1

Flowers, double fertilization, and fruits are key features of the angiosperm life cycle

The life cycles of all plants are characterized by an alternation of generations, in which multicellular haploid (*n*) and multicellular diploid (*2n*) generations alternately produce each other (see Figure 13.6b). The diploid plant, the *sporophyte*, produces haploid spores by meiosis. These spores divide by mitosis, giving rise to multicellular *gametophytes*, the male and female haploid plants that produce gametes (sperm and eggs). Fertilization, the fusion of gametes, results in a diploid zygote, which divides by mitosis and forms a new sporophyte. In angiosperms, the sporophyte is the dominant generation: It is larger, more conspicuous, and longer-lived

than the gametophyte. The key traits of the angiosperm life cycle can be remembered as the "three Fs"—*f*lowers, double *f*ertilization, and *f*ruits. We'll begin with flowers.

Flower Structure and Function

In angiosperms, the **flower** is the sporophytic structure specialized for sexual reproduction. A flower is typically composed of four types of floral organs: **carpels**, **stamens**, **petals**, and **sepals (Figure 38.2)**. When viewed from above, these organs take the form of concentric whorls. Carpels form the first (innermost) whorl, stamens the second, petals the third, and sepals the fourth (outermost) whorl. All are attached to a part of the stem called the **receptacle**. Flowers are determinate shoots; they cease growing after the flower and fruit are formed.

Carpels and stamens are sporophylls—modified leaves specialized for reproduction (see Concept 30.1); sepals and petals are sterile modified leaves. A carpel (megasporophyll) has an **ovary** at its base and a long, slender neck called the **style**. At the top of the style is a sticky structure called the **stigma** that captures pollen. Within the ovary are one or more **ovules**, which become seeds if fertilized; the number of ovules depends on the species. The flower in Figure 38.2 has a single carpel, but many species have multiple carpels. In most species, the carpels are fused, resulting in a compound ovary with two or more chambers, each containing one or more ovules. The term **pistil** is sometimes used to refer to a single carpel or two or more fused carpels **(Figure 38.3)**. A stamen (microsporophyll) consists of a stalk called the filament and a terminal structure called the **anther**; within the anther are chambers called microsporangia (pollen sacs) that produce

▼ **Figure 38.3 The relationship between the terms *carpel* and *pistil*.** A simple pistil consists of a single, unfused carpel. A compound pistil consists of two or more fused carpels. Some types of flowers have only a single pistil, while other types have many pistils. In either case, the pistils may be simple or compound.

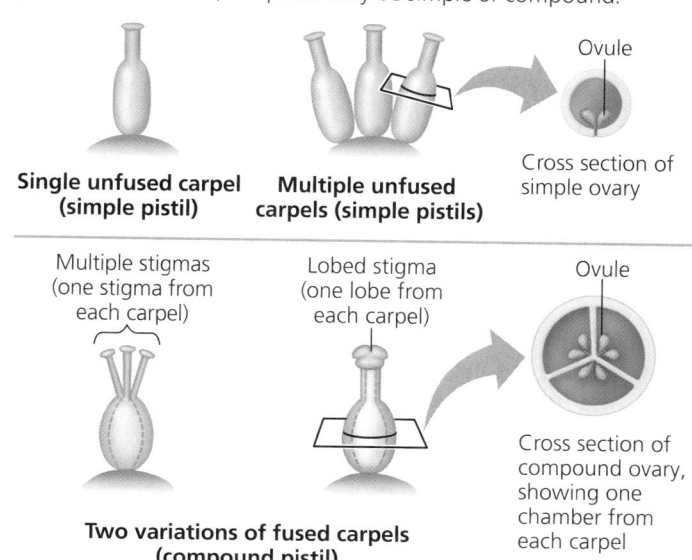

Single unfused carpel (simple pistil)

Multiple unfused carpels (simple pistils)

Cross section of simple ovary

Ovule

Multiple stigmas (one stigma from each carpel)

Lobed stigma (one lobe from each carpel)

Ovule

Cross section of compound ovary, showing one chamber from each carpel

Two variations of fused carpels (compound pistil)

▼ **Figure 38.2 The structure of an idealized flower.**

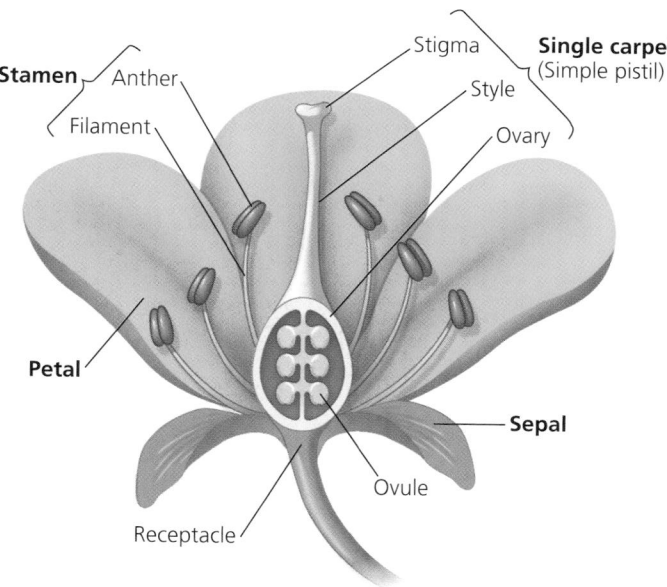

Stamen — Anther
Filament
Petal
Receptacle
Ovule
Sepal
Stigma
Style
Ovary
Single carpel (Simple pistil)

pollen. Petals are typically more brightly colored than sepals and advertise the flower to insects and other animal pollinators. Sepals, which enclose and protect unopened floral buds, usually resemble leaves more than the other floral organs do.

Complete flowers have all four basic floral organs (see Figure 38.2). Some species have **incomplete flowers**, lacking sepals, petals, stamens, or carpels. For example, most grass flowers lack petals. Some incomplete flowers

are sterile, lacking functional stamens and carpels; others are *unisexual* (sometimes called *imperfect*), lacking either stamens or carpels. Flowers also vary in size, shape, color, odor, organ arrangement, and time of opening. Some are borne singly, while others are arranged in showy clusters called **inflorescences**. For example, a sunflower consists of a central disk composed of hundreds of tiny incomplete flowers, surrounded by sterile, incomplete flowers that look

---

▼ Figure 38.4 **Exploring Flower Pollination**

Most angiosperm species rely on a living (biotic) or nonliving (abiotic) pollinating agent that can move pollen from the anther of a flower on one plant to the stigma of a flower on another plant. Approximately 80% of all angiosperm pollination is biotic, employing animal go-betweens. Among abiotically pollinated species, 98% rely on wind and 2% on water. (Some angiosperm species can self-pollinate, but such species are limited to inbreeding in nature.)

Abiotic Pollination by Wind

About 20% of all angiosperm species are wind-pollinated. Since their reproductive success does not depend on attracting pollinators, there has been no selective pressure favoring colorful or scented flowers. Accordingly, the flowers of wind-pollinated species are often small, green, and inconspicuous, and they produce neither scent nor the sugary solution called nectar. Most temperate trees and grasses are wind-pollinated. The flowers of hazel (*Corylus*

avellana) and many other temperate, wind-pollinated trees appear in the early spring, when there are no leaves to interfere with pollen movement. The relative inefficiency of wind pollination is compensated for by production of copious amounts of pollen grains. Wind tunnel studies reveal that wind pollination is often more efficient than it appears because floral structures can create eddy currents that aid in pollen capture.

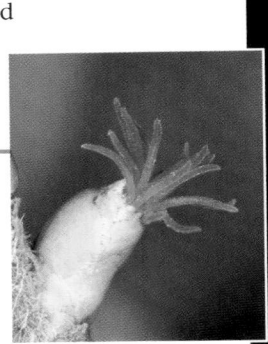

▲ Hazel carpellate flower (carpels only)

▲ Hazel staminate flowers (stamens only) releasing clouds of pollen

Pollination by Bees

▲ Common dandelion under normal light

▲ Common dandelion under ultraviolet light

About 65% of all flowering plants require insects for pollination; the percentage is even greater for major crops. Bees are the most important insect pollinators, and there is great concern in Europe and North America that honeybee populations have shrunk. Pollinating bees depend on nectar and pollen for food. Typically, bee-pollinated flowers have a delicate, sweet

fragrance. Bees are attracted to bright colors, primarily yellow and blue. Red appears dull to them, but they can see ultraviolet radiation. Many species of bee-pollinated flowers, such as the common dandelion (*Taraxacum vulgare*), have ultraviolet markings called "nectar guides" that help insects locate the nectaries (nectar-producing glands) but are only visible to human eyes under ultraviolet light.

> **Mastering Biology**
> Video: Bee Colony Decline

Pollination by Moths and Butterflies

Moths and butterflies detect odors, and the flowers they pollinate are often sweetly fragrant. Butterflies perceive many bright colors, but moth-pollinated flowers are usually white or yellow, which stand out at night when moths are active. A yucca plant (shown here) is typically pollinated by a single species of moth with appendages that pack pollen onto the stigma. The moth then

deposits eggs directly into the ovary. The larvae eat some developing seeds, but this cost is outweighed by the benefit of an efficient and reliable pollinator. If a moth deposits too many eggs, the flower aborts and drops off, selecting against individuals that overexploit the plant.

❓ *What are the benefits and dangers to a plant of having a highly specific animal pollinator?*

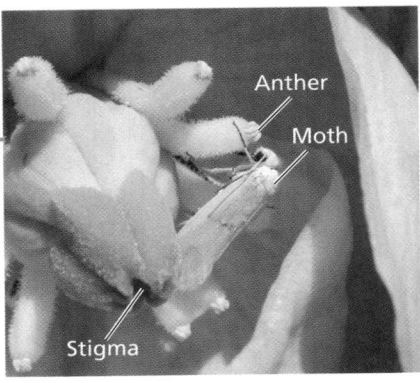

▲ Moth on yucca flower

like yellow petals (see Figure 40.23). Much of floral diversity represents adaptation to specific pollinators.

Methods of Pollination

Pollination is the transfer of pollen to the ovule-bearing structure of a seed plant. In angiosperms, this transfer is from an anther to a stigma. Pollination can occur by wind,

Pollination by Flies

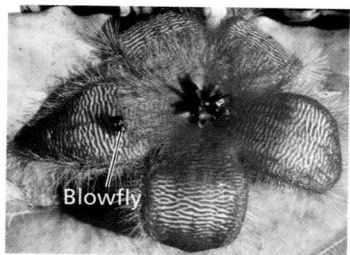

▲ Blowfly on carrion flower

Many fly-pollinated flowers are reddish and fleshy, with an odor like rotten meat. Blowflies visiting carrion flowers (*Stapelia* species) mistake the flower for a rotting corpse and lay their eggs on it. In the process, the blowflies become dusted with pollen that they carry to other flowers. When the eggs hatch, the larvae find no carrion to eat and die.

Pollination by Bats

Bat-pollinated flowers, like moth-pollinated flowers, are light-colored and aromatic, attracting their nocturnal pollinators. The lesser long-nosed bat (*Leptonycteris curasoae yerbabuenae*) feeds on the nectar and pollen of agave and cactus flowers in the southwestern United States and Mexico. In feeding, the bats transfer pollen from plant to plant. Long-nosed bats are an endangered species.

▲ Long-nosed bat feeding on agave flowers at night

Pollination by Birds

Bird-pollinated flowers, such as columbine flowers, are usually large and bright red or yellow, but they have little odor. Since birds often do not have a well-developed sense of smell, there has been no selective pressure favoring scent production. However, the flowers produce the sugary nectar that helps meet the high energy demands of the pollinating birds. The primary function of nectar, which is produced by nectaries at the base of many flowers, is to "reward" the pollinator. The petals of such flowers are often fused, forming a bent floral tube that fits the curved beak of the bird.

▶ Hummingbird drinking nectar of columbine flower

water, or animals **(Figure 38.4)**. In wind-pollinated species, including grasses and many trees, the release of clouds of typically small-sized pollen compensates for the randomness of dispersal by the wind. At certain times of the year, the air is loaded with pollen grains, as anyone plagued with pollen allergies can attest. Some species of aquatic plants rely on water to disperse pollen. Most angiosperm species, however, depend on insects, birds, or other animal pollinators to transfer pollen directly from one flower to another.

EVOLUTION Animal pollinators are drawn to flowers for the food they provide in the form of pollen and nectar. Attracting pollinators that are loyal to a given plant species is an efficient way to ensure that pollen is transferred to another flower of the same species. Natural selection, therefore, favors deviations in floral structure or physiology that make it more likely for a flower to be pollinated regularly by an effective animal species. If a plant species develops traits that make its flowers more prized by pollinators, there is a selective pressure for pollinators to become adept at harvesting food from these flowers. The joint evolution of two interacting species, each in response to selection imposed by the other, is called **coevolution**. For example, some species have fused flower petals that form long, tubelike structures bearing nectaries tucked deep inside. Charles Darwin suggested that coadaptations between flower and insect might lead to correspondences between the length of a floral tube and the length of an insect's proboscis, a straw-like mouthpart. Based on the length of a long, tubular flower endemic to Madagascar, Darwin predicted the existence of a pollinating moth with a 28-cm-long proboscis. Such a moth was discovered two decades after Darwin's death **(Figure 38.5)**.

Climate change may be affecting long-standing relationships between plants and animal pollinators. For example, flowers that require long-tongued pollinators have declined under the warmer conditions in the Rocky Mountains. As a result, there has been selective pressure favoring bumblebees

▼ **Figure 38.5 Coevolution of a flower and an insect pollinator.** The long floral tube of the Madagascar orchid *Angraecum sesquipedale* has coevolved with the 28-cm-long proboscis of its pollinator, the hawk moth *Xanthopan morganii praedicta*. The moth is named in honor of Darwin's prediction of its existence.

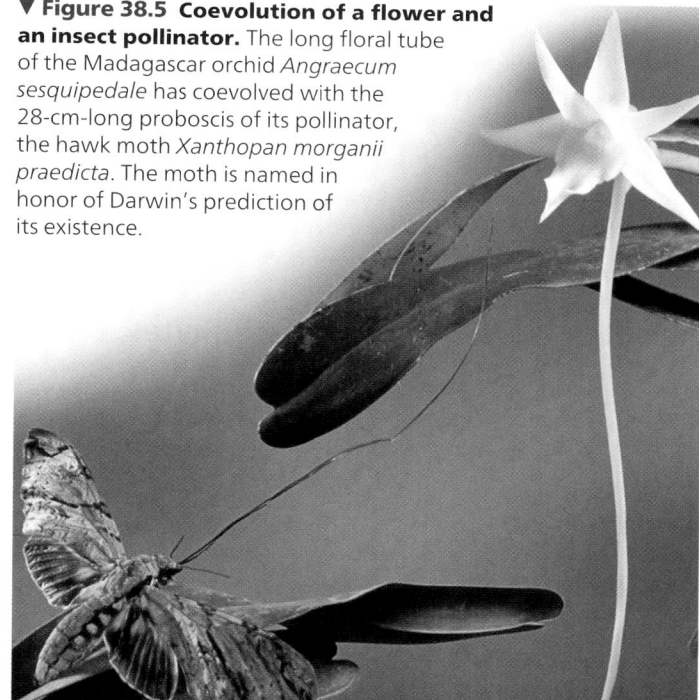

with shorter tongues. Two species of Rocky Mountain bumble-bees now have tongues that are about one-quarter shorter than those of bees of the same species 40 years ago.

The Angiosperm Life Cycle: An Overview

Pollination is one step in the angiosperm life cycle. **Figure 38.6** provides a complete overview of the life cycle, focusing on gametophyte development, sperm delivery by pollen tubes, double fertilization, and seed development.

Over the course of seed plant evolution, gametophytes became reduced in size and wholly dependent on the sporophyte for nutrients (see Figure 30.2). The gametophytes of angiosperms are the most reduced of all plants, consisting of only a few cells: They are microscopic, and their development is obscured by protective tissues.

Development of Female Gametophytes (Embryo Sacs)

As a carpel develops, one or more ovules form deep within its ovary, its swollen base. A female gametophyte, also known as an **embryo sac**, develops inside each ovule. The process of embryo sac formation occurs in a tissue called the megasporangium ❶ within each ovule. Two *integuments* (layers of protective sporophytic tissue that will develop into the seed coat) surround each megasporangium, except at a gap called the *micropyle*. Female gametophyte development begins when one cell in the megasporangium of each ovule, the *megasporocyte,* enlarges and undergoes meiosis, producing four haploid **megaspores**. Usually only one megaspore survives; the other three degenerate.

The nucleus of the surviving megaspore divides by mitosis three times without cytokinesis, resulting in one large cell with eight haploid nuclei. The multinucleate mass is then divided by membranes to form the embryo sac. Near the micropyle of the embryo sac, two synergid cells flank the egg and help attract and guide the pollen tube to the embryo sac. At the opposite end of the embryo sac are three antipodal cells of unknown function. The other two nuclei, called polar nuclei, are not partitioned into separate cells but share the cytoplasm of the large central cell of the embryo sac. The mature embryo sac thus consists of eight nuclei contained within seven cells. The ovule, which will become a seed if fertilized, now consists of the embryo sac, enclosed by the megasporangium (which eventually withers) and two surrounding integuments.

Development of Male Gametophytes in Pollen Grains

As the stamens are produced, each anther ❷ develops four microsporangia, also called pollen sacs. Within the microsporangia are many diploid cells called *microsporocytes*. Each microsporocyte undergoes meiosis, forming four haploid **microspores**, ❸ each of which eventually gives rise to a haploid male gametophyte. Each microspore then undergoes mitosis, producing a haploid

male gametophyte consisting of only two cells: the *generative cell* and the *tube cell*. Together, these two cells and the spore wall constitute a **pollen grain**. The spore wall, which consists of material produced by both the microspore and the anther, usually exhibits an elaborate pattern unique to the species. During maturation of the male gametophyte, the generative cell passes into the tube cell: The tube cell now has a completely free-standing cell inside it.

Sperm Delivery by Pollen Tubes

After the microsporangium breaks open and releases pollen, a pollen grain may be transferred to a receptive surface of a stigma—the act of pollination. There, it absorbs water and germinates by producing a **pollen tube**, a long cellular protuberance that delivers sperm to the female gametophyte.

A pollen grain typically consists of the spore wall and two cells: a tube cell and a second cell, the generative cell, contained within the tube cell. As the pollen tube elongates through the style, the nucleus of the generative cell divides by mitosis and produces two sperm, which remain inside the tube cell. The tube nucleus leads ahead of the two sperm as the pollen tube grows toward the micropyle in response to chemical attractants produced by the synergid cells. The arrival of the pollen tube initiates the death of one of the two synergids, thereby providing a passageway into the embryo sac. The two sperm are then released from the pollen tube ❹ in the vicinity of the female gametophyte.

Double Fertilization

Fertilization, the fusion of gametes, occurs after the two sperm reach the female gametophyte. One sperm fertilizes the egg, forming the zygote. The other sperm combines with the two polar nuclei, forming a triploid ($3n$) nucleus in the center of the large central cell of the female gametophyte. This cell will give rise to the **endosperm**, a multicellular, food-storing tissue of the seed. ❺ The union of the two sperm cells with different nuclei of the female gametophyte is called **double fertilization**. Double fertilization ensures that endosperm develops only in ovules where the egg has been fertilized, thereby preventing angiosperms from squandering nutrients on infertile ovules. Near the time of double fertilization, the tube nucleus, the other synergid cell, and the antipodal cells degenerate.

Seed Development

❻ After double fertilization, each ovule develops into a seed. Meanwhile, the ovary develops into a fruit, which encloses the seeds and aids in their dispersal by wind or animals. As the sporophyte embryo develops from the zygote, the seed stockpiles proteins, oils, and starch to varying degrees, depending on the species. This is why seeds are such a major nutrient drain. Initially, carbohydrates and other nutrients are stored in the seed's endosperm, but later, depending on the species, the swelling cotyledons (seed leaves) of the embryo may take over this function. When a seed germinates, ❼ the embryo develops into a new sporophyte. The mature

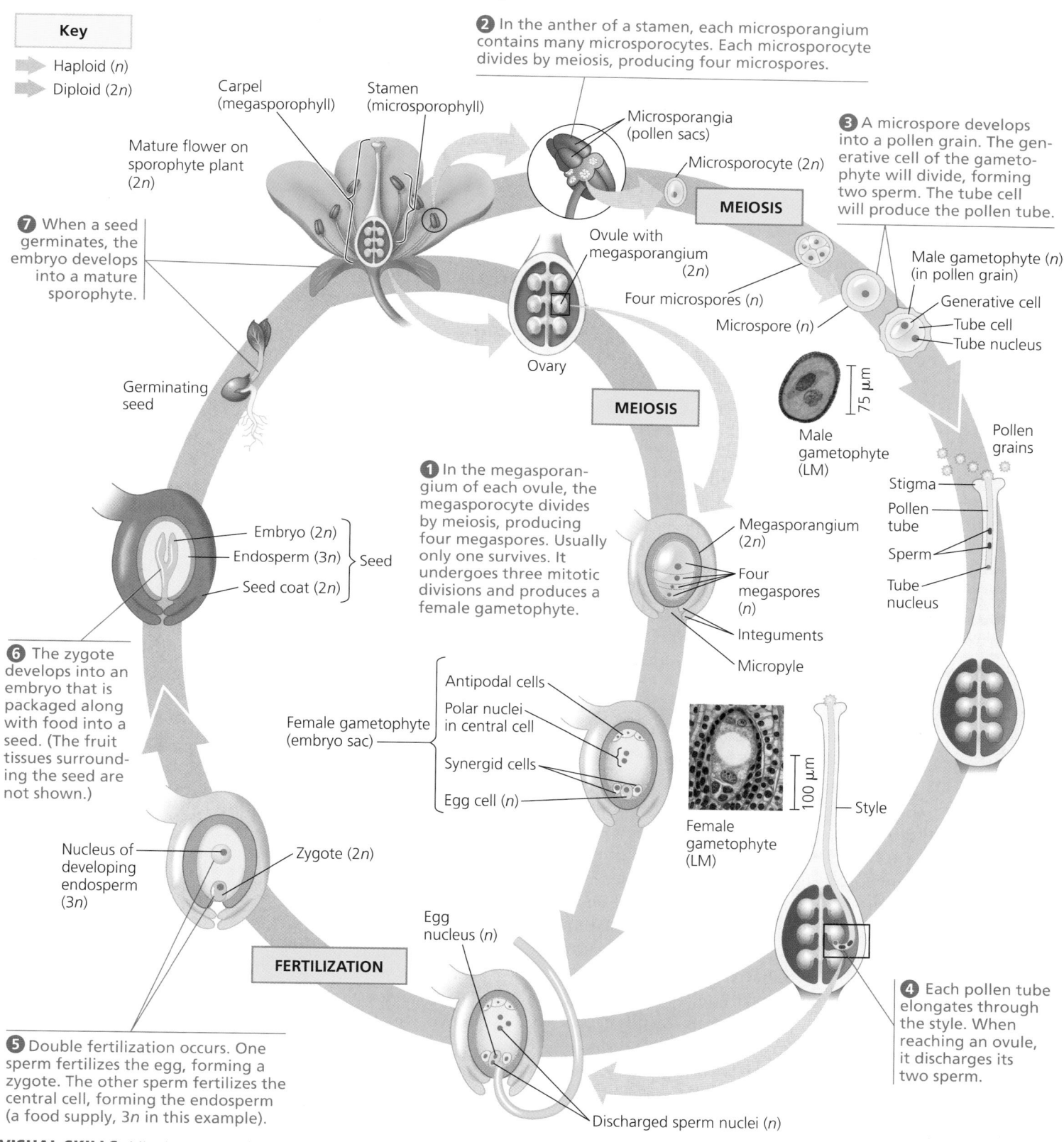

▼ **Figure 38.6 The life cycle of an angiosperm.** For simplicity, a flower with a single carpel (simple pistil) is shown. Many species have multiple carpels, either separate or fused.

Key

Haploid (*n*)
Diploid (2*n*)

Carpel (megasporophyll)
Stamen (microsporophyll)

Mature flower on sporophyte plant (2*n*)

2 In the anther of a stamen, each microsporangium contains many microsporocytes. Each microsporocyte divides by meiosis, producing four microspores.

Microsporangia (pollen sacs)
Microsporocyte (2*n*)

MEIOSIS

3 A microspore develops into a pollen grain. The generative cell of the gametophyte will divide, forming two sperm. The tube cell will produce the pollen tube.

7 When a seed germinates, the embryo develops into a mature sporophyte.

Ovule with megasporangium (2*n*)

Four microspores (*n*)
Microspore (*n*)

Male gametophyte (*n*) (in pollen grain)
Generative cell
Tube cell
Tube nucleus

Ovary

MEIOSIS

75 μm

Male gametophyte (LM)

Pollen grains

Stigma
Pollen tube
Sperm
Tube nucleus

Germinating seed

1 In the megasporangium of each ovule, the megasporocyte divides by meiosis, producing four megaspores. Usually only one survives. It undergoes three mitotic divisions and produces a female gametophyte.

Megasporangium (2*n*)
Four megaspores (*n*)
Integuments
Micropyle

Embryo (2*n*)
Endosperm (3*n*)   Seed
Seed coat (2*n*)

6 The zygote develops into an embryo that is packaged along with food into a seed. (The fruit tissues surrounding the seed are not shown.)

Antipodal cells
Polar nuclei in central cell
Synergid cells
Egg cell (*n*)

Female gametophyte (embryo sac)

Female gametophyte (LM)

100 μm

Style

Nucleus of developing endosperm (3*n*)
Zygote (2*n*)

Egg nucleus (*n*)

FERTILIZATION

4 Each pollen tube elongates through the style. When reaching an ovule, it discharges its two sperm.

5 Double fertilization occurs. One sperm fertilizes the egg, forming a zygote. The other sperm fertilizes the central cell, forming the endosperm (a food supply, 3*n* in this example).

Discharged sperm nuclei (*n*)

VISUAL SKILLS *Mitosis continues throughout the life cycle. Where in the cycle would the most mitotic divisions occur?*

→ Mastering Biology Figure Walkthrough
Animation: Angiosperm Life Cycle
Animation: Sexual Reproduction in Angiosperms

sporophyte produces its own flowers and fruits: The life cycle is now complete, but it is necessary to examine more closely how an ovule develops into a mature seed.

Seed Development and Structure

After successful pollination and double fertilization, a seed begins to form. During this process, both the endosperm and the embryo develop. When mature, a **seed** consists of a dormant embryo surrounded by stored food and protective layers.

Endosperm Development

Endosperm usually begins development before the embryo does. After double fertilization, the triploid nucleus of the ovule's central cell divides mitotically, forming a multinucleate "supercell" that has a milky consistency. This liquid mass, the endosperm, becomes multicellular when cytokinesis partitions the cytoplasm by forming membranes between the nuclei. Eventually, these "naked" cells produce cell walls, and the endosperm becomes solid. Coconut "milk" and "meat" are examples of liquid and solid endosperm. The white fluffy part of popcorn is another example of endosperm. The endosperms of just three grains—wheat, maize, and rice—provide much of the food energy for human sustenance.

In grains and most other species of monocots, as well as many eudicots, the endosperm stores nutrients that can be used by the seedling after germination. In other eudicot seeds, the food reserves of the endosperm are completely exported to the cotyledons before the seed completes its development; consequently, the mature seed lacks endosperm.

Embryo Development

The first mitotic division of the zygote is asymmetrical and splits the fertilized egg into a basal cell and a terminal cell **(Figure 38.7)**. The terminal cell eventually gives rise to most of the embryo. The basal cell continues to divide, producing a thread of cells called the suspensor, which anchors the embryo to the parent plant. The suspensor helps in transferring nutrients to the embryo from the parent plant and, in some species, from the endosperm. As the suspensor elongates, it pushes the embryo deeper into the nutritive and protective tissues. Meanwhile, the terminal cell divides several times and then forms a spherical proembryo (early embryo) that is attached to the suspensor. The cotyledons begin to form as bumps on the proembryo. A eudicot embryo, with its two cotyledons, is heart-shaped at this stage of development.

Soon after the rudimentary cotyledons appear, the embryo elongates. Cradled between the two cotyledons is the embryonic shoot apex. At the opposite end of the embryo's axis, where the suspensor attaches, an embryonic root apex forms. After the seed germinates—indeed, for the rest of the plant's life—the apical meristems at the apices of shoots and roots sustain primary growth (see Figure 35.11).

▼ **Figure 38.7 The development of a eudicot plant embryo.** By the time the ovule becomes a mature seed and the integuments harden and thicken into the seed coat, the zygote has given rise to an embryonic plant with rudimentary organs.

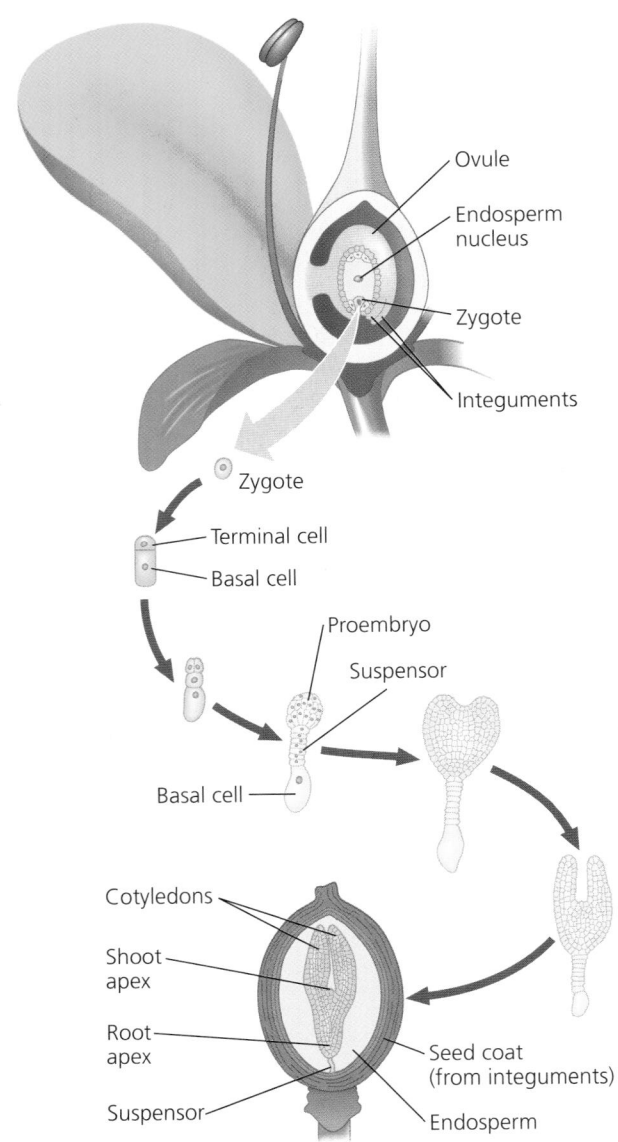

⊘ **Mastering Biology** Animation: Embryo and Endosperm Development

Structure of the Mature Seed

During the last stages of its maturation, the seed dehydrates until its water content is only about 5–15% of its weight. The embryo, which is surrounded by a food supply (cotyledons, endosperm, or both), enters **dormancy**; that is, it stops growing and its metabolism nearly ceases. The embryo and its food supply are enclosed by a hard, protective **seed coat** formed from the integuments of the ovule. In some species, dormancy is imposed by the presence of an intact seed coat rather than by the embryo itself.

If you split apart a seed of the garden bean, a type of eudicot, you can see that the embryo consists of an elongate structure, the embryonic axis, attached to two thick, fleshy cotyledons **(Figure 38.8a)**. Below where the cotyledons are attached, the embryonic axis is called the **hypocotyl** (from the Greek *hypo*, under). The hypocotyl terminates in the **radicle**, or embryonic root. The portion of the embryonic axis above where the cotyledons are attached and below the first pair of miniature leaves is the **epicotyl** (from the Greek *epi*, on, over). The epicotyl, young leaves, and shoot apical meristem are collectively called the *plumule*.

The cotyledons of the common garden bean are packed with starch before the seed germinates because they absorbed carbohydrates from the endosperm when the seed was developing. However, the seeds of some eudicot species, such as castor beans (*Ricinus communis*), retain their food supply in the endosperm and have very thin cotyledons. The cotyledons absorb nutrients from the endosperm and transfer them to the rest of the embryo when the seed germinates.

The embryos of monocots possess only a single cotyledon **(Figure 38.8b)**. Grasses, including maize and wheat, have a specialized cotyledon called a *scutellum* (from the Latin *scutella*, small shield, a reference to its shape). The scutellum, which has a large surface area, is pressed against the endosperm, from which it absorbs nutrients during germination. The embryo of a grass seed is enclosed within two protective sheaths: a **coleoptile**, which covers the young shoot, and a **coleorhiza**, which covers the young root. Both structures aid in soil penetration after germination.

Seed Dormancy: An Adaptation for Tough Times

The environmental conditions required to break seed dormancy vary among species. Some seed types germinate as soon as they are in a suitable environment. Others remain dormant, even if sown in a favorable place, until a specific environmental cue causes them to break dormancy.

The requirement for specific cues to break seed dormancy increases the chances that germination will occur at a time and place most advantageous to the seedling. The seeds of many desert plants, for instance, germinate only after a substantial rainfall. If they were to germinate after a mild drizzle, the soil might soon become too dry to support the seedlings. Where natural fires are common, many seeds require intense heat or smoke to break dormancy; seedlings are therefore most abundant after fire has cleared away competing vegetation. Where winters are harsh, seeds may require extended exposure to cold before they germinate; seeds sown during summer or fall will therefore not germinate until the following spring, ensuring a long growth season before the next winter. Many small seeds require light for germination, an adaptation that prevents them from germinating when they are buried so deeply in the soil that their energy reserves would be exhausted before their plumules could reach sunlight. Some seeds have coats that must be weakened by chemical attack as they pass through an animal's digestive

▼ **Figure 38.8 Seed structure.**

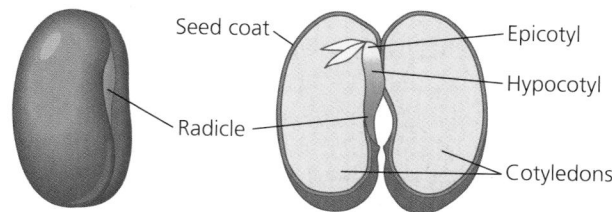

(a) Common garden bean, a eudicot with thick cotyledons. The fleshy cotyledons store food absorbed from the endosperm before the seed germinates.

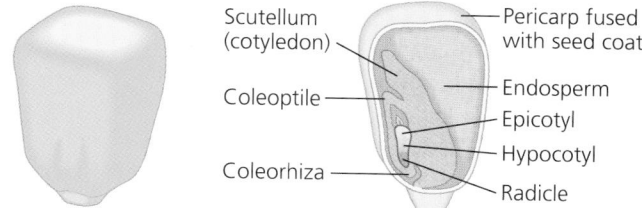

(b) Maize, a monocot. Like all monocots, maize has only one cotyledon. Maize and other grasses have a large cotyledon called a scutellum. The rudimentary shoot is sheathed in a structure called the coleoptile, and the coleorhiza covers the young root.

MAKE CONNECTIONS *In addition to cotyledon number, how do the structures of monocots and eudicots differ? (See Figure 30.16.)*

VISUAL SKILLS *Which mature seed lacks an endosperm? What happened to it?*

tract and thus are usually carried a long distance before germinating from feces.

The length of time a dormant seed remains viable and capable of germinating varies from a few days to decades or even longer, depending on the plant species and environmental conditions. The oldest carbon-14-dated seed that has grown into a viable plant was a 2,000-year-old date palm seed from Israel. Most seeds are durable enough to last a year or two until conditions are favorable for germinating. Thus, the soil has a bank of ungerminated seeds that may have accumulated for several years. This is one reason vegetation reappears so rapidly after an environmental disruption such as fire.

Sporophyte Development from Seed to Mature Plant

When environmental conditions are conducive for growth, seed dormancy is lost and germination proceeds. Germination is followed by growth of stems, leaves, and roots, and eventually by flowering.

Seed Germination

Seed germination is initiated by **imbibition**, the uptake of water due to the low water potential of the dry seed. Imbibition causes the seed to expand and rupture its coat and triggers changes in the embryo that enable it to resume growth. Following hydration, enzymes digest the storage materials of the endosperm or cotyledons, and the nutrients are transferred to the growing regions of the embryo.

The first organ to emerge from the germinating seed is the radicle, the embryonic root. The development of a root system anchors the seedling in the soil and supplies it with water necessary for cell expansion. A ready supply of water is a prerequisite for the next step, the emergence of the shoot tip into the drier conditions encountered above ground. In garden beans, for example, a hook forms in the hypocotyl, and growth pushes the hook above ground **(Figure 38.9a)**. In response to light, the hypocotyl straightens, the cotyledons separate, and the delicate epicotyl, now exposed, spreads its first true leaves (as distinct from the cotyledons, or seed leaves). These leaves expand, become green, and begin making food by photosynthesis. The shriveled, energy-depleted cotyledons that supplied food for the developing embryo are shed.

Some monocots, such as maize and other grasses, use a different method for breaking ground as they germinate **(Figure 38.9b)**. The coleoptile pushes up through the soil and into the air. The shoot tip grows through the tunnel provided by the coleoptile and breaks through the coleoptile's tip upon emergence.

▼ **Figure 38.9 Two common types of seed germination.**

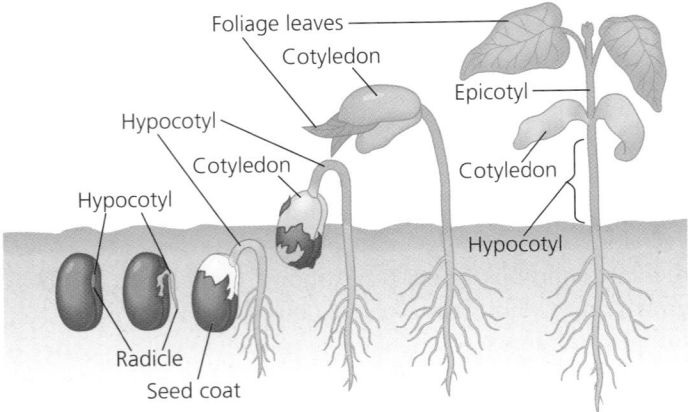

(a) Common garden bean. In common garden beans, straightening of a hook in the hypocotyl pulls the cotyledons from the soil.

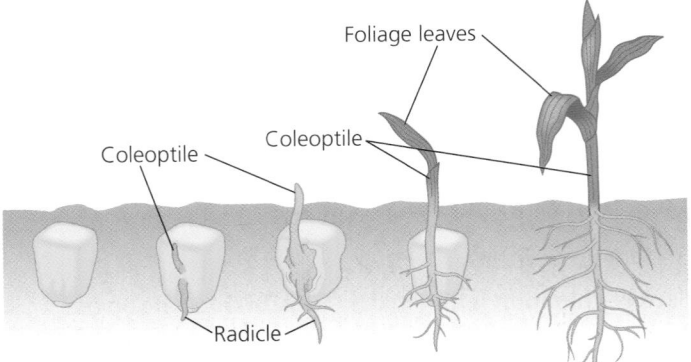

(b) Maize. In maize and other grasses, the shoot grows straight up through the tube of the coleoptile.

VISUAL SKILLS *How do bean and maize seedlings protect their shoot systems as they push through the soil?*

⊙ **Mastering Biology Animation: Seed Germination**

Growth and Flowering

Once a seed has germinated and started to photosynthesize, most of the plant's resources are devoted to primary and secondary growth of stems, leaves, and roots (also known as *vegetative growth*). This growth arises from the activity of meristematic cells (see Concept 35.2). During this stage, plants usually photosynthesize and grow as much as possible before flowering, the reproductive phase.

The flowers of a given plant species typically appear suddenly and simultaneously at a specific time of year. Such timing promotes outbreeding, the main advantage of sexual reproduction. Flower formation involves a developmental switch in the shoot apical meristem from a vegetative to a reproductive growth mode. This transition into a *floral meristem* is triggered by a combination of environmental cues (such as day length) and internal signals, as you'll learn in Concept 39.3. Once the transition to flowering has begun, the order of each organ's emergence from the floral meristem determines whether it will develop into a sepal, petal, stamen, or carpel (see Figure 35.33).

Fruit Structure and Function

Before a seed can germinate and develop into a mature plant, it must be deposited in suitable soil. Fruits play a key role in this process. A **fruit** is the mature ovary of a flower. While the seeds are developing from ovules, the flower develops into a fruit **(Figure 38.10)**. The fruit protects the enclosed seeds and, when mature, aids in their dispersal by wind or animals. Fertilization triggers hormonal changes that cause the ovary to begin its transformation into a fruit. If a flower has not been pollinated, fruit typically does not develop, and the flower usually withers and dies.

During fruit development, the ovary wall becomes the *pericarp*, the thickened wall of the fruit. In some fruits, such as soybean pods, the ovary wall dries out completely at maturity, whereas in other fruits, such as grapes, it remains fleshy. In still others, such as peaches, the inner part of the ovary

▼ **Figure 38.10 The flower-to-fruit transition.** After flowers are fertilized, stamens and petals fall off, stigmas and styles wither, and the ovary walls that house the developing seeds swell to form fruits. Developing seeds and fruits are major sinks for sugars and other carbohydrates. American pokeweed flowers and fruits are shown here.

▼ Figure 38.11 Developmental origin of different classes of fruits.

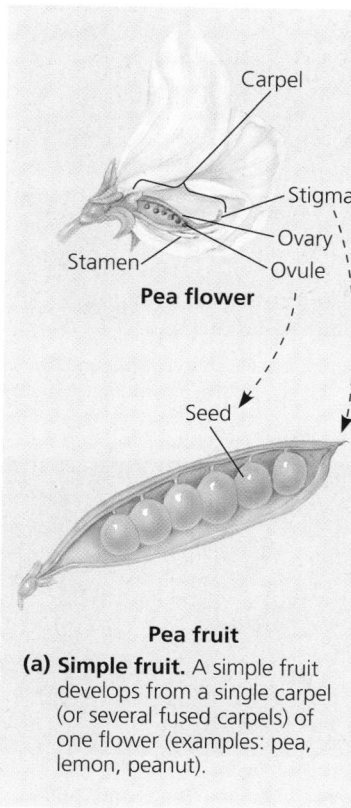

Pea flower

Carpel

Stigma
Ovary
Stamen
Ovule

Seed

Pea fruit

(a) Simple fruit. A simple fruit develops from a single carpel (or several fused carpels) of one flower (examples: pea, lemon, peanut).

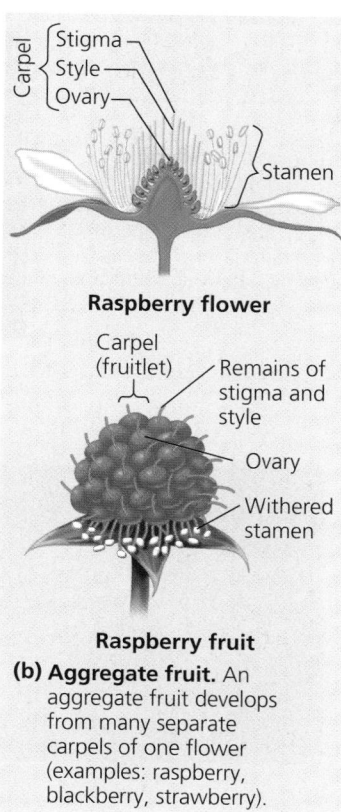

Carpel

Stigma
Style
Ovary
Stamen

Raspberry flower

Carpel (fruitlet)
Remains of stigma and style
Ovary
Withered stamen

Raspberry fruit

(b) Aggregate fruit. An aggregate fruit develops from many separate carpels of one flower (examples: raspberry, blackberry, strawberry).

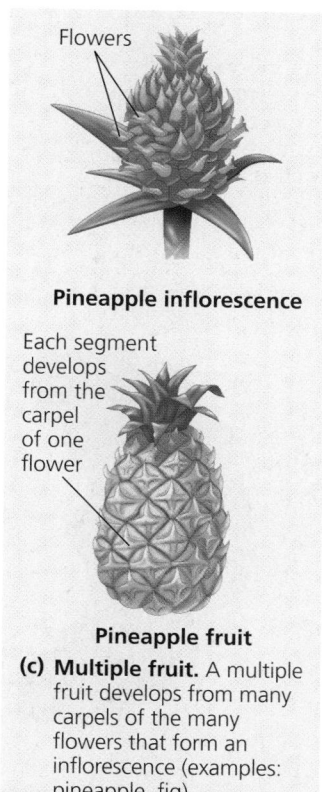

Flowers

Pineapple inflorescence

Each segment develops from the carpel of one flower

Pineapple fruit

(c) Multiple fruit. A multiple fruit develops from many carpels of the many flowers that form an inflorescence (examples: pineapple, fig).

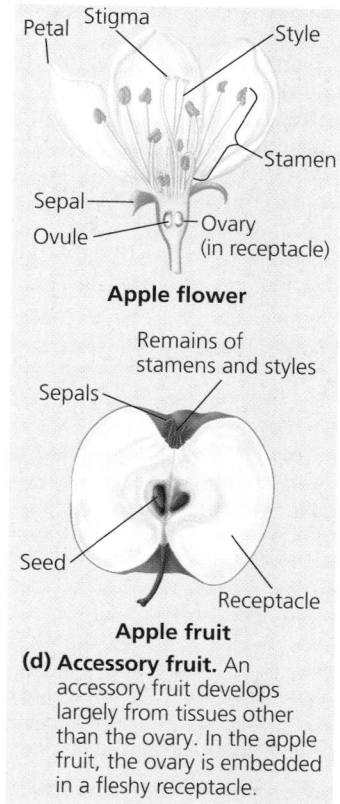

Petal
Stigma
Style
Stamen
Sepal
Ovule
Ovary (in receptacle)

Apple flower

Remains of stamens and styles
Sepals
Seed
Receptacle

Apple fruit

(d) Accessory fruit. An accessory fruit develops largely from tissues other than the ovary. In the apple fruit, the ovary is embedded in a fleshy receptacle.

becomes stony (the pit) while the outer parts stay fleshy. As the ovary grows, the other parts of the flower usually wither and are shed.

Fruits are classified into several types, depending on their developmental origin. Most fruits are derived from a single carpel or several fused carpels and are called **simple fruits** (Figure 38.11a). An **aggregate fruit** results from a single flower that has more than one separate carpel, each forming a small fruit (**Figure 38.11b**). These "fruitlets" are clustered together on a single receptacle, as in a raspberry. A **multiple fruit** develops from an inflorescence, a group of flowers tightly clustered together. When the walls of the many ovaries start to thicken, they fuse together and become incorporated into one fruit, as in a pineapple (**Figure 38.11c**).

In some angiosperms, other floral parts contribute to what we commonly call the fruit. Such fruits are called **accessory fruits**. In apple flowers, the ovary is embedded in the receptacle, and the fleshy part of this simple fruit is derived mainly from the enlarged receptacle; only the apple core develops from the ovary (**Figure 38.11d**). Another example is the strawberry, an aggregate fruit consisting of an enlarged receptacle studded with tiny, partially embedded fruits, each bearing a single seed.

A fruit usually ripens about the same time that its seeds complete their development. Whereas the ripening of a dry fruit, such as a soybean pod, involves the aging and drying out of fruit tissues, the process in a fleshy fruit is more elaborate. Complex interactions of hormones result in an edible fruit that entices animals that disperse the seeds. The fruit's "pulp" becomes softer as enzymes digest components of cell walls. The color usually changes from green to another color, making the fruit more visible among the leaves. The fruit becomes sweeter as organic acids or starch molecules are converted to sugar, which may reach a concentration of 20% in a ripe fruit. **Figure 38.12** examines some mechanisms of seed and fruit dispersal in more detail.

In this section, you have learned about the key features of sexual reproduction in angiosperms—flowers, double fertilization, and fruits. Next, we'll examine asexual reproduction.

CONCEPT CHECK 38.1

1. Distinguish between pollination and fertilization.
2. **WHAT IF?** If flowers had shorter styles, pollen tubes would more easily reach the embryo sac. Suggest an explanation for why very long styles have evolved in most species of flowering plants.
3. **MAKE CONNECTIONS** Does the life cycle of humans have any structures analogous to plant gametophytes? Explain your answer. (See Figures 13.5 and 13.6.)

For suggested answers, see Appendix A.

▼ Figure 38.12 Exploring Fruit and Seed Dispersal

A plant's life depends on finding fertile ground. But a seed that falls and sprouts beneath the parent plant will stand little chance of competing successfully for nutrients. To prosper, seeds must be widely dispersed. Plants use biotic dispersal agents as well as abiotic agents such as water and wind.

Dispersal by Water

▶ Some buoyant seeds and fruits can survive months or years at sea. In coconut, the seed embryo and fleshy white "meat" (endosperm) are within a hard layer (endocarp) surrounded by a thick and buoyant fibrous husk.

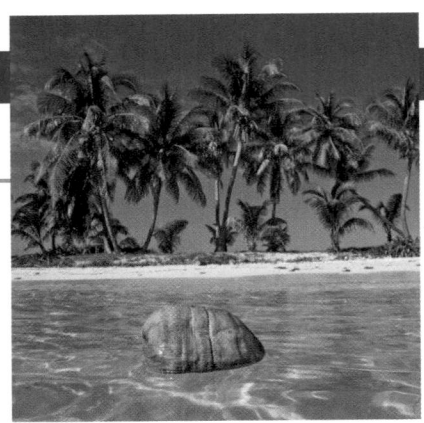

Dispersal by Wind

▶ With a wingspan of 12 cm, the giant seed of the tropical Asian climbing gourd *Alsomitra macrocarpa* glides through the air of the rain forest in wide circles when released.

▼ The winged fruit of a maple spins like a helicopter blade, slowing descent and increasing the chance of being carried farther by horizontal winds.

Dandelion fruit

▶ Tumbleweeds break off at the ground and tumble across the terrain, scattering their seeds.

▲ Some seeds and fruits are attached to umbrella-like "parachutes" that are made of intricately branched hairs and often produced in puffy clusters. These dandelion "seeds" (actually one-seeded fruits) are carried aloft by the slightest gust of wind.

Dispersal by Animals

◀ The sharp, tack-like spines on the fruits of puncture vine (*Tribulus terrestris*) can pierce bicycle tires and injure animals, including humans. When these painful "tacks" are removed and discarded, the seeds are dispersed.

◀ Some animals, such as squirrels, hoard seeds or fruits in underground caches. If the animal dies or forgets the cache's location, the buried seeds are well positioned to germinate.

▶ Seeds in edible fruits are often dispersed in feces, such as the black bear feces shown here. Such dispersal may carry seeds far from the parent plant.

▶ Ants are chemically attracted to seeds with "food bodies" rich in fatty acids, amino acids, and sugars. The ants carry the seed to their underground nest, where the food body (the lighter-colored portion shown here) is removed and fed to larvae. Due to the seed's size, unwieldy shape, or hard coating, the remainder is usually left intact in the nest, where it germinates.

Flowering plants reproduce sexually, asexually, or both

During **asexual reproduction**, offspring are derived from a single parent without any fusion of egg and sperm. The result is a clone, an individual genetically identical to its parent. Asexual reproduction is common in angiosperms, as well as in other plants, and for some species it is the main mode of reproduction.

Mechanisms of Asexual Reproduction

Asexual reproduction in plants is typically an extension of the capacity for indeterminate growth. Plant growth can be sustained or renewed indefinitely by meristems, regions of undifferentiated, dividing cells (see Concept 35.2). In addition, parenchyma cells throughout the plant can divide and differentiate into more specialized types of cells, enabling plants to regenerate lost parts. Detached root or stem fragments of some plants can develop into whole offspring; for example, pieces of a potato with an "eye" (bud) can each regenerate a whole plant. Such **fragmentation**, the separation of a parent plant into parts that develop into whole plants, is one of the most common modes of asexual reproduction. The plantlets on *Kalanchoë* leaves exemplify an unusual type of fragmentation (see Figure 35.7). In other cases, the root system of a single parent, such as an aspen tree, can give rise to many shoots that become separate shoot systems **(Figure 38.13)**. One aspen clone in Utah has been estimated to be composed of 47,000 stems of genetically identical trees. Although it is likely that some of the root system connections have been severed, making some of the trees isolated from the rest of the clone, each tree still shares a common genome.

A different mechanism of asexual reproduction has evolved in dandelions and some other plants. These plants can sometimes produce seeds without pollination or fertilization. This asexual production of seeds is called **apomixis** (from the Greek words meaning "away from the act of mixing") because there is no joining or, indeed, production of sperm and egg. Instead, a diploid cell in the ovule gives rise to the embryo, and the ovules mature into seeds, which in the dandelion are dispersed by windblown fruits. Thus, these plants clone themselves by an asexual process but have the advantage of seed dispersal, usually associated with sexual reproduction. Plant breeders are interested in introducing apomixis into hybrid crops because it would allow hybrid plants to pass desirable genomes intact to offspring.

Advantages and Disadvantages of Asexual and Sexual Reproduction

EVOLUTION An advantage of asexual reproduction in some cases is that there is no need for pollination. This may be beneficial in situations where plants of the same species are sparsely distributed and unlikely to be successfully pollinated. Asexual reproduction also allows the plant to pass on all its genetic legacy intact to its progeny. In contrast, when reproducing sexually, a plant passes on only half of its alleles. If a plant is superbly suited to its environment, asexual reproduction can be advantageous. A vigorous plant can potentially clone many copies of itself, and if the environmental circumstances remain stable, these offspring will also be genetically well adapted to the same environmental conditions under which the parent flourished.

Asexual plant reproduction based on the vegetative growth of stems, leaves, or roots is known as **vegetative reproduction**. Generally, the progeny produced by vegetative reproduction are stronger than seedlings produced by sexual reproduction. In contrast, seed germination is a precarious stage in a plant's life. The tough seed gives rise to a fragile seedling that may face exposure to predators, parasites, wind, and other hazards. In the wild, few seedlings survive to become parents themselves. Production of enormous numbers of seeds compensates for the odds against individual survival and gives natural selection ample genetic variations to screen. However, this is an expensive means of reproduction in terms of the resources consumed in flowering and fruiting.

Because sexual reproduction generates variation in offspring and populations, it can be advantageous in unstable environments where evolving pathogens and other fluctuating conditions affect survival and reproductive success. In contrast, the genotypic uniformity of asexually produced plants puts them at great risk of local extinction if there is a catastrophic environmental change, such as a new strain of disease. Moreover, seeds (which are almost always produced sexually) facilitate the dispersal of offspring to more distant locations. Finally, seed dormancy allows growth to be suspended until environmental conditions

▼ **Figure 38.13 Asexual reproduction in aspen trees.** Some aspen groves, such as those shown here, consist of thousands of trees descended by asexual reproduction. Each grove of trees derives from the root system of one parent. Thus, the grove is a clone. Notice that genetic differences between groves descended from different parents result in different timing for the development of fall color.

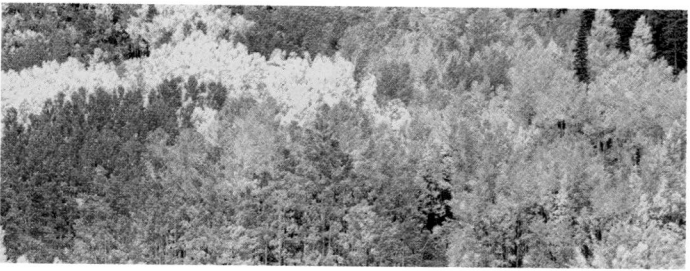

Using Positive and Negative Correlations to Interpret Data

Do Monkey Flower Species Differ in Allocating Their Energy to Sexual Versus Asexual Reproduction? Over the course of its life span, a plant captures only a finite amount of resources and energy, which must be allocated to best meet the plant's individual requirements for maintenance, growth, defense, and reproduction. Researchers examined how five species of monkey flower (genus *Mimulus*) use their resources for sexual and asexual reproduction.

How the Experiment Was Done After growing specimens of each species in separate pots in the open, the researchers determined averages for nectar volume, nectar concentration, seeds produced per flower, and the number of times the plants were

visited by broad-tailed hummingbirds (*Selasphorus platycercus*, shown here). Using greenhouse-grown specimens, they determined the average number of rooted branches per gram fresh shoot weight for each of the species. The phrase *rooted branches* refers to asexual reproduction through horizontal shoots that develop roots.

INTERPRET THE DATA

1. A correlation is a way to describe the relationship between two variables. In a positive correlation, as the values of one of the variables increase, the values of the second variable also increase. In a negative correlation, as the values of one of the variables increase, the values of the second variable decrease. Or there may be no correlation between two variables. If researchers know how two variables are correlated, they can make a prediction about one variable based on what they know about the other variable. **(a)** Which variable(s) is/are positively correlated with the volume of nectar production in this genus? **(b)** Which is/are negatively correlated? **(c)** Which show(s) no clear relationship?

2. **(a)** Which *Mimulus* species would you categorize as mainly asexual reproducers? Why? **(b)** Which species would you categorize as mainly sexual reproducers? Why?

3. **(a)** Which species would probably fare better in response to a pathogen that infects all *Mimulus* species? **(b)** Which species would fare better if a pathogen caused hummingbird populations to dwindle?

➡ **Instructors:** A version of this Scientific Skills Exercise can be assigned in **Mastering Biology**.

Data from the Experiment

Species	Nectar Volume (µL)	Nectar Concentration (% weight of sucrose/ total weight)	Seeds per Flower	Visits per Flower	Rooted Branches per Gram Shoot Weight
M. rupestris	4.93	16.6	2.2	0.22	0.673
M. eastwoodiae	4.94	19.8	25	0.74	0.488
M. nelsonii	20.25	17.1	102.5	1.08	0.139
M. verbenaceus	38.96	16.9	155.1	1.26	0.091
M. cardinalis	50.00	19.9	283.7	1.75	0.069

Data from S. Sutherland and R. K. Vickery, Jr., Trade-offs between sexual and asexual reproduction in the genus *Mimulus*, *Oecologia* 76:330–335 (1998).

become more favorable. In the **Scientific Skills Exercise**, you can use data to determine which species of monkey flower are mainly asexual reproducers and which are mainly sexual reproducers.

Although sexual reproduction involving two genetically different plants produces the most genetically diverse offspring, some plants, such as garden peas, usually self-fertilize. This process, called "selfing," is a desirable attribute in some crop plants because it heightens the probability that every ovule will develop into a seed. In many angiosperm species, however, mechanisms have evolved that make it difficult or impossible for a flower to fertilize itself, as we'll see next.

Mechanisms That Prevent Self-Fertilization

The various mechanisms that prevent self-fertilization contribute to genetic variety by ensuring that the sperm and egg come from different parents. Some species of plants cannot self-fertilize because different individuals have either staminate flowers (lacking carpels) or carpellate flowers (lacking stamens) **(Figure 38.14a)**. Other plants have flowers with functional stamens and carpels that mature at different times or are structurally arranged in such a way that it is unlikely that an animal pollinator could transfer pollen from an anther to a stigma of the same flower **(Figure 38.14b)**. However, the most common anti-selfing mechanism in flowering plants is **self-incompatibility**, the ability of a plant to reject its

▼ Figure 38.14 Some floral adaptations that prevent self-fertilization.

(a) Some species, such as *Sagittaria latifolia* (common arrowhead), have plants that produce only staminate flowers (left) or carpellate flowers (right).

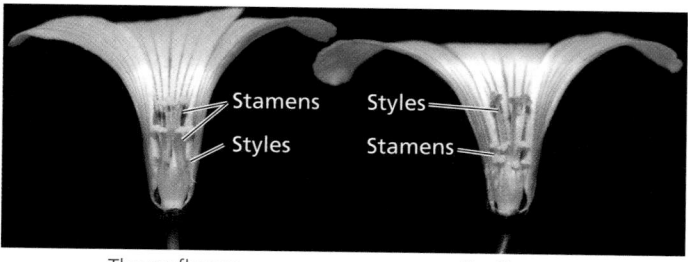

Thrum flower   Pin flower

(b) Some species, such as *Oxalis alpina* (alpine wood sorrel), produce two types of flowers on different individuals: "thrums," which have short styles and long stamens, and "pins," which have long styles and short stamens. An insect foraging for nectar would collect pollen on different parts of its body; thrum pollen would be deposited on pin stigmas, and vice versa.

own pollen and the pollen of closely related individuals. If a pollen grain lands on a stigma of a flower of the same plant or a closely related plant, a biochemical block prevents the pollen from completing its development and fertilizing an egg. This plant response is analogous to the immune response of animals because both are based on the ability to distinguish the cells of "self" from those of "nonself." The key difference is that the animal immune system rejects nonself, as when the immune system mounts a defense against a pathogen or rejects a transplanted organ (see Concept 43.3). In contrast, self-incompatibility in plants is a rejection of self.

Researchers are unraveling the molecular mechanisms of self-incompatibility. Recognition of "self" pollen is based on genes called *S*-genes. In the gene pool of a population, there can be dozens of alleles of an *S*-gene. If a pollen grain has an allele that matches an allele of the stigma on which it lands, the pollen tube either fails to germinate or fails to grow through the style to the ovary. There are two types of self-incompatibility: gametophytic and sporophytic.

In gametophytic self-incompatibility, the *S*-allele in the pollen genome governs the blocking of fertilization. For example, an S_1 pollen grain from an S_1S_2 parental sporophyte cannot fertilize eggs of an S_1S_2 flower but can fertilize an S_2S_3 flower. An S_2 pollen grain cannot fertilize either flower. In

some plants, this self-recognition involves the enzymatic destruction of RNA within a pollen tube. RNA-hydrolyzing enzymes are produced by the style and enter the pollen tube. If the pollen tube is a "self" type, they destroy its RNA.

In sporophytic self-incompatibility, fertilization is blocked by incompatibilities between *S*-allele gene products present in the sporophytic parental tissue adhering to the pollen wall and *S*-allele gene products secreted by the stigma of the receptive flower. For example, both the S_1 and S_2 pollen grains derived from an S_1S_2 parental sporophyte contain S_1-allele and S_2-allele gene products in the sporophytic tissue adhering to their pollen walls. The S_2-allele gene product in the pollen walls of these two pollen types would prevent either type from germinating on the stigma of a flower whose genotype includes either the S_1 or the S_2 allele. For example, an S_1 pollen grain, derived from an S_1S_2 parent, cannot germinate on the stigma of an S_2S_3 flower.

Research on self-incompatibility may have agricultural applications. Breeders often hybridize different genetic strains of a crop to combine the best traits of the two strains and to counter the loss of vigor that can result from excessive inbreeding. To prevent self-fertilization within the two strains, breeders must either laboriously remove the anthers from the parent plants that provide the seeds or use male-sterile strains of the crop plant, if they exist. If self-compatibility can be genetically engineered back into domesticated plant varieties, these limitations to the commercial hybridization of crop seeds could be overcome.

Totipotency, Vegetative Reproduction, and Tissue Culture

In a multicellular organism, any cell that can divide and asexually generate a clone of the original organism is said to be **totipotent**. Totipotency is found in many plants, particularly but not exclusively in their meristematic tissues. Plant totipotency underlies most of the techniques used by humans to clone plants.

Vegetative Propagation and Grafting

Vegetative reproduction occurs naturally in many plants, but it can often be facilitated or induced by humans, in which case it is called **vegetative propagation**. Most houseplants, landscape shrubs and bushes, and orchard trees are asexually reproduced from plant fragments called cuttings. In most cases, shoot cuttings are used. At the wounded end of the shoot, a mass of dividing, undifferentiated totipotent cells called a **callus** forms, and roots develop from the callus. If the shoot fragment includes a node, then roots form without a callus stage.

In grafting, a severed shoot from one plant is permanently joined to the truncated stem of another. This process, usually limited to closely related individuals, can combine the best qualities of different species or varieties into one plant. The plant that provides the roots is called the **stock**; the twig grafted onto the stock is known as the **scion**. For example, scions from varieties of vines that produce superior wine grapes are grafted onto rootstocks of varieties that produce inferior grapes but are more resistant to certain soil pathogens. The genes of the scion determine the quality of the fruit. During grafting, a callus first forms between the adjoining cut ends of the scion and stock; cell differentiation then completes the functional unification of the grafted individuals.

Test-Tube Cloning and Related Techniques

Plant biologists have adopted *in vitro* methods to clone plants for research or horticulture. Whole plants can be obtained by culturing small pieces of tissue from the parent plant on an artificial medium containing nutrients and hormones. The cells or tissues can come from any part of a plant, but growth may vary depending on the plant part, species, and artificial medium. In some media, the cultured cells divide and form a callus of undifferentiated totipotent cells **(Figure 38.15a)**. When the concentrations of hormones and nutrients are manipulated appropriately, a callus can sprout shoots and roots with fully differentiated cells **(Figure 38.15b and c)**. If desired, the cloned plantlets can then be transferred to soil, where they continue their growth.

Plant tissue culture is important in eliminating weakly pathogenic viruses from vegetatively propagated varieties. Although the presence of weak viruses may not be obvious, yield or quality may be substantially reduced as a result of infection. Strawberry plants, for example, are susceptible to more than 60 viruses, and typically the plants must be replaced each year because of viral infection. However, since the apical meristems are often virus-free, they can be excised and used to produce virus-free material for tissue culture.

Plant tissue culture also facilitates genetic engineering, the deliberate modification of the characteristics of an organism by manipulating its genetic material (see Concept 20.1). Many genetic engineering techniques require single plant cells as the starting material. Test-tube culture makes it possible to regenerate plants from a single genetically engineered plant cell. In the next section, we'll take a closer look at some of the promises and challenges surrounding the use of genetically modified plants in agriculture.

CONCEPT CHECK 38.2

1. What are three ways that flowering plants avoid self-fertilization?
2. The seedless banana, the world's most popular fruit, is losing the battle against two fungal epidemics. Why do such epidemics generally pose a greater risk to asexually propagated crops?
3. Self-fertilization, or selfing, seems to have obvious disadvantages as a reproductive "strategy" in nature, and it has even been called an "evolutionary dead end." So it is surprising that about 20% of angiosperm species primarily rely on selfing. Suggest a reason why selfing might be advantageous and yet still be an evolutionary dead end.

For suggested answers, see Appendix A.

CONCEPT 38.3

People modify crops by breeding and genetic engineering

People have intervened in the reproduction and genetic makeup of plants since the dawn of agriculture. Maize, for example, owes its existence to humans. Left on its own in nature, maize would soon become extinct for the simple reason that it cannot spread its seeds. Maize kernels are not only permanently attached to the central axis (the "cob") but also permanently protected by tough, overlapping leaf sheaths (the "husk") **(Figure 38.16)**. These attributes arose by artificial selection by humans. Despite having no understanding of the scientific principles underlying plant breeding, early farmers domesticated most of our crop species over a relatively short period about 10,000 years ago.

▼ **Figure 38.16 Maize: a product of artificial selection.**
Modern maize (bottom) was derived from teosinte (top). Teosinte kernels are tiny, and each row has a husk that must be removed to get at the kernel. The seeds are loose at maturity, allowing dispersal, which probably made harvesting difficult for early farmers. Neolithic farmers selected seeds from plants with larger cob and kernel size as well as the permanent attachment of seeds to the cob and the encasing of the entire cob by a tough husk.

 Mastering Biology HHMI Video: Popped Secret: The Mysterious Origin of Corn

▼ **Figure 38.15 Cloning a garlic plant. (a)** A root from a garlic clove gave rise to this callus culture, a mass of undifferentiated totipotent cells. **(b and c)** The differentiation of a callus into a plantlet depends on the nutrient levels and hormone concentrations in the artificial medium, as can be seen in these cultures grown for different lengths of time.

(a) (b) (c) Developing root

Plant Breeding

Plant breeding is the art and science of changing the traits of plants in order to produce desired characteristics. Breeders scrutinize their fields carefully and travel far and wide searching for domesticated varieties or wild relatives with desirable traits. Such traits occasionally arise spontaneously through mutation, but the natural rate of mutation is too slow and unreliable to produce all the mutations that breeders would like to study. Breeders sometimes hasten mutations by treating large batches of seeds or seedlings with radiation or chemicals.

In traditional plant breeding, when a desirable trait is identified in a wild species, the wild species is crossed with a domesticated variety. Generally, those progeny that have inherited the desirable trait from the wild parent have also inherited many traits that are not desirable for agriculture, such as small fruits or low yields. The progeny that express the desired trait are again crossed with members of the domesticated species and their progeny examined for the desired trait. This process is continued until the progeny with the desired wild trait resemble the original domesticated parent in their other agricultural attributes.

While most breeders cross-pollinate plants of a single species, some breeding methods rely on hybridization between two distant species of the same genus. Such crosses sometimes result in the abortion of the hybrid seed during development. Often in these cases the embryo begins to develop, but the endosperm does not. Hybrid embryos are sometimes rescued by surgically removing them from the ovule and culturing them *in vitro*.

It is important to note that the natural genetic modification of plants began long before humans started altering crops by artificial selection. For example, the wheat species we rely on for much of our food evolved by natural hybridization between different species of grasses. Such hybridization is common in plants and has long been exploited by breeders to introduce genetic variation for artificial selection and crop improvement. Another instance of natural genetic modification has been revealed by the genomic sequencing of sweet potato (*Ipomoea batatas*). Apparently an early ancestor of the modern sweet potato came into contact with the soil bacterium *Agrobacterium*, and a horizontal gene transfer event (see Concept 26.6) occurred. Thus, the introduction of a **transgene**, a gene that has been transferred from one organism into another, can occur in nature as well as in a genetic engineering laboratory.

Plant Biotechnology and Genetic Engineering

Plant biotechnology has two meanings. In the general sense, it refers to innovations in the use of plants (or substances obtained from plants) to make products of use to humans—an endeavor that began in prehistory. In a more specific sense, biotechnology refers to the use of genetically modified (GM) organisms in agriculture and industry. Indeed, in the last few decades, genetic engineering has become such a powerful force that the terms *genetic engineering* and *biotechnology* have become synonymous in the media. The revolutionary CRISPR-Cas9 gene-editing technology (see Figure 20.14) is changing plant biology as fast as it is breaking ground in other fields.

Unlike traditional plant breeders, modern plant biotechnologists, using techniques of genetic engineering, are not limited to the transfer of genes between closely related species or genera. For example, traditional breeding techniques could not be used to insert a desired gene from daffodil into rice because the many intermediate species between rice and daffodil and their common ancestor are extinct. In theory, if breeders had the intermediate species, over the course of several centuries they could probably introduce a daffodil gene into rice by traditional hybridization and breeding methods. With genetic engineering, however, such gene transfers can be done more quickly, more specifically, and without the need for intermediate species.

In the remainder of this chapter, we explore the prospects and controversies surrounding the agricultural use of crops that are genetically modified organisms (GMOs). Advocates for plant biotechnology believe that the genetic engineering of crop plants is the key to overcoming some of the most pressing problems of the 21st century, including world hunger and fossil fuel dependency.

> **Mastering Biology**
> Interview with Luis Herrera-Estrella:
> Using GMOs to help farmers

Reducing World Hunger and Malnutrition

Although global hunger affects nearly a billion people, there is much disagreement about its causes. Some argue that food shortages arise from inequities in distribution and that the most poverty-stricken simply cannot afford food. Others regard food shortages as evidence that the world is overpopulated—that the human species has exceeded the carrying capacity of the planet (see Concept 53.3). Whatever the causes of malnutrition, increasing food production is a humane objective. Because land and water are the most limiting resources, the best option is to increase yields on already existing farmland. Indeed, there is very little "extra" land that can be farmed, especially if the few remaining pockets of wilderness are to be preserved. Based on conservative estimates of population growth, farmers will have to produce 40% more grain per hectare to feed the human population in 2030. Plant biotechnology can help make these crop yields possible.

Crops that have been genetically modified to express transgenes from *Bacillus thuringiensis,* a soil bacterium, require less pesticide. The transgenes involved encode a protein (*Bt* toxin)

▼ Figure 38.17 Non-*Bt* versus *Bt* maize. Field trials reveal that non-*Bt* maize (left) is heavily damaged by insect feeding and *Fusarium* mold infection, whereas *Bt* maize (right) suffers little or no damage.

Non-*Bt* maize   *Bt* maize

that is toxic to many insect pests **(Figure 38.17)**. The *Bt* toxin used in crops is produced in the plant as a harmless protoxin that only becomes toxic if activated by alkaline conditions, such as in the guts of most insects. Because vertebrates have highly acidic stomachs, protoxin consumed by humans or livestock is rendered harmless by denaturation.

Biofortification, improving the nutritional quality of plants, is another strategy in the war on world hunger. For example, about 250,000 to 500,000 children go blind each year because of vitamin A deficiencies. More than half of these children die within a year of becoming blind. In response to this crisis, genetic engineers created "Golden Rice," a transgenic variety supplemented with transgenes that enable it to produce grain with increased levels of beta-carotene, a precursor of vitamin A. In 2018, after decades of safety testing and legal challenges, Golden Rice received positive food safety evaluations from the governments of Canada, New Zealand, and the United States. There are hopes that the governments of the Philippines and Bangladesh will soon follow suit.

Another major target for biofortification improvement is cassava (*Manihot esculenta*), a starchy root crop that provides carbohydrates for 800 million of the poorest people on our planet **(Figure 38.18)**. Researchers are also engineering plants with enhanced resistance to disease. In one case, a transgenic

> **Mastering Biology** Interview with Dennis Gonsalves: Saving the papaya industry (see the interview before Chapter 35)

▶ Figure 38.18 Engineering the perfect food? Cassava grows well in poor dry soils but must be supplemented by consumption of other crops. If those crops fail, people suffer malnutrition. What if cassava alone could supply a balanced diet? Transgenic, biofortified cassava plants have been developed with greatly increased levels of iron and beta-carotene (a vitamin A precursor). Other improvements have been made in the size of its storage roots, its ease of processing (less cyanide-producing chemicals), and its resistance to cassava mosaic virus disease.

papaya that is resistant to a ringspot virus was introduced into Hawaii, thereby saving its papaya industry.

Considerable controversy has arisen concerning transgenic crops that are resistant to the herbicide glyphosate. Glyphosate is lethal to a wide variety of plants because it inhibits a key enzyme in a biochemical pathway that is found in plants (and most bacteria) but not in animals. Researchers discovered a bacterial strain that had undergone a mutation in the gene encoding this enzyme that rendered it glyphosate resistant. When this mutated bacterial gene was spliced into the genome of various crops, these crops also became glyphosate resistant. Farmers achieved almost total weed control by spraying glyphosate over their fields of glyphosate-resistant crops. Unfortunately, the overuse of glyphosate created a huge selective pressure on weed species, with the result that many have evolved resistance to glyphosate. There has also been a growing appreciation in recent decades of the role that gut bacteria play in animal and human health, and claims have been made that glyphosate may be having negative effects on the health of humans and livestock by interfering with beneficial gut bacteria.

Reducing Fossil Fuel Dependency

Global sources of inexpensive fossil fuels, particularly oil, are rapidly being depleted. Moreover, most climatologists attribute global warming mainly to the rampant burning of fossil fuels, such as coal and oil, and the resulting release of the greenhouse gas CO_2. How can the world meet its energy demands in the 21st century in an economical and nonpolluting way? In certain localities, wind or solar power may become economically viable, but such alternative energy sources are unlikely to fill the global energy demands completely. Many scientists predict that **biofuels**—fuels derived from living biomass—could produce a sizable fraction of the world's energy needs in the not-too-distant future. **Biomass** is the total mass of organic matter in a group of organisms in a particular habitat. The use of biofuels from plant biomass would reduce the net emission of CO_2. Whereas burning fossil fuels increases atmospheric CO_2 concentrations, biofuel crops reabsorb by photosynthesis the CO_2 emitted when biofuels are burned, creating a cycle that is carbon neutral.

In working to create biofuel crops from wild precursors, scientists are focusing their domestication efforts on fast-growing plants, such as switchgrass (*Panicum virgatum*) and poplar (*Populus trichocarpa*), that can grow on soil that is too poor for food production. Scientists do not expect the plant biomass to be burned directly. Instead, the polymers in cell walls, such as cellulose and hemicellulose, which constitute the most abundant organic compounds on Earth, would be broken down into sugars by enzymatic reactions. These sugars, in turn, would be fermented into alcohol and distilled to yield biofuels. In addition to increasing plant polysaccharide content and overall biomass, researchers are trying to genetically engineer the cell walls of plants to increase the efficiency of the enzymatic conversion process.

The Debate over Plant Biotechnology

Much of the debate about GMOs in agriculture is political, social, economic, or ethical and therefore outside the scope of this book. But we *should* consider the biological concerns about GM crops. Some biologists, particularly ecologists, are concerned about the unknown risks associated with the release of GMOs into the environment. The debate centers on the extent to which GMOs could harm the environment or human health. Those who want to proceed more slowly with agricultural biotechnology (or end it) are concerned about the unstoppable nature of the "experiment." If a drug trial produces unanticipated harmful results, the trial is stopped. But we may not be able to stop the "trial" of introducing novel organisms into the biosphere. Here we examine some criticisms that have been leveled by opponents of GMOs, including the alleged effects on human health and nontarget organisms and the potential for transgene escape.

Issues of Human Health

Many GMO opponents worry that genetic engineering may inadvertently transfer allergens, molecules to which some people are allergic, from a species that produces an allergen to a plant used for food. However, biotechnologists are already removing genes that encode allergenic proteins from soybeans and other crops. So far, there is no credible evidence that GM plants designed for human consumption have allergenic effects on human health. In fact, some GM foods are potentially healthier than non-GM foods. For example, *Bt* maize (the transgenic variety with the *Bt* toxin) contains 90% less of a fungal toxin that causes cancer and birth defects than non-*Bt* maize. Called fumonisin, this toxin is highly resistant to degradation and has been found in alarmingly high concentrations in some batches of processed maize products, ranging from cornflakes to beer. Fumonisin is produced by a fungus (*Fusarium*) that infects insect-damaged maize. Because *Bt* maize generally suffers less insect damage than non-GM maize, it contains much less fumonisin.

Assessing the impact of GMOs on human health also involves considering the health of farmworkers, many of whom were commonly exposed to high levels of chemical insecticides prior to the adoption of *Bt* crops. In India, for example, the widespread adoption of *Bt* cotton has led to a 41% decrease in insecticide use and an 80% reduction in the number of acute poisoning cases involving farmers.

Possible Effects on Nontarget Organisms

Many ecologists are concerned that GM crops may have unforeseen effects on nontarget organisms. One laboratory study indicated that the larvae (caterpillars) of monarch butterflies responded adversely and even died after eating milkweed leaves (their preferred food) heavily dusted with pollen from transgenic *Bt* maize. This study has since been discredited, affording a good example of the self-correcting nature of science. As it turns out, when the original researcher shook the male maize inflorescences onto the milkweed leaves in the laboratory, the filaments of stamens, opened microsporangia, and other floral parts also rained onto the leaves. Subsequent research found that it was these other floral parts, *not* the pollen, that contained *Bt* toxin in high concentrations. Unlike pollen, these floral parts would not be carried by the wind to neighboring milkweed plants when shed under natural field conditions. Only one *Bt* maize line, accounting for less than 2% of commercial *Bt* maize production (and now discontinued), produced pollen with high *Bt* toxin concentrations.

In considering the negative effects of *Bt* pollen on monarch butterflies, one must also weigh the effects of an alternative to the cultivation of *Bt* maize—the spraying of non-*Bt* maize with chemical pesticides. Subsequent studies have shown that such spraying is much more harmful to nearby monarch populations than is *Bt* maize production. Although the effects of *Bt* maize pollen on monarch butterfly larvae appear to be minor, the controversy has emphasized the need for accurate field testing of all GM crops and the importance of targeting gene expression to specific tissues to improve safety.

Addressing the Problem of Transgene Escape

Perhaps the most serious concern raised about GM crops is the possibility of the introduced genes escaping from a transgenic crop into related weeds through crop-to-weed hybridization. The fear is that the spontaneous hybridization between a crop engineered for herbicide resistance and a wild relative might give rise to a "superweed" that would have a selective advantage over other weeds in the wild and would be much more difficult to control in the field. GMO advocates point out that the likelihood of transgene escape depends on the ability of the crop and weed to hybridize and on how the transgenes affect the overall fitness of the hybrids. A desirable crop trait—a dwarf phenotype, for example—might be disadvantageous to a weed growing in the wild. In other instances, there are no weedy relatives nearby with which to hybridize; soybean, for example, has no wild relatives in the United States. However, canola, sorghum, and many other crops do hybridize readily with weeds, and crop-to-weed transgene escape in a turfgrass has occurred. In 2003 a transgenic variety of creeping bentgrass (*Agrostis stolonifera*) genetically engineered to resist the herbicide glyphosate escaped from an experimental plot in Oregon following a windstorm. Despite efforts to eradicate the escapee, 62% of the *Agrostis* plants found in the vicinity three years later were glyphosate resistant. So far, the ecological impact of this event appears to be minor, but that may not be the case with future transgenic escapes.

Many strategies are being pursued with the goal of preventing transgene escape. For example, if male sterility could be engineered into plants, these plants would still produce seeds and fruit if pollinated by nearby nontransgenic plants, but they would produce no viable pollen. A second approach involves genetically engineering apomixis into transgenic crops. When a seed is produced by apomixis, the embryo and endosperm develop without fertilization. The transfer of this trait to transgenic crops would therefore minimize the possibility of transgene escape via pollen because plants could be male-sterile without compromising seed or fruit production. A third approach is to engineer the transgene into the chloroplast DNA of the crop. Chloroplast DNA in many plant species is inherited strictly from the egg, so transgenes in the chloroplast cannot be transferred by pollen. A fourth approach for preventing transgene escape is to genetically engineer flowers that develop normally but fail to open. Consequently, self-pollination would occur, but pollen would be unlikely to escape from the flower. This solution would require modifications to flower design. Several floral genes have been identified that could be manipulated to this end.

The continuing debate about GMOs in agriculture exemplifies one of this textbook's recurring ideas: the relationship of science and technology to society. Technological advances almost always involve some risk of unintended outcomes. In the case of genetically engineered crops, zero risk is probably unattainable. Therefore, scientists and the public must assess on a case-by-case basis the possible benefits of transgenic products versus the risks that society is willing to take. The best scenario is for these discussions and decisions to be based on sound scientific information and rigorous testing rather than on reflexive fear or blind optimism.

CONCEPT CHECK 38.3

1. Compare traditional plant-breeding methods with genetic engineering.
2. Why does *Bt* maize have less fumonisin than non-GM maize?
3. **WHAT IF?** In a few species, chloroplast genes are inherited only from sperm. How might this influence efforts to prevent transgene escape?

For suggested answers, see Appendix A.

---

38 Chapter Review

→ Go to **Mastering Biology** for Assignments, the eText, the Study Area, and Dynamic Study Modules.

SUMMARY OF KEY CONCEPTS

→ To review key terms, go to the **Vocabulary Self-Quiz** in the **Mastering Biology** eText or Study Area, or go to goo.gl/zkjz9t.

CONCEPT 38.1

Flowers, double fertilization, and fruits are key features of the angiosperm life cycle *(pp. 823–832)*

- Angiosperm reproduction involves an alternation of generations between a multicellular diploid sporophyte generation and a multicellular haploid gametophyte generation. **Flowers**, produced by the sporophyte, function in sexual reproduction.
- The four floral organs are sepals, petals, stamens, and carpels. **Sepals** protect the floral bud. **Petals** help attract pollinators. **Stamens** bear **anthers** in which haploid **microspores** develop into **pollen grains** containing a male gametophyte. **Carpels** contain **ovules** (immature seeds) in their swollen bases. Within the ovules, **embryo sacs** (female gametophytes) develop from **megaspores**.
- **Pollination**, which precedes **fertilization**, is the placing of pollen on the stigma of a carpel. After pollination, the **pollen tube** discharges two sperm into the female gametophyte. Two sperm are needed

Tube nucleus

One sperm will fuse with the egg, forming a zygote (2n).

One sperm cell will fuse with the two polar nuclei, forming an endosperm nucleus (3n).

for **double fertilization**, a process in which one sperm fertilizes the egg, forming a zygote and eventually an embryo, while the other sperm combines with the polar nuclei, giving rise to the food-storing endosperm.
- A **seed** consists of a dormant embryo along with a food supply stocked in either the **endosperm** or the **cotyledons**. Seed **dormancy** ensures that seeds germinate only when conditions for seedling survival are optimal. The breaking of dormancy often requires environmental cues, such as temperature or lighting changes.
- The **fruit** protects the enclosed seeds and aids in wind dispersal or in the attraction of seed-dispersing animals.

? *What changes occur to the four types of floral parts as a flower changes into a fruit?*

CONCEPT 38.2

Flowering plants reproduce sexually, asexually, or both *(pp. 833–836)*

- **Asexual reproduction**, also known as **vegetative reproduction**, enables successful plants to proliferate quickly. Sexual reproduction generates most of the genetic variation that makes evolutionary adaptation possible.
- Plants have evolved many mechanisms to avoid self-fertilization, including having male and female flowers on different individuals, nonsynchronous production of male and female parts within a single flower, and **self-incompatibility** reactions in which pollen grains that bear an allele identical to one in the female are rejected.
- Plants can be cloned from single cells, which can be genetically manipulated before being allowed to develop into a plant.

? *What are the advantages of asexual and sexual reproduction?*

People modify crops by breeding and genetic engineering (pp. 836–840)

- Hybridization of different varieties and even species of plants is common in nature and has been used by breeders, ancient and modern, to introduce new genes into crops. After two plants are successfully hybridized, plant breeders select those progeny that have the desired traits.
- In genetic engineering, genes from unrelated organisms are incorporated into plants. Genetically modified (GM) plants can increase the quality and quantity of food worldwide and may also become increasingly important as biofuels.
- There are concerns about the unknown risks of releasing GM organisms into the environment, but the potential benefits of crops that express **transgenes** will need to be considered.

? *Give two examples of how genetic engineering has improved or might potentially improve food quality.*

TEST YOUR UNDERSTANDING

➔ For more multiple-choice questions, go to the **Practice Test** in the **Mastering Biology** eText or Study Area, or go to goo.gl/GruWRg.

Levels 1-2: Remembering/Understanding

1. A fruit is
   (A) a mature ovary.
   (B) a mature ovule.
   (C) a seed plus its integuments.
   (D) an enlarged embryo sac.

2. Double fertilization means that
   (A) flowers must be pollinated twice to yield fruits and seeds.
   (B) every egg must receive two sperm to produce an embryo.
   (C) one sperm is needed to fertilize the egg, and a second sperm is needed to fertilize the polar nuclei.
   (D) every sperm has two nuclei.

3. *Bt* maize
   (A) is resistant to various herbicides, making it practical to weed maize fields with those herbicides.
   (B) contains transgenes that increase vitamin A content.
   (C) includes bacterial genes that produce a toxin that reduces damage from insect pests.
   (D) is a "boron (B)-tolerant" transgenic variety of maize.

4. Which statement concerning grafting is correct?
   (A) Stocks and scions refer to twigs of different species.
   (B) Stocks and scions must come from unrelated species.
   (C) Stocks provide root systems for grafting.
   (D) Grafting creates new species.

Levels 3-4: Applying/Analyzing

5. Some plant species produce male and female individuals in which the males have an XY genotype and the females have an XX genotype. After double fertilization, what would be the genotypes of the embryos and endosperm nuclei?
   (A) embryo XY/endosperm XXX or embryo XX/endosperm XXY
   (B) embryo XX/endosperm XX or embryo XY/endosperm XY
   (C) embryo XX/endosperm XXX or embryo XY/endosperm XYY
   (D) embryo XX/endosperm XXX or embryo XY/endosperm XXY

6. In a species showing sporophytic incompatibility, which type(s) of pollen could successfully fertilize an S_2S_3 flower?
   (A) S_1 pollen from an S_1S_3 flower
   (B) S_2 or S_3 pollen from an S_2S_3 flower
   (C) S_3 pollen from an S_1S_1 flower
   (D) S_1 pollen from an S_1S_1 flower

7. The black dots that cover strawberries are actually fruits formed from the separate carpels of a single flower. The fleshy and tasty portion of a strawberry derives from the receptacle of a flower with many separate carpels. Therefore, a strawberry is
   (A) a simple fruit with many seeds.
   (B) both a multiple fruit and an accessory fruit.
   (C) both a simple fruit and an aggregate fruit.
   (D) both an aggregate fruit and an accessory fruit.

8. **DRAW IT** Draw and label the parts of a flower.

Levels 5-6: Evaluating/Creating

9. **EVOLUTION CONNECTION** With respect to sexual reproduction, some plant species are fully self-fertile, others are fully self-incompatible, and some exhibit a "mixed strategy" with partial self-incompatibility. These reproductive strategies differ in their implications for evolutionary potential. How, for example, might a self-incompatible species fare as a small founder population or remnant population in a severe population bottleneck (see Concept 23.3), as compared with a self-fertile species?

10. **SCIENTIFIC INQUIRY** Critics of GM foods have argued that transgenes may disturb cellular functioning, causing unexpected and potentially harmful substances to appear inside cells. Toxic intermediary substances that normally occur in very small amounts may arise in larger amounts, or new substances may appear. The disruption may also lead to loss of substances that help maintain normal metabolism. If you were your nation's chief scientific advisor, how would you respond to these criticisms?

11. **SCIENCE, TECHNOLOGY, AND SOCIETY** Humans have engaged in genetic manipulation for millennia, producing plant and animal varieties through selective breeding and hybridization that significantly modify genomes of organisms. Why do you think modern genetic engineering, which often entails introducing or modifying only one or a few genes, has met with so much opposition? Should some forms of genetic engineering be of greater concern than others? Explain.

12. **WRITE ABOUT A THEME: ORGANIZATION** In a short essay (100–150 words), discuss how a flower's ability to reproduce with other flowers of the same species is an emergent property arising from floral parts and their organization.

13. **SYNTHESIZE YOUR KNOWLEDGE**

This colorized SEM shows pollen grains from six plant species. Explain how a pollen grain forms, how it functions, and how pollen grains contributed to the dominance of angiosperms and other seed plants.

For selected answers, see Appendix A.

39 Plant Responses to Internal and External Signals

KEY CONCEPTS

39.1 Signal transduction pathways link signal reception to response *p. 843*

39.2 Plants use chemicals to communicate *p. 845*

39.3 Responses to light are critical for plant success *p. 855*

39.4 Plants respond to a wide variety of stimuli other than light *p. 861*

39.5 Plants respond to attacks by pathogens and herbivores *p. 866*

Study Tip

Make a table: As you read the chapter, add specific examples for each of the general categories of responses shown in the diagram.

Factor	Example of plant response
Light	Seed germination in response to red light

→ Go to Mastering Biology

For Students (in eText and Study Area)
• Get Ready for Chapter 39
• Video: Gravitropism
• Video: *Mimosa* leaves

For Instructors to Assign (in Item Library)
• Activity: Leaf Abscission
• Activity: Plant Hormones

Figure 39.1 Sunflowers track the sun from east to west each day. After sunset, they reverse direction, facing the direction of the next sunrise. By facing the hot sun during the day, the floral heads become warmer and release greater amounts of chemicals that attract pollinators. Light is just one of the many factors to which a plant responds.

What are some factors that plants sense and respond to?

Light

Heat or cold stress

Hormones

Internal chemical signals

Touch by wind or objects

Drought or flooding

Time (daylength and seasons)

Gravity

Wounding by herbivores

Infection by pathogens

Signal transduction pathways link signal reception to response

The idea that plants are inert or passive is a common misconception. Nothing could be further from the truth. As shown in the examples of plant responses in **Figure 39.1**, plants must sense and integrate information about many facets of their environment. Although plant development is simpler than the development of multicellular animals, plant cells are just as complex as animal cells (see Figure 6.8). The molecular biology of plants is as complicated as that of animals: A Japanese plant (*Paris japonica*) currently has the largest genome on record, about fifty times greater than the human genome. Plants have among the largest genomes (see Table 21.1). At the levels of signal reception and signal transduction, your cells are not all that different from those of plants—the similarities far outweigh the differences. As an animal, however, your responses to environmental stimuli are generally quite different from those of plants. Animals commonly respond by movement; plants do so by altering growth and development.

As an example of a plant modifying its growth and development in response to environmental cues, consider a forgotten potato in the back corner of a kitchen cupboard. This modified underground stem, or tuber, has sprouted shoots from its "eyes" (axillary buds). These shoots, however, scarcely resemble those of a typical plant. Instead of sturdy stems and broad green leaves, this plant has ghostly pale stems and unexpanded leaves, as well as short, stubby roots **(Figure 39.2a)**. These physical adaptations for growing in darkness, collectively referred to as **etiolation**, make sense if we consider that a young potato plant in nature usually encounters continuous darkness when sprouting underground. Under these circumstances, expanded leaves would be a hindrance to soil penetration and would be damaged as the shoots pushed through the soil. Because the leaves are unexpanded and underground, there is little evaporative loss of water and little requirement for an extensive root system to replace the water lost by transpiration. Moreover, the energy expended in producing green chlorophyll would be wasted because there is no light for photosynthesis. Instead, a potato plant growing in the dark allocates as much energy as possible to elongating its stems. This adaptation enables the shoots to break ground before the nutrient reserves in the tuber are exhausted. The etiolation response is one example of how a plant's physical characteristics are tuned to its surroundings by complex interactions between environmental and internal signals.

When a shoot reaches light, the plant undergoes profound changes, collectively called **de-etiolation**

▼ **Figure 39.2 Light-induced de-etiolation (greening) of dark-grown potatoes.**

(a) Before exposure to light. A dark-grown potato has tall, spindly stems and nonexpanded leaves—morphological adaptations that enable the shoots to penetrate the soil. The roots are short, but there is little need for water absorption because little water is lost by the shoots.

(b) After a week's exposure to natural daylight. The potato plant begins to resemble a typical plant with broad green leaves, short sturdy stems, and long roots. This transformation begins with the reception of light by a specific pigment, phytochrome.

(informally known as greening). Stem elongation slows; leaves expand; roots elongate; and the shoot produces chlorophyll. In short, it begins to resemble a typical plant **(Figure 39.2b)**. In this section, we'll examine this de-etiolation response as an example of how a plant cell's reception of a signal—in this case, light—is transduced into a response (greening). Along the way, we will explore how studies of mutants provide insights into the molecular details of the stages of cell signal processing: reception, transduction, and response **(Figure 39.3)**.

▼ **Figure 39.3 Review of a general model for signal transduction pathways.** As discussed in Concept 11.1, a hormone or other kind of stimulus interacting with a specific receptor protein can trigger the sequential activation of relay proteins and also the production of second messengers that participate in the pathway. The signal is passed along, ultimately bringing about cellular responses. In this diagram, the receptor is on the surface of the target cell; in other cases, the stimulus interacts with receptors inside the cell.

Reception

Signals are first detected by receptors, proteins that undergo changes in shape in response to a specific stimulus. The receptor involved in de-etiolation is a type of *phytochrome*, a member of a class of photoreceptors that we'll explore more fully later in the chapter. Unlike most receptors, which are built into the plasma membrane, the type of phytochrome that functions in de-etiolation is located in the cytoplasm. Researchers demonstrated the requirement for phytochrome in de-etiolation through studies of the tomato, a close relative of the potato. The *aurea* mutant of tomato, which has reduced levels of phytochrome, greens less than wild-type tomatoes when exposed to light. (*Aurea* is Latin for "gold." In the absence of chlorophyll, the yellow and orange accessory pigments called carotenoids are more obvious.) Researchers produced a normal de-etiolation response in individual *aurea* leaf cells by injecting phytochrome from other plants and then exposing the cells to light. Such experiments indicated that phytochrome functions in light detection during de-etiolation.

Transduction

Receptors can be sensitive to very weak environmental or chemical signals. Some de-etiolation responses are triggered by extremely low levels of light, in certain cases as little as the equivalent of a few seconds of moonlight. The transduction of these extremely weak signals involves **second messengers**—small molecules and ions in the cell that amplify the signal and transfer it from the receptor to other proteins that carry out the response **(Figure 39.4)**. Concept 11.3 discussed several kinds of second messengers (see Figures 11.12 and 11.14). Here, we examine the particular roles of two types of second messengers in de-etiolation: calcium ions (Ca^{2+}) and cyclic GMP (cGMP).

Changes in cytosolic Ca^{2+} levels play an important role in phytochrome signal transduction. The concentration of cytosolic Ca^{2+} is generally very low (about 10^{-7} M), but phytochrome activation leads to the opening of Ca^{2+} channels and a transient 100-fold increase in cytosolic Ca^{2+} levels. In response to light, phytochrome undergoes a change in shape that leads to the activation of guanylyl cyclase, an enzyme that produces the second messenger cyclic GMP. Both Ca^{2+}

▼ **Figure 39.4 An example of signal transduction in plants: the role of phytochrome in the de-etiolation (greening) response.**

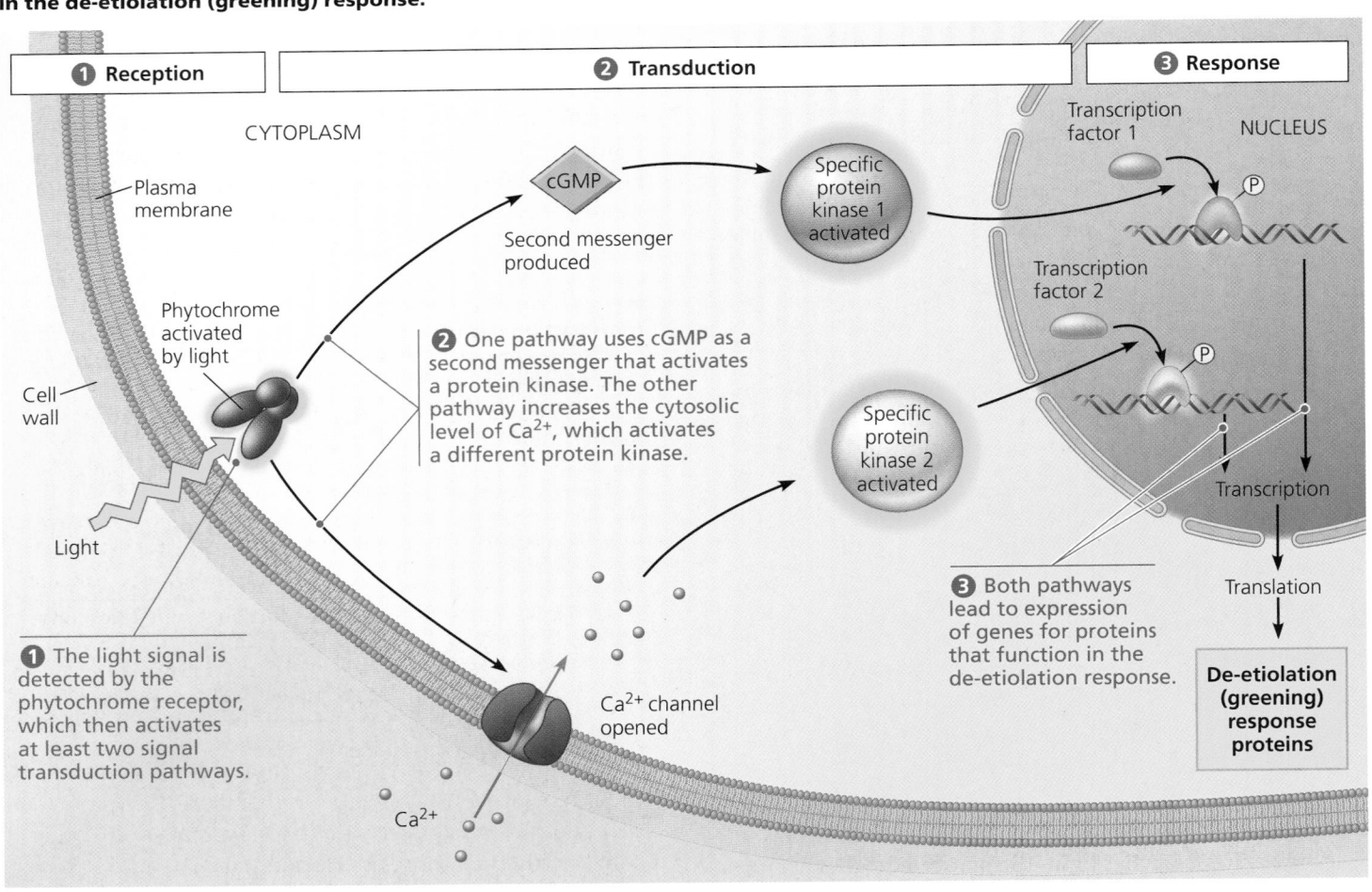

MAKE CONNECTIONS *Which panel in Figure 11.17 best exemplifies the phytochrome-dependent signal transduction pathway during de-etiolation? Explain.*

and cGMP must be produced for a complete de-etiolation response. The injection of cGMP into *aurea* tomato leaf cells, for example, induces only a partial de-etiolation response.

Response

Ultimately, second messengers regulate one or more cellular activities. In most cases, these responses involve the increased activity of particular enzymes. There are two main mechanisms by which a signaling pathway can enhance an enzymatic step in a biochemical pathway: transcriptional regulation and post-translational modification. Transcriptional regulation increases or decreases the synthesis of mRNA encoding a specific enzyme. Post-translational modification activates preexisting enzymes.

Post-translational Modification of Preexisting Proteins

In most signal transduction pathways, preexisting proteins are modified by the phosphorylation of specific amino acids, which alters the protein's hydrophobicity and activity. Many second messengers, including cGMP and Ca^{2+}, activate protein kinases directly. Often, one protein kinase will phosphorylate another protein kinase, which then phosphorylates another, and so on (see Figure 11.10). Such kinase cascades may link initial stimuli to responses at the level of gene expression, usually via the phosphorylation of transcription factors. As we'll see soon, many signal transduction pathways ultimately regulate the synthesis of new proteins by turning specific genes on or off.

Signal transduction pathways must also have a means for turning off when the initial signal is no longer present, such as when a sprouting potato is put back into the cupboard. Protein phosphatases, which are enzymes that dephosphorylate specific proteins, are important in these "switch-off" processes. At any particular moment, a cell's functioning depends on the balance of activity of many types of protein kinases and protein phosphatases.

Transcriptional Regulation

As discussed in Concept 18.2, the proteins we call *specific transcription factors* bind to specific regions of DNA and control the transcription of specific genes (see Figure 18.10). In the case of phytochrome-induced de-etiolation, several such transcription factors are activated by phosphorylation in response to the appropriate light conditions. The activation of some of these transcription factors depends on their phosphorylation by protein kinases activated by cGMP or Ca^{2+}.

The mechanism by which a signal promotes developmental changes may depend on transcription factors that are activators (which *increase* transcription of specific genes) or repressors (which *decrease* transcription) or both. For example, some *Arabidopsis* mutants, except for their pale color, have a light-grown appearance when grown in the dark; they have expanded leaves and short, sturdy stems but are not green because the final step in chlorophyll production requires light directly. These mutants have defects in a repressor that normally inhibits the expression of other genes that are activated by light. When the repressor is eliminated by mutation, the pathway that is normally blocked proceeds. Thus, these mutants appear to have been grown in the light, except for their pale color.

De-etiolation ("Greening") Proteins

What types of proteins are either activated by phosphorylation or newly transcribed during the de-etiolation process? Many are enzymes that function in photosynthesis directly; others are enzymes involved in supplying the chemical precursors necessary for chlorophyll production; still others affect the levels of plant hormones that regulate growth. For example, the levels of auxin and brassinosteroids, hormones that enhance stem elongation, decrease following the activation of phytochrome. That decrease explains the slowing of stem elongation that accompanies de-etiolation.

We have examined the signal transduction involved in the de-etiolation response of a potato plant in some detail to get a sense of the complexity of biochemical changes that underlie this one process. Every plant hormone and environmental stimulus will trigger one or more signal transduction pathways of comparable complexity. As in the studies on the *aurea* mutant tomato, the isolation of mutants (a genetic approach) and techniques of molecular biology are helping researchers identify these various pathways.

CONCEPT CHECK 39.1

1. What are the physical differences between dark- and light-grown plants? Explain how etiolation helps a seedling compete successfully.
2. Cycloheximide is a drug that inhibits protein synthesis. Predict what effect cycloheximide would have on de-etiolation.
3. **WHAT IF?** The sexual dysfunction drug Viagra inhibits an enzyme that breaks down cyclic GMP. If tomato leaf cells have a similar enzyme, would applying Viagra to these cells cause a normal de-etiolation of *aurea* mutant tomato leaves?

For suggested answers, see Appendix A.

CONCEPT 39.2

Plants use chemicals to communicate

As we have seen, plants sense their local environmental conditions using signal transduction pathways no less complicated than those used by animals. The activation of signal transduction pathways often changes the activity of enzymes and therefore the production of chemicals.

Plants use chemicals to communicate with the living world, as in the example of those released by the sunflowers in Figure 39.1. Plants also use chemicals for communicating between different parts of the plant, thereby optimizing the response of the whole organism. Chemical communication within plants is an exploding field of research. The discovery of new mobile information molecules, such as hundreds of small RNAs and dozens of small peptides, will keep researchers busy for decades. It has also been discovered that plants have an attribute that animals do not: The cytoplasmic bridges between plant cells (plasmodesmata), unlike the analogous structures in animals (gap junctions), can dilate enough that macromolecules such as proteins can move from cell to cell. This recent research builds on a long history of classic experiments that provided the first clues that mobile signaling molecules called hormones are internal regulators of plant growth.

General Characteristics of Plant Hormones

A **hormone**, in the original meaning of the term, is a signaling molecule that is produced in low concentrations by one part of an organism's body and transported to other parts, where it binds to a specific receptor and triggers responses in target cells and tissues. In animals, hormones are usually transported through the circulatory system, a criterion often included in definitions of the term. Many modern plant biologists, however, argue that the hormone concept, which originated from studies of animals, is too limiting to describe plant physiological processes. For example, plants don't have circulating blood to transport hormone-like signaling molecules. Moreover, some signaling molecules that are considered plant hormones act only locally. Finally, there are some signaling molecules in plants, such as glucose, that typically occur in plants at concentrations thousands of times greater than a

Table 39.1 Overview of Plant Hormones		
Hormone	**Where Produced or Found in Plant**	**Major Functions**
Auxin (IAA)	Shoot apical meristems and young leaves are the primary sites of auxin synthesis. Root apical meristems also produce auxin, although the root depends on the shoot for much of its auxin. Developing seeds and fruits contain high levels of auxin, but it is unclear whether it is newly synthesized or transported from maternal tissues.	Stimulates stem elongation (low concentration only); promotes the formation of lateral and adventitious roots; regulates development of fruit; enhances apical dominance; functions in phototropism and gravitropism; promotes vascular differentiation; retards leaf abscission
Cytokinins	These are synthesized primarily in roots and transported to other organs, although there are many minor sites of production as well.	Regulate cell division in shoots and roots; modify apical dominance and promote lateral bud growth; promote movement of nutrients into sink tissues; stimulate seed germination; delay leaf senescence
Gibberellins (GA)	Meristems of apical buds and roots, young leaves, and developing seeds are the primary sites of production.	Stimulate stem elongation, pollen development, pollen tube growth, fruit growth, and seed development and germination; regulate sex determination and the transition from juvenile to adult phases
Abscisic acid (ABA)	Almost all plant cells have the ability to synthesize abscisic acid, and its presence has been detected in every major organ and living tissue; it may be transported in the phloem or xylem.	Inhibits growth; promotes stomatal closure during drought stress; promotes seed dormancy and inhibits early germination; promotes leaf senescence; promotes desiccation tolerance
Ethylene	This gaseous hormone can be produced by most parts of the plant. It is produced in high concentrations during senescence, leaf abscission, and the ripening of some types of fruits. Synthesis is also stimulated by wounding and stress.	Promotes ripening of many types of fruit, leaf abscission, and the triple response in seedlings (inhibition of stem elongation, promotion of lateral expansion, and horizontal growth); enhances the rate of senescence; promotes root and root hair formation; promotes flowering in the pineapple family
Brassinosteroids	These compounds are present in all plant tissues, although different intermediates predominate in different organs. Internally produced brassinosteroids act near the site of synthesis.	Promote cell expansion and cell division in shoots; promote root growth at low concentrations; inhibit root growth at high concentrations; promote xylem differentiation and inhibit phloem differentiation; promote seed germination and pollen tube elongation
Jasmonates	These are a small group of related molecules derived from the fatty acid linolenic acid. They are produced in several parts of the plant and travel in the phloem to other parts of the plant.	Regulate a wide variety of functions, including fruit ripening, floral development, pollen production, tendril coiling, root growth, seed germination, and nectar secretion; also produced in response to herbivory and pathogen invasion
Strigolactones	These carotenoid-derived hormones and extracellular signals are produced in roots in response to low phosphate conditions or high auxin flow from the shoot.	Promote seed germination, control of apical dominance, and the attraction of mycorrhizal fungi to the root

typical hormone. Nevertheless, they activate signal transduction pathways that greatly alter the functioning of plants in a manner similar to a hormone. Thus, many plant biologists prefer the broader term *plant growth regulator* to describe organic compounds, either natural or synthetic, that modify or control one or more specific physiological processes within a plant. The terms *plant hormone* and *plant growth regulator* are used about equally, but for historical continuity this text will use the term *plant hormone* and adhere to the criterion that plant hormones are active at very low concentrations.

Although plant hormones are produced in very low concentrations, a tiny amount of hormone can have a profound effect on plant growth and development. Virtually every aspect of plant growth and development is under hormonal control to some degree. Each hormone has multiple effects, depending on its site of action, its concentration, and the developmental stage of the plant. Conversely, multiple hormones can influence a single process. Plant hormone responses commonly depend on both the amounts of the hormones involved and their relative concentrations. It is often the interactions between different hormones, rather than hormones acting in isolation, that control growth and development. These interactions will become apparent in the following survey of hormone function.

A Survey of Plant Hormones

Table 39.1 previews the major types and actions of plant hormones, including auxin, cytokinins, gibberellins, abscisic acid, ethylene, brassinosteroids, jasmonates, and strigolactones.

Auxin

The idea that chemical messengers exist in plants emerged from a series of classic experiments on how stems respond to light. As you know, the shoot of a houseplant on a windowsill grows toward light. Any growth response that results in plant organs curving toward or away from stimuli is called a **tropism** (from the Greek *tropos*, turn). The growth of a shoot toward light or away from it is called **phototropism**; the former is positive phototropism, and the latter is negative phototropism.

In natural ecosystems, where plants may be crowded, phototropism directs shoot growth toward the sunlight that powers photosynthesis. This response results from a differential growth of cells on opposite sides of the shoot; the cells on the darker side elongate faster than the cells on the brighter side.

Charles Darwin and his son Francis conducted some of the earliest experiments on phototropism in the late 1800s **(Figure 39.5)**. They observed that a grass seedling ensheathed in its coleoptile (see Figure 38.9b) could bend toward light only if the tip of the coleoptile was present. If the tip was removed, the coleoptile did not curve. The seedling also failed to grow toward light if the tip was covered with an opaque cap, but neither a transparent cap over the tip nor an opaque shield placed below the coleoptile tip prevented the phototropic response. It was the tip of the coleoptile, the Darwins concluded, that was responsible for sensing light. However, they noted that the differential growth response that led to curvature of the coleoptile occurred some distance below the tip. The Darwins postulated that some signal was transmitted downward from the tip to the elongating region of the coleoptile. A few decades later, the Danish scientist Peter Boysen-Jensen demonstrated that the signal was a mobile chemical substance. He separated the tip from the remainder of the coleoptile by a cube of gelatin, which prevented cellular contact but allowed chemicals to pass through. These seedlings responded normally, bending toward light. However, if the tip was experimentally separated from the lower coleoptile by an impermeable barrier, such as the mineral mica, no phototropic response occurred.

Subsequent research showed that a chemical was released from coleoptile tips and could be collected by means of diffusion into agar blocks. Little cubes of agar containing this chemical could induce "phototropic-like" curvatures even in complete darkness if the agar cubes were placed off-center atop the cut surface of decapitated coleoptiles. Coleoptiles curve toward light because of a higher concentration of this growth-promoting chemical on the darker side of the coleoptile. Since this chemical stimulated growth as it passed down the coleoptile, it was dubbed "auxin" (from the Greek *auxein*, to increase). Auxin was later purified, and its chemical structure determined to be indoleacetic acid (IAA). The term **auxin** is used for any chemical, synthetic or not, that promotes coleoptile elongation. The major natural auxin in plants is IAA, which has many additional effects. Unless noted otherwise, the terms *auxin* and *IAA* will be used interchangeably.

Auxin is produced predominantly in shoot tips and is transported from cell to cell down the stem at a rate of about 1 cm/hr. It moves only from tip to base, not in the reverse direction. This unidirectional transport of auxin is called *polar transport*. Polar transport is unrelated to gravity; experiments have shown that auxin travels upward when a stem or coleoptile segment is placed upside down. Rather, the polarity of auxin movement is attributable to the polar distribution of auxin transport protein in the cells. Concentrated at the basal end of a cell, the auxin transporters move the hormone out of the cell. The auxin can then enter the apical end of the neighboring cell **(Figure 39.6)**. Auxin has a variety of effects, including stimulating cell elongation and regulating plant architecture.

The Role of Auxin in Cell Elongation

One of auxin's chief functions is to stimulate elongation of cells within young developing shoots. As auxin from the shoot tip (see Figure 35.16) moves down to the region where cells are elongating, the hormone stimulates cell growth by binding to a receptor in the nucleus. Auxin stimulates growth only over a certain concentration range, from about 10^{-8} to 10^{-4} M. At higher concentrations, auxin may inhibit cell elongation by inducing production of ethylene, a hormone that generally

▼ **Figure 39.6** **Inquiry**

What causes polar movement of auxin from shoot tip to base?

Experiment To investigate how auxin is transported unidirectionally, Leo Gälweiler and colleagues designed an experiment to identify the location of the auxin transport protein. They used a greenish yellow fluorescent molecule to label antibodies that bind to the auxin transport protein. Then they applied the antibodies to longitudinally sectioned *Arabidopsis* stems.

Results The light micrograph on the left shows that auxin transport proteins are not found in all stem tissues, but only in the xylem parenchyma. In the light micrograph on the right, a higher magnification reveals that these proteins are primarily localized at the basal ends of the cells.

Data from L. Gälweiler et al., Regulation of polar auxin transport by AtPIN1 in *Arabidopsis* vascular tissue, *Science* 282:2226–2230 (1998).

Conclusion The results support the hypothesis that concentration of the auxin transport protein at the basal ends of cells mediates the polar transport of auxin.

WHAT IF? *If auxin transport proteins were equally distributed at both ends of the cells, would polar auxin transport still be possible? Explain.*

hinders growth. We will return to this hormonal interaction when we look at ethylene.

According to a model called the *acid growth hypothesis*, proton pumps play a major role in the growth response of cells to auxin. In a shoot's region of elongation, auxin stimulates the plasma membrane's proton (H^+) pumps. This pumping of H^+ increases the voltage across the membrane (membrane potential) and lowers the pH in the cell wall within minutes. Acidification of the wall activates proteins called **expansins** that break the cross-links (hydrogen bonds) between cellulose microfibrils and other cell wall constituents, loosening the wall's fabric **(Figure 39.7)**. Increasing the membrane potential enhances ion uptake into the cell, which causes osmotic uptake of water and increased turgor. Increased turgor and increased cell wall plasticity enable the cell to elongate.

▼ Figure 39.7 Cell elongation in response to auxin: the acid growth hypothesis. The cell expands in a direction mainly perpendicular to the main orientation of the microfibrils in the cell wall (see Figure 35.28).

Cell wall before auxin activates proton pumps

Cellulose microfibrils

Nucleus   Vacuoles

Cellulose microfibril   Cross-linking polysaccharide

PLANT CELL WALL

1 Auxin increases the activity of proton pumps, which pump H⁺ from the cytoplasm. The H⁺ reduces the pH of the cell wall.

2 The reduced pH activates wedge-shaped proteins called expansins (red) which separate the microfibrils (brown) from the polysaccharides (green).

Loosening of cell wall, enabling cell elongation

Elongation

3 The polysaccharides are cleaved by cell wall–loosening enzymes (purple). This process loosens the microfibrils, making the cell wall more flexible. Water uptake and increased turgor then cause the cell to elongate. At the same time, more microfibrils (not shown) are formed.

Auxin also rapidly alters gene expression, causing cells in the region of elongation to produce new proteins within minutes. Some of these proteins are short-lived transcription factors that repress or activate the expression of other genes. For sustained growth after this initial spurt, cells must make more cytoplasm and wall material. In addition, auxin stimulates this sustained growth response.

Auxin's Role in Plant Development   The polar transport of auxin plays a major role in controlling the spatial organization, or *pattern formation*, of a plant. The auxin synthesized in the shoot tips of a branch carries information about the growth potential of that branch. If a branch finds itself in an environment conducive for growth, it produces more auxin, and the plant diverts more resources to that branch. A reduced flow of auxin from a branch indicates that the branch is not being sufficiently productive: New branches are needed elsewhere. Thus, lateral buds below the branch are released from dormancy and begin to grow.

The transport of auxin also plays a key role in establishing the pattern of leaf emergence from the shoot apical meristem (see Figure 36.3). A leading model proposes that polar auxin transport in the shoot tip generates local peaks in auxin concentration that determine the site of leaf primordium formation and thereby the different leaf arrangements found in nature.

The polar transport of auxin from the leaf margin also directs the patterns of leaf veins. Inhibitors of polar auxin transport result in leaves that lack vascular continuity through the petiole and have broad, loosely organized main veins, an increased number of secondary veins, and a dense band of irregularly shaped vascular cells adjacent to the leaf margin.

The activity of the vascular cambium, the meristem that produces woody tissues, is also under the control of auxin transport. When a plant becomes dormant at the end of a growing season, there is a reduction in auxin transport capacity and the expression of genes encoding auxin transporters.

Auxin's effects on plant development are not limited to the familiar sporophyte plant that we see. Recent evidence suggests that the organization of the microscopic angiosperm female gametophytes is regulated by an auxin gradient.

Practical Uses for Auxins   Auxins, both natural and synthetic, have many commercial applications. For example, the natural auxin indolebutyric acid (IBA) is used in the vegetative propagation of plants by cuttings. Treating a detached leaf or stem with powder containing IBA often causes adventitious roots to form near the cut surface.

Certain synthetic auxins are widely used as herbicides, including 2,4-dichlorophenoxyacetic acid (2,4-D). Monocots, such as maize and turfgrass, can rapidly inactivate such synthetic auxins. However, eudicots cannot and therefore die from hormonal overdose. Spraying cereal fields or turf with 2,4-D eliminates eudicot (broadleaf) weeds.

Developing seeds produce auxin, which promotes fruit growth. In tomato plants grown in greenhouses, often fewer seeds are produced, resulting in poorly developed tomato fruits. However, spraying synthetic auxins on greenhouse-grown tomato vines induces normal fruit development, making the greenhouse-cultivated tomatoes commercially viable.

Cytokinins

Trial-and-error attempts to find chemical additives that would enhance the growth and development of plant cells in tissue culture led to the discovery of **cytokinins**. In the 1940s, researchers stimulated the growth of plant embryos in culture by adding coconut milk, the liquid endosperm of a coconut's giant seed. Subsequent researchers found that they could induce cultured tobacco cells to divide by adding degraded DNA samples. The active ingredients of both experimental additives were modified forms of adenine, a component of nucleic acids. These growth regulators were named cytokinins because they stimulate cytokinesis, or cell division. The most common natural cytokinin is zeatin, so named because it was discovered first in maize (*Zea mays*). Cytokinins influence cell division, cell differentiation, and apical dominance.

Control of Cell Division and Differentiation Cytokinins are produced in actively growing tissues, particularly in roots, embryos, and fruits. Cytokinins produced in roots reach their target tissues by moving up the plant in the xylem sap. Acting in concert with auxin, cytokinins stimulate cell division and influence the pathway of differentiation. The effects of cytokinins on cells growing in tissue culture provide clues about how this class of hormones may function in an intact plant. When a piece of parenchyma tissue from a stem is cultured in the absence of cytokinins, the cells grow very large but do not undergo mitosis. But if cytokinins are added along with auxin, the cells divide. Cytokinins alone have no effect. The ratio of cytokinins to auxin controls cell differentiation. When the concentrations of these two hormones are at certain levels, the mass of cells continues to grow, but it remains a cluster of undifferentiated cells called a callus (see Figure 38.15). If cytokinin levels increase, shoot buds develop from the callus. If auxin levels increase, roots form.

Control of Apical Dominance Apical dominance, the ability of the apical bud to suppress the development of axillary buds, is under the control of sugar and various plant hormones, including auxin, cytokinins, and strigolactones. The sugar demand of the shoot tip is critical for maintaining apical dominance. Cutting off the apical bud removes apical sugar demand and rapidly increases sugar (sucrose) availability to axillary buds. This increase of sugar is sufficient to initiate bud release. However, not all of the buds grow equally: Usually only one of the axillary buds closest to the cut surface will take over as the new apical bud.

Three plant hormones—auxin, cytokinins, and strigolactones—play a role

in determining the extent to which specific axillary buds elongate **(Figure 39.8)**. In an intact plant, auxin transported down the shoot from the apical bud *indirectly* inhibits axillary buds from growing, causing a shoot to lengthen at the expense of lateral branching. The polar flow of auxin down the shoot triggers the synthesis of strigolactones, which *directly* repress bud growth. Meanwhile, cytokinins entering the shoot system from roots counter the action of auxin and strigolactones by signaling axillary buds to begin growing. Thus, in an intact plant, the cytokinin-rich axillary buds closer to the base of the plant tend to be longer than the auxin-rich axillary buds closer to the apical bud. Mutants that overproduce cytokinins or plants treated with cytokinins tend to be bushier than normal.

Removing the apical bud, a major site of auxin biosynthesis, causes the auxin and strigolactone levels in the stem to wane, particularly in those regions close to the cut surface (see Figure 39.8). This causes the axillary buds closest to the cut surface to grow most vigorously, and one of these axillary buds will eventually take over as the new apical bud. Applying auxin to the cut surface of the shoot tip resuppresses the growth of the lateral buds.

Anti-aging Effects Cytokinins slow the aging of certain plant organs by inhibiting protein breakdown, stimulating RNA and protein synthesis, and mobilizing nutrients from surrounding tissues. If leaves removed from a plant are dipped in a cytokinin solution, they stay green much longer than otherwise.

Gibberellins

In the early 1900s, farmers in Asia noticed that some rice seedlings in their paddies grew so tall and spindly that they

▼ **Figure 39.8 Effects on apical dominance of removing the apical bud.** Apical dominance refers to the inhibition of the growth of axillary buds by the apical bud of a plant shoot. Removal of the apical bud enables lateral branches to grow. Multiple hormones play a role in this process, including auxin, cytokinin, and strigolactones.

The apical bud is a preferred sugar sink and a major site of auxin biosynthesis.

Auxin moving downward from the apical bud produces strigolactones that repress the growth of axillary buds.

Cytokinin coming from the root antagonizes the actions of auxin and strigolactone, allowing for a limited amount of axillary bud growth. Therefore, the axillary buds farthest from the apex are increasingly elongated.

Removal of the apical bud allows remaining buds to receive more sugars for growth. Auxin and strigolactone levels also decline, particularly near the cut surface. This decline allows the topmost axillary buds in particular to grow and take over as the new apical bud.

Plant with apical bud intact

Plant with apical bud removed

toppled over before they could mature. In 1926, it was discovered that a fungus of the genus *Gibberella* causes this "foolish seedling disease." By the 1930s, it was determined that the fungus causes hyperelongation of rice stems by secreting a chemical, which was given the name **gibberellin**, or gibberellic acid. In the 1950s, researchers discovered that plants also produce gibberellins. Since that time, scientists have identified more than 100 different gibberellins that occur naturally in plants, although a much smaller number occur in each plant species. "Foolish rice" seedlings, it seems, suffer from too much gibberellin. Gibberellins have a variety of effects, such as stem elongation, fruit growth, and seed germination.

Stem Elongation The major sites of gibberellin production are young roots and leaves. Gibberellins are best known for stimulating stem and leaf growth by enhancing cell elongation *and* cell division. One hypothesis proposes that they activate enzymes that loosen cell walls, facilitating entry of expansin proteins. Thus, gibberellins act in concert with auxin to promote stem elongation.

The effects of gibberellins in enhancing stem elongation are evident when certain dwarf (mutant) varieties of plants are treated with gibberellins. For instance, some dwarf pea plants (including the variety Mendel studied; see Concept 14.1) grow tall if treated with gibberellins. But there is often no response if the gibberellins are applied to wild-type plants. Apparently, these plants already produce an optimal dose of the hormone. The most dramatic example of gibberellin-induced stem elongation is *bolting*, rapid growth of the floral stalk **(Figure 39.9a)**.

Fruit Growth In many plants, both auxin and gibberellins must be present for fruit to develop. The most important commercial application of gibberellins is in the spraying of Thompson seedless grapes **(Figure 39.9b)**. The hormone makes the individual grapes grow larger, a trait valued by the consumer. The gibberellin sprays also make the internodes of the grape bunch elongate, allowing more space for the individual grapes. By enhancing air circulation between the grapes, this increase in space also makes it harder for yeasts and other microorganisms to infect the fruit.

Germination The embryo of a seed is a rich source of gibberellins. After water is imbibed, the release of gibberellins from the embryo signals the seed to break dormancy and germinate. Some seeds that normally require particular

▼ **Figure 39.9 Effects of gibberellins on stem elongation and fruit growth.**

(a) Some plants develop in a rosette form, low to the ground with very short internodes, as in the *Arabidopsis* plant shown at the left. As the plant switches to reproductive growth, a surge of gibberellins induces bolting: Internodes elongate rapidly, elevating floral buds that develop at stem tips (right).

(b) The Thompson seedless grape bunch on the left is from an untreated control vine. The bunch on the right is growing from a vine that was sprayed with gibberellin during fruit development.

environmental conditions to germinate, such as exposure to light or low temperatures, break dormancy if they are treated with gibberellins. Gibberellins support the growth of cereal seedlings by stimulating the synthesis of digestive enzymes such as α-amylase that mobilize stored nutrients **(Figure 39.10)**.

▼ **Figure 39.10 Mobilization of nutrients by gibberellins during the germination of grain seeds such as barley.**

1 After a seed imbibes water, the embryo releases gibberellin (GA), which sends a signal to the aleurone, the thin outer layer of the endosperm.

2 The aleurone responds to GA by synthesizing and secreting digestive enzymes that hydrolyze nutrients stored in the endosperm. One example is α-amylase, which hydrolyzes starch into sugars.

3 Sugars and other nutrients absorbed from the endosperm by the scutellum (cotyledon) are consumed during growth of the embryo into a seedling.

Seed coat
Aleurone
Endosperm
Embryo
GA
Water
GA
α-amylase
Sugar
Scutellum (cotyledon)

Abscisic Acid

In the 1960s, one research group studying the chemical changes that precede bud dormancy and leaf abscission in deciduous trees and another team investigating chemical changes preceding abscission of cotton fruits isolated the same compound, **abscisic acid (ABA)**. Ironically, ABA is no longer thought to play a primary role in leaf abscission, but it is very important in other functions. Unlike the growth-stimulating hormones discussed so far—auxin, cytokinins, gibberellins, and brassinosteroids—ABA *slows* growth. ABA often antagonizes the actions of growth hormones, and the ratio of ABA to one or more growth hormones determines the final physiological outcome. We will consider here two of ABA's many effects: seed dormancy and drought tolerance.

Seed Dormancy Seed dormancy increases the likelihood that seeds will germinate only when there are sufficient amounts of light, temperature, and moisture for the seedlings to survive (see Concept 38.1). What prevents seeds dispersed in autumn from germinating immediately, only to die in the winter? What mechanisms ensure that such seeds do not germinate until spring? For that matter, what prevents seeds from germinating in the dark, moist interior of the fruit? The answer to these questions is ABA. The levels of ABA may increase 100-fold during seed maturation. The high levels of ABA in maturing seeds inhibit germination and induce the production of proteins that help the seeds withstand the extreme dehydration that accompanies maturation.

Many types of dormant seeds germinate when ABA is removed or inactivated. The seeds of some desert plants break dormancy only when heavy rains wash ABA out of them. Other seeds require light or prolonged exposure to cold to inactivate ABA. Often, the ratio of ABA to gibberellins determines whether seeds remain dormant or germinate, and adding ABA to seeds that are primed to germinate makes them dormant again. Inactivated ABA or low levels of ABA can lead to precocious (early) germination **(Figure 39.11)**. For example, a maize mutant with grains that germinate while still on the cob lacks a functional transcription factor required for ABA to induce expression of certain genes. Precocious germination of red mangrove seeds, due to low ABA levels, is actually an adaptation that helps the young seedlings to plant themselves like darts in the soft mud below the parent tree.

Drought Tolerance ABA plays a major role in drought signaling. When a plant begins to wilt, ABA accumulates in the leaves and causes stomata to close rapidly, reducing transpiration and preventing further water loss. By affecting second messengers such as calcium, ABA causes potassium channels in the plasma membrane of guard cells to open, leading to a massive loss of potassium ions from the cells. The accompanying osmotic loss of water reduces guard cell turgor and leads to closing of the stomatal pores (see Figure 36.14).

▼ **Figure 39.11 Precocious germination of wild-type mangrove and mutant maize seeds.**

◀ Red mangrove (*Rhizophora mangle*) seeds produce only low levels of ABA, and the seeds germinate while still on the tree. In this case, early germination is a useful adaptation. When released, the radicle of the dart-like seedling deeply penetrates the soft mudflats in which the mangroves grow.

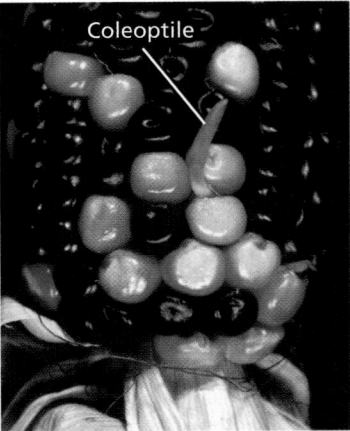

Coleoptile

▲ Precocious germination in this maize mutant is caused by lack of a functional transcription factor required for ABA action.

In some cases, water shortage stresses the root system before the shoot system, and ABA transported from roots to leaves may function as an "early warning system." Many mutants that are especially prone to wilting are deficient in ABA production.

Ethylene

During the 1800s, when coal gas was used as fuel for streetlights, leakage from gas pipes caused nearby trees to drop leaves prematurely. In 1901, the gas **ethylene** was demonstrated to be the active factor in coal gas. But the idea that it is a plant hormone was not widely accepted until the advent of a technique called gas chromatography simplified its identification.

Plants produce ethylene in response to stresses such as drought, flooding, mechanical pressure, injury, and infection. Ethylene is also produced during fruit ripening and programmed cell death and in response to high concentrations of externally applied auxin. Indeed, many effects previously ascribed to auxin, such as inhibition of root elongation, may be due to auxin-induced ethylene production. We focus here on four of ethylene's many effects: response to mechanical stress, senescence, leaf abscission, and fruit ripening.

The Triple Response to Mechanical Stress Imagine a pea seedling pushing upward through the soil, only to come up against a stone. As it pushes against the obstacle, the stress in its delicate tip induces the seedling to produce ethylene. The hormone then instigates a growth maneuver known as the **triple response** that enables the shoot to avoid the obstacle. The three parts of this response are a slowing of stem elongation, a thickening of the stem (which makes it stronger), and a curvature that causes the stem to start growing horizontally. As the effects of the initial ethylene pulse lessen, the stem resumes vertical growth. If it again contacts a barrier, another burst of ethylene is released, and horizontal growth resumes. However, if the upward touch detects no solid object, then ethylene production decreases, and the stem, now clear of the obstacle, resumes its normal upward growth. It is ethylene that induces the stem to grow horizontally rather than the physical obstruction itself; when ethylene is applied to normal seedlings growing free of physical impediments, they still undergo the triple response **(Figure 39.12)**.

Studies of *Arabidopsis* mutants with abnormal triple responses are an example of how biologists identify a signal transduction pathway. Scientists isolated ethylene-insensitive (*ein*) mutants, which fail to undergo the triple response after exposure to ethylene **(Figure 39.13a)**. Some types of *ein* mutants are insensitive to ethylene because they lack a functional ethylene receptor. Mutants of a different sort undergo the triple response even out of soil, in the air, where there are no physical obstacles. Some of these mutants have a regulatory defect that causes them to produce ethylene

▼ **Figure 39.12 The ethylene-induced triple response.** In response to ethylene, a gaseous plant hormone, germinating pea seedlings grown in the dark undergo the triple response—slowing of stem elongation, stem thickening, and horizontal stem growth. The response is greater with increased ethylene concentration.

Ethylene concentration (parts per million)

▼ **Figure 39.13 Ethylene triple-response *Arabidopsis* mutants.**

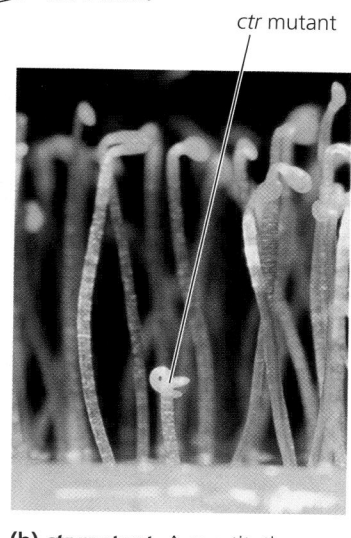

(a) *ein* mutant. An ethylene-insensitive (*ein*) mutant fails to undergo the triple response in the presence of ethylene.

(b) *ctr* mutant. A constitutive triple-response (*ctr*) mutant undergoes the triple response even in the absence of ethylene.

VISUAL SKILLS *If the* ein *single mutation is combined with an ethylene-overproducing* (eto) *mutation, would the phenotype of the double mutant be different from that of the single mutant? Explain.*

at rates 20 times normal. The phenotype of such ethylene-overproducing (*eto*) mutants can be restored to wild-type by treating the seedlings with inhibitors of ethylene synthesis. Other mutants, called constitutive triple-response (*ctr*) mutants, undergo the triple response in air but do not respond to inhibitors of ethylene synthesis **(Figure 39.13b)**. (Constitutive genes are genes that are continually expressed in all cells of an organism.) In *ctr* mutants, ethylene signal transduction is permanently turned on, even though ethylene is not present.

The affected gene in *ctr* mutants codes for a protein kinase. The fact that this mutation *activates* the ethylene response suggests that the normal kinase product of the wild-type allele is a *negative* regulator of ethylene signal transduction. Thus, binding of the hormone ethylene to the ethylene receptor normally leads to inactivation of the kinase, and the inactivation of this negative regulator allows synthesis of the proteins required for the triple response.

Senescence Consider the shedding of a leaf in autumn or the death of an annual after flowering. Or think about the final step in differentiation of a vessel element, when its living contents are destroyed, leaving a hollow tube behind. Such events involve **senescence**—the programmed death of certain cells or organs or the entire plant. Cells, organs, and plants genetically programmed to die on a schedule do not simply shut down cellular machinery and await death. Instead, at the molecular level, the onset of programmed cell death is a very busy time in a cell's life, requiring new gene expression. Newly formed enzymes break down many chemical components, including chlorophyll, DNA, RNA, proteins, and membrane lipids. The plant salvages many of the breakdown products. A burst of ethylene is almost always associated with the death of cells during senescence.

Leaf Abscission The loss of leaves from deciduous trees helps prevent desiccation during seasonal periods when the availability of water to the roots is severely limited. Before dying leaves abscise, many essential elements are salvaged from them and stored in stem parenchyma cells. These nutrients are recycled back to developing leaves during the following spring. The colors of autumn leaves are due to newly made red pigments as well as yellow and orange carotenoids (see Concept 10.3) that were already present in the leaves and are rendered visible by the breakdown of the dark green chlorophyll in autumn.

When an autumn leaf falls, it detaches from the stem at an abscission layer that develops near the base of the petiole **(Figure 39.14)**. The small parenchyma cells of this layer have very thin walls, and there are no fiber cells around the vascular tissue. The abscission layer is further weakened when enzymes hydrolyze polysaccharides in the cell walls. Finally, the weight of the leaf, with the help of the wind, causes a separation within the abscission layer. Even before the leaf falls, a layer of cork forms a protective scar on the twig side of the abscission layer, preventing pathogens from invading the plant.

▼ **Figure 39.14 Abscission of a maple leaf.** Abscission is controlled by a change in the ratio of ethylene to auxin. The abscission layer is seen in this longitudinal section as a vertical band at the base of the petiole. After the leaf falls, a protective layer of cork becomes the leaf scar that helps prevent pathogens from invading the plant (LM).

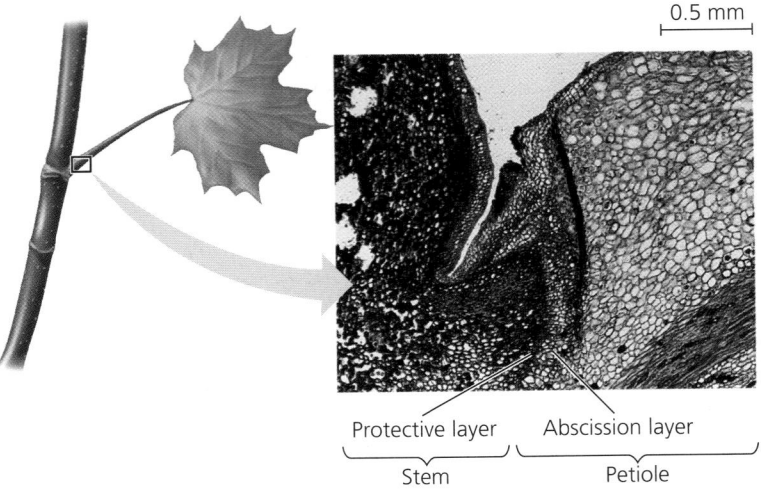

0.5 mm

Protective layer   Abscission layer

Stem   Petiole

→ **Mastering Biology** Animation: Leaf Abscission

A change in the ratio of ethylene to auxin controls abscission. An aging leaf produces less and less auxin, rendering the cells of the abscission layer more sensitive to ethylene. As the influence of ethylene on the abscission layer prevails, the cells produce enzymes that digest the cellulose and other components of cell walls.

Fruit Ripening Immature fleshy fruits are generally tart, hard, and green—features that help protect the developing seeds from herbivores. After ripening, the mature fruits help *attract* animals that disperse the seeds (see Figures 30.10 and 30.11). In many cases, a burst of ethylene production in the fruit triggers the ripening process. The enzymatic breakdown of cell wall components softens the fruit, and the conversion of starches and acids to sugars makes the fruit sweet. The production of new scents and colors helps advertise ripeness to animals, which eat the fruits and disperse the seeds.

A chain reaction occurs during ripening: Ethylene triggers ripening, and ripening triggers more ethylene production. The result is a huge burst in ethylene production. Because ethylene is a gas, the signal to ripen spreads from fruit to fruit. If you pick or buy green fruit, you may be able to speed ripening by storing the fruit in a paper bag, allowing ethylene to accumulate. On a commercial scale, many kinds of fruits are ripened in huge storage containers in which ethylene levels are enhanced. In other cases, fruit producers take measures to slow ripening caused by natural ethylene. Apples, for instance, are stored in bins flushed with carbon dioxide. Circulating the air prevents ethylene from accumulating, and carbon dioxide inhibits synthesis of new ethylene. Stored in this way, apples picked in autumn can still be shipped to grocery stores the following summer.

Given the importance of ethylene in the postharvest physiology of fruits, the genetic engineering of ethylene signal transduction pathways has potential commercial applications. For example, by engineering a way to block the transcription of one of the genes required for ethylene synthesis, molecular biologists have created tomato fruits that ripen on demand. These fruits are picked while green and will not ripen unless ethylene gas is added. As such methods are refined, they will reduce spoilage of fruits and vegetables, a problem that ruins almost half the produce harvested in the United States.

More Recently Discovered Plant Hormones

Auxin, gibberellins, cytokinins, abscisic acid, and ethylene are often considered the five "classic" plant hormones. However, more recently discovered hormones have swelled the list of important plant growth regulators.

Brassinosteroids are steroids similar to cholesterol and the sex hormones of animals. They induce cell elongation and division in stem segments and seedlings at concentrations as low as 10^{-12} M. They also slow leaf abscission (leaf drop) and promote xylem differentiation. These effects are so qualitatively similar to those of auxin that it took years for plant physiologists to determine that brassinosteroids were not types of auxins.

The identification of brassinosteroids as plant hormones arose from studies of an *Arabidopsis* mutant that even when grown in the dark exhibited physical features similar to plants grown in the light. The researchers discovered that the mutation affects a gene that normally codes for an enzyme similar to one involved in steroid synthesis in mammals. They also found that this brassinosteroid-deficient mutant could be restored to the wild-type phenotype by applying brassinosteroids.

Jasmonates, including *jasmonate* (JA) and *methyl jasmonate* (MeJA), are fatty acid–derived molecules that play important roles both in plant defense (see Concept 39.5) and, as discussed here, in plant development. Chemists first isolated MeJA as a key ingredient producing the enchanting fragrance of jasmine (*Jasminum grandiflorum*) flowers. Interest in jasmonates exploded when it was realized that jasmonates are produced by wounded plants and play a key role in controlling plant defenses against herbivores and pathogens. From studying jasmonate signal transduction mutants as well as the effects of applying jasmonates to plants, it soon became apparent that jasmonates and their derivatives regulate a wide variety of physiological processes in plants, including nectar secretion, fruit ripening, pollen production, flowering time, seed germination, root growth, tuber formation, mycorrhizal symbioses, and tendril coiling. In controlling plant processes, jasmonates also engage in cross-talk with phytochrome and various hormones, including GA, IAA, and ethylene.

Strigolactones are xylem-mobile chemicals that stimulate seed germination, suppress adventitious root formation, help establish mycorrhizal associations, and (as noted earlier) help control apical dominance. Their recent discovery relates back to studies of their namesake, *Striga*, a colorfully named genus of rootless parasitic plants that penetrate the roots of other plants, diverting essential nutrients from them and stunting their growth. (In Romanian legend, *Striga* is a vampire-like creature that lives for thousands of years, only needing to feed every 25 years or so.) Also known as witchweed, *Striga* may be the greatest obstacle to food production in Africa, infesting about two-thirds of the area devoted to cereal crops. Each *Striga* plant produces tens of thousands of tiny seeds that can remain dormant in the soil for many years until a suitable host begins to grow. Thus, *Striga* cannot be eradicated by growing non-grain crops for several years. Strigolactones, exuded by the host roots, were first identified as the chemical signals that stimulate the germination of *Striga* seeds.

CONCEPT CHECK 39.2

1. Fusicoccin is a fungal toxin that stimulates the plasma membrane H^+ pumps of plant cells. How may it affect the growth of isolated stem sections?
2. **WHAT IF?** If a plant has the double mutation *ctr* and *ein*, what is its triple-response phenotype? Explain your answer.
3. **MAKE CONNECTIONS** What type of feedback process is exemplified by the production of ethylene during fruit ripening? Explain. (See Figure 1.10.)

For suggested answers, see Appendix A.

CONCEPT 39.3

Responses to light are critical for plant success

Light is an especially important environmental factor in the lives of plants. In addition to being required for photosynthesis, light triggers many key events in plant growth and development, collectively known as **photomorphogenesis**. Light reception also allows plants to measure the passage of days and seasons.

Plants detect not only the presence of light signals but also their direction, intensity, and wavelength (color). A graph called an **action spectrum** depicts the relative effectiveness of different wavelengths of radiation in driving a particular process, such as photosynthesis (see Figure 10.9b). Action spectra are useful in studying *any* process that depends on light. By comparing action spectra of various plant responses, researchers determine which responses are mediated by the same photoreceptor (pigment). They also compare action spectra with absorption spectra of pigments; a close correspondence for a given pigment suggests that the pigment is the photoreceptor mediating the response. Action spectra reveal that red and blue light are the most important colors in regulating a plant's photomorphogenesis. These observations led researchers to two major classes of light receptors: blue-light photoreceptors and phytochromes.

Blue-Light Photoreceptors

Pigments that absorb blue light, known as **blue-light photoreceptors**, initiate a variety of responses in plants, including phototropism, the light-induced opening of stomata (see Figure 36.14), and the light-induced slowing of hypocotyl elongation that occurs when a seedling breaks ground. The biochemical identity of the blue-light photoreceptor was so elusive that in the 1970s, plant physiologists began to call this receptor "cryptochrome" (from the Greek *kryptos*, hidden, and *chrom*, pigment). In the 1990s, molecular biologists analyzing *Arabidopsis* mutants found that plants use different types of pigments to detect blue light. *Cryptochromes*, molecular relatives of DNA repair enzymes, are involved in the blue-light-induced inhibition of stem elongation that occurs, for example, when a seedling first emerges from the soil. *Phototropin* is a protein kinase involved in mediating blue-light-mediated stomatal opening, chloroplast movements in response to light, and phototropic curvatures **(Figure 39.15)**, such as those studied by the Darwins.

▼ **Figure 39.15 Action spectrum for blue-light-stimulated phototropism in maize coleoptiles.** Phototropic bending toward light is controlled by phototropin, a photoreceptor sensitive to blue and violet light, particularly blue light.

(a) This action spectrum illustrates that only light wavelengths below 500 nm (blue and violet light) induce curvature.

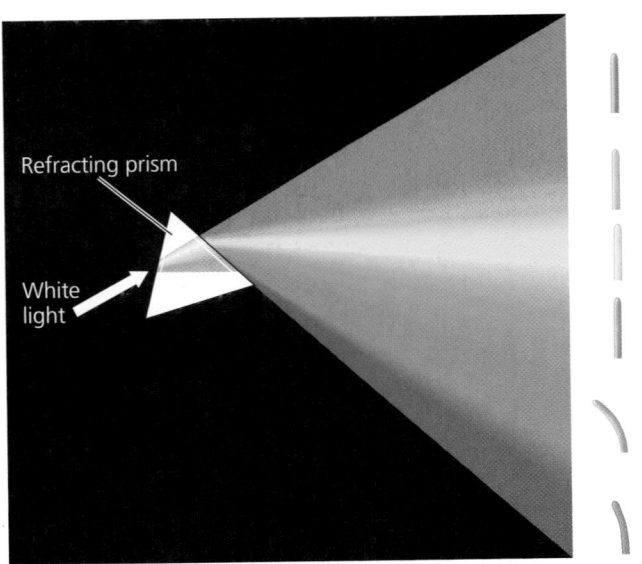

(b) When coleoptiles are exposed to light of various wavelengths as shown here, violet light induces slight curvature toward the light and blue light induces the most curvature. The other colors do not induce any curvature.

⊙ **Mastering Biology** Video: Phototropism

Phytochrome Photoreceptors

When exploring signal transduction in plants earlier in the chapter, we looked at the role of the plant pigments called phytochromes in the de-etiolation process. **Phytochromes**, pigments that absorb mostly red and far-red light, regulate many plant responses to light, including seed germination and shade avoidance.

Phytochromes and Seed Germination

Studies of seed germination led to the discovery of phytochromes. Because of limited nutrient reserves, many types of seeds, especially small ones, germinate only when the light environment and other conditions are near optimal. Such seeds often remain dormant for years until light conditions change. For example, the death of a shading tree or the plowing of a field may create a favorable light environment for germination.

In the 1930s, scientists determined the action spectrum for light-induced germination of lettuce seeds. They exposed water-swollen seeds to a few minutes of single-colored light of various wavelengths and then stored the seeds in the dark. After two days, the researchers counted the number of seeds that had germinated under each light regimen. They found that red light of wavelength 660 nm increased the germination percentage of lettuce seeds maximally, whereas far-red light—that is, light of wavelengths near the upper edge of human visibility (730 nm)—*inhibited* germination compared with dark controls **(Figure 39.16)**. What happens when the lettuce seeds are subjected to a flash of red light followed by a flash of far-red light or, conversely, to far-red light followed by red light? The *last* flash of light determines the seeds' response: The effects of red and far-red light are reversible.

The photoreceptors responsible for the opposing effects of red and far-red light are phytochromes. So far, researchers have identified five phytochromes in *Arabidopsis*, each with a slightly different polypeptide component. In most phytochromes, the light-absorbing portion is photoreversible, converting back and forth between two forms, depending on the color of light to which it is exposed. In its red-absorbing form (P_r), a phytochrome absorbs red (r) light maximally and is converted to its far-red-absorbing form (P_{fr}); in its P_{fr} form, it absorbs far-red (fr) light and is converted to its P_r form **(Figure 39.17)**. This $P_r \leftrightarrow P_{fr}$ interconversion is a switching mechanism that controls various light-induced events in the life of the plant. P_{fr} is the form of phytochrome that triggers many of a plant's developmental responses to light. For example, P_r in lettuce seeds exposed to red light is converted to P_{fr}, stimulating the cellular responses that lead to germination. When red-illuminated seeds are then exposed to far-red light, the P_{fr} is converted back to P_r, inhibiting the germination response.

How does the order of red and far-red illumination affect seed germination?

Experiment Scientists at the U.S. Department of Agriculture briefly exposed batches of lettuce seeds to red light or far-red light to test the effects on germination. After the light exposure, the seeds were placed in the dark, and the results were compared with control seeds that were not exposed to light.

Results The bar below each photo indicates the sequence of red light exposure, far-red light exposure, and darkness. The germination rate increased greatly in groups of seeds that were last exposed to red light (left). Germination was inhibited in groups of seeds that were last exposed to far-red light (right).

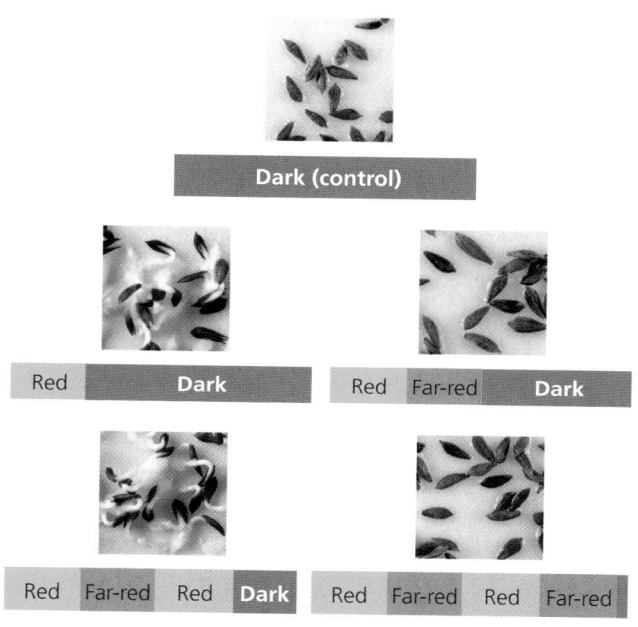

Dark (control)

| Red | Dark |

| Red | Far-red | Dark |

| Red | Far-red | Red | Dark |

| Red | Far-red | Red | Far-red |

Data from H. Borthwick et al., A reversible photoreaction controlling seed germination, *Proceedings of the National Academy of Sciences USA* 38:662–666 (1952).

Conclusion Red light stimulates germination, and far-red light inhibits germination. The final light exposure is the determining factor. The effects of red and far-red light are reversible.

WHAT IF? *Phytochrome responds faster to red light than to far-red light. If the seeds had been placed in white light instead of the dark after their red light and far-red light treatments, would the results have been different?*

How does phytochrome switching explain light-induced germination in nature? Plants synthesize phytochrome as P_r, and if seeds are kept in the dark, the pigment remains almost entirely in the P_r form (see Figure 39.17). Sunlight contains both red light and far-red light, but the conversion to P_{fr} is faster than the conversion to P_r. Therefore, the ratio of P_{fr} to P_r increases in the sunlight. When seeds are exposed to adequate sunlight, the production and accumulation of P_{fr} trigger their germination.

▼ **Figure 39.17 Phytochrome: a molecular switching mechanism.** The absorption of red light causes P_r to change to P_{fr}. Far-red light reverses this conversion. In most cases, it is the P_{fr} form of the pigment that switches on physiological and developmental responses in the plant.

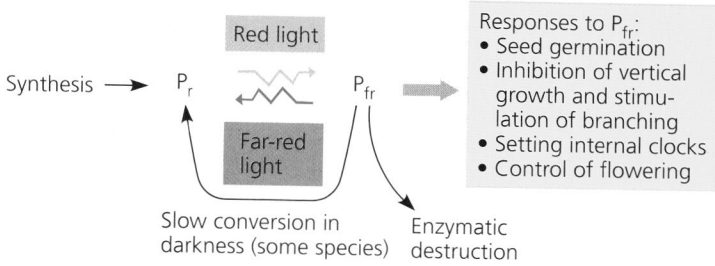

Responses to P_{fr}:
- Seed germination
- Inhibition of vertical growth and stimulation of branching
- Setting internal clocks
- Control of flowering

Phytochromes and Shade Avoidance

The phytochrome system also provides the plant with information about the *quality* of light. Because sunlight includes both red and far-red radiation, during the day the $P_r \leftrightarrow P_{fr}$ interconversion reaches a dynamic equilibrium, with the ratio of the two phytochrome forms indicating the relative amounts of red and far-red light. This sensing mechanism enables plants to adapt to changes in light conditions. Consider, for example, the "shade avoidance" response of a tree that requires relatively high light intensity. If other trees in a forest shade this tree, the phytochrome ratio shifts in favor of P_r because the forest canopy screens out more red light than far-red light. This is because the chlorophyll pigments in the leaves of the canopy absorb red light and allow far-red light to pass. The shift in the ratio of red to far-red light induces the tree to allocate more of its resources to growing taller. In contrast, direct sunlight increases the proportion of P_{fr}, which stimulates branching and inhibits vertical growth.

In addition to helping plants detect light, phytochrome helps a plant keep track of the passage of days and seasons. To understand phytochrome's role in these timekeeping processes, we must first examine the nature of the plant's internal clock.

Biological Clocks and Circadian Rhythms

Many plant processes, such as transpiration and the synthesis of certain enzymes, undergo a daily oscillation. Some of these cyclic variations are responses to the changes in light levels and temperature that accompany the 24-hour cycle of day and night. We can control these external factors by growing plants in growth chambers under rigidly maintained conditions of light and temperature. But even under artificially constant conditions, many physiological processes in plants, such as the opening and closing of stomata and the production of photosynthetic enzymes, continue to oscillate with a frequency of about 24 hours. For example, many

▼ **Figure 39.18 Sleep movements of a bean plant (*Phaseolus vulgaris*).** The movements are caused by reversible changes in the turgor pressure of cells on opposing sides of the pulvini, motor organs of the leaf.

Noon   10:00 PM

legumes lower their leaves in the evening and raise them in the morning **(Figure 39.18)**. A bean plant continues these "sleep movements" even if kept in constant light or constant darkness; the leaves are not simply responding to sunrise and sunset. Such cycles, with a frequency of about 24 hours and not directly controlled by any known environmental variable, are called **circadian rhythms** (from the Latin *circa*, approximately, and *dies*, day).

Recent research supports the idea that the molecular "gears" of the circadian clock really are internal and not a daily response to some subtle but pervasive environmental cycle, such as geomagnetism or cosmic radiation. Organisms, including plants and humans, continue their rhythms even after being placed in deep mine shafts or when orbited in satellites, conditions that alter these subtle geophysical periodicities. However, daily signals from the environment can entrain (set) the circadian clock to a period of precisely 24 hours.

If an organism is kept in a constant environment, its circadian rhythms deviate from a 24-hour period (a period is the duration of one cycle). These free-running periods, as they are called, vary from about 21 to 27 hours, depending on the particular rhythmic response. The sleep movements of bean plants, for instance, have a period of 26 hours when the plants are kept in the free-running condition of constant darkness. Deviation of the free-running period from exactly 24 hours does not mean that biological clocks drift erratically. Free-running clocks are still keeping perfect time, but they are not synchronized with the outside world. To understand the mechanisms underlying circadian rhythms, we must distinguish between the clock and the rhythmic processes it controls. For example, the leaves of the bean plant in Figure 39.18 are the clock's "hands" but are not the essence of the clock itself. If bean leaves are restrained for several hours and then released, they will reestablish the position appropriate for the time of day. We can interfere with a biological rhythm, but the underlying clockwork continues to tick.

At the heart of the molecular mechanisms underlying circadian rhythms are oscillations in the transcription of certain genes. Mathematical models propose that the 24-hour period arises from negative-feedback loops involving the transcription of a few central "clock genes." Some clock genes may encode transcription factors that inhibit, after a time delay, the transcription of the gene that encodes the transcription factor itself. Such negative-feedback loops, together with a time delay, are enough to produce oscillations.

Researchers have used a novel technique to identify clock mutants of *Arabidopsis*. One prominent circadian rhythm in plants is the daily production of certain photosynthesis-related proteins. Molecular biologists traced the source of this rhythm to the promoter that initiates the transcription of the genes for these photosynthesis proteins. To identify clock mutants, scientists inserted the gene for an enzyme responsible for the bioluminescence of fireflies, called luciferase, to the promoter. When the biological clock turned on the promoter in the *Arabidopsis* genome, it also turned on the production of luciferase. The plants began to glow with a circadian periodicity. Clock mutants were then isolated by selecting specimens that glowed for a longer or shorter time than normal. The genes altered in some of these mutants affect proteins that normally bind photoreceptors. Perhaps these particular mutations disrupt a light-dependent mechanism that sets the biological clock.

>
> ⊘ **Mastering Biology**
> **Interview with Ruth Satter:**
> **Investigating sleep movements in plants**

The Effect of Light on the Biological Clock

As previously noted, the free-running period of the circadian rhythm of bean leaf movements is 26 hours. Consider a bean plant placed at dawn in a dark cabinet for 72 hours: Its leaves would not rise again until 2 hours after natural dawn on the second day, 4 hours after natural dawn on the third day, and so on. Shut off from environmental cues, the plant becomes desynchronized. Desynchronization happens to humans when we fly across several time zones; when we reach our destination, the clocks on the wall are not synchronized with our internal clocks. Most organisms are probably prone to jet lag.

The factor that entrains the biological clock to precisely 24 hours every day is light. Both phytochromes and blue-light photoreceptors can entrain circadian rhythms in plants, but our understanding of how phytochromes do this is more complete. The mechanism involves turning cellular responses on and off by means of the $P_r \leftrightarrow P_{fr}$ switch.

Consider again the photoreversible system in Figure 39.17. In darkness, the phytochrome ratio shifts gradually in favor of the P_r form, partly as a result of turnover in the overall phytochrome pool. The pigment is synthesized in the P_r form, and enzymes destroy more P_{fr} than P_r. In some plant species, P_{fr} present at sundown slowly converts to P_r. In darkness, there is no means for the P_r to be reconverted to P_{fr}, but upon illumination, the P_{fr} level suddenly increases again as P_r is rapidly converted. This increase in P_{fr} each day at dawn resets the biological clock: Bean leaves reach their most extreme night position 16 hours after dawn.

In nature, interactions between phytochrome and the biological clock enable plants to measure the passage of night and day. The relative lengths of night and day, however, change over the course of the year (except at the equator). Plants use this change to adjust activities in timing with the seasons.

Photoperiodism and Responses to Seasons

Imagine the consequences if a plant produced flowers when pollinators were not present or if a deciduous tree produced leaves in the middle of winter. Seasonal events are of critical importance in the life cycles of most plants. Seed germination, flowering, and the onset and breaking of bud dormancy are all stages that usually occur at specific times of the year. The environmental cue that plants use to detect the time of year is the change in day length (*photoperiod*). A physiological response to specific night or day lengths, such as flowering, is called **photoperiodism**.

Photoperiodism and Control of Flowering

An early clue to how plants detect seasons came from a mutant variety of tobacco, Maryland Mammoth, that grew tall but failed to flower during summer. It finally bloomed in a greenhouse in December. After trying to induce earlier flowering by varying temperature, moisture, and mineral nutrition, researchers learned that the shortening days of winter stimulated this variety to flower. Experiments revealed that flowering occurred only if the photoperiod was 14 hours or shorter. This variety did not flower during summer because at Maryland's latitude the photoperiods were too long.

The researchers called Maryland Mammoth a **short-day plant** because it apparently required a light period *shorter* than a critical length to flower. Chrysanthemums, poinsettias, and some soybean varieties are also short-day plants, which generally flower in late summer, fall, or winter. Another group of plants flower only when the light period is *longer* than a certain number of hours. These **long-day plants** generally flower in late spring or early summer. Spinach, for example, flowers when days are 14 hours or longer. Radishes, lettuce, irises, and many cereal varieties are also long-day plants. **Day-neutral plants**, such as tomatoes, rice, and dandelions, are unaffected by photoperiod and flower when they reach a certain stage of maturity, regardless of photoperiod.

Critical Night Length   In the 1940s, researchers learned that flowering in short- and long-day plants is actually controlled by night length, not day length (photoperiod). Many of these scientists worked with cocklebur (*Xanthium strumarium*), a short-day plant that flowers only when days are 16 hours or shorter (and nights are at least 8 hours long). These researchers found that if the photoperiod is broken by a brief exposure to darkness, flowering proceeds. However, if the night length is interrupted by even a few minutes of dim light, cocklebur will not flower, and this turned out to be true for other short-day plants as well **(Figure 39.19a)**. Cocklebur is unresponsive to day length, but it requires at least 8 hours of continuous darkness to flower. Short-day plants are really long-night plants, but the older term is embedded firmly in the lexicon of plant physiology. Similarly, long-day plants are actually short-night plants. A long-day plant grown under long-night conditions that would not normally induce flowering will flower if the night length is interrupted by a few minutes of light **(Figure 39.19b)**.

Notice that long-day plants are *not* distinguished from short-day plants by an absolute night length. Instead, they are distinguished by whether the critical night length sets a *maximum* number of hours of darkness required for flowering (long-day plants) or a *minimum* number of hours of darkness required for flowering (short-day plants). In both cases, the actual number of hours in the critical night length is specific to each species of plant.

▼ **Figure 39.19 Photoperiodic control of flowering.**

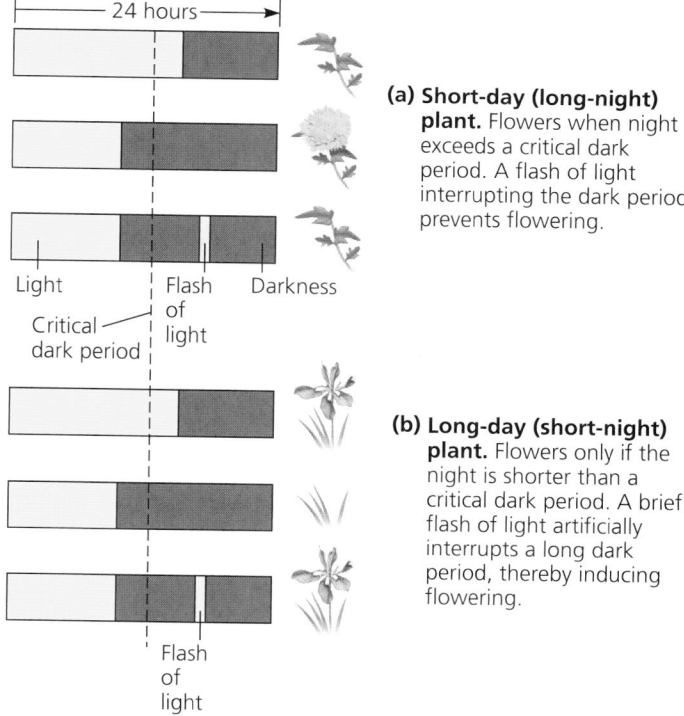

(a) Short-day (long-night) plant. Flowers when night exceeds a critical dark period. A flash of light interrupting the dark period prevents flowering.

(b) Long-day (short-night) plant. Flowers only if the night is shorter than a critical dark period. A brief flash of light artificially interrupts a long dark period, thereby inducing flowering.

▼ **Figure 39.20 Reversible effects of red and far-red light on photoperiodic response.** A flash of red (r) light shortens the dark period. A subsequent flash of far-red (fr) light cancels the red flash's effect.

24 hours

r
r fr
r fr r
r fr r fr

Critical dark period

Short-day (long-night) plant

Long-day (short-night) plant

VISUAL SKILLS *Under long-day conditions (as in the top panel) or under short-day conditions (as in the second panel), how would a single flash of far-red light during the dark period affect flowering?*

Red light is the most effective color in interrupting the night length. Action spectra and photoreversibility experiments show that phytochrome is the pigment that detects the red light **(Figure 39.20)**. For example, if a flash of red light during the night length is followed by a flash of far-red light, then the plant detects no interruption of night length. As in the case of phytochrome-mediated seed germination, red/far-red photoreversibility occurs.

Plants measure night lengths very precisely; some short-day plants will not flower if night is even 1 minute shorter than the critical length. Some plant species always flower on the same day each year. It appears that plants use their biological clock, entrained by night length with the help of phytochrome, to tell the season of the year. The floriculture (flower-growing) industry applies this knowledge to produce flowers out of season. Chrysanthemums, for instance, are short-day plants that normally bloom in fall, but their blooming can be stalled until Mother's Day in May by punctuating each long night with a flash of light, thus turning one long night into two short nights.

Some plants bloom after a single exposure to the photoperiod required for flowering. Other species need several successive days of the appropriate photoperiod. Still others respond to a photoperiod only if they have been previously exposed to some other environmental stimulus, such as a period of cold. Winter wheat, for example, will not flower unless it has been exposed to several weeks of temperatures below 10°C. The use of pretreatment with cold to induce flowering is called

vernalization (from the Latin for "spring"). Several weeks after winter wheat is vernalized, a long photoperiod (short night) induces flowering.

A Flowering Hormone?

Although flowers form from apical or axillary bud meristems, it is leaves that detect changes in photoperiod and produce signaling molecules that cue buds to develop as flowers. In many short-day and long-day plants, exposing just one leaf to the appropriate photoperiod is enough to induce flowering. Indeed, as long as one leaf is left on the plant, photoperiod is detected and floral buds are induced. If all leaves are removed, the plant is insensitive to photoperiod.

Classic experiments revealed that the floral stimulus could move across a graft from an induced plant to a noninduced plant and trigger flowering in the latter. Moreover, the flowering stimulus appears to be the same for short-day and long-day plants, despite the different photoperiodic conditions required for leaves to send this signal **(Figure 39.21)**. The hypothetical signaling molecule for flowering, called **florigen**, remained unidentified for over 70 years as scientists focused on small hormone-like molecules. However, large macromolecules, such as mRNA and proteins, can move by the symplastic route via plasmodesmata and regulate plant development. It now appears that florigen is a protein. A gene called *FLOWERING LOCUS T* (*FT*) is activated in leaf cells during conditions favoring flowering, and the FT protein travels through the symplasm to the shoot apical meristem, initiating the transition of a bud's meristem from a vegetative to a flowering state.

▼ **Figure 39.21 Experimental evidence for a flowering hormone.** If grown individually under short-day conditions, a short-day plant will flower and a long-day plant will not. However, both will flower if grafted together and exposed to short days. This result indicates that a flower-inducing substance (florigen) is transmitted across grafts and induces flowering in both short-day and long-day plants.

24 hours   24 hours   24 hours

Graft

Short-day plant

Long-day plant grafted to short-day plant

Long-day plant

WHAT IF? *If flowering were inhibited in both parts of the grafted plants, what would you conclude?*

CONCEPT 39.4

Plants respond to a wide variety of stimuli other than light

Although plants are immobile, some mechanisms have evolved by natural selection that enable them to adjust to a wide range of environmental circumstances by developmental or physiological means. Light is so important in the life of a plant that we devoted the entire previous section to the topic of a plant's reception of and response to this particular environmental factor. In this section, we will examine responses to some of the other environmental stimuli that a plant commonly encounters.

Gravity

Because plants are photoautotrophs, it is not surprising that mechanisms for growing toward light have evolved. But what environmental cue does the shoot of a young seedling use to grow upward when it is completely underground and there is no light for it to detect? Similarly, what environmental factor prompts the young root to grow downward? The answer to both questions is gravity.

Place a plant on its side, and it adjusts its growth so that the shoot bends upward and the root curves downward. In their responses to gravity, or **gravitropism**, roots display positive gravitropism **(Figure 39.22a)** and shoots exhibit negative gravitropism. Gravitropism occurs as soon as a seed germinates, ensuring that the root grows into the soil and the shoot grows toward sunlight, regardless of how the seed is oriented when it lands.

Plants may detect gravity by the settling of **statoliths**, dense cytoplasmic components that settle under the influence of gravity to the lower portions of the cell. The statoliths of vascular plants are specialized plastids containing dense starch grains **(Figure 39.22b)**. In roots, statoliths are located in certain cells of the root cap. According to one hypothesis, the aggregation of statoliths at the low points of these cells triggers a redistribution of calcium, which causes lateral transport of auxin within the root. The calcium and auxin accumulate on the lower side of the root's zone of elongation. At high concentration, auxin

▼ **Figure 39.22** Positive gravitropism in roots: the statolith hypothesis.

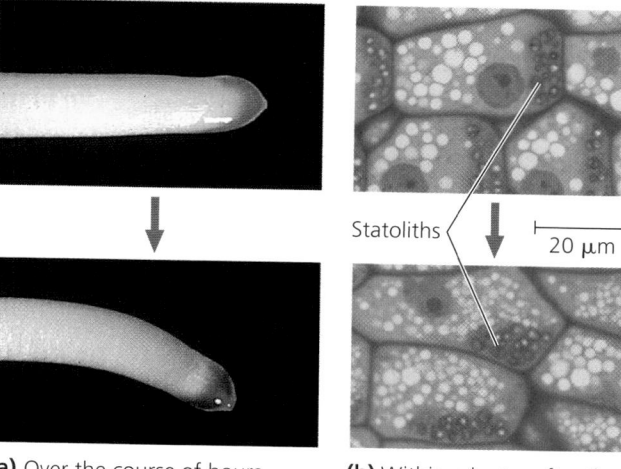

Statoliths

20 μm

(a) Over the course of hours, a horizontally oriented primary root of maize bends gravitropically until its growing tip becomes vertically oriented (LMs).

(b) Within minutes after the root is placed horizontally, plastids called statoliths begin settling to the lowest sides of root cap cells. This settling may be the gravity-sensing mechanism that leads to redistribution of auxin and differing rates of elongation by cells on opposite sides of the root (LMs).

➤ **Mastering Biology**
Video: Gravitropism

inhibits cell elongation, an effect that slows growth on the root's lower side. The more rapid elongation of cells on the upper side causes the root to grow straight downward.

Falling statoliths, however, may not be necessary for gravitropism. For example, there are mutants of *Arabidopsis* and tobacco that lack statoliths but are still capable of gravitropism, though the response is slower than in wild-type plants. It could be that the entire cell helps the root sense gravity by mechanically pulling on proteins that tether the protoplast to the cell wall, stretching the proteins on the "up" side and compressing the proteins on the "down" side of the root cells. Dense organelles, in addition to starch granules, may also contribute by distorting the cytoskeleton as they are pulled by gravity. Statoliths, because of their density, may enhance gravitational sensing by a mechanism that simply works more slowly in their absence.

Mechanical Stimuli

Trees in windy environments usually have shorter, stockier trunks than a tree of the same species growing in more sheltered locations. This stunted shape enables the plant to hold its ground against strong gusts of wind. The term **thigmomorphogenesis** (from the Greek *thigma*, touch) refers to the changes in form that result from mechanical perturbation. Plants are very sensitive to mechanical stress: Even the act of measuring the length of a leaf with a ruler alters its subsequent growth. Rubbing the stems of a young

▼ **Figure 39.23 Thigmomorphogenesis in *Arabidopsis*.** The shorter plant on the left was rubbed twice a day. The untouched plant (right) grew much taller.

plant a couple of times daily results in plants that are shorter than controls **(Figure 39.23)**.

Some plant species have become, over the course of their evolution, "touch specialists." Acute responsiveness to mechanical stimuli is an integral part of these plants' "life strategies." Most vines and other climbing plants have tendrils that coil rapidly around supports (see Figure 35.7). These grasping organs usually grow straight until they touch something; the contact stimulates a coiling response caused by differential growth of cells on opposite sides of the tendril. This directional growth in response to touch is called **thigmotropism**, and it allows the vine to take advantage of whatever mechanical supports it comes across as it climbs upward toward a forest canopy.

Other examples of touch specialists are plants that undergo rapid leaf movements in response to mechanical stimulation. For example, when the compound leaf of the sensitive plant *Mimosa pudica* is touched, it collapses and its leaflets fold together **(Figure 39.24)**. This response, which takes only a second or two, results from a rapid loss of turgor in cells within pulvini, specialized motor organs located at the joints of the leaf. The motor cells suddenly become flaccid after stimulation because they lose potassium ions, causing water to leave the cells by osmosis. It takes about 10 minutes for the cells to regain their turgor and restore the "unstimulated" form of the leaf. The function of the sensitive plant's behavior invites speculation. Perhaps the plant appears less leafy and appetizing to herbivores by folding its leaves and reducing its surface area when jostled.

A remarkable feature of rapid leaf movements is the mode of transmission of the stimulus through the plant. If one leaflet on a sensitive plant is touched, first that leaflet responds, then the adjacent leaflet responds, and so on, until all the leaflet pairs have folded together. From the point of stimulation, the signal that produces this response travels at a speed of about 1 cm/sec. An electrical impulse traveling at the same rate can be detected when electrodes are attached to the leaf. These impulses, called **action potentials**, resemble nerve impulses in animals, though the action potentials of plants are thousands of times slower. Action potentials have been discovered in many species of algae and plants and may be used as a form of internal communication. For example, in the Venus flytrap (*Dionaea muscipula*), action potentials are transmitted from sensory hairs in the trap to the cells that respond by closing the trap (see Figure 37.15). In the case of *Mimosa pudica*, a more violent stimulus, such as burning a leaf, causes *all* the leaves and leaflets to droop. This whole-plant response involves the spread of signaling molecules from the injured area to other parts of the shoot.

Environmental Stresses

Environmental stresses, such as flooding, drought, or extreme temperatures, can have devastating effects on plant survival, growth, and reproduction. In natural ecosystems, plants that cannot tolerate an environmental stress either die or are outcompeted by other plants. Thus, environmental stresses are an important factor in determining the geographic ranges of plants. In the last section of this chapter, we'll examine the defensive responses of plants to common **biotic** (living) stresses, such as herbivores and pathogens. Here we'll consider some of the more common **abiotic** (nonliving) stresses that plants encounter. Since these abiotic factors are major determinants of crop yields, there is currently much interest in trying to project how global climate change will impact crop production (see the **Problem-Solving Exercise**).

▼ **Figure 39.24 Rapid turgor movements by the sensitive plant (*Mimosa pudica*).**

(a) Unstimulated state (leaflets spread apart) **(b) Stimulated state (leaflets folded)**

→ **Mastering Biology** Video: *Mimosa* Leaves

How will climate change affect crop productivity?

Plant growth is significantly limited by air temperature, water availability, and solar radiation. A useful parameter for estimating crop productivity is the number of days per year when these three climate variables are suitable for plant growth. Camilo Mora (University of Hawaii at Manoa) and his colleagues analyzed global climate models to project the effect of climate change on suitable days for plant growth by the year 2100.

In this exercise, you will examine projected effects of climate change on crop productivity and identify the resulting human impacts.

Your Approach Analyze the map and table. Then answer the questions below.

Your Data The researchers projected the annual changes in suitable days for plant growth for all three climate variables: temperature, water availability, and solar radiation. They did so by subtracting recent averages (1996–2005) from projected future averages (2091–2100). The map shows the projected changes if no measures are taken to reduce climate change. The numbers identify locations of the 15 most populous nations. The table identifies their economies as either mainly industrial (🏭) or agricultural (🌱) and their annual per capita income category.

Nation	Map location	Estimated population in 2014 (millions)	Type of economy	Income category*
China	1	1,350	🏭	$$$
India	2	1,221	🌱	$$
United States	3	317	🏭	$$$$
Indonesia	4	251	🌱	$
Brazil	5	201	🌱	$$$
Pakistan	6	193	🌱	$$
Nigeria	7	175	🌱	$$
Bangladesh	8	164	🌱	$
Russia	9	143	🏭	$$$$
Japan	10	127	🏭	$$$$
Mexico	11	116	🌱	$$$
Philippines	12	106	🌱	$$
Ethiopia	13	94	🌱	$
Vietnam	14	92	🌱	$
Egypt	15	85	🌱	$$

Data from The World Bank
*Per capita income, based on World Bank categories: $ = low: < $1,035; $$ = lower middle: $1,036-$4,085; $$$ = upper middle: $4,086-$12,615; $$$$ = high: > $12,615.

300 200 100 0 −100 −200 −300
Annual change in suitable days for plant growth for all three climate variables

Map data from Camilo Mora, et al. Days for Plant Growth Disappear under Projected Climate Change: Potential Human and Biotic Vulnerability. *PLoS Biol* 13(6): e1002167 (2015).

➔ **Instructors:** A version of this Problem-Solving Exercise can be assigned in **Mastering Biology**.

Your Analysis

1. Camilo Mora began the study as a result of talking with someone who claimed that climate change improves plant growth because it increases the number of days above freezing. Based on the map data, how would you respond to this claim?

2. What do the map and the table data indicate about the impact of the projected changes on humans?

Drought

On a sunny, dry day, a plant may wilt because its water loss by transpiration exceeds water absorption from the soil. Prolonged drought, of course, will kill a plant, but plants have control systems that enable them to cope with less extreme water deficits.

Many of a plant's responses to water deficit help the plant conserve water by reducing the rate of transpiration. Water deficit in a leaf causes stomata to close, thereby slowing transpiration dramatically (see Figure 36.14). Water deficit stimulates increased synthesis and release of abscisic acid in the leaves; this hormone helps keep stomata closed by acting on guard cell membranes. Leaves respond to water deficit in several other ways. For example, when the leaves of grasses wilt, they roll into a tubelike shape that reduces transpiration by exposing less leaf surface to dry air and wind. Other plants, such as ocotillo (see Figure 36.15), shed their leaves in response to seasonal drought. Although these leaf responses conserve water, they also reduce photosynthesis, which is one reason why a drought diminishes crop yield. Plants can even take advantage of early warnings in the form of chemical signals from wilting neighbors and prime themselves to respond more readily and intensely to impending drought stress (see the **Scientific Skills Exercise**).

Interpreting Experimental Results from a Bar Graph

Do Drought-Stressed Plants Communicate Their Condition to Their Neighbors? Researchers wanted to learn if plants can communicate drought-induced stress to neighboring plants and, if so, whether they use aboveground or belowground signals. In this exercise, you will interpret a bar graph concerning widths of stomatal openings to investigate whether drought-induced stress can be communicated from plant to plant.

How the Experiment Was Done Eleven potted pea plants (*Pisum sativum*) were placed equidistantly in a row. The root systems of plants 6–11 were connected to those of their immediate neighbors by tubes, which allowed chemicals to move from the roots of one plant to the roots of the next plant without moving through the soil. The root systems of plants 1–6 were not connected. Osmotic shock was inflicted on plant 6 using a highly concentrated solution of mannitol, a natural sugar commonly used to mimic drought stress in vascular plants. Fifteen minutes following the osmotic shock to plant 6, researchers measured the width of stomatal openings in leaves from all the plants. A control experiment was also done in which water was added to plant 6 instead of mannitol.

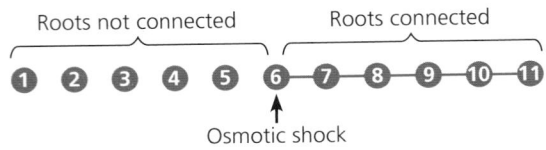

Roots not connected — Roots connected

① ② ③ ④ ⑤ ⑥—⑦—⑧—⑨—⑩—⑪

Osmotic shock

INTERPRET THE DATA

1. How do the widths of the stomatal openings of plants 6–8 and plants 9 and 10 compare with those of the other plants in the experiment? What does this indicate about the state of plants 6–8 and 9 and 10? (For information about reading graphs, see the Scientific Skills Review in Appendix D.)

2. Do the data support the idea that plants can communicate their drought-stressed condition to their neighbors? If so, do

Data from the Experiment

Data from O. Falik et al., Rumor has it…: Relay communication of stress cues in plants, *PLoS ONE* 6(11):e23625 (2011).

the data indicate that the communication is via the shoot system or the root system? Make specific reference to the data in answering both questions.

3. Why was it necessary to make sure that chemicals could not move through the soil from one plant to the next?

4. When the experiment was run for 1 hour rather than 15 minutes, the results were about the same except that the stomatal openings of plants 9–11 were comparable to those of plants 6–8. Suggest a reason why.

5. Why was water added to plant 6 instead of mannitol in the control experiment? What do the results of the control experiment indicate?

➡️ **Instructors:** A version of this Scientific Skills Exercise can be assigned in **Mastering Biology**.

Flooding

Too much water is also a problem for a plant. An overwatered houseplant may suffocate because the soil lacks the air spaces that provide oxygen for cellular respiration in the roots. Some plants are structurally adapted to very wet habitats. For example, the submerged roots of mangroves, which inhabit coastal marshes, are continuous with aerial roots exposed to oxygen (see Figure 35.4). But how do less specialized plants cope with oxygen deprivation in waterlogged soils? Oxygen deprivation stimulates the production of ethylene, which causes some cells in the root cortex to die. The destruction of these cells creates air tubes that function as "snorkels," providing oxygen to the submerged roots (**Figure 39.25**).

▼ **Figure 39.25 A developmental response of maize roots to flooding and oxygen deprivation. (a)** A cross section of a control root grown in an aerated hydroponic medium. **(b)** A root grown in a nonaerated hydroponic medium. Ethylene-stimulated programmed cell death creates the air tubes (SEMs).

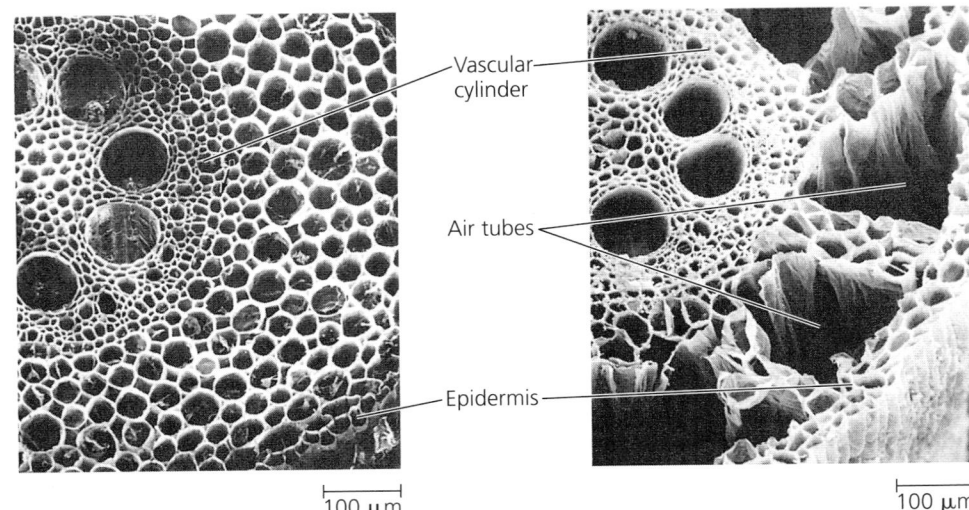

Vascular cylinder

Air tubes

Epidermis

100 μm

(a) Control root (aerated)

100 μm

(b) Experimental root (nonaerated)

Salt Stress

An excess of sodium chloride or other salts in the soil threatens plants for two reasons. First, by lowering the water potential of the soil solution, salt can cause a water deficit in plants even though the soil has plenty of water. As the water potential of the soil solution becomes more negative, the water potential gradient from soil to roots is lowered, thereby reducing water uptake (see Figure 36.12). Another problem with saline soil is that sodium and certain other ions are toxic to plants when their concentrations are too high. Many plants can respond to moderate soil salinity by producing solutes that are well tolerated at high concentrations: These mostly organic compounds keep the water potential of cells more negative than that of the soil solution without admitting toxic quantities of salt. However, most plants cannot survive salt stress for long. The exceptions are halophytes, salt-tolerant plants with adaptations such as salt glands that pump salts out across the leaf epidermis.

Heat Stress

Excessive heat may harm and even kill a plant by denaturing its enzymes. Transpiration helps cool leaves by evaporative cooling. On a warm day, for example, the temperature of a leaf may be 3–10°C below the ambient air temperature. Hot, dry weather also tends to dehydrate many plants; the closing of stomata in response to this stress conserves water but then sacrifices evaporative cooling. This dilemma is one reason why very hot, dry days take a toll on most plants.

Most plants have a backup response that enables them to survive heat stress. Above a certain temperature—about 40°C for most plants in temperate regions—plant cells begin synthesizing **heat-shock proteins**, which help protect other proteins from heat stress. This response also occurs in heat-stressed animals and microorganisms. Some heat-shock proteins function in unstressed cells as temporary scaffolds that help other proteins fold into their functional shapes. In their roles as heat-shock proteins, perhaps these molecules bind to other proteins and help prevent their denaturation.

Cold Stress

One problem plants face when the temperature of the environment falls is a change in the fluidity of cell membranes. When a membrane cools below a critical point, it loses its fluidity as the lipids become locked into crystalline structures. This alters solute transport across the membrane and also adversely affects the functions of membrane proteins. Plants respond to cold stress by altering the lipid composition of their membranes. For example, membrane lipids increase in their proportion of unsaturated fatty acids, which have shapes that help keep membranes more fluid at low temperatures. Such membrane modification requires from several hours to days, which is one reason why unseasonably cold temperatures are generally more stressful to plants than the more gradual seasonal drop in air temperature.

Freezing is another type of cold stress. At subfreezing temperatures, ice forms in the cell walls and intercellular spaces of most plants. The cytosol generally does not freeze at the cooling rates encountered in nature because it contains more solutes than the very dilute solution found in the cell wall, and solutes lower the freezing point of a solution. The reduction in liquid water in the cell wall caused by ice formation lowers the extracellular water potential, causing water to leave the cytoplasm. The resulting increase in the concentration of ions in the cytoplasm is harmful and can lead to cell death. Whether the cell survives depends largely on how well it resists dehydration. In regions with cold winters, native plants are adapted to cope with freezing stress. For example, before the onset of winter, the cells of many frost-tolerant species increase cytoplasmic levels of specific solutes, such as sugars, that are well tolerated at high concentrations and that help reduce the loss of water from the cell during extracellular freezing. The unsaturation of membrane lipids also increases, thereby maintaining proper levels of membrane fluidity.

EVOLUTION Many organisms, including certain vertebrates, fungi, bacteria, and many species of plants, have proteins that hinder ice crystals from growing, helping the organism escape freezing damage. First described in Arctic fish in the 1950s, these *antifreeze proteins* permit survival at temperatures below 0°C. They bind to small ice crystals and inhibit their growth or, in the case of plants, prevent the crystallization of ice. The five major classes of antifreeze proteins differ markedly in their amino acid sequences but have a similar three-dimensional structure, suggesting convergent evolution. Surprisingly, antifreeze proteins from winter rye are homologous to antifungal defense proteins, but they are produced in response to cold temperatures and shorter days, not fungal pathogens. Progress is being made in increasing the freezing tolerance of crop plants by engineering antifreeze protein genes into their genomes.

CONCEPT CHECK 39.4

1. Thermal images are photographs of the heat emitted by an object. Researchers have used thermal imaging of plants to isolate mutants that overproduce abscisic acid. Suggest a reason why they are warmer than wild-type plants under conditions that are normally nonstressful.

2. A greenhouse worker finds that potted chrysanthemums nearest the aisles are often shorter than those in the middle of the bench. Explain this "edge effect," a common problem in horticulture.

3. **WHAT IF?** If you removed the root cap from a root, would the root still respond to gravity? Explain.

For suggested answers, see Appendix A.

Plants respond to attacks by pathogens and herbivores

Through natural selection, plants have evolved many types of interactions with other species in their communities. Some interspecific interactions are mutually beneficial, such as the associations of plants with mycorrhizal fungi (see Figure 37.14) or with pollinators (see Figures 38.4 and 38.5). Many plant interactions with other organisms, however, do not benefit the plant. As primary producers, plants are at the base of most food webs and are subject to attack by a wide range of plant-eating (herbivorous) animals. A plant is also subject to infection by diverse viruses, bacteria, and fungi that can damage tissues or even kill the plant. Plants counter these threats with defense systems that deter animals and prevent infection or combat invading pathogens.

Defenses Against Pathogens

A plant's first line of defense against infection is the physical barrier presented by the epidermis and periderm of the plant body (see Figure 35.19). This line of defense, however, is not impenetrable. The mechanical wounding of leaves by herbivores, for example, opens up portals for invasion by pathogens. Even when plant tissues are intact, viruses, bacteria, and the spores and hyphae of fungi can still enter the plant through natural openings in the epidermis, such as stomata. Once the physical lines of defense are breached, a plant's next lines of defense are two types of immune responses: PAMP-triggered immunity and effector-triggered immunity.

PAMP-Triggered Immunity

When a pathogen succeeds in invading a host plant, the plant mounts the first of two lines of immune defense, which ultimately results in a chemical attack that isolates the pathogen and prevents its spread from the site of infection. This first line of immune defense, called *PAMP-triggered immunity*, depends on the plant's ability to recognize **pathogen-associated molecular patterns** (**PAMPs**; formerly called *elicitors*), molecular sequences that are specific to certain pathogens. For example, bacterial *flagellin*, a major protein found in bacterial flagella, is a PAMP. Many soil bacteria, including some pathogenic varieties, get splashed onto the shoots of plants by raindrops. If these bacteria penetrate the plant, a specific amino acid sequence within flagellin is perceived by a Toll-like receptor, a type of receptor also found in animals, where it plays a key role in the innate immune system (see Concept 43.1). The innate immune system is an evolutionarily old defense strategy and is the dominant immune system in plants, fungi, insects, and primitive multicellular organisms. Unlike vertebrates, plants do not have an adaptive immune system: Plants neither generate antibody or T cell responses nor possess mobile cells that detect and attack pathogens.

PAMP recognition in plants leads to a chain of signaling events that lead ultimately to the local production of broad-spectrum, antimicrobial chemicals called *phytoalexins*, which have fungicidal and bactericidal properties. The plant cell wall is also toughened, hindering further progress of the pathogen. Similar but even stronger defenses are initiated by the second plant immune response, effector-triggered immunity.

Effector-Triggered Immunity

EVOLUTION Over the course of evolution, plants and pathogens have engaged in an arms race. PAMP-triggered immunity can be overcome by the evolution of pathogens that can evade detection by the plant. These pathogens deliver **effectors**, pathogen-encoded proteins that cripple the plant's innate immune system, directly into plant cells. For example, some bacteria deliver effectors inside the plant cell that block the perception of flagellin. Thus, these effectors allow the pathogen to redirect the host's metabolism to the pathogen's advantage.

The suppression of PAMP-triggered immunity by pathogen effectors led to the evolution of *effector-triggered immunity*. Because there are thousands of effectors, this plant defense is typically made up of hundreds of disease resistance (*R*) genes. Each *R* gene codes for an R protein that can be activated by a specific effector. Signal transduction pathways then lead to an arsenal of defense responses, including a local defense called the *hypersensitive response* and a general defense called *systemic acquired resistance*. Local and systemic responses to pathogens require extensive genetic reprogramming and commitment of cellular resources. Therefore, a plant activates these defenses only after detecting a pathogen.

The Hypersensitive Response An important mechanism that restricts the spread of a pathogen is the **hypersensitive response**, the formation of a ring of local cell death around the infection site. As indicated in **Figure 39.26**, the hypersensitive response, which results from effector-triggered immunity, involves the production of enzymes and chemicals that impair the pathogen's cell wall integrity, metabolism, or reproduction. Effector-triggered immunity also stimulates the formation of lignin and the cross-linking of molecules within the plant cell wall, responses that hinder the spread of the pathogen to other parts of the plant. As shown in step 2 of the figure, the hypersensitive response results in localized lesions on a leaf. As "sick" as such a leaf appears, it will still survive, and its defensive response will help protect the rest of the plant.

Systemic Acquired Resistance

The hypersensitive response is localized and specific. However, pathogen invasions can also produce signaling molecules that "sound the alarm" of infection to the whole plant. The resulting **systemic acquired resistance** arises from the plant-wide expression of defense genes. It is nonspecific, providing protection against a diversity

▼ Figure 39.26 Defense responses against pathogens. Plants can often prevent the systemic spread of infection by instigating a hypersensitive response. This response helps isolate the pathogen by producing lesions that form "rings of death" around the sites of infection.

1 Pathogens often infect leaf cells and secrete effectors, proteins that bypass PAMP-triggered immunity.

2 In response to effectors, the hypersensitive response occurs in cells in and near the infected area: The cells produce antimicrobial molecules, seal off infected areas by modifying their cell walls, and then destroy themselves. This localized response results in lesions, regions of dead tissue that deprive the pathogen of nutrients, thereby helping to protect the rest of the infected leaf.

Pathogen
1 Effectors

2 Lesion formed as a result of hypersensitive response

Infected tobacco leaf with hypersensitive response lesions

4 In cells remote from the infected area, methylsalicylic acid is converted to salicylic acid, which induces systemic acquired resistance. This resistance consists of biochemical changes that protect the plant against a diversity of pathogens for several days.

Systemic acquired resistance

3 Before the infected cells die, they release the signaling molecule methylsalicylic acid, which is exported to the rest of the plant.

Infected area
Export of methylsalicylic acid to rest of plant

Infected area
Export of methylsalicylic acid to rest of plant

of pathogens that can last for days. A signaling molecule called methylsalicylic acid is produced around the infection site, carried by the phloem throughout the plant, and then converted to **salicylic acid** in areas remote from the sites of infection. Salicylic acid activates a signal transduction pathway that poises the defense system to respond rapidly to another infection (see step 4 of Figure 39.26).

Plant disease epidemics, such as the potato blight (see Concept 28.3) that caused the Irish potato famine of the 1840s, can lead to incalculable human misery. Other diseases, such as chestnut blight (see Concept 31.5) and sudden oak death (see Concept 54.5), can dramatically alter community structures. Plant epidemics are often the result of infected plants or timber being inadvertently transported around the world. As global commerce increases, such epidemics will become increasingly more common. To prepare for such outbreaks, plant biologists are stockpiling the seeds of wild relatives of crop plants in special storage facilities. Scientists hope that undomesticated relatives may have genes that will be able to curb the next plant epidemic.

Defenses Against Herbivores

Herbivory, animals eating plants, is a stress that plants face in any ecosystem. The mechanical damage caused by herbivores reduces the size of plants, hindering ability to acquire

resources. It can restrict growth because many species divert some energy to defend against herbivores. Also, it opens portals for infection by viruses, bacteria, and fungi. Plants prevent excessive herbivory through methods that span all levels of biological organization (**Figure 39.27**, before the Chapter Review), including physical defenses, such as thorns and trichomes (see Figure 35.9), and chemical defenses, such as distasteful or toxic compounds.

> **Mastering Biology**
> **Interview with Jeffery Dangl:**
> **Investigating plant defense responses**
>

CONCEPT CHECK 39.5

1. Why do pathogen-infected leaves often appear spotted?

2. Chewing insects mechanically damage plants and lessen the surface area of leaves for photosynthesis. In addition, these insects make plants more vulnerable to pathogen attack. Suggest a reason why.

3. Many fungal pathogens get food by causing plant cells to become leaky, releasing nutrients into the intercellular spaces. Would it benefit the fungus to kill the host plant in a way that results in all the nutrients leaking out? Explain.

4. **WHAT IF?** Suppose a scientist finds that a population of plants growing in a breezy location is more prone to herbivory by insects than a population of the same species growing in a sheltered area. Suggest an explanation.

For suggested answers, see Appendix A.

Levels of Plant Defenses Against Herbivores

Herbivory, animals eating plants, is ubiquitous in nature. Plant defenses against herbivores are examples of how biological processes can be observed at multiple levels of biological organization: molecular, cellular, tissue, organ, organism, population, and community. (See Figure 1.3.)

Opium poppy fruit

Molecular-Level Defenses

At the molecular level, plants produce chemical compounds that deter attackers. These compounds are typically terpenoids, phenolics, and alkaloids. Some terpenoids mimic insect hormones and cause insects to molt prematurely and die. Some examples of phenolics are tannins, which have an unpleasant taste and hinder the digestion of proteins. Their synthesis is often enhanced following attack. The opium poppy (*Papaver somniferum*) is the source of the narcotic alkaloids morphine, heroin, and codeine. These drugs accumulate in secretory cells called laticifers, which exude a milky-white latex (opium) when the plant is damaged.

Cellular-Level Defenses

Some plant cells are specialized for deterring herbivores. Trichomes on leaves and stems hinder the access of chewing insects. Laticifers and, more generally, the central vacuoles of plant cells may serve as storage depots for chemicals that deter herbivores. *Idioblasts* are specialized cells found in the leaves and stems of many species, including taro (*Colocasia esculenta*). Some idioblasts contain needle-shaped crystals of calcium oxalate called *raphides*. They penetrate the soft tissues of the tongue and palate, making it easier for an irritant produced by the plant, possibly a protease, to enter animal tissues and cause temporary swelling of the lips, mouth, and throat. The crystals act as a carrier for the irritant, enabling it to seep deeper into the herbivore's tissues. The irritant is destroyed by cooking.

Raphide crystals from taro plant

Tissue-Level Defenses

Some leaves deter herbivores by being especially tough to chew as a result of extensive growth of thick, hardened sclerenchyma tissue. The bright orange cells with thick cell walls seen in this cross section through the major vein of an olive leaf (*Olea europaea*) are tough sclerenchyma fibers.

Organ-Level Defenses

The shapes of plant organs may deter herbivores by causing pain or making the plant appear unappealing. Spines (modified leaves) and thorns (modified stems) provide mechanical defenses against herbivores. Bristles on the spines of some cacti have fearsome barbs that tear flesh during removal. The leaf of the snowflake plant (*Trevesia palmata*) looks as if it has been partially eaten, perhaps making it less attractive. Some plants mimic the presence of insect eggs on their leaves, dissuading insects from laying eggs there. For example, the leaf glands of some species of *Passiflora* (passion flowers) closely imitate the bright yellow eggs of *Heliconius* butterflies.

Bristles on cactus spines

Leaf of snowflake plant

Egg mimicry on leaf of passion flower plant

Organismal-Level Defenses

Mechanical damage by herbivores can greatly alter a plant's entire physiology, deterring further attack. For example, a species of wild tobacco called *Nicotiana attenuata* changes the timing of its flowering as a result of herbivory. It normally flowers at night, emitting the chemical benzyl acetone, which attracts hawk-moths as pollinators. Unfortunately for the plant, the moths often lay eggs on the leaves as they pollinate, and the larvae are herbivores. When the plants become too larvae-infested, they stop producing the chemical and instead open their flowers at dawn, when the moths are gone. They are then pollinated by hummingbirds. Research has shown that oral secretions from the munching larvae trigger the dramatic shift in the timing of flower opening.

Hummingbird pollinating wild tobacco plant

Population-Level Defenses

In some species, a coordinated behavior at the population level helps defend against herbivores. Some plants can communicate their distress from attack by releasing molecules that warn nearby plants of the same species. For example, lima bean (*Phaseolus lunatus*) plants infested with spider mites release a cocktail of chemicals that signal "news" of the attack to noninfested lima bean plants. In response, these neighbors instigate biochemical changes that make them less susceptible to attack. Another type of population-level defense is a phenomenon in some species called masting, in which a population synchronously produces a massive amount of seeds after a long interval. Regardless of environmental conditions, an internal clock signals each plant in the population that it is time to flower. Bamboo populations, for example, grow vegetatively for decades and suddenly flower en masse, set seed, and die. As much as 80,000 kg of bamboo seeds are released per hectare, much more than the local herbivores, mostly rodents, can eat. As a result, some seeds escape the herbivores' attention, germinate, and grow.

Flowering bamboo plants

Community-Level Defenses

Some plant species "recruit" predatory animals that help defend the plant against specific herbivores. Parasitoid wasps, for example, inject their eggs into caterpillars feeding on plants. The eggs hatch within the caterpillars, and the larvae eat through their organic containers from the inside out. The larvae then form cocoons on the surface of the host before emerging as adult wasps. The plant has an active role in this drama. A leaf damaged by caterpillars releases compounds that attract parasitoid wasps. The stimulus for this response is a combination of physical damage to the leaf caused by the munching caterpillar and a specific compound in the caterpillar's saliva.

Parasitoid wasp cocoons on caterpillar host

Adult wasp emerging from a cocoon

MAKE CONNECTIONS

As with plant adaptations against herbivores, other biological processes can involve multiple levels of biological organization (Figure 1.3). Discuss examples of specialized photosynthetic adaptations involving modifications at the molecular (Concept 10.5), tissue (Concept 36.4), and organismal (Concept 36.1) levels.

39 Chapter Review

→ Go to **Mastering Biology** for Assignments, the eText, the Study Area, and Dynamic Study Modules.

SUMMARY OF KEY CONCEPTS

→ To review key terms, go to the **Vocabulary Self-Quiz** in the **Mastering Biology** eText or Study Area, or go to goo.gl/zkjz9t.

CONCEPT 39.1

Signal transduction pathways link signal reception to response (pp. 843–845)

? *What are two common ways by which signal transduction pathways enhance the activity of specific enzymes?*

CONCEPT 39.2

Plants use chemicals to communicate (pp. 845–855)

- **Hormones** control plant growth and development by affecting the division, elongation, and differentiation of cells. Some also mediate the responses of plants to environmental stimuli.

Plant Hormone	Major Responses
Auxin (IAA)	Stimulates cell elongation; regulates branching and organ bending (phototropism and gravitropism)
Cytokinins	Stimulate plant cell division and differentiation; promote axillary bud growth
Gibberellins	Promote stem elongation; help seeds break dormancy and use stored reserves
Abscisic acid (ABA)	Promotes stomatal closure in response to drought; promotes seed dormancy
Ethylene	Mediates senescence, leaf abscission, fruit ripening, and obstacle avoidance by shoots (the triple response)
Brassinosteroids	Chemically similar to the sex hormones of animals; induce cell elongation and division
Jasmonates	Mediate plant defenses against insect herbivores; regulate a wide range of physiological processes
Strigolactones	Regulate apical dominance, seed germination, and mycorrhizal associations

? *Is there any truth to the old saying "One bad apple spoils the whole bunch?" Explain.*

CONCEPT 39.3

Responses to light are critical for plant success (pp. 855–861)

- **Blue-light photoreceptors** control hypocotyl elongation, stomatal opening, and phototropism.
- **Phytochromes** act like molecular "on-off" switches that regulate shade avoidance and germination of many seed types. Red light turns phytochrome "on," and far-red light turns it "off."

- Phytochrome conversion also provides information about the day length (photoperiod) and hence the time of year. **Photoperiodism** regulates the time of flowering in many species. **Short-day plants** require a night longer than a critical length to flower. **Long-day plants** need a night length shorter than a critical period to flower.
- Many daily rhythms in plant behavior are controlled by an internal circadian clock. Free-running **circadian rhythms** are approximately 24 hours long but are entrained to exactly 24 hours by dawn and dusk effects on phytochrome form.

? *Why did plant physiologists propose the existence of a mobile molecule (florigen) that triggers flowering?*

CONCEPT 39.4

Plants respond to a wide variety of stimuli other than light (pp. 861–865)

- **Gravitropism** is bending in response to gravity. Roots show positive gravitropism, and stems show negative gravitropism. **Statoliths**, starch-filled plastids, enable roots to detect gravity.
- **Thigmotropism** is a growth response to touch. Rapid leaf movements involve transmission of electrical impulses.
- Plants are sensitive to environmental stresses, including drought, flooding, high salinity, and extremes of temperature.

Environmental Stress	Major Response
Drought	ABA production, reducing water loss by closing stomata
Flooding	Formation of air tubes that help roots survive oxygen deprivation
Salt	Avoiding osmotic water loss by producing solutes tolerated at high concentrations
Heat	Synthesis of heat-shock proteins, which reduce protein denaturation at high temperatures
Cold	Adjusting membrane fluidity; avoiding osmotic water loss; producing antifreeze proteins

? *Plants that have acclimated to drought stress are often more resistant to freezing stress as well. Suggest a reason why.*

Plants respond to attacks by pathogens and herbivores (pp. 866–869)

- The **hypersensitive response** seals off an infection and destroys both pathogen and host cells in the region. **Systemic acquired resistance** is a generalized defense response in organs distant from the infection site.
- In addition to physical defenses such as thorns and trichomes, plants produce distasteful or toxic chemicals, as well as attractants that recruit animals that destroy herbivores.

? *How can insects make plants more susceptible to pathogens?*

TEST YOUR UNDERSTANDING

➡ For more multiple-choice questions, go to the **Practice Test** in the **Mastering Biology** eText or Study Area, or go to goo.gl/GruWRg.

Levels 1-2: Remembering/Understanding

1. The hormone that helps plants respond to drought is
   - (A) auxin.
   - (B) abscisic acid.
   - (C) cytokinin.
   - (D) ethylene.

2. A barley mutant lacking a gibberellic acid receptor would
   - (A) fail to make GA.
   - (B) catalyze starch more quickly.
   - (C) fail to make α-amylase.
   - (D) fail to take up water.

3. Charles and Francis Darwin discovered that
   - (A) auxin is responsible for phototropic curvature.
   - (B) red light is most effective in shoot phototropism.
   - (C) light destroys auxin.
   - (D) light is perceived by the tips of coleoptiles.

4. How may a plant respond to *severe* heat stress?
   - (A) by reorienting leaves to increase evaporative cooling
   - (B) by creating air tubes for ventilation
   - (C) by producing heat-shock proteins, which may protect the plant's proteins from denaturing
   - (D) by increasing the proportion of unsaturated fatty acids in cell membranes, reducing their fluidity

Levels 3-4: Applying/Analyzing

5. The signaling molecule for flowering might be released earlier than usual in a long-day plant exposed to flashes of
   - (A) far-red light during the night.
   - (B) red light during the night.
   - (C) red light followed by far-red light during the night.
   - (D) far-red light during the day.

6. If a long-day plant has a critical night length of 9 hours, which 24-hour cycle would prevent flowering?
   - (A) 16 hours light/8 hours dark
   - (B) 14 hours light/10 hours dark
   - (C) 4 hours light/8 hours dark/4 hours light/8 hours dark
   - (D) 8 hours light/8 hours dark/light flash/8 hours dark

7. A plant mutant that shows normal gravitropic bending but does not store starch in its plastids would require a reevaluation of the role of _____ in gravitropism.
   - (A) auxin
   - (B) calcium
   - (C) statoliths
   - (D) differential growth

8. **DRAW IT** Indicate the response to each condition by drawing a straight seedling or one with the triple response.

	Control	Ethylene added	Ethylene synthesis inhibitor
Wild-type			
Ethylene insensitive (*ein*)			
Ethylene overproducing (*eto*)			
Constitutive triple response (*ctr*)			

Levels 5-6: Evaluating/Creating

9. **EVOLUTION CONNECTION** In general, light-sensitive germination is more pronounced in small seeds compared with germination of large seeds. Suggest a reason why.

10. **SCIENTIFIC INQUIRY** A plant biologist observed a peculiar pattern when a tropical shrub was attacked by caterpillars. After a caterpillar ate a leaf, it would skip over nearby leaves and attack a leaf some distance away. Simply removing a leaf did not deter caterpillars from eating nearby leaves. The biologist suspected that an insect-damaged leaf sent out a chemical that signaled nearby leaves. How could the researcher test this hypothesis?

11. **SCIENCE, TECHNOLOGY, AND SOCIETY** Describe how our knowledge about the control systems of plants is being applied to agriculture or horticulture.

12. **WRITE ABOUT A THEME: INTERACTIONS** In a short essay (100–150 words), summarize phytochrome's role in altering shoot growth for the enhancement of light capture.

13. **SYNTHESIZE YOUR KNOWLEDGE**

 This mule deer is grazing on the shoot tips of a shrub. Describe how this event will alter the physiology, biochemistry, structure, and health of the plant and identify which hormones and other chemicals are involved in making these changes.

For selected answers, see Appendix A.

Meet the "phage wrangler," Steffanie Strathdee, Professor of Medicine in the Division of Infectious Diseases and Global Public Health at the University of California, San Diego. Born in Canada, she completed her M.Sc. and Ph.D. degrees in epidemiology at the University of Toronto. Since joining the faculty at UC San Diego in 2004, she has focused on HIV research and prevention in the underserved populations of Tijuana, just across the Mexican border. In 2017, extreme personal circumstances prompted a radical shift in her research, as detailed in *The Perfect Predator: A Scientist's Race to Save Her Husband from a Deadly Superbug*, a book co-authored by Dr. Strathdee and her husband. In 2018, Dr. Strathdee became co-director of the newly established Center for Innovative Phage Applications and Therapeutics at UC San Diego.

AN INTERVIEW WITH

Steffanie Strathdee

Tell us about your start in science.

I always had a natural curiosity as a young girl. I did not have an innate ability at science or math, even though I was interested in them. I struggled in math. Even in university, I got D's in calculus. I still don't like math, but I've learned to surround myself with people that do really well in math. I think it's important for students to realize that just because they're not good at everything doesn't mean that they can't pursue their dreams.

How did you become interested in epidemiology?

After both my course instructor and my Masters and Ph.D. advisor passed away from AIDS, I decided I wanted to focus on ending the HIV epidemic. I had started out in the laboratory and realized I was really lousy at tissue culture experiments. So I turned my attention to public health,

▼ Dr. Steffanie Strathdee's husband, Dr. Tom Patterson, holding an electron micrograph of his superbug (left), and Dr. Strathdee, with an artist's rendition of the bacteriophage that defeated it (right).

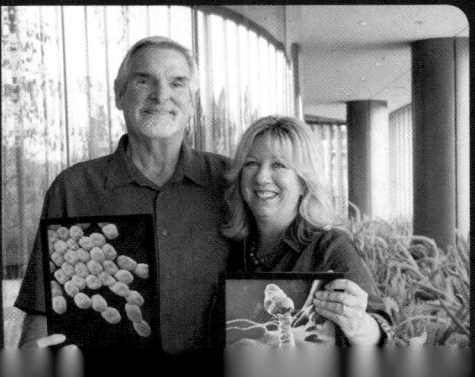

of which epidemiology is one part. Epidemiology involves studying risk factors and patterns, not just at an individual level, like what people eat and how they behave, but also the social, political, and economic forces that drive those behaviors.

What changed your research focus?

My husband became ill with a superbug infection—an infection caused by a bacterium that is resistant to many antibiotics. As an infectious disease epidemiologist, you would think that I would have understood the global crisis we are facing in multidrug-resistant organisms. But it wasn't until it hit me on a personal level that I fully understood the threat.

How did the story of your husband's illness unfold?

My husband and I were in Egypt, and he became very ill. The doctor gave him IV antibiotics and said, "He'll be fine," but he wasn't. It turned out that a gallstone had blocked his bile duct, causing an abscess (cavity) to form, and a superbug moved into that abscess and multiplied. Tom's particular bacterium had 51 different antibiotic-resistance genes. It was winning its battle against the immune system, none of the standard antibiotic treatments were working, and my husband was dying. I did a literature search and found a hundred-year-old, forgotten cure based on bacteriophages. These viruses, called phages for short, attack bacteria, but not human cells. Each phage is specific for a particular bacterium. Phages inject their DNA into bacteria, turning them into phage factories and killing the bacteria in the process. I emailed Tom's doctor, who is a colleague of Tom's and mine, and he thought the idea was worth a shot. So, I had to go out and find people who had phages that were active against Tom's bacterium, and that felt harder than looking for a needle in a haystack. Luckily, labs at the Naval Medical Research Center in Maryland and at Texas A&M University dedicated themselves to the search and found phages that matched. The phages were grown, purified, and injected into Tom's body, a billion phages per dose. Three days later, Tom woke up from his coma.

Tell us a bit more about your collaborators.

A Ph.D. student at Texas A&M worked around the clock and found the phages that made up the first infusion. This student was at a really low point in her studies, thinking, "I don't know what I'm going to do with my career, I don't know if what I do really matters." And then she ended up helping save the life of a total stranger, and it's jump-started a whole new field—phage therapy.

> *"This student . . . ended up helping save the life of a total stranger, and it's jump-started a whole new field."*

What advice do you have for students starting to study biology?

Students should look for mentors with whom they share the same values. For me, that was what really turned my D in calculus and some of the troubles that I had in science into success. I found mentors who were encouraging and helped me identify my strengths. Knowing what your weaknesses and your strengths are is, I think, one of the big components to making a successful career.

40 Basic Principles of Animal Form and Function

KEY CONCEPTS

40.1 Animal form and function are correlated at all levels of organization p. 874

40.2 Feedback control maintains the internal environment in many animals p. 881

40.3 Homeostatic processes for thermoregulation involve form, function, and behavior p. 884

40.4 Energy requirements are related to animal size, activity, and environment p. 889

Study Tip

Draw a diagram: When you encounter an example in the chapter of how an animal maintains a steady internal state, draw a simple circuit diagram (see example—illustrations are optional!). Label the variable being controlled, a perturbation that affects the variable, the response, and its effect in restoring the normal state.

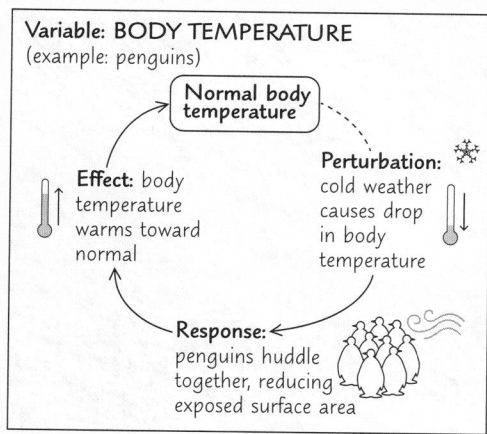

Variable: BODY TEMPERATURE (example: penguins)

Normal body temperature

Effect: body temperature warms toward normal

Perturbation: cold weather causes drop in body temperature

Response: penguins huddle together, reducing exposed surface area

→ Go to Mastering Biology

For Students (in eText and Study Area)
- Get Ready for Chapter 40
- Figure 40.17 Walkthrough: Thermoregulation in Humans

For Instructors to Assign (in Item Library)
- Everyday Biology: How to Keep Your Cool
- Tutorial: Thermoregulation

Figure 40.1 Emperor penguins (*Aptenodytes forsteri*) live in Antarctica, Earth's coldest and windiest continent. In summer, these birds catch fish by diving down 500 meters in water only 2°C above freezing. In winter, the females forage and the males incubate eggs while temperatures drop to –40°C and winds gust to 200 km/hr.

How do animals regulate their internal state even in changing or harsh environments?

Adaptations in **form, function, and behavior** help maintain an animal's internal environment. Adaptations that limit variation in temperature and other internal variables are widespread and diverse. Consider, for example, three adaptations that help an Emperor penguin stay warm:

Form (anatomy): An insulating layer of fat (blubber) reduces heat loss from most of the penguin's body (blue body areas in this thermal image).

Function (physiology): Rapid cycles of muscle contraction and relaxation during shivering produce heat at a cellular level.

Behavior: By packing together in groups of up to several thousand, Emperor penguins greatly reduce their exposure to wind and cold.

Animal form and function are correlated at all levels of organization

Over the course of its life, an Emperor penguin faces the same fundamental challenges as any other animal, whether hydra, hawk, or human. All animals must obtain nutrients and oxygen, fight off infection, and survive to produce offspring. Given that all animal species share these and other basic requirements, why does their form, including **anatomy**—biological structure—vary so widely? The answer lies in natural selection and adaptation. Natural selection favors those variations in a population that increase relative fitness (see Concept 23.4). The evolutionary adaptations that enable survival vary among environments and species but frequently result in a close match of form to function, as illustrated for the Emperor penguins in Figure 40.1. Because structure and function are correlated, examining anatomy often provides clues to **physiology**—biological function.

An animal's size and shape are fundamental aspects of form that significantly affect the way the animal interacts with its environment. Although we may refer to size and shape as elements of a "body plan" or "design," this does not imply a process of conscious invention. The body plan of an animal is the result of a pattern of development programmed by the genome, itself the product of millions of years of evolution.

Evolution of Animal Size and Shape

EVOLUTION Many different body plans have arisen during the course of evolution, but these variations fall within certain bounds. Physical laws that govern strength, diffusion, movement, and heat exchange limit the range of animal forms.

As an example of how physical laws constrain evolution, let's consider how some properties of water limit the possible shapes for animals that are fast swimmers. Water is about 1,000 times denser than air and also far more viscous. Therefore, any bump on an animal's body surface that causes drag impedes a swimmer more than it would a runner or flyer. Tuna and other fast ray-finned fishes can swim at speeds up to 80 km/hr (50 miles/hour). Sharks, penguins, dolphins, and seals are also relatively fast swimmers. As illustrated by the three examples in **Figure 40.2**, these animals all have a shape that is fusiform, meaning tapered on both ends. The similar streamlined shape found in these speedy vertebrates is an example of convergent evolution (see Concept 22.3). Natural selection often results in similar adaptations when diverse organisms face the same environmental challenge, such as overcoming drag during swimming.

Physical laws also influence animal body plans with regard to maximum size. As body dimensions increase, thicker skeletons are required to maintain adequate support. This limitation affects internal skeletons, such as those of vertebrates, as well as external skeletons, such as those of insects and other arthropods. In addition, as bodies increase in size, the muscles required for locomotion must represent an ever-larger fraction of the total body mass. At some point, mobility becomes limited. By considering the fraction of body mass in leg muscles and the effective force such muscles generate, scientists can estimate maximum speed for a wide range of body plans. In the case of the 6-meter-tall dinosaur *Tyrannosaurus rex*, there is controversy, with some scientists calculating a top running speed as fast as that of an Olympic sprinter—30 km/hr (19 miles/hour), but others inferring that *T. rex* was at best a fast walker.

Exchange with the Environment

Animals must exchange nutrients, waste products, and gases with their environment, and this requirement imposes an additional limitation on body plans. Exchange occurs as substances dissolved in an aqueous solution move across the plasma membrane of each cell. A single-celled organism, such as the amoeba in **Figure 40.3a**, has a sufficient membrane surface area in contact with its environment to carry out all necessary exchange. In contrast, an animal is composed of many cells, each with its own plasma membrane across which exchange must occur. The rate of exchange is proportional to the membrane surface area involved in exchange, whereas the amount of material that must be exchanged is proportional to the total body volume.

A multicellular organization therefore works only if every cell has access to a suitable aqueous environment, either inside or outside the animal's body.

Many animals with a simple internal organization have body plans that enable direct exchange between almost all their cells and the external environment. For example, a pond-dwelling hydra has a saclike body plan and a body wall only two cell layers thick **(Figure 40.3b)**. Because its gastrovascular cavity opens to the external environment, both the outer and inner layers of cells are constantly bathed by pond water. Another common body plan that maximizes exposure to the surrounding medium is a flat shape. Consider, for instance, a parasitic tapeworm, which can reach several meters in length (see Figure 33.11). A thin, flat shape places most cells of the worm in direct contact with its particular environment—the nutrient-rich intestinal fluid of a vertebrate host.

▼ Figure 40.2 Convergent evolution in fast swimmers.

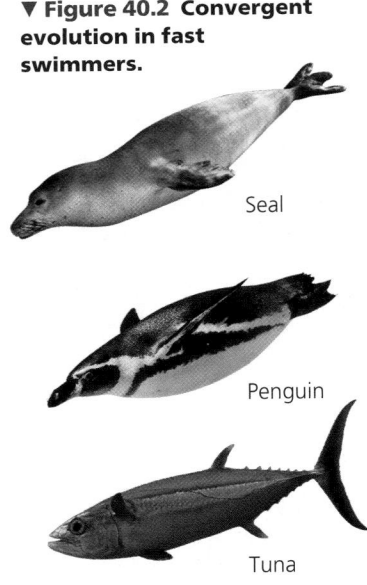

Seal

Penguin

Tuna

▼ Figure 40.3 Direct exchange with the environment.

Mouth

Gastrovascular cavity

Exchange

Exchange

Exchange

0.1 mm

1 mm

(a) An amoeba, a single-celled organism

(b) A hydra, an animal with two layers of cells

Our bodies and those of most other animals are composed of compact masses of cells, with an internal organization much more complex than that of a hydra or a tapeworm. For such a body plan, increasing the number of cells decreases the ratio of outer surface area to total volume. As an extreme comparison, the ratio of outer surface area to volume for a whale is hundreds of thousands of times smaller than that for a water flea. Nevertheless, every cell in the whale has access to oxygen, nutrients, and other resources. How is this accomplished?

In whales and most other animals, the evolutionary adaptations that enable sufficient exchange with the environment are specialized surfaces that are extensively branched or folded **(Figure 40.4)**. In almost all cases, these exchange surfaces lie within the body, an arrangement that protects their delicate tissues from abrasion or dehydration and allows for streamlined body contours. The branching or folding greatly increases surface area (see Figure 33.8). In humans, for example, the exchange surfaces for digestion, respiration, and circulation each have an area more than 25 times larger than that of the skin.

Internal body fluids link exchange surfaces to body cells. The spaces between cells are filled with fluid, known in many animals as **interstitial fluid** (from the Latin for "stand between"). Complex body plans also include a circulatory fluid, such as blood. Exchange between the interstitial fluid and the circulatory fluid enables cells throughout the body to obtain nutrients and get rid of wastes (see Figure 40.4).

Complex body plans offer numerous benefits. For example, an external skeleton can protect against predators, and sensory organs can provide detailed information on the animal's surroundings. Internal digestive organs can break down

▶ Figure 40.4 Internal exchange surfaces of complex animals.
Most animals have surfaces that are specialized for exchanging chemicals with the surroundings. These exchange surfaces are usually internal but are connected to the environment via openings on the body surface (the mouth, for example). The exchange surfaces are finely branched or folded, giving them a very large area. The digestive, respiratory, and excretory systems all have such exchange surfaces. Chemicals exchanged across these surfaces are transported throughout the body via the circulatory system.

VISUAL SKILLS *Using this diagram, explain how exchange carried out by animals can be described as both internal and external.*

EXTERNAL ENVIRONMENT

Food

CO_2 O_2

Mouth

ANIMAL BODY

Blood

Respiratory system

Heart

Cells

Nutrients

Circulatory system

Interstitial fluid

Digestive system

Excretory system

Anus

Unabsorbed matter (feces)

Metabolic waste products (nitrogenous waste)

250 μm

A microscopic view of the lung reveals that it is much more sponge-like than balloon-like. This construction provides an expansive wet surface for gas exchange with the environment (SEM).

100 μm

The lining of the small intestine has finger-like projections that expand the surface area for nutrient absorption (SEM).

50 μm

Within the kidney, blood is filtered across the surface of long, narrow blood vessels packed into ball-shaped structures (SEM).

food gradually, controlling the release of stored energy. In addition, specialized filtration systems can adjust the composition of the internal fluid that bathes the animal's body cells. In this way, an animal can maintain a relatively stable internal environment despite the fact that it is living in a changeable external environment. A complex body plan is especially advantageous for animals living on land, where the external environment may be highly variable.

Hierarchical Organization of Body Plans

Cells form a working animal body through their *emergent properties*, which arise from successive levels of structural and functional organization (see Concept 1.1). Cells are organized into **tissues**, groups of cells with a similar appearance and a common function. Different types of tissues are further organized into functional units called **organs**. (The simplest animals, such as sponges, lack organs or even true tissues.) Groups of organs that work together, providing an additional level of organization and coordination, make up an **organ system (Table 40.1)**. Thus, for example, the skin is an organ of the integumentary system, which protects against infection and helps regulate body temperature.

Many organs have more than one physiological role. If the roles are distinct enough, we consider the organ to belong to more than one organ system. The pancreas, for instance, produces enzymes critical to the function of the digestive system but also regulates the level of sugar in the blood as a vital part of the endocrine system.

Just as viewing the body's organization from the "bottom up" (from cells to organ systems) reveals emergent properties, a "top-down" view of the hierarchy reveals the multilayered basis of specialization. Organ systems include specialized organs made up of specialized tissues and cells. Consider the human digestive system: Each organ has specific roles. In the case of the stomach, one role is to initiate protein breakdown. This process requires a churning motion powered by stomach muscles, as well as digestive juices secreted by the stomach lining. Producing digestive juices, in turn, requires highly specialized cell types: One cell type secretes a protein-digesting enzyme, a second generates concentrated hydrochloric acid, and a third produces mucus, which protects the stomach lining.

The specialized and complex organ systems of animals are built from a limited set of cell and tissue types. For example, lungs and blood vessels have different functions but are lined by tissues that are of the same basic type and therefore share many properties.

There are four main types of animal tissues: epithelial, connective, muscle, and nervous. **Figure 40.5** explores the structure and function of each type. In later chapters, we'll discuss how these tissue types contribute to the functions of particular organ systems.

→ **Mastering Biology** Animation: Overview of Animal Tissues

Table 40.1 Organ Systems in Mammals		
Organ System	**Main Components**	**Main Functions**
Digestive	Mouth, pharynx, esophagus, stomach, intestines, liver, pancreas, anus (See Figure 41.8.)	Food processing (ingestion, digestion, absorption, elimination)
Circulatory	Heart, blood vessels, blood (See Figure 42.5.)	Internal distribution of materials
Respiratory	Lungs, trachea, other breathing tubes (See Figure 42.24.)	Gas exchange (uptake of oxygen; disposal of carbon dioxide)
Immune and lymphatic	Bone marrow, lymph nodes, thymus, spleen, lymph vessels (See Figure 43.6.)	Body defense (fighting infections and virally induced cancers)
Excretory	Kidneys, ureters, urinary bladder, urethra (See Figure 44.12.)	Disposal of metabolic wastes; regulation of osmotic balance of blood
Endocrine	Pituitary, thyroid, pancreas, adrenal, and other hormone-secreting glands (See Figure 45.8.)	Coordination of body activities (such as digestion and metabolism)
Reproductive	Ovaries or testes and associated organs (See Figures 46.9 and 46.10.)	Gamete production; promotion of fertilization; support of developing embryo
Nervous	Brain, spinal cord, nerves, sensory organs (See Figure 49.6.)	Coordination of body activities; detection of stimuli and formulation of responses to them
Integumentary	Skin and its derivatives (such as hair, claws, sweat glands) (See Figure 50.5.)	Protection against mechanical injury, infection, dehydration; thermoregulation
Skeletal	Skeleton (bones, tendons, ligaments, cartilage) (See Figure 50.37.)	Body support, protection of internal organs, movement
Muscular	Skeletal muscles (See Figure 50.26.)	Locomotion and other movement

Exploring Structure and Function in Animal Tissues

Epithelial Tissue

Occurring as sheets of cells, **epithelial tissues**, or **epithelia** (singular, *epithelium*), cover the outside of the body and line organs and cavities within the body. Because epithelial cells are closely packed, often with tight junctions, they function as a barrier against mechanical injury, pathogens, and fluid loss. Epithelia also form active interfaces with the environment. For example, the epithelium that lines the nasal passages is crucial for olfaction, the sense of smell. Note how different cell shapes and arrangements correlate with distinct functions.

 Mastering Biology
Animation: Epithelial Tissue

Stratified squamous epithelium

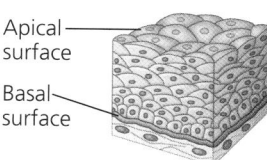
Apical surface
Basal surface

A stratified squamous epithelium is multilayered and regenerates rapidly. New cells formed by division near the basal surface push outward, replacing cells that are sloughed off. This epithelium is commonly found on surfaces subject to abrasion, such as the outer skin and the linings of the mouth, anus, and vagina.

Cuboidal epithelium

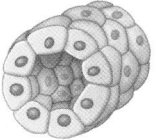

A cuboidal epithelium, with dice-shaped cells specialized for secretion, makes up the epithelium of kidney tubules and many glands, including the thyroid gland and salivary glands.

Simple columnar epithelium

The large, brick-shaped cells of simple columnar epithelia are often found where secretion or active absorption is important. For example, a simple columnar epithelium lines the intestines, secreting digestive juices and absorbing nutrients.

Simple squamous epithelium

The single layer of platelike cells that form a simple squamous epithelium functions in the exchange of material by diffusion. This type of epithelium, which is thin and leaky, lines blood vessels and the air sacs of the lungs, where diffusion of nutrients and gases is essential.

Pseudostratified columnar epithelium

A pseudostratified epithelium consists of a single layer of cells varying in height and the position of their nuclei. In many vertebrates, a pseudostratified epithelium of ciliated cells forms a mucous membrane that lines portions of the respiratory tract. The beating cilia sweep the film of mucus along the surface.

Lumen
Apical surface
Basal surface
10 μm

Polarity of epithelia

All epithelia are polarized, meaning that they have two different sides. The *apical* surface faces the lumen (cavity) or outside of the organ and is therefore exposed to fluid or air. Specialized projections often cover this surface. For example, the apical surface of the epithelium lining the small intestine is covered with microvilli, projections that increase the surface area available for absorbing nutrients. Opposite the apical surface of each epithelium is the *basal* surface.

Connective Tissue

Connective tissue, consisting of a sparse population of cells scattered through an extracellular matrix, holds many tissues and organs together and in place. The matrix generally consists of a web of fibers embedded in a liquid, jellylike, or solid foundation. Within the matrix are numerous cells called **fibroblasts**, which secrete fiber proteins, and **macrophages**, which engulf foreign particles and any cell debris by phagocytosis.

Connective tissue fibers are of three kinds: *Collagenous fibers* provide strength and flexibility,

reticular fibers join connective tissue to adjacent tissues, and *elastic fibers* make tissues elastic. If you pinch a fold of tissue on the back of your hand, the collagenous and reticular fibers prevent the skin from being pulled far from the bone, whereas the elastic fibers restore the skin to its original shape when you release your grip. Different mixtures of fibers and foundation form the major types of connective tissue shown below.

→ **Mastering Biology Animation: Connective Tissue**

Loose connective tissue

The most widespread connective tissue in the vertebrate body is *loose connective tissue*, which binds epithelia to underlying tissues and holds organs in place. Loose connective tissue gets its name from the loose weave of its fibers, which include all three types. It is found in the skin and throughout the body.

Collagenous fiber

120 μm

Elastic fiber

Fibrous connective tissue

Fibrous connective tissue is dense with collagenous fibers. It is found in **tendons**, which attach muscles to bones, and in **ligaments**, which connect bones at joints.

30 μm

Nuclei

Bone

The skeleton of most vertebrates is made of **bone**, a mineralized connective tissue. Bone-forming cells called *osteoblasts* deposit a matrix of collagen. Calcium, magnesium, and phosphate ions combine into a hard mineral within the matrix. The microscopic structure of hard mammalian bone consists of repeating units called *osteons*. Each osteon has concentric layers of the mineralized matrix, which are deposited around a central canal containing blood vessels and nerves.

Central canal

700 μm

Osteon

Adipose tissue

Adipose tissue is a specialized loose connective tissue that stores fat in adipose cells distributed throughout its matrix. Adipose tissue pads and insulates the body and stores fuel as fat molecules. Each adipose cell contains a large fat droplet that swells when fat is stored and shrinks when the body uses that fat as fuel.

Lipid droplets

150 μm

Blood

Blood has a liquid extracellular matrix called plasma, which consists of water, salts, and dissolved proteins. Suspended in plasma are erythrocytes (red blood cells), leukocytes (white blood cells), and cell fragments called platelets. Red cells carry oxygen, white cells function in defense, and platelets aid in blood clotting.

Plasma

White blood cells

50 μm

Red blood cells

Cartilage

Cartilage contains collagenous fibers embedded in a rubbery protein-carbohydrate complex called chondroitin sulfate. Cells called *chondrocytes* secrete the collagen and chondroitin sulfate, which together make cartilage a strong yet flexible support material. The skeletons of many vertebrate embryos contain cartilage that is replaced by bone as the embryo matures. Cartilage remains in some locations, such as the disks that act as cushions between vertebrae.

Chondrocytes

100 μm

Chondroitin sulfate

Muscle Tissue

The tissue responsible for nearly all types of body movement is **muscle tissue**. All muscle cells consist of filaments containing the proteins actin and myosin, which together enable muscles to contract. There are three types of muscle tissue in the vertebrate body: skeletal, smooth, and cardiac.

→ **Mastering Biology** Animation: Muscle Tissue

Skeletal muscle

Attached to bones by tendons, **skeletal muscle**, or *striated muscle*, is responsible for voluntary movements. Skeletal muscle consists of bundles of long cells that are called muscle fibers. During development, skeletal muscle fibers form by the fusion of many cells, resulting in multiple nuclei in each muscle fiber. The arrangement of contractile units, or sarcomeres, along the fibers gives the cells a striped (striated) appearance. In adult mammals, building muscle increases the size but not the number of muscle fibers.

Smooth muscle

Smooth muscle, which lacks striations, is found in the walls of the digestive tract, urinary bladder, arteries, and other internal organs. The cells are spindle-shaped. Smooth muscles are responsible for involuntary body activities, such as churning of the stomach and constriction of arteries.

Cardiac muscle

Cardiac muscle forms the contractile wall of the heart. It is striated like skeletal muscle and has similar contractile properties. Unlike skeletal muscle, however, cardiac muscle has branched fibers that interconnect via intercalated disks, which relay signals from cell to cell and help synchronize heart contraction.

Nuclei

Muscle fiber

Sarcomere

100 μm

Nucleus   Muscle fibers   25 μm

Nucleus   Intercalated disk   25 μm

Nervous Tissue

Nervous tissue functions in the receipt, processing, and transmission of information. Nervous tissue contains **neurons**, or nerve cells, which transmit nerve impulses, as well as support cells called **glial cells**, or simply **glia**. In many animals, a concentration of nervous tissue forms a brain, an information-processing center.

→ **Mastering Biology** Animation: Nervous Tissue

Neurons

Neurons are the basic units of the nervous system. A neuron receives nerve impulses from other neurons via its cell body and multiple extensions called dendrites. Neurons transmit impulses to neurons, muscles, or other cells via extensions called axons, which are often bundled together into nerves.

Neuron:

Dendrites

Cell body

Axon

45 μm

(Fluorescent LM)

Glia

The various types of glia help nourish, insulate, and replenish neurons, and in some cases, modulate neuron function.

Glia

15 μm

Axons of neurons

Blood vessel

(Confocal LM)

Coordination and Control

For an animal's tissues and organ systems to function effectively, they must act in concert with one another. For example, when the wolf shown in Figure 40.5 hunts, blood flow is regulated to bring adequate nutrients and gases to its leg muscles, which in turn are activated by the brain in response to cues detected by the nose. What signals coordinate activity? How do the signals move within the body?

Animals have two major systems for coordinating and controlling responses to stimuli: the endocrine and nervous systems **(Figure 40.6)**. In the **endocrine system**, signaling molecules released into the bloodstream by endocrine cells are carried to all locations in the body. In the **nervous system**, neurons transmit signals along dedicated routes connecting specific locations in the body. In each system, the type of pathway used is the same regardless of whether the signal's ultimate target is at the other end of the body or just a few cell diameters away.

The signaling molecules that are broadcast throughout the body by the endocrine system are called **hormones**. It takes seconds for hormones to be released into the bloodstream and carried throughout the body. The effects are often long-lasting, however, because hormones can remain in the bloodstream for minutes or even hours.

Different hormones cause distinct effects, and only cells that have receptors for a particular hormone respond (Figure 40.6a). Depending on which cells have receptors for that hormone, the hormone may have an effect in just a single location or in sites throughout the body. For example, thyroid-stimulating hormone (TSH) acts solely on cells in the thyroid gland. They in turn release thyroid hormone, which acts on nearly every body tissue to increase oxygen consumption and heat production.

In the nervous system, signals called nerve impulses travel to specific target cells along communication lines consisting mainly of axons (Figure 40.6b). Transmission in the nervous system is extremely fast; nerve impulses take only a fraction of a second to reach the target and last only a fraction of a second.

Nerve impulses can act on other neurons, on muscle cells, and on cells and glands that produce secretions. Unlike the endocrine system, the nervous system conveys information by the *pathway* the signal takes. For example, a person can distinguish different musical notes because each note's frequency activates neurons in the ear that connect to slightly different locations in the brain.

Communication in the nervous system usually involves more than one type of signal. Nerve impulses travel along axons, sometimes over long distances, as changes in voltage. In contrast, passing information from one neuron to another often involves very short-range chemical signals.

▼ **Figure 40.6 Signaling in the endocrine and nervous systems.**

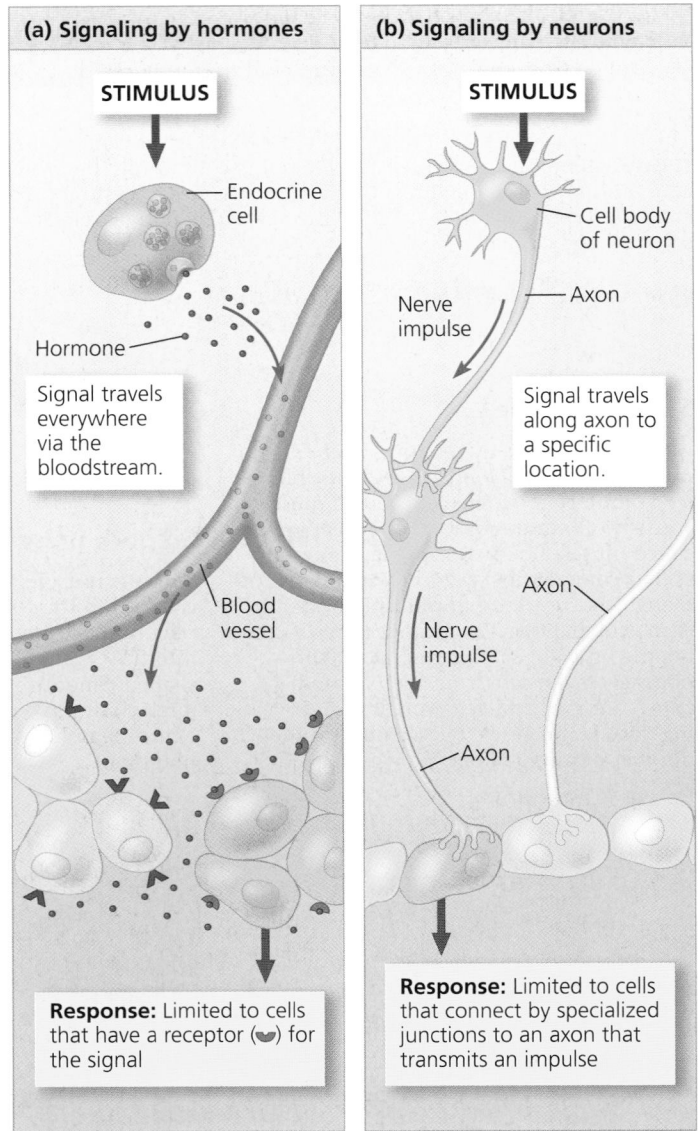

VISUAL SKILLS *After comparing the two diagrams, explain why a particular nerve impulse signal has only one physical pathway but a particular hormone molecule can have multiple physical pathways.*

Because the two major communication systems of the body differ in signal type, transmission, speed, and duration, it is not surprising that they are adapted to different functions. The endocrine system is especially well adapted for coordinating gradual changes that affect the entire body, such as growth, development, reproduction, metabolic processes, and digestion. The nervous system is well suited for directing immediate and rapid responses to the environment, such as reflexes and other rapid movements. Nevertheless, the two systems often work in close coordination. Both help maintain a stable internal environment, our next topic of discussion.

1. What properties do all types of epithelia share?

2. **VISUAL SKILLS** Consider the idealized animal in Figure 40.4. At which sites must oxygen cross a plasma membrane in traveling from the external environment to the cytoplasm of a body cell?

3. **WHAT IF?** Suppose you are standing at the edge of a cliff and suddenly slip, barely managing to keep your balance and avoid falling. As your heart races, you feel a burst of energy, due in part to a surge of blood into dilated (widened) vessels in your muscles and an upward spike in the level of glucose in your blood. Why might you expect that this "fight-or-flight" response requires both the nervous and endocrine systems?

For suggested answers, see Appendix A.

CONCEPT 40.2

Feedback control maintains the internal environment in many animals

Many organ systems play a role in managing an animal's internal environment, a task that can present a major challenge. Imagine if your body temperature soared every time you took a hot shower or slurped a steaming bowl of soup. Faced with environmental fluctuations, animals manage their internal environment by either regulating or conforming.

Regulating and Conforming

Compare the two sets of data in **Figure 40.7**. The river otter's body temperature is largely independent of that of the surrounding water, whereas the largemouth bass's body warms or cools when the water temperature changes. We can convey these two trends by labeling the otter a regulator and the bass a conformer with regard to body temperature. An animal is a **regulator** for an environmental variable if it uses internal mechanisms to control internal change in the face of external fluctuation. In contrast, an animal is a **conformer** if it allows its internal condition to change in accordance with external changes in the particular variable.

An animal may allow some internal conditions to conform to the environment but regulate others. For instance, the bass conforms to the temperature of the water in which it lives, but regulates the solute concentration in its blood and interstitial fluid. In addition, conforming need not involve changes in an internal variable: Many marine invertebrates, such as spider crabs (genus *Libinia*), let their internal solute concentration conform to the relatively stable salinity of their ocean environment.

Homeostasis

The steady body temperature of a river otter and the stable concentration of solutes in a freshwater bass are examples of **homeostasis**, which means the maintenance of internal balance. In achieving homeostasis, animals maintain a "steady state"—a relatively constant internal environment—even when the external environment changes significantly.

Many animals exhibit homeostasis for a range of physical and chemical properties. For example, humans maintain a fairly constant body temperature of about 37°C (98.6°F), a blood pH within 0.1 pH unit of 7.4, and a blood glucose concentration that is predominantly in the range of 70–110 mg of glucose per 100 mL of blood.

Mechanisms of Homeostasis

Homeostasis requires a control system. Before exploring homeostasis in animals, let's get a basic picture of how a control system works by considering a nonliving example: the regulation of room temperature. Let's assume you want to keep a room at 20°C (68°F), a comfortable temperature for normal activity. You set a control device—the thermostat—to 20°C. A thermometer in the thermostat monitors the room temperature. If the temperature falls

▼ **Figure 40.7 The relationship between body and environmental temperatures in an aquatic temperature regulator and an aquatic temperature conformer.** The river otter regulates its body temperature, keeping it stable across a wide range of environmental temperatures. The largemouth bass, meanwhile, allows its internal environment to conform to the water temperature.

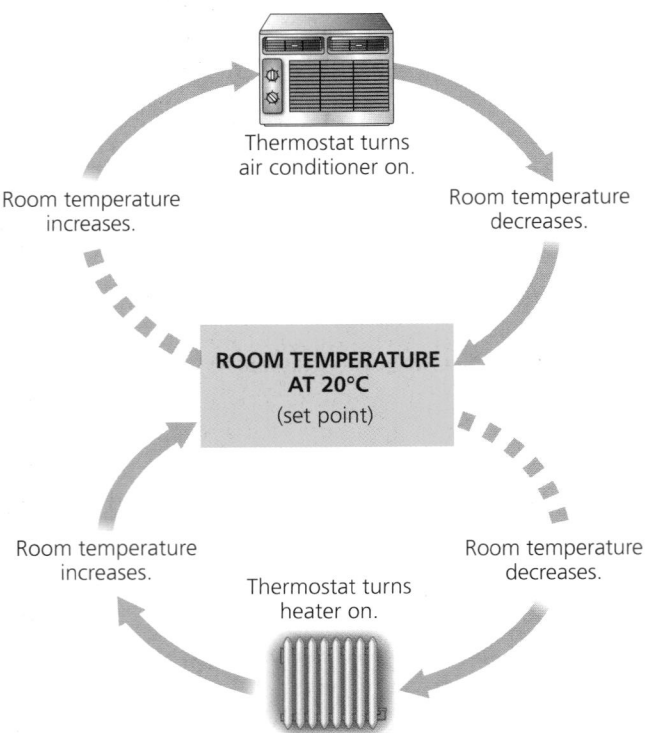

▼ Figure 40.8 A nonliving example of temperature regulation: control of room temperature. Regulating room temperature depends on a sensor/control center (a thermostat) that detects temperature change and activates mechanisms that reverse that change.

Thermostat turns air conditioner on.

Room temperature increases.

Room temperature decreases.

ROOM TEMPERATURE AT 20°C
(set point)

Room temperature increases.

Room temperature decreases.

Thermostat turns heater on.

DRAW IT *Label at least one stimulus, response, and sensor/control center in the above figure.*

below 20°C, the thermostat turns on a radiator, furnace, or other heater **(Figure 40.8)**. When the room exceeds 20°C, the thermostat switches off the heater. If the temperature then drifts below 20°C, the thermostat activates another heating cycle. If the temperature instead rises above 20°C, the thermostat activates a cooling mechanism, such as by turning on an air conditioner.

Like a home heating system, the homeostatic control system in animals maintains a variable, such as body temperature or solute concentration, at or near a particular value, or **set point**. A fluctuation in the variable above or below the set point serves as the **stimulus** detected by a **sensor**. The sensor signals a *control center*, which triggers a **response**, a physiological activity that helps return the variable to the set point. In the home heating example, a drop in temperature below the set point acts as a stimulus, the thermostat serves as the sensor and control center, and the heater produces the response.

Feedback Control in Homeostasis

If you examine the circuit in Figure 40.8, you can see that either response (heating or cooling) reduces the stimulus (the change in temperature) that triggered that response. The circuit thus displays **negative feedback**, a control mechanism

that "damps" its stimulus (see Figure 1.10). This type of feedback regulation plays a major role in homeostasis in animals. For example, when you exercise vigorously, you produce heat, which increases your body temperature. Your nervous system detects this increase and triggers sweating. The evaporation of sweat from your skin then cools your body, helping return body temperature to its set point and eliminating the stimulus.

Homeostasis is a dynamic equilibrium, an interplay between external factors that tend to change the internal environment and internal control mechanisms that oppose such changes. Note that physiological responses to stimuli are not instantaneous, just as switching on a furnace does not immediately warm a room. As a result, homeostasis moderates but doesn't eliminate changes in the internal environment. Fluctuation is greater if a variable has a *normal range*—an upper and lower limit—rather than a set point. This is equivalent to a heating system that is programmed to produce heat when the room temperature drops to 19°C (66°F) and to stop heating when the temperature reaches 21°C (70°F). Regardless of whether there is a set point or a normal range, homeostasis is enhanced by adaptations that reduce fluctuations, such as insulation in the case of temperature and physiological buffers in the case of pH.

Unlike negative feedback, **positive feedback** is a control mechanism that amplifies the stimulus. In animals, positive-feedback loops do not play a major role in homeostasis, but instead help drive processes to completion. During childbirth, for instance, the pressure of the baby's head against sensors near the opening of the mother's uterus stimulates the uterus to contract. These contractions result in greater pressure against the opening of the uterus, heightening the contractions and thereby causing even greater pressure, ultimately causing the baby to be born.

Alterations in Homeostasis

The set points and normal ranges for homeostasis can change under various circumstances. In fact, *regulated changes* in the internal environment are essential to normal body functions. Some regulated changes occur during a particular stage in life, such as the radical shift in hormone balance that occurs during puberty. Other regulated changes are cyclic, such as the variation in hormone levels responsible for a woman's menstrual cycle (see Figure 46.14).

In all animals (and plants), certain cyclic alterations in metabolism reflect a **circadian rhythm**, a set of physiological changes that occur roughly every 24 hours **(Figure 40.9)**. One way to observe this rhythm is to monitor body temperature, which in humans typically undergoes a cyclic rise and fall of more than 0.6°C (1°F) in every 24-hour period. Remarkably, a biological clock maintains this rhythm even when variations in human activity, room temperature, and light levels are minimized (see Figure 40.9a). A circadian rhythm is thus intrinsic to the body, although the biological

▼ **Figure 40.9 Human circadian rhythm.**

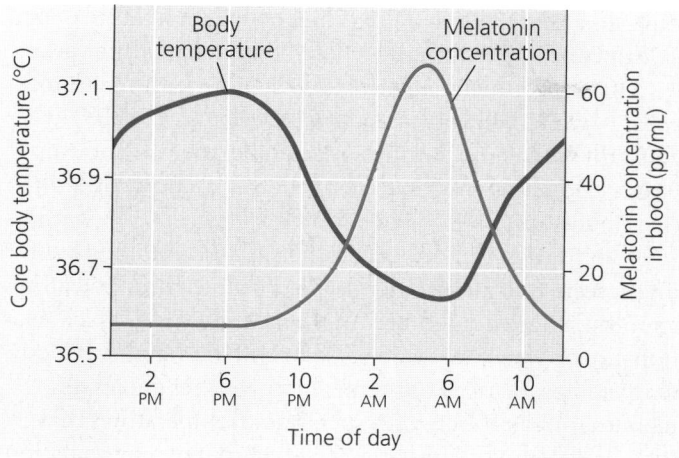

(a) Variation in core body temperature and melatonin concentration in blood. Researchers studied resting but awake volunteers in an isolation chamber with constant temperature and low light. (Melatonin is a hormone secreted by the pineal gland.)

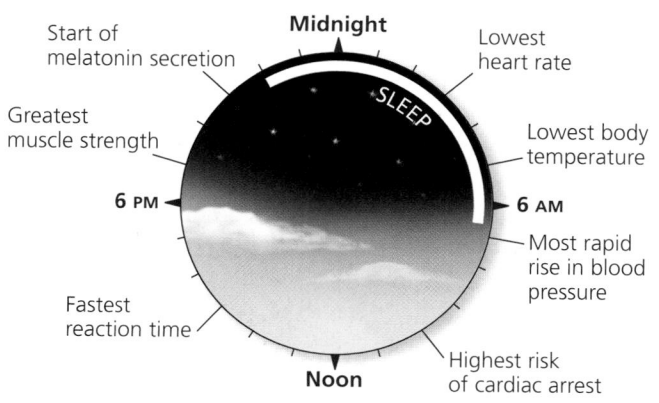

(b) The human circadian clock. Metabolic activities undergo daily cycles in response to the circadian clock. As illustrated for a typical individual who rises early in the morning, eats lunch around noon, and sleeps at night, these cyclic changes occur throughout a 24-hour day.

clock is normally coordinated with the cycle of light and darkness in the environment (see Figure 40.9b). For example, the hormone melatonin is secreted at night, and more is released during the longer nights of winter. External stimuli can reset the biological clock, but the effect is not immediate. That is why flying across several time zones results in jet lag, a mismatch between the circadian rhythm and local environment that persists until the clock fully resets.

Noting the importance of biological clocks to human health and disease, the Nobel Prize Committee awarded the 2017 Nobel Prize in Physiology or Medicine to Americans Jeffrey Hall, Michael Rosbash, and Michael Young, who studied the fruit fly *Drosophila* to map out the molecular mechanisms that underlie circadian rhythms.

▼ **Figure 40.10 Acclimatization by mountain climbers in the Himalayas.** To lessen the risk of altitude sickness when ascending a high peak, climbers acclimatize by camping partway up the mountain. Spending time at an intermediate altitude allows the circulatory and respiratory systems to become more efficient in capturing and distributing oxygen at a lower concentration.

Homeostasis is sometimes altered by **acclimatization**, an animal's physiological adjustment to changes in its external environment. For instance, when an elk moves up into the mountains from sea level, the lower oxygen concentration in the high mountain air stimulates the animal to breathe more rapidly and deeply. As a result, more CO_2 is lost through exhalation, raising blood pH above its normal range. As the animal acclimatizes over several days, changes in kidney function cause it to excrete urine that is more alkaline, returning blood pH to its normal range. Other mammals, including humans, are also capable of acclimatizing to dramatic altitude changes **(Figure 40.10)**, although health risks remain.

CONCEPT CHECK 40.2

1. **MAKE CONNECTIONS** How does negative feedback in thermoregulation differ from feedback inhibition in an enzyme-catalyzed biosynthetic process (see Figure 8.21)?

2. If you were deciding where to put the thermostat in a house, what factors would govern your decision? How do these factors relate to the fact that many homeostatic control sensors in humans are located in the brain?

3. **MAKE CONNECTIONS** Like animals, cyanobacteria have a circadian rhythm. By analyzing the genes that maintain biological clocks, scientists concluded that the 24-hour rhythms of humans and cyanobacteria reflect convergent evolution (see Concept 26.2). What evidence would have supported this conclusion? Explain.

For suggested answers, see Appendix A.

CHAPTER 40 Basic Principles of Animal Form and Function   **883**

Homeostatic processes for thermoregulation involve form, function, and behavior

In this section, we'll examine the regulation of body temperature as an example of how form and function work together in regulating an animal's internal environment. Later chapters in this unit will discuss other physiological systems involved in maintaining homeostasis.

Thermoregulation is the process by which animals maintain their body temperature within a normal range. Body temperatures outside the normal range can reduce the efficiency of enzymatic reactions, alter the fluidity of cellular membranes, and affect other temperature-sensitive biochemical processes, potentially with fatal results.

In talking about thermoregulation, we will need to talk about heat. Formally, heat is defined as thermal energy in transfer from one body of matter to another (see Concept 8.1). Here, however, we will use the term heat to refer simply to thermal energy.

Endothermy and Ectothermy

Heat for thermoregulation can come from either internal metabolism or the external environment. Humans and other mammals, as well as birds, are **endothermic**, meaning that they are warmed mostly by heat generated by metabolism. Some fishes and insect species and a few nonavian reptiles are also mainly endothermic. In contrast, amphibians, many nonavian reptiles and fishes, and most invertebrates are **ectothermic**, meaning that they gain most of their heat from external sources. Endothermy and ectothermy are not mutually exclusive, however. For example, a bird is mainly endothermic but may warm itself in the sun on a cold morning, much as an ectothermic lizard does.

Endotherms can maintain a stable body temperature even in the face of large fluctuations in the environmental temperature. In a cold environment, an endotherm generates enough heat to keep its body substantially warmer than its surroundings **(Figure 40.11a)**. In a hot environment, endothermic vertebrates have mechanisms for cooling their bodies, enabling them to withstand temperatures that are intolerable for most ectotherms.

Many ectotherms adjust their body temperature by behavioral means, such as seeking out shade or basking in the sun **(Figure 40.11b)**. Because their heat source is largely environmental, ectotherms generally need to consume much less food than endotherms of equivalent size—an advantage if food supplies are limited. Ectotherms also usually tolerate larger fluctuations in their internal temperature.

Variation in Body Temperature

Animals also differ in whether their body temperature is variable or constant. An animal whose body temperature varies with its environment is called a *poikilotherm* (from the Greek *poikilos*, varied). In contrast, a *homeotherm* has a relatively constant body temperature. For example, the largemouth bass is a poikilotherm, and the river otter is a homeotherm (see Figure 40.7).

From the descriptions of ectotherms and endotherms, it might seem that all ectotherms are poikilothermic and all endotherms are homeothermic. In fact, there is no fixed relationship between the source of heat and the stability of body temperature. Many ectothermic marine fishes and invertebrates inhabit waters with such stable temperatures that their body temperature varies less than that of mammals and other endotherms. Conversely, the body temperature of a few endotherms varies considerably. For example, the body temperature of some bats drops from 40°C to a few degrees above zero when they enter hibernation.

▼ **Figure 40.11 Thermoregulation by internal or external sources of heat.** Endotherms obtain heat from their internal metabolism, whereas ectotherms rely on heat from their external environment.

(a) King penguins (*Aptenodytes patagonicus*), endotherms

(b) Florida red-bellied turtles (*Pseudemys nelsoni*), ectotherms

It is a common misconception that ectotherms are "cold-blooded" and endotherms are "warm-blooded." Ectotherms do not necessarily have low body temperatures. On the contrary, when sitting in the sun, many ectothermic lizards have higher body temperatures than mammals. Thus, the terms *cold-blooded* and *warm-blooded* are misleading and are avoided in scientific communication.

Balancing Heat Loss and Gain

Thermoregulation depends on an animal's ability to control the exchange of heat with its environment. That exchange can occur by any of four processes: radiation, evaporation, convection, and conduction **(Figure 40.12)**. In each, heat is transferred from an object of higher temperature to one of lower temperature.

The essence of thermoregulation is maintaining a rate of heat gain that equals the rate of heat loss. Animals do this through mechanisms that either reduce heat exchange overall or favor heat exchange in a particular direction. In mammals, several of these mechanisms involve the **integumentary system**, the

▼ **Figure 40.12 Heat exchange between an organism and its environment.**

Radiation is the emission of electromagnetic waves by all objects warmer than absolute zero. Here, a lizard absorbs heat radiating from the distant sun and radiates a smaller amount of energy to the surrounding air.

Evaporation is the removal of heat from the surface of a liquid that is losing some of its molecules as gas. Evaporation of water from a lizard's moist surfaces that are exposed to the environment has a strong cooling effect.

Convection is the transfer of heat by the movement of air or liquid past a surface, as when a breeze contributes to heat loss from a lizard's dry skin or when blood moves heat from the body core to the extremities.

Conduction is the transfer of heat between molecules of objects in contact with each other, as when a lizard sits on a hot rock.

VISUAL SKILLS *If this figure showed a penguin (an endotherm) on an ice floe rather than an iguana (an ectotherm) on a rock, would any of the arrows point in a different direction? Explain.*

outer covering of the body, consisting of the skin, hair, and nails (claws or hooves in some species).

Insulation

Insulation, which reduces the flow of heat between an animal's body and its environment, is a major adaptation for thermoregulation in both mammals and birds. Insulation is found both at the body surface—hair and feathers—and beneath—layers of fat formed by adipose tissue. In addition, some animals secrete oily substances that repel water, protecting the insulating capacity of feathers or fur. Birds, for example, secrete oils that they apply to their feathers during preening.

Often, animals can adjust their insulating layers to further regulate body temperature. Most land mammals and birds, for example, react to cold by raising their fur or feathers. This action traps a thicker layer of air, thereby increasing the effectiveness of the insulation. Lacking feathers or fur, humans must rely primarily on fat for insulation. We do, however, get "goose bumps," a vestige of hair raising inherited from our furry ancestors.

Insulation is particularly important for marine mammals, such as whales and walruses. These animals swim in water colder than their body core, and many species spend at least part of the year in nearly freezing polar seas. Furthermore, the transfer of heat to water occurs 50 to 100 times more rapidly than heat transfer to air. Survival under these conditions is made possible by an evolutionary adaptation called blubber, a very thick layer of insulating fat just under the skin. The insulation that blubber provides is so effective that marine mammals can maintain body core temperatures of about 36–38°C (97–100°F) without requiring much more energy from food than land mammals of similar size.

Circulatory Adaptations

Circulatory systems provide a major route for heat flow between the interior and exterior of the body. Adaptations that regulate the extent of blood flow near the body surface or that trap heat within the body core play a significant role in thermoregulation.

In response to changes in the temperature of their surroundings, many animals alter the amount of blood (and hence heat) flowing between their body core and their skin. Nerve signals that relax the muscles of the vessel walls result in *vasodilation*, a widening of superficial blood vessels (those near the body surface). As a consequence of the increase in vessel diameter, blood flow in the skin increases. In endotherms, vasodilation usually increases the transfer of body heat to the environment by radiation, conduction, and convection (see Figure 40.12). The reverse process, *vasoconstriction*, reduces blood flow and heat transfer by decreasing the diameter of superficial vessels.

Like endotherms, some ectotherms control heat exchange by regulating blood flow. For example, when the marine

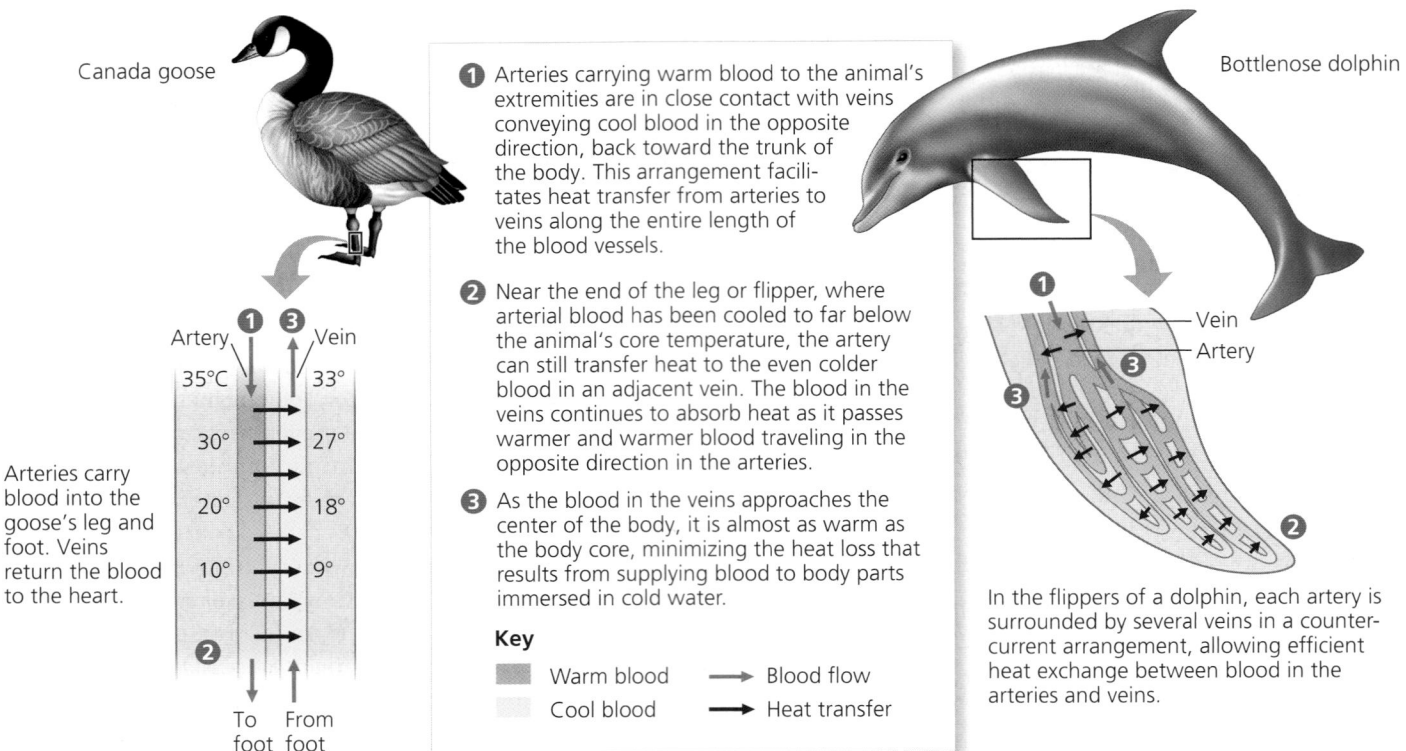

▼ **Figure 40.13 Countercurrent heat exchangers.** A countercurrent exchange system traps heat in the body core, thus reducing heat loss from the extremities, particularly when they are immersed in cold water or in contact with ice or snow. In essence, heat in the arterial blood emerging from the body core is transferred directly to the returning venous blood instead of being lost to the environment.

Canada goose

Arteries carry blood into the goose's leg and foot. Veins return the blood to the heart.

Artery ❶ ❸ Vein

35°C | 33°
30° | 27°
20° | 18°
10° | 9°

To foot   From foot

❶ Arteries carrying warm blood to the animal's extremities are in close contact with veins conveying cool blood in the opposite direction, back toward the trunk of the body. This arrangement facilitates heat transfer from arteries to veins along the entire length of the blood vessels.

❷ Near the end of the leg or flipper, where arterial blood has been cooled to far below the animal's core temperature, the artery can still transfer heat to the even colder blood in an adjacent vein. The blood in the veins continues to absorb heat as it passes warmer and warmer blood traveling in the opposite direction in the arteries.

❸ As the blood in the veins approaches the center of the body, it is almost as warm as the body core, minimizing the heat loss that results from supplying blood to body parts immersed in cold water.

Key

Warm blood   ⟶ Blood flow
Cool blood   ⟶ Heat transfer

Bottlenose dolphin

Vein
Artery

In the flippers of a dolphin, each artery is surrounded by several veins in a countercurrent arrangement, allowing efficient heat exchange between blood in the arteries and veins.

iguana of the Galápagos Islands swims in the cold ocean, its superficial blood vessels undergo vasoconstriction. This process routes more blood to the body core, conserving body heat.

In many birds and mammals, reducing heat loss from the body relies on **countercurrent exchange**, the transfer of heat (or solutes) between fluids that are flowing in opposite directions. In a countercurrent heat exchanger, arteries and veins are located adjacent to each other **(Figure 40.13)**. Because blood flows through the arteries and veins in opposite directions, this arrangement allows heat exchange to be remarkably efficient. As warm blood in the arteries moves outward from the body core, it transfers heat to the colder blood in the veins returning from the extremities. Most importantly, heat is transferred along the entire length of the exchanger, maximizing the rate of heat exchange and minimizing heat loss to the environment.

Although most sharks and fishes are temperature conformers, countercurrent heat exchangers are found in some large, powerful swimmers, including great white sharks, bluefin tuna, and swordfish. By keeping the main swimming muscles warm, this adaptation enables vigorous, sustained activity. Similarly, many endothermic insects (bumblebees, honeybees, and some moths) have a countercurrent exchanger that helps maintain a high temperature in their thorax, where flight muscles are located.

Cooling by Evaporative Heat Loss

Many mammals and birds live in places where regulating body temperature requires cooling at some times and warming at others. If environmental temperature is above body temperature, only evaporation can keep body temperature from rising. Water absorbs considerable heat when it evaporates (see Concept 3.2); this heat is carried away from the skin and respiratory surfaces with water vapor.

Some animals exhibit adaptations that greatly facilitate evaporative cooling. A few mammals, including horses and humans, have sweat glands. In many other mammals, as well as in birds, panting is important. Some birds have a pouch richly supplied with blood vessels in the floor of the mouth; fluttering the pouch increases evaporation. Pigeons can use this adaptation to keep their body temperature close to 40°C (104°F) in air temperatures as high as 60°C (140°F), as long as they have sufficient water.

Thermoregulatory behavior in a dragonfly. By orienting its body so that the narrow tip of its abdomen faces the sun, the dragonfly minimizes heating by solar radiation.

Behavioral Responses

Ectotherms, and sometimes endotherms, control body temperature through behavioral responses to changes in the environment. When cold, they seek warm places, orienting themselves toward heat sources and expanding the portion of their body surface exposed to the heat source (see Figure 40.11b). When hot, they bathe, move to cool areas, or turn in another direction, minimizing their absorption of heat from the sun. For example, a dragonfly's "obelisk" posture is an adaptation that minimizes the amount of body surface exposed to the sun and thus to heating **(Figure 40.14)**. Although these behaviors are relatively simple, they enable many ectotherms to maintain a nearly constant body temperature.

Social behavior contributes to thermoregulation in both endotherms and ectotherms. Among endotherms, for example, behavior contributes significantly to the winter survival of Emperor penguins (see Figure 40.1). Among ectotherms, honeybees are notable for their use of behavior in achieving homeostasis for temperature. In cold weather, they increase heat production and huddle together, thereby retaining heat. Individuals move between the cooler outer edges of the huddle and the warmer center, thus circulating and distributing the heat. In hot weather, honeybees cool the hive by transporting water to the hive and fanning with their wings, promoting evaporation and convection. Thus, a honeybee colony uses many of the mechanisms of thermoregulation characteristic of individual animals.

Adjusting Metabolic Heat Production

Because endotherms generally maintain a body temperature considerably higher than that of the environment, they must counteract continual heat loss. Endotherms can vary heat production—*thermogenesis*—to match changing rates of heat loss. Thermogenesis is increased by such muscle activity as moving or shivering. For example, shivering helps

chickadees, birds with a body mass of only 20 g, remain active and hold their body temperature nearly constant at 40°C (104°F) in environmental temperatures as low as −40°C (−40°F).

Flying insects such as bees and moths can also vary heat production. Many such endothermic insects warm up by shivering before taking off. As they contract their flight muscles in synchrony, only slight wing movements occur, but considerable heat is produced. Chemical reactions, and hence cellular respiration, accelerate in the warmed-up flight "motors," enabling flight even in cold air.

In some mammals, endocrine signals released in response to cold cause mitochondria to increase their metabolic activity and produce heat instead of ATP. This process, called *nonshivering thermogenesis*, takes place throughout the body. Some mammals also have a tissue called *brown fat* in their neck and between their shoulders that is specialized for rapid heat production. (The presence of extra mitochondria is what gives brown fat its characteristic color.) Brown fat is found in the infants of many mammals, representing about 5% of total body weight in human infants. Long known to be present in adult mammals that hibernate, brown fat has also recently been detected in human adults **(Figure 40.15)**. There, the amount has been found to vary, with individuals exposed to a cool environment for a month having increased amounts of brown fat.

Among the nonavian reptiles, endothermy has been observed in some large species in certain circumstances. For example, researchers found that a female Burmese python (*Python molurus bivittatus*) incubating eggs maintained a body temperature roughly 6°C (11°F) above that of the surrounding air. Where did the heat come from? Further studies showed that such pythons, like birds, can raise their body

▼ **Figure 40.15 Brown fat activity during cold stress.** This PET scan shows metabolically active brown fat deposits (indicated by the arrows) surrounding the neck.

▼ Figure 40.16 **Inquiry**

How does a Burmese python generate heat while incubating eggs?

Experiment Herndon Dowling and colleagues at the Bronx Zoo in New York observed that when a female Burmese python incubated eggs by wrapping her body around them, she raised her body temperature and frequently contracted the muscles in her coils. To learn if the contractions were elevating her body temperature, they placed the python and her eggs in a chamber. As they varied the chamber's temperature, they monitored the python's muscle contractions as well as her oxygen uptake, a measure of her rate of cellular respiration.

Results The python's oxygen consumption increased when the temperature in the chamber decreased. As shown in the graph, this increase in oxygen consumption paralleled an increase in the rate of muscle contraction.

Conclusion Because oxygen consumption, which generates heat through cellular respiration, was correlated with the rate of muscle contraction, the researchers concluded that the muscle contractions, a form of shivering, were the source of the Burmese python's elevated body temperature.

Data from V. H. Hutchison, H. G. Dowling, and A. Vinegar, Thermoregulation in a brooding female Indian python, *Python molurus bivittatus, Science* 151:694–696 (1966).

WHAT IF? *Suppose you varied air temperature and measured oxygen consumption for a female Burmese python without a clutch of eggs. Since she would not show shivering behavior, how would you expect the snake's oxygen consumption to vary with environmental temperature?*

temperature through shivering **(Figure 40.16)**. Whether certain groups of Mesozoic dinosaurs were similarly endothermic is a matter of active debate.

Acclimatization in Thermoregulation

Acclimatization contributes to thermoregulation in many animal species. In birds and mammals, acclimatization to seasonal temperature changes often includes adjusting insulation—growing a thicker coat of fur in the winter and shedding it in the summer, for example.

Acclimatization in ectotherms often includes adjustments at the cellular level. Cells may produce variants of enzymes that have the same function but different optimal temperatures. Also, the proportions of saturated and unsaturated lipids in membranes may change; unsaturated lipids help keep membranes fluid at lower temperatures (see Figure 7.5).

Remarkably, some ectotherms can survive subzero temperatures, producing "antifreeze" proteins that prevent ice formation in their cells. In the Arctic and Southern (Antarctic) Oceans, these proteins enable certain fishes to survive in water as cold as −2°C (28°F), a full degree Celsius below the freezing point of body fluids in other species.

Physiological Thermostats and Fever

In humans and other mammals, the sensors responsible for thermoregulation are concentrated in the **hypothalamus**, the brain region that also controls the circadian clock. Within the hypothalamus, a group of nerve cells functions as a thermostat, responding to body temperatures outside the normal range by activating mechanisms that promote heat loss or gain **(Figure 40.17)**.

At body temperatures above the normal range, the hypothalamic thermostat promotes cooling of the body by dilation of vessels in the skin, sweating, or panting. When body temperatures instead drop below the normal range, the thermostat inhibits heat loss mechanisms and activates mechanisms that either save heat, such as constricting vessels in the skin, or generate heat, such as shivering.

In the course of certain bacterial and viral infections, mammals and birds develop *fever*, an elevated body temperature. A variety of experiments have shown that fever reflects an increase in the normal range for the biological thermostat. For example, artificially *raising* the temperature of the hypothalamus in an infected animal *reduces* fever in the rest of the body.

Among certain ectotherms, an increase in body temperature upon infection reflects what is called a behavioral fever. For example, the desert iguana (*Dipsosaurus dorsalis*) responds to infection with certain bacteria by seeking a warmer environment and then maintaining a body temperature that is elevated by 2–4°C (4–7°F). Similar observations in fishes, amphibians, and even cockroaches indicate that fever is common to both endotherms and ectotherms.

Now that we have explored thermoregulation, we'll conclude our introduction to animal form and function by considering the different ways that animals allocate, use, and conserve energy.

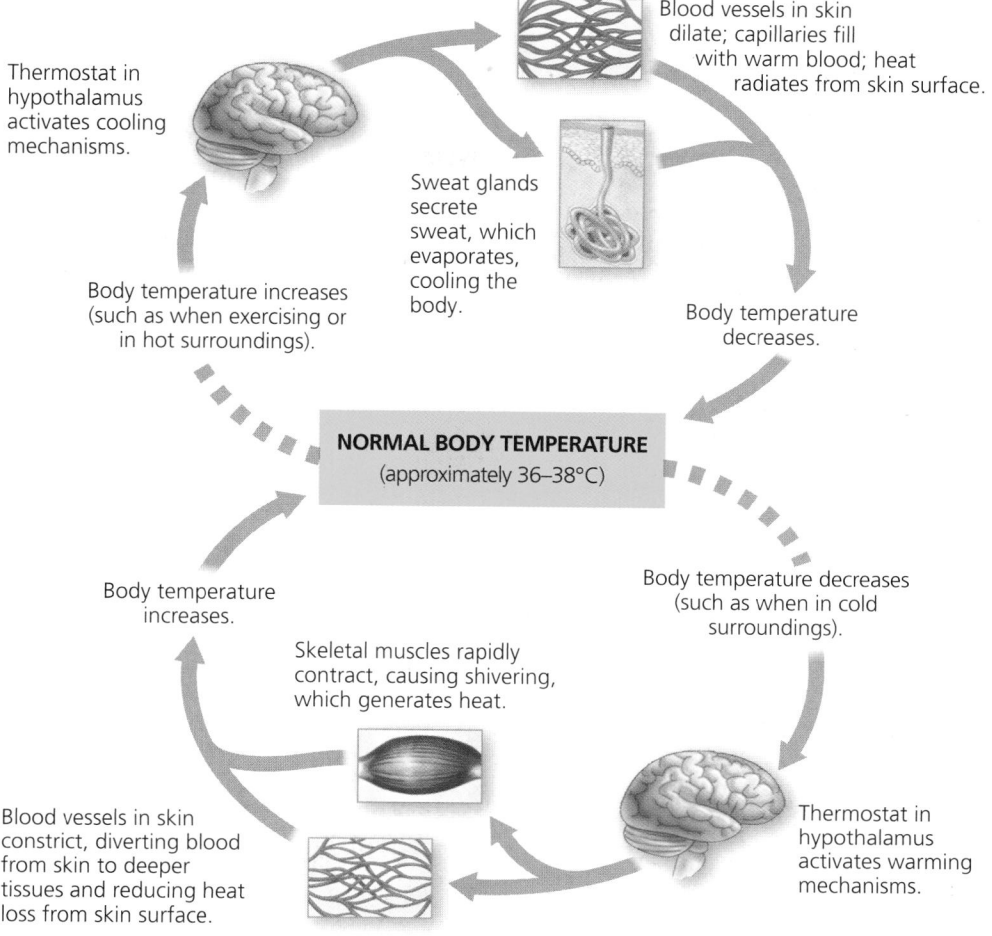

► **Figure 40.17 The thermostatic function of the hypothalamus in human thermoregulation.**

WHAT IF? *Suppose at the end of a hard run on a hot day you find that there are no drinks left in the cooler. If, out of desperation, you dunk your head into the cooler, how might the ice-cold water affect the rate at which your body temperature returns to normal?*

→ **Mastering Biology Figure Walkthrough**

Thermostat in hypothalamus activates cooling mechanisms.

Blood vessels in skin dilate; capillaries fill with warm blood; heat radiates from skin surface.

Sweat glands secrete sweat, which evaporates, cooling the body.

Body temperature increases (such as when exercising or in hot surroundings).

Body temperature decreases.

NORMAL BODY TEMPERATURE
(approximately 36–38°C)

Body temperature increases.

Body temperature decreases (such as when in cold surroundings).

Skeletal muscles rapidly contract, causing shivering, which generates heat.

Blood vessels in skin constrict, diverting blood from skin to deeper tissues and reducing heat loss from skin surface.

Thermostat in hypothalamus activates warming mechanisms.

CONCEPT CHECK 40.3

1. What mode of heat exchange is involved in "wind chill," when moving air feels colder than still air at the same temperature? Explain.

2. Flowers differ in how much sunlight they absorb. Why might this matter to a hummingbird seeking nectar on a cool morning?

3. **WHAT IF?** Why is shivering likely during the onset of a fever?

For suggested answers, see Appendix A.

CONCEPT 40.4

Energy requirements are related to animal size, activity, and environment

One of the unifying themes of biology, introduced in Concept 1.1, is that life requires energy transfer and transformation. Like other organisms, animals use chemical energy for growth, repair, activity, and reproduction. The overall flow and transformation of energy in an animal—its **bioenergetics**—determines nutritional needs and is related to the animal's size, activity, and environment.

Energy Allocation and Use

Organisms can be classified by how they obtain chemical energy. Most *autotrophs*, such as plants, harness light energy to build energy-rich organic molecules and then use those molecules for fuel. Most *heterotrophs*, such as animals, obtain their chemical energy from food, which contains organic molecules synthesized by other organisms.

Animals use chemical energy harvested from the food they eat to fuel metabolism and activity. Food is digested by enzymatic hydrolysis (see Figure 5.2b), and nutrients are absorbed by body cells. The ATP (adenosine triphosphate) produced by cellular respiration and fermentation powers cellular work, enabling cells, organs, and organ systems to perform the functions that keep an animal alive. Other uses of energy in the form of ATP include biosynthesis, which is needed for

▼ Figure 40.18 Bioenergetics of an animal: an overview.

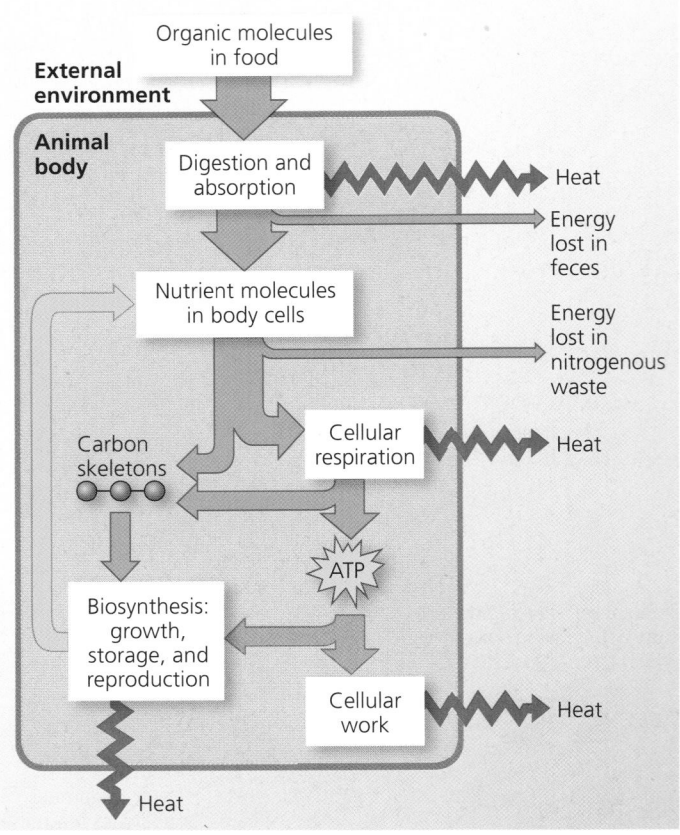

External environment

Organic molecules in food

Animal body

Digestion and absorption → Heat

→ Energy lost in feces

Nutrient molecules in body cells

→ Energy lost in nitrogenous waste

Carbon skeletons

Cellular respiration → Heat

ATP

Biosynthesis: growth, storage, and reproduction

Cellular work → Heat

→ Heat

MAKE CONNECTIONS *Use the idea of energy coupling to explain why heat is produced in the absorption of nutrients, in cellular respiration, and in the synthesis of biopolymers (see Concept 8.3).*

body growth and repair, synthesis of storage material such as fat, and production of gametes. The production and use of ATP generate heat, which the animal eventually gives off to its surroundings **(Figure 40.18)**.

Quantifying Energy Use

How much of the total energy an animal obtains from food does it need just to stay alive? How much energy must be expended to walk, run, swim, or fly from one place to another? What fraction of the energy intake is used for reproduction? Physiologists answer such questions by measuring the rate at which an animal uses chemical energy and how this rate changes in different circumstances.

The sum of all the energy an animal uses in a given time interval is called its **metabolic rate**. Energy is measured in joules (J) or in calories (cal) and

kilocalories (kcal). A kilocalorie equals 1,000 calories, or 4,184 joules. (The unit Calorie, with a capital C, as used by many nutritionists, is actually a kilocalorie.)

Metabolic rate can be determined in several ways. Because nearly all of the chemical energy used in cellular respiration eventually appears as heat, metabolic rate can be measured by monitoring an animal's rate of heat loss. For this approach, researchers use a calorimeter, which is a closed, insulated chamber equipped with a device that records the heat an animal gives off to its environment. Metabolic rate can also be determined from the amount of oxygen consumed or carbon dioxide produced by an animal's cellular respiration **(Figure 40.19)**. To calculate metabolic rate over longer periods, researchers record the rate of food consumption, the energy content of the food (about 4.5–5 kcal per gram of protein or carbohydrate and about 9 kcal per gram of fat), and the chemical energy lost in waste products (feces and urine or other nitrogenous wastes).

Minimum Metabolic Rate and Thermoregulation

Animals must maintain a minimum metabolic rate for basic functions such as cell maintenance, breathing, and circulation. Researchers measure this minimum metabolic rate differently for endotherms and ectotherms. The minimum metabolic rate of a nongrowing endotherm that is at rest, has an empty stomach, and is not experiencing stress is called the **basal metabolic rate (BMR)**. BMR is measured under a "comfortable" temperature range—a range that requires only the minimum generation or shedding of heat. The minimum metabolic rate of ectotherms is determined at a specific temperature because changes in the environmental temperature alter body temperature and therefore metabolic rate. The metabolic rate of a fasting, nonstressed

▼ Figure 40.19 Measuring the rate of oxygen consumption by a swimming shark.
A researcher monitors the decrease in oxygen level over time in the recirculating water of a juvenile hammerhead's tank.

ectotherm at rest at a particular temperature is called its **standard metabolic rate (SMR)**.

Comparisons of minimum metabolic rates reveal the different energy costs of endothermy and ectothermy. The BMR for humans averages 1,600–1,800 kcal per day for adult males and 1,300–1,500 kcal per day for adult females. These BMRs are about equivalent to the rate of energy use by a 75-watt lightbulb. In contrast, the SMR of an American alligator is only about 60 kcal per day at 20°C (68°F). As this represents less than $\frac{1}{20}$ the energy used by a comparably sized adult human, it is clear that ectothermy has a markedly lower energetic requirement than endothermy.

Influences on Metabolic Rate

Metabolic rate is affected by many factors other than an animal being an endotherm or an ectotherm. Some key factors are age, sex, size, activity, temperature, and nutrition. Here we'll examine the effects of size and activity.

Size and Metabolic Rate

Larger animals have more body mass and therefore require more chemical energy. Remarkably, the relationship between overall metabolic rate and body mass is constant across a wide range of sizes and forms, as illustrated for various mammals in **Figure 40.20a**. In fact, for even more varied organisms ranging in size from bacteria to blue whales, metabolic rate remains roughly proportional to body mass to the three-quarter power ($m^{3/4}$). Scientists are still researching the basis of this relationship, which applies to ectotherms as well as endotherms.

The relationship of metabolic rate to size profoundly affects energy consumption by body cells and tissues. As shown in **Figure 40.20b**, the energy it takes to maintain each gram of body mass is inversely related to body size. Each gram of a mouse, for instance, requires about 20 times as many calories as a gram of an elephant, even though the whole elephant uses far more calories than the whole mouse. The smaller animal's higher metabolic rate per gram demands a higher rate of oxygen delivery. To meet this demand, the smaller animal must have a higher breathing rate, blood volume (relative to its size), and heart rate.

Thinking about body size in bioenergetic terms reveals how trade-offs shape the evolution of body plans. As body size decreases, each gram of tissue increases in energy cost. As body size increases, energy costs per gram of tissue decrease, but an ever-larger fraction of body tissue is required for exchange, support, and locomotion.

Activity and Metabolic Rate

For both ectotherms and endotherms, activity greatly affects metabolic rate. Even a person reading quietly at a desk or an

▼ **Figure 40.20 The relationship of metabolic rate to body size.**

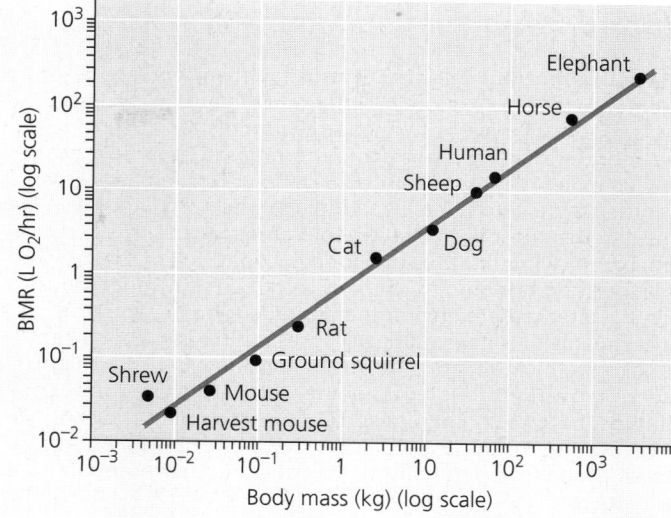

(a) Relationship of basal metabolic rate (BMR) to body size for various mammals. From shrew to elephant, size increases 1 millionfold.

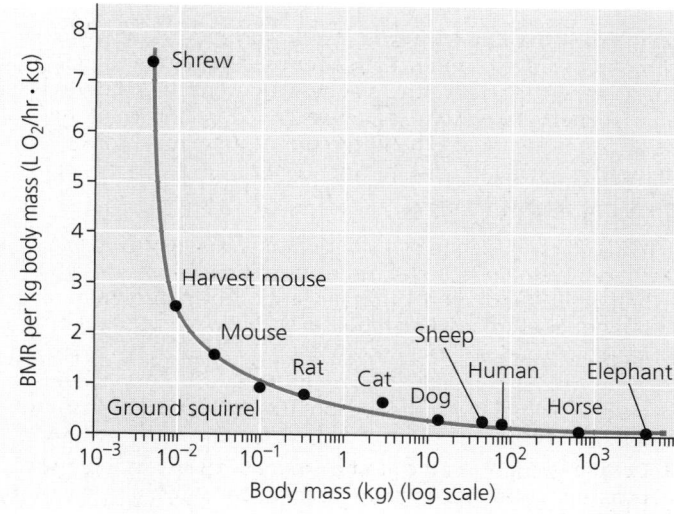

(b) Relationship of BMR per kilogram of body mass to body size for the same mammals as in (a).

INTERPRET THE DATA *Based on the graph in (a), one observer suggests that a group of 100 ground squirrels has the same basal metabolic rate as 1 dog. A second observer looking at the graph disagrees. Who is correct and why?*

insect twitching its wings consumes energy beyond the BMR or SMR. Maximum metabolic rates (the highest rates of ATP use) occur during peak activity, such as lifting a heavy object, sprinting, or swimming at high speed. In general, the maximum metabolic rate an animal can sustain is inversely related to the duration of activity.

For most terrestrial animals, the average daily rate of energy consumption is two to four times BMR (for endotherms) or SMR (for ectotherms). Humans in most developed countries have an unusually low average daily metabolic rate

Interpreting Pie Charts

How Do Energy Budgets Differ for Three Terrestrial Vertebrates? To explore bioenergetics in animal bodies, let's consider typical annual energy budgets for three terrestrial vertebrates that vary in size and thermoregulatory strategy: a 4-kg male Adélie penguin, a 25-g (0.025-kg) female deer mouse, and a 4-kg female ball python. The penguin is well-insulated against his Antarctic environment but must expend energy in swimming to catch food, incubating eggs laid by his partner, and bringing food to his chicks. The tiny deer mouse lives in a temperate environment where food may be read-ily available, but her small size causes rapid loss of body heat. Unlike the penguin and mouse, the python is ectothermic and keeps growing throughout her life. She produces eggs but does not incubate them. In this exercise, we'll compare the energy expenditures of these animals for five important functions: basal (standard) metabolism, reproduction, thermoregulation, activity, and growth.

How the Data Were Obtained Energy budgets were calculated for each of the animals based on measurements from field and laboratory studies.

Data from the Experiments Pie charts are a good way to compare *relative* differences in a set of variables. In the pie charts here, the sizes of the wedges represent the relative annual energy expenditures for the functions shown in the key. The total annual expenditure for each animal is given below its pie chart.

INTERPRET THE DATA

1. You can estimate the contribution of each wedge in a pie chart by remembering that the entire circle represents 100%, half is 50%, and so on. What percent of the mouse's energy budget goes to basal metabolism? What percent of the penguin's budget is for activity?

2. Without considering the sizes of the wedges, how do the three pie charts differ in which functions they include? Explain these differences.

3. Does the penguin or the mouse expend a greater proportion of its energy budget on thermoregulation? Why?

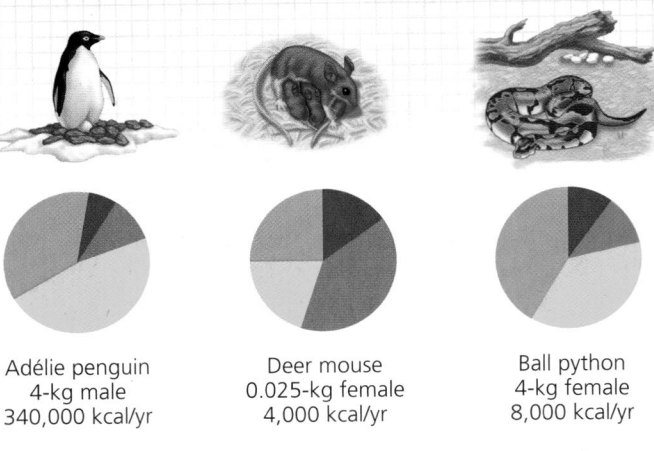

Adélie penguin	Deer mouse	Ball python
4-kg male	0.025-kg female	4-kg female
340,000 kcal/yr	4,000 kcal/yr	8,000 kcal/yr

Key

- Basal (standard) metabolism
- Reproduction
- Thermoregulation
- Activity
- Growth

Data from M. A. Chappell et al., Energetics of foraging in breeding Adélie pen-guins, *Ecology* 74:2450–2461 (1993); M. A. Chappell et al., Voluntary running in deer mice: speed, distance, energy costs, and temperature effects, *Journal of Experimental Biology* 207:3839–3854 (2004); T. M. Ellis and M. A. Chappell, Metabolism, temperature relations, maternal behavior, and reproductive en-ergetics in the ball python (*Python regius*), *Journal of Comparative Physiology B* 157:393–402 (1987).

4. Now look at the *total* annual energy expenditures for each animal. How much more energy does the penguin expend each year compared to the similarly sized python?

5. Which animal expends the most kilocalories per year on thermoregulation?

6. If you monitored energy allocation in the penguin for just a few months instead of an entire year, you might find the growth category to be a significant part of the pie chart. Given that adult penguins don't grow from year to year, how would you explain this finding?

➔ **Instructors:** A version of this Scientific Skills Exercise can be assigned in **Mastering Biology**.

---

of about 1.5 times BMR—an indication of a relatively seden-tary lifestyle.

The fraction of an animal's energy "budget" that is devoted to activity depends on many factors, including its environment, behavior, size, and thermoregulation. In the **Scientific Skills Exercise**, you'll interpret data on the annual energy budgets of three terrestrial vertebrates.

Torpor and Energy Conservation

Despite their many adaptations for homeostasis, animals may encounter conditions that severely challenge their abilities to balance their heat, energy, and materials budgets. For exam-ple, at certain times of the day or year, their surroundings may be extremely hot or cold, or food may be unavailable.

A major adaptation that enables animals to save energy in the face of such difficult conditions is **torpor**, a physiological state of decreased activity and metabolism.

Many birds and small mammals exhibit a daily torpor that is well adapted to feeding patterns. For instance, some bats feed at night and go into torpor in daylight. Similarly, chicka-dees and hummingbirds, which feed during the day, often go into torpor on cold nights.

All endotherms that exhibit daily torpor are relatively small; when active, they have high metabolic rates and thus very high rates of energy consumption. The changes in body temperature, and thus the energy savings, are often consid-erable: the body temperature of chickadees drops as much as 10°C (18°F) at night, and the core body temperature of a hummingbird can fall 25°C (45°F) or more.

Hibernation is long-term torpor that is an adaptation to winter cold and food scarcity. When a mammal enters hibernation, its body temperature declines as its body's thermostat is turned down **(Figure 40.21)**. Some hibernating mammals cool to as low as 1–2°C (34–36°F), and at least one, the Arctic ground squirrel (*Spermophilus parryii*), can enter a supercooled (unfrozen) state in which its body temperature dips below 0°C (32°F). Periodically, perhaps every two weeks or so, hibernating animals undergo arousal, raising their body temperature and becoming active briefly before resuming hibernation. Metabolic rates during hibernation can be 20 times lower than if the animal attempted to maintain normal body temperatures of 36–38°C (97–100°F). As a result, hibernators such as the ground squirrel can survive through the winter on limited supplies of energy stored in the body tissues or as food cached in a burrow. Similarly, the slow metabolism and inactivity of *estivation*, or summer torpor, enable animals to survive long periods of high temperatures and scarce water.

What happens to the circadian rhythm in hibernating animals? In the past, researchers reported detecting daily biological rhythms in hibernating animals. However, in some cases the animals were probably in a state of torpor from which they could readily arouse, rather than "deep" hibernation. More recently, a group of researchers in France addressed this question in a different way, examining the machinery of the biological clock rather than the rhythms it controls **(Figure 40.22)**. Working with the European hamster, they found that molecular components of the clock stopped oscillating during hibernation. These findings support the hypothesis that the circadian clock ceases operation during hibernation, at least in this species.

From tissue types to homeostasis, this chapter has focused on the whole animal. We also investigated how animals exchange materials with the environment and how size and activity affect metabolic rate. For much of the rest of this unit, we'll explore how specialized organs and organ systems enable animals to meet the basic challenges of life. In Unit 6, we investigated how plants meet the same challenges. **Figure 40.23**, on the next two pages, highlights some fundamental similarities and differences in the evolutionary adaptations of plants and animals. This figure is thus a review of Unit 6, an introduction to Unit 7, and, most importantly, an illustration of the connections that unify the myriad forms of life.

➔ **Mastering Biology Interview with George Bartholomew: Exploring connections between animal physiology and the environment**

▼ **Figure 40.21 A hazel dormouse (*Muscardinus avellanarius*) hibernating.**

▼ **Figure 40.22 Inquiry**

What happens to the circadian clock during hibernation?

Experiment To determine whether the 24-hour biological clock continues to run during hibernation, Paul Pévet and colleagues at the University of Louis Pasteur in Strasbourg, France, studied molecular components of the circadian clock in the European hamster (*Cricetus cricetus*). The researchers measured RNA levels for two clock genes—*Per2* and *Bmal1*—during normal activity (euthermia) and during hibernation in constant darkness. The RNA samples were obtained from the suprachiasmatic nuclei (SCN), a pair of structures in the mammalian brain that control circadian rhythms.

Results

Conclusion Hibernation disrupted circadian variation in the hamster's clock gene RNA levels. Further experiments demonstrated that this disruption was not simply due to the dark environment during hibernation, since for nonhibernating animals RNA levels during a darkened daytime were the same as in daylight. The researchers concluded that the biological clock stops running in hibernating European hamsters and, perhaps, in other hibernators as well.

Data from F. G. Revel et al., The circadian clock stops ticking during deep hibernation in the European hamster, *Proceedings of the National Academy of Sciences USA* 104:13816–13820 (2007).

WHAT IF? *Suppose you discovered a new hamster gene and found that the levels of RNA for this gene were constant during hibernation. What could you conclude about the day and night RNA levels for this gene during euthermia?*

CONCEPT CHECK 40.4

1. If a mouse and a small lizard of the same mass (both at rest) were placed in experimental chambers under identical environmental conditions, which animal would consume oxygen at a higher rate? Explain.

2. Which animal must eat a larger proportion of its weight in food each day: a house cat or an African lion caged in a zoo? Explain.

3. **WHAT IF?** Suppose the animals at a zoo were resting comfortably and remained at rest while the nighttime air temperature dropped. If the temperature change were sufficient to cause a change in metabolic rate, what changes would you expect for an alligator and a lion?

For suggested answers, see Appendix A.

MAKE CONNECTIONS

Life Challenges and Solutions in Plants and Animals

Multicellular organisms face a common set of challenges. Comparing the solutions that have evolved in plants and animals reveals both unity (shared elements) and diversity (distinct features) across these two lineages.

Nutritional Mode

All living things must obtain energy and carbon from the environment to grow, survive, and reproduce. Plants are autotrophs, obtaining their energy through photosynthesis and their carbon from inorganic sources, whereas animals are heterotrophs, obtaining their energy and carbon from food. Evolutionary adaptations in plants and animals support these different nutritional modes. The broad surface of many leaves enhances light capture for photosynthesis. When hunting, a bobcat relies on stealth, speed, and sharp claws. (See Figure 36.2 and Figure 41.16.)

Growth and Regulation

The growth and physiology of both plants and animals are regulated by hormones. In plants, hormones may act in a local area or be transported in the body. They control growth patterns, flowering, fruit development, and more. In animals, hormones circulate throughout the body and act in specific target tissues, controlling homeostatic processes and developmental events such as molting. (See Figure 39.10 and Figure 45.12.)

Environmental Response

All forms of life must detect and respond appropriately to conditions in their environment. Specialized organs sense environmental signals. For example, the floral head of a sunflower and an insect's eyes both contain photoreceptors that detect light. Environmental signals activate specific receptor proteins, triggering signal transduction pathways that initiate cellular responses coordinated by chemical and electrical communication. (See Figure 39.19 and Figure 50.15.)

→ **Mastering Biology BioFlix®**
Animation: Homeostasis:
Regulating Blood Sugar

Transport

All but the simplest multicellular organisms must transport nutrients and waste products between locations in the body. A system of tubelike vessels is the common evolutionary solution, while the mechanism of circulation varies. Plants harness solar energy to transport water, minerals, and sugars through specialized tubes (left). In animals, a pump (heart) moves circulatory fluid through vessels (right). (See Figure 35.10 and Figure 42.9.)

→ **Mastering Biology BioFlix® Animation:**
   Water Transport in Plants

Reproduction

In sexual reproduction, specialized tissues and structures produce and exchange gametes. Offspring are generally supplied with nutritional stores that facilitate rapid growth and development. For example, seeds have stored food reserves that supply energy to the young seedling, while milk provides sustenance for juvenile mammals. (See Figure 38.8 and Figure 34.40.)

Gas Exchange

The exchange of certain gases with the environment is essential for life. Respiration by plants and animals requires taking up oxygen (O_2) and releasing carbon dioxide (CO_2). In photosynthesis net exchange occurs in the opposite direction: CO_2 uptake and O_2 release. In both plants and animals, highly convoluted surfaces that increase the area available for gas exchange have evolved, such as the spongy mesophyll of leaves (left) and the alveoli of lungs (right). (See Figure 35.18 and Figure 42.24.)

→ **Mastering Biology BioFlix® Animation:**
   Gas Exchange

Absorption

Organisms need to absorb nutrients. The root hairs of plants (left) and the villi (projections) that line the intestines of vertebrates (right) increase the surface area available for absorption. (See Figure 35.3 and Figure 41.12.)

MAKE CONNECTIONS

Compare the adaptations that enable plants and animals to respond to the challenges of living in hot and cold environments. See Concept 39.4 and Concept 40.3.

40 Chapter Review

→ Go to **Mastering Biology** for Assignments, the eText, the Study Area, and Dynamic Study Modules.

SUMMARY OF KEY CONCEPTS

→ To review key terms, go to the **Vocabulary Self-Quiz** in the **Mastering Biology** eText or Study Area, or go to goo.gl/zkjz9t.

CONCEPT 40.1

Animal form and function are correlated at all levels of organization (pp. 874–881)

- Physical laws constrain the evolution of an animal's size and shape. These constraints contribute to convergent evolution in animal body forms.
- Each animal cell must have access to an aqueous environment. Simple two-layered sacs and flat shapes maximize exposure to the surrounding medium. More complex body plans have highly folded internal surfaces specialized for exchanging materials.
- Animal bodies are based on a hierarchy of cells, **tissues, organs**, and **organ systems. Epithelial tissue** forms active interfaces on external and internal surfaces; **connective tissue** binds and supports other tissues; **muscle tissue** contracts, moving body parts; and **nervous tissue** transmits nerve impulses throughout the body.
- The **endocrine** and **nervous systems** are the two means of communication between different locations in the body. The endocrine system broadcasts signaling molecules called **hormones** everywhere via the bloodstream, but only certain cells are responsive to each hormone. The nervous system uses dedicated cellular circuits involving electrical and chemical signals to send information to specific locations.

? *For a large animal, what challenges would a spherical shape pose for carrying out exchange with the environment?*

CONCEPT 40.2

Feedback control maintains the internal environment in many animals (pp. 881–883)

- An animal is a **regulator** if it controls an internal variable and a **conformer** if it allows an internal variable to vary with external changes. **Homeostasis** is the maintenance of a steady state despite internal and external changes.
- Homeostatic mechanisms are usually based on **negative feedback**, in which the **response** reduces the **stimulus**. In contrast, **positive feedback** involves amplification of a stimulus by the response and often brings about a change in state, such as the transition from pregnancy to childbirth.

- Regulated change in the internal environment is essential to normal function. **Circadian rhythms** are daily fluctuations in metabolism and behavior tuned to the cycles of light and dark in the environment. Other environmental changes may trigger **acclimatization**, a temporary shift in the steady state.

? *Is it accurate to define homeostasis as a constant internal environment? Explain.*

CONCEPT 40.3

Homeostatic processes for thermoregulation involve form, function, and behavior (pp. 884–889)

- An animal maintains its internal temperature within a tolerable range by **thermoregulation. Endothermic** animals are warmed mostly by heat generated by metabolism. **Ectothermic** animals get most of their heat from external sources. Endothermy requires a greater expenditure of energy. Body temperature may vary with environmental temperature, as in *poikilotherms*, or be relatively constant, as in *homeotherms*.
- In thermoregulation, physiological and behavioral adjustments balance heat gain and loss, which occur through radiation, evaporation, convection, and conduction. Insulation and **countercurrent exchange** reduce heat loss, whereas panting, sweating, and bathing increase evaporation, cooling the body. Many ectotherms and endotherms adjust their rate of heat exchange with their surroundings by vasodilation or vasoconstriction and by behavioral responses.
- Many mammals and birds adjust their amount of body insulation in response to changes in environmental temperature. Ectotherms undergo a variety of changes at the cellular level to acclimatize to shifts in temperature.
- The **hypothalamus** acts as the thermostat in mammalian regulation of body temperature. Fever reflects a resetting of this thermostat to a higher normal range in response to infection.

? *Given that humans thermoregulate, explain why your skin is cooler than your body core.*

CONCEPT 40.4

Energy requirements are related to animal size, activity, and environment (pp. 889–895)

- Animals obtain chemical energy from food, storing it for short-term use in ATP. The total amount of energy used in a unit of time defines an animal's **metabolic rate**.
- Under similar conditions and for animals of the same size, the **basal metabolic rate** of endotherms is substantially higher than the **standard metabolic rate** of ectotherms. Minimum metabolic rate per gram is inversely related to body size among similar animals. Animals allocate energy for basal (or standard) metabolism, activity, homeostasis, growth, and reproduction.
- **Torpor**, a state of decreased activity and metabolism, conserves energy during environmental extremes. Animals may enter torpor according to a circadian rhythm (daily torpor), in winter (**hibernation**), or in summer (estivation).

? *Why do small animals breathe more rapidly than large animals?*

➡ For more multiple-choice questions, go to the **Practice Test** in the **Mastering Biology** eText or Study Area, or go to goo.gl/GruWRg.

Levels 1–2: Remembering/Understanding

1. The body tissue that consists largely of material located outside of cells is
   (A) epithelial tissue.
   (B) connective tissue.
   (C) muscle tissue.
   (D) nervous tissue.

2. Which of the following would increase the rate of heat exchange between an animal and its environment?
   (A) feathers or fur
   (B) vasoconstriction
   (C) wind blowing across the body surface
   (D) countercurrent heat exchanger

3. Consider the energy budgets for a human, an elephant, a penguin, a mouse, and a snake. The _____ would have the highest total annual energy expenditure, and the _____ would have the highest energy expenditure per unit mass.
   (A) elephant; mouse
   (B) elephant; human
   (C) mouse; snake
   (D) penguin; mouse

Levels 3–4: Applying/Analyzing

4. Compared with a smaller cell, a larger cell of the same shape has
   (A) less surface area.
   (B) less surface area per unit of volume.
   (C) the same surface-area-to-volume ratio.
   (D) a smaller cytoplasm-to-nucleus ratio.

5. An animal's inputs of energy and materials would exceed its outputs
   (A) if the animal is an endotherm, which must always take in more energy because of its high metabolic rate.
   (B) if it is actively foraging for food.
   (C) if it is growing and increasing its mass.
   (D) never; due to homeostasis, these energy and material budgets always balance.

6. You are studying a large tropical reptile that has a high and relatively stable body temperature. How do you determine whether this animal is an endotherm or an ectotherm?
   (A) You know from its high and stable body temperature that it must be an endotherm.
   (B) You subject this reptile to various temperatures in the lab and find that its body temperature and metabolic rate change with the ambient temperature. You conclude that it is an ectotherm.
   (C) You note that its environment has a high and stable temperature. Because its body temperature matches the environmental temperature, you conclude that it is an ectotherm.
   (D) You measure the metabolic rate of the reptile, and because it is higher than that of a related species that lives in temperate forests, you conclude that this reptile is an endotherm and its relative is an ectotherm.

7. Which of the following animals uses the largest percentage of its energy budget for homeostatic regulation?
   (A) marine jelly (an invertebrate)
   (B) snake in a temperate forest
   (C) desert insect
   (D) desert bird

8. **DRAW IT** Draw a model of the control circuit(s) required for driving an automobile at a fairly constant speed over a hilly road. Indicate each feature that represents a sensor, stimulus, or response.

Levels 5–6: Evaluating/Creating

9. **EVOLUTION CONNECTION** In 1847, the German biologist Christian Bergmann noted that mammals and birds living at higher latitudes (farther from the equator) are on average larger and bulkier than related species found at lower latitudes. Suggest an evolutionary hypothesis to explain this observation.

10. **SCIENTIFIC INQUIRY** Eastern tent caterpillars (*Malacosoma americanum*) live in large groups in silk nests resembling tents, which they build in trees. They are among the first insects to be active in early spring, when daily temperature fluctuates from freezing to very hot. Over the course of a day, they display striking differences in behavior: Early in the morning, they rest in a tightly packed group on the tent's east-facing surface. In midafternoon, they are on its undersurface, each caterpillar hanging by a few of its legs. Propose a hypothesis to explain this behavior. How could you test it?

11. **SCIENCE, TECHNOLOGY, AND SOCIETY** Medical researchers are investigating artificial substitutes for various human tissues. Why might artificial blood or skin be useful? What characteristics would these substitutes need in order to function well in the body? Why do real tissues work better? Why not use the real tissues if they work better? What other artificial tissues might be useful? What problems do you anticipate in developing and applying them?

12. **WRITE ABOUT A THEME: ENERGY AND MATTER** In a short essay (about 100–150 words) focusing on energy transfer and transformation, discuss the advantages and disadvantages of hibernation.

13. **SYNTHESIZE YOUR KNOWLEDGE**

These macaques (*Macaca fuscata*) are partially immersed in a hot spring in a snowy region of Japan. What are some ways that form, function, and behavior contribute to homeostasis for these animals?

For selected answers, see Appendix A.

41 Animal Nutrition

KEY CONCEPTS

41.1 An animal's diet must supply chemical energy, organic building blocks, and essential nutrients *p. 899*

41.2 Food processing involves ingestion, digestion, absorption, and elimination *p. 902*

41.3 Organs specialized for sequential stages of food processing form the mammalian digestive system *p. 905*

41.4 Evolutionary adaptations of vertebrate digestive systems correlate with diet *p. 911*

41.5 Feedback circuits regulate digestion, energy storage, and appetite *p. 915*

Study Tip

Make a flowchart: Digestion occurs stepwise in the human alimentary canal. As you study the molecular details, keep track of the overall process by adding to this start of a flowchart listing the overall effect and location of each step.

1. Break food into smaller pieces and mix with saliva (oral cavity)

↓

2. Transfer boluses of food to stomach (esophagus)

↓

3. Denature proteins in food (stomach)

↓

⊙ Go to Mastering Biology

For Students (in eText and Study Area)
• Get Ready for Chapter 41
• Figure 41.10 Walkthrough: The Stomach and Its Secretions
• BioFlix® Animation: Homeostasis: Regulating Blood Sugar

For Instructors to Assign (in Item Library)
• Tutorial: Fat Absorption and Fat Structure
• Tutorial: What Role Do Genes Play in Appetite Regulation?

Ready-to-Go Teaching Module (in Instructor Resources)
• The Human Digestive System

Figure 41.1 For this herring gull, dinnertime has arrived. Once this meal has been taken in, the tissues of the sea star will be taken apart and its nutrients taken up. Paradoxically, the classes of nutrients in the sea star—largely proteins, fats, and carbohydrates—also make up the tissues of the gull.

How can animals extract the nutrients they need from food while not digesting their own tissues?

An animal digests food using **compartmentalized processing** in a tube-like system. Compartmentalization protects body tissues while allowing enzymes and acids to break down nutrients.

Uptake of **nutrients** and **water** into blood

Bloodstream

BODY TISSUES

Food

Waste

Secretions that promote digestion

To bloodstream

Nutrients are transported through epithelial cells and absorbed into the bloodstream.

Digestive secretions catalyze breakdown of food and release of nutrients

Protective secretions help separate body cells from digestive activity.

An animal's diet must supply chemical energy, organic building blocks, and essential nutrients

Although dining on sea stars, crabs, and fish is the herring gull's specialty, all animals eat other organisms—dead or alive, piecemeal or whole. Unlike plants, animals must consume food for both energy and the organic molecules used to assemble new molecules, cells, and tissues. Overall, an adequate diet must satisfy three needs: chemical energy for cellular processes, organic building blocks for macromolecules, and essential nutrients. The process by which an animal takes in and makes use of food to meet these needs constitutes **nutrition**.

The activities of cells, tissues, organs, and whole animals depend on sources of chemical energy in the diet. This energy is used to produce ATP, which powers processes ranging from DNA replication and cell division to vision and flight (see Concept 8.3). To meet the need for ATP, animals ingest and digest nutrients, including carbohydrates, proteins, and lipids, for use in cellular respiration and energy storage.

In addition to fuel for ATP production, an animal requires raw materials for biosynthesis. To build the complex molecules it needs to grow, maintain itself, and reproduce, an animal's food must provide a source of organic carbon (such as sugar) and a source of organic nitrogen (such as protein).

The third requirement of an animal's diet is to provide **essential nutrients**, substances that an animal requires but cannot assemble from simple organic molecules.

Essential Nutrients

Essential nutrients in the diet include certain amino acids and fatty acids, as well as vitamins and minerals. The key functions of essential nutrients include serving as substrates of enzymes, as coenzymes, and as cofactors in biosynthetic reactions (**Figure 41.2**).

In general, an animal can obtain all essential amino acids and fatty acids, as well as vitamins and minerals, by feeding on plants or other animals. Needs for particular nutrients vary among species. For instance, some animals (including humans) must get ascorbic acid (vitamin C) from their diet, whereas most animals can synthesize it from other nutrients.

Essential Amino Acids

All organisms require a standard set of 20 amino acids to make a complete set of proteins (see Figure 5.14). Plants and microorganisms normally can produce all 20. Most animals have the enzymes to synthesize about half of these amino acids, as long as their diet includes sulfur and organic nitrogen. The remaining amino acids must be obtained from the animal's food in prefabricated form and are therefore called **essential amino acids**. Many animals, including adult humans, require eight amino acids in their diet: isoleucine, leucine, lysine, methionine, phenylalanine, threonine, tryptophan, and valine. (Human infants also need a ninth, histidine.)

The proteins in animal products such as meat, eggs, and cheese are "complete," providing all the essential amino acids. In contrast, most plant proteins are "incomplete," being deficient in one or more essential amino acids. Corn (maize), for example, is deficient in tryptophan and lysine, whereas beans are lacking in methionine. However, vegetarians can easily obtain all of the essential amino acids by eating a varied diet of plant proteins.

Essential Fatty Acids

Animals require fatty acids to synthesize a variety of cellular components, including membrane phospholipids, signaling molecules, and storage fats. Although animals can synthesize many fatty acids, they lack the enzymes to form the double bonds found in certain required fatty acids. Instead, these molecules must be obtained from the diet and are considered **essential fatty acids**. In mammals, they include linoleic acid (see Figure 41.2). Animals typically obtain ample quantities of essential fatty acids from seeds, grains, and vegetables in their diet.

▼ **Figure 41.2 Roles of essential nutrients.** This example of a biosynthetic reaction illustrates some common functions for essential nutrients. The conversion of linoleic acid to γ-linoleic acid by the enzyme fatty acid desaturase involves all four classes of essential nutrients, as labeled in blue. Note that almost all enzymes and other proteins in animals contain some essential amino acids, as indicated in the partial sequence shown for fatty acid desaturase.

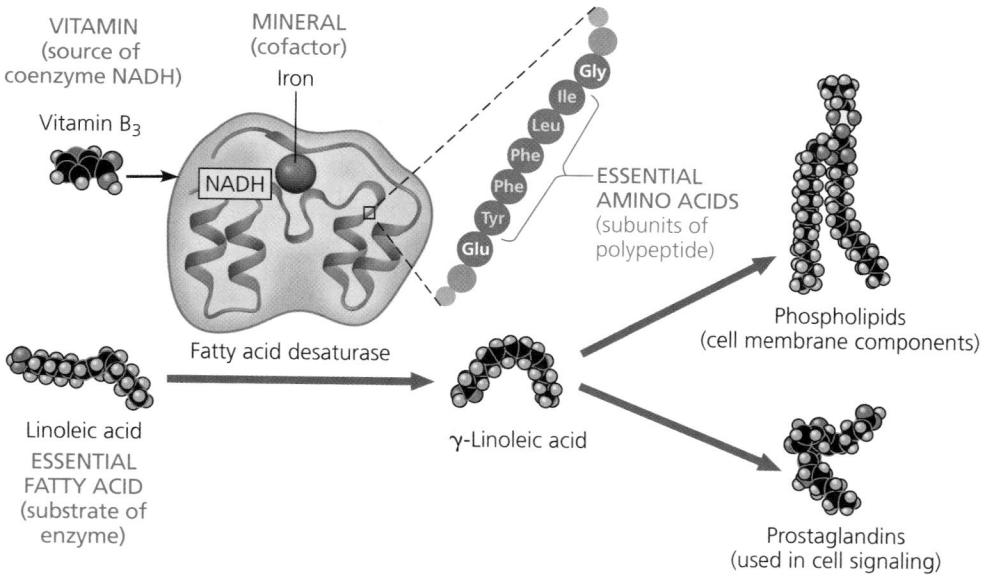

Vitamins

As Albert Szent-Györgyi, the discoverer of vitamin C, once quipped, "A vitamin is a substance that makes you ill if you *don't* eat it." **Vitamins** are organic molecules that are required in the diet in very small amounts (0.01–100 mg per day, depending on the vitamin).

The 13 vitamins required by humans vary in both chemical properties and function **(Table 41.1)**. Vitamin B_2, for example, is a water-soluble vitamin that is converted in the body to FAD, a coenzyme used in many metabolic processes, including cellular respiration (see Figure 9.12). Vitamin C, which is required for the production of connective tissue, is also water-soluble.

Fat-soluble vitamins include vitamin A, which is incorporated into visual pigments of the eye, and vitamin D, which aids in calcium absorption and bone formation. Our dietary requirement for vitamin D, unlike other vitamins, turns out to be variable. Why? When our skin is exposed to sunlight, our bodies synthesize vitamin D, reducing our dietary need.

For people with imbalanced diets, taking vitamin supplements at recommended daily levels is reasonable. It is far less clear that massive doses of vitamins confer any health benefits or are even safe. Moderate overdoses of water-soluble vitamins are probably harmless because excesses are excreted in urine. However, excesses of fat-soluble vitamins are deposited in body fat, so overconsumption may cause them to accumulate to toxic levels.

Minerals

Dietary **minerals** are inorganic nutrients, such as iron and sulfur, that are usually required in small amounts—from less than 1 mg to about 2,500 mg per day. As shown in **Table 41.2**, minerals have diverse functions in animal physiology. Some are assembled into the structure of proteins; iron, for example, is incorporated into the oxygen carrier hemoglobin as well as some enzymes (see Figure 41.2). Others, such as sodium, potassium, and chloride, are important in the functioning of nerves and muscles and in maintaining osmotic balance between cells and the surrounding body fluid. In vertebrates, the mineral iodine is incorporated into thyroid hormone, which regulates metabolic rate. Vertebrates also require relatively large quantities of calcium and phosphorus for building and maintaining bone.

Ingesting too much of some minerals can cause health problems. For example, excess sodium can contribute to

Table 41.1 Vitamin Requirements of Humans

Vitamin	Major Dietary Sources	Major Functions in the Body	Symptoms of Deficiency
Water-Soluble Vitamins			
B_1 (thiamine)	Pork, legumes, peanuts, whole grains	Coenzyme used in removing CO_2 from organic compounds	Beriberi (tingling, poor coordination, reduced heart function)
B_2 (riboflavin)	Dairy products, meats, enriched grains, vegetables	Component of coenzymes FAD and FMN	Skin lesions, such as cracks at corners of mouth
B_3 (niacin)	Nuts, meats, grains	Component of coenzymes NAD^+ and $NADP^+$	Skin and gastrointestinal lesions, delusions, confusion
B_5 (pantothenic acid)	Meats, dairy products, whole grains, fruits, vegetables	Component of coenzyme A	Fatigue, numbness, tingling of hands and feet
B_6 (pyridoxine)	Meats, vegetables, whole grains	Coenzyme used in amino acid metabolism	Irritability, convulsions, muscular twitching, anemia
B_7 (biotin)	Legumes, other vegetables, meats	Coenzyme in synthesis of fat, glycogen, and amino acids	Scaly skin inflammation, neuromuscular disorders
B_9 (folic acid)	Green vegetables, oranges, nuts, legumes, whole grains	Coenzyme in nucleic acid and amino acid metabolism	Anemia, neural tube malformation in fetus
B_{12} (cobalamin)	Meats, eggs, dairy products	Production of nucleic acids and red blood cells	Anemia, numbness, loss of balance
C (ascorbic acid)	Citrus fruits, broccoli, tomatoes	Used in collagen synthesis; antioxidant	Scurvy (degeneration of skin and teeth), delayed wound healing
Fat-Soluble Vitamins			
A (retinol)	Dark green and orange vegetables and fruits, dairy products	Component of visual pigments; maintenance of epithelial tissues	Blindness, skin disorders, impaired immunity
D	Dairy products, egg yolk	Aids in absorption and use of calcium and phosphorus	Rickets (bone deformities) in children, bone softening in adults
E (tocopherol)	Vegetable oils, nuts, seeds	Antioxidant; helps prevent damage to cell membranes	Nervous system degeneration
K (phylloquinone)	Green vegetables, tea; also made by colon bacteria	Important in blood clotting	Defective blood clotting

Table 41.2 Mineral Requirements of Humans*

Mineral		Major Dietary Sources	Major Functions in the Body	Symptoms of Deficiency
More than 200 mg per day required	Calcium (Ca)	Dairy products, dark green vegetables, legumes	Bone and tooth formation, blood clotting, nerve and muscle function	Impaired growth, loss of bone mass
	Phosphorus (P)	Dairy products, meats, grains	Bone and tooth formation, acid-base balance, nucleotide synthesis	Weakness, loss of minerals from bone, calcium loss
	Sulfur (S)	Proteins from many sources	Component of certain amino acids	Impaired growth, fatigue, swelling
	Potassium (K)	Meats, dairy products, many fruits and vegetables, grains	Acid-base balance, water balance, nerve function	Muscular weakness, paralysis, nausea, heart failure
	Chlorine (Cl)	Table salt	Acid-base balance, formation of gastric juice, nerve function, osmotic balance	Muscle cramps, reduced appetite
	Sodium (Na)	Table salt	Acid-base balance, water balance, nerve function	Muscle cramps, reduced appetite
	Magnesium (Mg)	Whole grains, green leafy vegetables	Enzyme cofactor; ATP bioenergetics	Nervous system disturbances
Iron (Fe)		Meats, eggs, legumes, whole grains, green leafy vegetables	Component of hemoglobin and of electron carriers; enzyme cofactor	Iron-deficiency anemia, weakness, impaired immunity
Fluorine (F)		Drinking water, tea, seafood	Maintenance of tooth structure	Higher frequency of tooth decay
Iodine (I)		Seafood, iodized salt	Component of thyroid hormones	Goiter (enlarged thyroid gland)

*Additional minerals required in trace amounts include cobalt (Co), copper (Cu), manganese (Mn), molybdenum (Mo), selenium (Se), and zinc (Zn). All of these minerals, as well as those in the table, can be harmful in excess.

high blood pressure. This is a particular problem in the United States, where the typical person consumes enough salt (sodium chloride) to provide about 20 times the required amount of sodium. Processed foods often contain large amounts of sodium chloride, even if they do not taste salty.

Variation in Diet

Despite many shared nutritional needs, animals have diverse diets. **Herbivores**, such as cattle, sea slugs, and caterpillars, dine mainly on plants or algae. **Carnivores**, such as sea otters, hawks, and spiders, mostly eat other animals. Rats and other **omnivores** (from the Latin *omnis*, all) don't in fact eat everything, but they do regularly consume animals as well as plants or algae. We humans are typically omnivores, as are cockroaches and crows.

The terms *herbivore, carnivore,* and *omnivore* represent the kinds of food an animal usually eats. However, most animals are opportunistic feeders, broadening their diet when their usual foods aren't available. For example, deer are herbivores, but occasionally eat insects, worms, or bird eggs. Similarly, herring gulls eat marine invertebrates, insects, and small fishes, but also human refuse. Note that microorganisms are an unavoidable "supplement" in every animal's diet.

Dietary Deficiencies

A diet that lacks one or more essential nutrients or consistently supplies less chemical energy than the body requires results in *malnutrition*, a failure to obtain adequate nutrition.

Malnutrition affects one out of four children worldwide, impairing health and often survival.

Deficiencies in Essential Nutrients

Insufficient intake of essential nutrients can cause deformities, disease, and even death. For example, deer and other herbivores can develop fragile bones if the plants they consume grew in phosphorus-deficient soil. In such environments, some grazing animals obtain missing nutrients by consuming concentrated sources of salt or other minerals **(Figure 41.3)**. Similarly, some birds supplement their diet with snail shells, and certain tortoises obtain minerals from stones they ingest.

Like other animals, humans sometimes have diets lacking in essential nutrients. A diet with insufficient amounts of one or more essential amino acids causes protein deficiency, the most common type of malnutrition among humans. In children, protein deficiency may arise if their diet shifts entirely from breast milk to foods that contain relatively little protein, such as rice. Such children, if they survive infancy, often have impaired physical and mental development.

▶ **Figure 41.3 Obtaining essential nutrients from an unusual source.** A juvenile chamois (*Rupicapra rupicapra*), an herbivore, licks salts from exposed rocks in its alpine habitat. This behavior is common among herbivores that live where soils and plants provide insufficient amounts of minerals.

Undernourishment

As mentioned earlier, malnutrition can also be caused by a diet that fails to provide enough chemical energy. In this situation, the body first uses up stored carbohydrates and fat. It then begins breaking down its own proteins for fuel: Muscles shrink, and the brain may become protein-deficient. If energy intake remains less than energy expenditures, the animal will eventually die. Even if a seriously undernourished animal survives, some of the damage may be irreversible.

Inadequate nourishment in humans is most common when drought, war, or other crisis severely disrupts the food supply. In sub-Saharan Africa, where the AIDS epidemic has crippled both rural and urban communities, approximately 200 million children and adults cannot obtain enough food.

Sometimes undernourishment occurs within well-fed human populations as a result of eating disorders. For example, anorexia nervosa involves weight loss to a level that is unhealthy for the individual's age and height and may be related to a distorted body image.

Assessing Nutritional Needs

Determining the ideal diet for the human population is an important but difficult problem for scientists. As objects of study, people present many challenges. Unlike laboratory animals, humans are genetically diverse. They also live in settings far more varied than the stable and uniform environment that scientists use in laboratory experiments. Ethical concerns present an additional barrier. For example, it is not acceptable to investigate the nutritional needs of children in a way that might harm a child's growth or development.

Many insights into human nutrition have come from *epidemiology*, the study of human health and disease at the population level. In the 1970s, for instance, researchers discovered that children born to women of low socioeconomic status were more likely to have neural tube defects, which occur when tissue fails to enclose the developing brain and spinal cord (see Concept 47.2). The English scientist Richard Smithells thought that malnutrition among these women might be responsible. As described in **Figure 41.4**, he found that vitamin supplementation greatly reduced the risk of neural tube defects. In other studies, he obtained evidence that folic acid (vitamin B$_9$) was the specific vitamin responsible, a finding confirmed by other researchers. Based on this evidence, the United States in 1998 began to require that folic acid be added to enriched grain products used to make bread, cereals, and other foods. Follow-up studies have documented the effectiveness of this program in reducing the frequency of neural tube defects. Thus, at a time when microsurgery and sophisticated diagnostic imaging dominate the headlines, a simple dietary change such as folic acid supplementation may be among the greatest contributors to human health.

▼ **Figure 41.4** Inquiry

Can diet influence the frequency of neural tube defects?

Experiment Richard Smithells, of the University of Leeds, in England, examined the effect of vitamin supplementation on the risk of neural tube defects. Women who had had one or more babies with such a condition were put into two study groups. The experimental group consisted of those who were planning a pregnancy and began taking a multivitamin at least four weeks before attempting conception. The control group, who were not given vitamins, included women who declined them and women who were already pregnant. The numbers of neural tube defects resulting from the pregnancies were recorded for each group.

Results

Group	Number of Infants/Fetuses Studied	Infants/Fetuses with a Neural Tube Defect
Vitamin supplements (experimental group)	141	1
No vitamin supplements (control group)	204	12

Data from R. W. Smithells et al., Possible prevention of neural-tube defects by periconceptional vitamin supplementation, *Lancet* 315:339–340 (1980).

Conclusion This controlled study provided evidence that vitamin supplementation protects against neural tube defects, at least after the first pregnancy. Follow-up trials demonstrated that folic acid alone provided an equivalent protective effect.

INTERPRET THE DATA *After folic acid supplementation became standard in the United States, the frequency of neural tube defects dropped to an average of just one in 5,000 live births. Propose two explanations of why the observed frequency was much higher in the experimental group of the Smithells study.*

INQUIRY IN ACTION Read and analyze the original paper in *Inquiry in Action: Interpreting Scientific Papers.*

CONCEPT CHECK 41.1

1. An animal requires 20 amino acids to make proteins. Why aren't all 20 essential to animal diets?

2. **MAKE CONNECTIONS** Considering how enzymes function (see Concept 8.4), explain why vitamins are required in very small amounts.

3. **WHAT IF?** If a zoo animal eating ample food shows signs of malnutrition, how might a researcher determine which nutrient is lacking in its diet?

For suggested answers, see Appendix A.

CONCEPT 41.2

Food processing involves ingestion, digestion, absorption, and elimination

Whatever their diet, animals need to process food. We can divide food processing into ingestion, digestion, absorption, and elimination. The first stage— **ingestion**—is the act of eating or feeding. As shown in **Figure 41.5**, four quite different categories describe the feeding mechanisms of most animal species.

Exploring Four Main Feeding Mechanisms of Animals

Filter Feeding

Baleen

Many aquatic animals are **filter feeders**, which strain small organisms or food particles from the surrounding medium. The humpback whale, shown above, is one example. Attached to the whale's upper jaw are comblike plates called baleen, which remove small invertebrates and fish from enormous volumes of water and sometimes mud. Filter feeding in water is a type of suspension feeding, which also includes removing suspended food particles from the surrounding medium by capture or trapping mechanisms.

Substrate Feeding

Substrate feeders are animals that live in or on their food source. This leaf miner caterpillar, the larva of a moth, is eating through the soft tissue of an oak leaf, leaving a dark trail of feces in its wake. Other substrate feeders include maggots (fly larvae), which burrow into animal carcasses.

Caterpillar   Feces

Fluid Feeding

Fluid feeders suck nutrient-rich fluid from a living host. This tsetse fly has pierced the skin of its human host with hollow, needlelike mouthparts and is consuming a blood meal. Similarly, aphids are fluid feeders that tap the phloem sap of plants. In contrast to such parasites, some fluid feeders actually benefit their hosts. For example, hummingbirds and bees move pollen between flowers as they fluid-feed on nectar.

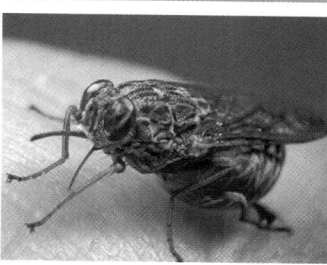

Bulk Feeding

Most animals, including humans, are **bulk feeders**, which eat relatively large pieces of food. Their adaptations include tentacles, pincers, claws, venomous fangs, jaws, and teeth that kill their prey or tear off pieces of meat or vegetation. In this amazing scene, a rock python is beginning to ingest a gazelle it has captured and killed. Snakes cannot chew their food into pieces and must swallow it whole—even if the prey is much bigger than the diameter of the snake. They can do so because the lower jaw is loosely hinged to the skull by an elastic ligament that permits the mouth and throat to open very wide. After swallowing its prey, which may take more than an hour, the python will spend two weeks or longer digesting its meal.

During **digestion**, the second stage of food processing, food is broken down into molecules small enough for the body to absorb. Both mechanical and chemical processes are typically required. Mechanical digestion, such as chewing or grinding, breaks food into smaller pieces, increasing surface area. The food particles then undergo chemical digestion, which cleaves large molecules into smaller components.

Chemical digestion is necessary because animals cannot directly use the proteins, carbohydrates, nucleic acids, fats, and phospholipids in food. These molecules are too large to pass through cell membranes and also are not all identical to those the animal needs for its particular tissues and functions. But when large molecules in food are broken down into their smaller components, the animal can use these products of digestion to assemble the large molecules it needs. For example, although the humpback whale and the tsetse fly in Figure 41.5 have very different diets, both break down proteins in their food to the same 20 amino acids from which they assemble all of the specific proteins in their bodies.

Enzyme-catalyzed synthesis of a fat or macromolecule links together smaller components, releasing a molecule of water for each new covalent bond formed. Chemical breakdown by digestive enzymes reverses this process, breaking bonds through the addition of water. This splitting process is called *enzymatic hydrolysis*. Polysaccharides and disaccharides are split into simple sugars, as shown here for sucrose and the enzyme sucrase:

▼ Enzymatic hydrolysis of a disaccharide

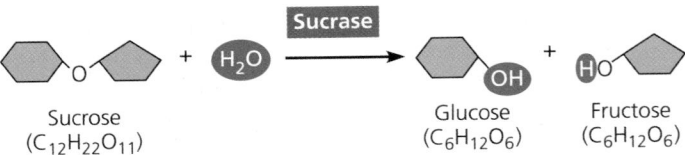

Sucrose $(C_{12}H_{22}O_{11})$	Glucose $(C_6H_{12}O_6)$	Fructose $(C_6H_{12}O_6)$

Similarly, proteins are broken down into small peptides and amino acids, and nucleic acids are cleaved into nucleotides and their components. Enzymatic hydrolysis also releases fatty acids and other components from fats and phospholipids. In many animals, bacteria living in the digestive system carry out some chemical digestion.

The last two stages of food processing occur after the food is digested. In the third stage, **absorption**, the animal's cells take up (absorb) small molecules such as amino acids and simple sugars. **Elimination**, in which undigested material passes out of the digestive system, completes the process.

Digestive Compartments

You have just read that digestive enzymes hydrolyze the same biological materials (such as proteins, fats, and carbohydrates) that make up the bodies of the animals themselves. How, then, are animals able to digest food without digesting their own cells and tissues? The evolutionary adaptation that allows animals to avoid self-digestion is the processing of food within specialized intracellular or extracellular compartments.

Intracellular Digestion

Food vacuoles—cellular organelles in which hydrolytic enzymes break down food—are the simplest digestive compartments. The hydrolysis of food inside vacuoles, called intracellular digestion, begins after a cell engulfs solid food by phagocytosis or liquid food by pinocytosis (see Figure 7.21). Newly formed food vacuoles fuse with lysosomes, organelles containing hydrolytic enzymes. This fusion of organelles brings food in contact with these enzymes, allowing digestion to occur safely within a compartment enclosed by a protective membrane. A few animals, such as sponges, digest all their food in this way (see Figure 33.4).

Extracellular Digestion

In most animal species, hydrolysis occurs largely by extracellular digestion, the breakdown of food in compartments that are continuous with the outside of the animal's body. Having one or more extracellular compartments for digestion enables an animal to devour much larger pieces of food than can be ingested by phagocytosis.

Animals with relatively simple body plans typically have a digestive compartment with a single opening. This pouch, called a **gastrovascular cavity**, functions in digestion as well as in the distribution of nutrients throughout the body (hence the *vascular* part of the term). Small freshwater cnidarians called hydras provide a good example **(Figure 41.6)**. The

▼ Figure 41.6 Digestion in a hydra. Digestion begins in the gastrovascular cavity and is completed intracellularly after small food particles are engulfed by specialized cells of the gastrodermis.

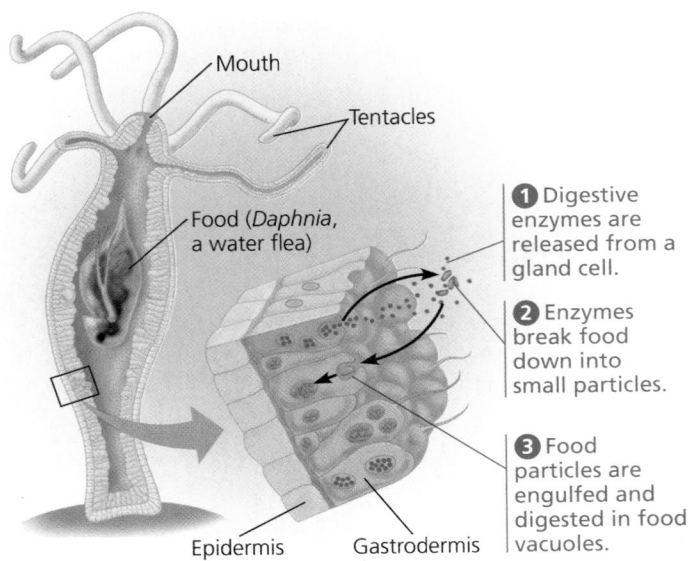

Mouth

Tentacles

Food (*Daphnia*, a water flea)

Epidermis   Gastrodermis

❶ Digestive enzymes are released from a gland cell.

❷ Enzymes break food down into small particles.

❸ Food particles are engulfed and digested in food vacuoles.

DRAW IT *Draw and label a simple diagram showing the pathway that nutrients follow from when food enters the hydra's mouth to when nutrients reach a cell on the outside of the tip of one of its tentacles.*

→ **Mastering Biology** Video: *Hydra* Eating *Daphnia*

hydra—a carnivore—uses its tentacles to stuff captured prey through its mouth into its gastrovascular cavity. Specialized gland cells of the hydra's gastrodermis, the tissue layer that lines the cavity, then secrete digestive enzymes that break the soft tissues of the prey into tiny pieces. Other cells of the gastrodermis engulf these food particles, and most of the hydrolysis of macromolecules occurs intracellularly. After the hydra has digested its meal, undigested materials that remain in its gastrovascular cavity, such as exoskeletons of small crustaceans, are eliminated through its mouth. Many flatworms also have a gastrovascular cavity (see Figure 33.9).

▼ **Figure 41.7 Variation in alimentary canals.** These examples illustrate how the organization and structure of compartments for digestion, storage, and absorption differ among animals.

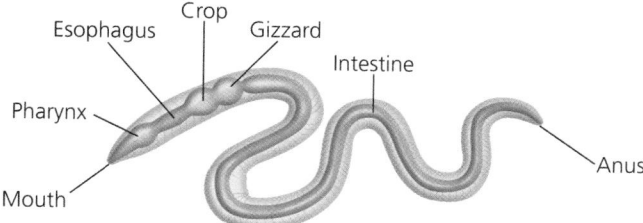

(a) Earthworm. The muscular pharynx of an earthworm sucks food in through the mouth. Food passes through the esophagus and is stored and moistened in the crop. Mechanical digestion occurs in the muscular gizzard, which pulverizes food with the aid of small bits of sand and gravel. Further digestion and absorption occur in the intestine before wastes are eliminated through the anus.

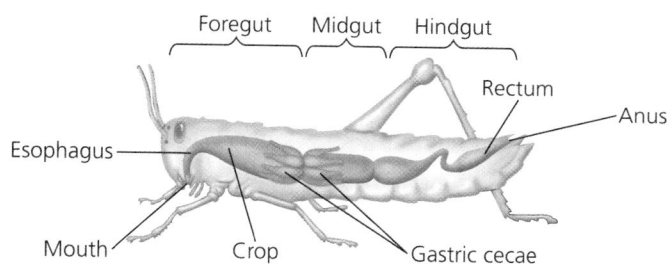

(b) Grasshopper. A grasshopper has several digestive chambers grouped into three main regions: a foregut, with an esophagus and crop; a midgut; and a hindgut. Food is moistened and stored in the crop, but most digestion occurs in the midgut. Pouches called gastric cecae (singular, *ceca*) extend from the beginning of the midgut and function in digestion and absorption.

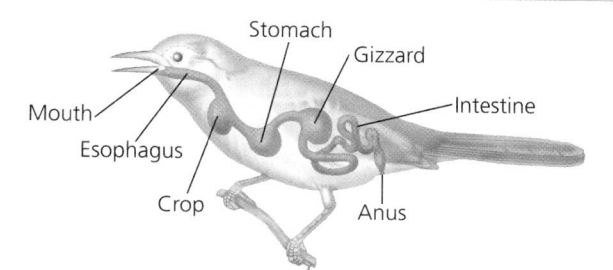

(c) Bird. Many birds have a crop for storing food and a stomach and gizzard for mechanically digesting it. Chemical digestion and absorption of nutrients occur in the intestine.

Rather than a gastrovascular cavity, animals with complex body plans have a digestive tube with two openings, a mouth and an anus **(Figure 41.7)**. Such a tube is called a *complete digestive tract* or, more commonly, an **alimentary canal**. Food moves along the alimentary canal in a single direction, encountering a series of specialized compartments that carry out stepwise digestion and nutrient absorption. An animal with an alimentary canal can ingest food while earlier meals are still being digested, a feat that is likely to be difficult or inefficient for an animal with a gastrovascular cavity.

Because most animals have an alimentary canal, we'll use the complete digestive system of mammals in the next section to illustrate the general principles of food processing.

CONCEPT CHECK 41.2

1. Distinguish the overall structure of a gastrovascular cavity from that of an alimentary canal.

2. In what sense are nutrients from a recently ingested meal not really "inside" your body prior to the absorption stage of food processing?

3. **WHAT IF?** Thinking in broad terms, what similarities can you identify between digestion in an animal body and the breakdown of gasoline in an automobile engine? (You don't have to know about auto mechanics.)

For suggested answers, see Appendix A.

CONCEPT 41.3

Organs specialized for sequential stages of food processing form the mammalian digestive system

In mammals, a number of accessory glands support food processing by secreting digestive juices through ducts into the alimentary canal. There are three pairs of salivary glands, as well as three individual glands: the pancreas, the liver, and the gallbladder. To explore the coordinated function of the accessory glands and alimentary canal, we'll consider the steps in food processing as a meal travels along the canal in a human.

The Oral Cavity, Pharynx, and Esophagus

As soon as food enters your mouth, or **oral cavity**, food processing begins **(Figure 41.8)**. Teeth with specialized shapes cut, mash, and grind, breaking the food into smaller pieces. This mechanical breakdown not only increases the surface area available for chemical breakdown but also facilitates swallowing. Meanwhile, the anticipation or arrival of food in the oral cavity triggers the release of saliva by the **salivary glands**.

▼ **Figure 41.8 The human digestive system.** After food is chewed and swallowed, it takes 5–10 seconds for it to pass down the esophagus and into the stomach, where it is stored for 2–6 hours during the first stages of processing. Complete digestion and nutrient absorption occur in the small intestine over a period of 5–6 hours. Processing is completed in the large intestine, and undigested material is expelled through the anus as feces.

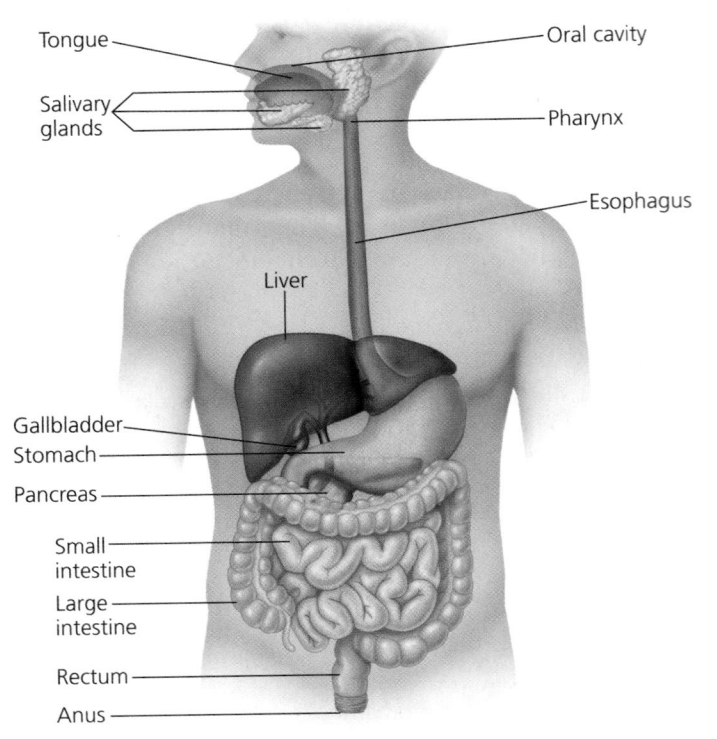

Tongue
Salivary glands
Oral cavity
Pharynx
Esophagus
Liver
Gallbladder
Stomach
Pancreas
Small intestine
Large intestine
Rectum
Anus

Mastering Biology Animation: Overview of the Human Digestive System

▼ **Figure 41.9 Intersection of the human airway and digestive tract.** In humans, the pharynx connects to the trachea and the esophagus. **(a)** At most times, a contracted sphincter seals off the esophagus while the trachea remains open. **(b)** When a food bolus arrives at the pharynx, the swallowing reflex is triggered. Movement of the larynx, the upper part of the airway, tips a flap of tissue called the epiglottis down, preventing food from entering the trachea. At the same time, the esophageal sphincter relaxes, allowing the bolus to pass into the esophagus. The trachea then reopens, and peristaltic contractions of the esophagus move the bolus to the stomach.

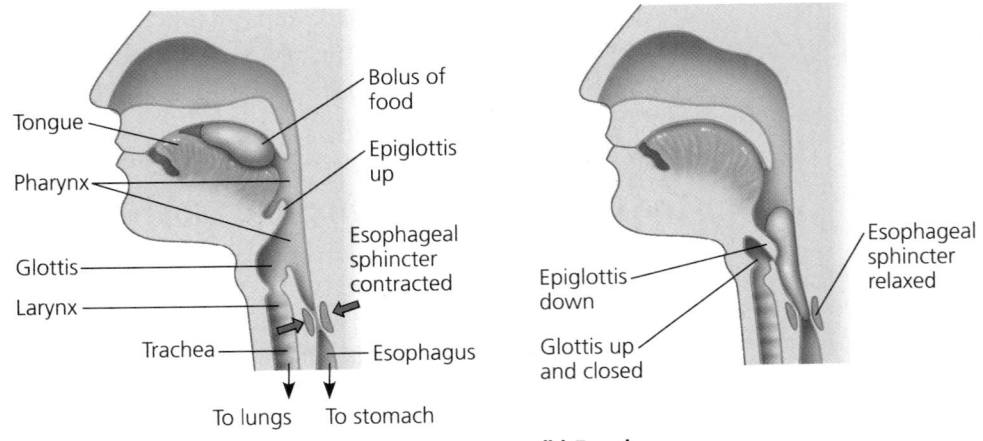

Tongue
Pharynx
Glottis
Larynx
Trachea
To lungs
To stomach
Bolus of food
Epiglottis up
Esophageal sphincter contracted
Esophagus

(a) Trachea open

Epiglottis down
Glottis up and closed
Esophageal sphincter relaxed

(b) Esophagus open

VISUAL SKILLS *If you laugh while drinking water, the liquid may be ejected from your nostrils. Use this diagram to explain why this happens, taking into account that laughing involves exhaling.*

Saliva is a complex mixture of materials with a number of vital functions. One major component is **mucus**, a viscous mixture of water, salts, cells, and slippery glycoproteins (carbohydrate-protein complexes). Mucus lubricates food for easier swallowing, protects the gums against abrasion, and facilitates taste and smell. Saliva also contains buffers, which help prevent tooth decay by neutralizing acid, and antimicrobial agents (such as lysozyme; see Figure 5.16), which protect against bacteria that enter the mouth with food.

Scientists have long been puzzled by the fact that saliva contains a large amount of the enzyme **amylase**, which breaks down starch (a glucose polymer from plants) and glycogen (a glucose polymer from animals). Most chemical digestion occurs not in the mouth but in the small intestine, where amylase is also present. Why, then, does saliva contain so much amylase? A current hypothesis is that amylase in saliva releases food particles that are stuck to the teeth, thereby reducing the nutrients available to microorganisms living in the mouth.

The tongue also has important roles in food processing. Much as a doorman screens and assists people entering a fancy hotel, the tongue aids digestive processes by evaluating ingested material, distinguishing which foods should be processed further and then enabling their passage. (See Concept 50.4 for a discussion of the sense of taste.) Once chewing begins, tongue movements manipulate the mixture of saliva and food, helping shape it into a ball called a *bolus* **(Figure 41.9)**. During swallowing, the tongue provides further assistance, pushing the bolus to the back of the oral cavity and into the pharynx.

Each bolus of food is received by the **pharynx**, or throat region, which leads to two passageways: the esophagus and the trachea. The **esophagus** is a muscular tube that connects to the stomach; the trachea (windpipe) leads to the lungs. Swallowing must therefore be carefully choreographed to keep food and liquids from entering the trachea and causing choking, a blockage of the trachea. The resulting lack of airflow into the lungs can be fatal if the material is not dislodged by vigorous coughing, a series of back slaps, or a forced upward thrust of the diaphragm (the Heimlich maneuver).

Within the esophagus, food is pushed along by **peristalsis**, alternating waves of smooth muscle contraction and relaxation. Upon reaching the end of the esophagus, the bolus encounters a **sphincter**, a ringlike valve of muscle **(Figure 41.10)**. Acting like a drawstring, the sphincter regulates passage of the ingested food into the next compartment, the stomach.

Digestion in the Stomach

The **stomach**, which is located just below the diaphragm, has two major roles in digestion. The first is storage. With a very elastic wall and accordion-like folds, the stomach can stretch to accommodate about 2 L of food and fluid. The second major function is to process food into a liquid suspension. The stomach secretes a digestive fluid called **gastric juice** and mixes it with the food through a churning action. This mixture of ingested food and gastric juice is called **chyme**.

Chemical Digestion in the Stomach

Two components of gastric juice help liquefy food in the stomach. First, hydrochloric acid (HCl) disrupts the extracellular matrix that binds cells together in meat and plant material. The concentration of HCl is so high that the pH of gastric juice is about 2, acidic enough to dissolve iron nails (and to kill most bacteria). This low pH denatures (unfolds) proteins in food, increasing exposure of their peptide bonds. The exposed bonds are then attacked by the second component of gastric juice—a **protease**, or protein-digesting enzyme, called **pepsin**. Unlike most enzymes, pepsin is adapted to work best in a very acidic environment. By breaking peptide bonds, it cleaves proteins into smaller polypeptides and further exposes the contents of ingested tissues.

Two types of cells in the gastric glands of the stomach produce the components of gastric juice (see Figure 41.10). *Parietal cells* use an ATP-driven pump to expel hydrogen ions into the lumen. At the same time, chloride ions diffuse into the lumen through specific membrane channels of the parietal cells. It is therefore only within the lumen that hydrogen and chloride ions combine to form HCl. Meanwhile, *chief cells* release pepsin into the lumen in an inactive form called **pepsinogen**. HCl converts pepsinogen to active pepsin by clipping off a small portion of the molecule and exposing its active site. Through these processes, both HCl and pepsin form in the lumen (cavity) of the stomach, not within the cells of the gastric glands. As a result, the parietal and chief cells produce gastric juice but are not digested from within by its components.

After hydrochloric acid converts a small amount of pepsinogen to pepsin, pepsin itself helps activate the remaining pepsinogen. Pepsin, like HCl, can clip pepsinogen to expose the enzyme's active site. This generates more pepsin, which activates more pepsinogen. This series of events is an example of positive feedback (see Concept 40.2).

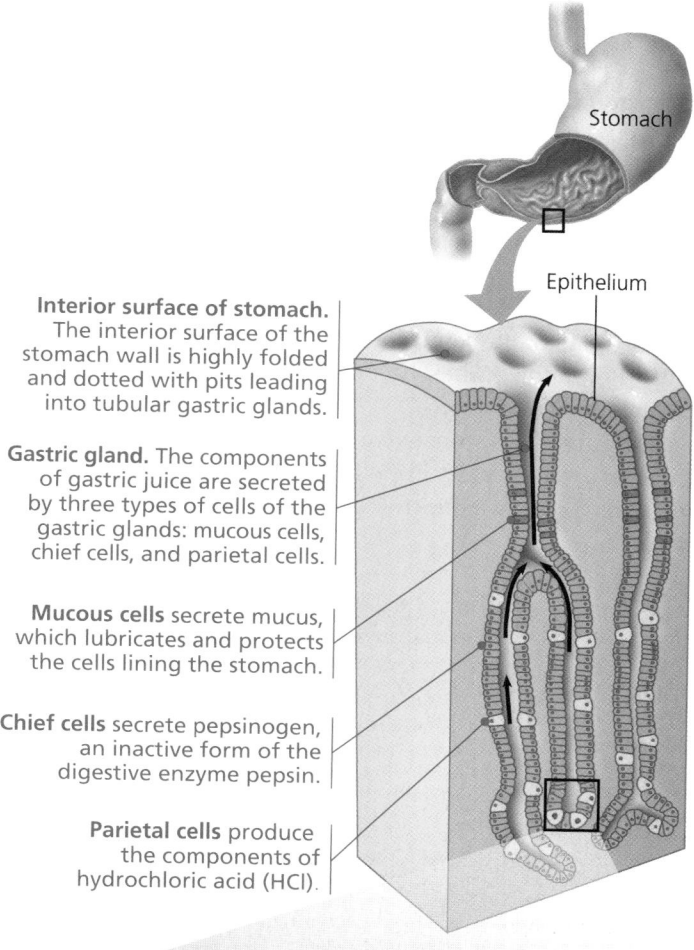

▼ **Figure 41.10** **The stomach and its secretions.**

Stomach

Epithelium

Interior surface of stomach. The interior surface of the stomach wall is highly folded and dotted with pits leading into tubular gastric glands.

Gastric gland. The components of gastric juice are secreted by three types of cells of the gastric glands: mucous cells, chief cells, and parietal cells.

Mucous cells secrete mucus, which lubricates and protects the cells lining the stomach.

Chief cells secrete pepsinogen, an inactive form of the digestive enzyme pepsin.

Parietal cells produce the components of hydrochloric acid (HCl).

Pepsinogen → Pepsin (active enzyme)

HCl

Chief cell

H⁺

Cl⁻

Parietal cell

The production of gastric juice

1 Pepsinogen and HCl are introduced into the lumen of the stomach.

2 HCl converts pepsinogen to pepsin.

3 Pepsin then activates more pepsinogen, starting a chain reaction. Pepsin begins the chemical digestion of proteins.

Mastering Biology **Figure Walkthrough**

Why don't HCl and pepsin eat through the lining of the stomach? For one thing, mucus secreted by cells in gastric glands protects against self-digestion (see Figure 41.10).

In addition, cell division adds a new epithelial layer every three days on average, replacing cells before the lining is fully eroded by digestive juices.

Stomach Dynamics

The breakdown of food by gastric juices is enhanced by muscular activity of the stomach. The coordinated series of muscle contractions and relaxations that we call "churning" mixes the stomach contents about every 20 seconds. Churning facilitates the action of HCl and pepsin by bringing all of the food into contact with the gastric juices secreted by the lining of the stomach. As a result, what began as a recently swallowed meal becomes the acidic, nutrient-rich broth known as chyme.

Contractions of stomach muscles also help move material through the alimentary canal. In particular, peristaltic contractions typically move the contents of the stomach into the small intestine within 2–6 hours after a meal. The sphincter located where the stomach opens to the small intestine helps regulate passage into the small intestine, allowing only one squirt of chyme to enter at a time.

Occasionally, the sphincter at the top of the stomach allows a movement, or flux, of chyme from the stomach back into the lower end of the esophagus. The painful irritation of the esophagus that results from this process of acid reflux is commonly called "heartburn."

Digestion in the Small Intestine

Although there is some chemical digestion in the oral cavity and stomach, most enzymatic hydrolysis of macromolecules from food occurs in the small intestine (**Figure 41.11**). This organ's name refers to its small diameter compared to the large intestine, not to its length. The **small intestine** is in

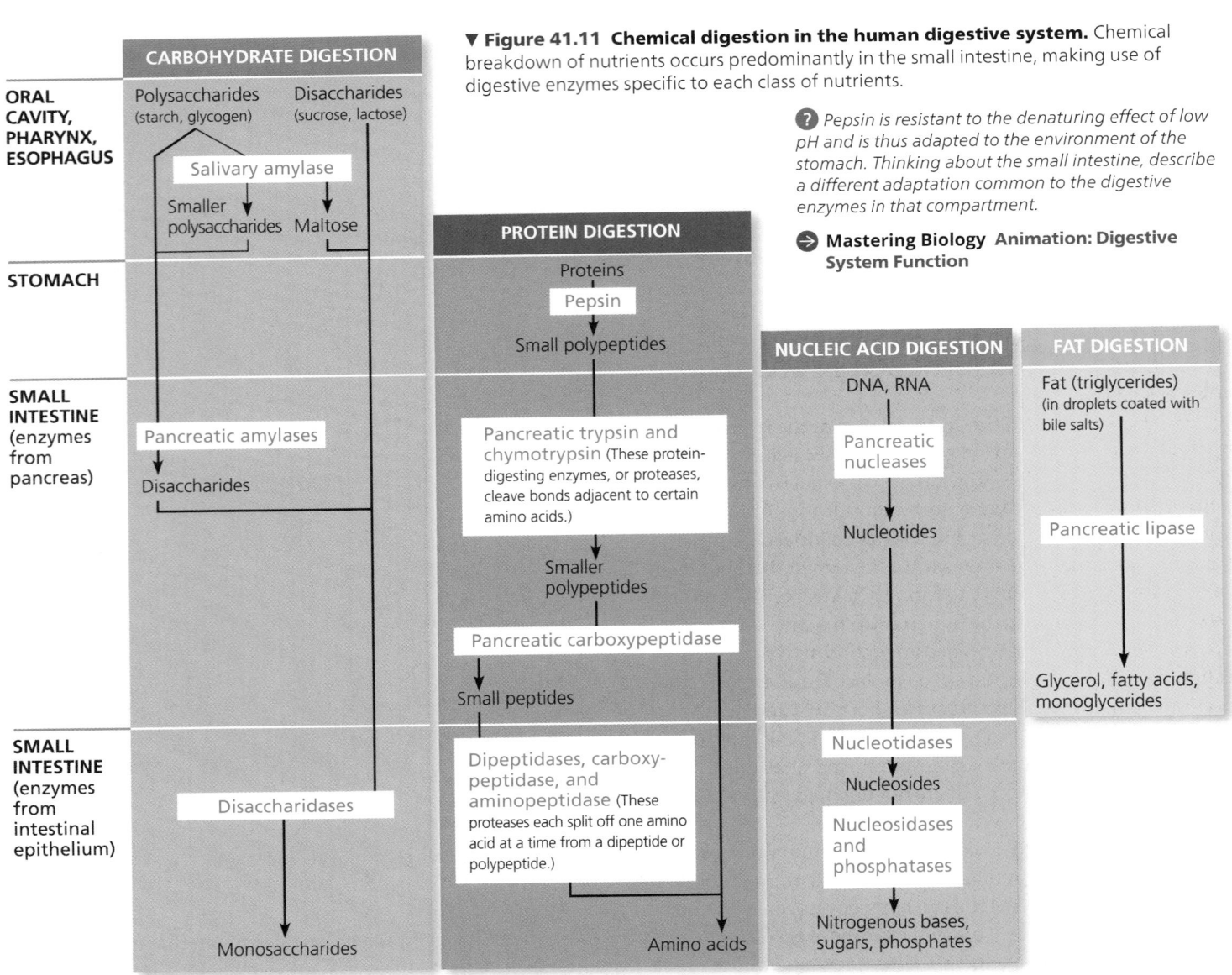

▼ **Figure 41.11 Chemical digestion in the human digestive system.** Chemical breakdown of nutrients occurs predominantly in the small intestine, making use of digestive enzymes specific to each class of nutrients.

? *Pepsin is resistant to the denaturing effect of low pH and is thus adapted to the environment of the stomach. Thinking about the small intestine, describe a different adaptation common to the digestive enzymes in that compartment.*

→ **Mastering Biology** Animation: Digestive System Function fact the alimentary canal's longest compartment—over 6 m (20 feet) long in humans. The first 25 cm (10 inches) or so of the small intestine forms the **duodenum**. It is here that chyme from the stomach mixes with digestive juices from the pancreas, liver, and gallbladder, as well as from gland cells of the intestinal wall itself. As you will read in Concept 41.5, hormones released by the stomach and duodenum control the digestive secretions into the alimentary canal.

The arrival of chyme in the duodenum triggers release of the hormone secretin, which stimulates the **pancreas** to secrete bicarbonate. Bicarbonate neutralizes the acidity of chyme and acts as a buffer for chemical digestion in the small intestine. The pancreas also secretes numerous digestive enzymes into the small intestine. These include the proteases trypsin and chymotrypsin, which are produced in inactive forms. In a chain reaction similar to that for pepsinogen, they are activated when safely located in the lumen of the duodenum.

The epithelial lining of the duodenum is the source of additional digestive enzymes. Some are secreted into the lumen of the duodenum, whereas others are bound to the surface of epithelial cells. Together with the enzymes from the pancreas, they complete most digestion in the duodenum.

Fats present a particular challenge for digestion. Insoluble in water, they form large globules that cannot be attacked efficiently by digestive enzymes. In humans and other vertebrates, fat digestion is facilitated by bile salts, which act as emulsifiers (detergents) that break apart fat and lipid globules. Bile salts are a major component of **bile**, a secretion of the **liver** that is stored and concentrated in the **gallbladder**.

Bile production is metabolically linked to another vital liver function: the destruction of red blood cells that are no longer fully functional. Pigments released during red blood cell disassembly are incorporated into bile pigments, which are eliminated from the body with the feces. In some liver and blood disorders, bile pigments accumulate in the skin, resulting in a yellowing called jaundice.

Absorption in the Small Intestine

With digestion largely complete, the contents of the duodenum move by peristalsis into the *jejunum* and *ileum*, the remaining regions of the small intestine. There, nutrient absorption occurs across the lining of the intestine **(Figure 41.12)**. Large folds in the lining encircle the intestine and are studded with finger-shaped projections called **villi** (singular, *villus*). Within the villi, each epithelial cell has many microscopic projections, or **microvilli**, that face the intestinal lumen. These epithelial microvilli have a brush-like appearance that is reflected in the name *brush border*. Together, the folds, villi, and microvilli have a surface area of 200–300 m^2, roughly the size of a tennis court. This enormous surface area is an evolutionary adaptation that greatly increases the rate of nutrient absorption (see Figure 33.9 for more discussion and examples of maximizing surface area in diverse organisms).

Depending on the nutrient, transport across the epithelial cells can be passive or active (see Concepts 7.3 and 7.4). The sugar fructose, for example, moves by facilitated diffusion down its concentration gradient from the lumen of the small

▼ **Figure 41.12 Nutrient absorption in the small intestine.** Water-soluble nutrients, such as amino acids and sugars, enter the bloodstream, whereas fats are transported to the lymphatic system.

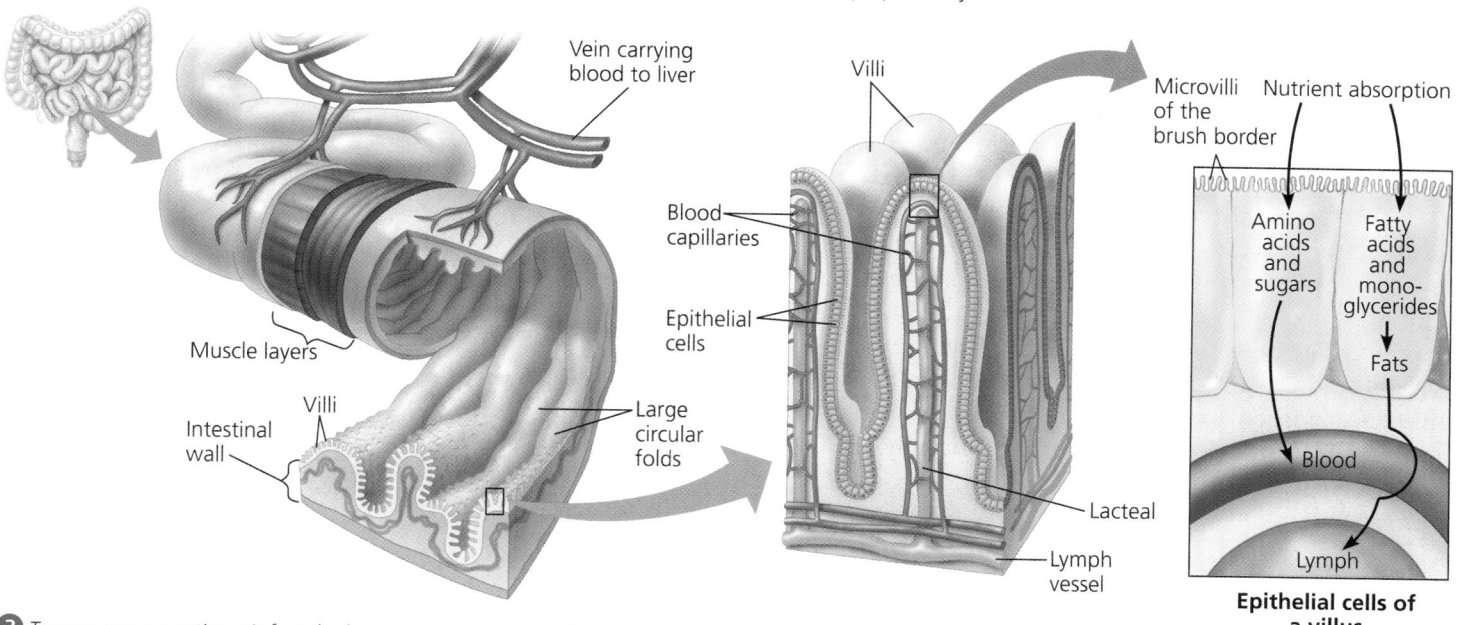

❓ *Tapeworms sometimes infect the human alimentary canal, anchoring themselves to the wall of the small intestine. Based on how digestion is compartmentalized along the mammalian alimentary canal, what digestive functions would you expect these parasites to have?*

→ **Mastering Biology** BioFlix® Animation: Nutrient Transport Across Membranes intestine into the epithelial cells. From there, fructose exits the basal surface and is absorbed into microscopic blood vessels, or capillaries, at the core of each villus. Other nutrients, including amino acids, small peptides, vitamins, and most glucose molecules, enter the epithelial cells of the villus by being pumped against concentration gradients. This active transport allows much more absorption of those nutrients than would be possible with passive diffusion alone.

The capillaries and veins that carry nutrient-rich blood away from the villi converge into the **hepatic portal vein**, a blood vessel that leads directly to the liver. From the liver, blood travels to the heart and then to other tissues and organs. This arrangement serves two major functions. First, it allows the liver to regulate the distribution of nutrients to the rest of the body. Because the liver converts many organic nutrients to different forms for use elsewhere, blood leaving the liver may have a very different nutrient balance than the blood that entered. Second, the arrangement allows the liver to remove toxic substances before they can circulate broadly. The liver is the primary site for detoxifying many organic molecules foreign to the body, such as drugs, and certain metabolic waste products.

Although many nutrients leave the small intestine through the bloodstream and pass through the liver for processing, some products of fat (triglyceride, also known as triacylglycerol) digestion take a different path **(Figure 41.13)**. Hydrolysis of a fat by lipase in the small intestine generates fatty acids and a monoglyceride (glycerol joined to a fatty acid). These products are absorbed by epithelial cells and recombined into triglycerides. They are then coated with phospholipids, cholesterol, and proteins, forming globules called **chylomicrons**.

In exiting the small intestine, chylomicrons first enter a **lacteal**, a vessel at the core of each villus. Lacteals are part of the vertebrate lymphatic system, which is a network of vessels filled with a clear fluid called lymph. Starting at the lacteals, lymph containing the chylomicrons passes into the larger vessels of the lymphatic system and eventually into large veins that return the blood directly to the heart.

In addition to absorbing nutrients, the small intestine recovers water and ions. Each day a person consumes about 2 L of water and secretes another 7 L in digestive juices into the alimentary canal. Typically all but 0.1 L of the water is reabsorbed, mostly in the small intestine. There is no mechanism for active transport of water. Instead, water is reabsorbed by osmosis when sodium and other ions are pumped out of the lumen of the small intestine.

Processing in the Large Intestine

The alimentary canal ends with the **large intestine**, which includes the colon, cecum, and rectum. The small intestine connects to the large intestine at a T-shaped junction **(Figure 41.14)**.

▼ **Figure 41.13 Digestion and absorption of fats.** Fats, which are insoluble in water, are broken down in the lumen of the small intestine, reassembled in epithelial cells, and then transported in water-soluble globules called chylomicrons. The chylomicrons enter the lymph via narrow vessels called lacteals and are later transferred to the blood in large veins leading to the liver and heart.

1 In the lumen of the small intestine, bile salts break up large fat globules into fat droplets, increasing exposure of triglycerides (fat molecules) on the surface to hydrolysis (step 2).

2 During enzymatic hydrolysis, the enzyme lipase breaks the exposed triglycerides down to fatty acids and monoglycerides.

3 After diffusing into epithelial cells, monoglycerides and fatty acids are re-formed into triglycerides. (Some glycerol and fatty acids pass directly into capillaries.)

4 The triglycerides are incorporated into particles called chylomicrons. Phospholipids and proteins on the surface make chylomicrons water-soluble.

5 Chylomicrons leave epithelial cells by exocytosis and enter lacteals, where they are carried away by the lymph and later pass into large veins that lead directly to the heart.

VISUAL SKILLS *Arrows at two locations in this figure indicate movement of materials between the epithelial cell and its surroundings. Identify and circle these arrows. At either of these locations, do the movements represented require an input of energy? Explain.*

→ **Mastering Biology** HHMI Animation: The Fate of Fat

One arm of the T is the 1.5-m-long **colon**, which leads to the rectum and anus. The other arm is a pouch called the **cecum**. In animals that eat large amounts of plant material, the cecum has an important role in fermenting ingested material. In humans, the cecum is small and has an **appendix**, a finger-shaped extension that acts as a reservoir for symbiotic microorganisms, which are discussed in Concept 41.4.

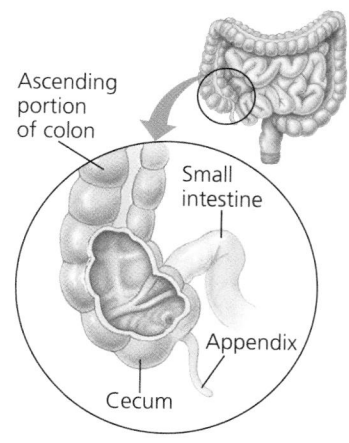

▲ **Figure 41.14 Junction of the small and large intestines.**

Ascending portion of colon

Small intestine

Appendix

Cecum

The colon completes the recovery of water that began in the small intestine. What remain are the **feces**, the wastes of the digestive system, which become increasingly solid as they are moved along the colon by peristalsis. It takes approximately 12–24 hours for material to travel the length of the colon. The undigested material in feces includes cellulose fiber. Although it provides no caloric value (energy) to humans, it helps move food along the alimentary canal.

If the lining of the colon is irritated—by a viral or bacterial infection, for instance—less water than normal may be reabsorbed, resulting in diarrhea. The opposite problem, constipation, occurs when the feces move along the colon too slowly. Too much water is reabsorbed, and the feces become compacted.

The community of bacteria living on unabsorbed organic material in the human colon contributes about one-third of the dry weight of feces. As by-products of their metabolism, many colon bacteria generate gases, including methane and hydrogen sulfide, the latter of which has an offensive odor. These gases and ingested air are expelled through the anus.

The terminal portion of the large intestine is the **rectum**, where the feces are stored before elimination. Two sphincters separate the rectum and the anus; the inner one is involuntary and the outer one is voluntary. Periodically, strong contractions of the colon create an urge to defecate. Because filling of the stomach triggers a reflex that increases the rate of contractions in the colon, the urge to defecate often follows a meal.

Having followed a meal through the alimentary canal, we'll look next at some adaptations of this general digestive plan in different animals.

CONCEPT CHECK 41.3

1. Explain why a proton pump inhibitor, such as the drug Prilosec, can relieve the symptoms of acid reflux.
2. The acids in bile salts have both lipid-soluble (hydrophobic) and water-soluble (hydrophilic) surfaces. How is such an organization beneficial for the role of bile salts in digestion?
3. **WHAT IF?** Predict what would happen if you mixed gastric juice with crushed food in a test tube.

For suggested answers, see Appendix A.

CONCEPT 41.4

Evolutionary adaptations of vertebrate digestive systems correlate with diet

EVOLUTION The digestive systems of vertebrates are varied, with many adaptations linked to an animal's diet. To highlight how form fits function, we'll examine a few of them.

Dental Adaptations

Dentition, an animal's assortment of teeth, is one example of structural variation reflecting diet **(Figure 41.15)**. The

▼ **Figure 41.15 Dentition and diet.**

Carnivore

Carnivores, such as members of the dog and cat families, generally have large, pointed incisors and canines that can be used to kill prey and rip or cut away pieces of flesh. The jagged premolars and molars crush and shred food.

Herbivore

Herbivores, such as horses and deer, usually have premolars and molars with broad, ridged surfaces that grind tough plant material. The incisors and canines are generally modified for biting off pieces of vegetation. In some herbivores, canines are absent.

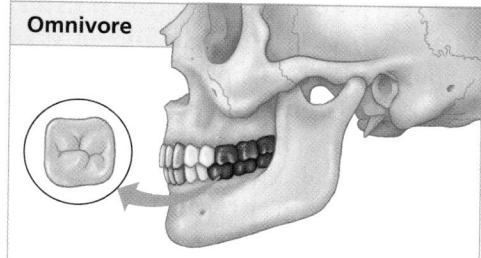

Omnivore

As omnivores, humans are adapted to eating both plants and meat. Adults have 32 teeth. From front to back along either side of the mouth are four bladelike incisors for biting, a pair of pointed canines for tearing, four premolars for grinding, and six molars for crushing (see inset, top view).

Key ▮ Incisors ▮ Canines ▮ Premolars ▮ Molars

evolutionary adaptation of teeth for processing different kinds of food is one of the major reasons that mammals have been so successful. For example, a sea otter uses its sharp canine teeth to tear apart prey such as crabs and its slightly rounded molars to crush their shells. Nonmammalian vertebrates generally have less specialized dentition, but there are interesting exceptions. Venomous snakes, such as rattlesnakes, have fangs, modified teeth that inject venom into prey. Some fangs are hollow, like syringes, whereas others drip the toxin along grooves on the surfaces of the teeth.

Stomach and Intestinal Adaptations

Evolutionary adaptations to differences in diet are sometimes apparent as variations in the dimensions of digestive organs. For example, large, expandable stomachs are common in carnivorous vertebrates, which may wait a long time between meals and must eat as much as they can when they do catch prey. An expandable stomach enables a rock python to ingest a whole gazelle (see Figure 41.5) and a 200-kg African lion to consume 40 kg of meat in one meal!

Adaptation is also apparent in the length of the digestive system in different vertebrates. In general, herbivores and omnivores have longer alimentary canals relative to their body size than do carnivores. Plant matter is more difficult to digest than meat because it contains cell walls. A longer digestive tract furnishes more time for digestion and more surface area for nutrient absorption. As an example, consider the lion and koala in **Figure 41.16**. The koala has much longer intestines, relative to its size, enhancing the processing of fibrous, protein-poor eucalyptus leaves from which it obtains nearly all of its nutrients and water.

Mutualistic Adaptations

An estimated 10–100 trillion bacteria live in the human digestive system. The coexistence of humans and many intestinal bacteria is an example of mutualism, an interaction between two species that benefits both of them (see Concept 54.1). For example, some intestinal bacteria produce vitamins, such as vitamin K, biotin, and folic acid, which are absorbed into the blood, supplementing our dietary intake. Intestinal bacteria also regulate the development of the intestinal epithelium and the function of the innate immune system. The bacteria in turn receive a steady supply of nutrients and a stable host environment.

Recently, we have greatly expanded our knowledge of the **microbiome**, the collection of microorganisms living in and on the body, along with their genetic material. To study the microbiome, scientists are using a DNA sequencing approach based on the polymerase chain reaction (PCR, see Figure 20.8). To date they have found more than 400 bacterial species in the human digestive

▼ **Figure 41.16 The alimentary canals of a carnivore (lion) and herbivore (koala).** The relatively short digestive tract of the lion is sufficient for digesting meat and absorbing its nutrients. In contrast, the koala's long alimentary canal is specialized for digesting eucalyptus leaves. Extensive chewing chops the leaves into tiny pieces, increasing exposure to digestive juices. In the long cecum and the upper portion of the colon, symbiotic bacteria further digest the shredded leaves, releasing nutrients that the koala can absorb.

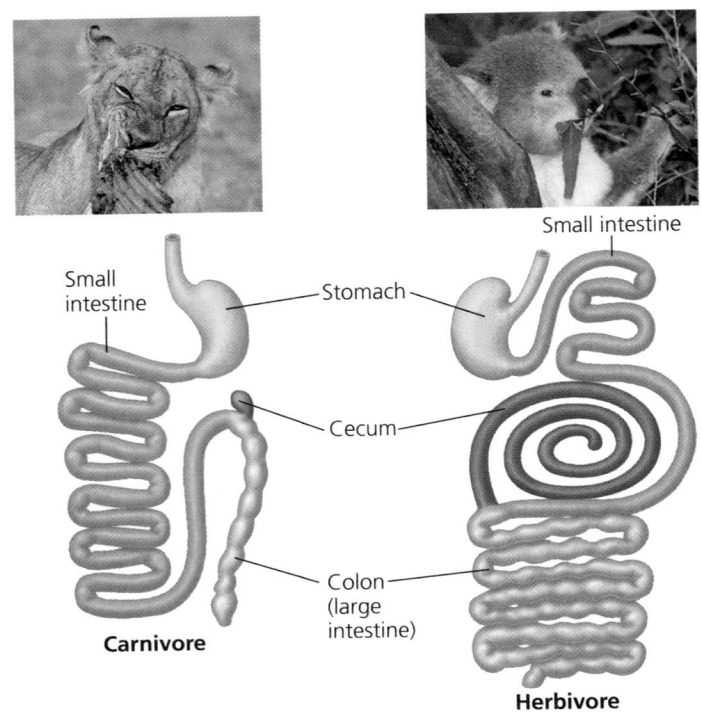

Carnivore

Herbivore

tract, a far greater number than had been identified through approaches relying on laboratory culture and characterization. Furthermore, researchers have found significant differences in the microbiome associated with diet, disease, and age **(Figure 41.17)**.

One example of the power of microbiome studies comes from research on gastric ulcers, a disease that damages the stomach lining. Experiments demonstrating that ulcers are caused by infection with the acid-tolerant bacterium *Helicobacter pylori* and can be cured with antibiotics earned Australian researchers Barry Marshall and Robin Warren a Nobel Prize in 2005.

Recently, scientists analyzed the microbiome in samples from human stomachs to learn how *H. pylori* infection leads to ulcers. Their findings were dramatic: *H. pylori* infection led to the near-complete elimination of all bacterial species commonly found in the stomach **(Figure 41.18)**.

Studies on the microbiome have already led to therapies for intestinal infections by antibiotic-resistant pathogens, a major public health problem. One such therapy is fecal microbial transplantation, in which the microbiome of a healthy individual is introduced into the patient's intestine. This approach has been used for untreatable diarrhea caused

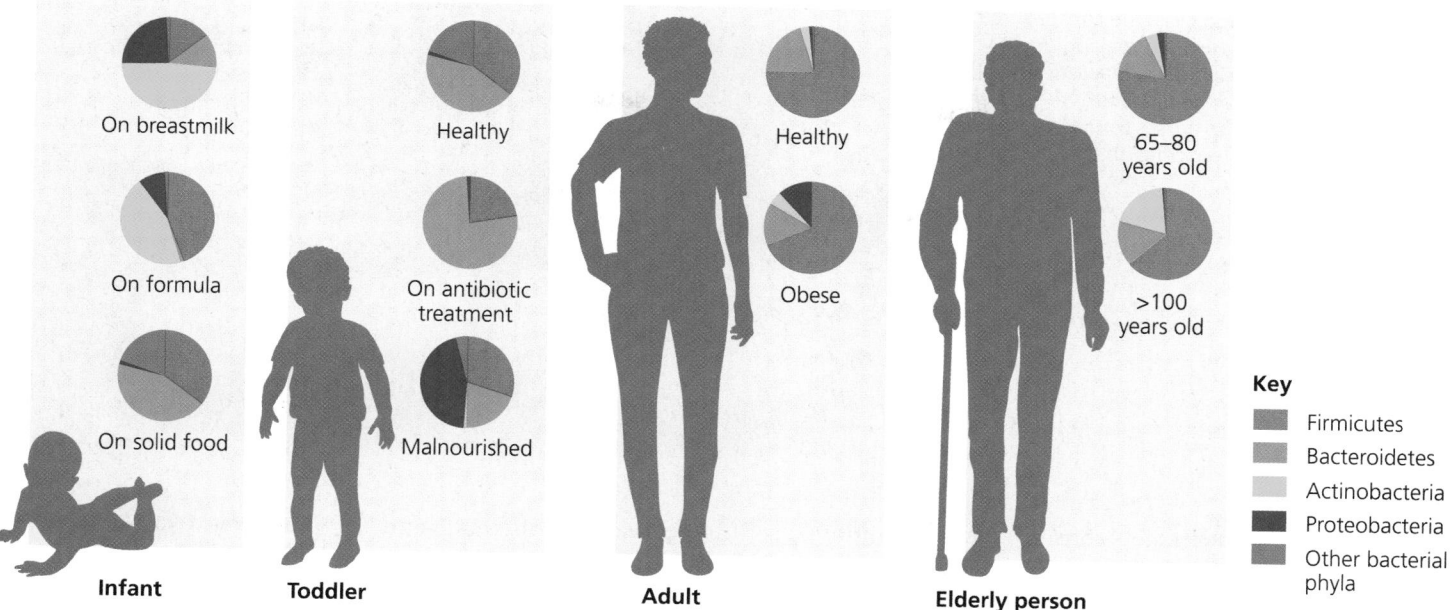

▼ **Figure 41.17 Variation in human gut microbiome at different life stages.** By copying and sequencing bacterial DNA in samples obtained from intestinal tracts of humans of different ages, researchers characterized the bacterial community that makes up the human gut and how it changes with age.

Infant
On breastmilk
On formula
On solid food

Toddler
Healthy
On antibiotic treatment
Malnourished

Adult
Healthy
Obese

Elderly person
65–80 years old
>100 years old

Key
Firmicutes
Bacteroidetes
Actinobacteria
Proteobacteria
Other bacterial phyla

INTERPRET THE DATA *Using data displayed in Figures 41.17 and 41.18, compare the relative abundance of Actinobacteria in the microbiome of a healthy adult's intestinal tract to that in a healthy stomach. Suggest a possible explanation for why the microbiome composition in the two organs is different even though the intestine and stomach are directly connected.*

▼ **Figure 41.18 The stomach microbiome and stomach health.** In samples from individuals infected with *Helicobacter pylori*, more than 95% of the sequences were from that species, which belongs to the phylum Proteobacteria. The stomach microbiome in uninfected individuals was much more diverse.

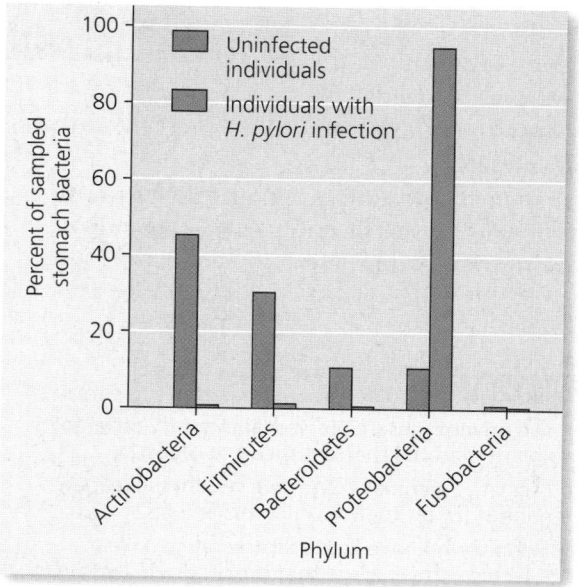

Percent of sampled stomach bacteria

■ Uninfected individuals
■ Individuals with *H. pylori* infection

Phylum: Actinobacteria, Firmicutes, Bacteroidetes, Proteobacteria, Fusobacteria

▶ **Figure 41.19**
Clostridium difficile.

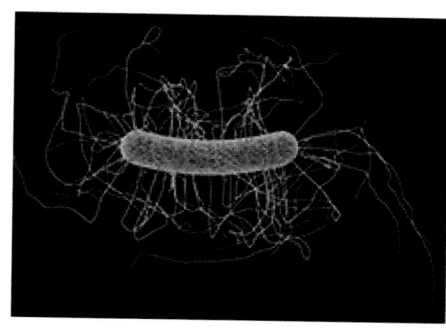

Another treatment against antibiotic-resistant pathogens makes use of bacteriophages, viruses that infect bacteria but not human cells (see Concept 19.1). In 2017, bacteriophages engineered to kill a multidrug-resistant bacterial pathogen, *Acinetobacter baumannii*, were successfully used to restore health to a severely affected patient.

> **Mastering Biology** Interview with Steffanie Strathdee: Harnessing phages to fight deadly infections (see the interview before Chapter 40)

Mutualistic Adaptations in Herbivores

by the bacterium *Clostridium difficile* (**Figure 41.19**). Such infections are most common after antibiotic treatment has killed off the resident microbiome. Clinical trials have shown the fecal transplantation approach to be effective, although there is a real risk of secondary infection.

Mutualistic relationships with microorganisms are also very important in herbivores. Herbivores get much of the chemical energy they need from the cellulose of plant cell walls, but, like other animals, herbivores do not produce enzymes that hydrolyze cellulose. Instead, many vertebrates (as well

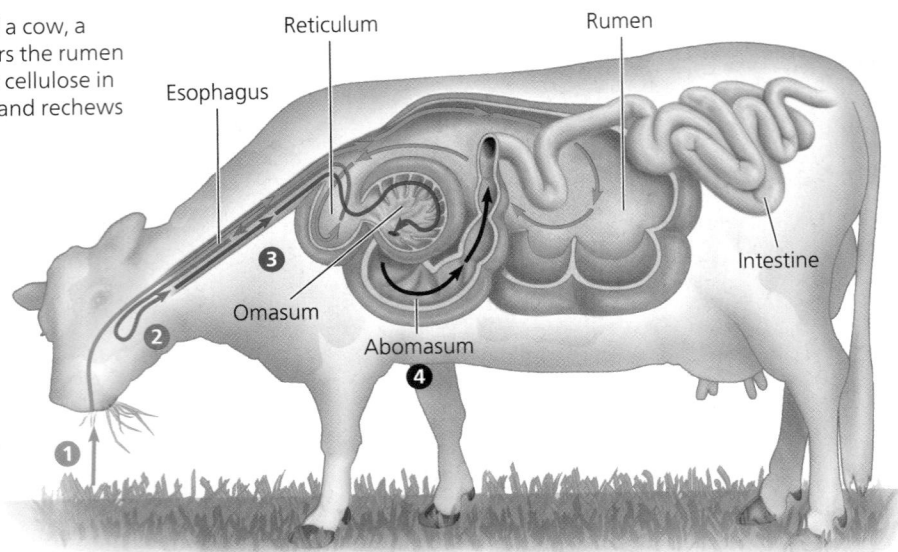

► **Figure 41.20 Ruminant digestion.** The stomach of a cow, a ruminant, has four chambers. **1** Chewed food first enters the rumen and reticulum, where mutualistic microorganisms digest cellulose in the plant material. **2** Periodically, the cow regurgitates and rechews "cud" from the reticulum, further breaking down fibers and thereby enhancing microbial action. **3** The reswallowed cud passes to the omasum, where some water is removed. **4** It then passes to the abomasum for digestion by the cow's enzymes. In this way, the cow obtains significant nutrients from both the grass and the mutualistic microorganisms, which maintain a stable population in the rumen.

as termites, whose wooden diets consist largely of cellulose) host large populations of mutualistic bacteria and protists in fermentation chambers in their alimentary canals. These symbiotic microorganisms have enzymes that can digest cellulose to simple sugars and other compounds that the animal can absorb. In many cases, the microorganisms also use the sugars from digested cellulose in the production of a variety of essential nutrients, such as vitamins and amino acids.

In horses, koalas, and elephants, symbiotic microorganisms are housed in a large cecum. In contrast, the hoatzin, an herbivorous bird found in South American rain forests, hosts microorganisms in a large, muscular crop (an esophageal pouch; see Figure 41.7). Hard ridges in the wall of the crop grind plant leaves into small fragments, and the microorganisms break down cellulose.

In rabbits and some rodents, mutualistic bacteria live in the large intestine as well as the cecum. Since most nutrients are absorbed in the small intestine, nourishing by-products of fermentation by bacteria in the large intestine are initially lost with the feces. Rabbits and rodents recover these nutrients by *coprophagy* (from the Greek, meaning "dung eating"), feeding on some of their feces and then passing the food through the alimentary canal a second time. The familiar rabbit "pellets," which are not reingested, are the feces eliminated after food has passed through the digestive tract twice.

The most elaborate adaptations for an herbivorous diet have evolved in the animals called *ruminants*, the cud-chewing animals that include deer, sheep, and cattle **(Figure 41.20)**.

Although we have focused our discussion on vertebrates, adaptations related to digestion are also widespread among other animals. Some of the most remarkable examples are the over 3-meter-long giant tubeworms **(Figure 41.21)** that live at pressures as high as 260 atmospheres around

deep-sea hydrothermal vents (see Figure 52.16). These worms have no mouth or digestive system. Instead, they obtain all of their energy and nutrients from mutualistic bacteria that live within their bodies. The bacteria carry out chemoautotrophy (see Concept 27.3) using the carbon dioxide, oxygen, hydrogen sulfide, and nitrate available at the vents. Thus, for invertebrates and vertebrates alike, the evolution of mutualistic relationships with symbiotic microorganisms is an adaptation that expands the sources of nutrition available to animals.

Having examined how animals optimize their extraction of nutrients from food, we'll next turn to the challenge of balancing the use of these nutrients.

▼ **Figure 41.21 Giant tubeworm, an animal without a digestive system.**

CONCEPT CHECK 41.4

1. What are two advantages of a longer alimentary canal for processing plant material that is difficult to digest?

2. What features of a mammal's digestive system make it an attractive habitat for mutualistic microorganisms?

3. **WHAT IF?** People who have lactose intolerance have a shortage of lactase, the enzyme that breaks down lactose in milk. As a result, they sometimes develop cramps, bloating, or diarrhea after consuming dairy products. Suppose such a person ate yogurt that contains bacteria that produce lactase. Why would eating yogurt likely provide at best only temporary relief of the symptoms?

For suggested answers, see Appendix A.

Feedback circuits regulate digestion, energy storage, and appetite

To complete our consideration of animal nutrition, we'll explore the ways that obtaining and using nutrients are matched to an animal's circumstances and need for energy.

Regulation of Digestion

Many animals face long gaps between meals. Under such circumstances, there is no need for their digestive systems to be active continuously. Instead, processing is activated step-wise. As food reaches each new compartment, it triggers the secretion of digestive juices for the next stage of processing. Muscular contractions then move the contents farther along the canal. For example, you learned earlier that nervous reflexes stimulate the release of saliva when food enters the oral cavity and orchestrate swallowing when a bolus of food reaches the pharynx. Similarly, the arrival of food in the stomach triggers churning and the release of gastric juices. These events, as well as peristalsis in the small and large intestines, are regulated by the *enteric nervous system*, a net-work of neurons dedicated to the digestive organs.

The endocrine system also plays a critical role in control-ling digestion. As described in **Figure 41.22**, a series of hor-mones released by the stomach and duodenum help ensure that digestive secretions are present only when needed. Like all hormones, they are transported through the bloodstream. This is true even for the hormone gastrin, which is secreted by the stomach and targets that same organ.

Regulation of Energy Storage

When an animal takes in more energy-rich molecules than it needs for metabolism and activity, it stores the excess energy (see Concept 40.4). In humans, liver and muscle cells serve as the primary sites for energy storage. In these cells, excess energy from the diet is stored in glycogen, a polysaccharide made up of many glucose units (see Figure 5.6b). Once glycogen depots are full, any additional excess energy is usually stored in fat in adipose cells.

At times when fewer calories are taken in than are expended—perhaps because of sustained heavy exercise or lack of food—the human body generally expends liver gly-cogen first and then draws on muscle glycogen and fat. Fats are especially rich in energy; oxidizing a gram of fat liberates about twice the energy liberated from a gram of carbohydrate or protein. For this reason, adipose tissue provides the most space-efficient way for the body to store large amounts of energy. Most healthy people have enough stored fat to sus-tain them through several weeks without food.

▼ **Figure 41.22 Hormonal control of digestion.**

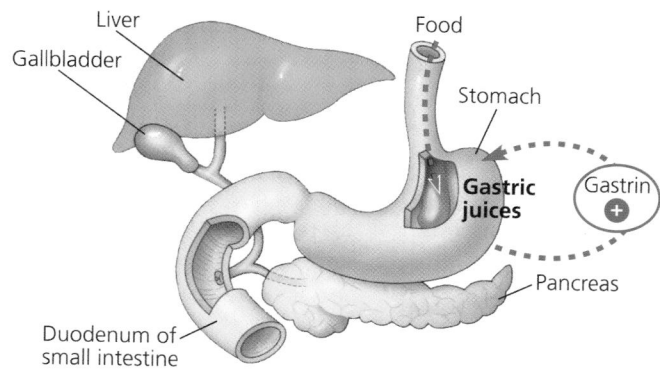

1 As food arrives at the stomach, it stretches the stomach walls, triggering release of the hormone *gastrin*. Gastrin circulates via the bloodstream back to the stomach, where it stimulates production of gastric juices.

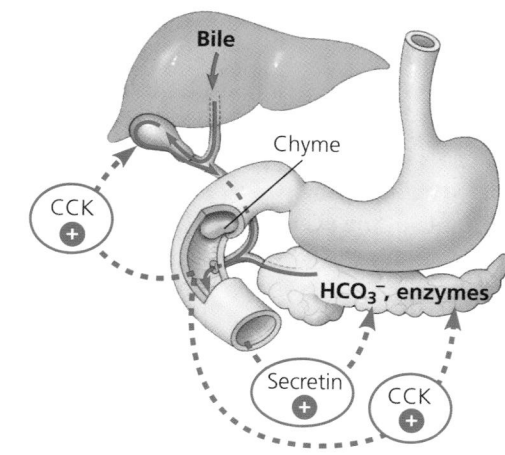

2 Chyme—an acidic mixture of partially digested food—eventually passes from the stomach to the duodenum. The duodenum responds by releasing the digestive hormones cholecystokinin and secretin. *Cholecystokinin (CCK)* stimulates the release of digestive enzymes from the pancreas and of bile from the gallbladder. *Secretin* stimulates the pancreas to release bicarbonate (HCO_3^-), which neutralizes chyme.

3 If the chyme is rich in fats, the high levels of secretin and CCK released inhibit peristalsis. This slows the movement of chyme and allows the more time-consuming digestion of fats to take place in the small intestine.

Key   Stimulation   Inhibition

Glucose Homeostasis

The synthesis and breakdown of glycogen are central not only to energy storage, but also to maintaining metabolic balance through glucose homeostasis. In humans, the normal range for the concentration of glucose in the blood is 70–110 mg/100 mL. Because glucose is a major fuel for cellular respiration and a key source of carbon skeletons for biosynthesis, maintaining blood glucose concentrations near this normal range is essential.

Glucose homeostasis relies predominantly on the antagonistic (opposing) effects of two hormones, insulin and glucagon **(Figure 41.23)**. When the blood glucose level rises above the normal range, the secretion of **insulin** triggers the uptake of glucose from the blood into body cells, decreasing the blood glucose concentration. When the blood glucose level drops below the normal range, the secretion of **glucagon** promotes the release of glucose into the blood from energy stores, such as liver glycogen, increasing the blood glucose concentration.

The liver is a key site of action for insulin and glucagon. After a carbohydrate-rich meal, for example, insulin secretion promotes biosynthesis of glycogen from glucose that enters the liver in the hepatic portal vein. Between meals, when blood in the hepatic portal vein has a much lower glucose concentration, glucagon stimulates the liver to break down glycogen, convert amino acids and glycerol to glucose, and release glucose into the blood. Together, these opposing effects of insulin and glucagon ensure that blood exiting the liver has a glucose concentration in the normal range at nearly all times.

Insulin also acts on nearly all body cells to stimulate glucose uptake from blood. A major exception is brain cells, which can take up glucose whether or not insulin is present. This evolutionary adaptation ensures that the brain almost always has access to circulating fuel, even if supplies are low.

Glucagon and insulin are both produced in the pancreas. Clusters of endocrine cells called pancreatic islets are scattered throughout this organ. Each islet has *alpha cells*, which make glucagon, and *beta cells*, which make insulin. Like other hormones, insulin and glucagon enter the interstitial fluid and then the circulatory system.

Overall, hormone-secreting cells make up only 1–2% of the mass of the pancreas. Other cells in the pancreas produce and secrete bicarbonate ions and the digestive enzymes active in the small intestine (see Figure 41.11). These secretions are released into small ducts that empty into the pancreatic duct, which leads to the small intestine. Thus, the pancreas has functions in both the endocrine and digestive systems.

Diabetes Mellitus

In discussing the role of insulin and glucagon in glucose homeostasis, we have focused exclusively on a healthy metabolic state. However, a number of disorders can disrupt glucose homeostasis with potentially serious consequences, especially for the heart, blood vessels, eyes, and kidneys. The best known and most prevalent of these disorders is diabetes mellitus.

▶ **Figure 41.23 Homeostatic regulation of cellular fuel.** After a meal is digested, glucose and other monomers are absorbed into the blood from the digestive tract. The human body regulates the use and storage of glucose, a major cellular fuel.

MAKE CONNECTIONS *What form of feedback control do these regulatory circuits reflect (see Concept 40.2)?*

➡ **Mastering Biology** BioFlix® **Animation: Homeostasis: Regulating Blood Sugar** • **Animation: Pancreatic Hormones Regulate Blood Glucose Level**

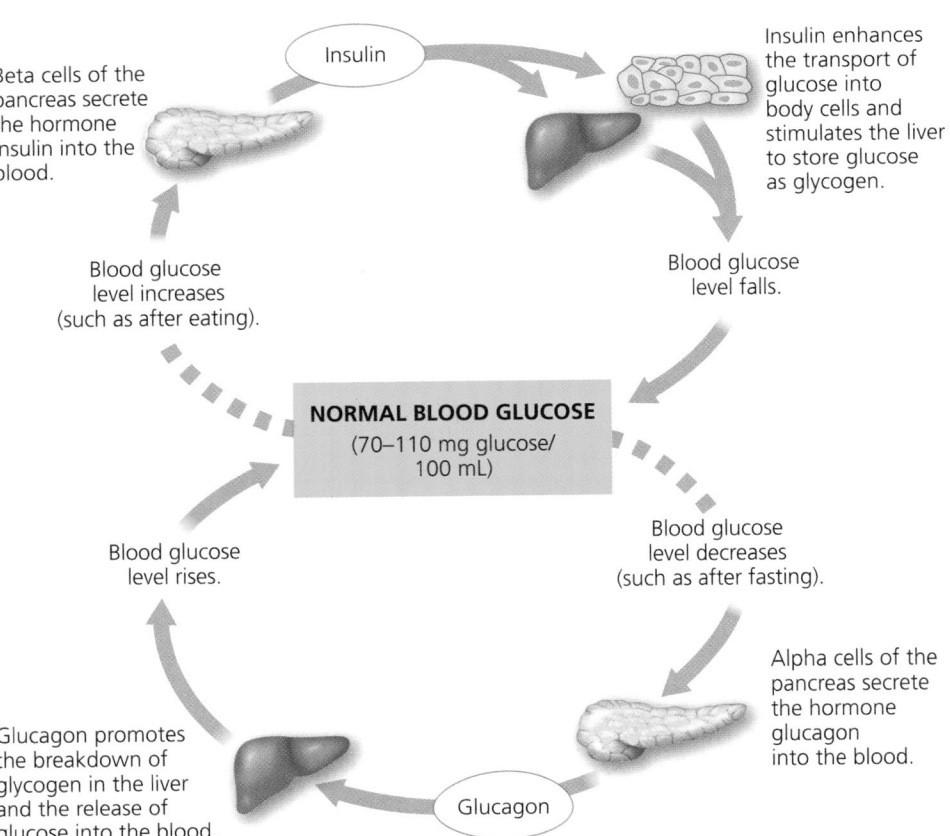

Beta cells of the pancreas secrete the hormone insulin into the blood.

Insulin

Insulin enhances the transport of glucose into body cells and stimulates the liver to store glucose as glycogen.

Blood glucose level falls.

Blood glucose level increases (such as after eating).

NORMAL BLOOD GLUCOSE (70–110 mg glucose/ 100 mL)

Blood glucose level decreases (such as after fasting).

Blood glucose level rises.

Glucagon promotes the breakdown of glycogen in the liver and the release of glucose into the blood.

Glucagon

Alpha cells of the pancreas secrete the hormone glucagon into the blood.

The disease **diabetes mellitus** is caused by a deficiency of insulin or a decreased response to insulin in target tissues. The blood glucose level rises, but cells are unable to take up enough glucose to meet metabolic needs. Instead, fat becomes the main substrate for cellular respiration. In severe cases, acidic metabolites formed during fat breakdown accumulate in the blood, threatening life by lowering blood pH and depleting sodium and potassium ions from the body.

In people with diabetes mellitus, the level of glucose in the blood may exceed the capacity of the kidneys to reabsorb this nutrient. Glucose that remains in the kidney filtrate is excreted. For this reason, the presence of sugar in urine is one test for this disorder. The presence of excess sugar in both urine and blood is the basis for the name *diabetes mellitus* (from the Greek *diabainein*, to pass through, and *meli*, honey).

There are two main types of diabetes mellitus: type 1 and type 2. Each is marked by high blood glucose levels, but with very different causes.

Type 1 Diabetes   Also called insulin-dependent diabetes, *type 1 diabetes* is an autoimmune disorder in which the immune system destroys the beta cells of the pancreas. Type 1 diabetes, which usually appears during childhood, destroys the person's ability to produce insulin. Treatment consists of insulin injections, typically given multiple times daily. In the past, insulin was extracted from animal pancreases, but now human insulin can be readily obtained at reasonable cost from genetically engineered bacteria (see Figure 20.4). Stem cell research may someday provide a cure for type 1 diabetes by generating replacement beta cells that restore insulin production by the pancreas.

Type 2 Diabetes   Non-insulin-dependent diabetes, or *type 2 diabetes*, is characterized by a failure of target cells to respond normally to insulin. Insulin is produced, but target cells fail to take up glucose from the blood, and the blood glucose level remains elevated. Although heredity can play a role in type 2 diabetes, excess body weight and lack of exercise significantly increase the risk of developing this disorder. This form of diabetes generally appears after age 40, but even children can develop the disease, particularly if they are overweight and sedentary. More than 90% of people with diabetes have type 2. Many can control their blood glucose levels with regular exercise and a healthy diet; some require medications. Nevertheless, type 2 diabetes is the seventh most common cause of death in the United States and a growing public health problem worldwide.

The resistance to insulin signaling in type 2 diabetes is sometimes due to a genetic defect in the insulin receptor or the insulin response pathway. In many cases, however, events in target cells suppress activity of an otherwise functional response pathway. One source of this suppression appears to be inflammatory signals generated by the innate immune system (see Concept 43.1). How obesity and inactivity relate to this suppression is being studied in both humans and laboratory animals.

→ **Mastering Biology** BioFlix® Animation: Diabetes

Regulation of Appetite and Consumption

Consuming more calories than the body needs for normal metabolism, or *overnourishment*, can lead to obesity, the excessive accumulation of fat. Obesity, in turn, contributes to a number of health problems, including type 2 diabetes, cancer of the colon and breast, and cardiovascular disease that can result in heart attacks and strokes. It is estimated that obesity is a factor in 300,000 deaths per year in the United States alone.

Researchers have discovered several homeostatic mechanisms that operate as feedback circuits controlling the storage and metabolism of fat. A network of neurons relays and integrates information from the digestive system to regulate secretion of hormones that regulate long-term and short-term appetite. The target for these hormones is a "satiety center" in the brain **(Figure 41.24)**. For example, *ghrelin*, a hormone secreted by the

▼ **Figure 41.24 A few of the appetite-regulating hormones.** Secreted by various organs and tissues, the hormones reach the brain via the bloodstream. These signals act on a region of the brain that in turn controls the "satiety center," which generates the nervous impulses that make us feel either hungry or satiated ("full"). The hormone ghrelin is an appetite stimulant; the other three hormones shown here are appetite suppressants.

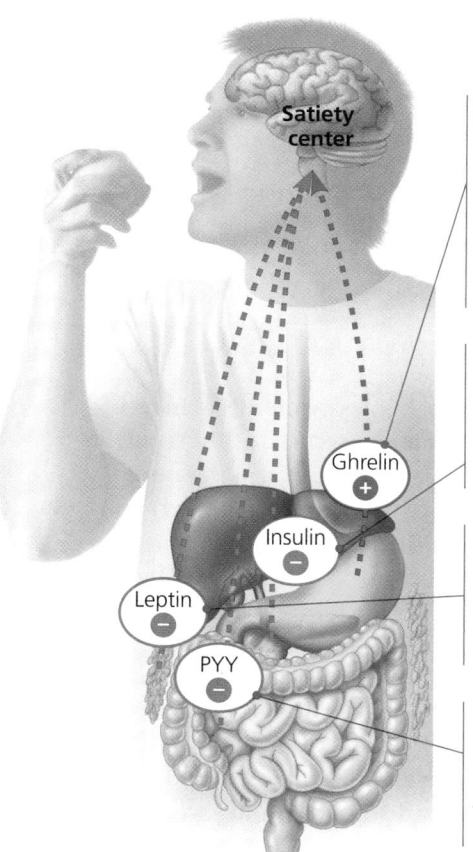

Satiety center

Ghrelin +

Insulin −

Leptin −

PYY −

Secreted by the stomach wall, **ghrelin** is one of the signals that triggers feelings of hunger as mealtimes approach. In dieters who lose weight, ghrelin levels increase, which may be one reason it's so hard to stay on a diet.

A rise in blood sugar level after a meal stimulates the pancreas to secrete **insulin**. In addition to its other functions, insulin suppresses appetite by acting on the brain.

Produced by adipose (fat) tissue, **leptin** suppresses appetite. When the amount of body fat decreases, leptin levels fall, and appetite increases.

The hormone **PYY**, secreted by the small intestine after meals, acts as an appetite suppressant that counters the appetite stimulant ghrelin.

Interpreting Data from an Experiment with Genetic Mutants

What Are the Roles of the *ob* and *db* Genes in Appetite Regulation? A mutation that disrupts a physiological process is often used to study the normal function of the mutated gene. Ideally, researchers use a standard set of conditions and compare animals that differ genetically only in whether a particular gene is mutant (nonfunctional) or wild-type (normal). In this way, a difference in phenotype, the physiological property being measured, can be attributed to a difference in genotype, the presence or absence of the mutation. To study the role of specific genes in regulating appetite, researchers used laboratory animals with known mutations in those genes.

Mice in which recessive mutations inactivate both copies of either the *ob* gene or the *db* gene eat voraciously and grow much more massive than wild-type mice.

▶ **The mouse on the right is wild-type, whereas the obese mouse on the left has an inactivating mutation in both copies of the *ob* gene.**

One hypothesis for the normal role of the *ob* and *db* genes is that they participate in a hormone pathway that suppresses appetite when caloric intake is sufficient. Before setting out to isolate the potential hormone, researchers explored this hypothesis genetically.

How the Experiment Was Done The researchers measured the mass of young mice of various genotypes (the "subjects") and surgically linked the circulatory system of each one to that of another mouse. This procedure ensured that any factor circulating in the bloodstream of either mouse would be transferred to the other in the pair. After eight weeks, they again measured the mass of each subject mouse.

Data from the Experiment

Genotype Pairing (red type indicates mutant genes)		Average Change in Body Mass of Subject (g)
Subject	**Paired with**	
(a) ob^+/ob^+, db^+/db^+	ob^+/ob^+, db^+/db^+	8.3
(b) ob/ob, db^+/db^+	ob/ob, db^+/db^+	38.7
(c) ob/ob, db^+/db^+	ob^+/ob^+, db^+/db^+	8.2
(d) ob/ob, db^+/db^+	ob^+/ob^+, db/db	−14.9*

*Due to pronounced weight loss and weakening, subjects in this pairing were remeasured after less than eight weeks.

Data from D. L. Coleman, Effects of parabiosis of obese mice with diabetes and normal mice, *Diabetologia* 9:294–298 (1973).

INTERPRET THE DATA

1. First, practice reading the genotype information given in the data table. For example, pairing (a) joined two mice that each had the wild-type version of both genes. Describe the two mice in pairing (b), pairing (c), and pairing (d). Explain how each pairing contributed to the experimental design.
2. Compare the results observed for pairing (a) and pairing (b) in terms of phenotype. If the results had been identical for these two pairings, what would that outcome have implied about the experimental design?
3. Compare the results observed for pairing (c) to those observed for pairing (b). Based on these results, does the ob^+ gene product appear to promote or suppress appetite? Explain your answer.
4. Describe the results observed for pairing (d). Note how these results differ from those for pairing (b). Suggest a hypothesis to explain this difference. How could you test your hypothesis using the kinds of mice in this study?

→ **Instructors:** A version of this Scientific Skills Exercise can be assigned in **Mastering Biology**.

---

stomach wall, triggers feelings of hunger before meals. In contrast, both insulin and *PYY*, a hormone secreted by the small intestine after meals, suppress appetite. *Leptin*, a hormone produced by adipose (fat) tissue, also suppresses appetite and appears to play a major role in regulating body fat levels. In the **Scientific Skills Exercise**, you'll interpret data from an experiment studying genes that affect leptin production and function in mice.

Obtaining food, digesting it, and absorbing nutrients are part of the larger story of how animals fuel their activities. Provisioning the body also involves circulating nutrients, and using nutrients for metabolism requires exchanging respiratory gases with the environment. These processes and the adaptations that facilitate such distribution and exchange are the focus of Chapter 42.

CONCEPT CHECK 41.5

1. Explain how people can become obese even if their intake of dietary fat is relatively low compared with carbohydrate intake.
2. **WHAT IF?** Suppose you were studying two groups of people who are obese and have genetic abnormalities in the leptin pathway. In one group, the leptin levels are abnormally high; in the other group, they are abnormally low. How would each group's leptin levels change if they ate a low-calorie diet for an extended period? Explain.
3. **WHAT IF?** An insulinoma is a cancerous mass of pancreatic beta cells that secrete insulin but do not respond to feedback mechanisms. How you would expect an insulinoma to affect blood glucose levels and liver activity?

For suggested answers, see Appendix A.

41 Chapter Review

→ Go to **Mastering Biology** for Assignments, the eText, the Study Area, and Dynamic Study Modules.

SUMMARY OF KEY CONCEPTS

→ To review key terms, go to the **Vocabulary Self-Quiz** in the **Mastering Biology** eText or Study Area, or go to goo.gl/zkjz9t.

CONCEPT 41.1

An animal's diet must supply chemical energy, organic building blocks, and essential nutrients (pp. 899–902)

- Food provides animals with energy for ATP production, carbon skeletons for biosynthesis, and **essential nutrients**—nutrients that must be supplied in preassembled form. Essential nutrients include certain amino acids and fatty acids that animals cannot synthesize; **vitamins**, which are organic molecules; and **minerals**, which are inorganic substances.
- Animals have diverse diets. **Herbivores** mainly eat plants; **carnivores** mainly eat other animals; and **omnivores** eat both. Animals must balance consumption, storage, and use of food.
- Malnutrition results from an inadequate intake of essential nutrients or a deficiency in sources of chemical energy. Studies of disease at the population level help researchers determine human dietary requirements.

? *How can an enzyme cofactor needed for an essential process be an essential nutrient for only some animals?*

CONCEPT 41.2

Food processing involves ingestion, digestion, absorption, and elimination (pp. 902–905)

- Animals differ in the ways they obtain and ingest food. Many animals are **bulk feeders**, eating large pieces of food. Other strategies include **filter feeding, substrate feeding,** and **fluid feeding**.
- Compartmentalization is necessary to avoid self-digestion. In intracellular digestion, food particles are engulfed by phagocytosis and digested within food vacuoles that have fused with lysosomes. In extracellular digestion, which is used by most animals, enzymatic hydrolysis occurs outside cells in a **gastrovascular cavity** or **alimentary canal**.

Gastrovascular cavity (jelly)

Alimentary canal (earthworm)

? *Propose an artificial diet that would eliminate the need for one of the first three steps in food processing.*

CONCEPT 41.3

Organs specialized for sequential stages of food processing form the mammalian digestive system (pp. 905–911)

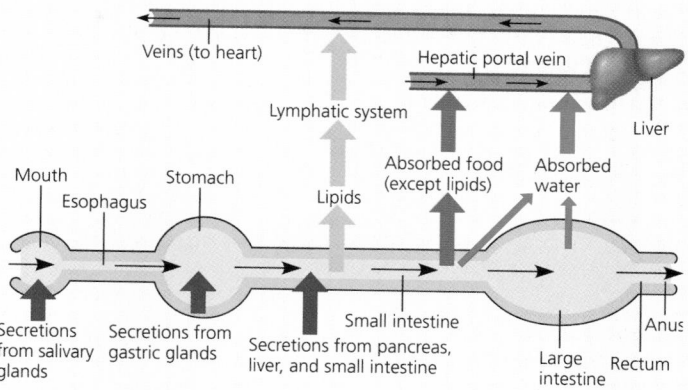

? *What structural feature of the small intestine makes it better suited for absorption of nutrients than the stomach?*

CONCEPT 41.4

Evolutionary adaptations of vertebrate digestive systems correlate with diet (pp. 911–914)

- Vertebrate digestive systems display many evolutionary adaptations associated with diet. For example, the assortment of teeth (dentition) generally correlates with diet. Also, many herbivores have fermentation chambers where mutualistic microorganisms digest cellulose. In addition, herbivores usually have longer alimentary canals than carnivores, reflecting the longer time needed to digest vegetation.

? *How does human anatomy indicate that our primate ancestors were not strict vegetarians?*

CONCEPT 41.5

Feedback circuits regulate digestion, energy storage, and appetite (pp. 915–918)

- Nutrition is regulated at multiple levels. Food intake triggers nervous and hormonal responses that cause secretion of digestive juices and promote movement of ingested material through the canal. The hormones **insulin** and **glucagon** control the synthesis and breakdown of glycogen, thereby regulating glucose availability.
- Vertebrates store excess calories in glycogen (in liver and muscle cells) and in fat (in adipose cells). These energy stores can be tapped when an animal expends more calories than it consumes. If, however, an animal consumes more calories than it needs for normal metabolism, the resulting overnourishment can cause obesity.
- Several hormones, including leptin and insulin, regulate appetite by affecting the brain's satiety center.

? *Explain why your stomach might make growling noises when you skip a meal.*

➡ For more multiple-choice questions, go to the **Practice Test** in the **Mastering Biology** eText or Study Area, or go to goo.gl/GruWRg.

Levels 1-2: Remembering/Understanding

1. Fat digestion yields fatty acids and glycerol. Protein digestion yields amino acids. Both digestive processes
   (A) occur inside cells in most animals.
   (B) add a water molecule to break bonds.
   (C) require a low pH resulting from HCl production.
   (D) consume ATP.

2. The mammalian trachea and esophagus both connect to the
   (A) pharynx.
   (B) stomach.
   (C) large intestine.
   (D) rectum.

3. In which organ does almost all enzymatic digestion of food occur?
   (A) stomach
   (B) small intestine
   (C) large intestine
   (D) pancreas

4. In which digestive system organ does nearly all nutrient absorption occur?
   (A) stomach
   (B) small intestine
   (C) large intestine
   (D) pancreas

Levels 3-4: Applying/Analyzing

5. If you put the following events in the order they occur in the human digestive system, which would be the third event in the series?
   (A) Cells in gastric pits secrete protons.
   (B) Pepsin activates pepsinogen.
   (C) HCl activates pepsinogen.
   (D) Partially digested food enters the small intestine.

6. After surgical removal of the gallbladder, a person might need to limit his or her dietary intake of
   (A) starch.   (C) sugar.
   (B) protein.   (D) fat.

7. If you were to jog 1 km a few hours after lunch, which stored fuel would you probably tap?
   (A) muscle proteins
   (B) muscle and liver glycogen
   (C) fat in the liver
   (D) fat in adipose tissue

Levels 5-6: Evaluating/Creating

8. **DRAW IT** Create a flowchart to summarize the events that occur after partially digested food leaves the stomach in a human. Use the following terms: bicarbonate secretion, circulation, decrease in acidity, increase in acidity, secretin secretion, signal detection. Next to each term, indicate the compartment(s) involved. You may use terms more than once.

9. **EVOLUTION CONNECTION** Lizards and snakes cannot breathe while chewing food because the connection between their external nostrils and their esophagus is in the mouth. In contrast, mammals can continue breathing through the nostrils while chewing food in the mouth. However, choking sometimes occurs when the paths of air and food cross each other. Thinking about the high oxygen demand of active endotherms, explain how the concept of descent with modification explains this "imperfect" anatomy of some amniotes (see Figure 34.2 to review vertebrate phylogeny).

10. **SCIENTIFIC INQUIRY** In human populations of northern European origin, the disorder called hemochromatosis causes excess iron uptake from food and affects one in 200 adults. Among individuals 15 to 50 years old, men are ten times as likely as women to suffer from iron overload. Taking into account this range of ages, propose a hypothesis that explains this difference.

11. **WRITE ABOUT A THEME: ORGANIZATION** Hair is largely made up of the protein keratin. In a short essay (100–150 words), explain why a shampoo containing protein cannot replace the protein in damaged hair.

12. **SYNTHESIZE YOUR KNOWLEDGE**

Owls, which have a varied diet, periodically cough up pellets composed of indigestible material, such as bones, feathers, and fur. (a) Compare this aspect of the owl's digestive process to those of the hydra, the rabbit, and the cow. (b) In which digestive compartment do you predict owl pellets form (see Figure 41.7c)? Explain.

For selected answers, see Appendix A.

42 Circulation and Gas Exchange

KEY CONCEPTS

42.1 Circulatory systems link exchange surfaces with cells throughout the body *p. 922*

42.2 Coordinated cycles of heart contraction drive double circulation in mammals *p. 926*

42.3 Patterns of blood pressure and flow reflect the structure and arrangement of blood vessels *p. 929*

42.4 Blood components function in exchange, transport, and defense *p. 934*

42.5 Gas exchange occurs across specialized respiratory surfaces *p. 939*

42.6 Breathing ventilates the lungs *p. 944*

42.7 Adaptations for gas exchange include pigments that bind and transport gases *p. 947*

Study Tip

Draw and annotate a diagram: Sketch a simplified double circulation system like the idealized one on this page. Label where each of the following is highest: blood pressure, CO_2 concentration, O_2 concentration, and total cross-sectional area of the blood vessels. Note that the left side of your diagram represents the right side of the heart in the body, and remember "**ReViTaLize**" to help remind you of the path of blood from the **R**ight **V**entricle **t**o the **L**ungs.

→ Go to Mastering Biology

For Students (in the eText and Study Area)
- Get Ready for Chapter 42
- Figure 42.10 Walkthrough: Interrelationship of Blood Vessel Area, Velocity, and Pressure
- Figure 42.22 Walkthrough: Structure and Function of Fish Gills
- BioFlix® Animation: Gas Exchange in the Human Body

For Instructors to Assign (in the Item Library)
- Animation: Gills Are Adapted for Gas Exchange in Aquatic Environments

Ready-to-Go Teaching Module
(in Instructor Resources)
- Cardiac Cycle and Heart Function

Figure 42.1 This animal may look like a creature from a science fiction film, but it's actually an axolotl, a salamander native to shallow ponds in central Mexico. The feathery red appendages jutting out from its head are gills. External gills, although uncommon in other adult animals, help axolotls carry out a process common to all organisms: the exchange of substances between body cells and the environment.

How are structure and function related in the exchange and circulation of oxygen and carbon dioxide?

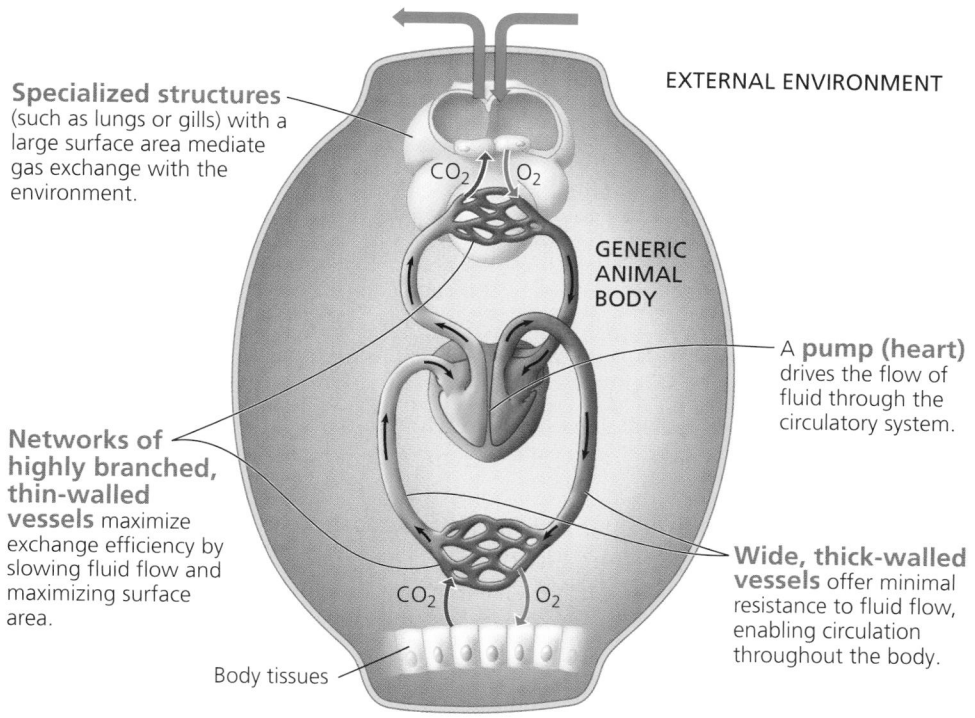

Specialized structures (such as lungs or gills) with a large surface area mediate gas exchange with the environment.

EXTERNAL ENVIRONMENT

GENERIC ANIMAL BODY

A **pump (heart)** drives the flow of fluid through the circulatory system.

Networks of highly branched, thin-walled vessels maximize exchange efficiency by slowing fluid flow and maximizing surface area.

Wide, thick-walled vessels offer minimal resistance to fluid flow, enabling circulation throughout the body.

Body tissues

Circulatory systems link exchange surfaces with cells throughout the body

The molecular trade that an animal carries out with its environment must ultimately involve every cell in the body. Required resources, such as nutrients and oxygen (O_2), enter the cytoplasm by crossing the plasma membrane. Waste products, such as carbon dioxide, (CO_2), exit the cell by crossing the same membrane.

Small molecules in and around cells, including O_2 and CO_2, undergo **diffusion**, which is random thermal motion (see Concept 7.3). When there is a difference in the concentration of a gas or other substance, such as between a cell and its immediate surroundings, diffusion can result in net movement.

Unicellular organisms exchange materials directly with the external environment via diffusion across the plasma membrane. For most multicellular organisms, however, direct exchange between every cell and the environment is not possible. Furthermore, net movement by diffusion is very slow for distances of more than a few millimeters. That's because the time it takes for a substance to diffuse from one place to another is proportional to the *square* of the distance. For example, a quantity of glucose that takes 1 second to diffuse 100 μm will take 100 seconds to diffuse 1 mm and almost 3 hours to diffuse 1 cm.

Given that net movement by diffusion is rapid only over very small distances, how does each cell of an animal participate in exchange? Natural selection has resulted in two basic adaptations that permit efficient exchange for all of an animal's cells.

One adaptation for efficient exchange is a simple body plan that places many or all cells in direct contact with the environment. Each cell can thus exchange materials directly with the surrounding medium. Such an arrangement is characteristic of certain invertebrates, including cnidarians and flatworms. Animals that lack a simple body plan display an alternative adaptation for efficient exchange: a circulatory system. Such systems move fluid between each cell's immediate surroundings and the body tissues. As a result, exchange with the environment and exchange with body tissues both occur over very short distances.

In most animals, the circulatory system is functionally linked to the exchange of gases with the environment and with body cells. For this reason, we'll discuss systems for circulation and gas exchange together in this chapter. By considering examples of these systems from a range of species, we'll explore their common elements as well as their extensive variation.

Gastrovascular Cavities

Let's begin by looking at some animals whose body shapes put many of their cells into contact with their environment, enabling them to live without a distinct circulatory system. In hydras, jellies, and other cnidarians, a central **gastrovascular cavity** functions in the distribution of substances throughout the body, as well as in digestion (see Figure 41.6). An opening at one end connects the cavity to the surrounding water. In a hydra, thin branches of the gastrovascular cavity extend into the animal's tentacles. In jellies and some other cnidarians, the gastrovascular cavity has a much more elaborate branching pattern **(Figure 42.2a)**.

In animals with a gastrovascular cavity, fluid bathes both the inner and outer tissue layers, facilitating exchange of gases and cellular waste. Only the cells lining the cavity have direct access to nutrients released by digestion. However, because the body wall is a mere two cells thick, nutrients only need to diffuse a short distance to reach the cells of the outer tissue layer.

▼ **Figure 42.2 Diversity of gastrovascular cavities.**

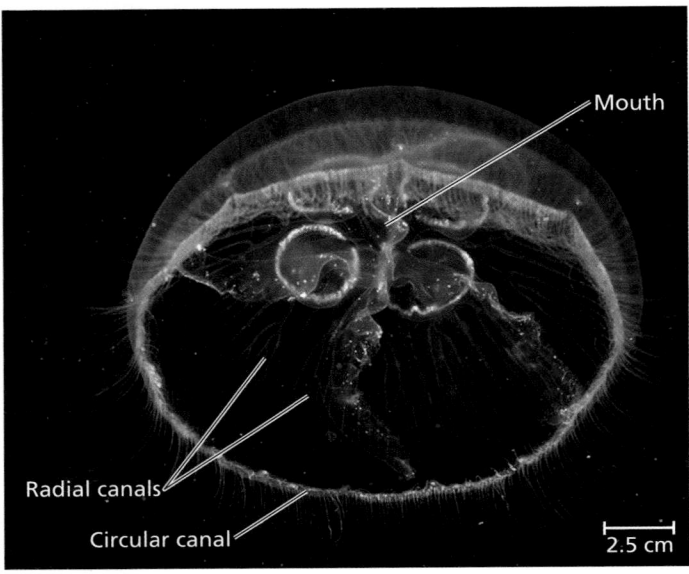

Mouth

Radial canals

Circular canal

2.5 cm

(a) The moon jelly *Aurelia*, a cnidarian. The jelly is viewed here from its underside (oral surface). The mouth leads to an elaborate gastrovascular cavity that consists of radial canals leading to and from a circular canal. Ciliated cells lining the canals circulate fluid within the cavity.

Mouth

Pharynx

Gastrovascular cavity

1 mm

(b) The planarian *Dugesia*, a flatworm. The mouth and pharynx on the ventral side lead to the highly branched gastrovascular cavity, stained dark red in this specimen (LM).

WHAT IF? *Suppose a gastrovascular cavity were open at two ends, with fluid entering one end and leaving the other. How would this affect the cavity's functions in gas exchange and digestion?*

Planarians and most other flatworms also survive without a circulatory system. Their combination of a gastrovascular cavity and a flat body is well suited for exchange with the environment **(Figure 42.2b)**. A flat body optimizes exchange by increasing surface area and minimizing diffusion distances.

Open and Closed Circulatory Systems

A circulatory system has three basic components: a circulatory fluid, a set of interconnecting vessels, and a muscular pump, the **heart**. The heart powers circulation by using metabolic energy to elevate the circulatory fluid's hydrostatic pressure, which is the pressure the fluid exerts on surrounding vessels. The fluid then flows through the vessels and back to the heart.

By transporting fluid throughout the body, the circulatory system functionally connects the aqueous environment of the body cells to the organs that exchange gases, absorb nutrients, and dispose of wastes. In mammals, for example, O_2 from inhaled air diffuses across only two layers of cells in the lungs before reaching the blood. The circulatory system then carries the oxygen-rich blood to all parts of the body. As the blood passes throughout the body tissues in tiny blood vessels, O_2 in the blood diffuses only a short distance before entering the fluid that directly bathes the cells.

Circulatory systems are either open or closed. In an **open circulatory system**, the circulatory fluid, called **hemolymph**, is also the *interstitial fluid* that bathes body cells. Arthropods, such as grasshoppers, and some molluscs, including clams, have open circulatory systems. Contraction of the heart pumps the hemolymph through the circulatory vessels into interconnected sinuses, spaces surrounding the organs **(Figure 42.3a)**. Within the sinuses, the hemolymph and body cells exchange gases and other chemicals. Relaxation of the heart draws hemolymph back in through pores, which have valves that close when the heart contracts. Body movements periodically squeeze the sinuses, helping circulate the hemolymph. The open circulatory system of larger crustaceans, such as lobsters and crabs, includes a more extensive system of vessels, as well as an accessory pump.

In a **closed circulatory system**, a circulatory fluid called **blood** is confined to vessels and is distinct from the interstitial fluid **(Figure 42.3b)**. One or more hearts pump blood into large vessels that branch into smaller ones, which infiltrate the tissues and organs. Chemical exchange occurs between the blood and the interstitial fluid, as well as between the interstitial fluid and body cells. Annelids (including earthworms), cephalopods (including squids and octopuses), and all vertebrates have closed circulatory systems.

The fact that both open and closed circulatory systems are widespread among animals suggests that each

▼ **Figure 42.3 Open and closed circulatory systems.**

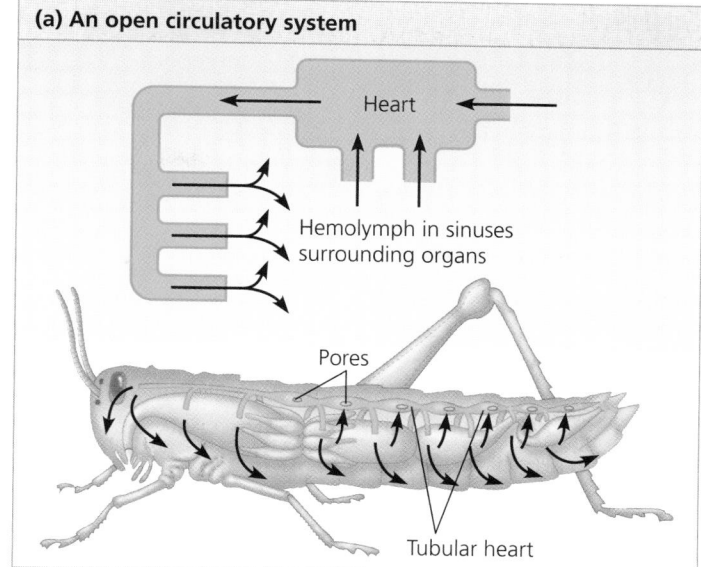

(a) An open circulatory system

Heart

Hemolymph in sinuses surrounding organs

Pores

Tubular heart

(b) A closed circulatory system

Heart

Interstitial fluid

Blood

Small branch vessels in each organ

Dorsal vessel (main heart)

Auxiliary hearts

Ventral vessels

system offers evolutionary advantages. The lower hydrostatic pressures typically associated with open circulatory systems allow them to use less energy than closed systems. In some invertebrates, open circulatory systems serve additional functions. For example, spiders use the hydrostatic pressure of their open circulatory system to extend their legs. The benefits of closed circulatory systems include blood pressure high enough to enable the effective delivery of O_2 and nutrients in larger and more active animals. Among molluscs, for instance, closed circulatory systems are found in the largest and most active species, the squids and octopuses. Closed systems are also particularly well suited to regulating the distribution of blood to different organs, as you'll learn later in this chapter. We'll now examine closed circulatory systems in more detail, focusing on vertebrates.

Organization of Vertebrate Circulatory Systems

The term **cardiovascular system** is often used to describe the heart and blood vessels in vertebrates. Blood circulates to and from the heart through an amazingly extensive network of vessels: The total length of blood vessels in an average human adult is twice Earth's circumference at the equator!

Arteries, veins, and capillaries are the three main types of blood vessels. Within each type, blood flows in only one direction. **Arteries** carry blood from the heart to organs throughout the body. Within organs, arteries branch into **arterioles**. These small vessels convey blood to **capillaries**, microscopic vessels with very thin, porous walls. Networks of capillaries, called **capillary beds**, infiltrate tissues, passing within a few cell diameters of every cell in the body. Across the thin walls of capillaries, dissolved gases and other chemicals are exchanged by diffusion between the blood and the interstitial fluid around the tissue cells. At their "downstream" end, capillaries converge into **venules**, and venules converge into **veins**, the vessels that carry blood back to the heart.

Note that arteries and veins are distinguished by the *direction* in which they carry blood, not by the O_2 content or other characteristics of the blood they contain. Arteries carry blood *away* from the heart toward capillaries, and veins return blood *toward* the heart from capillaries. The only exceptions are the portal veins, which carry blood between pairs of capillary beds. The hepatic portal vein, for example, carries blood from capillary beds in the digestive system to capillary beds in the liver.

The hearts of all vertebrates contain two or more muscular chambers. The chambers that receive blood entering the heart are called **atria** (singular, *atrium*). The chambers responsible for pumping blood out of the heart are called **ventricles**. The number of chambers and the extent to which they are separated from one another differ substantially among groups of vertebrates, as we'll discuss next. These important differences reflect the close fit of form to function that arises from natural selection.

Single Circulation

In sharks, rays, and bony fishes, blood travels through the body and returns to its starting point in a single circuit (loop), an arrangement called **single circulation (Figure 42.4a)**. These animals have a heart that consists of two chambers: an atrium and a ventricle. Blood entering the heart collects in the atrium before transfer to the ventricle. Contraction of the ventricle pumps blood to a capillary bed in the gills, where there is a net diffusion of O_2 into the blood and of CO_2 out of the blood. As blood leaves the gills, the capillaries converge into a vessel that carries oxygen-rich blood to capillary beds throughout the body. Following

▼ **Figure 42.4 Examples of vertebrate circulatory system organization.**

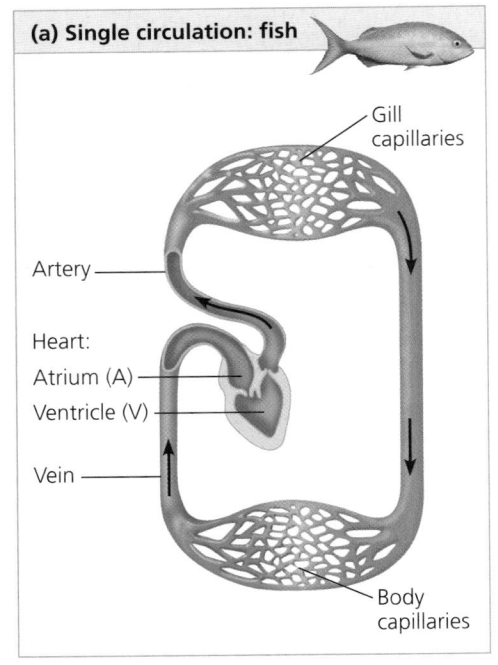

(a) Single circulation: fish

Gill capillaries

Artery

Heart:
Atrium (A)
Ventricle (V)

Vein

Body capillaries

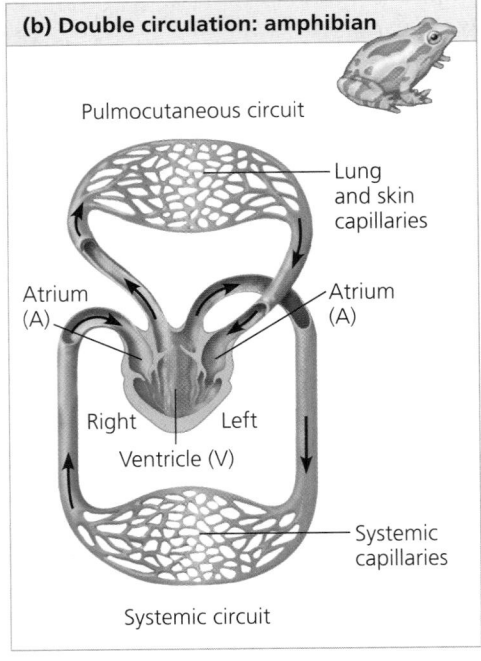

(b) Double circulation: amphibian

Pulmocutaneous circuit

Lung and skin capillaries

Atrium (A)

Atrium (A)

Right   Left

Ventricle (V)

Systemic capillaries

Systemic circuit

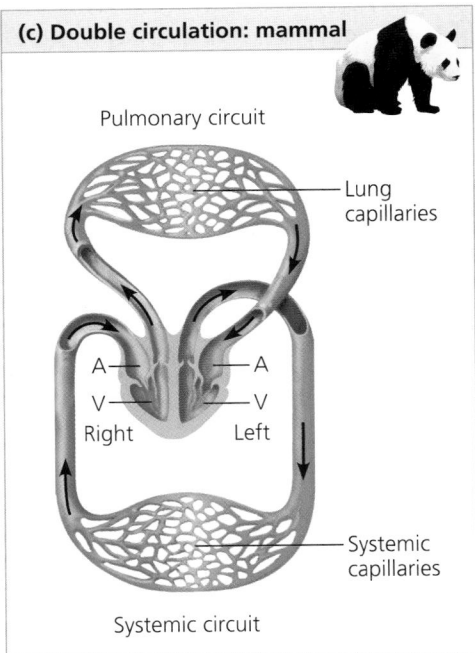

(c) Double circulation: mammal

Pulmonary circuit

Lung capillaries

A   A
V   V
Right   Left

Systemic capillaries

Systemic circuit

Key   ■ Oxygen-rich blood
          ■ Oxygen-poor blood

(Note that circulatory systems are shown as if the body were facing you: The right side of the heart is shown on the left, and vice versa.)

gas exchange in the capillary beds, blood enters veins and returns to the heart.

In single circulation, blood that leaves the heart passes through two capillary beds before returning to the heart. When blood flows through a capillary bed, blood pressure drops substantially, for reasons we'll explain later in the chapter. The blood pressure drop in the gills limits the rate of blood flow in the rest of the animal's body. As the animal swims, however, the contraction and relaxation of its muscles help accelerate the relatively sluggish pace of circulation.

Double Circulation

The circulatory systems of amphibians, reptiles, and mammals have two circuits of blood flow, an arrangement called **double circulation (Figure 42.4b** and **c)**. In animals with double circulation, the pumps for the two circuits are combined into a single organ, the heart. Having both pumps within a single heart simplifies coordination of the pumping cycles.

In one circuit, the right side of the heart pumps oxygen-poor blood to the capillary beds of the gas exchange tissues, where there is a net movement of O_2 into the blood and of CO_2 out of the blood. In most vertebrates, including reptiles and mammals, this is called the *pulmonary circuit* (from the Latin *pulmo*, lung) because gas exchange takes place in the lungs. For many amphibians, it is called the *pulmocutaneous circuit* because gas exchange takes place in capillaries in both the lungs and the skin.

The other circuit, called the *systemic circuit*, begins with the left side of the heart pumping oxygen-enriched blood from the gas exchange tissues to capillary beds in organs and tissues throughout the body. Following the exchange of O_2 and CO_2, as well as nutrients and waste products, the now oxygen-poor blood returns to the heart, completing the circuit.

Double circulation provides a vigorous flow of blood to the brain, muscles, and other organs because the heart repressurizes the blood after it passes through the capillary beds of the lungs or skin. Indeed, blood pressure is often much higher in the systemic circuit than in the gas exchange circuit. By contrast, in single circulation the blood flows under reduced pressure directly from the gas exchange organs to other organs.

Evolutionary Variation in Double Circulation

EVOLUTION Some vertebrates with double circulation are intermittent breathers. For example, amphibians and many reptiles fill their lungs with air periodically, passing long periods either without gas exchange or by relying on another gas exchange tissue, typically the skin. A variety of adaptations found among intermittent breathers enable their circulatory systems to temporarily bypass the lungs in part or in whole:

- Frogs and other amphibians have a heart with three chambers—two atria and one ventricle (see Figure 42.4b). A ridge within the ventricle diverts most (about 90%) of the oxygen-rich blood from the left atrium into the systemic circuit and most of the oxygen-poor blood from the right atrium into the pulmocutaneous circuit. When a frog is underwater, it takes advantage of the incomplete division of the ventricle, largely shutting off blood flow to its temporarily ineffective lungs. Blood flow continues to the skin, which acts as the sole site of gas exchange while the frog is submerged.

- In the three-chambered heart of turtles, snakes, and lizards, an incomplete septum partially divides the single ventricle into right and left chambers. Two major arteries, called aortas, lead to the systemic circulation. As with amphibians, the circulatory system enables control of the relative amount of blood flowing to the lungs and the rest of the body.

- In alligators, caimans, and other crocodilians, the ventricles are divided by a complete septum, but the pulmonary and systemic circuits connect where the arteries exit the heart. This connection allows arterial valves to shunt blood flow away from the lungs temporarily, such as when the animal is underwater.

Double circulation in birds and mammals, which for the most part breathe continuously, differs from double circulation in other vertebrates. As shown for a panda in Figure 42.4c, the heart has two atria and two completely divided ventricles. The left side of the heart receives and pumps only oxygen-rich blood, while the right side receives and pumps only oxygen-poor blood. Unlike amphibians and many reptiles, birds and mammals cannot vary blood flow to the lungs without varying blood flow throughout the body in parallel.

How has natural selection shaped the double circulation of birds and mammals? As endotherms, they use about ten times as much energy as equal-sized ectotherms (see Concept 40.4). Their circulatory systems therefore need to deliver about ten times as much fuel and O_2 to their tissues and remove ten times as much CO_2 and other wastes. This large-scale traffic of substances is made possible by the separate and independently powered systemic and pulmonary circuits and by large hearts. A powerful four-chambered heart arose independently in the distinct ancestors of birds and mammals and thus reflects convergent evolution (see Concept 22.3).

In the next section, we'll restrict our focus to circulation in mammals and to the anatomy and physiology of the key circulatory organ—the heart.

1. How is the flow of hemolymph through an open circulatory system similar to the flow of water through an outdoor fountain?

2. Three-chambered hearts with incomplete septa were once viewed as being less adapted to circulatory function than mammalian hearts. What advantage of such hearts did this viewpoint overlook?

3. **WHAT IF?** The heart of a normally developing human fetus has a hole between the left and right atria. In some cases, this hole does not close completely before birth. If the hole weren't surgically corrected, how would it affect the O_2 content of the blood entering the systemic circuit?

For suggested answers, see Appendix A.

CONCEPT 42.2

Coordinated cycles of heart contraction drive double circulation in mammals

The timely delivery of O_2 to the body's organs is critical. Some brain cells, for example, die if their O_2 supply is interrupted for even a few minutes. How does the mammalian cardiovascular system meet the body's continuous (although variable) demand for O_2? To answer this question, we must consider how the parts of the system are arranged and how each of these parts functions.

Mammalian Circulation

Let's first examine the overall organization of the mammalian cardiovascular system, beginning with the pulmonary circuit. (The circled numbers refer to labeled structures in **Figure 42.5**.) Contraction of ❶ the right ventricle pumps blood to the lungs via ❷ the pulmonary arteries. As the blood flows through ❸ capillary beds in the left and right lungs, it loads O_2 and unloads CO_2. Oxygen-rich blood returns from the lungs via the pulmonary veins to ❹ the left atrium of the heart. Next, the oxygen-rich blood flows into ❺ the heart's left ventricle, which pumps the oxygen-rich blood out to body tissues through the systemic circuit. Blood leaves the left ventricle via ❻ the aorta, which conveys blood to arteries leading throughout the body. The first branches leading from the aorta are the coronary arteries (not shown), which supply blood to the heart muscle itself. Branches further along the aorta lead to ❼ capillary beds in the head and arms (forelimbs). The aorta then descends into the abdomen, supplying oxygen-rich blood to arteries leading to ❽ capillary beds in the abdominal organs and legs (hind limbs). Within the capillary beds, there is a net diffusion of O_2 from the blood to the tissues and of CO_2 (produced by cellular respiration) into the blood. Capillaries rejoin, forming venules, which convey blood to veins. Oxygen-poor blood from the head, neck, and

▼ **Figure 42.5 The mammalian cardiovascular system: an overview.** Note that the dual circuits operate simultaneously, not in the serial fashion that the numbering in the diagram suggests. The two ventricles contract almost in unison and pump the same volume of blood. However, the total volume of blood in the systemic circuit is much greater than that in the pulmonary circuit.

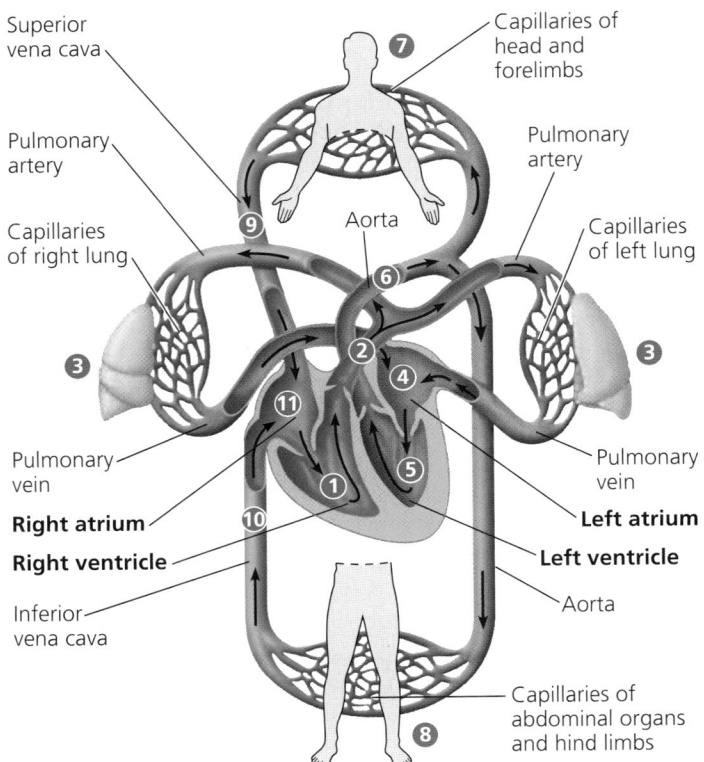

VISUAL SKILLS *If you trace the path of a molecule of carbon dioxide that starts in an arteriole in the right thumb and leaves the body in exhaled air, what is the minimum number of capillary beds the molecule encounters? Explain.*

→ **Mastering Biology Animation: The Human Heart and Circulation**

forelimbs is channeled into a large vein, ❾ the superior vena cava. Another large vein, ❿ the inferior vena cava, drains blood from the trunk and hind limbs. The two venae cavae empty their blood into ⓫ the right atrium, from which the oxygen-poor blood flows into the right ventricle.

The Mammalian Heart: *A Closer Look*

Using the human heart as an example, let's now take a closer look at how the mammalian heart works **(Figure 42.6)**. Located behind the sternum (breastbone), the human heart is about the size of a clenched fist and consists mostly of cardiac muscle (see Figure 40.5). The two atria have relatively thin walls and serve as collection chambers for blood returning to the heart from the lungs or other body tissues. Much of the blood that enters the atria flows into the ventricles while all four heart chambers are relaxed. The remainder is transferred by contraction of the atria before the ventricles

▼ Figure 42.6 The mammalian heart: a closer look. Notice the locations of the valves, which prevent backflow of blood within the heart. Also notice how the atria and left and right ventricles differ in the thickness of their muscular walls.

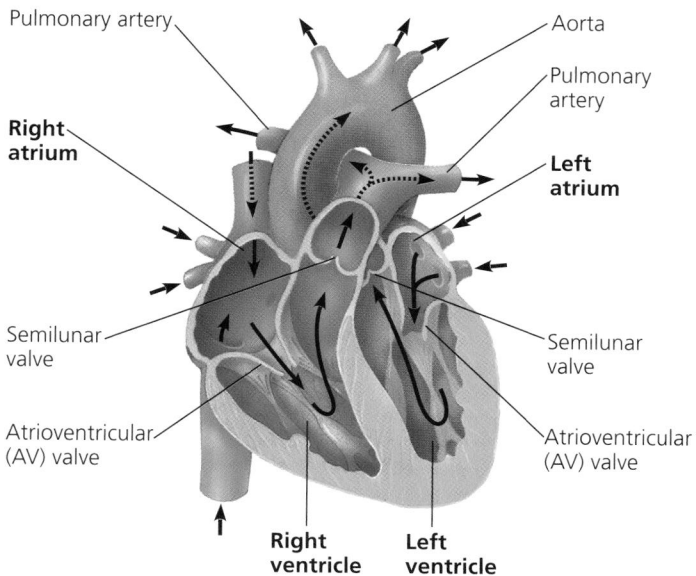

Pulmonary artery

Aorta

Pulmonary artery

Right atrium

Left atrium

Semilunar valve

Semilunar valve

Atrioventricular (AV) valve

Atrioventricular (AV) valve

Right ventricle

Left ventricle

➔ **Mastering Biology** Animation: Structure of the Human Heart

begin to contract. Compared with the atria, the ventricles have thicker walls and contract much more forcefully—especially the left ventricle, which pumps blood throughout the body via the systemic circuit. Although the left ventricle contracts with greater force than the right ventricle, it pumps the same volume of blood as the right ventricle during each contraction.

The heart contracts and relaxes in a rhythmic cycle. When it contracts, it pumps blood; when it relaxes, its chambers fill with blood. One complete sequence of pumping and filling is referred to as the **cardiac cycle**. The contraction phase of the cycle is called **systole**, and the relaxation phase is called **diastole (Figure 42.7)**.

The volume of blood each ventricle pumps per minute is the **cardiac output**. Two factors determine cardiac output: the rate of contraction, or **heart rate** (number of beats per minute), and the **stroke volume**, the amount of blood pumped by a ventricle in a single contraction. The average stroke volume in humans is about 70 mL. Multiplying this stroke volume by a typical resting heart rate of 72 beats per minute yields a cardiac output of 5 L/min—about equal to the total volume of blood in the human body. During heavy exercise, the increased demand for O_2 is met by an increase in cardiac output that can be as much as fivefold.

Four valves in the heart prevent backflow and keep blood moving in the correct direction (see Figures 42.6 and 42.7). Made of flaps of connective tissue, the valves open when pushed from one side and close when pushed from the other. An **atrioventricular (AV) valve** lies between each atrium

▼ Figure 42.7 The cardiac cycle. For an adult human at rest with a heart rate of about 72 beats per minute, one complete cardiac cycle takes about 0.8 second. Note that during all but 0.1 second of the cardiac cycle, the atria are relaxed and are filling with blood returning via the veins.

❶ Atrial and ventricular diastole. During a relaxation phase, blood returning from the large veins flows into the atria and then into the ventricles through the AV valves.

❷ Atrial systole and ventricular diastole. A brief period of atrial contraction then forces all blood remaining in the atria into the ventricles.

0.1 sec

0.3 sec

0.4 sec

❸ Ventricular systole and atrial diastole. During the remainder of the cycle, ventricular contraction pumps blood into the large arteries through the semilunar valves.

and ventricle. The AV valves are anchored by strong fibers that prevent them from turning inside out during ventricular systole. Pressure generated by the powerful contraction of the ventricles closes the AV valves, keeping blood from flowing back into the atria. **Semilunar valves** are located at the two exits of the heart: where the pulmonary artery leaves the right ventricle and where the aorta leaves the left ventricle. These valves are pushed open by the pressure generated during contraction of the ventricles. When the ventricles relax, blood pressure built up in the pulmonary artery and aorta closes the semilunar valves and prevents significant backflow.

You can follow the closing of the two sets of heart valves either with a stethoscope or by pressing your ear tightly against the chest of a friend (or a friendly dog). The sound pattern is "lub-dup, lub-dup, lub-dup." The first heart sound ("lub") is created by the recoil of blood against the closed AV valves. The second sound ("dup") is due to the vibrations caused by closing of the semilunar valves.

If blood squirts backward through a defective valve, it may produce an abnormal sound called a **heart murmur**. Some people are born with heart murmurs. In others, the valves may be damaged as a result of infection (for instance, from rheumatic fever, an inflammation caused by infection with certain bacteria). When a valve defect is severe enough to endanger health, surgeons may implant a mechanical replacement valve. However, not all heart murmurs are caused by a defect, and most valve defects do not reduce the efficiency of blood flow enough to warrant surgery.

Maintaining the Heart's Rhythmic Beat

In vertebrates, the heartbeat originates in the heart itself. Some cardiac muscle cells are autorhythmic, meaning they can contract and relax repeatedly without any signal from the nervous system. In fact, these rhythmic contractions continue in tissue removed from the heart and placed in a dish in the laboratory! Given that each of these cells has its own intrinsic contraction rhythm, how are their contractions coordinated in the intact heart? The answer lies in a group of autorhythmic cells located in the wall of the right atrium, near where the superior vena cava enters the heart. This cluster of cells, called the **sinoatrial (SA) node**, acts as a *pacemaker*, setting the rate and timing at which all cardiac muscle cells contract. (In contrast, some arthropods have pacemakers located in the nervous system, outside the heart.)

The SA node produces electrical impulses much like those produced by nerve cells. Because cardiac muscle cells are electrically coupled through gap junctions (see Figure 6.30), impulses from the SA node spread rapidly within heart tissue. These impulses generate currents that can be measured when they reach the skin via body fluids. In an **electrocardiogram** (**ECG** or **EKG**, from the German spelling), electrodes placed on the skin record the currents, thus measuring electrical activity of the heart. The graph of current against time has a shape that represents the stages in the cardiac cycle **(Figure 42.8)**.

Impulses from the SA node first spread rapidly through the walls of the atria, causing both atria to contract in unison. During atrial contraction, the impulses originating at the SA node reach other autorhythmic cells located in the wall between the left and right atria. These cells form a relay point called the **atrioventricular (AV) node**. Here the impulses are delayed for about 0.1 second before spreading to the heart apex. This delay allows the atria to empty completely before the ventricles contract. Then the signals from the AV node are conducted to the heart apex and throughout the ventricular walls by specialized structures called bundle branches and Purkinje fibers.

Physiological cues alter heart tempo by regulating the pacemaker function of the SA node. Two portions of the nervous system, the sympathetic and parasympathetic divisions, are largely responsible for this regulation. They function like the accelerator and brake in a car. For example, when you stand up and start walking, the sympathetic division speeds up your pacemaker. The resulting increase in heart rate provides the additional O_2 needed by the muscles that are powering your activity. If you then sit down and relax, the parasympathetic division slows down your pacemaker, decreasing your heart rate and thus conserving energy. Hormones secreted into the blood also influence the pacemaker. For instance, epinephrine, the "fight-or-flight" hormone secreted by the adrenal glands, speeds up the pacemaker. A third type of input that affects the pacemaker is body temperature. An increase of only 1°C raises the heart rate by about 10 beats per minute. This is the reason your heart beats faster when you have a fever.

Having examined the operation of the circulatory pump, we turn in the next section to the forces and structures that influence blood flow in the vessels of each circuit.

▼ **Figure 42.8 The control of heart rhythm.** Electrical signals follow a set path through the heart in establishing the heart rhythm. The diagrams at the top trace the movement of these signals (yellow) during the cardiac cycle; specialized muscle cells involved in controlling of the rhythm are indicated in orange. Under each step, the corresponding portion of an electrocardiogram (ECG) is highlighted (yellow). In step 4, the portion of the ECG to the right of the "spike" represents electrical activity that reprimes the ventricles for the next round of contraction.

1. Signals (yellow) from SA node spread through atria.
2. Signals are delayed at AV node.
3. Bundle branches pass signals to heart apex.
4. Signals spread throughout ventricles.

SA node (pacemaker)

AV node

Bundle branches   Heart apex

Purkinje fibers

ECG

WHAT IF? *If your doctor gave you a copy of your ECG recording, how could you determine what your heart rate had been during the test?*

1. Explain why blood has a higher O_2 concentration in the pulmonary veins than in the venae cavae, which are also veins.
2. Why is it important that the AV node delay the electrical impulse moving from the SA node and atria to the ventricles?
3. **WHAT IF?** Suppose that after you exercise regularly for several months, your resting heart rate decreases, but your cardiac output at rest is unchanged. Based on these observations, what other change in the function of your heart at rest likely occurred?

For suggested answers, see Appendix A.

CONCEPT 42.3

Patterns of blood pressure and flow reflect the structure and arrangement of blood vessels

To deliver oxygen and nutrients and remove wastes throughout the body, the vertebrate circulatory system relies on blood vessels that exhibit a close match of structure and function.

Blood Vessel Structure and Function

All blood vessels contain a central lumen (cavity) lined with an **endothelium**, a single layer of flattened epithelial cells. Like the polished surface of a copper pipe, the smooth endothelial layer minimizes resistance to fluid flow. Surrounding the endothelium are tissue layers that differ among capillaries, arteries, and veins, reflecting distinct adaptations to the particular functions of these vessels **(Figure 42.9)**.

Capillaries are the smallest blood vessels, having a diameter only slightly greater than that of a red blood cell. Capillaries also have very thin walls, which consist of just an endothelium and a surrounding extracellular layer called the *basal lamina*. The exchange of substances between the blood and interstitial fluid occurs only in capillaries because only there are the vessel walls thin enough to permit this exchange.

In contrast to capillaries, both arteries and veins have walls that consist of two layers of tissue surrounding the endothelium. The outer layer is formed by connective tissue

that contains elastic fibers, which allow the vessel to stretch and recoil, and collagen, which provides strength. The layer next to the endothelium contains smooth muscle and more elastic fibers.

Arterial walls are thick, strong, and elastic. They can thus accommodate blood pumped at high pressure by the heart, bulging outward as blood enters and recoiling as the heart relaxes between contractions. As we'll discuss shortly, this behavior of arterial walls has an essential role in maintaining blood pressure and flow to capillaries.

The smooth muscles in the walls of arteries and arterioles help regulate the path of blood flow. Signals from the nervous system and circulating hormones act on the smooth muscle of these vessels, causing dilation or constriction that modulates blood flow to different parts of the body.

Because veins convey blood back to the heart at a lower pressure, they do not require thick walls. For a given blood vessel diameter, a vein has a wall only about a third as thick as that of an artery. Unlike arteries, veins contain valves, which maintain a unidirectional flow of blood despite the low blood pressure in these vessels.

▼ **Figure 42.9 The structure of blood vessels.**

We consider next how blood vessel diameter, vessel number, and blood pressure influence the speed at which blood flows in different locations within the body.

Blood Flow Velocity

To understand how blood vessel diameter influences blood flow, consider how water flows through a thick-walled hose connected to a faucet. When the faucet is turned on, water flows at the same velocity at each point along the hose. What happens when a narrow nozzle is attached to the end of the hose? Because water doesn't compress under pressure, the volume of water moving through the nozzle in a given time must be the same as the volume moving through the rest of the hose. The cross-sectional area of the nozzle is smaller than that of the hose, so the water speeds up, exiting the nozzle at high velocity.

An analogous situation exists in the circulatory system, but blood *slows* as it moves from arteries to arterioles to the much narrower capillaries. Why? The number of capillaries is enormous, roughly 7 billion in a human body. Each artery conveys blood to so many capillaries that the *total* cross-sectional area is much greater in capillary beds than in the arteries or any other part of the circulatory system (**Figure 42.10**). This enormous increase in cross-sectional area results in a dramatic decrease in velocity from the arteries to the capillaries: Blood travels 500 times more slowly in the capillaries (about 0.1 cm/sec) than in the aorta (about 48 cm/sec). After passing through the capillaries, the blood speeds up as it enters the venules and veins and *total* cross-sectional areas decreases.

Blood Pressure

Blood, like all fluids, flows from areas of higher pressure to areas of lower pressure. Contraction of a heart ventricle generates blood pressure, which exerts a force in all directions. The part of the force directed lengthwise in an artery causes the blood to flow away from the heart, the site of highest pressure. The part of the force exerted sideways stretches the wall of the artery. Following ventricular contraction, the recoil of the elastic arterial walls plays a critical role in maintaining blood pressure, and hence blood flow, throughout the cardiac cycle. Once the blood enters the millions of tiny arterioles and capillaries, the narrow diameter of these vessels generates substantial resistance to flow. By the time the blood enters the veins, this resistance has dissipated much of the pressure generated by the pumping heart (see Figure 42.10).

Changes in Blood Pressure During the Cardiac Cycle

Arterial blood pressure is highest when the heart contracts during ventricular systole. The pressure at this time is called **systolic pressure** (see Figure 42.10). Each ventricular contraction causes a spike in blood pressure that stretches the walls

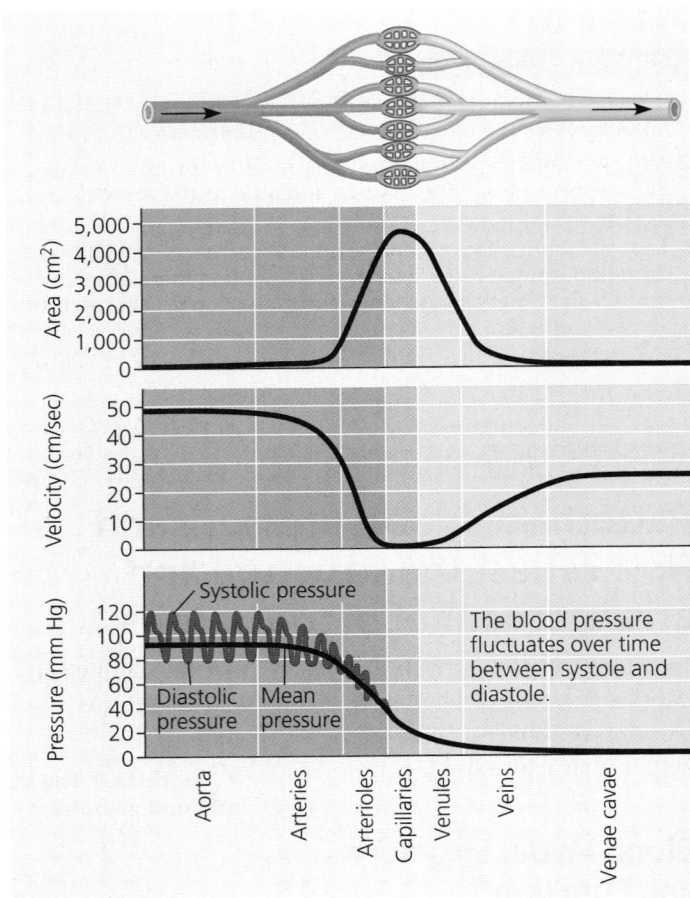

▲ **Figure 42.10 The interrelationship of cross-sectional area of blood vessels, blood flow velocity, and blood pressure.** As a result of an increase in total cross-sectional area, blood flow velocity decreases markedly in the arterioles and is lowest in the capillaries. Blood pressure, the main force driving blood from the heart to the capillaries, is highest in the aorta and other arteries.

⊕ **Mastering Biology** Figure Walkthrough

of the arteries. You can feel this **pulse**—the rhythmic bulging of the artery walls with each heartbeat—by placing the tips of your fingers on the inside of the opposite wrist. The pressure surge is partly due to the narrow openings of arterioles impeding the exit of blood from the arteries. When the heart contracts, blood enters the arteries faster than it can leave, and the vessels stretch to a wider diameter from the rise in pressure.

During diastole, the elastic walls of the arteries snap back. As a consequence, there is a lower but still substantial blood pressure when the ventricles are relaxed (**diastolic pressure**). Before enough blood has flowed into the arterioles to completely relieve pressure in the arteries, the heart contracts again. Because the arteries remain pressurized throughout the cardiac cycle (see Figure 42.10), blood continuously flows into arterioles and capillaries.

Regulation of Blood Pressure

Homeostatic mechanisms regulate arterial blood pressure by altering the diameter of arterioles. If the smooth muscles in arteriole walls contract, the arterioles narrow, a process called **vasoconstriction**. Vasoconstriction increases blood pressure upstream in the arteries. When the smooth muscles relax, the arterioles undergo **vasodilation**, an increase in diameter that causes blood pressure in the arteries to fall.

Researchers have identified nitric oxide (NO), a gas, as a major inducer of vasodilation and endothelin, a peptide, as the most potent inducer of vasoconstriction. Cues from the nervous and endocrine systems regulate production of NO and endothelin in blood vessels, where their opposing activities provide homeostatic regulation of blood pressure.

Vasoconstriction and vasodilation are often coupled to changes in cardiac output that also affect blood pressure. This coordination of regulatory mechanisms maintains adequate blood flow as the body's demands on the circulatory system change. During heavy exercise, for example, the arterioles in working muscles dilate, causing a greater flow of oxygen-rich blood to the muscles. By itself, this increased flow to the muscles would cause a drop in blood pressure (and therefore blood flow) in the body as a whole. However, cardiac output increases at the same time, maintaining blood pressure and supporting the necessary increase in blood flow.

> **Mastering Biology** Interview with Masashi Yanagisawa: Discovering the key inducer of vasoconstriction

flow. By causing your body to collapse to the ground, fainting effectively places your head at the level of your heart, quickly increasing blood flow to your brain.

For animals with very long necks, the blood pressure required to overcome gravity is particularly high. A giraffe, for example, requires a systolic pressure of more than 250 mm Hg near the heart to get blood to its head. When a giraffe lowers its head to drink, one-way valves and sinuses, along with feedback mechanisms that reduce cardiac output, reduce blood pressure in the head, preventing brain damage. A dinosaur with a neck nearly 10 m long would have required even greater systolic pressure—nearly 760 mm Hg—to pump blood to its brain when its head was fully raised. However, calculations based on anatomy and inferred metabolic rate suggest that dinosaurs did not have a heart powerful enough to generate such high pressure. Based on this evidence as well as studies of neck bone structure, some biologists have concluded that the long-necked dinosaurs fed close to the ground rather than on high foliage.

Gravity is also a consideration for blood flow in veins, especially those in the legs. When you stand or sit, gravity draws blood downward to your feet and impedes its upward return to the heart. Because blood pressure in veins is relatively low, valves inside the veins have an important function in maintaining the unidirectional flow of blood within these vessels. The return of blood to the heart is further enhanced by rhythmic contractions of smooth muscles in the walls of

Blood Pressure and Gravity

Blood pressure is generally measured for an artery in the arm at the same height as the heart **(Figure 42.11)**. For a healthy 20-year-old human at rest, arterial blood pressure in the systemic circuit is typically about 120 millimeters of mercury (mm Hg) at systole and 70 mm Hg at diastole, expressed as 120/70. (Arterial blood pressure in the pulmonary circuit is six to ten times lower.)

Gravity has a significant effect on blood pressure. When you are standing, for example, your head is roughly 0.35 m higher than your chest, and the arterial blood pressure in your brain is about 27 mm Hg less than that near your heart. This relationship of blood pressure and gravity is the key to understanding fainting. The fainting response is triggered when the nervous system detects that the blood pressure in your brain is below the level needed to provide adequate blood

▼ **Figure 42.11 Measurement of blood pressure.** Blood pressure is recorded as two numbers separated by a slash. The first number is the systolic pressure; the second is the diastolic pressure.

❶ A sphygmomanometer, an inflatable cuff attached to a pressure gauge, measures blood pressure in an artery. The cuff is inflated until the pressure closes the artery, so that no blood flows past the cuff. When this occurs, the pressure exerted by the cuff exceeds the pressure in the artery.

❷ The cuff is allowed to deflate gradually. When the pressure exerted by the cuff falls just below that in the artery, blood pulses into the forearm, generating sounds that can be heard with the stethoscope. The pressure measured at this point is the systolic pressure (120 mm Hg in this example).

❸ The cuff is allowed to deflate further, just until the blood flows freely through the artery and the sounds below the cuff disappear. The pressure at this point is the diastolic pressure (70 mm Hg in this example).

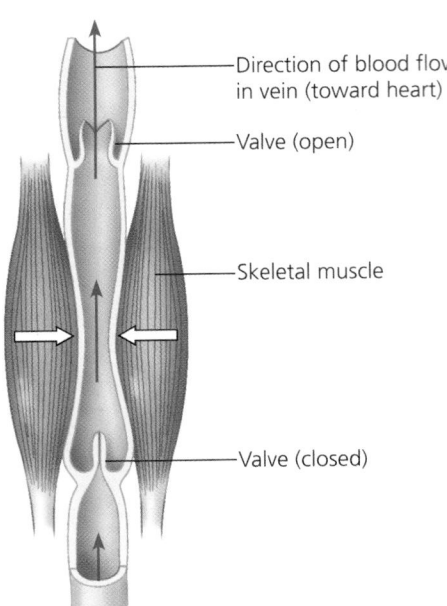

► **Figure 42.12 Blood flow in veins.** Skeletal muscle contraction squeezes and constricts veins. Flaps of tissue within the veins act as one-way valves that keep blood moving only toward the heart. If you sit or stand too long, the lack of muscular activity may cause your feet to swell as blood pools in your veins.

Direction of blood flow in vein (toward heart)

Valve (open)

Skeletal muscle

Valve (closed)

venules and veins and by the contraction of skeletal muscles during exercise **(Figure 42.12)**.

In rare instances, runners and other athletes can suffer heart failure if they stop vigorous exercise abruptly. When the leg muscles suddenly cease contracting and relaxing, less blood returns to the heart, which continues to beat rapidly. If the heart is weak or damaged, this inadequate blood flow may cause the heart to malfunction. To reduce the risk of stressing the heart excessively, athletes are encouraged to follow hard exercise with moderate activity, such as walking, to "cool down" until their heart rate approaches its resting level.

Capillary Function

At any given time, blood is flowing through only 5–10% of the body's capillaries. However, each tissue has many capillaries, so every part of the body is supplied with blood at all times. Capillaries in the brain, heart, kidneys, and liver usually remain at capacity, but at many other sites the blood supply varies over time as blood is diverted from one destination to another. For example, blood supply to the digestive tract increases after a meal. In contrast, blood is diverted from the digestive tract and supplied more generously to skeletal muscles during strenuous exercise.

Given that capillaries lack smooth muscle, how is blood flow in capillary beds altered? One mechanism is constriction or dilation of the arterioles that supply capillary beds. A second mechanism involves precapillary sphincters, rings of smooth muscle located at the entrance to capillary beds **(Figure 42.13)**. The opening and closing of these muscular rings regulate the passage of blood into particular sets of capillaries. The signals regulating blood flow by these mechanisms include nerve impulses, hormones traveling throughout the bloodstream, and chemicals produced locally. For example, the chemical histamine released by cells at a wound

▼ **Figure 42.13 Blood flow in capillary beds.** Precapillary sphincters regulate the passage of blood into capillary beds. Some blood flows directly from arterioles to venules through capillaries called thoroughfare channels, which are always open.

Precapillary sphincters

Thoroughfare channel

Capillaries

Arteriole

Venule

(a) Sphincters relaxed

Arteriole

Venule

(b) Sphincters contracted

site causes vasodilation. The result is increased blood flow and increased access of disease-fighting white blood cells to invading microorganisms.

As you have read, the critical exchange of substances between the blood and interstitial fluid takes place across the thin endothelial walls of the capillaries. How does exchange occur? A few macromolecules are carried across the endothelium in vesicles that form on one side by endocytosis and release their contents on the opposite side by exocytosis. Small molecules, such as O_2 and CO_2, simply diffuse across the endothelial cells or, in some tissues, through microscopic pores in the capillary wall. These openings also provide the route for transport of small solutes such as sugars, salts, and urea, as well as for bulk flow of fluid into tissues driven by blood pressure within the capillary.

Two opposing forces control the movement of fluid between the capillaries and the surrounding tissues: Blood pressure tends to drive fluid out of the capillaries, and the presence of blood proteins tends to pull fluid back **(Figure 42.14)**. Many blood proteins (and all blood cells) are too large to pass readily through the

▼ Figure 42.14 Fluid exchange between capillaries and the interstitial fluid. This diagram shows a hypothetical capillary in which blood pressure exceeds osmotic pressure throughout the entire length of the capillary. In other capillaries, blood pressure may be lower than osmotic pressure along all or part of the capillary.

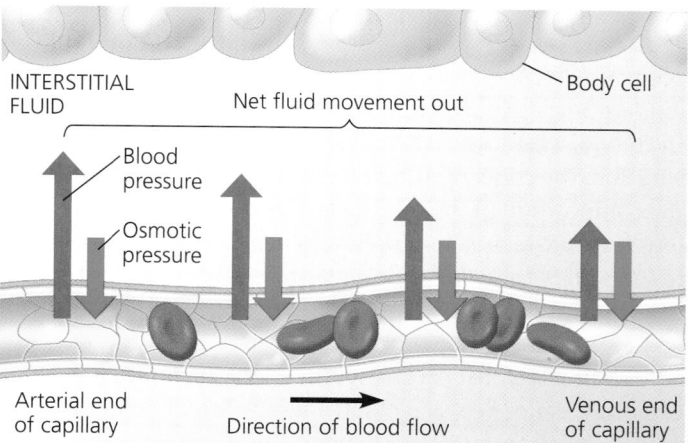

endothelium, so they remain in the capillaries. These dissolved proteins are responsible for much of the blood's *osmotic pressure* (the pressure produced by the difference in solute concentration across a membrane). The difference in osmotic pressure between the blood and the interstitial fluid opposes fluid movement out of the capillaries. On average, blood pressure is greater than the opposing forces, leading to a net loss of fluid from capillaries.

Fluid Return by the Lymphatic System

Each day the adult human body loses approximately 4–8 L of fluid from capillaries to the surrounding tissues. There is also some leakage of blood proteins, even though the capillary wall is not very permeable to large molecules. The lost fluid and the proteins within it are recovered and returned to the blood via the **lymphatic system**.

As shown in **Figure 42.15**, fluid diffuses into the lymphatic system via a network of tiny vessels intermingled with capillaries. The recovered fluid, called **lymph**, circulates within the lymphatic system before draining into a pair of large veins of the

▶ Figure 42.15 The close association of lymphatic vessels and blood capillaries. The lymphatic system, shown in green, extends throughout the body, terminating in narrow vessels intermingled with blood capillaries. The terminal lymphatic vessels have closed ends, but are permeable to interstitial liquid flowing in from surrounding tissue. Before reaching the heart, fluid in the lymphatic system undergoes filtering and monitoring by the immune system at small, bean-shaped structures called lymph nodes.

cardiovascular system at the base of the neck. This joining of the lymphatic and cardiovascular systems completes the recovery of fluid lost from capillaries as well as the transfer of lipids from the small intestine to the blood (see Figure 41.13).

The movement of lymph from peripheral tissues to the heart relies on many of the same mechanisms that assist blood flow in veins. Lymph vessels, like veins, have valves that prevent the backflow of fluid. Rhythmic contractions of the vessel walls help draw fluid into the small lymphatic vessels. In addition, skeletal muscle contractions play a role in moving lymph.

Disruptions in the movement of lymph often result in fluid accumulation, or edema, in affected tissues. In some circumstances, the consequence is severe. For example, certain species of parasitic worms that lodge in lymph vessels and thereby block lymph movement cause elephantiasis, a condition marked by extreme swelling in limbs or other body parts.

Along a lymph vessel are small, lymph-filtering organs called **lymph nodes**, which play an important role in the body's defense. Inside each lymph node is a honeycomb of connective tissue with spaces filled by white blood cells, which function in defense. When the body is fighting an infection, the white blood cells multiply rapidly, and the lymph nodes become swollen and tender. This is why your doctor may check for swollen lymph nodes in your neck, armpits, or groin when you feel sick. Because lymph nodes may also trap circulating cancer cells, examining the lymph nodes of patients with cancer may reveal the spread of the disease.

In recent years, evidence has surfaced demonstrating that the lymphatic system plays a role in harmful immune responses, such as those responsible for asthma. Because of these and other findings, the lymphatic system has become a very active area of biomedical research.

Fluid and proteins leak from the blood capillaries into the interstitial fluid.

Lymphatic vessels recover leaked fluid and proteins, carrying them to large veins at the base of the neck.

Blood capillary

Interstitial fluid

Tissue cells

Lymphatic vessel

Arteriole

Valves in larger lymphatic vessels prevent the backflow of fluid.

1. What is the primary cause of the low velocity of blood flow in capillaries?
2. What short-term changes in an animal's cardiovascular function might facilitate using skeletal muscles to escape from a dangerous situation?
3. **WHAT IF?** If you had additional hearts distributed throughout your body, what would be one likely advantage and one likely disadvantage?

For suggested answers, see Appendix A.

Blood components function in exchange, transport, and defense

As you read in Concept 42.1, the fluid transported by an open circulatory system is continuous with the fluid that surrounds all of the body cells and thus has the same composition. In contrast, the fluid in a closed circulatory system can be much more specialized, as is the case for the blood of vertebrates.

Blood Composition and Function

Vertebrate blood is a connective tissue consisting of cells suspended in a liquid matrix called **plasma**. Separating the components of blood using a centrifuge reveals that cellular elements (cells and cell fragments) occupy about 45% of the volume of blood (**Figure 42.16**). The remainder is plasma.

Plasma

Dissolved in the plasma are ions and proteins that, together with the blood cells, function in osmotic regulation, transport, and defense. Inorganic salts in the form of dissolved ions are an essential component of the blood. Some buffer the blood, while others help maintain osmotic balance. In addition, the concentration of ions in plasma directly affects the composition of the interstitial fluid, where many of these ions have a vital role in muscle and nerve activity. For plasma to serve all of these functions, plasma electrolytes must remain within narrow concentration ranges.

Like dissolved ions, plasma proteins such as albumins act as buffers against pH changes and help maintain the osmotic balance between blood and interstitial fluid. A number of plasma proteins have additional functions. Immunoglobulins, or antibodies, combat viruses and other foreign agents that invade the body (see Figure 43.10). Apolipoproteins escort lipids, which are insoluble in water and can travel in blood only when bound to proteins. Fibrinogens act as clotting factors that help plug leaks when blood vessels are injured. (The term *serum* refers to blood plasma from which these clotting factors have been removed.)

▼ **Figure 42.16 The composition of mammalian blood.** Centrifuged blood separates into three layers: plasma, leukocytes and platelets, and erythrocytes.

Plasma 55%			Cellular elements 45%		
Constituent	**Major functions**		**Cell type**	**Number per µL (mm³) of blood**	**Functions**
Water	Solvent		**Leukocytes (white blood cells)** Basophils, Lymphocytes, Eosinophils, Neutrophils, Monocytes	5,000–10,000	Defense and immunity
Ions (blood electrolytes) Sodium Potassium Calcium Magnesium Chloride Bicarbonate	Osmotic balance, pH buffering, and regulation of membrane permeability				
Plasma proteins Albumin	Osmotic balance, pH buffering				
Immunoglobulins (antibodies)	Defense and immunity		**Platelets**	250,000–400,000	Blood clotting
Apolipoproteins	Lipid transport				
Fibrinogen	Blood clotting		**Erythrocytes (red blood cells)**	5,000,000–6,000,000	Transport of O_2 and some CO_2
Substances transported by blood Nutrients (such as glucose, fatty acids, vitamins), waste products of metabolism, respiratory gases (O_2 and CO_2), and hormones					

Separated blood elements

Plasma also contains many other substances in transit, including nutrients, metabolic wastes, respiratory gases, and hormones. Plasma has a much higher protein concentration than interstitial fluid, although the two fluids are otherwise similar. (Capillary walls, remember, are not very permeable to proteins.)

Cellular Elements

Blood contains two classes of cells: red blood cells, which transport O_2, and white blood cells, which function in defense (see Figure 42.16). Also suspended in blood plasma are **platelets**, cell fragments that are involved in the clotting process.

Erythrocytes   Red blood cells, or **erythrocytes**, are by far the most numerous blood cells. Their main function is O_2 transport, and their structure is closely related to this function. Human erythrocytes are small disks (7–8 μm in diameter) that are biconcave—thinner in the center than at the edges. This shape increases surface area, enhancing the rate of diffusion of O_2 across the plasma membrane. Mature mammalian erythrocytes lack nuclei. This unusual characteristic leaves more space in these tiny cells for **hemoglobin**, the iron-containing protein that transports O_2 (see Figure 5.18). Erythrocytes also lack mitochondria and generate their ATP exclusively by anaerobic metabolism. Oxygen transport would be less efficient if erythrocytes were aerobic and consumed some of the O_2 they carry.

Despite its small size, an erythrocyte contains about 250 million molecules of hemoglobin (Hb). Because each molecule of hemoglobin binds up to four molecules of O_2, one erythrocyte can transport about 1 billion O_2 molecules. As erythrocytes pass through the capillary beds of lungs, gills, or other respiratory organs, O_2 diffuses into the erythrocytes and binds to hemoglobin. In the systemic capillaries, O_2 dissociates from hemoglobin and diffuses into body cells.

In **sickle-cell disease**, an abnormal form of hemoglobin (Hb^S) polymerizes into aggregates. Because the concentration of hemoglobin in erythrocytes is so high, these aggregates are large enough to distort the erythrocyte into an elongated, curved shape that resembles a sickle. This abnormality results from an alteration in the amino acid sequence of hemoglobin at a single position (see Figure 5.19).

Sickle-cell disease significantly impairs the function of the circulatory system. Sickled cells often lodge in arterioles and capillaries, preventing delivery of O_2 and nutrients and removal of CO_2 and wastes. Blood vessel blockage and resulting organ swelling can result in severe pain. In addition, sickled cells frequently rupture, reducing the number of red blood cells available for transporting O_2. The average life span of a sickled erythrocyte is only 20 days—one-sixth that of a normal erythrocyte. The rate of erythrocyte loss outstrips their production rate. Short-term therapy includes replacement of erythrocytes by blood transfusion; long-term treatments are generally aimed at inhibiting aggregation of Hb^S.

Leukocytes   The blood contains five major types of white blood cells, or **leukocytes**. Their function is to fight infections. Some are phagocytic, engulfing and digesting microorganisms and debris from the body's own dead cells. Other leukocytes, called lymphocytes, mount immune responses against foreign substances (as we'll discuss in Concepts 43.2 and 43.3). Normally, 1 μL of human blood contains about 5,000–10,000 leukocytes; their numbers increase temporarily whenever the body is fighting an infection. Unlike erythrocytes, leukocytes are also found outside the circulatory system, patrolling interstitial fluid and the lymphatic system.

Platelets   Platelets are pinched-off cytoplasmic fragments of specialized bone marrow cells. They are about 2–3 μm in diameter and have no nuclei. Platelets serve both structural and molecular functions in blood clotting.

Stem Cells and the Replacement of Cellular Elements

Erythrocytes, leukocytes, and platelets all develop from stem cells that are dedicated to replenishing the body's blood cell populations. As described in Concept 20.3, a **stem cell** can reproduce indefinitely, dividing mitotically to produce one daughter cell that remains a stem cell and another that adopts a specialized function. The stem cells that produce the cellular elements of blood cells are located in the red marrow inside bones, particularly the ribs, vertebrae, sternum, and pelvis. As they divide and self-renew, these stem cells give rise to two sets of progenitor cells with a more limited capacity for self-renewal **(Figure 42.17)**. One set, the lymphoid

▼ **Figure 42.17 Differentiation of blood cells.** Cell divisions of stem cells in bone marrow give rise to two specialized sets of cells. One set—the lymphoid progenitor cells—gives rise to immune cells called lymphocytes, primarily B and T cells. The second set—myeloid progenitor cells—gives rise to other immune cells, red blood cells (erythrocytes), and cell fragments called platelets.

progenitors, produces lymphocytes. The other set, the myeloid progenitors, produces all other white blood cells, red blood cells, and platelets.

Throughout a person's life, stem cells replace the worn-out cellular elements of blood. Erythrocytes are the shortest-lived, circulating for only 120 days on average before being replaced. A feedback mechanism sensitive to O_2 level controls erythrocyte production. If the O_2 level falls, the kidneys synthesize and secrete a hormone called **erythropoietin (EPO)** that stimulates the generation of more erythrocytes.

Today, EPO produced by recombinant DNA technology is used to treat disorders such as *anemia*, a condition of lower-than-normal erythrocyte or hemoglobin level that decreases the oxygen-carrying capacity of the blood. Some athletes inject themselves with EPO to increase their erythrocyte level. Because this practice is banned by most major sports organizations, runners, cyclists, and other athletes caught using EPO-related drugs have forfeited their records and been banned from future competitions.

Blood Clotting

When blood vessels are broken by an injury such as a small cut or scrape, a chain of events ensues that quickly seals the break, halting blood loss and exposure to infection. The key mechanical event in this response is coagulation, the conversion of the liquid components of blood into a solid—a blood clot.

In the absence of injury, the coagulant, or sealant, circulates in an inactive form called fibrinogen. Blood clotting begins when injury exposes the proteins in a broken blood vessel wall to blood constituents. The exposed proteins attract platelets, which gather at the site of injury and release clotting factors. These clotting factors trigger a cascade of reactions leading to the formation of an active enzyme, *thrombin*, from an inactive form, prothrombin **(Figure 42.18)**. Thrombin in turn converts fibrinogen to fibrin, which aggregates into threads that form the framework of the clot. Any mutation that blocks a step in the clotting process can cause hemophilia, a life-threatening disease characterized by excessive bleeding and bruising from even minor cuts and bumps (see Concept 15.2).

As shown in Figure 42.18, clotting involves a positive feedback loop. Initially, the clotting reactions convert only some of the prothrombin at the clot site to thrombin. However, thrombin itself stimulates the enzymatic cascade, leading to more conversion of prothrombin to thrombin and thus driving clotting to completion.

▼ **Figure 42.18 Blood clotting.**

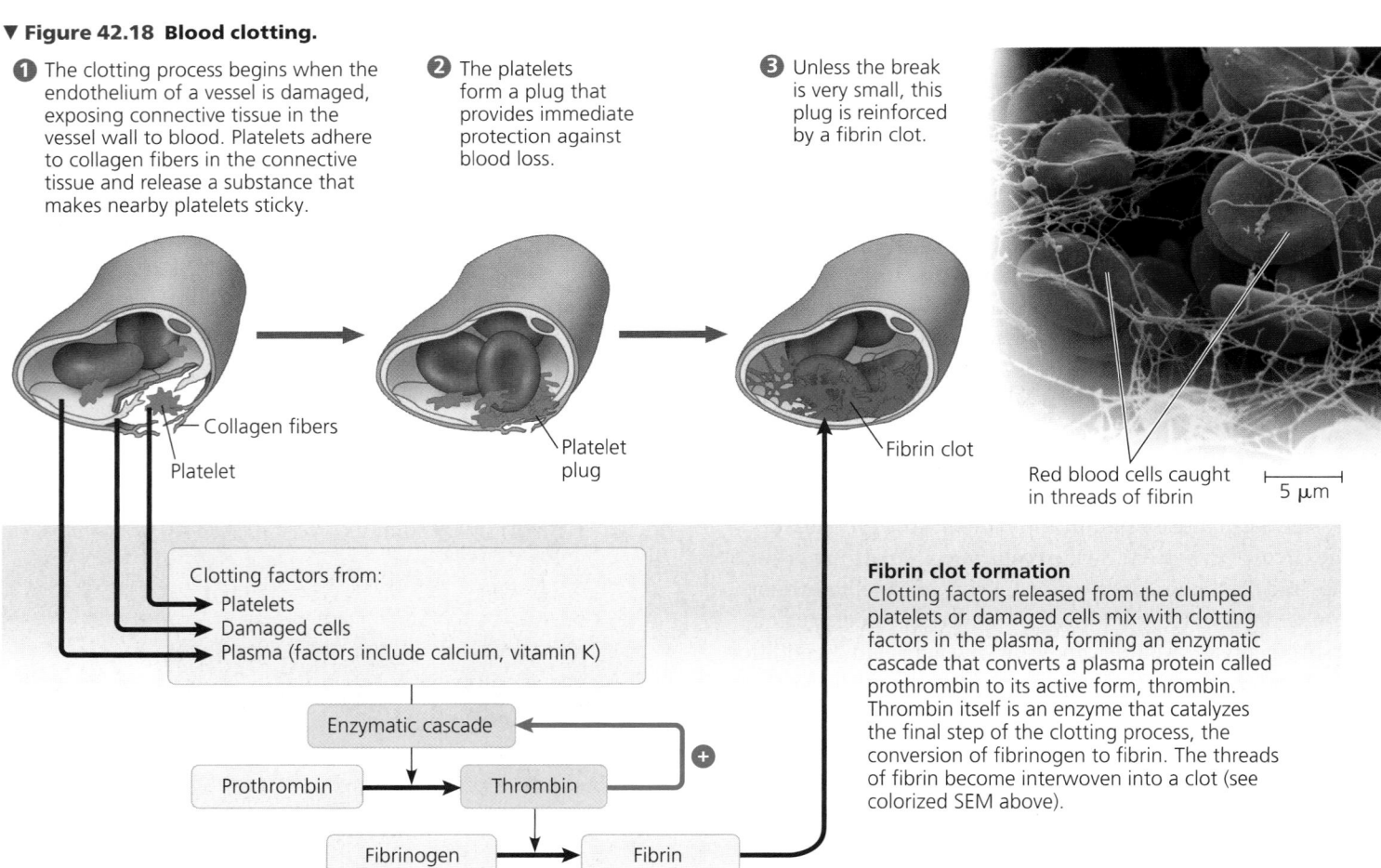

❶ The clotting process begins when the endothelium of a vessel is damaged, exposing connective tissue in the vessel wall to blood. Platelets adhere to collagen fibers in the connective tissue and release a substance that makes nearby platelets sticky.

❷ The platelets form a plug that provides immediate protection against blood loss.

❸ Unless the break is very small, this plug is reinforced by a fibrin clot.

Collagen fibers

Platelet

Platelet plug

Fibrin clot

Red blood cells caught in threads of fibrin

5 μm

Clotting factors from:
→ Platelets
→ Damaged cells
→ Plasma (factors include calcium, vitamin K)

Enzymatic cascade

Prothrombin → Thrombin

Fibrinogen → Fibrin

+

Fibrin clot formation
Clotting factors released from the clumped platelets or damaged cells mix with clotting factors in the plasma, forming an enzymatic cascade that converts a plasma protein called prothrombin to its active form, thrombin. Thrombin itself is an enzyme that catalyzes the final step of the clotting process, the conversion of fibrinogen to fibrin. The threads of fibrin become interwoven into a clot (see colorized SEM above).

Anticlotting factors in the blood normally prevent spontaneous clotting in the absence of injury. Sometimes, however, clots form within a blood vessel, blocking the flow of blood. Such a clot is called a **thrombus** (plural, *thrombi*). We'll explore shortly how thrombi form and the dangers they pose.

Cardiovascular Disease

Each year, cardiovascular diseases—disorders of the heart and blood vessels—kill more than 750,000 people in the United States. These diseases range from minor disturbances of vein or heart valve function to life-threatening disruptions of blood flow to the heart or brain.

Atherosclerosis, Heart Attacks, and Stroke

Healthy arteries have a smooth inner lining that reduces resistance to blood flow. However, damage or infection can roughen the lining and lead to **atherosclerosis**, the hardening of the arteries by accumulation of fatty deposits. A key player in the development of atherosclerosis is cholesterol, a steroid that is important for maintaining normal membrane fluidity in animal cells (see Figure 7.5).

In atherosclerosis, damage to the arterial lining results in *inflammation*, the body's reaction to injury. Leukocytes are attracted to the inflamed area and begin to take up lipids, including cholesterol. A fatty deposit, called a plaque, grows steadily, incorporating fibrous connective tissue and additional cholesterol. As the plaque grows, the walls of the artery become thick and stiff, and the obstruction of the artery increases. If the plaque ruptures, a thrombus can form in the artery **(Figure 42.19)**, potentially triggering a heart attack or a stroke.

▼ **Figure 42.19 Atherosclerosis.** In atherosclerosis, thickening of an arterial wall by plaque formation can restrict blood flow through the artery. If a plaque ruptures, a thrombus can form, further restricting blood flow. Fragments of a ruptured plaque can also travel via the bloodstream and become lodged in other arteries. If the blockage is in an artery that supplies the heart or brain, the result could be a heart attack or stroke, respectively.

A **heart attack**, also called a *myocardial infarction*, is the damage or death of cardiac muscle tissue resulting from blockage of one or more coronary arteries, which supply oxygen-rich blood to the heart muscle. The coronary arteries are small in diameter and therefore especially vulnerable to obstruction by atherosclerotic plaques or thrombi. Such blockage can destroy cardiac muscle quickly because the constantly beating heart muscle requires a steady supply of O_2. If a large enough portion of the heart is affected, the heart will stop beating. Such cardiac arrest causes death if a heartbeat is not restored within minutes by cardiopulmonary resuscitation (CPR) or some other emergency procedure.

A **stroke** is the death of nervous tissue in the brain due to a lack of O_2. Strokes usually result from rupture or blockage of arteries in the neck or head. The effects of a stroke and the individual's chance of survival depend on the extent and location of the damaged brain tissue. If a stroke results from arterial blockage by a thrombus, rapid administration of a clot-dissolving drug may help limit the damage.

Although atherosclerosis often isn't detected until critical blood flow is disrupted, there can be warning signs. Partial blockage of the coronary arteries may cause occasional chest pain, a condition known as angina pectoris. The pain is most likely to be felt when the heart is laboring under stress, and it signals that part of the heart is not receiving enough O_2. An obstructed artery may be treated surgically, either by inserting a mesh tube called a stent to expand the artery **(Figure 42.20)** or by transplanting a healthy blood vessel from the chest or a limb to bypass the blockage.

Risk Factors and Treatment of Cardiovascular Disease

Cholesterol travels in blood plasma mainly in particles that consist of thousands of cholesterol molecules and other lipids bound to a protein. One type of particle—**low-density lipoprotein (LDL)**—delivers cholesterol to cells for membrane production. Another type—**high-density lipoprotein (HDL)**—scavenges excess cholesterol

▼ **Figure 42.20 Inserting a stent to widen an obstructed artery.**

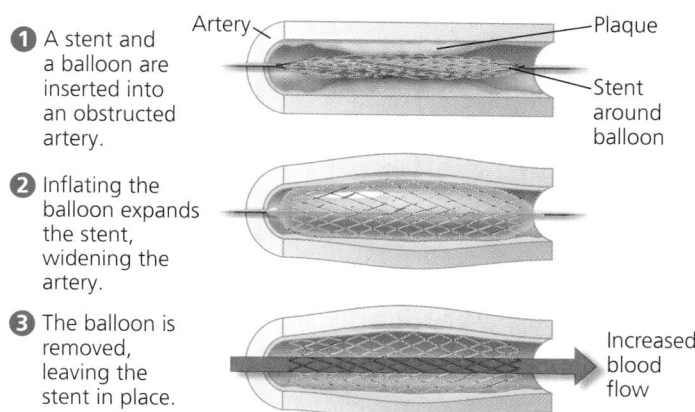

❶ A stent and a balloon are inserted into an obstructed artery.

❷ Inflating the balloon expands the stent, widening the artery.

❸ The balloon is removed, leaving the stent in place.

for return to the liver. Individuals with a high ratio of LDL to HDL are at substantially increased risk for atherosclerosis.

Although the tendency to develop particular cardiovascular diseases is inherited, it is also strongly influenced by lifestyle. For example, exercise decreases the LDL/HDL ratio, reducing the risk of cardiovascular disease. In contrast, consumption of certain processed vegetable oils called *trans fats*

and smoking increase the LDL/HDL ratio. For many individuals at high risk, treatment with drugs called statins can lower LDL levels and thereby reduce the risk of heart attacks. In the **Scientific Skills Exercise**, you can interpret the effect of a genetic mutation on blood LDL level.

⊙ **Mastering Biology** **BBC Video: The Impact of Smoking**

Scientific Skills Exercise

Making and Interpreting Histograms

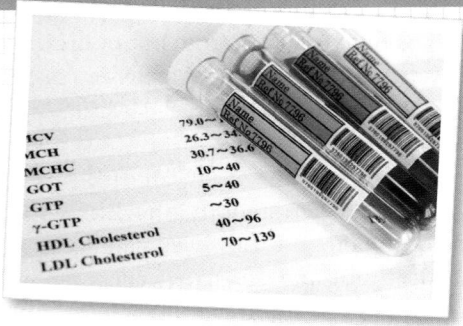

Does Inactivating the PCSK9 Enzyme Lower LDL Levels? Researchers interested in genetic factors affecting susceptibility to cardiovascular disease examined the DNA of 15,000 individuals. They found that 3% of the individuals had a mutation that inactivates one copy of the gene for PCSK9, a liver enzyme. Because mutations that *increase* the activity of PCSK9 are known to *increase* the level of LDL cholesterol in the blood, the researchers hypothesized that *inactivating* mutations in this gene would *lower* the LDL level. In this exercise, you will interpret the results of an experiment they carried out to test this hypothesis.

How the Experiment Was Done Researchers measured the LDL cholesterol level in blood plasma from 85 individuals with one copy of the *PCSK9* gene inactivated (the study group) and from 3,278 individuals with two functional copies of the gene (the control group).

Data from the Experiment

Individuals with an inactivating mutation in one copy of *PCSK9* gene (study group)

Individuals with two functional copies of *PCSK9* gene (control group)

INTERPRET THE DATA

1. The results are presented using a variant of a bar graph called a *histogram*. In a histogram, the variable on the *x*-axis is grouped into ranges. The height of each bar in this histogram reflects the percentage of samples that fall into the range specified on the *x*-axis for that bar. For example, in the top histogram, about 4% of individuals studied had plasma LDL cholesterol levels in the 25–50 mg/dL (milligrams per deciliter) range. Add the percentages for the relevant bars to calculate the percentage of individuals in the study and control groups that had an LDL level of 100 mg/dL or less. (For additional information about histograms, see the Scientific Skills Review in Appendix D.)

2. Compare the two histograms. Do you find support for the researchers' hypothesis? Explain.

3. What if instead of graphing the data, the researchers had compared the range of concentrations for plasma LDL cholesterol (low to high) in the control and study groups? How would their conclusions have differed?

4. What does the fact that the two histograms overlap as much as they do indicate about the extent to which PCSK9 determines plasma LDL cholesterol level?

5. Comparing these two histograms allowed researchers to draw a conclusion regarding the effect of *PCSK9* mutations on LDL cholesterol levels in blood. Consider two individuals with a plasma LDL level cholesterol of 160 mg/dL, one from the study group and one from the control group. **(a)** What do you predict regarding their relative risk of developing cardiovascular disease? **(b)** Explain how you arrived at your prediction. What role did the histograms play in making your prediction?

⊙ **Instructors:** A version of this Scientific Skills Exercise can be assigned in **Mastering Biology**.

Data from J. C. Cohen et al., Sequence variations in *PCSK9*, low LDL, and protection against coronary heart disease, *New England Journal of Medicine* 354:1264–1272 (2006).

The recognition that inflammation plays a central role in atherosclerosis and thrombus formation is also influencing the treatment of cardiovascular disease. For example, aspirin, which inhibits the inflammatory response, has been found to help prevent the recurrence of heart attacks and stroke.

Hypertension (high blood pressure) is yet another contributor to heart attack and stroke. According to one hypothesis, chronic high blood pressure damages the endothelium that lines the arteries, promoting plaque formation. The usual definition of hypertension in adults is a systolic pressure above 140 mm Hg or a diastolic pressure above 90 mm Hg. Fortunately, hypertension can often be prevented and controlled by quitting any use of tobacco products, changing the diet, increasing exercise, taking medication, or a combination of these approaches.

CONCEPT CHECK 42.4

1. Explain why a physician might order a white cell count for a patient with symptoms of an infection.

2. Clots in arteries can cause heart attacks and strokes. Why, then, does it make sense to treat people with hemophilia by introducing clotting factors into their blood?

3. **WHAT IF?** Nitroglycerin (the key ingredient in dynamite) is sometimes prescribed for heart disease patients. Within the body, nitroglycerin is converted to nitric oxide (see Concept 42.3). Why would you expect nitroglycerin to relieve chest pain caused by narrowing of the cardiac arteries?

4. **MAKE CONNECTIONS** How do stem cells from the bone marrow of an adult differ from embryonic stem cells (see Concept 20.3)?

For suggested answers, see Appendix A.

CONCEPT 42.5

Gas exchange occurs across specialized respiratory surfaces

In the remainder of this chapter, we will focus on the process of **gas exchange**. Although this process is often called respiratory exchange or respiration, it should not be confused with the energy transformations of cellular respiration. Gas exchange is the uptake of molecular O_2 from the environment and the discharge of CO_2 to the environment.

Partial Pressure Gradients in Gas Exchange

To understand the driving forces for gas exchange, we must consider **partial pressure**, which is simply the pressure exerted by a particular gas in a mixture of gases. Determining partial pressures enables us to predict the net movement of a gas at an exchange surface: A gas always undergoes net diffusion from a region of higher partial pressure to a region of lower partial pressure.

To calculate partial pressures, we need to know the pressure that a gas mixture exerts and the fraction of the mixture represented by a particular gas. Let's consider O_2 as an example. At sea level, the atmosphere exerts a downward force equal to that of a column of mercury (Hg) 760 mm high. Atmospheric pressure at sea level is thus 760 mm Hg. Since the atmosphere is 21% O_2 by volume, the partial pressure of O_2 is 0.21 × 760, or about 160 mm Hg. This value is called the *partial pressure* of O_2 (abbreviated P_{O_2}) because it is the part of atmospheric pressure contributed by O_2. The partial pressure of CO_2 (abbreviated P_{CO_2}) is much, much less, only 0.29 mm Hg at sea level.

Partial pressures also apply to gases dissolved in a liquid, such as water. When water is exposed to air, an equilibrium state is reached such that the partial pressure of each gas in the water equals the partial pressure of that gas in the air. Thus, water exposed to air at sea level has a P_{O_2} of 160 mm Hg, the same as in the atmosphere. However, the *concentrations* of O_2 in the air and water differ substantially because O_2 is much less soluble in water than in air **(Table 42.1)**. Furthermore, the warmer and saltier the water is, the less dissolved O_2 it can hold.

Table 42.1	Comparing Air and Water as Respiratory Media		
	Air (Sea Level)	Water (20°C)	Air-to-Water Ratio
O_2 partial pressure	160 mm	160 mm	1:1
O_2 concentration	210 ml/L	7 ml/L	30:1
Density	0.0013 kg/L	1 kg/L	1:770
Viscosity	0.02 cP	1 cP	1:50

Respiratory Media

The conditions for gas exchange vary considerably, depending on whether the respiratory medium—the source of O_2—is air or water. As already noted, O_2 is plentiful in air, making up about 21% of Earth's atmosphere by volume. As shown in Table 42.1, air is much less dense and less viscous than water, so it is easier to move and to force through small passageways. As a result, breathing air is relatively easy and the exchange does not need to not be particularly efficient. Humans, for example, extract only about 25% of the O_2 in inhaled air.

Water is a much more demanding gas exchange medium than air. The amount of O_2 dissolved in a given volume of water varies but is always less than in an equivalent volume of air. Water in many marine and freshwater habitats contains only about 7 mL of dissolved O_2 per liter, a concentration roughly 30 times less than in air. Water's lower O_2 content, greater density, and greater viscosity mean that aquatic animals such as fishes and lobsters must expend considerable energy to carry out gas exchange. In the context of these challenges, adaptations have evolved that enable most aquatic animals to be very efficient in gas exchange. Many of these adaptations involve the organization of the surfaces dedicated to exchange.

Respiratory Surfaces

Specialization for gas exchange is apparent in the structure of the respiratory surface, the part of an animal's body where gas exchange occurs. Like all living cells, the cells that carry out gas exchange have a plasma membrane that must be in contact with an aqueous solution. Respiratory surfaces are therefore always moist.

The movement of O_2 and CO_2 across respiratory surfaces takes place by diffusion. The rate of net diffusion is proportional to the surface area across which it occurs and inversely proportional to the square of the distance through which molecules must move. In other words, gas exchange is fast when the area for diffusion is large and the path for diffusion is short. As a result, respiratory surfaces tend to be large and thin.

In some relatively simple animals, such as sponges, cnidarians, and flatworms, every cell in the body is close enough to the external environment that gases can diffuse quickly between any cell and the environment. In many animals, however, the bulk of the body's cells lack immediate access to the environment. The respiratory surface in these animals is a thin, moist epithelium that constitutes a respiratory organ.

For earthworms, as well as some amphibians and other animals, the skin serves as a respiratory organ. A dense network of capillaries just below the skin facilitates the exchange of gases between the circulatory system and the environment. For most animals, however, the general body surface lacks sufficient area to exchange gases for the whole organism. The evolutionary solution to this limitation is a respiratory organ that is extensively folded or branched, thereby enlarging the available surface area for gas exchange. Gills, tracheae, and lungs are three such organs.

Gills in Aquatic Animals

Gills are outfoldings of the body surface that are suspended in the water. As illustrated in **Figure 42.21** (and in Figure 42.1), the distribution of gills over the body can vary considerably.

▼ Figure 42.21 **Diversity in the structure of gills, external body surfaces that function in gas exchange.**

(a) **Marine worm.** Many polychaetes (marine worms of the phylum Annelida) have a pair of flattened appendages called parapodia (singular, *parapodium*) on each body segment. The parapodia serve as gills and also function in crawling and swimming.

(b) **Crayfish.** Crayfish and other crustaceans have long, feathery gills covered by the exoskeleton. Specialized body appendages drive water over the gill surfaces.

(c) **Sea star.** The gills of a sea star are simple tubular projections of the skin. The hollow core of each gill is an extension of the coelom (body cavity). Gas exchange occurs by diffusion across the gill surfaces, and fluid in the coelom circulates in and out of the gills, aiding gas transport. The tube feet surfaces also function in gas exchange.

Regardless of their distribution, gills often have a total surface area much greater than that of the rest of the body's exterior.

Movement of the respiratory medium over the respiratory surface, a process called **ventilation**, maintains the partial pressure gradients of O_2 and CO_2 across the gill that are necessary for gas exchange. To promote ventilation, most gill-bearing animals either move their gills through the water or move water over their gills. For example, crayfish and lobsters have paddle-like appendages that drive a current of water over the gills, whereas mussels and clams move water with cilia. Octopuses and squids ventilate their gills by taking in and ejecting water, with the significant side benefit of getting about by jet propulsion. Fishes use the motion of swimming or coordinated movements of the mouth and gill covers to ventilate their gills. In both cases, a current of water enters the mouth of the fish, passes through slits in the pharynx, flows over the gills, and then exits the body **(Figure 42.22)**.

In fishes, the efficiency of gas exchange is maximized by **countercurrent exchange**, the exchange of a substance or heat between two fluids flowing in opposite directions. In a fish gill, the two fluids are blood and water. Because blood flows in the direction opposite to that of water passing over the gills, at each point in its travel blood is less saturated with O_2 than the water it meets (see Figure 42.22). As blood enters

a gill capillary, it encounters water that is completing its passage through the gill. Depleted of much of its dissolved O_2, this water nevertheless has a higher P_{O_2} than the incoming blood, and O_2 transfer takes place. As the blood continues its passage, its P_{O_2} steadily increases, but so does that of the water it encounters, since each successive position in the blood's travel corresponds to an earlier position in the water's passage over the gills. The result is a partial pressure gradient that favors the diffusion of O_2 from water to blood along the entire length of the capillary.

Countercurrent exchange mechanisms are remarkably efficient. In the fish gill, more than 80% of the O_2 dissolved in the water is removed as the water passes over the respiratory surface. In other settings, countercurrent mechanisms contribute to temperature regulation and to the functioning of the mammalian kidney (see Concepts 40.3 and 44.4).

Tracheal Systems in Insects

In most terrestrial animals, respiratory surfaces are enclosed within the body, exposed to the atmosphere only through narrow tubes. Although the most familiar example of such an arrangement is the lung, the most common is the insect **tracheal system**, a network of air tubes that branch throughout the body. The largest tubes, called tracheae,

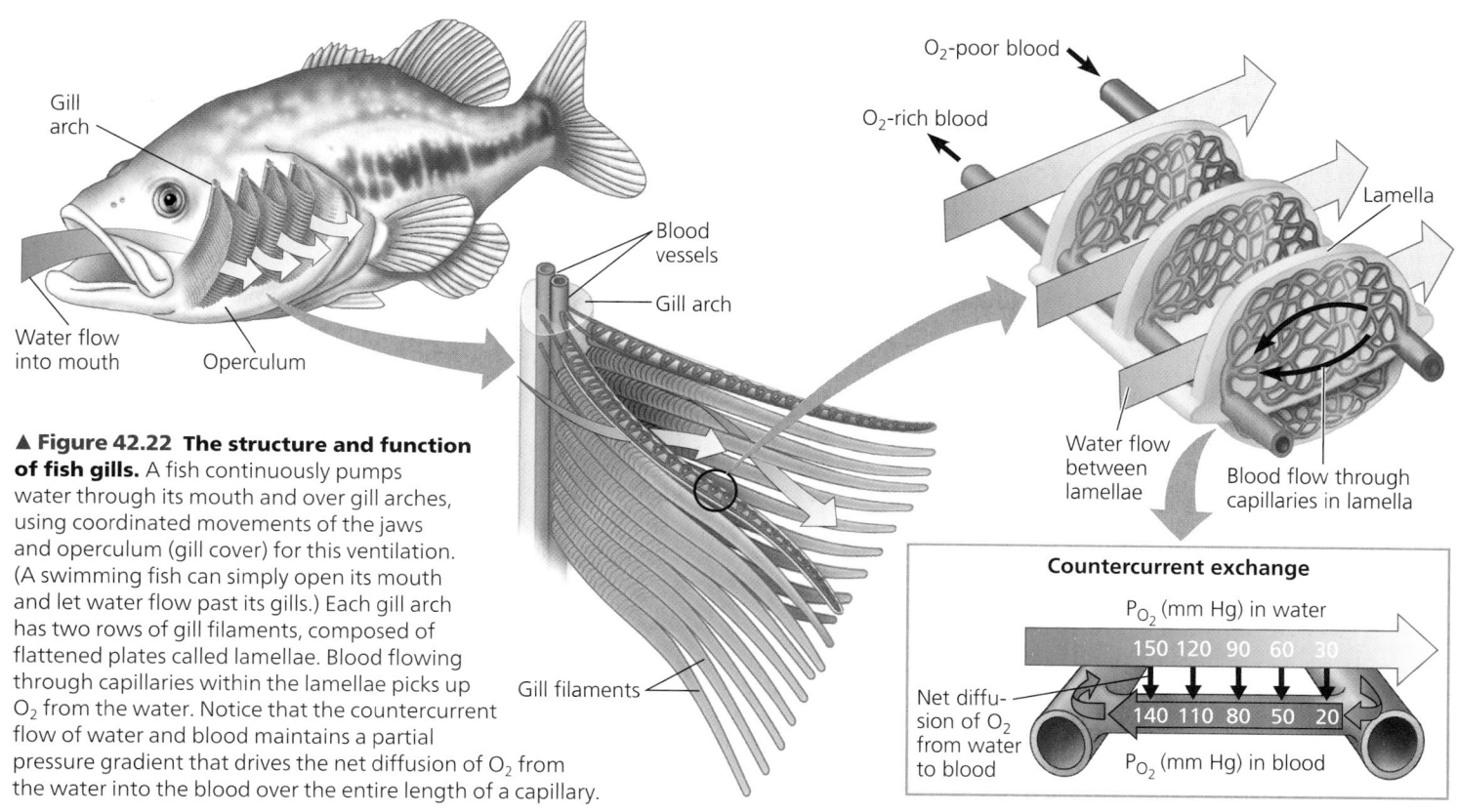

▲ **Figure 42.22 The structure and function of fish gills.** A fish continuously pumps water through its mouth and over gill arches, using coordinated movements of the jaws and operculum (gill cover) for this ventilation. (A swimming fish can simply open its mouth and let water flow past its gills.) Each gill arch has two rows of gill filaments, composed of flattened plates called lamellae. Blood flowing through capillaries within the lamellae picks up O_2 from the water. Notice that the countercurrent flow of water and blood maintains a partial pressure gradient that drives the net diffusion of O_2 from the water into the blood over the entire length of a capillary.

Gill arch
Water flow into mouth
Operculum
Blood vessels
Gill arch
Gill filaments

O_2-poor blood
O_2-rich blood
Lamella
Water flow between lamellae
Blood flow through capillaries in lamella

Countercurrent exchange
P_{O_2} (mm Hg) in water
150 120 90 60 30
Net diffusion of O_2 from water to blood
140 110 80 50 20
P_{O_2} (mm Hg) in blood

Mastering Biology Figure Walkthrough • Animation: Gas Exchange in Fish Gills

▼ **Figure 42.23 A tracheal system.**

(a) The respiratory system of an insect consists of branched internal tubes. The largest tubes, called tracheae, connect to external openings spaced along the insect's body surface. Air sacs formed from enlarged portions of the tracheae are found near organs that require a large supply of oxygen.

(b) Rings of chitin keep the tracheae open, allowing air to enter and pass into smaller tubes called tracheoles. The branched tracheoles deliver air directly to cells throughout the body. Tracheoles have closed ends filled with fluid (blue-gray). When the animal is active and using more O_2, most of the fluid is withdrawn into the body. This increases the surface area of air-filled tracheoles in contact with cells.

(c) This micrograph above (TEM) shows cross sections of tracheoles in a tiny piece of insect flight muscle. Each of the numerous mitochondria in the muscle cells lies within about 5 μm of a tracheole.

open to the outside **(Figure 42.23)**. At the tips of the finest branches, a moist epithelial lining enables gas exchange by diffusion. Because the tracheal system brings air within a very short distance of virtually every body cell in an insect, the efficient exchange of O_2 and CO_2 does not require the participation of the animal's open circulatory system.

Tracheal systems often exhibit adaptations directly related to bioenergetics. Consider, for example, a flying insect, which consumes 10 to 200 times more O_2 when in flight than it does at rest. In many flying insects, cycles of flight muscle contraction and relaxation pump air rapidly through the tracheal system. This pumping improves ventilation, bringing ample O_2 to the densely packed mitochondria that support the high metabolic rate of flight muscle (see Figure 42.23).

Lungs

Unlike tracheal systems, which branch throughout the insect body, **lungs** are localized respiratory organs. Representing an infolding of the body surface, they are typically subdivided into numerous pockets. Because the respiratory surface of a lung is not in direct contact with all other parts of the body, the gap must be bridged by the circulatory system, which transports gases between the lungs and the rest of the body. Lungs have evolved both in organisms with open circulatory systems, such as spiders and land snails, and in vertebrates.

Among vertebrates that lack gills, the use of lungs for gas exchange varies. Amphibians rely heavily on diffusion across external body surfaces, such as the skin, to carry out gas exchange; lungs, if present, are relatively small. In contrast, most reptiles (including all birds) and all mammals depend entirely on lungs for gas exchange. Turtles are an exception;

they supplement lung breathing with gas exchange across moist epithelial surfaces continuous with their mouth or anus. Lungs and air breathing also evolved in a few aquatic vertebrates as adaptations to living in oxygen-poor water or to spending part of their time exposed to air (for instance, when the water level of a pond recedes).

Mammalian Respiratory Systems: A Closer Look

In mammals, branching ducts convey air to the lungs, which are located in the *thoracic cavity*, enclosed by the ribs and diaphragm. Air enters through the nostrils and is then filtered by hairs, warmed, humidified, and sampled for odors as it flows through a maze of spaces in the nasal cavity. The nasal cavity leads to the pharynx, an intersection where the paths for air and food cross **(Figure 42.24)**. When food is swallowed, the **larynx** (the upper part of the respiratory tract) moves upward and tips the epiglottis over the glottis, which is the opening of the **trachea**, or windpipe. This allows food to go down the esophagus to the stomach (see Figure 41.9). The rest of the time, the glottis is open, enabling breathing.

From the larynx, air passes into the trachea. The cartilage that reinforces the walls of both the larynx and the trachea keeps this part of the airway open. Within the larynx of most mammals, the exhaled air rushes by a pair of elastic bands of muscle called vocal folds or, in humans, vocal cords. Sounds are produced when muscles in the larynx are tensed, stretching the cords so that they vibrate. High-pitched sounds result from tightly stretched cords vibrating rapidly; low-pitched sounds come from looser cords vibrating slowly.

The trachea branches into two **bronchi** (singular, *bronchus*), one leading to each lung. Within the lung, the

▼ **Figure 42.24 The mammalian respiratory system.** From the nasal cavity and pharynx, inhaled air passes through the larynx, trachea, and bronchi to the bronchioles, which end in microscopic alveoli lined by a thin, moist epithelium. Branches of the pulmonary arteries convey oxygen-poor blood to the alveoli; branches of the pulmonary veins transport oxygen-rich blood from the alveoli back to the heart.

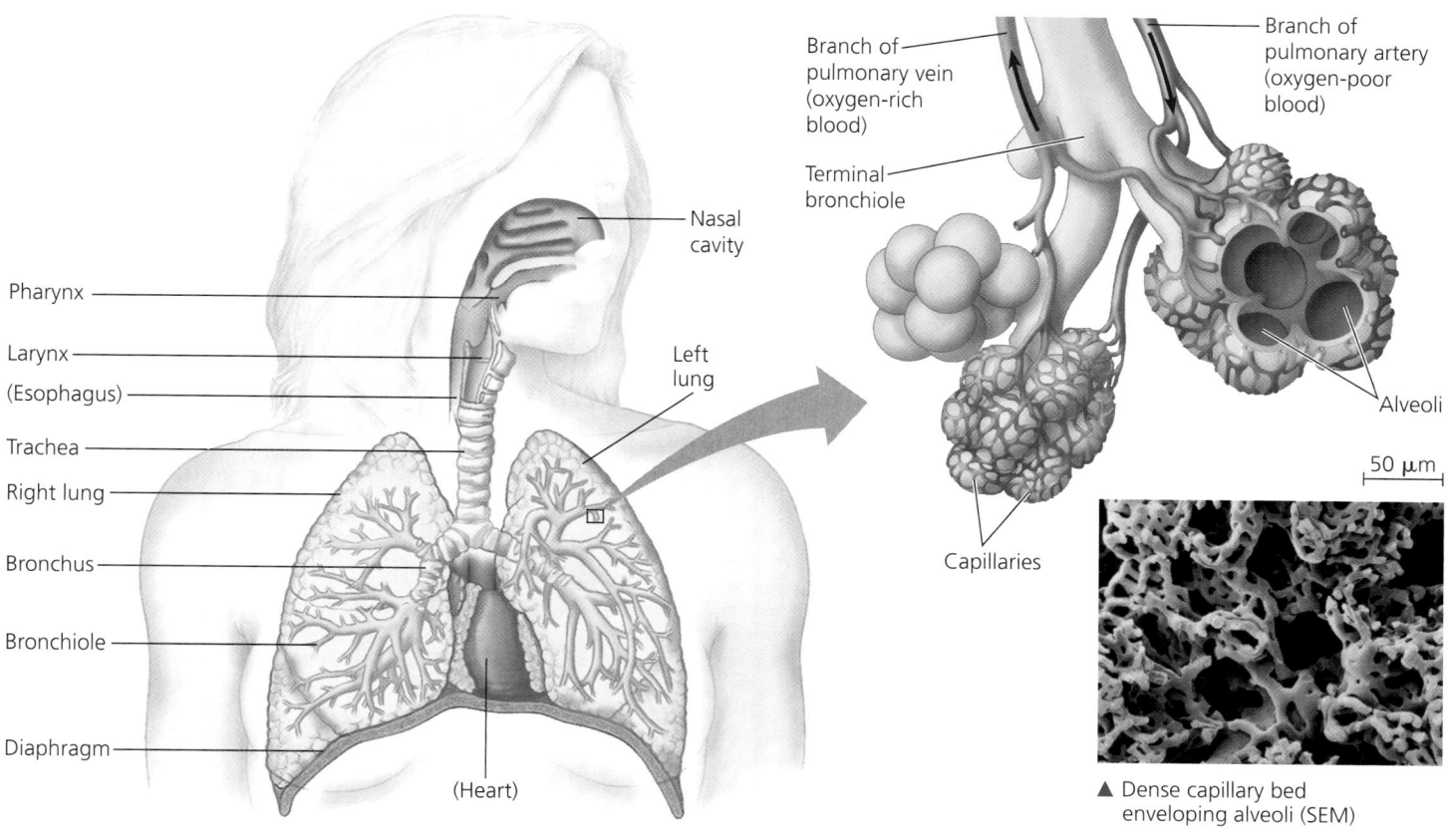

Branch of pulmonary vein (oxygen-rich blood)

Branch of pulmonary artery (oxygen-poor blood)

Terminal bronchiole

Nasal cavity

Pharynx

Larynx

(Esophagus)

Trachea

Right lung

Left lung

Bronchus

Alveoli

50 μm

Bronchiole

Capillaries

Diaphragm

(Heart)

▲ Dense capillary bed enveloping alveoli (SEM)

⊘ **Mastering Biology** Animation: The Human Respiratory System

bronchi branch repeatedly into finer and finer tubes called **bronchioles**. The entire system of air ducts has the appearance of an inverted tree, the trunk being the trachea. The epithelium lining the major branches of this respiratory tree is covered by cilia and a thin film of mucus. The mucus traps dust, pollen, and other particulate contaminants, and the beating cilia move the mucus upward to the pharynx, where it can be swallowed into the esophagus. This process, sometimes referred to as the "mucus escalator," plays a crucial role in cleansing the respiratory system.

Gas exchange in mammals occurs in **alveoli** (singular, *alveolus*; see Figure 42.24), air sacs clustered at the tips of the tiniest bronchioles. Human lungs contain millions of alveoli, which together have a surface area of about 100 m², 50 times that of the skin. Oxygen in the air entering the alveoli dissolves in the moist film lining their inner surfaces and rapidly diffuses across the epithelium into a web of capillaries that surrounds each alveolus. Net diffusion of carbon dioxide occurs in the opposite direction, from the capillaries across the epithelium of the alveolus and into the air space.

Lacking cilia or significant air currents to remove particles from their surface, alveoli are highly susceptible to contamination. White blood cells patrol the alveoli, engulfing foreign particles. However, if too much particulate matter reaches the alveoli, the defenses can be overwhelmed, leading to inflammation and irreversible damage. For example, particulates from smoking or vaping enter the alveoli and can cause a permanent reduction in lung capacity. For coal miners, inhalation of large amounts of coal dust can lead to silicosis, a disabling, irreversible, and sometimes fatal lung disease.

The film of liquid that lines alveoli is subject to surface tension, an attractive force that has the effect of minimizing a liquid's surface area (see Concept 3.2). Given their tiny diameter (about 0.25 mm), alveoli would be expected to collapse under high surface tension. It turns out, however, that these air sacs produce a mixture of phospholipids and proteins called **surfactant**, for *surf*ace-*act*ive agent, which coats the alveoli and reduces surface tension.

In the 1950s, Mary Ellen Avery did the first experiment linking surfactant deficiency to *respiratory distress syndrome* (RDS), a life-threatening disease of newborns that at the

▼ Figure 42.25 Inquiry

What causes respiratory distress syndrome?

Experiment Mary Ellen Avery, a research fellow at Harvard University, hypothesized that respiratory distress syndrome (RDS) in preterm infants is caused by the lack of a substance that reduces surface tension in the alveoli. To test this idea, she obtained autopsy samples of lungs from infants who had died of RDS or from other causes. She extracted material from each sample and let it form a film on water. Avery then measured the tension (in dynes per centimeter) across the water surface for each sample.

Results Avery noted a pattern when she separated out samples from infants with body mass less than 1,200 g, the average mass of a fetus at 29–30 weeks of pregnancy.

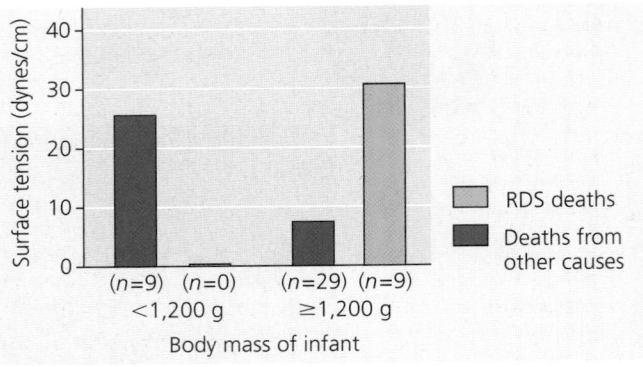

Data from M. E. Avery and J. Mead, Surface properties in relation to atelectasis and hyaline membrane disease, *American Journal of Diseases of Children* 97:517–523 (1959).

Conclusion For infants with a body mass of 1,200 g or greater, samples from those who had died of RDS exhibited much higher surface tension than samples from those who had died from other causes. Avery inferred that infants' lungs normally contain a surface-tension-reducing substance (now called surfactant) and that RDS occurs when this substance is lacking. The results from infants with a body mass less than 1,200 g were similar to those from infants who had died from RDS, suggesting that surfactant is not normally produced until a fetus reaches this size.

WHAT IF? *If the researchers had measured the amount of surfactant in lung samples from the infants, what relationship would you expect between the amount of surfactant and infant body mass?*

time killed 10,000 infants annually in the United States **(Figure 42.25)**. RDS is most common among infants born before 32 weeks of pregnancy. (The average full-term human pregnancy is 38 weeks.) Later studies revealed that surfactant typically appears in the lungs after 33 weeks of gestation. Artificial surfactants are now used to treat early preterm infants, and treated babies with a body mass over 900 g (2 pounds) at birth usually survive without long-term health problems. For her contributions, Avery received the National Medal of Science.

Having surveyed the route that air follows when we breathe, we'll turn next to the process of breathing itself.

CONCEPT CHECK 42.5

1. Why is an internal location for gas exchange tissues advantageous for terrestrial animals?

2. After a heavy rain, earthworms come to the surface. How would you explain this behavior in terms of an earthworm's requirements for gas exchange?

3. **MAKE CONNECTIONS** Describe similarities in the countercurrent exchange that facilitates respiration in fish and thermoregulation in geese (see Concept 40.3).

For suggested answers, see Appendix A.

CONCEPT 42.6

Breathing ventilates the lungs

Like fishes, terrestrial vertebrates rely on ventilation to maintain high O_2 and low CO_2 concentrations at the gas exchange surface. The process that ventilates lungs is **breathing**, the alternating inhalation and exhalation of air. A variety of mechanisms for moving air in and out of lungs have evolved, as we will see by considering breathing in amphibians, birds, and mammals.

How an Amphibian Breathes

An amphibian such as a frog ventilates its lungs by **positive pressure breathing**, inflating the lungs with forced airflow. Inhalation begins when muscles lower the floor of an amphibian's oral cavity, drawing in air through its nostrils. Next, with the nostrils and mouth closed, the floor of the oral cavity rises, forcing air down the trachea. Exhalation follows as air is expelled by the elastic recoil of the lungs and by compression of the muscular body wall. When male frogs puff themselves up in aggressive or courtship displays, they disrupt this breathing cycle, taking in air several times without allowing any release.

How a Bird Breathes

When a bird breathes, it passes air over the gas exchange surface in only one direction. Air sacs situated on either side of the lungs act as bellows that direct air flow through the lungs. Within the lungs, tiny channels called *parabronchi* serve as the sites of gas exchange. Passage of air through the entire system—air sacs and lungs—requires two cycles of inhalation and exhalation **(Figure 42.26)**.

Ventilation in birds is highly efficient. One reason is that birds pass air over the gas exchange surface in only one direction during breathing. In addition, incoming fresh air does not mix with air that has already carried out gas exchange, maximizing the partial pressure difference with blood flowing through the lungs.

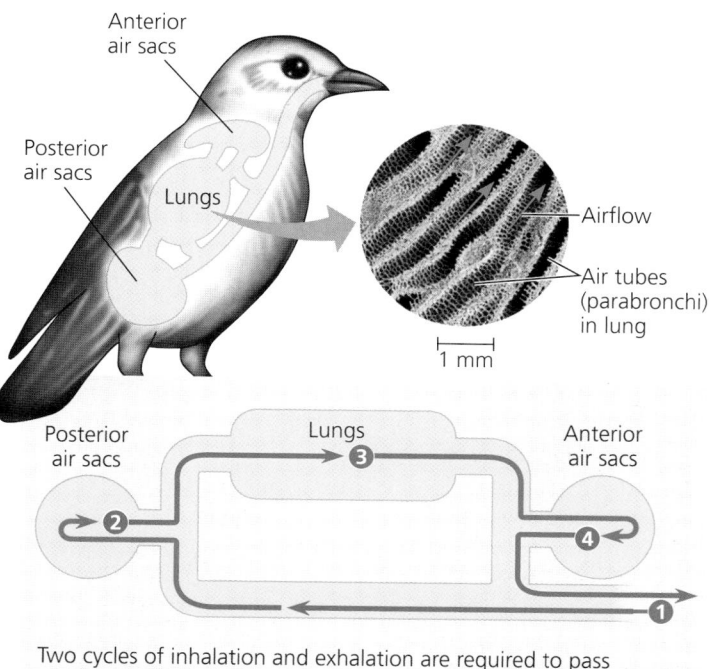

▼ **Figure 42.26 The avian respiratory system.** This diagram traces a breath of air through the respiratory system of a bird.

Anterior air sacs

Posterior air sacs

Lungs

Airflow

Air tubes (parabronchi) in lung

1 mm

Posterior air sacs

Lungs **③**

Anterior air sacs

②

④

①

Two cycles of inhalation and exhalation are required to pass one breath through the system:

① First inhalation: Air fills the posterior air sacs.

② First exhalation: Posterior air sacs contract, pushing air into lungs.

③ Second inhalation: Air passes through lungs and fills anterior air sacs.

④ Second exhalation: As anterior air sacs contract, air that entered body at first inhalation is pushed out of body.

How a Mammal Breathes

To understand how a mammal breathes, think about filling a syringe. By pulling back on the plunger, you lower the pressure in the syringe, drawing gas or fluid through the needle into the syringe chamber. Similarly, mammals employ **negative pressure breathing**—pulling, rather than pushing, air into their lungs **(Figure 42.27)**. Using muscle contraction to actively expand the thoracic cavity, mammals lower air pressure in their lungs below that of the air outside their body. Because gas flows from a region of higher pressure to a region of lower pressure, the lowered air pressure in the lungs causes air to rush through the nostrils and mouth and down the breathing tubes to the alveoli.

Expanding the thoracic cavity during inhalation involves the animal's rib muscles and the **diaphragm**, a sheet of skeletal muscle that forms the bottom wall of the cavity. Contracting the rib muscles pulls the ribs upward and the sternum outward, expanding the rib cage, the front wall of the thoracic cavity. At the same time, the diaphragm contracts, expanding the thoracic cavity downward. It is this descending movement of the diaphragm that is analogous to a plunger being drawn out of a syringe.

▼ **Figure 42.27 Negative pressure breathing.** A mammal breathes by changing the air pressure within its lungs relative to the pressure of the outside atmosphere.

Rib cage expands as rib muscles contract.

Rib cage gets smaller as rib muscles relax.

Lung

Diaphragm

① INHALATION: Diaphragm contracts (moves down).

② EXHALATION: Diaphragm relaxes (moves up).

WHAT IF? *The walls of alveoli contain elastic fibers that allow the alveoli to expand and contract with each breath. If the alveoli lost their elasticity, how would that affect gas exchange in the lungs?*

→ Mastering Biology BioFlix® **Animation: Mechanics of Breathing**

Whereas inhalation is always active and requires work, exhalation is usually passive. During exhalation, the muscles controlling the thoracic cavity relax, and the volume of the cavity is reduced. The increased air pressure in the alveoli forces air up the breathing tubes and out of the body.

Within the thoracic cavity, a double membrane surrounds the lungs. The inner layer of this membrane adheres to the outside of the lungs, and the outer layer adheres to the wall of the thoracic cavity. A thin space filled with fluid separates the two layers. Surface tension in the fluid causes the two layers to stick together like two plates of glass separated by a film of water: The layers can slide smoothly past each other, but they cannot be pulled apart easily. Consequently, the volume of the thoracic cavity and the volume of the lungs change in unison.

The rib muscles and diaphragm are sufficient to change lung volume when a mammal is at rest. During exercise, other muscles of the neck, back, and chest increase the volume of the thoracic cavity by raising the rib cage. In kangaroos and some other mammals, locomotion causes a rhythmic movement of organs in the abdomen, including the stomach and liver. The result is a piston-like pumping motion that pushes and pulls on the diaphragm, further increasing the volume of air moved in and out of the lungs.

The volume of air inhaled and exhaled with each breath, called **tidal volume**, averages about 500 mL in resting humans. The tidal volume during maximal inhalation and exhalation is the **vital capacity**, about 3.4 L and 4.8 L for college-age women and men, respectively. The air that remains after a forced exhalation is called the **residual volume**.

With age, the lungs lose their resilience, and residual volume increases at the expense of vital capacity.

Because the lungs in mammals do not completely empty with each breath, and because inhalation occurs through the same airways as exhalation, each inhalation mixes fresh air with oxygen-depleted residual air. As a result, the maximum P_{O_2} in alveoli is always considerably less than in the atmosphere. The maximum P_{O_2} in lungs is also less for mammals than for birds, which have a unidirectional flow of air through the lungs. This is one reason why mammals function less well than birds at high altitude. For example, humans have great difficulty obtaining enough O_2 when climbing at high elevations, such as those in the Himalayas. However, bar-headed geese and several other bird species easily fly through high Himalayan passes during their migrations.

Control of Breathing in Humans

Although you can voluntarily hold your breath or breathe faster and deeper, most of the time your breathing is regulated by involuntary mechanisms. These control mechanisms ensure that gas exchange is coordinated with blood circulation and with metabolic demand.

The neurons mainly responsible for regulating breathing are in the medulla oblongata, near the base of the brain **(Figure 42.28)**. Neural circuits in the medulla form a pair of *breathing control centers* that establish the breathing rhythm.

In regulating breathing, the medulla uses the pH of the fluid in which it is bathed as an indicator of blood CO_2 concentration. The pH can be used in this way because blood CO_2 is the main determinant of the pH of cerebrospinal fluid, the fluid surrounding the brain and spinal cord. Carbon dioxide diffuses from the blood to the cerebrospinal fluid, where it reacts with water and forms carbonic acid (H_2CO_3). The H_2CO_3 can then dissociate into a bicarbonate ion (HCO_3^-) and a hydrogen ion (H^+):

$$CO_2 + H_2O \rightleftharpoons H_2CO_3 \rightleftharpoons HCO_3^- + H^+$$

Consider what happens when metabolic activity increases, for example, during exercise. Increased metabolism raises the concentration of CO_2 in the blood and cerebrospinal fluid. Through the reactions shown above, the higher CO_2 concentration leads to an increase in the concentration of H^+, lowering pH. Sensors in the medulla as well as in major blood vessels detect this pH change. In response, the medulla's control circuits increase the depth and rate of breathing (see Figure 42.28). Both remain high until the excess CO_2 is eliminated in exhaled air and pH returns to a normal value.

The blood O_2 level usually has little effect on the breathing control centers. However, when the O_2 level drops very low (at high altitudes, for instance), O_2 sensors in the aorta and the carotid arteries in the neck send signals to the breathing control centers, which respond by increasing the breathing rate. The regulation of breathing is modulated by additional neural circuits, primarily in the pons, a part of the brain next to the medulla.

Breathing control is effective only if ventilation is matched to blood flow through alveolar capillaries. During exercise, for instance, such coordination couples an increased breathing rate, which enhances O_2 uptake and CO_2 removal, with an increase in cardiac output. When you breathe deeply, a negative-feedback mechanism prevents the lungs from over-expanding: During inhalation, sensors that detect stretching of the lung tissue send nerve impulses to the control circuits in the medulla, inhibiting further inhalation.

CONCEPT CHECK 42.6

1. How does an increase in the CO_2 concentration in the blood affect the pH of cerebrospinal fluid?

2. A drop in blood pH causes an increase in heart rate. What is the function of this control mechanism?

3. **WHAT IF?** If an injury tore a small hole in the membranes surrounding your lungs, what effect on lung function would you expect?

For suggested answers, see Appendix A.

▼ **Figure 42.28 Homeostatic control of breathing.**

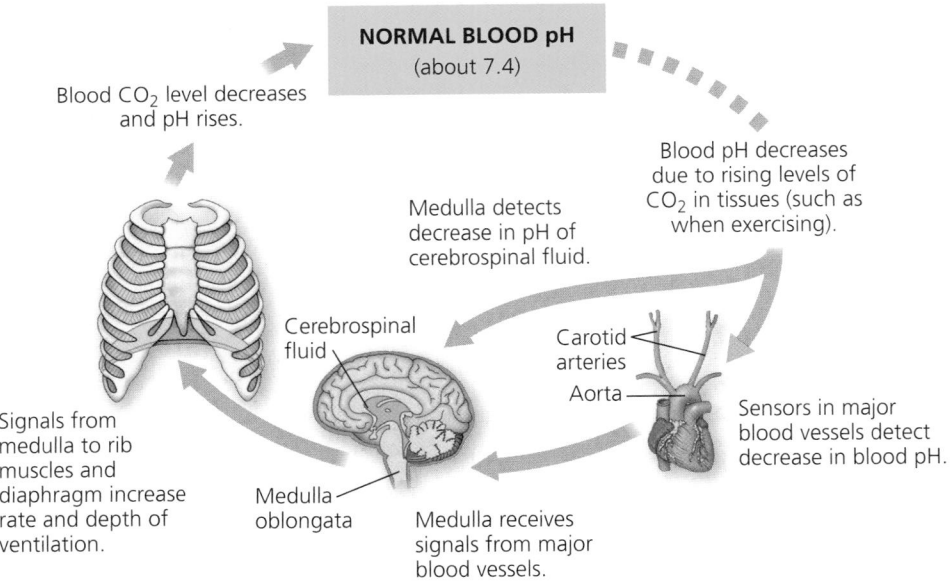

NORMAL BLOOD pH (about 7.4)

Blood CO_2 level decreases and pH rises.

Blood pH decreases due to rising levels of CO_2 in tissues (such as when exercising).

Medulla detects decrease in pH of cerebrospinal fluid.

Cerebrospinal fluid

Carotid arteries

Aorta

Sensors in major blood vessels detect decrease in blood pH.

Signals from medulla to rib muscles and diaphragm increase rate and depth of ventilation.

Medulla oblongata

Medulla receives signals from major blood vessels.

VISUAL SKILLS *Suppose a person began breathing very rapidly while resting. Tracing a path along this negative-feedback control circuit, describe the effect on the CO_2 level in the blood and the steps by which homeostasis would be restored.*

Adaptations for gas exchange include pigments that bind and transport gases

The high metabolic demands of many animals necessitate the exchange of large quantities of O_2 and CO_2. Here we'll examine how blood molecules called respiratory pigments facilitate this exchange through their interaction with O_2 and CO_2. We'll also investigate physiological adaptations that enable animals to be active under conditions of high metabolic load or very limiting P_{O_2}. As a basis for exploring these topics, let's summarize the basic gas exchange circuit in humans.

Coordination of Circulation and Gas Exchange

To appreciate how the gas exchange and circulatory systems function together, let's track the variation in partial pressure for O_2 and CO_2 across these systems **(Figure 42.29)**. ❶ During inhalation, fresh air mixes with air remaining in the lungs. ❷ The resulting mixture formed in the alveoli has a higher P_{O_2} than the blood flowing through the alveolar capillaries. Consequently, there is a net diffusion of O_2 down its partial pressure gradient from the air in the alveoli to the blood.

Meanwhile, the presence of a P_{CO_2} in the alveoli that is higher in the capillaries than in the air drives the net diffusion of CO_2 from blood to air. ❸ By the time the blood leaves the lungs in the pulmonary veins, its P_{O_2} and P_{CO_2} match the values for the air in alveoli. After returning to the heart, this blood is pumped through the systemic circuit.

❹ In the systemic capillaries, gradients of partial pressure favor the net diffusion of O_2 out of the blood and CO_2 into the blood. These gradients exist because cellular respiration in the mitochondria of cells near each capillary removes O_2 from and adds CO_2 to the surrounding interstitial fluid. ❺ Having unloaded O_2 and loaded CO_2, the blood is returned to the heart and pumped to the lungs again. ❻ There, exchange occurs across the alveolar capillaries, resulting in exhaled air enriched in CO_2 and partially depleted of O_2.

Respiratory Pigments

The low solubility of O_2 in water (and thus in blood) poses a problem for animals that rely on the circulatory system to deliver O_2. For example, a person requires almost 2 L of O_2 per minute during intense exercise, and all of it must be carried in the blood from the lungs to the active tissues. At normal body temperature and air pressure, however, only 4.5 mL of O_2 can dissolve into a liter of blood in the lungs. Even if 80% of the dissolved O_2 were delivered to the tissues, the heart would still need to pump 555 L of blood per minute!

In fact, animals transport most of their O_2 bound to proteins called **respiratory pigments**. Respiratory pigments circulate with the blood or hemolymph and are often contained within specialized cells. The pigments greatly increase the amount of O_2 that can be carried in the circulatory fluid (from 4.5 to about 200 mL of O_2 per liter in mammalian blood). In our example of an exercising human with an O_2 delivery rate of 80%, the presence of a respiratory pigment reduces the cardiac output necessary for O_2 transport to a manageable 12.5 L of blood per minute.

A variety of respiratory pigments have evolved in animals. With a few exceptions, these molecules have a distinctive color (hence the term *pigment*) and consist of a metal bound to a protein. One example is the blue pigment *hemocyanin*, which has copper as its oxygen-binding component and is found in arthropods and many molluscs.

The respiratory pigment of many invertebrate species and almost all species

▼ **Figure 42.29 Loading and unloading of respiratory gases.** The partial pressures of O_2 and CO_2 are given in mm Hg.

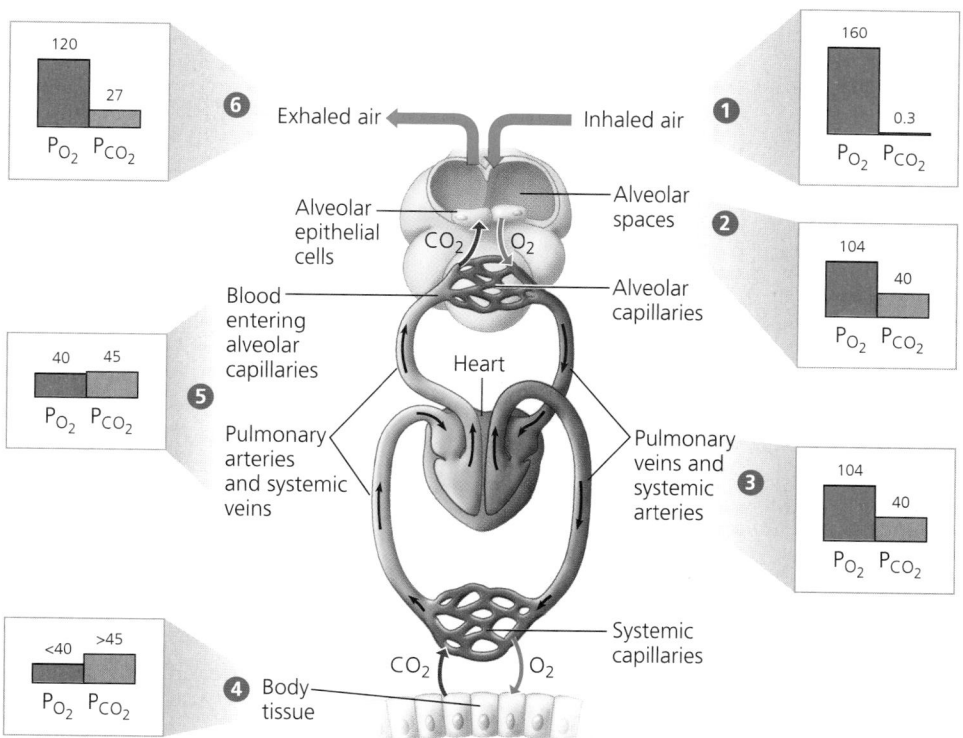

WHAT IF? *If you consciously forced more air out of your lungs each time you exhaled, how would that affect the values shown in the figure?*

➡ **Mastering Biology BioFlix® Animation: Gas Exchange in the Human Body** of vertebrates is hemoglobin (see Concept 42.4). In vertebrates, it is contained in erythrocytes and has four subunits. Each consists of a polypeptide and a heme group, a cofactor that has an iron atom at its center (**Figure 42.30**). Each iron atom binds one molecule of O_2, so a hemoglobin molecule can carry four O_2 molecules. Like all respiratory pigments, hemoglobin binds O_2 reversibly, loading O_2 in the lungs or gills and unloading it elsewhere in the body. This process is enhanced by cooperativity between the hemoglobin subunits (see Concept 8.5). When O_2 binds to one subunit, the others change shape slightly, increasing affinity for O_2. When four O_2 molecules are bound and one subunit unloads its O_2, the other three subunits more readily unload O_2, as an associated change in shape lowers their affinity for O_2.

Cooperativity in O_2 binding and release is evident in the dissociation curve for hemoglobin (**Figure 42.31a**). Over the range of P_{O_2} where the dissociation curve has a steep slope, even a slight change in P_{O_2} causes hemoglobin to load or unload a substantial amount of O_2. The steep part of the curve corresponds to the range of P_{O_2} found in body tissues. When cells in a particular location begin working harder—during exercise, for instance—P_{O_2} dips in their vicinity as the O_2 is consumed in cellular respiration. Because of subunit cooperativity, a slight drop in P_{O_2} causes a relatively large increase in the amount of O_2 the blood unloads.

Hemoglobin is especially efficient at delivering O_2 to tissues actively consuming O_2. However, this increased efficiency results not from O_2 consumption, but rather from CO_2 production. As tissues consume O_2 in cell respiration, they also produce CO_2. As we have seen, CO_2 reacts with water, forming carbonic acid, which lowers the pH of its surroundings. Low pH decreases the affinity of hemoglobin for O_2, an effect called the **Bohr shift** (**Figure 42.31b**). Thus, where CO_2 production is greater, hemoglobin releases more O_2, which can then be used to support more cellular respiration.

Hemoglobin also assists in buffering the blood—that is, preventing harmful changes in pH. In addition, it has a minor role in CO_2 transport, the topic we'll explore next.

Carbon Dioxide Transport

Only about 7% of the CO_2 released by respiring cells is transported in solution in blood plasma. The rest diffuses from plasma into erythrocytes and reacts with water (assisted by the enzyme carbonic anhydrase), forming H_2CO_3. The H_2CO_3 readily dissociates into H^+ and HCO_3^-. Most H^+ binds to hemoglobin and other proteins, minimizing change in blood pH. Most HCO_3^- diffuses out of the erythrocytes and is

▼ **Figure 42.30 Ribbon model of hemoglobin.**

Heme group with iron at center

▼ **Figure 42.31 Dissociation curves for hemoglobin at 37°C.**

(a) **P_{O_2} and hemoglobin dissociation at pH 7.4.** The curve shows the relative amounts of O_2 bound to hemoglobin exposed to solutions with different P_{O_2}. At a P_{O_2} of 100 mm Hg, typical in the lungs, hemoglobin is about 98% saturated with O_2. At a P_{O_2} of 40 mm Hg, common in resting tissues, hemoglobin is about 70% saturated, having unloaded nearly a third of its O_2. As shown in the above graph, hemoglobin can release much more O_2 to metabolically very active tissues, such as muscle tissue during exercise.

(b) **pH and hemoglobin dissociation.** In very active tissues, CO_2 from cellular respiration reacts with water to form carbonic acid, decreasing pH. Because hydrogen ions affect hemoglobin shape, a drop in pH shifts the O_2 dissociation curve toward the right (the Bohr shift). For a given P_{O_2}, hemoglobin releases more O_2 at a lower pH, supporting increased cellular respiration.

→ **Mastering Biology** Animation: Transport of Respiratory Gases

transported to the lungs in the plasma. The remaining HCO_3^-, representing about 5% of the CO_2, binds to hemoglobin and is transported in erythrocytes.

When blood flows through the lungs, the relative partial pressures of CO_2 favor the net diffusion of CO_2 out of the blood. As CO_2 diffuses into alveoli, the amount of CO_2 in the blood decreases. This decrease shifts the chemical equilibrium in favor of the conversion of HCO_3^- to CO_2, enabling further net diffusion of CO_2 into alveoli. Overall, the P_{CO_2} gradient is sufficient to drive a 10–15% reduction in P_{CO_2} during passage of blood through the lungs (see Figure 42.29).

Respiratory Adaptations of Diving Mammals

EVOLUTION Animals vary greatly in their ability to spend time in environments in which there is no access to their normal respiratory medium—for example, when an air-breathing mammal swims underwater. Whereas most humans cannot swim deeper than 20 m or hold their breath longer than 2–3 minutes, the Weddell seal of Antarctica routinely plunges to 200–500 m and remains there for a period ranging from 20 minutes to more than an hour **(Figure 42.32)**. Another diving mammal, the Cuvier's beaked whale, can reach depths of 2,900 m—nearly 2 miles—and stay submerged for more than 2 hours! What enables these amazing feats?

One evolutionary adaptation of diving mammals to prolonged stays underwater is a capacity to store large amounts of O_2 in their bodies. The volume of blood per kilogram of body mass in a Weddell seal is about twice that in a human. Furthermore, the muscles of seals and other diving mammals contain a high concentration of an oxygen-storing protein called **myoglobin** in their muscles. As a result, the Weddell seal can store about twice as much O_2 per kilogram of body mass as can a human.

Diving mammals not only have a relatively large O_2 stockpile but also have adaptations that conserve O_2. They swim with little muscular effort and glide passively for prolonged periods. During a dive, their heart rate and O_2 consumption rate decrease, and most blood is routed to vital tissues: the brain, spinal cord, eyes, adrenal glands, and, in pregnant seals, the placenta. Blood supply to the muscles is restricted or, during extended dives, shut off altogether. During these dives, a Weddell seal's muscles deplete the O_2 stored in myoglobin and then derive their ATP from fermentation instead of respiration (see Concept 9.5).

▼ **Figure 42.32 A Weddell seal, adapted to deep, long dives.**

How might such adaptations have arisen over the course of evolution? All mammals, including humans, have a diving reflex triggered by a plunge or fall into water: When the face contacts cold water, the heart rate immediately decreases and blood flow to body extremities is reduced. Genetic changes that strengthened this reflex would have provided a selective advantage to seal ancestors foraging underwater. Also, genetic variations that increased traits such as blood volume or myoglobin concentration would have improved diving ability and therefore been favored during selection over many generations.

CONCEPT CHECK 42.7

1. What determines whether O_2 and CO_2 undergo net diffusion into or out of capillaries? Explain.
2. How does the Bohr shift help deliver O_2 to very active tissues?
3. **WHAT IF?** A doctor might give bicarbonate (HCO_3^-) to a patient who is breathing very rapidly. What is the doctor assuming about the patient's blood chemistry?

For suggested answers, see Appendix A.

---

42 Chapter Review

→ Go to **Mastering Biology** for Assignments, the eText, the Study Area, and Dynamic Study Modules.

SUMMARY OF KEY CONCEPTS

→ To review key terms, go to the **Vocabulary Self-Quiz** in the **Mastering Biology** eText or Study Area, or go to goo.gl/zkjz9t.

CONCEPT 42.1

Circulatory systems link exchange surfaces with cells throughout the body (pp. 922–926)

- In animals with simple body plans, a **gastrovascular cavity** mediates exchange between the environment and cells that can be reached by **diffusion**. Because diffusion is slow over long distances, most complex animals have a circulatory system that moves fluid between cells and the organs that carry out exchange with the environment. Arthropods and most molluscs have an **open circulatory system**, in which **hemolymph** bathes organs directly. Vertebrates have a **closed circulatory system**, in which **blood** circulates in a closed network of pumps and vessels.

- The closed circulatory system of vertebrates consists of blood, blood vessels, and a two- to four-chambered **heart**. Blood pumped by a heart **ventricle** passes to **arteries** and then to the **capillaries**, sites of chemical exchange between blood and interstitial fluid. **Veins** return blood from capillaries to an **atrium**, which passes blood to a ventricle. Fishes, rays, and sharks have a single pump in their circulation. Air-breathing vertebrates have two pumps combined in a single heart. Variations in ventricle number and separation reflect adaptations to different environments and metabolic needs.

(?) *How does the flow of a fluid in a closed circulatory system differ from the movement of molecules between cells and their environment with regard to distance traveled, direction traveled, and driving force?*

CONCEPT **42.2**

Coordinated cycles of heart contraction drive double circulation in mammals *(pp. 926–929)*

- The **right ventricle** pumps blood to the lungs, where it loads O_2 and unloads CO_2. Oxygen-rich blood from the lungs entering the heart at the left atrium is pumped to the body tissues by the left ventricle. Blood then returns to the heart through the right atrium.

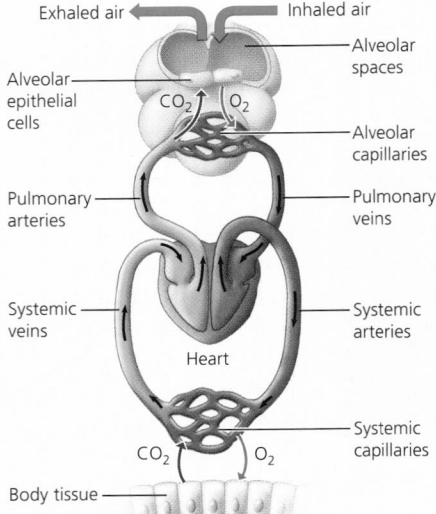

- The **cardiac cycle**, a complete sequence of the heart's pumping and filling, consists of a period of contraction, called **systole**, and a period of relaxation, called **diastole**. Heart function can be assessed by measuring the **pulse** (number of times the heart beats each minute) and **cardiac output** (volume of blood pumped by each ventricle per minute).
- The heartbeat originates with impulses at the **sinoatrial (SA) node** (pacemaker) of the right atrium. They trigger atrial contraction, are delayed at the **atrioventricular (AV) node**, and are then conducted along the bundle branches and Purkinje fibers, triggering ventricular contraction. The nervous system, hormones, and body temperature affect pacemaker activity.

(?) *What changes in cardiac function might you expect after surgical replacement of a defective heart valve?*

CONCEPT **42.3**

Patterns of blood pressure and flow reflect the structure and arrangement of blood vessels *(pp. 929–934)*

- Blood vessels have structures well adapted to function. Capillaries have narrow diameters and thin walls that facilitate exchange. Arteries contain thick elastic walls that maintain blood pressure. Veins contain one-way valves that contribute to the return of blood to the heart.
- The velocity of blood flow is lowest in the capillary beds as a result of their large total cross-sectional area. Blood pressure is altered by changes in cardiac output and by variable constriction of arterioles.

- Fluid leaks out of capillaries and is returned to blood by the **lymphatic system**, which also defends against infection.

(?) *If while standing you placed your forearm on your head, how, if at all, would the blood pressure in that arm change? Explain.*

CONCEPT **42.4**

Blood components function in exchange, transport, and defense *(pp. 934–939)*

- Whole blood consists of cells and cell fragments (**platelets**) suspended in a liquid matrix called **plasma**. Plasma proteins influence blood pH, osmotic pressure, and viscosity, and they function in lipid transport, immunity (antibodies), and blood clotting (fibrinogen). Red blood cells, or **erythrocytes**, transport O_2. Five types of white blood cells, or **leukocytes**, function in defense against microorganisms and foreign substances in the blood. Platelets function in blood clotting, a cascade of reactions that converts plasma fibrinogen to fibrin.
- A variety of diseases impair function of the circulatory system. In **sickle-cell disease**, an aberrant form of **hemoglobin** disrupts erythrocyte shape and function, leading to blockage of small blood vessels and a decrease in the oxygen-carrying capacity of the blood. In cardiovascular disease, inflammation of the arterial lining enhances deposition of lipids and cells, resulting in the potential for life-threatening damage to the heart or brain.

(?) *In the absence of infection, what percentage of cells in human blood are leukocytes?*

CONCEPT **42.5**

Gas exchange occurs across specialized respiratory surfaces *(pp. 939–944)*

- At all sites of **gas exchange**, a gas undergoes net diffusion from where its **partial pressure** is higher to where it is lower. Air is more conducive to gas exchange than water because air has a higher O_2 content, lower density, and lower viscosity.
- The structure and organization of respiratory surfaces differ among animal species. Gills are outfoldings of the body surface specialized for gas exchange in water. The effectiveness of gas exchange in some gills, including those of fishes, is increased by **ventilation** and **countercurrent exchange** between blood and water. Gas exchange in insects relies on a **tracheal system**, a branched network of tubes that bring O_2 directly to cells. Spiders, land snails, and most terrestrial vertebrates have internal **lungs**. In mammals, air inhaled through the nostrils passes through the pharynx into the **trachea, bronchi, bronchioles**, and dead-end **alveoli**, where gas exchange occurs.

(?) *Why does altitude have almost no effect on an animal's ability to rid itself of CO_2 through gas exchange?*

CONCEPT **42.6**

Breathing ventilates the lungs *(pp. 944–946)*

- **Breathing** mechanisms vary substantially among vertebrates. An amphibian ventilates its lungs by **positive pressure breathing**, which forces air down the trachea. Birds use a system of air sacs as bellows to keep air flowing through the lungs in one direction only, preventing the mixing of incoming and outgoing air. Mammals ventilate their lungs by **negative pressure breathing**, which pulls air into the lungs when the rib muscles and **diaphragm** contract. Incoming and outgoing air mix, decreasing the efficiency of ventilation.
- Sensors detect the pH of cerebrospinal fluid (reflecting CO_2 concentration in the blood), and a control center in the brain adjusts breathing

rate and depth to match metabolic demands. Additional input to the control center is provided by sensors in the aorta and carotid arteries that monitor blood levels of O_2 as well as CO_2 (via blood pH).

❓ *How does air in the lungs differ from the fresh air that enters the body during inspiration?*

CONCEPT 42.7

Adaptations for gas exchange include pigments that bind and transport gases *(pp. 947–949)*

- In the lungs, gradients of partial pressure favor the net diffusion of O_2 into the blood and CO_2 out of the blood. The opposite situation exists in the rest of the body. **Respiratory pigments** such as hemocyanin and hemoglobin bind O_2, greatly increasing the amount of O_2 transported by the circulatory system.
- Evolutionary adaptations enable some animals to satisfy extraordinary O_2 demands. Deep-diving mammals stockpile O_2 in blood and other tissues and deplete it slowly.

❓ *How are the roles of a respiratory pigment and an enzyme similar?*

TEST YOUR UNDERSTANDING

→ For more multiple-choice questions, go to the **Practice Test** in the **Mastering Biology** eText or Study Area, or go to goo.gl/GruWRg.

Levels 1-2: Remembering/Understanding

1. Which of the following respiratory systems is independent from a fluid-based circulatory system?
   (A) the lungs of a vertebrate
   (B) the gills of a fish
   (C) the tracheal system of an insect
   (D) the skin of an earthworm

2. Blood returning to the mammalian heart in a pulmonary vein drains first into the
   (A) left atrium.   (C) left ventricle.
   (B) right atrium.   (D) right ventricle.

3. Pulse is a direct measure of
   (A) blood pressure.   (C) cardiac output.
   (B) stroke volume.   (D) heart rate.

4. When you hold your breath, which of the following blood gas changes first leads to the urge to breathe?
   (A) rising O_2 (C) rising CO_2
   (B) falling O_2 (D) falling CO_2

5. One feature that amphibians and humans have in common is
   (A) the number of heart chambers.
   (B) a complete separation of circuits for circulation.
   (C) the number of circuits for circulation.
   (D) a low blood pressure in the systemic circuit.

Levels 3-4: Applying/Analyzing

6. A molecule of CO_2 released into the blood in your left toe can be exhaled from your nose without passing through which of the following structures?
   (A) the pulmonary vein.   (C) the right atrium.
   (B) the trachea.   (D) the right ventricle.

7. Compared with the interstitial fluid that bathes active muscle cells, blood reaching these cells in arterioles has a
   (A) higher P_{O_2}.
   (B) higher P_{CO_2}.
   (C) greater bicarbonate concentration.
   (D) lower pH.

Levels 5-6: Evaluating/Creating

8. **DRAW IT** Plot blood pressure against time for one cardiac cycle in humans, drawing separate lines for the pressure in the aorta, the left ventricle, and the right ventricle. Below the time axis, add a vertical arrow pointing to the time when you expect a peak in atrial blood pressure.

9. **EVOLUTION CONNECTION** One opponent of the movie monster Godzilla is Mothra, a mothlike creature with a wingspan of several dozen meters. The largest known insects were Paleozoic dragonflies with half-meter wingspans. Focusing on respiration and gas exchange, explain why giant insects are improbable.

10. **SCIENTIFIC INQUIRY • INTERPRET THE DATA**
The hemoglobin of a human fetus differs from adult hemoglobin. Compare the dissociation curves of the two hemoglobins in the graph at right. Describe how they differ, and propose a hypothesis to explain the benefit of this difference.

11. **SCIENCE, TECHNOLOGY, AND SOCIETY** Hundreds of studies have linked smoking with cardiovascular and lung disease. According to most health authorities, smoking is the leading cause of preventable, premature death in the United States. What are some arguments in favor of a total ban on cigarette advertising? What are arguments in opposition? Do you favor or oppose such a ban? Explain.

12. **WRITE ABOUT A THEME: INTERACTIONS** Some athletes prepare for competition at sea level by sleeping in a tent in which P_{O_2} is kept low. When climbing high peaks, some mountaineers breathe from bottles of pure O_2. In a short essay (100–150 words), relate these behaviors to the mechanism of O_2 transport in the human body and to physiological interactions with our gaseous environment.

13. **SYNTHESIZE YOUR KNOWLEDGE**

The diving bell spider (*Argyroneta aquatica*) stores air underwater in a net of silk. Explain why this adaptation could be more advantageous than having gills, taking into account differences in gas exchange media and gas exchange organs among animals.

For selected answers, see Appendix A.

43 The Immune System

KEY CONCEPTS

43.1 In innate immunity, recognition and response rely on traits common to groups of pathogens *p. 953*

43.2 In adaptive immunity, receptors provide pathogen-specific recognition *p. 957*

43.3 Adaptive immunity defends against infection of body fluids and body cells *p. 963*

43.4 Disruptions in immune system function can elicit or exacerbate disease *p. 970*

Study Tip

Make a table: As you study innate and adaptive immunity, compare and contrast these two forms of immune defense by continuing to fill in this table.

Defense mechanism	Key role in innate defense?	Key role in adaptive defense?	Effect on pathogen	Example of cell involved
Barrier	Yes	No	Entry into host tissues prevented	Epithelial cell
Phagocytosis				
Secreted molecules				
Destruction of infected host cells				

→ Go to Mastering Biology

For Students (in eText and Study Area)
• Get Ready for Chapter 43
• BioFlix® Animation: Activation of a Helper T cell
• BioFlix® Animation: Adaptive Defenses

For Instructors to Assign (in Item Library)
• Activity: The Inflammatory Response
• Tutorial: Acquired Immunity

Figure 43.1 Dedicated immune cells in most animals specifically interact with and attack pathogens, disease-causing agents such as bacteria or viruses. Here, an immune cell called a macrophage (brown) is engulfing bacteria (green).

How do immune cells recognize pathogens?

Innate recognition Each of a small set of receptors recognizes a molecule absent from animals, but common to a type of pathogen, as shown by these examples.

Pathogen	Virus	Bacterium	Fungus
Example of recognized molecule	dsRNA (nucleic acid in genome)	Flagellin (protein in flagella)	Mannan (oligosaccharide in cell wall)

Adaptive recognition Each of a vast number of receptors is specific for a particular part of a protein in one pathogen, such as a surface protein of the influenza (flu) virus. In this example, a different adaptive immune receptor would recognize each of the flu strains based on the proteins' amino acid sequences, which differ at the positions colored orange.

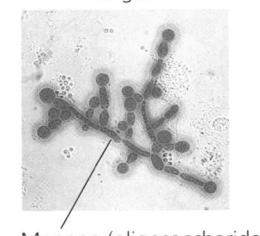

Influenza virus strain 1 — Ser Tyr Pro Lys Leu Lys Asn

Influenza virus strain 2 — Leu Tyr Pro Asn Leu Ser Asn

Influenza virus strain 3 — Ser Tyr Pro Asn Leu Lys Lys

Recognition by either type of immunity triggers a **response** that can eliminate or inactivate the pathogen, protecting the infected animal.

In innate immunity, recognition and response rely on traits common to groups of pathogens

For a **pathogen**—a bacterium, fungus, virus, or other disease-causing agent—the internal environment of an animal offers a source of nutrients, a protected setting, and a means of transport to new environments. From the animal's vantage point, the situation is not so ideal. Fortunately, adaptations have arisen over the course of evolution that protect animals against many pathogens. The body's defenses make up the **immune system**, which enables an animal to avoid or limit many infections. (Note that a foreign molecule or cell doesn't have to be pathogenic to elicit an immune response, but we'll focus on the immune system's role in defending against pathogens.)

The first lines of defense offered by immune systems help prevent pathogens from gaining entrance to the body. For example, an outer covering, such as skin or a shell, blocks entry by many pathogens. Sealing off the entire body surface is impossible, however, because gas exchange, nutrition, and reproduction require openings to the environment. Secretions that trap or kill pathogens guard the body's entrances and exits, while the linings of the digestive tract, airway, and other exchange surfaces provide additional barriers to infection.

If a pathogen breaches barrier defenses and enters the body, the problem of how to fend off attack changes substantially. Housed within body fluids and tissues, the invader is no longer an outsider. To fight infections, an animal's immune system must detect foreign particles and cells within the body. In other words, a properly functioning immune system distinguishes nonself from self. How is this accomplished? Immune cells produce receptor molecules that bind specifically to molecules from foreign cells or viruses and activate defense responses. The specific binding of immune receptors to foreign molecules is a type of *molecular recognition* and is the central event in identifying nonself molecules, particles, and cells.

Two types of immune defenses are found among animals. This concept focuses on **innate immunity**, the set of immune defenses common to all animals. The remainder of the chapter explores **adaptive immunity**, a set of molecular and cellular defense found only among vertebrates.

Innate Immunity of Invertebrates

The great success of insects in terrestrial and freshwater habitats teeming with diverse pathogens highlights the effectiveness of invertebrate innate immunity. One part of this defense system is a set of barrier defenses, including the insect exoskeleton. Composed largely of the polysaccharide chitin, the exoskeleton provides a physical barrier against most pathogens. Chitin also lines the insect intestine, where it blocks infection by many pathogens. In the digestive system, **lysozyme**, an enzyme that breaks down bacterial cell walls, acts as a chemical barrier against any pathogens ingested with food.

Any pathogen that breaches an insect's barrier defenses encounters internal immune defenses. Insect immune cells produce a set of recognition proteins, each of which binds to a molecule common to a broad class of pathogens. Many of these molecules are components of fungal or bacterial cell walls. Because such molecules are not normally found in animal cells, they function as "identity tags" for pathogen recognition. Once bound to a pathogen molecule, a recognition protein triggers an innate immune response.

In insects, the major immune cells are called *hemocytes*. Like amoebas, some hemocytes are phagocytic cells: They ingest and break down microorganisms by a process known as **phagocytosis (Figure 43.2)**. One class of hemocytes produces a type of defense molecule that helps entrap larger pathogens, such as *Plasmodium*, the single-celled parasite of mosquitoes that causes malaria in humans. Many other hemocytes release *antimicrobial peptides*, which circulate throughout the body of the insect and inactivate or kill bacteria or fungi by disrupting their plasma membranes.

The innate immune response of insects is specific for particular classes of pathogens. For example, if a fungus infects an insect, binding of recognition proteins to fungal cell wall molecules activates a transmembrane receptor called Toll. Toll in turn activates production and secretion of antimicrobial peptides that specifically kill fungal cells. Remarkably,

▼ **Figure 43.2 Phagocytosis.** This diagram depicts events in the ingestion and destruction of pathogens by a typical phagocytic cell.

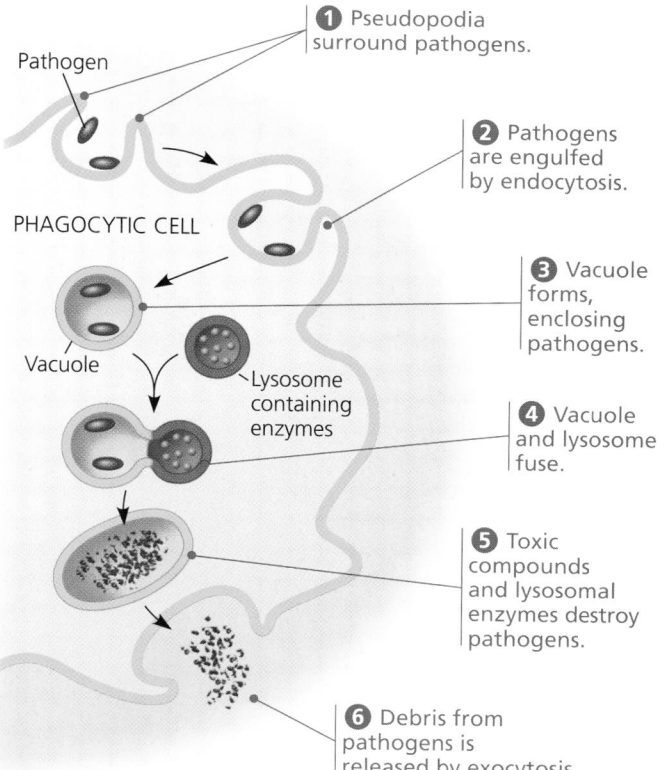

Pathogen

1 Pseudopodia surround pathogens.

2 Pathogens are engulfed by endocytosis.

PHAGOCYTIC CELL

3 Vacuole forms, enclosing pathogens.

Vacuole

Lysosome containing enzymes

4 Vacuole and lysosome fuse.

5 Toxic compounds and lysosomal enzymes destroy pathogens.

6 Debris from pathogens is released by exocytosis.

phagocytic mammalian cells use receptor proteins very similar to the Toll receptor to recognize viral, fungal, and bacterial components, a discovery that was recognized with the Nobel Prize in Physiology or Medicine in 2011.

Insects also have specific defenses that protect against infection by viruses. Many viruses that infect insects have a genome consisting of a single strand of RNA. When the virus replicates in the host cell, this RNA strand is the template for the synthesis of double-stranded RNA. Because animals do not produce double-stranded RNA, its presence can trigger a specific defense against the invading virus, as illustrated in **Figure 43.3**.

Innate Immunity of Vertebrates

In jawed vertebrates, innate immune defenses coexist with the more recently evolved system of adaptive immunity. Because most of the recent discoveries regarding vertebrate innate immunity have come from studies of mice and humans, we'll focus here on mammals. In this section, we'll consider the innate defenses that are similar to those found among invertebrates—barrier defenses, phagocytosis, and antimicrobial peptides—as well as some that are unique to vertebrates, such as natural killer cells, interferons, and the inflammatory response.

Barrier Defenses

The barrier defenses of mammals, which block the entry of many pathogens, include the mucous membranes and the skin. The mucous membranes that line the digestive, respiratory, urinary, and reproductive tracts produce *mucus*, a viscous fluid that traps pathogens and other particles. In the airway, ciliated epithelial cells sweep mucus and any entrapped material upward, helping prevent infection of the lungs. Saliva, tears, and mucous secretions that bathe various exposed epithelia provide a washing action that also inhibits colonization by fungi and bacteria.

Beyond their physical role in inhibiting microbial entry, body secretions create an environment that is hostile to many pathogens. Lysozyme in tears, saliva, and mucous secretions destroys the cell walls of susceptible bacteria as they enter the openings around the eyes or the upper respiratory tract. Pathogens in food or water and those in swallowed mucus must also contend with the acidic environment of the stomach (pH 2), which kills most of them before they can enter the intestines. Similarly, secretions from oil and sweat glands give human skin a pH ranging from 3 to 5, acidic enough to prevent the growth of many bacteria.

Cellular Innate Defenses

In mammals, as in insects, there are phagocytic innate immune cells dedicated to detecting, devouring, and destroying pathogens. In recognizing viral, fungal, or bacterial components, phagocytic mammalian cells rely on several types of receptors. As mentioned above, some are very similar to the insect innate immune receptor Toll. Each mammalian **Toll-like receptor (TLR)** binds to fragments of molecules

▼ **Figure 43.3 Antiviral defense in insects.** In defending against an infecting RNA virus, an insect cell turns the viral genome against itself, cutting the viral genome into small fragments that it then uses as guide molecules to find and destroy viral messenger RNAs (mRNAs).

VISUAL SKILLS *Compare and contrast the specificity of the Dicer-2 and Argo enzymes in terms of the size, number of strands, and sequence of the RNA molecules they each bind or act upon.*

Viral infection

During infection, the single-stranded viral RNA genome is replicated. A double strand of viral RNA forms briefly during replication.

RNA virus

HOST CELL

Viral RNA genome

Host defense

HOST CELL

Double-stranded viral RNA

❶ Dicer-2 enzyme

Single-stranded viral mRNA

❷ Argo protein complex

❸

❶ The host enzyme Dicer-2 recognizes the double-stranded RNA structure formed during the replicative cycle of an RNA virus. Dicer-2 cuts the viral RNA into fragments, each 21 nucleotides long.

❷ A protein complex containing the host enzyme Argo binds to an RNA fragment, displacing one of the two strands.

❸ The Argo complex uses the bound single-stranded fragment as a guide, matching it to the complementary sequence in a viral mRNA. Argo then cuts the viral mRNA, inactivating it and thus blocking synthesis of viral proteins.

▼ Figure 43.4 TLR signaling. Each mammalian Toll-like receptor (TLR) recognizes a molecular pattern shared by a group of pathogens. Lipopolysaccharide, flagellin, CpG DNA (DNA containing unmethylated CG sequences), and double-stranded (ds) RNA are found in bacteria, fungi, or viruses but not in animal cells. Once bound to such a pathogen molecule, TLR proteins trigger internal innate immune defenses, including production of cytokines and antimicrobial peptides.

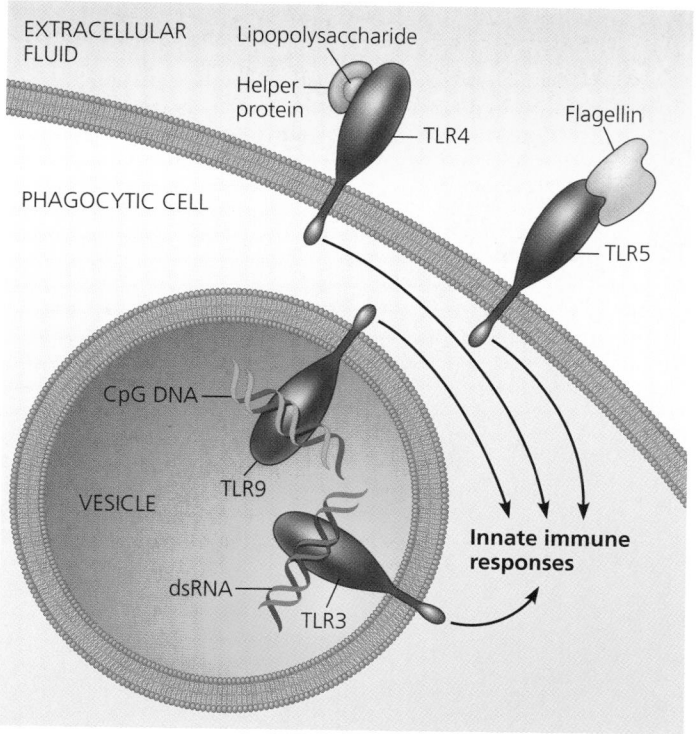

VISUAL SKILLS *Observe the locations of the TLR proteins and then suggest a possible benefit of their distribution.*

▼ Figure 43.5 Major events in a local inflammatory response.

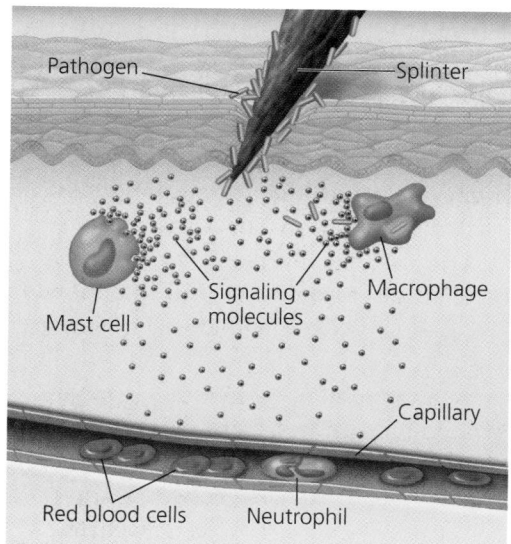

1 At the injury site, mast cells release histamines, and macrophages secrete cytokines. Some of these signaling molecules cause nearby capillaries to dilate, while others attract neutrophils.

2 Capillaries widen and become more permeable, allowing fluid containing antimicrobial peptides, as well as neutrophils, to enter the tissue. These immune cells digest pathogens and cell debris, allowing the tissue to heal.

? *From your experience with splinters, deduce whether the signals mediating an inflammatory response are short- or long-lived. Explain your answer.*

→ **Mastering Biology** Animation: Overview of the Inflammatory Response

characteristic of a set of pathogens **(Figure 43.4)**. For example, TLR3, on the inner surface of vesicles formed by endocytosis, binds to double-stranded RNA, a form of nucleic acid produced by certain viruses. Similarly, TLR4, located on immune cell plasma membranes, recognizes lipopolysaccharide, a type of molecule found on the surface of many bacteria, and TLR5 recognizes flagellin, the main protein of bacterial flagella.

The two main types of phagocytic cells in the mammalian body are neutrophils and macrophages. **Neutrophils**, which circulate in the blood, are attracted by signals from infected tissues and then engulf and destroy the infecting pathogens. **Macrophages** ("big eaters"), like the one shown in Figure 43.1, are larger phagocytic cells. Some migrate throughout the body, whereas others reside permanently in organs and tissues where they are likely to encounter pathogens. For example, some macrophages are located in the spleen, where pathogens in the blood are often trapped. Macrophages and neutrophils are both key components of the inflammatory response, as discussed shortly. In addition to these cells, a number of more specialized cell types also contribute to innate immune defenses:

- **Dendritic cells** mainly populate tissues, such as skin, that contact the environment. They stimulate adaptive immunity against pathogens that they engulf.
- **Eosinophils**, often found beneath an epithelium, are important in defending against multicellular invaders, such as parasitic worms. Upon encountering such parasites, eosinophils discharge destructive enzymes.
- **Natural killer cells** circulate through the body and detect the abnormal surface proteins found on some virus-infected and cancerous cells. Natural killer cells do not engulf stricken cells. Instead, they release chemicals that lead to cell death, inhibiting spread of the virus or cancer.
- **Mast cells** are found in connective tissue and make key contributions to the inflammatory response, described next, as well as to allergies, discussed later.

Local Inflammatory Response

When a splinter lodges under your skin, the surrounding area becomes swollen and warm. As **Figure 43.5** depicts, both changes reflect a local **inflammatory response**, a set of events triggered by signaling molecules released upon injury or infection.

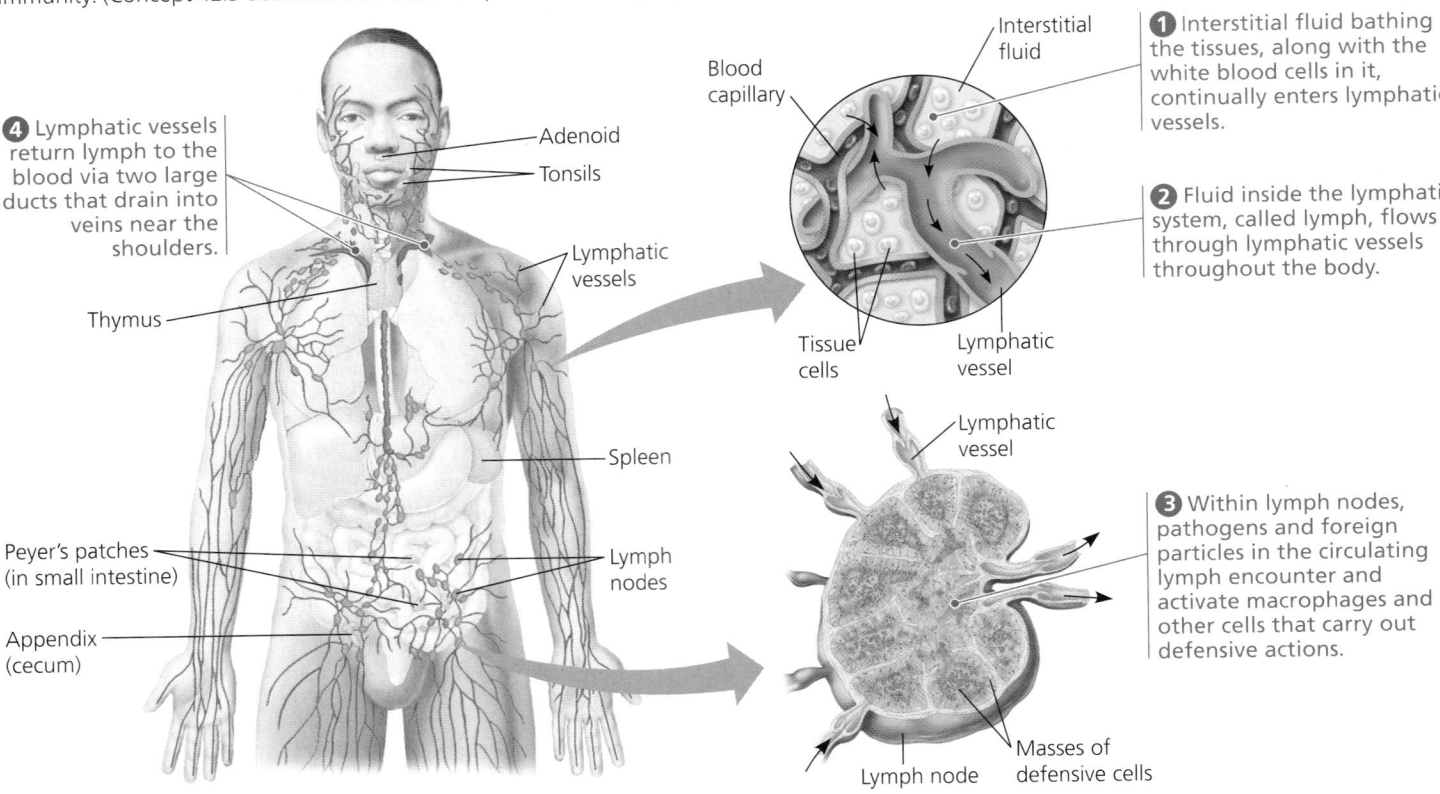

▼ Figure 43.6 The human lymphatic system. The lymphatic system consists of lymphatic vessels (shown in green), through which lymph travels, and structures that trap foreign substances. These structures include lymph nodes (orange) and lymphoid organs (yellow): the adenoids, tonsils, thymus, spleen, Peyer's patches, and appendix. Steps 1–4 trace the flow of lymph and illustrate the role of lymph nodes in activating adaptive immunity. (Concept 42.3 describes the relationship between the lymphatic and circulatory systems.)

④ Lymphatic vessels return lymph to the blood via two large ducts that drain into veins near the shoulders.

Adenoid

Tonsils

Lymphatic vessels

Thymus

Spleen

Peyer's patches (in small intestine)

Lymph nodes

Appendix (cecum)

Blood capillary

Interstitial fluid

① Interstitial fluid bathing the tissues, along with the white blood cells in it, continually enters lymphatic vessels.

Tissue cells

Lymphatic vessel

② Fluid inside the lymphatic system, called lymph, flows through lymphatic vessels throughout the body.

Lymphatic vessel

③ Within lymph nodes, pathogens and foreign particles in the circulating lymph encounter and activate macrophages and other cells that carry out defensive actions.

Lymph node

Masses of defensive cells

A local inflammatory response begins when activated macrophages discharge *cytokines*, signaling molecules that recruit neutrophils to the site of injury or infection. In addition, mast cells release the signaling molecule **histamine** at sites of damage. Histamine triggers nearby blood vessels to dilate and become more permeable. The resulting increase in local blood supply produces the redness and increased skin temperature typical of the inflammatory response (from the Latin *inflammare*, to set on fire).

Cycles of signaling and response continue the process of inflammation. Activated complement proteins promote further histamine release, attracting more phagocytic cells to the site of injury and infection. At the same time, enhanced blood flow to the site helps deliver antimicrobial peptides, which, as in insects, typically kill or inactivate pathogens by disrupting membrane integrity. The result is an accumulation of *pus*, a fluid rich in white blood cells, dead pathogens, and debris from damaged tissue.

At the end of the local inflammatory response, pus and excess fluid are taken up as lymph, the fluid transported in the network of vessels known as the lymphatic system **(Figure 43.6)**. Small organs called *lymph nodes* scattered within the lymphatic system contain macrophages, which engulf pathogens that enter the lymph from the interstitial fluid. Dendritic cells reside outside the lymphatic system but

migrate to the lymph nodes after interacting with pathogens. Within the lymph nodes, dendritic cells interact with other immune cells, stimulating adaptive immunity.

Systemic and Chronic Inflammation

A minor injury or infection causes a local inflammatory response, but more extensive tissue damage or infection may lead to a response that is systemic (throughout the body). Cells in injured or infected tissue often secrete molecules that stimulate the release of additional neutrophils from the bone marrow. In the case of a severe infection, such as meningitis or appendicitis, the number of white blood cells in the bloodstream may increase severalfold within only a few hours.

A systemic inflammatory response sometimes involves fever. In response to certain pathogens, substances released by activated macrophages cause the body's thermostat to reset to a higher temperature (see Concept 40.3). The benefits of the resulting fever are still a subject of debate. One hypothesis is that an elevated body temperature may enhance phagocytosis and, by speeding up chemical reactions, accelerate tissue repair.

Certain bacterial infections can induce an overwhelming systemic inflammatory response, leading to a life-threatening condition called *septic shock*. Characterized by very high fever, low blood pressure, and poor blood flow through capillaries, septic shock occurs most often in the very old and the very young. It is fatal in more than one-third of cases and kills more than 200,000 people each year in the United States.

Chronic (ongoing) inflammation can also threaten human health. For example, millions of individuals worldwide are affected by Crohn's disease and ulcerative colitis, often debilitating disorders in which an unregulated inflammatory response disrupts intestinal function.

Antimicrobial Peptides and Proteins

In mammals, pathogen recognition triggers the production and release of a variety of peptides and proteins that attack pathogens or impede their reproduction. A number of these, including the interferons and complement proteins, are unique to vertebrate immune systems.

Interferons are proteins that provide innate defense by interfering with viral infections. Virus-infected body cells secrete interferon proteins that induce nearby uninfected cells to produce substances that inhibit viral replication. In this way, these interferons limit the cell-to-cell spread of viruses in the body, helping control viral infections such as colds and influenza. Some white blood cells secrete a different type of interferon that helps activate macrophages, enhancing their phagocytic ability. Pharmaceutical companies now use recombinant DNA technology to mass-produce interferons to help treat certain viral infections, such as hepatitis C.

The infection-fighting **complement system** consists of roughly 30 proteins in blood plasma. These proteins circulate in an inactive state and are activated by substances on the surface of many pathogens. Activation results in a cascade of biochemical reactions that can lead to lysis (bursting) of invading cells. The complement system also functions in the inflammatory response as well as in the adaptive defenses discussed later in the chapter.

Evasion of Innate Immunity by Pathogens

Adaptations have evolved in some pathogens that enable them to avoid destruction by phagocytic cells. For example, the outer capsule that surrounds certain bacteria interferes with molecular recognition and phagocytosis. One such bacterium, *Streptococcus pneumoniae*, is a major cause of pneumonia and meningitis in humans (see Concept 16.1).

Some bacteria are recognized but resist breakdown after being engulfed by a host cell. One example is *Mycobacterium tuberculosis*, the bacterium shown in Figure 43.1. Rather than being destroyed, this bacterium grows and reproduces within host cells, effectively hidden from the body's immune defenses. The result of this infection is tuberculosis (TB), a disease that attacks the lungs and other tissues. Worldwide, TB kills more than 1 million people a year.

CONCEPT CHECK 43.1

1. Pus is both a sign of infection and an indicator of immune defenses in action. Explain.
2. **MAKE CONNECTIONS** How do the molecules that activate the vertebrate TLR signal transduction pathway differ from the ligands in most other signaling pathways (see Concept 11.2)?
3. **WHAT IF?** Parasitic wasps inject their eggs into host larvae of other insects. If the host immune system doesn't kill the wasp egg, the egg hatches and the wasp larva devours the host larva as food. Suggest an explanation for why some insect species initiate an innate immune response to a wasp egg, but others cannot.

For suggested answers, see Appendix A.

CONCEPT 43.2

In adaptive immunity, receptors provide pathogen-specific recognition

Vertebrates are unique in having adaptive as well as innate immunity. In adaptive immunity, molecular recognition relies on a vast arsenal of receptors, each of which recognizes a feature typically found only on a particular part of a particular molecule in a particular pathogen. As a result, recognition and response in adaptive immunity occur with remarkable specificity.

The adaptive immune response, also known as the acquired immune response, is activated after the innate immune response and develops more slowly. Unlike innate immunity, the adaptive response is enhanced by previous exposure to the infecting pathogen. **Figure 43.7** highlights

▼ **Figure 43.7 Overview of animal immunity.** Innate immunity offers a primary defense in all animals and sets the stage for adaptive immunity in vertebrates.

Pathogens (such as bacteria, fungi, and viruses)

INNATE IMMUNITY (all animals)
- Recognition of traits shared by broad ranges of pathogens, using a small set of receptors
- Rapid response

Barrier defenses:
Skin
Mucous membranes
Secretions

Internal defenses:
Phagocytic cells
Natural killer cells
Antimicrobial proteins
Inflammatory response

ADAPTIVE IMMUNITY (vertebrates only)
- Recognition of traits specific to particular pathogens, using a vast array of receptors
- Slower response

Humoral response:
Antibodies defend against infection in body fluids.

Cell-mediated response:
Cytotoxic cells defend against infection in body cells.

these and other fundamental similarities and differences between adaptive and innate immunity.

Adaptive immunity relies on T cells and B cells, which are types of white blood cells called **lymphocytes (Figure 43.8)**. Like all blood cells, lymphocytes originate from stem cells in the bone marrow. Some migrate from the bone marrow to the **thymus**, an organ in the thoracic cavity above the heart (see Figure 43.6). These lymphocytes mature into **T cells**. Lymphocytes that remain and mature in the bone marrow develop as **B cells**. (Lymphocytes of a third type remain in the blood and become the natural killer cells active in innate immunity.)

▼ **Figure 43.8 B and T lymphocytes.**

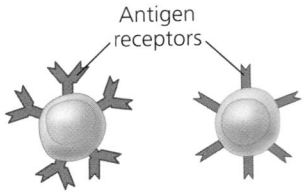

Antigen receptors

Mature B cell   **Mature T cell**

Antigens as the Trigger for Adaptive Immunity

Any substance that elicits a B or T cell response is an **antigen**. In adaptive immunity, recognition occurs when a B cell or T cell binds to an antigen, such as a bacterial or viral protein, via a protein called an **antigen receptor**. Each antigen receptor binds to just one part of one molecule from a particular pathogen, such as a species of bacteria or strain of virus.

The cells of the immune system produce millions of different antigen receptors. A given lymphocyte, however, produces just one variety; all of the antigen receptors made by a single B or T cell are identical. Infection by a virus, bacterium, or other pathogen triggers activation of B and T cells with antigen receptors specific for parts of that pathogen. Although drawings of B and T cells typically include just a few antigen receptors, a single B or T cell actually has about 100,000 antigen receptors on its surface.

Antigens are usually foreign and are typically large molecules, either proteins or polysaccharides. Many antigens protrude from the surface of foreign cells or viruses. Other antigens, such as toxins secreted by bacteria, are released into the extracellular fluid.

The small, accessible portion of an antigen that binds to an antigen receptor is called an **epitope**. An example is a group of amino acids in a particular protein. A single antigen usually has several epitopes, each binding a receptor with a different specificity. Because all antigen receptors produced by a single B cell or T cell are identical, they bind to the same epitope. Each B or T cell thus displays *specificity* for a particular epitope, enabling it to respond to any pathogen that produces molecules containing that epitope.

The antigen receptors of B cells and T cells have similar components, but they encounter antigens in different ways. We'll consider the two processes in turn.

▼ **Figure 43.9 The structure of a B cell antigen receptor.**

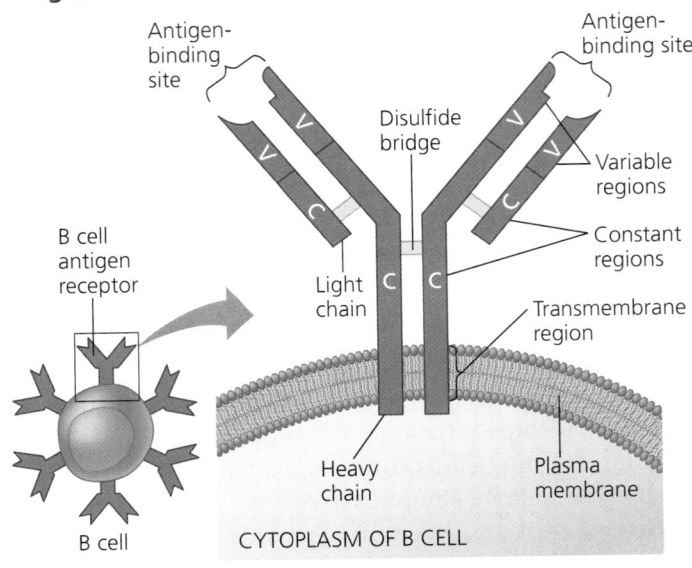

Antigen Recognition by B Cells and Antibodies

Each B cell antigen receptor is a Y-shaped protein consisting of four polypeptide chains: two identical **heavy chains** and two identical **light chains (Figure 43.9)**. Disulfide bridges link the chains together.

Each light chain or heavy chain has a *constant (C) region*, where amino acid sequences vary little among the receptors on different B cells. The constant region of heavy chains contains a transmembrane region, which anchors the receptor in the cell's plasma membrane. As shown in Figure 43.9, each light or heavy chain also has a *variable (V) region*, so named because its amino acid sequence varies extensively from one B cell to another. Together, parts of a heavy-chain V region and a light-chain V region form an asymmetric binding site for an antigen. Therefore, each B cell antigen receptor has two identical antigen-binding sites.

Binding of a B cell antigen receptor to an antigen is an early step in B cell activation, leading eventually to formation of cells that secrete a soluble form of the receptor **(Figure 43.10a)**. This secreted protein is called an **antibody**, also known as an **immunoglobulin (Ig)**. Antibodies have the same Y-shaped structure as B cell antigen receptors but lack a membrane anchor. As you'll see later, antibodies provide a direct defense against pathogens in body fluids.

The antigen-binding site of a membrane-bound receptor or antibody has a unique shape that provides a lock-and-key fit for a particular epitope. This stable interaction involves many noncovalent bonds between an epitope and the surface of the binding site. Differences in the amino acid sequences of variable regions provide the variation in binding surfaces that enables binding to be highly specific.

▼ Figure 43.10 Antigen recognition by B cells and antibodies.

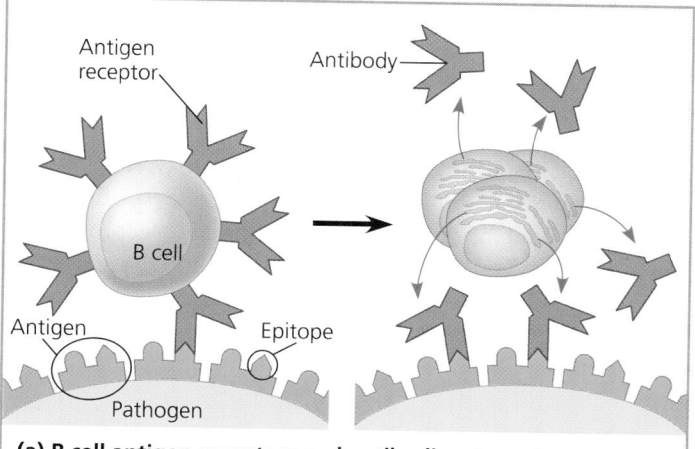

(a) B cell antigen receptors and antibodies. An antigen receptor of a B cell binds to an epitope, a particular part of an antigen. Following binding, the B cell gives rise to cells that secrete an antibody: a soluble form of the receptor that is specific for the same epitope as the original antigen receptor.

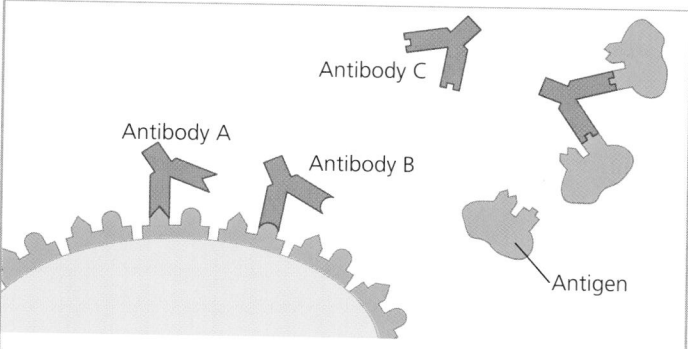

(b) Antigen receptor specificity. Different antibodies can recognize distinct epitopes on the same antigen. Furthermore, antibodies can recognize free antigens as well as antigens on a pathogen's surface.

MAKE CONNECTIONS *The interactions depicted here involve a highly specific binding between antigen and receptor (see also Figure 5.17). How is this type of interaction similar to an enzyme-substrate interaction (see Figure 8.15)?*

B cell antigen receptors and antibodies bind to intact antigens in the blood and lymph. As illustrated in **Figure 43.10b** for antibodies, they can bind to antigens on the surface of pathogens or free in body fluids.

Antigen Recognition by T Cells

For a T cell, the antigen receptor consists of two different polypeptide chains, an *α chain* and a *β chain*, linked by a disulfide bridge **(Figure 43.11)**. Near the base of the T cell antigen receptor (often called simply a T cell receptor) is a transmembrane region that anchors the molecule in the cell's plasma membrane. At the outer tip of the molecule, the variable (V) regions of the α and β chains together form a single antigen-binding site. The remainder of the molecule is made up of the constant (C) regions.

▼ Figure 43.11 The structure of a T cell antigen receptor.

Whereas the antigen receptors of B cells bind to epitopes of *intact* antigens protruding from pathogens or circulating free in body fluids, antigen receptors of T cells bind only to fragments of antigens that are displayed, or presented, on the surface of host cells. The host protein that displays the antigen fragment on the cell surface is called a **major histocompatibility complex (MHC) molecule**. By displaying antigen fragments, MHC molecules are essential for antigen recognition by T cells.

The display of protein antigens occurs when a pathogen infects a cell of the animal host or when an immune cell engulfs pathogen proteins or a whole pathogen **(Figure 43.12)**. Inside the animal cell, enzymes cleave each antigen into antigen fragments, which are small peptides. Each antigen fragment binds to an MHC molecule, which transports the bound peptide to the cell surface. The result is

▼ Figure 43.12 Antigen recognition by T cells. Inside the host cell, an antigen fragment from a pathogen binds to an MHC molecule and is brought up to the cell surface, where it is displayed. The combination of MHC molecule and antigen fragment is recognized by a T cell.

▼ **Figure 43.13 A closer look at antigen presentation.** As shown in this ribbon model, the top of the MHC molecule cradles an antigen fragment, like a bun holding a hot dog. An MHC molecule can display many different antigen fragments, but the antigen receptor of a T cell is specific for a single antigen fragment.

Top view

Antigen fragment

MHC molecule

Host cell

antigen presentation, the display of the antigen fragment in an exposed groove of the MHC protein.

Figure 43.13 shows a close-up view of antigen presentation. In effect, antigen presentations advertises the fact that a host cell contains a foreign substance. If the cell displaying an antigen fragment encounters a T cell with the right specificity, the antigen receptor on the T cell can bind to both the antigen fragment and the MHC molecule. This interaction of an MHC molecule, an antigen fragment, and an antigen receptor triggers an adaptive immune response, as we'll explore in Concept 43.3.

B Cell and T Cell Development

Now that you know how B cells and T cells recognize antigens, let's consider four major characteristics of adaptive immunity. First, the immense repertoire of lymphocytes and receptors enables detection of antigens and pathogens never before encountered. Second, adaptive immunity normally has self-tolerance, the lack of reactivity against an animal's own molecules and cells. Third, cell proliferation triggered by activation greatly increases the number of B and T cells specific for an antigen. Fourth, there is a stronger and more rapid response to an antigen encountered previously, due to a feature known as *immunological memory*, which we'll explore later in the chapter.

Receptor diversity and self-tolerance arise as a lymphocyte matures. Cell proliferation and the formation of immunological memory occur later, after a mature lymphocyte encounters and binds to a specific antigen. We'll consider these four characteristics in the order in which they develop.

The Basis of B Cell and T Cell Diversity

Each person makes more than 1 million different B cell antigen receptors and 10 million different T cell antigen receptors. Yet there are only about 20,000 protein-coding genes in the human genome. How, then, do we generate so many

different antigen receptors? The answer lies in combinations. Think of selecting a cell phone model that comes in three sizes and six colors. There are 18 (3×6) combinations to consider. Similarly, by combining variable elements, the immune system assembles millions of different receptors from a very small collection of parts.

To understand the origin of receptor diversity, let's consider an immunoglobulin (Ig) gene that encodes the light chain of both membrane-bound B cell antigen receptors and secreted antibodies (immunoglobulins). Although we'll analyze only a single Ig light-chain gene, all B and T cell antigen receptor genes undergo very similar transformations.

The capacity to generate diversity is built into the structure of Ig genes. A receptor light chain is encoded by three gene segments: a variable (V) segment, a joining (J) segment, and a constant (C) segment. The V and J segments together encode the variable region of the receptor chain, while the C segment encodes the constant region. The light-chain gene contains a single C segment, 40 different V segments, and 5 different J segments. The alternative copies of the V and J segments are arrayed along the gene in a series **(Figure 43.14)**. Because a functional gene is built from one copy of each type of segment, the pieces can be combined in 200 different ways (40 $V \times 5 J \times 1 C$). The number of different heavy-chain combinations is even greater, resulting in even more diversity.

Antigen Receptor Gene Rearrangement

Assembling a functional Ig gene requires rearranging the DNA. Early in B cell development, an enzyme complex called *recombinase* links one light-chain V gene segment to one J gene segment. This recombination event eliminates the long stretch of DNA between the segments, forming a single exon that is part V and part J.

Recombinase acts randomly, linking any one of the 40 V gene segments to any one of the 5 J gene segments. Heavy-chain genes undergo a similar rearrangement. In any given cell, however, only one allele of a light-chain gene and one allele of a heavy-chain gene are rearranged. Furthermore, the rearrangements are permanent and are passed on to the daughter cells when the lymphocyte divides.

After both a light-chain and a heavy-chain gene have been rearranged, antigen receptors can be synthesized. The rearranged genes are transcribed, and the transcripts are processed for translation. Following translation, the light chain and heavy chain assemble together, forming an antigen receptor (see Figure 43.14). Each pair of randomly rearranged heavy and light chains results in a different antigen-binding site. For the total population of B cells in a human body, the number of such combinations has been calculated as 3.5×10^6. Furthermore, mutations introduced during VJ recombination add additional

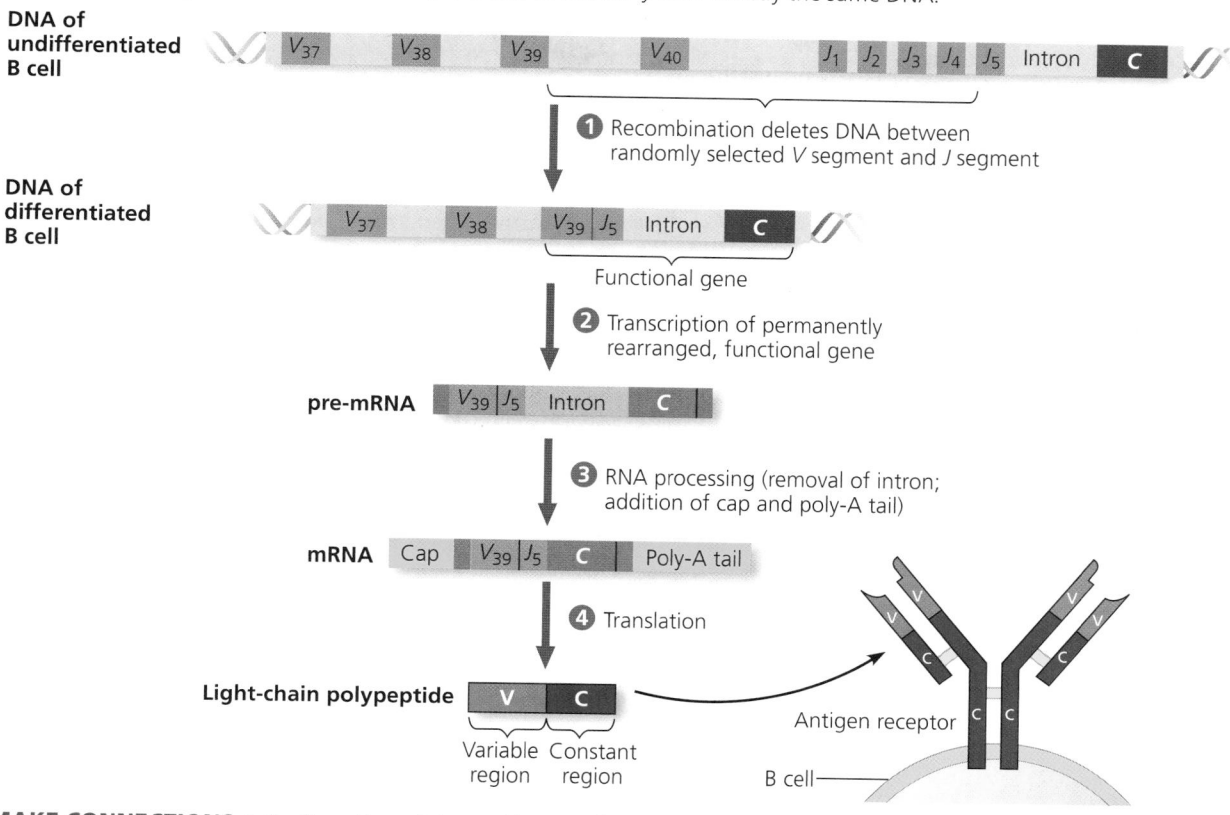

▼ **Figure 43.14 Immunoglobulin (antibody) gene rearrangement.** The joining of randomly selected *V* and *J* gene segments (V_{39} and J_5 in the example shown) results in a functional gene that encodes the light-chain polypeptide of a B cell antigen receptor. Transcription, splicing, and translation result in a light chain that combines with a polypeptide produced from an independently rearranged heavy-chain gene to form a functional receptor. Mature B cells (and T cells) are exceptions to the generalization that all diploid cells in the body have exactly the same DNA.

MAKE CONNECTIONS *Both alternative splicing and joining of V and J segments by recombination generate diverse gene products from a limited set of gene segments (see Figure 18.14). How do these processes differ?*

variation, making the number of antigen-binding specificities even greater.

Origin of Self-Tolerance

In adaptive immunity, how does the body distinguish self from nonself? Because antigen receptor genes are randomly rearranged, some immature lymphocytes produce receptors specific for epitopes on the organism's own molecules. If these self-reactive lymphocytes were not eliminated or inactivated, the immune system could not distinguish self from nonself and would attack body proteins, cells, and tissues. Instead, as lymphocytes mature in the bone marrow or thymus, their antigen receptors are tested for self-reactivity. Some B and T cells with receptors specific for the body's own molecules are destroyed by *apoptosis*, which is a programmed cell death (see Concept 11.5). The remaining self-reactive lymphocytes are typically rendered nonfunctional, leaving only those lymphocytes that react to foreign molecules. Since the body normally lacks mature lymphocytes that can react

against its own components, the immune system is said to exhibit *self-tolerance*.

Proliferation of B Cells and T Cells

Despite the enormous variety of antigen receptors, only a tiny fraction are specific for a given epitope. How then does an effective adaptive response develop? To begin with, an antigen is presented to a steady stream of lymphocytes in the lymph nodes (see Figure 43.6) until a match is made. A successful match between an antigen receptor and an epitope initiates events that activate the lymphocyte bearing the receptor.

Once activated, a B cell or T cell undergoes multiple cell divisions. For each activated cell, the result of this proliferation is a clone, a population of cells that are identical to the original cell. Some cells from this clone become **effector cells**, mostly short-lived cells that take effect immediately against the antigen and any pathogens producing that antigen.

For B cells, the effector forms are *plasma cells* (**Figure 43.15**), which secrete antibodies. For T cells, the effector forms are helper T cells and cytotoxic T cells, whose roles we'll explore in Concept 43.3. The remaining cells in the clone become **memory cells**, long-lived cells that can give rise to effector cells if the same antigen is encountered later in the animal's life.

Figure 43.16 summarizes the proliferation of lymphocyte into a clone of cells occurs in response to binding to an antigen, using B cells as an example. The process is called **clonal selection** because an encounter with an antigen *selects* which lymphocyte will divide to produce a *clonal* population of thousands of cells specific for a particular epitope.

Immunological Memory

Immunological memory is responsible for the long-term protection that a prior infection provides against many

▼ **Figure 43.15 A plasma cell.**

2 µm

diseases, such as chicken pox. This type of protection was noted almost 2,400 years ago by the Greek historian Thucydides. He observed that individuals who had recovered from the plague could safely care for those who were sick or dying, "for the same man was never attacked twice—never at least fatally."

Prior exposure to an antigen alters the speed, strength, and duration of the immune response. The effector cells formed by clones of lymphocytes after an initial exposure to an antigen produce a **primary immune response**. The primary response peaks about 10–17 days after the initial exposure. If the same antigen is encountered again later, there is a **secondary immune response**, a response that is faster (typically peaking only 2–7 days after exposure), of greater magnitude, and more prolonged. These differences between primary and secondary immune responses are readily apparent in a graph of the concentrations of specific antibodies in blood over time (**Figure 43.17**).

The secondary immune response relies on the reservoir of T and B memory cells generated upon initial exposure to an antigen. Because these cells are long-lived, they provide the basis for immunological memory, which can span many

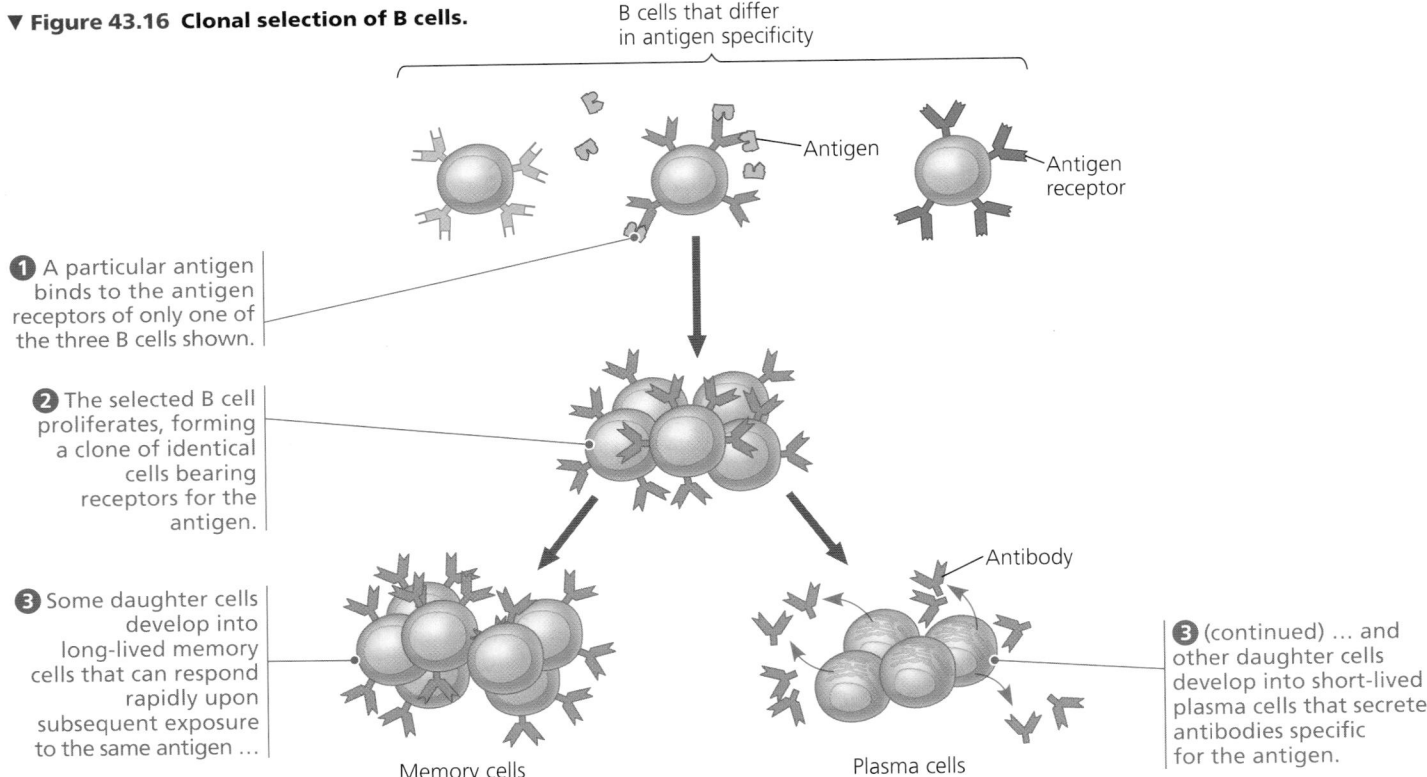

▼ **Figure 43.16 Clonal selection of B cells.**

B cells that differ in antigen specificity

Antigen

Antigen receptor

❶ A particular antigen binds to the antigen receptors of only one of the three B cells shown.

❷ The selected B cell proliferates, forming a clone of identical cells bearing receptors for the antigen.

❸ Some daughter cells develop into long-lived memory cells that can respond rapidly upon subsequent exposure to the same antigen …

Memory cells

Antibody

❸ (continued) … and other daughter cells develop into short-lived plasma cells that secrete antibodies specific for the antigen.

Plasma cells

VISUAL SKILLS *For the purpose of illustration, this figure shows only a few of each type of cell or molecule. Based on what you have read in Concept 43.2, provide estimates of the number of different B cells and number of antigen receptors on each B cell.*

→ **Mastering Biology Animation: Clonal Selection**

▼ **Figure 43.17 The specificity of immunological memory.**
Long-lived memory cells that are generated in the primary response
to antigen A give rise to a heightened secondary response to the
same antigen but don't affect the primary response to another
antigen (B).

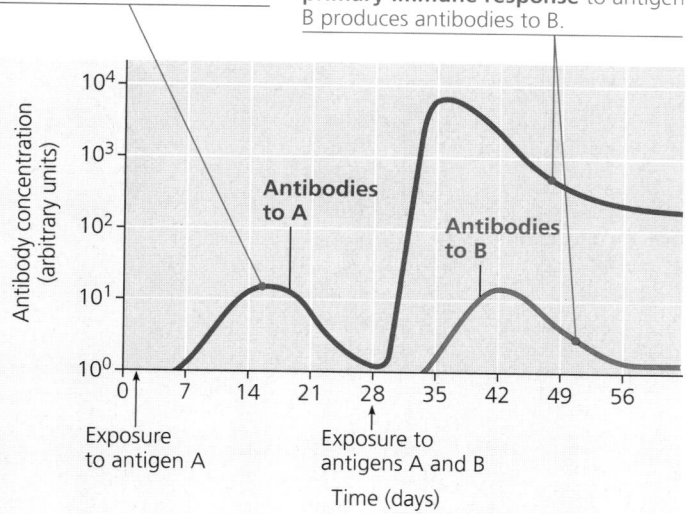

Primary immune response to antigen A produces antibodies to A.

Secondary immune response to antigen A produces antibodies to A; **primary immune response** to antigen B produces antibodies to B.

INTERPRET THE DATA *Assume that on average one out of every 10^5 B cells in the body is specific for antigen A on day 16 and that the number of B cells producing a specific antibody is proportional to the concentration of that antibody. What would you predict is the frequency of B cells specific for antigen A on day 36?*

decades. (Most effector cells have much shorter life spans, which is why the immune response diminishes after an infection is overcome.) If an antigen is encountered again, memory cells specific for that antigen enable the rapid formation of clones of thousands of effector cells that are also specific for that antigen, thus generating a greatly enhanced immune defense.

Although the processes for antigen recognition, clonal selection, and immunological memory are similar for B cells and T cells, these two classes of lymphocytes fight infection in different ways and in different settings, as we'll explore in Concept 43.3.

CONCEPT CHECK 43.2

1. **DRAW IT** Sketch a B cell antigen receptor. Label the V and C regions of the light and heavy chains. Where are the antigen-binding sites, disulfide bridges, and transmembrane region located relative to these regions?

2. Explain how memory cells strengthen the immune response when a pathogen is encountered for a second time.

3. **WHAT IF?** If both copies of a light-chain gene and a heavy-chain gene recombined in each (diploid) B cell, how would this affect B cell development and function?

For suggested answers, see Appendix A.

Adaptive immunity defends against infection of body fluids and body cells

Having considered how clones of lymphocytes arise, we now explore how these cells help fight infections and minimize damage by pathogens. The defenses provided by B and T lymphocytes can be divided into humoral and cell-mediated immune responses. The **humoral immune response** protects the blood and lymph (once called body *humors*, or fluids). In this response, antibodies help neutralize or eliminate toxins and pathogens in body fluids. In the **cell-mediated immune response**, specialized T cells destroy infected host cells. Both humoral and cellular immunity can include a primary and a secondary immune response, with memory cells enabling the secondary response.

Helper T Cells: Activating Adaptive Immunity

A type of T cell called a **helper T cell** activates humoral and cell-mediated immune responses. Before this can happen, however, two conditions must be met. First, a foreign molecule must be present that can bind specifically to the antigen receptor of the helper T cell. Second, this antigen must be displayed on the surface of an **antigen-presenting cell**. An antigen-presenting cell can be a dendritic cell, macrophage, or B cell.

Like immune cells, infected cells can display foreign antigens on their surface. What then distinguishes antigen-presenting cells from infected cells? The answer lies in two distinct types, or *classes*, of MHC molecules. Most body cells have only the class I MHC molecules, but antigen-presenting cells have both class I and class II MHC molecules. The class II molecules provide a molecular signature by which an antigen-presenting cell is recognized.

A helper T cell and the antigen-presenting cell displaying its specific epitope have a complex interaction. The antigen receptors on the surface of the helper T cell bind to the antigen fragment and to the class II MHC molecule displaying that fragment on the antigen-presenting cell **(Figure 43.18)**. At the same time, an accessory protein called CD4 on the helper T cell surface binds to the class II MHC molecule, helping keep the cells joined. As the two cells interact, signals are exchanged in the form of cytokines. For example, the cytokines secreted from a dendritic cell act in combination with the antigen to stimulate the helper T cell, causing it to produce its own set of cytokines. Also, extensive contact between the cell surfaces enables further information exchange.

▼ **Figure 43.18 The central role of helper T cells in humoral and cell-mediated immune responses.** Here, a helper T cell responds to a dendritic cell displaying an antigen.

Clone of activated helper T cells

B cell

Cytotoxic T cell

HUMORAL IMMUNITY (secretion of antibodies by plasma cells)

CELL-MEDIATED IMMUNITY (attack on infected cells)

1 An antigen-presenting cell engulfs a pathogen, degrades it, and displays antigen fragments complexed with class II MHC molecules on the cell surface. A specific helper T cell binds to this complex via its antigen receptor and an accessory protein called CD4.

2 Binding of the helper T cell promotes secretion of cytokines by the antigen-presenting dendritic cell. These cytokines, along with cytokines from the helper T cell itself, activate the helper T cell and stimulate its proliferation.

3 Cell proliferation produces a clone of activated helper T cells. All cells in the clone have receptors for the same antigen. These cells secrete other cytokines, which help activate B cells and cytotoxic T cells with the same antigen specificity.

→ **Mastering Biology** BioFlix® **Animation: Activation of a Helper T Cell by a Dendritic Cell (Example: Infection by a Rhinovirus)**

Antigen-presenting cells interact with helper T cells in several different contexts. Antigen presentation by a dendritic cell or macrophage activates a helper T cell, which then proliferates, forming a clone of activated cells. In contrast, B cells present antigens to *already* activated helper T cells, which in turn activate the B cells. Activated helper T cells also help stimulate cytotoxic T cells, as we'll discuss shortly.

B Cells and Antibodies: A Response to Extracellular Pathogens

Secretion of antibodies by clonally selected B cells is the hallmark of the humoral immune response. It begins with activation of the B cells.

Activation of B Cells

As illustrated in **Figure 43.19**, activation of B cells involves both helper T cells and proteins on the surface of pathogens. Stimulated by both an antigen and cytokines, the B cell proliferates and differentiates into memory B cells and antibody-secreting plasma cells.

The pathway for antigen processing and display in B cells differs from that in other antigen-presenting cells. A macrophage or dendritic cell can present fragments from a wide variety of protein antigens, whereas a B cell presents only the antigen to which it specifically binds. When an antigen first binds to receptors on the surface of a B cell, the cell takes in a few foreign molecules by receptor-mediated endocytosis (see Figure 7.19). The class II MHC protein of the B cell then presents an antigen fragment to a helper T cell. This direct cell-to-cell contact is usually critical to B cell activation (see step 2 in Figure 43.19).

B cell activation leads to a robust humoral immune response: A single activated B cell gives rise to thousands of identical plasma cells. These plasma cells stop expressing a membrane-bound antigen receptor and begin producing and secreting antibodies (see step 3 in Figure 43.19). Each plasma cell secretes approximately 2,000 antibodies every second during its four- to five-day life span, nearly a trillion antibody molecules in total. Furthermore, most antigens recognized by B cells contain multiple epitopes. An exposure to a single antigen therefore normally activates a variety of B cells, which give rise to different plasma cells producing antibodies directed against different epitopes on the common antigen.

Antibody Function

Antibodies do not directly kill pathogens, but by binding to antigens, they interfere with pathogen activity or mark

▼ **Figure 43.19 Activation of a B cell in the humoral immune response.** Most protein antigens require activated helper T cells to trigger a humoral response. A macrophage (shown here) or a dendritic cell can activate a helper T cell, which in turn can activate a B cell to give rise to antibody-secreting plasma cells.

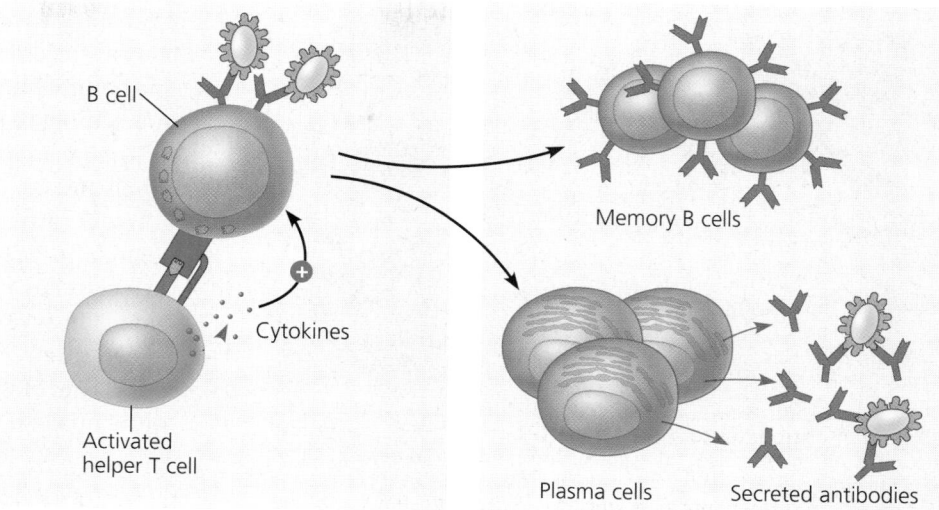

1 After an antigen-presenting cell engulfs and degrades a pathogen, it displays an antigen fragment complexed with a class II MHC molecule. A helper T cell that recognizes the complex is activated with the aid of cytokines secreted from the antigen-presenting cell.

2 When a B cell with receptors for the same epitope internalizes the antigen, it displays an antigen fragment on the cell surface in a complex with a class II MHC molecule. An activated helper T cell bearing receptors specific for the displayed fragment binds to and activates the B cell.

3 The activated B cell proliferates and differentiates into memory B cells and antibody-secreting plasma cells. The secreted antibodies are specific for the same antigen that initiated the response.

? *Looking at the steps in this figure, propose a function for the cell-surface antigen receptors of memory B cells.*

→ **Mastering Biology** BioFlix® **Animation: Adaptive Defenses: B Cells (Example: Infection by a Rhinovirus)**

pathogens in various ways for inactivation or destruction. Consider, for example, *neutralization*, a process in which antibodies bind to proteins on the surface of a virus **(Figure 43.20a)**. The bound antibodies prevent viral infection of a host cell, thus neutralizing the virus. Similarly, antibodies sometimes bind to toxins released in body fluids, preventing the toxins from entering body cells.

In *opsonization*, antibodies that are bound to antigens on bacteria do not block infection, but instead present a readily recognized structure for macrophages or neutrophils, thereby promoting phagocytosis **(Figure 43.20b)**. Because each antibody has two antigen-binding sites, antibodies can also facilitate phagocytosis by linking bacterial cells, viruses, or other foreign substances into aggregates.

When antibodies facilitate phagocytosis, as in opsonization, they also help fine-tune the humoral immune response. Recall that phagocytosis enables macrophages and dendritic cells to present antigens to and stimulate helper T cells, which in turn stimulate the very B cells whose antibodies contribute to phagocytosis. This positive feedback between innate and adaptive immunity contributes to a coordinated, effective response to infection.

▼ **Figure 43.20 Two mechanisms of antibody function.**

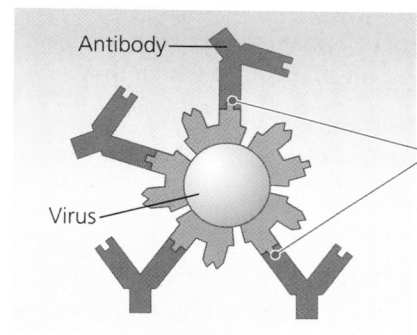

Antibodies bound to antigens on the surface of a virus neutralize it by blocking its ability to bind to a host cell.

(a) Neutralization

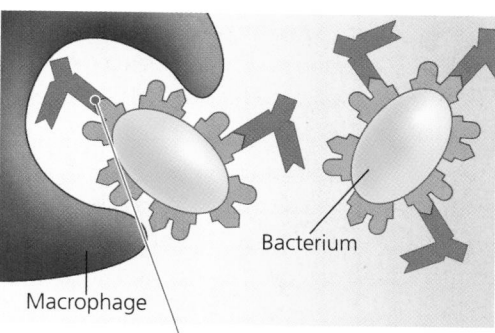

Binding of antibodies to antigens on the surface of bacteria promotes phagocytosis by macrophages and neutrophils.

(b) Opsonization

Figure 43.21 Activation of complement system and pore formation.

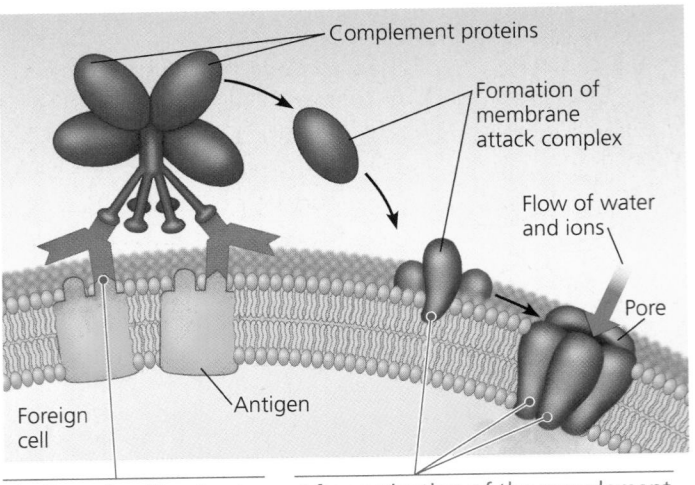

Complement proteins

Formation of membrane attack complex

Flow of water and ions

Pore

Foreign cell

Antigen

Binding of antibodies to antigens on the surface of a foreign cell activates the complement system.

After activation of the complement system, the membrane attack complex forms pores in the cell's membrane, allowing water and ions to rush in. The cell swells and lyses.

Antibodies sometimes work together with the proteins of the complement system **(Figure 43.21)**. (The name *complement* reflects the fact that these proteins add to the effectiveness of antibody-directed attacks on bacteria.) Binding of a complement protein to an antigen-antibody complex on a foreign cell triggers events leading to formation of a pore in the membrane of the cell. Ions and water rush into the cell, causing it to swell and lyse.

Although antibodies are the cornerstones of the response in body fluids, there is also a mechanism by which they

can bring about the death of infected body cells. When a virus uses a cell's biosynthetic machinery to produce viral proteins, these viral products can appear on the cell surface. If antibodies specific for epitopes on these viral proteins bind to the exposed proteins, the presence of bound antibody at the cell surface can recruit a natural killer cell. The natural killer cell then releases proteins that cause the infected cell to undergo apoptosis. Thus the activities of the innate and adaptive immune systems are once again closely linked.

B cells can express five different types, or *classes*, of immunoglobulin (IgA, IgD, IgE, IgG, and IgM). For a given B cell, each class has an identical antigen-binding specificity but a distinct heavy-chain C region. The B cell antigen receptor, known as IgD, is exclusively membrane bound. The other four classes of Ig have soluble forms, such as the antibodies found in blood, tears, saliva, and breast milk.

Cytotoxic T Cells: A Response to Infected Host Cells

In the absence of an immune response, pathogens can reproduce in and kill infected cells. In the cell-mediated immune response, **cytotoxic T cells** use toxic proteins to kill cells infected by viruses or other intracellular pathogens before pathogens fully mature **(Figure 43.22)**. To become active, cytotoxic T cells require signals from helper T cells and interaction with an antigen-presenting cell. Fragments of foreign proteins produced in infected host cells associate with class I MHC molecules and are displayed on the cell surface, where they can be recognized by activated cytotoxic

▼ Figure 43.22 The killing action of cytotoxic T cells on an infected host cell.
An activated cytotoxic T cell releases molecules that make pores in an infected cell's membrane and enzymes that break down proteins, promoting the cell's death.

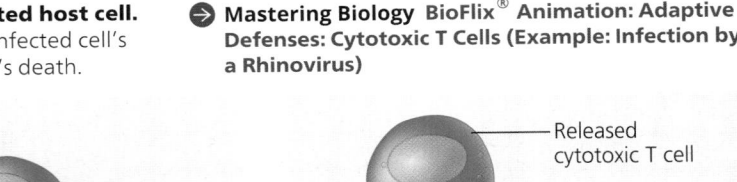

Mastering Biology BioFlix® Animation: Adaptive Defenses: Cytotoxic T Cells (Example: Infection by a Rhinovirus)

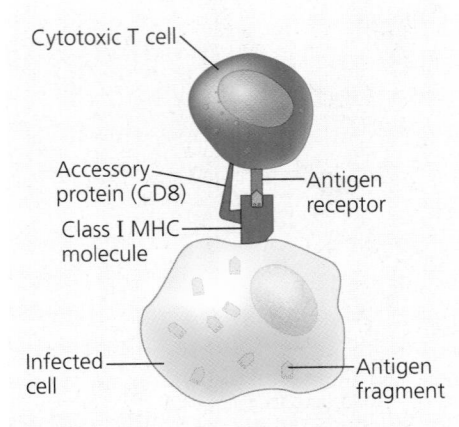

Cytotoxic T cell

Accessory protein (CD8)

Antigen receptor

Class I MHC molecule

Infected cell

Antigen fragment

❶ An activated cytotoxic T cell binds to a class I MHC–antigen fragment complex on an infected cell via the T cell's antigen receptor and an accessory protein called CD8.

Perforin

Granzymes

Pore

❷ The T cell releases perforin molecules, which form pores in the infected cell membrane, and granzymes, enzymes that break down proteins. Granzymes enter the infected cell via the perforin pores.

Released cytotoxic T cell

Dying infected cell

❸ The granzymes initiate apoptosis within the infected cell, leading to fragmentation of the nucleus and cytoplasm and eventual cell death. Once released, the cytotoxic T cell can attack other infected cells.

T cells. Like helper T cells, cytotoxic T cells have an accessory protein that can bind to an MHC molecule. The interaction of this accessory protein, called CD8, with a class I MHC molecule displaying antigen targets the infected cell for cytotoxic T cell activity.

The targeted destruction of an infected host cell by a cytotoxic T cell involves the secretion of proteins that disrupt membrane integrity and trigger cell death (apoptosis; see Figure 43.22). The death of the infected cell not only deprives the pathogen of a place to multiply but also exposes cell contents to circulating antibodies, which mark released antigens for disposal.

Summary of the Humoral and Cell-Mediated Immune Responses

As noted earlier, both humoral and cell-mediated immunity can include primary as well as secondary immune responses. Memory cells of each type—helper T cell, B cell, and cytotoxic T cell—enable the secondary response. For example, when body fluids are reinfected by a pathogen encountered previously, memory B cells and memory helper T cells initiate a secondary humoral response. **Figure 43.23** summarizes adaptive immunity, reviews the events that initiate humoral and cell-mediated immune responses, highlights the difference in

▼ **Figure 43.23 An overview of the adaptive immune response.**

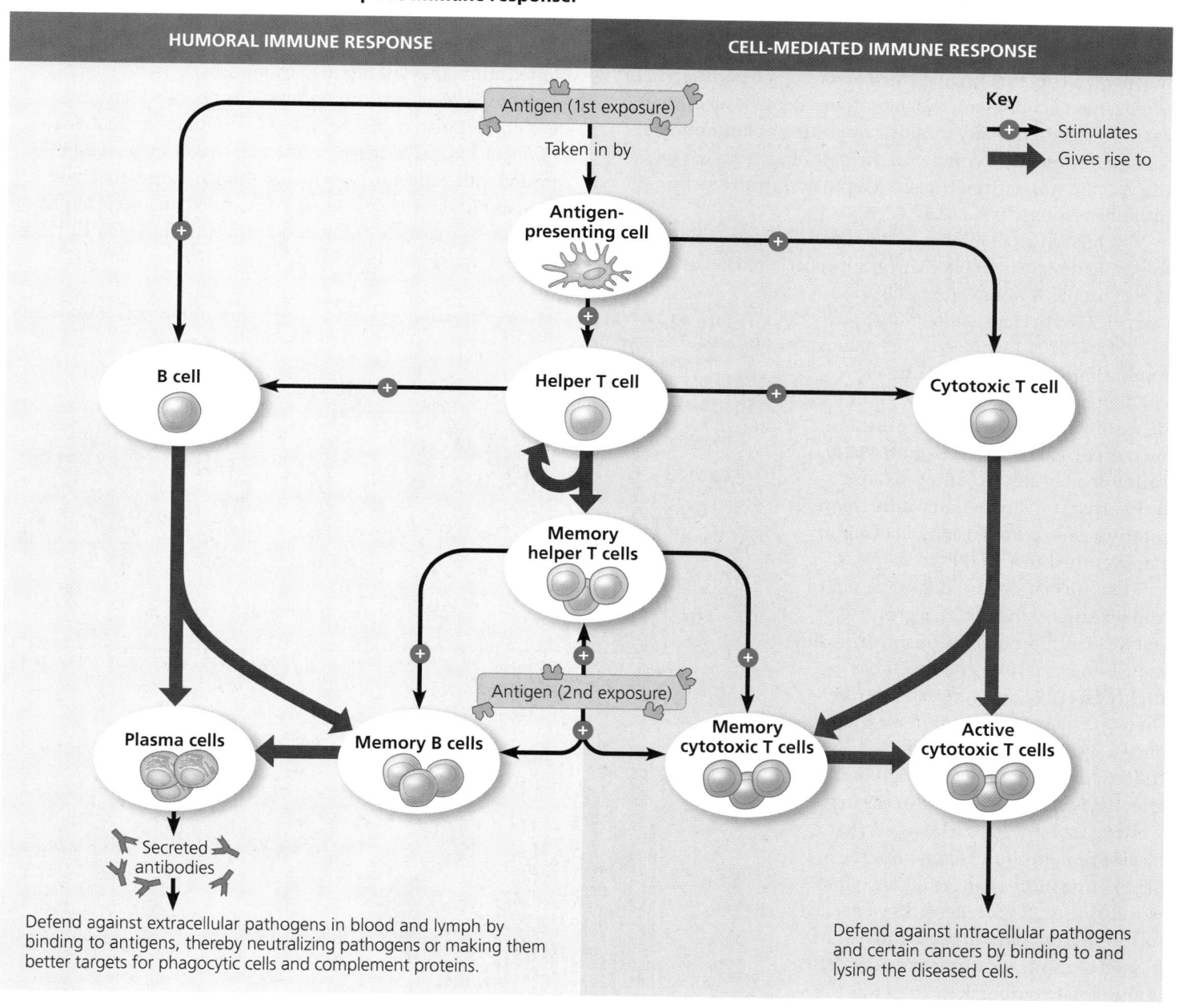

VISUAL SKILLS *Identify each arrow as representing part of the primary response or secondary response.*

Mastering Biology BioFlix® Animation: Summary of the Adaptive Immune Response (Example: Infection by a Rhinovirus)

response to pathogens in body fluids versus in body cells, and emphasizes the central role of the helper T cell.

Immunization

The protection provided by a second immune response provides the basis for **immunization**, the use of antigens artificially introduced into the body to generate an adaptive immune response and memory cell formation. In 1796, Edward Jenner noted that milkmaids who had cowpox, a mild disease usually seen only in cows, did not contract smallpox, a far more dangerous disease. In the first documented immunization (or *vaccination*, from the Latin *vacca*, cow), Jenner used the cowpox virus to induce adaptive immunity against the closely related smallpox virus. The *vaccines* used today for immunizations may be made from inactivated bacterial toxins, killed or weakened pathogens, or even genes encoding microbial proteins. Because all of these agents induce a primary immune response and immunological memory, an encounter with the pathogen from which the vaccine was derived triggers a rapid and strong secondary immune response (see Figure 43.17).

Vaccination programs have been successful against many infectious diseases that once killed or incapacitated large numbers of people. A worldwide vaccination campaign led to eradication of smallpox in the late 1970s. In industrialized nations, routine immunization of infants and children has dramatically reduced the incidence of sometimes devastating diseases, such as polio and measles **(Figure 43.24)**. Unfortunately, not all pathogens are easily managed by vaccination. Furthermore, some vaccines are not readily available in impoverished areas of the globe.

Misinformation about vaccine safety and disease risk has led to a growing public health problem. Consider measles as just one example. The vaccine is highly effective and very safe—fewer than one in a million children experience a significant allergic reaction. Indeed, a vaccination campaign had effectively eliminated the disease in the United States by 2000. However, the disease remains quite dangerous to this day, killing more than 200,000 people worldwide each year. Recently, a decline in vaccination rates in the United States has led to multiple outbreaks since 2010. A multi-state outbreak in 2019 was the worst in two decades, with more than 1,100 cases in the first six months of the year. Even with modern medical care, measles kills on average one out of every 1,000 infected individuals.

Active and Passive Immunity

Our discussion of adaptive immunity has to this point focused on **active immunity**, the defenses that arise when a pathogen infection or immunization prompts an immune response. A different type of immunity results when, for example, the IgG antibodies in the blood of a pregnant female cross the placenta to her fetus. This protection is called **passive immunity** because the antibodies in the recipient (in this case, the fetus) are produced by another individual (the mother). IgA antibodies present in breast milk provide additional passive immunity to the infant's digestive tract while the infant's immune system develops. Passive immunity persists only as long as the transferred antibodies last (up to a few months).

In artificial passive immunization, antibodies from an immune animal are injected into a nonimmune animal. For example, humans bitten by venomous snakes are sometimes treated with antivenin, serum from sheep or horses that have been immunized against a snake venom. When injected immediately after a snakebite, the antibodies in antivenin can neutralize toxins in the venom before the toxins do massive damage.

▼ **Figure 43.24 Vaccine-based protection against two life-threatening communicable diseases.** The graphs show deaths by year in the United States caused by polio and measles. The maps show examples of the global progress against these two diseases.

Antibodies as Tools

Antibodies that an animal produces after exposure to an antigen are the products of many different clones of plasma cells, each specific for a different epitope. However, antibodies can also be prepared from a single clone of B cells grown in culture. The **monoclonal antibodies** produced by such a culture are identical and specific for the same epitope on an antigen.

Monoclonal antibodies have provided the basis for many recent advances in medical diagnosis and treatment. For example, home pregnancy test kits use monoclonal antibodies to detect human chorionic gonadotropin (hCG). Because hCG is produced as soon as an embryo implants in the uterus (see Concept 46.5), the presence of this hormone in a woman's urine is a reliable indicator for a very early stage of pregnancy. Monoclonal antibodies are also being used as therapies for a growing number of human diseases, including many cancers.

One recently developed antibody tool uses a single drop of blood to identify every virus that a person has encountered through infection or vaccination **(Figure 43.25)**. To detect the antibodies formed against these viruses, researchers generated a set of nearly 100,000 bacteriophages, each of which displays a different peptide from one of the roughly 200 species of viruses that infect humans.

Immune Rejection

Like pathogens, cells from another person can be recognized as foreign and attacked by immune defenses. For example, skin transplanted from one person to a genetically nonidentical person will look healthy for a week or so but will then be destroyed (rejected) by the recipient's immune response. It turns out that MHC molecules are a primary cause of rejection. Why? Each of us expresses MHC proteins from more than a dozen different MHC genes. Furthermore, there are more than 100 different versions, or alleles, of human MHC genes. As a consequence, the sets of MHC proteins on cell surfaces are likely to differ between any two people, except identical twins. Such differences can stimulate an immune response in the recipient of a transplant or graft, causing rejection. To minimize rejection of a transplant

▼ **Figure 43.25 A comprehensive test for past viral encounters.** By combining the power of DNA sequencing with the specificity of antigen recognition by antibodies, researchers can identify every virus that a person's immune system has encountered during the person's lifetime.

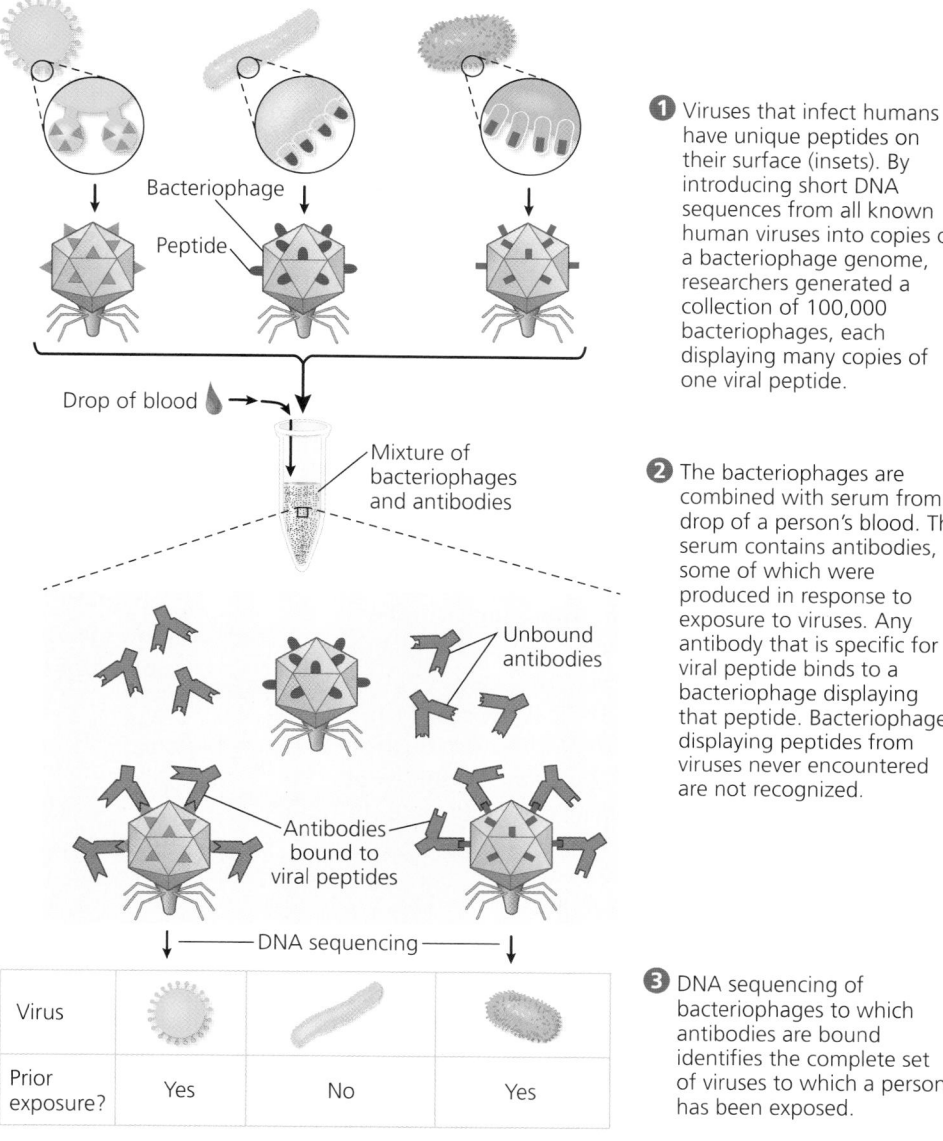

❶ Viruses that infect humans have unique peptides on their surface (insets). By introducing short DNA sequences from all known human viruses into copies of a bacteriophage genome, researchers generated a collection of 100,000 bacteriophages, each displaying many copies of one viral peptide.

❷ The bacteriophages are combined with serum from a drop of a person's blood. The serum contains antibodies, some of which were produced in response to exposure to viruses. Any antibody that is specific for a viral peptide binds to a bacteriophage displaying that peptide. Bacteriophages displaying peptides from viruses never encountered are not recognized.

❸ DNA sequencing of bacteriophages to which antibodies are bound identifies the complete set of viruses to which a person has been exposed.

WHAT IF? *All of the antibodies are shown with just one antigen-binding site occupied. If a single antibody bound to two bacteriophages, how would this affect the results?*

or graft, surgeons use donor tissue bearing MHC molecules that match those of the recipient as closely as possible. In addition, the recipient takes medicines that suppress immune responses (but as a result leave the recipient more susceptible to infections).

Blood Groups

In the case of blood transfusions, the recipient's immune system can recognize glycoproteins on the surface of blood cells as foreign, triggering an immediate and devastating reaction. To avoid this danger, the so-called ABO blood groups of the donor and recipient must be taken into account. Red blood cells are designated as type A if they have the A carbohydrate on their surface. Similarly, the B carbohydrate is found on the surface of type B red blood cells; both A and B carbohydrates are found on type AB red blood cells; and neither carbohydrate is found on type O red blood cells (see Figure 14.11).

Why does the immune system recognize particular sugars on red blood cells? It turns out that we are frequently exposed to certain bacteria that have epitopes very similar to the carbohydrates on blood cells. A person with type A blood will respond to the bacterial epitope similar to the B carbohydrate and make antibodies that will react with any B carbohydrate encountered upon a transfusion. However, that same person doesn't make antibodies against the bacterial epitope similar to the A carbohydrate because lymphocytes that would be reactive with the body's own cells and molecules were inactivated or eliminated during development.

To understand how ABO blood groups affect transfusions, let's consider further the example of a person with type A blood receiving a transfusion of type B blood. The person's anti-B antibodies would cause the transfused red blood cells to undergo lysis, triggering chills, fever, shock, and perhaps kidney malfunction. At the same time, anti-A antibodies in the donated type B blood would act against the recipient's red blood cells. Applying the same logic to a type O person, we can see that such interactions would cause a problem upon transfusion of any other blood type. Fortunately, the discovery of enzymes that can cleave the A and B carbohydrates from red blood cells may eliminate this problem in the future.

CONCEPT CHECK 43.3

1. If a child were born without a thymus gland, what cells and functions of the immune system would be deficient? Explain.

2. Treatment of antibodies with a particular protease clips the heavy chains in half, releasing the two arms of the Y-shaped molecule. How might the antibodies continue to function?

3. **WHAT IF?** Suppose that a snake handler bitten by a particular venomous snake species was treated with antivenin. Why might the same treatment for a second such bite have a harmful side effect?

For suggested answers, see Appendix A.

CONCEPT 43.4

Disruptions in immune system function can elicit or exacerbate disease

Although adaptive immunity protects against many pathogens, it is not fail-safe. Here we'll first examine the disorders and diseases that arise when adaptive immunity is blocked or misregulated. We'll then turn to some of the evolutionary adaptations of pathogens that diminish the effectiveness of adaptive immune responses in the host.

Exaggerated, Self-Directed, and Diminished Immune Responses

The highly regulated interplay among lymphocytes, other body cells, and foreign substances generates an immune response that provides extraordinary protection against many pathogens. When allergic, autoimmune, or immunodeficiency disorders disrupt this delicate balance, the effects are frequently severe.

Allergies

Allergies are exaggerated (hypersensitive) responses to certain antigens called *allergens*. The most common allergies involve antibodies of the IgE class. Hay fever, for instance, occurs when plasma cells secrete IgE antibodies specific for antigens on the surface of pollen grains **(Figure 43.26)**. Some IgE antibodies attach by their base to mast cells in connective tissues. Pollen grains that enter the body later attach to the antigen-binding sites of these IgE antibodies. Such attachment cross-links adjacent IgE molecules, inducing the mast cell to release histamine and other inflammatory chemicals. Acting on a variety of cell types, these chemicals bring about the typical allergy symptoms: sneezing, runny nose, teary eyes, and smooth muscle contractions in the lungs that can inhibit effective breathing.

▼ **Figure 43.26 Mast cells, IgE, and the allergic response.** In this example, pollen grains act as the allergen.

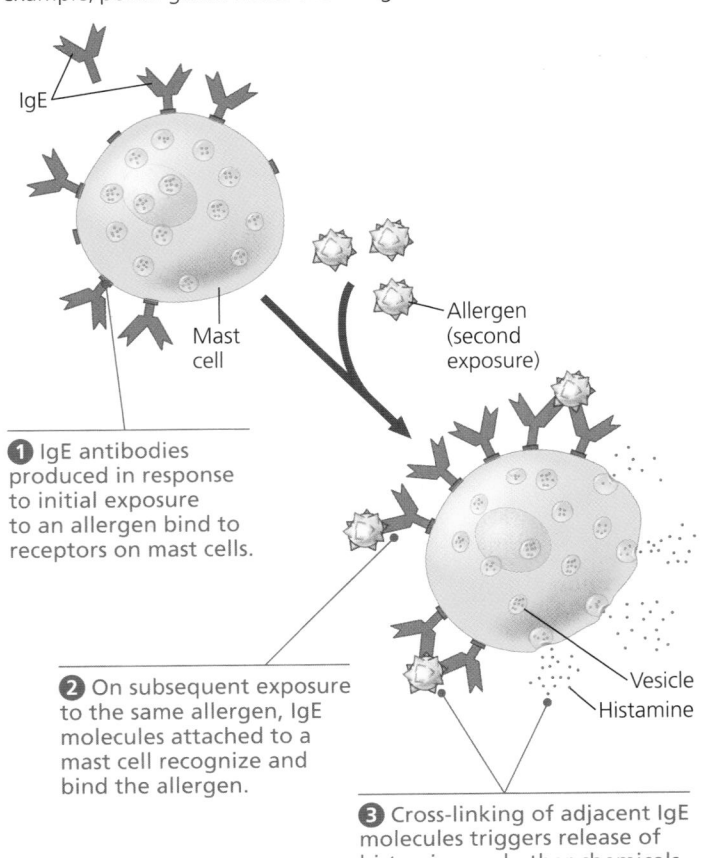

IgE

Mast cell

Allergen (second exposure)

❶ IgE antibodies produced in response to initial exposure to an allergen bind to receptors on mast cells.

❷ On subsequent exposure to the same allergen, IgE molecules attached to a mast cell recognize and bind the allergen.

Vesicle
Histamine

❸ Cross-linking of adjacent IgE molecules triggers release of histamine and other chemicals, leading to allergy symptoms.

Drugs known as antihistamines block receptors for histamine, diminishing allergy symptoms (and inflammation).

An acute allergic response sometimes leads to a life-threatening reaction called *anaphylactic shock*. Inflammatory chemicals released from immune cells trigger constriction of bronchioles and sudden dilation of peripheral blood vessels, which causes a precipitous drop in blood pressure. Death may occur within minutes due to the inability to breathe and lack of blood flow. Substances that can cause anaphylactic shock in allergic individuals include bee venom, penicillin, peanuts, and shellfish. People with severe hypersensitivities often carry an autoinjector containing the hormone epinephrine. An injection of epinephrine rapidly counteracts this allergic response, constricting peripheral blood vessels, reducing swelling in the throat, and relaxing muscles in the lungs to help breathing (see Figure 45.20b).

Autoimmune Diseases

In some people, the immune system is active against particular molecules of the body, causing an **autoimmune disease**. Such a loss of self-tolerance has many forms. In systemic lupus erythematosus, commonly called *lupus*, the immune system generates antibodies against histones and DNA released by the normal breakdown of body cells. These self-reactive antibodies cause skin rashes, fever, arthritis, and kidney dysfunction. Other targets of autoimmunity include the insulin-producing beta cells of the pancreas (in type 1 diabetes) and the myelin sheaths that encase many neurons (in multiple sclerosis).

Heredity, gender, and environment all influence susceptibility to autoimmune disorders. For example, members of certain families show an increased susceptibility to particular autoimmune disorders. In addition, many autoimmune diseases affect females more often than males. Women are nine times as likely as men to have lupus and two to three times as likely to develop *rheumatoid arthritis*, a damaging and painful inflammation of the cartilage and bone in joints **(Figure 43.27)**. The causes of this sex bias, as well as

▼ **Figure 43.27 X-ray of hands that are deformed by rheumatoid arthritis.**

the rise in autoimmune disease frequency in industrialized countries, are areas of active research and debate.

An additional focus of current research on autoimmune disorders is the activity of *regulatory T cells*, nicknamed Tregs. These specialized T cells help modulate immune system activity and prevent response to self-antigens.

Exertion, Stress, and the Immune System

Many forms of exertion and stress influence immune system function. For example, moderate exercise improves immune system function and significantly reduces susceptibility to the common cold and other infections of the upper respiratory tract. In contrast, exercise to the point of exhaustion leads to more frequent infections and more severe symptoms. Psychological stress likewise disrupts immune system regulation, altering the interplay of the hormonal, nervous, and immune systems (see Figure 45.20). Research has also shown that rest is important for immunity: Adults who averaged fewer than 7 hours of sleep got sick three times as often when exposed to a cold virus as those who averaged at least 8 hours.

Immunodeficiency Diseases

A disorder in which an immune system response to antigens is defective or absent is called an immunodeficiency. Whatever its cause and nature, an immunodeficiency can lead to frequent and recurrent infections and increased susceptibility to certain cancers.

An *inborn immunodeficiency* results from a genetic or developmental defect in the production of immune system cells or of specific proteins, such as antibodies or the proteins of the complement system. Depending on the specific defect, either innate or adaptive defenses—or both—may be impaired. In severe combined immunodeficiency (SCID), functional lymphocytes are rare or absent. Lacking an adaptive immune response, SCID patients are susceptible to infections that can cause death in infancy, such as pneumonia and meningitis. Treatments include bone marrow and stem cell transplantation.

Later in life, exposure to chemicals or biological agents can cause an *acquired immunodeficiency*. Drugs used to fight autoimmune diseases or prevent transplant rejection suppress the immune system, leading to an immunodeficient state. Certain cancers also suppress the immune system, especially Hodgkin's disease, which damages the lymphatic system. Acquired immunodeficiencies range from temporary states that may arise from physiological stress to the devastating disease AIDS (acquired immune deficiency syndrome), which we'll explore in the next section.

Evolutionary Adaptations of Pathogens That Underlie Immune System Avoidance

EVOLUTION Just as immune systems that ward off pathogens have evolved in animals, mechanisms that thwart immune responses have evolved in pathogens. Using human pathogens as examples, we'll examine some common mechanisms: antigenic variation, latency, and direct attack on the immune system.

Antigenic Variation

One mechanism for escaping the body's defenses involves a pathogen altering how it appears to the immune system. If a pathogen changes the epitopes that it expresses to ones that a host has not previously encountered, it can reinfect or remain in the host without triggering the rapid and robust response mediated by memory cells. Such changes in epitope expression are called *antigenic variation*. The parasite that causes sleeping sickness (trypanosomiasis) provides an extreme example, periodically switching at random among 1,000 versions of the protein found over its entire surface. In the **Scientific Skills Exercise**, you will interpret data on this form of antigenic variation and the body's response.

Antigenic variation is the main reason the influenza, or "flu," virus remains a major public health problem. As it

Scientific Skills Exercise

Comparing Two Variables on a Common x-Axis

How Does the Immune System Respond to a Changing Pathogen? Natural selection favors parasites that are able to maintain a low-level infection in a host for a long time. *Trypanosoma*, the unicellular parasite that causes sleeping sickness, is one example. The glycoproteins covering a trypanosome's surface are encoded by a gene that is duplicated more than 1,000 times in the organism's genome. Each copy is slightly different. By periodically switching among these genes, the trypanosome can display a series of surface glycoproteins with different molecular structures. In this exercise, you will interpret two data sets to explore hypotheses about the benefits to the trypanosome of its ever-shifting surface glycoproteins in avoiding the host's immune response.

Day of infection	Study A: Millions of parasites/mL blood	Study B: Antibody specific to Variant A	Study B: Antibody specific to Variant B
4	0.1	0	0
6	0.3	0	0
8	1.2	0.2	0
10	0.2	0.5	0
12	0.2	1	0
14	0.9	1	0.1
16	0.6	1	0.3
18	0.1	1	0.9
20	0.7	1	1
22	1.2	1	1
24	0.2	1	1

Data from L. J. Morrison et al., Probabilistic order in antigenic variation of *Trypanosoma brucei*, *International Journal for Parasitology* 35:961–972 (2005); and L. J. Morrison et al., Antigenic variation in the African trypanosome: molecular mechanisms and phenotypic complexity, *Cellular Microbiology* 1:1724–1734 (2009).

Data from the Studies Study A measured the abundance of parasites in the blood of one human patient during the first few weeks of a chronic infection. The results are shown in the second column of the data table. Many decades after scientists first observed the pattern of *Trypanosoma* abundance over the course of infection, researchers identified antibodies specific to different forms of the parasite's surface glycoprotein. The third and fourth columns of the table list the relative abundance of two such antibodies during the early period of chronic infection, using an index ranging from 0 (absent) to 1.

Part A: INTERPRET THE DATA

1. Which of the first two columns represents the independent variable and which the dependent variable? Plot the data from Study A as a line graph, putting the independent variable on the x-axis. (For additional information about graphs, see the Scientific Skills Review in Appendix D.)

2. Visually displaying data in a graph can help make patterns in the data more noticeable. Describe any patterns revealed by your graph.

3. Assume that a drop in parasite abundance reflects an effective immune response by the host. Formulate a hypothesis to explain the pattern you described in question 2.

Part B: INTERPRET THE DATA

4. Note that the data for Study B were collected over the same period of infection (days 4–24) as the parasite abundance data in Study A. Therefore, you can incorporate these new data into your first graph, using the same x-axis. However, since the antibody level data are measured in a different way than the parasite abundance data, add a second set of y-axis labels on the right side of your graph. Then, using different colors or sets of symbols, add the data for the two antibody types. Labeling the y-axis two different ways enables you to compare how two dependent variables change relative to a shared independent variable.

5. Describe any patterns you observe by comparing the two data sets over the same period. Do these patterns support your hypothesis from question 3? Do they prove it? Explain.

6. Scientists can now also distinguish the abundance of trypanosomes recognized specifically by antibodies type A and type B. How would incorporating such information change your graph?

➔ **Instructors:** A version of this Scientific Skills Exercise can be assigned in **Mastering Biology**.

replicates in one human host after another, the human flu virus undergoes frequent mutations. As a result, a new flu vaccine must be developed, produced, and distributed each year. In addition, the human flu virus occasionally forms new strains by exchanging genes with influenza viruses that infect domesticated animals, such as pigs or chickens. When this exchange of genes occurs, the new strain may not be recognized by any of the memory cells in the human population. The resulting outbreak can be deadly: The 1918–1919 influenza outbreak killed more than 20 million people.

Latency

Some viruses avoid an immune response by infecting cells and then entering a largely inactive state called *latency*. In latency, the production of most viral proteins and free viruses ceases; as a result, latent viruses do not trigger an adaptive immune response. Nevertheless, the viral genome persists in the nuclei of infected cells, either as a separate DNA molecule or as a copy integrated into the host genome. Latency typically persists until conditions arise that are favorable for viral transmission or unfavorable for host survival, such as when the host is infected by another pathogen. Such circumstances trigger the synthesis and release of free viruses that can infect new hosts.

Herpes simplex viruses provide a good example of latency. A number of these herpesviruses infect only humans and cause symptoms ranging from mild to life-threatening **(Table 43.1)**. The type 1 herpesvirus causes most oral herpes infections (often inaccurately called "cold" sores), whereas the sexually transmitted type 2 herpesvirus is responsible for most cases of genital herpes. People infected with either the type 1 or type 2 herpesvirus often have no symptoms. Instead, these viruses remain latent in certain neurons until a stimulus such as fever, emotional stress, or a hormonal change associated with the menstrual cycle reactivates the virus. Activation of the type 1 herpesvirus can result in

the appearance of blisters around the mouth. Infections of the type 2 herpesvirus pose a serious threat to the babies of infected mothers and can increase transmission of HIV.

Attack on the Immune System: HIV

The **human immunodeficiency virus (HIV)**, the pathogen that causes AIDS, both escapes and attacks the adaptive immune response. HIV infects helper T cells with high efficiency by binding specifically to the CD4 accessory protein. HIV also infects some cell types that have low levels of CD4, such as macrophages and brain cells. Inside cells, the HIV RNA genome is reverse-transcribed, and the product DNA is integrated into the host cell's genome (see Figure 19.8). In this form, the viral genome can direct the production of new viruses.

Although the body responds to HIV with an immune response sufficient to eliminate most viral infections, some HIV invariably escapes. One reason HIV persists is that it has a very high mutation rate. Altered proteins on the surface of some mutated viruses reduce interaction with antibodies and cytotoxic T cells. Such viruses replicate and mutate further. HIV thus evolves within the body.

Over time, an untreated HIV infection not only avoids the adaptive immune response but also abolishes it **(Figure 43.28)**. Viral replication and cell death triggered by the virus lead to loss of helper T cells, impairing both humoral and cell-mediated immune responses. The eventual result is **acquired immunodeficiency syndrome (AIDS)**, an impairment in immune responses that leaves the body susceptible to infections and cancers that a healthy immune system would usually defeat. For example, *Pneumocystis jirovecii*, a common fungus that does not cause disease in

▼ **Figure 43.28 The progress of an untreated HIV infection.**

Mastering Biology
Interview with Flossie Wong-Staal: Characterizing HIV

Table 43.1 Latency as a Shared Characteristic of Human Herpesviruses		
Human Herpesvirus	**Main Sites of Latency**	**Associated Diseases or Disorders**
Herpes simplex virus-1 (HSV-1)	Clusters of neurons in spinal nerves	Cold sores
Herpes simplex virus-2 (HSV-2)	Clusters of neurons in spinal nerves	Genital ulcers
Varicella-zoster virus (VZV)	Clusters of neurons in spinal nerves	Chicken pox, shingles
Epstein-Barr virus (EBV)	Memory B cells	Some lymphoma, mononucleosis
Cytomegalovirus (CMV)	Monocytes and lymphocytes	Abnormal fetal development
Human herpesvirus 8 (HHV-8)	B cells	Kaposi's sarcoma

healthy individuals, can result in severe pneumonia in people with AIDS. Such opportunistic diseases, as well as nerve damage and body wasting, are the primary causes of death from AIDS, rather than HIV itself.

Transmission of HIV requires the transfer of virus particles or infected cells from person to person via body fluids such as semen, blood, or breast milk. Unprotected sex (that is, without using a condom) and transmission via HIV-contaminated needles (often among intravenous drug users) cause the vast majority of HIV infections. The virus can enter the body through mucosal linings of the vagina, vulva, penis, or rectum during intercourse or via the mouth during oral sex. People infected with HIV can transmit the disease in the first few weeks of infection, *before* they express HIV-specific antibodies that can be detected in a blood test. Although no cure has been found for HIV infection, drugs that can significantly slow HIV replication and the progression to AIDS are available.

⊙ **Mastering Biology** Animation: HIV Reproductive Cycle

Cancer and Immunity

When adaptive immunity is inactivated, the frequency of certain cancers increases dramatically. For example, the risk of developing Kaposi's sarcoma is 20,000 times greater for patients with untreated AIDS than for people who are healthy. This observation was initially puzzling. If the immune system recognizes only nonself, it should fail to recognize the uncontrolled growth of self cells that is the hallmark of cancer. It turns out, however, that viruses are involved in about 15–20% of all human cancers. Because the immune system can recognize viral proteins as foreign, it can act as a defense against viruses that can cause cancer and against cancer cells that harbor viruses.

Scientists have identified six viruses that can cause cancer in humans. The Kaposi's sarcoma herpesvirus is one such virus (see Table 43.1). Hepatitis B virus, which can trigger liver cancer, is another. A vaccine introduced in 1986 for hepatitis B virus was the first vaccine shown to help prevent a specific human cancer. Rapid progress on developing vaccines for virus-induced cancers continues. In 2006, the release of a vaccine specific for human papillomavirus (HPV) marked a major victory against cervical cancer, as well as oral cancers that are becomingly increasingly common among men. The computer graphic image of an HPV particle in **Figure 43.29** illustrates the abundant copies of the capsid protein (yellow) that is used as the antigen in vaccination.

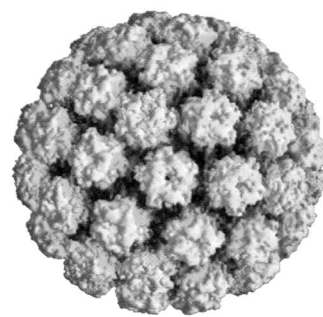
▼ **Figure 43.29 Human papillomavirus.**

⊙ **Mastering Biology**
Interview with Harald zur Hausen:
Discovering the cause of cervical cancer

CONCEPT CHECK 43.4

1. In the condition known as myasthenia gravis, antibodies bind to and block certain receptors on muscle cells, preventing muscle contraction. What type of disorder is myasthenia gravis?

2. People with herpes simplex type 1 viruses often get mouth sores when they have a cold or similar infection. How might this location benefit the virus?

3. **WHAT IF?** How would a macrophage deficiency likely affect a person's innate and adaptive defenses?

For suggested answers, see Appendix A.

43 Chapter Review

⊙ Go to **Mastering Biology** for Assignments, the eText, the Study Area, and Dynamic Study Modules.

SUMMARY OF KEY CONCEPTS

⊙ To review key terms, go to the **Vocab Self-Quiz** in the **Mastering Biology** eText or Study Area, or go to goo.gl/zkjz9t.

CONCEPT 43.1

In innate immunity, recognition and response rely on traits common to groups of pathogens
(pp. 953–957)

• **Innate immunity** is mediated by physical and chemical barriers as well as cell-based defenses. Recognition proteins in innate immunity are specific for broad classes of **pathogens**. Pathogens that penetrate barrier defenses are ingested by phagocytic cells,

which in vertebrates include **macrophages** and **dendritic cells**. Additional cellular defenses include **natural killer cells**, which can induce the death of virus-infected cells. **Complement system** proteins, **interferons**, and other antimicrobial peptides also act against pathogens. In the **inflammatory response**, **histamine** and other chemicals that are released at the injury site promote changes in blood vessels that enhance immune cell access and action.

• Pathogens sometimes evade innate immune defenses. For example, some bacteria have an outer capsule that prevents recognition, while others are resistant to breakdown within lysosomes.

❓ *In what ways does innate immunity protect the mammalian digestive tract?*

CONCEPT 43.2

In adaptive immunity, receptors provide pathogen-specific recognition (pp. 957–963)

- **Adaptive immunity** relies on two types of **lymphocytes: B cells** and **T cells**. Lymphocytes have cell-surface **antigen receptors** for foreign molecules (**antigens**). All receptor proteins on a single B or T cell are the same, but there are millions of B and T cells in the body that differ in the foreign molecules that their receptors recognize. Upon infection, B and T cells specific for the pathogen are activated. Some T cells help other lymphocytes; others kill infected host cells. B cells called **plasma cells** produce soluble proteins called **antibodies**, which bind to foreign molecules and cells. B and T **memory cells** defend against future infections by the same pathogen.

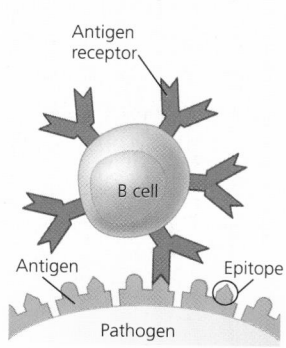

- Recognition of foreign molecules by B cells and T cells involves the binding of variable regions of receptors to an **epitope**, a small region of an antigen. B cells and antibodies recognize epitopes on the surface of antigens circulating in the blood or lymph. T cells recognize protein epitopes in small antigen fragments (peptides) that are presented on the surface of host cells by **major histocompatibility complex (MHC) molecules**. This interaction activates a T cell, enabling it to participate in adaptive immunity.

- The four major characteristics of B and T cell development are the generation of cell diversity, self-tolerance, proliferation, and immunological memory. Proliferation and memory are both based on **clonal selection**, illustrated here for B cells:

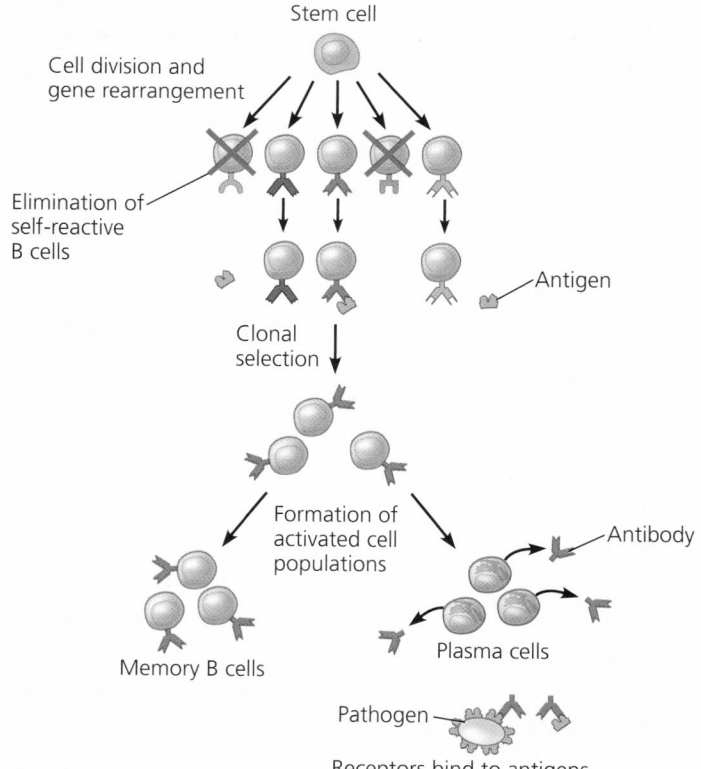

? *Why is the adaptive immune response to an initial infection slower than the innate response?*

CONCEPT 43.3

Adaptive immunity defends against infection of body fluids and body cells (pp. 963–970)

- **Helper T cells** interact with antigen fragments displayed by class II MHC molecules on the surface of **antigen-presenting cells**: dendritic cells, macrophages, and B cells. Activated helper T cells secrete cytokines that stimulate other lymphocytes. In the **humoral immune response**, antibodies help eliminate antigens by facilitating phagocytosis and complement-mediated lysis. In the **cell-mediated immune response**, activated **cytotoxic T cells** trigger destruction of infected cells.

- **Active immunity** develops in response to infection or to **immunization**. The transfer of antibodies in **passive immunity** provides immediate, short-term protection.

- Tissues or cells transferred from one person to another are subject to immune rejection. In tissue grafts and organ transplants, MHC molecules stimulate rejection. Lymphocytes in bone marrow transplants may cause a graft-versus-host reaction.

? *Is immunological memory after a natural infection fundamentally different from immunological memory after vaccination? Explain.*

CONCEPT 43.4

Disruptions in immune system function can elicit or exacerbate disease (pp. 970–974)

- In allergies, such as hay fever, the interaction of antibodies and allergens triggers immune cells to release histamine and other mediators that cause vascular changes and allergic symptoms. Loss of self-tolerance can lead to **autoimmune diseases**, such as multiple sclerosis. Inborn immunodeficiencies result from defects that interfere with innate, humoral, or cell-mediated defenses. **AIDS** is an acquired immunodeficiency caused by **HIV**.

- Antigenic variation, latency, and direct assault on the immune system allow some pathogens to thwart immune responses.

? *Is being infected with HIV the same as having AIDS? Explain.*

TEST YOUR UNDERSTANDING

→ For more multiple-choice questions, go to the **Practice Test** in the **Mastering Biology** eText or Study Area, or go to goo.gl/GruWRg.

Levels 1-2: Remembering/Understanding

1. Which of these is absent from insect immunity?
   (A) antibacterial digestive enzymes
   (B) activation of natural killer cells
   (C) phagocytosis by hemocytes
   (D) production of antimicrobial peptides

2. An epitope associates with which part of an antigen receptor or antibody?
   (A) the tail
   (B) the heavy-chain constant regions only
   (C) variable regions of a heavy chain and light chain combined
   (D) the light-chain constant regions only

3. Which statement best describes the difference between responses of effector B cells (plasma cells) and those of cytotoxic T cells?
   (A) B cells confer active immunity; cytotoxic T cells confer passive immunity.
   (B) B cells respond the first time a pathogen is present; cytotoxic T cells respond subsequent times.
   (C) B cells secrete antibodies against a pathogen; cytotoxic T cells kill pathogen-infected host cells.
   (D) B cells carry out the cell-mediated response; cytotoxic T cells carry out the humoral response.

Levels 3-4: Applying/Analyzing

4. Which of the following statements is true?
   (A) An antibody has one antigen-binding site.
   (B) A lymphocyte has receptors for a single antigen.
   (C) Every antigen has a single epitope.
   (D) A liver or muscle cell makes two classes of MHC molecule.

5. Which of the following should be the same in identical twins?
   (A) the set of antibodies produced
   (B) the set of MHC molecules produced
   (C) the set of T cell antigen receptors produced
   (D) the set of immune cells eliminated as self-reactive

Levels 5-6: Evaluating/Creating

6. Vaccination increases the number of
   (A) different receptors that recognize a pathogen.
   (B) lymphocytes with receptors that can bind to the pathogen.
   (C) epitopes that the immune system can recognize.
   (D) MHC molecules that can present an antigen.

7. Which of the following is least likely to help a virus avoid triggering an adaptive immune response?
   (A) having frequent mutations in genes for surface proteins
   (B) infecting cells that produce very few MHC molecules
   (C) producing proteins very similar to those of other viruses
   (D) infecting and killing helper T cells

8. **DRAW IT** Consider a pencil-shaped protein with two epitopes, Y (the "eraser" end) and Z (the "point" end). They are recognized by antibodies A1 and A2, respectively. Draw and label a picture showing the antibodies linking proteins into a complex that could trigger endocytosis by a macrophage.

9. **MAKE CONNECTIONS** Contrast clonal selection with Lamarck's idea for the inheritance of acquired characteristics (see Concept 22.1).

10. **EVOLUTION CONNECTION** Describe one invertebrate mechanism of defense against pathogens and discuss how it is an evolutionary adaptation retained in vertebrates.

11. **SCIENTIFIC INQUIRY** A major cause of septic shock is the presence in blood of lipopolysaccharide (LPS) from bacteria. Suppose you have available purified LPS and several strains of mice, each with a mutation that inactivates a particular TLR gene. Explain how you might use these mice to test the feasibility of treating septic shock with a drug that blocks TLR signaling.

12. **WRITE ABOUT A THEME: INFORMATION** Among all nucleated body cells, only B and T cells lose DNA during their development and maturation. In a short essay (100–150 words), discuss the relationship between this loss and DNA as heritable biological information, focusing on similarities between cellular and organismal generations.

13. **SYNTHESIZE YOUR KNOWLEDGE**

This child is receiving an oral vaccine against polio, a disease caused by a virus that infects neurons. Given that the body cannot replace most neurons, why must a polio vaccine stimulate both a cell-mediated and a humoral response?

For selected answers, see Appendix A.

Explore Scientific Papers with **Science in the Classroom** Ⲛ AAAS
How do immunizations affect deaths from infectious disease?
Go to "One Vaccine to Rule Them All"
at www.scienceintheclassroom.org.
→ **Instructors:** Questions can be assigned in **Mastering Biology**.

44 Osmoregulation and Excretion

KEY CONCEPTS

44.1 Osmoregulation balances the uptake and loss of water and solutes *p. 978*

44.2 An animal's nitrogenous wastes reflect its phylogeny and habitat *p. 982*

44.3 Diverse excretory systems are variations on a tubular theme *p. 983*

44.4 The nephron is organized for stepwise processing of blood filtrate *p. 987*

44.5 Hormonal circuits link kidney function, water balance, and blood pressure *p. 994*

Study Tip

Make a table: As you read Concept 44.2, complete a table like the one shown to help organize information about the major forms of nitrogenous waste. In the first three rows, use the terms "high," "medium," or "low" to describe the relevant properties of these forms.

Waste attribute	Waste product		
	Ammonia	Urea	Uric acid
Toxicity			
Energy cost to produce			
Water loss during excretion			
Example of organism excreting this product			

→ Go to Mastering Biology

For Students (in eText and Study Area)
• Get Ready for Chapter 44
• Figure 44.14 Walkthrough: How the Human Kidney Concentrates Urine
• BioFlix® Animation: Membrane Transport

For Instructors to Assign (in Item Library)
• Tutorial: Kidney Structure and Function
• Scientific Skills Exercise: Describing and Interpreting Quantitative Data

Figure 44.1 At 3.5 m, the wingspan of a wandering albatross (*Diomedea exulans*) is the largest of any living bird. This massive bird remains at sea day and night throughout the year, returning to land only to reproduce. Despite drinking only seawater, the albatross keeps its body salts and water in balance.

How do animals regulate water and salt concentrations in their tissues?

Lacking a means to pump water across cell membranes, animals transport salts and thereby direct water movement into or out of cells by **osmosis**, in which water undergoes net diffusion from an area of higher free H_2O concentration (lower solute concentration) to an area of lower free H_2O concentration (higher solute concentration).

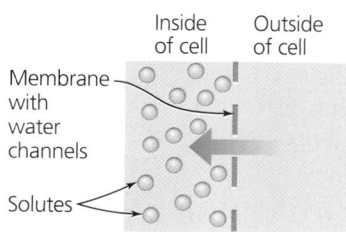

Cell is **hyperosmotic** to surroundings; **net flow of water** into cell

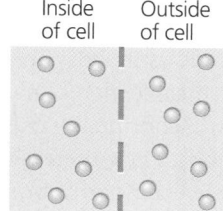

Cell is **isoosmotic** to surroundings; no net water movement

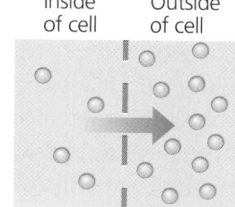

Cell is **hypoosmotic** to surroundings; **net flow of water** out of cell

Cell lyses

Cell in osmotic balance

Cell shrivels

As you'll read, excretory systems not only maintain salt and water balance, but also rid the body of **nitrogen-containing wastes**.

For both an albatross and a human, maintaining fluid balance requires that the relative concentrations of water and solutes be kept within narrow limits. In addition, ions such as sodium and calcium must be kept at concentrations that permit normal activity of muscles, neurons, and other body cells. Homeostasis thus requires **osmoregulation**, the general term for the processes by which animals control solute concentrations and balance water gain and loss.

In safeguarding their internal fluids, animals must also deal with ammonia, a toxic metabolite produced by the dismantling of *nitrogenous* (nitrogen-containing) molecules, chiefly proteins and nucleic acids. Several mechanisms have evolved for ridding the body of nitrogenous metabolites and other metabolic waste products, a process called **excretion**. Because systems for osmoregulation and excretion are structurally and functionally linked in many animals, we'll consider both of these processes in this chapter.

CONCEPT 44.1

Osmoregulation balances the uptake and loss of water and solutes

Osmoregulation requires controlled movement of water and solutes across plasma membrane. Ultimately, the driving force for the movement of both water and solutes is a concentration gradient of one or more solutes across the membrane.

Osmosis and Osmolarity

Water enters and leaves cells by osmosis, which occurs when two solutions separated by a membrane differ in total solute concentration. The unit of measurement for solute concentration is **osmolarity**, the number of moles of solute per liter of solution. The osmolarity of human blood is about 300 milliosmoles per liter (mOsm/L), whereas that of seawater is about 1,000 mOsm/L.

Two solutions with the same osmolarity are said to be *isoosmotic*. If a selectively permeable membrane separates the solutions, water molecules will continually cross the membrane at equal rates in both directions. Thus, there is no *net* movement of water by osmosis between isoosmotic solutions. When two solutions differ in osmolarity, the solution with the higher concentration of solutes is said to be *hyperosmotic*, and the more dilute solution is said to be *hypoosmotic*. Water flows by osmosis from a hypoosmotic solution to a hyperosmotic one, thus reducing the concentration difference for both solutes and free water (see the diagram on the first page of the chapter).

In this chapter, we use the terms *isoosmotic, hypoosmotic,* and *hyperosmotic*, which refer specifically to osmolarity, instead of *isotonic, hypotonic,* and *hypertonic*. The latter set of terms applies to the response of animal cells—whether they swell or shrink—in solutions of known solute concentrations.

Osmoregulatory Challenges and Mechanisms

An animal can maintain water balance in two ways. One is to be an **osmoconformer**: to be isoosmotic with its surroundings. All osmoconformers are marine animals. Because an osmoconformer's internal osmolarity is the same as that of its environment, there is no tendency to gain or lose water. Many osmoconformers live in water that has a stable composition and hence have a constant internal osmolarity.

The second way to maintain water balance is to be an **osmoregulator**: to control internal osmolarity independent of that of the external environment. In a hypoosmotic environment, an osmoregulator must discharge excess water. In a hyperosmotic environment, it must instead take in water to offset osmotic loss. Osmoregulation enables animals to live in environments that are uninhabitable for osmoconformers, such as freshwater and terrestrial habitats, or to move between marine and freshwater environments **(Figure 44.2)**.

Whether osmoconformers or osmoregulators, most animals cannot tolerate substantial changes in external osmolarity and are said to be *stenohaline* (from the Greek *stenos*, narrow, and *halos*, salt). In contrast, *euryhaline* animals (from the Greek *eurys*, broad) can survive large fluctuations in external osmolarity. Euryhaline osmoconformers include barnacles and mussels in estuaries that are alternately exposed to fresh and salt water; euryhaline osmoregulators include striped bass and the various species of salmon (see Figure 44.2).

Next, we'll examine adaptations for osmoregulation that have evolved in marine, freshwater, and terrestrial animals.

Marine Animals

Most marine invertebrates are osmoconformers. Their osmolarity is the same as that of seawater. Therefore, they face no

▼ **Figure 44.2 Sockeye salmon (*Oncorhynchus nerka*), osmoregulators that migrate between rivers and the ocean.**

substantial challenges in water balance. Nevertheless, they actively transport *specific* solutes that they maintain at levels different from those in the ocean. For example, homeostatic mechanisms in the Atlantic lobster (*Homarus americanus*) maintain a magnesium ion (Mg^{2+}) concentration in hemolymph (circulatory fluid) of less than 9 mM (millimolar, or 10^{-3} mol/L), far below the 50 mM concentration of Mg^{2+} in their environment.

Two osmoregulatory strategies evolved among marine vertebrates that address the challenges of a strongly dehydrating environment. One is found among marine "bony fishes," a group that includes ray-finned and lobe-finned fishes. The other is found in marine sharks and most other chondrichthyans (cartilaginous animals; see Concept 34.3).

Cod, shown in **Figure 44.3a**, and other marine bony fishes constantly lose water by osmosis. They balance water loss by drinking a lot of seawater. The excess salts ingested with seawater are eliminated through the gills and kidneys.

Like bony fishes, sharks have an internal salt concentration much lower than that of seawater. However, shark tissues have a high concentration of several other solutes, including urea and another organic molecule, trimethylamine oxide (TMAO). Because the total solute concentration is somewhat higher than 1,000 mOsm/L, water slowly *enters* the shark's body by osmosis and in food (sharks do not drink).

The small influx of water into the shark's body is disposed of in urine produced by the kidneys. The urine also removes some of the salt that diffuses into the shark's body; the rest is lost in feces or is secreted from a specialized gland.

Freshwater Animals

The osmoregulatory problems of freshwater animals are the opposite of those of marine animals. The body fluids of freshwater animals must be hyperosmotic because animal cells cannot tolerate salt concentrations as low as that of lake or river water. Since freshwater animals have internal fluids with an osmolarity higher than that of their surroundings, they face the problem of gaining water by osmosis. For the perch and many other freshwater animals, water balance relies on excreting large amounts of very dilute urine and drinking almost no water **(Figure 44.3b)**. In addition, salts lost by diffusion and in the urine are replenished by eating and by taking up salt across their gills.

Salmon and other euryhaline fishes that migrate between fresh water and seawater undergo dramatic changes in osmoregulatory status. When living in rivers and streams, salmon osmoregulate like other freshwater fishes, producing large amounts of dilute urine and taking up salt from the dilute environment through their gills. When they migrate to the ocean, salmon acclimatize (see Concept 40.2). They produce more of the steroid hormone cortisol, which increases the number and size of specialized salt-secreting cells. These and other physiological changes enable salmon in salt water to excrete excess salt from their gills and produce only small amounts of urine—just like bony fishes that spend their entire lives in salt water.

Animals That Live in Temporary Waters

Extreme dehydration, or *desiccation*, is fatal for most animals. However, a few aquatic invertebrates that live in temporary ponds and in films of water around soil particles can lose almost all their body water and survive. These animals enter a dormant state when their habitats dry up, an adaptation called **anhydrobiosis** ("life without water"). Among the most striking examples are the tardigrades, or water bears, tiny invertebrates less than

▼ Figure 44.3 **Osmoregulation in marine and freshwater bony fishes: a comparison.**

(a) Osmoregulation in a marine fish

Gain of water and salt ions from food

Excretion of salt ions from gills

Osmotic water **loss** through gills and other parts of body surface

Gain of water and salt ions from drinking seawater

Excretion of salt ions and small amounts of water in scanty urine from kidneys

SALT WATER

Key
⇨ Water
➡ Salt

(b) Osmoregulation in a freshwater fish

Gain of water and some ions in food

Uptake of salt ions by gills

Osmotic water **gain** through gills and other parts of body surface

FRESH WATER

Excretion of salt ions and large amounts of water in dilute urine from kidneys

▼ **Figure 44.4 Anhydrobiosis.** When hydrated, tardigrades (SEM images) inhabit temporary ponds as well as droplets of water in soil and on moist plants.

50 µm

Hydrated tardigrade

Dehydrated tardigrade

1 mm long **(Figure 44.4)**. In their active, hydrated state, they contain about 85% water by weight, but they can dehydrate to less than 2% water and survive in an inactive state, dry as dust, for a decade or more. Just add water, and within hours the rehydrated tardigrades are moving about and feeding.

Anhydrobiosis requires adaptations that keep cell membranes intact. Researchers are just beginning to learn how tardigrades survive drying out, but studies of anhydrobiotic roundworms (phylum Nematoda; see Concept 33.4) show that desiccated individuals contain large amounts of sugars. In particular, a disaccharide called trehalose seems to protect the cells by replacing the water that is normally associated with proteins and membrane lipids. Many insects that survive freezing in the winter also use trehalose as a membrane protectant, as do some plants resistant to desiccation.

Recently, scientists began applying lessons learned from the study of anhydrobiosis to the preservation of biological materials. Traditionally, samples of protein, DNA, and cells have been kept in ultracold freezers ($-80°C$), consuming large amounts of energy. Today, the manufacture of materials modeled after the protectants of anhydrobiotic species has enabled room temperature storage of such samples.

Land Animals

The threat of dehydration is a major regulatory problem for terrestrial plants and animals. Adaptations that reduce water loss are key to survival on land. Much as a waxy cuticle contributes to the success of land plants, the body coverings of most terrestrial animals help prevent dehydration. Examples are the waxy layers of insect exoskeletons, the shells of land snails, and the layers of dead, keratinized skin cells covering most terrestrial vertebrates, including humans.

Many terrestrial animals, especially desert-dwellers, are nocturnal, which reduces evaporative water loss because of the lower temperature and higher humidity of night air.

Despite anatomical and behavioral adaptations that conserve water, most terrestrial animals lose water through a variety of routes: in urine and feces, across the skin, and from the epithelial surfaces of gas exchange organs and airways. Land animals maintain water balance by drinking and eating moist foods and by producing water metabolically through cellular respiration.

A number of desert animals are well enough adapted for minimizing water loss that they can survive for long periods of time without drinking. Camels, for example, tolerate a 7°C rise in body temperature, greatly reducing the amount of water lost in sweat production. They can also lose 25% of their body water and survive. (In contrast, a human who loses half this fraction of body water will die from heart failure.) In the **Scientific Skills Exercise**, you can examine water balance in another desert species, the sandy inland mouse.

Energetics of Osmoregulation

Maintaining an osmolarity difference between an animal's body and its external environment carries an energy cost. Because diffusion tends to equalize concentrations in a system, osmoregulators must expend energy to maintain the osmotic gradients that cause water to move in or out. They do so by using active transport to manipulate solute concentrations in their body fluids.

The energy cost of osmoregulation depends on how different an animal's osmolarity is from its surroundings, how easily water and solutes can move across the animal's surface, and how much work is required to pump solutes across the membrane. Osmoregulation accounts for 5% or more of the resting metabolic rate of many fishes. For brine shrimp, small crustaceans that live in extremely salty lakes, the gradient between internal and external osmolarity is very large, and the cost of osmoregulation is correspondingly high—as much as 30% of the resting metabolic rate.

The energy cost to an animal of maintaining water and salt balance is minimized by having body fluids that are adapted to the salinity of the animal's habitat. Thus, the body fluids of most animals that live in fresh water (which has an osmolarity of 0.5–15 mOsm/L) have lower solute concentrations than the body fluids of their closest relatives that live in seawater (1,000 mOsm/L). For instance, whereas marine molluscs have body fluids with solute concentrations of approximately 1,000 mOsm/L, some freshwater molluscs maintain the osmolarity of their body fluids at just 40 mOsm/L. In each case, minimizing the osmotic difference between body fluids and the surrounding environment decreases the energy cost of osmoregulation.

Describing and Interpreting Quantitative Data

How Do Desert Mice Maintain Osmotic Homeostasis?

The sandy inland mouse (now known scientifically as *Pseudomys hermannsburgensis*) is an Australian desert mammal that can survive indefinitely on a diet of dried seeds without drinking water. To study this species' adaptations to its arid environment, researchers conducted a laboratory experiment in which they controlled access to water. In this exercise, you will analyze some of the data from the experiment.

How the Experiment Was Done Nine captured mice were kept in an environmentally controlled room and given birdseed (10% water by weight) to eat. In part A of the study, the mice had unlimited access to tap water for drinking; in part B of the study, the mice were not given any drinking water for 35 days, similar to conditions in their natural habitat. At the end of parts A and B, the researchers measured the osmolarity and urea concentration of the urine and blood of each mouse. The mice were also weighed three times a week.

Data from the Experiment

Access to Water	Mean Osmolarity (mOsm/L)		Mean Urea Concentration (m*M*)	
	Urine	Blood	Urine	Blood
Part A: Unlimited	490	350	330	7.6
Part B: None	4,700	320	2,700	11

In part A, the mice drank about 33% of their body weight each day. The change in body weight during the study was negligible for all mice.

INTERPRET THE DATA

1. In words, describe how the data differ between the unlimited water and no-water conditions for the following: **(a)** osmolarity of urine, **(b)** osmolarity of blood, **(c)** urea concentration in urine, **(d)** urea concentration in blood. **(e)** Does this data set provide evidence of homeostatic regulation? Explain.

2. **(a)** Calculate the ratio of urine osmolarity to blood osmolarity for mice with unlimited access to water. **(b)** Calculate this ratio for mice with no access to water. **(c)** What conclusion can you draw from these ratios?

3. If you learned that the amount of urine produced in part A were different from that in part B, how would that affect your calculation? Explain.

→ **Instructors:** A version of this Scientific Skills Exercise can be assigned in **Mastering Biology**.

Data from R. E. MacMillen et al., Water economy and energy metabolism of the sandy inland mouse, *Leggadina hermannsburgensis*, *Journal of Mammalogy* 53:529–539 (1972).

Transport Epithelia in Osmoregulation

Although the ultimate function of osmoregulation is to control solute concentrations in cells, most animals do this indirectly, managing instead the solute content of the body fluid that bathes the cells. In insects and other animals with an open circulatory system, the fluid surrounding cells is hemolymph. In vertebrates and other animals with a closed circulatory system, the cells are bathed in an interstitial fluid that contains a mixture of solutes controlled indirectly by the blood. Maintaining the composition of such fluids depends on structures ranging from individual cells that regulate solute movement to complex organs such as the vertebrate kidney.

In most animals, osmoregulation and metabolic waste disposal rely on **transport epithelia**—one or more layers of epithelial cells specialized for moving particular solutes in controlled amounts in specific directions. Transport epithelia are typically arranged into tubular networks with extensive surface areas. Some transport epithelia face the outside environment directly, whereas others line channels connected to the outside by an opening on the body surface.

The transport epithelium that enables the albatross and other marine birds to survive on seawater remained undiscovered for many years. To find it, researchers gave captive marine birds only seawater to drink. Although very little salt appeared in the birds' urine, tests revealed that fluid dripping from the tip of their beaks was a concentrated solution of salt (NaCl). The source of this solution turned out to be a pair of nasal salt glands packed with transport epithelia **(Figure 44.5)**. Salt glands, which are also found in sea turtles and marine iguanas, use active transport of ions to secrete a fluid much saltier than the ocean. Even though drinking seawater brings in a lot of salt, the salt gland enables these marine vertebrates to achieve a net gain of water. By contrast, humans who drink a given volume of seawater must use a *greater* volume of water to excrete the salt load, with the result that they become dehydrated.

Transport epithelia that function in maintaining water balance also often function in disposal of metabolic wastes. We'll see examples of this coordinated function in our upcoming consideration of earthworm and insect excretory systems as well as the vertebrate kidney.

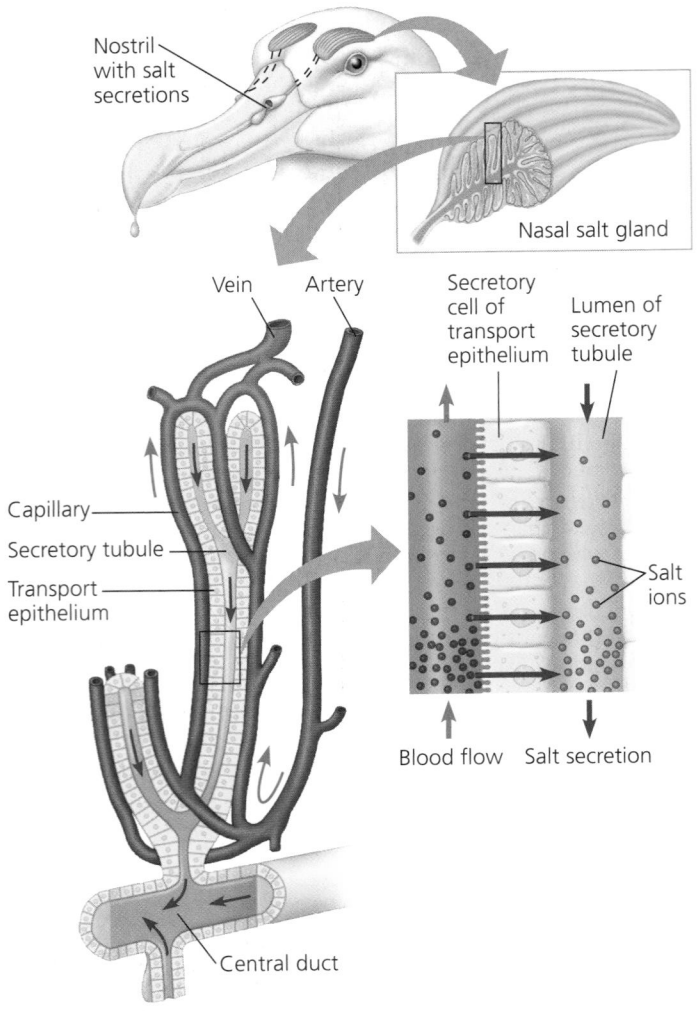

▼ **Figure 44.5 Salt secretion in the nasal glands of a marine bird.** A transport epithelium moves salt from the blood into secretory tubules, which drain into central ducts leading to the nostrils.

Nostril with salt secretions

Nasal salt gland

Vein   Artery

Secretory cell of transport epithelium

Lumen of secretory tubule

Capillary

Secretory tubule

Transport epithelium

Salt ions

Blood flow   Salt secretion

Central duct

CONCEPT CHECK 44.1

1. The movement of salt from the surrounding water to the blood of a freshwater fish requires the expenditure of energy in the form of ATP. Why?

2. Why aren't any freshwater animals osmoconformers?

3. **WHAT IF?** Researchers found that a camel in the sun required much more water when its fur was shaved off, although its body temperature was the same. What can you conclude about the relationship between osmoregulation and the insulation provided by fur?

For suggested answers, see Appendix A.

CONCEPT 44.2

An animal's nitrogenous wastes reflect its phylogeny and habitat

In regulating and safeguarding their internal fluids, animals must deal with **ammonia**, a toxic metabolite produced by the dismantling of *nitrogenous* (nitrogen-containing) molecules,

chiefly proteins and nucleic acids. Several mechanisms have evolved for ridding the body of ammonia and other metabolic waste products, a process called **excretion**. Because most metabolic wastes must be dissolved in water to be excreted from the body, the type and quantity of an animal's waste products may have a large impact on its water balance.

Forms of Nitrogenous Waste

Although some animal species excrete ammonia directly, others excrete alternative forms of nitrogenous waste, either urea or uric acid **(Figure 44.6)**. These different forms vary significantly in their toxicity, their solubility, and the energy costs of producing them.

Ammonia

Ammonia is very toxic, in part because its ion, ammonium (NH_4^+), can interfere with oxidative phosphorylation. Because ammonia can be tolerated only at very low concentrations, animals that excrete ammonia need access to lots of water. Therefore, ammonia excretion is most common in aquatic species. The highly soluble ammonia molecules, which interconvert between NH_3 and NH_4^+, easily pass through membranes and are readily lost by diffusion to the surrounding water. In many invertebrates, ammonia release occurs across the whole body surface.

▼ **Figure 44.6 Variations in forms of nitrogenous waste among animal species.**

Proteins

Nucleic acids

Breakdown of nitrogen-containing macromolecules

Amino acids

Nitrogenous bases

Removal of nitrogen-containing amino group

—NH_2 (amino groups)

Conversion to nitrogenous waste

NH_3

Ammonia

Most aquatic animals, including most bony fishes

Urea

Mammals, most amphibians, sharks, some bony fishes

Uric acid

Birds and many other reptiles, insects, land snails

Urea

Although ammonia excretion works well in many aquatic species, it is much less suitable for land animals. Ammonia is so toxic that it can be safely transported through and excreted from the body only in large volumes of very dilute solutions. Most terrestrial animals and many marine species simply do not have access to sufficient water to routinely excrete ammonia. Instead, they mainly excrete a different nitrogenous waste, **urea**. In vertebrates, urea is the product of an energy-consuming metabolic cycle that combines ammonia with carbon dioxide in the liver.

The main advantages of urea for nitrogenous waste excretion are its very low toxicity and its high solubility in water. The main disadvantage is its energy cost: Animals must expend energy to produce urea from ammonia. From a bioenergetic standpoint, we would predict that animals that spend part of their lives in water and part on land would switch between excreting ammonia (thereby saving energy) and excreting urea (reducing excretory water loss). Indeed, many amphibians excrete mainly ammonia when they are aquatic tadpoles and switch largely to urea excretion when they become land-dwelling adults.

Uric Acid

Insects, land snails, and many reptiles, including birds, excrete **uric acid** as their primary nitrogenous waste **(Figure 44.7)**. Uric acid is relatively nontoxic and does not readily dissolve in water. It therefore can be excreted as a semisolid paste with very little water loss. However, uric acid is even more energetically expensive than urea, requiring considerable ATP for synthesis from ammonia.

Although they are not primarily uric acid producers, humans and some other animals generate a small amount of uric acid from metabolism. Diseases that alter this process reflect problems that can arise when a metabolic product is insoluble. For example, a genetic defect predisposes Dalmatian dogs to form uric acid stones in their bladder. In humans, adult males are particularly susceptible to *gout*, a painful joint inflammation caused by deposits of uric acid crystals. Some dinosaurs appear to have been similarly affected: Fossilized bones of *Tyrannosaurus rex* exhibit joint damage characteristic of gout.

▼ Figure 44.7 Guano (bird excrement) is rich in uric acid.

The Influence of Evolution and Environment on Nitrogenous Wastes

EVOLUTION As a result of natural selection, the type and amount of nitrogenous waste a species produces are matched to its environment. One key factor in a habitat is the availability of water. For example, terrestrial turtles (which often live in dry areas) excrete mainly uric acid, whereas aquatic turtles excrete both urea and ammonia.

In some cases, an animal's egg is the immediate environment of relevance to the type of nitrogenous waste excreted. In an amphibian egg, which lacks a shell, ammonia or urea can simply diffuse out of the egg. Similarly, soluble wastes produced by a mammalian embryo can be carried away by the mother's blood. In the case of birds and other reptiles, however, the egg is surrounded by a shell that is permeable to gases but not to liquids. As a result, any soluble nitrogenous wastes released by the embryo would be trapped within the egg and could accumulate to dangerous levels. For this reason, using uric acid as an insoluble waste product conveys a selective advantage in reptiles. Stored within the egg as a harmless solid, the uric acid is left behind when the animal hatches.

Regardless of the type of nitrogenous waste, the amount produced is coupled to the animal's energy budget. Endotherms, which use energy at high rates, eat more food and produce more nitrogenous waste than ectotherms. The amount of nitrogenous waste is also linked to diet. Predators, which derive much of their energy from protein, excrete more nitrogen than animals that rely mainly on lipids or carbohydrates as energy sources.

Having surveyed the forms of nitrogenous waste and their interrelationship with habitat and energy consumption, we'll turn next to the processes and systems animals use to excrete these and other wastes.

CONCEPT CHECK 44.2

1. What advantage does uric acid offer as a nitrogenous waste in arid environments?
2. **WHAT IF?** Suppose a bird and a human both have gout. Why might reducing purine in their diets help the human much more than the bird?

For suggested answers, see Appendix A.

CONCEPT 44.3

Diverse excretory systems are variations on a tubular theme

By both disposing of metabolic wastes and controlling body fluid composition, excretory systems play a central role in homeostasis. Excretory processes begin when body fluid (blood, coelomic fluid, or hemolymph) is brought

▼ **Figure 44.8 Key steps of excretory system function: an overview.** Most excretory systems produce a filtrate by pressure-filtering body fluids and then modify the filtrate's contents. This diagram is modeled after the vertebrate excretory system.

Capillary

1 Filtration. The excretory tubule collects a filtrate from the blood. Water and solutes are forced by blood pressure across the selectively permeable membranes of a cluster of capillaries and into the excretory tubule.

Filtrate

Excretory tubule

2 Reabsorption. The transport epithelium reclaims valuable substances from the filtrate and returns them to the body fluids.

3 Secretion. Other substances, such as toxins and excess ions, are extracted from body fluids and added to the contents of the excretory tubule.

Urine

4 Excretion. The altered filtrate (urine) leaves the system and the body.

in contact with the selectively permeable membrane of a transport epithelium **(Figure 44.8)**. In most cases, hydrostatic pressure (blood pressure in many animals) drives a process of **filtration**. Cells, as well as proteins and other large molecules, cannot cross the epithelial membrane and remain in the body fluid. In contrast, water and small solutes, such as salts, sugars, amino acids, and nitrogenous wastes, cross the membrane, forming a solution called the **filtrate**.

The filtrate is converted to a waste fluid by the specific transport of materials into or out of the filtrate. The process of selective **reabsorption** recovers useful molecules and water from the filtrate and returns them to the body fluid. Valuable solutes—including glucose, certain salts, vitamins, hormones, and amino acids—are reabsorbed by active transport. Nonessential solutes and wastes are left in the filtrate or are added to it by selective **secretion**, which also occurs by active transport. The pumping of various solutes in turn determines whether water moves by osmosis into or out of the filtrate. In the last step—excretion—the processed filtrate containing nitrogenous wastes is released from the body as a fluid waste called urine.

Survey of Excretory Systems

The systems that perform the basic excretory functions outlined in Figure 44.8 vary widely among animal groups. However, they are generally built on a complex network of tubules that provide a large surface area for the exchange of water and solutes, including nitrogenous wastes. We'll examine the excretory systems of flatworms, earthworms, insects, and vertebrates as examples of evolutionary variations on tubule networks.

Protonephridia

As illustrated in **Figure 44.9**, flatworms (phylum Platyhelminthes), which lack a coelom or body cavity, have excretory systems called **protonephridia** (singular, *protonephridium*). Protonephridia consist of a network of dead-end tubules that branch throughout the body. Cellular units called flame bulbs cap each branch. Each flame bulb, consisting of a tubule cell and a cap cell, has a tuft of cilia projecting into the tubule.

During filtration, the beating of the cilia draws water and solutes from the interstitial fluid through the flame bulb, releasing filtrate into the tubule network. (The name *flame bulb* derives from the moving cilia's resemblance to a flickering flame.) The processed filtrate moves outward through the tubules and empties as urine via external openings. Because the urine excreted by freshwater flatworms is low in solutes, its production helps to balance the osmotic uptake of water from the environment.

▼ **Figure 44.9 Protonephridia in a planarian.**

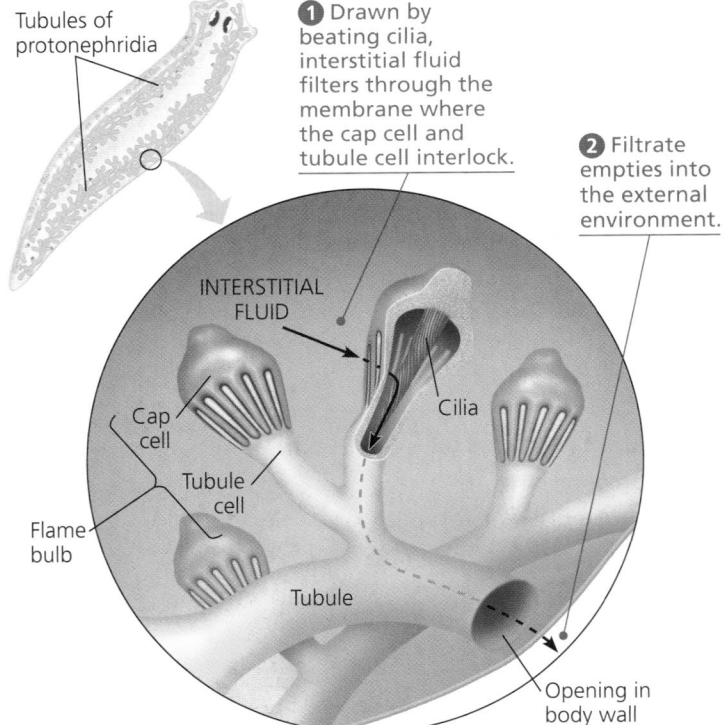

Tubules of protonephridia

1 Drawn by beating cilia, interstitial fluid filters through the membrane where the cap cell and tubule cell interlock.

2 Filtrate empties into the external environment.

INTERSTITIAL FLUID

Cap cell

Cilia

Tubule cell

Flame bulb

Tubule

Opening in body wall

Protonephridia are also found in rotifers, some annelids, mollusc larvae, and lancelets (see Figure 34.4). In the fresh-water flatworms, protonephridia serve chiefly in osmoregulation. Most metabolic wastes diffuse out of the animal across the body surface or are excreted into the gastrovascular cavity and eliminated through the mouth (see Figure 33.9). In contrast, parasitic flatworms that are isoosmotic to the surrounding fluids of their host organisms have protonephridia that primarily function in the disposal of nitrogenous wastes. Natural selection has thus adapted protonephridia to different tasks in different environments.

Metanephridia

Most annelids, such as earthworms, have **metanephridia** (singular, *metanephridium*), excretory organs that collect fluid directly from the coelom **(Figure 44.10)**. A pair of metanephridia are found in each segment of an annelid, where they are immersed in coelomic fluid and enveloped by a capillary network. A ciliated funnel surrounds the internal opening of each metanephridium. As the cilia beat, fluid is drawn into a collecting tubule, which includes a storage bladder that opens to the outside.

Earthworms inhabit damp soil and therefore usually experience a net uptake of water by osmosis through their skin. Their metanephridia balance the water influx by producing urine that is dilute (hypoosmotic to body fluids). In producing a hypoosmotic filtrate, the transport epithelium reabsorbs most solutes and returns them to the blood in the capillaries. Nitrogenous wastes, however, remain in the tubule and are excreted to the environment. The metanephridia of an earthworm thus serve both an excretory and an osmoregulatory function.

▼ **Figure 44.10 Metanephridia of an earthworm.** Each segment of the worm contains a pair of metanephridia, which collect coelomic fluid from the adjacent anterior segment. The region highlighted in yellow illustrates the organization of one metanephridium of a pair; the other would be behind it.

Coelom
Capillary network

Components of a metanephridium:

Collecting tubule
Internal opening
Bladder
External opening

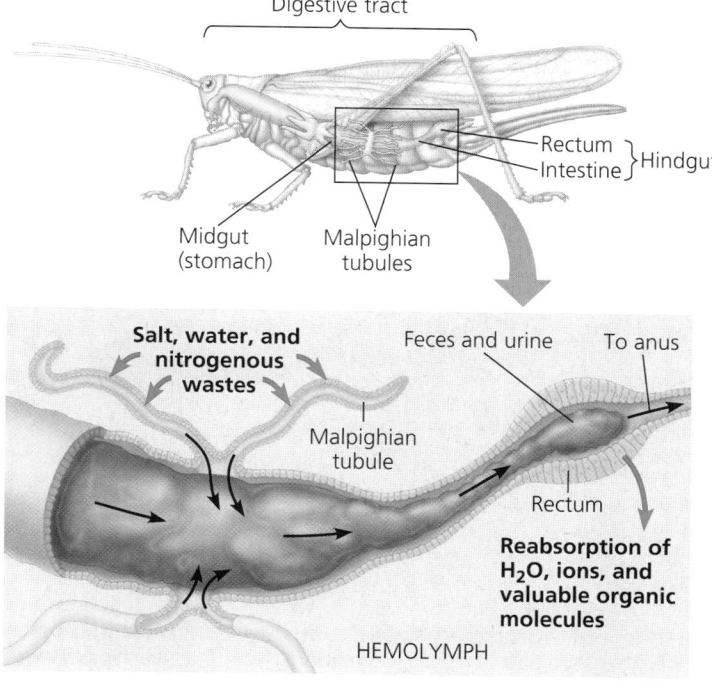

▼ **Figure 44.11 Malpighian tubules of insects.** Malpighian tubules are outpocketings of the digestive tract that remove nitrogenous wastes and function in osmoregulation.

Digestive tract

Rectum
Intestine } Hindgut
Midgut (stomach)
Malpighian tubules

Salt, water, and nitrogenous wastes
Feces and urine
To anus
Malpighian tubule
Rectum
Reabsorption of H$_2$O, ions, and valuable organic molecules
HEMOLYMPH

Malpighian Tubules

Insects and other terrestrial arthropods have organs called **Malpighian tubules** that remove nitrogenous wastes and that also function in osmoregulation **(Figure 44.11)**. The Malpighian tubules extend from dead-end tips immersed in hemolymph to openings into the digestive tract. The filtration step common to other excretory systems is absent. Instead, the transport epithelium that lines the tubules secretes certain solutes, including nitrogenous wastes, from the hemolymph into the lumen of the tubule. Water follows the solutes into the tubule by osmosis.

As fluid passes from the tubules into the rectum, most solutes are pumped back into the hemolymph; water reabsorption by osmosis follows. The nitrogenous wastes—mainly insoluble uric acid—are eliminated as nearly dry matter along with the feces. The insect excretory system is capable of conserving water very effectively, a key adaptation contributing to the tremendous success of insects on land.

Kidneys

In vertebrates and some other chordates, a compact organ called the **kidney** functions in both osmoregulation and excretion. Like the excretory organs of most animal phyla, kidneys consist of tubules. The tubules of kidneys are

Exploring the Mammalian Excretory System

Excretory Organs

Posterior vena cava

Renal artery and vein

Aorta

Ureter

Urinary bladder

Urethra

Kidney

In humans, the excretory system consists of **kidneys**, a pair of organs each about 10 cm in length, as well as organs for transporting and storing urine. Urine produced by each kidney exits through a duct called the **ureter**; the two ureters drain into a common sac called the **urinary bladder**. During urination, urine is expelled from the bladder through a tube called the **urethra**, which empties to the outside near the vagina in females and through the penis in males. Sphincter muscles near the junction of the urethra and bladder regulate urination.

Kidney Structure

Renal cortex

Renal medulla

Renal artery

Renal vein

Ureter

© Pearson Education, Inc.

Renal pelvis

Each kidney has an outer **renal cortex** and an inner **renal medulla**. Both regions are supplied with blood by a renal artery and drained by a renal vein. Within the cortex and medulla lie tightly packed excretory tubules and associated blood vessels. The excretory tubules carry and process a filtrate produced from the blood entering the kidney. Nearly all of the fluid in the filtrate is reabsorbed into the surrounding blood vessels and exits the kidney in the renal vein. The remaining fluid leaves the excretory tubules as urine, is collected in the inner **renal pelvis**, and exits the kidney via the ureter.

Nephron Types

Cortical nephron

Juxtamedullary nephron

Renal cortex

Renal medulla

Weaving back and forth across the renal cortex and medulla are the **nephrons**, the functional units of the vertebrate kidney. Of the roughly 1 million nephrons in a human kidney, 85% are **cortical nephrons**, which reach only a short distance into the medulla. The remainder, the **juxtamedullary nephrons**, extend deep into the medulla. Juxtamedullary nephrons are essential for production of urine that is hyperosmotic to body fluids, a key adaptation for water conservation in mammals.

---

arranged in a highly organized manner and are closely associated with a network of capillaries. The vertebrate excretory system also includes ducts and other structures that carry urine from the tubules out of the kidney and, eventually, the body.

Vertebrate kidneys are typically nonsegmented. However, hagfishes, which are jawless vertebrates (see Concept 34.2), have kidneys with segmentally arranged excretory tubules. Because hagfishes and other vertebrates share a common chordate ancestor, it is possible that the excretory structures of vertebrate ancestors were also segmented.

We conclude this introduction to excretory systems with an exploration of the anatomy of the mammalian kidney and associated structures **(Figure 44.12)**. Familiarizing yourself with the terms and diagrams in this figure will provide you

with a solid foundation for learning about filtrate processing in the kidney, the focus of the next section of this chapter.

CONCEPT CHECK 44.3

1. Compare and contrast the ways that metabolic waste products enter the excretory systems of flatworms, earthworms, and insects.

2. Where and how does filtrate originate in the vertebrate kidney, and by what two routes do the components of the filtrate exit the kidney?

3. **WHAT IF?** Kidney failure is often treated by hemodialysis, in which blood diverted out of the body is filtered and then allowed to flow on one side of a semipermeable membrane. Fluid called dialysate flows in the opposite direction on the other side of the membrane. In replacing the reabsorption and secretion of solutes in a functional kidney, the makeup of the starting dialysate is critical. What initial solute composition would work well?

For suggested answers, see Appendix A.

Nephron Organization

Afferent arteriole
from renal artery

Glomerulus

Bowman's
capsule

**Proximal
tubule**

**Distal
tubule**

Efferent
arteriole from
glomerulus

Peritubular
capillaries

Branch of
renal vein

Collecting
duct

Vasa
recta

Descending
limb

Ascending
limb

**Loop of
Henle**

Each nephron consists of a single long tubule as well as a ball of capillaries called the **glomerulus**. The blind end of the tubule forms a cup-shaped swelling, called **Bowman's capsule**, which surrounds the glomerulus. Filtrate is formed when blood pressure forces fluid from the blood in the glomerulus into the lumen of Bowman's capsule. Processing occurs as the filtrate passes through three major regions of the nephron: the **proximal tubule**, the **loop of Henle** (a hairpin turn with a descending limb and an ascending limb), and the **distal tubule**. A **collecting duct** receives processed filtrate from many nephrons and transports it to the renal pelvis.

Each nephron is supplied with blood by an *afferent arteriole*, an offshoot of the renal artery that branches and forms the capillaries of the glomerulus. The capillaries converge as they leave the glomerulus, forming an *efferent arteriole*. Branches of this vessel form the **peritubular capillaries**, which surround the proximal and distal tubules. Other branches extend downward and form the **vasa recta**, hairpin-shaped capillaries that serve the renal medulla, including the long loop of Henle of juxtamedullary nephrons.

▶ In this SEM of densely packed blood vessels from a human kidney, arterioles and peritubular capillaries appear pink; the glomeruli appear yellow.

200 µm

→ **Mastering Biology** Animation: Kidney Structure

CONCEPT 44.4

The nephron is organized for stepwise processing of blood filtrate

In the human kidney, filtrate forms when fluid passes from the bloodstream to the lumen of Bowman's capsule. The glomerular capillaries and specialized cells of Bowman's capsule retain blood cells and large molecules, such as plasma proteins, but are permeable to water and small solutes. Thus, the filtrate produced in the capsule contains salts, glucose, amino acids, vitamins, nitrogenous wastes, and other small molecules. Because such molecules pass freely between glomerular capillaries and Bowman's capsule, the

concentrations of these substances in the initial filtrate are the same as those in blood plasma.

Under normal conditions, roughly 1,600 L of blood flows through a pair of human kidneys each day, yielding about 180 L of initial filtrate. Both the volume and the composition of the filtrate are changed dramatically as processing occurs. About 99% of the water and nearly all of the sugars, amino acids, vitamins, and other organic nutrients are reabsorbed into the blood, leaving only about 1.5 L of urine to be transported to the bladder.

From Blood Filtrate to Urine: *A Closer Look*

To explore how filtrate is processed into urine, we'll follow the filtrate along its path in the nephron. Each circled number in the text and the figure refers to the processing

▼ **Figure 44.13 The nephron and collecting duct: regional functions of the transport epithelium.** The numbered regions in this diagram are keyed to the circled numbers in the text discussion of kidney function.

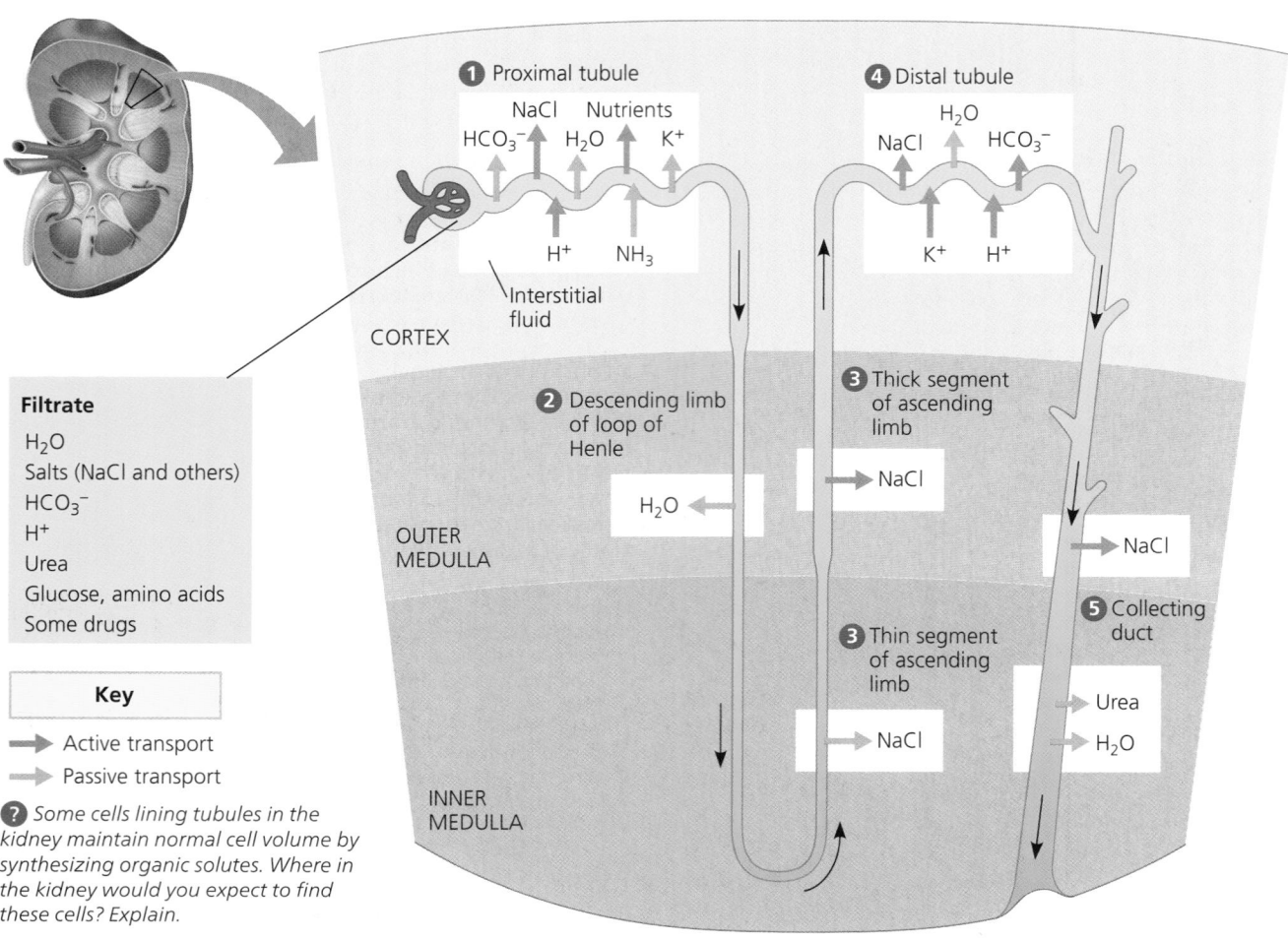

Filtrate

H₂O
Salts (NaCl and others)
HCO₃⁻
H⁺
Urea
Glucose, amino acids
Some drugs

Key

→ Active transport
→ Passive transport

? *Some cells lining tubules in the kidney maintain normal cell volume by synthesizing organic solutes. Where in the kidney would you expect to find these cells? Explain.*

⊙ **Mastering Biology** Animation: Nephron Function

in transport epithelia as the filtrate moves through the kidney cortex and medulla **(Figure 44.13)**.

❶ Proximal tubule. Reabsorption in the proximal tubule is critical for the recapture of ions, water, and valuable nutrients from the huge volume of initial filtrate. NaCl (salt) in the filtrate enters the cells of the transport epithelium by facilitated diffusion and cotransport mechanisms. There, Na⁺ ions are transferred to the interstitial fluid by active transport (see Concept 7.4). This transfer of positive charge out of the tubule drives the passive transport of Cl⁻.

As salt moves from the filtrate to the interstitial fluid, water follows by osmosis, reducing filtrate volume considerably. The salt and water that exit the filtrate diffuse from the interstitial fluid into the peritubular capillaries. Glucose, amino acids, potassium ions (K⁺), and other essential substances are also actively or passively transported from the filtrate to the interstitial fluid and then into the peritubular capillaries.

Processing of filtrate in the proximal tubule helps maintain a relatively constant pH in body fluids. Cells of the

transport epithelium secrete H⁺ into the lumen of the tubule but also synthesize and secrete ammonia, which acts as a buffer to trap H⁺ in the form of ammonium ions (NH₄⁺). The more acidic the filtrate is, the more ammonia the cells produce and secrete, and a mammal's urine usually contains some ammonia from this source (even though most nitrogenous waste is excreted as urea). The proximal tubules also reabsorb about 90% of the buffer bicarbonate (HCO₃⁻) from the filtrate, contributing further to pH balance in body fluids.

As the filtrate passes through the proximal tubule, materials to be excreted become concentrated. Many wastes leave the body fluids during the nonselective filtration process and remain in the filtrate while water and salts are reabsorbed. Urea, for example, is reabsorbed at a much lower rate than are salt and water. In addition, some materials are actively secreted into the filtrate from surrounding tissues. For example, drugs and toxins that have been processed in the liver pass from the peritubular capillaries into the interstitial fluid. These molecules are then actively secreted by the transport epithelium into the lumen of the proximal tubule.

② **Descending limb of the loop of Henle.** Upon leaving the proximal tubule, filtrate enters the loop of Henle, which further reduces filtrate volume via distinct stages of water and salt movement. In the first portion of the loop, the descending limb, numerous water channels formed by **aquaporin** proteins make the transport epithelium freely permeable to water. In contrast, there are almost no channels for salt and other small solutes, resulting in very low permeability for these substances.

For water to move out of the tubule by osmosis, the interstitial fluid bathing the tubule must be hyperosmotic to the filtrate. This condition is met along the entire length of the descending limb because the osmolarity of the interstitial fluid increases progressively from the cortex through the medulla. As a result, the filtrate loses water and increases in solute concentration all along its journey down the descending limb. The highest osmolarity (about 1,200 mOsm/L) occurs at the elbow of the loop of Henle.

③ **Ascending limb of the loop of Henle.** The filtrate reaches the tip of the loop and then returns to the cortex in the ascending limb. Unlike the descending limb, the ascending limb has a transport epithelium that lacks water channels. Consequently, the epithelial membrane that faces the filtrate in the ascending limb is impermeable to water.

The ascending limb has two specialized regions: a thin segment near the loop tip and a thick segment adjacent to the distal tubule. As filtrate ascends in the thin segment, NaCl, which became highly concentrated in the descending limb, diffuses out of the permeable tubule into the interstitial fluid. This movement of NaCl out of the tubule helps maintain the osmolarity of the interstitial fluid in the medulla.

In the thick segment of the ascending limb, the movement of NaCl out of the filtrate continues. Here, however, the epithelium actively transports NaCl into the interstitial fluid. As a result of losing salt but not water, the filtrate becomes progressively more dilute as it moves up to the cortex in the ascending limb of the loop.

Although the loop of Henle has a small net effect on filtrate composition, it is a major site for the recovery of water (descending loop) and salt (ascending loop) from the filtrate. It is this recovery that underlies water conservation in land-dwelling vertebrates, as we will explore shortly.

④ **Distal tubule.** The distal tubule plays a key role in regulating the K^+ and NaCl concentration of body fluids. This regulation involves variation in the amount of K^+ secreted into the filtrate as well as the amount of NaCl reabsorbed from the filtrate. The distal tubule also contributes to pH regulation by the controlled secretion of H^+ and reabsorption of HCO_3^-.

⑤ **Collecting duct.** The collecting duct processes the filtrate into urine, which it carries to the renal pelvis (see Figure 44.12). As filtrate passes along the transport epithelium of the collecting duct, hormonal control of permeability

and transport determines the extent to which the urine becomes concentrated.

When the kidneys are conserving water, aquaporin channels in the collecting duct allow water molecules to cross the epithelium. At the same time, the epithelium remains impermeable to salt and, in the renal cortex, to urea. As the collecting duct traverses the gradient of osmolarity in the kidney, the filtrate becomes increasingly concentrated, losing more and more water by osmosis to the hyperosmotic interstitial fluid. In the inner medulla, the duct becomes permeable to urea. Because of the high urea concentration in the filtrate at this point, some urea diffuses out of the duct and into the interstitial fluid. Along with NaCl, this urea contributes to the high osmolarity of the interstitial fluid in the medulla. The net result is urine that is hyperosmotic to the general body fluids.

When producing dilute rather than concentrated urine, the collecting duct actively absorbs salts without allowing water to follow by osmosis. At these times, the epithelium lacks aquaporin channels, and NaCl is actively transported out of filtrate. As we'll see, the presence of water channels in the collecting duct epithelium is controlled by hormones that regulate blood pressure, volume, and osmolarity.

Solute Gradients and Water Conservation

The ability of the mammalian kidney to conserve water is a key adaptation for terrestrial habitats. In humans, the osmolarity of blood is about 300 mOsm/L, but the kidney can excrete urine up to four times as concentrated—about 1,200 mOsm/L. Some mammals can do even better: Australian hopping mice, small marsupials that live in dry desert regions, can produce urine with an osmolarity of 9,300 mOsm/L, 25 times as concentrated as the animal's blood.

In a mammalian kidney, the production of hyperosmotic urine is possible only because considerable energy is expended for the active transport of solutes against concentration gradients. The nephrons—particularly the loops of Henle—can be thought of as energy-consuming machines that produce an osmolarity gradient suitable for extracting water from the filtrate in the collecting duct. The primary solutes affecting osmolarity are NaCl, which is concentrated in the renal medulla by the loop of Henle, and urea, which passes across the epithelium of the collecting duct in the inner medulla.

Concentrating Urine in the Mammalian Kidney

To better understand the physiology of the mammalian kidney as a water-conserving organ, let's retrace the flow of filtrate through the excretory tubule. This time, let's focus on how juxtamedullary nephrons maintain an osmolarity gradient in the tissues that surround the loop of Henle and how they use that gradient to excrete hyperosmotic urine

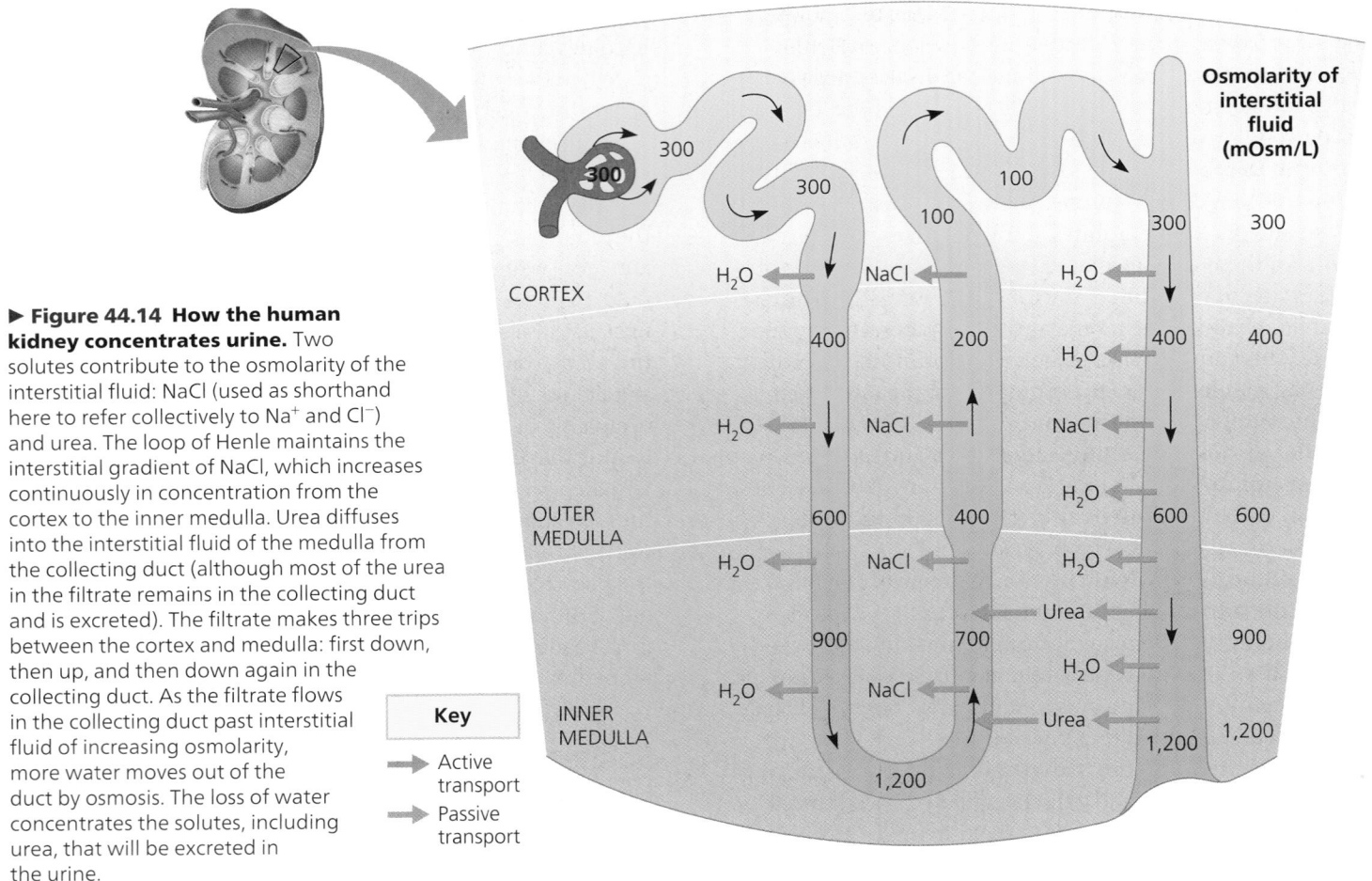

► **Figure 44.14 How the human kidney concentrates urine.** Two solutes contribute to the osmolarity of the interstitial fluid: NaCl (used as shorthand here to refer collectively to Na^+ and Cl^-) and urea. The loop of Henle maintains the interstitial gradient of NaCl, which increases continuously in concentration from the cortex to the inner medulla. Urea diffuses into the interstitial fluid of the medulla from the collecting duct (although most of the urea in the filtrate remains in the collecting duct and is excreted). The filtrate makes three trips between the cortex and medulla: first down, then up, and then down again in the collecting duct. As the filtrate flows in the collecting duct past interstitial fluid of increasing osmolarity, more water moves out of the duct by osmosis. The loss of water concentrates the solutes, including urea, that will be excreted in the urine.

Key

→ Active transport

→ Passive transport

WHAT IF? *The drug furosemide blocks the cotransporters for Na^+ and Cl^- in the ascending limb of the loop of Henle. What effect would you expect this drug to have on urine volume?*

→ **Mastering Biology** Figure Walkthrough

(Figure 44.14). Filtrate passing from Bowman's capsule to the proximal tubule has about the same osmolarity as blood. A large amount of water *and* salt is reabsorbed from the filtrate as it flows through the proximal tubule in the renal cortex. As a result, the filtrate's volume decreases substantially, but its osmolarity remains about the same.

As the filtrate flows from cortex to medulla in the descending limb of the loop of Henle, water leaves the tubule by osmosis. Solutes, including NaCl, become more concentrated, increasing the osmolarity of the filtrate. Diffusion of salt out of the tubule is maximal as the filtrate rounds the curve and enters the ascending limb, which is permeable to salt but not to water. NaCl diffusing from the ascending limb helps maintain a high osmolarity in the interstitial fluid of the renal medulla.

The loop of Henle and surrounding capillaries act as a type of countercurrent system to generate the steep osmotic gradient between the medulla and cortex. Recall that some endotherms have a countercurrent heat exchanger that reduces heat loss and that countercurrent gas exchange in fish gills maximizes oxygen absorption (see Figures 40.13

and 42.21). In those cases, the countercurrent mechanisms involve passive movement along either an oxygen concentration gradient or a heat gradient. In contrast, the countercurrent system of the loop of Henle involves active transport and thus an expenditure of energy. The active transport of NaCl from the filtrate in the upper part of the ascending limb of the loop maintains a high salt concentration in the interior of the kidney, enabling the kidney to form concentrated urine. Such countercurrent systems, which expend energy to create concentration gradients, are called **countercurrent multiplier systems**.

What prevents the capillaries of the vasa recta from dissipating the gradient by carrying away the high concentration of NaCl in the medulla's interstitial fluid? As shown in Figure 44.12, the descending and ascending vessels of the vasa recta carry blood in opposite directions through the kidney's osmolarity gradient. As the descending vessel conveys blood toward the inner medulla, water is lost from the blood and NaCl is gained by diffusion. These net fluxes are reversed as blood flows back toward the cortex in the ascending vessel of the vasa recta, with water reentering the blood and salt diffusing out. Thus, the vasa recta can supply the kidney with nutrients and other important substances carried by the blood without interfering with the osmolarity gradient in the inner and outer medulla.

The countercurrent-like characteristics of the loop of Henle and the vasa recta help to generate the steep osmotic gradient between the medulla and cortex. However, diffusion will eventually eliminate any osmotic gradient within animal tissue unless energy is expended to maintain the gradient. In the kidney, this expenditure largely occurs in the thick segment of the ascending limb of the loop of Henle, where NaCl is actively transported out of the tubule. Even with the benefits of countercurrent exchange, this process—along with other renal active transport systems—consumes considerable ATP. Thus, for its size, the kidney has one of the highest metabolic rates of any organ.

As a result of active transport of NaCl out of the thick segment of the ascending limb, the filtrate is actually hypoosmotic to body fluids by the time it reaches the distal tubule. Next, the filtrate descends again toward the medulla, this time in the collecting duct, which is permeable to water but not to salt. Therefore, osmosis extracts water from the filtrate as it passes from cortex to medulla and encounters interstitial fluid of increasing osmolarity. This process concentrates salt, urea, and other solutes in the filtrate. Some urea passes out of the lower portion of the collecting duct and contributes to the high interstitial osmolarity of the inner medulla. (This urea is recycled by diffusion into the loop of Henle, but continual leakage from the collecting duct maintains a high interstitial urea concentration.) When the kidney concentrates urine maximally, the urine reaches 1,200 mOsm/L, the osmolarity of the interstitial fluid in the inner medulla. Although *isoosmotic* to the inner medulla's interstitial fluid, the urine is *hyperosmotic* to blood and interstitial fluid elsewhere in the body. This high osmolarity allows the solutes remaining in the urine to be excreted from the body with minimal water loss.

Adaptations of the Vertebrate Kidney to Diverse Environments

EVOLUTION Vertebrates occupy habitats ranging from rain forests to deserts and from some of the saltiest bodies of water to the nearly pure waters of high mountain lakes. Comparing vertebrates across environments reveals adaptive variations in nephron structure and function. In the case of mammals, for example, the presence of juxtamedullary nephrons is a key adaptation that enables these terrestrial animals to shed salts and nitrogenous wastes without squandering water. Differences among species in the length of the loop of Henle in the juxtamedullary nephrons and in the relative numbers of juxtamedullary and cortical nephrons help to fine-tune osmoregulation to particular habitats.

Mammals

Mammals that excrete the most hyperosmotic urine, such as Australian hopping mice, North American kangaroo rats, and other desert mammals, have many juxtamedullary nephrons with loops of Henle that extend deep into the medulla. Long loops maintain steep osmotic gradients in the kidney, resulting in urine becoming very concentrated as it passes from cortex to medulla in the collecting ducts.

In contrast, beavers, muskrats, and other aquatic mammals that spend much of their time in fresh water and rarely face problems of dehydration have mostly cortical nephrons, resulting in a much lower ability to concentrate urine. Terrestrial mammals living in moist conditions have loops of Henle of intermediate length and the capacity to produce urine intermediate in concentration to that produced by freshwater and desert mammals.

Case Study: *Kidney Function in the Vampire Bat*

The South American vampire bat in **Figure 44.15** illustrates the versatility of the mammalian kidney. This species feeds at night on the blood of large birds and mammals. The bat uses its sharp teeth to make a small incision in the prey's skin and then laps up blood from the wound (the prey is typically not seriously harmed). Anticoagulants in the bat's saliva prevent the blood from clotting.

A vampire bat may search for hours and fly long distances to locate a suitable victim. When it does find prey, it benefits from consuming as much blood as possible. Often drinking more than half its body mass, the bat is at risk of becoming too heavy to fly. As the bat feeds, however, its kidneys excrete large volumes of dilute urine, up to 24% of body mass per hour. Having lost enough weight to take off, the bat can fly back to its roost in a cave or hollow tree, where it spends the day.

In the roost, the vampire bat faces a different regulatory problem. Most of the nutrition it derives from blood comes in the form of protein. Digesting proteins generates large

▶ **Figure 44.15 A vampire bat (*Desmodus rotundus*), a mammal with unique excretory challenges.**

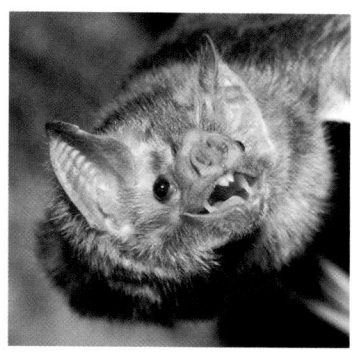

quantities of urea, but roosting bats lack access to the drinking water necessary to dilute it. Instead, their kidneys shift to producing small quantities of highly concentrated urine (up to 4,600 mOsm/L), an adjustment that disposes of the urea load while conserving as much water as possible. The vampire bat's ability to alternate rapidly between producing large amounts of dilute urine and small amounts of very hyperosmotic urine is an essential part of its adaptation to an unusual food source.

Birds and Other Reptiles

Most birds, including the albatross (see Figure 44.1) and the ostrich **(Figure 44.16)**, live in environments that are dehydrating. Like mammals but no other species, birds have kidneys with juxtamedullary nephrons. However, the nephrons of birds have loops of Henle that extend less far into the medulla than those of mammals. Thus, bird kidneys cannot concentrate urine to the high osmolarities achieved by mammalian kidneys. Although birds can produce hyperosmotic urine, their main water conservation adaptation is having uric acid as the nitrogenous waste molecule.

The kidneys of other reptiles have only cortical nephrons, and they produce urine that is isoosmotic or hypoosmotic to body fluids. However, the epithelium of the cloaca from which urine and feces leave the body conserves fluid by reabsorbing water from these wastes. Like birds, most other reptiles excrete their nitrogenous wastes as uric acid.

▼ **Figure 44.16 An ostrich (*Struthio camelus*), an animal well adapted to its dry environment.**

Freshwater Fishes and Amphibians

Hyperosmotic to their surroundings, freshwater fishes produce large volumes of very dilute urine. Their kidneys, which are packed with cortical nephrons, produce filtrate at a high rate. Salt conservation relies on the reabsorption of ions from the filtrate in the distal tubules.

Amphibian kidneys function much like those of freshwater fishes. When frogs are in fresh water, their kidneys excrete dilute urine while their skin accumulates certain salts from the water by active transport. On land, where dehydration is the most pressing problem of osmoregulation, frogs conserve body fluid by reabsorbing water across the epithelium of the urinary bladder.

Marine Bony Fishes

Compared with freshwater fishes, marine fishes have fewer and smaller nephrons, and their nephrons lack a distal tubule. In addition, their kidneys have small glomeruli or lack glomeruli entirely. In keeping with these features, filtration rates are low and very little urine is excreted.

The main function of kidneys in marine bony fishes is to get rid of divalent ions (those with a charge of 2+ or 2−) such as calcium (Ca^{2+}), magnesium (Mg^{2+}), and sulfate (SO_4^{2-}). Marine fishes take in divalent ions by incessantly drinking seawater. They rid themselves of these ions by secreting them into the proximal tubules of the nephrons and excreting them in urine. Osmoregulation in marine bony fishes also relies on specialized *chloride cells* in the gills. By establishing ion gradients that enable secretion of salt (NaCl) into seawater, the chloride cells maintain proper levels of monovalent ions (charge of 1+ or 1−) such as Na^+ and Cl^-.

The generation of ion gradients and the movement of ions across membranes are central to salt and water balance in marine bony fishes. These events, however, are by no means unique to these organisms nor to homeostasis. As illustrated by the examples in **Figure 44.17**, osmoregulation by chloride cells is but one of many diverse physiological processes that are driven by the movement of ions across a membrane.

CONCEPT CHECK 44.4

1. What do the number and length of nephrons in a fish's kidney indicate about the fish's habitat? How do they correlate with urine production?

2. Many medications make the epithelium of the collecting duct less permeable to water. How would taking such a medication affect kidney output?

3. **WHAT IF?** If blood pressure in the afferent arteriole leading to a glomerulus decreased, how would the rate of blood filtration within Bowman's capsule be affected? Explain.

For suggested answers, see Appendix A.

Ion Movement and Gradients

The transport of ions across the plasma membrane of a cell is a fundamental activity of all animals, and indeed of all living things. By generating ion gradients, ion transport provides the potential energy that powers processes ranging from an organism's regulation of salts and gases in internal fluids to its perception of and locomotion through its environment.

Osmoregulation

In marine bony fishes, ion gradients drive secretion of salt (NaCl), a process essential to avoid dehydration. Within gills, the pumps, cotransporters, and channels of specialized chloride cells function together to drive salt from the blood across the gill epithelium and into the surrounding salt water. (See Figure 44.3.)

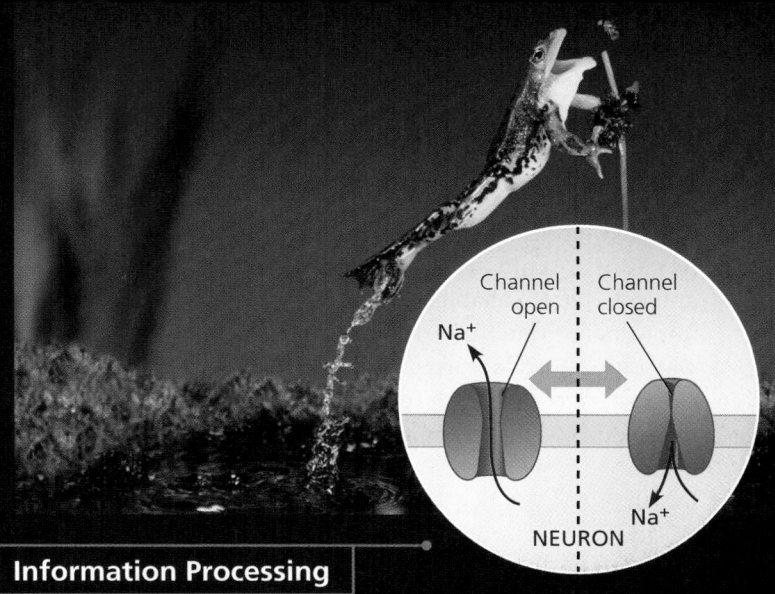

Information Processing

In neurons, transmission of information as nerve impulses is made possible by the opening and closing of channels selective for sodium or other ions. These signals enable nervous systems to receive and process input and to direct appropriate output, such as this leap of a frog capturing prey. (See Concept 48.3 and Concept 50.5.)

Gas Exchange

Ion gradients provide the basis for the opening of a plant stoma by surrounding guard cells. Active transport of H^+ out of a guard cell generates a voltage (membrane potential) that drives inward movement of K^+. This uptake of K^+ by guard cells triggers an osmotic influx of water that changes cell shape, bowing the guard cells outward and thereby opening the stoma. (See

Locomotion

A gradient of H^+ powers the bacterial flagellum. An electron transport chain generates this gradient, establishing a higher concentration of H^+ outside the bacterial cell. Protons reentering the cell provide a force that causes the flagellar motor to rotate. The rotating motor turns the curved hook, causing the attached filament to propel the cell. (See Concept 9.4 and Figure 27.7.)

MAKE CONNECTIONS

Explain why the set of forces driving ion movement across the plasma membrane of a cell is described as an electrochemical (electrical and chemical) gradient (see Concept 7.4).

→ **Mastering Biology BioFlix® Animation: Membrane Transport**

Hormonal circuits link kidney function, water balance, and blood pressure

In mammals, both the volume and osmolarity of urine are adjusted according to an animal's water and salt balance and its rate of urea production. In situations of high salt intake and low water availability, a mammal can excrete urea and salt in small volumes of hyperosmotic urine with minimal water loss. If salt is scarce and fluid intake is high, the kidney can instead eliminate the excess water with little salt loss by producing large volumes of hypoosmotic urine. At such times, the urine can be as dilute as 70 mOsm/L, less than one-fourth the osmolarity of human blood.

How are urine volume and osmolarity regulated so effectively? As we'll explore in this final portion of the chapter, two major control circuits that respond to different stimuli together restore and maintain normal water and salt balance.

Homeostatic Regulation of the Kidney

A combination of nervous and hormonal controls manages the osmoregulatory function of the mammalian kidney. Through their effects on the amount and osmolarity of urine, these controls contribute to homeostasis for both blood pressure and blood volume.

Antidiuretic Hormone

One key hormone in of the kidney is **antidiuretic hormone (ADH)**, also called *vasopressin*. ADH molecules released from the posterior pituitary bind to and activate membrane receptors on the surface of collecting duct cells. The activated receptors initiate a signal transduction cascade that directs insertion of aquaporin proteins into the membrane lining the collecting duct **(Figure 44.18)**. More aquaporin channels result in more water recapture, reducing urine volume. (A high level of urine production is called diuresis; ADH is therefore called *anti*diuretic hormone.)

To understand the regulatory circuitry based on ADH, let's consider first what occurs when blood osmolarity rises, such as after eating salty food or losing water through sweating **(Figure 44.19)**. When osmolarity rises above the normal range (275–295 mOsm/L), osmoreceptor cells in the hypothalamus trigger increased

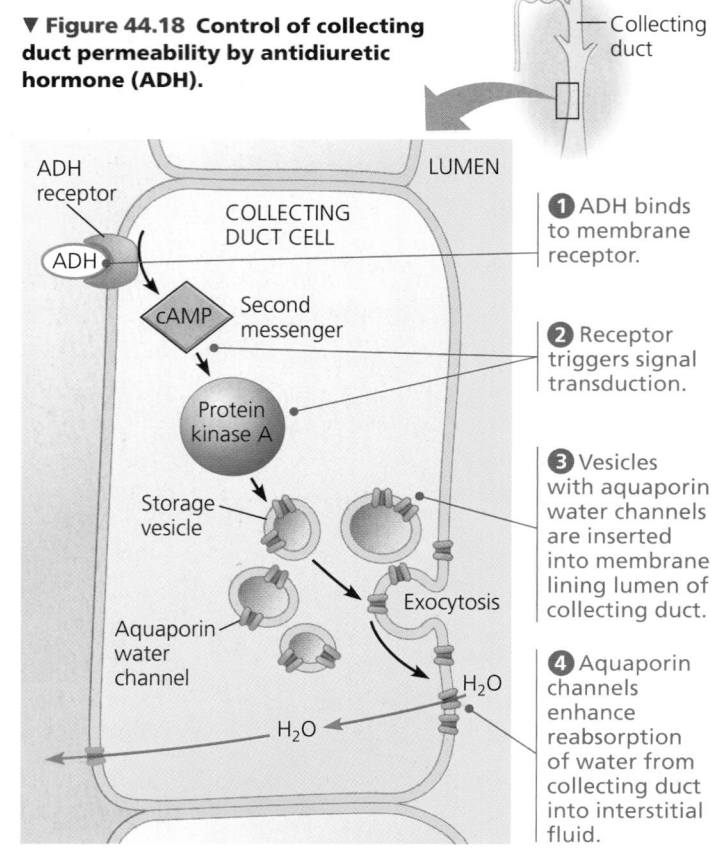

▼ **Figure 44.18 Control of collecting duct permeability by antidiuretic hormone (ADH).**

Collecting duct

LUMEN

ADH receptor

COLLECTING DUCT CELL

ADH

cAMP — Second messenger

Protein kinase A

Storage vesicle

Aquaporin water channel

Exocytosis

H_2O

H_2O

❶ ADH binds to membrane receptor.

❷ Receptor triggers signal transduction.

❸ Vesicles with aquaporin water channels are inserted into membrane lining lumen of collecting duct.

❹ Aquaporin channels enhance reabsorption of water from collecting duct into interstitial fluid.

▼ **Figure 44.19 Regulation of fluid retention in the kidney.** Osmoreceptors in the hypothalamus monitor blood osmolarity via its effect on the net diffusion of water into or out of the receptor cells. When blood osmolarity increases, signals from the osmoreceptors trigger a release of ADH from the posterior pituitary and generate thirst. Water reabsorption in the collecting duct and water intake restore normal blood osmolarity, inhibiting further ADH secretion.

Osmoreceptors in hypothalamus trigger release of ADH from posterior pituitary.

Hypothalamus

Specific neurons of hypothalamus generate thirst.

Posterior pituitary

Distal tubule

ADH

Blood osmolarity increases (such as after sweating profusely).

Collecting duct

Drinking water reduces blood osmolarity.

H_2O reabsorption reduces blood osmolarity.

NORMAL BLOOD OSMOLARITY (275–295 mOsm/L)

→ **Mastering Biology** Animation: Control of Water Reabsorption release of ADH from the posterior pituitary. The resulting increase in water reabsorption in the collecting duct concentrates urine, reduces urine volume, and lowers blood osmolarity back toward the normal range. As the osmolarity of the blood falls, a negative-feedback mechanism reduces the osmoreceptor cell activity in the hypothalamus, and ADH secretion is reduced.

What happens if, instead of ingesting salt or sweating profusely, you drink a large amount of water? Blood osmolarity falls below the normal range, causing a drop in ADH secretion to a very low level. The resulting decrease in permeability of the collecting ducts reduces water reabsorption, resulting in discharge of large volumes of dilute urine.

Contrary to common belief, caffeinated drinks increase urine production to no greater degree than water of comparable volume: Numerous studies of coffee and tea drinkers have found no diuretic effect for caffeine.

Blood osmolarity, ADH release, and water reabsorption in the kidney are normally linked in a feedback circuit that contributes to homeostasis. Anything that disrupts this circuit can interfere with water balance. For example, alcohol inhibits ADH release, leading to excessive urinary water loss and dehydration (which may cause some of the symptoms of a hangover).

Mutations that prevent ADH production or that inactivate the ADH receptor gene disrupt homeostasis by blocking the insertion of additional aquaporin channels in the collecting duct membrane. The resulting disorder can cause severe dehydration and solute imbalance due to production of copious dilute urine. These symptoms give the disorder its name: *diabetes insipidus* (from the Greek for "to pass through" and "having no flavor"). Could mutations in an aquaporin gene have a similar effect? **Figure 44.20** describes an experimental approach that addressed this question.

The Renin-Angiotensin-Aldosterone System

The release of ADH is a response to an increase in blood osmolarity, as when the body is dehydrated from excessive water loss or inadequate water intake. However, an excessive loss of both salt and body fluids—caused, for example, by a major wound or severe diarrhea—will reduce blood volume *without* increasing osmolarity. Given that this will not affect ADH release, how does the body respond? It turns out that an endocrine circuit called the **renin-angiotensin-aldosterone system (RAAS)** also regulates kidney function. The RAAS responds to the drop in blood volume and pressure by increasing water and Na^+ reabsorption.

The RAAS involves the **juxtaglomerular apparatus (JGA)**, a specialized tissue consisting of cells of and around the afferent arteriole, which supplies blood to the glomerulus. When blood pressure or volume drops in the afferent

▼ **Figure 44.20** **Inquiry**

Can aquaporin mutations cause diabetes?

Experiment Researchers studied a patient with diabetes insipidus who had a normal ADH receptor gene but two mutant alleles (A and B) of the aquaporin-2 gene. The resulting changes are shown below in an alignment of protein sequences that includes other species.

Source of Aquaporin-2 Gene Sequence	Amino Acids 183–191* in Encoded Protein	Amino Acids 212–220* in Encoded Protein
Frog (*Xenopus laevis*)	MNPARSFAP	GIFASLIYN
Lizard (*Anolis carolinensis*)	MNPARSFGP	AVVASLLYN
Chicken (*Gallus gallus*)	MNPARSFAP	AAAASIIYN
Human (*Homo sapiens*)	MNPARSLAP	AILGSLLYN
Conserved residues	MNPARS–P	–S–YN
Patient's gene: allele A	MNPACSLAP	AILGSLLYN
Patient's gene: allele B	MNPARSLAP	AILGPLLYN

*The numbering is based on the human aquaporin-2 protein sequence.

Each mutation changed the protein sequence at a highly conserved position. To test the hypothesis that the changes affect function, researchers used frog oocytes, cells that will express foreign messenger RNA and can be readily collected from adult female frogs.

❶ Messenger RNA transcribed from wild-type and mutant aquaporin genes is injected into frog oocytes, where it directs the synthesis of aquaporin proteins.

❷ The oocytes are transferred from a 200-mOsm to a 10-mOsm solution. The rate of oocyte swelling is measured as an indicator of water permeability.

Results

Source of Injected mRNA	Rate of Swelling (μm/sec)
Human wild type	196
Patient's allele A	17
Patient's allele B	18
None (H_2O control)	20

Conclusion Because each mutation renders aquaporin inactive as a water channel, researchers concluded that these mutations cause the disorder common to the patients.

Data from P. M. Deen et al., Requirement of human renal water channel aquaporin-2 for vasopressin-dependent concentration of urine, *Science* 264:92–95 (1994).

WHAT IF? *If you measured ADH levels in patients with ADH receptor mutations and in patients with aquaporin mutations, what would you expect to find, compared with wild-type subjects?*

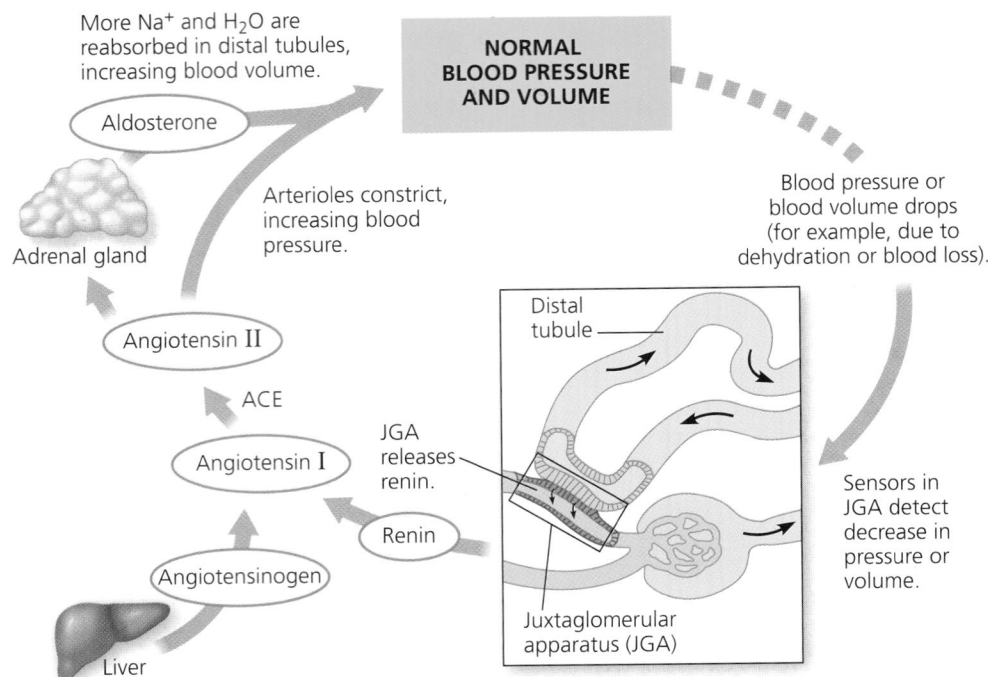

▶ **Figure 44.21 Regulation of blood volume and blood pressure by the renin-angiotensin-aldosterone system (RAAS).**

VISUAL SKILLS *Label each arrow that represents the secretion of a hormone.*

Labels in figure:

More Na$^+$ and H$_2$O are reabsorbed in distal tubules, increasing blood volume.

NORMAL BLOOD PRESSURE AND VOLUME

Aldosterone

Adrenal gland

Arterioles constrict, increasing blood pressure.

Blood pressure or blood volume drops (for example, due to dehydration or blood loss).

Angiotensin II

ACE

Angiotensin I

Renin

Angiotensinogen

Liver

Distal tubule

JGA releases renin.

Juxtaglomerular apparatus (JGA)

Sensors in JGA detect decrease in pressure or volume.

---

arteriole (for instance, as a result of dehydration), the JGA releases the enzyme renin. Renin initiates a sequence of steps that cleave a plasma protein called angiotensinogen, ultimately yielding a peptide called *angiotensin II* **(Figure 44.21)**.

Functioning as a hormone, angiotensin II triggers vasoconstriction, increasing blood pressure and decreasing blood flow to capillaries in the kidney (and elsewhere). Angiotensin II also stimulates the adrenal glands to release a hormone called *aldosterone*. Aldosterone causes the nephrons' distal tubules and collecting duct to reabsorb more Na$^+$ and water, increasing blood volume and pressure.

Because angiotensin II results in increased blood pressure, drugs that block angiotensin II production are widely used to treat hypertension (chronic high blood pressure). Many of these drugs are specific inhibitors of angiotensin converting enzyme (ACE), which catalyzes one of the steps in the production of angiotensin II.

The RAAS operates as a feedback circuit. A drop in blood pressure and blood volume triggers renin release. The resulting production of angiotensin II and release of aldosterone cause a rise in blood pressure and volume, reducing the release of renin from the JGA.

Coordinated Regulation of Salt and Water Balance

Both ADH and RAAS increase water reabsorption in the kidney. However, whereas ADH alone would lower blood Na$^+$ concentration via water reabsorption in the kidney, the RAAS helps maintain body fluid osmolarity within the normal range by stimulating Na$^+$ reabsorption.

Another hormone, **atrial natriuretic peptide (ANP)**, opposes the RAAS. The walls of the atria of the heart release ANP in response to an increase in blood volume and pressure. ANP inhibits the release of renin from the JGA, inhibits NaCl reabsorption by the collecting ducts, and reduces aldosterone release from the adrenal glands. These actions lower blood volume and pressure. Thus, ADH, the RAAS, and ANP provide an elaborate system of checks and balances that regulate the kidney's ability to control blood osmolarity, salt concentration, volume, and pressure.

Thirst plays an essential role in the control of water and salt balance. Recently, researchers have identified neurons in the hypothalamus dedicated to regulating thirst. Stimulating one set of neurons in mice causes intense drinking behavior, even if the animal is fully hydrated. Stimulating a second set causes an immediate halt in water consumption, even in dehydrated animals. Follow-up studies are focused on identifying the cellular and molecular pathways linking these neurons to the behavioral responses.

CONCEPT CHECK 44.5

1. How does alcohol affect regulation of water balance in the body?

2. Why could it be dangerous to drink a very large amount of water in a short period of time?

3. **WHAT IF?** Conn's syndrome is a condition caused by tumors of the adrenal cortex that secrete high amounts of aldosterone in an unregulated manner. What would you expect to be the major symptom of this disorder?

For suggested answers, see Appendix A.

44 Chapter Review

→ Go to **Mastering Biology** for Assignments, the eText, the Study Area, and Dynamic Study Modules.

SUMMARY OF KEY CONCEPTS

→ To review key terms, go to the **Vocabulary Self-Quiz** in the **Mastering Biology** eText or Study Area, or go to goo.gl/zkjz9t.

CONCEPT 44.1

Osmoregulation balances the uptake and loss of water and solutes (pp. 978–982)

Animal	Inflow/Outflow	Urine
Freshwater fish. Lives in water less concentrated than body fluids; fish tends to gain water, lose salt	Does not drink water Salt in (active transport by gills) H₂O in Salt out H₂O out	▶ Large volume of urine ▶ Urine is less concentrated than body fluids
Marine bony fish. Lives in water more concentrated than body fluids; fish tends to lose water, gain salt	Drinks water Salt in H₂O out Salt out (active transport by gills)	▶ Small volume of urine ▶ Urine is slightly less concentrated than body fluids
Terrestrial vertebrate. Terrestrial environment; tends to lose body water to air	Drinks water Salt in (by mouth) H₂O and salt out	▶ Moderate volume of urine ▶ Urine is more concentrated than body fluids

- Cells balance water gain and loss through **osmoregulation**, a process based on the controlled movement of solutes between internal fluids and the external environment and on the movement of water, which follows by osmosis.
- **Osmoconformers** are isoosmotic with their marine environment and do not regulate their **osmolarity**. In contrast, **osmoregulators** control water uptake and loss in a hypoosmotic or hyperosmotic environment, respectively. Water-conserving excretory organs help terrestrial animals avoid desiccation, which can be life-threatening. Animals that live in temporary waters may enter a dormant state called **anhydrobiosis** when their habitats dry up.
- **Transport epithelia** contain specialized epithelial cells that control the solute movements required for waste disposal and osmoregulation.

? *Under what environmental conditions does water move into a cell by osmosis?*

CONCEPT 44.2

An animal's nitrogenous wastes reflect its phylogeny and habitat (pp. 982–983)

- Protein and nucleic acid metabolism generates **ammonia**. Most aquatic animals excrete ammonia. Mammals and most adult amphibians convert ammonia to the less toxic **urea**, which is excreted with a minimal loss of water. Insects and many reptiles, including birds, convert ammonia to **uric acid**, a mostly insoluble waste excreted in a paste-like urine.
- The kind of nitrogenous waste excreted depends on an animal's habitat, whereas the amount excreted is coupled to the animal's energy budget and dietary protein intake.

MAKE CONNECTIONS *Metabolism of carbohydrates and fats requires several nitrogen-containing molecules, such as NAD⁺/NADH (see Figure 9.12), but is not a significant source of nitrogen-containing waste. Why?*

CONCEPT 44.3

Diverse excretory systems are variations on a tubular theme (pp. 983–987)

- Most excretory systems carry out **filtration, reabsorption, secretion**, and **excretion**. Invertebrate excretory systems include the **protonephridia** of flatworms, the **metanephridia** of earthworms, and the **Malpighian tubules** of insects. **Kidneys** function in both excretion and osmoregulation in vertebrates.
- Excretory tubules (consisting of **nephrons** and **collecting ducts**) and blood vessels pack the mammalian kidney. Blood pressure forces fluid from blood in the **glomerulus** into the lumen of **Bowman's capsule**. Following reabsorption and secretion, **filtrate** flows into a collecting duct. The **ureter** conveys urine from the **renal pelvis** to the **urinary bladder**.

? *What is the function of the filtration step in excretory systems?*

CONCEPT 44.4

The nephron is organized for stepwise processing of blood filtrate (pp. 987–993)

- Within the nephron, selective secretion and reabsorption in the **proximal tubule** alter filtrate volume and composition. The descending limb of the **loop of Henle** is permeable to water but not salt; water moves by osmosis into the interstitial fluid. The ascending limb is permeable to salt but not water; salt leaves by diffusion and by active transport. The **distal tubule** and collecting duct regulate K⁺ and NaCl levels in body fluids.
- In mammals, **a countercurrent multiplier system** involving the loop of Henle maintains the gradient of salt concentration in the kidney interior. Urea exiting the collecting duct contributes to the osmotic gradient of the kidney.
- Natural selection has shaped the form and function of nephrons in various vertebrates to the osmoregulatory challenges of the animals' habitats. For example, desert mammals, which excrete the most hyperosmotic urine, have loops of Henle that extend deep into the **renal medulla,** whereas mammals in moist habitats have shorter loops and excrete more dilute urine.

? *How do cortical and juxtamedullary nephrons differ with respect to reabsorbing nutrients and concentrating urine?*

Hormonal circuits link kidney function, water balance, and blood pressure (pp. 994–996)

- The posterior pituitary gland releases **antidiuretic hormone (ADH)** when blood osmolarity rises above the normal range, such as when water intake is inadequate. ADH increases the permeability to water of the collecting ducts by increasing the number of epithelial **aquaporin** channels.
- When blood pressure or blood volume in the afferent arteriole drops, the **juxtaglomerular apparatus** releases renin. Angiotensin II formed in response to renin constricts arterioles and triggers release of the hormone aldosterone, raising blood pressure and reducing the release of renin. This **renin-angiotensin-aldosterone system** has functions that overlap with those of ADH and are opposed by **atrial natriuretic peptide**.

? *Why can only some patients with diabetes insipidus be treated effectively with ADH?*

TEST YOUR UNDERSTANDING

⟶ For more multiple-choice questions, go to the **Practice Test** in the **Mastering Biology** eText or Study Area, or go to goo.gl/GruWRg.

Levels 1-2: Remembering/Understanding

1. *Unlike* an earthworm's metanephridia, a mammalian nephron
  (A) is intimately associated with a capillary network.
  (B) functions in both osmoregulation and excretion.
  (C) receives filtrate from blood instead of coelomic fluid.
  (D) has a transport epithelium.

2. Which process in the nephron is *least* selective?
  (A) filtration   (C) active transport
  (B) reabsorption   (D) secretion

3. Which of the following animals generally has the lowest volume of urine production?
  (A) vampire bat   (C) marine bony fish
  (B) salmon in fresh water (D) freshwater flatworm

Levels 3-4: Applying/Analyzing

4. The high osmolarity of the renal medulla is maintained by which of the following?
  (A) active transport of salt from the upper region of the descending limb.
  (B) the loose packing of juxtamedullary nephrons.
  (C) diffusion of urea into the collecting duct.
  (D) diffusion of salt from the descending limb of the loop of Henle.

5. In which of the following species should natural selection favor the highest proportion of juxtamedullary nephrons?
  (A) a river otter
  (B) a mouse species living in a temperate broadleaf forest
  (C) a mouse species living in a desert
  (D) a beaver

6. African lungfish, which are often found in small, stagnant pools of fresh water, produce urea as a nitrogenous waste. What is an advantage of this adaptation?
  (A) Urea takes less energy to synthesize than ammonia.
  (B) Small, stagnant pools do not provide enough water to dilute ammonia, which is toxic.
  (C) Urea forms an insoluble precipitate.
  (D) Urea makes lungfish tissue hypoosmotic to the pool.

Levels 5-6: Evaluating/Creating

7. INTERPRET THE DATA (a) Use the data below to draw four pie charts for the average daily water gain and loss in a kangaroo rat and a human.

	Kangaroo Rat	Human
Average Water Gain (mL/day)		
Ingested in food	0.2	750
Ingested in liquid	0	1,500
Derived from metabolism	1.8	250
Average Water Loss (mL/day)		
Urine	0.45	1,500
Feces	0.09	100
Evaporation	1.46	900

(b) Which routes of water gain and loss make up a much larger share of the total in a kangaroo rat than in a human?

8. EVOLUTION CONNECTION Merriam's kangaroo rats (*Dipodomys merriami*) live in North American habitats ranging from moist, cool woodlands to hot deserts. Based on the hypothesis that there are adaptive differences in water conservation between *D. merriami* populations, predict how the rates of evaporative water loss would differ for populations that live in moist versus dry environments. Propose a test of your prediction, using a humidity sensor to detect evaporative water loss by kangaroo rats.

9. SCIENTIFIC INQUIRY You are exploring kidney function in kangaroo rats. You measure urine volume and osmolarity, as well as the amount of chloride (Cl⁻) and urea in the urine. If the water source provided to the animals were switched from tap water to a 2% NaCl solution, indicate what change in urine osmolarity you would expect. How would you determine if this change was more likely due to a change in the excretion of Cl⁻ or urea?

10. WRITE ABOUT A THEME: ORGANIZATION In a short essay (100–150 words), compare how membrane structures in the loop of Henle and collecting duct of the mammalian kidney enable water to be recovered from filtrate in the process of osmoregulation.

11. SYNTHESIZE YOUR KNOWLEDGE

The marine iguana (*Amblyrhynchus cristatus*), which spends long periods under water feeding on seaweed, relies on both salt glands and kidneys for homeostasis of its internal fluids. Describe how these organs together meet the particular osmoregulatory challenges of this animal's environment.

For selected answers, see Appendix A.

45 Hormones and the Endocrine System

KEY CONCEPTS

45.1 Hormones and other signaling molecules bind to target receptors, triggering specific response pathways *p. 1000*

45.2 Feedback regulation and coordination with the nervous system are common in hormone pathways *p. 1004*

45.3 Endocrine glands respond to diverse stimuli in regulating homeostasis, development, and behavior *p. 1011*

Study Tip

Make a flowchart: Many hormones, such as insulin, parathyroid hormone, and epinephrine, have multiple physiological effects in a single organism. To keep track of the action and function of each such hormone, make a flowchart like this example. Use arrows to indicate how the hormone's diverse effects contribute to an overall outcome for the organism.

High level of testosterone in male elephant seals

Enlarged proboscis: Enables mating roar

Increased body size: Helps defend territory

Stimulates testes: Initiates sperm production

Reproductive success

→ Go to Mastering Biology

For Students (in eText and Study Area)
• Get Ready for Chapter 45
• Animation: Water-Soluble Hormone Pathway
• Animation: Steroid Hormone Pathway

For Instructors to Assign (in Item Library)
• Scientific Skills Exercise: Designing a Controlled Experiment
• Problem-Solving Exercise: Is Thyroid Regulation Normal in this Patient?

Figure 45.1 Male and female elephant seals (*Mirounga angustirostris*) differ greatly in appearance and behavior. The male is much larger, and only he has the prominent proboscis for which the species is named. The male is also far more territorial, using the proboscis to emit loud roars during mating season. Underlying each of these differences is a single hormone—testosterone. Like all hormones, testosterone is an endocrine signaling molecule that circulates in the blood throughout the body.

What variables shape a hormone's effect on an animal's body and behavior?

Concentration of the hormone in the body:
Testosterone is present in both male and female mammals, but typically at a much higher concentration in males.

Testosterone in urine (g/mL) — Male seals, Female seals

Presence of the hormone receptor in a cell:
A hormone circulates throughout the bloodstream, but cells only respond to a hormone if they have a receptor that binds that hormone specifically.

Blood vessel
Testosterone
Target cells

Response of the cell when the receptor binds the hormone:
Cells in different tissues may respond differently to the same hormone.

In juvenile proboscis | **In testis**

Testosterone

Cells elongate, enlarging proboscis. | Cells produce sperm.

▼ **Male elephant seals sparring**

Hormones and other signaling molecules bind to target receptors, triggering specific response pathways

A **hormone** (from the Greek *horman*, to excite) is a secreted molecule that circulates throughout the body and stimulates specific cells. Although a given hormone reaches all cells of the body, it only elicits a response—such as a change in metabolism—in specific *target cells*, those that have a receptor that binds the hormone specifically. Cells lacking a receptor for that hormone are unaffected.

Chemical signaling by hormones is the function of the **endocrine system**, one of the two basic systems for communication and regulation in the animal body. The other major communication and control system is the **nervous system**, a network of specialized cells—neurons—that transmit signals along dedicated pathways. These signals in turn regulate neurons, muscle cells, and endocrine cells. Because signaling by neurons can regulate the release of hormones, the nervous and endocrine systems often overlap in function.

As a background to our further exploration of the endocrine system, we'll begin with an overview of the diverse ways that animal cells use chemical signals to communicate.

Intercellular Information Flow

Communication between animal cells via secreted signals is often classified by two criteria: the type of secreting cell and the route taken by the signal in reaching its target. **Figure 45.2** illustrates five forms of signaling distinguished in this manner.

Endocrine Signaling

In endocrine signaling (see Figure 45.2a), hormones secreted into extracellular fluid by endocrine cells reach target cells via the bloodstream (or hemolymph). One function of endocrine signaling is to maintain homeostasis. Hormones regulate properties that include blood pressure and volume, energy metabolism and allocation, and solute concentrations in body fluids. Endocrine signaling also mediates responses to environmental stimuli, regulates growth and development, and triggers physical and behavioral changes underlying sexual maturity and reproduction (see Figure 45.1).

Paracrine and Autocrine Signaling

Many types of cells produce and secrete **local regulators**, molecules that act over short distances, reach their target cells solely by diffusion, and act on their target cells within seconds or even milliseconds. Local regulators play roles in many physiological processes, including blood pressure regulation, nervous system function, and reproduction.

▼ **Figure 45.2 Intercellular communication by secreted molecules.** In each type of signaling, secreted molecules (•) bind to a specific receptor protein (◡) expressed by target cells. Some receptors are located inside cells, but for simplicity, here all are drawn on the cell surface.

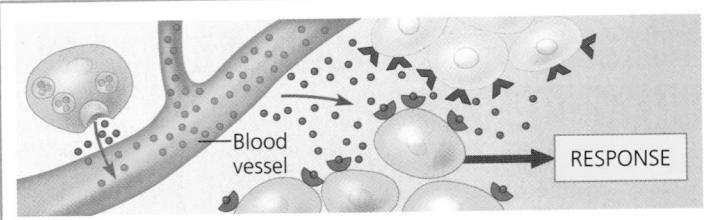

(a) In **endocrine signaling**, secreted molecules diffuse into the bloodstream and trigger responses in target cells anywhere in the body.

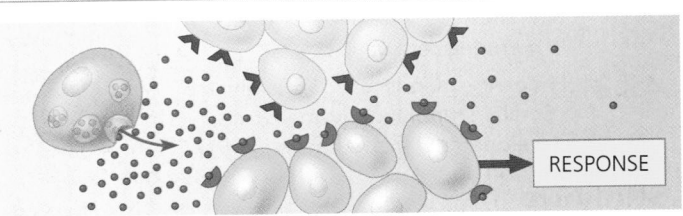

(b) In **paracrine signaling**, secreted molecules diffuse locally and trigger a response in neighboring cells.

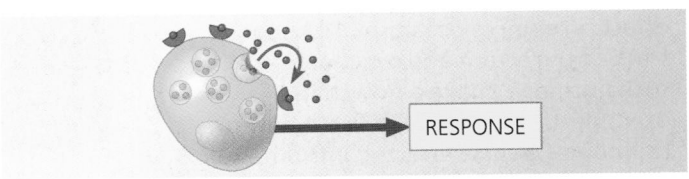

(c) In **autocrine signaling**, secreted molecules diffuse locally and trigger a response in the cells that secrete them.

(d) In **synaptic signaling**, neurotransmitters diffuse across synapses and trigger responses in target tissues (neurons, muscles, or glands).

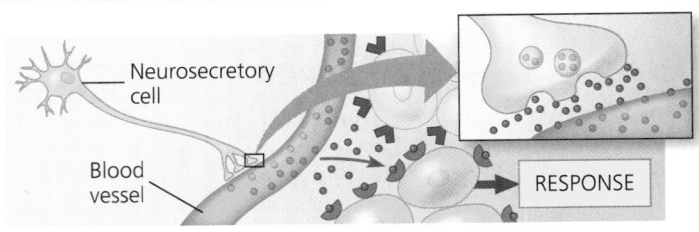

(e) In **neuroendocrine signaling**, neurohormones diffuse into the bloodstream and trigger responses in target cells anywhere in the body.

Depending on the target cell, signaling by local regulators is in general either paracrine or autocrine. In **paracrine** signaling (from the Greek *para*, to one side of), target cells lie near the secreting cell (see Figure 45.2b). In **autocrine** signaling (from the Greek *auto*, self), the secreting cells themselves are the target cells (see Figure 45.2c).

One group of local regulators are the **prostaglandins**, which are produced throughout the body and have diverse functions. In the immune system, for example, prostaglandins promote inflammation and the sensation of pain in response to injury. Drugs that block prostaglandin synthesis, such as aspirin and ibuprofen, prevent these activities, producing both anti-inflammatory and pain-relieving effects.

Prostaglandins are modified fatty acids. Many other local regulators are polypeptides, including cytokines, which enable immune cell communication (see Figure 43.16 and Figure 43.17), and growth factors, which promote cell growth, division, and development.

Some local regulators, such as **nitric oxide (NO)**, are gases. When the level of oxygen in the blood falls, endothelial cells in blood vessel walls synthesize and release NO. After diffusing into the surrounding smooth muscle cells, NO activates an enzyme that relaxes the cells. The result is vasodilation, which increases blood flow to tissues.

In human males, NO's ability to promote vasodilation enables sexual function by increasing blood flow into the penis, producing an erection. The drug Viagra (sildenafil citrate), a treatment for male erectile dysfunction, sustains an erection by prolonging activity of the NO response pathway.

Synaptic and Neuroendocrine Signaling

Secreted molecules are essential for the function of the nervous system. Neurons communicate with target cells, such as other neurons and muscle cells, via specialized junctions called synapses. At most synapses, neurons secrete molecules called **neurotransmitters** that diffuse a very short distance (a fraction of a cell diameter) and bind to receptors on the target cells (see Figure 45.2d). Such *synaptic signaling* is central to sensation, memory, cognition, and movement (as we'll explore in Chapters 48–50).

In *neuroendocrine signaling*, neurons called neurosecretory cells secrete **neurohormones**, which diffuse from nerve cell endings into the bloodstream (see Figure 45.2e). One example of a neurohormone is antidiuretic hormone, which functions in kidney function and water balance as well as courtship behavior. Many neurohormones regulate endocrine signaling, as we'll discuss later in this chapter.

▼ **Figure 45.3 Signaling by pheromones.** Using their lowered antennae, these Asian army ants (*Leptogenys distinguenda*) carry pupae and larvae along a pheromone-marked trail to a new nest site.

Signaling by Pheromones

Not all secreted signaling molecules act within the body. Members of a particular animal species sometimes communicate with each other via **pheromones**, chemicals that are released into the external environment. For example, when a foraging ant discovers a new food source, it marks its path back to the nest with a pheromone. Ants also use pheromones for guidance when a colony migrates to a new location **(Figure 45.3)**.

Pheromones serve a wide range of functions that include defining territories, warning of predators, and attracting potential mates. The polyphemus moth (*Antheraea polyphemus*) provides a noteworthy example: The sex pheromone released into the air by a female enables her to attract a male of the species from up to 4.5 km away. You'll read more about pheromone function when we take up the topic of animal behavior in Chapter 51.

Chemical Classes of Hormones

Hormones fall into three major chemical classes: polypeptides, steroids, and amines **(Figure 45.4)**. The hormone insulin, for example, is a polypeptide that contains two chains in its active form. Steroid hormones, such as cortisol, are lipids

▼ **Figure 45.4 Variation in hormone solubility and structure.**

Water-soluble (hydrophilic)	Lipid-soluble (hydrophobic)
Polypeptides	**Steroids**

Insulin 0.8 nm

Cortisol

Amines

Epinephrine

Thyroxine

MAKE CONNECTIONS *Cells synthesize epinephrine from the amino acid tyrosine. On the structure of epinephrine shown above, draw a circle around the portion of the molecule corresponding to the R group of tyrosine (see Figure 5.14).*

that contain four fused carbon rings; all are derived from the steroid cholesterol (see Figure 5.12). Epinephrine and thyroxine are amine hormones, each synthesized from a single amino acid, either tyrosine or tryptophan.

As Figure 45.4 indicates, hormones vary in their solubility in aqueous and lipid-rich environments. Polypeptides and most amine hormones are water-soluble, whereas steroid hormones and other largely nonpolar (hydrophobic) hormones, such as thyroxine, are lipid-soluble.

Cellular Hormone Response Pathways

Water-soluble and lipid-soluble hormones differ in their response pathways. One key difference is the location of the receptor proteins in target cells. Water-soluble hormones are secreted by exocytosis and travel freely in the bloodstream. Being insoluble in lipids, they cannot diffuse through the plasma membranes of target cells. Instead, these hormones bind to cell-surface receptors, inducing changes in cytoplasmic molecules and sometimes altering gene transcription **(Figure 45.5a)**. In contrast, lipid-soluble hormones exit endocrine cells by diffusing out across the membranes. They then bind to transport proteins, which keep them soluble in blood. After circulating in the blood, they diffuse into target cells and typically bind to receptors in the cytoplasm or nucleus **(Figure 45.5b)**. The hormone-bound receptor then triggers changes in gene transcription.

To explore further the distinct cellular responses to water-soluble and lipid-soluble hormones, we'll examine the two response pathways in turn.

Response Pathway for Water-Soluble Hormones

The binding of a water-soluble hormone to a cell-surface receptor protein triggers a cellular response. The response may be the activation of an enzyme, a change in the uptake or secretion of specific molecules, or a rearrangement of the cytoskeleton. In some cases, cell-surface receptors cause proteins in the cytoplasm to move into the nucleus and alter the transcription of specific genes.

The chain of events that converts the extracellular chemical signal to a specific intracellular response is called **signal transduction**. As an example, we'll consider one response to short-term stress. When you are in a stressful situation, perhaps running to catch a bus, the adrenal glands that lie atop your kidneys secrete the water-soluble hormone **epinephrine**, also known as *adrenaline*. Epinephrine regulates many organs, including the liver, where it binds to a G protein-coupled receptor in the plasma membrane of target cells. As shown in **Figure 45.6**, this interaction triggers a cascade of events involving synthesis of cyclic AMP (cAMP) as a short-lived *second messenger*. Activation of protein kinase A by cAMP leads to activation of an enzyme required for breakdown of

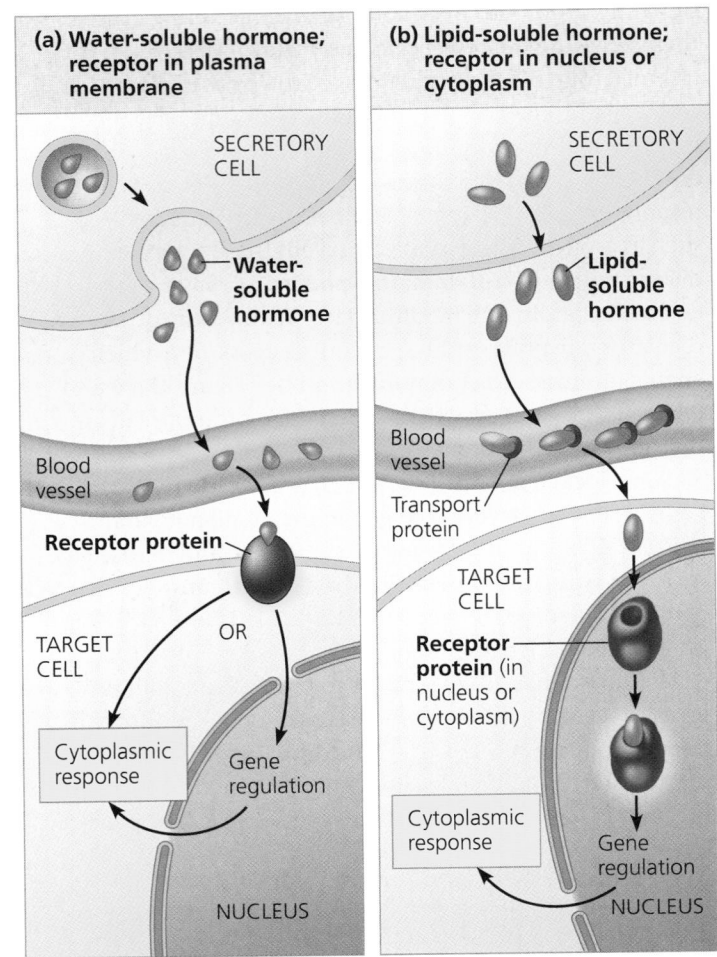

▼ **Figure 45.5 Variation in hormone receptor location.**

(a) Water-soluble hormone; receptor in plasma membrane

SECRETORY CELL

Water-soluble hormone

Blood vessel

Receptor protein

TARGET CELL

OR

Cytoplasmic response

Gene regulation

NUCLEUS

(b) Lipid-soluble hormone; receptor in nucleus or cytoplasm

SECRETORY CELL

Lipid-soluble hormone

Blood vessel

Transport protein

TARGET CELL

Receptor protein (in nucleus or cytoplasm)

Cytoplasmic response

Gene regulation

NUCLEUS

WHAT IF? *Suppose you are studying a cell's response to a particular hormone. You observe that the cell produces the same response to the hormone whether or not the cell is treated with a chemical that blocks transcription. What can you surmise about the hormone and its receptor?*

→ **Mastering Biology** Animation: Binding of Hormones

glycogen into glucose, as well as inactivation of an enzyme needed for glycogen synthesis.

Note that there are three enzymes in this signal transduction cascade—adenylyl cyclase (which converts AMP to its cyclic form), protein kinase A, and, for example, the enzyme that breaks down glycogen into glucose. Each enzyme-catalyzed step in the cascade provides an opportunity for signal amplification: One enzyme molecule can catalyze many reactions, thereby generating multiple signals at that step in the cascade. Furthermore, because the three enzymes act at different steps in the same pathway, the net effect can be enormous. If, for instance, each enzyme carried out 1,000 reactions, the binding of one molecule of epinephrine to its receptor would trigger cleavage of a billion ($10^3 \times 10^3 \times 10^3$) glycogen molecules. The net result is that the liver releases a substantial quantity of glucose into the bloodstream, quickly providing the body with extra fuel.

▼ **Figure 45.6 Signal transduction triggered by a cell-surface hormone receptor.**

VISUAL SKILLS *A series of arrows represents the steps linking epinephrine to protein kinase A. How does the event represented by the arrow between ATP and cAMP differ from the other four?*

→ **Mastering Biology** Animation: Water-Soluble Hormone Pathway

Response Pathway for Lipid-Soluble Hormones

Intracellular receptors for lipid-soluble hormones perform the entire task of transducing a signal within a target cell. The hormone activates the receptor, which then directly triggers the cell's response. In most cases, the response to a lipid-soluble hormone is a change in gene expression.

Most steroid hormone receptors are located in the cytosol prior to binding to a hormone. Binding of a steroid hormone to its cytosolic receptor forms a complex that moves into the nucleus (see Figure 11.9). There, the receptor portion of the complex interacts with a specific DNA-binding protein or response element in the DNA, altering transcription of particular genes. (In some cell types, steroid hormones trigger additional responses by interacting with other kinds of receptor proteins located at the cell surface).

Among the best-characterized steroid hormone receptors are those that bind to estrogens, steroid hormones necessary for female reproductive function in vertebrates. For example, in female birds and frogs, estradiol, a form of estrogen, binds to a cytoplasmic receptor in liver cells. Binding of estradiol to this receptor activates transcription of the vitellogenin gene **(Figure 45.7)**. Following translation of the messenger RNA, vitellogenin protein is secreted and transported in the blood to the reproductive system, where it is used to produce egg yolk.

▼ **Figure 45.7 Direct regulation of gene expression by a steroid hormone receptor.**

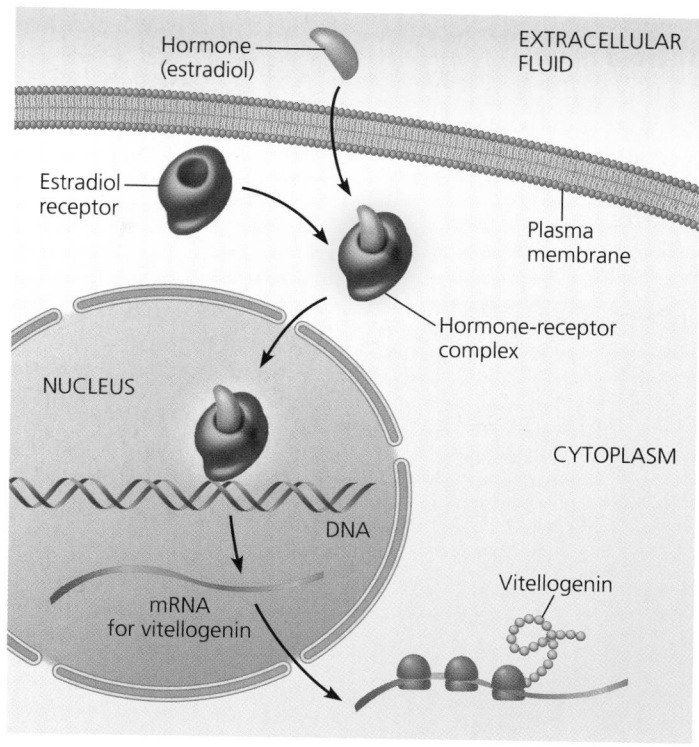

→ **Mastering Biology** Animation: Steroid Hormone Pathway

Thyroxine, vitamin D, and other lipid-soluble hormones that are not steroids typically have receptors in the nucleus. These receptors bind to hormone molecules that diffuse from the bloodstream across both the plasma membrane and nuclear envelope. Once bound to a hormone, the receptor binds to specific sites in the cell's DNA and stimulates the transcription of specific genes.

Multiple Responses to a Single Hormone

Although hormones bind to specific receptors, a particular hormone can vary in its effects. A hormone can elicit distinct responses in particular target cells if those cells differ in receptor type or in the molecules that produce the response. In this way a single hormone can trigger a range of activities that together bring about a coordinated response to a stimulus. For example, the multiple effects of epinephrine form the basis for the "fight-or-flight" response, a rapid response to stress that you'll read about in Concept 45.3.

Endocrine Tissues and Organs

Some endocrine cells are found in organs that are part of other organ systems. For example, the stomach contains isolated endocrine cells that help regulate digestive processes by secreting the hormone gastrin. More often, endocrine cells are grouped in ductless organs called **endocrine glands**,

▼ **Figure 45.8 Human endocrine glands and their hormones.** This figure highlights the location and primary functions of the major human endocrine glands. Endocrine tissues and cells are also located in the thymus, heart, liver, stomach, kidneys, and small intestine.

Thyroid gland
• *Thyroid hormone (T_3 and T_4):* Stimulates and maintains metabolic processes
• *Calcitonin:* Lowers blood calcium level

Parathyroid glands
• *Parathyroid hormone (PTH):* Raises blood calcium level

Adrenal glands (atop kidneys)
  Adrenal medulla
  • *Epinephrine* and *norepinephrine:* Raise blood glucose level; increase metabolic activity; constrict or dilate blood vessels
  Adrenal cortex
  • *Glucocorticoids:* Raise blood glucose level; control inflammation
  • *Mineralocorticoids:* Promote reabsorption of Na^+ and excretion of K^+ in kidneys

Ovaries (in females)
• *Estrogens*:* Stimulate uterine lining growth; promote development and maintenance of female secondary sex characteristics
• *Progesterone*:* Promotes uterine lining growth

Testes (in males)
• *Androgens*:* Support sperm formation; promote development and maintenance of male secondary sex characteristics

*Found in both males and females, but with a major role in one sex

Pineal gland
• *Melatonin:* Participates in regulation of biological rhythms

Hypothalamus
Hormones released from posterior pituitary (see below)
• *Releasing and inhibiting hormones:* Regulate anterior pituitary

Pituitary gland
  Posterior pituitary
  • *Oxytocin:* Stimulates contraction of uterus and mammary gland cells; modulates behavior
  • *Vasopressin,* also called *antidiuretic hormone (ADH):* Promotes retention of water by kidneys; modulates behavior
  Anterior pituitary
  • *Follicle-stimulating hormone (FSH)* and *luteinizing hormone (LH):* Stimulate gonads (ovaries or testes)
  • *Thyroid-stimulating hormone (TSH):* Stimulates thyroid gland
  • *Adrenocorticotropic hormone (ACTH):* Stimulates adrenal cortex
  • *Prolactin:* Stimulates mammary gland growth and milk synthesis in mammals
  • *Growth hormone (GH):* Stimulates growth and metabolic functions
  • *Melanocyte-stimulating hormone (MSH):* Affects color of melanocytes, a type of skin cell

Pancreas
• *Insulin:* Lowers blood glucose level
• *Glucagon:* Raises blood glucose level

such as the thyroid and parathyroid glands and the gonads, either testes in males or ovaries in females **(Figure 45.8)**.

Note that endocrine glands secrete hormones directly into the surrounding fluid. In contrast, *exocrine glands* have ducts that carry secreted substances, such as sweat or saliva, onto body surfaces or into body cavities. This distinction is reflected in the glands' names: The Greek *endo* (within) and *exo* (out of) refer to secretion into or out of body fluids, while *crine* (from the Greek word meaning "separate") refers to movement away from the secreting cell. In the case of the pancreas, endocrine and exocrine tissues are found in the same gland: Ductless tissues secrete hormones, whereas tissues with ducts secrete enzymes and bicarbonate.

CONCEPT CHECK 45.1

1. How do response mechanisms in target cells differ for water-soluble and lipid-soluble hormones?

2. What type of gland would you expect to secrete pheromones? Explain.

3. **WHAT IF?** Predict what would happen if you injected a water-soluble hormone into the cytosol of a target cell.

For suggested answers, see Appendix A.

Feedback regulation and coordination with the nervous system are common in hormone pathways

Having explored hormone structure, recognition, and response, we now consider how regulatory pathways controlling hormone secretion are organized.

Simple Endocrine Pathways

In a *simple endocrine pathway,* endocrine cells respond directly to an internal or environmental stimulus by secreting a particular hormone. The hormone travels in the bloodstream to target cells, where it interacts with its specific receptors. Signal transduction within target cells brings about a physiological response.

▼ **Figure 45.9 A simple endocrine pathway.** Endocrine cells respond to a change in some internal or external variable—the stimulus—by secreting hormone molecules that binds to a specific receptor protein expressed by target cells, triggering a particular response. In the case of secretin signaling, the simple endocrine pathway is self-limiting because the response to secretin (bicarbonate release) reduces the stimulus (low pH) through negative feedback.

▼ **Figure 45.10 A simple neuroendocrine pathway.** Sensory neurons respond to a stimulus by sending nerve impulses to a neurosecretory cell, triggering secretion of a neurohormone. Upon reaching its target cells, the neurohormone binds to its receptor, triggering a specific response. In oxytocin signaling, the response increases the stimulus, forming a positive-feedback loop that amplifies signaling.

Simple endocrine pathway	Example: secretin signaling
STIMULUS	Low pH in duodenum
Endocrine cell	S cells of duodenum
Hormone	Secretin (•)
Circulation throughout body via blood	
Target cells	Pancreatic cells
RESPONSE	Bicarbonate release

Negative feedback

Simple neuroendocrine pathway	Example: oxytocin signaling
STIMULUS	Suckling
Hypothalamus — Neurosecretory cell	
Posterior pituitary — Hormone	
Blood vessel	
Circulation throughout body via blood	Oxytocin (■)
Target cells	Smooth muscle in mammary glands
RESPONSE	Milk release

Positive feedback

The activity of endocrine cells in the duodenum, the first part of the small intestine, provides a useful example of a simple endocrine pathway. During digestion, the partially processed food that enters the duodenum contains highly acidic digestive juices secreted by the stomach. Before further digestion can occur, this acidic mixture must be neutralized. **Figure 45.9** outlines the simple endocrine pathway that ensures neutralization takes place.

The low pH of partially digested food entering the small intestine is detected by S cells, which are endocrine cells in the lining of the duodenum. In response, the S cells secrete the hormone *secretin*, which diffuses into the blood. Traveling throughout the circulatory system, secretin reaches the pancreas. Target exocrine cells in the pancreas have receptors for secretin and respond by releasing bicarbonate into ducts that lead to the duodenum. In the last step of the pathway, the bicarbonate released into the duodenum raises the pH, neutralizing the stomach acid.

Simple Neuroendocrine Pathways

In a *simple neuroendocrine pathway*, the stimulus is received by a sensory neuron rather than endocrine tissue. The sensory

neuron in turn stimulates a neurosecretory cell. In response, the neurosecretory cell secretes a neurohormone. Like other hormones, the neurohormone diffuses into the bloodstream and travels in the circulation to target cells.

As an example of a simple neuroendocrine pathway, consider the regulation of milk release during nursing in mammals **(Figure 45.10)**. When an infant suckles, it stimulates sensory neurons in the nipples, generating nerve impulses that reach the hypothalamus. This input triggers the secretion of the neurohormone **oxytocin** from the posterior pituitary gland. Oxytocin then causes contraction of mammary gland cells, forcing milk from reservoirs in the gland.

Feedback Regulation

A feedback loop linking a response back to an initial stimulus is a feature of many control pathways. Often, this loop involves **negative feedback**, in which the response reduces the initial stimulus. For instance, bicarbonate released in response to secretin increases pH in the intestine, eliminating the stimulus and thereby shutting off secretin release (see Figure 45.9). By decreasing hormone signaling, negative-feedback regulation prevents excessive pathway activity.

Whereas negative feedback dampens a stimulus, **positive feedback** reinforces a stimulus, driving a process to completion. For example, in the pathway outlined in Figure 45.10, the mammary glands secrete milk in response to circulating oxytocin. The released milk in turn leads to more suckling and thus more stimulation. Activation of the pathway is sustained until the baby is full and stops suckling. Other functions of oxytocin, such as stimulating contractions of the uterus during birthing, also exhibit positive feedback.

Comparing negative and positive feedback, we see that only negative feedback helps restore a preexisting state. It is not surprising, therefore, that hormone pathways involved in homeostasis typically exhibit negative feedback. Often such pathways are paired, providing even more balanced control. For example, the blood glucose level is tightly regulated by the opposing, or antagonistic, effects of insulin and glucagon (see Figure 41.23).

Coordination of the Endocrine and Nervous Systems

In a wide range of animals, endocrine organs in the brain integrate function of the endocrine system with that of the nervous system. We'll explore the basic principles of such integration in invertebrates and vertebrates.

Invertebrates

The control of development in a moth illustrates neuroendocrine coordination in invertebrates. A moth larva, such as the caterpillar of the giant silk moth (*Hyalophora cecropia*) shown in **Figure 45.11**, grows in stages. Because its exoskeleton cannot stretch, the larva must periodically molt, shedding the old exoskeleton and secreting a new one. The endocrine pathway that controls molting originates in the larval brain **(Figure 45.12)**. Neurosecretory cells in the brain produce PTTH, a polypeptide neurohormone. When PTTH in body fluids reaches an endocrine organ called the prothoracic gland, it directs release of a second hormone, *ecdysteroid*. Bursts of ecdysteroid trigger each successive molt.

Ecdysteroid also controls a remarkable change in form called metamorphosis. Within the larva lie islands of tissues that will become the eyes, wings, brain, and other adult structures. Once a plump, crawling larva becomes a stationary pupa, these islands of cells take over. They complete their program of development, while many larval tissues undergo programmed cell death. The end result is the transformation of the crawling caterpillar into a free-flying moth.

▼ **Figure 45.11 Larva of the giant silk moth.**

❶ Neurosecretory cells in the brain produce prothoracicotropic hormone (PTTH), which is stored in the corpora cardiaca until release.

❷ PTTH signals its main target organ, the prothoracic gland, to produce the hormone ecdysteroid.

❸ Ecdysteroid secretion from the prothoracic gland is episodic, with each release stimulating a molt.

❹ Juvenile hormone (JH), secreted by the corpora allata, determines the result of the molt. At relatively high concentrations, JH suppresses metamorphosis. Under these conditions, molting stimulated by ecdysteroid produces another larval stage. When JH drops below a certain concentration, a pupa forms at the next ecdysteroid-induced molt. The adult insect emerges from the pupa.

Brain
Neurosecretory cells
Corpora cardiaca
Corpora allata
PTTH
Prothoracic gland
High JH
Low JH
Ecdysteroid

EARLY LARVA — LATER LARVA — PUPA — ADULT

▶ **Figure 45.12 Regulation of insect development and metamorphosis.** As shown here for a giant silk moth, most insects go through a series of larval stages, with each molt (shedding of the old exoskeleton) leading to a larger larva. Molting of the final larval stage gives rise to a pupa, in which metamorphosis produces the adult form of the insect. Neurohormones and hormones control the progression of stages.

Given that ecdysteroid can cause either molting or metamorphosis, what determines which process takes place? The answer is another signal, juvenile hormone (JH), secreted by a pair of endocrine glands behind the brain. JH modulates ecdysteroid activity. When the level of JH in body fluids is high, ecdysteroid stimulates molting (and thus maintains the "juvenile" larval state). When the JH level drops, ecdysteroid instead induces formation of a pupa, within which metamorphosis occurs.

Knowledge of the coordination between the nervous system and endocrine system in insects has provided a basis for novel methods of agricultural pest control. For example, one tool to control insect pests is a chemical that binds to the ecdysteroid receptor, causing insect larvae to molt prematurely and die.

Vertebrates

In vertebrates, coordination of endocrine signaling relies heavily on the **hypothalamus (Figure 45.13)**. The hypothalamus receives information from nerves throughout the body and, in response, initiates neuroendocrine signaling appropriate to environmental conditions. In many vertebrates, for example, nerve signals from the brain pass sensory information to the hypothalamus about seasonal changes. The hypothalamus, in turn, regulates the release of reproductive hormones required during the breeding season.

Signals from the hypothalamus travel to the **pituitary gland**, a gland located at the base of the hypothalamus

▼ **Figure 45.13 Endocrine glands in the human brain.** This side view of the brain indicates the position of the hypothalamus, the pituitary gland, and the pineal gland. (The pineal gland plays a role in regulating biological rhythms.)

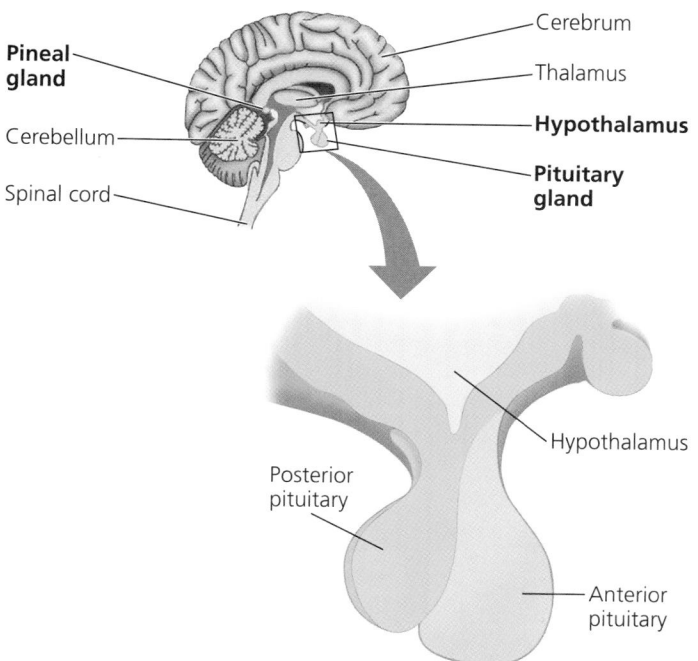

(see Figure 45.13). Roughly the size and shape of a lima bean, the pituitary is made up of two glands that fused during development but remain as discrete posterior and anterior parts, or lobes, that perform very different functions. The **posterior pituitary** is an extension of the neural tissue of the hypothalamus. Hypothalamic axons that reach into the posterior pituitary secrete neurohormones synthesized in the hypothalamus. In contrast, the **anterior pituitary** is an endocrine gland that synthesizes and secretes hormones in response to hormones from the hypothalamus.

Posterior Pituitary Hormones   Neurosecretory cells of the hypothalamus synthesize the two posterior pituitary hormones: antidiuretic hormone (ADH) and oxytocin. After traveling to the posterior pituitary within the long axons of the neurosecretory cells, these neurohormones are stored, to be released into the bloodstream in response to nerve impulses transmitted by the hypothalamus **(Figure 45.14)**.

Antidiuretic hormone (ADH), or *vasopressin*, regulates kidney function. Circulating ADH increases water retention in the kidneys, helping maintain normal blood osmolarity (see Concept 44.5). ADH also has an important role in social behavior (see Concept 51.4).

Oxytocin has multiple functions related to reproduction. As we have seen, in female mammals oxytocin controls milk secretion by the mammary glands and regulates uterine contractions during birthing. In addition, oxytocin has targets in

▼ **Figure 45.14 Production and release of posterior pituitary hormones.** The posterior pituitary gland is an extension of the hypothalamus. Certain neurosecretory cells in the hypothalamus make antidiuretic hormone (ADH) and oxytocin, which are transported to the posterior pituitary, where they are stored. Nerve signals from the brain trigger release of these neurohormones.

the brain, where it influences behaviors related to maternal care, pair bonding, and sexual activity.

Anterior Pituitary Hormones   Hormones secreted by the anterior pituitary control diverse processes in the human body, including metabolism, osmoregulation, and reproduction. As illustrated in **Figure 45.15**, many anterior pituitary hormones, but not all, regulate endocrine glands or tissues.

Hormones secreted by the hypothalamus control the release of all anterior pituitary hormones. Each hypothalamic hormone that regulates release of one or more hormones by the anterior pituitary is called a *releasing* or *inhibiting* hormone. *Prolactin-releasing hormone*, for example, is a hypothalamic hormone that stimulates the anterior pituitary to secrete **prolactin**, which has activities that include stimulating milk production. Each anterior pituitary hormone is controlled by at least one releasing hormone. Some, such as prolactin, have both a releasing hormone and an inhibiting hormone.

The hypothalamic releasing and inhibiting hormones are secreted near capillaries at the base of the hypothalamus. The capillaries drain into short blood vessels, called portal vessels, which subdivide into a second capillary bed within the anterior pituitary. Releasing and inhibiting hormones thus have direct access to the gland they control.

In neuroendocrine pathways, sets of hormones from the hypothalamus, the anterior pituitary, and a target endocrine gland are often organized into a *hormone cascade*, a form of regulation in which multiple endocrine organs and signals act in series. Signals to the brain stimulate the hypothalamus to secrete a hormone that stimulates or inhibits release of a specific anterior pituitary hormone. The anterior pituitary hormone in turn stimulates another endocrine organ to secrete yet another hormone, which affects specific target tissues. In reproduction, for example, the hypothalamus signals the anterior pituitary to release the hormones FSH and LH, which in turn regulate hormone secretion by the gonads (ovaries or testes).

In a sense, hormone cascade pathways redirect signals from the hypothalamus to other endocrine glands. For this reason, the anterior pituitary hormones in such pathways are called *tropic* hormones, or tropins, and are said to have a *tropic* effect (from the Greek *trope*, to turn). Thus, FSH and LH are gonadotropins because they convey signals from the hypothalamus to the gonads. To learn more about tropic hormones and hormone cascade pathways, we'll turn next to thyroid gland function and regulation.

Thyroid Regulation: A Hormone Cascade Pathway

In mammals, **thyroid hormone** regulates bioenergetics; helps maintain normal blood pressure, heart rate, and muscle tone; and regulates digestive and reproductive functions. **Figure 45.16** provides an overview of the hormone cascade

▶ **Figure 45.15 Production and release of anterior pituitary hormones.** The release of hormones synthesized in the anterior pituitary gland is controlled by hypothalamic releasing and inhibiting hormones. The hypothalamic hormones are secreted by neurosecretory cells and enter a capillary network within the hypothalamus. These capillaries drain into portal vessels that connect with a second capillary network in the anterior pituitary.

Neurosecretory cells of the hypothalamus

Hypothalamic releasing and inhibiting hormones

Portal vessels

Posterior pituitary

Endocrine cells of the anterior pituitary

Anterior pituitary hormones

HORMONE	FSH and LH	TSH	ACTH	Prolactin	MSH	GH
TARGET	Testes or ovaries	Thyroid	Adrenal cortex	Mammary glands	Melanocytes	Liver, bones, other tissues

Tropic effects only   **Nontropic effects only**   **Tropic and nontropic effects**

▼ Figure 45.16 Regulation of thyroid hormone secretion: a hormone cascade pathway.

STIMULUS

Neurosecretory cell in the hypothalamus

Negative feedback

TRH

TSH

Anterior pituitary

Circulation throughout body via blood

Thyroid gland

Thyroid hormone

Circulation throughout body via blood

RESPONSE

1 Thyroid hormone level drops below the normal range. Sensory neurons respond by sending nerve impulses to neurosecretory cells in the hypothalamus.

2 Neurosecretory cells secrete thyrotropin-releasing hormone (TRH ●) into the blood, which carries it to the anterior pituitary.

3 TRH causes the anterior pituitary to secrete thyroid-stimulating hormone (TSH, also known as thyrotropin ▲) into the circulatory system.

4 TSH stimulates endocrine cells in the thyroid gland to secrete thyroid hormone (T_3 and T_4 ■) into the circulatory system.

5 Thyroid hormone level increases in the blood and body tissues. Thyroid hormone acts on target cells throughout the body to control bioenergetics; help maintain normal blood pressure, heart rate, and muscle tone; and regulate digestive and reproductive functions.

6 As its level returns to the normal range, thyroid hormone blocks TRH release from the hypothalamus and TSH release from the anterior pituitary, forming a negative-feedback loop that prevents overproduction of thyroid hormone.

pathway that regulates thyroid hormone release. If the level of thyroid hormone in the blood drops, the hypothalamus secretes thyrotropin-releasing hormone (TRH), causing the anterior pituitary to secrete thyrotropin, a tropic hormone also known as thyroid-stimulating hormone (TSH). TSH in turn stimulates the **thyroid gland**, an organ in the neck consisting of two lobes on the ventral surface of the trachea. The thyroid gland responds by secreting thyroid hormone, which increases metabolic rate.

As with other hormone cascade pathways, feedback regulation often occurs at multiple levels. For example, thyroid hormone exerts negative feedback on the hypothalamus and on the anterior pituitary, in each case blocking release of the hormone that promotes its production (see Figure 45.16).

Disorders of Thyroid Function and Regulation

Disruption of thyroid hormone production and regulation can result in serious disorders. One such disorder reflects the unusual chemical makeup of thyroid hormone, the only iodine-containing molecule synthesized in the body. *Thyroid hormone* is actually a pair of very similar molecules derived from the amino acid tyrosine. *Triiodothyronine* (T_3) contains three iodine atoms, whereas tetraiodothyronine, or *thyroxine* (T_4), contains four (see Figure 45.4).

Although iodine is readily obtained from seafood or iodized salt, people in many parts of the world lack enough iodine in their diet to synthesize adequate amounts of thyroid hormone. With only a low blood level of thyroid hormone, the pituitary receives no negative feedback and continues to secrete TSH. An elevated TSH level in turn causes the thyroid gland to enlarge, resulting in goiter, a marked swelling of the neck.

Hormonal Regulation of Growth

Growth hormone (GH), which is secreted by the anterior pituitary, stimulates growth through both tropic and non-tropic effects. A major target, the liver, responds to GH by releasing *insulin-like growth factors* (*IGFs*), which circulate in the blood and directly stimulate bone and cartilage growth. (IGFs also appear to play a key role in aging in many animal species.) In the absence of GH, the skeleton of an immature animal stops growing. GH also exerts diverse metabolic effects that tend to raise the blood glucose level, thus opposing the effects of insulin.

Abnormal production of GH in humans can result in several disorders, depending on when the problem occurs and whether it involves hypersecretion (too much) or hyposecretion (too little). Hypersecretion of GH during childhood can lead to gigantism, in which the person grows unusually tall but retains relatively normal body

PROBLEM-SOLVING EXERCISE

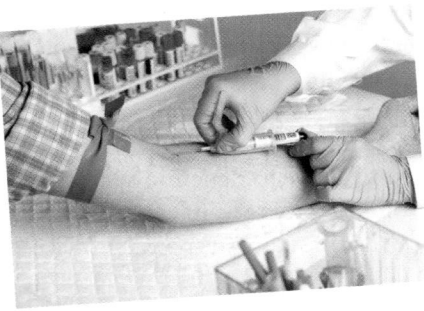

Is thyroid regulation normal in this patient?

Normal health requires proper regulation of the thyroid gland. Hypothyroidism, the secretion of too little thyroid hormone (T_3 and T_4), can cause weight gain, lethargy, and intolerance to cold in adults. In contrast, excessive secretion of thyroid hormone, known as hyperthyroidism, can lead to high body temperature, profuse sweating, weight loss, muscle weakness, irritability, and high blood pressure. Thyroid-stimulating hormone (TSH) stimulates the thyroid to release thyroid hormone. Testing for levels of T_3, T_4, and TSH in the blood can help diagnose various medical conditions.

Instructors: A version of this Problem-Solving Exercise can be assigned in **Mastering Biology**.

In this exercise, you will determine whether a 35-year-old man who came to the emergency room with episodes of paralysis has thyroid problems.

Your Approach As the emergency physician, you order a set of blood tests, including four that measure thyroid function. To determine whether the thyroid activity of your patient is normal, you will compare his blood test results with the normal range, as determined from a large set of healthy people.

Your Data

#	Test*	Patient	Normal Range	Comments
1	Total triiodothyronine (T_3)	2.93 nmol/L	0.89–2.44 nmol/L	
2	Free thyroxine (T_4)	27.4 pmol/L	9.0–21.0 pmol/L	
3	TSH	5.55 mU/L	0.35–4.94 mU/L	
4	TSH receptor autoantibody	0.2 U/mL	0–1.5 U/mL	

*T_3 and T_4 levels are measured as the number of molecules per unit volume: here, nanomoles (nmol, 10^{-9} moles) or picomoles (pmol, 10^{-12} moles) per liter (L). The levels of TSH and the autoantibody for its receptor are measured as activity, expressed in units (U) or milliunits (mU) per unit volume.

Your Analysis

1. For each test, determine whether the patient's test value is high, low, or normal relative to the normal range. Then write *High, Low*, or *Normal* in the comments column of the table.

2. Based on tests 1–3, is your patient hypothyroid or hyperthyroid?

3. Test 4 measures the level of autoantibodies (self-reactive antibodies) that bind to and activate the body's receptor for TSH. A high level of autoantibodies causes sustained thyroid hormone production and the autoimmune disorder called Graves' disease. Is it likely that your patient has this disease? Explain.

4. A thyroid tumor increases the mass of cells producing T_3 and T_4, whereas a tumor in the anterior pituitary increases the mass of TSH-secreting cells. Would you expect either condition to result in the observed blood test values? Explain.

---

proportions **(Figure 45.17)**. Excessive GH production in adulthood stimulates bony growth in the few body parts that are still responsive to the hormone—predominantly the face, hands, and feet. The result is an overgrowth of the extremities called acromegaly (from the Greek *acros*, extreme, and *mega*, large).

Hyposecretion of GH in childhood retards long-bone growth and can lead to pituitary dwarfism. People with this disorder are for the most part properly proportioned but generally reach a height of only about 1.2 m (4 feet). If diagnosed before puberty, pituitary dwarfism can be treated with human GH (also called HGH). Treatment with HGH produced by recombinant DNA technology is common.

Whereas the effects of altered growth hormone levels are readily related to a change in adult height, disrupting some endocrine pathways can have effects that appear unrelated to normal pathway function. The **Problem-Solving Exercise** explores one such example of a medical mystery difficult to diagnose based on symptoms alone.

▼ **Figure 45.17 Effect of growth hormone overproduction.** Shown here surrounded by his family, Robert Wadlow grew to a height of 2.7 m (8 feet 11 inches) by age 22, making him the tallest man in history. His height was due to excess secretion of growth hormone by his pituitary gland.

1. What are the roles of oxytocin and prolactin in regulating the mammary glands?

2. How do the two fused glands of the pituitary gland differ in function?

3. **WHAT IF?** Propose an explanation for why defects in a particular hormone cascade pathway observed in patients typically affect the final gland in the pathway rather than the hypothalamus or pituitary.

4. **WHAT IF?** Lab tests of two patients, each diagnosed with excessive thyroid hormone production, revealed an elevated level of TSH in one but not the other. Was the diagnosis of one patient necessarily incorrect? Explain.

For suggested answers, see Appendix A.

CONCEPT 45.3

Endocrine glands respond to diverse stimuli in regulating homeostasis, development, and behavior

In the remainder of this chapter, we'll focus on endocrine function in homeostasis, development, and behavior. We'll begin with another example of a simple hormone pathway, the regulation of calcium ion concentration in the circulatory system.

Parathyroid Hormone and Vitamin D: Control of Blood Calcium

Because calcium ions (Ca^{2+}) are essential to the normal functioning of all cells, homeostatic control of the level of calcium in the blood is vital. If the blood Ca^{2+} level falls substantially, skeletal muscles begin to contract convulsively, a potentially fatal condition. If the blood Ca^{2+} level rises substantially, calcium phosphate can form precipitates in body tissues, leading to widespread organ damage.

In mammals, the **parathyroid glands**, a set of four small structures embedded in the posterior surface of the thyroid (see Figure 45.8), play a major role in blood Ca^{2+} regulation. When the blood Ca^{2+} level falls below a set point of about 10 mg/100 mL, these glands release **parathyroid hormone (PTH)**.

PTH raises the level of blood Ca^{2+} through direct effects in bones and the kidneys and an indirect effect on the intestines (**Figure 45.18**). In bones, PTH causes the mineralized matrix to break down, releasing Ca^{2+} into the blood. In the kidneys, PTH directly stimulates reabsorption of Ca^{2+} through the renal tubules. In addition, PTH indirectly raises the blood Ca^{2+} level by promoting production of vitamin D. A precursor form of vitamin D is obtained from food or synthesized by skin exposed to sunlight. Conversion of this precursor to active vitamin D begins in the liver. PTH acts in the kidney to stimulate completion of the conversion process. Vitamin D in turn acts on the intestines, stimulating the uptake of Ca^{2+} from food. As the blood Ca^{2+} level rises, a negative-feedback loop inhibits further release of PTH from the parathyroid glands (not shown in Figure 45.18).

The thyroid gland can also contribute to calcium homeostasis. If the blood Ca^{2+} level rises above the set point, the thyroid gland releases **calcitonin**, a hormone that inhibits bone breakdown and enhances Ca^{2+} excretion by the kidneys. In fishes, rodents, and some other animals, calcitonin is required for Ca^{2+} homeostasis. In humans, however, calcitonin is apparently needed only during the extensive bone growth of childhood.

▶ **Figure 45.18 The roles of parathyroid hormone (PTH) in regulating the blood calcium level in mammals.**

NORMAL BLOOD Ca²⁺ LEVEL (about 10 mg/100 mL)

Blood Ca²⁺ level rises.

Blood Ca²⁺ level falls (such as when diet provides less calcium than is excreted in urine).

In the intestines, active vitamin D increases Ca²⁺ absorption from food into bloodstream.

In the kidney, PTH stimulates Ca²⁺ reabsorption from filtrate and promotes activation of vitamin D.

In bones, PTH stimulates Ca²⁺ release.

PTH

Parathyroid glands release PTH.

Adrenal Hormones: Response to Stress

The adrenal glands of vertebrates play a major role in the response to *stress*, a state of threatened homeostasis. Located atop the kidneys (the *renal* organs), each **adrenal gland** is actually made up of two glands with different cell types, functions, and embryonic origins: the adrenal *cortex*, the outer portion, and the adrenal *medulla*, the central portion **(Figure 45.19)**. The adrenal cortex consists of true endocrine cells, whereas the secretory cells of the adrenal medulla develop from neural tissue. Thus, like the pituitary gland, each adrenal gland is a fused endocrine and neuroendocrine gland.

The Role of the Adrenal Medulla

Imagine that while walking in the woods at night you hear a growling noise nearby. "A bear?" you wonder. Your heart beats faster, your breathing quickens, your muscles tense, and your thoughts speed up. These and other rapid responses to perceived danger comprise the "fight-or-flight" response. This coordinated set of physiological changes is triggered by two hormones of the adrenal medulla, epinephrine (adrenaline) and **norepinephrine** (also known as noradrenaline). Both are *catecholamines*, a class of amine hormones synthesized from the amino acid tyrosine. Both molecules also function as neurotransmitters, as you'll read in Concept 48.4.

As hormones, epinephrine and norepinephrine increase the amount of chemical energy available for immediate use (see Figure 45.19a). Both catecholamines increase the rate of glycogen breakdown in the liver and skeletal muscles and promote the release of glucose by liver cells and of fatty acids from fat cells. The released glucose and fatty acids circulate in the blood and can be used by body cells as fuel.

Catecholamines also exert profound effects on the cardiovascular and respiratory systems. For example, they increase heart rate and stroke volume and dilate the bronchioles in the lungs, actions that raise the rate of oxygen delivery to body cells. For this reason, doctors may prescribe epinephrine as a heart stimulant or to open the airways during an asthma attack. Catecholamines also alter blood flow, causing constriction of some blood vessels and dilation of others. The overall effect is to shunt blood away from the skin, digestive organs, and kidneys while increasing the blood supply to the heart, brain, and skeletal muscles.

▼ **Figure 45.19 Stress and the adrenal gland.**

(a) Stress response and the adrenal medulla

❶ The stimulus of physical threat, exercise, or cold exposure causes the hypothalamus to activate the adrenal medulla via nerve impulses.

❷ The adrenal medulla secretes epinephrine and norepinephrine.

Spinal cord (cross section)
Nerve impulses
Neuron
Adrenal medulla
Neuron
Stress
Releasing hormone
Hypothalamus
Anterior pituitary
ACTH
Circulation throughout body via blood
Adrenal gland
Kidney
Adrenal cortex

(b) Stress response and the adrenal cortex

❶ The stimulus of low blood sugar, decreased blood volume and pressure, or shock causes the hypothalamus to activate the adrenal cortex via hormonal signals.

❷ The adrenal cortex secretes mineralocorticoids and glucocorticoids.

Effects of epinephrine and norepinephrine:
- Glycogen broken down to glucose; increased blood glucose
- Increased blood pressure
- Increased breathing rate
- Increased metabolic rate
- Change in blood flow patterns, leading to increased alertness and decreased digestive, excretory, and reproductive system activity

Effects of mineralocorticoids:
- Retention of sodium ions and water by kidneys
- Increased blood volume and blood pressure

Effects of glucocorticoids:
- Proteins and fats broken down and converted to glucose, leading to increased blood glucose
- Partial suppression of immune system

Mastering Biology Animation: Hormonal Response to Stress

Epinephrine's Multiple Effects: *A Closer Look*   How can epinephrine coordinate a response to stress that involves widely varying effects in individual tissues? We can answer that question by examining different response pathways **(Figure 45.20)** in a range of target cells:

- In liver cells, epinephrine binds to a β-type receptor in the plasma membrane. This receptor activates the enzyme protein kinase A, which in turn regulates enzymes of glycogen metabolism, causing release of glucose into the blood (see Figure 45.20a). Note that this is the signal transduction pathway illustrated in Figure 45.6.
- In the smooth muscle cells that line blood vessels supplying skeletal muscle, the same kinase activated by the same epinephrine receptor inactivates a muscle-specific enzyme. The result is smooth muscle relaxation, leading to vasodilation and hence increased blood flow to skeletal muscles (see Figure 45.20b).
- In the smooth muscle cells lining blood vessels of the intestines, epinephrine binds to an α-type receptor (see Figure 45.20c). This receptor triggers a signaling pathway that involves enzymes other than protein kinase A and that causes smooth muscle contraction rather than

relaxation. The resulting vasoconstriction reduces blood flow to the intestines, facilitating the redirection of blood to active skeletal muscle.

Thus, epinephrine elicits multiple responses if its target cells differ in the receptor protein they express or in the molecules activated by the receptor upon hormone binding. As illustrated in these examples, such variation in response plays a key role in enabling epinephrine to trigger a range of activities that together bring about a coordinated rapid response to stressful stimuli.

The Role of the Adrenal Cortex

Like the adrenal medulla, the adrenal cortex mediates an endocrine response to stress (see Figure 45.19b). The two portions of the adrenal gland differ, however, in both the types of stress that trigger a response and the targets of the hormones that are released.

The adrenal cortex becomes active under stressful conditions that include low blood sugar, decreased blood volume and pressure, and shock. Such stimuli cause the hypothalamus to secrete a releasing hormone that stimulates the anterior pituitary to release adrenocorticotropic hormone (ACTH), a tropic hormone. When ACTH reaches the adrenal cortex via the bloodstream, it stimulates the endocrine cells to synthesize and secrete a family of steroids called *corticosteroids*. The two main types of corticosteroids in humans are glucocorticoids and mineralocorticoids.

Glucocorticoids, such as cortisol (see Figure 45.4), make more glucose available as fuel by promoting glucose synthesis from noncarbohydrate sources, such as proteins. Glucocorticoids also act on skeletal muscle, causing the breakdown of muscle proteins into amino acids. These are transported to the liver and kidneys, converted to glucose, and released into the blood. The synthesis of glucose upon the breakdown of muscle proteins provides circulating fuel when the body requires more glucose than the liver can mobilize from its glycogen stores.

If glucocorticoids are introduced into the body at a level above that normally present, they suppress certain components of the body's immune system. For this reason, glucocorticoids are sometimes used to treat inflammatory diseases such as arthritis. However, their long-term use can have serious side effects on metabolism. Nonsteroidal anti-inflammatory drugs (NSAIDs), such as

▼ **Figure 45.20 One hormone, different effects.** Epinephrine, the primary "fight-or-flight" hormone, produces different responses in different target cells. Target cells with the same receptor exhibit different responses if they have different signal transduction pathways or effector proteins; compare **(a)** with **(b)**. Target cells with different receptors for the hormone often exhibit different responses; compare **(b)** with **(c)**.

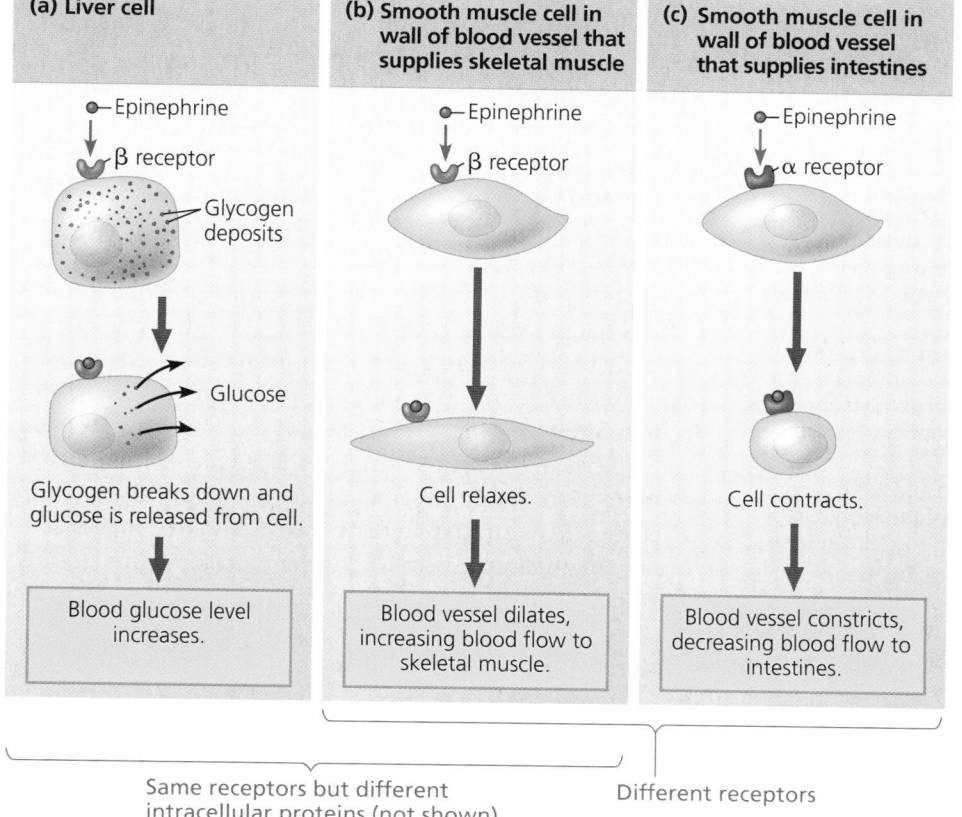

(a) Liver cell

Epinephrine
β receptor
Glycogen deposits
Glucose

Glycogen breaks down and glucose is released from cell.

Blood glucose level increases.

(b) Smooth muscle cell in wall of blood vessel that supplies skeletal muscle

Epinephrine
β receptor

Cell relaxes.

Blood vessel dilates, increasing blood flow to skeletal muscle.

(c) Smooth muscle cell in wall of blood vessel that supplies intestines

Epinephrine
α receptor

Cell contracts.

Blood vessel constricts, decreasing blood flow to intestines.

Same receptors but different intracellular proteins (not shown)

Different receptors

Designing a Controlled Experiment

How Is Nighttime ACTH Secretion Related to Expected Sleep Duration? Humans secrete increasing amounts of adrenocorticotropic hormone (ACTH) during the late stages of normal sleep, with the peak secretion occurring at the time of spontaneous waking. Because ACTH is released in response to stressful stimuli, scientists hypothesized that ACTH secretion prior to waking might be an anticipatory response to the stress associated with transitioning from sleep to a more active state. If so, an individual's expectation of waking at a particular time might influence the timing of ACTH secretion. How can such a hypothesis be tested? In this exercise, you will examine how researchers designed a controlled experiment to study the role of expectation.

How the Experiment Was Done Researchers studied 15 healthy volunteers in their mid-20s over three nights. Each night, each subject was told when he or she would be awakened: 6:00 or 9:00 AM. The subjects went to sleep at midnight. Subjects in the "short" or "long" protocol group were awakened at the expected time (6:00 or 9:00 AM, respectively). Subjects in the "surprise" protocol group were told they would be awakened at 9:00 AM, but were actually awakened 3 hours early, at 6:00 AM. At set times, blood samples were drawn to determine plasma levels of ACTH. To determine the change (Δ) in ACTH concentration post-waking, the researchers compared samples drawn at waking and 30 minutes later.

Data from the Experiment

Sleep Protocol	Expected Wake Time	Actual Wake Time	Mean Plasma ACTH Level (pg/mL)		
			1:00 AM	6:00 AM	Δ in the 30 Minutes Post-waking
Short	6:00 AM	6:00 AM	9.9	37.3	10.6
Long	9:00 AM	9:00 AM	8.1	26.5	12.2
Surprise	9:00 AM	6:00 AM	8.0	25.5	22.1

Data from J. Born et al., Timing the end of nocturnal sleep, *Nature* 397:29–30 (1999).

INTERPRET THE DATA

1. Describe the role of the "surprise" protocol in the experimental design.
2. Each subject was given a different protocol on each of the three nights, and the order of the protocols was varied among the subjects that so that one-third had each protocol each night. What factors were the researchers attempting to control for with this approach?
3. For subjects in the short protocol, what was the mean ACTH level at waking? Using the data in the last two columns, calculate the mean level 30 minutes later. Was the rate of change faster or slower in that 30-minute period than during the interval from 1:00 to 6:00 AM?
4. How does the change in ACTH level between 1:00 and 6:00 AM for the surprise protocol compare to that for the short and long protocols? Does this result support the hypothesis being tested? Explain.
5. Using the data in the last two columns, calculate the mean ACTH concentration 30 minutes post-waking for the surprise protocol and compare to your answer for question 3. What do your results suggest about a person's physiological response immediately after waking?
6. What are some variables that weren't controlled for in this experiment that could be explored in a follow-up study?

➔ **Instructors:** A version of this Scientific Skills Exercise can be assigned in **Mastering Biology**.

---

aspirin and ibuprofen, are therefore generally preferred for treating chronic inflammatory conditions.

Mineralocorticoids act principally in maintaining salt and water balance. For example, the mineralocorticoid *aldosterone* functions in ion and water homeostasis of the blood (see Figure 44.21). Like glucocorticoids, mineralocorticoids not only mediate stress responses, but also participate in homeostatic regulation of metabolism. In the **Scientific Skills Exercise**, you can explore an experiment investigating changes in ACTH secretion as humans awaken from sleep.

Sex Hormones

Sex hormones affect growth, development, reproductive cycles, and sexual behavior. Although the adrenal glands secrete small quantities of these hormones, the gonads (testes of males and ovaries of females) are their principal sources. The gonads produce and secrete three major types of steroid sex hormones: androgens, estrogens, and progesterone. All three types are found in both males and females but in different proportions.

The testes primarily synthesize **androgens**, the main one being **testosterone**. In humans, testosterone first functions in male (XY) embryos, promoting development of male reproductive structures **(Figure 45.21)**. In female (XX) embryos, the absence of testosterone allows the development of female reproductive structures. You can learn more about this role of hormones in the development of an embryo as male or female in the Scientific Skills Exercise in Chapter 46.

Androgens play a major role again at puberty, when they are responsible for the development of male secondary sex characteristics. High concentrations of androgens lead to lengthening and thickening of the vocal cords that lower the voice, male patterns of hair growth, and increases in muscle and bone mass. The muscle-building, or anabolic, action of testosterone and related steroids has enticed some athletes to take them as supplements, despite prohibitions against their use in nearly all sports. Use of anabolic steroids, while effective in increasing muscle mass, can cause severe acne outbreaks and liver damage, as well as significant decreases in sperm count and testicular size.

Estrogens, of which the most important is **estradiol**, are responsible for the maintenance of the female reproductive system and for the development of female secondary sex characteristics. In contrast, **progesterone** is involved in preparing and maintaining tissues of the mammalian uterus required to support the growth and development of an embryo.

Gonadal androgens, estrogens, and progesterone are components of hormone cascade pathways. Synthesis of these hormones is primarily controlled by two gonadotropins from the anterior pituitary, follicle-stimulating hormone and luteinizing hormone (see Figure 45.15). Gonadotropin secretion is in turn controlled by GnRH (gonadotropin-releasing hormone) from the hypothalamus.

Figure 45.21 Sex hormones regulate formation of internal reproductive structures in human development.

▼ **Figure 45.21 Sex hormones regulate formation of internal reproductive structures in human development.** In a male (XY) embryo, the bipotential gonads (gonads that can develop into either of two forms) become the testes, which secrete testosterone and anti-Müllerian hormone (AMH). Testosterone directs formation of sperm-carrying ducts (vas deferens and seminal vesicles), while AMH causes the female ducts to degenerate. In the absence of these testis hormones, the male ducts degenerate and female structures form, including the oviduct, uterus, and vagina.

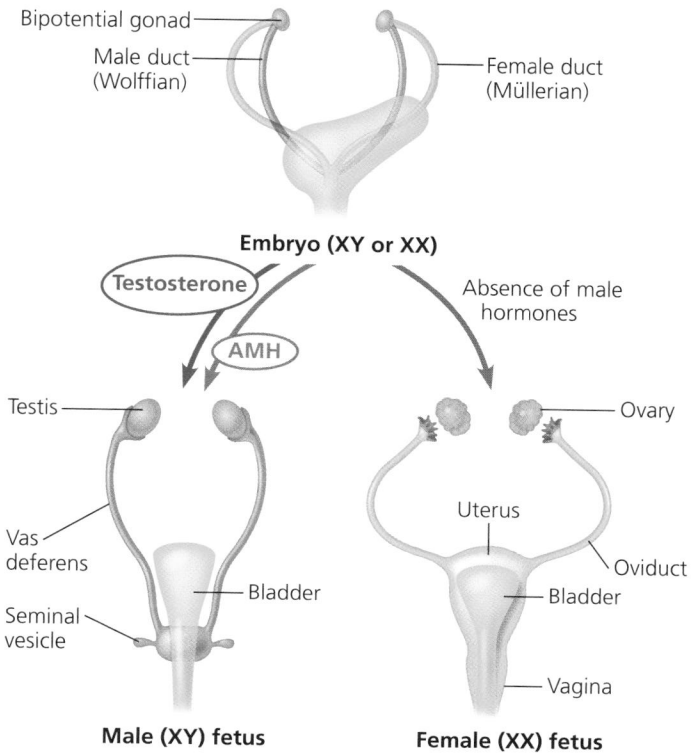

VISUAL SKILLS *Looking at this figure, explain why the adjective* bipotential *is only used to describe the gonad.*

→ **Mastering Biology** BBC Video: Male, Female, or Intersex?

We'll examine the feedback relationships that regulate gonadal hormone secretion in detail in Chapter 46.

Endocrine Disruptors

Between 1938 and 1971, some pregnant women at risk for pregnancy complications were prescribed a synthetic estrogen called diethylstilbestrol (DES). What was not known until 1971 was that exposure to DES can alter reproductive system development in the fetus. Daughters of women who took DES more frequently developed certain reproductive abnormalities, including vaginal and cervical cancer, structural changes in the reproductive organs, and increased risk of miscarriage (spontaneous abortion). DES is now recognized as an *endocrine disruptor*, a foreign molecule that interrupts the normal function of a hormone pathway.

In recent years, some scientists have hypothesized that molecules in the environment also act as endocrine disruptors. For example, bisphenol A, a chemical used in making some plastics,

has been studied for potential interference with normal reproduction and development. In addition, it has been suggested that some estrogen-like molecules, such as those present in soybeans and other edible plant products, have the beneficial effect of lowering breast cancer risk. Sorting out such effects, whether harmful or beneficial, has proven quite difficult, in part because enzymes in the liver change the properties of any such molecules entering the body through the digestive system.

Hormones and Biological Rhythms

There is still much to be learned about the hormone **melatonin**, a modified amino acid that regulates functions related to light and the seasons. Melatonin is produced by the **pineal gland**, a small mass of tissue near the center of the mammalian brain (see Figure 45.13).

Although melatonin affects skin pigmentation in many vertebrates, its primary effects relate to biological rhythms associated with reproduction and with the daily activity level (see Figure 40.9). Melatonin is secreted at night, and the amount released depends on the length of the night. In winter, for example, when days are short and nights are long, more melatonin is secreted. There is also good evidence that nightly increases in the level of melatonin play a significant role in promoting sleep.

The release of melatonin by the pineal gland is controlled by a group of neurons in the hypothalamus called the suprachiasmatic nucleus (SCN). The SCN functions as a biological clock and receives input from specialized light-sensitive neurons in the retina of the eye. Although the SCN regulates melatonin production during the 24-hour light/dark cycle, melatonin also influences SCN activity. We'll consider biological rhythms further in Concept 49.2, where we analyze experiments on SCN function.

Evolution of Hormone Function

EVOLUTION Over the course of evolution, the functions of a given hormone often diverge between species. An example is thyroid hormone, which across many evolutionary lineages plays a role in regulating metabolism (see Figure 45.16). In frogs, however, the thyroid hormone thyroxine (T_4) has taken on an apparently unique function: stimulating resorption of the tadpole's tail during metamorphosis **(Figure 45.22)**.

▼ **Figure 45.22 Specialized role of a hormone in frog metamorphosis.** The hormone thyroxine is responsible for the resorption of the tadpole's tail as the frog develops into its adult form.

▲ **Tadpole**

▲ **Adult frog**

The hormone *prolactin* has an especially broad range of activities. Prolactin stimulates mammary gland growth and milk synthesis in mammals, regulates fat metabolism and reproduction in birds, delays metamorphosis in amphibians, and regulates salt and water balance in freshwater fishes. These varied roles indicate that prolactin is an ancient hormone with functions that have diversified during the evolution of vertebrate groups.

Melanocyte-stimulating hormone (MSH), secreted by the anterior pituitary, provides another example of a hormone with distinct functions in different evolutionary lineages. In amphibians, fishes, and reptiles, MSH regulates skin color by controlling pigment distribution in skin cells called melanocytes. In mammals, MSH functions in hunger and metabolism in addition to skin coloration.

The specialized action of MSH that has evolved in the mammalian brain may prove to be of particular medical importance. Many patients with late-stage cancer, AIDS, tuberculosis, and certain aging disorders develop a devastating wasting condition called cachexia. Characterized by weight loss, muscle atrophy, and loss of appetite, cachexia responds poorly to existing therapies. However, it turns out that activation of a brain receptor for MSH produces some of the same changes seen in cachexia. Moreover, in experiments on mice with mutations that cause cancer and consequently cachexia, treatment with drugs that blocked the brain receptor for MSH prevented cachexia. Whether such drugs can be used to treat cachexia in humans is an area of active study.

45 Chapter Review

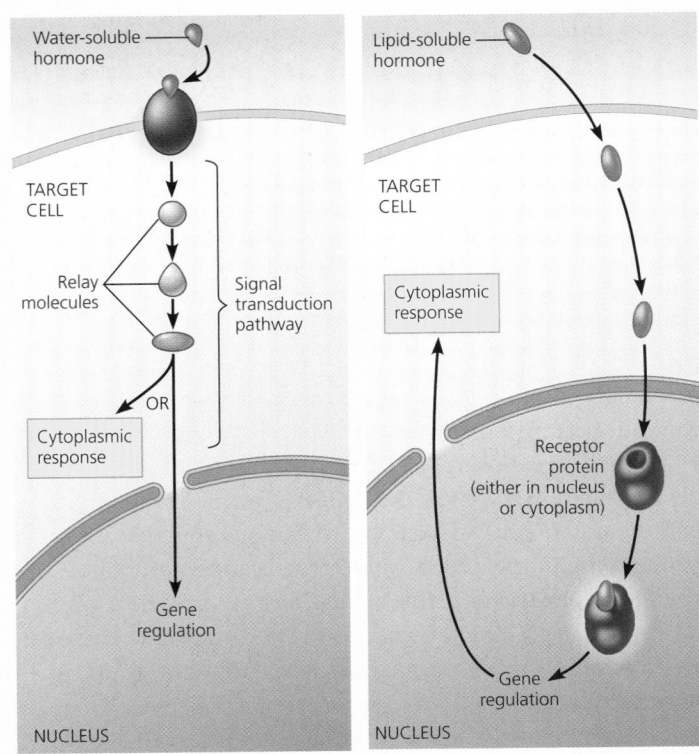

→ Go to **Mastering Biology** for Assignments, the eText, the Study Area, and Dynamic Study Modules.

SUMMARY OF KEY CONCEPTS

→ To review key terms, go to the **Vocabulary Self-Quiz** in the **Mastering Biology** eText or Study Area, or go to goo.gl/zkjz9t.

CONCEPT 45.1

Hormones and other signaling molecules bind to target receptors, triggering specific response pathways (pp. 1000–1004)

- The forms of signaling between animal cells differ in the type of secreting cell and the route taken by the signal to its target. **Endocrine** signals, or **hormones**, are secreted into the extracellular fluid by endocrine cells or ductless glands and reach target cells via circulatory fluids. There the binding of a hormone to a receptor specific for that particular hormone triggers a cellular response. **Paracrine** signals act on neighboring cells, whereas **autocrine** signals act on the secreting cell itself. **Neurotransmitters** also act locally, but **neurohormones** can act throughout the body. **Pheromones** are released into the environment for communication between animals of the same species.

- **Local regulators**, which carry out paracrine and autocrine signaling, include cytokines and growth factors (polypeptides), **prostaglandins** (modified fatty acids), and **nitric oxide** (a gas).

- Polypeptides, steroids, and amines comprise the major classes of animal hormones. Depending on whether they are water-soluble or lipid-soluble, hormones activate different response pathways. The endocrine cells that secrete hormones are often located in glands dedicated in part or in whole to endocrine signaling.

MAKE CONNECTIONS *What forms of signaling activate a helper T cell in immune responses (see Figure 43.18)?*

Feedback regulation and coordination with the nervous system are common in hormone pathways (pp. 1004–1011)

- In a simple endocrine pathway, endocrine cells respond directly to a stimulus. By contrast, in a simple neuroendocrine pathway, a sensory neuron receives the stimulus.
- Hormone pathways may include **negative feedback**, which dampens the stimulus and thus limits the response, or **positive feedback**, which amplifies the stimulus and drives the response to completion.

- In insects, molting and development are controlled by three hormones: PTTH; ecdysteroid, whose release is triggered by PTTH; and juvenile hormone. Coordination of signals from the nervous and endocrine systems and modulation of one hormone activity by another bring about the sequence of developmental stages that lead to an adult form.
- In vertebrates, neurosecretory cells in the **hypothalamus** produce two hormones that are secreted by the **posterior pituitary** and that act directly on nonendocrine tissues: **oxytocin**, which induces uterine contractions and release of milk from mammary glands, and **antidiuretic hormone (ADH)**, which enhances water reabsorption in the kidneys.
- Other hypothalamic cells produce hormones that are transported to the **anterior pituitary**, where they stimulate or inhibit the release of particular hormones.
- Often, anterior pituitary hormones act in a cascade. For example, the secretion of thyroid-stimulating hormone (TSH) is regulated by thyrotropin-releasing hormone (TRH). TSH in turn induces the **thyroid gland** to secrete **thyroid hormone**, a combination of the iodine-containing hormones T_3 and T_4. Thyroid hormone stimulates metabolism and influences development and maturation.

- Most anterior pituitary hormones are tropic hormones, acting on endocrine tissues or glands to regulate hormone secretion. Tropic hormones of the anterior pituitary include TSH, follicle-stimulating hormone (FSH), luteinizing hormone (LH), and adrenocorticotropic hormone (ACTH). **Growth hormone (GH)** has both tropic and nontropic effects. It promotes growth directly, affects metabolism, and stimulates the production of growth factors by other tissues.

? *Which major endocrine organs described in Figure 45.8 are regulated independently of the hypothalamus and pituitary?*

Endocrine glands respond to diverse stimuli in regulating homeostasis, development, and behavior (pp. 1011–1016)

- **Parathyroid hormone (PTH)**, secreted by the **parathyroid glands**, causes bone to release Ca^{2+} into the blood and stimulates reabsorption of Ca^{2+} in the kidneys. PTH also stimulates the kidneys to activate vitamin D, which promotes intestinal uptake of Ca^{2+} from food. **Calcitonin**, secreted by the thyroid, has the opposite effects in bones and kidneys as PTH. Calcitonin is important for calcium homeostasis in adults of some vertebrates, but not humans.
- In response to stress, neurosecretory cells in the adrenal medulla release **epinephrine** and **norepinephrine**, which mediate various fight-or-flight responses. The adrenal cortex releases **glucocorticoids**, such as cortisol, which influence glucose metabolism and the immune system. It also releases **mineralocorticoids**, primarily aldosterone, which help regulate salt and water balance.
- Sex hormones regulate growth, development, reproduction, and sexual behavior. Although the adrenal cortex produces small amounts of these hormones, the gonads (testes and ovaries) serve as the major source. All three types—**androgens, estrogens**, and **progesterone**—are produced in males and females, but in different proportions.
- The **pineal gland**, located within the brain, secretes **melatonin**, which functions in biological rhythms related to reproduction and sleep. Release of melatonin is controlled by the SCN, the region of the brain that functions as a biological clock.
- Hormones have acquired distinct roles in different species over the course of evolution. **Prolactin** stimulates milk production in mammals but has diverse effects in other vertebrates. **Melanocyte-stimulating hormone (MSH)** influences fat metabolism in mammals and skin pigmentation in other vertebrates.

? *ADH and epinephrine act as hormones when released into the bloodstream and as neurotransmitters when released in synapses between neurons. What is similar about the endocrine glands that produce these two molecules?*

TEST YOUR UNDERSTANDING

→ For more multiple-choice questions, go to the **Practice Test** in the **Mastering Biology** eText or Study Area, or go to goo.gl/GruWRg.

Levels 1-2: Remembering/Understanding

1. Which statement is accurate?
   (A) Hormones that differ in effect reach their target cells by different routes through the body.
   (B) Pairs of hormones that have the same effect are said to have antagonistic functions.
   (C) Hormones are often regulated through feedback loops.
   (D) Hormones of the same chemical class usually have the same function.

2. The hypothalamus
   (A) synthesizes all of the hormones produced by the pituitary gland.
   (B) influences the function of only one lobe of the pituitary gland.
   (C) produces only inhibitory hormones.
   (D) regulates both reproduction and body temperature.

3. Growth factors are local regulators that
   (A) are produced by the anterior pituitary.
   (B) are modified fatty acids that stimulate bone and cartilage growth.
   (C) are found on the surface of cancer cells and stimulate abnormal cell division.
   (D) bind to cell-surface receptors and stimulate growth and development of target cells.

4. Which hormone is *correctly* paired with its action?
   (A) oxytocin—stimulates uterine contractions during childbirth
   (B) thyroxine—inhibits metabolic processes
   (C) ACTH—inhibits the release of glucocorticoids by the adrenal cortex
   (D) melatonin—raises blood calcium level

Levels 3-4: Applying/Analyzing

5. What do steroid and peptide hormones typically have in common?
   (A) their solubility in cell membranes
   (B) their requirement for travel through the bloodstream
   (C) the location of their receptors
   (D) their reliance on signal transduction in the cell

6. Which of the following is the most likely explanation for hypothyroidism in a patient whose iodine level is normal?
   (A) greater production of T_3 than of T_4
   (B) hyposecretion of TSH
   (C) hypersecretion of MSH
   (D) a decrease in the thyroid secretion of calcitonin

7. The relationship between the insect hormones ecdysteroid and PTTH is an example of
   (A) an interaction of the endocrine and nervous systems.
   (B) homeostasis achieved by positive feedback.
   (C) homeostasis maintained by antagonistic hormones.
   (D) competitive inhibition of a hormone receptor.

8. **DRAW IT** In mammals, milk production by mammary glands is controlled by prolactin and prolactin-releasing hormone. Draw a simple sketch of this pathway, including glands, tissues, hormones, routes for hormone movement, and effects.

Levels 5-6: Evaluating/Creating

9. **EVOLUTION CONNECTION** The intracellular receptors used by all the steroid and thyroid hormones are similar enough in structure that they are all considered members of one "superfamily" of proteins. Propose a hypothesis for how the genes encoding these receptors may have evolved. (Hint: See Figure 21.13.) Explain how you could test your hypothesis using DNA sequence data.

10. **SCIENTIFIC INQUIRY • INTERPRET THE DATA** A chronically high level of glucocorticoids can result in obesity, muscle weakness, and depression, a combination of symptoms called Cushing's syndrome. Excessive activity of either the pituitary or the adrenal gland can be the cause. To determine which gland has abnormal activity in a particular patient, doctors use the drug dexamethasone, a synthetic glucocorticoid that blocks ACTH release. Based on the graph, identify which gland is affected in patient X.

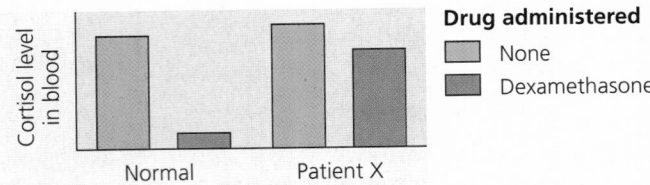

11. **WRITE ABOUT A THEME: INTERACTIONS** In a short essay (100–150 words), discuss the role of hormones in an animal's responses to changes in its environment. Use specific examples.

12. **SYNTHESIZE YOUR KNOWLEDGE**

The frog on the left was injected with MSH, causing a change in skin color within minutes due to a rapid redistribution of pigment granules in specialized skin cells. Using what you know about neuroendocrine signaling, explain how a frog could use MSH to match its skin coloration to that of its surroundings.

For selected answers, see Appendix A.

46 Animal Reproduction

KEY CONCEPTS

46.1 Both asexual and sexual reproduction occur in the animal kingdom *p. 1020*

46.2 Fertilization depends on mechanisms that bring together sperm and eggs of the same species *p. 1022*

46.3 Reproductive organs produce and transport gametes *p. 1025*

46.4 The interplay of tropic and sex hormones regulates reproduction in mammals *p. 1030*

46.5 In placental mammals, an embryo develops fully within the mother's uterus *p. 1034*

Study Tip

Make a table: To help you keep track of the roles of the tropic hormones in mammalian reproductive systems, make a table like the one shown here. Complete the last cell for GNRH, and fill in the rows for FSH and LH.

Hormone	Source	In males: Target/effect	In females: Target/effect
GNRH= gonado-tropin-releasing hormone	Hypo-thalamus	Anterior pituitary/ promote FSH and LH release	
FSH=			
LH=			

Go to Mastering Biology

For Students (in eText and Study Area)
- Get Ready for Chapter 46
- Animation: Human Spermatogenesis
- Animation: Human Oogenesis

For Instructors to Assign (in Item Library)
- Scientific Skills Exercise: Making Inferences and Designing an Experiment
- Tutorial: Sex Hormones and Mammalian Reproduction

Figure 46.1 This colony of coral polyps is reproducing. Tiny yellow orbs packed with eggs and sperm emerge from the polyps, rise to the sea surface, and burst. There, the eggs and sperm form embryos that become larvae, eventually drifting down and establishing new coral colonies.

In what different ways do animals reproduce?

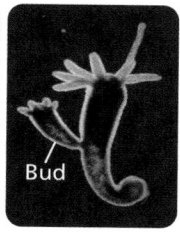

Bud

Asexual reproduction
No fusion of a sperm and an egg.

Sexual reproduction
Fusion of a sperm and an egg forms a zygote.

Hermaphroditism
An individual animal produces both male and female gametes.

Separate sexes
Male produces sperm and female produces eggs.

External fertilization
An egg encounters sperm outside of the female reproductive tract.

Internal fertilization
An egg encounters sperm inside the female reproductive tract. In some animals, including most mammals, the embryo develops internally and is born from the mother's body.

Both asexual and sexual reproduction occur in the animal kingdom

There are two modes of animal reproduction—sexual and asexual. In **sexual reproduction**, the fusion of haploid gametes forms a diploid cell, the **zygote**. The animal that develops from a zygote can in turn give rise to gametes by meiosis (see Figure 13.8). The female gamete, the **egg**, is large and nonmotile, whereas the male gamete, the **sperm**, is generally much smaller and motile. In **asexual reproduction**, new individuals are generated without the fusion of egg and sperm. For most asexual animals, reproduction relies entirely on mitotic cell division. Asexual and sexual reproduction are both common among animals.

Mechanisms of Asexual Reproduction

Among animals, several simple forms of asexual reproduction are found exclusively in invertebrates. One of these is *budding*, in which new individuals arise from outgrowths of existing ones (see Figure 13.2). In stony corals, for example, buds form and remain attached to the parent. The eventual result is a colony more than 1 m across, consisting of thousands of connected individuals. Also common among invertebrates is **fission**, the splitting and separation of a parent organism into two individuals of approximately equal size.

Invertebrate asexual reproduction can also occur by *fragmentation*, the breaking of the body into several pieces, followed by *regeneration*, regrowth of lost body parts. If more than one piece grows and develops into a complete animal, the effect is reproduction. For example, certain annelid worms can split into several fragments, each regenerating a complete worm. Numerous corals, sponges, cnidarians, and tunicates also reproduce by fragmentation and regeneration.

A wide range of animal species reproduce asexually by **parthenogenesis**, in which an egg develops without being fertilized. Among invertebrates, parthenogenesis occurs in certain species of bees, wasps, and ants. The offspring can be either haploid or diploid. In the case of honeybees, males (drones) are fertile haploid adults that arise by parthenogenesis. In contrast, female honeybees, including both the sterile workers and the fertile queens, are diploid adults that develop from fertilized eggs.

Among vertebrates, parthenogenesis is thought to be a rare response to low population density. For example, female Komodo dragons, hammerhead sharks, and zebra sharks have been observed to produce offspring when kept in captivity apart from males of their species. In 2015, DNA analysis of a group of female sawfish from a Florida river identified specimens that had two identical copies of all loci tested, providing evidence of vertebrate parthenogenesis in the wild.

Variation in Patterns of Sexual Reproduction

In many animal species, including humans, sexual reproduction involves the mating of a female and a male. In certain circumstances, however, finding a partner for reproduction can be challenging. Adaptations that arose during the evolution of some species meet this challenge by blurring the distinction between male and female. One such adaptation is particularly common among sessile (stationary) animals, such as barnacles, burrowing animals, such as clams, and some parasites, including tapeworms. That adaptation is **hermaphroditism**, in which each individual has both male and female reproductive systems (the term *hermaphrodite* merges the names of Hermes and Aphrodite, a Greek god and goddess). Sessile animals have a very limited opportunity to find a mate, but because each hermaphrodite reproduces as both a male and a female, *any* two individuals can mate. Each animal donates and receives sperm during mating, as shown for a pair of sea slugs in **Figure 46.2**. In some species, including many corals, hermaphrodites can also self-fertilize, allowing a form of sexual reproduction that doesn't require any partner.

▼ **Figure 46.2 Reproduction among hermaphrodites.** In this mating of sea slugs, or nudibranchs (*Nembrotha chamberlaini*), each hermaphrodite is providing sperm to fertilize the eggs of the other.

The bluehead wrasse (*Thalassoma bifasciatum*) provides an example of a quite different variation in sexual reproduction. These coral reef fish live in harems, each consisting of a single male and several females. When the lone male dies, the opportunity for sexual reproduction would appear to be lost. Instead, within a week the largest female in the harem transforms into a male and begins to produce sperm instead of eggs. What selective pressure in the evolution of the bluehead wrasse resulted in sex reversal for the female with the largest body? Because it is the male wrasse that defends a harem against intruders, a larger size may be particularly important for a male in ensuring successful reproduction.

Certain oyster species also undergo sex reversal. In this case, individuals reproduce as males and then later as females, when their size is greatest. Since the number of gametes produced generally increases with size much more for females than for males, sex reversal in this direction maximizes gamete production. The result is enhanced reproductive success: Because oysters are sedentary animals and release their gametes into the surrounding water rather than mating directly, releasing more gametes tends to result in more offspring.

Reproductive Cycles

Most animals, whether asexual or sexual, exhibit cycles in reproductive activity, often related to changing seasons. These cycles are controlled by hormones, whose secretion is in turn regulated by environmental cues. In this way, animals expend resources to reproduce only when sufficient energy sources are available and when environmental conditions favor the survival of offspring. For example, ewes (female sheep) have a reproductive cycle lasting 15–17 days. **Ovulation**, the release of mature eggs, occurs at the midpoint of each cycle. For ewes, reproductive cycles generally occur only during fall and early winter, and the length of any pregnancy is five months. Thus, most lambs are born in the early spring, when their chances of survival are optimal.

Because seasonal temperature is often an important cue for reproduction, climate change can decrease reproductive success. Researchers have discovered such an effect on caribou (wild reindeer) in Greenland. In spring, caribou migrate to calving grounds to eat sprouting plants, give birth, and care for their calves. Prior to 1993, the arrival of the caribou at the calving grounds coincided with the brief period during which the plants were nutritious and digestible. By 2006, however, average spring temperatures in the calving grounds had increased by more than 4°C, and the plants sprouted two weeks earlier. Because caribou migration is triggered by day length, not temperature, a mismatch arose between the timing of new plant growth and caribou birthing. Without adequate nutrition for the nursing females, production of caribou offspring has declined by more than 75% since 1993. To learn more about the effects of climate change on caribou and other organisms, see Make Connections Figure 56.31.

Reproductive cycles are also found among animals that can reproduce both sexually and asexually. Consider, for instance, the water flea (genus *Daphnia*). A female *Daphnia* can produce eggs of two types. One type of egg requires fertilization to develop, but the other type does not and develops instead by parthenogenesis. *Daphnia* reproduce asexually when environmental conditions are favorable and sexually during times of environmental stress. As a result, the switch between sexual and asexual reproduction is roughly linked to season.

For some asexual animal species, a cycle of reproductive behavior appears to reflect a sexual evolutionary past. In the parthenogenetic lizard species *Aspidoscelis uniparens*, reproduction is asexual, and all individuals are female. Nevertheless, these lizards have courtship and mating behaviors very similar to those of sexual species of *Aspidoscelis*. One member of each mating pair undergoes **ovulation**, the production and release of mature eggs. The other female mimics a male **(Figure 46.3a)**. Over the course of the breeding season, the two lizards alternate roles two or three times. An individual adopts female behavior prior to ovulation, when the concentration of the hormone estradiol is high, and then switches to male-like behavior after ovulation, when the concentration of the hormone progesterone is high **(Figure 46.3b)**. A female is more likely to ovulate if she is mounted at a critical time of

▼ **Figure 46.3 Sexual behavior in parthenogenetic lizards.** The desert grassland whiptail lizard (*Aspidoscelis uniparens*) is an all-female species. These reptiles reproduce by parthenogenesis, the development of an unfertilized egg, but ovulation is stimulated by mating behavior.

(a) Both lizards in this photograph are *A. uniparens* females. The one on top is playing the role of a male. Individuals switch sex roles two or three times during the breeding season.

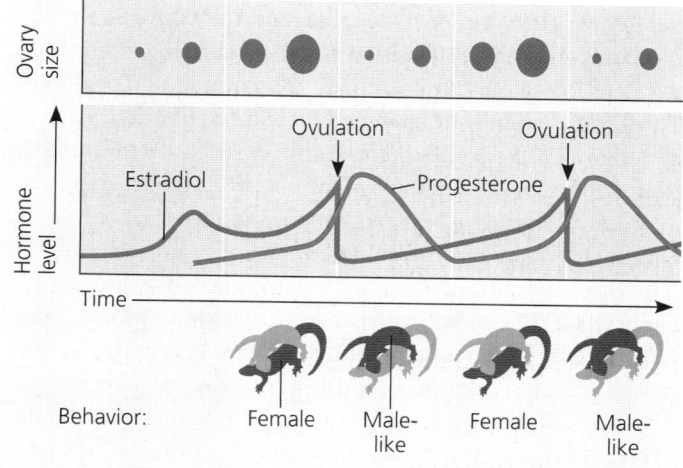

(b) The changes in sexual behavior of *A. uniparens* individuals are correlated with the cycles of ovulation and changing levels of the sex hormones estradiol and progesterone. These drawings track the changes in ovary size, hormone levels, and sexual behavior of one female lizard (shown in brown).

INTERPRET THE DATA *If you plotted hormone levels for the lizard shown in gray, how would your graph differ from the graph in part (b)?*

the hormone cycle; isolated lizards lay fewer eggs than those that go through the motions of sex. These findings support the hypothesis that these parthenogenetic lizards evolved from species having two sexes and still require certain sexual stimuli for maximum reproductive success.

Sexual Reproduction: An Evolutionary Enigma

EVOLUTION Although our species and many others reproduce sexually, the existence of sexual reproduction is actually puzzling. To see why, imagine an animal population in which half the females reproduce sexually and half reproduce asexually. We'll assume that the number of offspring per female is a constant, two in this case. The two offspring of an

▼ **Figure 46.4 The "reproductive handicap" of sex.** These diagrams contrast asexual versus sexual reproduction over four generations, assuming two surviving offspring per female.

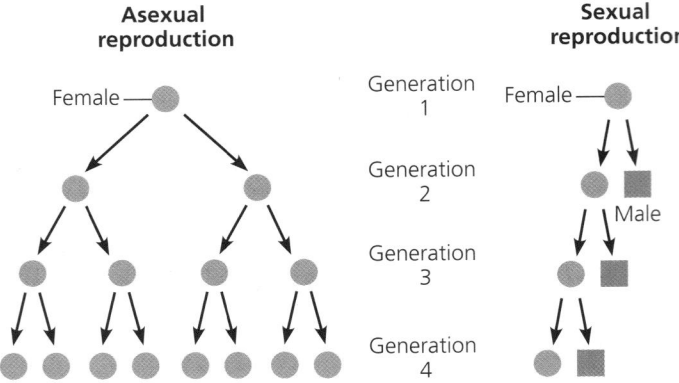

asexual female will both be daughters that will each give birth to two more reproductive daughters. In contrast, half of a sexual female's offspring will be male **(Figure 46.4)**. The number of sexual offspring will remain the same at each generation because both a male and a female are required to reproduce. Thus, the asexual condition will increase in frequency at each generation. Yet despite this "twofold cost," sex is maintained, even in animal species that can also reproduce asexually.

What advantage does sexual reproduction provide that counteracts its twofold cost? The answer remains uncertain. Most hypotheses focus on the unique combinations of parental genes formed during meiotic recombination and fertilization. By producing offspring of varied genotypes, sexual reproduction may enhance the reproductive success of parents when environmental factors, such as pathogens, change relatively rapidly. In contrast, asexual reproduction is expected to be most advantageous in stable, favorable environments because it can perpetuate successful genotypes precisely.

There are a number of reasons why the unique gene combinations formed during sexual reproduction might be advantageous. One is that beneficial gene combinations arising through recombination might speed up adaptation. Although this idea appears straightforward, the theoretical advantage is significant only when the rate of beneficial mutations is high and population size is small. Another idea is that the shuffling of genes during sexual reproduction might allow a population to rid itself of sets of harmful genes more readily.

CONCEPT CHECK 46.1

1. Compare and contrast the outcomes of asexual and sexual reproduction.

2. Parthenogenesis is the most common form of asexual reproduction in animals that at other times reproduce sexually. What characteristic of parthenogenesis might explain this observation?

3. **WHAT IF?** If a hermaphrodite self-fertilizes, will the offspring be identical to the parent? Explain.

4. **MAKE CONNECTIONS** What examples of plant reproduction are most similar to asexual reproduction in animals? (See Concept 38.2.)

For suggested answers, see Appendix A.

Fertilization depends on mechanisms that bring together sperm and eggs of the same species

The union of sperm and egg—**fertilization**—can be external or internal. In species with *external fertilization*, the female releases eggs into the environment, where the male fertilizes them **(Figure 46.5)**. In species with *internal fertilization*, sperm deposited in or near the female reproductive tract fertilize eggs within the tract. (We'll discuss cellular and molecular details of fertilization in Concept 47.1.)

A moist habitat is almost always required for external fertilization, both to prevent the gametes from drying out and to allow the sperm to swim to the eggs. Many aquatic invertebrates simply shed their eggs and sperm into the surroundings, and fertilization occurs without the parents making physical contact. However, timing is crucial to ensure that mature sperm and eggs encounter one another.

Among some species with external fertilization, individuals clustered in the same area release their gametes into the water at the same time, a process known as *spawning*. In some cases, chemical signals that one individual generates in releasing gametes trigger others to release gametes. In other cases, environmental cues, such as temperature or day length, cause a whole population to release gametes at one time. For example, the palolo worm of the South Pacific, like the coral in Figure 46.1, times its spawning to both the season and the lunar cycle. In spring, when the moon is in its last quarter, palolo worms break in half, releasing tail segments engorged with sperm or eggs. These packets rise to the ocean surface and burst

▼ **Figure 46.5 External fertilization.** Many species of amphibians reproduce by external fertilization. In most of these species, behavioral adaptations ensure that a male is present when the female releases eggs. Here, a female frog (on bottom) has released a mass of eggs in response to being clasped by a male. The male released sperm (not visible) at the same time, and external fertilization has already occurred in the water.

in such vast numbers that the sea appears milky with gametes. The sperm quickly fertilize the floating eggs, and within hours the palolo's once-a-year reproductive frenzy is complete.

When external fertilization is not synchronous across a population, individuals may exhibit specific "courtship" behaviors leading to the fertilization of the eggs of one female by one male (see Figure 46.5). By triggering the release of both sperm and eggs, these behaviors increase the probability of successful fertilization.

Internal fertilization is an adaptation that enables sperm to reach an egg even when the external environment is dry. It typically requires sophisticated and compatible reproductive systems, as well as cooperative behavior that leads to copulation. The male copulatory organ delivers sperm, and the female reproductive tract often has receptacles for storage and delivery of sperm to mature eggs.

No matter how fertilization occurs, the mating animals may make use of *pheromones*, chemicals released by one organism that influence the physiology and behavior of other individuals of the same species. Pheromones are small, volatile or water-soluble molecules that disperse into the environment and are active at very low concentrations (see Concept 45.1). Many pheromones function as mate attractants. For example, pheromones allow some female insects to be detected by males more than a kilometer away.

Evidence for human pheromones remains controversial. It was once argued that female roommates produce pheromones that trigger synchrony in menstrual cycles, but further statistical analyses have failed to support this finding.

Ensuring the Survival of Offspring

Typically, animals that fertilize eggs internally produce fewer gametes than species with external fertilization, but a higher fraction of their zygotes survive. Better zygote survival is due in part to the fact that eggs fertilized internally are sheltered from potential predators. However, internal fertilization is also more often associated with mechanisms that provide greater protection of the embryos and parental care of the young. For example, the internally fertilized eggs of birds and other reptiles have shells and internal membranes that protect against water loss and physical damage during the eggs' external development (see Figure 34.26). In contrast, the eggs of fishes and amphibians have only a gelatinous coat and lack internal membranes.

Rather than secreting a protective eggshell, some animals retain the embryo for a portion of its development within the female's reproductive tract. The offspring of marsupial mammals, such as kangaroos and opossums, spend only a short period in the uterus as embryos; they then crawl out and complete development attached to a mammary gland in the mother's pouch. Embryos of eutherian (placental) mammals, such as zebras and humans, remain in the uterus throughout fetal development. There they are nourished by the mother's blood supply through a temporary organ, the placenta. The embryos of some fishes and sharks also complete development internally.

▼ **Figure 46.6 Parental care in an invertebrate.** Compared with many other insects, giant water bugs of the genus *Belostoma* produce relatively few offspring but offer much greater parental protection. Following internal fertilization, the female glues her fertilized eggs to the back of the male. The male (shown here) carries the eggs for days, frequently fanning water over them to keep them moist, aerated, and free of parasites.

When a caribou or kangaroo is born or when a baby eagle hatches out of an egg, the newborn is not yet capable of independent existence. Instead, mammals nurse their offspring, and adult birds feed their young. Parental care of eggs or offspring is in fact widespread among animals, even including some invertebrates **(Figure 46.6)**.

Gamete Production and Delivery

Sexual reproduction in animals relies on sets of cells that are precursors for eggs and sperm. Cells dedicated to this function are often established early in the formation of the embryo and remain inactive while the body plan takes shape. Cycles of growth and mitosis then increase, or *amplify*, the number of cells available for making gametes—eggs or sperm.

In producing gametes from the precursor cells and making them available for fertilization, animals employ a variety of reproductive systems. **Gonads**, organs that produce gametes, are found in many but not all animals. Exceptions include the palolo, discussed above. The palolo and most other polychaete worms (phylum Annelida) have separate sexes but lack distinct gonads; rather, the eggs and sperm develop from undifferentiated cells lining the coelom (body cavity). As the gametes mature, they are released from the body wall and fill the coelom. Depending on the species, mature gametes in these worms may be shed through the excretory opening, or the swelling mass of eggs may split open a portion of the body, spilling the eggs into the environment.

More elaborate reproductive systems include sets of accessory tubes and glands that carry, nourish, and protect the gametes and sometimes the developing embryos. For example, fruit flies and most other insects have separate sexes with

▼ Figure 46.7 An example of insect reproductive anatomy.

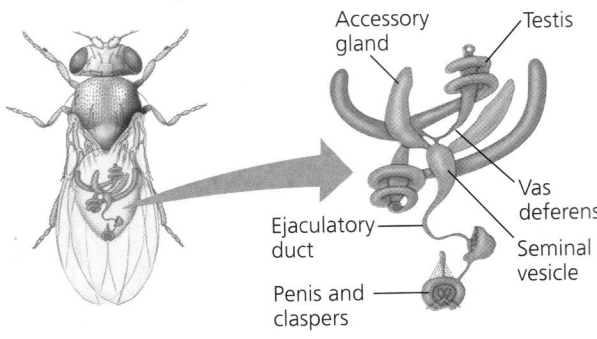

Accessory gland · Testis · Vas deferens · Seminal vesicle · Ejaculatory duct · Penis and claspers

(a) Male fruit fly. Sperm form in the testes, pass through a sperm duct (vas deferens), and are stored in the seminal vesicles. The male ejaculates sperm along with fluid from the accessory glands. (Males of some species of insects and other arthropods have appendages called claspers that grasp the female during copulation.)

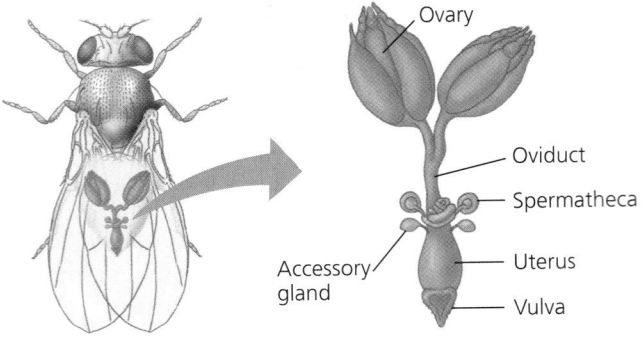

Ovary · Oviduct · Spermatheca · Uterus · Accessory gland · Vulva

(b) Female fruit fly. Eggs develop in the ovaries and then travel through the oviducts to the uterus. After mating, sperm are stored in the spermathecae, which are connected to the uterus by short ducts. The female uses a stored sperm to fertilize each egg as it enters the uterus before she passes the egg out through the vulva.

VISUAL SKILLS *Study the two drawings, and then describe the movement of fruit fly sperm from formation to fertilization.*

complex reproductive systems **(Figure 46.7)**. In many insect species, the female reproductive system includes one or more **spermathecae** (singular, *spermatheca*), sacs in which sperm may be kept alive and stored for extended periods, a year or more in some cases. Because the female releases male gametes from the spermathecae and thus fertilizes her eggs only in response to the appropriate stimuli, fertilization occurs under conditions likely to be well suited to survival of offspring.

Vertebrate reproductive systems display limited but significant variations. In some vertebrates, the uterus is divided into two chambers; in others, including humans and birds, it is a single structure. In many nonmammalian vertebrates, the digestive, excretory, and reproductive systems have a common opening to the outside, the **cloaca**, a structure probably present in the ancestors of all vertebrates. Lacking a well-developed penis, males of these species instead release sperm by turning the cloaca inside out. In contrast, mammals generally lack a cloaca and have a separate opening for the digestive tract. In addition, most female mammals have separate openings for the excretory and reproductive systems.

Although fertilization involves the union of a single egg and sperm, animals often mate with more than one member

of the other sex. Monogamy, the sustained sexual partnership of two individuals, is rare among animals, including most mammals. Mechanisms have evolved, however, that enhance the reproductive success of a male with a particular female and diminish the chance of that female mating successfully with another partner. For example, some male insects transfer secretions that make a female less receptive to courtship, reducing the likelihood of her mating again.

Can females also influence the relative reproductive success of their mates? This question intrigued two scientific collaborators working in Europe. Studying female fruit flies that copulated with one male and then another, the researchers traced the fate of sperm transferred in the first mating. As shown in **Figure 46.8**, females play a major role in determining the outcome of multiple matings. The processes by which gametes and individuals compete during reproduction remain a vibrant research area.

▼ Figure 46.8 Inquiry

Why is sperm usage biased when female fruit flies mate twice?

Experiment When a female fruit fly mates twice, 80% of the offspring result from the second mating. Scientists had hypothesized that ejaculate from the second mating displaces sperm from the first mating. To test this hypothesis, Rhonda Snook, at the University of Sheffield, and David Hosken, at the University of Zurich, used mutant males with altered reproductive systems. "No-ejaculate" males mate but do not transfer sperm or fluid to females. "No-sperm" males mate and ejaculate but make no sperm. The researchers allowed females to mate first with wild-type males and then with wild-type males, no-sperm males, or no-ejaculate males. As a control, some females were mated only once (to wild-type males). The scientists then dissected each female under a microscope and recorded whether sperm were absent from the spermathecae, the major sperm storage organs.

Results

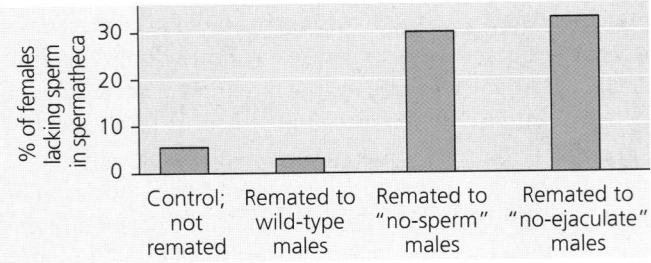

Conclusion Because remating reduces sperm storage when no sperm or fluids are transferred, the hypothesis that ejaculate from a second mating displaces stored sperm is incorrect. Instead, it appears females may get rid of stored sperm in response to remating, perhaps allowing for replacing stored sperm, possibly of diminished fitness, with fresh sperm.

Data from R. R. Snook and D. J. Hosken, Sperm death and dumping in *Drosophila, Nature* 428:939–941 (2004).

WHAT IF? *Suppose males in the first mating had a mutant allele that resulted in smaller eyes as a dominant trait. What fraction of the females would produce some offspring with smaller eyes?*

1. How does internal fertilization facilitate life on land?

2. What mechanisms have evolved in animals with (a) external fertilization and (b) internal fertilization that help offspring survive to adulthood?

3. **MAKE CONNECTIONS** What are the shared and distinct functions of the uterus of an insect and the ovary of a flowering plant? (See Figure 38.6.)

For suggested answers, see Appendix A.

Reproductive organs produce and transport gametes

Having surveyed some of the general features of animal reproduction, we'll focus in the rest of the chapter on humans, beginning with the reproductive anatomy of each sex.

Human Male Reproductive Anatomy

The human male's external reproductive organs are the scrotum and penis. The internal reproductive organs consist of gonads that produce both sperm and reproductive hormones, accessory glands that secrete products essential to sperm movement, and ducts that carry the sperm and glandular secretions (**Figure 46.9**).

▼ **Figure 46.9 Reproductive anatomy of the human male.** Labels in parentheses identify nonreproductive structures shown for orientation purposes.

⊖ **Mastering Biology** Animation: Reproductive System of the Human Male

Testes

The male gonads, or **testes** (singular, *testis*), produce sperm in highly coiled tubes called **seminiferous tubules**. Most mammals produce sperm properly only when the testes are cooler than the rest of the body. In humans and many other mammals, testis temperature is maintained about 2°C below the core body temperature by the **scrotum**, a fold of the body wall.

The testes develop in the abdominal cavity and descend into the scrotum just before birth (a testis within a scrotum is a *testicle*). In many rodents, the testes are drawn back into the cavity between breeding seasons, interrupting sperm maturation. Some mammals whose body temperature is low enough to allow sperm maturation—such as whales and elephants—retain the testes in the abdominal cavity at all times.

Ducts

From the seminiferous tubules of a testis, the sperm pass into the coiled duct of an **epididymis**. In humans, it takes three weeks for sperm to travel the 6-m length of this duct, during which time the sperm complete maturation and become motile. During **ejaculation**, the sperm are propelled from each epididymis through a muscular duct, the **vas deferens**. Each vas deferens (one from each epididymis) extends around and behind the urinary bladder, where it joins a duct from the seminal vesicle, forming a short *ejaculatory duct*. The ejaculatory ducts open into the **urethra**, the outlet tube for both the excretory system and the reproductive system. The urethra runs through the penis and opens to the outside at the tip of the penis.

Accessory Glands

Three sets of accessory glands—the seminal vesicles, the prostate gland, and the bulbourethral glands—produce secretions that combine with sperm to form **semen**, the fluid that is ejaculated. Two **seminal vesicles** contribute about 60% of the volume of semen. The fluid from the seminal vesicles is thick, yellowish, and alkaline. It contains mucus, the sugar fructose (which provides most of the sperm's energy), a coagulating enzyme, ascorbic acid, and local regulators called prostaglandins (see Concept 45.1).

The **prostate gland** secretes its products directly into the urethra through small ducts. Thin and milky, the fluid from this gland contains anticoagulant enzymes and citrate (a sperm nutrient). The *bulbourethral glands* are a pair of small glands along the urethra below the prostate. Before ejaculation, they secrete clear mucus that neutralizes any acidic urine remaining in the urethra. There is evidence that bulbourethral fluid carries some sperm released before ejaculation, which may contribute to the high failure rate of the withdrawal method of birth control (coitus interruptus).

Penis

The human **penis** contains the urethra as well as three cylinders of spongy erectile tissue. During sexual arousal, the erectile tissue fills with blood from the arteries. As this tissue fills, the increasing pressure seals off the veins that drain the penis, causing it to engorge with blood. The resulting erection enables the penis to be inserted into the vagina. Alcohol consumption, certain drugs, emotional issues, and aging all can cause an inability to achieve an erection (erectile dysfunction). For individuals with long-term erectile dysfunction, drugs such as Viagra promote the vasodilating action of the local regulator nitric oxide (NO; see Concept 45.1); the resulting relaxation of smooth muscles in the blood vessels of the penis enhances blood flow into the erectile tissues. Although all mammals rely on penile erection for mating, the penis of dogs, raccoons, walruses, and several other mammals also contains a bone, the baculum, which is thought to further stiffen the penis for mating.

The main shaft of the penis is covered by relatively thick skin. The head of the penis, the male **glans**, has a much thinner outer layer and is much more sensitive to stimulation. A fold of skin called the *prepuce* covers the glans of humans. The male prepuce, or foreskin, is removed if a male is circumcised.

Human Female Reproductive Anatomy

The human female's external reproductive structures are the clitoris and two sets of labia, which surround the clitoris and vaginal opening. The internal organs consist of gonads, which produce eggs and reproductive hormones, and a system of ducts and chambers, which receive and carry gametes and house the embryo and fetus **(Figure 46.10)**.

Ovaries

The female gonads are a pair of **ovaries** that flank the uterus and are held in place in the abdominal cavity by ligaments. The outer layer of each ovary is packed with **follicles**, each consisting of an **oocyte**, a partially developed egg, surrounded by support cells. The surrounding cells nourish and protect the oocyte during much of its formation and development.

Oviducts and Uterus

An **oviduct**, or fallopian tube, extends from the uterus toward a funnel-like opening at each ovary. The dimensions of this tube vary along its length, with the inside diameter near the uterus being as narrow as a human hair. Upon ovulation, cilia on the epithelial lining of the oviduct begin beating. This motion draws fluid from the body cavity into the oviduct, bringing along the egg. Further motion of the cilia, together with wavelike contractions of the oviduct, conveys the egg down the duct to the uterus.

The **uterus**, also known as the womb, is a thick, muscular organ that can expand during pregnancy to accommodate a 4-kg fetus. The inner lining of the uterus, the **endometrium**, is richly supplied with blood vessels. The neck of the uterus, called the **cervix**, opens into the vagina.

Vagina and Vulva

The **vagina** is a muscular but elastic chamber that is the site for insertion of the penis and deposition of sperm during copulation. The vagina, which also serves as the birth canal through which a baby is born, opens to the outside at the **vulva**, the collective term for the external female genitalia.

A pair of thick, fatty ridges called the **labia majora** enclose and protect the rest of the vulva. The vaginal opening and the separate opening of the urethra are located within a cavity bordered by a pair of slender skin folds, the **labia minora**. A thin piece of tissue called the *hymen* partly covers the vaginal opening in humans at birth, but becomes thinner over time and typically wears away through physical activity. Located at the top of the labia minora, the **clitoris** consists of erectile tissue supporting a rounded glans, or head, covered by a small hood of skin, the prepuce. During sexual arousal, the clitoris, vagina, and labia minora all engorge with blood and enlarge. Richly supplied with nerve endings, the clitoris is one of the most sensitive points of sexual stimulation. Sexual arousal also induces the vestibular glands near the vaginal opening to secrete lubricating mucus, thereby facilitating intercourse.

Mammary Glands

The **mammary glands** are present in both sexes, but they normally produce milk only in females. Though not part of

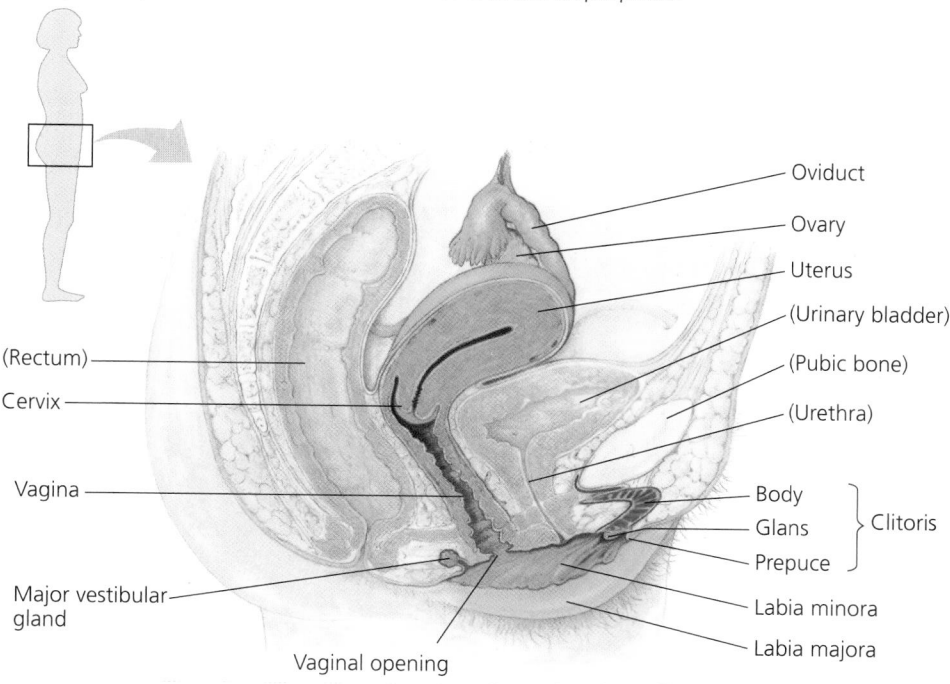

▼ **Figure 46.10 Reproductive anatomy of the human female.** Labels in parentheses identify nonreproductive structures shown for orientation purposes.

Oviduct
Ovary
Uterus
(Urinary bladder)
(Pubic bone)
(Urethra)
(Rectum)
Cervix
Vagina
Body
Glans
Prepuce
} Clitoris
Major vestibular gland
Labia minora
Labia majora
Vaginal opening

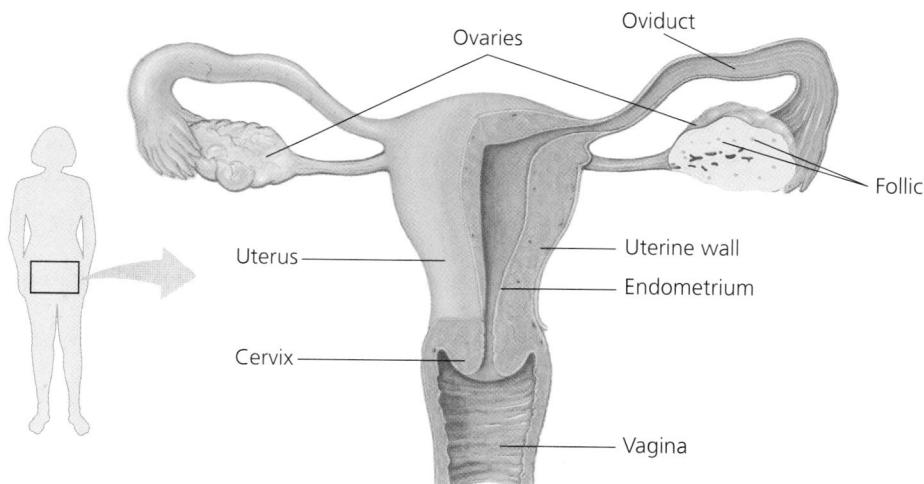

Ovaries
Oviduct
Follicles
Uterus
Uterine wall
Endometrium
Cervix
Vagina

⊙ **Mastering Biology** Animation: Reproductive System of the Human Female

the reproductive system, the female mammary glands are important to reproduction. Within the glands, small sacs of epithelial tissue secrete milk, which drains into a series of ducts that open at the nipple. The breasts contain connective and fatty (adipose) tissue in addition to the mammary glands.

Gametogenesis

With this overview of reproductive anatomy in mind, we turn now to **gametogenesis**, the production of gametes. In both males and females, there is a close relationship between the gonads' structure and their function. As shown in **Figure 46.11**, there are many parallels between **spermatogenesis**—the production of sperm—and **oogenesis**—the production of oocytes (eggs). Both processes generate haploid gametes via meiotic divisions of a set of dedicated diploid cells. In addition, support cells in the gonad play an essential role in both spermatogenesis and oogenesis. However, there are also several significant differences in gametogenesis between human males and females:

- Spermatogenesis is continuous in adult males. In producing hundreds of millions of sperm each day, cell division and maturation occur throughout the seminiferous tubules. For a single sperm, the process takes about seven weeks. In contrast, oogenesis is a prolonged process in the human female. Immature eggs form in the ovary of the female embryo but do not complete their development until years, and often decades, later.

- In spermatogenesis, the four products of meiosis develop into mature gametes. In oogenesis, cytokinesis during meiosis is unequal, with almost all the cytoplasm segregated to a single daughter cell. This large cell is destined to become the egg; the other products of meiosis, smaller cells known as polar bodies, degenerate.

- Spermatogenesis occurs throughout adolescence and adulthood. In contrast, the mitotic divisions that occur in oogenesis in human females are thought to be complete before birth, and the production of mature gametes ceases at about age 50.

- Spermatogenesis produces mature sperm from precursor cells in a continuous sequence, whereas oogenesis has long interruptions.

CONCEPT CHECK 46.3

1. Why might frequent use of a hot tub make it harder for a couple to conceive a child?

2. The process of oogenesis is often described as the production of a haploid egg by meiosis, but in some animals, including humans, this description is not entirely accurate. Explain.

3. **WHAT IF?** If each vas deferens in a male were surgically sealed off, what changes would you expect in sexual response and ejaculate composition?

For suggested answers, see Appendix A.

Spermatogenesis

Stem cells that give rise to sperm are situated near the outer edge of the seminiferous tubules. Their progeny move inward as they pass through the spermatocyte and spermatid stages, and sperm are released into the lumen (fluid-filled cavity) of the tubule. The sperm travel along the tubule into the epididymis, where they become motile.

⊘ **Mastering Biology** **Animation: Human Spermatogenesis**

The stem cells arise from division and differentiation of primordial germ cells. In mature testes, the stem cells divide mitotically to form **spermatogonia**, which in turn generate spermatocytes by mitosis. Each spermatocyte gives rise to four spermatids through meiosis, reducing the chromosome number from diploid ($2n = 46$ in humans) to haploid ($n = 23$). Spermatids undergo extensive changes as they differentiate into sperm.

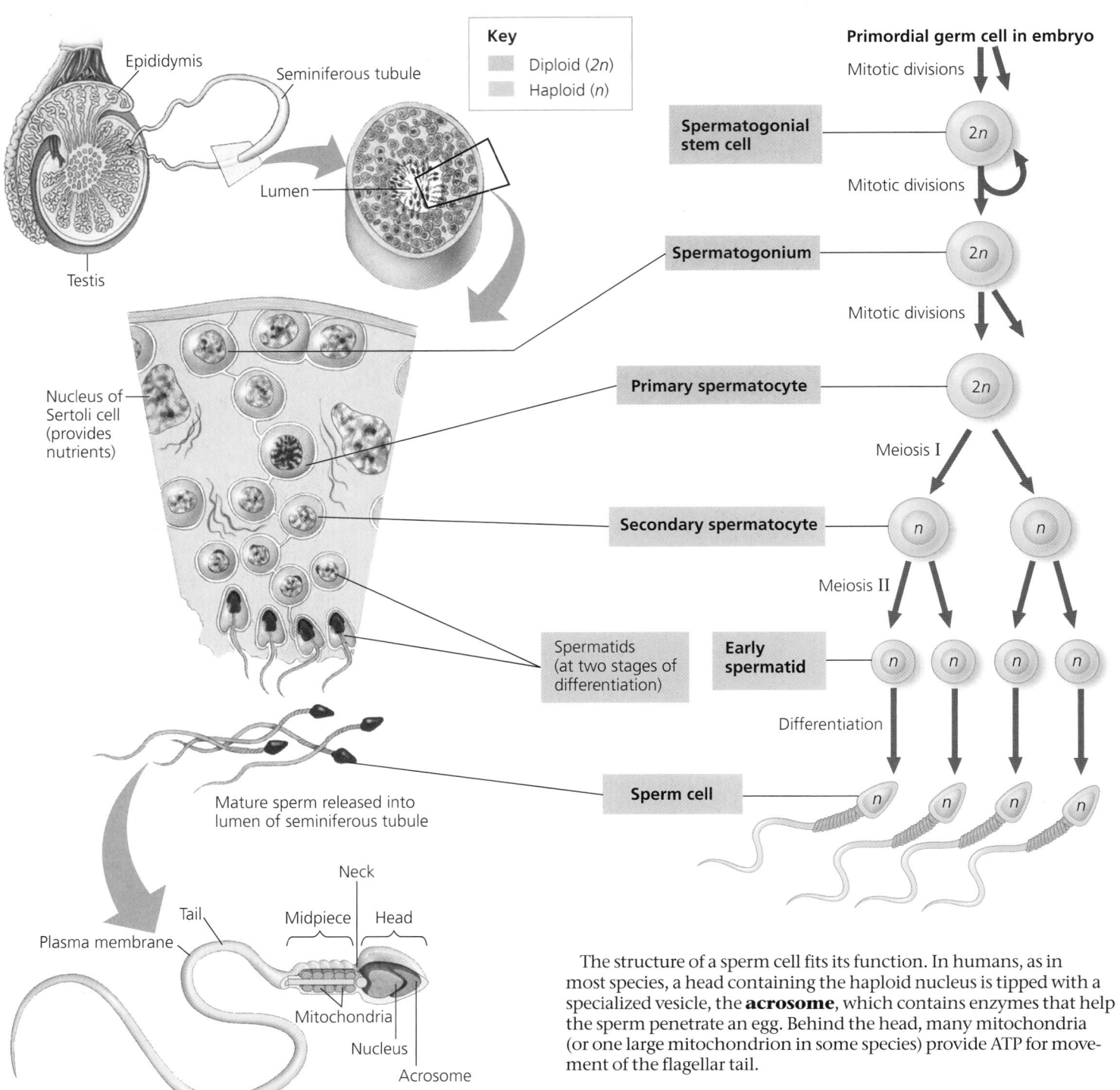

The structure of a sperm cell fits its function. In humans, as in most species, a head containing the haploid nucleus is tipped with a specialized vesicle, the **acrosome**, which contains enzymes that help the sperm penetrate an egg. Behind the head, many mitochondria (or one large mitochondrion in some species) provide ATP for movement of the flagellar tail.

Oogenesis

Oogenesis begins in the female embryo with the production of **oogonia** from primordial germ cells. The oogonia divide by mitosis to form cells that begin meiosis, but stop the process at prophase I before birth. These developmentally arrested cells, which are **primary oocytes**, each reside within a small follicle, a cavity lined with protective cells. At birth, the ovaries together contain about 1–2 million primary oocytes, of which about 500 fully mature between puberty and menopause.

→ **Mastering Biology** Animation: Human Oogenesis

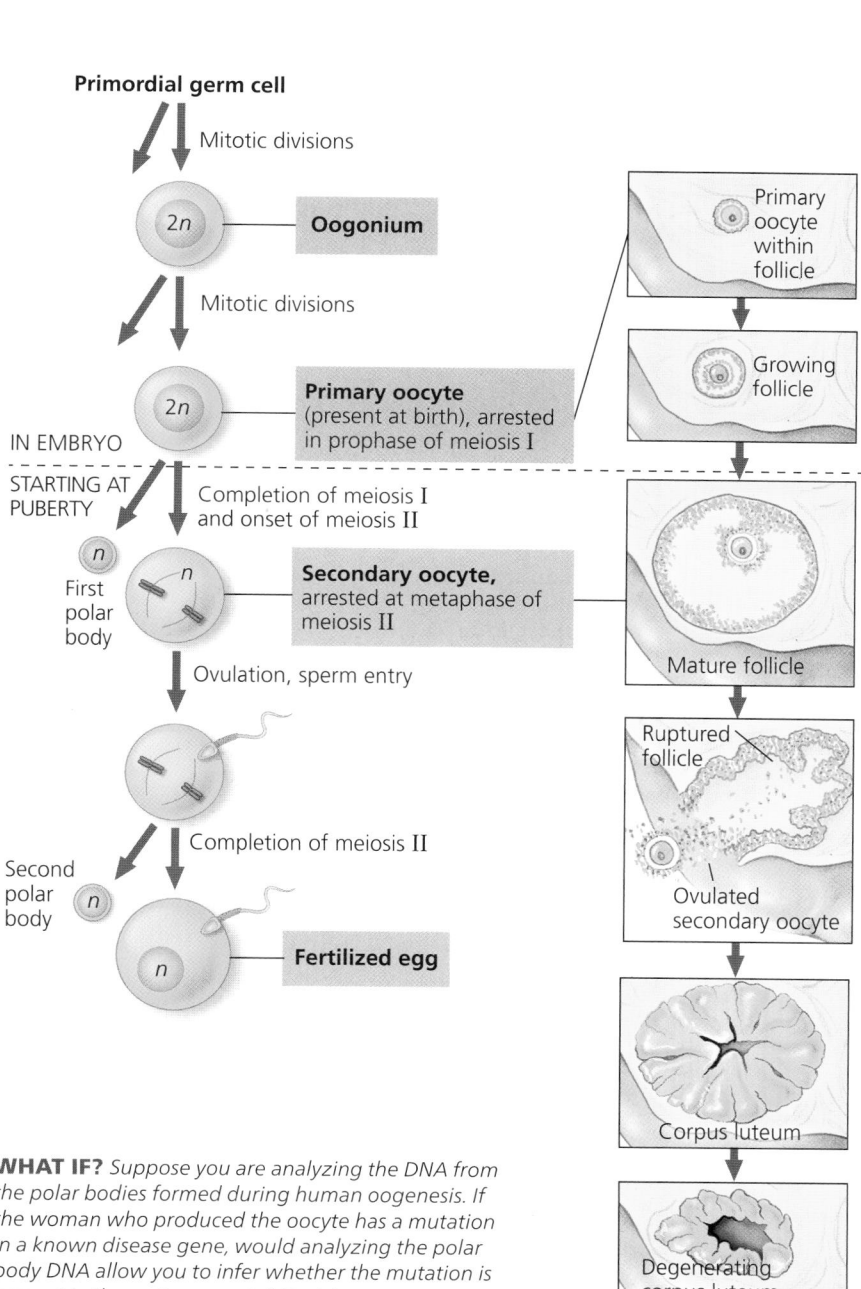

Beginning at puberty, follicle-stimulating hormone (FSH) periodically stimulates a small number of follicles to resume growth and development. Typically, only one follicle fully matures each month, with its primary oocyte completing meiosis I. The second meiotic division begins, but stops at metaphase. Thus arrested in meiosis II, the **secondary oocyte** is released at ovulation, when its follicle breaks open. Only if a sperm penetrates the oocyte does meiosis II resume. (In some other animal species, the sperm enters the oocyte earlier or later.) Each of the two meiotic divisions involves unequal cytokinesis, with the smaller cells becoming polar bodies that eventually degenerate (the first polar body may or may not divide again). As a result, the functional product of complete oogenesis is a single mature egg containing a sperm head. Fertilization is defined strictly as the fusion of the haploid nuclei of the sperm and secondary oocyte, although the term is often used loosely to mean the entry of the sperm head into the egg.

The ruptured follicle left behind after ovulation develops into the **corpus luteum**. The corpus luteum secretes estradiol as well as progesterone, a hormone that helps maintain the uterine lining during pregnancy. If the egg is not fertilized, the corpus luteum degenerates, and a new follicle matures during the next cycle.

WHAT IF? *Suppose you are analyzing the DNA from the polar bodies formed during human oogenesis. If the woman who produced the oocyte has a mutation in a known disease gene, would analyzing the polar body DNA allow you to infer whether the mutation is present in the mature oocyte? Explain.*

The interplay of tropic and sex hormones regulates reproduction in mammals

Mammalian reproduction is governed by the coordinated actions of hormones from the hypothalamus, anterior pituitary, and gonads. Endocrine control of reproduction begins with the hypothalamus, which secretes *gonadotropin-releasing hormone (GnRH)*. This hormone directs the anterior pituitary to secrete the gonadotropins **follicle-stimulating hormone (FSH)** and **luteinizing hormone (LH)** (see Figure 45.15). Both are tropic hormones, meaning that they regulate the activity of endocrine cells or glands. They are called *gonadotropins* because they act on the male and female gonads. FSH and LH support gametogenesis, in part by stimulating sex hormone production by the gonads.

The gonads produce and secrete three major types of steroid sex hormones: *androgens*, principally **testosterone**; *estrogens*, principally **estradiol**; and **progesterone**. All three hormones are found in both males and females, but at quite different concentrations. The concentration of testosterone in the blood is roughly ten times higher in males than in females. In contrast, the blood estradiol level is about ten times higher in females than in males; the peak blood progesterone level is also much higher in females. Although the gonads are the major source of sex hormones, the adrenal glands also secrete sex hormones in small amounts.

In mammals, sex hormone function in reproduction begins in the embryo. In particular, androgens produced in male embryos direct the appearance of the primary sex characteristics of males, the structures directly involved in reproduction. These include the seminal vesicles and associated ducts, as well as external reproductive structures. In the **Scientific Skills Exercise**, you can interpret the results of an experiment investigating the development of reproductive structures in mammals.

During sexual maturation, sex hormones in human males and females induce formation of secondary sex characteristics, the physical and behavioral differences between males and females that are not directly related to the reproductive system. Secondary sex characteristics often lead to sexual dimorphism, the difference in appearance between the male and female adults of a species **(Figure 46.12)**. When human males enter puberty, androgens cause the voice to deepen, facial and pubic hair to develop, and muscles to grow (by stimulating protein synthesis). Androgens also

---

Scientific Skills Exercise

Making Inferences and Designing an Experiment

What Role Do Hormones Play in Making a Mammal Male or Female? In non-egg-laying mammals, females have two X chromosomes, whereas males have one X chromosome and one Y chromosome. In the 1940s, French physiologist Alfred Jost wondered whether development of mammalian embryos as female or male in accord with their chromosome set requires instructions in the form of hormones produced by the gonads. In this exercise, you will interpret the results of an experiment that Jost performed to answer this question.

How the Experiment Was Done Working with rabbit embryos still in the mother's uterus at a stage before sex differences are observable, Jost surgically removed the portion of each embryo that would form the ovaries or testes. When the baby rabbits were born, he made note of their chromosomal sex and whether their genital structures were male or female.

Data from the Experiment

Chromosome Set	Appearance of Genitalia	
	No Surgery	Embryonic Gonad Removed
XY (male)	Male	Female
XX (female)	Female	Female

Data from A. Jost, Recherches sur la differenciation sexuelle de l'embryon de lapin (Studies on the sexual differentiation of the rabbit embryo), *Archives d'Anatomie Microscopique et de Morphologie Experimentale* 36:271–316 (1947).

► Human sex chromosomes in metaphase (duplicated)

INTERPRET THE DATA

1. This experiment is an example of a research approach in which scientists infer how something works normally based on what happens when the normal process is blocked. **(a)** What normal process was blocked in Jost's experiment? **(b)** From the results, what inference can you make about the role of the gonads in controlling the development of mammalian genitalia?

2. The data in Jost's experiment could be explained if some aspect of the surgery other than gonad removal caused female genitalia to develop. If you were to repeat Jost's experiment, how might you test the validity of such an explanation?

3. What result would Jost have obtained if female development also required a signal from the gonad?

4. Design another experiment to determine whether the signal that controls male development is a hormone. Make sure to identify your hypothesis, prediction, data collection plan, and controls.

➡ **Instructors:** A version of this Scientific Skills Exercise can be assigned in **Mastering Biology**.

▼ Figure 46.12 Androgen-dependent male anatomy and behavior in a moose. The male and female in this mating pair of moose (*Alces alces*) differ in both anatomy and physiology. High levels of testosterone in the male are responsible for the appearance of secondary sex characteristics, such as antlers, and for male courtship and territorial behavior.

promote specific sexual behaviors and sex drive, as well as an increase in general aggressiveness. Estrogens similarly have multiple effects in females. At puberty, estradiol stimulates breast and pubic hair development. Estradiol also influences female sexual behavior, induces fat deposition in the breasts and hips, increases water retention, and alters calcium metabolism.

Biological Sex, Gender Identity, and Sexual Orientation in Human Sexuality

A newborn baby is assigned a "biological sex" that typically reflects the genitals present at birth and the child's chromosomes. In mammals, the Y chromosome carries a gene called *SRY* that directs development of the gonad into a testis. In XX embryos, which lack the Y and hence lack *SRY*, the gonad becomes an ovary. Although most individuals are born male or female, roughly one in 100 is intersex, having both male and female biological characteristics. For example, intersex individuals may have a nonstandard chromosome set (such as XXY) or may differ in the hormone-directed pathways that control sexual development.

Though often confused with assigned sex, gender identity is distinct and refers to a person's internal sense of being male, female, some combination, or neither. The term *cisgender* describes a person having a gender identity in line with their assigned sex. In contrast, a *transgender* person experiences a mismatch between their gender identity and their assigned sex. Thus, for example, an individual may have an assigned sex of female, but a male gender identity.

Whereas gender identity is about a person's self, sexual orientation identifies the gender of people to whom an individual is attracted romantically, emotionally, and sexually. Members of a human population may have a sexual orientation that is heterosexual (straight), homosexual (lesbian or gay), bisexual, or asexual. As these descriptions make clear, human sexuality varies considerably.

Hormonal Control of the Male Reproductive System

When mammals reach sexual maturity, the sex hormones and gonadotropins have essential roles in gametogenesis. In exploring this hormonal control of reproduction, we'll begin with the relatively simple system found in males.

In directing spermatogenesis, FSH and LH act on two types of cells in the testis **(Figure 46.13)**. FSH stimulates *Sertoli cells*, located within the seminiferous tubules, to nourish developing sperm (see Figure 46.11). LH causes *Leydig cells*, scattered in connective tissue between the tubules, to produce testosterone and other androgens, which promote spermatogenesis in the tubules.

Two negative-feedback mechanisms control sex hormone production in males (see Figure 46.13). Testosterone regulates the blood concentration of GnRH, FSH, and LH through inhibitory effects on the hypothalamus and anterior pituitary. In addition, *inhibin*, a hormone that in males is produced by Sertoli cells, acts on the anterior pituitary gland to reduce FSH secretion. Together, these negative-feedback circuits maintain the level of androgen in the normal range.

Leydig cells have other roles besides producing testosterone. They in fact secrete small quantities of many other hormones and local regulators, including oxytocin, renin, angiotensin, corticotropin-releasing factor, growth factors, and prostaglandins. These signals coordinate the activity of reproduction with growth, metabolism, homeostasis, and behavior.

▼ Figure 46.13 Hormonal control of the testes.

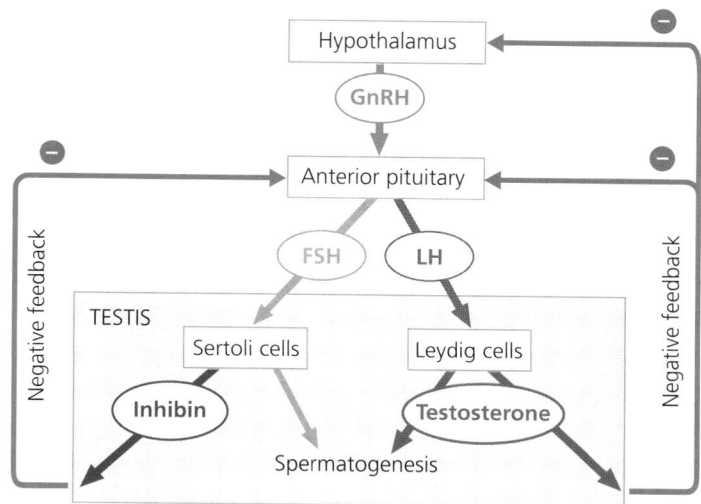

Hormonal Control of Female Reproductive Cycles

Whereas sperm are produced continuously in human males, there are two closely linked reproductive cycles in human females. Both are controlled by cyclic patterns of endocrine signaling.

Cyclic events in the ovaries define the **ovarian cycle**: Once per cycle a follicle matures and an oocyte is released. Changes in the uterus define the **uterine cycle**, which in humans and some other primates is a menstrual cycle. In each **menstrual cycle**, the endometrium (lining of the uterus) thickens and develops a rich blood supply before being shed through the cervix and vagina if pregnancy does not occur. By linking the ovarian and uterine cycles, hormone activity synchronizes ovulation with the establishment of a uterine lining that can support embryo implantation and development.

If an oocyte is not fertilized and pregnancy does not occur, the uterine lining is sloughed off, and another pair of ovarian and uterine cycles begins. The cyclic shedding of the blood-rich endometrium from the uterus, a process that occurs in a flow through the cervix and vagina, is called **menstruation**. Menstrual (and ovarian) cycles average 28 days but can range from about 20 to 40 days. **Figure 46.14** uses the 28-day average to highlight the major events of these cycles for one ovarian and uterine cycle, illustrating the close coordination across different tissues of the reproductive system.

The Ovarian Cycle

In human females, as in males, the hypothalamus has a central role in regulating reproduction. The ovarian cycle begins when the hypothalamus releases GnRH, which stimulates the anterior pituitary to secrete small amounts of FSH and LH.

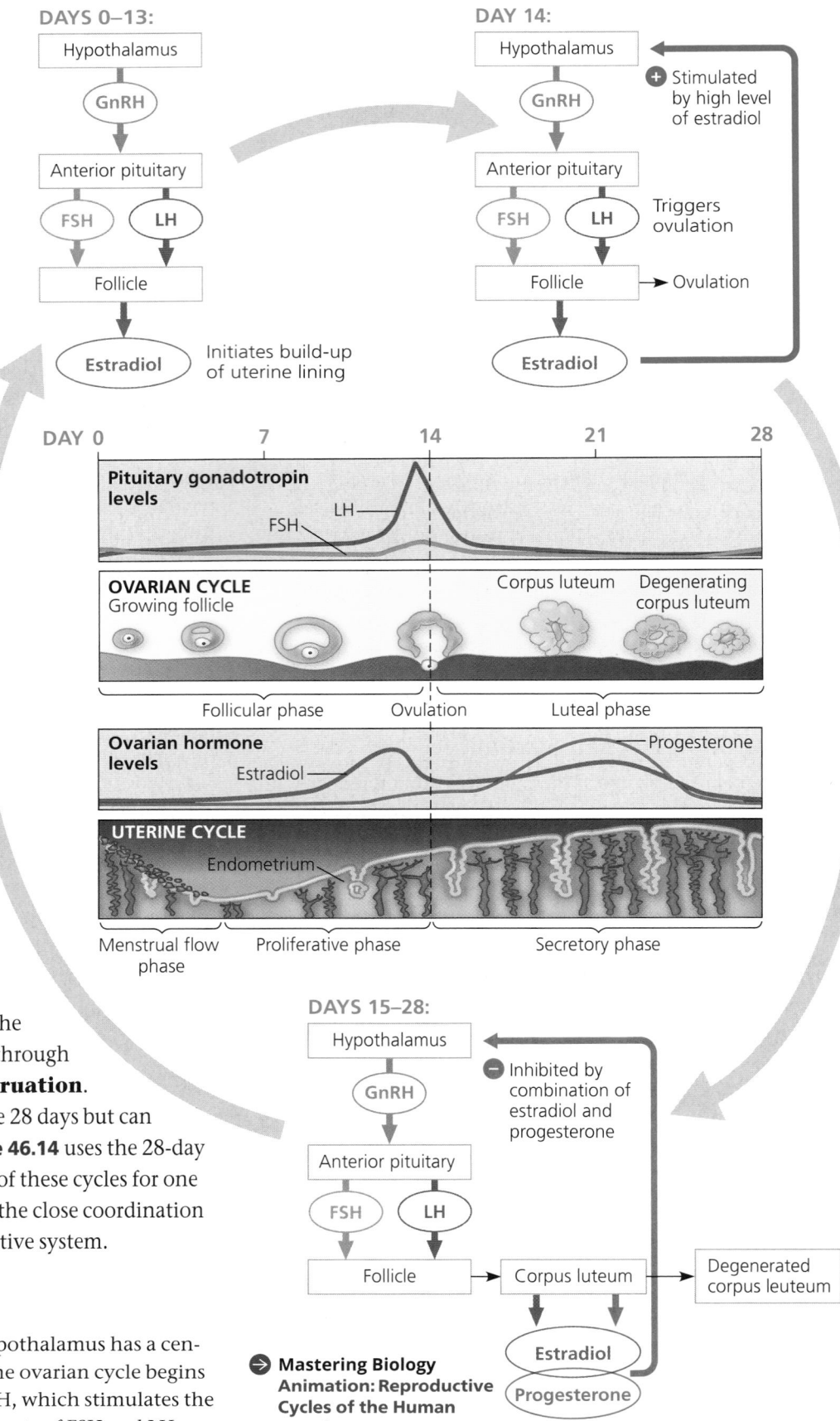

▼ **Figure 46.14 The reproductive cycles of the human female.** This figure shows how the ovarian cycle and the uterine (menstrual) cycle are regulated by changing hormone levels in the blood.

Mastering Biology Animation: Reproductive Cycles of the Human Female

Follicle-stimulating hormone (as its name implies) stimulates follicle growth, aided by LH, and the cells of the growing follicles start to make estradiol. Estradiol concentration slowly rises during most of the *follicular phase* (days 0–14), the part of the ovarian cycle during which follicles grow and oocytes mature. (Several follicles begin to grow with each cycle, but usually only one matures; the others disintegrate.) A low concentration of estradiol inhibits secretion of pituitary hormones, keeping the concentration of FSH and LH relatively low. In this portion of the cycle, regulation of the reproductive hormones closely parallels the regulation in males.

When estradiol secretion by the follicle begins to rise steeply, the levels of FSH and LH increase markedly. Why? Whereas a low level of estradiol inhibits secretion of pituitary gonadotropins, a high concentration has the opposite effect: It stimulates gonadotropin secretion by causing the hypothalamus to increase output of GnRH. A high estradiol concentration also increases the GnRH sensitivity of LH-releasing cells in the pituitary, further increasing the LH level.

The maturing follicle, containing a fluid-filled cavity, enlarges to form a bulge at the surface of the ovary. The follicular phase ends at ovulation (day 14), about a day after the LH surge. In response to FSH and the peak in LH level, the follicle and adjacent wall of the ovary rupture, releasing the secondary oocyte. At or near the time of ovulation, women may feel a pain in the lower abdomen, on the same side as the ovary that released the oocyte.

The *luteal phase* (days 15–28) of the ovarian cycle follows ovulation. LH stimulates the remaining follicular tissue to form the corpus luteum, a glandular structure. Stimulated by LH, the corpus luteum secretes progesterone and estradiol, which in combination exert negative feedback on the hypothalamus and pituitary. This feedback greatly reduces LH and FSH secretion, preventing maturation of another egg when a pregnancy may be under way.

If pregnancy does not occur, the low gonadotropin concentration at the end of the luteal phase cause the corpus luteum to disintegrate, triggering a sharp decline in estradiol and progesterone concentrations. This decline liberates the hypothalamus and pituitary from negative feedback. The pituitary can then secrete enough FSH to stimulate the growth of new follicles, initiating the next ovarian cycle.

The Uterine (Menstrual) Cycle

Prior to ovulation, ovarian steroid hormones stimulate the uterus to prepare for support of an embryo. Estradiol secreted in increasing amounts by growing follicles signals the endometrium to thicken. In this way, the follicular phase of the ovarian cycle is coordinated with the *proliferative phase* (days 6–14) of the uterine cycle. After ovulation, the estradiol and progesterone secreted by the corpus luteum stimulate maintenance and further development of the uterine lining, including enlargement of arteries and growth of endometrial glands. These glands secrete a nutrient fluid that can sustain

an early embryo even before it implants in the uterine lining. Thus, the luteal phase of the ovarian cycle is coordinated with the *secretory phase* (days 15–28) of the uterine cycle.

Once the corpus luteum has disintegrated, the rapid drop in ovarian hormone concentration causes arteries in the endometrium to constrict. Deprived of its circulation, the uterine lining largely disintegrates, releasing blood that is shed along with endometrial tissue and fluid. The result is menstruation—the *menstrual flow phase* (days 1–5) of the uterine cycle. During this phase, which usually lasts a few days, a new set of ovarian follicles begin to grow. By convention, the first day of flow is designated day 1 of the new uterine (and ovarian) cycle.

About 7% of women of reproductive age suffer from a disorder called **endometriosis**, in which some cells of the uterine lining migrate to an abdominal location that is abnormal, or **ectopic** (from the Greek *ektopos*, away from a place). Having migrated to a location such as an oviduct, ovary, or large intestine, the ectopic tissue responds to hormones in the bloodstream. Like the uterine endometrium, the ectopic tissue swells and breaks down during each ovarian cycle, resulting in pelvic pain and bleeding into the abdomen. Researchers have not yet determined why endometriosis occurs, but hormonal therapy or surgery can be used to lessen discomfort.

Menopause

After about 500 cycles, a woman undergoes **menopause**, the cessation of ovulation and menstruation. Menopause usually occurs between the ages of 46 and 54. During this interval, the ovaries lose their responsiveness to FSH and LH, resulting in a decline in estradiol production.

Menopause is an unusual phenomenon. In most other species, females and males can reproduce throughout life. Is there an evolutionary explanation for menopause? One intriguing hypothesis proposes that during early human evolution, undergoing menopause after bearing several children allowed a mother to provide better care for her children and grandchildren, thereby increasing the chances for survival of individuals who share much of her genetic makeup.

Menstrual Versus Estrous Cycles

In all female mammals, the endometrium thickens before ovulation, but only humans and some other primates have menstrual cycles. In other mammals, both domesticated and wild, the uterus reabsorbs the endometrium in the absence of a pregnancy, and no extensive fluid flow occurs. For these animals, the cyclic changes in the uterus occur as part of an **estrous cycle** that also controls the sexual receptivity of females: Whereas human females may engage in sexual activity throughout the menstrual cycle, mammals with estrous cycles usually copulate only during the period surrounding ovulation. This period, called estrus (from the Latin *oestrus*, frenzy, passion), is the only time the female is receptive to mating. It is often called "heat," and the female's temperature does increase slightly.

The length, frequency, and nature of estrous cycles vary. Bears and wolves have one estrous cycle per year; elephants have several. Rats have estrous cycles throughout the year, each lasting just five days. The rat's nemesis, the household cat, ovulates only upon mating.

Human Sexual Response

In humans, the arousal of sexual interest is complex, involving a variety of psychological as well as physical factors. Although reproductive structures in the male and female differ in appearance, a number serve similar functions in arousal, reflecting their shared developmental origin. For example, the same embryonic tissues give rise to the scrotum and the labia majora, to the skin on the penis shaft and the labia minora, and to the glans of the penis and the clitoris. Furthermore, the general pattern of human sexual response is similar in males and females. Two types of physiological reactions predominate in both sexes: *vasocongestion*, the filling of a tissue with blood, and *myotonia*, increased muscle tension.

The sexual response cycle can be divided into four phases: excitement, plateau, orgasm, and resolution. An important function of the excitement phase is to prepare the vagina and penis for *coitus* (sexual intercourse). During this phase, vasocongestion is particularly evident in erection of the penis and clitoris and in enlargement of the testicles, labia, and breasts. The vagina becomes lubricated, and myotonia may occur, as evident in nipple erection or tension of the limbs.

In the plateau phase, sexual responses continue as a result of direct stimulation of the genitalia. In females, the outer third of the vagina becomes vasocongested, while the inner two-thirds slightly expands. This change, coupled with the elevation of the uterus, forms a depression for receiving sperm at the back of the vagina. Breathing quickens and heart rate rises, sometimes to 150 beats per minute—not only in response to the physical effort of sexual activity, but also as an involuntary reaction to stimulation by the autonomic nervous system (see Figure 49.9).

Orgasm is characterized by rhythmic, involuntary contractions of the reproductive structures in both sexes. Male orgasm has two stages. The first, emission, occurs when the glands and ducts of the reproductive tract contract, forcing semen into the urethra. Expulsion, or ejaculation, occurs when the urethra contracts and the semen is expelled. During female orgasm, the uterus and outer vagina contract, but the inner two-thirds of the vagina does not. Orgasm is the shortest phase of the sexual response cycle, usually lasting only a few seconds. In both sexes, contractions occur at about 0.8-second intervals and may also involve the anal sphincter and several abdominal muscles.

The resolution phase completes the cycle and reverses the responses of the earlier stages. Vasocongested organs return to normal size and color, and muscles relax. Most of these changes are completed within 5 minutes, but some may take as long as an hour. Following orgasm, the male typically enters a refractory period, lasting from a few minutes to hours, when erection and orgasm cannot be achieved. Females do not have a refractory period, making possible multiple orgasms within a short period of time.

CONCEPT CHECK 46.4

1. How are the functions of FSH and LH in females and males similar?
2. How does an estrous cycle differ from a menstrual cycle? In what animals are the two types of cycles found?
3. **WHAT IF?** If a human female begins taking estradiol and progesterone immediately after the start of a new menstrual cycle, how will ovulation be affected? Explain.
4. **MAKE CONNECTIONS** A coordination of events is characteristic of the reproductive cycle of a human female and the replicative cycle of an enveloped RNA virus (see Figure 19.8). What is the nature of the coordination in each of these cycles?

For suggested answers, see Appendix A.

CONCEPT 46.5

In placental mammals, an embryo develops fully within the mother's uterus

Having surveyed the ovarian and uterine cycles of human females, we turn now to reproduction itself, beginning with the events that transform an egg into a developing embryo.

Conception, Embryonic Development, and Birth

During human copulation, hundreds of millions of sperm are transferred in 2–5 mL of semen. When first ejaculated, the semen coagulates, which likely keeps the ejaculate in place until sperm reach the cervix. Soon after, anticoagulants liquefy the semen, and the sperm swim through the cervix and oviducts. Fertilization—also called **conception** in humans—occurs when a sperm fuses with an egg (mature oocyte) in an oviduct **(Figure 46.15)**.

The zygote begins a series of cell divisions called cleavage about 24 hours after fertilization and after an additional 4 days produces a **blastocyst**, a sphere of cells surrounding a central cavity. A few days later, the embryo implants into the endometrium of the uterus. The condition of carrying one or more embryos in the uterus is called **pregnancy**, or **gestation**. Human pregnancy averages 266 days (38 weeks) from fertilization of the egg, or 40 weeks from the start of the last menstrual cycle. In comparison, gestation averages 21 days in many

▼ Figure 46.15 Formation of a human zygote and early postfertilization events.

❸ Cleavage (cell division) begins in the oviduct as the embryo is moved toward the uterus by peristalsis and the movements of cilia.

❷ Fertilization occurs: A sperm enters the oocyte, the oocyte completes meiosis, and the nuclei of the oocyte and the sperm fuse, producing a zygote.

❹ Cleavage continues. The embryo reaches the uterus as a ball of cells. There it floats for several days, nourished by endometrial secretions as it becomes a blastocyst.

❶ Ovulation releases a secondary oocyte, which enters the oviduct.

❺ The blastocyst implants in the endometrium about 7 days after conception.

Ovary

Uterus

Endometrium

VISUAL SKILLS *If a woman's eggs need to be fertilized in vitro, they can be readily introduced into the uterus but not the extremely narrow oviduct. Based on this drawing, propose conditions for culturing a fertilized egg that you predict will optimize the chance of a successful pregnancy.*

rodents, 280 days in cows, and more than 600 days in elephants. The roughly nine months of human gestation are divided into three *trimesters* of equal length.

First Trimester

During the first trimester, the implanted embryo secretes hormones that signal its presence and regulate the mother's reproductive system. One embryonic hormone, *human chorionic gonadotropin* (*hCG*), acts like pituitary LH in maintaining secretion of progesterone and estrogens by the corpus luteum through the first few months of pregnancy. Some hCG passes from the maternal blood to the urine, where it can be detected by the most common early pregnancy tests.

During its first 2–4 weeks of development, the embryo obtains nutrients directly from the endometrium. Meanwhile, the outer layer of the blastocyst, which is called the **trophoblast**, grows outward and mingles with the endometrium, eventually helping form the **placenta (Figure 46.16)**.

▶ Figure 46.16 Placental circulation. Maternal blood enters the placenta in arteries, flows through blood pools in the endometrium, and leaves via veins. Embryonic or fetal blood, which remains in vessels, enters the placenta through arteries and passes through capillaries in finger-like chorionic villi, where oxygen and nutrients are acquired. Fetal blood leaves the placenta through veins leading back to the fetus.

WHAT IF? *In a rare genetic disorder, the absence of a particular enzyme leads to increased testosterone production. When the fetus has this disorder, the mother develops a male-like pattern of body hair during the pregnancy. Explain why.*

Placenta

Uterus

Umbilical cord

Maternal arteries

Maternal veins

Maternal portion of placenta

Chorionic villus, containing fetal capillaries

Maternal blood pool

Fetal portion of placenta (chorion)

Fetal arteriole   Fetal venule

Umbilical arteries

Umbilical vein

Umbilical cord

▼ Figure 46.17 Some stages of human development during the first and second trimesters.

(a) **5 weeks.** Limb buds, eyes, the heart, the liver, and rudiments of all other organs have started to develop in the embryo, which is only about 1 cm long.

(b) **14 weeks.** Growth and development of the offspring, now called a fetus, continue during the second trimester. This fetus is about 6 cm long.

This disk-shaped organ contains both embryonic and maternal blood vessels. Exchange between the maternal and embryonic circulatory systems supplies nutrients, provides immune protection, exchanges respiratory gases, and disposes of metabolic wastes for the embryo. Blood from the embryo travels to the placenta through the arteries of the umbilical cord and returns via the umbilical vein.

Occasionally, an embryo splits during the first month of development, resulting in identical, or *monozygotic* (one-egg), twins. In contrast, fraternal, or *dizygotic*, twins arise when two follicles mature in a single cycle, followed by independent fertilization and implantation of two genetically distinct embryos.

The first trimester is the main period of **organogenesis**, the development of the body organs **(Figure 46.17a)**. During organogenesis, the embryo is particularly susceptible to damage. For example, alcohol that passes through the placenta and reaches the developing central nervous system of the embryo can cause developmental and intellectual disabilities and other serious symptoms of fetal alcohol syndrome. The heart begins beating by the 4th week; a heartbeat can be detected at 8–10 weeks. At 8 weeks, all the major structures of the adult are present in rudimentary form, and the embryo is called a **fetus**.

At the end of the first trimester, the fetus is well differentiated, but only 5 cm long. Meanwhile, a high progesterone level brings about rapid changes in the mother: Mucus in the cervix forms a plug that protects against infection, the maternal part of the placenta grows, the mother's breasts and uterus get larger, and both ovulation and menstrual cycling stop. About three-fourths of all pregnant women experience nausea, misleadingly called "morning sickness," during the first trimester.

Second and Third Trimesters

During the second trimester, the fetus grows to about 30 cm in length. Development continues, including formation of fingernails, external sex organs, and outer ears **(Figure 46.17b)**. The mother may feel fetal movements as early as one month into the second trimester, and fetal activity is typically visible through the abdominal wall one to two months later. Hormone concentrations stabilize as hCG secretion declines; the corpus luteum deteriorates; and the placenta completely takes over the production of progesterone, the hormone that maintains the pregnancy.

During the third trimester, the fetus grows to about 3–4 kg in weight and 50 cm in length. Fetal activity may decrease as the fetus fills the available space. As the fetus grows and the uterus expands around it, the mother's abdominal organs become compressed and displaced, leading to digestive blockages and a need for frequent urination.

Childbirth begins with *labor*, a series of strong, rhythmic uterine contractions that push the fetus and placenta out of the body. Once labor begins, local regulators (prostaglandins) and hormones (chiefly estradiol and oxytocin) induce and regulate further contractions of the uterus **(Figure 46.18)**. Central

▼ Figure 46.18 Positive feedback in labor.

VISUAL SKILLS *Based on the feedback circuits shown, predict the effect of a single dose of oxytocin on a pregnant woman at the end of 39 weeks gestation.*

to this regulation is a positive-feedback loop (see Concept 45.2) in which uterine contractions stimulate secretion of oxytocin, which in turn stimulates further contractions.

Labor is typically described as having three stages (**Figure 46.19**). The first stage is the thinning and opening up (dilation) of the cervix. The second stage is the expulsion, or delivery, of the baby. Continuous strong contractions force the fetus out of the uterus and through the vagina. The final stage of labor is the delivery of the placenta.

▼ **Figure 46.19 The three stages of labor.**

❶ Dilation of the cervix

Placenta
Umbilical cord
Uterus
Cervix

❷ Expulsion: delivery of the infant

❸ Delivery of the placenta

Uterus
Placenta (detaching)
Umbilical cord

The newborn human or other mammal receives nourishment in the form of mother's milk. In response to suckling by the newborn and changes in estradiol concentration after birth, the hypothalamus signals the anterior pituitary to secrete prolactin, which stimulates the mammary glands to produce milk. Suckling also stimulates the secretion of oxytocin from the posterior pituitary, which triggers release of milk from the mammary glands (see Figure 45.15).

Maternal Immune Tolerance of the Embryo and Fetus

Pregnancy is an immunological puzzle. Because half of the embryo's genes are inherited from the father, many of the chemical markers present on the surface of the embryo are foreign to the mother. Why, then, does the mother not reject the embryo as a foreign body, as she would a tissue or organ graft from another person? One intriguing clue comes from the relationship between certain autoimmune disorders and pregnancy. For example, the symptoms of rheumatoid arthritis, an autoimmune disease of the joints, become less severe during pregnancy. Such observations suggest that the overall regulation of the immune system changes during pregnancy. Sorting out these changes and how they might protect the developing fetus is an active area of research for immunologists.

Contraception and Abortion

Contraception, the deliberate prevention of pregnancy, can be achieved in a number of ways. Some contraceptive methods prevent gamete development or release from female or male gonads; others prevent fertilization by keeping sperm and egg apart; and still others prevent implantation of an embryo. For complete information on contraceptive methods, you should consult a health-care provider. The following brief introduction to the biology of the most common methods and the corresponding diagram in **Figure 46.20** make no pretense of being a contraception manual.

Fertilization can be prevented by abstinence from sexual intercourse or by any of several kinds of barriers that keep live sperm from contacting the egg. Temporary abstinence, sometimes called *natural family planning*, depends on refraining from intercourse when conception is most likely. Because the egg can survive in the oviduct for 24–48 hours and sperm for up to 5 days, a couple practicing temporary abstinence should not engage in intercourse for a number of days before and after ovulation. Furthermore, because the timing of ovulation can vary significantly, the

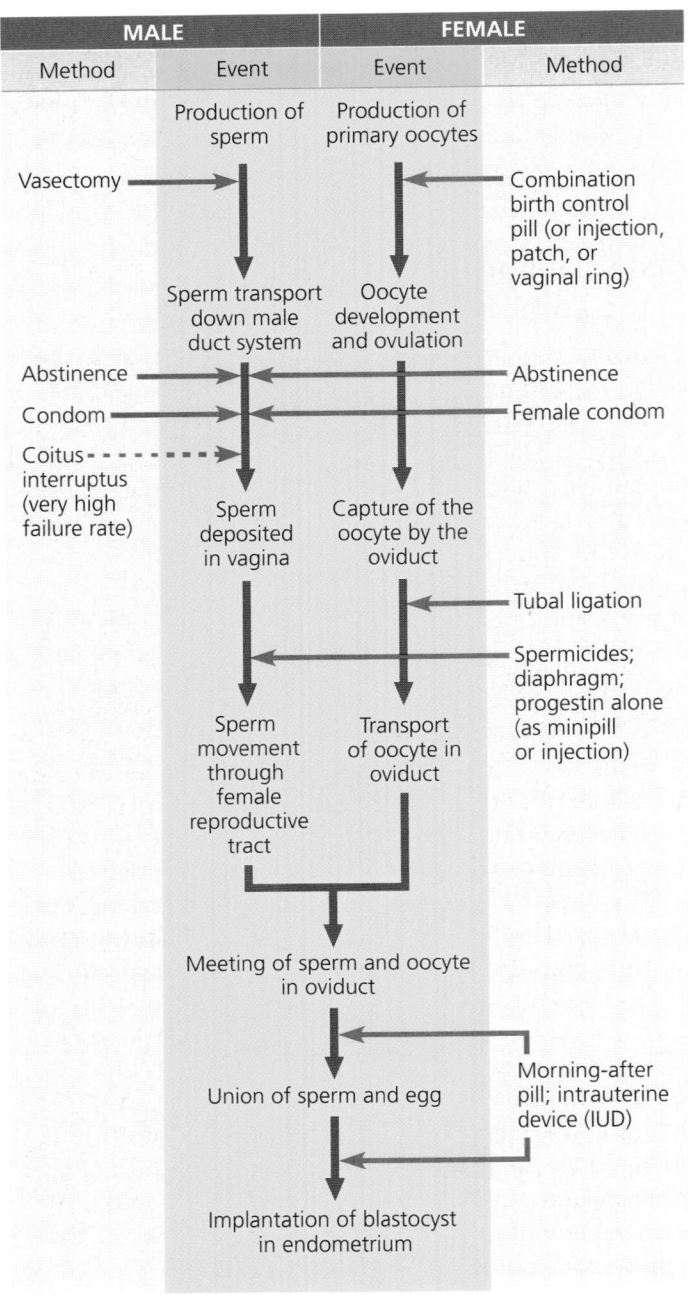

▼ **Figure 46.20 Mechanisms of several contraceptive methods.** Red arrows indicate where these methods, devices, or products interfere with events from the production of sperm and primary oocytes to implantation of a developing embryo.

MALE		FEMALE	
Method	Event	Event	Method

Production of sperm → Sperm transport down male duct system → Sperm deposited in vagina → Sperm movement through female reproductive tract → Meeting of sperm and oocyte in oviduct → Union of sperm and egg → Implantation of blastocyst in endometrium

Production of primary oocytes → Oocyte development and ovulation → Capture of the oocyte by the oviduct → Transport of oocyte in oviduct

Male methods: Vasectomy, Abstinence, Condom, Coitus interruptus (very high failure rate)

Female methods: Combination birth control pill (or injection, patch, or vaginal ring), Abstinence, Female condom, Tubal ligation, Spermicides; diaphragm; progestin alone (as minipill or injection), Morning-after pill; intrauterine device (IUD)

As a method of preventing fertilization, *coitus interruptus,* or withdrawal (removal of the penis from the vagina before ejaculation), is unreliable. Sperm from a previous ejaculate may be transferred in secretions that precede ejaculation. Furthermore, a split-second lapse in timing or willpower can result in tens of millions of sperm being transferred before withdrawal.

Used properly, several methods of contraception that block sperm from meeting the egg have pregnancy rates of less than 10%. The *condom* is a thin, latex rubber or natural membrane sheath that fits over the penis to collect the semen. For sexually active individuals, latex condoms are the only contraceptives that are highly effective in preventing the spread of AIDS and other *sexually transmitted infections* (*STIs*), also known as *sexually transmitted diseases* (*STDs*). This protection is not absolute, however. Another common barrier device is the *diaphragm,* a dome-shaped rubber cap inserted into the upper portion of the vagina before intercourse. Both of these devices have lower pregnancy rates when used in conjunction with a spermicidal (sperm-killing) foam or jelly. Other barrier devices include the vaginal pouch, also called a "female condom."

Except for complete abstinence from sexual intercourse or sterilization (discussed later), the most effective means of birth control are the intrauterine device (IUD) and hormonal contraceptives. The IUD has a pregnancy rate of 1% or less and is the most commonly used reversible method of birth control outside the United States. Placed in the uterus by a doctor, the IUD interferes with fertilization and implantation. Hormonal contraceptives, most often in the form of **birth control pills**, also have pregnancy rates of 1% or less.

The most commonly prescribed hormonal contraceptives contain a synthetic estrogen and a synthetic progesterone-like hormone called progestin. This combination mimics negative feedback in the ovarian cycle, stopping the release of GnRH by the hypothalamus and thus of FSH and LH by the pituitary. The prevention of LH release blocks ovulation. In addition, the inhibition of FSH secretion by the low dose of estrogens in the pills prevents follicles from developing.

Another hormonal contraceptive with a very low pregnancy rate contains only progestin. Progestin causes thickening of a woman's cervical mucus so that it blocks sperm from entering the uterus. Progestin also decreases the frequency of ovulation and causes changes in the endometrium that may interfere with implantation if fertilization occurs. This contraceptive can be administered as injections that last for three months or as a tablet ("minipill") taken daily.

Hormonal contraceptives have both harmful and beneficial side effects. They increase the risk of some cardiovascular disorders slightly for nonsmokers and quite substantially (three- to tenfold) for women who smoke regularly. At the same time, oral contraceptives eliminate

couple needs knowledge of physiological indicators associated with ovulation, such as changes in cervical mucus. Note also that a pregnancy rate of 10–20% is typical for couples practicing natural family planning. (In this context, pregnancy rate is the percentage of women who become pregnant in one year while using a particular pregnancy prevention method.)

the dangers of pregnancy; women on birth control pills have mortality rates about one-half those of pregnant women. Birth control pills also decrease the risk of ovarian and endometrial cancers. No hormonal contraceptives are available for men.

Sterilization is the permanent prevention of gamete production or release. For women, the most common method is **tubal ligation**, the sealing shut or tying off (ligating) of a section of each oviduct to prevent eggs from traveling into the uterus. Similarly, **vasectomy** in men is the cutting and tying off of each vas deferens to prevent sperm from entering the urethra. Sex hormone secretion and sexual function are unaffected by both procedures, with no change in menstrual cycles in females or ejaculate volume in males. Although tubal ligation and vasectomy are considered permanent, both procedures can in many cases be reversed by microsurgery.

The termination of a pregnancy in progress is called **abortion**. Spontaneous abortion, or *miscarriage*, is very common; it occurs in as many as one-third of all pregnancies, often before the woman is even aware she is pregnant. In addition, each year about 700,000 women in the United States choose to have an abortion performed by a physician.

A drug called mifepristone, or RU486, can terminate a pregnancy nonsurgically within the first 7 weeks. RU486 blocks progesterone receptors in the uterus, thus preventing progesterone from maintaining the pregnancy. It is taken with a small amount of prostaglandin to induce uterine contractions.

Modern Reproductive Technologies

Recent scientific and technological advances have made it possible to address many reproductive problems, including genetic diseases and infertility.

Infertility and in Vitro Fertilization

Infertility—an inability to conceive offspring—is quite common, affecting about one in ten couples in the United States and worldwide. The likelihood of infertility is nearly the same for men and women and causes vary. For women, however, the risk of reproductive difficulties, as well as genetic abnormalities of the fetus, increases steadily past age 35. Evidence suggests that the prolonged period of time oocytes spend in meiosis is largely responsible for this increased risk.

Among preventable causes of infertility, sexually transmitted infections (STIs) are the most significant. In women 15–24 years old, approximately 830,000 cases of chlamydia and gonorrhea are reported annually in the United States. The actual number of women infected with the chlamydia or gonorrhea bacterium is considerably higher because most women with these infections have no symptoms and are therefore unaware of their infection.

▼ **Figure 46.21** *In vitro* **fertilization (IVF).** In this form of IVF, a technician holds the egg in place with a pipette (left) and uses a very fine needle (right) to inject one sperm into the egg cytoplasm (colorized LM).

Up to 40% of women who remain untreated for either chlamydia or gonorrhea develop an inflammatory disorder that can scar the oviduct, substantially increasing the risk of a tubal, or ectopic, pregnancy. Rather than implanting in the uterus, the embryo becomes lodged in the oviduct (fallopian tube) where fertilization occurred. Such pregnancies cannot be sustained and may rupture the oviduct, resulting in serious internal bleeding.

Some forms of infertility are treatable. Hormone therapy can sometimes increase sperm or egg production, and surgery can often correct ducts that formed improperly or have become blocked. In some cases, doctors recommend *in vitro* **fertilization (IVF)**, which involves combining oocytes and sperm in the laboratory. Fertilized eggs are incubated until they form eight or more cells and are then transferred to the woman's uterus for implantation. If mature sperm are defective or low in number, a whole sperm or a spermatid nucleus may be injected directly into an oocyte **(Figure 46.21)**. Though costly and not always successful, IVF procedures have enabled more than a million couples to conceive children.

Detecting Disorders During Pregnancy

Many developmental problems and genetic diseases can now be diagnosed while the fetus is in the uterus. Ultrasound imaging, which generates images using sound frequencies above the normal hearing range, is commonly used to analyze the fetus's size and condition. In amniocentesis and chorionic villus sampling, a needle is used to obtain fetal cells from either fluid or tissue, respectively, surrounding the embryo; these cells then provide the basis for genetic analysis (see Figure 14.19).

A new reproductive technology makes use of a pregnant mother's blood to analyze the genome of her fetus. As discussed in Concept 14.4, a pregnant woman's blood contains

DNA from the growing embryo. How does it get there? The mother's blood reaches the embryo through the placenta. When cells produced by the embryo grow old, die, and break open within the placenta, the released DNA enters the mother's circulation. Although the blood also contains pieces of DNA from the mother, about 10–15% of the DNA circulating in the blood is from the fetus. Both the polymerase chain reaction (PCR) and high-throughput sequencing can convert the bits of fetal DNA into useful information.

Unfortunately, almost all detectable disorders remain untreatable in the uterus, and many cannot be corrected even after birth. Genetic testing may leave parents faced with difficult decisions about whether to terminate a pregnancy or to raise a child who may have profound developmental abnormalities and a short life expectancy. These are complex issues that demand careful, informed thought and competent genetic counseling.

Parents will be receiving even more genetic information and confronting further questions in the near future. Indeed, in 2012 the first infant was born whose entire genome had been sequenced before the baby was born.

Nevertheless, completing a genome sequence does not ensure complete information. Consider, for example, Klinefelter syndrome, in which males have an extra X chromosome. This disorder is quite common, affecting one in 1,000 men, and can cause reduced testosterone, a feminized appearance, and infertility. However, while some men with an extra X chromosome have a debilitating disorder, others have symptoms so mild that they are unaware of the condition. For other disorders, such as diabetes, heart disease, or cancer, a genome sequence may only indicate the degree of risk. How parents will use this and other information in having and raising children is a question with no clear answers.

CONCEPT CHECK 46.5

1. Why does testing for hCG (human chorionic gonadotropin) work as a pregnancy test early in pregnancy but not late in pregnancy? What is the function of hCG in pregnancy?
2. In what ways are tubal ligation and vasectomy similar?
3. **WHAT IF?** If a sperm nucleus is injected into an oocyte, what steps of gametogenesis and conception are bypassed?

For suggested answers, see Appendix A.

46 Chapter Review

→ Go to **Mastering Biology** for Assignments, the eText, the Study Area, and Dynamic Study Modules.

SUMMARY OF KEY CONCEPTS

→ To review key terms, go to the **Vocabulary Self-Quiz** in the **Mastering Biology** eText or Study Area, or go to goo.gl/zkjz9t.

CONCEPT 46.1

Both asexual and sexual reproduction occur in the animal kingdom (pp. 1020–1022)

- **Sexual reproduction** requires the fusion of male and female gametes, forming a diploid **zygote**. Examples of **asexual reproduction**—the production of offspring without gamete fusion—include budding, **fission**, and fragmentation with regeneration. Variations on the mode of reproduction are achieved through **parthenogenesis, hermaphroditism**, and sex reversal. Hormones and environmental cues control reproductive cycles.

? *Would a pair of haploid offspring produced by parthenogenesis be genetically identical? Explain.*

CONCEPT 46.2

Fertilization depends on mechanisms that bring together sperm and eggs of the same species (pp. 1022–1025)

- **Fertilization** occurs externally, when sperm and eggs are both released outside the body, or internally, when sperm deposited by the male fertilize an egg in the female reproductive system. In either case, fertilization requires coordinated timing, which may

be mediated by environmental cues, pheromones, or courtship behavior. Internal fertilization is often associated with relatively fewer offspring and greater protection of offspring by the parents.
- Systems for gamete production and delivery range from undifferentiated cells in the body cavity to complex systems that include **gonads**, which produce gametes, and accessory tubes and glands that protect or transport gametes and embryos. Although sexual reproduction involves a partnership, it also provides an opportunity for competition between individuals and between gametes.

Complex reproductive systems in fruit flies

Male fruit fly — Testis, Vas deferens, Ejaculatory duct, Seminal vesicle, Penis and claspers

Female fruit fly — Ovary, Oviduct, Spermatheca, Accessory gland, Uterus, Vulva

Key to labels:
Gamete production
Gamete protection and transport

? *Identify which of the following, if any, are unique to mammals: a female uterus, a male vas deferens, extended internal development, and parental care of newborns.*

Reproductive organs produce and transport gametes
(pp. 1025–1029)

- In human males, **sperm** are produced in **testes**, which are suspended outside the body in the **scrotum**. Ducts connect the testes to internal accessory glands and to the **penis**. The reproductive system of the human female consists principally of the **labia** and the **glans** of the **clitoris** externally and the **vagina, uterus, oviducts,** and **ovaries** internally. **Eggs** are produced in the ovaries and upon fertilization develop in the uterus.

- **Gametogenesis**, or gamete production, consists of the processes of **spermatogenesis** in males and **oogenesis** in females. Human spermatogenesis is continuous and produces four sperm per meiosis. Human oogenesis is discontinuous and cyclic, generating one egg per meiosis.

Human gametogenesis

Spermatogenesis   Oogenesis

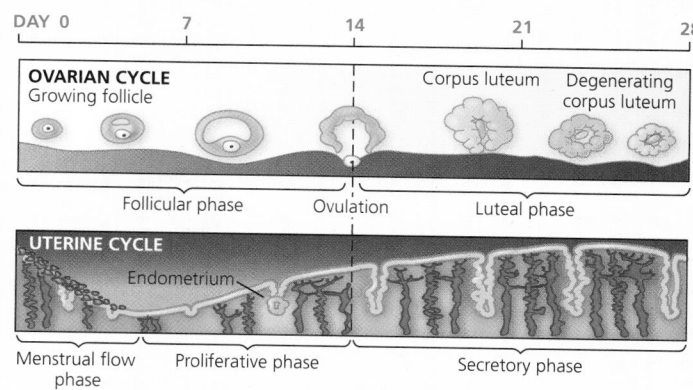

? How does the difference in size and cellular contents between sperm and eggs relate to their specific functions in reproduction?

The interplay of tropic and sex hormones regulates reproduction in mammals *(pp. 1030–1034)*

- Human sexuality, which includes biological sex, gender identity, and sexual orientation, exhibits considerable variation.

- In mammals, GnRH from the hypothalamus regulates the release of two hormones, **FSH** and **LH**, from the anterior pituitary. In males, FSH and LH control the secretion of androgens (chiefly **testosterone**) and sperm production. In females, cyclic secretion of FSH and LH orchestrates the **ovarian** and **uterine cycles** via estrogens (primarily **estradiol**) and **progesterone**. The developing **follicle** and the **corpus luteum** also secrete hormones, which help coordinate the uterine and ovarian cycles through positive and negative feedback.

- In **estrous cycles**, the lining of the **endometrium** is reabsorbed, and sexual receptivity is limited to a heat period. Reproductive structures with a shared origin in development underlie many features of human sexual arousal and orgasm common to males and females.

? Why do anabolic steroids lead to reduced sperm counts?

In placental mammals, an embryo develops fully within the mother's uterus *(pp. 1034–1040)*

- After fertilization and the completion of meiosis in the oviduct, the zygote undergoes a series of cell divisions and develops into a **blastocyst** before implantation in the endometrium. All major organs start developing by 8 weeks. A pregnant woman's acceptance of her "foreign" offspring likely reflects partial suppression of the maternal immune response.

- **Contraception** may prevent release of mature gametes from the gonads, fertilization, or embryo implantation. **Abortion** is the termination of a pregnancy in progress.

- Reproductive technologies can help detect problems before birth and can assist infertile couples. Infertility may be treated through hormone therapy or ***in vitro* fertilization**.

? What route would oxygen in the mother's blood follow to arrive at a body cell of the fetus?

TEST YOUR UNDERSTANDING

→ For more multiple-choice questions, go to the **Practice Test** in the **Mastering Biology** eText or Study Area, or go to goo.gl/GruWRg.

Levels 1–2: Remembering/Understanding

1. Which of the following characterizes parthenogenesis?
   (A) An individual may change its sex during its lifetime.
   (B) Specialized groups of cells grow into new individuals.
   (C) An organism is first a male and then a female.
   (D) An egg develops without being fertilized.

2. In male mammals, excretory and reproductive systems share
   (A) the vas deferens.   (C) the seminal vesicle.
   (B) the urethra.   (D) the prostate.

3. Which of the following is properly paired?
   (A) seminiferous tubule—cervix
   (B) vas deferens—oviduct
   (C) corpus luteum—Sertoli cell
   (D) scrotum—clitoris

4. Peaks of LH and FSH production occur during
(A) the menstrual flow phase of the uterine cycle.
(B) the beginning of the follicular phase of the ovarian cycle.
(C) the period just before ovulation.
(D) the secretory phase of the uterine cycle.

5. During human gestation, rudiments of all organs develop
(A) in the first trimester.
(B) in the second trimester.
(C) in the third trimester.
(D) during the blastocyst stage.

Levels 3-4: Applying/Analyzing

6. Which of the following is a true statement?
(A) All mammals have menstrual cycles.
(B) The endometrial lining is shed in menstrual cycles but reabsorbed in estrous cycles.
(C) Estrous cycles are more frequent than menstrual cycles.
(D) Ovulation occurs before the endometrium thickens in estrous cycles.

7. For which of the following is the number the same in human males and females?
(A) interruptions in meiotic divisions
(B) functional gametes produced by meiosis
(C) meiotic divisions required to produce each gamete
(D) different cell types produced by meiosis

8. Which statement about human reproduction is true?
(A) Fertilization occurs in the vagina.
(B) Spermatogenesis and oogenesis both require normal body temperature.
(C) An oocyte completes meiosis after a sperm penetrates it.
(D) The earliest stages of spermatogenesis occur closest to the lumen of the seminiferous tubules.

Levels 5-6: Evaluating/Creating

9. DRAW IT In human spermatogenesis, mitosis of a stem cell gives rise to one cell that remains a stem cell and one cell that becomes a spermatogonium. (a) Draw four rounds of mitosis for a stem cell, and label the daughter cells. (b) For one spermatogonium, draw the cells it would produce from one round of mitosis followed by meiosis. Label the cells, and label mitosis and meiosis. (c) Explain what would happen if stem cells divided like spermatogonia.

10. EVOLUTION CONNECTION Hermaphroditism is often found in animals that are fixed to a surface. Motile species are less often hermaphroditic. Explain why.

11. SCIENTIFIC INQUIRY Suppose that you discover a new egg-laying worm species. You dissect four adults and find both oocytes and sperm in each. Cells outside the gonad contain five chromosome pairs. Lacking genetic variants, explain how you would determine whether the worms can self-fertilize.

12. WRITE ABOUT A THEME: ENERGY AND MATTER In a short essay (100–150 words), discuss how different types of energy investment by females contribute to the reproductive success of a frog, a chicken, and a human.

13. SYNTHESIZE YOUR KNOWLEDGE

A female Komodo dragon (*Varanus komodoensis*) kept in isolation in a zoo had offspring. Each of the offspring had two identical copies of every gene in its genome. However, the offspring were not identical to one another. Based on your understanding of parthenogenesis and meiosis, propose a hypothesis to explain these observations.

For selected answers, see Appendix A.

47 Animal Development

KEY CONCEPTS

47.1 Fertilization and cleavage initiate embryonic development *p. 1044*

47.2 Morphogenesis in animals involves specific changes in cell shape, position, and survival *p. 1049*

47.3 Cytoplasmic determinants and inductive signals regulate cell fate *p. 1057*

Study Tip

Use word roots to learn vocabulary: Word roots often provide clues to the meaning of biological terms. Study this list of word roots and their meanings to help you infer and remember the definitions of vocabulary words from this chapter and others.

Word root—meaning		
blast—group of immature cells	poly—many	endo—internal
tropho—change	derm—tissue	meso—middle
coel—cavity	genesis—origin	ecto—external
morpho—shape	diplo—two	
	triplo—three	

Term—definition
blastocoel—cavity in ball of immature cells
diploblast
triploblast
trophoblast
endoderm
morphogenesis

→ Go to Mastering Biology

For Students (in eText and Study Area)
- Get Ready for Chapter 47
- Video: Sea Urchin Fertilization and Cleavage
- HHMI Video: Differentiation and the Fate of Cells

For Instructors to Assign (in Item Library)
- Tutorial: Visualizing Gastrulation
- Experimental Inquiry: How Do Calcium Ions Help to Prevent Polyspermy During Egg Fertilization?

1 mm

Chick embryo

Figure 47.1 This human embryo (main photo) and chick embryo (inset) are quite different in age (41 vs. 3.5 days old), but remarkably alike in appearance. The developing eyes, heart, and digestive tract are recognizable. So too are the repeated blocks of tissue that will form the vertebrae. These similarities reflect the fact that the early stages of embryonic development are shared among many animals, giving rise to a basic body plan that is later transformed into distinct and specialized forms.

What processes transform an egg into an embryo with recognizable structures?

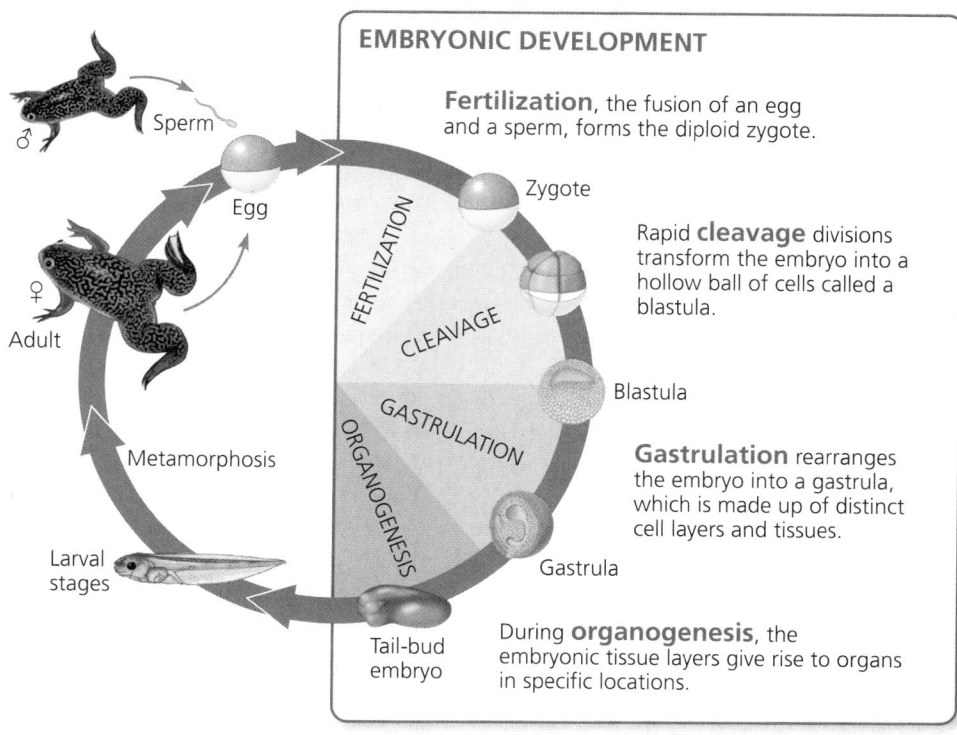

EMBRYONIC DEVELOPMENT

Fertilization, the fusion of an egg and a sperm, forms the diploid zygote.

Rapid **cleavage** divisions transform the embryo into a hollow ball of cells called a blastula.

Gastrulation rearranges the embryo into a gastrula, which is made up of distinct cell layers and tissues.

During **organogenesis**, the embryonic tissue layers give rise to organs in specific locations.

Sperm / Egg / Zygote / Adult / Metamorphosis / Larval stages / Tail-bud embryo / Gastrula / Blastula

FERTILIZATION / CLEAVAGE / GASTRULATION / ORGANOGENESIS

Fertilization and cleavage initiate embryonic development

Development occurs at multiple points in the animal life cycle. In a frog, for example, the larva (tadpole) undergoes sweeping changes in anatomy in becoming an adult. Development occurs in adult animals too, as when stem cells in the gonads produce sperm and eggs (gametes). In this chapter, however, our focus is on development in the embryonic stage.

Embryonic development in many animal species involves common stages in a set order. The first is fertilization, the fusion of sperm and egg. Next is cleavage, a series of cell divisions that divide, or cleave, the embryo into many cells. These cleavage divisions, which typically are rapid and lack accompanying cell growth, generate a hollow ball of cells called a blastula. In gastrulation, this blastula folds in on itself, rearranging into a multilayered embryo, the gastrula. Then, during organogenesis, local changes in cell shape and large-scale changes in cell location generate the rudimentary organs.

Because embryonic development has many common features across the animal kingdom, lessons learned from studying one animal can often be applied quite broadly. For this reason, studying development lends itself well to the use of **model organisms**, species chosen for the ease with which they can be studied. For example, the sea urchin (phylum *Echinodermata*) is useful for studying cell-surface events that take place during fertilization. Sea urchin gametes are easy to collect and fertilization occurs outside the body. As a result, researchers can observe fertilization and development in the laboratory simply by combining eggs and sperm in seawater.

In this chapter, we will concentrate on the sea urchin and four other model organisms: the fruit fly, the frog, the chick, and the nematode (roundworm). We will also explore some aspects of human embryonic development. We'll begin with the events surrounding **fertilization**, the formation of a diploid zygote from a haploid egg and sperm.

Fertilization

When sea urchins release their gametes into the water, the jelly coat of the egg exudes soluble molecules that attract the sperm, which swim toward the egg.

The Acrosomal Reaction

As detailed in **Figure 47.2**, the moment a sperm head contacts the egg surface, molecules in the jelly coat trigger

▼ Figure 47.2 The acrosomal and cortical reactions during sea urchin fertilization.

This icon is a simplified drawing of an adult sea urchin. Throughout the chapter, this icon and others of an adult frog, chicken, nematode, and human indicate the source of embryos in figures.

2 Acrosomal reaction. Hydrolytic enzymes released from the acrosome make a hole in the jelly coat. Growing actin filaments form the acrosomal process, which protrudes from the sperm head and through the jelly coat. Proteins on the surface of the acrosomal process bind to specific partner proteins in the egg plasma membrane.

3 Contact and fusion of sperm and egg membranes. Fusion triggers depolarization of the membrane, which acts as a fast block to polyspermy.

4 Cortical reaction. Cortical granules in the egg fuse with the plasma membrane. The secreted contents clip off sperm-binding receptors and cause the fertilization envelope to form. This acts as a slow block to polyspermy.

1 Contact. The sperm contacts the egg's jelly coat, triggering exocytosis of the sperm's acrosome.

5 Entry of sperm nucleus.

Sperm plasma membrane
Sperm nucleus
Acrosomal process
Basal body (centriole)
Sperm head
Actin filament
Acrosome
Jelly coat
Sperm-binding receptors
Vitelline layer
Egg plasma membrane
Hydrolytic enzymes
Fused plasma membranes
Cortical granule
Perivitelline space
Fertilization envelope
EGG CYTOPLASM

the **acrosomal reaction** in the sperm. This reaction begins with the discharge of hydrolytic enzymes from the **acrosome**, a specialized vesicle at the tip of the sperm. These enzymes partially digest the jelly coat, enabling a sperm structure called the *acrosomal process* to form, elongate, and penetrate the coat. Protein molecules on the tip of the acrosomal process bind to specific proteins in the egg plasma membrane. This "lock-and-key" recognition is especially important for sea urchins and other species with external fertilization because the water into which sperm and eggs are released may contain gametes of other species (see, for example, Figure 24.3h).

The recognition event between the sperm and egg triggers fusion of their plasma membranes. The sperm nucleus enters the egg cytoplasm as ion channels open in the egg's plasma membrane. Sodium ions diffuse into the egg and cause depolarization, a decrease in the charge difference, or potential, across the plasma membrane (see Concept 7.4).

Once depolarization takes place, additional sperm cannot fuse with the egg's plasma membrane. In this way, depolarization acts as a barrier to **polyspermy**, the entry of multiple sperm nuclei into the egg. If polyspermy were to occur, the resulting abnormal number of chromosomes would be lethal for the embryo. Because depolarization occurs 1–3 seconds after a sperm binds to an egg, depolarization acts as a *fast block to polyspermy*.

The Cortical Reaction

Although membrane depolarization in sea urchins lasts for only a minute or so, there is a longer-lasting change that also prevents polyspermy. This *slow block to polyspermy* is established by vesicles in the outer rim, or cortex, of the cytoplasm. Within seconds after a sperm binds to the egg, these vesicles, called cortical granules, fuse with the egg plasma membrane (see Figure 47.2). Contents of the cortical granules are released into the space between the plasma membrane and the surrounding vitelline layer, a structure formed by the egg's extracellular matrix. Enzymes and

other granule contents then trigger a *cortical reaction*, which lifts the vitelline layer away from the egg and hardens the layer into a protective fertilization envelope.

Formation of the fertilization envelope requires a high concentration of calcium ions (Ca^{2+}) in the egg. As described in **Figure 47.3**, researchers wondered whether a change in the Ca^{2+} concentration triggers the cortical reaction. Using a

---

▼ Figure 47.3 Inquiry

Does the distribution of Ca^{2+} in an egg correlate with formation of the fertilization envelope?

Experiment Investigators mixed sea urchin eggs with sperm, waited 10–60 seconds, and then added a chemical fixative, freezing cellular structures in place. When photomicrographs of each sample are ordered according to the time of fixation, they illustrate the stages in the formation of a fertilization membrane for a single egg.

Fertilization envelope

10 sec after fertilization | 25 sec | 35 sec | 1 min   500 μm

Calcium ion (Ca^{2+}) signaling controls fusion of vesicles with the plasma membrane during neurotransmitter release, insulin secretion, and plant pollen tube formation. Researchers hypothesized that Ca^{2+} signaling plays a similar role in forming the fertilization envelope. To test this hypothesis, they tracked the release of free Ca^{2+} in sea urchin eggs after sperm binding. A fluorescent dye that glows when it binds free Ca^{2+} was injected into unfertilized eggs. The scientists then added sea urchin sperm and used fluorescence to produce the results shown here.

Results A rise in the concentration of Ca^{2+} in the cytosol began near where the sperm entered and spread in a wave. Soon after the wave passed, the fertilization envelope rose from the egg surface.

Point of sperm nucleus entry   Spreading wave of Ca^{2+}

1 sec before fertilization | 10 sec after fertilization | 20 sec | 30 sec   500 μm

Conclusion Ca^{2+} release correlates with formation of the fertilization envelope, supporting the researchers' hypothesis that an increase in the Ca^{2+} level triggers cortical granule fusion.

Data from R. Steinhardt et al., Intracellular calcium release at fertilization in the sea urchin egg, *Developmental Biology* 58:185–197 (1977); M. Hafner et al., Wave of free calcium at fertilization in the sea urchin egg visualized with Fura-2, *Cell Motility and the Cytoskeleton* 9:271–277 (1988).

→ **Instructors:** A related Experimental Inquiry Tutorial can be assigned in **Mastering Biology**.

WHAT IF? *Suppose a particular molecule could enter the egg and bind to Ca^{2+}, blocking its function. How would you use this molecule to further test the hypothesis that a rise in the Ca^{2+} level triggers cortical granule fusion?*

calcium-sensitive dye, they assessed how Ca²⁺ is distributed in the egg before and during fertilization. They found that Ca²⁺ spread across the egg in a wave that correlated with the appearance of the fertilization envelope.

Further studies demonstrated that the binding of a sperm to the egg activates a signal transduction pathway that triggers release of Ca²⁺ into the cytosol from the endoplasmic reticulum. The resulting increase in Ca²⁺ level causes cortical granules to fuse with the plasma membrane. A cortical reaction triggered by Ca²⁺ also occurs in vertebrates such as fishes and mammals.

Egg Activation

Fertilization initiates and speeds up metabolic reactions that bring about the onset of embryonic development, "activating" the egg. For example, cellular respiration and protein synthesis in the egg markedly accelerate following the entry of the sperm nucleus. Soon thereafter, the egg and sperm nuclei fully fuse, and cycles of DNA synthesis and cell division begin.

What triggers activation of the egg? A major clue came from experiments demonstrating that the unfertilized eggs of sea urchins and many other species can be activated by an injection of Ca²⁺. Based on this discovery, researchers concluded that the rise in Ca²⁺ concentration that causes the cortical reaction also causes egg activation. Further experiments revealed that artificial activation is possible even if the nucleus has been removed from the egg. This further finding indicates that the proteins and mRNAs required for activation are already present in the cytoplasm of the unfertilized egg.

Not until about 20 minutes after the sperm nucleus enters the sea urchin egg do the sperm and egg nuclei fuse. DNA synthesis is then initiated. The first cell division, which occurs after about 90 minutes, marks the end of the fertilization stage.

Fertilization in other species shares many features with the process in sea urchins. However, there are differences, such as the stage of meiosis the egg has reached by the time it is fertilized. Sea urchin eggs have already completed meiosis when they are released from the female. In many other species, eggs are arrested at a specific stage of meiosis and do not complete the meiotic divisions until a sperm head enters. Human eggs, for example, are arrested at metaphase of meiosis II until sperm entry (see Figure 46.11).

Fertilization in Mammals

Unlike sea urchins and most other marine invertebrates, terrestrial animals, including mammals, fertilize their eggs internally. Support cells of the developing follicle surround the mammalian egg before and after ovulation. As shown in **Figure 47.4**, a sperm must travel through this

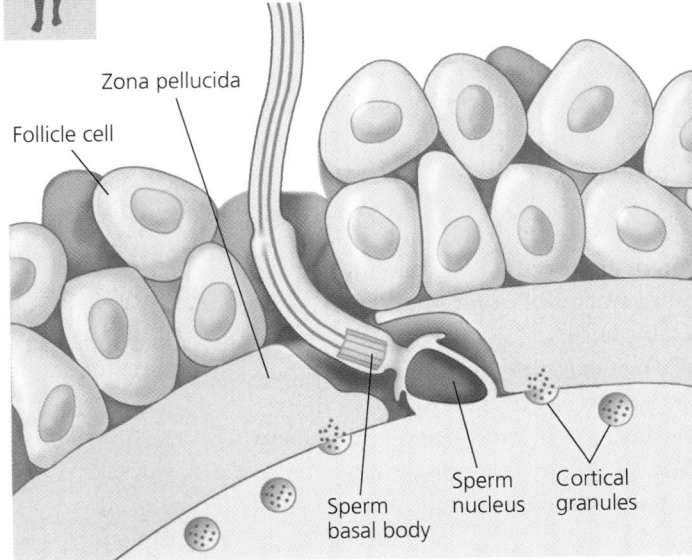

▼ **Figure 47.4 Fertilization in mammals.** The sperm shown here has traveled through the follicle cells and zona pellucida and has fused with the egg. The cortical reaction has begun, initiating events that ensure that only one sperm nucleus enters the egg.

Zona pellucida

Follicle cell

Sperm basal body

Sperm nucleus

Cortical granules

layer of follicle cells before it reaches the **zona pellucida**, the extracellular matrix of the egg. There, the binding of a sperm to a sperm receptor induces an acrosomal reaction, facilitating sperm entry.

As in sea urchins, sperm binding triggers a cortical reaction, the release of enzymes from cortical granules to the outside of the cell. These enzymes catalyze changes in the zona pellucida, which then acts as the slow block to polyspermy. (No fast block to polyspermy is known in mammals.)

Overall, the process of fertilization is much slower in mammals than in sea urchins: The first cell division occurs within 12–36 hours after sperm binding in mammals, compared with about 1.5 hours in sea urchins. This cell division marks the end of fertilization and the beginning of the next stage of development, cleavage.

Cleavage

The single nucleus in a newly fertilized egg has too little DNA to produce the amount of mRNA required to meet the cell's need for new proteins. Instead, initial development is carried out by mRNA and proteins deposited in the egg during oogenesis. There is still a need, however, to restore a balance between the cell's size and its DNA content. The process that addresses this challenge is **cleavage**, a series of rapid cell divisions during early development **(Figure 47.5)**.

▼ **Figure 47.5 Variation in mitotic cell division in embryonic development.** Circles represent embryonic cells, with letters and colors indicating different cell types (fates).

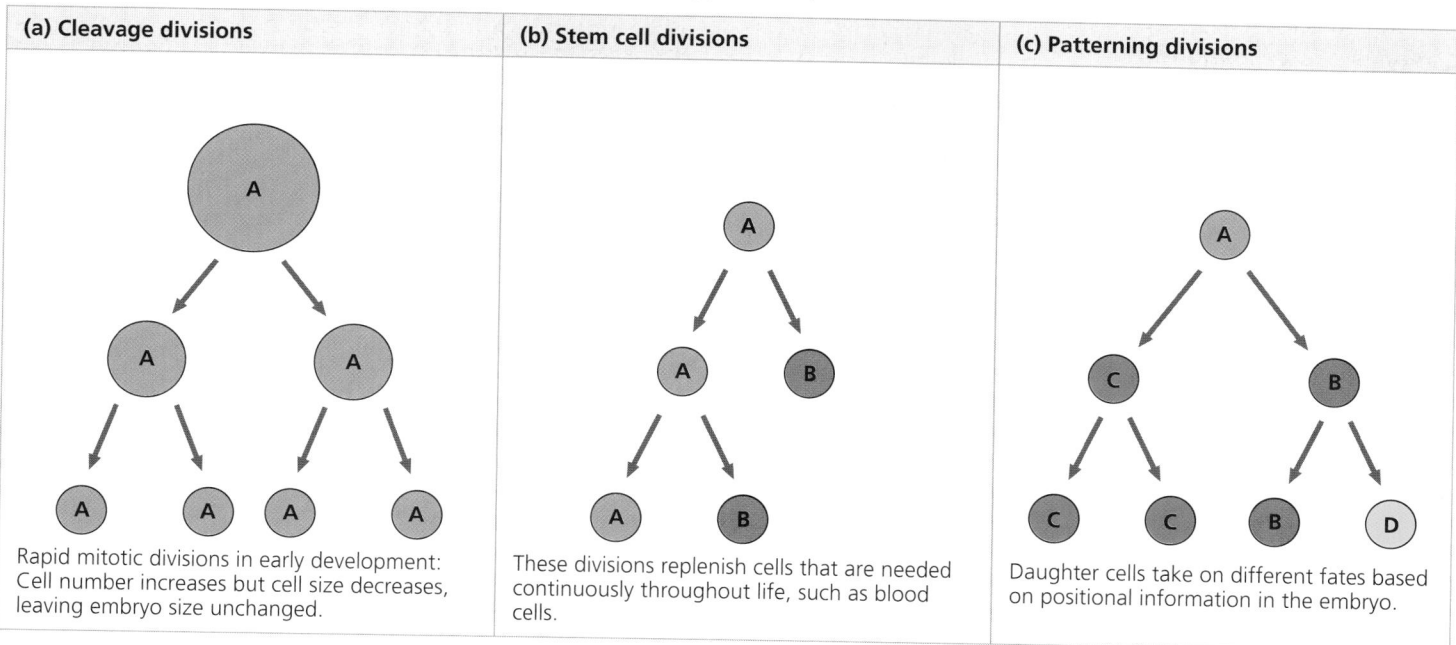

(a) Cleavage divisions

Rapid mitotic divisions in early development: Cell number increases but cell size decreases, leaving embryo size unchanged.

(b) Stem cell divisions

These divisions replenish cells that are needed continuously throughout life, such as blood cells.

(c) Patterning divisions

Daughter cells take on different fates based on positional information in the embryo.

During cleavage, the cell cycle consists primarily of the S (DNA synthesis) and M (mitosis) phases (see Figure 12.6 for a review of the cell cycle). The G_1 and G_2 (gap) phases are essentially skipped, and there is no increase in mass. Instead, cleavage partitions the cytoplasm of the large fertilized egg into many smaller cells called **blastomeres**. Because each blastomere is much smaller than the entire egg, its nucleus can make enough RNA to program further development.

The first five to seven cleavage divisions produce a hollow ball of cells, the **blastula**, surrounding a fluid-filled cavity called the **blastocoel**. In some species, including sea urchins and other echinoderms, the division pattern is uniform across the embryo **(Figure 47.6)**. In others, including frogs, the

▼ **Figure 47.6 Cleavage in an echinoderm embryo.** Cleavage is a series of mitotic cell divisions that transform the fertilized egg into a blastula, a hollow ball composed of cells called blastomeres. These light micrographs show the cleavage stages of a sand dollar embryo, which are virtually identical to those of a sea urchin. Each image is taken from above the embryo, with the focal plane, and hence the cells that are visible, at the equator.

(a) Fertilized egg. Shown here is the zygote shortly before the first cleavage division, surrounded by the fertilization envelope.

(b) Four-cell stage. Remnants of the mitotic spindle can be seen between the two pairs of cells that have just completed the second cleavage division.

(c) Early blastula. After further cleavage divisions, the embryo is a multicellular ball. A cavity—the blastocoel—has begun to form in the center.

(d) Later blastula. A single layer of cells surrounds the blastocoel. Morphogenesis follows, after which the embryo hatches from the fertilization envelope and begins swimming.

VISUAL SKILLS *If the embryo in (c) or (d) were photographed at a focal plane midway from the equator to one of the poles, what visible features would change?*

Mastering Biology Video: Sea Urchin Fertilization and Cleavage pattern is asymmetric, with regions of the embryo differing in both the number and size of newly formed cells.

Cleavage Pattern in Frogs

In the eggs of frogs (and many other animals), stored nutrients called **yolk** are concentrated toward one pole, called the **vegetal pole**, and away from the opposite or **animal pole**. This asymmetric distribution of yolk not only gives the two halves of the egg—the animal and vegetal hemispheres—different colors, but also influences the pattern of cleavage divisions, for reasons we will explore next.

When an animal cell divides, an indentation called a *cleavage furrow* forms in the cell surface as cytokinesis divides the cell in half **(Figure 47.7)**. In the frog embryo, the first two cleavage furrows form parallel to the line (or meridian) connecting the two poles. During these divisions, the dense yolk slows completion of cytokinesis. As a result, the first cleavage furrow is still dividing the yolky cytoplasm in the vegetal hemisphere when the second cell division begins. Eventually, four blastomeres of equal size extend from the animal pole to the vegetal pole.

During the third division, the yolk begins to affect the relative size of cells produced in the two hemispheres. This division is equatorial (perpendicular to the line connecting the poles) and produces an eight-cell embryo. However, as each of the four blastomeres begins this division, yolk near the vegetal pole displaces the mitotic apparatus and the cleavage furrow from the egg equator toward the animal pole. The result is a smaller blastomere size in the animal hemisphere than in the vegetal hemisphere. The displacing effect of the yolk persists in subsequent divisions, causing the blastocoel to form entirely in the animal hemisphere (see Figure 47.7).

Cleavage Patterns in Other Animals

Although yolk affects where division occurs in the eggs of frogs and other amphibians, the cleavage furrow still passes entirely through the egg. Cleavage in amphibian development is therefore said to be *holoblastic* (from the Greek *holos*, complete). Holoblastic cleavage is also seen in many other groups of animals, including echinoderms, mammals, and annelids. In those animals whose eggs contain relatively little yolk, the blastocoel forms centrally and the blastomeres are often of similar size, particularly during the first few divisions of cleavage (see Figure 47.6). This is the case for humans.

Yolk is most plentiful and has its most pronounced effect on cleavage in the eggs of birds and other reptiles, many fishes, and insects. In these animals, the volume of yolk is so great that cleavage furrows cannot pass through it, and only the region of the egg lacking yolk undergoes cleavage. This incomplete cleavage of a yolk-rich egg is said to be *meroblastic* (from the Greek *meros*, partial).

For chickens and other birds, the part of the egg that we commonly call the yolk is actually the entire egg cell. Cell divisions are limited to a small whitish area at the animal pole. These divisions produce a cap of cells that sort into upper and lower layers. The cavity between these two layers is the avian version of the blastocoel.

In *Drosophila* and most other insects, yolk is found throughout the egg. Early in development, multiple rounds of mitosis occur without cytokinesis. In other words, no cell membranes form around the early nuclei. The first several hundred nuclei spread throughout the yolk and later migrate to the outer edge of the embryo. After several more rounds of mitosis, a plasma membrane forms around each nucleus, and the embryo, now the equivalent of a blastula, consists of a single layer of about 6,000 cells surrounding a mass of yolk (see Figure 18.22). Given that the number of cleavage divisions varies among animals, what determines the end of the cleavage stage? The **Scientific Skills Exercise** explores one of the landmark studies addressing this question.

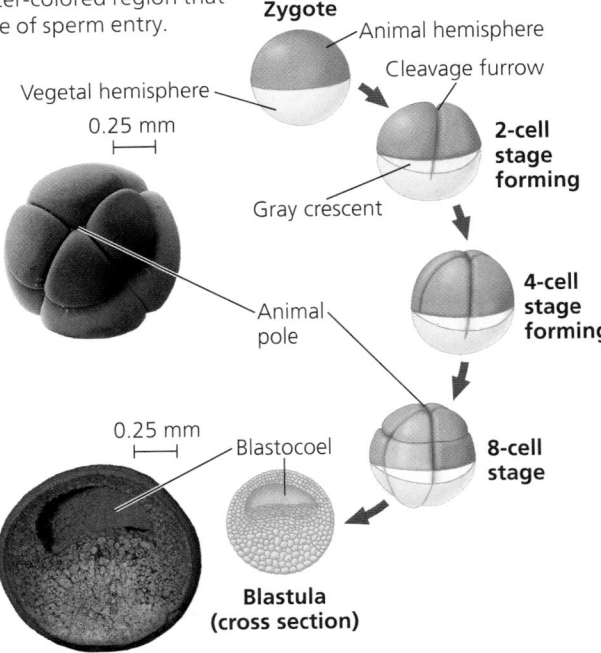

▼ **Figure 47.7 Cleavage in a frog embryo.** The cleavage planes in the first and second divisions extend from the animal pole to the vegetal pole, but the third cleavage is perpendicular to the polar axis. In some species, the first division bisects the gray crescent, a lighter-colored region that appears opposite the site of sperm entry.

Zygote
Animal hemisphere
Cleavage furrow
Vegetal hemisphere
0.25 mm
Gray crescent
2-cell stage forming
4-cell stage forming
Animal pole
8-cell stage
0.25 mm
Blastocoel
Blastula (cross section)

Eight-cell stage (viewed from the animal pole). The large amount of yolk displaces the third cleavage toward the animal pole, forming two tiers of cells. The four cells near the animal pole (closer, in this view) are smaller than the other four cells (colorized SEM).

Blastula (at least 128 cells). As cleavage continues, a fluid-filled cavity, the blastocoel, forms within the embryo. Because of unequal cell division, the blastocoel is located in the animal hemisphere. Both the drawing and the micrograph (assembled from fluorescence images) show cross sections of a blastula with about 4,000 cells.

Interpreting a Change in Slope

What Causes the End of Cleavage in a Frog Embryo? For a frog embryo in the cleavage stage, the cell cycle consists mainly of the S (DNA synthesis) and M (mitosis) phases. However, after the 12th cell division, G_1 and G_2 phases appear, and the cells grow, producing proteins and cytoplasmic organelles. What triggers this change?

How the Experiments Were Done Researchers tested the hypothesis that a mechanism for counting cell divisions determines when cleavage ends. They let frog embryos take up radioactively labeled nucleosides, either thymidine (to measure DNA synthesis) or uridine (to measure RNA synthesis). They then repeated the experiments in the presence of cytochalasin B, a chemical that prevents cell division by blocking cleavage furrow formation and cytokinesis.

Data from the Experiments

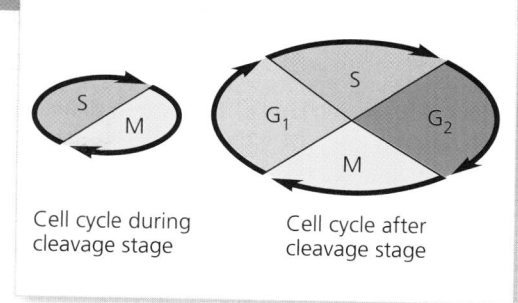

Cell cycle during cleavage stage

Cell cycle after cleavage stage

INTERPRET THE DATA

1. How does the use of particular labeled nucleosides allow independent measurement of DNA and RNA synthesis?
2. Describe the changes in synthesis that occur at the end of cleavage (time point 5 corresponds to cell division 12).
3. Comparing the rate of DNA synthesis with and without cytochalasin B, the researchers hypothesized that the toxin increases diffusion of thymidine into the embryos. Explain their logic.
4. Do the data support the hypothesis that the timing of the end of cleavage depends on counting cell divisions? Explain.
5. In a separate experiment, researchers disrupted the block to polyspermy, generating embryos with seven to ten sperm nuclei. At the end of cleavage, these embryos had the same nucleus-to-cytoplasm ratio as the wild-type embryos, but cleavage ended at the 10th cell division rather than the 12th cell division. What do these results indicate about the timing of the end of cleavage?

Data from J. Newport and M. Kirschner, A major developmental transition in early Xenopus embryos: I. Characterization and timing of cellular changes at the mid-blastula stage, *Cell* 30:675–686 (1982).

➜ **Instructors:** A version of this Scientific Skills Exercise can be assigned in **Mastering Biology**.

CONCEPT CHECK 47.1

1. How does the fertilization envelope form in sea urchins? What is its function?
2. **WHAT IF?** Predict what would happen if Ca^{2+} was injected into an unfertilized sea urchin egg.
3. **MAKE CONNECTIONS** Review Figure 12.16 on cell cycle control. Would you expect MPF (maturation-promoting factor) activity to remain steady during cleavage? Explain.

For suggested answers, see Appendix A.

CONCEPT 47.2

Morphogenesis in animals involves specific changes in cell shape, position, and survival

The cellular and tissue-based processes that are called **morphogenesis** and by which the animal body takes shape occur over the last two stages of embryonic development.

During **gastrulation**, a set of cells at or near the surface of the blastula moves to an interior location, cell layers are established, and a primitive digestive tube is formed. Further transformation occurs during **organogenesis**, the formation of organs. We will discuss these two stages in turn.

Gastrulation

Gastrulation is a dramatic reorganization of the hollow blastula into a two-layered or three-layered embryo called a **gastrula**. Cells move during gastrulation, taking up new positions and often acquiring new neighbors. **Figure 47.8** will help you visualize these complex three-dimensional changes. The cell layers produced are collectively called the embryonic **germ layers** (from the Latin *germen*, to sprout or germinate). In the late gastrula, **ectoderm** forms the outer layer and **endoderm** forms the lining of the digestive tract. In a few radially symmetric animals, only these two germ layers form during gastrulation. Such animals are called

VISUALIZING GASTRULATION

The embryos of all animals, and only animals, gastrulate. Cells change position: Some are brought inside the embryo, and others spread over the surface. The net result is an embryo with two, or more commonly three, cell layers. This figure will help you visualize the basic choreography of gastrulation before you explore the specific steps in different types of animals.

→ **Instructors:** Additional questions related to this Visualizing Figure can be assigned in **Mastering Biology**.

Reorganizing the animal embryo in three dimensions

Gastrulation typically begins with invagination, the infolding of a sheet of cells, shown here in both surface view and cross section. The resulting changes to the epithelium covering the embryo resemble what happens if you push a finger into one end of a lightly inflated balloon.

Surface view:

Plane of section

1. *In the gastrula, how many compartments are formed from the blastocoel?*

Cross section:

Blastula

Blastocoel

Archenteron

Gastrula

Blastopore

Primitive gut

At the end of cleavage, a single sheet of cells covers the surface of the blastula.

A group of cells buckle into the blastocoel, forming a shallow depression.

Continued invagination forms a blind-ended tube, the **archenteron.**

The open end of the archenteron, the first opening formed in the embryo, is called the **blastopore.**

The tip of the archenteron reaches the embryo surface, completing formation of the primitive gut of the embryo, now a gastrula.

Forming the primary cell layers of the animal body

In diploblasts, gastrulation forms two germ layers: ectoderm and endoderm. Triploblasts—both protostomes and deuterostomes—gain a third layer: mesoderm. At the end of embryogenesis, each germ layer gives rise to specific tissues and organs. To visualize how this occurs, trace the fate of each germ layer in each row of drawings (all views are cross sections except the sea urchin larva, which is a transparent surface view).

Key
- Ectoderm
- Mesoderm
- Endoderm

Diploblast: *Hydra*

In diploblasts, the blastopore becomes the open end of the gastrovascular cavity.

Blastopore

Open end of gastro-vascular cavity

Adult polyp

Triploblast: *Planaria* (protostome)

In protostomes, the mouth forms from the blastopore.

Blastopore

Adult worm

Mouth (on underside)

Gastrovascular cavity

Triploblast: Sea urchin (deuterostome)

In deuterostomes, including all vertebrates as well as some inverte-brates, the mouth forms *opposite* the blastopore.

Blastopore

Larva

Mouth

Anus

2. *What generalizations can you make about the location of mesoderm, when it is present, during and after development?*

→ **Mastering Biology** Video: **Sea Urchin Gastrulation**

diploblasts. In contrast, vertebrates and other bilaterally symmetric animals are triploblasts: In these animals, a third germ layer, the **mesoderm**, forms between the ectoderm and the endoderm.

Gastrulation in Frogs

Each embryonic germ layer contributes to a distinct set of structures in the adult animal, as shown in **Figure 47.9**. The embryonic organization of the germ layers is often reflected in the adult: The ectoderm forms the nervous system and outer body layer, the mesoderm gives rise to muscles and skeleton, and the endoderm lines many organs and ducts. There are, however, numerous exceptions.

Figure 47.10 details gastrulation in a frog. The blastula of frogs and other triploblasts has a dorsal (top) and ventral

▼ **Figure 47.9 Major derivatives of the three embryonic germ layers in vertebrates.**

ECTODERM (outer layer)	MESODERM (middle layer)	ENDODERM (inner layer)
• Epidermis of skin and its derivatives (including sweat glands, hair follicles) • Nervous and sensory systems • Pituitary gland, adrenal medulla • Jaws and teeth	• Skeletal and muscular systems • Circulatory and lymphatic systems • Excretory and reproductive systems (except germ cells) • Dermis of skin • Adrenal cortex	• Epithelial lining of digestive tract and associated organs • Epithelial lining of respiratory, excretory, and reproductive tracts and ducts • Thymus, thyroid, and parathyroid glands

→ **Mastering Biology** HHMI Video: Differentiation and the Fate of Cells

 ▼ **Figure 47.10 Gastrulation in a frog embryo.** In the frog blastula, the blastocoel is displaced toward the animal pole and is surrounded by a wall several cells thick.

❶ Gastrulation begins when cells on the dorsal side invaginate to form a small indented crease, the blastopore. The part above the crease is called the **dorsal lip**. As the blastopore is forming, a sheet of cells begins to spread out of the animal hemisphere, rolls inward over the dorsal lip (involution), and moves into the interior (shown by the dashed arrow). In the interior, these cells will form endoderm and mesoderm, with the endodermal layer on the inside. Meanwhile, cells at the animal pole change shape and begin spreading over the outer surface.

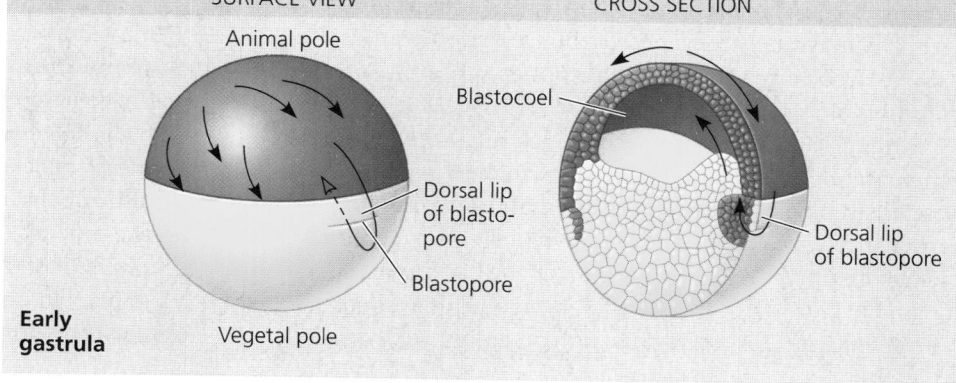

❷ The blastopore extends around both sides of the embryo as more cells invaginate. When the ends meet, the blastopore forms a circle that becomes smaller as ectoderm spreads downward over the surface. Internally, continued involution expands the endoderm and mesoderm; an archenteron forms and grows as the blastocoel shrinks and eventually disappears.

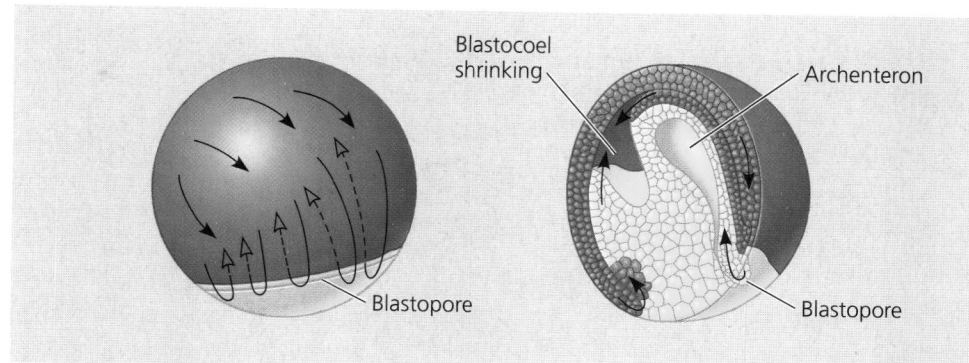

❸ Late in gastrulation, the cells remaining on the surface make up the ectoderm. The endoderm is the innermost layer, and the mesoderm lies between the ectoderm and endoderm. The circular blastopore surrounds a plug of yolk-filled cells.

Key

■ Future ectoderm
■ Future mesoderm
□ Future endoderm

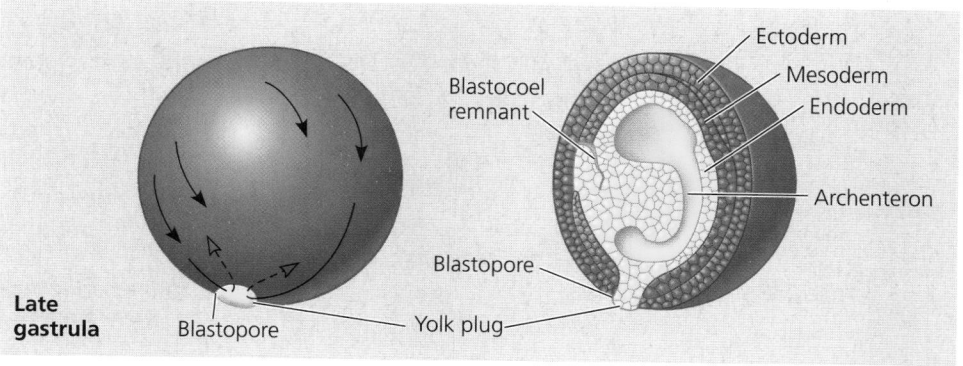

(bottom) side, a left and right side, and an anterior (front) and posterior (back) end. The cell movements that begin gastrulation occur on the dorsal side, opposite where the sperm entered the egg. The frog's anus develops from the blastopore, and the mouth eventually breaks through at the opposite end of the archenteron.

Gastrulation in Chicks

At the onset of gastrulation in chicks, an upper and a lower layer of cells—the *epiblast* and *hypoblast*—lie atop a yolk mass. All the cells that will form the embryo come from the epiblast. During gastrulation, some epiblast cells move toward the midline, detach, and move inward toward the yolk **(Figure 47.11)**. The pileup of cells moving inward at the midline produces a visible thickening called the *primitive streak*. Some of these cells move downward and form endoderm, pushing aside the hypoblast cells, while others migrate laterally (sideways) and form mesoderm. The cells left behind on the surface at the end of gastrulation will become ectoderm. The hypoblast cells later segregate from the endoderm and eventually form part of the sac that surrounds the yolk and also part of the stalk that connects the yolk mass to the embryo.

Although different terms describe gastrulation in different vertebrate species, the rearrangements and movements of cells exhibit a number of fundamental similarities. In particular, the primitive streak, shown in Figure 47.11 for the chick embryo, is the counterpart of the blastopore lip, shown in Figure 47.10 for the frog embryo. Formation of a primitive streak is also central to human embryo gastrulation, our next topic.

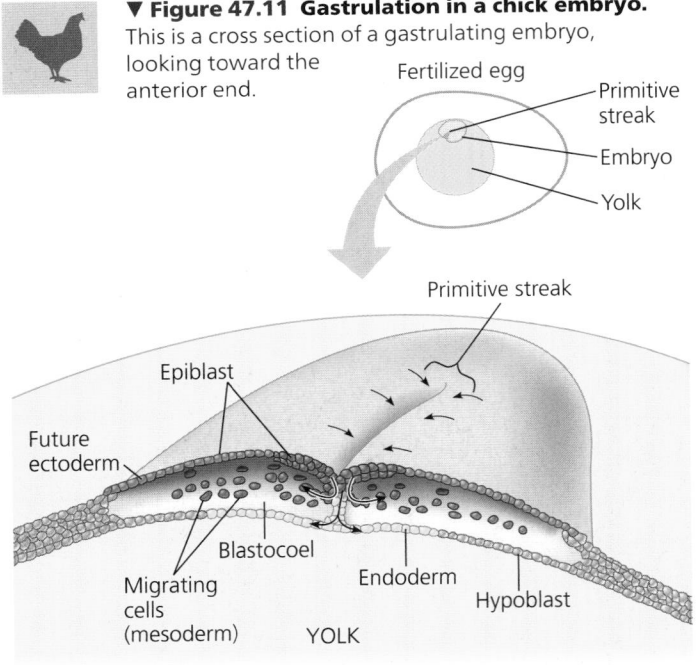

▼ **Figure 47.11 Gastrulation in a chick embryo.** This is a cross section of a gastrulating embryo, looking toward the anterior end.

Fertilized egg

Primitive streak

Embryo

Yolk

Primitive streak

Epiblast

Future ectoderm

Migrating cells (mesoderm)

Blastocoel

Endoderm

Hypoblast

YOLK

Gastrulation in Humans

Unlike the large, yolky eggs of many vertebrates, human eggs are quite small, storing little in the way of food reserves. Fertilization takes place in the oviduct, and development begins while the embryo completes its journey down the oviduct to the uterus (see Figure 46.15).

Figure 47.12 outlines the development of the human embryo, starting about 6 days after fertilization. This depiction is largely based on observations of embryos from other mammals, such as the mouse, and of very early human embryos following *in vitro* fertilization.

❶ At the end of cleavage, the embryo has more than 100 cells arranged around a central cavity and has reached the uterus. At this stage, the embryo is called a **blastocyst**, the mammalian version of a blastula. Clustered at one end of the blastocyst cavity is a group of cells called the **inner cell mass**, which will develop into the embryo proper. It is the cells of the inner cell mass that are the source of embryonic stem cell lines (see Concept 20.3).

❷ Implantation of the embryo is initiated by the **trophoblast**, the outer epithelium of the blastocyst. Enzymes secreted by the trophoblast during implantation break down molecules of the endometrium, the lining of the uterus, allowing invasion by the blastocyst. The trophoblast also extends finger-like projections that cause capillaries in the endometrium to spill out blood that can be captured by trophoblast tissues. Around the time the embryo undergoes implantation, the inner cell mass of the blastocyst forms a flat disk with an inner layer of cells, the *epiblast*, and an outer layer, the *hypoblast*. As is true for a bird embryo, the human embryo develops almost entirely from epiblast cells.

❸ Following implantation, the trophoblast continues to expand into the endometrium, and four new membranes appear. Although these **extraembryonic membranes** arise from the embryo, they enclose specialized structures located outside the embryo. As implantation is completed, gastrulation begins. Some epiblast cells remain as ectoderm on the surface, while others move inward through a primitive streak and form mesoderm and endoderm, just as in the chick (see Figure 47.11).

❹ By the end of gastrulation, the embryonic germ layers have formed. Extraembryonic mesoderm and four distinct extraembryonic membranes now surround the embryo. As development proceeds, cells of the invading trophoblast, the epiblast, and the adjacent endometrial tissue all contribute to the formation of the placenta. This vital organ mediates the exchange of nutrients, gases, and nitrogenous wastes between the developing embryo and the mother (see Figure 46.16).

➔ **Mastering Biology** HHMI Video: Human Embryonic Development

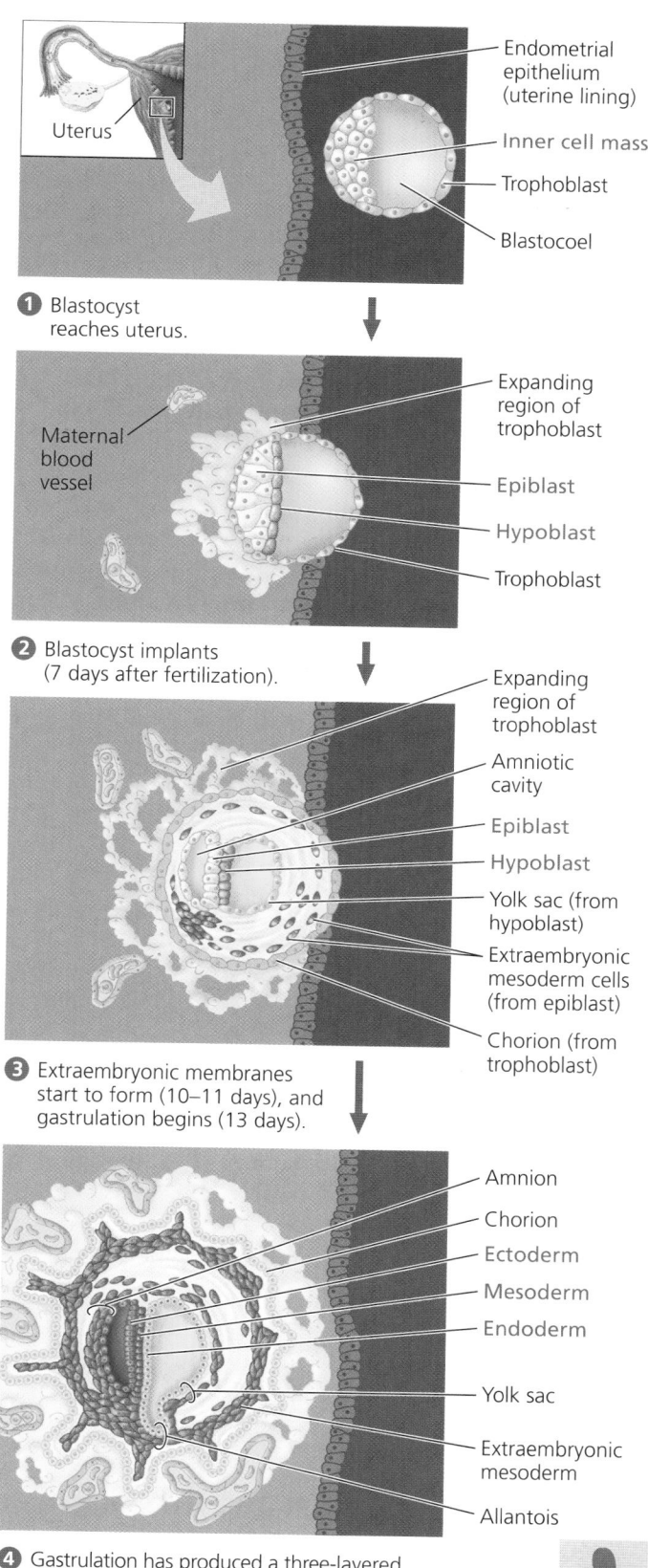

① Blastocyst reaches uterus.

- Endometrial epithelium (uterine lining)
- Inner cell mass
- Trophoblast
- Blastocoel

② Blastocyst implants (7 days after fertilization).

- Expanding region of trophoblast
- Epiblast
- Hypoblast
- Trophoblast

Maternal blood vessel

③ Extraembryonic membranes start to form (10–11 days), and gastrulation begins (13 days).

- Expanding region of trophoblast
- Amniotic cavity
- Epiblast
- Hypoblast
- Yolk sac (from hypoblast)
- Extraembryonic mesoderm cells (from epiblast)
- Chorion (from trophoblast)

④ Gastrulation has produced a three-layered embryo with four extraembryonic membranes: the amnion, chorion, yolk sac, and allantois.

- Amnion
- Chorion
- Ectoderm
- Mesoderm
- Endoderm
- Yolk sac
- Extraembryonic mesoderm
- Allantois

▲ **Figure 47.12 Four stages in the early embryonic development of a human.** The names of the tissues that develop into the embryo proper are printed in blue.

Developmental Adaptations of Amniotes

EVOLUTION During embryonic development, mammals and reptiles (including birds) form four extraembryonic membranes: the chorion, allantois, amnion, and yolk sac **(Figure 47.13)**. In all these groups, such membranes provide a "life-support system" for further development. Why did this adaptation appear in the evolutionary history of reptiles and mammals, but not other vertebrates, such as fishes and amphibians? We can formulate a reasonable hypothesis by considering a few basic facts about embryonic development. All vertebrate embryos require an aqueous environment for their development. The embryos of fishes and amphibians usually develop in the surrounding sea or pond and need no specialized water-filled enclosure. However, the extensive colonization of land by vertebrates was possible only after the evolution of structures that would allow reproduction in dry environments. Two such structures exist today: (1) the shelled egg of birds and other reptiles as well as a few mammals (the monotremes) and (2) the uterus of marsupial and eutherian mammals. Inside the shell or uterus, the embryos of these animals are surrounded by fluid within a sac formed by one of the extraembryonic membranes, the amnion. Mammals and reptiles, including birds, are therefore called **amniotes** (see Concept 34.5).

For the most part, the extraembryonic membranes have similar functions in mammals and reptiles, consistent with a common evolutionary origin (see Figure 34.25). The chorion is the site of gas exchange, and the fluid within the amnion physically protects the developing embryo. (This amniotic fluid is released from the vagina when a pregnant woman's

▼ **Figure 47.13 The shelled egg of reptiles.**

(a) The four extraembryonic membranes in a reptile egg

- Chorion
- Allantois
- Amnion
- Yolk sac

(b) A baby red-tailed racer snake (*Gonyosoma oxycephala*) hatching from its protective egg

"water breaks" before childbirth.) The allantois, which disposes of wastes in the reptilian egg, is incorporated into the umbilical cord in mammals. There, it forms blood vessels that transport oxygen and nutrients from the placenta to the embryo and rid the embryo of carbon dioxide and nitrogenous wastes. The fourth extraembryonic membrane, the yolk sac, encloses yolk in the eggs of reptiles. In mammals it is a site of early formation of blood cells, which later migrate into the embryo proper. Thus, the extraembryonic membranes common to reptiles and mammals exhibit adaptations specific to development within a shelled egg or a uterus.

After gastrulation is complete and any extraembryonic membranes are formed, the next stage of embryonic development begins: organogenesis, the formation of organs.

Organogenesis

During organogenesis, regions of the three embryonic germ layers develop into the rudiments of organs. Often, cells from two or three germ layers participate in forming a single organ, with interactions between cells of different germ layers helping to specify cell fates. Adopting particular developmental fates may in turn cause cells to change shape or, in certain circumstances, migrate to another location in the body. To see how these processes contribute to organogenesis, we'll consider *neurulation*, the early steps in the formation of the brain and spinal cord in vertebrates.

Neurulation

Neurulation begins as cells from the dorsal mesoderm form the **notochord**, a rod that extends along the dorsal side of the chordate embryo, as seen for the frog in **Figure 47.14a**. Signaling molecules secreted by these mesodermal cells and other tissues cause the ectoderm above the notochord to become the *neural plate*. Formation of the neural plate is thus an example of **induction**, a process in which a group of cells or tissues influences the development of another group through close-range interactions (see Figure 18.17b).

 ▼ **Figure 47.14 Neurulation in a frog embryo.**

⊘ **Mastering Biology** Video: Frog Development

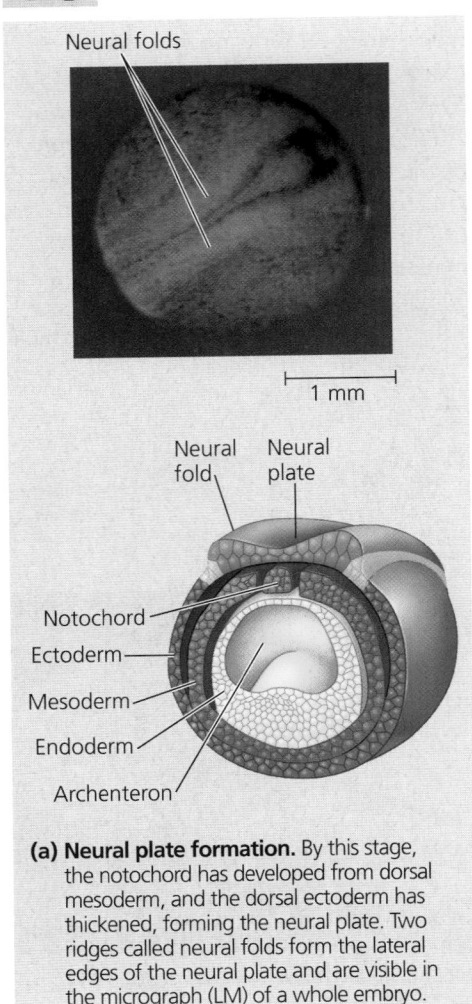

(a) Neural plate formation. By this stage, the notochord has developed from dorsal mesoderm, and the dorsal ectoderm has thickened, forming the neural plate. Two ridges called neural folds form the lateral edges of the neural plate and are visible in the micrograph (LM) of a whole embryo.

(b) Neural tube formation. Infolding and pinching off of the neural plate generates the neural tube.

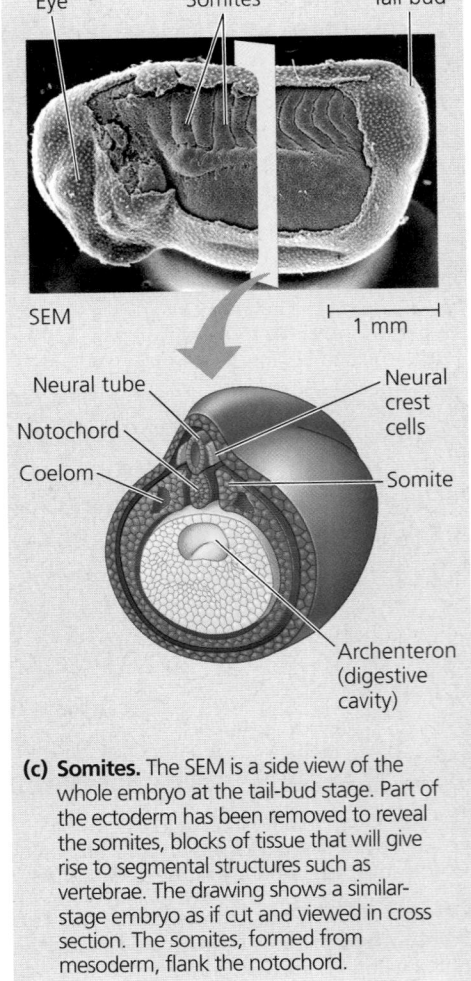

(c) Somites. The SEM is a side view of the whole embryo at the tail-bud stage. Part of the ectoderm has been removed to reveal the somites, blocks of tissue that will give rise to segmental structures such as vertebrae. The drawing shows a similar-stage embryo as if cut and viewed in cross section. The somites, formed from mesoderm, flank the notochord.

After the neural plate is formed, its cells change shape, curving the structure inward. In this way, the neural plate rolls itself into the **neural tube**, which runs along the anterior-posterior axis of the embryo **(Figure 47.14b)**. The neural tube will become the brain in the head and the spinal cord along the rest of the body. In contrast, the notochord disappears before birth, although parts persist as the inner portions of the disks in the adult spine. (These are the disks that can herniate or rupture, causing back pain.)

Neurulation, like other stages of development, is sometimes imperfect. For example, *spina bifida*, the most common congenital disability in the United States, occurs when a portion of the neural tube fails to develop or close properly. The opening in the spinal column that remains causes nerve damage, resulting in varying degrees of leg paralysis. Although the opening can be surgically repaired shortly after birth, the nerve damage is permanent.

Cell Migration in Organogenesis

During organogenesis, some cells undergo long-range migration, including two sets of cells that develop near the vertebrate neural tube. The first set is called the **neural crest**, a set of cells that develops along the borders where the neural tube pinches off from the ectoderm (see Figure 47.14b). Neural crest cells subsequently migrate to many parts of the embryo, forming a variety of tissues that include peripheral nerves as well as parts of the teeth and skull bones.

A second set of migratory cells is formed when groups of mesodermal cells lateral to the notochord separate into blocks called **somites (Figure 47.14c)**. Somites play a significant role in organizing the segmented structure of the vertebrate body. Parts of the somites dissociate into mesenchyme

cells. Some form the vertebrae; others form the muscles associated with the vertebral column and the ribs.

By contributing to the formation of vertebrae, ribs, and associated muscles, somites form repeated structures in the adult. Chordates, including ourselves, are thus segmented, although in the adult form the segmentation is much less obvious than in shrimp and other segmented invertebrates.

Organogenesis in Chicks and Insects

Early organogenesis in the chick is quite similar to that in the frog. For example, the borders of the chick blastoderm fold downward and come together, pinching the embryo into a three-layered tube joined under the middle of the body to the yolk **(Figure 47.15a)**. By the time the chick embryo is 56 hours old, rudiments of the major organs, including the brain, eyes, and heart, are readily apparent **(Figure 47.15b)**.

Comparing organogenesis in invertebrates with that in vertebrates often reveals fundamental similarities in mechanism that are masked by differences in pattern and appearance. For example, consider neurulation. In insects, tissues of the nervous system form on the ventral, not dorsal, side of the embryo. However, ectoderm along the anterior-posterior axis then rolls into a tube inside the embryo, just as in vertebrate neurulation. Furthermore, the molecular signaling pathways that bring about these similar events in different locations have many steps in common, underscoring a shared evolutionary history.

Like gastrulation, organogenesis in vertebrates and invertebrates relies substantially on changes in cell shape and location. We turn now to an exploration of how these changes take place.

▶ **Figure 47.15 Organogenesis in a chick embryo.**

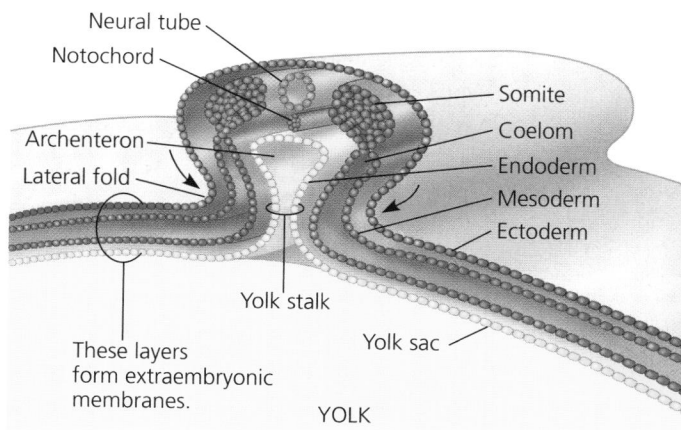

(a) **Early organogenesis.** The archenteron forms when lateral folds pinch the embryo away from the yolk. The embryo remains open to the yolk, attached by the yolk stalk, as shown in this cross section.

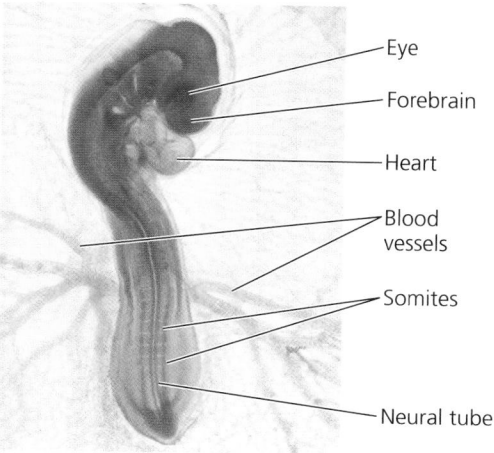

(b) **Late organogenesis.** Rudiments of most major organs have already formed in this 56-hour-old chick embryo. Blood vessels extending from the embryo supply the extraembryonic membranes, as seen in this light micrograph (LM).

The Cytoskeleton in Morphogenesis

In animals, movement of parts of a cell can bring about changes in cell shape or enable a cell to migrate from one place to another within the embryo. One set of cellular components essential to these events is the collection of microtubules and microfilaments that make up the cytoskeleton (see Table 6.1).

Cell Shape Changes in Morphogenesis

Reorganization of the cytoskeleton is a major force in changing cell shape during development. As an example, let's return to the topic of neurulation. At the onset of neural tube formation, microtubules oriented from dorsal to ventral in a sheet of ectodermal cells help lengthen the cells along that axis **(Figure 47.16)**. At the apical end of each cell is a bundle of actin filaments (microfilaments) oriented crosswise. These actin filaments contract, giving the cells a wedge shape that bends the ectoderm layer inward.

The generation of wedge-shaped cells by apical constriction of actin filaments is a common mechanism in development for invaginating a cell layer. For instance, during gastrulation in the fruit fly *Drosophila melanogaster*, the formation of wedge-shaped cells along the ventral surface drives generation of the tube of cells that forms the mesoderm.

▼ **Figure 47.16 Change in cell shape during morphogenesis.** Reorganization of the cytoskeleton is associated with morphogenetic changes in embryonic tissues, as shown here.

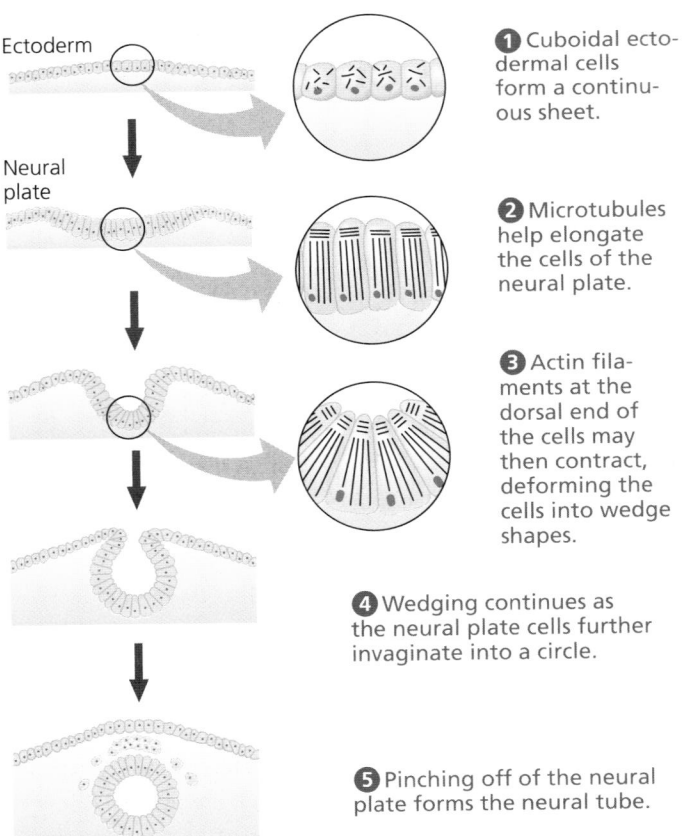

Ectoderm

Neural plate

1 Cuboidal ectodermal cells form a continuous sheet.

2 Microtubules help elongate the cells of the neural plate.

3 Actin filaments at the dorsal end of the cells may then contract, deforming the cells into wedge shapes.

4 Wedging continues as the neural plate cells further invaginate into a circle.

5 Pinching off of the neural plate forms the neural tube.

The cytoskeleton also directs a morphogenetic movement called **convergent extension**, a rearrangement that causes a sheet of cells to become narrower (converge) while it becomes longer (extends). This type of lengthening and narrowing of cells occurs often in gastrulation, including formation of the primitive streak in the fertilized chick egg (see Figure 47.11) and elongation of the archenteron in the sea urchin embryo (see Figure 47.8). Convergent extension is also important during involution in the frog gastrula. There, convergent extension changes the gastrulating embryo from a spherical shape to the rounded rectangular shape seen in Figure 47.14c.

The cell movements during convergent extension are quite simple: The cells elongate, with their ends pointing in the direction they will move, and then wedge between each other, forming fewer columns of cells **(Figure 47.17)**. It's like a crowd of people about to enter a theater moving forward while also merging into a single-file line.

Cell Migration in Morphogenesis

The cytoskeleton is responsible not only for cell shape changes but also for cell migration. During organogenesis in vertebrates, cells from the neural crest and from somites migrate to locations throughout the embryo. Cells "crawl" within the embryo by using cytoskeletal fibers to extend and retract cellular protrusions. This type of motility is akin to amoeboid movement (see Figure 6.26b). Transmembrane glycoproteins called *cell adhesion molecules* play a key role in cell migration by promoting interaction between pairs of cells. Cell migration also involves the *extracellular matrix* (*ECM*), the meshwork of secreted glycoproteins and other macromolecules lying outside the plasma membranes of cells (see Figure 6.28).

The ECM helps to guide cells in many types of movements, such as migration of individual cells and shape changes of cell sheets. Cells that line migration pathways regulate the movement of migrating cells by secreting specific molecules into the ECM. For these reasons, researchers are attempting to generate an artificial ECM that can serve as a scaffold for the repair or replacement of damaged tissues or organs. One promising approach involves the use of nanofiber fabrication to produce materials that mimic the essential properties of the natural ECM.

Programmed Cell Death

Just as certain cells of the embryo are programmed to change shape or location, others are programmed to die. At various times in development, individual cells, sets of cells, or whole tissues cease to develop, die, and are engulfed by neighboring cells. Thus, *programmed cell death*, also called **apoptosis**, is a common feature of animal development.

One circumstance for programmed cell death is when a structure functions only in a larval or other immature form of the organism. One familiar example is the tail of a tadpole, the free-swimming larval stage of a frog or toad. The tail forms during early development, enables locomotion during

▶ **Figure 47.17 Convergent extension of a sheet of cells.** In this simplified diagram, the cells elongate coordinately in a particular direction and crawl between each other (convergence) as the sheet becomes longer and narrower (extension).

Convergence
Cells elongate and crawl between each other.

Extension
The sheet of cells becomes longer and narrower.

larval growth, and is then eliminated during metamorphosis into the adult form (see Figure 45.22).

Apoptosis may also occur when a large set of cells is formed, but only a subset has the properties required for further function. Such is the case in both nervous and immune system development. In the vertebrate nervous system, for instance, many more neurons are produced during development than exist in the adult. Neurons that make functional connections with other neurons typically survive; many of the rest undergo apoptosis. Similarly, in the adaptive immune system, self-reactive cells—cells with the potential to attack the animal itself rather than invading pathogens—are often eliminated by apoptosis.

Some cells that undergo apoptosis don't seem to have any function. Why do such cells form? The answer can be found by considering the evolution of amphibians, birds, and mammals. When these groups began to diverge during evolution, the basic developmental program for making a vertebrate body was already in place. The differences in present-day body forms arose through modification of that common developmental program. For example, the shared developmental program generates webbing between the embryonic digits, but in many birds and mammals, including humans, the webbing is eliminated by apoptosis (see Figure 11.21). This is one reason why such different adult forms arise from early vertebrate embryos that look so much alike.

As you have seen, cell behavior and the molecular mechanisms underlying it are crucial to the morphogenesis of the embryo. In the next section, you'll learn some ways in which shared cellular and genetic processes ensure that particular types of cells each end up in the right place.

CONCEPT CHECK 47.2

1. In the frog embryo, convergent extension elongates the notochord. Explain how the words *convergent* and *extension* apply to this process.

2. **WHAT IF?** Predict what would happen if, just before neural tube formation, you treated frog embryos with a drug that enters all the cells of the embryo and blocks the function of microfilaments.

3. **MAKE CONNECTIONS** Unlike some other types of developmental abnormalities, neural tube defects are largely preventable. Explain (see Figure 41.4).

For suggested answers, see Appendix A.

CONCEPT 47.3

Cytoplasmic determinants and inductive signals regulate cell fate

During embryonic development, cells arise by division, take up particular locations in the body, and become specialized in structure and function. Where a cell resides, how it appears, and what it does define its development *fate*. Developmental biologists use the terms **determination** to refer to the process by which a cell or group of cells becomes committed to a particular fate and **differentiation** to refer to the resulting specialization in structure and function. You may find it a useful analogy to think about determination being equivalent to declaring a major in college and differentiation being comparable to taking the courses required by your major.

Every diploid cell formed during an animal's development has the same genome. With the exception of certain mature immune cells, the collection of genes present in a given cell is the same throughout the cell's life. How, then, do cells acquire different fates? As discussed in Concept 18.4, particular tissues, and often cells within a tissue, differ from one another by expressing distinct sets of genes from their shared genome.

Even animals that display widely differing body plans share many basic mechanisms of development and often use a common set of regulatory genes. For example, the gene that specifies where eyes form in a vertebrate embryo has a close counterpart with a nearly identical function in the fruit fly *Drosophila melanogaster*. Indeed, when the gene from a mouse is experimentally introduced into a fly embryo, the mouse gene directs eye formation wherever it is expressed in the fly.

A major focus of developmental biology is to uncover the mechanisms that direct the differences in gene expression underlying developmental fates. As one step toward this goal, scientists often seek to trace tissues and cell types back to their origins in the early embryo.

▼ **Figure 47.18 Fate mapping for two chordates.**

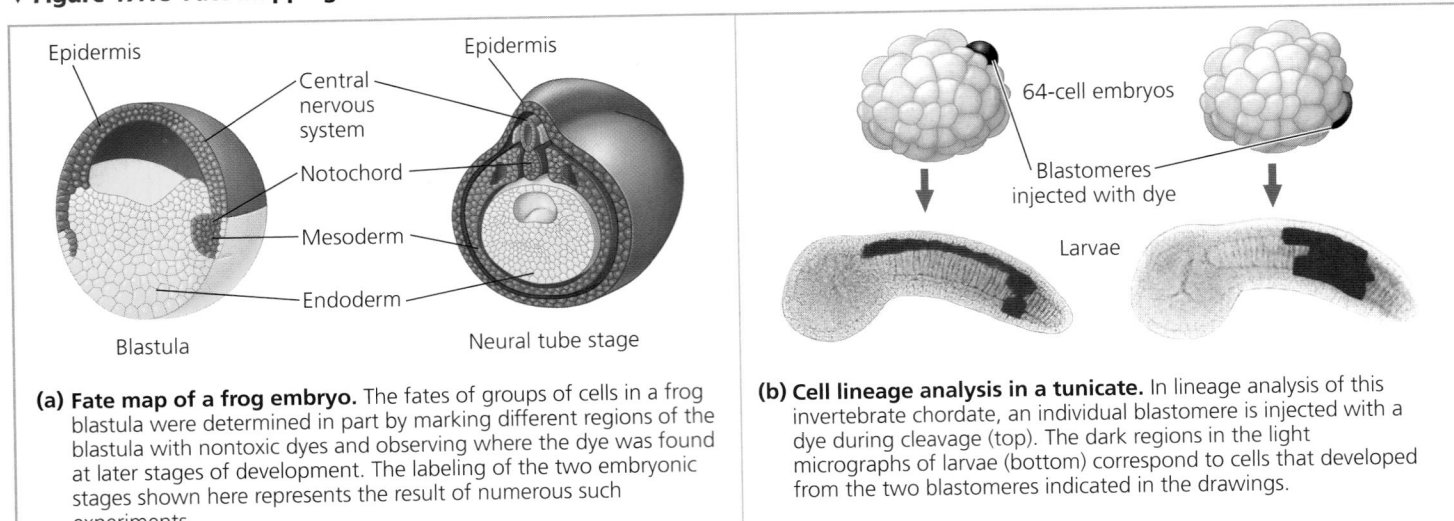

(a) Fate map of a frog embryo. The fates of groups of cells in a frog blastula were determined in part by marking different regions of the blastula with nontoxic dyes and observing where the dye was found at later stages of development. The labeling of the two embryonic stages shown here represents the result of numerous such experiments.

(b) Cell lineage analysis in a tunicate. In lineage analysis of this invertebrate chordate, an individual blastomere is injected with a dye during cleavage (top). The dark regions in the light micrographs of larvae (bottom) correspond to cells that developed from the two blastomeres indicated in the drawings.

Fate Mapping

One way to trace the ancestry of embryonic cells is direct observation through the microscope. Such studies produced the first **fate maps**, diagrams showing the structures arising from each region of an embryo. In the 1920s, German embryologist Walther Vogt used this approach to determine where groups of cells from the blastula end up in the gastrula **(Figure 47.18a)**. Later, researchers developed techniques that allowed them to mark an individual blastomere during cleavage and then follow the marker as it was distributed to all the mitotic descendants of that cell **(Figure 47.18b)**.

A much more comprehensive approach to fate mapping has been carried out on the soil-dwelling nematode *Caenorhabditis elegans*, as shown in **Figure 47.19**. This roundworm is about 1 mm long, has a simple, transparent body with only a few types of cells, and develops into a mature adult hermaphrodite worm in only 3.5 days in the laboratory. These attributes allowed Sydney Brenner, Robert Horvitz, and John Sulston to determine the complete developmental history, or *lineage*, of every cell in *C. elegans*. They found that every adult hermaphrodite worm has exactly 959 somatic cells, which arise

▼ **Figure 47.19 Cell lineage in *Caenorhabditis elegans*.** The *C. elegans* embryo is transparent, making it possible for researchers to trace the lineage of every cell, from the zygote to the adult worm (LM). The diagram shows a detailed lineage only for the intestine, which is derived exclusively from one of the first four cells formed from the zygote.

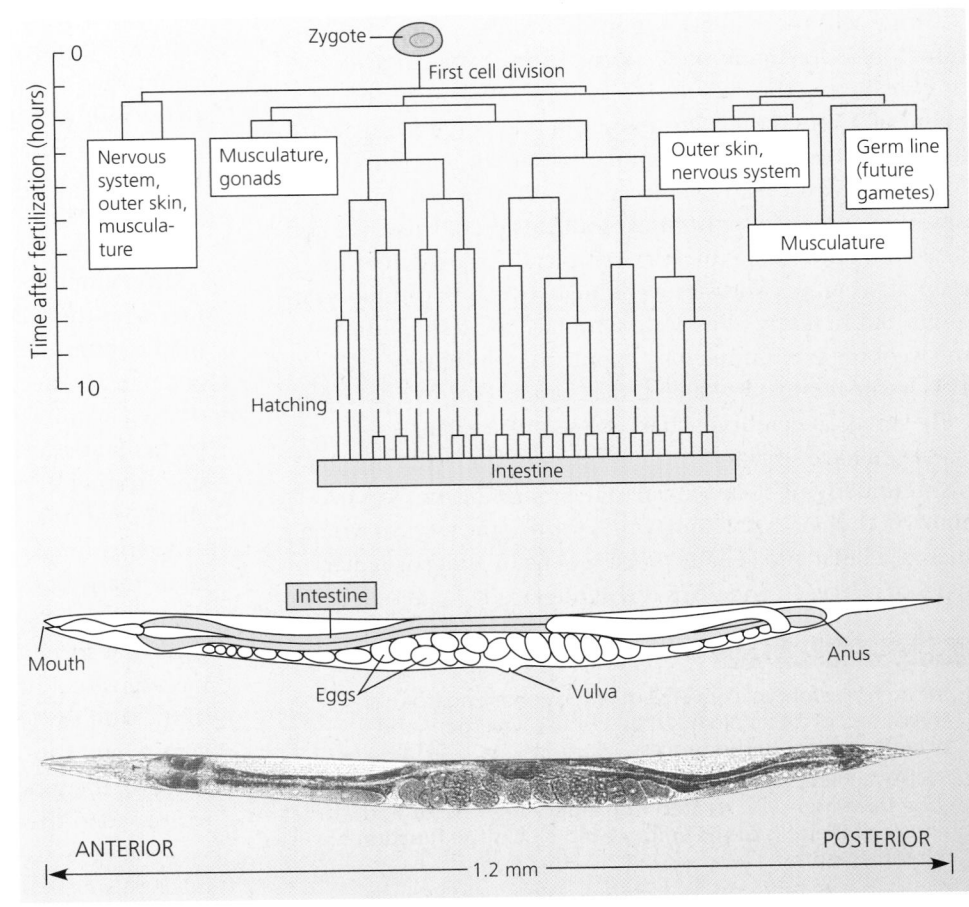

VISUAL SKILLS *The pattern of divisions is exactly the same in every* C. elegans *embryo. How many divisions of the fertilized egg give rise to the intestinal cell closest to the worm's mouth?*

▼ **Figure 47.20 Determination of germ cell fate in C. elegans.** Labeling with a fluorescent antibody that is specific for a C. elegans P granule protein (green) reveals the incorporation of P granules into four cells of the newly hatched larva (two of the four cells are visible in this view).

100 μm

Cells with P granules

from the fertilized egg in virtually the same way for every individual. Careful microscopic observations of worms at all stages of development, coupled with experiments in which particular cells or groups of cells were destroyed by a laser beam or through mutations, resulted in the cell lineage diagram shown in Figure 47.19. Using this cell lineage diagram, you can identify all of the progeny of a single cell, just as you would use a family history to trace the descendants of, for example, one great-great-grandparent.

As an example of a particular cell fate, let's consider *germ cells*, the specialized cells that give rise to eggs or sperm. In all animals studied, complexes of RNA and protein direct particular cells to become germ cells. In *C. elegans*, such complexes, called *P granules*, can be detected in four cells of the newly hatched larva **(Figure 47.20)** and, later, in the cells of the adult gonad that produce sperm or eggs.

Tracing the position of the P granules provides a dramatic illustration of how cells acquire a specific fate, during development. As shown in **Figure 47.21**, the P granules are distributed throughout the newly fertilized egg but move to the posterior end of the zygote before the first cleavage division. As a result, only the posterior of the two cells formed by the first division contains P granules. The P granules continue to be asymmetrically partitioned during subsequent divisions. Thus, the P granules act as cytoplasmic determinants (see Concept 18.4), fixing germ cell fate at the earliest stage of *C. elegans* development.

Fate mapping in *C. elegans* paved the way for major discoveries about programmed cell death. Lineage analysis demonstrated that exactly 131 cells die during normal *C. elegans* development. In the 1980s, researchers found that a mutation inactivating a single gene allows all 131 cells to live. Further research revealed that this gene is part of a pathway that controls and carries out apoptosis in a wide range of animals, including humans. In 2002, Brenner, Horvitz, and Sulston shared a Nobel Prize for their use of the *C. elegans* fate map in studies of programmed cell death and organogenesis.

Having established fate maps for early development, scientists were positioned to answer questions about underlying

▼ **Figure 47.21 Partitioning of P granules during C. elegans development.** The differential interference contrast micrographs (left) highlight the boundaries of nuclei and cells through the first two cell divisions. The fluorescence micrographs (right) show identically staged embryos labeled with a fluorescent antibody specific for a P granule protein.

20 μm

❶ Newly fertilized egg

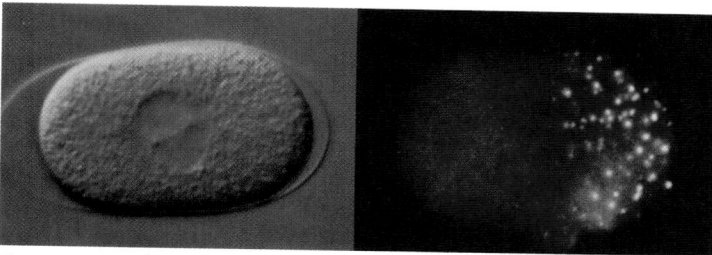

❷ Zygote prior to first division

❸ Two-cell embryo

❹ Four-cell embryo

mechanisms, such as how the basic axes of the embryo are established, a process known as axis formation.

Axis Formation

A body plan with bilateral symmetry is found across a range of animals, including nematodes, echinoderms, and vertebrates (see Concept 32.4). This body plan exhibits asymmetry along the dorsal-ventral and anterior-posterior axes, as shown for a frog tadpole in **Figure 47.22a**. The right-left axis is largely symmetrical, as the two sides are roughly mirror images. When and how are these three axes established? We'll begin answering this question by considering the frog.

Axis Formation in the Frog

In the frog, the future position of the anterior-posterior axis is determined during oogenesis. Asymmetry in the egg is apparent in the formation of two distinct hemispheres: Dark melanin granules are embedded in the cortex of the animal hemisphere, whereas a yellow yolk fills the vegetal hemisphere. This animal-vegetal asymmetry dictates where the anterior-posterior axis forms in the embryo. Note, however, that the anterior-posterior and animal-vegetal axes are not the same; that is, the head of the embryo does not coincide with the animal pole.

Surprisingly, the dorsal-ventral axis of the frog embryo is determined at random. Specifically, wherever the sperm enters in the animal hemisphere determines where the dorsal-ventral axis forms. Once the sperm and egg have fused, the egg surface—the plasma membrane and associated cortex—rotates with respect to the inner cytoplasm, a movement called *cortical rotation*. From the perspective of the animal pole, this rotation is always toward the point of sperm entry **(Figure 47.22b)**. The resulting interactions between molecules in the vegetal cortex and in the inner cytoplasm of the animal hemisphere activate regulatory proteins. Once activated, these proteins direct expression of one set of genes in dorsal regions and another set of genes in ventral regions.

Axis Formation in Birds, Mammals, and Insects

It turns out that there are many different processes by which animal embryos establish their body axes. In mammals, the sperm appears to contribute to axis formation, but not in the same manner as in frogs. In particular, the orientation of the egg and sperm nuclei before they fuse influences the location of the first cleavage plane. In chicks, the anterior-posterior axis is established by the pull of gravity during the time when the soon-to-be-laid egg is traveling down the hen's oviduct. In zebrafish, signals within the embryo gradually establish the anterior-posterior axis over the course of a day. Still other mechanisms occur in insects, where gradients of active transcription factors across the body establish both the anterior-posterior and dorsal-ventral axes (see Concept 18.4).

Once the anterior-posterior and dorsal-ventral axes are established, the position of the left-right axis is fixed. Nevertheless, specific molecular mechanisms must establish which side is left and which is right. In vertebrates, there are marked left-right differences in the location of internal organs as well as in the organization and structure of the heart and brain. Recent research has revealed a key role for cilia in setting up this left-right asymmetry, as we will discuss toward the end of this chapter.

 ▼ **Figure 47.22 The body axes and their establishment in an amphibian.** All three axes are established before the zygote begins to undergo cleavage.

(a) The three axes of the fully developed embryo

Dorsal
Right
Anterior
Posterior
Left
Ventral

(b) Establishing the axes. The polarity of the egg and cortical rotation are critical in setting up the body axes.

❶ The polarity of the egg determines the anterior-posterior axis before fertilization.

Animal pole
Animal hemisphere
Vegetal hemisphere
Vegetal pole

❷ At fertilization, the pigmented cortex slides over the underlying cytoplasm toward the point of sperm nucleus entry. This rotation (black arrows) exposes a region of lighter-colored cytoplasm, the gray crescent, which is a marker of the future dorsal side.

Point of sperm nucleus entry
Pigmented cortex
Future dorsal side
Gray crescent

❸ The first cleavage division bisects the gray crescent. Once the anterior-posterior and dorsal-ventral axes are defined, so is the position of the left-right axis.

First cleavage

WHAT IF? *When researchers allowed normal cortical rotation to occur and then forced the opposite rotation, the result was a two-headed embryo. How might you explain this finding, thinking about how cortical rotation influences body axis formation?*

Restricting Developmental Potential

Earlier we described determination in terms of commitment to a particular cell fate. The fertilized egg gives rise to all cell fates. How long during development do cells retain this ability? The German zoologist Hans Spemann addressed this question in 1938. By manipulating embryos to perturb normal development and then examining cell fate after the manipulation, he was able to assay a cell's *developmental potential*, the range of structures to which it can give rise

(Figure 47.23). The work of Spemann and others demonstrated that the first two blastomeres of the frog embryo are **totipotent**, meaning that they can each develop into all the different cell types of that species.

In mammals, embryonic cells remain totipotent through the eight-cell stage, much longer than in many other animals. Recent work, however, indicates that the very early cells (even the first two) are not actually equivalent in a normal embryo. Rather, their totipotency when isolated likely means that the cells can regulate their fate in response to their embryonic environment. Once the 16-cell stage is reached, mammalian cells are determined to form the trophoblast or the inner cell mass. Although the cells have a limited developmental potential from this point onward, their nuclei remain totipotent, as demonstrated in transplantation and cloning experiments (see Figures 20.17 and 20.18).

The totipotency of cells early in human embryogenesis is the reason why you or a classmate may have an identical twin. Identical (monozygotic) twins result when cells or groups of cells from a single embryo become separated. If the separation occurs before the trophoblast and inner cell mass become differentiated, two embryos grow, each with its own chorion and amnion. This is the case for about a third of identical twins. For the rest, the two embryos that develop share a chorion and, in very rare cases where separation is particularly late, an amnion as well.

Regardless of how uniform or varied early embryonic cells are in a particular species, the progressive restriction of developmental potential is a general feature of development in all animals. In general, the tissue-specific fates of cells are fixed in a late gastrula, but not always so in an early gastrula. For example, if the dorsal ectoderm of an early amphibian gastrula is experimentally replaced with ectoderm from some other location in the same gastrula, the transplanted tissue forms a neural plate. But if the same experiment is performed on a late-stage gastrula, the transplanted ectoderm does not respond to its new environment and does not form a neural plate.

▼ **Figure 47.23** **Inquiry**

How does distribution of the gray crescent affect the developmental potential of the first two daughter cells?

Experiment Hans Spemann, at the University of Freiburg, Germany, carried out the following experiment in 1938 to test whether substances were located asymmetrically in the gray crescent.

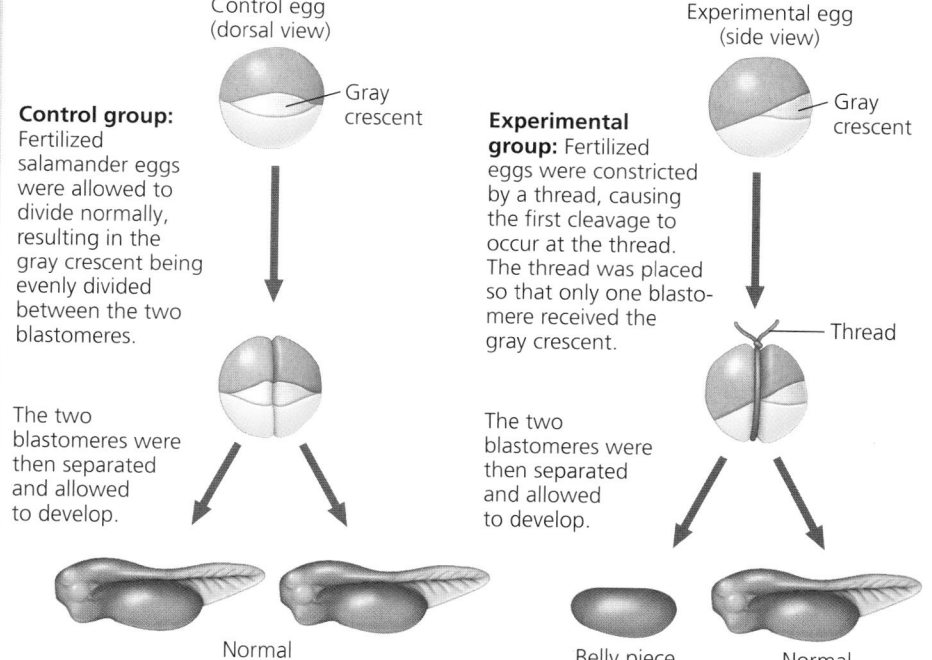

Control group: Fertilized salamander eggs were allowed to divide normally, resulting in the gray crescent being evenly divided between the two blastomeres.

The two blastomeres were then separated and allowed to develop.

Experimental group: Fertilized eggs were constricted by a thread, causing the first cleavage to occur at the thread. The thread was placed so that only one blastomere received the gray crescent.

The two blastomeres were then separated and allowed to develop.

Control egg (dorsal view) — Gray crescent

Experimental egg (side view) — Gray crescent

Thread

Normal   Normal

Belly piece

Results Blastomeres that received half or all of the material in the gray crescent developed into normal embryos, but a blastomere that received none of the gray crescent gave rise to an abnormal embryo without dorsal structures. Spemann called it a "belly piece."

Conclusion The developmental potential of the two blastomeres normally formed during the first cleavage division depends on their acquisition of cytoplasmic determinants localized in the gray crescent.

Data from H. Spemann, *Embryonic Development and Induction,* Yale University Press, New Haven, CT (1938).

DRAW IT *Draw lines to show the plane of the first cell division that would occur in the fertilized eggs above if no manipulations were done.*

WHAT IF? *In a similar experiment 40 years earlier, embryologist Wilhelm Roux allowed the first cleavage to occur and then used a needle to kill just one blastomere. The embryo that developed from the remaining blastomere (plus remnants of the dead cell) was abnormal, resembling a half-embryo. How might the presence of molecules in the dead cell explain why Roux's result differed from the control result in Spemann's experiment?*

Cell Fate Determination and Pattern Formation by Inductive Signals

As embryonic development continues, cells influence each other's fates by induction. At the molecular level, the response to an inductive signal is usually to switch on a set of genes that make the receiving cells differentiate into a specific cell type or tissue. Here we will examine examples of this important developmental process in organizing the basic body plan of an embryo and in directing the three-dimensional development of a vertebrate limb.

The "Organizer" of Spemann and Mangold

Before his studies of totipotency in the fertilized frog egg, Spemann had investigated cell fate determination during gastrulation. In these experiments, he and his student Hilde Mangold transplanted tissues between early gastrulas. In their most famous such experiment, summarized in **Figure 47.24**,

▼ **Figure 47.24** Inquiry

Can the dorsal lip of the blastopore induce cells in another part of the amphibian embryo to change their developmental fate?

Experiment In 1924, Hans Spemann and Hilde Mangold, at the University of Freiburg, Germany, investigated the inductive ability of the dorsal lip of the gastrula. Using newts, they transplanted a piece of the dorsal lip from one gastrula to the ventral side of a second gastrula. Because the donor embryo was albino and thus lacked pigmentation, the researchers could visually follow how the transplanted material altered the fate of the recipient embryo.

Results The photograph in this figure documents a repeat of this classic experiment, using the frog *Xenopus laevis*. The tadpole at the top developed from a control gastrula. When an experimental gastrula received the transplant of a dorsal lip from an albino donor (lower left), the recipient embryo formed a second notochord and neural tube in the region of the transplant. Eventually most of a second embryo developed, producing a twinned tadpole (lower right).

Dorsal lip from albino gastrula

Conclusion The transplanted dorsal lip was able to induce cells in a different region of the recipient to form structures different from their normal fate. In effect, the transplanted dorsal lip "organized" the later development of an entire extra embryo.

Data from H. Spemann and H. Mangold, Induction of embryonic primordia by implantation of organizers from a different species, Trans. V. Hamburger (1924). Reprinted in *International Journal of Developmental Biology* 45:13–38 (2001); and E. M. De Robertis and H. Kuroda, Dorsal-ventral patterning and neural induction in *Xenopus* embryos, *Annual Review of Cell and Developmental Biology* 20:285–308 (2004).

WHAT IF? *Because the transplant caused the recipient tissue to become something it would not have otherwise, a signal must have passed from the dorsal lip. If you identified a protein candidate for the signaling molecule, how would injecting it into ventral cells of a gastrula test its function?*

they made a remarkable discovery. Not only did a transplanted dorsal lip of the blastopore continue to be a blastopore lip, but it also triggered gastrulation of the surrounding tissue. They concluded that the dorsal lip of the blastopore in the early gastrula functions as an "organizer" of the embryo's body plan, inducing changes in surrounding tissue that direct formation of the notochord, the neural tube, and other organs.

Nearly a century later, developmental biologists are still studying the basis of induction by what is now called *Spemann's organizer*. An important clue has come from studies of a growth factor called bone morphogenetic protein 4 (BMP-4). One major function of the organizer seems to be to inactivate BMP-4 on the dorsal side of the embryo. Inactivating BMP-4 allows cells on the dorsal side to make dorsal structures, such as the notochord and neural tube. Proteins related to BMP-4 and its inhibitors are found as well in invertebrates such as the fruit fly, where they also function in regulating the dorsal-ventral axis.

Formation of the Vertebrate Limb

Inductive signals play a major role in **pattern formation**, the process governing the arrangement of organs and tissues in their characteristic places in three-dimensional space. The molecular cues that control pattern formation, called **positional information**, tell a cell where it is with respect to the animal's body axes and help to determine how the cell and its descendants will respond to molecular signaling during embryonic development.

In Concept 18.4, we discussed pattern formation in the development of *Drosophila*. For the study of pattern formation in vertebrates, a classic model system has been limb development in the chick. The wings and legs of chicks, like all vertebrate limbs, begin as limb buds (see Figure 47.1). These bumps consist of mesodermal tissue covered by a layer of ectoderm (see Figure 47.1, inset). Each component of a chick limb, such as a specific bone or muscle, develops with a precise location and orientation relative to three axes: proximal-distal (shoulder to fingertip), anterior-posterior (thumb to little finger), and dorsal-ventral (knuckle to palm), as shown in **Figure 47.25**.

Two regions in a limb bud have profound effects on its development. One such region is the **apical ectodermal ridge (AER)**, a thickened area of ectoderm at the tip of the bud (see Figure 47.25a). Surgically removing the AER blocks outgrowth of the limb along the proximal-distal axis. Why? The AER secretes a protein signal called fibroblast growth factor (FGF) that promotes limb-bud outgrowth. If the AER is replaced with beads soaked with FGF, a nearly normal limb develops.

The second major limb-bud regulatory region is the **zone of polarizing activity (ZPA)**, a specialized block of mesodermal tissue (see Figure 47.25a). The ZPA regulates development along the anterior-posterior axis of the limb.

▼ Figure 47.25 Regulation of vertebrate limb development by organizer regions.

50 μm

(a) Organizer regions. Vertebrate limbs develop from protrusions called limb buds. Two regions in each limb bud, the apical ectodermal ridge (AER, shown in this SEM) and the zone of polarizing activity (ZPA), play key roles as organizers in limb pattern formation.

Apical ectodermal ridge (AER)

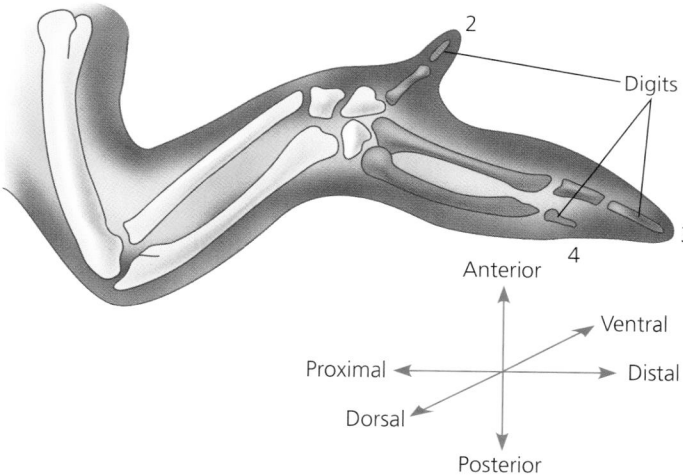

(b) Wing of chick embryo. Each embryonic cell receives positional information indicating location along the three axes of the limb. The AER and ZPA secrete molecules that help provide this information. (Numbers are assigned to the digits based on a convention established for vertebrate limbs. The chicken wing has only four digits; the first digit points backward and is not shown.)

Cells nearest the ZPA form posterior structures, such as the most posterior of the chick's digits (equivalent to our little finger); cells farthest from the ZPA form anterior structures, including the most anterior digit (like our thumb).

Like the AER, the ZPA influences development by secreting a protein signal. The signal secreted by the ZPA is called Sonic hedgehog, named after both a video game character and a similar protein in *Drosophila* that also regulates development. Implanting cells genetically engineered

---

▼ Figure 47.26 Inquiry

What role does the zone of polarizing activity (ZPA) play in limb pattern formation in vertebrates?

Experiment In 1985, researchers were eager to investigate the nature of the zone of polarizing activity. They transplanted ZPA tissue from a donor chick embryo under the ectoderm in the anterior margin of a limb bud in another chick (the host).

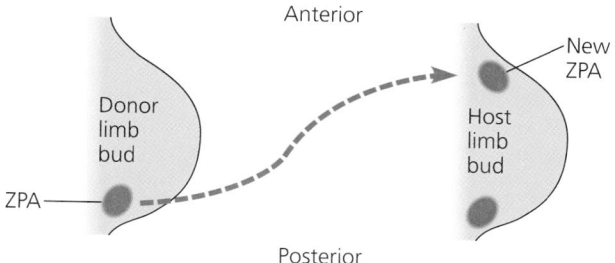

Results The host limb bud developed extra digits from host tissue in a mirror-image arrangement to the normal digits, which also formed (compare with Figure 47.25b, which shows a normal chick wing).

Conclusion The mirror-image duplication observed in this experiment suggests that ZPA cells secrete a signal that diffuses from its source and conveys positional information indicating "posterior." As the distance from the ZPA increases, the signal concentration decreases, and hence more anterior digits develop.

Data from L. S. Honig and D. Summerbell, Maps of strength of positional signaling activity in the developing chick wing bud, *Journal of Embryology and Experimental Morphology* 87:163–174 (1985).

WHAT IF? *Suppose you learned that the ZPA forms after the AER, leading you to develop the hypothesis that the AER is necessary for formation of the ZPA. If you removed the AER and looked for expression of Sonic hedgehog, how would that test your hypothesis?*

to produce Sonic hedgehog into the anterior region of a normal limb bud causes formation of a mirror-image limb—just as if a ZPA had been grafted there **(Figure 47.26)**. Furthermore, experiments with mice reveal that production of Sonic hedgehog in part of the limb bud where it is normally absent can result in extra toes.

The AER and ZPA regulate the axes of a limb bud, but what determines whether the bud develops into a forelimb or hind limb? That information is provided by spatial patterns of *Hox* genes, which specify different developmental fates in particular body regions (see Figure 21.20).

BMP-4, FGF, Sonic hedgehog, and Hox proteins are examples of a much larger set of molecules that govern cell fates in animals. Having mapped out many of the basic functions of these molecules in embryonic development, researchers are now addressing their role in organogenesis, focusing in particular on the development of the brain.

Cilia and Cell Fate

Researchers have found that the cellular organelles known as cilia are essential for specifying cell fate in human embryos. Like other mammals, humans have stationary and motile cilia (see Figure 6.24). Stationary primary cilia, or *monocilia*, jut from the surface of nearly all cells, one per cell. In contrast, motile cilia are restricted to cells that propel fluid over their surface, such as the epithelial cells of airways, and sperm (as flagella that propel sperm movement). Both stationary and motile cilia have crucial roles in development.

Genetic studies provided vital clues to the developmental role of monocilia. In 2003, researchers discovered that certain mutations disrupting development of the mouse nervous system affect genes that function in the assembly of monocilia. Other geneticists found that mutations responsible for a severe kidney disease in mice alter a gene important for the transport of materials up and down monocilia. In addition, mutations in humans that block the function of monocilia were linked to cystic kidney disease.

How do monocilia function in development? Evidence indicates that monocilia act as antennae on the cell surface, receiving signals from multiple signaling proteins, including Sonic hedgehog. Mechanisms that regulate which types of receptor proteins are present tune the cilium to particular signals. When the monocilia are defective, signaling is disrupted.

Insight into the role of motile cilia in development grew from studies of Kartagener's syndrome, a set of medical conditions that often appear together. These conditions include male infertility due to immotile sperm and infections of the nasal sinuses and bronchi in both males and females. A distinctive feature of Kartagener's syndrome is *situs inversus*, a reversal of the normal left-right asymmetry of the organs in the chest and abdomen **(Figure 47.27)**. The heart, for example, is on the right side rather than the left. (By itself, situs inversus causes no significant medical problems.)

Scientists studying Kartagener's syndrome came to realize that all of the associated conditions result from a defect that makes cilia immotile. Without motility, sperm tails cannot beat and airway cells cannot sweep mucus and microbes out

▼ **Figure 47.27 Situs inversus, a reversal of normal left-right asymmetry in the chest and abdomen.**

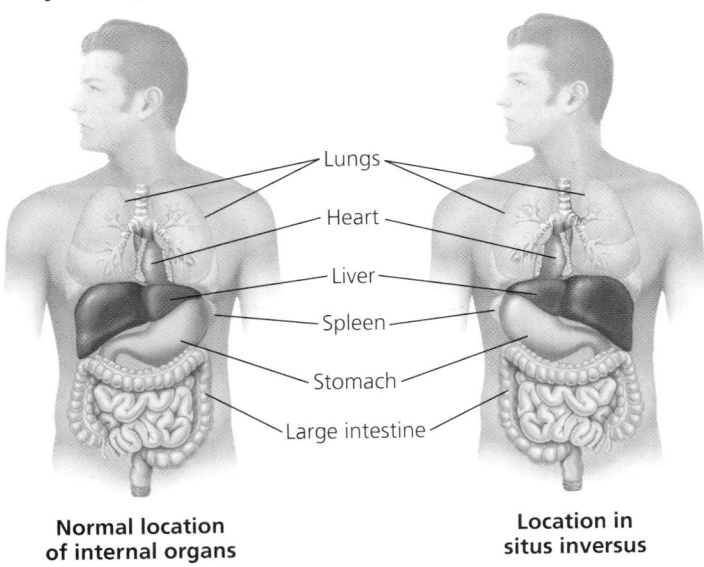

Lungs
Heart
Liver
Spleen
Stomach
Large intestine

Normal location of internal organs

Location in situs inversus

of the airway. But what causes situs inversus in these individuals? The current model proposes that ciliary motion in a particular part of the embryo is essential for normal development. Evidence indicates that movement of the cilia generates a leftward fluid flow, breaking the symmetry between left and right sides. Without that flow, asymmetry along the left-right axis arises randomly, and half of the affected embryos develop situs inversus.

If we consider development as a whole, we see a sequence of events marked by cycles of signaling and differentiation. Initial cell asymmetries allow different types of cells to influence each other, resulting in the expression of specific sets of genes. The products of these genes then direct cells to differentiate into specific types. Through pattern formation and morphogenesis, differentiated cells ultimately produce a complex arrangement of tissues and organs, each functioning in its appropriate location and in coordination with other cells, tissues, and organs throughout the organism.

CONCEPT CHECK 47.3

1. How do axis formation and pattern formation differ?
2. **MAKE CONNECTIONS** How does a morphogen gradient differ from cytoplasmic determinants and inductive interactions with regard to the set of cells it affects (see Concept 18.4)?
3. **WHAT IF?** If the ventral cells of an early frog gastrula were experimentally induced to express large amounts of a protein that inhibits BMP-4, could a second embryo develop? Explain.
4. **WHAT IF?** If you removed the ZPA from a limb bud and then placed a bead soaked in Sonic hedgehog in the middle of the bud, what would be the most likely result?

For suggested answers, see Appendix A.

47 Chapter Review

→ Go to **Mastering Biology** for Assignments, the eText, the Study Area, and Dynamic Study Modules.

SUMMARY OF KEY CONCEPTS

→ To review key terms, go to the **Vocabulary Self-Quiz** in the **Mastering Biology** eText or Study Area, or go to goo.gl/zkjz9t.

CONCEPT 47.1

Fertilization and cleavage initiate embryonic development (pp. 1044–1049)

- **Fertilization** forms a diploid zygote and initiates embryonic development. The **acrosomal reaction** releases hydrolytic enzymes from the sperm head that digest material surrounding the egg.

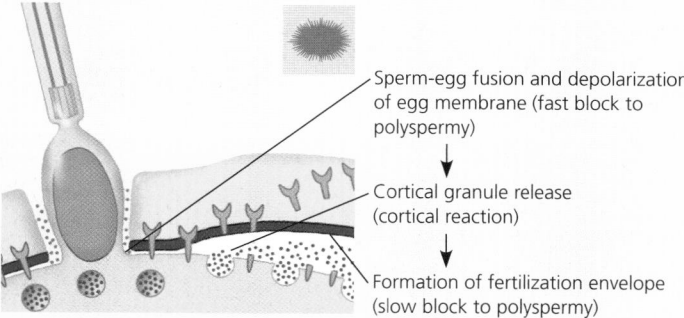

Sperm-egg fusion and depolarization of egg membrane (fast block to polyspermy)

↓

Cortical granule release (cortical reaction)

↓

Formation of fertilization envelope (slow block to polyspermy)

In mammalian fertilization, the cortical reaction modifies the **zona pellucida** as a slow block to polyspermy.

- Fertilization is followed by **cleavage**, a period of rapid cell division without growth, producing a large number of cells called **blastomeres**. The amount and distribution of **yolk** strongly influence the pattern of cleavage. In many species, the completion of the cleavage stage generates a **blastula** containing a fluid-filled cavity, the **blastocoel**.

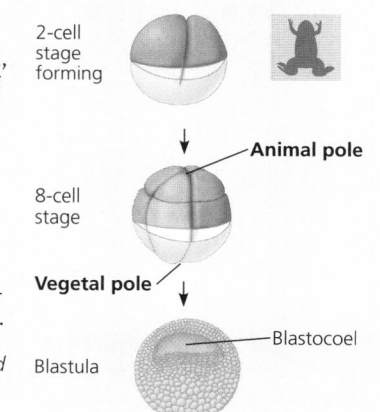

2-cell stage forming

8-cell stage

Animal pole

Vegetal pole

Blastula

Blastocoel

? *What cell-surface event would likely fail if a sperm contacted an egg of another species?*

CONCEPT 47.2

Morphogenesis in animals involves specific changes in cell shape, position, and survival (pp. 1049–1057)

- **Gastrulation** converts the blastula to a **gastrula**, which has a primitive digestive cavity and three **germ layers: ectoderm** (blue), which forms the outer layer of the embryo, **mesoderm** (red), which forms the middle layer, and **endoderm** (yellow), which gives rise to the innermost tissues.

- Gastrulation and organogenesis in mammals resemble the processes in reptiles, including birds. After fertilization and early cleavage in the oviduct, the **blastocyst** implants in the uterus. The **trophoblast** initiates formation of the fetal portion of the placenta, and the embryo proper develops from a cell layer, the epiblast, within the blastocyst.
- The embryos of birds, other reptiles, and mammals develop within a fluid-filled sac that is contained within a shell or the uterus. In these organisms, the three germ layers produce four **extraembryonic membranes**: the amnion, chorion, yolk sac, and allantois.
- The organs of the animal body develop from specific portions of the three embryonic germ layers. Early events in **organogenesis** in vertebrates include neurulation: formation of the **notochord** by cells of the dorsal mesoderm and development of the **neural tube** from infolding of the ectodermal neural plate.

Neural tube

Notochord

Coelom

Neural tube

Notochord

Coelom

- Cytoskeletal rearrangements cause changes in the shape of cells that underlie cell movements in gastrulation and organogenesis, including invaginations and **convergent extension**. The cytoskeleton is also involved in cell migration, which relies on cell adhesion molecules and the extracellular matrix to help cells reach specific destinations. Migratory cells arise both from the neural crest and from **somites**.
- Some processes in animal development require **apoptosis**, programmed cell death.

? *What are some functions of apoptosis in development?*

CONCEPT 47.3

Cytoplasmic determinants and inductive signals regulate cell fate (pp. 1057–1064)

- Experimentally derived **fate maps** of embryos show that specific regions of the zygote or blastula develop into specific parts of older embryos. The complete cell lineage has been worked out for *C. elegans*, revealing that programmed cell death contributes to animal development. In all species, the developmental potential of cells becomes progressively more limited as embryonic development proceeds.
- Cells in a developing embryo receive and respond to **positional information** that varies with location. This information is often in the form of signaling molecules secreted by cells in specific regions of the embryo, such as the dorsal lip of the blastopore in the amphibian gastrula and the **apical ectodermal ridge** and **zone of polarizing activity** of the vertebrate limb bud.

? *Suppose you found two classes of mouse mutations, one that affected limb development only and one that affected both limb and kidney development. Which class would be more likely to alter the function of monocilia? Explain.*

TEST YOUR UNDERSTANDING

→ For more multiple-choice questions, go to the **Practice Test** in the **Mastering Biology** eText or Study Area, or go to goo.gl/GruWRg.

Levels 1-2: Remembering/Understanding

1. The cortical reaction of sea urchin eggs functions directly in
   (A) the formation of a fertilization envelope.
   (B) the production of a fast block to polyspermy.
   (C) the generation of an electrical impulse by the egg.
   (D) the fusion of egg and sperm nuclei.

2. Which of the following is common to the development of both birds and mammals?
   (A) holoblastic cleavage
   (B) epiblast and hypoblast
   (C) trophoblast
   (D) gray crescent

3. The archenteron develops into
   (A) the mesoderm.
   (B) the endoderm.
   (C) the placenta.
   (D) the lumen of the digestive tract.

4. What structural adaptation in chickens allows them to lay their eggs in arid environments rather than in water?
   (A) extraembryonic membranes
   (B) yolk
   (C) cleavage
   (D) gastrulation

Levels 3-4: Applying/Analyzing

5. If an egg cell were treated with EDTA, a chemical that binds calcium and magnesium ions,
   (A) the acrosomal reaction would be blocked.
   (B) the fusion of sperm and egg nuclei would be blocked.
   (C) the fast block to polyspermy would not occur.
   (D) the fertilization envelope would not form.

6. In humans, identical twins are possible because
   (A) extraembryonic cells interact with the zygote nucleus.
   (B) convergent extension occurs.
   (C) early blastomeres can form a complete embryo if isolated.
   (D) the gray crescent divides the dorsal-ventral axis into new cells.

7. Cells transplanted from the neural tube of a frog embryo to the ventral part of another embryo develop into nervous system tissues. This result indicates that the transplanted cells were
   (A) totipotent.   (C) differentiated.
   (B) determined.   (D) mesenchymal.

8. **DRAW IT** Each blue circle in the figure below represents a cell in a cell lineage. Draw two modified versions of the cell lineage so that each version produces three cells. Use apoptosis in one of the versions, marking any dead cells with an X.

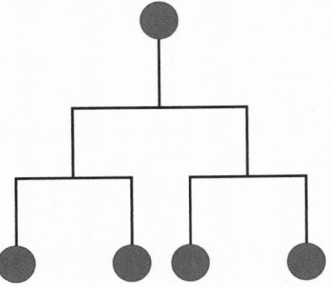

Levels 5-6: Evaluating/Creating

9. **EVOLUTION CONNECTION** Evolution in insects and vertebrates has involved the repeated duplication of body segments, followed by fusion of some segments and specialization of their structure and function. In vertebrates, what anatomical features reflect segmentation?

10. **SCIENTIFIC INQUIRY** The "snout" region of a salamander has a mustache-shaped structure called a balancer, whereas that of a frog tadpole does not. When you transplant tissue from the side of a young salamander embryo to the snout of a frog embryo, the tadpole that develops has a balancer. If you use a slightly older salamander embryo as the donor, no balancer forms. Propose a hypothesis to explain these results and explain how you might test your hypothesis.

11. **SCIENCE, TECHNOLOGY, AND SOCIETY** Scientists can now make identical copies, or clones, of animals ranging from dairy cows to pet cats. Propose a few arguments for and against this application of discoveries about embryonic development.

12. **WRITE ABOUT A THEME: ORGANIZATION** In a short essay (100–150 words), describe how the emergent properties of the cells of the gastrula direct embryonic development.

13. **SYNTHESIZE YOUR KNOWLEDGE**

Occasionally, two-headed animals such as this turtle are born. Thinking about the occurrence of identical twins and the property of totipotency, explain how this might happen.

For selected answers, see Appendix A.

48 Neurons, Synapses, and Signaling

KEY CONCEPTS

48.1 Neuron structure and organization reflect function in information transfer *p. 1068*

48.2 Ion pumps and ion channels establish the resting potential of a neuron *p. 1069*

48.3 Action potentials are the signals conducted by axons *p. 1072*

48.4 Neurons communicate with other cells at synapses *p. 1077*

Study Tip

Make a table: The term *potential* is central to understanding neuronal signaling, as it provides the driving force for both electrical and chemical signaling. Fill in a table like this one as you read to help you distinguish characteristics of a *resting potential, graded potential*, and *action potential*, as well as chemical potentials for sodium and potassium ion concentrations.

Membrane potentials of a neuron		Occurs in cell body?	Occurs in axon?	Variable strength?
Electrical	Resting	Yes	Yes	No
Electrical	Graded			
Electrical	Action			
Chemical	[Na⁺]			
Chemical	[K⁺]			

→ Go to Mastering Biology

For Students (in eText and Study Area)
- Get Ready for Chapter 48
- BioFlix® Animation: How Neurons Work
- BioFlix® Animation: How Synapses Work

For Instructors to Assign (in Item Library)
- Scientific Skills Exercise: Interpreting Data Values Expressed in Scientific Notation

Ready-to-Go Teaching Module
(in Instructor Resources)
- Resting and Action Potentials

Figure 48.1 This cone snail (*Conus geographus*) is slow-moving, but a dangerous hunter. Delivering venom with a hollow, harpoon-like tooth, it can paralyze a fish almost instantaneously. Scuba divers picking up a cone snail have died from a single injection. Underlying the fast-acting, lethal nature of the venom is an ability to block information transfer by neurons, specialized cells of the nervous system.

How does a neuron transmit information?

A neuron receives information, transmits it along an extension called an **axon**, and transmits the information to other cells via specialized junctions called **synapses**.

Input to neuron
Sensory information or output from other neurons

Dendrites

Nucleus

Cell body

Electrical signaling along axon
Electrical impulses travel in one direction only. Every impulse is identical, but the frequency of signaling varies.

Axon

SYNAPSE

Synaptic terminal

Another neuron or other cell

Neurotransmitter

Output from neuron
Received by muscles, glands, or other neurons

CONCEPT 48.1

Neuron structure and organization reflect function in information transfer

Our starting point for exploring the nervous system is the **neuron**, a cell type exemplifying the close fit of form and function that often arises over the course of evolution.

Neuron Structure and Function

The ability of a neuron to receive and transmit information is based on the highly specialized cellular organization shown in Figure 48.1. Most of a neuron's organelles, including its nucleus, are located in the **cell body**. In a typical neuron, the cell body is studded with numerous highly branched extensions called **dendrites** (from the Greek *dendron*, tree). Together with the cell body, the dendrites *receive* signals from other neurons.

A typical neuron has a single **axon**, the extension that *transmits* signals to other cells. Axons are often much longer than dendrites, and some, such as those that reach from the spinal cord of a giraffe to the muscle cells in its feet, are over a meter long. The specialized structure of axons allows them to use pulses of electrical current to transmit information, even over long distances. The cone-shaped base of an axon, called the axon hillock, is typically where signals that travel down the axon are generated. Near its other end, an axon usually divides into many branches.

Each branched end of an axon transmits information to another cell at a junction called a **synapse (Figure 48.2)**. The part of each axon branch that forms this specialized junction is a *synaptic terminal*. At most synapses, chemical messengers called **neurotransmitters** pass information from the transmitting neuron to the receiving cell. Cone snail venom is particularly potent because it interferes not only with electrical signaling along axons but also with chemical signaling across synapses.

▼ **Figure 48.2 Synaptic terminals on the cell body of a postsynaptic neuron.** (colorized SEM)

Postsynaptic neuron

Synaptic terminals of pre-synaptic neurons

5 μm

▼ **Figure 48.3 Summary of information processing.** The cone snail's siphon acts as a sensor, transferring information to the neuronal circuits in the snail's head for processing. If prey is detected, these circuits issue motor commands—signals that control muscle activity. In this example, motor commands trigger release of a harpoon-like tooth from the proboscis, spearing the prey.

Siphon

Sensory input

Sensor

Integration

Motor output

Proboscis

Processing center

Effector

> **Mastering Biology**
> Interview with Baldomero Olivera: Developing drugs from research on cone snail venom

Introduction to Information Processing

Information processing by a nervous system occurs in three stages: sensory input, integration, and motor output. As an example, let's consider the cone snail, focusing on the steps involved in identifying and attacking its prey. To generate sensory input to the nervous system, the snail surveys its environment with sensors in its tubelike siphon, sampling scents that might reveal a nearby fish **(Figure 48.3)**. During the integration stage, networks of neurons in the snail brain process this information to determine if a fish is in fact present and, if so, where the fish is located. Motor output from the processing center then initiates attack, activating neurons that trigger release of the harpoon-like tooth toward the prey.

In all but the simplest animals, specialized populations of neurons handle each stage of information processing.

- **Sensory neurons**, like those in the snail's siphon, transmit information about external stimuli (such as light, touch, or smell), and internal conditions (such as blood pressure or muscle tension).
- **Interneurons** form the local circuits connecting neurons in the brain or ganglia. Interneurons are responsible for the integration (analysis and interpretation) of sensory input.
- **Motor neurons** transmit signals to muscle cells, causing them to contract. Additional neurons that extend out of the processing centers trigger gland activity.

All neurons transmit electrical signals within the cell in an identical manner. Thus a neuron that detects an odor transmits information along its length in the same way as a neuron that controls the movement of a body part. The particular connections made by the active neuron are what distinguish the type of information being transmitted. Interpreting nerve impulses therefore involves sorting neuronal paths and connections.

As shown in **Figure 48.4**, the shape of a neuron can vary from simple to quite complex, depending on its role in

▼ **Figure 48.4 Structural diversity of neurons.** In these drawings of neurons, cell bodies and dendrites are black and axons are red.

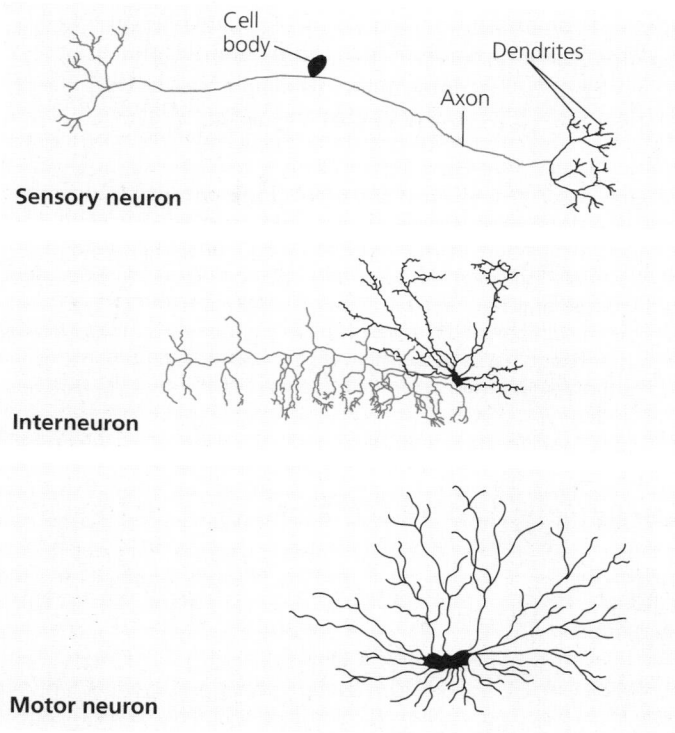

Sensory neuron

Interneuron

Motor neuron

▼ **Figure 48.5 Glia in the mammalian brain.** This fluorescently labeled laser confocal micrograph shows a region of the rat brain packed with glia and interneurons. The glia are labeled red, the DNA in nuclei is labeled blue, and the dendrites of neurons are labeled green.

80 μm

Glia   Cell bodies of neurons

information processing. Neurons that have highly branched dendrites, such as some interneurons, can receive input through tens of thousands of synapses. Similarly, neurons that transmit information to many target cells do so through highly branched axons. When grouped together, the axons of neurons form the bundles we call **nerves**.

In many animals, the neurons that carry out sorting, processing, and integration are organized in a **central nervous system (CNS)**, which may include a **brain** or simpler clusters called **ganglia**. The neurons that carry information into and out of the CNS constitute the **peripheral nervous system (PNS)**. Neurons of both the CNS and PNS require supporting cells called glial cells, or **glia** (from a Greek word meaning "glue") **(Figure 48.5)**.

CONCEPT CHECK 48.1

1. Compare and contrast the structure and function of axons and dendrites.

2. Describe the basic pathway of information flow through neurons that causes you to turn your head when someone calls your name.

3. **WHAT IF?** How might increased branching of an axon help coordinate responses to signals communicated by the nervous system?

For suggested answers, see Appendix A.

CONCEPT 48.2

Ion pumps and ion channels establish the resting potential of a neuron

We turn now to the essential role of ions in neuronal signaling. In neurons, as in other cells, ions are unequally distributed between the interior of cells and the surrounding fluid (see Concept 7.4). As a result, the inside of a cell is negatively charged relative to the outside. This charge difference, or *voltage*, across the plasma membrane is called the **membrane potential**, reflecting the fact that the attraction of opposite charges across the plasma membrane is a source of potential energy. For a resting neuron—one that is not sending a signal—the membrane potential is called the **resting potential** and is typically between −60 and −80 millivolts (mV).

When a neuron receives a stimulus, the membrane potential changes. Rapid shifts in membrane potential are what enable us to see the intricate structure of a spiderweb, hear a song, or ride a bicycle. These changes, which are known as *action potentials*, will be discussed in Concept 48.3. To understand how they convey information, we need to explore the ways in which membrane potentials are formed, maintained, and altered.

Formation of the Resting Potential

Potassium ions (K^+) and sodium ions (Na^+) play an essential role in the formation of the resting potential. These ions each have a concentration gradient across the plasma membrane of a neuron **(Table 48.1)**. In most neurons, the concentration of K^+ is higher inside the cell, while the concentration of Na^+ is higher outside. The Na^+ and K^+ concentration gradients are maintained by the **sodium-potassium pump**. This pump uses the energy of ATP hydrolysis to actively transport Na^+ out of the cell and K^+ into the cell. (There are also concentration gradients for chloride ions (Cl^-) and other anions, as shown in Table 48.1, but we can ignore these for now.)

Table 48.1 Ion Concentrations Inside and Outside of Mammalian Neurons

Ion	Intracellular Concentration (mM)	Extracellular Concentration (mM)
Potassium (K^+)	140	5
Sodium (Na^+)	15	150
Chloride (Cl^-)	10	120
Large anions (A^-), such as proteins, inside cell	100	Not applicable

The sodium-potassium pump transports three Na^+ out of the cell for every two K^+ that it transports in **(Figure 48.6)**. Although this pumping generates a net export of positive charge, the pump acts slowly. The resulting change in the membrane potential is therefore quite small—only a few millivolts. Why, then, is there a membrane potential of -60 to -80 mV in a resting neuron? The answer lies in ion movement through **ion channels**, pores formed by clusters of specialized proteins that span the membrane. Ion channels allow ions to diffuse back and forth across the membrane. As ions diffuse through channels, they carry with them units of electrical

charge. Furthermore, ions can move quite rapidly through ion channels. When this occurs, the resulting current—a *net* movement of positive or negative charge—generates a membrane potential, or voltage across the membrane.

Concentration gradients of ions across a plasma membrane represent a chemical form of potential energy that can be harnessed for cellular processes (see Figure 44.17). In neurons, the ion channels that convert this chemical potential energy to electrical potential energy can do so because they have *selective permeability*, allowing only certain ions to pass. For example, a potassium channel allows K^+ to diffuse freely across the membrane, but not other ions, such as Na^+ or Cl^-.

Diffusion of K^+ through potassium channels that are always open (sometimes called *leak channels*) is critical for establishing the resting potential **(Figure 48.7)**. The K^+

▼ **Figure 48.7 The basis of the membrane potential.** The sodium-potassium pump generates and maintains the concentration gradients of Na^+ and K^+ shown in Table 48.1. The [Na^+] gradient results in very little net diffusion of Na^+ in a resting neuron because very few sodium channels are open. In contrast, the many open potassium channels allow a significant net outflow of K^+. Because the membrane is only weakly permeable to chloride and other anions, this outflow of K^+ results in a net negative charge inside the cell.

Key

 Na^+
 K^+

 Sodium-potassium pump

 Potassium channel

 Sodium channel

❓ *The potassium and sodium channels have the same rough overall structure, as shown. How must these proteins differ to allow passage of only a particular ion?*

→ **Mastering Biology BioFlix® Animation: Resting Potential**

◀ **Figure 48.6 Summary of active transport by the sodium-potassium pump.** You can find a step-by-step description of the pump's activity in Figure 7.15.

OUTSIDE OF CELL

Na^+/K^+ pump

3 Na^+

ATP

2 K^+

INSIDE OF CELL

concentration is 140 millimolar (m*M*) inside the cell, but only 5 m*M* outside. The chemical concentration gradient thus favors a net outflow of K⁺. Furthermore, a resting neuron has many open potassium channels, but very few open sodium channels. Because Na⁺ and other ions can't readily cross the membrane, K⁺ outflow leads to a net negative charge inside the cell. This buildup of negative charge within the neuron is the major source of the membrane potential.

What stops the buildup of negative charge? The excess negative charges inside the cell exert an attractive force that opposes the flow of additional positively charged potassium ions out of the cell. The separation of charge (voltage) thus results in an electrical gradient that counterbalances the chemical concentration gradient of K⁺.

Modeling the Resting Potential

The net flow of K⁺ out of a neuron proceeds until the chemical and electrical forces are in balance. We can model this process by considering a pair of chambers separated by an artificial membrane. To begin, imagine that the membrane contains many open ion channels, all of which allow only K⁺ to diffuse across **(Figure 48.8a)**. To produce a K⁺ concentration gradient like that of a mammalian neuron, we place a solution of 140 m*M* potassium chloride (KCl) in the inner chamber and 5 m*M* KCl in the outer chamber. The K⁺ will diffuse down its concentration gradient into the outer chamber. But because the chloride ions (Cl⁻) lack a means of crossing the membrane, there will be an excess of negative charge in the inner chamber.

When our model neuron reaches equilibrium, the electrical gradient will exactly balance the chemical gradient so that no further net diffusion of K⁺ occurs across the membrane.

The magnitude of the membrane voltage at equilibrium for a particular ion is called that ion's **equilibrium potential** (*E_{ion}*). For a membrane permeable to a single type of ion, E_{ion} can be calculated using a formula called the *Nernst equation*. At human body temperature (37°C) and for an ion with a net charge of 1+, such as K⁺ or Na⁺, the Nernst equation is

$$E_{ion} = 62\,\text{mV}\left(\log\frac{[\text{ion}]_{\text{outside}}}{[\text{ion}]_{\text{inside}}}\right)$$

Plugging the K⁺ concentrations into the Nernst equation reveals that the equilibrium potential for K⁺ (E_K) is −90 mV (see Figure 48.8a). The minus sign indicates that K⁺ is at equilibrium when the inside of the membrane is 90 mV more negative than the outside.

Whereas the equilibrium potential for K⁺ is −90 mV, the resting potential of a mammalian neuron is somewhat less negative. This difference reflects the small but steady movement of Na⁺ across the few open sodium channels in a resting neuron. The concentration gradient of Na⁺ has a direction opposite to that of K⁺ (see Table 48.1). Na⁺ therefore diffuses into the cell, making the inside of the cell less negative. If we model a membrane in which the only open channels are selectively permeable to Na⁺, we find that a tenfold higher concentration of Na⁺ in the outer chamber results in an equilibrium potential (E_{Na}) of +62 mV **(Figure 48.8b)**. In an actual neuron, the resting potential (−60 to −80 mV) is much closer to E_K than to E_{Na} because there are many open potassium channels but only a small number of open sodium channels.

Because neither K⁺ nor Na⁺ is at equilibrium in a resting neuron, there is a net flow of each ion across the membrane. The resting potential remains steady, which means that these K⁺ and Na⁺ currents are equal and opposite. Ion

▶ **Figure 48.8 Modeling a mammalian neuron.** In this model of the membrane potential of a resting neuron, an artificial membrane divides each container into two chambers. Ion channels allow free diffusion for particular ions, resulting in the net ion flow represented by arrows. **(a)** The presence of open potassium channels makes the membrane selectively permeable to K⁺, and the inner chamber contains a 28-fold higher concentration of K⁺ than the outer chamber; at equilibrium, the inside of the membrane is −90 mV relative to the outside. **(b)** The membrane is selectively permeable to Na⁺, and the inner chamber contains a tenfold lower concentration of Na⁺ than the outer chamber; at equilibrium, the inside of the membrane is +62 mV relative to the outside.

WHAT IF? *How would adding potassium or chloride channels to the membrane in (b) affect the membrane potential?*

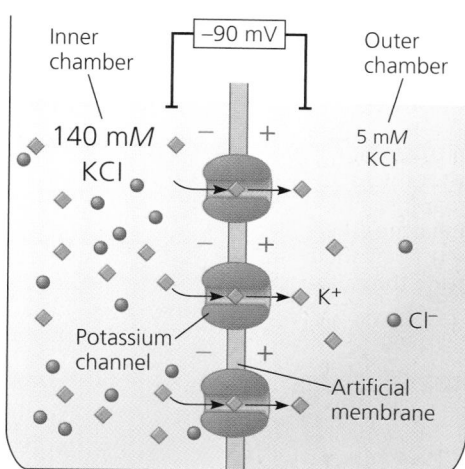

(a) Membrane selectively permeable to K⁺

Nernst equation for K⁺ equilibrium potential at 37°C:

$$E_K = 62\,\text{mV}\left(\log\frac{5\,\text{m}M}{140\,\text{m}M}\right) = -90\,\text{mV}$$

(b) Membrane selectively permeable to Na⁺

Nernst equation for Na⁺ equilibrium potential at 37°C:

$$E_{Na} = 62\,\text{mV}\left(\log\frac{150\,\text{m}M}{15\,\text{m}M}\right) = +62\,\text{mV}$$

concentrations on either side of the membrane also remain steady. Why? The resting potential arises from the net movement of far fewer ions than would be required to alter the concentration gradients.

If Na^+ is allowed to cross the membrane more readily, the membrane potential will move toward E_{Na} and away from E_K. As you'll see, this happens in generation of a nerve impulse.

→ **Mastering Biology** Animation: Membrane Potentials

CONCEPT CHECK 48.2

1. Under what circumstances could ions flow through an ion channel from a region of lower ion concentration to a region of higher ion concentration?

2. **WHAT IF?** Suppose a cell's membrane potential shifts from -70 mV to -50 mV. What changes in the cell's permeability to K^+ or Na^+ could cause such a shift?

3. **MAKE CONNECTIONS** Review Figure 7.11, which illustrates the diffusion of dye molecules across a membrane. Could diffusion eliminate the concentration gradient of a dye that has a net charge? Explain.

For suggested answers, see Appendix A.

CONCEPT 48.3

Action potentials are the signals conducted by axons

When a neuron responds to a stimulus, the membrane potential changes. Using intracellular recording, researchers can monitor these changes as a function of time **(Figure 48.9)**. As

▼ **Figure 48.9** **Research Method**

Intracellular Recording

Application Electrophysiologists use intracellular recording to measure the membrane potential of neurons and other cells.

Technique A microelectrode is made from a glass capillary tube filled with an electrically conductive salt solution. One end of the tube tapers to an extremely fine tip (diameter < 1 μm). While looking through a microscope, the experimenter uses a micropositioner to insert the tip of the microelectrode into a cell. A voltage recorder (usually an oscilloscope or a computer-based system) measures the voltage between the microelectrode tip inside the cell and a reference electrode placed in the solution outside the cell.

▼ **Figure 48.10** **Voltage-gated ion channel.** A change in the membrane potential in one direction (solid arrow) opens the voltage-gated channel. The opposite change (dotted arrow) closes the channel.

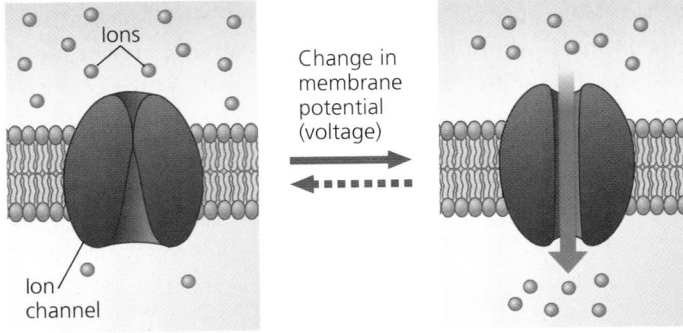

Gate closed: No ions flow across membrane.

Gate open: Ions flow through channel.

VISUAL SKILLS *Gated ion channels allow ion flow in either direction. Using visual information in this figure, explain why there is net ion movement when the channel opens.*

you will see, such recordings have been central to the study of information transfer by neurons.

How does a stimulus alter the membrane potential? Certain ion channels in a neuron, called **gated ion channels**, open or close in response to stimuli. When a gated ion channel opens or closes, it alters the membrane's permeability to particular ions **(Figure 48.10)**. The result is a rapid flow of ions across the membrane, altering the membrane potential.

Particular types of gated channels respond to different stimuli. For example, Figure 48.10 illustrates a **voltage-gated ion channel**, a channel that opens or closes in response to a shift in the voltage across the plasma membrane of the neuron. Later in this chapter, we will discuss gated channels that are located in neurons and are regulated by chemical signals.

Hyperpolarization and Depolarization

Let's consider now what happens in a neuron when a stimulus causes closed voltage-gated ion channels to open. If gated potassium channels in a resting neuron open, the membrane's permeability to K^+ increases. As a result, net diffusion of K^+ out of the neuron increases, shifting the membrane potential toward E_K (-90 mV at 37°C). This increase in the magnitude of the membrane potential, called a **hyperpolarization**, makes the inside of the membrane more negative **(Figure 48.11a)**. In a resting neuron, hyperpolarization results from any stimulus that increases the outflow of positive ions or the inflow of negative ions.

Although opening potassium channels in a resting neuron causes hyperpolarization, opening some other types of ion channels has an opposite effect, making the inside of the membrane less negative **(Figure 48.11b)**. A reduction in the

▼ **Figure 48.11 Graded potentials and an action potential in a neuron.**

(a) Graded hyperpolarizations produced by two stimuli that increase membrane permeability to K⁺. The larger stimulus produces a larger hyperpolarization.

(b) Graded depolarizations produced by two stimuli that increase membrane permeability to Na⁺. The larger stimulus produces a larger depolarization.

(c) Action potential triggered by a depolarization that reaches the threshold.

DRAW IT *Redraw the graph in (c), extending the y-axis. Then label the positions of E_K and E_{Na}.*

magnitude of the membrane potential is a **depolarization**. In neurons, depolarization often involves gated sodium channels. If a stimulus causes gated sodium channels to open, the membrane's permeability to Na⁺ increases. Na⁺ diffuses into the cell along its concentration gradient, causing a depolarization as the membrane potential shifts toward E_{Na} (+62 mV at 37°C).

Graded Potentials and Action Potentials

Sometimes, the response to hyperpolarization or depolarization is simply a shift in the membrane potential. This shift, called a **graded potential**, has a magnitude that varies with the strength of the stimulus: A larger stimulus causes a greater change in the membrane potential (see Figure 48.11a and 48.11b). Graded potentials induce a small electrical current that dissipates as it flows along the membrane. Graded potentials thus decay with time and with distance from their source.

If a depolarization shifts the membrane potential sufficiently, the result is a massive change in membrane voltage called an **action potential**. Unlike graded potentials, action potentials have a constant magnitude and can regenerate in adjacent regions of the membrane. Action potentials can therefore spread along axons, making them well suited for transmitting a signal over long distances.

Action potentials arise because some of the ion channels in neurons are voltage-gated (see Figure 48.10). If a depolarization increases the membrane potential to a level

called **threshold**, the voltage-gated sodium channels open. The resulting flow of Na⁺ into the neuron results in further depolarization. Because the sodium channels are voltage gated, the increased depolarization causes more sodium channels to open, leading to an even greater flow of current. The result is a process of positive feedback that triggers a very rapid opening of many voltage-gated sodium channels and the marked temporary change in membrane potential that defines an action potential **(Figure 48.11c)**.

The positive-feedback loop of channel opening and depolarization triggers an action potential whenever the membrane potential reaches threshold, about −55 mV for many mammals. Once initiated, the action potential has a magnitude that is independent of the strength of the triggering stimulus. Because action potentials either occur fully or do not occur at all, they represent an *all-or-none* response to stimuli.

Generation of Action Potentials: *A Closer Look*

The characteristic shape of the action potential graph in Figure 48.11c reflects changes in membrane potential resulting from ion movement through voltage-gated sodium and potassium channels. Depolarization opens both types of channels, but they respond independently and sequentially. Sodium channels open first, initiating the action potential. As the action potential proceeds, sodium channels remain open but become *inactivated*: A portion of the channel protein called an inactivation loop blocks ion flow through the open channel. Sodium channels remain inactivated until after the membrane returns to the resting potential and the channels close. Potassium channels open more slowly than sodium channels, but remain open and functional until the end of the action potential.

To understand further how voltage-gated channels shape the action potential, consider the process as a series of stages,

as depicted in **Figure 48.12**. ❶ When the membrane of the axon is at the resting potential, most voltage-gated sodium channels are closed. Some potassium channels are open, but most voltage-gated potassium channels are closed. ❷ When a stimulus depolarizes the membrane, some gated sodium channels open, allowing more Na⁺ to diffuse into the cell. If the stimulus is sufficiently strong, the Na⁺ inflow persists, causing further depolarization, which opens more gated sodium channels, allowing even more Na⁺ to diffuse into the

▼ **Figure 48.12 The role of voltage-gated ion channels in the generation of an action potential.**
The circled numbers on the graph in the center and the colors of the action potential phases correspond to the five diagrams showing voltage-gated sodium and potassium channels in a neuron's plasma membrane.

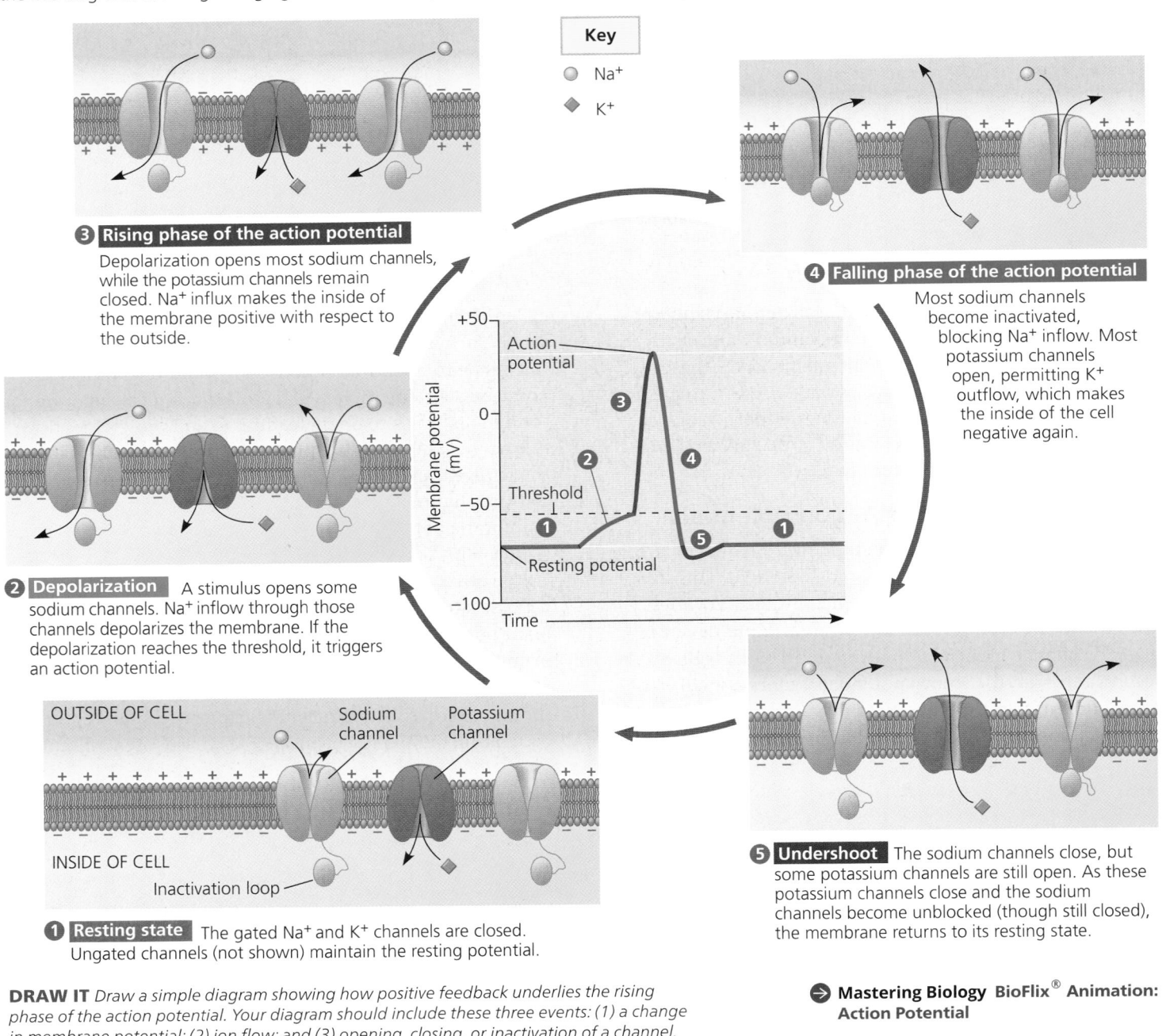

❸ **Rising phase of the action potential**
Depolarization opens most sodium channels, while the potassium channels remain closed. Na⁺ influx makes the inside of the membrane positive with respect to the outside.

❷ **Depolarization** A stimulus opens some sodium channels. Na⁺ inflow through those channels depolarizes the membrane. If the depolarization reaches the threshold, it triggers an action potential.

❶ **Resting state** The gated Na⁺ and K⁺ channels are closed. Ungated channels (not shown) maintain the resting potential.

❹ **Falling phase of the action potential**
Most sodium channels become inactivated, blocking Na⁺ inflow. Most potassium channels open, permitting K⁺ outflow, which makes the inside of the cell negative again.

❺ **Undershoot** The sodium channels close, but some potassium channels are still open. As these potassium channels close and the sodium channels become unblocked (though still closed), the membrane returns to its resting state.

DRAW IT *Draw a simple diagram showing how positive feedback underlies the rising phase of the action potential. Your diagram should include these three events: (1) a change in membrane potential; (2) ion flow; and (3) opening, closing, or inactivation of a channel.*

→ **Mastering Biology** BioFlix® Animation: **Action Potential** cell. ❸ Once the threshold is crossed, the positive-feedback cycle then rapidly brings the membrane potential close to E_{Na}. This stage of the action potential is called the *rising phase*. ❹ Two events prevent the membrane potential from actually reaching E_{Na}: Voltage-gated sodium channels inactivate soon after opening, halting the flow of Na^+ into the cell, and most voltage-gated potassium channels open, causing a rapid outflow of K^+. Both events quickly bring the membrane potential back toward E_K. This stage is called the *falling phase*. ❺ In the final phase of an action potential, called the *undershoot*, the membrane's permeability to K^+ is higher than at rest, so the membrane potential is closer to E_K than it is at the resting potential. The gated potassium channels eventually close, and the membrane potential returns to the resting potential.

Why is inactivation of channels required during an action potential? Because they are voltage gated, the sodium channels open when the membrane potential reaches the threshold of -55 mV and don't close until the resting potential is restored. They are therefore open throughout the action potential. However, the resting potential cannot be restored unless the flow of Na^+ into the cell stops. This is accomplished by inactivation. The sodium channels remain in the "open" state, but sodium ions cease flowing once inactivation occurs. The end of Na^+ inflow allows K^+ outflow to repolarize the membrane.

The sodium channels remain inactivated during the falling phase and the early part of the undershoot. As a result, if a second depolarizing stimulus occurs during this period, it will be unable to trigger an action potential. The "downtime" when a second action potential cannot be initiated is called the **refractory period**. Note that the refractory period is due to the inactivation of sodium channels, not to a change in the ion concentration gradients across the plasma membrane. The flow of charged particles during an action potential involves far too few ions to change the concentration on either side of the membrane significantly.

Conduction of Action Potentials

Having described the events of a single action potential, we'll explore next how a series of action potentials moves a signal along an axon. At the site where an action potential is initiated (usually the axon hillock), Na^+ inflow during the rising phase creates an electrical current that depolarizes the neighboring region of the axon membrane **(Figure 48.13)**. The depolarization is large enough to reach threshold, causing an action potential in the neighboring region. This process is repeated many times along the length of the axon. Because an action potential is an all-or-none event, the magnitude and duration of the action potential are the same at each position along the axon. The net result is the movement of a nerve impulse from the cell body to the synaptic terminals, much like the cascade of events triggered by knocking over the first domino in a line.

▼ **Figure 48.13 Conduction of an action potential.** This figure shows events at three successive times as an action potential passes from left to right. At each point along the axon, voltage-gated ion channels go through the sequence of changes shown in Figure 48.9. Membrane colors correspond to the action potential phases in that figure.

❶ An action potential is generated as Na^+ flows inward across the membrane in one region.

❷ The depolarization of the action potential spreads to the neighboring region of the membrane, reinitiating the action potential there. To the left of this region, the membrane is repolarizing as K^+ flows outward.

❸ The depolarization-repolarization process is repeated in the next region of the membrane. In this way, local currents of ions *across* the plasma membrane cause the action potential to be propagated *along* the length of the axon.

DRAW IT *For the axon segment shown, consider a point at the left end, a point in the middle, and a point at the right end. Draw a graph for each point showing the change in membrane potential over time at that point as a single nerve impulse moves from left to right across the segment.*

⮕ **Mastering Biology BioFlix® Animation: Conduction of an Action Potential**

An action potential that starts at the axon hillock moves along the axon only toward the synaptic terminals. Why? Immediately behind the traveling zone of depolarization, the sodium channels remain inactivated, making the membrane temporarily refractory (not responsive) to further input. Consequently, the inward current that depolarizes the axon membrane *ahead* of the action potential cannot produce another action potential *behind* it. This is why action potentials do not travel back toward the cell body.

After the refractory period is complete, depolarization of the axon hillock to threshold will trigger a new action potential. In many neurons, action potentials last less than 2 milliseconds (msec), and the firing rate can thus reach hundreds of action potentials per second.

The frequency of action potentials conveys information: The rate at which action potentials are produced in a particular neuron is proportional to input signal strength. In hearing, for example, louder sounds result in more frequent action potentials in neurons connecting the ear to the brain. Similarly, increased frequency of action potentials in a neuron that stimulates skeletal muscle tissue will increase the tension in the contracting muscle. Differences in the number of action potentials in a given time are in fact the only variable in how information is encoded and transmitted along an axon.

Gated ion channels and action potentials have a central role in nervous system activity. As a consequence, mutations in genes that encode ion channel proteins can cause disorders affecting the nerves or brain—or the muscles or heart, depending largely on where in the body the gene for the ion channel protein is expressed. For example, mutations affecting voltage-gated sodium channels in skeletal muscle cells can cause myotonia, a periodic spasming of those muscles. Mutations affecting sodium channels in the brain can cause epilepsy, in which groups of nerve cells fire simultaneously and excessively, producing seizures.

Evolutionary Adaptations of Axon Structure

EVOLUTION The rate at which the axons within nerves conduct action potentials governs how rapidly an animal can react to danger or opportunity. As a consequence, natural selection often results in anatomical adaptations that increase conduction speed. One such adaptation is a wider axon. In the same way that a wide hose offers less resistance to the flow of water than does a narrow hose, a wide axon provides less resistance to the current associated with an action potential than does a narrow axon.

In invertebrates, conduction speed varies from several centimeters per second in very narrow axons to approximately 30 m/sec in the giant axons of some arthropods and molluscs. These giant axons (up to 1 mm in diameter) function in rapid behavioral responses, such as the muscle contraction that propels a hunting squid toward its prey.

Vertebrate axons have narrow diameters but can still conduct action potentials at high speed. How is this possible? The evolutionary adaptation that enables fast conduction in vertebrate axons is electrical insulation, analogous to the plastic insulation that encases many electrical wires. Insulation causes the depolarizing current associated with an action potential to travel farther along the axon interior, bringing more distant regions to the threshold sooner.

The electrical insulation that surrounds vertebrate axons is called a **myelin sheath (Figure 48.14)**. Myelin sheaths are produced by glia: **oligodendrocytes** in the CNS and **Schwann cells** in the PNS. During development, these specialized glia wrap axons in many layers of membrane. The membranes forming these layers are mostly lipid, which is a poor conductor of electrical current and thus a good insulator.

In myelinated axons, voltage-gated sodium channels are restricted to gaps in the myelin sheath called **nodes of Ranvier** (see Figure 48.14). Furthermore, the extracellular fluid is in contact with the axon membrane only at the nodes. As a result, action potentials are not generated in the regions between the nodes. Rather, the inward current produced during the rising phase of the action potential at a node travels within the axon all the way to the next node. There, the current

▼ **Figure 48.14 Schwann cells and the myelin sheath.** In the PNS, glia called Schwann cells wrap themselves around axons, forming layers of myelin. Gaps between adjacent Schwann cells are called nodes of Ranvier. The TEM shows a cross section through a myelinated axon.

0.1 μm

▼ **Figure 48.15 Propagation of action potentials in myelinated axons.** In a myelinated axon, the depolarizing current during an action potential at one node of Ranvier spreads along the interior of the axon to the next node (blue arrows), where voltage-gated sodium channels enable reinitiation. Thus, the action potential appears to jump from node to node as it travels along the axon (red arrows).

➔ **Mastering Biology** Animation: Propagation of an Action Potential in Unmyelinated and Myelinated Axons

depolarizes the membrane and regenerates the action potential (**Figure 48.15**).

Action potentials propagate more rapidly in myelinated axons because the time-consuming process of opening and closing of ion channels occurs at only a limited number of positions along the axon. This mechanism for propagating action potentials is called **saltatory conduction** (from the Latin *saltare*, to leap) because the action potential appears to jump from node to node along the axon.

The major selective advantage of myelination is its space efficiency. A myelinated axon 20 μm in diameter has a conduction speed faster than that of a squid giant axon with a diameter 40 times greater. Consequently, more than 2,000 of those myelinated axons can be packed into the space occupied by just one giant axon.

For any axon, myelinated or not, the conduction of an action potential to the end of the axon sets the stage for the next step in neuronal signaling—the transfer of information to another cell. This information handoff occurs at synapses, our next topic.

CONCEPT CHECK 48.3

1. How do action potentials and graded potentials differ?
2. In multiple sclerosis (from the Greek *skleros*, hard), a person's myelin sheaths harden and deteriorate. How would this affect nervous system function?
3. How do both negative and positive feedback contribute to the changes in membrane potential during an action potential?
4. **WHAT IF?** Suppose a mutation caused gated sodium channels to remain inactivated longer after an action potential. How would this affect the frequency at which action potentials could be generated? Explain.

For suggested answers, see Appendix A.

CONCEPT 48.4

Neurons communicate with other cells at synapses

Transmission of information from neurons to other cells occurs at synapses. Synapses are either electrical or chemical.

Electrical synapses contain gap junctions (see Figure 6.30) that allow electrical current to flow directly from one neuron to another. Such synapses often play a role in synchronizing the activity of neurons that direct rapid, unvarying behaviors. For example, electrical synapses associated with the giant axons of squids and lobsters facilitate swift escapes from danger. Electrical synapses are also found in the vertebrate heart and brain.

The majority of synapses are *chemical synapses*, which rely on the release of a chemical neurotransmitter by the presynaptic neuron to transfer information to the target cell. While at rest, the presynaptic neuron synthesizes the neurotransmitter at each synaptic terminal, packaging it in multiple membrane-enclosed compartments called *synaptic vesicles*. When an action potential arrives at a chemical synapse, it depolarizes the plasma membrane at the synaptic terminal, opening voltage-gated channels that allow Ca^{2+} to diffuse in. The Ca^{2+} concentration in the terminal rises, causing synaptic vesicles to fuse with the terminal membrane and release the neurotransmitter.

Neurotransmitter released from the synaptic terminus diffuses across the *synaptic cleft*, the gap that separates the presynaptic neuron from the postsynaptic cell. Diffusion time is very short because the gap is less than 50 nm across. Upon reaching the postsynaptic membrane, the neurotransmitter binds to and activates a specific receptor in the membrane. This series of events at the synapse is summarized in **Figure 48.16**.

▼ **Figure 48.16 A chemical synapse.** This figure illustrates the sequence of events that transmits a signal across a chemical synapse. In response to binding of neurotransmitter, ligand-gated ion channels in the postsynaptic membrane open (as shown here) or, less commonly, close. Synaptic transmission ends when the neurotransmitter diffuses out of the synaptic cleft, is taken up by the synaptic terminal or by another cell, or is degraded by an enzyme.

WHAT IF? *If all the Ca^{2+} in the fluid surrounding a neuron were removed, how would this affect the transmission of information within and between neurons?*

→ **Mastering Biology** BioFlix® Animation: **How Synapses Work**
HHMI Videos: Baldomero Olivera on the Deadly *Conus geographicus* • Blockage of Motor Neuron Synapses by Cone Snail Toxins

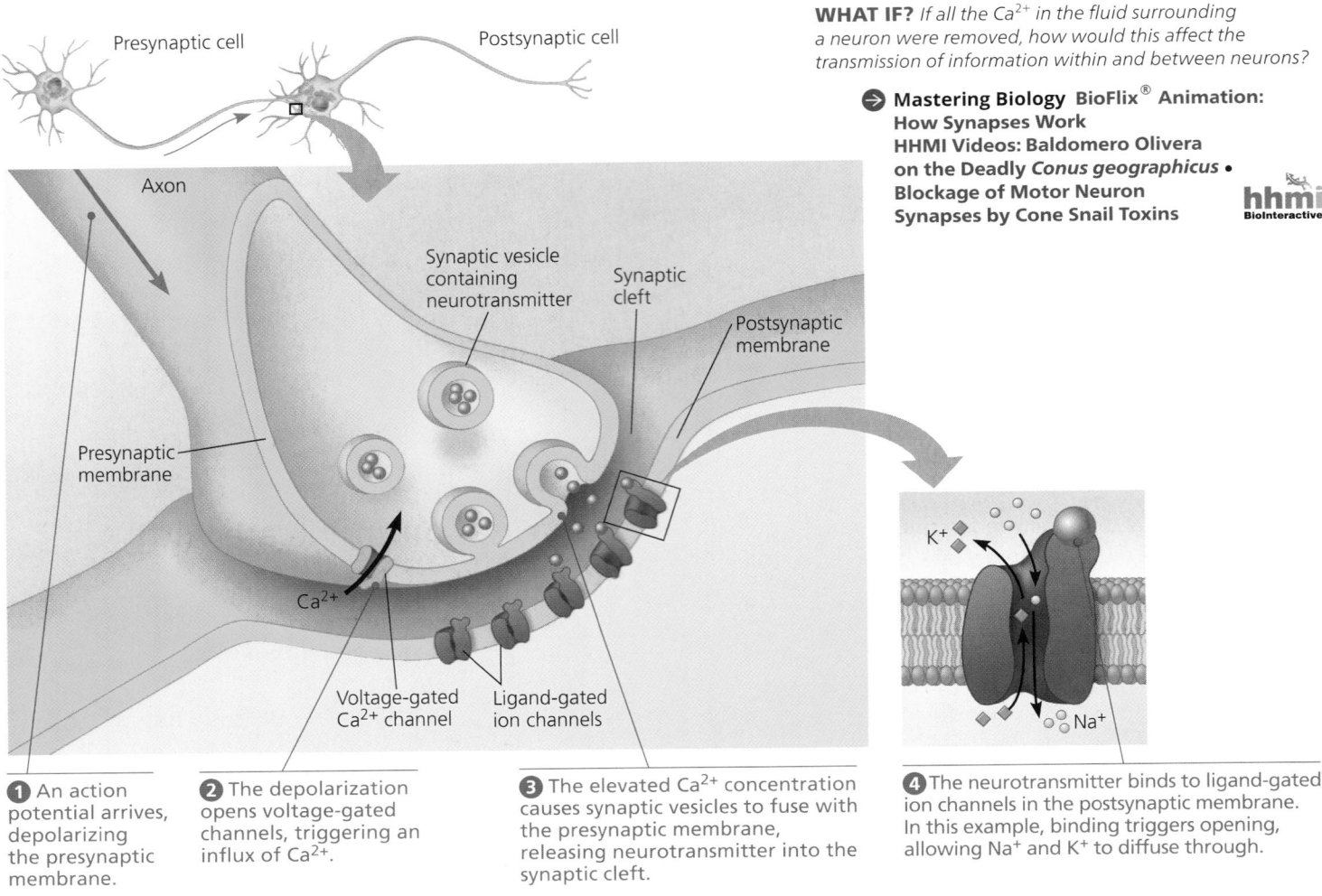

❶ An action potential arrives, depolarizing the presynaptic membrane.

❷ The depolarization opens voltage-gated channels, triggering an influx of Ca^{2+}.

❸ The elevated Ca^{2+} concentration causes synaptic vesicles to fuse with the presynaptic membrane, releasing neurotransmitter into the synaptic cleft.

❹ The neurotransmitter binds to ligand-gated ion channels in the postsynaptic membrane. In this example, binding triggers opening, allowing Na^+ and K^+ to diffuse through.

Information transfer at chemical synapses can be modified by altering either the amount of neurotransmitter that is released or the responsiveness of the postsynaptic cell. Such modifications underlie an animal's ability to alter its behavior in response to change and also form the basis for learning and memory, as you will read in Concept 49.4.

Generation of Postsynaptic Potentials

At many chemical synapses, the receptor protein that binds and responds to neurotransmitters is a **ligand-gated ion channel**, often called an *ionotropic receptor*. These receptors are clustered in the membrane of the postsynaptic cell, directly opposite the synaptic terminal. Binding of the neurotransmitter (the receptor's ligand) to a particular part of the receptor opens the channel and allows specific ions to diffuse across the postsynaptic membrane. The result is a *postsynaptic potential*, a graded potential in the postsynaptic cell.

At some chemical synapses, the ligand-gated ion channels are permeable to both K^+ and Na^+ (see Figure 48.16). When these channels open, the membrane potential depolarizes toward a value roughly midway between E_K and E_{Na}. Because such a depolarization brings the membrane potential toward threshold, it is called an **excitatory postsynaptic potential (EPSP)**.

At other chemical synapses, the ligand-gated ion channels are selectively permeable for only K^+ or Cl^-. When such channels open, the postsynaptic membrane hyperpolarizes. A hyperpolarization produced in this manner is an **inhibitory postsynaptic potential (IPSP)** because it moves the membrane potential farther from threshold.

Summation of Postsynaptic Potentials

The interplay between multiple excitatory and inhibitory inputs is the essence of integration in the nervous system. The cell body and dendrites of a given postsynaptic neuron may receive inputs from chemical synapses formed with hundreds or even thousands of synaptic terminals (see Figure 48.2). How do so many synapses contribute to information transfer?

The input from an individual synapse is typically insufficient to trigger a response in a postsynaptic neuron. To see why, consider an EPSP arising at a single synapse. As a graded potential, the EPSP becomes smaller as it spreads from the synapse. Therefore, by the time a particular EPSP reaches the axon hillock, it is usually too small to trigger an action potential **(Figure 48.17a)**.

On some occasions, individual postsynaptic potentials combine to produce a larger postsynaptic potential, a process called **summation**. For instance, two EPSPs may occur at a single synapse in rapid succession. If the second EPSP arises before the postsynaptic membrane potential returns to its resting value, the EPSPs add together through *temporal summation*. If the summed postsynaptic potentials depolarize the membrane at the axon hillock to threshold, the result is an action potential **(Figure 48.17b)**. Summation can also involve multiple synapses on the same postsynaptic

neuron. If such synapses are active at the same time, the resulting EPSPs can add together through *spatial summation* **(Figure 48.17c)**.

Summation applies as well to IPSPs: Two or more IPSPs occurring nearly simultaneously at synapses in the same region or in rapid succession at the same synapse have a larger hyperpolarizing effect than a single IPSP. Through summation, an IPSP can also counter the effect of an EPSP **(Figure 48.17d)**.

The axon hillock is the neuron's integrating center, the region where the membrane potential at any instant represents the summed effect of all EPSPs and IPSPs. Whenever the membrane potential at the axon hillock reaches threshold, an action potential is generated and travels along the axon to its synaptic terminals. After the refractory period, the neuron may produce another action potential, provided the membrane potential at the axon hillock once again reaches threshold.

Termination of Neurotransmitter Signaling

After a response is triggered, the chemical synapse returns to its resting state. How does this happen? The key step is clearing the neurotransmitter molecules from the synaptic cleft. Some neurotransmitters are inactivated by enzymatic

▼ **Figure 48.17 Summation of postsynaptic potentials.** These graphs trace changes in the membrane potential at a postsynaptic neuron's axon hillock. The red arrows indicate times when postsynaptic potentials occur at two excitatory synapses (E_1 and E_2, green in the diagrams above the graphs) and at one inhibitory synapse (I, purple). Like most EPSPs, those produced at E_1 or E_2 do not reach the threshold at the axon hillock without summation.

Mastering Biology
Animation: Action Potentials

VISUAL SKILLS *Using these drawings, propose an argument for all summation being in some sense temporal.*

Figure 48.18 Two mechanisms of terminating neurotransmission.

(a) Enzymatic breakdown of neurotransmitter in the synaptic cleft

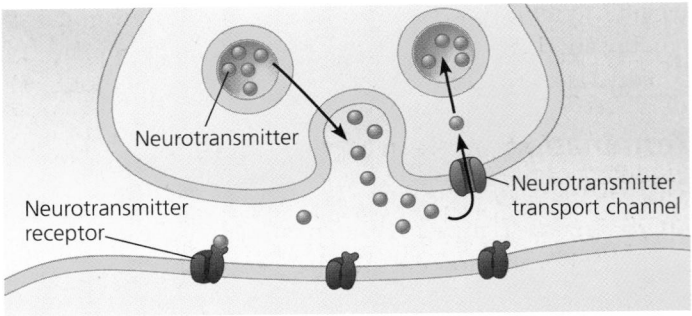

(b) Reuptake of neurotransmitter by presynaptic neuron

hydrolysis **(Figure 48.18a)**. Other neurotransmitters are recaptured into the presynaptic neuron **(Figure 48.18b)**. After this reuptake occurs, neurotransmitters are repackaged in synaptic vesicles or transferred to glia for metabolism or recycling to neurons.

Clearing neurotransmitter from the synaptic cleft is an essential step in the transmission of information through the nervous system. Indeed, blocking this process can have severe consequences. For example, the nerve gas sarin triggers paralysis and death because it inhibits the enzyme that breaks down the neurotransmitter controlling skeletal muscles.

Modulated Signaling at Synapses

So far, we have focused on synapses where a neurotransmitter binds directly to an ion channel, causing the channel to open. However, there are also chemical synapses in which the receptor for the neurotransmitter is *not* part of an ion channel. At these synapses, the neurotransmitter binds to a G protein-coupled receptor, activating a signal transduction pathway in the postsynaptic cell involving a second messenger (see Concept 11.3). Because the resulting opening or closing of ion channels depends on one or more metabolic steps, these G protein-coupled receptors are also called *metabotropic receptors*.

G protein-coupled receptors modulate the responsiveness and activity of postsynaptic neurons in diverse ways.

Consider, for example, the metabotropic receptor for the neurotransmitter norepinephrine. Binding of norepinephrine to its G protein-coupled receptor activates a G protein, which in turn activates adenylyl cyclase, the enzyme that converts ATP to cAMP (see Figure 11.11). Cyclic AMP activates protein kinase A, which phosphorylates specific ion channel proteins in the postsynaptic membrane, causing them to open or close. Because of the amplifying effect of the signal transduction pathway, the binding of one norepinephrine molecule can trigger the opening or closing of many channels.

Many neurotransmitters have both ionotropic and metabotropic receptors. Compared with the postsynaptic potentials produced by ligand-gated channels, the effects of G protein pathways typically have a slower onset but last longer.

Neurotransmitters

Signaling at a synapse brings about a response that depends on both the neurotransmitter released from the presynaptic membrane and the receptor produced at the postsynaptic membrane. A single neurotransmitter may bind specifically to more than a dozen different receptors. Indeed, a particular neurotransmitter can excite postsynaptic cells expressing one receptor and inhibit postsynaptic cells expressing a different receptor. As an example, let's examine **acetylcholine**, a common neurotransmitter in both invertebrates and vertebrates.

Acetylcholine

Acetylcholine is vital for nervous system functions that include muscle stimulation, memory formation, and learning. In vertebrates, there are two major classes of acetylcholine receptor. One is a ligand-gated ion channel. We know the most about its function at the vertebrate *neuromuscular junction*, the site where a motor neuron forms a synapse with a skeletal muscle cell. When acetylcholine released by motor neurons binds this receptor, the ion channel opens, producing an EPSP. This excitatory activity is soon terminated by acetylcholinesterase, an enzyme in the synaptic cleft that hydrolyzes the neurotransmitter.

A G protein-coupled receptor for acetylcholine is found at locations that include the vertebrate CNS and heart. In heart muscle, acetylcholine released by neurons activates a signal transduction pathway. The G proteins in the pathway inhibit adenylyl cyclase and open potassium channels in the muscle cell membrane. Both effects reduce the rate at which the heart pumps. Thus, the effect of acetylcholine in heart muscle is inhibitory rather than excitatory.

Several chemicals with profound effects on the nervous system mimic or alter the function of acetylcholine. Nicotine, a chemical found in tobacco and tobacco smoke, acts as a stimulant by binding to an ionotropic acetylcholine receptor in the CNS. As discussed earlier, the nerve gas sarin blocks enzymatic cleavage of acetylcholine. A third example is botulinum toxin, which inhibits presynaptic release of acetylcholine. The result is a form of food poisoning called botulism. Because muscles required for breathing fail to contract when acetylcholine release is blocked, untreated botulism is typically fatal. Today, injections of the botulinum toxin, known by the trade name Botox, are used cosmetically to minimize wrinkles around the eyes or mouth by inhibiting synaptic transmission to particular facial muscles.

Although acetylcholine has many roles, it is just one of more than 100 known neurotransmitters. As shown by the examples in **Table 48.2**, the rest fall into four classes: amino acids, biogenic amines, neuropeptides, and gases.

Amino Acids

Glutamate is one of several amino acids that can act as a neurotransmitter. In invertebrates, glutamate, rather than acetylcholine, is the neurotransmitter at the neuromuscular junction. In vertebrates, glutamate is the most common neurotransmitter in the CNS. Synapses at which glutamate is the neurotransmitter have a key role in the formation of long-term memory, as you will see in Concept 49.4.

Two amino acids act as inhibitory neurotransmitters in the CNS. *Glycine* acts at inhibitory synapses in parts of the CNS that lie outside of the brain. Within the brain, the amino acid *gamma-aminobutyric acid (GABA)* is the neurotransmitter at most inhibitory synapses. Binding of GABA to receptors in postsynaptic cells increases membrane permeability to Cl^-, resulting in an IPSP. The widely prescribed drug diazepam (Valium) reduces anxiety through binding to a site on a GABA receptor, increasing the response to GABA.

Biogenic Amines

The neurotransmitters grouped as *biogenic amines* are synthesized from amino acids and include *norepinephrine*, which is made from tyrosine. Norepinephrine is an excitatory neurotransmitter in the autonomic nervous system, a branch of the PNS. Outside the nervous system, norepinephrine has distinct but related functions as a hormone, as does the chemically similar biogenic amine *epinephrine* (see Concept 45.3).

The biogenic amines *dopamine* (made from tyrosine) and *serotonin* (made from tryptophan) are released at many sites in the brain and affect sleep, mood, attention, and learning. Some psychoactive drugs, including LSD and mescaline, apparently produce their hallucinatory effects by binding to brain receptors for these neurotransmitters.

Biogenic amines have a central role in a number of nervous system disorders and treatments (see Concept 49.5). The degenerative illness Parkinson's disease is associated with a lack of dopamine in the brain. In addition, depression is often treated with drugs that increase the brain concentrations of biogenic amines. Prozac, for instance, enhances the effect of serotonin by inhibiting its reuptake after it is released by presynaptic neurons.

Neuropeptides

Several **neuropeptides**, relatively short chains of amino acids, serve as neurotransmitters that operate via G protein-coupled receptors. Such peptides are typically produced by cleavage of much larger protein precursors. The neuropeptide *substance P* is a key excitatory neurotransmitter that mediates our perception of pain. Other neuropeptides, called **endorphins**, function as natural analgesics, decreasing pain perception.

Endorphins are produced in the brain during times of physical or emotional stress, such as childbirth. In addition

Table 48.2 Major Neurotransmitters

Neurotransmitter	Structure
Acetylcholine	
Amino Acids	
Glutamate	
GABA (gamma-aminobutyric acid)	
Glycine	
Biogenic Amines	
Norepinephrine	
Dopamine	
Serotonin	
Neuropeptides (a very diverse group, only two of which are shown)	
Substance P	Arg—Pro—Lys—Pro—Gln—Gln—Phe—Phe—Gly—Leu—Met
Met-enkephalin (an endorphin)	Tyr—Gly—Gly—Phe—Met
Gases	
Nitric oxide	$N{=}O$

Interpreting Data Values Expressed in Scientific Notation

Does the Brain Have Specific Protein Receptors for Opiates?

Researchers were looking for opiate receptors in the mammalian brain. Knowing that the drug naloxone blocks the analgesic effect of opiates, they hypothesized that naloxone acts by binding tightly to brain opiate receptors without activating them. In this exercise, you will interpret the results of an experiment to test this hypothesis.

How the Experiment Was Done The researchers incubated radioactive naloxone with a protein mixture prepared from rodent brains. If the mixture contained opiate receptors or other proteins that could bind naloxone, the radioactivity would stably associate with the mixture. To determine whether the binding was due to specific opiate receptors, they tested other drugs, opiate and non-opiate, for their ability to block naloxone binding.

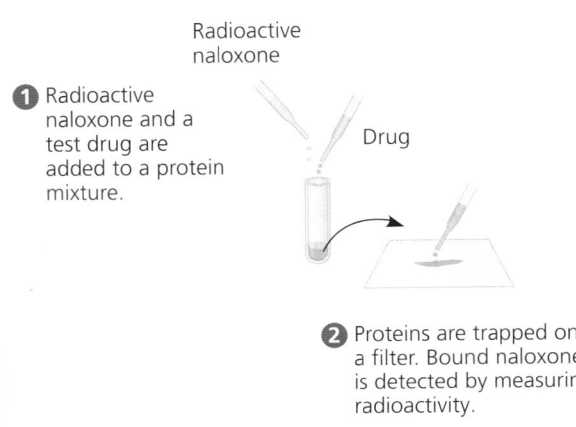

① Radioactive naloxone and a test drug are added to a protein mixture.

Radioactive naloxone

Drug

② Proteins are trapped on a filter. Bound naloxone is detected by measuring radioactivity.

Data from the Experiment

Drug	Opiate	Lowest Concentration That Blocked Naloxone Binding
Morphine	Yes	$6 \times 10^{-9} M$
Methadone	Yes	$2 \times 10^{-8} M$
Levorphanol	Yes	$2 \times 10^{-9} M$
Phenobarbital	No	No effect at $10^{-4} M$
Atropine	No	No effect at $10^{-4} M$
Serotonin	No	No effect at $10^{-4} M$

Data from C. B. Pert and S. H. Snyder, Opiate receptor: demonstration in nervous tissue, *Science* 179:1011–1014 (1973).

INTERPRET THE DATA

1. The data above are expressed in scientific notation: a numerical factor times a power of 10. Remember that a negative power of 10 means a number less than 1. For example, $10^{-1} M$ (molar) can also be written as 0.1 M. Write the concentrations in the table for morphine and atropine in this alternative format.

2. Compare the concentrations listed in the table for methadone and phenobarbital. Which concentration is higher? By how much?

3. Would phenobarbital, atropine, or serotonin have blocked naloxone binding at a concentration of $10^{-5} M$? Explain.

4. **(a)** Which drugs blocked naloxone binding in this experiment? **(b)** What do these results indicate about the brain receptors for naloxone?

5. When the researchers instead used tissue from intestinal muscles rather than brains, they found no naloxone binding. What does that suggest about opiate receptors in mammalian muscle?

→ **Instructors:** A version of this Scientific Skills Exercise can be assigned in **Mastering Biology**.

---

to relieving pain, they reduce urine output, decrease respiration, and produce euphoria as well as other emotional effects. Because opiates (drugs such as morphine and heroin) bind to the same receptor proteins as endorphins, opiates mimic endorphins and produce many of the same physiological effects (see Figure 2.16). In the **Scientific Skills Exercise**, you can interpret data from an experiment designed to search for opiate receptors in the brain.

Gases

Some vertebrate neurons release dissolved gases as neurotransmitters. In human males, for example, certain neurons release nitric oxide (NO) into the erectile tissue of the penis during sexual arousal. The resulting relaxation of smooth muscle in the blood vessel walls of the spongy erectile tissue allows the tissue to fill with blood, producing an erection. The erectile dysfunction drug Viagra works by inhibiting an enzyme that terminates the action of NO.

Unlike most neurotransmitters, NO is not stored in cytoplasmic vesicles but is instead synthesized on demand. NO diffuses into neighboring target cells, produces a change, and is broken down—all within a few seconds. In many of its targets, including smooth muscle cells, NO works like many hormones, stimulating an enzyme to synthesize a second messenger that directly affects cellular metabolism.

Although the gas carbon monoxide (CO) is a deadly poison if inhaled, vertebrates produce small amounts of CO as a neurotransmitter. For example, CO synthesized in the brain regulates the release of hormones from the hypothalamus.

In the next chapter, we'll consider how the cellular and biochemical mechanisms we have discussed contribute to nervous system function on the system level.

→ **Mastering Biology** The Visual Brain: Neural Conduction and Synaptic Transmission

48 Chapter Review

→ Go to **Mastering Biology** for Assignments, the eText, the Study Area, and Dynamic Study Modules.

SUMMARY OF KEY CONCEPTS

→ To review key terms, go to the **Vocabulary Self-Quiz** in the **Mastering Biology** eText or Study Area, or go to goo.gl/zkjz9t.

CONCEPT 48.1

Neuron structure and organization reflect function in information transfer *(pp. 1068–1069)*

- Most **neurons** have branched **dendrites** that receive signals from other neurons and an **axon** that transmits signals to other cells at **synapses**. Neurons rely on **glia** for functions that include nourishment, insulation, and regulation.

- A **central nervous system (CNS)** and a **peripheral nervous system (PNS)** together process information in three stages: sensory input, integration, and motor output to effector cells.

? *How would severing an axon affect the flow of information in a neuron?*

CONCEPT 48.2

Ion pumps and ion channels establish the resting potential of a neuron *(pp. 1069–1072)*

- Ion concentration gradients generate a voltage difference, or **membrane potential**, across the plasma membrane of cells. The concentration of Na^+ is higher outside than inside; the reverse is true for K^+. In resting neurons, the plasma membrane has many open potassium channels but few open sodium channels. Diffusion of ions, principally K^+, through channels generates a **resting potential**, with the inside more negative than the outside.

? *Suppose you placed an isolated neuron in a solution similar to extracellular fluid and later transferred the neuron to a solution lacking any Na^+. What change would you expect in the resting potential?*

CONCEPT 48.3

Action potentials are the signals conducted by axons *(pp. 1072–1077)*

- Neurons have **gated ion channels** that open or close in response to stimuli, leading to changes in the membrane potential. An increase in the magnitude of the membrane potential is a **hyperpolarization**; a decrease is a **depolarization**. Changes in membrane potential that vary continuously with the strength of a stimulus are known as **graded potentials**.

- An **action potential** is a brief, all-or-none depolarization of a neuron's plasma membrane. When a graded depolarization brings the membrane potential to **threshold**, many **voltage-gated ion channels** open, triggering an inflow of Na^+ that rapidly brings the membrane potential to a positive value. A negative membrane potential is restored by the inactivation of sodium channels and by the opening of many voltage-gated potassium channels, which increases K^+ outflow. A **refractory period** follows, corresponding to the interval when the sodium channels are inactivated.

- A nerve impulse travels from the axon hillock to the synaptic terminals by propagating a series of action potentials along the axon. The speed of conduction increases with the diameter of the axon and, in many vertebrate axons, with myelination. Action potentials in myelinated axons appear to jump from one **node of Ranvier** to the next, a process called **saltatory conduction**.

INTERPRET THE DATA *Assuming a refractory period equal in length to the action potential (see graph above), what is the maximum frequency per unit time at which a neuron could fire action potentials?*

CONCEPT 48.4

Neurons communicate with other cells at synapses
(pp. 1077–1083)

- In an electrical synapse, electrical current flows directly from one cell to another. In a chemical synapse, depolarization causes synaptic vesicles to fuse with the terminal membrane and release **neurotransmitter** into the synaptic cleft.
- At many synapses, the neurotransmitter binds to **ligand-gated ion channels** in the postsynaptic membrane, producing an **excitatory** or **inhibitory postsynaptic potential** (**EPSP** or **IPSP**). The neurotransmitter then diffuses out of the cleft, is taken up by surrounding cells, or is degraded by enzymes. A single neuron has many synapses on its dendrites and cell body. Temporal and spatial **summation** of EPSPs and IPSPs at the axon hillock determine whether a neuron generates an action potential.
- Different receptors for the same neurotransmitter produce different effects. Some neurotransmitter receptors activate signal transduction pathways, which can produce long-lasting changes in postsynaptic cells. Major neurotransmitters include **acetylcholine**; the amino acids GABA, glutamate, and glycine; biogenic amines; **neuropeptides**; and gases such as NO.

? *Why are many drugs that are used to treat nervous system diseases or to affect brain function targeted to specific receptors rather than particular neurotransmitters?*

TEST YOUR UNDERSTANDING

➜ For more multiple-choice questions, go to the **Practice Test** in the **Mastering Biology** eText or Study Area, or go to goo.gl/GruWRg.

Levels 1-2: Remembering/Understanding

1. What happens when a resting neuron's membrane depolarizes?
   (A) There is a net diffusion of Na^+ out of the cell.
   (B) The equilibrium potential for K^+ (E_K) becomes more positive.
   (C) The neuron's membrane voltage becomes more positive.
   (D) The cell's inside becomes more negative than the outside.

2. A common feature of action potentials is that they
   (A) cause the membrane to hyperpolarize and then depolarize.
   (B) can undergo temporal and spatial summation.
   (C) are triggered by a depolarization that reaches threshold.
   (D) move at the same speed along all axons.

3. Where are neurotransmitter receptors located?
   (A) the nuclear membrane
   (B) the nodes of Ranvier
   (C) the postsynaptic membrane
   (D) synaptic vesicle membranes

Levels 3-4: Applying/Analyzing

4. Why are action potentials usually conducted in one direction?
   (A) Ions can flow along the axon in only one direction.
   (B) The brief refractory period prevents reopening of voltage-gated Na^+ channels.
   (C) The axon hillock has a higher membrane potential than the terminals of the axon.
   (D) Voltage-gated channels for both Na^+ and K^+ open in only one direction.

5. Which of the following is the most *direct* result of depolarizing the presynaptic membrane of an axon terminal?
   (A) Voltage-gated calcium channels in the membrane open.
   (B) Synaptic vesicles fuse with the membrane.
   (C) Ligand-gated channels open, allowing neurotransmitters to enter the synaptic cleft.
   (D) An EPSP or IPSP is generated in the postsynaptic cell.

6. Suppose a particular neurotransmitter causes an IPSP in postsynaptic cell X and an EPSP in postsynaptic cell Y. A likely explanation is that
   (A) the threshold value in the postsynaptic membrane for cell X is different from that for cell Y.
   (B) the axon of cell X is myelinated, but that of cell Y is not.
   (C) only cell Y produces an enzyme that terminates the activity of the neurotransmitter.
   (D) cells X and Y express different receptor molecules for this particular neurotransmitter.

Levels 5-6: Evaluating/Creating

7. **WHAT IF?** Ouabain, a plant substance used in some cultures to poison hunting arrows, disables the sodium-potassium pump. What change in the resting potential would you expect to see if you treated a neuron with ouabain? Explain.

8. **WHAT IF?** If a drug mimicked the activity of GABA in the CNS, what general effect on behavior might you expect? Explain.

9. **DRAW IT** Suppose a researcher inserts electrodes at two positions along the middle of an axon dissected out of a squid. By applying a depolarizing stimulus at both positions to threshold, the researcher brings the plasma membrane at both positions to threshold. Using the drawing below as a model, create one or more drawings that illustrate where each action potential would terminate.

Electrode

Squid axon

10. **EVOLUTION CONNECTION** An action potential is an all-or-none event. This on/off signaling is an evolutionary adaptation of animals that must sense and act in a complex environment. Imagine a nervous system in which the action potentials are graded, with the amplitude depending on the size of the stimulus. Describe what evolutionary advantage on/off signaling might have over this continuously variable kind of signaling.

11. **SCIENTIFIC INQUIRY** From what you know about action potentials and synapses, propose two hypotheses for how various anesthetics might block pain.

12. **WRITE ABOUT A THEME: ORGANIZATION** In a short essay (100–150 words), describe how the structure and electrical properties of vertebrate neurons reflect similarities and differences with other animal cells.

13. **SYNTHESIZE YOUR KNOWLEDGE**

This diamond-back rattlesnake (*Crotalus atrox*) alerts enemies to its presence with a rattle—a set of modified scales at the tip of its tail. Describe the roles of gated ion channels in initiating and moving a signal along the nerve from the snake's head to its tail and then to the muscle that shakes the rattle.

For selected answers, see Appendix A.

49 Nervous Systems

KEY CONCEPTS

49.1 Nervous systems consist of circuits of neurons and supporting cells *p. 1086*

49.2 The vertebrate brain is regionally specialized *p. 1091*

49.3 The cerebral cortex controls voluntary movement and cognitive functions *p. 1096*

49.4 Changes in synaptic connections underlie memory and learning *p. 1099*

49.5 Many nervous system disorders can now be explained in molecular terms *p. 1102*

Study Tip

Think in pairs: Regional specialization involves many examples of paired structures or circuits. Fill in the complementary or reciprocal functions of these pairs to help you understand their roles in the brain or nervous system.

Structure A/*Function*	Structure B/*Function*
CNS (brain and spinal cord) *Integration of information*	PNS (ganglia and peripheral nerves) *Transfer of information to/from CNS*
Sympathetic division of autonomic nervous system	Parasympathetic division of autonomic nervous system
Left hemisphere of brain	Right hemisphere of brain
Hippocampus (role in memory)	Cerebral cortex (role in memory)

→ Go to Mastering Biology

For Students (in eText and Study Area)
- Get Ready for Chapter 49
- The Visual Brain: Nervous System
- HHMI Video: The Human Suprachiasmatic Nucleus

For Instructors to Assign (in Item Library)
- Scientific Skills Exercise: Designing an Experiment Using Genetic Mutants
- Tutorial: The Vertebrate Nervous System

Figure 49.1 Neuroscientists engineered mice to express a random combination of four fluorescent proteins in each brain cell. The result, shown here, is a "brainbow," with each neuron displaying one of 90 different color combinations. This brainbow technology holds promise for studying particular pathways within the mouse brain. Ultimately, however, we want to understand our own brains, which contain an estimated 10^{11} (100 billion) neurons and 10^{14} (100 trillion) connections.

How are billions of neurons organized to perform complex tasks?

Regional specialization: Complex tasks, such as responding to a spoken question, involve the stepwise functions of different brain regions.

Memory formation: Information is stored by reinforcing patterns of active connections (synapses) among particular neurons.

Synapses that are active in synchrony (A and B) are strengthened. Synapses that are not part of an active circuit (C) are weakened or lost.

Nervous systems consist of circuits of neurons and supporting cells

The ability to sense and react originated billions of years ago in prokaryotes, enhancing survival and reproductive success in changing environments. Later in evolution, modification of simple recognition and response processes provided a basis for communication between cells in an animal body. By the time of the Cambrian explosion, more than 500 million years ago (see Concept 32.2), specialized nervous systems had appeared that enable animals to sense their surroundings and respond rapidly.

Hydras, jellies, and other cnidarians are the simplest animals with nervous systems. In most cnidarians, interconnected neurons form a diffuse *nerve net* **(Figure 49.2a)**, which controls the contraction and expansion of the gastrovascular cavity. In more complex animals, the axons of multiple neurons are often bundled together, forming **nerves**. These fibrous structures channel information flow along specific routes through the nervous system. For example, sea stars have a set of radial nerves connecting to a central nerve ring **(Figure 49.2b)**. Within each arm of a sea star, the radial nerve is linked to a nerve net from which it receives input and to which it sends signals that control muscle contraction.

Animals with elongated, bilaterally symmetrical bodies have even more specialized nervous systems. In particular, they exhibit *cephalization*, an evolutionary trend toward a clustering of sensory neurons and interneurons at the anterior (front) end of the body. Nerves that extend toward the posterior (rear) end enable these anterior neurons to communicate with cells elsewhere in the body.

In many animals, neurons that carry out integration form a **central nervous system (CNS)**, and neurons that carry information into and out of the CNS form a **peripheral nervous system (PNS)**. In nonsegmented worms, such as the planarian in **Figure 49.2c**, a small brain and longitudinal nerve cords constitute the simplest clearly defined CNS. In certain nonsegmented worms, the entire nervous system is constructed from only a small number of cells, as in the case of the nematode *Caenorhabditis elegans*. In this species, an adult worm (hermaphrodite) has exactly 302 neurons, no more and no fewer. More complex invertebrates, such as segmented worms **(Figure 49.2d)** and arthropods **(Figure 49.2e)**, have many more neurons. Their behavior is regulated by more complicated brains and by ventral nerve cords containing **ganglia**, segmentally arranged clusters of neurons that act as relay points in transmitting information.

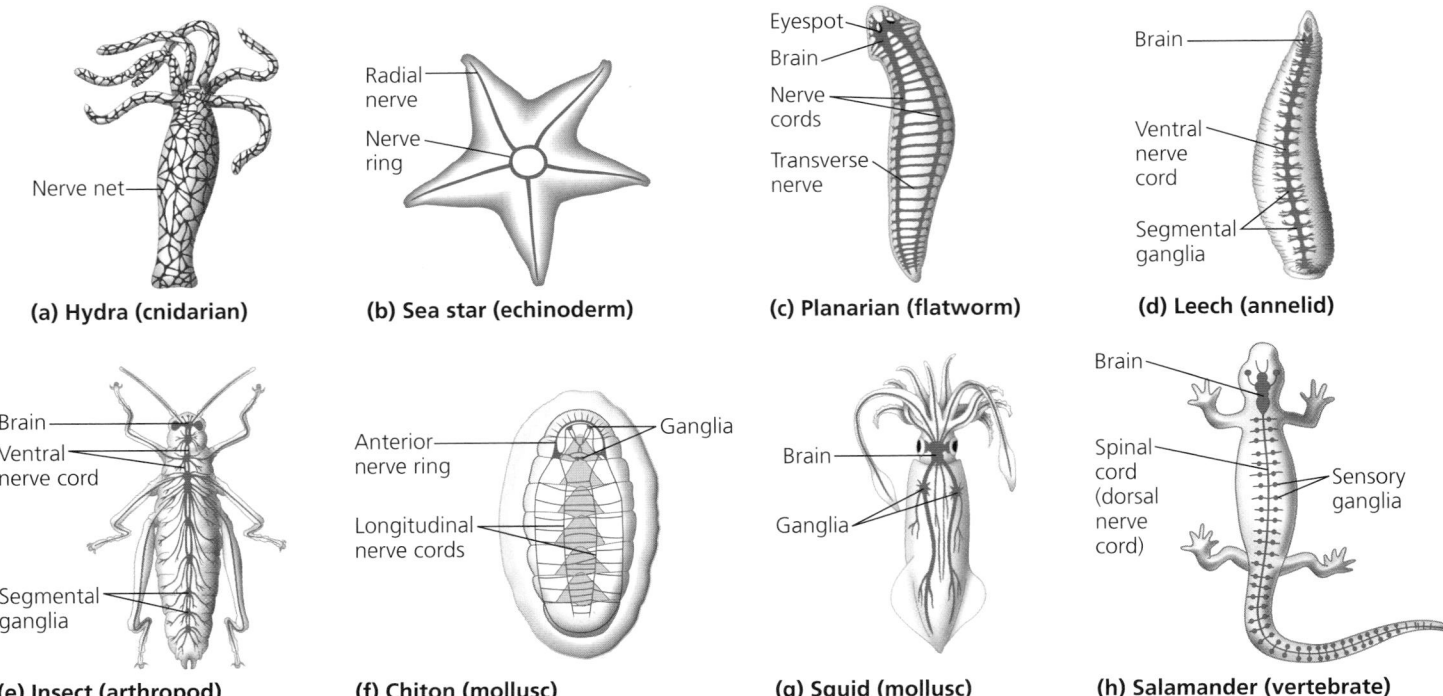

▼ **Figure 49.2 Nervous system organization. (a)** A hydra contains individual neurons (purple) organized in a diffuse nerve net. **(b–h)** Animals with more sophisticated nervous systems contain groups of neurons (blue) organized into nerves and often ganglia and a brain.

Nerve net

(a) Hydra (cnidarian)

Radial nerve
Nerve ring

(b) Sea star (echinoderm)

Eyespot
Brain
Nerve cords
Transverse nerve

(c) Planarian (flatworm)

Brain
Ventral nerve cord
Segmental ganglia

(d) Leech (annelid)

Brain
Ventral nerve cord
Segmental ganglia

(e) Insect (arthropod)

Anterior nerve ring
Ganglia
Longitudinal nerve cords

(f) Chiton (mollusc)

Brain
Ganglia

(g) Squid (mollusc)

Brain
Spinal cord (dorsal nerve cord)
Sensory ganglia

(h) Salamander (vertebrate)

Within an animal group, nervous system organization often correlates with lifestyle. Among the molluscs, for example, sessile and slow-moving species, such as clams and chitons, have relatively simple sense organs and little or no cephalization (**Figure 49.2f**). In contrast, active predatory molluscs, such as octopuses and squids (**Figure 49.2g**), have the most sophisticated nervous systems of any invertebrate. With their large, image-forming eyes and a brain containing millions of neurons, octopuses can learn to discriminate between visual patterns and to perform complex tasks, such as unscrewing a jar to feed on its contents.

In vertebrates such as a salamander (**Figure 49.2h**) or human (**Figure 49.3**), the brain and the spinal cord form the CNS; nerves and ganglia are the key elements of the PNS. Regional specialization is a hallmark of both systems, as we will see throughout this chapter.

Organization of the Vertebrate Nervous System

During embryonic development in vertebrates, the central nervous system develops from the hollow dorsal nerve cord—a hallmark of chordates (see Figure 34.3). The cavity of the nerve cord gives rise to the narrow *central canal* of the spinal cord as well as the *ventricles* of the brain. Both the canal and ventricles fill with *cerebrospinal fluid*, which is formed in the brain by filtering arterial blood (**Figure 49.4**). The cerebrospinal fluid supplies the CNS with nutrients and hormones and carries away wastes, circulating through the ventricles and central canal before draining into the veins.

In addition to fluid-filled spaces, the brain and spinal cord have gray and white matter. **Gray matter** is primarily made up of neuron cell bodies. **White matter** consists of bundled axons. In the spinal cord, white matter forms the outer layer, reflecting its role in linking the CNS to sensory and motor neurons of the PNS. In the brain, white matter is predominantly in the interior, where signaling between neurons functions in learning, feeling emotions, processing sensory information, and generating commands.

In vertebrates, the spinal cord runs lengthwise inside the vertebral column, known as the spine. The spinal cord conveys information to and from the brain and generates basic patterns of locomotion. It also acts independently of the brain as part of the simple nerve circuits that produce **reflexes**, the body's automatic responses to certain stimuli.

Reflexes protect the body by providing rapid, involuntary responses to particular stimuli. Reflexes are rapid because sensory information is used to activate motor neurons without the information first having to travel from the spinal cord to the brain and back. If you accidentally put your hand on a hot burner, a reflex jerks your hand back even before your brain processes pain. Similarly, the knee-jerk reflex provides an immediate protective response when you pick up an unexpectedly heavy object. If your legs buckle, the tension across your knees triggers contraction of your thigh muscle (quadriceps), helping you stay upright and support the load.

▼ **Figure 49.3 The vertebrate nervous system.** The central nervous system consists of the brain and spinal cord (yellow). Left-right pairs of cranial nerves, spinal nerves, and ganglia make up most of the peripheral nervous system (dark gold).

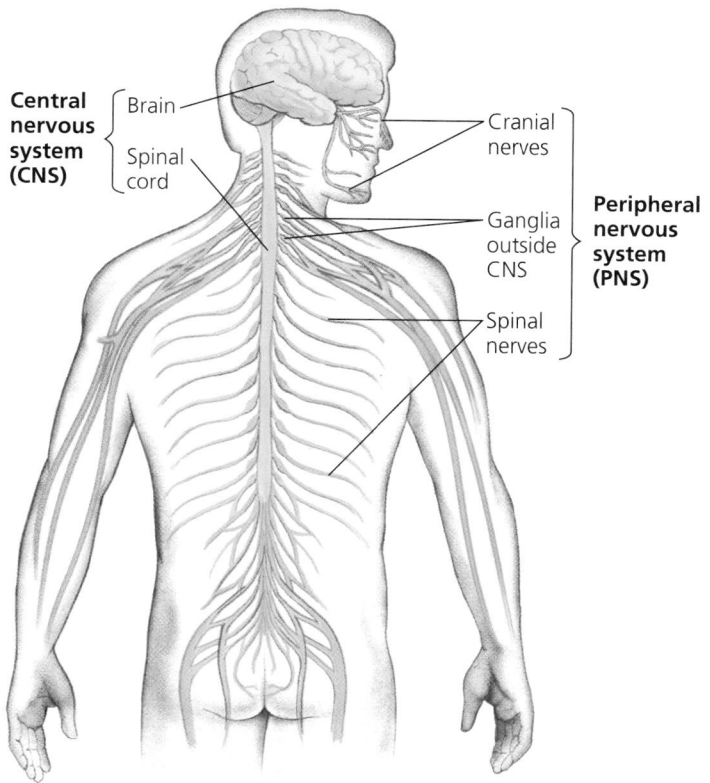

▼ **Figure 49.4 Ventricles, gray matter, and white matter.** Ventricles deep in the brain's interior contain cerebrospinal fluid. In the cerebrum, most of the gray matter is on the surface, surrounding the white matter.

▶ **Figure 49.5 The knee-jerk reflex.** Many neurons are involved in this reflex, but for simplicity only a few neurons are shown.

MAKE CONNECTIONS *Using the nerve signals to the hamstring and quadriceps in this reflex as an example, propose a model for regulation of smooth muscle activity in the esophagus during the swallowing reflex (see Figure 41.9).*

① The reflex (shown here in the movement of the right leg) is initiated artificially by tapping the tendon connected to the quadriceps muscle.

② Sensors detect a sudden stretch in the quadriceps, and **sensory neurons** convey the information to the spinal cord.

③ In response to signals from the sensory neurons, **motor neurons** convey signals to the quadriceps, causing it to contract and jerking the lower leg forward.

Cell body of sensory neuron in dorsal root ganglion

Gray matter   White matter

Quadriceps muscle

Spinal cord (cross section)

Hamstring muscle

During a physical exam, your doctor may trigger the knee-jerk reflex with a triangular mallet to help assess nervous system function **(Figure 49.5)**.

The Peripheral Nervous System

The PNS transmits information to and from the CNS and plays a large role in regulating an animal's movement and its internal environment **(Figure 49.6)**. Sensory information reaches the CNS along PNS neurons designated as *afferent* (from the Latin, meaning "to carry toward"). Following information processing within the CNS, instructions then travel to muscles, glands, and endocrine cells along PNS neurons designated as *efferent* (from the Latin, meaning "to carry away"). Note that most nerves are bundles of both afferent and efferent neurons.

The PNS has two efferent components: the motor system and the autonomic nervous system (see Figure 49.8). The neurons of the **motor system** carry signals to skeletal muscles. Motor control can be voluntary, as when you raise your hand to ask a question, or involuntary, as in the knee-jerk reflex controlled by the spinal cord. In contrast, regulation of smooth and cardiac muscles by the **autonomic nervous system** is generally involuntary. The sympathetic and parasympathetic divisions of the autonomic nervous system regulate organs of the cardiovascular, excretory, and endocrine systems. A distinct network of neurons now known as the **enteric nervous system** exerts direct and partially independent control over the digestive tract, pancreas, and gallbladder.

Homeostasis often relies on cooperation between the motor and autonomic nervous systems. In response to a

⑤ Motor neurons that lead to the hamstring muscle are inhibited by the interneurons. This inhibition prevents contraction of the hamstring, which would resist the action of the quadriceps.

④ Interneurons in the spinal cord also receive signals from sensory neurons.

Key ━● Sensory neuron   ━● Motor neuron   ━● Interneuron

▼ **Figure 49.6 Functional hierarchy of the vertebrate peripheral nervous system.**

► **Figure 49.7 The parasympathetic and sympathetic divisions of the autonomic nervous system.** Most pathways in each division involve two neurons connecting the CNS to target organs. The axon of the first neuron extends from a cell body in the CNS to a set of PNS neurons whose cell bodies are clustered into a ganglion (plural, *ganglia*). The axons of these PNS neurons transmit instructions to internal organs, where they form synapses with smooth muscle, cardiac muscle, or gland cells.

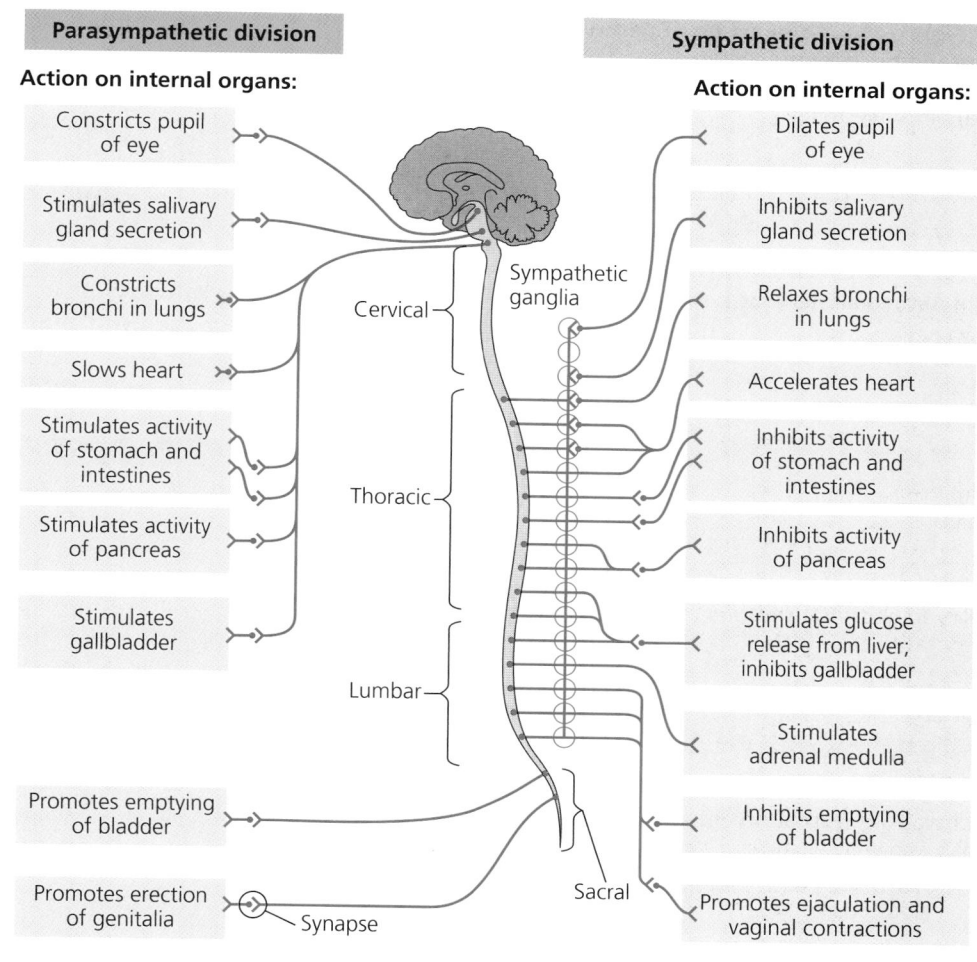

Parasympathetic division

Action on internal organs:

- Constricts pupil of eye
- Stimulates salivary gland secretion
- Constricts bronchi in lungs
- Slows heart
- Stimulates activity of stomach and intestines
- Stimulates activity of pancreas
- Stimulates gallbladder
- Promotes emptying of bladder
- Promotes erection of genitalia

Synapse

Sympathetic division

Action on internal organs:

- Dilates pupil of eye
- Inhibits salivary gland secretion
- Relaxes bronchi in lungs
- Accelerates heart
- Inhibits activity of stomach and intestines
- Inhibits activity of pancreas
- Stimulates glucose release from liver; inhibits gallbladder
- Stimulates adrenal medulla
- Inhibits emptying of bladder
- Promotes ejaculation and vaginal contractions

Cervical

Thoracic

Lumbar

Sacral

Sympathetic ganglia

drop in body temperature, for example, the hypothalamus signals the motor system to cause shivering, which increases heat production. At the same time, the hypothalamus signals the autonomic nervous system to constrict surface blood vessels, reducing heat loss.

The sympathetic and parasympathetic divisions of the autonomic nervous system have largely antagonistic (opposite) functions in regulating organ function **(Figure 49.7)**. Activation of the **sympathetic division** corresponds to arousal and energy generation (the "fight-or-flight" response). For example, the heart beats faster, digestion is inhibited, the liver converts glycogen to glucose, and the adrenal medulla increases secretion of epinephrine (adrenaline). Activation of the **parasympathetic division** generally causes opposite responses that promote calming and a return to self-maintenance functions ("rest and digest"). Thus, heart rate decreases, digestion is enhanced, and glycogen production increases. However, in regulating reproductive activity, a function that is not homeostatic, the parasympathetic division complements rather than antagonizes the sympathetic division, as shown at the bottom of Figure 49.7.

The two divisions differ not only in overall function but also in organization and signals released. Parasympathetic nerves exit the CNS at the base of the brain or spinal cord and form synapses in ganglia near or within an internal organ. In contrast, sympathetic nerves typically exit the CNS midway along the spinal cord and form synapses in ganglia located just outside of the spinal cord.

In both the sympathetic and parasympathetic divisions, the pathway for information flow typically involves a preganglionic and a postganglionic neuron. The *preganglionic neurons* have cell bodies in the CNS and release acetylcholine as a neurotransmitter (see Concept 48.4). In the case of the *postganglionic neurons*, those of the parasympathetic division release acetylcholine, whereas nearly all their counterparts in the sympathetic division release norepinephrine. It is this difference in neurotransmitters that enables the sympathetic and parasympathetic divisions to bring about opposite effects in organs such as the lungs, heart, intestines, and bladder. A comparison of these pathways in the autonomic nervous system, along with a motor system pathway, is shown in **Figure 49.8**.

▼ Figure 49.8 Comparison of pathways in the motor and autonomic nervous systems.

(a) Motor system

Target: skeletal muscle cell

Single neuron

(b) Autonomic nervous system

Sympathetic division

Preganglionic neurons   Ganglia   Postganglionic neurons

Parasympathetic division

Target: smooth muscle, cardiac muscle, or gland

Key to neurotransmitters • Acetylcholine • Norepinephrine

Glia

The nervous systems of vertebrates and most invertebrates include not only neurons but also **glial cells**, or **glia**. The Schwann cells that myelinate axons in the PNS are an example of glia, as are oligodendrocytes, their counterparts in the CNS. **Figure 49.9** provides an overview of the major types

of glia in the adult vertebrate and the ways in which they nourish, support, and regulate the functioning of neurons.

In embryos, two types of glia play essential roles in the development of the nervous system: radial glia and astrocytes. *Radial glia* form tracks along which newly formed neurons migrate from the neural tube, the structure that gives rise to the CNS (see Figure 47.14). Later, *astrocytes* that are adjacent to brain capillaries participate in the formation of the *blood-brain barrier*, a filtering mechanism that prevents many substances in the blood from entering the CNS. Both radial glia and astrocytes can also act as stem cells, which undergo unlimited cell divisions to self-renew and to form more specialized cells.

CONCEPT CHECK 49.1

1. Which division of the autonomic nervous system would likely be activated if a student learned that an exam she had forgotten about would start in 5 minutes? Explain your answer.

2. **WHAT IF?** Suppose a person had an accident that severed a small nerve required to move some of the fingers of the right hand. Would you also expect an effect on sensation from those fingers?

3. **MAKE CONNECTIONS** Most tissues regulated by the autonomic nervous system receive both sympathetic and parasympathetic input from postganglionic neurons. Responses are typically local. In contrast, the adrenal medulla receives input only from the sympathetic division and only from preganglionic neurons, yet responses are observed throughout the body. Explain why (see Figure 45.19).

For suggested answers, see Appendix A.

▼ Figure 49.9 Glia in the vertebrate nervous system.

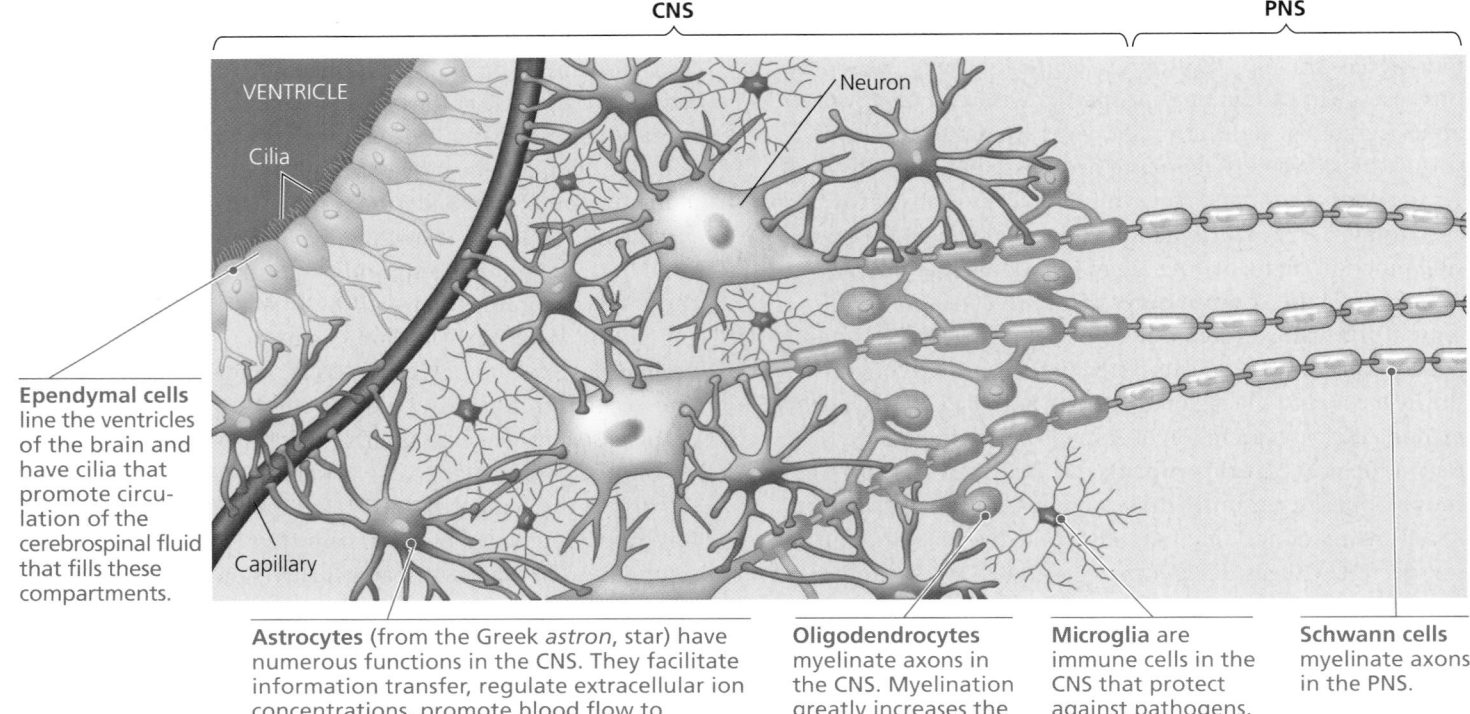

CNS

PNS

VENTRICLE

Cilia

Neuron

Ependymal cells line the ventricles of the brain and have cilia that promote circulation of the cerebrospinal fluid that fills these compartments.

Capillary

Astrocytes (from the Greek *astron*, star) have numerous functions in the CNS. They facilitate information transfer, regulate extracellular ion concentrations, promote blood flow to neurons, help form the blood-brain barrier, and act as stem cells to replenish certain neurons.

Oligodendrocytes myelinate axons in the CNS. Myelination greatly increases the conduction speed of action potentials.

Microglia are immune cells in the CNS that protect against pathogens.

Schwann cells myelinate axons in the PNS.

CONCEPT 49.2

The vertebrate brain is regionally specialized

We turn now to the vertebrate brain, which has three major regions: the forebrain, midbrain, and hindbrain (shown here for a ray-finned fish).

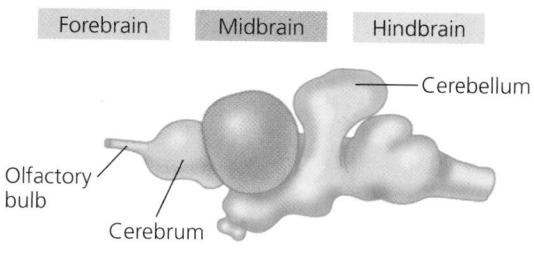

Each region is specialized in function. The **forebrain**, which contains the *olfactory bulb* and *cerebrum*, has activities that include processing of olfactory input (smells), regulation of sleep, learning, and any complex processing. The **midbrain**, located centrally in the brain, coordinates routing of sensory input. The **hindbrain**, part of which forms the *cerebellum*, controls involuntary activities, such as blood circulation, and coordinates motor activities, such as locomotion.

EVOLUTION Comparing vertebrates across a phylogenetic tree, we see that the relative sizes of particular brain regions vary **(Figure 49.10)**. Furthermore, these size differences reflect differences in the importance of particular brain functions. Consider, for example, ray-finned fishes, which explore their environment using olfaction, vision, and a lateral line system that detects water currents, electrical stimuli, and body position. The olfactory bulb, which detects scents in the water, is relatively large in these fishes. So is the midbrain, which processes input from the visual and lateral line systems. In contrast, the cerebrum, required for complex processing and learning, is relatively small.

The correlation between the size and function of brain regions can also be observed by considering the cerebellum. Free swimming ray-finned fishes, such as the tuna, control movement in three dimensions in the open water and have a relatively large cerebellum. In comparison, the cerebellum is much smaller in species that don't swim actively, such as the lamprey. Evolution has thus resulted in a close match of structure to function, with the size of particular brain regions correlating with their importance for that species in nervous system function and, hence, species survival and reproduction.

If one compares birds and mammals with groups that diverged from the common vertebrate ancestor earlier in evolution, two trends are apparent. First, the forebrain of birds and mammals occupies a larger fraction of the brain than it does in amphibians, fishes, and other vertebrates. Second, birds and mammals have much larger brains relative to body size than do other groups. Indeed, the ratio of brain size to body weight for birds and mammals is ten times larger than that for their evolutionary ancestors. These differences in both overall brain size and the relative size of the forebrain reflect the greater capacity of birds and mammals for cognition and higher-order reasoning, traits we will return to later in this chapter.

In the case of humans, the 100 billion neurons in the brain make 100 trillion connections. How are so many cells and links organized into circuits and networks that can perform highly sophisticated information processing, storage, and retrieval? In addressing this question, let's begin with **Figure 49.11**, which explores the overall architecture of the human brain. You can use this figure to trace how brain structures arise during embryonic development; as a reference for their size, shape, and location in the adult brain; and as an introduction to their best-understood functions.

To learn more about how particular brain structure and brain organization overall relate to brain function in humans, we'll first consider activity cycles of the brain and the physiological basis of emotion. Then, in Concept 49.3, we'll shift our attention to regional specialization within the cerebrum.

▼ **Figure 49.10 Vertebrate brain structure and evolution.** These examples of vertebrate brains are drawn to the same overall dimensions to highlight differences in the relative size of major structures. These differences in relative size, which arose over the course of vertebrate evolution, correlate with the importance of particular brain functions for particular vertebrate groups.

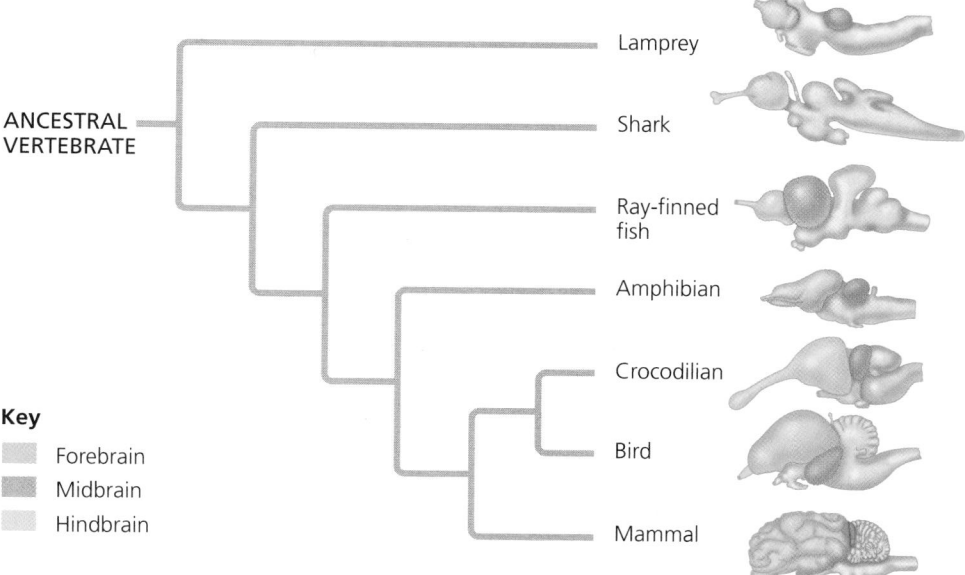

The brain is the most complex organ in the human body. Surrounded by the thick bones of the skull, the brain is divided into a set of distinctive structures, some of which are visible in the magnetic resonance image (MRI) of an adult's head shown at right. The diagram below traces the development of these structures in the embryo. Their major functions are explained in the main text of the chapter.

Human Brain Development

As a human embryo develops, the neural tube forms three anterior bulges—the forebrain, midbrain, and hindbrain—that together produce the adult brain. The midbrain and portions of the hindbrain give rise to the **brainstem,** a stalk that joins with the spinal cord at the base of the brain. The rest of the hindbrain gives rise to the **cerebellum**, which lies behind the brainstem. Meanwhile, the forebrain develops into the diencephalon, including the neuroendocrine tissues of the brain, and the telencephalon, which becomes the **cerebrum**. Rapid, expansive growth of the telencephalon during the second and third months causes the outer portion, or cortex, of the cerebrum to extend over and around much of the rest of the brain.

Embryonic brain regions		Brain structures in child and adult
Forebrain	Telencephalon	Cerebrum (includes cerebral cortex, basal nuclei)
	Diencephalon	Diencephalon (thalamus, hypothalamus, epithalamus)
Midbrain	Mesencephalon	Midbrain (part of brainstem)
Hindbrain	Metencephalon	Pons (part of brainstem), cerebellum
	Myelencephalon	Medulla oblongata (part of brainstem)

Midbrain

Hindbrain

Forebrain

Embryo at 1 month

Mesencephalon

Metencephalon

Diencephalon

Myelencephalon

Spinal cord

Telencephalon

Embryo at 5 weeks

Cerebrum

Diencephalon

Midbrain

Pons

Medulla oblongata

Cerebellum

Spinal cord

Brainstem

Child

The Cerebrum

The cerebrum controls skeletal muscle contraction and is the center for learning, emotion, memory, and perception. It is divided into right and left *cerebral hemispheres*. The outer layer of the cerebrum is called the **cerebral cortex** and is vital for perception, voluntary movement, and learning. The left side of the cerebral cortex receives information from, and controls the movement of, the right side of the body, and vice versa. A thick band of axons known as the **corpus callosum** enables the right and left cerebral cortices to communicate. Deep within the white matter, clusters of neurons called *basal nuclei* serve as centers for planning and learning movement sequences. Damage to these sites during fetal development can result in cerebral palsy, a disorder resulting from a disruption in the transmission of motor commands to the muscles.

The Cerebellum

The cerebellum coordinates movement and balance and helps in learning and remembering motor skills. The cerebellum receives sensory information about the positions of the joints and the lengths of the muscles, as well as input from the auditory (hearing) and visual systems. It also monitors motor commands issued by the cerebrum. The cerebellum integrates this information as it carries out coordination and error checking during motor and perceptual functions.

Hand-eye coordination is an example of cerebellar control; if the cerebellum is damaged, the eyes can follow a moving object, but they will not stop at the same place as the object. Hand movement toward the object will also be erratic.

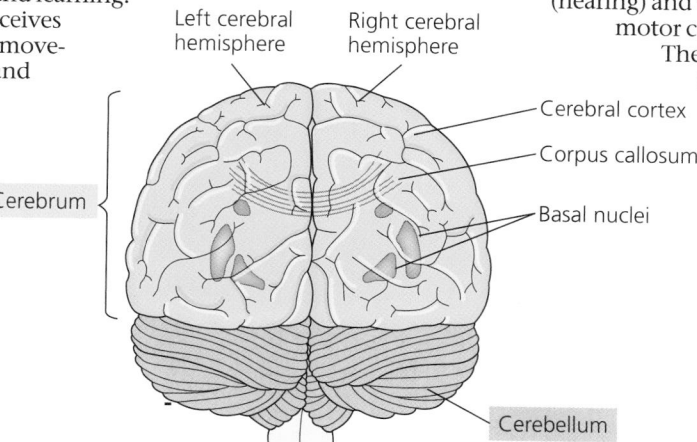

Left cerebral hemisphere

Right cerebral hemisphere

Cerebral cortex

Corpus callosum

Basal nuclei

Cerebrum

Cerebellum

Adult brain viewed from the rear

The Diencephalon

The diencephalon gives rise to the thalamus, hypothalamus, and epithalamus. The **thalamus** is the main input center for sensory information going to the cerebrum. Incoming information from all the senses, as well as from the cerebral cortex, is sorted in the thalamus and sent to the appropriate cerebral centers for further processing. The thalamus is formed by two masses, each roughly the size and shape of a walnut. A much smaller structure, the **hypothalamus**, constitutes a control center that includes the body's thermostat as well as the central biological clock. Through its regulation of the pituitary gland, the hypothalamus regulates hunger and thirst, plays a role in sexual and mating behaviors, and initiates the fight-or-flight response. The hypothalamus is also the source of posterior pituitary hormones and of releasing hormones that act on the anterior pituitary. The *epithalamus* includes the pineal gland, the source of melatonin.

Diencephalon
Thalamus
Pineal gland
Hypothalamus
Pituitary gland

Brainstem
Midbrain
Pons
Medulla oblongata

Spinal cord

The Brainstem

The brainstem consists of the midbrain, the **pons**, and the **medulla oblongata** (commonly called the *medulla*). The midbrain receives and integrates several types of sensory information and sends it to specific regions of the forebrain. All sensory axons involved in hearing either terminate in the midbrain or pass through it on their way to the cerebrum. In addition, the midbrain coordinates visual reflexes, such as the peripheral vision reflex: The head turns toward an object approaching from the side without the brain having formed an image of the object.

A major function of the pons and medulla is to transfer information between the PNS and the midbrain and forebrain. The pons and medulla also help coordinate large-scale body movements, such as running and climbing. Most axons that carry instructions about these movements cross from one side of the CNS to the other as they pass through the medulla. As a result, the right side of the brain controls much of the movement of the left side of the body, and vice versa. An additional function of the medulla is the control of several automatic, homeostatic functions, including breathing, heart and blood vessel activity, swallowing, vomiting, and digestion. The pons also participates in some of these activities; for example, it regulates the breathing centers in the medulla.

→ **Mastering Biology The Visual Brain: Nervous System**

Arousal and Sleep

If you've ever drifted off to sleep while listening to a lecture (or reading a book), you know that your attentiveness and mental alertness can change rapidly. Such transitions are regulated by the brainstem and cerebrum, which control arousal and sleep. Arousal is a state of awareness of the external world. Sleep is a state in which external stimuli are received but not consciously perceived.

Contrary to appearances, sleep is an active state, at least for the brain. By placing electrodes at multiple sites on the scalp, we can record patterns of electrical activity called brain waves in an electroencephalogram (EEG). These recordings reveal that brain wave frequencies change as the brain progresses through distinct stages of sleep.

Some animals have evolutionary adaptations that allow for substantial activity during sleep. Bottlenose dolphins, for example, swim while sleeping, rising to the surface to breathe air on a regular basis. How is this possible? As in other mammals, the forebrain is physically and functionally divided into two halves, the right and left hemispheres. Noting that dolphins sleep with one eye open and one closed, researchers hypothesized that only one side of the brain is asleep at a time. EEG recordings from each hemisphere of sleeping dolphins support this hypothesis **(Figure 49.12)**.

Although sleep is essential for survival, we still know very little about its function. One hypothesis is that sleep and dreams are involved in consolidating learning and memory. Evidence supporting this hypothesis includes the finding that test subjects who are kept awake for 36 hours have a reduced ability to remember when particular events occurred, even if they first "perk up" with caffeine. Other experiments show that regions of the brain that are activated during a learning task can become active again during sleep.

Arousal and sleep are controlled in part by the *reticular formation*, a diffuse network formed primarily by neurons in

▼ **Figure 49.12 Dolphins can be asleep and awake at the same time.** EEG recordings were made separately for the two sides of a dolphin's brain. At each time point, low-frequency activity characteristic of sleep was recorded in one hemisphere while higher-frequency activity typical of being awake was recorded in the other hemisphere.

Location	Time: 0 hours	Time: 1 hour
Left hemisphere	⩓⩓⩓⩓	⩗⩗⩗
Right hemisphere	⩗⩗⩗	⩓⩓⩓⩓

Key

⩓ Low-frequency waves characteristic of sleep

⩗ High-frequency waves characteristic of wakefulness

▼ **Figure 49.13 The reticular formation.** Once thought to consist of a single diffuse network of neurons, the reticular formation is now recognized as many distinct clusters of neurons. These clusters function in part to filter sensory input (blue arrows), blocking familiar and repetitive information that constantly enters the nervous system before sending the filtered input to the cerebral cortex (green arrows).

Eye

Reticular formation

Input from touch, pain, and temperature receptors

Input from nerves of ears

the midbrain and pons **(Figure 49.13)**. These neurons control the timing of sleep periods characterized by rapid eye movements (REMs) and by vivid dreams. Sleep is also regulated by the biological clock, discussed next, and by regions of the forebrain that regulate sleep intensity and duration.

⊙ **Mastering Biology** The Visual Brain: Sleep and Waking

Biological Clock Regulation

Cycles of sleep and wakefulness are an example of a circadian rhythm, a daily cycle of biological activity. Such cycles, which occur in organisms ranging from bacteria to humans, rely on a **biological clock**, a molecular mechanism that directs periodic gene expression and cellular activity. Although biological clocks are typically synchronized to the cycles of light and dark in the environment, they can maintain a roughly 24-hour cycle even in the absence of environmental cues (see Figure 40.9). For example, in a constant environment humans exhibit a sleep/wake cycle of 24.2 hours, with very little variation among individuals.

What normally links the biological clock to environmental cycles of light and dark in an animal's surroundings? In mammals, circadian rhythms are coordinated by clustered neurons in the hypothalamus (see Figure 49.11). These neurons form a structure called the **SCN**, which stands for **suprachiasmatic nucleus**. (Certain clusters of neurons in the CNS are referred to as "nuclei.") In response to sensory information from the eyes, the SCN acts as a pacemaker, synchronizing the biological clock in cells throughout the body to the natural cycles of day length. In the **Scientific Skills Exercise**, you can interpret

Designing an Experiment Using Genetic Mutants

Does the SCN Control the Circadian Rhythm in Hamsters? By surgically removing the SCN from laboratory mammals, scientists demonstrated that the SCN is required for circadian rhythms. Those experiments did not, however, reveal whether circadian rhythms originate in the SCN. To answer this question, researchers performed an SCN transplant experiment on wild-type and mutant hamsters (*Mesocricetus auratus*). Whereas wild-type hamsters have a circadian cycle lasting about 24 hours in the absence of external cues, hamsters that are homozygous for the τ (tau) mutation have a cycle lasting only about 20 hours. In this exercise, you will evaluate the design of this experiment and propose additional experiments to gain further insight.

How the Experiment Was Done The researchers surgically removed the SCN from wild-type and τ hamsters. Several weeks later, each of these hamsters received a transplant of an SCN from a hamster of the opposite genotype. To determine the periodicity of rhythmic activity for the hamsters before the surgery and after the transplants, the researchers measured activity levels over a three-week period. They plotted the data collected for each day in the manner shown in Figure 40.9a and then calculated the circadian cycle period.

Data from the Experiment In 80% of the hamsters in which the SCN had been removed, transplanting an SCN from another hamster restored rhythmic activity. For hamsters in which an SCN transplant restored a circadian rhythm, the net effect of the two procedures (SCN removal and replacement) on the circadian cycle period is graphed at the upper right. Each red line connects the two data points for an individual hamster.

Legend:
- ○ Wild-type hamster
- ▲ Wild-type hamster with SCN from τ hamster
- ● τ hamster
- ◐ τ hamster with SCN from wild-type hamster

Data from M. R. Ralph et al., Transplanted suprachiasmatic nucleus determines circadian period, *Science* 247:975–978 (1990).

INTERPRET THE DATA

1. In a controlled experiment, researchers manipulate one variable at a time. **(a)** What was the variable manipulated in this study? **(b)** Why did the researchers use more than one hamster for each procedure? **(c)** What traits of the individual hamsters would likely have been held constant among the treatment groups?

2. For the wild-type hamsters that received τ SCN transplants, what would have been an appropriate experimental control?

3. **(a)** What general trends does the graph above reveal about the circadian cycle period of the transplant recipients? **(b)** Do the trends differ for the wild-type and τ recipients? Based on these data, what can you conclude about the role of the SCN in determining the period of the circadian rhythm?

4. **(a)** In 20% of the hamsters, there was no restoration of rhythmic activity following the SCN transplant. What are some possible reasons for this finding? **(b)** How confident are you in your conclusion about the role of the SCN based on data from 80% of the hamsters?

5. Suppose that researchers identified a mutant hamster that lacked rhythmic activity; that is, its circadian activity cycle had no regular pattern. Propose SCN transplant experiments using such a mutant along with **(a)** wild-type and **(b)** τ hamsters. Predict the results of those experiments in light of your conclusion in question 3(b).

→ **Instructors:** A version of this Scientific Skills Exercise can be assigned in **Mastering Biology**.

data from an experiment and propose experiments to test the role of the SCN in hamster circadian rhythms.

→ **Mastering Biology** HHMI Video: The Human Suprachiasmatic Nucleus

Emotions

Whereas a single structure in the brain controls the biological clock, the generation and experience of emotions depend on many brain structures, including the amygdala, hippocampus, and parts of the thalamus. As shown in **Figure 49.14**, these structures border the brainstem in mammals and are therefore called the *limbic system* (from the Latin *limbus*, border).

One way the limbic system contributes to our emotions is by storing emotional experiences as memories that can be recalled by similar circumstances. This is why, for example, a situation that causes you to remember a frightening event

◄ **Figure 49.14 The limbic system of the human brain.** This diagram shows the brain and brainstem, with the left cerebral hemisphere of the brain removed.

Labels: Thalamus, Olfactory bulbs, Hypothalamus, Hippocampus, Amygdala can trigger a faster heart rate, sweating, or fear, even if there is currently nothing scary or threatening in your surroundings. Such storage and recall of emotional memory are especially dependent on function of the **amygdala**, an almond-shaped brain structure near the base of the cerebrum.

Often, generating emotion and experiencing emotion require interactions between different regions of the brain. For example, both laughing and crying involve the limbic system interacting with sensory areas of the forebrain. Similarly, structures in the forebrain attach emotional "feelings" to survival-related functions controlled by the brainstem, including aggression, feeding, and sexuality.

To study the function of the human amygdala, researchers sometimes present adult subjects with an image followed by an unpleasant experience, such as a mild electrical shock. After several trials, study participants experience *autonomic arousal*—as measured by increased heart rate or sweating—if they see the image again. Subjects with brain damage confined to the amygdala can recall the image because their explicit memory is intact. However, they do not exhibit autonomic arousal, indicating that damage to the amygdala has resulted in a reduced capacity for emotional memory.

Functional Imaging of the Brain

In recent years, scientists have begun studying the amygdala and other brain structures with functional imaging techniques. By scanning the brain while the subject performs a particular function, such as forming a mental image of a person's face, researchers are able to match particular functions with activity in specific brain areas.

Several approaches are available for functional imaging. The first widely used technique was positron-emission tomography (PET), in which injection of radioactive glucose enables a display of metabolic activity. Today, the most commonly used approach is functional magnetic resonance imaging (fMRI). In fMRI, a subject lies with his or her head in the center of a large, doughnut-shaped magnet. Brain activity is detected by an increase in the flow of oxygen-rich blood into a particular region.

In one experiment using fMRI, researchers mapped brain activity while subjects listened to music that they described as sad or happy **(Figure 49.15)**. The findings were striking: Different regions of the brain were associated with the experience of each of these contrasting emotions. Subjects who heard sad music had increased activity in the amygdala. In contrast, listening to happy music led to increased activity in the *nucleus accumbens*, a brain structure important for the perception of pleasure.

Functional imaging has an ever-increasing number of applications. Hospitals use fMRI to, for example, monitor recovery from stroke, map abnormalities in migraine headaches, and increase the effectiveness of brain surgery.

▼ **Figure 49.15 Functional imaging in the working brain.** These images resulted from using fMRI to reveal brain activity associated with music that listeners described as happy or sad. (Each view shows activity in a single plane of the brain, as seen from above.)

Nucleus accumbens   Amygdala

Happy music   Sad music

VISUAL SKILLS *The two images reveal activity in different horizontal planes through the brain. How can you tell this from the two photographs? What can you conclude about the location of the nucleus accumbens and the amygdala?*

CONCEPT CHECK 49.2

1. When you wave your right hand, what part of your brain initiates the action?

2. People who are inebriated have difficulty touching their nose with their eyes closed. Which brain region does this observation indicate is one of those impaired by alcohol?

3. **WHAT IF?** Suppose you examine two groups of individuals with CNS damage. In one group, the damage has resulted in a coma (a prolonged state of unconsciousness). In the other group, it has caused paralysis (a loss of skeletal muscle function throughout the body). Relative to the position of the midbrain and pons, where is the likely site of damage in each group? Explain.

For suggested answers, see Appendix A.

CONCEPT 49.3

The cerebral cortex controls voluntary movement and cognitive functions

We turn now to the cerebrum, the part of the brain essential for language, cognition, memory, consciousness, and awareness of our surroundings. As shown in Figure 49.11, the cerebrum is the largest structure in the human brain. Like the brain overall, it exhibits regional specialization. For the most part, cognitive functions reside in the cortex, the outer layer of the cerebrum. Within this cortex, *sensory areas* receive and process sensory information, *association areas* integrate the information, and *motor areas* transmit instructions to other parts of the body.

In discussing the location of particular functions in the cerebral cortex, neurobiologists often use four regions, or *lobes*, as physical landmarks. Each lobe—frontal, temporal, occipital, and parietal—is named for a nearby bone of the skull, and each is the focus of specific brain activities **(Figure 49.16)**.

▼ **Figure 49.16 The human cerebral cortex.** Each of the four lobes of the cerebral cortex has specialized functions, some of which are listed here. Some areas on the left side of the brain (shown here) have different functions from those on the right side (not shown).

Motor cortex (control of skeletal muscles)

Somatosensory cortex (sense of touch)

Frontal lobe

Parietal lobe

Prefrontal cortex (decision making, planning)

Sensory association cortex (integration of sensory information)

Broca's area (forming speech)

Visual association cortex (combining images and object recognition)

Temporal lobe

Occipital lobe

Auditory cortex (hearing)

Visual cortex (processing visual stimuli and pattern recognition)

Wernicke's area (comprehending language)

Cerebellum

Information Processing

Broadly speaking, the human cerebral cortex receives sensory information from two sources. Some sensory input originates in individual receptors in the hands, scalp, and elsewhere in the body. These somatic sensory, or *somatosensory*, receptors (from the Greek *soma*, body) provide information about touch, pain, pressure, temperature, and the position of muscles and limbs. Other sensory input comes from groups of receptors clustered in dedicated sensory organs, such as the eyes and nose.

Most sensory information coming into the cortex is directed via the thalamus to primary sensory areas within the brain lobes. Information received at the primary sensory areas is passed along to nearby association areas, which process particular features in the sensory input. In the occipital lobe, for instance, some groups of neurons in the primary visual area are specifically sensitive to rays of light oriented in a particular direction. In the visual association area, information related to such features is

combined in a region dedicated to recognizing complex images, such as faces.

Once processed, sensory information passes to the prefrontal cortex, which helps plan actions and movement. The cerebral cortex may then generate motor commands that cause particular behaviors—moving a limb or saying hello, for example. These commands consist of action potentials produced by neurons in the motor cortex, which lies at the rear of the frontal lobe (see Figure 49.16). The action potentials travel along axons to the brainstem and spinal cord, where they excite motor neurons, which in turn excite skeletal muscle cells.

In the somatosensory cortex and motor cortex, neurons are arranged according to the part of the body that generates the sensory input or receives the motor commands **(Figure 49.17)**. For example, neurons that process sensory information from the legs and feet lie in the region of the somatosensory cortex closest to the midline. Neurons that control muscles in the legs and feet are located in the corresponding region of the motor cortex. Notice in Figure 49.17 that the cortical surface area

▼ **Figure 49.17 Body part representation in the primary motor and primary somatosensory cortices.** In these cross-sectional maps of the cortices, the cortical surface area devoted to each body part is represented by the relative size of that part in the cartoons.

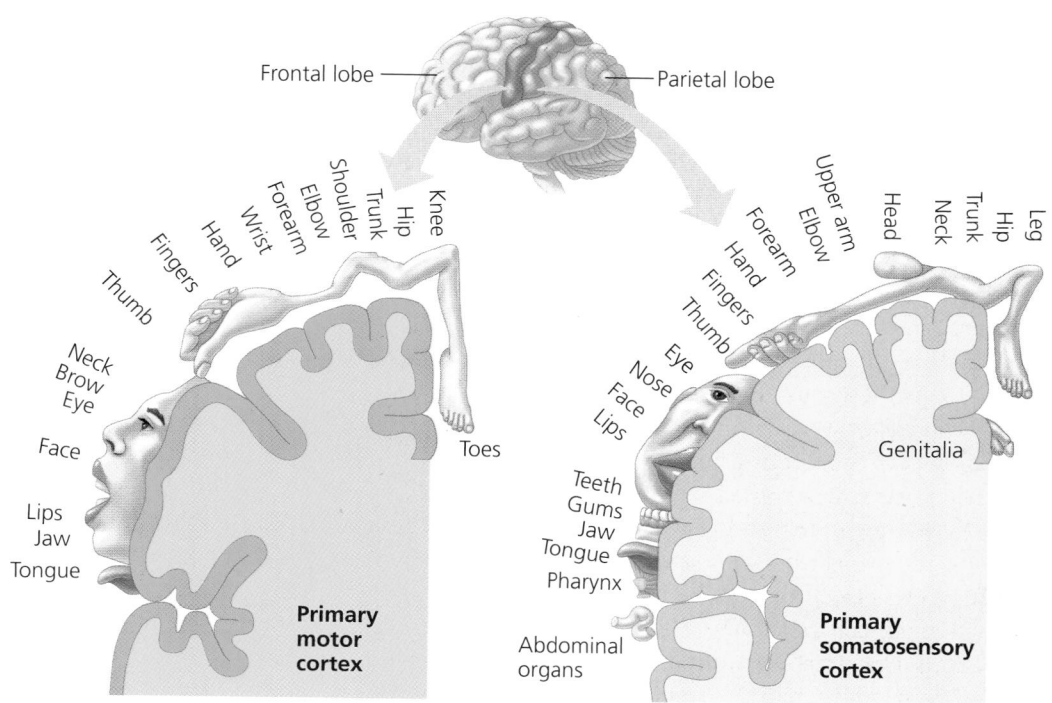

Frontal lobe

Parietal lobe

Shoulder
Trunk
Elbow
Hip
Forearm
Knee
Wrist
Hand
Fingers
Thumb

Neck
Brow
Eye
Face

Lips
Jaw
Tongue

Toes

Primary motor cortex

Upper arm
Elbow
Forearm
Trunk
Hand
Head
Fingers
Neck
Thumb
Leg
Hip

Eye
Nose
Face
Lips

Genitalia

Teeth
Gums
Jaw
Tongue
Pharynx

Abdominal organs

Primary somatosensory cortex

VISUAL SKILLS *Why is the hand larger than the forearm in both parts of this figure?*

devoted to each body part is not proportional to the size of the part. Instead, surface area correlates with the extent of neuronal control needed (for the motor cortex) or with the number of sensory neurons that extend axons to that part (for the somatosensory cortex). Thus, the surface area of the motor cortex devoted to the face is proportionately quite large, reflecting the extensive involvement of facial muscles in communication.

Language and Speech

The mapping of cognitive functions within the cortex began in the 1800s when physicians studied the effects of damage to particular regions of the cortex by injuries, strokes, or tumors. Pierre Broca conducted postmortem (after death) examinations of patients who had been able to understand language but unable to speak. He discovered that many had defects in a small region of the left frontal lobe, now known as *Broca's area*. Karl Wernicke found that damage to a posterior portion of the left temporal lobe, now called *Wernicke's area*, abolished the ability to comprehend speech but not the ability to speak. PET studies have now confirmed activity in Broca's area during speech generation and Wernicke's area when speech is heard **(Figure 49.18)**.

Lateralization of Cortical Function

Both Broca's area and Wernicke's area are located in the left cerebral hemisphere, reflecting a greater role in language for the left side of the cerebrum than for the right side. The left hemisphere is also more adept at math and logical operations. In contrast, the right hemisphere appears to dominate in the recognition of faces and patterns, spatial relations, and nonverbal thinking. This difference in function between the right and left hemispheres is called **lateralization**.

The two cerebral hemispheres normally exchange information through the fibers of the corpus callosum (see Figure 49.11). Severing this connection (a treatment of last resort for the most extreme forms of epilepsy, a seizure disorder) results in a "split-brain" effect. In such patients, the two hemispheres function independently. For example, they cannot read even a familiar word that appears only in their left field of vision: The sensory information travels from the left field of vision to the right hemisphere, but cannot then reach the language centers in the left hemisphere.

Frontal Lobe Function

In 1848, a horrific accident pointed to the role of the prefrontal cortex in temperament and decision making. Phineas Gage was the foreman of a railroad construction crew when an explosion drove an iron rod through his head. The rod, which was more than 3 cm in diameter at one end, entered his skull just below his left eye and exited through the top of his head, damaging large portions of his frontal lobe **(Figure 49.19)**. Gage recovered, but his personality changed dramatically. He became emotionally detached, impatient, and erratic in his behavior, providing evidence of the role of the prefrontal cortex in temperament and decision making.

▼ **Figure 49.19 Phineas Gage's skull injury.**

Two further sets of observations support the hypothesis that Gage's brain injury and personality change inform us about frontal lobe function. First, frontal lobe tumors cause similar symptoms: Intellect and memory seem intact, but decision making is flawed and emotional responses are diminished. Second, the same problems arise when the connection between the prefrontal cortex and the limbic system is surgically severed. (This procedure, called a frontal lobotomy, was once a common treatment for severe behavioral disorders but is no longer in use.) Together, these observations provide evidence that the frontal lobes have a substantial influence on what are called "executive functions."

Evolution of Cognition in Vertebrates

EVOLUTION In nearly all vertebrates, the brain has the same basic structures (see Figure 49.10). Given this uniform organization, how did a capacity for advanced cognition, the perception and reasoning that constitute knowledge,

▼ **Figure 49.18 Mapping language areas in the cerebral cortex.** These PET images show activity levels on the left side of one person's brain during four activities, all related to speech. Increases in activity are seen in Wernicke's area when hearing words, Broca's area when speaking words, the visual cortex when seeing words, and the prefrontal cortex when generating words (without reading them).

Hearing words

Seeing words

Speaking words

Generating words

Max

Min

evolve in certain species? For many years researchers favored the hypothesis that higher-order reasoning in vertebrates required evolution of an extensively convoluted cerebral cortex, as is found in humans, other primates, and cetaceans (whales, dolphins, and porpoises). In humans, for example, the cerebral cortex accounts for about 80% of total brain mass.

Birds lack a convoluted cerebral cortex and were long thought to have much lower intellectual capacity than primates and cetaceans. However, recent experiments have refuted this idea: Western scrub jays (*Aphelocoma californica*) can remember which food items they hid first. New Caledonian crows (*Corvus moneduloides*) are highly skilled at making and using tools, an ability otherwise well documented only for humans and some other apes. African grey parrots (*Psittacus erithacus*) understand numerical and abstract concepts, such as "same" and "different" and "none."

What brain structures enable some birds to have such sophisticated information processing? The answer appears to be a nuclear (clustered) organization of neurons within the *pallium*, the top or outer portion of the brain **(Figure 49.20a)**. Note that this arrangement is different from that in the human cerebral cortex **(Figure 49.20b)**, where six parallel layers of neurons are arranged tangential to the surface. Thus, vertebrate evolution has resulted in two different types of outer brain organization that can support complex and flexible function.

▼ **Figure 49.20 Comparison of regions for higher cognition in avian and human brains.** Although structurally different, the **(a)** pallium of a songbird brain and the **(b)** cerebral cortex of the human brain play similar roles in higher cognitive activities and make many similar connections with other brain structures.

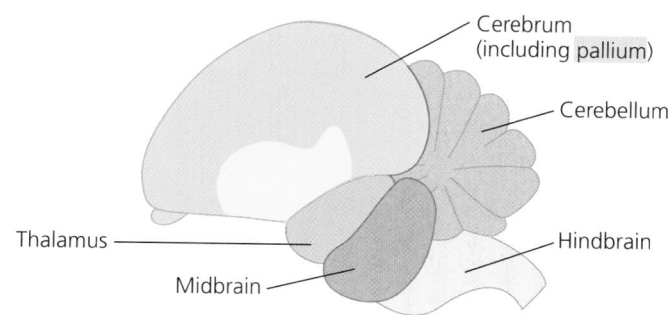

(a) Songbird brain (cross section)

(b) Human brain (cross section)

How did the bird pallium and human cerebral cortex arise during evolution? The current consensus is that the common ancestor of birds and mammals had a pallium in which neurons were organized into nuclei, as is still found in birds. Early in mammalian evolution, this clustered organization was transformed into a layered one. However, connectivity was maintained such that, for example, the thalamus relays sensory input relating to sights, sounds, and touch to the pallium in birds and to the cerebral cortex in mammals.

Sophisticated information processing depends not only on the overall organization of a brain but also on the very small-scale changes that enable learning and encode memory. We'll turn to these changes in the context of humans in the next section.

→ **Mastering Biology**
Interview with Erich Jarvis: Studying how songbirds learn melodies

---

CONCEPT CHECK 49.3

1. How can studying individuals with damage to a particular brain region provide insight into the normal function of that region?

2. How do the functions of Broca's area and Wernicke's area each relate to the activity of the surrounding cortex?

3. **WHAT IF?** If a woman with a severed corpus callosum viewed a photograph of a familiar face, first in her left field of vision and then in her right field, why would she find it difficult to put a name to the face?

For suggested answers, see Appendix A.

CONCEPT 49.4

Changes in synaptic connections underlie memory and learning

Formation of the nervous system occurs stepwise. First, regulated gene expression and signal transduction determine where neurons form in the developing embryo. Next, neurons compete for survival. Every neuron requires growth-supporting factors, which are produced in limited quantities by tissues that direct neuron growth. Neurons that don't reach the proper locations fail to receive such factors and undergo programmed cell death. The net effect is the preferential survival of neurons that are in a proper location. The competition is so severe that half of the neurons formed in the embryo are eliminated.

In the final phase of organizing the nervous system, synapse elimination takes place. During development, each neuron forms numerous synapses, more than are required for its proper function. Once a neuron begins to function, its activity stabilizes some synapses and destabilizes others. By the time the embryo completes development, more than half of all synapses have been eliminated. In humans, this elimination of unnecessary connections, a process called synaptic pruning, continues after birth and throughout childhood.

Together, neuron development, neuron death, and synapse elimination establish the basic network of cells and connections within the nervous system required throughout life.

Neuronal Plasticity

Although the overall organization of the CNS is established during embryonic development, the connections between neurons can be modified. This capacity for the nervous system to be remodeled, especially in response to its own activity, is called **neuronal plasticity**.

Much of the reshaping of the nervous system occurs at synapses. Synapses belonging to circuits that link information in useful ways are maintained, whereas those that convey bits of information lacking any context may be lost. Specifically, when the activity of a synapse coincides with that of other synapses, changes may occur that reinforce that synaptic connection. Conversely, when the activity of a synapse fails to coincide with that of other synapses, the synaptic connection sometimes becomes weaker.

Figure 49.21a illustrates how activity-dependent events can trigger the gain or loss of a synapse. If you think of signals in the nervous system as traffic on a highway, such changes are comparable to adding or removing an entrance ramp. The net effect is to increase signaling between particular pairs of neurons and decrease signaling between other pairs. Signaling at a synapse can also be strengthen or weakened, as

▼ **Figure 49.21 Neuronal plasticity.** Synaptic connections can change over time, strengthening or weakening in response to the level of activity at the synapse.

(a) High-level activity at the synapse of neuron N_1 with the postsynaptic neuron leads to recruitment of additional axon terminals from that neuron. Lack of activity at the synapse with neuron N_2 leads to loss of functional connections with that neuron.

(b) If two synapses on a postsynaptic cell are often active at the same time, the strength of both responses may increase.

shown in **Figure 49.21b**. In our traffic analogy, this would be equivalent to widening or narrowing an entrance ramp.

A defect in neuronal plasticity may underlie *autism spectrum disorder*, which results in impaired communication and social interaction, as well as stereotyped and repetitive behaviors beginning in early childhood. There is now growing evidence that autism spectrum disorder involves a disruption of activity-dependent remodeling at synapses. At the same time, extensive research has ruled out any link to vaccine preservatives, once proposed as a potential risk factor based on fraudulent data.

Although the underlying causes of autism are unknown, there is a strong genetic contribution to this and related disorders. Further understanding of the autism-associated disruption in synaptic plasticity may help efforts to better understand and treat this disorder.

Memory and Learning

Neuronal plasticity is essential to the formation of memories. We are constantly checking what is happening against what just happened. We hold information for a time in **short-term memory** and then release it if it becomes irrelevant. If we wish to retain knowledge of a name, phone number, or other fact, the mechanisms of **long-term memory** are activated. If we later need to recall the name or number, we fetch it from long-term memory and return it to short-term memory.

Short-term and long-term memory both involve the storage of information in the cerebral cortex. In short-term memory, this information is accessed via temporary links formed in the hippocampus. When memories are made long-term, the links in the hippocampus are replaced by connections within the cerebral cortex itself. As discussed earlier, some of this consolidation of memory is thought to occur during sleep. Furthermore, the reactivation of the hippocampus that is required for memory consolidation likely forms the basis for at least some of our dreams.

According to our current understanding of memory, the hippocampus is essential for acquiring new long-term memories but not for maintaining them. This hypothesis readily explains the symptoms of some individuals who suffer damage to the hippocampus: They cannot form any new lasting memories but can freely recall events from before their injury. In effect, their lack of normal hippocampal function traps them in their past. Hippocampal damage and memory loss are common in the early stages of Alzheimer's disease (see Concept 49.5).

What evolutionary advantage might be offered by organizing short-term and long-term memories differently? One hypothesis is that the delay in forming connections in the cerebral cortex allows long-term memories to be integrated gradually into the existing store of knowledge and experience, providing a basis for more meaningful associations.

Consistent with this hypothesis, the transfer of information from short-term to long-term memory is enhanced by the association of new data with data previously learned and stored in long-term memory. For example, it's easier to learn a new card game if you already have "card sense" from playing other card games.

Motor skills, such as tying your shoes or writing, are usually learned by repetition. You can perform these skills without consciously recalling the individual steps required to do these tasks correctly. Learning skills and procedures, such as those required to ride a bicycle, appears to involve cellular mechanisms very similar to those responsible for brain growth and development. In such cases, neurons actually make new connections. In contrast, memorizing phone numbers, facts, and places—which can be very rapid and may require only one exposure to the relevant item—may rely mainly on changes in the strength of existing neuronal connections. Next we will consider one way that such changes in strength can take place.

Long-Term Potentiation

In searching for the physiological basis of memory, researchers have concentrated their attention on processes that can alter a synaptic connection, making the flow of communication either more efficient or less efficient. We will focus here on **long-term potentiation (LTP)**, a lasting increase in the strength of synaptic transmission. Data indicate that LTP represents a fundamental process for memory storage and learning.

First characterized in tissue slices from the hippocampus, LTP involves a presynaptic neuron that releases the excitatory neurotransmitter glutamate. LTP involves two types of glutamate receptors, each named for a molecule—NMDA or AMPA—that can be used to artificially activate that particular receptor. As shown in **Figure 49.22**, the set of receptors present on the postsynaptic membrane changes when two conditions are met: a rapid series of action potentials at the presynaptic neuron and a depolarizing stimulus elsewhere on the postsynaptic cell. The result is LTP—a stable increase in the size of the postsynaptic potentials at a synapse whose activity coincides with that of another input.

→ **Mastering Biology** The Visual Brain: Learning and Memory

CONCEPT CHECK 49.4

1. Outline two mechanisms by which information flow between two neurons in an adult can increase.

2. Individuals with localized brain damage have been very useful in the study of many brain functions. Why is this unlikely to be true for consciousness?

3. **WHAT IF?** Suppose that a person with damage to the hippocampus is unable to acquire new long-term memories. Why might the acquisition of short-term memories also be impaired?

For suggested answers, see Appendix A.

▼ **Figure 49.22 Long-term potentiation in the brain.**

(a) Synapse prior to long-term potentiation (LTP). The NMDA glutamate receptors open in response to glutamate but are blocked by Mg^{2+}.

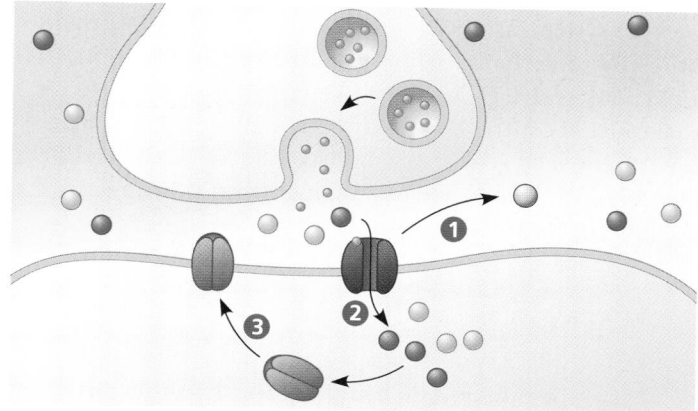

(b) Establishing LTP. Activity at nearby synapses (not shown) depolarizes the postsynaptic membrane, causing ❶ Mg^{2+} release from NMDA receptors. The unblocked receptors respond to glutamate by allowing ❷ an influx of Na^+ and Ca^{2+}. The Ca^{2+} influx triggers ❸ insertion of stored AMPA glutamate receptors into the postsynaptic membrane.

(c) Synapse exhibiting LTP. Glutamate release activates ❶ AMPA receptors that trigger ❷ depolarization. The depolarization unblocks ❸ NMDA receptors. Together, the AMPA and NMDA receptors trigger postsynaptic potentials strong enough to initiate ❹ action potentials without input from other synapses. Additional mechanisms (not shown) contribute to LTP, including receptor modification by protein kinases.

Many nervous system disorders can now be explained in molecular terms

Disorders of the nervous system, including schizophrenia, depression, drug addiction, Alzheimer's disease, and Parkinson's disease, are a major public health problem. Together, they result in more hospitalizations in the United States than do heart disease or cancer. Until recently, hospitalization was typically the only available treatment, and many affected individuals were institutionalized for the rest of their lives. Today, many disorders that alter mood or behavior can be treated with medication, reducing average hospital stays for these disorders to only a few weeks. Nevertheless, many challenges remain with regard to preventing or treating nervous system disorders, especially Alzheimer's disease and other disorders that lead to nervous system degeneration.

Major research efforts are under way to identify genes that cause or contribute to disorders of the nervous system. Identifying such genes offers hope for identifying causes, predicting outcomes, and developing effective treatments. For most nervous system disorders, however, genetic contributions only partially account for which individuals are affected. The other significant contribution to disease comes from environmental factors. Unfortunately, such environmental contributions are typically very difficult to identify.

To distinguish between genetic and environmental variables, scientists often carry out family studies. In these studies, researchers track how family members are related genetically, which individuals are affected, and which family members grew up in the same household. These studies are especially informative when one of the affected individuals has either an adopted sibling who is genetically unrelated or an identical twin, as we'll see for the disorder schizophrenia, our next topic.

Schizophrenia

Approximately 1% of the world's population suffers from **schizophrenia**, a severe mental disturbance characterized by psychotic episodes in which patients have a distorted perception of reality. People with schizophrenia typically experience hallucinations (such as "voices" that only they can hear) and delusions (for example, the idea that others are plotting to harm them). Family studies have revealed a very strong genetic component for schizophrenia. However, as shown in **Figure 49.23**, the disease is also subject to environmental influences, since an individual who shares 100% of his or her genes with a twin with schizophrenia has only a 48% chance of developing the disorder. Despite the commonly held notion, schizophrenia does not necessarily result in multiple personalities. Rather, the name *schizophrenia* (from the Greek

▼ Figure 49.23 Genetic contribution to schizophrenia. First cousins, uncles, and aunts of a person with schizophrenia have twice the risk of unrelated members of the population of developing the disease. The risks for closer relatives are many times greater.

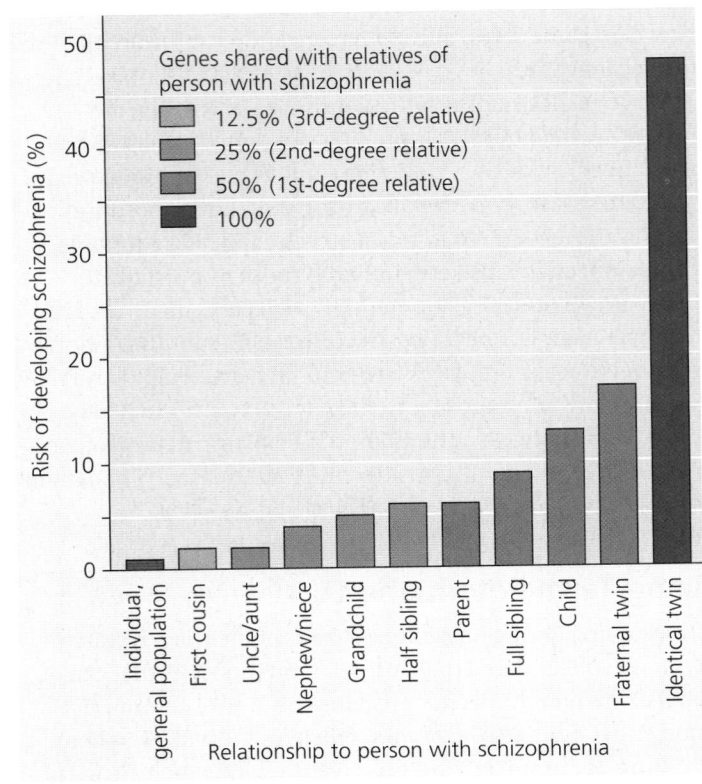

INTERPRET THE DATA *What is the likelihood of a person developing schizophrenia if the disorder affects his or her fraternal twin? How would the likelihood change if DNA sequencing revealed that the twins shared the genetic variants that contribute to the disorder?*

schizo, split, and *phren*, mind) refers to the fragmentation of what are normally integrated brain functions.

One current hypothesis is that neuronal pathways that use dopamine as a neurotransmitter are disrupted in schizophrenia. Supporting evidence comes from the fact that many drugs that alleviate the symptoms of schizophrenia block dopamine receptors. In addition, the drug amphetamine ("speed"), which stimulates dopamine release, can produce the same set of symptoms as schizophrenia. Recent genetic studies suggest a link between schizophrenia and particular forms of the complement protein C4, an immune system component.

Depression

Depression is a disorder characterized by depressed mood, as well as abnormalities in sleep, appetite, and energy level. Two broad forms of depressive illness are known: major depressive disorder and bipolar disorder. Individuals affected by **major depressive disorder** undergo periods—often lasting many months—during which once enjoyable activities provide no pleasure and provoke no interest. One of the most common nervous system disorders, major depression affects about one in every seven adults at some point, and twice as many women as men.

Bipolar disorder, or manic-depressive disorder, involves extreme swings of mood and affects about 1% of the world's population. The manic phase is characterized by high self-esteem, increased energy, a flow of ideas, overtalkativeness, and increased risk taking. In its milder forms, this phase is sometimes associated with great creativity, and some well-known artists, musicians, and literary figures (Vincent van Gogh, Robert Schumann, Virginia Woolf, and Ernest Hemingway, to name a few) have had productive periods during manic phases. The depressive phase comes with lessened motivation, sense of worth, and ability to feel pleasure, as well as sleep disturbances. These symptoms can be so severe that affected individuals attempt suicide.

Major depressive and bipolar disorders are among the nervous system disorders for which therapies are available. Many drugs used to treat depressive illness, including fluoxetine (Prozac), increase activity of biogenic amines in the brain.

The Brain's Reward System and Drug Addiction

Emotions are strongly influenced by a neuronal circuit in the brain called the *reward system*. The reward system provides motivation for activities that enhance survival and reproduction, such as eating in response to hunger, drinking when thirsty, and engaging in sexual activity when aroused. Inputs to the reward system are received by neurons in the *ventral tegmental area* (*VTA*), a region located within the midbrain (see Figure 49.11). When activated, these neurons release the neurotransmitter dopamine from their synaptic terminals **(Figure 49.24)**. Targets of this dopamine signaling include the nucleus accumbens and the prefrontal cortex.

The brain's reward system is dramatically affected by drug addiction, a disorder characterized by compulsive consumption of a drug and loss of control in limiting intake. Addictive drugs range from sedatives to stimulants and include alcohol, cocaine, and nicotine, as well as opioids, such as heroin, fentanyl and oxycodone. All enhance the activity of the dopamine pathway (see Figure 49.24). As addiction develops, there are also long-lasting changes in the reward circuitry. The result is a craving for the drug independent of any pleasure associated with consumption. In 2018, the Centers for Disease Control and Prevention reported that an average of 130 Americans died from opioid overdose every day.

Laboratory animals are highly valuable in modeling and studying addiction. Rats, for example, will provide themselves with heroin, cocaine, or amphetamine when given a dispensing system linked to a lever in their cage. Furthermore, they exhibit addictive behavior, continuing to self-administer the drug rather than seek food, even to the point of starvation.

As scientists expand their knowledge about the brain's reward system and the various forms of addiction, there is hope that the insights will lead to more effective prevention and treatment.

▼ **Figure 49.24 Effects of addictive drugs on the reward system of the mammalian brain.** Addictive drugs alter the transmission of signals in the pathway formed by neurons of the ventral tegmental area (VTA), a region located near the base of the brain.

Nicotine stimulates dopamine-releasing VTA neuron.

Inhibitory neuron

Opioids decrease activity of inhibitory neuron.

Dopamine-releasing VTA neuron

Cocaine and amphetamines block removal of dopamine from synaptic cleft.

Cerebral neuron of reward pathway

Reward system response

MAKE CONNECTIONS *Review depolarization in Concept 48.3. What effect would you expect if you depolarized the neurons in the VTA? Explain.*

Mastering Biology
Interview with Ulrike Heberlein: Research with drunk flies

Alzheimer's Disease

The condition now known as **Alzheimer's disease** is a mental deterioration, or dementia, characterized by confusion and memory loss. Its incidence is age related, rising from about 10% at age 65 to about 35% at age 85. Overall, Alzheimer's disease accounts for about two of every three cases of dementia. It is also the sixth most common cause of death among adults in the United States, affecting individuals including former President Ronald Reagan, author E.B. White, and civil rights heroine Rosa Parks.

Alzheimer's disease is progressive; patients gradually become less able to function and eventually need to be dressed, bathed, and fed by others. Individuals with Alzheimer's disease often lose their ability to recognize people and may treat even immediate family members with suspicion and hostility.

Examining the brains of individuals who have died of Alzheimer's disease reveals two characteristic

▼ **Figure 49.25 Microscopic signs of Alzheimer's disease.** A hallmark of Alzheimer's disease is the presence in brain tissue of neurofibrillary tangles surrounding plaques made of β-amyloid (LM).

Amyloid plaque   Neurofibrillary tangle   20 μm

⊙ **Mastering Biology** BBC Video: Searching for a Cure for Alzheimer's

features: amyloid plaques and neurofibrillary tangles, as shown in **Figure 49.25**. There is also often massive shrinkage of brain tissue, reflecting the death of neurons in many areas of the brain, including the hippocampus and cerebral cortex.

The plaques are aggregates of β-amyloid, an insoluble peptide that is cleaved from the extracellular portion of a membrane protein found in neurons. Membrane enzymes, called secretases, catalyze the cleavage, causing β-amyloid to accumulate in plaques outside the neurons. It is these plaques that appear to trigger the death of surrounding neurons.

The neurofibrillary tangles observed in Alzheimer's disease are primarily made up of the tau protein. (This protein is unrelated to the tau mutation that affects circadian rhythm in hamsters.) The tau protein normally helps assemble and maintain microtubules that transport nutrients along axons. In Alzheimer's disease, tau undergoes changes that cause it to bind to itself, resulting in neurofibrillary tangles. There is evidence that changes in tau are associated with the appearance of early-onset Alzheimer's disease, a much less common disorder that affects relatively young individuals. At present, there are no treatments available that block the progression of Alzheimer's disease.

Tau protein accumulation is also characteristic of a degenerative brain disease found in athletes, military veterans, and others with a history of concussion or other repetitive brain trauma. (A concussion is a brain injury caused by a blow or jolt to the head or a hit to the body that shakes the brain.) Known as *chronic traumatic encephalopathy (CTE)*, this disease was first described in the early 2000s and has now been detected in postmortem diagnosis of more than 100 men who had played professional football in the National Football League, among others.

Parkinson's Disease

Symptoms of **Parkinson's disease**, a motor disorder, include muscle tremors, poor balance, a flexed posture, and a shuffling gait. Facial muscles become rigid, limiting the ability of patients to vary their expressions. Cognitive defects may also develop. Like Alzheimer's disease, Parkinson's disease is a progressive brain illness and is more common with advancing age. The incidence of Parkinson's disease is about 1% at age 65 and about 5% at age 85. In the United States population, approximately 1 million people have Parkinson's disease.

Parkinson's disease involves the death of neurons in the midbrain that normally release dopamine at synapses in the basal nuclei. As with Alzheimer's disease, protein aggregates accumulate. Most cases of Parkinson's disease lack an identifiable cause; however, a rare form of the disease that appears in relatively young adults has a clear genetic basis. Molecular studies of mutations linked to this early-onset Parkinson's disease reveal disruption of genes required for certain mitochondrial functions. Researchers are investigating whether mitochondrial defects also contribute to the more common, later-onset form of the disease.

At present, Parkinson's disease can be treated, but not cured. Approaches used to manage the symptoms include brain surgery, deep-brain stimulation, and a dopamine-related drug, L-dopa. Unlike dopamine, L-dopa crosses the blood-brain barrier. Within the brain, the enzyme dopa decarboxylase converts the drug to dopamine, reducing the severity of Parkinson's disease symptoms:

$$HO—\bigcirc—CH_2—CH(NH_2)—CO_2H \xrightarrow{\text{Dopa decarboxylase}} HO—\bigcirc—CH_2—CH_2—NH_2$$

L-dopa → Dopamine

Future Directions in Brain Research

In 2014, the United States government launched a bold 12-year project, the BRAIN (Brain Research through Advancing Innovative Neurotechnologies) Initiative. The aim is to develop innovative technologies and drive scientific advances, similar to the major projects that accomplished landing a person on the moon and mapping the human genome. Specific BRAIN goals are to map brain circuits, measure activity within those circuits, and discover how this activity is translated into thought and behavior.

CONCEPT CHECK 49.5

1. Compare Alzheimer's disease and Parkinson's disease.
2. How is dopamine activity related to schizophrenia, drug addiction, and Parkinson's disease?
3. **WHAT IF?** If you could detect early-stage Alzheimer's disease, would you expect to see brain changes that were similar to, although less extensive than, those seen in patients who have died of this disease? Explain.

For suggested answers, see Appendix A.

49 Chapter Review

→ Go to **Mastering Biology** for Assignments, the eText, the Study Area, and Dynamic Study Modules.

SUMMARY OF KEY CONCEPTS

→ To review key terms, go to the **Vocabulary Self-Quiz** in the **Mastering Biology** eText or Study Area, or go to goo.gl/zkjz9t.

CONCEPT 49.1

Nervous systems consist of circuits of neurons and supporting cells (pp. 1086–1090)

- Invertebrate nervous systems range in complexity from simple nerve nets to highly centralized nervous systems having complicated brains and ventral nerve cords.

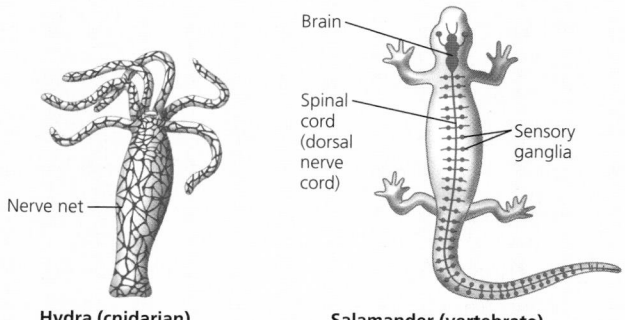

Hydra (cnidarian)   **Salamander (vertebrate)**

- In vertebrates, the **central nervous system (CNS)**, consisting of the brain and the spinal cord, integrates information, while the **nerves** of the **peripheral nervous system (PNS)** transmit sensory and motor signals between the CNS and the rest of the body. The simplest circuits control **reflex** responses, in which sensory input is linked to motor output without involvement of the brain.

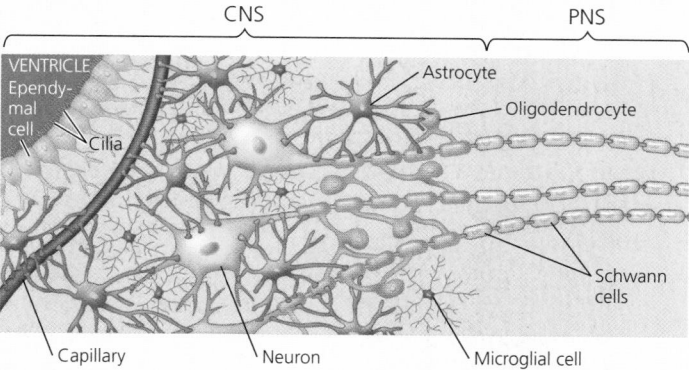

- Afferent neurons carry sensory signals to the CNS. Efferent neurons function in either the **motor system**, which carries signals to skeletal muscles, or the **autonomic nervous system**, which regulates smooth and cardiac muscles. The **sympathetic** and **parasympathetic divisions** of the autonomic nervous system have antagonistic effects on a diverse set of target organs, while the **enteric nervous system** controls the activity of many digestive organs.
- Vertebrate neurons are supported by **glia**, including astrocytes, oligodendrocytes, and Schwann cells. Some glia serve as stem cells that can differentiate into mature neurons.

? *How does the circuitry of a reflex facilitate a rapid response?*

CONCEPT 49.2

The vertebrate brain is regionally specialized (pp. 1091–1096)

- The cerebrum has two hemispheres, each of which consists of cortical **gray matter** overlying **white matter** and basal nuclei. The basal nuclei are important in planning and learning movements. The **pons** and **medulla oblongata** are relay stations for information traveling between the PNS and the cerebrum. The reticular formation, a network of neurons within the **brainstem**, regulates sleep and arousal. The **cerebellum** helps coordinate motor, perceptual, and cognitive functions. The **thalamus** is the main center through which sensory information passes to the **cerebrum**. The **hypothalamus** regulates homeostasis and basic survival behaviors. Within the hypothalamus, a group of neurons called the **suprachiasmatic nucleus (SCN)** acts as the pacemaker for circadian rhythms. The **amygdala** plays a key role in recognizing and recalling a number of emotions.

? *What roles do the midbrain, cerebellum, thalamus, and cerebrum play in vision and responses to visual input?*

CONCEPT 49.3

The cerebral cortex controls voluntary movement and cognitive functions (pp. 1096–1099)

- Each side of the **cerebral cortex** has four lobes—frontal, temporal, occipital, and parietal—that contain primary sensory areas and association areas. Association areas integrate information from different sensory areas. Broca's area and Wernicke's area are essential for generating and understanding language. These functions are concentrated in the left cerebral hemisphere, as are math and logic operations. The right hemisphere appears to be stronger at pattern recognition and nonverbal thinking.
- In the somatosensory cortex and the motor cortex, neurons are distributed according to the part of the body that generates sensory input or receives motor commands.
- Primates and cetaceans, which are capable of higher cognition, have an extensively convoluted cerebral cortex. In birds, a brain region called the pallium contains clustered nuclei that carry out functions similar to those performed by the cerebral cortex of mammals. Some birds can solve problems and understand abstractions in a manner indicative of higher cognition.

? *A patient has trouble with language and has paralysis on one side of the body. Which side would you expect to be paralyzed? Why?*

CONCEPT 49.4

Changes in synaptic connections underlie memory and learning (pp. 1099–1101)

- During development, more neurons and synapses form than will exist in the adult. The programmed death of neurons and elimination of synapses in embryos establish the basic structure of the nervous system. In the adult, reshaping of the nervous system can involve the loss or addition of synapses or the strengthening or weakening of signaling at synapses. This capacity for remodeling is termed **neuronal plasticity.** Our **short-term memory** relies on temporary links in the hippocampus. In **long-term memory**, these links are replaced by connections within the cerebral cortex.

? *Learning multiple languages is typically easier early in childhood than later in life. How does this fit with our understanding of neural development?*

CONCEPT 49.5

Many nervous system disorders can now be explained in molecular terms (pp. 1102–1104)

- **Schizophrenia,** which is characterized by hallucinations, delusions, and other symptoms, affects neuronal pathways that use dopamine as a neurotransmitter. Drugs that increase the activity of biogenic amines in the brain can be used to treat **bipolar disorder** and **major depressive disorder**. The compulsive drug use that characterizes addiction reflects altered activity of the brain's reward system, which normally provides motivation for actions that enhance survival or reproduction.
- **Alzheimer's disease** and **Parkinson's disease** are neurodegenerative and typically age related. Alzheimer's disease is a dementia in which neurofibrillary tangles and amyloid plaques form in the brain. Parkinson's disease is a motor disorder caused by the death of dopamine-secreting neurons and associated with the presence of protein aggregates.

? *The fact that both amphetamine and PCP have effects similar to the symptoms of schizophrenia suggests a potentially complex basis for this disease. Explain.*

TEST YOUR UNDERSTANDING

→ For more multiple-choice questions, go to the **Practice Test** in the **Mastering Biology** eText or Study Area, or go to goo.gl/GruWRg.

Levels 1-2: Remembering/Understanding

1. Activation of the parasympathetic branch of the autonomic nervous system
   (A) increases heart rate.
   (B) enhances digestion.
   (C) triggers release of epinephrine.
   (D) causes conversion of glycogen to glucose.

2. Which of the following structures or regions is correctly paired with its function?
   (A) limbic system—motor control of speech
   (B) medulla oblongata—homeostatic control
   (C) cerebrum—coordination of movement and balance
   (D) amygdala—short-term memory

3. Patients with damage to Wernicke's area have difficulty
   (A) coordinating limb movement.
   (B) generating speech.
   (C) recognizing faces.
   (D) understanding language.

4. The cerebral cortex plays a major role in
   (A) emotional memory.
   (B) hand-eye coordination.
   (C) circadian rhythm.
   (D) breath holding.

Levels 3-4: Applying/Analyzing

5. After suffering a stroke, a patient can see objects anywhere in front of him but pays attention only to objects in his right field of vision. When asked to describe these objects, he has difficulty judging their size and distance. What part of the brain was likely damaged by the stroke?
   (A) the left frontal lobe
   (B) the right frontal lobe
   (C) the right parietal lobe
   (D) the corpus callosum

6. Injury localized to the hypothalamus would most likely disrupt
   (A) regulation of body temperature.
   (B) short-term memory.
   (C) executive functions, such as decision making.
   (D) sorting of sensory information.

7. **DRAW IT** The reflex that pulls your hand away when you prick your finger on a sharp object relies on a neuronal circuit with two synapses in the spinal cord. (a) Using a circle to represent a cross section of the spinal cord, draw the circuit. Label the types of neurons, the direction of information flow in each, and the locations of synapses. (b) Draw a simple diagram of the brain indicating where pain would eventually be perceived.

Levels 5-6: Evaluating/Creating

8. **EVOLUTION CONNECTION** Scientists often use measures of "higher-order thinking" to assess intelligence in other animals. For example, birds are judged to have sophisticated thought processes because they can use tools and make use of abstract concepts. Identify problems you see in defining intelligence in these ways.

9. **SCIENTIFIC INQUIRY** Consider an individual who had been fluent in American Sign Language before suffering an injury to his left cerebral hemisphere. After the injury, he could still understand that sign language but could not readily generate sign language that represented his thoughts. Propose *two* hypotheses that could explain this finding. How might you distinguish between them?

10. **SCIENCE, TECHNOLOGY, AND SOCIETY** With increasingly sophisticated methods for scanning brain activity, scientists are developing the ability to detect an individual's particular emotions and thought processes from outside the body. What benefits and problems do you envision when such technology becomes readily available? Explain.

11. **WRITE ABOUT A THEME: INFORMATION** In a short essay (100–150 words), explain how specification of the adult nervous system by the genome is incomplete.

12. **SYNTHESIZE YOUR KNOWLEDGE**

Imagine you are standing at a microphone in front of a crowd. Checking your notes, you begin speaking. Using the information in this chapter, describe the series of events in particular regions of the brain that enabled you to say the very first word.

For selected answers, see Appendix A.

50 Sensory and Motor Mechanisms

KEY CONCEPTS

50.1 Sensory receptors transduce stimulus energy and transmit signals to the central nervous system *p. 1108*

50.2 In hearing and equilibrium, mechanoreceptors detect moving fluid or settling particles *p. 1112*

50.3 The diverse visual receptors of animals depend on light-absorbing pigments *p. 1117*

50.4 The senses of taste and smell rely on similar sets of sensory receptors *p. 1123*

50.5 The physical interaction of protein filaments is required for muscle function *p. 1125*

50.6 Skeletal systems transform muscle contraction into locomotion *p. 1132*

Study Tip

Diagram a process: To help you understand the sliding-filament mechanism of muscle contraction, fill in the remaining steps of the process. For each step, draw the position of the myosin head and list 1) the form of ATP or its components bound; 2) the energy state (low/high); and 3) whether a cross-bridge forms at that step.

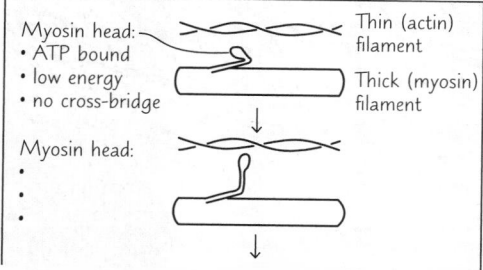

Myosin head:
• ATP bound
• low energy
• no cross-bridge

Thin (actin) filament

Thick (myosin) filament

Myosin head:
•
•
•

→ Go to Mastering Biology

For Students (in eText and Study Area)
• Get Ready for Chapter 50
• Figure 50.18 Walkthrough: Response of a Photoreceptor Cell to Light
• BioFlix® Animation: Muscle Contraction

For Instructors to Assign (in Item Library)
• Activity: Structure and Function of the Eye
• Tutorial: Energy Costs of Locomotion

Figure 50.1 Tunneling beneath North American wetlands, the star-nosed mole (*Condylura cristata*) finds its prey in almost total darkness. To do so, the mole relies on a star-shaped cluster of appendages, each with 25,000 touch-sensitive receptors. Relaying signals to the mole's brain, the receptors initiate information processing that enables the animal to capture and eat its prey in a fraction of a second.

What steps link sensory stimuli to animal activity?

Mechanoreceptors
Sound, touch, motion

Electromagnetic receptors
Light, electricity

Pain receptors
Noxious chemicals, temperatures

Chemoreceptors
Solutes, tastes, smells

Thermoreceptors
Heat, cold

Sensory receptors respond to specific stimuli by transmitting signals...

Sensory information encoded as action potentials is sent along specific paths (axons).

...that the central nervous system decodes, processes, and responds to...

Processing center

Motor commands encoded as action potentials are sent along specific paths (axons).

...in the form of motor output.

Muscles
Locomotion, posture

Glands
Secretions

Sensory receptors transduce stimulus energy and transmit signals to the central nervous system

All sensory processes begin with stimuli, and all stimuli represent forms of energy. A sensory receptor converts stimulus energy to a change in membrane potential, thereby regulating the output of action potentials to the central nervous system (CNS). Decoding of this information within the CNS results in sensation.

When a stimulus is received and processed by the nervous system, a motor response may be generated. One of the simplest stimulus-response circuits is a reflex, such as the knee-jerk reflex shown in Figure 49.5. For many other behaviors, sensory input undergoes more elaborate processing. As an example, consider how the star-nosed mole in Figure 50.1 searches for food, or forages, in a tunnel. When the mole's nose contacts an object, touch receptors in the nose are activated **(Figure 50.2)**. These receptors transmit sensory information about the object, such as whether the object is moving, to the mole's brain. Circuits in the brain integrate the input and initiate one of two response pathways. If prey

or other food is detected, the brain sends motor output commands to skeletal muscles that cause the jaws to bite down. If no food is detected, the brain sends instructions to skeletal muscles to continue movement along the tunnel.

With this overview in mind, let's examine the general organization and activity of animal sensory systems. We'll focus on four basic functions common to sensory pathways: sensory reception, transduction, transmission, and perception.

Sensory Reception and Transduction

A sensory pathway begins with **sensory reception**, the detection of a stimulus by specialized sensory cells. Each sensory cell is either a neuron or a cell that regulates a neuron **(Figure 50.3)**. Some sensory cells exist singly; others are collected in sensory organs, such as the star-shaped nose of the mole in Figure 50.1.

The term **sensory receptor** describes a sensory cell or organ, as well as the subcellular structure that detects stimuli. Some sensory receptors respond to stimuli from within the body, such as blood pressure and body position. Other receptors detect stimuli from outside the body, such as heat, light, pressure, or chemicals. Some of these receptors are sensitive to the smallest possible unit of stimulus. Most light receptors, for example, can detect a single quantum (photon) of light.

Although animals use a range of sensory receptors to detect widely varying stimuli, the effect in all cases is to open or close ion channels. The resulting change in the flow of ions across

▼ **Figure 50.2 A simple response pathway: foraging by a star-nosed mole.**

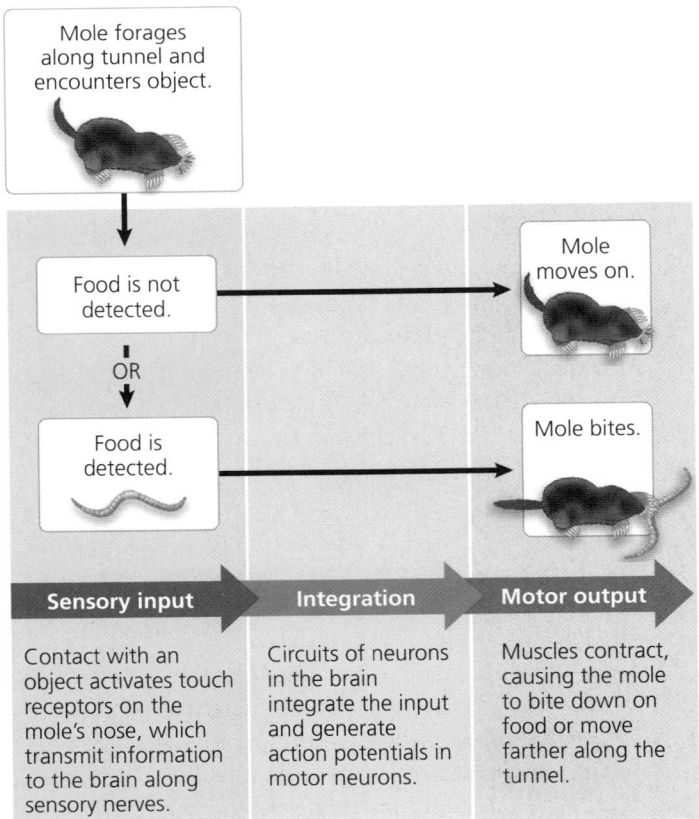

▼ **Figure 50.3 Neuronal and non-neuronal sensory receptors.**

the membrane alters the membrane potential. The change in membrane potential is called a **receptor potential**, and conversion of the stimulus to a receptor potential is known as **sensory transduction**. Note that receptor potentials are graded potentials: Their magnitude varies with the strength of the stimulus.

Transmission

Sensory information travels through the nervous system as action potentials. A sensory receptor that is also a neuron generates action potentials that travel along an axon extending into the CNS (see Figure 50.3). In contrast, a non-neuronal sensory receptor does not itself generate action potentials, but instead conveys information to an afferent neuron via a chemical synapse. Because this chemical signaling alters the rate at which the afferent neurons produce action potentials, the sensory information generated also enters the CNS in the form of action potentials.

The size of a receptor potential increases with the intensity of the stimulus. If the receptor is a sensory neuron, a larger receptor potential results in more frequent action potentials **(Figure 50.4)**. If the receptor is not a sensory neuron, a larger receptor potential usually causes the receptor to release more neurotransmitter.

Many sensory neurons spontaneously generate action potentials at a low rate. In these neurons, a stimulus does not switch the production of action potentials on or off, but instead changes *how often* an action potential is produced, alerting the nervous system to changes in stimulus intensity.

Processing of sensory information can occur before, during, and after transmission of action potentials to the CNS. In many cases, the *integration* of sensory information begins as soon as the information is received. Receptor potentials produced by stimuli delivered to different parts of a sensory receptor cell are integrated through summation, as are postsynaptic potentials in neurons that receive input from multiple sensory receptors. As you'll see shortly, sensory structures such as eyes also provide higher levels of integration, and the brain further processes all incoming signals.

Perception

When action potentials reach the brain via afferent neurons, circuits of neurons process this input, generating the **perception** of the stimulus. An action potential triggered by light striking the eye has the same properties as an action potential triggered by air vibrating in the ear. How, then, do we distinguish sights, sounds, and other stimuli? The answer lies in the connections that link sensory receptors to the brain. Action potentials from sensory receptors travel along neurons that are dedicated to a particular stimulus; these dedicated neurons form synapses with particular neurons in the brain or spinal cord. As a result, the brain distinguishes stimuli such as sight and sound solely by the path along which the action potentials have arrived.

Perceptions—such as colors, smells, sounds, and tastes—are constructions formed in the brain and do not exist outside it. So if a tree falls and no animal is present to hear it, is there a sound? The falling tree certainly produces pressure waves in the air, but if sound is defined as a perception, then there is no sound unless an animal senses the waves and its brain perceives them.

Amplification and Adaptation

The transduction of stimuli by sensory receptors is subject to two types of modification—amplification and adaptation. **Amplification** refers to the strengthening of a sensory signal during transduction. The effect of amplification can be considerable. For example, an action potential conducted from the eye to the human brain has about 100,000 times as much energy as the few photons of light that triggered it.

Amplification that occurs in sensory receptor cells often requires signal transduction that involves enzyme-catalyzed reactions (see Concept 11.3). Because a single enzyme molecule catalyzes the formation of many product molecules, these pathways amplify signal strength considerably. Amplification may also take place in accessory structures of a sense organ. For example, the lever system formed by three small bones in the ear enhances the pressure associated with sound waves more than 20-fold before the stimulus reaches receptors in the innermost part of the ear.

Upon continued stimulation, many receptors undergo a decrease in responsiveness known as **sensory adaptation** (not to be confused with the evolutionary term *adaptation*). Sensory adaptation has very important roles in our perception of ourselves and our surroundings. Without it, you

▼ **Figure 50.4 Coding of stimulus intensity by a single sensory receptor.**

Gentle pressure

Sensory receptor

Low frequency of action potentials per receptor

More pressure

High frequency of action potentials per receptor

would feel each beat of your heart and be constantly aware of every bit of clothing on your body. Furthermore, adaptation is essential for you to see, hear, and smell changes in external stimuli that vary widely in intensity.

Types of Sensory Receptors

Sensory receptors fall into five categories based on the nature of the stimuli they transduce: mechanoreceptors, chemoreceptors, electromagnetic receptors, thermoreceptors, and pain receptors.

Mechanoreceptors

Our senses of hearing and balance, as well as our responses to pressure, touch, stretch, and motion, rely on sensory receptors called **mechanoreceptors**, which sense physical deformation caused by forms of mechanical energy. Mechanoreceptors typically consist of ion channels that are linked to structures that extend outside the cell, such as "hairs" (cilia), and also anchored to internal cell structures, such as the cytoskeleton. Bending or stretching of the external structure generates tension that alters ion channel permeability. This change in turn alters the membrane potential, resulting in a receptor potential—a depolarization or hyperpolarization.

The vertebrate stretch receptor, a mechanoreceptor that detects muscle movement, triggers the familiar knee-jerk reflex (see Figure 49.5). Stretch receptors in vertebrates are dendrites of sensory neurons that spiral around the middle of certain small skeletal muscle fibers. When the muscle fibers are stretched, the sensory neurons depolarize, triggering nerve impulses that reach the spinal cord, activate motor neurons, and generate a reflex response.

Mechanoreceptors that are the dendrites of sensory neurons are also responsible for the mammalian sense of touch. Touch receptors are often embedded in layers of connective tissue. The structure of the connective tissue and the location of the receptors dramatically affect the type of mechanical energy (light touch, vibration, or strong pressure) that best stimulates them **(Figure 50.5)**. Receptors that detect a light touch or vibration are close to the surface of the skin; they transduce very slight inputs of mechanical energy into receptor potentials. Receptors that respond to stronger pressure and vibrations are in deep skin layers.

Some animals use mechanoreceptors to literally get a feel for their environment. For example, cats and many rodents have extremely sensitive mechanoreceptors at the base of their whiskers. Like the appendages on the face of the star-nosed mole, whiskers act as touch organs. Deflection of different whiskers triggers action potentials that reach different cells in the brain. As a result, an animal's whiskers enable the brain to assemble a "touch map" detailing the location of nearby objects such as food or obstacles.

▼ **Figure 50.5 Sensory receptors in human skin.** Most receptors in the dermis are encapsulated by connective tissue. Receptors in the epidermis are naked dendrites, as are hair movement receptors that wind around the base of hairs in the dermis.

Receptors that detect gentle pressure and vibration are close to the surface of the skin.

Naked dendrites in the epidermis respond to temperature or noxious stimuli.

Connective tissue

Hair

Epidermis

Dermis

Hypodermis

Nerve

Naked dendrites wind around the base of hairs and detect hair movement.

Receptors that respond to strong pressure are in deep skin layers.

Chemoreceptors

Chemoreceptors that monitor the internal environment fall into two broad categories. Some transmit information about overall solute concentration. For example, osmoreceptors in the mammalian brain detect changes in the total solute concentration of the blood and stimulate thirst when osmolarity increases (see Figure 44.19). Other chemoreceptors respond to specific molecules in body fluids, including glucose, oxygen, carbon dioxide, and amino acids.

Animals also use chemoreceptors to detect stimuli in their diet and in the environment they occupy. The antennae of the male silkworm moth contain two of the most sensitive and specific chemoreceptors known **(Figure 50.6)**; these receptors can detect components of the sex pheromone released by a female moth several kilometers away. For pheromones and other molecules detected by chemoreceptors, the stimulus molecule binds to the specific receptor on the membrane of the sensory cell and initiates changes in ion permeability.

Electromagnetic Receptors

An **electromagnetic receptor** detects a form of electromagnetic energy, such as light, electricity, and magnetism. For instance, the platypus has electroreceptors on its bill

▼ Figure 50.6 Chemoreceptors in an insect. The antennae of the male silkworm moth *Bombyx mori* are covered with sensory hairs, visible in the SEM enlargement. The hairs have chemoreceptors that are highly sensitive to the sex pheromone released by the female.

▼ Figure 50.7 Examples of electromagnetic reception and thermoreception.

(a) Some migrating animals, such as these beluga whales, apparently sense Earth's magnetic field and use the information, along with other cues, for orientation.

(b) This rattlesnake and other pit vipers have a pair of heat-sensing pit organs, one anterior to and just below each eye. These organs are sensitive enough to detect the infrared radiation emitted by a warm prey a meter away. The snake moves its head from side to side until the radiation is detected equally by the two pit organs, indicating that the prey is straight ahead.

that can detect the electric field generated by the muscles of crustaceans, small fishes, and other prey. In a few cases, the animal detecting the stimulus is also its source: Some fishes generate electric currents and then use electroreceptors to locate prey or other objects that disturb those currents.

Many animals can use Earth's magnetic field lines to orient themselves as they migrate **(Figure 50.7a)**. In 2015, researchers identified a pair of proteins that appear to act as a sensor for the Earth's magnetic field in many animals that can orient to it, including monarch butterflies, pigeons, and minke whales. One of these proteins binds iron; the other is a receptor that is sensitive to electromagnetic radiation.

Thermoreceptors

Thermoreceptors detect heat and cold. For example, certain venomous snakes rely on thermoreceptors to detect the infrared radiation emitted by warm prey. These thermoreceptors are located in a pair of pit organs on the snake's head **(Figure 50.7b)**. In humans, thermoreceptors are located in the skin and in the anterior hypothalamus.

Recently, our understanding of thermoreception has increased substantially, thanks to scientists with an appreciation for fiery foods. Jalapeno and cayenne peppers that we describe as "hot" contain a substance called capsaicin. Applying capsaicin to a sensory neuron causes an influx of calcium ions. When scientists identified the receptor protein in neurons that binds capsaicin, they made a fascinating discovery: The receptor opens a calcium channel in response not only to capsaicin but also to high temperatures (42°C or

higher). In essence, spicy foods taste "hot" because they activate the same receptors as hot soup and coffee.

Mammals have a variety of thermoreceptors, each specific for a particular temperature range. The capsaicin receptor and at least five other types of thermoreceptors belong to the TRP (transient receptor potential) family of ion channel proteins. Just as the TRP-type receptor specific for high temperature is sensitive to capsaicin, the receptor for low temperatures (below 28°C) can be activated by menthol, a plant product that we perceive to have a "cool" flavor.

Pain Receptors

Extreme pressure or temperature, as well as certain chemicals, can damage animal tissues. To detect stimuli that reflect such noxious (harmful) conditions, animals rely on **nociceptors** (from the Latin *nocere*, to hurt), also called **pain receptors**. By triggering defensive reactions, such as withdrawal from danger, the perception of pain serves an important function. The capsaicin receptor of mammals can detect dangerously high temperatures, so it also functions as a pain receptor.

Chemicals produced in an animal's body sometimes enhance the perception of pain. For example, damaged tissues produce prostaglandins, which act as local regulators of inflammation (see Concept 45.1). Prostaglandins worsen pain by increasing nociceptor sensitivity to noxious stimuli. Aspirin and ibuprofen reduce pain by inhibiting the synthesis of prostaglandins.

Next we'll turn our focus to sensory systems, beginning with systems for maintaining balance and detecting sound.

CONCEPT CHECK 50.1

1. Which one of the five categories of sensory receptors is primarily dedicated to external stimuli?
2. Why can eating "hot" peppers cause a person to sweat?
3. **WHAT IF?** If you stimulated a sensory neuron electrically, how would that stimulation be perceived?

For suggested answers, see Appendix A.

CONCEPT 50.2

In hearing and equilibrium, mechanoreceptors detect moving fluid or settling particles

For most animals, the sense of hearing is closely related to the sense of balance, the perception of body equilibrium. For both senses, mechanoreceptor cells produce receptor potentials in response to deflection of cell-surface structures by settling particles or moving fluid.

Sensing of Gravity and Sound in Invertebrates

To sense gravity and maintain equilibrium, most invertebrates rely on mechanoreceptors located in organs called **statocysts (Figure 50.8)**. In a typical statocyst, **statoliths**,

▼ **Figure 50.8 The statocyst of an invertebrate.** The settling of granules called statoliths to the low point in the chamber bends cilia on receptor cells in that location, providing the brain with information about the orientation of the body with respect to gravity.

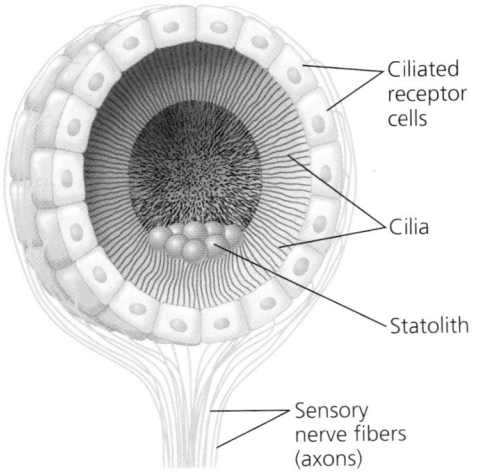

Ciliated receptor cells

Cilia

Statolith

Sensory nerve fibers (axons)

▼ **Figure 50.9 An insect's "ear"—on its leg.** The tympanic membrane, visible in this SEM of a cricket's front leg, vibrates in response to sound waves. The vibrations stimulate mechanoreceptors attached to the inside of the tympanic membrane.

Tympanic membrane

1 mm

granules formed by grains of sand or other dense materials, sit freely in a chamber lined with ciliated cells. Each time an animal repositions itself, the statoliths resettle, stimulating mechanoreceptors at the low point in the chamber.

How did researchers test the hypothesis that resettling of statoliths provides information about body position relative to Earth's gravity? In one key experiment, statoliths were replaced with metal shavings. Researchers then "tricked" crayfish into swimming upside down by using magnets to pull the shavings to the upper end of statocysts at the base of their antennae.

Many (perhaps most) insects have body hairs that vibrate in response to sound waves. Hairs differing in stiffness and length vibrate at different frequencies. For example, fine hairs on the antennae of a male mosquito vibrate in a specific way in response to the hum produced by the beating wings of flying females. The importance of this sensory system in attracting males to a potential mate can be readily demonstrated: A tuning fork vibrating at the same frequency as that of a female's wings by itself attracts males.

Many insects also detect sound by means of vibration-sensitive organs, which consist in some species of a tympanic membrane (eardrum) stretched over an internal air chamber **(Figure 50.9)**. Cockroaches lack such a tympanic membrane, but instead have vibration-sensitive organs that sense air movement, such as that caused by a descending human foot.

Hearing and Equilibrium in Mammals

In mammals, as in most other terrestrial vertebrates, the sensory organs for hearing and equilibrium are closely associated. **Figure 50.10** explores the structure and function of these organs in the human ear.

1 Overview of Ear Structure

The **outer ear** consists of the external pinna and the auditory canal, which collect sound waves and channel them to the **tympanic membrane** (eardrum), a thin tissue that separates the outer ear from the **middle ear**. In the middle ear, three small bones—the malleus (hammer), incus (anvil), and stapes (stirrup)—transmit vibrations to the **oval window**, which is a membrane beneath the stapes. The middle ear also opens into the **Eustachian tube**, a passage that connects to the pharynx and equalizes pressure between the middle ear and the atmosphere. The **inner ear** consists of fluid-filled chambers, including the **semicircular canals**, which function in equilibrium, and the coiled **cochlea** (from the Latin meaning "snail"), a bony chamber that is involved in hearing.

⊙ **Mastering Biology** Animation: **The Human Ear**

2 The Cochlea

The cochlea, shown here in cross section, has two large canals—an upper vestibular canal and a lower tympanic canal—separated by a smaller cochlear duct. Both canals are filled with fluid.

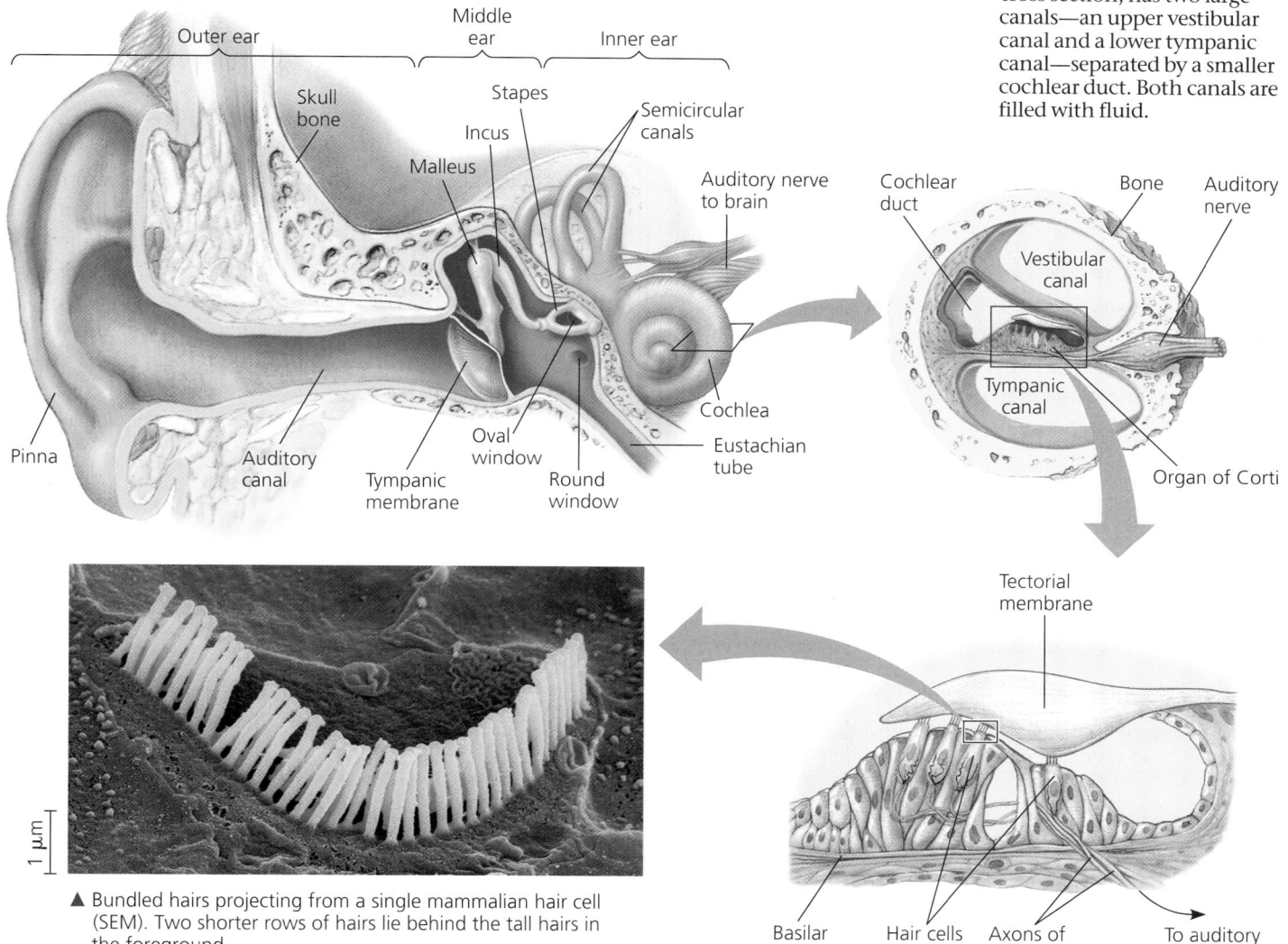

▲ Bundled hairs projecting from a single mammalian hair cell (SEM). Two shorter rows of hairs lie behind the tall hairs in the foreground.

4 Hair Cell

Within the bundle of hairs projecting from each hair cell lies a core of actin filaments. Vibration of the basilar membrane in response to sound raises and lowers the hair cells, bending the hairs against the surrounding fluid and the tectorial membrane. Displacing the hairs causes a change in the membrane potential of the hair cell.

3 The Organ of Corti

The floor of the cochlear duct, the basilar membrane, is the base of the **organ of Corti**, which contains the mechanoreceptors of the ear—hair cells with bundles of rod-shaped "hairs" projecting into the cochlear duct. Many of the hairs are attached to the tectorial membrane, which hangs over the organ of Corti like an awning.

Hearing

Vibrating objects, such as a plucked guitar string or the vocal cords of a person who is speaking, create pressure waves in the surrounding air. In *hearing*, the ear transduces this mechanical stimulus (pressure waves) into nerve impulses that the brain perceives as sound. To hear music, speech, or other sounds in our environment, we rely on **hair cells**, sensory cells with hairlike projections that detect motion.

Before vibration waves reach hair cells, they are amplified and transformed by accessory structures. The first steps involve structures in the ear that convert the vibrations of moving air to pressure waves in fluid. Moving air that reaches the outer ear causes the tympanic membrane to vibrate. The three bones of the middle ear transmit these vibrations to the oval window, a membrane on the cochlea's surface. When one of those bones, the stapes, vibrates against the oval window, it creates pressure waves in the fluid inside the cochlea.

Upon entering the vestibular canal, fluid pressure waves push down on the cochlear duct and basilar membrane. In response, the basilar membrane and attached hair cells vibrate up and down. The hairs projecting from the hair cells are deflected by the fixed tectorial membrane, which lies above (see Figure 50.10). With each vibration, the hairs bend first in one direction and then the other, causing ion channels in the hair cells to open or close. Bending in one direction depolarizes hair cells, increasing neurotransmitter release and the frequency of action potentials directed to the brain along the auditory nerve **(Figure 50.11)**. Bending the hairs in the other direction hyperpolarizes hair cells,

reducing neurotransmitter release and the frequency of auditory nerve sensations.

What prevents pressure waves from reverberating within the ear and causing prolonged sensation? After propagating through the vestibular canal, pressure waves pass around the apex (tip) of the cochlea and dissipate as they strike the **round window (Figure 50.12a)**. This damping of sound waves resets the apparatus for the next vibrations that arrive.

The ear captures information about two important sound variables: volume and pitch. *Volume* (loudness) is determined by the amplitude, or height, of the sound wave. A large-amplitude wave causes more vigorous vibration of the basilar membrane, greater bending of the hairs on hair cells, and more action potentials in the afferent neurons that transmit information to the brain. *Pitch* is determined by a sound wave's frequency, the number of vibrations per unit time. The detection of sound wave frequency takes place in the cochlea and relies on the asymmetric structure of that organ.

The cochlea can distinguish pitch because the basilar membrane is not uniform along its length: It is relatively narrow and stiff near the oval window and wider and more flexible at the apex at the base of the cochlea. Each region of the basilar membrane is tuned to a different vibration frequency **(Figure 50.12b)**. Furthermore, each region is connected by axons to a different location in the cerebral cortex. Consequently, when a sound wave causes vibration of a particular region of the basilar membrane, nerve impulses are transduced to a specific site in our cortex and we perceive sound of a particular pitch.

▼ **Figure 50.11 Sensory reception by hair cells.** In hearing or balance, each hair cell forms a synapse with an afferent neuron that conducts action potentials to the CNS. Bending of the hairs of the hair cell in one direction depolarizes the cell. Depolarization increases release of excitatory neurotransmitter, resulting in more frequent action potentials in the afferent neuron. Bending the hairs in the other direction decreases neurotransmitter release, reducing action potential frequency in the afferent neuron.

(a) No bending of hairs

(b) Bending of hairs in one direction

(c) Bending of hairs in other direction

▼ **Figure 50.12 Sensory transduction in the cochlea.**

(a) Vibrations of the stapes against the oval window produce pressure waves (black arrows) in the fluid (perilymph; blue) of the cochlea. (For purposes of illustration, the cochlea on the right is drawn partially uncoiled.) The waves travel to the apex via the vestibular canal and back towards the base via the tympanic canal. The energy in the waves causes the basilar membrane (pink) to vibrate, stimulating hair cells (not shown). Because the basilar membrane varies in stiffness along its length, each point along the membrane vibrates maximally in response to waves of a particular frequency.

(b) These graphs show the patterns of vibration along the basilar membrane for three different frequencies, high (top), medium (middle), and low (bottom). The higher the frequency, the closer the vibration to the oval window.

INTERPRET THE DATA *A musical chord consists of several notes, each formed by a sound wave of different frequency. If a chord had notes with frequencies of 100, 1,000, and 6,000 Hz, what would happen to the basilar membrane? How would this result in your hearing a chord?*

➔ **Mastering Biology Animation: Detecting Pitch**

Equilibrium

Several organs in the inner ear of humans and most other mammals detect body movement, position, and equilibrium. For example, the chambers called the *utricle* and *saccule* allow us to perceive position with respect to gravity as well as linear movement **(Figure 50.13)**. Situated in a vestibule behind the oval window, each of these chambers contains hair cells that project into a gelatinous material. Embedded in this gel are small calcium carbonate particles called *otoliths* ("ear stones"). When you tilt your head, the otoliths shift position, contacting a different set of hairs protruding into the gel. The hair cell receptors transform this deflection into a change in neurotransmitter output. This alters the activity of afferent neurons, signaling the brain that your head is at an angle. The otoliths are also responsible for your ability to perceive acceleration, such as when a stationary car in which you are sitting moves forward.

▼ **Figure 50.13 Organs of equilibrium in the inner ear.**

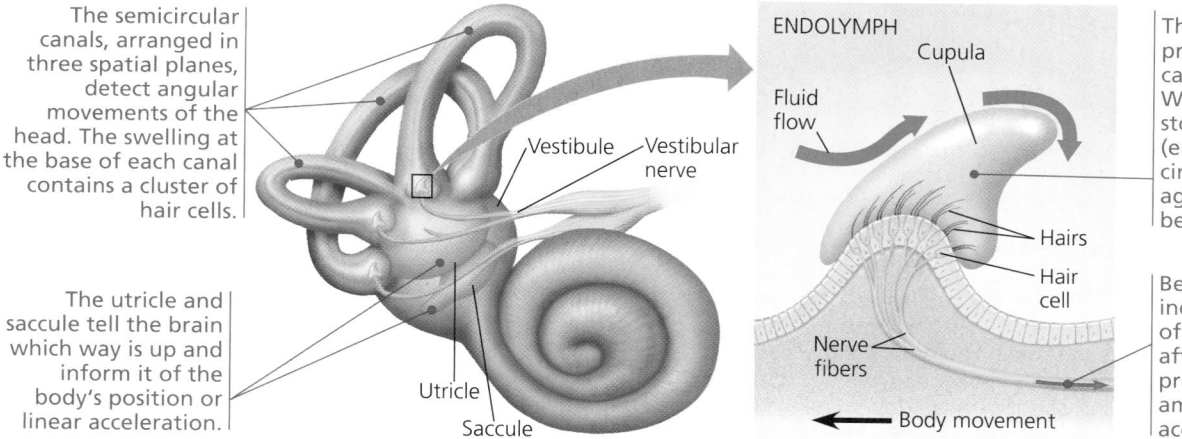

The semicircular canals, arranged in three spatial planes, detect angular movements of the head. The swelling at the base of each canal contains a cluster of hair cells.

The utricle and saccule tell the brain which way is up and inform it of the body's position or linear acceleration.

The hairs of the hair cells project into a gelatinous cap called the cupula. When the head starts or stops rotating, the fluid (endolymph) in the semicircular canals presses against the cupula, bending the hairs.

Bending of the hairs increases the frequency of action potentials in afferent neurons in direct proportion to the amount of rotational acceleration.

Three fluid-filled semicircular canals connected to the utricle detect turning of the head and other rotational acceleration. Within each canal, the hair cells form a cluster, with the hairs projecting into a gelatinous cap called a cupula (see Figure 50.13). Because the three canals are arranged in the three spatial planes, they can detect angular motion of the head in any direction. If you spin in place, the fluid in each canal eventually comes to equilibrium and remains in that state until you stop. At that point, the moving fluid encounters a stationary cupula, triggering the false sensation of angular motion that we call dizziness.

Hearing and Equilibrium in Other Vertebrates

Fishes rely on several systems for detecting movement and vibrations in their aquatic environment. One system involves a pair of inner ears that contain otoliths and hair cells. Unlike the ears of mammals, these ears have no eardrum, cochlea, or opening to the outside of the body. Instead, the vibrations of the water caused by sound waves are conducted to the inner ear through the skeleton of the head. Some fishes also have a series of bones that conduct vibrations from the swim bladder to the inner ear.

Most fishes and aquatic amphibians are able to detect low-frequency waves by means of a **lateral line system** along each side of their body **(Figure 50.14)**. As in our semicircular canals, receptors are formed from a cluster of hair cells whose hairs are embedded in a cupula. Water entering the lateral line system through numerous pores bends the cupula, leading to depolarization of the hair cells and production of action potentials. In this way, the fish perceives its movement through water or the direction and velocity of water currents flowing over its body. The lateral line system also detects water movements or vibrations generated by prey, predators, and other moving objects.

In the ear of a frog or toad, sound vibrations in the air are conducted to the inner ear by a tympanic membrane on the body surface and a single middle ear bone. The same is true in birds and other reptiles, although they, like mammals, have a cochlea.

CONCEPT CHECK 50.2

1. How are otoliths adaptive for burrowing mammals, such as the star-nosed mole?

2. **WHAT IF?** Suppose a series of pressure waves in your cochlea caused a vibration of the basilar membrane that moved gradually from the apex toward the base. How would your brain interpret this stimulus?

3. **WHAT IF?** If the stapes became fused to the other middle ear bones or to the oval window, how would this condition affect hearing? Explain.

4. **MAKE CONNECTIONS** Plants use statoliths to detect gravity (see Figure 39.22). How do plants and animals differ with regard to the type of compartment in which statoliths are found and the physiological mechanism for detecting their response to gravity?

For suggested answers, see Appendix A.

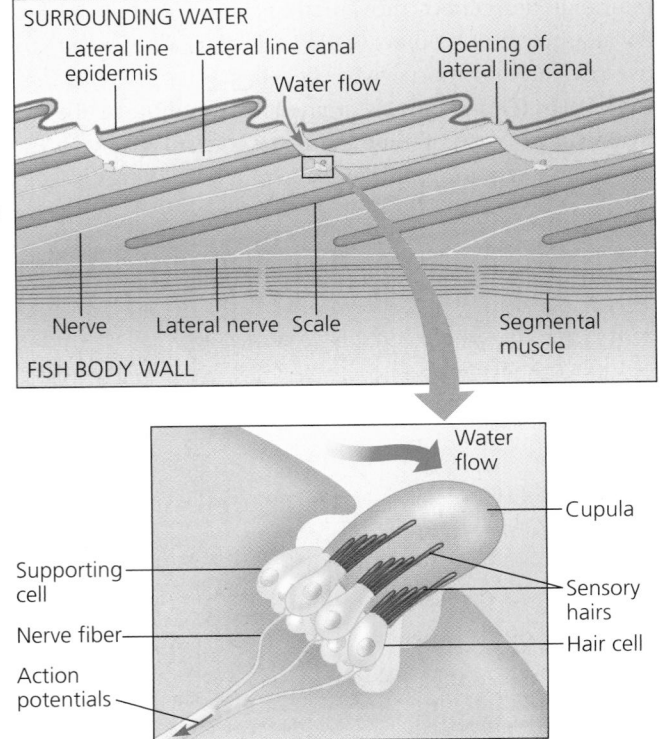

▲ **Figure 50.14 The lateral line system in a fish.** The sensory organs of the lateral line stretch from head to tail along each side of the fish. Water movement into and through the lateral line canals pushes on the gelatinous cupula, bending the hair cells within. In response, the hair cells generate receptor potentials, triggering action potentials that are conveyed to the brain. This information enables a fish to monitor water currents, any pressure waves produced by moving objects, and any low-frequency sounds conducted through the water.

The diverse visual receptors of animals depend on light-absorbing pigments

The ability to detect light has a central role in the interaction of nearly all animals with their environment. Although the organs used for vision vary considerably among animals, the underlying mechanism for capturing light is the same, suggesting a common evolutionary origin.

Evolution of Visual Perception

EVOLUTION Light detectors in the animal kingdom range from simple clusters of cells that detect only the direction and intensity of light to complex organs that form images. These diverse light detectors all contain **photoreceptors**, sensory cells that contain light-absorbing pigment molecules. Furthermore, the genes that specify where and when photoreceptors arise during embryonic development are shared among flatworms, annelids, arthropods, and vertebrates. It is thus very probable that the genetic underpinnings of all photoreceptors were already present in the earliest bilaterian animals.

→ **Mastering Biology** BBC Video: How Did Eyes Evolve?

Light-Detecting Organs

Most invertebrates have some kind of light-detecting organ. One of the simplest is that of planarians **(Figure 50.15)**.

▼ **Figure 50.15 Ocelli and orientation behavior of a planarian.**

(a) The planarian's brain directs the body to turn until the sensations from the two ocelli are equal and minimal, causing the animal to move away from light.

(b) Whereas light striking the front of an ocellus excites the photoreceptors, light striking the back is blocked by the screening pigment. In this way, the ocelli indicate the direction of a light source, enabling the light avoidance behavior.

A pair of ocelli (singular, *ocellus*), sometimes called eye-spots, are located in the head region. Photoreceptors in each ocellus receive light only through an opening where there are no pigmented cells. By comparing the rate of action potentials coming from the two ocelli, the planarian can move away from a light source until it reaches a shaded location, where the object providing the shade may hide it from predators.

Compound Eyes

Insects, crustaceans, and some polychaete worms have **compound eyes**, each consisting of up to several thousand light detectors called **ommatidia (Figure 50.16)**. Each of these "facets" of the eye has its own light-focusing lens that captures light from a tiny portion of the visual field (the total area seen when the eyes point forward). A compound eye is very effective at detecting movement, an important adaptation for flying insects and small animals constantly threatened with predation. Many compound eyes, including those of the fly in Figure 50.16, offer a very wide field of view.

Insects have excellent color vision, and some, such as bees, can see into the ultraviolet (UV) range of the electromagnetic spectrum. Because UV light is invisible to humans, we don't see differences in the environment that bees and

▼ **Figure 50.16 Compound eyes.**

(a) The faceted eyes on the head of a fly form a repeating pattern visible in this photomicrograph.

(b) The cornea and crystalline cone of each ommatidium together function as a lens that focuses light on the rhabdom, an organelle formed by and extending inward from a circle of photoreceptors. The rhabdom traps light, serving as the photosensitive part of the ommatidium. Information gathered from different intensities of light entering the many ommatidia from different angles is used to form a visual image.

other insects detect. In studying animal behavior, we cannot simply extrapolate our sensory world to other species; different animals have different sensitivities and different brain organizations.

Single-Lens Eyes

Among invertebrates, **single-lens eyes** are found in some jellies and polychaete worms, as well as in spiders and many molluscs. A single-lens eye works somewhat like a camera.

The eye of an octopus or squid, for example, has a small opening, the **pupil**, through which light enters. Like a camera's adjustable aperture, the **iris** expands or contracts, changing the diameter of the pupil to let in more or less light. Behind the pupil, a single lens directs light on a layer of photoreceptors. Similar to a camera's focusing action, muscles in an invertebrate's single-lens eye move the lens forward or backward, focusing on objects at different distances.

▼ **Figure 50.17** **Exploring the Structure of the Human Eye**

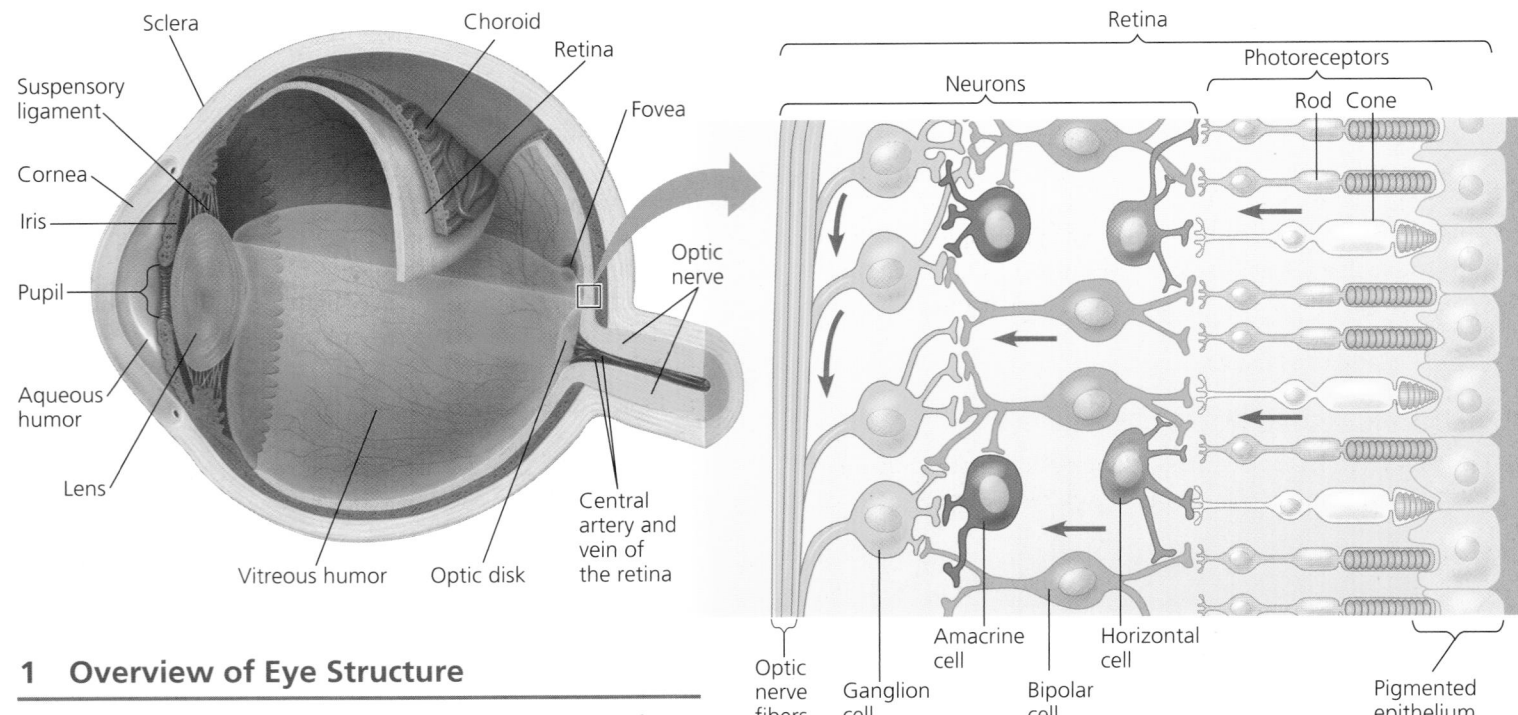

1   Overview of Eye Structure

Starting from the outside, the human eye is surrounded by the conjunctiva, a mucous membrane (not shown); the sclera, a connective tissue; and the choroid, a thin, pigmented layer. At the front, the sclera forms the transparent *cornea* and the choroid forms the colored *iris*. By changing size, the iris regulates the amount of light entering the pupil, the hole in the center of the iris. Just inside the choroid, the neurons and photoreceptors of the **retina** form the innermost layer of the eyeball. The optic nerve exits the eye at the optic disk.

The **lens**, a transparent disk of protein, divides the eye into two cavities. In front of the lens lies the *aqueous humor*, a clear watery substance. Blockage of ducts that drain this fluid can produce glaucoma, a condition in which increased pressure in the eye damages the optic nerve, causing vision loss. Behind the lens lies the jellylike *vitreous humor* (illustrated here in the lower portion of the eyeball).

2   The Retina

Light (coming from the left in this diagram) strikes the retina, passing through largely transparent layers of neurons before reaching the rods and cones, two types of photoreceptors that differ in shape and in function. The neurons of the retina then relay visual information captured by the photo receptors to the optic nerve and brain along the pathways shown with red arrows. Each *bipolar cell* receives information from several rods or cones, and each *ganglion cell* gathers input from several bipolar cells. *Horizontal* and *amacrine cells* integrate information across the retina.

The optic disk, where the optic nerve exits the retina, lacks photoreceptors. As a result, this region forms a "blind spot" where light is not detected.

The eyes of all vertebrates have a single lens. In fishes, focusing occurs as in invertebrates, with the lens moving forward or backward. In other species, including mammals, focusing is achieved by changing the shape of the lens.

The Vertebrate Visual System

The human eye will serve as our model of vision in vertebrates. As described in **Figure 50.17**, vision begins when photons of light enter the eye and strike the rods and cones. There the energy of each photon is captured in retinal, the light-absorbing molecule in the visual pigment rhodopsin.

Although light detection in the eye is the first stage in vision, remember that it is actually the brain that "sees." Thus, to understand vision, we must examine how the capture of light by retinal changes the production of action potentials and then follow these signals to the visual centers of the brain, where images are perceived.

Retinal ⎤
Opsin ⎦ **Rhodopsin**

CYTOSOL

INSIDE OF DISK

3   Photoreceptor Cells

Humans have two main types of photoreceptor cells: rods and cones. Within the outer segment of a rod or cone is a stack of membranous disks in which *visual pigments* are embedded. **Rods** are more sensitive to light than cones are, but they do not distinguish colors; rods enable us to see at night, but only in black and white. **Cones** provide color vision, but, being less sensitive, contribute very little to night vision. There are three types of cones. Each has a different sensitivity across the visible spectrum, providing an optimal response to red, green, or blue light.

In the colorized SEM shown above, cones (green), rods (light tan), and adjacent neurons (red) are visible. The pigmented epithelium, which was removed in this preparation, would be to the right.

4   Visual Pigments

Vertebrate visual pigments consist of a light-absorbing molecule called **retinal** (a derivative of vitamin A) bound to a membrane protein called an **opsin**. Seven α helices of each opsin molecule span the disk membrane. The visual pigment of rods, shown here, is called **rhodopsin**.

Retinal exists as two isomers. Absorption of light shifts one bond in retinal from a *cis* to a *trans* arrangement, converting the molecule from an angled shape to a straight shape. This change in configuration activates the opsin protein to which retinal is bound.

Retinal: *cis* isomer

Light   Enzymes

Retinal: *trans* isomer

→ **Mastering Biology** Animations: **Structure and Function of the Eye** • **Function of the Retina** • **Photoreception**

VISUAL SKILLS *The isomers of retinal have the same number of atoms and bonds but differ in the spatial arrangement at one carbon-carbon double bond (C═C). In each isomer, circle that (C═C). Looking at the atoms around that bond, to what atoms do the terms cis (same side) and trans (opposite side) refer?*

Sensory Transduction in the Eye

The transduction of visual information to the nervous system begins with the light-induced conversion of *cis*-retinal to *trans*-retinal in rods and cones. Like other *cis-trans* pairs, these isomers of retinal differ only in the spatial arrangement of atoms at a carbon-carbon double bond (see Figure 4.7).

As shown in Figure 50.17, *trans*-retinal and *cis*-retinal differ in shape. This shift in shape activates the visual pigment (in rods, rhodopsin), which activates a G protein, which in turn activates the enzyme phosphodiesterase. The substrate for this enzyme in rods and cones is cyclic GMP (cGMP), which in the dark binds to sodium ion (Na^+) channels and keeps them open **(Figure 50.18a)**. When the enzyme hydrolyzes cGMP, Na^+ channels close, and the cell becomes hyperpolarized **(Figure 50.18b)**. The signal transduction pathway then shuts off as enzymes convert retinal back to the *cis* form, inactivating the visual pigment.

In bright light, rhodopsin remains active, and the response in the rods becomes saturated. If the amount of light entering the eyes abruptly decreases, the rods do not regain full responsiveness for several minutes. This is why you are briefly blinded if you pass quickly from bright sunshine into a dark movie theater. (Because light activation changes the color of rhodopsin from purple to yellow, rods in which the light response is saturated are often described as "bleached.")

Processing of Visual Information in the Retina

The processing of visual information begins in the retina itself, where both rods and cones form synapses with bipolar cells (see Figure 50.17). In the dark, rods and cones are depolarized and continually release the neurotransmitter glutamate at these synapses **(Figure 50.19)**. When light strikes the rods and cones, they hyperpolarize, shutting off their release of glutamate. This decrease triggers a change in the membrane potential of the bipolar cells, altering their regulation of action potential transmission to the brain.

Processing of signals from rods and cones occurs via several different pathways in the retina. Some information passes directly from photoreceptors to bipolar cells to ganglion cells. In other cases, horizontal cells carry signals from one rod or cone to other photoreceptors and to several bipolar cells.

How is it adaptive for visual information to follow several paths? We'll consider one example. When an illuminated rod or cone stimulates a horizontal cell, the horizontal cell inhibits more distant photoreceptors and bipolar cells that are not illuminated. The result is that the region receiving light appears lighter and the dark surroundings even darker. This form of integration, called *lateral inhibition*, sharpens edges and enhances contrast in the image. An essential part of visual processing, lateral inhibition occurs in the brain as well as the retina.

▼ **Figure 50.18 Response of a photoreceptor cell to light.** Light triggers a receptor potential in a rod (shown here) or cone. Note that for photoreceptors, this change in membrane potential is a hyperpolarization.

(a) In the dark, Na^+ flows through channels that are open when bound to cyclic GMP (cGMP), which is present at high levels in the cytosol. This flow of Na^+ keeps the plasma membrane of the photoreceptor cell depolarized.

Rod cell

DARK — Inactive rhodopsin — INSIDE OF DISK — Disk membrane — EXTRACELLULAR FLUID — Plasma membrane

Transducin — cGMP — cGMP — cGMP — cGMP — Na+ — CYTOSOL — cGMP — Inactive phosphodiesterase

Membrane potential (mV) across plasma membrane

Depolarized state

In the dark, the depolarized cell releases the neurotransmitter glutamate at the synaptic terminals.

(b) Light triggers a cascade of events that activate rhodopsin, transducin, and the enzyme phosphodiesterase, which converts cGMP to GMP. When the cGMP level falls, cGMP no longer occupies the binding sites on the channels, causing the channels to close and the cell to hyperpolarize.

LIGHT — Active rhodopsin — INSIDE OF DISK — Disk membrane — EXTRACELLULAR FLUID — Plasma membrane

GMP — Transducin — GMP — GMP — Activated phosphodiesterase — cGMP — GMP — GMP — Na+ — CYTOSOL

Membrane potential (mV) across plasma membrane

Light strikes retina

Hyperpolarization

In the light, the cell becomes hyperpolarized and ceases to release glutamate.

Synaptic terminals

→ **Mastering Biology** Figure Walkthrough

▼ **Figure 50.19 Synaptic activity of rod cells in light and dark.**

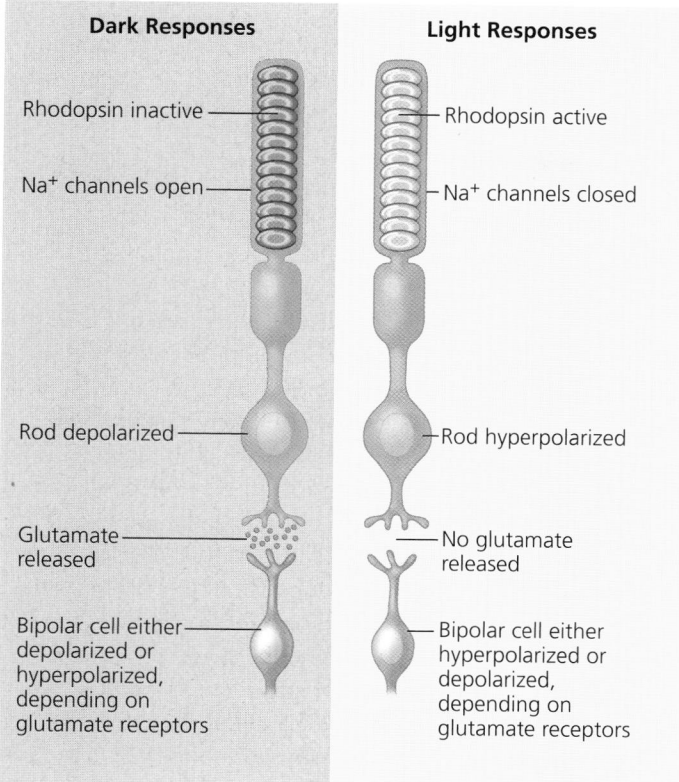

Dark Responses	Light Responses
Rhodopsin inactive	Rhodopsin active
Na+ channels open	Na+ channels closed
Rod depolarized	Rod hyperpolarized
Glutamate released	No glutamate released
Bipolar cell either depolarized or hyperpolarized, depending on glutamate receptors	Bipolar cell either hyperpolarized or depolarized, depending on glutamate receptors

❓ *Like rods, cone cells are depolarized when their opsin molecules are inactive. In the case of a cone, why might it be misleading to call this a dark response?*

A single ganglion cell receives information from an array of rods and cones, each of which responds to light coming from a particular location. Together, the rods and cones that are feeding information to one ganglion cell define a *receptive field*—the part of the visual field to which that ganglion cell can respond. The fewer rods or cones that supply a single ganglion cell, the smaller the receptive field. A smaller receptive field typically results in a sharper image because the information about where light has struck the retina is more precise.

Processing of Visual Information in the Brain

Axons of ganglion cells form the optic nerves that transmit action potentials from the eyes to the brain (**Figure 50.20**). The two optic nerves meet at the *optic chiasm* near the center of the base of the cerebral cortex. Axons in the optic nerves are routed at the optic chiasm such that sensations from the left visual field are transmitted to the right side of the brain, and sensations from the right visual field are transmitted to the left side of the brain. (Note that each visual field, whether right or left, involves input from both eyes.)

Within the brain, most ganglion cell axons lead to the *lateral geniculate nuclei*, which have axons that reach the *primary visual cortex* in the cerebrum. Additional neurons carry the information to higher-order visual processing and

▼ **Figure 50.20 Neural pathways for vision.** Each optic nerve contains about a million axons that synapse with interneurons in the brain. The nuclei relay sensations to the primary visual cortex, one of many brain centers that cooperate in constructing visual perceptions.

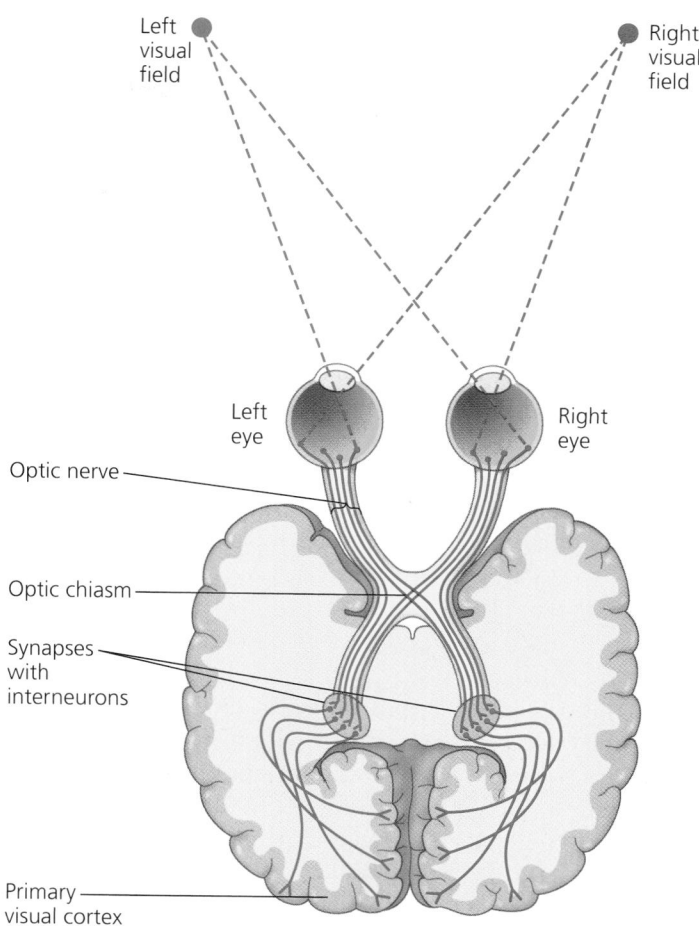

Left visual field
Right visual field
Left eye
Right eye
Optic nerve
Optic chiasm
Synapses with interneurons
Primary visual cortex

integrating centers elsewhere in the cortex. Researchers estimate that at least 30% of the cerebral cortex, comprising hundreds of millions of neurons in perhaps dozens of integrating centers, takes part in formulating what we actually "see." Determining how these centers integrate such components of our vision as color, motion, depth, shape, and detail is the focus of much exciting research.

Color Vision

Among vertebrates, most fishes, amphibians, and reptiles, including birds, have very good color vision. Humans and other primates also see color well, but are among the minority of mammals with this ability. For cats and other mammals that are most active at night, a high proportion of rods in the retina is an adaptation that provides keen night vision. Color vision among these nocturnal animals is limited, and they probably see a pastel world during the day.

In humans, the perception of color is based on three types of cones, each with a different visual pigment—red, green, or blue. The three visual pigments, called *photopsins*, are formed from the binding of retinal to three distinct opsin proteins. Slight differences in the opsin proteins cause each photopsin to absorb light optimally at a different wavelength. Although the visual pigments are designated as red, green, and blue, their absorption spectra in fact overlap. For this reason, the brain's perception of intermediate hues depends on the differential stimulation of two or more classes of cones. For example, if both red and green cones are stimulated, we see either yellow or orange, depending on which class of cones is more strongly stimulated.

Abnormal color vision typically results from mutations in the genes for one or more photopsin proteins. In humans, color blindness nearly always affects perception of red or green and is far more common in one gender than the other: 5–8% of males, fewer than 1% of females. Why? The human genes for both the red and green pigments are X-linked. Thus, males are affected by a single mutation, whereas females are color-blind only if both copies are mutant. (The human gene for the blue pigment is on chromosome 7.)

Experiments on color vision in the squirrel monkey (*Saimiri sciureus*) enabled an early breakthrough in the field of gene therapy. These monkeys have only two opsin genes, one sensitive to blue light and the other sensitive to either red or green light, depending on the allele. Because the red/green opsin gene is X-linked, all males have only the red- or green-sensitive version and are red-green color-blind. When researchers injected a virus containing the gene for the missing version into the retina of adult male monkeys, evidence of full color vision was apparent after 20 weeks (**Figure 50.21**).

The squirrel monkey gene therapy studies demonstrate that the neural circuits required to process visual information can be generated or activated even in adults, making it possible to treat a range of vision disorders. Indeed, gene therapy

▼ **Figure 50.21 Gene therapy for vision.** Once color-blind, this adult male monkey treated with gene therapy demonstrates his ability to distinguish red from green.

MAKE CONNECTIONS *Red-green color blindness is X-linked in squirrel monkeys and in humans (see Figure 15.7). Why is the inheritance pattern in humans not apparent in squirrel monkeys?*

has been used to treat Leber's congenital amaurosis (LCA), an inherited retinal degenerative disease that causes severe loss of vision. After using gene therapy to restore vision in dogs and mice with LCA, researchers successfully treated the disease in humans by injecting the functional LCA gene in a viral vector (see Figure 20.22).

→ **Mastering Biology** HHMI Video: Genes as Medicine

The Visual Field

The brain not only processes visual information but also controls what information is captured. One important control is focusing, which in humans occurs by changing the shape of the lens, as noted earlier and illustrated in **Figure 50.22**. When you focus your eyes on a close object, your lenses become almost spherical. When you view a distant object, your lenses are flattened.

Although our peripheral vision allows us to see objects over a nearly 180° range, the distribution of photoreceptors across the eye limits both what we see and how well we see it. Overall, the human retina contains about 125 million rods and 6 million cones. At the **fovea**, the center of the visual field, there are no rods but a very high density of

▼ **Figure 50.22 Focusing in the mammalian eye.** Ciliary smooth muscles control the shape of the lens, which bends light and focuses it on the retina. The thicker the lens, the more sharply the light is bent.

→ **Mastering Biology** Animation: Near and Distance Vision

cones—about 150,000 cones per square millimeter. The ratio of cones to rods falls with distance from the fovea, with the peripheral regions having only rods. In daylight, you achieve your sharpest vision by looking directly at an object, such that light shines on the tightly packed cones in your fovea. At night, looking directly at a dimly lit object is ineffective, since the rods—the more sensitive light receptors—are absent from the fovea. For this reason, you see a dim star best by focusing on a point just to one side of it.

CONCEPT CHECK 50.3

1. Contrast the light-detecting organs of planarians and flies. How is each organ adaptive for the lifestyle of the animal?

2. In a condition called presbyopia, the eyes' lenses lose much of their elasticity and maintain a flat shape. How would you expect this condition to affect a person's vision?

3. **WHAT IF?** Our brain receives more action potentials when our eyes are exposed to light even though our photoreceptors release more neurotransmitter in the dark. Propose an explanation.

4. **MAKE CONNECTIONS** Compare the function of retinal in the eye with that of the pigment chlorophyll in a plant photosystem (see Concept 10.2).

For suggested answers, see Appendix A.

CONCEPT 50.4

The senses of taste and smell rely on similar sets of sensory receptors

Animals use their chemical senses for a wide range of purposes, including finding mates, recognizing marked territories, and helping navigate during migration. In addition, animals such as ants and bees that live in large social groups rely extensively on chemical "conversation."

In all animals, chemical senses are important for feeding behavior. The perceptions of **gustation** (taste) and **olfaction** (smell) both depend on chemoreceptors. In the case of terrestrial animals, taste is the detection of chemicals called **tastants** that are present in a solution, and smell is the detection of **odorants** that are carried through the air. There is no distinction between taste and smell in aquatic animals.

In insects, taste receptors are located within sensory hairs located on the feet and in mouthparts, where they are used to select food. A tasting hair contains several chemoreceptors, each especially responsive to a particular class of tastant, such as sugar or salt. Insects are also capable of smelling airborne odorants using olfactory hairs, usually located on their antennae (see Figure 50.6). The chemical DEET (*N,N*-diethyl-meta-toluamide), marketed as an insect "repellant," actually protects against bites by blocking the olfactory receptor in mosquitoes that detects human scent.

Taste in Mammals

Humans and other mammals perceive five tastes: sweet, sour, salty, bitter, and umami. Umami (which means "pleasant savory taste" in Japanese) is elicited by the amino acid glutamate. Sometimes used as a flavor enhancer in the form of monosodium glutamate, or MSG, glutamate occurs naturally in meat, aged cheese, and other foods, to which it imparts a "savory" quality.

For decades, many researchers assumed that a taste cell could have more than one type of receptor. An alternative idea is that each taste cell has a single receptor type, programming the cell to recognize only one of the five tastes. To test this hypothesis, scientists used a cloned bitter taste receptor to genetically reprogram gustation in a mouse **(Figure 50.23)**. This reprogramming experiment, together with follow-up studies, revealed that an individual taste cell does in fact express a single receptor type and detects tastants representing only one of the five tastes.

▼ Figure 50.23 | Inquiry

How do mammals detect different tastes?

Experiment To investigate the basis of mammalian taste perception, researchers used a chemical called phenyl-β-D-glucopyranoside (PBDG). Humans find PBDG extremely bitter. Mice, however, lack a receptor for PBDG. Mice avoid drinking water containing other bitter tastants but show no aversion to water that contains PBDG.

Using a molecular cloning strategy, the researchers generated mice that made the human PBDG receptor in cells that normally make either a sweet receptor or a bitter receptor. The mice were given a choice of two bottles, one filled with pure water and one filled with water containing PBDG at varying concentrations. The researchers then observed whether the mice had an attraction or an aversion to PBDG.

Results

Conclusion The researchers found that the presence of a bitter receptor in sweet taste cells is sufficient to cause mice to be attracted to a bitter chemical. They concluded that the mammalian brain must therefore perceive sweet or bitter taste solely on the basis of which afferent neurons are activated.

Data from K. L. Mueller et al., The receptors and coding logic for bitter taste, *Nature* 434:225–229 (2005).

WHAT IF? *Suppose that instead of the PBDG receptor, the researchers had used a receptor specific for a sweetener that humans crave but mice ignore. How would the results of the experiment have differed?*

The receptor cells for taste in mammals are modified epithelial cells organized into **taste buds**, which are scattered in several areas of the tongue and mouth **(Figure 50.24)**. Most taste buds on the tongue are associated with nipple-shaped projections called papillae. Any region of the tongue with taste buds can detect any of the five types of taste. (The frequently reproduced "taste maps" of the tongue are thus not accurate.)

Researchers have identified the receptor proteins for all five tastes. The sensations of sweet, umami, and bitter tastes each require one or more genes encoding a G protein-coupled receptor, or GPCR (see Figures 11.7 and 11.8). Humans have one type of sweet receptor and one type of umami receptor, each assembled from a different pair of GPCR proteins. In contrast, humans have more than 30 different receptors for bitter taste, and each receptor is able to recognize multiple bitter tastants. GPCR proteins are also critical for the sense of smell, as will be discussed shortly.

The receptor for sour tastants belongs to the TRP family and is similar to the capsaicin receptor and other thermoreceptor proteins. In taste buds, the TRP proteins of the sour receptor assemble into an ion channel in the plasma membrane of the taste cell. Binding of an acid or other sour-tasting substance to the receptor triggers a change in the ion channel. Depolarization occurs, activating an afferent neuron that relays information to the CNS, triggering the perception of a sour taste.

The taste receptor for salt turns out to be a sodium channel. Not surprisingly, it specifically detects sodium salts, such as the NaCl that we use in cooking and flavoring.

Smell in Humans

In olfaction, unlike gustation, the sensory cells are neurons. Olfactory receptor cells line the upper portion of the nasal cavity and send impulses along their axons to the olfactory bulb of the brain **(Figure 50.25)**. The receptive ends of the cells contain cilia that extend into the layer of mucus coating the nasal cavity. When an odorant diffuses into this region, it binds to a specific GPCR protein called an olfactory receptor (OR) on the plasma membrane of the olfactory cilia. These events trigger signal transduction leading to the production of cyclic AMP. In olfactory cells, cyclic AMP opens channels in the plasma membrane that are permeable to both Na^+ and Ca^{2+}. The flow of these ions into the receptor cell leads to depolarization of the membrane, generating action potentials.

Mammals can distinguish thousands of different odors, each caused by a structurally distinct odorant. How is this remarkable sensory discrimination possible? American researchers Linda Buck and Richard Axel found the answer—a very large gene family. Working with mice, Buck and Axel discovered 1,200 different OR genes—and were honored with a Nobel Prize in 2004. Humans have just 380 OR genes, far fewer than mice, but still nearly 2% of all the genes in our genome. Identification of particular odors relies on two basic properties of the olfactory system. First, each olfactory receptor cell expresses one OR gene. Second, those cells that express the same OR gene transmit action potentials to the same small region of the olfactory bulb.

After odorants are detected, information from olfactory receptors is collected and integrated. Genetic studies on mice, worms, and flies have shown that signals from the nervous system regulate this process, dialing the response to particular odorants up or down. As a result, animals can detect the location of food sources even if the concentration of a key odorant is particularly low or high.

▼ Figure 50.24 Human taste receptors.

(a) The tongue. Small raised structures called papillae cover the tongue surface. The enlarged cross section shows the side walls of a papilla lined with taste buds.

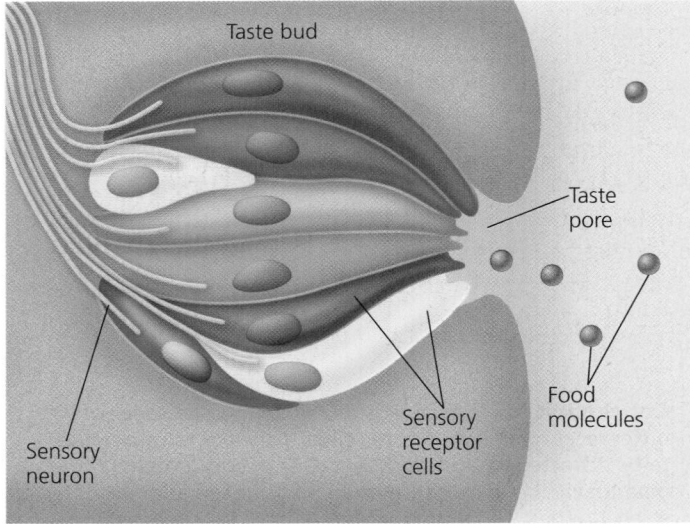

Key

▮ Sweet ▮ Bitter ▮ Umami
▮ Salty   ▮ Sour

(b) A taste bud. Taste buds in all regions of the tongue contain sensory receptor cells specific for each of the five taste types.

▼ **Figure 50.25 Smell in humans.** Odorant molecules bind to specific chemoreceptor proteins in the plasma membrane of olfactory receptor cells, triggering action potentials. Each olfactory receptor cell has just one type of chemoreceptor. As shown, cells that express different chemoreceptors detect different odorants.

WHAT IF? *If you spray an air freshener in a musty room, would you be affecting detection, transmission, or perception of the odorants responsible for the musty smell?*

Studies of model organisms also reveal that complex mixtures of odorants are not processed as the simple sum of each input. Rather, the brain integrates olfactory information from different receptors into single sensations. These sensations contribute to the perception of the environment in the present and to the memory of events and emotions.

Although the receptors and neuronal pathways for taste and smell are independent, the two senses do interact. Indeed, much of the complex flavor humans experience when eating is due to our sense of smell. If the olfactory system is blocked, as occurs when you have a head cold, the perception of flavor is sharply reduced.

CONCEPT CHECK 50.4

1. Explain why some taste receptor cells and all olfactory receptor cells use G protein-coupled receptors, yet only olfactory receptor cells produce action potentials.

2. Pathways involving G proteins provide an opportunity for an increase in signal strength in the course of signal transduction, a change referred to as amplification. How might this be beneficial in olfaction?

3. **WHAT IF?** If you discovered a mutation in mice that disrupted the ability to taste sweet, bitter, and umami but not sour or salty, what might you predict about where this mutation acts in the signaling pathways used by these receptors?

For suggested answers, see Appendix A.

CONCEPT 50.5

The physical interaction of protein filaments is required for muscle function

In discussing sensory mechanisms, we have seen how sensory inputs to the nervous system can result in specific behaviors: the touch-guided foraging of a star-nosed mole, the upside-down swimming of a crayfish with manipulated statocysts, and the light-avoiding maneuvers of planarians. Underlying these diverse behaviors are common fundamental mechanisms: Feeding, swimming, and crawling all require muscle activity in response to nervous system motor output.

Muscle cell contraction relies on the interaction between protein structures called thin and thick filaments. The major component of **thin filaments** is the globular protein actin. In thin filaments, two strands of polymerized actin are coiled around one another; similar actin structures called microfilaments function in cell motility. The **thick filaments** are staggered arrays of myosin molecules. Muscle contraction is the result of filament movement powered by chemical energy; muscle extension occurs only passively. To understand how filaments bring about muscle contraction, we will begin by examining the skeletal muscle of vertebrates.

Vertebrate Skeletal Muscle

Vertebrate **skeletal muscle**, which moves bones and body, has a hierarchy of smaller and smaller units **(Figure 50.26)**. Within a typical skeletal muscle is a bundle of long fibers running along the length of the muscle. Each individual fiber is a single cell. Within are multiple nuclei, each derived from one of the embryonic cells that fused to form the fiber.

▼ **Figure 50.26 The structure of skeletal muscle.**

VISUAL SKILLS *Looking at this figure, would you say that there are multiple sarcomeres per myofibril or multiple myofibrils per sarcomere? Explain.*

Surrounding these nuclei are longitudinal **myofibrils**, which consist of bundles of thin and thick filaments.

The myofibrils in muscle fibers are made up of repeating sections called **sarcomeres**, which are the basic contractile units of skeletal muscle. The borders of the sarcomere line up in adjacent myofibrils, forming a pattern of light and dark bands (striations) visible with a light microscope. For this reason, skeletal muscle is also called *striated muscle*. Thin filaments attach at the Z lines at the sarcomere ends, while thick filaments are anchored in the middle of the sarcomere (M line).

In a resting (relaxed) myofibril, thick and thin filaments partially overlap. Near the edge of the sarcomere there are only thin filaments, whereas the zone in the center contains only thick filaments. This partially overlapping arrangement is the key to how the sarcomere, and hence the whole muscle, contracts.

The Sliding-Filament Model of Muscle Contraction

A contracting muscle shortens, but the filaments that bring about contraction stay the same length. To explain this apparent paradox, we'll focus first on a single sarcomere. As shown in **Figure 50.27**, the filaments slide past each other, much like the segments of a telescoping support pole. According to the well-accepted **sliding-filament model**, the thin and thick filaments ratchet past each other, powered by myosin molecules.

Figure 50.28 illustrates the cycles of change in the myosin molecule that convert the chemical energy of ATP into the longitudinal sliding of thick and thin filaments.

As shown in the figure, each myosin molecule has a long "tail" region and a globular "head" region. The tail adheres to the tails of other myosin molecules, binding together the thick filament. The head, jutting to the side, can bind ATP. Hydrolysis of bound ATP converts myosin to a high-energy form that binds to actin, forming a cross-bridge between the myosin and the thin filament. The myosin head then returns to its low-energy form as it helps to pull the thin filament toward the center of the sarcomere. When a new ATP molecule binds to the myosin head, the cross-bridge is disrupted, releasing the myosin head from the actin filament.

Muscle contraction requires repeated cycles of binding and release. During each cycle of each myosin head, the head is freed from a cross-bridge, cleaves the newly bound ATP, and binds again to actin. Because the thin filament moves toward the center of the sarcomere in each cycle, the myosin head now attaches to a binding site farther along the thin filament than in the previous cycle. Each end of a thick filament contains approximately 300 heads, each of which forms and re-forms about five cross-bridges per second, driving the thick and thin filaments past each other.

► **Figure 50.27 The sliding-filament model of muscle contraction.** The drawings on the left show that the lengths of the thick (myosin) filaments (purple) and thin (actin) filaments (orange) remain the same as a muscle fiber contracts.

⊙ **Mastering Biology** BioFlix®
Animation: From Muscle Cells to Movement

▼ **Figure 50.28 Myosin-actin interactions underlying muscle fiber contraction.**

Thin filaments

Thick filament

1 Starting here, the myosin head is bound to ATP and is in its low-energy configuration.

Thin filament

Myosin head (low-energy configuration)

Thick filament

5 Binding of a new molecule of ATP releases the myosin head from actin, and a new cycle begins.

ATP

2 The myosin head hydrolyzes ATP to ADP and P_i (inorganic phosphate) and is in its high-energy configuration.

Actin

Myosin-binding sites

The thin filament moves toward the center of the sarcomere.

Myosin head (low-energy configuration)

ADP
P_i
Myosin head (high-energy configuration)

4 The cross-bridge couples the release of ADP and P_i to a power stroke that slides the thin filament along the myosin and returns the myosin head to a low-energy state.

ADP + P_i

ADP
P_i
Cross-bridge

3 The myosin head binds to actin, forming a cross-bridge with the thin filament.

❓ *When ATP binds the myosin head, what prevents the filaments from sliding back into their original positions?*

⊙ **Mastering Biology** BioFlix® **Animation: Sliding-Filament Model of Muscle Contraction**

At rest, most muscle fibers contain only enough ATP for a few contractions. Powering repetitive contractions requires two other storage compounds: creatine phosphate and glycogen. The transfer of a phosphate group from creatine phosphate to ADP in an enzyme-catalyzed reaction synthesizes additional ATP. In this way, the resting supply of creatine phosphate can sustain contractions for about 15 seconds. ATP stores are also replenished when glycogen is broken down to glucose. During light or moderate muscle activity, this glucose is metabolized by aerobic respiration. This highly efficient metabolic process yields enough power to sustain contractions for nearly an hour. During intense muscle activity, oxygen becomes limiting and ATP is instead generated by lactic acid fermentation (see Concept 9.5). This anaerobic pathway, although very rapid, generates much less ATP per glucose molecule and can sustain contraction for only about 1 minute.

The Role of Calcium and Regulatory Proteins

Proteins bound to actin play crucial roles in controlling muscle contraction. In a muscle fiber at rest, **tropomyosin**, a regulatory protein, and the **troponin complex**, a set of additional regulatory proteins, are bound to the actin strands of thin filaments. Tropomyosin covers the myosin-binding sites along the thin filament, preventing actin and myosin from interacting **(Figure 50.29a)**.

Motor neurons enable actin and myosin to interact by triggering a release of calcium ions (Ca^{2+}) into the cytosol. Once in the cytosol, Ca^{2+} binds to the troponin complex, causing the myosin-binding sites on actin to be exposed **(Figure 50.29b)**. Note that the effect of Ca^{2+}

is indirect: Binding to Ca^{2+} causes the troponin complex to change shape, dislodging tropomyosin from the myosin-binding sites.

When the Ca^{2+} concentration rises in the cytosol, the cycle of cross-bridge formation begins, the thin and thick filaments slide past each, and the muscle fiber contracts. When the Ca^{2+} concentration falls, the binding sites are covered, and contraction stops.

Motor neurons cause muscle contraction through a multi-step process triggering the movement of Ca^{2+} into the cytosol of muscle cells. This process is summarized in **Figure 50.30** and detailed step by step in **Figure 50.31**. First, the arrival of an action potential at the synaptic terminal of a motor neuron ❶ causes release of the neurotransmitter acetylcholine. Binding of acetylcholine to receptors on the muscle fiber leads to a depolarization that initiates an action potential. Within the muscle fiber, the action potential spreads deep into the interior, following infoldings of the plasma membrane called **transverse (T) tubules**. ❷ These make close contact with the **sarcoplasmic reticulum (SR)**, a specialized endoplasmic reticulum. As the action potential spreads along the T tubules, it triggers changes in the SR, opening Ca^{2+} channels ❸. Calcium ions stored in the interior of the SR flow through open channels into the cytosol ❹ and bind to the troponin complex, ❺ initiating the muscle fiber contraction.

When motor neuron input stops, the filaments slide back to their starting position as the muscle relaxes. Relaxation

▼ **Figure 50.29 The role of regulatory proteins and calcium in muscle fiber contraction.** Each thin filament consists of two strands of actin and associated regulatory proteins: two long molecules of tropomyosin and multiple copies of the troponin complex.

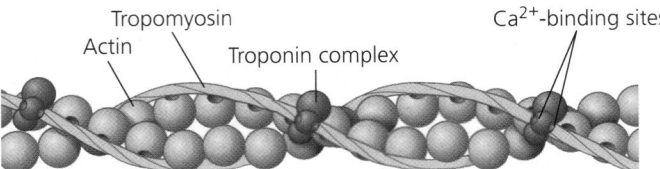

(a) Myosin-binding sites blocked by tropomyosin

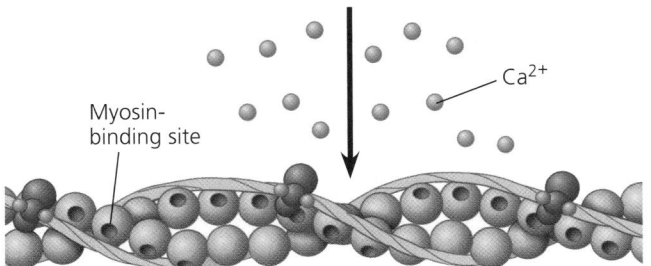

(b) Myosin-binding sites exposed

▼ **Figure 50.30 The roles of the sarcoplasmic reticulum and T tubules in muscle fiber contraction.** The synaptic terminal of a motor neuron releases acetylcholine, which depolarizes the plasma membrane of the muscle fiber. The depolarization causes action potentials (red arrows) to sweep across the muscle fiber and deep into it along the transverse (T) tubules. The action potentials trigger the release of calcium (green dots) from the sarcoplasmic reticulum into the cytosol. Calcium initiates the sliding of filaments by allowing myosin to bind to actin.

⊘ **Mastering Biology BioFlix® Animation: Muscle Structure**

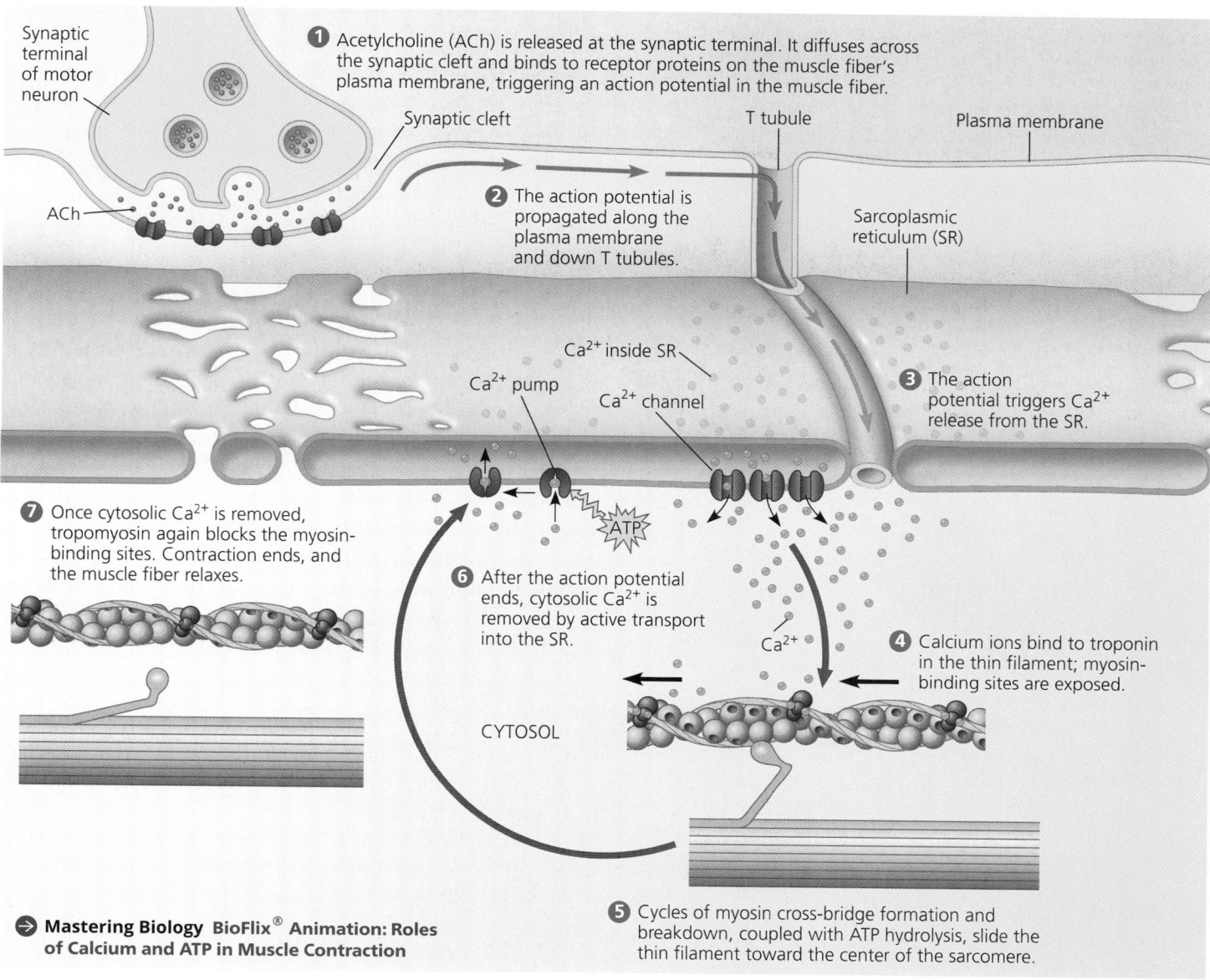

▼ Figure 50.31 Summary of contraction in a skeletal muscle fiber.

Synaptic terminal of motor neuron

ACh

1 Acetylcholine (ACh) is released at the synaptic terminal. It diffuses across the synaptic cleft and binds to receptor proteins on the muscle fiber's plasma membrane, triggering an action potential in the muscle fiber.

Synaptic cleft

T tubule

Plasma membrane

2 The action potential is propagated along the plasma membrane and down T tubules.

Sarcoplasmic reticulum (SR)

Ca^{2+} inside SR

Ca^{2+} pump

Ca^{2+} channel

3 The action potential triggers Ca^{2+} release from the SR.

7 Once cytosolic Ca^{2+} is removed, tropomyosin again blocks the myosin-binding sites. Contraction ends, and the muscle fiber relaxes.

ATP

6 After the action potential ends, cytosolic Ca^{2+} is removed by active transport into the SR.

Ca^{2+}

4 Calcium ions bind to troponin in the thin filament; myosin-binding sites are exposed.

CYTOSOL

Mastering Biology BioFlix® Animation: Roles of Calcium and ATP in Muscle Contraction

5 Cycles of myosin cross-bridge formation and breakdown, coupled with ATP hydrolysis, slide the thin filament toward the center of the sarcomere.

begins as proteins in the SR pump Ca^{2+} back into the SR **6** from the cytosol. When the Ca^{2+} concentration in the cytosol drops to a low level, the regulatory proteins bound to the thin filament shift back to their starting position, **7** once again blocking the myosin-binding sites. At the same time, the Ca^{2+} pumped from the cytosol accumulates in the SR, providing the stores needed to respond to the next action potential.

Several diseases cause paralysis by interfering with the excitation of skeletal muscle fibers by motor neurons. In amyotrophic lateral sclerosis (ALS), motor neurons in the spinal cord and brainstem degenerate, and muscle fibers atrophy. ALS is progressive and usually fatal within five years after symptoms appear. In myasthenia gravis, a person produces antibodies to the acetylcholine receptors of skeletal

muscle. As the disease progresses and the number of receptors decreases, transmission between motor neurons and muscle fibers declines. Myasthenia gravis can generally be controlled with drugs that inhibit acetylcholinesterase or suppress the immune system.

Nervous Control of Muscle Tension

Whereas contraction of a single skeletal muscle fiber is a brief all-or-none twitch, contraction of a whole muscle, such as the biceps in your upper arm, is graded; you can voluntarily alter the extent and strength of its contraction. The nervous system produces graded contractions of whole muscles by varying (1) the number of muscle fibers that contract and (2) the rate at which muscle fibers are stimulated. Let's consider each mechanism in turn.

▼ **Figure 50.32 Motor units in a vertebrate skeletal muscle.**
Each muscle fiber (cell) forms synapses with only one motor neuron, but each motor neuron typically synapses with many muscle fibers. A motor neuron and all the muscle fibers it controls constitute a motor unit.

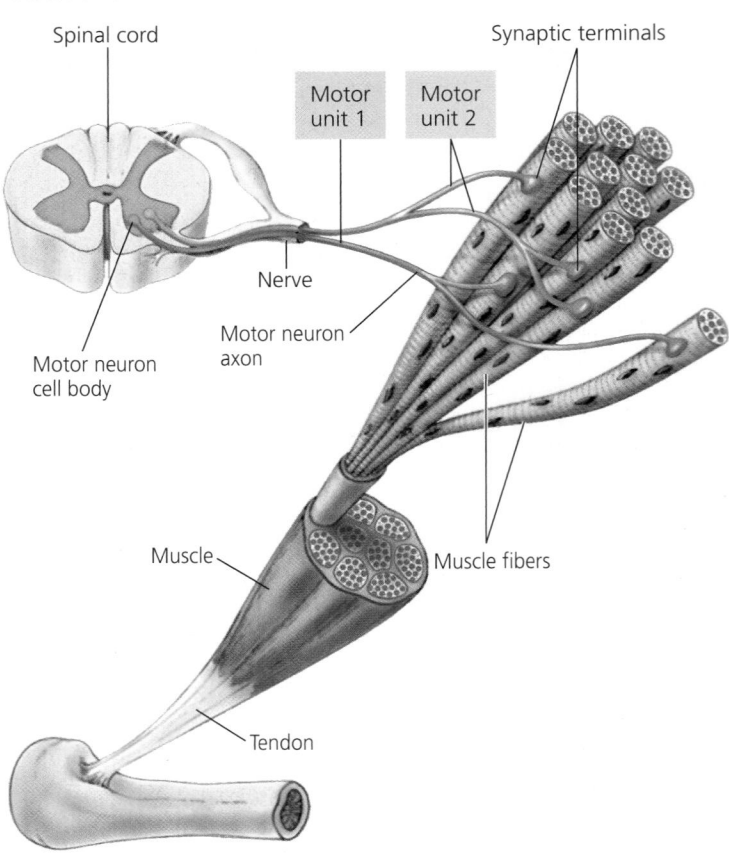

In vertebrates, each branched motor neuron may form synapses with many muscle fibers, although each fiber is controlled by only one motor neuron. A **motor unit** consists of a single motor neuron and all the muscle fibers it controls **(Figure 50.32)**. When a motor neuron produces an action potential, all the muscle fibers in its motor unit contract as a group. The strength of the resulting contraction depends on how many muscle fibers the motor neuron controls.

In the whole muscle, there may be hundreds of motor units. As more and more of the corresponding motor neurons are activated, a process called *recruitment*, the force (tension) developed by a muscle progressively increases. Depending on the number of motor neurons your brain recruits and the size of their motor units, you can lift a fork or something much heavier, like your biology textbook. Some muscles, especially those that hold up the body and maintain posture, are almost always partially contracted. In such muscles, the nervous system may alternate activation among the motor units, reducing the length of time any one set of fibers is contracted.

▼ **Figure 50.33 Summation of twitches.** This graph illustrates how the number of action potentials during a short period of time influences the tension developed in a muscle fiber.

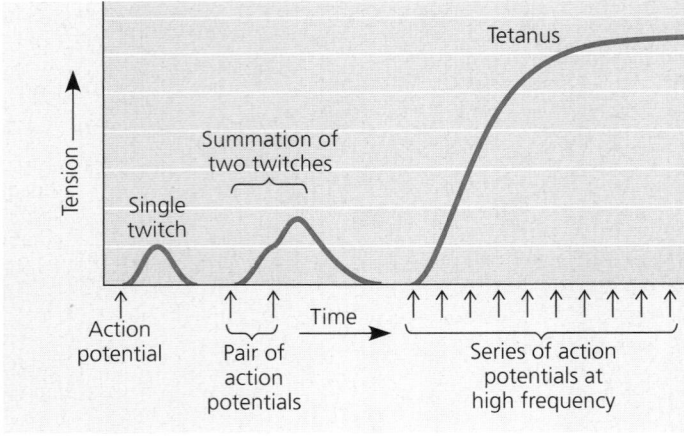

? *How could the nervous system cause a skeletal muscle to produce the most forceful contraction it is capable of?*

The nervous system regulates muscle contraction not only by controlling which motor units are activated but also by varying the rate of muscle fiber stimulation. A single action potential produces a twitch lasting about 100 milliseconds or less. If a second action potential arrives before the muscle fiber has completely relaxed, the two twitches add together, resulting in greater tension **(Figure 50.33)**. Further summation occurs as the rate of stimulation increases. When the rate is so high that the muscle fiber cannot relax at all between stimuli, the twitches fuse into one smooth, sustained contraction called **tetanus**. (Note that tetanus is also the name of a disease of uncontrolled muscle contraction caused by a bacterial toxin.)

Types of Skeletal Muscle Fibers

Our discussion to this point has focused on the general properties of vertebrate skeletal muscles. There are, however, several distinct types of skeletal muscle fibers, each of which is adapted to a particular set of functions. We typically classify these varied fiber types both by the source of ATP used to power their activity and by the speed of their contraction **(Table 50.1)**.

Oxidative and Glycolytic Fibers Fibers that rely mostly on aerobic respiration are called oxidative fibers. Such fibers are specialized in ways that enable them to make use of a steady energy supply: They have many mitochondria, a rich blood supply, and a large amount of an oxygen-storing protein called **myoglobin**. A brownish red pigment, myoglobin binds oxygen more tightly than does hemoglobin, enabling oxidative fibers to extract oxygen from the blood efficiently. In contrast, glycolytic fibers have a larger diameter and less myoglobin. Also, glycolytic fibers use glycolysis

Table 50.1 Types of Skeletal Muscle Fibers

	Slow-Twitch	Fast-Twitch	
	Oxidative	Oxidative	Glycolytic
Contraction speed	Slow	Fast	Fast
Major ATP source	Aerobic respiration	Aerobic respiration	Glycolysis
Rate of fatigue	Slow	Intermediate	Fast
Mitochondria	Many	Many	Few
Myoglobin content	High (red muscle)	High (red muscle)	Low (white muscle)

as their primary source of ATP and fatigue more readily than oxidative fibers. These two fiber types are readily apparent in the muscle of poultry and fish: The dark meat (red muscle) is made up of oxidative fibers rich in myoglobin, and the light meat (white muscle) is composed of glycolytic fibers.

Fast-Twitch and Slow-Twitch Fibers   Muscle fibers vary in the speed with which they contract: **Fast-twitch fibers** develop tension two to three times faster than **slow-twitch fibers**. Fast fibers enable brief, rapid, powerful contractions. Compared with a fast fiber, a slow fiber has less sarcoplasmic reticulum and pumps Ca^{2+} more slowly. Because Ca^{2+} remains in the cytosol longer, a muscle twitch in a slow fiber lasts about five times as long as one in a fast fiber.

The difference in contraction speed between slow-twitch and fast-twitch fibers mainly reflects the rate at which their myosin heads hydrolyze ATP. However, there isn't a one-to-one relationship between contraction speed and ATP source. Whereas all slow-twitch fibers are oxidative, fast-twitch fibers can be either glycolytic or oxidative.

Most human skeletal muscles contain both fast-twitch and slow-twitch fibers, although the muscles of the eye and hand are exclusively fast-twitch. In a muscle that has a mixture of fast and slow fibers, the relative proportions of each are genetically determined. However, if such a muscle is used repeatedly for activities requiring high endurance, some fast glycolytic fibers can develop into fast oxidative fibers. Because fast oxidative fibers fatigue more slowly than fast glycolytic fibers, the result will be a muscle that is more resistant to fatigue.

Some vertebrates have skeletal muscle fibers that twitch at rates far faster than any human muscle. For example, superfast muscles produce a rattlesnake's rattle and a dove's coo. Even faster are the muscles surrounding the gas-filled swim bladder of the male oyster toadfish **(Figure 50.34)**. In producing its "boat whistle" mating call, the toadfish can contract and relax these muscles more than 200 times per second!

▼ **Figure 50.34 Specialization of skeletal muscles.** The male oyster toadfish (*Opsanus tau*) uses superfast muscles to produce its mating call.

Other Types of Muscle

Although all muscles share the same fundamental mechanism of contraction—actin and myosin filaments sliding past each other—there are many different types of muscle. Vertebrates, for example, have cardiac muscle and smooth muscle in addition to skeletal muscle (see Figure 40.5).

Vertebrate **cardiac muscle** is found only in the heart and, like skeletal muscle, is striated. Unlike the cells of skeletal muscle fibers, some cardiac muscle cells can initiate rhythmic depolarization and contraction without nervous system input. Normally, however, cells in one part of the heart act as a pacemaker to initiate contraction. Signals from the pacemaker spread throughout the heart because specialized regions called *intercalated disks* electrically couple each cardiac muscle cell to the adjacent cells. It is this coupling that enables action potentials generated in one part of the heart to trigger contraction throughout the organ. Although these action potentials last up to 20 times longer than those of skeletal muscle fibers, a long refractory period prevents summation and tetanus.

Smooth muscle in vertebrates is found in the walls of hollow organs, such as vessels and tracts of the circulatory, digestive, and reproductive systems. In these locations it controls blood flow in the arteries, moves food through the digestive system, carries out uterine contractions during labor, and assists in regulating the temperature of the testicles. Smooth muscle is also found in the eye, where its action controls focusing and pupil diameter.

Smooth muscle cells lack striations because their actin and myosin filaments are not regularly arrayed along the length of the cell. Instead, the thick filaments are scattered throughout the cytoplasm, and the thin filaments are attached to structures called dense bodies, some of which are tethered to the plasma membrane. There is less myosin than in striated muscle fibers, and the myosin is not associated with specific actin strands. Some smooth muscle cells contract only when stimulated by neurons of the autonomic nervous system. Others are electrically coupled to one another and can generate action potentials without input from neurons. Smooth muscles contract and relax more slowly than striated muscles.

Although Ca^{2+} regulates smooth muscle contraction, smooth muscle cells have no troponin complex or T tubules, and their sarcoplasmic reticulum is not well developed. During an action potential, Ca^{2+} enters the cytosol mainly through the plasma membrane. Calcium ions cause contraction by binding to the protein calmodulin, which activates an enzyme that phosphorylates the myosin head, enabling cross-bridge activity.

Invertebrates have muscle cells similar to vertebrate skeletal and smooth muscle cells; in fact, arthropod skeletal muscles are nearly identical to those of vertebrates. However, because the flight muscles of insects contract in response to stretching, the wings of some insects can actually beat up and down faster than action potentials can arrive from the central nervous system. Another interesting evolutionary adaptation has been discovered in the muscles that hold a clam's shell closed. Modification of certain proteins in these muscles allows them to remain contracted for as long as a month with only a low rate of energy consumption.

CONCEPT CHECK 50.5

1. Contrast the role of Ca^{2+} in the contraction of a skeletal muscle fiber and a smooth muscle cell.

2. **WHAT IF?** Why are the muscles of an animal that has recently died likely to be stiff?

3. **MAKE CONNECTIONS** How does the activity of tropomyosin and troponin in muscle contraction compare with the activity of a competitive inhibitor in enzyme action? (See Figure 8.18b.)

For suggested answers, see Appendix A.

CONCEPT 50.6

Skeletal systems transform muscle contraction into locomotion

Converting muscle contraction to movement requires a skeleton—a rigid structure to which muscles can attach. An animal changes its shape or location by contracting muscles connecting two parts of its skeleton. Often muscles are anchored to bone indirectly via connective tissue formed into a tendon.

Because muscles exert force only during contraction, moving a body part back and forth typically requires two

▼ **Figure 50.35 The interaction of muscles and skeletons in movement.** Back-and-forth movement of a body part is generally accomplished by antagonistic muscles. This arrangement works with either an internal skeleton, as in mammals, or an external skeleton, as in insects.

	Human forearm (internal skeleton)	Grasshopper tibia (external skeleton)
Flexion	Biceps / Triceps	Extensor muscle / Flexor muscle
Extension	Biceps / Triceps	Extensor muscle / Flexor muscle

Key ■ Contracting muscle ▨ Relaxing muscle

muscles attached to the same section of the skeleton. We can see such an arrangement in the upper portion of a human arm or grasshopper leg **(Figure 50.35)**. Although we call such muscles an antagonistic pair, their function is actually cooperative, coordinated by the nervous system. For example, when you extend your arm, motor neurons trigger your triceps muscle to contract while the absence of neuronal input allows your biceps to relax.

Vital for movement, animal skeletons also function in support and protection. Most land animals would collapse if they had no skeleton to support their mass. Furthermore, an animal living in water would be formless without a framework to maintain its shape. In many animals, a hard skeleton also protects soft tissues. For example, the vertebrate skull protects the brain, and the ribs of terrestrial vertebrates form a cage around the heart, lungs, and other internal organs.

Types of Skeletal Systems

Although we tend to think of skeletons only as interconnected sets of bones, skeletons come in many different forms. Hardened support structures can be external (as in exoskeletons), internal (as in endoskeletons), or even absent (as in fluid-based, or hydrostatic, skeletons).

Hydrostatic Skeletons

A **hydrostatic skeleton** consists of fluid held under pressure in a closed body compartment. This is the main type of skeleton in most cnidarians, flatworms, nematodes, and annelids (see Figure 33.2). These animals control their form and movement by using muscles to change the shape of fluid-filled compartments. Among the cnidarians, for example, a hydra elongates by closing its mouth and constricting its central gastrovascular cavity using contractile cells in its body wall. Because water maintains its volume under pressure, the cavity must elongate when its diameter is decreased.

Worms carry out locomotion in a variety of ways. In planarians and other flatworms, body movement results mainly from muscles in the body wall exerting localized forces against the interstitial fluid. In nematodes (roundworms), longitudinal muscles contracting around the fluid-filled body cavity move the animal forward by wavelike motions called undulations. In earthworms and many other annelids, circular and longitudinal muscles act together to change the shape of individual fluid-filled segments, which are divided by septa. These shape changes bring about **peristalsis**, a movement produced by rhythmic waves of muscle contractions passing from front to back **(Figure 50.36)**.

Hydrostatic skeletons are well suited for life in aquatic environments. On land, they provide support for crawling and burrowing and may cushion internal organs from shocks. However, a hydrostatic skeleton cannot support walking or running, in which an animal's body is held off the ground.

Exoskeletons

The clam shell you find on a beach once served as an **exoskeleton**, a hard covering deposited on an animal's surface. The shells of clams and most other molluscs are made of calcium carbonate secreted by the mantle, a sheet-like extension of the body wall (see Figure 33.15). Clams and other bivalves close their hinged shell using muscles attached to the inside of this exoskeleton. As the animal grows, it enlarges its shell by adding to the outer edge.

Insects and other arthropods have a jointed exoskeleton called a *cuticle*, a coat secreted by the epidermis. About 30–50% of the arthropod cuticle consists of **chitin**, a polysaccharide similar to cellulose (see Figure 5.8). Fibrils of chitin are embedded in a protein matrix, forming a composite material that combines strength and flexibility. The cuticle may be hardened with organic compounds and, in some cases, calcium salts. In body parts that must be flexible, such as leg joints, the cuticle remains unhardened. Muscles are attached to knobs and plates of the cuticle that extend into the interior of the body. With each growth spurt,

▼ **Figure 50.36 Crawling by peristalsis.** Contraction of the longitudinal muscles thickens and shortens the earthworm; contraction of the circular muscles constricts and elongates it.

1 At the moment depicted, body segments at the earthworm's head end and just in front of the rear end are short and thick (longitudinal muscles contracted; circular muscles relaxed) and are anchored to the ground by bristles. The other segments are thin and elongated (circular muscles contracted; longitudinal muscles relaxed).

2 The head has moved forward because circular muscles in the head segments have contracted. Segments behind the head and at the rear are now thick and anchored, thus preventing the worm from slipping backward.

3 The head segments are thick again and anchored in their new positions. The rear segments have released their hold on the ground and have been pulled forward.

→ **Mastering Biology** Video: Earthworm Locomotion

an arthropod must shed its exoskeleton (molt) and produce a larger one.

Endoskeletons

Animals ranging from sponges to mammals have a hardened internal skeleton, or **endoskeleton**, buried within their soft tissues. In sponges, the endoskeleton consists of hard needle-like structures of inorganic material or fibers made of protein. Echinoderms' bodies are reinforced by ossicles, hard plates composed of magnesium carbonate and calcium carbonate crystals. Whereas the ossicles of sea urchins are tightly bound, the ossicles of sea stars are more loosely linked, allowing a sea star to change the shape of its arms.

Vertebrates have an endoskeleton consisting of cartilage, bone, or some combination of these materials (see Figure 40.5). The mammalian skeleton contains more than 200 bones, some fused together and others connected at joints by ligaments that allow freedom of movement (**Figure 50.37** and **Figure 50.38**). Cells called *osteoblasts* secrete bone matrix and thereby build and repair bone (see Figure 40.5). *Osteoclasts* have an opposite function, resorbing bone components in remodeling of the skeleton.

How thick does an endoskeleton need to be? We can begin to answer this question by applying ideas from civil engineering. The weight of a building increases with the cube of its dimensions. However, the strength of a support depends on its cross-sectional area, which only increases with the square of its diameter. We can thus predict that if we scaled up a mouse to the size of an elephant, the legs of the giant mouse would be too thin to support its weight. Indeed, the body proportions of large animals are very different from those of small ones.

In applying the building analogy, we might also predict that the size of leg bones should be directly proportional to the strain imposed by body weight. Animal bodies, however, are complex and nonrigid. In supporting body weight, it turns out that body posture—the position of the legs relative to the main body—is more important than leg size, at least in mammals and birds. In addition, muscles and tendons, which hold the legs of large mammals relatively straight and positioned under the body, actually bear most of the load.

▼ **Figure 50.37 Bones and joints of the human skeleton.**

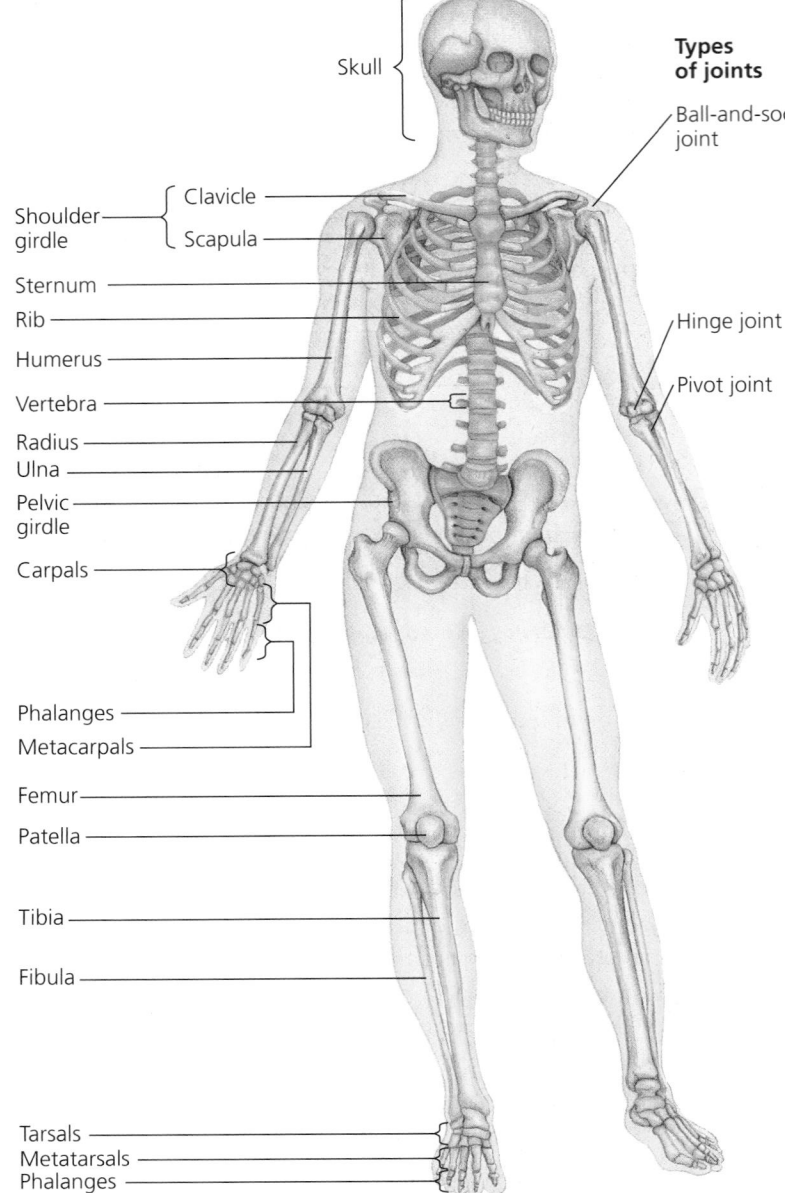

▼ **Figure 50.38 Types of joints.**

Ball-and-socket joint

Ball-and-socket joints are found where the humerus contacts the shoulder girdle and where the femur contacts the pelvic girdle. These joints enable the arms and legs to rotate and move in several planes.

Hinge joint

Hinge joints, such as between the humerus and the head of the ulna, restrict movement to a single plane.

Pivot joint

Pivot joints enable rotating the forearm at the elbow and turning the head from side to side.

Types of Locomotion

Movement is a hallmark of animals. Even animals fixed to a surface move their body parts: Sponges use beating flagella to generate water currents that draw and trap small food particles, and sessile cnidarians wave tentacles that capture prey. Most animals, however, are mobile and spend a considerable portion of their time and energy actively searching for food, escaping from danger, and seeking mates. These activities involve **locomotion**—active travel from place to place.

Friction and gravity tend to keep an animal stationary and therefore oppose locomotion. To move, an animal must expend energy to overcome these two forces. As we will see next, the amount of energy required to oppose friction or gravity is often reduced by an animal body plan adapted for movement in a particular environment.

Locomotion on Land

On land, a walking, running, hopping, or crawling animal must be able to support itself and move against gravity, but air poses relatively little resistance, at least at moderate speeds. When a land animal walks, runs, or hops, its leg muscles expend energy both to propel it and to keep it from falling down. With each step, the animal's leg muscles must overcome inertia by accelerating a leg from a standing start. For moving on land, powerful muscles and strong skeletal support are therefore more important than a streamlined shape.

Diverse adaptations for traveling on land have evolved in various vertebrates. For example, kangaroos have large, powerful muscles in their hind legs, suitable for locomotion by hopping **(Figure 50.39)**. As a kangaroo lands after each leap, tendons in its hind legs momentarily store energy. Much like the energy in a compressed spring, the energy stored in the tendons from one jump is released in the next jump, reducing the energy the animal must expend to travel. The legs of an insect, dog, or human also retain some energy during walking or running, although a considerably smaller share than those of a kangaroo.

Maintaining balance is another prerequisite for walking, running, or hopping. A cat, dog, or horse maintains balance by keeping three feet on the ground when walking. Illustrating the same principle, bipedal animals, such as humans and birds, keep part of at least one foot on the ground when walking. At running speeds, momentum more than foot contact keeps the body upright, enabling all the feet to be off the ground briefly. A kangaroo's large tail forms a stable tripod with its hind legs when the animal sits or moves slowly and helps balance its body during leaps. Recent studies reveal that the tail also generates significant force during hopping, helping to propel the kangaroo's forward movement.

Crawling poses a very different challenge. Having much of its body in contact with the ground, a crawling animal must exert considerable effort to overcome friction. As you have read, earthworms crawl by peristalsis. In contrast, many

▲ **Figure 50.39 Energy-efficient locomotion on land.** Members of the kangaroo family travel from place to place mainly by leaping on their large hind legs. Kinetic energy momentarily stored in tendons after each leap provides a boost for the next leap. In fact, a large kangaroo hopping at 30 km/hr uses no more energy per minute than it does at 6 km/hr. The large tail provides extra force for leaping and helps to balance the kangaroo when it leaps as well as when it sits.

> **Mastering Biology**
> **Interview with Terry Dawson: Kangaroos are "rule" breakers!**

snakes crawl by undulating their entire body from side to side. Waves of bending propagate from head to tail, with each portion of the body eventually following the same undulating path as the head and neck. Some other snakes, such as boa constrictors and pythons, creep in a straight line, driven by muscles that lift belly scales off the ground, tilt the scales forward, and then push them backward against the ground.

Swimming

Because most animals are reasonably buoyant in water, overcoming gravity is less of a problem for swimming animals than for species that move on land or through the air. On the other hand, water is a much denser and more viscous medium than air, and thus drag (friction) is a major problem for aquatic animals. A sleek, fusiform (torpedo-like) shape is a common adaptation of fast swimmers (see Figure 40.2).

Swimming occurs in diverse ways. Many insects and four-legged vertebrates use their legs as oars to push against the water. Sharks and bony fishes swim by moving their body and tail from side to side, while whales and dolphins move by undulating their body and tail up and down. Squids, scallops, and some cnidarians are jet-propelled, taking in water and squirting it out in bursts. In contrast, jellies appear to generate a region of low pressure in the water ahead of them such that they are pulled forward rather than pushed.

Flying

Active flight (in contrast to gliding downward from a tree) has evolved in only a few animal groups: insects, reptiles (including birds), and, among the mammals, bats. One group of flying reptiles, the pterosaurs, died out millions of years ago, leaving birds and bats as the only flying vertebrates.

> **Mastering Biology Video: Soaring Hawk**

Interpreting a Graph with Log Scales

What Are the Energy Costs of Locomotion? In the 1960s, animal physiologist Knut Schmidt-Nielsen, at Duke University, wondered whether general principles govern the energy costs of different forms of locomotion among diverse animal species. To answer this question, he drew on his own experiments as well as those of other researchers. In this exercise you will analyze the combined results of these studies and evaluate the rationale for plotting the experimental data on a graph with logarithmic scales.

How the Experiments Were Done Researchers measured the rate of oxygen consumption or carbon dioxide production in animals that ran on treadmills, swam in water flumes, or flew in wind tunnels. For example, a tube connected to a plastic face mask collected gases exhaled by a parakeet during flight (see photo). From these measurements, Schmidt-Nielsen calculated the amount of energy each animal used to transport a given amount of body mass over a given distance [calories/(kilogram · meter)].

Data from the Experiments Schmidt-Nielsen plotted the cost of running, flying, and swimming versus body mass on a single graph with logarithmic (log) scales for the axes. He then drew a best-fit straight line through the data points for each form of locomotion. (On the graph here, the individual data points are not shown.)

Data from K. Schmidt-Nielsen, Locomotion: Energy cost of swimming, flying, and running, *Science* 177:222–228 (1972). Reprinted with permission from AAAS.

INTERPRET THE DATA

1. The body masses of the animals used in these experiments ranged from about 0.001 g to 1,000,000 g, and their rates of energy use ranged from about 0.1 cal/(kg · m) to 100 cal/(kg · m). If you were to plot these data on a graph with linear instead of log scales for the axes, how would you draw the axes so that all of the data would be visible? What is the advantage of using log scales for plotting data with a wide range of values? (For additional information about graphs, see the Scientific Skills Review in Appendix D.)

2. Based on the graph, how much greater is the energy cost of flying for an animal that weighs 10^{-3} g than for an animal that weighs 1 g? For any given form of locomotion, which travels more efficiently, a larger animal or smaller animal?

3. The slopes of the flying and swimming lines are very similar. Based on your answer to question 2, if the energy cost of a 2-g swimming animal is 1.2 cal/(kg · m), what is the estimated energy cost of a 2-kg swimming animal?

4. Considering animals with a body mass of about 100 g, rank the three forms of locomotion from highest energy cost to lowest energy cost. Were these the results you expected, based on your own experience? What could explain the energy cost of running compared to that of flying or swimming?

5. Schmidt-Nielson calculated the swimming cost in a mallard duck and found that it was nearly 20 times as high as the swimming cost in a salmon of the same body mass. What could explain the greater swimming efficiency of salmon?

➜ **Instructors:** A version of this Scientific Skills Exercise can be assigned in **Mastering Biology**.

To fly, an animal's wings must develop enough lift to overcome gravity's downward force. The key to meeting this challenge is wing shape. All wings act as airfoils—structures whose shape alters air currents in a way that helps animals or airplanes stay aloft. As for the body to which the wings attach, a fusiform shape helps reduce drag in air as it does in water.

Flying animals are relatively light, with body masses ranging from less than a gram for some insects to about 20 kg for the largest flying birds. The low body mass of many flying animals is due to specialized structural adaptations. Birds, for example, have no urinary bladder or teeth and have relatively large bones with air-filled regions that help lessen the bird's weight (see Figure 34.30).

Flying, running, and swimming each impose different energetic demands on animals. In the **Scientific Skills Exercise**, you can interpret a graph that compares the relative energy costs of these three forms of locomotion.

CONCEPT CHECK 50.6

1. Contrast swimming and flying in terms of the main problems they pose and the adaptations that allow animals to overcome those problems.

2. **MAKE CONNECTIONS.** Peristalsis contributes to the locomotion of many annelids and to the movement of food in the digestive tract (see Concept 41.3). Using the muscles of your hand and a toothpaste tube as a model of peristalsis, how would your demonstration differ for the two processes?

3. **WHAT IF?** When using your arms to lower yourself into a chair, you bend your arms without using your biceps. Explain how this is possible. (Hint: Think about gravity as an antagonistic force.)

For suggested answers, see Appendix A.

50 Chapter Review

→ Go to **Mastering Biology** for Assignments, the eText, the Study Area, and Dynamic Study Modules.

SUMMARY OF KEY CONCEPTS

→ To review key terms, go to the **Vocabulary Self-Quiz** in the **Mastering Biology** eText or Study Area, or go to goo.gl/zkjz9t.

CONCEPT 50.1

Sensory receptors transduce stimulus energy and transmit signals to the central nervous system (pp. 1108–1112)

- The detection of a stimulus precedes **sensory transduction**, the change in the membrane potential of a **sensory receptor** in response to a stimulus. The resulting **receptor potential** controls transmission of action potentials to the CNS, where sensory information is integrated to generate **perceptions**. The frequency of action potentials in an axon and the number of axons activated reflect stimulus strength. The identity of the axon carrying the signal encodes the nature or quality of the stimulus.
- **Mechanoreceptors** respond to stimuli such as pressure, touch, stretch, motion, and sound. **Chemoreceptors** detect either total solute concentrations or specific molecules. **Electromagnetic receptors** detect different forms of electromagnetic radiation. **Thermoreceptors** signal surface and core temperatures of the body. Pain is detected by a group of **nociceptors** that respond to excess heat, pressure, or specific classes of chemicals.

? *To simplify sensory receptor classification, why might it make sense to eliminate nociceptors as a distinct class?*

CONCEPT 50.2

In hearing and equilibrium, mechanoreceptors detect moving fluid or settling particles (pp. 1112–1116)

- Most invertebrates sense their orientation with respect to gravity by means of **statocysts**. Specialized **hair cells** form the basis for hearing and balance in mammals and for detection of water movement in fishes and aquatic amphibians. In mammals, the **tympanic membrane** (eardrum) transmits sound waves to bones of the **middle ear**, which transmit the waves through the **oval window** to the fluid in the coiled **cochlea** of the **inner ear**. Pressure waves in the fluid vibrate the basilar membrane, depolarizing hair cells and triggering action potentials that travel via the auditory nerve to the brain. Receptors in the inner ear function in balance and equilibrium.

? *How are music volume and pitch encoded in signals to the brain?*

CONCEPT 50.3

The diverse visual receptors of animals depend on light-absorbing pigments (pp. 1117–1123)

- Invertebrates have varied light detectors, including simple light-sensitive eyespots, image-forming **compound eyes**, and **single-lens eyes**. In the vertebrate eye, a single **lens** is used to focus light on **photoreceptors** in the **retina**. Both **rods** and **cones** contain a pigment, **retinal**, chemically bonded to a protein (**opsin**). Absorption of light by retinal triggers a signal transduction pathway that hyperpolarizes the photoreceptors, causing them to release less neurotransmitter. Synapses transmit information from photoreceptors to cells that integrate information and convey it to the brain along axons that form the optic nerve.

? *How does processing of sensory information sent to the vertebrate brain in vision differ from that in hearing or olfaction?*

CONCEPT 50.4

The senses of taste and smell rely on similar sets of sensory receptors (pp. 1123–1125)

- Taste (**gustation**) and smell (**olfaction**) depend on stimulation of chemoreceptors by small dissolved molecules. In humans, sensory cells in **taste buds** express a receptor type specific for one of the five taste perceptions: sweet, sour, salty, bitter, and umami (elicited by glutamate). Olfactory receptor cells line the upper part of the nasal cavity. Hundreds of different genes code for membrane proteins that bind to specific classes of **odorants**, and each receptor cell appears to express only one of those genes.

? *Why is the flavor of food less intense when you have a head cold?*

CONCEPT 50.5

The physical interaction of protein filaments is required for muscle function (pp. 1125–1132)

- The muscle cells (fibers) of vertebrate **skeletal muscle** contain **myofibrils** composed of **thin filaments** of (mostly) actin and **thick filaments** of myosin. These filaments are organized into repeating units called **sarcomeres**. Myosin heads, energized by the hydrolysis of ATP, bind to the thin filaments, form cross-bridges, and then release upon binding ATP anew. As this cycle repeats, the thick and thin filaments slide past each other, shortening the sarcomere and contracting the muscle fiber.

- Motor neurons release acetylcholine, triggering action potentials in muscle fibers that stimulate the release of Ca^{2+} from the **sarcoplasmic reticulum**. When the Ca^{2+} binds the **troponin complex**, **tropomyosin** moves, exposing the myosin-binding sites on actin and thus initiating cross-bridge formation. A **motor unit** consists of a motor neuron and the muscle fibers it controls. A twitch results from one action potential. Skeletal muscle fibers are **slow-twitch** or **fast-twitch** and oxidative or glycolytic.
- **Cardiac muscle**, found in the heart, consists of striated cells electrically connected by intercalated disks. Nervous system input controls the rate at which the heart contracts, but is not strictly required for cardiac muscle contraction. In **smooth muscles**, contractions are initiated by the muscles or by stimulation from neurons in the autonomic nervous system.

? *What are two major functions of ATP hydrolysis in skeletal muscle activity?*

CONCEPT 50.6

Skeletal systems transform muscle contraction into locomotion (pp. 1132–1136)

- Skeletal muscles, often in antagonistic pairs, contract and pull against the skeleton. Skeletons may be **hydrostatic** and maintained by fluid pressure, as in worms; hardened into **exoskeletons**, as in insects; or in the form of **endoskeletons**, as in vertebrates.
- Each form of **locomotion**—swimming, movement on land, or flying—presents a particular challenge. For example, swimmers need to overcome substantial friction, but face less of a challenge from gravity than do animals that move on land or fly.

? *Explain how microscopic and macroscopic anchoring of muscle filaments enables you to bend your elbow.*

TEST YOUR UNDERSTANDING

→ For more multiple-choice questions, go to the **Practice Test** in the **Mastering Biology** eText or Study Area, or go to goo.gl/GruWRg.

Levels 1-2: Remembering/Understanding

1. Which sensory receptor—category pair is correct?
   (A) hair cell—nociceptor
   (B) snake pit organ—mechanoreceptor
   (C) taste receptor—chemoreceptor
   (D) olfactory receptor—electromagnetic receptor

2. The middle ear converts
   (A) air pressure waves to fluid pressure waves.
   (B) air pressure waves to nerve impulses.
   (C) fluid pressure waves to nerve impulses.
   (D) pressure waves to hair cell movements.

3. During vertebrate skeletal muscle contraction, calcium ions
   (A) break cross-bridges as a cofactor in hydrolysis of ATP.
   (B) bind to troponin, exposing myosin-binding sites.
   (C) transmit action potentials to the muscle fiber.
   (D) spread action potentials through the T tubules.

Levels 3-4: Applying/Analyzing

4. The human brain differentiates tastes from smells because action potentials for the two sensations differ in
   (A) magnitude and shape.
   (B) threshold potential.
   (C) where they are received in the brain.
   (D) how long they take to reach the brain.

5. The transduction of sound waves into action potentials occurs
   (A) in the tectorial membrane as it is stimulated by hair cells.
   (B) when hair cells are bent against the tectorial membrane, causing them to depolarize and release neurotransmitter that stimulates afferent neurons.
   (C) as the basilar membrane vibrates at different frequencies in response to the varying volume of sounds.
   (D) within the middle ear as the vibrations are amplified by the malleus, incus, and stapes.

Levels 5-6: Evaluating/Creating

6. Metal objects cause sharks to misdirect their bites. Sharks also can find batteries buried under sand. These facts suggest that sharks track their prey just before they bite in the same way that
   (A) a rattlesnake finds a mouse in its burrow.
   (B) an insect avoids being stepped on.
   (C) a star-nosed mole locates its prey in tunnels.
   (D) a platypus locates its prey in a muddy river.

7. **DRAW IT** Based on the information in the text, fill in the following graph. Use one line for rods and another line for cones.

Position along retina (in degrees away from fovea)

8. **EVOLUTION CONNECTION** In general, locomotion on land will require more energy than locomotion in water. By integrating what you learned about animal form and function in Unit 7, discuss some of the evolutionary adaptations of mammals that support the high energy requirements for moving on land.

9. **SCIENTIFIC INQUIRY • INTERPRET THE DATA** To demonstrate how energy is stored in tendons during hopping, an instructor asked student volunteers to hop at a frequency that felt "natural" and then, after resting, to hop at exactly half that frequency. Hopping was done at a standard height and volunteers' mass, O_2 consumption, and CO_2 production were measured. Here is a representative set of results calculated for one student.

Frequency (hops/sec)	Energy used (joules/sec)
1.85	735
0.92	716

The student consumed 159 joules/sec when standing. For each hop frequency, subtract this standing value from the energy used during hopping. Then divide by the hop frequency to calculate the energy cost per hop. How does the energy cost per hop differ at the two frequencies, and how might this be related to energy storage in tendons?

10. **WRITE ABOUT A THEME: ORGANIZATION** In a short essay (100–150 words), describe three ways in which the structure of the lens of the human eye is well adapted to its function in vision.

11. **SYNTHESIZE YOUR KNOWLEDGE**

Bloodhounds, which are adept at following a scent trail even days old, have no more olfactory receptor genes than other dogs. Predict how the sensory and nervous systems of bloodhounds differ from those of other dogs in ways that contribute to their tracking ability.

For selected answers, see Appendix A.

51 Animal Behavior

KEY CONCEPTS

51.1 Discrete sensory inputs can stimulate both simple and complex behaviors *p. 1140*

51.2 Learning establishes specific links between experience and behavior *p. 1143*

51.3 Selection for individual survival and reproductive success can explain diverse behaviors *p. 1148*

51.4 Genetic analyses and the concept of inclusive fitness provide a basis for studying the evolution of behavior *p. 1154*

Study Tip

Identifying components of experimental design: Animal behavior is studied in both the wild and the laboratory and involves approaches ranging from passive observation to active intervention. As you study this chapter, note how scientific methods are applied to answer questions in diverse contexts, such as foraging in honeybees (Figure 51.5; see example started here) and fruit flies (Figure 51.13).

1) **Make observations**—returning bees do a number of different but set dances
   ↓
2) **Form a hypothesis**—the dances convey information about food location
   ↓
3) **Generate predictions**
   ↓
4) **Gather and analyze data**
   ↓
5) **Draw conclusions**

→ Go to Mastering Biology

For Students (in eText and Study Area)
• Get Ready for Chapter 51
• HHMI Videos: Studying Elephant Communication; How Lizards Find Their Way Home

For Instructors to Assign (in Item Library)
• Interpret the Data: Evolution of Foraging Behavior
• Tutorial: Animal Behavior and Learning
• Scientific Skills Exercise: Crow Foraging Behavior

Figure 51.1 A male magnificent frigatebird (*Fregata magnificens*) lives mostly at sea. He often feeds by piracy, harassing diving birds into coughing up their catch for his consumption. Periodically, however, he alights on an island, points his bill to the sky, and inflates an enormous red throat pouch. Clacking his beak, he produces a drumming sound that resonates in the pouch. If a curious female takes notice and lands nearby, a mating ritual ensues.

What questions do biologists seek to answer in studying animal behavior?

1. What event or signal triggers the behavior?

Males inflate their pouch only if females are nearby. The intensity of the display increases when a female circles overhead.

2. How does experience during growth and development influence the behavior?

Older males have larger pouches that produce lower drumming frequencies, which attract more females.

3. How does the behavior aid survival and reproduction?

After mating, males and females cooperate in nest building and feeding the chicks.

4. How has the behavior been shaped by natural selection?

The throat pouch can be inflated prominently for mating, and deflated to maximize flight efficiency.

Answering these questions can provide fundamental insights into **how** behaviors occur and **why** they arise.

Discrete sensory inputs can stimulate both simple and complex behaviors

A **behavior** is an action carried out by muscles under control of the nervous system. Examples include an animal using its throat muscles to produce a song or releasing a scent to mark its territory. Behavior is an essential part of acquiring nutrients and finding a partner for sexual reproduction. Behavior also contributes to homeostasis, as when honeybees huddle to conserve heat (see Concept 40.3). In short, all of animal physiology contributes to behavior, and behavior influences all of physiology.

Many behaviors, especially those involved in recognition and communication, rely on specialized body structures or form. For instance, the throat pouch of a male frigatebird (see Figure 51.1) is essential for courtship. Inflating the pouch exposes its vibrantly colored surface and provides a resonating chamber that amplifies the male's drumming. As this example illustrates, the process of natural selection that shapes behaviors also influences the evolution of animal anatomy.

What approach do biologists use to determine how behaviors arise and what functions they serve? The Dutch scientist Niko Tinbergen, a pioneer in the study of animal behavior, suggested that understanding any behavior requires answering four questions, which were highlighted in Figure 51.1:

1. What stimulus triggers the behavior and how do the various body systems bring it about?
2. How does the animal's experience during growth and development influence the response to the stimulus?
3. How does the behavior aid survival and reproduction?
4. What is the behavior's evolutionary history?

The first two questions ask about *proximate causation—how* a behavior occurs or is modified. The last two questions ask about *ultimate causation—why* a behavior occurs in the context of natural selection.

Experiments on proximate causation by Tinbergen earned him a share of a Nobel Prize awarded in 1973. Such studies defined the field of *ethology*, the study of animal behavior observed in a natural environment. We'll consider those and related experiments in the early part of the chapter. The concept of ultimate causation is central to **behavioral ecology**, the study of the ecological and evolutionary basis for animal behavior. We'll explore this vibrant area of modern biological research in the rest of the chapter.

Fixed Action Patterns

In addressing Tinbergen's first question, the nature of the stimuli that trigger behavior, we'll begin with behavioral responses to well-defined stimuli, starting with an example from one of Tinbergen's own experiments.

As part of his research, Tinbergen kept fish tanks containing three-spined sticklebacks (*Gasterosteus aculeatus*), a species in which males, but not females, have red bellies. Male sticklebacks attack other males that invade their nesting territories **(Figure 51.2)**. Tinbergen noticed that his male sticklebacks also behaved aggressively when a red truck passed within view of their tank. Inspired by this chance observation, he carried out experiments showing that the red color of an intruder's underside is the proximate cause of the attack behavior. A male stickleback will not attack a fish lacking red coloration, but will attack even unrealistic models if they contain areas of red color. The territorial response of male sticklebacks is an example of a **fixed action pattern**, a sequence of unlearned acts directly linked to a simple stimulus. The fixed action pattern of certain moths takes the form of evasive flight maneuvers—loops and spirals—performed instantaneously upon hearing the sounds of an echolocating bat or an ultrasonic whistle in the same frequency range. Fixed action patterns are essentially unchangeable and, once initiated, are usually carried to completion. The trigger for the behavior is an external cue called a **sign stimulus**, such as a red object that prompts a male stickleback's aggressive behavior.

Migration

Environmental stimuli not only trigger behaviors but also provide cues that animals use to carry out those behaviors. For example, a wide variety of birds, fishes, and other animals use environmental cues to guide **migration**—a regular, long-distance change in location. In the course of migration, many animals pass through environments they have not

▶ **Figure 51.2 Sign stimuli in a classic fixed action pattern. (a)** A male stickleback fish attacks other male sticklebacks that invade its nesting territory. **(b)** Models, whether realistic or not, can trigger the aggressive behavior, but only if the underside is red. Researchers concluded that the red belly of the intruding male is a sign stimulus that releases the aggressive behavior.

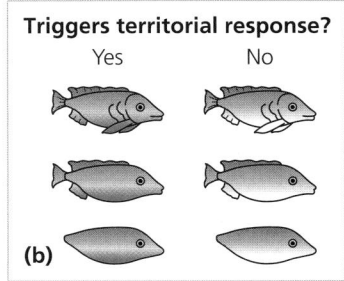

? *Suggest an explanation for why this behavior might have evolved (its ultimate causation).*

previously encountered. How, then, do they find their way in these foreign settings?

Some migrating animals track their position relative to the sun, even though the sun's position relative to Earth changes throughout the day. Animals can adjust for these changes by means of a *circadian clock*, an internal mechanism that has a 24-hour periodicity (see Concept 49.2). For example, experiments have shown that migrating birds orient differently relative to the sun at distinct times of the day. Nocturnal animals can instead use the North Star, which has a constant position in the night sky.

Although the sun and stars can provide useful clues for navigation, clouds can obscure these landmarks. How do migrating animals overcome this problem? A simple experiment with homing pigeons provided one answer. On an overcast day, placing a small magnet on the head of a homing pigeon prevented it from returning efficiently to its roost. Researchers concluded that pigeons sense their position relative to Earth's magnetic field and can thereby navigate without solar or celestial cues.

Behavioral Rhythms

Although the circadian clock plays a small but significant role in navigation by some migrating species, it has a major role in the daily activity of all animals. As discussed in Concepts 40.2 and 49.2, the clock is responsible for a circadian rhythm, a daily cycle of rest and activity. The clock is normally synchronized with the light and dark cycles of the environment but can maintain rhythmic activity even under constant environmental conditions, such as during hibernation.

Some behaviors, such as migration and reproduction, reflect biological rhythms with a longer cycle, or period, than the circadian rhythm. Behavioral rhythms linked to the yearly cycle of seasons are called *circannual rhythms*. Although migration and reproduction typically correlate with food availability, these behaviors are not a direct response to changes in food intake. Instead, circannual rhythms, like circadian rhythms, are influenced by the periods of daylight and darkness in the environment. For example, studies with several bird species have shown that an artificial environment with extended daylight can induce out-of-season migratory behavior.

Not all biological rhythms are linked to the light and dark cycles in the environment. Consider, for instance, fiddler crab courtship behavior, which is linked to the lunar cycle. Possessed with front claws of very different size, the male fiddler crab (genus *Uca*) waves the larger claw in front of him to attract potential mates **(Figure 51.3)**. Timing this claw-waving courtship behavior to the new or full moon helps the development of offspring. How? Fiddler crabs begin their lives as larvae settling in the mudflats. The tides disperse larvae to deeper waters, where they complete early development

▼ **Figure 51.3 Courtship display of male fiddler crab.**

in relative safety before returning to the tidal flats. By courting at the time of the new or full moon, crabs link their reproduction to the times of greatest tidal movement.

Animal Signals and Communication

Claw waving by fiddler crabs during courtship is an example of one animal (the male crab) generating the stimulus that guides the behavior of another animal (the female crab). A stimulus transmitted from one organism to another is called a **signal**. The transmission and reception of signals between animals constitute **communication**, which often has a role in the proximate causation of behavior.

 Mastering Biology Video: Albatross Courtship Ritual • Video: Giraffe Courtship Ritual HHMI Video: Studying Elephant Communication **hhmi** BioInteractive

Forms of Animal Communication

Let's consider the courtship behavior of the fruit fly, *Drosophila melanogaster*, as an introduction to the four common modes of animal communication: visual, chemical, tactile, and auditory.

Fruit fly courtship constitutes a *stimulus-response chain*, in which the response to each stimulus is itself a stimulus for the next behavior. In the first step, a male detects a female in his field of vision and orients his body toward hers. To confirm she belongs to his species, he uses his olfactory system to detect chemicals she releases into the air. The male then approaches and touches the female with a foreleg **(Figure 51.4)**. This touching, or tactile communication, alerts the female to the male's presence. In the third stage, the male extends and vibrates one wing, producing a courtship song. This auditory

▼ **Figure 51.4 Male fruit fly tapping female with foreleg.**

communication informs the female whether he is of the same species. Only if all of these forms of communication are successful will the female allow the male to attempt copulation.

In general, the form of communication that evolves is closely related to an animal's lifestyle and environment. For example, most terrestrial mammals are nocturnal, which makes visual displays relatively ineffective. Instead, these species use olfactory and auditory signals, which work as well in the dark as in the light. In contrast, most birds are diurnal (active mainly in daytime) and communicate primarily by visual and auditory signals. Humans are also diurnal and, like birds, use primarily visual and auditory communication. We can thus detect and appreciate the songs and bright colors used by birds to communicate but miss many chemical cues on which other mammals base their behavior.

The information content of animal communication varies considerably. One of the most remarkable examples is the symbolic language of the European honeybee (*Apis mellifera*), discovered in the early 1900s by Austrian researcher Karl von Frisch. Using glass-walled observation hives, he and his students spent several decades observing honeybees. Methodical recordings of bee movements enabled von Frisch to decipher a "dance language" that returning foragers use to inform other bees about the distance and direction of travel to a source of nectar.

When a successful forager returns to the hive, its movements, as well as sounds and odors, quickly become the center of attention for other bees, called followers. Moving along the vertical wall of the honeycomb, the forager performs a "waggle dance" that communicates to the follower bees both the direction and distance of the food source in relation to the hive **(Figure 51.5)**. In performing the dance, the bee follows a half-circle swing in one direction, a straight run during which it waggles its abdomen, and a half-circle swing in the other direction. What von Frisch and colleagues deduced was that the angle of the straight run relative to the hive's vertical surface indicates the horizontal angle of the food in relation to the sun. Thus, if the returning bee runs at a 30° angle to the right of vertical, the follower bees leaving the hive fly 30° to the right of the horizontal direction of the sun.

How does the waggle dance communicate distance to the nectar source? It turns out that a dance with a longer straight run, and therefore more abdominal waggles per run, indicates a greater distance to the food found by the forager. As follower bees exit the hive, they fly almost directly to the area indicated by the waggle dance. By using flower odor and other clues, they locate the source of nectar within this area.

If the food source is close to the hive (less than 50 m away), the waggle dance takes a slightly different form that primarily advertises the availability of nectar nearby. In this form of the waggle dance, which von Frisch called "round," the returning bee moves in tight circles while moving its abdomen from side to side. In response, the follower bees leave the hive and search in all directions for nearby flowers rich in nectar.

Pheromones

Animals that communicate through odors or tastes emit chemical substances called **pheromones**. Pheromones are especially common among mammals and insects and often relate to reproductive behavior. For example, pheromones are the basis for the chemical communication in fruit fly

▼ **Figure 51.5 Honeybee dance language.** Honeybees returning to the hive communicate the location of food sources through the symbolic language of a dance.

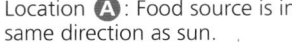

Location **A**: Food source is in same direction as sun.

Location **B**: Food source is in direction opposite sun.

Location **C**: Food source is 30° to right of sun.

The waggle dance, performed when food is distant, resembles a figure eight. Distance is indicated by the number of abdominal waggles performed in the straight-run part of the dance. Direction is indicated by the angle of the straight run relative to the vertical surface of the hive. The dance communicates this information to workers clustered around the dancing bee, directing them to food at a particular location.

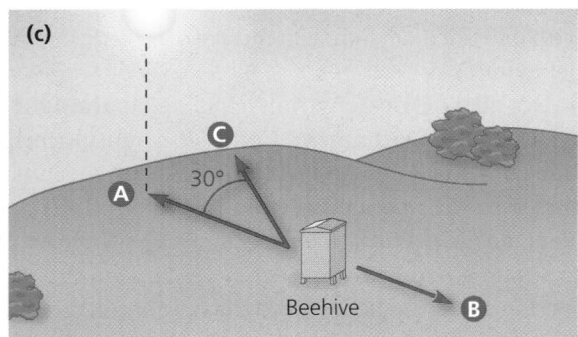

VISUAL SKILLS *What information, if any, might be conveyed by the portions of the waggle dance between the straight runs? Explain.*

courtship (see Figure 51.4). Pheromones are not limited to short-distance signaling, however. Male silkworm moths have receptors that can detect the pheromone from a female moth from several kilometers away (see Figure 50.6).

In a honeybee colony, pheromones produced by the queen and her daughters, the workers, maintain the hive's complex social order. One pheromone (once called the queen substance) has a particularly wide range of effects. It attracts workers to the queen, inhibits development of ovaries in workers, and attracts males (drones) to the queen during her mating flights out of the hive.

Pheromones can also serve as alarm signals. For example, when a minnow is injured, a substance released from the fish's skin disperses in the water, inducing a fright response in other minnows in the area. These nearby fish become more vigilant and often form tightly packed schools near the river or lake bottom, where they are safer from attack (**Figure 51.6**). Pheromones can be very effective at remarkably low concentrations. For instance, just 1 cm² of skin from a fathead minnow contains sufficient alarm substance to induce a reaction in 58,000 L of water.

So far in this chapter, we have explored the types of stimuli that elicit behaviors—the first part of Tinbergen's first question. The second part of that question—the physiological mechanisms that mediate responses—involves the nervous, muscular, and skeletal systems explored elsewhere in this unit: Stimuli activate sensory systems, are processed in the central nervous system, and result in motor outputs that constitute

▼ **Figure 51.6 Fathead minnows (*Pimephales promelas*) responding to the presence of an alarm substance.**

① Minnows are widely dispersed in an aquarium before an alarm substance is introduced.

② Within seconds of the alarm substance being introduced, minnows aggregate near the bottom of the aquarium and reduce their movement.

behavior. Thus, we are ready to focus on Tinbergen's second question—how experience influences behavior.

CONCEPT CHECK 51.1

1. If an egg rolls out of a greylag goose's nest, the mother goose will retrieve it by nudging it with her beak and head. If researchers remove the egg or substitute a ball during this process, the goose continues to bob her beak and head while she moves back to the nest. Explain how and why this behavior occurs.

2. **WHAT IF?** Suppose you exposed various fish species from the minnows' environment to the alarm substance from minnows. Thinking about natural selection, suggest why some species might respond like minnows, some might increase their activity, and some might show no change.

3. **MAKE CONNECTIONS** How is the lunar-linked rhythm of fiddler crab courtship similar in mechanism and function to the seasonal timing of plant flowering? (See Concept 39.3.)

For suggested answers, see Appendix A.

CONCEPT 51.2

Learning establishes specific links between experience and behavior

Some behaviors—such as a fixed action pattern, a courtship stimulus-response chain, or pheromone signaling—are performed by all individuals the same way each time. Behavior that is developmentally fixed in this way is known as **innate behavior**. Consider, for example, the construction of a spider's web. Although a spider's web is quite complex, web building is innate, not learned. Many other behaviors, however, vary considerably with experience.

▶ **A spider in family Theridiidae spinning a web**

Experience and Behavior

Tinbergen's second question asks how an animal's experiences during growth and development influence the response to stimuli. One informative approach to this question is a *cross-fostering study*, in which the young of one species are placed in the care of adults from another species in the same or a similar environment. The extent to which the offspring's behavior changes in such a situation provides a measure of how the social and physical environment influences behavior.

Certain mouse species have behaviors well suited for cross-fostering studies. Male California mice (*Peromyscus californicus*)

Table 51.1	Influence of Cross-Fostering on Male Mice*		
Species	Aggression Toward an Intruder	Aggression in Neutral Situation	Paternal Behavior
California mice fostered by white-footed mice	Reduced	No difference	Reduced
White-footed mice fostered by California mice	No difference	Increased	No difference

*Comparisons are with mice raised by parents of their own species.

are highly aggressive toward other mice and provide extensive parental care. In contrast, male white-footed mice (*Peromyscus leucopus*) are less aggressive and engage in little parental care. Cross-fostering—placing pups in the nests of the other species—altered some behaviors of both species **(Table 51.1)**. For instance, male California mice raised by white-footed mice were less aggressive toward intruders. Thus, experience during development can strongly influence aggressive behavior in these rodents.

One of the most important findings of the cross-fostering experiments with mice was that the influence of experience on behavior can be passed on to progeny: When the cross-fostered California mice became parents, they spent less time retrieving offspring who wandered off than did California mice raised by their own species. Thus, experience during development can modify physiology in a way that alters parental behavior, extending the influence of environment to a subsequent generation.

For humans, the influence of genetics and environment on behavior can be explored by a *twin study*, in which researchers compare the behavior of identical twins raised apart with the behavior of those raised in the same household. Twin studies have been instrumental in studying disorders that alter human behavior, such as anxiety disorders, schizophrenia, and alcoholism.

▼ **Identical twins who were raised separately**

Learning

One powerful way that an animal's environment can influence its behavior is through **learning**, the modification of behavior as a result of specific experiences. The capacity for learning depends on nervous system organization established during development following instructions encoded in the genome. Learning itself involves the formation of memories by specific changes in neuronal connectivity (see Concept 49.4). Therefore, the essential challenge for research

into learning is not to decide between nature and nurture, but rather to explore the contributions of *both* nature (genes) and nurture (environment) in shaping learning and, more generally, behavior.

Imprinting

In some species, the ability of offspring to recognize and be recognized by a parent is essential for survival. In the young, this learning often takes the form of **imprinting**, the establishment of a long-lasting behavioral response to a particular individual or object. Imprinting can take place only during a specific time period in development, called the **sensitive period**. Among gulls, for instance, the sensitive period for a parent to bond with its young lasts one to two days. During the sensitive period, the young imprint on their parent and learn basic behaviors, while the parent learns to recognize its offspring. If bonding does not occur, the parent will not care for the offspring, leading to the death of the offspring and a decrease in the reproductive success of the parent.

How do the young know on whom—or what—to imprint? Experiments with many species of waterfowl indicate that young birds have no innate recognition of "mother." Rather, they identify with the first object they encounter that has certain key characteristics. In the 1930s, experiments showed that the principal imprinting stimulus in greylag geese (*Anser anser*) is a nearby object that is moving away from the young. When incubator-hatched goslings spent their first few hours with a person rather than with a goose, they imprinted on the human and steadfastly followed that person from then on **(Figure 51.7)**. Furthermore, they showed no recognition of their biological mother.

Imprinting has become an important component of efforts to save endangered species, such as the whooping crane (*Grus americana*). Scientists tried raising whooping cranes in captivity by using sandhill cranes (*Grus canadensis*) as foster parents. However, because the whooping cranes imprinted on their foster parents, none formed a *pair-bond* (strong attachment) with a whooping crane mate. To avoid such problems, captive breeding programs now isolate young cranes, exposing them to the sights and sounds of members of their own species.

Until recently, scientists made further use of imprinting to teach cranes born in captivity to migrate along safe routes. Young whooping cranes were imprinted on humans in "crane suits" and then allowed to follow these "parents" as they flew ultralight aircraft along selected migration routes. Beginning in 2016, efforts shifted to a focus on minimizing human intervention as part of an overall strategy aimed at fostering self-sustaining crane populations.

Spatial Learning and Cognitive Maps

Every natural environment has spatial variation, as in locations of nest sites, hazards, food, and prospective mates.

▼ **Figure 51.7 Imprinting.** Young graylag geese imprinted on a man.

WHAT IF? *Suppose the geese were bred to each other. How might their imprinting on a human affect their offspring? Explain.*

→ **Mastering Biology** Video: Ducklings

Therefore, an organism's fitness may be enhanced by the capacity for **spatial learning**, the establishment of a memory that reflects the environment's spatial structure.

The idea of spatial learning intrigued Tinbergen while he was a graduate student in the Netherlands. At that time, he was studying females of a digger wasp species (*Philanthus triangulum*) that nests in small burrows dug into sand dunes. When a wasp leaves her nest to go hunting, she hides the entrance from potential intruders by covering it with sand. When she returns, however, she flies directly to her hidden nest, despite the presence of hundreds of other burrows in the area. How does she accomplish this feat? Tinbergen hypothesized that a wasp locates her nest by learning its position relative to visible landmarks. To test his hypothesis, he carried out an experiment in the wasps' natural habitat **(Figure 51.8)**. By manipulating objects around nest entrances, he demonstrated that digger wasps engage in spatial learning. This experiment was so simple and informative that it could be summarized very concisely. In fact, at 32 pages, Tinbergen's Ph.D. thesis from 1932 is still the shortest ever approved at Leiden University.

In some animals, spatial learning involves formulating a **cognitive map**, a representation in an animal's nervous

▼ **Figure 51.8 Inquiry**

Does a digger wasp use landmarks to find her nest?

Experiment A female digger wasp covers the entrance to her nest while foraging for food, but finds the correct wasp nest reliably upon her return 30 minutes or more later. Niko Tinbergen wanted to test the hypothesis that a wasp learns visual landmarks that mark her nest before she leaves on hunting trips. First, he marked one nest with a ring of pinecones while the wasp was in the burrow. After leaving the nest to forage, the wasp returned to the nest successfully.

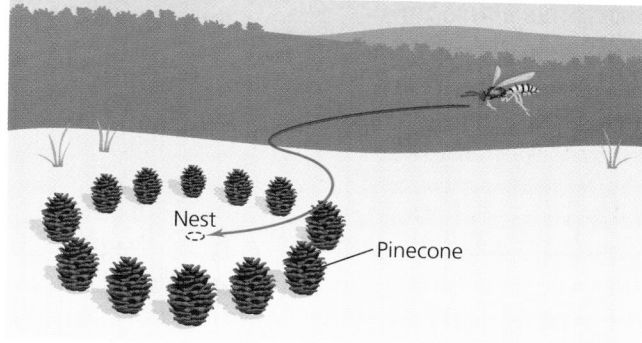

Two days later, after the wasp had again left, Tinbergen shifted the ring of pinecones away from the nest. Then he waited to observe the wasp's behavior.

Results When the wasp returned, she flew to the center of the pinecone circle instead of to the nearby nest. Repeating the experiment with many wasps, Tinbergen obtained the same results.

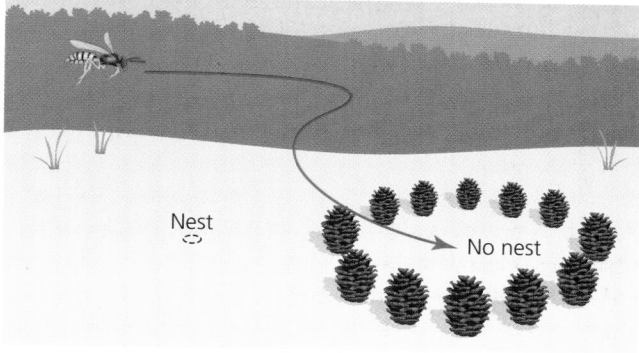

Conclusion The experiment supported the hypothesis that digger wasps use visual landmarks to keep track of their nests.

Data from N. Tinbergen, *The Study of Instinct,* Clarendon Press, Oxford (1951).

WHAT IF? Suppose the digger wasp had returned to her original nest site, despite the pinecones having been moved. What alternative hypotheses might you propose regarding how the wasp finds her nest and why the pinecones didn't misdirect the wasp?

→ **Mastering Biology** Animation: Digger Wasps and Landmarks system of the spatial relationships between objects in its surroundings. One striking example is found in the Clark's nutcracker (*Nucifraga columbiana*), a relative of ravens, crows, and jays. In the fall, nutcrackers hide pine seeds for retrieval during the winter. By experimentally varying the distance between landmarks in the birds' environment, researchers discovered that the birds kept track of the halfway point between landmarks, rather than a fixed distance, to find their hidden food stores.

⊙ **Mastering Biology** HHMI Video: How Lizards Find Their Way Home

hhmi
BioInteractive

Associative Learning

Learning often involves making associations between experiences. Consider, for example, a blue jay (*Cyanocitta cristata*) that ingests a brightly colored monarch butterfly (*Danaus plexippus*). Substances that the monarch accumulates from milkweed plants cause the blue jay to vomit almost immediately **(Figure 51.9)**. Following such experiences, blue jays avoid attacking monarchs and similar-looking butterflies. The ability to associate one environmental feature (such as a color) with another (such as a foul taste) is called **associative learning**.

Associative learning is well suited to study in the laboratory. In the 1890s, Russian physiologist Ivan Pavlov demonstrated that if he always rang a bell just before feeding a dog, the dog would eventually salivate when the bell sounded, anticipating food. This form of learning, in which an arbitrary stimulus becomes associated with a particular outcome, is known as *classical conditioning*. Four decades later, the American researcher B. F. Skinner explored how a rat learns through trial and error to obtain food by pressing a lever. Such learning, in which a behavior becomes associated with a reward or punishment, is known as trial-and-error learning or operant conditioning (see Figure 51.9).

▼ **Figure 51.9 Associative learning.** Having ingested and vomited a monarch butterfly, a blue jay has probably learned to avoid this species.

Studies reveal that animals can learn to link many pairs of features of their environment, but not all. For example, pigeons can learn to associate danger with a sound but not with a color. However, they can learn to associate a color with food. What does this mean? The development and organization of the pigeon's nervous system apparently restrict the associations that can be formed. Moreover, such restrictions are not limited to birds. Rats, for example, can learn to avoid illness-inducing foods on the basis of smells, but not on the basis of sights or sounds.

If we consider how behavior evolves, the fact that some animals can't learn to make particular associations appears logical. The associations an animal can readily form typically reflect relationships likely to occur in nature. Conversely, associations that can't be formed are those unlikely to be of selective advantage in a native environment. In the case of a rat's diet in the wild, for example, a harmful food is far more likely to have a certain odor than to be associated with a particular sound.

Cognition and Problem Solving

The most complex forms of learning involve **cognition**—the process of knowing that involves awareness, reasoning, recollection, and judgment. Although it was once argued that only primates and certain marine mammals have high-level thought processes, many other groups of animals, including insects, appear to exhibit cognition in controlled laboratory studies. For example, an experiment using Y-shaped mazes provided evidence for abstract thinking in honeybees. One maze had different colors, and one had different black-and-white striped patterns, either vertical or horizontal bars. Two groups of honeybees were trained in the color maze. Upon entering, a bee would see a sample color and could then choose between an arm of the maze with the same color or an arm with a different color. Only one arm contained a food reward. The first group of bees were rewarded for flying into the arm with the *same* color as the sample **(Figure 51.10,)**; the second group were rewarded for choosing the arm with the *different* color. Next, bees from each group were tested in the bar maze, which had no food reward. After encountering a sample black-and-white pattern of bars, a bee could choose an arm with the same pattern or an arm with a different pattern. The bees in the first group most often chose the arm with the same pattern **(Figure 51.10, ❷)**, whereas those in the second group typically chose the arm with the different pattern.

The maze experiments provide strong experimental support for the hypothesis that honeybees can distinguish on the basis of "same" and "different." Remarkably, research published in 2010 indicates that honeybees can also learn to distinguish human faces.

The information-processing ability of a nervous system can also be revealed in **problem solving**, the cognitive

▼ **Figure 51.10 A maze test of abstract thinking by honeybees.** These mazes are designed to test whether honeybees can distinguish "same" from "different."

Decision chamber

Food

Lid

Stimulus

Entrance

1 **Bees were trained in a color maze.** As shown here, one group was rewarded for choosing the same color as the stimulus.

2 **Bees were tested in a pattern maze.** If previously rewarded for choosing the same color, bees most often chose lines oriented the same way as the stimulus.

VISUAL SKILLS *Describe how you would set up the pattern maze to control for an inherent preference for or against a particular orientation of the black bars.*

activity of devising a method to proceed from one condition to another in the face of real or apparent obstacles. For example, if a chimpanzee is placed in a room with several boxes on the floor and a banana hung high out of reach, the chimpanzee can assess the situation and stack the boxes, enabling it to reach the food. Problem-solving behavior is highly developed in some mammals, especially primates and dolphins. Notable examples have also been observed in some bird species. In one study, ravens were confronted with food hanging from a branch by a string. After failing to grab the food in flight, one raven flew to the branch and alternately pulled up and stepped on the string until the food was within reach. A number of other ravens eventually arrived at similar solutions. Nevertheless, some ravens failed to solve the problem, indicating that problem-solving success in this species, as in others, varies with individual experience and abilities.

Development of Learned Behaviors

Most of the learned behaviors we have discussed develop over a relatively short time. Some behaviors develop more gradually. For example, some bird species learn songs in stages. In the case of the white-crowned sparrow (*Zonotrichia leucophrys*), the first stage of song learning takes place early in life, when the fledgling sparrow first hears the song. If a fledgling is prevented from hearing real sparrows or recordings of sparrow songs during the first 50 days of its life, it fails to develop the adult song of its species.

Although a young sparrow does not sing during the sensitive period, it memorizes the song of its species by listening to other white-crowned sparrows sing. During the sensitive period, fledglings chirp more in response to songs of their own species than to songs of other species. Thus, when young white-crowned sparrows learn the songs they will sing later on, that learning appears to be bounded by genetically controlled preferences.

The sensitive period when a white-crowned sparrow memorizes its species' song is followed by a second learning phase when the juvenile bird sings tentative notes called a subsong. The juvenile bird hears its own singing and compares it with the song memorized during the sensitive period. Once a sparrow's own song matches the one it memorized, the song "crystallizes" as the final song, and the bird sings only this adult song for the rest of its life.

Song learning is one of many examples of how animals learn from other members of their species. In finishing our exploration of learning, we'll look at several more examples that reflect the more general phenomenon of social learning.

Social Learning

Many animals learn to solve problems by observing the behavior of other individuals. Learning through observing and interpreting behaviors and their consequences is called **social learning**. Young wild chimpanzees, for example, learn how to crack open oil palm nuts with two stones by copying experienced chimpanzees **(Figure 51.11)**.

▼ **Figure 51.11 A young chimpanzee (*Pan troglodytes*) learning to crack oil palm nuts by observing an elder.**

Mastering Biology **Video: Chimp Cracking Nut**

Another example of how social learning can modify behavior comes from studies of the vervet monkeys (*Chlorocebus pygerythrus*) in Amboseli National Park, Kenya. Vervet monkeys, which are about the size of a domestic cat, produce a complex set of alarm calls. Amboseli vervets give distinct alarm calls for leopards, eagles, and snakes. When a vervet sees a leopard, it gives a loud barking sound; when it sees an eagle, it gives a short double-syllable cough; and the snake alarm call is a "chutter." Upon hearing a particular alarm call, other vervets in the group behave in an appropriate way: They run up a tree on hearing the alarm for a leopard (vervets are nimbler than leopards in the trees); look up on hearing the alarm for an eagle; and look down on hearing the alarm for a snake (**Figure 51.12**).

Infant vervet monkeys give alarm calls, but in a relatively undiscriminating way. For example, they give the "eagle" alarm on seeing any bird, including harmless birds such as bee-eaters. With age, the monkeys improve their accuracy. In fact, adult vervet monkeys give the eagle alarm only on seeing an eagle belonging to either of the two species that eat vervets. Infants probably learn how to give the right call by observing other members of the group and receiving social confirmation. For instance, if the infant gives the call on the right occasion—say, an eagle alarm when there is an eagle overhead—another member of the group will also give the eagle call. But if the infant gives the call when a bee-eater flies by, the adults in the group are silent. Thus, vervet monkeys have an initial, unlearned tendency to give calls upon seeing potentially threatening objects in the environment. Learning fine-tunes the call so that adult vervets give calls only in response to genuine danger and can fine-tune the alarm calls of the next generation.

Social learning forms the roots of **culture**, a system of information transfer through social learning or teaching that influences the behavior of individuals in a population. Cultural transfer of information can alter behavioral phenotypes and thereby influence the fitness of individuals.

Changes in behavior that result from natural selection occur on a much longer time scale than does learning. In Concept 51.3, we'll examine the relationship between particular behaviors and the processes of selection related to survival and reproduction.

CONCEPT CHECK 51.2

1. How might associative learning explain why different species of distasteful or stinging insects have similar colors?
2. **WHAT IF?** How might you position and manipulate a few objects in a lab to test whether an animal can use a cognitive map to remember the location of a food source?
3. **MAKE CONNECTIONS** How might a learned behavior contribute to speciation? (See Concept 24.1.)

For suggested answers, see Appendix A.

CONCEPT 51.3

Selection for individual survival and reproductive success can explain diverse behaviors

EVOLUTION We turn now to Tinbergen's third question—how behavior enhances survival and reproduction in a population. The focus thus shifts from proximate causation—the "how" questions—to ultimate causation—the "why" questions. We'll begin by considering the activity of gathering food. Food-obtaining behavior, or **foraging**, includes not only eating but also any activities an animal uses to search for, recognize, and capture food items.

Evolution of Foraging Behavior

The fruit fly allows us to examine one way that foraging behavior might have evolved. Variation in a gene called *forager* (*for*) dictates how far *Drosophila* larvae travel when foraging. On average, larvae carrying the for^R ("Rover") allele travel nearly twice as far while foraging as do larvae with the for^S ("sitter") allele.

▼ **Figure 51.12 Vervet monkeys learning correct use of alarm calls.** On seeing a python (foreground), vervet monkeys give a distinct "snake" alarm call (inset), and the members of the group stand upright and look down.

▼ **Figure 51.13 Evolution of foraging behavior by laboratory populations of *Drosophila melanogaster*.** After 74 generations of living at low population density, *Drosophila* larvae (populations R1–R3) followed foraging paths significantly shorter than those of *Drosophila* larvae that had lived at high density (populations K1–K3).

INTERPRET THE DATA *What alternative hypothesis is made far less likely by having three R and K lines, rather than one of each?*

Both the *for^R* and *for^s* alleles are present in natural populations. What circumstances might favor one or the other allele? An answer became apparent in experiments that maintained flies at either low or high population densities for many generations. Larvae in populations kept at a low density foraged over shorter distances than those in populations kept at high density **(Figure 51.13)**. Furthermore, the *for^s* allele increased in frequency in the low-density populations, whereas the *for^R* allele increased in frequency in the high-density group. These changes make sense. At a low population density, short-distance foraging yields sufficient food, while long-distance foraging would result in unnecessary energy expenditure. Under crowded conditions, long-distance foraging could enable larvae to move beyond areas depleted of food. Thus, an interpretable evolutionary change in behavior occurred in the course of the experiment.

Optimal Foraging Model

To study the ultimate causation of foraging strategies, biologists sometimes apply a type of cost-benefit analysis used in economics. This idea proposes that foraging behavior is a compromise between the benefits of nutrition and the costs of obtaining food. These costs might include the energy expenditure of foraging as well as the risk of being eaten while foraging. According to this **optimal foraging model**, natural selection should favor a foraging behavior that minimizes the costs of foraging and maximizes the benefits. The **Scientific Skills Exercise** provides an example of how this model can be applied to animals in the wild.

Balancing Risk and Reward

One of the most significant potential costs to a forager is risk of predation. Maximizing energy gain and minimizing energy costs are of little benefit if the behavior makes the forager a likely meal for a predator. It seems logical, therefore, that predation risk would influence foraging behavior. Such appears to be the case for the mule deer (*Odocoileus hemionus*), which lives in the mountains of western North America. Researchers found that the food available for mule deer was fairly uniform across the potential foraging areas, although somewhat lower in open, nonforested areas. In contrast, the risk of predation differed greatly; mountain lions (*Puma concolor*), the major predator, killed large numbers of mule deer at forest edges and only a small number in open areas and forest interiors.

How does mule deer foraging behavior reflect the differences in predation risk in particular areas? Mule deer feed predominantly in open areas. Thus, it appears that mule deer foraging behavior reflects the large variation in predation risk and not the smaller variation in food availability. This result underscores the point that behavior typically reflects a compromise between competing selective pressures.

Mating Behavior and Mate Choice

Just as foraging is crucial for individual survival, mating behavior and mate choice play a major role in determining reproductive success. These behaviors include seeking or attracting mates, choosing among potential mates, competing for mates, and caring for offspring.

Mating Systems and Sexual Dimorphism

For humans and other animal species, there is considerable variation in patterns of sexual contact. In the context of reproduction, biologists describe differences in *mating systems*, the length and number of relationships between males and females. In some animal species, mating is *promiscuous*, with no strong pair-bonds. In others, mates form a relationship of some duration that is **monogamous** (one male mating with one female) or **polygamous** (an individual of one sex mating with several of the other). Polygamous relationships involve either *polygyny*, a single male and many females, or *polyandry*, a single female and multiple males.

The extent to which males and females differ in appearance, a characteristic known as *sexual dimorphism*, typically varies with the type of mating system. Among monogamous species, males and females often look very similar. In contrast, among polygamous species, the sex that attracts multiple mating partners is typically showier and larger than

Testing a Hypothesis with a Quantitative Model

Do Crows Display Optimal Foraging Behavior? On islands off British Columbia, Canada, Northwestern crows (*Corvus caurinus*) search rocky tide pools for sea snails called whelks. After spotting a whelk, the crow picks it up in its beak, flies upward, and drops the whelk onto the rocks. If the drop is successful, the shell breaks and the crow can dine on the whelk's soft parts. If not, the crow flies up and drops the whelk again and again until the shell breaks. What determines how high the crow flies? If energetic considerations dominated selection for the crow's foraging behavior, the average drop height might reflect a trade-off between the cost of flying higher and the benefit of more frequent success. In this exercise you'll test how well this optimal foraging model predicts the average drop height observed in nature.

How the Experiments Were Done There were two parts to the experiment. First, the researcher measured the height of drops made by crows in the wild by referring to a marked pole erected nearby. Second, the researcher built a device that dropped a whelk onto the rocks from a fixed platform. For each whelk, he kept track of the number of drops carried out before the whelk broke open. Averaging over many trials with the device and combining the data for each platform height, he calculated a predicted total "flight" height: the height of the platform times the average number of drops required to break open the whelk.

Data from the Experiment The graph summarizes the results of the experiment, comparing the behavior of crows in the wild with the measurements using the whelk-dropping device.

Data from R. Zach, Shell-dropping: Decision-making and optimal foraging in northwestern crows, *Behavior* 68:106–117 (1979).

INTERPRET THE DATA

1. How does the average number of drops required to break open a whelk depend on platform height for a drop of 5 meters or less? For drops of more than 5 meters?

2. Total flight height can be considered to be a measure of the total energy required to break open a whelk. Why is this value lower for a platform set at 5 meters than for one at 2 or 15 meters?

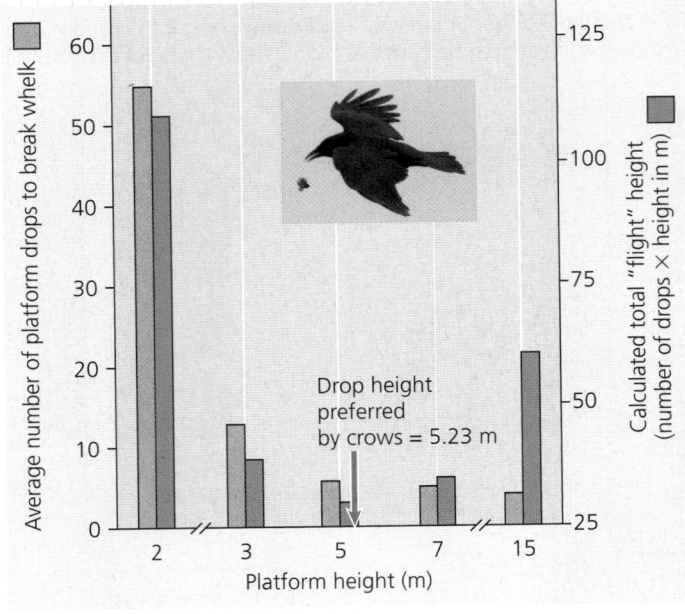

Drop height preferred by crows = 5.23 m

3. Compare the drop height preferred by crows with the graph of total flight height for the platform drops. Are the data consistent with the hypothesis of optimal foraging? Explain.

4. In testing the optimal foraging model, it was assumed that changing the height of the drop only changed the total energy required. Do you think this is a realistic limitation, or might other factors than total energy be affected by height?

5. Researchers observed that the crows only gather and drop the largest whelks. What are some reasons crows might favor larger whelks?

6. It turned out that the probability of a whelk breaking was the same for a whelk dropped for the first time as for an unbroken whelk dropped several times previously. If the probability of breaking instead increased, what change might you predict in the crow's behavior?

→ **Instructors:** A version of this Scientific Skills Exercise can be assigned in **Mastering Biology**.

---

the opposite sex **(Figure 51.14)**. We'll discuss the evolutionary basis of these differences shortly.

Mating Systems and Parental Care

The needs of the young are an important factor constraining the evolution of mating systems. Most newly hatched birds, for instance, cannot care for themselves. Rather, they require a large, continuous food supply, a need that is difficult for a single parent to meet. In such cases, a male that stays with and helps a single mate may ultimately have more viable offspring than it would by going off to seek additional mates. This may explain why many bird species are monogamous. In contrast, for birds with young that can feed and care for themselves almost

immediately after hatching, the males derive less benefit from staying with their partner. Males of these species, such as pheasants and quail, can maximize their reproductive success by seeking other mates, and polygyny is relatively common in such birds. In the case of mammals, the lactating female is often the only food source for the young, and males usually play no role in raising the young. In mammalian species where males protect the females and young, such as lions, a male or small group of males typically cares for a harem of many females.

Another factor influencing mating behavior and parental care is *certainty of paternity*. Young born to or eggs laid by a female definitely contain that female's genes. However, even if there is a strong pair-bond, a male other than the female's

▼ Figure 51.14 Relationship between mating system and male and female forms.

(a) Monogamy (one male, one female)

In monogamous species, such as these western gulls (*Larus occidentalis*), males and females are difficult to distinguish using external characteristics only.

(b) Polygyny (one male, multiple females)

Among polygynous species, such as elk (*Cervus canadensis*), the male (right) is often highly ornamented.

(c) Polyandry (one female, multiple males)

In polyandrous species, such as these red-necked phalaropes (*Phalaropus lobatus*), females (right) are generally more ornamented than males.

usual mate may on occasion father offspring with that female. The certainty of paternity is relatively low in most species with internal fertilization because the acts of mating and birth (or mating and egg laying) are separated over time. This could be one reason that exclusively male parental care is rare in bird and mammal species. However, the males of many species with internal fertilization engage in behaviors that appear to increase their certainty of paternity. These behaviors include guarding females, removing any sperm from the female reproductive tract before copulation, and introducing large quantities of sperm that displace the sperm of other males.

Certainty of paternity is high when egg laying and mating occur together, as in external fertilization. This may explain why parental care in aquatic invertebrates, fishes, and amphibians, when it occurs at all, is at least as likely to be by males as by females (**Figure 51.15**; see also Figure 46.6). Among fishes and amphibians, parental care occurs in fewer than 10% of species with internal fertilization but in more than half of species with external fertilization.

It is important to point out that certainty of paternity does not mean that animals are aware of those factors when they behave a certain way. Parental behavior correlated with certainty of paternity exists because it has been reinforced over generations by natural selection. The intriguing relationship between certainty of paternity and male parental care remains an area of active research.

Sexual Selection and Mate Choice

Sexual dimorphism results from sexual selection, a form of natural selection in which differences in reproductive success among individuals are a consequence of differences in mating success (see Concept 23.4). Sexual selection can

▼ Figure 51.15 Paternal care by a male yellowhead jawfish (*Opistognathus aurifrons*). The male jawfish, which lives in tropical marine environments, holds the eggs it has fertilized in its mouth, keeping them aerated and protecting them from predators until the young fish hatch.

take the form of *intersexual selection*, in which members of one sex choose mates on the basis of characteristics of the other sex, such as courtship songs, or *intrasexual selection*, which involves competition between members of one sex for mates.

Mate Choice by Females   Mate preferences of females may play a central role in the evolution of male behavior and anatomy through intersexual selection. Consider, for example, the courtship behavior of stalk-eyed flies. The eyes of these insects are at the tips of stalks, which are longer in males than in females. During courtship, a male approaches the female headfirst. Researchers have shown that females are more likely to mate with males that have relatively long eyestalks. Why would females favor this seemingly arbitrary trait? Ornaments such as long eyestalks in these flies and bright coloration in birds correlate in general with health and vitality. A female whose mate choice is a healthy male is likely to produce more offspring that survive to reproduce. As a result, males may compete with each other in ritualized contests to attract female attention **(Figure 51.16)**.

▼ **Figure 51.18 Sexual selection influenced by imprinting.** Experiments demonstrated that female zebra finch chicks that had imprinted on artificially ornamented fathers preferred ornamented males as adult mates. For all experimental groups, male offspring showed no preference for either ornamented or non-ornamented female mates.

▶ **Figure 51.16 A face-off between male stalk-eyed flies (*Cyrtodiopsis whitei*) competing for female attention.**

▶ **Figure 51.17 Appearance of zebra finches in nature.** The male zebra finch (right) is more highly patterned and colorful than the female zebra finch (left).

In face-offs between male stalk-eyed flies, the male whose eyestalk length is smaller usually retreats peacefully.

Mate choice can also be influenced by imprinting, as revealed by experiments carried out with zebra finches (*Taeniopygia guttata*). Both male and female zebra finches normally lack any feather crest on their head **(Figure 51.17)**. To explore whether parental appearance affects mate preference in offspring independent of any genetic influence, researchers provided zebra finches with artificial ornamentation. A 2.5-cm-long red feather was taped to the forehead feathers of either or both zebra finch parents when their chicks were 8 days old, approximately 2 days before they opened their eyes. A control group of zebra finches was raised by unadorned parents. When the chicks matured, they were presented with prospective mates that were either artificially ornamented with a red feather or non-ornamented **(Figure 51.18)**. Males showed no preference. Females raised by a male parent that was not ornamented also showed no preference. However, females raised by an ornamented male parent preferred ornamented males as their own mates. Thus, female finches apparently take cues from their fathers in choosing mates.

Mate-choice copying, a behavior in which individuals in a population copy the mate choice of others, has been studied in the guppy, *Poecilia reticulata*. When a female guppy chooses between males with no other females present, the female almost always chooses the male with more orange coloration. To explore if the behavior of other females could influence this preference, an experiment was set up using both living females and artificial model females **(Figure 51.19)**. If a female guppy observed the model "courting" a male with less extensive orange markings, she often copied the preference of the model female. That is, the female chose the male that had been presented in association with a model female rather than a more orange alternative. The exceptions were also informative. Mate-choice copying typically did not change when the difference in coloration was particularly large; that is, the female guppy chose the male with the much stronger coloration even if a model female was associated with the less orange male. Mate-choice copying can thus mask genetically controlled female preference below a certain threshold of difference, in this case for male color.

Mate-choice copying, a form of social learning, has also been observed in several other fish and bird species. What is the selective pressure for such a mechanism? One possibility is that a female that mates with males that are attractive to other females increases the probability that her male offspring will also be attractive and have high reproductive success.

Male Competition for Mates The previous examples show how female choice can select for one best type of male in a given situation, resulting in low variation among males. Similarly, male competition for mates can reduce variation among males. Such competition may involve *agonistic behavior*, a contest, typically ritualized, that determines which competitor gains access to a resource, such as food or a mate **(Figure 51.20**; see also Figure 51.16).

Despite the potential for male competition to select for reduced variation, behavioral and morphological variation in males is extremely high in some vertebrate species, including species of fish and deer, as well as in a wide variety of invertebrates. In some species, sexual selection has led to the evolution of alternative male mating behavior and morphology. How do scientists analyze situations where more than one mating behavior can result in

▼ **Figure 51.19 Mate choice copying by female guppies (*Poecilia reticulata*).** In the absence of other females (control group), female guppies generally choose males with more orange coloration. However, when a female model is placed near one of the males (experimental group), female guppies often copy the apparent mate choice of the model, even if the male is less colorful than others. Guppy females ignored the mate choice of the model only if an alternative male had much more orange coloration.

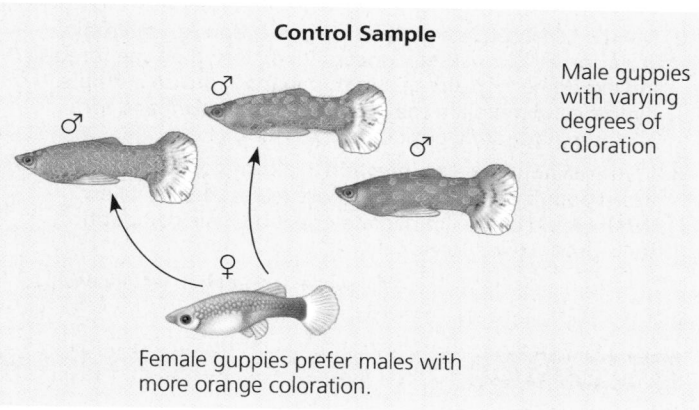

Control Sample

Male guppies with varying degrees of coloration

Female guppies prefer males with more orange coloration.

Experimental Sample

Female model in mock courtship with less orange male

Female guppies prefer males that are associated with another female.

▼ **Figure 51.20 Agonistic interaction.** Male eastern grey kangaroos (*Macropus giganteus*) often "box" in contests that determine which male is most likely to mate with an available female. Typically, one male snorts loudly and strikes the other with his forelimbs. If the male under attack does not retreat, the fight may escalate into grappling or the two males balancing on their tails while attempting to kick each other with the sharp toenails of their hind feet.

→ **Mastering Biology** Video: Agonistic Behavior in Wolves • Video: Snake Ritual Wrestling

successful reproduction? One approach relies on the rules that govern games.

Applying Game Theory

Often, the fitness of a particular behavioral phenotype is influenced by other behavioral phenotypes in the population. In studying such situations, behavioral ecologists use a range of tools, including game theory. Developed by American mathematician John Nash and others to model human economic behavior, **game theory** evaluates alternative strategies in situations where the outcome depends on the strategies of all the individuals involved.

As an example of applying game theory to mating behavior, let's consider the common side-blotched lizard (*Uta stansburiana*) of California. Genetic variations give rise to males with orange, blue, or yellow throats **(Figure 51.21)**. One would expect that natural selection would favor one of the three color types, yet all three persist. Why? The answer appears to lie in the fact that each throat color is associated with a different pattern of behavior: Orange-throat males are the most aggressive and defend large territories that contain many females. Blue-throat males are also territorial but defend smaller territories and fewer females. Yellow-throats are nonterritorial males that mimic females and use "sneaky" tactics to gain the chance to mate.

Evidence indicates that the mating success of each male lizard type is influenced by the relative abundance of the other types, an example of frequency-dependent selection. In one study population, the most frequent throat coloration changed over a period of several years from blue to orange to yellow and back to blue.

By comparing the competition between common side-blotched lizard males to the children's game of rock-paper-scissors, scientists devised an explanation for the cycles of variation in the lizard population. In the game, paper defeats rock, rock defeats scissors, and scissors defeats paper. Each hand symbol thus wins one matchup but loses

the other. Similarly, each type of male lizard has an advantage over one of the other two types:

- When blue-throats are abundant, they can defend the few females in their territories from the advances of the sneaky yellow-throat males. However, blue-throats cannot defend their territories against the hyperaggressive orange-throats.
- Once the orange-throats become the most abundant, the larger number of females in each territory provides the opportunity for the yellow-throats to have greater mating success.
- The yellow-throats become more frequent, but then give way to the blue-throats, whose tactic of guarding small territories once again allows them the most success.

Thus, following the population over time, one sees a persistence of all three color types and a periodic shift in which type is most prevalent.

Game theory provides a way to think about complex evolutionary problems in which relative performance (reproductive success relative to other phenotypes), not absolute performance, is the key to understanding the evolution of behavior. This makes game theory an important tool because the relative performance of one phenotype compared with others is a measure of Darwinian fitness.

CONCEPT CHECK 51.3

1. Why does the mode of fertilization correlate with the presence or absence of male parental care?

2. **MAKE CONNECTIONS** Balancing selection can maintain variation at a locus (see Concept 23.4). Based on the foraging experiments described in this chapter, devise a simple hypothesis to explain the presence of both for^R and for^s alleles in natural fly populations.

3. **WHAT IF?** Suppose an infection in a common side-blotched lizard population killed many more males than females. What would be the immediate effect on male competition for reproductive success?

For suggested answers, see Appendix A.

CONCEPT 51.4

Genetic analyses and the concept of inclusive fitness provide a basis for studying the evolution of behavior

EVOLUTION We'll now explore issues related to Tinbergen's fourth question—the evolutionary history of behaviors. We will first look at the genetic control of a behavior. Next, we will examine the genetic variation underlying the evolution of particular behaviors. Finally, we will see how expanding the definition of fitness beyond individual survival can help explain "selfless" behavior.

▼ **Figure 51.21 Male polymorphism in the common side-blotched lizard (*Uta stansburiana*).** An orange-throat male, left; a blue-throat male, center; a yellow-throat male, right.

Genetic Basis of Behavior

In exploring the genetic basis of behavior, we'll begin with the courtship behavior of the male fruit fly (see Figure 51.4). During courtship, the male fly carries out a complex series of actions in response to multiple sensory stimuli. Genetic studies have revealed that a single gene called *fru* controls this entire courtship ritual. If the *fru* gene is mutated to an inactive form, males do not court or mate with females. (The name *fru* is short for *fruitless*, reflecting the absence of offspring from the mutant males.) Normal male and female flies express distinct forms of the *fru* gene. When females are genetically manipulated to express the male form of *fru*, they court other females, performing the role normally played by the male.

How does the *fru* gene control so many different actions? Experiments carried out cooperatively in several laboratories demonstrated that *fru* is a master regulatory gene that directs the expression and activity of many genes with narrower functions. Together, genes that are controlled by the *fru* gene bring about sex-specific development of the fly nervous system. In effect, *fru* programs the fly for male courtship behavior by overseeing a male-specific wiring of the central nervous system.

In many cases, differences in behavior arise not from gene inactivation, but from variation in the activity or amount of a gene product. One striking example comes from the study of two related species of voles, which are small, mouse-like rodents. Male meadow voles (*Microtus pennsylvanicus*) are solitary and do not form lasting relationships with mates. Following mating, they pay little attention to their pups. In contrast, male prairie voles (*Microtus ochrogaster*) form a pair-bond with a single female after they mate **(Figure 51.22)**. Male prairie voles provide care for their young pups, hovering over them, licking them, and carrying them, while acting aggressively toward intruders.

A neurotransmitter released during mating is critical for the partnering and parental behavior of male voles. Known as **antidiuretic hormone (ADH) or vasopressin** (see Concept 44.5), this peptide is released during mating and binds to a specific receptor in the central nervous system. When male prairie voles are given a drug that inhibits the receptor in the brain that detects vasopressin, the male voles fail to form pair-bonds after mating.

The vasopressin receptor gene is much more highly expressed in the brain of prairie voles than in the brain of meadow voles. Testing the hypothesis that the level of vasopressin receptor in the brain regulates postmating behavior, researchers inserted the vasopressin receptor gene from prairie voles into the genome of meadow voles. The male meadow voles carrying this gene not only developed brains with higher levels of the vasopressin receptor but also showed many of the same mating behaviors as male prairie voles,

▼ **Figure 51.22 A pair of prairie voles (*Microtus ochrogaster*) huddling.** Among North American prairie voles, males associate closely with their mates, as shown here, and contribute substantially to the care of young.

such as pair-bonding. Thus, although many genes influence pair-bonding and parenting in male voles, a change in the level of expression of the vasopressin receptor is sufficient to alter the development of these behaviors.

Genetic Variation and the Evolution of Behavior

Behavioral differences between closely related species, such as meadow and prairie voles, are common. Significant differences in behavior can also be found *within* a species but are often less obvious. When behavioral variation between populations of a species correlates with variation in environmental conditions, it may reflect natural selection.

Case Study: *Variation in Prey Selection*

An example of genetically based behavioral variation within a species involves prey selection by the western garter snake (*Thamnophis elegans*). The natural diet of this species differs widely across its range in California. Coastal populations feed predominantly on banana slugs (*Ariolimax californicus*). Inland populations feed on frogs, leeches, and fish, but not banana slugs. In fact, banana slugs are rare or absent in the inland habitats.

▼ **Figure 51.23 Western garter snake from a coastal habitat eating a banana slug.** Experiments indicate that the preference of these snakes for banana slugs may be influenced more by genetics than by environment.

When researchers offered banana slugs to snakes collected from each wild population, most coastal snakes readily ate them **(Figure 51.23)**. In contrast, inland snakes tended to refuse. To what extent does genetic variation among snake species contribute to a fondness for banana slugs? To answer this question, researchers collected pregnant snakes from the wild coastal and inland populations and then housed these females in separate cages in the laboratory. While the offspring were still very young, they were each offered a small piece of banana slug on ten successive days. More than 60% of the young snakes from coastal mothers ate banana slugs on eight or more of the ten days. In contrast, fewer than 20% of the young snakes from inland mothers ate a piece of banana slug even once. Perhaps not surprisingly, banana slugs thus appear to be a genetically acquired taste.

How did a genetically determined difference in feeding preference come to match the snakes' habitats so well? It turns out that the two populations also vary with respect to their ability to recognize and respond to odor molecules produced by banana slugs. Researchers hypothesize that when inland snakes colonized coastal habitats more than 10,000 years ago, some of them could recognize banana slugs by scent. Because these snakes took advantage of this food source, they had higher fitness than snakes in the population that ignored the slugs. Over hundreds or thousands of generations, the capacity to recognize the slugs as prey increased in frequency in the coastal population. The marked variation in behavior observed today between the coastal and inland populations may be evidence of this past evolutionary change.

Case Study: *Variation in Migratory Patterns*

Another species suited to the study of behavioral variation is the blackcap (*Sylvia atricapilla*), a small migratory warbler. Blackcaps that breed in Germany generally migrate southwest to Spain and then south to Africa for the winter. In the 1950s, a few blackcaps began to spend their winters in Britain, and over time the population of blackcaps wintering in Britain grew to many thousands. Leg bands showed that some of these birds had migrated westward from central Germany. Was this change in the pattern of migration the outcome of natural selection? If so, the birds wintering in Britain must have a heritable difference in migratory behavior. To test this hypothesis, researchers at the Max Planck Institute for Ornithology in Radolfzell, Germany, devised a strategy to study migratory orientation in the laboratory **(Figure 51.24)**. The results demonstrated that the two patterns of migration—to the west and to the southwest—do in fact reflect genetic differences between the two populations.

The study of western European blackcaps indicated that the change in their migratory behavior occurred both recently and rapidly. Before the year 1950, there were no known westward-migrating blackcaps in Germany. By the 1990s, westward migrants made up 7–11% of the blackcap populations of Germany. Once westward migration began, it persisted and increased in frequency, perhaps due to the widespread use of winter bird feeders in Britain, as well as shorter migration distances.

Altruism

We typically assume that behaviors are selfish; that is, they benefit the individual at the expense of others, especially competitors. For example, superior foraging ability by one individual may leave less food for others. The problem comes with "unselfish" behaviors. How can such behaviors arise through natural selection? To answer this question, let's look more closely at some examples of unselfish behavior and consider how they might arise.

In discussing selflessness, we will use the term **altruism** to describe a behavior that reduces an animal's individual fitness but increases the fitness of other individuals in the population. Consider, for example, the Belding's ground squirrel (*Urocitellus beldingi*), which lives in the western United States and is vulnerable to predators such as coyotes and hawks. A squirrel that sees a predator approach often gives a high-pitched alarm call that alerts unaware individuals to retreat to their burrows. In warning others, however, this squirrel brings attention to its location and thus increases its own risk of being killed.

Another example of altruistic behavior occurs in honeybee societies, in which the workers are sterile. The workers themselves never reproduce, but they labor on behalf of a single fertile queen. Furthermore, the workers sting intruders, a behavior that helps defend the hive but results in the death of those workers.

Are differences in migratory orientation within a species genetically determined?

Experiment Birds known as blackcaps that live in Germany winter elsewhere. Most migrate to Spain and Africa, but a few fly to Britain, where they find food left out by city dwellers. German scientist Peter Berthold and colleagues wondered if this change had a genetic basis. To test this hypothesis, they captured blackcaps wintering in Britain and bred them in Germany in an outdoor cage. They also collected young blackcaps from nests in Germany and raised them in cages. In the autumn, the blackcaps captured in Britain and the birds raised in cages were placed in large, glass-covered funnel cages. When the funnels were lined with carbon-coated paper and placed outside at night, the birds moved around, making marks on the paper that indicated the direction in which they were trying to "migrate."

Scratch marks

Results The wintering adult birds captured in Britain and their laboratory-raised offspring both attempted to migrate to the west. In contrast, the young birds collected from nests in southern Germany attempted to migrate to the southwest.

Adults from Britain and offspring of British adults

Young from SW Germany

Conclusion The young of the British blackcaps and the young birds from Germany (the control group) were raised under similar conditions but showed very different migratory orientations, indicating that their migratory orientation has a genetic basis.

Data from P. Berthold et al., Rapid microevolution of migratory behavior in a wild bird species, *Nature* 360:668–690 (1992).

WHAT IF? *Suppose the birds had not shown a difference in orientation in these experiments. Could you conclude that the behavior was not genetically based? Explain.*

▼ **Figure 51.25 Naked mole rats, a species of colonial mammal that exhibits altruistic behavior.** Pictured here is a queen nursing offspring while surrounded by other members of the colony.

Altruism is also observed in naked mole rats (*Heterocephalus glaber*), highly social rodents that live in underground chambers and tunnels in southern and northeastern Africa. The naked mole rat, which is almost hairless and nearly blind, lives in colonies of 20 to 300 individuals **(Figure 51.25)**. Each colony has only one reproducing female, the queen, who mates with one to three males, called kings. The rest of the colony consists of nonreproductive females and males who at times sacrifice themselves to protect the queen or kings from snakes or other predators that invade the colony.

Inclusive Fitness

With these examples from ground squirrels, honeybees, and mole rats in mind, let's return to the question of how altruistic behavior arises during evolution. The easiest case to consider is that of parents sacrificing for their offspring. When parents sacrifice their own well-being to produce and aid offspring, this act actually increases the fitness of the parents because it maximizes their genetic representation in the population. By this logic, altruistic behavior can be maintained by evolution even though it does not enhance the survival and reproductive success of the self-sacrificing individuals.

What about circumstances when individuals help others who are not their offspring? By considering a broader group of relatives than just parents and offspring, biologist William Hamilton found an answer. He began by proposing that an animal could increase its genetic representation in the next generation by helping close relatives other than its own offspring. Like parents and offspring, full siblings have half their genes in common. Therefore, selection might also favor helping siblings or helping one's parents produce more siblings. This thinking led Hamilton to the idea of **inclusive fitness**, the total effect an individual has on proliferating its genes by producing its own offspring *and* by providing aid that enables other close relatives to produce offspring.

Hamilton's Rule and Kin Selection

The power of Hamilton's hypothesis was that it provided a way to measure, or quantify, the effect of altruism on fitness. According to Hamilton, the three key variables in an act of altruism are the benefit to the recipient, the cost to the altruist, and the coefficient of relatedness. The benefit, B, is the average number of *extra* offspring that the recipient of an altruistic act produces. The cost, C, is how many *fewer* offspring the altruist produces. The **coefficient of relatedness**, r, equals the fraction of genes that, on average, are shared. Natural selection favors altruism when the benefit to the recipient multiplied by the coefficient of relatedness exceeds the cost to the altruist—in other words, when $rB > C$. This statement is called **Hamilton's rule**.

To better understand Hamilton's rule, let's apply it to a human population in which the average individual has two children. We'll imagine that a young man is close to drowning in heavy surf, and his sister risks her life to swim out and pull her sibling to safety. If the young man had drowned, his reproductive output would have been zero; but now, if we use the average, he can father two children. The benefit to the man is thus two offspring ($B = 2$). What cost does his sister incur? Let's say that she has a 25% chance of drowning in attempting the rescue. The cost of the altruistic act to the sister is then 0.25 times 2, the number of offspring she would be expected to have if she had stayed on shore ($C = 0.25 \times 2 = 0.5$). Finally, we note that a brother and sister share half their genes on average ($r = 0.5$). One way to see this is in terms of the separation of homologous chromosomes that occurs during meiosis of gametes **(Figure 51.26;** see also Figure 13.7).

We can use our values of B, C, and r to evaluate whether natural selection would favor the altruistic act in our imaginary scenario. For the surf rescue, $rB = 0.5 \times 2 = 1$, whereas $C = 0.5$. Because rB is greater than C, Hamilton's rule is satisfied; thus, natural selection would favor this altruistic act.

Averaging over many individuals and generations, any particular gene in a sister faced with the situation described will be passed on to more offspring if she risks the rescue than if she does not. Among the genes propagated in this way may be some that contribute to altruistic behavior. Natural selection that thus favors altruism by enhancing the reproductive success of relatives is called **kin selection**.

Kin selection weakens with hereditary distance. Siblings have an r of 0.5, but between an aunt and her niece, $r = 0.25$ (¼), and between first cousins, $r = 0.125$ (⅛). Notice that as the degree of relatedness decreases, the rB term in the Hamilton inequality also decreases. Would natural selection favor rescuing a cousin? Not unless the surf were less treacherous. For the original conditions, $rB = 0.125 \times 2 = 0.25$, which is only half the value of C (0.5). British geneticist J. B. S. Haldane appears to have

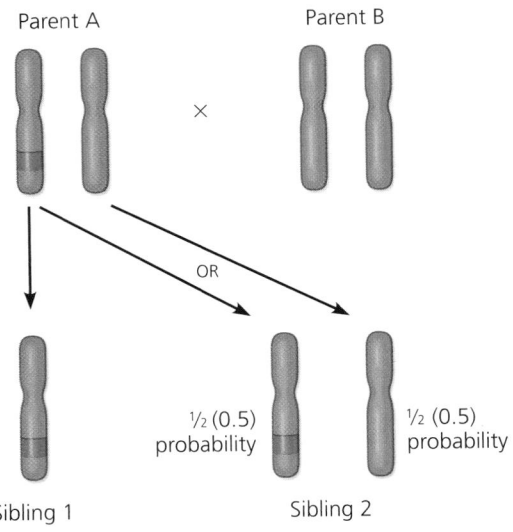

▼ Figure 51.26 The coefficient of relatedness between siblings. The red band indicates a particular allele (version of a gene) present on one chromosome, but not its homolog, in parent A. Sibling 1 has inherited the allele from parent A. There is a probability of ½ that sibling 2 will also inherit this allele from parent A. Any allele present on one chromosome of either parent will behave similarly. The coefficient of relatedness between the two siblings is thus ½, or 0.5.

Parent A   Parent B

×

OR

½ (0.5) probability

½ (0.5) probability

Sibling 1   Sibling 2

WHAT IF? *The coefficient of relatedness of an individual to a full (nontwin) sibling or to either parent is the same: 0.5. Does this value also hold true in cases of polyandry and polygyny?*

anticipated these ideas when he jokingly stated that he would not lay down his life for one brother, but would do so for two brothers or eight cousins.

If kin selection explains altruism, then the examples of unselfish behavior we observe among diverse animal species should involve close relatives. This is apparently the case, but often in complex ways. Like most mammals, female Belding's ground squirrels settle close to their site of birth, whereas males settle at distant sites **(Figure 51.27)**. Since nearly all alarm calls are given by females, they are most likely aiding close relatives. In the case of worker bees, who are all sterile, anything they do to help the entire hive benefits the only permanent member who is reproductively active—the queen, who is their mother.

In the case of naked mole rats, DNA analyses have shown that all the individuals in a colony are closely related. Genetically, the queen appears to be a sibling, daughter, or mother of the kings, and the nonreproductive mole rats are the queen's direct descendants or her siblings. Therefore, when a nonreproductive individual enhances a queen's or king's chances of reproducing, the altruist increases the chance that some genes identical to its own will be passed to the next generation.

Reciprocal Altruism

Some animals occasionally behave altruistically toward others who are not relatives. A baboon may help an unrelated companion in a fight, or a wolf may offer food to another wolf

▼ **Figure 51.27 Kin selection and altruism in Belding's ground squirrels.** This graph helps explain the male-female difference in altruistic behavior of ground squirrels. Once weaned (pups are nursed for about one month), females are more likely than males to live near close relatives. Alarm calls that warn these relatives increase the inclusive fitness of the female altruist.

even though they share no kinship. Such behavior can be adaptive if the aided individual returns the favor in the future. This sort of exchange of aid, called **reciprocal altruism**, is commonly invoked to explain altruism that occurs between unrelated humans. Reciprocal altruism is rare in other animals; it is limited largely to species (such as chimpanzees) with social groups stable enough that individuals have many chances to exchange aid. It is generally thought to occur when individuals are likely to meet again and when there would be negative consequences associated with not returning favors to individuals who had been helpful in the past, a pattern of behavior that behavioral ecologists refer to as "cheating."

Since cheating may benefit the cheater substantially, how could reciprocal altruism evolve? Game theory provides a possible answer in the form of a behavioral strategy called *tit for tat*. In the tit-for-tat strategy, an individual treats another in the same way it was treated the last time they met. Individuals adopting this behavior are always altruistic, or cooperative, on the first encounter with another individual and will remain so as long as their altruism is reciprocated. When their cooperation is not reciprocated, however, individuals employing tit for tat will retaliate immediately but return to cooperative behavior as soon as the other individual becomes cooperative. The tit-for-tat strategy has been used to explain the few apparently reciprocal altruistic interactions observed in animals—ranging from blood sharing between nonrelated vampire bats to social grooming in primates.

Evolution and Human Culture

As animals, humans behave (and, sometimes, misbehave). Just as humans vary extensively in anatomical features, we display substantial variations in behavior. Environment

intervenes in the path from genotype to phenotype for physical traits, but does so much more profoundly for behavioral traits. Furthermore, as a consequence of our marked capacity for learning, humans are probably more able than any other animal to acquire new behaviors and skills **(Figure 51.28)**.

Some human activities have a less easily defined function in survival and reproduction than do, for example, foraging or courtship. One of these activities is play, which is sometimes defined as behavior that appears purposeless. We recognize play in children and what we think is play in the young of other vertebrates. Behavioral biologists describe "object play," such as chimpanzees playing with leaves, "locomotor play," such as the acrobatics of an antelope, and "social play," such as the interactions and antics of lion cubs. These categories, however, do little to inform us about the function of play. One idea is that, rather than generating specific skills or experience, play serves as preparation for unexpected events and for circumstances that cannot be controlled.

Human behavior and culture are related to evolutionary theory in the discipline of *sociobiology*. The main premise of sociobiology is that certain behavioral characteristics exist because they are expressions of genes that have been perpetuated by natural selection. In his seminal 1975 book *Sociobiology: The New Synthesis,* E. O. Wilson speculated about the evolutionary basis of certain kinds of social behavior. By including a few examples from human culture, he sparked a debate that continues today.

Over our recent evolutionary history, we have built up structured societies with governments, laws, cultural values, and religions that define what is acceptable behavior and what is not, even when unacceptable behavior might enhance an individual's Darwinian fitness. Perhaps it is our social and cultural

▶ **Figure 51.28 Learning a new behavior.**

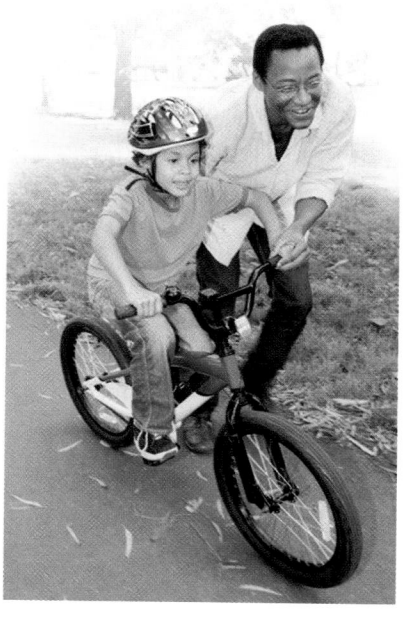

institutions that make us distinct and that provide those qualities that at times make less apparent the continuum between humans and other animals. One such quality, our considerable capacity for reciprocal altruism, will be essential as we tackle current challenges, including global climate change, in which individual and collective interests often appear to be in conflict.

> **Mastering Biology**
> Interview with E. O. Wilson: Pioneering the field of sociobiology

CONCEPT CHECK 51.4

1. Explain why geographic variation in garter snake prey choice might indicate that the behavior evolved by natural selection.
2. Suppose an individual organism aids the survival and reproductive success of the offspring of its sibling. How might this behavior result in indirect selection for certain genes carried by that individual?
3. **WHAT IF?** Suppose you applied Hamilton's logic to a situation in which one individual is past reproductive age. Could there still be selection for an altruistic act?

For suggested answers, see Appendix A.

51 Chapter Review

> Go to **Mastering Biology** for Assignments, the eText, the Study Area, and Dynamic Study Modules.

SUMMARY OF KEY CONCEPTS

> To review key terms, go to the **Vocabulary Self-Quiz** in the **Mastering Biology** eText or Study Area, or go to goo.gl/zkjz9t.

CONCEPT 51.1

Discrete sensory inputs can stimulate both simple and complex behaviors *(pp. 1140–1143)*

- **Behavior** is the sum of an animal's responses to external and internal stimuli. In behavior studies, proximate, or "how," questions focus on the stimuli that trigger a behavior and on genetic, physiological, and anatomical mechanisms underlying a behavioral act. Ultimate, or "why," questions address evolutionary significance.
- A **fixed action pattern** is a largely invariant behavior triggered by a simple cue known as a **sign stimulus**.
- **Migratory** movements involve navigation, which can be based on orientation relative to the sun, the stars, or Earth's magnetic field. Animal behavior is often synchronized to the circadian cycle of light and dark in the environment or to the seasons.
- The transmission and reception of **signals** constitute animal **communication**. Animals use visual, auditory, chemical, and tactile signals. Chemical substances called **pheromones** transmit species-specific information between members of a species in behaviors ranging from foraging to courtship.

? *How is migration based on circannual rhythms poorly suited for adaptation to global climate change?*

CONCEPT 51.2

Learning establishes specific links between experience and behavior *(pp. 1143–1148)*

- Cross-fostering studies can be used to measure the influence of social environment and experience on behavior.
- **Learning**, the modification of behavior as a result of experience, can take many forms, as summarized in the diagram that follows.

Imprinting

Cognition

Forms of learning and problem solving

Spatial learning

Associative learning

Social learning

? *How do imprinting in geese and song development in sparrows differ with regard to the resulting behavior?*

CONCEPT 51.3

Selection for individual survival and reproductive success can explain diverse behaviors *(pp. 1148–1154)*

- Controlled experiments in the laboratory can give rise to interpretable evolutionary changes in behavior.
- An **optimal foraging model** is based on the idea that natural selection should favor **foraging** behavior that minimizes the costs of foraging and maximizes the benefits.
- Sexual dimorphism correlates with the types of mating relationship, which include **monogamous** and **polygamous** mating

systems. Variations in mating system and mode of fertilization affect certainty of paternity, which in turn, through evolution, influences mating behavior and parental care.

- **Game theory** provides a way of thinking about evolution in situations where the fitness of a particular behavioral phenotype is influenced by other behavioral phenotypes in the population.

? *In some spider species, the female eats the male immediately after copulation. How might you explain this behavior from an evolutionary perspective?*

CONCEPT 51.4

Genetic analyses and the concept of inclusive fitness provide a basis for studying the evolution of behavior
(pp. 1154–1160)

- Genetic studies in insects have revealed the existence of master regulatory genes that control complex behaviors. Within the underlying hierarchy, multiple genes influence specific behaviors, such as a courtship song. Research on voles illustrates how variation in a single gene can determine differences in complex behaviors.
- Behavioral variation within a species that corresponds to environmental variation may be evidence of past evolution.
- **Altruism** can be explained by the concept of **inclusive fitness**, the effect an individual has on proliferating its genes by producing its own offspring *and* by providing aid that enables close relatives to reproduce. The **coefficient of relatedness** and **Hamilton's rule** provide a way of measuring the strength of the selective forces favoring altruism against the potential cost of the "selfless" behavior. **Kin selection** favors altruistic behavior by enhancing the reproductive success of relatives.

? *What insight about the genetic basis of behavior emerges from studying the effects of courtship mutations in fruit flies and of pair-bonding in voles?*

TEST YOUR UNDERSTANDING

→ For more multiple-choice questions, go to the **Practice Test** in the **Mastering Biology** eText or Study Area, or go to goo.gl/GruWRg.

Levels 1-2: Remembering/Understanding

1. Which of the following is true of innate behaviors?
   (A) Their expression is only weakly influenced by genes.
   (B) They occur with or without environmental stimuli.
   (C) They are expressed in most individuals in a population.
   (D) They occur in invertebrates and some vertebrates but not mammals.

2. According to Hamilton's rule,
   (A) natural selection does not favor altruistic behavior that causes the death of the altruist.
   (B) natural selection favors altruistic acts when the resulting benefit to the recipient, corrected for relatedness, exceeds the cost to the altruist.
   (C) natural selection is more likely to favor altruistic behavior that benefits an offspring than altruistic behavior that benefits a sibling.
   (D) the effects of kin selection are larger than the effects of direct natural selection on individuals.

3. Female spotted sandpipers aggressively court males and, after mating, leave the clutch of young for the male to incubate. This sequence may be repeated several times with different males until no available males remain, forcing the female to incubate her last clutch. Which of the following terms best describes this behavior?
   (A) polygyny
   (B) polyandry
   (C) promiscuity
   (D) certainty of paternity

Levels 3-4: Applying/Analyzing

4. A region of the canary forebrain shrinks during the nonbreeding season and enlarges when breeding season begins. This change is probably associated with the annual
   (A) addition of new syllables to a canary's song repertoire.
   (B) crystallization of subsong into adult songs.
   (C) sensitive period in which canary parents imprint on new offspring.
   (D) elimination of the memorized template for songs sung the previous year.

5. Although many chimpanzees live in environments with oil palm nuts, members of only a few populations use stones to crack open the nuts. The likely explanation is that
   (A) the behavioral difference is caused by genetic differences between populations.
   (B) members of different populations have different nutritional requirements.
   (C) the cultural tradition of using stones to crack nuts has arisen in only some populations.
   (D) members of different populations differ in learning ability.

6. Which of the following is *not* required for a behavioral trait to evolve by natural selection?
   (A) In each individual, the form of the behavior is determined entirely by genes.
   (B) The behavior varies among individuals.
   (C) An individual's reproductive success depends in part on how the behavior is performed.
   (D) Some component of the behavior is genetically inherited.

Levels 5-6: Evaluating/Creating

7. **DRAW IT** You are considering two optimal foraging models for the behavior of a mussel-feeding shorebird, the oystercatcher. In model A, the energetic reward increases solely with mussel size. In model B, you take into consideration that larger mussels are more difficult to open. Draw a graph of reward (energy benefit on a scale of 0–10) versus mussel length (scale of 0–70 mm) for each model. Assume that mussels under 10 mm provide no benefit and are ignored by the birds. Also assume that mussels start becoming difficult to open when they reach 40 mm in length and impossible to open when 70 mm long. Considering the graphs you have drawn, indicate what observations and measurements you would want to make in this shorebird's habitat to help determine which model is more accurate.

8. **EVOLUTION CONNECTION** We often explain our behavior in terms of subjective feelings, motives, or reasons, but evolutionary explanations are based on reproductive fitness. Can both kinds of explanation be valid? For instance, is an explanation for behavior such as "falling in love" incompatible with an evolutionary explanation?

9. **SCIENTIFIC INQUIRY** Scientists studying scrub jays found that "helpers" often assist mated pairs of birds by gathering food for their offspring. (a) Propose a hypothesis to explain what advantage there might be for the helpers to engage in this behavior instead of seeking their own territories and mates. (b) Explain how you would test your hypothesis. If it is correct, what results would you expect your tests to yield?

10. **SCIENCE, TECHNOLOGY, AND SOCIETY** Researchers are very interested in studying identical twins separated at birth and raised apart. So far, the data reveal that such twins frequently have similar personalities, mannerisms, habits, and interests. What general question do you think researchers hope to answer by studying such twins? Why do identical twins make good subjects for this research? What are the potential pitfalls of this research? What abuses might occur if the studies are not evaluated critically? Explain your thinking.

11. **WRITE ABOUT A THEME: INFORMATION** Learning is defined as a change in behavior as a result of experience. In a short essay (100–150 words), describe how heritable information contributes to the acquisition of learning, using some examples from imprinting and associative learning.

12. **SYNTHESIZE YOUR KNOWLEDGE**

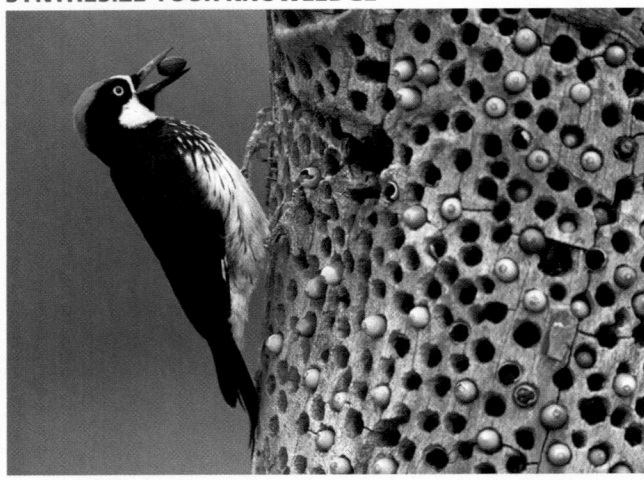

Acorn woodpeckers (*Melanerpes formicivorus*) stash acorns in storage holes they drill in trees. When these woodpeckers breed, the offspring from previous years often help with parental duties. Activities of these nonbreeding helpers include incubating eggs and defending stashed acorns. Propose some questions that a behavioral biologist could ask about the proximate and ultimate causation of these behaviors.

For selected answers, see Appendix A.

Meet Chelsea Rochman, a pioneer and internationally recognized researcher on the ecological effects of microplastic pollution. After receiving a B.S. in Biology from University of California (UC) San Diego, Dr. Rochman completed a joint Ph.D. program in Ecology at UC Davis and San Diego State University. An Assistant Professor in the Department of Ecology and Evolutionary Biology at University of Toronto, Canada, Dr. Rochman and her students are studying how plastic waste and other contaminants affect individual organisms, populations, communities, and ecosystems. Dr. Rochman has also worked with industry and government leaders to shape science-based policy, including the Microbead-Free Waters Act in the United States and the G7 Ocean Plastics Charter led by Canada.

AN INTERVIEW WITH

Chelsea Rochman

How did you become interested in ecology and the impact of plastic pollutants?

I began my undergraduate studies at Santa Monica College, a community college near Los Angeles. Initially, I was really interested in the performing arts and the fine arts. But then I took a marine biology class, and I absolutely loved it. I switched to a biology major and after one year, I, transferred to UC San Diego. While there, I was able to study abroad, at the University of Queensland, Australia. While taking a course in ecology, we went to an island that was basically uninhabited. Even though there were so few people, there was debris all over the beach, and there were sea turtles harmed by plastic they had eaten. It was the first time I realized that there is a lot of plastic in the ocean and it was harming wildlife. I've never looked back—I've been working on this topic ever since.

What are microplastics, and what drew you to working on them?

Microplastics are small pieces of plastic debris, up to 5 mm in size. They are either

▼ Collecting samples of plastic waste.

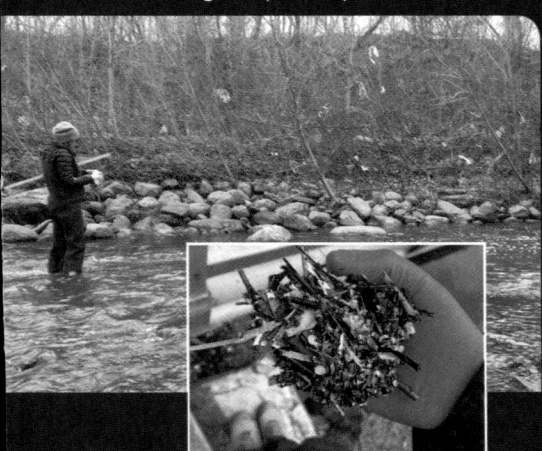

produced as microplastics, such as microbeads in face wash, or they are fragments of larger plastic items, such as a bottle or bag. In 2009, I took my first trip to the Great Pacific Garbage Patch. But it didn't look like an island of trash the way it is described in the media. Instead, it is a soup of tiny plastic particles, the size of a pea or smaller. I realized that these small particles of plastic could infiltrate every level of the food chain. And so as I began my graduate work, I was focused on the small stuff because of what I had seen.

How do plastic pollutants affect life in the ocean?

Plastic pollution is everywhere and affects marine life at every level of biological organization. Effects of microplastics can be complex, depending on their size and the type of polymer used to make them. Some microplastics cause liver tumors, others disrupt the endocrine system, and still others affect the number of offspring produced or their chances of survival. Large pieces of plastic also cause lots of problems. An animal can die after eating a big piece of plastic or being entangled by it. Large pieces also can spread diseases or smother sections of a coral reef, harming the community of organisms that live there. In addition, animals (or their eggs) attached to large pieces of plastic can be transported long distances, bringing those animals to new locations where they may harm existing communities.

What is the most surprising discovery you've made?

The biggest surprise came gradually over my career: Microplastic waste is literally everywhere. People would ask, are microplastics in the food we eat or the water we drink? I was skeptical. The first time someone asked me about plastic in drinking water, I said, "We're not going to find anything." To check, I filtered some tap water and put the filter under a microscope. And of course we found it. I live near the Great Lakes, and we can find more than a hundred pieces of microplastic in one fish. It's in the air, tap water, bottled water, dust, the food we eat. And recently, we've learned that microplastics are in humans—they're present in human stool (feces). Now we're trying to learn how the microplastics we eat, drink, and breathe affect human health. Currently, we just don't know.

What can people do to reduce plastic pollution?

The first step is to grapple with the fact that plastic is a valuable resource. It's made from oil. It is not something that should be used as a throwaway material. Once we realize that, there are many things we can do. As individuals, we vote with our wallet when we buy things. We can refuse to buy single-use plastics, such as plastic straws, cups, or disposable razor blades. At the government level, we need to create a materials management stream where the plastic we use truly is recycled. Right now we use a lot of plastic products, and they go right into the trash, much of which ends up contaminating the environment. To fix this, we need local, national, and international agreements where we agree to reduce the amount of plastic that contaminates the environment by a certain amount, and we provide support and incentives to make that happen.

> *"Plastic pollution is everywhere and affects marine life at every level of biological organization."*

52 An Introduction to Ecology and the Biosphere

KEY CONCEPTS

52.1 Earth's climate varies by latitude and season and is changing rapidly *p. 1167*

52.2 The distribution of terrestrial biomes is controlled by climate and disturbance *p. 1171*

52.3 Aquatic biomes are diverse and dynamic systems that cover most of Earth *p. 1177*

52.4 Interactions between organisms and the environment limit the distribution of species *p. 1178*

52.5 Ecological change and evolution affect one another over long and short periods of time *p. 1187*

Study Tip

Make a table: As you read the chapter, build a table listing factors that influence species distribution in terrestrial and aquatic environments. Add figure numbers or page numbers that give examples of these factors in each type of environment.

Type of environment	Factor affecting what species live there	Example
Lake	Amount of sunlight that reaches organisms	Photosynthetic plants and algae live at shallow edges and sunlit surface (Figs. 52.14 and 52.16)

→ Go to Mastering Biology

For Students (in eText and Study Area)
- Get Ready for Chapter 52
- Figure 52.4 Walkthrough: Seasonal Variation in Sunlight Intensity

For Instructors to Assign (in Item Library)
- Interpreting Data: Cholera and Sea Surface Temperature
- Scientific Thinking: Carbon Dioxide and the Growth of Forest Stands

Figure 52.1 Dwarfed by a dime, this tiny frog (*Paedophryne swiftorum*) was discovered on a 2008 expedition to Papua New Guinea. The entire genus *Paedophryne* is known only from a single peninsula in the eastern part of the country. Adult frogs in this genus are about 8 mm long and are among the smallest adult vertebrates on Earth.

What determines where a species such as this tiny frog lives?

The climate

has the strongest effect on

where terrestrial organisms live.

Light and nutrient availability

are two factors that have a strong effect on

where aquatic organisms live.

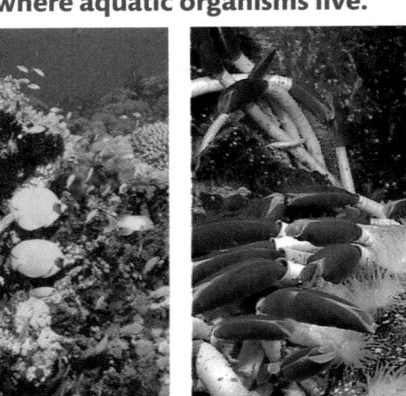

Dispersal and **interactions among organisms**, such as competition, also affect where species are found.

When studying a species in nature, scientists often begin by asking, what environmental factors limit where it is found? How do variations in an organism's food supply or interactions with other species, such as predators, affect the size of their populations?

Questions like these are the subject of **ecology**, the scientific study of the interactions between organisms and the environment. (Note that here and throughout this text,

the term *environment* refers to other organisms as well as the physical aspects of an organism's surroundings.) The interactions studied by ecologists can be organized into a hierarchy that ranges in scale from single organisms to the planet **(Figure 52.2)**. We'll begin our study of ecology by considering how Earth's climate and other factors determine the location of major life zones on land and in the oceans.

▼ Figure 52.2 **Exploring the Scope of Ecological Research**

Ecologists work at different levels of the biological hierarchy, from individual organisms to the planet. Here we present a sample research question for each level of the hierarchy.

Organismal Ecology

Organismal ecology, which includes the subdisciplines of physiological, evolutionary, and behavioral ecology, is concerned with how an organism's structure, physiology, and behavior meet the challenges posed by its environment.

◀ How do flamingos select a mate?

Population Ecology

A **population** is a group of individuals of the same species living in an area. **Population ecology** analyzes factors that affect population size and how and why it changes through time.

◀ What environmental factors affect the reproductive rate of flamingos?

Community Ecology

A **community** is a group of populations of different species in an area. **Community ecology** examines how interactions between species, such as predation and competition, affect community structure and organization.

◀ What factors influence the diversity of species that interact at this African lake?

Ecosystem Ecology

An **ecosystem** is the community of organisms in an area and the physical factors with which those organisms interact. **Ecosystem ecology** emphasizes energy flow and chemical cycling between organisms and the environment.

◀ What factors control photosynthetic productivity in this aquatic ecosystem?

Landscape Ecology

A **landscape** (or seascape) is a mosaic of connected ecosystems. Research in **landscape ecology** focuses on the factors controlling exchanges of energy, materials, and organisms across multiple ecosystems.

◀ To what extent do nutrients from terrestrial ecosystems affect organisms in the lake?

Global Ecology

The **biosphere** is the global ecosystem—the sum of all the planet's ecosystems and landscapes. **Global ecology** examines how the regional exchange of energy and materials influences the functioning and distribution of organisms across the biosphere.

◀ How do global patterns of air circulation affect the distribution of organisms?

Latitudinal Variation in Sunlight Intensity

Earth's curved shape causes latitudinal variation in the intensity of sunlight. Because sunlight strikes the tropics (those regions that lie between 23.5° north latitude and 23.5° south latitude) most directly, more heat and light per unit of surface area are delivered there. At higher latitudes, sunlight strikes Earth at an oblique angle, and thus the light energy is more diffuse on Earth's surface.

Global Air Circulation and Precipitation Patterns

Intense solar radiation near the equator initiates a global pattern of air circulation and precipitation. High temperatures in the tropics evaporate water from Earth's surface and cause warm, wet air masses to rise (blue arrows) and flow toward the poles. As the rising air masses expand and cool, they release much of their water content, creating abundant precipitation in tropical regions. The high-altitude air masses, now dry, descend (tan arrows) toward Earth around 30° north and south, absorbing moisture from the land and creating an arid climate conducive to the development of the deserts that are common at those latitudes. Some of the descending air then flows toward the poles. At latitudes around 60° north and south, the air masses again rise and release abundant precipitation (though less than in the tropics). Some of the cold, dry rising air then flows to the poles, where it descends and flows back toward the equator, absorbing moisture and creating the comparatively rainless and bitterly cold climates of the polar regions.

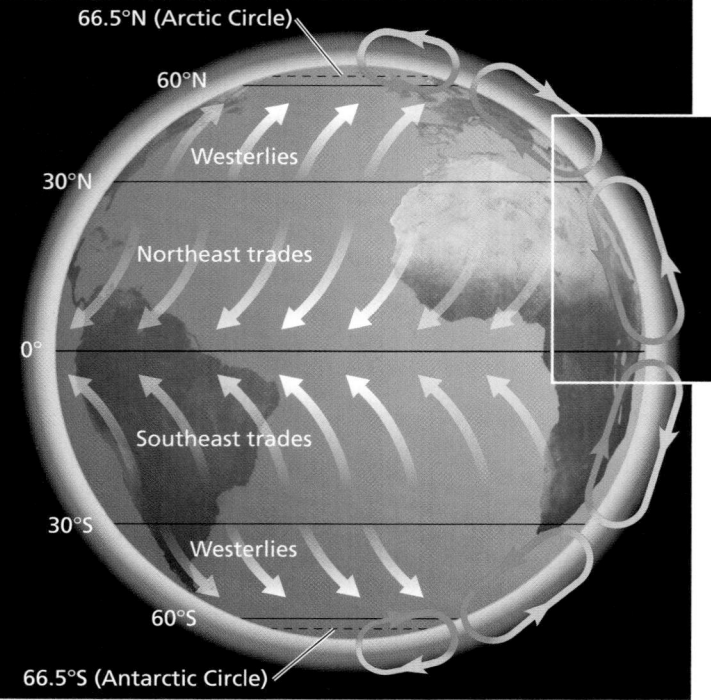

Air flowing close to Earth's surface creates predictable global wind patterns. As Earth rotates on its axis, land near the equator moves faster than that at the poles, deflecting the winds from the vertical paths shown above and creating the more easterly and westerly flows shown at left. Cooling trade winds blow from east to west in the tropics; prevailing westerlies blow from west to east in the temperate zones, defined as the regions between the Tropic of Cancer and the Arctic Circle and between the Tropic of Capricorn and the Antarctic Circle.

Earth's climate varies by latitude and season and is changing rapidly

The most significant influence on the distribution of organisms on land is **climate**, the long-term prevailing weather conditions in a given area. Four physical factors—temperature, precipitation, sunlight, and wind—are key components of climate. To set the stage for understanding how climate—and climate change—affect life on Earth, we'll begin by examining patterns in climate at the global, regional, and local levels.

Global Climate Patterns

Global climate patterns are determined largely by the input of solar energy and Earth's movement in space. The sun warms the atmosphere, land, and water. This warming establishes the temperature variations, movements of air and water, and evaporation of water that cause dramatic latitudinal variations in climate. **Figure 52.3** summarizes Earth's climate patterns and how they are formed.

Regional and Local Effects on Climate

Climate varies seasonally and can be modified by other factors, such as large bodies of water and mountain ranges. We will examine each of these factors in more detail.

Seasonality

In middle to high latitudes, Earth's tilted axis of rotation and its annual passage around the sun cause strong seasonal cycles in day length, solar radiation, and temperature **(Figure 52.4)**. The changing angle of the sun over the course of the year also affects local environments. For example, the belts of wet and dry air on either side of the equator move slightly northward and southward as the sun's angle changes; this produces marked wet and dry seasons around 20° north and 20° south latitude, where many tropical deciduous forests grow. In addition, seasonal changes in wind patterns alter ocean currents, sometimes causing the upwelling of cold water from deep ocean layers. This nutrient-rich water stimulates the growth of surface-dwelling phytoplankton and the organisms that feed on them. Though these upwelling zones make up only a few percent of the ocean's area, they are the source of more than 25% of all fish caught globally.

▼ **Figure 52.4 Seasonal variation in sunlight intensity.** Because Earth is tilted on its axis relative to its plane of orbit around the sun, the intensity of solar radiation varies seasonally. This variation is smallest in the tropics and increases toward the poles.

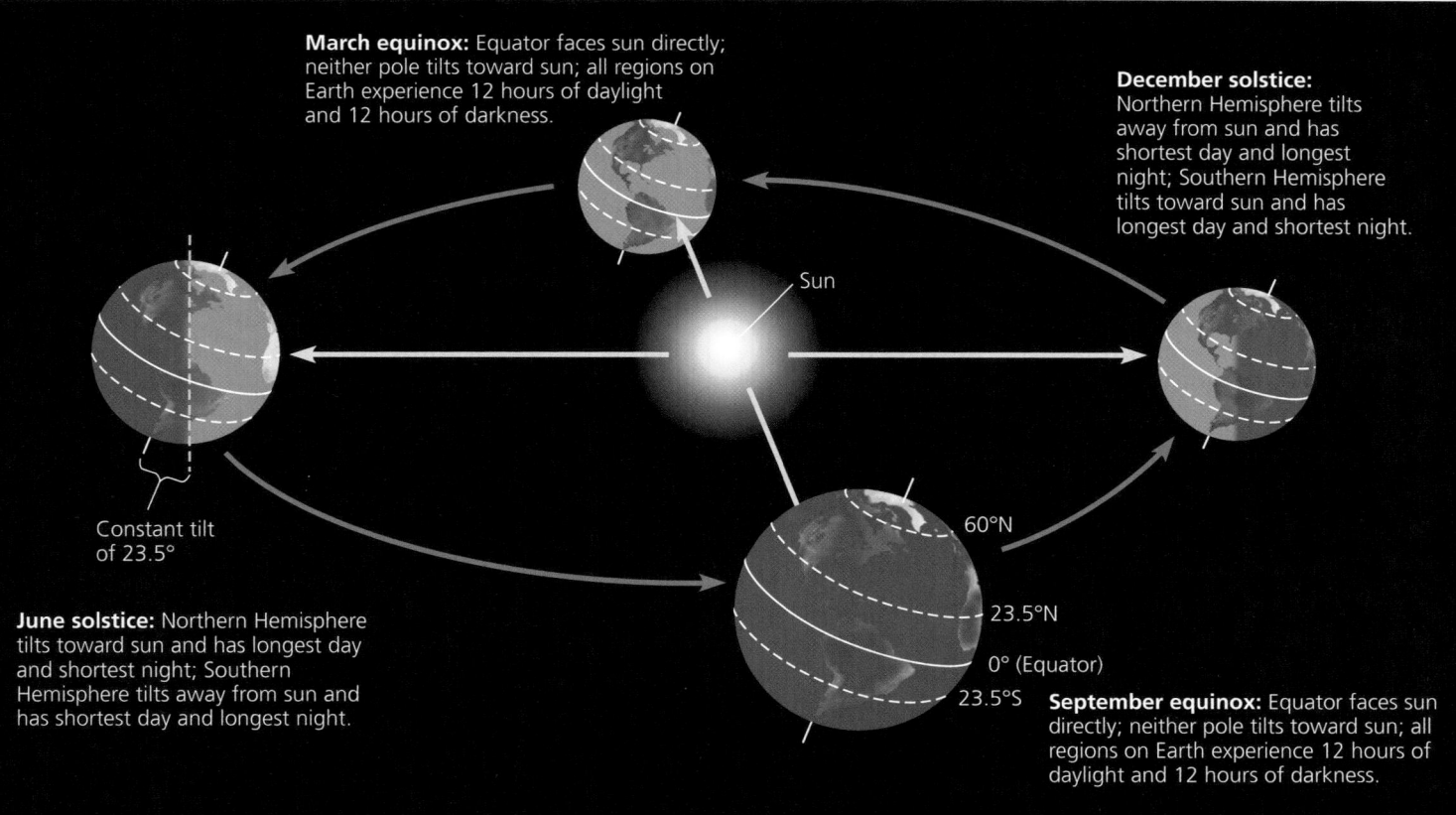

March equinox: Equator faces sun directly; neither pole tilts toward sun; all regions on Earth experience 12 hours of daylight and 12 hours of darkness.

December solstice: Northern Hemisphere tilts away from sun and has shortest day and longest night; Southern Hemisphere tilts toward sun and has longest day and shortest night.

Sun

June solstice: Northern Hemisphere tilts toward sun and has longest day and shortest night; Southern Hemisphere tilts away from sun and has shortest day and longest night.

Constant tilt of 23.5°

60°N

23.5°N

0° (Equator)

23.5°S

September equinox: Equator faces sun directly; neither pole tilts toward sun; all regions on Earth experience 12 hours of daylight and 12 hours of darkness.

⟶ **Mastering Biology Figure Walkthrough**

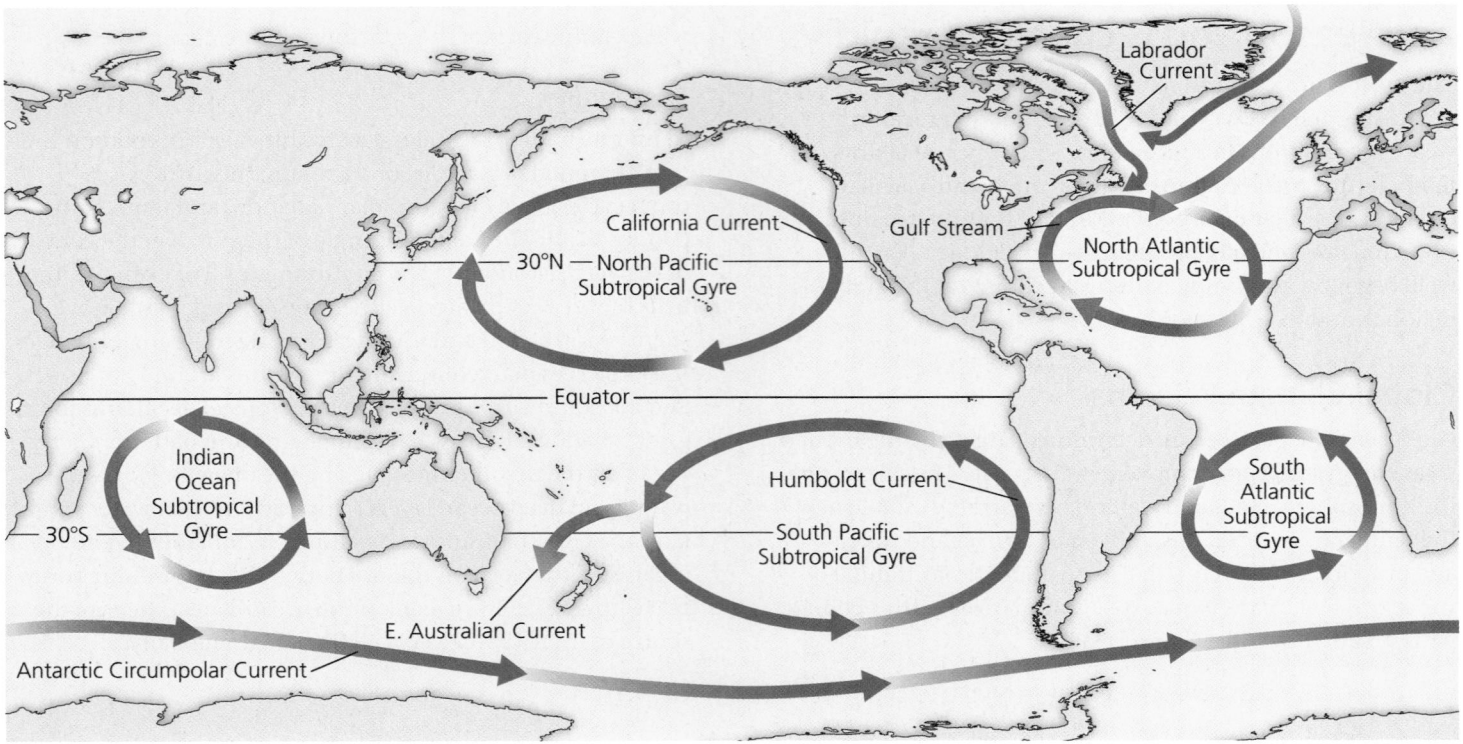

▼ **Figure 52.5 Global circulation of surface water in the oceans.** Water is warmed at the equator and flows north and south toward the poles, where it cools. Note the similarities between the direction of water circulation in the gyres and the direction of the trade winds in Figure 52.3.

Bodies of Water

Ocean currents influence climate along the coasts of continents by heating or cooling overlying air masses that pass across the land. Coastal regions are also generally wetter than inland areas at the same latitude. The cool, misty climate produced by the cold California Current that flows southward along western North America supports a coniferous rain forest ecosystem along much of the continent's Pacific coast and large redwood groves farther south. Conversely, the west coast of northern Europe has a mild climate because the Gulf Stream carries warm water from the equator to the North Atlantic **(Figure 52.5)**. As a result, northwestern Europe is warmer during winter than southeastern Canada, which is farther south but is cooled by the Labrador Current flowing south from the coast of Greenland.

Because of the high specific heat of water (see Concept 3.2), oceans and large lakes tend to moderate the climate of nearby land. During a hot day, when land is warmer than the water, air over the land heats up and rises, drawing a cool breeze from the water across the land **(Figure 52.6)**. In contrast, because temperatures drop more quickly over land than over water at night, air over the now warmer water rises, drawing cooler air from the land back out over the water and replacing it with warmer air from offshore. This local moderation of climate can be limited to the coast itself, however. In southern California and southwestern Australia, cool, dry ocean breezes in summer are warmed when they contact the land, absorbing moisture and creating a hot, arid climate just a few kilometers inland (see Figure 3.5). This climate pattern also occurs around the Mediterranean Sea, which gives it the name *Mediterranean climate*.

Mountains

Like large bodies of water, mountains influence air flow over land. When warm, moist air approaches a mountain, the air rises and cools, releasing moisture on the windward side of the peak (see Figure 52.6). On the leeward side, cooler, dry air descends, absorbing moisture and producing a "rain shadow." Such leeward rain shadows determine where many deserts are found, including the Mojave Desert of western North America and the Gobi Desert of Asia.

Mountains also affect the amount of sunlight reaching an area and thus the local temperature and rainfall. South-facing slopes in the Northern Hemisphere receive more sunlight than north-facing slopes and are therefore warmer and drier. These physical differences influence species distributions locally. On many mountains in western North America, spruce and other conifers grow on the cooler north-facing slopes, but shrubby, drought-resistant plants inhabit the south-facing slopes. In addition, every 1,000-m increase in elevation produces an average temperature drop of 6°C, equivalent to that produced by an 880-km increase in latitude. This is one reason that high-elevation plant communities near the equator, for example, can be similar to lower-elevation plant communities that are far from the equator.

▼ Figure 52.6 How large bodies of water and mountains affect climate. This figure illustrates what can happen on a hot summer day.

1 Cool air flows inland from the water, moderating temperatures near the shore.

2 Air that encounters mountains flows upward, cools at higher altitudes, and releases water as rain and snow.

3 Less moisture is left in the air reaching the leeward side, which therefore has little precipitation. This rain shadow can create a desert on the back side of the mountain range.

Windward side of mountains

Leeward side of mountains

Mountain range

Ocean

Effects of Vegetation on Climate

Climate affects where terrestrial organisms can live, but organisms can also affect climate. This is especially true of forests, which can alter the climate at local and even regional scales.

When viewed from above, a forest is darker in color than a desert or grassland. As a result, a forest absorbs more (and reflects less) solar energy than does a desert or grassland, thereby contributing to a warming of Earth's surface in forested areas. This warming effect, however, is more than offset by *transpiration*, the evaporative loss of water from a plant that cools the plant's surface—similar to how your body is cooled when you sweat. The evaporative loss of water is much greater in forests than in other ecosystems, causing forests to affect climate in two key ways: Forests reduce Earth's surface temperature and they increase precipitation rates **(Figure 52.7)**. As documented in regions around the world, the climate becomes hotter and drier in areas where humans have cut down large forests and becomes cooler and wetter where humans have restored large forests.

Microclimate

At an even smaller scale is the **microclimate**, very fine, localized patterns in climatic conditions. Many features in the environment influence microclimate by casting shade, altering evaporation from soil, or changing wind patterns. For example, forest trees often moderate the microclimate below them. Cleared areas therefore typically experience greater temperature extremes than the forest interior because of greater solar radiation and wind currents that arise from the rapid heating and cooling of open land. Within a forest, low-lying ground is usually wetter than higher ground and tends to be occupied by different tree species. A log or large stone can shelter organisms such as salamanders, worms, and insects, buffering them from the extremes of temperature and moisture.

Every environment on Earth exhibits small-scale differences in chemical and physical attributes, such as temperature, light, water, and nutrients. Later, we'll examine how these **abiotic**, or nonliving, factors influence the distribution and abundance

▼ Figure 52.7 How cutting down a forest affects regional climate.

(a) Forested land surface

1 More vegetation means more solar radiation is absorbed, fueling higher rates of photosynthesis and transpiration.

3 Overall effect: Lower temperatures, more precipitation.

2 Higher transpiration rates return more water to the atmosphere; the surface cools.

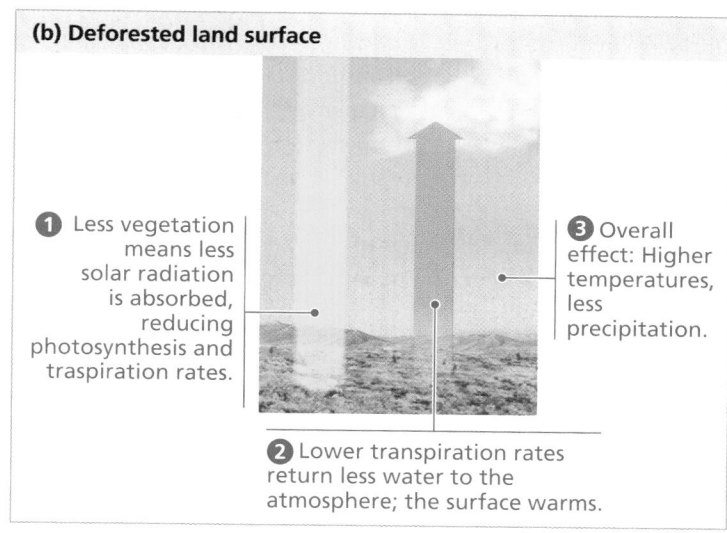

(b) Deforested land surface

1 Less vegetation means less solar radiation is absorbed, reducing photosynthesis and transpiration rates.

3 Overall effect: Higher temperatures, less precipitation.

2 Lower transpiration rates return less water to the atmosphere; the surface warms.

of organisms. Similarly, all of the **biotic**, or living, factors—the other organisms that are part of an individual's environment—also influence the distribution and abundance of life on Earth.

Global Climate Change

Because climatic variables affect the geographic ranges of most plants and animals, any large-scale change in Earth's climate profoundly affects the biosphere. In fact, a large-scale climate "experiment" is under way: The burning of fossil fuels and deforestation are increasing the concentrations of carbon dioxide and other greenhouse gases in the atmosphere. This has caused **climate change**, a directional change to the global climate that lasts three decades or more (as opposed to short-term changes in the weather). As we'll explore in more detail in Concept 56.4, Earth has warmed an average of 0.9°C (1.6°F) since 1900 and is projected to warm 1–6°C (2–11°F) more by the year 2100. Wind and precipitation patterns also are shifting, and extreme weather events (such as major storms and droughts) are occurring more frequently.

How will such changes affect the distribution of organisms? One way to answer this question is to look at the changes that have occurred since the last ice age ended. Until about 16,000 years ago, glaciers covered much of North America and Eurasia. As the climate warmed and the glaciers retreated, tree species distributions expanded northward. Fossil pollen shows that while some species moved northward rapidly, others moved more slowly, with their range expansion lagging several thousand years behind the shift in suitable habitat.

Will plants and other species be able to keep up with the much more rapid warming projected for this century? Consider the American beech, *Fagus grandifolia*. Ecological models predict that the northern limit of the beech's range may move 700–900 km northward in the next century, and its southern range limit will shift even more. The current and predicted geographic ranges of this species under two different climate-change scenarios are illustrated in **Figure 52.8**. If these predictions are even approximately correct, the beech's range must shift 7–9 km northward per year to keep pace with the warming climate. However, since the end of the last ice age, the beech has moved at a rate of only 0.2 km per year. Without human help in moving to new habitats, species such as the American beech may have much smaller ranges or even become extinct.

Indeed, the climate change that has *already* occurred has affected the geographic ranges of hundreds of terrestrial, marine, and freshwater organisms. For example, as the climate has warmed, 22 of 35 European butterfly species studied have shifted their ranges farther north by 35–240 km in recent decades. In western North America, nearly 200 plant species have moved to lower elevations, most likely in response to decreased rain and snow at higher elevations. Other research shows that a Pacific diatom species, *Neodenticula seminae*, recently colonized the Atlantic Ocean for the first time in 800,000 years. In these and many other

▼ **Figure 52.8 Current range and predicted ranges for the American beech under two climate-change scenarios.**

(a) Current range **(b) 4.5°C warming over next century** **(c) 6.5°C warming over next century**

 The predicted range in each scenario is based on climate factors alone. What other factors might alter the distribution of this species?

→ **Mastering Biology**
Interview with Margaret Davis: Using fossil pollen to track tree species migration

such cases, when climate change enables or causes a species to move into a new geographic area, other organisms living there may be harmed (see Figure 56.31).

Furthermore, as the climate changes, some species are facing a shortage of suitable habitat while others cannot migrate quickly enough. For example, a recent study found that on average, the geographic ranges of 67 bumblebee species **(Figure 52.9)** in the Northern Hemisphere were shrinking: They were retreating from the southern edges of their distributions, but failing to expand their ranges to the north. Overall, climate change is causing the populations of many species to decrease in size or even disappear (see Figure 1.12).

▼ **Figure 52.9 The rusty-patched bumblebee (*Bombus affinis*).**

CONCEPT CHECK 52.1

1. Explain how the sun's unequal heating of Earth's surface results in deserts near 30° north and south of the equator.

2. What are some of the differences in microclimate between an unplanted agricultural field and a nearby stream corridor with trees?

3. **WHAT IF?** Changes in Earth's climate at the end of the last ice age happened gradually, taking centuries to thousands of years. If the planet continues to warm at its current, rapid rate, how might this affect the evolution of long-lived trees compared with that of annual plants, which have much shorter generation times?

4. **MAKE CONNECTIONS** Focusing just on the effects of temperature, would you expect the global distribution of C_4 plants to expand or contract as Earth becomes warmer? Why? (See Concept 10.4.)

For suggested answers, see Appendix A.

▼ **Figure 52.10 The distribution of major terrestrial biomes.**

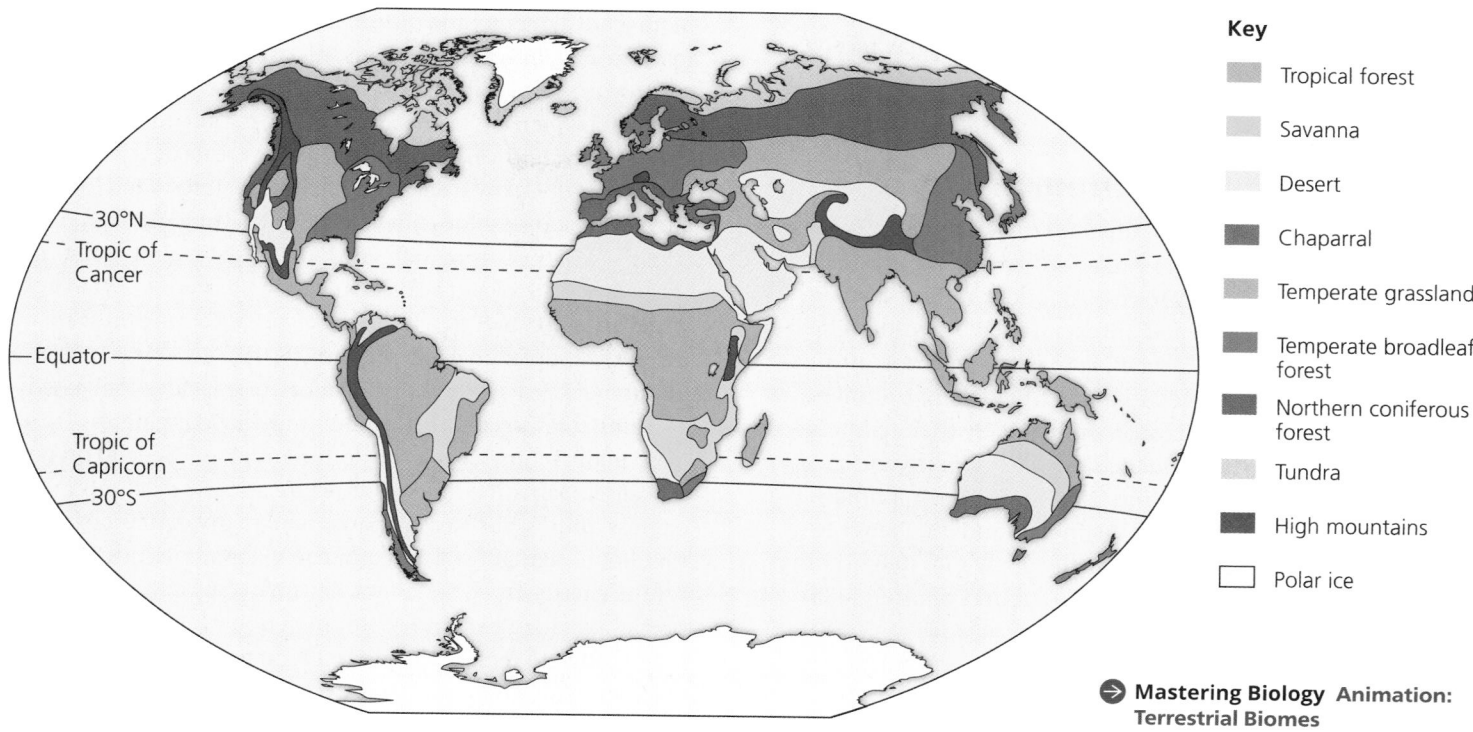

Key

- Tropical forest
- Savanna
- Desert
- Chaparral
- Temperate grassland
- Temperate broadleaf forest
- Northern coniferous forest
- Tundra
- High mountains
- Polar ice

→ **Mastering Biology** Animation: Terrestrial Biomes

CONCEPT 52.2

The distribution of terrestrial biomes is controlled by climate and disturbance

Earth's life is distributed on a grand scale in **biomes**, major life zones characterized by vegetation type in terrestrial biomes (or by the physical environment in aquatic biomes). On land, what determines where biomes are located?

Climate and Terrestrial Biomes

Because climate has a strong influence on the distribution of plant species, it is a major factor in determining the locations of terrestrial biomes **(Figure 52.10)**. The importance of climate on the distribution of biomes can be shown in a **climograph**, a plot of the annual mean temperature and precipitation in a particular region. **Figure 52.11** is a climograph for some North American biomes. Notice, for instance, that the range of precipitation in northern coniferous and temperate forests is similar but that temperate forests are generally warmer. Grasslands are typically drier than either kind of forest, and deserts are drier still.

Factors other than mean temperature and precipitation also play a role in determining where biomes exist. Some areas in North America with a particular combination of temperature and precipitation support a temperate broadleaf forest, but other areas with similar values support a coniferous forest (see the overlap in Figure 52.11). One reason for

▼ **Figure 52.11 A climograph for some major biomes in North America.** The areas plotted here encompass the ranges of annual mean temperature and precipitation in the biomes.

Desert

Temperate grassland

Temperate broadleaf forest

Tropical forest

Northern coniferous forest

Tundra

Annual mean temperature (°C)

30

15

0

−15

0   100   200   300   400

Annual mean precipitation (cm)

INTERPRET THE DATA *Some arctic tundra ecosystems receive as little rainfall as deserts but have much more dense vegetation. What climatic factor could cause this difference? Explain.*

this variation is that climographs are based on annual *averages*, but the *pattern* of climatic variation is often as important as the average climate. For example, some areas may receive regular precipitation throughout the year, whereas other areas may have distinct wet and dry seasons.

General Features of Terrestrial Biomes

Most terrestrial biomes are named for major climatic features and for their predominant vegetation. Temperate grasslands, for instance, are generally found in middle latitudes, where the climate is moderate, and are dominated by various grass species. Each biome is also characterized by microorganisms, fungi, and animals adapted to that particular environment. Temperate grasslands are usually more likely than temperate broadleaf forests to be populated by arbuscular mycorrhizal fungi (see Figure 37.14) and by large grazing mammals. Although Figure 52.10 shows distinct boundaries between the biomes, terrestrial biomes usually grade into neighboring biomes, sometimes over large areas. The area of intergradation, called an **ecotone**, may be wide or narrow.

Vertical layering of vegetation is an important feature of terrestrial biomes. In many forests, the layers from top to bottom consist of the upper **canopy**, the low-tree layer, the shrub understory, the ground layer of herbaceous plants, the forest floor (litter layer), and the root layer. Nonforest biomes have similar, though usually less pronounced, layers. Layering of vegetation provides many different habitats for animals, which sometimes exist in well-defined feeding groups, from the insectivorous birds and bats that feed above canopies to the worms, arthropods, and small mammals that search for food in the litter and root layers below.

The species composition of each kind of biome varies from place to place. For instance, in the northern coniferous forest (taiga) of North America, red spruce is common in the east but does not occur in most other areas, where black spruce and white spruce are abundant. As **Figure 52.12** shows, cacti living in deserts of North and South America resemble plants called euphorbs found in African deserts. But since cacti and euphorbs belong to different evolutionary lineages, their similarities are due to convergent evolution rather than shared ancestry.

Disturbance and Terrestrial Biomes

Biomes are dynamic, and disturbance rather than stability tends to be the rule. In ecological terms, **disturbance** is an event such as a storm, fire, or human activity that changes a community, removing organisms from it and altering resource availability. For example, hurricanes and other storms create openings for new species in many tropical and temperate forests and can alter forest composition. After Hurricane Katrina struck the Gulf coast of the United States in 2005, mixed swamp forests in the area shifted toward a dominance of bald cypress (*Taxodium distichum*) and water tupelo (*Nyssa aquatica*) because these species are less susceptible to wind damage than other tree species found there. As a result of disturbances, biomes are often patchy, containing several different communities in a single area.

In many biomes, even the dominant plants depend on periodic disturbance. Natural wildfires are an integral component of grasslands, savannas, chaparral, and many coniferous forests. Before agricultural and urban development, much of the southeastern United States was dominated by a single conifer species, the longleaf pine. Without periodic burning, broadleaf trees tended to replace the pines. Forest managers now use fire as a tool to help maintain many coniferous forests.

Figure 52.13 summarizes the major features of terrestrial biomes. As you read about the characteristics of each biome, remember that humans have altered much of Earth's surface, replacing natural communities with urban and agricultural ones. The central United States, for example, is classified as grassland and once contained extensive areas of tallgrass prairie. Very little of the original prairie remains today, however, having been converted to farmland.

CONCEPT CHECK 52.2

1. Based on the climograph in Figure 52.11, what mainly differentiates temperate grassland from temperate broadleaf forest?

2. Using Figure 52.13, identify the natural biome in which you live, and summarize its abiotic and biotic characteristics. Do these reflect your actual surroundings? Explain.

3. **WHAT IF?** If average temperatures on Earth increase by 4°C in this century, predict which biome would be most likely to replace tundra in some locations as a result. Explain.

For suggested answers, see Appendix A.

▼ **Figure 52.12 Convergent evolution in a cactus and a euphorb.** Cacti in the genus *Cereus* are found in the Americas; *Euphorbia canariensis*, a euphorb, is native to the Canary Islands, off the northwest coast of Africa.

Euphorbia ▶
canariensis

◀ *Cereus* sp.

▼ Figure 52.13 Exploring Terrestrial Biomes

Tropical Forest

Distribution Tropical forest occurs in equatorial and subequatorial regions.

Precipitation In **tropical rain forests**, rainfall is relatively constant, about 200–400 cm annually. In **tropical dry forests**, precipitation is highly seasonal, about 150–200 cm annually, with a six- to seven-month dry season.

Temperature High year-round, averaging 25–29°C with little seasonal variation.

Plants Tropical forests are vertically layered, and competition for light is intense. Layers in rain forests include trees that grow above a closed canopy, the canopy trees, one or two layers of subcanopy trees, and layers of shrubs and herbs (small, nonwoody plants). There are generally fewer layers in tropical dry forests. Broadleaf evergreen trees are dominant in tropical rain forests, whereas many tropical dry forest trees drop their leaves during the dry season. Epiphytes such as bromeliads and orchids generally cover tropical forest trees but are less abundant in dry forests. Thorny shrubs and succulent plants are common in some tropical dry forests.

Animals Earth's tropical forests are home to millions of species, including an estimated 5–30 million still undescribed species of insects, spiders, and other arthropods. In fact, animal diversity is

A tropical rain forest in Costa Rica

higher in tropical forests than in any other terrestrial biome. The animals, including amphibians, birds and other reptiles, mammals, and arthropods, are adapted to the vertically layered environment and are often inconspicuous.

Human Impact Humans have long had thriving communities in tropical forests. Many tropical forests are now being cut down and converted to farmland, urban areas, and other types of land use.

Desert

Distribution Deserts occur in bands near 30° north and south latitude or at other latitudes in the interior of continents (for instance, the Gobi Desert of north-central Asia).

Precipitation Precipitation is low and highly variable, generally less than 30 cm per year.

Temperature Temperature is variable seasonally and daily. Maximum air temperature in hot deserts may exceed 50°C; in cold deserts air temperature may fall below –30°C.

Plants Deserts are dominated by low, widely scattered vegetation; the proportion of bare ground is high compared with other biomes. The plants include succulents such as cacti or euphorbs, deeply rooted shrubs, and herbs that grow during the infrequent moist periods. Desert plant adaptations include tolerance of heat and dessication, water storage, and reduced leaf surface area. Physical defenses, such as spines, and chemical defenses, such as toxins in the leaves of shrubs, are common. Many of the plants exhibit C_4 or CAM photosynthesis.

Animals Common desert animals include snakes and lizards, scorpions, ants, beetles, migratory and resident birds, and seed-eating rodents. Many species are nocturnal. Water conservation is a common adaptation, with some species surviving solely on water obtained from breaking down carbohydrates in seeds.

Human Impact Long-distance transport of water and deep groundwater wells have allowed humans to maintain substantial populations in deserts. Urbanization and conversion to irrigated agriculture have reduced the natural biodiversity of some deserts.

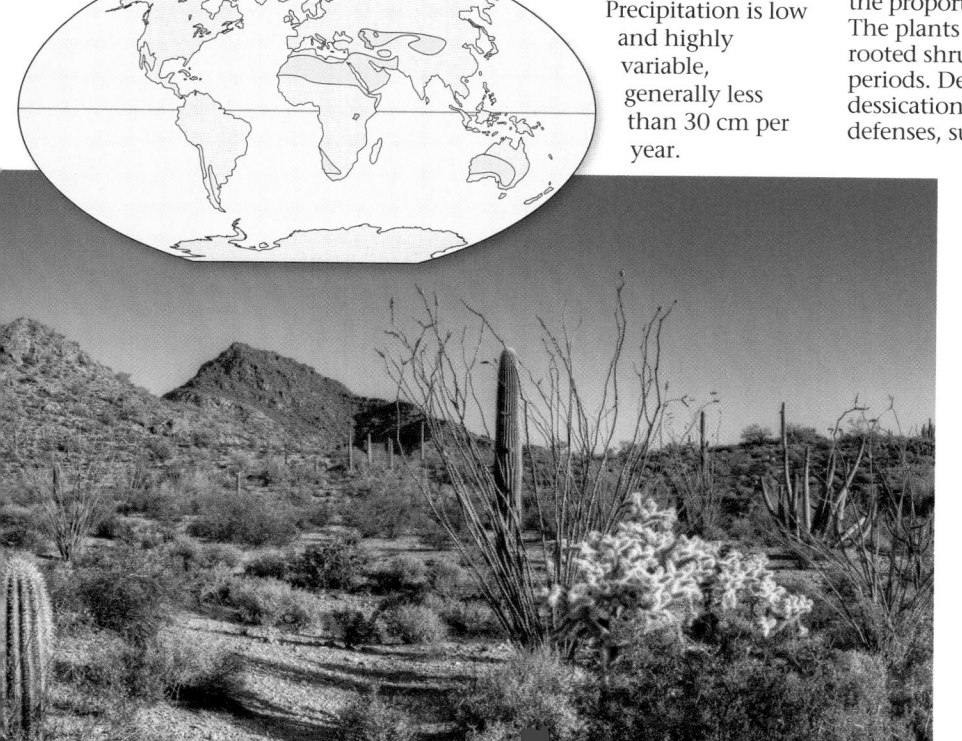

Organ Pipe Cactus National Monument, Arizona

Continued on next page

Savanna

Distribution Savanna occurs in equatorial and subequatorial regions.

Precipitation Seasonal rainfall averages 30–50 cm per year. The dry season can last up to eight or nine months.

Temperature The **savanna** is warm year-round, averaging 24–29°C, but with somewhat more seasonal variation than in tropical forests.

Plants The scattered trees found at different densities in the savanna often are thorny and have small leaves, an apparent adaptation to the relatively dry conditions. Fires are common in the dry season, and the dominant plant species are fire-adapted and tolerant of seasonal drought. Grasses and small nonwoody plants called forbs, which make up most of the ground cover, grow rapidly in response to seasonal rains and are tolerant of grazing by large mammals and other herbivores.

Animals Large plant-eating mammals, such as wildebeests and zebras, and predators, including lions and hyenas, are common inhabitants. However, the dominant herbivores are actually insects, especially termites. During seasonal droughts, grazing mammals often migrate to parts of the savanna with more forage and scattered watering holes.

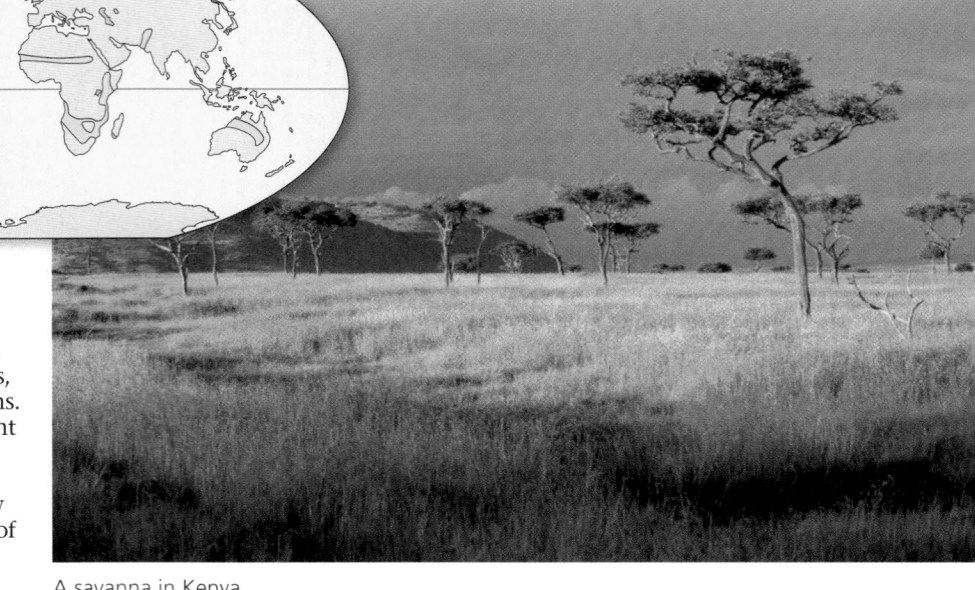

A savanna in Kenya

Human Impact The earliest humans may have lived in savannas. Fires set by humans may help maintain this biome, though overly frequent fires reduce tree regeneration by killing the seedlings and saplings. Cattle ranching and overhunting have led to declines in large-mammal populations.

Chaparral

Distribution This biome occurs in midlatitude coastal regions on several continents, and its many names reflect its far-flung distribution: **chaparral** in North America, *matorral* in Spain and Chile, *garigue* and *maquis* in southern France, and *fynbos* in South Africa.

Precipitation Precipitation is highly seasonal, with rainy winters and dry summers. Annual precipitation generally falls within the range of 30–50 cm.

Temperature Fall, winter, and spring are cool, with average temperatures in the range of 10–12°C. Average summer temperature can reach 30°C, and daytime maximum temperature can exceed 40°C.

Plants Chaparral is dominated by shrubs and small trees, along with many kinds of grasses and herbs. Plant diversity is high, with many species confined to a specific, relatively small geographic area. Adaptations of the woody plants to drought include their tough evergreen leaves, which reduce water loss. Adaptations to fire are also prominent. Some of the shrubs produce seeds that will germinate only after a hot fire; food reserves stored in their fire-resistant roots enable them to resprout quickly and use nutrients released by the fire.

Animals Native mammals include browsers, such as deer and goats, that feed on twigs and buds of woody vegetation, and a high diversity of small mammals. Chaparral areas also support many species of amphibians, birds and other reptiles, and insects.

Human Impact Chaparral areas have been heavily settled and reduced through conversion to agriculture and urbanization. Humans contribute to the fires that sweep across the chaparral.

An area of chaparral in California

Temperate Grassland

Distribution The veldts of South Africa, the *puszta* of Hungary, the pampas of Argentina and Uruguay, the steppes of Russia, and the plains and prairies of central North America are examples of **temperate grasslands**.

Precipitation Precipitation is often highly seasonal, with relatively dry winters and wet summers. Annual precipitation generally averages between 30 and 100 cm. Periodic drought is common.

Temperature Winters are generally cold, with average temperatures falling below –10°C. Summers, with average temperatures often approaching 30°C, are hot.

Plants The dominant plants are grasses and forbs, which vary in height from a few centimeters to 2 m in tallgrass prairie. Many grassland plants have adaptations that help them survive periodic, protracted droughts and fire. For example, grasses can sprout quickly following fire. Grazing by large mammals helps prevent establishment of woody shrubs and trees.

Animals Native mammals include large grazers such as bison and wild horses. Temperate grasslands are also inhabited by a wide variety of burrowing mammals, such as prairie dogs in North America.

A grassland in Mongolia

Human Impact Deep, fertile soils make temperate grasslands ideal places for agriculture, especially for growing grains. As a consequence, most grassland in North America and much of Eurasia has been converted to farmland. In some drier grasslands, cattle and other grazers have turned parts of the biome into desert.

Northern Coniferous Forest

Distribution Extending in a broad band across northern North America and Eurasia to the edge of the arctic tundra, the **northern coniferous forest**, or *taiga*, is the largest terrestrial biome on Earth.

Precipitation Annual precipitation generally ranges from 30 to 70 cm, and periodic droughts are common. However, some coastal coniferous forests of the U.S. Pacific Northwest are temperate rain forests that may receive over 300 cm of annual precipitation.

Temperature Winters are usually cold; summers may be hot. Some areas of coniferous forest in Siberia typically range in temperature from –50°C in winter to over 20°C in summer.

Plants Northern coniferous forests are dominated by cone-bearing trees, such as pine, spruce, fir, and hemlock, some of which depend on fire to regenerate. The conical shape of many conifers prevents too much snow from accumulating and breaking their branches, and their needle- or scale-like leaves reduce water loss. The diversity of plants in the shrub and herb layers of these forests is lower than in temperate broadleaf forests.

Animals While many migratory birds nest in northern coniferous forests, other species reside there year-round. The mammals of this biome, which include moose, brown bears, and Siberian tigers, are diverse. Periodic outbreaks of insects that feed on the dominant trees can kill vast tracts of trees.

Human Impact Although they have not been heavily settled by human populations, northern coniferous forests are being logged at an alarming rate, and the old-growth stands of these trees may soon disappear.

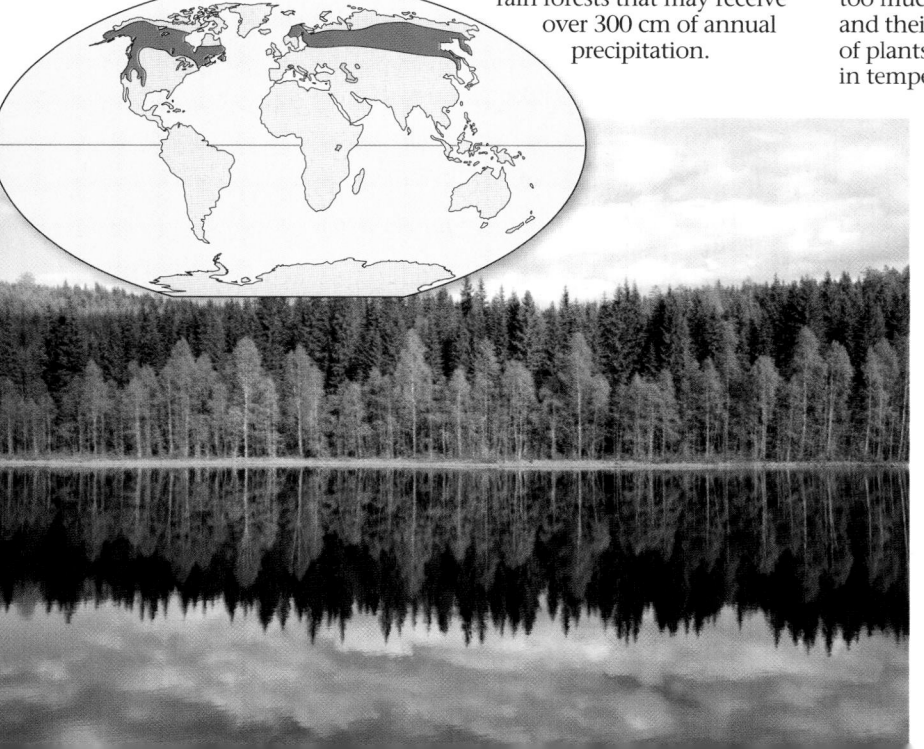

A coniferous forest in Norway

Continued on next page

Temperate Broadleaf Forest

Distribution Temperate broadleaf forest is found mainly at midlatitudes in the Northern Hemisphere, with smaller areas in Chile, South Africa, Australia, and New Zealand.

Precipitation Precipitation can average from about 70 to over 200 cm annually. Significant amounts fall during all seasons, including summer rain and, in some forests, winter snow.

Temperature Winter temperatures average 0°C. Summers, with temperatures up to 35°C, are hot and humid.

Plants A mature **temperate broadleaf forest** has distinct vertical layers, including a closed canopy, one or two strata of understory trees, a shrub layer, and an herb layer. There are few epiphytes. The dominant plants in the Northern Hemisphere are deciduous trees, which drop their leaves before winter, when low temperatures would reduce photosynthesis and make water uptake from frozen soil difficult. In Australia, evergreen eucalyptus trees dominate these forests.

Animals In the Northern Hemisphere, many mammals hibernate in winter, while many bird species migrate to warmer climates. Mammals, birds, and insects make use of all the vertical layers of the forest.

A temperate broadleaf forest in New Jersey

Human Impact Temperate broadleaf forest has been heavily settled on all continents. Logging and land clearing for agriculture and urban development cleared virtually all the original deciduous forests in North America. However, owing to their capacity for recovery, these forests are returning over much of their former range.

Tundra

Distribution Tundra covers expansive areas of the Arctic, amounting to 20% of Earth's land surface. High winds and low temperatures produce similar plant communities, called *alpine tundra*, on very high mountaintops at all latitudes, including the tropics.

Precipitation Precipitation averages from 20 to 60 cm annually in arctic tundra but may exceed 100 cm in alpine tundra.

Temperature Winters are cold, with averages in some areas below –30°C. Summer temperatures generally average less than 10°C.

Plants The vegetation of tundra is mostly herbaceous, consisting of a mixture of mosses, grasses, and forbs, along with some dwarf shrubs and trees and lichens. A permanently frozen layer of soil called permafrost restricts the growth of plant roots.

Animals Large grazing musk oxen are resident, while caribou and reindeer are migratory. Predators include bears, wolves, and foxes. Many bird species migrate to the tundra for summer nesting.

Human Impact Tundra is sparsely settled but has become the focus of significant mineral and oil extraction in recent years.

Dovrefjell-Sunndalsfjella National Park, Norway, in autumn

Aquatic biomes are diverse and dynamic systems that cover most of Earth

Unlike terrestrial biomes, aquatic biomes are characterized primarily by their physical and chemical environment. They also show far less latitudinal variation, with all types found across the globe. For example, marine biomes generally have salt concentrations that average 3%, whereas freshwater biomes are usually characterized by a salt concentration of less than 0.1%.

The oceans make up the largest marine biome, covering about 75% of Earth's surface. Because of their vast size, they greatly impact the biosphere. Water evaporated from the oceans provides most of the planet's rainfall. Marine algae and photosynthetic bacteria supply much of the world's oxygen and consume large amounts of atmospheric carbon dioxide. Ocean temperatures have a major effect on global climate and wind patterns (see Figure 52.3), and along with large lakes, oceans tend to moderate the climate of nearby land.

Freshwater biomes are closely linked to the soils and biotic components of the surrounding terrestrial biome. The particular characteristics of a freshwater biome are also influenced by the patterns and speed of water flow and the climate to which the biome is exposed.

Zonation in Aquatic Biomes

Many aquatic biomes are physically and chemically stratified (layered), vertically and horizontally, as illustrated for both a lake and a marine environment in **Figure 52.14**. Light is absorbed by water and by photosynthetic organisms, so its intensity decreases rapidly with depth. The upper **photic zone** is the region where there is sufficient light for photosynthesis, while the lower **aphotic zone** is the region where little light penetrates. The photic and aphotic zones together make up the **pelagic zone**. Deep in the aphotic zone lies the **abyssal zone**, the part of the ocean 2,000–6,000 m below the surface. At the bottom of all of these aquatic zones, deep or shallow, is the **benthic zone**. Made up of sand and organic and inorganic sediments, the benthic zone is occupied by communities of organisms collectively called the **benthos**. A major source of food for many benthic species is dead organic matter called **detritus**, which "rains" down from the productive surface waters of the photic zone.

Thermal energy from sunlight warms surface waters to whatever depth the sunlight penetrates, but the deeper waters remain quite cold. In the ocean and in most lakes, a narrow

▼ **Figure 52.14 Zonation in aquatic environments.**

(a) Zonation in a lake

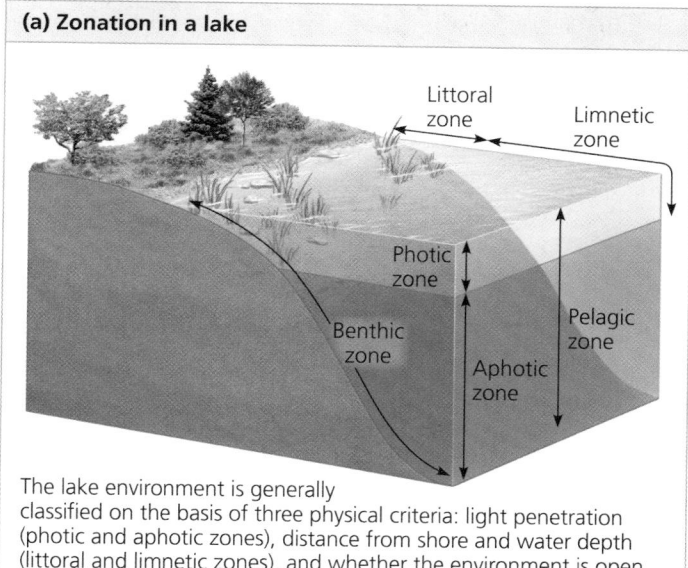

The lake environment is generally classified on the basis of three physical criteria: light penetration (photic and aphotic zones), distance from shore and water depth (littoral and limnetic zones), and whether the environment is open water (pelagic zone) or on the bottom (benthic zone).

(b) Marine zonation

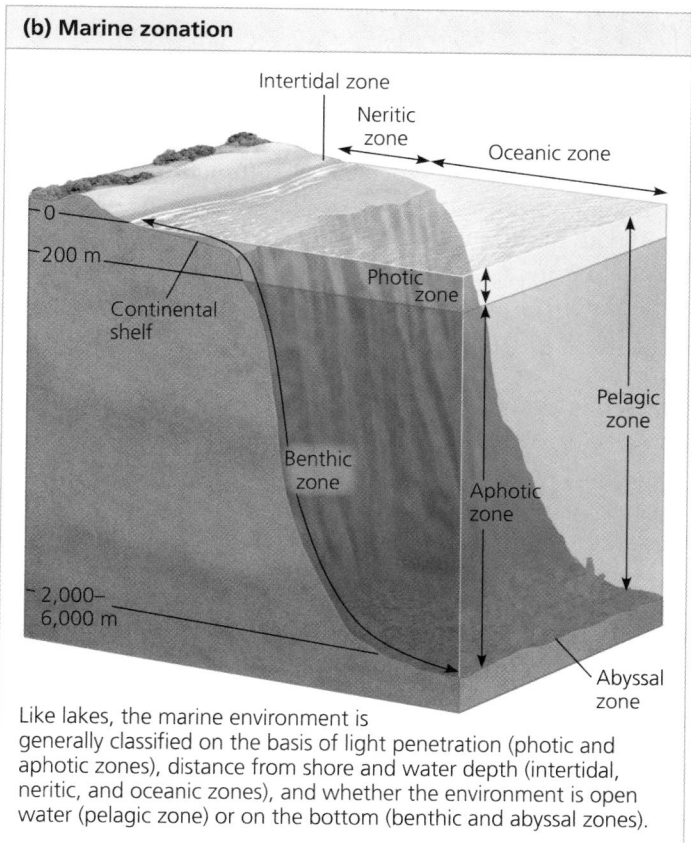

Like lakes, the marine environment is generally classified on the basis of light penetration (photic and aphotic zones), distance from shore and water depth (intertidal, neritic, and oceanic zones), and whether the environment is open water (pelagic zone) or on the bottom (benthic and abyssal zones).

layer of abrupt temperature change called a **thermocline** separates the more uniformly warm upper layer from more uniformly cold deeper waters. Lakes tend to be particularly layered with respect to temperature, especially during summer and winter, but many temperate lakes undergo a

▼ **Figure 52.15 Seasonal turnover in lakes with winter ice cover.** Seasonal turnover causes the lake waters to be well oxygenated at all depths in spring and autumn; in winter and summer, when the lake is stratified by temperature, the oxygen concentration decreases with depth.

Winter   **Spring**   **Summer**   **Autumn**

❶ In winter, the coldest water in the lake (0°C) lies just below the surface ice; water becomes progressively warmer at deeper levels of the lake, typically 4°C at the bottom.

❷ In spring, the surface water warms to 4°C and mixes with the layers below, eliminating thermal stratification. Spring winds help mix the water, bringing oxygen to the bottom and nutrients to the surface.

❸ In summer, the lake regains a distinctive thermal profile, with warm surface water separated from cold bottom water by a narrow vertical zone of abrupt temperature change, called a thermocline.

❹ In autumn, as surface water cools rapidly, it sinks beneath the underlying layers, remixing the water until the surface begins to freeze and the winter temperature profile is reestablished.

semiannual mixing of their waters as a result of changing temperature profiles **(Figure 52.15)**. This **turnover**, as it is called, sends oxygenated water from a lake's surface to the bottom and brings nutrient-rich water from the bottom to the surface in both spring and autumn.

In both freshwater and marine environments, communities are distributed according to water depth, degree of light penetration, distance from shore, and whether they are found in open water or near the bottom. Marine communities, in particular, illustrate the limitations on species distribution that result from these abiotic factors. Plankton and many fish species live in the relatively shallow photic zone (see Figure 52.14b). Because water absorbs light so well and the ocean is so deep, most of the ocean volume is dark (the aphotic zone) and harbors relatively little life.

Figure 52.16 explores the main characteristics of Earth's major aquatic biomes.

CONCEPT CHECK 52.3

1. Why are phytoplankton, and not benthic algae or rooted aquatic plants, the dominant photosynthetic organisms of the oceanic pelagic zone? (See Figure 52.16.)

2. **MAKE CONNECTIONS** Many organisms living in estuaries experience freshwater and saltwater conditions each day with the rising and falling of tides. Explain how these changing conditions challenge the survival of these organisms (see Concept 44.1).

3. **MAKE CONNECTIONS** As noted in Figure 52.16, the addition of nutrients to a lake can cause an algal bloom. When these algae die, complex molecules in their bodies are broken down by decomposers using aerobic respiration. Explain why this could reduce the lake's oxygen levels (see Concept 9.1).

For suggested answers, see Appendix A.

CONCEPT 52.4

Interactions between organisms and the environment limit the distribution of species

Broadly speaking, species distributions are a consequence of both ecological factors and evolutionary history. Consider kangaroos, which are found in Australia and nowhere else in the world. Fossil evidence indicates that kangaroos and their close relatives originated in Australia, roughly 5 million years ago. By that time, Australia had moved close to its present location (by continental drift; see Concept 25.4), and it was not connected to other landmasses. Thus, kangaroos occur only in Australia in part because of an accident of history: The kangaroo lineage originated there at a point in time when the continent was geographically isolated.

But ecological factors are also important. To date, kangaroos have not dispersed (on their own) to other continents; hence, they are restricted to the continent on which they originated. And within Australia, kangaroos are found in some habitats but not in others. The red kangaroo, for example, occurs in the arid grasslands of central Australia, but not in the tall, open forests of eastern Australia. Moreover, kangaroos are not unusual in this respect—all species are found in some habitats but not others. Hence, ecologists ask not only *where* species occur, but also *why* species occur where they do: What ecological factors—biotic and abiotic—determine their distribution?

In many cases, both biotic and abiotic factors affect the distribution of a species, as is true of the saguaro cactus (*Carnegiea gigantea*). Saguaros are found almost exclusively in the Sonoran Desert of the southwestern United States and

Lakes

Physical Environment Standing bodies of water range from ponds a few square meters in area to lakes covering thousands of square kilometers. Light decreases with depth, creating stratification. Temperate lakes may have a seasonal thermocline; tropical lowland lakes have a thermocline year-round.

Chemical Environment The salinity, oxygen concentration, and nutrient content differ greatly among lakes and can vary with season. **Oligotrophic lakes** tend to be nutrient-poor and generally oxygen-rich; **eutrophic lakes** are nutrient-rich and often depleted of oxygen in the deepest zone in summer and if covered with ice in winter. The amount of decomposable organic matter in bottom sediments is low in oligotrophic lakes and high in eutrophic lakes; high rates of decomposition in deeper layers of eutrophic lakes cause periodic oxygen depletion.

Geologic Features Oligotrophic lakes tend to have less surface area relative to their depth than eutrophic lakes. Oligotrophic lakes may become more eutrophic over time as runoff adds sediments and nutrients.

Photosynthetic Organisms Rooted and floating aquatic plants in lakes live in the

An oligotrophic lake in Jasper National Park, Alberta

littoral zone, the shallow, well-lit waters close to shore. Farther from shore, where water is too deep to support rooted aquatic plants, the **limnetic zone** is inhabited by a variety of phytoplankton, including cyanobacteria.

Heterotrophs In the limnetic zone, small drifting heterotrophs, or zooplankton, graze on the phytoplankton. The benthic zone is inhabited by assorted invertebrates whose species composition depends partly on oxygen levels. Fishes live in all zones with sufficient oxygen.

A eutrophic lake in the Okavango Delta, Botswana

Human Impact Runoff from fertilized land and dumping of wastes lead to nutrient enrichment, which can produce large numbers of algae (an algal "bloom") oxygen depletion, and fish kills.

Wetlands

Physical Environment A **wetland** is a habitat that is inundated by water at least some of the time and that supports plants adapted to water-saturated soil. Some wetlands are inundated at all times, whereas others flood infrequently.

Chemical Environment Because of high organic production by plants and decomposition by microbes and other organisms, both the water and the soils are periodically low in dissolved oxygen. Wetlands have a high capacity to filter dissolved nutrients and chemical pollutants.

Geologic Features *Basin wetlands* develop in shallow basins, ranging from upland depressions to filled-in lakes and ponds. *Riverine wetlands* develop along shallow and periodically flooded banks of rivers and streams. *Fringe wetlands* occur along the coasts of large lakes and seas, where water flows back and forth because of rising lake levels or tidal action. Thus, fringe wetlands include both freshwater and marine biomes.

Photosynthetic Organisms Wetlands are among the most productive biomes on Earth. Their water-saturated soils favor the growth of plants such as pond lilies and cattails, many sedges, bald cypress, and black spruce, which have adaptations enabling them to grow in water or in soil that is periodically anaerobic owing to the presence of unaerated water. Woody plants dominate the vegetation of swamps, while bogs are dominated by sphagnum mosses.

Heterotrophs Wetlands are home to a diverse community of invertebrates, birds, and many other organisms. Herbivores, from crustaceans and aquatic insect larvae to muskrats, consume algae, detritus, and plants. Carnivores are also varied and may include dragonflies, otters, frogs, alligators, and herons.

Human Impact Wetlands help purify water and reduce peak flooding. Draining and filling have destroyed up to 90% of wetlands in Europe.

A basin wetland in the United Kingdom

Continued on next page

Streams and Rivers

Physical Environment The most prominent physical characteristic of streams and rivers is the speed and volume of their flow. Head-water streams are generally cold, clear, swift, and turbulent. Farther downstream, where numerous tributaries may have joined, forming a river, the water is generally warmer and more turbid because of suspended sediment. Streams and rivers are stratified into vertical zones.

Chemical Environment The salt and nutrient content of streams and rivers increases from the headwaters to the mouth. Headwaters are generally rich in oxygen. Downstream water may also contain substantial oxygen, except where there has been organic enrichment. A large fraction of the organic matter in rivers consists of dissolved or highly fragmented material that is carried by the current from forested streams.

Geologic Features Headwater stream channels are often narrow, have a rocky bottom, and alternate between shallow sections and deeper pools. The downstream stretches of rivers are generally wide and meandering. River bottoms are often silty from sediments deposited over long periods of time.

A headwater stream in Washington

Photosynthetic Organisms Headwater streams that flow through grasslands or deserts may be rich in phytoplankton or rooted aquatic plants.

Heterotrophs A great diversity of fishes and invertebrates inhabit unpolluted rivers and streams, distributed according to, and throughout, the vertical zones. In streams flowing through temperate or tropical forests, organic matter from terrestrial vegetation is the primary source of food for aquatic consumers.

The Loire River in France, far from its headwaters

Human Impact Municipal, agricultural, and industrial pollution degrade water quality and kill aquatic organisms. Damming and flood control impair the natural functioning of stream and river ecosystems and threaten migratory species such as salmon.

Estuaries

Physical Environment An **estuary** is a transition area between river and sea. Seawater flows up the estuary channel during a rising tide and flows back down during the falling tide. Often, higher-density seawater occupies the bottom of the channel and mixes little with the lower-density river water at the surface.

Chemical Environment Salinity varies spatially within estuaries, from nearly that of fresh water to that of seawater. Salinity also varies with the rise and fall of the tides. Nutrients from the river make estuaries, like wetlands, among the most productive biomes.

Geologic Features Estuarine flow patterns combined with the sediments carried by river and tidal waters create a complex network of tidal channels, islands, natural levees, and mudflats.

Photosynthetic Organisms Saltmarsh grasses and algae, including phytoplankton, are the major producers in estuaries.

Heterotrophs Estuaries support an abundance of worms, oysters, crabs, and many fish species that humans consume. Many marine invertebrates and fishes use estuaries as a breeding ground or migrate through them to freshwater habitats upstream. Estuaries are also crucial feeding areas for waterfowl and some marine mammals.

Human Impact Filling, dredging, and pollution from upstream have disrupted estuaries worldwide.

An estuary in southern Spain

Intertidal Zones

Physical Environment An **intertidal zone** is periodically submerged and exposed by the tides, twice daily on most marine shores. Upper zones experience longer exposures to air and greater variations in temperature and salinity. Changes in physical conditions from the upper to the lower intertidal zones limit the distributions of many organisms to particular strata, as shown in the photograph.

Chemical Environment Oxygen and nutrient levels are generally high and are renewed with each turn of the tides.

Geologic Features The substrates of intertidal zones, which are generally either rocky or sandy, select for particular behavior and anatomy among intertidal organisms. The configuration of bays or coastlines influences the magnitude of tides and the relative exposure of intertidal organisms to wave action.

Photosynthetic Organisms A high diversity and biomass of attached marine algae inhabit rocky intertidal zones, especially in the lower zone. Sandy intertidal zones exposed to vigorous wave action generally lack attached plants or algae, while sandy intertidal zones in protected bays or lagoons often support rich beds of seagrass and algae.

Heterotrophs Many of the animals in rocky intertidal environments have structural adaptations that enable them to attach to the hard substrate. The composition, density, and diversity of animals change markedly from the upper to the lower intertidal zones. Many of the animals in sandy or muddy intertidal zones, such as worms, clams, and predatory crustaceans, bury themselves and feed as the tides bring sources of food. Other common animals are sponges, sea anemones, echinoderms, and small fishes.

Human Impact Oil pollution has disrupted many intertidal areas. The construction of rock walls and barriers to reduce erosion from waves and storm surges has disrupted this zone in some locations.

A rocky intertidal zone on the Oregon coast

Oceanic Pelagic Zone

Physical Environment The **oceanic pelagic zone** is a vast realm of open blue water, constantly mixed by wind driven oceanic currents. Because of higher water clarity, the photic zone extends to greater depths than in coastal marine waters.

Chemical Environment Oxygen levels are generally high. Nutrient concentrations are generally lower than in coastal waters. Because they are thermally stratified year-round, some tropical areas of the oceanic pelagic zone have lower nutrient concentrations than temperate oceans. Turnover between fall and spring renews nutrients in the photic zones of temperate and high-latitude ocean areas.

Geologic Features This biome covers approximately 70% of Earth's surface and has an average depth of nearly 4,000 m. The deepest point in the ocean is more than 10,000 m beneath the surface.

Photosynthetic Organisms The dominant photosynthetic organisms are phytoplankton, including photosynthetic bacteria, that drift with the oceanic currents. Spring turnover renews nutrients in temperate oceans, producing a surge of phytoplankton growth. Because of this biome's large size, photosynthetic plankton account for about half of the photosynthetic activity on Earth.

Heterotrophs The most abundant heterotrophs in this biome are zooplankton. These protists, worms, copepods, shrimp-like krill, jellies, and small larvae of invertebrates and fishes graze on photosynthetic plankton. The oceanic pelagic zone also includes numerous free-swimming animals, such as large squids, fishes, sea turtles, and marine mammals.

Human Impact Overfishing has depleted fish stocks in all Earth's oceans; marine life has also been harmed by pollution, ocean acidification, and global warming.

Continued on next page

Open ocean near Iceland

Coral Reefs

Physical Environment Coral reefs are formed largely from the calcium carbonate skeletons of corals. Shallow reef-building corals live in the photic zone of relatively stable tropical marine environments with high water clarity, primarily near islands and along the edge of some continents. They are sensitive to temperatures below about 18–20°C and above 30°C. Deep-sea coral reefs, found between 200 and 1,500 m deep, are less known than their shallow counterparts but harbor as much diversity as many shallow reefs do.

Chemical Environment Corals require high oxygen levels and are excluded by high inputs of fresh water and nutrients.

Geologic Features Corals require a solid substrate for attachment. A typical coral reef begins as a *fringing reef* on a young, high island, forming an offshore *barrier reef* later in the history of the island and becoming a *coral atoll* as the older island submerges.

Photosynthetic Organisms Unicellular algae live within the tissues of the corals, forming a mutualistic relationship that provides the corals with organic molecules. Diverse multicellular red and green algae growing on the reef also contribute substantial amounts of photosynthesis.

A coral reef in the Red Sea

Heterotrophs Corals, a diverse group of cnidarians, are themselves the predominant animals on coral reefs. However, fish and invertebrate diversity is exceptionally high. Overall animal diversity on coral reefs rivals that of tropical forests.

Human Impact Collecting of coral skeletons and overfishing have reduced populations of corals and reef fishes. Global warming and pollution may be contributing to large-scale coral death. Development of coastal mangroves for aquaculture has also reduced spawning grounds for many species of reef fishes.

Marine Benthic Zone

Physical Environment The **marine benthic zone** consists of the seafloor below the surface waters of the coastal, or **neritic**, zone and the offshore pelagic zone. Except for shallow, near-coastal areas, the marine benthic zone receives no sunlight. Water temperature declines with depth, while pressure increases. As a result, organisms in the very deep benthic, or abyssal, zone are adapted to continuous cold (about 3°C) and very high water pressure.

Chemical Environment Except in areas of organic enrichment, oxygen is usually present at sufficient concentrations to support diverse animal life.

Geologic Features Soft sediments cover most of the benthic zone. However, there are areas of rocky substrate on reefs, submarine mountains, and new oceanic crust.

Autotrophs Photosynthetic organisms, mainly algae, are limited to shallow benthic areas with sufficient light to support them. Unique assemblages of organisms live near deep-sea **hydrothermal vents** on mid-ocean ridges. In these dark, hot environments, the food producers are chemoautotrophic prokaryotes that obtain energy by oxidizing H_2S formed by a reaction of the hot water with dissolved sulfate (SO_4^{2-}).

Heterotrophs Neritic benthic communities include numerous invertebrates and fishes. Beyond the photic zone, most consumers depend entirely on organic matter raining down from above. Among the animals of the deep-sea hydrothermal vent communities are giant tube worms (pictured), some more than 1 m long. They are nourished by chemoautotrophic prokaryotes that live as symbionts within their bodies. Many other invertebrates, including arthropods and echinoderms, are also abundant around the vents.

Human Impact Overfishing has decimated important benthic fish populations, such as the cod of the Grand Banks off Newfoundland. Dumping of organic wastes has created oxygen-deprived benthic areas.

A deep-sea hydrothermal vent community

→ **Mastering Biology** Video: Tubeworms

northwestern Mexico (**Figure 52.17**). To the north, their range is limited by an abiotic factor: temperature. Saguaros tolerate freezing temperatures only briefly, typically for less than a day, and generally cannot survive at temperatures below –4°C (25°F). For the same reason, saguaros are rarely found at elevations above 1,200 m (4,000 feet).

However, temperature alone does not fully explain the distribution of saguaros, including why they are missing from the western portion of the Sonoran Desert. Water availability is critical because seedling survival typically requires consecutive years of moist conditions, something that may occur only a few times each century. Biotic factors almost certainly influence their distribution as well. Mice and grazers such as goats eat the seedlings, and bats pollinate the large, white flowers that open at night. Saguaros are also vulnerable to a deadly bacterial disease. Thus, for the saguaro, as for most other species, ecologists need to consider multiple factors and alternative hypotheses when attempting to explain the distribution of a species.

To see how ecologists might arrive at such an explanation, let's examine the ecological factors highlighted by the questions in the flowchart in **Figure 52.18**.

Dispersal and Distribution

One factor that contributes greatly to the global distribution of organisms is **dispersal**, the movement of individuals or gametes away from their area of origin or from centers of high population density. For example, while land-bound kangaroos have not reached Africa under their own power, other organisms that disperse more readily, such as some birds, have. The dispersal of organisms is critical to understanding the role of geographic isolation in evolution (see Concept 24.2) as well as the patterns of species distribution we see today, including that of the Pacific diatom discussed earlier in this chapter.

Natural Range Expansions and Adaptive Radiation

The importance of dispersal is most evident when organisms reach an area where they did not exist previously, called a *range expansion*. For instance, 200 years ago, the cattle egret (*Bubulcus ibis*) was found only in Africa and southwestern Europe. But in the late 1800s, some of these birds managed to cross the Atlantic Ocean and colonize northeastern

▲ Figure 52.17 Distribution of the saguaro cactus in North America. Freezing temperatures strongly limit where saguaros are found, but other abiotic and biotic factors are also important.

- ▢ Sonoran desert
- • Saguaro cacti observed

100 km

South America. From there, cattle egrets gradually spread southward and also northward through Central America and into North America, reaching Florida by 1960 (**Figure 52.19**). Today they have breeding populations as far west as the Pacific coast of the United States and as far north as southern Canada.

In rare cases, such long-distance dispersal can lead to adaptive radiation, the rapid evolution of an ancestral species into new species that fill many ecological niches. The diversity of Hawaiian silverswords is one example of adaptive radiation that was possible only with the long-distance dispersal of an ancestral tarweed from North America (see Figure 25.22).

Natural range expansions clearly show the influence of dispersal on distribution. However, opportunities to observe such dispersal directly are rare, so ecologists often turn to experimental methods to better understand the role of dispersal in limiting the distribution of species.

▼ Figure 52.18 Flowchart of factors limiting geographic distribution. An ecologist studying factors limiting a species' distribution might consider questions like these. As suggested by the arrows leading from the "Yes" responses, the ecologist would answer all of these questions because more than one factor can limit a species' distribution.

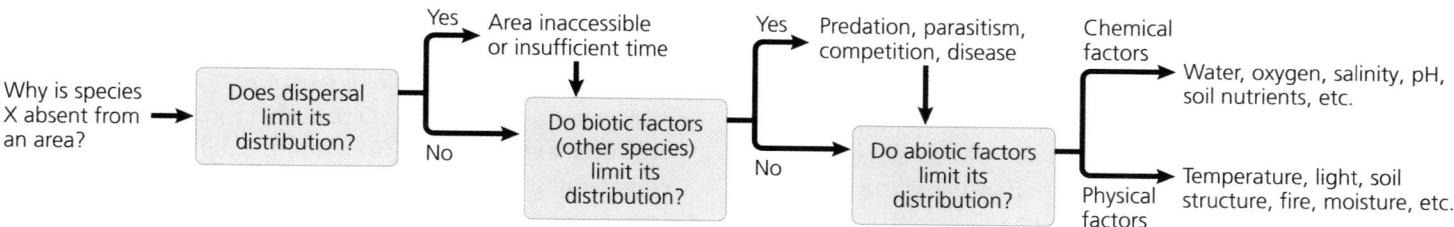

? *How might the importance of various abiotic factors differ for aquatic and terrestrial ecosystems?*

▲ Figure 52.19 Dispersal of the cattle egret in the Americas. Native to Africa, cattle egrets were first reported in South America in 1877. The color-coding and dates show the regions to which the birds had spread at different points in time.

Species Transplants

To determine if dispersal is a key factor limiting the distribution of a species, ecologists may observe the results of intentional or accidental transplants of the species to areas where it was previously absent. For a transplant to be successful, some of the organisms must not only survive in the new area but also reproduce there sustainably. If a transplant is successful, then we can conclude that the *potential* range of the species is larger than its *actual* range; in other words, the species *could* live in certain areas where it currently does not.

In some cases, species introduced to new geographic locations have disrupted the communities and ecosystems to which they have been introduced (see Concept 56.1). Consequently, ecologists rarely move species to new regions. Instead, they document the outcome when a species reaches a new region in other ways, such as when a predator is introduced to control a pest species or when a species has been moved accidentally to the area.

Biotic Factors

Our next question is whether biotic factors—other species—limit the distribution of a species. Often, the ability of a species to survive and reproduce is reduced by its interactions with other species, such as predators (organisms that kill their prey) or herbivores (organisms that eat plants or algae). **Figure 52.20** shows how an herbivore, the long-spined sea urchin (*Centrostephanus rodgersii*), has affected the distribution of a food species. In certain marine ecosystems, there is often an inverse relationship between the abundance of sea urchins and seaweeds (multicellular algae, such as kelp). Where urchins that graze on seaweeds and other algae are common, large stands of seaweeds do not become established. As shown in Figure 52.20, Australian

▼ Figure 52.20 Effects of feeding by sea urchins on seaweed distribution. Seaweed abundance in areas from which the long-spined sea urchin (*Centrostephanus rodgersii*) had been removed was much higher than seaweed abundance in adjacent control sites from which the urchin was not removed.

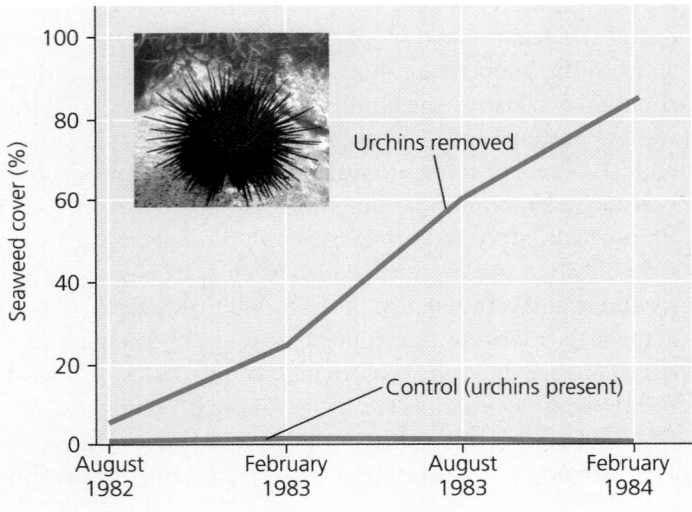

researchers tested whether *C. rodgersii* is a biotic factor limiting seaweed distribution. When this urchin was removed from experimental plots, seaweed cover increased dramatically, showing that *C. rodgersii* limited the distribution of seaweeds.

In addition to predation and herbivory, the presence or absence of pollinators, food resources, parasites, pathogens, and competing organisms can act as a biotic limitation on species distribution. Such biotic limitations are common in nature.

Abiotic Factors

The last question in the flowchart in Figure 52.18 considers whether abiotic factors, such as temperature, water, oxygen, salinity, sunlight, or soil, might be limiting a species' distribution. If the physical conditions at a site do not allow a species to survive and reproduce, then the species will not be found there. Throughout this discussion, keep in mind that most abiotic factors vary substantially over space and time. Daily and annual fluctuations of abiotic factors may either blur or accentuate regional distinctions. Furthermore, organisms can avoid some stressful conditions temporarily through behaviors such as dormancy or hibernation (see Concept 40.4).

Temperature

Environmental temperature is an important factor in the distribution of organisms because of its effect on biological processes. Cells may rupture if the water they contain freezes (at temperatures below 0°C), and the proteins of most organisms denature at temperatures above 45°C. Organisms typically function best within a specific range of environmental temperature. Temperatures outside that range may force some animals to expend energy regulating their internal temperature, as mammals and birds do (see Figure 40.16). Extraordinary adaptations

▼ Figure 52.21 A sea urchin's expanding range. Since the 1950s, water temperatures along the coast of Tasmania have increased, allowing the sea urchin *C. rodgersii* to expand its range to the south. The bold orange type indicates the years when *C. rodgersii* was first observed colonizing those locations. Once its population has been established in a new location, *C. rodgersii* has eliminated the local kelp community.

AUSTRALIA

TASMANIA

AUSTRALIA

East Australian Current

1 Southern portion of *C. rodgersii* native range

1960s

2 Expanded range of *C. rodgersii*; kelp communities decimated

1978

TASMANIA

1980s

1980s

1990s

2005

3 Continuation of expanded range; kelp communities threatened

100 km

enable certain organisms, such as thermophilic prokaryotes, to live outside the temperature range habitable by other life.

As mentioned earlier, climate change has already caused hundreds of species to alter their geographic ranges. A shift in the range of one species can also have profound effects on the distribution of other species. Consider how rising sea temperatures have affected the geographic range of the sea urchin *C. rodgersii*. Since 1950, water temperatures along the coast of Tasmania, an island south of mainland Australia, have increased from 11.5°C to 12.5°C. This has enabled *C. rodgersii*—whose larvae fail to develop properly if temperatures drop below 12°C—to expand its range to the south **(Figure 52.21)**. The urchin is a voracious consumer of kelp and other algae. As a result, algal communities that once harbored a rich diversity of other species have been completely destroyed in regions where the urchin has become well established (denoted by a solid orange line in Figure 52.21).

Water and Oxygen

The dramatic variation in water availability among habitats is another important factor in species distribution. Species living at the seashore or in tidal wetlands can desiccate (dry out) as the tide recedes. Terrestrial organisms face a nearly constant threat of desiccation, and the distribution of terrestrial species reflects their ability to obtain and conserve water. Many amphibians, such as the tiny frog in Figure 52.1, are particularly vulnerable to drying because they use their moist, delicate skin for gas exchange. Desert organisms exhibit a variety of adaptations for acquiring and conserving water in dry environments, as described in Concept 44.4.

Water affects oxygen availability in aquatic environments and in flooded soils, where the slow diffusion of oxygen in water can limit cellular respiration and other physiological processes. Oxygen concentrations can be particularly low in deep ocean and deep lake waters, as well as in sediments where organic matter is abundant. Flooded wetland soils may also have low oxygen content. Mangroves and other trees have specialized roots that project above the water and help the root system obtain oxygen (see Figure 35.4). Unlike many flooded wetlands, the surface waters of streams and rivers tend to be well oxygenated because of rapid exchange of gases with the atmosphere.

Salinity

The salt concentration of water in the environment affects the water balance of organisms through osmosis. Most aquatic organisms are restricted to either freshwater or saltwater habitats by their limited ability to osmoregulate (see Concept 44.1). Most terrestrial organisms can excrete excess salts from specialized glands or in feces or urine. However, the salt concentrations in some habitats (such as salt flats) are so high that few species of plants or animals can survive there **(Figure 52.22)**. In the **Scientific Skills Exercise**, you can interpret data from an experiment investigating the influence of salinity on plant distributions.

Salmon that migrate between freshwater streams and the ocean use both behavioral and physiological mechanisms to osmoregulate. They balance their salt content by adjusting the amount of water they drink and by switching their gills from taking up salt in fresh water to excreting salt in the ocean.

Sunlight

Sunlight provides the energy that drives most ecosystems, and too little sunlight can limit the distribution of photosynthetic species. In forests, shading by leaves makes competition for light especially intense, particularly for seedlings growing on the forest floor. In aquatic environments, every meter of water depth absorbs about 45% of the red light and about 2% of the blue light passing through it. Thus, most photosynthesis occurs relatively near the water surface.

▼ Figure 52.22 Salar de Uyuni, in Bolivia, the largest salt flat in the world. Aside from the flamingos that breed there annually, very few animals or plants inhabit the vast expanse of crusty white salt.

Making a Bar Graph and a Line Graph to Interpret Data

How Do Salinity and Competition Affect the Distribution of Plants in an Estuary? Field observations show that *Spartina patens* (salt marsh hay) is a dominant plant in salt marshes and *Typha angustifolia* (cattail) is a dominant plant in freshwater marshes. In this exercise, you will graph and interpret data from an experiment that examined the influence of an abiotic factor, salinity, and a biotic factor, competition, on the growth of these two species.

How the Experiment Was Done Researchers planted *S. patens* and *T. angustifolia* in salt marshes and freshwater marshes with and without neighboring plants. After two growing seasons (1.5 years), they measured the biomass of each species in each treatment. The researchers also grew both species in a greenhouse at six salinity levels and measured the biomass at each level after eight weeks.

Data from the Field Experiment (averages of 16 replicate samples)

	Average Biomass (g/100 cm²)			
	Spartina patens		*Typha angustifolia*	
	Salt Marshes	Freshwater Marshes	Salt Marshes	Freshwater Marshes
With neighbors	8	3	0	18
Without neighbors	10	20	0	33

Data from the Greenhouse Experiment

Salinity (parts per thousand)	0	20	40	60	80	100
% maximum biomass (*S. patens*)	77	40	29	17	9	0
% maximum biomass (*T. angustifolia*)	80	20	10	0	0	0

Data from C. M. Crain et al., Physical and biotic drivers of plant distribution across estuarine salinity gradients, *Ecology* 85:2539–2549 (2004).

▲ *Spartina patens*

▲ *Typha angustifolia*

INTERPRET THE DATA

1. Make a bar graph of the data from the field experiment. (For additional information about graphs, see the Scientific Skills Review in Appendix D.) What do these data indicate about the salinity tolerances of *S. patens* and *T. angustifolia*?

2. What do the data from the field experiment indicate about the effect of competition on the growth of these two species? Which species was limited more by competition?

3. Make a line graph of the data from the greenhouse experiment. Decide which values constitute the dependent and independent variables, and use these values to set up the axes of your graph.

4. **(a)** In the field, *S. patens* is typically absent from freshwater marshes. Based on the data, does this appear to be due to salinity or competition? Explain your answer. **(b)** *T. angustifolia* does not grow in salt marshes. Does this appear to be due to salinity or competition? Explain your answer.

→ **Instructors:** A version of this Scientific Skills Exercise can be assigned in **Mastering Biology**.

Too much light can also limit the survival of organisms. In some ecosystems, such as deserts, high light levels can increase temperature stress if animals and plants are unable to avoid the light or to cool themselves through evaporation (see Figure 40.12). At high elevations, the sun's rays are more likely to damage DNA and proteins because the atmosphere is thinner, absorbing less ultraviolet (UV) radiation. Damage from UV radiation, combined with other abiotic stresses, prevents trees from surviving above a certain elevation, resulting in the appearance of a tree line on mountain slopes **(Figure 52.23)**.

Rocks and Soil

In terrestrial environments, the pH, mineral composition, and physical structure of rocks and soil limit the distribution of plants and thus of the animals that feed on them, contributing to the patchiness of terrestrial ecosystems. The pH of soil can limit the distribution of organisms directly, through extreme acidic or basic conditions, or indirectly, by affecting the solubility of toxins and nutrients. Soil phosphorus, for instance, is relatively insoluble in basic soils and precipitates into forms unavailable to plants.

In a river, the composition of rocks and soil that make up the substrate (riverbed) can affect water chemistry, which in turn influences the resident organisms. In freshwater and marine environments, the structure of the substrate determines the organisms that can attach to it or burrow into it.

▼ **Figure 52.23 Alpine tree line in Banff National Park, Canada.** Organisms living at high elevations are exposed not only to high levels of ultraviolet radiation but also to freezing temperatures, moisture deficits, and strong winds. Above the tree line, the combination of such factors restricts the growth and survival of trees.

1. Describe human actions that could expand a species' range by changing its (a) dispersal or (b) biotic interactions.

2. **WHAT IF?** You suspect that deer are restricting the distribution of a tree species by preferentially eating the seedlings of the tree. How might you test this hypothesis?

3. **MAKE CONNECTIONS** Hawaiian silverswords underwent a remarkable adaptive radiation after their ancestor reached Hawaii, while the islands were still young (see Figure 25.23). Would you expect the cattle egret to undergo a similar adaptive radiation in the Americas (see Figure 52.19)? Explain.

For suggested answers, see Appendix A.

CONCEPT 52.5

Ecological change and evolution affect one another over long and short periods of time

Biologists have long recognized that ecological interactions can cause evolutionary change, and vice versa **(Figure 52.24)**. The history of life includes many examples of these reciprocal effects occurring over long periods of time. Consider the origin and diversification of plants. As described in Concept 29.3, the evolutionary origin of plants altered the chemical cycling of carbon, leading to the removal of large quantities of carbon dioxide from the atmosphere. As the adaptive radiation of plants continued over time, the appearance of new plant species provided new habitats and new sources of food for insects and other animals. In turn, the availability of new habitats and new food sources stimulated bursts of speciation in animals, leading to further ecological changes. Here, as in many other such examples, ecological and evolutionary changes had ongoing and major effects upon one another.

The interplay between ecological and evolutionary change illustrated by the origin of plants occurred over millions of years. Reciprocal "eco-evolutionary" effects occurring over centuries or thousands of years are also well documented, as in the examples of the mosquitofish and the apple maggot fly

▼ **Figure 52.24 Reciprocal effects of ecological and evolutionary change.** An ecological change, such as the expansion of a predator's range, can alter the selective pressures faced by prey populations. This could cause evolutionary change, such as an increase in the frequency of a new defensive mechanism in a prey population; that change, in turn, could alter the outcome of ecological interactions.

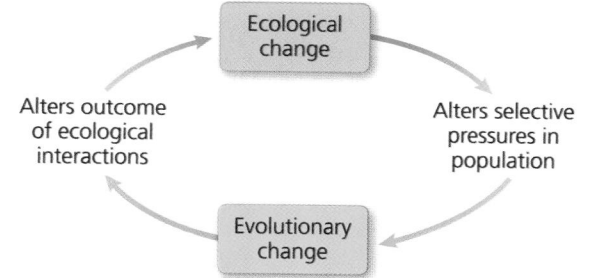

▼ **Figure 52.25 Rapid evolution leading to rapid ecological change.** Dotted lines connect algal abundance at control sites (with no guppies) to algal abundance at nearby sites in the same streams inhabited by guppies that had evolved under either low or high levels of predation.

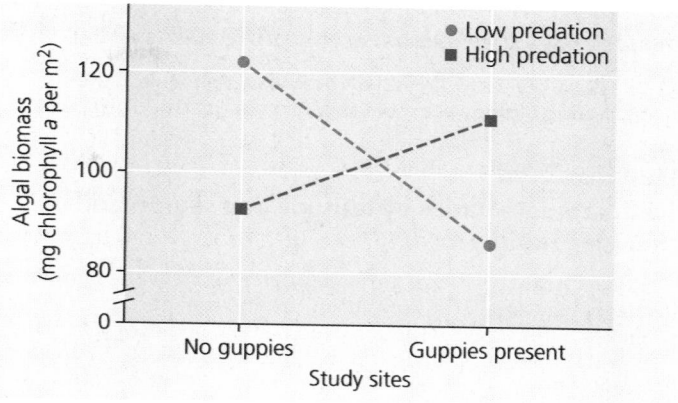

discussed in Concept 24.2. But are such joint effects common over even shorter periods of time? As we've seen in previous chapters, ecological change can cause evolutionary change over the course of a few years to decades; examples include beak length evolution in soapberry bugs (see Figure 22.13) and the formation of new sunflower species (see Figure 24.18).

Recent studies show that the causation can run both ways: Rapid evolution can also cause rapid ecological change. For example, Trinidadian guppy (*Poecilia reticulata*) populations evolve rapidly when predators are removed: Guppy color patterns, jaw morphology, and feeding preferences all change within a few years. In 2017, researchers showed that these rapid evolutionary changes affect the stream ecosystems in which the guppies live. For instance, guppies that had evolved under different levels of predation had contrasting effects on algal abundance **(Figure 52.25)**. Guppies that had evolved under low predation fed primarily on algae (thereby reducing algal abundance), while guppies that had evolved under high predation fed primarily on invertebrates (thereby tending to increase algal abundance because some invertebrates eat algae). As producers, algae are key components of the ecosystem: Other organisms depend on them for food, either directly (by eating them) or indirectly (by eating an organism that ate algae). Overall, this and other studies show that ecological change and evolution have the potential to exert rapid feedback effects on each other.

CONCEPT CHECK 52.5

1. Describe a scenario showing how ecological change and evolution can affect one another.

2. **MAKE CONNECTIONS** Fisheries target older, larger cod fish, causing cod that reproduce at a younger age and smaller size to be favored by natural selection. Younger, smaller cod have fewer offspring than do older cod. Predict how evolution in response to fishing would affect the ability of a cod population to recover from overfishing. What other reciprocal eco-evolutionary effects might occur? (See Concept 23.3.)

For suggested answers, see Appendix A.

52 Chapter Review

→ Go to **Mastering Biology** for Assignments, the eText, the Study Area, and Dynamic Study Modules.

SUMMARY OF KEY CONCEPTS

→ To review key terms, go to the **Vocabulary Self-Quiz** in the **Mastering Biology** eText or Study Area, or go to goo.gl/zkjz9t.

CONCEPT 52.1

Earth's climate varies by latitude and season and is changing rapidly (pp. 1167–1170)

- Global **climate** patterns are largely determined by the input of solar energy and Earth's revolution around the sun.
- The changing angle of the sun over the year, bodies of water, and mountains exert seasonal, regional, and local effects on climate.
- Climate affects where plants can live, but these effects run both ways: Vegetation can alter local and regional climate.
- Fine-scale differences in certain **abiotic** (nonliving) factors, such as sunlight and temperature, determine **microclimate**.
- Increasing greenhouse gas concentrations in the air are warming Earth and altering the distributions of many species. Some species will not be able to shift their ranges quickly enough to reach suitable habitat in the future.

❓ *Suppose global air circulation suddenly reversed, with moist air ascending at 30° north and south latitude and descending at the equator. At what latitude would you most likely find deserts in this scenario?*

CONCEPT 52.2

The distribution of terrestrial biomes is controlled by climate and disturbance (pp. 1171–1176)

- **Climographs** show that temperature and precipitation are correlated with **biomes**. Because other factors also play roles in biome location, biomes overlap.
- Terrestrial biomes are often named for major physical or climatic factors and for their predominant vegetation. Vertical layering is an important feature of terrestrial biomes.
- **Disturbance**, both natural and human-induced, influences the type of vegetation found in biomes. Humans have altered much of Earth's surface, replacing the natural terrestrial communities described and depicted in Figure 52.13 with urban and agricultural ones.
- The pattern of climatic variation is as important as the average climate in determining where biomes occur.

❓ *In what ways are disturbances important for savanna ecosystems and the plants in them?*

CONCEPT 52.3

Aquatic biomes are diverse and dynamic systems that cover most of Earth (pp. 1177–1178)

- Aquatic biomes are characterized primarily by their physical environment rather than by climate and are often layered with regard to light penetration, temperature, and community structure. Marine biomes have a higher salt concentration than freshwater biomes.
- In the ocean and in most lakes, an abrupt temperature change called a **thermocline** separates a more uniformly warm upper layer from more uniformly cold deeper waters.
- Many temperate lakes undergo a **turnover,** or mixing of water in spring and fall, that sends deep, nutrient-rich water to the surface and shallow, oxygen-rich water to deeper layers.

❓ *In which aquatic biomes might you find an aphotic zone?*

CONCEPT 52.4

Interactions between organisms and the environment limit the distribution of species (pp. 1178–1187)

Ecologists want to know not only *where* species occur but also *why* those species occur where they do.

VISUAL SKILLS *If you were an ecologist studying the chemical and physical limits to the distributions of species, how might you rearrange the flowchart preceding this question?*

CONCEPT 52.5

Ecological change and evolution affect one another over long and short periods of time (p. 1187)

- Ecological interactions can cause evolutionary change, as when predators cause natural selection in a prey population.
- Likewise, an evolutionary change, such as an increase in the frequency of a new defensive mechanism in a prey population, can alter the outcome of ecological interactions.

❓ *Suppose humans introduced a species to a new continent where it had few predators or parasites. How might this lead to eco-evolutionary feedback effects?*

TEST YOUR UNDERSTANDING

→ For more multiple-choice questions, go to the **Practice Test** in the **Mastering Biology** eText or Study Area, or go to goo.gl/GruWRg.

Levels 1-2: Remembering/Understanding

1. Which of the following areas of study focuses on the exchange of energy, organisms, and materials between ecosystems?
   (A) organismal ecology   (C) ecosystem ecology
   (B) landscape ecology   (D) community ecology

2. Which lake zone would be absent in a very shallow lake?
   (A) benthic zone   (C) pelagic zone
   (B) aphotic zone   (D) littoral zone

Levels 3-4: Applying/Analyzing

3. Which of the following is characteristic of most terrestrial biomes?
   (A) a distribution predicted almost entirely by rock and soil patterns
   (B) clear boundaries between adjacent biomes
   (C) vegetation demonstrating vertical layering
   (D) cold winter months

4. The oceans affect the biosphere by
(A) absorbing a substantial amount of the biosphere's oxygen.
(B) regulating the pH of freshwater biomes and terrestrial groundwater.
(C) decreasing winter temperatures of coastal terrestrial biomes.
(D) removing carbon dioxide from the atmosphere.

5. Which statement about dispersal is true?
(A) Dispersal is not a component of the life cycles of most plants and animals.
(B) Dispersal occurs only on an evolutionary time scale.
(C) Colonization of devastated areas after floods or volcanic eruptions depends on dispersal.
(D) The ability to disperse does not affect the geographic distribution of a species.

6. When climbing a mountain, we can observe transitions in biological communities that are analogous to the changes
(A) in biomes at different latitudes.
(B) in different depths in the ocean.
(C) in a community through different seasons.
(D) in an ecosystem as it evolves over time.

7. Suppose that the number of bird species is determined mainly by the number of vertical strata found in the environment. If so, in which of the following biomes would you find the greatest number of bird species?
(A) tropical rain forest
(B) savanna
(C) desert
(D) temperate broadleaf forest

Levels 5-6: Evaluating/Creating

8. WHAT IF? If the direction of Earth's rotation reversed, the most predictable effect would be
(A) a big change in the length of the year.
(B) winds blowing from west to east along the equator.
(C) a loss of seasonal variation at high latitudes.
(D) the elimination of ocean currents.

9. INTERPRET THE DATA After examining Figure 52.20, you decide to study feeding relationships among sea otters, sea urchins, and kelp. You know that sea otters prey on sea urchins and that urchins eat kelp. At four coastal sites, you measure kelp abundance. Then you spend one day at each site and mark whether otters are present or absent every 5 minutes during the day. Graph kelp abundance (on the *y*-axis) versus otter density (on the *x*-axis), using the data below. Then formulate a hypothesis to explain any pattern you observe.

Site	Otter Density (# sightings per day)	Kelp Abundance (% cover)
1	98	75
2	18	15
3	85	60
4	36	25

10. EVOLUTION CONNECTION Discuss how the distribution of a species can be affected both by its evolutionary history and by ecological factors. Could ongoing evolutionary change also affect its distribution? Explain.

11. SCIENTIFIC INQUIRY Jens Clausen and colleagues, at the Carnegie Institution of Washington, studied how the size of yarrow plants (*Achillea lanulosa*) growing on the slopes of the Sierra Nevada varied with elevation. They found that plants from low elevations were generally taller than plants from high elevations, as shown in the diagram.

Seed collection sites

Data from J. Clausen et al., Experimental studies on the nature of species. III. Environmental responses of climatic races of *Achillea*, Carnegie Institution of Washington Publication No. 581 (1948).

Clausen and colleagues proposed two hypotheses to explain this variation within a species: (1) There are genetic differences between populations of plants found at different elevations. (2) The species has developmental flexibility and can assume tall or short growth forms, depending on local abiotic factors. If you had seeds from yarrow plants found at low and high elevations, how would you test these hypotheses?

12. WRITE ABOUT A THEME: INTERACTIONS Global warming is occurring rapidly in arctic marine and terrestrial ecosystems. The reflective white snow and ice cover are melting quickly and extensively, uncovering darker-colored ocean water, plants, and rocks. In a short essay (100–150 words), explain how this process might exemplify positive feedback.

13. SYNTHESIZE YOUR KNOWLEDGE

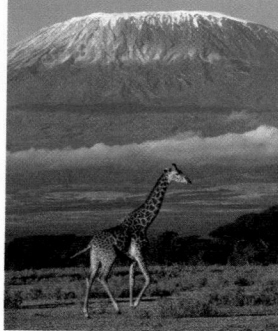

If you were to hike up Mount Kilimanjaro in Tanzania, you would pass through several habitats, including savanna at the base, forest on the slopes, and alpine tundra near the top. Explain how such diverse habitats can be found at one location near the equator.

For selected answers, see Appendix A.

53 Population Ecology

KEY CONCEPTS

53.1 Biotic and abiotic factors affect population density, dispersion, and demographics *p. 1191*

53.2 The exponential model describes population growth in an idealized, unlimited environment *p. 1196*

53.3 The logistic model describes how a population grows more slowly as it nears its carrying capacity *p. 1197*

53.4 Life history traits are products of natural selection *p. 1200*

53.5 Density-dependent factors regulate population growth *p. 1202*

53.6 The human population is no longer growing exponentially but is still increasing extremely rapidly *p. 1207*

Study Tip

Annotate a graph: This chapter contains several *scatter plots*, graphs in which each measurement or observation is plotted as a single point. To make sure you understand a scatter plot, select one point and describe what it represents, as shown here for Figure 53.8. Do this for Figures 53.4, 53.8, 53.10, 53.16, and 53.18.

Represents the total number of elephants in the park in 1965

→ Go to Mastering Biology

For Students (in eText and Study Area)
- Get Ready for Chapter 53
- Table 53.1 Walkthrough: Life Table for Female Belding's Ground Squirrels
- BioFlix® Animation: Population Ecology

For Instructors to Assign (in Item Library)
- Interpreting Data: Survivorship Curves
- Interpreting Data: Human Population Growth

Figure 53.1 Although it was once common along the Atlantic coast, by 1985 prospects looked grim for the piping plover (*Charadrius melodus*). Since then, nesting site protection has enabled the piping plover's numbers to grow, with recent surveys finding 1,400 to 4,000 adults in recent years. Such variation is typical of natural populations, all of which vary in number (size) over time.

What affects the size of a population and how it changes over time?

Adding individuals to or **removing** individuals from a population **changes the population size:**

Births

Deaths

Immigration

Emigration

Over time, a population may…

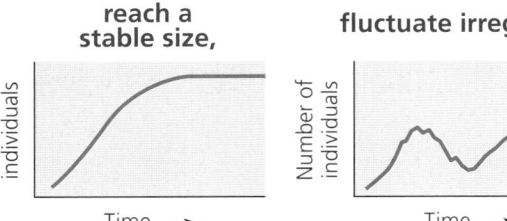

reach a stable size,

fluctuate irregularly,

or rise and fall in regular cycles.

Biotic and abiotic factors affect population density, dispersion, and demographics

A **population** is a group of individuals of a single species living in the same general area. Members of a population rely on the same resources, are influenced by similar environmental factors, and are likely to interact and breed with one another.

Populations are often described by their boundaries and size (the number of individuals living within those boundaries). Ecologists usually begin investigating a population by defining boundaries appropriate to the organism under study and to the questions being asked. A population's boundaries may be natural ones, as in the case of an island or a lake, or they may be arbitrarily defined by an investigator—for example, a specific county in Minnesota for a study of oak trees.

In this chapter, we will explore how ecologists describe and analyze populations and the factors that can determine the abundance of organisms. We will then examine recent trends in the size and makeup of the human population.

Density and Dispersion

The **density** of a population is the number of individuals per unit area or volume: the number of oak trees per square kilometer in the Minnesota county or the number of *Escherichia coli* bacteria per milliliter in a test tube. **Dispersion** is the pattern of spacing among individuals within the boundaries of the population.

Density: A Dynamic Perspective

In some cases, population size and density can be determined by counting all individuals within the boundaries of the population. We could count all the sea stars in a tide pool, for instance. Large mammals that live in herds, such as elephants, can sometimes be counted accurately from airplanes.

In most cases, however, it is impractical or impossible to count all individuals in a population. Instead, ecologists use various sampling techniques to estimate densities and total population sizes. They might count the number of oak trees in several randomly located 100 × 100 m plots, calculate the average density in the plots, and then extend the estimate to the population size in the entire area. Such estimates are most accurate when there are many sample plots and when the habitat is fairly homogeneous. In other cases, instead of counting single organisms, ecologists estimate density from an indicator of population size, such as the number of nests, burrows, tracks, or fecal droppings. Ecologists also use the **mark-recapture method** to estimate the size of wildlife populations **(Figure 53.2)**.

▼ **Figure 53.2** **Research Method**

Determining Population Size Using the Mark-Recapture Method

Application Ecologists cannot count all the individuals in a population if the organisms move too quickly or are hidden from view. In such cases, researchers often use the mark-recapture method to estimate population size. Andrew Gormley and his colleagues at the University of Otago applied this method to a population of endangered Hector's dolphins (*Cephalorhynchus hectori*) near Banks Peninsula, in New Zealand.

▲ **Hector's dolphins**

Technique Scientists typically begin by capturing a random sample of individuals in a population. They tag, or "mark," each individual and then release it. With some species, researchers can identify individuals without physically capturing them. For example, Gormley and colleagues identified 180 Hector's dolphins by photographing their distinctive dorsal fins from boats.

After waiting for the marked or otherwise identified individuals to mix back into the population, usually a few days or weeks, scientists capture or sample a second set of individuals. At Banks Peninsula, Gormley's team encountered 44 dolphins in their second sampling, 7 of which they had photographed before. The number of marked animals captured in the second sampling (x) divided by the total number of animals captured in the second sampling (n) should equal the number of individuals marked and released in the first sampling (s) divided by the estimated population size (N):

$$\frac{x}{n} = \frac{s}{N} \text{ or, solving for population size, } N = \frac{sn}{x}$$

The method assumes that marked and unmarked individuals have the same probability of being captured or sampled, that the marked organisms have mixed completely back into the population, and that no individuals are born, die, immigrate, or emigrate during the resampling interval.

Results Based on these initial data, the estimated population size of Hector's dolphins at Banks Peninsula would be 180 × 44/7 = 1,131 individuals. Repeated sampling by Gormley and colleagues suggested a true population size closer to 1,100.

Data from A. M. Gormley et al., Capture-recapture estimates of Hector's dolphin abundance at Banks Peninsula, New Zealand, *Marine Mammal Science* 21:204–216 (2005).

INTERPRET THE DATA *Suppose that none of the 44 dolphins encountered in the second sampling had been photographed before. Would you be able to solve the equation for N? What might you conclude about population size in this case?*

Density is not a static property but can increase or decrease over time as individuals are added to or removed from a population. As seen on the opening page of this chapter, additions occur through birth (which we define here to include all forms of reproduction) and **immigration**, the influx of new individuals from other areas. The factors that remove individuals from a population are death (mortality) and **emigration**, the movement of individuals out of a population and into other locations.

While birth and death rates influence the density of all populations, immigration and emigration can also have substantial effects. For example, studies of a population of Hector's dolphins (see Figure 53.2) in New Zealand showed that immigration accounted for about 15% of the total population size each year. Emigration was also important, typically occurring during the winter season when the dolphins move farther from shore. In general, both immigration and emigration represent key factors affecting the populations of many species. Emigration, for example, not only reduces the density of the original population, it also can result in the establishment of new populations in regions of favorable habitat, as we'll discuss later in this chapter.

⊙ **Mastering Biology Animation: Techniques for Estimating Population Density and Size**

Patterns of Dispersion

Within the boundaries of a population, the spacing among individuals may differ substantially, creating contrasting patterns of dispersion. Such differences in spacing can provide insight into the biotic and abiotic factors that affect individuals in the population.

The most common pattern of dispersion is *clumped*, in which individuals are aggregated in patches. Plants and fungi are often clumped where soil conditions and other environmental factors favor germination and growth. Mushrooms, for instance, may be clumped within and on top of a rotting log. Insects and salamanders may be clumped under the same log because of the higher humidity there. Clumping of animals may also be associated with mating behavior. Sea stars group together in tide pools, where food is readily available and where they can breed successfully **(Figure 53.3a)**. The aggregation of individuals into groups may also increase the effectiveness of predation or defense; for example, a wolf pack is more likely than a single wolf to subdue a moose, and a flock of birds is more likely than a single bird to warn of a potential attack.

A *uniform*, or evenly spaced, pattern of dispersion may result from direct interactions between individuals in the population. Some plants secrete chemicals that inhibit the germination and growth of nearby individuals that could compete for resources. Animals often exhibit uniform dispersion as a result of antagonistic social

▼ **Figure 53.3 Patterns of dispersion within a population's geographic range.**

(a) Clumped. Sea stars group together where food is abundant.

(b) Uniform. King penguins and other birds that nest on small islands often exhibit uniform spacing, maintained by aggressive interactions between neighbors.

(c) Random. Dandelions grow from windblown seeds that land at random and later germinate.

❓ *Patterns of dispersion can depend on scale. How might the penguin dispersion look from an airplane flying high above the ocean?*

interactions, such as **territoriality**—the defense of a bounded physical space against encroachment by other individuals **(Figure 53.3b)**.

In *random* dispersion (unpredictable spacing), the position of each individual in a population is independent of other individuals. This pattern occurs in the absence of strong attractions or repulsions among individuals or where key physical or chemical factors are relatively constant across the study area. Plants established by windblown seeds, such as dandelions, may be randomly distributed in a fairly uniform habitat (Figure 53.3c).

Demographics

The biotic and abiotic factors that influence population density and dispersion patterns also influence other characteristics of populations, including birth, death, and migration rates. **Demography** is the study of these key characteristics of populations and how they change over time. A useful way to summarize demographic information for a population is to make a life table.

Life Tables

A **life table** summarizes the survival and reproductive rates of individuals in specific age-groups within a population. To construct a life table, researchers often follow the fate of a **cohort**, a group of individuals of the same age, from birth until all of the individuals are dead. Building the life table requires determining the proportion of the cohort that survives from one age-group to the next. It is also necessary to keep track of the number of offspring produced by females in each age-group.

Demographers who study sexually reproducing species often ignore the males and concentrate on the females in a population because only females produce offspring. Using this approach, a population is viewed in terms of females giving rise to new females. **Table 53.1** is a life table built in this way for female Belding's ground squirrels (*Urocitellus beldingi*) from a population located in the Sierra Nevada mountains of California. Next, we'll take a closer look at some of the data presented in a life table.

Survivorship Curves

The survival rate data in a life table can be represented graphically as a **survivorship curve**, a plot of the proportion or numbers in a cohort still alive at each age. As an example, let's use the data for female Belding's ground squirrels in Table 53.1 to draw a survivorship curve. Often, a survivorship curve begins with a cohort of a convenient size—say, 1,000 individuals. To obtain the other points in the curve for the ground squirrel population, we multiply the proportion alive at the start of each year (the third column of Table 53.1) by 1,000 (the hypothetical beginning cohort). The result is the number alive at the start of each year. Plotting these numbers versus age for female Belding's ground squirrels

Table 53.1 Life Table for Female Belding's Ground Squirrels (Tioga Pass, in the Sierra Nevada Mountains of California)

Age (years)	Number Alive at Start of Year	Proportion Alive at Start of Year*	Death Rate[†]	Average Number of Female Offspring per Female
0–1	653	1.000	0.614	0.00
1–2	252	0.386	0.496	1.07
2–3	127	0.197	0.472	1.87
3–4	67	0.106	0.478	2.21
4–5	35	0.054	0.457	2.59
5–6	19	0.029	0.526	2.08
6–7	9	0.014	0.444	1.70
7–8	5	0.008	0.200	1.93
8–9	4	0.006	0.750	1.93
9–10	1	0.002	1.00	1.58

Data from P. W. Sherman and M. L. Morton, Demography of Belding's ground squirrel, *Ecology* 65:1617–1628 (1984).

*Indicates the proportion of the original cohort of 653 individuals that are still alive at the start of a time interval.

[†]The death rate is the proportion of individuals alive at the start of a time interval that die during that time interval.

▲ Researchers working with a Belding's ground squirrel

● **Mastering Biology** Table Walkthrough

▼ **Figure 53.4 Survivorship curve for female Belding's ground squirrels.** The logarithmic scale on the *y*-axis allows the number of survivors to be visible across the entire range (2–1,000 individuals) on the graph.

? *Based on this graph, what percentage of the female ground squirrels survive to be 3 years old?*

yields **Figure 53.4**. The approximately straight line of the plot indicates a relatively constant rate of death.

Figure 53.4 represents just one of many patterns of survivorship exhibited by natural populations. Though diverse, survivorship curves can be classified into three general types **(Figure 53.5)**. A Type I curve is flat at the start, reflecting low death rates during early and middle life, and then drops steeply as death rates increase among older age-groups. Many large mammals, including humans and elephants, that produce few offspring but provide them with good care exhibit this kind of curve.

▼ **Figure 53.5 Idealized survivorship curves: Types I, II, and III.** The *y*-axis is logarithmic and the *x*-axis is on a relative scale so that species with widely varying life spans can be presented together on the same graph.

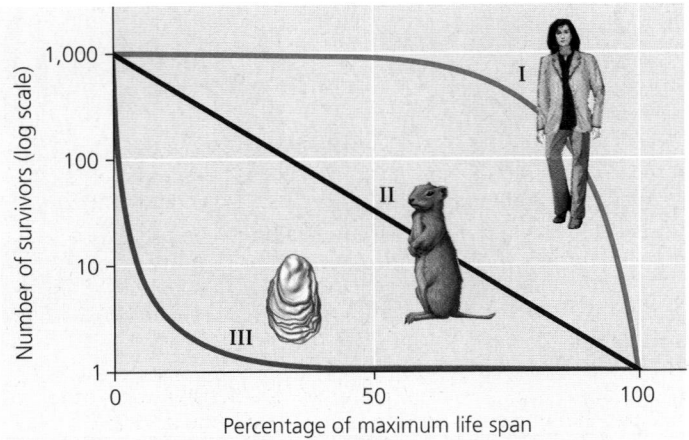

→ **Mastering Biology** **Animation: Investigating the Survivorship Curve of Oysters**

In contrast, a Type III curve drops sharply at the start, reflecting very high death rates for the young, but flattens out as death rates decline for those few individuals that survive the early period of die-off. This type of curve is usually associated with organisms that produce very large numbers of offspring but provide little or no care, such as long-lived plants, many fishes, and most marine invertebrates. An oyster, for example, may release millions of eggs, but most larvae hatched from fertilized eggs die from predation or other causes. Those few offspring that survive long enough to attach to a suitable substrate and begin growing a hard shell tend to survive for a relatively long time. Type II curves are intermediate, with a constant death rate over the organism's life span. This kind of survivorship occurs in Belding's ground squirrels and some other rodents, many invertebrates, lizards, and annual plants.

Many species fall somewhere between these basic types of survivorship or show more complex patterns. In birds, mortality is often high among the youngest individuals (as in a Type III curve) but fairly constant among adults (as in a Type II curve). Some invertebrates, such as crabs, may show a "stair-stepped" curve, with brief periods of increased mortality during molts, followed by periods of lower mortality when their protective exoskeleton is hard. In addition to such variation among species, survivorship curves also can differ among the populations of a single species.

In populations experiencing low levels of immigration and emigration, survivorship is one of the two key factors determining changes in population size. The other key factor that affects how population size changes over time is reproductive rate.

Reproductive Rates

As mentioned above, demographers often ignore the males and concentrate on the females in a population because only females produce offspring. Therefore, demographers view populations in terms of females giving rise to new females. The simplest way to describe the reproductive pattern of a population is to identify how reproductive output varies with the number of breeding females and their ages.

How do ecologists estimate the number of breeding females in a population? Possible approaches include direct counts and the mark-recapture method (see Figure 53.2). Increasingly, ecologists also use molecular tools. For example, scientists working in the state of Georgia collected skin samples from 198 female loggerhead turtles between 2005 and 2009. From these samples, they amplified nuclear short tandem repeats at 14 loci using the polymerase chain reaction (PCR) and produced a genetic profile for each female **(Figure 53.6)**. They then extracted DNA from an eggshell from each turtle nest on the beaches they studied and, using their database of genetic profiles, matched the nest to a

▼ Figure 53.6 Using genetic profiles from loggerhead turtle eggshells to identify which female laid the eggs.

Part 1: Developing the Database:

Skin samples are collected from female loggerhead turtles.

In a lab, DNA is extracted from each skin sample, and short tandem repeats at 14 loci are amplified by PCR.

A genetic profile is determined for each turtle and stored in a database.

Breeding female #108   Breeding female #109

Part 2: Comparing Samples to the Database:

An eggshell is collected from a loggerhead turtle nest.

In a lab, DNA is extracted from the eggshell, and short tandem repeats at 14 loci are amplified by PCR.

A genetic profile is determined for each eggshell sample.

Eggshell sample #74

The genetic profile of the eggshell is compared with an established database containing genetic profiles of adult female loggerhead turtles.

A match identifies the female that laid the eggs in the nest.

VISUAL SKILLS *Use the profiles displayed in the figure to determine which breeding female laid the eggs in the nest from which eggshell sample #74 was taken.*

---

specific female. This approach allowed them to determine how many of the 198 females were breeding—and how many offspring each female produced—without having to disturb the females during egg laying.

Reproductive output for sexual organisms such as birds and mammals is typically measured as the average number of female offspring produced by the females in a given age-group. For some organisms, the number of offspring for each female can be counted directly; alternatively, molecular methods can be used (see Figure 53.6). Researchers directly counted the offspring of the Belding's ground squirrels, which begin to reproduce at age 1 year. The squirrels' reproductive output rises to a peak at 4–5 years of age and then gradually falls off in older females (see Table 53.1).

Age-specific reproductive rates vary considerably by species. Squirrels, for example, have a litter of two to six young once a year for less than a decade, whereas oak trees drop thousands of acorns each year for tens or hundreds of years.

Mussels and other invertebrates may release millions of eggs and sperm in a spawning cycle. However, a high reproductive rate will not lead to rapid population growth unless conditions are near ideal for the growth and survival of offspring, as you'll learn in the next section.

CONCEPT CHECK 53.1

1. **DRAW IT** Each female of a particular fish species produces millions of eggs per year. Draw and label the most likely survivorship curve for this species, and explain your choice.

2. **WHAT IF?** Suppose you are constructing a life table for a population of Belding's ground squirrels (see Table 53.1). If 485 individuals are alive at the start of year 0–1 and 218 are still alive at the start of year 1–2, what is the proportion alive at the start of each of these years (see column 3 in Table 53.1)?

3. **MAKE CONNECTIONS** A male stickleback fish attacks other males that invade its nesting territory (see Figure 51.2a). Predict the likely pattern of dispersion for male sticklebacks, and explain your reasoning.

For suggested answers, see Appendix A.

The exponential model describes population growth in an idealized, unlimited environment

Populations of all species have the potential to expand greatly when resources are abundant. To appreciate the potential for population increase, consider a bacterium that can reproduce by fission every 20 minutes under ideal laboratory conditions. There would be two bacteria after 20 minutes, four after 40 minutes, and eight after 60 minutes. If reproduction continued at this rate for a day and a half without mortality, there would be enough bacteria to form a layer 30 cm deep over the entire globe! But unlimited growth does not occur for long in nature, where individuals typically have access to fewer resources as a population grows. Nonetheless, ecologists study population growth in ideal, unlimited environments to reveal how fast populations are capable of growing and the conditions under which rapid growth might actually occur.

Changes in Population Size

Imagine a population consisting of a few individuals living in an ideal, unlimited environment. Under these conditions, there are no external limits on the abilities of individuals to harvest energy, grow, and reproduce. The population will increase in size with every birth and with the immigration of individuals from other populations, and it will decrease in size with every death and with the emigration of individuals out of the population. We can thus define a change in population size during a fixed time interval with the following verbal equation:

Change in population size = Births + Immigrants entering population − Deaths − Emigrants leaving population

For now, we will simplify our discussion by ignoring the effects of immigration and emigration.

We can use mathematical notation to express this simplified relationship more concisely. If N represents population size and t represents time, then ΔN is the change in population size and Δt is the time interval (appropriate to the life span or generation time of the species) over which we are evaluating population growth. (The Greek letter delta, Δ, indicates change, such as change in time.) Using B for the number of births in the population during the time interval and D for the number of deaths, we can rewrite the verbal equation:

$$\frac{\Delta N}{\Delta t} = B - D$$

Population ecologists often are most interested in changes in population size—the number of individuals that are added to or subtracted from a population during a given time interval, symbolized by R. Here, R represents the *difference* between the number of births (B) and the number of deaths (D) that

occur in the time interval. Thus, $R = B - D$, and we can simplify our equation by writing the following:

$$\frac{\Delta N}{\Delta t} = R$$

Next, we can convert our model to one in which changes in population size are expressed on a per individual (per capita) basis. The *per capita* change in population size ($r_{\Delta t}$) represents the contribution that an average member of the population makes to the number of individuals added to or subtracted from the population during the time interval Δt. If, for example, a population of 1,000 individuals increases by 16 individuals per year, then on a per capita basis, the annual change in population size is 16/1,000, or 0.016. If we know the annual per capita change in population size, we can use the formula $R = r_{\Delta t} N$ to calculate how many individuals will be added to (or subtracted from) a population each year. For example, if $r_{\Delta t}$ is 0.016 and the population size is 500,

$$R = r_{\Delta t} N = 0.016 \times 500 = 8 \text{ per year}$$

Since the number of individuals added to (or subtracted from) the population (R) can be expressed on a per capita basis as $R = r_{\Delta t} N$, we can revise our population growth equation to take this into account:

$$\frac{\Delta N}{\Delta t} = r_{\Delta t} N$$

Remember that our equation is for a specific time interval (often one year). However, many ecologists prefer to use differential calculus to express population growth as a rate of change *at each instant in time*:

$$\frac{dN}{dt} = rN$$

In this case, r represents the per capita change in population size that occurs at each instant in time (whereas $r_{\Delta t}$ represented the per capita change that occurred during the time interval Δt). If you have not yet studied calculus, don't be intimidated by the last equation; it is similar to the previous one, except that the time intervals Δt are very short and are expressed in the equation as dt. In fact, as Δt becomes shorter, $r_{\Delta t}$ and r become increasingly close to one another in value.

Exponential Growth

Earlier we described a population whose members all have access to abundant food and can reproduce at their physiological capacity. In some cases, a population that experiences such ideal conditions increases in size by a constant proportion at each instant in time. When this occurs, the pattern of growth that results is called **exponential population growth**. The equation for exponential growth is the one presented at the end of the previous section, namely:

$$\frac{dN}{dt} = rN$$

▼ Figure 53.7 Population growth predicted by the exponential model.
This graph compares growth in a population with $r = 1.0$ (blue curve) to growth in a population with $r = 0.5$ (red curve).

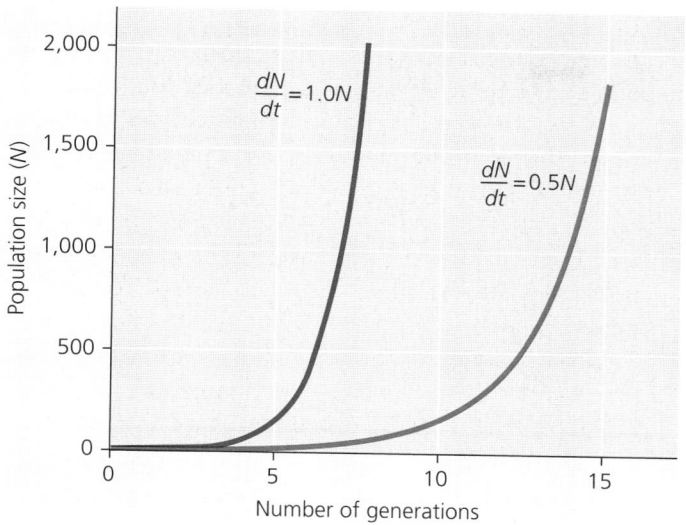

$$\frac{dN}{dt} = 1.0N$$

$$\frac{dN}{dt} = 0.5N$$

❓ *How many generations does it take these populations to reach a size of 1,500 individuals?*

→ **Mastering Biology** **BioFlix® Animation: Exponential Growth**

In this equation, *dN/dt* represents the rate at which the population is increasing in size at each moment in time, akin to how a glance at the speedometer of a car reveals the speed at that instant in time. As seen in the equation, *dN/dt* equals the current population size, *N*, multiplied by a constant, *r*. Ecologists refer to *r* as the **intrinsic rate of increase**, the per capita rate at which an exponentially growing population increases in size at each instant in time.

The size of a population that is growing exponentially increases at a constant rate per individual, resulting eventually in a J-shaped growth curve when population size is plotted over time **(Figure 53.7)**. Although the per capita rate of population growth is constant (and equals *r*), more new individuals are added per unit of time when the population is large than when it is small; thus, the curves in Figure 53.7 get progressively steeper over time. This occurs because population growth depends on *N* as well as *r*, and hence more individuals are added to larger populations than to small ones growing at the same per capita rate. It is also clear from Figure 53.7 that a population with a higher intrinsic rate of increase (*dN/dt* = 1.0*N*) will grow faster than one with a lower intrinsic rate of increase (*dN/dt* = 0.5*N*).

The J-shaped curve of exponential growth is characteristic of some populations that are introduced into a new environment or whose numbers have been drastically reduced by a catastrophic event and are rebounding. For example, the population of elephants in Kruger National Park, South Africa, grew exponentially for approximately 60 years after they were first protected from hunting **(Figure 53.8)**. The increasingly

▼ Figure 53.8 Exponential growth in the African elephant population of Kruger National Park, South Africa.

large number of elephants eventually caused enough damage to vegetation in the park that a collapse in their food supply was likely. To protect other species and the park ecosystem before that happened, park managers began limiting the elephant population by giving birth control medication to females and by exporting elephants to other countries.

CONCEPT CHECK 53.2

1. Explain why a constant per capita rate of growth (*r*) for a population produces a curve that is J-shaped.

2. Where is exponential growth by a plant population more likely—in an area where a forest was destroyed by fire or in a mature, undisturbed forest? Why?

3. **WHAT IF?** In 2018, the United States had a population of 327 million people. If the (annual) per capita change in population size $(r_{\Delta t})$ was 0.005, how many people were added to the population that year (ignoring immigration and emigration)? What would you need to know to determine whether the United States is currently experiencing exponential growth?

For suggested answers, see Appendix A.

CONCEPT 53.3

The logistic model describes how a population grows more slowly as it nears its carrying capacity

The exponential growth model assumes that resources remain abundant, which is rarely the case in the real world. Instead, as the size of a population increases, each individual has access to fewer resources. Ultimately, there is a limit to the number of individuals that can occupy a habitat. Ecologists define the **carrying capacity**, symbolized by *K*, as the maximum population size that a particular environment can sustain. Carrying capacity varies over space and time with the abundance of limiting resources. Energy, shelter, refuge from predators, nutrient availability, water, and suitable nesting sites can all be limiting factors. For example, the carrying capacity for bats may be high in a habitat with abundant flying insects and roosting sites but lower where there is abundant food but fewer suitable shelters.

Crowding and resource limitation can have a profound effect on population growth rate. If individuals cannot obtain sufficient resources to reproduce, then the per capita birth rate will decline. Similarly, if starvation or disease increases with density, then the per capita death rate may increase. Falling per capita birth rates or rising per capita death rates will cause the per capita rate of population growth to drop, a very different situation from the constant per capita growth rate (r) seen in a population that is growing exponentially.

The Logistic Growth Model

We can modify our mathematical model so that the per capita population growth rate decreases as N increases. In the **logistic population growth** model, the per capita rate of population growth approaches zero as the population size nears the carrying capacity (K).

To construct the logistic model, we start with the exponential population growth model and add an expression that reduces the per capita rate of population growth as N increases. If the carrying capacity is K, then $K - N$ is the number of additional individuals the environment can support, and $(K - N)/K$ is the fraction of K that is still available for population growth. By multiplying the exponential rate of population growth rN by $(K - N)/K$, we modify the change in population size as N increases:

$$\frac{dN}{dt} = rN\frac{(K - N)}{K}$$

When N is small compared to K, the term $(K - N)/K$ is close to 1. In this scenario, the per capita rate of population growth, $r[(K - N)/K]$, will be close to (but slightly less than) r, the intrinsic rate of increase seen in exponential population growth. But when N is large and resources are limiting, then $(K - N)/K$ is close to 0, and the per capita population growth rate is small. When N equals K, the population stops growing. **Table 53.2** shows calculations of population growth rate for a hypothetical population growing according to the logistic model, with $r = 1.0$ per individual per year. Notice that the overall population growth rate is highest, +375 individuals per year, when the population size is 750, or half the carrying capacity. At a population size of 750, the per capita population growth rate remains relatively high (one-half the value of r), and there are more reproducing individuals (N) in the population than at lower population sizes.

As shown in **Figure 53.9**, the logistic model of population growth produces a sigmoid (S-shaped) growth curve when N is plotted over time (the red line). New individuals

Table 53.2 Logistic Growth of a Hypothetical Population ($K = 1,500$)

Population Size (N)	Intrinsic Rate of Increase (r)	$\dfrac{K - N}{K}$	Per Capita Population Growth Rate, $r\dfrac{(K - N)}{K}$	Population Growth Rate,* $rN\dfrac{(K - N)}{K}$
25	1.0	0.983	0.983	+25
100	1.0	0.933	0.933	+93
250	1.0	0.833	0.833	+208
500	1.0	0.667	0.667	+333
750	1.0	0.500	0.500	+375
1,000	1.0	0.333	0.333	+333
1,500	1.0	0.000	0.000	0

*Rounded to the nearest whole number.

▼ **Figure 53.9 Population growth predicted by the logistic model.** The rate of population growth decreases as population size (N) approaches the carrying capacity (K) of the environment. The red line shows logistic growth in a population where $r = 1.0$ and $K = 1,500$ individuals. For comparison, the blue line illustrates a population continuing to grow exponentially with the same r.

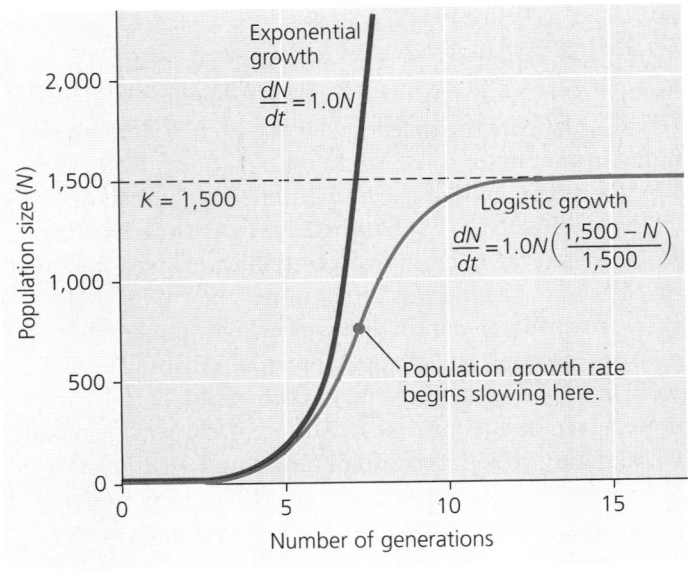

> Mastering Biology BioFlix® Animation: Logistic Growth

are added to the population most rapidly at intermediate population sizes, when there is not only a breeding population of substantial size, but also lots of available space and other resources in the environment. The number of individuals added to the population decreases dramatically as N approaches K. As a result, the population growth rate (dN/dt) also decreases as N approaches K.

Note that we haven't said anything yet about *why* the population growth rate decreases as N approaches K. For a population's growth rate to decrease, the birth rate must decrease, the death rate must increase, or both. Later in the chapter, we'll consider some of the factors affecting these rates, including the presence of disease, predation, and limited amounts of food and other resources.

The Logistic Model and Real Populations

The growth of laboratory populations of some small animals, such as beetles and crustaceans, and of some microorganisms, such as bacteria, *Paramecium*, and yeasts, fits an S-shaped curve fairly well under conditions of limited resources **(Figure 53.10a)**. These populations are grown in a constant environment lacking predators and competing species that may reduce growth of the populations, conditions that rarely occur in nature.

Populations in nature rarely match the predictions of the logistic model as closely as do some laboratory populations, such as the *Paramecium* population in Figure 53.10a. This is not surprising, since some of the assumptions built into the logistic model clearly do not apply to all populations. The logistic model assumes that populations adjust instantaneously to growth and approach carrying capacity smoothly. In reality, there is often a delay before the negative effects of an increasing population are realized. If food becomes limiting for a population, for instance, reproduction will decline eventually, but females may use their energy reserves to continue reproducing for a short time. This may cause the population to overshoot its carrying capacity temporarily, as shown for the water fleas in **Figure 53.10b**. In the **Scientific Skills Exercise**, you can model what can happen to such a population when *N* becomes greater than *K*. Other populations fluctuate greatly, making it difficult even to define carrying capacity. We will examine some possible reasons for such fluctuations later in the chapter.

The logistic model provides a useful starting point for thinking about how populations grow and for constructing more complex models. As such, its role is similar to that played by the Hardy-Weinberg equation for thinking about the evolution of populations. The logistic model is also important in conservation biology for predicting how rapidly a particular population might increase in numbers after it has been reduced to a small size and for estimating sustainable harvest rates for wildlife populations. Conservation biologists can also use the model to estimate the critical size below which populations of certain organisms, such as the northern subspecies of the white rhinoceros (*Ceratotherium simum*), may become extinct **(Figure 53.11)**.

▼ **Figure 53.11 White rhinoceros mother and calf.** The two animals pictured here are members of the southern subspecies, which has a population of more than 20,000 individuals. The northern subspecies is critically endangered, with just a few known individuals.

▶ **Figure 53.10 How well do these populations fit the logistic growth model?** In each graph, the black dots plot the measured growth of the population, and the red curve is the growth predicted by the logistic model.

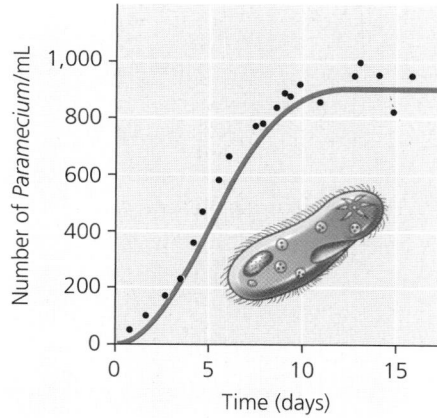

(a) A *Paramecium* population in the lab. The growth of *Paramecium aurelia* in a small culture closely approximates logistic growth if the researcher maintains a constant environment.

(b) A *Daphnia* population in the lab. The growth of a population of water fleas (*Daphnia*) in a small laboratory culture does not correspond well to the logistic model. This population overshoots the carrying capacity of its artificial environment before it settles down to an approximately stable population size.

Using the Logistic Equation to Model Population Growth

What Happens to the Size of a Population When It Overshoots Its Carrying Capacity? In the logistic population growth model, the per capita rate of population increase approaches zero as the population size (N) approaches the carrying capacity (K). Under some conditions, however, a population in the laboratory or the field can overshoot K, at least temporarily. If food becomes limiting to a population, for instance, there may be a delay before reproduction declines, and N may briefly exceed K. In this exercise, you will use the logistic equation to model the growth of the hypothetical population in Table 53.2 when N > K.

► *Daphnia*

INTERPRET THE DATA

1. Assuming that r = 1.0 and K = 1,500, calculate the population growth rate for four cases where population size (N) is greater than carrying capacity (K): N = 1,510, 1,600, 1,750 and 2,000 individuals. To do this, first write the equation for population growth rate given in Table 53.2. Plug in the values for each of the four cases, starting with N = 1,510, and solve the equation for each one. Which population size has the highest growth rate?

2. If r is doubled, predict how the population growth rates will change for the four population sizes given in question 1. Now calculate the population growth rate for the same four cases, this time assuming that r = 2.0 (and with K still = 1,500).

3. Now let's see how the growth of a real-world population of *Daphnia* corresponds to this model. At what times in Figure 53.10b is the *Daphnia* population changing in ways that correspond to the values you calculated? Hypothesize why the population drops below the carrying capacity briefly late in the experiment.

→ **Instructors:** A version of this Scientific Skills Exercise can be assigned in **Mastering Biology**.

CONCEPT CHECK 53.3

1. Explain why a population that fits the logistic growth model increases more rapidly at an intermediate size than at relatively small and large sizes.

2. **WHAT IF?** Given the latitudinal differences in sunlight intensity (see Figure 52.3), how might you expect the carrying capacity of plant species found at the equator to compare with that of plant species found at high latitudes?

3. **MAKE CONNECTIONS** Suppose that a sudden change in environmental conditions caused a substantial drop in a population's carrying capacity. Predict how natural selection and genetic drift might affect this population. (See Concept 23.3.)

For suggested answers, see Appendix A.

CONCEPT 53.4

Life history traits are products of natural selection

EVOLUTION Natural selection favors traits that improve an organism's chances of survival and reproductive success. In every species, there are trade-offs between survival and reproductive traits such as frequency of reproduction, number of offspring (number of seeds produced by plants; litter or clutch size for animals), or investment in parental care. The traits that affect an organism's schedule of reproduction and survival make up its **life history**. Life history traits of an organism are evolutionary outcomes reflected in its development, physiology, and behavior.

Diversity of Life Histories

We'll focus on three key components of an organism's life history: when reproduction begins (the age at first reproduction or age at maturity), how often the organism reproduces, and how many offspring are produced per reproductive episode. The fundamental idea that evolution accounts for the diversity of life is manifest in the broad range of these life history characteristics in nature. For example, the age at which reproduction begins varies considerably across species. A typical loggerhead turtle is about 30 years old when it first crawls onto a beach to lay eggs. In contrast, the coho salmon (*Oncorhynchus kisutch*) is often only three or four years old when it spawns.

Organisms also vary in how often they reproduce. The coho salmon is an example of organisms that undergo a "one-shot" pattern of big-bang reproduction, or **semelparity** (from the Latin *semel*, once, and *parere*, to beget). It hatches in the headwaters of a freshwater stream and then migrates to the Pacific Ocean, where it typically requires a few years to mature. The salmon eventually returns to the same stream to spawn, producing thousands of eggs in a single reproductive opportunity before it dies. Semelparity also occurs in some plants, such as the agave, or "century plant" **(Figure 53.12a)**. Agaves generally grow in arid climates with unpredictable rainfall and poor soils. An agave grows for years, accumulating nutrients in its tissues, until there is an unusually wet year. It then sends up a large flowering stalk, produces seeds, and dies. This life history appears to be an adaptation to the agave's harsh desert environment.

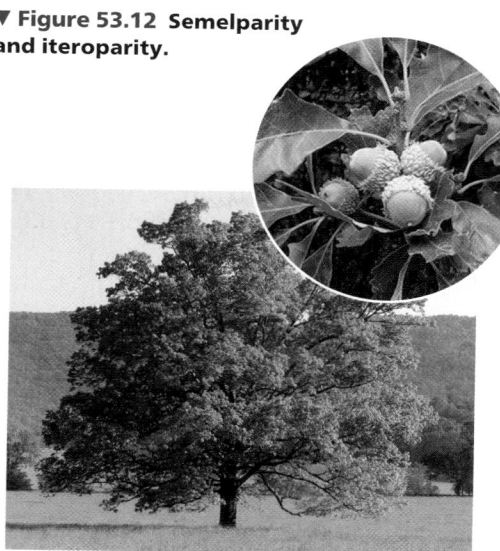

▼ **Figure 53.12 Semelparity and iteroparity.**

(a) Semelparity: one-time reproduction. An agave (*Agave americana*) is an example of semelparity. The leaves of the plant are visible at the base of the giant flowering stalk, which is produced only at the end of the agave's life.

(b) Iteroparity: repeat reproduction. Organisms that reproduce repeatedly, such as the bur oak (*Quercus macrocarpa*), undergo iteroparity. One oak tree can produce thousands of acorns (inset) per year over the course of many decades.

In contrast to semelparity is **iteroparity** (from the Latin *iterare*, to repeat), or repeated reproduction. As an example, a female loggerhead turtle produces four clutches totaling approximately 300 eggs in a year. It then typically waits two to three years before laying more eggs; presumably, the turtles lack sufficient resources to produce that many eggs every year. A mature turtle may lay eggs for 30 years after the first clutch. Horses and other large mammals also reproduce repeatedly, as do many fish, sea urchins, and long-lived trees, such as maples and oaks **(Figure 53.12b)**.

Finally, organisms also vary in how many offspring they produce. Some species, such as the white rhinoceros (see Figure 53.11), produce a single calf when they reproduce, while most insects and many plants produce large numbers of offspring. Such variation in offspring number has other consequences as well; as you'll read, a species that produces one or a few offspring may provision them better than does a species that produces many offspring.

"Trade-offs" and Life Histories

No organism could produce thousands of offspring yet provision each of them as well as does a white rhinoceros caring for its single calf. There is a trade-off between the number of offspring and the amount of resources a parent can devote to each offspring. Such trade-offs occur because organisms do not have access to unlimited amounts of resources. As a result, the use of resources for one function (such as reproduction) can reduce

the resources available for supporting another function (such as survival). In Eurasian kestrels, for example, caring for a larger number of young lowered the survival rates of the parents **(Figure 53.13)**. In another study, in Scotland, researchers found that female red deer that reproduced in a given summer were more likely to die the next winter than were females that did not reproduce.

Selective pressures also influence trade-offs between the number and size of offspring. Plants and animals whose young have a low change of survival often produce many small offspring. For example, plants that colonize disturbed environments usually produce many small seeds, only a few of which may reach a suitable habitat. Small size may also increase the chance of seedling establishment by enabling the seeds to be carried longer

▼ **Figure 53.13 Inquiry**

How does caring for offspring affect parental survival in kestrels?

Experiment Cor Dijkstra and colleagues in the Netherlands studied the effects of parental caregiving in Eurasian kestrels over five years. The researchers transferred chicks among nests to produce reduced broods (three or four chicks), normal broods (five or six), and enlarged broods (seven or eight). They then measured the percentage of male and female parent birds that survived the following winter. (Both males and females provide care for chicks.)

Results

Conclusion The lower survival rates of kestrels with larger broods indicate that caring for more offspring negatively affects survival of the parents.

Data from C. Dijkstra et al., Brood size manipulations in the kestrel (*Falco tinnunculus*): effects on offspring and parent survival, *Journal of Animal Ecology* 59:269–285 (1990).

INTERPRET THE DATA *The males of some bird species provide no parental care. If this were true for the Eurasian kestrel, predict how the experimental results would differ from those shown here.*

▼ Figure 53.14 Variation in the number and size of seeds in plants.

(a) Dandelions grow quickly and release a large number of tiny fruits, each containing a single seed. Producing numerous seeds increases the chance that at least some will grow into plants that eventually produce seeds themselves.

(b) Some plants, such as the Brazil nut tree (right), produce a moderate number of large seeds in pods (above). Each seed's large endosperm provides nutrients for the embryo, an adaptation that helps a relatively large fraction of offspring survive.

distances to a broader range of habitats **(Figure 53.14a)**. Animals that suffer high predation rates, such as quail, sardines, and mice, also tend to produce many offspring.

In other organisms, extra investment on the part of the parent greatly increases the offspring's chances of survival. Brazil nut and walnut trees produce large seeds packed with nutrients that help the seedlings become established **(Figure 53.14b)**. Primates generally bear only one or two offspring at a time; parental care and an extended period of learning in the first several years of life are very important to offspring fitness. Such provisioning and extra care can be especially important in habitats with high population densities.

One way to categorize variation in life history traits is related to the logistic growth model discussed in Concept 53.3. Selection for traits that are advantageous at high densities is referred to as **K-selection**. In contrast, selection for traits that maximize reproductive success in uncrowded environments (low densities) is called **r-selection**. These names follow from the variables of the logistic equation. K-selection is said to operate in populations living at a density near the limit imposed by their resources (the carrying

capacity, K), where competition among individuals is stronger. Mature trees growing in an old-growth forest are an example of K-selected organisms. In contrast, r-selection is said to maximize r, the intrinsic rate of increase, and occurs in environments in which population densities are well below carrying capacity or individuals face little competition. Such conditions are often found in disturbed habitats that are being recolonized. Weeds growing in an abandoned agricultural field are an example of r-selected organisms.

The concepts of K- and r-selection represent two extremes in a range of actual life histories. The framework of K- and r-selection, grounded in the idea of carrying capacity, also relates to the important question we alluded to earlier: *Why* does population growth rate decrease as population size approaches carrying capacity? Answering this question is the focus of the next section.

CONCEPT CHECK 53.4

1. Identify three key life history traits, and give examples of organisms that vary widely in each of these traits.

2. In the fish called the peacock wrasse (*Symphodus tinca*), females disperse some of their eggs widely and lay other eggs in a nest. Only the latter receive parental care. Explain the trade-offs in reproduction that this behavior illustrates.

3. **WHAT IF?** Mice that experience stress such as a food shortage will sometimes abandon their young. Explain how this behavior might have evolved in the context of reproductive trade-offs and life history.

For suggested answers, see Appendix A.

CONCEPT 53.5

Density-dependent factors regulate population growth

What environmental factors keep populations from growing indefinitely? Why are some populations fairly stable in size, while others are not?

Answers to such questions can be important in practical applications. Farmers may want to reduce the abundance of insect pests or stop the growth of a weed that is spreading rapidly. Conservation ecologists need to know what environmental factors create favorable feeding or breeding habitats for endangered species, such as the white rhinoceros and the whooping crane. Overall, whether seeking to reduce the size of an unwanted population or increase the size of one that is endangered, it is helpful to understand factors that affect population abundance.

Population Change and Population Density

To understand why a population stops growing when it reaches a certain size, ecologists study how the rates of birth, death, immigration, and emigration change as population density rises. If immigration and emigration offset each other, then a population grows when the birth rate exceeds the death rate and declines when the death rate exceeds the birth rate.

A birth rate or death rate that does *not* change with population density is said to be **density independent**. For example, researchers found that the mortality of dune fescue grass (*Vulpia fasciculata*) is mainly due to physical factors that kill similar proportions of a local population, regardless of its density. Drought stress that arises when the roots of the grass are uncovered by shifting sands is a density-independent factor that can kill these plants. In contrast, a death rate that increases with population density or a birth rate that falls with rising density is said to be **density dependent**. Researchers found that reproduction by dune fescue declines as population density increases, in part because water or nutrients become more scarce. Thus, the key factors affecting birth rate in this population are density dependent, while death rate is largely determined by density-independent factors. For situations in which immigration and emigration offset each other, **Figure 53.15** shows how the combination of density-dependent reproduction and density-independent mortality can stop population growth (as in species such as dune fescue).

Variation in density-independent factors such as temperature and precipitation can cause dramatic changes in population size. For example, a drought or heat wave can cause a sharp increase in mortality rates, causing the abundance of a population to plummet. Note, however, that a density-independent factor cannot consistently cause a population to decrease in size when it is large or increase in size when it is small—only a density-dependent factor can consistently cause such changes. With this in mind, a population is said to be *regulated* when one or more density-dependent factors cause its size to decrease when large (or increase when small).

→ **Mastering Biology** BioFlix® Animation: Density Dependence

Mechanisms of Density-Dependent Population Regulation

Without some type of negative feedback between population density and the rates of birth and death, a population would never stop growing. But no population can increase in size indefinitely. Ultimately, at large population sizes, negative feedback is provided by density-dependent regulation, which halts population growth through mechanisms that reduce birth rates or increase death rates. For example, a study of kelp perch (*Brachyistius frenatus*) populations showed that the fish's death rate rose proportionally as its density increased **(Figure 53.16)**. This occurred because at high densities, the fish ran out of safe spaces in the kelp where they could hide from predators. Several other mechanisms that can cause density-dependent regulation are described in **Figure 53.17**.

These various examples of population regulation by negative feedback show how increased densities cause population growth rates to decline by affecting reproduction, growth, and survival. Although negative feedback helps explain why populations stop growing, it does not address why some populations fluctuate dramatically while others remain relatively stable.

▼ **Figure 53.15 Determining equilibrium for population density.** In this example, the birth rate changes with population density, while the death rate is constant; immigration and emigration rates are assumed to be equal. At the equilibrium density (*Q*), the birth and death rates are equal. As a result, the number of individuals added to the population equals the number removed from it, and the population stops changing in size.

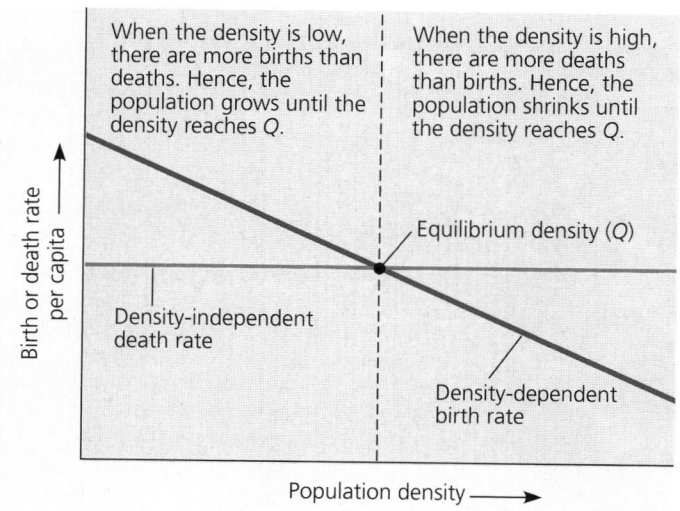

DRAW IT *Redraw this figure for the case where the birth and death rates are both density dependent, as occurs for many species.*

▼ **Figure 53.16 An example of density-dependent regulation.** At high population densities, the death rate of kelp perch increased because they ran out of spaces to hide from predators.

Exploring Mechanisms of Density-Dependent Regulation

As population density increases, density-dependent mechanisms operating within a species can slow or stop population growth by decreasing birth rates or increasing death rates.

Competition for Resources

Increasing population density leads to competition among members of a population for nutrients and other resources, reducing reproductive rates. Farmers minimize the effect of competition on the growth of wheat (*Triticum aestivum*) and other crops by applying fertilizers to reduce nutrient limitations on crop yield.

Territoriality

Territoriality can limit population density when space becomes the resource for which individuals compete. Cheetahs (*Acinonyx jubatus*) use a chemical marker in urine to warn other cheetahs of their territorial boundaries. The presence of surplus, or nonbreeding, individuals is a good indication that territoriality is restricting population growth.

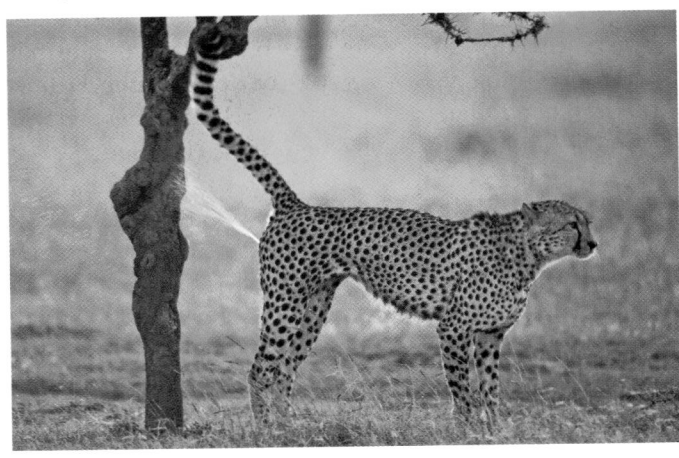

Disease

If the transmission rate of a disease increases as a population becomes more crowded, then the disease's impact is density dependent. In humans, the respiratory diseases influenza (flu) and tuberculosis are spread through the air when an infected person sneezes or coughs. Both diseases strike a greater percentage of people in densely populated cities than in rural areas.

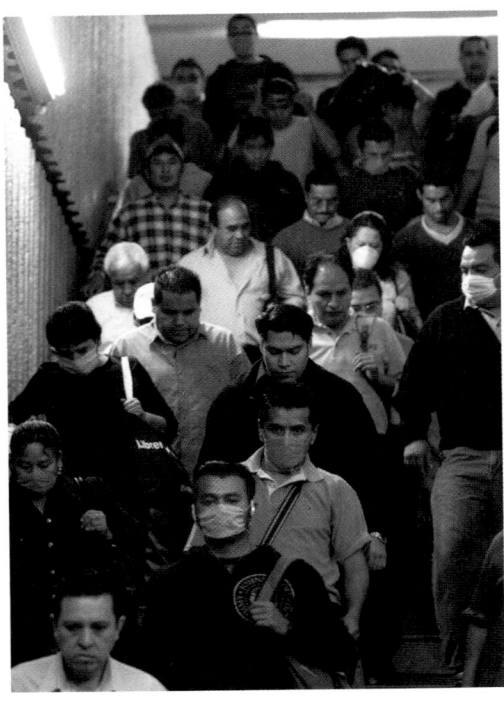

Intrinsic Factors

Intrinsic physiological factors (those operating within an individual organism) sometimes regulate population size. Reproductive rates of white-footed mice (*Peromyscus leucopus*) in a field enclosure can drop even when food and shelter are abundant. This drop in reproduction at high population density is associated with aggressive interactions and hormonal changes within individual mice that delay sexual maturation and depress the immune system.

Toxic Wastes

Yeasts, such as the brewer's yeast *Saccharomyces cerevisiae*, are used to convert carbohydrates to ethanol in winemaking. The ethanol that accumulates in the wine is toxic to yeasts and contributes to density-dependent regulation of yeast population size. The alcohol content of wine is usually less than 13% because that is the maximum concentration of ethanol that most wine-producing yeast cells can tolerate.

5 µm

Population Dynamics

Like the piping plover (see Figure 53.1), all populations show some fluctuation in size. Such population fluctuations from year to year or place to place, called **population dynamics**, are influenced by many factors and in turn affect other species. For example, fluctuations in fish populations affect populations of seabirds that eat fish. The study of population dynamics focuses on the complex interactions between biotic and abiotic factors that cause variation in population sizes.

Stability and Fluctuation

Populations of large mammals were once thought to remain relatively stable, but long-term studies have challenged that idea. For instance, the moose population on Isle Royale in Lake Superior has fluctuated substantially since around 1900. At that time, moose from the Ontario mainland (25 km away) colonized the island, perhaps by walking across the lake when it was frozen. Wolves, which rely on moose for most of their food, reached the island around 1950 by walking across the frozen lake. The lake has not frozen over since the early 1950s, and both populations appear to have been isolated from immigration and emigration since then. Despite this isolation, the moose population experienced two major increases and collapses during the last 50 years **(Figure 53.18)**.

What factors cause the size of the moose population to change so dramatically? Harsh weather, particularly cold winters with heavy snowfall, can weaken moose and reduce food availability, decreasing the population size. When moose numbers are low and the weather is mild, food is readily available and the population grows quickly. Conversely, when moose numbers are high, factors such as predation and

an increase in the density of ticks and other parasites cause the population to shrink. The effects of some of these factors can be seen in Figure 53.18. The first major collapse coincided with a peak in the numbers of wolves from 1975 to 1980. The second major collapse, around 1995, coincided with harsh winter weather, which increased the energy needs of the moose and made it harder for them to find food under the deep snow.

Population Cycles: Scientific Inquiry

While many populations fluctuate at unpredictable intervals, others undergo regular boom-and-bust cycles. Some small herbivorous mammals, such as voles and lemmings, tend to have 3- to 4-year cycles, while some birds, such as ruffed grouse and ptarmigans, have 9- to 11-year cycles.

One striking example of population cycles is the roughly 10-year cycling of snowshoe hares (*Lepus americanus*) and lynx (*Lynx canadensis*) in the far northern forests of Canada and Alaska. Lynx are predators that feed predominantly on snowshoe hares, so lynx numbers might be expected to rise and fall with the numbers of hares **(Figure 53.19)**. But why do hare numbers rise and fall in approximately 10-year cycles? Two main hypotheses have been proposed. First, the cycles may be caused by food shortage during winter. Hares eat the

▼ **Figure 53.19 Population cycles in the snowshoe hare and lynx.** Population counts are based on the number of pelts sold by trappers to the Hudson Bay Company.

INTERPRET THE DATA *What do you observe about the relative timing of the peaks in lynx numbers and hare numbers? What might explain this observation?*

▼ **Figure 53.18 Fluctuations in moose and wolf population sizes on Isle Royale, 1959–2011.**

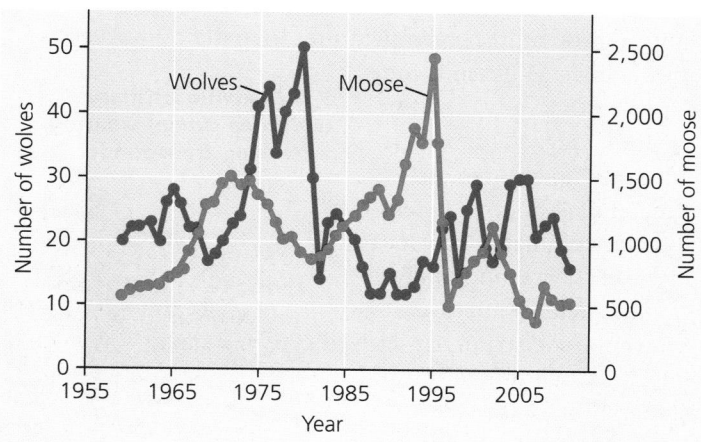

⊝ **Mastering Biology** BioFlix® Animation: Predator-Prey Interactions

terminal twigs of small shrubs such as willow and birch in winter, although why this food supply might cycle in 10-year intervals is uncertain. Second, the cycles may be due to predator-prey interactions. Many predators other than lynx eat hares, and they may overexploit their prey.

Let's consider the evidence for the two hypotheses. If hare cycles are due to winter food shortage, then the cycles should stop if extra food is provided to a field population. Researchers conducted such experiments in the Yukon for 20 years—over two hare cycles. They found that hare populations in the areas with extra food increased about threefold in density but continued to cycle in the same way as the unfed control populations. Therefore, food supplies alone do not cause the hare cycles shown in Figure 53.19, so we can reject the first hypothesis.

To study the effects of predation, ecologists used radio collars to track individual hares to determine why they died. Predators, including lynx, coyotes, hawks, and owls, killed 95% of the hares in such studies. None of the hares appeared to have died of starvation. These data support the second hypothesis. When ecologists set up electric fences to exclude predators from certain areas, the collapse in survival that normally occurs in the decline phase of the cycle was nearly eliminated. Overexploitation by predators thus seems to be an essential part of snowshoe hare cycles; without predators, it is unlikely that hare populations would cycle in northern Canada.

Immigration, Emigration, and Metapopulations

So far, our discussion of population dynamics has focused mainly on the contributions of births and deaths. However, immigration and emigration also influence populations. When a population becomes crowded and resource competition increases, emigration often increases.

Immigration and emigration are particularly important when a number of local populations are linked, forming a **metapopulation**. Local populations in a metapopulation can be thought of as occupying discrete patches of suitable habitat in a sea of otherwise unsuitable habitat. Such patches vary in size, quality, and isolation from other patches, factors that influence how many individuals move among the populations. If one population becomes extinct, the patch it occupied may be recolonized by immigrants from another population.

The Glanville fritillary (*Melitaea cinxia*) illustrates the movement of individuals between populations. On average, this butterfly is found in about 500 meadows across the Åland Islands of Finland, but its potential habitat in the islands is much larger, approximately 4,000 suitable patches. New populations of the butterfly regularly appear and existing populations become extinct, constantly shifting the locations of the 500 colonized patches **(Figure 53.20)**. The species persists in a balance of local extinctions and recolonizations.

An individual's ability to move between populations depends on a number of factors, including its genetic

▼ **Figure 53.20 The Glanville fritillary: a metapopulation.** On the Åland Islands, local populations of this butterfly (filled circles) are found in only a fraction of the suitable habitat patches at any given time. Individuals can move between local populations and colonize unoccupied patches (open circles).

• Occupied patch
○ Unoccupied patch

5 km

makeup. One gene that has a strong effect on the Glanville fritillary's ability to move is *Pgi*, which codes for the enzyme phosphoglucoisomerase. This enzyme catalyzes the second step of glycolysis (see Figure 9.8), and its activity correlates with the rate of CO_2 production from respiration by the butterflies. Ecologists studied butterflies known to be heterozygous or homozygous for a single nucleotide polymorphism in *Pgi*. They tracked the movements of individual butterflies using radar and transponders attached to the butterflies that emit an identifying signal. Butterfly movements ranged widely, from 10 m to 4 km, in 2-hour periods. Heterozygous individuals flew more than twice as far in the morning and at lower ambient temperatures than did homozygous individuals. The results indicated a fitness advantage to the heterozygous genotype in low temperatures and a greater likelihood of heterozygotes colonizing new locations in the metapopulation.

The metapopulation concept underscores the significance of immigration and emigration for the Glanville fritillary and many other species. It also helps ecologists understand population dynamics and gene

▼ **A Glanville fritillary (*Melitaea cinxia*) wearing a tracking transponder.**

flow in patchy habitats, providing a framework for the conservation of species living in a network of habitat fragments and reserves.

CONCEPT CHECK 53.5

1. Describe three attributes of habitat patches that could affect population density and rates of immigration and emigration.
2. **DRAW IT** Circle the portion of the red curve in Figure 53.9 in which density-dependent population regulation would likely have the most pronounced effects.
3. **WHAT IF?** Suppose you were studying a species that has a population cycle of about ten years. How long would you need to study the species to determine if its population size were declining? Explain.
4. **MAKE CONNECTIONS** Negative feedback is a process that regulates biological systems (see Concept 40.2). Explain how the density-dependent birth rate of dune fescue grass exemplifies negative feedback.

For suggested answers, see Appendix A.

CONCEPT 53.6

The human population is no longer growing exponentially but is still increasing extremely rapidly

In the last few centuries, the human population has grown at an unprecedented rate, more like the elephant population in Kruger National Park (see Figure 53.8) than the fluctuating populations we considered in Concept 53.5. No population can grow indefinitely, however. In this section of the chapter, we'll apply the concepts of population dynamics to the specific case of the human population.

The Global Human Population

The human population has grown explosively over the last four centuries **(Figure 53.21)**. In 1650, about 500 million people inhabited Earth. Our population doubled to 1 billion within the next two centuries, doubled again to 2 billion by 1930, and doubled still again to 4 billion by 1975. Notice that the time it took our population to double in size decreased from 200 years in 1650 to just 45 years in 1930. Thus, historically our population has grown even *faster* than exponential growth, which has a constant rate of increase and hence a constant doubling time.

The global population is now more than 7.6 billion people and is increasing by about 80 million each year. This translates into more than 200,000 people each day, the equivalent of adding a city the size of Amarillo, Texas. At this rate, it takes only about four years to add the equivalent of another United States to the world population. Ecologists

▼ **Figure 53.21 Human population growth (data as of 2018).** The global human population has grown almost continuously throughout history, but it skyrocketed after the Industrial Revolution.

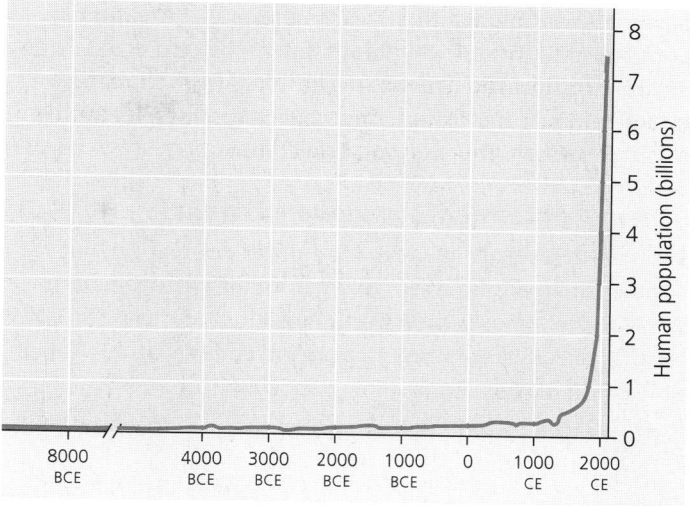

▼ **Figure 53.22 Annual percent increase and recent human population growth.** The sharp dip in the rate of increase around 1960 is due mainly to a famine in China in which about 60 million people died. (Data as of 2018.)

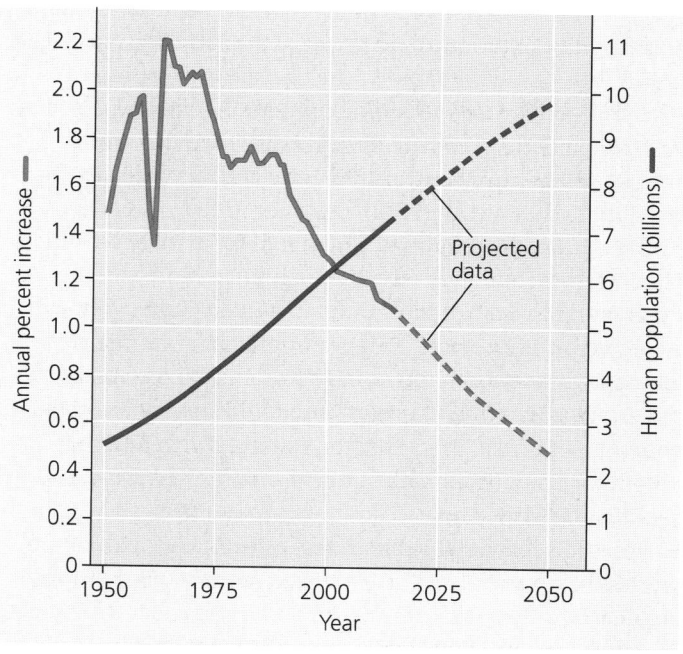

VISUAL SKILLS *The blue curve in this figure appears to tell a different story about human population growth than does the curve in Figure 53.21, yet both curves are accurate. Summarize what each curve conveys about recent human population growth and explain how both can be accurate.*

predict there will be about 9.8 billion people on Earth by the year 2050.

Though the global population is still growing, the *rate* of growth began to slow during the 1960s **(Figure 53.22)**. The annual rate of increase in the global population peaked at 2.2% in 1962 but was only 1.1% in 2018. Current models project a growth rate of 0.5% by 2050, which would add 49 million more people per year if the population climbs to a projected 9.8 billion. The reduction in annual growth rate observed over the past five decades shows that the human population is now growing more slowly than expected in exponential growth. This change resulted from fundamental shifts in population dynamics due to diseases, including AIDS, and to voluntary population control.

Regional Patterns of Population Change

We have described changes in the global population, but population dynamics vary widely from region to region. In a stable regional population, birth rate equals death rate (disregarding the effects of immigration and emigration). Two possible configurations for a stable population are

Zero population growth = High birth rate − High death rate

or

Zero population growth = Low birth rate − Low death rate

The movement from high birth and death rates toward low birth and death rates, which tends to accompany industrialization and improved living conditions, is called the **demographic transition**. In Sweden, this transition took about 150 years, from 1810 to 1975, when birth rates finally approached death rates. In Mexico, where the human population is still growing rapidly, the transition is projected to take until at least 2050. Demographic transition is associated with an increase in the quality of health care and sanitation as well as improved access to education, especially for women.

After 1950, death rates declined rapidly in most developing countries, but birth rates have declined more variably. Birth rates have fallen most dramatically in China. The expected number of children per woman per lifetime in China decreased from 5.9 in 1970 to 1.6 in 2011, largely due to the government's strict "one-child" policy. In some countries of Africa, the transition to lower birth rates has also been rapid, though birth rates remain high in most of sub-Saharan Africa.

How do such variable birth rates affect the growth of the world's population? In industrialized nations, populations are near equilibrium, with reproductive rates near the replacement level of 2.1 children per female (over their lifetime). In many industrialized countries—including the United States, Canada, Germany, Japan, and the United Kingdom—total reproductive rates are in fact *below* the replacement level. These populations will eventually decline if there is no immigration and if the birth rate does not change. In fact, the population is already declining in many eastern and central European countries. Most of the current global population growth occurs in less industrialized countries, where about 80% of the world's people now live.

A unique feature of human population growth is our ability to control family sizes through planning and voluntary contraception. Social change and the rising educational and career aspirations of women in many cultures encourage women to delay marriage and postpone reproduction. Delayed reproduction helps to decrease population growth rates and to move a society toward zero population growth under conditions of low birth rates and low death rates. However, there is a great deal of disagreement as to how much support should be provided for global family planning efforts.

Age Structure

Another important factor that can affect population growth is a country's **age structure**, the relative number of individuals of each age in the population. Age structure is commonly graphed as "pyramids" like those in **Figure 53.23**. For Zambia, the pyramid is bottom heavy, skewed toward young individuals who will grow up and perhaps sustain the explosive growth with their own reproduction. The age structure for the United States is relatively even until the older, postreproductive ages. Although the current total reproductive rate in the United States is about 1.8 children per woman—below the replacement rate—the population is projected to grow slowly through 2050 as a result of immigration. For Italy, the pyramid has a small base, indicating that individuals younger than reproductive age are relatively underrepresented in the population. This situation contributes to the projection of a future population decrease in Italy.

Age-structure diagrams not only predict a population's growth trends but also can illuminate social conditions. Based on the diagrams in Figure 53.23, we can predict that employment and education opportunities will continue to be a problem for Zambia in the foreseeable future. In the United States and Italy, a decreasing proportion of younger working-age people will soon be supporting an increasing population of retired "boomers." This demographic feature has made the future of Social Security and Medicare a major political issue in the United States. Understanding age structures is necessary to plan for the future.

Infant Mortality and Life Expectancy

Infant mortality, the number of infant deaths per 1,000 live births, and *life expectancy at birth*, the predicted average length of life at birth, vary widely in different countries. In 2015, for example, the infant mortality rate was 71 (7.1%) in Angola but only 2 (0.2%) in Japan. Life expectancy at birth was 59 years in Angola but 85 years in Japan. These differences reflect the quality of life faced by children at birth and influence the reproductive choices parents make. If infant

▼ Figure 53.23 Age-structure pyramids for the human population of three countries. The annual growth rate was approximately 2.8% in Zambia, 1.0% in the United States, and 0.0% in Italy. Note that Zambia's pyramid has a different scale for its *x*-axis from that of the other two pyramids. (Data as of 2010.)

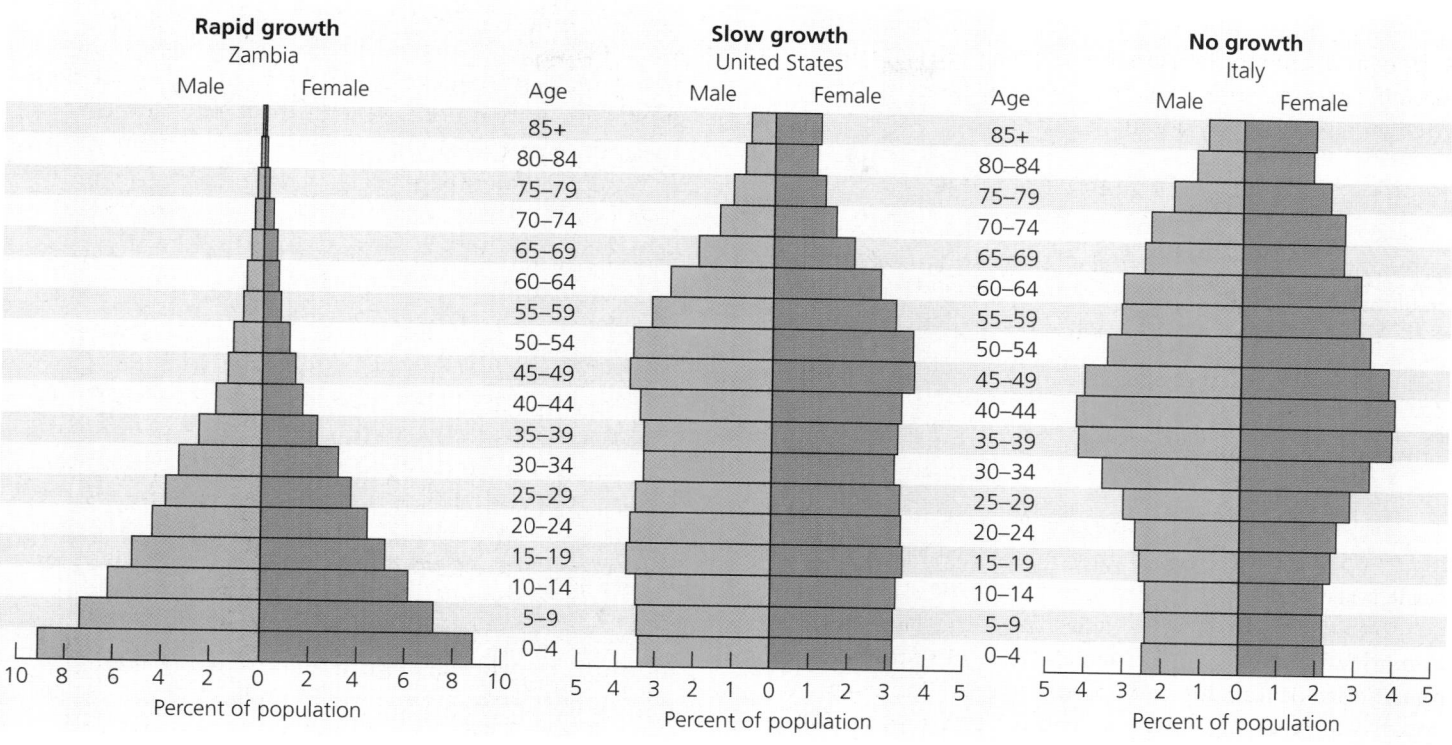

→ **Mastering Biology** Animation: Analyzing Age-Structure Pyramids

mortality is high, then parents may choose to have more children to ensure that some reach adulthood.

Although global life expectancy has been increasing since about 1950, it has recently dropped in a number of regions, including countries of the former Soviet Union and in sub-Saharan Africa. In these regions, social upheaval, decaying infrastructure, and infectious diseases such as AIDS and tuberculosis are reducing life expectancy.

Global Carrying Capacity

No ecological question is more important than the future size of the human population. As noted earlier, population ecologists project a global population of about 9.8 billion people in 2050. That means that over 2 billion people will be added to the population in the next four decades because of the momentum of population growth. But just how many humans can the biosphere support? Will the world be overpopulated in 2050? Is it *already* overpopulated?

Estimates of Carrying Capacity

For over three centuries, scientists have attempted to estimate the human carrying capacity of Earth. The first known estimate, 13.4 billion people, was made in 1679 by Anton van Leeuwenhoek (a Dutch scientist who also discovered protists). Since then, estimates have varied from less than 1 billion to more than 1,000 billion (1 trillion).

Carrying capacity is difficult to estimate, and scientists use different methods to produce their estimates. Some researchers use curves like that produced by the logistic equation (see Figure 53.9) to predict the future maximum of the human population. Others generalize from existing "maximum" population density and multiply this number by the area of habitable land. Still others base their estimates on a single limiting factor, such as food, and consider variables such as the amount of farmland, the average yield of crops, the prevalent diet—vegetarian or meat based—and the number of calories needed per person per day.

Limits on Human Population Size

A more comprehensive approach to estimating the carrying capacity of Earth is to recognize that humans have multiple constraints: We need food, water, fuel, building materials, and other resources, such as clothing and transportation. The **ecological footprint** concept summarizes the aggregate land and water area required by each person, city, or nation

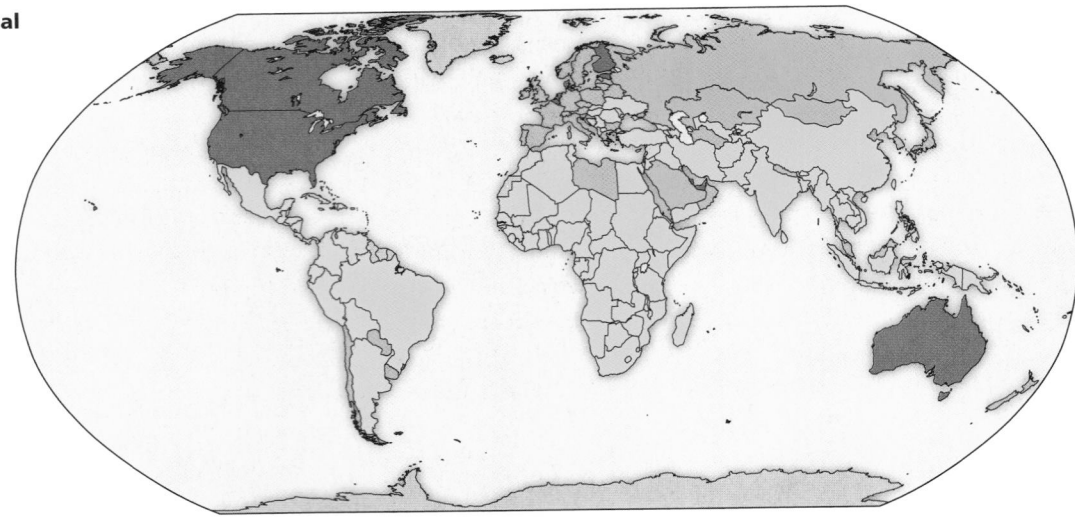

▶ **Figure 53.24 Per capita ecological footprint by country.**

Per capita ecological footprint in global hectares (gha)

⬜ 0–3	⬛ >6
⬜ 3–6	⬜ Insufficient data

❓ *Earth has a total of 11.9 billion gha of productive land. How many people could Earth support sustainably if the average ecological footprint were 8 gha per person (as in the United States)?*

to produce all the resources it consumes and to absorb all the waste it generates.

What is a sustainable ecological footprint for the entire human population? One way to estimate this footprint is to add up all the ecologically productive land on the planet and divide by the size of the human population. Typically, this estimate is made using *global hectares*, where a global hectare (gha) represents a hectare of land or water with a productivity equal to the average of all biologically productive areas on Earth (1 hectare = 2.47 acres). This calculation yields an allotment of 1.7 gha per person—the benchmark for comparing actual ecological footprints. Anyone who consumes resources that require more than 1.7 gha to produce is using an unsustainable share of Earth's

resources, as is the case for the citizens of many countries **(Figure 53.24)**. For example, a typical ecological footprint for a person in the United States is 8 gha. Globally, the average footprint is 2.7 gha per person, more than a 50% overshoot of the sustainable use (1.7 gha per person) of Earth's resources.

Our impact on the planet can also be assessed using currencies other than area, such as energy use. Average energy use differs greatly across different regions of the world **(Figure 53.25)**. A typical person in the United States, Canada, or Norway consumes roughly 30 times the energy that a person in central Africa does. Moreover, fossil fuels, such as oil, coal, and natural gas, are the source of 80% or more of the energy used in most developed nations. As Concept 56.4

▶ **Figure 53.25 The uneven electrification of the planet.** This composite image taken from space of Earth's surface at night illustrates the varied density of electric lights worldwide, one aspect of energy use by humans.

discusses in more detail, this unsustainable reliance on fossil fuels is changing Earth's climate and increasing the amount of waste that humans produce. Ultimately, the combination of resource use per person and population density determines our global ecological footprint.

What factors will eventually limit the growth of the human population? Perhaps food will be the main limiting factor. Malnutrition and famine are common in some regions, but they result mainly from the unequal distribution of food rather than from inadequate production. So far, technological improvements in agriculture have allowed food supplies to keep up with global population growth. In contrast, the demands of many populations have already far exceeded the local and even regional supplies of one renewable resource—fresh water. More than 1 billion people do not have access to sufficient water to meet their basic sanitation needs. The human population may also be limited by the capacity of the environment to absorb its wastes. If so, then Earth's current human occupants could lower the planet's long-term carrying capacity for future generations.

Technology has substantially increased Earth's carrying capacity, but no population can grow indefinitely. After reading this chapter, you should realize that there is no single carrying capacity. How many people our planet can sustain depends on the quality of life each of us has and the distribution of wealth across people and nations, topics of great concern and political debate. We can decide whether zero population growth will be attained through social changes based on human choices or, instead, through increased mortality due to resource limitation, plagues, war, and environmental degradation.

CONCEPT CHECK 53.6

1. How does a human population's age structure affect its growth rate?
2. How have the rate and number of people added to the human population each year changed in recent decades?
3. **WHAT IF?** Type "personal ecological footprint calculator" into a search engine and use one of the resulting calculators to estimate your footprint. Is your current lifestyle sustainable? If not, what choices can you make to influence your own ecological footprint?

For suggested answers, see Appendix A.

53 Chapter Review

→ Go to **Mastering Biology** for Assignments, the eText, the Study Area, and Dynamic Study Modules.

SUMMARY OF KEY CONCEPTS

→ To review key terms, go to the **Vocabulary Self-Quiz** in the **Mastering Biology** eText or Study Area, or go to goo.gl/zkjz9t.

CONCEPT 53.1

Biotic and abiotic factors affect population density, dispersion, and demographics (pp. 1191–1195)

- Population **density**—the number of individuals per unit area or volume—reflects the interplay of births, deaths, immigration, and emigration. Environmental and social factors influence the **dispersion** of individuals.

Patterns of dispersion

  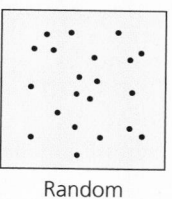

| Clumped | Uniform | Random |

- Populations increase from births and **immigration** and decrease from deaths and **emigration. Life tables** and **survivorship curves** summarize specific trends in **demography**.

? *Gray whales (Eschrichtius robustus) gather each winter near Baja California to give birth. How might such behavior make it easier for ecologists to estimate birth and death rates for the species?*

CONCEPT 53.2

The exponential model describes population growth in an idealized, unlimited environment (pp. 1196–1197)

- If immigration and emigration are ignored, a population's per capita growth rate equals its birth rate minus its death rate.
- The **exponential growth** equation $dN/dt = rN$ represents a population's growth when resources are relatively abundant, where r is the **intrinsic rate of increase** and N is the number of individuals in the population.

$$\frac{dN}{dt} = rN$$

Population size (N)

Number of generations

? *Suppose one population has an r that is twice as large as the r of another population. What is the maximum size that both populations will reach over time, based on the exponential model?*

CONCEPT 53.3

The logistic model describes how a population grows more slowly as it nears its carrying capacity (pp. 1197–1200)

- Exponential growth cannot be sustained in any population. A more realistic population model limits growth by incorporating **carrying capacity** (*K*), the maximum population size the environment can support.
- According to the **logistic growth** equation $dN/dt = rN(K - N)/K$, growth levels off as population size approaches the carrying capacity.

$$\frac{dN}{dt} = rN\left(\frac{K - N}{K}\right)$$

K = carrying capacity

Population size (*N*) — vertical axis

Number of generations — horizontal axis

- The logistic model fits few real populations perfectly, but it is useful for estimating possible growth.

? *As an ecologist who manages a wildlife preserve, you want to increase the preserve's carrying capacity for a particular endangered species. How might you go about accomplishing this?*

CONCEPT 53.4

Life history traits are products of natural selection (pp. 1200–1202)

- **Life history** traits are evolutionary outcomes reflected in the development, physiology, and behavior of organisms.
- Big-bang, or **semelparous**, organisms reproduce once and die. **Iteroparous** organisms produce offspring repeatedly.
- Life history traits such as brood size, age at maturity, and parental caregiving represent trade-offs between conflicting demands for time, energy, and nutrients. Two hypothetical life history patterns are *K*-selection and *r*-selection.

? *Explain why ecological trade-offs are common.*

CONCEPT 53.5

Density-dependent factors regulate population growth (pp. 1202–1207)

- In **density-dependent** population regulation, death rates rise and birth rates fall with increasing density. A birth or death rate that does not vary with density is said to be **density independent**.
- Density-dependent changes in birth and death rates curb population increase through negative feedback and can eventually stabilize a population near its carrying capacity. Density-dependent limiting factors include intraspecific competition for limited food or space, increased predation, disease, and intrinsic physiological factors.
- Because changing environmental conditions periodically disrupt them, all populations exhibit some size fluctuations. Many populations undergo regular boom-and-bust cycles that are influenced by complex interactions between biotic and abiotic factors. A **metapopulation** is a group of populations linked by immigration and emigration.

? *Give an example of one biotic and one abiotic factor that contribute to yearly fluctuations in the size of the human population.*

CONCEPT 53.6

The human population is no longer growing exponentially but is still increasing extremely rapidly (pp. 1207–1211)

- Since about 1650, the global human population has grown exponentially, but within the last 50 years, the rate of growth has fallen by half. Differences in **age structure** show that while some nations' populations are growing rapidly, those of others are stable or declining in size. Infant mortality rates and life expectancy at birth vary widely in different countries.
- **Ecological footprint** is the aggregate land and water area needed to produce all the resources a person or group of people consume and to absorb all of their waste. It is one measure of how close we are to the carrying capacity of Earth, which is uncertain. With a world population of more than 7.6 billion people, we are already using many resources in an unsustainable manner.

? *How do humans differ from other species in the ability to "choose" a carrying capacity for their environment?*

TEST YOUR UNDERSTANDING

→ For more multiple-choice questions, go to the **Practice Test** in the **Mastering Biology** eText or Study Area, or go to goo.gl/GruWRg.

Levels 1-2: Remembering/Understanding

1. Population ecologists follow the fate of same-age cohorts to
   (A) determine a population's carrying capacity.
   (B) determine the birth rate and death rate of each group in a population.
   (C) determine if a population is regulated by density-dependent processes.
   (D) determine the factors that affect the size of a population.

2. A population's carrying capacity
   (A) may change as environmental conditions change.
   (B) can be accurately calculated using the logistic growth model.
   (C) increases as the per capita population growth rate decreases.
   (D) can never be exceeded.

3. Scientific study of the population cycles of the snowshoe hare and its predator, the lynx, has revealed that
   (A) predation is the dominant factor affecting prey population cycling.
   (B) hares and lynx are so mutually dependent that each species cannot survive without the other.
   (C) both hare and lynx population sizes are affected mainly by abiotic factors.
   (D) the hare population is *r*-selected and the lynx population is *K*-selected.

4. Analyzing ecological footprints reveals that
(A) Earth's carrying capacity would increase if per capita meat consumption increased.
(B) current demand by industrialized countries for resources is much smaller than the ecological footprint of those countries.
(C) it is not possible for technological improvements to increase Earth's carrying capacity for humans.
(D) the ecological footprint of the United States is large because per capita resource use is high.

5. Based on current growth rates, Earth's human population in 2019 will be closest to
(A) 2.5 million.   (C) 7.8 billion.
(B) 4.5 billion.   (D) 10.5 billion.

Levels 3-4: Applying/Analyzing

6. The observation that members of a population are uniformly distributed suggests that
(A) resources are distributed unevenly.
(B) the members of the population are competing for access to a resource.
(C) the members of the population are neither attracted to nor repelled by one another.
(D) the density of the population is low.

7. According to the logistic growth equation

$$\frac{dN}{dt} = rN\frac{(K - N)}{K}$$

(A) the number of individuals added per unit time is greatest when N is close to zero.
(B) the per capita population growth rate increases as N approaches K.
(C) population growth is zero when N equals K.
(D) the population grows exponentially when K is small.

8. During exponential growth, a population always
(A) has a constant per capita population growth rate.
(B) quickly reaches its carrying capacity.
(C) cycles through time.
(D) loses some individuals to emigration.

9. Which of the following statements about human populations in industrialized countries is correct?
(A) Birth rates and death rates are high.
(B) Average family size is relatively large.
(C) The population has undergone the demographic transition.
(D) The survivorship curve is Type II.

Levels 5-6: Evaluating/Creating

10. INTERPRET THE DATA To estimate which age cohort in a population of females produces the most female offspring, you need information about the number of offspring produced per capita within that cohort and the number of individuals alive in the cohort. Make this estimate for Belding's ground squirrels by multiplying the number of females alive at the start of the year (column 2 in Table 53.1) by the average number of female offspring produced per female (column 5 in Table 53.1). Draw a bar graph with female age in years on the *x*-axis (0–1, 1–2, and so on) and total number of female offspring produced for each age cohort on the *y*-axis. Which cohort of female Belding's ground squirrels produces the most female young?

11. EVOLUTION CONNECTION Contrast the selective pressures operating in high-density populations (those near the carrying capacity, K) versus low-density populations.

12. SCIENTIFIC INQUIRY You are testing the hypothesis that increased population density of a particular plant species increases the rate at which a pathogenic fungus infects the plant. Because the fungus causes visible scars on the leaves, you can easily determine whether a plant is infected. Design an experiment to test your hypothesis. Describe your experimental and control groups, how you would collect data, and what results you would see if your hypothesis is correct.

13. SCIENCE, TECHNOLOGY, AND SOCIETY Some people regard the rapid population growth of less industrialized countries as our most serious environmental problem. Others think that the population growth in industrialized countries, though smaller, is actually a greater environmental threat. What problems result from population growth in (a) less industrialized countries and (b) industrialized nations? Which do you think is a greater threat, and why?

14. WRITE ABOUT A THEME: INTERACTIONS In a short essay (100–150 words), identify the factor or factors in Figure 53.17 that you think may ultimately be most important for density-dependent population regulation in humans, and explain your reasoning.

15. SYNTHESIZE YOUR KNOWLEDGE

Locusts (grasshoppers in the family Acrididae) undergo cyclic population outbreaks, leading to massive swarms such as this one in the Canary Islands, off the west coast of Africa. Of the mechanisms of density-dependent regulation shown in Figure 53.17, choose the two that you think most apply to locust swarms, and explain why.

For selected answers, see Appendix A.

54 Community Ecology

KEY CONCEPTS

54.1 Interactions between species can help, harm, or have no effect on the individuals involved *p. 1215*

54.2 Diversity and trophic structure characterize biological communities *p. 1222*

54.3 Disturbance influences species diversity and composition *p. 1228*

54.4 Biogeographic factors affect community diversity *p. 1231*

54.5 Pathogens alter community structure locally and globally *p. 1234*

Study Tip

Make a table: As you read the chapter, build a table listing factors that can affect community structure. Add in figure numbers or page numbers that give examples of these factors.

Factor affecting community structure	What aspect of community structure is affected?	Example
Competition	Locations where a species can live in a community	Competition between two barnacle species (Figure 54.3)

→ Go to Mastering Biology

For Students (in eText and Study Area)
- Get Ready for Chapter 54
- Figure 54.17 Walkthrough: An Antarctic Marine Food Web
- HHMI Videos: Moth Mimicry: Using Ultrasound to Avoid Bats; Some Animals Are More Equal Than Others: Keystone Species and Trophic Cascades; Trophic Cascades in Salt Marsh Ecosystems; Riverine Food Webs: How Flow Rates Affect Biomass

For Instructors to Assign (in Item Library)
- Tutorial: Make Connections: Character Displacement and Modes of Selection
- Interpreting Data: Bird Populations During Old-Field Succession

Figure 54.1 Although this moray eel could easily eat the cleaner wrasse in its mouth, the eel remains still, allowing the wrasse to eat parasites living in the eel's mouth. The wrasse and the eel—and the parasites in the eel's mouth—live together in a **community**, a group of populations of different species living in close enough proximity to interact.

What are some factors that influence the structure of a community?

Foundation species
Species that are large or abundant, such as the corals that form this reef, may affect community structure by providing habitat and food for other organisms, like the parrotfish below.

Interactions between species
Predation and other interactions affect the number of species in a community and the particular species that are present. Interactions can be classified in three categories:
- **Competition**: negatively affects both organisms involved
- **Exploitation**: benefits one organism while harming the other
- **Positive interactions**: benefit one or both organisms without harming either

Disturbances
Marine heat waves (resulting in coral bleaching, above), storms, human activities, and other disturbances can affect a community by removing organisms or altering resource availability.

In the interaction shown in **Figure 54.1**, both organisms benefit: The cleaner wrasse gains access to a supply of food, and the moray eel is freed of parasites that might weaken it or spread disease. However, other interactions between members of different species can harm one of the participants, and still others can reduce the reproduction and survival of both participants. We'll begin this chapter by examining ecological interactions between the members of different species that live together in a community. We'll then turn to factors that affect **community structure**—the number of species found in a community, the particular species that are present, and the relative abundance of these species.

CONCEPT 54.1

Interactions between species can help, harm, or have no effect on the individuals involved

Some key relationships in the life of an organism are its interactions with individuals of other species in the community. These **interspecific interactions** include competition, predation, herbivory, parasitism, mutualism, and commensalism. In this section, we'll define and describe each of these interactions, grouping them according to whether they have positive (+) or negative (−) effects on the survival and reproduction of individuals engaged in the interaction.

For example, predation is a +/− interaction, with a positive effect on the survival and reproduction of members of the predator population and a negative effect on members of the prey population. Mutualism is a +/+ interaction in which the survival and reproduction of individuals of each species is increased in the presence of the other. A zero (0) indicates that the members of a species are not affected by the interaction.

We'll consider three broad categories of ecological interactions: competition (−/−), exploitation (+/−), and positive interactions (+/+ or +/0). As you study examples of these interactions, bear in mind that their effects can change over time. For example, an interaction that typically benefits individuals of both species may at times continue to benefit one of them, while harming or having no effect on the other.

Competition

Competition is a −/− interaction that occurs when individuals of different species each use a resource that limits the survival and reproduction of both individuals. (As described in Figure 53.18, competition for resources can also occur between members of the same species; this case is sometimes distinguished as *intraspecific competition*.) Weeds growing in a garden compete with garden plants for soil nutrients and water. Lynx and foxes in the northern forests of Alaska and Canada compete for prey such as snowshoe hares. In contrast, some resources, such as oxygen, are rarely in short supply on land; most terrestrial species use this resource but do not usually compete for it.

Competitive Exclusion

What happens in a community when individuals of two species compete for limited resources? In 1934, Russian ecologist G. F. Gause studied this question using laboratory experiments with two closely related ciliate species, *Paramecium aurelia* and *Paramecium caudatum* (see Figure 28.19a). He cultured the species under stable conditions, adding a constant amount of food each day. When Gause grew the two species separately, each population increased rapidly in number and then leveled off at the apparent carrying capacity of the culture (see Figure 53.10a for an illustration of the logistic growth of a *Paramecium* population). But when Gause grew the two species together, *P. caudatum* became extinct in the culture. Gause inferred that *P. aurelia* had a competitive edge in obtaining food. More generally, his results led him to conclude that two species whose members compete for the same limiting resources cannot coexist permanently in the same place. In the absence of disturbance, one species will use the resources more efficiently and reproduce more rapidly than the other. Even a slight reproductive advantage will eventually lead to local elimination of the inferior competitor, an outcome called **competitive exclusion**.

Ecological Niches and Natural Selection

EVOLUTION Competition for limited resources can cause evolutionary change in populations. One way to examine how this occurs is to focus on an organism's **ecological niche**, the specific set of biotic and abiotic resources that an organism uses in its environment. The niche of a tropical tree lizard, for instance, includes the temperature range it tolerates, the size of branches on which it perches, the time of day when it is active, and the sizes and kinds of insects it eats. Such factors define the lizard's niche, or ecological role—how it fits into an ecosystem.

We can use the niche concept to restate the principle of competitive exclusion: Two species cannot coexist permanently in a community if their niches are identical. However, ecologically similar species *can* coexist in a community if one or more significant differences in their niches arise through time. Evolution by natural selection can result in one of the species using a different set of resources or similar resources at different times of the day or year. The differentiation of niches that enables similar species to coexist in a community is called **resource partitioning (Figure 54.2)**.

As a result of competition, a species' *fundamental niche*, which is the niche potentially occupied by that species, is often different from its *realized niche*, the portion of its fundamental niche that it actually occupies. Ecologists can identify the fundamental niche of a species by testing the range of conditions in which it grows and reproduces in the absence of competitors. They can also test whether a potential

▼ Figure 54.2 Resource partitioning among Dominican Republic lizards. Seven species of *Anolis* lizards live in close proximity, and all feed on insects and other small arthropods. However, competition for food is reduced because each lizard species has a different preferred perch, thus occupying a distinct niche.

A. distichus perches on fence posts and other sunny surfaces.

A. insolitus usually perches on shady branches.

A. ricordii

A. insolitus
A. christophei

A. aliniger
A. distichus

A. cybotes
A. etheridgei

competitor limits a species' realized niche by removing the competitor and seeing if the first species expands into the newly available space. The classic experiment depicted in **Figure 54.3** clearly showed that competition between individuals of two barnacle species kept one species from occupying part of its fundamental niche.

Species can partition their niches not just in space, as lizards and barnacles do, but in time as well. The common spiny mouse (*Acomys cahirinus*) and the golden spiny mouse (*A. russatus*; **Figure 54.4**) live in rocky habitats of the Middle East and Africa, using similar microhabitats and food sources. Where they coexist, *A. cahirinus* is nocturnal (active at night), while *A. russatus* is diurnal (active during the day). Surprisingly, laboratory research showed that *A. russatus* is naturally nocturnal. To be active during the day, it must override its biological clock in the presence of *A. cahirinus*. When researchers in Israel removed all *A. cahirinus* individuals from a site in the species' natural habitat, *A. russatus* individuals at that site became nocturnal, consistent with the laboratory results. This change in behavior suggests that these individuals were competing for resources and that partitioning of their active time helps them coexist.

◀ **Figure 54.4 The golden spiny mouse (*Acomys russatus*).**

Can a species' niche be influenced by competition?

Experiment Ecologist Joseph Connell studied two barnacle species—*Chthamalus stellatus* and *Balanus balanoides*—that have a stratified distribution on rocks along the coast of Scotland. *Chthamalus* is usually found higher on the rocks than *Balanus*. To determine whether the distribution of *Chthamalus* is the result of competition with *Balanus*, Connell removed *Balanus* from the rocks at several sites.

Chthamalus

Balanus

High tide

Chthamalus realized niche

Balanus realized niche

Ocean

Low tide

Results *Chthamalus* spread into the region formerly occupied by *Balanus*.

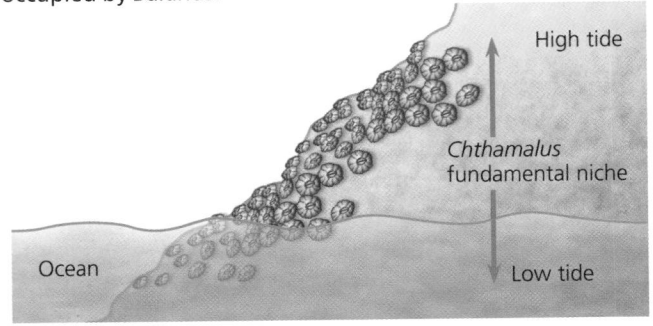

High tide

Chthamalus fundamental niche

Ocean

Low tide

Conclusion Competition makes the realized niche of *Chthamalus* much smaller than its fundamental niche.

Data from J. H. Connell, The influence of interspecific competition and other factors on the distribution of the barnacle *Chthamalus stellatus*, *Ecology* 42:710–723 (1961).

→ **Instructors:** A related Experimental Inquiry Tutorial can be assigned in **Mastering Biology**.

WHAT IF? *Other observations showed that Balanus cannot survive high on the rocks because it dries out during low tides. How would Balanus's realized niche compare with its fundamental niche?*

Character Displacement

Closely related species whose populations are sometimes allopatric (geographically separate; see Concept 24.2) and sometimes sympatric (geographically overlapping) provide additional evidence of how competition affects communities. In some cases, the allopatric populations of such species are morphologically similar and use similar resources. By contrast, sympatric populations, which would potentially compete for resources, show differences in body structures and in the resources they use. This tendency for characteristics to diverge more in sympatric than in allopatric populations of two species is called **character displacement**. An example of character

▼ **Figure 54.5 Character displacement: indirect evidence of past competition.** Allopatric populations of *Geospiza fuliginosa* and *Geospiza fortis* on Los Hermanos and Daphne Islands have similar beak morphologies (top two graphs) and presumably eat similarly sized seeds. However, where the two species are sympatric on Floreana and San Cristóbal, *G. fuliginosa* has a shallower, smaller beak and *G. fortis* a deeper, larger one (bottom graph), adaptations that favor eating different-sized seeds.

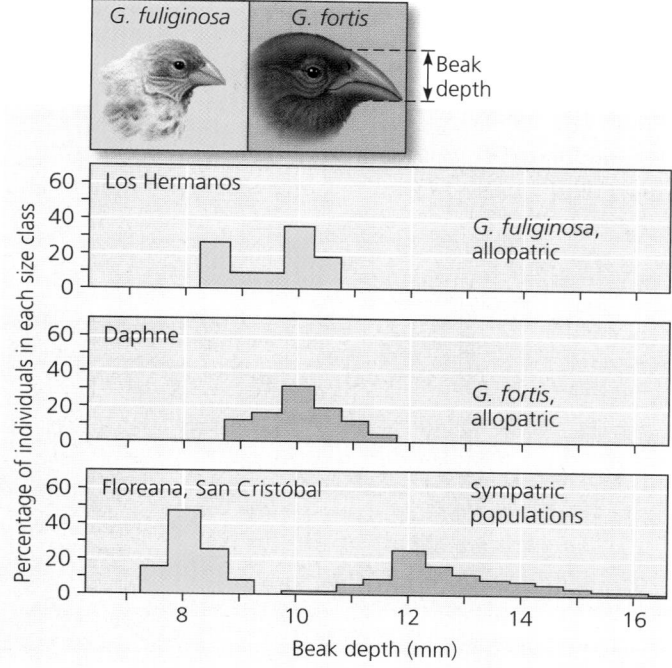

WHAT IF? *If* G. fuliginosa *became extinct on Floreana, predict how beak depth would change over time in Floreana's* G. fortis *population. Explain your reasoning.*

displacement can be seen in two species of Galápagos finches, *Geospiza fuliginosa* and *Geospiza fortis*: Beak depths in these species are similar in allopatric populations but have diverged considerably in sympatric populations **(Figure 54.5)**.

Exploitation

All nonphotosynthetic organisms must eat, and all organisms are at risk of being eaten. As a result, much of the drama in nature involves **exploitation**, a term for any type of +/− interaction in which individuals of one species benefit by feeding on (and thereby harming) individuals of the other species. Exploitative interactions include predation, herbivory, and parasitism.

Predation

Predation is a +/− interaction in which an individual of one species, the predator, kills and eats an individual of the other species, the prey. Though the term *predation* generally elicits such images as a lion attacking and eating an antelope, it applies to a wide range of interactions. A rotifer (a tiny aquatic animal that is smaller than many unicellular protists) that kills a protist by eating it can also be considered a predator. Because eating and avoiding being eaten are prerequisites to reproductive success, the adaptations of both predators and prey tend to be refined through natural selection. In the **Scientific Skills Exercise**,

---

Making a Bar Graph and a Scatter Plot

Can a Native Predator Species Adapt Rapidly to an Introduced Prey Species? Cane toads (*Bufo marinus*) were introduced to Australia in 1935 in a failed attempt to control an insect pest. Since then, the toads have spread across northeastern Australia, with a population of over 200 million today. Cane toads have glands that produce a toxin that is poisonous to snakes and other potential predators. In this exercise, you will graph and interpret data from a two-part experiment conducted to determine whether native Australian predators have developed resistance to the cane toad toxin.

How the Experiment Was Done In part 1, researchers collected 12 red-bellied black snakes (*Pseudechis porphyriacus*) from areas where cane toads had existed for 40–60 years and another 12 from areas free of cane toads. They recorded the percentage of snakes from each area that ate either a freshly killed native frog (*Limnodynastes peronii*, a species the snakes commonly eat) or a freshly killed cane toad from which the toxin gland had been removed (making the toad nonpoisonous). In part 2, researchers collected snakes from areas where cane toads had been present for 5–60 years. To assess how cane toad toxin affected the physiological activity of these snakes, they injected small amounts of the toxin into the snakes' stomachs and measured the snakes' swimming speed in a small pool.

Data from the Experiment, Part 1

Type of Prey Offered	% of Snakes from Each Area That Ate Each Type of Prey	
	Area with Cane Toads Present for 40–60 Years	Area with No Cane Toads
Native frog	100	100
Cane toad	0	50

Data from the Experiment, Part 2

Number of Years Cane Toads Had Been Present in the Area	5	10	10	20	50	60	60	60	60	60
% Reduction in Snake Swimming Speed	52	19	30	30	5	5	9	11	12	22

Data from B. L. Phillips and R. Shine, An invasive species induces rapid adaptive change in a native predator: cane toads and black snakes in Australia, *Proceedings of the Royal Society B* 273:1545–1550 (2006).

INTERPRET THE DATA

1. Make a bar graph of the data in part 1. (For additional information about graphs, see the Scientific Skills Review in Appendix D.)
2. What do the data represented in the graph suggest about the effects of cane toads on the predatory behavior of red-bellied black snakes in areas where the toads are and are not currently found?
3. Suppose a novel enzyme that deactivates the cane toad toxin evolved in a snake population exposed to cane toads. If the researchers repeated part 1 of this study, predict how the results would change.
4. Identify the dependent and independent variables in part 2 and make a scatter plot. What conclusion would you draw about whether exposure to cane toads is having a selective effect on the snakes? Explain.
5. Explain why a bar graph is appropriate for presenting the data in part 1 and a scatter plot is appropriate for the data in part 2.

➡ **Instructors:** A version of this Scientific Skills Exercise can be assigned in **Mastering Biology**.

you can interpret data on the impact of natural selection for a specific predator-prey interaction.

Many important feeding adaptations of predators are obvious and familiar. Most predators have acute senses that enable them to find and identify potential prey. Rattlesnakes and other pit vipers, for example, find their prey with a pair of heat-sensing organs located between their eyes and nostrils (see Figure 50.7b). Owls have characteristically large eyes that help them see prey at night. Many predators also have adaptations such as claws, fangs, or poison that help them catch and subdue their food. Predators that pursue their prey are generally fast and agile, whereas those that lie in ambush are often disguised in their environments.

Just as predators possess adaptations for capturing prey, potential prey have adaptations that help them avoid being eaten. In animals, these adaptations include behavioral defenses such as hiding, fleeing, and forming herds or schools. Active self-defense is less common, though some large grazing mammals vigorously defend their young from predators.

Animals also display a variety of morphological and physiological defensive adaptations. Mechanical or chemical defenses protect species such as porcupines and skunks (**Figure 54.6a** and **b**). Some animals, such as the European fire salamander, can synthesize toxins; others accumulate toxins passively from the plants they eat. Animals with effective chemical defenses often exhibit bright

▼ **Figure 54.6 Examples of defensive adaptations in animals.**

(a) Mechanical defense

▶ Porcupine

(b) Chemical defense

▶ Skunk

(c) Aposematic coloration: warning coloration

▶ Poison dart frog

(d) Cryptic coloration: camouflage

▶ Canyon tree frog

(e) Batesian mimicry: A harmless species mimics a harmful one.

▲ Venomous green parrot snake

◀ Nonvenomous hawkmoth larva

(f) Müllerian mimicry: Two unpalatable species mimic each other.

▲ Yellow jacket

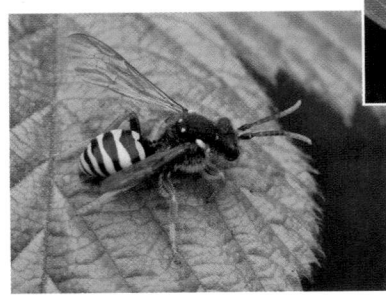

◀ Cuckoo bee

MAKE CONNECTIONS *Explain how natural selection could increase the resemblance of a harmless species to a distantly related harmful species. In addition to selection, what else could account for a harmless species resembling a closely related harmful species? (See Concept 22.2.)*

→ **Mastering Biology** HHMI Video: **Moth Mimicry: Using Ultrasound to Avoid Bats**

 hhmi BioInteractive

aposematic coloration, or warning coloration, such as that of poison dart frogs **(Figure 54.6c)**. Such coloration seems to be adaptive because predators often avoid brightly colored prey. **Cryptic coloration**, or camouflage, makes prey difficult to see **(Figure 54.6d)**.

Some prey species are protected by their resemblance to other species. For example, in **Batesian mimicry**, a palatable or harmless species mimics an unpalatable or harmful species to which it is not closely related. The larva of the hawkmoth *Hemeroplanes ornatus* puffs up its head and thorax when disturbed, looking like the head of a small venomous snake **(Figure 54.6e)**. In this case, the mimicry even involves behavior; the larva weaves its head back and forth and hisses like a snake. Such cases of Batesian mimicry are thought to result from natural selection, as individuals in the harmless species that happen to more closely resemble the harmful one are avoided by predators who have learned not to eat the harmful ones. Over time, closer and closer resemblance to the harmful species evolves. In **Müllerian mimicry**, two or more unpalatable species, such as the cuckoo bee and yellow jacket, resemble each other **(Figure 54.6f)**. Presumably, the more unpalatable prey there are, the faster predators learn to avoid prey with that particular appearance. Mimicry has also evolved in many predators. The mimic octopus *Thaumoctopus mimicus* **(Figure 54.7)** can take on the appearance and movement of

▼ **Figure 54.7 The mimic octopus. (a)** After hiding six of its tentacles in a hole in the seafloor, the octopus waves its other two tentacles to mimic a sea snake. **(b)** Flattening its body and arranging its arms to trail behind, the octopus mimics a flounder (a flat fish). **(c)** It can mimic a stingray by flattening most of its tentacles alongside its body while allowing one tentacle to extend behind it.

(a) Mimicking a sea snake

(b) Mimicking a flounder

(c) Mimicking a stingray

more than a dozen marine animals, including crabs, sea stars, sea snakes, fish, and stingrays. This octopus's ability to mimic other animals enables it to approach prey—for example, imitating a crab to approach another crab and eat it. The octopus also can defend itself from predators through mimicry. When attacked by a damselfish, the octopus quickly mimics a banded sea snake, a known predator of the damselfish.

> → **Mastering Biology Interview with Tracy Langkilde: Studying the evolution of novel defensive adaptations in lizards**

Herbivory

Herbivory is an exploitative (+/−) interaction in which an organism—an herbivore—eats parts of a plant or alga, thereby harming it but usually not killing it. While large mammalian herbivores such as cattle, sheep, and water buffalo may be most familiar, most herbivores are actually invertebrates, such as grasshoppers, caterpillars, and beetles. In the ocean, herbivores include sea urchins, some tropical fishes, and certain mammals **(Figure 54.8)**.

Like predators, herbivores have many specialized adaptations. Many herbivorous insects have chemical sensors on their feet that enable them to distinguish between plants based on their toxicity or nutritional value. Some mammalian herbivores, such as goats, use their sense of smell to examine plants, rejecting some and eating others. They may also eat just a specific part of a plant, such as the flowers. Many herbivores also have specialized teeth or digestive systems adapted for processing vegetation (see Concept 41.4).

Unlike animals, plants cannot run away to avoid being eaten. Instead, a plant's arsenal against herbivores may feature chemical toxins or structures such as spines and thorns.

▼ **Figure 54.8 An herbivorous marine mammal.** This West Indian manatee (*Trichechus manatus*) in Florida is grazing on *Hydrilla*, an introduced plant species.

Among the plant compounds that serve as chemical defenses are the poison strychnine, produced by the tropical vine *Strychnos toxifera*; nicotine, from the tobacco plant; and tannins, from a variety of plant species. Compounds that are not toxic to humans but may be distasteful to many herbivores are responsible for the familiar flavors of cinnamon, cloves, and peppermint. Certain plants produce chemicals that cause abnormal development in some insects that eat them. For more examples of how plants defend themselves, see Make Connections Figure 39.27, "Levels of Plant Defenses Against Herbivores."

Parasitism

Parasitism is a +/− exploitative interaction in which one organism, the **parasite**, derives its nourishment from another organism, its **host**, which is harmed in the process. Parasites that live within the body of their host, such as tapeworms, are called **endoparasites**; parasites that feed on the external surface of a host, such as ticks and lice, are called **ectoparasites**. Some ecologists have estimated that at least one-third of all species on Earth are parasites. In one particular type of parasitism, parasitoid insects—usually small wasps—lay eggs on or in living hosts. The larvae then feed on the body of the host, eventually killing it.

Many parasites have complex life cycles involving multiple hosts. The blood fluke, which currently infects approximately 200 million people around the world, requires two hosts at different times in its development: humans and freshwater snails (see Figure 33.10). Some parasites change the behavior of their current host in ways that increase the likelihood that the parasite will reach its next host. For instance, crustaceans that are parasitized by acanthocephalan (spiny-headed) worms leave protective cover and move into the open, where they are more likely to be eaten by the birds that are the second host in the worm's life cycle.

Parasites can significantly affect the survival, reproduction, and density of their host population, either directly or indirectly. For example, ticks that feed as ectoparasites on moose can weaken their hosts by withdrawing blood and causing hair breakage and loss. In their weakened condition, the moose have a greater chance of dying from cold stress or predation by wolves.

Positive Interactions

While nature abounds with dramatic and gory examples of exploitative interactions, ecological communities are also heavily influenced by **positive interactions**, a term that refers to a +/+ or +/0 interaction between members of two species in which at least one individual benefits and neither is harmed. Positive interactions include mutualism and commensalism. As we'll see, positive interactions can affect the diversity of species found in an ecological community.

Mutualism

Mutualism is a +/+ interaction that benefits individuals of both of the interacting species. Mutualisms are common in nature, as illustrated by examples seen in previous chapters, including cellulose digestion by microorganisms in the digestive systems of termites and ruminant mammals, animals that pollinate flowers or disperse seeds, nutrient exchange between fungi and plant roots in mycorrhizae, and photosynthesis by unicellular algae in corals. In some mutualisms, such as the acacia-ant example shown in **Figure 54.9**, each of the interacting individuals depends on the other for their survival and reproduction. In other mutualisms, however, both individuals can survive on their own.

Typically, both partners in a mutualism incur costs as well as benefits. In mycorrhizae, for example, the plant often

▼ **Figure 54.9 Mutualism between acacia trees and ants.**

(a) Certain species of acacia trees (genus *Acacia*) in Central and South America have hollow thorns (not shown) that are home to stinging ants of the genus *Pseudomyrmex*. The ants feed on nectar produced by the tree and on protein-rich swellings (yellow in the photograph) at the tips of leaflets.

(b) The acacia benefits because the pugnacious ants, which attack anything that touches the tree, remove fungal spores, small herbivores, and debris. They also clip vegetation that grows close to the acacia.

transfers carbohydrates to the fungus, while the fungus transfers limiting nutrients, such as phosphorus, to the plant. Each partner benefits, but each partner also experiences a cost: It transfers materials that it could have used to support its own growth and metabolism. For an interaction to be considered a mutualism, the benefits to each partner must exceed the costs. When this is not the case, the mutualism may break down, at least temporarily. In some mycorrhizae, for example, the plant may cease to supply carbohydrates to its fungal partner when soil nutrients are plentiful—a change that occurs because the cost of supporting the fungus has become greater than the benefits the fungus can provide.

Commensalism

An interaction that benefits the individuals of one of the interacting species but neither harms nor helps the individuals of the other species is called **commensalism** (+/0). Like mutualism, commensalisms are common in nature. For instance, many wildflowers that grow best in low light levels are found only in shaded, forest floor environments. Such shade-tolerant "specialists" depend entirely on the trees that tower above them—the trees provide their dim habitat. Yet the survival and reproduction of the trees are not affected by these wildflowers. Thus, these species are involved in a +/0 interaction in which the wildflowers benefit and the trees are not affected.

In another example of a commensalism, cattle egrets feed on insects flushed out of the grass by grazing bison, cattle, horses, and other herbivores **(Figure 54.10)**. Because the birds increase their feeding rates when following the herbivores, they clearly benefit from the association. Much of the time, the herbivores are not affected by the birds. At times, however, the herbivores too may derive some benefit;

▼ **Figure 54.10 Commensalism between cattle egrets and African buffalo.**

▼ **Figure 54.11 Positive interactions in New England salt marshes.** When black rush (*Juncus gerardii*) is present, soil salt concentrations drop and soil oxygen levels rise, increasing the number of plant species that can live in the marsh.

(a) Salt marsh with *Juncus* (foreground)

(b)

for example, the birds may remove and eat ticks and other ectoparasites from the herbivores' skin, or they may warn the herbivores of a predator's approach. This example provides another illustration of a key point about ecological interactions: Their effects can change over time. In this case, an interaction whose effects are typically +/0 (commensalism) may at times become +/+ (mutualism).

Positive interactions can have major effects on ecological communities. For instance, the black rush *Juncus gerardii* alters soil conditions in ways that benefit other plant species in New England salt marshes **(Figure 54.11a)**. *Juncus* shades the soil surface, which reduces evaporation and thus lowers the concentration of salt in the soil. The presence of *Juncus* also increases soil oxygen levels; this occurs because some oxygen leaks into the soil as *Juncus* transports oxygen to its belowground tissues. In one study, when *Juncus* was removed from areas of the marsh, those areas supported 50% fewer plant species **(Figure 54.11b)**.

Like positive interactions, competition and exploitation (predation, herbivory, and parasitism) also can strongly affect ecological communities, as examples throughout the rest of this chapter will show.

CONCEPT CHECK 54.1

1. Explain how competition, predation, and mutualism differ in their effects on members of the two interacting species.

2. According to the principle of competitive exclusion, what outcome is expected when two species with identical niches compete for a resource? Why?

3. **MAKE CONNECTIONS** Figure 24.14 illustrates how a hybrid zone can change over time. Imagine that two finch species colonize a new island and are capable of hybridizing (mating and producing viable offspring). The island contains two plant species, one with large seeds and one with small seeds, growing in isolated habitats. If the two finch species specialize in eating different plant species, would reproductive barriers be reinforced, weakened, or unchanged in this hybrid zone? Explain.

For suggested answers, see Appendix A.

Diversity and trophic structure characterize biological communities

Ecological communities can be characterized by certain general attributes, including how diverse they are and the feeding relationships of their species. In some cases, as you'll read, a few species exert strong control on a community's *structure*—the number, identity, and relative abundance of its species.

Species Diversity

The **species diversity** of a community—the variety of different kinds of organisms that make up the community—has two components. One is **species richness**, the number of different species in the community. The other is the **relative abundance** of the different species, the proportion each species represents of all individuals in the community.

Imagine two small forest communities, each with 100 individuals distributed among four tree species (A, B, C, and D) as follows:

> Community 1: 25A, 25B, 25C, 25D
> Community 2: 80A, 5B, 5C, 10D

The species richness is the same for both communities because they both contain four species of trees, but the relative abundance is very different **(Figure 54.12)**. You would easily notice the four types of trees in community 1, but

▼ **Figure 54.12 Which forest is more diverse?** Ecologists would say that community 1 has greater species diversity, a measure that includes both species richness and relative abundance.

Community 1
A: 25% B: 25% C: 25% D: 25%

Community 2
A: 80% B: 5% C: 5% D: 10%

unless you looked carefully, you might see only the abundant species A in the second forest. Most observers would intuitively describe community 1 as the more diverse of the two communities.

Ecologists use many tools to compare the diversity of communities. They often calculate indexes of diversity based on species richness and relative abundance. One widely used index is the **Shannon diversity index** (H):

$$H = -(p_A \ln p_A + p_B \ln p_B + p_C \ln p_C + \ldots)$$

where A, B, C ... are the species in the community, p is the relative abundance of each species, and ln is the natural logarithm; the ln of each value of p can be determined using the "ln" key on a scientific calculator. A higher value of H indicates a more diverse community. Let's use this equation to calculate the Shannon diversity index of the two communities in Figure 54.12. For community 1, $p = 0.25$ for each species, so

$$H = -4(0.25 \ln 0.25) = 1.39.$$

For community 2,

$$H = -[0.8 \ln 0.8 + 2(0.05 \ln 0.05) + 0.1 \ln 0.1] = 0.71.$$

These calculations confirm our intuitive description of community 1 as more diverse.

Determining the number and relative abundance of species in a community can be challenging. Because most species in a community are relatively rare, it may be hard to obtain a sample size large enough to be representative. It can also be difficult to identify some of the species in the community. If an unknown organism cannot be identified on the basis of morphology alone, it is useful to compare all or part of its genome to a reference database of DNA sequences from known organisms. For example, although the two samples of red algae shown in **Figure 54.13** might appear to be two different species, comparing their sequences of a short standardized section of DNA—a *DNA "barcode"*—to a reference database shows that they belong to the same species. Researchers are increasingly using DNA sequencing for species identification as it becomes cheaper and as DNA sequences from more organisms are added to comparative databases.

It can also be difficult to census the highly mobile or less visible members of communities, such as microorganisms, deep-sea creatures, and nocturnal species. The small size of microorganisms makes them particularly difficult to sample, so ecologists now commonly use molecular tools to help determine microbial diversity **(Figure 54.14)**.

▼ **Figure 54.13 Two samples of one species, the red alga *Chondracanthus harveyanus*.**

Determining Microbial Diversity Using Molecular Tools

Application Ecologists are increasingly using molecular techniques to determine microbial diversity and richness in environmental samples. One such technique produces a DNA profile for microbial taxa based on sequence variations in the DNA that encodes the small subunit of ribosomal RNA. Researchers used this method to compare the diversity of soil bacteria in 98 habitats across North and South America to help identify environmental variables associated with high bacterial diversity.

Technique Researchers first extracted and purified DNA from the microbial community in each sample. They used the polymerase chain reaction (PCR; see Figure 20.7) to amplify the ribosomal DNA and labeled it with a fluorescent dye.

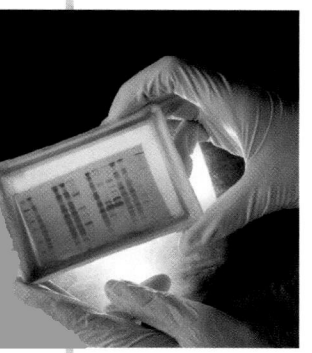

Restriction enzymes then cut the amplified, labeled DNA into fragments of different lengths, which are separated by gel electrophoresis. (A gel is shown here; see also Figure 20.6.) The number and abundance of these fragments characterize the sample's DNA profile, which was then used to calculate the Shannon diversity index (*H*) of each sample. Researchers then looked for a correlation between *H* and several environmental variables, including vegetation type, mean annual temperature and rainfall, and soil acidity.

Results The diversity of bacterial communities was related to soil pH, with the Shannon diversity index being highest in neutral soils (pH 7) and lowest in very acidic soils (pH < 5). Amazonian rain forests, which have extremely high plant and animal diversity, had the most acidic soils and the lowest bacterial diversity of the samples tested.

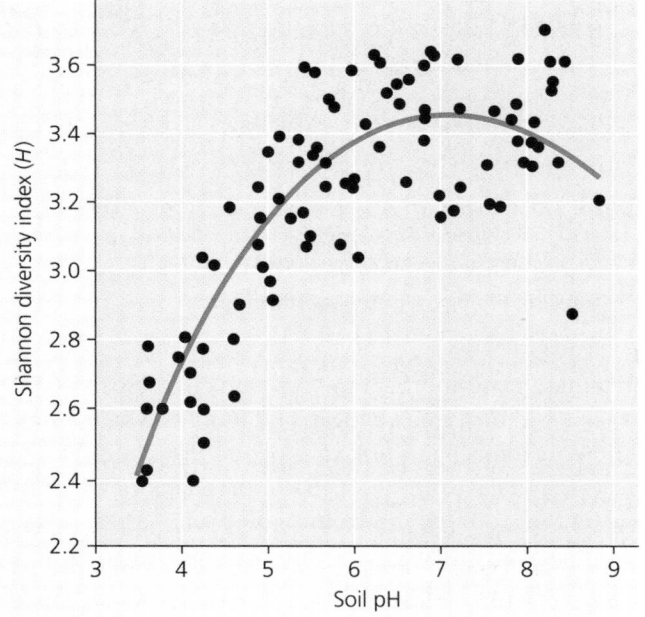

Data from N. Fierer and R. B. Jackson, The diversity and biogeography of soil bacterial communities, *Proceedings of the National Academy of Sciences USA* 103:626–631 (2006).

▶ **Figure 54.15 Study plots at the Cedar Creek Ecosystem Science Reserve, site of long-term experiments on manipulating plant diversity.**

Diversity and Community Stability

In addition to measuring species diversity, ecologists manipulate diversity in experimental communities in nature and in the laboratory. Many experiments examine the potential benefits of diversity, including increased productivity and stability of biological communities.

Researchers at the Cedar Creek Ecosystem Science Reserve, in Minnesota, have been manipulating plant diversity in experimental communities for more than three decades **(Figure 54.15)**. Higher-diversity communities generally are more productive and are better able to withstand and recover from environmental stresses, such as droughts. More diverse communities are also more stable year to year in their productivity. In one decade-long experiment, for instance, researchers at Cedar Creek created 168 plots, each containing 1, 2, 4, 8, or 16 perennial grassland species. The most diverse plots consistently produced more **biomass** (the total mass of all organisms in a habitat) each year than did the single-species plots.

Higher-diversity communities are often more resistant to *introduced species*, which are organisms that humans have moved to regions outside of the species' native range. In research conducted off the coast of Connecticut, scientists created communities of different levels of diversity consisting of sessile marine invertebrates, including tunicates (see Figure 34.5). They then examined how vulnerable these experimental communities were to an introduced tunicate. They found that the introduced tunicate was four times more likely to survive in lower-diversity communities than in higher-diversity ones. The researchers concluded that relatively diverse communities captured more of the resources available in the system, leaving fewer resources for the introduced species and thus decreasing its survival.

Trophic Structure

In addition to species diversity, the structure and dynamics of a community also depend on the feeding relationships between organisms—the **trophic structure** of the community. The transfer of chemical energy from its source in plants and other autotrophs (primary producers) through herbivores (primary consumers) to carnivores (secondary, tertiary, and

▼ Figure 54.16 Examples of terrestrial and marine food chains. The arrows trace the transfer of the chemical energy in food through the trophic levels of a community when organisms feed on one another. Decomposers, which consume the remains of organisms from all trophic levels, are not shown here.

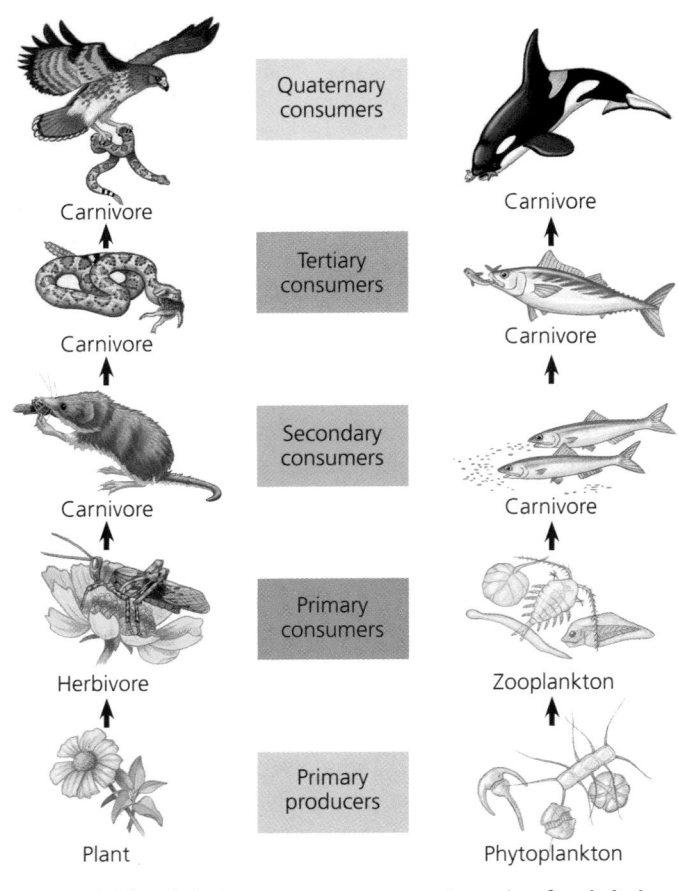

A terrestrial food chain   A marine food chain

VISUAL SKILLS *Suppose the abundance of carnivores that eat zooplankton increased greatly. Use this diagram to infer how that might affect phytoplankton abundance.*

quaternary consumers) and eventually to decomposers is referred to as a **food chain (Figure 54.16)**. The position an organism occupies in a food chain is called its **trophic level**.

Food Webs

A food chain is not an isolated unit, separate from other feeding relationships in a community. Instead, a group of food chains are linked together to form a **food web**. Ecologists diagram the trophic relationships of a community using arrows that link species according to who eats whom. In an Antarctic pelagic community, for example, the primary producers are phytoplankton, which serve as food for the dominant grazing zooplankton, especially krill and copepods, both of which are crustaceans **(Figure 54.17)**. These zooplankton species are in turn eaten by various carnivores, including other plankton, penguins, seals, fishes, and baleen whales. Squids, which are carnivores that feed on fishes and

▼ Figure 54.17 An Antarctic marine food web. Arrows follow the transfer of the chemical energy in food from the producers (phytoplankton) up through the trophic levels. For simplicity, this diagram omits decomposers. At various times over the last two centuries, humans have also played a role in the Antarctic food web as consumers of fish, krill, and whales.

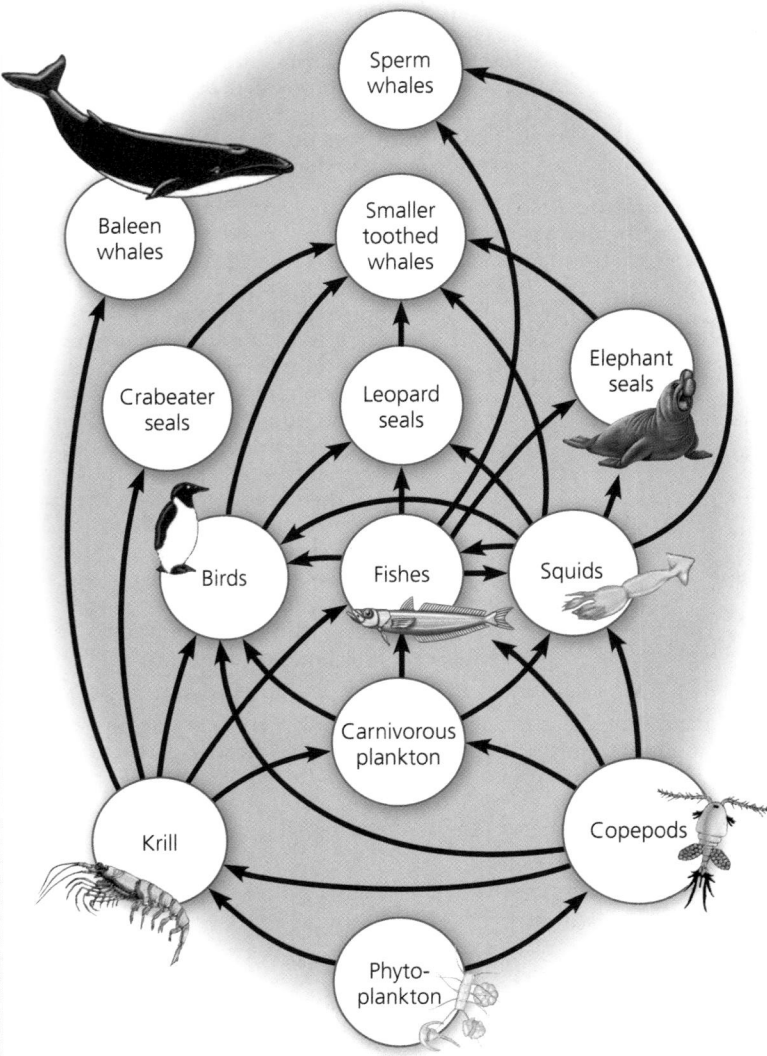

VISUAL SKILLS *For each organism in this food web, indicate the number of other kinds of organisms that it eats. Which two organisms are both predator and prey for each other?*

→ **Mastering Biology** Figure Walkthrough

zooplankton, are another important link in these food webs, as they are in turn eaten by seals and toothed whales.

How are food chains linked into food webs? A given species may weave into the web at more than one trophic level. In the food web shown in Figure 54.17, krill feed on phytoplankton as well as on other grazing zooplankton, such as copepods. Such "nonexclusive" consumers are also found in terrestrial communities. For instance, foxes are omnivores whose diet includes berries and other plant materials, herbivores such as mice, and other predators, such as weasels. Humans are among the most versatile of omnivores.

▼ **Figure 54.18 Partial food web for Chesapeake Bay estuary.**
The sea nettle (*Chrysaora quinquecirrha*) and the juvenile striped bass (*Morone saxatilis*) are the main predators of fish larvae.

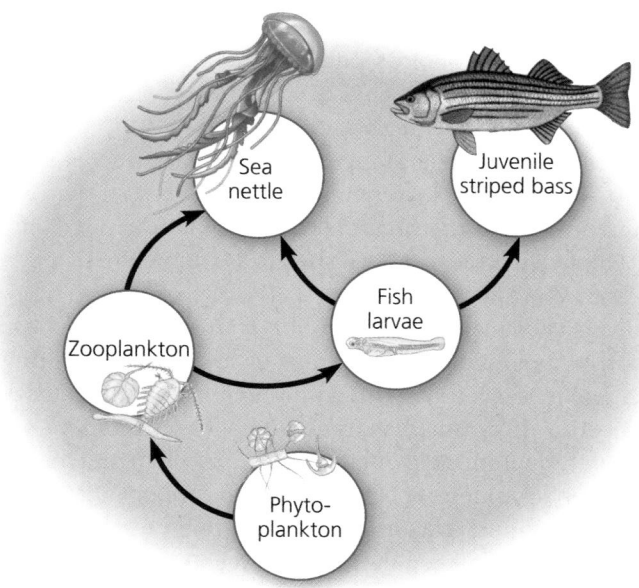

VISUAL SKILLS *Based on this food web, identify organisms that function as primary, secondary, or tertiary consumers.*

Complicated food webs can be simplified in two ways for easier study. First, species with similar trophic relationships in a given community can be grouped into broad functional groups. In Figure 54.17, more than 100 phytoplankton species are grouped as the primary producers in the food web. A second way to simplify a food web is to isolate a portion of the web that interacts very little with the rest of the community. **Figure 54.18** illustrates a partial food web for sea nettles (a type of cnidarian) and juvenile striped bass in the Chesapeake Bay estuary on the Atlantic coast of the United States.

→ **Mastering Biology** HHMI Video: Riverine Food Webs: How Flow Rates Affect Biomass

Limits on Food Chain Length

Each food chain within a food web is usually only a few links long. In the Antarctic web of Figure 54.17, there are rarely more than seven links from the producers to any top-level predator, and most chains in this web have fewer links. In fact, most food webs studied to date have chains consisting of five or fewer links.

Why do food chains tend to be relatively short? The most common explanation, known as the **energetic hypothesis**, suggests that the length of a food chain is limited by the inefficiency of energy transfer along the chain. On average, only about 10% of the energy stored in the organic matter of each trophic level is converted to organic matter at the next trophic level (see Concept 55.3). Thus, a producer level consisting of 100 kg of plant material can support about 10 kg of herbivore biomass and 1 kg of carnivore biomass. The energetic hypothesis predicts that food chains should be relatively

▼ **Figure 54.19 Test of the energetic hypothesis for the restriction of food chain length.** Researchers manipulated the productivity of experimental tree-hole communities in Queensland, Australia, by providing leaf litter input at three levels. Reducing energy input reduced food chain length, a result consistent with the energetic hypothesis.

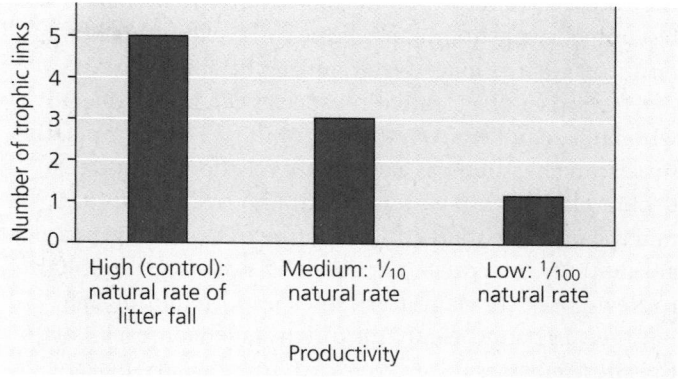

longer in habitats characterized by higher photosynthetic production, since the amount of energy stored in primary producers is greater than in habitats with lower photosynthetic production.

Ecologists tested the energetic hypothesis in experiments that mimicked the "tree-hole" communities found in tropical forests. Many trees have small branch scars that rot, forming holes in the tree trunk. These tree holes hold water and provide a habitat for tiny communities consisting of microorganisms and insects that feed on leaf litter, as well as predatory insects. **Figure 54.19** shows the results of experiments in which researchers manipulated productivity by varying the amount of leaf litter in an experiment using artificial tree holes (water-filled pots placed around the trees); previous studies had shown that the communities that colonized these pots were similar to those in natural tree holes. As predicted by the energetic hypothesis, the pots with the most leaf litter, and hence the greatest total food supply at the producer level, supported the longest food chains.

Another factor that may limit food chain length is that carnivores tend to be larger at higher trophic levels. The size of a carnivore puts an upper limit on the size of food it can take into its mouth. And except in a few cases, large carnivores cannot live on very small food items because they cannot obtain enough food in a given time to meet their metabolic needs. Among the exceptions are baleen whales, huge filter feeders with adaptations that enable them to consume enormous quantities of krill and other small organisms (see Figure 41.5).

Species with a Large Impact

Certain species have an especially large impact on the structure of entire communities because they are highly abundant or play a pivotal role in community dynamics. The impact of these species occurs through trophic interactions and their influence on the physical environment.

Foundation species have strong effects on their communities as a result of their large size or high abundance. Examples of foundation species include trees, desert shrubs such as creosote, and certain marine algae, such as kelp. Such species often have community-wide effects because they provide habitat or food. Foundation species may also be competitively dominant—superior in exploiting key resources such as space, water, nutrients, or light.

The impact of a foundation species can be discovered when it is removed from the community. For example, the American chestnut was a competitively dominant tree in deciduous forests of eastern North America before 1910, making up more than 40% of mature trees. Then humans accidentally introduced the fungal disease chestnut blight to New York City via nursery stock imported from Asia. Between 1910 and 1950, this fungus killed almost all of the chestnut trees in eastern North America. In this case,

removing the foundation species had a relatively small impact on some species but large effects on others. Oaks, hickories, beeches, and red maples that were already present in the forest increased in abundance and replaced the chestnuts. No mammals or birds seemed to have been harmed by the loss of the chestnut, but seven species of moths and butterflies that fed on the tree became extinct.

In contrast to foundation species, **keystone species** are not usually abundant in a community. They exert strong control on community structure not by numerical might but by their pivotal ecological roles. **Figure 54.20** highlights the importance of a keystone species, a sea star, in maintaining the diversity of an intertidal community. In this case, the sea star affects its community by feeding on and thereby limiting the abundance of a competitively dominant species, a mussel.

Still other organisms exert their influence on a community not through trophic interactions but by changing the physical environment. Species that create or dramatically alter their environment are called **ecosystem engineers**. A familiar ecosystem engineer is the beaver **(Figure 54.21)**. The effects of ecosystem engineers on other species can be positive or negative, depending on the needs of the other species. Some foundation species, such as trees, can also be considered ecosystem engineers because their presence modifies the physical environment in ways that create habitats on which other species depend.

▼ **Figure 54.20** Inquiry

Is *Pisaster ochraceus* a keystone species?

Experiment In rocky intertidal communities of western North America, the relatively uncommon sea star *Pisaster ochraceus* preys on mussels such as *Mytilus californianus*, a dominant competitor for space. Robert Paine, of the University of Washington, removed *Pisaster* from an area in the intertidal zone and examined the effect on species richness.

Results In the absence of *Pisaster*, species richness declined as mussels monopolized the rock face and eliminated most other invertebrates and algae. In a control area where *Pisaster* was not removed, species richness changed very little.

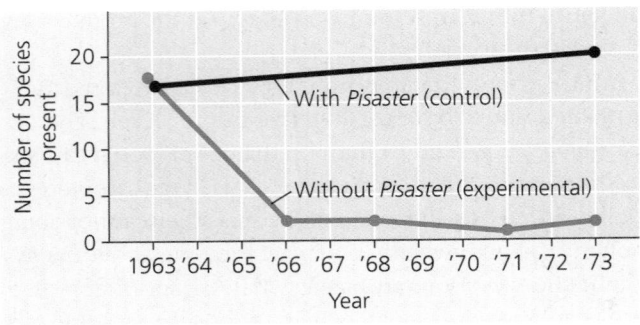

Conclusion *Pisaster* acts as a keystone species, exerting an influence on the community that is not reflected in its abundance.

Data from R. T. Paine, Food web complexity and species diversity, *American Naturalist* 100:65–75 (1966).

WHAT IF? *Suppose that an invasive fungus killed most individuals of* Mytilus *at these sites. Predict how species richness would be affected if* Pisaster *were then removed.*

▼ **Figure 54.21 Beavers as ecosystem engineers.** By felling trees, building dams, and creating ponds, beavers can transform large areas of forest into flooded wetlands.

Bottom-Up and Top-Down Controls

The ways in which adjacent trophic levels affect one another can be useful for describing community organization. These effects occur in two general ways: organisms can be controlled by what they eat ("bottom-up" control) or by what eats them ("top-down" control).

In **bottom-up control**, the abundance of organisms at each trophic level is limited by nutrient supply or the

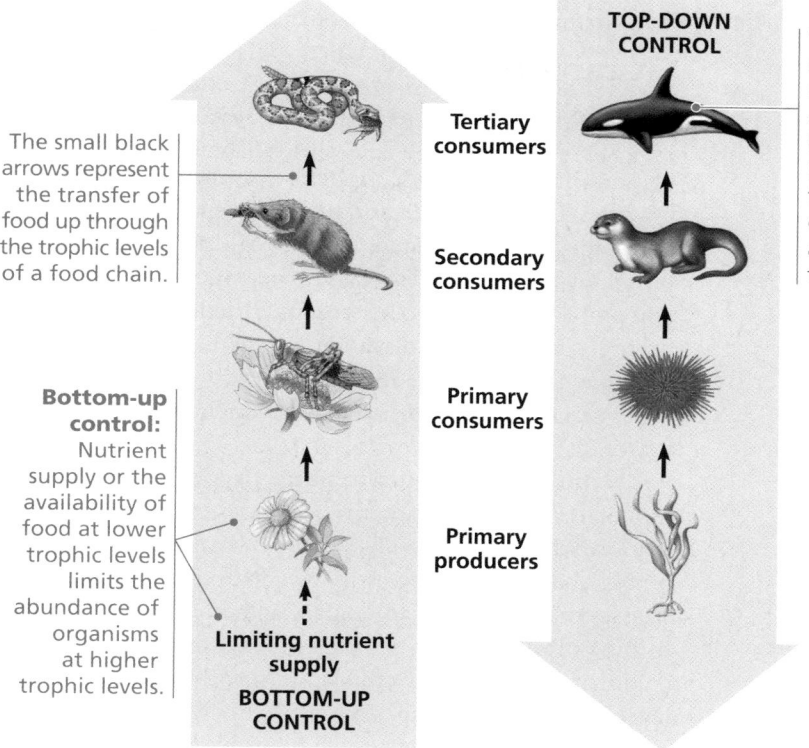

▼ **Figure 54.22 Bottom-up and top-down control.** The abundance of organisms at each trophic level can be can be controlled from the bottom up (by the availability of food or nutrients) or from the top down (by rates of consumption at higher trophic levels). Large gray arrows indicate the direction of control. Small black arrows show the transfer of food through a food chain.

The small black arrows represent the transfer of food up through the trophic levels of a food chain.

Bottom-up control: Nutrient supply or the availability of food at lower trophic levels limits the abundance of organisms at higher trophic levels.

Limiting nutrient supply

BOTTOM-UP CONTROL

Tertiary consumers

Secondary consumers

Primary consumers

Primary producers

TOP-DOWN CONTROL

Top-down control: The abundance of organisms at higher trophic levels controls the abundance of organisms at lower trophic levels.

availability of food at lower trophic levels. In this case, the supply of nutrients controls plant numbers, which control herbivore numbers, which in turn control predator numbers (**Figure 54.22**, left side). To change the structure of a bottom-up community, you would need to alter the biomass or abundance of organisms at lower trophic levels, allowing those changes to propagate up through the food web. For example, if you add nutrients to stimulate plant growth, then each of the higher trophic levels should also increase in biomass. If you change predator abundance, however, the effect should not extend down to the lower trophic levels.

In **top-down control**, the abundance of organisms at each trophic level is controlled by the abundance of consumers at higher trophic levels. Thus, in top-down control, predators limit herbivores, and herbivores limit plants (Figure 54.22, right side). In a lake community with four trophic levels, removing the top carnivores should increase the abundance of primary carnivores, decrease the number of herbivores, and increase phytoplankton abundance. The effects thus move down the trophic structure as alternating +/− effects.

Ecologists have applied top-down control to improve water quality in lakes with a high abundance of algae. For example, in lakes with three trophic levels, removing fish should improve water quality by increasing zooplankton density, thereby decreasing algal populations **(Figure 54.23)**. In lakes

with four trophic levels, adding top predators should have the same effect.

Ecologists in Finland used this approach to help purify Lake Vesijärvi, a large lake that was polluted with city sewage and industrial waste until 1976. After pollution controls reduced these inputs, the lake's water quality began to improve. In 1986, however, massive blooms of cyanobacteria began occurring. These blooms coincided with an increase in the population of roach, a fish species that eats zooplankton, which otherwise keep cyanobacteria in check. To reverse these changes, ecologists removed nearly a million kilograms of fish from the lake between 1989 and 1993, reducing roach abundance by about 80%. At the same time, they added a fourth trophic level by stocking the lake with pike perch, a predatory fish that eats roach. The water cleared up, and the last cyanobacterial bloom was in 1989. Ecologists continue to monitor the lake for cyanobacterial blooms and low oxygen availability, but the lake has remained clear, even though roach removal ended in 1993.

 Mastering Biology HHMI Videos: Some Animals Are More Equal Than Others: Keystone Species and Trophic Cascades • Trophic Cascades in Salt Marsh Ecosystems

hhmi BioInteractive

▼ **Figure 54.23 Restoring water quality in a lake with top-down control.** Decreasing the abundance of fish that ate zooplankton results in a decrease in the biomass of algae, improving water quality. Arrow thickness indicates the relative strength of each top-down control.

Lake Condition

Before Restoration	After Restoration
Fish abundant	Fish rare
Zooplankton rare	Zooplankton abundant
Algae abundant	Algae rare

1. What two components contribute to species diversity? Explain how two communities with the same number of species can differ in species diversity.

2. How is a food chain different from a food web?

3. **WHAT IF?** Consider a grassland with five trophic levels: grasses, mice, snakes, raccoons, and bobcats. If you released additional bobcats into the grassland, how would grass biomass change if bottom-up control applied? If top-down control applied?

4. **MAKE CONNECTIONS** Rising atmospheric CO_2 levels lead to ocean acidification (see Figure 3.12) and higher ocean temperatures, both of which can reduce krill abundance. Predict how a drop in krill abundance might affect other organisms in the food web shown in Figure 54.17. Which organisms are particularly at risk? Explain.

For suggested answers, see Appendix A.

Disturbance influences species diversity and composition

Decades ago, most ecologists thought that biological communities are at equilibrium, a more or less stable balance, unless seriously disturbed by human activities. This "balance of nature" view focused on competition as a key factor determining the composition and stability of communities. *Stability* in this context refers to a community's tendency to reach and maintain a relatively constant composition of species.

Proponents of this view thought that the community of plants at a site had only one stable equilibrium, a *climax community* controlled solely by climate. They also argued that biotic interactions caused the species in the community to function as an integrated unit—in effect, as a superorganism. This argument was based on the observation that certain species of plants are often found together, such as the oaks, maples, birches, and beeches in deciduous forests of the northeastern United States.

Other ecologists questioned whether most communities were at equilibrium and challenged the concept of a single climax community. They thought that differences in soils, topography, and other factors created many potential communities that were stable within a region. Rather than functioning as an integrated unit, communities could be viewed as chance assemblages of species found together because they have similar abiotic requirements, such as for temperature, rainfall, and soil type. Moreover, evidence shows that disturbance keeps many communities from reaching a state of equilibrium in species diversity or composition. A **disturbance** is an event, such as a storm, fire, flood, drought, or human activity, that changes a community by removing organisms from it or altering resource availability.

This emphasis on change in communities led to the formulation of the **nonequilibrium model**, which describes most communities as constantly changing after disturbance. Even relatively stable communities can be rapidly transformed into nonequilibrium communities. Let's examine some of the ways that disturbances influence community structure.

Characterizing Disturbance

The types of disturbances and their frequency and severity vary among communities. Storms disturb almost all communities, even those in the oceans through the action of waves. Fire is a significant disturbance; in fact, chaparral and some grassland biomes require regular burning to maintain their structure and species composition. Many streams and ponds are disturbed by seasonal flooding and drying. A high level of disturbance is generally the result of frequent *and* intense disturbance, while low disturbance levels can result from either a low frequency or low intensity of disturbance.

The **intermediate disturbance hypothesis** states that moderate levels of disturbance foster greater species diversity than do high or low levels of disturbance. High levels of disturbance reduce diversity by creating environmental stresses that exceed the tolerances of many species or by disturbing the community so often that slow-growing or slow-colonizing species are excluded. At the other extreme, low levels of disturbance can reduce species diversity by allowing competitively dominant species to exclude less competitive ones. Meanwhile, intermediate levels of disturbance can foster greater species diversity by opening up habitats for occupation by less competitive species. Such intermediate disturbance levels rarely create conditions so severe that they exceed the environmental tolerances or recovery rates of potential community members.

The intermediate disturbance hypothesis is supported by many terrestrial and aquatic studies. In one study, ecologists in New Zealand compared the richness of invertebrates living in the beds of streams exposed to different frequencies and intensities of flooding **(Figure 54.24)**. When floods occurred either very frequently or rarely, invertebrate richness was low. Frequent floods made it difficult for some species to become established in the streambed, while rare floods resulted in species being displaced by superior competitors. Invertebrate richness peaked in streams that had an intermediate frequency or intensity of flooding, as predicted by the hypothesis.

Although moderate levels of disturbance appear to maximize species diversity in some cases, small and large disturbances also can have important effects on community structure. Small-scale disturbances can create patches of different habitats across a landscape, which help maintain diversity in a community. Large-scale disturbances are also a natural part of many communities. Much of Yellowstone National Park, for example, is dominated by lodgepole pine, a tree species that requires the rejuvenating influence

▼ Figure 54.24 Testing the intermediate disturbance hypothesis. Researchers identified the taxa (species or genera) of invertebrates at two locations in each of 27 New Zealand streams. They assessed the intensity of flooding at each location using an index of streambed disturbance. The number of invertebrate taxa peaked where the intensity of flooding was at intermediate levels.

of periodic fires. Lodgepole pine cones remain closed until exposed to intense heat. When a forest fire burns the trees, the cones open and the seeds are released. The new generation of lodgepole pines can then thrive on nutrients released from the burned trees and in the sunlight that is no longer blocked by taller trees.

In the summer of 1988, extensive areas of Yellowstone burned during a severe drought **(Figure 54.25a)**. By 1989, many burned areas in the park were largely covered with new vegetation, suggesting that the species in this community are adapted to rapid recovery after fire **(Figure 54.25b)**. In fact, large-scale fires have periodically swept through the lodgepole pine forests of Yellowstone and other northern areas for thousands of years. In contrast, more southerly pine forests were historically affected by frequent but low-intensity fires. In these forests, a century of human intervention to suppress small fires has allowed an unnatural buildup of fuels in some places and elevated the risk of large, severe fires to which the species are not adapted.

Studies of the Yellowstone forest community and many others indicate that they are nonequilibrium communities, changing continually because of natural disturbances and the internal processes of growth and reproduction. Mounting evidence suggests that nonequilibrium conditions are in fact the norm for most communities.

Ecological Succession

Changes in the composition of terrestrial communities are most apparent after a severe disturbance, such as a volcanic eruption or a glacier, strips away all the existing vegetation. The disturbed area may be colonized by a variety of species, which are gradually replaced by other species, which are in turn replaced by still other species—a process called **ecological succession**. When this process begins in a virtually lifeless area, such as on a new volcanic island or on

▼ Figure 54.25 Recovery following a large-scale disturbance. The 1988 Yellowstone National Park fires burned large areas of forests dominated by lodgepole pines.

(a) Soon after fire. While all trees in the foreground of this photograph were killed by the fire, unburned trees can be seen in other locations.

(b) One year after fire. The community has begun to recover. Herbaceous plants, different from those in the former forest, cover the ground.

> **Mastering Biology** Interview with Monica Turner: **Studying the consequences of fire and other disturbances on ecosystems**

the rubble (moraine) left by a retreating glacier, it is called **primary succession**.

During primary succession, the only life-forms initially present are often prokaryotes and protists. Lichens and mosses, which grow from windblown spores, are commonly the first macroscopic photosynthesizers to colonize such areas. Soil develops gradually as rocks weather and organic matter accumulates from the decomposed remains of the early colonizers. Once soil is present, the lichens and mosses are usually overgrown by grasses, shrubs, and trees that sprout from seeds blown in from nearby areas or carried in by animals. Eventually, an area is colonized by plants that become the community's dominant form of vegetation.

① Pioneer stage

② *Dryas* stage

④ Spruce stage

③ Alder stage

▲ **Figure 54.26 Glacial retreat and primary succession near the shores of Glacier Bay, Alaska.** The different shades of blue on the map are based on historical information of how the area covered by Glacier Bay has increased over time due to retreat of the glaciers since 1760.

Producing such a community through primary succession may take hundreds or thousands of years.

Early-arriving species and later-arriving ones may be linked by one of three key processes. The early arrivals may *facilitate* the appearance of the later species by making the environment more favorable—for example, by increasing the fertility of the soil. Alternatively, the early species may *inhibit* establishment of the later species, so that successful colonization by later species occurs in spite of, rather than because of, the activities of the early species. Finally, the early species may be completely independent of the later species, which *tolerate* conditions created early in succession but are neither helped nor hindered by early species.

Ecologists have conducted some of the most extensive research on primary succession at Glacier Bay in southeastern Alaska, where glaciers have retreated more than 100 km since 1760 **(Figure 54.26)**. By studying the communities at different distances from the mouth of the bay—with the oldest exposed areas being closest to the mouth and the more recently exposed areas being farthest—ecologists can examine different stages in succession. As the glaciers have retreated, the following sequence of events occurred on areas above the waterline of the bay: ① The exposed glacial moraine is colonized first by pioneering species that include liverworts, mosses, fireweed, scattered *Dryas* (a mat-forming shrub), and willows. ② After about three decades, *Dryas* dominates the plant community. ③ A few decades later, the area is invaded by alder, which forms dense thickets up to 9 m tall. ④ In the next two centuries, these alder stands are overgrown first by Sitka spruce and later by western hemlock and mountain hemlock. In areas of poor drainage, the forest floor of this spruce-hemlock forest is invaded by sphagnum moss, which holds water and acidifies the soil, eventually killing the trees. Thus, by about 300 years after glacial retreat, the vegetation consists of sphagnum bogs on the poorly drained flat areas and spruce-hemlock forest on the well-drained slopes.

Succession on glacial moraines is related to changes in soil nutrients and other environmental factors caused by transitions in the vegetation. Because the bare soil after glacial retreat is low in nitrogen content, almost all the pioneer plant species begin succession with poor growth and yellow leaves due to limited nitrogen supply. The exceptions are *Dryas* and alder, whose roots host symbiotic bacteria that fix atmospheric nitrogen (see Figure 37.12). Soil nitrogen content increases quickly during the alder stage of succession and keeps increasing during the spruce stage **(Figure 54.27)**. By altering soil properties, pioneer plant

▼ Figure 54.27 Changes in soil nitrogen content during succession at Glacier Bay.

MAKE CONNECTIONS *Figure 37.11 illustrates two types of atmospheric nitrogen fixation by prokaryotes. At the earliest stages of primary succession, before any plants are present at a site, which type of nitrogen fixation would occur, and why?*

species can facilitate colonization by new plant species during succession.

In contrast to primary succession, **secondary succession** involves the recolonization of an area after a major disturbance has removed most but not all of the organisms in a community, as in Yellowstone following the 1988 fires (see Figure 54.25). Following the disturbance, the area may return to something like its original state. For instance, in a forested area that has been cleared for farming and later abandoned, the earliest plants to recolonize are often herbaceous species that grow from windblown or animal-borne seeds. If the area has not been burned or heavily grazed, woody shrubs may in time replace most of the herbaceous species, and forest trees may eventually replace most of the shrubs.

Human Disturbance

Ecological succession is a response to disturbance of the environment, and some of the strongest disturbances result from human activities. Agricultural development has disrupted what were once the vast grasslands of the North American prairie. Tropical rain forests are quickly disappearing as a result of clear-cutting for lumber, cattle grazing, and farmland. Centuries of overgrazing and agricultural disturbance have contributed to famine in parts of Africa by turning seasonal grasslands into vast barren areas.

Human actions have disturbed marine ecosystems as well as terrestrial ones. The effects of ocean trawling, in which boats drag weighted nets across the seafloor, are similar to those of clear-cutting a forest or plowing a field. The trawls scrape and scour corals and other life on the seafloor **(Figure 54.28)**. In a typical

▼ Figure 54.28 Disturbance of the ocean floor by trawling. These photos show the seafloor off northwestern Australia before and after deep-sea trawlers have passed through.

◄ Before trawling

After ► trawling

year, ships trawl an area about the size of South America, 150 times larger than the area of forests that are clear-cut annually.

Because disturbance by human activities is often severe, it reduces species diversity in many communities. In Chapter 56, we'll take a closer look at how human-caused disturbance is affecting the diversity of life.

CONCEPT CHECK 54.3

1. Why do high and low levels of disturbance usually reduce species diversity? Why does an intermediate level of disturbance promote species diversity?
2. During succession, how might the early species facilitate the arrival of other species?
3. **WHAT IF?** Most prairies experience regular fires, typically every few years. If these disturbances were relatively modest, how would the species diversity of a prairie likely be affected if no burning occurred for 100 years? Explain.

For suggested answers, see Appendix A.

CONCEPT 54.4

Biogeographic factors affect community diversity

So far, we have examined relatively small-scale or local factors that influence the diversity of communities, including the effects of species interactions, foundation species, and many types of disturbances. Large-scale biogeographic factors also contribute to the tremendous range of diversity observed in biological communities.

Two biogeographic factors in particular—the latitude of a community and the area it occupies—have been investigated for more than a century.

Latitudinal Gradients

In the 1850s, both Charles Darwin and Alfred Wallace pointed out that plant and animal life was generally more abundant and diverse in the tropics than in other parts of the globe. Since that time, many researchers have confirmed this observation. One study found that a 6.6-hectare (1 ha = 10,000 m²) plot in tropical Malaysia contained 711 tree species, while a 2-ha plot of deciduous forest in Michigan typically contained just 10 to 15 tree species. Many groups of animals show similar latitudinal gradients. For instance, there are more than 200 species of ants in Brazil but only 7 in Alaska.

Two key factors that can affect latitudinal gradients of species richness are evolutionary history and climate. Over the course of evolution, a series of speciation events may lead to increased species richness in a community (see Concept 24.2). Tropical communities are generally older than temperate or polar communities, which have repeatedly "started over" after major disturbances such as glaciations. As a result, species diversity may be highest in the tropics simply because there has been more time for speciation to occur in tropical communities than in temperate or polar communities.

Climate is another key factor thought to affect latitudinal gradients of richness and diversity. In terrestrial communities, the two main climatic factors correlated with diversity are sunlight and precipitation, both of which occur at high levels in the tropics. These factors can be considered together by measuring a community's rate of **evapotranspiration**, the evaporation of water from soil and plants. Evapotranspiration, a function of solar radiation, temperature, and water availability, is much higher in hot areas with abundant rainfall than in areas with low temperatures or low precipitation. The species richness of plants and animals correlates with measures of evapotranspiration, as shown for vertebrates in **Figure 54.29**.

Area Effects

In 1807, naturalist and explorer Alexander von Humboldt described one of the first patterns of species richness to be recognized, the **species-area curve**: All other factors being equal, the larger the geographic area of a community, the more species it has. One explanation for this relationship is that larger areas offer a greater diversity of habitats and microhabitats.

Predictions of the species-area relationship have been tested by examining the number of animals and plants in many different regions. As one example, in the Sunda Islands of Malaysia, the number of bird species increased with island size **(Figure 54.30)**. Different species-area curves vary in how rapidly species richness increases with area. Even so, the basic concept of diversity increasing with increasing area applies in many situations, from surveys of ant diversity in New Guinea to studies of plant species richness on islands of different sizes.

▼ **Figure 54.30 Species richness and island area.** The number of bird species on the Sunda Islands of Malaysia increases with island size.

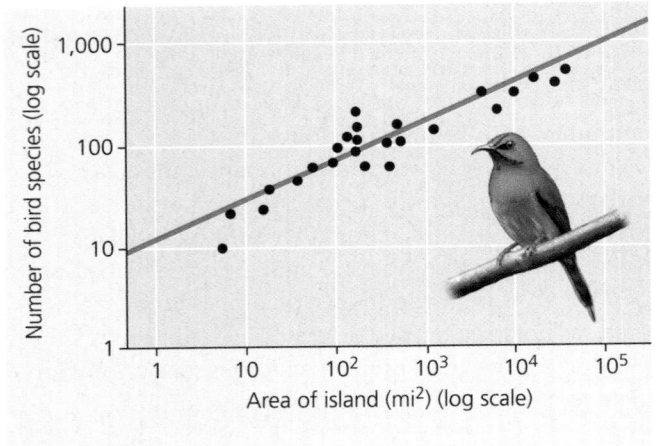

Island Equilibrium Model

Because of their isolation and limited size, islands provide excellent opportunities for studying how area and other factors affect the species diversity of communities. By "islands," we mean not only oceanic islands, but also habitat islands on land, such as lakes, mountain peaks, or habitat fragments—any patch surrounded by an environment not suitable for the "island" species.

Recognizing the opportunities that studying islands provides, Robert MacArthur and E. O. Wilson developed a method for predicting the species diversity of islands **(Figure 54.31)**. In their approach, the number of species on an island represents a balance between the immigration of new species to the island and the extinction of species already there.

In Figure 54.31, note that the immigration rate *decreases* as the number of species on the island gets larger, while the

▼ **Figure 54.29 Energy, water, and species richness.** Vertebrate species richness in North America increases with an index of evapotranspiration, expressed as rainfall equivalents (mm/yr).

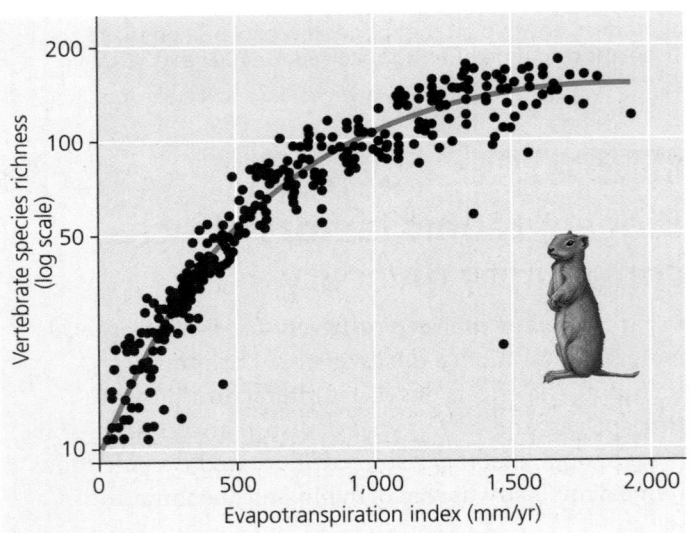

▼ Figure 54.31 MacArthur and Wilson's island equilibrium model. The equilibrium number of species on an island represents a balance between the immigration of new species (red curve) and the extinction of species already there (black curve). The point at which these curves intersect shows the predicted equilibrium number of species (Q).

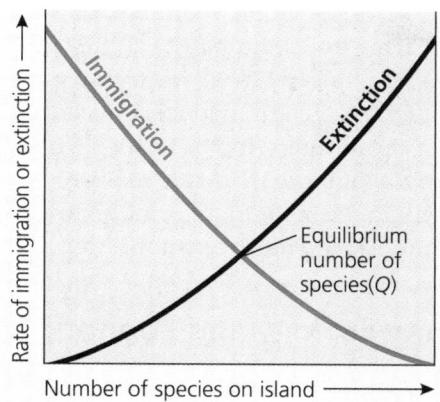

WHAT IF? *Suppose rising sea levels substantially decreased the size of the island. How would that affect (a) the population sizes of species already on the island, (b) the extinction curve shown above, and (c) the predicted equilibrium number of species?*

extinction rate *increases*. To see why this is so, consider a newly formed oceanic island that receives colonizing species from a distant mainland. At any given time, an island's immigration and extinction rates are affected by the number of species currently present. As the number of species already on the island increases, the immigration rate of new species decreases, because any individual reaching the island is less likely to represent a species that is not already present. Moreover, as more species inhabit an island, extinction rates on the island increase because of the greater likelihood of competitive exclusion.

An island's size and its distance from the mainland are also important. Small islands generally have lower immigration rates because potential colonizers are less likely to reach a small island than a large one. Small islands also have higher extinction rates because they generally contain fewer resources, have less diverse habitats, and have smaller population sizes. With respect to distance, an island that is closer to the mainland generally has a higher immigration rate and a lower extinction rate than one farther away. Arriving colonists increase the abundance of a species on a near island, thereby reducing the species' chance of extinction.

MacArthur and Wilson's model is called the *island equilibrium model* because an equilibrium will eventually be reached where the rate of species immigration equals the rate of species extinction. The number of species at this equilibrium point is correlated with the island's size and distance from the mainland. Although the number of species ultimately should stabilize at a constant level, immigration and extinction continue and hence the exact species composition on the island may change over time.

Researchers tested the island equilibrium model in an experiment on six small mangrove islands in the Florida Keys

▶ Figure 54.32 Mangrove island. The islands that researchers studied were small, each consisting of one or a few mangrove trees.

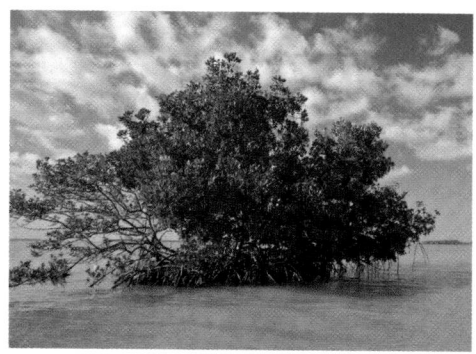

(Figure 54.32). To do this, they first painstakingly identified and counted all of the arthropod species on each island. As predicted by the model, they found more species on islands that were larger and closer to the mainland. They then fumigated four of the islands with methyl bromide to kill all of the arthropods. As also predicted by the model, the number of arthropod species on the islands increased over time to near their pre-fumigation values **(Figure 54.33).** The island closest to the mainland recovered first, while the island farthest from the mainland was the slowest to recover. The number of arthropod species on the remaining two islands—which were not fumigated and hence served as controls—remained approximately constant during the study.

⊘ **Mastering Biology** Animation: Exploring Island Biogeography

▼ Figure 54.33 Testing the island equilibrium model. The graph shows results for one of the islands studied. The number of arthropod species increased over time; by 140 days, it had reached levels similar to those found on the island prior to the start of the experiment.

CONCEPT CHECK 54.4

1. Describe two hypotheses as to why species diversity is greater in the tropics than in temperate and polar regions.
2. Describe how an island's size and distance from the mainland affect the island's species richness.
3. **WHAT IF?** Based on MacArthur and Wilson's island equilibrium model, predict how the richness of birds on an island would compare with that of snakes and lizards there.

For suggested answers, see Appendix A.

Pathogens alter community structure locally and globally

Now that we have examined several important factors that structure biological communities, we'll finish the chapter by examining community interactions involving **pathogens**, disease-causing microorganisms, viruses, viroids, or prions. (Viroids and prions are infectious RNA molecules and proteins, respectively; see Concept 19.3.) As you'll see, pathogens have strong effects on ecological communities.

Pathogens produce especially clear effects when they are introduced into new habitats, as in the case of the fungus that causes chestnut blight (see Concept 54.2). A pathogen can be particularly virulent in a new habitat because new host populations have not had a chance to become resistant to the pathogen through natural selection. The invasive chestnut blight fungus had far stronger effects on the American chestnut, for instance, than it had on Asian chestnut species in the fungus's native habitat.

Effects on Community Structure

The ecological importance of disease can be highlighted by how pathogens have affected coral reef communities. White-band disease, caused by an unknown pathogen, has resulted in dramatic changes in Caribbean reefs. The disease kills corals by causing their tissue to slough off in a band from the base to the tip of the branches. Because of the disease, staghorn coral (*Acropora cervicornis*) has virtually disappeared from the Caribbean. Populations of elkhorn coral (*Acropora palmata*) have also been decimated. Such corals provide key habitat for lobsters as well as snappers and other fish species. When the corals die, they become overgrown by algae. Surgeonfish and other herbivores that feed on algae come to dominate the fish community. Eventually, the corals topple because of damage from storms and other disturbances. The complex, three-dimensional structure of the reef disappears, and diversity plummets.

Pathogens also influence community structure in terrestrial ecosystems. Consider sudden oak death (SOD), a recently discovered disease that is caused by the protist *Phytophthora ramorum* (see Concept 28.6). SOD was first described in California in 1995, when hikers noticed trees dying around San Francisco Bay. By 2014, it had spread more than 1,000 km, from the central California coast to southern Oregon, and it had killed more than a million oaks and other trees. The loss of the oaks has led to the decreased abundance of at least five bird species, including the acorn woodpecker and the oak titmouse, that rely on oaks for food and habitat. Although there is currently no cure for SOD, scientists recently sequenced the genome of *P. ramorum* in hopes of finding a way to fight the pathogen.

Human activities are transporting pathogens around the world at unprecedented rates. Genetic analyses using DNA sequencing suggest that *P. ramorum* likely came to North America from Europe through the horticulture trade. Similarly, the pathogens that cause human diseases are spread by our global economy. H1N1, the virus that causes "swine flu" in humans, was first detected in Veracruz, Mexico, in early 2009. It quickly spread around the world when infected individuals traveled on airplanes. By the end of the outbreak in 2011, this flu had a confirmed death toll of more than 18,000 people. The actual number may have been far higher since many people with flu-like symptoms who died were not tested for H1N1.

Community Ecology and Zoonotic Diseases

Three-quarters of emerging human diseases and many of the most devastating established diseases are caused by **zoonotic pathogens**—those that are transferred to humans from other animals, either through direct contact with an infected animal or via an intermediate species, called a **vector**. The vectors that spread zoonotic diseases are often parasites, including ticks, lice, and mosquitoes.

Identifying the community of hosts and vectors for a pathogen can help prevent illnesses such as Lyme disease, which is spread by ticks. For years, scientists thought that the primary host for the Lyme pathogen was the white-footed mouse because mice are heavily parasitized by young ticks. When researchers vaccinated mice against Lyme disease and released them into the wild, however, the number of infected ticks hardly changed. Further investigation in New York revealed that two inconspicuous shrew species were the source for more than half the infected ticks collected in the field **(Figure 54.34)**. Identifying the most important hosts for a pathogen provides information that may be used to control the hosts most responsible for spreading diseases.

Ecologists also use their knowledge of community interactions to track the spread of zoonotic diseases. One example, avian flu, is caused by highly contagious viruses transmitted through the saliva and feces of birds (see Concept 19.3). Most of these viruses affect wild birds mildly, but they often cause stronger symptoms in domesticated birds, the most common source of human infections. Since 2003, one particular viral strain, called H5N1, has killed hundreds of millions of poultry and more than 400 people.

Control programs that quarantine domestic birds or monitor their transport may be ineffective if avian flu spreads naturally through the movements of wild birds. From 2003 to 2006, the H5N1 strain spread rapidly from southeast Asia into Europe and Africa. The most likely place for infected wild birds to enter the Americas is Alaska, the entry point for ducks, geese, and shorebirds that migrate across the Bering Sea from Asia every year. Ecologists are studying the spread of

▼ **Figure 54.34 Unexpected hosts of the Lyme disease pathogen.** A combination of ecological data and genetic analyses enabled scientists to show that more than half of ticks carrying the Lyme pathogen became infected by feeding on the northern short-tailed shrew (*Blarina brevicauda*) or the masked shrew (*Sorex cinereus*).

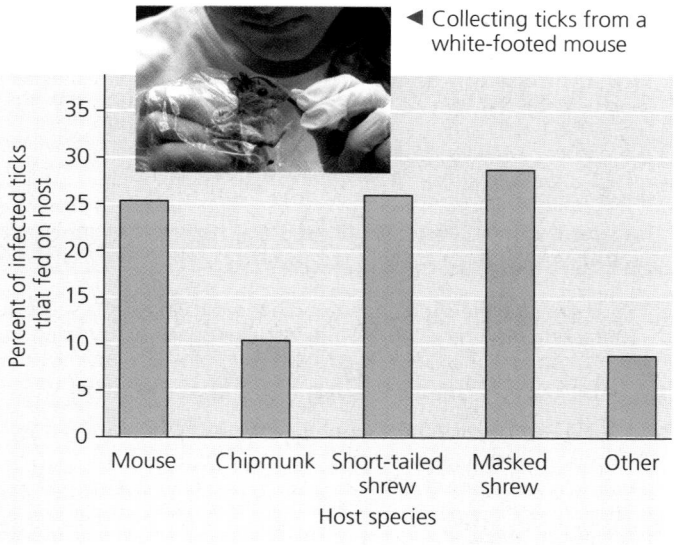

◄ Collecting ticks from a white-footed mouse

MAKE CONNECTIONS *Concept 23.1 discusses the significance of genetic variation between populations. How might genetic variation between shrew populations in different locations affect the number of infected ticks?*

► **Figure 54.35 Tracking avian flu.** A graduate student bands a young gyrfalcon as part of a project to monitor the spread of avian flu.

the virus by trapping and testing migrating and resident birds in Alaska **(Figure 54.35)**.

While our emphasis here has been on community ecology, pathogens are also greatly influenced by changes in the physical environment. To control pathogens and the diseases they cause, scientists need an ecosystem perspective—an intimate knowledge of how the pathogens interact with other species and with all aspects of their environment. Ecosystems are the subject of Chapter 55.

CONCEPT CHECK 54.5

1. What are pathogens?
2. **WHAT IF?** Rabies, a viral disease in mammals, is not currently found in the British Isles. If you were in charge of disease control there, what practical approaches might you employ to keep the rabies virus from reaching these islands?

For suggested answers, see Appendix A.

54 Chapter Review

→ Go to **Mastering Biology** for Assignments, the eText, the Study Area, and Dynamic Study Modules.

SUMMARY OF KEY CONCEPTS

→ To review key terms, go to the **Vocabulary Self-Quiz** in the **Mastering Biology** eText or Study Area, or go to goo.gl/zkjz9t.

CONCEPT 54.1

Interactions between species can help, harm, or have no effect on the individuals involved *(pp. 1215–1221)*

- **Interspecific interactions** affect the survival and reproduction of the individuals that engage in them. As shown in the table, these interactions can be grouped into three broad categories: competition, exploitation, and positive interactions.
- **Competitive exclusion** states that two species whose members compete for the same resource cannot coexist permanently in the same place. **Resource partitioning** is the differentiation of **ecological niches** that enables species to coexist in a **community**.

❓ *For each interaction listed in the table, give an example of a pair of species that exhibit the interaction.*

Interaction	Description
Competition (−/−)	Individuals of different species each use a limited resource, reducing the survival or reproduction of both individuals.
Exploitation (+/−)	Members of one species benefit by feeding upon (and thereby harming) members of the other species. Exploitation includes the following:
Predation	An individual of one species, the predator, kills and eats an individual of the other, the prey.
Herbivory	An herbivore eats part of a plant or alga.
Parasitism	The parasite derives its nourishment from a second organism, its host.
Positive interactions (+/+ or 0/+)	Members of one species benefit, while members of the other benefit or are not harmed. Positive interactions include the following:
Mutualism (+/+)	Members of both species benefit from the interaction.
Commensalism (+/0)	Members of one species benefit, while members of the other are not affected.

Diversity and trophic structure characterize biological communities (pp. 1222–1228)

- **Species diversity** is affected by both the number of species in a community—its **species richness**—and their **relative abundance**.
- More diverse communities typically produce more **biomass** and show less year-to-year variation in growth than less diverse communities and are more resistant to introduced species.
- **Trophic structure** is a key factor in community dynamics. **Food chains** link the trophic levels from producers to top carnivores. Branching food chains and complex trophic interactions form **food webs**.
- **Foundation species** are large or abundant members of a community that provide food or habitat. **Keystone species** are usually less abundant species that exert a disproportionate influence on community structure. **Ecosystem engineers** influence community structure through their effects on the physical environment.
- In **bottom-up control**, the abundance of organisms at each trophic level is limited by nutrient supply or food availability. In **top-down control**, each trophic level is controlled by the abundance of consumers at higher trophic levels.

? *Based on indexes such as the Shannon diversity index, is a community of higher species richness always more diverse than a community of lower species richness? Explain.*

Disturbance influences species diversity and composition (pp. 1228–1231)

- Increasing evidence suggests that **disturbance** and lack of equilibrium, rather than stability and equilibrium, are the norm for most communities. According to the **intermediate disturbance hypothesis**, moderate levels of disturbance can foster higher species diversity than can low or high levels of disturbance.
- **Ecological succession** is the sequence of community and ecosystem changes after a disturbance. **Primary succession** occurs in an area that is virtually lifeless, while **secondary succession** occurs after a disturbance has removed most but not all of the organisms in a community.

? *Is the disturbance pictured in Figure 54.28 more likely to initiate primary or secondary succession? Explain.*

Biogeographic factors affect community diversity (pp. 1231–1233)

- Species richness generally declines along a latitudinal gradient from the tropics to the poles. Climate influences the diversity gradient through energy (heat and light) and water. The greater age of tropical environments also may contribute to their greater species richness.
- Species richness is directly related to a community's geographic size, a principle formalized in the **species-area curve**.
- Species richness on islands depends on island size and distance from the mainland. The island equilibrium model maintains that species richness on an ecological island reaches an equilibrium where new immigrations are balanced by extinctions.

? *How have periods of glaciation influenced latitudinal patterns of diversity?*

Pathogens alter community structure locally and globally (pp. 1234–1235)

- **Pathogens** play a key in structuring terrestrial and marine communities.
- **Zoonotic pathogens** are transferred from other animals to humans and cause most emerging human diseases. Community ecology provides the framework for identifying key species interactions associated with such pathogens and for helping us track and control their spread.

? *Suppose a pathogen attacks a keystone species. Explain how this could alter the structure of a community.*

TEST YOUR UNDERSTANDING

➡️ For more multiple-choice questions, go to the **Practice Test** in the **Mastering Biology** eText or Study Area, or go to goo.gl/GruWRg.

Levels 1-2: Remembering/Understanding

1. The feeding relationships among the species in a community determine the community's
   - (A) secondary succession.
   - (B) ecological niche.
   - (C) species richness.
   - (D) trophic structure.

2. The principle of competitive exclusion states that
   - (A) two species cannot coexist in the same habitat.
   - (B) competition between two species always causes extinction or emigration of one species.
   - (C) two species that have exactly the same niche cannot coexist in a community.
   - (D) two species will stop reproducing until one species leaves the habitat.

3. Based on the intermediate disturbance hypothesis, a community's species diversity is increased by
   - (A) frequent massive disturbance.
   - (B) stable conditions with no disturbance.
   - (C) moderate levels of disturbance.
   - (D) human intervention to eliminate disturbance.

4. According to the island equilibrium model, species richness would be greatest on an island that is
   - (A) large and remote.
   - (B) small and remote.
   - (C) large and close to a mainland.
   - (D) small and close to a mainland.

Levels 3-4: Applying/Analyzing

5. Predators that are keystone species can maintain species diversity in a community if they
   - (A) competitively exclude other predators.
   - (B) prey on the community's dominant competitors.
   - (C) reduce the number of disruptions in the community.
   - (D) prey on the least abundant species in the community.

6. Food chains tend to be short because
   - (A) only a single species of herbivore feeds on each plant species.
   - (B) local extinction of a species causes extinction of the other species in its food chain.
   - (C) most of the energy in a trophic level is lost as energy passes to the next higher level.
   - (D) most producers are inedible.

7. Which of the following could qualify as a top-down control on a grassland community?
   (A) limitation of plant biomass by rainfall amount
   (B) influence of temperature on competition among plants
   (C) influence of soil nutrients on the abundance of grasses versus wildflowers
   (D) effect of grazing by bison on plant species diversity

8. The most plausible hypothesis to explain why species richness is higher in tropical than in temperate regions is that
   (A) tropical communities are younger.
   (B) tropical regions generally have more available water and higher levels of solar radiation.
   (C) higher temperatures cause more rapid speciation.
   (D) diversity increases as evapotranspiration decreases.

9. Community 1 contains 100 individuals distributed among four species: 5A, 5B, 85C, and 5D. Community 2 contains 100 individuals distributed among three species: 30A, 40B, and 30C. Calculate the Shannon diversity index (H) for each community. Which community is more diverse?

Levels 5-6: Evaluating/Creating

10. **DRAW IT** In the Chesapeake Bay estuary, the blue crab is an omnivore that eats eelgrass and other primary producers as well as clams. It is also a cannibal. In turn, the crabs are eaten by humans and by the endangered Kemp's Ridley sea turtle. Based on this information, draw a food web that includes the blue crab. Assuming that top-down control occurs in this system, describe what would happen to the abundance of eelgrass if humans stopped eating blue crabs.

11. **EVOLUTION CONNECTION** Explain why adaptations of particular organisms to competition may not necessarily represent instances of character displacement. What would a researcher have to demonstrate about two competing species to make a convincing case for character displacement?

12. **SCIENTIFIC INQUIRY** An ecologist studying desert plants performed the following experiment. She staked out two identical plots, containing sagebrush plants and small annual wildflowers. She found the same five wildflower species in roughly equal numbers on both plots. She then enclosed one plot with a fence to keep out kangaroo rats, the most common grain-eaters of the area. After two years, four of the wildflower species were no longer present in the fenced plot, but one species had become much more abundant. The control plot had not changed in species diversity. Using the principles of community ecology, propose a hypothesis to explain her results. What additional evidence would support your hypothesis?

13. **WRITE ABOUT A THEME: INTERACTIONS** In Batesian mimicry, a palatable species gains protection by mimicking an unpalatable one. Imagine that individuals of a palatable, brightly colored fly species are blown to three remote islands. The first island has no predators of that species; the second has predators but no similarly colored, unpalatable species; and the third has both predators and a similarly colored, unpalatable species. In a short essay (100–150 words), predict what might happen to the coloration of the palatable species on each island through time if coloration is a genetically controlled trait. Explain your predictions.

14. **SYNTHESIZE YOUR KNOWLEDGE**

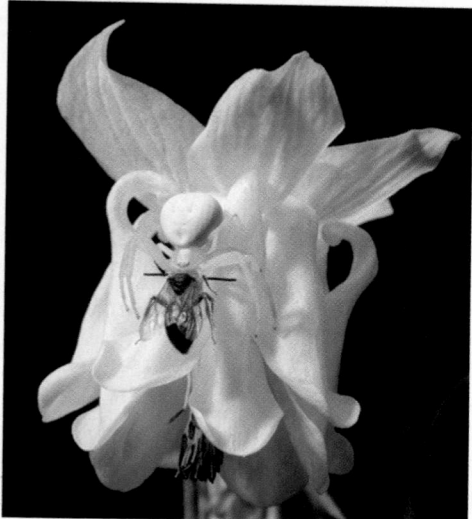

Describe two types of ecological interactions that appear to be occurring between the three species shown in this photo. (Look closely to see the three species: a flower, a bee, and a spider!) What morphological adaptation can be seen in the species that is at the highest trophic level in this scene?

For selected answers, see Appendix A.

55 Ecosystems and Restoration Ecology

KEY CONCEPTS

55.1 Physical laws govern energy flow and chemical cycling in ecosystems *p. 1239*

55.2 Energy and other limiting factors control primary production in ecosystems *p. 1241*

55.3 Energy transfer between trophic levels is typically only 10% efficient *p. 1246*

55.4 Biological and geochemical processes cycle nutrients and water in ecosystems *p. 1248*

55.5 Restoration ecologists return degraded ecosystems to a more natural state *p. 1253*

Study Tip

Make a flowchart: To connect the processes shown in Figures 55.4 and 55.14 to actual events in an ecosystem, make a flowchart like the one begun here. It should show how a carbon atom exhaled as CO_2 by a squirrel could eventually enter another squirrel after passing through decomposers, a small bird, grass, a hawk, an oak tree, and a grasshopper (not in that order).

Carbon atom pathway

Squirrel

CO_2 released by respiration

Grass (CO_2 used in photosynthesis)

→ Go to Mastering Biology

For Students (in eText and Study Area)
- Get Ready for Chapter 55
- Figure 55.4 Walkthrough: Overview of Energy and Nutrient Dynamics
- BioFlix® Animation: The Carbon Cycle

For Instructors to Assign (in Item Library)
- Scientific Thinking: Restoring Native Oysters in the Chesapeake Bay
- Interpreting Data: Net Primary Production

Ready-to-Go Teaching Module (in Instructor Resources)
- Trophic Relationships (Concept 55.3)

Figure 55.1 This arctic fox (*Vulpes lagopus*) will obtain chemical nutrients and energy when it eats the seabird it has killed. Before its death, the seabird ate fish from the sea and excreted waste that provided key nutrients for the plants on this Alaskan island. These interactions illustrate a few of the many ways that chemicals and energy move through ecosystems.

What are the dynamics of energy and chemical nutrients in an ecosystem?

There is a one-way flow of **energy** through an ecosystem, whereas **chemicals** cycle within the ecosystem.

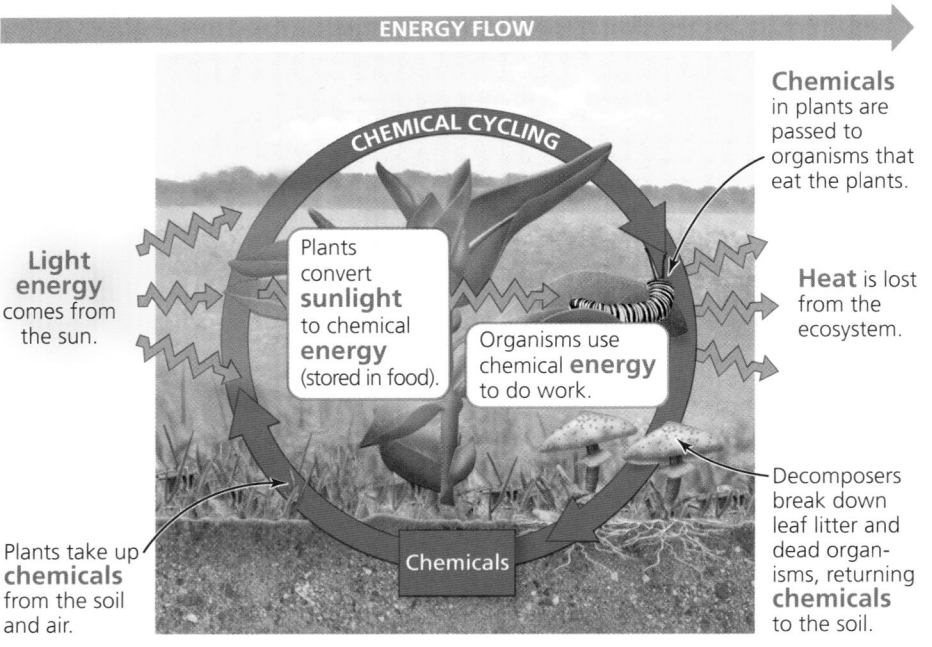

ENERGY FLOW

CHEMICAL CYCLING

Light energy comes from the sun.

Plants convert **sunlight** to chemical **energy** (stored in food).

Organisms use chemical **energy** to do work.

Plants take up **chemicals** from the soil and air.

Chemicals

Chemicals in plants are passed to organisms that eat the plants.

Heat is lost from the ecosystem.

Decomposers break down leaf litter and dead organisms, returning **chemicals** to the soil.

Physical laws govern energy flow and chemical cycling in ecosystems

The "fox island" in Figure 55.1 and its community of organisms are an example of an **ecosystem**, the sum of all the organisms living in a given area and the abiotic factors with which they interact. An ecosystem can encompass a large area, such as a lake, forest, or island, or a microcosm, such as the space under a fallen log or a small desert spring **(Figure 55.2)**. As with populations and communities, the boundaries of ecosystems are not always discrete. Many ecologists view the entire biosphere as a global ecosystem, a composite of all the local ecosystems on Earth.

Energy Flow and Chemical Cycling

An ecosystem, regardless of its size, has two key emergent properties: energy flow and chemical cycling (see Figure 55.1). Energy enters most ecosystems as sunlight. This light energy is converted to chemical energy by autotrophs, passed to heterotrophs in the organic compounds of food, and dissipated as heat.

As for chemical cycling, elements such as carbon and nitrogen are passed between the biotic and abiotic components of the ecosystem. Photosynthetic and chemosynthetic organisms take up these elements in inorganic form from the air, soil, and water and incorporate them into organic compounds, some of which are consumed by animals. The elements are returned in inorganic form to the environment by the metabolism of organisms and by the breakdown of organic wastes and dead organisms by decomposers.

Both energy and chemicals are transformed in ecosystems through photosynthesis and feeding relationships. But unlike chemicals, energy cannot be recycled. An ecosystem requires a continuous influx of energy from an external source—in most cases, the sun. As we'll see, energy has a one-way flow through ecosystems, whereas chemicals cycle within them.

▼ **Figure 55.2 A desert spring ecosystem.**

Conservation of Energy

Cells transform energy and matter, subject to the laws of thermodynamics (see Concept 8.1). Cell biologists study these transformations within organelles and cells, and measure the amounts of energy and chemical compounds that cross the cells' boundaries. Ecosystem ecologists do the same thing, except in their case the "cell" is an entire ecosystem. By studying feeding relationships among organisms and how organisms interact with their physical environment, ecologists can follow the transformations of energy in an ecosystem and map the movements of chemical elements.

To study energy flow and chemical cycling, ecosystem ecologists use approaches based on laws of physics and chemistry. The first law of thermodynamics states that energy cannot be created or destroyed but only transferred or transformed (see Concept 8.1). Plants and other photosynthetic organisms convert solar energy to chemical energy, but the total amount of energy does not change: The amount of energy stored in organic molecules must equal the total solar energy intercepted by the plant minus the amounts reflected and dissipated as heat. Ecosystem ecologists measure energy transfers within and across ecosystems, in part to understand how many organisms a habitat can support and the amount of food humans can harvest from a site.

The second law of thermodynamics states that every exchange of energy increases the entropy of the universe. One implication of this law is that energy conversions are inefficient. Some energy is always lost as heat. As a result, each unit of energy that enters an ecosystem eventually exits as heat. Thus, energy flows through ecosystems—it does not cycle within them for long periods of time. Because energy flowing through ecosystems is ultimately lost as heat, most ecosystems would vanish if the sun were not continuously providing energy to Earth.

Conservation of Mass

Matter, like energy, cannot be created or destroyed. This **law of conservation of mass** is as important for ecosystems as are the laws of thermodynamics. Because mass is conserved, we can determine how much of a chemical element cycles within an ecosystem or is gained or lost by that ecosystem over time.

Unlike energy, chemical elements are continually recycled within ecosystems. For example, a carbon atom in CO_2 might be released from the soil by a decomposer, taken up by a blade of grass through photosynthesis, consumed by a grazing animal, and returned to the soil in the animal's waste.

In addition to cycling within ecosystems, elements can also be gained or lost by an ecosystem. For example, a forest gains mineral nutrients—the essential elements that plants obtain from soil—that enter as dust or as solutes dissolved in rainwater or leached from rocks in the ground. Nitrogen is also supplied through the biological process of nitrogen fixation (see Figure 37.11). In terms of losses, some elements return to the atmosphere as gases, while others are carried out of the ecosystem by moving water or by wind. Like organisms, ecosystems are open systems, absorbing energy and mass and releasing heat and waste products.

Most gains and losses to ecosystems are small compared to the amounts that cycle within them. Even so, the balance between inputs and outputs is important because it determines whether an ecosystem stores or loses a given element. In particular, if a nutrient's outputs exceed its inputs, that nutrient will eventually limit production in that ecosystem. Human activities often change the balance of inputs and outputs considerably, as we'll see later in this chapter and in Concept 56.4.

Energy, Mass, and Trophic Levels

Ecologists group species in an ecosystem into trophic levels based on feeding relationships (see Concept 54.2). The trophic level that ultimately supports all others consists of autotrophs, also called the **primary producers** of the ecosystem. Most autotrophs are photosynthetic organisms that use light energy to synthesize sugars and other organic compounds, which they use as fuel for cellular respiration and as building material for growth. The most common autotrophs are plants, algae, and photosynthetic prokaryotes. However, chemosynthetic prokaryotes are the primary producers in some ecosystems, such as deep-sea hydrothermal vents (see Figure 52.16) and places deep under the ground or ice.

Organisms in trophic levels above the primary producers are heterotrophs (consumers), which depend directly or indirectly on the outputs of primary producers for their source of energy. Herbivores, which eat plants and other primary producers, are **primary consumers**. Carnivores that eat herbivores are **secondary consumers**, and carnivores that eat other carnivores are **tertiary consumers**.

Another group of heterotrophs is the **decomposers**, consumers that get their energy from detritus. **Detritus** is nonliving

▼ **Figure 55.3 Decomposers.**

▼ Fungi decomposing a dead tree

▲ **Rod-shaped and spherical bacteria in compost (colorized SEM)**

organic material, such as the remains of dead organisms, feces, and fallen leaves. Although some animals (such as earthworms) feed on detritus, the main decomposers are prokaryotes and fungi **(Figure 55.3)**. These organisms secrete enzymes that digest organic material; they then absorb the breakdown products. Many decomposers are in turn eaten by secondary and tertiary consumers. In a forest, for instance, birds eat earthworms that have been feeding on leaf litter and its associated prokaryotes and fungi. As a result, chemicals originally synthesized by plants pass from the plants to leaf litter to decomposers to birds.

By recycling chemical elements to producers, decomposers also play a key role in the trophic relationships of an ecosystem **(Figure 55.4)**. Decomposers convert organic matter from all trophic levels to inorganic compounds usable by primary producers. When the decomposers excrete waste products or die, those inorganic compounds are returned to the soil. Producers can then absorb these elements and use them to synthesize

▶ **Figure 55.4 An overview of energy and nutrient dynamics in an ecosystem.** Energy enters, flows through, and exits an ecosystem, whereas chemical nutrients cycle within it. Energy (dark orange arrows) entering from the sun as radiation is transferred as chemical energy through the food web; each of these units of energy ultimately exits as heat radiated into space. Most transfers of nutrients (blue arrows) through the food web lead eventually to detritus; the nutrients then cycle back to the primary producers.

VISUAL SKILLS In this diagram, one blue arrow leads to the box labeled "Primary consumers," and three blue arrows come out of it. For each of these arrows, describe an example of nutrient transfer that the arrow could represent, using specific organisms and other components of an ecosystem in your answer.

⊖ Mastering Biology Figure Walkthrough

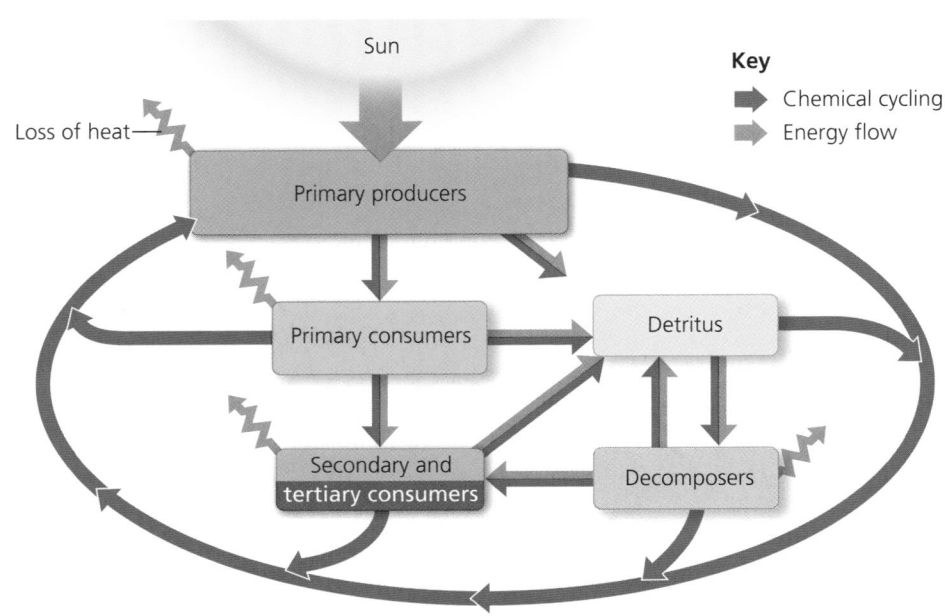

Sun

Loss of heat

Key
➡ Chemical cycling
➡ Energy flow

Primary producers

Primary consumers

Detritus

Secondary and tertiary consumers

Decomposers

organic compounds. If decomposition stopped, life as we know it would cease as detritus piled up and the supply of ingredients needed to synthesize organic matter was exhausted.

→ **Mastering Biology** **Animation: Energy Flow and Chemical Cycling**

CONCEPT 55.2

Energy and other limiting factors control primary production in ecosystems

The theme of energy transfer underlies all biological interactions (see Concept 1.1). In most ecosystems, the amount of light energy converted to chemical energy—in the form of organic compounds—by autotrophs during a given time period is the ecosystem's **primary production**. In ecosystems where the primary producers are chemoautotrophs, the initial energy input is chemical, and the initial products are the organic compounds synthesized by the microorganisms.

Ecosystem Energy Budgets

In most ecosystems, primary producers use light energy to synthesize energy-rich organic molecules, and consumers acquire their organic fuels secondhand (or even third- or fourth-hand) through food webs (see Figure 54.17). Therefore, the total amount of photosynthetic production sets the "spending limit" for the entire ecosystem's energy budget.

The Global Energy Budget

Each day, Earth's atmosphere is bombarded by approximately 10^{22} joules of solar radiation ($1 J = 0.239 \, cal$). This is enough energy to supply the demands of the entire human population for 19 years at 2013 energy consumption levels. The intensity of the solar energy striking Earth varies with latitude, with the tropics receiving the greatest input (see Figure 52.3). About 50% of incoming solar radiation is absorbed, scattered, or reflected by clouds and dust in the atmosphere. The amount of solar radiation that ultimately reaches Earth's surface limits the possible photosynthetic output of ecosystems.

However, only a small fraction of the sunlight that reaches Earth's surface is actually used in photosynthesis. Much of the radiation strikes materials that don't photosynthesize, such as ice and soil. Of the radiation that does reach photosynthetic organisms, only certain wavelengths are absorbed by photosynthetic pigments (see Figure 10.9); the rest is transmitted, reflected, or lost as heat. As a result, only about 1% of the visible light that strikes photosynthetic organisms is converted to chemical energy. Nevertheless, Earth's primary producers create about 150 billion metric tons ($1.50 \times 10^{14} kg$) of organic material each year.

Gross and Net Production

Total primary production in an ecosystem is known as that ecosystem's **gross primary production (GPP)**—the amount of energy from light (or chemicals, in chemoautotrophic systems) converted to the chemical energy of organic molecules per unit time. Not all of this production is stored as organic material in the primary producers because they use some of the molecules as fuel for their own cellular respiration. **Net primary production (NPP)** is equal to gross primary production minus the energy used by the primary producers (autotrophs) for their cellular respiration (R_a, where "a" stands for autotrophs):

$$NPP = GPP - R_a$$

On average, NPP is about one-half of GPP. To ecologists, NPP is the key measurement because it represents the storage of chemical energy that will be available to consumers in the ecosystem. Using the analogy of a paycheck, you can think of net primary production (NPP) as the take-home pay, which equals gross primary production (GPP), the gross pay, minus respiration (R_a), the taxes.

Net primary production can be expressed as energy per unit area per unit time [$J/(m^2 \cdot yr)$] or as biomass (mass of vegetation) added per unit area per unit time [$g/(m^2 \cdot yr)$]. (Note that biomass is usually expressed in terms of the dry mass of organic material.) An ecosystem's NPP should not be confused with the total biomass of photosynthetic autotrophs present. The net primary production is the amount of *new* biomass added in a given period of time. Although the total biomass of a forest is large, its NPP may actually be less than that of some grasslands; grasslands do not accumulate as much biomass as forests because animals consume the plants rapidly and because grasses and herbs decompose more quickly than trees do.

Satellites provide a powerful tool for studying global patterns of primary production. Images produced from satellite data show that different ecosystems vary considerably in their NPP. For example, tropical rain forests are among the most productive terrestrial ecosystems and contribute a large portion of the planet's

► **Figure 55.5 Global net primary production.** The map is based on satellite-collected data, such as amount of sunlight absorbed by vegetation. Note that tropical land areas have the highest rates of production (yellow to red).

VISUAL SKILLS *Does this map accurately reflect the significance of wetlands, coral reefs, and coastal zones, all highly productive habitats? Explain.*

Net primary production [kg carbon/(m² · yr)]

NPP **(Figure 55.5)**. Estuaries and coral reefs also have very high NPP, but their contribution to the global total is smaller because these ecosystems cover only about one-tenth the area covered by tropical rain forests. In contrast, while the open oceans are relatively unproductive, their vast size means that together they contribute as much global NPP as terrestrial systems do.

Whereas NPP can be expressed as the amount of new biomass added by producers in a given period of time, **net ecosystem production (NEP)** is a measure of the biomass accumulation by producers *and* consumers during that time. NEP is defined as gross primary production minus the total respiration of all organisms in the system (R_T)—not just primary producers, as for the calculation of NPP, but decomposers and other heterotrophs as well:

$$NEP = GPP - R_T$$

NEP is useful to ecologists because its value determines whether an ecosystem is gaining or losing carbon over time. A forest may have a positive NPP but still lose carbon if heterotrophs release it as CO_2 more quickly than primary producers incorporate it into organic compounds.

The most common way to estimate NEP is to measure the net flux (flow) of CO_2 or O_2 entering or leaving the ecosystem. If more CO_2 enters than leaves, the system is storing carbon. Because O_2 release is directly coupled to photosynthesis and respiration (see Figure 9.1), a system that is giving off O_2 is also storing carbon. On land, ecologists typically measure only the net flux of CO_2 from ecosystems because detecting small changes in O_2 flux in a large atmospheric O_2 pool is difficult.

Next, we'll examine factors that limit production in ecosystems, focusing first on aquatic ecosystems.

Primary Production in Aquatic Ecosystems

In aquatic (marine and freshwater) ecosystems, both light and nutrients are important in controlling primary production.

Light Limitation

Because solar radiation drives photosynthesis, you would expect light to be a key variable in controlling primary production in oceans. Indeed, the depth of light penetration affects primary production throughout the photic zone of an ocean or lake (see Figure 52.14). About half of the solar radiation is absorbed in the first 15 m of water. Even in "clear" water, only 5–10% of the radiation may reach a depth of 75 m.

If light were the main variable limiting primary production in the ocean, you would expect production to increase along a gradient from the poles toward the equator, which receives the greatest intensity of light. However, you can see in Figure 55.5 that there is no such gradient. What other factor strongly influences primary production in the ocean?

Nutrient Limitation

More than light, nutrients limit primary production in most oceans and lakes. A **limiting nutrient** is the element that must be added for production to increase. The nutrients that most often limit marine production are nitrogen and phosphorus. Concentrations of these nutrients are typically low in the photic zone because they are rapidly taken up by phytoplankton and because detritus tends to sink.

In one study, detailed in **Figure 55.6**, nutrient enrichment experiments found that nitrogen was limiting phytoplankton growth off the south shore of Long Island, New York. Moreover, as also seen in Figure 55.6, primary production can increase dramatically when the nutrient status of an ecosystem changes from nutrient-poor to nutrient-rich, a process known as **eutrophication** (from the Greek *eutrophos*, well nourished). One practical application of this work is in preventing algal blooms caused by excess nitrogen runoff that fertilizes the phytoplankton. Preventing such blooms is critical because their occurrence can lead to the formation of large marine "dead zones," regions in which oxygen concentrations drop to levels that are fatal to many organisms (see Figure 56.24).

The macronutrients nitrogen and phosphorus are not the only nutrients that limit aquatic production. Several large areas of the ocean have low phytoplankton densities despite relatively high nitrogen concentrations. The Sargasso Sea, a subtropical region of the Atlantic Ocean, has some of the clearest water in the world because of its low phytoplankton density. Nutrient enrichment experiments have revealed that the availability of the micronutrient iron limits primary production there **(Table 55.1)**. Windblown dust from land supplies most of the iron to the oceans but is relatively scarce in the Sargasso Sea and certain other regions compared to the oceans as a whole.

Which nutrient limits phytoplankton production along the coast of Long Island?

Experiment Pollution from duck farms concentrated near Moriches Bay adds both nitrogen and phosphorus to the coastal water off Long Island, New York. To determine which nutrient limits phytoplankton growth in this area, John Ryther and William Dunstan, of the Woods Hole Oceanographic Institution, cultured the phytoplankton *Nannochloris atomus* with water collected from several sites, identified as A through G. They added either ammonium (NH_4^+) or phosphate (PO_4^{3-}) to some of the cultures.

Results The addition of ammonium caused heavy phytoplankton growth in the cultures, but the addition of phosphate did not.

Conclusion Nitrogen is the nutrient that limits phytoplankton growth in this ecosystem because adding phosphorus did not increase *Nannochloris* growth, whereas adding nitrogen increased phytoplankton density dramatically.

Data from J. H. Ryther and W. M. Dunstan, Nitrogen, phosphorus, and eutrophication in the coastal marine environment, *Science* 171:1008–1013 (1971).

WHAT IF? *Predict how the results would change if water samples were drawn from areas where new duck farms had greatly increased the amount of pollution in the water. Explain.*

On the flip side, areas of *upwelling*, where deep, nutrient-rich waters circulate to the ocean surface, have exceptionally high primary production. This fact supports the hypothesis that nutrient availability determines marine primary production. Because upwelling stimulates growth of the phytoplankton that form the base of marine food webs, upwelling areas typically host highly productive, diverse ecosystems and are prime fishing locations. The largest areas of upwelling occur in the Southern Ocean (also called the Antarctic Ocean), along the equator, and in the coastal waters off Peru, California, and parts of western Africa.

Nutrient limitation is also common in freshwater lakes. During the 1970s, scientists showed that the sewage and fertilizer runoff from farms and lawns adds considerable nutrients to lakes, promoting the growth of primary producers.

Table 55.1 Nutrient Enrichment Experiment for Sargasso Sea	
Nutrients Added to Experimental Culture	**Relative Uptake of ^{14}C by Cultures***
None (controls)	1.00
Nitrogen (N) + phosphorus (P) only	1.10
N + P + metals, excluding iron (Fe)	1.08
N + P + metals, including Fe	12.90
N + P + Fe	12.00

*^{14}C uptake by cultures measures primary production.

Data from D. W. Menzel and J. H. Ryther, Nutrients limiting the production of phytoplankton in the Sargasso Sea, with special reference to iron, *Deep Sea Research* 7:276–281 (1961).

INTERPRET THE DATA *The element molybdenum (Mo) is another micronutrient that can limit primary production in the oceans. If the researchers found the following results for additions of Mo, what would you conclude about its relative importance for growth?*

N + P + Mo: 6.0   N + P + Fe + Mo: 72.0

When the primary producers die, their bodies are broken down by aerobic decomposers. This depletes the water of much or all of its oxygen, killing large numbers of fish and causing the loss of many fish species from the lakes.

To prevent such fish kills, scientists need to know which nutrient is responsible. While nitrogen rarely limits primary production in lakes, many whole-lake experiments showed that phosphorus availability limited cyanobacterial growth. This and other ecological research led to the use of phosphate-free detergents and other water quality reforms.

Primary Production in Terrestrial Ecosystems

At regional and global scales, temperature and moisture are the main factors controlling primary production in terrestrial ecosystems. Primary production is greater in wetter ecosystems, as shown for the plot of NPP and annual precipitation in **Figure 55.7**. NPP also increases with temperature and the

▼ **Figure 55.7 A global relationship between NPP and mean annual precipitation for terrestrial ecosystems.**

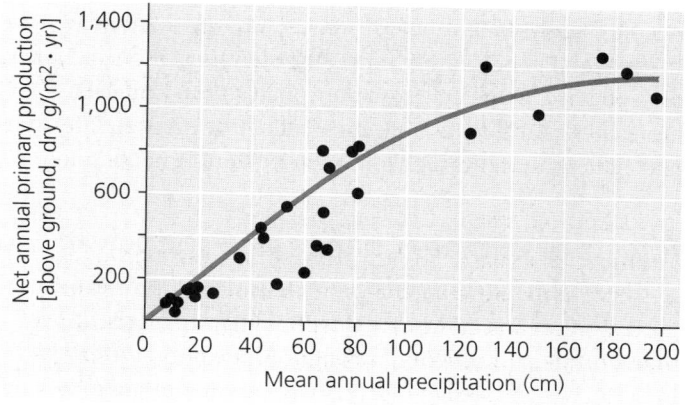

amount of solar energy available to drive evaporation and transpiration. As a result, tropical rain forests, with their warm, wet conditions that promote plant growth, are the most productive terrestrial ecosystems. In contrast, low-productivity systems are generally hot and dry, like many deserts, or cold and dry, like arctic tundra. Between these extremes lie the temperate forest and grassland ecosystems, with moderate climates and intermediate productivity.

Nutrients can also affect NPP in terrestrial ecosystems. A dramatic example occurred after the arctic fox (see Figure 55.1) was introduced onto islands near Alaska. The presence of foxes had the surprising effect of converting grasslands to tundra, thereby lowering NPP. The foxes fed voraciously on the islands' seabirds, decreasing their density almost 100-fold. Fewer seabirds meant less bird guano (waste), a primary source of essential nutrients for plants on the islands. Researchers suspected that the scarcity of nutrients reduced the growth of nutrient-hungry grasses, favoring instead the slower-growing plants typical of tundra. To test this explanation, they added fertilizer to plots of tundra on one of the fox-infested islands. Three years later, the fertilized plots had reverted back to grassland.

Nutrient Limitations and Adaptations That Reduce Them

EVOLUTION As in aquatic systems, nitrogen and phosphorus are the nutrients that most commonly limit terrestrial production. Globally, nitrogen limits plant growth most. Phosphorus limitations are common in older soils where phosphate molecules have been leached away by water, such as in many tropical ecosystems. Note that adding a nonlimiting nutrient, even one that is scarce, will not stimulate production. Conversely, adding more of the limiting nutrient will increase production until some other nutrient becomes limiting.

Various adaptations have evolved in plants that can increase their uptake of limiting nutrients. One important adaptation is the mutualism between plant roots and nitrogen-fixing bacteria. Another is the mycorrhizal association between plant roots and fungi that supply phosphorus and other limiting elements to plants (see Figure 37.14). Plant roots also have hairs and other anatomical features that increase the area of soil in contact with the roots (see Figures 33.8 and 35.3). Many plants release enzymes and other substances into the soil that increase the availability of limiting nutrients; such substances include phosphatases, which cleave a phosphate group from larger molecules, and certain molecules (called chelating agents) that make micronutrients such as iron more soluble in the soil.

Effects of Climate Change on Production

As we've seen, climatic factors such as temperature and precipitation affect terrestrial NPP. Thus, we might expect that climate change could affect production in terrestrial

ecosystems—and it does. For example, satellite data showed that from 1982 to 1999, NPP increased by 6% in terrestrial ecosystems. Nearly half of this increase occurred in the tropical forests of the Amazon, where changing climate patterns had caused cloud cover to decrease, thereby increasing the amount of solar energy available to primary producers. Since 2000, however, these gains in NPP have been erased. This reversal was affected by another aspect of climate change: a series of major droughts in the Southern Hemisphere.

Effects of climate change on NPP can also be seen in the impact of "hotter droughts" on wildfires and insect outbreaks. Consider forests in the American Southwest. In recent decades, the forests of this region have experienced droughts driven by climate warming and changing patterns of precipitation. These ongoing droughts, in turn, have led to increases in the area burned by wildfires and the area affected by outbreaks of bark beetles such as the mountain pine beetle *Dendroctonus ponderosae* **(Figure 55.8)**. As a result, tree mortality has increased and NPP has decreased in these forests.

▼ **Figure 55.8 Climate change, wildfires, and insect outbreaks.** Forests in the American southwest are experiencing hotter droughts caused by rising temperatures in the summer and reduced snowfall in the winter. The drought-stress index indicates how greatly trees are stressed by these conditions; rising values of this index correspond to increasing drought. Higher drought stress correlates with increasing area burned by wildfires (top) and affected by bark beetles (bottom), which target drought-stressed trees with weakened defenses.

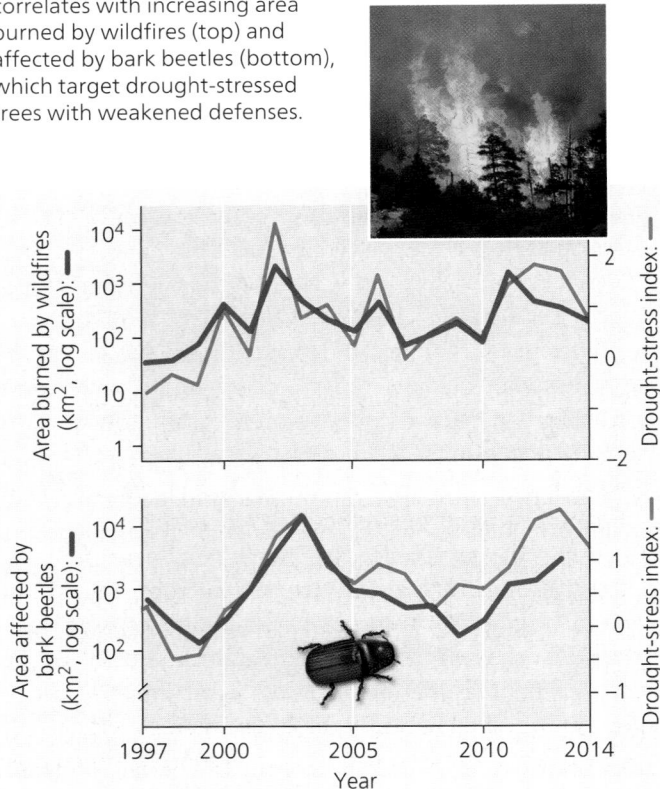

Can an insect outbreak threaten a forest's ability to absorb CO_2 from the atmosphere?

One way to combat climate change is to plant trees, since trees absorb large amounts of CO_2 from the atmosphere, converting it to biomass through photosynthesis. But what happens to the carbon stored as biomass in trees when an insect population explodes in number? Such insect outbreaks have become more frequent with climate change.

A tree with dozens of "pitch tubes," indications of a damaging outbreak of mountain pine beetles (inset).

In this exercise, you will test whether an outbreak of the mountain pine beetle (*Dendroctonus ponderosae*) alters the amount of CO_2 that a forest ecosystem absorbs from and releases to the atmosphere.

Your Approach The principle guiding your investigation is that every ecosystem both absorbs and releases CO_2. Net ecosystem production (NEP) indicates whether an ecosystem is a carbon sink (absorbing more CO_2 from the atmosphere than it releases; this occurs when NEP > 0) or a carbon source (releasing more CO_2 than it absorbs; NEP < 0). To find out if the mountain pine beetle affects NEP, you will determine a forest's NEP before and after a recent outbreak of this insect.

Your Data From 2000 to 2006, an outbreak of the mountain pine beetle killed millions of trees in British Columbia, Canada. The impact of such outbreaks on whether forests gain carbon (NEP > 0) or lose carbon (NEP < 0) was poorly understood. To find out, ecologists estimated net primary production (NPP) and cellular respiration by decomposers and other heterotrophs (R_h), before and after the outbreak. These data allow forest NEP to be calculated from the equation NEP = NPP − R_h.

	NPP $[g/(m^2 \cdot yr)]$	R_h $[g/(m^2 \cdot yr)]$	NEP $[g/(m^2 \cdot yr)]$
Before outbreak	440	408	
After outbreak	400	424	

Your Analysis

1. Complete the table by calculating the values of NEP before and after the outbreak. Before the outbreak, was the forest a carbon sink or a carbon source? After the outbreak?
2. NEP is often defined as NEP = GPP − R_T, where GPP is gross primary production and R_T equals cellular respiration by autotrophs (R_a) *plus* cellular respiration by heterotrophs (R_h). Use the relation NPP = GPP − R_a to show that the two equations for NEP introduced in this exercise are equivalent.
3. Based on your results in question 1, predict whether the mountain pine beetle outbreak could have feedback effects on the global climate. Explain.

→ **Instructors:** A version of this Problem-Solving Exercise can be assigned in **Mastering Biology**.

Climate change can also affect whether an ecosystem stores or loses carbon over time. As discussed earlier, net ecosystem production, or NEP, reflects the total biomass accumulation that occurs during a given period of time. When NEP > 0, the ecosystem gains more carbon than it loses; such ecosystems store carbon and are said to be a carbon *sink*. In contrast, when NEP < 0, the ecosystem loses more carbon than it gains; such ecosystems are a carbon *source*.

Recent research shows that climate change can cause an ecosystem to switch from a carbon sink to a carbon source. For example, in some arctic ecosystems, climate warming has increased the metabolic activities of soil microorganisms, causing an uptick in the amount of CO_2 produced in cellular respiration. In these ecosystems, the total amount of CO_2 produced in cellular respiration now exceeds what is absorbed in photosynthesis. As a result, these ecosystems—which once were carbon sinks—are now carbon sources. When this happens, an ecosystem may contribute to climate change

by releasing more CO_2 than it absorbs. Indeed, a 2017 study found that with climate warming, large regions of tundra in Alaska now release more CO_2 than they absorb—and for the years 2013 and 2014, the entire state of Alaska released more CO_2 than it absorbed. In the **Problem-Solving Exercise**, you can examine how outbreaks of an insect population may affect the NEP of forest ecosystems.

CONCEPT CHECK 55.2

1. Why is only a small portion of the solar energy that strikes Earth's atmosphere stored by primary producers?
2. How can ecologists experimentally determine the factor that limits primary production in an ecosystem?
3. **WHAT IF?** Suppose a forest was heavily burned by a wildfire. Predict how NEP of this forest would change over time.
4. **MAKE CONNECTIONS** Explain how nitrogen and phosphorus, the nutrients that most often limit primary production, are necessary for the Calvin cycle to function in photosynthesis (see Concept 10.3).

For suggested answers, see Appendix A.

Energy transfer between trophic levels is typically only 10% efficient

The amount of chemical energy in consumers' food that is converted to their own new biomass during a given period is called the **secondary production** of the ecosystem. Consider the transfer of organic matter from primary producers to herbivores, the primary consumers. In most ecosystems, herbivores eat only a small fraction of plant material produced; globally, they consume only about one-sixth of total plant production. Moreover, they cannot digest all the plant material that they *do* eat, as anyone who has walked through a field where cattle have been grazing will attest. Most of an ecosystem's production is eventually consumed by decomposers. Let's analyze the process of energy transfer more closely.

Production Efficiency

We'll begin by examining secondary production in one organism—a caterpillar. When a caterpillar feeds on a leaf, only about 33 J out of 200 J, or one-sixth of the potential energy in the leaf, is used for secondary production, or growth **(Figure 55.9)**. The caterpillar stores some of the remaining energy in organic compounds that will be used for cellular respiration and passes the rest in its feces. The energy in the feces remains in the ecosystem temporarily, but most of it is lost as heat after the feces are consumed by decomposers. The energy used for the caterpillar's respiration is also eventually lost from the ecosystem as heat. Only the chemical energy stored by herbivores as biomass, through growth or the production of offspring, is available as food to secondary consumers.

▼ **Figure 55.9 Energy partitioning within a link of the food chain.**

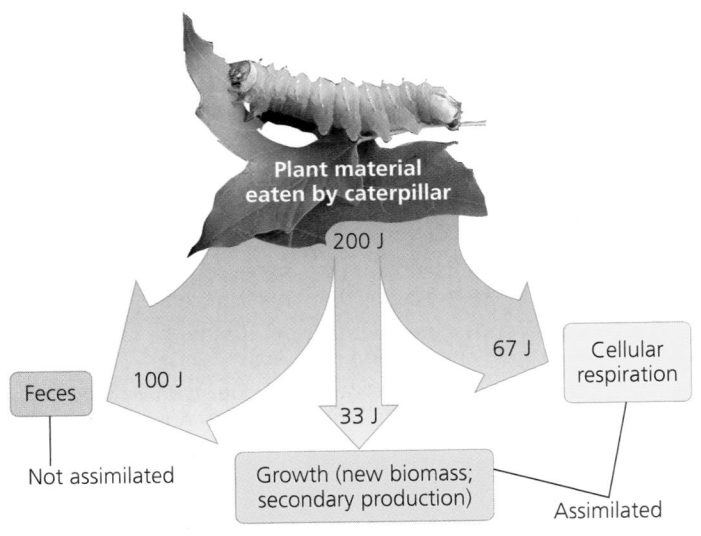

INTERPRET THE DATA *What percentage of the energy in the caterpillar's food is actually used for secondary production (growth)?*

We can measure the efficiency of animals as energy transformers using the following equation:

$$\text{Production efficiency} = \frac{\text{Net secondary production} \times 100\%}{\text{Assimilation of primary production}}$$

Net secondary production is the amount of energy an organism has consumed and used for growth and reproduction. Assimilation consists of the total amount of energy an organism has consumed and used for growth, reproduction, and respiration. **Production efficiency**, therefore, is the percentage of energy stored in assimilated food that is used for growth and reproduction, *not* respiration. For the caterpillar in Figure 55.9, production efficiency is 33%; 67 J of the 100 J of assimilated energy is used for respiration. (The 100 J of energy lost as undigested material in feces does not count toward assimilation.) Birds and mammals typically have low production efficiencies, in the range of 1–3%, because they use so much energy in maintaining a constant, high body temperature. Fishes, which are mainly ectothermic (see Concept 40.3), have production efficiencies around 10%. Insects and microorganisms are even more efficient, with production efficiencies averaging 40% or more.

Trophic Efficiency and Ecological Pyramids

Let's scale up now from the production efficiencies of individual consumers to the flow of energy through trophic levels.

Trophic efficiency is the percentage of production transferred from one trophic level to the next. Trophic efficiencies must always be less than production efficiencies because they take into account not only the energy lost through respiration and contained in feces, but also the energy in organic material in a lower trophic level that is not consumed by the next trophic level. Trophic efficiencies range from roughly 5% to 20% in different ecosystems, but on average are only about 10%. In other words, 90% of the energy available at one trophic level typically is *not* transferred to the next. This loss is multiplied over the length of a food chain. If 10% of available energy is transferred from primary producers to primary consumers, such as caterpillars, and 10% of that energy is transferred to secondary consumers (carnivores), then only 1% of net primary production is available to secondary consumers (10% of 10%). In the Scientific Skills Exercise, you can calculate trophic efficiency and other measures of energy flow in a salt marsh ecosystem.

The loss of energy along a food chain limits the abundance of top-level carnivores that an ecosystem can support. Only about 0.1% of the chemical energy fixed by photosynthesis can flow through a food web to a tertiary consumer, such as a snake or a shark. This helps explain why most food webs include only about four or five trophic levels (see Figure 54.17).

The loss of energy with each transfer in a food chain can be represented by an *energy pyramid*, in which the net productions of different trophic levels are arranged in tiers **(Figure 55.10)**. The width of each tier is proportional to the net production,

Interpreting Quantitative Data

How Efficient Is Energy Transfer in a Salt Marsh Ecosystem? In a classic experiment, John Teal studied the flow of energy through the producers, consumers, and decomposers in a salt marsh. In this exercise, you will use the data from this study to calculate some measures of energy transfer between trophic levels in this ecosystem.

How the Study Was Done Teal measured the amount of solar radiation entering a salt marsh in Georgia over a year. He also measured the aboveground biomass of the dominant primary producers, which were grasses, as well as the biomass of the dominant consumers, including insects, spiders, and crabs, and of the detritus that flowed out of the marsh to the surrounding coastal waters. To determine the amount of energy in each unit of biomass, he dried the biomass, burned it in a calorimeter, and measured the amount of heat produced.

INTERPRET THE DATA

1. What percentage of solar energy is incorporated into the marsh as gross primary production? As net primary production?

Data from the Study

Form of Energy	Amount [kcal/(m² · yr)]
Solar radiation	600,000
Gross grass production	34,580
Net grass production	6,585
Gross insect production	305
Net insect production	81
Detritus leaving marsh	3,671

Data from J. M. Teal, Energy flow in the salt marsh ecosystem of Georgia, *Ecology* 43:614–624 (1962).

2. How much energy is lost by primary producers as respiration in this ecosystem? How much is lost as respiration by the insect population?

3. If all of the detritus leaving the marsh is plant material, what percentage of all net primary production leaves the marsh as detritus each year?

⮕ **Instructors:** A version of this Scientific Skills Exercise can be assigned in **Mastering Biology.**

▼ **Figure 55.10 An idealized pyramid of energy.** This example assumes a trophic efficiency of 10% for each link in the food chain. Notice that primary producers convert only about 1% of the energy available to them to net primary production.

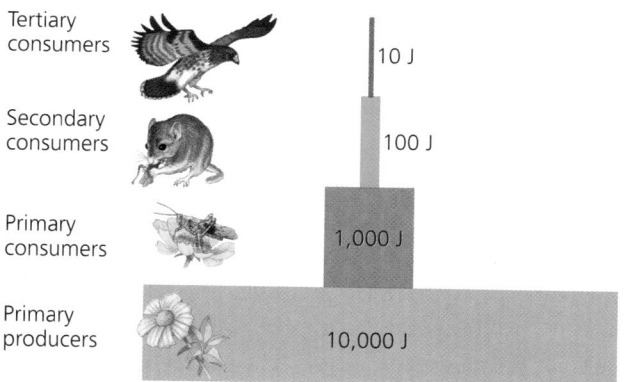

Tertiary consumers — 10 J
Secondary consumers — 100 J
Primary consumers — 1,000 J
Primary producers — 10,000 J

1,000,000 J of sunlight

expressed in joules, of each trophic level. The highest level, which represents top-level predators, contains relatively few individuals. The small population size typical of top predators is one reason they tend to be vulnerable to extinction (and to the evolutionary consequences of small population size; see Concept 23.3).

One ecological consequence of low trophic efficiencies is represented in a *biomass pyramid*, in which each tier represents the total dry mass of all organisms in one trophic level. Most biomass pyramids narrow sharply from primary producers at the base to top-level carnivores at the apex because energy transfers between trophic levels are so inefficient **(Figure 55.11a)**. Certain aquatic ecosystems, however, have

▼ **Figure 55.11 Pyramids of biomass.** Numbers denote the dry mass of all organisms at each trophic level.

Trophic level	Dry mass (g/m²)
Tertiary consumers	1.5
Secondary consumers	11
Primary consumers	37
Primary producers	809

(a) Most biomass pyramids show a sharp decrease in biomass at successively higher trophic levels, as illustrated by data from a Florida bog.

Trophic level	Dry mass (g/m²)
Primary consumers (zooplankton)	21
Primary producers (phytoplankton)	4

(b) In some aquatic ecosystems, such as the English Channel, a small biomass of primary producers (phytoplankton) supports a larger biomass of primary consumers (zooplankton).

inverted biomass pyramids, in which primary consumers outweigh the producers **(Figure 55.11b)**.

Such inverted biomass pyramids occur because the producers—phytoplankton—grow, reproduce, and are consumed so quickly by the zooplankton that their total biomass remains at comparatively low levels. However, because the phytoplankton continually replace their biomass at such a rapid rate, they can support a biomass of zooplankton bigger than their own biomass. Likewise, because phytoplankton reproduce so quickly and have much higher production than zooplankton, the pyramid of *energy* for this ecosystem is still bottom-heavy, like the one in Figure 55.10.

The dynamics of energy flow through ecosystems have implications for human consumers. For example, eating meat is a relatively inefficient way of tapping photosynthetic production. The same pound of soybeans that a person could eat for protein produces only a fifth of a pound of beef or less when fed to a cow. Agriculture worldwide could, in fact, feed many more people and require less land if we all fed more efficiently—as primary consumers, eating plant material.

CONCEPT CHECK 55.3

1. If an insect that eats plant seeds containing 100 J of energy uses 30 J of that energy for respiration and excretes 50 J in its feces, what is the insect's net secondary production? What is its production efficiency?
2. Tobacco leaves contain nicotine, a poisonous compound that is energetically expensive for the plant to make. What advantage might the plant gain by using some of its resources to produce nicotine?
3. **WHAT IF?** Decomposers are consumers that obtain their energy from detritus. How many joules of energy are potentially available to decomposers in the ecosystem represented in Figure 55.10?

For suggested answers, see Appendix A.

CONCEPT 55.4

Biological and geochemical processes cycle nutrients and water in ecosystems

Although most ecosystems receive abundant solar energy, chemical elements are available only in limited amounts. Life therefore depends on the recycling of essential chemical elements. Much of an organism's chemical stock is replaced continuously as nutrients are assimilated and waste products are released. When the organism dies, the atoms in its body are returned to the atmosphere, water, or soil by decomposers. By liberating nutrients from organic matter, decomposition replenishes the pools of inorganic nutrients that plants and other autotrophs use to build new organic matter.

Decomposition and Nutrient Cycling Rates

Decomposers are heterotrophs that get their energy from detritus. Their growth is controlled by the same factors that limit primary production in ecosystems, including temperature, moisture, and nutrient availability. Decomposers usually grow faster and decompose material more quickly in warmer ecosystems **(Figure 55.12)**. In tropical rain forests, most organic material decomposes in a few months to a few years, whereas in temperate forests, decomposition takes four to six years, on average. The difference is largely the result of the higher temperatures and more abundant precipitation in tropical rain forests.

▼ **Figure 55.12** Inquiry

How does temperature affect litter decomposition in an ecosystem?

Experiment Researchers with the Canadian Forest Service placed identical samples of organic material—litter—on the ground in 21 sites across Canada. Three years later, they returned to see how much of each sample had decomposed.

Results The mass of litter in the warmest ecosystem decreased four times faster than in the coldest ecosystem.

Conclusion Decomposition rate increases with temperature across much of Canada.

Data from J. A. Trofymow and the CIDET Working Group, *The Canadian Intersite Decomposition Experiment: Project and Site Establishment Report* (Information Report BC-X-378), Natural Resources Canada, Canadian Forest Service, Pacific Forestry Centre (1998) and T. R. Moore et al., Litter decomposition rates in Canadian forests, *Global Change Biology* 5:75–82 (1999).

WHAT IF? *What factors other than temperature might also have varied across these 21 sites? How might this variation have affected the interpretation of the results?*

Because decomposition in a tropical rain forest is rapid, relatively little organic material accumulates as leaf litter on the forest floor; about 75% of the ecosystem's nutrients is present in the woody trunks of trees, and only about 10% is contained in the soil. Thus, the relatively low concentrations of some nutrients in the soil of tropical rain forests result from a short cycling time, not from a lack of these elements in the ecosystem. In temperate forests, where decomposition is much slower, the soil may contain as much as 50% of all the organic material in the ecosystem. The nutrients that are present in temperate forest detritus and soil may remain there for years before plants assimilate them.

Decomposition on land is also slower when conditions are either too dry for decomposers to thrive or too wet to supply them with enough oxygen. Decomposers grow especially poorly in ecosystems that are both cold and wet, such as peatlands. As a result, net primary production greatly exceeds the rate of decomposition in such ecosystems, causing them to store large amounts of organic matter.

In aquatic ecosystems, decomposition in anaerobic muds can take 50 years or longer. Bottom sediments are comparable to the detritus layer in terrestrial ecosystems, but algae and aquatic plants usually assimilate nutrients directly from the water. Thus, the sediments often constitute a nutrient sink, and aquatic ecosystems are very productive only when there is exchange between the bottom layers of water and surface waters (as occurs in the upwelling regions described earlier).

Biogeochemical Cycles

Because nutrient cycles involve both biotic and abiotic components, they are called **biogeochemical cycles**. We can recognize two general scales of biogeochemical cycles: global and local. Gaseous forms of carbon, oxygen, sulfur, and nitrogen occur in the atmosphere, and these elements' cycles are essentially global. For example, some of the carbon atoms a plant acquires from the air as CO_2 may have been released into the atmosphere by the respiration of an organism in a distant locale. Other elements—including phosphorus, potassium, and calcium—are too heavy to occur as gases at Earth's surface, although they are transported in dust. In terrestrial ecosystems, these elements cycle more locally, absorbed from the soil by plant roots and eventually returned

to the soil by decomposers. In aquatic systems, however, they cycle more broadly as dissolved forms carried in currents.

Let's first look at a general model of nutrient cycling that includes reservoirs where elements exist and processes that transfer elements between them **(Figure 55.13)**. The nutrients in living organisms and detritus (reservoir A) are available to other organisms when consumers feed and when decomposers consume nonliving organic matter. The low pH and low oxygen levels found in the waterlogged sediments of swamps can inhibit decomposition, leading to the formation of peat. When this occurs, organic materials from dead organisms can be transferred from reservoir A to reservoir B; eventually, peat may be converted to fossil fuels such as coal or oil. Inorganic materials that are dissolved in water or present in soil or air (reservoir C) are available for use. Although most organisms cannot directly tap into the inorganic elements tied up in rocks (reservoir D), these nutrients may slowly become available through weathering and erosion.

Figure 55.14 provides a detailed look at the cycling of water, carbon, nitrogen, and phosphorus. When you study each cycle, consider which steps are driven primarily by biological

▼ **Figure 55.13** VISUALIZING BIOGEOCHEMICAL CYCLES

A biogeochemical cycle diagram summarizes the movements of a chemical element between living and nonliving components of the biosphere. This figure uses a generic model of nutrient cycling to show what types of information may be represented. Then, when you study Figure 55.14, notice how this information is depicted in the specific nutrient cycles.

Cycle diagrams include a representation of the **reservoirs** in which the chemical element is found. A reservoir can consist of either organic or inorganic materials, and those materials may be either available for direct use by organisms or unavailable.

A cycle diagram also represents the movements of the chemical element between reservoirs, usually with arrows. These movements occur through biological, chemical, or geologic processes.

Nutrient cycle diagrams often do not represent the time scale of the different processes. Thus, it's important to keep in mind that some processes, such as photosynthesis, occur on a short time scale within an individual's life span, while others, such as rock formation, occur on a long geologic time scale.

Reservoir A
Organic materials *available* as nutrients

Living organisms, detritus

Nutrient uptake, photosynthesis

Respiration, decomposition, excretion

Waterlogging, fossilization

Reservoir C
Inorganic materials *available* as nutrients

Atmosphere

Water

Soil

Reservoir B
Organic materials *unavailable* as nutrients

Peat

Coal

Oil

Burning of fossil fuels

Weathering, erosion

Reservoir D
Inorganic materials *unavailable* as nutrients

Minerals in rocks

Formation of sedimentary rock

2. *Compare and contrast the time scales at which materials move into and out of reservoirs A and B.*

1. *Suppose decomposition rates slowed greatly. Use the diagram to hypothesize how that might affect the transfer of materials from reservoir A to reservoirs B and C.*

➔ **Instructors:** Additional questions related to this Visualizing Figure can be assigned in **Mastering Biology**.

Examine each cycle closely, considering the major reservoirs of water, carbon, nitrogen, and phosphorus and the processes that drive each cycle. The widths of the arrows in the diagrams approximately reflect the relative contribution of each process to the movement of water or a nutrient in the biosphere.

The Water Cycle

Biological importance Water is essential to all organisms, and its availability influences the rates of ecosystem processes, particularly primary production and decomposition in terrestrial ecosystems.

Forms available to life All organisms are capable of exchanging water directly with their environment. Liquid water is the primary physical phase in which water is used, though some organisms can harvest water vapor. Freezing of soil water can limit water availability to terrestrial plants.

Reservoirs The oceans contain 97% of the water in the biosphere. Approximately 2% is bound in glaciers and polar ice caps, and the remaining 1% is in lakes, rivers, and groundwater, with a negligible amount in the atmosphere.

Key processes The main processes driving the water cycle are evaporation of liquid water by solar energy, condensation of water vapor into clouds, and precipitation. Transpiration by terrestrial plants also moves large volumes of water into the atmosphere. Surface and groundwater flow can return water to the oceans, completing the water cycle.

The Carbon Cycle

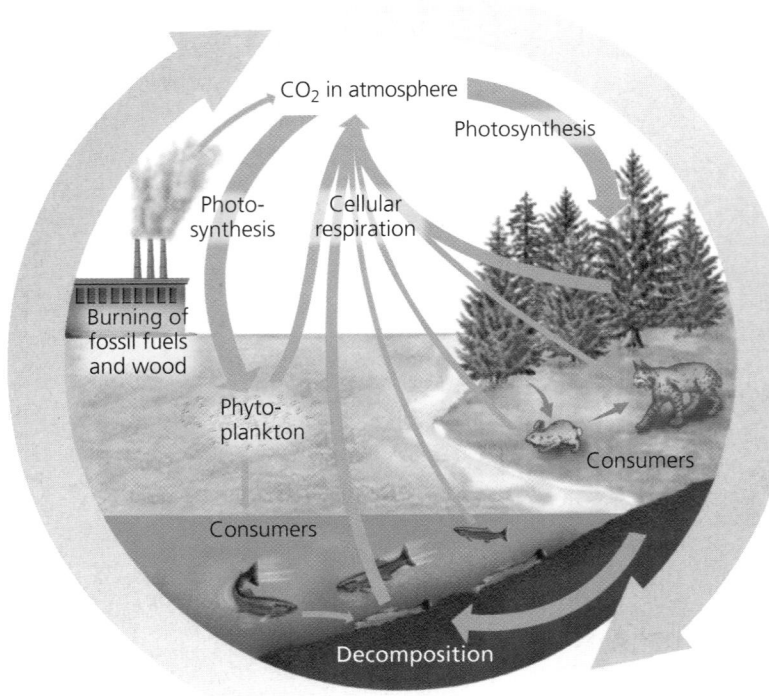

Biological importance Carbon forms the framework of the organic molecules essential to all organisms.

Forms available to life Photosynthetic organisms utilize CO_2 during photosynthesis and convert the carbon to organic forms that are used by consumers, including animals, fungi, and heterotrophic protists and prokaryotes.

Reservoirs The major reservoirs of carbon include fossil fuels, soils, the sediments of aquatic ecosystems, the oceans (dissolved carbon compounds), plant and animal biomass, and the atmosphere (CO_2). The largest reservoir is sedimentary rocks such as limestone; however, carbon remains in this pool for long periods of time. All organisms are capable of returning carbon directly to their environment in its original form (CO_2) through respiration.

Key processes Photosynthesis by plants and phytoplankton removes substantial amounts of atmospheric CO_2 each year. This quantity is approximately equal to the CO_2 added to the atmosphere through cellular respiration by producers and consumers. The burning of fossil fuels and wood is adding significant amounts of additional CO_2 to the atmosphere. Over geologic time, volcanoes are also a substantial source of CO_2.

→ **Mastering Biology** BioFlix® Animation: The Carbon Cycle

The Nitrogen Cycle

Biological importance Nitrogen is part of amino acids, proteins, and nucleic acids and is often a limiting plant nutrient.

Forms available to life Plants can assimilate (use) two inorganic forms of nitrogen—ammonium (NH_4^+) and nitrate (NO_3^-)—and some organic forms, such as amino acids. Various bacteria can use all of these forms as well as nitrite (NO_2^-). Animals can use only organic forms of nitrogen.

Reservoirs The main reservoir of nitrogen is the atmosphere, which is 80% free nitrogen gas (N_2). The other reservoirs of inorganic and organic nitrogen compounds are soils and the sediments of lakes, rivers, and oceans; surface water and groundwater; and the biomass of living organisms.

Key processes The major pathway for nitrogen to enter an ecosystem is via *nitrogen fixation*, the conversion of N_2 to forms that can be used to synthesize organic nitrogen compounds. Certain bacteria, as well as lightning and volcanic activity, fix nitrogen naturally. Nitrogen inputs from human activities now outpace natural inputs on land. Two major contributors are industrially produced fertilizers and legume crops that fix nitrogen via bacteria in their root nodules. Other bacteria in soil convert nitrogen to different forms. Examples include nitrifying bacteria, which convert ammonium to nitrate, and denitrifying bacteria, which convert nitrate to nitrogen gas. Human activities also release large quantities of reactive nitrogen gases, such as nitrogen oxides, to the atmosphere.

→ **Mastering Biology** Animation: The Nitrogen Cycle

The Phosphorus Cycle

Biological importance Organisms require phosphorus as a major constituent of nucleic acids, phospholipids, and ATP and other energy-storing molecules and as a mineral constituent of bones and teeth.

Forms available to life The most biologically important inorganic form of phosphorus is phosphate (PO_4^{3-}), which plants absorb and use in the synthesis of organic compounds.

Reservoirs The largest accumulations of phosphorus are in sedimentary rocks of marine origin. There are also large quantities of phosphorus in soil, in the oceans (in dissolved form), and in organisms. Because soil particles bind PO_4^{3-}, the recycling of phosphorus tends to be quite localized in ecosystems.

Key processes Weathering of rocks gradually adds PO_4^{3-} to soil; some leaches into groundwater and surface water and may eventually reach the sea. Phosphate taken up by producers and incorporated into biological molecules may be eaten by consumers. Phosphate is returned to soil or water by either decomposition of biomass or excretion by consumers. Because there are no significant phosphorus-containing gases, only relatively small amounts of phosphorus move through the atmosphere, usually in the forms of dust and sea spray.

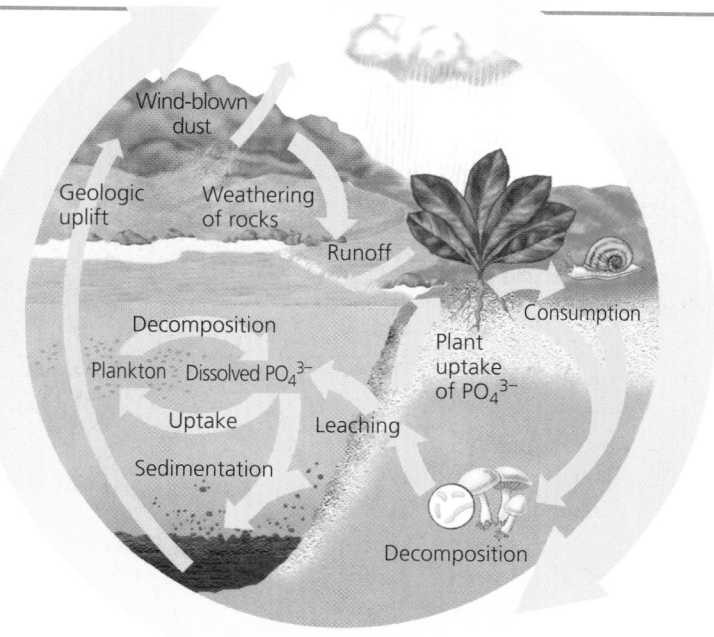

processes. For the carbon cycle, for instance, plants, animals, and other organisms control most of the key steps, including photosynthesis and decomposition. For the water cycle, however, purely physical processes control many key steps, such as evaporation from the oceans. Note also that human actions, such as the burning of fossil fuels and the production of fertilizers, have had major effects on the global cycling of carbon and nitrogen.

Case Study: Nutrient Cycling in the Hubbard Brook Experimental Forest

Since 1963, ecologist Gene Likens and colleagues have been studying nutrient cycling at the Hubbard Brook Experimental Forest in the White Mountains of New Hampshire. Their research site is a deciduous forest that grows in six small valleys, each drained by a single creek. Impermeable bedrock underlies the soil of the forest.

The research team first determined the mineral budget for each of six valleys by measuring the input and outflow of several key nutrients. They collected rainfall at several sites to measure the amount of water and dissolved minerals added to the ecosystem. To monitor the loss of water and minerals, they constructed a small concrete dam with a V-shaped spillway across the creek at the bottom of each valley **(Figure 55.15a)**. They found that about 60% of the water added to the ecosystem as rainfall and snow exits through the stream, and the remaining 40% is lost by evapotranspiration.

Preliminary studies confirmed that internal cycling conserved most of the mineral nutrients in the system. For example, only about 0.3% more calcium (Ca^{2+}) leaves a valley via its creek than is added by rainwater, and this small net loss is probably replaced by chemical decomposition of the bedrock. During most years, the forest even registers small net gains of a few mineral nutrients, including nitrogen.

Experimental deforestation of a watershed dramatically increased the flow of water and minerals leaving the watershed **(Figure 55.15b)**. Over three years, water runoff from the newly deforested watershed was 30–40% greater than in a control watershed, apparently because there

were no plants to absorb and transpire water from the soil. Most remarkable was the loss of nitrate, whose concentration in the creek increased 60-fold, reaching levels considered unsafe for drinking water **(Figure 55.15c)**. The Hubbard Brook deforestation study showed that the amount of nutrients leaving an intact forest ecosystem is controlled mainly by the plants. Retaining nutrients in an ecosystem helps to maintain the productivity of the system as well as to avoid problems elsewhere, such as algal blooms caused by excess nutrient runoff that enters a downstream ecosystem.

→ **Mastering Biology** Interview with Eugene Likens: Co-founder of the Hubbard Brook Forest Study

▼ **Figure 55.15 Nutrient cycling in the Hubbard Brook Experimental Forest.**

(a) Concrete dams and weirs built across streams at the bottom of watersheds enabled researchers to monitor the outflow of water and nutrients from the ecosystem.

(b) One watershed was clear-cut to study the effects of the loss of vegetation on drainage and nutrient cycling. All of the original plant material was left in place to decompose.

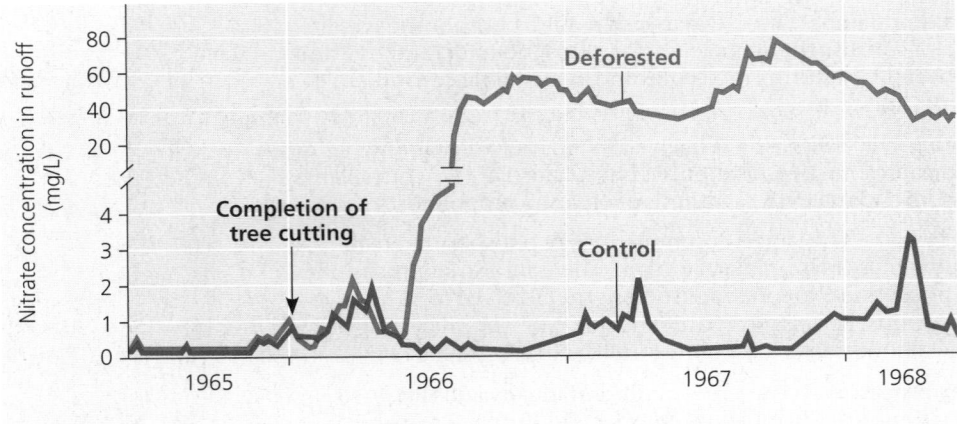

(c) The concentration of nitrate in runoff from the deforested watershed was 60 times greater than in a control (unlogged) watershed.

VISUAL SKILLS *If this graph were redrawn so that the y-axis had a consistent scale, what information would be emphasized compared to the above graph? What information might be lost or harder to see? Explain.*

→ **Instructors:** A related Experimental Inquiry Tutorial can be assigned in **Mastering Biology**.

CONCEPT 55.5

Restoration ecologists return degraded ecosystems to a more natural state

Ecosystems can recover naturally from many disturbances (including the experimental deforestation at Hubbard Brook) through the stages of ecological succession (see Concept 54.3). Sometimes, however, that recovery takes centuries, particularly when humans have degraded the environment. Tropical areas that are cleared for farming may quickly become unproductive because of nutrient losses. Mining activities may last for several decades, and the lands are often abandoned in a degraded state. Ecosystems can also be damaged by salts that build up in soils from irrigation and by toxic chemicals or oil spills. Biologists increasingly are called on to help restore and repair damaged ecosystems.

Restoration ecologists seek to initiate or speed up the recovery of degraded ecosystems. One of the basic assumptions is that environmental damage is at least partly reversible. This optimistic view must be balanced by a second assumption—that ecosystems are not infinitely resilient. Restoration ecologists therefore work to identify and manipulate the processes that most limit recovery of ecosystems

from disturbances. Where disturbance is so severe that restoring all of a habitat is impractical, ecologists try to reclaim as much of a habitat or ecological process as possible, within the limits of the time and money available to them.

In extreme cases, the physical structure of an ecosystem may need to be restored before biological restoration can occur. If a stream was straightened to channel water quickly through a suburb, ecologists may reconstruct a meandering channel to slow down the flow of water eroding the stream bank. To restore an open-pit mine, engineers may grade the site with heavy equipment to reestablish a gentle slope, spreading topsoil when the slope is in place **(Figure 55.16)**.

After any physical reconstruction of the ecosystem is complete, the next step is biological restoration. The long-term objective of restoration is to return an ecosystem as closely as possible to its predisturbance state. **Figure 55.17** explores four ambitious and successful restoration projects. These and the many other such projects throughout the world often employ two key strategies: bioremediation and biological augmentation.

 Mastering Biology HHMI Video: Response of Corals to Ocean Warming and Efforts to Restore Damaged Reefs

Bioremediation

Using organisms—usually prokaryotes, fungi, or plants—to detoxify polluted ecosystems is known as **bioremediation**. Some plants and lichens adapted to soils containing heavy metals can accumulate high concentrations of toxic metals such as lead and cadmium in their tissues. Restoration ecologists can introduce such species to sites polluted by mining and other human activities and then harvest these organisms to remove the metals from the ecosystem. For instance, researchers in the United Kingdom have discovered a lichen species that grows

▼ **Figure 55.16 A gravel and clay mine site in New Jersey before and after restoration.**

(a) In 1991, before restoration

(b) In 2010, several years after restoration

The examples highlighted on this page are just a few of the many restoration ecology projects taking place around the world.

Kissimmee River, Florida

In the 1960s, the Kissimmee River was converted from a meandering river to a 90-km canal to control flooding. This channelization diverted water from the floodplain, causing the wetlands to dry up, threatening many fish and wetland bird populations. Kissimmee River restoration has filled 12 km of drainage canal and reestablished 24 km of the original 167 km of natural river channel. Pictured here is a section of the Kissimmee canal that has been plugged (wide, light strip on the right side of the photo), diverting flow into remnant river channels (center of the photo). The project will also restore natural flow patterns, which will foster self-sustaining populations of wetland birds and fishes.

Succulent Karoo, South Africa

In the Succulent Karoo desert region of southern Africa, as in many arid regions, overgrazing by livestock has damaged vast areas. Private landowners and government agencies in South Africa are restoring large areas of this unique region, revegetating the land and employing more sustainable resource management. The photo shows a small sample of the exceptional plant diversity of the Succulent Karoo; its 5,000 plant species include the highest diversity of succulent plants in the world.

Maungatautari, New Zealand

Weasels, rats, pigs, and other introduced species pose a serious threat to New Zealand's native plants and animals, including kiwis, a group of flightless, ground-dwelling bird species. The goal of the Maungatautari restoration project is to exclude all exotic mammals from a 3,400-ha reserve located on a forested volcanic cone. A specialized fence around the reserve eliminates the need to continue setting traps and using poisons that can harm native wildlife. In 2006, a pair of critically endangered takahe (a species of flightless rail) were released into the reserve with the hope of reestablishing a breeding population of this colorful bird on New Zealand's North Island.

Coastal Indonesia

Seaweed and seagrass beds are important nursery grounds for a wide variety of fishes and shellfish. Once extensive but now reduced by development, these beds are being restored in the coastal areas of Indonesia. Techniques include constructing suitable seafloor habitat, hand seeding, and transplanting seaweeds and seagrasses from natural beds (shown in this photograph).

on soil polluted with uranium dust left over from mining. The lichen concentrates uranium in a dark pigment, making it useful as a biological monitor and potentially as a remediator.

Ecologists already utilize many prokaryotes to carry out bioremediation of soils and water. Scientists have sequenced the genomes of at least ten prokaryotic species specifically for their bioremediation potential. One of the species, the bacterium *Shewanella oneidensis*, appears particularly promising. It can metabolize a dozen or more elements under aerobic and anaerobic conditions. In doing so, it converts soluble forms of uranium, chromium, and nitrogen to insoluble forms that are less likely to leach into streams or groundwater. Researchers at Oak Ridge National Laboratory, in Tennessee, stimulated the growth of *Shewanella* and other uranium-reducing bacteria by adding ethanol to groundwater contaminated with uranium; the bacteria can use ethanol as an energy source. In just five months, the concentration of soluble uranium in the ecosystem dropped by 80% **(Figure 55.18)**.

Biological Augmentation

In contrast to bioremediation, which is a strategy for removing harmful substances from an ecosystem, **biological augmentation** uses organisms to *add* essential materials to a degraded ecosystem. To augment ecosystem processes, restoration ecologists need to determine which factors, such as chemical nutrients, have been lost from a system and are limiting its recovery.

Encouraging the growth of plants that thrive in nutrient-poor soils often speeds up succession and ecosystem recovery. In alpine ecosystems of the western United States, nitrogen-fixing plants such as lupines are often planted to raise nitrogen concentrations in soils disturbed by mining and other activities. Once these nitrogen-fixing plants become established, other native species are better able to obtain enough soil nitrogen to survive. In other systems where the soil has been severely disturbed or where topsoil is missing entirely, plant roots may lack the mycorrhizal symbionts that help them meet their nutritional needs (see Concept 31.1). Ecologists restoring a tallgrass prairie in Minnesota recognized this limitation and enhanced the recovery of native species by adding mycorrhizal symbionts to the soil they seeded.

Restoring the physical structure and plant community of an ecosystem does not always ensure that animal species will recolonize a site and persist there. Because animals provide critical ecosystem services, including pollination and seed dispersal, restoration ecologists sometimes help wildlife to reach and use restored ecosystems. They might release animals at a site or establish habitat corridors that connect a restored site to places where the animals are found. They may build perches for birds to use. These and other efforts can increase the biodiversity of restored ecosystems and help the community persist.

▼ **Figure 55.18 Bioremediation of groundwater contaminated with uranium at Oak Ridge National Laboratory, Tennessee.**

(a) Wastes containing uranium were dumped in these four unlined pits for more than 30 years, contaminating soils and groundwater.

(b) After ethanol was added, microbial activity decreased the concentration of soluble uranium in groundwater near the pits.

Ecosystems: *A Review*

Figure 55.19 illustrates energy transfer, nutrient cycling, and other key processes for an arctic tundra ecosystem. Note the conceptual similarities between this figure and Make Connections Figure 10.22, "The Working Cell." The scale of the two figures is different, but the physical laws and biological rules that govern life apply equally to both systems.

CONCEPT CHECK 55.5

1. Identify the main goal of restoration ecology.
2. How do bioremediation and biological augmentation differ?
3. **WHAT IF?** In what way is the Kissimmee River project a more complete ecological restoration than the Maungatautari project (see Figure 55.17)?

For suggested answers, see Appendix A.

MAKE CONNECTIONS

The Working Ecosystem

This arctic tundra ecosystem teems with life during the short two-month growing season each summer. In ecosystems, organisms interact with each other and with the environment around them in diverse ways, including those illustrated here.

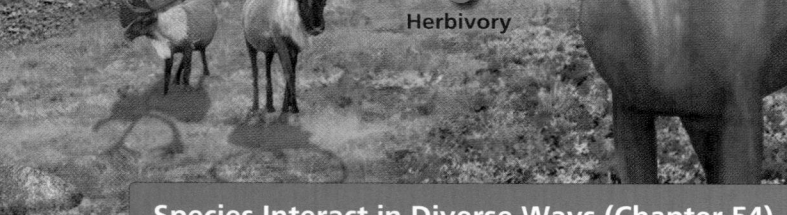

1 Caribou

2 Snow geese

Populations Are Dynamic (Chapter 53)

1 Caribou migrate across the tundra to give birth at their calving grounds each year. In general, populations change in size through births and deaths and through immigration and emigration. (See Figure 53.1 and Concept 53.1.)

2 Snow geese and many other species migrate to the Arctic each spring for the abundant food found there in summer. (See Concept 53.5.)

3 Birth and death rates influence the density of all populations. Death in the tundra comes from many causes, including predation, competition for resources, and lack of food in winter. (See Figure 53.17.)

→ **Mastering Biology** BioFlix® Animation: **Population Ecology**

5 Herbivory

Arctic fox

3

4 Predation

Snow goose

Species Interact in Diverse Ways (Chapter 54)

4 In predation, an individual of one species kills and eats another. (See Concept 54.1.)

5 In herbivory, an individual of one species eats part of a plant or other primary producer, such as a caribou eating a lichen. (See Concept 54.1.)

6 In mutualism, individuals of two species interact in ways that benefit each other. In some mutualisms, the partners live in direct contact, forming a symbiosis; for example, a lichen is a symbiotic mutualism between a fungus and an alga or cyanobacterium. (See Concept 54.1 and Figures 31.22 and 31.23.)

7 In competition, individuals of two species seek to acquire the same limiting resources. For example, snow geese and caribou both eat cottongrass. (See Concept 54.1.)

Nitrogen cycle

N₂

Denitrification

Organisms — N fixation

12

Carbon cycle

CO₂

Cellular respiration

Photosynthesis

Organisms Transfer Energy and Matter in Ecosystems (Chapter 55)

8 Primary producers convert the energy in sunlight to chemical energy through photosynthesis. Their growth is often limited by abiotic factors such as low temperatures, scarce soil nutrients, and lack of light in winter. (See Figures 10.6, 52.13, and 55.4.)

9 Food chains are typically short in the tundra because primary production is lower than in most other ecosystems. (See Figure 54.16.)

10 Energy flows through ecosystems. When one organism eats another, only 10% of the energy transfers from one trophic level to the next. (See Figure 55.10.)

11 Decomposers recycle chemical elements back to primary producers. (See Figures 55.3 and 55.4.)

12 Chemical elements such as carbon and nitrogen move in cycles between the physical environment and organisms. (See Figures 55.13 and 55.14.)

→ **Mastering Biology** BioFlix® Animation: **The Carbon Cycle**

Secondary consumers (wolves)

9

10

Primary consumers (caribou)

Chemical elements

Primary producers (plants and lichens)

7 Competition

8

6 Mutualism

Algal cell

Fungal hyphae

Lichen

Decomposers (soil fungi and prokaryotes)

11

MAKE CONNECTIONS

Human actions are causing climate change, thereby affecting Earth's ecosystems, few of which have been affected as greatly as those in the Arctic. Hypothesize how climate change might cause evolution in arctic tundra populations. Explain. (See Concepts 1.1, 22.2, and 25.4.)

55 Chapter Review

→ Go to **Mastering Biology** for Assignments, the eText, the Study Area, and Dynamic Study Modules.

SUMMARY OF KEY CONCEPTS

→ To review key terms, go to the **Vocabulary Self-Quiz** in the **Mastering Biology** eText or Study Area, or go to goo.gl/zkjz9t.

CONCEPT 55.1

Physical laws govern energy flow and chemical cycling in ecosystems (pp. 1239–1241)

- An **ecosystem** consists of all the organisms in a community and all the abiotic factors with which they interact. Energy is conserved but released as heat during ecosystem processes. As a result, energy flows through ecosystems (rather than being recycled).
- Chemical elements enter and leave an ecosystem and cycle within it, subject to the **law of conservation of mass**. Inputs and outputs are generally small compared to recycled amounts, but their balance determines whether the ecosystem gains or loses an element over time.

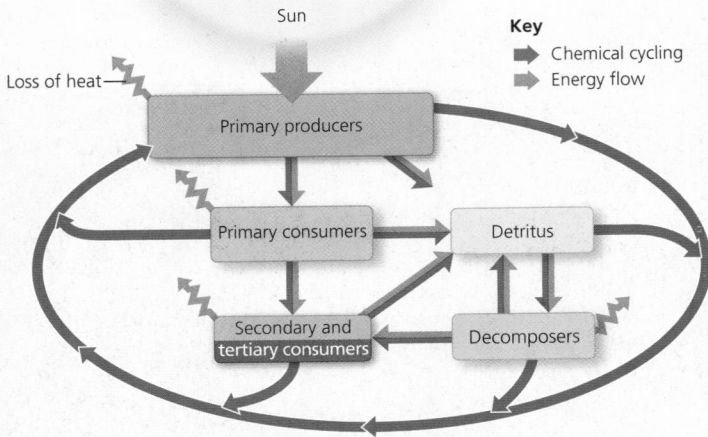

? Considering the second law of thermodynamics, would you expect the typical biomass of primary producers in an ecosystem to be greater than or less than the biomass of consumers in the system? Explain your reasoning.

CONCEPT 55.2

Energy and other limiting factors control primary production in ecosystems (pp. 1241–1245)

- **Primary production** sets the spending limit for the global energy budget. **Gross primary production** is the total energy assimilated by an ecosystem in a given period. **Net primary production**, the energy accumulated in autotroph biomass, equals gross primary production minus the energy used by the primary producers for respiration. **Net ecosystem production** is the total biomass accumulation of an ecosystem, defined as the difference between gross primary production and total ecosystem respiration.
- In aquatic ecosystems, light and nutrients limit primary production. In terrestrial ecosystems, climatic factors such as temperature and moisture affect primary production at large scales, but a soil nutrient is often the limiting factor locally.

? If you know NPP for an ecosystem, what additional variable do you need to know to estimate NEP? Why might measuring this variable be difficult, for instance, in a sample of ocean water?

CONCEPT 55.3

Energy transfer between trophic levels is typically only 10% efficient (pp. 1246–1248)

- The amount of energy available to each trophic level is determined by the net primary production and the **production efficiency**, the efficiency with which chemical energy (stored in food) is converted to biomass at each link in the food chain.
- The percentage of energy transferred from one trophic level to the next, called **trophic efficiency**, is typically 10%. Pyramids of energy and biomass reflect low trophic efficiency.

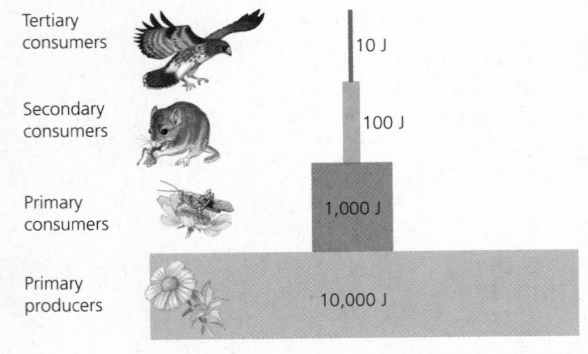

? Why would runners have a lower production efficiency when running a long-distance race than when they are sedentary?

CONCEPT 55.4

Biological and geochemical processes cycle nutrients and water in ecosystems (pp. 1248–1253)

- Water moves in a global cycle driven by solar energy. The carbon cycle primarily reflects the reciprocal processes of photosynthesis and cellular respiration. Nitrogen enters ecosystems through atmospheric deposition and nitrogen fixation by prokaryotes.
- The proportion of a nutrient in a particular form varies among ecosystems, largely because of differences in the rate of decomposition.
- Nutrient cycling is strongly regulated by vegetation. The Hubbard Brook case study showed that logging increases water runoff and can cause large losses of minerals.

? If decomposers usually grow faster and decompose material more quickly in warmer ecosystems, why is decomposition in hot deserts relatively slow?

CONCEPT 55.5

Restoration ecologists return degraded ecosystems to a more natural state (pp. 1253–1257)

- Restoration ecologists harness organisms to detoxify polluted ecosystems through the process of **bioremediation**.
- In **biological augmentation**, ecologists use organisms to add essential materials to ecosystems.

? In preparing a site for surface mining and later restoration, why would engineers separate the topsoil from the deeper soil, rather than removing all soil at once and mixing it in a single pile?

TEST YOUR UNDERSTANDING

➡️ For more multiple-choice questions, go to the **Practice Test** in the **Mastering Biology** eText or Study Area, or go to goo.gl/GruWRg.

Levels 1-2: Remembering/Understanding

1. Which of the following organisms is correctly paired with a trophic level?
   (A) cyanobacterium—secondary consumer
   (B) grasshopper—primary consumer
   (C) zooplankton—primary producer
   (D) grass—decomposer

2. Which of these ecosystems has the *lowest* net primary production per square meter?
   (A) a salt marsh   (C) a coral reef
   (B) an open ocean   (D) a tropical rain forest

3. The discipline that applies ecological principles to returning degraded ecosystems to a more natural state is known as
   (A) restoration ecology.
   (B) thermodynamics.
   (C) eutrophication.
   (D) biogeochemistry.

Levels 3-4: Applying/Analyzing

4. Nitrifying bacteria participate in the nitrogen cycle mainly by
   (A) converting nitrogen gas to ammonia.
   (B) releasing ammonium from organic compounds, thus returning it to the soil.
   (C) converting ammonium to nitrate, which plants absorb.
   (D) incorporating nitrogen into amino acids and organic compounds.

5. Which of the following has the greatest effect on the rate of chemical cycling in an ecosystem?
   (A) the rate of decomposition in the ecosystem
   (B) the production efficiency of the ecosystem's consumers
   (C) the trophic efficiency of the ecosystem
   (D) the location of the nutrient reservoirs in the ecosystem

6. Which of the following was a result of the Hubbard Brook watershed deforestation experiment?
   (A) Most minerals were not recycled within the intact forest ecosystem.
   (B) Calcium levels remained high in the soil of deforested areas.
   (C) Deforestation decreased water runoff.
   (D) The nitrate concentration in waters draining the deforested area became dangerously high.

7. Which of the following is an example of bioremediation?
   (A) adding nitrogen-fixing microorganisms to a degraded ecosystem to increase nitrogen availability
   (B) using a bulldozer to regrade a strip mine
   (C) reconfiguring the channel of a river
   (D) adding seeds of a chromium-accumulating plant to soil contaminated by chromium

8. If you applied a fungicide to a cornfield, what would you expect to happen to the rate of decomposition and net ecosystem production (NEP)?
   (A) Both decomposition rate and NEP would decrease.
   (B) Neither would change.
   (C) Decomposition rate would increase and NEP would decrease.
   (D) Decomposition rate would decrease and NEP would increase.

Levels 5-6: Evaluating/Creating

9. **DRAW IT** (a) Draw a simplified global water cycle showing ocean, land, atmosphere, and runoff from the land to the ocean. Label your drawing with these annual water fluxes:
   - ocean evaporation, 425 km^3
   - ocean evaporation that returns to the ocean as precipitation, 385 km^3
   - ocean evaporation that falls as precipitation on land, 40 km^3
   - evapotranspiration from plants and soil that falls as precipitation on land, 70 km^3
   - runoff to the oceans, 40 km^3

   (b) What is the ratio of ocean evaporation that falls as precipitation on land compared with runoff from land to the oceans? (c) How would this ratio change during an ice age, and why?

10. **EVOLUTION CONECTION** Some biologists have suggested that ecosystems are emergent, "living" systems capable of evolving. One manifestation of this idea is environmentalist James Lovelock's Gaia hypothesis, which views Earth itself as a living, homeostatic entity—a kind of superorganism. Are ecosystems capable of evolving? If so, would this be a form of Darwinian evolution? Why or why not? Explain.

11. **SCIENTIFIC INQUIRY** Using two neighboring ponds in a forest as your study site, design a controlled experiment to measure the effect of falling leaves on net primary production in a pond.

12. **WRITE ABOUT A THEME: ENERGY AND MATTER** Decomposition typically occurs quickly in moist tropical forests. However, waterlogging in the soil of some moist tropical forests results over time in a buildup of organic matter called "peat." In a short essay (100–150 words), discuss the relationship of net primary production, net ecosystem production, and decomposition for such an ecosystem. Are NPP and NEP likely to be positive? What do you think would happen to NEP if a landowner drained the water from a tropical peatland, exposing the organic matter to air?

13. **SYNTHESIZE YOUR KNOWLEDGE**

This dung beetle (genus *Scarabaeus*) is burying a ball of dung it has collected from a large mammalian herbivore in Kenya. Explain why this process is important for the cycling of nutrients and for primary production.

For selected answers, see Appendix A.

56 Conservation Biology and Global Change

KEY CONCEPTS

56.1 Human activities threaten Earth's biodiversity *p. 1261*

56.2 Population conservation focuses on population size, genetic diversity, and critical habitat *p. 1266*

56.3 Landscape and regional conservation help sustain biodiversity *p. 1270*

56.4 Earth is changing rapidly as a result of human actions *p. 1274*

56.5 Sustainable development can improve human lives while conserving biodiversity *p. 1284*

Study Tip

Make a table: As you read the chapter, fill in a table like the one shown to organize what you learn about human activities that have major effects on the environment. Describe the activity; summarize its impact; and list actions at personal and government levels that could reduce harmful effects of the activity.

Activity	Harmful effects	Personal choice/ governmental action
Cutting down tropical forests (often for cattle ranches)	Destroys habitat for tropical species; contributes to rising CO_2 levels	Eat less beef to reduce demand/ establish conservation areas

→ Go to Mastering Biology

For Students (in eText and Study Area)
- Get Ready for Chapter 56
- Animation: The Global Carbon Cycle and the Greenhouse Effect

For Instructors to Assign (in Item Library)
- Interpreting Data: Recent Global Temperature Changes
- Scientific Thinking: Using Forensics to Uncover Illegal Whaling

Ready-to-Go Teaching Module (in Instructor Resources)
- Interpreting Data on Introduced Species

Figure 56.1 Researchers discovered this colorful lizard, the psychedelic rock gecko (*Cnemaspis psychedelica*), during an expedition to the Greater Mekong region of Southeast Asia. In fact, between 2000 and 2010, more than 1,000 new species were discovered in this region alone. Unfortunately, the psychedelic rock gecko and many other newly discovered species' survival is threatened by deforestation and other human activities.

How can we protect the many species threatened by human activities?

Ways to protect the diversity of life include:

Restore or preserve habitat that threatened species need to survive

End the overharvesting of species whose populations are in decline

Combat climate change and other types of human-caused, global environmental change

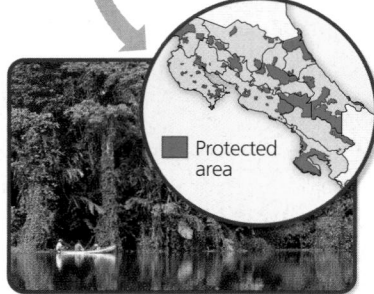
Protected area

Establish regional networks of protected areas

Protect "hot spots," relatively small areas with high biodiversity

Human activities threaten earth's biodiversity

Throughout the biosphere, human activities are altering natural disturbances, trophic structures, energy flow, and chemical cycling—ecosystem processes on which we and all other species depend. We have physically altered nearly half of Earth's land surface, and we use over half of all accessible surface fresh water. In the oceans, stocks of most major fisheries are shrinking because of overharvesting. By some estimates, we may be pushing more species toward extinction than did the asteroid that triggered the Cretaceous mass extinction 66 million years ago (see Figure 25.18).

In this chapter, we'll examine changes happening across Earth, focusing on **conservation biology**, a discipline that integrates ecology, physiology, molecular biology, genetics, and evolutionary biology to conserve the diversity of life on Earth. Efforts to sustain ecosystem processes and stem the extinction of species also connect the life sciences with the social sciences, economics, and humanities.

Extinction is a natural phenomenon that has been occurring since life first evolved; it is the high *rate* of extinction that is of concern. More than 1,000 species have become extinct in the last 400 years, a rate that is 100 to 1,000 times the "background," or typical, extinction rate seen in the fossil record (see Concept 25.4). This comparison indicates that the extinction rate today is high and that human activities threaten Earth's biological diversity at all levels.

Three Levels of Biodiversity

Biodiversity—short for biological diversity—can be considered at three main levels: genetic diversity, species diversity, and ecosystem diversity **(Figure 56.2)**.

Genetic Diversity

Genetic diversity comprises not only the individual genetic variation *within* a population, but also genetic differences *between* populations—differences that often are associated with adaptations to local conditions. If one population becomes extinct, then a species may have lost some of the genetic diversity that makes microevolution possible. This erosion of genetic diversity in turn reduces the adaptive potential of the species.

Species Diversity

Public awareness of the biodiversity crisis centers on species diversity—the number of species in an ecosystem or across the biosphere. To date, scientists have described and formally named about 1.8 million species of organisms. In addition to these named species, many others remain to be discovered:

▼ **Figure 56.2 Three levels of biodiversity.** The oversized chromosomes in the top diagram symbolize the genetic variation within the population.

Genetic diversity in a vole population

Species diversity in a coastal redwood **ecosystem**

Community and ecosystem diversity across the landscape of an entire region

Estimates for the number of species that currently exist range from 5 million to 100 million.

Of particular concern are species that are endangered or threatened. An **endangered species** is one that is in danger of extinction throughout all or much of its range **(Figure 56.3)**, while a **threatened species** is considered likely to become endangered in the near future. The following are just a few statistics that illustrate the problem of species loss:

- According to the International Union for Conservation of Nature and Natural Resources (IUCN), 13% of the 10,000 known species of birds and 22% of the 5,500 known species of mammals are threatened.
- A survey by the Center for Plant Conservation showed that of the nearly 20,000 known plant species in the United States, 200 have become extinct since such records have been kept, and 730 are endangered or threatened.

▼ **Figure 56.3 A hundred heartbeats from extinction.** These are two members of what Harvard biologist E. O. Wilson calls the Hundred Heartbeat Club, species with fewer than 100 individuals remaining on Earth. The Yangtze River dolphin may be extinct, but a few individuals were reportedly sighted in 2007.

Philippine eagle

Yangtze River dolphin

 To document that a species has actually become extinct, what factors would you need to consider?

- In North America, at least 123 freshwater animal species have become extinct since 1900, and hundreds more species are threatened. The extinction rate for North American freshwater fauna is about five times as high as that for terrestrial animals.

The local populations of a species can also be driven to extinction; for example, populations of a species may be lost in one river system but survive in an adjacent one. Global extinction of a species means that it is lost from *all* the ecosystems in which it lived, leaving them permanently impoverished.

➡ **Mastering Biology** HHMI Video: Surveying Gorongosa's Biodiversity

hhmi BioInteractive

Ecosystem Diversity

The variety of ecosystems on Earth is a third level of biological diversity. Because of the many interactions between members of different species in an ecosystem, the extinction of populations of one species can have a negative impact on other species in the ecosystem (see Figure 54.20). For instance, bats called "flying foxes" are important pollinators and seed dispersers in the Pacific Islands, where they are increasingly hunted as a luxury food **(Figure 56.4)**. Conservation biologists fear that the extinction of flying foxes would also harm the native plants of the Samoan Islands, where four-fifths of the tree species depend on flying foxes for pollination or seed dispersal.

▼ **Figure 56.4 The endangered Marianas "flying fox" bat (*Pteropus mariannus*), an important pollinator.**

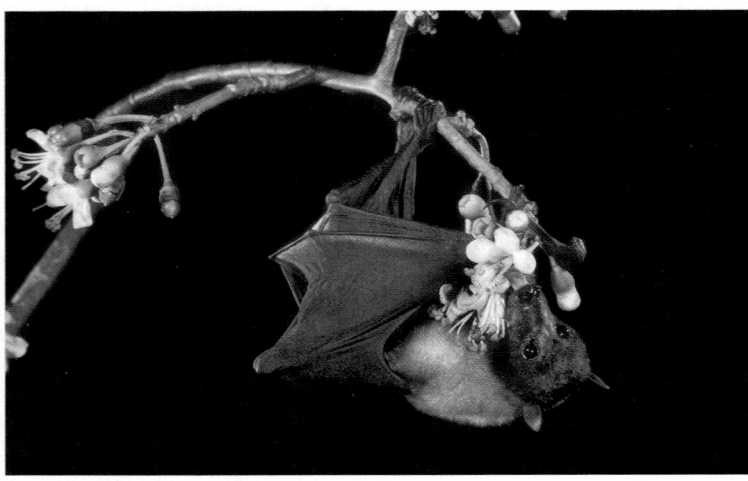

Some ecosystems have already been heavily affected by humans, and others are being altered at a rapid pace. Since European colonization, more than half of the wetlands in the contiguous United States have been drained and converted to agricultural and other uses. In California, Arizona, and New Mexico, roughly 90% of native riparian (streamside) communities have been affected by overgrazing, flood control, water diversions, lowering of water tables, and invasion by non-native plants.

Biodiversity and Human Welfare

Why should we care about the loss of biodiversity? One basic reason concerns our human sense of connection to nature and all forms of life, termed *biophilia*. Moreover, the belief that other species are entitled to life is a pervasive theme of many religions and the basis of a moral argument that we should protect biodiversity. There is also a concern for future human generations. Paraphrasing an old proverb, G. H. Brundtland, a former prime minister of Norway, said, "We must consider our planet to be on loan from our children, rather than being a gift from our ancestors." In addition to such philosophical and moral justifications, species and genetic diversity bring us many practical benefits.

Benefits of Species and Genetic Diversity

Many threatened species could potentially provide medicines, food, and fibers for human use, making biodiversity a crucial natural resource. Products from aspirin to antibiotics were derived originally from natural sources. If we lose wild populations of plants closely related to agricultural species, we lose genetic resources that could be used to improve crop qualities, such as disease resistance. For instance, in the 1970s, plant breeders responded to devastating outbreaks of the grassy stunt virus in rice (*Oryza sativa*) by screening

7,000 populations of this species and its close relatives for resistance to the virus. One population of a single relative, Indian rice (*Oryza nivara*), was found to be resistant to the virus, and scientists succeeded in breeding the resistance trait into commercial rice varieties. Today, the original disease-resistant population has apparently become extinct in the wild.

In the United States, about 25% of the prescriptions dispensed from pharmacies contain substances originally derived from plants. For example, researchers discovered that the rosy periwinkle, which grows in Madagascar, contains alkaloids that inhibit cancer cell growth **(Figure 56.5)**. This discovery led to treatments for two deadly forms of cancer, Hodgkin's lymphoma and childhood leukemia, resulting in remission in most cases. Madagascar is also home to five other species of periwinkles, one of which is approaching extinction. Losing these species would mean the loss of any possible medicinal benefits they might offer.

▼ **Figure 56.5 The rosy periwinkle (*Catharanthus roseus*), a plant that saves lives.**

Each species lost means the loss of unique genes, some of which may code for enormously useful proteins. The enzyme Taq polymerase was first extracted from a bacterium, *Thermus aquaticus*, found in hot springs at Yellowstone National Park. This enzyme is essential for the polymerase chain reaction (PCR) because it is stable at the high temperatures required for automated PCR (see Figure 20.8). DNA from many other species of prokaryotes, living in a variety of environments, is used in the mass production of proteins for new medicines, foods, petroleum substitutes, industrial chemicals, and other products. However, because many species of prokaryotes and other organisms may become extinct before we discover them, we stand to lose the valuable genetic potential held in their unique libraries of genes.

Ecosystem Services

The benefits that individual species provide to humans are substantial, but saving individual species is only part of the reason for preserving ecosystems. Humans evolved in Earth's ecosystems, and we rely on these systems and their inhabitants for our survival. **Ecosystem services** encompass all the processes through which natural ecosystems help sustain human life. Ecosystems purify our air and water. They detoxify and decompose our wastes and reduce the impacts of extreme weather and flooding. The organisms in ecosystems pollinate our crops, control pests, and

create and preserve our soils. Moreover, these diverse services are provided for free.

If we had to pay for them, how much would the services of natural ecosystems be worth? In 1997, scientists estimated the value of Earth's ecosystem services at $33 trillion per year, nearly twice the gross national product of all the countries on Earth at the time ($18 trillion). It may be more realistic to do the accounting on a smaller scale. For example, New York City invested $1.5 billion to buy land and restore habitat in the Catskill Mountains, the source of much of the city's fresh water. This investment was spurred by increasing pollution of the water by sewage, pesticides, and fertilizers. By harnessing ecosystem services to purify its water naturally, the city avoided spending $2 billion to build new water treatment plants and $300 million a year to run those plants.

There is growing evidence that the functioning of ecosystems, and hence their capacity to perform services, is linked to biodiversity. As human activities reduce biodiversity, we are reducing the capacity of the planet's ecosystems to perform processes critical to our own survival.

⊙ **Mastering Biology**
Interview with E. O. Wilson: Advocate for protecting biodiversity

Threats to Biodiversity

Many different human activities threaten biodiversity on local, regional, and global scales. The threats posed by these activities include four major types: habitat loss, introduced species, overharvesting, and global change.

Habitat Loss

Human alteration of habitat is the single greatest threat to biodiversity throughout the biosphere. Habitat loss has been brought about by factors such as agriculture, urban development, forestry, mining, and pollution. As discussed in more detail later in this chapter, global climate change is also altering habitats today and will have an even larger effect later this century. When no alternative habitat is available or a species is unable to move, habitat loss may mean extinction. The IUCN implicates destruction of habitat as a contributing cause for 73% of the species that have become extinct, endangered, vulnerable, or rare in the last few hundred years.

Habitat loss and fragmentation may occur over immense regions. Approximately 98% of the tropical dry forests of Central America and Mexico have been cut down. The clearing of tropical rain forest in the state of Veracruz, Mexico, mostly for cattle ranching, has resulted in the loss of more than 90% of the original forest, leaving relatively small,

▼ **Figure 56.6 Habitat fragmentation in the foothills of Los Angeles.** Development in the valleys may confine the organisms that inhabit the narrow strips of hillside.

▼ **Figure 56.7 Kudzu, an introduced species, thriving in the southeastern United States.**

isolated patches of forest. Many other natural habitats have also been fragmented by human activities **(Figure 56.6)**.

Habitat fragmentation often reduces the number of species in a region because the smaller populations in habitat fragments have a higher probability of extinction. Prairie covered about 800,000 hectares of southern Wisconsin when Europeans first arrived in North America but occupies only 800 hectares today; most of the original prairie in this area is now used to grow crops. Plant diversity surveys of 54 Wisconsin prairie remnants conducted in 1948–1954 and 1987–1988 showed that the remnants lost 8–60% of their plant species in the time between the two surveys.

Habitat loss is also a major threat to aquatic biodiversity. About 70% of coral reefs, among Earth's most species-rich aquatic communities, have been damaged by human activities. At the current rate of destruction, 40–50% of the reefs, home to one-third of marine fish species, could disappear in the next 30 to 40 years. Freshwater habitats are also being lost, often as a result of the dams, reservoirs, channel modification, and flow regulation now affecting most of the world's rivers. For example, the more than 30 dams and locks built along the Mobile River basin in the southeastern United States changed river depth and flow. While providing the benefits of hydroelectric power and increased ship traffic, these dams and locks also helped drive more than 40 species of mussels and snails to extinction.

Introduced Species

Introduced species—sometimes called invasive, non-native, or exotic species—are those that humans move intentionally or accidentally from the species' native locations to new geographic regions. Human travel by ship and airplane has accelerated the transplant of species. Free from the predators, herbivores, pathogens, or competitors that limit their populations in their native habitats, such transplanted species may spread rapidly through a new region.

Some introduced species disrupt their new community, often by preying on native organisms or outcompeting native organisms for resources. For example, the brown tree snake was accidentally introduced to the island of Guam from other parts of the South Pacific after World War II, as a "stowaway" in military cargo. Since then, 12 species of birds and 6 species of lizards that the snakes ate have become extinct on Guam. In 1988, the devastating zebra mussel, a suspension-feeding mollusc, was discovered in the Great Lakes of North America, most likely introduced in the ballast water of ships arriving from Europe. Zebra mussels form dense colonies and have disrupted freshwater ecosystems, threatening native aquatic species. They have also clogged water intake structures, causing billions of dollars in damage to domestic and industrial water supplies.

Humans have deliberately introduced many species with good intentions but disastrous effects. An Asian plant called kudzu, which the U.S. Department of Agriculture once introduced in the southern United States to help control erosion, has taken over large areas of the landscape there **(Figure 56.7)**. The European starling was brought intentionally into New York's Central Park in 1890 by a citizens' group intent on introducing all the plants and animals mentioned in Shakespeare's plays. It quickly spread across North America, where its population now exceeds 100 million, displacing many native songbirds.

Introduced species are a worldwide problem, contributing to approximately 40% of the extinctions recorded since 1750 and costing billions of dollars each year in damage and control efforts. There are more than 50,000 introduced species in the United States alone.

Overharvesting

The term *overharvesting* refers generally to the harvesting of wild organisms at rates exceeding the ability of their populations to rebound. Species with restricted habitats, such as those that inhabit small islands, are particularly vulnerable to overharvesting. One such species was the great auk, a large, flightless seabird found on islands in the North Atlantic. By the 1840s, humans had hunted the great auk to extinction to satisfy demand for its feathers, eggs, and meat.

Also susceptible to overharvesting are large organisms with low reproductive rates, such as elephants, whales, and rhinoceroses. The decline of Earth's largest terrestrial animals, the African elephants, is a classic example of the impact of overhunting. Largely because of the trade in ivory, elephant populations have been declining in most of Africa for the last 50 years. An international ban on the sale of new ivory resulted in increased poaching (illegal hunting), so the ban had little effect in much of central and eastern Africa. From 2006 to 2015, the number of elephants dropped by 110,000, a 22% decline (from 525,000 to 415,000). Only in South Africa, where once-decimated herds have been well protected for nearly a century, have elephant populations been stable or increasing (see Figure 53.8). If such protections are not extended to other regions, elephants could become extinct in the wild across most of Africa.

Conservation biologists are increasingly using the tools of molecular genetics to track the origins of tissues harvested from endangered species. For example, researchers have used DNA isolated from samples of elephant dung to construct a DNA reference map for the African elephant (*Loxodonta africana*). By comparing this reference map with DNA isolated from ivory harvested legally or by poachers, they can determine to within a few hundred kilometers where the elephants were killed (**Figure 56.8**). Such work in Zambia suggested that poaching rates were 30 times higher than previously estimated, a finding that has stimulated improved anti-poaching efforts by the Zambian government. Similarly, biologists using phylogenetic analyses of mitochondrial DNA (mtDNA) showed that some whale meat sold in Japanese fish markets

▼ **Figure 56.8 Ecological forensics and elephant poaching.** These severed tusks were part of an illegal shipment of ivory intercepted on its way from Africa to Singapore in 2002. DNA-based evidence showed that the thousands of elephants killed for the tusks came from a relatively narrow east-west band centered in Zambia rather than from across Africa.

MAKE CONNECTIONS *Figure 26.6 describes a similar example in which conservation biologists used DNA analyses to compare harvested samples of whale meat with a reference DNA database. How are these examples similar, and how are they different? What limitations might there be to using such forensic methods in other suspected cases of poaching?*

▼ **Figure 56.9 Overharvesting.** Atlantic bluefin tuna are auctioned in a Japanese fish market.

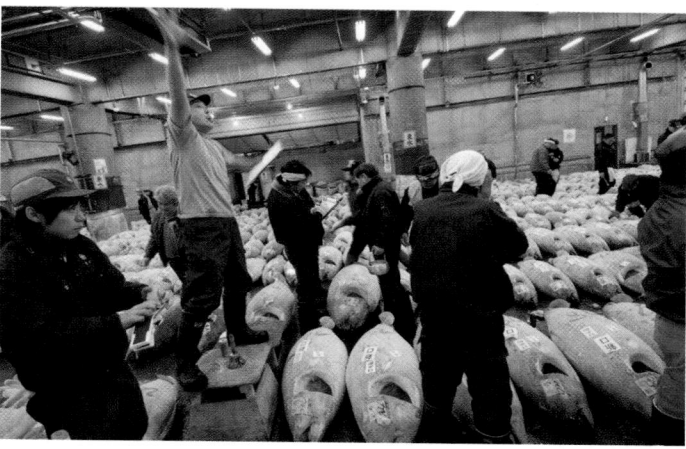

came from illegally harvested species, including fin and humpback whales, which are endangered (see Figure 26.6).

Many commercially important fish populations, once thought to be inexhaustible, have been decimated by overfishing. Demands from an increasing human population for protein-rich food, coupled with new harvesting technologies, such as long-line fishing and modern trawlers, have reduced these fish populations to levels that cannot sustain further exploitation. Until the past few decades, the Atlantic bluefin tuna was considered a sport fish of little commercial value—just a few cents per pound for use in cat food. In the 1980s, however, wholesalers began flying fresh, iced bluefin to Japan for sushi and sashimi. In that market, the fish can now bring over $500 per pound (**Figure 56.9**). With increased harvesting spurred by such high prices, it took just ten years to reduce the western Atlantic bluefin population to less than 20% of its 1980 size.

Global Change

The fourth threat to biodiversity, global change, alters the fabric of Earth's ecosystems at regional to global scales. Global change includes alterations in climate, atmospheric chemistry, and broad ecological systems that reduce the capacity of Earth to sustain life.

One of the first types of global change to cause concern was *acid precipitation*, which is rain, snow, or fog with a pH less than 5.2. The burning of wood and fossil fuels releases oxides of sulfur and nitrogen that react with water in air, forming sulfuric and nitric acids. The acids eventually fall to Earth's surface, where they cause chemical reactions that decrease nutrient supplies and increase concentrations of toxic metals. These changes to the soil and water harm some terrestrial and aquatic organisms.

In the 1960s, ecologists determined that lake-dwelling organisms in eastern Canada were dying because of acid precipitation caused by air pollution from factories in the Midwestern United States. Newly hatched lake trout, for instance, die when the pH drops below 5.4. Similarly, lakes and streams in southern Norway and Sweden were losing fish

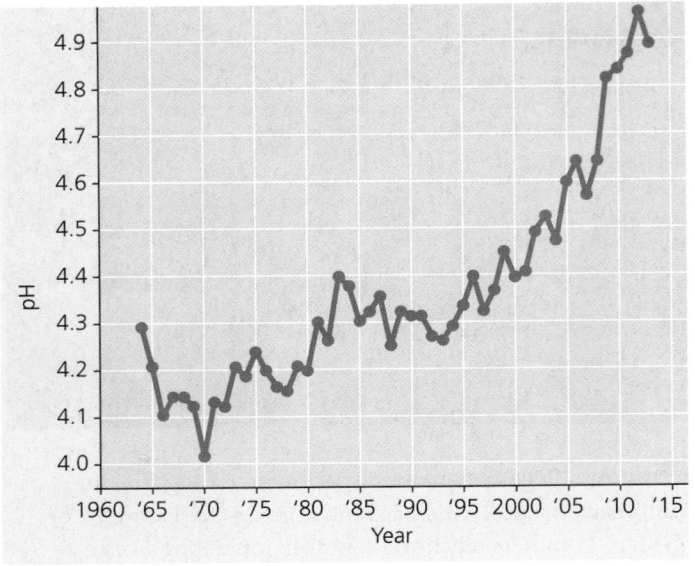

▼ **Figure 56.10 Changes in the pH of precipitation at the Hubbard Brook Experimental Forest, New Hampshire.**

MAKE CONNECTIONS *Describe the relationship between pH and acidity. (See Concept 3.3.) Overall, is the precipitation in this forest becoming more acidic or less acidic?*

because of pollution generated in Great Britain and central Europe. By 1980, the pH of precipitation in large areas of North America and Europe averaged 4.0–4.5 and sometimes dropped as low as 3.0. (To review pH, see Concept 3.3.)

Environmental regulations and new technologies have enabled many countries to reduce sulfur dioxide emissions in recent decades. In the United States, sulfur dioxide emissions decreased more than 75% between 1990 and 2013, gradually reducing the acidity of precipitation **(Figure 56.10)**. However, ecologists estimate that it will take decades for aquatic ecosystems to recover. Meanwhile, emissions of nitrogen oxides are increasing in the United States, and emissions of sulfur dioxide and acid precipitation continue to damage forests in central and eastern Europe.

We'll explore the importance of global change for Earth's biodiversity in more detail in Concept 56.4, where we examine such factors as climate change and ozone depletion.

Can Extinct Species Be Resurrected?

To the best of our knowledge, extinction has always been forever. Some scientists are nevertheless trying to use cloning techniques (see Concept 20.3) to resurrect species that have become extinct. For example, researchers used this approach in an attempt to reverse the extinction of the Pyrenean ibex, *Capra pyrenaica pyrenaica*. In 1999, researchers removed a skin sample from the last living individual, a female, and froze it. When that individual died a year later, scientists used the frozen cells to try to produce more ibex. Eventually, one ibex was born in 2009. Sadly, it lived for only 7 minutes before succumbing to lung defects similar to those observed in other cloned animals.

More recently, researchers completed an early step toward the possible use of cloning to prevent the extinction of the northern white rhinoceros (*Ceratotherium simum cottoni*). The last *C. simum cottoni* male died in 2018—dooming the northern white rhinoceros to eventual extinction. That same year, researchers successfully created hybrid embryos using sperm from northern white rhinoceroses and eggs from closely related southern white rhinoceroses, *C. simum simum* (the few living females from the northern subspecies all had severe reproductive issues).

Many steps remain before the northern white rhinoceros could be considered resurrected from extinction. Moreover, even if scientists eventually can resurrect one or a few species, that approach will not solve the extinction crisis we face today. To do that, we must take steps to clean up the environment and protect natural habitat for the thousands of species currently threatened by extinction.

CONCEPT CHECK 56.1

1. Explain why it is too narrow to define the biodiversity crisis as simply a loss of species.
2. Identify the four main threats to biodiversity and explain how each damages diversity.
3. **WHAT IF?** Imagine two populations of a fish species, one in the Mediterranean Sea and one in the Caribbean Sea. Now imagine two scenarios: (1) The populations breed separately, and (2) adults of both populations migrate yearly to the North Atlantic to interbreed. Which scenario would result in a greater loss of genetic diversity if the Mediterranean population were harvested to extinction? Explain your answer.

For suggested answers, see Appendix A.

CONCEPT 56.2

Population conservation focuses on population size, genetic diversity, and critical habitat

Biologists who work on conservation at the population and species levels use two main approaches. One approach focuses on populations that are small and hence often vulnerable, while the other emphasizes critical habitat.

Extinction Risks in Small Populations

Small populations are particularly vulnerable to overharvesting, habitat loss, and the other threats to biodiversity that you read about in Concept 56.1. After such factors have reduced a population's size to a small number of individuals, the small size itself can push the population to extinction.

The Extinction Vortex: Evolutionary Implications of Small Population Size

EVOLUTION A small population is vulnerable to inbreeding and genetic drift, which can draw the population down an **extinction vortex** toward smaller and smaller population

▼ Figure 56.11 Processes driving an extinction vortex.

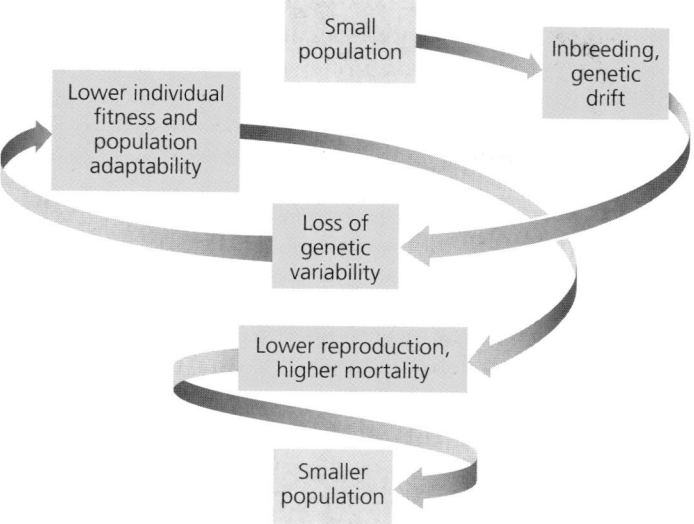

size until no individuals survive (**Figure 56.11**). A key factor driving the extinction vortex is the loss of genetic variation that can enable evolutionary responses to environmental change, such as the appearance of new strains of pathogens. Both inbreeding and genetic drift can cause a loss of genetic variation (see Concept 23.3), and their effects become more harmful as a population shrinks. Inbreeding often reduces fitness because offspring are more likely to be homozygous for harmful recessive traits.

Not all small populations are doomed by low genetic diversity, and low genetic variability does not automatically lead to permanently small populations. For instance, overhunting of northern elephant seals in the 1890s reduced the species to only 20 individuals—clearly a bottleneck with reduced genetic variation. Since that time, however, the northern elephant seal populations have rebounded to about 150,000 individuals today, though their genetic variation remains relatively low.

Case Study: *The Greater Prairie Chicken and the Extinction Vortex*

When Europeans arrived in North America, the greater prairie chicken (*Tympanuchus cupido*) was common from New England to Virginia and across the western prairies of the continent. Land cultivation for agriculture fragmented the populations of this species, and its abundance decreased rapidly (see Figure 23.11). Illinois had millions of greater prairie chickens in the 19th century but fewer than 50 by 1993. Researchers found that the decline in the Illinois population was associated with reduced genetic variation and a decrease in fertility. As a test of the extinction vortex hypothesis, scientists increased genetic variation by importing 271 birds from larger populations elsewhere (**Figure 56.12**). The Illinois population rebounded, indicating that it had been on its way to extinction until rescued by the transfusion of genetic variation.

▼ Figure 56.12 Inquiry

What caused the drastic decline of the Illinois greater prairie chicken population?

Experiment Researchers had observed that the population collapse of the greater prairie chicken was mirrored in a reduction in fertility, as measured by the hatching rate of eggs. Comparison of DNA samples from the Jasper County, Illinois, population with DNA from feathers in museum specimens showed that genetic variation had declined in the study population (see Figure 23.11). In 1992, researchers began translocating prairie chickens from neighboring states in an attempt to increase genetic variation.

Results After translocation (black arrow), the viability of eggs rapidly increased, and the population rebounded.

(a) Population dynamics

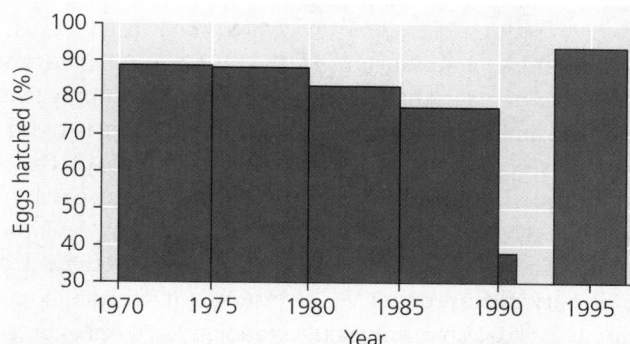

(b) Hatching rate. Bar height represents the average rate for the years spanned by the bar.

Conclusion Reduced genetic variation had started the Jasper County population of prairie chickens down the extinction vortex.

Data from R. L. Westemeier et al., Tracking the long-term decline and recovery of an isolated population, *Science* 282:1695–1698 (1998). © 1998 by AAAS. Reprinted with permission.

INQUIRY IN ACTION Read and analyze the original paper in *Inquiry in Action: Interpreting Scientific Papers.*

WHAT IF? *Given the success of using transplanted birds as a tool for increasing the percentage of hatched eggs in Illinois, would you support transplanting additional birds to Illinois? Why or why not?*

Minimum Viable Population Size

How small does a population have to be before it starts down an extinction vortex? The answer depends on the type of organism and other factors. Large predators that feed high on the food chain usually require extensive individual ranges, resulting in low population densities. Such species may be rare yet of little concern to conservation biologists. All populations, however, require some minimum size to remain viable.

The minimal population size at which a species is able to sustain its numbers is known as the **minimum viable population (MVP)**. MVP is usually estimated for a given species using computer models that integrate many factors. The calculation may include, for instance, an estimate of how many individuals in a small population are likely to be killed by a natural catastrophe such as a storm. Once in the extinction vortex, two or three consecutive years of bad weather could finish off a population that is already below its MVP.

Effective Population Size

Genetic variation can be of considerable importance in small populations. The *total* size of a population may be misleading because only certain members of the population breed successfully and pass their alleles on to offspring. Therefore, a meaningful estimate of MVP requires the researcher to determine the **effective population size**, which is based on the breeding potential of the population.

The following formula illustrates one way to estimate the effective population size, abbreviated N_e:

$$N_e = \frac{4N_f N_m}{N_f + N_m}$$

where N_f and N_m are, respectively, the number of females and the number of males that successfully breed. If we apply this formula to an idealized population whose total size is 1,000 individuals, N_e will also be 1,000 if every individual breeds and the sex ratio is 500 females to 500 males. In this case, $N_e = (4 \times 500 \times 500)/(500 + 500) = 1,000$. Any deviation from these conditions (not all individuals breed or there is not a 1:1 sex ratio) reduces N_e. For instance, if the total population size is 1,000 but only 400 females and 400 males breed, then $N_e = (4 \times 400 \times 400)/(400 + 400) = 800$, or 80% of the total population size. Numerous factors can influence N_e. As a result, alternative formulas for estimating N_e have been developed to account for factors such as age at maturation, genetic relatedness among population members, the effects of gene flow, and population fluctuations.

In actual study populations, N_e is always some fraction of the total population. Thus, simply determining the total number of individuals in a small population does not necessarily provide a good measure of whether the population is large enough to avoid extinction. Whenever possible, conservation programs attempt to sustain total population sizes that include at least the minimum viable number of *reproductively active* individuals. The conservation goal of sustaining effective population size (N_e) above MVP stems from the concern that populations retain enough genetic diversity to adapt as their environment changes.

The MVP of a population is often used in population viability analysis. The objective of this analysis is to predict a population's chances for survival, usually expressed as a specific probability of survival, such as a 95% chance, over a particular time interval, perhaps 100 years. Such modeling approaches allow conservation biologists to explore the potential consequences of alternative management plans.

Case Study: *Analysis of Grizzly Bear Populations*

One of the first population viability analyses was conducted in 1978 by Mark Shaffer as part of a long-term study of grizzly bears in Yellowstone National Park and its surrounding areas **(Figure 56.13)**. A threatened species in the United States, the grizzly bear (*Ursus arctos horribilis*) is currently found in only 4 of the 48 contiguous states. Its populations in those states have been drastically reduced and fragmented. In 1800, an estimated 100,000 grizzlies ranged over about 500 million hectares of habitat, while today only 1,000 individuals in six relatively isolated populations range over less than 5 million hectares.

Shaffer attempted to determine viable sizes for the Yellowstone grizzly population. Using life history data obtained for individual Yellowstone bears over a 12-year period, he simulated the effects of environmental factors on survival and reproduction. His models predicted that, given a suitable habitat, a Yellowstone grizzly bear population of 70–90 individuals would have about a 95% chance of surviving for 100 years. A slightly larger population of only 100 bears would have a 95% chance of surviving for twice as long, about 200 years.

How does the actual size of the Yellowstone grizzly population compare with Shaffer's predicted MVP? A current estimate puts the total grizzly bear population in the greater Yellowstone

▼ **Figure 56.13 Monitoring a grizzly bear population.** A scientist measures a tranquilized bear that has been fitted with a radio collar so that the bear's movements can be compared with those of other grizzlies in Yellowstone National Park.

ecosystem at about 700 individuals. The relationship of this estimate to the effective population size (N_e) depends on several factors. Usually, only a few dominant males breed, and it may be difficult for them to locate females, since individuals inhabit such large areas. Moreover, females may reproduce only when there is abundant food. As a result, N_e is only about 25% of the total population size, or about 175 bears.

Because small populations tend to lose genetic variation over time, researchers have used proteins, mitochondrial DNA, and short tandem repeats (see Concept 21.4) to assess genetic variability in the Yellowstone grizzly bear population. All results to date indicate that the Yellowstone population has less genetic variability than other grizzly bear populations in North America.

How might conservation biologists increase the effective size and genetic variation of the Yellowstone grizzly bear population? Migration between isolated populations of grizzlies could increase both effective and total population sizes. Computer models predict that introducing only two unrelated bears each decade into a population of 100 individuals would reduce the loss of genetic variation by about half. For the grizzly bear, and probably for many other species with small populations, finding ways to promote dispersal among populations may be one of the most urgent conservation needs.

This case study and that of the greater prairie chicken show how an understanding of the ecological and evolutionary consequences of small populations can have practical applications in conservation. Next, we look at an alternative approach to understanding the biology of extinction.

Critical Habitat

A loss of critical habitat can cause threatened and endangered populations to show a downward trend, even if the population is far above its minimum viable population size. A focus on the importance of critical habitat emphasizes the environmental factors that can cause a population decline in the first place. If, for instance, an area is deforested, then species that depend on trees will decline in abundance, whether or not they retain genetic variation. The following case study illustrates how a critical-habitat approach has been applied to the conservation of an endangered species.

Case Study: *Decline of the Red-Cockaded Woodpecker*

The red-cockaded woodpecker (*Picoides borealis*) is found only in the southeastern United States. It requires mature pine forests, preferably ones dominated by the longleaf pine, for its habitat. A critical-habitat factor for the red-cockaded woodpecker is that the undergrowth of plants around the pine trunks must be low **(Figure 56.14a)**. Breeding birds tend to abandon nests when vegetation among the pines is thick and higher than about 4.5 m **(Figure 56.14b)**; the birds appear to need a clear flight path between their home trees and the neighboring feeding grounds. Periodic fires have

▼ **Figure 56.14 A habitat requirement of the red-cockaded woodpecker.**

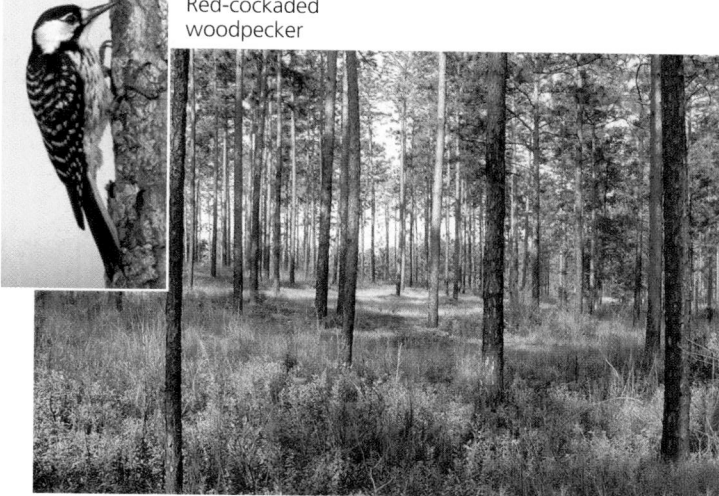

Red-cockaded woodpecker

(a) Forests that can sustain red-cockaded woodpeckers have low undergrowth.

(b) Forests that cannot sustain red-cockaded woodpeckers have high, dense undergrowth that interferes with the woodpeckers' access to feeding grounds.

? *How is habitat disturbance absolutely necessary for the long-term survival of the woodpecker?*

historically swept through longleaf pine forests, keeping the undergrowth low.

In addition, although most woodpeckers nest in dead trees, the red-cockaded woodpecker drills its nest holes in mature, living pine trees. It also drills small holes around the entrance to its nest cavity, which causes resin from the tree to ooze down the trunk. The resin seems to repel predators, such as corn snakes, that eat bird eggs and nestlings.

One factor leading to decline of the red-cockaded woodpecker has been the destruction or fragmentation of suitable habitats by logging and agriculture **(Figure 56.15)**. By recognizing key habitat factors, protecting some longleaf

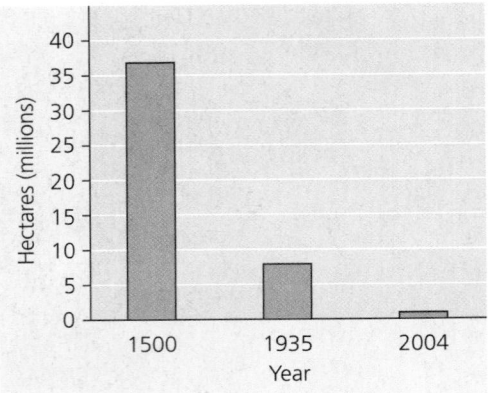

▼ **Figure 56.15 Decline of longleaf pine forests.** Fire suppression, logging, urban development, and other human activities have reduced longleaf pine forests to 3% of their original area; the covered area has not increased since 2004.

pine forests, and using controlled fires to reduce forest undergrowth, conservation managers have helped restore habitat that can support viable populations.

Sometimes conservation managers also help species colonize restored habitats. Because red-cockaded woodpeckers take months to excavate nesting cavities, researchers performed an experiment to see whether providing cavities for the birds would make them more likely to use a site. The researchers constructed cavities in pine trees at 20 sites. The results were dramatic. Cavities in 18 of the 20 sites were colonized by red-cockaded woodpeckers, and new breeding groups formed only in those sites. Based on this experiment, conservationists initiated a habitat maintenance program that included controlled burning and excavation of new nesting cavities, enabling this endangered species to begin to recover.

Weighing Conflicting Demands

Determining population numbers and habitat needs is only part of a strategy to save species. Scientists also need to weigh a species' needs against other conflicting demands. Conservation biology often highlights the relationship between science, technology, and society. For example, an ongoing, sometimes bitter debate in the western United States pits habitat preservation for wolf, grizzly bear, and bull trout populations against job opportunities in the grazing and resource extraction industries. Programs that restocked wolves in Yellowstone National Park remain controversial for people concerned about human safety and for many ranchers concerned with potential loss of livestock outside the park.

Large, high-profile vertebrates are not always the focal point in such conflicts, but habitat use is almost always the issue. Should work proceed on a new highway bridge that benefits the local economy but destroys the only remaining habitat of a species of freshwater mussel? If you were the owner of a coffee plantation growing varieties that thrive in bright sunlight, would you be willing to change to shade-tolerant varieties that produce less coffee per hectare but can grow beneath trees that support large numbers of songbirds? Successful conservation efforts often strive for a "win-win" outcome: to protect both endangered species *and* jobs or other aspects of a region's economy.

Another important consideration is the ecological role of a species. Because we cannot save every endangered species, we must determine which species are most important for conserving biodiversity as a whole. Identifying keystone species and finding ways to sustain their populations can be central to maintaining communities and ecosystems. In most situations, conservation biologists must also look beyond single species and consider the whole community and ecosystem as an important unit of biodiversity.

CONCEPT CHECK 56.2

1. How does the reduced genetic diversity of small populations make them more vulnerable to extinction?
2. If there were 100 greater prairie chickens in a population, and 30 females and 10 males bred, what would be the effective population size (N_e)?
3. **WHAT IF?** In 2005, at least ten grizzly bears in the greater Yellowstone ecosystem were killed through contact with people. Most of these deaths resulted from three things: collisions with automobiles, hunters (of other animals) shooting when charged by a female grizzly bear with cubs nearby, and conservation managers killing bears that attacked livestock repeatedly. If you were a conservation manager, what steps might you take to minimize such encounters in Yellowstone?

For suggested answers, see Appendix A.

CONCEPT 56.3

Landscape and regional conservation help sustain biodiversity

Although conservation efforts have historically focused on saving individual species, efforts today often seek to sustain the biodiversity of entire communities, ecosystems, and landscapes. Such a broad view requires applying not just ecological principles, but aspects of human population dynamics and economics as well.

Landscape Structure and Biodiversity

The biodiversity of a given landscape is heavily influenced by its physical features, or *structure*. Understanding landscape structure is critically important in conservation because many species use more than one kind of ecosystem, and many live on the borders between ecosystems.

Fragmentation and Edges

The boundaries, or *edges*, between ecosystems—such as between a lake and the surrounding forest or between cropland and suburban housing tracts—are defining features of

▼ Figure 56.16 Natural edges between ecosystems in Siberia.

VISUAL SKILLS *What edges between ecosystems do you observe in this photo?*

▼ Figure 56.17 Amazon rain forest fragments created as part of the Biological Dynamics of Forest Fragments Project.

landscapes **(Figure 56.16)**. An edge has its own set of physical conditions, which differ from those on either side of it. The soil surface of an edge between a forest patch and a burned area receives more sunlight and is usually hotter and drier than the forest interior, but it is cooler and wetter than the soil surface in the burned area.

Some organisms thrive in edge communities because they gain resources from both adjacent areas. The ruffed grouse (*Bonasa umbellus*) is a bird that needs forest habitat for nesting, winter food, and shelter, but it also needs forest openings with dense shrubs and herbs for summer food.

Ecosystems in which edges arise from human alterations often have reduced biodiversity and a preponderance of edge-adapted species. For example, white-tailed deer (*Odocoileus virginianus*) thrive in edge habitats, where they can browse on woody shrubs; deer populations often expand when forests are logged and more edges are generated. The brown-headed cowbird (*Molothrus ater*) is an edge-adapted species that lays its eggs in the nests of other birds, often migratory songbirds. Cowbirds need forests, where they can parasitize the nests of other birds, and open fields, where they forage on seeds and insects. Consequently, their populations are growing where forests are being cut and fragmented, creating more edge habitat and open land. Increasing cowbird parasitism and habitat loss are correlated with declining populations of several of the cowbird's host species.

The influence of fragmentation on the structure of communities has been explored since 1979 in the long-term Biological Dynamics of Forest Fragments Project. Located in the heart of the Amazon River basin, the study area consists of isolated fragments of tropical rain forest separated from surrounding continuous forest by distances of 80–1,000 m **(Figure 56.17)**. Numerous researchers working on this project have clearly documented the effects of this fragmentation on organisms ranging from bryophytes to beetles to birds. They have consistently found that species adapted to forest interiors show the greatest declines when patches are the smallest, suggesting that landscapes dominated by small fragments will support fewer species.

Corridors That Connect Habitat Fragments

In fragmented habitats, the presence of a **movement corridor**, a narrow strip or series of small clumps of habitat connecting otherwise isolated patches, can be important for conserving biodiversity. Riparian habitats often serve as corridors, and in some nations, government policy prohibits altering these habitats. In areas of heavy human use, artificial corridors are sometimes constructed. Bridges or tunnels, for instance, can reduce the number of animals killed trying to cross highways **(Figure 56.18)**.

Movement corridors can also promote dispersal and reduce inbreeding in declining populations. Corridors have been shown to increase the exchange of individuals among populations of many organisms, including butterflies, voles, and aquatic plants. Corridors are especially important to species

▼ Figure 56.18 An artificial corridor. This highway overpass in the Netherlands helps animals cross a human-created barrier.

that migrate between different habitats seasonally. But corridors can also be harmful—for example, by allowing the spread of disease. In a 2003 study, a scientist at the University of Zaragoza, Spain, showed that habitat corridors facilitate the movement of disease-carrying ticks among forest patches in northern Spain. All the effects of corridors are not yet understood, and their impact is an area of active research.

 Mastering Biology HHMI Video: Seed Dispersal and Forest Fragmentation

Establishing Protected Areas

Currently, governments have set aside about 7% of the world's land in various forms of reserves. Choosing where to place nature reserves and how to design them poses many challenges. Should the reserve be managed to minimize the risks of fire and predation to a threatened species? Or should the reserve be left as natural as possible, with such processes as fires ignited by lightning allowed to play out on their own? This is just one of the debates that arise among people who share an interest in the health of national parks and other protected areas.

Preserving Biodiversity Hot Spots

In deciding which areas are of highest conservation priority, biologists often focus on hot spots of biodiversity. A **biodiversity hot spot** is a relatively small area with numerous endemic species (species found nowhere else in the world) and a large number of endangered and threatened species **(Figure 56.19)**. Nearly 30% of all bird species can be found in hot spots that make up only about 2% of Earth's land area. Together, the "hottest" of the terrestrial biodiversity hot spots total less than 1.5% of Earth's land but are home to more than

a third of all species of plants, amphibians, reptiles (including birds), and mammals. Aquatic ecosystems also have hot spots, such as coral reefs and certain river systems.

Biodiversity hot spots are good choices for nature reserves, but identifying them is not always simple. One problem is that a hot spot for one taxonomic group, such as butterflies, may not be a hot spot for some other taxonomic group, such as birds. Designating an area as a biodiversity hot spot is often biased toward saving vertebrates and plants, with less attention paid to invertebrates and microorganisms. Some biologists are also concerned that the hot-spot strategy places too much emphasis on such a small fraction of Earth's surface.

Climate change makes the task of preserving hot spots even more challenging because the conditions that favor a particular community may not occur there in the future. The biodiversity hot spot in the southwest corner of Australia (see Figure 56.19) holds thousands of species of endemic (unique to that area) plants and many endemic vertebrates. Researchers recently concluded that between 5% and 25% of the plant species they examined may become extinct by 2080 because the plants will be unable to tolerate the increased dryness predicted for this region.

Philosophy of Nature Reserves

Nature reserves are protected "islands" of biodiversity in a sea of habitat altered or degraded by human activity. An earlier policy—that protected areas should be set aside to remain unchanged forever—was based on the concept that ecosystems are balanced, self-regulating units. However, disturbance is common in all ecosystems (see Concept 54.3). Management policies that ignore disturbances or attempt to prevent them have generally failed. For instance, setting aside an area of a fire-dependent community, such as a portion of a tallgrass prairie, chaparral, or dry pine forest, with the intention of saving it is unrealistic if periodic burning is excluded. Without the dominant disturbance, the fire-adapted species are usually outcompeted and biodiversity is reduced.

An important conservation question is whether to create numerous small reserves or fewer large reserves. Small, unconnected reserves may slow the spread of disease between populations. One argument for large reserves is that large, far-ranging animals with low-density populations, such as the grizzly bear, require extensive habitats. Large reserves also have proportionately smaller perimeters than small reserves and are therefore less affected by edges.

As conservation biologists have learned more about the requirements for achieving minimum viable populations for endangered species, they have realized that most national parks and other reserves are far too small. The area needed for the long-term survival of

▼ **Figure 56.19 Earth's terrestrial (■) and marine (▲) biodiversity hot spots.**

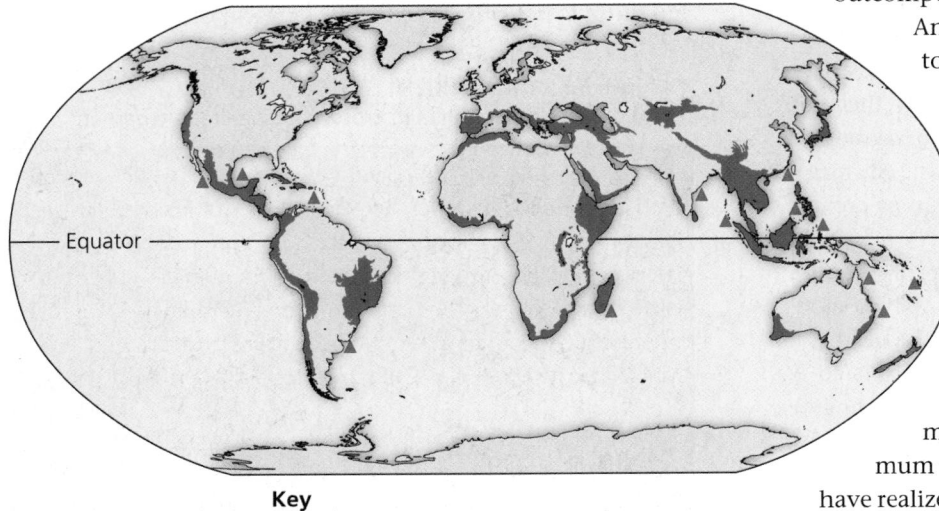

Key
■ Terrestrial hot spot ▲ Marine hot spot

the Yellowstone grizzly bear population, for instance, is more than 11 times the area of Yellowstone National Park. Areas of private and public land surrounding reserves will likely have to contribute to biodiversity conservation.

Zoned Reserves

Several nations have adopted a zoned reserve approach to landscape management. A **zoned reserve** is an extensive region that includes areas relatively undisturbed by humans surrounded by areas that have been changed by human activity and are used for economic gain. The key challenge of the zoned reserve approach is to develop a social and economic climate in the surrounding lands that is compatible with the long-term viability of the protected core. These surrounding areas continue to support human activities, but regulations prevent the types of extensive alterations likely to harm the protected area. As a result, the surrounding habitats serve as buffer zones against further intrusion into the undisturbed area.

The Central American nation of Costa Rica has become a world leader in establishing zoned reserves. An agreement initiated in 1987 reduced Costa Rica's international debt in return for land preservation there. The country is now divided into 11 Conservation Areas, which include national parks and other protected areas, both on land and in the ocean **(Figure 56.20)**. Costa Rica is making progress toward managing its zoned reserves, and the buffer zones provide a steady, lasting supply of forest products, water, and hydro-electric power while also supporting sustainable agriculture and tourism, both of which employ local people.

Costa Rica relies on its zoned reserve system to maintain at least 80% of its native species, but the system is not without problems. A 2003 analysis of land cover change between 1960

▼ **Figure 56.20 Protected areas in Costa Rica.** The boundaries of the 11 Conservation Areas are indicated by black outlines.

▼ **Figure 56.21 A diver measuring coral in the Florida Keys National Marine Sanctuary.**

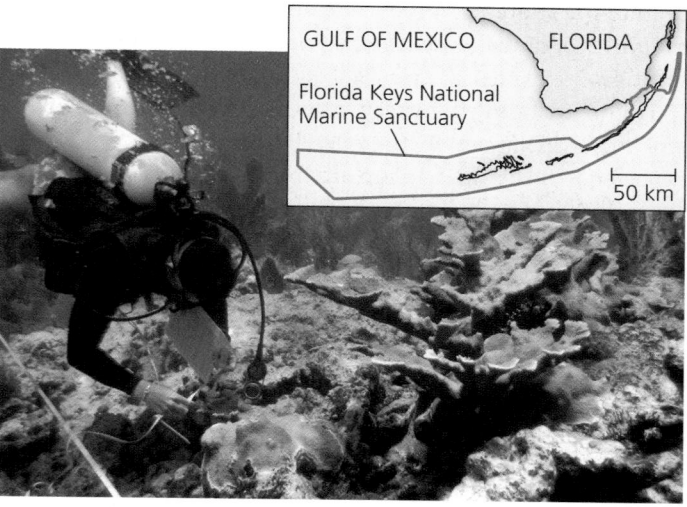

and 1997 showed negligible deforestation within Costa Rica's national parks and a gain in forest cover in the 1-km buffer around the parks. However, significant losses in forest cover were discovered in the 10-km buffer zones around all national parks, threatening to turn the parks into isolated habitat islands.

Although marine ecosystems have also been heavily affected by human exploitation, reserves in the ocean are far less common than reserves on land. Many fish populations around the world have collapsed as increasingly sophisticated equipment puts nearly all potential fishing grounds within human reach. In response, scientists have proposed establishing marine reserves around the world that would be off-limits to fishing. They present strong evidence that a patchwork of marine reserves can serve as a means of both increasing fish populations within the reserves and improving fishing success in nearby areas. Their proposed system is a modern application of a centuries-old practice in the Fiji Islands in which some areas have historically remained closed to fishing—a traditional example of the zoned reserve concept.

The United States adopted such a system in creating a set of 13 national marine sanctuaries, including the Florida Keys National Marine Sanctuary, which was established in 1990 **(Figure 56.21)**. Populations of marine organisms, including fishes and lobsters, recovered quickly after harvests were banned in the 9,500-km^2 reserve. Larger and more abundant fish now produce larvae that help repopulate reefs and improve fishing outside the sanctuary. The increased marine life within the sanctuary also makes it a favorite for recreational divers, increasing the economic value of this zoned reserve.

Urban Ecology

The zoned reserves that you just read about combine habitats that are relatively undisturbed by human activity with those that are used extensively by people for economic gain. Increasingly, ecologists are looking at species preservation even in the context of cities. The field of **urban ecology** examines organisms and their environment in urban settings.

For the first time in history, more than half of the people on Earth live in cities. By the year 2030, 5 billion people are expected to be living in urban environments. As cities expand in number and size, protected areas that were once outside city boundaries become incorporated into urban landscapes. Ecologists are now studying cities as ecological laboratories, seeking to balance species preservation and other ecological needs with the needs of people.

One critical area of research centers on urban streams, including the quality and flow of their water and the organisms living in them. Urban streams tend to rise and fall more quickly after rain than natural streams. This rapid change in water level occurs because of the concrete and other impervious surfaces in cities as well as the drainage systems that route water out of cities as quickly as possible to avoid flooding. Urban streams also tend to have higher concentrations of nutrients and contaminants and are often straightened or even channeled underground.

Near Vancouver, Canada, ecologists and volunteers have worked to restore a degraded urban stream, Guichon Creek. They stabilized its banks, removed introduced plants, and planted native trees and shrubs along the creek **(Figure 56.22)**. Their efforts returned the water flow and the communities of invertebrates and fish in the stream much closer to what they had been 50 years before the stream became degraded. A few years ago, ecologists successfully reestablished cutthroat trout in the stream. The trout are now thriving.

As cities continue to expand into the landscapes around them, understanding the ecological effects of this expansion will only increase in importance. Research on and conservation of urban habitats will continue to grow.

▼ **Figure 56.22 Volunteers working to remove invasive species along the urban Guichon Creek.**

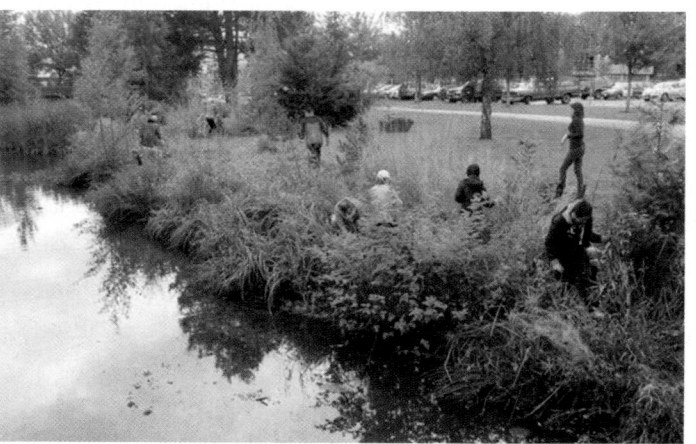

1. What is a biodiversity hot spot?
2. How do zoned reserves provide economic incentives for long-term conservation of protected areas?
3. **WHAT IF?** Suppose a developer proposes to clear-cut a forest that serves as a corridor between two parks. To compensate, the developer also proposes to add the same area of forest to one of the parks. As a professional ecologist, how might you argue for retaining the corridor?

For suggested answers, see Appendix A.

CONCEPT 56.4

Earth is changing rapidly as a result of human actions

As we've discussed, landscape and regional conservation help protect habitats and preserve species. However, environmental changes that result from human activities are creating new challenges. As a consequence of human-caused climate change, for example, the place where a vulnerable species is found today may not be the same place that is needed for its preservation in the future. What would happen if *many* habitats on Earth changed so quickly that the locations of preserves today were unsuitable for their species in 10, 50, or 100 years? Such a scenario is increasingly possible.

The rest of this section describes four types of environmental change that humans are bringing about: nutrient enrichment, toxin accumulation, climate change, and ozone depletion. The impacts of these and other changes are evident not just in human-dominated ecosystems, such as cities and farms, but also in the most remote ecosystems on Earth.

Nutrient Enrichment

Human activity often removes nutrients from one part of the biosphere and adds them to another. Someone eating broccoli in Washington, DC, consumes nutrients that only days before were in the soil in California; a short time later, some of these nutrients will be in the Potomac River, having passed through the person's digestive system and a local sewage treatment facility. Likewise, nutrients in farm soil may run off into streams and lakes, depleting nutrients in one area, increasing them in another, and altering chemical cycles in both.

Farming illustrates how human activities can lead to nutrient enrichment. After vegetation is cleared from an area, the existing reserve of nutrients in the soil is depleted over time because many of these nutrients are exported from the area in crop biomass. How long this takes varies greatly. When some of the early North American prairie lands were first tilled, good crops could be produced for decades because the large store of organic materials in the soil continued to decompose and provide nutrients. By contrast, some cleared land in the tropics can be farmed for only one or two years because so little of the ecosystems' nutrient load is contained in the soil. Despite such variations, the natural store of nutrients eventually becomes exhausted.

Consider nitrogen, the main nutrient lost through agriculture (see Figure 55.14). Plowing mixes the soil and speeds up decomposition of organic matter, releasing nitrogen that is then removed when crops are harvested. Fertilizers containing nitrates and other forms of nitrogen that plants can absorb are used to replace the nitrogen that is lost. However, after crops are harvested, few plants remain to take up nitrates from the soil. As shown in Figure 55.15, without plants to absorb them, nitrates are often leached from the ecosystem.

Recent studies indicate that human activities have more than doubled Earth's supply of fixed nitrogen available to primary producers. Industrial fertilizers provide the largest additional nitrogen source. Fossil fuel combustion also releases nitrogen oxides, which enter the atmosphere and dissolve in rainwater; the nitrogen ultimately enters ecosystems as nitrate. Increased cultivation of legumes, with their nitrogen-fixing symbionts, is a third way in which humans increase the amount of fixed nitrogen in the soil.

A problem arises when the nutrient level in an ecosystem exceeds the **critical load**, the amount of added nutrient, usually nitrogen or phosphorus, that can be absorbed by plants without damaging ecosystem integrity. For example, nitrogenous minerals in the soil that exceed the critical load eventually leach into groundwater or run off into freshwater and marine ecosystems, contaminating water supplies and causing eutrophication in aquatic ecosystems (see Concept 55.2). Nitrate concentrations in groundwater are increasing in most agricultural regions, sometimes reaching levels that are unsafe for drinking.

Many rivers contaminated with nitrates and ammonium from agricultural runoff and sewage drain into the Atlantic Ocean, with the highest inputs coming from northern Europe and the central United States. The Mississippi River carries nitrogen pollution to the Gulf of Mexico, fueling a phytoplankton bloom each summer. When the phytoplankton die, their decomposition by oxygen-using organisms results in an extensive "dead zone" of low oxygen levels along the coast **(Figure 56.23)**. When this occurs, fish and other marine animals disappear from some of the most economically important waters in the United States. To reduce the size of the dead zone, farmers have begun using fertilizers more efficiently, and managers are restoring wetlands in the Mississippi watershed.

Nutrient runoff can also lead to the eutrophication of lakes. The bloom and subsequent die-off of algae and cyanobacteria and the ensuing depletion of oxygen are similar to what occurs in a marine dead zone. Such conditions threaten the survival of many organisms. For example, eutrophication of Lake Erie coupled with overfishing wiped out commercially important fishes such as blue pike, whitefish, and lake trout by the 1960s. Since then, tighter regulations on the dumping of sewage into the lake have enabled some fish populations to rebound, but many native species of fish and invertebrates have not recovered.

> **Mastering Biology** Animation: Water Pollution from Nitrates

Toxins in the Environment

Humans release an immense variety of toxic chemicals, including thousands of synthetic compounds previously unknown in nature, with little regard for the ecological consequences. Organisms acquire toxic substances from the environment along with nutrients and water. Some of the poisons are metabolized or excreted, but others accumulate in specific tissues, often fat. One of the reasons accumulated toxins are particularly harmful is that they become more concentrated in successive trophic levels of a food web. This phenomenon, called **biological magnification**, occurs because the biomass at any given trophic level is produced from a much larger biomass ingested from the level below (see Concept 55.3). Thus, top-level carnivores tend to be most severely affected by toxic compounds in the environment. We'll discuss three types of environmental toxins: industrial compounds and pesticides, pharmaceuticals, and plastic waste.

Industrial Compounds and Pesticides

Chlorinated hydrocarbons are a class of industrially synthesized compounds that have demonstrated biological magnification. Chlorinated hydrocarbons include the

▶ **Figure 56.23 A dead zone arising from nitrogen pollution in the Mississippi basin.** Each year, a large dead zone (a region where dissolved O_2 levels are less than 2.0 mg/L) forms in the Gulf of Mexico.

(a) Nutrients drain from agricultural land (green) and cities (red) through the vast Mississippi watershed to the Gulf of Mexico.

(b) The 2017 dead zone, represented here, was the largest yet measured. It occupied 22,730 km^2 (8,776 mi^2), an area slightly larger than New Jersey.

▼ Figure 56.24 Biological magnification of PCBs in a Great Lakes food web. (ppm = parts per million)

Concentration of PCBs

Herring
gull eggs
124 ppm

Lake trout
4.83 ppm

Smelt
1.04 ppm

Zooplankton
0.123 ppm

Phytoplankton
0.025 ppm

❓ *Calculate how much the PCB concentration increased at each step in the food web.*

▶ Figure 56.25 Rachel Carson. Biologist and author Rachel Carson helped promote a new environmental ethic through her writing and her testimony before the U.S. Congress. Her efforts led to a ban on DDT use in the United States and stronger controls on the use of other chemicals.

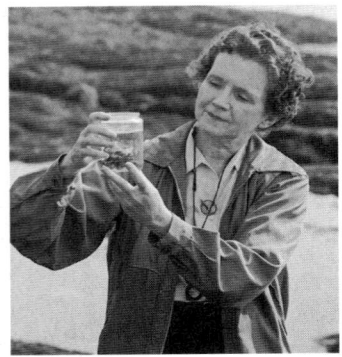

industrial chemicals called PCBs (polychlorinated biphenyls) and many pesticides, such as DDT. Current research implicates many of these compounds in endocrine system disruption in numerous animal species, including humans. Biological magnification of PCBs has been found in the food web of the Great Lakes, where the concentration of PCBs in herring gull eggs, at the top of the food web, is nearly 5,000 times that in phytoplankton, at the base **(Figure 56.24)**.

An infamous case of biological magnification that harmed top-level carnivores involved DDT, a chemical used to control insects such as mosquitoes and agricultural pests. In the decade after World War II, the use of DDT grew rapidly. By the 1950s, scientists were learning that DDT persists in the environment and is transported by water to areas far from where it is applied. One of the first signs that DDT was a serious environmental problem was a decline in the populations of pelicans, ospreys, and eagles, birds that feed at the top of food webs. The accumulation of DDT (and DDE, a product of its breakdown) in the tissues of these birds interfered with the deposition of calcium in their eggshells. When the birds tried to incubate their eggs, the weight of the parents broke the shells of affected eggs, resulting in catastrophic declines

in the birds' reproduction rates. Rachel Carson's book *Silent Spring* helped bring the problem to public attention in the 1960s **(Figure 56.25)**, and DDT was banned in the United States in 1971. A dramatic recovery in populations of the affected bird species followed.

In much of the tropics, DDT is still used to control the mosquitoes that spread malaria and other diseases. Societies there face a trade-off between saving human lives and protecting other species. Moreover, resistance to DDT has evolved in many insect species, further complicating its use (for a specific example, see Chapter Review question 7 in Chapter 22). Overall, the best approach seems to be to apply DDT sparingly and to couple its use with mosquito netting and other low-technology solutions.

Pharmaceuticals

Pharmaceuticals make up another group of toxins in the environment, one that is a growing concern among ecologists. The use of over-the-counter and prescription drugs has risen in recent years, particularly in industrialized nations. People who consume such products excrete residual chemicals in their waste and may also dispose of unused drugs improperly, such as in their toilets or sinks. Drugs that are not broken down in sewage treatment plants may then enter rivers and lakes with the material discharged from these plants. Growth-promoting drugs given to farm animals can also enter rivers and lakes with agricultural runoff. As a consequence, many pharmaceuticals are spreading in low concentrations across the world's freshwater ecosystems **(Figure 56.26)**.

Among the pharmaceuticals that ecologists are studying are the sex steroids, including forms of estrogen used for birth control. Some fish species are so sensitive to certain estrogens that concentrations of a few parts per trillion in their water can alter sexual differentiation and shift the female-to-male sex ratio toward females. Researchers in Ontario, Canada, conducted a seven-year experiment in which they applied the synthetic estrogen used in contraceptives to a lake in very low concentrations (5–6 ng/L). They found that chronic exposure of the fathead minnow (*Pimephales promelas*) to the estrogen led to feminization of males and a near extinction of the population of this species from the lake.

▼ Figure 56.26 Sources and movements of pharmaceuticals in the environment.

Pharmaceuticals

Humans

Toilet

Farm animals

Manure

Sludge

Agricultural runoff

Farms

Treated effluent

Sewage treatment plant

Lakes and rivers

Plastic Waste

Plastics are synthetic compounds that typically are made from petroleum products, such as oil or natural gas. An estimated 4.8 to 12.7 million metric tons (11 to 28 billion pounds) of plastic waste enters the ocean each year, making plastics the most common type of marine debris.

Over time, larger pieces of plastic waste are broken into successively smaller pieces by the actions of waves and by UV light, which breaks the bonds that hold plastic compounds together. As a result, the oceans are contaminated by large pieces of plastic and by **microplastics**, plastic particles less than 5 mm in size. Data suggest that plastic waste can persist in the environment for hundreds to thousands of years.

The enormous quantities of plastic waste in the oceans have wide-ranging effects. For example, birds, marine mammals, turtles, and fish can die after mistaking plastic debris for food **(Figure 56.27)**. A 2018 study documented another effect: Coral reefs contaminated by plastic waste had much higher rates of disease than did reefs without plastic waste **(Figure 56.28)**. What causes this dramatic rise in disease levels? Bacterial pathogens can "hitchhike" on plastic waste—so pieces of

plastic waste may introduce new pathogens to a reef or increase the abundance of pathogens already present. In addition, plastic waste that becomes entangled on a reef can damage corals or deprive them of light and oxygen, making them more susceptible to pathogens.

As for microplastics, these small particles now contaminate all of the world's oceans. Microplastics have been found in organisms from every type of marine food chain, and harmful effects of microplastics have been documented in some fish and invertebrates. Microplastics also may pose a threat to human health. Recent studies have found that 25% of supermarket fish in Indonesia and the United States contained microplastics in their guts. Shellfish sold in supermarkets also contained microplastics. While less is known about the extent to which microplastics contaminate terrestrial and freshwater ecosystems, current evidence indicates they are widespread there as well—they have been found in lakes, streams, soils, invertebrates, birds, and freshwater fish. Some evidence indicates that microplastics can be transported in the atmosphere, enabling them to be deposited in remote locations.

Research into the extent and effects of plastic waste is still in the early stages. Even so, it is clear that plastic waste is a major and growing environmental problem—one that will require new and innovative ways to reduce both the amount of plastic waste we generate and how much of that waste is mismanaged, thereby contaminating natural ecosystems.

> ⊘ **Mastering Biology** Interview with Chelsea Rochman: Studying the ecological effects of plastic waste (see the interview before Chapter 52)

▶ Figure 56.27 A victim of plastic waste. Found in a wildlife refuge in Hawaii, this Laysan albatross (*Phoebastria immutabilis*) chick died after consuming large amounts of plastic debris.

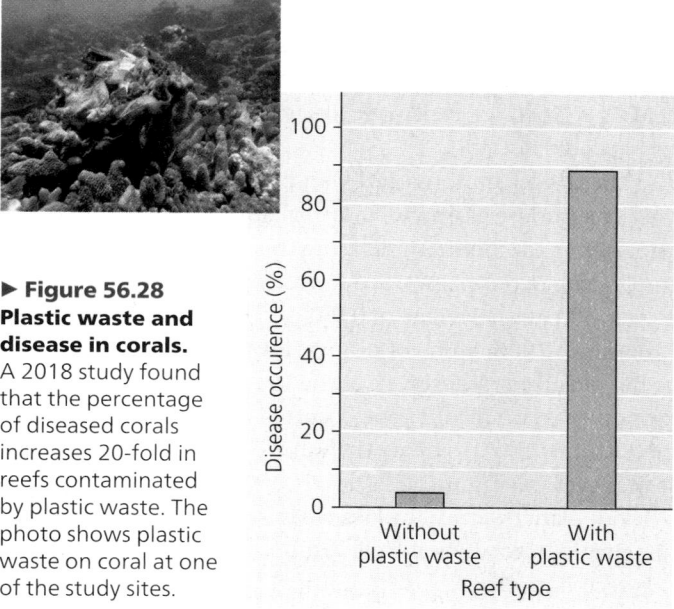

▶ Figure 56.28 Plastic waste and disease in corals. A 2018 study found that the percentage of diseased corals increases 20-fold in reefs contaminated by plastic waste. The photo shows plastic waste on coral at one of the study sites.

Disease occurence (%)

Without plastic waste

With plastic waste

Reef type

Greenhouse Gases and Climate Change

Human activities release a variety of gaseous waste products. People once thought that the vast atmosphere could absorb these materials indefinitely, but we now know that our actions can lead to **climate change**, a directional change to the global climate that lasts for three decades or more (as opposed to short-term changes in the weather).

Rising Atmospheric CO₂ Levels

To see how human actions can cause climate change, consider atmospheric CO_2 levels. Over the past 170 years, the concentration of CO_2 in the atmosphere has been increasing as a result of the burning of fossil fuels and deforestation. Scientists estimate that the average CO_2 concentration in the atmosphere before 1850 was about 274 ppm. In 1958, a monitoring station began taking accurate measurements on Hawaii's Mauna Loa peak, a location far from cities and high enough for the atmosphere to be well mixed. At that time, the average CO_2 concentration was 315 ppm **(Figure 56.29)**. By 2018, the CO_2 level exceeded 410 ppm, an increase of more than 50% since the mid-19th century. In the **Scientific Skills Exercise**, you can graph and interpret changes in CO_2 concentration that occur over three ten-year periods of time.

The increase in the concentration of atmospheric CO_2 over the last 170 years concerns scientists because of its link to increased global temperature. Much of the solar radiation that strikes the planet is emitted toward space as infrared radiation (known informally as "heat radiation"). Although CO_2, methane, water vapor, and other greenhouse gases in the atmosphere are transparent to visible light, they intercept and absorb much of the infrared radiation that Earth emits, radiating most of it back toward Earth. This process, called the **greenhouse effect**, retains some of the solar heat **(Figure 56.30)**. If it were not for this greenhouse effect, the average air temperature at Earth's surface would be a frigid −18°C (−0.4°F), and most life as we know it could not exist.

As the concentrations of CO_2 and other greenhouse gases rise, more solar heat is retained, thereby increasing the temperature of our planet. So far, Earth has warmed by an average of 0.9°C (1.6°F) since 1900. Much of this warming has occurred recently: 18 of the 19 warmest years on record have occurred since 2001, with 2016 being the warmest year ever (see Figure 56.29).

As our planet warms, the climate is changing in other ways as well: Wind and precipitation patterns are shifting,

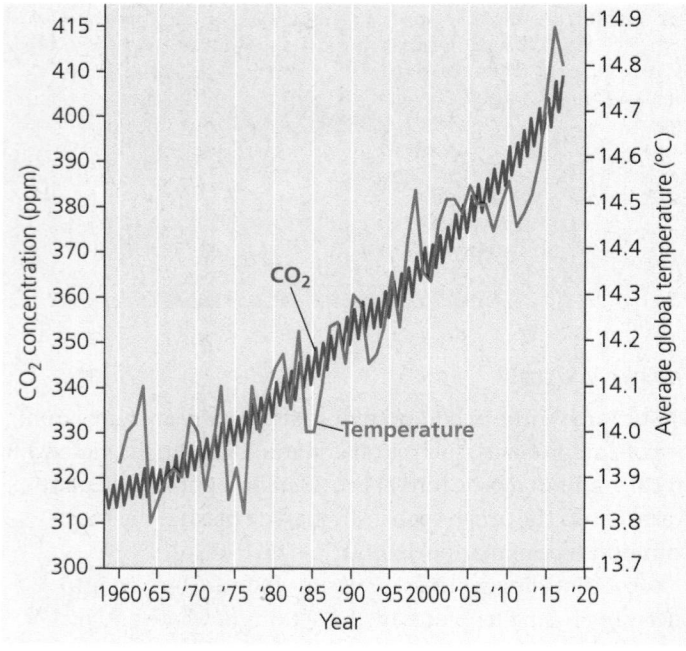

▼ **Figure 56.29 Increase in atmospheric carbon dioxide concentration at Mauna Loa, Hawaii, and average global temperatures.** Aside from normal seasonal fluctuations, the CO_2 concentration (blue curve) increased steadily since 1958. Though average global temperatures (red curve) fluctuated a great deal over time, there is a clear warming trend.

→ **Mastering Biology** MP3 Tutor: Global Warming

▼ **Figure 56.30 The greenhouse effect.** Carbon dioxide and other greenhouse gases in the atmosphere absorb heat emitted from Earth's surface and then radiate much of that heat back to Earth.

❶ Some of the incoming solar radiation is reflected back to space, but most passes through the atmosphere to reach Earth's surface.

❹ Of the heat emitted from Earth, some escapes to space. Much of the rest is absorbed by greenhouse gases and radiated back toward Earth, thereby trapping heat.

SPACE

ATMOSPHERE

❷ Of the radiation that reaches Earth's surface, some is reflected back to space. Much of the rest is absorbed, warming Earth's surface.

❸ Some of the energy that warms the surface is then emitted from Earth as heat.

→ **Mastering Biology** HHMI Animation: Greenhouse Effect
Animation: The Global Carbon Cycle and the Greenhouse Effect

Graphing Data and Evaluating Evidence

How Has the Rise in Atmospheric CO₂ Concentration Changed over Time? The blue curve in Figure 56.29 shows how the concentration of CO_2 in Earth's atmosphere has changed over a span of more than 50 years. For each year in that span, scientists collected daily CO_2 concentration data and used those data to calculate the average CO_2 level for that year. In this exercise, you'll graph average CO_2 concentrations for three ten-year periods and analyze how those concentrations have changed over time.

Data from the Study The data in the table below are average CO_2 concentrations (in parts per million, or ppm) at the Mauna Loa monitoring station for each of the indicated years.

1969	324.6	1989	353.1	2009	387.4
1970	325.7	1990	354.4	2010	389.9
1971	326.3	1991	355.6	2011	391.7
1972	327.5	1992	356.5	2012	393.9
1973	329.7	1993	357.1	2013	396.5
1974	330.2	1994	358.8	2014	398.7
1975	331.1	1995	360.8	2015	400.8
1976	332.0	1996	362.6	2016	404.2
1977	333.8	1997	363.7	2017	406.6
1978	335.4	1998	366.7	2018	408.5

Data from National Oceanic & Atmospheric Administration, Earth System Research Laboratory, Global Monitoring Division.

INTERPRET THE DATA

1. Plot the data for each ten-year period on one graph, where the x-axis runs from year 1 to year 10 (producing three curves); this approach can help you to compare and contrast how the CO_2 level has changed during the three decades. Select a type of graph that is appropriate for these data, and choose

▶ A researcher sampling the air at the Mauna Loa monitoring station, Hawaii.

a vertical axis scale that allows you to clearly see how the CO_2 concentration has changed, both from year to year and from one ten-year period to the next. (For additional information about graphs, see the Scientific Skills Review in Appendix D.)

2. For each ten-year period, what is the pattern of change in CO_2 concentration? Do the curves you have drawn suggest that the pattern has changed from decade to decade?

3. Calculate the annual rate at which the CO_2 concentration increased during each ten-year period. Perform these calculations as shown here for data from 1959 to 1968 (not in the table): in 1959 the CO_2 concentration was 316 ppm, while in 1968 it was 323. Thus, from 1959 to 1968, the CO_2 concentration increased at an average annual rate of

$$\frac{323 - 316}{9}$$

or 0.8 ppm per year (we divide by 9 because during a 10-year period of time, there are 9 year-to-year changes in CO_2 concentration).

4. **(a)** If the annual increase in CO_2 concentration were to remain at the level observed for 2009–2018, predict the CO_2 concentration in 2100. **(b)** Use your results to evaluate the effectiveness of efforts to reduce CO_2 emissions.

➔ **Instructors:** A version of this Scientific Skills Exercise can be assigned in **Mastering Biology**.

and extreme weather events (such as droughts, heat waves, and hurricanes) are occurring more often. How are such changes to the global climate affecting life on Earth?

Biological Effects of Climate Change

Many organisms, especially those such as plants that cannot disperse rapidly over long distances, may not be able to survive the rapid warming and changing precipitation patterns projected to occur over the next century. Furthermore, many habitats today are more fragmented than ever, further limiting the ability of many organisms to migrate. Indeed, the climate change that has occurred to date has *already* altered the geographic ranges of hundreds of species, in some cases leading to declining population sizes and shrinking geographic ranges (see Concept 52.1). For example, a 2015 study of 67 species of bumblebees found that on average, the geographic ranges of these important pollinators have decreased in size as the climate has warmed since 1900. Several other observed effects of climate change are discussed in **Figure 56.31**.

The ecosystems where the climate has changed the most include those in the far north, particularly northern coniferous forests and tundra. As snow and ice melt and uncover darker, more absorptive surfaces, these systems reflect less radiation back to the atmosphere and warm further (see Figure 56.30). Satellite data show that the area covered by Arctic sea ice in the summer has been declining since 1979 (the first year the data were available). At current rates of decline, there may be no summer ice there within a few decades, decreasing habitat for polar bears, seals, and seabirds. In addition, as discussed in Concept 55.2, rising temperatures have caused some Arctic regions to switch from being a CO_2 *sink* (absorbing more CO_2 from the atmosphere than they release to the atmosphere) to a CO_2 *source* (releasing more CO_2 than they absorb)—a worrisome change that could contribute to further climate warming.

Coniferous forests in western North America have also been hard hit, in this case by a combination of higher temperatures,

Climate Change Has Effects at All Levels of Biological Organization

The burning of fossil fuels by humans has caused atmospheric concentrations of carbon dioxide and other greenhouse gases to rise dramatically (see Figure 56.29). This, in turn, is changing Earth's climate: The planet's average temperature has increased by about 1°C since 1900, and extreme weather events are occurring more often in some regions of the globe. How are these changes affecting life on Earth today?

Effects on Cells

Temperature affects the rates of enzymatic reactions (see Figure 8.17), and as a result, the rates of DNA replication, cell division, and other key processes in cells are affected by rising temperatures.

Global warming and other aspects of climate change have also impaired some organisms' defense responses at the cellular level. For example, in the vast coniferous forests of western North America, climate change has reduced the ability of pine trees to defend themselves against attack by the mountain pine beetle (*Dendroctonus ponderosae*).

Resin canal

Resin cells   100 μm

◀ Pine defenses include specialized resin cells that secrete a sticky substance (resin) that can entrap and kill mountain pine beetles. Resin cells produce less resin in trees that are stressed by rising temperatures and drought conditions.

When beetles overwhelm a tree's cellular defenses, they produce large numbers of offspring that tunnel through the wood, causing extensive damage. Rising temperatures have shortened how long it takes beetles to mature and reproduce, resulting in even more beetles. The beetles can also infect the tree with a harmful fungus, which appears as blue stains on the wood.

◀ This aerial view shows the scope of destruction in one North American forest due to mountain pine beetles; dead trees appear orange and red.

Effects on Individual Organisms

Organisms must maintain relatively constant internal conditions (see Concept 40.3); for example, an individual will die if its body temperature becomes too high. Global warming has increased the risk of overheating in some species, leading to reduced food intake and reproductive failure.

For instance, an American pika (*Ochotona princeps*) will die if its body temperature rises just 3°C above its resting temperature—this can happen quickly in regions where climate change has caused significant warming.

▶ As summer temperatures have risen, American pikas are spending more time in their burrows to escape the heat. Thus, they have less time to forage for food. Lack of food has caused mortality rates to increase and birth rates to drop. Pika populations have dwindled, some to the point of extinction. (See Figure 1.11 for another example.)

▲ This graph represents conditions in 2015 at 67 sites that previously supported a pika population; the population at 10 of these sites had become extinct. Most extinctions occurred at sites with high summer temperatures and a small area of pika habitat. As temperatures continue to increase, more extinctions are expected.

Climate change has caused some populations to increase in size, while others have declined (see Concept 1.1 and Concept 46.1). In particular, as the climate has changed, some species have adjusted when they grow, reproduce, or migrate—but others have not, causing their populations to face food shortages and reduced survival or reproductive success.

In one example, researchers have documented a link between rising temperatures and declining populations of caribou (*Rangifer tarandus*) in the Arctic.

Caribou populations migrate north in the spring to give birth and to eat sprouting plants.

► Alpine chickweed is an early-flowering plant on which caribou depend.

▲ As the climate has warmed, the plants on which caribou depend have emerged earlier in the spring. Caribou have not made similar changes in the timing of when they migrate and give birth. As a result, there is a shortage of food, and caribou offspring production has dropped fourfold.

Climate affects where species live (see Figure 52.8). Climate change has caused hundreds of species to move to new locations, in some cases leading to dramatic changes in ecological communities. Climate change has also altered primary production (see Figure 28.25) and nutrient cycling in ecosystems.

In the example we discuss here, rising temperatures have enabled a sea urchin to invade southern regions along the coast of Australia, causing catastrophic changes to marine communities there.

▲ The sea urchin *Centrostephanus rodgersii* requires water temperatures above 12°C to reproduce successfully, as shown in this graph. As ocean waters rise above this critical temperature, the urchin has been able to expand its range to the south, destroying kelp beds as it moves into new regions.

▲ As the urchin has expanded its range to the south, it has destroyed high-diversity kelp communities, leaving bare regions called "urchin barrens" in its wake.

MAKE CONNECTIONS

In addition to contributing to climate change, rising concentrations of CO_2 also contribute to ocean acidification (see Figure 3.12). Explain how ocean acidification can affect individual organisms, and how that, in turn, can cause dramatic changes in ecological communities.

decreased winter snowfall, and a lengthening of the summer dry period. Since the latter half of the 20th century, otherwise healthy forests have experienced a steady increase in the percentage of trees that die each year. Higher temperatures and more frequent droughts also increase the likelihood of fires (see Figure 55.8). In boreal forests of western North America and Russia, for example, fires have burned twice the usual area in recent decades, again leading to widespread tree mortality. As the climate continues to warm, other changes in the geographic distribution of precipitation are likely, such as agricultural areas of the central United States becoming much drier.

Climate change has already affected many other ecosystems as well. In Europe and Asia, for example, plants are producing leaves earlier in the spring, while in tropical regions, the growth and survival of some species of coral have declined as water temperatures have risen. Recently, Australia's Great Barrier Reef had a catastrophic decline: The combined effects of two consecutive marine heat waves (one in 2016, another in 2017) killed so many corals that two-thirds of the reef was severely degraded—so degraded that some scientists worry it may never recover. A key take-home message from these examples is that a given effect of climate change may, in turn, cause a series of other biological changes. The exact nature of such cascading biological effects can be hard to predict, but it is clear that the more our planet warms, the more severely its ecosystems will be affected.

Modeling Climate Change

At the current rates that human activities are adding CO_2 and other greenhouse gases to the atmosphere, global models predict that Earth's temperature will rise by an additional 3°C (5°F) by the end of the 21st century. What data are used to construct such models, and are their predictions accurate?

Global models are constructed using data on factors that affect the absorption of solar radiation at Earth's surface. These data are crucial because global temperatures rise when more solar radiation is absorbed by Earth's surface, and they fall when less solar radiation is absorbed by the surface. Natural factors that affect the absorption of solar radiation include an 11-year solar cycle and volcanic explosions. During each round of the solar cycle, the amount of energy emitted by the sun increases and decreases in a regular way, thereby contributing to a well-understood rise or fall in global temperatures at different points in time. Volcanic explosions emit gases and tiny particles that block incoming solar radiation, thereby contributing to a temporary drop in global temperatures. (Changes in the shape of Earth's orbit around the sun also contribute to increases or decreases in global temperatures, but those changes occur gradually over tens of thousands of years and hence are not a cause of recent climate change.)

Scientists also have collected vast amounts of data on how human activities affect absorption of solar radiation. When we burn fossil fuels for electricity, heat, and transportation,

we cause CO_2 and other greenhouse gases to be released to the atmosphere. As shown in Figure 56.30, adding CO_2 and other greenhouse gases to the atmosphere increases the absorption of solar radiation, leading to global warming. Other human activities decrease the absorption of solar radiation, thereby tending to reduce global temperatures. For example, dust released by plowing fields and the emission of gases such as sulfur dioxide (SO_2) both decrease the absorption of solar energy, thus reducing global temperatures.

Each of these and other types of data provides a single, well-understood piece of a grand puzzle—how Earth's temperature is affected by a combination of natural and human factors. Scientists use computer models to put the pieces of this puzzle together. Each year, large amounts of data on natural and human factors that affect the absorption of solar radiation are entered into the model, which then sums their effects to predict Earth's temperature. Over time, predictions from these models have become increasingly accurate: Global climate models can now reproduce observed changes such as the increase in global warming that has occurred since the 1950s **(Figure 56.32)** and the global cooling that results from volcanic eruptions.

But why go to all this trouble developing models when Earth's temperature can be measured each year? The reason is that we have only one Earth and hence cannot perform

▼ **Figure 56.32 Factors contributing to rising global temperatures.** Scientists have measured the temperature of our planet since 1880. These observed temperatures (red) can be compared to results from climate models that incorporate only natural factors that affect global temperatures (blue) and from models that incorporate both natural and human factors (black).

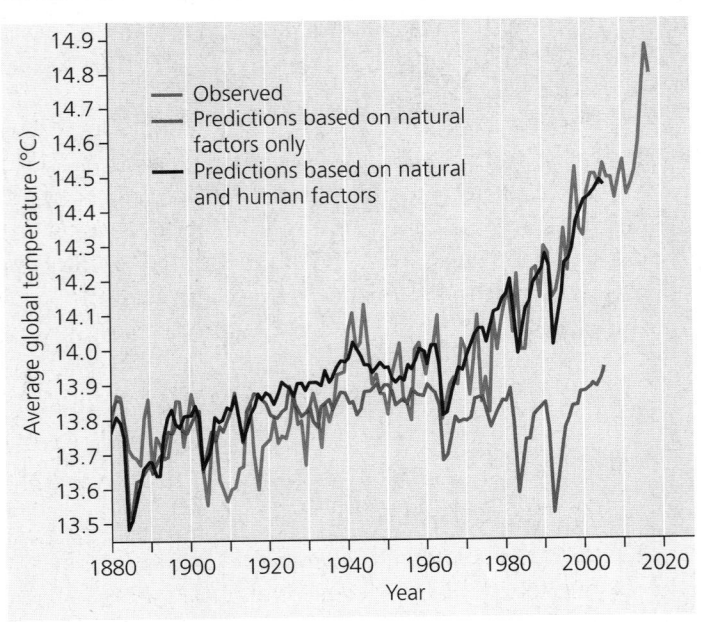

VISUAL SKILLS *Describe how well the results shown in the blue and black curves match observed temperature changes over time. Use those results to evaluate whether human activities such as burning fossil fuels have contributed to observed temperature changes over time.*

experiments to determine how emitting different amounts of CO_2 affects the temperature of our planet. Instead, because we know that global climate models yield accurate predictions, we can use the models to perform "if-then" thought experiments: *If* we add a certain amount of CO_2 to the atmosphere, *then* global temperatures will rise by a certain number of degrees.

The predictions are sobering. If we continue with "business as usual," where emissions of CO_2 continue at current levels, by 2100 Earth's temperature will likely be 4°C higher than it was in 1900—an increase that would have dire effects for all of Earth's species, including our own. Moreover, even if all CO_2 emissions were to stop immediately, the temperature of our planet would continue to rise by an additional 0.6°C, making it 1.5°C higher in 2100 than it was in 1900. Earth will continue to warm long after emissions stop because it takes decades for the CO_2 already emitted to warm the oceans, and thus that long for the full effect of past emissions to be realized.

Finding Solutions to Address Climate Change

We will need many approaches to slow global warming and other aspects of climate change. Quick progress can be made by using energy more efficiently and by replacing fossil fuels with renewable solar and wind power and, more controversially, with nuclear power. Today, coal, gasoline, wood, and other organic fuels remain central to industrialized societies and cannot be burned without releasing CO_2. Stabilizing CO_2 emissions will require concerted international effort and changes in both personal lifestyles and industrial processes.

Progress toward finding solutions to address climate change was made in 2015 when all nations agreed—for the first time—to take steps to reduce CO_2 emissions and limit the extent to which global temperatures ultimately rise. This international effort, known as the Paris Agreement, has been ratified by 169 nations, including China, the United States, and all other nations that emit substantial quantities of CO_2 and other greenhouse gases. The effectiveness of the agreement was recently called into question, however, when the United States announced its intention to withdraw from the agreement by 2020. This setback highlights a potential difference between what we know and what we choose to do. An overwhelming amount of evidence indicates that climate change is real, that the rise in global temperatures since the 1950s was caused primarily by human actions, and that there will be negative consequences for human societies and all life on Earth unless we act now. What we choose to do with this information is up to us.

Depletion of Atmospheric Ozone

Like carbon dioxide and other greenhouse gases, atmospheric ozone (O_3) has also changed in concentration because of human activities. Life on Earth is protected from the damaging effects of ultraviolet (UV) radiation by a layer of ozone located in the stratosphere 17–25 km above Earth's surface. However, satellite studies of the atmosphere show that the springtime ozone

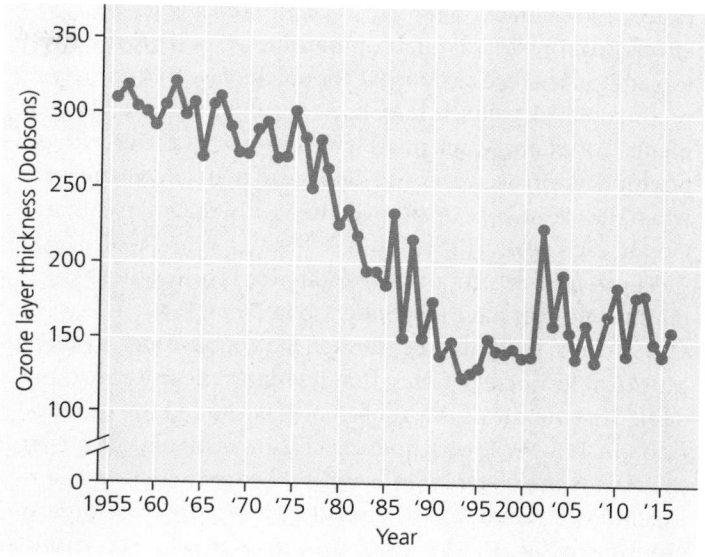

▼ **Figure 56.33 Thickness of the October ozone layer over Antarctica, in units called Dobsons.**

layer over Antarctica has thinned substantially since the mid-1970s **(Figure 56.33)**. The destruction of atmospheric ozone results primarily from the accumulation of chlorofluorocarbons (CFCs), chemicals once widely used in refrigeration and manufacturing. In the stratosphere, chlorine atoms released from CFCs react with ozone, reducing it to molecular O_2. Subsequent chemical reactions liberate the chlorine, allowing it to react with other ozone molecules in a catalytic chain reaction.

The thinning of the ozone layer is most apparent over Antarctica in spring, where cold, stable air allows the chain reaction to continue **(Figure 56.34)**. The magnitude of ozone depletion and the size of the ozone hole have been slightly smaller in recent years than the average for the last 20 years, but the hole still sometimes extends as far as the southernmost portions of Australia, New Zealand, and South America. At the more heavily populated middle latitudes, ozone levels have decreased 2–10% during the past 20 years.

Decreased ozone levels in the stratosphere increase the intensity of UV rays reaching Earth's surface. The consequences

▼ **Figure 56.34 Erosion of Earth's ozone shield.** The ozone hole over Antarctica is visible as the dark blue patch in these images based on atmospheric data.

September 1979   **September 2018**

of ozone depletion may be severe for plants, animals, and microorganisms. Some scientists expect increases in skin cancer and cataracts in humans, as well as unpredictable effects on crops and natural communities, especially the phytoplankton that are responsible for much of Earth's primary production. Such effects have been documented in field experiments. For example, scientists have shown that the growth of phytoplankton decreases and damage to their DNA increases when the ozone hole opens over the Southern Ocean (Antarctic Ocean) each year.

The good news about the ozone hole is how quickly many countries have responded to it. Since 1987, at least 197 nations, including the United States, have signed the Montreal Protocol, a treaty that regulates the use of ozone-depleting chemicals. Most nations have ended the production of CFCs. As a consequence, chlorine concentrations in the stratosphere have stabilized and ozone depletion is slowing. But even though CFC emissions today are close to zero, chlorine molecules already in the atmosphere will continue to influence stratospheric ozone levels for at least 50 years.

The partial destruction of Earth's ozone shield is one more example of how greatly humans can disrupt the dynamics of ecosystems and the biosphere. It also highlights our ability to solve environmental problems when we set our minds to it. However, in 2018 researchers detected an unexpected rise in atmospheric CFC levels, suggesting that unreported new production of these compounds may be occurring. Thus, once environmental regulations are in place, we cannot take them for granted—ongoing monitoring and enforcement are needed.

> ### CONCEPT CHECK 56.4
>
> 1. How can the addition of excess mineral nutrients to a lake threaten its fish population?
> 2. **MAKE CONNECTIONS** There are vast stores of carbon-containing organic matter in the soils of northern coniferous forests and tundra around the world. Suggest why scientists who study global warming are closely monitoring these carbon reservoirs (see Figure 55.14).
> 3. **MAKE CONNECTIONS** Mutagens are chemical and physical agents that induce mutations in DNA (see Concept 17.5). How does reduced ozone concentration in the atmosphere increase the likelihood of mutations in various organisms?
>
> *For suggested answers, see Appendix A.*

CONCEPT 56.5

Sustainable development can improve human lives while conserving biodiversity

With the increasing loss and fragmentation of habitats, changes in Earth's physical environment and climate, and increasing human population (see Concept 53.6), we face difficult trade-offs in managing the world's resources. Preserving all habitat patches isn't feasible, so biologists must help societies set conservation priorities by identifying which habitat patches are most crucial, while likewise improving the quality of life for local people. Ecologists use the concept of *sustainability* as a tool to establish long-term conservation priorities.

Sustainable Development

We need to understand the interconnections of the biosphere to protect species from extinction and improve the quality of human life. To this end, many nations, scientific societies, and other groups have embraced the concept of **sustainable development**, economic development that meets the needs of people today without limiting the ability of future generations to meet their needs.

Achieving sustainable development is an ambitious goal. To sustain ecosystem processes and stem the loss of biodiversity, we must connect life science with the social sciences, economics, and the humanities. We must also reassess our personal values. Those of us living in wealthier nations have a larger ecological footprint than do people living in developing nations (see Concept 53.6). By including the long-term costs of consumption in our decision-making processes, we can learn to value the ecosystem services that sustain us.

Case Study: *Sustainable Development in Costa Rica*

The success of conservation in Costa Rica discussed earlier has required a partnership between the government, nongovernment organizations (NGOs), and private citizens. Many nature reserves established by individuals have been recognized as national wildlife reserves and given significant tax benefits. However, conservation and restoration are only one facet of sustainable development; the other is improving the human condition.

How have the living conditions of the Costa Rican people changed as the country has pursued its conservation goals? As represented in **Figure 56.35**, two fundamental indicators

▼ **Figure 56.35 Infant mortality and life expectancy at birth in Costa Rica.**

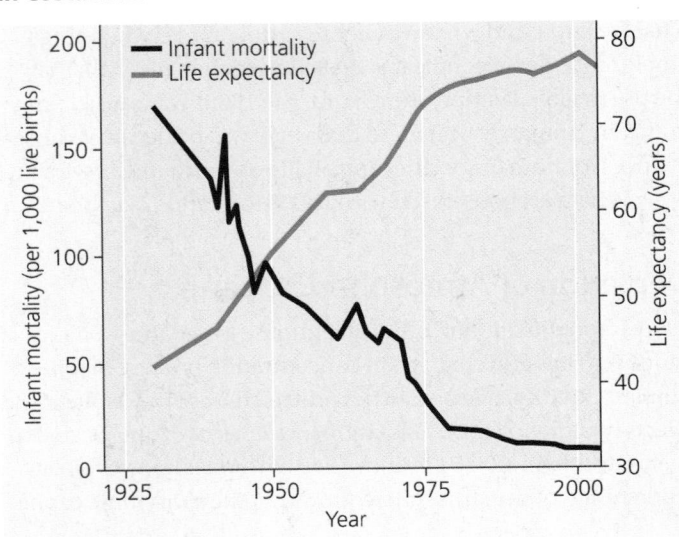

of living conditions are infant mortality rate and life expectancy (see Concept 53.6). From 1930 to 2010, the infant mortality rate in Costa Rica declined from 170 to 9 per 1,000 live births; over the same period, life expectancy increased from about 43 years to 79 years. Another indicator of living conditions is the literacy rate. The 2011 literacy rate in Costa Rica was 96%, compared to an average of 82% in the other six Central American countries. Such statistics show that living conditions in Costa Rica have improved greatly in the time in which the country has dedicated itself to conservation and restoration. While this does not prove that conservation *causes* an improvement in human welfare, we can say with certainty that development in Costa Rica has attended to both nature *and* people.

▼ **Figure 56.36 Biophilia, past and present.**

(a) Detail of animals in a 17,000-year-old cave painting, Lascaux, France

(b) A 30,000-year-old ivory carving of a water bird, found in Germany

(c) Nature lovers on a wildlife-watching expedition

The Future of the Biosphere

Our modern lives are very different from those of early humans, who hunted and gathered to survive. Their reverence for nature is evident in the early cave paintings of wildlife **(Figure 56.36a)** and in the visions of life they sculpted from bone and ivory **(Figure 56.36b)**.

Our lives reflect remnants of our ancestral attachment to nature and the diversity of life—the concept of *biophilia* introduced early in this chapter. We evolved in natural environments rich in biodiversity, and we still have an affinity for such settings **(Figure 56.36c** and **d)**. Indeed, our biophilia may be innate, an evolutionary product of natural selection acting on a brainy species whose survival depended on a close connection to the environment and a practical appreciation of plants and animals.

Our appreciation of life guides the field of biology today. We celebrate life by deciphering the genetic code that makes each species unique. We embrace life by using fossils and DNA to chronicle evolution through time. We preserve life through our efforts to classify and protect the millions of species on Earth. We respect life by using nature responsibly and reverently to improve human welfare.

Biology is the scientific expression of our desire to know nature. We are most likely to protect what we appreciate and are most likely to appreciate what we understand. By learning about the processes and diversity of life, we become more aware of our place in the biosphere. We hope this text has served you well in this lifelong adventure.

CONCEPT CHECK 56.5

1. What is meant by the term *sustainable development*?

2. How might biophilia influence us to conserve species and restore ecosystems?

3. **WHAT IF?** Suppose a new fishery is discovered, and you are put in charge of developing it sustainably. What ecological data might you want on the fish population? What criteria would you apply for the fishery's development?

For suggested answers, see Appendix A.

(d) A young biologist holding a songbird

→ Go to **Mastering Biology** for Assignments, the eText, the Study Area, and Dynamic Study Modules.

56 Chapter Review

SUMMARY OF KEY CONCEPTS

→ To review key terms, go to the **Vocabulary Self-Quiz** in the **Mastering Biology** eText or Study Area, or go to goo.gl/zkjz9t.

CONCEPT 56.1

Human activities threaten earth's biodiversity (pp. 1261–1266)

- Biodiversity can be considered at three main levels:

Genetic diversity: enables adaptation to environmental change

Species diversity: maintains communities and food webs

Ecosystem diversity: provides life-sustaining services

- Our biophilia enables us to recognize the value of biodiversity for its own sake. Other species also provide humans with food, fiber, medicines, and **ecosystem services**.
- Four major threats to biodiversity are habitat loss, **introduced species**, overharvesting, and global change.

? *Give at least three examples of key ecosystem services that nature provides for people.*

CONCEPT 56.2

Population conservation focuses on population size, genetic diversity, and critical habitat (pp. 1266–1270)

- When a population drops below a **minimum viable population (MVP)** size, its loss of genetic variation due to inbreeding and genetic drift can trap it in an **extinction vortex**.
- A loss of critical habitat can cause threatened populations to show a downward trend, even if the population is far above its minimum viable population size.
- Conserving species often requires resolving conflicts between the habitat needs of **endangered species** and human demands.

? *Why is the minimum viable population size smaller for a genetically diverse population than for a less genetically diverse population?*

CONCEPT 56.3

Landscape and regional conservation help sustain biodiversity (pp. 1270–1274)

- The structure of a landscape can strongly influence biodiversity. As habitat fragmentation increases and edges become more extensive, biodiversity tends to decrease. **Movement corridors** can promote dispersal and help sustain populations.
- **Biodiversity hot spots** are also hot spots of extinction and thus prime candidates for protection. Sustaining biodiversity in parks and reserves requires management to ensure that human activities in the surrounding landscape do not harm the protected habitats. The **zoned reserve** model recognizes that conservation efforts often involve working in landscapes that are greatly affected by human activity.
- **Urban ecology** is the study of organisms and their environment in primarily urban settings.

? *Give two examples that show how habitat fragmentation can harm species in the long term.*

CONCEPT 56.4

Earth is changing rapidly as a result of human actions (pp. 1274–1284)

- Agriculture removes plant nutrients from ecosystems, so large supplements are usually required. The nutrients in fertilizer can pollute groundwater and surface-water aquatic ecosystems (eutrophication), where they can stimulate excess algal growth.
- The release of industrial compounds, pesticides, pharmaceuticals, and plastic waste has polluted the environment with harmful substances that often persist for long periods and become increasingly concentrated in successively higher trophic levels of food webs (**biological magnification**).
- Because of the burning of fossil fuels and other human activities, the atmospheric concentration of CO_2 and other greenhouse gases has been steadily increasing. These increases have caused **climate change**, including significant global warming and changing patterns of precipitation. Climate change has already affected many ecosystems.
- The ozone layer reduces the penetration of UV radiation through the atmosphere. Human activities, notably the release of chlorine-containing pollutants, have eroded the ozone layer, but government policies are helping to solve the problem.

? *Thinking about biological magnification of toxins, is it healthier to feed at a lower or higher trophic level? Explain.*

CONCEPT 56.5

Sustainable development can improve human lives while conserving biodiversity (pp. 1284–1285)

- **Sustainable development** refers to economic development that meets the needs of people today without limiting the ability of future generations to meet their needs.
- Costa Rica's success in conserving tropical biodiversity has involved a partnership among the government, other organizations, and private citizens. Human living conditions in Costa Rica have improved along with ecological conservation.

? *Why is sustainability such an important goal for conservation biologists?*

TEST YOUR UNDERSTANDING

→ For more multiple-choice questions, go to the **Practice Test** in the **Mastering Biology** eText or Study Area, or go to goo.gl/GruWRg.

Levels 1-2: Remembering/Understanding

1. One characteristic that distinguishes a population in an extinction vortex from most other populations is that
   (A) its members are rare, top-level predators.
   (B) its effective population size is lower than its total population size.
   (C) its genetic diversity is very low.
   (D) it is not well adapted to edge conditions.

2. The main cause of the increase in the amount of CO_2 in Earth's atmosphere over the past 170 years is
   (A) increased worldwide primary production.
   (B) increased worldwide fertilizer production.
   (C) increased infrared radiation absorption by the atmosphere.
   (D) the burning of fossil fuels and deforestation.

3. What is the single greatest threat to biodiversity?
   (A) overharvesting of commercially important species
   (B) habitat alteration, fragmentation, and destruction
   (C) introduced species that compete with native species
   (D) novel pathogens

Levels 3-4: Applying/Analyzing

4. Which of these is a consequence of biological magnification?
   (A) Toxic chemicals in the environment pose greater risk to top-level predators than to primary consumers.
   (B) Populations of top-level predators are generally smaller than populations of primary consumers.
   (C) The biomass of producers in an ecosystem is generally higher than the biomass of primary consumers.
   (D) Only a small portion of the energy captured by producers is transferred to consumers.

5. Which of the following strategies would most rapidly increase the genetic diversity of a population in an extinction vortex?
   (A) Establish a reserve that protects the population's habitat.
   (B) Introduce new individuals transported from other populations of the same species.
   (C) Sterilize the least fit individuals in the population.
   (D) Control populations of the endangered population's predators and competitors.

6. Which of the following statements about protected areas established to preserve biodiversity is true?
   (A) About 25% of Earth's land area is now protected.
   (B) National parks are the only type of protected area.
   (C) Management of a protected area does not need to be coordinated with management of the surrounding area.
   (D) It is especially important to protect biodiversity hot spots.

Levels 5-6: Evaluating/Creating

7. **DRAW IT** Suppose you are managing a forest reserve, and one of your goals is to protect local populations of woodland birds from parasitism by the brown-headed cowbird. You know that female cowbirds don't venture more than about 100 m into a forest and that nest parasitism is reduced when woodland birds nest away from forest edges. The reserve extends about 6,000 m from east to west and 3,000 m from north to south. It is surrounded by a deforested pasture on the west, an agricultural field for 500 m in the southwest corner, and intact forest everywhere else. You must build a road, 10 m by 3,000 m, from the north to the south side of the reserve and construct a maintenance building that will take up 100 m². Draw a map of the reserve and explain where you would put the road and building to minimize cowbird intrusion along edges.

8. **EVOLUTION CONNECTION** The fossil record indicates that there have been five mass extinction events in the past 500 million years (see Concept 25.4). Many ecologists think we are on the verge of another. Briefly discuss the history of mass extinctions and the length of time it typically takes for species diversity to recover through evolution. Explain why this should motivate us to slow the loss of biodiversity today.

9. **WRITE ABOUT A THEME: INTERACTIONS** One factor favoring rapid population growth by an introduced species is the absence of the predators, parasites, and pathogens that controlled its population in the region where it evolved. In a short essay (100–150 words), explain how evolution by natural selection would influence the rate at which native predators, parasites, and pathogens attack an introduced species.

10. **SYNTHESIZE YOUR KNOWLEDGE**

Big cats, such as the Siberian tiger (*Panthera tigris altaica*), are one of the most endangered groups of mammals in the world. Based on what you've learned in this chapter, discuss some approaches you would use to help preserve them.

For selected answers, see Appendix A.

Explore Scientific Papers with Science in the Classroom ⋀ AAAS
How can genetic analyses help in the fight against poaching?
Go to "CSI Africa: Tracking Ivory Poachers" at
www.scienceintheclassroom.org.
→ **Instructors:** Questions can be assigned in Mastering Biology.

APPENDIX A Answers

Chapter 1

Figure Questions

Figure 1.4 Dividing the length of the prokaryotic cell by the length of the scale bar, the length of the prokaryotic cell is about 1.4 scale bars. Each scale bar represents 1 μm, so the prokaryotic cell is about 1.4 μm long. Dividing the diameter of the eukaryotic cell by the length of the scale bar, the diameter of the eukaryotic cell is about 8.2 scale bars, which is 8.2 μm. **Figure 1.10** The response to insulin is glucose uptake by cells and glucose storage in liver cells. The initial stimulus is a high glucose level, which is reduced when glucose is taken up by cells.

Figure 1.18

As the soil gradually becomes lighter gray, beetles that match the color of the soil will not be seen by birds and therefore will not be eaten. For example, when the soil is medium-colored, it will be easiest for birds to see and eat the darkest and lightest of the beetles in a population. (Most or all of the lighter beetles will have been eaten earlier, but new light beetles will arise due to variation in new generations of the population.) Thus, over time, the population will become lighter as the soil becomes lighter.

⑤ Environmental change resulting in survival of organisms with different traits

Concept Check 1.1

1. Examples: A molecule consists of *atoms* bonded together. Each organelle has an orderly arrangement of *molecules*. Photosynthetic plant cells contain *organelles* called chloroplasts. A tissue consists of a group of similar *cells*. Organs such as the heart are constructed from several *tissues*. A complex multicellular organism, such as a plant, has several types of *organs*, such as leaves and roots. A population is a set of *organisms* of the same species. A community consists of *populations* of the various species inhabiting a specific area. An ecosystem consists of a biological *community* along with the nonliving factors important to life, such as air, soil, and water. The biosphere is made up of all of Earth's *ecosystems*. **2.** (a) New properties emerge at successive levels of biological organization: Structure and function are correlated. (b) Life's processes involve the expression and transmission of genetic information. (c) Life requires the transfer and transformation of energy and matter. **3.** Sample answers: *Organization (Emergent properties)*: The ability of a human heart to pump blood requires an intact heart; it is not a capability of any of the heart's tissues or cells working alone. *Organization (Structure and function)*: The strong, sharp teeth of a wolf are well suited to grasping and dismembering its prey. *Information*: Human eye color is determined by the combination of genes inherited from the two parents. *Energy and Matter*: A plant, such as a grass, absorbs energy from the sun and transforms it into molecules that act as stored fuel. Animals can eat parts of the plant and use the food for energy to carry out their activities. *Interactions (Molecules)*: When your stomach is full, it signals your brain to decrease your appetite. *Interactions (Ecosystems)*: A mouse eats food, such as nuts or grasses, and deposits some of the food material as wastes (feces and urine). Construction of a nest rearranges the physical environment and may hasten degradation of some of its components. The mouse may also act as food for a predator.

Concept Check 1.2

1. The naturally occurring heritable variation in a population is "edited" by natural selection because individuals with traits better suited to the environment survive and reproduce more successfully than others, and these traits are passed on to the next generation more often. Over time, better-suited individuals persist and their percentage in the population increases, while less well-suited individuals become less prevalent—a type of population editing. **2.** Here is one possible explanation: The ancestor species of the green warbler finch lived on an island where insects were a plentiful food source. Among individuals in the ancestor population, there was likely variation in beak shape and size. Individuals with slender, sharp beaks were likely more successful at picking up insects for food. Being well-nourished, they gave rise to more offspring than birds with thick, short beaks. Their many offspring inherited slender, sharp beaks (because of genetic information being passed from generation to generation, although Darwin didn't know this). In each generation, the offspring birds with the beaks of a shape best at picking up insects would eat more and have more offspring. Therefore, the green warbler finch of today has a slender beak that is very well matched (adapted) to its food source, insects.

3.

Concept Check 1.3

1. Mouse coat color matches the environment for both beach and inland populations. **2.** Inductive reasoning derives generalizations from specific cases; deductive reasoning predicts specific outcomes from general premises. **3.** Compared to a hypothesis, a scientific theory is usually more general and substantiated by a much greater amount of evidence. Natural selection is an explanatory idea that applies to all kinds of organisms and is supported by vast amounts of evidence of various kinds. **4.** Based on the mouse coloration in Figure 1.25, you might expect that the mice that live on the sandy soil would be lighter in color and those that live on the lava rock would be much darker. And in fact, that is what researchers have found. You would predict that each color of mouse would be less preyed upon in its native habitat than it would be in the other habitat. (Research results also support this prediction.) You could repeat the Hoekstra experiment with colored models, painted to resemble these two types of mouse. Or you could try transplanting some of each population to its non-native habitat and counting how many you can recapture over the next few days, then comparing the four samples as was done in Hoekstra's experiment. (The painted models are easier to recapture, of course!) In the live mouse transplantation experiment, you would have to do controls to eliminate the variable represented by the transplanted mice being in a new, unknown territory. You could control for the transplantation process by transplanting some dark mice from one area of lava rock to one far distant, and some light mice from one area of sandy soil to a distant area.

Concept Check 1.4

1. Science aims to understand natural phenomena and the underlying mechanisms that affect them, while technology involves application of scientific discoveries for a particular purpose or to solve a specific problem. **2.** Natural selection could be operating. Malaria is present in sub-Saharan Africa, so there might be an advantage to people with the sickle-cell disease form of the gene that makes them more able to survive and pass on their genes to offspring. Among those of African descent living in the United States, where malaria is absent, there would be no advantage, so they would be selected against more strongly, resulting in fewer individuals with the sickle-cell disease form of the gene.

Summary of Key Concepts Questions

1.1 Finger movements rely on the coordination of the many structural components of the hand (muscles, nerves, bones, etc.), each of which is composed of elements from lower levels of biological *organization* (cells, molecules). The development of the hand relies on the genetic *information* encoded in chromosomes found in cells throughout the body. To power the finger movements that result in a text message, muscle and nerve cells require chemical *energy* that they transform in powering muscle contraction or in propagating nerve impulses. Texting is in essence communication, an *interaction* that conveys information between organisms, in this case of the same species. **1.2** Ancestors of the beach mouse may have exhibited variations in their coat color. Because of the prevalence of visual predators, the better-camouflaged (lighter) mice in the beach habitat may have survived longer and been able to produce more offspring. Over time, a higher and higher proportion of individuals in the population would have had the adaptation of lighter fur that acted to camouflage the mouse in the beach habitat. **1.3** Gathering and interpreting data are core activities in the scientific process, and they are affected by, and affect in turn, three other arenas of the scientific process: exploration and discovery, community analysis and feedback, and societal benefits and outcomes. **1.4** Different approaches taken by scientists studying natural phenomena at different levels complement each other, so more is learned about each problem being studied. A diversity of backgrounds among scientists may lead to fruitful ideas in the same way that important innovations have often arisen where a mix of cultures coexist, due to multiple different viewpoints.

Test Your Understanding

1. B **2.** C **3.** C **4.** B **5.** C **6.** A **7.** D **8.** Your figure should show the following: (1) for the biosphere, the Earth with an arrow coming out of a tropical ocean; (2) for the ecosystem, a distant view of a coral reef; (3) for the community, a collection of reef animals and algae, with corals, fishes, some seaweed, and any other organisms you can think of; (4) for the population, a group of fish of the same species; (5) for the organism, one fish from your population; (6) for the organ, the fish's stomach; (7) for a tissue, a group of similar cells from the stomach; (8) for a cell, one cell from the tissue, showing its nucleus and a few other organelles; (9) for an organelle, the nucleus, where most of the cell's DNA is located; and (10) for a molecule, a DNA double helix. Your sketches can be very rough!

Chapter 2

Figure Questions

Figure 2.7 Atomic number = 12; 12 protons, 12 electrons; three electron shells; 2 valence electrons

Figure 2.14 One possible answer:

Figure 2.17

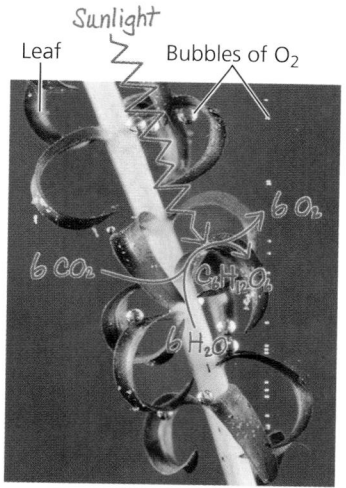

Concept Check 2.1

1. Table salt (sodium chloride) is made up of sodium and chlorine. We are able to eat the compound, showing that it has different properties from those of a metal (sodium) and a poisonous gas (chlorine). **2.** Yes. An organism requires trace elements, though only in small amounts. **3.** A person with an iron deficiency will probably show fatigue and other effects of a low oxygen level in the blood. (The condition is called anemia and can also result from too few red blood cells or abnormal hemoglobin.) **4.** Variant ancestral plants that could tolerate elevated levels of the elements in serpentine soils could grow and reproduce there. (Plants that were well adapted to nonserpentine soils would not be expected to survive in serpentine areas.) The offspring of the variants would also vary, with those most capable of thriving under serpentine conditions growing best and reproducing most. Over many generations, this probably led to the serpentine-adapted species we see today.

Concept Check 2.2

1. 7 **2.** $^{15}_{7}N$ **3.** 9 electrons; two electron shells; $1s$, $2s$, $2p$ (three orbitals); 1 electron is needed to fill the valence shell. **4.** The elements in a row all have the same number of electron shells. All the elements in a column have the same number of electrons in their valence shells.

Concept Check 2.3

1. In this structure, each carbon atom has only three covalent bonds instead of the required four. **2.** The attraction between oppositely charged ions, forming ionic bonds **3.** If you could synthesize molecules that mimic these shapes, you might be able to treat diseases or conditions caused by the inability of affected individuals to synthesize such molecules—or to block the function of such molecules if overproduction is the cause of the disorder.

Concept Check 2.4

1.

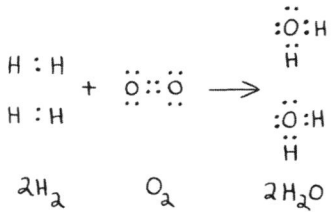

2. At equilibrium, the forward and reverse reactions occur at the same rate. **3.** $C_6H_{12}O_6 + 6\,O_2 \rightarrow 6\,CO_2 + 6\,H_2O + Energy$. Glucose and oxygen react to form carbon dioxide and water, releasing energy. We breathe in oxygen because we need it for this reaction to occur, and we breathe out carbon dioxide because it is a by-product of this reaction. (This reaction is called cellular respiration, and you will learn more about it in Chapter 9.)

Summary of Key Concepts Questions

2.1 A compound is made up of two or more elements combined in a fixed ratio, while an element is a substance that cannot be broken down to other substances.
2.2

Neon ($_{10}$Ne)   Argon ($_{18}$Ar)

Both neon and argon have completed valence shells, containing 8 electrons. They do not have unpaired electrons that could participate in chemical bonds. **2.3** Electrons are shared equally between the two atoms in a nonpolar covalent bond. In a polar covalent bond, the electrons are drawn closer to the more electronegative atom. In the formation of ions, an electron is completely transferred from one atom to a much more electronegative atom. **2.4** The concentration of

products would increase as the added reactants were converted to products. Eventually, an equilibrium would again be reached in which the forward and reverse reactions were proceeding at the same rate and the relative concentrations of reactants and products returned to where they were before the addition of more reactants.

Test Your Understanding

1. D **2.** A **3.** B **4.** A **5.** D **6.** B **7.** C **8.** D

9. a.

This structure makes sense because all valence shells are complete, and all bonds have the correct number of electrons.

b.

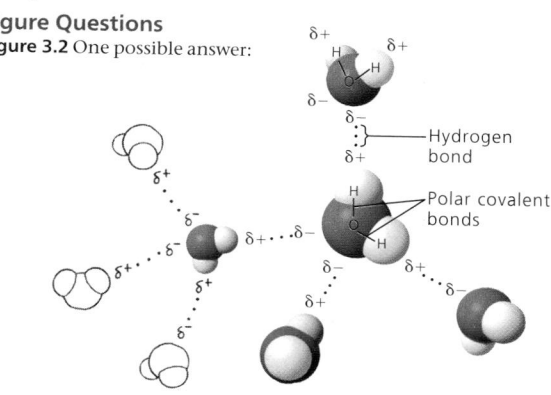

This structure doesn't make sense because H has only 1 electron to share, so it cannot form bonds with 2 atoms.

Chapter 3

Figure Questions

Figure 3.2 One possible answer:

Hydrogen bond
Polar covalent bonds

Figure 3.8 Heating the solution would cause the water to evaporate faster than it is evaporating at room temperature. At a certain point, there wouldn't be enough water molecules to dissolve the salt ions. The salt would start coming out of solution and re-forming crystals. Eventually, all the water would evaporate, leaving behind a pile of salt like the original pile. **Figure 3.12** Adding excess CO_2 to the oceans ultimately reduces the rate at which calcification (by organisms) can occur.

Concept Check 3.1

1. Electronegativity is the attraction of an atom for the electrons of a covalent bond. Because oxygen is more electronegative than hydrogen, the oxygen atom in H_2O pulls electrons toward itself, resulting in two partial negative charges on the oxygen atom and a partial positive charge on each hydrogen atom. Atoms in neighboring water molecules with opposite partial charges are attracted to each other, forming a hydrogen bond. **2.** Due to its two polar covalent bonds, a water molecule has regions of partial negative charge on the O atom that allow it to form hydrogen bonds with hydrogen atoms on neighboring water molecules, and regions of partial positive charge on the H atoms that allow them to form hydrogen bonds with oxygen atoms on neighboring water molecules. **3.** The hydrogen atoms of one molecule, with their partial positive charges, would repel the hydrogen atoms of the adjacent molecule. **4.** The covalent bonds of water molecules would not be polar, so no regions of the molecule would carry partial charges and water molecules would not form hydrogen bonds with each other.

Concept Check 3.2

1. Hydrogen bonds hold neighboring water molecules together. This cohesion helps chains of water molecules move upward against gravity in water-conducting cells as water evaporates from the leaves. Adhesion between water molecules and the walls of the water-conducting cells also helps counter gravity. **2.** High humidity hampers cooling by suppressing the evaporation of sweat. **3.** As water freezes, it expands because water molecules move farther apart in forming ice crystals. When there is water in a crevice of a boulder, expansion due to freezing may crack the boulder. **4.** The hydrophobic substance repels water, perhaps helping to keep the ends of the legs from becoming coated with water and breaking through the surface. If the legs were coated with a hydrophilic substance, water would be drawn up them, possibly making it more difficult for the water strider to walk on water.

Concept Check 3.3

1. 10^5, or 100,000 **2.** $[H^+] = 0.01\ M = 10^{-2}\ M$, so pH = 2. **3.** $CH_3COOH \rightarrow CH_3COO^- + H^+$. CH_3COOH is the acid (the H^+ donor), and CH_3COO^- is the base (the H^+ acceptor). **4.** The pH of the water should decrease from 7 to about 2 (as mentioned in the text); the pH of the acetic acid solution will decrease only a small amount, because as a weak acid, it acts (like carbonic acid) as a buffer. The reaction shown for question 3 will shift to the left, with CH_3COO^- accepting the influx of H^+ and becoming CH_3COOH molecules.

Summary of Key Concepts Questions
3.1

No. A covalent bond is a strong bond in which electrons are shared between two atoms. A hydrogen bond is a weak bond, which does not involve electron sharing, but is simply an attraction between two partial charges on neighboring atoms. **3.2** Ions dissolve in water when polar water molecules form a hydration shell around them, with partially charged regions of water molecules being attracted to ions of the opposite charge. Polar molecules dissolve as water molecules form hydrogen bonds with them and surround them. Solutions are homogeneous mixtures of solute and solvent. **3.3** The concentration of hydrogen ions (H^+) would be 10^{-11}, and the pH of the solution would be 11.

Test Your Understanding
1. C **2.** D **3.** C **4.** A **5.** D
6.

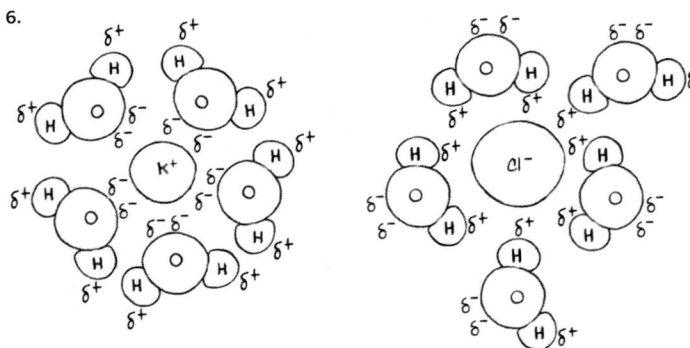

7. Due to intermolecular hydrogen bonds, water has a high specific heat (the amount of heat required to increase the temperature of water by 1°C). When water is heated, much of the heat is absorbed in breaking hydrogen bonds before the water molecules increase their motion and the temperature increases. Conversely, when water is cooled, many H bonds are formed, which releases a significant amount of heat. This release of heat can provide some protection against freezing of the plants' leaves, thus protecting the cells from damage. **8.** Both global warming and ocean acidification are caused by increasing levels of carbon dioxide in the atmosphere, the result of burning fossil fuels.

Chapter 4

Figure Questions
Figure 4.2 Because the concentration of the reactants influences the equilibrium (as discussed in Concept 2.4), there might have been more HCN relative to CH_2O since there would have been a higher concentration of the reactant gas containing nitrogen.
Figure 4.4

Na· ·Si· ·P· ·S: ·Cl:

Figure 4.6 The tails of fats contain only carbon-hydrogen bonds, which are relatively nonpolar. Because the tails occupy the bulk of a fat molecule, they make the molecule as a whole nonpolar and therefore incapable of forming hydrogen bonds with water.

Figure 4.7

```
        H
        |
    H - C - H
    H   |   H
    |   |   |
H - C — C — C - H
    |   |   |
    H   |   H
    H - C - H
        |
        H
```

Concept Check 4.1
1. The sparks provided energy needed for the inorganic molecules in the atmosphere to react with each other. (You'll learn more about energy and chemical reactions in Chapter 8.)

Concept Check 4.2
1.

a.
```
H         H
 \       /
  C = C
 /       \
H         H
```
b.
```
H         Cl
 \       /
  C = C
 /       \
Cl        H
```

2. The forms of C_4H_{10} in (b) are structural isomers, as are the butenes (forms of C_4H_8) in (c). **3.** Both consist largely of hydrocarbon chains, which provide fuel—gasoline for engines and fats for plant embryos and animals. Reactions of both types of molecules release energy. **4.** No. There is not enough diversity in propane's atoms. It can't form structural isomers because there is only one way for three carbons to attach to each other (in a line). There are no double bonds, so *cis-trans* isomers are not possible. Each carbon has at least two hydrogens attached to it, so the molecule is symmetrical and cannot have enantiomers.

Concept Check 4.3
1. An amino acid has both an amino group (—NH_2), which makes it an amine, and a carboxyl group (—COOH), which makes it a carboxylic acid. **2.** The ATP molecule loses a phosphate, becoming ADP.
3.

```
 O      H      O
  \\    |    //
   C —  C  — C
  /     |      \
HO     CH₂      OH
        |
        SH
```

A chemical group that can act as a base has been replaced with a group that can act as an acid, increasing the acidic properties of the molecule. The shape of the molecule would also change, likely changing the molecules with which it can interact. The original cysteine molecule has an asymmetric carbon in the center. After replacement of the amino group with a carboxyl group, this carbon is no longer asymmetric.

Summary of Key Concepts Questions
4.1 Miller showed that organic molecules could form under the physical and chemical conditions estimated to have been present on early Earth. This abiotic synthesis of organic molecules would have been a first step in the origin of life. **4.2** Acetone and propanal are structural isomers. Acetic acid and glycine have no asymmetric carbons, whereas glycerol phosphate has one. Therefore, glycerol phosphate can exist as forms that are enantiomers, but acetic acid and glycine cannot. **4.3** The methyl group is nonpolar and not reactive. The other six groups are called functional groups because they can participate in chemical reactions. Also, all except the sulfhydryl group are hydrophilic, increasing the solubility of organic compounds in water.

Test Your Understanding
1. B **2.** B **3.** C **4.** C **5.** A **6.** B **7.** A
8. The molecule on the right; the middle carbon is asymmetric.
9.

·Si·

Silicon has 4 valence electrons, the same number as carbon. Therefore, silicon would be able to form long chains, including branches, that could act as skeletons for large molecules. It would clearly do this much better than neon (with no valence electrons) or aluminum (with 3 valence electrons).

Chapter 5

Figure Questions
Figure 5.3 Glucose and fructose are structural isomers.
Figure 5.4

Four carbons are in the fructose ring, and two are not. (The latter two carbons are attached to carbons 2 and 5, which are in the ring.) The fructose ring differs from the glucose ring, which has five carbons in the ring and one that is not. (Note that the orientation of this fructose molecule is flipped horizontally relative to that of the one in Figure 5.5b; also, note that the oxygen on carbon 5 lost its proton and that the oxygen on carbon 2, which used to be the carbonyl oxygen, gained a proton.

Figure 5.5

Glucose + Glucose → Maltose

Glucose + Fructose → Sucrose

(a) In maltose, the linkage is called a 1–4 glycosidic linkage because the number 1 carbon in the left monosaccharide (glucose) is linked to the number 4 carbon in the right monosaccharide (also glucose). (b) In sucrose, the linkage is called a 1–2 glycosidic linkage because the number 1 carbon in the left monosaccharide (glucose) is linked to the number 2 carbon in the right monosaccharide (fructose). (Note that the fructose molecule is oriented differently from glucose in Figures 5.4 and 5.5b, where carbon 2 is on the right. In fructose in Figure 5.5b and here, carbon 2 of fructose is on the left.)

Figure 5.11   **Figure 5.12**

Figure 5.15

Met (M)   Tyr (Y)   Cys (C)

Peptide bond   Carboxyl group   Amino group

Figure 5.16 (1) The polypeptide backbone is most easily followed in the ribbon model.

(2)

(3) The point of this diagram is to show that a pancreas cell secretes insulin proteins, so the shape is not important to the process being illustrated.   **Figure 5.17** We can see that their complementary shapes allow the two proteins to fit together quite precisely.   **Figure 5.19** The R group on glutamic acid is acidic and hydrophilic, whereas that on valine is nonpolar and hydrophobic. Therefore, it is unlikely that valine and glutamic acid participate in the same intramolecular interactions. A change in these interactions could (and does) cause a disruption of molecular structure.   **Figure 5.26** Using a genomics approach allows us to use gene sequences to identify species and to learn about evolutionary relationships among any two species. This is because all species are related by their evolutionary history, and the evidence is in the DNA sequences. Proteomics—looking at proteins that are expressed—allows us to learn about how organisms or cells are functioning at a given time or in an association with another species.

Concept Check 5.1
1. The four main classes are proteins, carbohydrates, lipids, and nucleic acids. Lipids are not polymers.   **2.** Nine, with one water molecule required to hydro-lyze each connection between adjacent monomers   **3.** The amino acids in the fish protein must be released in hydrolysis reactions and incorporated into other proteins in dehydration reactions.

Concept Check 5.2
1. $C_3H_6O_3$ **2.** $C_{12}H_{22}O_{11}$ **3.** The antibiotic treatment is likely to have killed the cellulose-digesting prokaryotes in the cow's gut. The absence of these prokary-otes would hamper the cow's ability to obtain energy from food and could lead to weight loss and possibly death. Thus, prokaryotic species are reintroduced, in appropriate combinations, in the gut culture given to treated cows.

Concept Check 5.3
1. Both have a glycerol molecule attached to fatty acids. The glycerol of a fat has three fatty acids attached, whereas the glycerol of a phospholipid is attached to two fatty acids and one phosphate group.   **2.** Human sex hor-mones are steroids, a type of compound that is hydrophobic and thus clas-sified as a lipid.   **3.** The oil droplet membrane could consist of a single layer of phospholipids rather than a bilayer, because an arrangement in which the hydrophobic tails of the membrane phospholipids were in contact with the hydrocarbon regions of the oil molecules would be more stable.

Concept Check 5.4
1. Secondary structure involves hydrogen bonds between atoms of the polypep-tide backbone. Tertiary structure involves interactions between atoms of the side chains of the amino acid subunits.   **2.** The two ring forms of glucose are called α and β, depending on how the glycosidic bond dictates the position of a hydroxyl group. Proteins have α helices and β pleated sheets, two types of repeating structures found in polypeptides due to interactions between the repeating con-stituents of the chain (not the side chains). The hemoglobin molecule is made up of two types of polypeptides: It contains two molecules each of α-globin and β-globin.   **3.** During formation of a polypeptide by polymerization of amino acids, the amino group of one amino acid reacts with the carboxyl group of the next, forming a peptide bond. Therefore, in a polypeptide, there is only one amino group on the N-terminus and one carboxyl group on the C-terminus, along with any carboxyl groups or amino groups located in the amino acid side chains (R groups).   **4.** These are all nonpolar, hydrophobic amino acids, so you would expect this region to be located in the interior of the folded polypeptide, where it would not contact the aqueous environment inside the cell.

Concept Check 5.5
1.

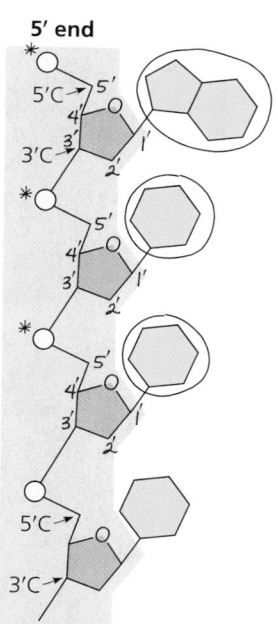

5' end

3' end

2.

5'–T A G G C C T–3'
3'–A T C C G G A–5'

Concept Check 5.6
1. The DNA of an organism encodes all of its proteins, and proteins are the molecules that carry out the work of cells, whether an organism is unicellular or multicellular. By knowing the DNA sequence of an organism, scientists would be able to catalog the protein sequences as well.   **2.** Ultimately, the DNA sequence carries the information necessary to make the proteins that determine the traits of a particular species. Because the traits of the two species are similar, you would expect the proteins to be similar as well, and therefore the gene sequences should also have a high degree of similarity.

Summary of Key Concepts Questions

5.1 The polymers of large carbohydrates (polysaccharides), proteins, and nucleic acids are built from three different types of monomers (monosaccharides, amino acids, and nucleotides, respectively). **5.2** Both starch and cellulose are polymers of glucose, but the glucose monomers are in the α configuration in starch and the β configuration in cellulose. The glycosidic linkages thus have different geometries, giving the polymers different shapes and thus different properties. Starch is an energy-storage compound in plants; cellulose is a structural component of plant cell walls. Humans can hydrolyze starch to provide energy but cannot hydrolyze cellulose. Cellulose aids in the passage of food through the digestive tract. **5.3** Lipids are not polymers because they do not exist as a chain of linked monomers. They are not considered macromolecules because they do not reach the giant size of many polysaccharides, proteins, and nucleic acids. **5.4** A polypeptide, which may consist of hundreds of amino acids in a specific sequence (primary structure), has regions of coils and pleats (secondary structure), which are then folded into irregular contortions (tertiary structure) and may be noncovalently associated with other polypeptides (quaternary structure). The linear order of amino acids, with the varying properties of their side chains (R groups), determines what secondary and tertiary structures will form to produce a protein. The resulting unique three-dimensional shapes of proteins are key to their specific and diverse functions. **5.5** The complementary base pairing of the two strands of DNA makes possible the precise replication of DNA every time a cell divides, ensuring that genetic information is faithfully transmitted. In some types of RNA, complementary base pairing enables RNA molecules to assume specific three-dimensional shapes that facilitate diverse functions. **5.6** You would expect the human gene sequence to be most similar to that of the mouse (another mammal), then to that of the fish (another vertebrate), and least similar to that of the fruit fly (an invertebrate).

Test Your Understanding

1. D **2.** A **3.** B **4.** A **5.** B **6.** B **7.** C

8.

	Monomers or Components	Polymer or larger molecule	Type of linkage
Carbohydrates	Monosaccharides	Polysaccharides	Glycosidic linkages
Fats	Fatty acids	Triacylglycerols	Ester linkages
Proteins	Amino acids	Polypeptides	Peptide bonds
Nucleic acids	Nucleotides	Polynucleotides	Phosphodiester linkages

9.

Original Strand   Complementary strand

Chapter 6

Figure Questions

Figure 6.3 The cilia in the lower portion of the TEM were oriented lengthwise in the plane of the slice, while those in the upper portion of the TEM were oriented perpendicular to the plane of the slice. Therefore, the cilia in the lower portion were cut in longitudinal section, and the cilia in the upper portion were cut in cross section.   **Figure 6.4** You would use the pellet from the final fraction, which is rich in ribosomes. These are the sites of protein translation.
Figure 6.6 The dark bands in the TEM correspond to the hydrophilic heads of the phospholipids, while the light band corresponds to the hydrophobic fatty acid tails of the phospholipids.   **Figure 6.9** The DNA in a chromosome dictates synthesis of a messenger RNA (mRNA) molecule, which then moves out to the cytoplasm. There, the information is used for the production, on ribosomes, of proteins that carry out cellular functions.   **Figure 6.10** Any of the bound

ribosomes (attached to the endoplasmic reticulum) could be circled, because any could be making a protein that will be secreted.   **Table 6.1** Three dimers
Figure 6.22 Each centriole has nine sets of 3 microtubules, so the entire centrosome (two centrioles) has 54 microtubules. Each microtubule consists of a helical array of tubulin dimers (as shown in Table 6.1).

Figure 6.24 The two central microtubules terminate above the basal body, so they aren't present at the level of the cross section through the basal body, indicated by the lower red rectangle shown in the EM on the left.

Central pair of microtubules

They terminate here

Figure 6.32 (1) nuclear pore, ribosome, proton pump, cyt *c*. (2) As shown in the figure, the enzyme RNA polymerase moves along the DNA, transcribing the genetic information into an mRNA molecule. Given that RNA polymerase is somewhat larger than a nucleosome, the enzyme would not be able to fit between the histone proteins of the nucleosome and the DNA itself. Thus, the group of histone proteins must be separated from or moved along the DNA somehow in order for the RNA polymerase enzyme to access the DNA. (3) A mitochondrion.

Concept Check 6.1

1. Stains used for light microscopy are colored molecules that bind to cell components, affecting the light passing through, while stains used for electron microscopy involve heavy metals that affect the beams of electrons.
2. (a) Light microscope, (b) scanning electron microscope.

Concept Check 6.2

1. See Figure 6.8.
2.

This cell would have the same volume as the cells in columns 2 and 3 in Figure 6.7 but proportionally more surface area than that in column 2 and less than that in column 3. Thus, the surface-to-volume ratio should be greater than 1.2 but less than 6. To obtain the surface area, you would add the area of the six sides (the top, bottom, sides, and ends): 125 + 125 + 125 + 125 + 1 + 1 = 502. The surface-to-volume ratio equals 502 divided by a volume of 125, or roughly 4.0.

Concept Check 6.3

1. Ribosomes in the cytoplasm translate the genetic message, carried from the DNA in the nucleus by mRNA, into a polypeptide chain.   **2.** Nucleoli consist of DNA and the ribosomal RNAs (rRNAs) made according to its genes in the DNA, as well as proteins imported from the cytoplasm. Together, the rRNAs and proteins are assembled into large and small ribosomal subunits. (These are exported through nuclear pores to the cytoplasm, where they will participate in polypeptide synthesis.)   **3.** Each chromosome consists of one long DNA molecule attached to numerous protein molecules, a combination called chromatin. As a cell begins division, each chromosome becomes "condensed" as its diffuse mass of chromatin coils up.

Concept Check 6.4

1. The primary distinction between rough and smooth ER is the presence of bound ribosomes on the rough ER. Both types of ER make phospholipids, but membrane proteins and secretory proteins are all produced by the ribosomes on the rough ER. The smooth ER also functions in detoxification, carbohydrate metabolism, and storage of calcium ions.   **2.** Transport vesicles move membranes and the substances they enclose between other components of the endomembrane system.   **3.** The mRNA is synthesized in the nucleus and then passes out through a nuclear pore to the cytoplasm, where it is translated on a bound ribosome, attached to the rough ER. The protein is synthesized into the lumen of the ER and may be modified there. A transport vesicle carries the protein to the Golgi apparatus. After further modification in the Golgi, another transport vesicle carries it back to the ER, where it will perform its cellular function.

Concept Check 6.5

1. Both organelles are involved in energy transformation, mitochondria in cellular respiration and chloroplasts in photosynthesis. They both have multiple membranes that separate their interiors into compartments. In both organelles, the innermost membranes—cristae, or infoldings of the inner membrane, in mitochondria and the thylakoid membranes in chloroplasts—have large surface areas with embedded enzymes that carry out their main functions.   **2.** Yes. Plant cells are able to make their own sugar by photosynthesis, but mitochondria in plant cells (which are, of course, eukaryotic) are the organelles that are able to generate ATP molecules to be used for energy generation from sugars, a function required in all cells.   **3.** Mitochondria and chloroplasts are not derived from the ER, nor are they connected physically or via transport vesicles to organelles of the endomembrane system. Mitochondria and chloroplasts are structurally quite different from vesicles derived from the ER, which are bounded by a single membrane.

Concept Check 6.6

1. Dynein arms, powered by ATP, move neighboring doublets of microtubules relative to each other. Because they are anchored within the flagellum or cilium and with respect to one another, the doublets bend instead of sliding past each other. Synchronized bending of the nine microtubule doublets brings about bending of both cilia and flagella.   **2.** Such individuals have defects in the microtubule-based movement of cilia and flagella. Thus, the sperm can't move because of malfunctioning or nonexistent flagella, and the airways are compromised because cilia that line the trachea malfunction or don't exist, and so mucus cannot be cleared from the lungs.

Concept Check 6.7

1. The most obvious difference is the presence of direct cytoplasmic connections between cells of plants (plasmodesmata) and animals (gap junctions). These connections result in the cytoplasm being continuous between adjacent cells. **2.** The cell would not be able to function properly and would probably soon die, as the cell wall or ECM must be permeable to allow the exchange of matter between the cell and its external environment. Molecules involved in energy production and use must be allowed entry, as well as those that provide information about the cell's environment. Other molecules, such as products synthesized by the cell for export and the by-products of cellular respiration, must be allowed to exit. **3.** The parts of the protein that face aqueous regions would be expected to have polar or charged (hydrophilic) amino acids, while the parts that go through the membrane would be expected to have nonpolar (hydrophobic) amino acids. You would predict polar or charged amino acids at each end (tail), in the region of the cytoplasmic loop, and in the regions of the two extracellular loops. You would predict nonpolar amino acids in the four regions that go through the membrane between the tails and loops.

Concept Check 6.8

1. *Colpidium colpoda* moves around in freshwater using cilia, projections from the plasma membrane that enclose microtubules in a "9 + 2" arrangement. The interactions between motor proteins and microtubules cause the cilia to bend synchronously, propelling the cell through the water. This is powered by ATP, obtained via breaking down sugars from food in a process that occurs in mitochondria. *C. colpoda* obtains bacteria as their food source, maybe via the same process (involving filopodia) the macrophage uses in Figure 6.31. This process uses actin filaments and other elements of the cytoskeleton to ingest the bacteria. Once ingested, the bacteria are broken down by enzymes in lysosomes. The proteins involved in all of these processes are encoded by genes on DNA in the nucleus of the *C. colpoda*.

Summary of Key Concepts Questions

6.1 Both light and electron microscopy allow cells to be studied visually, thus helping us understand internal cellular structure and the arrangement of cell components. Cell fractionation techniques separate out different groups of cell components, which can then be analyzed biochemically to determine their function. Performing microscopy on the same cell fraction helps to correlate the biochemical function of the cell with the cell component responsible.
6.2 The separation of different functions in different organelles has several advantages. Reactants and enzymes can be concentrated in one area instead of spread throughout the cell. Reactions that require specific conditions, such as a lower pH, can be compartmentalized. And enzymes for specific reactions are often embedded in the membranes that enclose or partition an organelle.
6.3 The nucleus contains the genetic material of the cell in the form of DNA, which specifies messenger RNA, which in turn provides instructions for the synthesis of proteins (including the proteins that make up part of the ribosomes). DNA also codes for ribosomal RNAs, which are combined with proteins in the nucleolus into the subunits of ribosomes. Within the cytoplasm, ribosomes join with mRNA to build polypeptides, using the genetic information in the mRNA.   **6.4** Transport vesicles move proteins and membranes synthesized by the rough ER to the Golgi for further processing and then to the plasma membrane, lysosomes, or other locations in the cell, including back to the ER.
6.5 According to the endosymbiont theory, mitochondria originated from an oxygen-using prokaryotic cell that was engulfed by a cell that was ancestral to eukaryotic cells. Over time, the host and endosymbiont evolved into a single unicellular organism containing a mitochondrion. Chloroplasts originated when at least one of these eukaryotic cells containing mitochondria engulfed and then retained a photosynthetic prokaryote, which eventually evolved into a chloroplast.   **6.6** Inside the cell, motor proteins interact with components of the cytoskeleton to move cellular parts. Motor proteins "walk" vesicles along microtubules. The movement of cytoplasm within a cell involves interactions of the motor protein myosin and microfilaments (actin filaments). Whole cells can be moved by the rapid bending of flagella or cilia, which is caused by the motor protein–powered sliding of microtubules within these structures. Cell movement can also occur when pseudopodia form at one end of a cell (caused by actin polymerization into a filamentous network), followed by contraction of the cell toward that end; this amoeboid movement is powered by interactions of microfilaments with myosin. Interactions of motor proteins and microfilaments in muscle cells causes muscle contraction that can propel whole organisms (for example, by walking or swimming).   **6.7** A plant cell wall is primarily composed of microfibrils of cellulose embedded in other polysaccharides and proteins. The ECM of animal cells is primarily composed of collagen and other protein fibers, such as fibronectin and other glycoproteins. These fibers are embedded in a network of carbohydrate-rich proteoglycans. A plant cell wall provides structural support for the cell and, collectively, for the plant body. In addition to giving support, the ECM of an animal cell also allows for communication of environmental changes into the cell.   **6.8** The nucleus houses the chromosomes; each is made up of proteins and a single DNA molecule. The genes that exist along the DNA carry the genetic information necessary to make the proteins involved in ingesting a bacterial cell, such as the actin of microfilaments that form pseudopodia (filopodia), the proteins in the mitochondria responsible for providing the necessary ATP, and the enzymes present in the lysosomes that will digest the bacterial cell.

Test Your Understanding

1. B   **2.** C   **3.** B   **4.** A   **5.** D   **6.** See Figure 6.8.

Chapter 7

Figure Questions

Figure 7.2

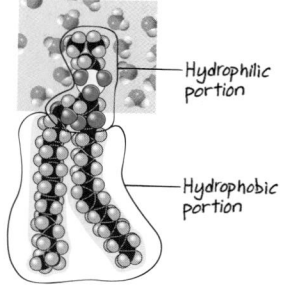

Hydrophilic portion

Hydrophobic portion

The hydrophilic portion is in contact with an aqueous environment (cytosol or extracellular fluid), and the hydrophobic portion is in contact with the hydrophobic portions of other phospholipids in the interior of the bilayer.   **Figure 7.4** You couldn't rule out movement of proteins within membranes of the same species. You might propose that the membrane lipids and proteins from one species weren't able to mingle with those from the other species because of some incompatibility.   **Figure 7.7** A transmembrane protein like the dimer in (f) might change its shape upon binding to a particular extracellular matrix (ECM) molecule. The new shape might enable the interior portion of the protein to bind to a second, cytoplasmic protein that would relay the message to the inside of the cell, as shown in (c).   **Figure 7.8** The shape of a protein on the HIV surface is likely to be complementary to the shape of the receptor (CD4) and also to that of the co-receptor (CCR5). A molecule with a shape similar to that of the HIV surface protein could bind CCR5, blocking HIV binding. (Another answer would be a molecule that bound to CCR5 and changed the shape of CCR5 so it could no longer bind HIV; in fact, this is how maraviroc works.)

Figure 7.9

The protein would contact the extracellular fluid. (Because one end of the protein is in the ER membrane, no part of the protein extends into the cytoplasm.) The part of the protein not in the membrane extends into the ER lumen. Once the vesicle fuses with the plasma membrane, the "inside" of the ER membrane, facing the lumen, will become the "outside" of the plasma membrane, facing the extracellular fluid. **Figure 7.12** The orange dye would be evenly distributed throughout the solution on both sides of the membrane. The solution levels would not be affected because the orange dye can diffuse through the membrane and equalize its concentration. Thus, no additional osmosis would take place in either direction. In this experiment, the membrane is meant to represent the plasma membrane of a cell. **Figure 7.13** The cells take up water and become turgid, causing the stalk to lose its limpness and become crisp. **Figure 7.16** The sodium ion concentration ($[Na^+]$) is low inside the cell and high outside, while potassium ion concentration ($[K^+]$) is low outside the cell and high inside. Three Na^+ ions are moved out of the cell and two K^+ ions into the cell for each cycle. **Figure 7.17** The diamond solutes are moving into the cell (downward), and the round solutes are moving out of the cell (upward); each is moving against its concentration gradient. **Figure 7.21** (a) In the micrograph of the algal cell, the diameter of the algal cell is about 2.3 times longer than the scale bar, which represents 5 μm, so the diameter of the algal cell is about 11.5 μm. (b) In the micrograph of the coated vesicle, the diameter of the coated vesicle is about 1.2 times longer than the scale bar, which represents 0.25 μm, so the diameter of the coated vesicle is about 0.3 μm. (c) Therefore, the food vacuole around the algal cell will be about 40 × larger than the coated vesicle.

Concept Check 7.1
1. They are on the inside of the transport vesicle membrane. **2.** The grasses living in the cooler region would be expected to have more unsaturated fatty acids in their membranes because those fatty acids remain fluid at lower temperatures. The grasses living immediately adjacent to the hot springs would be expected to have more saturated fatty acids, which would allow the fatty acids to "stack" more closely, making the membranes less fluid and therefore helping them to stay intact at higher temperatures. (In plants, cholesterol is generally not used to moderate the effects of temperature on membrane fluidity because it is found at vastly lower levels in membranes of plant cells than in those of animal cells.)

Concept Check 7.2
1. O_2 and CO_2 are both small, nonpolar molecules that can easily pass through the hydrophobic interior of a membrane. **2.** Water is a polar molecule, so it cannot pass very rapidly through the hydrophobic region in the middle of a phospholipid bilayer. **3.** The hydronium ion is charged, while glycerol is not. Charge is probably more significant than size as a basis for exclusion by the aquaporin channel.

Concept Check 7.3
1. CO_2 is a nonpolar molecule that can diffuse through the plasma membrane. As long as it diffuses away so that the concentration remains low outside the cell, it will continue to exit the cell in this way. (This is the opposite of the case for O_2, described in this section of the text.) **2.** *Paramecium*'s contractile vacuole will become less active. The vacuole pumps out excess water that accumulates in the cell; this accumulation occurs only in a hypotonic environment.

Concept Check 7.4
1. These pumps use ATP. To establish a voltage, ions have to be pumped against their gradients, which requires energy. **2.** Each ion is being transported against its electrochemical gradient. If either ion were transported down its electrochemical gradient, this *would* be considered cotransport. **3.** The internal environment of a lysosome is acidic, so it has a higher concentration of H^+ than does the cytoplasm. Therefore, you might expect the membrane of the lysosome to have a proton pump such as that shown in Figure 7.18 to pump H^+ into the lysosome.

Concept Check 7.5
1. Exocytosis. When a transport vesicle fuses with the plasma membrane, the vesicle membrane becomes part of the plasma membrane.

2.

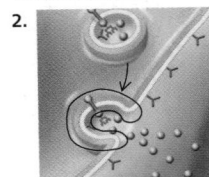

3. The glycoprotein is synthesized in the ER lumen, moves through the Golgi apparatus, and then travels in a vesicle to the plasma membrane, where it undergoes exocytosis and becomes part of the ECM.

Summary of Key Concepts Questions
7.1 Plasma membranes define the cell by separating the cellular components from the external environment. This allows conditions inside cells to be controlled by membrane proteins, which regulate entry and exit of molecules and even cell function (see Figure 7.7). The processes of life can be carried out inside the controlled environment of the cell, so membranes are crucial. In eukaryotes, membranes also function to subdivide the cytoplasm into different compartments where distinct processes can occur, even under differing conditions such as low or high pH. **7.2** Aquaporins are channel proteins that greatly increase the permeability of a membrane to water molecules, which are polar and therefore do not readily diffuse through the hydrophobic interior of the membrane. **7.3** There

will be a net diffusion of water out of a cell into a hypertonic solution. The free water concentration is higher inside the cell than in the solution (where not as many water molecules are free, because many are clustered around the numerous solute particles). **7.4** One of the solutes moved by the cotransporter is actively transported against its concentration gradient. The energy for this transport comes from the concentration gradient of the other solute, which was established by an electrogenic pump that used energy to transport the other solute across the membrane. Because energy is required overall to drive this process (because ATP is used to establish the concentration gradient), it is considered active transport. **7.5** In receptor-mediated endocytosis, specific molecules bind to receptors on the plasma membrane in a region where a coated pit develops. The cell can acquire bulk quantities of those specific molecules when the coated pit forms a vesicle and carries the bound molecules into the cell.

Test Your Understanding
1. B **2.** C **3.** A **4.** C **5.** B
6. (a)

(b) The solution outside is hypotonic. It has less sucrose, which is a nonpenetrating solute. (c) See answer for (a). (d) The artificial cell will become more turgid. (e) Eventually, the two solutions will have the same solute concentrations. Even though sucrose can't move through the membrane, water flow (osmosis) will lead to isotonic conditions.

Chapter 8

Figure Questions
Figure 8.5 With a proton pump (Figure 7.18), the energy stored in ATP is used to pump protons across the membrane and build up a higher (nonrandom) concentration outside of the cell, so this process results in higher free energy. When solute molecules (analogous to hydrogen ions) are uniformly distributed, similar to the random distribution in the bottom of (b), the system has less free energy than it does in the top of (b). The system in the bottom can do no work. Because the concentration gradient created by a proton pump (Figure 7.18) represents higher free energy, this system has the potential to do work once there is a higher concentration of protons on one side of the membrane (as you will see in Figure 9.15). **Figure 8.10** Glutamic acid (Glu) has a carboxyl group at the end of its R group. Glutamine (Gln) has exactly the same structure as glutamic acid, except that there is an amino group in place of the $-O^-$ on the R group. (This O atom on the R group leaves during the synthesis reaction.) Thus, in this figure, Gln is drawn as a Glu with an attached NH_2.
Figure 8.13

Figure 8.16

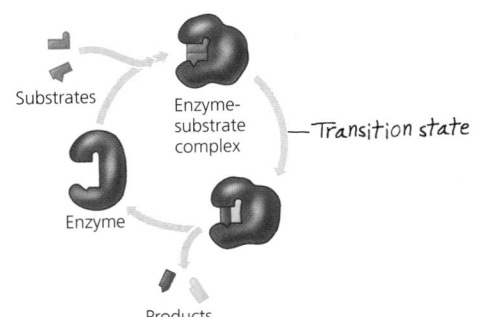

Concept Check 8.1
1. The second law is the trend toward randomization, or increasing entropy. When the concentrations of a substance on both sides of a membrane are equal, the distribution is more random than when they are unequal. Diffusion of a substance to a region where it is initially less concentrated increases entropy,

making it an energetically favorable (spontaneous) process as described by the second law. This explains the process seen in Figure 7.11.   **2.** The apple has potential energy in its position hanging on the tree, and the sugars and other nutrients it contains have chemical energy. The apple has kinetic energy as it falls from the tree to the ground. Finally, when the apple is digested and its molecules broken down, some of the chemical energy is used to do work, and the rest is lost as thermal energy.   **3.** The sugar crystals become less ordered (entropy increases) as they dissolve and become randomly spread out in the water. Over time, the water evaporates, and the crystals form again because the water volume is insufficient to keep them in solution. While the reappearance of sugar crystals may represent a "spontaneous" increase in order (decrease in entropy), it is balanced by the decrease in order (increase in entropy) of the water molecules, which changed from a relatively compact arrangement as liquid water to a much more dispersed and disordered form as water vapor.

Concept Check 8.2

1. Cellular respiration is a spontaneous and exergonic process. The energy released from glucose is used to do work in the cell or is lost as heat.   **2.** Catabolism breaks down organic molecules, releasing their chemical energy and resulting in smaller products with more entropy, as when moving from the top to the bottom of Figure 8.5c. Anabolism consumes energy to synthesize larger molecules from simpler ones, as when moving from the bottom to the top of part (c).   **3.** The reaction is exergonic because it releases energy—in this case, in the form of light. (This is a nonbiological version of the bioluminescence seen in Figure 8.1.)

Concept Check 8.3

1. ATP usually transfers energy to an endergonic process by phosphorylating (adding a phosphate group to) another molecule. (Exergonic processes, in turn, phosphorylate ADP to regenerate ATP.)   **2.** A set of coupled reactions can transform the first combination into the second. Since this is an exergonic process overall, ΔG is negative and the first combination must have more free energy (see Figure 8.10).   **3.** Active transport: The solute is being transported against its concentration gradient, which requires energy, provided by ATP hydrolysis.

Concept Check 8.4

1. A spontaneous reaction is a reaction that is exergonic. However, if it has a high activation energy that is rarely attained, the rate of the reaction may be low.   **2.** O_2 is required as a substrate to react with the gas. There is O_2 in the lab air, which is why the Bunsen burner has a flame above the gas delivery opening, but there is no O_2 in the rubber tubing or the gas supply.   **3.** In the presence of malonate, increase the concentration of the normal substrate (succinate) and see whether the rate of reaction increases. If it does, malonate is a competitive inhibitor.   **4.**

Concept Check 8.5

1. The activator binds in such a way that it stabilizes the active form of an enzyme, whereas the inhibitor stabilizes the inactive form.   **2.** A catabolic pathway breaks down organic molecules, generating energy that is stored in ATP molecules. In feedback inhibition of such a pathway, ATP (one product) would act as an allosteric inhibitor of an enzyme catalyzing an early step in the catabolic process. When ATP is plentiful, the pathway would be turned off and no more would be made.

Summary of Key Concepts Questions

8.1 The process of "ordering" a cell's structure is accompanied by an increase in the entropy (disorder) of the universe. For example, an animal cell takes in highly ordered organic molecules as the source of matter and energy used to build and maintain its structures. In the same process, however, the cell releases heat and the simple molecules of CO_2 and H_2O to the surroundings. The increase in entropy of the latter process offsets the entropy decrease in the former.   **8.2** A spontaneous reaction has a negative ΔG and is exergonic. For a chemical reaction to proceed with a net release of free energy ($-\Delta G$), the enthalpy or total energy of the system must decrease ($-\Delta H$), and/or the entropy or disorder must increase (yielding a more negative term, $-T\Delta S$). Spontaneous reactions supply the energy to perform cellular work.   **8.3** The free energy released from the hydrolysis of ATP may drive endergonic reactions through the transfer of a phosphate group to a reactant molecule, forming a more reactive phosphorylated intermediate. ATP hydrolysis also powers the mechanical and transport work of a cell, often by powering shape changes in the relevant motor proteins. Cellular respiration, the catabolic breakdown of glucose, provides the energy for the endergonic regeneration of ATP from ADP and Ⓟᵢ.   **8.4** Activation energy barriers prevent the complex molecules of the cell, which are rich in free energy, from spontaneously breaking down to less ordered, more stable molecules. Enzymes permit a regulated metabolism by binding to specific substrates and forming enzyme-substrate complexes that selectively lower the E_A for the chemical reactions in a cell.   **8.5** A cell tightly regulates its metabolic

pathways in response to fluctuating needs for energy and materials. The binding of activators or inhibitors to regulatory sites on allosteric enzymes stabilizes either the active or the inactive form of the subunits. For example, the binding of ATP to a catabolic enzyme in a cell with excess ATP would inhibit that pathway. Such types of feedback inhibition preserve chemical resources within a cell. If ATP supplies are depleted, binding of ADP to the regulatory site of catabolic enzymes would activate that pathway, generating more ATP.

Test Your Understanding

1. B   **2.** C   **3.** B   **4.** A   **5.** C   **6.** D   **7.** C

9.

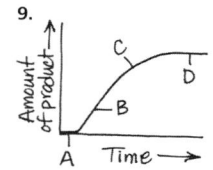

A. The substrate molecules are entering the pancreatic cells, so no product is made yet.
B. There is sufficient substrate, so the reaction is proceeding at a maximum rate.
C. As the substrate is used up, the rate decreases (the slope is less steep).
D. The line is flat because no new substrate remains and thus no new product appears.

Chapter 9

Figure Questions

Figure 9.2 The C atom is oxidized. The electrons that used to be equally shared with H atoms in methane are now much closer to the O atoms in CO_2 than they are to the C.   **Figure 9.3** The reduced form has an extra hydrogen, along with 2 electrons, bound to the carbon shown at the top of the nicotinamide (opposite the N). There are different numbers and positions of double bonds in the two forms: The oxidized form has three double bonds in the ring, while the reduced form has only two. (In organic chemistry you may have learned, or will learn, that three double bonds in a ring are able to "resonate," or act as a ring of electrons. Having three resonant double bonds is more "oxidized" than having only two double bonds in the ring.) In the oxidized form there is a + charge on the N (because it is sharing 4 electron pairs), whereas in the reduced form it is only sharing 3 electron pairs (having a pair of electrons to itself).   **Figure 9.6** Because there is no external source of energy for the reaction, it must be exergonic, and the reactants must be at a higher energy level than the products.   **Figure 9.8** The removal would probably stop glycolysis, or at least slow it down, since it would push the equilibrium for step 5 toward DHAP (toward the bottom in this figure). If less (or no) glyceraldehyde 3-phosphate were available, step 6 would slow down (or be unable to occur).   **Figure 9.12** The electrons in NADH are in a C—H bond (right side of Figure 9.3). In H_2O, the electrons are in an O—H bond. Because the electronegativities of C and H are similar, the electrons are equally shared in NADH. The electronegativity of O is much higher than that of C or H, so the electrons are much closer to O in H_2O and have "fallen" in potential energy.   **Figure 9.14** At first, some ATP could be made, since electron transport could proceed as far as complex III, and a small H^+ gradient could be built up. Soon, however, no more electrons could be passed to complex III because it could not be reoxidized by passing its electrons to complex IV.   **Figure 9.15** First, there are 2 NADH from the oxidation of pyruvate plus 6 NADH from the citric acid cycle (CAC); $8\,NADH \times 2.5\,ATP/NADH = 20\,ATP$. Second, there are 2 $FADH_2$ from the CAC; $2\,FADH_2 \times 1.5\,ATP/FADH_2 = 3\,ATP$. Third, the 2 NADH from glycolysis enter the mitochondrion through one of two types of shuttle. They pass their electrons either to 2 FAD, which become $FADH_2$ and result in 3 ATP, or to 2 NAD^+, which become NADH and result in 5 ATP. Thus, $20 + 3 + 3 = 26$ ATP, or $20 + 3 + 5 = 28$ ATP from all NADH and $FADH_2$.

Concept Check 9.1

1. Both processes include glycolysis, the citric acid cycle, and oxidative phosphorylation. In aerobic respiration, the final electron acceptor is molecular oxygen (O_2); in anaerobic respiration, the final electron acceptor is a different substance.   **2.** $C_4H_6O_5$ would be oxidized and NAD^+ would be reduced.

Concept Check 9.2

1. NAD^+ acts as the oxidizing agent in step 6, accepting electrons from glyceraldehyde 3-phosphate (G3P), which thus acts as the reducing agent.

Concept Check 9.3

1. NADH and $FADH_2$; one ATP is produced during substrate-level phosphorylation in step 5.   **2.** The CO_2 that we exhale is produced by pyruvate oxidation and the citric acid cycle.   **3.** In both cases, the precursor molecule loses a CO_2 molecule and then donates electrons to an electron carrier in an oxidation step. Also, the product has been activated due to the attachment of a CoA group by its S atom.

Concept Check 9.4

1. Oxidative phosphorylation would eventually stop entirely, resulting in no ATP production by this process. Without oxygen to "pull" electrons down the electron transport chain, H^+ would not be pumped into the mitochondrion's intermembrane space and chemiosmosis would not occur.   **2.** Decreasing the pH means addition of H^+. This would establish a proton gradient even without the function of the electron transport chain, and we would expect ATP synthase to function and synthesize ATP. (In fact, it was experiments like this that provided support for chemiosmosis as an energy-coupling mechanism.) **3.** One of the components of the electron transport chain, ubiquinone (Q), must be able to diffuse within the membrane. It could not do so if the membrane components were locked rigidly into place.

Concept Check 9.5

1. A derivative of pyruvate, such as acetaldehyde during alcohol fermentation, or pyruvate itself during lactic acid fermentation; O_2; another electron acceptor at the end of an electron transport chain, such as sulfate (SO_4^{2-})   **2.** The cell would need to consume glucose at a rate about 16 times the consumption rate in the aerobic environment (2 ATP are generated by fermentation versus up to 32 ATP by cellular respiration).

Concept Check 9.6

1. The fat is much more reduced; it has many $-CH_2-$ units, and in all these bonds the electrons are equally shared. The electrons present in a carbohydrate molecule are already somewhat oxidized (shared unequally in bonds; there are more C—O and O—H bonds), as quite a few of them are bound to oxygen. Electrons that are equally shared, as in fat, have a higher energy level than electrons that are unequally shared, as in carbohydrates. Thus, fat is a much better fuel than carbohydrate.   **2.** When you consume more food than necessary for metabolic processes, your body synthesizes fat as a way of storing energy for later use.   **3.** AMP will accumulate, stimulating phosphofructokinase, and thus increasing the rate of glycolysis. Since oxygen is not present, the cell will convert more pyruvate to lactate, providing a supply of ATP.   **4.** When O_2 is present, the fatty acid chains containing most of the energy of a fat are oxidized and fed into the citric acid cycle and the electron transport chain. During intense exercise, however, O_2 is scarce in muscle cells, so ATP must be generated by glycolysis alone. A very small part of the fat molecule, the glycerol backbone, can be oxidized via glycolysis, but the amount of energy released by this portion is insignificant compared to that released by the fatty acid chains. (This is why moderate exercise, staying below 70% maximum heart rate, is better for burning fat—because enough O_2 remains available to the muscles.)

Summary of Key Concepts Questions

9.1 Most of the ATP produced in cellular respiration comes from oxidative phosphorylation, in which the energy released from redox reactions in an electron transport chain is used to produce ATP. In substrate-level phosphorylation, an enzyme directly transfers a phosphate group to ADP from an intermediate substrate. All ATP production in glycolysis occurs by substrate-level phosphorylation; this form of ATP production also occurs at one step in the citric acid cycle.   **9.2** The oxidation of the three-carbon sugar, glyceraldehyde 3-phosphate, yields energy. In this oxidation, electrons and H^+ are transferred to NAD^+, forming NADH, and a phosphate group is attached to the oxidized substrate. ATP is then formed by substrate-level phosphorylation when this phosphate group is transferred to ADP.   **9.3** The release of six molecules of CO_2 represents the complete oxidation of glucose. During the processing of two pyruvates to acetyl CoA, the fully oxidized carboxyl groups ($-COO^-$) are given off as 2 CO_2. The remaining four carbons are released as CO_2 in the citric acid cycle as citrate is oxidized back to oxaloacetate.   **9.4** The flow of H^+ through the ATP synthase complex causes the rotor and attached rod to rotate, exposing catalytic sites in the knob portion that produce ATP from ADP and ℗$_i$. ATP synthases are found in the inner mitochondrial membrane, the plasma membrane of prokaryotes, and membranes within chloroplasts.   **9.5** Anaerobic respiration yields more ATP. The 2 ATP produced by substrate-level phosphorylation in glycolysis represent the total energy yield of fermentation. NADH passes its "high-energy" electrons to pyruvate or a derivative of pyruvate, recycling NAD^+ and allowing glycolysis to continue. In anaerobic respiration, the NADH produced during glycolysis, as well as additional molecules of NADH produced when pyruvate is oxidized, are used to generate ATP molecules. An electron transport chain captures the energy of the electrons in NADH via a series of redox reactions; ultimately, the electrons are transferred to an electronegative atom in a molecule other than oxygen.   **9.6** The ATP produced by catabolic pathways is used to drive anabolic pathways. Also, many of the intermediates of glycolysis and the citric acid cycle are used in the biosynthesis of a cell's molecules.

Test Your Understanding

1. C   **2.** C   **3.** A   **4.** B   **5.** D   **6.** A   **7.** B
8. Since the overall process of glycolysis results in net production of ATP, it would make sense for the process to slow down when ATP levels have increased substantially. Thus, we would expect ATP to allosterically inhibit phosphofructokinase.   **9.** The proton pump in Figures 7.18 and 7.19 is carrying out active transport, using ATP hydrolysis to pump protons against their concentration gradient. Because ATP is required, this is active transport of protons. The ATP synthase in Figure 9.13 is using the flow of protons down their concentration gradient to power ATP synthesis. Because the protons are moving down their concentration gradient, no energy is required, and this is passive transport.
10.

Rotor — Stator
Internal rod
Catalytic knob

12.

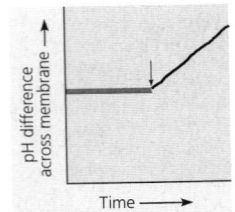

H^+ would continue to be pumped across the membrane into the intermembrane space, increasing the difference between the matrix pH and the intermembrane space pH. H^+ would not be able to flow back through ATP synthase, since the enzyme is inhibited by the poison, so rather than maintaining a constant difference across the membrane, the difference would continue to increase. (Over a longer period of time, the H^+ concentration in the intermembrane space would be so high that no more H^+ would be able to be pumped against the gradient. This wasn't asked for in the question, but if your graph levels off at the right end and this is your reasoning, your answer is correct.)

Chapter 10

Figure Questions

Figure 10.11 In the leaf, most of the chlorophyll electrons excited by photon absorption are used to power the reactions of photosynthesis.   **Figure 10.15** The person at the top of the photosystem I tower would not turn to his left and throw his electron into the NADPH bucket. Instead, he would throw it onto the top of the ramp at his right, next to the photosystem II tower. The electron would then roll down the ramp, get energized by a photon, and return to him. This cycle would continue as long as light was available. (This is why it's called cyclic electron flow.)   **Figure 10.16** You would (a) decrease the pH outside the mitochondrion (thus increasing the H^+ concentration) and (b) increase the pH in the chloroplast stroma (thus decreasing the H^+ concentration). In both cases, this would generate an H^+ gradient across the membrane that would cause ATP synthase to synthesize ATP.   **Figure 10.17** Steps that increase the [H^+] in the thylakoid space or decrease [H^+] in the stroma contribute to the [H^+] concentration gradient across the thylakoid membrane. In step 2, water is split in the thylakoid space, releasing 2 H^+ and increasing [H^+]. In step 3, as electrons travel down the electron transport chain, 4 H^+ are pumped into the thylakoid space, increasing [H^+]. In step 5, NADPH formation uses an H^+ from the stroma, decreasing [H^+] in the stroma.   **Figure 10.22** The gene encoding hexokinase is part of the DNA of a chromosome in the nucleus. There, the gene is transcribed into mRNA, which is transported to the cytoplasm where it is translated on a free ribosome into a polypeptide. The polypeptide folds into a functional protein with secondary and tertiary structure. Once functional, it carries out the first reaction of glycolysis in the cytoplasm.

Concept Check 10.1

1. Because heterotrophs cannot photosynthesize, they cannot capture the energy of sunlight and produce energy-rich compounds like sugars, as autotrophs can. Sugars are oxidized by cellular respiration, providing energy (in the form of ATP) for cellular processes. Without this ability, heterotrophs depend on autotrophs to provide sugars as the food molecules that fuel their vital processes.   **2.** The main product of fossil fuel combustion (for example, by cars) is CO_2. Placing containers of algae near sources of CO_2 emission makes sense because the algae need CO_2 to carry out photosynthesis. The higher the CO_2 concentration, the higher will be the rate of algal photosynthesis. (At the same time, algae would be reducing the CO_2 concentration in those areas, which would otherwise contribute to climate change; see Concept 1.1.)

Concept Check 10.2

1. CO_2 enters the leaves via stomata, and, being a nonpolar molecule, can cross the leaf cell membrane and the chloroplast membranes to reach the stroma of the chloroplast.   **2.** Using ^{18}O, a heavy isotope of oxygen, as a label, researchers were able to confirm van Niel's hypothesis that the oxygen atoms in O_2 produced during photosynthesis comes from H_2O, not from CO_2.   **3.** The light reactions could *not* keep producing NADPH and ATP without the $NADP^+$, ADP, and ℗$_i$ that the Calvin cycle generates. The two cycles are interdependent.

Concept Check 10.3

1. Green, because green light is mostly transmitted and reflected—not absorbed—by photosynthetic pigments   **2.** Water (H_2O) is the initial electron donor; $NADP^+$ accepts electrons at the end of the electron transport chain, becoming reduced to NADPH.   **3.** In this experiment, the rate of ATP synthesis would slow and eventually stop. Because the added compound would not allow a proton gradient to build up across the membrane, ATP synthase could not catalyze ATP production.

Concept Check 10.4

1. 6, 18, 12   **2.** The more potential energy and reducing power a molecule stores, the more energy and reducing power are required for the formation of that molecule. Glucose is a valuable energy source because it is highly reduced (has lots of C—H bonds), storing lots of potential energy in its electrons. To reduce CO_2 to glucose, a large amount of energy and a lot of reducing power are required in the form of large numbers of ATP and NADPH molecules, respectively.   **3.** Yes, it would inhibit the dark reactions. The light reactions require ADP and $NADP^+$, which would not be formed in sufficient quantities from ATP and NADPH if the Calvin cycle stopped.

4.

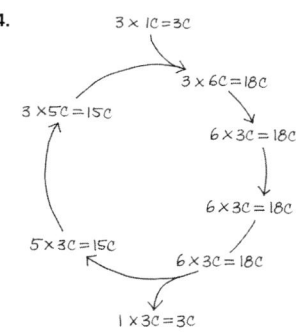

$3 \times 1C = 3C$
$3 \times 6C = 18C$
$3 \times 5C = 15C$
$6 \times 3C = 18C$
$6 \times 3C = 18C$
$5 \times 3C = 15C$
$6 \times 3C = 18C$
$1 \times 3C = 3C$

For every three turns of the Calvin cycle, the total number of carbon atoms remains constant because three carbons enter as part of CO_2 molecules, replacing the three that left as part of a G3P molecule.   **5.** In glycolysis, G3P acts as an intermediate. The six-carbon sugar fructose 1,6-bisphosphate is cleaved into two three-carbon sugars, one of which is G3P. The other is an isomer called dihydroxyacetone phosphate (DHAP), which can be converted to G3P by an isomerase. Because G3P is the substrate for the next enzyme, it is constantly removed, and the reaction reaches equilibrium is pulled in the direction of conversion of DHAP to more G3P. In the Calvin cycle, G3P acts as both an intermediate and a product. For every three CO_2 molecules that enter the cycle, six G3P molecules are formed, five of which must remain in the cycle and become rearranged to regenerate three five-carbon RuBP molecules. The one remaining G3P is a product, which can be thought of as the result of "reducing" the three CO_2 molecules that entered the cycle into a three-carbon sugar that can later be used to generate energy.

Concept Check 10.5

1. Photorespiration decreases photosynthetic output by adding O_2, instead of CO_2, to the Calvin cycle. As a result, no sugar is generated (no carbon is fixed), and O_2 is used rather than generated.   **2.** Without PS II, no O_2 is generated in bundle-sheath cells. This avoids the problem of O_2 competing with CO_2 for binding to rubisco in these cells.   **3.** Both problems are caused by a drastic change in Earth's atmosphere due to burning of fossil fuels. The increase in CO_2 concentration affects ocean chemistry by decreasing pH, thus affecting calcification by marine organisms. On land, CO_2 concentration and air temperature are conditions that plants have become adapted to, and changes in these characteristics have a strong effect on photosynthesis by plants. Thus, alteration of these two fundamental factors could have critical effects on organisms all around the planet, in all different habitats.   **4.** You would expect that C_4 and CAM species would replace many of the C_3 species.

Concept Check 10.6

1. Plants can break down the sugar they make (in the form of glucose) by cellular respiration, producing ATPs for various cellular processes such as endergonic chemical reactions, transport of substances across membranes, and movement of molecules in the cell. ATPs are also used for the movement of chloroplasts during cellular streaming in some plant cells (see Figure 6.26).

Summary of Key Concepts Questions

10.1 Photosynthesizing autotrophs are called producers because they use sunlight as energy to make their own organic molecules (including sugars) from CO_2 and H_2O. Heterotrophs are called consumers because they cannot make their own food and therefore must feed on other organisms, either autotrophs or other heterotrophs. Heterotrophs that consume the remains of dead organisms are called decomposers. Photosynthesis enables almost all organisms to survive because it makes food for photosynthesizing autotrophs, which in turn are food for most heterotrophs.   **10.2** CO_2 and H_2O are the products of cellular respiration; they are the reactants in photosynthesis. In respiration, glucose is oxidized to CO_2 and electrons are passed through an electron transfer chain from glucose to O_2, producing H_2O. In photosynthesis, H_2O is the source of electrons, which are energized by light, temporarily stored in NADPH, and used to reduce CO_2 to carbohydrate.   **10.3** The action spectrum of photosynthesis shows that some wavelengths of light that are not absorbed by chlorophyll *a* are still effective at promoting photosynthesis. The light-harvesting complexes of photosystems contain accessory pigments such as chlorophyll *b* and carotenoids, which absorb different wavelengths and pass the energy to chlorophyll *a*, broadening the spectrum of light usable for photosynthesis.

10.4

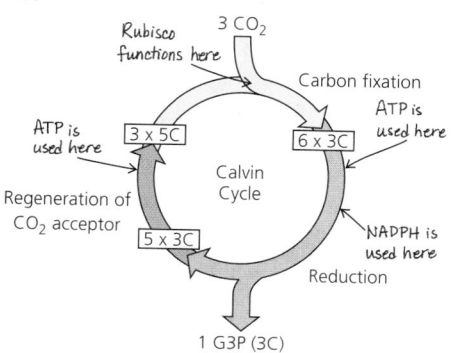

Rubisco functions here

3 CO_2

Carbon fixation

ATP is used here

$3 \times 5C$ $6 \times 3C$

ATP is used here

Calvin Cycle

Regeneration of CO_2 acceptor

$5 \times 3C$

NADPH is used here

Reduction

1 G3P (3C)

In the reduction phase of the Calvin cycle, ATP phosphorylates a three-carbon compound, and NADPH then reduces this compound to G3P. ATP is also used in the regeneration phase, when five molecules of G3P are converted to three molecules of the five-carbon compound RuBP. Rubisco catalyzes the first step of carbon fixation—the addition of CO_2 to RuBP.

10.5 Both C_4 photosynthesis and CAM photosynthesis involve initial fixation of CO_2 to produce a four-carbon compound (in mesophyll cells in C_4 plants and at night in CAM plants). These compounds are then broken down to release CO_2 (in the bundle-sheath cells in C_4 plants and during the day in CAM plants). ATP is required for recycling the molecule that is used initially to combine with CO_2. These pathways avoid the photorespiration that consumes ATP

and reduces the photosynthetic output of C_3 plants when they close stomata on hot, dry, bright days. Thus, hot, arid climates would favor C_4 and CAM plants.   **10.6** Sucrose made in the leaves of plants is transported through veins to nonphotosynthetic parts of the plant, where some of it is oxidized by cellular respiration, producing ATP for cellular processes. Other sugar molecules enter anabolic pathways, where they are used for synthesis of proteins, lipids, and polysaccharides such as cellulose, the main component of cell walls. Excess sugar is stockpiled as glucose subunits of the polysaccharide starch.

Test Your Understanding

1. D   **2.** B   **3.** C   **4.** A   **5.** A   **6.** B   **7.** C

10.

High H^+ concentration (pH=4)

H^+

Low H^+ concentration (pH=8)

ADP + P_i H^+ ATP

The ATP would end up outside the thylakoid. The thylakoids were able to make ATP in the dark because the researchers set up an artificial proton concentration gradient across the thylakoid membrane; thus, the light reactions were not necessary to establish the H^+ gradient required for ATP synthesis by ATP synthase.

Chapter 11

Figure Questions

Figure 11.6 Epinephrine is a signaling molecule outside the cell; presumably, it binds to a cell-surface receptor protein and thus is part of the signal reception step.   **Figure 11.8** This is an example of passive transport. The ion is moving down its concentration gradient, and no energy is required.   **Figure 11.9** The aldosterone molecule, a steroid, doesn't need a receptor protein—it is hydrophobic and can therefore pass directly through the hydrophobic lipid bilayer of the plasma membrane into the cell. (Hydrophilic molecules cannot do this.)   **Figure 11.10** The entire phosphorylation cascade wouldn't operate. Regardless of whether or not the signaling molecule was bound, protein kinase 2 would always be inactive and would not be able to activate the purple-colored protein leading to the cellular response.   **Figure 11.11** The signaling molecule (cAMP) would remain in its active form and would continue to signal; the pathway would remain active even in the absence of ligand because the cAMP would persist.
Figure 11.12

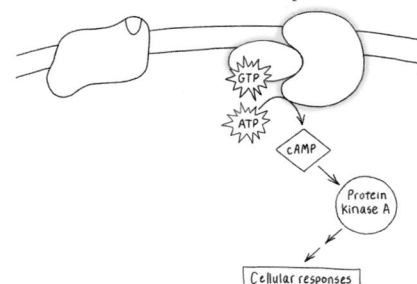

GTP

ATP

cAMP

Protein Kinase A

Cellular responses

Figure 11.14 The Ca^{2+} pump shown in Figure 11.13 is carrying out active transport of Ca^{2+} ions against their concentration gradient, using ATP for energy. These pumps help maintain a significantly lower [Ca^{2+}] in the cytoplasm than outside the cell and inside the ER. The Ca^{2+} channel proteins seen in Figure 11.14 are facilitating diffusion of Ca^{2+} down its concentration gradient; here, the Ca^{2+} is moving from inside the ER, where it is more concentrated, to the cytoplasm. Because the ion is moving down its concentration gradient, no energy is required.   **Figure 11.16** 100,000,000 (one hundred million, or 10^8) glucose molecules are released just from one epinephrine binding to the receptor. The first step results in $100\times$ amplification (one epinephrine activates 100 G proteins); the next step does not amplify the response; the next step is a $100\times$ amplification (10^2 active adenylyl cyclase molecules to 10^4 cyclic AMPs); the next step does not amplify; the next two steps are each $10\times$ amplifications, and the final step is a $100\times$ amplification.   **Figure 11.17** The signaling pathway shown in Figure 11.14 leads to the splitting of PIP_2 into the second messengers DAG and IP_3, which produce different responses. (The response elicited by DAG is mentioned but not shown.) The pathway shown for cell B is similar in that it branches and leads to two responses.

Concept Check 11.1

1. The two cells of opposite mating type (**a** and **α**) each secrete a unique signaling molecule, which can only be bound by receptors carried on cells of the opposite mating type. Thus, the **a** mating factor cannot bind to another **a** cell and cause it to grow toward the first **a** cell. Only an **α** cell can "receive" the signaling molecule and respond by directed growth.   **2.** Glycogen phosphorylase acts in the third stage, the cellular response to epinephrine signaling.   **3.** Glucose 1-phosphate would not be generated because the activation of the enzyme requires an intact cell, with an intact receptor in the membrane and an intact signal transduction pathway. The enzyme cannot be activated directly by interaction with the signaling molecule in the cell-free mixture.

Concept Check 11.2

1. NGF is water-soluble (hydrophilic), so it cannot pass through the lipid membrane to reach intracellular receptors, as steroid hormones can. Therefore, you'd expect the NGF receptor to be in the plasma membrane—which is, in fact, the case. **2.** The cell with the faulty receptor would not be able to respond appropriately to the signaling molecule when it was present. This would most likely have dire consequences for the cell since regulation of the cell's activities by this receptor would not occur appropriately. **3.** Binding of a ligand to a receptor changes the shape of the receptor, altering the ability of the receptor to transmit a signal. Binding of an allosteric regulator to an enzyme changes the shape of the enzyme, either promoting or inhibiting enzyme activity.

Concept Check 11.3

1. A protein kinase is an enzyme that transfers a phosphate group from ATP to a protein, usually activating that protein (often a second type of protein kinase). Many signal transduction pathways include a series of such interactions, in which each phosphorylated protein kinase in turn phosphorylates the next protein kinase in the series. Such phosphorylation cascades carry a signal from outside the cell to the cellular protein(s) that will carry out the response. **2.** Protein phosphatases reverse the effects of the kinases by dephosphorylation, and unless the signaling molecule is at a high enough concentration that it is continuously rebinding the receptor, the kinase molecules will all be returned to their inactive states by phosphatases. **3.** The signal that is being transduced is the *information* that a signaling molecule is bound to the cell-surface receptor. Information is transduced by way of sequential protein-protein interactions that change protein shapes, causing them to function in a way that passes the signal (the information) along. **4.** The IP$_3$-gated channel would open, allowing calcium ions to flow out of the ER and into the cytoplasm, which would raise the cytosolic Ca^{2+} concentration.

Concept Check 11.4

1. At each step in a cascade of sequential activations, one molecule or ion may activate numerous molecules functioning in the next step. This causes the response to be amplified at each such step and overall results in a large amplification of the original signal. **2.** Scaffolding proteins hold molecular components of signaling pathways in a complex with each other. Different scaffolding proteins would assemble different collections of proteins, facilitating different molecular interactions and leading to different cellular responses in the two cells. **3.** A malfunctioning protein phosphatase would not be able to dephosphorylate a particular receptor or relay protein. As a result, the signaling pathway, once activated, would not be able to be terminated. (In fact, one study found altered protein phosphatases in cells from 25% of colorectal tumors.) **4.** The proteins in the two cells are different, so the cellular response is different. In heart muscle cells, the pathway shown in Figure 11.16 allows glucose to fuel faster muscle contractions and heart rate. In respiratory muscles, the relay proteins must be different, so that the effect is to block muscle contraction. (In fact, the steps are the same through protein kinase A (PKA), but in respiratory muscle cells, PKA phosphorylates a protein that is required for muscle contraction—and in this case, phosphorylation *inactivates* that protein. So muscle contraction cannot occur.)

Concept Check 11.5

1. In formation of the hand or paw in mammals, cells in the regions between the digits are programmed to undergo apoptosis. This serves to shape the digits of the hand or paw so that they are not webbed. (A lack of apoptosis in these regions in water birds results in webbed feet.) **2.** If a receptor protein for a death-signaling molecule were defective such that it was activated even in the absence of the death signal, this would lead to apoptosis when it wouldn't normally occur. Similar defects in any of the proteins in the signaling pathway would have the same effect if the defective proteins activated relay or response proteins in the absence of interaction with the previous protein or second messenger in the pathway. Conversely, if any protein in the pathway were defective in its ability to respond to an interaction with an early protein or other molecule or ion, apoptosis would not occur when it normally should. For example, a receptor protein for a death-signaling ligand might not be able to be activated, even when ligand was bound. This would stop the signal from being transduced into the cell.

Summary of Key Concepts Questions

11.1 A cell is able to respond to a hormone only if it has a receptor protein on the cell surface or inside the cell that can bind to the hormone. The response to a hormone depends on the specific signal transduction pathway within the cell, which will lead to the specific cellular response. The response can vary for different types of cells. **11.2** Both GPCRs and RTKs have an extracellular binding site for a signaling molecule (ligand) and one or more α-helical regions of the polypeptide that spans the membrane. A GPCR functions singly, while RTKs tend to dimerize or form larger groups of RTKs. GPCRs usually trigger a single transduction pathway, whereas the multiple activated tyrosines on an RTK dimer may trigger several different transduction pathways at the same time. **11.3** A protein kinase is an enzyme that adds a phosphate group to another protein. Protein kinases are often part of a phosphorylation cascade that transduces a signal. A second messenger is a small, nonprotein molecule or ion that rapidly diffuses and relays a signal throughout a cell. Both protein kinases and second messengers can operate in the same pathway. For example, the second messenger cAMP often activates protein kinase A, which then phosphorylates other proteins. **11.4** In G protein-coupled pathways, the GTPase portion of a G protein converts GTP to

GDP, inactivating the G protein. Protein phosphatases remove phosphate groups from activated proteins, thus stopping a phosphorylation cascade of protein kinases. Phosphodiesterase converts cAMP to AMP, thus reducing the effect of cAMP in a signal transduction pathway. **11.5** The basic mechanism of controlled cell suicide evolved early in eukaryotic evolution, and the genetic basis for these pathways has been conserved during animal evolution. Such a mechanism is essential to the development and maintenance of all animals.

Test Your Understanding

1. D **2.** A **3.** B **4.** A **5.** C **6.** C **7.** C **8.** This is one possible drawing of the pathway. (Similar drawings would also be correct.)

Chapter 12

Figure Questions

Figure 12.4

One sister chromatid

Circling the other chromatid instead would also be correct. **Figure 12.5** The chromosome has four arms. The single (duplicated) chromosome in ❷ becomes two (unduplicated) chromosomes in ❸. The duplicated chromosome in step 2 is considered one single chromosome.
Figure 12.7 12; 2; 2; 1
Figure 12.8

Chromosome movement

Metaphase plate

Figure 12.9 The mark would have moved toward the pole closer to it. The lengths of fluorescent microtubules between that pole and the mark would have decreased, while the lengths between the chromosomes and the mark would have remained the same. **Figure 12.14** In both cases, the G$_1$ nucleus would have remained in G$_1$ until the time it normally would have entered the S phase. Chromosome condensation and spindle formation would not have occurred until the S and G$_2$ phases had been completed. **Figure 12.16** Passing the G$_2$ checkpoint in the diagram corresponds to the beginning of the "Time" axis of the graph, and entry into the mitotic phase (yellow background on the diagram) corresponds to the peaks of MPF activity and cyclin concentration on the graph (see the yellow M banner over the peaks). During G$_1$ and S phase in the diagram, Cdk is present without cyclin, so on the graph both cyclin concentration and MPF activity are low. The curved purple gradient arrow in the diagram shows increasing cyclin concentration, seen on the graph during the end of S phase and throughout G$_2$ phase. Then the cell cycle begins again. **Figure 12.17** The cell would divide under conditions where it was inappropriate to do so. If the daughter cells and their descendants also ignored either of the checkpoints and divided, there would soon be an abnormal mass of cells. (This type of inappropriate cell division can contribute to the development of cancer.) **Figure 12.18** The cells in the vessel with PDGF would not be able to respond to the growth factor signal and thus would not divide. The culture would resemble that without the added PDGF.

Concept Check 12.1

1. 1; 1; 2 **2.** 39; 39; 78

Concept Check 12.2

1. 6 chromosomes; they are duplicated; 12 chromatids   **2.** Following mitosis, cytokinesis results in two genetically identical daughter cells in both plant cells and animal cells. However, the mechanism of dividing the cytoplasm is different in animals and plants. In an animal cell, cytokinesis occurs by cleavage, which divides the parent cell in two with a contractile ring of actin filaments. In a plant cell, a cell plate forms in the middle of the cell and grows until its membrane fuses with the plasma membrane of the parent cell. A new cell wall grows inside the cell plate, thus eventually between the two new cells.   **3.** From the end of S phase in interphase through the end of metaphase in mitosis   **4.** During eukaryotic cell division, tubulin is involved in spindle formation and chromosome movement, while actin functions during cytokinesis. In bacterial binary fission, it's the opposite: Actin-like molecules are thought to move the daughter bacterial chromosomes to opposite ends of the cell, and tubulin-like molecules are thought to act in daughter cell separation.   **5.** A kinetochore connects the spindle (a motor; note that it has motor proteins) to a chromosome (the cargo it will move). **6.** Microtubules made up of tubulin in the cell provide "rails" along which vesicles and other organelles can travel, based on interactions of motor proteins with tubulin in the microtubules. In muscle cells, actin in microfilaments interacts with myosin filaments to cause muscle contraction.

Concept Check 12.3

1. The nucleus on the right was originally in the G_1 phase; therefore, it had not yet duplicated its chromosomes. The nucleus on the left was in the M phase, so it had already duplicated its chromosomes.   **2.** A sufficient amount of MPF has to exist for a cell to pass the G_2 checkpoint; this occurs through the accumulation of cyclin proteins, which combine with Cdk to form (active) MPF. MPF then phosphorylates other proteins, initiating mitosis.   **3.** The intracellular receptor (for example, an estrogen receptor), once activated, would be able to act as a transcription factor in the nucleus, turning on genes that may cause the cell to pass a checkpoint and divide. The RTK receptor, when activated by a ligand, would form a dimer, and each subunit of the dimer would phosphorylate the other. This would lead to a series of signal transduction steps, ultimately turning on genes in the nucleus. As in the case of the estrogen receptor, the genes would code for proteins necessary to cause the cell to pass a checkpoint and divide.

Summary of Key Concepts Questions

12.1 The DNA of a eukaryotic cell is packaged into structures called *chromosomes*. Each chromosome is a long molecule of DNA, which carries hundreds to thousands of genes, with associated proteins that maintain chromosome structure and help control gene activity. This DNA-protein complex is called *chromatin*. The chromatin of each chromosome is long and thin when the cell is not dividing. Prior to cell division, each chromosome is duplicated, and the resulting sister *chromatids* are attached to each other by proteins at the centromeres and, for many species, all along their lengths (a phenomenon called sister chromatid cohesion).   **12.2** A chromosome exists as a single DNA molecule in G_1 of interphase and in anaphase and telophase of mitosis. During S phase, DNA replication produces two sister chromatids per chromosome, which persist during G_2 of interphase and through prophase, prometaphase, and metaphase of mitosis. **12.3** Checkpoints allow cellular surveillance mechanisms to determine whether the cell is prepared to go to the next stage. Internal and external signals move a cell past these checkpoints. The G_1 checkpoint determines whether a cell will proceed forward in the cell cycle or switch into the G_0 phase. The signals to pass this checkpoint often are external, such as growth factors. Passing the G_2 checkpoint requires sufficient numbers of active MPF complexes, which in turn orchestrate several mitotic events. MPF also initiates degradation of its cyclin component, terminating the M phase. The M phase will not begin again until sufficient cyclin is produced during the next S and G_2 phases. The signal to pass the M phase checkpoint is not activated until all chromosomes are attached to kinetochore fibers and are aligned at the metaphase plate. Only then will sister chromatid separation occur.

Test Your Understanding

1. B **2.** A **3.** C **4.** C **5.** A **6.** B **7.** A **8.** D **9.** See Figure 12.7 for a description of major events. Only one cell is indicated for each stage, but other correct answers are also present in this micrograph.

10.

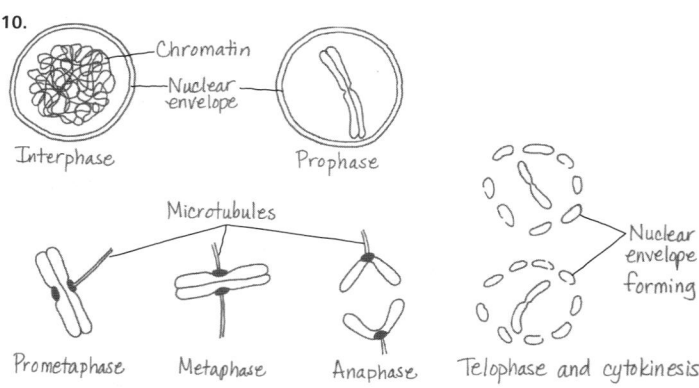

Chapter 13

Figure Questions

Figure 13.3 The chromosomes in a karyotype are photographed at their most condensed, at metaphase, which occurs during M phase of the cell cycle. At this point, the chromosomes have already been duplicated during S phase, so each chromosome exists as two sister chromatids.   **Figure 13.4** Two sets of chromosomes are present. Three pairs of homologous chromosomes are present. One long chromosome would be red (maternal), and one would be blue (paternal); one medium chromosome would be red, and one would be blue; one short chromosome would be red, and one would be blue.   **Figure 13.6** In (a), haploid cells do not undergo mitosis. In (b), haploid spores undergo mitosis to form the gametophyte, and haploid cells of the gametophyte undergo mitosis to form gametes. In (c), haploid cells undergo mitosis to form either a multicellular haploid organism or a new unicellular haploid organism, and these haploid cells undergo mitosis to form gametes.

Figure 13.7

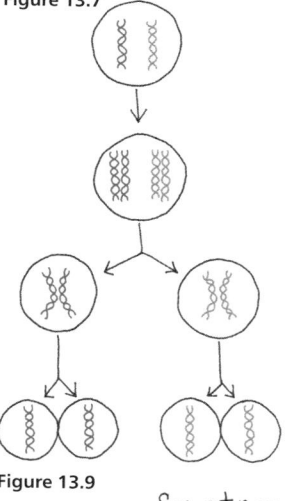

(A short strand of DNA is shown here for simplicity, but each chromosome or chromatid contains a very long coiled and folded DNA molecule.)   **Figure 13.8** If a cell with six chromosomes undergoes two rounds of mitosis, each of the four resulting cells will have six chromosomes, while the four cells resulting from meiosis in Figure 13.8 each have three chromosomes. In mitosis, DNA replication (and thus chromosome duplication) precedes each prophase, ensuring that daughter cells have the same number of chromosomes as the parent cell. In contrast, in meiosis, DNA replication occurs only before prophase I (not before prophase II). Thus, in two rounds of mitosis, the chromosomes duplicate twice and divide twice, while in meiosis, the chromosomes duplicate once and divide twice.

Figure 13.9

Figure 13.10 Yes. Each of the six chromosomes (three per cell) shown in telophase I has one nonrecombinant chromatid and one recombinant chromatid. Therefore, eight (2 possibilities for the first chromosome × 2 for the second × 2 for the third) possible sets of chromosomes can be generated for the cell on the left and eight for the cell on the right.

Concept Check 13.1

1. Parents pass genes to their offspring; by dictating the production of messenger RNAs (mRNAs), the genes program cells to make specific enzymes and other proteins, whose cumulative action produces an individual's inherited traits. **2.** Such organisms reproduce by mitosis, which generates offspring whose genomes are exact copies of the parent's genome (in the absence of mutation).   **3.** She should clone (generate a clone of) it. Crossbreeding it with another plant would generate offspring that have additional variation, which she no longer desires now that she has obtained her ideal orchid.

Concept Check 13.2

1. Each of the six chromosomes is duplicated, so each contains two DNA molecules (double helices), so there are 12 DNA molecules in the cell. The haploid number, n, is 3. One set is always haploid. **2.** There are 23 pairs of chromosomes and two sets. **3.** A human diploid cell would have 23 "pairs of shoes." A haploid cell would have 23 "shoes," one of each pair. **4.** This organism has the life cycle shown in Figure 13.6c, since the zygote doesn't undergo mitosis but immediately undergoes meiosis. Therefore, it must be a fungus or a protist, perhaps an alga.

Concept Check 13.3

1. The chromosomes are similar in that each is composed of two sister chromatids, and the individual chromosomes are positioned similarly at the metaphase plate. The chromosomes differ in that in a mitotically dividing cell, sister chromatids of each chromosome are genetically identical, but in a meiotically dividing cell, sister chromatids are genetically distinct because of crossing over in meiosis I. Moreover, the chromosomes in metaphase of mitosis can be a diploid set or a haploid set, but the chromosomes in metaphase of meiosis II always consist of a haploid set. **2.** If crossing over did not occur, the two homologs would not be associated in any way. This is because each sister chromatid would be either all maternal or all paternal DNA, and the single DNA molecule would therefore not have been joined to the DNA of a nonsister chromatid, holding the complex together. The lack of association of homologs could easily result in the incorrect arrangement of homologs during metaphase I (both might go toward the same pole, for instance) and ultimately in formation of gametes with an abnormal number of chromosomes.

Concept Check 13.4

1. Mutations in a gene lead to the different versions (alleles) of that gene. **2.** Without crossing over, independent assortment of chromosomes during meiosis I theoretically can generate 2^n possible haploid gametes, and random fertilization can produce $2^n \times 2^n$ possible diploid zygotes. Because the haploid number (n) of grasshoppers is 23 and that of fruit flies is 4, two grasshoppers would be expected to produce a greater variety of zygotes than would two fruit flies. **3.** If the segments of the maternal and paternal chromatids that undergo crossing over are genetically identical and thus have the same two alleles for every gene, then the recombinant chromosomes will be genetically equivalent to the parental chromosomes. Crossing over contributes to genetic variation only when it involves the rearrangement of different alleles.

Summary of Key Concepts Questions

13.1 Genes program specific traits, and offspring inherit their genes from each parent, accounting for similarities in their appearance to one or the other parent. Humans reproduce sexually, which ensures new combinations of genes (and thus traits) in the offspring. Consequently, the offspring are not clones of their parents (which would be the case if humans reproduced asexually). **13.2** Animals and plants both reproduce sexually, alternating meiosis with fertilization. Both have haploid gametes that unite to form a diploid zygote, which then goes on to divide mitotically, forming a diploid multicellular organism. In animals, haploid cells become gametes and don't undergo mitosis, while in plants, the haploid cells resulting from meiosis undergo mitosis to form a haploid multicellular organism, the gametophyte. This organism then goes on to generate haploid gametes. (In plants such as trees, the gametophyte is quite reduced in size and not obvious to the casual observer.) **13.3** At the end of meiosis I, the two members of a homologous pair end up in different cells, so they cannot pair up and undergo crossing over during prophase II. **13.4** First, during independent assortment in metaphase I, each pair of homologous chromosomes lines up independently of each other pair at the metaphase plate, so a daughter cell of meiosis I randomly inherits either a maternal or a paternal chromosome of each pair. Second, due to crossing over, each chromosome is not exclusively maternal or paternal, but includes regions at the ends of the chromatid from a nonsister chromatid (a chromatid of the other homolog). (The nonsister segment can also be in an internal region of the chromatid if a second crossover occurs beyond the first one before the end of the chromatid.) This provides much additional diversity in the form of new combinations of alleles. Third, random fertilization ensures even more variation since any sperm of a large number containing many possible genetic combinations can fertilize any egg of a similarly large number of possible combinations.

Test Your Understanding

1. A **2.** B **3.** A **4.** D **5.** C
6. (a) One possible answer:

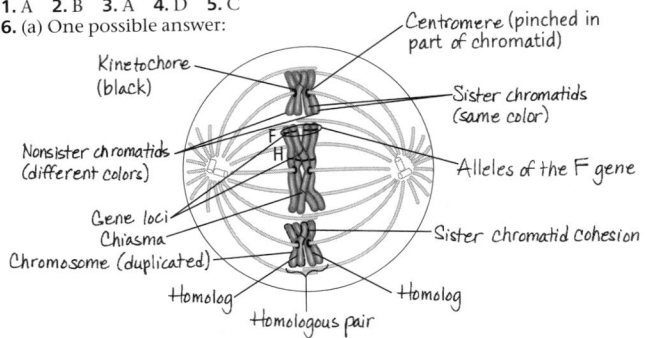

(b) A haploid set is made up of one long, one medium, and one short chromosome, no matter what combination of colors. For example, one red long, one blue medium, and one red short chromosome make up a haploid set. (In

cases where crossovers have occurred, a haploid set of one color may include segments of chromatids of the other color.) All red and blue chromosomes together make up a diploid set. (c) Metaphase I **7.** This cell must be undergoing meiosis because the two homologs of a homologous pair are associated with each other at the metaphase plate; this does not occur in mitosis. Also, chiasmata are clearly present, meaning that crossing over has occurred, another process unique to meiosis.

Chapter 14

Figure Questions

Figure 14.3 All offspring would have purple flowers. (The ratio of purple to white flowers would be 1 purple: 0 white.) The P generation plants are true-breeding, so mating two purple-flowered plants produces the same result as self-pollination: All the offspring have the same trait. If Mendel had stopped after the F_1 generation, he could have concluded that the white factor had disappeared entirely and would not ever reappear.
Figure 14.8

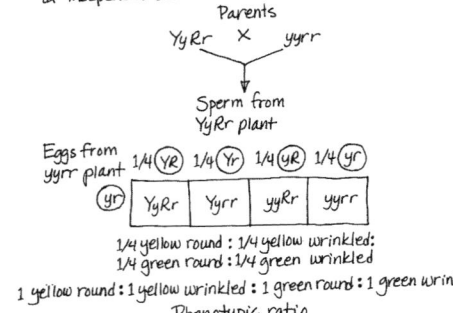

Yes, this cross would also have allowed Mendel to make different predictions for the two hypotheses, thereby allowing him to distinguish the correct one.
Figure 14.10 Your classmate would probably point out that the F_1 generation hybrids show an intermediate phenotype between those of the homozygous parents, which supports the blending hypothesis. You could respond that crossing the F_1 hybrids results in the reappearance of the white phenotype, rather than identical pink offspring, which fails to support the idea of traits blending during inheritance. The blending hypothesis predicts that the white trait would have been lost after the F_1 generation. **Figure 14.11** Both the I^A and I^B alleles are dominant to the i allele because the i allele results in no attached carbohydrate. The I^A and I^B alleles are codominant; both are expressed in the phenotype of $I^A I^B$ heterozygotes, who have type AB blood. **Figure 14.12** In this cross, the final "3" and "1" of a standard cross are lumped together as a single phenotype. This occurs because in dogs that are ee, no pigment is deposited, thus the three dogs that have a B in their genotype (normally black) can no longer be distinguished from the dog that is bb (normally brown). **Figure 14.16** In the Punnett square, two of the three individuals with normal coloration are carriers, so the probability is $2/3$. (Note that you must take into account everything you know when you calculate probability: You know she is not aa, so there are only three possible genotypes to consider.)

Concept Check 14.1

1. According to the law of independent assortment, 25 plants ($1/16$ of the offspring) are predicted to be $aatt$, or recessive for both characters. The actual result is likely to differ slightly from this value. **2.** The plant could make eight different gametes (*YRI, YRi, YrI, Yri, yRI, yRi, yrI,* and *yri*). To fit all the possible gametes in a self-pollination, a Punnett square would need 8 rows and 8 columns. It would have spaces for the 64 possible unions of gametes in the offspring. **3.** Self-pollination is sexual reproduction because meiosis is involved in forming gametes, which unite during fertilization. As a result, the offspring in self-pollination are genetically different from the parent. (As mentioned in the footnote near the beginning of Concept 14.1, we have simplified the explanation in referring to the single pea plant as a parent. Technically, the gametophytes in the flower are the two "parents.")

Concept Check 14.2

1. $\frac{1}{2}$ homozygous dominant (*AA*), 0 homozygous recessive (*aa*), and $\frac{1}{2}$ heterozygous (*Aa*) **2.** $\frac{1}{4}$ *BBDD*; $\frac{1}{4}$ *BbDD*; $\frac{1}{4}$ *BBDd*; $\frac{1}{4}$ *BbDd* **3.** The genotypes that fulfill this condition are *ppyyII, ppyyIi, ppYYii, ppYyii, PpyyiI,* and *ppyyii.* Use the multiplication rule to find the probability of getting each genotype, and then use the addition rule to find the overall probability of meeting the conditions of this problem:

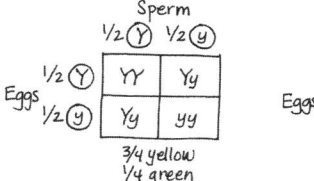

Concept Check 14.3

1. Incomplete dominance describes the relationship between two alleles of a single gene, whereas epistasis relates to the genetic relationship between two genes (and the respective alleles of each). **2.** Half of the children would be expected to have type A blood and half type B blood. **3.** The black and white alleles are incompletely dominant, with heterozygotes being gray in color. A cross between a gray rooster and a black hen should yield approximately equal numbers of gray and black offspring.

Concept Check 14.4

1. $\frac{1}{9}$ (Since cystic fibrosis is caused by a recessive allele, Lucia and Jared's siblings who have CF must be homozygous recessive. Therefore, each parent must be a carrier of the recessive allele. Since neither Lucia nor Jared has CF and are thus not *cc* (using *C/c* as the alleles for the CF gene), this means they each have a $\frac{2}{3}$ chance of being a carrier. If they are both carriers, there is a $\frac{1}{4}$ chance that they will have a child with CF; $\frac{2}{3} \times \frac{2}{3} \times \frac{1}{4} = \frac{1}{9}$); virtually 0 (Both Lucia and Jared would have to be carriers to produce a child with the disease, unless a very rare mutation (change) occurred in the DNA of cells making eggs or sperm in a noncarrier that resulted in the CF allele.) **2.** In normal hemoglobin, the sixth amino acid is glutamic acid (Glu), which is acidic (has a negative charge on its side chain). In sickle-cell hemoglobin, Glu is replaced by valine (Val), which is a nonpolar amino acid, very different from Glu. The primary structure of a protein (its amino acid sequence) ultimately determines the shape of the protein and thus its function. The substitution of Val for Glu enables the hemoglobin molecules to interact with each other and form long fibers, leading to the protein's deficient function and the deformation of the red blood cell. **3.** Juanita's genotype is *Dd.* Because the allele for polydactyly (*D*) is dominant to the allele for five digits per appendage (*d*), the trait is expressed in people with either the *DD* or *Dd* genotype. But because Juanita's father does not have polydactyly, his genotype must be *dd*, which means that Juanita inherited a *d* allele from him. Therefore, Juanita, who does have the trait, must be heterozygous. **4.** In the monohybrid cross involving flower color, the ratio is 3.15 purple : 1 white, while in the human family in the pedigree, the ratio in the third generation is 1 taster of PTC : 1 nontaster of PTC. The difference is due to the small sample size (two offspring) in the human family. If the second-generation couple in this pedigree were able to have 929 offspring as in the pea plant cross, the ratio would likely be closer to 3:1. (Note that none of the pea plant crosses in Table 14.1 yielded *exactly* a 3:1 ratio.)

Summary of Key Concepts Questions

14.1 Alternative versions of genes, called alleles, are passed from parent to offspring during sexual reproduction. In a cross between purple- and white-flowered homozygous parents, the F_1 offspring are all heterozygous, each inheriting a purple allele from one parent and a white allele from the other. Because the purple allele is dominant, it determines the phenotype of the F_1 offspring to be purple, and the expression of the recessive white allele is masked. Only in the F_2 generation is it possible for some of the offspring to be homozygous recessive, which causes the white trait to be expressed.

14.2

14.3 The ABO blood group is an example of multiple alleles because this single gene has more than two alleles (I^A, I^B, and i). Two of the alleles, I^A and I^B, exhibit codominance since both carbohydrates (A and B) are present when these two alleles exist together in a genotype. I^A and I^B each exhibit complete dominance over the *i* allele. This situation is not an example of incomplete dominance because each allele affects the phenotype in a distinguishable way, so the result is not intermediate between the two phenotypes. Because this situation involves a single gene, it is not an example of epistasis or polygenic inheritance. **14.4** The chance of the fourth child having cystic fibrosis is $\frac{1}{4}$, as it was for each of the other children, because each birth is an independent event. We already know both parents are carriers, so whether their first three children are carriers or not has no bearing on the probability that their next child will have the disease. The parents' genotypes provide the only relevant information.

Test Your Understanding

1. A cross of *Ii* × *ii* would yield offspring with a genotypic ratio of 1 *Ii* : 1 *ii* (2:2 is an equivalent answer) and a phenotypic ratio of 1 inflated : 1 constricted (2:2 is equivalent).

2. Man $I^A i$; woman $I^B i$; child *ii*. Genotypes for future children are predicted to be $\frac{1}{4}$ $I^A I^B$, $\frac{1}{4}$ $I^A i$, $\frac{1}{4}$ $I^B i$, $\frac{1}{4}$ *ii*. **3.** $\frac{1}{2}$

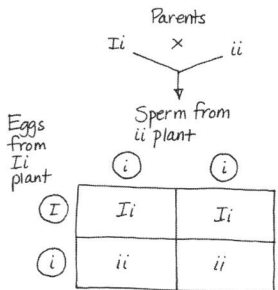

4. The green pod trait is dominant, so the green pod allele is *G* and the yellow pod allele is *g*; the inflated pod trait is dominant, so the inflated pod allele is *I* and the constricted pod allele is *i*. The cross described is *GgIi* × *GgIi*.

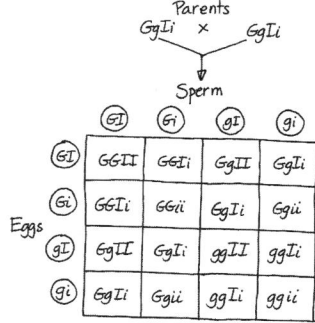

5. (a) $\frac{1}{64}$; (b) $\frac{1}{64}$; (c) $\frac{1}{8}$; (d) $\frac{1}{32}$ **6.** (a) $\frac{3}{4} \times \frac{3}{4} \times \frac{3}{4} = \frac{27}{64}$; (b) $1 - \frac{27}{64} = \frac{37}{64}$; (c) $\frac{1}{4} \times \frac{1}{4} \times \frac{1}{4} = \frac{1}{64}$; (d) $1 - \frac{1}{64} = \frac{63}{64}$ **7.** (a) $\frac{1}{256}$; (b) $\frac{1}{16}$; (c) $\frac{1}{256}$; (d) $\frac{1}{64}$; (e) $\frac{1}{128}$ **8.** (a) 1; (b) $\frac{1}{32}$; (c) $\frac{1}{8}$; (d) $\frac{1}{2}$ **9.** $\frac{1}{9}$
10. Matings of the original mutant cat with true-breeding noncurl cats will produce both curl and noncurl offspring if the curl allele is dominant, but only noncurl offspring if the curl allele is recessive. Whether the curl trait is dominant or recessive, you would obtain some true-breeding offspring homozygous for the curl allele from matings between the F_1 cats resulting from the original curl × noncurl crosses. If dominant, you wouldn't be able to tell true-breeding, homozygous offspring from heterozygotes without further crosses. You know that cats are true-breeding when curl × curl matings produce only curl offspring. As it turns out, the allele that causes curled ears is dominant. **11.** 25%, or $\frac{1}{4}$, will be cross-eyed; all (100%) of the cross-eyed offspring will also be white. **12.** The dominant allele *I* is epistatic to the *P/p* locus, and thus the genotypic ratio for the F_1 generation will be 9 *I–P–* (colorless) : 3 *I–pp* (colorless) : 3 *iiP–* (purple) : 1 *iipp* (red). Overall, the phenotypic ratio is 12 colorless : 3 purple : 1 red. **13.** Recessive.

All individuals with alkaptonuria (Ariana, Benito, Elena, and Carlota) are homozygous recessive *aa*. Jorge is *Aa*, since some of his children with Ariana have alkaptonuria. Diego, Carmen, Hector, and Julio are each *Aa*, since they are all unaffected children with one affected parent. Miguel also is *Aa*, since he has an affected child (Carlota) with his heterozygous wife Carmen. Mariposa, Paloma, and Roberto can each have either the *AA* or *Aa* genotype.   **14.** 1/6

Chapter 15

Figure Questions

Figure 15.3 About 3/4 of the F2 offspring would have red eyes and about 1/4 would have white eyes. Since the sex chromosomes are not involved in determining eye color in this hypothetical cross, about half of the white-eyed flies would be female and half would be male; similarly, about half of the red-eyed flies would be female and half would be male. (Note that the autosomes with the eye color alleles would be the same shape in the Punnett square—unlike the X and Y chromosomes—and each offspring would inherit two alleles. The sex of the flies would be determined separately by inheritance of the sex chromosomes. Thus your F2 Punnett square would have four possible combinations in sperm and two in eggs; it would have eight squares altogether.)   **Figure 15.4** The ratio would be 1 yellow round : 1 green round : 1 yellow wrinkled : 1 green wrinkled. (This is similar to a test cross.)
Figure 15.7 All the males would be color-blind, and all the females would be carriers. (Another way to say this is that 1/2 the offspring would be color-blind males, and 1/2 the offspring would be carrier females.)   **Figure 15.9** The two largest classes would still be the offspring with the phenotypes of the true-breeding P generation flies, but now they would be gray vestigial and black normal, which is now the "parental type" because those were the specific allele combinations in the P generation (the linked alleles on their chromosomes).   **Figure 15.10** The two chromosomes on the left side of the sketch below are like the two chromosomes inherited by the F1 female, one from each P generation fly. They are passed by the F1 female intact to the offspring and thus could be called "parental" chromosomes. The other two chromosomes result from crossing over during meiosis in the F1 female's meiosis. Because they have combinations of alleles not seen in either of the F1 female's chromosomes, they can be called "recombinant" chromosomes. (Note that in this example, the allele combinations on the recombinant chromosomes, *b+ vg+* and *b vg*, are the allele combinations that were on the parental chromosomes in the cross shown in Figures 15.9 and 15.10. The basis for calling them parental chromosomes is that they have the combination of alleles that was present on the P generation chromosomes.)

Concept Check 15.1

1. To show the mutant phenotype, a male needs to possess only one mutant allele. If this gene had been on a pair of autosomes, an individual would show the recessive phenotype only if both alleles were mutant, a much less probable situation.   **2.** The law of segregation describes the inheritance of alleles for a single character. The law of independent assortment of alleles describes the inheritance of alleles for two characters.   **3.** The physical basis for the law of segregation is the separation of homologs in anaphase I. The physical basis for the law of independent assortment is the alternative arrangements of all the different homologous chromosome pairs in metaphase I.

Concept Check 15.2

1. Because the gene for this eye color character is located on the X chromosome, all female offspring will be red-eyed and heterozygous ($X^{w+}X^w$); all male offspring will inherit a Y chromosome from the father and be white-eyed (X^wY). (Another way to say this is that 1/2 the offspring will be red-eyed heterozygous [carrier] females, and 1/2 will be white-eyed males.)   **2.** 1/4 (1/2 chance that the child will inherit a Y chromosome from the father and be male × 1/2 chance that he will inherit the X carrying the disease allele from his mother). If the child is a boy, there is a 1/2 chance he will have the disease; a female would have zero chance (but 1/2 chance of being a carrier).   **3.** In a disorder caused by a dominant allele, there is no such thing as a "carrier," since those with the allele have the disorder. Because the allele is dominant, the females lose any "advantage" in having two X chromosomes, since one disorder-associated allele is sufficient to result in the disorder. All fathers who have the dominant allele will pass it along to *all* their daughters, who will also have the disorder. A mother who has the allele (and thus the disorder) will pass it to half of her sons and half of her daughters.

Concept Check 15.3

1. Crossing over during meiosis I in the heterozygous parent produces some gametes with recombinant genotypes for the two genes. Offspring with a recombinant phenotype arise from fertilization of the recombinant gametes by homozygous recessive gametes from the double-mutant parent.   **2.** In each case, the alleles contributed by the female parent (in the egg) determine the phenotype of the offspring because the male in this cross contributes only recessive alleles. Thus, identifying the phenotype of the offspring tells you what alleles were in the mother's (the dihybrid female's) egg.   **3.** No. The order could be *A-C-B* or *C-A-B*. To determine which possibility is correct, you need to know the recombination frequency between *B* and *C*.

Concept Check 15.4

1. In meiosis, a combined 14-21 chromosome will behave as one chromosome. If a gamete receives the combined 14-21 chromosome and a normal copy of chromosome 21, trisomy 21 will result when this gamete combines with a normal gamete (with its own chromosome 21) during fertilization.   **2.** No. The child can be either I^AI^Ai or I^Aii. A sperm of genotype I^AI^A could result from nondisjunction in the father during meiosis II, while an egg with the genotype *ii* could result from nondisjunction in the mother during either meiosis I or meiosis II.   **3.** Activation of this gene could lead to the production of too much of this kinase. If the kinase is involved in a signaling pathway that triggers cell division, too much of it could trigger unrestricted cell division, which in turn could contribute to the development of a cancer (in this case, a cancer of one type of white blood cell).

Concept Check 15.5

1. Inactivation of an X chromosome in females and genomic imprinting. Because of X inactivation, the effective dose of genes on the X chromosome is the same in males and females. As a result of genomic imprinting, only one allele of certain genes is phenotypically expressed.   **2.** The genes for leaf coloration are located in plastids within the cytoplasm. Normally, only the maternal parent transmits plastid genes to offspring. Since variegated offspring are produced only when the female parent is of the B variety, we can conclude that variety B contains both the wild-type and mutant alleles of pigment genes, producing variegated leaves. (Variety A must contain only the wild-type allele of pigment genes.)   **3.** Each cell contains numerous mitochondria, and in affected individuals, most cells contain a variable mixture of normal and mutant mitochondria. The normal mitochondria carry out enough cellular respiration for survival. (The situation is similar for chloroplasts.)

Summary of Key Concepts Questions

15.1 Because the sex chromosomes are different from each other and because they determine the sex of the offspring, Morgan could use the sex of the offspring as a phenotypic character to follow the parental chromosomes. (He could also have followed them under a microscope, as the X and Y chromosomes look different.) At the same time, he could record eye color to follow the eye color alleles.   **15.2** Males have only one X chromosome, along with a Y chromosome, while females have two X chromosomes. The Y chromosome has very few genes on it, while the X has about 1,000. When a recessive X-linked allele that causes a disorder is inherited by a male on the X from his mother, there isn't a second allele present on the Y (males are hemizygous), so the male has the disorder. Because females have two X chromosomes, they must inherit two recessive alleles in order to have the disorder, a rarer occurrence.   **15.3** Crossing over results in new combinations of alleles. Crossing over is a random occurrence, and the more distance there is between two genes, the more chances there are for crossing over to occur, leading to new allele combinations.   **15.4** In inversions and reciprocal translocations, the same genetic material is present in the same relative amount but just organized differently. In aneuploidy, duplications, deletions, and nonreciprocal translocations, the balance of genetic material is upset, as large segments are either missing or present in more than one copy. Apparently, this type of imbalance is very damaging to the organism. (Although it isn't lethal in the developing embryo, the reciprocal translocation that produces the Philadelphia chromosome can lead to a serious condition, cancer, by altering the expression of important genes.)   **15.5** In these cases, the sex of the parent contributing an allele affects the inheritance pattern. For imprinted genes, either the paternal or the maternal allele is expressed, depending on the imprint. For mitochondrial and chloroplast genes, only the maternal contribution will affect offspring phenotype because the offspring inherit these organelles from the mother, via the egg cytoplasm.

Test Your Understanding

1. 0; 1/2; 1/16   **2.** Recessive; if the disorder were dominant, it would affect at least one parent of a child born with the disorder. The disorder's inheritance is sex-linked because it is seen only in boys. For a girl to have the disorder, she would have to inherit recessive alleles from *both* parents. This would be very rare, since males with the recessive allele on their X chromosome die in their early teens.   **3.** 17%; yes, it is consistent. In Figure 15.9, the recombination frequency was also 17%. (You'd expect this to be the case since these are the very same two genes, and their distance from each other wouldn't change from one experiment to another.)

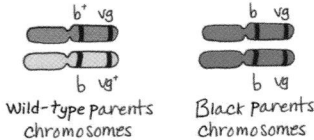

4. Between *T* and *A*, 12%; between *A* and *S*, 5%   **5.** Between *T* and *S*, 16%; sequence of genes is *T-A-S*   **6.** 6%; wild-type heterozygous for normal wings and red eyes × recessive homozygous for vestigial wings and purple eyes   **7.** Fifty percent of the offspring will show phenotypes resulting from crossovers. These results would be the same as those from a cross where *A* and *B* were *not* on the same chromosome, and you would interpret the results to mean that the genes are unlinked. (Further crosses involving other genes between *A* and *B* on the same chromosome would reveal genetic linkage and map distances.)
8. 450 each of blue/oval and white/round (parentals) and 50 each of blue/round and white/oval (recombinants)   **9.** About one-third of the distance along the

chromosome from the vestigial wing locus toward the brown eye locus
10. Because bananas are triploid, homologous pairs cannot line up during meiosis. Therefore, it is not possible to generate gametes that can fuse to produce a zygote with the triploid number of chromosomes.   **12.** (a) For each pair of genes, you had to generate an F₁ dihybrid fly; let's use the *A* and *B* genes as an example. You obtained homozygous parental flies, either the first with dominant alleles of the two genes (*AABB*) and the second with recessive alleles (*aabb*), or the first with dominant alleles of gene *A* and recessive alleles of gene *B* (*AAbb*) and the second with recessive alleles of gene *A* and dominant alleles of gene *B* (*aaBB*). Breeding either of these pairs of P generation flies gave you an F₁ dihybrid, which you then testcrossed with a doubly homozygous recessive fly (*aabb*). You classed the offspring as parental or recombinant, based on the genotypes of the P generation parents (either of the two pairs described above). You added up the number of recombinant types and then divided by the total number of offspring. This gave you the recombination percentage (in this case, 8%), which you can translate into map units (8 map units) to construct your map.

(b)

Chapter 16

Figure Questions
Figure 16.2 The living S cells found in the blood sample were able to reproduce to yield more S cells, indicating that the S trait is a permanent, heritable change, rather than just a one-time use of the dead S cells' capsules.   **Figure 16.4** The radioactivity would have been found in the pellet when proteins were labeled (batch 1) because proteins would have had to enter the bacterial cells to program them with genetic instructions. It's hard for us to imagine now, but the DNA might have played a structural role that allowed some of the proteins to be injected while it remained outside the bacterial cell (thus no radioactivity would be found in the pellet in batch 2).   **Figure 16.7** (1) The nucleotides in a single DNA strand are held together by covalent bonds between an oxygen on the 3′ carbon of one nucleotide and the phosphate group on the 5′ carbon of the next nucleotide in the chain. Instead of covalent bonds, the bonds that hold the two strands together are hydrogen bonds between a nitrogenous base on one strand and the complementary nitrogenous base on the other strand. (Hydrogen bonds are weaker than covalent bonds, but there are so many hydrogen bonds in a DNA double helix that, together, they are enough to hold the two strands together.) (2) One end, the 5′ end, has a phosphate group, which is attached to the 5′ carbon of the sugar, the one that is not in the ring. The other end, the 3′ end, has an —OH group attached to the 3′ carbon of the sugar; this carbon is in the ring. (3) The left diagram shows the most detail. It shows that each sugar-phosphate backbone is made up of sugars (blue pentagons) and phosphates (yellow circles) joined by covalent bonds (black lines). The middle diagram doesn't show any detail in the backbone. Both the left and middle diagrams label the bases and represent their complementarity by the complementary shapes at the ends of the bases (curves/indents for G/C or V's/notches for T/A). The diagram on the right is the least detailed, implying that the base pairs pair up, but showing all bases as the same shape so not including the information about specificity and complementarity visible in the other two diagrams. The left and right diagrams show that the strand on the left was synthesized most recently, as indicated by the light blue color. All three diagrams show the 5′ and 3′ ends of the strands.
Figure 16.12 The tube from the first replication would look the same, with a middle band of hybrid ¹⁵N–¹⁴N DNA, but the second tube would not have the upper band of DNA molecules made up of two light blue strands. Instead, it would have a bottom band of DNA made up of two dark blue strands, like the bottom band in the result predicted after one replication in the conservative model.   **Figure 16.13** In the bubble at the top of the micrograph in (b), arrows should be drawn pointing left and right to indicate the two replication forks.   **Figure 16.15**

Figure 16.18

Figure 16.19

Figure 16.24 The two members of a homologous pair (which would be the same color) would be associated tightly together at the metaphase plate during metaphase I of meiosis I. In metaphase of mitosis, however, each chromosome would be lined up individually, so the two chromosomes of the same color would be in different places at the metaphase plate.

Concept Check 16.1
1. In order to tell which end is the 5′ end, you need to know which end has a phosphate group on the 5′ carbon (the 5′ end) and/or which end has an —OH group on the 3′ carbon (the 3′ end).   **2.** Griffith expected that the mouse injected with the mixture of heat-killed S cells and living R cells would survive, since neither type of cell alone would kill the mouse.

Concept Check 16.2
1. Complementary base pairing ensures that the two daughter molecules are exact copies of the parental molecule. When the two strands of the parental molecule separate, each serves as a template on which nucleotides are arranged by the base-pairing rules; they will then be polymerized by enzymes into new complementary strands.   **2.** DNA pol III covalently adds nucleotides to new DNA strands and proofreads each added nucleotide for correct base pairing.   **3.** In the cell cycle, DNA synthesis occurs during the S phase, between the G₁ and G₂ phases of interphase. DNA replication is therefore complete before the mitotic phase begins.
4. Synthesis of the leading strand is initiated by an RNA primer, which must be removed and replaced with DNA, a task that could not be performed if the cell's DNA pol I were nonfunctional. In the overview box in Figure 16.18, just to the left of the top origin of replication, a functional DNA pol I would replace the RNA primer of the leading strand (shown in red) with DNA nucleotides (blue). The nucleotides would be added on to the 3′ end of the first Okazaki fragment of the upper lagging strand (the right half of the replication bubble).

Concept Check 16.3
1. A nucleosome is made up of eight histone proteins, two each of four different types, around which DNA is wound. Linker DNA runs from one nucleosome to the next.   **2.** The 10-nm fiber of euchromatin is less compacted during interphase than in mitosis and is accessible to the cellular proteins responsible for gene expression. In contrast, the 10-nm fiber of heterochromatin is relatively compacted (densely arranged) during interphase, and genes in heterochromatin are largely inaccessible to proteins necessary for gene expression.   **3.** The nuclear lamina is a netlike array of protein filaments that provides mechanical support just inside the nuclear envelope and thus maintains the shape of the nucleus. Considerable evidence also supports the existence of a nuclear matrix, a framework of protein fibers extending throughout the nuclear interior.

Summary of Key Concepts Questions

16.1 Each strand in the double helix has polarity; the end with a phosphate group on the 5′ carbon of the sugar is called the 5′ end, and the end with an —OH group on the 3′ carbon of the sugar is called the 3′ end. The two strands run in opposite directions, one running 5′ → 3′ and the other alongside it running 3′ → 5′. Thus, each end of the molecule has both a 5′ and a 3′ end, one on each strand of the double helix. This arrangement is called antiparallel. If the strands were parallel, they would both run 5′ → 3′ in the same direction, so an end of the molecule would have either two 5′ ends or two 3′ ends.   **16.2** On both the leading and lagging strands, DNA polymerase adds onto the 3′ end of an RNA primer synthesized by primase, synthesizing DNA in the 5′ → 3′ direction. Because the parental strands are antiparallel, however, only on the leading strand does synthesis proceed continuously into the replication fork. The lagging strand is synthesized bit by bit in the direction away from the fork as a series of shorter Okazaki fragments, which are later joined together by DNA ligase. Each fragment is initiated by synthesis of an RNA primer by primase as soon as a given stretch of single-stranded template strand is opened up. Although both strands are synthesized at the same rate, synthesis of the lagging strand is delayed because initiation of each fragment begins only when sufficient template strand is available.   **16.3** The chromatin in an interphase nucleus is present as the 10-nm fiber, either fairly loosely arranged in euchromatin or more densely arranged in heterochromatin (such as at the centromeres and telomeres). The euchromatin is also subdivided into larger compartments and smaller looped domains. This organization may reflect differences in gene expression occurring in these regions.

Test Your Understanding

1. C   **2.** C   **3.** B   **4.** D   **5.** A   **6.** D   **7.** B   **8.** A   **9.** Like histones, the *E. coli* proteins would be expected to contain many basic (positively charged) amino acids, such as lysine and arginine, which can form weak bonds with the negatively charged phosphate groups on the sugar-phosphate backbone of the DNA molecule.
11.

New DNA strand (olive)   Parental DNA strand (purple)

Sliding clamp   DNA pol III   Single-strand binding protein

Direction of replication

Chapter 17

Figure Questions

Figure 17.3 The previously presumed pathway would have been wrong. The new results would support this pathway: precursor → citrulline → ornithine → arginine. They would also indicate that class I mutants have a defect in the second step and class II mutants have a defect in the first step.   **Figure 17.5** The mRNA sequence (5′-UGGUUUGGCUCA-3′) is the same as the nontemplate DNA strand sequence (5′-TGGTTTGGCTCA-3′), except there is a U in the mRNA wherever there is a T in the DNA. The nontemplate strand is probably used to represent a DNA sequence because it so closely resembles the mRNA sequence, containing codons. (This is why it's called the coding strand.)   **Figure 17.6** Arg (or R)–Glu (or E)–Pro (or P)–Arg (or R)   **Figure 17.8** The processes are similar in that polymerases form polynucleotides complementary to an antiparallel DNA template strand. In replication, however, both strands act as templates, whereas in transcription, only one DNA strand acts as a template. This reflects, of course, the fact that replication results in a double-stranded product (DNA) while transcription results in a single-stranded product (an RNA).   **Figure 17.9** The RNA polymerase would bind directly to the promoter, rather than being dependent on the previous binding of transcription factors.
Figure 17.12

Start codon   Stop codon

5′ Cap
mRNA   Poly-A tail

5′ UTR   1–146   3′ UTR
Coding segment

Figure 17.16 The anticodon on the tRNA is 3′-AAG-5′, so it would bind to the mRNA codon 5′-UUC-3′. This codon codes for phenylalanine (Phe, or F), which is the amino acid this tRNA would carry.   **Figure 17.22** It would be packaged in a vesicle, transported to the Golgi apparatus for further processing, and then transported via a vesicle to the plasma membrane. The vesicle would fuse with the membrane, releasing the protein outside the cell.   **Figure 17.24** The mRNA farthest to the right (the longest one) started transcription first. The ribosome at the top, closest to the DNA, started translating first and thus has the longest polypeptide.

Concept Check 17.1

1. Recessive   **2.** A polypeptide made up of 10 Gly (glycine) amino acids
3.

Nontemplate sequence (from template sequence given in problem),
written 3′ → 5′:   3′-ACGACTGAA-5′

If above used as template, mRNA sequence: 5′-UGCUGACUU-3′

Translated: Cys-STOP

If the nontemplate sequence could have been used as a template for transcribing the mRNA, the protein translated from the mRNA would have a completely different amino acid sequence, so it would not be able to function as the original protein (translated from an mRNA transcribed from the template strand). It would also be shorter because of the UGA stop signal shown in the mRNA sequence above—and possibly others earlier in the mRNA sequence.

Concept Check 17.2

1. A promoter is the region of DNA to which RNA polymerase binds to begin transcription. It is at the upstream end of the gene (transcription unit).
2. In a bacterial cell, part of the RNA polymerase recognizes the gene's promoter and binds to it. In a eukaryotic cell, transcription factors must bind to the promoter first, then the RNA polymerase binds to them. In both cases, sequences in the promoter determine the precise binding of RNA polymerase so the enzyme is in the right location and orientation.   **3.** The transcription factor that recognizes the TATA sequence would be unable to bind, so RNA polymerase could not bind and transcription of that gene most likely would not occur.

Concept Check 17.3

1. Due to alternative splicing of exons, each gene can result in multiple different mRNAs and can thus direct synthesis of multiple different proteins.   **2.** In watching a pre-recorded show, you watch segments of the show itself (exons) and fast-forward through the commercials, which are thus like introns. However, unlike introns, commercials remain in the recording, while the introns are cut out of the RNA transcript during RNA processing.   **3.** Once the mRNA has exited the nucleus, the cap prevents it from being degraded by hydrolytic enzymes and facilitates its attachment to ribosomes. If the cap were removed from all mRNAs, the cell would no longer be able to synthesize any proteins and would probably die.

Concept Check 17.4

1. First, each aminoacyl-tRNA synthetase specifically recognizes a single amino acid and attaches it only to an appropriate tRNA. Second, a tRNA charged with its specific amino acid binds only to an mRNA codon for that amino acid.   **2.** A signal peptide on the leading end (amino end, or N-terminus) of the polypeptide being synthesized is recognized by a signal-recognition particle that brings the ribosome to the ER membrane. There, the ribosome attaches and continues to synthesize the polypeptide, depositing it in the ER lumen.   **3.** Because of wobble, the tRNA could bind to either 5′-GCA-3′ or 5′-GCG-3′, both of which code for alanine (Ala, or A). Alanine would be attached to the tRNA (see diagram, upper right).   **4.** When one ribosome terminates translation and dissociates, the two subunits would be very close to the cap. This could facilitate their rebinding and initiating synthesis of a new polypeptide, thus increasing the efficiency of translation.

tRNA   Ala

3′ CGU 5′

GCA   GCG
5′ ⊔⊔⊔ 3′   5′ ⊔⊔⊔ 3′

Concept Check 17.5

1. In the mRNA, the reading frame downstream from the deletion is shifted, leading to a long string of incorrect amino acids in the polypeptide, and in most cases, a stop codon will occur, leading to premature termination. The polypeptide will most likely be nonfunctional.   **2.** Heterozygous individuals, said to have sickle-cell trait, have a copy each of the wild-type allele and the sickle-cell allele. Both alleles will be expressed, so these individuals will have both normal and sickle-cell hemoglobin molecules. Apparently, having a mix of the two forms of β-globin has no effect under most conditions, but during prolonged periods of low blood oxygen (such as at higher altitudes), these individuals can show some signs of sickle-cell disease.
3.

Normal DNA sequence
(template strand is on top):   3′-TACTTGTCCGATATC-5′
                              5′-ATGAACAGGCTATAG-3′

mRNA sequence: 5′-AUGAACAGGCUAUAG-3′

Amino acid sequence: Met-Asn-Arg-Leu-STOP

Mutated DNA sequence
(template strand is on top):   3′-TACTTGTCCAATATC-5′
                              5′-ATGAACAGGTTATAG-3′

mRNA sequence: 5′-AUGAACAGGUUAUAG-3′

Amino acid sequence: Met-Asn-Arg-Leu-STOP

No effect: The amino acid sequence is Met-Asn-Arg-Leu both before and after the mutation because the mRNA codons 5′-CUA-3′ and 5′-UUA-3′ both code for Leu. (The fifth codon is a stop codon.)   **4.** A Cas9–guide RNA could be synthesized complementary to the mutated sequence, and the Cas9–guide RNA complex injected into a cell with the mutation. A region of double-stranded DNA with the correct sequence would also be provided. Cas9 would cut the mutated sequence, and the DNA repair system in the cell would repair the DNA, using the correct sequence as a template. Based on the answer to question 3, however, this would not be worthwhile because there is no amino acid change in the encoded protein, so the function of the protein encoded by the mutated gene is normal.

Summary of Key Concepts Questions

17.1 A gene contains genetic information in the form of a nucleotide sequence. The gene is first transcribed into an RNA molecule, and a messenger RNA molecule is ultimately translated into a polypeptide. The polypeptide makes up part or all of a protein, which performs a function in the cell and contributes to the phenotype of the organism.   **17.2** Both bacterial and eukaryotic genes have promoters, regions where RNA polymerase ultimately binds and begins transcription. In bacteria, RNA polymerase binds directly to the promoter; in eukaryotes, transcription factors bind first to the promoter, and then RNA polymerase binds to the transcription factors and promoter together.
17.3 Both the 5′ cap and the 3′ poly-A tail help the mRNA exit from the nucleus and then, in the cytoplasm, help ensure mRNA stability and allow it to bind to ribosomes.   **17.4** In the context of the ribosome, tRNAs function as translators between the nucleotide-based language of mRNA and the amino-acid-based language of polypeptides. A tRNA carries a specific amino acid, and the anticodon on the tRNA is complementary to the codon on the mRNA that codes for that amino acid. In the ribosome, the tRNA binds to the A site. Then the polypeptide being synthesized (currently on the tRNA in the P site) is joined to the new amino acid, which becomes the new (C-terminal) end of the polypeptide. Next, the tRNA in the A site moves to the P site. After the polypeptide is transferred to the new tRNA, thus adding the new amino acid, the now empty tRNA moves from the P site to the E site, where it exits the ribosome.   **17.5** When a nucleotide base is altered chemically, its base-pairing characteristics may be changed. When that happens, an incorrect nucleotide is likely to be incorporated into the complementary strand during the next replication of the DNA, and successive rounds of replication will perpetuate the mutation. Once the gene is transcribed, the mutated codon may code for a different amino acid that inhibits or changes the function of a protein. If, however, the chemical change in the base is detected and repaired by the DNA repair system before the next replication, no mutation will result.

Test Your Understanding

1. B   **2.** C   **3.** A   **4.** B   **5.** B   **6.** C   **7.** D   **8.** No. Transcription and translation are separated in space and time in a eukaryotic cell, as a result of the eukaryotic cell's nuclear membrane
9.

Type of RNA	Functions
Messenger RNA (mRNA)	Carries information specifying amino acid sequences of polypeptides from DNA to ribosomes
Transfer RNA (tRNA)	Serves as translator molecule in protein synthesis; translates mRNA codons into amino acids
Ribosomal RNA (rRNA)	In a ribosome, plays a structural role; as a ribozyme, plays a catalytic role (catalyzes peptide bond formation)
Primary transcript	Is a precursor to mRNA, rRNA, or tRNA, before being processed; some intron RNA acts as a ribozyme, catalyzing its own splicing
Small RNAs in spliceosome	Play structural and catalytic roles in spliceosomes, the complexes of protein and RNA that splice pre-mRNA

Chapter 18

Figure Questions

Figure 18.3 As the concentration of tryptophan in the cell falls, eventually there will be none bound to *trp* repressor molecules. These will then change into their inactive shapes and dissociate from the operator, allowing transcription of the operon to resume. The enzymes for tryptophan synthesis
will be made, and they will again synthesize tryptophan in the cell.
Figure 18.10 Each of the two polypeptides has two regions—one that makes up part of MyoD's DNA-binding domain and one that makes up part of MyoD's activation domain. Each functional domain in the complete MyoD protein is made up of parts of both polypeptides.   **Figure 18.12** In both types of cell, the albumin gene enhancer has the three control elements colored yellow, gray, and red. The sequences in the liver and lens cells would be identical, since the cells are in the same organism.   **Figure 18.18** Even if the mutant MyoD protein couldn't activate the *myoD* gene, it could still turn on genes for the other proteins in the pathway (other transcription factors, which would turn on the genes for muscle-specific proteins, for example). Therefore, some differentiation would occur. But unless there were other activators that could compensate for the loss of the MyoD protein's activation of the *myoD* gene, the cell would not be able to maintain its differentiated state.   **Figure 18.22** Normal Bicoid protein would be made in the anterior end and compensate for the presence of mutant *bicoid* mRNA put into the egg by the mother. Development should be normal, with a head present. (This is what was observed.)   **Figure 18.25** The mutation is likely to be recessive

because it is more likely to have an effect if both copies of the gene are mutated and code for nonfunctional proteins. If one normal copy of the gene is present, its product could inhibit the cell cycle. (However, there are also known cases of dominant *p53* mutations, and the HNPCC gene, discussed later, is a dominant mutation in a tumor-suppressor pathway.)   **Figure 18.27** Cancer is a disease in which cell division occurs without its usual regulation. Cell division can be stimulated by growth factors (see Figure 12.18), which bind to cell-surface receptors (see Figure 11.8). Cancer cells evade these normal controls and can often divide in the absence of growth factors (see Figure 12.19). This suggests that the receptor proteins or some other components in a signaling pathway are abnormal in some way (see, for example, the mutant Ras protein in Figure 18.24) or are expressed at abnormal levels, as seen for the receptors in this figure. Under some circumstances in the mammalian body, steroid hormones such as estrogen and progesterone can also promote cell division. These molecules also use cell-signaling pathways, as described in Concept 11.2 (see Figure 11.9). Because signaling receptors are involved in triggering cells to undergo cell division, it is not surprising that altered genes encoding these proteins might play a significant role in the development of cancer. Genes might be altered through either a mutation that changes the function of the protein product or a mutation that causes the gene to be expressed at abnormal levels that disrupt the overall regulation of the signaling pathway.

Concept Check 18.1

1. Binding by the *trp* corepressor (tryptophan) activates the *trp* repressor, which binds to the *trp* operator, shutting off transcription of the *trp* operon. Binding by the *lac* inducer (allolactose) inactivates the *lac* repressor, so that it can no longer bind to the *lac* operator, leading to transcription of the *lac* operon.   **2.** When glucose is scarce, cAMP is bound to CRP and CRP is bound to the *lac* promoter, favoring the binding of RNA polymerase. However, in the absence of lactose, the *lac* repressor is bound to the *lac* operator, blocking RNA polymerase from transcribing the *lac* operon genes.   **3.** The cell would continuously produce β-galactosidase and the two other enzymes for using lactose, even in the absence of lactose, thus wasting cell resources.

Concept Check 18.2

1. Histone acetylation is generally associated with gene expression, while DNA methylation is generally associated with lack of expression.   **2.** The same enzyme could not methylate both a histone and a DNA base. Enzymes are very specific in structure, and an enzyme that could methylate an amino acid of a protein would not be able to fit the base of a DNA nucleotide into the same active site.   **3.** General transcription factors function in assembling the transcription initiation complex at the promoters for all genes. Specific transcription factors bind to control elements associated with a particular gene and, once bound, either increase (activators) or decrease (repressors) transcription of that gene.   **4.** Regulation of translation initiation, degradation of the mRNA, activation of the protein (by chemical modification, for example), and protein degradation   **5.** The three genes should have some similar or identical sequences in the control elements of their enhancers. Because of this similarity, the same specific transcription factors that are present in muscle cells could bind to the enhancers of all three genes and stimulate their expression coordinately.

Concept Check 18.3

1. Both miRNAs and siRNAs are small, single-stranded RNAs that associate with a complex of proteins and then can base-pair with mRNAs that have a complementary sequence. This base pairing leads to either degradation of the mRNA or blockage of its translation. In some yeasts, siRNAs associated with proteins in a different complex can bind back to centromeric chromatin, recruiting enzymes that cause condensation of that chromatin into heterochromatin. Both miRNAs and siRNAs are processed from double-stranded RNA precursors but have subtle variations in the structure of those precursors.   **2.** The mRNA would persist and be translated into the cell division–promoting protein, and the cell would probably divide. If the intact miRNA is necessary for inhibition of cell division, then division of this cell might be inappropriate. Uncontrolled cell division could lead to formation of a mass of cells (tumor) that prevents proper functioning of the organism and could contribute to the development of cancer.   **3.** The *XIST* RNA is transcribed from the *XIST* gene on the X chromosome that will be inactivated. It then binds to that chromosome and induces heterochromatin formation. A likely model is that *XIST* RNA somehow recruits chromatin modification enzymes that lead to formation of heterochromatin.

Concept Check 18.4

1. Cells undergo differentiation during embryonic development, becoming different from each other. Therefore, the adult organism is made up of many highly specialized cell types that are different from each other.   **2.** By binding to a receptor on the receiving cell's surface and triggering a signal transduction pathway, involving intracellular molecules such as second messengers and transcription factors that affect gene expression   **3.** The products of maternal effect genes, made and deposited into the egg by the mother, determine the head and tail ends, as well as the back and belly, of the egg and embryo (and eventually the adult fly).   **4.** The lower cell is synthesizing signaling molecules because the gene encoding them is activated, meaning that the appropriate specific transcription factors are binding to the gene's enhancer. The genes encoding these specific transcription factors are also being expressed in this cell because the transcription factor activators that can turn them on were expressed in the precursor to this cell. A similar explanation also applies to the cells expressing the receptor proteins. This scenario began with specific cytoplasmic determinants localized in specific regions of the egg. These cytoplasmic determinants were distributed unevenly to daughter cells, resulting in cells going down different developmental pathways.

Concept Check 18.5

1. A cancer-causing mutation in a proto-oncogene usually makes the gene product overactive, whereas a cancer-causing mutation in a tumor-suppressor gene usually makes the gene product nonfunctional.   **2.** When an individual has inherited an oncogene or a mutant allele of a tumor-suppressor gene   **3.** Apoptosis is signaled by the p53 protein when a cell has extensive DNA damage, so apoptosis plays a protective role in eliminating a cell that might contribute to cancer. If mutations in the genes in the apoptotic pathway blocked apoptosis, a cell with such damage could continue to divide and might lead to tumor formation.

Summary of Key Concepts Questions

18.1 A corepressor and an inducer are both small molecules that bind to the repressor protein in an operon, causing the repressor to change shape. In the case of a corepressor (like tryptophan), this shape change allows the repressor to bind to the operator, blocking transcription. In contrast, an inducer causes the repressor to dissociate from the operator, allowing transcription to begin. **18.2** In that specific type of cell, the chromatin must not be tightly condensed because it must be accessible to transcription factors. The appropriate specific transcription factors (activators), which are made in that type of cell, must bind to the control elements in the enhancer of the gene, while repressors must not be bound. The DNA must be bent by a bending protein so the activators can contact the mediator proteins and form a complex with general transcription factors at the promoter. Then RNA polymerase must bind and begin transcription. **18.3** miRNAs do not "code" for the amino acids of a protein—they are never translated. Each miRNA associates with a group of proteins to form a complex. Binding of the complex to an mRNA with a complementary sequence causes that mRNA to be degraded or blocks its translation. This is considered gene regulation because it controls the amount of a particular mRNA that can be translated into a functional protein.   **18.4** The first process involves cytoplasmic determinants, including mRNAs and proteins, placed into specific locations in the egg by maternal cells. The embryonic cells that are formed from different regions of the egg during early cell divisions will have different proteins in them, which will direct different programs of gene expression. The second process involves the cell in question responding to signaling molecules secreted by neighboring cells (induction). The signaling pathway in the responding cell also leads to a different pattern of gene expression. The coordination of these two processes results in each cell following a unique pathway in the developing embryo. **18.5** The protein product of a proto-oncogene is usually involved in a pathway that stimulates cell division. The protein product of a tumor-suppressor gene is usually involved in a pathway that inhibits cell division.

Test Your Understanding

1. C   **2.** A   **3.** B   **4.** C   **5.** C   **6.** D   **7.** A   **8.** C   **9.** B   **10.** D
11. (a)

The purple, blue, and red activator proteins would be present.
(b)

Only gene 4 would be transcribed.
(c) In nerve cells, the yellow, blue, green, and black activators would have to be present, thus activating transcription of genes 1, 2, and 4. In skin cells, the red, black, purple, and blue activators would have to be present, thus activating genes 3 and 5.

Chapter 19

Figure Questions

Figure 19.2 Beijerinck might have concluded that the agent was a toxin produced by the plant that was able to pass through a filter but that became more and more dilute. In this case, he would have concluded that the infectious agent could not replicate.
Figure 19.4

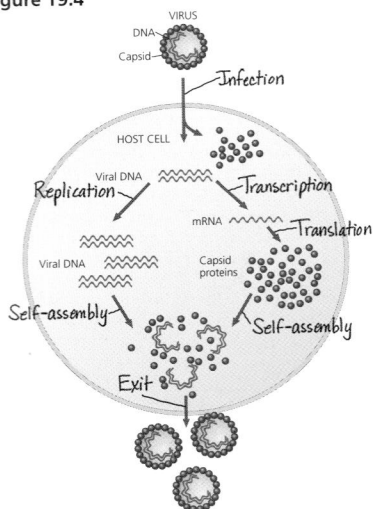

Figure 19.9 The main protein on the cell surface that HIV binds to is called CD4. However, HIV also requires a "co-receptor," which in many cases is a protein called CCR5. HIV binds to both of these proteins together and then is taken into the cell. Researchers discovered this requirement by studying individuals who seemed to be resistant to HIV infection despite multiple exposures. These individuals turned out to have mutations in the gene that encodes CCR5 such that the protein apparently cannot act as a co-receptor, and so HIV can't enter and infect cells.

Concept Check 19.1

1. Both viruses consist of RNA as the genetic material, associated with proteins. However, TMV consists of one molecule of RNA surrounded by a helical array of proteins, while the influenza virus has eight molecules of RNA, each associated with proteins and wound into a double helix. Another difference between the viruses is that the influenza virus has an outer envelope and TMV does not.   **2.** The T2 phages were an excellent choice for use in the Hershey-Chase experiment because they consist of only DNA surrounded by a protein coat, and DNA and protein were the two candidates for macromolecules that carried genetic information. Hershey and Chase were able to radioactively label each type of molecule alone and follow it during separate infections of E. coli cells with T2. Only the DNA entered the bacterial cell during infection, and only labeled DNA showed up in some of the progeny phage. Hershey and Chase concluded that the DNA must carry the genetic information necessary for the phage to reprogram the cell and produce progeny phages.

Concept Check 19.2

1. Lytic phages can only carry out lysis of the host cell, whereas lysogenic phages may either lyse the host cell or integrate into the host chromosome. In the latter case, the viral DNA (prophage) is simply replicated along with the host chromosome. Under certain conditions, a prophage may exit the host chromosome and initiate a lytic cycle.   **2.** Both the CRISPR-Cas system and miRNAs involve RNA molecules bound in a protein complex and acting as "homing devices" that enable the complex to bind a complementary sequence. However, miRNAs are involved in regulating gene expression (by affecting mRNAs) and the CRISPR-Cas system protects bacterial cells from foreign invaders—infecting phages. Thus the CRISPR-Cas system is more like an immune system than is the miRNA system.   **3.** Both the viral RNA polymerase and the RNA polymerase in Figure 17.10 synthesize an RNA molecule complementary to a template strand. However, the RNA polymerase in Figure 17.10 uses one of the strands of the DNA double helix as a template, whereas the viral RNA polymerase uses the RNA of the viral genome as a template.   **4.** HIV is called a retrovirus because it synthesizes DNA using its RNA genome as a template. This is the reverse ("retro") of the usual DNA → RNA information flow.   **5.** There are many steps that could be interfered with: binding of the virus to the cell, reverse transcriptase function, integration into the host cell chromosome, genome synthesis (in this case, transcription of RNA from the integrated provirus), assembly of the virus inside the cell, and budding of the virus. (Many of these, if not all, are targets of actual medical strategies to block progress of the infection in HIV-infected people.)

Concept Check 19.3

1. Mutations can lead to a new strain of a virus that can no longer be effectively fought by the immune system, even if an animal had been exposed to the original strain; a virus can jump from one species to a new host; and a rare virus can spread if a host population becomes less isolated.   **2.** In horizontal transmission, a plant is infected from an external source of virus, which enters through a break in the plant's epidermis due to damage by herbivores or other agents. In vertical transmission, a plant inherits viruses from its parent either via infected seeds (sexual reproduction) or via an infected cutting (asexual reproduction). **3.** Humans are not within the host range of TMV, so they can't be infected by the virus. (TMV can't bind to receptors on human cells and infect them.)

Summary of Key Concepts Questions

19.1 Viruses are generally considered nonliving because they are not capable of replicating outside of a host cell and are unable to carry out the energy-transforming reactions of metabolism. To replicate and carry out metabolism, they depend completely on host enzymes and resources. **19.2** Single-stranded RNA viruses require an RNA polymerase that can make RNA using an RNA template. (Cellular RNA polymerases make RNA using a DNA template.) Retroviruses require reverse transcriptases to make DNA using an RNA template. (Once the first DNA strand has been made, the same enzyme can promote synthesis of the second DNA strand.) **19.3** The mutation rate of RNA viruses is higher than that of DNA viruses because RNA polymerase has no proofreading function, so errors in replication are not corrected. Their higher mutation rate is one reason that RNA viruses change faster than DNA viruses, leading to their being able to have an altered host range and to evade immune defenses in possible hosts.

Test Your Understanding

1. C **2.** D **3.** C **4.** D **5.** B
6. As shown below, the viral genome would be translated into capsid proteins and envelope glycoproteins directly, rather than after a complementary RNA copy was made. A complementary RNA strand would still be made, however, that could be used as a template for many new copies of the viral genome.

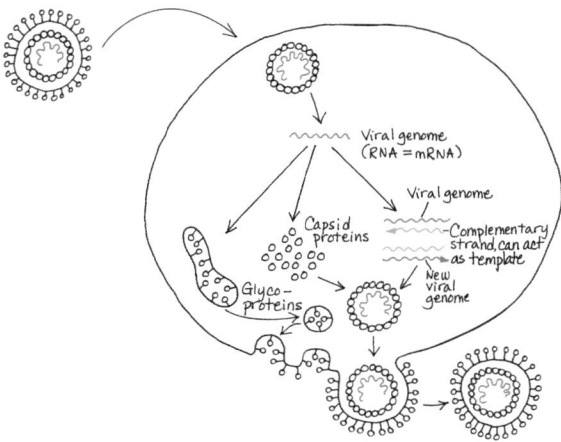

Chapter 20

Figure Questions
Figure 20.5

Figure 20.16 None of the eggs with the transplanted nuclei from the four-cell embryo at the upper left would have developed into a tadpole. Also, the result might include only some of the tissues of a tadpole, which might differ, depending on which nucleus was transplanted. (This assumes that there was some way to tell the four cells apart, as one can in some frog species.) **Figure 20.21** Using converted iPS cells would not carry the same risk, which is its major advantage. Because the donor cells would come from the patient, they would be perfectly matched. The patient's immune system would recognize them as "self" cells and would not mount an attack (which is what leads to rejection). On the other hand, cells that are rapidly dividing might carry a risk of inducing some type of tumor or contributing to development of cancer.

Concept Check 20.1
1. The covalent sugar-phosphate bonds of the DNA strands **2.** Yes, *Pvu*I will cut the molecule (at the position indicated by the dashed red line).

3. Some eukaryotic genes are too large to be incorporated into bacterial plasmids. Bacterial cells lack the means to process RNA transcripts into mRNA, and even if the need for RNA processing is avoided by using cDNA, bacteria lack enzymes to catalyze the post-translational processing that many eukaryotic proteins require to function properly. (This is often the case for human proteins, which are a focus of biotechnology.) **4.** During the replication of the ends of linear DNA molecules (see Figure 16.21), an RNA primer is used at the 5′ end of

each new strand. The RNA must be replaced by DNA nucleotides, but DNA polymerase is incapable of starting from scratch at the 5′ end of a new DNA strand. During PCR, the primers are made of DNA nucleotides already, so they don't need to be replaced—they just remain as part of each new strand. Therefore, there is no problem with end replication during PCR, and the fragments don't shorten with each replication.

Concept Check 20.2
1. Complementary base pairing is involved in cDNA synthesis, which is required for all three techniques: RT-PCR, DNA microarray analysis, and RNA sequencing. Reverse transcriptase uses mRNA as a template to synthesize the first strand of cDNA, adding nucleotides complementary to those on the mRNA. Complementary base pairing is also involved when DNA polymerase synthesizes the second strand of the cDNA. Furthermore, in RT-PCR, the primers must base-pair with their target sequences in the DNA mixture, locating one specific region among many. Also, the DNA polymerase (for example, Taq polymerase) used in PCR relies on complementary base pairing to the template strand to add new nucleotides during synthesis of the fragments. In DNA microarray analysis, the labeled cDNA probe binds only to the specific target sequence due to complementary nucleic acid hybridization (DNA-DNA hybridization). In RNA-seq, when sequencing the cDNAs, base complementarity plays a role in the sequencing process. **2.** As a researcher interested in how cancer develops, you would want to study genes represented by spots that are green or red because these are genes for which the expression level differs between the two types of tissue. Some of these genes may be expressed differently as a result of cancer, while others might play a role in causing cancer, so both would be of interest.

Concept Check 20.3
1. The state of chromatin modification in the nucleus from the intestinal cell was undoubtedly less similar to that of a nucleus from a fertilized egg, explaining why many fewer of these nuclei were able to be reprogrammed. In contrast, the chromatin in a nucleus from a cell at the four-cell stage would have been much more like that of a nucleus in a fertilized egg and therefore much more easily programmed to direct development. **2.** No, primarily because of subtle (and perhaps not so subtle) differences in the environment in which the clone develops and lives compared with that in which the original pet lived (see the differences noted in Figure 20.18). This does provoke ethical questions. To produce Dolly, also a mammal, several hundred embryos were cloned, but only one survived to adulthood. If any of the "reject" dog embryos survived to birth as defective dogs, would they be killed? Is it ethical to produce living animals that may be defective? You can probably think of other ethical issues as well. **3.** Given that muscle cell differentiation involves a master regulatory gene (*MyoD*), you might start by introducing either the MyoD protein or an expression vector carrying the *MyoD* gene into stem cells. (This is not likely to work, because the embryonic precursor cell in Figure 18.18 is more differentiated than the stem cells you are working with, and some other changes would need to be introduced as well. But it's a good way to start! And you may be able to think of others.)

Concept Check 20.4
1. Stem cells continue to reproduce themselves, ensuring that the corrective gene product will continue to be made. **2.** Herbicide resistance, pest resistance, disease resistance, drought resistance, and delayed ripening **3.** Because hepatitis A is an RNA virus, you could isolate RNA from the blood and try to detect copies of hepatitis A RNA by RT-PCR. You would first reverse-transcribe the blood mRNA into cDNA and then use PCR to amplify the cDNA, using primers specific to hepatitis A sequences. If you then ran the products on an electrophoretic gel, the presence of a band of the appropriate size would support your hypothesis. Alternatively, you could use RNA-seq to sequence all the RNAs in your patient's blood and see whether any of the sequences match up with that of hepatitis A. (Since you are only seeking one sequence, though, RT-PCR is probably a better choice.)

Summary of Key Concepts Questions
20.1 A plasmid vector and a source of foreign DNA to be cloned are both cut with the same restriction enzyme, generating restriction fragments with sticky ends. These fragments are mixed together, ligated, and reintroduced into bacterial cells. The plasmid has a gene for resistance to an antibiotic. That antibiotic is added to the host cells, and only cells that have taken up a plasmid will grow. (Another technique allows researchers to select only the cells that have a recombinant plasmid, rather than the original plasmid without an inserted gene.) **20.2** The genes that are expressed in a given tissue or cell type determine the proteins (and noncoding RNAs) that are the basis of the structure and functions of that tissue or cell type. Understanding which groups of interacting genes establish particular structures and carry out certain functions will help us learn how the parts of an organism work together. We will also be better able to treat diseases that occur when faulty gene expression leads to malfunctioning tissues. **20.3** (1) Cloning a mouse involves transplanting a nucleus from a differentiated mouse cell into a mouse egg cell that has had its own nucleus removed. Activating the egg cell and promoting its development into an embryo in a surrogate mother results in a mouse that is genetically identical to the mouse that donated the nucleus. In this case, the differentiated nucleus has been reprogrammed by factors in the egg cytoplasm. (2) Mouse ES cells are generated from inner cells in mouse blastocysts, so in this case the cells are "naturally" reprogrammed by the process of reproduction and development. (Cloned mouse embryos can also be used as a source of ES cells.) (3) iPS cells can be generated without the use of embryos from a differentiated adult mouse cell by adding certain transcription factors into the cell. In this case, the transcription factors are reprogramming the cells to become pluripotent. **20.4** First, the disease must be caused by a single gene, and the molecular basis of the problem must be understood. Second, the cells that are going to be introduced

into the patient must be cells that will integrate into body tissues and continue to multiply (and provide the needed gene product). Third, the gene must be able to be introduced into the cells in question in a safe way, as there have been instances of cancer resulting from some gene therapy trials. (Note that this will require testing the procedure in mice; moreover, the factors that determine a safe vector are not yet well understood. Maybe one of you will go on to solve this problem!)

Test Your Understanding

1. D **2.** B **3.** C **4.** B **5.** C **6.** A **7.** B **8.** You would use PCR to amplify the gene. This could be done from genomic DNA. Alternatively, mRNA could be isolated from lens cells and reverse-transcribed by reverse transcriptase to make cDNA. This cDNA could then be used for PCR. In either case, the gene would then be inserted into an expression vector so you could produce the protein and study it. **9.** Crossing over, which causes recombination, is a random event. The chance of crossing over occurring between two loci increases as the distance between them increases. If a SNP is located very close to a disease-associated allele, it is said to be genetically linked. Crossing over will rarely occur between the SNP and the allele, so the SNP can be used as a genetic marker indicating the presence of the particular allele.
10.

Chapter 21

Figure Questions

Figure 21.2 In step 2 of this figure, the order of the fragments relative to each other is not known and will be determined later by computer. The unordered nature of the fragments is reflected by their scattered arrangement in the diagram.   **Figure 21.8** The transposon would be cut out of the DNA at the original site rather than copied, so the figure would show the original stretch of DNA without the transposon after the mobile transposon had been cut out.
Figure 21.10 The RNA transcripts extending from the DNA in each transcription unit are shorter on the left and longer on the right. This means that RNA polymerase must be starting on the left end of the unit and moving toward the right.
Figure 21.13

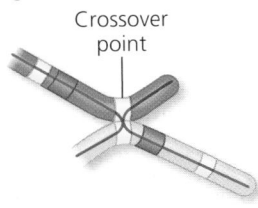

Crossover point

Figure 21.14 Pseudogenes are nonfunctional. They could have arisen by any mutations in the second copy that made the gene product unable to function. Examples are base changes that introduce stop codons in the sequence, alter amino acids, or change a region of the gene promoter so that the gene can no longer be expressed.   **Figure 21.15** At position 5, there is an R (arginine) in lysozyme and a K (lysine) in α-lactalbumin; both of these are basic amino acids.   **Figure 21.16** Let's say a transposable element (TE) existed in the intron to the left of the indicated EGF exon in the EGF gene, and the same TE was present in the intron to the right of the indicated F exon in the fibronectin gene. During meiotic recombination, these TEs could cause nonsister chromatids on homologous chromosomes to pair up incorrectly, as seen in Figure 21.13. One gene might end up with an F exon next to an EGF exon. Further mistakes in pairing over many generations might result in these two exons being separated from the rest of the gene and placed next to a single or duplicated K exon. In general, the presence of repeated sequences in introns and between genes facilitates these processes because it allows incorrect pairing of nonsister chromatids, leading to novel exon combinations.   **Figure 21.18** Since you know that chimpanzees do not speak but humans do, you'd probably want to know how many amino acid differences there are between the human wild-type FOXP2 protein and that of the chimpanzee and whether these changes affect the function of the protein. (As we explain later in the text, there are two amino acid differences.) You know that humans with mutations in this gene have severe language impairment. You would want to learn more about the human mutations by checking whether they affect the same amino acids in the gene product that the chimpanzee sequence differences affect. If so, those amino acids might play an important role in the function of the protein in language. Going further, you could analyze the differences between the chimpanzee and mouse FOXP2 proteins. You might ask: Are they more similar than the chimpanzee and human proteins? (It turns out that the chimpanzee and mouse proteins have only one amino acid difference and thus are more similar than the chimpanzee and human proteins, which have two differences, and also are more similar than the human and mouse proteins, which have three differences.)

Concept Check 21.1

1. In the whole-genome shotgun approach, short fragments are generated by cutting the genome with multiple restriction enzymes. These fragments are cloned, sequenced, and then ordered by computer programs that identify overlapping regions.

Concept Check 21.2

1. The Internet allows centralization of databases such as GenBank and software resources such as BLAST, making them freely accessible. Having all the data in a central database, easily accessible on the Internet, minimizes the possibility of errors and of researchers working with different data. It streamlines the process of science, since all researchers are able to use the same software programs, rather than each having to obtain their own, possibly different, software. It speeds up dissemination of data and ensures as much as possible that errors are corrected in a timely fashion. These are just a few answers; you can probably think of more.
2. Cancer is a disease caused by multiple factors. Focusing on a single gene or a single defect would mean ignoring other factors that may influence the cancer and even the behavior of the single gene being studied. The systems approach, because it takes into account many factors at the same time, is more likely to lead to an understanding of the causes and most useful treatments for cancer.
3. Some of the transcribed region is accounted for by introns. The rest is transcribed into noncoding RNAs, including small RNAs, such as microRNAs (miRNAs), siRNAs, and piRNAs. Some of these RNAs help regulate gene expression by blocking translation, binding to the promoter and repressing transcription, or causing remodeling of chromatin structure. The long noncoding RNAs (lncRNAs) may also contribute to gene regulation or to remodeling of chromatin structure.   **4.** Genome-wide association studies use the systems biology approach in that they consider the correlation of many single nucleotide polymorphisms (SNPs) with particular diseases, such as heart disease and diabetes, in an attempt to find patterns of SNPs that correlate with each disease.

Concept Check 21.3

1. Alternative splicing of RNA transcripts from a gene and post-translational processing of polypeptides   **2.** At the top of the web page, you can see the number of genomes completed and those considered permanent drafts in a bar graph by year. Scrolling down, you can see the number of complete and incomplete sequencing projects by year, the number of projects by domain by year (the genomes of viruses and metagenomes are counted too, even though these are not "domains"), the phylogenetic distribution of bacterial genome projects, and projects by sequencing center. Finally, near the bottom, you can see a pie chart of the "Project Relevance of Bacterial Genome Projects," which shows that about 58.6% have medical relevance. (This number may vary over time.) The web page ends with another pie chart showing the sequencing centers for archaeal and bacterial projects.   **3.** A prokaryotic cell is generally smaller than a eukaryotic cell, and prokaryotes reproduce by binary fission. The evolutionary process involved is natural selection for more quickly reproducing cells: The faster they can replicate their DNA and divide, the more likely they will be able to dominate a population of prokaryotes. The less DNA they have to replicate, then, the faster they will reproduce.

Concept Check 21.4

1. The number of genes is higher in mammals, and the amount of noncoding DNA is greater. Also, the presence of introns in mammalian genes makes them larger, on average, than prokaryotic genes.   **2.** The copy-and-paste transposon mechanism and retrotransposition   **3.** In the rRNA gene family, identical transcription units, each containing genes for all three different RNA products, are present in long arrays, repeated one after the other. The large number of copies of the rRNA genes enable organisms to produce the rRNA for enough ribosomes to carry out active protein synthesis, and the single transcription unit for the three rRNAs ensures that the relative amounts of the different rRNA molecules produced are correct—every time one rRNA is made, a copy of each of the other two is made as well. Rather than numerous identical units, each globin gene family consists of a relatively small number of nonidentical genes. The differences in the globin proteins encoded by these genes result in production of hemoglobin molecules adapted to particular developmental stages of the organism.   **4.** The exons would be classified as exons (1.5%); the enhancer region containing the distal control elements, the region closer to the promoter containing the proximal control elements, and the promoter itself would be classified as regulatory sequences (5%); and the introns would be classified as introns (20%).

Concept Check 21.5

1. If meiosis is faulty, two copies of the entire genome can end up in a single cell. Errors in crossing over during meiosis can lead to one segment being duplicated while another is deleted. During DNA replication, slippage backward along the template strand can result in segment duplication. Also, a "copy-and-paste" transposable element could copy and paste a segment of DNA, resulting in a duplication of that segment (and of the transposable element itself.)   **2.** For either gene, a mistake in crossing over during meiosis could have occurred between the two copies of that gene, such that one ended up with a duplicated exon. (The other copy would have ended up with a deleted exon.) This could have happened several times, resulting in the multiple copies of a particular exon in each gene.
3. Homologous transposable elements scattered throughout the genome provide sites where recombination can occur between different chromosomes. Movement of these elements into coding or regulatory sequences may change expression of genes, which can affect the phenotype in a way that is subject to natural selection.

Transposable elements also can carry genes with them, leading to dispersion of genes and in some cases different patterns of expression. Transport of an exon during transposition and its insertion into a gene may add a new functional domain to the originally encoded protein, a type of exon shuffling. (For any of these changes to be heritable, they must happen in germ cells, cells that will give rise to gametes.) **4.** Because more offspring are born to women who have this inversion, it must provide some advantage during the process of reproduction and development. Because proportionally more offspring have this inversion, we would expect it to persist and spread in the population. (In fact, evidence in the study allowed the researchers to conclude that it has been increasing in proportion in the population. You'll learn more about population genetics in the next unit.)

Concept Check 21.6

1. Because both humans and macaques are primates, their genomes are expected to be more similar than the macaque and mouse genomes are. The mouse lineage diverged from the primate lineage before the human and macaque lineages diverged.
2. Homeotic genes differ in their *non*homeobox sequences, which determine the interactions of homeotic gene products with other transcription factors and hence which genes are regulated by the homeotic genes. These nonhomeobox sequences differ in the two species, as do the expression patterns of the homeobox genes.
3. *Alu* elements underwent transposition more actively in the human genome for some reason. Their increased sites of insertion may have then allowed more recombination errors in the human genome, resulting in more or different duplications. The divergence of the organization and content of the two genomes presumably made the chromosomes of each genome less similar to those of the other, thus accelerating divergence of the two species by making matings less and less likely to result in fertile offspring due to the mismatch of genetic information.

Summary of Key Concepts Questions

21.1 One focus of the Human Genome Project was to improve sequencing technology in order to speed up the process. During the project, many advances in sequencing technology allowed faster reactions and detection of products, which were therefore less expensive. **21.2** The most significant finding is that more than 75% of the human genome appears to be transcribed at some point in at least one of the cell types studied. Also, at least 80% of the genome contains an element that is functional, participating in gene regulation or maintaining chromatin structure in some way. The project was expanded to include other species to further investigate the functions of these transcribed DNA elements. It is necessary to carry out this type of analysis on the genomes of species that can be used in laboratory experiments. **21.3** (a) In general, bacteria and archaea have smaller genomes, lower numbers of genes, and higher gene density than eukaryotes. (b) Among eukaryotes, there is no apparent systematic relationship between genome size and phenotype. The number of genes is often lower than would be expected from the size of the genome—in other words, the gene density is often lower in larger genomes. (Humans are a good example.) **21.4** Transposable element–related sequences can move from place to place in the genome, and some of these sequences make a new copy of themselves when they do so. Thus, it is not surprising that they make up a significant percentage of the genome, and this percentage might be expected to increase over evolutionary time. **21.5** Chromosomal rearrangements within a species lead to some individuals having different chromosomal arrangements. Each of these individuals could still undergo meiosis and produce gametes, and fertilization involving gametes with different chromosomal arrangements could result in viable offspring. However, during meiosis in the offspring, the maternal and paternal chromosomes might not be able to pair up, causing gametes with incomplete sets of chromosomes to form. Most often, when zygotes are produced from such gametes, they do not survive. Ultimately, a new species could form if two different chromosomal arrangements became prevalent within a population and individuals could mate successfully only with other individuals having the same arrangement. **21.6** Comparing the genomes of two closely related species can reveal information about more recent evolutionary events, perhaps events that resulted in the distinguishing characteristics of the two species. Comparing the genomes of very distantly related species can tell us about evolutionary events that occurred a very long time ago. For example, genes that are shared between two distantly related species must have arisen before the two species diverged.

Test Your Understanding

1. B **2.** A **3.** C **4.** Answers for (a) through (c):

(Chimpanzee)	PKSSD ... TSSTT ... NARRD
Mouse	PKSS**E** ... TSS**TT** ... **N**ARRD
(Gorilla)	PKSSD ... TSSTT ... NARRD
Human	PKSSD ... TSS**N**T ... **S**ARRD
(Rhesus monkey)	PKSSD ... TSSTT ... NARRD

(d) There is one difference between the sequence for the mouse and the sequence for the chimpanzee, gorilla, and rhesus monkey. There are two differences between the human sequence and the sequence for the chimpanzee, gorilla, and rhesus monkey. These facts might lead to the hypothesis that the *FOXP2* gene has been evolving faster in the human lineage than in other primates: Two differences between humans and other primates occurred during the 6 million years since they diverged, but only one difference occurred during the much longer period of 65 million years since rodents and primates diverged. However, as described in the text, later analysis that included more genomes that were more diverse failed to support this hypothesis.

Chapter 22

Figure Questions

Figure 22.6 You should have circled the branch located at the far left of Figure 1.20. Although three of the descendants (*Certhidea olivacea, Camarhynchus pallidus,* and *Camarhynchus parvulus*) of this common ancestor ate insects, the other three species that descended from this ancestor did not eat insects. **Figure 22.8** The common ancestor lived about 5.5 million years ago. **Figure 22.13** These results show that being reared from the egg stage on one plant species or the other did not result in the adult having a beak length appropriate for that host; instead, adult beak lengths were determined primarily by the population from which the eggs were obtained. Because an egg from a balloon vine population likely had long-beaked parents, while an egg from a golden rain tree population likely had short-beaked parents, these results indicate that beak length is an inherited trait. **Figure 22.14** Both strategies should increase the time that it takes *S. aureus* to become resistant to a new drug. If a drug that harms *S. aureus* does not harm other bacteria, natural selection will not favor resistance to that drug in the other species. This would decrease the chance that *S. aureus* would acquire resistance genes from other bacteria—thus slowing the evolution of resistance. Similarly, selection for resistance to a drug that slows the growth but does not kill *S. aureus* is much weaker than selection for resistance to a drug that kills *S. aureus*—again slowing the evolution of resistance. **Figure 22.17** Based on this evolutionary tree, crocodiles are more closely related to birds than to lizards because they share a more recent common ancestor with birds (ancestor ⑤) than with lizards (ancestor ④). **Figure 22.20** Hind limb structure changed first. *Rodhocetus* lacked flukes, but its pelvic bones and hind limbs had changed substantially from how those bones were shaped and arranged in *Pakicetus*. For example, in *Rodhocetus*, the pelvis and hind limbs appear to be oriented for paddling, whereas they were oriented for walking in *Pakicetus*.

Concept Check 22.1

1. Hutton and Lyell proposed that geologic events in the past were caused by the same processes operating today, at the same gradual rate. This principle suggested that Earth must be much older than a few thousand years, the age that was widely accepted in the early 19th century. Hutton's and Lyell's ideas also stimulated Darwin to reason that the slow accumulation of small changes could ultimately produce the profound changes documented in the fossil record. In this context, the age of Earth was important to Darwin, because unless Earth was very old, he could not envision how there would have been enough time for evolution to occur. **2.** By this criterion, Cuvier's explanation of the fossil record and Lamarck's hypothesis of evolution are both scientific. Cuvier thought that species did not evolve over time. He also suggested that sudden, catastrophic events caused extinctions in particular areas and that such regions were later repopulated by a different set of species that immigrated from other areas. These assertions can be tested against the fossil record. Lamarck's principle of use and disuse can be used to make testable predictions for fossils of groups such as whale ancestors as they adapted to a new habitat. Lamarck's principle of use and disuse and his associated principle of the inheritance of acquired characteristics can also be tested directly in living organisms.

Concept Check 22.2

1. Organisms share characteristics (the unity of life) because they share common ancestors; the great diversity of life occurs because new species have repeatedly formed when descendant organisms gradually adapted to different environments, thereby becoming different from their ancestors. **2.** The fossil mammal species (or its ancestors) would most likely have colonized the Andes from within South America, whereas ancestors of mammals currently found in Asian mountains would most likely have colonized those mountains from other parts of Asia. As a result, the Andes fossil species would share a more recent common ancestor with South American mammals than with mammals in Asia. Thus, for many of its traits, the fossil mammal species would probably more closely resemble mammals that live in South American jungles than mammals that live on Asian mountains. It is also possible that for some of its features, the Andean fossil mammal species could closely resemble a mammal from the mountains of Asia. This could happen because similar environments had selected for similar adaptations (even though the fossil and Asian species were only distantly related to one another). **3.** As long as the white phenotype (encoded by the genotype *pp*) continues to be favored by natural selection, the proportion of white individuals in the population should increase over time relative to the proportion of purple individuals (encoded by the genotypes *PP* and *Pp*). As a result, the frequency of the *p* allele in the population would likely increase over time.

Concept Check 22.3

1. An environmental factor such as a drug does not create new traits, such as drug resistance, but rather selects for traits among those that are already present in the population. **2.** (a) Despite their different functions, the forelimbs of different mammals are structurally similar because they all represent modifications of a structure found in the common ancestor; thus, they are homologous structures. (b) In this case, the similar features of these mammals represent analogous features that arose by convergent evolution. The similarities between the sugar glider and flying squirrel indicate that similar environments selected for similar adaptations despite different ancestry. **3.** At the time that dinosaurs originated, Earth's landmasses formed a single large continent, Pangaea. Because many dinosaurs were large and mobile, it is likely that early members of these groups lived on many different parts of Pangaea. When Pangaea broke apart, fossils of these organisms would have moved with the rocks in which they were deposited. As a result, we would predict that fossils of early dinosaurs would have a broad geographic distribution (this prediction has been upheld).

Summary of Key Concepts Questions

Concept 22.1 Darwin thought that descent with modification occurred as a gradual, steplike process. The age of Earth was important to him because if Earth were only a few thousand years old (as conventional wisdom suggested), there wouldn't have been sufficient time for major evolutionary change. **Concept 22.2** All species have the potential to overreproduce—that is, to produce more offspring than can be supported by the environment. This ensures that there will be what Darwin called a "struggle for existence" in which many of the offspring are eaten, starved, diseased, or unable to reproduce for a variety of other reasons. Members of a population exhibit a range of heritable variations, some of which make it likely that their bearers will leave more offspring than other individuals (for example, the bearer may escape predators more effectively or be more tolerant of the physical conditions of the environment). Over time, natural selection resulting from factors such as predators, lack of food, or the physical conditions of the environment can increase the proportion of individuals with favorable traits in a population (evolutionary adaptation). **Concept 22.3** The hypothesis that cetaceans originated from a terrestrial mammal and are closely related to even-toed ungulates is supported by several lines of evidence. For example, fossils document that early cetaceans had hind limbs, as expected for organisms that descended from a land mammal; these fossils also show that cetacean hind limbs became reduced over time. Other fossils show that early cetaceans had a type of ankle bone that is otherwise found only in even-toed ungulates, providing strong evidence that even-toed ungulates are the land mammals to which cetaceans are most closely related. DNA sequence data also indicate that even-toed ungulates are the land mammals to which cetaceans are, most closely related.

Test Your Understanding

1. C **2.** D **3.** C **4.** A **5.** B

7. (a)

(b) The rapid rise in the percentage of mosquitoes resistant to DDT was most likely caused by natural selection in which mosquitoes resistant to DDT could survive and reproduce while other mosquitoes could not. (c) In India—where DDT resistance first appeared—natural selection would have caused the frequency of resistant mosquitoes to increase over time. If resistant mosquitoes then migrated from India (for example, transported by wind or in planes, trains, or ships) to other parts of the world, the frequency of DDT resistance would increase there as well. In addition, if resistance to DDT were to arise independently in mosquito populations outside of India, those populations would also experience an increase in the frequency of DDT resistance.

Chapter 23

Figure Questions

Figure 23.4 The genetic code is redundant, meaning that more than one codon can specify the same amino acid. As a result, a substitution at a particular site in a coding region of the *Adh* gene might change the codon but not the translated amino acid, and thus not the resulting protein encoded by the gene. One way an insertion in an exon would not affect the gene produced is if it occurs in an untranslated region of the exon. (This is the case for the insertion at location 1,703.) **Figure 23.7** There should be 24 red balls. **Figure 23.8** The predicted frequencies are 36% $C^R C^R$, 48% $C^R C^W$, and 16% $C^W C^W$. **Figure 23.9** Overall, by chance the frequency of the C^W allele first increases in generation 2 and then falls to zero in generation 3—causing the C^R allele to become fixed (reach a frequency of 100%). **Figure 23.12** The frequency of banded color patterns in island populations would probably increase in the year immediately after the storm. Since mainland populations did not decline in size, the number of individuals migrating from the mainland to the islands would probably not decline either. As a result, after island populations had decreased in size (because of the storm), alleles encoding banded coloration that were transferred from the mainland would comprise a larger proportion of the gene pool in island populations. This would cause the frequency of banded color patterns in island populations to increase. **Figure 23.13** Directional selection. The seeds of golden rain tree are buried less deeply than are the seeds of the native host, balloon vine. Thus, in soapberry bug populations feeding on golden rain tree, bugs with shorter beaks had an advantage, resulting in directional selection for shorter beak length. **Figure 23.16** Crossing a single female's eggs with both an SC and an LC male's sperm allowed the researchers to directly compare the effects of the males' contribution to the next generation since both batches of offspring had the same maternal contribution. This isolation of the male's impact enabled researchers to draw conclusions about differences in genetic "quality" between the SC and LC males. **Figure 23.18** Under prolonged low-oxygen conditions, some of the red blood cells of a heterozygote may sickle, leading to harmful effects. This does not occur in individuals with two wild-type hemoglobin alleles, suggesting that there may be selection against heterozygotes in malaria-free regions (where heterozygote advantage does not occur). However, since heterozygotes are healthy under most conditions, selection against them is unlikely to be strong.

Concept Check 23.1

1. Within a population, genetic differences among individuals provide the raw material on which natural selection and other mechanisms can act. Without such differences, allele frequencies could not change over time—and hence the population could not evolve. **2.** Many mutations occur in somatic cells, which do not produce gametes and so are lost when the organism dies. Of the mutations that do occur in cell lines that produce gametes, many do not have a phenotypic effect on which natural selection can act. Others have a harmful effect and are thus unlikely to increase in frequency because they decrease the reproductive success of their bearers. **3.** Its genetic variation (whether measured at the level of the gene or at the level of nucleotide sequences) would probably drop over time. During meiosis, crossing over and the independent assortment of chromosomes produce many new combinations of alleles. In addition, a population contains a vast number of possible mating combinations, and fertilization brings together the gametes of individuals with different genetic backgrounds. Thus, via crossing over, independent assortment of chromosomes, and fertilization, sexual reproduction reshuffles alleles into fresh combinations each generation. Without sexual reproduction, the rate of forming new combinations of alleles would be vastly reduced, causing the overall amount of genetic variation to drop.

Concept Check 23.2

1. There are 700 individuals in the population: 85 of genotype *AA*, 320 of genotype *Aa*, and 295 of genotype *aa*. The genotype frequencies are thus 0.12 (85/700) for genotype *AA*, 0.46 (320/700) for genotype *Aa*, and 0.42 (295/700) for genotype *aa*. Each individual has two alleles, so the total number of alleles is 1,400. To calculate the frequency of allele *A*, note that each of the 85 individuals of genotype *AA* has two *A* alleles, each of the 320 individuals of genotype *Aa* has one *A* allele, and each of the 295 individuals of genotype *aa* has zero *A* alleles. Thus, the frequency (*p*) of allele *A* is

$$p = \frac{(2 \times 85) + (1 \times 320) + (0 \times 295)}{1,400} = 0.35$$

There are only two alleles (*A* and *a*) in our population, so the frequency of allele *a* must be $q = 1 - p = 0.65$. **2.** Because the frequency of allele *a* is 0.45, the frequency of allele *A* must be 0.55. Thus, the expected genotype frequencies are $p^2 = 0.3025$ for genotype *AA*, $2pq = 0.495$ for genotype *Aa*, and $q^2 = 0.2025$ for genotype *aa*. **3.** There are 120 individuals in the population, so there are 240 alleles. Of these, there are 124 *V* alleles—32 from the 16 *VV* individuals and 92 from the 92 *Vv* individuals. Thus, the frequency of the *V* allele is $p = 124/240 = 0.52$; hence, the frequency of the *v* allele is $q = 0.48$. Based on the Hardy-Weinberg equation, if the population were not evolving, the frequency of genotype *VV* should be $p^2 = 0.52 \times 0.52 = 0.27$; the frequency of genotype *Vv* should be $2pq = 2 \times 0.52 \times 0.48 = 0.5$; and the frequency of genotype *vv* should be $q^2 = 0.48 \times 0.48 = 0.23$. In a population of 120 individuals, these expected genotype frequencies lead us to predict that there would be 32 *VV* individuals (0.27×120), 60 *Vv* individuals (0.5×120), and 28 *vv* individuals (0.23×120). The actual numbers for the population (16 *VV*, 92 *Vv*, 12 *vv*) deviate from these expectations (fewer homozygotes and more heterozygotes than expected). This indicates that the population is not in Hardy-Weinberg equilibrium and hence may be evolving at this locus.

Concept Check 23.3

1. Natural selection is more "predictable" in that it alters allele frequencies in a nonrandom way: It tends to increase the frequency of alleles that increase the organism's reproductive success in its environment and decrease the frequency of alleles that decrease the organism's reproductive success. Alleles subject to genetic drift increase or decrease in frequency by chance alone, whether or not they are advantageous. **2.** Genetic drift results from chance events that cause allele frequencies to fluctuate at random from generation to generation; within a population, this process tends to decrease genetic variation over time. Gene flow is the transfer of alleles between populations, a process that can introduce new alleles to a population and hence may increase its genetic variation (albeit slightly, since rates of gene flow are often low). **3.** Selection is not important at this locus; furthermore, the populations are not small, and hence the effects of genetic drift should not be pronounced. Gene flow is occurring via the movement of pollen and seeds. Thus, allele and genotype frequencies in these populations should become more similar over time as a result of gene flow.

Concept Check 23.4

1. The relative fitness of a mule is zero, because fitness includes reproductive contribution to the next generation, and a sterile mule cannot produce offspring. **2.** Although both gene flow and genetic drift can increase the frequency of advantageous alleles in a population, they can also decrease the frequency of advantageous alleles or increase the frequency of harmful alleles. Only natural selection *consistently* results in an increase in the frequency of alleles that enhance survival or reproduction. Thus, natural selection is the only mechanism that consistently leads to adaptive evolution. **3.** The three modes of natural selection (directional, stabilizing, and disruptive) are defined in terms of the selective advantage of different *phenotypes*, not different genotypes. Thus, the type of selection represented by heterozygote advantage depends on the phenotype of the heterozygote. In this question, because heterozygous individuals have a more extreme phenotype than either homozygote, heterozygote advantage represents directional selection.

Summary of Key Concepts Questions

23.1 Much of the nucleotide variability at a genetic locus occurs within introns. Nucleotide variation at these sites typically does not affect the phenotype because introns do not code for the protein product of the gene. (Note: In certain circumstances, it is possible that a change in an intron could affect RNA splicing and ultimately have some phenotypic effect on the organism, but such mechanisms are not covered in this introductory text.) There are also many variable nucleotide sites within exons. However, most of the variable sites within exons reflect changes to the DNA sequence that do not change the sequence of amino acids encoded by the gene (and hence may not affect the phenotype). **23.2** No, this is not an example of circular reasoning. Calculating p and q from observed genotype frequencies does not imply that those genotype frequencies must be in Hardy-Weinberg equilibrium. For example, consider a population that has 195 individuals of genotype AA, 10 of genotype Aa, and 195 of genotype aa. Calculating p and q from these values yields $p = q = 0.5$. Using the Hardy-Weinberg equation, the predicted equilibrium frequencies are $p^2 = 0.25$ for genotype AA, $2pq = 0.5$ for genotype Aa, and $q^2 = 0.25$ for genotype aa. Since there are 400 individuals in the population, these predicted genotype frequencies indicate that there should be 100 AA individuals, 200 Aa individuals, and 100 aa individuals—numbers that differ greatly from the values that we used to calculate p and q. **23.3** It is unlikely that two such populations would evolve in similar ways. Since their environments are very different, the alleles favored by natural selection would probably differ between the two populations. Although genetic drift may have important effects in each of these small populations, drift causes unpredictable changes in allele frequencies, so it is unlikely that drift would cause the populations to evolve in similar ways. Both populations are geographically isolated, suggesting that little gene flow would occur between them (again making it less likely that they would evolve in similar ways). **23.4** Compared to males, it is likely that the females of such species would be larger, more colorful, endowed with more elaborate ornamentation (for example, a large morphological feature such as the peacock's tail), and more apt to engage in behaviors intended to attract mates or prevent other members of their sex from obtaining mates.

Test Your Understanding
1. D **2.** C **3.** B **4.** A **5.** C

Chapter 24

Figure Questions
Figure 24.7 If this had not been done, the strong preference of "starch flies" and "maltose flies" to mate with like-adapted flies could have occurred simply because the flies could detect (for example, by sense of smell) what their potential mates had eaten as larvae—and preferred to mate with flies that had a similar smell to their own. **Figure 24.11** *Tragopogon dubius* and *T. pratensis* are the parent species of the polyploid species *T. miscellus*. *T. dubius* and *T. porrifolius* are the parent species of the other polyploid species, *T. mirus*. **Figure 24.12** In murky waters where females distinguish colors poorly, females of each species might mate often with males of the other species. Hence, since hybrids between these species are viable and fertile, the gene pools of the two species might become more similar over time. **Figure 24.13** The graph indicates that there has been gene flow of some fire-bellied toad alleles into the range of the yellow-bellied toad. Otherwise, all individuals located to the left of the hybrid zone portion of the graph would have allele frequencies equal to 1. **Figure 24.15** Because the populations had only just begun to diverge from one another at this point in the process, it is likely that any existing barriers to reproduction would weaken over time. **Figure 24.19** Over time, the chromosomes of the experimental hybrids came to resemble those of *H. anomalus*. This occurred even though conditions in the laboratory differed greatly from conditions in the field, where *H. anomalus* is found, suggesting that selection for laboratory conditions was not strong. Thus, it is unlikely that the observed rise in the fertility of the experimental hybrids was due to selection for life under laboratory conditions. **Figure 24.20** The presence of *M. cardinalis* plants that carry the *M. lewisii yup* allele would make it more likely that bumblebees would transfer pollen between the two monkey flower species. As a result, we would expect the number of hybrid offspring to increase.

Concept Check 24.1
1. (a) All except the biological species concept can be applied to both asexual and sexual species because they define species on the basis of characteristics other than the ability to reproduce. In contrast, the biological species concept can be applied only to sexual species. (b) The easiest species concept to apply in the field would be the morphological species concept because it is based only on the appearance of the organism. Additional information about its ecological habits or reproduction is not required. **2.** Because these birds live in fairly similar environments and can breed successfully in captivity, the reproductive barrier in nature is probably prezygotic; given the species' differences in habitat preference, this barrier could result from habitat isolation.

Concept Check 24.2
1. In allopatric speciation, a new species forms while in geographic isolation from its parent species; in sympatric speciation, a new species forms in the absence of geographic isolation. Geographic isolation greatly reduces gene flow between populations, whereas ongoing gene flow is more likely in sympatric populations. As a result, allopatric speciation is more common than sympatric speciation. **2.** Gene flow between subsets of a population that live in the same area can be reduced in a variety of ways. In some species—especially plants—changes in chromosome number can block gene flow and establish reproductive isolation in a single generation. Gene flow can also be reduced in sympatric populations by habitat differentiation (as seen in the apple maggot fly, *Rhagoletis*) and sexual selection (as seen in Lake Victoria cichlids). **3.** Allopatric speciation would be less likely to occur on an island near a mainland than on a more isolated island of the same size. We expect this result because continued gene flow between mainland populations and those on a nearby island reduces the chance that enough genetic divergence will take place for allopatric speciation to occur. **4.** If all of the homologs failed to separate during anaphase I of meiosis, some gametes would end up with an extra set of chromosomes (and others would end

up with no chromosomes). If a gamete with an extra set of chromosomes fused with a normal gamete, a triploid would result; if two gametes with an extra set of chromosomes fused with each other, a tetraploid would result.

Concept Check 24.3
1. Hybrid zones are regions in which members of different species meet and mate, producing some offspring of mixed ancestry. Such regions can be viewed as "natural laboratories" in which to study speciation because scientists can directly observe factors that cause (or fail to cause) reproductive isolation. **2.** (a) If hybrids consistently survived and reproduced poorly compared with the offspring of intraspecific matings, reinforcement could occur. If it did, natural selection would cause prezygotic barriers to reproduction between the parent species to strengthen over time, decreasing the production of unfit hybrids and leading to a completion of the speciation process. If reinforcement did not occur, hybrids may continue to be produced even though they are selected against (as in the *Bombina* hybrid zone). (b) If hybrid offspring survived and reproduced as well as the offspring of intraspecific matings, indiscriminate mating between the parent species would lead to the production of large numbers of hybrid offspring. As these hybrids mated with each other and with members of both parent species, the gene pools of the parent species could fuse over time, reversing the speciation process.

Concept Check 24.4
1. The time between speciation events includes (1) the length of time that it takes for populations of a newly formed species to begin diverging reproductively from one another and (2) the time it takes for speciation to be complete once this divergence begins. Although speciation can occur rapidly once populations have begun to diverge from one another, it may take millions of years for that divergence to begin. **2.** Investigators transferred alleles at the *yup* locus (which influences flower color) from each parent species to the other. *M. lewisii* plants with an *M. cardinalis yup* allele received many more visits from hummingbirds than usual; hummingbirds usually pollinate *M. cardinalis* but avoid *M. lewisii*. Similarly, *M. cardinalis* plants with an *M. lewisii yup* allele received many more visits from bumblebees than usual; bumblebees usually pollinate *M. lewisii* and avoid *M. cardinalis*. Thus, alleles at the *yup* locus can influence pollinator choice, which in these species provides the primary barrier to interspecific mating. Nevertheless, the experiment does not prove that the *yup* locus alone controls barriers to reproduction between *M. lewisii* and *M. cardinalis*; other genes might enhance the effect of the *yup* locus (by modifying flower color) or cause entirely different barriers to reproduction (for example, gametic isolation or a postzygotic barrier). **3.** Crossing over. If crossing over did not occur, each chromosome in an experimental hybrid would remain as in the F_1 generation: composed entirely of DNA from one parent species or the other.

Summary of Key Concepts Questions
24.1 According to the biological species concept, a species is a group of populations whose members interbreed and produce viable, fertile offspring; thus, gene flow occurs between populations of a species. In contrast, members of different species do not interbreed and hence no gene flow occurs between their populations. Overall, then, in the biological species concept, species can be viewed as designated by the *absence* of gene flow—making gene flow of central importance to the biological species concept. **24.2** Sympatric speciation can be promoted by factors such as polyploidy, sexual selection, and habitat shifts, all of which can reduce gene flow between the subpopulations of a larger population. Of these factors, sexual selection and habitat shifts can also occur in allopatric populations and hence can also promote allopatric speciation. **24.3** If the hybrids are selected against, the hybrid zone could persist if individuals from the parent species regularly travel into the zone, where they mate to produce hybrid offspring. If hybrids are not selected against, there is no cost to the continued production of hybrids, and large numbers of hybrid offspring may be produced. However, natural selection for life in different environments may keep the gene pools of the two parent species distinct, thus preventing the loss (by fusion) of the parent species and once again causing the hybrid zone to be stable over time. **24.4** As the goatsbeard plant, Bahamas mosquitofish, and apple maggot fly illustrate, speciation continues to happen today. A new species can begin to form whenever gene flow is reduced between populations of the parent species. Such reductions in gene flow can occur in many ways: A new, geographically isolated population may be founded by a few colonists; some members of the parent species may begin to utilize a new habitat; or sexual selection may isolate formerly connected populations or subpopulations. These and many other such events are happening today.

Test Your Understanding
1. B **2.** C **3.** B **4.** A **5.** D **6.** C
8. Here is one possibility:

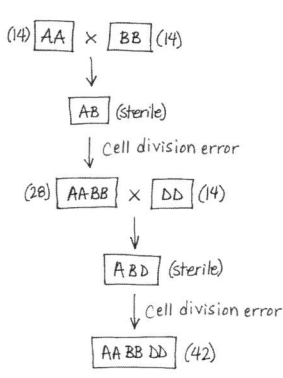

Chapter 25

Figure Questions

Figure 25.2 Proteins are almost always composed of the same 20 amino acids shown in Figure 5.14. However, many other amino acids could potentially form in this or any other experiment. For example, any molecule that had an R group that differed from those listed in Figure 5.14 would still be an amino acid as long as it also contained an α carbon, an amino group, and a carboxyl group—but that molecule would not be one of the 20 amino acids commonly found in nature. **Figure 25.4** The hydrophobic regions of such molecules are attracted to one another and excluded from water, whereas the hydrophilic regions have an affinity for water. As a result, the molecules can form a bilayer in which the hydrophilic regions are on the outside of the bilayer (facing water on each side of the bilayer) and the hydrophobic regions point toward each other (that is, toward the inside of the bilayer). **Figure 25.6** Because uranium-238 has a half-life of 4.5 billion years, the *x*-axis would be relabeled (in billions of years) as 4.5, 9, 13.5, and 18. **Figure 25.8** (1) The earliest direct evidence of life comes from fossils of prokaryotes that date to 3.5 billion years ago. Fossil evidence also shows that for the next 2 billion years (3.5 to 1.5 billion years ago), life on Earth consisted entirely of unicellular organisms. In fact, from 3.5 billion years ago to 1.8 billion years ago, all of Earth's organisms were prokaryotes; around 1.8 billion years ago, these unicellular prokaryotes were joined by unicellular eukaryotes (multicellular eukaryotes emerged about 1.3 billion years ago). (2) Sample answer: In Figure 25.12, there are two hatch marks on the *x*-axis. These hatch marks represent large time spans. Unless the graph was very wide, showing these entire time spans would obscure key details for the figure, such as when selected animal groups first appear in the fossil record. (3) The horizontal time scale indicates that prokaryotes originated 3.5 billion years ago and that the colonization of land took place 500 million years ago. On a 1-hour time scale, this indicates that prokaryotes appeared about 46 minutes ago, while the colonization of land took place less than 7 minutes ago. **Figure 25.12** You should have circled the node, shown in the tree diagram at approximately 635 million years ago (mya), that leads to the echinoderm/chordate lineage and to the lineage that gave rise to brachiopods, annelids, molluscs, and arthropods. To determine a minimum estimate of the age of the ancestor represented by this node, note that the most recent common ancestor of chordates and annelids must be at least as old as any of its descendants. Since fossil molluscs date to about 560 mya, the common ancestor represented by the circled branch point must be at least 560 million years old. **Figure 25.14** There are two speciation events and five extinctions in lineage A, while there are five speciation events and one extinction in lineage B during the last 2 million years. **Figure 25.17** The Australian plate's current direction of movement is roughly similar to the northeasterly direction the continent traveled over the past 66 million years. **Figure 25.27** The coding sequence of the *Pitx1* gene would differ between the marine and lake populations, but patterns of gene expression would not.

Concept Check 25.1

1. The hypothesis that conditions on early Earth could have permitted the synthesis of organic molecules from inorganic ingredients **2.** In contrast to random mingling of molecules in an open solution, segregation of molecular systems by membranes could concentrate organic molecules, assisting biochemical reactions. **3.** Today, genetic information usually flows from DNA to RNA, as when the DNA sequence of a gene is used as a template to synthesize the mRNA encoding a particular protein. However, the life cycle of retroviruses such as HIV shows that genetic information can flow in the reverse direction (from RNA to DNA). In these viruses, the enzyme reverse transcriptase uses RNA as a template for DNA synthesis, suggesting that a similar enzyme could have played a key role in the transition from an RNA world to a DNA world.

Concept Check 25.2

1. The fossil record shows that different groups of organisms dominated life on Earth at different points in time and that many organisms once alive are now extinct; specific examples of these points can be found in Figure 25.5. The fossil record also indicates that new groups of organisms can arise via the gradual modification of previously existing organisms, as illustrated by fossils that document the origin of mammals from their cynodont ancestors (see Figure 25.7). **2.** 22,920 years (four half-lives: $5,730 \times 4$)

Concept Check 25.3

1. Free oxygen attacks chemical bonds and can inhibit enzymes and damage cells. As a result, the appearance of oxygen in the atmosphere probably caused many prokaryotes that had thrived in anaerobic environments to survive and reproduce poorly, ultimately driving many of these species to extinction. **2.** All eukaryotes have mitochondria or remnants of these organelles, but not all eukaryotes have plastids. **3.** A fossil record of life today would include many organisms with hard body parts (such as vertebrates and many marine invertebrates), but might not include some species we are very familiar with, such as those that have small geographic ranges and/or small population sizes (for example, endangered species such as the giant panda, tiger, and several rhinoceros species).

Concept Check 25.4

1. The theory of plate tectonics describes the movement of Earth's continental plates, which alters the physical geography and climate of Earth, as well as the extent to which organisms are geographically isolated. Because these factors affect extinction and speciation rates, plate tectonics has a major impact on life on Earth. **2.** Mass extinctions; major evolutionary innovations; the diversification of another group of organisms (which can provide new sources of food);

migration to new locations where few competitor species exist **3.** Evidence from previous mass extinctions indicates that the diversity of life on Earth would not recover, for millions of years, to what it had been before the mass extinction—a much greater period of time than our species has been in existence (about 200,000 years). Although new speciation events would eventually cause the total number of species on Earth to recover, the many species and evolutionary lineages driven to extinction would be gone forever, thus forever changing the course of evolution on our planet. In addition, previous evidence suggests that a sixth mass extinction would reduce thriving and complex ecological communities (such as forests and coral reefs) so greatly that they might hardly resemble what they are like now. A sixth mass extinction would also change the types of organisms that live in ecological communities and how those organisms interact with one another. Finally, a sixth mass extinction would pave the way for new adaptive radiations in some of the groups that survive the extinction.

Concept Check 25.5

1. Heterochrony can cause a variety of morphological changes. For example, if the timing of the onset of sexual maturity changes, retention of juvenile characteristics (paedomorphosis) may result. Paedomorphosis can be caused by small genetic changes that result in large changes in morphology, as seen in the axolotl salamander. **2.** In animal embryos, *Hox* genes influence the development of structures such as limbs and feeding appendages. As a result, changes in these genes—or in the regulation of these genes—are likely to have major effects on morphology. **3.** From genetics, we know that gene regulation is altered by how well transcription factors bind to noncoding DNA sequences called control elements. Thus, if changes in morphology are often caused by changes in gene regulation, portions of noncoding DNA that contain control elements are likely to be strongly affected by natural selection.

Concept Check 25.6

1. Complex structures do not evolve all at once, but in increments, with natural selection selecting for adaptive variants of the earlier versions. **2.** Although the myxoma virus is highly lethal, initially some of the rabbits are resistant (0.2% of infected rabbits are not killed). Thus, assuming resistance is an inherited trait, we would expect the rabbit population to show a trend for increased resistance to the virus. We would also expect the virus to show an evolutionary trend toward reduced lethality. We would expect this trend because a rabbit infected with a less lethal virus would be more likely to live long enough for a mosquito to bite it and hence potentially transmit the virus to another rabbit. (A virus that kills its rabbit host before a mosquito transmits the virus to another rabbit dies with its host.)

Summary of Key Concepts Questions

Concept 25.1 Particles of montmorillonite clay may have provided surfaces on which organic molecules became concentrated and hence were more likely to react with one another. Montmorillonite clay particles may also have facilitated the transport of key molecules, such as short strands of RNA, into vesicles. These vesicles can form spontaneously from simple precursor molecules, "reproduce" and "grow" on their own, and maintain internal concentrations of molecules that differ from those in the surrounding environment. These features of vesicles represent key steps in the emergence of protocells and (ultimately) the first living cells. **Concept 25.2** One challenge is that radioisotopes with very long half-lives are not used by organisms to build their bones or shells. As a result, fossils older than 75,000 years cannot be dated directly. Fossils are often found in sedimentary rock, but those rocks typically contain sediments of different ages, again posing a challenge when trying to date old fossils. To circumvent these challenges, geologists use radioisotopes with long half-lives to date layers of volcanic rock that surround old fossils. This approach provides minimum and maximum estimates for the ages of fossils sandwiched between two layers of volcanic rock. **Concept 25.3** The "Cambrian explosion" refers to a relatively short interval of time (535–525 million years ago) during which large forms of many present-day animal phyla first appear in the fossil record. The evolutionary changes that occurred during this time, such as the appearance of large predators and well-defended prey, were important because they set the stage for many of the key events in the history of life over the last 500 million years. **Concept 25.4** The broad evolutionary changes documented by the fossil record reflect the rise and fall of major groups of organisms. In turn, the rise or fall of any particular group results from a balance between speciation and extinction rates: A group increases in size when the rate at which its members produce new species is greater than the rate at which its member species are lost to extinction, while a group shrinks in size if extinction rates are greater than speciation rates. **Concept 25.5** A change in the sequence or regulation of a developmental gene can produce major morphological changes. In some cases, such morphological changes may enable organisms to perform new functions or live in new environments—thus potentially leading to an adaptive radiation and the formation of a new group of organisms. **Concept 25.6** Evolutionary change results from interactions between organisms and their current environments. No goal is involved in this process. As environments change over time, the features of organisms favored by natural selection may also change. When this happens, what once may have seemed like a "goal" of evolution (for example, improvements in the function of a feature previously favored by natural selection) may cease to be beneficial or may even be harmful.

Test Your Understanding

1. B **2.** A **3.** D **4.** B **5.** D **6.** C **7.** A

Chapter 26

Figure Questions

Figure 26.5 (1) In this tree, frogs are most closely related to a group consisting of lizards, chimps, and humans. (2) You should have circled the branch point splitting the frog lineage from the lineage leading to lizards, chimps, and humans. (3) Four: chimps–humans, lizards–chimps/humans; frogs–lizards/chimps/humans; and fishes–frogs/lizards/chimps/humans.

(4)

(5)

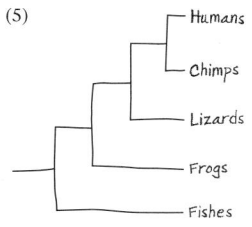

Each of the three trees identifies chimps and lizards as the two closest relatives of humans in these trees because they are the groups shown with whom we share the two most recent common ancestors.

Figure 26.6 Unknown 1b (a portion of sample 1) and unknowns 9–13 all would have to be located on the branch of the tree that currently leads to Minke (Southern Hemisphere) and unknowns 1a and 2–8. **Figure 26.9** There are four possible bases (A, C, G, T) at each nucleotide position. If the base at each position depends on chance, not common descent, we would expect roughly one out of four (25%) of them to be the same. **Figure 26.11** You should have circled the branch point that is drawn farthest to the left (the common ancestor of all taxa shown). Both cetaceans and seals descended from terrestrial lineages of mammals, indicating that the cetacean–seal common ancestor lacked a streamlined body form and hence would not be part of the cetacean–seal group. **Figure 26.12** Hinged jaws are a shared ancestral character for the group that includes frogs, turtles, and leopards. Thus, you should have circled the frog, turtle, and leopard lineages, along with their most recent common ancestor. **Figure 26.16** Crocodilians are the sister taxon to the dinosaur clade (which includes birds) because crocodilians and the dinosaur clade share an immediate common ancestor that is not shared by any other group. **Figure 26.21** This tree indicates that the sequences of rRNA and other genes in mitochondria are most closely related to those of proteobacteria, while the sequences of chloroplast genes are most closely related to those of cyanobacteria. These gene sequence relationships are what would be predicted from endosymbiont theory, which posits that both mitochondria and chloroplasts originated as engulfed prokaryotic cells.

Concept Check 26.1

1. In Figure 26.4, leopards are the sister taxon to a group consisting of the family Mustelidae (which includes badgers) and the family Canidae (which includes wolves). Since members of a sister group are each other's closest relatives, leopards are equally related to badgers and to wolves. **2.** The tree in (c) shows a different pattern of evolutionary relationships. In (c), C and B are sister taxa, whereas C and D are sister taxa in (a) and (b). **3.** The redrawn version of Figure 26.4 is shown below.

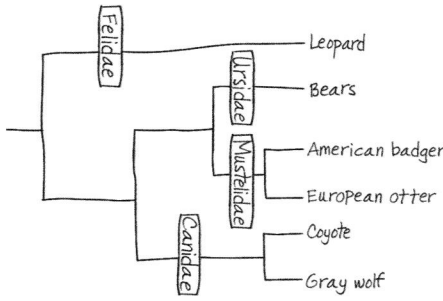

Concept Check 26.2

1. (a) Analogy, since porcupines and cacti are not closely related and since most other animals and plants do not have similar structures; (b) homology, since cats and humans are both mammals and have homologous forelimbs, of which the hand and paw are the lower part; (c) analogy, since owls and hornets are not closely related and since the structure of their wings is very different. **2.** Species B and C are more likely to be closely related. Small genetic changes (as between species B and C) can produce divergent physical appearances, but if many genes have diverged greatly (as in species A and B), then the lineages have probably been separate for a long time.

Concept Check 26.3

1. No; hair is a shared ancestral character common to all mammals and thus is not helpful in distinguishing different mammalian subgroups. **2.** The principle of maximum parsimony states that the hypothesis about nature we investigate first should be the simplest explanation found to be consistent with the facts. Actual evolutionary relationships may differ from those inferred by parsimony owing to complicating factors such as convergent evolution. **3.**

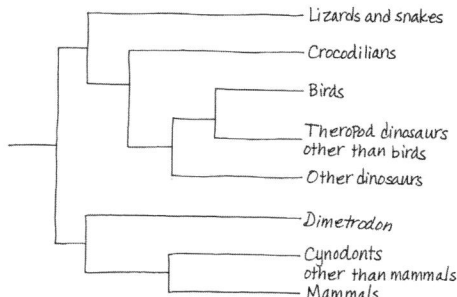

The traditional classification provides a poor match to evolutionary history, thus violating the basic principle of cladistics—that classification should be based on common descent. Both birds and mammals originated from groups traditionally designated as reptiles, making reptiles (as traditionally delineated) a paraphyletic group. These problems can be addressed by removing *Dimetrodon* and cynodonts from the reptiles and by regarding birds as a group of reptiles (specifically, as a group of dinosaurs).

Concept Check 26.4

1. Proteins are gene products. Their amino acid sequences are determined by the nucleotide sequences of the DNA that codes for them. Thus, differences between comparable proteins in two species reflect underlying genetic differences that have accumulated as the species diverged from one another. As a result, differences between the proteins can reflect the evolutionary history of the species. **2.** Orthologous genes are homologous genes in which the homology results from a speciation event and hence occurs between genes found in different species. Paralogous genes are homologous genes in which the homology results from gene duplication. Orthologous genes should be used to infer phylogeny since differences among them reflect the history of speciation events. **3.** In RNA processing, the exons or coding regions of a gene can be spliced together in different ways, yielding different mRNAs and hence different protein products. As a result, different proteins could potentially be produced from the same gene in different tissues, thereby enabling the gene to perform different functions in these different tissues.

Concept Check 26.5

1. A molecular clock is a method of estimating the actual time of evolutionary events based on numbers of base changes in orthologous genes. It is based on the assumption that the regions of genomes being compared evolve at constant rates. **2.** There are many portions of the genome that do not code for genes; mutations that alter the sequence of bases in such regions could accumulate through drift without affecting an organism's fitness. Even in coding regions of the genome, some mutations may not have a critical effect on genes or proteins.

Concept Check 26.6

1. The kingdom Monera included bacteria and archaea, but we now know that these organisms are in separate domains. Kingdoms are subsets of domains, so a single kingdom (like Monera) that includes taxa from different domains is not valid. **2.** Because of horizontal gene transfer, some genes in eukaryotes are more closely related to bacteria, while others are more closely related to archaea; thus, depending on which genes are used, phylogenetic trees constructed from DNA data can yield conflicting results. **3.** Eukaryotes are hypothesized to have originated when a heterotrophic prokaryote (an archaeal host cell) engulfed a bacterium that would later become an organelle found in all eukaryotes—the mitochondrion. Over time, a fusion of organisms occurred as the archaeal host cell and its bacterial endosymbiont evolved to become a single organism. As a result, we would expect the cell of a eukaryote to include both archaeal DNA and bacterial DNA, making the origin of eukaryotes an example of horizontal gene transfer.

Summary of Key Concepts Questions

26.1 The fact that humans and chimpanzees are sister species indicates that we share a more recent common ancestor with chimpanzees than we do with any other living primate species. But that does not mean that humans evolved from chimpanzees, or vice versa; instead, it indicates that both humans and chimpanzees are descendants of that common ancestor. **26.2** Homologous characters result from shared ancestry. As organisms diverge over time, some of their homologous characters will also diverge. The homologous characters of organisms that diverged long ago typically differ more than do the homologous characters of organisms that diverged more recently. As a result, differences in homologous characters can be used to infer phylogeny. In contrast, analogous characters result from convergent evolution, not shared ancestry, and hence can give misleading estimates of phylogeny. **26.3** All features of organisms arose at some point in the history of life. In the group in which a new feature first arose, that feature is a shared derived character that is unique to that clade. The group in which each shared derived character first appeared can be determined, and the resulting nested pattern can be used to infer evolutionary history. **26.4** Orthologous genes should be used; for such genes, the homology results from speciation and hence reflects evolutionary history. **26.5** A key assumption of molecular clocks is that nucleotide substitutions occur at fixed rates, and hence the number of nucleotide differences between two DNA sequences is proportional to the time since the sequences diverged from each other. Some limitations of molecular clocks: No gene marks time with complete precision; natural selection can favor certain DNA changes over others; nucleotide substitution rates can change over long periods of time (causing molecular clock estimates of when events in the distant past occurred to be highly uncertain); and the same gene can evolve at different rates in different organisms. **26.6** Genetic data indicated that many prokaryotes differed as much from each other as they did from eukaryotes. This indicated that organisms should be grouped into three "super-kingdoms," or domains (Archaea, Bacteria, Eukarya). These data also indicated that the previous kingdom Monera (which had contained all the prokaryotes) did not make biological sense and should be abandoned. Later genetic and morphological data also indicated that the former kingdom Protista (which had primarily contained single-celled organisms) should be abandoned because some protists are more closely related to plants, fungi, or animals than they are to other protists.

Test Your Understanding

1. A **2.** C **3.** B **4.** C **5.** B **6.** A **7.** D
9.

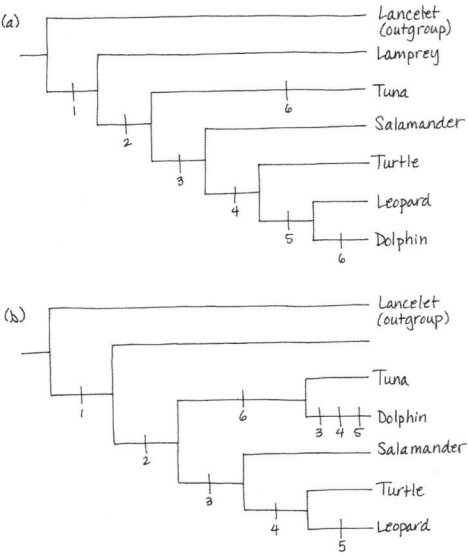

(c) The tree in (a) requires seven evolutionary changes, while the tree in (b) requires nine evolutionary changes. Thus, the tree in (a) is more parsimonious since it requires fewer evolutionary changes.

Chapter 27

Figure Questions

Figure 27.7 The top ring, to which the hook is attached, is embedded within the interior, hydrophobic portion of the lipid bilayer of the outer membrane, suggesting that the top ring is hydrophobic. Likewise, the third ring down is embedded within the hydrophobic portion of the plasma membrane's lipid bilayer, suggesting that this ring also is hydrophobic. **Figure 27.9** The third plasmid is the separate, small twisted loop located just above and to the left of the line pointing to the label "Chromosome." **Figure 27.10** It is likely that the expression or sequence of genes that affect glucose metabolism may have changed; genes for metabolic processes no longer needed by the cell also may have changed.
Figure 27.11 Transduction results in horizontal gene transfer when the host and recipient cells are members of different species. **Figure 27.16** Eukarya
Figure 27.18 Thermophiles live in very hot environments, so it is likely that their enzymes can continue to function normally at much higher temperatures than can the enzymes of other organisms. At low temperatures, however, the enzymes of thermophiles may not function as well as the enzymes of other organisms.
Figure 27.19 From the graph, plant uptake can be estimated as 0.72, 0.62, and 0.96 mg K^+ for strains 1, 2, and 3, respectively. These values average to 0.77 mg K^+. If bacteria had no effect, the average plant uptake of K^+ for strains 1, 2, and 3 should be close to 0.51 mg K^+, the value observed for plants grown in bacteria-free soil.
Figure 27.22 Penicillin

Concept Check 27.1

1. Adaptations include the capsule (shields prokaryotes from the host's immune system) and endospores (enable cells to survive harsh conditions and to revive when the environment becomes favorable). **2.** Prokaryotic cells lack the complex compartmentalization associated with the membrane-enclosed organelles of eukaryotic cells. Prokaryotic genomes have much less DNA than eukaryotic genomes, and most of this DNA is contained in a single ring-shaped chromosome located in the nucleoid rather than within a true membrane-enclosed nucleus. In addition, many prokaryotes also have plasmids, small ring-shaped DNA molecules containing a few genes. **3.** Plastids such as chloroplasts are thought to have evolved from an endosymbiotic photosynthetic prokaryote. More specifically, the phylogenetic tree shown in Figure 26.21 indicates that plastids are closely related to cyanobacteria. Hence, we can hypothesize that the thylakoid membranes of chloroplasts resemble those of cyanobacteria because chloroplasts evolved from an endosymbiotic cyanobacterium.

Concept Check 27.2

1. Prokaryotes can have extremely large population sizes, in part because they often have short generation times. The large number of individuals in prokaryotic populations makes it likely that in each generation there will be many individuals that have new mutations at any particular gene, thereby adding considerable genetic diversity to the population. **2.** In transformation, naked, foreign DNA from the environment is taken up by a bacterial cell. In transduction, phages carry bacterial genes from one bacterial cell to another. In conjugation, a bacterial cell directly transfers plasmid or chromosomal DNA to another cell via a mating bridge that temporarily connects the two cells. **3.** Yes. Genes for antibiotic resistance could be transferred (by transformation, transduction, or conjugation) from the nonpathogenic bacterium to a pathogenic bacterium; this could make the pathogen an even greater threat to human health. In general, transformation, transduction, and conjugation tend to increase the spread of resistance genes.

Concept Check 27.3

1. A phototroph derives its energy from light, while a chemotroph gets its energy from chemical sources. An autotroph derives its carbon from CO_2, HCO_3^-, or related compounds, while a heterotroph gets its carbon from organic nutrients such as glucose. Thus, there are four nutritional modes: photoautotrophic, photoheterotrophic (unique to prokaryotes), chemoautotrophic (unique to prokaryotes), and chemoheterotrophic. **2.** Chemoheterotrophy; the bacterium must rely on chemical sources of energy, since it is not exposed to light, and it must be a heterotroph if it requires a source of carbon other than CO_2 (or a related compound, such as HCO_3^-).
3. If humans could fix nitrogen, we could build proteins using atmospheric N_2 and hence would not need to eat high-protein foods such as meat, fish, or soy. Our diet would, however, need to include a source of carbon, along with minerals and water. Thus, a typical meal might consist of carbohydrates as a carbon source, along with fruits and vegetables to provide essential minerals (and additional carbon).

Concept Check 27.4

1. Molecular systematic studies indicate that some organisms once classified as bacteria are more closely related to eukaryotes and belong in a domain of their own: Archaea. Such studies have also shown that horizontal gene transfer is common and plays an important role in the evolution of prokaryotes. By not requiring that organisms be cultured in the laboratory, metagenomic studies have revealed an immense diversity of previously unknown prokaryotic species. Over time, the ongoing discovery of new species by metagenomic analyses may alter our understanding of prokaryotic phylogeny greatly. **2.** The three domains shown in Figure 27.16 are not valid under this assumption because domain Eukarya would be nested within domain Archaea. As such, Eukarya would be a subset of Archaea, not a separate domain of its own.

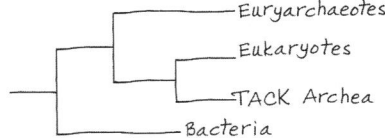

Concept Check 27.5

1. Although prokaryotes are small, their large numbers and metabolic abilities enable them to play key roles in ecosystems by decomposing wastes, recycling chemicals, and affecting the concentrations of nutrients available to other organisms. **2.** Cyanobacteria produce oxygen when water is split in the light reactions of photosynthesis. The Calvin cycle incorporates CO_2 from the air into organic molecules, which are then converted to sugars.

Concept Check 27.6

1. Sample answers: eating fermented foods such as yogurt, sourdough bread, or cheese; receiving clean water from sewage treatment; taking medicines produced by bacteria **2.** No. If the poison is secreted as an exotoxin, live bacteria could be transmitted to another person. But the same is true if the poison is an endotoxin—only in this case, the live bacteria that are transmitted may be descendants of the (now-dead) bacteria that produced the poison. **3.** Some of the many different species of prokaryotes that live in the human gut compete with one another for resources (from the food that you eat). Because different prokaryotic species have different adaptations, a change in diet may alter which species can grow most rapidly, thus altering species abundance.

Summary of Key Concepts Questions

27.1 Specific structural features that enable prokaryotes to thrive in diverse environments include their cell walls (which provide shape and protection), flagella (which function in directed movement), and ability to form capsules or endospores (both of which can protect against harsh conditions). Prokaryotes also possess biochemical adaptations for growth in varied conditions, such as those that enable them to tolerate extremely hot or salty environments. **27.2** Many prokaryotic species can reproduce extremely rapidly, and their populations can number in the trillions. As a result, even though mutations are rare, every day many offspring are produced that have new mutations at particular gene loci. In addition, even though prokaryotes reproduce asexually and hence the vast majority of offspring are genetically identical to their parent, the genetic variation of their populations can be increased by transduction, transformation, and conjugation. Each of these (nonreproductive) processes can increase genetic variation by transferring DNA from one cell to another—even among cells that are of different species. **27.3** Prokaryotes have an exceptionally broad range of metabolic adaptations. As a group, prokaryotes perform all four modes of nutrition (photoautotrophy, chemoautotrophy, photoheterotrophy, and chemoheterotrophy), whereas eukaryotes perform only two of these (photoautotrophy and chemoheterotrophy). Prokaryotes are also able to metabolize nitrogen in a wide variety of forms (again unlike eukaryotes), and they frequently cooperate with other prokaryotic cells of the same or different species. **27.4** Phenotypic criteria such as shape, motility, and nutritional mode do not provide a clear picture of the evolutionary history of the prokaryotes. In contrast, molecular data have elucidated relationships among major groups of prokaryotes. Molecular data have also allowed researchers to sample genes directly from the environment; using such genes to construct phylogenies has led to the discovery of major new groups of prokaryotes **27.5** Prokaryotes play key roles in the chemical cycles on which life depends. For example, prokaryotes are important decomposers, breaking down corpses and waste materials, thereby releasing nutrients to the environment where they can be used by other organisms. Prokaryotes also convert inorganic compounds to forms that other organisms can use. With respect to their ecological interactions, many prokaryotes form life-sustaining mutualisms with other species. In some cases, such as hydrothermal vent communities, the metabolic activities of prokaryotes provide an energy source on which hundreds of other species depend; in the absence of the prokaryotes, the community would collapse. **27.6** Human well-being depends on our associations with mutualistic prokaryotes, such as the many species that live in our intestines and digest food that we cannot. Humans also can harness the remarkable metabolic capabilities of prokaryotes to produce a wide range of useful products and to perform key services such as bioremediation. Negative effects of prokaryotes result primarily from bacterial pathogens that cause disease.

Test Your Understanding

1. D **2.** A **3.** B **4.** C **5.** D **6.** A

Chapter 28

Figure Questions

Figure 28.3 The diagram shows that a single secondary endosymbiosis event gave rise to the stramenopiles and alveolates—thus, these groups can trace their ancestry back to a single heterotrophic protist (shown in yellow) that ingested a red alga. In contrast, euglenids and chlorarachniophytes each descended from a different heterotrophic protist (one of which is shown in gray, the other in brown). Hence, it is likely that stramenopiles and alveolates are more closely related than are euglenids and chlorarachniophytes.
Figure 28.5

Simplified tree that shows 4 supergroups:

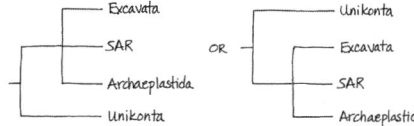

Simplified tree that shows Unikonta as sister group to all other eukaryotes:

Figure 28.14 The sperm cells in the diagram are produced by the asexual (mitotic) division of cells in a single male gametophyte, which was itself produced by the asexual (mitotic) division of a single zoospore. Thus, the sperm cells are all derived from a single zoospore and so are genetically identical to one another. **Figure 28.18** Merozoites are produced by the asexual (mitotic) cell division of haploid sporozoites; similarly, gametocytes are produced by the asexual cell division of merozoites. Hence, it is likely that individuals in these three stages have the same complement of genes and that morphological differences between them result from changes in gene expression.
Figure 28.19 These events have a similar overall effect to fertilization. In both cases, haploid nuclei that were originally from two genetically different cells fuse to form a diploid nucleus. **Figure 28.25** The following stage should be

circled: step 6, where a mature cell undergoes mitosis and forms four or more daughter cells. In step 7, the zoospores eventually grow into mature haploid cells, but they do not produce new daughter cells. Likewise, in step 2, a mature cell develops into a gamete, but it does not produce new daughter cells.
Figure 28.26

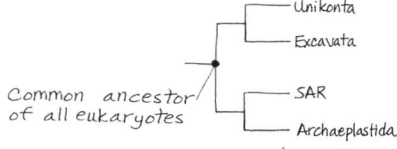

Figure 28.28 They would be haploid because originally each of these cells was a haploid, solitary amoeba.

Concept Check 28.1

1. Sample response: Protists include unicellular, colonial, and multicellular organisms; photoautotrophs, heterotrophs, and mixotrophs; species that reproduce asexually, sexually, or both ways; and organisms with diverse physical forms and adaptations. **2.** Strong evidence shows that eukaryotes acquired mitochondria after a host cell (either an archaean or a cell closely related to the archaea) first engulfed and then formed an endosymbiotic association with an alpha proteobacterium. Similarly, chloroplasts in red and green algae appear to have descended from a photosynthetic cyanobacterium that was engulfed by an ancient heterotrophic eukaryote. Secondary endosymbiosis also played an important role: Various protistan lineages acquired plastids by engulfing unicellular red or green algae. **3.** Four. The first (and primary) genome is the DNA located in the chlorarachniophyte nucleus. A chlorarachniophyte also contain remnants of a green alga's nuclear DNA, located in the nucleomorph. Finally, mitochondria and chloroplasts contain DNA from the (different) bacteria from which they evolved. These two prokaryotic genomes comprise the third and fourth genomes contained within a chlorarachniophyte.

Concept Check 28.2

1. Their mitochondria do not have an electron transport chain and so cannot function in aerobic respiration. **2.** Since the unknown protist is more closely related to diplomonads than to euglenids, it must have originated after the lineage leading to the diplomonads and parabasalids diverged from the euglenozoans. In addition, since the unknown species has fully functional mitochondria—yet both diplomonads and parabasalids do not—it is likely that the unknown species originated *before* the last common ancestor of the diplomonads and parabasalids.

Concept Check 28.3

1. Because foram tests are hardened with calcium carbonate, they form long-lasting fossils in marine sediments and sedimentary rocks. **2.** The plastid DNA would likely be more similar to the chromosomal DNA of cyanobacteria based on the well-supported hypothesis that eukaryotic plastids (such as those found in the eukaryotic groups listed) originated by an endosymbiosis event in which a eukaryote engulfed a cyanobacterium. If the plastid is derived from the cyanobacterium, its DNA would be derived from the bacterial DNA. **3.** Figure 13.6b. Algae and plants with alternation of generations have a multicellular haploid stage *and* a multicellular diploid stage. In the other two life cycles, either the haploid stage or the diploid stage is unicellular. **4.** During photosynthesis, aerobic algae produce O_2 and use CO_2. O_2 is produced as a by-product of the light reactions, while CO_2 is used as an input to the Calvin cycle (the end products of which are sugars). Aerobic algae also perform cellular respiration, which uses O_2 as an input and produces CO_2 as a waste product.

Concept Check 28.4

1. Many red algae contain a photosynthetic pigment called phycoerythrin, which gives them a reddish color and allows them to carry out photosynthesis in relatively deep coastal water. Also unlike brown algae, red algae have no flagellated stages in their life cycle and must depend on water currents to bring gametes together for fertilization. **2.** *Ulva* contains many cells and its body is differentiated into leaflike blades and a rootlike holdfast. *Caulerpa*'s body is composed of multinucleate filaments without cross-walls, so it is essentially one large cell. **3.** Red algae have no flagellated stages in their life cycle and hence must depend on water currents to bring their gametes together. This feature of their biology might increase the difficulty of reproducing on land. In contrast, the gametes of green algae are flagellated, making it possible for them to swim in thin films of water. In addition, a variety of green algae contain compounds in their cytoplasm, cell wall, or zygote coat that protect against intense sunlight and other terrestrial conditions. Such compounds may have increased the chance that descendants of green algae could survive on land.

Concept Check 28.5

1. Amoebozoans have lobe- or tube-shaped pseudopodia, whereas forams have threadlike pseudopodia. **2.** Slime molds are fungus-like in that they produce fruiting bodies that aid in the dispersal of spores, and they are animal-like in that they are motile and ingest food. However, slime molds are more closely related to tubulinids and entamoebas than to fungi or animals. **3.** The genes used to estimate the tree shown in Figure 28.26 were transferred from an alpha proteobacterium to an early eukaryote. Based on the sequences of these genes, the eukaryotes should be more closely related to alpha proteobacteria than they are to any other lineage of prokaryotes. Thus, alpha proteobacteria are well suited as an outgroup to the eukaryotes (the group of species whose relationships we are trying to determine).

Concept Check 28.6

1. Because photosynthetic protists constitute the base of aquatic food webs, many aquatic organisms depend on them for food, either directly or indirectly. (In addition, a substantial percentage of the oxygen produced by photosynthesis is made by photosynthetic protists.)   **2.** Protists form mutualistic and parasitic associations with other organisms. Examples include photosynthetic dinoflagellates that form a mutualistic symbiosis with coral polyps; parabasalids that form a mutualistic symbiosis with termites; and the stramenopile *Phytophthora ramorum*, a parasite of oak trees. **3.** Corals depend on their dinoflagellate symbionts for nourishment, so coral bleaching could cause the corals to die. As the corals die, less food would be available for fishes and other species that eat coral. As a result, populations of these species might decline, and that, in turn, might cause populations of their predators to decline. **4.** The two approaches differ in the evolutionary changes they may bring about. A strain of *Wolbachia* that confers resistance to infection by *Plasmodium* and does not harm mosquitoes would spread rapidly through the mosquito population. In this case, natural selection would favor any *Plasmodium* individuals that could overcome the resistance to infection conferred by *Wolbachia*. If insecticides are used, mosquitoes that are resistant to the insecticide would be favored by natural selection. Hence, use of *Wolbachia* could cause evolution in *Plasmodium* populations, while using insecticides could cause evolution in mosquito populations.

Summary of Key Concepts Questions

28.1 Sample response: Protists, plants, animals, and fungi are similar in that their cells have a nucleus and other membrane-enclosed organelles, unlike the cells of prokaryotes. These membrane-enclosed organelles make the cells of eukaryotes more complex than the cells of prokaryotes. Protists and other eukaryotes also differ from prokaryotes in having a well-developed cytoskeleton that enables them to have asymmetric forms and to change in shape as they feed, move, or grow. With respect to differences between protists and other eukaryotes, most protists are unicellular, unlike animals, plants, and most fungi. Protists also have greater nutritional diversity than other eukaryotes.   **28.2** Unique cytoskeletal features are shared by many excavates. In addition, some members of Excavata have an "excavated" feeding groove for which the group was named. Moreover, recent genomic studies support the monophyly of the excavate supergroup.   **28.3** Stramenopiles and alveolates are hypothesized to have originated by secondary endosymbiosis. Under this hypothesis, we can infer that the common ancestor of these two groups had a plastid, in this case of red algal origin. Thus, we would expect that apicomplexans (and alveolate or stramenopile protists) either would have plastids or would have lost their plastids over the course of evolution.   **28.4** Red algae, green algae, and plants are placed in the same supergroup because considerable evidence indicates that these organisms all descended from the same ancestor, an ancient heterotrophic protist that acquired a cyanobacterial endosymbiont. **28.5** The unikonts are a diverse group of eukaryotes that includes many protists, along with animals and fungi. Most of the protists in Unikonta are amoebozoans, a clade of amoebas that have lobe- or tube-shaped pseudopodia (as opposed to the threadlike pseudopodia of rhizarians). Other protists in Unikonta include several groups that are closely related to fungi and several other groups that are closely related to animals.   **28.6** Sample response: Ecologically important protists include photosynthetic dinoflagellates that provide essential sources of energy to their symbiotic partners, the corals that build coral reefs. Other important protistan symbionts include those that enable termites to digest wood and *Plasmodium*, the pathogen that causes malaria. Photosynthetic protists such as diatoms are among the most important producers in aquatic communities; as such, many other species in aquatic environments depend on them for food.

Test Your Understanding

1. D   **2.** B   **3.** B   **4.** D   **5.** D   **6.** C
7.

Pathogens that share a relatively recent common ancestor with humans will likely also share metabolic and structural characteristics with humans. Because drugs target the pathogen's metabolism or structure, developing drugs that harm the pathogen but not the patient should be most difficult for pathogens with whom we share the most recent evolutionary history. Working backward in time, we can use the phylogenetic tree to determine the order in which humans shared a common ancestor with pathogens in different taxa. This process leads to the prediction that it should be hardest to develop drugs to combat animal pathogens, followed by choanoflagellate pathogens, fungal and nucleariid pathogens, amoebozoans, other protists, and finally prokaryotes.

Chapter 29

Figure Questions

Figure 29.5 The life cycles of plants and some algae, shown in Figure 13.6b, have alternation of generations; other life cycles do not. Unlike in the animal life cycle (Figure 13.6a), in the plant/algal life cycle, meiosis produces spores, not gametes. These haploid spores then divide repeatedly by mitosis, ultimately forming a multicellular haploid individual that produces gametes. There is no multicellular haploid stage in the animal life cycle. An alternation of generations life cycle also has a multicellular diploid stage, whereas the life cycle of most fungi and

some protists shown in Figure 13.6c does not.   **Figure 29.10** Plants, vascular plants, and seed plants are monophyletic because each of these groups includes the common ancestor of the group and all of the descendants of that common ancestor. The other two categories of plants, the nonvascular plants and the seedless vascular plants, are paraphyletic: These groups do not include all of the descendants of the group's most recent common ancestor.   **Figure 29.11** Yes. As shown in the diagram, the sperm cell and the egg cell that fuse each resulted from the mitotic division of spores produced by the same sporophyte. However, these spores would differ genetically from one another because they were produced by meiosis, a cell division process that generates genetic variation among the offspring cells.   **Figure 29.14** Because the moss reduces nitrogen loss from the ecosystem, species that typically colonize the soils after the moss probably experience higher soil nitrogen levels than they otherwise would. The resulting increased availability of nitrogen may benefit these species because nitrogen is an essential nutrient that often is in short supply.   **Figure 29.17** A fern that had wind-dispersed sperm would not require water for fertilization, thus removing a difficulty that ferns face when they live in arid environments. The fern would also be under strong selection to produce sperm above ground (as opposed to the current situation, where some fern gametophytes are located below ground).

Concept Check 29.1

1. Plants share some key traits only with charophytes: rings of cellulose-synthesizing complexes and similarity in sperm structure. Comparisons of nuclear, chloroplast, and mitochondrial DNA sequences also indicate that certain groups of charophytes (such as *Zygnema*) are the closest living relatives of plants.   **2.** Possible answers include walls toughened by sporopollenin (protects against harsh environmental conditions); multicellular, dependent embryos (provide nutrients and protection to the developing embryo); cuticle (reduces water loss); stomata (control gas exchange and reduce water loss)   **3.** The multicellular diploid stage of the life cycle would not produce gametes. Instead, both males and females would produce haploid spores by meiosis. These spores would give rise to multicellular male and female haploid stages—a major change from the single-celled haploid stages (sperm and eggs) that we actually have. The multicellular haploid stages would produce gametes and reproduce sexually. An individual at the multicellular haploid stage of the human life cycle might look like us, or it might look completely different.

Concept Check 29.2

1. Most bryophytes do not have a vascular transport system, and their life cycle is dominated by gametophytes rather than sporophytes.   **2.** Answers may include the following: Large surface area of protonema enhances absorption of water and minerals; the vase-shaped archegonia protect eggs during fertilization and transport nutrients to the embryos via placental transfer cells; the stalk-like seta conducts nutrients from the gametophyte to the capsule, where spores are produced; the peristome enables gradual spore discharge; stomata enable CO_2/O_2 exchange while minimizing water loss; lightweight spores are readily dispersed by wind.   **3.** Effects of global warming on peatlands could result in positive feedback, which occurs when an end product of a process increases its own production. In this case, global warming is expected to lower the water levels of some peatlands. This would expose peat to air and cause it to decompose, thereby releasing stored CO_2 to the atmosphere. The release of more stored CO_2 to the atmosphere could cause additional global warming, which in turn could cause further drops in water levels, the release of still more CO_2 to the atmosphere, additional warming, and so on: an example of positive feedback.

Concept Check 29.3

1. Lycophytes have microphylls, whereas seed plants and monilophytes (ferns and their relatives) have megaphylls. Monilophytes and seed plants also share other traits not found in lycophytes, such as the initiation of new root branches at various points along the length of an existing root.   **2.** Both seedless vascular plants and bryophytes have flagellated sperm that require moisture for fertilization; this shared similarity poses challenges for these species in arid regions. With respect to key differences, seedless vascular plants have lignified, well-developed vascular tissue, a trait that enables the sporophyte to grow tall and that has transformed life on Earth (via the formation of forests). Seedless vascular plants also have true leaves and roots, which, when compared with bryophytes, provide increased surface area for photosynthesis and improve their ability to extract nutrients from soil.   **3.** Three mechanisms contribute to the production of genetic variation in sexual reproduction: independent assortment of chromosomes, crossing over, and random fertilization. If fertilization were to occur between gametes from the same gametophyte, all of the offspring would be genetically identical. This would be the case because all of the cells produced by a gametophyte—including its sperm and egg cells—are the descendants of a single spore and hence are genetically identical. Although crossing over and the independent assortment of chromosomes would continue to generate genetic variation during the production of spores (which ultimately develop into gametophytes), overall the amount of genetic variation produced by sexual reproduction would drop.

Summary of Key Concepts Questions

29.1

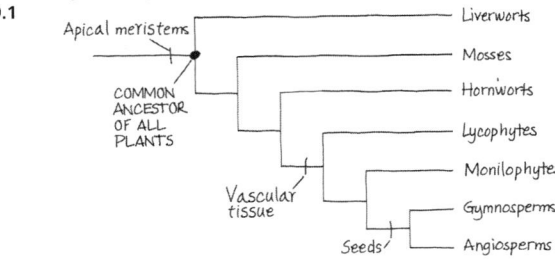

29.2 Some mosses colonize bare, sandy soils, leading to the increased retention of nitrogen in these otherwise low-nitrogen environments. Other mosses harbor nitrogen-fixing cyanobacteria that increase the availability of nitrogen in the ecosystem. The moss *Sphagnum* is often a major component of deposits of peat (partially decayed organic material). Boggy regions with thick layers of peat, known as peatlands, cover broad geographic regions and contain large reservoirs of carbon. By storing large amounts of carbon—in effect, removing CO_2 from the atmosphere—peatlands affect the global climate, making them of considerable ecological importance. **29.3** Lignified vascular tissue provided the strength needed to support a tall plant against gravity, as well as a means to transport water and nutrients to plant parts located high above ground. Roots were another key trait, anchoring the plant to the ground and providing additional structural support for plants that grew tall. Tall plants could shade shorter plants, thereby outcompeting them for light. Because the spores of a tall plant disperse farther than the spores of a short plant, it is also likely that tall plants could colonize new habitats more rapidly than short plants.

Test Your Understanding

1. B **2.** D **3.** C **4.** A **5.** C
6. (a) diploid; (b) haploid; (c) haploid; (d) diploid
7. Based on our current understanding of the evolution of major plant groups, the phylogeny has the four branch points shown here:

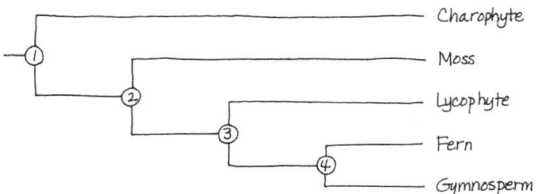

Derived characters unique to the charophyte and plant clade (indicated by branch point 1) include rings of cellulose-synthesizing complexes and flagellated sperm structure. Derived characters unique to the plant clade (branch point 2) include alternation of generations; multicellular, dependent embryos; walled spores produced in sporangia; and apical meristems. Derived characters unique to the vascular plant clade (branch point 3) include life cycles with dominant sporophytes, complex vascular systems (xylem and phloem), and well-developed roots and leaves. Derived characters unique to the monilophyte and seed plant clade (branch point 4) include megaphylls and roots that can branch at various points along the length of an existing root.

Chapter 30

Figure Questions

Figure 30.2 Retaining the gametophyte within the sporophyte shields the egg-containing gametophyte from UV radiation. UV radiation is a mutagen. Hence, we would expect fewer mutations to occur in the egg cells produced by a gametophyte retained within the body of a sporophyte. Most mutations are harmful. Thus, the fitness of embryos should increase because fewer embryos would carry harmful mutations. **Figure 30.3** The seed contains cells from three generations: (1) the current sporophyte (cells of ploidy $2n$, found in the seed coat and in the megasporangium remnant that surrounds the spore wall), (2) the female gametophyte (cells of ploidy n, found in the food supply), and (3) the sporophyte of the next generation (cells of ploidy $2n$, found in the embryo). **Figure 30.4** Mitosis. A single haploid megaspore divides by mitosis to produce a multicellular, haploid female gametophyte. (Likewise, a single haploid microspore divides by mitosis to produce a multicellular male gametophyte.)

Figure 30.9

Figure 30.12 The largest number of seeds this flower could produce is six. We can infer this because the flower is shown as having six ovules, each of which can develop into a single seed. At least six pollen grains would have to germinate to produce six seeds. **Figure 30.14** No. The branching order shown could still be correct if species on the lineages leading to basal angiosperms and magnoliids had originated prior to 150 million years ago, but fossils of that age from those lineages had not yet been discovered. In such a situation, the 140-million-year-old date for the origin of the angiosperms shown on the phylogeny would be incorrect.

Concept Check 30.1

1. To reach the eggs, the flagellated sperm of seedless plants must swim through a film of water, usually over a distance of no more than a few centimeters. In contrast, the sperm of seed plants do not require water because they are produced within pollen grains that can be transported long distances by wind or by animal pollinators. Although flagellated in some species, the sperm of seed plants do not require mobility because pollen tubes convey them from the point at which the pollen grain is deposited (near the ovules) directly to the eggs.
2. The reduced gametophytes of seed plants are nurtured by sporophytes and protected from stress, such as drought conditions and UV radiation. Pollen grains, with walls containing sporopollenin, provide protection during transport by wind or animals. Seeds have one or two layers of protective tissue, the seed coat, that improve survival by providing more protection from environmental stresses than do the walls of spores. Seeds also contain a stored supply of food, which provides nourishment for growth after dormancy is broken and the embryo emerges as a seedling. **3.** If a seed could not enter dormancy, the

embryo would continue to grow after it was fertilized. As a result, the embryo might rapidly become too large to be dispersed, thus limiting its transport. The embryo's chance of survival might also be reduced because it could not delay growth until conditions become favorable.

Concept Check 30.2

1. The pine life cycle illustrates heterospory, as ovulate cones produce megaspores and pollen cones produce microspores. The reduced gametophytes are evident in the form of the microscopic pollen grains that develop from microspores and the microscopic female gametophyte that develops from the megaspore. The egg is shown developing within an ovule, and a pollen tube is shown conveying the sperm. The figure also shows the protective and nutritive features of a seed. **2.** In pines, pollination occurs when a pollen grain is transferred to an ovulate cone scale, the part of a pine that contains ovules. Fertilization occurs when a sperm nucleus unites with an egg nucleus, forming a diploid zygote. Although pollination is required for fertilization, they are separate events: In pines, pollination usually occurs more than a year before fertilization.
3.

Concept Check 30.3

1. In the oak's life cycle, the tree (the sporophyte) produces flowers, which contain gametophytes in pollen grains and ovules; the eggs in ovules are fertilized; the mature ovaries develop into dry fruits called acorns. We can view the oak's life cycle as starting when the acorn seeds germinate, resulting in embryos giving rise to seedlings and finally to mature trees, which produce flowers—and then more acorns. **2.** Pine cones and flowers both have sporophylls, modified leaves that produce spores. Pine trees have separate pollen cones (with pollen grains) and ovulate cones (with ovules inside cone scales). In flowers, pollen grains are produced by the anthers of stamens, and ovules are within the ovaries of carpels. Unlike pine cones, many flowers produce both pollen and ovules. **3.** The fact that the clade with bilaterally symmetrical flowers had more species establishes a correlation between flower shape and the rate of plant speciation. Flower shape is not necessarily responsible for the result because the shape (that is, bilateral or radial symmetry) may have been correlated with another factor that was the actual cause of the observed result. Note, however, that flower shape was associated with increased speciation rates when averaged across 19 different pairs of plant lineages. Since these 19 lineage pairs were independent of one another, this association suggests—but does not establish—that differences in flower shape cause differences in speciation rates. In general, strong evidence for causation can come from controlled, manipulative experiments, but such experiments are usually not possible for studies of past evolutionary events.

Concept Check 30.4

1. Plant diversity can be considered a resource because plants provide many important benefits to humans; as a resource, plant diversity is nonrenewable because if a species is lost to extinction, that loss is permanent. **2.** A detailed phylogeny of the seed plants would identify many different monophyletic groups of seed plants. Using this phylogeny, researchers could look for clades that contained species in which medicinally useful compounds had already been discovered. Identification of such clades would allow researchers to concentrate their search for new medicinal compounds among clade members—as opposed to searching for new compounds in species that were selected at random from the more than 290,000 existing species of seed plants.

Summary of Key Concepts Questions

30.1 The integument of an ovule develops into the protective coat of a seed. The ovule's megaspore develops into a haploid female gametophyte, and two parts of the seed are related to that gametophyte: The food supply of the seed is derived from haploid gametophyte cells, and the embryo of the seed develops after the female gametophyte's egg cell is fertilized by a sperm cell. A remnant of the ovule's megasporangium surrounds the spore wall that encloses the seed's food supply and embryo. **30.2** Gymnosperms arose about 305 million years ago, making them a successful group in terms of their evolutionary longevity. Gymnosperms have the five derived traits common to all seed plants (reduced gametophytes, heterospory, ovules, pollen, and seeds), making them well adapted for life on land. Finally, because gymnosperms dominate immense geographic regions today, the group is also highly successful in geographic distribution. **30.3** Based on fossils known during his lifetime, Darwin was troubled by the relatively sudden and geographically widespread appearance of angiosperms in the fossil record. Recent fossil evidence shows that angiosperms arose and began to diversify over a period of 20–30 million years, a less rapid event than was suggested by the fossils known during Darwin's lifetime. Fossil discoveries have also uncovered extinct lineages of woody seed plants thought to have been more closely related to angiosperms than to gymnosperms; one such group, the Bennettitales, had flowerlike structures that may have been pollinated by insects. In addition, phylogenetic analyses have identified a woody species, *Amborella*, as the most basal lineage of extant angiosperms. The fact that both the extinct seed plant ancestors of angiosperms and the most basal taxon of extant angiosperms were woody suggests that the common ancestor of angiosperms also was woody. **30.4** The loss of tropical forests could contribute to global warming (which would have negative effects on many human societies).

People also depend on Earth's biodiversity for many products and services and hence would be harmed by the loss of species that would occur if the world's remaining tropical forests were cut down. With respect to a possible mass extinction, tropical forests harbor at least 50% of the species on Earth. If the remaining tropical forests were destroyed, large numbers of these species could be driven to extinction, thus rivaling the losses that occurred in the five mass extinction events documented in the fossil record.

Test Your Understanding
1. C **2.** B **3.** A **4.** D **5.** C
6.

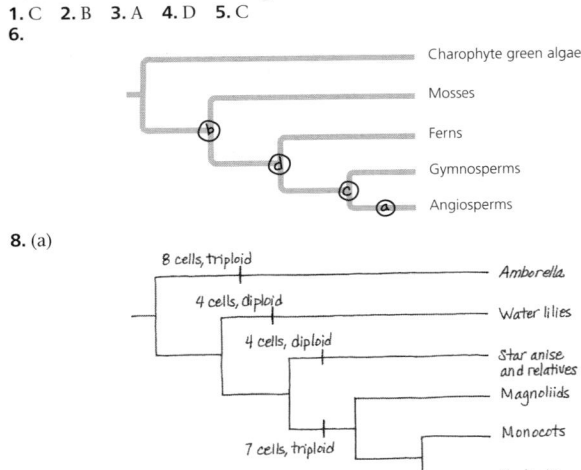

8. (a)

(b) The phylogeny indicates that basal angiosperms differed from other angiosperms in terms of the number of cells in female gametophytes and the ploidy of the endosperm. The ancestral state of the angiosperms cannot be determined from these data alone. It is possible that the common ancestor of angiosperms had seven-celled female gametophytes and triploid endosperm and hence that the eight-celled and four-celled conditions found in basal angiosperms represent derived traits for those lineages. Alternatively, either the eight-celled or four-celled condition may represent the ancestral state.

Chapter 31

Figure Questions
Figure 31.2 DNA from each of these mushrooms would be identical if each mushroom is part of a single hyphal network, as is likely. **Figure 31.5** The haploid spores produced in the sexual portion of the life cycle develop from haploid nuclei that were produced by meiosis; because genetic recombination occurs during meiosis, these spores will differ genetically from one another. In contrast, the haploid spores produced in the asexual portion of the life cycle develop from nuclei that were produced by mitosis; as a result, these spores are genetically identical to one another. **Figure 31.17** One or both of the following would apply to each species: DNA analyses would reveal that it is a member of the ascomycete clade, or aspects of its sexual life cycle would indicate that it is an ascomycete (for example, it would produce asci and ascospores). **Figure 31.18** The hypha is composed of cells that are haploid (n), as indicated by the teal-colored arrow behind it. **Figure 31.20** The mushroom is a basidiocarp, or fruiting body, of the dikaryotic mycelium, and so a cell from its stalk would be dikaryotic ($n + n$). **Figure 31.22** Two possible controls would be E−P− and E+P−. Results from an E−P− control could be compared with results from the E−P+ experiment, and results from an E+P− control could be compared with results from the E+P+ experiment. Together, these two comparisons would indicate whether the addition of the pathogen causes an increase in leaf mortality. Results from an E−P− control could also be compared with results from the second control (E+P−) to determine whether adding the fungal endophytes has a negative effect on the plant.

Concept Check 31.1
1. Both a fungus and a human are heterotrophs. Many fungi digest their food externally by secreting enzymes into the food and then absorbing the small molecules that result from digestion. Other fungi absorb such small molecules directly from their environment. In contrast, humans (and most other animals) ingest relatively large pieces of food and digest the food within their bodies. **2.** The ancestors of such a mutualist most likely secreted powerful enzymes to digest the body of their insect host. Since such enzymes would harm a living host, it is likely that the mutualist would not produce such enzymes or would restrict their secretion and use. **3.** Carbon that enters the plant through stomata is fixed into sugar through photosynthesis. Some of these sugars are absorbed by the fungus that partners with the plant to form mycorrhizae; others are transported within the plant body and used in the plant. Thus, the carbon may be deposited in either the body of the plant or the body of the fungus.

Concept Check 31.2
1. The majority of the fungal life cycle is spent in the haploid stage, whereas the majority of the human life cycle is spent in the diploid stage. **2.** The two mushrooms might be reproductive structures of the same mycelium (the same

organism). Or they might be parts of two separate organisms that have arisen from a single parent organism through asexual reproduction (for example, from two genetically identical asexual spores) and thus carry the same genetic information.

Concept Check 31.3
1. DNA evidence indicates that fungi, animals, and their protistan relatives form a clade, the opisthokonts. Furthermore, chytrids and other fungi thought to be members of basal lineages have posterior flagella, as do most other opisthokonts. This suggests that other fungal lineages lost their flagella after diverging from ancestors that had flagella. **2.** Mycorrhizae form extensive networks of hyphae through the soil, enabling nutrients to be absorbed more efficiently than a plant can do on its own; this is true today, and similar associations were probably very important for the earliest plants (which lacked roots). Evidence for the antiquity of mycorrhizal associations includes fossils showing arbuscular mycorrhizae in the early plant *Aglaophyton* and molecular results showing that genes required for the formation of mycorrhizae are present in liverworts and other basal plant lineages. **3.** Fungi are heterotrophs. Prior to the colonization of land by plants, terrestrial land would have lived where other organisms (or their remains) were present and provided a source of food. Thus, if fungi colonized land before plants, they could have fed on prokaryotes or protists that lived on land or by the water's edge—but not on the plants or animals on which many fungi feed today.

Concept Check 31.4
1. Flagellated spores; molecular evidence also suggests that chytrids include species that belong to lineages that diverged from other fungi early in the history of the group. **2.** Possible answers include the following: In mucoromycetes, the sturdy, thick-walled zygosporangium can withstand harsh conditions and then undergo karyogamy and meiosis when the environment is favorable for reproduction. In one group of mucoromycetes, the glomeromycetes, the hyphae have a specialized morphology that enables the fungi to form arbuscular mycorrhizae with plant roots. In ascomycetes, the asexual spores (conidia) are often produced in chains or clusters at the tips of conidiophores, where they are easily dispersed by wind. The often cup-shaped ascocarps house the sexual spore-forming asci. In basidiomycetes, the basidiocarp supports and protects a large surface area of basidia, from which spores are dispersed. **3.** Such a change to the life cycle of an ascomycete would reduce the number and genetic diversity of ascospores that result from a mating event. Ascospore number would drop because a mating event would lead to the formation of only one ascus. Ascospore genetic diversity would also drop because in ascomycetes, one mating event leads to the formation of asci by many different dikaryotic cells. As a result, genetic recombination and meiosis occur independently many different times—which could not happen if only a single ascus was formed. It is also likely that if such an ascomycete formed an ascocarp, the shape of the ascocarp would differ considerably from that found in its close relatives.

Concept Check 31.5
1. A suitable environment for growth, retention of water and minerals, protection from intense sunlight, and protection from being eaten **2.** A hardy spore stage enables dispersal to host organisms through a variety of mechanisms; their ability to grow rapidly in a favorable new environment enables them to capitalize on the host's resources. **3.** Many different outcomes might have occurred. Organisms that currently form mutualisms with fungi might have gained the ability to perform the tasks currently done by their fungal partners, or they might have formed similar mutualisms with other organisms (such as bacteria). Alternatively, organisms that currently form mutualisms with fungi might be less effective at living in their present environments. For example, the colonization of land by plants might have been more difficult. And if plants did eventually colonize land without fungal mutualists, natural selection might have favored plants that formed more highly divided and extensive root systems (in part replacing mycorrhizae).

Summary of Key Concepts Questions
31.1 The body of a multicellular fungus typically consists of thin filaments called hyphae. These filaments form an interwoven mass (mycelium) that penetrates the substrate on which the fungus grows and feeds. Because the individual filaments are thin, the surface-to-volume ratio of the mycelium is maximized, making nutrient absorption highly efficient.
31.2

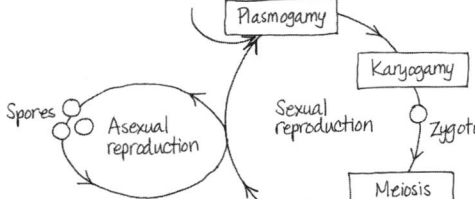

31.3 Phylogenetic analyses show that fungi and animals are more closely related to each other than either is to other multicellular eukaryotes (such as plants or multicellular algae). These analyses also show that fungi are more closely related to single-celled protists called nucleariids than they are to animals, whereas animals are more closely related to a different group of single-celled protists, the choanoflagellates, than they are to fungi. In combination, these results indicate that multicellularity evolved in fungi and animals independently, from different single-celled ancestors.

Cryptomycetes
Microsporidians
Chytrids
Zoopagomycetes
Mucoromycetes
Acomycetes
Basidiomycetes

31.5 As decomposers, fungi break down the bodies of dead organisms, thereby recycling elements between the living and nonliving environments. Without the activities of fungi and bacterial decomposers, essential nutrients would remain tied up in organic matter, and life as we know it would cease. As an example of their key role as mutualists, fungi form mycorrhizal associations with plants. These associations improve the growth and survival of plants, thereby indirectly affecting the many other species (humans included) that depend on plants. As pathogens, fungi harm other species. In some cases, fungal pathogens have caused their host populations to decline across broad geographic regions, as in the case of the American chestnut.

Test Your Understanding

1. B **2.** D **3.** A **4.** D

Chapter 32

Figure Questions

Figure 32.3 As described in ❶ and ❷, choanoflagellates and a broad range of animals have collar cells. Since collar cells have never been observed in plants, fungi, or non-choanoflagellate protists, this suggests that choanoflagellates may be more closely related to animals than to other eukaryotes. If choanoflagellates are more closely related to animals than to any other group of eukaryotes, choanoflagellates and animals should share other traits that are not found in other eukaryotes. The data described in ❸ are consistent with this prediction.
Figure 32.8 (1) Any imaginary slice through the central axis of a radial animal divides its body into mirror images. As a result, a radial animal has no front and back sides and no left and right sides.
(2)

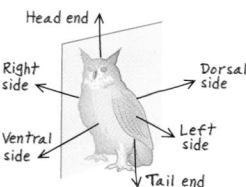

Head end
Right side
Dorsal side
Ventral side
Left side
Tail end

Figure 32.11 Cnidaria is the sister phylum in this tree.

Concept Check 32.1

1. In most animals, the zygote undergoes cleavage, which leads to the formation of a blastula. Next, in gastrulation, one end of the embryo folds inward, producing layers of embryonic tissue. As the cells of these layers differentiate, a wide variety of animal forms are produced. Despite the diversity of animal forms, animal development is controlled by a similar set of *Hox* genes across a broad range of taxa.   **2.** The imaginary plant would require tissues composed of cells that were analogous to the muscle and nerve cells found in animals: "Muscle" tissue would be necessary for the plant to chase prey, and "nerve" tissue would be required for the plant to coordinate its movements when chasing prey. To digest captured prey, the plant would need to either secrete enzymes into one or more digestive cavities (which could be modified leaves, as in a Venus flytrap) or secrete enzymes outside of its body and feed by absorption. To extract nutrients from the soil—yet be able to chase prey—the plant would need something other than fixed roots, perhaps retractable "roots" or a way to ingest soil. To conduct photosynthesis, the plant would require chloroplasts. Overall, such an imaginary plant would be very similar to an animal that had chloroplasts and retractable roots.

Concept Check 32.2

1. c, b, a, d **2.** The red-colored portion of the tree represents ancestors of animals that lived between 1 billion years ago and 770 million years ago. Although these ancestors are more closely related to animals than to fungi, they would not be classified as animals. One example of an ancestor represented by the red-colored portion of this tree is the most recent common ancestor shared by choanoflagellates and animals (see Figure 32.3). That common ancestor was not an animal (or a choanoflagellate), but it was a direct ancestor of the animals.   **3.** In descent with modification, an organism shares characteristics with its ancestors (due to their shared ancestry), yet it also differs from its ancestors (because organisms accumulate differences over time as they adapt to their surroundings). As an example, consider the evolution of animal cadherin proteins, a key step in the origin of multicellular animals. These proteins illustrate both

of these aspects of descent with modification: Animal cadherin proteins share many protein domains with a cadherin-like protein found in their choanoflagellate ancestors, yet they also have a unique "CCD" domain that is not found in choanoflagellates.

Concept Check 32.3

1. A snail has a spiral and determinate cleavage pattern; a human has radial, indeterminate cleavage. In a snail, the coelomic cavity is formed by splitting of mesoderm masses; in a human, the coelom forms from folds of archenteron. In a snail, the mouth forms from the blastopore; in a human, the anus develops from the blastopore.   **2.** Animals that lack a body cavity tend to have thin, flat bodies. Such animals don't require an internal transport system: With bodies that are only a few cells thick, the exchange of nutrients, gases, and wastes can occur across the entire body surface.   **3.** Most triploblasts have two openings to their digestive tract, a mouth and an anus. As such, their bodies have a structure that is analogous to that of a doughnut: The digestive tract (the hole of the doughnut) runs from the mouth to the anus and is surrounded by various tissues (the solid part of the doughnut). The doughnut analogy is most obvious at early stages of development (see Figure 32.10c).

Concept Check 32.4

1. Cnidarians possess tissues, while sponges do not. Also unlike sponges, cnidarians exhibit body symmetry, though it is radial and not bilateral as in most other animal phyla.
2.

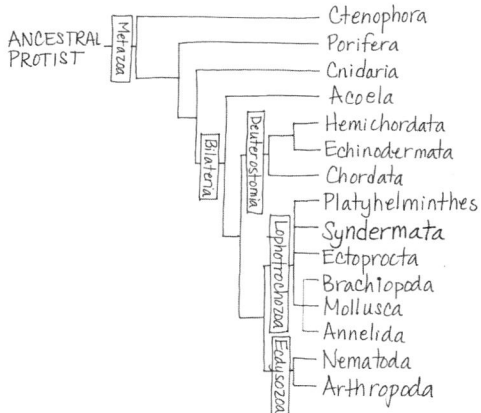

ANCESTRAL PROTIST — Metazoa
Ctenophora
Porifera
Cnidaria
Acoela
Hemichordata
Echinodermata
Chordata
Platyhelminthes
Syndermata
Ectoprocta
Brachiopoda
Mollusca
Annelida
Nematoda
Arthropoda

Under the hypothesis that ctenophores are basal metazoans, sponges (which lack tissues) would be nested within a clade whose other members all have tissues. As a result, a group composed of animals with tissues would not form a clade.   **3.** The phylogeny in Figure 32.11 indicates that molluscs are members of Lophotrochozoa, one of the three main groups of bilaterians (the others being Deuterostomia and Ecdysozoa). As seen in Figure 25.11, the fossil record shows that molluscs were present tens of millions of years before the Cambrian explosion. Thus, long before the Cambrian explosion, the lophotrochozoan clade had formed and was evolving independently of the evolutionary lineages leading to Deuterostomia and Ecdysozoa. Based on the phylogeny in Figure 32.11, we can also conclude that the lineages leading to Deuterostomia and Ecdysozoa were independent of one another before the Cambrian explosion. Since the lineages leading to the three main clades of bilaterians were evolving independently of one another prior to the Cambrian explosion, that explosion could be viewed as consisting of three "explosions," not one.

Summary of Key Concepts Questions

32.1 Unlike animals, which are heterotrophs that ingest their food, plants are autotrophs, and fungi are heterotrophs that grow on their food and feed by absorption. Animals lack cell walls, which are found in both plants and fungi. Animals also have muscle tissue and nerve tissue, which are not found in either plants or fungi. In addition, the sperm and egg cells of animals are produced by meiotic division, unlike what occurs in plants and fungi (where reproductive cells such as sperm and eggs are produced by mitotic division). Finally, animals regulate the development of body form with *Hox* genes, a unique group of genes that is not found in either plants or fungi.   **32.2** Current hypotheses about the cause of the Cambrian explosion include new predator-prey relationships, an increase in atmospheric oxygen, and an increase in developmental flexibility provided by the origin of *Hox* genes and other genetic changes.   **32.3** Body plans provide a helpful way to compare and contrast key features of organisms. However, phylogenetic analyses show that similar body plans have arisen independently in different groups of organisms. As such, similar body plans may have arisen by convergent evolution and hence may not be informative about evolutionary relationships.   **32.4** Listed in order from the most to the least inclusive clade, humans belong to Metazoa, Eumetazoa, Bilateria, Deuterostomia, and Chordata.

Test Your Understanding

1. A **2.** D **3.** C **4.** C

Chapter 33

Figure Questions

Figure 33.7 *Obelia* is an animal, and its life cycle is indeed most similar to the generalized life cycle of animals (Figure 13.6a). In *Obelia*, both the polyp and the medusa are diploid organisms. As in other animals, in *Obelia* only the single-celled gametes are haploid. By contrast, plants and some algae (Figure 13.6b) have a multicellular haploid generation and a multicellular diploid generation. *Obelia* also differs from fungi and some protists (Figure 13.6c) in that the diploid stage of those organisms is unicellular. **Figure 33.8** Possible examples include the Golgi apparatus (flattening; increases area for receiving and transporting proteins), the cristae of mitochondria (folding; increases the surface area available for cellular respiration), the cardiovascular system (branching; increase area for materials exchange in tissues), and root hairs (projections; increase area for absorption). **Figure 33.10** Adding fertilizer to the water supply would probably increase the abundance of algae, and that, in turn, would likely increase the abundance of snails (which eat algae). If the water was also contaminated with infected human feces or urine, an increase in the number of snails would likely lead to an increase in the abundance of blood flukes (which require snails as an intermediate host). As a result, the occurrence of schistosomiasis might increase. **Figure 33.21** The extinction of freshwater bivalves might lead to an increase in the abundance of photosynthetic protists and bacteria. Because these organisms are at the base of aquatic food webs, increases in their abundance could have major effects on aquatic communities (including both increases and decreases in the abundance of other species). **Figure 33.29** Such a result would be consistent with the origin of the *Ubx* and *abd-A Hox* genes having played a major role in the evolution of increased body segment diversity in arthropods. However, note that such a result would simply show that the presence of the *Ubx* and *abd-A Hox* genes was *correlated with* an increase in body segment diversity in arthropods; it would not provide direct experimental evidence that the origin of the *Ubx* and *abd-A* genes *caused* an increase in arthropod body segment diversity. **Figure 33.35** You should have circled the clade that includes the insects, remipedians, and other crustaceans, along with the branch point that represents their most recent common ancestor.

Concept Check 33.1

1. The flagella of choanocytes draw water through their collars, which trap food particles. The particles are engulfed by phagocytosis and digested, either by choanocytes or by amoebocytes. **2.** The collar cells of sponges bear a striking resemblance to a choanoflagellate cell. This suggests that the last common ancestor of animals and their protist sister group may have resembled a choanoflagellate. Nevertheless, mesomycetozoans could still be the sister group of animals. If this is the case, the lack of collar cells in mesomycetozoans would indicate that over time their structure evolved in ways that caused it to no longer resemble a choanoflagellate cell. It is also possible that choanoflagellates and sponges share similar looking collar cells as a result of convergent evolution.

Concept Check 33.2

1. Both the polyp and the medusa are composed of an outer epidermis and an inner gastrodermis separated by a gelatinous layer, the mesoglea. The polyp is a cylindrical form that adheres to the substrate by its aboral end; the medusa is a flattened, mouth-down form that moves freely in the water. **2.** Both a feeding polyp and a medusa are diploid, as indicated by the pink arrow in the diagram. The medusa stage produces haploid gametes. **3.** Evolution is not goal oriented; hence, it would not be correct to argue that cnidarians are not "highly evolved" simply because their form had changed relatively little over the past 560 million years. Instead, the fact that cnidarians have persisted for hundreds of millions of years indicates that the cnidarian body plan is a highly successful one.

Concept Check 33.3

1. Tapeworms can absorb food from their environment and release ammonia into their environment through their body surface because their body is very flat, due in part to the lack of a body cavity. **2.** The inner tube is the alimentary canal, which runs the length of the body. The outer tube is the body wall. The two tubes are separated by the coelom. **3.** All molluscs have inherited a foot from their common ancestor. However, in different groups of molluscs, the structure of the foot has been modified over time by natural selection. In gastropods, the foot is used as a holdfast or to move slowly on the substrate. In cephalopods, the foot has been modified into part of the tentacles and into an excurrent siphon, through which water is propelled (resulting in movement in the opposite direction).

Concept Check 33.4

1. Nematodes lack body segments and a coelom; annelids have both. **2.** The arthropod exoskeleton, which had already evolved in the ocean, allows terrestrial species to retain water and support their bodies on land. Wings allow insects to disperse quickly to new habitats and to find food and mates. The tracheal system allows for efficient gas exchange despite the presence of an exoskeleton. **3.** Yes. Under the traditional hypothesis, we would expect body segmentation to be controlled by similar *Hox* genes in annelids and arthropods. However, if annelids are in Lophotrochozoa and arthropods are in Ecdysozoa (as current evidence suggests), body segmentation may have evolved independently in these two groups. In such a case, we might expect that different *Hox* genes would control the development of body segmentation in the two clades.

Concept Check 33.5

1. Each tube foot consists of an ampulla and a podium. When the ampulla squeezes, it forces water into the podium, which causes the podium to expand and contact the substrate. Adhesive chemicals are then secreted from the base of the podium, thereby attaching the podium to the substrate. **2.** Both insects and nematodes are members of Ecdysozoa, one of the three major clades of

bilaterians. Therefore, a characteristic shared by *Drosophila* and *Caenorhabditis* may be informative for other members of their clade—but not necessarily for members of Deuterostomia. Instead, Figure 33.2 suggests that a species within Echinodermata or Chordata might be a more appropriate invertebrate model organism from which to draw inferences about humans and other vertebrates. **3.** Echinoderms include species with a wide range of body forms. However, even echinoderms that look very different from one another, such as sea stars and sea cucumbers, share characteristics unique to their phylum, including a water vascular system and tube feet. The differences between echinoderm species illustrate the diversity of life, while the characteristics they share illustrate the unity of life. The match between organisms and their environments can be seen in such echinoderm features as the eversible stomachs of sea stars (enabling them to digest prey that are larger than their mouth) and the complex, jaw-like structure that sea urchins use to eat seaweed.

Summary of Key Concepts Questions

33.1 The sponge body consists of two layers of cells, both of which are in contact with water. As a result, gas exchange and waste removal occur as substances diffuse into and out of the cells of the body. Choanocytes and amoebocytes ingest food particles from the surrounding water. Choanocytes also release food particles to amoebocytes, which then digest the food particles and deliver nutrients to other cells. **33.2** The cnidarian body plan consists of a sac with a central digestive compartment, the gastrovascular cavity. The single opening to this compartment serves as both a mouth and an anus. The two main variations on this body plan are sessile polyps (which adhere to the substrate at the end of the body opposite to the mouth/anus) and motile medusae (which move freely through the water and resemble flattened, mouth-down versions of polyps). **33.3** No. Some lophotrochozoans have a crown of ciliated tentacles that function in feeding (called a lophophore), while others go through a distinctive developmental stage known as trochophore larvae. Many other lophotrochozoans do not have either of these features. As a result, the clade is defined primarily by DNA similarities, not morphological similarities. **33.4** Many nematode species live in soil and in sediments on the bottom of bodies of water. These free-living species play important roles in decomposition and nutrient cycling. Other nematodes are parasites, including many species that attack the roots of plants and some that attack animals (including humans). Arthropods have profound effects on all aspects of ecology. In aquatic environments, crustaceans play key roles as grazers (of algae), scavengers, and predators and some species, such as krill, are important sources of food for whales and other vertebrates. On land, it is difficult to think of features of the natural world that are not affected in some way by insects and other arthropods, such as spiders and ticks. There are more than 1 million species of insects, many of which have enormous ecological effects as herbivores, predators, parasites, decomposers, and vectors of disease. Insects are also key sources of food for many organisms, including humans in some regions of the world. **33.5** Echinoderms and chordates are both members of Deuterostomia, one of the three main clades of bilaterian animals. As such, chordates (including humans) are more closely related to echinoderms than we are to animals in any of the other phyla covered in this chapter. Nevertheless, echinoderms and chordates have evolved independently for over 500 million years. This statement does not contradict the close relationship of echinoderms and chordates, but it does make clear that "close" is a relative term indicating that these two phyla are more closely related to each other than either is to animal phyla not in Deuterostomia.

Test Your Understanding

1. A   **2.** C   **3.** B   **4.** C   **5.** C   **6.** D

Chapter 34

Figure Questions
Figure 34.2

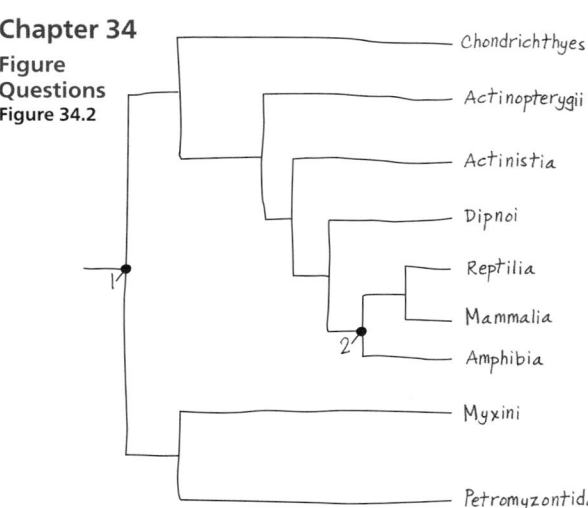

The redrawn tree shows mammals (including humans) as nested near the middle of the evolutionary tree of vertebrates. Showing the vertebrate tree in this way provides a visual illustration of the fact that the evolutionary history of vertebrates did not consist of a series of steps "leading to"

humans. **Figure 34.6** The patterns in these figures suggest that specific *Hox* genes, as well as the order in which they are expressed, have been highly conserved over the course of evolution. **Figure 34.20** *Tiktaalik* was a lobe-fin fish that had both fish and tetrapod characters. Like a fish, *Tiktaalik* had fins, scales, and gills. As described by Darwin's concept of descent with modification, such shared characters can be attributed to descent from ancestral species—in this case, *Tiktaalik*'s descent from fish ancestors. *Tiktaalik* also had traits that were unlike a fish but like a tetrapod, including a flat skull, a neck, a full set of ribs, and the skeletal structure of its fin. These characters illustrate the second part of descent with modification, showing how ancestral features had become modified over time. **Figure 34.21** Sometime between 350 mya and 340 mya. We can infer this because amphibians must have originated after the most recent common ancestor of amphibians and amniotes (and that ancestor is shown as having lived 350 mya), but no later than the date of the earliest known fossils of amphibians (shown in the figure as 340 mya). **Figure 34.25** Pterosaurs did not descend from the common ancestor of all dinosaurs; hence, pterosaurs are not dinosaurs. However, birds are descendants of the common ancestor of the dinosaurs. As a result, a monophyletic clade of dinosaurs must include birds. In that sense, birds are dinosaurs. **Figure 34.37** In a catabolic pathway, like the aerobic processes of cellular respiration, water is released as a by-product when an organic compound such as glucose is mixed with oxygen. The kangaroo rat can retain and use that water, decreasing its need to drink water. **Figure 34.38** In general, the process of exaptation occurs as a structure that had one function acquires a different function via a series of intermediate stages. Each of these intermediate stages typically has some function in the organism in which it is found. The incorporation of articular and quadrate bones into the mammalian ear illustrates exaptation because these bones originally evolved as part of the jaw, where they functioned as the jaw hinge, but over time they became co-opted for another function, namely, the transmission of sound. **Figure 34.44** As shown in this phylogeny, chimpanzees and humans represent the tips of separate branches of evolution. As such, the human and chimpanzee lineages have evolved independently after they diverged from their common ancestor—an event that took place about 8 million years ago. Hence, it is incorrect to say that humans evolved from chimpanzees (or vice versa). If humans had descended from chimpanzees, for example, the human lineage would be nested within the chimpanzee lineage, much as birds are nested within the reptile clade (see Figure 34.25).

Concept Check 34.1

1. The four characters are a notochord; a dorsal, hollow nerve chord; pharyngeal slits or clefts; and a muscular, post-anal tail. **2.** In humans, these characters are present only in the embryo. The notochord becomes disks between the vertebrae; the dorsal, hollow nerve cord develops into the brain and spinal cord; the pharyngeal clefts develop into various adult structures, and the tail is almost completely lost. **3.** You would expect the vertebrate groups Actinopterygii, Actinistia, Dipnoi, Amphibia, Reptilia, and Mammalia to have lungs or lung derivatives. All of these groups originate to the right of (evolved after) the hatch mark indicating the appearance of this derived character in their lineage.

Concept Check 34.2

1. Parasitic lampreys have a round, rasping mouth, which they use to attach to fish. Non-parasitic lampreys feed only as larvae; these larvae resemble lancelets and like them, are suspension feeders. Conodonts had two sets of mineralized dental elements, which may have been used to impale prey and cut it into smaller pieces. **2.** Such a finding suggests that early organisms with a head were favored by natural selection in several different evolutionary lineages. However, while a logical argument can be made that having a head was advantageous, fossils alone do not constitute proof. **3.** In armored jawless vertebrates, bone served as external armor that may have provided protection from predators. Some species also had mineralized mouthparts, which could be used for either predation or scavenging.

Concept Check 34.3

1. Both are gnathostomes and have jaws, four clusters of *Hox* genes, enlarged forebrains, and lateral line systems. Shark skeletons consist mainly of cartilage, whereas tuna have bony skeletons. Sharks also have a spiral valve. Tuna have an operculum and a swim bladder, as well as flexible rays supporting their fins. **2.** Aquatic gnathostomes have jaws (an adaptation for feeding) and paired fins and a tail (adaptations for swimming). Aquatic gnathostomes also typically have streamlined bodies for efficient swimming and swim bladders or other mechanisms (such as oil storage in sharks) for buoyancy. **3.**

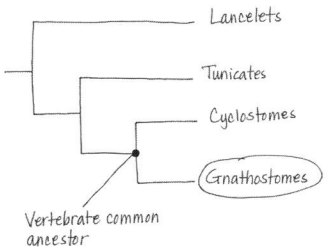

4. Yes, that could have happened. The paired appendages of aquatic gnathostomes other than the lobe-fins could have served as a starting point for the evolution of limbs. The colonization of land by aquatic gnathostomes other than the lobe-fins might have been facilitated in lineages that possessed lungs, as that would have enabled those organisms to breathe air.

Concept Check 34.4

1. Tetrapods are thought to have originated about 365 million years ago when the fins of some lobe-fins evolved into the limbs of tetrapods. In addition to their four limbs with digits—a key derived trait for which the group is named—other derived traits of tetrapods include a neck (consisting of vertebrae that separate the head from the rest of the body) and a pelvic girdle that is fused to the backbone. **2.** Some fully aquatic species are paedomorphic, retaining larval features for life in water as adults. Species that live in dry environments may avoid dehydration by burrowing or living under moist leaves, and they protect their eggs with foam nests, viviparity, and other adaptations. **3.** Many amphibians spend part of their life cycle in aquatic environments and part on land. Thus, they may be exposed to a wide range of environmental problems, including water and air pollution and the loss or degradation of aquatic and/or terrestrial habitats. In addition, amphibians have highly permeable skin, providing relatively little protection from external conditions, and their eggs do not have a protective shell.

Concept Check 34.5

1. The amniotic egg provides protection to the embryo and allows the embryo to develop on land, eliminating the necessity of a watery environment for reproduction. Another key adaptation is rib cage ventilation, which improves the efficiency of air intake and may have allowed early amniotes to dispense with breathing through their skin. Finally, not breathing through their skin allowed amniotes to develop relatively impermeable skin, thereby conserving water. **2.** Yes. Although snakes lack limbs, they descended from lizards with legs. Some snakes retain vestigial pelvic and leg bones, providing evidence of their descent from an ancestor with legs. **3.** Birds have weight-saving modifications, including the absence of teeth, a urinary bladder, and a second ovary in females. The wings and feathers are adaptations that facilitate flight, as do efficient respiratory and circulatory systems that support a high metabolic rate. **4.** (a) synapsids, (b) tuataras, (c) turtles

Concept Check 34.6

1. Monotremes lay eggs. Marsupials give birth to very small live young that attach to a nipple in the mother's pouch, where they complete development. Eutherians give birth to more developed live young. **2.** Hands and feet adapted for grasping, flat nails, large brain, forward-looking eyes on a flat face, parental care, and movable big toe and thumb **3.** Mammals are endothermic, enabling them to live in a wide range of habitats. Milk provides young with a balanced set of nutrients, and hair and a layer of fat under the skin help mammals retain heat. Mammals have differentiated teeth, enabling them to eat many different kinds of food. Mammals also have relatively large brains, and many species are capable learners. Following the mass extinction at the end of the Cretaceous period, the absence of large terrestrial dinosaurs may have opened many new ecological niches to mammals, promoting an adaptive radiation. Continental drift also isolated many groups of mammals from one another, promoting the formation of many new species.

Concept Check 34.7

1. Hominins are a clade within the ape clade that includes humans and all species more closely related to humans than to other apes. The derived characters of hominins include bipedal locomotion and relatively larger brains. **2.** In hominins, bipedal locomotion evolved long before large brain size. *Homo ergaster*, for example, was fully upright, bipedal, and as tall as modern humans, but its brain was significantly smaller than that of modern humans. **3.** Yes, both can be correct. *Homo sapiens* may have established populations outside of Africa as early as 180,00 years ago, as indicated by the fossil record. However, those populations may have left few or no descendants today. Instead, all living humans may have descended from Africans that spread from Africa roughly 50,000 years ago, as indicated by genetic data.

Summary of Key Concepts Questions

34.1 Lancelets are the most basal group of living chordates, and as adults they have key derived characters of chordates. This suggests that the chordate common ancestor may have resembled a lancelet in having an anterior end with a mouth along with the following four derived characters: a notochord; a dorsal, hollow nerve cord; pharyngeal slits or clefts; and a muscular, post-anal tail. **34.2** Conodonts, among the earliest vertebrates in the fossil record, were very abundant for 300 million years. While jawless, their well-developed teeth provide early signs of bone formation. Other species of jawless vertebrates developed armor on the outside of their bodies, which probably helped protect them from predators. Like lampreys, these species had paired fins for locomotion and an inner ear with semicircular canals that provided a sense of balance. There were many species of these armored jawless vertebrates, but they all became extinct by the close of the Devonian period, 359 million years ago. **34.3** The origin of jaws altered how fossil gnathostomes obtained food, which in turn had large effects on ecological interactions. Predators could use their jaws to grab prey or remove chunks of flesh, stimulating the evolution of increasingly sophisticated means of defense in prey species. Evidence for these changes can be found in the fossil record, which includes fossils of 10-m-long predators with remarkably powerful jaws, as well as lineages of well-defended prey species whose bodies were covered by armored plates. **34.4** Amphibians require water for reproduction; their bodies can lose water rapidly through their moist, highly permeable skin; and amphibian eggs do not have a shell and hence are vulnerable to desiccation. **34.5** Birds are descended from theropod dinosaurs, and dinosaurs are nested within the archosaur lineage, one of the two main reptile lineages. Thus, the other living archosaur reptiles, the crocodilians, are more closely related to birds than they are to non-archosaur reptiles such as lizards. As a result, birds are considered reptiles. (Note that if reptiles were defined as excluding birds, the reptiles would not form a clade; instead, the reptiles would be a paraphyletic group.) **34.6** Mammals are members of a group of amniotes called synapsids. Early (nonmammalian) synapsids laid eggs and had a sprawling gait. Fossil evidence shows that mammalian features arose gradually over a period of more than 100 million years. For example, the jaw was modified over time in nonmammalian synapsids, eventually coming to resemble that of a mammal. By 180 million years ago, the first mammals had appeared. There were many species of early mammals, but most of them were small, and they were not abundant or dominant members of their community. Mammals did not rise to ecological dominance until after the extinction of the dinosaurs. **34.7** The fossil record shows that from 4.5 to 2.5 million years ago, a wide range of hominin species walked upright but had relatively small brain sizes. About 2.5 million years ago, the first members of genus *Homo* emerged. These species used tools and had larger brains than those of earlier hominins. Fossil evidence indicates that multiple members of our genus were alive at any given point in time. Furthermore, until about 1.3 million years ago, these various *Homo* species also coexisted with members of earlier hominin lineages, such as *Paranthropus*. The different hominins alive at the same periods of time varied in body size, body shape, brain size, dental morphology, and the capacity for tool use. Ultimately, except for *Homo sapiens*, all of these species became extinct. Overall, human evolution can be viewed as an evolutionary tree with many branches—the only surviving lineage of which is our own.

Test Your Understanding

1. D **2.** C **3.** B **4.** C **5.** D **6.** A

8. (a) Because brain size tends to increase consistently in such lineages, we can conclude that natural selection favored the evolution of larger brains and hence that the benefits outweighed the costs. (b) As long as the benefits of brains that are large relative to body size are greater than the costs, large brains can evolve. Natural selection might favor the evolution of brains that are large relative to body size because such brains confer an advantage in obtaining mates and/or an advantage in survival.

(c)

Mortality tends to be lower in birds with larger brains.

Chapter 35

Figure Questions

Figure 35.5 All three examples have nodes because they're all stems. The nodes are the locations where the axillary buds arise.

Figure 35.11

(1)

(2)

As a result of the addition of secondary xylem cells, the vascular cambium is pushed farther to the outside.

Figure 35.15

Figure 35.17 Pith and cortex are defined, respectively, as ground tissue that is internal and ground tissue that is external to vascular tissue. Since vascular bundles of monocot stems are scattered throughout the ground tissue, there is no clear distinction between internal and external relative to the vascular tissue. **Figure 35.19** The vascular cambium produces growth that increases the diameter of a stem or root. The tissues that are exterior to the vascular cambium cannot keep pace with the growth because their cells no longer divide. As a result, these tissues rupture. **Figure 35.23** Periderm (mainly cork and cork cambium), primary phloem, secondary phloem, vascular cambium, secondary xylem (sapwood and heartwood), primary xylem, and pith. At the base of ancient redwood that is many centuries old, the remnants of primary growth (primary phloem, primary xylem, and pith) would be quite insignificant. **Figure 35.30** Every root epidermal cell would develop a root hair. **Figure 35.32** Another example of homeotic gene mutation is the mutation in a *Hox* gene that causes legs to form in place of antennae in *Drosophila* (see Figure 18.20).

Figure 35.33

(b)

Concept Check 35.1

1. The vascular tissue system connects leaves and roots, allowing sugars to move from leaves to roots in the phloem and allowing water and minerals to move to the leaves in the xylem. **2.** To get sufficient energy from photosynthesis, we would need lots of surface area exposed to the sun. This large surface-to-volume ratio, however, would create a new problem—evaporative water loss. We would have to be permanently connected to a water source—the soil, also our source of minerals. In short, we would probably look and behave very much like plants. **3.** As plant cells enlarge, they typically form a huge central vacuole that contains a dilute, watery sap. Central vacuoles enable plant cells to become large with only a minimal investment of new cytoplasm. The orientation of the cellulose microfibrils in plant cell walls affects the growth pattern of cells.

Concept Check 35.2

1. Yes. In a woody plant, secondary growth is occurring in the older parts of the stem and root, while primary growth is occurring at the root and shoot tips. **2.** The largest, oldest leaves would be lowest on the shoot. Since they would probably be heavily shaded, they would not photosynthesize much regardless of their size. Determinate growth benefits the plant by keeping it from investing an ever-increasing amount of resources into organs that provide little photosynthetic product. **3.** No. The carrot roots will probably be smaller at the end of the second year because the food stored in the roots will be used to produce flowers, fruits, and seeds.

Concept Check 35.3

1. In roots, primary growth occurs in three successive stages, moving away from the tip of the root: the zones of cell division, elongation, and differentiation. In shoots, it occurs at the tip of apical buds, with leaf primordia arising along the sides of an apical meristem. Most growth in length occurs in older internodes below the shoot tip. **2.** The fossil probably came from a floating leaf because having stomata exclusively on the upper surface would be a poor adaptation for a desert plant since it would lead to water loss. A floating leaf, on the other hand, would benefit from having stomata on its upper surface since only that surface is in contact with the gaseous environment. **3.** Root hairs are cellular extensions that increase the surface area of the root epidermis, thereby enhancing the absorption of minerals and water. Microvilli are extensions that increase the absorption of nutrients by increasing the surface area of the gut.

Concept Check 35.4

1. The sign will still be 2 m above the ground because this part of the tree is no longer growing in length (primary growth); it is now growing only in thickness (secondary growth). **2.** Stomata must be able to close because evaporation is much more intensive from leaves than from the trunks of woody trees as a result of the higher surface-to-volume ratio in leaves. **3.** Since there is little seasonal temperature variation in the tropics, the growth rings of a tree from the tropics would be difficult to discern unless the tree came from an area that had pronounced wet and dry seasons. **4.** The tree would die slowly. Girdling removes an entire ring of secondary phloem (part of the bark), completely preventing transport of sugars and starches from the shoots to the roots. After several weeks, the roots would have used all of their stored carbohydrate reserves and would die.

Concept Check 35.5

1. Although all the living vegetative cells of a plant have the same genome, they develop different forms and functions because of differential gene expression. **2.** Plants show indeterminate growth; juvenile and mature phases are found on the same individual plant; and cell differentiation in plants is more dependent on final position than on lineage. **3.** One hypothesis is that tepals arise if *B* gene activity is present in all three of the outer whorls of the flower.

Summary of Key Concepts Questions

35.1 Here are a few examples: The cuticle of leaves and stems protects these structures from desiccation. Collenchyma and sclerenchyma cells have thick walls that provide support for plants. Strong, branching root systems help anchor plants in the soil. **35.2** Primary growth arises from apical meristems and involves production and elongation of organs. Secondary growth arises from lateral meristems and adds to the diameter of roots and stems. **35.3** Lateral roots emerge from the pericycle and destroy plant cells as they emerge. In stems, branches arise from axillary buds and do not destroy any cells. **35.4** With the evolution of secondary growth, plants were able to grow taller and shade competitors. **35.5** The orientation of cellulose microfibrils in the innermost layers of the cell wall causes growth along one axis. Microtubules in the cell's outermost cytoplasm play a key role in regulating the axis of cell expansion because it is their orientation that determines the orientation of cellulose microfibrils.

1. D **2.** C **3.** C **4.** A **5.** D **6.** C **7.** D **8.** B **9.** D **10.** D
11.

12. The tea and iris leaves differ in the arrangement of photosynthetic mesophyll cells. In the tea leaf, the mesophyll cells are divided into two layers, with the lower part of the leaf having a loosely arranged spongy layer with large air spaces, and the upper part having a palisade layer of tightly packed cells. In contrast, the mesophyll cells in the iris leaf are evenly distributed with no large air spaces. These differences are associated with the natural orientation of the leaves of these two species. Tea leaves are horizontally oriented and are more likely to receive light on the upper surface, so most of the mesophyll cells are concentrated in a palisade layer to absorb light efficiently. In contrast, iris leaves are vertically oriented with both sides illuminated about equally during the course of the day, so the mesophyll cells are evenly distributed. In both leaves, structure fits function because the arrangement of mesophyll cells maximizes photosynthesis.
13. Pranksters must have been carried the bicycle high into the tree and threaded it over a young woody branch: Primary growth did not raise the bicycle off the ground. Over time, secondary growth in the region of the tree notch enveloped the bicycle.

Chapter 36

Figure Questions

Figure 36.2 There would be no reference to photosynthesis because it ceases at night. Also, the directions of the CO_2 and O_2 arrows associated with the leaves would be reversed because at night only the gas exchange related to cellular respiration is occurring.   **Figure 36.3** The leaves are being produced in a counterclockwise spiral. The next leaf primordium will emerge approximately between and to the inside of leaves 8 and 13.   **Figure 36.4** A higher leaf area index will not necessarily increase photosynthesis because of upper leaves shading lower leaves.   **Figure 36.6** A proton pump inhibitor would depolarize (increase) the membrane potential because fewer hydrogen ions would be pumped out across the plasma membrane. The immediate effect of an inhibitor of the H^+/sucrose transporter would be to hyperpolarize (decrease) the membrane potential because fewer hydrogen ions would be leaking back into the cell through these cotransporters. An inhibitor of the H^+/NO_3^- cotransporter would have no effect on the membrane potential because the simultaneous cotransport of a positively charged ion and a negatively charged ion has no net effect on charge difference across the membrane. An inhibitor of the potassium ion channels would decrease the membrane potential because additional positively charged ions would not be accumulating outside the cell.   **Figure 36.8** Few, if any, mesophyll cells are more than three cells from a vein.   **Figure 36.9** The Casparian strip blocks water and minerals from moving between endodermal cells or moving around an endodermal cell via the cell's wall. Therefore, water and minerals must pass through an endodermal cell's plasma membrane.   **Figure 36.13** A section 500 μm by 300 μm is equal to 150,000 μm² or 0.0015 cm². Since there are five stomata visible in 0.0015 cm², the stomatal density of this bean leaf is approximately 3,333 stomata per square centimeter of leaf surface.   **Figure 36.18** Because the xylem is under negative pressure (tension), excising a stylet that had been inserted into a tracheid or vessel element would probably introduce air into the cell. No xylem sap would exude unless positive root pressure was predominant.

Concept Check 36.1

1. Vascular plants must transport minerals and water absorbed by the roots to all the other parts of the plant. They must also transport sugars from sites of production to sites of use.   **2.** Increased stem elongation would raise the plant's upper leaves. Erect leaves and reduced lateral branching would make the plant less subject to shading by the encroaching neighbors.   **3.** Pruning shoot tips removes apical dominance, resulting in lateral shoots (branches) growing from axillary buds (see Concept 35.3). This branching produces a bushier plant with a higher leaf area index.

Concept Check 36.2

1. The cell's Ψ_P is 0.7 MPa. In a solution with a Ψ of −0.4 MPa, the cell's Ψ_P at equilibrium would be 0.3 MPa.   **2.** The cell would still adjust to changes in its osmotic environment, but its responses would be slower. Although aquaporins do not affect the water potential gradient across membranes, they allow for more rapid osmotic adjustments.   **3.** If tracheids and vessel elements were alive at maturity, their cytoplasm would impede water movement, preventing rapid long-distance transport.   **4.** The protoplasts would burst. Because the cytoplasm has many dissolved solutes, water would enter the protoplast continuously without reaching equilibrium. (When present, the cell wall prevents rupturing by limiting expansion of the protoplast.)

Concept Check 36.3

1. At dawn, a drop is exuded from the rooted stump because the xylem is under positive pressure due to root pressure. At noon, the xylem is under negative pressure (tension) when it is cut, and the xylem sap is pulled back into the rooted stump. Root pressure cannot keep pace with the increased rate of transpiration at noon.   **2.** Perhaps greater root mass helps compensate for the lower water permeability of the plasma membranes.   **3.** The Casparian strip and tight junctions both prevent movement of fluid between cells.

Concept Check 36.4

1. Stomatal opening at dawn is controlled mainly by light, CO_2 concentration, and a circadian rhythm. Environmental stresses such as drought, high temperature, and wind can stimulate stomata to close during the day. Water deficiency during the peak of the day can trigger release of the plant hormone abscisic acid, which signals guard cells to close stomata.   **2.** The activation of the proton pumps of stomatal cells would cause the guard cells to take up K^+. The increased turgor of the guard cells would lock the stomata open and lead to extreme evaporation from the leaf.   **3.** After the flowers are cut, transpiration from any leaves and from the petals (which are modified leaves) will continue to draw water up the xylem. If cut flowers are transferred directly to a vase, air pockets in xylem vessels prevent delivery of water from the vase to the flowers. Cutting stems again underwater, a few centimeters from the original cut, will sever the xylem above the air pocket. The water droplets prevent another air pocket from forming while the flowers are transferred to a vase.   **4.** Water molecules are in constant motion, traveling at different speeds. If water molecules gain enough energy, the most energetic molecules near the liquid's surface will have sufficient speed, and therefore sufficient kinetic energy, to leave the liquid in the form of gaseous molecules (water vapor). As the molecules with the highest kinetic energy leave the liquid, the average kinetic energy of the remaining liquid decreases. Because a liquid's temperature is directly related to the average kinetic energy of its molecules, the temperature drops as evaporation proceeds.

Concept Check 36.5

1. In both cases, the long-distance transport is a bulk flow driven by a pressure difference at opposite ends of tubes. Pressure is generated at the source end of a sieve tube by the loading of sugar and resulting osmotic flow of water into the phloem, and this pressure *pushes* sap from the source end to the sink end of the tube. In contrast, transpiration generates a negative pressure potential (tension) that *pulls* the ascent of xylem sap.   **2.** The main sources are fully grown leaves (producing sugar by photosynthesis) and fully developed storage organs (producing sugar by breakdown of starch). Roots, buds, stems, expanding leaves, and fruits are powerful sinks because they are actively growing. A storage organ may be a sink in the summer when accumulating carbohydrates but a source in the spring when breaking down starch into sugar for growing shoot tips.   **3.** Positive pressure, whether it be in the xylem when root pressure predominates or in the sieve-tube elements of the phloem, requires active transport. Most long-distance transport in the xylem depends on bulk flow driven by the negative pressure potential generated ultimately by the evaporation of water from the leaf and does not require living cells.   **4.** The spiral slash prevents optimal bulk flow of the phloem sap to the root sinks. Therefore, more phloem sap can move from the source leaves to the fruit sinks, making them sweeter.

Concept Check 36.6

1. Plasmodesmata, unlike gap junctions, have the ability to pass RNA, proteins, and viruses from cell to cell.   **2.** Long-distance signaling is critical for the integrated functioning of all large organisms, but the speed of such signaling is much less critical to plants because their responses to the environment, unlike those of animals, do not typically involve rapid movements.   **3.** Although this strategy would eliminate the systemic spread of viral infections, it would also severely impact the development of the plants.

Summary of Key Concepts Questions

36.1 Plants with tall shoots and elevated leaf canopies generally had an advantage over shorter competitors. A consequence of the selective pressure for tall shoots was the further separation of leaves from roots. This separation created problems for the transport of materials between root and shoot systems. Plants with xylem cells were more successful at supplying their shoot systems with soil resources (water and minerals). Similarly, those with phloem cells were more successful at supplying sugar sinks with carbohydrates.   **36.2** Xylem sap is pulled up the plant by transpiration much more often than it is pushed up the plant by root pressure.   **36.3** Hydrogen bonds are necessary for the cohesion of water molecules to each other and for the adhesion of water to other materials, such as cell walls. Both adhesion and cohesion of water molecules are involved in the ascent of xylem sap under conditions of negative pressure.   **36.4** Although stomata account for most of the water lost from plants, they are necessary for exchange of gases—for example, for the uptake of carbon dioxide needed for photosynthesis. The loss of water through stomata also drives the long-distance transport of water that brings soil nutrients from roots to the rest of the plant.   **36.5** Although the movement of phloem sap depends on bulk flow, the pressure gradient that drives phloem transport depends on the osmotic uptake of water in response to the loading of sugars into sieve-tube elements at sugar sources. Phloem loading depends on H^+ cotransport processes that ultimately depend on H^+ gradients established by active K^+ pumping.   **36.6** Electrical signaling, cytoplasmic pH, cytoplasmic Ca^{2+} concentration, and viral movement proteins all affect symplastic communication, as do developmental changes in the number of plasmodesmata.

Test Your Understanding

1. A **2.** B **3.** B **4.** C **5.** B **6.** C **7.** A **8.** D

Chapter 37

Figure Questions

Figure 37.3 Cations. At low pH, there would be more protons (H^+) to displace mineral cations from negatively charged soil particles into the soil solution.
Figure 37.4 The A horizon, which consists of the topsoil **Table 37.1** During photosynthesis, CO_2 is fixed into carbohydrates, which contribute to the dry mass. In cellular respiration, O_2 is reduced to H_2O and does not contribute to the dry mass. **Figure 37.9** Some other examples of mutualism are the following relationships. *Flashlight fish and bioluminescent bacteria:* The bacteria gain nutrients and protection from the fish, while the bioluminescence attracts prey and mates for the fish. *Flowering plants and pollinators:* Animals distribute the pollen and are rewarded by a meal of nectar or pollen. *Vertebrate herbivores and some bacteria in the digestive system:* Microorganisms in the alimentary canal break down cellulose to glucose and, in some cases, provide the animal with vitamins or amino acids. Meanwhile, the microorganisms have a steady supply of food and a warm environment. *Humans and some bacteria in the digestive system:* Some bacteria provide humans with vitamins, while the bacteria get nutrients from the digested food. **Figure 37.11** Both ammonium and nitrate. A decomposing animal would release amino acids into the soil that would be converted into ammonium by ammonifying bacteria. Some of this ammonium could be used directly by the plant. A large part of the ammonium, however, would be converted by nitrifying bacteria to form nitrate ions that could also be absorbed by the plant root system. **Figure 37.12** The legume plants benefit because the bacteria fix nitrogen that is absorbed by their roots. The bacteria benefit because they acquire photosynthetic products from the plants. **Figure 37.13** All three plant tissue systems are affected. Root hairs (dermal tissue) are modified to allow *Rhizobium* penetration. The cortex (ground tissue) and pericycle (vascular tissue) proliferate during nodule formation. The vascular tissue of the nodule connects to the vascular cylinder of the root to allow for efficient nutrient exchange.

Concept Check 37.1

1. Overwatering deprives roots of oxygen. Overfertilizing is wasteful and can lead to soil salinization and water pollution. **2.** As lawn clippings decompose, they restore mineral nutrients to the soil. If they are removed, the minerals lost from the soil must be replaced by fertilization. **3.** Because of their small size and negative charge, clay particles would increase the number of binding sites for cations and water molecules and would therefore increase cation exchange and water retention in the soil. **4.** Due to hydrogen bonding between water molecules, water expands when it freezes, and this causes mechanical fracturing of rocks. Water also coheres to many objects, and this cohesion combined with other forces, such as gravity, can help tug particles from rock. Finally, water, because it is polar, is an excellent solvent that allows many substances, including ions, to become dissolved in solution.

Concept Check 37.2

1. No. Even though macronutrients are required in greater amounts, all essential elements are necessary for a plant to complete its life cycle. **2.** No. The fact that the addition of an element results in an increase in the growth rate of a crop does not mean that the element is strictly required for the plant to complete its life cycle. **3.** Inadequate aeration of the roots of hydroponically grown plants would promote alcohol fermentation, which uses more energy and may lead to the accumulation of ethanol, a toxic by-product of fermentation.

Concept Check 37.3

1. The rhizosphere is the zone in the soil immediately adjacent to living roots. It harbors many rhizobacteria with which the root systems form beneficial mutualisms. Some rhizobacteria produce antibiotics that protect roots from disease. Others absorb toxic metals or make nutrients more available to roots. Still others convert gaseous nitrogen into forms usable by the plant or produce chemicals that stimulate plant growth. **2.** Soil bacteria and mycorrhizae enhance plant nutrition by making certain minerals more available to plants. For example, many types of soil bacteria are involved in the nitrogen cycle, and the hyphae of mycorrhizae provide a large surface area for the absorption of nutrients, particularly phosphate ions. **3.** Mixotrophy refers to the strategy of using photosynthesis and heterotrophy for nutrition. Euglenids are well-known mixotrophic protists. **4.** Saturating rainfall may deplete the soil of oxygen. A lack of soil oxygen would inhibit nitrogen fixation by the peanut root nodules and decrease the nitrogen available to the plants. Alternatively, heavy rain may leach nitrate from the soil. A symptom of nitrogen deficiency is yellowing of older leaves.

Summary of Key Concepts Questions

37.1 The term *ecosystem* refers to the communities of organisms within a given area and their interactions with the physical environment around them. Soil is teeming with many communities of organisms, including bacteria, fungi, animals, and the root systems of plants. The vigor of these individual communities depends on nonliving factors in the soil environment, such as minerals, oxygen, and water, as well as on interactions, both positive and negative, between different communities of organisms. **37.2** No. Plants can complete their life cycle when grown hydroponically, that is, in aerated salt solutions containing the proper ratios of all the minerals needed by plants. **37.3** No. Some parasitic plants obtain their energy by siphoning off carbon nutrients from other organisms.

Test Your Understanding

1. B **2.** B **3.** A **4.** D **5.** B **6.** B **7.** D **8.** C **9.** D
10.

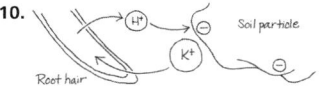

Chapter 38

Figure Questions

Figure 38.4 Having a specific pollinator is more efficient because less pollen gets delivered to flowers of the wrong species. However, it is also a risky strategy: If the pollinator population suffers to an unusual degree from predation, disease, or climate change, then the plant may not be able to produce seeds.
Figure 38.6 The part of the angiosperm life cycle characterized by the most mitotic divisions is the step between seed germination and the mature sporophyte.
Figure 38.8 Make Connections In addition to having a single cotyledon, monocots have leaves with parallel vein venation, scattered vascular bundles in their stems, a fibrous root system, floral parts in threes or multiples of threes, and pollen grains with only one opening. In contrast, eudicots have two cotyledons, netlike leaf venation, vascular bundles in a ring, taproots, floral parts in fours or fives or multiples thereof, and pollen grains with three openings. **Visual Skills** The mature garden bean seed lacks an endosperm at maturity. Its endosperm was consumed during seed development, and its nutrients were stored anew in the cotyledons. **Figure 38.9** Beans use a hypocotyl hook to push through the soil. The delicate leaves and shoot apical meristem are also protected by being sandwiched between two large cotyledons. The coleoptile of maize seedlings helps protect the emerging leaves.

Concept Check 38.1

1. In angiosperms, pollination is the transfer of pollen from an anther to a stigma. Fertilization is the fusion of the egg and sperm to form the zygote; it cannot occur until after the growth of the pollen tube from the pollen grain. **2.** Long styles help to weed out pollen grains that are genetically inferior and not capable of successfully growing long pollen tubes. **3.** No. The haploid (gametophyte) generation of plants is multicellular and arises from spores. The haploid phase of the animal life cycles is a single-celled gamete (egg or sperm) that arises directly from meiosis: There are no spores.

Concept Check 38.2

1. Flowering plants can avoid self-fertilization by self-incompatibility, having male and female flowers on separate plants (dioecious species), or having stamens and styles of different heights on separate plants ("pin" and "thrum" flowers). **2.** Asexually propagated crops lack genetic diversity. Genetically diverse populations are less likely to become extinct in the face of an epidemic because there is a greater likelihood that a few individuals in the population are resistant. **3.** In the short term, selfing may be advantageous in a population that is so dispersed and sparse that pollen delivery is unreliable. In the long term, however, selfing is an evolutionary dead end because it leads to a loss of genetic diversity that may preclude adaptive evolution.

Concept Check 38.3

1. Traditional breeding and genetic engineering both involve artificial selection for desired traits. However, genetic engineering techniques facilitate faster gene transfer and are not limited to transferring genes between closely related varieties or species. **2.** *Bt* maize suffers less insect damage; therefore, *Bt* maize plants are less likely to be infected by fumonisin-producing fungi that infect plants through wounds. **3.** In such species, engineering the transgene into the chloroplast DNA would not prevent its escape in pollen; such a method requires that the chloroplast DNA be found only in the egg. An entirely different method of preventing transgene escape would therefore be needed, such as male sterility, apomixis, or self-pollinating closed flowers.

Summary of Key Concepts Questions

38.1 After pollination and fertilization, a flower changes into a fruit. The petals, sepals, and stamens typically fall off the flower. The stigma of the pistil withers, and the ovary begins to swell. The ovules (embryonic seeds) inside the ovary begin to mature. **38.2** Asexual reproduction can be advantageous in a stable environment because individual plants that are well suited to that environment pass on all their genes to offspring. Also, asexual reproduction generally results in offspring that are less fragile than the seedlings produced by sexual reproduction. However, sexual reproduction offers the advantage of dispersal of tough seeds. Moreover, sexual reproduction produces genetic variety, which may be advantageous in an unstable environment. The likelihood is better that at least one offspring of sexual reproduction will survive in a changed environment. **38.3** "Golden Rice," although not yet in commercial production, has been engineered to produce more vitamin A, thereby raising the nutritional value of rice. A protoxin gene from a soil bacterium has been engineered into *Bt* maize. This protoxin is lethal to invertebrates but harmless to vertebrates. *Bt* crops require less pesticide spraying and have lower levels of fungal infection and fungal toxins. The nutritional value of cassava is being increased in many ways by genetic engineering. Enriched levels of iron and beta-carotene (a vitamin A precursor) have been achieved, and cyanide-producing chemicals have been almost eliminated from the roots.

Test Your Understanding

1. A **2.** C **3.** C **4.** C **5.** D **6.** D **7.** D
8.

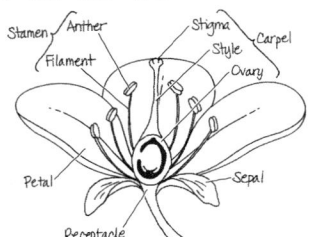

Chapter 39

Figure Questions
Figure 39.4 Panel B in Figure 11.17 shows a branching signal transduction pathway that resembles the branching phytochrome-dependent pathway involved in de-etiolation.   **Figure 39.5** To determine which wavelengths of light are most effective in phototropism, you could use a glass prism to split white light into its component colors and see which colors cause the quickest bending (the answer is blue; see Figure 39.15).   **Figure 39.6** No. Polar auxin transport depends on the distribution of auxin transport proteins at the basal ends of cells.   **Figure 39.13** No. Since the *ein* mutation renders the seedling "blind" to ethylene, enhancing ethylene production by adding an *eto* mutation would have no effect on phenotype compared with the *ein* mutation alone. **Figure 39.16** Yes. The white light, which contains red light, would stimulate seed germination in all treatments.   **Figure 39.20** Since far-red light, like darkness, causes an accumulation of the red-absorbing form (P_r) of phytochrome, single flashes of far-red light at night would have no effect on flowering beyond what the dark periods alone would have.   **Figure 39.21** If this were true, then florigen would be an inhibitor of flowering, not an inducer.   **Figure 39.27** Photosynthetic adaptations can occur at the molecular level, as is apparent in the fact that C_3 plants use rubisco to fix carbon dioxide initially, whereas C_4 and CAM plants use PEP carboxylase. An adaptation at the tissue level is that plants have different stomatal densities based on their genotype and environmental conditions. At the organismal level, plants alter their shoot architectures to make photosynthesis more efficient. For example, self-pruning removes branches and leaves that respire more than they photosynthesize.

Concept Check 39.1
1. Dark-grown seedlings are etiolated: They have long stems, underdeveloped root systems, and unexpanded leaves, and their shoots lack chlorophyll. Etiolated growth is beneficial to seeds sprouting under the dark conditions they would encounter underground. By devoting more energy to stem elongation and less to leaf expansion and root growth, a plant increases the likelihood that the shoot will reach the sunlight before its stored foods run out.   **2.** Cycloheximide should inhibit de-etiolation by preventing the synthesis of new proteins necessary for de-etiolation.   **3.** No. Applying Viagra, like injecting cyclic GMP as described in the text, should cause only a partial de-etiolation response. Full de-etiolation would require activation of the calcium branch of the signal transduction pathway.

Concept Check 39.2
1. Fusicoccin's ability to cause an increase in plasma H^+ pump activity has an auxin-like effect and promotes stem cell elongation.   **2.** The plant will exhibit a constitutive triple response. Because the kinase that normally prevents the triple response is dysfunctional, the plant will undergo the triple response regardless of whether ethylene is present or the ethylene receptor is functional.   **3.** Since ethylene often stimulates its own synthesis, it is under positive-feedback regulation.

Concept Check 39.3
1. Not necessarily. Many environmental factors, such as temperature and light, change over a 24-hour period in the field. To determine whether the enzyme is under circadian control, a scientist would have to demonstrate that its activity oscillates even when environmental conditions are held constant.   **2.** It is impossible to say. To establish that this species is a short-day plant, it would be necessary to establish the critical night length for flowering and that this species only flowers when the night is longer than the critical night length. **3.** According to the action spectrum of photosynthesis, red and blue light are the most effective in photosynthesis. Thus, it is not surprising that plants assess their light environment using blue- and red-light-absorbing photoreceptors.

Concept Check 39.4
1. A plant that overproduces ABA would undergo less evaporative cooling because its stomata would not open as widely.   **2.** Plants close to the aisles may be more subject to mechanical stresses caused by passing workers and air currents. The plants nearer the center of the bench may also be taller as a result of shading and less evaporative stress.   **3.** No. Because root caps are involved in sensing gravity, roots that have their root caps removed are almost completely insensitive to gravity.

Concept Check 39.5
1. An infection can trigger the hypersensitive response, which causes a ring of cell death around the infection site, thereby isolating the pathogen from living host cells and preventing systemic infection.   **2.** Mechanical damage breaches a plant's first line of defense against infection, its protective dermal tissue. **3.** No. Pathogens that kill their hosts would soon run out of victims and might themselves go extinct.   **4.** Perhaps the breeze dilutes the local concentration of a volatile defense compound that the plants produce.

Summary of Key Concepts Questions
39.1 Signal transduction pathways often activate protein kinases, enzymes that phosphorylate other proteins. Protein kinases can directly activate certain preexisting enzymes by phosphorylating them, or they can regulate gene transcription (and enzyme production) by phosphorylating specific transcription factors.
39.2 Yes, there is truth to the old adage that one bad apple spoils the whole bunch. Ethylene, a gaseous hormone that stimulates ripening, is produced by damaged, infected, or overripe fruits. Ethylene can diffuse to healthy fruit in the "bunch" and stimulate their rapid ripening.   **39.3** Plant physiologists proposed the existence of a floral-promoting factor (florigen) based on the fact that a plant induced to flower could induce flowering in a second plant to which it was grafted, even though the second plant was not in an environment that would normally induce flowering in that species.   **39.4** Plants subjected to drought stress are often more resistant to freezing stress because the two types of stress are quite similar. Freezing of water in

the extracellular spaces causes free water concentrations outside the cell to decrease. This, in turn, causes free water to leave the cell by osmosis, leading to the dehydration of cytoplasm, much like what is seen in drought stress.   **39.5** Chewing insects make plants more susceptible to pathogen invasion by disrupting the waxy cuticle of shoots, thereby creating an opening for infection. Moreover, substances released from damaged cells can serve as nutrients for the invading pathogens.

Test Your Understanding
1. B   **2.** C   **3.** D   **4.** C   **5.** B   **6.** B   **7.** C
8.

Chapter 40

Figure Questions
Figure 40.4 Such exchange surfaces are internal in the sense that they are inside the body. However, they are also continuous with openings on the external body surface that contact the environment.   **Figure 40.6** Signals in the nervous system always travel on a direct route between the sending and receiving cell. In contrast, hormones that reach target cells can have an effect regardless of the path by which they arrive or how many times they travel through the circulatory system.   **Figure 40.8** The stimuli (gray boxes) are the room temperature increasing in the top loop and decreasing in the bottom loop. The responses could include the heater turning off and the temperature decreasing in the top loop and the heater turning on and the temperature increasing in the bottom loop. The sensor/control center is the thermostat. The air conditioner would form a second control circuit, cooling the house when air temperature exceeded the set point. Such opposing, or antagonistic, pairs of control circuits increase the effectiveness of a homeostatic mechanism.   **Figure 40.12** The conduction arrows would be in the opposite direction, transferring heat from the penguin to the ice because the penguin is warmer than the ice.   **Figure 40.16** If a female Burmese python were not incubating eggs, her oxygen consumption would decrease with decreasing temperature, as for any other ectotherm.   **Figure 40.17** The ice water would cool tissues in your head, including blood that would then circulate throughout your body. This effect would accelerate the return to a normal body temperature. If, however, the ice water reached the eardrum and cooled the blood vessel that supplies the hypothalamus, the hypothalamic thermostat would respond by inhibiting sweating and constricting blood vessels in the skin, slowing cooling elsewhere in the body.   **Figure 40.18** The transport of nutrients across membranes and the synthesis of RNA and protein are coupled to ATP hydrolysis. These processes proceed spontaneously because there is an overall drop in free energy, with the excess energy given off as heat. Similarly, less than half of the free energy in glucose is captured in the coupled reactions of cellular respiration. The remainder of the energy is released as heat.   **Figure 40.22** Nothing. Although genes that show a circadian variation in expression during euthermia exhibit constant RNA levels during hibernation, a gene that shows constant expression during hibernation might also show constant expression during euthermia.   **Figure 40.23** In hot environments, both plants and animals experience evaporative cooling as a result of transpiration (in plants) or bathing, sweating, and panting (in animals); both plants and animals synthesize heat-shock proteins, which protect other proteins from heat stress; and animals also use various behavioral responses to minimize heat absorption. In cold environments, both plants and animals increase the proportion of unsaturated fatty acids in their membrane lipids and use antifreeze proteins that prevent or limit the formation of intracellular ice crystals; plants increase cytoplasmic levels of specific solutes that help reduce the loss of intracellular water during extracellular freezing; and animals increase metabolic heat production and use insulation, circulatory adaptations such as countercurrent exchange, and behavioral responses to minimize heat loss.

Concept Check 40.1
1. All types of epithelia consist of cells that line a surface, are tightly packed, are situated on top of a basal lamina, and form an active and protective interface with the external environment.   **2.** An oxygen molecule must cross a plasma membrane when entering the body at an exchange surface in the respiratory system, in both entering and exiting the circulatory system, and in moving from the interstitial fluid to the cytoplasm of the body cell.   **3.** You need the nervous system to perceive the danger and provoke a split-second muscular response to keep from falling. The nervous system, however, does not make a direct connection with blood vessels or glucose-storing cells in the liver. Instead, the nervous system triggers the release of a hormone (called epinephrine, or adrenaline) by the endocrine system, bringing about a change in these tissues in just a few seconds.

Concept Check 40.2
1. In thermoregulation, the product of the pathway (a change in temperature) decreases pathway activity by reducing the stimulus. In an enzyme-catalyzed biosynthetic process, the product of the pathway (in this case, isoleucine) inhibits the pathway that generated it.   **2.** You would want to put the thermostat close to

where you would be spending time, where it would be protected from environmental perturbations, such as direct sunshine, and not right in the path of the output of the heating system. Similarly, the sensors for homeostasis located in the human brain are separated from environmental influences and can monitor conditions in a vital and sensitive tissue. **3.** In convergent evolution, the same biological trait arises independently in two or more species. Gene analysis can provide evidence for an independent origin. In particular, if the genes responsible for the trait in one species lack significant sequence similarity to the corresponding genes in another species, scientists conclude that there is a separate genetic basis for the trait in the two species and thus an independent origin. In the case of circadian rhythms, the clock genes in cyanobacteria appear unrelated to those in humans.

Concept Check 40.3

1. "Wind chill" involves heat loss through convection, as the moving air contributes to heat loss from the skin surface. **2.** The hummingbird, being a very small endotherm, has a very high metabolic rate. If by absorbing sunlight certain flowers warm their nectar, a hummingbird feeding on these flowers is saved the metabolic expense of warming the nectar to its body temperature. **3.** To raise body temperature to the higher range of fever, the hypothalamus triggers heat generation by muscular contractions, or shivering. The person with a fever may in fact say that they feel cold, even though their body temperature is above normal.

Concept Check 40.4

1. The mouse would consume oxygen at a higher rate because it is an endotherm, so its basal metabolic rate is higher than the ectothermic lizard's standard metabolic rate. **2.** The house cat; smaller animals have a higher metabolic rate per unit body mass and a greater demand for food per unit body mass. **3.** The alligator's body temperature would decrease along with the air temperature. Its metabolic rate would therefore also decrease as chemical reactions slowed. In contrast, the lion's body temperature would not change. Its metabolic rate would increase as it shivered and produced heat to keep its body temperature constant.

Summary of Key Concepts Questions

40.1 Animals exchange materials with their environment across their body surface, and a spherical shape has the minimum surface area per unit volume. As body size increases, the ratio of surface area to body volume decreases. **40.2** No; an animal's internal environment fluctuates slightly around set points or within normal ranges. Homeostasis is a dynamic state. Furthermore, there are sometimes programmed changes in set points, such as those resulting in radical increases in hormone levels at particular times in development. **40.3** Heat exchange across the skin is a primary mechanism for the regulation of body core temperature, with the result that the skin is cooler than the body core. **40.4** Small animals have a higher BMR per unit mass and therefore consume more oxygen per unit mass than large animals. A higher breathing rate is required to support this increased oxygen consumption.

Test Your Understanding

1. B **2.** C **3.** A **4.** B **5.** C **6.** B **7.** D
8.

```
                    SENSOR:
                    Cruise control OR
                    driver reads
                    speedometer.

  STIMULUS:              RESPONSE:
  Car speeds up, such    Brake applied and
  as when going          car slows down.
  downhill.

              STEADY STATE:
              Car drives at
              desired speed.

  RESPONSE:              STIMULUS:
  Gas pedal applied      Car speeds up, such
  and car speeds up.     as when going
                         downhill.

                    SENSOR:
                    Cruise control
                    OR driver reads
                    speedometer.
```

Chapter 41

Figure Questions

Figure 41.6 Your diagram should show food entering through the hydra's mouth and being digested into nutrients in the large portion of the gastrovascular cavity. The nutrients then diffuse into the extensions of that cavity that reach into the tentacles. There, nutrients would be absorbed by cells of the gastrodermis and transported to cells of the epidermis of a tentacle. **Figure 41.9** The airway must be open for exhaling to occur. If the epiglottis is up, water that entered the throat from the mouth encounters air forced out of the lungs and is carried along into the nasal cavity and out the nose. **Figure 41.11** Since enzymes are proteins, and proteins are hydrolyzed in the small intestine, the digestive enzymes in that compartment need to be resistant to enzymatic cleavage other than the cleavage required to activate them. **Figure 41.12** None. Since digestion is completed in the small intestine, tapeworms simply absorb predigested nutrients through their large body surface. **Figure 41.13** Yes. The exit of the chylomicrons involves exocytosis, an active process that consumes energy in the form of ATP. In contrast, the entry of monoglycerides and fatty acids into the cell by diffusion is a passive process that does not consume energy. **Figure 41.23** Both insulin and glucagon are involved in negative feedback circuits.

Concept Check 41.1

1. The only essential amino acids are those that an animal cannot synthesize from other molecules. **2.** Many vitamins serve as enzyme cofactors, which, like enzymes themselves, are unchanged by the chemical reactions in which they participate. Therefore, only very small amounts of vitamins are needed. **3.** To identify the essential nutrient missing from an animal's diet, a researcher could supplement the diet with individual nutrients one at a time and determine which nutrient eliminates the signs of malnutrition.

Concept Check 41.2

1. A gastrovascular cavity is a digestive pouch with a single opening that functions in both ingestion and elimination; an alimentary canal is a digestive tube with a separate mouth and anus at opposite ends. **2.** As long as nutrients are within the cavity of the alimentary canal, they are in a compartment that is continuous with the outside environment via the mouth and anus and have not yet crossed a membrane to enter the body. **3.** In both cases, high-energy fuels are consumed, complex molecules are broken down into simpler ones, and waste products are eliminated. In addition, gasoline, like food, is broken down in a specialized compartment, so that surrounding structures are protected from disassembly. Finally, just as food and wastes remain outside the body in a digestive tract, neither gasoline nor its waste products enter the passenger compartment of the automobile.

Concept Check 41.3

1. Because parietal cells in the stomach pump hydrogen ions into the stomach lumen where they combine with chloride ions to form HCl, a proton pump inhibitor reduces the acidity of chyme and thus the irritation that occurs when chyme enters the esophagus. **2.** Bile acids act as emulsifiers, dispersing large fat globules into smaller fat droplets. The lipid-soluble surfaces of the bile acids bind to the fat droplets and the water-soluble surfaces of the bile acids interact with the digestive fluids. As a result, surface tension is reduced and the droplets are stabilized, facilitating enzymatic digestion of fats by lipases. **3.** Proteins would be denatured and digested into peptides. Further digestion, to individual amino acids, would require enzymatic secretions found in the small intestine. No digestion of carbohydrates or lipids would occur.

Concept Check 41.4

1. The increased time for transit through the alimentary canal allows for more extensive processing, and the increased surface area of the canal provides greater opportunity for absorption. **2.** A mammal's digestive system provides mutualistic microorganisms with an environment that is protected against other microorganisms by saliva and gastric juice, that is held at a constant temperature conducive to enzyme action, and that provides a steady source of nutrients. **3.** For the yogurt treatment to be effective, the bacteria from yogurt would have to establish a mutualistic relationship with the small intestine, where disaccharides are broken down and sugars are absorbed. Conditions in the small intestine are likely to be very different from those in a yogurt culture. The bacteria might be killed before they reach the small intestine, or they might not be able to grow there in sufficient numbers to aid in digestion.

Concept Check 41.5

1. Over the long term, the body stores excess calories in fat, whether those calories come from fat, carbohydrate, or protein in food. **2.** In most individuals, leptin levels decline during fasting. Individuals in the group with low levels of leptin are likely to be unable to produce leptin, so leptin levels would remain low regardless of food intake. Individuals in the group with high leptin levels are likely to be unable to respond to leptin, but they still should shut off leptin production as fat stores are used up. **3.** Excess insulin production will cause the blood glucose level to decrease below normal physiological levels. It will also trigger glycogen synthesis in the liver, further decreasing the blood glucose level. However, a low blood glucose level will stimulate the release of glucagon from alpha cells in the pancreas, which will trigger glycogen breakdown. Thus, there will be antagonistic effects in the liver.

Summary of Key Concepts Questions

41.1 Since the cofactor is necessary in all animals, those animals that do not require it in their diet must be able to synthesize it from other organic molecules. **41.2** A liquid diet containing glucose, amino acids, and other building blocks could be ingested and absorbed without the need for mechanical or chemical digestion. **41.3** The small intestine has a much larger surface area than the stomach. **41.4** The assortment of teeth in our mouth and the short length of our cecum suggest that our ancestors' digestive systems were not specialized for digesting plant material. **41.5** When mealtime arrives, nervous inputs from the brain signal the stomach to prepare to digest food through secretions and churning.

Test Your Understanding

1. B **2.** A **3.** B **4.** B **5.** B **6.** D **7.** B

8.

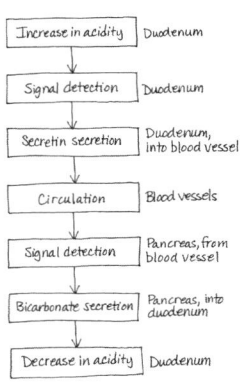

```
Increase in acidity    | Duodenum

Signal detection       | Duodenum

Secretin secretion     | Duodenum,
                       | into blood vessel

Circulation            | Blood vessels

Signal detection       | Pancreas, from
                       | blood vessel

Bicarbonate secretion  | Pancreas, into
                       | duodenum

Decrease in acidity    | Duodenum
```

Chapter 42

Figure Questions

Figure 42.2 Although gas exchange might be improved by a steady, one-way flow of fluid, there would likely be inadequate time for food to be digested and nutrients absorbed if fluid flowed through the cavity in this manner.

Figure 42.5 Two capillary beds. The molecule of carbon dioxide would need to enter a capillary bed in the thumb before returning to the right atrium and ventricle, then travel to the lung and enter a capillary from which it could diffuse into an alveolus and be available to be exhaled. **Figure 42.8** Each feature of the ECG recording, such as the sharp upward spike, occurs once per cardiac cycle. Using the x-axis to measure the time in seconds between successive spikes and dividing that number by 60 would yield the heart rate as the number of cycles per minute.

Figure 42.25 The reduction in surface tension results from the presence of surfactant. Therefore, for all the infants who had died of RDS, you would expect the amount of surfactant to be near zero. For infants who had died of other causes, you would expect the amount of surfactant to be near zero for body masses less than 1,200 g but much greater than zero for body masses above 1,200 g.

Figure 42.27 Since exhalation is largely passive, the recoil of the elastic fibers in alveoli helps force air out of the lungs. When alveoli lose their elasticity, as occurs in the disease emphysema, less air is exhaled. Because more air is left in the lungs, less fresh air can be inhaled. With a smaller volume of air exchanged, there is a decrease in the partial pressure gradient that drives gas exchange.

Figure 42.28 Breathing at a rate greater than that needed to meet metabolic demand (hyperventilation) would lower the blood CO_2 level. Sensors in major blood vessels and the medulla would signal the breathing control center to decrease the rate of contraction of the diaphragm and rib muscles, decreasing the breathing rate and restoring normal CO_2 level in the blood and other tissues.

Figure 42.29 The resulting increase in tidal volume would enhance ventilation within the lungs, increasing P_{O_2} and decreasing P_{CO_2} in the alveoli.

Concept Check 42.1

1. In both an open circulatory system and a fountain, fluid is pumped through a tube and then returns to the pump after collecting in a pool. **2.** The ability to shut off blood supply to the lungs when the animal is submerged **3.** The O_2 content would be abnormally low because some oxygen-depleted blood returned to the right atrium from the systemic circuit would mix with the oxygen-rich blood in the left atrium.

Concept Check 42.2

1. The pulmonary veins carry blood that has just passed through capillary beds in the lungs, where it accumulated O_2. The venae cavae carry blood that has just passed through capillary beds in the rest of the body, where it lost O_2 to the tissues. **2.** The delay allows the atria to empty completely, filling ventricles fully before they contract. **3.** The heart, like any other muscle, becomes stronger through regular exercise. You would expect a stronger heart to have a greater stroke volume, which would allow for the decrease in heart rate.

Concept Check 42.3

1. The large total cross-sectional area of the capillaries **2.** An increase in blood pressure and cardiac output combined with the diversion of more blood to the skeletal muscles would increase the capacity for action by increasing the rate of blood circulation and delivering more O_2 and nutrients to the skeletal muscles. **3.** Additional hearts could be used to improve blood return from the legs. However, it might be difficult to coordinate the activity of multiple hearts and to maintain adequate blood flow to hearts far from the gas exchange organs.

Concept Check 42.4

1. An increase in the number of white blood cells (leukocytes) may indicate that the person is combating an infection. **2.** Clotting factors do not initiate clotting but are essential steps in the clotting process. **3.** The chest pain results from inadequate blood flow in coronary arteries. Vasodilation promoted by nitric oxide from nitroglycerin increases blood flow, providing the heart muscle with additional oxygen and thus relieving the pain. **4.** Embryonic stem cells are pluripotent rather than multipotent, meaning they can give rise to many rather than a few different cell types.

Concept Check 42.5

1. Their interior position helps gas exchange tissues stay moist. If the respiratory surfaces of lungs extended into the terrestrial environment, they would quickly dry out, and diffusion of O_2 and CO_2 across these surfaces would stop. **2.** Earthworms need to keep their skin moist for gas exchange, but they need air outside this moist layer. If they stay in their waterlogged tunnels after a heavy rain, they will suffocate because they cannot get as much O_2 from water as from air. **3.** In fish, water passes over the gills in the direction opposite to that of blood flowing through the gill capillaries, maximizing the extraction of oxygen from the water along the length of the exchange surface. Similarly, in the extremities of some vertebrates, blood flows in opposite directions in neighboring veins and arteries; this countercurrent arrangement maximizes the recapture of heat from blood leaving the body core in arteries, which is important for thermoregulation in cold environments.

Concept Check 42.6

1. An increase in blood CO_2 concentration causes an increase in the rate of CO_2 diffusion into the cerebrospinal fluid, where the CO_2 combines with water to form carbonic acid. Dissociation of carbonic acid releases hydrogen ions, decreasing the pH of the cerebrospinal fluid. **2.** Increased heart rate increases the rate at which CO_2-rich blood is delivered to the lungs, where CO_2 is removed. **3.** A hole would allow air to enter the space between the inner and outer layers of the double membrane, resulting in a condition called a pneumothorax. The two layers would no longer stick together, and the lung on the side with the hole would collapse and cease functioning.

Concept Check 42.7

1. Differences in partial pressure between the capillaries and the surrounding tissues or medium; the net diffusion of a gas occurs from a region of higher partial pressure to a region of lower partial pressure. **2.** The Bohr shift causes hemoglobin to release more O_2 at a lower pH, such as is found in the vicinity of tissues with high rates of cellular respiration and CO_2 release. **3.** The doctor is assuming that the rapid breathing is the body's response to low blood pH. Metabolic acidosis, the lowering of blood pH as a result of metabolism, can have many causes, including complications of certain types of diabetes, shock (extremely low blood pressure), and poisoning.

Summary of Key Concepts Questions

42.1 In a closed circulatory system, an ATP-driven muscular pump generally moves fluids in one direction on a scale of millimeters to meters. Exchange between cells and their environment relies on diffusion, which involves random movements of molecules. Concentration gradients of molecules across exchange surfaces can drive rapid net diffusion on a scale of 1 mm or less. **42.2** Replacement of a defective valve should increase stroke volume. A lower heart rate would therefore be sufficient to maintain the same cardiac output. **42.3** Blood pressure in the arm would fall by 25–30 mm Hg, the same difference as is normally seen between your heart and your brain. **42.4** One microliter of blood contains about 5 million erythrocytes and 5,000 leukocytes, so leukocytes make up only about 0.1% of the cells in the absence of infection. **42.5** Because CO_2 is such a small fraction of atmospheric gas (0.29 mm Hg/760 mm Hg, or less than 0.04%), the partial pressure gradient of CO_2 between the respiratory surface and the environment always strongly favors the release of CO_2 to the atmosphere. **42.6** Because the lungs do not empty completely with each breath, incoming and outgoing air mix. Lungs thus contain a mixture of fresh and stale air. **42.7** An enzyme speeds up a reaction without changing the equilibrium and without being consumed. Similarly, a respiratory pigment speeds up the exchange of gases between the body and the external environment without changing the equilibrium state and without being consumed.

Test Your Understanding

1. C **2.** A **3.** D **4.** C **5.** C **6.** A **7.** A
8.

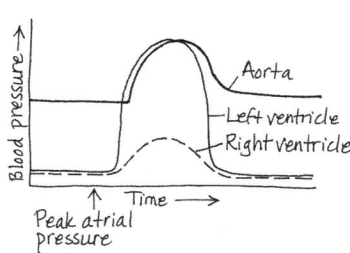

Chapter 43

Figure Questions

Figure 43.3 Dicer-2 binds double-stranded RNA without regard to size or sequence and then cuts that RNA into fragments, each 21 base pairs long. The Argo complex binds to double-stranded RNA fragments that are each 21 base pairs long, displaces one strand, and then uses the remaining strand to match to a particular target sequence in a single-stranded mRNA. **Figure 43.4** Cell-surface TLRs recognize molecules on the surface of pathogens, whereas TLRs in vesicles recognize internal molecules of pathogens after the pathogens are broken down.

Figure 43.5 Because the pain of a splinter stops almost immediately when you remove it from the skin, you can correctly deduce that the signals that mediate the inflammatory response are quite short-lived. **Figure 43.10** Part of the enzyme or antigen receptor provides a structural "backbone" that maintains overall shape, while interaction occurs at a surface with a close fit to the substrate or antigen. The combined effect of multiple noncovalent interactions at the active site or binding site is a high-affinity interaction of tremendous specificity.

Figure 43.14 After gene rearrangement, a lymphocyte and its daughter cells make a single version of the antigen receptor. In contrast, alternative splicing is not heritable and can give rise to diverse gene products in a single cell. **Figure 43.16** A single B cell has more than 100,000 identical antigen receptors on its surface, not four, and there are more than 1 million B cells differing in their antigen specificity, not three. **Figure 43.19** These receptors enable memory cells to present antigen on their cell surface to a helper T cell. This presentation of antigen is required to activate memory cells in a secondary immune response. **Figure 43.23** Primary response: arrows extending from Antigen (1st exposure), Antigen-presenting cell, Helper T cell, B cell, Plasma cells, Cytotoxic T cell, and Active cytotoxic T cells; secondary response: arrows extending from Antigen (2nd exposure), Memory helper T cell, Memory B cells, Memory cytotoxic T cells, Plasma cells, and Active cytotoxic T cells **Figure 43.25** There would be no change in the results. Because the two antigen-binding sites of an antibody have identical specificity, the two bacteriophages bound would have to display the same viral peptide.

Concept Check 43.1

1. Because pus contains white blood cells, fluid, and cell debris, it indicates an active and at least partially successful inflammatory response against invading pathogens. **2.** Whereas the ligand for the TLR receptor is a foreign molecule, the ligand for many signal transduction pathways is a molecule produced by the organism itself. **3.** Mounting an immune response would require recognition of some molecular feature of the wasp egg not found in the host. It might be that only some potential hosts have a receptor with the necessary specificity.

Concept Check 43.2

1. See Figure 43.9. The transmembrane regions lie within the C regions, which also form the disulfide bridges. In contrast, the antigen-binding sites are in the V regions. **2.** Generating memory cells ensures both that a receptor specific for a particular epitope will be present and that there will be more lymphocytes with this specificity than in a host that had never encountered the antigen. **3.** If each B cell produced two different light and heavy chains for its antigen receptor, different combinations would make four different receptors. If any one were self-reactive, the lymphocyte would be eliminated in the generation of self-tolerance. For this reason, many more B cells would be eliminated, and those that could respond to a foreign antigen would be less effective at doing so due to the variety of receptors (and antibodies) they express.

Concept Check 43.3

1. A child lacking a thymus would have no functional T cells. Without helper T cells to help activate B cells, the child would be unable to produce antibodies against extracellular bacteria. Furthermore, without cytotoxic T cells or helper T cells, the child's immune system would be unable to kill virus-infected cells. **2.** Since the antigen-binding site is intact, the antibody fragments could neutralize viruses and opsonize bacteria. **3.** If the handler developed immunity to proteins in the antivenin, another injection could provoke a severe immune response.

Concept Check 43.4

1. Myasthenia gravis is considered an autoimmune disease because the immune system produces antibodies against self molecules (certain receptors on muscle cells). **2.** A person with a cold is likely to produce oral and nasal secretions that facilitate viral transfer. In addition, since sickness can cause incapacitation or death, a virus that is programmed to exit the host when there is a physiological stress has the opportunity to find a new host at a time when the current host may cease to function. **3.** A person with a macrophage deficiency would have frequent infections. The causes would be poor innate responses, due to diminished phagocytosis and inflammation, and poor adaptive responses, due to the lack of macrophages to present antigens to helper T cells.

Summary of Key Concepts Questions

43.1 Lysozyme in saliva destroys bacterial cell walls; the viscosity of mucus helps trap bacteria; acidic pH in the stomach kills many bacteria; and the tight packing of cells lining the gut provides a physical barrier to infection. **43.2** Sufficient numbers of cells to mediate an innate immune response are always present, whereas an adaptive response requires selection and proliferation of an initially very small cell population specific for the infecting pathogen. **43.3** No. Immunological memory after a natural infection and that after vaccination are very similar. There may be minor differences in the particular antigens that can be recognized in a subsequent infection. **43.4** No. AIDS refers to a loss of immune function that can occur over time in an individual infected with HIV. However, certain multidrug combinations ("cocktails") or rare genetic variations usually prevent progression to AIDS in people infected with HIV.

Test Your Understanding

1. B **2.** C **3.** C **4.** B **5.** B **6.** B **7.** C
8. One possible answer:

9. Lamarck's discredited idea was that organisms changed their form to fit challenges and then somehow passed those changes on to their descendants. In clonal selection, heritable differences that give rise to variation arise prior to any challenge. An encounter with a particular antigen results in proliferation of the variants best suited to recognize and respond to that challenge.

Chapter 44

Figure Questions

Figure 44.13 You would expect to find these cells lining tubules where they pass through the renal medulla. Because the extracellular fluid of the renal medulla has a very high osmolarity, production of organic solutes by tubule cells in this region keeps intracellular osmolarity high, with the result that these cells maintain normal volume. **Figure 44.14** Furosemide increases urine volume. The absence of ion transport in the ascending limb leaves the filtrate too concentrated for substantial volume reduction in the distal tubule and collecting duct. **Figure 44.17** When the concentration of an ion differs across a plasma membrane, the difference in the concentration of ions inside and outside represents chemical potential energy, while the resulting difference in charge inside and outside represents electrical potential energy. **Figure 44.20** The ADH levels would likely be elevated in both sets of patients with mutations because either

defect prevents the recapture of water that restores blood osmolarity to normal levels. **Figure 44.21** Arrows that would be labeled "secretion" are the arrows indicating secretion of aldosterone, angiotensinogen, and renin.

Concept Check 44.1

1. Because the salt is moved against its concentration gradient, from low concentration (fresh water) to high concentration (blood) **2.** A freshwater osmoconformer would have body fluids too dilute to carry out life's processes. **3.** Without a layer of insulating fur, the camel must use the cooling effect of evaporative water loss to maintain body temperature, thus linking thermoregulation and osmoregulation.

Concept Check 44.2

1. Because uric acid is largely insoluble in water, it can be excreted as a semisolid paste, thereby reducing an animal's water loss. **2.** Humans produce uric acid from purine breakdown, and reducing purines in the diet often lessens the severity of gout. Birds, however, produce uric acid as a waste product of general nitrogen metabolism. They would therefore need a diet low in all nitrogen-containing compounds, not just purines.

Concept Check 44.3

1. In flatworms, ciliated cells draw interstitial fluids containing waste products into protonephridia. In earthworms, waste products pass from interstitial fluids into the coelom. From there, cilia move the wastes into metanephridia via a funnel surrounding an internal opening to the metanephridia. In insects, the Malpighian tubules pump fluids from the hemolymph, which receives waste products during exchange with cells in the course of circulation. **2.** Filtrate is formed when the glomerulus filters blood from the renal artery within Bowman's capsule. Some of the filtrate contents are recovered, enter capillaries, and exit in the renal vein; the rest remain in the filtrate and pass out of the kidney in the ureter. **3.** The presence of Na^+ and other ions (electrolytes) in the dialysate would limit the extent to which they would be removed from the filtrate during dialysis. Adjusting the electrolytes in the starting dialysate can thus lead to the restoration of proper electrolyte concentrations in the plasma. Similarly, the absence of urea and other waste products in the starting dialysate facilitates their removal from the filtrate.

Concept Check 44.4

1. The numerous nephrons and well-developed glomeruli of freshwater fishes produce urine at a high rate, while the small numbers of nephrons and smaller glomeruli of marine fishes produce urine at a low rate. **2.** The kidney medulla would absorb less water; thus, the drug would increase the amount of water lost in the urine. **3.** A decline in blood pressure in the afferent arteriole would reduce the rate of filtration by moving less material through the vessels.

Concept Check 44.5

1. Alcohol inhibits the release of ADH, causing an increase in urinary water loss and increasing the chance of dehydration. **2.** The consumption of a very large amount of water in a short period of time, coupled with an absence of solute intake, can reduce sodium levels in the blood below tolerable levels. This condition, called hyponatremia, leads to disorientation and, sometimes, respiratory distress. It has occurred in some marathon runners who drink water rather than sports drinks. (It has also caused the death of a fraternity pledge as a consequence of a water hazing ritual and the death of a contestant in a water-drinking competition.) **3.** High blood pressure

Summary of Key Concepts Questions

44.1 Water moves into a cell by osmosis when the fluid outside the cells is hypoosmotic (has a lower solute concentration than the cytosol). **44.2** As cofactors for the enzymes that catalyze metabolism, nitrogen-containing molecules such as NAD^+/NADH are "recycled" during cellular respiration. They thus are not broken down and their components are not absorbed or excreted. **44.3** Filtration produces a fluid for exchange processes that is free of cells and large molecules, which are of benefit to the animal and could not readily be reabsorbed. **44.4** Both types of nephrons have proximal tubules that can reabsorb nutrients, but only juxtamedullary nephrons have loops of Henle that extend deep into the renal medulla. Thus, only kidneys containing juxtamedullary nephrons can produce urine that is more concentrated than the blood. **44.5** Patients who don't produce ADH have symptoms relieved by treatment with the hormone, but many patients with diabetes insipidus lack functional receptors for ADH.

Test Your Understanding

1. C **2.** A **3.** C **4.** D **5.** C **6.** B

Chapter 45

Figure Questions
Figure 45.4

Epinephrine

Figure 45.5 The hormone is water-soluble and has a cell-surface receptor. Such receptors, unlike those for lipid-soluble hormones, can cause observable changes in cells without hormone-dependent gene transcription. **Figure 45.6** ATP is enzymatically converted to cAMP. The other steps represent binding reactions. **Figure 45.21** The embryonic gonad can become either a testis or an ovary. In contrast, the ducts either form a particular structure or degenerate, and the bladder forms in both males and females.

Concept Check 45.1

1. Water-soluble hormones, which cannot penetrate the plasma membrane, bind to cell-surface receptors. This interaction triggers an intracellular signal transduction pathway that ultimately alters the activity of a preexisting protein in the cytoplasm and/or changes transcription of specific genes in the nucleus. Steroid hormones are lipid-soluble and can cross the plasma membrane into the cell interior, where they bind to receptors located in the cytosol or nucleus. The hormone-receptor complex then functions directly as a transcription factor that changes transcription of specific genes. **2.** An exocrine gland, because pheromones are not secreted into interstitial fluid but instead are typically released onto a body surface or into the environment **3.** Because receptors for water-soluble hormones are located on the cell surface, facing the extracellular space, injecting the hormone into the cytosol would not trigger a response.

Concept Check 45.2

1. Prolactin regulates milk production, and oxytocin regulates milk release. **2.** The posterior pituitary, an extension of the hypothalamus that contains the axons of neurosecretory cells, is the storage and release site for two neurohormones, oxytocin and antidiuretic hormone (ADH). The anterior pituitary contains endocrine cells that make at least six different hormones. Secretion of anterior pituitary hormones is controlled by hypothalamic hormones that travel via blood vessels to the anterior pituitary. **3.** The hypothalamus and pituitary glands function in many different endocrine pathways. Many defects in these glands, such as those affecting growth or organization, would therefore disrupt many hormone pathways. Only a very specific defect, such as a mutation affecting a particular hormone receptor, would alter just one endocrine pathway. The situation is quite different for the final gland in a pathway, such as the thyroid gland. In this case, a wide range of defects that disrupt gland function would disrupt only the one pathway or small set of pathways in which that gland functions. **4.** Both diagnoses could be correct. In one case, the thyroid gland may produce excess thyroid hormone despite normal hormonal input from the hypothalamus and anterior pituitary. In the other, abnormally elevated hormonal input (an elevated TSH level) may be the cause of the overactive thyroid gland.

Concept Check 45.3

1. If the function of the pathway is to provide a transient response, a short-lived stimulus would be less dependent on negative feedback. **2.** You would be exploiting the anti-inflammatory activity of glucocorticoids. Local injection avoids the effects on glucose metabolism that would occur if glucocorticoids were taken orally and transported throughout the body in the bloodstream. **3.** Both hormones produce opposite effects in different target tissues. In the fight-or-flight response, epinephrine increases blood flow to skeletal muscles and reduces blood flow to smooth muscles in the digestive system. In establishing apical dominance, auxin promotes the growth of apical buds and inhibits the growth of lateral buds.

Summary of Key Concepts Questions

45.1 As shown in Figure 43.18, helper T cell activation by cytokines acting as local regulators involves both autocrine and paracrine signaling. **45.2** The pancreas, parathyroid glands, and pineal gland **45.3** Both the pituitary and the adrenal glands are formed by fusion of neural and nonneural tissue. ADH is secreted by the neurosecretory portion of the pituitary gland, and epinephrine is secreted by the neurosecretory portion of the adrenal gland.

Test Your Understanding

1. C **2.** D **3.** D **4.** A **5.** B **6.** B **7.** A
8.

Prolactin-releasing hormone circulates in body via blood

↓

Anterior pituitary secretes prolactin (o)

Prolactin circulates in body via blood

↓

Mammary glands

↓

Milk production

Chapter 46

Figure Questions

Figure 46.7 Newly formed sperm enter the seminal vesicle from the testis and exit via the ejaculatory duct during intercourse. Sperm enter the spermatheca after intercourse and, after storage, are released into the oviduct to fertilize an egg moving into the uterus. **Figure 46.8** When successfully courted by a second male, regardless of his genotype, about one-third of the females rid themselves of all sperm from the first mating. Thus, two-thirds retained some sperm from the first mating. We would therefore predict that two-thirds of those females would have some offspring exhibiting the small-eye phenotype of the dominant mutation carried by the males with which the females mated first. **Figure 46.11** The

analysis would be informative because the polar bodies contain all of the maternal chromosomes that don't end up in the mature egg. For example, finding two copies of the disease gene in the polar bodies would indicate its absence in the egg. This method of genetic testing is sometimes carried out when oocytes collected from a female are fertilized with sperm in a laboratory dish. **Figure 46.15** The embryo normally implants about a week after conception, but it spends several days in the uterus before implanting, receiving nutrients from the endometrium. Therefore, the fertilized egg should be cultured for several days in liquid that is at normal body temperature and contains the same nutrients as those provided by the endometrium before implantation. **Figure 46.16** Testosterone can pass from fetal blood to maternal blood via the placental circulation, temporarily upsetting the hormonal balance in the mother. **Figure 46.18** Oxytocin would most likely induce labor, starting a positive-feedback loop that would direct labor to completion. Synthetic oxytocin is in fact frequently used to induce labor when prolonged pregnancy might endanger the mother or fetus.

Concept Check 46.1

1. The offspring of sexual reproduction are more genetically diverse. However, asexual reproduction can produce more offspring over multiple generations. **2.** Unlike other forms of asexual reproduction, parthenogenesis involves gamete production. By controlling whether or not haploid eggs are fertilized, species such as honeybees can readily switch between asexual and sexual reproduction. **3.** No. Owing to random assortment of chromosomes during meiosis, the offspring may receive the same copy or different copies of a particular parental chromosome from the sperm and the egg. Furthermore, genetic recombination during meiosis will result in reassortment of genes between pairs of parental chromosomes. **4.** Fragmentation occurs in both plants and animals. Also, budding in animals and the growth of adventitious plant roots both involve emergence of new individuals from outgrowths of the parent.

Concept Check 46.2

1. Internal fertilization allows sperm to reach the egg without either gamete drying out. **2.** (a) Animals with external fertilization tend to release many gametes at once, resulting in the production of enormous numbers of zygotes. This increases the chances that some will survive to adulthood. (b) Animals with internal fertilization produce fewer offspring but generally exhibit greater care of the embryos and the young. **3.** Like the uterus of an insect, the ovary of a plant is the site of fertilization. Unlike the plant ovary, the uterus is not the site of egg production, which occurs in the insect ovary. In addition, the fertilized insect egg is expelled from the uterus, whereas the plant embryo develops within a seed in the ovary.

Concept Check 46.3

1. Spermatogenesis occurs normally only when the testicles are cooler than normal body temperature. Extensive use of a hot tub (or of very tight-fitting underwear) can cause a decrease in sperm quality and number. **2.** In humans, the secondary oocyte combines with a sperm before it finishes the second meiotic division. Thus, oogenesis is completed after, not before, fertilization. **3.** The only effect of sealing off each vas deferens is an absence of sperm in the ejaculate. Sexual response and ejaculate volume are unchanged. The cutting and sealing off of these ducts, a *vasectomy*, is a common surgical procedure for men who do not wish to produce any (more) offspring.

Concept Check 46.4

1. In the testis, FSH stimulates the Sertoli cells, which nourish developing sperm. LH stimulates the production of androgens (mainly testosterone), which in turn stimulate sperm production. In both females and males, FSH encourages the growth of cells that support and nourish developing gametes (follicle cells in females and Sertoli cells in males), and LH stimulates the production of sex hormones that promote gametogenesis (estrogens, primarily estradiol, in females and androgens, especially testosterone, in males). **2.** In estrous cycles, which occur in most female mammals, the endometrium is reabsorbed (rather than shed) if fertilization does not occur. Estrous cycles often occur just once or a few times a year, and the female is usually receptive to copulation only during the period around ovulation. Menstrual cycles are found only in humans and some other primates. They control the buildup and breakdown of the uterine lining, but not sexual receptivity. **3.** The combination of estradiol and progesterone would have a negative-feedback effect on the hypothalamus, blocking release of GnRH. This would interfere with LH secretion by the pituitary, thus preventing ovulation. This is in fact one basis of action of the most common hormonal contraceptives. **4.** In the viral replicative cycle, the production of new viral genomes is coordinated with capsid protein expression and with the production of phospholipids for viral coats. In the reproductive cycle of a human female, there is hormonally based coordination of egg maturation with the development of support tissues of the uterus.

Concept Check 46.5

1. The secretion of hCG by the early embryo stimulates the corpus luteum to make progesterone, which helps maintain the pregnancy. During the second trimester, however, hCG production drops, the corpus luteum disintegrates, and the placenta completely takes over progesterone production. **2.** Both tubal ligation and vasectomy block the movement of gametes from the gonads to a site where fertilization could take place. **3.** The introduction of a sperm nucleus directly into an oocyte bypasses the sperm's acquisition of motility in the epididymis, its swimming to meet the egg in the oviduct, and its fusion with the egg.

Summary of Key Concepts Questions

46.1 No. Because parthenogenesis involves meiosis, the mother would pass on to each offspring a random and therefore typically distinct combination of the chromosomes she inherited from her mother and father. **46.2** None **46.3** The small size and lack of cytoplasm characteristic of a sperm are adaptations well suited to its function as a delivery vehicle for DNA. The large size and rich cytoplasmic contents of eggs support the growth and development of the embryo. **46.4** Circulating anabolic steroids mimic the feedback regulation of testosterone, turning off pituitary signaling to the testes and thereby blocking the release of signals required for spermatogenesis. **46.5** Oxygen in maternal blood diffuses from pools in the endometrium into fetal capillaries in the chorionic villi of the placenta and from there travels throughout the circulatory system of the fetus.

Test Your Understanding

1. D **2.** B **3.** B **4.** C **5.** A **6.** B **7.** C **8.** C
9.

(a)

(b) Sperm

(c) The supply of stem cells would be used up, and spermatogenesis would not be able to continue.

Chapter 47

Figure Questions

Figure 47.3 You could inject the compound into an unfertilized egg, expose the egg to sperm, and see whether the fertilization envelope forms.
Figure 47.6 There would be fewer cells, and they would be closer together.
Figure 47.8 (1) The blastocoel forms a single compartment that surrounds the gut, much like a doughnut surrounds a hole. (2) Ectoderm forms the outer covering of the animal, and endoderm lines the internal organs, such as the digestive tract. Mesoderm fills much of the space between these two layers.
Figure 47.19 Eight cell divisions are required to give rise to the intestinal cell closest to the mouth. **Figure 47.22** When the researchers allowed normal cortical rotation to occur, the "back-forming" determinants were activated. When they then forced the opposite rotation to occur, the back was established on the opposite side as well. Because the molecules on the normal side were already activated, forcing the opposite rotation apparently did not "cancel out" the establishment of the back side by the first rotation.
Figure 47.23 Draw It

Control egg (dorsal view)   Plane of separation of the blastomeres   Experimental egg (side view)

Gray crescent   Gray crescent

What If? In Spemann's control, the two blastomeres were physically separated, and each grew into a whole embryo. In Roux's experiment, remnants of the dead blastomere were still contacting the live blastomere, which developed into a half-embryo. Therefore, molecules present in the dead cell's remnants may have been signaling to the live cell, inhibiting it from making all the embryonic structures.
Figure 47.24 You could inject the isolated protein (or an mRNA encoding it) into ventral cells of an earlier gastrula. If dorsal structures form on the ventral side, that would support the idea that the protein is the signaling molecule secreted or presented by the dorsal lip. You should also do a control experiment to make sure the injection process alone did not cause dorsal structures to form. **Figure 47.26** Either Sonic hedgehog mRNA or protein can serve as a marker of the zone of polarizing activity (ZPA). The absence of either one after removal of the apical ectodermal ridge would support your hypothesis. You could also block fibroblast growth factor function and see whether the ZPA formed (by looking for Sonic hedgehog).

Concept Check 47.1

1. The fertilization envelope forms after cortical granules release their contents outside the egg, causing the vitelline membrane to rise and harden. The fertilization envelope serves as a barrier to fertilization by more than one sperm.   **2.** The increased Ca^{2+} concentration in the egg would cause the cortical granules to fuse with the plasma membrane, releasing their contents and causing a fertilization envelope to form, even though no sperm had entered. This would prevent fertilization.   **3.** You would expect it to fluctuate. The fluctuation of MPF drives the transition between DNA replication (S phase) and mitosis (M phase), which is still required in the abbreviated cleavage cell cycle.

Concept Check 47.2

1. The cells of the notochord migrate toward the midline of the embryo (converge), rearranging themselves so there are fewer cells across the notochord, which thus becomes longer overall (extends; see Figure 47.17).   **2.** Because microfilaments would not be able to contract and decrease the size of one end of the cell, both the inward bending in the middle of the neural tube and the outward bending of the hinge regions at the edges would be blocked. Therefore, the neural tube probably would not form.   **3.** Dietary intake of the vitamin folic acid dramatically reduces the frequency of neural tube defects.

Concept Check 47.3

1. Axis formation establishes the location and polarity of the three axes that provide the coordinates for development. Pattern formation positions particular tissues and organs in the three-dimensional space defined by those coordinates.   **2.** Morphogen gradients act by specifying cell fates across a field of cells through variation in the level of a determinant. Morphogen gradients thus act more globally than cytoplasmic determinants or inductive interactions between pairs of cells.   **3.** Yes, a second embryo could develop because inhibiting BMP-4 activity would have the same effect as transplanting an organizer.   **4.** The limb that developed probably would have a mirror-image duplication, with the most posterior digits in the middle and the most anterior digits at either end.

Summary of Key Concepts Questions

47.1 The binding of a sperm to a receptor on the egg surface is very specific and likely would not occur if the two gametes were from different species. Without sperm binding, the sperm and egg membranes would not fuse.   **47.2** Apoptosis functions to eliminate structures required only in an immature form, nonfunctional cells from a pool larger than the number required, and tissues formed by a developmental program that is not adaptive for the organism as it has evolved.
47.3 Mutations that affected both limb and kidney development would be more likely to alter the function of monocilia because these organelles are important in several signaling pathways. Mutations that affected limb development but not kidney development would more likely alter a single pathway, such as Hedgehog signaling.

Test Your Understanding

1. A **2.** B **3.** D **4.** A **5.** D **6.** C **7.** B
8.

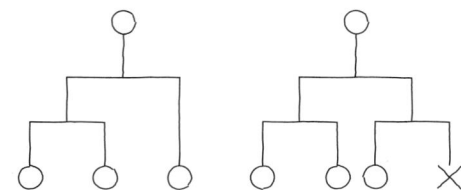

Chapter 48

Figure Questions

Figure 48.7 Potassium and sodium channels must differ in the structure of the channel through which the ions pass. The channel could differ in the size of the opening, the distribution of charge, or other properties that would allow one type of ion but not others to diffuse through the channel.   **Figure 48.8** Adding chloride channels would make the membrane potential less positive. Adding potassium channels would have no effect because there are no potassium ions present.   **Figure 48.10** In the absence of other forces, chemical concentration gradients govern net diffusion. In this case, ions are more concentrated outside of the cell and move in when the channel opens.
Figure 48.11

Figure 48.12

Figure 48.13

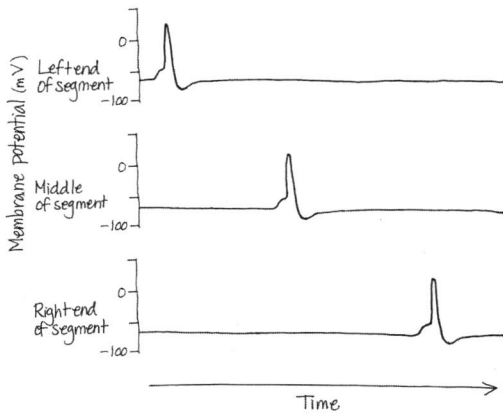

Figure 48.16 The production and transmission of action potentials would be unaffected. However, action potentials arriving at chemical synapses would be unable to trigger release of neurotransmitter. Signaling at such synapses would thus be blocked. **Figure 48.17** Summation only occurs if inputs occur simultaneously or nearly so. Thus, spatial summation, in which input is received from two different sources, is in effect also temporal summation.

Concept Check 48.1

1. A typical neuron has multiple dendrites and one axon. Dendrites transfer information to the cell body, whereas axons transmit information from the cell body. Both axons and dendrites extend from the cell body and function in information flow. **2.** Sensors in your ear transmit information to your brain. There, the activity of interneurons in processing centers enables you to recognize your name. In response, signals transmitted via motor neurons cause contraction of muscles that turn your neck. **3.** Increased branching would allow control of a greater number of postsynaptic cells, enhancing coordination of responses to nervous system signals.

Concept Check 48.2

1. Ions can flow against a chemical concentration gradient if there is an opposing electrical gradient of greater magnitude. **2.** A decrease in permeability to K^+, an increase in permeability to Na^+, or both **3.** Charged dye molecules could equilibrate only if other charged molecules could also cross the membrane. If not, a membrane potential would develop that would counterbalance the chemical gradient.

Concept Check 48.3

1. A graded potential has a magnitude that varies with stimulus strength, whereas an action potential has an all-or-none magnitude that is independent of stimulus strength. **2.** Loss of the insulation provided by myelin sheaths leads to a disruption of action potential propagation along axons. Voltage-gated sodium channels are restricted to the nodes of Ranvier, and without the insulating effect of myelin, the inward current produced at one node during an action potential cannot depolarize the membrane to the threshold at the next node. **3.** Positive feedback is responsible for the rapid opening of many voltage-gated sodium channels, causing the rapid outflow of sodium ions responsible for the rising phase of the action potential. As the membrane potential becomes positive, voltage-gated potassium channels open in a form of negative feedback that helps bring about the falling phase of the action potential. **4.** The maximum frequency would decrease because the refractory period would be extended.

Concept Check 48.4

1. It can bind to different types of receptors, each triggering a specific response in postsynaptic cells. **2.** These toxins would prolong the EPSPs that acetylcholine produces because the neurotransmitter would remain longer in the synaptic cleft. **3.** Membrane depolarization, exocytosis, and membrane fusion each occur in fertilization and in neurotransmission.

Summary of Key Concepts Questions

48.1 It would prevent information from being transmitted away from the cell body along the axon. **48.2** There are very few open sodium channels in a resting neuron, so the resting potential either would not change or would become slightly more negative (hyperpolarization). **48.4** A given neurotransmitter can have many receptors that differ in their location and activity. Drugs that target receptor activity rather than neurotransmitter release or stability are therefore likely to exhibit greater specificity and potentially have fewer undesirable side effects.

Test Your Understanding

1. C **2.** C **3.** C **4.** B **5.** A **6.** D
7. The activity of the sodium-potassium pump is essential to maintain the resting potential. With the pump inactivated, the sodium and potassium concentration gradients would gradually disappear, resulting in a greatly reduced resting potential. **8.** Since GABA is an inhibitory neurotransmitter in the CNS, this drug would be expected to decrease brain activity. A decrease in brain activity

might be expected to slow down or reduce behavioral activity. Many sedative drugs act in this fashion. **9.** As shown in this pair of drawings, a pair of action potentials would move outward in both directions from each electrode. (Action potentials are unidirectional only if they begin at one end of an axon.) However, because of the refractory period, the two action potentials between the electrodes both stop where they meet. Thus, only one action potential reaches the synaptic terminals.

Chapter 49

Figure Questions

Figure 49.5 During swallowing, muscles along the esophagus alternately contract and relax, resulting in peristalsis. One model to explain this alternation is that each section of muscle receives nerve impulses that alternate between excitation and inhibition, just as the quadriceps and hamstring receive opposing signals in the knee-jerk reflex. **Figure 49.15** The gray areas have a different shape and pattern, indicating different planes through the brain. This fact indicates that the nucleus accumbens and the amygdala are in different planes. **Figure 49.17** The hand is shown larger than the forearm because the hand receives more innervation than the forearm for sensory input to the brain and motor output from the brain. **Figure 49.24** If the depolarization brings the membrane potential to or past threshold, it should initiate action potentials that cause dopamine release from the VTA neurons. This should mimic natural stimulation of the brain reward system, resulting in positive and perhaps pleasurable sensations.

Concept Check 49.1

1. The sympathetic division would likely be activated. It mediates the "fight-or-flight" response in stressful situations. **2.** Nerves contain bundles of axons, some that belong to motor neurons, which send signals outward from the CNS, and some that belong to sensory neurons, which bring signals into the CNS. Therefore, you would expect effects on both motor control and sensation. **3.** Neurosecretory cells of the adrenal medulla secrete the hormones epinephrine and norepinephrine in response to preganglionic input from sympathetic neurons. These hormones travel in the circulation throughout the body, triggering responses in many tissues.

Concept Check 49.2

1. The cerebral cortex on the left side of the brain initiates voluntary movement of the right side of the body. **2.** Alcohol diminishes function of the cerebellum. **3.** A coma reflects a disruption in the cycles of sleep and arousal regulated by communication between the midbrain and pons (reticular formation) and the cerebrum. You would expect this group to have damage to the midbrain, the pons, the cerebrum, or any part of the brain between these structures. Paralysis reflects an inability to carry out motor commands transmitted from the cerebrum to the spinal cord. You would expect this group to have damage to the portion of the CNS extending from the spinal cord up to but not including the midbrain and pons.

Concept Check 49.3

1. Brain damage that disrupts behavior, cognition, memory, or other functions provides evidence that the portion of the brain affected by the damage is important for the normal activity that is blocked or altered. **2.** Broca's area, which is active during the generation of speech, is located near the motor cortex, which controls skeletal muscles, including those in the face. Wernicke's area, which is active when speech is heard, is located in the posterior part of the temporal lobe, which is involved in hearing. **3.** Each cerebral hemisphere is specialized for different parts of this task—the right for face recognition and the left for language. Without an intact corpus callosum, neither hemisphere can take advantage of the other's processing abilities.

Concept Check 49.4

1. There can be an increase in the number of synapses between the neurons or an increase in the strength of existing synaptic connections. **2.** If consciousness is an emergent property resulting from the interaction of many different regions of the brain, then it is unlikely that localized brain damage will have a discrete effect on consciousness. **3.** The hippocampus is responsible for organizing newly acquired information. Without hippocampal function, the links necessary to retrieve information from the cerebral cortex will be lacking, and no functional memory, short- or long-term, will be formed.

Concept Check 49.5

1. Both are progressive brain diseases whose risk increases with advancing age. Both result from the death of brain neurons and are associated with the accumulation of peptide or protein aggregates. **2.** The symptoms of schizophrenia can be mimicked by a drug that stimulates dopamine-releasing neurons. The brain's reward system, which is involved in drug addiction, is composed of dopamine-releasing neurons that connect the ventral tegmental area to regions in the cerebrum. Parkinson's disease results from the death of dopamine-releasing neurons. **3.** Not necessarily. It might be that the plaques, tangles, and missing regions of the brain seen at death reflect secondary effects, the consequence of other unseen changes that are actually responsible for the alterations in brain function.

Summary of Key Concepts Questions

49.1 Because reflex circuits involve only a few neurons—the simplest consist of a sensory neuron and a motor neuron—the path for information transfer is short and simple, increasing the speed of the response.   **49.2** The midbrain coordinates visual reflexes; the cerebellum controls coordination of movement that depends on visual input; the thalamus serves as a routing center for visual information; and the cerebrum is essential for converting visual input to a visual image.   **49.3** You would expect the right side of the body to be paralyzed because it is controlled by the left cerebral hemisphere, where language generation and interpretation are localized.   **49.4** Learning a new language likely requires the maintenance of synapses that are formed during early development but are otherwise lost prior to adulthood.   **49.5** Whereas amphetamine stimulates dopamine release, PCP blocks glutamate receptors, suggesting that schizophrenia does not reflect a defect in the function of just one neurotransmitter.

Test Your Understanding

1. B   **2.** A   **3.** D   **4.** D   **5.** C   **6.** A
7.

Chapter 50

Figure Questions

Figure 50.17

Figure 50.19 Each of the three types of cones is most sensitive to a different wavelength of light. A cone might be fully depolarized when there is light present if the light is of a wavelength far from its optimum.   **Figure 50.21** In humans, an X chromosome with a defect in the red or green opsin gene is much less common than a wild-type X chromosome. Color blindness therefore typically skips a generation as the defective allele passes from an affected male to a carrier daughter and back to an affected grandson. In squirrel monkeys, no X chromosome can confer full color vision. As a result, all males are color-blind and no unusual inheritance pattern is observed.   **Figure 50.23** The results of the experiment would have been identical. What matters is the activation of particular sets of neurons, not the manner in which they are activated. Any signal from a bitter cell will be interpreted by the brain as a bitter taste, regardless of the nature of the compound and the receptor involved.   **Figure 50.25** Only perception. Binding of an odorant to its receptor will cause action potentials to be sent to the brain. Although an excess of that odorant might cause a diminished response through adaptation, another odorant can mask the first only at the level of perception in the brain.   **Figure 50.26** Both. A muscle fiber contains many myofibrils bundled together and divided lengthwise into many sarcomeres. A sarcomere is a contractile unit made up of portions of many myofibrils, and each myofibril is a part of many sarcomeres.   **Figure 50.28** Hundreds of myosin heads participate in sliding each pair of thick and thin filaments past each other. Because cross-bridge formation and breakdown are not synchronized, many myosin heads are exerting force on the thin filaments at all times during muscle contraction.   **Figure 50.33** By causing all of the motor neurons that control the muscle to generate action potentials at a rate high enough to produce tetanus in all of the muscle fibers

Concept Check 50.1

1. Electromagnetic receptors in general detect only external stimuli. Nonelectromagnetic receptors, such as chemoreceptors or mechanoreceptors, can act as either internal or external sensors.   **2.** The capsaicin present in the peppers activates the thermoreceptor for high temperatures. In response to the perceived high temperature, the nervous system triggers sweating to achieve evaporative cooling.   **3.** You would perceive the electrical stimulus as if the sensory receptors that regulate that neuron had been activated. For example, electrical stimulation of the sensory neuron controlled by the thermoreceptor activated by menthol would likely be perceived as a local cooling.

Concept Check 50.2

1. Otoliths detect the animal's orientation with respect to gravity, providing information that is essential in environments such as the tunnel habitat of the star-nosed mole, where light cues are absent.   **2.** As a sound that changes gradually from a very low to a very high pitch   **3.** The stapes and the other middle ear bones transmit vibrations from the tympanic membrane to the oval window. Fusion of these bones (as occurs in a disease called otosclerosis) would block this transmission and result in hearing loss.   **4.** In animals, the statoliths are extracellular. In contrast, the statoliths of plants are found within an intracellular organelle. The methods for detecting their location also differ. In animals, detection is by means of mechanoreceptors on ciliated cells. In plants, the mechanism appears to involve calcium signaling.

Concept Check 50.3

1. Planarians have ocelli that cannot form images but can sense the intensity and direction of light, providing enough information to enable the animals to find protection in shaded places. Flies have compound eyes that form images and excel at detecting movement.   **2.** The person can focus on distant objects but not close objects (without glasses) because close focusing requires the lens to become almost spherical. This problem is common after age 50.   **3.** The signal produced by rod and cone cells is glutamate, and their release of glutamate decreases upon exposure to light. However, a decrease in glutamate production causes other retinal cells to increase the rate at which action potentials are sent to the brain, so that the brain receives more action potentials in light than in dark.   **4.** Absorption of light by retinal converts retinal from its *cis* isomer to its *trans* isomer, initiating the process of light detection. In contrast, a photon absorbed by chlorophyll does not bring about isomerization, but instead boosts an electron to a higher energy orbital, initiating the electron flow that generates ATP and NADPH.

Concept Check 50.4

1. Both taste cells and olfactory cells have receptor proteins in their plasma membrane that bind certain substances, leading to membrane depolarization through a signal transduction pathway involving a G protein. However, olfactory cells are sensory neurons, whereas taste cells are not.   **2.** Since animals rely on chemical signals for behaviors that include finding mates, marking territories, and avoiding dangerous substances, it is adaptive for the olfactory system to have a robust response to a very small number of molecules of a particular odorant.   **3.** Because the sweet, bitter, and umami tastes involve GPCR proteins but the sour taste does not, you might predict that the mutation is in a molecule that acts in the signal transduction pathway common to the different GPCRs.

Concept Check 50.5

1. In a skeletal muscle fiber, Ca^{2+} binds to the troponin complex, which moves tropomyosin away from the myosin-binding sites on actin and allows cross-bridges to form. In a smooth muscle cell, Ca^{2+} binds to calmodulin, which activates an enzyme that phosphorylates the myosin head and thus enables cross-bridge formation.   **2.** *Rigor mortis*, a Latin phrase meaning "stiffness of death," results from the complete depletion of ATP in skeletal muscle. Since ATP is required to release myosin from actin and to pump Ca^{2+} out of the cytosol, muscles become chronically contracted beginning about 3–4 hours after death.   **3.** A competitive inhibitor binds to the same site as the substrate for the enzyme. In contrast, the troponin and tropomyosin complex masks, but does not bind to, the myosin-binding sites on actin.

Concept Check 50.6

1. The main problem in swimming is drag; a fusiform body minimizes drag. The main problem in flying is overcoming gravity; wings shaped like airfoils provide lift, and adaptations such as air-filled bones reduce body mass.   **2.** In modeling peristalsis you would constrict the toothpaste tube at different points along its length, using your hand to encircle the tube and squeeze concentrically. To demonstrate movement of food through the digestive tract you would want the cap off the toothpaste tube, whereas you would want the cap on to show how peristalsis contributes to worm locomotion.   **3.** When you grasp the sides of the chair, you are using a contraction of the triceps to keep your arms extended against the pull of gravity on your body. As you lower yourself slowly into the chair, you gradually decrease the number of motor units in the triceps that are contracted. Contracting your biceps would jerk you down, since you would no longer be opposing gravity.

Summary of Key Concepts Questions

50.1 Nociceptors overlap with other classes of receptors in the type of stimulus they detect. They differ from other receptors only in how a particular stimulus is perceived.   **50.2** Volume is encoded by the frequency of action potentials transmitted to the brain; pitch is encoded by which axons are transmitting action potentials.   **50.3** The major difference is that neurons in the retina integrate information from multiple sensory receptors (photoreceptors) before transmitting information to the central nervous system.   **50.4** Our olfactory sense is responsible for most of what we describe as distinct tastes. A head cold or other source of congestion blocks odorant access to receptors lining portions of the nasal cavity.   **50.5** Hydrolysis of ATP is required to convert myosin to a high-energy configuration for binding to actin and to power the Ca^{2+} pump that removes cytosolic Ca^{2+} during muscle relaxation.   **50.6** Human body movements rely on the contraction of muscles anchored to a rigid endoskeleton. Tendons attach muscles to bones, which in turn are composed of fibers built up from a basic organizational unit, the sarcomere. The thin and thick filaments have separate points of attachment within the sarcomere. In response to nervous system motor output, the formation and breakdown of cross-bridges between myosin heads and actin ratchet the thin and thick filaments past each other. Because the filaments are anchored, this sliding movement shortens the muscle fibers. Furthermore, because the fibers themselves are part of the muscles attached at each end to bones, muscle contraction moves bones of the body relative to each other. In this way, the structural anchoring of muscles and filaments enables muscle function, such as the bending of an elbow by contraction of the biceps.

Test Your Understanding
1. C 2. A 3. B 4. C 5. B 6. D
7.

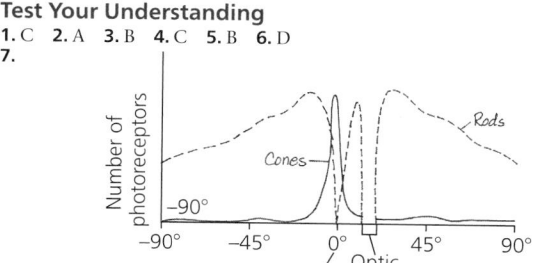

Position along retina (in degrees away from fovea)

The answer shows the actual distribution of rods and cones in the human eye. Your graph may differ, but should have the following properties: only cones at the fovea; fewer cones and more rods at both ends of the *x*-axis; no photoreceptors in the optic disk.

Chapter 51

Figure Questions
Figure 51.2 The fixed action pattern based on the sign stimulus of a red belly ensures that the male will chase away any invading males of his species. By chasing away such males, the defender decreases the chance that another male will fertilize eggs laid in his nesting territory.   **Figure 51.5** The straight-run portion conveys two pieces of information: direction, via the angle of that run relative to the wall of the hive, and distance, via the number of waggles performed during the straight run. At a minimum, the portions between the straight runs identify the activity as a waggle dance. Since they also provide contact with workers to one side and then the other, they may ensure transmission of information to a larger number of other bees.   **Figure 51.7** There should be no effect. Imprinting is an innate behavior that is carried out anew in each generation. Assuming the nest was not disturbed, the offspring of the geese imprinted on a human would imprint on the mother goose. **Figure 51.8** Perhaps the wasp doesn't use visual cues. It might also be that wasps recognize objects native to their environment, but not foreign objects, such as the pinecones. Tinbergen addressed these ideas before carrying out the pinecone study. When he swept away the pebbles and sticks around the nest, the wasps could no longer find their nests. If he shifted the natural objects in their natural arrangement, the shift in the landmarks caused a shift in the site to which the wasps returned. Finally, if natural objects around the nest site were replaced with pinecones while the wasp was in the burrow, the wasp nevertheless found her way back to the nest site.   **Figure 51.10** Switching the orientations of all three grids would control for an inherent preference for or against a particular orientation. If there were no inherent preference or bias, the experiment should work equally well after the switch.   **Figure 51.24** It might be that the birds require stimuli during flight to exhibit their migratory preference. If this were true, the birds would show the same orientation in the funnel experiment despite their distinct genetic programming.   **Figure 51.26** It holds true for some, but not all individuals. If a parent has more than one reproductive partner, the offspring of different partners will have a coefficient of relatedness less than 0.5.

Concept Check 51.1
1. The proximate explanation for this fixed action pattern might be that nudging and rolling are released by the sign stimulus of an object outside the nest, and the behavior is carried to completion once initiated. The ultimate explanation might be that ensuring that eggs remain in the nest increases the chance of producing healthy offspring. **2.** There might be selective pressure for other prey fish to detect an injured fish because the source of the injury might threaten them as well. Among predators, there might be selection for those that are attracted to the alarm substance because they would be more likely to encounter crippled prey. Fish with adequate defenses might show no change because they have a selective advantage if they do not waste energy responding to the alarm substance. **3.** In both cases, the detection of periodic variation in the environment results in a reproductive cycle timed to environmental conditions that optimize the opportunity for success.

Concept Check 51.2
1. Natural selection would tend to favor convergence in color pattern because a predator learning to associate a pattern with a sting or bad taste would avoid all other individuals with that same color pattern, regardless of species. **2.** You might move objects around to establish an abstract rule, such as "past landmark A, the same distance as A is from the starting point," while maintaining a minimum of fixed metric relationships, that is, avoiding having the food directly

adjacent to or a set distance from a landmark. As you might surmise, designing an informative experiment of this kind is not easy.   **3.** Learned behavior, just like innate behavior, can contribute to reproductive isolation and thus to speciation. For example, learned bird songs contribute to species recognition during courtship, thereby helping ensure that only members of the same species mate.

Concept Check 51.3
1. Certainty of paternity is higher with external fertilization.   **2.** Balancing selection could maintain the two alleles at the *forager* locus if population density fluctuated from one generation to another. At times of low population density, the energy-conserving sitter larvae (carrying the *for*^{s} allele) would be favored, while at higher population density, the more mobile Rover larvae (*for*^{R} allele) would have a selective advantage.   **3.** Because females would now be present in much larger numbers than males, all three types of males should have some reproductive success. Nevertheless, since the advantage that the blue-throats rely on—a limited number of females in their territory—will be absent, the yellow-throats are likely to increase in frequency in the short term.

Concept Check 51.4
1. Because this geographic variation corresponds to differences in prey availability between two garter snake habitats, it seems likely that snakes with characteristics enabling them to feed on the abundant prey in their locale would have had increased survival and reproductive success. In this way, natural selection would have resulted in the divergent foraging behaviors.   **2.** The fact that the individual shares some genes with the offspring of its sibling (in the case of humans, with the individual's niece or nephew) means that the reproductive success of that niece or nephew increases the representation of those genes in the population (selects for them).   **3.** The older individual cannot be the beneficiary because he or she cannot have extra offspring. However, the cost is low for an older individual performing the altruistic act because that individual has already reproduced (but perhaps is still caring for a child or grandchild). There can therefore be selection for an altruistic act by a postreproductive individual that benefits a young relative.

Summary of Key Concepts Questions
51.1 Circannual rhythms are typically based on the cycles of light and dark in the environment. As the global climate changes, animals that migrate in response to these rhythms may shift to a location before or after local environmental conditions are optimal for reproduction and survival.   **51.2** For the goose, all that is acquired is an object at which the behavior is directed. In the case of the sparrow, learning takes place that will give shape to the behavior itself.   **51.3** Because feeding the female is likely to improve her reproductive success, the genes from the sacrificed male are likely to appear in a greater number of progeny.   **51.4** Studying the genetic basis of these behaviors reveals that changes in a single gene can have large-scale effects on even complex behaviors.

Test Your Understanding
1. C 2. B 3. B 4. A 5. C 6. A
7.

You could measure the size of mussels that oystercatchers successfully open and compare that with the size distribution in the habitat.

Chapter 52

Figure Questions
Figure 52.8 The species' distribution could be altered by dispersal limitations, the activities of people (such as a broad-scale conversion of forests to agriculture or selective harvesting), or many other factors, including those discussed later in the chapter (see Figure 52.18).   **Figure 52.18** Some factors, such as fire, are relevant only for terrestrial systems. At first glance, water availability is primarily a terrestrial factor, too. However, species living along the intertidal zone of oceans or along the edge of lakes also suffer desiccation. Salinity stress is important for species in some aquatic and terrestrial systems. Oxygen availability is an important factor primarily for species in some aquatic systems and in soils and sediments.

Concept Check 52.1
1. In the tropics, high temperatures evaporate water and cause warm, moist air to rise. The rising air cools and releases much of its water as rain over the tropics. The remaining dry air descends at approximately 30° north and south, causing deserts to occur in those regions.   **2.** The microclimate around the stream will be cooler, moister, and shadier than that around the unplanted agricultural field.   **3.** Trees that require a long time to reach reproductive age are likely to evolve more slowly than annual plants in

response to climate change, constraining the potential ability of such trees to respond to rapid climate change. **4.** Plants with C_4 photosynthesis are likely to expand their range globally as Earth's climate warms. C_4 photosynthesis minimizes photorespiration and enhances sugar production, an advantage that is especially useful in warmer regions where C_4 plants are found today.

Concept Check 52.2

1. The biggest difference between the two biomes is the higher amounts of precipitation that the forest receives. **2.** Answers will vary by location but should be based on the information and maps in Figure 52.13. How much your local area has been altered from its natural state will influence how much it reflects the expected characteristics of your biome, particularly the expected plants and animals. **3.** Northern coniferous forest is likely to replace tundra along the boundary between these biomes. To see why, note that northern coniferous forest is adjacent to tundra throughout North America, northern Europe, and Asia (see Figure 52.10) and that the temperature range for northern coniferous forest is just above that for tundra (see Figure 52.11).

Concept Check 52.3

1. In the oceanic pelagic zone, the ocean bottom lies below the photic zone, so there is too little light to support benthic algae or rooted plants. **2.** Aquatic organisms either gain or lose water by osmosis if the osmolarity of their environment differs from their internal osmolarity. Water gain can cause cells to swell, and water loss can cause them to shrink. To avoid excessive changes in cell volume, organisms that live in estuaries must be able to compensate for both water gain (under freshwater conditions) and water loss (under saltwater conditions). **3.** Oxygen serves as a reactant when decomposers break down the bodies of dead algae using aerobic respiration. Following an algal bloom, there are many dead algae; hence, decomposers may use a lot of oxygen to break down the bodies of dead algae, causing the lake's oxygen levels to drop.

Concept Check 52.4

1. (a) Humans might transplant a species to a new area that it could not previously reach because of a geographic barrier. (b) Humans might eliminate a predator or herbivore species, such as sea urchins, from an area. **2.** One test would be to build a fence around a plot of land in an area that has trees of that species, excluding all deer from the plot. You could then compare the abundance of tree seedlings inside and outside the fenced plot over time. **3.** Because the ancestor of the silverswords reached isolated Hawaii early in the islands' existence, it likely faced little competition and was able to occupy many unfilled niches. The cattle egret, in contrast, arrived in the Americas only recently and has to compete with a well-established group of species. Thus, its opportunities for adaptive radiation have probably been much more limited.

Concept Check 52.5

1. Changes in how organisms interact with one another and their environment can cause evolutionary change. In turn, an evolutionary change, such as an improvement in the ability of a predator to detect its prey, can alter ecological interactions. **2.** As cod adapt to the pressure of commercial fishing by reproducing at younger ages and smaller sizes, the number of offspring they produce each year will be lower. This may cause the population to decline as time goes on, thereby further reducing the population's ability to recover. If that happened, as the population becomes smaller over time, effects of genetic drift might become increasingly important. Drift could, for example, lead to the fixation of harmful alleles, which would further hinder the ability of the cod population to recover from overfishing.

Summary of Key Concepts Questions

52.1 Because dry air would descend at the equator instead of at 30° north and south latitude (where deserts exist today), deserts would be more likely to exist along the equator (see Figure 52.3). **52.2** The dominant plants in savanna ecosystems tend to be adapted to fire and tolerant of seasonal droughts. The savanna biome is maintained by periodic fires, both natural and set by humans, but humans are also clearing savannas for agriculture and other uses. **52.3** An aphotic zone is most likely to be found in the deep waters of a lake, the oceanic pelagic zone, or the marine benthic zone. **52.4** You might arrange a flowchart that begins with abiotic limitations—first determining the physical and chemical conditions under which a species could survive—and then moves through the other factors listed in the flowchart. **52.5** Because the introduced species had few predators or parasites, it might outcompete native species and thereby increase in number and expand its range in the new location. As the introduced species increased in abundance, natural selection might cause evolution in populations of competing species, favoring individuals with traits that made them more effective competitors with the introduced species. Selection could also cause evolution in populations of potential predator or parasite species, in this case favoring individuals with traits that enabled them to take advantage of this new potential source of food. Such evolutionary changes could modify the outcome of ecological interactions, potentially leading to further evolutionary changes, and so on.

Test Your Understanding

1. B **2.** B **3.** C **4.** D **5.** C **6.** A **7.** A **8.** B

Chapter 53

Figure Questions

Figure 53.3 The dispersion of the penguins would likely appear clumped as you flew over densely populated islands and sparsely populated ocean.
Figure 53.4 Ten percent (100/1,000) of the females survive to be 3 years old.
Figure 53.6 #109 **Figure 53.7** The population with $r = 1.0$ (blue curve) reaches 1,500 individuals in about 7.5 generations, whereas the population with $r = 0.5$ (red curve) reaches 1,500 individuals in about 14.5 generations.

Figure 53.15

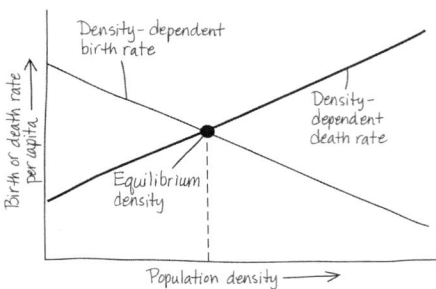

Figure 53.22 Based on Figure 53.21, which highlights the long-term, explosive growth of the human population, one might conclude (mistakenly) that the growth rate of the human population has not decreased in recent decades. However, the growth rate of the human population *has* decreased in recent decades—a slowdown that is evident in the blue curve shown in Figure 53.22. Both curves are accurate, but they convey different messages because they differ in the time scale over which human population size is represented. The time period covered by Figure 53.21 is so long (over 6,000 years in the "unbroken" portion of the x-axis lying to the right of the hatch mark) that the recent slowdown in how fast the human population is growing is not visually apparent. In contrast, Figure 53.22 covers only 100 years, a time period that is short enough to show the recent decrease in the growth rate of the human population.
Figure 53.24 If the average ecological footprint were 8 gha per person, Earth could support about 1.5 billion people in a sustainable fashion. This estimate is obtained by dividing the total amount of Earth's productive land (11.9 billion gha) by the number of global hectares used per person (8 gha/person), which yields 1.49 billion people.

Concept Check 53.1

1.

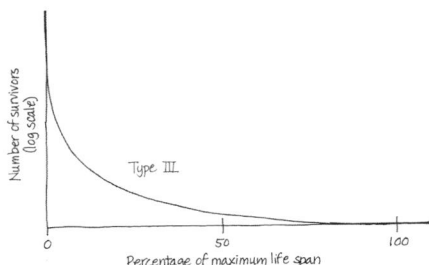

A Type III survivorship curve is most likely because very few of the young probably survive.
2. The proportion alive at the start of year 0–1 is $485/485 = 1.0$. The proportion alive at the start of year 1–2 is $218/485 = 0.449$. **3.** Male sticklebacks would likely have a uniform pattern of dispersion, with antagonistic interactions maintaining a relatively constant spacing between them.

Concept Check 53.2

1. Although r is constant, the population size (N) is increasing. As r is applied to an increasingly large N, population growth (rN) accelerates, producing the J-shaped curve. **2.** Exponential growth is more likely in the area where a forest was destroyed by fire. The first plants that found suitable habitat there would encounter an abundance of space, nutrients, and light. In the undisturbed forest, competition among plants for these resources would be intense. **3.** The equation for the number of people added to the population each year is $\Delta N/\Delta t = r_{\Delta t}N$. Therefore, the net population growth in 2018 was

$$\Delta N/\Delta t = 0.005 \times 327,000,000 = 1,635,000$$

or roughly 1.6 million people. To determine whether the population is growing exponentially, you would need to determine whether $r > 0$ and if it is constant through time (across multiple years).

Concept Check 53.3

1. When N (population size) is small, there are relatively few individuals producing offspring. When N is large, near the carrying capacity, the per capita growth rate is relatively small because it is limited by available resources. The steepest part of the logistic growth curve corresponds to a population with a number of reproducing individuals that is substantial but not yet near carrying capacity. **2.** All else being equal, you would expect a plant species to have a larger carrying capacity at the equator than at high latitudes because there is more incident sunlight near the equator. **3.** The sudden change in environmental conditions might alter the phenotypic traits favored by natural selection. Assuming the newly favored traits were encoded at least in part by genes, natural selection might alter gene frequencies in this population. In addition, a substantial drop in the carrying capacity of the population could cause the size of the population to drop considerably. If this occurred, effects of genetic drift could become more pronounced—and that in turn could lead to the fixation of harmful alleles, hindering the ability of the population to rebound in size.

Concept Check 53.4

1. Three key life history traits are when reproduction begins, how often reproduction occurs, and how many offspring are produced per reproductive episode. Organisms differ widely for each of these traits. For example, the age of first reproduction is typically 3–4 years in coho salmon compared to 30 years in loggerhead turtles. Similarly, an agave reproduces only once during its lifetime, whereas an oak tree reproduces many times. Finally, the white rhinoceros produces a single calf when it reproduces, while most insects produce many offspring each time they reproduce. **2.** By preferentially investing in the eggs it lays in the nest, the peacock wrasse increases the chance those eggs will survive. The eggs it disperses widely and does not provide care for are less likely to survive, at least some of the time, but require a lower investment by the adults. (In this sense, the adults avoid the risk of placing all their eggs in one basket.) **3.** If a parent's survival is compromised greatly by bearing young during times of stress, the animal's fitness may increase if it abandons its current young and survives to produce healthier young at a later time.

Concept Check 53.5

1. Three attributes are the size, quality, and isolation of patches. A patch that is larger or of higher quality is more likely to attract individuals and to be a source of individuals for other patches. A patch that is relatively isolated will undergo fewer exchanges of individuals with other patches. **2.** You should have circled the portion of the curve where it is close to K (after generation 10). **3.** You would need to study the population for more than one cycle (longer than 10 years and probably at least 20) before having sufficient data to examine changes through time. Otherwise, it would be impossible to know whether an observed decrease in the population size reflected a long-term trend or was part of the normal cycle. **4.** In negative feedback, the output, or product, of a process slows that process. In populations that have a density-dependent birth rate, such as dune fescue grass, an accumulation of product (more individuals, resulting in a higher population density) slows the process (population growth) by decreasing the birth rate.

Concept Check 53.6

1. A bottom-heavy age structure, with a disproportionate number of young people, portends continuing growth of the population as these young people begin reproducing. In contrast, a more evenly distributed age structure predicts a more stable population size, and a top-heavy age structure predicts a decrease in population size because relatively fewer young people are reproducing. **2.** The growth rate of Earth's human population has dropped by half since the 1960s, from 2.2% in 1962 to 1.1% today. Nonetheless, the yearly increase in population size has not slowed as much because the smaller growth rate is counterbalanced by increased population size; hence, the number of additional people on Earth each year remains enormous—approximately 80 million. **3.** Each student will calculate his or her own ecological footprint. Each of us influences our ecological footprint by how we live—what we eat, how much energy we use, and the amount of waste we generate—as well as by how many children we have. Making choices that reduce our demand for resources makes our ecological footprint smaller.

Summary of Key Concepts Questions

53.1 Ecologists can potentially estimate birth rates by counting the number of young born each year, and they can estimate death rates by seeing how the number of adults changes each year. **53.2** Under the exponential model, both populations will continue to grow to infinite size, regardless of the specific value of r (see Figure 53.7). **53.3** There are many things you can do to increase the carrying capacity of the species, including increasing its food supply, protecting it from predators, and providing more sites for nesting or reproduction. **53.4** Ecological trade-offs are common because organisms do not have access to unlimited amounts of energy and resources. As a result, the use of energy or resources for one function (such as reproduction) can decrease the energy or resources available to support other functions (such as growth or survival). **53.5** An example of a biotic factor is disease caused by a pathogen; natural disasters, such as earthquakes and floods, are examples of abiotic factors. **53.6** Humans are unique in our potential ability to reduce global population through contraception and family planning. Humans also are capable of consciously choosing their diet and personal lifestyle, and these choices influence the number of people Earth can support.

Test Your Understanding

1. B **2.** A **3.** A **4.** D **5.** C **6.** B **7.** C **8.** A **9.** C

Chapter 54

Figure Questions

Figure 54.3 Its realized and fundamental niches would be similar, unlike those of *Chthamalus*. **Figure 54.5** Beak depths in the *G. fortis* population would likely decrease over time. With the extinction of *G. fuliginosa*, the small seeds eaten by that species would increase in abundance. As a result, natural selection would favor *G. fortis* individuals with smaller beaks because those individuals can eat small seeds more efficiently than could *G. fortis* individuals with larger beaks. **Figure 54.6** Individuals of a harmless species that resembled a distantly related harmful species might be attacked by predators less often than were other individuals that did not resemble the harmful species. As a result, individuals of the harmless species that resembled a harmful species would tend to contribute more offspring to the next generation than would other individuals of the harmless species. Over time, as natural selection by predators continued to favor those individuals of the harmless species that most closely resembled the harmful species, the resemblance of the harmless species to the harmful species would increase. However, selection is not the only process that could cause

a harmless species to resemble a closely related harmful species. In this case, the two species could also resemble each other because they descended from a recent common ancestor and hence share many traits (including a resemblance to one another). **Figure 54.16** An increase in the abundance of carnivores that eat zooplankton might cause zooplankton abundance to drop, thereby causing phytoplankton abundance to increase. **Figure 54.17** The number of types of organisms eaten is zero for phytoplankton; one for copepods, crab-eater seals, and baleen whales; two for krill, carnivorous plankton, elephant seals, and sperm whales; three for squids, fishes, and leopard seals; and five for birds and smaller toothed whales. The two groups that both consume and are consumed by each other are fishes and squids. **Figure 54.18** Zooplankton are primary consumers of phytoplankton. Fish larvae are secondary consumers of zooplankton. Sea nettles function as secondary consumers when they eat zooplankton, but as tertiary consumers when they eat fish larvae. Juvenile striped bass are tertiary consumers of fish larvae. **Figure 54.20** The death of individuals of *Mytilus*, a competitively dominant species, should open up space for individuals of other species and thereby increase species richness even in the absence of *Pisaster*. **Figure 54.27** At the earliest stages of primary succession, free-living prokaryotes in the soil would reduce atmospheric N_2 to NH_3. Symbiotic nitrogen fixation could not occur until plants were present at the site. **Figure 54.31** We would expect that (a) population sizes would decrease because there would be fewer resources and less suitable habitat; (b) the extinction curve would rise more rapidly as the number of species on the island increased because small islands generally have fewer resources, less diverse habitats, and smaller population sizes; and (c) the predicted equilibrium species number would be smaller than shown in Figure 54.31. **Figure 54.34** Shrew populations in different locations and habitats might show substantial genetic variation in their susceptibility to the Lyme pathogen. As a result, there might be fewer infected ticks where shrew populations are less susceptible to the Lyme pathogen and more infected ticks where shrews are more susceptible.

Concept Check 54.1

1. Competition has negative effects on individuals of both species $(-/-)$. In predation, members of the predator population benefit by killing and eating members of the prey population; this is an example of exploitation $(+/-)$. Mutualism is an interaction in which individuals of both species benefit $(+/+)$. **2.** One of the competing species will become locally extinct because of the greater reproductive success of the more efficient competitor. **3.** By specializing in eating seeds of different plant species, individuals of the two finch species may be less likely to come into contact in the separate habitats, thereby reinforcing a reproductive barrier to hybridization.

Concept Check 54.2

1. Species richness, the number of species in the community, and relative abundance, the proportions of the community represented by the various species, both contribute to species diversity. Compared to a community with a very high proportion of one species, one with a more even proportion of species is considered more diverse. **2.** A food chain presents a set of one-way transfers of food energy up to successively higher trophic levels. A food web documents how food chains are linked together, with many species weaving into the web at more than one trophic level. **3.** In bottom-up control, adding extra predators would have little effect on lower trophic levels, particularly vegetation. If top-down control applied, increased bobcat numbers would decrease raccoon numbers, increase snake numbers, decrease mouse numbers, and increase grass biomass. **4.** A decrease in krill abundance might increase the abundance of organisms that krill eat (phytoplankton and copepods), while decreasing the abundance of organisms that eat krill (baleen whales, crabeater seals, birds, fishes, and carnivorous plankton); baleen whales and crabeater seals might be particularly at risk because they eat only krill. However, many of these possible changes could lead to other changes as well, making the overall outcome hard to predict. For example, a decrease in krill abundance could cause an increase in copepod abundance—but an increase in copepod abundance could counteract some of the other effects of decreased krill abundance (since like krill, copepods eat phytoplankton and are eaten by carnivorous plankton and fishes).

Concept Check 54.3

1. High levels of disturbance are generally so disruptive that they eliminate many species from communities, leaving the community dominated by a few tolerant species. Low levels of disturbance permit competitively dominant species to exclude other species from the community. On the other hand, moderate levels of disturbance can facilitate coexistence of a greater number of species in a community by preventing competitively dominant species from becoming abundant enough to eliminate other species from the community. **2.** Early successional species can facilitate the arrival of other species in many ways, including increasing the fertility or water-holding capacity of soils or providing shelter to seedlings from wind and intense sunlight. **3.** The absence of fire for 100 years would represent a change to a low level of disturbance. According to the intermediate disturbance hypothesis, this change should cause diversity to decline as competitively dominant species gain sufficient time to exclude less competitive species.

Concept Check 54.4

1. Ecologists propose that the greater species richness of tropical regions is the result of their longer evolutionary history and the greater solar energy input and water availability in tropical regions. **2.** Immigration of species to islands declines with distance from the mainland and increases with island area. Extinction of species is lower on larger islands and on less isolated islands. Since the number of species on islands is largely determined by the difference between rates of immigration and extinction, the number of species will be highest on large islands near the mainland and lowest on small islands far from the mainland.   **3.** Because of their greater mobility, birds disperse to islands more often than snakes and lizards, so birds should have greater richness.

Concept Check 54.5
1. Pathogens are microorganisms, viruses, viroids, or prions that cause disease. **2.** To keep the rabies virus out, you could ban imports of all mammals, including pets. Potentially, you could also attempt to vaccinate all dogs in the British Isles against the virus. A more practical approach might be to quarantine all pets brought into the country that are potential carriers of the disease, the approach the British government actually takes.

Summary of Key Concepts Questions
54.1 Note: Sample answers follow; other answers could also be correct. Competition: a fox and a bobcat competing for prey. Predation: an orca eating a sea otter. Herbivory: a bison eating grass. Parasitism: a parasitoid wasp that lays its eggs on a caterpillar. Mutualism: a fungus and an alga that make up a lichen. Commensalism: a wildflower that grows in a maple forest and a maple tree.   **54.2** Not necessarily if the more species-rich community is dominated by only one or a few species **54.3** Similar to clear-cutting a forest or plowing a field, some species would be present initially. As a result, the disturbance would initiate secondary succession in spite of its severe appearance.   **54.4** Glaciations are major disturbances that can completely destroy communities found in temperate and polar regions. As a result, tropical communities may be older than temperate or polar communities. This can cause species diversity to be high in the tropics simply because there has been more time for speciation to occur.   **54.5** A keystone species is one with a pivotal ecological role. Hence, a pathogen that reduces the abundance of (or otherwise harms) a keystone species could greatly alter the structure of the community. For example, if a novel pathogen drove a keystone species to local extinction, drastic changes in species diversity could occur.

Test Your Understanding
1. D   **2.** C   **3.** C   **4.** C   **5.** B   **6.** C   **7.** D   **8.** B
9. Community 1: $H = -(0.05 \ln 0.05 + 0.05 \ln 0.05 + 0.85 \ln 0.85 + 0.05 \ln 0.05) = 0.59$. Community 2: $H = -(0.30 \ln 0.30 + 0.40 \ln 0.40 + 0.30 \ln 0.30) = 1.1$. Community 2 is more diverse.   **10.** Crab numbers should increase, reducing the abundance of eelgrass.

Chapter 55

Figure Questions
Figure 55.4 The blue arrow leading to *Primary consumers* could represent a grasshopper feeding on a plant. The blue arrow leading from *Primary consumers* to *Detritus* could represent the remains of a dead primary consumer (such as a grasshopper) becoming part of the detritus found in the ecosystem. The blue arrow leading from *Primary consumers* to *Secondary and tertiary consumers* could represent a bird (the secondary consumer) eating a grasshopper (the primary consumer). Finally, the blue arrow leading from *Primary consumers* to *Primary producers* could represent CO_2 released by a grasshopper in cellular respiration.   **Figure 55.5** The map does not accurately reflect the productivity of wetlands, coral reefs, and coastal zones because these habitats cover areas that are too small to show up clearly on global maps.   **Figure 55.6** New duck farms would add extra nitrogen and phosphorus to the water samples used in the experiment. We would expect that the extra phosphorus from these new duck farms would not alter the results (because in the original experiment, phosphorus levels were *already* so high that adding phosphorus did not increase phytoplankton growth). However, the new duck farms might increase nitrogen levels to the point where adding extra nitrogen in an experiment would not increase phytoplankton density.   **Figure 55.12** The availability of water and exposure to light are other factors that may have varied across the sites. Factors such as these that are not included in the experimental design could make the results more difficult to interpret. Multiple factors can also be correlated to each other in nature, so ecologists must be careful that the factor they are studying is actually causing the observed response and is not just correlated with it.   **Figure 55.13** (1) If the rate of decomposition slowed, more organic materials would be transferred from reservoir A to reservoir B; eventually, this might lead to more organic material becoming fossilized into fossil fuels. In addition, a decrease in decomposition rate would cause fewer inorganic materials to become available as nutrients in reservoir C, which would ultimately slow the rates of nutrient uptake and photosynthesis by living organisms. (2) Materials move into and out of reservoir A on a much shorter time scale than they move into reservoir B. Materials may remain in reservoir B for a very long time, or humans may remove them at a rapid pace by excavating and burning fossil fuels.   **Figure 55.15** If the *y*-axis had a consistent scale with no break, the change in nitrate concentration in runoff from one tick mark to the next would remain constant throughout the entire axis. For example, if the *y*-axis were redrawn to have a consistent scale that ran from 0 to 80 mg/L with nine evenly spaced tick marks, the nitrate concentration would increase by 10 mg/L from each tick mark to the next. Drawing the graph with a consistent scale would emphasize the dramatic increase in nitrate concentration that occurred in 1966,

but it would be harder to see other, comparatively small changes that occurred from 1965 to 1968 in both control and deforested areas.   **Figure 55.19** Populations evolve as organisms interact with each other and with the physical and chemical conditions of their environment. As a result, any human action that alters the environment has the potential to cause evolutionary change. In particular, since climate change has greatly affected arctic ecosystems, we would expect that climate change will cause evolution in arctic tundra populations.

Concept Check 55.1
1. Energy passes through an ecosystem, entering as sunlight and leaving as heat. It is not recycled within the ecosystem.   **2.** You would need to know how much biomass the wildebeests ate from your plot and how much nitrogen was contained in that biomass. You would also need to know how much nitrogen they deposited in urine or feces.   **3.** The second law states that in any energy transfer or transformation, some of the energy is dissipated to the surroundings as heat. For the ecosystem to remain intact, this "escape" of energy from the ecosystem must be offset by the continuous influx of solar radiation.

Concept Check 55.2
1. Only a fraction of solar radiation strikes plants or algae, only a portion of that fraction is of wavelengths suitable for photosynthesis, and much energy is reflected or lost as heat.   **2.** By manipulating the level of the factors of interest, such as phosphorus availability or soil moisture, and measuring responses by primary producers   **3.** It is likely that NEP would decline after the fire. To see why, recall that NEP = GPP − R_T, where GPP is gross primary production and R_T is the total amount of cellular respiration in the ecosystem. By killing trees and other plants, the fire would cause GPP to decline from its pre-fire level. In addition, as decomposers broke down the remains of trees killed by fire, the overall amount of cellular respiration (R_T) in the ecosystem could increase (because of increased cellular respiration by decomposers).   **4.** The enzyme rubisco, which catalyzes the first step in the Calvin cycle, is the most abundant protein on Earth. Like all proteins, rubisco contains nitrogen, and because photosynthetic organisms require so much rubisco, they also require considerable nitrogen to make it. Phosphorus is also needed as a component of several metabolites in the Calvin cycle and as a component of both ATP and NADPH (see Figure 10.19).

Concept Check 55.3
1. 20 J; 40%   **2.** Nicotine protects the plant from herbivores.   **3.** Total net primary production is $10,000 + 1,000 + 100 + 10$ J = 11,110 J. This is the amount of energy theoretically available to decomposers.

Concept Check 55.4
1. For example, for the carbon cycle:

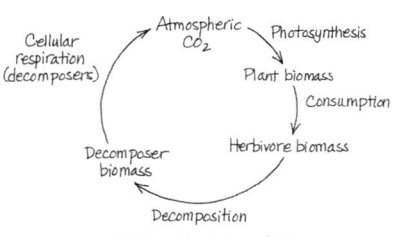

Cycling of a carbon atom

2. Removal of the trees stops nitrogen uptake from the soil, allowing nitrate to accumulate in the soil. The nitrate is washed away by precipitation and enters the streams.   **3.** Most of the nutrients in a tropical rain forest are contained in the trees, so removing the trees by logging rapidly depletes nutrients from the ecosystem. The nutrients that remain in the soil are quickly carried away into streams and groundwater by the abundant precipitation.

Concept Check 55.5
1. The main goal is to restore degraded ecosystems to a more natural state. **2.** Bioremediation uses organisms—generally prokaryotes, fungi, or plants—to detoxify or remove pollutants from ecosystems. Biological augmentation uses organisms, such as nitrogen-fixing plants, to add essential materials to degraded ecosystems.   **3.** The Kissimmee River project returns the flow of water to the original channel and restores natural flow, a self-sustaining outcome. Ecologists at the Maungatautari reserve will need to maintain the integrity of the fence indefinitely, an outcome that is not self-sustaining in the long term.

Summary of Key Concepts Questions
55.1 Because energy conversions are inefficient, with some energy inevitably lost as heat, you would expect that a given mass of primary producers would support a smaller biomass of consumers.   **55.2** If you know NPP and want to estimate NEP, you must be able to determine how much of the total respiration (R_T) results from heterotrophs and how much results from autotrophs. In a sample of ocean water, primary producers and other organisms are usually mixed together, making their respective respirations hard to separate.   **55.3** Runners use much more energy in respiration when they are running than when they are sedentary, reducing their production efficiency.   **55.4** Factors other than temperature, including a shortage of water and nutrients, slow decomposition in hot deserts.   **55.5** If the topsoil and deeper soil are kept separate, the engineers could return the deeper soil to the site first and then apply the more fertile topsoil to improve the success of revegetation and other restoration efforts.

1. B **2.** B **3.** A **4.** C **5.** A **6.** D **7.** D **8.** D
9. (a)

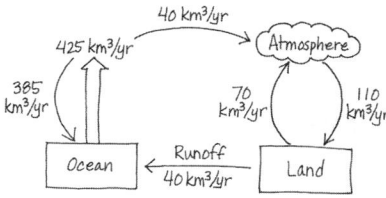

(b) On average, the ratio is 1, with equal amounts of water moving from the ocean to land as precipitation and moving from land to ocean in runoff. (c) During an ice age, the amount of ocean evaporation falling on land as precipitation would be greater than the amount returning to the oceans in runoff; thus, the ratio would be 71. The difference would build up on land as ice.

Chapter 56

Figure Questions

Figure 56.3 You would need to know the complete range of the species and that it is missing across all of that range. You would also need to be certain that the species isn't hidden, as might be the case for an animal that is hibernating underground or a plant that is present in the form of seeds or spores. **Figure 56.8** The two examples are similar in that segments of DNA from the harvested samples were analyzed and compared with segments from specimens of known origin. One difference is that the whale researchers investigated relatedness at species and population levels to determine whether illegal activity had occurred, whereas the elephant investigators determined relatedness at the population level to determine the precise location of the poaching. Another difference is that mtDNA was used for the whale study, whereas nuclear DNA was used for the elephant study. The primary limitations of such approaches are the need to have (or generate) a reference database and the requirement that the organisms have sufficient variation in their DNA to reveal the relatedness of samples. **Figure 56.10** The higher the pH, the lower the acidity. Thus, the precipitation in this forest is becoming less acidic. **Figure 56.12** Answers may vary, but there are two reasons not to support transplanting additional birds. First, the Illinois population has a different genetic makeup than birds in other regions, and you would want to maintain to the greatest extent possible the frequency of beneficial genes or alleles found only in the Illinois population. Second, the translocation of birds from other states already caused the percentage of hatched eggs to increase dramatically, indicating that the transplantation of additional birds is not necessary. **Figure 56.14** The natural disturbance regime in this habitat includes frequent fires that clear undergrowth but do not kill mature pine trees. Without these fires, the undergrowth quickly fills in and the habitat becomes unsuitable for red-cockaded woodpeckers. **Figure 56.16** The photo shows edges between forest and grassland ecosystems, and grassland and river ecosystems. **Figure 56.24** The PCB concentration increased by a factor of 4.9 from phytoplankton to zooplankton, 41.6 from phytoplankton to smelt, 8.5 from zooplankton to smelt, 4.6 from smelt to lake trout, 119.2 from smelt to herring gull eggs, and 25.7 from lake trout to herring gull eggs. **Figure 56.31** Ocean acidification reduces the availability of carbonate ions (CO_3^{2-}). Corals and many other marine organisms require carbonate ions to build their shells. Since shell-building organisms depend upon their shells for survival, scientists have predicted that ocean acidification will cause many shell-building organisms to die. In turn, increased mortality rates of organisms that build shells would cause many other changes to ecological communities. For example, increased mortality rates of corals would harm the many other species that seek protection in coral reefs or that feed upon the species living there. **Figure 56.32** The model results in the blue curve (natural factors only) and the results in the purple curve (natural *and* human factors) both provide a good match to observed temperature changes until about 1960. After 1960, however, the results in the blue curve are a poor match to observed temperature changes, whereas the results in the purple curve continue to provide a good match. These results suggest that human activities such as burning fossil fuels have contributed to the observed rise in global temperatures, especially for the period 1960 to the present.

Concept Check 56.1

1. In addition to species loss, the biodiversity crisis includes the loss of genetic diversity within populations and species and the degradation of entire ecosystems. **2.** Habitat destruction, such as deforestation, channelizing of rivers, or conversion of natural ecosystems to agriculture or cities, deprives species of places to live. Introduced species, which are transported by humans to regions outside their native range, often reduce the population sizes of native species through competition or by feeding on them (as predators, herbivores, or pathogens). Overharvesting has reduced populations of plants and animals or driven them to extinction. Finally, global change is altering the environment to the extent that it reduces the capacity of Earth to sustain life. **3.** If both populations breed separately, then gene flow between the populations would not occur and genetic differences between them would be greater. As a result, the loss of genetic diversity would be greater than if the populations interbreed.

Concept Check 56.2

1. Reduced genetic variation decreases the capacity of a population to evolve in the face of change. **2.** The effective population size, N_e, would be

$4(30 \times 10)/(30 + 10) = 30$ birds. **3.** Because millions of people use the greater Yellowstone ecosystem each year, it would be impossible to eliminate all contact between people and bears. Instead, you might try to reduce the kinds of encounters where bears are killed. You might recommend lower speed limits on roads in the park, adjust the timing or location of hunting seasons (where hunting is allowed outside the park) to minimize contact with mother bears and cubs, and provide financial incentives for livestock owners to try alternative means of protecting livestock, such as using guard dogs.

Concept Check 56.3

1. A small area supporting numerous endemic species as well as a large number of endangered and threatened species **2.** Zoned reserves may provide sustained supplies of forest products, water, hydroelectric power, educational opportunities, and income from tourism. **3.** Habitat corridors can increase the rate of movement or dispersal of organisms between habitat patches and thus the rate of gene flow between subpopulations. They thus help prevent a decrease in fitness attributable to inbreeding. They can also minimize interactions between organisms and humans as the organisms disperse; in cases involving potential predators, such as bears or large cats, minimizing such interactions is desirable.

Concept Check 56.4

1. Adding nutrients causes population explosions of algae and the organisms that feed on them. Increased respiration by algae and consumers, including decomposers, depletes the lake's oxygen, which the fish require. **2.** Decomposers are consumers that use nonliving organic matter as fuel for cellular respiration, which releases CO_2 as a by-product. Because higher temperatures lead to faster decomposition, organic matter in these soils could be decomposed to CO_2 more rapidly, thereby speeding up global warming. **3.** Reduced concentrations of ozone in the atmosphere increase the amount of UV radiation that reaches Earth's surface and the organisms living there. UV radiation can cause mutations by producing disruptive thymine dimers in DNA.

Concept Check 56.5

1. Sustainable development is an approach to development that works toward the long-term prosperity of human societies and the ecosystems that support them, which requires linking the biological sciences with the social sciences, economics, and humanities. **2.** Biophilia, our sense of connection to nature and all forms of life, may act as a significant motivation for the development of an environmental ethic that resolves not to allow species to become extinct or ecosystems to be destroyed. Such an ethic is necessary if we are to become more attentive and effective custodians of the environment. **3.** At a minimum, you would want to know the size of the population and the average reproductive rate of the individuals in it. To develop the fishery sustainably, you would seek a harvest rate that maintains the population near its original size and maximizes its harvest in the long term rather than the short term.

Summary of Key Concepts Questions

56.1 Nature provides us with many beneficial services, including a supply of reliable, clean water, the production of food and fiber, and the dilution and detoxification of our pollutants. **56.2** A more genetically diverse population is better able to withstand pressures from disease or environmental change, making it less likely to become extinct over a given period of time. **56.3** Habitat fragmentation can isolate populations, leading to inbreeding and genetic drift, and it can make populations more susceptible to local extinctions resulting from edge effects, including a change in physical conditions and an increase in competition or predation with edge-adapted species. **56.4** It's healthier to feed at a lower trophic level because biological magnification increases the concentration of toxins at higher levels. **56.5** One goal of conservation biology is to preserve as many species as possible. Sustainable approaches that maintain the quality of habitats are required for the long-term survival of organisms.

Test Your Understanding

1. C **2.** D **3.** B **4.** A **5.** B **6.** D
7.

To minimize the area of forest into which the cowbirds penetrate, you should locate the road along the west edge of the reserve (since that edge abuts deforested pasture and an agricultural field). Any other location would increase the area of affected habitat. Similarly, the maintenance building should be in the southwest corner of the reserve to minimize the area susceptible to cowbirds.

APPENDIX B Classification of Life

Appendix B Classification of Life

This appendix presents a taxonomic classification for the major extant groups of organisms discussed in this text; not all phyla are included. The classification presented here is based on the three-domain system, which assigns the two major groups of prokaryotes, bacteria and archaea, to separate domains (with eukaryotes making up the third domain).

Various alternative classification schemes are discussed in Unit Five of the text. The taxonomic turmoil includes debates about the number and boundaries of kingdoms and about the alignment of the Linnaean classification hierarchy with the findings of modern cladistic analysis.

DOMAIN BACTERIA

- **Proteobacteria**
- **Chlamydia**
- **Spirochetes**
- **Cyanobacteria**
- **Gram-positive Bacteria**

DOMAIN ARCHAEA

- **Euryarchaeota**
- **Thaumarchaeota**
- **Aigarchaeota**
- **Crenarchaeota**
- **Korarchaeota**

DOMAIN EUKARYA

In the phylogenetic hypothesis we present in Chapter 28, major clades of eukaryotes are grouped together in the four "supergroups" listed in blue type. Formerly, all the eukaryotes generally called protists were assigned to a single kingdom, Protista. However, advances in systematics have made it clear that some protists are more closely related to plants, fungi, or animals than they are to other protists. As a result, the kingdom Protista has been abandoned.

Excavata
- Diplomonadida (diplomonads)
- Parabasala (parabasalids)
- Euglenozoa (euglenozoans)
  - Kinetoplastida (kinetoplastids)
  - Euglenophyta (euglenids)

SAR
- Stramenopila (stramenopiles)
  - Oomycota (oomycetes)
  - Phaeophyta (brown algae)
  - Bacillariophyta (diatoms)

- Alveolata (alveolates)
  - Dinoflagellata (dinoflagellates)
  - Apicomplexa (apicomplexans)
  - Ciliophora (ciliates)
- Rhizaria (rhizarians)
  - Radiolaria (radiolarians)
  - Foraminifera (forams)
  - Cercozoa (cercozoans)

Archaeplastida
- Rhodophyta (red algae)
- Chlorophyta (green algae: chlorophytes)
- Charophyta (green algae: charophytes)
- Plantae
  - Phylum Hepatophyta (liverworts) ⎫
  - Phylum Bryophyta (mosses) ⎬ Nonvascular plants (bryophytes)
  - Phylum Anthocerophyta (hornworts) ⎭
  - Phylum Lycophyta (lycophytes) ⎫ Seedless vascular plants
  - Phylum Monilophyta (ferns, horsetails, whisk ferns) ⎭
  - Phylum Ginkgophyta (ginkgo) ⎫
  - Phylum Cycadophyta (cycads) ⎬ Gymnosperms ⎫
  - Phylum Gnetophyta (gnetophytes) ⎪ ⎬ Seed plants
  - Phylum Coniferophyta (conifers) ⎭ ⎪
  - Phylum Anthophyta (flowering plants) ⎬ Angiosperms ⎭

DOMAIN EUKARYA, continued

Unikonta (also called Amorphea)
- Amoebozoa (amoebozoans)
  - Tubulinea (tubulinids)
  - Myxogastrida (plasmodial slime molds)
  - Dictyostelida (cellular slime molds)
  - Entamoeba (entamoebas)
- Nucleariida (nucleariids)
- Fungi
  - Phylum Cryptomycota (cryptomycetes)
  - Phylum Microsporidia (microsporidians)
  - Phylum Chytridiomycota (chytrids)
  - Phylum Zoopagomycota (zoopagomycetes)
  - Phylum Mucoromycota (mucoromycetes)
  - Phylum Ascomycota (ascomycetes)
  - Phylum Basidiomycota (basidiomycetes)

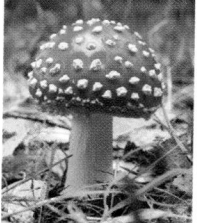

- Choanoflagellata (choanoflagellates)
- Animalia
  - Phylum Porifera (sponges)
  - Phylum Ctenophora (comb jellies)
  - Phylum Cnidaria (cnidarians)
    - Medusozoa (hydrozoans, jellies, box jellies)
    - Anthozoa (sea anemones and most corals)
  - Phylum Acoela (acoel flatworms)
  - Phylum Placozoa (placozoans)
  - Lophotrochozoa (lophotrochozoans)
  - Phylum Platyhelminthes (flatworms)
    - Catenulida (chain worms)
    - Rhabditophora (planarians, flukes, tapeworms)
  - Phylum Nemertea (ribbon worms)
  - Phylum Ectoprocta (ectoprocts)
  - Phylum Brachiopoda (brachiopods)
  - Phylum Syndermata (rotifers and spiny-headed worms)
  - Phylum Gastrotricha (gastrotrichs)
  - Phylum Cycliophora (cycliophorans)
  - Phylum Mollusca (molluscs)
    - Polyplacophora (chitons)
    - Gastropoda (gastropods)
    - Bivalvia (bivalves)
    - Cephalopoda (cephalopods)

Phylum Annelida (segmented worms)
  Errantia (errantians)
  Sedentaria (sedentarians)
Ecdysozoa (ecdysozoans)
  Phylum Loricifera (loriciferans)
  Phylum Priapula (priapulans)
  Phylum Nematoda (roundworms)
  Phylum Arthropoda (This survey groups arthropods into a single phylum, but some zoologists now split the arthropods into multiple phyla.)
    Chelicerata (horseshoe crabs, arachnids)
    Myriapoda (millipedes, centipedes)
    Pancrustacea (crustaceans, insects)
  Phylum Tardigrada (tardigrades)
  Phylum Onychophora (velvet worms)
Deuterostomia (deuterostomes)
  Phylum Hemichordata (hemichordates)
  Phylum Echinodermata (echinoderms)
    Asteroidea (sea stars, sea daisies)
    Ophiuroidea (brittle stars)
    Echinoidea (sea urchins, sand dollars)
    Crinoidea (sea lilies)
    Holothuroidea (sea cucumbers)
  Phylum Chordata (chordates)
    Cephalochordata (cephalochordates: lancelets)
    Urochordata (urochordates: tunicates)

    Cyclostomata (cyclostomes)
      Myxini (hagfishes)
      Petromyzontida (lampreys)
    Gnathostomata (gnathostomes)
      Chondrichthyes (sharks, rays, chimaeras)
      Actinopterygii (ray-finned fishes)
      Actinistia (coelacanths)
      Dipnoi (lungfishes)
      Amphibia (amphibians: frogs, salamanders, caecilians)
      Reptilia (reptiles: tuataras, lizards, snakes, turtles, crocodilians, birds)
      Mammalia (mammals)

} Vertebrates

Light Microscope

In light microscopy, light is focused on a specimen by a glass condenser lens; the image is then magnified by an objective lens and an ocular lens for projection on the eye, digital camera, digital video camera, or photographic film.

Electron Microscope

In electron microscopy, a beam of electrons (top of the microscope) is used instead of light, and electromagnets are used instead of glass lenses. The electron beam is focused on the specimen by a condenser lens; the image is magnified by an objective lens and a projector lens for projection on a digital detector, fluorescent screen, or photographic film.

Graphs

Graphs provide a visual representation of numerical data. They may reveal patterns or trends in the data that are not easy to recognize in a table. A graph is a diagram that shows how one variable in a data set is related (or perhaps not related) to another variable. The **independent variable** is the factor that is manipulated or changed by the researchers. The **dependent variable** is the factor that the researchers are measuring in relation to the independent variable. The independent variable is typically plotted on the *x*-axis and the dependent variable on the *y*-axis. Types of graphs that are frequently used in biology include scatter plots, line graphs, bar graphs, and histograms.

➤ A **scatter plot** is used when the data for all variables are numerical and continuous. Each piece of data is represented by a point. In a **line graph**, each data point is connected to the next point in the data set with a straight line, as in the graph to the right. (To practice making and interpreting scatter plots and line graphs, see the Scientific Skills Exercises in Chapters 2, 3, 7, 8, 10, 13, 19, 24, 34, 43, 47, 49, 50, 52, 54, and 56.)

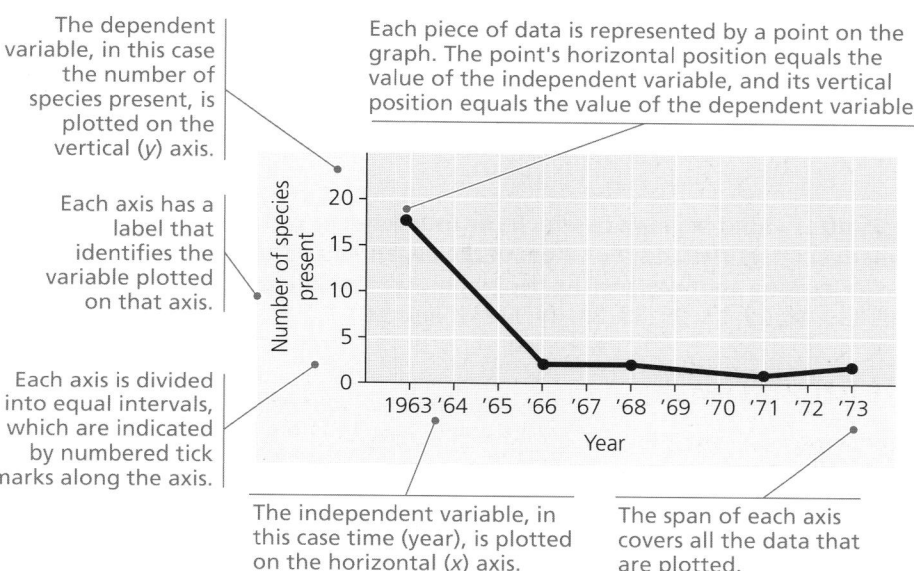

The dependent variable, in this case the number of species present, is plotted on the vertical (*y*) axis.

Each axis has a label that identifies the variable plotted on that axis.

Each axis is divided into equal intervals, which are indicated by numbered tick marks along the axis.

Each piece of data is represented by a point on the graph. The point's horizontal position equals the value of the independent variable, and its vertical position equals the value of the dependent variable.

The independent variable, in this case time (year), is plotted on the horizontal (*x*) axis.

The span of each axis covers all the data that are plotted.

▼ Two or more data sets can be plotted on the same line graph to show how two dependent variables are related to the same independent variable. (To practice making and interpreting line graphs with two or more data sets, see the Scientific Skills Exercises in Chapters 7, 43, 47, 49, 50, 52, and 56.)

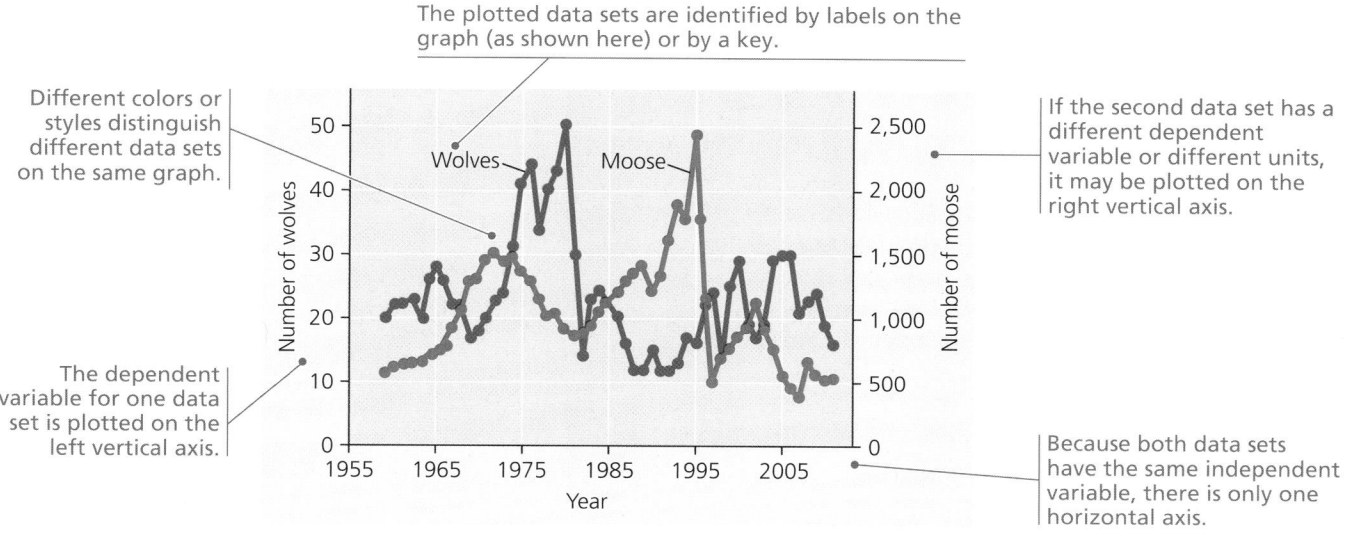

The plotted data sets are identified by labels on the graph (as shown here) or by a key.

Different colors or styles distinguish different data sets on the same graph.

The dependent variable for one data set is plotted on the left vertical axis.

If the second data set has a different dependent variable or different units, it may be plotted on the right vertical axis.

Because both data sets have the same independent variable, there is only one horizontal axis.

▼ In some scatter plot graphs, a straight or curved line is drawn through the entire data set to show the general trend in the data. A straight line that mathematically best fits the data is called a *regression line*. Alternatively, a mathematical function that best fits the data may describe a curved line, often termed a *best-fit curve*. (To practice making and interpreting regression lines, see the Scientific Skills Exercises in Chapters 3, 10, and 34.)

The regression line can be expressed as a mathematical equation. It allows you to predict the value of the dependent variable for any value of the independent variable within the span of the data set and, less commonly, beyond the span of the data.

Dependent variable

Independent variable

▼ A **bar graph** is a kind of graph in which the independent variable represents groups or nonnumerical categories and the values of the dependent variable(s) are shown by bars. (To practice making and interpreting bar graphs, see the Scientific Skills Exercises in Chapters 1, 9, 18, 22, 25, 29, 33, 35, 39, 51, 52, and 54.)

Each piece of data is represented by a bar on the graph. The top of the bar aligns with the value of the dependent variable.

If multiple data sets are plotted on the same bar graph, they are distinguished by bars of different colors or styles and identified by labels or a key.

As in a line graph or scatter plot, the vertical axis is usually used for the dependent variable.

The axis for the dependent variable is labeled and divided into equal intervals indicated by numbered tick marks.

The groups or categories of the independent variable are usually spaced equally along the horizontal axis. (In some bar graphs, the horizontal axis is used for the dependent variable and the vertical axis for the independent variable.)

➤ A variant of a bar graph called a **histogram** can be made for numeric data by first grouping, or "binning," the variable plotted on the *x*-axis into intervals of equal width. The "bins" may be integers or spans of numbers. In the histogram at right, the intervals are 25 mg/dL wide. The height of each bar shows the percent (or, alternatively, the number) of experimental subjects whose characteristics can be described by one of the intervals plotted on the *x*-axis. (To practice making and interpreting histograms, see the Scientific Skills Exercises in Chapters 12, 14, and 42.)

The height of this bar shows the percent of individuals (about 4%) whose plasma LDL cholesterol levels are in the interval indicated on the *x*-axis.

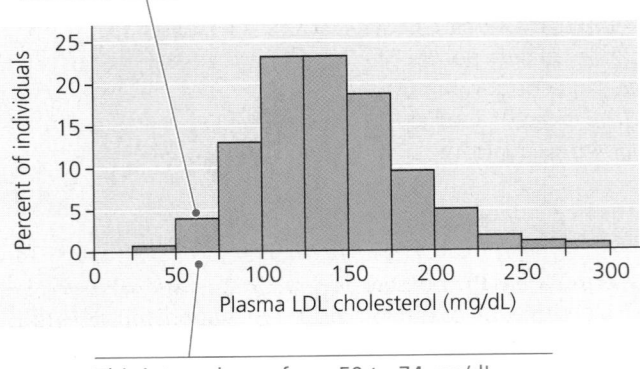

This interval runs from 50 to 74 mg/dL.

Glossary of Scientific Inquiry Terms

See Concept 1.3 for more discussion of the process of scientific inquiry.

control group In a controlled experiment, a set of subjects that lacks (or does not receive) the specific factor being tested. Ideally, the control group is identical to the experimental group in other respects.

controlled experiment An experiment designed to compare an experimental group with a control group; ideally, the two groups differ only in the factor being tested.

data Recorded observations.

deductive reasoning A type of logic in which specific results are predicted from a general premise.

dependent variable A factor whose value is measured during an experiment to see whether it is influenced by changes in another factor (the independent variable).

experiment A scientific test. Often carried out under controlled conditions that involve manipulating one factor in a system in order to see the effects of changing that factor.

experimental group A set of subjects that has (or receives) the specific factor being tested in a controlled experiment. Ideally, the experimental group is identical to the control group for all other factors.

hypothesis A testable explanation for a set of observations based on the available data and guided by inductive reasoning. A hypothesis is narrower in scope than a theory.

independent variable A factor whose value is manipulated or changed during an experiment to reveal possible effects on another factor (the dependent variable).

inductive reasoning A type of logic in which generalizations are based on a large number of specific observations.

inquiry The search for information and explanation, often focusing on specific questions.

model A physical or conceptual representation of a natural phenomenon.

prediction In deductive reasoning, a forecast that follows logically from a hypothesis. By testing predictions, experiments may allow certain hypotheses to be rejected.

theory An explanation that is broader in scope than a hypothesis, generates new hypotheses, and is supported by a large body of evidence.

variable A factor that varies during an experiment.

Chi-Square (χ^2) Distribution Table

To use the table, find the row that corresponds to the degrees of freedom in your data set. (The degrees of freedom is the number of categories of data minus 1.) Move along that row to the pair of values that your calculated χ^2 value lies between. Move up from those numbers to the probabilities at the top of the columns to find the probability range for your χ^2 value. A probability of 0.05 or less is generally considered significant. (To practice using the chi-square test, see the Scientific Skills Exercise in Chapter 15.)

Degrees of Freedom (df)	Probability										
	0.95	0.90	0.80	0.70	0.50	0.30	0.20	0.10	0.05	0.01	0.001
1	0.004	0.02	0.06	0.15	0.45	1.07	1.64	2.71	3.84	6.64	10.83
2	0.10	0.21	0.45	0.71	1.39	2.41	3.22	4.61	5.99	9.21	13.82
3	0.35	0.58	1.01	1.42	2.37	3.66	4.64	6.25	7.82	11.34	16.27
4	0.71	1.06	1.65	2.19	3.36	4.88	5.99	7.78	9.49	13.28	18.47
5	1.15	1.61	2.34	3.00	4.35	6.06	7.29	9.24	11.07	15.09	20.52
6	1.64	2.20	3.07	3.83	5.35	7.23	8.56	10.64	12.59	16.81	22.46
7	2.17	2.83	3.82	4.67	6.35	8.38	9.80	12.02	14.07	18.48	24.32
8	2.73	3.49	4.59	5.53	7.34	9.52	11.03	13.36	15.51	20.09	26.12
9	3.33	4.17	5.38	6.39	8.34	10.66	12.24	14.68	16.92	21.67	27.88
10	3.94	4.87	6.18	7.27	9.34	11.78	13.44	15.99	18.31	23.21	29.59

Mean and Standard Deviation

The **mean** is the sum of all data points in a data set divided by the number of data points. The mean (or average) represents a "typical" or central value around which the data points are clustered. The mean of a variable x (denoted by $\bar{x}$) is calculated from the following equation:

$$\bar{x} = \frac{\sum\limits_{i=1}^{n} x_i}{n}$$

In this formula, n is the number of observations, and x_i is the value of the ith observation of variable x; the "Σ" symbol indicates that the n values of x_i are to be summed. (To practice calculating the mean, see the Scientific Skills Exercises in Chapters 27, 32, and 34.)

The **standard deviation** provides a measure of the variation found in a set of data points. The standard deviation (s) of a variable x is calculated from the following equation:

$$s = \sqrt{\frac{\sum\limits_{i=1}^{n} (x_i - \bar{x})^2}{n - 1}}$$

In this formula, n is the number of observations, x_i is the value of the ith observation of variable x, and $\bar{x}$ is the mean of x; the "Σ" symbol indicates that the n values of $(x_i - \bar{x})^2$ are to be summed. (To practice calculating standard deviation, see the Scientific Skills Exercises in Chapters 27, 32, and 34.)

Appendix D Scientific Skills Review

Performing a *t*-Test

One way to assess whether the results of an experiment are statistically significant is to perform a *t*-test. Consider an experiment in which one group of bean plants was treated with fertilizer, whereas a control group was not. Before beginning the experiment, the researchers hypothesized that fertilizer would not affect plant height (the fertilizer would neither increase plant height nor decrease it).

When the experiment was completed, the fertilized plants appeared overall to have grown taller than the unfertilized ones—that is, the mean height of the fertilized plants was greater than the mean height of the unfertilized ones. This result suggests that the fertilizer did, in fact, have an effect and, hence, that the two means are not equal. However, it is also possible that the different means in the two study groups resulted from natural variation of plant heights within the two groups, particularly if the total number of plants is small. How can we determine the likelihood that the observed differences are meaningful and therefore indicate that the fertilizer had an effect? The *t*-test provides a standardized way to decide whether the fertilizer had a significant effect on mean plant height.

To perform a *t*-test, the first step is to calculate the value T (named for "*t*-test"):

$$T = \frac{\bar{x}_1 - \bar{x}_2}{\sqrt{\dfrac{(s_1^2 + s_2^2)}{n}}}$$

In this equation, $\bar{x}_1$ is the mean for the experimental group (fertilized plants), $\bar{x}_2$ is the mean for the control group (unfertilized plants), s_1 is the standard deviation for the experimental group, and s_2 is the standard deviation for the control group. Finally, n is the number of observations in each of the groups. [Note: The formula shown here is valid when the experimental and control groups have the same number of observations (n). A different formula would be used if the two groups had different numbers of observations.]

To calculate T, plug in the values for $\bar{x}_1$, $\bar{x}_2$, s_1, s_2, and n. T will be close to zero when the means $\bar{x}_1$ and $\bar{x}_2$ are nearly equal, and T will be farther from zero (will differ more greatly from zero) when the means are considerably different.

Is the calculated value T different enough from zero to reject the hypothesis that the two means are equal? This decision is based on the probability (p) that an observed difference between two means could have occurred simply by chance (assuming the initial hypothesis was correct, that is, that the two means are equal). The value of p can be determined using a *t* distribution that has $2(n-1)$ degrees of freedom, where n is the number of observations. When p is small (typically, when it is less than 0.05), we reject the hypothesis that the means $\bar{x}_1$ and $\bar{x}_2$ are equal. The value of p can be obtained from an online calculator or looked up in tables for *t* distributions in a statistics textbook.

Credits

Photo Credits

Front Endpapers and Front Matter **laptop** Lemberg Vector studio/Shutterstock; **iPhone, iPad** Pixedon; **Brief Contents, Students page, pp. i, iii, xiv, xvi water lily** Rohrbaugh Photography; **p. iv donkeys** AFP/Getty images; **p. v sunflowers** Christopher Ison/Alamy Stock Photo Image; **p. vi CRISPR video** HHMI BioInteractive; **breathing animation** BioFlix® 3-D Animation; **p. xii human** lanych/Shutterstock; **monkey** David Bagnall/Alamy Stock Photo; **gibbon** Eric Isselee/Shutterstock; **frogs** Joel Sartore/ Getty Images; **p. xiii students** YanLev/Shutterstock; **p. xv authors** Josh Frost, Pearson Education, **Neil Campbell** Pearson Education; **p. xviii bioluminescence** Biosphoto/Alamy Stock Photo; **mice** Permission for use granted by Randy Jirtle, Professor of Epigenetics, NC State University, Raleigh, NC; **Dutch Hunger Winter** History and Art Collection/ Alamy Stock Photo; **p. xix snowshoe hares (a)** From: Camouflage mismatch in seasonal coat color due to decreased snow duration. L. S. Mills, et al. *Proc Natl Acad Sci U S A*. 2013 Apr 30;110(18):7360-5. Fig. 1; **(b)** L. Scott Mills; **p. xix graph** Data from K. Kupferschmidt, Resistance Fighters. *Science* 352(6287):758-761. 13 May 2016; **p. xxii space-filling model, ribbon model** Data from PDB ID 2LYZ: R. Diamond. Real-Space Refinement of the Structure of Hen Egg-white Lysozyme. *Journal of Molecular Biology* 82(3):371–91 (Jan. 25, 1974); **wireframe model** Clive Freeman, The Royal Institution/Science Source; **p. xxiv sunflower** © Radius Images/Getty Images; **p. xxv seal** Hemis Morales/agefotostock; **cryo-EM** Veronica Falconieri and Siriam Subramaniam, National Cancer Institute; **students** Pearson Education; **p. xxvi Olden** CUNY; **Bautista** Mark Joseph Hanson; **Mojica** University of Alicante; **Extavour** Kris Snibbe; **Chisholm** Gretchen Ertl; **Gonsalves** Courtesy of Dennis Gonsalves; **Strathdee** UCSD Health; **Rochman** Courtesy of Chelsea Rochman; **p. xxxii seal** Hemis Morales/agefotostock; **p. xxxiii researchers** CDC; **p. xxxiv marmot** Tierfotoagentur/Alamy Stock Photo; **caterpillar** blickwinkel/Alamy Stock Photo Image; **p. xxxv impala** Nature Picture Library/Alamy Stock Photo; **pea blossom** John Swithinbank/ Agefotostock; **p. xxxvi DNA** 4X-image/E+/Getty Images; **p. xxxvii cell** Thomas Deerinck, NCMIR/Science Source; **sea horse** Rich Cary/Shutterstock; **p. xxxviii hare** L. Scott Mills; **amber** Iolanda Astor/AGE fotostock/Alamy Stock Photo; **p. xxxix dulse** Andrew J. Martinez/Science Source; **p. xl seed** Howard Sandler/shutterstock; **fungi** Sava Krstic; **p. xli ladybug** André Skonieczny/F1online digitale Bildagentur GmbH/Alamy Stock Photo; **shark** Gino Santa Maria/shutterstock; **tree** Raimund Linke/Photodisc/Getty Images; **p. xlii farmer** Visuals Stock/Alamy Stock Photo; **pollen** Dartmouth College Electron Microscope Facility; **p. xliii penguins** Paul Nicklen/National Geographic Image Collection/ Getty Images; **p. xliv virus** Kateryna Kon/Shutterstock; **p. xlv zebras** Mike Taylor/Alamy Stock Photo; **p. xlvi plover** Joel Sartore/National Geographic Images; **p. xlvii reef** Matthew Banks/Alamy Stock Photo Image; **tiger** Edwin Giesbers/Nature Picture Library.

Chapter 1 **1.1 top** J.B. Miller/Florida Department of Environmental Protection; **bottom left** Hopi Hoekstra/Harvard University; **bottom right** Shawn P. Carey/Migration Productions; **1.2 sunflower** John Foxx/ImageState Media Partners; **seahorse** R. Dirscherl/ OceanPhoto/Frank Lane Picture Agency; **rabbit** Joe McDonald/Encyclopedia/Corbis; **butterfly** Louise Docker Sydney Australia/Moment/Getty Images; **seedling** Frederic Didillon/Garden Picture Library/Getty Images; **venus flytrap** Maximilian Weinzierl/Alamy Stock Photo; **giraffe** Malcolm Schuyl/Frank Lane Picture Agency; **1.3 biosphere** Leonello Calvetti/Stocktrek Images/Getty Images; **ecosystems** Terry Donnelly/Alamy Stock Photo; **communities, populations** Floris van Breugel/naturepl.com; **organisms** Greg Vaughn/ Alamy Stock Photo; **organs** Pat Burner/Pearson Education; **tissues** Photo Researchers/Science Source; **cells** Andreas Holzenburg, University of Texas Rio Grande Valley; **organelles** Jeremy Burgess/Science Source; **p. 6 hummingbird** Jim Zipp/Science Source; **1.4 left** Steve Gschmeissner/Science Source; **1.5** Conly L. Rieder, Wadsworth Center, Albany, NY; **1.6** Gelpi/Fotolia; **1.8a** Carol Yepes/Moment/Getty Images; **inset** Ralf Dahm/Max Planck Institute of Neurobiology; **1.11** James Balog/Aurora/Getty Images; **1.12** Kefca/Shutterstock; **1.13a, b** Eye of Science/Science Source; **1.13c plantae** John Delapp/Design Pics/Getty Images; **fungi** daksel/Fotolia; **animalia** Anup Shah/naturepl.com; **protists** M. I. Walker/ Science Source; **1.14 lake** Basel101658/Shutterstock; **paramecium** SPL/Science Source; **cilium** Dartmouth College Electron Microscope Facility; **cilia** Steve Gschmeissner/Science Source; **1.15** Robert Clark/National Geographic Image Collection; **1.16 left** G. Richmond/FineArt/Alamy Stock Photo; **1.16 right** Origin of Species/Charles Darwin, 1859. Murray edition; **1.17 hawk** jhayes44/E+/Getty Images; **robin** Sebastian Knight/Shutterstock; **flamingo** zhaoyan/Shutterstock; **penguin** Volodymyr Goinyk/Shutterstock; **1.19** Dorling Kindersley ltd/Alamy Stock Photo; **1.21** Michael Nichols/National Geographic Image Collection; **p. 17 Goodall** Jim Dallas/Alamy Stock Photo; **1.23 top** Martin Shields/ Alamy Stock Photo; **center** xpacifica/Getty Images; **bottom left** Rolf Hicker Photography/All Canada Photos/Alamy Stock Photo; **bottom right** Maureen Spuhler/Pearson Education; **1.24** HildeAnna/Shutterstock; **inset** Courtesy of Hopi Hoekstra/Harvard University; **right** Sacha Vignieri; **inset** Shawn P. Carey, Migration Productions; **1.25** From: The selective advantage of cryptic coloration in mice. Vignieri, S. N., J. Larson, and H. E. Hoekstra. 2010. *Evolution* 64:2153–2158. Fig. 1; **p. 21 Hoekstra** Josh Frost/Pearson Education; **Scientific Skills Exercise** Rolf Nussbaumer Photography/Alamy Stock Photo; **1.26 left** John Amis/AP Images; **top** Don Heupel/AP Images; **right** LBJ Presidential Library/ Alamy Stock Photo; **bottom** McClatchy-Tribune/Tribune Content Agency LLC/Alamy Stock Photo; **p. 26 gecko** Chris Mattison/Alamy Stock Photo.

Unit One Interview CUNY.

Chapter 2 **2.1 top** www.pqpictures.co.uk/Alamy Stock Photo; **bottom** Nature Picture Library/Alamy Stock Photo; **2.2 left** sciencephotos/Alamy Stock Photo; **center, right** Stephen Frisch/Pearson Education; **p. 29 Olden** CUNY; **2.3 community** Richard Wong/ Alamy Stock Photo; **lily** Tom Hilton; **rock** Andrew Alden; **2.5** National Library of Medicine; **Scientific Skills Exercise** Pascal Goetgheluck/Science Source; **2.13** Stephen

Frisch/Pearson Education; **p. 39 gecko** nico99/Shutterstock; **hairs** Andrew Syred/Science Source; **p. 40 Pert** Photo 12/Alamy Stock Photo; **2.17** Nigel Cattlin/Science Source; **p. 43 top** Rolf Nussbaumer Photography/Alamy Stock Photo; **bottom** From: Spray aiming in the bombardier beetle: photographic evidence. T. Eisner et al. *Proc Natl Acad Sci U S A* 1999 Aug 17;96(17):9705-9. Fig. 1.

Chapter 3 **3.1 top** Hemis Morales/agefotostock; **bottom** Paul Nicklen/National Geographic Image Collection; **3.3** Alasdair James/E+/Getty Images; **3.4** N.C Brown Center for Ultrastructure Studies, SUNY-ESF, Syracuse, NY; **3.6** Four Oaks/Shutterstock; **p. 48 Solomon** Justina Thorsen; **3.10** JPL/University of Arizona/NASA; **3.11 lemon** Paulista/Fotolia; **ant** Nature Picture Library/Alamy Stock Photo; **blood cells** SCIEPRO/SPL/AGE Fotostock; **bleach** Beth Van Trees/Shutterstock; **Scientific Skills Exercise** Vlad61/Shutterstock; **p. 55 cat** Eric Guilloret/Biosphoto/Science Source.

Chapter 4 **4.1** Florian Möllers/Nature Picture Library; **p. 57 Miller** Robin Heyden, Pearson Education; **Scientific Skills Exercise** Mandeville Special Collections Library; **vials** Jeffrey Bada/Scripps Institution of Oceanography/University of California San Diego; **p. 60 Gordon** Stuart Brinin; **4.6a** David M. Phillips/Science Source; **p. 65 lions** George Sanker/ Nature Picture Library.

Chapter 5 **5.1** Mark J. Winter/Science Source; **5.6a arm** Dougal Waters/Getty Images; **plastids** Omikron/Science Source; **5.6b** Mediscan/Alamy Stock Photo; **5.6c cell** John Durham/Science Source; **microfibrils** Biophoto Associates/Science Source; **5.8** blickwinkel/ Alamy Stock Photo; **5.10a** Vincent Giordano Photo/Shutterstock; **5.10b** Bamorgan91/ Shutterstock; **p. 75 Jones** Courtesy of Lovell Jones; **5.13 eggs** Andrey Stratilatov/Shutterstock; **muscle, collagen** Nina Zanetti/Pearson Education; **5.16 wireframe model** Clive Freeman, The Royal Institution/Science Source; **5.17** Peter M. Colman; **5.18 spider** Dieter Hopf/imageBROKER/Getty Images; **blood cells** SCIEPRO/SPL/AGE Fotostock; **5.19** Eye of Science/Science Source; **p. 82 Pauling** James W. Behnke/Pearson Education; **5.21 top** Dsrjsr; **bottom** Laguna Design/Science Source; **5.25** Centers for Disease Control and Prevention (CDC); **5.26 DNA** Alfred Pasieka/Science Source; **neanderthal** Mark Thiessen/ National Geographic; **hippo** Frontline Photography/Alamy Stock Photo; **whale** WaterFrame_mus/Alamy Stock Photo; **5.26 doctor** Chassenet/BSIP/Alamy Stock Photo; **elephants** Villiers Steyn/Shutterstock; **roots** D.J. Read, Department of Animal and Plant Sciences, University of Sheffield; **Scientific Skills Exercise human** lanych/ Shutterstock; **monkey** David Bagnall/Alamy Stock Photo; **gibbon** Eric Isselee/Shutterstock; **Problem-Solving Exercise** Cindy Hopkins/Alamy Stock Photo; **p. 91 chick** Africa Studio/Shutterstock.

Unit Two Interview **top** Mark Joseph Hansen; **bottom** Robin Heyden.

Chapter 6 **6.1** M. I. Walker/Science Source; **6.3 brightfield, phase-contrast, Nomarski** Elisabeth Pierson/Pearson Education; **fluorescence** Michael W. Davidson/The Florida State University Research Foundation; **confocal** Karl Garsha; **deconvolution** Hans van der Voort SVI; **super-resolution** Muthugapatti K. Kandasamy, Biomedical Microscopy Core, University of Georgia; **SEM, TEM** Steve Gschmeissner/Science Source; **cryo-EM** Veronica Falconieri and Siriam Subramaniam, National Cancer Institute; **6.5b** CNRI/Science Source; **6.6a** Don W. Fawcett/Science Source; **Scientific Skills Exercise** Kelly Tatchell; **6.8 human cells** S. Cinti/Science Source; **yeast SEM** SPL/Science Source; **yeast TEM** A. Barry Dowsett/Science Source; **duckweed** Biophoto Associates/Science Source; **alga SEM** SPL/ Science Source; **alga TEM** From: Flagellar microtubule dynamics in *Chlamydomonas*: cytochalasin D induces periods of microtubule shortening and elongation; and colchicine induces disassembly of the distal, but not proximal, half of the flagellum. W. L. Dentler et al. *J Cell Biol*. 1992 Jun;117(6):1289-98. Fig. 10d; **p. 102 nucleus** Thomas Deerinck/Mark Ellisman/NCMIR; **6.9 nuclear envelope** Biophoto Associates/Science Source; **pore complex** Don W. Fawcett/Science Source; **nuclear lamina** Ueli Aebi; **6.10 left** Don W. Fawcett/Science Source; **right** Harry Noller; **p. 103 Ramakrishnan** Courtesy of Venki Ramakrishnan; **6.11** R. W. Bolender; Don W. Fawcett/Science Source; **6.12** Don W. Fawcett/Science Source; **6.13a** Steve Gschmeissner/Science Source; **6.13b** Don W. Fawcett/Science Source; **6.14** Eldon H. Newcomb; **6.17a** Keith R. Porter/Science Source; **6.17c** From: The shape of mitochondria and the number of mitochondrial nucleoids during the cell cycle of *Euglena gracilis*. Y. Hayashi and K. Ueda. *Journal of Cell Science*, 93:565-570, fig. 3. Copyright © 1989 by Company of Biologists; **6.18a** Jeremy Burgess/Mary Martin/Science Source; **6.18b** Ed Reschke/Photolibrary/Getty Images; **6.19** Eldon H. Newcomb; **6.20** Albert Tousson; **p. 113 Langford** Syracuse University; **6.21b** Bruce J. Schnapp; **Table 6.1 left to right** Gopal Murti/ Science Source, Nikon MicroscopyU (www.microscopyu.com), Mark Ladinsky; **6.22** Kent L. McDonald; **6.23a** Biophoto Associates/Science Source; **6.23b** Oliver Meckes, Nicole Ottawa/Eye of Science/Science Source; **6.24a** Omikron/Science Source; **6.24b** Dartmouth College Electron Microscope Facility; **6.24c** Richard W. Linck; **6.25** From: Organization of actin, myosin, and intermediate filaments in the brush border of intestinal epithelial cells. Hirokawa et al. *J Cell Biol*. 1982 Aug;94(2):425-43. Fig. 1. The Rockefeller University Press; **6.26a** Clara Franzini-Armstrong/University of Pennsylvania; **6.26b** M. I. Walker/Science Source; **6.26c** Michael Clayton/University of Wisconsin; **6.27** G. F. Leedale/Science Source; **6.29** Eldon H. Newcomb, University of Wisconsin, Department of Botany; **6.30 top** Reproduced with permission from: *Freeze-Etch Histology*, by L. Orci and A. Perrelet, Springer-Verlag, Heidelberg, 1975. Plate 32. Page 68. Copyright 1975 by Springer-Verlag GmbH & Co KG; **6.30 center** From: Fine structure of desmosomes, hemidesmosomes, and an adepidermal globular layer in developing newt epidermis. DE Kelly. *J Cell Biol*. 1966 Jan; 28(1):51-72. Fig. 7. Reproduced by permission of Rockefeller University Press; **bottom** From: Low resistance junctions in crayfish. Structural changes with functional uncoupling. C. Peracchia and A. F. Dulhunty, *The Journal of Cell Biology*. 1976 Aug; 70(2 pt 1):419-39. Fig. 6. Reproduced by permission of Rockefeller University Press; **6.31** Eye of Science/ Science Source; **p. 125 epithelial cell** Susumu Nishinaga/Science Source.

Alamy Stock Photo; **bottom left** Amar and Isabelle Guillen - Guillen Photo LLC/Alamy Stock Photo; **top right** Eldon H. Newcomb; **bottom right** Science Photo Library/Alamy Stock Photo; **33.10** CDC; **33.11** Eye of Science/Science Source; **33.11** Eye of Science/Science Source; **33.12** M. I. Walker/Science Source; **33.13** Holger Herlyn, University of Mainz, Germany; **33.14a** blickwinkel/Alamy Stock Photo; **33.14b** Image Quest Marine; **33.16** Image Quest Marine; **33.17a** Lubos Chlubny/Fotolia; **33.17b** Terry Moore/Stocktrek Images/Alamy Stock Photo; **Scientific Skills Exercise** Christophe Courteau/Water Rights/Alamy Stock Photo; **33.18** Andrew J. Martinez/Science Source; **33.20 top** Mark Conlin/VWPics/Alamy Stock Photo; **center** Photonimo/Shutterstock; **bottom** SeaTops/Alamy Stock Photo; **33.21 left** Dave Clarke/Zoological Society of London; **right** The U.S. Bureau of Fisheries; **33.22** Fredrik Pleijel; **33.23** Wolcott Henry/National Geographic; **33.24** Astrid Michler; Hanns-Frieder Michler/Science Source; **33.25** Wayne Taylor/The AGE/Fairfax Media via Getty Images; **33.26** London Scientific Films/Oxford Scientific/Getty Images; **33.27** Power and Syred/Science Source; **33.28** Dan Cooper; **33.29b** Courtesy of Sean B. Carroll; **33.31** Mark Newman/Frank Lane Picture Agency; **33.32 top** Tim Flach/The Image Bank/Getty Images; **center** Andrew Syred/Science Source; **bottom** Reinhard Hölzl/ImageBROKER/AGE Fotostock; **33.34a** Premaphotos/Nature Picture Library; **33.34b** Tom McHugh/Science Source; **33.36** Maximilian Weinzierl/Alamy Stock Photo; **33.37** Peter Herring/Image Quest Marine; **33.38** Peter Parks/Image Quest Marine; **33.40** André Skonieczny/F1online digitale Bildagentur GmbH/Alamy Stock Photo; **33.41a, b, d, e** Cathy Keifer/Shutterstock; **33.41c** Jim Zipp/Science Source; **33.42** Archaeognatha Kevin Murphy; **Zygentoma** Denis Crawford/Alamy Stock Photo; **Coleoptera** Premaphotos/Nature Picture Library; **Diptera** Bruce Marlin; **Hymenoptera** John Cancalosi/Nature Picture Library; **Lepidoptera** Hans Christoph Kappel/Nature Picture Library; **Hemiptera** Dante Fenolio/Science Source; **Orthoptera** Chris Mattison/Alamy Stock Photo; **33.43** Andrey Nekrasov/Image Quest Marine; **33.44** Daniel Janies; **33.45** Jeff Rotman/Science Source; **33.46** Louise Murray/robertharding/Alamy Stock Photo; **33.47** Jurgen Freund/Nature Picture Library; **33.48** Hal Beral/Corbis; **p. 717** beetles Lucy Arnold.

Chapter 34 34.1 top to bottom Derek Siveter, Tom McHugh/Science Source, Gino Santa Maria/Shutterstock; Digital Vision/Photodisc/Getty Images, Arnaz Mehta, Tom McHugh/Science Source, Rolf Nussbaumer Photography/Alamy Stock Photo, Visceralimage/Fotolia; **34.4** Natural Visions/Alamy Stock Photo; **34.5c** Ethan Daniels/Stocktrek Images/Alamy Stock Photo; **34.8** Tom McHugh/Science Source; **34.9** Marevision/AGE Fotostock; **inset** Hartl/blickwinkel/Alamy Stock Photo; **34.10** Junyuan Chen/Nanjing Institute of Geology and Palaeontology, Chinese Academy of Sciences; **34.14** Field Museum Library/Premium Archive/Getty Images; **34.15a** Gino Santa Maria/Shutterstock; **34.15b** Masa Ushioda/Image Quest Marine; **34.15c** RGB Ventures/SuperStock/Alamy Stock Photo; **34.17** tuna James D. Watt/Image Quest Marine; **lionfish** Teigler/blickwinkel/Alamy Stock Photo; **sea horse** George Grall/National Geographic; **eel** Fred McConnaughey/Science Source; **34.18** Reprinted by permission from Macmillan Publishers Ltd: From: The oldest articulated osteichthyan reveals mosaic gnathostome characters. M. Zhu. *Nature.* 2009 Mar 26;458(7237):469–74. doi: 10.1038/nature07855. Fig. 2; **34.19** Arnaz Mehta/SeaPics; **34.20** fossil, ribs, scales Ted Daeschler/Academy of Natural Sciences/Vireo; **fin** Kalliopi Monoyios Studio; **34.22a** Alberto Fernández/AGE Fotostock/Alamy Stock Photo; **34.22b** Anneka/Shutterstock; **34.22c** Zeeshan Mirza/ephotocorp/Alamy Stock Photo; **34.23a** DP Wildlife Vertebrates/Alamy Stock Photo; **34.23b** FLPA/Alamy Stock Photo; **34.23c** John Cancalosi/Photolibrary/Getty Images; **34.24** Hinrich Kaiser, Victor Valley College; **Problem-Solving Exercise** Joel Sartore/National Geographic; **34.27** Nobumichi Tamura; **34.28** Chris Mattison/Alamy Stock Photo; **p. 736 Sereno** Paul Sereno Fossil Lab; **34.29a** Natural Visions/Alamy Stock Photo; **34.29b** Lee T. Matt; **34.29c** Nick Garbutt/Nature Picture Library; **34.29d** Juniors Bildarchiv/AGE Fotostock; **34.29e** Carl & Ann Purcell/Corbis NX/Getty Images; **34.30a** Visceralimage/Fotolia; **34.30b** The Natural History Museum/Alamy Stock Photo; **34.32** Boris Karpinski/Alamy Stock Photo; **34.33** DLILLC/Corbis/VCG/Getty Images; **34.34** Mariusz Blach/Fotolia; **34.35** The Africa Image Library/Alamy Stock Photo; **inset** mychicport/Shutterstock; **34.36** Gianpiero Ferrari/Frank Lane Picture Agency Limited; **34.39** Clearviewstock/Shutterstock; **inset** Commonwealth Scientific and Industrial Research Organization; **34.40a** John Cancalosi/Alamy Stock Photo; **34.40b** Martin Harvey/Alamy Stock Photo; **34.40c** Rick & Nora Bowers/Alamy Stock Photo; **34.43** ImageBroker/Alamy Stock Photo; **34.45a** Kevin Schafer/AGE Fotostock; **34.45b** J & C Sohns/Picture Press/Getty Images; **34.46a** Morales/AGE Fotostock; **34.46b** Tim Laman/NaturePL; **34.46c** T.J. Rich/Nature Picture Library; **34.46d** E.A. Janes/AGE Fotostock; **34.46e** Martin Harvey/Photolibrary/Getty Images; **34.48** T. White/David L. Brill Photography; **34.49a** John Reader/Science Source; **34.49b** Mauricio Anton/Science Source; **Scientific Skills Exercise** Golfx/Shutterstock; **34.50** Danita Delimont/Alamy Stock Photo; **34.52a** Erik Trinkaus; **34.52b** Tom Higham, University of Oxford; **34.53** David L. Brill Photography; **34.54** From: *Homo naledi,* a new species of the genus *Homo* from the Dinaledi Chamber, South Africa. L. R. Berger et al. *eLife* 2015;4:e09560. Fig. 6; **34.55** C. Henshilwood; **p. 756 animal** Tony Heald/Nature Picture Library.

Unit Six Interview Courtesy of Dennis Gonsalves.

Chapter 35 35.1 tree Raimund Linke/Photodisc/Getty Images; **seedling** Beata Becia/Shutterstock; **leaf cross section** P&R Fotos/AGE Fotostock/Alamy Stock Photo; **chloroplasts** John Durham/Science Source; **tube-shaped cells** Science Photo Library/Alamy Stock Photo; **root hairs** Scenics & Science/Alamy Stock Photo; **35.3** Jeremy Burgess/Science Source; **35.4 buttress roots** Karl Weidmann/Science Source; **prop roots** Natalie Bronstein; **beet** Rob Walls/Alamy Stock Photo; **pneumatophores** Bjorn Svensson/AGE Fotostock/Alamy Stock Photo; **strangling roots** Dana Tezarr/Photodisc/Getty Images; **35.5 top** Maureen Spuhler/Seelevel.com; **center** Dorling Kindersley ltd/Alamy Stock Photo; **bottom** Toshihiko Watanabe/Aflo/Alamy Stock Photo; **35.7 tendrils** Neil Cooper/Alamy Stock Photo; **spines** Martin Ruegner/Photodisc/Getty Images; **storage leaves** Godunova Tatiana/Shutterstock; **reproductive leaves** Dmytro Skorobogatov/123RF; **Scientific Skills Exercise** Dorling Kindersley ltd/Alamy Stock Photo; **35.9** Steve Gschmeissner/SPL/AGE Fotostock; **35.10 parenchyma** M I (Spike) Walker/Alamy Stock Photo; **35.10 collenchyma** Keith Wheeler/Science Source; **sclereid** Graham Kent/Pearson Education; **fiber** Graham Kent/Pearson Education; **tracheids and vessels** N.C Brown Center for Ultrastructure Studies; **sieve-tube element (TEM)** From: *Plant Cell*

Biology on DVD: Information for students and a resource for teachers Springer-Verlag 2009, by B Gunning; **sieve-tube element (LM)** Ray F. Evert; **sieve plate** Graham Kent/Pearson Education; **35.13** From: ABA-mediated ROS in mitochondria regulate root meristem activity by controlling PLETHORA expression in *Arabidopsis*. Yang L. *PLoS Genet.* 2014 Dec 18;10(12):e1004791. doi: 10.1371/journal.pgen.1004791. eCollection 2014 Dec. Figure 6G; **35.14a, b** Ed Reschke; **35.14a center** Chuck Brown/Science Source; **35.15, 35.16** Michael Clayton; **35.17, 35.18** Ed Reschke; **35.20 left** Michael Clayton; **right** Alison W. Roberts; **35.23** University of Southern California; **35.25** Edu Boer, NVWA, NL; **35.26** From: Natural variation in *Arabidopsis*. Plant genetics to ecological genomics. D. Weigel. *Plant Physiol.* 2012 Jan;158(1):2-22. doi: 10.1104/pp.111.189845. Epub 2011 Dec 6. Fig. 1; **p. 776 Chory** Micheline Pelletier/Courtesy of Joanne Chory; **35.28** From: Microtubule plus-ends reveal essential links between intracellular polarization and localized modulation of endocytosis during division-plane establishment in plant cells. P. Dhonukshe. *BMC Biol.* 2005 Apr 14;3:11. Fig4B; **35.28 (inset)** B. Wells and K. Roberts; **35.29** From: The making of a compound leaf: genetic manipulation of leaf architecture in tomato. D. Hareven. *Cell.* 1996 Mar 8;84(5):735-44. Fig. 1; **35.30** From: A common position-dependent mechanism controls cell-type patterning and GLABRA2 regulation in the root and hypocotyl epidermis of *Arabidopsis*. C. Y. Hung et al. *Plant Physiol.* 1998 May;117(1):73-84. Fig. 2g; **p. 778 Benfey** Jie Huang/Duke University; **35.31** Lawrence Jensen; **35.32** From: Genetic interactions among floral homeotic genes of *Arabidopsis*. JL Bowman, DR Smyth, EM Meyerowitz. *Development.* 1991 May;112(1):1-20; Fig. 1A; **p. 780 Walbot** Courtesy of Virginia Walbot; **p. 782 woody eudicot** From: Anatomy of the vessel network within and between tree rings of *Fraxinus lanuginosa* (Oleaceae). P. B. Kitin et al. *American Journal of Botany.* 2004;91:779-788. Fig. 1; **p. 783 tea leaves** Volodymyr Burdiak/Shutterstock; **tea leaf cross section** Keith Wheeler/Science Source; **iris leaves** Rob Stark/Shutterstock; **iris leaf cross section** www.willemsmicroscope.com; **bicycle** Janet Horton/Alamy Stock Photo; **Hakea purpurea** Biophoto Associates/Science Source.

Chapter 36 36.1 velislava/Alamy Stock Photo; **36.3** Rolf Rutishauser and Evelin Pfeifer; **p. 790 plant** Nigel Cattlin/Alamy Stock Photo; **36.8** Benjamin Blonder and David Elliott; **36.10** Scott Camazine/Science Source; **36.13** AGE Fotostock/Alamy Stock Photo; **36.14** Power and Syred/Science Source; **36.15** leafless ocotillo mike lane/Alamy Stock Photo; **green ocotillo** Rick & Nora Bowers/Alamy Stock Photo; **ocotillo close-up** Mint Images/SuperStock; **oleander cross section** Natalie Bronstein; **blossom** SutidaS/Shutterstock; **cactus** Danita Delimont/Alamy Stock Photo; **36.18** M. H. Zimmerman/Harvard Forest; **36.19** From: A coiled-coil interaction mediates cauliflower mosaic virus cell-to-cell movement. L. Stavolone et al. *Proc Natl Acad Sci U S A.* 2005 Apr 26;102(17):6219-24. Epub 2005 Apr 18. Fig. 5c; **p. 802 Zambryski** Noah Berger Photography; **p. 804 forest** Catalin Petolea/Alamy Stock Photo.

Chapter 37 37.1 Visuals Stock/Alamy Stock Photo; **37.2** ARS/USDA; **37.4** National Oceanic and Atmospheric Administration (NOAA); **37.5** Menlo Park/USGS; **37.6** Kevin Horan/The Image Bank/Getty Images; **p. 811 Coruzzi** Courtesy of Gloria M. Coruzzi; **37.8 healthy** View Stock RF/AGE Fotostock; **nitrogen-deficient** Guillermo Roberto Pugliese/International Plant Nutrition Institute (IPNI); **phosphorus-deficient** C. Witt/IPNI; **potassium-deficient** M.K. Sharma and P. Kumar/IPNI; **Scientific Skills Exercise** Nigel Cattlin/Science Source; **37.9** lichen David T. Webb, University of Montana; **lichen section** Courtesy of Ralf Wagner; **puffer** Andrey Nekrasov/Pixtal/AGE Fotostock; **Azolla** Daniel L Nickrent; **37.9** ant Juan Carlos Vindas/Moment Open/Getty Images; **fungal garden** Martin Dohrn/Nature Picture Library; **root** Yoshihiro Kobae; **nectar** Oxford Scientific/Getty Images; **37.10** Sarah Lydia Lebeis; **37.12** Scimat/Science Source; **37.14 sheath** Hugues B. Massicotte/University of Northern British Columbia Ecosystem and Management Program, Prince George, BC, Canada; **cells, arbuscules** Mark Brundrett; **37.15 fern** David Wall/Alamy Stock Photo; **mistletoe** Peter Lane/Alamy Stock Photo; **dodder** Emilio Ereza/Alamy Stock Photo; **Indian pipe** Martin Shields/Alamy Stock Photo; **pitcher plants** Dorling Kindersley ltd/Alamy Stock Photo; **ant on pitcher plant** Paul Zahl/Science Source; **sundew** Fritz Polking/Frank Lane Picture Agency Limited W. Rolfes/Arco Images GmbH/Alamy Stock Photo; **Venus flytrap** Chris Mattison/Nature Picture Library; **p. 821 footprint** Mode Images/Alamy Stock Photo.

Chapter 38 38.1 blickwinkel/Alamy Stock Photo; **inset** Nicolas J. Vereecken; **38.4 hazel carpellate** Friedhelm Adam/imageBROKER/Getty Images; **hazel staminate** Wildlife GmbH/Alamy Stock Photo; **dandelions** © Bjørn Rørslett/NN/Samfoto/Sipa USA; **moth** Doug Backlund/WildPhotosPhotography.com; **blowfly** Kjell B. Sandved/Science Source; **bat** Rolf Nussbaumer/imageBROKER/AGE Fotostock; **hummingbird** Rolf Nussbaumer/Nature Picture Library; **38.5** W. Barthlott and W.Rauh/Nees Institute for Biodiversity of Plants; **38.6 top** Michael Clayton/Botany Dept., University of Wisconsin; **bottom** Ed Reschke/Photolibrary/Getty Images; **38.10** Blickwinkel/Alamy Stock Photo; **38.12 coconut** Kevin Schafer/Alamy Stock Photo; **Alsomitra macrocarpa** Aquiya/Fotolia; **dandelion** Steve Bloom Images/Alamy Stock Photo; **maple seed** Chrispo/Fotolia; **tumbleweed** Nurlan Kalchinov/Alamy Stock Photo; **Tribulus terrestris** California Department of Food and Agriculture's Plant Health and Pest Prevention Services; **squirrel** Alan Williams/Alamy Stock Photo; **feces** Kim A Cabrera; **ant** Benoit Guénard; **38.13** Dennis Frates/Alamy Stock Photo; **Scientific Skills Exercise** monkeyflower Ken Barber/Alamy Stock Photo; **hummingbird** Dec Hogan/Shutterstock; **38.14a** Marcel Dorken; **38.14b** Nobumitsu Kawakubo; **38.15** Meriel G. Jones, University of Liverpool School of Biological Sciences; **38.16** Dorling Kindersley ltd/Alamy Stock Photo; **38.17** Gary P. Munkvold; **p. 837 Herrera-Estrella** Michael Starghill, Michael Starghill Photography; **38.17** Gary P. Munkvold; **p. 838 Gonsalves** Courtesy of Dennis Gonsalves; **38.18** ton koene/Alamy Stock Photo; **p. 841 pollen** Dartmouth College Electron Microscope Facility.

Chapter 39 39.1 Christopher Ison/Alamy Stock Photo; **39.2** Natalie Bronstein; **39.6** From: Regulation of polar auxin transport by AtPIN1 in *Arabidopsis* vascular tissue. L. Gälweiler et al. *Science.* 1998 Dec 18;282(5397):2226-30; Fig. 4; **39.9a** Richard Amasino; **39.9b** Fred Jensen, Kearney Agricultural Center; **39.11 left** Mia Molvray; **right** Karen E. Koch; **39.13a** Kurt Stepnitz; **39.13b** Joseph J. Kieber; **39.14** Ed Reschke; **39.16** Nigel Cattlin/Alamy Stock Photo; **39.18** Martin Shields/Alamy Stock Photo; **p. 858 Satter** Robin Heyden/Pearson Education; **39.22** Michael L. Evans/Ohio State University; **39.23** From the cover of *Cell,* Volume 60, Issue 3, 9 February 1990. Janet Braam, Ronald W. Davis. Used by permission, Copyright ©1990 Cell Press. Image courtesy of Elsevier Sciences, Ltd; **39.24** Martin Shields/Alamy Stock Photo; **39.25** J. L. Basq/M. C. Drew; **39.26** New York State Agricultural Experiment Station/Cornell University College of Agriculture and Life Sciences; **p. 867 Dangl** Courtesy of Jeff Dangl; **39.27 poppy seed** De Meester Johan/Arterra Picture Library/Alamy Stock Photo; **taro plant** David T. Webb; **olive leaf** Science Photo Library/Alamy Stock Photo; **cactus spines** Susumu Nishinaga/Science Source; **snowflake plant** Giuseppe Mazza; **passion flower** Lawrence E. Gilbert/University of Texas-Austin; **hummingbird** Danny Kessler; **bamboo plants** Kim Jackson/Mode Images/Alamy Stock Photo; **wasp, cocoons** Custom Life Science Images/Alamy Stock Photo; **p. 871 deer** Gary Crabbe/Alamy Stock Photo.

Unit Seven Interview UCSD Health.

Chapter 40 40.1 top Paul Nicklen/National Geographic/Getty Images; **center left** CNRS/IPEV/IPHC, France; **bottom** Nature Picture Library/Alamy Stock Photo; **40.2 seal** Dave Fleetham/Robert Harding World Imagery; **penguin** WILDLIFE GmbH/Alamy Stock Photo; **tuna** Andre Seale/Image Quest Marine; **40.4 intestine** Eye of Science/Science Source; **lung, kidney** Susumu Nishinaga/Science Source; **40.5 epithelia** Steve Downing/Pearson Education; **loose connective tissue, adipose tissue, bone, skeletal muscle** Nina Zanetti/Pearson Education; **40.5 blood** Jarun Ontakrai/Shutterstock; **cartilage** Chuck Brown/Science Source; **fibrous connective tissue, smooth muscle, cardiac muscle** Ed Reschke/Photolibrary/Getty Images; **neuron** James Cavallini/Science Source; **glia** Thomas Deerinck; **40.7 otter** Kaufung Agency/blickwinkel/Alamy Stock Photo; **bass** Roderick Paul Walker/Alamy Stock Photo; **40.10** Meiqianbao/Shutterstock; **40.11a** Paul Souders/Danita Delimont Creative/Alamy Stock Photo; **40.11b** Bill Gozansky/Alamy Stock Photo; **40.14** Mirko Graul/Shutterstock; **40.15** From: Assessment of oxidative metabolism in brown fat using PET imaging. Otto Muzik, Thomas J. Mangner and James G. Granneman. *Front. Endocrinol.*, 08 February 2012 | http://dx.doi.org/10.3389/fendo.2012.00015 Fig. 2; **40.19** Jeff Rotman/Alamy Stock Photo; **40.21** FLPA/Alamy Stock Photo; **p. 893 Bartholomew** Robin Heyden **40.23 plants** Irin-K/Shutterstock; **bobcat** Thomas Kitchin/Victoria Hurst/All Canada Photos/AGE Fotostock; **sunflowers** Phil_Good/Fotolia; **fly** WildPictures/Alamy Stock Photo; **sprouts** Bogdan Wankowicz/Shutterstock; **molting** Nature's Images/Science Source; **plant vessels** Last Refuge/Robert Harding Picture Library Ltd/Alamy Stock Photo; **blood vessels** Susumu Nishinaga/Science Source; **peas** Scott Rothstein/Shutterstock; **pigs** steven goodier/Alamy Stock Photo; **intestinal lining** David M. Martin/Science Source; **root hairs, mesophyll** Rosanne Quinnell © The University of Sydney. eBot http://hdl.handle.net/10.100.100/1463, http://hdl.handle.net/102.100.100/2574; **alveoli** David M. Phillips/Science Source; **p. 897 macaques** Yoshiteru Takahashi/Sebun Photo/amana images/Getty Images.

Chapter 41 41.1 Milo Burcham/First Light/Getty Images; **41.3** Stefan Huwiler/Rolf Nussbaumer Photography/Alamy Stock Photo; **41.5 baleen** Vicki Beaver/Alamy Stock Photo; **caterpillar** Stuart Wilson/Science Source; **fly** Peter Parks/Image Quest Marine; **python** Gunter Ziesler/Photolibrary/Getty Images; **41.16 left** McPhoto/INS Agency/blickwinkel/Alamy Stock Photo; **right** Tom Brakefield/Stockbyte/Getty Images; **41.19** James Archer, CDC; **p. 913 Strathdee** UCSD Health; **41.21** Peter Batson/Image Quest Marine; **Scientific Skills Exercise** ORNL/Science Source; **p. 920 owl** Stefan Huwiler/imageBROKER RF/AGE Fotostock.

Chapter 42 42.1 John Cancalosi/Alamy Stock Photo; **42.2a** Reinhard Dirscherl/WaterFrame/Getty Images; **42.2b** Eric Grave/Science Source; **42.9 top** Indigo Instruments; **bottom** Ed Reschke/Photolibrary/Getty Images; **p. 933 Yanagisawa** Skeeter Hagler; **42.18** Eye of Science/Science Source; **42.19** Image Source Plus/Alamy Stock Photo; **Scientific Skills Exercise** cassis/Fotolia; **42.21a** Peter Batson/Image Quest Marine; **42.21b** Olgysha/Shutterstock; **42.21c** Greg Amptman/Shutterstock; **42.23c** Prepared by Dr. Hong Y. Yan, University of Kentucky and Dr. Peng Chai, University of Texas; **42.24** Motta and Macchiarelli, Anatomy Dept., Univ. La Sapienza, Rome/Science Source; **42.26** Hans-Rainer Duncker, Institute of Anatomy and Cell Biology, Justus-Liebig-University Giessen; **42.32** Doug Allan/Nature Picture Library; **p. 951 spider** CB2/ZOB/WENN.com/Newscom.

Chapter 43 43.1 macrophage SPL/Science Source; **virus** James Cavallini/BSIP SA/Alamy Stock Photo; **bacterium** Chris Bjornberg/Science Source; **fungus** Callista Images/Cultura Creative (RF)/Alamy Stock Photo; **influenza** Kateryna Kon/Shutterstock; **43.15** Steve Gschmeissner/Science Source; **Scientific Skills Exercise** Eye of Science/Science Source; **p. 973 Wong-Staal** Bill Branson/NIH; **43.29** Stephen C. Harrison/The Laboratory of Structural Cell Biology/Harvard Medical School; **p. 974 zur Hausen** Tobias Schwerdt; **p. 976 vaccine** Tatan Yuflana/AP Images.

Chapter 44 44.1 David Wall/Alamy Stock Photo; **44.2** Mark Conlin/VWPICS/Visual&Written SL/Alamy Stock Photo; **Scientific Skills Exercise** Jiri Lochman/Lochman Transparencies; **44.6 left** GeorgePeters/E+/Getty Images; **center** Eric Isselée/Fotolia; **right** Maksym Gorpenyuk/Shutterstock; **44.7** Stephane Bidouze/Shutterstock; **44.12** Steve Gschmeissner/Science Source; **44.15** Michael Lynch/Shutterstock; **44.16** v_blinov/Fotolia; **44.17 fish** Image Quest Marine; **stomata** Eye of Science/Science Source; **frog** F1online digitale Bildagentur GmbH/Alamy Stock Photo; **bacterium** Power and Syred/Science Source; **p. 998 iguana** Steven A. Wasserman.

Chapter 45 45.1 top Phillip Colla/Oceanlight.com; **bottom** Craig K. Lorenz/Science Source; **45.3** Volker Witte/Ludwig-Maximilians-Universitat Munchen; **45.11** Cathy Keifer/123rf; **Problem-Solving Exercise** angellodeco/Fotolia; **45.17** AP Images; **45.22 left** Blickwinkel/Alamy Stock Photo; **right** Jurgen and Christine Sohns/Frank Lane Picture Agency; **p. 1018 frogs** Eric Roubos.

Chapter 46 46.1 coral Auscape/UIG/Getty Images; **hydra** Roland Birke/Okapia/Science Source; **nudibranchs** Colin Marshall/Frank Lane Picture Agency; **frogs** Andy Sands/Nature Picture Library; **sperm** Don W. Fawcett/Science Source; **cardinals** William Leaman/Alamy Stock Photo; **zebras** Mike Taylor/Alamy Stock Photo; **46.2** Colin Marshall/Frank Lane Picture Agency; **46.3a** P. de Vries/Crews, David; **46.5** Andy Sands/Nature Picture Library; **46.6** John Cancalosi/Alamy Stock Photo; **Scientific Skills Exercise** Andrew Syred/Science Source; **46.12** Design Pics Inc/Alamy Stock Photo; **46.17** Tidningarnas Telelgrambyra AB; **46.21** M.I. Walker/Science Source; **p. 1042 Komodo dragon** Dave Thompson/AP Images.

Chapter 47 47.1 Brad Smith/Stamps School of Art & Design, University of Michigan; **inset** Oxford Scientific/Getty Images; **47.3 top** Victor D. Vacquier; **bottom** From: Wave of free calcium at fertilization in the sea urchin egg visualized with fura-2. M. Hafner et al, *Cell Motil Cytoskeleton*. 1988;9(3):271-7. Fig. 1; **47.6** George von Dassow; **47.7 top** Jürgen Berger/Max Planck Institute for Developmental Biology, Tübingen Germany; **bottom** Andrew J. Ewald, Johns Hopkins Medical School; **47.13b** Alejandro Díaz Díez/AGE Fotostock/Alamy Stock Photo; **47.14a** P. Huw Williams and Jim Smith, The Wellcome Trust/Cancer Research UK Gurdon Institute; **47.14c** Thomas Poole, SUNY Health Science Center; **47.15b** Keith Wheeler/Science Source; **47.18b** From: Cell lineage analysis in ascidian embryos by intracellular injection of a tracer enzyme. III. Up to the tissue restricted stage. H. Nishida. *Dev Biol*. 1987 Jun;121(2):526-41. Fig. 1. Reprinted by permission of Academic Press; **47.19** From: Post-embryonic cell lineages of the nematode, *Caenorhabditis elegans*. E. Sulston et al. *Dev Biol*. 1977 Mar;56(1):110-56. Fig. 1; **47.20** Susan Strome; **47.21** Susan Strome; **47.24** From: Dorsal-ventral patterning and neural induction in *Xenopus* embryos. E. M. De Robertis and H. Kuroda. *Annu Rev Cell Dev Biol*. 2004;20:285-308. Fig. 1; **47.25a** Kathryn Tosney, University of Michigan; **47.26** Based on Honig and Summerbell, courtesy of Lawrence S. Honig; **p. 1066 turtle** James Gerholdt/Getty Images.

Chapter 48 48.1 Franco Banfi/Science Source; **48.2** Edwin R. Lewis; **p. 1068 Olivera** From: QnAs with Baldomero M. Olivera. Beth Azar. *PNAS* August 21, 2012 109 (34) 13470; https://doi.org/10.1073/pnas.1211581109. Fig. 1; **48.5** Thomas Deerinck/National Center for Microscopy and Imaging Research, University of California, San Diego; **48.14** Alan Peters; **p. 1084** B.A.E./Alamy Stock Photo.

Chapter 49 49.10 Tamily Weissman; **49.11** Larry Mulvehill/Corbis; **49.15** From: A functional MRI study of happy and sad affective states induced by classical music. M. T. Mitterschiffthaler et al. *Hum Brain Mapp*. 2007 Nov. 28(11):1150-62. Fig. 1; **49.18** Marcus E. Raichle, Washington University Medical Center. From research based on "Positron emission tomographic studies of the cortical anatomy of single-word processing". S.E. Petersen et al. *Nature* 331:585-589 (1988); **49.19** National Library of Medicine (NLM); **p. 1099 Jarvis** Walter Oleksy/Alamy Stock Photo; **p. 1103 Heberlein** Matthew Staley; **49.25** Martin M. Rotker/Science Source; **p. 1106 microphone** Eric Delmar/Getty Images.

Chapter 50 50.1 Kenneth Catania; **50.6** CSIRO Publishing; **50.7a** Michael Nolan/Robert Harding World Imagery; **50.7b** Grischa Georgiew/Panther Media/AGE Fotostock; **50.9** From: Richard Elzinga, *Fundamentals of Entomology*, 3rd ed. ©1987, p. 185. Reprinted by permission of Prentice-Hall, Upper Saddle River, NJ; **50.10** SPL/Science Source; **50.16a** APHIS Animal and Plant Health Inspection Service/USDA; **50.17** Steve Gschmeissner/Science Source; **50.21** Neitz Laboratories; **50.26** Clara Franzini-Armstrong; **50.27** H. E. Huxley; **50.34** Joe Quinn/Alamy Stock Photo; **50.39** Dave Watts/NHPA/Science Source; **p. 1135** Terence Dawson; **Scientific Skills Exercise** Vance A. Tucker; **p. 1138 hound** Dogs/Fotolia.

Chapter 51 51.1 Robert Koss/Shutterstock; **51.3** Manamana/Shutterstock; **51.5b** Scott Camazine/Alamy Stock Photo; **p. 1143** Ian Fletcher/Shutterstock; **p. 1144** Dustin Finkelstein/Getty Images for SXSW; **51.7** Thomas D. McAvoy/The LIFE Picture Collection/Getty Images; **51.9** Lincoln Brower/Sweet Briar College; **51.11** Dr Clive Bromhall/Oxford Scientific/Getty Images; **51.12** Richard Wrangham; **inset** Mike Korostelev www.mkorostelev.com/Moment Open/Getty Images; **Scientific Skills Exercise** Matt Goff; **51.14a** Matt T. Lee; **51.14b** David Osborn/Alamy Stock Photo; **51.14c** David Tipling/Frank Lane Picture Agency Limited; **51.15** Fotograferen.net/Alamy Stock Photo; **51.16** Gerald S. Wilkinson; **51.17** Juniors Bildarchiv/F300/Alamy Stock Photo; **51.20** Martin Harvey/Photolibrary/Getty Images; **51.21** Erik Svensson/Lund University, Sweden; **51.22** Lowell Getz; **51.23** Rory Doolin; **51.25** Jennifer Jarvis, University of Cape Town; **51.27** Fred van Wijk/Alamy Stock Photo; **51.28** Jupiterimages/Creatas/Thinkstock/Getty Images; **p. 1160 Wilson** Michael Dwyer/Alamy Stock Photo; **p. 1162 woodpecker** William Leaman/Alamy Stock Photo.

Unit Eight Interview Courtesy of Chelsea Rochman.

Chapter 52 52.1 top Christopher Austin; **bottom, left to right:** Siepmann/imageBROKER/Alamy Stock Photo, Anton Foltin/123RF, Digital Vision/Photodisc/Getty Images, NOAA Okeanos Explorer Program; **52.2 top to bottom:** Luca Nichetti/Shutterstock, Barrie Britton/Nature Picture Library, Oleg Znamenskiy/Fotolia, Bluegreen Pictures/Alamy Stock Photo, Juan Carlos Muñoz, AGE Fotostock/Alamy Stock Photo, 1Xpert/Fotolia; **p. 1170 Davis** Courtesy of Margaret Davis; **52.9** Susan Carpenter; **52.11 desert** Anton Foltin/123RF; **grassland** David Halbakken/AGE Fotostock; **broadleaf forest** Gary718/Shutterstock; **tropical forest** Siepmann/ImageBroker/Alamy Stock Photo; **coniferous forest** Bent G. Nordeng/Shutterstock; **tundra** Juan Carlos Munoz/Nature Picture Library; **52.12 left** JTB Media Creation, Inc./Alamy Stock Photo; **right** Krystyna Szulecka/Alamy Stock Photo; **52.13 tropical forest** Siepmann/imageBROKER/Alamy Stock Photo; **desert** Anton Foltin/123RF; **savanna** Robert Harding Picture Library/Alamy Stock Photo; **chaparral** blickwinkel/Alamy Stock Photo; **grassland** David Halbakken/AGE Fotostock; **coniferous forest** Bent Nordeng/Shutterstock; **broadleaf forest** Gary718/Shutterstock; **tundra** Juan Carlos Munoz/Nature Picture Library; **52.16 oligotrophic lake** Susan Lee Powell; **eutrophic lake** AfriPics.com/Alamy Stock Photo; **wetland** David Tipling/Nature Picture Library; **headwater stream** scubaluna/Shutterstock; **Loire river** Photononstop/SuperStock; **estuaries** Juan Carlos Munoz/AGE Fotostock; **intertidal zone** Stuart Westmorland/Danita Delimont/Alamy Stock Photo; **ocean** Tatonka/Shutterstock; **coral reef** Digital Vision/Photodisc/Getty Images; **benthic zone** NOAA Okeanos Explorer Program; **52.17** JLV Image Works/Fotolia; **52.19** Sylvain Oliveira/Alamy Stock Photo; **52.20** Scott Ling; **52.22** Sabastien Lecocq/Alamy Stock Photo; **Scientific Skills Exercise** *Spartina* John W. Bova/Science Source; **Typha** Dave Bevan/Alamy Stock Photo; **52.23** Harold Stiver/Alamy Stock Photo; **p. 1189 giraffe** Daryl Balfour/The Image Bank/Getty Images.

Chapter 53 53.1 top Joel Sartore/National Geographic Image Collection; **bottom** Villiers Steyn/Shutterstock; **53.2** Todd Pusser/Nature Picture Library; **53.3a** Bernard Castelein/Nature Picture Library/Alamy Stock Photo; **53.3b** Michael S Nolan/AGE Fotostock; **53.3c** Alexander Chaikin/Shutterstock; **Table 53.1 top** Kevin Ebi/Alamy Stock Photo; **bottom** Jennifer A. Dever; **53.8** Villiers Steyn/Shutterstock; **53.11** Bence Mate/

Nature Picture Library/Alamy Stock Photo; **Scientific Skills Exercise** Lebendkulturen. de/Shutterstock; **53.12a** Stone Nature Photography/Alamy Stock Photo; **53.12b** Kent Foster/Science Source; **inset** Robert D. and Jane L. Dorn; **53.13** Dietmar Nill/Nature Picture Library; **53.14a** Steve Bloom Images/Alamy Stock Photo; **53.14b left** Fernanda Preto/Alamy Stock Photo; **right** Edward Parker/Alamy Stock Photo; **53.16** National Geographic/Alamy Stock Photo; **53.17 wheat** FotoVoyager/E+/Getty Images; **cheetah** Ian Cumming/Axiom/Design Pics/Alamy Stock Photo; **humans** Jorge Dan/Reuters; **mice** Nicholas Bergkessel Jr./Science Source; **yeast** Andrew Syred/Science Source; **53.19** Alan & Sandy Carey/Science Source; **53.20** Robert Pickett/Papilio/Alamy Stock Photo; **p. 1206 transponder** From: Tracking butterfly movements with harmonic radar reveals an effect of population age on movement distance. O. Ovaskainen et al. *Proc Natl Acad Sci U S A.* 2008 Dec 9;105(49):19090-5. doi: 10.1073/pnas.0802066105. Epub 2008 Dec 5. Fig. 1; **53.25** NASA; **p. 1213 locusts** Carlos Guevara/Reuters/Newscom.

Chapter 54 54.1 eel Jeremy Brown/123RF; **coral reef** Jan Wlodarczyk/Alamy Stock Photo; **triggerfish** imageBROKER/Alamy Stock Photo; **shark** Andrey Armyagov/Alamy Stock Photo; **coral bleaching** Reinhard Dirscherl/Alamy Stock Photo; **54.2 left** Joseph T. Collins/Science Source; **right** National Museum of Natural History/Smithsonian Institution; **54.4** Frank W Lane/Frank Lane Picture Agency Limited; **Scientific Skills Exercise** Johan Larson/Shutterstock; **54.6a** Tony Heald/Nature Picture Library; **54.6b** Tom Brakefield/Getty Images; **54.6c** Dirk Ercken/Shutterstock; **54.6d** Barry Mansell/Nature Picture Library; **54.6e left** Daniel Janzen/JANZEN.UPENN.EDU/Caters News; **right** Robert Pickett/Papilio/Alamy Stock Photo; **54.6f left** David J Martin/Shutterstock; **right** Lightwriter1949/Alamy Stock Photo; **54.7** Roger Steene/Image Quest Marine; **p. 1219 Langkilde** Patrick Mansell/PennState; **54.8** Doug Perrine/NaturePL.com; **54.9a** Bazzano Photography/Alamy Stock Photo; **54.9b** Nicholas Smythe/Science Source; **54.10** Daryl Balfour/Gallo Images/Getty Images; **54.11a** Sally D. Hacker; **54.13** Gary W. Saunders; **54.14** Dung Vo Trung/Science Source; **54.15** Cedar Creek Ecosystem Science Reserve, University of Minnesota; **54.20a** Genny Anderson; **54.21** Adam Welz; **54.25** National Park Service (NPS.gov); **p. 1229 Turner** From: Tracking butterfly movements with harmonic radar reveals an effect of population age on movement distance. O. Ovaskainen et al. *Proc Natl Acad Sci U S A.* 2008 Dec 9;105(49):19090-5. doi: 10.1073/ pnas.0802066105. Epub 2008 Dec 5. Fig. 1; **54.26a** Charles D. Winters/Science Source; **54.26b** Keith Boggs; **54.26c** Terry Donnelly/Mary Liz Austin; **54.26d** Glacier Bay National Park and Preserve; **54.27 left to right** Charles D. Winters/Science Source, Keith Boggs, Terry Donnelly/Mary Liz Austin, Glacier Bay National Park/Preserve; **54.28 top** R. Grant Gilmore/NOAA; **54.28 bottom** Lance Horn/National Undersea Research Center/University of North Carolina-Wilmington/NOAA; **54.32** Tim Laman/National Geographic/Getty Images; **54.34** Nelish Pradhan/Bates College/Lewiston, ME; **54.35** Josh Spice; **p. 1237 flower** Jim Holden/Alamy Stock Photo.

Chapter 55 55.1 Steven Kazlowski/RGB Ventures/SuperStock/Alamy Stock Photo; **55.2** Stone Nature Photography/Alamy Stock Photo; **55.3 left** Scimat/Science Source; **right** Justus de Cuveland/imageBROKER/AGE Fotostock; **55.5** MODIS Science Team/ Earth Observatory/NASA; **55.8** A. T. Willett/Alamy Stock Photo; **Problem-Solving Exercise tree** Steven Katovich/USDA Forest Service; **mountain pine beetles** British Columbia Ministry of Forests, Lands and Natural Resource Operations; **55.9** Matt Meadows/Photolibrary/Getty Images; **Scientific Skills Exercise** David R. Frazier Photolibrary/Science Source; **p. 1252 Likens** Courtesy of Gene Likens; **55.15** Hubbard Brook Research Foundation/USDA Forest Service; **55.16** Mark Gallagher/Princeton Hydro, LLC/Ringoes, NJ; **55.17 top to bottom** Kissimmee Division/South Florida Water Management District, Jean Hall/FLPA/Science Source, Tim Day/Xcluder Pest Proof Fencing Company, From: Species richness accelerates marine ecosystem restoration in the Coral Triangle. S. L. Williams et al. *Proc Natl Acad Sci U S A.* 2017 Nov 7;114(45):11986-11991. doi: 10.1073/pnas.1707962114. Epub 2017 Oct 24. Fig. 1a. Photos courtesy of D. Trockel, University of California, Davis, CA; **55.18** U.S. Department of Energy; **p. 1259 beetle** Dr Eckart Pott/NHPA/Photoshot.

Chapter 56 56.1 gecko Phung My Trung, vncreatures.net; **clearcutting** Mason Vranish/Alamy Stock Photo; **tusks, 56.8** Benezeth M. Mutayoba; **energy** Ververidis Vasilis/ Shutterstock; **park** Edwin Giesbers/Nature Picture Library; **reef** Matthew Banks/Alamy Stock Photo Image; **56.3 top** Alaz/Shutterstock; **bottom** Mark Carwardine/Photolibrary/Getty Images; **56.4** Merlin D. Tuttle/Science Source; **56.5** Scott Camazine/Science Source; **p. 1263 Wilson** Michael Dwyer/Alamy Stock Photo; **56.6** Michael Edwards/ The Image Bank/Getty Images; **56.7** Bruce Coleman/Alamy Stock Photo; **56.9** Travel Pictures/Alamy Stock Photo; **56.12** Bruce Montagne/Dembinsky Photo Associates/ Alamy Stock Photo; **56.13** Courtesy Interagency Grizzly Bear Study Team, USGS; **56.14a** Chuck Bargeron; **inset** William Leaman/Alamy Stock Photo; **56.14b** William D. Boyer/USDA; **56.16** Vladimir Melnikov/Shutterstock; **56.17** Richard O. Bierregaard, Jr.; **56.18** Frans Lemmens/Alamy Stock Photo; **56.21** Mark Chiappone; **56.22** Lyda Bergman, Green Teams of Canada; **56.25** Alfred Eisenstaedt/The LIFE Picture Collection/ Getty Images; **56.27** Claire Fackler, NOAA National Marine Sanctuaries; **p. 1277 Rochman** Courtesy of Chelsea Rochman; **56.28** Courtesy of Bette Willis and Joleah Lamb; **Scientific Skills Exercise** Hank Morgan/Science Source; **56.31 resin canal** Biophoto Associates/Science Source; **tunnels** Ladd Livingston, Idaho Department of Lands, Bugwood.org; **dead trees** Dezene Huber; **pika** Chris Ray; **caribou** E.A. Janes/Robert Harding World Imagery; **chickweed** Gilles Delacroix/Garden World Images/AGE Fotostock; **urchin** Scott Ling; **56.34** NASA Ozone Watch; **56.36a** Serge de Sazo/Science Source; **56.36b** Javier Trueba/MSF/Science Source; **56.36c** Gabriel Rojo/Nature Picture Library; **56.36d** Titus Lacoste/The Image Bank/Getty Images; **p. 1287 tiger** Edwin Giesbers/ Nature Picture Library.

Appendix A Figure 2.17 Nigel Cattlin/Science Source; **Figure 6.24 left** Omikron/ Science Source; **6.24 right** Dartmouth College Electron Microscope Facility; **Ch. 9 Test Your Understanding 10** Medical Research Council; **Figure 12.4** Biophoto/Science Source; **Figure 12.8** Jane Stout and Claire Walczak, Indiana University; **Ch. 12 Test Your Understanding 9** Scenics & Science/Alamy Stock Photo; **Ch. 16 Test Your Understanding 11** Thomas A. Steitz, Yale University, New Haven; **Figure 30.9** Paul Atkinson/Shutterstock; **Ch. 35 Test Your Understanding 11** From: Anatomy of the vessel network within and between tree rings of *Fraxinus lanuginosa* (Oleaceae). Peter B. Kitin, Tomoyuki Fujii, Hisashi Abe and Ryo Funada. *American Journal of Botany.* 2004;91:779-788.

Appendix B Bacteria, Archaea Eye of Science/Science Source; **diatoms** M I Walker/ NHPA/Photoshot/Newscom; **lily** Howard Rice/Dorling Kindersley, Ltd./Alamy Stock Photo; **fungus** daksel/fotolia; **chimpanzees** E.A. Janes/AGE Fotostock.

Illustration and Text Credits

Chapter 1 1.23 Adapted from *The Real Process of Science* (2013), Understanding Science website. The University of California Museum of Paleontology, Berkeley, and the Regents of the University of California. Retrieved from http://undsci.berkeley.edu/ article/howscienceworks_02; **1.25** Data from S. N. Vignieri, J. G. Larson, and H. E. Hoekstra, The Selective Advantage of Crypsis in Mice, *Evolution* 64:2153–2158 (2010); **Scientific Skills Exercise** Data from D. W. Kaufman, Adaptive Coloration in *Peromyscus polionotus*: Experimental Selection by Owls, *Journal of Mammalogy* 55:271–283 (1974).

Chapter 2 Scientific Skills Exercise Data from R. Pinhasi et al., Revised Age of late Neanderthal Occupation and the End of the Middle Paleolithic in the Northern Caucasus, *Proceedings of the National Academy of Sciences USA* 147:8611–8616 (2011). doi 10.1073/pnas.1018938108.

Chapter 3 3.7 map based on NOAA Fisheries, Bowhead Whale (*Balaena mysticetus*); sea ice extent from National Snow and Ice Data Center (https://nsidc.org/arcticseaicenews/); **3.9** Based on Simulating Water and the Molecules of Life by Mark Gerstein and Michael Levitt, from *Scientific American*, November 1998; **Scientific Skills Exercise** Data from C. Langdon et al., Effect of Calcium Carbonate Saturation State on the Calcification Rate of an Experimental Coral Reef, *Global Biogeochemical Cycles* 14:639–654 (2000).

Chapter 4 4.2 Data from S. L. Miller, A Production of Amino Acids Under Possible Primitive Earth Conditions, *Science* 117:528–529 (1953); **Scientific Skills Exercise** Data from E. T. Parker et al., Primordial Synthesis of Amines and Amino Acids in a 1958 Miller H_2S-rich Spark Discharge Experiment, *Proceedings of the National Academy of Sciences USA* 108:5526–5531 (2011). www.pnas.org/cgi/doi/10.1073/pnas.1019191108; **4.7** Adapted from Becker, Wayne M.; Reece, Jane B.; Poenie, Martin F., *The World of the Cell*, 3rd Ed., ©1996. Reprinted and electronically reproduced by permission of Pearson Education, Inc., Upper Saddle River, New Jersey.

Chapter 5 5.11 Adapted from Wallace/Sanders/Ferl, *Biology: The Science of Life*, 3rd Ed., ©1991. Reprinted and electronically reproduced by permission of Pearson Education, Inc., Upper Saddle River, New Jersey; **5.13** Collagen Data from Protein Data Bank ID 1CGD: "Hydration Structure of a Collagen Peptide" by Jordi Bella et al., from *Structure*, September 1995, Volume 3(9); **5.16 Space-filling model, ribbon model** Data from PDB ID 2LYZ: R. Diamond. Real-Space Refinement of the Structure of Hen Egg-white Lysozyme. *Journal of Molecular Biology* 82(3):371–91 (Jan. 25, 1974); **5.18 Transthyretin** Data from PDB ID 3GS0: S.K. Palaninathan, N.N. Mohamedmohaideen, E. Orlandini, G. Ortore, S. Nencetti, A. Lapucci, A. Rossello, J.S. Freundlich, J.C. Sacchettini. Novel Transthyretin Amyloid Fibril Formation Inhibitors: Synthesis, Biological Evaluation, and X-ray Structural Analysis. *Public Library of Science ONE* 4:e6290–e6290 (2009); **5.18 Collagen** Data from PDB ID 1CGD: J. Bella, B. Brodsky, and H.M. Berman. Hydration Structure of a Collagen Peptide, *Structure* 3:893–906 (1995); **5.18 Hemoglobin** Data from PDB ID 2HHB: G. Fermi, M.F. Perutz, B. Shaanan, R. Fourme. The Crystal Structure of Human Deoxyhaemoglobin at 1.74 Å resolution. *J. Mol. Biol.* 175:159–174 (1984).

Chapter 6 6.6 Adapted from Becker, Wayne M.; Reece, Jane B.; Poenie, Martin F., *The World of the Cell*, 3rd Ed., ©1996. Reprinted and electronically reproduced by permission of Pearson Education, Inc. Upper Saddle River, New Jersey; **6.8 Animal cell** Adapted from Marieb, Elaine N.; Hoehn, Katja, *Human Anatomy and Physiology*, 8th Ed., © 2010. Printed and electronically reproduced by permission of Pearson Education, Inc., Upper Saddle River, New Jersey; **6.9–6.13, 6.17, 6.22, 6.24 Small cell** Adapted from Marieb, Elaine N.; Hoehn, Katja, *Human Anatomy and Physiology*, 8th Ed., ©2010. Printed and electronically reproduced by permission of Pearson Education, Inc., Upper Saddle River, New Jersey; **6.14** Adapted from Marieb, Elaine N.; Hoehn, Katja, *Human Anatomy and Physiology*, 8th Ed., ©2010. Printed and electronically reproduced by permission of Pearson Education, Inc., Upper Saddle River, New Jersey; **Table 6.1** Adapted from Hardin Jeff; Bertoni Gregory Paul, Kleinsmith, Lewis J., *Becker's World of the Cell*, 8th Edition, © 2012, p. 423. Reprinted and electronically reproduced by permission of Pearson Education, Inc. Upper Saddle River, New Jersey; **6.32** Data from: Proton pump: PDB ID 3B8C: Crystal Structure of the Plasma Membrane Proton Pump, Pedersen, B.P., Buch-Pedersen, M. J., Morth, J.P., Palmgren, M.G., Nissen, P. (2007) *Nature* 450: 1111–1114; calcium channel: PDB ID 5E1J: Structure of the Voltage-Gated Two-Pore Channel TPC1 from *Arabidopsis thaliana*, Guo, J., Zeng, W., Chen, Q., Lee, C., Chen, L., Yang, Y., Cang, C., Ren, D., Jiang, Y. (2016) *Nature* 531: 196–201; aquaporin: PDB ID 5I32: Crystal Structure of an Ammonia-Permeable Aquaporin, Kirscht, A., Kaptan, S.S., Bienert, G.P., Chaumont, F., Nissen, P., de Groot, B.L., Kjellbom, P., Gourdon, P., Johanson, U. (2016) *Plos Biol.* 14: e1002411–e1002411; BRI1 and SERK1 co-receptors: PDB ID 4LSX: Molecular mechanism for plant steroid receptor activation by somatic embryogenesis co-receptor kinases, Santiago, J., Henzler, C., Hothorn, M. (2013) *Science* 341: 889–892; BRI1 kinase domain: PDB ID 4OAC: Crystal structures of the phosphorylated BRI1 kinase domain and implications for brassinosteroid signal initiation, Bojar, D., Martinez, J., Santiago, J., Rybin, V., Bayliss, R., Hothorn, M. (2014) *Plant J.* 78: 31–43; BAK1 kinase domain: PDB ID 3UIM: Structural basis for the impact of phosphorylation on the activation of plant receptor-like kinase BAK1, Yan, L., Ma, Y., Liu, D., Wei, X., Sun, Y., Chen, X., Zhao, H., Zhou, J., Wang, Z., Wang, J., Shui, W., Lou, Z.Y. (2012) *Cell Res.* 22: 1304–1308; BSK8 pseudokinase: PDB ID: 4I92 Structural Characterization of the RLCK Family Member BSK8: A Pseudokinase with an Unprecedented Architecture, Grutter, C., Sreeramulu, S., Sessa, G., Rauh, D. (2013) *J. Mol. Biol.* 425: 4455–4467; ATP synthase PDB ID 1E79: The Structure of the Central Stalk in Bovine F(1)-ATPase at 2.4 Å Resolution, Gibbons, C., Montgomery, M.G., Leslie, A.G.W., Walker, J.E. (2000) *Nat. Struct. Biol.* 7: 1055; ATP synthase PDB ID 1C17: Structural changes linked to proton translocation by subunit c of the ATP synthase, Rastogi, V.K., Girvin, M.E. (1999) *Nature* 402: 263–268; ATP synthase PDB ID 1L2P: The "Second Stalk" of *Escherichia coli* ATP Synthase: Structure of the Isolated Dimerization Domain, Del Rizzo, P.A., Bi, Y., Dunn,

S.D., Shilton, B.H. (2002) *Biochemistry* 41: 6875–6884; ATP synthase PDB ID 2A7U: Structural Characterization of the Interaction of the Delta and Alpha Sub-units of the *Escherichia coli* F(1)F(0)-ATP Synthase by NMR Spectroscopy, Wilkens, S., Borchardt, D., Weber, J., Senior, A.E. (2005) *Biochemistry* 44: 11786–11794; Phosphofructokinase: PDB ID 1PFK: Crystal Structure of the Complex of Phosphofructokinase from *Escherichia coli* with Its Reaction Products, Shirakihan, Y., Evans, P.R. (1988) *J. Mol. Biol.* 204: 973–994; Hexokinase: PDB ID 4QS8: Biochemical and Structural Study of *Arabidopsis* Hexokinase 1, Feng, J., Zhao, S., Chen, X., Wang, W., Dong, W., Chen, J., Shen, J.-R., Liu, L., Kuang, T. (2015) *Acta Crystallogr.*, Sect. D 71: 367–375; Isocitrate dehydrogenase: PDB ID 3BLW: Allosteric Motions in Structures of Yeast NAD+-specific Isocitrate Dehydrogenase, Taylor, A.B., Hu, G., Hart, P.J., McAlister-Henn, L. (2008) *J. Biol. Chem.* 283:10872–10880; NADH-quinone oxidoreduc-tase: PDB ID 3M9S: The architecture of respiratory complex I, Efremov, R.G., Barada-ran, R., Sazanov, L.A. (2010) *Nature* 465: 441–445; NADH-quinone oxidore-ductase: PDB ID 3RKO: Structure of the membrane domain of respiratory complex I, Efremov, R.G., Sazanov, L.A. (2011) *Nature* 476: 414–420; Succinate dehydrogenase: PDB ID 1NEK: Architecture of Succinate Dehydrogenase and Reactive Oxygen Species Generation, Yankovskaya, V., Horsefield, R., Tornroth, S., Luna-Chavez, C., Miyoshi, H., Leger, C., Byrne, B., Cecchini, G., Iwata, S. (2003) *Science* 299: 700–704; Ubiquinone: http://www.proteopedia.org/wiki/index.php/Image: Co-enzyme_Q10.pdb; Cytochrome bc1: PDB ID 1BGY: Complete structure of the 11-subunit bovine mitochondrial cytochrome bc1 complex, Iwata, S., Lee, J.W., Okada, K., Lee, J.K., Iwata, M., Rasmussen, B., Link, T.A., Ramaswamy, S., Jap, B.K. (1998) *Science* 281: 64–71; Cytochrome *c*: PDB ID 3CYT: Redox Conformation Changes in Refined Tuna Cytochrome c, Takano, T., Dickerson, R.E. (1980) *Proc. Natl. Acad. Sci. USA* 77: 6371–6375; Cytochrome *c* oxidase: PDB ID 1OCO: Redox-Coupled Crystal Structural Changes in Bovine Heart Cytochrome c Oxidase, Yoshikawa, S., Shinzawa-Itoh, K., Nakashima, R., Yaono, R., Yamashita, E., Inoue, N., Yao, M., Fei, M.J., Libeu, C.P., Mizushima, T., Yamaguchi, H., Tomizaki, T., Tsukihara, T. (1998) *Science* 280: 1723–1729; Rubisco: PDB ID 1RCX: The Structure of the Complex between Rubisco and its Natural Substrate Ribulose 1,5-Bisphosphate, Taylor, T., Andersson, I. (1997) *J. Mol. Biol.* 265: 432–444; Photosystem II: PDB ID 1S5L: Architecture of the Photosynthetic Oxygen-Evolving Center, Ferreira, K.N., Iverson, T.M., Maghlaoui, K., Barber, J., Iwata, S. (2004) *Science* 303: 1831–1838; Plastoquinone: http://www.rcsb.org/pdb/ligand/ligandsummary.do?hetId=PL9; Photosystem I: PDB ID 1JB0: Three-Dimensional Structure of Cyanobacterial Photosystem I at 2.5 Å Resolution, Jordan, P., Fromme, P., Witt, H.T., Klukas, O., Saenger, W., Krauss, N. (2001) *Nature* 411: 909–917; Ferredoxin-NADP+ reductase: PDB ID 3W5V: Concentration-Dependent Oligomerization of Cross-Linked Complexes between Ferredoxin and Ferredoxin-NADP(+) Reductase; DNA:PDB ID 1BNA: Structure of a B-DNA Dodecamer: Conformation and Dynamics, Drew, H.R., Wing, R.M., Takano, T., Broka, C., Tanaka, S., Itakura, K., Dickerson, R.E. (1981) *Proc. Natl. Acad. Sci. USA* 78: 2179–2183; RNA polymerase: PDB ID 2E2I: Structural basis of transcription: role of the trigger loop in substrate specificity and catalysis, Wang, D., Bushnell, D.A., Westover, K.D., Kaplan, C.D., Kornberg, R.D. (2006) *Cell* (Cambridge, Mass.) 127: 941–954; Nucleosome: PDB ID 1AOI: Crystal Structure of the Nucleosome Core Particle at 2.8 Å Resolution, Luger, K., Mader, A.W., Richmond, R.K., Sargent, D.F., Richmond, T.J. (1997) *Nature* 389: 251–260; tRNA: PDB ID 4TNA: Further refinement of the structure of yeast tRNAPhe, Hingerty, B., Brown, R.S., Jack, A. (1978) *J. Mol. Biol.* 124: 523–534; Ribosome: PDB ID 1FJF: Structure of the 30S Ribosomal Subunit, Wimberly, B.T., Brodersen, D.E., Clemons Jr., W.M., Morgan-Warren, R.J., Carter, A.P., Vonrhein, C., Hartsch, T., Ramakrishnan, V. (2000) *Nature* 407: 327–339; Ribosome: PDB ID 1JJ2: The Kink-Turn: A New RNA Secondary Structure Motif, Klein, D.J., Schmeing, T.M., Moore, P.B., Steitz, T.A. (2001) *EMBO J.* 20: 4214–4221; Microtubule: PDB ID 3J2U: Structural Model for Tubulin Recognition and Deformation by Kinesin-13 Microtubule Depolymerases, Asenjo, A.B., Chatterjee, C., Tan, D., Depaoli, V., Rice, W.J., Diaz-Avalos, R., Silvestry, M., Sosa, H. (2013) *Cell Rep.* 3: 759–768; Actin microfilament: PDB ID 1ATN: Atomic Structure of the Actin:DNase I Complex. Kabsch, W., Mannherz, H.G., Suck, D., Pai, E.F., Holmes, K.C. (1990) *Nature* 347: 37–44; Myosin: PDB ID 1M8Q: Molecular Modeling of Averaged Rigor Crossbridges from Tomograms of Insect Flight Muscle, Chen, L.F., Winkler, H., Reedy, M.K., Reedy, M.C., Taylor, K.A. (2002) *J. Struct. Biol.* 138: 92–104; Phosphoglucose Isomerase: PDB ID 1IAT: The Crystal Structure of Human Phosphoglucose Isomerase at 1.6 Å Resolution: Implications for Catalytic Mechanism, Cytokine Activity and Haemolytic Anaemia, Read, J., Pearce, J., Li, X., Muirhead, H., Chirgwin, J., Davies, C. (2001) *J. Mol. Biol.* 309: 447–463; Aldolase: PDB ID 1ALD: Activity and Specificity of Human Aldolases, Gamblin, S.J., Davies, G.J., Grimes, J.M., Jackson, R.M., Littlechild, J.A., Watson, H.C. (1991) *J. Mol. Biol.* 219: 573–576; Triosephosphate Isomerase: PDB ID 7TIM: Structure of the Triose-Phosphate Isomerase-Phosphoglycolohydroxamate Complex: An Analogue of the Intermediate on the Reaction Pathway, Davenport, R.C., Bash, P.A., Seaton, B.A., Karplus, M., Petsko, G.A., Ringe, D. (1991) *Biochemistry* 30: 5821–5826; Glyceraldehyde-3-Phosphate Dehydrogenase: PDB ID 3GPD: Twinning in Crystals of Human Skeletal Muscle D-Glyceraldehyde-3-Phosphate Dehydrogenase, Mercer, W.D., Winn, S.I., Watson, H.C. (1976) *J. Mol. Biol.* 104: 277–283; Phosphoglycerate Kinase: PDB ID 3PGK: Sequence and Structure of Yeast Phosphoglycerate Kinase, Watson, H.C., Walker, N.P., Shaw, P.J., Bryant, T.N., Wendell, P.L., Fothergill, L.A., Perkins, R.E., Conroy, S.C., Dobson, M.J., Tuite, M.F. (1982) *EMBO J.* 1: 1635–1640; Phosphoglycerate Mutase: PDB ID 3PGM: Structure and Activity of Phosphoglycerate Mutase, Winn, S.I., Watson, H.C., Harkins, R.N., Fothergill, L.A. (1981) *Philos. Trans. R. Soc. London, Ser. B* 293: 121–130; Enolase: PDB ID 5ENL: Inhibition of Enolase: The Crystal Structures of Enolase-Ca2(+)-2-Phosphoglycerate and Enolase-Zn2(+)-Phosphoglycolate Complexes at 2.2-Å Resolution, Lebioda, L., Stec, B., Brewer, J.M., Tykarska, E. (1991) *Biochemistry* 30: 2823–2827; Pyruvate Kinase: PDB ID 1A49: Structure of the Bis(Mg2+)-ATP-Oxalate Complex of the Rabbit Muscle Pyruvate Kinase at 2.1 Å Resolution: ATP Binding over a Barrel, Larsen, T.M., Benning, M.M., Rayment, I., Reed, G.H. (1998) *Biochemistry* 37: 6247–6255; Citrate Synthase: PDB ID 1CTS: Crystallographic Refinement and Atomic Models of Two Different Forms of Citrate Synthase at 2.7 and 1.7 Å Resolution, Remington, S., Wiegand, G., Huber, R. (1982) *J. Mol. Biol.* 158: 111–152; Succinyl-CoA Synthetase: PDB ID 2FP4: Interactions of GTP with the ATP-Grasp Domain of GTP-Specific Succinyl-CoA Synthetase, Fraser, M.E., Hayakawa, K., Hume, M.S., Ryan, D.G., Brownie, E.R. (2006) *J. Biol. Chem.* 281: 11058-11065; Malate Dehydrogenase: PDB ID

4WLE: Crystal Structure of Citrate Bound MDH2, Eo, Y.M., Han, B.G., Ahn, H.C. To Be Published; **Summary art nucleus, Golgi apparatus and endoplasmic reticulum** Adapted from Marieb, Elaine N.; Hoehn, Katja, *Human Anatomy and Physiology*, 8th Ed., ©2010. Printed and electronically reproduced by permission of Pearson Education, Inc., Upper Saddle River, New Jersey.

Chapter 7 **7.4** Data from L. D. Frye and M. Edidin, The Rapid Intermixing of Cell Surface Antigens after Formation of Mouse-human Heterokaryons, *Journal of Cell Science* 7:319 (1970); **7.6** Based on Similar Energetic Contributions of Packing in the Core of Membrane and Water-Soluble Proteins by Nathan H. Joh et al., from *Journal of the American Chemical Society*, Volume 131(31); **Scientific Skills Exercise** Data from Figure 1 in T. Kondo and E. Beutler, Developmental Changes in Glucose Transport of Guinea Pig Erythrocytes, *Journal of Clinical Investigation* 65:1–4 (1980).

Chapter 8 **8 Scientific Skills Exercise** Data from S. R. Commerford et al., Diets Enriched in Sucrose or Fat Increase Gluconeogenesis and G-6-pase but not Basal Glucose Production in Rats, *American Journal of Physiology—Endocrinology and Metabolism* 283:E545–E555 (2002); **8.19** Data from Protein Data Bank ID 3e1f: "Direct and Indirect Roles of His-418 in Metal Binding and in the Activity of Beta-Galactosidase (*E. coli*)" by Douglas H. Juers et al., from *Protein Science*, June 2009, Volume 18(6); **8.20** Data from Protein Data Bank ID 1MDYO: "Crystal Structure of MyoD bHLH Domain-DNA Complex: Perspectives on DNA Recognition and Implications for Transcriptional Activation" from *Cell*, May 1994, Volume 77(3); **8.22 Small cell** Adapted from Marieb, Elaine N.; Hoehn, Katja, *Human Anatomy and Physiology*, 8th Ed., ©2010. Printed and electronically reproduced by permission of Pearson Education, Inc., Upper Saddle River, New Jersey.

Chapter 9 **9.4** Adaptation of Figure 2.69 from *Molecular Biology of the Cell*, 4th Edition, by Bruce Alberts et al. Garland Science/Taylor & Francis LLC; **9.8** Figure adapted from *Biochemistry*, 4th Edition, by Christopher K. Mathews et al. Pearson Education, Inc.; **Scientific Skills Exercise** Data from M. E. Harper and M. D. Brand, The Quantitative Contributions of Mitochondrial Proton Leak and ATP Turnover Reactions to the Changed Respiration Rates of Hepatocytes from Rats of Different Thyroid Status, *Journal of Biological Chemistry* 268:14850–14860 (1993).

Chapter 10 **10.9** Data from T. W. Engelmann, Bacterium Photometricum. Ein Beitrag zur Vergleichenden Physiologie des Lichtund Farbensinnes, *Archiv. für Physiologie* 30:95–124 (1883); **10.12b** Data from Architecture of the Photosynthetic Oxygen-Evolving Center by Kristina N. Ferreira et al., from *Science*, March 2004, Volume 303(5665); **10.14** Adaptation of Figure 4.1 from *Energy, Plants, and Man*, by Richard Walker and David Alan Walker. © 1992 by Richard Walker and David Alan Walker. Reprinted with permission of Richard Walker; **Scientific Skills Exercise** Data from D. T. Patterson and E. P. Flint, Potential Effects of Global Atmospheric CO2 Enrichment on the Growth and Competitiveness of C3 and C4 Weed and Crop Plants, *Weed Science* 28(1):71–75 (1980).

Chapter 11 **Problem-Solving Exercise** Data from N. Balaban et al., Treatment of *Staphylococcus aureus* Biofilm Infection by the Quorum-Sensing Inhibitor RIP, *Antimicrobial Agents and Chemotherapy*, 51:2226–2229 (2007); **11.8, 11.12** Adapted from Becker, Wayne M.; Reece, Jane B.; Poenie, Martin F., *The World of the Cell*, 3rd Edition, © 1996. Reprinted and electronically reproduced by permission of Pearson Education, Inc., Upper Saddle River, New Jersey.

Chapter 12 **12.9** Data from G. J. Gorbsky, P. J. Sammak, and G. G. Borisy, Chromosomes Move Poleward in Anaphase along Stationary Microtubules that Coordinately Disassemble from their Kinetochore Ends, *Journal of Cell Biology* 104:9–18 (1987); **12.13** Adaptation of Figure 18.41 from *Molecular Biology of the Cell*, 4th Edition, by Bruce Alberts et al. Garland Science/Taylor & Francis LLC; **12.14** Data from R. T. Johnson and P. N. Rao, Mammalian Cell Fusion: Induction of Premature Chromosome Condensation in Interphase Nuclei, *Nature* 226:717–722 (1970); **Scientific Skills Exercise** Data from K. K. Velpula et al., Regulation of Glioblastoma Progression by Cord Blood Stem Cells is Mediated by Downregulation of Cyclin D1, *PLoS ONE* 6(3): e18017 (2011).

Chapter 14 **14.3, 14.8** Data from G. Mendel, Experiments in Plant Hybridization, *Proceedings of the Natural History Society of Brünn* 4:3–47 (1866).

Chapter 15 **15.3** Data from T. H. Morgan, Sex-limited inheritance in *Drosophila*, *Science* 32:120–122 (1910); **15.9** Based on the data from "The Linkage of Two Factors in *Drosophila* That Are Not Sex-Linked" by Thomas Hunt Morgan and Clara J. Lynch, from *Biological Bulletin*, August 1912, Volume 23(3).

Chapter 16 **16.2** Data from F. Griffith, The Significance of Pneumococcal Types, *Journal of Hygiene* 27:113–159 (1928); **16.4** Data from A. D. Hershey and M. Chase, Independent Functions of Viral Protein and Nucleic Acid in Growth of Bacteriophage, *Journal of General Physiology* 36:39–56 (1952); **Scientific Skills Exercise** Data from several papers by Chargaff: for example, E. Chargaff et al., Composition of the Desoxypentose Nucleic Acids of Four Genera of Sea-urchin, *Journal of Biological Chemistry* 195:155–160 (1952); **pp. 320–321 quote** J. D. Watson and F. H. C. Crick, Genetical Implications of the Structure of Deoxyribonucleic Acid, *Nature* 171:964–967 (1953); **16.11** Data from M. Meselson and F. W. Stahl, The Replication of DNA in *Escherichia coli*, *Proceedings of the National Academy of Sciences USA* 44:671–682 (1958).

Chapter 17 **17.3** Data from A. M. Srb and N. H. Horowitz, The Ornithine Cycle in Neurospora and Its Genetic Control, *Journal of Biological Chemistry* 154:129–139 (1944); **17.12** Adapted from Becker, Wayne M.; Reece, Jane B.; Poenie, Martin F., *The World of the Cell*, 3rd Edition, © 1996. Reprinted and electronically reproduced by permission of Pearson Education, Inc., Upper Saddle River, New Jersey; **17.14** Adapted from Klein-Smith, Lewis J., Kish, Valerie M.; *Principles of Cell and Molecular Biology*. Reprinted and electronically reproduced by permissions of Pearson Education, Inc., Upper Saddle River, New Jersey; **17.18** Adapted from Mathews, Christopher K.; Van Holde, Kensal E., *Biochemistry*, 2nd ed., ©1996. Reprinted and electronically reproduced by permission of Pearson Education, Inc. Upper Saddle River, New Jersey; **Scientific Skills Exercise** Material provided courtesy of Dr. Thomas Schneider, National Cancer Institute, National Institutes of Health, 2012; **Problem-Solving Exercise** Data from N. Nishi and K. Nanjo, Insulin Gene Mutations and Diabetes, *Journal of Diabetes Investigation* Vol. 2: 92–100 (2011).

Chapter 18 **18.10** Data from PDB ID 1MDY: P. C. Ma et al. Crystal structure of MyoD bHLH Domain-DNA Complex: Perspectives on DNA Recognition and Implications for Transcriptional Activation, *Cell* 77:451–459 (1994); **Scientific Skills Exercise** Data from J. N. Walters et al., Regulation of Human Microsomal Prostaglandin E Synthase-1 by IL-1b Requires a Distal Enhancer Element with a Unique Role for C/EBPb, *Biochemical Journal* 443:561–571 (2012); **18.26** Adapted from Becker, Wayne M.; Reece, Jane B.; Poenie, Martin F., *The World of the Cell*, 3rd Edition, © 1996. Reprinted and electronically reproduced by permission of Pearson Education, Inc., Upper Saddle River, New Jersey.

Chapter 19 **19.2** Data from M. J. Beijerinck, Concerning a Contagium Vivum Fluidum as Cause of the Spot Disease of Tobacco Leaves, *Verhandelingen der Koninkyke Akademie Wettenschappen te Amsterdam* 65:3–21 (1898). Translation published in English as Phytopathological Classics Number 7 (1942), American Phytopathological Society Press, St. Paul, MN; **Scientific Skills Exercise** Data from J.-R. Yang et al., New Variants and Age Shift to High Fatality Groups Contribute to Severe Successive Waves in the 2009 Influenza Pandemic in Taiwan, *PLoS ONE* 6(11): e28288 (2011).

Chapter 20 **20.7** Adapted from Becker, Wayne M.; Reece, Jane B.; Poenie, Martin F., *The World of the Cell*, 3rd Edition, © 1996. Reprinted and electronically reproduced by permission of Pearson Education, Inc., Upper Saddle River, New Jersey; **20.16** Data from J. B. Gurdon et al., The Developmental Capacity of Nuclei Transplanted from Keratinized Cells of Adult Frogs, *Journal of Embryology and Experimental Morphology* 34:93–112 (1975); **20.21** Data from K. Takahashi et al., Induction of pluripotent stem cells from adult human fibroblasts by defined factors, *Cell* 131:861–872 (2007).

Chapter 21 **21.3** Simulated screen shots based on Mac OS X and from data found at NCBI, U.S. National Library of Medicine using Conserved Domain Database, Sequence Alignment Viewer, and Cn3D; **21.8, 21.9** Adapted from Becker, Wayne M.; Reece, Jane B.; Poenie, Martin F., *The World of the Cell*, 3rd Edition, © 1996. Reprinted and electronically reproduced by permission of Pearson Education, Inc., Upper Saddle River, New Jersey; **21.10 Hemoglobin** Data from PDB ID 2HHB: G. Fermi, M.F. Perutz, B. Shaanan, and R. Fourme. The Crystal Structure of Human Deoxyhaemoglobin at 1.74 Å resolution. *J. Mol. Biol.* 175:159–174 (1984); **21.15a** Drawn from data in Protein Data Bank ID 1LZ1: "Refinement of Human Lysozyme at 1.5 Å Resolution Analysis of Nonbonded and Hydrogen-bond Interactions" by P. J. Artymiuk and C. C. Blake, from *Journal of Molecular Biology*, 1981, 152:737–762; **21.15b** Drawn from data in Protein Data Bank ID 1A4V: "Structural Evidence for the Presence of a Secondary Calcium Binding Site in Human Alpha-Lactalbumin" by N. Chandra et al., from *Biochemistry*, 1998, 37:4767–4772; **Hemoglobin in Scientific Skills Exercise** PDB ID 2HHB: G. Fermi, M.F. Perutz, B. Shaanan, and R. Fourme. The Crystal Structure of Human Deoxyhaemoglobin at 1.74 Å Resolution. *J. Mol. Biol.* 175:159–174 (1984); **Scientific Skills Exercise** Compiled using data from NCBI; **21.18** Data from W. Shu et al., Altered ultrasonic vocalization in mice with a disruption in the *Foxp2* gene, *Proceedings of the National Academy of Sciences USA* 102:9643–9648 (2005); **21.19** Adapted from *The Homeobox: Something Very Precious That We Share with Flies, From Egg to Adult* by Peter Radetsky, © 1992. Reprinted by permission from William McGinnis; **21.20** Adaptation from "Hox Genes and the Evolution of Diverse Body Plans" by Michael Akam, from *Philosophical Transactions of the Royal Society B: Biological Sciences*, September 29, 1995, Volume 349(1329): 313–319. Reprinted by permission from The Royal Society.

Chapter 22 **22.8** Artwork by Utako Kikutani (as appeared in "What Can Make a Four-Ton Mammal a Most Sensitive Beast?" by Jeheskel Shoshani, from *Natural History*, November 1997, Volume 106(1), 36–45). Copyright © 1997 by Utako Kikutani. Reprinted with permission of the artist; **22.13** Data from "Host Race Radiation in the Soapberry Bug: Natural History with the History" by Scott P. Carroll and Christin Boyd, from *Evolution*, 1992, Volume 46(4); **22.14** Figure created by Dr. Binh Diep on request of Michael Cain. Copyright © 2011 by Binh Diep. Reprinted with permission; **Scientific Skills Exercise** Data from J. A. Endler, Natural Selection on Color Patterns in *Poecilia reticulata*, *Evolution* 34:76–91 (1980); **Test Your Understanding Question 7** Data from C. F. Curtis et al., Selection for and Against Insecticide Resistance and Possible Methods of Inhibiting the Evolution of Resistance in Mosquitoes, *Ecological Entomology* 3:273–287 (1978).

Chapter 23 **23.4** Based on the data from *Evolution*, by Douglas J. Futuyma. Sinauer Associates, 2006; and Nucleotide Polymorphism at the Alcohol Dehydrogenase Locus of *Drosophila melanogaster* by Martin Kreitman, from *Nature*, August 1983, Volume 304(5925); **23.11a** Maps adapted from Figure 20.6 from *Discover Biology*, 2nd Edition, edited by Michael L. Cain, Hans Damman, Robert A. Lue, and Carol Kaesuk Loon. W. W. Norton & Company, Inc.; **23.12** Data from Joseph H. Camin and Paul R. Ehrlich, Natural Selection in Water Snakes (*Natrix sipedon* L.) on Islands in Lake Erie, *Evolution* 12:504–511 (1958); **23.14** Based on many sources: *Evolution* by Douglas J. Futuyma. Sinauer Associates 2005; and *Vertebrate Paleontology and Evolution* by Robert L. Carroll. W.H. Freeman & Co., 1988; **23.16** Data from A. M. Welch et al., Call Duration as an Indicator of Genetic Quality in Male Gray Tree Frogs, *Science* 280:1928–1930 (1998); **23.17** Adapted from Frequency-Dependent Natural Selection in the Handedness of Scale-Eating Cichlid Fish by Michio Hori, from *Science*, April 1993, Volume 260(5105); **Test Your Understanding Question 7** Data from R. K. Koehn and T. J. Hilbish, The Adaptive Importance of Genetic Variation, *American Scientist* 75:134–141 (1987).

Chapter 24 **24.6** Original unpublished graph created by Brian Langerhans; **24.7** Data from D. M. B. Dodd, Reproductive Isolation as a Consequence of Adaptive Divergence in *Drosophila pseudoobscura*, *Evolution* 43:1308–1311 (1989); **Scientific Skills Exercise** Data from S. G. Tilley, A. Verrell, and S. J. Arnold, Correspondence between Sexual Isolation and Allozyme Differentiation: A Test in the Salamander *Desmognathus ochrophaeus*, *Proceedings of the National Academy of Sciences USA* 87:2715–2719 (1990); **24.12** Data from O. Seehausen and J. J. M. van Alphen, The Effect of Male Coloration on Female Mate Choice in Closely Related Lake Victoria Cichlids (*Haplochromis nyererei* complex), *Behavioral Ecology and Sociobiology* 42:1–8 (1998); **24.13d** Based on *Hybrid Zone and the Evolutionary Process*, edited by Richard G. Harrison. Oxford University Press; **24.19b** Data from Role of Gene Interactions in Hybrid Speciation: Evidence from Ancient and Experimental Hybrids by Loren H. Rieseberg et al., from *Science*, May 1996, Volume 272(5262).

Chapter 25 **25.2** Based on data from The Miller Volcanic Spark Discharge Experiment by Adam P. Johnson et al., from *Science*, October 2008, Volume 322(5900); **25.4** Based on "Experimental Models of Primitive Cellular Compartments: Encapsulation, Growth, and Division" by Martin M. Hanczyc, Shelly M. Fujikawa, and Jack W. Szostak, from *Science*, October 2003, Volume 302(5645); **25.6** Eicher, D. L, *Geologic Time*, 2nd Ed., ©1976, p. 119. Adapted and electronically reproduced by permission of Pearson Education, Inc., Upper Saddle River, New Jersey; **25.7 First four skulls** Adapted from many sources including D.J. Futuyma, *Evolution*, Fig. 4.10, Sunderland, MA: Sinauer Associates, Sunderland, MA (2005) and from R.L. Carroll, *Vertebrate Paleontology and Evolution*. W.H. Freeman & Co. (1988); **25.7 Last skull** Adapted from Z. Luo et al., A New Mammaliaform from the Early Jurassic and Evolution of Mammalian Characteristics, *Science* 292:1535 (2001); **25.8** Adapted from When Did Photosynthesis Emerge on Earth? by David J. Des Marais, from *Science*, September 2000, Volume 289(5485). **25.9** Adapted from The Rise of Atmospheric Oxygen by Lee R. Kump, from *Nature*, January 2008, Volume 451(7176); **Scientific Skills Exercise** Data from T. A. Hansen, Larval Dispersal and Species Longevity in Lower Tertiary Gastropods, *Science* 199:885–887 (1978); **25.16** Data from *Earthquake Information Bulletin*, December 1977, Volume 9(6), edited by Henry Spall; **25.18** Based on many sources: D.M. Raup and J. J. Sepkoski, Jr., Mass Extinctions in the Marine Fossil Record, *Science* 215:1501–1503 (1982); J. J. Sepkoski, Jr., A Kinetic Model of Phanerozoic Taxonomic Diversity. III. Post-Paleozoic Families and Mass Extinctions, *Paleobiology* 10:246–267 (1984); and D. J. Futuyma, *The Evolution of Biodiversity*, p. 143, Fig. 7.3a and p. 145, Fig. 7.6, Sinauer Associates, Sunderland, MA; **25.20** Based on data from A Long-Term Association between Global Temperature and Biodiversity, Origination and Extinction in the Fossil Record by P.J. Mayhew, G.B. Jenkins and T.G. Benton, *Proceedings of the Royal Society B: Biological Sciences* 275(1630):47–53. The Royal Society, 2008; **25.21** Adapted from Anatomical and Ecological Constraints on Phanerozoic Animal Diversity in the Marine Realm by Richard K. Bambach et al., from *Proceedings of the National Academy of Sciences USA*, May 2002, Volume 99(10); **25.26** Based on data from The Miller Volcanic Spark Discharge Experiment by Adam P. Johnson et al., from *Science*, October 2008, Volume 322(5900); **25.27** Data from Genetic and Developmental Basis of Evolutionary Pelvic Reduction in Threespine Sticklebacks by Michael D. Shapiro et al., from *Nature*, April 2004, Volume 428(6984); **25.29** Adaptations of Figure 3-1 (a–d, f) from *Evolution*, 3rd Edition, by Monroe W. Strickberger. Jones & Bartlett Learning, Burlington, MA.

Chapter 26 **26.6** Data from C. S. Baker and S. R. Palumbi, Which Whales Are Hunted? A Molecular Genetic Approach to Monitoring Whaling, *Science* 265:1538–1539 (1994); **26.13** Based on The Evolution of the Hedgehog Gene Family in Chordates: Insights from Amphioxus Hedgehog by Sebastian M. Shimeld, from *Developmental Genes and Evolution*, January 1999, Volume 209(1); **26.19** Based on *Molecular Markers, Natural History, and Evolution*, 2nd ed., by J.C. Advise. Sinauer Associates, 2004; **26.20** Adapted from Timing the Ancestor of the HIV-1 Pandemic Strains by B. Korber et al., *Science* 288(5472):1789–1796 (6/9/00); **Scientific Skills Exercise** Data from Nancy A. Moran, Yale University. See N. A. Moran and T. Jarvik, Lateral transfer of genes from fungi underlies carotenoid production in aphids, *Science* 328:624–627 (2010); **26.23** Adapted from Phylogenetic Classification and the Universal Tree by W.F. Doolittle, *Science* 284(5423):2124–2128 (6/25/99).

Chapter 27 **27.10** Graph Data from V. S. Cooper and R. E. Lenski, The Population Genetics of Ecological Specialization in Evolving *Escherichia coli* Populations, *Nature* 407:736–739 (2000); **27.19** Data from Root-Associated Bacteria Contribute to Mineral Weathering and to Mineral Nutrition in Trees: A Budgeting Analysis by Christophe Calvaruso et al., *Applied and Environmental Microbiology*, February 2006, Volume 72(2); **27.22** Data from K. Kupferschmidt, Resistance Fighters. *Science* 352(6287):758-761. 13 May 2016; **Scientific Skills Exercise** Data from L. Ling et al. A New Antibiotic Kills Pathogens without Detectable Resistance, *Nature* 517:455–459 (2015); **Test Your Understanding Question 8** Data from J. J. Burdon et al., Variation in the Effectiveness of Symbiotic Associations between Native Rhizobia and Temperate Australian Acacia: Within Species Interactions, *Journal of Applied Ecology* 36:398–408 (1999).

Chapter 28 **Scientific Skills Exercise** Data from D. Yang et al., Mitochondrial Origins, *Proceedings of the National Academy of Sciences USA* 82:4443–4447 (1985); **28.19** Adaptation of illustration by Kenneth X. Probst, from *Microbiology* by R.W. Bauman. Copyright © 2004 by Kenneth X. Probst; **28.26** Data from R. Derelle et al., Bacterial proteins pinpoint a single eukaryotic root, *Proceedings of the National Academy of Sciences USA* 112:E693–699 (2015); **28.32** Based on Global Phytoplankton Decline over the Past Century by Daniel G. Boyce et al., from *Nature*, July 29, 2010, Volume 466(7306); and authors' personal communications.

Chapter 29 **29.14** Data from "Inputs, Outputs, and Accumulation of Nitrogen in an Early Successional Moss (Polytrichum) Ecosystem" by Richard D. Bowden, from *Ecological Monographs*, June 1991, Volume 61(2); **Scientific Skills Exercise** Data from T.M. Lenton et al, First Plants Cooled the Ordovician. *Nature Geoscience* 5:86–89 (2012); **Test Your Understanding Question 8** Data from O. Zackrisson et al., Nitrogen Fixation Increases with Successional Age in Boreal Forests, *Ecology* 85:3327–3334 (2006).

Chapter 30 **Scientific Skills Exercise** Data from S. Sallon et al, Germination, Genetics, and Growth of an Ancient Date Seed. *Science* 320:1464 (2008); **30.14a** Adapted from "A Revision of Williamsoniella" by T. M. Harris, from *Proceedings of the Royal Society B: Biological Sciences*, October 1944, Volume 231(583): 313–328; **30.14b** Adaptation of Figure 2.3, *Phylogeny and Evolution of Angiosperm*, 2nd Edition, by Douglas E. Soltis et al. (2005). Sinauer Associates, Inc.

Chapter 31 **Scientific Skills Exercise** Data from F. Martin et al., The genome of *Laccaria bicolor* provides insights into mycorrhizal symbiosis, *Nature* 452:88–93 (2008); **31.22** Data from A. E. Arnold et al., Fungal Endophytes Limit Pathogen Damage in a Tropical Tree, *Proceedings of the National Academy of Sciences USA* 100:15649–15654 (2003); **31.27** Adaption of Figure 1 from "Reversing Introduced Species Effects: Experimental Removal of Introduced Fish Leads to Rapid Recovery of a Declining Frog" by Vance T. Vredenburg, from *Proceedings of the National Academy of Sciences USA*, May 2004, Volume 101(20). Copyright (2004) National Academy of Sciences, U.S.A.;

Test Your Understanding Question 5 Data from R. S. Redman et al., Thermotolerance Generated by Plant/Fungal Symbiosis, *Science* 298:1581 (2002).

Chapter 32 Scientific Skills Exercise Data from Bradley Deline, University of West Georgia, and Kevin Peterson, Dartmouth College, 2013.

Chapter 33 Scientific Skills Exercise Data from R. Rochette et al., Interaction between an Invasive Decapod and a Native Gastropod: Predator Foraging Tactics and Prey Architectural Defenses, *Marine Ecology Progress Series* 330:179–188 (2007); **33.21** Adaptation of Figure 3 from "The Global Decline of Nonmarine Mollusks" by Charles Lydeard et al., from *Bioscience*, April 2004, Volume 54(4). American Institute of Biological Sciences. Oxford University Press; **33.29** Tree Data from J. K. Grenier et al., Evolution of the Entire Arthropod Hox Gene Set Predated the Origin and Radiation of the Onychophoran/Arthropod Clade, *Current Biology* 7:547–553 (1997).

Chapter 34 34.10 Adaptation of Figure 1a from "Fossil Sister Group of Craniates: Predicted and Found" by Jon Mallatt and Jun-yuan Chen, from *Journal of Morphology*, May 15, 2003, Volume 258(1). John Wiley & Sons, Inc.; **34.12** Adapted from *Vertebrates: Comparative Anatomy, Function, Evolution* (2002) by Kenneth Kardong. The McGraw-Hill Companies, Inc.; **34.18** Adaptation of Figure 3 from "The Oldest Articulated Osteichthyan Reveals Mosaic Gnathostome Characters" by Min Zhu et al., from *Nature*, March 26, 2009, Volume 458(7237); **34.21** Adaptation of Figure 4 from "The Pectoral Fin of *Tiktaalik roseae* and the Origin of the Tetrapod Limb" by Neil H. Shubin et al., from *Nature*, April 6, 2006, Volume 440(7085). Macmillan Publishers Ltd.; **34.21** *Acanthostega* adaptation of Figure 27 from "The Devonian Tetrapod *Acanthostega gunnari* Jarvik: Postcranial Anatomy, Basal Tetrapod Relationships and Patterns of Skeletal Evolution" by Michael I. Coates, from *Transactions of the Royal Society of Edinburgh: Earth Sciences*, Volume 87: 398; **34.38a** Based on many sources including Figure 4.10 from *Evolution*, by Douglas J. Futuyma. Sinauer Associates, 2005; and *Vertebrate Paleontology and Evolution* by Robert L. Carroll. W.H. Freeman & Co., 1988; **34.47** Based on many photos of fossils. Some sources are *O. tugenensis* photo in "Early Hominid Sows Division" by Michael Balter, from *Science Now*, Feb. 22, 2001; *A. garhi* and *H. neanderthalensis* based on *The Human Evolution Coloring Book* by Adrienne L. Zihlman and Carla J. Simmons. Harper Collins, 2001; *K. platyops* based on photo in "New Hominin Genus from Eastern Africa Shows Diverse Middle Pliocene Lineages" by Meave Leakey et al., from *Nature*, March 2001, Volume 410(6827); *P. boisei* based on a photo by David Brill; *H. ergaster* based on a photo at www.museumsinhand.com; *S. tchadensis* based on Figure 1b from "A New Hominid from the Upper Miocene of Chad, Central Africa" by Michel Brunet et al., from *Nature*, July 2002, Volume 418(6894); **Scientific Skills Exercise** Data from Dean Falk, Florida State University, 2013; **Test Your Understanding Question 8** Data from D. Sol et al., Big-Brained Birds Survive Better in Nature, *Proceedings of the Royal Society B* 274:763–769 (2007).

Chapter 35 Scientific Skills Exercise Data from D. L. Royer et al., Phenotypic Plasticity of Leaf Shape Along a Temperature Gradient in Acer rubrum, *PLOS ONE* 4(10):e7653 (2009); **35.21** Data from "Mongolian Tree Rings and 20th-Century Warming" by Gordon C. Jacoby, et al., from *Science,* August 9, 1996, Volume 273(5276): 771–773.

Chapter 36 Scientific Skills Exercise Data from J. D. Murphy and D. L. Noland, Temperature Effects on Seed Imbibition and Leakage Mediated by Viscosity and Membranes, *Plant Physiology* 69:428–431 (1982); **36.18** Data from S. Rogers and A. J. Peel, Some Evidence for the Existence of Turgor Pressure in the Sieve Tubes of Willow (*Salix*), *Planta* 126:259–267 (1975).

Chapter 37 37.10b Data from D.S. Lundberg et al., Defining the Core *Arabidopsis thaliana* Root Microbiome, *Nature* 488:86–94 (2012).

Chapter 38 Scientific Skills Exercise Data from S. Sutherland and R. K. Vickery, Jr. Trade-offs between Sexual and Asexual Reproduction in the Genus *Mimulus*. *Oecologia* 76:330–335 (1998).

Chapter 39 39.5 Data from C. R. Darwin, *The Power of Movement in Plants*, John Murray, London (1880). P. Boysen-Jensen, Concerning the Performance of Phototropic Stimuli on the Avenacoleoptile, *Berichte der Deutschen Botanischen Gesellschaft (Reports of the German Botanical Society)* 31:559–566 (1913); **39.6** Data from L. Gälweiler et al., Regulation of Polar Auxin Transport by AtPIN1 in *Arabidopsis* Vascular Tissue, *Science* 282:2226–2230 (1998); **39.15a** Based on *Plantwatching: How Plants Remember, Tell Time, Form Relationships and More* by Malcolm Wilkins. Facts on File, 1988; **39.16** Data from H. Borthwick et al., A Reversible Photo Reaction Controlling Seed Germination, *Proceedings of the National Academy of Sciences USA* 38:662–666 (1952); **Problem-Solving Exercise** Map data from Camilo Mora et al. Days for Plant Growth Disappear under Projected Climate Change: Potential Human and Biotic Vulnerability. *PLoS Biol.* 13(6): e1002167 (2015); **Scientific Skills Exercise** Data from O. Falik et al., Rumor Has It …: Relay Communication of Stress Cues in Plants, *PLoS ONE* 6(11):e23625 (2011).

Chapter 40 40.16 Data from V. H. Hutchison, H. G. Dowling, and A. Vinegar, Thermoregulation in a Brooding Female Indian Python, *Python molurus bivittatus*, *Science* 151:694–696 (1966); **Scientific Skills Exercise** Based on the data from M. A. Chappell et al., Energetics of Foraging in Breeding Adélie Penguins, *Ecology* 74:2450–2461 (1993); M. A. Chappell et al., Voluntary Running in Deer Mice: Speed, Distance, Energy Costs, and Temperature Effects, *Journal of Experimental Biology* 207:3839–3854 (2004); T. M. Ellis and M. A. Chappell, Metabolism, Temperature Relations, Maternal Behavior, and Reproductive Energetics in the Ball Python (*Python regius*), *Journal of Comparative Physiology* B 157:393–402 (1987); **40.22** Data from F. G. Revel et al., The Circadian Clock Stops Ticking During Deep Hibernation in the European Hamster, *Proceedings of the National Academy of Sciences USA* 104:13816–13820 (2007).

Chapter 41 41.4 Data from R. W. Smithells et al., Possible Prevention of Neural-Tube Defects by Periconceptional Vitamin Supplementation, *Lancet* 315:339–340 (1980); **41.8** Adapted from Marieb, Elaine; Hoehn, Katja, *Human Anatomy and Physiology*, 8th Edition, 2010, p. 852, Reprinted and electronically reproduced by permission of Pearson Education, Upper Saddle River, New Jersey; **41.17** Adapted from Ottman N., Smidt H., de Vos W.M. and Belzer C. (2012) The function of our microbiota: who is out there and what do they do? *Front. Cell. Inf. Microbiol.* 2:104. doi: 10.3389/fcimb.2012.00104; **41.24** Republished with permission of American Association for the Advancement of

Science, from Cellular Warriors at the Battle of the Bulge by Kathleen Sutliff and Jean Marx, from *Science,* February 2003, Volume 299(5608); **Scientific Skills Exercise** Based on the data from D. L. Coleman, Effects of Parabiosis of Obese Mice with Diabetes and Normal Mice, Diabetologia 9:294–298 (1973).

Chapter 42 Scientific Skills Exercise Data from J. C. Cohen et al., Sequence Variations in PCSK9, Low LDL, and Protection Against Coronary Heart Disease, *New England Journal of Medicine* 354:1264–1272 (2006); **42.25** Data from M. E. Avery and J. Mead, Surface Properties in Relation to Atelectasis and Hyaline Membrane Disease, *American Journal of Diseases of Children* 97:517–523 (1959).

Chapter 43 43.5 Adapted from *Microbiology: An Introduction*, 11th Edition, by Gerard J. Tortora, Berdell R. Funke, and Christine L. Case. Pearson Education, Inc.; **43.6** Adapted from Marieb, Elaine N.; Hoehn, Katja, *Human Anatomy and Physiology*, 8th Ed., © 2010. Reprinted and electronically reproduced by permission of Pearson Education, Inc., Upper Saddle River, New Jersey; **43.24** Based on multiple sources: WHO/UNICEF Coverage Estimates 2014 Revision. July 2015. Map Production: Immunization Vaccines and Biologicals (IVB). World Health Organization, 16 July 2015; Our Progress Against Polio, May 1, 2014. CDC; **Scientific Skills Exercise** Data from sources: L. J. Morrison et al., Probabilistic Order in Antigenic Variation of *Trypanosoma brucei*, *International Journal for Parasitology* 35:961-972 (2005); and L. J. Morrison et al., Antigenic Variation in the African Trypanosome: Molecular Mechanisms and Phenotypic Complexity, *Cellular Microbiology* 1: 1724–1734 (2009).

Chapter 44 Scientific Skills Exercise Data from R. E. MacMillen et al., Water Economy and Energy Metabolism of the Sandy Inland Mouse, *Leggadina hermannsburgensis*, *Journal of Mammalogy* 53:529–539 (1972); **44.6** Adapted from Mitchell, Lawrence G., *Zoology*, © 1998. Reprinted and electronically reproduced by permission of Pearson Education, Inc., Upper Saddle River, New Jersey; **44.12 Kidney structure** Adapted from Marieb, Elaine N.; Hoehn, Katja, *Human Anatomy and Physiology*, 8th Ed., 2010. Reprinted and electronically reproduced by permission of Pearson Education, Inc., Upper Saddle River, New Jersey; **44.13a Kidney structure** Adapted from Marieb, Elaine N.; Hoehn, Katja, *Human Anatomy and Physiology,* 8th Ed., 2010. Reprinted and electronically reproduced by permission of Pearson Education, Inc., Upper Saddle River, New Jersey; **44.20** Data in tables from P. M. Deen et al., Requirement of Human Renal Water Channel Aquaporin-2 for Vasopressin-Dependent Concentration of Urine, *Science* 264:92–95 (1994); **Summary Figure** Adapted from Beck, *Life: An Introduction to Biology*, 3rd Ed., ©1991, p. 643. Reprinted and electronically reproduced by permission of Pearson Education, Inc., Upper Saddle River, New Jersey; **Test Your Understanding Question 7** Data for kangaroo rat from *Animal Physiology: Adaptation and Environment* by Knut Schmidt-Nielsen. Cambridge University Press, 1991.

Chapter 45 Scientific Skills Exercise Data from J. Born et al., Timing the End of Nocturnal Sleep, *Nature* 397:29–30 (1999).

Chapter 46 46.8 Data from R. R. Snook and D. J. Hosken, Sperm Death and Dumping in *Drosophila*, *Nature* 428:939–941 (2004); **Scientific Skills Exercise** Data from A. Jost, Recherches Sur la Differenciation Sexuelle de l'embryon de Lapin (Studies on the Sexual Differentiation of the Rabbit Embryo), *Archives d'Anatomie Microscopique et de Morphologie Experimentale* 36:271–316 (1947); **46.16** Adapted from Marieb, Elaine N., Hoehn, Katja, *Human Anatomy and Physiology*, 8th Ed., 2010. Reprinted and electronically reproduced by permission of Pearson Education, Inc., Upper Saddle River, New Jersey.

Chapter 47 47.3 Data from "Intracellular Calcium Release at Fertilization in the Sea Urchin Egg" by R. Steinhardt et al., from *Developmental Biology*, July 1977, Volume 58(1); **Scientific Skills Exercise** Data from J. Newport and M. Kirschner, A Major Developmental Transition in Early *Xenopus* Embryos: I. Characterization and Timing of Cellular Changes at the Midblastula Stage, *Cell* 30:675–686 (1982); **47.10** Adapted from Keller, R. E. 1986. The Cellular Basis of Amphibian Gastrulation. In L. Browder (ed.), *Developmental Biology: A Comprehensive Synthesis*, Vol. 2. Plenum, New York, pp. 241–327; **47.14** Based on "Cell Commitment and Gene Expression in the Axolotl Embryo" by T. J. Mohun et al., from *Cell*, November 1980, Volume 22(1); **47.17** *Principles of Development*, 2nd Edition by Wolpert (2002), Fig. 8.26, p. 275. By permission of Oxford University Press; **47.19** Republished with permission of Garland Science, Taylor & Francis Group, from *Molecular Biology of the Cell*, Bruce Alberts et al., 4th Edition, © 2002; permission conveyed through Copyright Clearance Center, Inc.; **47.23** Data from H. Spemann, *Embryonic Development and Induction*, Yale University Press, New Haven, CT (1938); **47.24** Data from H. Spemann and H. Mangold, Induction of Embryonic Primordia by Implantation of Organizers from a Different Species, Trans. V. Hamburger (1924). Reprinted in *International Journal of Developmental Biology* 45:13–38 (2001); **47.26** Data from L. S. Honig and D. Summerbell, Maps of strength of positional signaling activity in the developing chick wing bud, *Journal of Embryology and Experimental Morphology* 87:163–174 (1985); **47.27** Adapted from Marieb, Elaine N.; Hoehn, Katja, *Human Anatomy and Physiology*, 8th Edition, 2010. Reprinted and electronically reproduced by permission of Pearson Education, Inc., Upper Saddle River, New Jersey.

Chapter 48 48.12 Graph Based on Figure 6-2d from *Cellular Physiology of Nerve and Muscle*, 4th Edition, by Gary G. Matthews. Wiley-Blackwell, 2003; **Scientific Skills Exercise** Data from C. B. Pert and S. H. Snyder, Opiate Receptor: Demonstration in Nervous Tissue, *Science* 179:1011–1014 (1973).

Chapter 49 49.7 Adapted from Marieb, Elaine N.; Hoehn, Katja, *Human Anatomy and Physiology*, 8th Ed., © 2010. Reprinted and electronically reproduced by permission of Pearson Education, Inc., Upper Saddle River, New Jersey; **49.12** Based on "Sleep in Marine Mammals" by L. M. Mukhametov, from *Sleep Mechanisms*, edited by Alexander A. Borberly and J. L. Valatx. Springer; **Scientific Skills Exercise** Data from M. R. Ralph et al., Transplanted Suprachiasmatic Nucleus Determines Circadian Period, *Science* 247:975–978 (1990); **49.20** Adaptation of Figure 1c from "Avian Brains and a New Understanding of Vertebrate Brain Evolution" by Erich D. Jarvis et al., from *Nature Reviews Neuroscience*, February 2005, Volume 6(2); **49.23** Adaptation of Figure 10 from *Schizophrenia Genesis: The Origins of Madness* by Irving I. Gottesman. Worth Publishers.

Chapter 50 50.12a, 50.13, 50.17 eye structure, 50.24a, 50.26, 50.31 Adapted from Marieb, Elaine N; Hoehn, Katja, *Human Anatomy and Physiology*, 8th Ed., © 2010 Reprinted and electronically reproduced by Permission of Pearson Education, Inc., Upper Saddle River, New Jersey; **50.23** Data from K. L. Mueller et al., The receptors and coding logic for bitter taste, *Nature* 434:225–229 (2005); **50.35 grasshopper** Based on Hickman et al., *Integrated Principles of Zoology*, 9th ed., p. 518, Fig. 22.6, McGraw-Hill Higher Education, NY (1993); **Scientific Skills Exercise** Data from K. Schmidt-Nielsen, Locomotion: Energy Cost of Swimming, Flying, and Running, *Science* 177:222–228 (1972).

Chapter 51 51.4 Based on "*Drosophila*: Genetics Meets Behavior" by Marla B. Sokolowski, from *Nature Reviews: Genetics*, November 2001, Volume 2(11); **51.8** Data from *The Study of Instinct*, N. Tinbergen, Clarendon Press, Oxford (1951); **51.10** Adapted from "Prospective and Retrospective Learning in Honeybees" by Martin Giurfa and Julie Bernard, from *International Journal of Comparative Psychology*, 2006, Volume 19(3); **51.13** Adapted from Evolution of Foraging Behavior in *Drosophila* by Density Dependent Selection by Maria B. Sokolowski et al., from *Proceedings of the National Academy of Sciences USA*, July 8, 1997, Volume 94(14); **Scientific Skills Exercise** Data from Shell Dropping: Decision-Making and Optimal Foraging in Northwestern Crows by Reto Zach, from *Behaviour*, 1979, Volume 68(1–2); 51; **51.18** Reprinted by permission from Klaudia Witte; **51.24 Illustration** Adaptations of photograph by Jonathan Blair, Figure/PhotoID: 3.14, as appeared in *Animal Behavior: An Evolutionary Approach*, 8th Edition, Editor: John Alcock, p. 88. Reprinted by permission; **51.24 Map** data from "Rapid Microevolution of Migratory Behaviour in a Wild Bird Species" by P. Berthold et al., from *Nature*, December 1992, Volume 360(6405); **Art for Concept 51.2 Summary**: Data from *The Study of Instinct*, N. Tinbergen, Clarendon Press, Oxford (1951).

Chapter 52 52.19 Based on the data from *Ecology and Field Biology* by Robert L. Smith. Pearson Education, 1974; and *Sibley Guide to Birds* by David Allen Sibley. Random House, 2000; **52.20** Based on the data from W.J. Fletcher, Interactions among Subtidal Australian Sea Urchins, Gastropods and Algae: Effects of Experimental Removals, *Ecological Monographs* 57:89–109 (1987); **52.21** Based on S. D. Ling et al. Climate-Driven Range Extension of a Sea Urchin: Inferring Future Trends by Analysis of Recent Population Dynamics, *Global Change Biology* (2009) 15, 719–731, doi: 10.1111/j.1365-2486.2008.01734.x; **Scientific Skills Exercise** Based on the data from C. M. Crain et al., Physical and Biotic Drivers of Plant Distribution Across Estuarine Salinity Gradients, *Ecology* 85:2539–2549 (2004); **Graph for Test Your Understanding Question 11** Based on the data from J. Clausen et al., *Experimental Studies on the Nature of Species. III. Environmental Responses of Climatic Races of Achillea*, Carnegie Institution of Washington Publication No. 581 (1948).

Chapter 53 53.2 Data from A. M. Gormley et al., Capture-Recapture Estimates of Hector's Dolphin Abundance at Banks Peninsula, New Zealand, *Marine Mammal Science* 21:204–216 (2005); **Table 53.1** Data from P. W. Sherman and M. L. Morton, Demography of Belding's Ground Squirrel, *Ecology* 65:1617–1628 (1984); **53.4** Based on Demography of Belding's Ground Squirrels by Paul W. Sherman and Martin L. Morton, from *Ecology*, October 1984, Volume 65(5); **53.13** Data from Brood Size Manipulations in the Kestrel (*Falco tinnunculus*): Effects on Offspring and Parent Survival by C. Dijkstra et al., from *Journal of Animal Ecology*, 1990, Volume 59(1); **53.15** Based on Climate and Population Regulation: The Biogeographer's Dilemma by J. T. Enright, from *Oecologia*, 1976, Volume 24(4); **53.16** Based on the data from Predator Responses, Prey Refuges, and Density-Dependent Mortality of a Marine Fish by T.W. Anderson, *Ecology* 82(1):245–257 (2001); **53.18** Based on the data provided by Dr. Rolf O. Peterson; **53.21** Based on U.S. Census Bureau International Data Base; **53.22** Based on U.S. Census Bureau International Data Base; **53.23** Based on U.S. Census Bureau International Data Base; **53.24** Based on Ewing B., D. Moore, S. Goldfinger, A. Oursler, A. Reed, and M. Wackernagel. 2010. *The Ecological Footprint Atlas 2010*. Oakland: Global Footprint Network, p. 33 (www.footprintnetwork.org).

Chapter 54 54.1, 54.2 Based on A. Stanley Rand and Ernest E. Williams. The Anoles of La Palma: Aspects of Their Ecological Relationships, *Breviora*, Volume 327: 1–19. Museum of Comparative Zoology, Harvard University; **54.3** Data from J. H. Connell, The Influence of Interspecific Competition and Other Factors on the Distribution of the Barnacle *Chthamalus stellatus*, *Ecology* 42:710–723 (1961); **Scientific Skills Exercise** Based on the data from B. L. Phillips and R. Shine, An Invasive Species Induces Rapid Adaptive Change in a Native Predator: Cane Toads and Black Snakes in Australia, *Proceedings of the Royal Society B* 273:1545–1550 (2006); **54.11** Based on the data from Sally D. Hacker and Mark D. Bertness, Experimental Evidence for Factors Maintaining Plant Species Diversity in a New England Salt Marsh. *Ecology*, September 1999, Volume

80(6); **54.14 Graph** Data from N. Fierer and R. B. Jackson, The Diversity and Biogeography of Soil Bacterial Communities, *Proceedings of the National Academy of Sciences USA* 103:626–631 (2006); **54.17** Based on George A. Knox. Antarctic Marine Ecosystems, from *Antarctic Ecology*, Volume 1, edited by Martin W. Holdgate. Academic Press, 1970; **54.18** Adapted from Denise L. Breitburg et al., Varying Effects of Low Dissolved Oxygen on Trophic Interactions in an Estuarine Food Web. *Ecological Monographs*, November 1997, Volume 67(4). Used by permission of the Ecological Society of America; **54.19** Based on B. Jenkins et al., Productivity, Disturbance and Food Web Structure at a Local Spatial Scale in Experimental Container Habitats. *OIKOS*, November 1992, Volume 65(2); **54.20 Graph** Data from R. T. Paine, Food web complexity and species diversity, *American Naturalist* 100:65–75 (1966); **54.24** Based on the data from C.R. Townsend, M.R. Scarsbrook, and S. Doledec, The Intermediate Disturbance Hypothesis, Refugia, and Biodiversity in Streams, *Limnology and Oceanography* 42:938–949 (1997); **54.26** Based on Robert L. Crocker and Jack Major. Soil Development in Relation to Vegetation and Surface Age at Glacier Bay, Alaska. *Journal of Ecology*, July 1955, Volume 43(2); **54.27** Adapted from F. Stuart Chapin et al., Mechanisms of Primary Succession Following Deglaciation at Glacier Bay. *Ecological Monographs*, May 1994, Volume 64(2). Ecological Society of America; **54.29** Adapted from D. J. Currie. Energy and Large-Scale Patterns of Animal-and Plant-Species Richness. *American Naturalist*, January 1991, Volume 137(1): 27–49; **54.30** Adapted from Robert H. MacArthur and Edward O. Wilson, An Equilibrium Theory of Insular Zoogeography. *Evolution*, December 1963, Volume 17(4). Society for the Study of Evolution; **54.33** Daniel S. Simberloff and Edward O. Wilson. 1969. Experimental Zoogeography of Islands: The Colonization of Empty Islands. *Ecology*, Vol. 50, No. 2 (Mar., 1969), pp. 278–296.

Chapter 55 55.4 Based on Figure 1.2 from Donald L. DeAngelis (1992), *Dynamics of Nutrient Cycling and Food Webs*. Taylor & Francis; **55.6** Data from J. H. Ryther and W. M. Dunstan, Nitrogen, Phosphorus, and Eutrophication in the Coastal Marine Environment, *Science* 171:1008–1013 (1971); **Table 55.1** Data from D. W. Menzel and J. H. Ryther, Nutrients Limiting the Production of Phytoplankton in the Sargasso Sea, with Special Reference to Iron, *Deep Sea Research* 7:276–281 (1961); **55.7** Based on the data from Fig. 4.1, p. 82, in R.H. Whittaker (1970), *Communities and Ecosystems*. Macmillan, New York; **55.8** Based on Fig. 3c and 3d from Temperate Forest Health in an Era of Emerging Megadisturbance, Constance I. Millar and Nathan L. Stephenson, *Science* 349, 823 (2015); doi: 10.1126/science.aaa9933; **Scientific Skills Exercise** Data from J. M. Teal, Energy Flow in the Salt Marsh Ecosystem of Georgia, *Ecology* 43:614–624 (1962); **55.12** Data from J. A. Trofymow and the CIDET Working Group, The Canadian Intersite Decomposition Experiment: Project and Site Establishment Report (Information Report BC-X-378), Natural Resources Canada, Canadian Forest Service, Pacific Forestry Centre (1998) and T. R. Moore et al., Litter decomposition rates in Canadian forests, *Global Change Biology* 5:75–82 (1999); **55.14** Adapted from Figure 7.4 from Robert E. Ricklefs (2001), *The Economy of Nature*, 5th edition. W.H. Freeman and Company; **55.18b** Based on the data from Wei-Min Wu et al. (2006), Pilot-Scale in Situ Bioremediation of Uranium in a Highly Contaminated Aquifer. 2. Reduction of U(VI) and Geochemical Control of U(VI) Bioavailability. *Environmental Science Technology* 40 (12):3986–3995 (5/13/06); **Art for Concept 55.1 Summary** Based on Figure 1.2 from Donald L. DeAngelis (1992). *Dynamics of Nutrient Cycling and Food Webs*. Taylor & Francis.

Chapter 56 56.10 Based on data from Gene Likens; **56.11** Krebs, Charles J., *Ecology: The Experimental Analysis of Distribution and Abundance*, 5th Ed., © 2001. Reprinted and electronically reproduced by permission of Pearson Education, Inc., Upper Saddle River, New Jersey; **56.13** Data from "Tracking the Long-Term Decline and Recovery of an Isolated Population" by R.L. Westemeier et al., *Science* Volume 282(5394):1695–1698 (11/27/98), AAAS; **56.19** Adapted from Norman Myers et al. (2000). Biodiversity Hotspots for Conservation Priorities, *Nature*, February 24, 2000, Volume 403(6772); **56.28** Based on CO_2 data from www.esrl.noaa.gov/gmd/ccgg/trends. Temperature data from www.giss.nasa.gov/gistemps/graphs/Fig. A.lrg.gif; **Scientific Skills Exercise** Based on data from National Oceanic & Atmospheric Administration, Earth System Research Laboratory, Global Monitoring Division; **56.32** Based on the data from "History of the Ozone Hole," from NASA website, February 26, 2013; and "Antarctic Ozone," from British Antarctic Society website, June 7, 2013; **56.35** Based on the data from Instituto Nacional de Estadistica y Censos de Costa Rica and Centro Centroamericano de Poblacion, Universidad de Costa Rica.

Appendix A Figure 5.11 Wallace/Sanders/Ferl, *Biology: The Science of Life*, 3rd Ed., © 1991. Reprinted and electronically reproduced by permission of Pearson Education, Inc., Upper Saddle River, New Jersey.

Glossary

Pronunciation Key

ā	ace
a/ah	ash
ch	chose
ē	meet
e/eh	bet
g	game
ī	ice
i	hit
ks	box
kw	quick
ng	song
ō	robe
o	ox
oy	boy
s	say
sh	shell
th	thin
ū	boot
u/uh	up
z	zoo
' = primary accent	
' = secondary accent	

5′ cap A modified form of guanine nucleotide added onto the 5′ end of a pre-mRNA molecule.

ABC hypothesis A model of flower formation identifying three classes of organ identity genes that direct formation of the four types of floral organs.

abiotic (ā'-bī-ot'-ik) Nonliving; referring to the physical and chemical properties of an environment.

abortion The termination of a pregnancy in progress.

abscisic acid (ABA) (ab-sis'-ik) A plant hormone that slows growth, often antagonizing the actions of growth hormones. Two of its many effects are to promote seed dormancy and facilitate drought tolerance.

absorption The third stage of food processing in animals: the uptake of small nutrient molecules by an organism's body.

absorption spectrum The range of a pigment's ability to absorb various wavelengths of light; also a graph of such a range.

abyssal zone (uh-bis'-ul) The part of the ocean's benthic zone between 2,000 and 6,000 m deep.

acanthodian (ak'-an-thō'-dē-un) Any of a group of ancient jawed aquatic vertebrates from the Silurian and Devonian periods.

accessory fruit A fruit, or assemblage of fruits, in which the fleshy parts are derived largely or entirely from tissues other than the ovary.

acclimatization (uh-klī'-muh-tī -zā'-shun) Physiological adjustment to a change in an environmental factor.

acetyl CoA Acetyl coenzyme A; the entry compound for the citric acid cycle in cellular respiration, formed from a two-carbon fragment of pyruvate attached to a coenzyme.

acetylcholine (as'-uh-til-kō'-lēn) One of the most common neurotransmitters; functions by binding to receptors and altering the permeability of the postsynaptic membrane to specific ions, either depolarizing or hyperpolarizing the membrane.

acid A substance that increases the hydrogen ion concentration of a solution.

acquired immunodeficiency syndrome (AIDS) The symptoms and signs present during the late stages of HIV infection, defined by a specified reduction in the number of T cells and the appearance of characteristic secondary infections.

acrosomal reaction (ak'-ruh-sōm'-ul) The discharge of hydrolytic enzymes from the acrosome, a vesicle in the tip of a sperm, when the sperm approaches or contacts an egg.

acrosome (ak'-ruh-sōm) A vesicle in the tip of a sperm containing hydrolytic enzymes and other proteins that help the sperm reach the egg.

actin (ak'-tin) A globular protein that links into chains, two of which twist helically about each other, forming microfilaments (actin filaments) in muscle and other kinds of cells.

action potential An electrical signal that propagates (travels) along the membrane of a neuron or other excitable cell as a nongraded (all-or-none) depolarization.

action spectrum A graph that profiles the relative effectiveness of different wavelengths of radiation in driving a particular process.

activation energy The amount of energy that reactants must absorb before a chemical reaction will start; also called free energy of activation.

activator A protein that binds to DNA and stimulates gene transcription. In prokaryotes, activators bind in or near the promoter; in eukaryotes, activators generally bind to control elements in enhancers.

active immunity Long-lasting immunity conferred by the action of B cells and T cells and the resulting B and T memory cells specific for a pathogen. Active immunity can develop as a result of natural infection or immunization.

active site The specific region of an enzyme that binds the substrate and that forms the pocket in which catalysis occurs.

active transport The movement of a substance across a cell membrane against its concentration or electrochemical gradient, mediated by specific transport proteins and requiring an expenditure of energy.

adaptation Inherited characteristic of an organism that enhances its survival and reproduction in a specific environment.

adaptive evolution A process in which traits that enhance survival or reproduction tend to increase in frequency over time, resulting in a better match between organisms and their environment.

adaptive immunity A vertebrate-specific defense that is mediated by B lymphocytes (B cells) and T lymphocytes (T cells) and that exhibits specificity, memory, and self-nonself recognition; also called acquired immunity.

adaptive radiation Period of evolutionary change in which groups of organisms form many new species whose adaptations allow them to fill different ecological roles in their communities.

addition rule A rule of probability stating that the probability of any one of two or more mutually exclusive events occurring can be determined by adding their individual probabilities.

adenosine triphosphate *See* ATP (adenosine triphosphate).

adenylyl cyclase (uh-den'-uh-lil) An enzyme that converts ATP to cyclic AMP in response to an extracellular signal.

adhesion The clinging of one substance to another, such as water to plant cell walls, in this case by means of hydrogen bonds.

adipose tissue A connective tissue that insulates the body and serves as a fuel reserve; contains fat-storing cells called adipose cells.

adrenal gland (uh-drē'-nul) One of two endocrine glands located adjacent to the kidneys in mammals. Endocrine cells in the outer portion (cortex) respond to adrenocorticotropic hormone (ACTH) by secreting steroid hormones that help maintain homeostasis during long-term stress. Neurosecretory cells in the central portion (medulla) secrete epinephrine and norepinephrine in response to nerve signals triggered by short-term stress.

aerobic respiration A catabolic pathway for organic molecules, using oxygen (O_2) as the final electron acceptor in an electron transport chain and ultimately producing ATP. This is the most efficient catabolic pathway and is carried out in most eukaryotic cells and many prokaryotic organisms.

age structure The relative number of individuals of each age in a population.

aggregate fruit A fruit derived from a single flower that has more than one carpel.

AIDS (acquired immunodeficiency syndrome) The symptoms and signs present during the late stages of HIV infection, defined by a specified reduction in the number of T cells and the appearance of characteristic secondary infections.

alcohol fermentation Glycolysis followed by the reduction of pyruvate to ethyl alcohol, regenerating NAD^+ and releasing carbon dioxide.

alga (plural, **algae**) A general term for any species of photosynthetic protist, including both unicellular and multicellular forms. Algal species are included in three eukaryote supergroups (Excavata, SAR, and Archaeplastida).

alimentary canal (al′-uh-men′-tuh-rē) A complete digestive tract, consisting of a tube running between a mouth and an anus.

alkaline vent A deep-sea hydrothermal vent that releases water that is warm (40–90°C) rather than hot and that has a high pH (is basic). These vents consist of tiny pores lined with iron and other catalytic minerals that some scientists hypothesize might have been the location of the earliest abiotic synthesis of organic compounds.

allele (uh-lē′-ul) Any of the alternative versions of a gene that may produce distinguishable phenotypic effects.

allopatric speciation (al′-uh-pat′-rik) The formation of new species in populations that are geographically isolated from one another.

allopolyploid (al′-ō-pol′-ē-ployd) A fertile individual that has more than two chromosome sets as a result of two different species interbreeding and combining their chromosomes.

allosteric regulation The binding of a regulatory molecule to a protein at one site that affects the function of the protein at a different site.

alpha (α) helix (al′-fuh hē′-liks) A coiled region constituting one form of the secondary structure of proteins, arising from a specific pattern of hydrogen bonding between atoms of the polypeptide backbone (not the side chains).

alternation of generations A life cycle in which there is both a multicellular diploid form, the sporophyte, and a multicellular haploid form, the gametophyte; characteristic of plants and some algae.

alternative RNA splicing A type of eukaryotic gene regulation at the RNA-processing level in which different mRNA molecules are produced from the same primary transcript, depending on which RNA segments are treated as exons and which as introns.

altruism (al′-trū-iz-um) Selflessness; behavior that reduces an individual's fitness while increasing the fitness of another individual.

alveolates (al-vē′-uh-lets) One of the three major subgroups for which the SAR eukaryotic supergroup is named. This clade arose by secondary endosymbiosis, and its members have membrane-enclosed sacs (alveoli) located just under the plasma membrane.

alveolus (al-vē′-uh-lus) (plural, **alveoli**) One of the dead-end air sacs where gas exchange occurs in a mammalian lung.

Alzheimer's disease (alts′-hī-merz) An age-related dementia (mental deterioration) characterized by confusion and memory loss.

amino acid (uh-mēn′-ō) An organic molecule possessing both a carboxyl and an amino group. Amino acids serve as the monomers of polypeptides.

amino group (uh-mēn′-ō) A chemical group consisting of a nitrogen atom bonded to two hydrogen atoms; can act as a base in solution, accepting a hydrogen ion and acquiring a charge of 1+.

aminoacyl-tRNA synthetase An enzyme that joins each amino acid to the appropriate tRNA.

ammonia A small, toxic molecule (NH_3) produced by nitrogen fixation or as a metabolic waste product of protein and nucleic acid metabolism.

ammonite A member of a group of shelled cephalopods that were important marine predators for hundreds of millions of years until their extinction at the end of the Cretaceous period (65.5 million years ago).

amniocentesis (am′-nē-ō-sen-tē′-sis) A technique associated with prenatal diagnosis in which amniotic fluid is obtained by aspiration from a needle inserted into the uterus. The fluid and the fetal cells it contains are analyzed to detect certain genetic and congenital defects in the fetus.

amniote (am′-nē-ōt) A member of a clade of tetrapods named for a key derived character, the amniotic egg, which contains specialized membranes, including the fluid-filled amnion, that protect the embryo. Amniotes include mammals as well as birds and other reptiles.

amniotic egg An egg that contains specialized membranes that function in protection, nourishment, and gas exchange. The amniotic egg was a major evolutionary innovation, allowing embryos to develop on land in a fluid-filled sac, thus reducing the dependence of tetrapods on water for reproduction.

amoeba (uh-mē′-buh) A protist characterized by the presence of pseudopodia.

amoebocyte (uh-mē′-buh-sīt′) An amoeba-like cell that moves by pseudopodia and is found in most animals. Depending on the species, it may digest and distribute food, dispose of wastes, form skeletal fibers, fight infections, or change into other cell types.

amoebozoan (uh-mē′-buh-zō′-an) A protist in a clade that includes many species with lobe- or tube-shaped pseudopodia.

amphibian A member of the clade of tetrapods that includes salamanders, frogs, and caecilians.

amphipathic (am′-fē-path′-ik) Having both a hydrophilic region and a hydrophobic region.

amplification The strengthening of stimulus energy during transduction.

amygdala (uh-mig′-duh-luh) A structure in the temporal lobe of the vertebrate brain that has a major role in the processing of emotions.

amylase (am′-uh-lās′) An enzyme that hydrolyzes starch (a glucose polymer from plants) and glycogen (a glucose polymer from animals) into smaller polysaccharides and the disaccharide maltose.

anabolic pathway (an′-uh-bol′-ik) A metabolic pathway that consumes energy to synthesize a complex molecule from simpler molecules.

anaerobic respiration (an-er-ō′-bik) A catabolic pathway in which inorganic molecules other than oxygen accept electrons at the "downhill" end of electron transport chains.

analogous Having characteristics that are similar because of convergent evolution, not homology.

analogy (an-al′-uh-jē) Similarity between two species that is due to convergent evolution rather than to descent from a common ancestor with the same trait.

anaphase The fourth stage of mitosis, in which the chromatids of each chromosome have separated and the daughter chromosomes are moving to the poles of the cell.

anatomy The structure of an organism.

anchorage dependence The requirement that a cell must be attached to a substratum in order to initiate cell division.

androgen (an′-drō-jen) Any steroid hormone, such as testosterone, that stimulates the development and maintenance of the male reproductive system and secondary sex characteristics.

aneuploidy (an′-yū-ploy′-dē) A chromosomal aberration in which one or more chromosomes are present in extra copies or are deficient in number.

angiosperm (an′-jē-ō-sperm) A flowering plant, which forms seeds inside a protective chamber called an ovary.

anhydrobiosis (an-hī′-drō-bī-ō′-sis) A dormant state involving loss of almost all body water.

animal pole The point at the end of an egg in the hemisphere where the least yolk is concentrated; opposite of vegetal pole.

anion (an′-ī-on) A negatively charged ion.

anterior Pertaining to the front, or head, of a bilaterally symmetrical animal.

anterior pituitary A portion of the pituitary gland that develops from nonneural tissue; consists of endocrine cells that synthesize and secrete several tropic and nontropic hormones.

anther In an angiosperm, the terminal pollen sac of a stamen, where pollen grains containing sperm-producing male gametophytes form.

antheridium (an-thuh-rid′-ē-um) (plural, **antheridia**) In plants, the male gametangium, a moist chamber in which gametes develop.

anthropoid (an′-thruh-poyd) A member of a primate group made up of the monkeys and the apes (gibbons, orangutans, gorillas, chimpanzees, bonobos, and humans).

antibody A protein secreted by plasma cells (differentiated B cells) that binds to a particular antigen; also called immunoglobulin. All antibodies have the same Y-shaped structure and in their monomer form consist of two identical heavy chains and two identical light chains.

anticodon (an′-tī-kō′-don) A nucleotide triplet at one end of a tRNA molecule that base-pairs with a particular complementary codon on an mRNA molecule.

antidiuretic hormone (ADH) (an′-tī-dī-yū-ret′-ik) A peptide hormone, also called vasopressin, that promotes water retention by the kidneys. Produced in the hypothalamus and released from the posterior pituitary, ADH also functions in the brain.

antigen (an′-ti-jen) A substance that elicits an immune response by binding to receptors of B or T cells.

antigen presentation (an′-ti-jen) The process by which an MHC molecule binds to a fragment of an intracellular protein antigen and carries it to the cell surface, where it is displayed and can be recognized by a T cell.

antigen-presenting cell (an'-ti-jen) A cell that upon ingesting pathogens or internalizing pathogen proteins generates peptide fragments that are bound by class II MHC molecules and subsequently displayed on the cell surface to T cells. Macrophages, dendritic cells, and B cells are the primary antigen-presenting cells.

antigen receptor (an'-ti-jen) The general term for a surface protein, located on B cells and T cells, that binds to antigens, initiating adaptive immune responses. The antigen receptors on B cells are called B cell receptors, and the antigen receptors on T cells are called T cell receptors.

antiparallel Referring to the arrangement of the sugar-phosphate backbones in a DNA double helix (they run in opposite $5' \rightarrow 3'$ directions).

aphotic zone (ā'-fō'-tik) The part of an ocean or lake beneath the photic zone, where light does not penetrate sufficiently for photosynthesis to occur.

apical bud (ā'-pik-ul) A bud at the tip of a plant stem; also called a terminal bud.

apical dominance (ā'-pik-ul) Tendency for growth to be concentrated at the tip of a plant shoot because the apical bud partially inhibits axillary bud growth.

apical ectodermal ridge (AER) (ā'-pik-ul) A thickened area of ectoderm at the tip of a limb bud that promotes outgrowth of the limb bud.

apical meristem (ā'-pik-ul mār'-uh-stem) A localized region at a growing tip of a plant body where one or more cells divide repeatedly. The dividing cells of an apical meristem enable the plant to grow in length.

apicomplexan (ap'-ē-kom-pleks'-un) A group of alveolate protists, this clade includes many species that parasitize animals. Some apicomplexans cause human disease.

apomixis (ap'-uh-mik'-sis) The ability of some plant species to reproduce asexually through seeds without fertilization by a male gamete.

apoplast (ap'-ō-plast) Everything external to the plasma membrane of a plant cell, including cell walls, intercellular spaces, and the space within dead structures such as xylem vessels and tracheids.

apoptosis (ā-puh-tō'-sus) A type of programmed cell death, which is brought about by activation of enzymes that break down many chemical components in the cell.

aposematic coloration (ap'-ō-si-mat'-ik) The bright warning coloration of many animals with effective physical or chemical defenses.

appendix A small, finger-like extension of the vertebrate cecum; contains a mass of white blood cells that contribute to immunity.

aquaporin A channel protein in a cellular membrane that specifically facilitates osmosis, the diffusion of free water across the membrane.

aqueous solution (ā'-kwē-us) A solution in which water is the solvent.

arachnid A member of a subgroup of the major arthropod clade Chelicerata. Arachnids have six pairs of appendages, including four pairs of walking legs, and include spiders, scorpions, ticks, and mites.

arbuscular mycorrhiza (ar-bus'-kyū-lur mī'-kō-rī'-zuh) Association of a fungus with a plant root system in which the fungus causes the invagination of the host (plant) cells' plasma membranes; also called endomycorrhiza.

arbuscular mycorrhizal fungus (ar-bus'-kyū-lur) A symbiotic fungus whose hyphae grow through the cell wall of plant roots and extend into the root cell (enclosed in tubes formed by invagination of the root cell plasma membrane).

arbuscules Specialized branching hyphae that are found in some mutualistic fungi and exchange nutrients with living plant cells.

Archaea (ar'-kē'-uh) One of two prokaryotic domains, the other being Bacteria.

Archaeplastida (ar'-kē-plas'-tid-uh) One of four supergroups of eukaryotes proposed in a current hypothesis of the evolutionary history of eukaryotes. This monophyletic group, which includes red algae, green algae, and plants, descended from an ancient protistan ancestor that engulfed a cyanobacterium. *See also* Excavata, SAR, and Unikonta.

archegonium (ar-ki-gō'-nē-um) (plural, **archegonia**) In plants, the female gametangium, a moist chamber in which gametes develop.

archenteron (ar-ken'-tuh-ron) The endoderm-lined cavity, formed during gastrulation, that develops into the digestive tract of an animal.

archosaur (ar'-kō-sōr) A member of the reptilian group that includes crocodiles, alligators and dinosaurs, including birds.

arteriole (ar-ter'-ē-ōl) A vessel that conveys blood between an artery and a capillary bed.

artery A vessel that carries blood away from the heart to organs throughout the body.

arthropod A segmented ecdysozoan with a hard exoskeleton and jointed appendages. Familiar examples include insects, spiders, millipedes, and crabs.

artificial selection The selective breeding of domesticated plants and animals to encourage the occurrence of desirable traits.

ascocarp The fruiting body of a sac fungus (ascomycete).

ascomycete (as'-kuh-mī'-sēt) A member of the fungal phylum Ascomycota, commonly called sac fungus. The name comes from the saclike structure in which the spores develop.

ascus (plural, **asci**) A saclike spore capsule located at the tip of a dikaryotic hypha of a sac fungus.

asexual reproduction The generation of offspring from a single parent that occurs without the fusion of gametes. In most cases, the offspring are genetically identical to the parent.

A site One of a ribosome's three binding sites for tRNA during translation. The A site holds the tRNA carrying the next amino acid to be added to the polypeptide chain. (A stands for aminoacyl tRNA.)

assisted migration The translocation of a species to a favorable habitat beyond its native range for the purpose of protecting the species from human-caused threats.

associative learning The acquired ability to associate one environmental feature (such as a color) with another (such as danger).

atherosclerosis A cardiovascular disease in which fatty deposits called plaques develop in the inner walls of the arteries, obstructing the arteries and causing them to harden.

atom The smallest unit of matter that retains the properties of an element.

atomic mass The total mass of an atom, numerically equivalent to the mass in grams of 1 mole of the atom. (For an element with more than one isotope, the atomic mass is the average mass of the naturally occurring isotopes, weighted by their abundance.)

atomic nucleus An atom's dense central core, containing protons and neutrons.

atomic number The number of protons in the nucleus of an atom, unique for each element and designated by a subscript.

ATP (adenosine triphosphate) (a-den'-ō-sēn trī-fos'-fāt) An adenine-containing nucleoside triphosphate that releases free energy when its phosphate bonds are hydrolyzed. This energy is used to drive endergonic reactions in cells.

ATP synthase A complex of several membrane proteins that functions in chemiosmosis with adjacent electron transport chains, using the energy of a hydrogen ion (proton) concentration gradient to make ATP. ATP synthases are found in the inner mitochondrial membranes of eukaryotic cells and in the plasma membranes of prokaryotes.

atrial natriuretic peptide (ANP) (ā'-trē-ul na'-trē-yū-ret'-ik) A peptide hormone secreted by cells of the atria of the heart in response to high blood pressure. ANP's effects on the kidney alter ion and water movement and reduce blood pressure.

atrioventricular (AV) node A region of specialized heart muscle tissue between the left and right atria where electrical impulses are delayed for about 0.1 second before spreading to both ventricles and causing them to contract.

atrioventricular (AV) valve A heart valve located between each atrium and ventricle that prevents a backflow of blood when the ventricle contracts.

atrium (ā'-trē-um) (plural, **atria**) A chamber of the vertebrate heart that receives blood from the veins and transfers blood to a ventricle.

autocrine Referring to a secreted molecule that acts on the cell that secreted it.

autoimmune disease An immunological disorder in which the immune system turns against self.

autonomic nervous system (ot'-ō-nom'-ik) An efferent branch of the vertebrate peripheral nervous system that regulates the internal environment; consists of the sympathetic and parasympathetic divisions and the enteric nervous system.

autopolyploid (ot'-ō-pol'-ē-ployd) An individual that has more than two chromosome sets that are all derived from a single species.

autosome (ot'-ō-sōm) A chromosome that is not directly involved in determining sex; not a sex chromosome.

autotroph (ot'-ō-trōf) An organism that obtains organic food molecules without eating other organisms or substances derived from other organisms. Autotrophs use energy from the sun or from oxidation of inorganic substances to make organic molecules from inorganic ones.

auxin (ok'-sin) A term that primarily refers to indoleacetic acid (IAA), a natural plant hormone that has a variety of effects, including cell elongation, root formation, secondary growth, and fruit growth.

axillary bud (ak′-sil-ār-ē) A structure that has the potential to form a lateral shoot, or branch. The bud appears in the angle formed between a leaf and a stem.

axon (ak′-son) A typically long extension, or process, of a neuron that carries nerve impulses away from the cell body toward target cells.

B cells The lymphocytes that complete their development in the bone marrow and become effector cells for the humoral immune response.

Bacteria One of two prokaryotic domains, the other being Archaea.

bacteriophage (bak-tēr′-ē-ō-fāj) A virus that infects bacteria; also called a phage.

bacteroid A form of the bacterium *Rhizobium* contained within the vesicles formed by the root cells of a root nodule.

balancing selection Natural selection that maintains two or more phenotypic forms in a population.

bar graph A graph in which the independent variable represents groups or nonnumerical categories and the values of the dependent variable(s) are shown by bars.

bark All tissues external to the vascular cambium, consisting mainly of the secondary phloem and layers of periderm.

Barr body A dense object lying along the inside of the nuclear envelope in cells of female mammals, representing a highly condensed, inactivated X chromosome.

basal angiosperm A member of one of three clades of early-diverging lineages of extant flowering plants. Examples are *Amborella*, water lilies, and star anise and its relatives.

basal body (bā′-sul) A eukaryotic cell structure consisting of a "9 + 0" arrangement of microtubule triplets. The basal body may organize the microtubule assembly of a cilium or flagellum and is structurally very similar to a centriole.

basal metabolic rate (BMR) The metabolic rate of a resting, fasting, and nonstressed endotherm at a comfortable temperature.

basal taxon In a specified group of organisms, a taxon whose evolutionary lineage diverged early in the history of the group.

base A substance that reduces the hydrogen ion concentration of a solution.

basidiocarp Elaborate fruiting body of a dikaryotic mycelium of a club fungus.

basidiomycete (buh-sid′-ē-ō-mī′-sēt) A member of the fungal phylum Basidiomycota, commonly called club fungus. The name comes from the club-like shape of the basidium.

basidium (plural, **basidia**) (buh-sid′-ē-um, buh-sid′-ē-ah) A reproductive appendage that produces sexual spores on the gills of mushrooms (club fungi).

Batesian mimicry (bāt′-zē-un mim′-uh-krē) A type of mimicry in which a harmless species resembles an unpalatable or harmful species to which it is not closely related.

behavior Individually, an action carried out by muscles or glands under control of the nervous system in response to a stimulus; collectively, the sum of an animal's responses to external and internal stimuli.

behavioral ecology The study of behavioral interactions between individuals within populations and communities, usually in an evolutionary context.

benign tumor A mass of abnormal cells with specific genetic and cellular changes such that the cells are not capable of surviving at a new site and generally remain at the site of the tumor's origin.

benthic zone The bottom surface of an aquatic environment.

benthos (ben′-thōz) The communities of organisms living in the benthic zone of an aquatic biome.

beta (β) pleated sheet One form of the secondary structure of proteins in which the polypeptide chain folds back and forth. Two regions of the chain lie parallel to each other and are held together by hydrogen bonds between atoms of the polypeptide backbone (not the side chains).

beta oxidation A metabolic sequence that breaks fatty acids down to two-carbon fragments that enter the citric acid cycle as acetyl CoA.

bilateral symmetry Body symmetry in which a central longitudinal plane divides the body into two equal but opposite halves.

bilaterian (bī′-luh-ter′-ē-uhn) A member of a clade of animals with bilateral symmetry and three germ layers.

bile A mixture of substances that is produced in the liver and stored in the gallbladder; enables formation of fat droplets in water as an aid in the digestion and absorption of fats.

binary fission A method of asexual reproduction in single-celled organisms in which the cell grows to roughly double its size and then divides into two cells. In prokaryotes, binary fission does not involve mitosis, but in single-celled eukaryotes that undergo binary fission, mitosis is part of the process.

binomial A common term for the two-part, latinized format for naming a species, consisting of the genus and specific epithet; also called a binomen.

biodiversity hot spot A relatively small area with numerous endemic species and a large number of endangered and threatened species.

bioenergetics (1) The overall flow and transformation of energy in an organism. (2) The study of how energy flows through organisms.

biofilm A surface-coating colony of one or more species of unicellular organisms that engage in metabolic cooperation; most known biofilms are formed by prokaryotes.

biofuel A fuel produced from biomass.

biogeochemical cycle Any of the various chemical cycles, which involve both biotic and abiotic components of ecosystems.

biogeography The scientific study of the past and present geographic distributions of species.

bioinformatics The use of computers, software, and mathematical models to process and integrate biological information from large data sets.

biological augmentation An approach to restoration ecology that uses organisms to add essential materials to a degraded ecosystem.

biological clock An internal timekeeper that controls an organism's biological rhythms. The biological clock marks time with or without environmental cues but often requires signals from the environment to remain tuned to an appropriate period. *See also* circadian rhythm.

biological magnification A process in which retained substances become more concentrated at each higher trophic level in a food chain.

biological species concept Definition of a species as a group of populations whose members have the potential to interbreed in nature and produce viable, fertile offspring but do not produce viable, fertile offspring with members of other such groups.

biology The scientific study of life.

biomass The total mass of organic matter comprising a group of organisms in a particular habitat.

biome (bī′-ōm) Any of the world's major ecosystem types, often classified according to the predominant vegetation for terrestrial biomes and the physical environment for aquatic biomes and characterized by adaptations of organisms to that particular environment.

bioremediation The use of organisms to detoxify and restore polluted and degraded ecosystems.

biosphere The entire portion of Earth inhabited by life; the sum of all the planet's ecosystems.

biotechnology The manipulation of organisms or their components to produce useful products.

biotic (bī-ot′-ik) Pertaining to the living factors—the organisms—in an environment.

bipolar disorder A depressive mental illness characterized by swings of mood from high to low; also called manic-depressive disorder.

birth control pill A hormonal contraceptive that inhibits ovulation, retards follicular development, or alters a woman's cervical mucus to prevent sperm from entering the uterus.

blade (1) A leaflike structure of a seaweed that provides most of the surface area for photosynthesis. (2) The flattened portion of a typical leaf.

blastocoel (blas′-tuh-sēl) The fluid-filled cavity that forms in the center of a blastula.

blastocyst (blas′-tuh-sist) The blastula stage of mammalian embryonic development, consisting of an inner cell mass, a cavity, and an outer layer, the trophoblast. In humans, the blastocyst forms 1 week after fertilization.

blastomere An early embryonic cell arising during the cleavage stage of an early embryo.

blastopore (blas′-tō-pōr) In a gastrula, the opening of the archenteron that typically develops into the anus in deuterostomes and the mouth in protostomes.

blastula (blas′-tyū-luh) A hollow ball of cells that marks the end of the cleavage stage during early embryonic development in animals.

blood A connective tissue with a fluid matrix called plasma in which red blood cells, white blood cells, and cell fragments called platelets are suspended.

blue-light photoreceptor Any of several classes of light-absorbing molecules that have physiological effects when activated by blue light.

body cavity A fluid- or air-filled space between the digestive tract and the body wall.

body plan In multicellular eukaryotes, a set of morphological and developmental traits that are integrated into a functional whole—the living organism.

Bohr shift A lowering of the affinity of hemoglobin for oxygen, caused by a drop in pH. It facilitates the release of oxygen from hemoglobin in the vicinity of active tissues.

bolus A lubricated ball of chewed food.

bone A connective tissue consisting of living cells held in a rigid matrix of collagen fibers embedded in calcium salts.

book lung An organ of gas exchange in spiders, consisting of stacked plates contained in an internal chamber.

bottleneck effect Genetic drift that occurs when the size of a population is reduced, as by a natural disaster or human actions. Typically, the surviving population is no longer genetically representative of the original population.

bottom-up control A situation in which the abundance of organisms at each trophic level is limited by nutrient supply or the availability of food at lower trophic levels; thus, the supply of nutrients controls plant numbers, which control herbivore numbers, which in turn control predator numbers.

Bowman's capsule (bō'-munz) A cup-shaped receptacle in the vertebrate kidney that is the initial, expanded segment of the nephron, where filtrate enters from the blood.

brachiopod (bra'-kē-uh-pod') A marine lophotrochozoan with a shell divided into dorsal and ventral halves; also called lamp shells.

brain Organ of the central nervous system where information is processed and integrated.

brainstem A collection of structures in the vertebrate brain, including the midbrain, the pons, and the medulla oblongata; functions in homeostasis, coordination of movement, and conduction of information to higher brain centers.

branch point The representation on a phylogenetic tree of the divergence of two or more taxa from a common ancestor. A branch point is usually shown as a dichotomy in which a branch representing the ancestral lineage splits (at the branch point) into two branches, one for each of the two descendant lineages.

brassinosteroid A steroid hormone in plants that has a variety of effects, including inducing cell elongation, retarding leaf abscission, and promoting xylem differentiation.

breathing Ventilation of the lungs through alternating inhalation and exhalation.

bronchiole (brong'-kē-ōl') A fine branch of the bronchi that transports air to alveoli.

bronchus (brong'-kus) (plural, **bronchi**) One of a pair of breathing tubes that branch from the trachea into the lungs.

brown alga A multicellular, photosynthetic protist with a characteristic brown or olive color that results from carotenoids in its plastids. Most brown algae are marine, and some have a plantlike body.

bryophyte (brī'-uh-fīt) An informal name for a moss, liverwort, or hornwort; a nonvascular plant that lives on land but lacks some of the terrestrial adaptations of vascular plants.

buffer A solution that contains a weak acid and its corresponding base. A buffer minimizes changes in pH when acids or bases are added to the solution.

bulk feeder An animal that eats relatively large pieces of food.

bulk flow The movement of a fluid due to a difference in pressure between two locations.

bundle-sheath cell In C_4 plants, a type of photosynthetic cell arranged into tightly packed sheaths around the veins of a leaf.

C_3 plant A plant that uses the Calvin cycle for the initial steps that incorporate CO_2 into organic material, forming a three-carbon compound as the first stable intermediate.

C_4 plant A plant in which the Calvin cycle is preceded by reactions that incorporate CO_2 into a four-carbon compound, the end product of which supplies CO_2 for the Calvin cycle.

calcitonin (kal'-si-tō'-nin) A hormone secreted by the thyroid gland that lowers the blood calcium level by promoting calcium deposition in bone and calcium excretion from the kidneys; nonessential in adult humans.

callus A mass of dividing, undifferentiated cells growing at the site of a wound or in culture.

calorie (cal) The amount of heat energy required to raise the temperature of 1 g of water by 1°C; also the amount of heat energy that 1 g of water releases when it cools by 1°C. The Calorie (with a capital C), usually used to indicate the energy content of food, is a kilocalorie.

Calvin cycle The second of two major stages in photosynthesis (following the light reactions), involving fixation of atmospheric CO_2 and reduction of the fixed carbon into carbohydrate.

Cambrian explosion A relatively brief time in geologic history when many present-day phyla of animals first appeared in the fossil record. This burst of evolutionary change occurred about 535–525 million years ago and saw the emergence of the first large, hard-bodied animals.

cAMP (cyclic AMP) Cyclic adenosine monophosphate, named because of its ring structure, is a common chemical signal that has a diversity of roles, including as a second messenger in many eukaryotic cells, and as a regulator of some bacterial operons.

CAM plant A plant that uses crassulacean acid metabolism, an adaptation for photosynthesis in arid conditions. In this process, CO_2 entering open stomata during the night is converted to organic acids, which release CO_2 for the Calvin cycle during the day, when stomata are closed.

canopy The uppermost layer of vegetation in a terrestrial biome.

capillary (kap'-il-ār'-ē) A microscopic blood vessel that penetrates the tissues and consists of a single layer of endothelial cells that allows exchange between the blood and interstitial fluid.

capillary bed (kap'-il-ār'-ē) A network of capillaries in a tissue or organ.

capsid The protein shell that encloses a viral genome. It may be rod-shaped, polyhedral, or more complex in shape.

capsule (1) In many prokaryotes, a dense and well-defined layer of polysaccharide or protein that surrounds the cell wall and is sticky, protecting the cell and enabling it to adhere to substrates or other cells. (2) The sporangium of a bryophyte (moss, liverwort, or hornwort).

carbohydrate (kar'-bō-hī'-drāt) A sugar (monosaccharide) or one of its dimers (disaccharides) or polymers (polysaccharides).

carbon fixation The initial incorporation of carbon from CO_2 into an organic compound by an autotrophic organism (a plant, another photosynthetic organism, or a chemoautotrophic prokaryote).

carbonyl group (kar'-buh-nil) A chemical group present in aldehydes and ketones and consisting of a carbon atom double-bonded to an oxygen atom.

carboxyl group (kar-bok'-sil) A chemical group present in organic acids and consisting of a single carbon atom double-bonded to an oxygen atom and also bonded to a hydroxyl group.

cardiac cycle (kar'-dē-ak) The alternating contractions and relaxations of the heart.

cardiac muscle (kar'-dē-ak) A type of striated muscle that forms the contractile wall of the heart. Its cells are joined by intercalated disks that relay the electrical signals underlying each heartbeat.

cardiac output (kar'-dē-ak) The volume of blood pumped per minute by each ventricle of the heart.

cardiovascular system A closed circulatory system with a heart and branching network of arteries, capillaries, and veins. The system is characteristic of vertebrates.

carnivore An organism that consumes animals for nutrition.

carotenoid (kuh-rot'-uh-noyd') An accessory pigment, either yellow or orange, in the chloroplasts of plants and in some prokaryotes. By absorbing wavelengths of light that chlorophyll cannot, carotenoids broaden the spectrum of colors that can drive photosynthesis.

carpel (kar'-pul) The ovule-producing reproductive organ of a flower, consisting of the stigma, style, and ovary.

carrier In genetics, an individual who is heterozygous at a given genetic locus for a recessively inherited disorder. The heterozygote is generally phenotypically normal for the disorder but can pass on the recessive allele to offspring.

carrying capacity The maximum population size that can be supported by the available resources, symbolized as K.

cartilage (kar'-til-ij) A flexible connective tissue with an abundance of collagenous fibers embedded in chondroitin sulfate.

Casparian strip (ka-spār'-ē-un) A water-impermeable ring of wax in the endodermal cells of plants that blocks the passive flow of water and solutes into the stele by way of cell walls.

catabolic pathway (kat'-uh-bol'-ik) A metabolic pathway that releases energy by breaking down complex molecules to simpler molecules.

catalysis (kuh-ta'-luh-sis) A process by which a chemical agent called a catalyst selectively increases the rate of a reaction without being consumed by the reaction.

catalyst (kat'-uh-list) A chemical agent that selectively increases the rate of a reaction without being consumed by the reaction.

cation (cat'-ī'-on) A positively charged ion.

cation exchange (cat'-ī'-on) A process in which positively charged minerals are made available to a plant when hydrogen ions in the soil displace mineral ions from the clay particles.

cecum (sē'-kum) (plural, **ceca**) The blind pouch forming one branch of the large intestine.

cell Life's fundamental unit of structure and function; the smallest unit of organization that can perform all activities required for life.

cell body The part of a neuron that houses the nucleus and most other organelles.

cell cycle An ordered sequence of events in the life of a cell, from its origin in the division of a parent cell until its own division into two. The eukaryotic cell cycle is composed of interphase (including G_1, S, and G_2 phases) and M phase (including mitosis and cytokinesis).

cell cycle control system A cyclically operating set of molecules in the eukaryotic cell that both triggers and coordinates key events in the cell cycle.

cell division The reproduction of cells.

cell fractionation The disruption of a cell and separation of its parts by centrifugation at successively higher speeds.

cell-mediated immune response The branch of adaptive immunity that involves the activation of cytotoxic T cells, which defend against infected cells.

cell plate A membrane-bounded, flattened sac located at the midline of a dividing plant cell, inside which the new cell wall forms during cytokinesis.

cellular respiration The catabolic pathways of aerobic and anaerobic respiration, which break down organic molecules and use an electron transport chain for the production of ATP.

cellulose (sel'-yū-lōs) A structural polysaccharide of plant cell walls, consisting of glucose monomers joined by β glycosidic linkages.

cell wall A protective layer external to the plasma membrane in the cells of plants, prokaryotes, fungi, and some protists. Polysaccharides such as cellulose (in plants and some protists), chitin (in fungi), and peptidoglycan (in bacteria) are important structural components of cell walls.

central nervous system (CNS) The portion of the nervous system where signal integration occurs; in vertebrate animals, the brain and spinal cord.

central vacuole In a mature plant cell, a large membranous sac with diverse roles in growth, storage, and sequestration of toxic substances.

centriole (sen'-trē-ōl) A structure in the centrosome of an animal cell composed of a cylinder of microtubule triplets arranged in a "9 + 0" pattern. A centrosome has a pair of centrioles.

centromere (sen'-trō-mēr) In a duplicated chromosome, the region on each sister chromatid where it is most closely attached to its sister chromatid by proteins that bind to the centromeric DNA. Other proteins condense the chromatin in that region, so it appears as a narrow "waist" on the duplicated chromosome. (An unduplicated chromosome has a single centromere, identified by the proteins bound there.)

centrosome (sen'-trō-sōm) A structure present in the cytoplasm of animal cells that functions as a microtubule-organizing center and is important during cell division. A centrosome has two centrioles.

cercozoan An amoeboid or flagellated protist that feeds with threadlike pseudopodia.

cerebellum (sār'-ruh-bel'-um) Part of the vertebrate hindbrain located dorsally; functions in unconscious coordination of movement and balance.

cerebral cortex (suh-rē'-brul) The surface of the cerebrum; the largest and most complex part of the mammalian brain, containing nerve cell bodies of the cerebrum; the part of the vertebrate brain most changed through evolution.

cerebrum (suh-rē'-brum) The dorsal portion of the vertebrate forebrain, composed of right and left hemispheres; the integrating center for memory, learning, emotions, and other highly complex functions of the central nervous system.

cervix (ser'-viks) The neck of the uterus, which opens into the vagina.

chaparral A scrubland biome of dense, spiny evergreen shrubs found at midlatitudes along coasts where cold ocean currents circulate offshore; characterized by mild, rainy winters and long, hot, dry summers.

chaperonin (shap'-er-ō'-nin) A protein complex that assists in the proper folding of other proteins.

character An observable heritable feature that may vary among individuals.

character displacement The tendency for characteristics to be more divergent in sympatric populations of two species than in allopatric populations of the same two species.

checkpoint A control point in the cell cycle where stop and go-ahead signals can regulate the cycle.

chelicera (kē-lih'-suh-ruh) (plural, **chelicerae**) One of a pair of clawlike feeding appendages characteristic of chelicerates.

chelicerate (kē-lih-suh'-rāte) An arthropod that has chelicerae and a body divided into a cephalothorax and an abdomen. Living chelicerates include sea spiders, horseshoe crabs, scorpions, ticks, and spiders.

chemical bond An attraction between two atoms, resulting from a sharing of outer-shell electrons or the presence of opposite charges on the atoms. The bonded atoms gain complete outer electron shells.

chemical energy Energy available in molecules for release in a chemical reaction; a form of potential energy.

chemical equilibrium In a chemical reaction, the state in which the rate of the forward reaction equals the rate of the reverse reaction, so that the relative concentrations of the reactants and products do not change with time.

chemical reaction The making and breaking of chemical bonds, leading to changes in the composition of matter.

chemiosmosis (kem'-ē-oz-mō'-sis) An energy-coupling mechanism that uses energy stored in the form of a hydrogen ion gradient across a membrane to drive cellular work, such as the synthesis of ATP. Under aerobic conditions, most ATP synthesis in cells occurs by chemiosmosis.

chemoautotroph (kē'-mō-ot'-ō-trōf) An organism that obtains energy by oxidizing inorganic substances and needs only carbon dioxide as a carbon source.

chemoheterotroph (kē-mō-het'-er-ō-trōf) An organism that requires organic molecules for both energy and carbon.

chemoreceptor A sensory receptor that responds to a chemical stimulus, such as a solute or an odorant.

chiasma (plural, **chiasmata**) (kī-az'-muh, kī-az'-muh-tuh) The X-shaped, microscopically visible region where crossing over has occurred earlier in prophase I between homologous nonsister chromatids. Chiasmata become visible after synapsis ends, with the two homologs remaining associated due to sister chromatid cohesion.

chitin (kī'-tin) A structural polysaccharide, consisting of amino sugar monomers, found in many fungal cell walls and in the exoskeletons of all arthropods.

chlorophyll (klōr'-ō-fil) A green pigment located in membranes within the chloroplasts of plants and algae and in the membranes of certain prokaryotes. Chlorophyll *a* participates directly in the light reactions, which convert solar energy to chemical energy.

chlorophyll *a* (klōr'-ō-fil) A photosynthetic pigment that participates directly in the light reactions, which convert solar energy to chemical energy.

chlorophyll *b* (klōr'-ō-fil) An accessory photosynthetic pigment that transfers energy to chlorophyll *a*.

chloroplast (klōr'-ō-plast) An organelle found in plants and photosynthetic protists that absorbs sunlight and uses it to drive the synthesis of organic compounds from carbon dioxide and water.

choanocyte (kō-an'-uh-sīt) A flagellated feeding cell found in sponges. Also called a collar cell, it has a collar-like ring that traps food particles around the base of its flagellum.

cholesterol (kō-les'-tuh-rol) A steroid that forms an essential component of animal cell membranes and acts as a precursor molecule for the synthesis of other biologically important steroids, such as many hormones.

chondrichthyan (kon-drik'-thē-an) A member of the clade Chondrichthyes, vertebrates with skeletons made mostly of cartilage, such as sharks and rays.

chordate A member of the phylum Chordata, animals that at some point during their development have a notochord; a dorsal, hollow nerve cord; pharyngeal slits or clefts; and a muscular, post-anal tail.

chorionic villus sampling (CVS) (kōr'-ē-on'-ik vil'-us) A technique associated with prenatal diagnosis in which a small sample of the fetal portion of the placenta is removed for analysis to detect certain genetic and congenital defects in the fetus.

chromatin (krō'-muh-tin) The complex of DNA and proteins that makes up eukaryotic chromosomes. When the cell is not dividing, chromatin exists in its dispersed form, as a mass of very long, thin fibers that are not visible with a light microscope.

chromosome (krō'-muh-sōm) A cellular structure consisting of one DNA molecule and associated protein molecules. A duplicated chromosome has two DNA molecules. (In some contexts, such as genome sequencing, the term may refer to the DNA alone.) A eukaryotic cell typically has multiple, linear chromosomes, which are located in the nucleus. A prokaryotic cell often has a single, circular chromosome, which is found in the nucleoid, a region that is not enclosed by a membrane. *See also* chromatin.

chromosome theory of inheritance (krō'-muh-sōm) A basic principle in biology stating that genes are located at specific positions (loci) on chromosomes and that the behavior of chromosomes during meiosis accounts for inheritance patterns.

chylomicron (kī'-lō-mī'-kron) A lipid transport globule composed of fats mixed with cholesterol and coated with proteins.

chyme (kīm) The mixture of partially digested food and digestive juices formed in the stomach.

chytrid (kī'-trid) A member of the fungal phylum Chytridiomycota, mostly aquatic fungi with flagellated zoospores that represent an early-diverging fungal lineage.

ciliate (sil'-ē-it) A type of protist that moves by means of cilia.

cilium (sil'-ē-um) (plural, **cilia**) A short appendage containing microtubules in eukaryotic cells. A motile cilium is specialized for locomotion or moving fluid past the cell; it is formed from a core of nine outer doublet microtubules and two inner single microtubules (the "9 + 2" arrangement) ensheathed in an extension of the plasma membrane. A primary cilium is usually nonmotile and plays a sensory and signaling role; it lacks the two inner microtubules (the "9 + 0" arrangement).

circadian rhythm (ser-kā'-dē-un) A physiological cycle of about 24 hours that persists even in the absence of external cues.

***cis-trans* isomer** One of several compounds that have the same molecular formula and covalent bonds between atoms but differ in the spatial arrangements of their atoms owing to the inflexibility of double bonds; also called a geometric isomer.

citric acid cycle A chemical cycle involving eight steps that completes the metabolic breakdown of glucose molecules begun in glycolysis by oxidizing acetyl CoA (derived from pyruvate) to carbon dioxide; occurs within the mitochondrion in eukaryotic cells and in the cytosol of prokaryotes; together with pyruvate oxidation, the second major stage in cellular respiration.

clade (klād) A group of species that includes an ancestral species and all of its descendants. A clade is equivalent to a monophyletic group.

cladistics (kluh-dis'-tiks) An approach to systematics in which organisms are placed into groups called clades based primarily on common descent.

class In Linnaean classification, the taxonomic category above the level of order.

cleavage (1) The process of cytokinesis in animal cells, characterized by pinching of the plasma membrane. (2) The succession of rapid cell divisions without significant growth during early embryonic development that converts the zygote to a ball of cells.

cleavage furrow The first sign of cleavage in an animal cell; a shallow groove around the cell in the cell surface near the old metaphase plate.

climate The long-term prevailing weather conditions at a given place.

climate change A directional change in temperature, precipitation, or other aspect of the global climate that lasts for three decades or more.

climograph A plot of the temperature and precipitation in a particular region.

clitoris (klit'-uh-ris) An organ at the upper intersection of the labia minora that engorges with blood and becomes erect during sexual arousal.

cloaca (klō-ā'-kuh) A common opening for the digestive, urinary, and reproductive tracts found in many nonmammalian vertebrates but in few mammals.

clonal selection The process by which an antigen selectively binds to and activates only those lymphocytes bearing receptors specific for the antigen. The selected lymphocytes proliferate and differentiate into a clone of effector cells and a clone of memory cells specific for the stimulating antigen.

clone (1) A group of genetically identical individuals or cells. (2) In popular usage, an individual that is genetically identical to another individual. (3) As a verb, to make one or more genetic replicas of an individual or cell. *See also* gene cloning.

cloning vector In genetic engineering, a DNA molecule that can carry foreign DNA into a host cell and replicate there. Cloning vectors include plasmids and bacterial artificial chromosomes (BACs), which move recombinant DNA from a test tube back into a cell, and viruses that transfer recombinant DNA by infection.

closed circulatory system A circulatory system in which blood is confined to vessels and is kept separate from the interstitial fluid.

cnidocyte (nī'-duh-sīt) A specialized cell unique to the phylum Cnidaria; contains a capsule-like organelle housing a coiled thread that, when discharged, explodes outward and functions in prey capture or defense.

cochlea (kok'-lē-uh) The complex, coiled organ of hearing that contains the organ of Corti.

coding strand Nontemplate strand of DNA, which has the same sequence as the mRNA except it has thymine (T) instead of uracil (U).

codominance The situation in which the phenotypes of both alleles are exhibited in the heterozygote because both alleles affect the phenotype in separate, distinguishable ways.

codon (kō'-don) A three-nucleotide sequence of DNA or mRNA that specifies a particular amino acid or termination signal; the basic unit of the genetic code.

coefficient of relatedness The fraction of genes that, on average, are shared by two individuals.

coelom (sē'-lōm) A body cavity lined by tissue derived only from mesoderm.

coenocytic fungus (sē'-no-si'-tic) A fungus that lacks septa and hence whose body is made up of a continuous cytoplasmic mass that may contain hundreds or thousands of nuclei.

coenzyme (kō-en'-zīm) An organic molecule serving as a cofactor. Most vitamins function as coenzymes in metabolic reactions.

coevolution The joint evolution of two interacting species, each in response to selection imposed by the other.

cofactor Any nonprotein molecule or ion that is required for the proper functioning of an enzyme. Cofactors can be permanently bound to the active site or may bind loosely and reversibly, along with the substrate, during catalysis.

cognition The process of knowing that may include awareness, reasoning, recollection, and judgment.

cognitive map A neural representation of the abstract spatial relationships between objects in an animal's surroundings.

cohesion The linking together of like molecules, often by hydrogen bonds.

cohesion-tension hypothesis The leading explanation of the ascent of xylem sap. It states that transpiration exerts pull on xylem sap, putting the sap under negative pressure, or tension, and that the cohesion of water molecules transmits this pull along the entire length of the xylem from shoots to roots.

cohort A group of individuals of the same age in a population.

coleoptile (kō'-lē-op'-tul) The covering of the young shoot of the embryo of a grass seed.

coleorhiza (kō'-lē-uh-rī'-zuh) The covering of the young root of the embryo of a grass seed.

collagen A glycoprotein in the extracellular matrix of animal cells that forms strong fibers, found extensively in connective tissue and bone; the most abundant protein in the animal kingdom.

collecting duct The location in the kidney where processed filtrate, called urine, is collected from the renal tubules.

collenchyma cell (kō-len'-kim-uh) A flexible plant cell type that occurs in strands or cylinders that support young parts of the plant without restraining growth.

colon (kō'-len) The largest section of the vertebrate large intestine; functions in water absorption and formation of feces.

commensalism (kuh-men'-suh-lizm) A +/0 ecological interaction that benefits the individuals of one species but neither harms nor helps the individuals of the other species.

communication (1) In behavior, a process involving the transmission and reception of signals between organisms. (2) Transfer of information from one cell or molecule to another by means of chemical or physical signals.

community All the organisms that inhabit a particular area; an assemblage of populations of different species living close enough together for potential interaction.

community ecology The study of how interactions between species affect community structure and organization.

community structure The number of species found in an ecological community, the particular species that are present, and the relative abundance of these species.

companion cell A type of plant cell that is connected to a sieve-tube element by many plasmodesmata and whose nucleus and ribosomes may serve one or more adjacent sieve-tube elements.

competition A −/− interaction that occurs when individuals of different species both use a resource that limits the survival and reproduction of each species.

competitive exclusion The concept that when populations of two similar species compete for the same limited resources, one population will use the resources more efficiently and have a reproductive advantage that will eventually lead to the elimination of the other population.

competitive inhibitor A substance that reduces the activity of an enzyme by entering the active site in place of the substrate, whose structure it mimics.

complement system A group of about 30 blood proteins that may amplify the inflammatory response, enhance phagocytosis, or directly lyse extracellular pathogens.

complementary DNA (cDNA) A double-stranded DNA molecule made *in vitro* using mRNA as a template and the enzymes reverse transcriptase and DNA polymerase. A cDNA molecule corresponds to the exons of a gene.

complete dominance The situation in which the phenotypes of the heterozygote and dominant homozygote are indistinguishable.

complete flower A flower that has all four basic floral organs: sepals, petals, stamens, and carpels.

complete metamorphosis The transformation of a larva into an adult that looks very different, and often functions very differently in its environment, than the larva.

compound A substance consisting of two or more different elements combined in a fixed ratio.

compound eye A type of multifaceted eye in insects and crustaceans consisting of up to several thousand light-detecting, focusing ommatidia.

concentration gradient A region along which the density of a chemical substance increases or decreases.

conception The fertilization of an egg by a sperm in humans.

cone A cone-shaped cell in the retina of the vertebrate eye, sensitive to color.

conformer An animal for which an internal condition conforms to (changes in accordance with) changes in an environmental variable.

conidium (plural, **conidia**) A haploid spore produced at the tip of a specialized hypha in ascomycetes during asexual reproduction.

conifer A member of the largest gymnosperm phylum. Most conifers are cone-bearing trees, such as pines and firs.

conjugation (kon′-jū-gā′-shun) (1) In prokaryotes, the direct transfer of DNA between two cells that are temporarily joined. When the two cells are members of different species, conjugation results in horizontal gene transfer. (2) In ciliates, a sexual process in which two cells exchange haploid micronuclei but do not reproduce.

connective tissue Animal tissue that functions mainly to bind and support other tissues, having a sparse population of cells scattered through an extracellular matrix.

conodont An early, soft-bodied vertebrate with prominent eyes and dental elements.

conservation biology The integrated study of ecology, evolutionary biology, physiology, molecular biology, and genetics to sustain biological diversity at all levels.

consumer An organism that feeds on producers, other consumers, or nonliving organic material.

contraception The deliberate prevention of pregnancy.

contractile vacuole A membranous sac that helps move excess water out of certain freshwater protists.

control element A segment of noncoding DNA that helps regulate transcription of a gene by serving as a binding site for a transcription factor. Multiple control elements are present in a eukaryotic gene's enhancer.

control group In a controlled experiment, a set of subjects that lacks (or does not receive) the specific factor being tested. Ideally, the control group is identical to the experimental group in other respects.

controlled experiment An experiment designed to compare an experimental group with a control group; ideally, the two groups differ only in the factor being tested.

convergent evolution The evolution of similar features in independent evolutionary lineages.

convergent extension A process in which the cells of a tissue layer rearrange themselves in such a way that the sheet of cells becomes narrower (converges) and longer (extends).

cooperativity A kind of allosteric regulation whereby a shape change in one subunit of a protein caused by substrate binding is transmitted to all the other subunits, facilitating binding of additional substrate molecules to those subunits.

coral reef Typically a warm-water, tropical ecosystem dominated by the hard skeletal structures secreted primarily by corals. Some coral reefs also exist in cold, deep waters.

corepressor A small molecule that binds to a bacterial repressor protein and changes the protein's shape, allowing it to bind to the operator and switch an operon off.

cork cambium (kam′-bē-um) A cylinder of meristematic tissue in woody plants that replaces the epidermis with thicker, tougher cork cells.

corpus callosum (kor′-pus kuh-lō′-sum) The thick band of nerve fibers that connects the right and left cerebral hemispheres in mammals, enabling the hemispheres to process information together.

corpus luteum (kor′-pus lū′-tē-um) A secreting tissue in the ovary that forms from the collapsed follicle after ovulation and produces progesterone.

cortex (1) The outer region of cytoplasm in a eukaryotic cell, lying just under the plasma membrane, that has a more gel-like consistency than the inner regions due to the presence of multiple microfilaments. (2) In plants, ground tissue that is between the vascular tissue and dermal tissue in a root or eudicot stem.

cortical nephron In mammals and birds, a nephron with a loop of Henle located almost entirely in the renal cortex.

cotransport The coupling of the "downhill" diffusion of one substance to the "uphill" transport of another against its own concentration gradient.

cotyledon (kot′-uh-lē′-dun) A seed leaf of an angiosperm embryo. Some species have one cotyledon, others two.

countercurrent exchange The exchange of a substance or heat between two fluids flowing in opposite directions. For example, blood in a fish gill flows in the opposite direction of water passing over the gill, maximizing diffusion of oxygen into and carbon dioxide out of the blood.

countercurrent multiplier system A countercurrent system in which energy is expended in active transport to facilitate exchange of materials and generate concentration gradients.

covalent bond (kō-vā′-lent) A type of strong chemical bond in which two atoms share one or more pairs of valence electrons.

crassulacean acid metabolism (CAM) (crass-yū-lā′-shen) An adaptation for photosynthesis in arid conditions, first discovered in the family Crassulaceae. In this process, a plant takes up CO_2 and incorporates it into a variety of organic acids at night; during the day, CO_2 is released from organic acids for use in the Calvin cycle.

CRISPR-Cas9 system A technique for editing genes in living cells, involving a bacterial protein called Cas9 associated with a guide RNA complementary to a gene sequence of interest.

crista (plural, **cristae**) (kris′-tuh, kris′-tē) An infolding of the inner membrane of a mitochondrion. The inner membrane houses electron transport chains and molecules of the enzyme catalyzing the synthesis of ATP (ATP synthase).

critical load The amount of added nutrient, usually nitrogen or phosphorus, that can be absorbed by plants without damaging ecosystem integrity.

crop rotation The practice of growing different crops in succession on the same land chiefly to preserve the productive capacity of the soil.

cross-fostering study A behavioral study in which the young of one species are placed in the care of adults from another species.

crossing over The reciprocal exchange of genetic material between nonsister chromatids during prophase I of meiosis.

cross-pollination In angiosperms, the transfer of pollen from an anther of a flower on one plant to the stigma of a flower on another plant of the same species.

cryptic coloration Camouflage that makes a potential prey difficult to spot against its background.

cryptomycete A member of the fungal phylum Cryptomycota, unicellular fungi that have flagellated spores; cryptomycetes and their sister taxon (microsporidians) are a basal fungal lineage.

culture A system of information transfer through social learning or teaching that influences the behavior of individuals in a population.

cuticle (kyū'-tuh-kul) Any of a variety of tough but flexible, non-mineral outer coverings of an organism, or parts of an organism, which provide protection.

cyclic AMP (cAMP) Cyclic adenosine monophosphate, named because of its ring structure, is a common chemical signal that has a diversity of roles, including as a second messenger in many eukaryotic cells, and as a regulator of some bacterial operons.

cyclic electron flow A route of electron flow during the light reactions of photosynthesis that involves only one photosystem and that produces ATP but not NADPH or O_2.

cyclin (sī'-klin) A cellular protein that occurs in a cyclically fluctuating concentration and that plays an important role in regulating the cell cycle.

cyclin-dependent kinase (Cdk) (sī'-klin) A protein kinase that is active only when attached to a particular cyclin.

cyclostome (sī'-cluh-stōm) Member of one of the two main clades of vertebrates; cyclostomes lack jaws and include lampreys and hagfishes. *See also* gnathostome.

cystic fibrosis (sis'-tik fī-brō'-sis) A human genetic disorder caused by a recessive allele for a chloride channel protein; characterized by an excessive secretion of mucus and consequent vulnerability to infection; fatal if untreated.

cytochrome (sī'-tō-krōm) An iron-containing protein that is a component of electron transport chains in the mitochondria and chloroplasts of eukaryotic cells and the plasma membranes of prokaryotic cells.

cytokine (sī'-tō-kīn') Any of a group of small proteins secreted by a number of cell types, including macrophages and helper T cells, that regulate the function of other cells.

cytokinesis (sī'-tō-kuh-nē'-sis) The division of the cytoplasm to form two separate daughter cells immediately after mitosis, meiosis I, or meiosis II.

cytokinin (sī'-tō-kī'-nin) Any of a class of related plant hormones that retard aging and act in concert with auxin to stimulate cell division, influence the pathway of differentiation, and control apical dominance.

cytoplasm (sī'-tō-plaz-um) The contents of the cell bounded by the plasma membrane; in eukaryotes, the portion exclusive of the nucleus.

cytoplasmic determinant A maternal substance, such as a protein or RNA, that when placed into an egg influences the course of early development by regulating the expression of genes that affect the developmental fate of cells.

cytoplasmic streaming A circular flow of cytoplasm, involving interactions of myosin and actin filaments, that speeds the distribution of materials within cells.

cytoskeleton A network of microtubules, microfilaments, and intermediate filaments that extend throughout the cytoplasm and serve a variety of mechanical, transport, and signaling functions.

cytosol (sī'-tō-sol) The semifluid portion of the cytoplasm.

cytotoxic T cell A type of lymphocyte that, when activated, kills infected cells as well as certain cancer cells and transplanted cells.

dalton A measure of mass for atoms and subatomic particles; the same as the atomic mass unit, or amu.

data Recorded observations.

day-neutral plant A plant in which flower formation is not controlled by photoperiod or day length.

decomposer An organism that absorbs nutrients from nonliving organic material such as corpses, fallen plant material, and the wastes of living organisms and converts them to inorganic forms; a detritivore.

deductive reasoning A type of logic in which specific results are predicted from a general premise.

de-etiolation The changes a plant shoot undergoes in response to sunlight; also known informally as greening.

dehydration reaction A chemical reaction in which two molecules become covalently bonded to each other with the removal of a water molecule.

deletion (1) A deficiency in a chromosome resulting from the loss of a fragment through breakage. (2) A mutational loss of one or more nucleotide pairs from a gene.

demographic transition In a stable population, a shift from high birth and death rates to low birth and death rates.

demography The study of changes over time in the vital statistics of populations, especially birth rates and death rates.

denaturation (dē-nā'-chur-ā'-shun) In proteins, a process in which a protein loses its native shape due to the disruption of weak chemical bonds and interactions, thereby becoming biologically inactive; in DNA, the separation of the two strands of the double helix. Denaturation occurs under extreme (noncellular) conditions of pH, salt concentration, or temperature.

dendrite (den'-drīt) One of usually numerous, short, highly branched extensions of a neuron that receive signals from other neurons.

dendritic cell An antigen-presenting cell, located mainly in lymphatic tissues and skin, that is particularly efficient in presenting antigens to helper T cells, thereby initiating a primary immune response.

density The number of individuals per unit area or volume.

density dependent Referring to any characteristic that varies with population density.

density-dependent inhibition The phenomenon observed in normal animal cells that causes them to stop dividing when they come into contact with one another.

density independent Referring to any characteristic that is not affected by population density.

deoxyribonucleic acid (DNA) (dē-ok'-sē-rī'-bō-nū-klā'-ik) A nucleic acid molecule, usually a double-stranded helix, in which each polynucleotide strand consists of nucleotide monomers with a deoxyribose sugar and the nitrogenous bases adenine (A), cytosine (C), guanine (G), and thymine (T); capable of being replicated and determining the inherited structure of a cell's proteins.

deoxyribose (dē-ok'-si-rī'-bōs) The sugar component of DNA nucleotides, having one fewer hydroxyl group than ribose, the sugar component of RNA nucleotides.

dependent variable A factor whose value is measured during an experiment or other test to see whether it is influenced by changes in another factor (the independent variable).

depolarization A change in a cell's membrane potential such that the inside of the membrane is made less negative relative to the outside. For example, a neuron membrane is depolarized if a stimulus decreases its voltage from the resting potential of -70 mV in the direction of zero voltage.

dermal tissue The outer protective covering of plants.

desert A terrestrial biome characterized by very low precipitation.

desmosome A type of intercellular junction in animal cells that functions as a rivet, fastening cells together.

determinate cleavage A type of embryonic development in protostomes that rigidly casts the developmental fate of each embryonic cell very early.

determinate growth A type of growth characteristic of most animals and some plant organs, in which growth stops after a certain size is reached.

determination The progressive restriction of developmental potential in which the possible fate of each cell becomes more limited as an embryo develops. At the end of determination, a cell is committed to its fate.

detritus (di-trī'-tus) Dead organic matter.

deuteromycete (dū'-tuh-rō-mī'-sēt) Traditional classification for a fungus with no known sexual stage.

deuterostome development (dū'-tuh-rō-stōm') In animals, a developmental mode distinguished by the development of the anus from the blastopore; often also characterized by radial cleavage and by the body cavity forming as outpockets of mesodermal tissue.

Deuterostomia (dū'-tuh-rō-stōm'-ē-uh) One of the three main lineages of bilaterian animals. *See also* Ecdysozoa and Lophotrochozoa.

development The events involved in an organism's changing gradually from a simple to a more complex or specialized form.

diabetes mellitus (dī'-uh-bē'-tis mel'-uh-tus) An endocrine disorder marked by an inability to maintain glucose homeostasis. The type 1 form results from autoimmune destruction of insulin-secreting cells; treatment usually requires daily insulin injections. The type 2 form most commonly results from reduced responsiveness of target cells to insulin; obesity and lack of exercise are risk factors.

diacylglycerol (DAG) (dī-a'-sil-glis'-er-ol) A second messenger produced by the cleavage of the phospholipid PIP_2 in the plasma membrane.

diaphragm (dī'-uh-fram') (1) A sheet of muscle that forms the bottom wall of the thoracic cavity in mammals. Contraction of the diaphragm pulls air into the lungs. (2) A dome-shaped rubber cup fitted into the upper portion of the vagina before sexual intercourse. It serves as a physical barrier to the passage of sperm into the uterus.

diapsid (dī-ap′-sid) A member of an amniote clade distinguished by a pair of holes on each side of the skull. Diapsids include the lepidosaurs and archosaurs.

diastole (dī-as′-tō-lē) The stage of the cardiac cycle in which a heart chamber is relaxed and fills with blood.

diastolic pressure Blood pressure in the arteries when the ventricles are relaxed.

diatom Photosynthetic protist in the stramenopile clade; diatoms have a unique glass-like wall made of silicon dioxide embedded in an organic matrix.

dicot A term traditionally used to refer to flowering plants that have two embryonic seed leaves, or cotyledons. Recent molecular evidence indicates that dicots do not form a clade; species once classified as dicots are now grouped into eudicots, magnoliids, and several lineages of basal angiosperms.

differential gene expression The expression of different sets of genes by cells with the same genome.

differentiation The process by which a cell or group of cells becomes specialized in structure and function.

diffusion The random thermal motion of particles of liquids, gases, or solids. In the presence of a concentration or electrochemical gradient, diffusion results in the net movement of a substance from a region where it is more concentrated to a region where it is less concentrated.

digestion The second stage of food processing in animals: the breaking down of food into molecules small enough for the body to absorb.

dihybrid (dī′-hī′-brid) An organism that is heterozygous with respect to two genes of interest. All the offspring from a cross between parents doubly homozygous for different alleles are dihybrids. For example, parents of genotypes *AABB* and *aabb* produce a dihybrid of genotype *AaBb*.

dihybrid cross (dī′-hī′-brid) A cross between two organisms that are each heterozygous for both of the characters being followed (or the self-pollination of a plant that is heterozygous for both characters).

dikaryotic (dī′-kār-ē-ot′-ik) Referring to a fungal mycelium with two haploid nuclei per cell, one from each parent.

dinoflagellate (dī′-nō-flaj′-uh-let) A member of a group of mostly unicellular photosynthetic algae with two flagella situated in perpendicular grooves in cellulose plates covering the cell.

dinosaur A member of an extremely diverse clade of reptiles varying in body shape, size, and habitat. Birds are the only extant dinosaurs.

dioecious (dī-ē′-shus) In plant biology, having the male and female reproductive parts on different individuals of the same species.

diploblastic Having two germ layers.

diploid cell (dip′-loyd) A cell containing two sets of chromosomes (*2n*), one set inherited from each parent.

diplomonad A protist that has modified mitochondria, two equal-sized nuclei, and multiple flagella.

directional selection Natural selection in which individuals at one end of the phenotypic range survive or reproduce more successfully than do other individuals.

disaccharide (dī-sak′-uh-rīd) A double sugar, consisting of two monosaccharides joined by a glycosidic linkage formed by a dehydration reaction.

dispersal The movement of individuals or gametes away from their parent location. This movement sometimes expands the geographic range of a population or species.

dispersion The pattern of spacing among individuals within the boundaries of a population.

disruptive selection Natural selection in which individuals on both extremes of a phenotypic range survive or reproduce more successfully than do individuals with intermediate phenotypes.

distal tubule In the vertebrate kidney, the portion of a nephron that helps refine filtrate and empties it into a collecting duct.

disturbance A natural or human-caused event that changes a biological community and usually removes organisms from it. Disturbances, such as fires and storms, play a pivotal role in structuring many communities.

disulfide bridge A strong covalent bond formed when the sulfur of one cysteine monomer bonds to the sulfur of another cysteine monomer.

DNA (deoxyribonucleic acid) (dē-ok′-sē-rī′-bō-nū-klā′-ik) A nucleic acid molecule, usually a double-stranded helix, in which each polynucleotide strand consists of nucleotide monomers with a deoxyribose sugar and the nitrogenous bases adenine (A), cytosine (C), guanine (G), and thymine (T); capable of being replicated and determining the inherited structure of a cell's proteins.

DNA cloning The production of multiple copies of a specific DNA segment.

DNA ligase (lī′-gās) A linking enzyme essential for DNA replication; catalyzes the covalent bonding of the 3′ end of one DNA fragment (such as an Okazaki fragment) to the 5′ end of another DNA fragment (such as a growing DNA chain).

DNA methylation The presence of methyl groups on the DNA bases (usually cytosine) of plants, animals, and fungi. (The term also refers to the process of adding methyl groups to DNA bases.)

DNA microarray assay A method to detect and measure the expression of thousands of genes at one time. Tiny amounts of a large number of single-stranded DNA fragments representing different genes are fixed to a glass slide and tested for hybridization with samples of labeled cDNA.

DNA polymerase (puh-lim′-er-ās) An enzyme that catalyzes the elongation of new DNA (for example, at a replication fork) by the addition of nucleotides to the 3′ end of an existing chain. There are several different DNA polymerases; DNA polymerase III and DNA polymerase I play major roles in DNA replication in *E. coli*.

DNA replication The process by which a DNA molecule is copied; also called DNA synthesis.

DNA sequencing Determining the complete nucleotide sequence of a gene or DNA segment.

DNA technology Techniques for sequencing and manipulating DNA.

domain (1) A taxonomic category above the kingdom level. The three domains are Archaea, Bacteria, and Eukarya. (2) A discrete structural and functional region of a protein.

dominant allele An allele that is fully expressed in the phenotype of a heterozygote.

dormancy A condition typified by extremely low metabolic rate and a suspension of growth and development.

dorsal In an animal with bilateral symmetry, pertaining to the top (in most animals) or back (in animals with upright posture) of the body.

dorsal lip The region above the blastopore on the dorsal side of the amphibian embryo.

double bond A double covalent bond; the sharing of two pairs of valence electrons by two atoms.

double circulation A circulatory system consisting of separate pulmonary and systemic circuits, in which blood passes through the heart after completing each circuit.

double fertilization A mechanism of fertilization in angiosperms in which two sperm cells unite with two cells in the female gametophyte (embryo sac) to form the zygote and endosperm.

double helix The form of native DNA, referring to its two adjacent antiparallel polynucleotide strands wound around an imaginary axis into a spiral shape.

Down syndrome A human genetic disease usually caused by the presence of an extra chromosome 21; characterized by developmental delays and heart and other defects that are generally treatable or non-life-threatening.

Duchenne muscular dystrophy (duh-shen′) A human genetic disease caused by a sex-linked recessive allele; characterized by progressive weakening and a loss of muscle tissue.

duodenum (dū′-uh-dēn′-um) The first section of the small intestine, where chyme from the stomach mixes with digestive juices from the pancreas, liver, and gallbladder as well as from gland cells of the intestinal wall.

duplication An aberration in chromosome structure due to fusion with a fragment from a homologous chromosome, such that a portion of a chromosome is duplicated.

dynein (dī′-nē-un) In cilia and flagella, a large motor protein extending from one microtubule doublet to the adjacent doublet. ATP hydrolysis drives changes in dynein shape that lead to bending of cilia and flagella.

E site One of a ribosome's three binding sites for tRNA during translation. The E site is the place where discharged tRNAs leave the ribosome. (E stands for exit.)

Ecdysozoa (ek′-dē-sō-zō′-uh) One of the three main lineages of bilaterian animals; many ecdysozoans are molting animals. *See also* Deuterostomia and Lophotrochozoa.

echinoderm (i-kī′-nō-derm) A slow-moving or sessile marine deuterostome with a water vascular system and, in larvae, bilateral symmetry. Echinoderms include sea stars, brittle stars, sea urchins, feather stars, and sea cucumbers.

ecological footprint The aggregate land and water area required by a person, city, or nation to produce all of the resources it consumes and to absorb all of the waste it generates.

ecological niche (nich) The sum of a species' use of the biotic and abiotic resources in its environment.

ecological species concept Definition of a species in terms of ecological niche, the sum of how members of the species interact with the nonliving and living parts of their environment.

ecological succession Transition in the species composition of a community following a disturbance; establishment of a community in an area virtually barren of life.

ecology The study of how organisms interact with each other and their environment.

ecosystem All the organisms in a given area as well as the abiotic factors with which they interact; one or more communities and the physical environment around them.

ecosystem ecology The study of energy flow and the cycling of chemicals among the various biotic and abiotic components in an ecosystem.

ecosystem engineer An organism that influences community structure by causing physical changes in the environment.

ecosystem service A function performed by an ecosystem that directly or indirectly benefits humans.

ecotone The transition from one type of habitat or ecosystem to another, such as the transition from a forest to a grassland.

ectoderm (ek'-tō-durm) The outermost of the three primary germ layers in animal embryos; gives rise to the outer covering and, in some phyla, the nervous system, inner ear, and lens of the eye.

ectomycorrhiza (plural, **ectomycorrhizae**) (ek'-tō-mī'-kō-rī'-zuh, ek'-tō-mī'-kō-rī'-zē) Association of a fungus with a plant root system in which the fungus surrounds the roots but does not cause invagination of the host (plant) cell's plasma membrane.

ectomycorrhizal fungus A symbiotic fungus that forms sheaths of hyphae over the surface of plant roots and also grows into extracellular spaces of the root cortex.

ectoparasite A parasite that feeds on the external surface of a host.

ectopic Occurring in an abnormal location.

ectoproct A sessile, colonial lophotrochozoan; also called a bryozoan.

ectothermic Referring to organisms for which external sources provide most of the heat for temperature regulation.

Ediacaran biota (ē'-dē-uh-keh'-run bī-ō'-tuh) An early group of macroscopic, mostly soft-bodied, multicellular eukaryotes known from fossils that range in age from 635 million to 541 million years old.

effective population size An estimate of the size of a population based on the numbers of females and males that successfully breed; generally smaller than the total population.

effector A pathogen-encoded protein that cripples the host's innate immune system.

effector cell (1) A muscle cell or gland cell that carries out the body's response to stimuli as directed by signals from the brain or other processing center of the nervous system. (2) A lymphocyte that has undergone clonal selection and is capable of mediating an adaptive immune response.

egg The female gamete.

ejaculation The propulsion of sperm from the epididymis through the muscular vas deferens, ejaculatory duct, and urethra.

electrocardiogram (ECG or EKG) A record of the electrical impulses that travel through heart muscle during the cardiac cycle.

electrochemical gradient The diffusion gradient of an ion, which is affected by both the concentration difference of an ion across a membrane (a chemical force) and the ion's tendency to move relative to the membrane potential (an electrical force).

electrogenic pump An active transport protein that generates voltage across a membrane while pumping ions.

electromagnetic receptor A receptor of electromagnetic energy, such as visible light, electricity, or magnetism.

electromagnetic spectrum The entire spectrum of electromagnetic radiation, ranging in wavelength from less than a nanometer to more than a kilometer.

electron A subatomic particle with a single negative electrical charge and a mass about 1/2,000 that of a neutron or proton. One or more electrons move around the nucleus of an atom.

electron microscope (EM) A microscope that uses magnets to focus an electron beam on or through a specimen, resulting in a practical resolution that is 100-fold greater than that of a light microscope using standard techniques. A transmission electron microscope (TEM) is used to study the internal structure of thin sections of cells. A scanning electron microscope (SEM) is used to study the fine details of cell surfaces.

electron shell An energy level of electrons at a characteristic average distance from the nucleus of an atom.

electron transport chain A sequence of electron carrier molecules (membrane proteins) that shuttle electrons down a series of redox reactions that release energy used to make ATP.

electronegativity The attraction of a given atom for the electrons of a covalent bond.

electroporation A technique to introduce recombinant DNA into cells by applying a brief electrical pulse to a solution containing the cells. The pulse creates temporary holes in the cells' plasma membranes, through which DNA can enter.

element Any substance that cannot be broken down to any other substance by chemical reactions.

elimination The fourth and final stage of food processing in animals: the passing of undigested material out of the body.

embryo sac (em'-brē-ō) The female gametophyte of angiosperms, formed from the growth and division of the megaspore into a multicellular structure that typically has eight haploid nuclei.

embryonic lethal A mutation with a phenotype leading to death of an embryo or larva.

embryophyte Alternate name for land plants that refers to their shared derived trait of multicellular, dependent embryos.

emergent properties New properties that arise with each step upward in the hierarchy of life, owing to the arrangement and interactions of parts as complexity increases.

emigration The movement of individuals out of a population.

enantiomer (en-an'-tē-ō-mer) One of two compounds that are mirror images of each other and that differ in shape due to the presence of an asymmetric carbon.

endangered species A species that is in danger of extinction throughout all or a significant portion of its range.

endemic (en-dem'-ik) Referring to a species that is confined to a specific geographic area.

endergonic reaction (en'-der-gon'-ik) A nonspontaneous chemical reaction in which free energy is absorbed from the surroundings.

endocrine gland (en'-dō-krin) A ductless gland that secretes hormones directly into the interstitial fluid, from which they diffuse into the bloodstream.

endocrine system (en'-dō-krin) In animals, the internal system of communication involving hormones, the ductless glands that secrete hormones, and the molecular receptors on or in target cells that respond to hormones; functions in concert with the nervous system to effect internal regulation and maintain homeostasis.

endocytosis (en'-dō-sī-tō'-sis) Cellular uptake of biological molecules and particulate matter via formation of vesicles from the plasma membrane.

endoderm (en'-dō-durm) The innermost of the three primary germ layers in animal embryos; lines the archenteron and gives rise to the liver, pancreas, lungs, and the lining of the digestive tract in species that have these structures.

endodermis In plant roots, the innermost layer of the cortex that surrounds the vascular cylinder.

endomembrane system The collection of membranes inside and surrounding a eukaryotic cell, related either through direct physical contact or by the transfer of membranous vesicles; includes the plasma membrane, the nuclear envelope, the smooth and rough endoplasmic reticulum, the Golgi apparatus, lysosomes, vesicles, and vacuoles.

endometriosis (en'-dō-mē-trē-ō'-sis) The condition resulting from the presence of endometrial tissue outside of the uterus.

endometrium (en'-dō-mē'-trē-um) The inner lining of the uterus, which is richly supplied with blood vessels.

endoparasite A parasite that lives within a host.

endophyte A harmless fungus, or occasionally another organism, that lives between cells of a plant part or multicellular alga.

endoplasmic reticulum (ER) (en'-dō-plaz'-mik ruh-tik'-yū-lum) An extensive membranous network in eukaryotic cells, continuous with the outer nuclear membrane and composed of ribosome-studded (rough) and ribosome-free (smooth) regions.

endorphin (en-dōr'-fin) Any of several hormones produced in the brain and anterior pituitary that inhibit pain perception.

endoskeleton A hard skeleton buried within the soft tissues of an animal.

endosperm In angiosperms, a nutrient-rich tissue formed by the union of a sperm with two polar nuclei during double fertilization. The endosperm provides nourishment to the developing embryo in angiosperm seeds.

endospore A thick-coated, resistant cell produced by some bacterial cells when they are exposed to harsh conditions.

endosymbiont theory The theory that mitochondria and plastids originated as prokaryotic cells engulfed by a host cell. The engulfed cell and its host cell then evolved into a single organism. *See also* endosymbiosis.

endosymbiosis A relationship between two species in which one organism lives inside the cell or cells of another organism. *See also* endosymbiont theory.

endothelium (en′-dō-thē′-lē-um) The simple squamous layer of cells lining the lumen of blood vessels.

endothermic Referring to organisms that are warmed by heat generated by their own metabolism. This heat usually maintains a relatively stable body temperature higher than that of the external environment.

endotoxin A toxic component of the outer membrane of certain gram-negative bacteria that is released only when the bacteria die.

energetic hypothesis The concept that the length of a food chain is limited by the inefficiency of energy transfer along the chain.

energy The capacity to cause change, especially to do work (to move matter against an opposing force).

energy coupling In cellular metabolism, the use of energy released from an exergonic reaction to drive an endergonic reaction.

enhancer A segment of eukaryotic DNA containing multiple control elements, usually located far from the gene whose transcription it regulates.

enteric nervous system Within the autonomic nervous system, a distinct network of neurons that exerts partially independent control over the digestive tract, pancreas, and gallbladder.

entropy A measure of molecular disorder, or randomness.

enzyme (en′-zīm) A macromolecule serving as a catalyst, a chemical agent that increases the rate of a reaction without being consumed by the reaction. Most enzymes are proteins.

enzyme-substrate complex (en′-zīm) A temporary complex formed when an enzyme binds to its substrate molecule(s).

eosinophil Immune system cell that secretes destructive enzymes and helps defend against multicellular pathogens.

epicotyl (ep′-uh-kot′-ul) In an angiosperm embryo, the embryonic axis above the point of attachment of the cotyledon(s) and below the first pair of miniature leaves.

epidemic A widespread outbreak of a disease.

epidermis (1) The dermal tissue of nonwoody plants, usually consisting of a single layer of tightly packed cells. (2) The outermost layer of cells in an animal.

epididymis (ep′-uh-did′-uh-mus) A coiled tubule located adjacent to the mammalian testis where sperm are stored.

epigenetics The study of the inheritance of traits transmitted by mechanisms that do not involve the nucleotide sequence.

epinephrine (ep′-i-nef′-rin) A catecholamine that, when secreted as a hormone by the adrenal medulla, mediates "fight-or-flight" responses to short-term stresses; also released by some neurons as a neurotransmitter; also called adrenaline.

epiphyte (ep′-uh-fīt) A plant that nourishes itself but grows on the surface of another plant for support, usually on the branches or trunks of trees.

epistasis (ep′-i-stā′-sis) A type of gene interaction in which the phenotypic expression of one gene alters that of another independently inherited gene.

epithelial tissue (ep′-uh-thē′-lē-ul) Sheets of tightly packed cells that line organs and body cavities as well as external surfaces.

epithelium (plural, **epithelia**) An epithelial tissue.

epitope A small, accessible region of an antigen to which an antigen receptor or antibody binds.

equilibrium potential (E_{ion}) The magnitude of a cell's membrane voltage at equilibrium; calculated using the Nernst equation.

erythrocyte (eh-rith′-ruh-sīt) A blood cell that contains hemoglobin, which transports oxygen; also called a red blood cell.

erythropoietin (EPO) (eh-rith′-rō-poy′-uh-tin) A hormone that stimulates the production of erythrocytes. It is secreted by the kidney when body tissues do not receive enough oxygen.

esophagus (eh-sof′-uh-gus) A muscular tube that conducts food, by peristalsis, from the pharynx to the stomach.

essential amino acid An amino acid that an animal cannot synthesize itself and must be obtained from food in prefabricated form.

essential element A chemical element required for an organism to survive, grow, and reproduce.

essential fatty acid An unsaturated fatty acid that an animal needs but cannot make.

essential nutrient A substance that an organism cannot synthesize from any other material and therefore must absorb in preassembled form.

estradiol (es′-truh-dī′-ol) A steroid hormone that stimulates the development and maintenance of the female reproductive system and secondary sex characteristics; the major estrogen in mammals.

estrogen (es′-trō-jen) Any steroid hormone, such as estradiol, that stimulates the development and maintenance of the female reproductive system and secondary sex characteristics.

estrous cycle (es′-trus) A reproductive cycle characteristic of female mammals except humans and certain other primates, in which the endometrium is reabsorbed in the absence of pregnancy and sexual response occurs only during a mid-cycle point known as estrus.

estuary The area where a freshwater stream or river merges with the ocean.

ethylene (eth′-uh-lēn) A gaseous plant hormone involved in responses to mechanical stress, programmed cell death, leaf abscission, and fruit ripening.

etiolation Plant morphological adaptations for growing in darkness.

euchromatin (yū-krō′-muh-tin) The less condensed form of eukaryotic chromatin that is available for transcription.

eudicot (yū-dī′-kot) A member of a clade that contains the vast majority of flowering plants that have two embryonic seed leaves, or cotyledons.

euglenid (yū′-glen-id) A protist, such as *Euglena* or its relatives, characterized by an anterior pocket from which one or two flagella emerge.

euglenozoan A member of a diverse clade of flagellated protists that includes predatory heterotrophs, photosynthetic autotrophs, and pathogenic parasites.

Eukarya (yū-kār′-ē-uh) The domain that includes all eukaryotic organisms.

eukaryote A single-celled or multicellular organism comprised of eukaryotic cells; eukaryotes include protists, plants, fungi, and animals.

eukaryotic cell (yū′-kār-ē-ot′-ik) A type of cell with a membrane-enclosed nucleus and membrane-enclosed organelles. Organisms with eukaryotic cells (protists, plants, fungi, and animals) are called eukaryotes.

eumetazoan (yū′-met-uh-zō′-un) A member of a clade of animals with true tissues. All animals except sponges and a few other groups are eumetazoans.

eurypterid (yur-ip′-tuh-rid) An extinct carnivorous chelicerate; also called a water scorpion.

Eustachian tube (yū-stā′-shun) The tube that connects the middle ear to the pharynx.

eutherian (yū-thēr′-ē-un) Placental mammal; mammal whose young complete their embryonic development within the uterus, joined to the mother by the placenta.

eutrophic lake (yū-trōf′-ik) A lake that has a high rate of biological productivity supported by a high rate of nutrient cycling.

eutrophication A process by which nutrients, particularly phosphorus and nitrogen, become highly concentrated in a body of water, leading to increased growth of organisms such as algae or cyanobacteria.

evaporative cooling The process in which the surface of an object becomes cooler during evaporation, a result of the molecules with the greatest kinetic energy changing from the liquid to the gaseous state.

evapotranspiration The total evaporation of water from an ecosystem, including water transpired by plants and evaporated from a landscape, usually measured in millimeters and estimated for a year.

evo-devo Evolutionary developmental biology; a field of biology that compares developmental processes of different multicellular organisms to understand how these processes have evolved and how changes can modify existing organismal features or lead to new ones.

evolution Descent with modification; the process by which species accumulate differences from their ancestors as they adapt to different environments over time; also defined as a change in the genetic composition of a population from generation to generation.

evolutionary lineage The sequence of ancestral organisms leading to a particular taxon;

represented by a branch (line) in a phylogenetic tree.

evolutionary tree A branching diagram that reflects a hypothesis about evolutionary relationships among groups of organisms.

Excavata (ex'-kuh-vah'-tuh) One of four supergroups of eukaryotes proposed in a current hypothesis of the evolutionary history of eukaryotes. Excavates have unique cytoskeletal features, and some species have an "excavated" feeding groove on one side of the cell body. *See also* SAR, Archaeplastida, and Unikonta.

excitatory postsynaptic potential (EPSP) An electrical change (depolarization) in the membrane of a postsynaptic cell caused by the binding of an excitatory neurotransmitter from a presynaptic cell to a postsynaptic receptor; makes it more likely for a postsynaptic cell to generate an action potential.

excretion The disposal of nitrogen-containing metabolites and other waste products.

exergonic reaction (ek'-ser-gon'-ik) A spontaneous chemical reaction in which there is a net release of free energy.

exocytosis (ek'-sō-sī-tō'-sis) The cellular secretion of biological molecules by the fusion of vesicles containing them with the plasma membrane.

exon A sequence within a primary transcript that remains in the RNA after RNA processing; also refers to the region of DNA from which this sequence was transcribed.

exoskeleton A hard encasement on the surface of an animal, such as the shell of a mollusc or the cuticle of an arthropod, that provides protection and points of attachment for muscles.

exotoxin (ek'-sō-tok'-sin) A toxic protein that is secreted by a prokaryote or other pathogen and that produces specific symptoms, even if the pathogen is no longer present.

expansin Plant enzyme that breaks the cross-links (hydrogen bonds) between cellulose microfibrils and other cell wall constituents, loosening the wall's fabric.

experiment A scientific test. Often carried out under controlled conditions that involve manipulating one factor in a system in order to see the effects of changing that factor.

experimental group A set of subjects that has (or receives) the specific factor being tested in a controlled experiment. Ideally, the experimental group is identical to the control group for all other factors.

exploitation A +/− ecological interaction in which individuals of one species benefit by feeding on (and thereby harming) individuals of the other species. Exploitative interactions include predation, herbivory, and parasitism.

exponential population growth Growth of a population in an ideal, unlimited environment, represented by a J-shaped curve when population size is plotted over time.

expression vector A cloning vector that contains a highly active bacterial promoter just upstream of a restriction site where a eukaryotic gene can be inserted, allowing the gene to be expressed in a bacterial cell. Expression vectors are also available that have been genetically engineered for use in specific types of eukaryotic cells.

extinction vortex A downward population spiral in which inbreeding and genetic drift combine to cause a small population to shrink and, unless the spiral is reversed, become extinct.

extracellular matrix (ECM) The meshwork surrounding animal cells, consisting of glycoproteins, polysaccharides, and proteoglycans synthesized and secreted by cells.

extraembryonic membrane One of four membranes (yolk sac, amnion, chorion, and allantois) located outside the embryo that support the developing embryo in reptiles and mammals.

extreme halophile An organism that lives in a highly saline environment, such as the Great Salt Lake or the Dead Sea.

extreme thermophile An organism that thrives in hot environments (often 60–80°C or hotter).

extremophile An organism that lives in environmental conditions so extreme that few other species can survive there. Extremophiles include extreme halophiles ("salt lovers") and extreme thermophiles ("heat lovers").

F_1 generation The first filial, hybrid (heterozygous) offspring arising from a parental (P generation) cross.

F_2 generation The offspring resulting from interbreeding (or self-pollination) of the hybrid F_1 generation.

facilitated diffusion The passage of molecules or ions down their electrochemical gradient across a biological membrane with the assistance of specific transmembrane transport proteins, requiring no energy expenditure.

facultative anaerobe (fak'-ul-tā'-tiv an'-uh-rōb) An organism that makes ATP by aerobic respiration if oxygen is present but that switches to anaerobic respiration or fermentation if oxygen is not present.

family In Linnaean classification, the taxonomic category above genus.

fast-twitch fiber A muscle fiber used for rapid, powerful contractions.

fat A lipid consisting of three fatty acids linked to one glycerol molecule; also called a triacylglycerol or triglyceride.

fate map A territorial diagram of embryonic development that displays the future derivatives of individual cells and tissues.

fatty acid A carboxylic acid with a long carbon chain. Fatty acids vary in length and in the number and location of double bonds; three fatty acids linked to a glycerol molecule form a fat molecule, also called triacylglycerol or triglyceride.

feces (fē'-sēz) The wastes of the digestive tract.

feedback inhibition A method of metabolic control in which the end product of a metabolic pathway acts as an inhibitor of an enzyme within that pathway.

feedback regulation The regulation of a process by its output or end product.

fermentation A catabolic process that makes a limited amount of ATP from glucose (or other organic molecules) without an electron transport chain and that produces a characteristic end product, such as ethyl alcohol or lactic acid.

fertilization (1) The union of haploid gametes to produce a diploid zygote. (2) The addition of mineral nutrients to the soil.

fetus (fē'-tus) A developing mammal that has all the major structures of an adult. In humans, the fetal stage lasts from the 9th week of gestation until birth.

F factor In bacteria, the DNA segment that confers the ability to form pili for conjugation and associated functions required for the transfer of DNA from donor to recipient. The F factor may exist as a plasmid or be integrated into the bacterial chromosome.

fibroblast (fī'-brō-blast) A type of cell in loose connective tissue that secretes the protein ingredients of the extracellular fibers.

fibronectin An extracellular glycoprotein secreted by animal cells that helps them attach to the extracellular matrix.

filament In an angiosperm, the stalk portion of the stamen, the pollen-producing reproductive organ of a flower.

filter feeder An animal that feeds by using a filtration mechanism to strain small organisms or food particles from its surroundings.

filtrate Cell-free fluid extracted from the body fluid by the excretory system.

filtration In excretory systems, the extraction of water and small solutes, including metabolic wastes, from the body fluid.

fimbria (plural, **fimbriae**) A short, hairlike appendage of a prokaryotic cell that helps it adhere to the substrate or to other cells.

first law of thermodynamics The principle of conservation of energy: Energy can be transferred and transformed, but it cannot be created or destroyed.

fission The separation of an organism into two or more individuals of approximately equal size.

fixed action pattern In animal behavior, a sequence of unlearned acts that is essentially unchangeable and, once initiated, usually carried to completion.

flaccid (flas'-id) Limp. Lacking turgor (stiffness or firmness), as in a plant cell in surroundings where there is a tendency for water to leave the cell. (A walled cell becomes flaccid if it has a higher water potential than its surroundings, resulting in the loss of water.)

flagellum (fluh-jel'-um) (plural, **flagella**) A long cellular appendage specialized for locomotion. Like motile cilia, eukaryotic flagella have a core with nine outer doublet microtubules and two inner single microtubules (the "9 + 2" arrangement) ensheathed in an extension of the plasma membrane. Prokaryotic flagella have a different structure.

florigen A flowering signal, probably a protein, that is made in leaves under certain conditions and that travels to the shoot apical meristems, inducing them to switch from vegetative to reproductive growth.

flower In an angiosperm, a specialized shoot with up to four sets of modified leaves, bearing structures that function in sexual reproduction.

fluid feeder An animal that lives by sucking nutrient-rich fluids from another living organism.

fluid mosaic model The currently accepted model of cell membrane structure, which envisions the membrane as a mosaic of protein molecules drifting laterally in a fluid bilayer of phospholipids.

follicle (fol'-uh-kul) A microscopic structure in the ovary that contains the developing oocyte and secretes estrogens.

follicle-stimulating hormone (FSH) (fol'-uh-kul) A tropic hormone that is produced and secreted by the anterior pituitary and that stimulates the production of eggs by the ovaries and sperm by the testes.

food chain The pathway along which food energy is transferred from trophic level to trophic level, beginning with producers.

food vacuole A membranous sac formed by phagocytosis of microorganisms or particles to be used as food by the cell.

food web The interconnected feeding relationships in an ecosystem.

foot (1) The portion of a bryophyte sporophyte that gathers sugars, amino acids, water, and minerals from the parent gametophyte via transfer cells. (2) One of the three main parts of a mollusc; a muscular structure usually used for movement. *See also* mantle and visceral mass.

foraging The seeking and obtaining of food.

foram (foraminiferan) An aquatic protist that secretes a hardened shell containing calcium carbonate and extends pseudopodia through pores in the shell.

forebrain One of three ancestral and embryonic regions of the vertebrate brain; develops into the thalamus, hypothalamus, and cerebrum.

fossil A preserved remnant or impression of an organism that lived in the past.

foundation species A species that has strong effects on its community as a result of its large size, high abundance, or pivotal role in community dynamics. Foundation species may provide significant habitat or food for other species; they may also be competitively dominant in exploiting key resources.

founder effect Genetic drift that occurs when a few individuals become isolated from a larger population and form a new population whose gene pool composition is not reflective of that of the original population.

fovea (fō'-vē-uh) The place on the retina at the eye's center of focus, where cones are highly concentrated.

F plasmid The plasmid form of the F factor.

fragmentation A means of asexual reproduction whereby a single parent breaks into parts that regenerate into whole new individuals.

frameshift mutation A mutation occurring when nucleotides are inserted in or deleted from a gene and the number inserted or deleted is not a multiple of three, resulting in the improper grouping of the subsequent nucleotides into codons.

free energy The portion of a biological system's energy that can perform work when temperature and pressure are uniform throughout the system. The change in free energy of a system (ΔG) is calculated by the equation $\Delta G = \Delta H - T\Delta S$, where ΔH is the change in enthalpy (in biological systems, equivalent to total energy), ΔT is the absolute temperature, and ΔS is the change in entropy.

frequency-dependent selection Selection in which the fitness of a phenotype depends on how common the phenotype is in a population.

fruit A mature ovary of a flower. The fruit protects dormant seeds and often functions in their dispersal.

functional group A specific configuration of atoms commonly attached to the carbon skeletons of organic molecules and involved in chemical reactions.

fusion In evolutionary biology, a process in which gene flow between two species that can form hybrid offspring weakens barriers to reproduction between the species. This process causes their gene pools to become increasingly alike and can cause the two species to fuse into a single species.

G_0 phase A nondividing state occupied by cells that have left the cell cycle, sometimes reversibly.

G_1 phase The first gap, or growth phase, of the cell cycle, consisting of the portion of interphase before DNA synthesis begins.

G_2 phase The second gap, or growth phase, of the cell cycle, consisting of the portion of interphase after DNA synthesis occurs.

gallbladder An organ that stores bile and releases it as needed into the small intestine.

game theory An approach to evaluating alternative strategies in situations where the outcome of a particular strategy depends on the strategies used by other individuals.

gametangium (gam'-uh-tan'-jē-um) (plural, **gametangia**) Multicellular plant structure in which gametes are formed. Female gametangia are called archegonia, and male gametangia are called antheridia.

gamete (gam'-ēt) A haploid reproductive cell, such as an egg or sperm, that is formed by meiosis or is the descendant of cells formed by meiosis. Gametes unite during sexual reproduction to produce a diploid zygote.

gametogenesis (guh-mē'-tō-gen'-uh-sis) The process by which gametes are produced.

gametophyte (guh-mē'-tō-fīt) In organisms (plants and some algae) that have alternation of generations, the multicellular haploid form that produces haploid gametes by mitosis. The haploid gametes unite and develop into sporophytes.

ganglion (gan'-glē-uhn) (plural, **ganglia**) A cluster (functional group) of nerve cell bodies.

gap junction A type of intercellular junction in animal cells, consisting of proteins surrounding a pore that allows the passage of materials between cells.

gas exchange The uptake of molecular oxygen from the environment and the discharge of carbon dioxide to the environment.

gastric juice A digestive fluid secreted by the stomach.

gastrovascular cavity A central cavity with a single opening in the body of certain animals, including cnidarians and flatworms, that functions in both the digestion and distribution of nutrients.

gastrula (gas'-trū-luh) An embryonic stage in animal development encompassing the formation of three layers: ectoderm, mesoderm, and endoderm.

gastrulation (gas'-trū-lā'-shun) In animal development, a series of cell and tissue movements in which the blastula-stage embryo folds inward, producing a three-layered embryo, the gastrula.

gated channel A transmembrane protein channel that opens or closes in response to a particular stimulus.

gated ion channel A gated channel for a specific ion. The opening or closing of such channels may alter a cell's membrane potential.

gel electrophoresis (ē-lek'-trō-fōr-ē'-sis) A technique for separating nucleic acids or proteins on the basis of their size and electrical charge, both of which affect their rate of movement through an electric field in a gel made of agarose or another polymer.

gene A discrete unit of hereditary information consisting of a specific nucleotide sequence in DNA (or RNA, in some viruses).

gene annotation Analysis of genomic sequences to identify protein-coding genes and determine the function of their products.

gene cloning The production of multiple copies of a gene.

gene drive A process that biases inheritance such that a particular allele is more likely to be inherited than are other alleles, causing the favored allele to spread (be "driven") through the population.

gene editing Altering genes in a specific, predictable way.

gene expression The process by which information encoded in DNA directs the synthesis of proteins or, in some cases, RNAs that are not translated into proteins and instead function as RNAs.

gene flow The transfer of alleles from one population to another, resulting from the movement of fertile individuals or their gametes.

gene pool The aggregate of all copies of every type of allele at all loci in every individual in a population. The term is also used in a more restricted sense as the aggregate of alleles for just one or a few loci in a population.

gene therapy The introduction of genes into an afflicted individual for therapeutic purposes.

genetic drift A process in which chance events cause unpredictable fluctuations in allele frequencies from one generation to the next. Effects of genetic drift are most pronounced in small populations.

genetic engineering The direct manipulation of genes for practical purposes.

genetic map An ordered list of genetic loci (genes or other genetic markers) along a chromosome.

genetic profile An individual's unique set of genetic markers, detected most often today by PCR or, previously, by electrophoresis and nucleic acid probes.

genetic recombination General term for the production of offspring with combinations of traits that differ from those found in either parent.

genetic variation Differences among individuals in the composition of their genes or other DNA sequences.

genetically modified organism (GMO) An organism that has acquired one or more genes by artificial means.

genetics The scientific study of heredity and hereditary variation.

genome (jē'-nōm) The genetic material of an organism or virus; the complete complement of an organism's or virus's genes along with its noncoding nucleic acid sequences.

genome-wide association study (jē'-nōm) A large-scale analysis of the genomes of many people having a certain phenotype or disease, with the aim of finding genetic markers that correlate with that phenotype or disease.

genomic imprinting (juh-nō'-mik) A phenomenon in which expression of an allele in offspring depends on whether the allele is inherited from the male or female parent.

genomics (juh-nō'-miks) The systematic study of whole sets of genes (or other DNA) and their interactions within a species, as well as genome comparisons between species.

genotype (jē'-nō-tīp) The genetic makeup, or set of alleles, of an organism.

genus (jē'-nus) (plural, **genera**) A taxonomic category above the species level, designated by the first word of a species' two-part scientific name.

geologic record A standard time scale dividing Earth's history into time periods, grouped into four eons—Hadean, Archaean, Proterozoic, and Phanerozoic—and further subdivided into eras, periods, and epochs.

germ layer One of the three main layers in a gastrula that will form the various tissues and organs of an animal body.

gestation (jes-tā'-shun) Pregnancy; the condition of carrying one or more embryos in the uterus.

gibberellin (jib'-uh-rel'-in) Any of a class of related plant hormones that stimulate growth in the stem and leaves, trigger the germination of seeds and breaking of bud dormancy, and (with auxin) stimulate fruit development.

glans In humans, the rounded head of the penis (in males) or of the clitoris (in females); the glans is highly sensitive to stimulation.

glia (glial cells) Cells of the nervous system that support, regulate, and augment the functions of neurons.

global ecology The study of the functioning and distribution of organisms across the biosphere and how the regional exchange of energy and materials affects them.

glomeromycete (glō'-mer-ō-mī'-sēt) A member of the fungal phylum Glomeromycota, characterized by a distinct branching form of mycorrhizae called arbuscular mycorrhizae.

glomerulus (glō-mār'-yū-lus) A ball of capillaries surrounded by Bowman's capsule in the nephron and serving as the site of filtration in the vertebrate kidney.

glucocorticoid A steroid hormone that is secreted by the adrenal cortex and that influences glucose metabolism and immune function.

glucagon (glū'-kuh-gon) A hormone secreted by the pancreas that raises blood glucose levels. It promotes glycogen breakdown and release of glucose by the liver.

glyceraldehyde 3-phosphate (G3P) (glis'-er-al'-de-hīd) A three-carbon carbohydrate that is the direct product of the Calvin cycle; it is also an intermediate in glycolysis.

glycogen (glī'-kō-jen) An extensively branched glucose storage polysaccharide found in the liver and muscle of animals; the animal equivalent of starch.

glycolipid A lipid with one or more covalently attached carbohydrates.

glycolysis (glī-kol'-uh-sis) A series of reactions that ultimately splits glucose into pyruvate. Glycolysis occurs in almost all living cells, serving as the starting point for fermentation or cellular respiration.

glycoprotein A protein with one or more covalently attached carbohydrates.

glycosidic linkage A covalent bond formed between two monosaccharides by a dehydration reaction.

gnathostome (na'-thu-stōm) Member of one of the two main clades of vertebrates; gnathostomes have jaws and include sharks and rays, ray-finned fishes, coelacanths, lungfishes, amphibians, reptiles, and mammals. *See also* cyclostome.

Golgi apparatus (gol'-jē) An organelle in eukaryotic cells consisting of stacks of flat membranous sacs that modify, store, and route products of the endoplasmic reticulum and synthesize some products, notably noncellulose carbohydrates.

gonad (gō'-nad) A male or female gamete-producing organ.

G protein A GTP-binding protein that relays signals from a plasma membrane signal receptor, known as a G protein-coupled receptor, to other signal transduction proteins inside the cell.

G protein-coupled receptor (GPCR) A signal receptor protein in the plasma membrane that responds to the binding of a signaling molecule by activating a G protein. Also called a G protein-linked receptor.

graded potential An electrical response of a cell to a stimulus, consisting of a change in voltage across the membrane proportional to the stimulus strength.

Gram stain A staining method that distinguishes between two different kinds of bacterial cell walls; may be used to help determine medical response to an infection.

gram-negative Describing the group of bacteria that have a cell wall that is structurally more complex and contains less peptidoglycan than the cell wall of gram-positive bacteria. Gram-negative bacteria are often more toxic than gram-positive bacteria.

gram-positive Describing the group of bacteria that have a cell wall that is structurally less complex and contains more peptidoglycan than the cell wall of gram-negative bacteria. Gram-positive bacteria are usually less toxic than gram-negative bacteria.

granum (gran'-um) (plural, **grana**) A stack of membrane-bounded thylakoids in the chloroplast. Grana function in the light reactions of photosynthesis.

gravitropism (grav'-uh-trō'-pizm) A response of a plant or animal to gravity.

gray matter Regions of clustered neuron cell bodies within the CNS.

green alga A photosynthetic protist, named for green chloroplasts that are similar in structure and pigment composition to the chloroplasts of plants. Green algae are a paraphyletic group; some members are more closely related to plants than they are to other green algae.

greenhouse effect The warming of Earth due to the atmospheric accumulation of carbon dioxide and certain other gases, which absorb reflected infrared radiation and reradiate some of it back toward Earth.

gross primary production (GPP) The total primary production of an ecosystem.

ground tissue Plant tissue that is neither vascular nor dermal, fulfilling a variety of functions, such as storage, photosynthesis, and support.

growth factor (1) A protein that must be present in the extracellular environment (culture medium or animal body) for the growth and normal development of certain types of cells. (2) A local regulator that acts on nearby cells to stimulate cell proliferation and differentiation.

growth hormone (GH) A hormone that is produced and secreted by the anterior pituitary and that has both direct (nontropic) and tropic effects on a wide variety of tissues.

guard cells The two cells that flank the stomatal pore and regulate the opening and closing of the pore.

gustation The sense of taste.

guttation The exudation of water droplets from leaves, caused by root pressure in certain plants.

gymnosperm (jim'-nō-sperm) A vascular plant that bears naked seeds—seeds not enclosed in protective chambers.

hagfish Marine jawless vertebrates that have highly reduced vertebrae and a skull made of cartilage; most hagfishes are bottom-dwelling scavengers.

hair cell A mechanosensory cell that alters output to the nervous system when hairlike projections on the cell surface are displaced.

half-life The amount of time it takes for 50% of a sample of a radioactive isotope to decay.

halophile See extreme halophile.

Hamilton's rule The principle that for natural selection to favor an altruistic act, the benefit to the recipient, devalued by the coefficient of relatedness, must exceed the cost to the altruist.

haploid cell (hap'-loyd) A cell containing only one set of chromosomes (n).

Hardy-Weinberg equilibrium The state of a population in which frequencies of alleles and genotypes remain constant from generation to generation, provided that only Mendelian segregation and recombination of alleles are at work.

heart A muscular pump that uses metabolic energy to elevate the hydrostatic pressure of the circulatory fluid (blood or hemolymph). The fluid then flows down a pressure gradient through the body and eventually returns to the heart.

heart attack The damage or death of cardiac muscle tissue resulting from prolonged blockage of one or more coronary arteries.

heart murmur A hissing sound that most often results from blood squirting backward through a leaky valve in the heart.

heart rate The frequency of heart contraction (in beats per minute).

heat Thermal energy in transfer from one body of matter to another.

heat of vaporization The quantity of heat a liquid must absorb for 1 g of it to be converted from the liquid to the gaseous state.

heat-shock protein A protein that helps protect other proteins during heat stress.

Heat-shock proteins are found in plants, animals, and microorganisms.

heavy chain One of the two types of polypeptide chains that make up an antibody molecule and B cell receptor; consists of a variable region, which contributes to the antigen-binding site, and a constant region.

helicase An enzyme that untwists the double helix of DNA at replication forks, separating the two strands and making them available as template strands.

helper T cell A type of T cell that, when activated, secretes cytokines that promote the response of B cells (humoral response) and cytotoxic T cells (cell-mediated response) to antigens.

hemocoel A body cavity lined by tissue derived from mesoderm and by tissue derived from endoderm.

hemoglobin (hē′-mō-glō′-bin) An iron-containing protein in red blood cells that reversibly binds oxygen.

hemolymph (hē′-mō-limf′) In invertebrates with an open circulatory system, the body fluid that bathes tissues.

hemophilia (hē′-muh-fil′-ē-uh) A human genetic disease caused by a sex-linked recessive allele resulting in the absence of one or more blood-clotting proteins; characterized by excessive bleeding following injury.

hepatic portal vein A large vessel that conveys nutrient-laden blood from the small intestine to the liver, which regulates the blood's nutrient content.

herbivore (hur′-bi-vōr′) An animal that mainly eats plants or algae.

herbivory A +/− ecological interaction in which an organism eats part of a plant or alga.

heredity The transmission of traits from one generation to the next.

hermaphrodite (hur-maf′-ruh-dīt′) An individual that functions as both male and female in sexual reproduction by producing both sperm and eggs.

hermaphroditism (hur-maf′-rō-dī-tizm) A condition in which an individual has both female and male gonads and functions as both a male and a female in sexual reproduction by producing both sperm and eggs.

heterochromatin (het′-er-ō-krō′-muh-tin) Eukaryotic chromatin that remains highly compacted during interphase and is generally not transcribed.

heterochrony (het′-uh-rok′-ruh-nē) Evolutionary change in the timing or rate of an organism's development.

heterocyst (het′-er-ō-sist) A specialized cell that engages in nitrogen fixation in some filamentous cyanobacteria; also called a heterocyte.

heterokaryon (het′-er-ō-kār′-ē-un) A fungal mycelium that contains two or more haploid nuclei per cell.

heteromorphic (het′-er-ō-mōr′-fik) Referring to a condition in the life cycle of plants and certain algae in which the sporophyte and gametophyte generations differ in morphology.

heterosporous (het-er-os′-pōr-us) Referring to a plant species that has two kinds of spores: microspores, which develop into male gametophytes, and megaspores, which develop into female gametophytes.

heterotroph (het′-er-ō-trōf) An organism that obtains organic food molecules by eating other organisms or substances derived from them.

heterozygote An organism that has two different alleles for a gene (encoding a character).

heterozygote advantage Greater reproductive success of heterozygous individuals compared with homozygotes; tends to preserve variation in a gene pool.

heterozygous (het′-er-ō-zī′-gus) Having two different alleles for a given gene.

hibernation A long-term physiological state in which metabolism decreases, the heart and respiratory system slow down, and body temperature is maintained at a lower level than normal.

high-density lipoprotein (HDL) A particle in the blood made up of thousands of cholesterol molecules and other lipids bound to a protein. HDL scavenges excess cholesterol.

hindbrain One of three ancestral and embryonic regions of the vertebrate brain; develops into the medulla oblongata, pons, and cerebellum.

histamine (his′-tuh-mēn) A substance released by mast cells that causes blood vessels to dilate and become more permeable in inflammatory and allergic responses.

histogram A variant of a bar graph that is made for numeric data by first grouping, or "binning," the variable plotted on the x-axis into intervals of equal width. The "bins" may be integers or ranges of numbers. The height of each bar shows the percent or number of experimental subjects whose characteristics can be described by one of the intervals plotted on the x-axis.

histone (his′-tōn) A small protein with a high proportion of positively charged amino acids that binds to the negatively charged DNA and plays a key role in chromatin structure.

histone acetylation (his′-tōn) The attachment of acetyl groups to certain amino acids of histone proteins.

HIV (human immunodeficiency virus) The infectious agent that causes AIDS. HIV is a retrovirus.

holdfast A rootlike structure that anchors a seaweed.

homeobox (hō′-mē-ō-boks′) A 180-nucleotide sequence within homeotic genes and some other developmental genes that is widely conserved in animals. Related sequences occur in plants and yeasts.

homeostasis (hō′-mē-ō-stā′-sis) The steady-state physiological condition of the body.

homeotic gene (hō-mē-o′-tik) Any of the master regulatory genes that control placement and spatial organization of body parts in animals, plants, and fungi by controlling the developmental fate of groups of cells.

hominin (hō′-mi-nin) A group consisting of humans and the extinct species that are more closely related to us than to chimpanzees.

homologous chromosomes (or homologs) (hō-mol′-uh-gus) A pair of chromosomes of the same length, centromere position, and staining pattern that possess genes for the same characters at corresponding loci. One homologous chromosome is inherited from the organism's father, the other from the mother. Also called a homologous pair.

homologous pair See homologous chromosomes.

homologous structures (hō-mol′-uh-gus) Structures in different species that are similar because of common ancestry.

homologs See homologous chromosomes.

homology (hō-mol′-ō-jē) Similarity in characteristics resulting from a shared ancestry.

homoplasy (hō′-muh-play′-zē) A similar (analogous) structure or molecular sequence that has evolved independently in two species.

homosporous (hō-mos′-puh-rus) Referring to a plant species that has a single kind of spore, which typically develops into a bisexual gametophyte.

homozygote An organism that has a pair of identical alleles for a gene (encoding a character).

homozygous (hō′-mō-zī′-gus) Having two identical alleles for a given gene.

horizontal gene transfer The transfer of genes from one genome to another through mechanisms such as transposable elements, plasmid exchange, viral activity, and perhaps fusions of different organisms.

hormone In multicellular organisms, one of many types of secreted chemicals that are formed in specialized cells, travel in body fluids, and act on specific target cells in other parts of the organism, changing the target cells' functioning.

hornwort A small, herbaceous, nonvascular plant that is a member of the phylum Anthocerophyta.

host The larger participant in a symbiotic relationship, often providing a home and food source for the smaller symbiont.

host range The limited number of species whose cells can be infected by a particular virus.

Human Genome Project An international collaborative effort to map and sequence the DNA of the entire human genome.

human immunodeficiency virus (HIV) The infectious agent that causes AIDS (acquired immunodeficiency syndrome). HIV is a retrovirus.

humoral immune response (hyū′-mer-ul) The branch of adaptive immunity that involves the activation of B cells and that leads to the production of antibodies, which defend against bacteria and viruses in body fluids.

humus (hyū′-mus) Decomposing organic material that is a component of topsoil.

Huntington's disease A human genetic disease caused by a dominant allele; characterized by uncontrollable body movements and degeneration of the nervous system; usually fatal 10 to 20 years after the onset of symptoms.

hybrid Offspring that results from the mating of individuals from two different species or from two true-breeding varieties of the same species.

hybrid zone A geographic region in which members of different species meet and mate, producing at least some offspring of mixed ancestry.

hybridization In genetics, the mating, or crossing, of two true-breeding varieties.

hydration shell The sphere of water molecules around a dissolved ion.

hydrocarbon An organic molecule consisting only of carbon and hydrogen.

hydrogen bond A type of weak chemical bond that is formed when the slightly positive hydrogen atom of a polar covalent bond in one molecule is attracted to the slightly negative atom of a polar covalent bond in another molecule or in another region of the same molecule.

hydrogen ion A single proton with a charge of 1+. The dissociation of a water molecule (H_2O) leads to the generation of a hydroxide ion (OH^-) and a hydrogen ion (H^+); in water, H^+ is not found alone but associates with a water molecule to form a hydronium ion.

hydrolysis (hī-drol´-uh-sis) A chemical reaction that breaks bonds between two molecules by the addition of water; functions in disassembly of polymers to monomers.

hydronium ion A water molecule that has an extra proton bound to it; H_3O^+, commonly represented as H^+.

hydrophilic (hī´-drō-fil´-ik) Having an affinity for water.

hydrophobic (hī´-drō-fō´-bik) Having no affinity for water; tending to coalesce and form droplets in water.

hydrophobic interaction (hī´-drō-fō´-bik) A type of weak chemical interaction caused when molecules that do not mix with water coalesce to exclude water.

hydroponic culture A method in which plants are grown in mineral solutions rather than in soil.

hydrostatic skeleton A skeletal system composed of fluid held under pressure in a closed body compartment; the main skeleton of most cnidarians, flatworms, nematodes, and annelids.

hydrothermal vent An area on the seafloor where heated water and minerals from Earth's interior gush into the seawater, producing a dark, hot, oxygen-deficient environment. The producers in a hydrothermal vent community are chemoautotrophic prokaryotes.

hydroxide ion A water molecule that has lost a proton; OH^-.

hydroxyl group (hī-drok´-sil) A chemical group consisting of an oxygen atom joined to a hydrogen atom. Molecules possessing this group are soluble in water and are called alcohols.

hyperpolarization A change in a cell's membrane potential such that the inside of the membrane becomes more negative relative to the outside. Hyperpolarization reduces the chance that a neuron will transmit a nerve impulse.

hypersensitive response A plant's localized defense response to a pathogen, involving the death of cells around the site of infection.

hypertension A disorder in which blood pressure remains abnormally high.

hypertonic Referring to a solution that, when surrounding a cell, will cause the cell to lose water.

hypha (plural, **hyphae**) (hī´-fuh, hī´-fē) One of many connected filaments that collectively make up the mycelium of a fungus.

hypocotyl (hī´-puh-cot´-ul) In an angiosperm embryo, the embryonic axis below the point of attachment of the cotyledon(s) and above the radicle.

hypothalamus (hī´-pō-thal´-uh-mus) The ventral part of the vertebrate forebrain; functions in maintaining homeostasis, especially in coordinating the endocrine and nervous systems; secretes hormones of the posterior pituitary and releasing factors that regulate the anterior pituitary.

hypothesis (hī-poth´-uh-sis) A testable explanation for a set of observations based on the available data and guided by inductive reasoning. A hypothesis is narrower in scope than a theory.

hypotonic Referring to a solution that, when surrounding a cell, will cause the cell to take up water.

imbibition The uptake of water by a seed or other structure, resulting in swelling.

immigration The influx of new individuals into a population from other areas.

immune system An organism's system of defenses against agents that cause disease.

immunization The process of generating a state of immunity by artificial means. In vaccination, an inactive or weakened form of a pathogen is administered, inducing B and T cell responses and immunological memory. In passive immunization, antibodies specific for a particular pathogen are administered, conferring immediate but temporary protection.

immunoglobulin (Ig) (im´-yū-nō-glob´-yū-lin) See antibody.

imprinting In animal behavior, the formation at a specific stage in life of a long-lasting behavioral response to a specific individual or object. See also genomic imprinting.

inclusive fitness The total effect an individual has on proliferating its genes by producing its own offspring and by providing aid that enables other close relatives to increase production of their offspring.

incomplete dominance The situation in which the phenotype of heterozygotes is intermediate between the phenotypes of individuals homozygous for either allele.

incomplete flower A flower in which one or more of the four basic floral organs (sepals, petals, stamens, or carpels) are either absent or nonfunctional.

incomplete metamorphosis A type of development in certain insects, such as grasshoppers, in which the young (called nymphs) resemble adults but are smaller and have different body proportions. The nymph goes through a series of molts, each time looking more like an adult, until it reaches full size.

independent variable A factor whose value is manipulated or changed during an experiment to reveal possible effects on another factor (the dependent variable).

indeterminate cleavage A type of embryonic development in deuterostomes in which each cell produced by early cleavage divisions retains the capacity to develop into a complete embryo.

indeterminate growth A type of growth characteristic of plants, in which the organism continues to grow as long as it lives.

induced fit Caused by entry of the substrate, the change in shape of the active site of an enzyme so that it binds more snugly to the substrate.

inducer A specific small molecule that binds to a bacterial repressor protein and changes the repressor's shape so that it cannot bind to an operator, thus switching an operon on.

induction A process in which a group of cells or tissues influences the development of another group through close-range interactions.

inductive reasoning A type of logic in which generalizations are based on a large number of specific observations.

inflammatory response An innate immune defense triggered by physical injury or infection of tissue involving the release of substances that promote swelling, enhance the infiltration of white blood cells, and aid in tissue repair and destruction of invading pathogens.

inflorescence A group of flowers tightly clustered together.

ingestion The first stage of food processing in animals: the act of eating.

ingroup A species or group of species whose evolutionary relationships are being examined in a given analysis.

inhibitory postsynaptic potential (IPSP) An electrical change (usually hyperpolarization) in the membrane of a postsynaptic neuron caused by the binding of an inhibitory neurotransmitter from a presynaptic cell to a postsynaptic receptor; makes it more difficult for a postsynaptic neuron to generate an action potential.

innate behavior Animal behavior that is developmentally fixed and under strong genetic control. Innate behavior is exhibited in virtually the same form by all individuals in a population despite internal and external environmental differences during development and throughout their lifetimes.

innate immunity A form of defense common to all animals that is active immediately upon exposure to a pathogen and that is the same whether or not the pathogen has been encountered previously.

inner cell mass An inner cluster of cells at one end of a mammalian blastocyst that subsequently develops into the embryo proper and some of the extraembryonic membranes.

inner ear One of the three main regions of the vertebrate ear; includes the cochlea (which in turn contains the organ of Corti) and the semicircular canals.

inositol trisphosphate (IP$_3$) (in-ō´-suh-tol) A second messenger that functions as an intermediate between certain signaling molecules and a subsequent second messenger, Ca^{2+}, by causing a rise in cytoplasmic Ca^{2+} concentration.

inquiry The search for information and explanation, often focusing on specific questions.

insertion A mutation involving the addition of one or more nucleotide pairs to a gene.

in situ hybridization A technique using nucleic acid hybridization with a labeled probe to detect the location of a specific mRNA in an intact organism.

insulin (in´-suh-lin) A hormone secreted by pancreatic beta cells that lowers blood glucose levels. It promotes the uptake of glucose by most body cells and the synthesis and storage of glycogen in the liver and also stimulates protein and fat synthesis.

integral protein A transmembrane protein with hydrophobic regions that extend into and often completely span the hydrophobic interior of the membrane and with hydrophilic regions in contact with the aqueous solution on one or both sides of the membrane (or lining the channel in the case of a channel protein).

integrin (in'-tuh-grin) In animal cells, a transmembrane receptor protein with two subunits that interconnects the extracellular matrix and the cytoskeleton.

integument (in-teg'-yū-ment) Layer of sporophyte tissue that contributes to the structure of an ovule of a seed plant.

integumentary system The outer covering of a mammal's body, including skin, hair, and nails, claws, or hooves.

interferon (in'-ter-fēr'-on) A protein that has antiviral or immune regulatory functions. For example, interferons secreted by virus-infected cells help nearby cells resist viral infection.

intermediate disturbance hypothesis The concept that moderate levels of disturbance can foster greater species diversity than low or high levels of disturbance.

intermediate filament A component of the cytoskeleton that includes filaments intermediate in size between microtubules and microfilaments.

interneuron An association neuron; a nerve cell within the central nervous system that forms synapses with sensory and/or motor neurons and integrates sensory input and motor output.

internode A segment of a plant stem between the points where leaves are attached.

interphase The period in the cell cycle when the cell is not dividing. During interphase, cellular metabolic activity is high, chromosomes and organelles are duplicated, and cell size may increase. Interphase often accounts for about 90% of the cell cycle.

intersexual selection A form of natural selection in which individuals of one sex (usually the females) are choosy in selecting their mates from the other sex; also called mate choice.

interspecific interaction A relationship between individuals of two or more species in a community.

interstitial fluid The fluid filling the spaces between cells in most animals.

intertidal zone The shallow zone of the ocean adjacent to land and between the high- and low-tide lines.

intrasexual selection A form of natural selection in which there is direct competition among individuals of one sex for mates of the opposite sex.

intrinsic rate of increase (r) In population models, the per capita rate at which an exponentially growing population increases in size at each instant in time.

introduced species A species moved by humans, either intentionally or accidentally, from its native location to a new geographic region; sometimes called a non-native species, exotic species, or invasive species.

intron (in'-tron) A noncoding, intervening sequence within a primary transcript that is removed from the transcript during RNA processing; also refers to the region of DNA from which this sequence was transcribed.

inversion An aberration in chromosome structure resulting from reattachment of a chromosomal fragment in a reverse orientation to the chromosome from which it originated.

invertebrate An animal without a backbone. Invertebrates make up 95% of animal species.

in vitro fertilization (IVF) (vē'-trō) Fertilization of oocytes in laboratory containers followed by artificial implantation of the early embryo in the mother's uterus.

in vitro mutagenesis A technique used to discover the function of a gene by cloning it, introducing specific changes into the cloned gene's sequence, reinserting the mutated gene into a cell, and studying the phenotype of the mutant.

ion (ī'-on) An atom or group of atoms that has gained or lost one or more electrons, thus acquiring a charge.

ion channel (ī'-on) A transmembrane protein channel that allows a specific ion to diffuse across the membrane down its concentration or electrochemical gradient.

ionic bond (ī-on'-ik) A chemical bond resulting from the attraction between oppositely charged ions.

ionic compound (ī-on'-ik) A compound resulting from the formation of an ionic bond; also called a salt.

iris The colored part of the vertebrate eye, formed by the anterior portion of the choroid.

isomer (ī'-sō-mer) One of two or more compounds that have the same numbers of atoms of the same elements but different structures and hence different properties.

isomorphic Referring to alternating generations in plants and certain algae in which the sporophytes and gametophytes look alike, although they differ in chromosome number.

isotonic (ī'-sō-ton'-ik) Referring to a solution that, when surrounding a cell, causes no net movement of water into or out of the cell.

isotope (ī'-sō-tōp') One of several atomic forms of an element, each with the same number of protons but a different number of neutrons, thus differing in atomic mass.

iteroparity Reproduction in which adults produce offspring over many years; also called repeated reproduction.

jasmonate Any of a class of plant hormones that regulate a wide range of developmental processes in plants and play a key role in plant defense against herbivores.

joule (J) A unit of energy: 1 J = 0.239 cal; 1 cal = 4.184 J.

juxtaglomerular apparatus (JGA) (juks'-tuh-gluh-mār'-yū-ler) A specialized tissue in nephrons that releases the enzyme renin in response to a drop in blood pressure or volume.

juxtamedullary nephron In mammals and birds, a nephron with a loop of Henle that extends far into the renal medulla.

karyogamy (kār'-ē-og'-uh-mē) In fungi, the fusion of haploid nuclei contributed by the two parents; occurs as one stage of sexual reproduction, preceded by plasmogamy.

karyotype (kār'-ē-ō-tīp) A display of the chromosome pairs of a cell arranged by size and shape.

keystone species A species that is not necessarily abundant in a community yet exerts strong control on community structure by the nature of its ecological role or niche.

kidney In vertebrates, one of a pair of excretory organs where blood filtrate is formed and processed into urine.

kilocalorie (kcal) A thousand calories; the amount of heat energy required to raise the temperature of 1 kg of water by 1°C.

kinetic energy (kuh-net'-ik) The energy associated with the relative motion of objects. Moving matter can perform work by imparting motion to other matter.

kinetochore (kuh-net'-uh-kōr) A structure of proteins attached to the centromere that links each sister chromatid to the mitotic spindle.

kinetoplastid A protist, such as a trypanosome, that has a single large mitochondrion that houses an organized mass of DNA.

kingdom A taxonomic category, the second broadest after domain.

kin selection Natural selection that favors altruistic behavior by enhancing the reproductive success of relatives.

K-selection Selection for life history traits that are sensitive to population density.

labia majora A pair of thick, fatty ridges that enclose and protect the rest of the vulva.

labia minora A pair of slender skin folds that surround the openings of the vagina and urethra.

lacteal (lak'-tē-ul) A tiny lymph vessel extending into the core of an intestinal villus and serving as the destination for absorbed chylomicrons.

lactic acid fermentation Glycolysis followed by the reduction of pyruvate to lactate, regenerating NAD^+ with no release of carbon dioxide.

lagging strand A discontinuously synthesized DNA strand that elongates by means of Okazaki fragments, each synthesized in a $5' \rightarrow 3'$ direction away from the replication fork.

lamprey Any of the jawless vertebrates with highly reduced vertebrae that live in freshwater and marine environments. Almost half of extant lamprey species are parasites that feed by clamping their round, jawless mouth onto the flank of a live fish; nonparasitic lampreys are suspension feeders that feed only as larvae.

lancelet A member of the clade Cephalochordata, small blade-shaped marine chordates that lack a backbone.

landscape An area containing several different ecosystems linked by exchanges of energy, materials, and organisms.

landscape ecology The study of how the spatial arrangement of habitat types affects the distribution and abundance of organisms and ecosystem processes.

large intestine The portion of the vertebrate alimentary canal between the small intestine and the anus; functions mainly in water absorption and the formation of feces.

larva (lar'-vuh) (plural, **larvae**) A free-living, sexually immature form in some animal life cycles that may differ from the adult animal in morphology, nutrition, and habitat.

larynx (lār'-inks) The portion of the respiratory tract containing the vocal cords; also called the voice box.

lateralization Segregation of functions in the cortex of the left and right cerebral hemispheres.

lateral line system A mechanoreceptor system consisting of a series of pores and receptor units along the sides of the body in fishes and aquatic amphibians; detects water movements made by the animal itself and by other moving objects.

lateral meristem (mār′-uh-stem) A meristem that thickens the roots and shoots of woody plants. The vascular cambium and cork cambium are lateral meristems.

lateral root A root that arises from the pericycle of an established root.

law of conservation of mass A physical law stating that matter can change form but cannot be created or destroyed. In a closed system, the mass of the system is constant.

law of independent assortment Mendel's second law, stating that each pair of alleles segregates, or assorts, independently of each other pair during gamete formation; applies when genes for two characters are located on different pairs of homologous chromosomes or when they are far enough apart on the same chromosome to behave as though they are on different chromosomes.

law of segregation Mendel's first law, stating that the two alleles in a pair segregate (separate from each other) into different gametes during gamete formation.

leading strand The new complementary DNA strand synthesized continuously along the template strand toward the replication fork in the mandatory 5′ → 3′ direction.

leaf The main photosynthetic organ of vascular plants.

leaf primordium (plural, **primordia**) A finger-like projection along the flank of a shoot apical meristem, from which a leaf arises.

learning The modification of behavior as a result of specific experiences.

lens The structure in an eye that focuses light rays onto the photoreceptors.

lenticel (len′-ti-sel) A small raised area in the bark of stems and roots that enables gas exchange between living cells and the outside air.

lepidosaur (leh-pid′-uh-sōr) A member of the reptilian group that includes lizards, snakes, and two species of New Zealand animals called tuataras.

leukocyte (lū′-kō-sīt′) A blood cell that functions in fighting infections; also called a white blood cell.

lichen The mutualistic association between a fungus and a photosynthetic alga or cyanobacterium.

life cycle The generation-to-generation sequence of stages in the reproductive history of an organism.

life history The traits that affect an organism's schedule of reproduction and survival.

life table A summary of the age-specific survival and reproductive rates of individuals in a population.

ligament A fibrous connective tissue that joins bones together at joints.

ligand (lig′-und) A molecule that binds specifically to another molecule, usually a larger one.

ligand-gated ion channel (lig′-und) A transmembrane protein containing a pore that opens or closes as it changes shape in response to a signaling molecule (ligand), allowing or blocking the flow of specific ions; also called an ionotropic receptor.

light chain One of the two types of polypeptide chains that make up an antibody molecule and B cell receptor; consists of a variable region, which contributes to the antigen-binding site, and a constant region.

light-harvesting complex A complex of proteins associated with pigment molecules (including chlorophyll *a*, chlorophyll *b*, and carotenoids) that captures light energy and transfers it to reaction-center pigments in a photosystem.

light microscope (LM) An optical instrument with lenses that refract (bend) visible light to magnify images of specimens.

light reactions The first of two major stages in photosynthesis (preceding the Calvin cycle). These reactions, which occur on the thylakoid membranes of the chloroplast or on membranes of certain prokaryotes, convert solar energy to the chemical energy of ATP and NADPH, releasing oxygen in the process.

lignin (lig′-nin) A strong polymer embedded in the cellulose matrix of the secondary cell walls of vascular plants that provides structural support in terrestrial species.

limiting nutrient An element that must be added for production to increase in a particular area.

limnetic zone In a lake, the well-lit, open surface waters far from shore.

linear electron flow A route of electron flow during the light reactions of photosynthesis that involves both photosystems (I and II) and produces ATP, NADPH, and O_2. The net electron flow is from H_2O to $NADP^+$.

line graph A graph in which each data point is connected to the next point in the data set with a straight line.

linkage map A genetic map based on the frequencies of recombination between markers during crossing over of homologous chromosomes.

linked genes Genes located close enough together on a chromosome that they tend to be inherited together.

lipid (lip′-id) Any of a group of large biological molecules, including fats, phospholipids, and steroids, that mix poorly, if at all, with water.

littoral zone In a lake, the shallow, well-lit waters close to shore.

liver A large internal organ in vertebrates that performs diverse functions, such as producing bile, maintaining blood glucose level, and detoxifying poisonous chemicals in the blood.

liverwort A small, herbaceous, nonvascular plant that is a member of the phylum Hepatophyta.

loam The most fertile soil type, made up of roughly equal amounts of sand, silt, and clay.

lobe-fin Member of a clade of osteichthyans having rod-shaped muscular fins. The group includes coelacanths, lungfishes, and tetrapods.

local regulator A secreted molecule that influences cells near where it is secreted.

locomotion Active motion from place to place.

locus (plural, **loci**), (lō′-kus), (lō′-sī) A specific place along the length of a chromosome where a given gene is located.

logistic population growth Population growth that levels off as population size approaches carrying capacity.

long-day plant A plant that flowers (usually in late spring or early summer) only when the light period is longer than a critical length.

long noncoding RNA (lncRNA) An RNA between 200 and hundreds of thousands of nucleotides in length that does not code for protein but is expressed at significant levels.

long-term memory The ability to hold, associate, and recall information over one's lifetime.

long-term potentiation (LTP) An enhanced responsiveness to an action potential (nerve signal) by a receiving neuron.

loop of Henle (hen′-lē) The hairpin turn, with a descending and ascending limb, between the proximal and distal tubules of the vertebrate kidney; functions in water and salt reabsorption.

lophophore (lof′-uh-fōr) In some lophotrochozoan animals, including brachiopods, a crown of ciliated tentacles that surround the mouth and function in feeding.

Lophotrochozoa (lo-phah′-truh-kō-zō′-uh) One of the three main lineages of bilaterian animals; lophotrochozoans include organisms that have lophophores or trochophore larvae. *See also* Deuterostomia and Ecdysozoa.

low-density lipoprotein (LDL) A particle in the blood made up of thousands of cholesterol molecules and other lipids bound to a protein. LDL transports cholesterol from the liver for incorporation into cell membranes.

lung An infolded respiratory surface of a terrestrial vertebrate, land snail, or spider that connects to the atmosphere by narrow tubes.

luteinizing hormone (LH) (lū′-tē-uh-nī′-zing) A tropic hormone that is produced and secreted by the anterior pituitary and that stimulates ovulation in females and androgen production in males.

lycophyte (lī′-kuh-fīt) An informal name for a member of the phylum Lycophyta, which includes club mosses, spike mosses, and quillworts.

lymph The colorless fluid, derived from interstitial fluid, in the lymphatic system of vertebrates.

lymph node An organ located along a lymph vessel. Lymph nodes filter lymph and contain cells that attack viruses and bacteria.

lymphatic system A system of vessels and nodes, separate from the circulatory system, that returns fluid, proteins, and cells to the blood.

lymphocyte A type of white blood cell that mediates immune responses. The two main classes are B cells and T cells.

lysogenic cycle (lī′-sō-jen′-ik) A type of phage replicative cycle in which the viral genome becomes incorporated into the bacterial host chromosome as a prophage, is replicated along with the chromosome, and does not kill the host.

lysosome (lī′-suh-sōm) A membrane-enclosed sac of hydrolytic enzymes found in the cytoplasm of animal cells and some protists.

lysozyme (lī′-sō-zīm) An enzyme that destroys bacterial cell walls; in mammals, it is found in sweat, tears, and saliva.

lytic cycle (lit'-ik) A type of phage replicative cycle resulting in the release of new phages by lysis (and death) of the host cell.

macroevolution Evolutionary change above the species level. Examples of macroevolutionary change include the origin of a new group of organisms through a series of speciation events and the impact of mass extinctions on the diversity of life and its subsequent recovery.

macromolecule A giant molecule formed by the joining of smaller molecules. Polysaccharides, proteins, and nucleic acids are macromolecules.

macronutrient An essential element that an organism must obtain in relatively large amounts. *See also* micronutrient.

macrophage (mak'-rō-fāj) A phagocytic cell present in many tissues that functions in innate immunity by destroying microorganisms and in acquired immunity as an antigen-presenting cell.

magnoliid A member of the angiosperm clade that is most closely related to the combined eudicot and monocot clades. Extant examples are magnolias, laurels, and black pepper plants.

major depressive disorder A mood disorder characterized by feelings of sadness, lack of self-worth, emptiness, or loss of interest in nearly all things.

major histocompatibility complex (MHC) molecule A host protein that functions in antigen presentation. Foreign MHC molecules on transplanted tissue can trigger T cell responses that may lead to rejection of the transplant.

malignant tumor A cancerous tumor containing cells that have significant genetic and cellular changes and are capable of invading and surviving in new sites. Malignant tumors can impair the functions of one or more organs.

Malpighian tubule (mal-pig'-ē-un) A unique excretory organ of insects that empties into the digestive tract, removes nitrogenous wastes from the hemolymph, and functions in osmoregulation.

mammal A member of the clade Mammalia, amniotes that have hair and mammary glands (glands that produce milk).

mammary gland An exocrine gland that secretes milk for nourishing the young. Mammary glands are characteristic of mammals.

mantle One of the three main parts of a mollusc; a fold of tissue that drapes over the mollusc's visceral mass and may secrete a shell. *See also* foot and visceral mass.

mantle cavity A water-filled chamber that houses the gills, anus, and excretory pores of a mollusc.

map unit A unit of measurement of the distance between genes. One map unit is equivalent to a 1% recombination frequency.

marine benthic zone The ocean floor.

mark-recapture method A sampling technique used to estimate the size of animal populations.

marsupial (mar-sū'-pē-ul) A mammal, such as a koala, kangaroo, or opossum, whose young complete their embryonic development inside a maternal pouch called the marsupium.

mass extinction The elimination of a large number of species throughout Earth, the result of global environmental changes.

mass number The total number of protons and neutrons in an atom's nucleus.

mast cell Immune system cell that secretes histamine; plays role in inflammatory response and allergies.

mate-choice copying Behavior in which individuals in a population copy the mate choice of others, apparently due to social learning.

maternal effect gene A gene that, when mutant in the mother, results in a mutant phenotype in the offspring, regardless of the offspring's genotype. Maternal effect genes, also called egg-polarity genes, were first identified in *Drosophila melanogaster*.

matter Anything that takes up space and has mass.

maximum likelihood As applied to DNA sequence data, a principle that states that when considering multiple phylogenetic hypotheses, one should take into account the hypothesis that reflects the most likely sequence of evolutionary events, given certain rules about how DNA changes over time.

maximum parsimony The principle that when considering multiple explanations for an observation, one should first investigate the simplest explanation that is consistent with the facts.

mean The sum of all data points in a data set divided by the number of data points.

mechanoreceptor A sensory receptor that detects physical deformation in the body's environment associated with pressure, touch, stretch, motion, or sound.

medulla oblongata (meh-dul'-uh ob'-long-go'-tuh) The lowest part of the vertebrate brain, commonly called the medulla; a swelling of the hindbrain anterior to the spinal cord that controls autonomic, homeostatic functions, including breathing, heart and blood vessel activity, swallowing, digestion, and vomiting.

medusa (muh-dū'-suh) (plural, **medusae**) The floating, mouth-down form of the cnidarian body plan. The alternate form is the polyp.

megapascal (MPa) (meg'-uh-pas-kal') A unit of pressure equivalent to about 10 atmospheres of pressure.

megaphyll (meh'-guh-fil) A leaf with a highly branched vascular system, found in almost all vascular plants other than lycophytes. *See also* microphyll.

megaspore A spore from a heterosporous plant species that develops into a female gametophyte.

meiosis (mī-ō'-sis) A modified type of cell division in sexually reproducing organisms consisting of two rounds of cell division but only one round of DNA replication. It results in cells with half the number of chromosome sets as the original cell.

meiosis I (mī-ō'-sis) The first division of a two-stage process of cell division in sexually reproducing organisms that results in cells with half the number of chromosome sets as the original cell.

meiosis II (mī-ō'-sis) The second division of a two-stage process of cell division in sexually reproducing organisms that results in cells with half the number of chromosome sets as the original cell.

melanocyte-stimulating hormone (MSH) A hormone produced and secreted by the anterior pituitary with multiple activities, including regulating the behavior of pigment-containing cells in the skin of some vertebrates.

melatonin A hormone that is secreted by the pineal gland and that is involved in the regulation of biological rhythms and sleep.

membrane potential The difference in electrical charge (voltage) across a cell's plasma membrane due to the differential distribution of ions. Membrane potential affects the activity of excitable cells and the transmembrane movement of all charged substances.

memory cell One of a clone of long-lived lymphocytes, formed during the primary immune response, that remains in a lymphoid organ until activated by exposure to the same antigen that triggered its formation. Activated memory cells mount the secondary immune response.

menopause The cessation of ovulation and menstruation marking the end of a human female's reproductive years.

menstrual cycle (men'-strū-ul) In humans and certain other primates, the periodic growth and shedding of the uterine lining that occurs in the absence of pregnancy.

menstruation The shedding of portions of the endometrium during a uterine (menstrual) cycle.

meristem (mār'-uh-stem) Plant tissue that remains embryonic as long as the plant lives, allowing for indeterminate growth.

mesoderm (mez'-ō-derm) The middle primary germ layer in a triploblastic animal embryo; develops into the notochord, the lining of the coelom, muscles, skeleton, gonads, kidneys, and most of the circulatory system in species that have these structures.

mesohyl (mez'-ō-hīl) A gelatinous region between the two layers of cells of a sponge.

mesophyll (mez'-ō-fil) Leaf cells specialized for photosynthesis. In C_3 and CAM plants, mesophyll cells are located between the upper and lower epidermis; in C_4 plants, they are located between the bundle-sheath cells and the epidermis.

messenger RNA (mRNA) A type of RNA, synthesized using a DNA template, that attaches to ribosomes in the cytoplasm and specifies the primary structure of a protein. (In eukaryotes, the primary RNA transcript must undergo RNA processing to become mRNA.)

metabolic pathway A series of chemical reactions that either builds a complex molecule (anabolic pathway) or breaks down a complex molecule to simpler molecules (catabolic pathway).

metabolic rate The total amount of energy an animal uses in a unit of time.

metabolism (muh-tab'-uh-lizm) The totality of an organism's chemical reactions, consisting of catabolic and anabolic pathways, which manage the material and energy resources of the organism.

metagenomics The collection and sequencing of DNA from a group of species, usually an environmental sample of microorganisms. Computer software sorts partial sequences and assembles them into genome sequences of individual species making up the sample.

metamorphosis (met'-uh-mōr'-fuh-sis) A developmental transformation that turns an animal larva into either an adult or an adult-like stage that is not yet sexually mature.

metanephridium (met'-uh-nuh-frid'-ē-um) (plural, **metanephridia**) An excretory organ found in many invertebrates that typically consists of tubules connecting ciliated internal openings to external openings.

metaphase The third stage of mitosis, in which the spindle is complete and the chromosomes, attached to microtubules at their kinetochores, are all aligned at the metaphase plate.

metaphase plate An imaginary structure located at a plane midway between the two poles of a cell in metaphase on which the centromeres of all the duplicated chromosomes are located.

metapopulation A group of spatially separated populations of one species that interact through immigration and emigration.

metastasis (muh-tas'-tuh-sis) The spread of cancer cells to locations distant from their original site.

methanogen (meth-an'-ō-jen) An organism that produces methane as a waste product of the way it obtains energy. All known methanogens are in domain Archaea.

methyl group A chemical group consisting of a carbon bonded to three hydrogen atoms. The methyl group may be attached to a carbon or to a different atom.

microbiome The collection of microorganisms living in or on an organism's body, along with their genetic material.

microclimate Climate patterns on a very fine scale, such as the specific climatic conditions underneath a log.

microevolution Evolutionary change below the species level; change in the allele frequencies in a population over generations.

microfilament A cable composed of actin proteins in the cytoplasm of almost every eukaryotic cell, making up part of the cytoskeleton and acting alone or with myosin to cause cell contraction; also called an actin filament.

micronutrient An essential element that an organism needs in very small amounts. *See also* macronutrient.

microphyll (mī'-krō-fil) A small, usually spine-shaped leaf supported by a single strand of vascular tissue, found only in lycophytes.

microplastic A plastic particle less than 5 mm in size; microplastics have contaminated all of the world's oceans as well as freshwater and terrestrial ecosystems.

micropyle A pore in the integuments of an ovule.

microRNA (miRNA) A small, single-stranded RNA molecule, generated from a double-stranded RNA precursor. The miRNA associates with one or more proteins in a complex that can degrade or prevent translation of an mRNA with a complementary sequence.

microspore A spore from a heterosporous plant species that develops into a male gametophyte.

microsporidian A member of the fungal phylum Microsporidia, unicellular parasites of protists and animals; microsporidians and their sister taxon (cryptomycetes) are a basal fungal lineage.

microtubule A hollow rod composed of tubulin proteins that makes up part of the cytoskeleton in all eukaryotic cells and is found in cilia and flagella.

microvillus (plural, **microvilli**) One of many fine, finger-like projections of the epithelial cells in the lumen of the small intestine that increase its surface area.

midbrain One of three ancestral and embryonic regions of the vertebrate brain; develops into sensory integrating and relay centers that send sensory information to the cerebrum.

middle ear One of three main regions of the vertebrate ear; in mammals, a chamber containing three small bones (the malleus, incus, and stapes) that convey vibrations from the eardrum to the oval window.

middle lamella (luh-mel'-uh) In plants, a thin layer of adhesive extracellular material, primarily pectins, found between the primary walls of adjacent young cells.

migration A regular, long-distance change in location.

mineral In nutrition, a simple nutrient that is inorganic and therefore cannot be synthesized in the body.

mineralocorticoid A steroid hormone secreted by the adrenal cortex that regulates salt and water homeostasis.

minimum viable population (MVP) The smallest population size at which a species is able to sustain its numbers and survive.

mismatch repair The cellular process that uses specific enzymes to remove and replace incorrectly paired nucleotides.

missense mutation A nucleotide-pair substitution that results in a codon that codes for a different amino acid.

mitochondrial matrix The compartment of the mitochondrion enclosed by the inner membrane and containing enzymes and substrates for the citric acid cycle, as well as ribosomes and DNA.

mitochondrion (mī'-tō-kon'-drē-un) (plural, **mitochondria**) An organelle in eukaryotic cells that serves as the site of cellular respiration; uses oxygen to break down organic molecules and synthesize ATP.

mitosis (mī-tō'-sis) A process of nuclear division in eukaryotic cells conventionally divided into five stages: prophase, prometaphase, metaphase, anaphase, and telophase. Mitosis conserves chromosome number by allocating replicated chromosomes equally to each of the daughter nuclei.

mitotic (M) phase The phase of the cell cycle that includes mitosis and cytokinesis.

mitotic spindle An assemblage of microtubules and associated proteins that is involved in the movement of chromosomes during mitosis.

mixotroph An organism that is capable of both photosynthesis and heterotrophy.

model A physical or conceptual representation of a natural phenomenon.

model organism A particular species chosen for research into broad biological principles because it is representative of a larger group and usually easy to grow in a lab.

molarity A common measure of solute concentration, referring to the number of moles of solute per liter of solution.

mold Informal term for a fungus that grows as a filamentous fungus, producing haploid spores by mitosis and forming a visible mycelium.

mole (mol) The number of grams of a substance that equals its molecular or atomic mass in daltons; a mole contains Avogadro's number of the molecules or atoms in question.

molecular clock A method for estimating the time required for a given amount of evolutionary change, based on the observation that some regions of genomes evolve at constant rates.

molecular mass The sum of the masses of all the atoms in a molecule; sometimes called molecular weight.

molecule Two or more atoms held together by covalent bonds.

molting A process in ecdysozoans in which the exoskeleton is shed at intervals, allowing growth by the production of a larger exoskeleton.

monilophyte An informal name for a member of the phylum Monilophyta, which includes ferns, horsetails, and whisk ferns and their relatives.

monoclonal antibody (mon'-ō-klōn'-ul) Any of a preparation of antibodies that have been produced by a single clone of cultured cells and thus are all specific for the same epitope.

monocot A member of a clade consisting of flowering plants that have one embryonic seed leaf, or cotyledon.

monogamous (muh-nog'-uh-mus) Referring to a type of relationship in which one male mates with just one female.

monohybrid An organism that is heterozygous with respect to a single gene of interest. All the offspring from a cross between parents homozygous for different alleles are monohybrids. For example, parents of genotypes *AA* and *aa* produce a monohybrid of genotype *Aa*.

monohybrid cross A cross between two organisms that are heterozygous for the character being followed (or the self-pollination of a heterozygous plant).

monomer (mon'-uh-mer) The subunit that serves as the building block of a polymer.

monophyletic (mon'-ō-fī-let'-ik) Pertaining to a group of taxa that consists of a common ancestor and all of its descendants. A monophyletic taxon is equivalent to a clade.

monosaccharide (mon'-ō-sak'-uh-rīd) The simplest carbohydrate, active alone or serving as a monomer for disaccharides and polysaccharides. Also called simple sugars, monosaccharides have molecular formulas that are generally some multiple of CH_2O.

monosomic Referring to a diploid cell that has only one copy of a particular chromosome instead of the normal two.

monotreme An egg-laying mammal, such as a platypus or echidna. Like all mammals, monotremes have hair and produce milk, but they lack nipples.

morphogen A substance, such as Bicoid protein in *Drosophila*, that provides positional information in the form of a concentration gradient along an embryonic axis.

morphogenesis (mōr′-fō-jen′-uh-sis) The development of the form of an organism and its structures.

morphological species concept Definition of a species in terms of measurable anatomical criteria.

moss A small, herbaceous, nonvascular plant that is a member of the phylum Bryophyta.

motor neuron A nerve cell that transmits signals from the brain or spinal cord to muscles or glands.

motor protein A protein that interacts with cytoskeletal elements and other cell components, producing movement of the whole cell or parts of the cell.

motor system An efferent branch of the vertebrate peripheral nervous system composed of motor neurons that carry signals to skeletal muscles in response to external stimuli.

motor unit A single motor neuron and all the muscle fibers it controls.

movement corridor A series of small clumps or a narrow strip of quality habitat (usable by organisms) that connects otherwise isolated patches of quality habitat.

MPF Maturation-promoting factor (or M-phase-promoting factor); a protein complex required for a cell to progress from late interphase to mitosis. The active form consists of cyclin and a protein kinase.

mucoromycete A member of the fungal phylum Mucoromycota, characterized by the formation of a sturdy structure called a zygosporangium during sexual reproduction.

mucus A viscous and slippery mixture of glycoproteins, cells, salts, and water that moistens and protects the membranes lining body cavities that open to the exterior.

Müllerian mimicry (myū-lār′-ē-un mim′-uh-krē) Reciprocal mimicry by two unpalatable species.

multifactorial Referring to a phenotypic character that is influenced by multiple genes and environmental factors.

multigene family A collection of genes with similar or identical sequences, presumably of common origin.

multiple fruit A fruit derived from an entire inflorescence.

multiplication rule A rule of probability stating that the probability of two or more independent events occurring together can be determined by multiplying their individual probabilities.

muscle tissue Tissue consisting of long muscle cells that can contract, either on its own or when stimulated by nerve impulses.

mutagen (myū′-tuh-jen) A chemical or physical agent that interacts with DNA and can cause a mutation.

mutation (myū-tā′-shun) A change in the nucleotide sequence of an organism's DNA or in the DNA or RNA of a virus.

mutualism (myū′-chū-ul-izm) A +/+ ecological interaction that benefits individuals of both interacting species.

mycelium (mī-sē′-lē-um) (plural, **mycelia**) The densely branched network of hyphae in a fungus.

mycorrhiza (plural, **mycorrhizae**) (mī′-kō-rī′-zuh, mī′-kō-rī′-zē) A mutualistic association of plant roots and fungus.

mycosis (mī-kō′-sis) General term for a fungal infection.

myelin sheath (mī′-uh-lin) Wrapped around the axon of a neuron, an insulating coat of cell membranes from Schwann cells or oligodendrocytes. It is interrupted by nodes of Ranvier, where action potentials are generated.

myofibril (mī′-ō-fī′-bril) A longitudinal bundle in a muscle cell (fiber) that contains thin filaments of actin and regulatory proteins and thick filaments of myosin.

myoglobin (mī′-uh-glō′-bin) An oxygen-storing, pigmented protein in muscle cells.

myosin (mī′-uh-sin) A type of motor protein that associates into filaments that interact with actin filaments to cause cell contraction.

myriapod (mir′-ē-uh-pod′) A terrestrial arthropod with many body segments and one or two pairs of legs per segment. Millipedes and centipedes are the two major groups of living myriapods.

NAD$^+$ The oxidized form of nicotinamide adenine dinucleotide, a coenzyme that can accept electrons, becoming NADH. NADH temporarily stores electrons during cellular respiration.

NADH The reduced form of nicotinamide adenine dinucleotide that temporarily stores electrons during cellular respiration. NADH acts as an electron donor to the electron transport chain.

NADP$^+$ The oxidized form of nicotinamide adenine dinucleotide phosphate, an electron carrier that can accept electrons, becoming NADPH. NADPH temporarily stores energized electrons produced during the light reactions.

NADPH The reduced form of nicotinamide adenine dinucleotide phosphate; temporarily stores energized electrons produced during the light reactions. NADPH acts as "reducing power" that can be passed along to an electron acceptor, reducing it.

natural killer cell A type of white blood cell that can kill tumor cells and virus-infected cells as part of innate immunity.

natural selection A process in which individuals that have certain inherited traits tend to survive and reproduce at higher rates than other individuals because of those traits.

negative feedback A form of regulation in which accumulation of an end product of a process slows the process; in physiology, a primary mechanism of homeostasis, whereby a change in a variable triggers a response that counteracts the initial change.

negative pressure breathing A breathing system in which air is pulled into the lungs.

nematocyst (nem′-uh-tuh-sist′) In a cnidocyte of a cnidarian, a capsule-like organelle containing a coiled thread that when discharged can penetrate the body wall of the prey.

nephron (nef′-ron) The tubular excretory unit of the vertebrate kidney.

neritic zone The shallow region of the ocean overlying the continental shelf.

nerve A fiber composed primarily of the bundled axons of neurons.

nervous system In animals, the fast-acting internal system of communication involving sensory receptors, networks of nerve cells, and connections to muscles and glands that respond to nerve signals; functions in concert with the endocrine system to effect internal regulation and maintain homeostasis.

nervous tissue Tissue made up of neurons and supportive cells.

net ecosystem production (NEP) The gross primary production of an ecosystem minus the energy used by all autotrophs and heterotrophs for respiration.

net primary production (NPP) The gross primary production of an ecosystem minus the energy used by the producers for respiration.

neural crest In vertebrates, a region located along the sides of the neural tube where it pinches off from the ectoderm. Neural crest cells migrate to various parts of the embryo and form pigment cells in the skin and parts of the skull, teeth, adrenal glands, and peripheral nervous system.

neural tube A tube of infolded ectodermal cells that runs along the anterior-posterior axis of a vertebrate, just dorsal to the notochord. It will give rise to the central nervous system.

neurohormone A molecule that is secreted by a neuron, travels in body fluids, and acts on specific target cells, changing their functioning.

neuron (nyūr′-on) A nerve cell; the fundamental unit of the nervous system, having structure and properties that allow it to conduct signals by taking advantage of the electrical charge across its plasma membrane.

neuronal plasticity The capacity of a nervous system to change with experience.

neuropeptide A relatively short chain of amino acids that serves as a neurotransmitter.

neurotransmitter A molecule that is released from the synaptic terminal of a neuron at a chemical synapse, diffuses across the synaptic cleft, and binds to the postsynaptic cell, triggering a response.

neutral variation Genetic variation that does not provide a selective advantage or disadvantage.

neutron A subatomic particle having no electrical charge (electrically neutral), with a mass of about 1.7×10^{-24} g, found in the nucleus of an atom.

neutrophil The most abundant type of white blood cell. Neutrophils are phagocytic and tend to self-destruct as they destroy foreign invaders, limiting their life span to a few days.

nitric oxide (NO) A gas produced by many types of cells that functions as a local regulator and as a neurotransmitter.

nitrogen cycle The natural process by which nitrogen, either from the atmosphere or from decomposed organic material, is converted by soil bacteria to compounds assimilated by plants. This incorporated nitrogen is then taken in by other organisms and subsequently released, acted on by bacteria, and made available again to the nonliving environment.

nitrogen fixation The conversion of atmospheric nitrogen (N_2) to ammonia (NH_3). Biological nitrogen fixation is carried out by certain prokaryotes, some of which have mutualistic relationships with plants.

nociceptor (nō′-si-sep′-tur) A sensory receptor that responds to noxious or painful stimuli; also called a pain receptor.

node A point along the stem of a plant at which leaves are attached.

node of Ranvier (ron′-vē-ā′) Gap in the myelin sheath of certain axons where an action potential may be generated. In saltatory conduction, an action potential is regenerated at each node, appearing to "jump" along the axon from node to node.

nodule A swelling on the root of a legume. Nodules are composed of plant cells that contain nitrogen-fixing bacteria of the genus *Rhizobium*.

noncompetitive inhibitor A substance that reduces the activity of an enzyme by binding to a location remote from the active site, changing the enzyme's shape so that the active site no longer effectively catalyzes the conversion of substrate to product.

nondisjunction An error in meiosis or mitosis in which members of a pair of homologous chromosomes or a pair of sister chromatids fail to separate properly from each other.

nonequilibrium model A model that maintains that communities change constantly after being buffeted by disturbances.

nonpolar covalent bond A type of covalent bond in which electrons are shared equally between two atoms of similar electronegativity.

nonsense mutation A mutation that changes an amino acid codon to one of the three stop codons, resulting in a shorter and usually nonfunctional protein.

norepinephrine (nor-ep′-i-nef′-rin) A catecholamine that is chemically and functionally similar to epinephrine and acts as a hormone or neurotransmitter; also called noradrenaline.

northern coniferous forest A terrestrial biome characterized by long, cold winters and dominated by cone-bearing trees.

no-till agriculture A plowing technique that minimally disturbs the soil, thereby reducing soil loss.

notochord (nō′-tuh-kord′) A longitudinal, flexible rod made of tightly packed mesodermal cells that runs along the anterior-posterior axis of a chordate in the dorsal part of the body.

nuclear envelope In a eukaryotic cell, the double membrane that surrounds the nucleus, perforated with pores that regulate traffic with the cytoplasm. The outer membrane is continuous with the endoplasmic reticulum.

nuclear lamina A netlike array of protein filaments that lines the inner surface of the nuclear envelope and helps maintain the shape of the nucleus.

nucleariid A member of a group of unicellular, amoeboid protists that are more closely related to fungi than they are to other protists.

nuclease An enzyme that cuts DNA or RNA, either removing one or a few bases or hydrolyzing the DNA or RNA completely into its component nucleotides.

nucleic acid (nū-klā′-ik) A polymer (polynucleotide) consisting of many nucleotide monomers; serves as a blueprint for proteins and, through the actions of proteins, for all cellular activities. The two types are DNA and RNA.

nucleic acid hybridization (nū-klā′-ik) The base pairing of one strand of a nucleic acid to the complementary sequence on a strand from *another* nucleic acid molecule.

nucleic acid probe (nū′-klā′-ik) In DNA technology, a labeled single-stranded nucleic acid molecule used to locate a specific nucleotide sequence in a nucleic acid sample. Molecules of the probe hydrogen-bond to the complementary sequence wherever it occurs; radioactive, fluorescent, or other labeling of the probe allows its location to be detected.

nucleoid (nū′-klē-oyd) A non-membrane-enclosed region in a prokaryotic cell where its chromosome is located.

nucleolus (nū-klē′-ō-lus) (plural, **nucleoli**) A specialized structure in the nucleus, consisting of chromosomal regions containing ribosomal RNA (rRNA) genes along with ribosomal proteins imported from the cytoplasm; site of rRNA synthesis and ribosomal subunit assembly. *See also* ribosome.

nucleosome (nū′-klē-ō-sōm′) The basic, bead-like unit of DNA packing in eukaryotes, consisting of a segment of DNA wound around a protein core composed of two copies of each of four types of histone.

nucleotide (nū′-klē-ō-tīd′) The building block of a nucleic acid, consisting of a five-carbon sugar covalently bonded to a nitrogenous base and one to three phosphate groups.

nucleotide excision repair (nū′-klē-ō-tīd′) A repair system that removes and then correctly replaces a damaged segment of DNA using the undamaged strand as a guide.

nucleotide-pair substitution (nū′-klē-ō-tīd′) A type of point mutation in which one nucleotide in a DNA strand and its partner in the complementary strand are replaced by another pair of nucleotides.

nucleus (1) An atom's central core, containing protons and neutrons. (2) The organelle of a eukaryotic cell that contains the genetic material in the form of chromosomes, made up of chromatin. (3) A cluster of neurons.

nutrition The process by which an organism takes in and makes use of food substances.

obligate aerobe (ob′-lig-et ār′-ōb) An organism that requires oxygen for cellular respiration and cannot live without it.

obligate anaerobe (ob′-lig-et an′-uh-rōb) An organism that carries out only fermentation or anaerobic respiration. Such organisms cannot use oxygen and in fact may be poisoned by it.

ocean acidification The process by which the pH of the ocean is lowered (made more acidic) when excess CO_2 dissolves in seawater and forms carbonic acid (H_2CO_3).

oceanic pelagic zone Most of the ocean's waters far from shore, constantly mixed by ocean currents.

odorant A molecule that can be detected by sensory receptors of the olfactory system.

Okazaki fragment (ō′-kah-zah′-kē) A short segment of DNA synthesized away from the replication fork on a template strand during DNA replication. Many such segments are joined together to make up the lagging strand of newly synthesized DNA.

olfaction The sense of smell.

oligodendrocyte A type of glial cell that forms insulating myelin sheaths around the axons of neurons in the central nervous system.

oligotrophic lake A nutrient-poor, clear lake with few phytoplankton.

ommatidium (ōm′-uh-tid′-ē-um) (plural, **ommatidia**) One of the facets of the compound eye of arthropods and some polychaete worms.

omnivore An animal that regularly eats animals as well as plants or algae.

oncogene (on′-kō-jēn) A gene found in viral or cellular genomes that is involved in triggering molecular events that can lead to cancer.

oocyte (ō′-uh-sīt) A cell in the female reproductive system that differentiates to form an egg.

oogenesis (ō′-uh-jen′-uh-sis) The process in the ovary that results in the production of female gametes.

oogonium (ō′-uh-gō′-nē-em) (plural, **oogonia**) A cell that divides mitotically to form oocytes.

open circulatory system A circulatory system in which fluid called hemolymph bathes the tissues and organs directly and there is no distinction between the circulating fluid and the interstitial fluid.

operator In bacterial and phage DNA, a sequence of nucleotides near the start of an operon to which an active repressor can attach. The binding of the repressor prevents RNA polymerase from attaching to the promoter and transcribing the genes of the operon.

operculum (ō-per′-kyuh-lum) In aquatic osteichthyans, a protective bony flap that covers and protects the gills.

operon (op′-er-on) A unit of genetic function found in bacteria and phages, consisting of a promoter, an operator, and a coordinately regulated cluster of genes whose products function in a common pathway.

opisthokont (uh-pis′-thuh-kont′) A member of an extremely diverse clade of eukaryotes that includes fungi, animals, and several closely related groups of protists.

opposable thumb A thumb that can touch the ventral surface (fingerprint side) of the fingertip of all four fingers of the same hand with its own ventral surface.

opsin A membrane protein bound to a light-absorbing pigment molecule.

optimal foraging model The basis for analyzing behavior as a compromise between feeding costs and feeding benefits.

oral cavity The mouth of an animal.

orbital The three-dimensional space where an electron is found 90% of the time.

order In Linnaean classification, the taxonomic category above the level of family.

organ A specialized center of body function composed of several different types of tissues.

organelle (ōr-guh-nel′) Any of several membrane-enclosed structures with specialized functions, suspended in the cytosol of eukaryotic cells.

organic chemistry The study of carbon compounds (organic compounds).

organism An individual living thing, consisting of one or more cells.

organismal ecology The branch of ecology concerned with the morphological, physiological, and behavioral ways in which individual organisms meet the challenges posed by their biotic and abiotic environments.

organ of Corti (kor′-tē) The actual hearing organ of the vertebrate ear, located in the floor of the cochlear duct in the inner ear; contains the receptor cells (hair cells) of the ear.

organogenesis (ōr-gan′-ō-jen′-uh-sis) The process in which organ rudiments develop from the three germ layers after gastrulation.

organ system A group of organs that work together in performing vital body functions.

origin of replication Site where the replication of a DNA molecule begins, consisting of a specific sequence of nucleotides.

orthologous genes Homologous genes that are found in different species because of speciation.

osculum (os′-kyuh-lum) A large opening in a sponge that connects the spongocoel to the environment.

osmoconformer An animal that is isoosmotic with its environment.

osmolarity (oz′-mō-lār′-uh-tē) Solute concentration expressed as molarity.

osmoregulation Regulation of solute concentrations and water balance by a cell or organism.

osmoregulator An animal that controls its internal osmolarity independent of the external environment.

osmosis (oz-mō′-sis) The diffusion of free water across a selectively permeable membrane.

osteichthyan (os′-tē-ik′-thē-an) A member of a vertebrate clade with jaws and mostly bony skeletons.

outer ear One of the three main regions of the ear in reptiles (including birds) and mammals; made up of the auditory canal and, in many birds and mammals, the pinna.

outgroup A species or group of species from an evolutionary lineage that is known to have diverged before the lineage that contains the group of species being studied. An outgroup is selected so that its members are closely related to the group of species being studied, but not as closely related as any study-group members are to each other.

oval window In the vertebrate ear, a membrane-covered gap in the skull bone, through which sound waves pass from the middle ear to the inner ear.

ovarian cycle (ō-vār′-ē-un) The cyclic recurrence of the follicular phase, ovulation, and the luteal phase in the mammalian ovary, regulated by hormones.

ovary (ō′-vuh-rē) (1) In flowers, the portion of a carpel in which the egg-containing ovules develop. (2) In animals, the structure that produces female gametes and reproductive hormones.

oviduct (ō′-vuh-duct) A tube passing from the ovary to the vagina in invertebrates or to the uterus in vertebrates, where it is also called a fallopian tube.

oviparous (ō-vip′-uh-rus) Referring to a type of development in which young hatch from eggs laid outside the mother's body.

ovoviviparous (ō′-vō-vī-vip′-uh-rus) Referring to a type of development in which young hatch from eggs that are retained in the mother's uterus.

ovulation The release of an egg from an ovary. In humans, an ovarian follicle releases an egg during each uterine (menstrual) cycle.

ovule (o′-vyūl) A structure that develops within the ovary of a seed plant and contains the female gametophyte.

oxidation The complete or partial loss of electrons from a substance involved in a redox reaction.

oxidative phosphorylation (fos′-fōr-uh-lā′-shun) The production of ATP using energy derived from the redox reactions of an electron transport chain; the third major stage of cellular respiration.

oxidizing agent The electron acceptor in a redox reaction.

oxytocin (ok′-si-tō′-sen) A hormone produced by the hypothalamus and released from the posterior pituitary. It induces contractions of the uterine muscles during labor and causes the mammary glands to eject milk during nursing.

p53 gene A tumor-suppressor gene that codes for a specific transcription factor that promotes the synthesis of proteins that inhibit the cell cycle.

paedomorphosis (pē′-duh-mōr′-fuh-sis) The retention in an adult organism of the juvenile features of its evolutionary ancestors.

pain receptor A sensory receptor that responds to noxious or painful stimuli; also called a nociceptor.

paleoanthropology The study of human origins and evolution.

paleontology (pā′-lē-un-tol′-ō-jē) The scientific study of fossils.

pancreas (pan′-krē-us) A gland with exocrine and endocrine tissues. The exocrine portion functions in digestion, secreting enzymes and an alkaline solution into the small intestine via a duct; the ductless endocrine portion functions in homeostasis, secreting the hormones insulin and glucagon into the blood.

pancrustacean A member of a diverse arthropod clade that includes lobsters, crabs, barnacles and other crustaceans, as well as insects and their six-legged terrestrial relatives.

pandemic A global epidemic.

Pangaea (pan-jē′-uh) The supercontinent that formed near the end of the Paleozoic era, when plate movements brought all the landmasses of Earth together.

parabasalid A protist, such as a trichomonad, with modified mitochondria.

paracrine Referring to a secreted molecule that acts on a neighboring cell.

paralogous genes Homologous genes that are found in the same genome as a result of gene duplication.

paraphyletic (pār′-uh-fī-let′-ik) Pertaining to a group of taxa that consists of a common ancestor and some, but not all, of its descendants.

parareptile A basal group of reptiles, consisting mostly of large, stocky quadrupedal herbivores. Parareptiles died out in the late Triassic period.

parasite (pār′-uh-sīt) An organism that feeds on the cell contents, tissues, or body fluids of another species (the host) while in or on the host organism. Parasites harm but usually do not kill their host.

parasitism (pār′-uh-sit-izm) A +/− ecological interaction in which one organism, the parasite, benefits by feeding upon another organism, the host, which is harmed; some parasites live within the host (feeding on its tissues), while others feed on the host's external surface.

parasympathetic division A division of the autonomic nervous system; generally enhances body activities that gain and conserve energy, such as digestion and reduced heart rate.

parathyroid gland One of four small endocrine glands, embedded in the surface of the thyroid gland, that secrete parathyroid hormone.

parathyroid hormone (PTH) A hormone secreted by the parathyroid glands that raises blood calcium level by promoting calcium release from bone and calcium retention by the kidneys.

parenchyma cell (puh-ren′-ki-muh) A relatively unspecialized plant cell type that carries out most of the metabolism, synthesizes and stores organic products, and develops into a more differentiated cell type.

parental type An offspring with a phenotype that matches one of the true-breeding parental (P generation) phenotypes; also refers to the phenotype itself.

Parkinson's disease A progressive brain disease characterized by difficulty in initiating movements, slowness of movement, and rigidity.

parthenogenesis (par′-thuh-nō′-jen′-uh-sis) A form of asexual reproduction in which females produce offspring from unfertilized eggs.

partial pressure The pressure exerted by a particular gas in a mixture of gases (for instance, the pressure exerted by oxygen in air).

passive immunity Short-term immunity conferred by the transfer of antibodies, as occurs in the transfer of maternal antibodies to a fetus or nursing infant.

passive transport The diffusion of a substance across a biological membrane with no expenditure of energy.

pathogen An organism or virus that causes disease.

pathogen-associated molecular pattern (PAMP) A molecular sequence that is specific to a certain pathogen.

pattern formation The development of a multicellular organism's spatial organization, the arrangement of organs and tissues in their characteristic places in three-dimensional space.

peat Extensive deposits of partially decayed organic material often formed primarily from the wetland moss *Sphagnum*.

pedigree A diagram of a family tree with conventional symbols, showing the occurrence of heritable characters in parents and offspring over multiple generations.

pelagic zone The open-water component of aquatic biomes.

penis The copulatory structure of male mammals.

PEP carboxylase An enzyme that adds CO_2 to phosphoenolpyruvate (PEP) to form oxaloacetate in mesophyll cells of C_4 plants. It acts prior to photosynthesis.

pepsin An enzyme present in gastric juice that begins the hydrolysis of proteins.

pepsinogen The inactive form of pepsin secreted by chief cells located in gastric pits of the stomach.

peptide bond The covalent bond between the carboxyl group on one amino acid and the amino group on another, formed by a dehydration reaction.

peptidoglycan (pep'-tid-ō-glī'-kan) A type of polymer in bacterial cell walls consisting of modified sugars cross-linked by short polypeptides.

perception The interpretation of sensory system input by the brain.

pericycle The outermost layer in the vascular cylinder, from which lateral roots arise.

periderm (pār'-uh-derm') The protective coat that replaces the epidermis in woody plants during secondary growth, formed of the cork and cork cambium.

peripheral nervous system (PNS) The sensory and motor neurons that connect to the central nervous system.

peripheral protein A protein loosely bound to the surface of a membrane or to part of an integral protein and not embedded in the lipid bilayer.

peristalsis (pār'-uh-stal'-sis) (1) Alternating waves of contraction and relaxation in the smooth muscles lining the alimentary canal that push food along the canal. (2) A type of movement on land produced by rhythmic waves of muscle contractions passing from front to back, as in many annelids.

peristome (pār'-uh-stōme') A ring of interlocking, tooth-like structures on the upper part of a moss capsule (sporangium), often specialized for gradual spore discharge.

peritubular capillary One of the tiny blood vessels that form a network surrounding the proximal and distal tubules in the kidney.

peroxisome (puh-rok'-suh-sōm') An organelle containing enzymes that transfer hydrogen atoms from various substrates to oxygen (O_2), producing and then degrading hydrogen peroxide (H_2O_2).

personalized medicine A type of medical care in which each person's specific genetic profile can provide information about diseases or conditions for which the person is especially at risk and help make health-care decisions.

petal A modified leaf of a flowering plant. Petals are the often colorful parts of a flower that advertise it to insects and other pollinators.

petiole (pet'-ē-ōl) The stalk of a leaf, which joins the leaf to a node of the stem.

P generation The true-breeding (homozygous) parent individuals from which F_1 hybrid offspring are derived in studies of inheritance. (P stands for parental.)

pH A measure of hydrogen ion concentration equal to $-\log[H^+]$ and ranging in value from 0 to 14.

phage (fāj) A virus that infects bacteria; also called a bacteriophage.

phagocytosis (fag'-ō-sī-tō'-sis) A type of endocytosis in which large particulate substances or small organisms are taken up by a cell. It is carried out by some protists and by certain immune cells of animals (in mammals, mainly macrophages, neutrophils, and dendritic cells).

pharyngeal cleft (fuh-rin'-jē-ul) In chordate embryos, one of the grooves that separate a series of arches along the outer surface of the pharynx and may develop into a pharyngeal slit.

pharyngeal slit (fuh-rin'-jē-ul) In chordate embryos, one of the slits that form from the pharyngeal clefts and open into the pharynx, later developing into gill slits in many vertebrates.

pharynx (fār'-inks) (1) An area in the vertebrate throat where air and food passages cross. (2) In flatworms, the muscular tube that protrudes from the ventral side of the worm and ends in the mouth.

phase change (1) A shift from one developmental phase to another. (2) In plants, a morphological change that arises from a transition in shoot apical meristem activity.

phenotype (fē'-nō-tīp) The observable physical and physiological traits of an organism, which are determined by its genetic makeup.

pheromone (fār'-uh-mōn) In animals and fungi, a small molecule released into the environment that functions in communication between members of the same species. In animals, it acts much like a hormone in influencing physiology and behavior.

phloem (flō'-em) Vascular plant tissue consisting of living cells arranged into elongated tubes that transport sugar and other organic nutrients throughout the plant.

phloem sap (flō'-em) The sugar-rich solution carried through a plant's sieve tubes.

phosphate group A chemical group consisting of a phosphorus atom bonded to four oxygen atoms; important in energy transfer.

phospholipid (fos'-fō-lip'-id) A lipid made up of glycerol joined to two fatty acids and a phosphate group. The hydrocarbon chains of the fatty acids act as nonpolar, hydrophobic tails, while the rest of the molecule acts as a polar, hydrophilic head. Phospholipids form bilayers that function as biological membranes.

phosphorylated intermediate (fos'-fōr-uh-lā'-ted) A molecule (often a reactant) with a phosphate group covalently bound to it, making it more reactive (less stable) than the unphosphorylated molecule.

phosphorylation cascade (fos'-fōr-uh-lā'-shun) A series of chemical reactions during cell signaling mediated by enzymes (kinases), in which each kinase in turn phosphorylates and activates another, ultimately leading to phosphorylation of many proteins.

photic zone (fō'-tic) The narrow top layer of an ocean or lake, where light penetrates sufficiently for photosynthesis to occur.

photoautotroph (fō'-tō-ot'-ō-trōf) An organism that harnesses light energy to drive the synthesis of organic compounds from carbon dioxide.

photoheterotroph (fō'-tō-het'-er-ō-trōf) An organism that uses light to generate ATP but must obtain carbon in organic form.

photomorphogenesis Effects of light on plant morphology.

photon (fō'-ton) A quantum, or discrete quantity, of light energy that behaves as if it were a particle.

photoperiodism (fō'-tō-pēr'-ē-ō-dizm) A physiological response to photoperiod, the interval in a 24-hour period during which an organism is exposed to light. An example of photoperiodism is flowering.

photophosphorylation (fō'-tō-fos'-fōr-uh-lā'-shun) The process of generating ATP from ADP and phosphate by means of chemiosmosis, using a proton-motive force generated across the thylakoid membrane of the chloroplast or the membrane of certain prokaryotes during the light reactions of photosynthesis.

photoreceptor An electromagnetic receptor that detects the radiation known as visible light.

photorespiration A metabolic pathway that consumes oxygen and ATP, releases carbon dioxide, and decreases photosynthetic output. Photorespiration generally occurs on hot, dry, bright days, when the stomata close and the $O_2:CO_2$ ratio in the leaf increases, favoring the binding of O_2 rather than CO_2 by rubisco.

photosynthesis (fō'-tō-sin'-thi-sis) The conversion of light energy to chemical energy that is stored in sugars or other organic compounds; occurs in plants, algae, and certain prokaryotes.

photosystem A light-capturing unit located in the thylakoid membrane of the chloroplast or in the membrane of some prokaryotes, consisting of a reaction-center complex surrounded by numerous light-harvesting complexes. There are two types of photosystems, I and II; they absorb light best at different wavelengths.

photosystem I (PS I) A light-capturing unit in a chloroplast's thylakoid membrane or in the membrane of some prokaryotes; it has two molecules of P700 chlorophyll *a* at its reaction center.

photosystem II (PS II) One of two light-capturing units in a chloroplast's thylakoid membrane or in the membrane of some prokaryotes; it has two molecules of P680 chlorophyll *a* at its reaction center.

phototropism (fō'-tō-trō'-pizm) The bending of a plant or other organism in response to light, either toward the source of light (positive phototropism) or away from it (negative phototropism).

phylogenetic tree A branching diagram that represents a hypothesis about the evolutionary history of a group of organisms.

phylogeny (fī-loj'-uh-nē) The evolutionary history of a species or group of related species.

phylum (fī'-lum) (plural, **phyla**) In Linnaean classification, the taxonomic category above class.

physiology The processes and functions of an organism.

phytochromes (fī'-tuh-krōm) Plant pigments that absorb mostly red and far-red light and regulate many plant responses, such as seed germination and shade avoidance.

phytoremediation An emerging technology that seeks to reclaim contaminated areas by taking advantage of some plant species' ability to extract heavy metals and other pollutants from the soil and to concentrate them in easily harvested portions of the plant.

pilus (plural, **pili**) (pī'-lus, pī'-lī) In bacteria, a structure that links one cell to another at the start of conjugation; also called a sex pilus or conjugation pilus.

pineal gland (pī'-nē-ul) A small gland on the dorsal surface of the vertebrate forebrain that secretes the hormone melatonin.

pinocytosis (pī'-nō-sī-tō'-sis) A type of endocytosis in which the cell ingests extracellular fluid and its dissolved solutes.

pistil A single carpel (a simple pistil) or a group of fused carpels (a compound pistil).

pith Ground tissue that is internal to the vascular tissue in a stem; in many monocot roots, parenchyma cells that form the central core of the vascular cylinder.

pituitary gland (puh-tū'-uh-tār'-ē) An endocrine gland at the base of the hypothalamus; consists of a posterior lobe, which stores and releases two hormones produced by the hypothalamus, and an anterior lobe, which produces and secretes many hormones that regulate diverse body functions.

placenta (pluh-sen'-tuh) A structure in the uterus of a pregnant eutherian mammal that nourishes the fetus with the mother's blood supply; formed from the uterine lining and embryonic membranes.

placoderm A member of an extinct group of fishlike vertebrates that had jaws and were enclosed in a tough outer armor.

planarian A free-living flatworm found in ponds and streams.

plasma (plaz'-muh) The liquid matrix of blood in which the blood cells are suspended.

plasma membrane (plaz'-muh) The membrane at the boundary of every cell that acts as a selective barrier, regulating the cell's chemical composition.

plasmid (plaz'-mid) A small, circular, double-stranded DNA molecule that carries accessory genes separate from those of a bacterial chromosome; in DNA cloning, plasmids are used as vectors carrying up to about 10,000 base pairs (10 kb) of DNA. Plasmids are also found in some eukaryotes, such as yeasts.

plasmodesma (plaz'-mō-dez'-muh) (plural, **plasmodesmata**) An open channel through the cell wall that connects the cytoplasm of adjacent plant cells, allowing water, small solutes, and some larger molecules to pass between the cells.

plasmogamy (plaz-moh'-guh-mē) In fungi, the fusion of the cytoplasm of cells from two individuals; occurs as one stage of sexual reproduction, followed later by karyogamy.

plasmolysis (plaz-mol'-uh-sis) A phenomenon in walled cells in which the cytoplasm shrivels and the plasma membrane pulls away from the cell wall; occurs when the cell loses water to a hypertonic environment.

plastid One of a family of closely related organelles that includes chloroplasts, chromoplasts, and amyloplasts. Plastids are found in cells of photosynthetic eukaryotes.

plate tectonics The theory that the continents are part of great plates of Earth's crust that float on the hot, underlying portion of the mantle. Movements in the mantle cause the continents to move slowly over time.

platelet A pinched-off cytoplasmic fragment of a specialized bone marrow cell. Platelets circulate in the blood and are important in blood clotting.

pleiotropy (plī'-o-truh-pē) The ability of a single gene to have multiple effects.

pluripotent Describing a cell that can give rise to many, but not all, parts of an organism.

point mutation A change in a single nucleotide pair of a gene.

polar covalent bond A covalent bond between atoms that differ in electronegativity. The shared electrons are pulled closer to the more electronegative atom, making it slightly negative and the other atom slightly positive.

polar molecule A molecule (such as water) with an uneven distribution of charges in different regions of the molecule.

polarity A lack of symmetry; structural differences in opposite ends of an organism or structure, such as the root end and shoot end of a plant.

pollen grain In seed plants, a structure consisting of the male gametophyte enclosed within a pollen wall.

pollen tube A tube that forms after germination of the pollen grain and that functions in the delivery of sperm to the ovule.

pollination (pol'-uh-nā'-shun) The transfer of pollen to the part of a seed plant containing the ovules, a process required for fertilization.

poly-A tail A sequence of 50–250 adenine nucleotides added onto the 3' end of a pre-mRNA molecule.

polygamous Referring to a type of relationship in which an individual of one sex mates with several of the other.

polygenic inheritance (pol'-ē-jen'-ik) An additive effect of two or more genes on a single phenotypic character.

polymer (pol'-uh-mer) A long molecule consisting of many similar or identical monomers linked together by covalent bonds.

polymerase chain reaction (PCR) (puh-lim'-uh-rās) A technique for amplifying DNA *in vitro* by incubating it with specific primers, a heat-resistant DNA polymerase, and nucleotides.

polynucleotide (pol'-ē-nū'-klē-ō-tīd) A polymer consisting of many nucleotide monomers in a chain. The nucleotides can be those of DNA or RNA.

polyp The sessile variant of the cnidarian body plan. The alternate form is the medusa.

polypeptide (pol'-ē-pep'-tīd) A polymer of many amino acids linked together by peptide bonds.

polyphyletic (pol'-ē-fī-let'-ik) Pertaining to a group of taxa that includes distantly related organisms but does not include their most recent common ancestor.

polyploidy (pol'-ē-ploy'-dē) A chromosomal alteration in which the organism possesses more than two complete chromosome sets. It is the result of an accident of cell division.

polyribosome (polysome) (pol'-ē-rī'-buh-sōm') A group of several ribosomes attached to, and translating, the same messenger RNA molecule.

polysaccharide (pol'-ē-sak'-uh-rīd) A polymer of many monosaccharides, formed by dehydration reactions.

polyspermy The fertilization of an egg by more than one sperm.

polytomy (puh-lit'-uh-mē) In a phylogenetic tree, a branch point from which more than two descendant taxa emerge. A polytomy indicates that the evolutionary relationships between the descendant taxa are not yet clear.

pons A portion of the brain that participates in certain automatic, homeostatic functions, such as regulating the breathing centers in the medulla.

population A group of individuals of the same species that live in the same area and interbreed, producing fertile offspring.

population dynamics The study of how complex interactions between biotic and abiotic factors influence variations in population size.

population ecology The study of populations in relation to their environment, including environmental influences on population density and distribution, age structure, and variations in population size.

positional information Molecular cues that control pattern formation in an animal or plant embryonic structure by indicating a cell's location relative to the organism's body axes. These cues elicit a response by genes that regulate development.

positive feedback A form of regulation in which an end product of a process speeds up that process; in physiology, a control mechanism in which a change in a variable triggers a response that reinforces or amplifies the change.

positive interaction A +/+ or +/0 ecological interaction between individuals of two species in which at least one individual benefits and neither is harmed; positive interactions include mutualism and commensalism.

positive pressure breathing A breathing system in which air is forced into the lungs.

posterior Pertaining to the rear, or tail end, of a bilaterally symmetrical animal.

posterior pituitary An extension of the hypothalamus composed of nervous tissue that secretes oxytocin and antidiuretic hormone made in the hypothalamus; a temporary storage site for these hormones.

postzygotic barrier (pōst-zī-got'-ik) A reproductive barrier that prevents hybrid zygotes produced by two different species from developing into viable, fertile adults.

potential energy The energy that matter possesses as a result of its location or spatial arrangement (structure).

predation An interaction in which an individual of one species, the predator, kills and eats an individual of the other species, the prey.

prediction In deductive reasoning, a forecast that follows logically from a hypothesis. By testing predictions, experiments may allow certain hypotheses to be rejected.

pregnancy The condition of carrying one or more embryos in the uterus; also called gestation.

pressure potential (Ψ_P) A component of water potential that consists of the physical pressure on a solution, which can be positive, zero, or negative.

prezygotic barrier (prē'-zī-got'-ik) A reproductive barrier that impedes mating between species or hinders fertilization if interspecific mating is attempted.

primary cell wall In plants, a relatively thin and flexible layer that surrounds the plasma membrane of a young cell.

primary consumer An herbivore; an organism that eats plants or other autotrophs.

primary electron acceptor In the thylakoid membrane of a chloroplast or in the membrane of some prokaryotes, a specialized molecule that shares the reaction-center complex with a pair of chlorophyll *a* molecules and that accepts an electron from them.

primary growth Growth produced by apical meristems, lengthening stems and roots.

primary immune response The initial adaptive immune response to an antigen, which appears after a lag of about 10–17 days.

primary meristems The three meristematic derivatives (protoderm, procambium, and ground meristem) of an apical meristem.

primary oocyte (ō´-uh-sīt) An oocyte prior to completion of meiosis I.

primary producer An autotroph, usually a photosynthetic organism. Collectively, autotrophs make up the trophic level of an ecosystem that ultimately supports all other levels.

primary production The amount of light energy converted to chemical energy (organic compounds) by the autotrophs in an ecosystem during a given time period.

primary structure The level of protein structure referring to the specific linear sequence of amino acids.

primary succession A type of ecological succession that occurs in an area where there were originally no organisms present and where soil has not yet formed.

primary transcript An initial RNA transcript from any gene; also called pre-mRNA when transcribed from a protein-coding gene.

primase An enzyme that joins RNA nucleotides to make a primer during DNA replication, using the parental DNA strand as a template.

primer A short polynucleotide with a free 3′ end, bound by complementary base pairing to the template strand and elongated with DNA nucleotides during DNA replication.

prion An infectious agent that is a misfolded version of a normal cellular protein. Prions appear to increase in number by converting correctly folded versions of the protein to more prions.

problem solving The cognitive activity of devising a method to proceed from one state to another in the face of real or apparent obstacles.

producer An organism that produces organic compounds from CO_2 by harnessing light energy (in photosynthesis) or by oxidizing inorganic chemicals (in chemosynthetic reactions carried out by some prokaryotes).

product A material resulting from a chemical reaction.

production efficiency The percentage of energy stored in assimilated food that is not used for respiration or eliminated as waste.

progesterone A steroid hormone that contributes to the menstrual cycle and prepares the uterus for pregnancy; the major progestin in mammals.

progestin Any steroid hormone with progesterone-like activity.

prokaryote A single-celled organism of the domain Bacteria or Archaea.

prokaryotic cell (prō´-kār´-ē-ot´-ik) A type of cell lacking a membrane-enclosed nucleus and membrane-enclosed organelles. Organisms with prokaryotic cells (bacteria and archaea) are called prokaryotes.

prolactin A hormone produced and secreted by the anterior pituitary with a great diversity of effects in different vertebrate species. In mammals, it stimulates growth of and milk production by the mammary glands.

prometaphase The second stage of mitosis, in which the nuclear envelope fragments and the spindle microtubules attach to the kinetochores of the chromosomes.

promoter A specific nucleotide sequence in the DNA of a gene that binds RNA polymerase, positioning it to start transcribing RNA at the appropriate place.

prophage (prō´-fāj) A phage genome that has been inserted into a specific site on a bacterial chromosome.

prophase The first stage of mitosis, in which the chromatin condenses into discrete chromosomes visible with a light microscope, the mitotic spindle begins to form, and the nucleolus disappears but the nucleus remains intact.

prostaglandin (pros´-tuh-glan´-din) One of a group of modified fatty acids that are secreted by virtually all tissues and that perform a wide variety of functions as local regulators.

prostate gland (pros´-tāt) A gland in human males that secretes an acid-neutralizing component of semen.

protease (prō´-tē-āz) An enzyme that digests proteins by hydrolysis.

protein (prō´-tēn) A biologically functional molecule consisting of one or more polypeptides folded and coiled into a specific three-dimensional structure.

protein kinase (prō´-tēn kī´-nās) An enzyme that transfers phosphate groups from ATP to a protein, thus phosphorylating the protein.

protein phosphatase (prō´-tēn fos´-fuh-tās) An enzyme that removes phosphate groups from (dephosphorylates) proteins, often functioning to reverse the effect of a protein kinase.

proteoglycan (prō´-tē-ō-glī´-kan) A large molecule consisting of a small core protein with many carbohydrate chains attached, found in the extracellular matrix of animal cells. A proteoglycan may consist of up to 95% carbohydrate.

proteome The entire set of proteins expressed by a given cell, tissue, or organism.

proteomics (prō´-tē-ō´-miks) The systematic study of sets of proteins and their properties, including their abundance, chemical modifications, and interactions.

protist An informal term applied to any eukaryote that is not a plant, animal, or fungus. Most protists are unicellular, though some are colonial or multicellular.

protocell An abiotic precursor of a living cell that had a membrane-like structure and that maintained an internal chemistry different from that of its surroundings.

proton (prō´-ton) A subatomic particle with a single positive electrical charge, with a mass

of about 1.7×10^{-24} g, found in the nucleus of an atom.

protonema (prō´-tuh-nē´-muh) (plural, **protonemata**) A mass of green, branched, one-cell-thick filaments produced by germinating moss spores.

protonephridium (prō´-tō-nuh-frid´-ē-um) (plural, **protonephridia**) An excretory system, such as the flame bulb system of flatworms, consisting of a network of tubules lacking internal openings.

proton-motive force (prō´-ton) The potential energy stored in the form of a proton electrochemical gradient, generated by the pumping of hydrogen ions (H^+) across a biological membrane during chemiosmosis.

proton pump (prō´-ton) An active transport protein in a cell membrane that uses ATP to transport hydrogen ions out of a cell against their concentration gradient, generating a membrane potential in the process.

proto-oncogene (prō´-tō-on´-kō-jēn) A normal cellular gene that has the potential to become an oncogene.

protoplast The living part of a plant cell, which also includes the plasma membrane.

protostome development In animals, a developmental mode distinguished by the development of the mouth from the blastopore; often also characterized by spiral cleavage and by the body cavity forming when solid masses of mesoderm split.

provirus A viral genome that is permanently inserted into a host genome.

proximal tubule In the vertebrate kidney, the portion of a nephron immediately downstream from Bowman's capsule that conveys and helps refine filtrate.

pseudogene (sū´-dō-jēn) A DNA segment that is very similar to a real gene but does not yield a functional product; a DNA segment that formerly functioned as a gene but has become inactivated in a particular species because of mutation.

pseudopodium (sū´-dō-pō´-dē-um) (plural, **pseudopodia**) A cellular extension of amoeboid cells used in moving and feeding.

P site One of a ribosome's three binding sites for tRNA during translation. The P site holds the tRNA carrying the growing polypeptide chain. (P stands for peptidyl tRNA.)

pterosaur Winged reptile that lived during the Mesozoic era.

pulse The rhythmic bulging of the artery walls with each heartbeat.

punctuated equilibria In the fossil record, long periods of apparent stasis, in which a species undergoes little or no morphological change, interrupted by relatively brief periods of sudden change.

Punnett square A diagram used in the study of inheritance to show the predicted genotypic results of random fertilization in genetic crosses between individuals of known genotype.

pupil The opening in the iris, which admits light into the interior of the vertebrate eye. Muscles in the iris regulate its size.

purine (pyū´-rēn) One of two types of nitrogenous bases found in nucleotides, characterized by a six-membered ring fused to a five-membered ring. Adenine (A) and guanine (G) are purines.

pyrimidine (puh-rim'-uh-dēn) One of two types of nitrogenous bases found in nucleotides, characterized by a six-membered ring. Cytosine (C), thymine (T), and uracil (U) are pyrimidines.

quantitative character A heritable feature that varies continuously over a range rather than in an either-or fashion.

quaternary structure (kwot'-er-nār'-ē) The particular shape of a complex, aggregate protein, defined by the characteristic three-dimensional arrangement of its constituent subunits, each a polypeptide.

radial cleavage A type of embryonic development in deuterostomes in which the planes of cell division that transform the zygote into a ball of cells are either parallel or perpendicular to the vertical axis of the embryo, thereby aligning tiers of cells one above the other.

radial symmetry Symmetry in which the body is shaped like a pie or barrel (lacking a left side and a right side) and can be divided into mirror-imaged halves by any plane through its central axis.

radicle An embryonic root of a plant.

radioactive isotope An isotope (an atomic form of a chemical element) that is unstable; the nucleus decays spontaneously, giving off detectable particles and energy.

radiolarian A protist, usually marine, with a shell generally made of silica and pseudopodia that radiate from the central body.

radiometric dating A method for determining the absolute age of rocks and fossils, based on the half-life of radioactive isotopes.

radula A straplike scraping organ used by many molluscs during feeding.

ras gene A gene that codes for Ras, a G protein that relays a growth signal from a growth factor receptor on the plasma membrane to a cascade of protein kinases, ultimately resulting in stimulation of the cell cycle.

ratite (rat'-īt) A member of the group of flightless birds.

ray-finned fish A member of the clade Actinopterygii, aquatic osteichthyans with fins supported by long, flexible rays, including tuna, bass, and herring.

reabsorption In excretory systems, the recovery of solutes and water from filtrate.

reactant A starting material in a chemical reaction.

reaction-center complex A complex of proteins associated with a special pair of chlorophyll *a* molecules and a primary electron acceptor. Located centrally in a photosystem, this complex triggers the light reactions of photosynthesis. Excited by light energy, the pair of chlorophylls donates an electron to the primary electron acceptor, which passes an electron to an electron transport chain.

reading frame On an mRNA, the triplet grouping of ribonucleotides used by the translation machinery during polypeptide synthesis.

receptacle The base of a flower; the part of the stem that is the site of attachment of the floral organs.

reception In cellular communication, the first step of a signaling pathway in which a signaling molecule is detected by a receptor molecule on or in the cell.

receptor-mediated endocytosis (en'-dō-sī-tō'-sis) The movement of specific molecules into a cell by the infolding of vesicles containing proteins with receptor sites specific to the molecules being taken in; enables a cell to acquire bulk quantities of specific substances.

receptor potential A graded potential occurring in a receptor cell.

receptor tyrosine kinase (RTK) A receptor protein spanning the plasma membrane, the cytoplasmic (intracellular) part of which can catalyze the transfer of a phosphate group from ATP to a tyrosine on another protein. Receptor tyrosine kinases often respond to the binding of a signaling molecule by dimerizing and then phosphorylating a tyrosine on the cytoplasmic portion of the other receptor in the dimer.

recessive allele An allele whose phenotypic effect is not observed in a heterozygote.

reciprocal altruism Altruistic behavior between unrelated individuals, whereby the altruistic individual benefits in the future when the beneficiary reciprocates.

recombinant chromosome A chromosome created when crossing over combines DNA from two parents into a single chromosome.

recombinant DNA molecule A DNA molecule made *in vitro* with segments from different sources.

recombinant type (recombinant) An offspring whose phenotype differs from that of the true-breeding P generation parents; also refers to the phenotype itself.

rectum The terminal portion of the large intestine, where the feces are stored prior to elimination.

red alga A photosynthetic protist, named for its color, which results from a red pigment that masks the green of chlorophyll. Most red algae are multicellular and marine.

redox reaction (rē'-doks) A chemical reaction involving the complete or partial transfer of one or more electrons from one reactant to another; short for **red**uction-**ox**idation reaction.

reducing agent The electron donor in a redox reaction.

reduction The complete or partial addition of electrons to a substance involved in a redox reaction.

reference genome A complete sequence that researchers agree best represents the genome of a given species, arrived at by sequencing multiple individuals.

reflex An automatic reaction to a stimulus, mediated by the spinal cord or lower brain.

refractory period (rē-frakt'-ōr-ē) A period, immediately following a response to stimulation, during which a cell or organ is unresponsive to further stimulation.

regulator An animal for which mechanisms of homeostasis moderate internal changes in a particular variable in the face of external fluctuation of that variable.

regulatory gene A gene that codes for a protein, such as a repressor, that controls the transcription of another gene or group of genes.

reinforcement In evolutionary biology, a process in which natural selection strengthens prezygotic barriers to reproduction, thus reducing the chances of hybrid formation. Such a process is likely to occur only if hybrid offspring are less fit than members of the parent species.

relative abundance The proportional abundance of different species in a community.

relative fitness The contribution an individual makes to the gene pool of the next generation, relative to the contributions of other individuals in the population.

renal cortex The outer portion of the vertebrate kidney.

renal medulla The inner portion of the vertebrate kidney, beneath the renal cortex.

renal pelvis The funnel-shaped chamber that receives processed filtrate from the vertebrate kidney's collecting ducts and is drained by the ureter.

renin-angiotensin-aldosterone system (RAAS) A hormone cascade pathway that helps regulate blood pressure and blood volume.

repetitive DNA Nucleotide sequences, usually noncoding, that are present in many copies in a eukaryotic genome. The repeated units may be short and arranged tandemly (in series) or long and dispersed in the genome.

replication fork A Y-shaped region on a replicating DNA molecule where the parental strands are being unwound and new strands are being synthesized.

repressor A protein that inhibits gene transcription. In prokaryotes, repressors bind to the DNA in or near the promoter. In eukaryotes, repressors may bind to control elements within enhancers, to activators, or to other proteins in a way that blocks activators from binding to DNA.

reproductive isolation The existence of biological factors (barriers) that impede members of two species from producing viable, fertile offspring.

reptile A member of the clade of amniotes that includes tuataras, lizards and snakes, turtles, crocodilians, and birds.

reservoir In biogeochemical cycles, location of a chemical element, consisting of either organic or inorganic materials that are either available for direct use by organisms or unavailable as nutrients.

residual volume The amount of air that remains in the lungs after forceful exhalation.

resource partitioning The division of environmental resources by coexisting species such that the niche of each species differs by one or more significant factors from the niches of all coexisting species.

respiratory pigment A protein that transports oxygen in blood or hemolymph.

response (1) In cellular communication, the change in a specific cellular activity brought about by a transduced signal from outside the cell. (2) In feedback regulation, a physiological activity triggered by a change in a variable.

resting potential The membrane potential characteristic of a nonconducting excitable cell, with the inside of the cell more negative than the outside.

restriction enzyme An endonuclease (type of enzyme) that recognizes and cuts DNA molecules foreign to a bacterium (such as phage genomes). The enzyme cuts at specific nucleotide sequences (restriction sites).

restriction fragment A DNA segment that results from the cutting of DNA by a restriction enzyme.

restriction site A specific sequence on a DNA strand that is recognized and cut by a restriction enzyme.

retina (ret'-i-nuh) The innermost layer of the vertebrate eye, containing photoreceptor cells (rods and cones) and neurons; transmits images formed by the lens to the brain via the optic nerve.

retinal The light-absorbing pigment in rods and cones of the vertebrate eye.

retrotransposon (re'-trō-trans-pō'-zon) A transposable element that moves within a genome by means of an RNA intermediate, a transcript of the retrotransposon DNA.

retrovirus (re'-trō-vī'-rus) An RNA virus that replicates by transcribing its RNA into DNA and then inserting the DNA into a cellular chromosome; an important class of cancer-causing viruses.

reverse transcriptase (tran-skrip'-tās) An enzyme encoded by certain viruses (retroviruses) that uses RNA as a template for DNA synthesis.

reverse transcriptase–polymerase chain reaction (RT-PCR) A technique for determining expression of a particular gene. It uses reverse transcriptase and DNA polymerase to synthesize cDNA from all the mRNA in a sample and then subjects the cDNA to PCR amplification using primers specific for the gene of interest.

rhizarians (rī-za'-rē-uhns) One of the three major subgroups for which the SAR eukaryotic supergroup is named. Many species in this clade are amoebas characterized by threadlike pseudopodia.

rhizobacterium A soil bacterium whose population size is much enhanced in the rhizosphere, the soil region close to a plant's roots.

rhizoid (rī'-zoyd) A long, tubular single cell or filament of cells that anchors bryophytes to the ground. Unlike roots, rhizoids are not composed of tissues, lack specialized conducting cells, and do not play a primary role in water and mineral absorption.

rhizosphere The soil region close to plant roots and characterized by a high level of microbiological activity.

rhodopsin (rō-dop'-sin) A visual pigment consisting of retinal and opsin. Upon absorbing light, the retinal changes shape and dissociates from the opsin.

ribonucleic acid (RNA) (rī'-bō-nū-klā'-ik) A type of nucleic acid consisting of a polynucleotide made up of nucleotide monomers with a ribose sugar and the nitrogenous bases adenine (A), cytosine (C), guanine (G), and uracil (U); usually single-stranded; functions in protein synthesis, in gene regulation, and as the genome of some viruses.

ribose The sugar component of RNA nucleotides.

ribosomal RNA (rRNA) (rī'-buh-sō'-mul) RNA molecules that, together with proteins, make up ribosomes; the most abundant type of RNA.

ribosome (rī'-buh-sōm) A complex of rRNA and protein molecules that functions as a site of protein synthesis in the cytoplasm; consists of a large and a small subunit. In eukaryotic cells, each subunit is assembled in the nucleolus. *See also* nucleolus.

ribozyme (rī'-buh-zīm) An RNA molecule that functions as an enzyme, such as an intron that catalyzes its own removal during RNA splicing.

RNA interference (RNAi) A mechanism for silencing the expression of specific genes. In RNAi, double-stranded RNA molecules that match the sequence of a particular gene are processed into siRNAs that either block translation or trigger the degradation of the gene's messenger RNA. This happens naturally in some cells and can be carried out in laboratory experiments as well.

RNA polymerase An enzyme that links ribonucleotides into a growing RNA chain during transcription, based on complementary binding to nucleotides on a DNA template strand.

RNA processing Modification of RNA primary transcripts, including splicing out of introns, joining together of exons, and alteration of the 5′ and 3′ ends.

RNA sequencing (RNA-seq) (RNA-sēk) A method of analyzing large sets of RNAs that involves making cDNAs and sequencing them.

RNA splicing After synthesis of a eukaryotic primary RNA transcript, the removal of portions of the transcript (introns) that will not be included in the mRNA and the joining together of the remaining portions (exons).

rod A rodlike cell in the retina of the vertebrate eye, sensitive to low light intensity.

root An organ in vascular plants that anchors the plant and enables it to absorb water and minerals from the soil.

root cap A cone of cells at the tip of a plant root that protects the apical meristem.

root hair A tiny extension of a root epidermal cell, growing just behind the root tip and increasing surface area for absorption of water and minerals.

root pressure Pressure exerted in the roots of plants as the result of osmosis, causing exudation from cut stems and guttation of water from leaves.

root system All of a plant's roots, which anchor it in the soil, absorb and transport minerals and water, and store food.

rooted Describing a phylogenetic tree that contains a branch point (often, the one farthest to the left) representing the most recent common ancestor of all taxa in the tree.

rough ER That portion of the endoplasmic reticulum with ribosomes attached.

round window In the mammalian ear, the point of contact where vibrations of the stapes create a traveling series of pressure waves in the fluid of the cochlea.

R plasmid A bacterial plasmid carrying genes that confer resistance to certain antibiotics.

r-selection Selection for life history traits that maximize reproductive success in uncrowded environments.

rubisco (rū-bis'-kō) Ribulose bisphosphate (RuBP) carboxylase-oxygenase, the enzyme that normally catalyzes the first step of the Calvin cycle (the addition of CO_2 to RuBP). When excess O_2 is present or CO_2 levels are low, rubisco can bind oxygen, resulting in photorespiration.

ruminant (rūh'-muh-nent) A cud-chewing animal, such as a cow or sheep, with multiple stomach compartments specialized for an herbivorous diet.

salicylic acid (sal'-i-sil'-ik) A signaling molecule in plants that may be partially responsible for activating systemic acquired resistance to pathogens.

salivary gland A gland associated with the oral cavity that secretes substances that lubricate food and begin the process of chemical digestion.

salt A compound resulting from the formation of an ionic bond; also called an ionic compound.

saltatory conduction (sol'-tuh-tōr'-ē) Rapid transmission of a nerve impulse along an axon, resulting from the action potential jumping from one node of Ranvier to another, skipping the myelin-sheathed regions of membrane.

SAR One of four supergroups of eukaryotes proposed in a current hypothesis of the evolutionary history of eukaryotes. This supergroup contains a large, extremely diverse collection of protists from three major subgroups: stramenopiles, alveolates, and rhizarians. *See also* Excavata, Archaeplastida, and Unikonta.

sarcomere (sar'-kō-mēr) The fundamental, repeating unit of striated muscle, delimited by the Z lines.

sarcoplasmic reticulum (SR) (sar'-kō-plaz'-mik ruh-tik'-yū-lum) A specialized endoplasmic reticulum that regulates the calcium concentration in the cytosol of muscle cells.

saturated fatty acid A fatty acid in which all carbons in the hydrocarbon tail are connected by single bonds, thus maximizing the number of hydrogen atoms that are attached to the carbon skeleton.

savanna A tropical grassland biome with scattered individual trees and large herbivores and maintained by occasional fires and drought.

scaffolding protein A type of large relay protein to which several other relay proteins are simultaneously attached, increasing the efficiency of signal transduction.

scanning electron microscope (SEM) A microscope that uses an electron beam to scan the surface of a sample, coated with metal atoms, to study details of its topography.

scatter plot A graph in which each piece of data is represented by a point. A scatter plot is used when the data for all variables are numerical and continuous.

schizophrenia (skit'-suh-frē'-nē-uh) A severe mental disturbance characterized by psychotic episodes in which patients have a distorted perception of reality.

Schwann cell A type of glial cell that forms insulating myelin sheaths around the axons of neurons in the peripheral nervous system.

science An approach to understanding the natural world.

scion (sī'-un) The twig grafted onto the stock when making a graft.

sclerenchyma cell (skluh-ren′-kim-uh) A rigid, supportive plant cell type usually lacking a protoplast and possessing thick secondary walls strengthened by lignin at maturity.

scrotum A pouch of skin outside the abdomen that houses the testes; functions in maintaining the testes at the lower temperature required for spermatogenesis.

second law of thermodynamics The principle stating that every energy transfer or transformation increases the entropy of the universe. Usable forms of energy are at least partly converted to heat.

second messenger A molecule that relays messages in a cell from a receptor to a target where an action within the cell takes place.

secondary cell wall In plant cells, a strong and durable matrix that is often deposited in several laminated layers around the plasma membrane and provides protection and support.

secondary consumer A carnivore that eats herbivores.

secondary endosymbiosis A process in eukaryotic evolution in which a heterotrophic eukaryotic cell engulfed a photosynthetic eukaryotic cell, which survived in a symbiotic relationship inside the heterotrophic cell.

secondary growth Growth produced by lateral meristems, thickening the roots and shoots of woody plants.

secondary immune response The adaptive immune response elicited on second or subsequent exposures to a particular antigen. The secondary immune response is more rapid, of greater magnitude, and of longer duration than the primary immune response.

secondary oocyte (ō′-uh-sīt) An oocyte that has completed meiosis I.

secondary production The amount of chemical energy in consumers' food that is converted to their own new biomass during a given time period.

secondary structure Regions of repetitive coiling or folding of the polypeptide backbone of a protein due to hydrogen bonding between constituents of the backbone (not the side chains).

secondary succession A type of succession that occurs where an existing community has been cleared by some disturbance that leaves the soil or substrate intact.

secretion (1) The discharge of molecules synthesized by a cell. (2) The active transport of wastes and certain other solutes from the body fluid into the filtrate in an excretory system.

seed An adaptation of some terrestrial plants consisting of an embryo packaged along with a store of food within a protective coat.

seed coat A tough outer covering of a seed, formed from the outer coat of an ovule. In a flowering plant, the seed coat encloses and protects the embryo and endosperm.

seedless vascular plant An informal name for a plant that has vascular tissue but lacks seeds. Seedless vascular plants form a paraphyletic group that includes the phyla Lycophyta (club mosses and their relatives) and Monilophyta (ferns and their relatives).

selective permeability A property of biological membranes that allows them to regulate the passage of substances across them.

self-incompatibility The ability of a seed plant to reject its own pollen and sometimes the pollen of closely related individuals.

semelparity (seh′-mel-pār′-i-tē) Reproduction in which an organism produces all of its offspring in a single event; also called big-bang reproduction.

semen (sē′-mun) The fluid that is ejaculated by the male during orgasm; contains sperm and secretions from several glands of the male reproductive tract.

semicircular canals A three-part chamber of the inner ear that functions in maintaining equilibrium.

semiconservative model Type of DNA replication in which the replicated double helix consists of one old strand, derived from the parental molecule, and one newly made strand.

semilunar valve A valve located at each exit of the heart, where the aorta leaves the left ventricle and the pulmonary artery leaves the right ventricle.

seminal vesicle (sem′-i-nul ves′-i-kul) A gland in males that secretes a fluid component of semen that lubricates and nourishes sperm.

seminiferous tubule (sem′-i-nif′-er-us) A highly coiled tube in the testis in which sperm are produced.

senescence (se-nes′-ens) The programmed death of certain cells or organs or the entire organism.

sensitive period A limited phase in an animal's development when learning of particular behaviors can take place; also called a critical period.

sensor In homeostasis, a receptor that detects a stimulus.

sensory adaptation The tendency of sensory neurons to become less sensitive when they are stimulated repeatedly.

sensory neuron A nerve cell that receives information from the internal or external environment and transmits signals to the central nervous system.

sensory reception The detection of a stimulus by sensory cells.

sensory receptor A specialized structure or cell that responds to a stimulus from an animal's internal or external environment.

sensory transduction The conversion of stimulus energy to a change in the membrane potential of a sensory receptor cell.

sepal (sē′-pul) A modified leaf in angiosperms that helps enclose and protect a flower bud before it opens.

septum (plural, **septa**) One of the cross-walls that divide a fungal hypha into cells. Septa generally have large pores large enough to allow ribosomes, mitochondria, and even nuclei to flow from cell to cell.

serial endosymbiosis A hypothesis for the origin of eukaryotes consisting of a sequence of endosymbiotic events in which mitochondria, chloroplasts, and perhaps other cellular structures were derived from small prokaryotes that had been engulfed by larger cells.

set point In homeostasis in animals, a value maintained for a particular variable, such as body temperature or solute concentration.

seta (sē′-tuh) (plural, **setae**) The elongated stalk of a bryophyte sporophyte.

sex chromosome A chromosome responsible for determining the sex of an individual.

sex-linked gene A gene located on either sex chromosome. Most sex-linked genes are on the X chromosome and show distinctive patterns of inheritance; there are very few genes on the Y chromosome.

sexual dimorphism (dī-mōr′-fizm) Differences between the secondary sex characteristics of males and females of the same species.

sexual reproduction Reproduction arising from fusion of two gametes.

sexual selection A process in which individuals with certain inherited characteristics are more likely than other individuals of the same sex to obtain mates.

Shannon diversity index An index of community diversity symbolized by H and represented by the equation $H = -(p_A \ln p_A + p_B \ln p_B + p_C \ln p_C + \cdots)$, where A, B, C … are species, p is the relative abundance of each species, and ln is the natural logarithm.

shared ancestral character A character, shared by members of a particular clade, that originated in an ancestor that is not a member of that clade.

shared derived character An evolutionary novelty that is unique to a particular clade.

shoot system The aerial portion of a plant body, consisting of stems, leaves, and (in angiosperms) flowers.

short tandem repeat (STR) Simple sequence DNA containing multiple tandemly repeated units of two to five nucleotides. Variations in STRs act as genetic markers in STR analysis, used to prepare genetic profiles.

short-day plant A plant that flowers (usually in late summer, fall, or winter) only when the light period is shorter than a critical length.

short-term memory The ability to hold information, anticipations, or goals for a time and then release them if they become irrelevant.

sickle-cell disease A recessively inherited human blood disorder in which a single nucleotide change in the α-globin gene causes hemoglobin to aggregate, changing red blood cell shape and causing multiple symptoms in afflicted individuals.

sieve plate An end wall in a sieve-tube element, which facilitates the flow of phloem sap in angiosperm sieve tubes.

sieve-tube element A living cell that conducts sugars and other organic nutrients in the phloem of angiosperms; also called a sieve-tube member. Connected end to end, they form sieve tubes.

sign stimulus An external sensory cue that triggers a fixed action pattern by an animal.

signal Any kind of information sent from one organism to another, or from one place in an organism to another place.

signal peptide A sequence of about 20 amino acids at or near the leading (amino) end of a polypeptide that targets it to the endoplasmic reticulum or other organelles in a eukaryotic cell.

signal-recognition particle (SRP) A protein-RNA complex that recognizes a signal peptide as it emerges from a ribosome and helps direct the ribosome to the endoplasmic reticulum (ER) by binding to a receptor protein on the ER.

signal transduction The linkage of a mechanical, chemical, or electromagnetic stimulus to a specific cellular response.

signal transduction pathway A series of steps linking a mechanical, chemical, or electrical stimulus to a specific cellular response.

silent mutation A nucleotide-pair substitution that has no observable effect on the phenotype; for example, within a gene, a mutation that results in a codon that codes for the same amino acid.

simple fruit A fruit derived from a single carpel or several fused carpels.

simple sequence DNA A DNA sequence that contains many copies of tandemly repeated short sequences.

single bond A single covalent bond; the sharing of a pair of valence electrons by two atoms.

single circulation A circulatory system consisting of a single pump and circuit, in which blood passes from the sites of gas exchange to the rest of the body before returning to the heart.

single-lens eye The camera-like eye found in some jellies, polychaete worms, spiders, and many molluscs.

single nucleotide polymorphism (SNP) (snip) A single base-pair site in a genome where nucleotide variation is found in at least 1% of the population.

single-strand binding protein A protein that binds to the unpaired DNA strands during DNA replication, stabilizing them and holding them apart while they serve as templates for the synthesis of complementary strands of DNA.

sinoatrial (SA) node (sī′-nō-ā′-trē-uhl) A region in the right atrium of the heart that sets the rate and timing at which all cardiac muscle cells contract; the pacemaker.

sister chromatids Two copies of a duplicated chromosome attached to each other by proteins at the centromere and, sometimes, along the arms. While joined, two sister chromatids make up one chromosome. Chromatids are eventually separated during mitosis or meiosis II.

sister taxa Groups of organisms that share an immediate common ancestor and hence are each other's closest relatives.

skeletal muscle A type of striated muscle that is generally responsible for the voluntary movements of the body.

sliding-filament model The idea that muscle contraction is based on the movement of thin (actin) filaments along thick (myosin) filaments, shortening the sarcomere, the basic unit of muscle organization.

slow-twitch fiber A muscle fiber that can sustain long contractions.

small interfering RNA (siRNA) One of multiple small, single-stranded RNA molecules generated by cellular machinery from a long, linear, double-stranded RNA molecule. The siRNA associates with one or more proteins in a complex that can degrade or prevent translation of an mRNA with a complementary sequence.

small intestine The longest section of the alimentary canal, so named because of its small diameter compared with that of the large intestine; the principal site of the enzymatic hydrolysis of food macromolecules and the absorption of nutrients.

smooth ER That portion of the endoplasmic reticulum that is free of ribosomes.

smooth muscle A type of muscle lacking the striations of skeletal and cardiac muscle because of the uniform distribution of myosin filaments in the cells; responsible for involuntary body activities.

social learning Modification of behavior through the observation of other individuals.

sociobiology The study of social behavior based on evolutionary theory.

sodium-potassium pump A transport protein in the plasma membrane of animal cells that actively transports sodium out of the cell and potassium into the cell.

soil horizon A soil layer with physical characteristics that differ from those of the layers above or beneath.

solute (sol′-yūt) A substance that is dissolved in a solution.

solute potential (Ψ_S) A component of water potential that is proportional to the molarity of a solution and that measures the effect of solutes on the direction of water movement; also called osmotic potential, it can be either zero or negative.

solution A liquid that is a homogeneous mixture of two or more substances.

solvent The dissolving agent of a solution. Water is the most versatile solvent known.

somatic cell (sō-mat′-ik) Any cell in a multicellular organism except a sperm or egg or their precursors.

somite One of a series of blocks of mesoderm that exist in pairs just lateral to the notochord in a vertebrate embryo.

soredium (suh-rē′-dē-um) (plural, **soredia**) In lichens, a small cluster of fungal hyphae with embedded algae.

sorus (plural, **sori**) A cluster of sporangia on a fern sporophyll. Sori may be arranged in various patterns, such as parallel lines or dots, which are useful in fern identification.

spatial learning The establishment of a memory that reflects the environment's spatial structure.

speciation (spē-sē-ā′-shun) An evolutionary process in which one species splits into two or more species.

species (spē′-sēz) A population or group of populations whose members have the potential to interbreed in nature and produce viable, fertile offspring but do not produce viable, fertile offspring with members of other such groups.

species-area curve (spē′-sēz) The biodiversity pattern that shows that the larger the geographic area of a community is, the more species it has.

species diversity (spē′-sēz) The number and relative abundance of species in a biological community.

species richness (spē′-sēz) The number of species in a biological community.

specific heat The amount of heat that must be absorbed or lost for 1 g of a substance to change its temperature by 1°C.

spectrophotometer An instrument that measures the proportions of light of different wavelengths absorbed and transmitted by a pigment solution.

sperm The male gamete.

spermatheca (sper′-muh-thē′-kuh) (plural, **spermathecae**) In many insects, a sac in the female reproductive system where sperm are stored.

spermatogenesis (sper-ma′-tō-gen′-uh-sis) The continuous and prolific production of mature sperm in the testis.

spermatogonium (sper-ma′-tō-gō′-nē-um) (plural, **spermatogonia**) A cell that divides mitotically to form spermatocytes.

S phase The synthesis phase of the cell cycle; the portion of interphase during which DNA is replicated.

sphincter (sfink′-ter) A ringlike band of muscle fibers that controls the size of an opening in the body, such as the passage between the esophagus and the stomach.

spiral cleavage A type of embryonic development in protostomes in which the planes of cell division that transform the zygote into a ball of cells are diagonal to the vertical axis of the embryo. As a result, the cells of each tier sit in the grooves between cells of adjacent tiers.

spliceosome (splī′-sō-sōm) A large complex made up of proteins and RNA molecules that splices RNA by interacting with the ends of an RNA intron, releasing the intron and joining the two adjacent exons.

spongocoel (spon′-jō-sēl) The central cavity of a sponge.

spontaneous process A process that occurs without an overall input of energy; a process that is energetically favorable.

sporangium (spōr-an′-jē-um) (plural, **sporangia**) A multicellular organ in fungi and plants in which meiosis occurs and haploid cells develop.

spore (1) In the life cycle of a plant or alga undergoing alternation of generations, a haploid cell produced in the sporophyte by meiosis. A spore can divide by mitosis to develop into a multicellular haploid individual, the gametophyte, without fusing with another cell. (2) In fungi, a haploid cell, produced either sexually or asexually, that produces a mycelium after germination.

sporocyte (spō′-ruh-sīt) A diploid cell within a sporangium that undergoes meiosis and generates haploid spores; also called a spore mother cell.

sporophyll (spō′-ruh-fil) A modified leaf that bears sporangia and hence is specialized for reproduction.

sporophyte (spō′-ruh-fīt) In organisms (plants and some algae) that have alternation of generations, the multicellular diploid form that results from the union of gametes. Meiosis in the sporophyte produces haploid spores that develop into gametophytes.

sporopollenin (spōr-uh-pol′-eh-nin) A durable polymer that covers exposed zygotes of charophyte algae and forms the walls of plant spores, preventing them from drying out.

stability In evolutionary biology, a term referring to a hybrid zone in which hybrids continue to be produced; this causes the hybrid zone to be "stable" in the sense of persisting over time.

stabilizing selection Natural selection in which intermediate phenotypes survive or reproduce more successfully than do extreme phenotypes.

stamen (stā'-men) The pollen-producing reproductive organ of a flower, consisting of an anther and a filament.

standard deviation A measure of the variation found in a set of data points.

standard metabolic rate (SMR) Metabolic rate of a resting, fasting, and nonstressed ectotherm at a particular temperature.

starch A storage polysaccharide in plants, consisting entirely of glucose monomers joined by glycosidic linkages.

start point In transcription, the nucleotide position on the promoter where RNA polymerase begins synthesis of RNA.

statocyst (stat'-uh-sist') A type of mechanoreceptor that functions in equilibrium in invertebrates by use of statoliths, which stimulate hair cells in relation to gravity.

statolith (stat'-uh-lith') (1) In plants, a specialized plastid that contains dense starch grains and may play a role in detecting gravity. (2) In invertebrates, a dense particle that settles in response to gravity and is found in sensory organs that function in equilibrium.

stele (stēl) The vascular tissue of a stem or root.

stem A vascular plant organ consisting of an alternating system of nodes and internodes that support the leaves and reproductive structures.

stem cell Any relatively unspecialized cell that can produce, during a single division, two identical daughter cells or two more specialized daughter cells that can undergo further differentiation, or one cell of each type.

steroid A type of lipid characterized by a carbon skeleton consisting of four fused rings with various chemical groups attached.

sticky end A single-stranded end of a double-stranded restriction fragment.

stigma (plural, **stigmata**) The sticky part of a flower's carpel, which receives pollen grains.

stimulus In feedback regulation, a fluctuation in a variable that triggers a response.

stipe A stemlike structure of a seaweed.

stock The plant that provides the root system when making a graft.

stoma (stō'-muh) (plural, **stomata**) A microscopic pore surrounded by guard cells in the epidermis of leaves and stems that allows gas exchange between the environment and the interior of the plant.

stomach An organ of the digestive system that stores food and performs preliminary steps of digestion.

stramenopiles (strah'-men-ō'-pē-lēs) One of the three major subgroups for which the SAR eukaryotic supergroup is named. This clade arose by secondary endosymbiosis and includes diatoms and brown algae.

stratum (strah'-tum) (plural, **strata**) A rock layer formed when new layers of sediment cover older ones and compress them.

strigolactone Any of a class of plant hormones that inhibit shoot branching, trigger the germination of parasitic plant seeds, and stimulate the association of plant roots with mycorrhizal fungi.

strobilus (strō-bī'-lus) (plural, **strobili**) The technical term for a cluster of sporophylls known commonly as a cone, found in most gymnosperms and some seedless vascular plants.

stroke The death of nervous tissue in the brain, usually resulting from rupture or blockage of arteries in the neck or head.

stroke volume The volume of blood pumped by a heart ventricle in a single contraction.

stroma (strō'-muh) The dense fluid within the chloroplast surrounding the thylakoid membrane and containing ribosomes and DNA; involved in the synthesis of organic molecules from carbon dioxide and water.

stromatolite Layered rock that results from the activities of prokaryotes that bind thin films of sediment together.

structural isomer One of two or more compounds that have the same molecular formula but differ in the covalent arrangements of their atoms.

style The stalk of a flower's carpel, with the ovary at the base and the stigma at the top.

substrate The reactant on which an enzyme works.

substrate feeder An animal that lives in or on its food source, eating its way through the food.

substrate-level phosphorylation (fos'-fōr-uh-lā'-shun) The enzyme-catalyzed formation of ATP by direct transfer of a phosphate group to ADP from an intermediate substrate in catabolism.

sugar sink A plant organ that is a net consumer or storer of sugar. Growing roots, buds, stems, and fruits are examples of sugar sinks supplied by phloem.

sugar source A plant organ in which sugar is being produced by either photosynthesis or the breakdown of starch. Mature leaves are the primary sugar sources of plants.

sulfhydryl group A chemical group consisting of a sulfur atom bonded to a hydrogen atom.

summation A phenomenon of neural integration in which the membrane potential of the postsynaptic cell is determined by the combined effect of EPSPs or IPSPs produced in rapid succession at one synapse or simultaneously at different synapses.

suprachiasmatic nucleus (SCN) (sūp'-ruh-kē'-as-ma-tik) A group of neurons in the hypothalamus of mammals that functions as a biological clock.

surface tension A measure of how difficult it is to stretch or break the surface of a liquid. Water has a high surface tension because of the hydrogen bonding of surface molecules.

surfactant A substance secreted by alveoli that decreases surface tension in the fluid that coats the alveoli.

survivorship curve A plot of the number of members of a cohort that are still alive at each age; one way to represent age-specific mortality.

suspension feeder An animal that feeds by removing suspended food particles from the surrounding medium by a capture, trapping, or filtration mechanism.

sustainable agriculture Long-term productive farming methods that are environmentally safe.

sustainable development Development that meets the needs of people today without limiting the ability of future generations to meet their needs.

swim bladder In aquatic osteichthyans, an air sac that enables the animal to control its buoyancy in the water.

symbiont (sim'-bē-ont) The smaller participant in a symbiotic relationship, living in or on the host.

symbiosis An ecological relationship between organisms of two different species that live together in direct and intimate contact.

sympathetic division A division of the autonomic nervous system; generally increases energy expenditure and prepares the body for action.

sympatric speciation (sim-pat'-rik) The formation of new species in populations that live in the same geographic area.

symplast In plants, the continuum of cytosol connected by plasmodesmata between cells.

synapse (sin'-aps) The junction where a neuron communicates with another cell across a narrow gap via a neurotransmitter or an electrical coupling.

synapsid (si-nap'-sid) A member of an amniote clade distinguished by a single hole on each side of the skull. Synapsids include the mammals.

synapsis (si-nap'-sis) The pairing and physical connection of one duplicated chromosome to its homolog during prophase I of meiosis.

synaptonemal complex (si-nap'-tuh-nē'-muhl) A zipper-like structure composed of proteins, which connects a chromosome to its homolog tightly along their lengths during part of prophase I of meiosis.

systematics A scientific discipline focused on classifying organisms and determining their evolutionary relationships.

systemic acquired resistance A defensive response in infected plants that helps protect healthy tissue from pathogenic invasion.

systemic circuit The branch of the circulatory system that supplies oxygenated blood to and carries deoxygenated blood away from organs and tissues throughout the body.

systems biology An approach to studying biology that aims to model the dynamic behavior of whole biological systems based on a study of the interactions among the system's parts.

systole (sis'-tō-lē) The stage of the cardiac cycle in which a heart chamber contracts and pumps blood.

systolic pressure Blood pressure in the arteries during contraction of the ventricles.

taproot A main vertical root that develops from an embryonic root and gives rise to lateral (branch) roots.

tastant Any chemical that stimulates the sensory receptors in a taste bud.

taste bud A collection of modified epithelial cells on the tongue or in the mouth that are receptors for taste in mammals.

TATA box A DNA sequence in eukaryotic promoters crucial in forming the transcription initiation complex.

taxis (tak'-sis) An oriented movement toward or away from a stimulus.

taxon (plural, **taxa**) A named taxonomic unit at any given level of classification.

taxonomy (tak-son'-uh-mē) A scientific discipline concerned with naming and classifying the diverse forms of life.

Tay-Sachs disease A human genetic disease caused by a recessive allele for a dysfunctional enzyme, leading to accumulation of certain lipids in the brain. Seizures, blindness, and degeneration of motor and mental performance usually become manifest a few months after birth, followed by death within a few years.

T cells The class of lymphocytes that mature in the thymus; they include both effector cells for the cell-mediated immune response and helper cells required for both branches of adaptive immunity.

technology The application of scientific knowledge for a specific purpose, often involving industry or commerce but also including uses in basic research.

telomere (tel'-uh-mēr) The tandemly repetitive DNA at the end of a eukaryotic chromosome's DNA molecule. Telomeres protect the organism's genes from being eroded during successive rounds of replication. *See also* repetitive DNA.

telophase The fifth and final stage of mitosis, in which daughter nuclei are forming and cytokinesis has typically begun.

temperate broadleaf forest A biome located throughout midlatitude regions where there is sufficient moisture to support the growth of large, broadleaf deciduous trees.

temperate grassland A terrestrial biome that exists at midlatitude regions and is dominated by grasses and forbs.

temperate phage A phage that is capable of replicating by either a lytic or lysogenic cycle.

temperature A measure in degrees of the average kinetic energy (thermal energy) of the atoms and molecules in a body of matter.

template strand The DNA strand that provides the pattern, or template, for ordering, by complementary base pairing, the sequence of nucleotides in an RNA transcript.

tendon A fibrous connective tissue that attaches muscle to bone.

terminator In bacteria, a sequence of nucleotides in DNA that marks the end of a gene and signals RNA polymerase to release the newly made RNA molecule and detach from the DNA.

territoriality A behavior in which an animal defends a bounded physical space against encroachment by other individuals, usually of its own species.

tertiary consumer (ter'-shē-ār'-ē) A carnivore that eats other carnivores.

tertiary structure (ter'-shē-ār'-ē) The overall shape of a protein molecule due to interactions of amino acid side chains, including hydrophobic interactions, ionic bonds, hydrogen bonds, and disulfide bridges.

test In foram protists, a porous shell that consists of a single piece of organic material hardened with calcium carbonate.

testcross Breeding an organism of unknown genotype with a homozygous recessive individual to determine the unknown genotype. The ratio of phenotypes in the offspring reveals the unknown genotype.

testis (plural, **testes**) The male reproductive organ, or gonad, in which sperm and reproductive hormones are produced.

testosterone A steroid hormone required for development of the male reproductive system, spermatogenesis, and male secondary sex characteristics; the major androgen in mammals.

tetanus (tet'-uh-nus) The maximal, sustained contraction of a skeletal muscle, caused by a very high frequency of action potentials elicited by continual stimulation.

tetrapod A vertebrate clade whose members have limbs with digits. Tetrapods include mammals, amphibians, and birds and other reptiles.

thalamus (thal'-uh-mus) An integrating center of the vertebrate forebrain. Neurons with cell bodies in the thalamus relay neural input to specific areas in the cerebral cortex and regulate what information goes to the cerebral cortex.

theory An explanation that is broader in scope than a hypothesis, generates new hypotheses, and is supported by a large body of evidence.

thermal energy Kinetic energy due to the random motion of atoms and molecules; energy in its most random form. *See also* heat.

thermocline A narrow stratum of abrupt temperature change in the ocean and in many temperate-zone lakes.

thermodynamics (ther'-mō-dī-nam'-iks) The study of energy transformations that occur in a collection of matter. *See also* first law of thermodynamics and second law of thermodynamics.

thermophile *See* extreme thermophile.

thermoreceptor A receptor stimulated by either heat or cold.

thermoregulation The maintenance of internal body temperature within a tolerable range.

theropod A member of a group of dinosaurs that were bipedal carnivores.

thick filament A filament composed of staggered arrays of myosin molecules; a component of myofibrils in muscle fibers.

thigmomorphogenesis (thig'-mō-mor'-phō-gen'-uh-sis) A response in plants to chronic mechanical stimulation, resulting from increased ethylene production. An example is thickening stems in response to strong winds.

thigmotropism (thig-mō'-truh-pizm) A directional growth of a plant in response to touch.

thin filament A filament consisting of two strands of actin and two strands of regulatory protein coiled around one another; a component of myofibrils in muscle fibers.

threatened species A species that is considered likely to become endangered in the foreseeable future.

threshold The potential that an excitable cell membrane must reach for an action potential to be initiated.

thrombus (plural, **thrombi**) A fibrin-containing clot that forms in a blood vessel and blocks the flow of blood.

thylakoid (thī'-luh-koyd) A flattened, membranous sac inside a chloroplast. Thylakoids often exist in stacks called grana that are interconnected; their membranes contain molecular "machinery" used to convert light energy to chemical energy.

thymus (thī'-mus) A small organ in the thoracic cavity of vertebrates where maturation of T cells is completed.

thyroid gland An endocrine gland, located on the ventral surface of the trachea, that secretes two iodine-containing hormones, triiodothyronine (T_3) and thyroxine (T_4), as well as calcitonin.

thyroid hormone Either of two iodine-containing hormones (triiodothyronine and thyroxine) that are secreted by the thyroid gland and that help regulate metabolism, development, and maturation in vertebrates.

thyroxine (T_4) One of two iodine-containing hormones that are secreted by the thyroid gland and that help regulate metabolism, development, and maturation in vertebrates.

tidal volume The volume of air a mammal inhales and exhales with each breath.

tight junction A type of intercellular junction between animal cells that prevents the leakage of material through the space between cells.

tissue An integrated group of cells with a common structure, function, or both.

tissue system One or more tissues organized into a functional unit connecting the organs of a plant.

Toll-like receptor (TLR) A membrane receptor on a phagocytic white blood cell that recognizes fragments of molecules common to a set of pathogens.

tonicity The ability of a solution surrounding a cell to cause that cell to gain or lose water.

top-down control A situation in which the abundance of organisms at each trophic level is controlled by the abundance of consumers at higher trophic levels; thus, predators limit herbivores, and herbivores limit plants.

topoisomerase A protein that breaks, swivels, and rejoins DNA strands. During DNA replication, topoisomerase helps to relieve strain in the double helix ahead of the replication fork.

topsoil A mixture of particles derived from rock, living organisms, and decaying organic material (humus).

torpor A physiological state in which activity is low and metabolism decreases.

totipotent (tō'-tuh-pōt'-ent) Describing a cell that can give rise to all parts of the embryo and adult, as well as extraembryonic membranes in species that have them.

trace element An element indispensable for life but required in extremely minute amounts.

trachea (trā'-kē-uh) The portion of the respiratory tract that passes from the larynx to the bronchi; also called the windpipe.

tracheal system In insects, a system of branched, air-filled tubes that extends throughout the body and carries oxygen directly to cells.

tracheid (trā'-kē-id) A long, tapered water-conducting cell found in the xylem of nearly all vascular plants. Functioning tracheids are no longer living.

trait One of two or more detectable variants in a genetic character.

***trans* fat** An unsaturated fat, formed artificially during hydrogenation of oils, containing one or more *trans* double bonds.

transcription The synthesis of RNA using a DNA template.

transcription factor A regulatory protein that binds to DNA and affects transcription of specific genes.

transcription initiation complex The completed assembly of transcription factors and RNA polymerase bound to a promoter.

transcription unit A region of DNA that is transcribed into an RNA molecule.

transduction A process in which phages (viruses) carry bacterial DNA from one bacterial cell to another. When these two cells are members of different species, transduction results in horizontal gene transfer. *See also* signal transduction pathway.

transfer RNA (tRNA) An RNA molecule that functions as a translator between nucleic acid and protein languages by picking up a specific amino acid and carrying it to the ribosome, where the tRNA recognizes the appropriate codon in the mRNA.

transformation (1) The process by which a cell in culture acquires the ability to divide indefinitely, similar to the division of cancer cells. (2) A change in genotype and phenotype due to the assimilation of external DNA by a cell. When the external DNA is from a member of a different species, transformation results in horizontal gene transfer.

transgene A gene that has been transferred naturally or by a genetic engineering technique from one organism to another.

transgenic Pertaining to an organism whose genome contains DNA introduced from another organism of the same or a different species.

translation The synthesis of a polypeptide using the genetic information encoded in an mRNA molecule. There is a change of "language" from nucleotides to amino acids.

translocation (1) An aberration in chromosome structure resulting from attachment of a chromosomal fragment to a nonhomologous chromosome. (2) During protein synthesis, the third stage in the elongation cycle, when the RNA carrying the growing polypeptide moves from the A site to the P site on the ribosome. (3) The transport of organic nutrients in the phloem of vascular plants.

transmembrane protein A type of integral protein that spans the entire membrane.

transmission electron microscope (TEM) A microscope that passes an electron beam through very thin sections stained with metal atoms and is primarily used to study the internal structure of cells.

transpiration The evaporative loss of water from a plant.

transport epithelium One or more layers of specialized epithelial cells that carry out and regulate solute movement.

transport protein A transmembrane protein that helps a certain substance or class of closely related substances to cross the membrane.

transport vesicle A small membranous sac in a eukaryotic cell's cytoplasm carrying molecules produced by the cell.

transposable element A segment of DNA that can move within the genome of a cell by means of a DNA or RNA intermediate; also called a transposable genetic element.

transposon A transposable element that moves within a genome by means of a DNA intermediate.

transverse (T) tubule An infolding of the plasma membrane of skeletal muscle cells.

triacylglycerol (trī-as'-ul-glis'-uh-rol) A lipid consisting of three fatty acids linked to one glycerol molecule; also called a fat or triglyceride.

trichome An epidermal cell that is a highly specialized, often hairlike outgrowth on a plant shoot.

triple response A plant growth maneuver in response to mechanical stress, involving slowing of stem elongation, thickening of the stem, and a curvature that causes the stem to start growing horizontally.

triplet code A genetic information system in which a series of three-nucleotide-long words specifies a sequence of amino acids for a polypeptide chain.

triploblastic Possessing three germ layers: the endoderm, mesoderm, and ectoderm. All bilaterian animals are triploblastic.

trisomic Referring to a diploid cell that has three copies of a particular chromosome instead of the normal two.

trochophore larva (trō'-kuh-fōr) Distinctive larval stage observed in some lophotrochozoan animals, including some annelids and molluscs.

trophic efficiency The percentage of production transferred from one trophic level to the next higher trophic level.

trophic level The position an organism occupies in a food chain.

trophic structure The different feeding relationships in an ecosystem, which determine the route of energy flow and the pattern of chemical cycling.

trophoblast The outer epithelium of a mammalian blastocyst. It forms the fetal part of the placenta, supporting embryonic development but not forming part of the embryo proper.

tropical dry forest A terrestrial biome characterized by relatively high temperatures and precipitation overall but with a pronounced dry season.

tropical rain forest A terrestrial biome characterized by relatively high precipitation and temperatures year-round.

tropics Latitudes between 23.5° north and south.

tropism A growth response that results in the curvature of whole plant organs toward or away from stimuli due to differential rates of cell elongation.

tropomyosin The regulatory protein that blocks the myosin-binding sites on actin molecules.

troponin complex The regulatory proteins that control the position of tropomyosin on the thin filament.

true-breeding Referring to organisms that produce offspring of the same variety over many generations of self-pollination.

tubal ligation A means of sterilization in which a woman's two oviducts (fallopian tubes) are tied closed and a segment of each is removed to prevent eggs from reaching the uterus.

tube foot One of numerous extensions of an echinoderm's water vascular system. Tube feet function in locomotion and feeding.

tumor-suppressor gene A gene whose protein product inhibits cell division, thereby preventing the uncontrolled cell growth that contributes to cancer.

tundra A terrestrial biome at the extreme limits of plant growth. At the northernmost limits, it is called arctic tundra, and at high altitudes, where plant forms are limited to low shrubby or matlike vegetation, it is called alpine tundra.

tunicate A member of the clade Urochordata, sessile marine chordates that lack a backbone.

turgid (ter'-jid) Swollen or distended, as in plant cells. (A walled cell becomes turgid if it has a lower water potential than its surroundings, resulting in entry of water.)

turgor pressure The force directed against a plant cell wall after the influx of water and swelling of the cell due to osmosis.

turnover The mixing of waters as a result of changing water-temperature profiles in a lake.

tympanic membrane Another name for the eardrum, the membrane between the outer and middle ear.

Unikonta (yū'-ni-kon'-tuh) One of four supergroups of eukaryotes proposed in a current hypothesis of the evolutionary history of eukaryotes. This clade, which is supported by studies of myosin proteins and DNA, consists of amoebozoans and opisthokonts. *See also* Excavata, SAR, and Archaeplastida.

unsaturated fatty acid A fatty acid that has one or more double bonds between carbons in the hydrocarbon tail. Such bonding reduces the number of hydrogen atoms attached to the carbon skeleton.

urban ecology The study of organisms and their environment in urban and suburban settings.

urea A soluble nitrogenous waste produced in the liver by a metabolic cycle that combines ammonia with carbon dioxide.

ureter (yū-rē'-ter) A duct leading from the kidney to the urinary bladder.

urethra (yū-rē'-thruh) A tube that releases urine from the mammalian body near the vagina in females and through the penis in males; also serves in males as the exit tube for the reproductive system.

uric acid A product of protein and purine metabolism and the major nitrogenous waste product of insects, land snails, and many reptiles. Uric acid is relatively nontoxic and largely insoluble in water.

urinary bladder The pouch where urine is stored prior to elimination.

uterine cycle The cyclic changes in the endometrium (uterine lining) of mammals that occur in the absence of pregnancy. In certain primates, including humans, the uterine cycle is a menstrual cycle.

uterus A female organ where eggs are fertilized and/or development of the young occurs.

vaccine A harmless variant or derivative of a pathogen that stimulates a host's immune system to mount defenses against the pathogen.

vacuole (vak'-yū-ōl') A membrane-bounded vesicle whose specialized function varies in different kinds of cells.

vagina Part of the female reproductive system between the uterus and the outside opening; the birth canal in mammals. During copulation, the vagina accommodates the male's penis and receives sperm.

valence The bonding capacity of a given atom; the number of covalent bonds that an atom can form, which usually equals the number of unpaired electrons in its outermost (valence) shell.

valence electron An electron in the outermost electron shell.

valence shell The outermost energy shell of an atom, containing the valence electrons involved in the chemical reactions of that atom.

van der Waals interactions Weak attractions between molecules or parts of molecules that result from transient local partial charges.

variable A factor that varies in an experiment.

variation Differences between members of the same species.

vas deferens In mammals, the tube in the male reproductive system in which sperm travel from the epididymis to the urethra.

vasa recta The capillary system in the kidney that serves the loop of Henle.

vascular cambium A cylinder of meristematic tissue in woody plants that adds layers of secondary vascular tissue called secondary xylem (wood) and secondary phloem.

vascular plant A plant with vascular tissue. Vascular plants include all living plant species except liverworts, mosses, and hornworts.

vascular tissue Plant tissue consisting of cells joined into tubes that transport water and nutrients throughout the plant body.

vasectomy The cutting and sealing of each vas deferens to prevent sperm from entering the urethra.

vasoconstriction A decrease in the diameter of blood vessels caused by contraction of smooth muscles in the vessel walls.

vasodilation An increase in the diameter of blood vessels caused by relaxation of smooth muscles in the vessel walls.

vasopressin *See* antidiuretic hormone (ADH).

vector An organism that transmits pathogens from one host to another.

vegetal pole The point at the end of an egg in the hemisphere where most yolk is concentrated; opposite of animal pole.

vegetative propagation Asexual reproduction in plants that is facilitated or induced by humans.

vegetative reproduction Asexual reproduction in plants.

vein (1) In animals, a vessel that carries blood toward the heart. (2) In plants, a vascular bundle in a leaf.

ventilation The flow of air or water over a respiratory surface.

ventral In an animal with bilateral symmetry, pertaining to the underside (in most animals) or front (in animals with upright posture) of the body.

ventricle (ven'-tri-kul) (1) A heart chamber that pumps blood out of the heart. (2) A space in the vertebrate brain, filled with cerebrospinal fluid.

venule (ven'-yūl) A vessel that conveys blood between a capillary bed and a vein.

vernalization The use of cold treatment to induce a plant to flower.

vertebrate A chordate animal with vertebrae, the series of bones that make up the backbone.

vesicle (ves'-i-kul) A membrane-bound sac in or outside a cell.

vessel A continuous water-conducting micropipe found in most angiosperms and a few nonflowering vascular plants.

vessel element A short, wide, water-conducting cell found in the xylem of most angiosperms and a few nonflowering vascular plants. Dead at maturity, vessel elements are aligned end to end to form micropipes called vessels.

vestigial structure A feature of an organism that is a historical remnant of a structure that served a function in the organism's ancestors.

villus (plural, **villi**) (1) A finger-like projection of the inner surface of the small intestine. (2) A finger-like projection of the chorion of the mammalian placenta. Large numbers of villi increase the surface areas of these organs.

viral envelope A membrane, derived from membranes of the host cell, that cloaks the capsid, which in turn encloses a viral genome.

virulent phage A phage that replicates only by a lytic cycle.

virus An infectious particle incapable of replicating outside of a cell, consisting of an RNA or DNA genome surrounded by a protein coat (capsid) and, for some viruses, a membranous envelope.

visceral mass One of the three main parts of a mollusc; the part containing most of the internal organs. *See also* foot and mantle.

visible light That portion of the electromagnetic spectrum that can be detected as various colors by the human eye, ranging in wavelength from about 380 nm to about 740 nm.

vital capacity The maximum volume of air that a mammal can inhale and exhale with each breath.

vitamin An organic molecule required in the diet in very small amounts. Many vitamins serve as coenzymes or parts of coenzymes.

viviparous (vī-vip'-uh-rus) Referring to a type of development in which the young are born alive after having been nourished in the uterus by blood from the placenta.

voltage-gated ion channel A specialized ion channel that opens or closes in response to changes in membrane potential.

vulva Collective term for the female external genitalia.

water potential (ψ) The physical property predicting the direction in which water will flow, governed by solute concentration and applied pressure.

water vascular system A network of hydraulic canals unique to echinoderms that branches into extensions called tube feet, which function in locomotion and feeding.

wavelength The distance between crests of waves, such as those of the electromagnetic spectrum.

wetland A habitat that is inundated by water at least some of the time and that supports plants adapted to water-saturated soil.

white matter Tracts of axons within the CNS.

whole-genome shotgun approach Procedure for genome sequencing in which the genome is randomly cut into many overlapping short segments that are sequenced; computer software then assembles the complete sequence.

wild type The phenotype most commonly observed in natural populations; also refers to the individual with that phenotype.

wilting The drooping of leaves and stems as a result of plant cells becoming flaccid.

wobble Flexibility in the base-pairing rules in which the nucleotide at the 5′ end of a tRNA anticodon can form hydrogen bonds with more than one kind of base in the third position (3′ end) of a codon.

xerophyte (zir'-ō-fīt') A plant adapted to an arid climate.

X-linked gene A gene located on the X chromosome; such genes show a distinctive pattern of inheritance.

X-ray crystallography A technique used to study the three-dimensional structure of molecules. It depends on the diffraction of an X-ray beam by the individual atoms of a crystallized molecule.

xylem (zī'-lum) Vascular plant tissue consisting mainly of tubular dead cells that conduct most of the water and minerals upward from the roots to the rest of the plant.

xylem sap (zī'-lum) The dilute solution of water and minerals carried through vessels and tracheids.

yeast Single-celled fungus. Yeasts reproduce asexually by binary fission or by the pinching of small buds off a parent cell. Many fungal species can grow both as yeasts and as a network of filaments; relatively few species grow only as yeasts.

yolk Nutrients stored in an egg.

zero population growth (ZPG) A period of stability in population size, when additions to the population through births and immigration are balanced by subtractions through deaths and emigration.

zona pellucida The extracellular matrix surrounding a mammalian egg.

zoned reserve An extensive region that includes areas relatively undisturbed by humans surrounded by areas that have been changed by human activity and that are used for economic gain.

zone of polarizing activity (ZPA) A block of mesoderm located just under the ectoderm where the posterior side of a limb bud is attached to the body; required for proper pattern formation along the anterior-posterior axis of the limb.

zoonotic pathogen A disease-causing agent that is transmitted to humans from other animals.

zoopagomycete A member of the fungal phylum Zoopagomycota, multicellular parasites or commensal symbionts of animals; sexual reproduction, where known, involves the formation of a sturdy structure called a zygosporangium.

zoospore Flagellated spore found in chytrid fungi and some protists.

zygomycete (zī'-guh-mī'-sēt) A member of the fungal phylum Zygomycota, characterized by the formation of a sturdy structure called a zygosporangium during sexual reproduction.

zygosporangium (zī'-guh-spōr-an'-jē-um) (plural, **zygosporangia**) In zygomycete fungi, a sturdy multinucleate structure in which karyogamy and meiosis occur.

zygote (zī'-gōt) The diploid cell produced by the union of haploid gametes during fertilization; a fertilized egg.

Index

NOTE: A page number in regular type indicates where a topic is discussed in the text; a **bold** page number indicates where a term is bold and defined; an *f* following a page number indicates a figure (the topic may also be discussed in the text on that page); a *t* following a page number indicates a table (the topic may also be discussed in the text on that page).

2,4-dichlorophenoxyacetic acid (2,4-D), 849
3′ end (sugar-phosphate backbone), 85*f*, 345*f*, 348*f*
3-phosphoglycerate, 201, 202*f*
5′ cap, **345***f*, 348*f*
5′ end (sugar-phosphate backbone), 85*f*
5-methylcytosine, 63*f*
10-nm fibers, DNA, 330*f*–331*f*, 332
30-nm fibers, DNA, 330*f*

A

a (yeast mating type), 213*f*, 214, 217
ABC hypothesis, flower formation, **780***f*
abd-A gene, 706*f*
Abdomen, insect, 710*f*
Abiotic factors, **1169**
 microclimate and, 1169–1170
 in pollination, 824*f*
 in species distributions, 1178, 1183*f*–1186*f*
Abiotic stresses, plant, **862**–863, 864*f*, 865
Abiotic synthesis, organic molecule, 57*f*, 58, 526*f*–527*f*
Abnormal chromosome number disorders, 307*f*
ABO blood groups, 280*f*, 292, 969–970
Abomasum, 914*f*
Abortion, **1039**
 spontaneous, 306
Abscisic acid (ABA), **798**, 846*t*, **852***f*, 863
Abscission, leaf, 852, 854*f*
Absorption, **904**
 in animal food processing, 904, 905*f*
 of carbon dioxide, 1245
 in large intestine, 910
 of light energy and primary production, 1242*f*
 of nutrients, 895*f*
 plant and animal, 895*f*
 in small intestine, 909*f*–910*f*
 of water and mineral by root cells, 792
Absorption spectrum, **193***f*
Abstinence, 1037, 1038*f*
Abstract thinking, 1146
Abyssal zone, **1177***f*
Acacia trees, 10*f*, 1220*f*
Acanthocephalans, 687*f*, 698*f*, 1220
Acanthodians, **726**
Accessory fruits, **831***f*
Accessory glands, male reproductive, 1025*f*, 1026
Acclimatization, **883***f*, 888
Accommodation, visual, 1122*f*
Acetic acid, 63*f*
Acetone, 63*f*
Acetylation, histone, **371***f*
Acetylcholine, **1080**, 1081*t*, 1128, 1129*f*
Acetylcholinesterase, 1080
Acetyl CoA (acetyl coenzyme A), **171***f*–173*f*
Acetylsalicylic acid, 651
Achillea lanulosa, 1189
Achondroplasia, 287*f*
Acid bogs, 529*f*
Acid growth hypothesis, 848, 849*f*
Acidic rain, 55
Acidification, ocean, 53*f*, 54
Acid precipitation, 821, 1265, 1266*f*
Acid reflux, 908
Acids, **51**
 acid precipitation and, 821, 1265, 1266*f*
 amino acids as, 76, 77*f*
 buffers and, 52–53
 hydrogen ions, bases, and, 51

 ocean acidification and, 53*f*, 54
 pH scale and, 51, 52*f*
Acinetobacter baumannii, 913
Acoela, 683*f*, 687*f*
Acorn woodpeckers (*Melanerpes formicivorus*), 1162
Acorn worms, 689*f*
Acquired immunodeficiency, 971, 972*f*, 973. *See also* AIDS
Acquired traits, noninheritance of, 471*f*
Acromegaly, 1010
Acrosomal reactions, 1044*f*, **1045**
Acrosomes, **1028***f*, 1044*f*, **1045**
Actias luna, 43
Actin, 76*f*, **115**, 241, 879*f*
Actin filaments, 113*t*, 1056*f*, 1125, 1126*f*–1127*f*.
 See also Microfilaments
Actinistia, 719*f*, 729*f*
Actinomycetes, 584*f*
Actinopterygii (ray-finned fishes), 719*f*, 728*f*–729*f*
Action potentials, neuron, **1073**
 adaptations of axon structure for, 1076*f*–1077*f*
 conduction of, 1075*f*, 1076
 evolution of, 1084
 generation of, 1073, 1074*f*, 1075
 graded potentials, voltage-gated ion channels, and, 1073, 1074*f*
 hyperpolarization and depolarization of membrane potentials and, 1072*f*–1073*f*
 in long-term potentiation, 1101*f*
 in sensory systems, 1109*f*
Action potentials, plant, **862**
Action spectrum, **193**, 194*f*, **855**
Activation, allosteric, 160*f*, 161
Activation, egg, 1046
Activation energy, 153, **154***f*–155*f*
Activators, 160*f*, **369***f*, 374*f*–375*f*, 397
Active immunity, **968**
Active sites, enzyme, **155***f*–156*f*
Active transport, 126, **136**
 ATP as energy for, 137*f*
 cotransport in, 138*f*, 139
 maintenance of membrane potential by ion pumps in, 137, 138*f*
 passive transport vs., 137*f*
 in plant cells, 209*f*
 of solutes across plant plasma membranes, 788, 789*f*
Activity, animal metabolic rate and, 891–892
Actual range, 1184
Acyclovir, 409
Adaptation, sensory, 1109–1110
Adaptations, **472**. *See also* Evolution; Natural selection
 in amniote development, 1053*f*, 1054
 axon width and myelination as, 1076*f*–1077*f*
 evolution and, 2*f*, 14*f*–16*f*, 472, 473*f*–476*f*
 as evolutionary compromises, 501*f*
 floral, to prevent self-fertilization, 834, 835*f*
 gas exchange, 947*f*–949*f*
 for heat exchange in animals, 873*f*, 885*f*–889*f*
 against herbivory, 1219–1220
 mycorrhizae as plant, 817
 of pathogens to evade immune systems, 957, 971–972, 973*f*, 973*t*, 974, 976
 of plants and animals to life challenges, 894*f*–895*f*
 of plants to global climate change, 204–205
 of plants to terrestrial life, 203
 of plants to toxic elements, 30*f*
 predator and prey, 1217, 1218*f*–1219*f*, 1237
 prokaryotic, 573*f*–577*f*, 578, 581*t*, 582*f*
 as property of life, 3*f*
 to reduce terrestrial nutrient limitations, 1244
 of seed plants, 653
 sexual reproduction patterns as, 1022
 of smooth muscles, 1132
 terrestrial, of fungi and plants, 619*f*, 621, 660
 terrestrial, of seed plants, 637*f*–638*f*, 639
 of vertebrate digestive systems, 911*f*–914*f*
Adaptive evolution, **494**, 498, 499*f*, 501. *See also* Evolution; Natural selection

Adaptive immunity, **953**
 active and passive immunity in, 968
 antigen recognition by B cells and antibodies of, 958*f*–959*f*
 antigen recognition by T cells, 959*f*–960*f*
 B cell and T cell development of, 960, 961*f*–963*f*
 B cell and T cell proliferation in, 961, 962*f*
 B cells and antibodies in responses of, 964, 965*f*–966*f*
 cytotoxic T cells in responses of, to infected cells, 966*f*, 967
 helper T cells in responses of, 963, 964*f*
 immune rejection and, 969–970
 immunization and, 968*f*, 976
 immunological memory of, 960, 962, 963*f*
 medical tools using antibodies of, 969*f*
 molecular recognition by, 957*f*
 overview of, 957*f*, 967*f*
 pathogen recognition in, 952*f*, 957*f*–958*f*
 self-tolerance in, 961
Adaptive radiations, 15*f*–16*f*, 525, **542**, 543*f*, 544, 736, 1183, 1184*f*
Addition rule, 277*f*, 278
Adélie penguin, 892
Adenine, 84, 85*f*, 86, 317*f*, 318, 341*f*, 850
Adenomatous polyposis coli (APC), 391*f*, 394
Adenosine diphosphate. *See* ADP
Adenosine triphosphate. *See* ATP
Adenoviruses, 400*f*
Adenylyl cyclase, **223***f*–224*f*
Adhesion, 46*f*, 795–796
Adhesive chemicals, echinoderm, 713, 714*f*
Adipose tissue, **878***f*, 915, 918
ADP (adenosine diphosphate)
 as enzyme activator, 160
 hydrolysis of ATP to, 151*f*
 in sliding-filament model of muscle contraction, 1126, 1127*f*
 synthesis of ATP from, 153*f*, 169, 175, 176*f*, 177–178
Adrenal cortex, 1004*f*, 1013–1014
Adrenal glands, **1012**
 epinephrine and. *See* Epinephrine (adrenaline)
 in human endocrine system, 1004*f*
 responses of, to stress, 1012*f*–1013*f*, 1014
 smooth ER and, 104
Adrenaline. *See* Epinephrine
Adrenal medulla, 1004*f*, 1012*f*–1013*f*
Adrenocorticotropic hormone (ACTH), 1004*f*, 1008*f*, 1012*f*, 1013–1014
Adult stem cells, 430, 431*f*
Adventitious roots, 760*f*
Adventitious shoots, 833*f*
Aerial roots, 760*f*
Aerobic prokaryotes, 577*f*
Aerobic respiration, **165**, 179–181
Afferent arteriole, 987*f*
Afferent neurons, 1088*f*, 1108*f*
Africa, human population in, 1208, 1209*f*
African-Americans, sickle-cell disease in, 286*f*, 287
African buffalo, 1221*f*
African elephants (*Loxodonta africana*), 474*f*, 1197*f*, 1265*f*
African golden mole, 558*f*, 559
African grey parrots, 1099
Africans
 genomes of, 462–463
 sickle-cell disease in, 286*f*, 287
African sleeping sickness, 713
Agave, 1200, 1201*f*
Age-related macular degeneration (AMD), 432
Age structure, human population, **1208**, 1209*f*
Aggregate fruits, **831***f*
Aging, 233, 289
Aglaophyton major, 628*f*, 660
Agonistic behavior, 1153*f*
agouti gene, 372*f*
Agriculture. *See also* Crop plants
 allopolyploidy in, 514*f*–515*f*

biotechnology in, 437–439
C₃ plants in, 203
C₄ plants in, 203
climate change and, 863
community disturbances by, 1231
effects of atmospheric carbon dioxide on productivity of, 205
fertilizing in. *See* Fertilization, soil
fungal food products of, 670
global human population size and, 1211
importance of insects to, 713
importance of mycorrhizae to, 818
nematode pests in, 705
nitrogen fixation and, 817
nutrient pollution from, 1274, 1275f
plant biotechnology and genetic engineering in, 836f–838f, 839–840
plant breeding in, 823, 835–837
plant cloning in, 428
plant control systems in, 871
seed plants in, 651
soil conservation and sustainable, 807f–809f, 1273
vegetative propagation of plants in, 835–836
Agrobacterium tumefaciens, 422, 595, 837
Aguirre-Jarquin, Clemente, 24f
AIDS (acquired immunodeficiency syndrome), **406**, **973**f, 974. *See also* HIV
    cell-surface proteins and blocking HIV to prevent, 130f
    drug cocktails in treatment of, 489
    emergence of, 410
    G protein-coupled receptors and, 217f
    HIV and, 398f, 406, 407f
    host range of, 401
Aigarchaeota clade, 586
Ailuropoda melanoleuca, 448t
Ain, Michael C., 287f
Air circulation patterns, global, 1166f
Air roots, 760f
Air sacs, 942f–945f
Alanine, 77f
Alarm calls, 1148f
Albatross, 977f, 1277f
Albinism, 285f, 335f, 336–337
Albumen, 735f
Albumin, 377f, 934f
Albuterol, 62f
Alcohol compounds, 63f
Alcohol dehydrogenase, 66f
Alcohol fermentation, **180**f
Alcoholic beverages, 670
Aldehyde compounds, 63f
Alder, 1230f–1231f
Aldoses, 68f
Aldosterone, 220, 221f, 995, 996f, 1014
Algae, **596**
    alternation of generations in, 258f, 259
    blooms of, 1227f, 1242, 1243f
    brown, 602, 603f, 604
    chloroplasts in, 110, 111f
    DNA identification of, 1222
    as earliest multicellular eukaryotes, 535f
    evolution of plants from green, 617, 618, 619f, 621
    evolution of photosynthetic, 596f–597f
    fossils of, 529f, 533f
    fungi and, as lichens, 663, 668f–669f
    green, 101f, 596f–597f, 610f–611f, 614, 617, 618, 619f, 621
    horizontal gene transfer in, 569f
    photosynthesis in, 188f
    as protists, 599f
    red, 596f–597f, 609f, 610
    structure and organelles of cells of, 101f
Algin, 603
Alimentary canals, 688f, **697**f, 704f, **905**f, 912f
Alkaline vents, **526**f
Alkaloids, 868f
Alkaptonuria, 293, 336
Allantois, 735f, 1053f, 1054
Alleles, **272**. *See also* Genes
    combination predictions of, 268
    correlating behavior of chromosome pairs with, 295, 296f–297f
    degrees of dominance of, and phenotypes, 279f, 280
    dominant, in genetic disorders, 287f
    dominant vs. recessive, in Mendelian inheritance, 272, 273f–276f, 293

frequencies of, in gene pools of populations, 490f
    genetic markers for disease-causing, 427f
    genetic variation from recombination of, 305. *See also* Recombination
    genetic variation preserved in recessive, 500
    genomic imprinting of maternal or paternal, 372
    as information, 293
    microevolution as alteration in frequencies of, in populations, 487f, 493, 494f–497f
    multiple, and pleiotropy, 280f
    mutations as sources of, 265, 488–489
    recessive, in genetic disorders, 285f–286f, 287
    in sexual life cycles, 259
    sickle-cell, 502f–503f
    testing populations for frequencies of, 490, 491f, 492–493
    transmission of, 269
Allergens, 839, 970f, 971
Allergies, 438, 970f, 971
Alliaria petiolata, 818
Alligator mississippiensis. See American alligator
Alligators, 737f, 738, 925
Allohexaploid, 514–515, 524
Allolactose, 368f–396f, 370
Allopatric populations, character displacement in, 1216, 1217f
Allopatric speciation, **511**f–512f, 513, 539–540
Allopolyploids, **514**f–515f
All-or-none responses, 1073
Allosteric regulation, **160**f–161f
α (yeast mating type), 213f, 214, 217
α chain, 959f
α-globin gene family, 453f
α helix, **80**f
α-lactalbumin, 456, 457f
Alpha (α) carbon, 75
Alpha cells, 916f
Alpha proteobacteria, 584f, 596
Alpheus species, 512f, 513
Alpine chickweed, 1281f
Alpine pennycress (*Thlaspi caerulescens*), 809
Alpine tundra, 1176f
Alpine wood sorrel, 835f
Alternate phyllotaxy, 786f
Alternation of generations, **258**f, **603**f, 604, **620**f, 621
Alternative RNA splicing, **347**f, **378**f, 449
Altruism, **1156**f–1159f, 1162
Alu elements, 452, 459
Aluminum toxicity, plant, 809
Alveolates, 599f, **604**f–606f
Alveoli, 604f, **943**f–944f
Alzheimer's disease, 83, 231, 311, 361, 413, 433, 435, **1103**, 1104f
Amacrine cells, 1118f
Amazon rain forest, 1271f
Amber, fossils in, 529f
Amborella trichopoda, 648f, 650f, 653
Amebic dysentery, 613
American alligator (*Alligator mississippiensis*), 737f
American beech trees (*Fagus grandifolia*), 1170f
American chestnut trees, 1226, 1234
American Dust Bowl, 807f
American flamingo, 14f
American pika (*Ochotona princeps*), 1280f
American pokeweed, 830f
Amine compounds, 63f
Amines, 1001f
Amino acids
    abiotic synthesis of, 526f–527f
    activation of, in eukaryotic cells, 356f
    deamination of, for catabolism, 182
    in DNA-binding proteins, 91
    in enzymatic catalysis, 156
    essential, 899f
    in evolution of enzymes, 159f
    in genetic code, 340f–342f
    human dietary deficiencies in, 901
    as neurotransmitters, 1081t
    in nitrogen cycle, 1251f
    in polypeptides and proteins, 67, 75, 76f–78f. *See also* Polypeptides; Proteins
    in proteins, 75, 77f
    sequences of, 444, 445f, 446, 458
    sickle-cell disease and, 82f
    specified by nucleotide triplets in translation, 347f–350f
    structure of, 91

using sequences of, to test hypothesis on horizontal gene transfer, 570
Amino acid sequence identity tables, reading, in Scientific Skills Exercise, 458
Amino acid sequences, 444, 445f, 446, 458
Aminoacyl tRNA, 349f
Aminoacyl-tRNA synthetases, **349**f
Amino end. *See* N-terminus
Amino group, 63f
Amitochondriate protists, 597
Ammonia, **982**
    as base, 51
    hydrogen bonds and, 39f
    as nitrogenous waste, 982f, 983, 988
Ammonites, **702**
Ammonium, in nitrogen cycle, 1251f
Amniocentesis, **288**, 289f, 1039–1040
Amnion, 480, 735f, 1053f
Amniotes, **734**, **1053**
    derived characters of, 734, 735f
    developmental adaptations of, 1053f, 1054
    evolution of, 678
    fossils and early evolution of, 735f
    mammals as, 741
    phylogeny of, 734f
    reptiles, 735f–740f
Amniotic egg, 718, **734**, 735f
Amoebas, 117f, 235f, 449, 599f, **606**
Amoebocytes, **690**f, 691
Amoeboid movement, 117f
Amoebozoans, **612**f–613f
Amorphea, 599, 611
AMP (adenosine monophosphate), 183f, 184, 349f
AMPA receptors, 1101f
Amphetamines, 1102, 1103f
Amphibians, **731**
    adaptations of kidneys of, 992
    axis formation in, 1060f
    brains of, 1091f
    breathing in, 925, 944
    cell developmental potential in, 1060, 1061f
    cell fate and pattern formation by inductive signals in, 1062f
    cleavage in, 1048f, 1049
    diversity of, 732f–733f
    double circulation in, 924f, 925
    evolution of, 678
    external fertilization in, 1022f
    fungal parasites of, 669, 670f
    gastrulation in, 1051f, 1052
    hearing and equilibrium in, 1116f
    neurulation in, 1054f, 1055
    parental care in, 1151
    phylogeny of, 719f
    species distributions of, 1164f, 1185
    vaccine for, 733
Amphipathic molecules, **127**
Amplification, sensory, **1109**–1110
Amplification, signal, 221, 226–227
Ampulla, sea star, 714f
Amygdala, 1095f–**1096**f
Amylase, **906**
Amyloid plaques, 1104f
Amylopectin, 70f
Amyloplasts, 111
Amylose, 70f, 111
Amyotrophic lateral sclerosis (ALS), 1129
Anabaena, 582f
Anableps anableps, 365f
Anabolic pathways, **144**, 183
Anabolic steroids, 1014
Anabrus simplex, 449, 712f, 1112f
Anaerobic respiration, 165, 179–182, **582**
Analogies, **558**f
Analogous structures, **481**
Anaphase, 252
Anaphase (mitosis), **237**, 239f, 241f, 243f, 263f
Anaphase I, 260f, 263f, 266f
Anaphase II, 261f, 266f
Anaphylactic shock, 971
Anatomical homologies, 479f, 480
Anatomy, 873f–**874**. *See also* Animal form and function; Morphology; Plant structure
Ancestral characters, shared, **560**, 561f
Ancestry, common, 15f–16f, 479f–481f, 555, 556f–560f, 648
Anchorage dependence, **247**, 248f

Anchoring junctions, 120f
Androgens, 1004f, **1014**, 1015f, 1030, 1031f
Anemia, 936
Anesthetics, 1084
Aneuploidies, **307**, 308f, 309
Angiosperm reproduction
 asexual, 833f–836f
 breeding and genetic engineering in, 836f–838f, 839–840
 flower pollination in, 824f–825f
 flower structure and function in, 823f–824f, 830
 fruit structure and function in, 830f–831f
 life cycle of, 646f, 647, 653, 826, 827f, 828
 mutualisms in, 825f
 seeds in, 828f–830f
Angiosperms, **623**f
 bulk flow in sugar translocation in, 800, 801f
 characteristics of, 644f–646f, 647
 evolutionary mystery of, for C. Darwin, 477
 evolution of, 647f–649f
 evolution of seeds in, 639
 flowers of, 644f
 fruits of, 644, 645f
 gametophyte-sporophyte relationships in, 637f, 638
 life cycle of, 646f, 647, 653, 826, 827f, 828
 G. Mendel's techniques of crossing, 270f–271f
 origin of, 533f
 phylogeny of, 622t, 648f–650f, 653
 reproduction of. See Angiosperm reproduction
Angiotensin II, 995, 996f
Anhydrobiosis, **979**, 980f
Animal behavior, 1140
 cerebral cortex information processing and, 1097f, 1098
 genetics, altruism, and inclusive fitness in evolution of, 1154, 1155f–1159f, 1160
 hormones and, 999f
 learning and, 1144t, 1145f–1148f
 stimuli for simple and complex, 1140f–1143f
 survival and reproductive success in evolution of, 1148, 1149f–1154f, 1161
 in thermoregulation, 873f, 887f
Animal cells. See also Eukaryotic cells
 apoptosis in, 230f–231f
 cell junctions of, 120f
 cellular respiration in. See Cellular respiration
 cytokinesis in, 241, 242f
 endocytosis in, 140f
 extracellular matrix of, 118, 119f
 local and long-distance cell signaling in, 215f, 216
 meiosis in, 260f–261f
 nuclear transplantation of differentiated, 428, 429f–430f
 stem cells, 430f–431f
 structure and organelles of, 100f
 structure and specialization of, 674
 water balance of, 134f–135f
Animal development. See also Embryonic development
 adaptations of, 894f
 animal phylogeny and, 682, 683f, 684
 cell fate specification in, 1057, 1058f–1064f
 comparing processes of, 463f–464f
 developmental biology and, 1057
 fertilization and cleavage in, 1044f–1048f, 1049
 morphogenesis in, 1049, 1050f–1057f
 protostome vs. deuterostome, 681, 682f
 reproduction and embryonic, 674f, 675
Animal form and function
 anatomy and physiology as, 873f–874
 bioenergetics of, 889, 890f––893f
 body plans, 679, 680f–682f
 correlation of, at all levels of organization, 874f–880f, 876t
 evolution of body size and shape in, 874f
 exchange with environment in, 874, 875f, 876
 feedback regulation of homeostasis in, 881f–883f
 hierarchical organization of body plans in, 876t, 877f–879f
 mammalian organ systems in, 876t
 regulation of, by endocrine and nervous systems, 880f
 thermoregulation in, 884f–889f
Animal hormones, **880**, **1000**. See also Endocrine systems; Hormones
 behavior and, 999f
 birth control, **1038**f, 1039
 cascade pathways of, 1008, 1009f, 1010, 1014

cellular response pathways of, 1002f–1003f
 chemical classes of, 1002f, 1003
 as chemical signals of endocrine system, 1000f–1004f
 in childbirth and labor, 1037f
 embryonic, 1035
 endocrine system glands and, 1003, 1004f
 endocrine systems and, in cell signaling, 880f
 erythropoietin, 936
 evolution of, 1015f, 1016
 feedback regulation of, 1005–1006
 in fight-or-flight responses, 212
 multiple effects of single, 1003
 in neuroendocrine signaling, 1000f, 1001, 1006f–1008f. See also Neuroendocrine signaling
 plant hormones vs., 216, 894f
 receptors for, 1000f–1004f, 1018
 regulation of appetite and consumption by, 917f, 918
 in regulation of digestive systems, 915f
 in regulation of mammalian reproduction, 1030, 1031f–1032f, 1033–1034
 in regulatory functions of endocrine glands, 1011f–1015f, 1016
 in sex determination, 1030
 in simple endocrine pathway, 1004, 1005f
 in simple neuroendocrine pathway, 1005f
Animalia kingdom, **12**f, 568
Animal nutrition. See also Human nutrition
 adaptations of vertebrate digestive systems for diets in, 911f–914f
 compartmentalized processing in, 898
 diets and requirements for, 899f–902f, 900t–901t
 digestive systems and, 905, 906f–911f
 feedback regulation of digestion, energy storage, and appetite in, 914, 915f–917f, 918
 feeding mechanisms in, 902, 903f
 food processing stages in, 902, 903f–905f
 nutritional modes in, 673, 674, 894f
Animal pole, **1048**f
Animal reproduction. See also Human reproduction
 amphibian, 732f–733f
 asexual, 266, 267f, 1020
 development and, 674f, 675
 different ways of, 1019f
 evolution of sexual, 1021, 1022f
 fertilization mechanisms in, 1022f–1024f
 of fish, 728
 hormonal regulation of mammalian, 1030, 1031f–1032f, 1033–1034
 reproductive cycles in, 1021f
 sexual life cycles in, 258f. See also Sexual life cycles
 of sharks and rays, 727
 variations in patterns of sexual, 1020f
Animals. See also Animal behavior; Animal development; Animal form and function; Animal hormones; Animal nutrition; Animal reproduction
 aquatic. See Aquatic animals
 catabolic pathways in, 182f, 183
 cells of, 6f. See also Animal cells
 cell structure and specialization of tissues of, 674
 cellular respiration in hibernating, 178
 circulatory and gas exchange systems of. See Cardiovascular systems; Circulatory systems; Gas exchange
 climate change and, 11f, 1280f–1281f
 cloning of, 428, 429f–430f
 as consumers and predators, 673f
 correlation of diversity of miRNAs with complexity of, 678
 defense mechanisms of, 28f
 in domain Eukarya, 12f
 in ecosystem interaction, 10f, 11
 endangered or threatened, 1261, 1262f, 1288
 in energy flow and chemical cycling, 9f
 evolutionary history of, 675f–677f, 678–679
 evolutionary links between plants and, 648, 649f
 extinctions of, 702f
 flower pollination by, 824f–825f
 fruit and seed dispersal by, 832f
 fungal mutualisms with, 668f
 fungal parasites of, 669f–670f
 glycogen as storage polysaccharide for, 70f, 71
 herbivore adaptations in, 477f, 478
 immune systems of. See Immune systems
 kingdom and domain mutualisms with, 813f

land colonization by, 533f, 536, 537f
 latitude and size of, 897
 life challenges and solutions for, 894f–895f
 maximizing body surface area of, 695f
 microevolution of populations of. See Microevolution
 neurons and nervous systems of. See Nervous systems; Neurons
 nutrition of, 12f, 13
 as opisthokonts, 613
 osmoregulation and excretion of. See Excretory systems; Osmoregulation
 osmoregulation of, 978f–980f, 981
 phylogeny of, 682, 683f, 684, 686
 plant recruitment of, as herbivore defense, 869f
 protein production by "pharm" transgenic, 436f
 relationship of, to unikont protists, 611–612
 reproduction and development of, 674f, 675
 seed dispersal by, 645f
 sensory and motor systems of. See Motor systems; Sensory systems
 tropical deforestation and extinctions of, 651f, 652
 zoonotic pathogens and, 1234, 1235f
Animal viruses
 classes of, 406t
 as pathogens, 407f–413f
 replicative cycles of, 404, 405f, 406t, 407f
Anions, **37**, 38f, 806, 807f, 1070t
Ankle bones, 481f
Annelids, 688f, 703f–704f, 705, 923f, 985f, 1086f
Annual human population growth rates, 1207f, 1208
Annuals, 766
Antagonistic functions, autonomic nervous system, 1089f
Antagonistic muscle pairs, 1132f
Antarctica, 1224f
Antarctic Circumpolar Current, 1168f
Antennae, 43, 1110, 1111f, 1123
Anterior pituitary gland, 1004f, **1007**f–1008f, 1016, 1030, 1031f–1032f
Anterior-posterior axis, 1055, 1060f
Antheraea polyphemus, 1001
Antheridia, 624f, **625**, 630f
Anthers, **644**f, **823**f
Anthocerophyta (hornworts), 623, 626f
Anthophyta. See Angiosperms
Anthozoans, 692f–693f
Anthrax, 575f, 584f
Anthropoids, **746**f–747f
Anti-aging effects, 850
Antibiotic drugs
 bacteria and, 349
 bacterial resistance to, 213, 581, 588f–589f
 for cystic fibrosis, 286
 as enzyme inhibitors, 158–159
 evolution of resistance to, 478f, 592
 fungal, 670
 gram-positive bacteria and, 584f
 peptidoglycan and, 574–575
 prokaryotic ribosomes and, 577
 from soil bacteria, 590
 sponges and, 691
 viruses and, 409
Antibodies, **958**
 in allergic reactions, 970f, 971
 antigen recognition by, 958f–959f
 gene rearrangement by, 960, 961f
 in humoral immune response, 963, 964f–967f
 immunological memory, 963f
 as medical tools, 969f
 as proteins, 76f, 79f
 in responses to extracellular pathogens, 964, 965f–966f
Anticodons, 348f–349f
Antidiuretic hormone (ADH), 994f–995f, 1004f, **1007**f, **1155**f
Antifreeze proteins, 865, 888
Antigen fragments, 959f–960f
Antigenic variation, 972–973
Antigen presentation, **959**f–960f
Antigen-presenting cells, **963**, 964f
Antigen receptors, **958**f–961f
Antigens, **958**, 963f–967f
 in allergic reactions, 970f, 971
 B cell recognition of, 958f–959f
 T cell recognition of, 959f–960f

Antihistamines, 971
Anti-inflammatory drugs, 1013–1014
Antimicrobial peptides, 953–954, 955*f*, 956–957
Antioxidants, 195
Antiparallel DNA sugar-phosphate backbones, **86***f*, 317*f*–320*f*, **318**, 324, 325*f*–327*f*
Antithrombin, 436*f*
Antivenin, 968
Antiviral drugs, 409
Ants, 28*f*, 298*f*, 485, 668*f*, 712*f*, 813*f*, 832*f*, 1001*f*, 1220*f*
Anurans, 732*f*–733*f*
Anus, 911
Aorta, 926*f*, 930*f*
Apes, 746*f*–747*f*
Aphids, 570, 801*f*
Aphotic zone, **1177***f*
Apical buds, **761**
Apical dominance, **770**, 850*f*
Apical ectodermal ridge (AER), **1062***f*, 1063*f*
Apical meristems, 621*f*, 766*f*–767*f*
Apical surface, epithelial, 877*f*
Apicomplexans, **605**, 606*f*
Apicoplasts, 606*f*
Apodans, 732*f*
Apomixis, **833**
Apoplast, **787**, 788*f*, 800*f*
Apoplastic route, 788*f*, 793*f*
Apoptosis, **229**
    cell-signaling pathways of, 230, 231*f*
    cytotoxic T cell response and, 966*f*, 967
    emergent properties of, 233
    ethylene in response to senescence and, 853–854
    in immune cells, 233
    molecular mechanisms of, 230*f*
    in morphogenesis, 1056–1057
    p53 gene and, 390*f*
    plant response to flooding with, 864*f*
    as programmed cell death, 229*f*, 230
    self-tolerance vs., 961
Aposematic coloration, 1218*f*, **1219**
Appendages, arthropod, 706, 707*f*, 708–709
Appendix, **911***f*
Appetite regulation, 917*f*, 918
Apple fruit, 831*f*
Apple maggot fly (*Rhagoletis pomonella*), 508*f*, 515–516
Aptenodytes forsteri, 873*f*
Aptenodytes patagonicus, 740*f*, 884*f*, 1192*f*
Aquaporins, **132**, **790**, **989**
    cellular membrane selective permeability and, 132*f*
    facilitated diffusion and, 135*f*
    in kidney regulation, 994*f*–995*f*
    mutations in, as causes of diabetes insipidus, 995*f*
    water diffusion and role of, 790–791
Aquatic animals
    adaptations of kidneys of, 992, 993*f*
    gills for gas exchange in, 921*f*, 940*f*–941*f*
    ice and, 44
    nitrogenous wastes of, 982*f*
    osmoregulation in, 979*f*–980*f*
Aquatic biomes
    acid precipitation in, 1265, 1266*f*
    biodiversity hot spots in, 1272*f*
    coral reefs, 1182*f*
    decomposition in, 1248–1249
    estuaries, 1180*f*
    eutrophication of, 1275*f*
    food chains in, 1224*f*
    habitat loss in, 1264
    intertidal zones, 1181*f*
    lakes, 1179*f*
    locomotion in, 1135
    marine benthic zones, 1182*f*
    nutrient cycling in, 1250*f*–1251*f*
    oceanic pelagic zones, 1181*f*
    plastic waste in, 1277*f*
    primary production in, 1242, 1243*f*, 1243*t*
    protists as producers in, 614, 615*f*
    streams and rivers, 1180*f*
    thermoregulation in, 881*f*
    wetlands, 1179*f*
    zonation in, 1177*f*–1178*f*
Aqueous humor, 1118*f*
Aqueous solutions, **49**
    acidic and basic conditions of, 51, 52*f*–53*f*, 54
    solvents, solutes, and, 49*f*, 50
Aquifers, 808

Arabidopsis thaliana (mustard plant)
    altering gene expression of, with touch, 862*f*
    genome size of, 448*t*
    as model organism, 22, 774, 776*f*, 783
    triple response in, 853*f*
Arachnids, 689*f*, **708***f*
Arbuscular mycorrhizae, **656**, 660, 663, **817**, 818*f*
Arbuscule, 537*f*, **656***f*
Archaea, **12***f*
    domain, 12*f*, 568, 569*f*, 585*t*, 586
    genome sizes and number of genes for, 448*t*, 449
    prokaryotic cells of, 6*f*, 97. *See also* Prokaryotes; Prokaryotic cells
Archaean eon, 530*f*, 532*f*–533*f*
Archaefructus sinensis, 533*f*, 647*f*
Archaeoglobus fulgidus, 448*t*
Archaeognatha (bristletails), 712*f*
Archaeology, peat moss and, 627*f*
Archaeopteris, 641
Archaeopteryx, 739*f*
Archaeplastida, 599*f*, **609***f*–610*f*
Archegonia, 624*f*, **625**, 630*f*
Archenterons, 682*f*, **1050***f*
Architeuthis dux, 702
Archosaurs, **736**
Arctic, 48*f*
Arctic fox, 1238*f*, 1244, 1256*f*
Arctic ground squirrel (*Spermophilus parryii*), 893
Arctic sea ice, 1279
Arctic tundra ecosystem, 1238*f*, 1244, 1256*f*–1257*f*
Ardipithecus ramidus, 748*f*–749*f*
Area effects, community diversity and, 1232*f*
Arginine, 77*f*, 336*f*–338*f*
Argyroneta aquatica, 951
Arid conditions, 142, 203, 204*f*–206*f*, 1166*f*, 1168
Aristotle, 470
Armenian oak, 650*f*
Arms, chromatid, 236, 264
Arousal
    autonomic, 1096
    brain functions in sleep and, 1094*f*
    human sexual, 1034, 1081
Arsenic, 29
Art, humans and, 754*f*
Arteries, **924***f*–926*f*, 929*f*–931*f*, 937*f*, 939
Arterioles, **924**, 929*f*–932*f*
Arthrophytes, 632*f*
Arthropods, **706**. *See also* Ecdysozoans
    chelicerates, 707, 708*f*
    chitin as structural polysaccharide for, 72*f*
    compound eyes of, 1117*f*, 1118
    crustaceans and insects as pancrustaceans, 709*f*–712*f*, 713
    evolution of, 678
    exoskeletons of, 1133
    general characteristics of, 707*f*, 708
    Hox genes and body plan of, 706*f*, 707
    land colonization by, 536–537
    Malpighian tubules of, 985*f*
    myriapods, 707–708, 709*f*
    nervous systems of, 1086*f*
    origins of, 706*f*, 707
    phylogeny of, 689*f*, 709*f*
Artificial corridors, 1271*f*, 1272
Artificial selection, 474, **475***f*, 476, 651, 823, 836*f*, 837
Artiodactyls, 745*f*
Ascending limb, loop of Henle, 988*f*, 989
Asci, **663**
Ascocarps, **663***f*
Ascomycetes, 660*f*, **663***f*–664*f*, 665*t*, 671*f*
Asexual reproduction, **255**, **833**, **1020**
    in angiosperms, 833*f*–836*f*
    in bryophytes, 625
    evolution of, in animals, 266, 267*f*
    in fungi, 658*f*–659*f*
    inheritance in, 255*f*
    in lichens, 668*f*–669*f*
    mechanisms of, 1020
    in protists, 594
    in rotifers, 697–698
    sexual reproduction vs., 255*f*, 833–834, 1021, 1022*f*. *See also* Sexual reproduction
    switch to, 268
    of triploid organisms, 268
Asian army ants (*Leptogenys distinguenda*), 1001*f*
Asian elephant, 474*f*
Asian ladybird beetles, 475*f*

A site (aminoacyl-tRNA binding site), **350***f*, 352*f*–353*f*
A soil horizon, 806*f*
Asparagine, 77*f*
Aspartic acid, 77*f*
Aspen trees (*Populus tremuloides*), 791*f*, 833*f*
Aspirin, 651, 1001, 1013–1014, 1112
Assassin bugs, 712*f*
Assembly stage, phage lytic cycle, 402*f*
Association areas, cerebral cortex, 1096
Associative learning, **1146***f*
Asteroidea, 713, 714*f*
Asters, 238*f*, 240*f*
Asthma, 62, 217, 933, 1012
Astragalus bones, 481*f*
Astrobiologists, 50
Astrocytes, 1090*f*
Asymmetrical cell division, 777*f*
Asymmetric carbon, 61*f*–62*f*, 68, 75
Asymmetry, body, 1059–1060, 1061*f*, 1064*f*
Atherosclerosis, 74, 75, 139, 141, **937***f*, 938–939
Athletes
    abuse of anabolic steroids by, 1014
    blood doping by, 936
Athlete's foot, 670
Atlantic lobster (*Homarus americanus*), 979
Atlantic salmon (*Salmo salar*), 89
Atmosphere
    animal evolution and oxygen in, 677
    carbon dioxide in, 11
    Earth's early, 526*f*–527*f*
    global climate change and carbon dioxide in, 1278*f*, 1279, 1282, 1283*f*
    ozone in, 1283*f*, 1284
    photosynthesis and oxygen in, 533, 534*f*
Atomic mass, 31
Atomic mass unit (amu), 31
Atomic nucleus, **30***f*, 31
Atomic number, **31**
Atoms, **30***f*–36*f*, 37
ATP (adenosine triphosphate), **64**, **150**
    in bioenergetics, 890*f*
    catabolic pathways and production of, 165
    conversion of, to cyclic AMP, 223*f*–224*f*
    in DNA replication, 324
    energy coupling and, 150, 151*f*–153*f*
    as energy for active transport, 137*f*
    as energy for membrane proteins, 130*f*
    as energy source for cellular processes, 64
    in feedback regulation of cellular respiration, 183*f*, 184
    in muscle fiber contraction speed, 1131
    phosphofructokinase and, 186
    in photosynthesis, 191*f*, 192, 197*f*–200*f*, 201, 206, 207*f*
    regeneration of, in ATP cycle, 153*f*
    regulation of regeneration of, 160
    in sliding-filament model of muscle contraction, 1126, 1127*f*, 1128
    synthesis of, by cellular respiration, 164*f*, 165, 169. *See also* Cellular respiration
    synthesis of, by fermentation and anaerobic respiration, 179, 180*f*, 181
    thylakoids and production of, 211
    in translation, 349*f*
    work as hydrolysis of, 151*f*, 152
    yield of, at each stage of cellular respiration, 177*f*, 178
ATP synthase, **175***f*–176*f*, 177, 186, 311
Atria, heart, **924***f*–927*f*
Atrial natriuretic peptide (ANP), **996**
Atrioventricular (AV) nodes, **928***f*
Atrioventricular (AV) valves, **927***f*
Attachment function, membrane protein, 130*f*
Attachment stage, phage lytic cycle, 402*f*
Auditory communication, 1141–1142
Aurea mutant tomato, 844*f*, 845
Australia, 481*f*, 539–540, 605*f*, 742–744
Australian moles, 558*f*, 559
Australian thorny devil lizard (*Moloch horridus*), 737*f*
Australopiths, 749, 750*f*
Autism spectrum disorder, 1100
Autocrine signaling, 1000*f*, **1001**
Autoimmune diseases, **971***f*
Autonomic arousal, 1096
Autonomic nervous system, **1088***f*–1089*f*
Autophagy, 107*f*, 108
Autopolyploids, **514***f*
Autosomes, **257**

Autotrophs, **188**f, 581t, 586, 889, 894f, 1240f
Auxin, 845, 846t, 847f, **848**f, 849f, 850, 854f, 861
Avery, Mary Ellen, 943, 944f
Avery, Oswald, 315
Avian flu, 410, 1234, 1235f
Avogadro's number, 50
Axel, Richard, 1124
Axillary buds, **761**
Axis formation, 385f–386f, 387, 1059, 1060f
Axolotls, 544, 545f, 921f
Axons, 879f, 1067, **1068**f, 1076f–1077f, 1084,
     1086–1087, 1089f, 1090
Azidothymidine (AZT), 409
Azolla, 813f, 817

B
Bacillus anthracis, 575f, 584f
Bacillus thuringiensis, 837, 838f
Backbone, nucleic acid, 86f
Backbone, polypeptide, 78f
Bacteria, **12**f
    alcohol fermentation by, 180f
    anaerobic respiration in, 179–180
    antibiotic drugs and, 349
    antibiotic resistance in, 478f, 581
    binary fission in, 242, 243f
    bioremediation using, 1255f
    Cas9 protein of, 360, 361f
    cell signaling in, 213f, 214
    cell structure of, 97f
    cellular integration and, 121f
    cholera and, 224
    conjugation in, 579, 580f, 581
    as decomposers, 1240f
    in digestive systems, 912, 913f–914f
    diversity of, 1223f
    as DNA cloning vectors, 418f, 419
    DNA packing in chromosomes of, 330–331
    DNA replication in, 322f–327f
    domain, 12f, 568, 569f, 583f–584f, 585t
    evidence for DNA in research on, 315f–316f, 317
    evolution of cell division in, 242, 243f
    evolution of glycolysis in, 182
    expressing cloned eukaryotic genes in, 422–423
    flagellum, movement of, 993f
    genome sizes and number of genes for, 448t, 449
    G proteins and infections by, 218f
    Gram staining of, 574, 575f
    immune system recognition of, 952f
    land colonization by, 536
    as model organism. *See Escherichia coli*
    mutualistic and pathogenic, 587, 588f, 813f
    nitrogen-fixing, and plants, 814f–816f, 817, 1244
    nutrient limitations and, 1244
    origin of photosynthesis in, 533, 534f
    origins of mitochondria and chloroplasts, 109,
         110f
    in Permian mass extinction, 540
    photosynthesis of, 188f, 189, 199
    phylogeny of, 583f–584f, 585
    plant defenses against, 866
    in polymerase chain reaction, 421, 422f
    prokaryotic cells of, 6f, 97. *See also* Prokaryotes;
         Prokaryotic cells
    regulation of transcription in, 365f–369f, 370
    relatedness of, to mitochondria, 595
    reproduction rate of, 592
    root mutualism with, 592
    soil, 592
    transcription and translation in, 339f, 355f
    transcription in, 342, 343f–344f
    translation in. *See* Translation
    viral infections of. *See* Phages
Bacteriophages (phages), **315**, **401**
    capsids of, 400f, 401
    Cas9 protein and, 360, 361f
    defense against, 404f
    in DNA research, 315f–316f, 317
    in fighting infections, 872, 931
    prophages and temperate, 402, 403f
    replicative cycles of, 402f–404f
    in transduction, 579f
    virulent, 402f
Bacteriorhodopsin, 129f
Bacteroides thetaiotaomicron, 587
Bacteroids, **816**f, 817
Bada, Jeffrey, 58

Bait-and-switch defense, 600–601
Baker, C. S., 557f
Baker's yeast. *See Saccharomyces cerevisiae*
Balance, body, 1112f–1116f, 1135
Balancer, 1066
Balancing selection, **500**, 501f–503f
Balanus balanoides, 1216f
Baleen, 903f
Ball-and-socket joints, 1134f
Ball-and-stick models, 39f, 59f
Ballooning, spider, 708
Ball pythons, 892
Banana, 268, 313
Banana slugs, 1155, 1156f
Barbiturates, 104–105
Barbs, 645f
Bar graphs in Scientific Skills Exercises, 23, 179, 376,
     483, 590, 629, 700, 762, 864, 1150, 1186, 1217
Bark, **774**
Bark beetles, 1244f, 1245, 1280f
Barley, 650f, 851f
Barnacles, 708–709, 710f, 1216f
Barr, Murray, 300
Barr body, **300**f, 309
Barrier contraceptives, 1038f
Barrier defenses, 953–954
Barrier reefs, 1182f
Basal angiosperms, **649**, 650f
Basal animals, 682, 683f, 684, 690
Basal body, **115**, 116f
Basal cells, 828f
Basal lamina, 929f
Basal-like breast cancer, 393f
Basal metabolic rate (BMR), **890**f–891f, 892
Basal nuclei, 1093f
Basal surface, epithelial, 877f
Basal taxon, 556f, **557**
Base pairing, DNA and RNA, 86f, 319f–321f
Bases, **51**
    amino acids as, 76, 77f
    buffers and, 52–53
    hydrogen ions, acids, and, 51
    pH scale and, 51, 52f
Basidiocarps, **666**f
Basidiomycetes, 657, 660f, **665**f–667f
Basidiospores, 666f
Basidium, **665**
Basilar membrane, 1113f
Basin wetlands, 1179f
Basking, reptilian, 735–736
Basophils, 934f–935f
Bass fishes, 881f
Bassham, James, 191
Batesian mimicry, 1218f, **1219**, 1237
Batrachochytrium dendrobatidis, 669, 670f
Bats, 15f, 717, 825f, 991f, 992, 1262f
Bautista, Diana, 92
B cells, **958**f–960f, 961f–966f
    DNA of, 976
Bdelloid rotifers, 266, 267f, 698
Beach mouse (*Peromyscus polionotus*), 2f, 20f–21f
Beadle, George, 336f–338f
Beagle, HMS, C. Darwin's voyage on, 471, 472f
Beaks
    finch, 473f, 486f–487f
    shapes of bird, 740f
    soapberry bug, 477f, 478
Beans, 829f–830f, 858f, 899
Bears, 145f, 510f, 1268f, 1269
Beavers, 1226f
Bed bugs, 712f
Beech trees, 1170f
Bees, 298f, 712f, 824f, 887
Beetles, 475f, 712f, 717, 1244f, 1245, 1259, 1280f
Behavior, **1140**. *See also* Animal behavior
Behavioral ecology, **1140**
Behavioral isolation, 508f
Beijerinck, Martinus, 399
Belding's ground squirrels, 1156, 1158, 1159f, 1193t,
     1194f, 1195, 1213
Beluga whales, 1111f
Benign tumors, **249**
Bennettitales, 648f
Benson, Andrew, 191
Benthic zone, **1177**f
Benthos, **1177**
Bergmann, Christian, 897

Berthold, Peter, 1157f
β_2-adrenergic receptor, 217f, 220
β-amyloid, 1104
β chain, 959f
β-galactosidase, 159f, 368f
β-globin, 87, 89, 357f, 358, 364
β-globin gene family, 453f
β-keratin, bird feathers and, 738f, 739
β pleated sheet, **80**f
Beta-carotene, 838f
Beta cells, 916f
Beta oxidation, **183**
BGI (formerly Beijing Genome Institute), 444
bicoid gene, 386f–387f, 463
Biennials, 766
Big-bang reproduction, 1200
Bilateral symmetry
    animal, 679, 680f, 683
    axis formation and, 1059, 1060f
    flower, 644f, 649f
Bilaterians, **677**
    chordates, 719f
    Deuterostomia, 713
    Ecdysozoans, 705
    invertebrates, 686f
    Lophotrochozoans, 694
    origin of, 677
    phylogeny of, 683f, 717
Bilayers, phospholipid. *See* Phospholipid bilayers
Bile, **909**, 915f
Binary fission, **242**, 243f, 577–578, 606, 607f
Binding sites, ribosome, 350f
Binomials (nomenclature), 470, **554**, 555f
Biochemical pathways. *See* Metabolic pathways
Biochemistry, 96
Biodiversity
    angiosperm, 649f–650f. *See also* Angiosperms
    animal. *See* Animals
    of bacteria and archaea. *See* Archaea; Bacteria
    cellular structures and, 125
    classification of, 470
    climate change effects on, 1279
    conservation biology and, 1261. *See also*
         Conservation biology
    current crisis of, 1260f–1262f
    effects of mass extinctions on, 540f–542f
    evolution and, 14f–16f, 473. *See also* Evolution
    of fungi. *See* Fungi
    gymnosperm, 641, 642f–643f
    habitat loss and fragmentation and, 1263, 1264f
    human welfare and, 1262, 1263f
    invertebrate. *See* Invertebrates
    landscape ecology and regional conservation to
         sustain, 1270, 1271f–1274f
    levels of, 1261f–1262f
    plant. *See* Plants
    protection of, 1260
    protist. *See* Protists
    resurrection of, 1266
    within species, 507f
    sustainable development and, 1284,
         1285f, 1286
    taxonomy and classification of, 12f–13f
    threats to, 1260f, 1261, 1263, 1264f–1266f
    tree of life and, 15f–16f. *See also* Phylogenetic
         trees; Phylogenies; Tree of life
    tropical deforestation as threat to, 651f, 652
    unity in, 13f–14f, 26, 469, 473
    vertebrate. *See* Vertebrates
Biodiversity hot spots, **1272**f
Bioenergetics, **144**, **889**. *See also* Metabolism
    energy allocation and use in, 889, 890f
    energy budgets in, 892
    energy costs of foraging in, 1149
    influences on metabolic rate in, 891f, 892
    of locomotion, 1136
    metabolic rates and thermoregulation in, 890–891
    of osmoregulation, 980
    principles of, 163
    thyroid regulation of, 1009
    torpor, hibernation, and energy conservation in,
         892, 893f
    of urea and uric acid wastes, 983
Biofilms, 213, **582**f
Biofortification, 838f
Biofuels, 671f, **838**
Biogenic amines, 1081t

Biogeochemical cycles, **1249**f–1252f. *See also* Energy flow and chemical cycling
Biogeographic factors, community diversity and, 1231, 1232f–1234f
Biogeography, **482**f, 483
Bioinformatics, **9**, **87**, **443**
  in analysis of protein structure, 83
  centralized resources of, for genome analysis, 444, 445f
  genome analysis using genomics and, 443
  genomics, proteomics, and, 86, 87f
  identifying protein-coding genes using gene annotation in, 445–446
  in study of genomes, 9
  systems biology and proteomics in study of genes and gene expression in, 446, 447f–448f
Biological augmentation, **1255**
Biological clocks, 857, 858f–860f, **1094**–1095. *See also* Circadian rhythms
Biological Dynamics of Forest Fragments Project, 1271f
Biological magnification, **1275**, 1276f
Biological molecules. *See also* Organic compounds
  analyzing polypeptide sequence data of, 87, 89
  carbohydrates as. *See* Carbohydrates
  four classes of, 66
  genomics and proteomics in study of, 86, 87f
  lipids as, 72, 73f–75f
  macromolecules as polymers of monomers and, 66f–67f
  as measures of evolution, 87, 89
  nucleic acids as, 84, 85f–86f
  proteins as. *See* Proteins
Biological species concept, **507**f–510f
Biology, **3**
  astrobiology in, 50f
  biodiversity in. *See* Animals; Biodiversity; Plants
  biophilia and, 1285f
  cells in. *See* Cells
  classification in. *See* Cladistics; Phylogenies; Systematics; Taxonomy
  connection of, to chemistry. *See* Chemistry
  conservation biology in. *See* Conservation biology
  developmental biology in, 1057
  ecology in. *See* Ecology
  emergent properties at levels of biological organization in, 4–6, 4f–5f
  evolutionary developmental (evo-devo) biology in, 387, 463, 544
  evolution as theme of, 2, 3, 11, 12f–16f. *See also* Evolution
  expression and transmission of genetic information in, 6, 7f–8f. *See also* Genetics
  genomics and proteomics in, 86, 87f
  interactions in biological systems in, 10f–11f. *See also* Interactions
  molecular biology in. *See* Molecular biology
  organization in, 4f–5f, 6
  science of, 3, 16, 17f–24f. *See also* Case studies; Inquiry Figures; Research Method Figures; Scientific Skills Exercises
  sociobiology in, 1159–1160
  systems biology in, 5–6, 446, 447f–448f
  transfer and transformation of energy and matter in, 9f. *See also* Energy
  unifying themes of, 2f–11f
Bioluminescence, 143f
Biomass, **838**, **1223**, 1241–1242
Biomass pyramid, 1247f, 1248
Biomes, **1171**. *See also* Aquatic biomes; Biosphere; Global ecology; Terrestrial biomes
  aquatic, 1177f–1182f
  terrestrial, 1171f–1176f, 1189
Biophilia, 1262, 1285f
Bioremediation, **591**f, 809, **1253**
  ecosystem restoration using, 1253, 1255f
Biorhythms, melatonin and, 1015
Biosphere, 4f, **1165**f. *See also* Earth
  aquatic biomes of, 1177f–1182f
  biomes of. *See* Aquatic biomes; Terrestrial biomes
  climate of, 1166f–1170f
  ecological role of prokaryotes in, 586f–587f
  future of, 1285f, 1286
  global climate change of. *See* Climate change
  global ecology of, 1165f. *See also* Global ecology
  human population, and carrying capacity of, 1209, 1210f, 1211

importance of seedless vascular plants to, 633–634
as level of biological organization, 4f
photosynthesis as process that feeds, 188f. *See also* Photosynthesis
terrestrial biomes of, 1171f–1176f, 1189
Biosynthetic pathways, 144, 183
Biotechnology, **433**
  DNA technology. *See* DNA technology
  evolution and, 441
  genetic code and, 342f, 441
  genetic engineering of plants, 837, 838f, 839–840
  in genetic testing, 288, 289f
  genome-sequencing, 9
  phytoremediation, 809
  practical applications of, 433, 434f–437f, 438–439
  prokaryotes in, 589, 590f–591f
  science, society, and, 23, 24f
Biotic factors, **1170**
  microclimate and, 1169–1170
  in pollination, 824f–825f
  in species distributions, 1178, 1183f–1184f
Biotic stresses, plant, **862**, 866, 867f–869f
Bipedal animals, 750f, 1135
Bipolar cells, 1118f, 1120, 1121f
Bipolar disorder, **1103**
Birds
  adaptations of kidneys of, 992f
  alimentary canals in, 905f
  avian flu in, 1234, 1235f
  axis formation in, 1060
  brains of, 1091f
  breathing by, 925, 944, 945f
  cleavage in, 1048
  DDT and, 1276f
  derived characters of, 738f, 739
  as descended from dinosaurs, 564f, 565
  endangered or threatened, 1261, 1262f
  evolution of, 679
  evolution of cognition and brains of, 1098, 1099f
  evolution of finches, 473f, 486f–487f
  evolution of genes in, 456, 457f
  flight adaptations of, 1135–1136
  flower pollination by, 825f
  gastrulation in, 1052f
  genetic variation in migration patterns of, 1156, 1157f
  greater prairie chicken, 495f, 496, 1267f
  hearing and equilibrium in, 1116
  limb formation in, 1062, 1063f, 1064
  living, 739, 740f
  migration behaviors of, 1140, 1141f
  nitrogenous wastes of, 982f, 983f
  organogenesis in, 1055f
  origin of, 739f
  phylogeny of, 719f
  plastic waste and, 1277f
  as pollinators, 522
  populations of, 1190f
  problem solving of, 1147
  red-cockaded woodpecker decline, 1269f–1270f
  salt excretion in marine, 982f
  sex determination of, 298f
  species-area curves for, 1232f
  thermoregulation in, 886f
  unity and diversity of, 14f
  wings of, 739, 740f
Birth control, human, 1037, 1038f, 1039
Birth control pills, **1038**f, 1039
Birth defects, human, 902f, 1039–1040
Births
  demographics of, 1193t, 1194f–1195f
  in density-dependent population growth, 1203f–1205f
  in exponential population growth, 1196, 1197f
  in population dynamics, 1192
Births, human
  birth defects in, 902f
  effects of vitamin supplementation on neural tube defects in, 902f
  life-expectancy at, 1208–1209
  newborn screening and, 289–290
  stages of labor in, 1036, 1037f
  zero population growth and, 1208
Bisphenol A, 1015
Bitter tastes, 1123f–1124f

Bivalves, 699, 701f–702f
Black-bellied seedcracker finches, 498
Black bread mold, 662f, 663
Black-breasted hill turtle (*Geoemyda spengleri*), 736f
Black-capped chickadees (*Poecile atricapillus*), 517f
Blackcap warblers, 1156, 1157f
Black rush plants, 1221f
Blacktip reef sharks (*Carcharhinus melanopterus*), 727f
Bladderwort, 448t, 449
Blades, **602**, 603f, **761**
Blastocoel, **1047**
Blastocysts, 430, 431f, **1034**, 1035f, **1052**, 1053f
Blastomeres, **1047**f
Blastopores, **682**f, **1050**f, 1062f
BLAST program, 444, 445f
Blastula, **674**f, 1044, **1047**f
Blebbing, 229f–230f
Blending hypothesis on inheritance, 271
Blindness, 311, 495, 584f, 1122
Blind spot, eye, 1118f
Blood, **878**f, **923**. *See also* Blood pressure; Blood vessels
  ABO blood groups for human, 280f, 292
  animal thermoregulation and, 885, 886f
  apoptosis of human white blood cells, 229f
  blood groups of, 969–970
  cell division of bone marrow cells and, 235f
  cholesterol in, and atherosclerosis, 75
  in closed circulatory systems, 707, 923, 923f. *See also* Cardiovascular systems; Closed circulatory systems
  clotting of, 10, 300, 418f, 436, 703f, 704, 936f, 937
  components of, 934f–936f, 937
  diffusion from, to interstitial fluid across capillary walls, 932, 933f
  enzymes and glucose levels in, 157
  filtration of, by nephrons, 987, 988f, 989
  flow velocity of, 930f
  gas exchange systems and components of, 947f–949f
  glucose level regulation in, 10f
  glycoproteins and types of human, 131
  hormones in, 999
  immune system rejection of transfusions of, 969–970
  melatonin concentrations in human, 883f
  pH of human, 52f, 53
  regulation of calcium levels in, 1011f
  sickle-cell disease and. *See* Sickle-cell disease
  small intestine and, 125
  vampire bat digestion of, 991f, 992
Blood-brain barrier, 1090
Blood doping, 936
Blood flow velocity, 930f
Blood flukes, 696f, 1220
Blood groups, 280f, 969–970
Bloodhounds, 1138
Bloodletting, 703f, 704
Blood pressure
  cardiac cycle changes in, 930, 951
  in closed circulatory systems, 925
  gravity and, 931f–932f, 951
  hormonal regulation of, 994f–996f
  hypertension and, 939
  regulation of, 930–931
Blood vessels
  blood flow velocity in, 930f
  blood pressure in, 930, 931f–932f, 951
  capillary function, 932f–933f
  lymphatic systems and, 933f
  structure and function of, 929f, 930
  Viagra and, 224
Blooms
  algal, 1227f, 1242, 1243f
  diatom, 602
  dinoflagellate, 605f
  nitrogen pollution and phytoplankton, 1275f
Blowfly, 825f
Blueberry maggot fly (*Rhagoletis mendax*), 508f
Blue crab, 1237
Blue dragons (*Glaucus atlanticus*), 686f
Bluefin tuna, 1265f
Blue-footed boobies, 508f
Bluehead wrasse (*Thalassoma bifasciatum*), 1020
Blue jays (*Cyanocitta cristata*), 1146f
Blue-light photoreceptors, **855**, 856f, 858–859

Blue whales, 719
Bobcat, 894*f*
Body axes, 1060*f*
Body cavities, **680**, 681*f*, 717
Body hairs, insect, 1112*f*
Body plans, **679**
　angiosperm, 648*f*
　animal, 679, 680*f*–682*f*
　arthropod, 706*f*, 707
　cell fate and. *See* Cell fate
　correlation of diversity of miRNAs with complexity of animal, 678
　fungal, 655*f*–656*f*
　hierarchical organization of animal, 876*t*, 877*f*–879*f*. *See also* Animal form and function
　homeotic genes and, 463*f*–464*f*
　human. *See* Human body
　lichen, 668*f*–669*f*
　macroevolution of, 544*f*–546*f*, 547
　maximizing body surface area in animal, 695*f*
　mollusc, 699*f*
　morphogenesis and. *See* Morphogenesis
　pattern formation of, 384, 385*f*–387*f*
　plant, as herbivore defense, 868*f*
Body size, metabolic rate and animal, 891*f*
Body temperature regulation. *See* Thermoregulation
Bog mummy, 627*f*
Bohr shift, **948***f*
Bolting, 851
Bolus, 906*f*
Bombardier beetle, 43
Bombina toads, 516, 517*f*, 520
Bombus affinis, 1170*f*
Bonasa umbellus, 1271
Bonds, chemical. *See* Chemical bonds
Bone marrow, 235*f*, 434*f*
Bone morphogenetic protein 4 (BMP-4), 1062
Bones, 725, **878***f*
　of human skeleton, 1134*f*
　of mammalian ear, 741, 742*f*
Bonobos, 460, 746*f*–747*f*
Bony fishes, 979*f*, 992, 993*f*
Book lungs, **708***f*
Boom-and-bust population cycles, 1205*f*, 1206
Borisy, Gary, 241*f*
Borrelia burgdorferi, 584*f*
Botox, 1080
Bottleneck effect, **495***f*, 496
Bottlenose dolphins, 886*f*, 1094*f*
Bottom-up control, 1226, **1227***f*
Botulism, 218*f*, 403, 584*f*, 588, 1080
Boundaries
　ecosystem, 1270, 1271*f*, 1287
Bound ribosomes, 103*f*, 104, 354*f*
Boveri, Theodor, 295
Bowden, Richard, 627*f*
Bowman's capsule, **987***f*
Boysen-Jensen, Peter, 847*f*, 848
Brachiopods, 688*f*, **698***f*, 699, 713
Brachyistius frenatus, 1203*f*
Brainbow technology, 1085*f*
Brain cancer, 250
Brain cells, 126*f*
Brain Research through Advancing Innovative Neurotechnologies (BRAIN), 1106
Brains, **1069**. *See also* Nervous systems
　arousal and sleep functions of, 1094*f*
　biological clock regulation by, 1094–1095
　breathing control centers in human, 946*f*
　in central nervous systems, 1086*f*, 1087
　cerebral cortex and cognitive functions of, 1096, 1097*f*–1099*f*
　development of, 722*f*, 1092*f*, 1099–1100
　disorders of, 1103, 1104*f*
　drug addiction and reward system of, 1103*f*
　emotional functions of, 1095*f*, 1096
　evolution of chordate and vertebrate, 722*f*
　evolution of cognition in avian pallium and human, 1098, 1099*f*
　evolution of vertebrate, 1091*f*
　frontal lobe function of, 1098
　glia in mammalian, 1069*f*
　glioblastoma cancer of, 447
　human, 748, 1092*f*–1093*f*, 1095*f*–1104*f*
　hypothalamus of human, in thermoregulation, 888, 889*f*

imaging of, 1085*f*, 1096*f*
information processing by, 1096, 1097*f*–1098*f*
language and speech functions of, 1098*f*
lateralization of cortical function of, 1098
mammalian, 741
Neanderthal, 752
nervous tissue of, 879*f*
in neuroendocrine signaling, 1007*f*–1008*f*
neurons in, 1068, 1085*f*, 1091. *See also* Neurons
opiate receptors in mammalian, 1082
of primates, 747–784
regions of, 1091*f*–1093*f*
in sensory systems, 1109
size of, 756
strokes in, 937
visual information processing in, 1121*f*
Zika virus and, 409
Brainstem, **1092***f*–1093*f*, 1094–1095
Brain waves, 1094*f*
Branching, body surface area and, 695*f*
Branching, carbon skeleton, 60*f*
Branching, plant, 770
Branching evolution, 548–549
Branch length, phylogenetic tree, 561, 562*f*
Branch points, **555**, **556***f*
Brassinosteroids, 845, 846*t*, **854**–855
Brazil nut tree, 1202*f*
BRCA1 and *BRCA2* genes, 392*f*–393*f*, 394
Bread mold. *See Neurospora crassa*
Breakdown pathways, 144
Breast cancer, 220, 249*f*, 391, 392*f*–393*f*, 394, 433
Breasts, 1026–1027
Breathing, 756, 920, 925, **944**, 945*f*–946*f*
Breathing control centers, 946*f*
Breeding
　as artificial selection, 474, 475*f*
　plant, 823, 835–837
Brenner, Sydney, 1058–1059
Brewer's yeast. *See Saccharomyces cerevisiae*
Briggs, Robert, 428
Brightfield microscopy, 95*f*
Brine shrimp, 464*f*, 545*f*, 546
Bristlecone pine tree, 643*f*
Bristletails, 712*f*
Brittle stars, 714*f*, 715
Broca, Pierre, 1098
Broca's area, 1098*f*
Bronchi, **942**, 943*f*
Bronchioles, **943***f*
Brood bodies, 625*f*
Brown algae, 598*f*, **602**, 603*f*, 604
Brown bears, 145*f*
Brown fat, 178, 887*f*
Brown-headed cowbird (*Molothrus ater*), 1271, 1287
Brown tree snake, 1264
Brundtland, G. H., 1262
Brush border, 909*f*
Brushtail possum, 743*f*
Bryophytes, **622**–623
　ecological and economic importance of, 627*f*, 628
　evolution of, 635
　gametangia of, 625, 626*f*
　gametophytes of, 624*f*–626*f*
　gametophyte-sporophyte relationships in, 637*f*, 638
　mosses, liverworts, and hornworts as, 623, 626*f*
　as nonvascular plants, 622, 622*t*
　phylogeny of, 623*f*
　sporophytes of, 625, 626*f*
Bryozoans, 687*f*, 698*f*
B soil horizon, 806*f*
Bt toxin, 837, 838*f*, 839
Bubulcus ibis, 1183, 1184*f*, 1221*f*
Buck, Linda, 1124
Budding, 100*f*, 255*f*, 659*f*, 1020
Buff-end moth (*Phalera bucephala*), 476*f*
Buffers, **52**–53
Bugs, 712*f*
Bulbourethral glands, 1025*f*, 1026
Bulk feeders, **903***f*
Bulk flow, **791**
　as long-distance transport, 791*f*
　as translocation mechanism in angiosperms, 800, 801*f*
　of water and minerals from roots to shoots, 792, 793*f*–795*f*, 796

Bulk transport, 126
Bumblebees, 522, 1170*f*, 1279
Bundle sheath, 771*f*
Bundle-sheath cells, **203**, 204*f*
Burkholderia glathei, 586*f*
Burkitt's lymphoma, 394
Burmese pythons (*Python molurus bivittatus*), 887, 888*f*
Bush babies, 746*f*
Butterflies, 313, 551, 711*f*–712*f*, 824*f*, 839
Buttress roots, 760*f*
Buxbaum, Joseph, 462*f*

C

C_3 plants, **203**, 205
C_4 plants, **203**, 204*f*–206*f*
C_4 Rice Project, 438
Cabomba aquatica, 775*f*
Cacao tree, 667*f*, 668
Cachexia, 1016
Cactus, 799*f*, 825*f*, 1172*f*, 1178, 1183*f*
Caddisflies, 551
Cadherin proteins, 676*f*
Caecilians, 732*f*
Caenorhabditis elegans (soil nematode)
　apoptosis in, 230*f*
　fate mapping for, 1058*f*–1059*f*
　genome size and number of genes of, 448*t*, 449
　as model organism, 22, 705*f*
　nervous system of, 1086*f*, 1087
Caffeine, 233
Caimans, 925
Calcitonin, 1004*f*, **1011***f*
Calcium, 29*t*
Calcium homeostasis, 1011*f*
Calcium ions
　diffusion of, across synapses, 1077, 1078*f*
　in formation of fertilization envelope, 1045*f*, 1046
　in regulation of muscle contraction, 1128, 1129*f*, 1131–1132
　in signal transduction pathways, 224, 225*f*, 844*f*, 845
California Current, 1168*f*
California mice, 1143, 1144*t*
Callus, **835**–836
Calmodulin, 1132
Calochortus tiburonensis, 30*f*
Calories (cal), **46**, 890
Calorimeters, 890
Calvin, Melvin, 191
Calvin cycle, **191***f*, 192, 201, 202*f*, 206, 207*f*
Cambrian explosion, 533*f*, **536***f*, **677***f*, 678, 685
Camels, 980
Camouflage, 20*f*–21*f*, 23, 468*f*, 476*f*, 501*f*, 1218*f*, 1219
cAMP (cyclic AMP), **223***f*–224*f*, 233, **369***f*, 1002, 1003*f*, 1080
CAM (crassulacean acid metabolism) plants, 205, **206***f*, 798, 799*f*
cAMP receptor protein (CRP), 369*f*
Canada goose, 886*f*
Canary Islands, 1213
Cancer
　abnormal cell cycle control systems in, 248, 249*f*, 250
　abnormal protein kinases in, 222–223
　biotechnology in treatment of, 434
　carcinogen screening and, 360
　chromosomal translocations and, 309
　DNA microarray detection of, 426
　endocrine disruptors and, 1015
　faulty apoptosis in, 231
　faulty cell-surface receptors in, 217, 220
　faulty growth factors in, 226
　genetic markers for, 428
　genomics, cell signaling, and breast, 391, 392*f*–393*f*, 394
　genomics and proteomics in study and treatment of, 88*f*
　HIV and, 974
　immune system and, 974*f*
　inherited disposition and environmental factors in, 394
　interferences with normal cell-signaling pathways in development of, 389, 390*f*, 391
　lymph nodes and, 933

mismatch repair and colon, 327
multistep model of development of, 391f, 394
ozone depletion, UV radiation, and, 1284
personal genome analysis and, 433
personalized medicine and, 433
PET scanners and, 31, 32f
skin, 328, 1284
species and genetic diversity and treatments for, 1263f
sponges and, 691
stem cells and, 431
systems biology approach to, 447, 448f
telomeres and prevention of, 329
treatment of, with cell cycle inhibitor, 250
tumor-suppressor genes and, 388
types of genes associated with, 388, 389f
viruses in, 394–395
Cancer Genome Atlas, 392f–393f, 394, 447
Candida albicans, 670
Cane toads, 1217
Canine parvovirus, 409
Cannibalism, 413, 1237
Canopy, **1172**
Canyon tree frog, 1218f
Capillaries, **924**f–926f, 929f–930f, 932f–933f
Capillary beds, **924**, 932f–933f
*Capra pyrenaica pyrenaic*a, 1266
Capsaicin, 1111
Capsids, **400**f–401f
Capsomeres, 400f, 402
Capsule, 97f, **575**f
Capsule, sporangium, **625**
Carbohydrates, **68**
cell-cell recognition role of membrane, 130f–131f
digestion and absorption of, 908f–909f
as fuel for catabolism, 182f, 183
glycoproteins and, 400–401. *See also* Glycoproteins
as macromolecules, 66
monosaccharides and disaccharides, 68f–69f, 70
oxidation of, during cellular respiration, 166–167
in plant composition, 809
polysaccharides, 70f–72f
as product of photosynthesis, 206, 207f
Carbon
in amino acids, 75
as essential element, 29t, 64
half-life of, 33
isotopes of, 31–33
in organic compounds, 58, 59f, 60. *See also* Organic compounds
in peatlands, 628
in plant composition, 809
Carbon-12, 530
Carbon-14, 530
Carbonate ions, 53f, 54
Carbon cycle, 1250f, 1257f
Carbon dioxide (CO_2)
atmospheric, 205, 1278f, 1279, 1282, 1283f
in carbon cycle, 1250f
in carbon fixation, 191f, 201, 202f–206f
covalent bonding of carbon atoms in, 59–60
diatom capture of, 602
diffusion of, across capillary walls, 932
in ecosystem interaction, 10f, 11
food quality and levels of, 812
fossil fuels, ocean acidification, and, 53f, 54
in gas exchange, 921, 939t, 940, 947f–949f
in global climate change, 204–205
global climate change and, 11, 48, 1170
inhibition of fruit ripening with, 854
insect effects on forest absorption of, 1245
in mammalian circulation, 926f
metabolic rate and, 890
net ecosystem production and, 1242, 1245
nonvascular plants and, in Ordovician period climate change, 629
in Permian mass extinction, 540
in photosynthesis, 41f
photosynthetic processing of, by marine protists, 615f
in plant cells, 209f
prokaryotic chemical recycling of, 586
in regulation of human breathing, 946f
rubisco as acceptor for, 201, 202f
seedless vascular plants and, 633
in *Sphagnum* peat moss, 628

as stimulus for stomatal opening and closing, 797–798
tropical rain forest deforestation and, 651f, 652
Carbon fixation, **191**f, 192, 201, 202f–206f
Carbonic acid, 51, 53f, 54, 946
Carbon monoxide, 1082
Carbon sink, 1245
Carbon skeletons, 56, 60f–61f
Carbon source, 1245
Carbonyl group, 63f
Carboxyl end. *See* C-terminus
Carboxyl group, 63f
Carboxylic acid, 63f
Carcharhinus melanopterus, 727f
Carcinogens, 360, 388. *See also* Cancer
Cardiac cycle, **927**f–928f, 930, 951
Cardiac muscle, **879**f, **1131**
Cardiac output, **927**
Cardiomyopathy, familial, 357
Cardiovascular diseases, 74, 311, 937f, 938–939, 951
Cardiovascular systems, **924**
adaptations of, 947f–949f
blood composition and function in, 934f–937f, 938–939
blood vessels of, 929f–933f, 951
closed circulatory systems. *See* Circulatory systems
coordination of gas exchange systems and. *See* Gas exchange
diseases of, 74, 937f, 938–939, 951
effects of adrenal hormones on, 1012
evolutionary variations in double circulation of, 925
hearts and blood vessels in single and double circulation of, 924f, 925
hearts in mammalian, 926f–928f
lymphatic systems and, 933f
single and double circulation in, 924f–926f
Caribou (*Rangifer tarandus*), 490f, 1021, 1256f–1257f, 1281f
Carnegiea gigantea, 1178, 1183f
Carnivora, phylogenetic tree of, 555f, 745f
Carnivores, **901**
alimentary canals of, 912f
dentition and diet in, 911f, 912
energetic hypothesis and biomass of, 1225f
Carnivorous plants, 818, 819f, 820
Carolina chickadees (*Poecile carolinensis*), 517f
Carotenoids, 193, 194f, **195**, 570, 605f
Carpellate flowers, 834
Carpels, 270f, 271, **644**f, **823**f
Carrier proteins, 132, 135f–137f
Carriers, genetic disorder, **285**f, 286, 288
Carrion flower, 825f
Carrots, 428
Carrying capacity (*K*), **1197**
global, for human population, 1209, 1210f, 1211
in logistic population growth model, 1198f–1199f, 1198t, 1200
Carson, Rachel, 1276f
Cartilage, **878**
Cartilage skeleton, 723–726
Cas9 protein, 360, 361f. *See also* CRISPR-Cas9 system
in gene editing, 426, 427f, 434–435
prokaryotes and, 589, 590f
Cascade
hormonal, 1008, 1009f, 1010, 1014
Casein, 76f
Case studies
on decline of red-cockaded woodpecker, 1269f–1270f
on evolution of tolerance to toxic elements, 30f
on greater prairie chicken extinction vortex, 1267f
on grizzly bear populations, 1268f, 1269
on kidney function in vampire bats, 991f, 992
on nutrient cycling in Hubbard Brook Experimental Forest, 1252f
on predation and mouse coat coloration, 20f–21f
sustainable development in Costa Rica, 1284, 1285f
on variation in migratory patterns, 1156, 1157f
on variation in prey selection, 1155, 1156f
Casparian strip, **792**, 793f
Caspases, 230, 233
Cas protein, 404f
Cassava, 651, 838f
Castor beans, 829
Catabolic pathways, **144**, 165, 166f–168f, 182f–184

Catalysis, **154**, 155f–156f. *See also* Enzymatic catalysis
Catalysts, **75**, **153**. *See also* Enzymatic catalysis
Catalytic cycle, 156f
Catalytic knob, ATP synthase, 175f
Cataracts, 1284
Catecholamines, 1012f
Catenulida, 694
Caterpillars, 488f, 871, 897, 903f, 1006f, 1007, 1246f
Catharanthus roseus, 1263f
Cation exchange, **806**, 807f
Cations, **37**, 38f, 806, 807f
Cats, 293, 300f, 364, 430f
Catskill Mountains, 1263
Cattails (*Typha angustifolia*), 1186
Cattle, 914f
Cattle egrets (*Bubulcus ibis*), 1183, 1184f, 1221f
Caulerpa, 610f
Causation, behavioral, 1140
Cavendish banana, 268
CC (Carbon Copy, cloned cat), 430f
CCR5 gene, 435
CCR5 protein, 130f
Cdks (cyclin-dependent kinases), **245**, 246f
Cecum, **911**f
Cedar Creek Ecosystem Science Reserve, 1223f
Celera Genomics, 443f
Cell adhesion molecules, 1056
Cell body, 879f, **1068**f, 1079f
Cell-cell recognition
by cellular membranes, 130f, 131
in local cell signaling, 215f
Cell cycle, **237**. *See also* Cell cycle control system; Cell division
binary fission in bacterial, 242, 243f
cell division roles in, 234f–235f
cellular organization of chromosomes in, 235f, 236
cytokinesis in, 241, 242f–243f
distribution of chromosomes during eukaryotic, 236f, 237
evolution of, 252
evolution of mitosis of, 243, 244f
interpreting histograms on, 250
mitosis stages of, in animal cells, 238f–239f
mitotic phases and interphases of, 237
mitotic spindle in mitotic phase of, 240f–241f
regulation of eukaryotic, by cell cycle control system, 244, 245f–249f, 250
treating cancer by inhibiting, 250
Cell cycle control system, **245**. *See also* Cell cycle
in cancer development, 248, 249f, 250, 388, 389f–393f, 394
checkpoints in, 245f
cyclins and cyclin-dependent kinases in, 245, 246f
cytoplasmic signals in, 244, 245f
internal and external signals at checkpoints of, as stop and go signs, 246, 247f–248f
interpreting histograms on, 250
Cell cycle–inhibiting pathway, 389, 390f
Cell cycle–stimulating pathway, 389, 390f
Cell differentiation. *See* Differentiation, cell
Cell division, **235**
bacterial, 242, 243f
cancer and interference with cell-signaling pathways of, 389, 390f, 391
in cell cycle, 234f–235f. *See also* Cell cycle
cytokinins in, 850
cytoplasmic determinants and induction in, 382f, 383
distribution of chromosomes during eukaryotic, 235f–236f, 237
as embryonic development process, 381f, 382. *See also* Embryonic development
evolution of, 243, 244f
of HeLa cancer cells, 252
key roles of, 235f
in meiosis, 259f–262f. *See also* Meiosis
in mitosis vs. in meiosis, 262, 263f, 264
newt lung cell, 7f
in plant growth, 776, 777f
prokaryotic, 577–578
Cell expansion, 777f
Cell fate, 1057–1059f, 1061–1064f
Cell fractionation, **96**f, 97, 125
Cell-free fetal DNA, 288

Cell junctions
  in local cell signaling, 215f
  plasmodesmata in plants, 119f, 120
  tight junctions, desmosomes, and gap junctions in animals, 120f
Cell-mediated immune response, 957f, **963**, 966f–967f
Cell migration
  in organogenesis, 1054f, 1055
Cell motility, 112, 113t, 114, 115f–117f, 1056f
Cell plate, **241**, 242f–243f
Cells, 5f, **759**
  animal. See Animal cells
  auxin in differentiation of, 850
  auxin in elongation of, 848, 849f
  blood, 934f–936f, 937
  calculating volume and surface area of, 99
  cell fractionation in study of, 96f, 97
  cellular integration of, 121f
  cellular membranes of. See Cellular membranes
  cellular respiration and. See Cellular respiration
  climate change effects on, 1280f
  communication between. See Cell signaling
  cytokinins in division and differentiation of, 850
  differentiation of. See Differentiation, cell
  division of, as fundamental to life, 235. See also Cell cycle; Cell division
  eukaryotic vs. prokaryotic, 97f–98f, 99. See also Eukaryotic cells; Prokaryotic cells
  as fundamental units of life, 5f–6f, 93f
  locations of enzymes in, 161f
  metabolism of. See Metabolism
  microscopy in study of, 94f–95f, 96
  molecular machinery in, 122f–123f
  photosynthesis and. See Photosynthesis
  plant. See Plant cells
  programmed death of. See Apoptosis
  programming of, by viral DNA, 315f–316f, 317
  protein folding in, 83f
  protocells as first, 528f
  sequential gene regulation in differentiation of, 383, 384f
  shape changes in animal morphogenesis, 1056f–1057f
  in sickle-cell disease, 502f
  size range of, 94f
  of small intestine, 125
  stem cells. See Stem cells
  transcription specific to type of, 374–375, 377f
  unity in, 13f
  water regulation of, 142
Cell sap, 108
Cell signaling. See also Signal transduction pathways
  by animal endocrine and nervous systems, 880f
  in apoptosis, 229f–231f
  cancer and interference with normal, 389, 390f, 391
  in cell cycle control system, 244, 245f–248f. See also Cell cycle control system
  cellular membrane selective permeability and, 130f
  cilia in, 114
  endocrine system, 1000f–1004f
  evolution of, 213f, 214, 233
  feedback regulation of, 1005–1006
  fight-or-flight responses in, 212f
  local and long-distance, 215f, 216
  mechanical, 119
  pathways of. See Endocrine signaling; Neuroendocrine signaling
  reception stage of, 217f–221f
  response stage of, 226f–228f, 229
  in simple endocrine pathway, 1004, 1005f
  in simple neuroendocrine pathway, 1005f
  symplastic, 801–802
  three stages of, 216f, 217
  transduction stage of, 221, 222f–225f
Cell-surface proteins, 119f, 130f
Cell-surface receptor tyrosine kinases (RTKs), 250
Cell-surface transmembrane receptors, 217f–220f
Cellular hormone response pathways, 1002f–1003f
Cellular innate immune defenses, 954f, 955f
Cellular-level herbivore defenses, plant, 868f
Cellular membranes. See also Plasma membranes
  active transport across, 136, 137f–138f, 139
  animal, 674
  bulk transport across, by exocytosis and endocytosis, 139, 140f, 141

  evolution of differences in lipid composition of, 129
  fluidity of, 128f, 129
  as fluid mosaics of lipids and proteins, 127f, 128
  interpreting scatter plots on glucose uptake across, 136
  membrane carbohydrates in cell-cell recognition by, 130f–131f
  membrane proteins of, 129f–130f, 132
  of mitochondria, 110, 111f
  movement across plant cell, 209f
  nuclear envelopes, 102, 103f
  organelles and internal, 98f, 99
  passive transport as diffusion across, 132, 133f–135f
  phospholipids in, 74f, 75
  in plant response to cold stress, 865
  selective permeability of, 127f, 131, 132f
  specialized prokaryotic, 577f
  synthesis and sidedness of, 131f
Cellular respiration, **165**
  ATP production by catabolic pathways and, 165
  ATP yield at each stage of, 177f, 178
  bar graphs of, 179
  biosynthesis in anabolic pathways and, 183
  in carbon cycle, 1250f
  as catabolic, 144, 165
  in energy flow and chemical cycling, 9f, 164f, 165
  enzymes for, in mitochondria, 161f
  evolutionary significance of glycolysis in, 182
  fermentation vs., 165, 179, 180f–181f, 182
  glycolysis in, 170f–171f
  metabolic rate and, 890f
  mitochondria in, 109–110, 111f
  monosaccharides in, 69
  origin of, 534, 535f
  overall reaction for, 149
  oxidative phosphorylation in, 174f–177f
  oxygen diffusion and, 133
  photosynthesis vs., 190. See also Photosynthesis
  in plant cells, 209f
  pyruvate oxidation and citric acid cycle in, 171f–173f
  redox reactions and, 165, 166f–168f
  regulation of, via feedback mechanisms, 183f, 184
  scale of, 122f
  stages of, 168, 169f
  using cell fractionation to study, 97
  versatility of catabolic pathways and, 182f, 183
Cellular slime molds, 612, 613f
Cellulose, **71**
  microfibrils, 777f
  in plant cell walls, 70f–71f, 72, 118
  as product of photosynthesis, 206, 207f
  proteins synthesizing, 619f
  water and, 49
Cellulose synthase, 118
Cell walls, **118**
  cellulose in plant, 619
  fungal cell, 100f, 661
  osmosis, water balance, and, 134f–135f
  plant cell, 101f, 118f
  prokaryotic, 97f, 574, 575f
Cenozoic era, 530f, 533f, 539f, 679
Center for Plant Conservation, 1261
Centipedes, 708, 709f
Central axis, 680f
Central canal, 1087f
Central disk, sea star, 714f
Central dogma, DNA, 339
Central nervous system (CNS), **1069**, **1086**. See also Brains
  neuronal plasticity of, **1100**f
  neurons of, 1069
  neurotransmitters and, 1080
  peripheral nervous system and, 1086, 1087f–1090f
  in sensory systems, 1108–1109
  structure and function of vertebrate, 1087f–1088f
Central vacuoles, 101f, **108**
Centrifuge, 96f, 97
Centrioles, **114**f
Centromeres, **236**f, 264
Centromeric DNA, 452
Centrosomes, 100f, **114**f, 238f, **240**f
Centrostephanus rodgersii, 1281f
Century plants, 1200, 1201f
Cephalization, 1086–1087
Cephalochordata (lancelets), 719f, **720**, 721f–722f
Cephalopods, 699, 701f, 702

Ceratotherium simum cottoni, 1266
Cercozoans, **608**f, 609
Cerebellum, 1091f, **1092**f–1093f
Cerebral cortex, **1093**f, 1096, 1097f–1099f, 1100, 1114
Cerebral ganglia, 704f–710f
Cerebral hemispheres, 1093f
Cerebrospinal fluid, 946f, 1087f
Cerebrum, 1091f, **1092**f–1093f, 1094
Certainty of paternity, 1150, 1151f
Cervical cancer, 974f
Cervix, **1026**, 1027f
Cetaceans, 481f–482f, 745f
Cetartiodactyla, 745f
cGMP (cyclic GMP), 224, 844f, 845
Chaetae, 703, 704f
Chagas' disease, 600
Chain worms, 694
Chambered nautilus, 701f, 702
Chameleons, 673f, 735f
Chamerion angustifolium, 636f
Chamois (Rupicapra rupicapra), 901f
Change, global. See Global change
Change, spontaneous, 148f
Channel proteins, 132, 135f, 136
Chaparral biomes, **1174**f
Character displacement, **1216**, 1217f, 1237
Characters, **270**
  dominant vs. recessive traits and, 271f, 272t
  multifactorial, 282f
  shared ancestral and shared derived, 560, 561f
  taxonomy and, 554
  traits and, 270–271
Character tables, 561f
Charadrius melodus, 1190f
Chargaff, Edwin, 317f, 318
Chargaff's rules, 317f, 318, 320
Charged tRNA, 349f
Charophytes, 610, 619f–620f
Charpentier, Emmanuelle, 426
Chase, Martha, 316f, 317
Cheating behavior, 1159
Checkpoints, cell cycle control system, **245**f–247f
Cheetahs, 212f, 1204f
Chelicerae, **708**
Chelicerates, **707**, 708f
Chemical bonds, **36**
  with carbon, 56, 58, 59f, 60
  covalent, 36f–37f
  hydrogen bonds, 38, **39**f
  ionic, 37, 38f
  van der Waals interactions, 39
Chemical cycling
  in ecosystems, 1238f–1240f, 1241
Chemical cycling, energy flow and. See Biogeochemical cycles; Energy flow and chemical cycling
Chemical defense, prey, 1218f
Chemical digestion, 907f–908f, 909
Chemical energy, 143–**144**, 145f, 889, 890f. See also Cellular respiration; Photosynthesis
Chemical equilibrium, **41**
  buffers and, 52–53
  in chemical reactions, 41
  free energy change and, 147, 148f
  metabolism and, 148, 149f–150f
Chemical mutagens, 360
Chemical reactions, **40**
  activation energy barrier of, 153, 154f–155f
  chemical energy in, 144, 145f
  endergonic, 148, 149f
  enzymatic catalysis of. See Enzymatic catalysis
  exergonic, 148, 149f, 154f, 165
  free energy change and, 147, 148f
  making and breaking of chemical bonds by, 40f–41f
  metabolism and. See Metabolism
  in photosynthesis, 190f, 191. See also Light reactions
Chemical signals, 1068. See also Animal hormones; Hormones
Chemical structure, DNA, 319f
Chemical synapses, 1077, 1078f. See also Synapses
Chemical work, 150–151, 152f
Chemiosmosis, 169, 175, **176**f–177f, 191f, 199f–200f, 201
Chemistry
  atoms in, 30f–36f
  biological molecules in. See Biological molecules

calculating standard radioactive isotope decay curves in, 33
chemical bonding between atoms in, 36*f*–41*f*
chemical bonding with carbon in, 58, 59*f*, 60
connection to biology. *See* Biology
matter as elements and compounds in, 29*f*–30*f*. *See also* Compounds; Molecules
organic, as study of carbon compounds, 57*f*, 58. *See also* Organic compounds
of water. *See* Water
Chemoautotrophs, 581*t*
Chemoheterotrophs, 581*t*
Chemoreceptors, **1110**, 1111*f*, 1123*f*–1125*f*
Chemosynthetic organisms, 1240
Chemotaxis, 576
Chemotherapy, 249, 393*f*
Chemotrophs, 581*t*
Chesapeake Bay estuary food web, 1225*f*, 1237
Chestnut blight, 669, 867, 1226, 1234
Chewing, 920
Chiasmata, **260***f*, 262*f*–263*f*, 264
Chickadees, 517*f*
Chicken pox, 962
Chicks
embryo image, 1043*f*
gastrulation in, 1052*f*
limb formation in, 1062, 1063*f*, 1064
organogenesis in, 1055*f*
Chicxulub crater, 541*f*
Chief cells, 907*f*
Chikungunya virus, 409*f*, 412
Childbirth, human, 1036, 1037*f*. *See also* Births, human
Chimaeras, 719*f*, 726, 727*f*
Chimpanzees (*Pan troglodytes*)
comparison of human genome with genome of, 87, 89, 454*f*–455*f*, 460*f*, 461
heterochrony and differential growth rates in skulls of, 544*f*
humans vs., 748
J. Goodall's research on, 17*f*
as primates, 746*f*–747*f*
problem solving of, 1147
skulls of humans and, 558
social learning in, 1147*f*
tool use by, 750
China, 1208
Chips, human gene microarray, 447, 448*f*
Chiroptera, 745*f*
Chi-square (χ^2) test in Scientific Skills Exercise, 304
Chisholm, Penny, 552
Chitin, **72***f*, **656**, 661, 707, **1133**
Chitons, 699*f*, 1086*f*
Chlamydias, 584*f*, 1039
Chlamydomonas, 101*f*, 610, 611*f*. *See also* Green algae
Chlamydomonas nivalis, 211
Chlorarachniophytes, 596*f*, 609
Chloride cells, 992, 993*f*
Chloride ions, 1069*f*, 1070*t*, 1071*f*
Chloride transport channels, 286
Chlorinated hydrocarbons, 1275, 1276*f*
Chlorine, 29*t*, 1283–1284
Chlorofluorocarbons (CFCs), 1283*f*, 1284
Chlorophyll, 5*f*, **189***f*, 190
chemiosmosis, 199, 200*f*, 201
cyclic electron flow in, 198*f*, 199
light excitation of, 195*f*
linear electron flow in, 197*f*–198*f*
in photosystems, 195, 196*f*, 197
structure of, 194*f*
Chlorophyll *a*, **193**, 194*f*, 195–198
Chlorophyll *b*, **193**, 194*f*, 196
Chlorophyll *d*, 194
Chlorophyll *f*, 194–195
Chlorophytes, 610*f*
Chloroplasts, **109**, **189**
chemiosmosis in, 177
chemiosmosis in mitochondria vs. in, 199*f*–200*f*, 201
evolutionary origins of, 109, 110*f*
folding of, 695*f*
light reactions in. *See* Light reactions
as organelles, 5*f*, 6
photosynthesis by, 109–110, 111*f*, 187
in plant cells, 101*f*, 209*f*
as sites of photosynthesis, 189*f*, 190, 206, 207*f*
transgenic crops and DNA in, 840

Chlorosis, 810, 812
Choanocytes, **690***f*
Choanoflagellates, 614, 675*f*–676*f*
Cholecystokinin (CCK), 915*f*
Cholera, 218*f*, 224, 584*f*, 588
Cholesterol, **75***f*
in cellular membranes, 127*f*–128*f*
DNA methylation and, 372*f*
in egg yolks, 91
G protein-coupled receptors and, 217*f*
receptor-mediated endocytosis and, 139, 141
types of, in blood, 937–938
Chondracanthus harveyanus, 1222
Chondrichthyans, 719*f*, **726**, 727*f*
Chondrocytes, 878*f*
Chondroitin sulfate, 878*f*
Chordates, **719**
endoskeletons of, 1133, 1134*f*
evolution of, 722*f*
hagfishes and lampreys, 723*f*, 724
invertebrate, 689*f*, 715, 719*f*–722*f*
lancelets, 720, 721*f*–722*f*
phylogeny and derived characters of, 719*f*–720*f*
phylogeny of, 683*f*
tunicates, 721*f*, 722
vertebrates as, 689*f*, 715, 719*f*, 722, 723*f*–725*f*
Chorion, 735*f*, 1053*f*
Chorionic villus sampling (CVS), **289***f*, 1039
Choroid, 1118*f*
Christmas tree worm, 703*f*
Chromatin, **102**, **235**, **331**
animal cell, 100*f*
in cell division, 235*f*–236*f*, 237
in eukaryotic cell nucleus, 102, 103*f*
in eukaryotic chromosomes, 330*f*–332*f*
plant cell, 101*f*
regulation of structure of eukaryotic, 371*f*–372*f*, 373
remodeling of, by siRNAs, 380–381
Chromatin loops, 377, 378*f*
ChromEMT, 330*f*
Chromoplasts, 111
Chromosomal basis of inheritance
as basis for Mendelian inheritance, 294*f*, 295
chromosomal alterations and genetic disorders in, 306, 307*f*–309*f*
evolution of gene concept from, 361
exceptions to Mendelian inheritance in, 310*f*, 311
genomic imprinting in, 310*f*, 311
inheritance of organelle genes in, 311*f*
linked genes and linkage in, 301*f*–306*f*
sex-linked genes in, 298*f*–300*f*
T. H. Morgan's experimental discovery of, 295*f*–296*f*
Chromosomal breakage points, 454–455
Chromosome conformation capture (3C) techniques, 377
Chromosomes, **102**, **235**. *See also* DNA; Genes
alleles on, 272, 273*f*. *See also* Alleles
alterations of, 306, 307*f*–309*f*, 454*f*–455*f*
bacterial, 242, 243*f*
in cancer cells, 249–250
in cell division, 234*f*–235*f*, 244*f*
in chromosomal basis of Mendelian inheritance, 294*f*, 295. *See also* Chromosomal basis of inheritance
correlating behavior of alleles with pairs of, 295, 296*f*–297*f*
crossing over and recombinant, 265, 266*f*
distribution of, during eukaryotic cell division, 235*f*–236*f*, 237
DNA, genes, and, 7*f*–8*f*
DNA and chromatin packing in, 330*f*–332*f*
in eukaryotic cell nucleus, 102, 103*f*
gene expression and interaction of, in interphase nucleus, 377, 378*f*
genetic variation due to mutations in, 488–489
in genome evolution, 454*f*–455*f*
homologous, 256*f*–257*f*
human, 102, 235, 236*f*, 237, 256*f*–257*f*, 258
independent assortment of, 265*f*
as information, 268, 313
inheritance of genes and, 255
in interphase, 252
karyotypes of, 256*f*
locating genes along, 295, 296*f*–297*f*
mapping distance between genes on, 305*f*–306*f*
in meiosis, 254, 259*f*–263*f*, 264
molecular tags and karyotypes of human, 332*f*

movement of, on kinetochore microtubules, 240, 241*f*
in prokaryotic and eukaryotic cells, 97*f*, 98, 577*f*
prokaryotic conjugation and gene transfer between, 579, 580*f*, 581
Chromosome theory of inheritance, **295**
Chronic inflammation, 956–957
Chronic myelogenous leukemia (CML), 309*f*, 435
Chronic traumatic encephalopathy (CTE), 1104
Chroococcidiopsis thermalis, 194*f*, 195
Chrysanthemums, 860
Chthamalus stellatus, 1216*f*
Chum salmon (*Oncorhynchus keta*), 89
Chylomicrons, **910***f*
Chyme, **907**, 915*f*
Chytrids, 660*f*, **661***f*, 669, 670*f*, 733
Cichlid fish, 515*f*, 519*f*, 520
Cigarette smoke, cardiovascular and lung disease and, 394, 951
Cilia, **114**
architecture of eukaryotic, and unity, 13*f*
bronchial, 943
cell fate and, 1064*f*
ciliate, 607*f*
flagella vs., 115*f*
as microtubules, 114, 115*f*–116*f*
Ciliates, **606**, 607*f*, 617
Cilium-based signaling, 114
Cinnabar (*cn*) gene, 305*f*–306*f*
Circadian rhythms, **798**, **858**, **882**. *See also* Biological clocks
in animal behavior, 1141
in animal homeostasis, 882, 883*f*
brain regulation of, 1094–1095
hibernation and, 893*f*
melatonin and, 1015
in plant responses to light, 857, 858*f*–860*f*
in stomatal opening and closing, 798
Circannual rhythms, 1141
Circulatory systems
cardiovascular systems. *See* Cardiovascular systems
gas exchange and, 921*f*–924*f*, 925, 947*f*. *See also* Gas exchange
gas exchange systems and. *See* Gas exchange
gastrovascular cavities as, 922*f*, 923
internal exchange surfaces and, 875*f*
invertebrate, 699*f*, 702, 704*f*, 707
open and closed, 923*f*
organization of vertebrate, 924*f*, 925. *See also* Cardiovascular systems
thermoregulatory adaptations of animal, 885, 886*f*
cis face, Golgi apparatus, 106*f*, 107
cis isomers, 61*f*
Cisternae, 104, 106*f*, 107
Cisternal maturation model, 106
Cisternal space, 104
cis-trans isomers, 61*f*
Citric acid cycle, **168**, 169*f*, 172*f*–173*f*, 177*f*
Citrulline, 338*f*
Clades, **559**, 560*f*, 561, 622–623
Cladistics, **559**, 560*f*, 572. *See also* Systematics; Taxonomy
Clams, 688*f*, 699, 701*f*
Clark's nutcracker (*Nucifraga columbiana*), 1146
Classes, taxonomy, **554**, 555*f*
Classical conditioning, 1146
Clausen, Jens, 1189
Claw-waving behavior of male fiddler crab, 1141*f*
Cleaner wrasse, 1214*f*
Cleanup, environmental, 418*f*, 437
Cleavage, **241**, **674**, **1046**
in animal embryonic development, 674*f*, 1044, 1046, 1047*f*–1048*f*, 1049
in cell cycle, 241, 242*f*–243*f*
in human embryonic development, 1035*f*
in protostome and deuterostome development, 681, 682*f*
Cleavage furrows, 239*f*, **241**, 242*f*, 1048*f*
Click beetle (*Pyrophorus nyctophanus*), 143*f*
Climate, **1167**. *See also* Climate change
continental drift and changes in, 539
effect of large bodies of water on, 47*f*
global patterns of, 1166*f*, 1167
greenhouse gases and, 1278*f*–1283*f*
latitudinal gradients and, affecting community diversity, 1232*f*

Climate (continued)
    microclimate and, 1169–1170
    nonvascular plants in Ordovician period changes of, 629
    Permian mass extinction and changes in, 540
    regional and local effects on, 1167*f*–1169*f*
    seedless vascular plants and ancient, 633
    species distributions and, 1164, 1170*f*
    terrestrial biomes and, 1171*f*
    using dendrochronology to study, 773*f*
    vegetation effects on, 1169*f*
Climate change, **11**, 1170*f*, **1278**
    biodiversity hot spot preservation and, 1272
    biological effects of, 1279, 1280*f*–1281*f*, 1282
    coral reefs and, 693
    crop productivity and, 863
    ecological footprints, fossil fuels, and, 1210*f*, 1211
    as ecosystem interaction, 11
    effects of, on photosynthetic marine protists, 615*f*
    extinction rates and, 541*f*, 542
    food quality and, 812
    fossil fuel burning and, 11
    greenhouse gases and, 1278*f*–1283*f*
    habitat loss from, 1263
    hybrid zones and, 517*f*
    lizards and, 11*f*
    melting of Arctic sea ice and, 48*f*
    models of, 1282, 1283*f*
    nonvascular plants in Ordovician period, 629
    ocean acidification and, 53*f*, 54–55
    overharvesting of peat moss and, 628
    in Permian mass extinction, 540
    plant adaptations to, 204–205
    positive feedbacks of, 1189
    primary production response to, 1244*f*, 1245
    ringed seals and, 44*f*
    seedless vascular plants and, 633–634
    snowshoe hares and, 501*f*
    solutions for, 1282–1283
    tropical rain forest deforestation and, 651*f*, 652
    tropical rain forest photosynthesis and, 211
    using dendrochronology to study, 773*f*
    viral transmission and, 412
Climax communities, 1228
Climographs, **1171***f*
Clitoris, **1026**, 1027*f*
Cloaca, **727**, **1024**
Clock, cell cycle, 245, 246*f*
Clock genes, 858
Clonal selection, **962***f*, 963, 976
Clone (term). *See* Organismal cloning
Cloned genes. *See also* DNA cloning; Gene cloning
    of crystallin, 441
    expressing eukaryotic, 422–423
    in gene therapy, 434*f*
    uses for, 418*f*
Clones, **255**, 1066. *See also* Organismal cloning
    asexual reproduction of, 255*f*
    of extinct species, 1266
    fragmentation and, 833*f*
    from plant cuttings, 835–836, 849
    test-tube or *in vitro*, 836*f*
Cloning vectors, **418**. *See also* Recombinant DNA
Closed circulatory systems, 702, 704*f*, **923***f*. *See also* Cardiovascular systems
Clostridium botulinum, 584*f*, 588
Clostridium difficile, 912
Clotting, blood, 10, 300, 418*f*, 436, 703*f*, 704, 936*f*, 937
Cloudina, 677*f*
Club fungi, 665*f*–667*f*
Club mosses, 631, 632*f*, 633
Clumped dispersion, 1192*f*
Clutch size, 1200, 1201*f*–1202*f*
Cnemaspis psychedelica, 1260*f*
Cnidarians, 687*f*, 691*f*–693*f*, 1086*f*
Cnidocytes, **691***f*–692*f*
Coal, 633–634. *See also* Fossil fuels
Coal gas, 852
Coastal Indonesia restoration project, 1254*f*
Coat coloration case studies, 20*f*–21*f*, 23
Cocaine, 1103*f*
Coccidioidomycosis, 670
Coccosteus cuspidatus, 533*f*
Cochlea, **1113***f*–1115*f*
Cocklebur, 859
Cocktails, drug, 409, 489

Coconut, 828, 832*f*
Cod (*Gadus morhua*), 728, 979*f*
Coding strand, 340*f*
Codominance, **279**
Codon recognition, 352, 353*f*
Codons, **340**
    evolution of, 364
    in genetic code, 340*f*–342*f*
    in translation, 347*f*–350*f*, 352, 353*f*
Coefficient of relatedness (r), **1158***f*–1159*f*
Coefficients, correlation, 678, 751
Coelacanths, 719*f*, 729*f*
Coelom, **680**, 681*f*, 698, 704*f*, 717
Coelomates, 681
Coenocytic fungi, **656***f*
coenzyme Q (CoQ), 175, 186
Coenzymes, **158**
Coevolution, **825***f*
Cofactors, **158**
Coffee, 651
Cognition, 1096, 1098, 1099*f*, **1146**, 1147*f*
Cognitive maps, **1146**
Cohesins, 236, 240, 262*f*
Cohesion, **45**, 46*f*, 795–796
Cohesion-tension hypothesis, **794***f*–795*f*, 796
Cohorts, **1193**
Coho salmon (*Oncorhynchus kisutch*), 89, 1200
Coitus, human, 1034
Coitus interruptus, 1038*f*
Cold
    plant response to stress of, 865
    thermoreceptors and, 1111*f*
Cold viruses, 401, 404–405
Coleoptera (beetles), 712*f*
Coleoptiles, **829***f*–830*f*, 847*f*, 848
Coleorhiza, **829***f*
Colistin, 588
Collagen, 76*f*, 81*f*, **118**, 119*f*, 674
Collagenous fibers, 878*f*
Collar cells, 675*f*
Collared flycatchers (*Ficedula albicollis*), 519
Collecting duct, **987***f*–988*f*, 989
Collenchyma cells, **764***f*, 770
Colon, **911***f*
Colon cancer, 327
colony collapse disorder, 661
Coloration
    case studies on mouse, 20*f*–21*f*, 23
    of chromosomes, 332*f*
    as prey defensive adaptation, 1218*f*, 1219
    skin, 1016, 1018
Color blindness, 299*f*, 1122*f*
Colorectal cancer, 391*f*, 394
Color vision, 1117–1118, 1121, 1122*f*
Colossal squid, 702
Columbine flowers, 825*f*
Columnar epithelium, 877*f*
Combinatorial control elements, 375–376, 377*f*
Comb jellies, 687*f*
Comet collision, mass extinction by, 541*f*
Commensalism, **587**, 662, **1221***f*
Commercial value
    of fungi, 670, 671*f*
    of mosses, 627*f*
Common arrowhead flower, 835*f*
Common juniper, 643*f*
Communicating junctions, 120*f*
Communication, animal, 880*f*, **1141***f*–1143*f*
Communication, cellular. *See* Cell signaling
Communities, 4f, **1165**f
    biogeographic factors affecting, 1231, 1232*f*–1234*f*
    climate change effects on, 1281*f*
    disturbances of, 1214, 1228, 1229*f*–1231*f*
    diversity in, 1261*f*–1262*f*
    interspecific interactions in, 1214*f*–1221*f*, 1237
    as level of biological organization, 4*f*
    pathogen alteration of structure of, 1234, 1235*f*
    scientific, 19*f*, 24
    species diversity and stability of, 1222*f*–1223*f*. *See also* Species diversity
    structure of, 1214*f*, 1215, 1234, 1235*f*
    study of, by community ecology, 1165*f*. *See also* Community ecology
    trophic structure of, 1223, 1224*f*–1227*f*. *See also* Trophic structure
Community diversity, 1261*f*–1262*f*

Community ecology, **1165***f*. *See also* Ecology
    biogeographic factors in, 1231, 1232*f*–1234*f*
    community boundaries in. *See* Communities
    disturbances in, 1214, 1228, 1229*f*–1231*f*
    influences on community structure and, 1214*f*, 1215
    interspecific interactions in, 1214*f*–1221*f*, 1237
    pathogens in, 1234, 1235*f*
    species diversity and trophic structure in, 1222*f*–1227*f*. *See also* Species diversity; Trophic structure
Community-level herbivore defenses, plant, 869*f*
Community structure, 1214*f*, **1215**
    pathogen alteration of, 1234, 1235*f*
Compact body cavity, 681*f*
Companion cells, **765***f*
Compartmentalized processing, **898**
Competition, **1215**, 1256*f*–1257*f*
    density-dependent population regulation by, 1204*f*
    interspecific, 1215, 1216*f*–1217*f*, 1237
    sexual, 499*f*–500*f*, 1153*f*
    in species distributions of plants, 1186
Competitive exclusion, **1215**
Competitive inhibitors, **158**, 159*f*
Complementary base pairing, DNA and RNA, 86*f*
Complementary DNA (cDNA), **424**, 425*f*–426*f*
Complement systems, **957**, 966*f*
Complete digestive tracts, 905*f*
Complete dominance, **279**
Complete flowers, **824**
Complete growth medium, 336*f*–338*f*
Complete metamorphosis, **711***f*, 712*f*
Complex eyes, 547, 548*f*
Compound eyes, 712*f*, 894*f*, **1117***f*, 1118
Compound leaves, 761*f*
Compounds, **29**. *See also* Molecules
    biological. *See* Biological molecules
    emergent properties of, 29*f*
    ionic, 38*f*
    organic. *See* Organic compounds
    properties of, 28
    pure elements vs., 37
Compromises, evolutionary, 501*f*
Computational tools, 9, 444, 445*f*–448*f*, 776. *See also* Bioinformatics
Concentration gradients, **133***f*, 136, 137*f*–138*f*, 139
Concentrations, chemical reactions and, 41
Conception, human, **1034**, **1035***f*
Condensation reaction, 67
Condensins, 331*f*
Condoms, 1038*f*
Conduction, animal heat exchange and, **885***f*
Conduction, of action potential, 1075*f*, 1076
Cones (photoreceptor), **1119***f*–1121*f*, 1122–1123, 1138
Cones, gymnosperm, 640, 643*f*
Cone snails (*Conus geographus*), 1067*f*–1068
Confocal microscopy, 95*f*, 96
Conformer animals, **881***f*
Congenital disorders, 256*f*
Conidia, **664***f*
Conifers, 623, **640***f*, 641, 643*f*
Conjugation, **579**, 580*f*, 581, **606**, 607*f*
Connective tissue, animal, 76*f*, **878***f*
Connell, Joseph, 1216*f*
Conodonts, **724***f*, 725
Consanguineous mating, human, 285–286
Conservation Areas, Costa Rican, 1273*f*
Conservation biology, **1261**. *See also* Ecology
    biodiversity and, 1260*f*–1266*f*
    conservation of mollusc species, 702*f*
    genomics and proteomics in, 88*f*
    global change and, 1265, 1266*f*, 1274, 1275*f*–1283*f*, 1284
    landscape and regional conservation in, 1270, 1271*f*–1274*f*
    population conservation in, 1266, 1267*f*–1270*f*
    species-area curves of species richness in, 1232*f*
    sustainable development in, 1284, 1285*f*, 1286
Conservation of energy, 145*f*, 146, 1239
Conservation of mass, 1239–1240
Conservative model, DNA replication, 321*f*–322*f*
Conserved Domain Database (CDD), 445*f*
Constant (C) region, light and heavy chain, 958*f*–961*f*
Constipation, 911
Consumers, **9**, 188, 673*f*

in carbon cycle, 1250*f*
primary, 1240*f*, 1246, 1247*f*, 1248
secondary, 1240*f*, 1246, 1247*f*, 1248
tertiary, 1240*f*, 1246, 1247*f*, 1248
Consumption, regulation of animal, 917*f*, 918
Continental drift, 482–483, 525, 538*f*–539*f*, 540
Contour tillage, 809*f*
Contraception, **1037**, 1038*f*, 1039, 1208
Contraceptives, as environmental toxins, 1276
Contractile proteins, 76*f*
Contractile vacuoles, **108**, 135*f*
Contraction, muscle, 1126, 1127*f*, 1128. *See also* Muscle
Contrast, 94
Control
animal, 880*f*
Control center, homeostatic, 882*f*
Control elements, **373***f*–377*f*
Control groups, 20, 21*f*
Controlled experiments, **20**–21
designing, in Scientific Skills Exercise, 1014
Conus geographus, 1067*f*–1068
Convection, animal heat exchange and, **885***f*
Convergent evolution, **481**
analogies and, 558*f*
of cacti and euphorbs, 1172*f*
of fast swimmers, 874*f*
of homologies, 481*f*
of marsupials, 744*f*
in phylogenies, 553*f*–554*f*
Convergent extension, **1056**, 1057*f*
Conversion, data, 264
Cooksonia sporangium, 622*f*
Cooling, evaporative, **47**, 48*f*, 886
Cooper, Vaughn, 578*f*
Cooperativity, **160**
in allosteric activation, 160*f*
prokaryotic metabolic, 582*f*
science and, 22–23, 24*f*
Coordinately controlled genes, 366, 375–376
Coordination
animal, 880*f*
Coordination, cell-signaling response, 227, 228*f*
Copepods, 710*f*
Coprophagy, 914
Copy-number variants (CNVs), 461–462
CoQ (coenzyme Q), 175, 186
Coral atolls, 1182*f*
Coral polyps, 1019*f*
Coral reefs, 53*f*, 54, 614, **1182***f*, 1234, 1264, 1277*f*, 1282
Corals, 692*f*–693*f*
Corepressors, **367***f*
Cork cambium, **766***f*, 774
Cork cells, 774, 783
Cormorant, flightless (*Phalacrocorax harrisi*), 506*f*, 507, 511
Corn (*Zea mays*), 205, 448*t*, 651. *See also* Maize
Cornea, 1118*f*
Corn smut, 669*f*
Corpus callosum, **1093***f*, 1098
Corpus luteum, **1029***f*, 1032*f*, 1033
Correlation coefficients in Scientific Skills Exercise, 678, 751
Correlations, positive and negative, in Scientific Skills Exercise, 834
Correns, Carl, 311*f*
Corridors, movement, 1271*f*, 1272
Cortex, **116**, **763**, 770*f*
Cortical microfilaments, 116
Cortical nephrons, **986***f*, 991
Cortical reactions, 1045*f*, 1046
Cortical rotation, 1060*f*
Cortisol, 1001*f*
Corynebacterium diphtheriae, 97*f*
Costa Rica, 1273*f*, 1284, 1285*f*
Cost-benefit behavior analysis, 1149
Cotransport, **138***f*, 139
Cotton, 49, 804
Cottongrass, 1256*f*
Cotyledons, **647**, 649*f*, 828*f*–830*f*
Counseling, genetic, 287–288
Countercurrent exchange, **886***f*, **941***f*
Countercurrent multiplier systems, **990**–991
Courtship rituals. *See also* Mating behavior
behavioral isolation and, 508*f*
external fertilization and, 1023

forms of animal communication in, 1141*f*–1142*f*
genetic basis of, 1155*f*
reproductive cycles and, 1021*f*
sexual selection and, 499*f*–500*f*, 1151, 1152*f*–1153*f*, 1154
Covalent bonds, **36**
of disaccharides, 69*f*
formation of, 36*f*
in organic compounds, 58, 59*f*, 60, 64
in protein tertiary structure, 81*f*
types of, 36*f*–37*f*
Cowbirds, 1271, 1287
Crabs, 709*f*
"Crank" drug, 62
Crassulacean acid metabolism (CAM) plants, 205, **206***f*, 798, 799*f*
Crawling, 1133*f*, 1135
Crayfish, 709, 940*f*, 1112
Creatine phosphate, 1128
Crenarchaeota clade, 586–587
Cretaceous mass extinction, 540, 541*f*
Creutzfeldt-Jakob disease, 412
Cricetus cricetus, 893*f*
Crick, Francis
central dogma of, 339
discovery of DNA molecular structure by, 4, 23–24, 314*f*, 317, 318*f*–320*f*
model of DNA replication by, 320, 321*f*
Crickets (*Anabrus simplex*), 449, 712*f*, 1112*f*
Cri du chat, 309
Crinoidea, 715*f*
CRISPR-Cas9 system, **360**, 361*f*, 380, 404*f*, **589**
in gene editing, 426, 427*f*, 434–435
in gene identification, 446
plant mutations and, 776
prokaryotes in development of, 589, 590*f*
Cristae, **110**
Critical habitat, 1269*f*–1270*f*
Critical load, **1275**
Crocodiles, 564*f*, 719*f*, 737*f*, 738, 925, 1091*f*
Crop (esophageal pouch), 905*f*, 913
Crop plants. *See also* Agriculture; Plants
artificial selection and breeding of, 835–837
biotechnology and genetic engineering of, 837, 838*f*, 839–840
climate change and, 863
effects of atmospheric carbon dioxide on, 205
as polyploids, 514–515
seed plants as, 651
transgenic and genetically modified, 438–439
Crop rotation, **817**
Cross-fostering studies, 1143, 1144*t*
Crossing over, **260***f*, **302**
chromosomal alterations during, 307, 308*f*
evolution and, 313
gene duplication due to unequal, 455*f*
genetic variation from, 265, 266*f*
in meiosis, 260*f*, 262*f*
recombination of linked genes in, 302, 303*f*
Cross-pollination, **646**
angiosperm, 646*f*, 647
G. Mendel's techniques of, 270*f*–271*f*
of plants, 837
Cross-talk, cell-signaling, 228*f*
Crotalus atrox, 1084
Crows, 1150
CRP (cAMP receptor protein), 369*f*
Crustaceans, 464*f*, 545*f*, 546, 689*f*, 708, 709*f*–710*f*
Crustose lichens, 668*f*, 672
Cryo-electron microscopy (cryo-EM), 95*f*, 96
Cryptic coloration, 1218*f*, **1219**
Cryptochromes, 856
Cryptomycetes, 660*f*–**661***f*
Cryptophyte, 597*f*
Crypts, 799*f*
Crystallin, 377*f*, 441
Crystallin proteins, 8*f*
Crystals, ice, 48*f*
C soil horizon, 806*f*
Ctenophora, 687*f*
Ctenophores, 683*f*, 684
C-terminus, 78*f*, 352
Cuatro ojos fish, 365*f*
Cuboidal epithelium, 877*f*
Cubozoans, 692*f*–693*f*
Cuckoo bee, 1218*f*, 1219
Cud, 914*f*

Culex pipiens, 496
Culture, **1148**, 1159*f*, 1160
Cupula, 1116*f*
Curl cat, 293
Curvularia, 672
Cuscuta, 819*f*
Cushing's syndrome, 1018
Cuticle, ecdysozoan, **705**, 707
Cuticle, exoskeleton, 1133
Cuticle, leaf, **763**
Cuticle, plant, **621**
Cuttings, plant, 428, 835–836, 849
Cutworms, 705
Cuvier, Georges, 469*f*, 470–471
Cyanobacteria
blooms of, 1227*f*
bryophyte symbiosis with, 626*f*, 627
chemical recycling by, 586*f*
evolution of glycolysis in, 182
fungi and, as lichens, 663, 668*f*–669*f*
metabolic cooperation in, 582*f*
mutualism with, 813*f*
origin of photosynthesis in, 533, 534*f*
photosynthesis by, 188*f*, 194*f*, 195, 199, 584*f*
protist endosymbiosis and photosynthetic, 596*f*–597*f*, 608*f*, 609
Cyanocitta cristata, 1146*f*
Cycads, 642*f*
Cyclic AMP (cyclic adenosine monophosphate, cAMP), **223***f*–224*f*, 233, **369***f*, 1002, 1003*f*, 1080
Cyclic electron flow, **198***f*, 199
Cyclic GMP (cGMP), 224, 844*f*, 845
Cyclin, **245**, 246*f*
Cyclin-dependent kinases (Cdks), **245**, 246*f*
Cycliophora, 688*f*
Cyclosporine, 670
Cyclostomes, **723**
Cynodonts, 531*f*, 532
Cysteine, 63*f*, 77*f*
Cystic fibrosis, 83, 280, **286**, 433
Cystic kidney disease, 1064
Cytochromes, **175**, 199, 230, 566
Cytogenetic maps, 306
Cytokines, 955–956, 1001
Cytokinesis, **236**
in meiosis, 260*f*–261*f*
mitosis and, 234, 239*f*, 241, 242*f*–243*f*
nuclear envelope during, 252
Cytokinins, 846*t*, 850*f*
Cytology, 96
Cytoplasm, **98**
cell cycle control signals in, 244, 245*f*
cell-signaling responses in, 226, 227*f*
cytokinesis and division of, 236, 239*f*, 241, 242*f*–243*f*
of prokaryotic and eukaryotic cells, 98–99
Cytoplasmic determinants, **382***f*, 383
Cytoplasmic genes, 311
Cytoplasmic responses, cell-signaling, 226, 227*f*
Cytoplasmic streaming, **117***f*, 608
Cytosine, 84, 85*f*, 86, 317*f*, 318, 341*f*
Cytoskeletons, **112**
actin microfilaments of, 113*t*, 115, 116*f*–117*f*
animal cell, 100*f*
ATP in mechanical work of, 152*f*
intermediate filaments of, 113*t*, 117
membrane proteins and attachment to, 130*f*
microtubules of, 113*t*, 114*f*–116*f*
in morphogenesis, 1056*f*–1057*f*
plant cell, 101*f*
scale of, 123*f*
structure and function of, 113*t*
support and motility roles of, 112*f*–113*f*
Cytosol, **97**
Cytosolic calcium ions, 844*f*, 845
Cytotoxic chemotherapy, 393*f*
Cytotoxic T cells, 962, **966***f*, 967

D
DAG (diacylglycerol), **225***f*
Dalton (atomic mass unit), **31**, 50
Dalton, John, 31
Danaus plexippus, 839, 1146*f*
Dance language, honeybee, 1142*f*
Dandelions, 824*f*, 832*f*, 1192*f*, 1202*f*
Dangl, Jeffery, 814*f*
Danio rerio (zebrafish) model organism, 22

Daphnia pulex, 448t, 1021, 1199f, 1200
Darkness
  flowering in long-night plants and, 859f, 860
  plant etiolation response to, 843f
Dark responses, rod cell, 1121f
D'Arrigo, Rosanne, 773f
Darwin, Charles. *See also* Evolution
  on barnacles, 710
  Beagle voyage and field research of, 471, 472f–473f
  on coevolution of flower-pollinator mutualism, 825
  on earthworms, 704
  evidence supporting theory of, 476, 477f–482f, 483
  on grandeur of evolutionary process, 484
  historical context of life and ideas of, 469f–471f
  on island species, 483
  on lung evolution from swim bladders, 728
  on mystery of flowering plants, 477, 647, 825f
  on mystery of speciation, 506f, 507
  on natural selection, 266, 267f, 487
  On the Origin of Species by Means of Natural Selection by, 469, 473, 483–484, 487
  speciation theory of, 473, 474f–476f
  on species diversity of tropics, 1232
  study by, of phototropism in grass coleoptiles, 847f
  theory of descent with modification by, 14f–16f, 469–470, 473f–476f, 483–484
  timeline of work of, 469f
Darwin, Francis, 847f
Dasyatis americana, 727f
Data, **17**, 19f, 21–22. *See also* Scientific Skills Exercises
Databases
  in estimating reproductive rates, 1194, 1195f
  genome-sequence, 444, 445f
Dating, radiometric, **32**–33, 529, 530f
dATP, 324
Daughter cells, 234, 235f–236f, 237, 266f
Day-neutral plants, **859f**
db gene, 918
DDT pesticide, 158, 485, 518, 1276f
Dead-leaf moth (*Oxytenis modestia*), 476f
Dead Sea, 585
Dead zone, 1242, 1275f
Deamination, amino acid, 182
Deaths
  demographics of, 1193t, 1194f–1195f
  in density-dependent population growth, 1203f–1205f
  in exponential population growth, 1196, 1197f
  in human population dynamics, 1208–1209
  in population dynamics, 1192, 1256f
Death signals, apoptosis, 230, 231f. *See also* Apoptosis
Decapods, 709f, 710
Decay curves, radioactive isotope, 33
December solstice, 1167f
Deciduous forest, nutrient cycling in, 1252f
Decomposers, 188, **586**, **1240f**
  in energy flow and chemical recycling, 9f, 1240f, 1241
  fungi as, 654–655, 658f, 661f, 663, 665, 667
  lichens as, 669
  prokaryotic, 586
Decomposition, 1240f, 1241, 1248f–1251f, 1259
Deconvolution microscopy, 95f, 96
Deductive reasoning, **18**, 617
Deep-sea hydrothermal vents, 527f, 1182f
Deer, 871, 1149, 1271
Deer mice, 498f, 892
DEET insect repellant, 1123
De-etiolation (greening), **843f**–844f, 845
Defensive adaptations, 28f, 43, 1217, 1218f–1219f, 1220, 1237
Defensive proteins, 76f
Deficiencies, plant, 810, 811f, 818
Deforestation
  climate effects of, 1169f
  experimental, and nutrient cycling, 1252f
  as human community disturbance, 1231
  loss of species from, 1260f, 1263
  rising atmospheric carbon dioxide levels from, 1278
  of tropical rain forests, 651f, 652, 1263
Degradation, protein, 379
Dehydration
  animal, 980
  plant, 203, 204f–206f, 536

Dehydration reactions, **67f**
Dehydrogenases, 167f, 168
DeJac, Lynn, 24f
Delayed reproduction, 1208
Deletions (mutations), 358f, **359**
Deletions, chromosome, **307**, 308f, 309
Dementia, 1103, 1104f
Demographics, population, 1193t, 1194f–1195f
Demographic transition, **1208**
Demography, **1193**. *See also* Demographics, population
Denaturation, **82f**, 83
Dendrites, 879f, **1068f**, 1078, 1110f
Dendritic cells, **955**
Dendrobates pumilio, 524
Dendrochronology, 773f
Dendroctonus ponderosae, 1244f, 1245, 1280f
Dengue fever, 412
Denitrifying bacteria, 1251f
Density, population, **1191f**–1192f, 1193, 1202, 1203f, 1256f
Density-dependent inhibition, **247**, 248f
Density-dependent population change, 1202, **1203f**–1206f, 1207, 1213
Density-independent population change, **1203**
Dentition, 530, 531f, 532, 911f, 912
Deoxyribonucleic acid. *See* DNA
Deoxyribose, **85f**, 317f, 324
Dependent assortment hypothesis, 275, 276f
Dependent variables, **21**, 513
Dephosphorylation, protein, 223
Depolarization, **1072f**–1074f
  in fertilization, 1044f, 1045
Depolymerization, protein, 240
Depression, 1081, 1102–1103
Derelle, Romain, 612f
Derived characters, shared, 560, 561f
Derived traits, plant, 620f–621f
Dermal tissue system, plant, 758, **762**, 763f
DES (diethylstilbestrol), 1015
Descending limb, loop of Henle, 988f, 989
Descent with modification theory, 14f–16f, 469–470, 473f–476f, 483–484. *See also* Evolution
Desert animals, 980–981
Desert iguana (*Dipsosaurus dorsalis*), 888
Desert mice, 981
Deserts, 798, 799f, 1168, 1169f, **1173f**
Desert spring, 1239f
Desiccation, 979
Desmognathus ochrophaeus, 513
Desmosomes, 120f
Desynchronization, 858–859
Determinate cleavage, **681**, 682f
Determinate growth, **766f**–767f
Determination, **383**, 384f, **1057**
Detoxification, 104–105, 112, 809
Detritus, **1177**, **1240f**, 1241
Deuteromycetes, **659**
Deuterostome development, **681**, 682f
Deuterostomes, **683f**, 689f, 713, 714f–715f, 719f. *See also* Chordates; Echinoderms
Development, **775**
  of brain, 722f, 1092f, 1099–1100
  as cell division function, 235f
  embryonic. *See* Embryonic development
  in human life cycle, 257f, 258. *See also* Human embryonic development
  macroevolution of, 545f–546f, 547
  plant and animal, 894f. *See also* Animal development; Plant development
  postzygotic barriers and, 509f, 510
  as property of life, 3f
  sustainable. *See* Sustainable development
Developmental biology, 1057
Developmental genes. *See* Homeotic genes; *Hox* genes
Developmental plasticity, 775f
Developmental potential, cell fate and, 1060, 1061f
Diabetes
  aquaporin mutations as causes of diabetes insipidus, 995f
  autoimmunity and, 971
  disruption of glucose homeostasis in, 916–917
  genetic engineering of insulin to treat, 436
  genetic markers for, 427
  mitochondrial mutations in, 311
  neonatal, insulin mutations and, 359
  stem cells for, 431

Diabetes insipidus, 995f
Diabetes mellitus, **916**–917
Diacodexis, 482f
Diacylglycerol (DAG), **225f**
Diagnosis
  antibodies as tools in, 969f, 434f
  biotechnology in, 433, 434f, 435
Diamond-back rattlesnake (*Crotalus atrox*), 1084
Diaphragm in breathing, **945f**
Diaphragm, birth control, 1038f
Diapsids, **736**
Diarrhea, 139, 224, 588, 598f, 911
Diastole, **927f**
Diastolic pressure, **930**, 931f
Diatomaceous earth, 602
Diatoms, 244f, 599f, **602f**, 617, 1170
Diazepam, 1081
Dicer-2, 954f
Dicots, **649**
Dictyostelium discoideum, 613f
Dideoxy DNA sequencing, 416
Dideoxyribonucleotide (dideoxy) chain termination DNA sequencing, 443
Didinium, 617
Diencephalon, 1093f
Diets. *See also* Food
  adaptations of vertebrate digestive systems for, 911f–914f
  assessing nutritional needs in, 902f
  catabolism and human, 182f, 183
  deficiencies in, 901f, 902
  essential nutrients for, 899f, 900t–901t, 901f
  genetic variation in prey selection and, 1155, 1156f
  nonheritable variation and, 488f
  phenylketonuria and, 492
  typical and opportunistic, 901
Differential centrifugation, 96f, 97
Differential gene expression, **370**. *See also* Gene regulation
  cytoplasmic determinants and induction in, 382f, 383
  in eukaryotic gene regulation, 370f, 371
  in gene regulation, 365f
  in pattern formation of body plans, 384, 385f–387f
  in processes of embryonic development, 381f, 382. *See also* Embryonic development
  sequential regulation of, during cellular differentiation, 383, 384f
Differential-interference microscopy, 95f
Differential reproductive success, 266, 267f
Differential speciation success, 548–549
Differentiation, **1057**
Differentiation, cell, **381**
  cytokinins in, 850
  as embryonic development process, 381f, 382. *See also* Embryonic development
  plant development and, 763, 775–776, 778f
  sequential gene regulation in, 383, 384f
  stem cells and, 430f, 431. *See also* Stem cells
Diffusion, **132**, **922**
  body surface area and, 695f
  effects on water balance of osmosis as, 133f–135f
  extracellular, 768
  free energy change and, 148f
  interpreting scatter plots on glucose uptake as, 136
  as passive transport, 132, 133f
  in plant cells, 209f
  proteins and facilitated, 135f, 136
  of water across plant plasma membranes, 788, 789f–791f
  of water and minerals into root cells, 792
Digestion, **904**
  in animal food processing, 904f–905f
  compartmentalized processing in, 898
  digestive compartments in, 902, 903f–905f
  digestive systems and, 875f
  extracellular, 904f–905f
  feeding mechanisms and, 902, 903f
  fungal, 655, 668
  hydrolysis in, 67
  intracellular, 904
  lysosomes in intracellular, 107f, 108
  regulation of animal, 915f
  sea star, 713, 714f

in small intestine, 908f, 909
in stomach, 907f–908f
vertebrate adaptations for, 911f–914f
Digestive enzymes, 76f
Digestive secretions, 898, 907f–908f, 909–910, 915f
Digestive systems
adaptations of, 911f–914f
alimentary canals, 905f, 912f
bacteria in, 911–912, 913f–914f
feedback regulation of, 915f
internal exchange surfaces and, 875f
of invertebrates, 717
large intestine in, 910, 911f
oral cavity, pharynx, and esophagus in, 905, 906f–908f
small intestine in, 908f–910f
stomach in, 907f–908f
Digger wasps, 1145f
Dihybrid crosses, **275**, 276f
Dihybrids, **274**–275, 276f
Dihydroxyacetone, 68f
Dijkstra, Cor, 1201f
Dikaryotic mycelia, **658**
Dimers, tubulin, 114
Dimorphism, sexual, **499f**, 752, 1149–1150, 1151f
Dinitrophenol (DNP), 186
Dinoflagellates, 244f, 594f, **604**, 605f
Dinosaurs, **736**
blood pressure of, 931
disappearance of, 679
as early reptiles, 736
flying, 1135
in fossil record, 529f, 564f, 565
mass extinction of, 541f, 542
in Mesozoic era, 533
Diomedea exulans, 977f
Dionaea muscipula, 802, 819f, 862
Diphtheria, 403
Diploblastic animals, **680**
Diploid cells, **257**
genetic variation preserved in recessive alleles of, 500
haploid cells vs., 257
mitosis in, 262, 263f, 264
in sexual life cycles, 258f, 259
Diplomonads, **600f**
Dipnoi (lungfishes), 719f, 729
Dipodomys merriami, 741f, 998
Dipsosaurus dorsalis, 888
Diptera, 712f
Direct contact, cell signaling by, 215f
Direct inhibition hypothesis, 850
Directionality, DNA replication, 324, 325f–326f
Directional selection, **497**, 498f
Disaccharides, 68, **69f**, 70
Diseases, animal
density-dependent population regulation by, 1204f
movement corridors and spread of, 1271f, 1272
viral, 400f, 407f, 409f–412f
Diseases, plant
community structure and pathogens in, 1234
density-dependent population regulation by, 1204f
disease-resistant genes and, 866, 867f
viral, 399f–400f, 412f
Diseases and disorders, human
adenoviruses and, 400f
alkaptonuria, 336
allergies, 970f, 971
Alzheimer's disease, 413, 1103, 1104f
amebic dysentery, 613
amyotrophic lateral sclerosis (ALS), 1129
antibiotic resistance and, 478f
asthma, 62, 217
atherosclerosis, 74, 75, 139, 141, 937f, 938–939
autism spectrum disorder, 1100
autoimmune, 971f
bacterial, 403, 574, 575f–576f, 584f, 587, 588f
biotechnology in diagnosis and treatment of, 433, 434f–436f
cachexia, 1016
cancer. See Cancer
cardiovascular diseases, 74, 311, 937f, 938–939, 951
chikungunya virus, 409f
cholera, 224, 584f, 588

color blindness, 299f, 1122f
community ecology, pathogens, and, 1234, 1235f
cri du chat and chronic myelogenous leukemia, 309f
CRISPR-Cas9 system for, 360, 361f
cystic fibrosis, 286, 293
cystic kidney disease, 1064
density-dependent population regulation by, 1204f
detecting fetal, during pregnancy, 1039–1040
diabetes. See Diabetes
diabetes insipidus, 995f
diarrhea, 139
dominantly inherited, 287f
Down syndrome, 307, 308f, 309
drug addiction, 1103f
Duchenne muscular dystrophy, 299
due to chromosomal alterations, 306, 307f–309f
Ebola virus, 409f
emerging viral diseases and, 409f–412f
endocrine disruptors and, 1015
endometriosis, 1033
epilepsy, 1098
erectile dysfunction, 1081–1082
familial cardiomyopathy, 357
faulty apoptosis in nervous system and, 231
faulty cell-surface receptors and, 218f, 220
fetal screening for, 288, 289f
flatworm parasites and, 696f–697f
flesh-eating disease, 478f
fungal, 670
gastric ulcers and acid reflux, 908
genetic. See Genetic disorders
genomics and proteomics in, 88f
gonorrhea, 576, 584f
gout, 983
growth-related, 1009, 1010f
heart murmurs, 928
hemophilia, 300
HIV/AIDS. See AIDS; HIV
Hodgkin's disease, 971
Huntington's disease, 287
hypercholesterolemia, 139
hypertension, 939
immune system disruptions and, 970f–974f, 973t
immunization against, 968f, 976
immunodeficiency, 971, 972f, 973
influenza, 400f, 410
insects as carriers of, 713
iodine deficiency and goiter, 29
Kartagener's syndrome, 1064
karyotypes and, 256f
Klinefelter syndrome, 309
lactose intolerance, 70
Leber's congenital amaurosis, 1122
lymphatic system and, 933
major depressive disorder and bipolar disorder, 1102–1103
malaria. See Malaria
mitochondrial, 311
mosaicism, 300f
multifactorial, 287
mutations and, 357f–358f, 359
myasthenia gravis, 1129
myotonia and epilepsy, 1076
nematode parasites and trichinosis, 705f, 706
of nervous system, 1102f–1104f
neurodegenerative, 413
neurotransmitters and, 1080–1082
from ozone depletion, 1284
parasitic. See Parasites
Parkinson's disease, 413, 1081, 1104
phenylketonuria, 492–493
from plastic waste, 1277
pleiotropy and inherited, 280
pneumonia, 315f, 579
polydactyly, 280
protein misfolding and, 83
protists and, 598f–599f
recessively inherited, 285f–286f, 287
respiratory distress syndrome, 943, 944f
retinitis pigmentosa, 495
schizophrenia, 1102f
sexually transmitted diseases. See Sexually transmitted diseases
sickle-cell disease. See Sickle-cell disease

sleeping sickness, 600, 601f
spina bifida, 1055
Tay-Sachs disease and lysosomal storage, 108, 280, 285, 288, 289f
testing genetic markers for, 427f
thyroid, 1009
Turner syndrome, 309
Wiskott-Aldrich syndrome, 229
xeroderma pigmentosum, 328
X-linked disorders, 299f, 300
Zika virus, 409f
Disease-suppressive soil, 590
Disorder, entropy and, 146
Disparity, vertebrate, 719
Dispersal, **1183**
fruit and seed, 832f
movement corridors and, 1271f, 1272
seed, 645f
species, 1183, 1184f
Dispersion, population, **1191f**–1192f, 1193
Dispersive model, DNA replication, 321f–322f
Disruptive selection, **498f**
Dissociation, water, 51
Distal control elements, 374
Distal tubule, **987f**–988f, 989
Distance vision, 1122f
Distribution patterns, analyzing, 283
Distributions, species. See Species distributions
Disturbances, **1172**, 1214, **1228**, 1229f–1231f
Disulfide bridges, **81f**
Divergence
allopatric speciation and, 506, 511f–512f, 513
of angiosperms, 647f, 648
of closely related species, 460–461, 462f
of fungi, 659f, 660
in phylogenetic trees, 555, 556f, 557
of unikonts from other eukaryotes, 612f
Diversity. See also Biodiversity
B cell and T cell, 960, 961f
biological. See Biodiversity
cellular structures and, 125
evolution and, 11, 468f, 469, 473
in science, 24
within a species, 507f
Diving bell spider (*Argyroneta aquatica*), 951
Diving mammals, respiratory adaptations of, 949f
Division, cell. See Cell division
Dixon, Henry, 794
Dizygotic twins, 1036
Dizziness, 1116
DNA (deoxyribonucleic acid), **6**, **84**. See also Chromosomes; Genes; Genetics; Nucleic acids
amplification of, using polymerase chain reaction, 420, 421f, 422
analyzing DNA deletion experiments, 376
in bacterial binary fission, 242, 243f
of B and T cells, 976
bending of, 374, 375f, 451
in cancer cells, 249–250
cell-free fetal, 288
changes of, in meiosis of yeast cells, 264
Chargaff's rules on structure of, 317f, 318
complementary DNA. See Complementary DNA
components of, 84, 85f
CRISPR-Cas9 system and, 360, 361f
discovery of structure of, 4, 18–19, 23–24, 314f, 317, 318f–320f
diseases of mitochondrial, 311f
distribution of, during eukaryotic cell division, 235f–236f, 237
in ecological forensics, 1265f
elevation and UV damage to, affecting species distributions, 1186f
eukaryotic, 6f, 97f, 98, 102, 103f
evidence for, as genetic material, 315f–317f, 318
evolutionary significance of mutations of, 328
evolution of genomes from changes in, 454f–457f, 458–459
in expression and transmission of genetic information, 6, 7f–8f
gene density and noncoding, 449–450
genetic code for, 340f–342f
genetic variation due to mutations in, 488–489
genomics, bioinformatics, and proteomics in study of, 9, 86, 87f
homeoboxes in, 463f–464f

DNA (deoxyribonucleic acid) (continued)
  human gene microarray chips containing, 447, 448f
  in hybrids, 524
  inheritance of, in chromosomes and genes, 255
  as measure of evolution, 87, 89, 397
  methylation of, 371, 372f, 373, 391, 430
  molecular homologies and, 479–480, 558, 559f
  ozone depletion and damage to, 1284
  p53 gene and repair of, 390–391
  packing of proteins and, into chromosomes, 330f–332f
  phylogenies based on, 554f
  in plant cells, 208f
  programming of cells by viral, 315f–316f, 317
  prokaryotic, 6f, 97f, 98, 577f, 579f–580f, 581
  proofreading and repairing of, 327, 328f
  recombinant. See Recombinant DNA
  repetitive noncoding, 450f–451f, 452
  replication of. See DNA replication
  roles of, in gene expression, 84f
  sequencing of. See DNA sequencing
  simple sequence and short-tandem repeat, 452
  species identity in mitochondrial, 557f
  structure of, 86f
  structure of, and inheritance, 334, 364
  technology of. See DNA technology
  testing of, in forensic science, 436, 437f
  in transcription, 337, 339f
  viral, 400f–407f, 406t
DNA amplification, 420, 421f, 422
DNA "barcode," 1222
DNA-binding domain, 374f
DNA-binding proteins, 91
DNA chips, 426
DNA cloning, 418
  amplification of DNA using polymerase chain reaction in, 420, 421f, 422
  copying DNA with gene cloning and, 418f, 419
  expressing cloned eukaryotic genes in, 422–423
  in gene therapy, 434f
  using restriction enzymes to make recombinant DNA plasmids for, 419f–420f
DNA Data Bank of Japan, 444
DNA deletion experiments in Scientific Skills Exercise, 376
DNA ligases, 325f–326f, 327t, 328f, 419f, 420
DNA methylation, 371, 372f, 373, 391, 430
DNA microarray assays, 289, 426f. See also Microarray chips, human genome
DNA pol I and pol III, 325, 326f, 327t
DNA polymerases, 324f
DNA profiles, 1223f
DNA replication, 320
  base pairing to template strands in semiconservative model of, 320, 321f–322f
  errors in, and genome evolution, 455f–457f
  evolutionary significance of mutations during, 328
  genetic information and, 314
  inheritance and, 7f–8f
  in molecular basis of inheritance, 314–315. See also Molecular basis of inheritance
  proofreading and repairing of DNA during, 327, 328f
  steps of, 322f–327f
  of telomeres, 328, 329f
DNA replication complex, 326f–327f, 334
DNA sequences
  analyzing phylogenetic trees based on, to understand viral evolution, 411
  animal phylogeny and, 682, 683f, 684
  constructing phylogenetic trees using, 562f, 563f, 564
  evaluating molecular homologies in, 558f–559f
  exon and intron, 345f–347f
  genes as, 361–362
  interpreting sequence logos for, 351
  on medical records, 448
  noncoding, 449–450
  promoter and terminator, 342, 343f
  types of, in human genome, 450f
DNA sequencing, 87, 416
  of complementary DNA, 426f
  in genetic testing for cancer predisposition, 394
  genome sequencing and, 443f, 444
  genomics, bioinformatics, proteomics, and, 86, 87f

human genome sequencing by, 415f, 416
  standard vs. next-generation techniques for, 416f–417f
  systems biology, medicine, and, 447, 448f
  three-domain taxonomy system and, 12f, 13
DNA strands, 7f–8f, 317, 318f–320f, 340f, 419f, 420. See also DNA replication
DNA technology, 416
  amplifying DNA with polymerase chain reaction in, 420, 421f, 422
  in analyzing gene expression and function, 423, 424f–428f
  bioinformatics and, 444, 445f–448f
  in biotechnology applications, 416, 433, 434f–437f, 438–439. See also Biotechnology
  in breast cancer treatment, 250
  creating recombinant DNA plasmids using restriction enzymes and gel electrophoresis in, 419f–420f
  DNA cloning and gene cloning in, 418f, 419
  DNA sequencing in, 415f–417f
  in ecological forensics, 1265f
  in eukaryotic gene regulation, 371
  evolution and, 441
  in expressing cloned eukaryotic genes, 422–423
  genetic code and gene transplantation in, 341, 342f
  organismal cloning in, 428, 429f–432f
  science, society, and, 23, 24f
DNA testing, forensic, 436, 437f
DNA viruses, 400f–407f, 406t
DNP (dinitrophenol), 186
Dobzhansky, Theodosius, 11
Dodder (Cuscuta), 819f
Dog rose, 650f
Dolly (cloned lamb), 429f
Dolphins, 481f–482f, 886f, 1094f, 1191f, 1192, 1262f
Domains, protein, 347f, 444, 445f, 446, 676f
Domains, taxonomy, 12f, 13, 460f, 554, 555f, 568, 569f, 585t. See also Archaea, domain; Bacteria, domain; Eukarya, domain
Domestication, plant, 651
Dominance, degrees of, 279f, 280
Dominant alleles, 272, 273f–276f, 279f, 280, 287f, 293. See also Alleles
Dominantly inherited disorders, human, 287f
Dominant traits, 271f, 272t
Donkeys, 335f, 337
L-Dopa, 65
Dopamine, 56, 1081t, 1102, 1103f, 1104
Doppler, Christian, 270
Dormancy, endospore, 575f
Dormancy, seed, 639, 828–829, 833–834, 852
Dormouse, 893f
Dorsal lip, 1051f, 1062f
Dorsal sides, 680f
Dorsal-ventral axis, 1060f
Double bonds, 36, 37f, 60f
Double circulation, 924f–926f, 925
Double fertilization, 646f, 647, 826, 827f
Double helix, DNA, 7f, 86f, 314f, 318f–320f, 330f
Double-stranded DNA (dsDNA) viruses, 406f, 408
Double-stranded RNA (dsRNA) viruses, 406t
Doudna, Jennifer, 42f, 361, 426
Douglas fir, 643f
Doushantuophyton, 535f
Dowling, Herndon, 888f
Down syndrome, 256f, 289, 307, 308f, 309, 432
Dragonflies, 887f
Draw It Questions, 15f, 16, 26, 39f, 41f, 42–43, 45f, 54–55, 61f, 62, 64–65, 69f, 74f, 78f, 86, 91, 102, 103f, 116f, 125, 131f, 141–142, 154f, 159, 163, 186, 202, 210, 224f, 233, 236f, 240f, 252, 259f, 263f, 268, 276, 290, 292, 303f, 313, 323f, 326f–327f, 334, 342, 345f, 355, 362, 397, 414, 419f, 423, 441, 466, 485, 491f, 524, 530f, 558f, 560f–561f, 572, 598f, 617, 635, 644f, 653, 709f, 719f, 730, 756, 769f, 780f, 782, 821, 841, 871, 904f, 920, 951, 963, 976, 1018, 1042, 1066, 1073f–1075f, 1084, 1106, 1138, 1161, 1195, 1203f, 1237, 1253, 1259, 1287
Drip irrigation, 808
Dromaius novaehollandiae, 739, 740f
Drosophila melanogaster (fruit fly). See also Fruit flies
  alternative RNA splicing in, 378f
  analyzing single-gene expression in, 423, 424f–425f

changes in developmental genes of, 545f, 546
  circadian rhythm molecular mechanisms in, 883
  courtship behaviors of, 1141f
  diploid and haploid numbers of, 257
  female bias in sperm usage of, 1024f
  foraging genes of, 1148, 1149f
  gene density of fungi vs., 665t
  genetic basis of behavior in, 1155
  genetic variation of, 487, 488f
  genome size and number of genes of, 448t, 449
  homeotic genes in, 463f–464f
  linkage maps of, 305f–306f
  linked genes and, 301f–303f
  as model organism, 22, 295f, 1044
  natural selection and adaptive evolution of, 494
  one gene–one enzyme hypothesis on, 336
  pattern formation of body plan of, 384, 385f–387f
  phylogenetic trees of, 562f
  reproductive anatomy of, 1024f
Drosophila species (fruit flies), 512f, 522–523, 567
Drought, 1228, 1244f
  abscisic acid in plant tolerance to, 852
  American Dust Bowl, 807f
  climate change and, 11
  plant responses to, 863–864
Drugs. See also Medicine; Pharmaceutical products
  addiction to, 1103f
  antibiotic. See Antibiotic drugs
  antiviral, 409
  biotechnology in production of, 435, 436f
  cocktails of, in AIDS treatment, 489
  enantiomers in, 62f
  as environmental toxins, 1276, 1277f
  fungi and, 670–671
  molecular shape and, 40f, 78
  plant-derived, 651t, 652
  resistance to, 478f, 592
  species and genetic diversity and, 1262, 1263f
  from sponges, 691
  tolerance of, 104–105
Dryas, 1230f–1231f
Dry fruits, 645f
Dubautia species, 543f
Duchenne muscular dystrophy, 299, 433
Duckweed (Spirodela oligorrhiza), 101f
Ducts, male reproductive, 1025f
Dulse, 609f
Dune fescue grass, 1203
Dung beetle, 1259
Dunstan, William, 1243f
Duodenum, 909
Duplications, chromosome, 307, 308f, 454f–456f
Duplications, gene, 489, 565f, 566
Dusky salamanders (Desmognathus ochrophaeus), 513
Dust Bowl, American, 807f
Dust mites, 708f
Dutch Hunger Winter, 372f
Dwarfism, 287f, 436, 1010
Dynamics, population, 1191f, 1192, 1205f–1206f, 1207, 1256f. See also Population growth
Dyneins, 115, 116f
Dysentery, 613

E
Eagles, 1262f
Ear bones, mammalian, 530, 531f, 532, 741, 742f
Eardrum, 1112, 1113f
Ears. See also Hearing
  bones of mammalian, 530, 531f, 532, 741, 742f
  human, 1113f
  insect, 1112f
Earth. See also Biosphere; Global ecology
  development of photosynthesis and atmospheric oxygen on, 533, 534f
  importance of seed plants to ecosystems of, 636f
  mass extinctions of life on, 540f–542f
  origins of life on, 57f, 58, 526f–528f
  plate tectonics of, 538f–539f, 540
Earthworms, 688f, 704f, 705, 807, 905f, 923f, 985f
East Australian Current, 1168f
Eastern tent caterpillars (Malacosoma americanum), 897
Ebola virus, 409f
Ecdysis, 683–684, 705, 707, 894f
Ecdysozoans, 683f, 684
  arthropods, 706f–712f, 713. See also Arthropods
  nematodes, 705f, 706
  phylogeny of, 688f–689f

Ecdysteroid, 1006f, 1007
Echidnas, 742, 743f
Echinoderms, 689f, **713**, 714f–715f, 719f, 1044f–1047f, 1086f, 1133
Echinoidea, 715f
ECM. *See* Extracellular matrix
Ecological footprint, **1209**, 1210f, 1211
Ecological forensics, 1265f
Ecological niches, **1215**, 1216f
Ecological pyramids, 1246, 1247f, 1248
Ecological species concept, **510**
Ecological succession, **1229**, 1230f–1231f
Ecology, **1165**. *See also* Community ecology; Conservation biology; Ecosystem ecology; Global ecology; Landscape ecology; Organismal ecology; Population ecology
  evolution and, 1187f
  factors of, in evolutionary rates, 538
  of fungi, 667f–670f
  genomics and proteomics in, 88f
  as interactions between organisms and environment, 1165
  of mosses, 627f, 628
  population growth and, 1213
  prokaryotes in, 586f–591f
  scope and fields of, 1165f
  of seedless vascular plants, 633–634
  urban, 1273, 1274f
Ecosystem diversity, 1261f–1262f
Ecosystem ecology, **1165**f. *See also* Ecology; Ecosystems
Ecosystem engineers, **1226**f
Ecosystems, **4**f, **1165**f, **1239**, 1256f–1257f
  biogeochemical cycles in, 1248f–1252f
  climate change effects on, 1279, 1281f
  decomposition in, 1240f, 1241, 1248f, 1249
  diversity of, 1261f–1262f
  edges between, 1270, 1271f, 1287
  effects of mass extinctions on, 542f
  energy budgets of, 1241, 1242f
  energy flow and chemical cycling in, 9f, 164–165, 1238f–1240f, 1241, 1256f–1257f
  evolution of, 1259
  fungi in, 667f–670f
  importance of mosses to, 627f, 628
  importance of seedless vascular plants to, 633–634
  importance of seed plants to, 636f
  interactions in, 10f–11f
  as level of biological organization, 4f
  metagenomics and genome sequencing of species in, 444
  primary production in, 1241, 1242f–1244f, 1243t, 1245
  prokaryotes in, 586f, 587
  protists in, 614f–615f
  restoration of degraded, 1253f–1255f
  secondary production in, 1246f–1247f, 1248
Ecosystem services, **1263**
Ecotones, **1172**
Ectoderm, **680**, **1049**, 1050f–1051f
Ectomycorrhizae, **817**, 818f
Ectomycorrhizal fungi, **656**
Ectoparasites, **1220**
Ectopic cells and tissue, **1033**
Ectopic pregnancies, 1039
Ectoprocts, 688f, **698**f, 713
Ectothermic organisms, **736**, **884**f, 885f–887f, 888, 890–892
Edema, 933
Edges, ecosystem, 1270, 1271f, 1287
Ediacaran biota, 535f, 536, **676**f–677f
Edidin, Michael, 128f
Effective population size, **1268**
Effector cells, **961**, 962f
Effectors, **866**
Effector-triggered immunity, 866
Efferent arteriole, 987f
Efferent neurons, 1088f
Efficiency, cell-signaling, 228f, 229
Egg-polarity genes, 386
Eggs, **1020**
  amniotic, 718, 734, 735f
  amphibian, 732, 733f
  of birds and dinosaurs, 564f, 565
  Burmese python thermogenesis for incubating, 887, 888f

in embryonic development, 381f, 382
in fertilization, 1022f–1024f, 1044f, 1045–1046
  human, 236–237, 1027, 1035f
  human oogenesis and, 1029f
  monotremes, 742
  nitrogenous wastes and, 983
  ovules and production of, in seed plants, 638f
Egg yolk, 91
Egrets, 1183, 1184f
Ejaculation, **1025**, 1034
Ejaculatory duct, 1025f
Elastic fibers, 878f
Elastin, 76f
Electrically charged side chains, 76, 77f
Electrical signals
  neuron, 880f, 1067–1068. *See also* Neurons
  phloem and symplastic, 802
Electrical synapses, 1077. *See also* Synapses
Electrocardiogram (ECG or EKG), **928**f
Electrochemical gradients, 137, **138**f
Electroencephalogram (EEG), 1094f
Electrogenic pump, **138**f
Electrolytes, blood plasma, 934f
Electromagnetic receptors, **1110**, 1111f
Electromagnetic spectrum, **192**f
Electron distribution diagrams, 34f–35f, 37f
Electron microscope (EM), **94**f–95f
Electrons, **30**
  chemical properties and, 34f, 35
  cyclic flow of, in light reactions of photosynthesis, 198f, 199
  distribution of, 34f–35f
  in electron transport chains, 167f–168f
  energy levels of, 32f, 33–34
  linear flow of, in light reactions of photosynthesis, 197f–198f
  orbitals of, 35f, 36, 39f, 40
  in organic compounds, 58, 59f, 60
  in redox reactions, 165, 166f
  as subatomic particles, 30f, 31
Electron shells, 32f, **34**, 35f, 36
Electron transfer, 37, 38f
Electron transport chains, **168**
  in anaerobic respiration and fermentation, 179–180
  in cellular respiration, 167f–168f
  in chemiosmosis, 175, 176f, 177
  in oxidative phosphorylation, 174f, 175
Electrophysiologists, 1072f
Electroporation, **422**
Elements, **29**, 809
  atomic number and atomic mass, 31
  electron distribution in, 34f–35f
  electron orbitals in, 35f, 36
  energy levels of, 32f, 33–34
  isotopes of, 31
  of life, 29t
  organisms ratio of, 43
  subatomic particles in, 30f, 31
  tolerance to toxic, 30f
Elephantiasis, 933
Elephants, 10f, 88f, 474f, 1197f, 1265f
Elephant seals (*Mirounga angustirostris*), 999f, 1267
Elephant shark (*Callorhinchus milii*), 442f
Elevation
  climate and, 1169f
  genetics and, 1189
  ultraviolet (UV) light damage at, 1186f
Elicitors, 866
Elimination, **904**, 905f, 911
Elk, 1151f
Elkhorn coral, 1234
Elkinsia, 641f
Elodea, 55
Elongation, antiparallel DNA strand, 324, 325f–326f, 327t
Elongation factors, 352, 353f
Elongation stage
  transcription, 343f–344f
  translation, 352, 353f
Embryo development, plant, 828f–829f
Embryonic development. *See also* Differential gene expression
  analyzing single-gene expression in, 423, 424f–425f

  animal, 674f, 675. *See also* Animal development
  cell division, cell differentiation, and morphogenesis in, 381f, 382
  cell fate specification in, 1057, 1058f–1064f
  cleavage in, 1044, 1046, 1047f–1048f, 1049
  cytoplasmic determinants and induction in, 382f, 383
  fertilization in, 1044f–1046f
  gene conservation in, 466
  genomic imprinting and, 310f, 311
  human. *See* Human embryonic development
  morphogenesis in, 1049
  pattern formation of body plans in, 384, 385f–387f
  processes of, 1043f
  sequential gene regulation in, 383, 384f
Embryonic lethals, **386**
Embryonic stem (ES) cells, 431f
Embryophytes, 619f–**620**f, 621
Embryos. *See also* Embryonic development
  anatomical homologies of vertebrate, 479f, 480
  plant, 620f
  maternal immune tolerance of, 1037
  monocot vs. eudicot, 649f
  survival of, 1023f
Embryo sacs, **646**f, **826**, 827f
Emergent properties, **5**
  of apoptosis, 233
  chemical waste and, 43
  of compounds, 29f
  of flowers, 841
  of gastrula, 1066
  hierarchical organization of animal body plans and, 876
  integration of Mendelian inheritance with, 282–283
  levels of biological organization and, 4f–6
  life as, 125
  of protein function, 78
  of sickle-cell disease, 505
  of water. *See* Water
  weak chemical bonds and, 38
Emerging diseases, 409f–412f, 1234
Emigration, **1192**, 1196, 1197f, 1206f, 1207, 1256f
Emotions, 1095f, 1096, 1106
Emperor penguins (*Aptenodytes forsteri*), 873f
Emu (*Dromaius novaehollandiae*), 739, 740f
Enantiomers, **61**f–62f, 65
Encephalitis, 401, 409
ENCODE (Encyclopedia of DNA Elements), 446, 450
Endangered species, 1199f, **1261**, 1262f, 1272f, 1288
Endergonic reactions, 148, **149**f
Endocarp, 832f
Endocrine disruptors, 1015, 1276
Endocrine glands, **1003**. *See also specific glands*
  of human endocrine system, 1004f
  in neuroendocrine signaling, 1007f–1008f
  regulatory functions of, 1011f–1015f, 1016
Endocrine signaling, 1000f. *See also* Endocrine systems
  cascade pathways in, 1008, 1009f, 1010, 1014
  in cell signaling, 215, 216f
  cellular response pathways for, 1002f–1003f
  feedback regulation of, 1005–1006
  simple pathway of, 1004, 1005f
Endocrine systems, **880**, **1000**. *See also* Animal hormones
  coordination of nervous systems and. *See* Neuroendocrine signaling
  disruption of, 1015, 1276
  evolution of hormone function in, 1015f, 1016
  feedback regulation of, 1005–1006
  glands and hormones of human, 1003, 1004f
  hormone cascade pathways of, 1008, 1009f, 1010, 1014
  hormones and cell signaling in, 880f
  nervous system coordination with, 1000f, 1001, 1006f–1008f
  in regulation of blood pressure, 931
  in regulation of digestive systems, 915f
  regulatory functions of endocrine glands in, 1011f–1015f, 1016
  signaling molecules, target receptors, and response pathways in, 1000f–1004f, 1018
  simple endocrine pathway of, 1004, 1005f
  simple neuroendocrine pathway of, 1005f
Endocytosis, 126, **139**, 140f, 141, 208f–209f

Endoderm, **680**, **1049**, 1050*f*–1051*f*
Endodermis, **768**, **792**, 793*f*
Endomembrane system, **104**
  bound ribosomes and, 354*f*
  endoplasmic reticulum of, 104, 105*f*
  Golgi apparatus, 105, 106*f*, 107
  lysosomes, 107*f*, 108
  signal mechanism for targeting polypeptides to, 354*f*
  vacuoles, 108*f*
Endometriosis, **1033**
Endometrium, **1026**, 1027*f*
Endomycorrhizae, **656**, 660, 663, **817**, 818*f*
Endoparasites, **1220**
Endophytes, **667***f*, 668, **814**
Endoplasmic reticulum (ER), **104**, 109*f*
  animal cell, 100*f*
  as biosynthetic factory, 104, 105*f*
  cellular membrane synthesis and, 131*f*
  ribosomes and, 103*f*, 104
  in RNA virus replicative cycle, 405*f*
  targeting polypeptides to, 354*f*, 355
Endorphins, 40*f*, 78, **1081***t*
Endoskeletons, 1132*f*–1134*f*, **1133**
Endosperm, **647**, 826, **826**, 827*f*–829*f*
Endospores, **575***f*
Endosymbiont theory, **109**, 110*f*, 189, 534, 535*f*
Endosymbiosis, 109, 110*f*, **534**, 535*f*, **594**–595, 596*f*–597*f*, 608*f*, 609, 617
Endothelin, 931
Endothelium, blood vessel, **929***f*
Endothermic organisms, **736**, **884***f*, 885*f*–889*f*, 890–892
Endotoxins, **588**
Energetic hypothesis on food chain length, **1225***f*
Energy, **32**, **144**
  allocation and use of, 889, 890*f*
  animal. *See* Bioenergetics
  in arctic tundra ecosystem, 1256*f*–1257*f*
  ATP energy coupling and, 150, 151*f*–153*f*
  biofuel technology to reduce dependence on fossil fuels for, 838
  chemical, 144, 890, 890*f*
  chemiosmosis as energy-coupling mechanism, 175, 176*f*, 177
  conservation of, 1239
  in ecosystems, 1238*f*–1240*f*, 1241
  electrons and levels of, 32*f*, 33–34
  in energy flow and chemical cycling, 164*f*, 165
  enzymatic catalysis and. *See* Enzymatic catalysis
  forms of, 144, 145*f*
  free-energy change and. *See* Free-energy change
  global human use of, 1210*f*, 1211
  heat as, 46
  hydrothermal vents, prokaryotes, and, 592
  kinetic, 46, 144, 145*f*, 163
  laws of transformation of, 145*f*–146*f*, 147
  light. *See* Light energy
  locomotion and, 1135–1136
  metabolism and. *See* Metabolism
  in photosynthesis, 41*f*
  potential, 32, 144, 145*f*, 163
  primary production of, 1241, 1242*f*–1244*f*, 1243*t*, 1245
  processing of, as theme of biology, 2, 3*f*
  regulation of storage of, in animal nutrition, 915, 916*f*, 917
  secondary production of, 1246*f*–1247*f*, 1248
  storage of, in fats, 74
  thermal, 46, 132, 144
  transfer and transformation of, as theme of biology, 9*f*, 143
  transformation of, 109, 110*f*–111*f*, 209*f*
  trophic levels and, 1240*f*, 1241, 1246, 1247*f*, 1248
Energy and matter
  bioenergetics, 163
  decomposition and, 1259
  eukaryotic cells and, 93*f*
  hibernation, 897
  photosynthesis and, 211
  reproductive success and, 1042
  as theme of biology, 2, 3
  transfer and transformation of, 9*f*
Energy budgets
  animal, 892
  ecosystem, 1241, 1242*f*
Energy conservation, animal, 892, 893*f*

Energy coupling, **150**, 151*f*–153*f*
Energy flow
  in ecosystems, 1238*f*–1240*f*, 1241, 1256*f*–1257*f*
  between trophic levels, 1246*f*–1247*f*, 1248
Energy flow and chemical cycling, 9*f*, 164*f*, 165.
    See also Biogeochemical cycles; Primary production; Secondary production
Energy pyramids, 1246, 1247*f*, 1248
Engelmann, Theodor W., 193, 194*f*
English ivy, 784*f*
Enhancers, **374***f*–375*f*, 397
Entamoebas, 613
Enteric division, peripheral nervous system, 915, **1088***f*–1089*f*
Enthalpy, 147
Entomopathogenic fungus, 672
Entomophthora muscae, 662*f*
Entropy, 143, **146**–147, 1239
Entry stage, phage lytic cycle, 402*f*
Enveloped viruses, 401, 405*f*
Environment
  adaptation to, 2*f*
  adaptive evolution as fitness to, through natural selection, 494, 498, 499*f*. *See also* Adaptive evolution
  animal and plant responses to, 894*f*
  behavior and stimuli from, 1140*f*–1141*f*
  biotechnology in cleanup of, 418*f*, 437
  bottleneck effect from changes in, 495*f*, 496
  cancer and factors of, 394–395
  C. Darwin on natural selection and adaptations to, 14*f*–16*f*. *See also* Adaptations; Evolution; Natural selection
  cell cycle control system and factors of, 247, 248*f*
  cellular membranes and factors of, 128*f*, 129
  Earth's early, and origin of life, 526*f*–528*f*
  ecology as interactions between organisms and. *See* Ecology
  enzymatic catalysis and factors of, 157, 158*f*, 159
  evolution and, 485
  exchanges with, in animals, 874, 875*f*, 876
  as factor in red maple leaf structure, 762
  gene regulation in response to, 377
  heat exchange with, in animals, 873*f*, 885*f*–889*f*
  hormones and, 1018
  human impacts on. *See* Human environmental impacts
  hybrid zones and, 517*f*
  impact of, on phenotypes, 282*f*
  induction in differential gene expression as response to, 382*f*, 383
  influence of evolution and, on nitrogenous wastes, 983
  interaction with, as theme of biology, 10*f*–11*f*
  metagenomics and genome sequencing of groups of species in, 444
  nervous system disorders and, 1102*f*
  organism interactions with, 1165, 1178, 1183*f*–1186*f*
  as organisms and their surroundings, 469
  plant responses to abiotic stresses from, 797–798, 862–863, 864*f*, 865
  population ecology as study of populations in, 1190*f*, 1191. *See also* Population ecology
  population growth in unlimited, 1196, 1197*f*
  prokaryotic adaptations to extreme conditions in, 573*f*, 574
  protein structure and factors of, 82*f*, 83
  reproductive cycles and cues from, 1021
  response to, as property of life, 3*f*
  strength of ionic bonds and factors in, 38
  vertebrate kidney adaptations to diverse, 991*f*–993*f*
Environmental issues. *See also* Climate change; Ecology; Pollution
  decline in amphibian populations, 733
  density-dependent population regulation by toxic waste, 1204*f*
  prokaryotes, biotechnology, and, 589, 590*f*–591*f*
  threats to biodiversity, 1260*f*, 1261, 1263, 1264*f*–1266*f*
  threats to seed plant diversity, 651*f*, 652
Environmental toxins, 1275, 1276*f*–1277*f*
Enzymatic catalysis. *See also* Enzymes
  activation energy barrier and, 153, 154*f*
  allosteric regulation of, 160*f*–161*f*
  of cellular respiration, 183*f*, 184

cofactors and, 158
  effects of temperature and pH on, 157, 158*f*
  in enzyme active sites, 156*f*
  enzyme inhibitors and, 158, 159*f*
  evolution of enzymes and, 159*f*
  graphing, for blood glucose levels, 157
  lowering of activation energy by enzymes in, 154, 155*f*
  in plant cells, 208*f*
  regulation of, 366*f*–369*f*, 370
  by ribozymes, 346
  substrate specificity of enzymes in, 155*f*, 156
Enzymatic hydrolysis, 904, 908*f*
  neurotransmission and, 1079, 1080*f*
  in vacuoles, 108
Enzymatic proteins, 76*f*
Enzymes, **67**, **153**. *See also* Enzymatic catalysis
  active sites of, 155*f*–156*f*
  activity measurement of, 163
  as catalysts, 153
  in cell-signaling nuclear responses, 226, 227*f*
  in chemical digestion, 907*f*–908*f*, 909
  evolution of, 159*f*
  facilitation of synthesis and breakdown of polymers by, 67*f*
  fungal, 654–655
  in gastric juice, 907*f*–908*f*
  gene relationship with, in protein synthesis, 336*f*–338*f*
  inducible and repressible, 367*f*–368*f*, 369
  locations of, in cells, 161*f*
  lowering plasma LDL levels by inactivating liver, 938
  lysosomes and, 107*f*, 108
  as proteins, 75, 76*f*, 130*f*
  restriction. *See* Restriction enzymes
  ribozymes as, 346
  in saliva, 906
  smooth ER and rough ER, 104–105
  substrate specificity of, 155*f*, 156
  from *Thermus aquaticus*, 441
Enzyme-substrate complexes, **155***f*, 156
Eosinophils, 934*f*–935*f*, **955**
Ependymal cells, 1090*f*
Ephedra, 642*f*
Ephrussi, Boris, 336
Epiblast, 1052*f*–1053*f*
Epicotyl, **829***f*–830*f*
Epidemics, **409***f*, 410, 867
Epidemiology, 902
Epidermis, 5*f*, **763***f*
Epididymis, **1025***f*
Epigenetic inheritance, **372***f*, 373
Epigenetic modifications, cancer gene, 388, 389*f*
Epigenetics, 430
Epiglottis, 906*f*
Epilepsy, 1076, 1098
Epinephrine (adrenaline), **1002**, 1003*f*
  adrenal glands and, 1004*f*, 1012*f*–1013*f*
  in fight-or-flight responses, 212*f*, 928
  multiple effects of, 1004*f*, 1013*f*
  second messenger of, 223*f*–224*f*
  signal transduction pathway of, 216, 226, 227*f*, 233
  as water-soluble hormone, 1001*f*
Epiphytes, 632*f*, 818, **819***f*, 820
Epistasis, **281***f*
Epithalamus, 1093*f*
Epithelial milk-secreting cell, 392*f*
Epithelial tissue, **877***f*
  as barrier defense, 954
  cell junctions in, 120*f*
  in small intestine, 909*f*
  transport, 981, 982*f*, 988*f*, 989
Epitopes, **958**, 959*f*, 969, 976
Epstein-Barr virus, 394
Equilibrium
  chemical. *See* Chemical equilibrium
  mechanoreceptors for sensing, 1112*f*–1116*f*
  population, 1203*f*
Equilibrium potential (Eion), **1071***f*
Equine encephalitis virus, 401
Equisetum, 632*f*, 633
Equus, 548, 549*f*
ER. *See* Endoplasmic reticulum
Erectile dysfunction, 1026, 1081–1082
Ergots, 669*f*

ER lumen, 104
Erosion, soil, 809f
Errantians, 703f
Errors, DNA replication, 327, 328f
Erythrocytes (red blood cells), 82f, 500–501, 502f–503f, 878f, 934f–**935**f
Erythropoietin (EPO), **936**
Erythropsidinium, 594f
Escherichia coli (*E. coli*) bacteria, 595
    binary fission in, 242, 243f
    DNA cloning and gene cloning of, 418f, 419, 589
    DNA packing in chromosomes of, 330–331
    DNA replication and repair in, 322f–327f
    genetic recombination and conjugation in, 579, 580f, 581
    genome size and number of genes of, 448t, 449
    as model organism, 22
    pathogenic strains of, 588
    rapid adaptive evolution of, 578f, 579
    regulation of gene expression in, 366f–369f, 370
    viral infection of, 401, 402f–403f
    viral phages in DNA research on, 316f, 317
E site (exit site), **350**f, 352f–353f
Esophagus, **906**f, 908f
Essential amino acids, **899**f
Essential elements, **29**t, 64, 809, **810**f, 811t
Essential fatty acids, **899**f
Essential nutrients, animal, **899**f, 900t–901t, 901f
Estivation, 893
Estradiol, 62f, 1003f, **1014**, 1015f, **1030**, 1032f, 1037f
Estrogen receptor alpha (ERα), 392f–393f
Estrogens, 62f, 376, 397, 1003f–1004f, **1014**, 1015f, 1030, 1032f, 1038, 1276
Estrous cycles, **1033**–1034
Estuaries, **1180**f, 1186
Ethane, 59f
Ethanol, 63f, 180f, 590, 1204f
Ethene (ethylene), 59f
Ethical issues
    on biotechnology, 438–439, 839–840
    on diagnosing fetal genetic diseases, 1039–1040
    on DNA technology, 24
    on gene therapy, 435
    on seed plant extinctions, 652
    on silencing gene expression in humans, 427
Ethylene, 216, 846t, **852**, 853f–854f, 864f
Etiolation, **843**f
Euchromatin, 330f, **332**
Eudicots, **649**f–650f, 761, 769f–770f, 828f–830f
Eugenics, 435
Euglenids, **601**f
Euglenozoans, **600**f–601f
Euhadra species, 522
Eukarya, domain, **12**f, 568, 569f, 585t, 594
Eukaryotes. *See also* Animals; Fungi; Plants
    cells of. *See* Eukaryotic cells
    chemiosmosis in, 175, 176f, 177
    cilia in, 13f
    electron transport chains in, 168–169
    endosymbiosis in evolution of, 594–595, 596f–597f
    Eukarya, domain, 12f, 568, 569f, 585t, 594
    genome sizes and number of genes for, 448t, 449
    origins of multicellular, 533f, 535f–536f
    origins of single-celled, 533f–535f
    photosynthesis of unicellular, 188f
    phylogenetic tree of, 612f
    protists as single-celled, 594. *See also* Protists
    taxonomy of, 568, 569f
    unikonts as first to diverge, 612f
Eukaryotic cells, **6**, **97**. *See also* Cells
    animal and plant, 100f–101f. *See also* Animal cells; Plant cells
    cellular integration of, 121f
    chromatin packing in chromosomes of, 330f–332f
    combinatorial transcription control for types of, 374–375, 377f
    cytoskeletons of, 112f–117f
    distribution of chromosomes in cell division of, 235f–236f, 237
    DNA replication in, 322, 323f
    in embryonic development, 381f, 382
    endomembrane system of, 104, 105f–109f
    evolution of cell division of, 243, 244f
    expressing cloned eukaryotic genes in, 422–423

extracellular components and connections between, 118f–120f
    genetic instructions in nucleus of, 102, 103f
    internal membranes and organelles of, 93f, 98f, 99, 100f–101f
    mitochondria, chloroplasts, and peroxisomes of, 109, 110f–112f
    organization of typical genes in, 373f
    origins of, 534, 535f
    plasma membranes of, 98f, 99
    prokaryotic cells vs., 6f, 97f–98f, 99. *See also* Prokaryotic cells
    regulation of gene expression in. *See* Eukaryotic gene regulation
    replication of DNA telomeres of, 328, 329f
    ribosomes as protein factories of, 102, 103f, 104
    RNA processing in, 345f–347f
    transcription and translation in, 339f, 342, 343f–344f, 356f. *See also* Translation volume and surface area of, 98f, 99
Eukaryotic gene regulation. *See also* Gene regulation
    analyzing DNA deletion experiments on, 376
    differential gene expression in, 365f, 370f, 371. *See also* Differential gene expression
    post-transcriptional regulation in, 377, 378f, 379
    regulation of chromatin structure in, 371f–372f, 373
    regulation of transcription initiation in, 373f–378f, 379
Eukaryotic genomes
    evolution of, from DNA changes, 454f–457f, 458–459
    genes and multigene families in, 452, 453f
    horizontal gene transfer in, 569, 570f
    noncoding repetitive DNA sequences in, 450f–451f, 452
    sizes and number of genes for, 448t, 449
Eumetazoans, **683**f, 686, 691
Euphorbs, 1172f
European flycatchers, 519
European hamster (*Cricetus cricetus*), 893f
European honeybees, 1142f
European kestrels, 1201f
European larch, 643f
European Molecular Biology Laboratory, 444
European robin, 14f
European starling, 1264
Euryarchaeota clade, 586
Euryhaline animals, 978
Eurypterids, **708**
Eustachian tube, **1113**f
Eutherians (placental mammals), 481f, 543f, **744**f–747f. *See also* Primates
Eutrophication, 1227f, **1242**, 1275f
Eutrophic lakes, **1179**f
Evaporation, 46f, 47, **885**f, 886
    in water cycle, 1250f, 1259
Evaporative cooling, **47**, 48f
Evapotranspiration, **1232**f
    in water cycle, 1250f
Even-toed ungulates, 481f–482f
Evergreen trees, 643f
Evidence
    fossils as, for evolution, 13f
    scientific data as, 17f, 19f
    theories and, 21–22
Evo-devo. *See* Evolutionary developmental biology
Evolution, **11**, **469**. *See also* Adaptations; Natural selection
    abiotic synthesis of organic molecules as origin of life on Earth, 57f, 58
    of action potentials, 1084
    altered DNA nucleotides as mutations in, 328
    of alternative carbon fixation mechanisms in plants, 203
    of amniotes, 735f
    of anatomical and molecular features, 485
    of angiosperms, 647f–649f
    of animals, 675f–677f, 678–679
    of animal size and shape, 874f
    of antifreeze proteins, 865
    of arthropods, 706f, 707
    of ATP synthase, 186
    of axon width and myelination, 1076f–1077f
    of behaviors by associative learning, 1146
    bilaterian phylogenetic tree, 717
    of biochemical pathways, 163

    of biological diversity. *See* Biodiversity
    of biological order, 147
    of bryophytes, 635
    of cell cycle, 252
    of cell signaling, 213f, 214, 233
    cellular structures in, 125
    of cellular water regulation, 142
    of chewing while breathing, 920
    of chordates, 722f
    cladistic analysis and, 572
    classification of diversity of life and, 12f–13f
    of codons, 364
    coevolution of flowers and pollinators in, 825f
    of cognition in vertebrates, 1098, 1099f
    comparing genome sequences to study, 459, 460f–464f
    convergent. *See* Convergent evolution
    crossing over and, 313
    cross-species gene expression in, 423
    Darwinian theory of, as descent with modification by natural selection, 14f–16f, 469–470, 473f–476f, 483–484
    of differences in cellular membrane lipid composition, 129
    divergence of human globin genes during, 458
    DNA and proteins as measures of, 87, 89, 397
    DNA-based technologies and, 441
    of double circulation, 925
    of drug resistance, 435, 592
    early vertebrate, 724f–725f
    ecological change and, 1187f
    of ecological niches, 1215, 1216f
    ecosystem, 1259
    elemental makeup of organisms and, 43
    endosymbiosis in eukaryotic, 594–595, 596f–597f, 608f, 609
    environment and, 485
    of enzymes, 159f
    evidence supporting Darwinian theory of, 476, 477f–482f, 483
    evolutionary developmental biology "evo-devo" in study of, 387
    exaptation, 783
    of extraembryonic membranes in amniote development, 1053f, 1054
    extremophiles and, 55
    field research by C. Darwin on, 471, 472f–473f
    of fishes, 730
    of flowers, 825f
    of foraging behaviors, **1148**, 1149f
    of fungi, 659f, 660
    of fungus-alga symbiosis, 672
    of gas exchange, 947f–949f
    of gecko appearance, 26
    gene conservation in, 466
    of genes and genomes, 454f–457f, 458–459, 565f, 566
    genes shared between organisms, 26
    of genetic code, 341, 342f
    genetic variation and, of behavior, 1155f–1157f
    of genetic variation in populations, 266, 267f
    genomics and proteomics in study of, 88f
    of glycolysis, 182
    of gnathostomes and jaws, 725f, 726
    of gymnosperms, 641f
    of hermaphroditism, 1042
    of high-density populations, 1213
    historical context of Darwinian theory of, 469f–471f
    hominin and human, 748f–754f
    of hormone function, 1015f, 1016
    of human genome, 524, 566
    imperfections of organisms and, 505
    influence of environment and, on nitrogenous wastes, 983
    of insect size, 951
    of intelligence, 1106
    of intracellular receptors, 1018
    of introns, 346, 347f
    of kidneys, 991, 998
    of plants, 617, 618, 619f, 621, 622t
    late-acting dominant lethal alleles, 293
    of life history traits, 1200, 1201f–1202f
    of life on planets with water, 50f
    light-sensitive germination, 871
    of lignin, 783
    of locomotion, 1138

Evolution (continued)
  making and testing predictions of Darwinian, 483
  of mammals, 741, 742f
  mass extinctions and, 653, 1287
  of mitochondria and chloroplasts, 109, 110f
  of mitosis, 243, 244f
  molecular clocks and rates of, 566, 567f–568f
  mutation rate and, 334
  of mycorrhizae in plants, 817
  natural selection and genetic variation from recombination of alleles in, 305
  of nitrogen recycling in bacteria, 821
  of pathogen detection by plants, 866
  of pathogens that evade immune systems, 957, 971–972, 973f, 973t, 974, 976
  pattern and process aspects of, 470, 483
  photorespiration and, 211
  phylogenetic bracketing and, 572
  phylogenies as history of, 554f–556f. See also Phylogenies
  of plant secondary growth, 774
  of plant self-compatibility, 841
  of prokaryotic flagella, 576f, 577
  as property of life, 3f
  protein divergence and, 91
  radiometric dating and, 32–33
  rapid prokaryotic, 573, 578f, 579
  of reproductive method switch, 268
  of reptiles, 736
  of resource acquisition adaptations in vascular plants, 784f–786f, 787
  reversals of, 756
  of RNA interference, 380
  of seeds in seed plant, 639
  of segmentation, 1066
  of sex reversal, 1020
  sexual reproduction as enigma of, 1021, 1022f
  of short-term and long-term memory, 1100, 1101f
  sickle-cell disease and, 286f, 287
  silicon-based life and, 65
  small population size and extinction vortex in, 1266, 1267f
  of smooth muscles, 1132
  speciation as conceptual bridge between macroevolution and microevolution, 507, 523. See also Macroevolution; Microevolution; Speciation
  species distributions and, 1183, 1189
  of tetrapods, 730
  as theme of biology, 2, 3, 11, 12f–16f
  theoretical aspects of Darwinian, 483–484
  of tolerance to toxic elements, 30f
  of transport cells in kelp, 804
  tree of life and, 15f–16f
  unity and diversity in, 13f, 26
  using protein data to test hypotheses on horizontal gene transfer in, 570
  of vascular plants, 628f–631f
  of vertebrate brain structure, 1091f
  of viruses, 406, 408, 411, 414
  of visual perception, 1117f–1119f
Evolutionary developmental biology (evo-devo), 387, 463, 467, 544
Evolutionary lineage, 555, 556f
Evolutionary trees, 474f, 480f. See also Phylogenetic trees
Exaptations, 548, 577, 783
Excavates, 597, 598f, 600
Exchange surfaces, animal, 874, 875f, 876
Excitatory postsynaptic potential (EPSP), 1078–1079
Excitement phase, sexual, 1034
Excretion, 978, 982, 984f. See also Excretory systems; Osmoregulation
Excretory systems, 983
  hormonal regulation of, 994f–996f
  internal exchange surfaces and, 875f
  kidneys in mammalian and human, 985, 986f–993f
  Malpighian tubules in, 985f
  metanephridia in, 985f
  nitrogenous wastes and, 977–978, 982f–983f
  osmoregulation and, 977f–982f
  processes of, 984f
  processing of blood filtrate by nephrons in, 987, 988f, 989
  protonephridia in, 984f, 985
Executive functions, brain, 1098

Exercise, immune systems and, 971
Exercises, Scientific Skills. See Scientific Skills Exercises
Exergonic reactions, 148, 149f, 154f, 165
Exhalation, 944, 945f
Exit tunnel, ribosome, 350f
Exocrine glands, 1004
Exocytosis, 126, 139f, 208f–209f
Exome, 434
Exons, 345f–347f, 346, 449, 456, 457f, 488f
Exon shuffling, 347f, 457f
Exoskeletons, 693, 1133
  animal, 683
  anthozoan, 693
  arthropod, 707f
  chitin as structural polysaccharide in, 72f
  ectoproct, 698f
  in locomotion, 1132f, 1133
Exotic species, 1264f
Exotoxins, 588
Expansins, 848, 849f
Experimental groups, 20, 21f
Experiments, 17–21. See also Case studies; Inquiry Figures; Research Method Figures; Scientific Skills Exercises
Exploitation, 1217, 1218f–1219f, 1220
Exploration, 17f
Explosive seed dispersal, 645f
Exponential population growth, 1196, 1197f, 1207f, 1208
Expressed sequence tags (ESTs), 445–446
Expression vectors, 422
Extavour, Cassandra, 467
Extension, muscle, 1132f
Extensively drug-resistant tuberculosis (XDR-TB), 589
External factors, cell cycle control system, 246, 247f–248f
External fertilization, 1022f, 1023, 1151
Extinctions
  of amphibians, 733
  climate change and, 11
  current rate of, 1261, 1262f
  of dinosaurs, 736
  ecological factors affecting rates of, 538
  in fossil record, 481, 529f
  global and local, 1262
  global temperature and, 541f, 542
  human impacts on, 1261, 1262f
  introduced species and, 1264f, 1287
  island equilibrium model and, 1232, 1233f
  mass, 540f–542f, 1287
  of molluscs, 702f
  resurrection of species after, 1266
  of seed plant species, 651f, 652
  small population risks for, 1266, 1267f–1268f, 1269
  speciation and, 520–523
Extinction vortex, 1266, 1267f
Extracellular diffusion, 768
Extracellular digestion, 904f–905f
Extracellular matrix (ECM), 118, 119f, 130f, 1056
Extraembryonic membranes, 734, 735f, 1052, 1053f, 1054
Extranuclear genes, 311
Extreme halophiles, 585–586
Extreme thermophiles, 585f, 586
Extremophiles, 55, 585f, 586
Eyes. See also Visual systems
  euglenid eyespot as, 601f
  evolution of, 547f–548f
  evolution of light detecting organs and, 1117f–1119f
  insect compound, 712f, 894f
  vertebrate, 1118f–1122f, 1123
Eyespots, 601f, 1117f

F

F₁ (first filial) generations, 271f, 524
F₂ (second filial) generations, 271f
Faceted eyes, arthropod, 1117f, 1118
Facilitated diffusion, 135f, 136, 209f
Facultative anaerobes, 181, 582
FAD (flavin adenine dinucleotide), 172f–173f, 178
FADH₂, 172f–173f, 175, 178
Fagus grandifolia, 1170f
Fairy ring, mushroom, 666, 667f

Falling phase, action potential, 1074f, 1075
Familial cardiomyopathy, 357
Families, taxonomy, 554, 555f
Family planning, 1208
Family resemblance, genetics and, 254f
Family studies, 1102f
Family trees, 284f, 285
Far-red light, plant response to, 857f
Fas molecule, 233
Fast block to polyspermy, 1045
Fast-twitch fibers, 1131
Fate maps, 1058f–1059f
Fathead minnows (Pimephales promelas), 1276
Fatigue, muscle, 1131
Fats, 72
  absorption of, in small intestine, 910f
  animal heat exchange and, 885, 887f
  blood cholesterol and, 938
  digestion and absorption of, 908f–910f
  in egg yolks, 91
  as fuel for catabolism, 182f, 183
  in glyoxysomes, 112
  hydrocarbons in, 60, 61f
  as lipids, 72, 73f, 74
  oxidation of, during cellular respiration, 166–167
  trans fats, 61
Fat-soluble vitamins, 900t
Fatty acids, 72
  beta oxidation of, for catabolism, 182f, 183
  essential, 899f
  fats and, 72, 73f, 74
Fd, 198f, 199
Feathers, 738f, 739
Feather stars, 715f
Fecal microbial transplantation, 912
Feces, 697, 832f, 911, 1246f
Feedback, scientific, 19f
Feedback inhibition, 161f, 183f, 184, 366f
Feedback mechanisms, 397
Feedback regulation, 10. See also Regulation
  of animal digestion, energy storage, and appetite, 914, 915f–917f, 918
  of animal homeostasis, 881f–883f
  Cambrian explosion and, 685
  in density-dependent population growth, 1203
  of endocrine systems, 1005–1006
  of molecular interactions, 10f
Feeding mechanisms, 655, 902, 903f
Feet, bird adaptation to perching, 740f
Feline panleukopenia virus, 409
Female condoms, 1038
Female gametophytes, angiosperm, 826, 827f
Females
  autoimmune diseases in human, 971
  fruit fly bias in sperm usage by, 1024f
  hormonal regulation of reproductive systems of human, 1032f, 1033–1034
  hormones of, 999f, 1004f, 1007f, 1008, 1014, 1015f
  inactivation of X-linked genes in mammalian, 300f
  mate choice by, 499f–500f, 1151, 1152f–1153f, 1154
  maternal immune tolerance of embryo and fetus during pregnancy, 1037
  oogenesis in human, 1027, 1029f
  parental care by, 1151
  parthenogenesis by, 697–698
  reproductive anatomy of human, 1026, 1027f
  reproductive rates and, 1194, 1195f, 1204f, 1208, 1213
  sex chromosomes of, 298f, 299
Fermentation, 165
  anaerobic and aerobic respiration vs., 179–180
  cellular respiration vs., 165, 181f, 182
  types of, 180f, 181
Fern galls, 678
Ferns, 630f–632f, 633, 635, 637f
Ferredoxin (Fd), 198f, 199
Fertilization, reproductive, 258, 826, 1022, 1044. See also Reproductive isolation
  angiosperm double, 646f, 647, 826, 827f. See also Pollination
  in animal embryonic development, 1044f–1046f
  external versus internal, 1022f, 1023
  gamete production and delivery in, 1023f, 1024
  genetic variation from random, 266, 305
  in human life cycle, 254, 257f, 258, 1035f

mechanisms preventing angiosperm self-fertilization, 834, 835*f*
meiosis and, 237
G. Mendel's techniques of plant, 270*f*–271*f*
offspring survival following, 1023*f*
parental care and internal vs. external, 1151
parthenogenic self-fertilization, 1020, 1021*f*
prezygotic barriers and, 508*f*–509*f*, 510
in varieties of sexual life cycles, 258*f*, 259
in vitro, 1039*f*
Fertilization, soil, **807**–808, 983, 1244, 1275*f*
Fertilization envelope, 1045*f*, 1046*f*
Fertilizers, in nitrogen cycle, 1251*f*
Fescue grass, 1203
Fetal alcohol syndrome, 1036
Fetal testing, 288, 289*f*
Fetus, **1036**
   detecting disorders of, during pregnancy, 1039–1040
   gestation and birth of, 1035*f*–1037*f*
   maternal immune tolerance of, 1037
   Zika virus and, 409
Fever, 888, 889*f*, 956–957
F factor, **580***f*, 581
Fiber cells, 764*f*
Fibers, muscle, 1130, 1131*t*
Fibrillanosema crangonycis, 661*f*
Fibrin, 936*f*
Fibrinogen, 936*f*
Fibroblast growth factor (FGF), 1062
Fibroblasts, 113*t*, 247, **878***f*
Fibronectin, **119**
Fibrous connective tissue, 878*f*
Fibrous proteins, 78
Fibrous root systems, 760
Ficedula species, 519
Fiddler crabs, 1141*f*
Fierer, Noah, 1223*f*
Fight-or-flight responses, 212*f*, 928, 1003, 1012*f*–1013*f*, 1089
Filamentous fungi, 659
Filaments, 879*f*
Filaments, flagellum, 576*f*, 577
Filaments, flower, **644***f*
Filter feeders, **690***f*, **903***f*
Filtrates, **984***f*, 987, 988*f*, 989
Filtration, **984***f*
Fimbriae, 97*f*, **575**, 576*f*
Finches, 15, 16*f*, 21, 473*f*, 486*f*–487*f*, 498, 519, 1217*f*
Fingerprints, DNA, 436
Finland, 1227*f*
Fire, 1172, 1228, 1229*f*, 1244*f*
Firefly gene, 341, 342*f*
Fireweed, 636*f*
First law of thermodynamics, 143, **145***f*, 146, 1239
Fishapod discovery, 730*f*, 731
Fishes
   adaptations of kidneys of, 992, 993*f*
   allopatric speciation in, 511*f*, 512
   in biomanipulation, 1227*f*
   changes in gene regulation of, 546*f*, 547
   differential gene expression in, 365*f*
   discovery of "fishapod" *Tiktaalik*, 730*f*, 731
   endangered or threatened, 1262
   flashlight, 397, 587*f*
   frequency-dependent selection in, 500, 501*f*
   gills for gas exchange in, 941*f*
   hearing and equilibrium in, 1116*f*
   hybrid zones of, 515*f*, 519*f*, 520
   lux genes in, 397
   osmoregulation in, 979*f*
   parental care in, 1151*f*
   parthenogenesis in, 1020
   plastic waste and, 1277
   protists as pathogens of, 614
   ray-finned, and lobe-finned, 728*f*–729*f*, 730
   sex determination of, 298*f*
   sex reversal in, 1020
   single circulation in, 924*f*
   thermoregulation in, 881*f*
Fish fraud, 89
Fission, **1020**
Fitness, relative, **497**
FitzRoy, Robert, 472
Fixed action patterns, **1140***f*
Fixed alleles, 490, 496
Flaccid cells, **135**, **790**, 791*f*, 797*f*

Flagella, **114**
   in animal cell, 100*f*
   bacterial, movement of, 993*f*
   cilia vs., 115*f*
   dinoflagellate, 605*f*
   euglenozoan, 600*f*
   flagellated sperm, 619
   as microtubules, 114, 115*f*–116*f*
   in prokaryotic cell, 97*f*, 576*f*, 577
   in protistan cell, 101*f*
   stramenopile, 602*f*
Flagellated sperm, 619
Flagellin, 866
Flame bulbs, 694, 984
Flamingo (*Phoenicopterus ruber*), 14*f*, 740*f*
Flashlight fish, 397, 587*f*
Flattening, body surface area and, 695*f*
Flatworms
   characteristics of, 687*f*, 694, 695*f*–697*f*
   gastrovascular cavities of, 922*f*, 923
   hydrostatic skeletons of, 1133*f*
   nervous systems of, 1086*f*
   protonephridia of, 984*f*, 985
Flavin adenine dinucleotide (FAD), 172*f*–173*f*, 178
Flavin mononucleotide (FMN), 175
Flavoproteins, 175
Flemming, Walther, 237
Flesh-eating disease, 478*f*
Fleshy fruits, 645*f*
Flexion, muscle, 1132*f*
Flies, 712*f*, 825*f*
Flight, 710, 711*f*–712*f*, 736, 738*f*, 739, 1135–1136
Flightless birds, 506*f*, 507, 511, 739, 740*f*
Floating bladderwort (*Utricularia gibba*), 448*t*, 449
Floating fern, 813*f*
Floating of ice, 48*f*
Flooding, 1228
   plant responses to, 804, 864*f*
Floral meristems, 830
Florida, restoration projects in, 1254*f*
Florida Keys mangrove islands, 1233*f*
Florida Keys National Marine Sanctuary, 1273*f*
Florida red-bellied turtles (*Pseudemys nelsoni*), 884*f*
Florigen, **860***f*
Flowering, 830
FLOWERING LOCUS T (*FT*) gene, 860*f*
Flowers, **644**, **823**
   adaptations of, that prevent self-fertilization, 834, 835*f*
   coevolution of pollinators and, 825*f*
   as emergent property, 841
   genetic control of formation and flowering of, 779*f*–780*f*
   hormonal control of flowering of, 860*f*
   monocot vs. eudicot, and pollination of, 649*f*
   photoperiodism and flowering of, 859*f*–860*f*
   pollination of, 822*f*–825*f*
   preventing transgene escape with genetically engineered, 839–840
   structure and function of, 644*f*, 645, 823*f*–824*f*
   trends in evolution of, 825*f*, 826
Flu. *See* Influenza viruses
Fluctuation, population, 1205*f*
Fluid feeders, **903***f*
Fluid mosaic model, **127***f*
Flukes, 687*f*, 696*f*
Fluorescence, 195*f*, 250
Fluorescence microscopy, 95*f*, 96
Fluoxetine, 1103
Flycatchers, European, 519
Fly death fungus, 662*f*
Flying fox bats (*Pteropus mariannus*), 1262*f*
Flying squirrels, 481*f*, 517
FMN (flavin mononucleotide), 175
Focusing, visual, 1122*f*, 1123
Folding, body surface area and, 695*f*
Folding, protein, 83
Folic acid, 372*f*
Foliose lichens, 668*f*, 672
Follicles, **1026**, 1027*f*
Follicle-stimulating hormone (FSH), 1004*f*, 1008*f*, **1030**, 1031*f*–1032*f*, 1033
Follicular phase, 1032*f*, 1033
Food. *See also* Animal nutrition; Trophic structure
   animal processing of, 902, 903*f*–905*f*
   in bioenergetics and metabolic rates, 889, 890*f*
   brown algae as human, 603

calories of, 46
climate change and quality of, 812
digestion of. *See* Digestion; Digestive systems
as fuel for catabolism, 182*f*, 183
as fuel for cellular respiration, 164*f*, 165
fungi as human, 670
genetically modified organisms (GMOs) as, 438–439, 837, 838*f*, 839–840
as limiting factor for human population size, 1211
plants as, for animals, 617, 619
population cycles and shortages of, 1205*f*, 1206
ray-finned fishes as human, 728
red algae as human, 609*f*, 610
seed plants as human, 651
Food chains, **1224***f*–1225*f*, 1257*f*
   biological magnification in, 1275, 1276*f*
   energy loss along, 1246, 1247*f*, 1248
Food poisoning, 403, 588, 1080
Food vacuoles, 107*f*, **108**, 607*f*, 904
Food webs, 614, 615*f*, **1224***f*–1225*f*, 1237
Foolish seedling disease, 851
Foot, mollusc, **699***f*
Foot, sporophyte, **625**
forager gene, 1148, 1149*f*
Foraging behaviors, **1148**, 1149*f*, 1150, 1161
Foraminiferans (forams), 599*f*, **608***f*
Forebrain, **1091***f*–1092*f*
Forelimbs, mammalian, 479*f*
Forensic ecology, 1265*f*
Forensic science, 24*f*, 436, 437*f*
Foreskin, penile, 1026
Forest fires, 1228, 1229*f*, 1244*f*
Forests. *See also* Deforestation
   climate change effects on, 1245, 1282
   climate effects of, 1169*f*
   decomposition in, 1248
   northern coniferous, 1175*f*
   temperate broadleaf, 1176*f*
   tropical, 1173*f*
Formic acid, 28*f*
Fossil fuels
   biofuel technology to reduce dependence on, 838
   in carbon cycle, 1250*f*
   ecological footprints and, 1210*f*, 1211
   global climate change and, 204–205, 1278, 1280*f*–1281*f*, 1282, 1283*f*
   humans and burning of, 11
   hydrocarbons as, 60
   nutrient cycling and, 1249*f*
   ocean acidification and, 53*f*, 54
   photosynthesis as source of, 188
   seedless vascular plants and, 633–634
Fossil record. *See also* Fossils
   angiosperms in, 647*f*, 648
   animals in, 675*f*–677*f*, 678–679
   arthropods in, 706*f*, 707, 710
   colonization of land in, 536, 537*f*
   ecological factors affecting evolutionary rates in, 538
   evidence in, for dinosaurs as ancestors of birds, 564*f*, 565
   evidence in, for evolutionary change, 13*f*
   evolutionary trends in, 548, 549*f*
   forams in, 608*f*
   fungi in, 659*f*, 660
   geologic record and, 532*f*–533*f*
   gymnosperms in, 641*f*
   as history of life, 525, 529*f*
   molecular clocks and, 567
   origin of mammals in, 530, 531*f*, 532
   origin of multicellular organisms in, 533*f*, 535*f*–536*f*
   origin of single-celled organisms in, 532*f*–535*f*
   plant origin and diversification in, 621, 622*f*
   seedless vascular plants in, 628*f*
   whales in, 525*f*, 526
Fossils, **470**. *See also* Fossil record
   amniote, 735*f*
   biogeography and, 482–483
   bird, 739*f*
   dating of, 529, 530*f*
   early *Homo*, 750–751, 752*f*
   of early vertebrates, 724*f*–725*f*
   as evidence for Darwinian evolution, 481*f*–482*f*
   evolutionary theories and, 470*f*
   of gnathostomes, 726*f*
   hominin, 748*f*–749*f*, 750

Fossils (continued)
Homo sapiens, 753f, 754
horseshoe crabs as living, 708f
radiometric dating of, 32–33
reptile and dinosaur, 736
speciation patterns of, 520f, 521
tetrapod, 730f–731f
whisk ferns as living, 633
Foundation species, 1214, **1226**
Founder effect, **494**–495
Fovea, **1122**f, 1123
Fox, 1238f, 1244, 1256f
FOXP2 gene, 461, 462f
F plasmids, 580f, **581**
Fragmentation, habitat, 1263, 1264f, 1270, 1271f
Fragmentation, reproductive, **833**, 1020
Frameshift mutations, 358f, **359**
Franklin, Rosalind, 317, 318f, 320
Fraternal twins, 1036
Free energy, **147**
Free-energy change, 149f–150f, 152f
Free energy of activation (activation energy), 153, 154f
Free ribosomes, 103f, 104, 354f
Freezing, 865
Fregata magnificens, 1139f
Frequency-dependent selection, 500, 501f
Freshwater animals, 979f, 992
Freshwater biomes, 1177f–1178f
habitat loss in, 1264
Freshwater lakes, primary production in, 1242–1243
Freshwater marshes, 1186
Friction, locomotion and, 1135
Fringe wetlands, 1179f
Fringing reefs, 1182f
Fritillaries, 1206f, 1207
Frogs
as amphibians, 719f, 732f–733f
axis formation in, 1060f
breathing in, 925
cell developmental potential in, 1060, 1061f
cell fate and pattern formation by inductive signals in, 1062f
cleavage in, 1048f, 1049
coloration of, 1218f
discovery of new species of, 1164f, 1185
embryonic development of, 381f
external fertilization of, 1022f
fungal parasites of, 669f, 670f
gastrulation in, 1051f, 1052
mate choice among tree, 500f
metamorphosis of, 1015f
neurulation in, 1054f, 1055
nuclear transplantation in, 428, 429f
polyploidy in, 514
reproductive isolation in, 524
Fronds, 632f
Frontal lobe, 1085, 1097f, 1098
Frontal lobotomy, 1098
Frost-tolerant plants, 865
Fructose, 68f–69f, 153, 155
Fructose 6-phosphate, 186
fru gene, 1155
Fruit flies
courtship behavior of, 1155
hybrid sterility of, 522–523
model organism. *See Drosophila melanogaster*
molecular clock of, 567
reproductive isolation of allopatric populations of, 512f
Fruiting bodies, 213f, 612f–613f, 663f–664f, 665t
Fruitlets, 830, 831f
Fruits, **645**, **830**
angiosperm seeds as, 644, 645f
auxin in growth of, 849
dispersal of, 832f
in ecosystem interaction, 10f, 11
ethylene in ripening of, 854
gibberellins in growth of, 851f
structure and function of, 830f–831f
Fruticose lichens, 668f, 672
Frye, Larry, 128f
Fuels
bacteria in production of ethanol, 590
fossil. *See* Fossil fuels
genetic engineering of, using fungi, 671f

peat and coal, 627f, 633–634
seed plants as, 651
Fumonisin, 839
Function, structure and, 6
Functional groups, **62**f–63f
Functional magnetic resonance imaging (fMRI), 1096f
Fundamental niches, 1215, 1216f
Fungi
amphibian population declines due to, 733
ascomycetes, 663f–664f, 665t, 671f
basidiomycetes, 665f–667f
body structure of, 655f–656f
cells of, 100f
chitin in, 72
chytrids, 661f
cryptomycetes, 661f
as decomposers, 188, 1240f
in domain Eukarya, 12f
expressing cloned eukaryotic genes in, 422–423
fungal plant toxin, 839
immune system recognition of, 952f
kingdom and domain mutualisms with, 813f
land colonization by, 536, 537f
maximizing surface area by, 695f
microsporidians, 661f
mucoromycetes, 663f
mycorrhizal, 621, 656f, 657, 787, 817, 818f
nutrient limitations and, 1244
nutritional and ecological roles of, 12f, 13, 655, 667f–670f
as opisthokonts, 613
origin and evolution of, 659f, 660
phylogeny and diversity of, 660f–667f
practical uses of, for humans, 670, 671f
relationship of, to unikont protists, 611–612
sexual and asexual life cycles of, 258f, 259, 657, 658f–659f
symbiotic associations of, 672
terrestrial adaptations of, 660
zoopagomycetes, 662f, 663
Fungi kingdom, 12f, 568
Furcifer pardalis, 735f
Fusarium, 839
Fusion, hybrid zone, 519f, 520
Fynbos, 1174f

G

G₀ phase, **246**, 247f

G$_0$ phase, **246**, 247f
G$_1$ checkpoint, 245f–247f
G$_1$ phase, **237**f
G$_2$ checkpoint, 245f–247f
G$_2$ phase, **237**f–238f
G3P, **201**, 202f, 206, 207f
Gadus morhua, 728, 979f
Gage, Phineas, 1098f
Gaia hypothesis, 1259
Galactose, 68f, 368f
Galápagos Islands, 15, 16f, 21, 472f–473f, 506f, 519
Galdieria sulphuraria, 569f
Gallbladders, **909**
Gälweiler, Leo, 848f
Gambusia hubbsi, 511f, 512
Gametangia, **625**, 626f
Gametes, **236**, **255**
formation of, 269
human gametogenesis and, 1027, 1028f–1029f
production and delivery of, in animal reproduction, 1023f, 1024
production of, by meiosis, 237
in sexual life cycles, 255, 257
Game theory, **1154**f, 1159
Gametic isolation, 509f
Gametogenesis, human, **1027**, 1028f–1029f
Gametophytes, **620**f, 823
in alternation of generations, 258f
angiosperm, 826, 827f
brown algae, 603f, 604
of plants, 620f–621f, 624f–626f
sporophyte relationships with, in plants, 637f, 638
Gamma-aminobutyric acid (GABA), 1081t, 1084
Ganglia, **1069**, **1086**f
Ganglia, planarian, 696f. *See also* Cerebral ganglia
Ganglion cells, 1118f, 1120, 1121f
Gap junctions, 120f, 215f
Garden peas, G. Mendel's experiments with, 269f–276f

Garigue, 1174f
Garlic, 836f
Garlic mustard (*Alliaria petiolata*), 818
Garrod, Archibald, 336
Garter snakes, 1155, 1156f
Gas chromatography, 852
Gases
greenhouse, 11, 48, 53f, 1170
as neurotransmitters, 1081t, 1082
Gas exchange, **939**
adaptations for, 947f–949f
amphibian, 732
arthropod, 707–708, 710f
breathing in, 925, 944, 945f–946f
chordate, 720
circulatory systems and, 921f–924f, 925, 947f.
See also Circulatory systems
coordination of cardiovascular systems and. *See* Cardiovascular systems
fish, 728
gills for, in aquatic animals, 921f, 940f–941f
of insects, 951
ion movements and gradients and, 993f
in lungs, 924f, 925, 942, 943f–946f
mammalian respiratory systems, 942, 943f–946f.
See also Respiratory systems
partial pressure gradients in, 939
plant and animal, 895f
respiratory media in, 939t, 940
respiratory surfaces in, 940
shark, 726
tracheal systems for, in insects, 941, 942f
Gasterosteus aculeatus, 481, 546f, 547, 1140f
Gastric glands, 907f
Gastric juices, **907**f, 908, 915f
Gastric ulcers, 912, 913f
Gastrin, 915f
Gastrodermis, 691f
Gastropods, 542, 699, 700f–701f, 702
Gastrotrichs, 688f
Gastrovascular cavities, **691**f, **922**f, 923
Gastrovascular cavity, 696f, **904**f
Gastrula, 674f, 1044, **1049**, 1050f–1053f, 1066
Gastrulation, 674f, 680, 1044, **1049**, 1050f–1053f
Gated channels, **135**
Gated ion channels, **1072**f–1077f
Gause, G. F., 1215
Geckos, 26, 39, 1260f
Geese, 886f, 1256f–1257f
Gel electrophoresis, **420**f
GenBank, 444, 445f
Gender, 298
Genealogy, molecular, 87, 89
Gene amplification, cancer gene, 388, 389f
Gene annotation, **445**–446
Gene cloning, **418**f, 419, 421, 422f
Gene drive, **427**
Gene editing, **360**, 361f
Gene expression, **8**, **84**, **336**. *See also* Genes; Genetics
auxin and, 848
basic principles of transcription and translation in, 337, 339f
of brain development genes in lancelets and vertebrates, 722f
changes in, in macroevolution, 545f–546f, 547
of cloned eukaryotic genes, 422–423
control of plant cell differentiation and, 778f
differential. *See* Differential gene expression
DNA, RNA, and genes in, 8f
DNA technology in analysis of, 423, 424f–428f
DNA technology in silencing, 427
evidence for, in study of metabolic defects, 336f–338f
evolutionary significance of cross-species, 423
experiment on, for red maple leaf structure, 762
flowering and, 860f
as flow of genetic information, 335–336
gene concept and, 361–362
genetic code in, 340f–342f
interacting groups of, 425f–426f
interpreting sequence logos to identify ribosome binding sites in, 351
mutations in, 357f–358f, 359
nematode parasite control of host, 706
nucleic acids in, 84f
polypeptide synthesis via RNA-directed translation in, 347f–356f

regulation of. *See* Gene regulation
RNA modification after transcription by eukaryotic cells in, 345*f*–347*f*
RNA synthesis via DNA-directed transcription in, 342, 343*f*–344*f*
stages of, that can be regulated, 370*f*
study of, by systems biology, 446, 447*f*–448*f*
summary of eukaryotic transcription and translation in, 356*f*
as transcription, 371. *See also* Transcription
Gene families, 565*f*, 566
Gene flow, **496**
  biological species concept and, 506–507, 508*f*–509*f*, 510*f*
  as cause of microevolution, 496, 497*f*
  Hardy-Weinberg equilibrium and, 492*t*
  in macroevolution, 551
  between Neanderthals and humans, 752*f*–753*f*
  speciation and, 522
Gene pools, **490***f*
General transcription factors, 373–374
Generative cells, 646*f*, 826
Gene regulation
  analyzing DNA deletion experiments on, 376
  in bacterial transcription, 366*f*–369*f*, 370
  cancer due to faulty cell cycle control in, 388, 389*f*–393*f*, 394–395. *See also* Cancer
  changes in, in macroevolution, 546*f*, 547
  differential gene expression and, 365*f*. *See also* Differential gene expression
  in eukaryotic cells. *See* Eukaryotic gene regulation
  faulty, in cloned animals, 430
  noncoding RNAs in, 379, 380*f*, 381
  of plant cell differentiation, 778*f*
  in plant signal transduction pathways, 845
  steroid hormones and, 1003*f*
  in treehopper, 466
Genes, 7, **84**, **255**. *See also* Chromosomes; DNA; Genetics; Genomes
  activator binding to, 397
  alleles as alternative versions of, 272. *See also* Alleles
  animal developmental. *See Hox* genes
  apoptosis, 230*f*
  appetite regulation, 918
  associated with cancer, 388, 389*f*
  B cell and T cell diversity and rearrangement of, 960, 961*f*
  calibrating molecular clocks of, 566, 567*f*
  for color vision, 1122*f*
  conservation of, 466
  coordinately controlled, 366, 375–376
  density of, in genomes, 449
  divergence of human globin, during evolution, 458
  DNA technology in determining functions of, 426, 427*f*, 428
  duplication of, due to unequal crossing over, 455*f*
  editing of, 426, 427*f*, 434–435
  effects of developmental, 544*f*–545*f*
  enzyme relationship with, in protein synthesis, 336*f*–338*f*
  evolution of, 455–457*f*, 461, 462*f*, 565*f*, 566
  extending Mendelian inheritance for, 278, 279*f*–282*f*. *See also* Mendelian inheritance
  flower formation, 779*f*–780*f*
  foraging, 1148, 1149*f*
  gene expression and, 7*f*–8*f*, 361–362. *See also* Gene expression
  genetic diversity and, 1263
  genetic variation due to alterations of number or position of, 489
  genomic imprinting and, 310*f*, 311
  genomics, bioinformatics, and proteomics in study of, 9, 86, 87*f*
  as hereditary units, 255
  homeotic and *Hox*, 385*f*, 463*f*–464*f*, 545*f*, 546
  homologous, 479–480
  horizontal gene transfer of, 568, 569*f*–570*f*
  Hox genes, 1064
  identifying linked or unlinked, 304
  identifying protein-coding, 445*f*, 446
  inheritance of organelle, 311*f*
  jumping. *See* Transposable elements
  linked, 301*f*–303*f*. *See also* Linked genes
  locating, along chromosomes, 295, 296*f*–297*f*
  in macroevolution of development, 545*f*–546*f*, 547

mapping distance between, on chromosomes, 305*f*–306*f*
maternal effect, egg-polarity, and *bicoid*, 386*f*–387*f*
G. Mendel's hereditary factors as, 269–270
meristem identity and organ identity, 779*f*–780*f*
molecular homologies and, 558, 559*f*
multigene families and, in eukaryotic genomes, 452, 453*f*
notation system for, 295
nucleic acids and, 84. *See also* DNA; Nucleic acids; RNA
number of, in genomes, 448*t*, 449
olfactory, 1124
organization of typical eukaryotic, 373*f*
pseudogenes, 450
rearrangement of parts of, 456, 457*f*
regulation of. *See* Gene regulation
regulatory, 367*f*
sex-linked, 298*f*–300*f*
speciation and, 522*f*, 523
split, 345*f*–347*f*
study of, by systems biology, 446, 447*f*–448*f*
transcription factors for, 221
transcription of, during cell-signal response stage, 226*f*
transgenes, 436*f*
transplanting, into different species, 341, 342*f*
variability of, in genetic variation, 488
Gene therapy, 426, **434***f*, 1122*f*
Genetically modified organisms (GMOs), **438**
  for fossil fuel dependency reduction, 838
  gene cloning and, 419
  for hunger and malnutrition reduction, 837, 838*f*
  issues about, 438–439, 839–840
Genetically modified (GM) organisms. *See also* Transgenic animals; Transgenic plants
  fungi as, 671
  issues about, 839–841
  plant biotechnology and genetic engineering of, 837, 838*f*
Genetic code
  biotechnology and, 441
  codons and triplet code of nucleotides as, 340*f*, 341
  cracking of, 341*f*
  of DNA, 7*f*–8*f*
  evolution of, 341, 342*f*
  as molecular homology, 479–480
  in mutations, 357*f*–358*f*, 359
  neutral variation and redundancy of, 488–489
  as sequence of nitrogenous bases, 85
  universality of, 16
Genetic counseling, 287–288, 293
Genetic disorders
  alkaptonuria, 336
  biotechnology in diagnosis and treatment of, 433, 434*f*, 435
  from chromosomal alterations, 306, 307*f*–309*f*
  counseling for, 287–288, 293
  diagnosing fetal, 1039–1040
  dominantly inherited, 287*f*
  ethics of testing for, 24
  multifactorial, 287
  mutations and, 357*f*–358*f*, 359
  recessively inherited, 285*f*–286*f*, 287
  sickle-cell disease. *See* Sickle-cell disease
  testing for, 288, 289*f*, 290
Genetic diversity. *See also* Genetic variation
  in biodiversity, 1261*f*
  as factor in extinction vortex, 1266, 1267*f*
  in human welfare, 1262, 1263*f*
  prokaryotic, 578*f*–580*f*, 581, 583*f*
  in small populations, 1266, 1267*f*–1268*f*, 1269
Genetic drift, 492, **494***f*–495*f*, 496, 551
Genetic engineering, **416**
  of antifreeze proteins, 865
  CRISPR-Cas9 system, 360, 361*f*
  DNA technology tools for, 416, 418*f*–422*f*, 423
  of ethylene signal transduction pathways, 854
  fungi in, 671
  of genetically modified organisms, 438–439. *See also* Genetically modified organisms
  opposition to, 839–841
  of "pharm" animals, 436*f*
  plant biotechnologists, 837

of plants. *See* Transgenic plants
plant tissue culture and, 836
prokaryotes in, 589, 590*f*–591*f*
Genetic Information Nondiscrimination Act, 288
Genetic maps, 305*f*–306*f*
Genetic markers , 427*f*, 434, 436, 437*f*, 441
Genetic mutants in Scientific Skills Exercises, 918, 1095
Genetic profiles, 434, **436**, 1194, 1195*f*
Genetic prospecting, 583
Genetic recombination, **302**
  identifying linked genes for, 304
  of linked genes in crossing over, 302, 303*f*
  natural selection and genetic variation from, 305
  in prokaryotes, 579*f*–580*f*, 581
  transposable elements and, 459
  of unlinked genes in independent assortment of chromosomes, 302*f*
Genetics, **255**. *See also* Genetic variation; Inheritance; Mendelian inheritance
  of blood cholesterol levels, 938
  designing experiments using genetic mutants, 1095
  in ecological forensics, 1265*f*
  in estimating reproductive rates, 1194, 1195*f*
  flow of genetic information in plant cells, 208*f*
  of foraging behaviors, 1148, 1149*f*
  genetic basis of animal behavior, 1155*f*
  genomics and proteomics in, 86, 87*f*
  inheritance of chromosomes and genes in, 255
  interpreting data from experiments with genetic mutants, 918
  of nervous system disorders, 1102*f*
  Punnett square as tool in, 273*f*, 274
  of sickle-cell disease, 502*f*
  solving complex problems of, with rules of probability, 277–278
  of speciation, 522*f*, 523
  as study of heredity and hereditary variation, 255
  vocabulary of, 274*f*
Genetic sequences in Scientific Skills Exercise, 595
Genetic testing
  biotechnology in, 433–434
  for breast cancer predisposition, 394
  fetal, 288, 289*f*, 1039–1040
  identifying carriers, 288
  newborn, 289–290
  personal genome analysis, 433
Genetic variation, **255**, **487**
  chromosome activity and, 268
  from crossing over and recombinant chromosomes, 265, 266*f*
  evolutionary significance of, within populations, 266, 267*f*
  extinction vortex and loss of, 1266, 1267*f*
  in genetic diversity, 1261*f*. *See also* Genetic diversity
  genetics as study of, 255. *See also* Genetics
  from independent assortment of chromosomes, 265*f*
  microevolution of populations and sources of, 487*f*–488*f*, 489
  in migratory patterns, 1156, 1157*f*
  molecular clocks and rates of, 566, 567*f*
  natural selection and, from recombinant chromosomes, 305
  phylogenetic tree branch lengths and, 561, 562*f*
  preservation of, 500
  in prey selection, 1155, 1156*f*
  from random fertilization, 266
  in sexual reproduction, 255
  in small populations, 1266, 1267*f*–1268*f*, 1269
Gene trees, 557*f*
Genomes, **8**, **235**. *See also* Genes
  analyzing phylogenetic trees based on, to understand viral evolution, 411
  animal phylogeny and, 682, 683*f*, 684
  bioinformatics in analysis of, 444, 445*f*–448*f*
  in cell division, 235
  comparing, 459, 460*f*–464*f*, 676*f*
  complete, 665, 722, 776
  differential gene expression for identical, 370*f*, 371. *See also* Differential gene expression
  evolutionary history in, 565*f*, 566
  evolution of, from DNA duplication, rearrangement, and mutation, 454*f*–457*f*, 458–459

Genomes (continued)
  gene density of, 665t
  genetic testing of fetal, 1039–1040
  genome-wide association studies of, 427f, 433
  genomics, proteomics, and bioinformatics in study of, 9, 86, 87f, 443
  horizontal gene transfer between, 568, 569f–570f
  Human Genome Project and development of DNA-sequencing techniques for, 443f, 444
  interpreting data from, and generating hypotheses, 657
  nervous systems and, 1106
  noncoding DNA and multigene families in eukaryotic, 450f–453f
  p53 gene as guardian angel of, 390–391
  personal analysis of, 433
  prokaryotic, 577f
  reference, 443
  species with complete sequences of, 448t
  variations in size, number of genes, gene density, and noncoding DNA of, 448t, 449
  viral, 400, 401f–407f, 406t
  widespread conservation of developmental genes in animal, 463f–464f
Genome sequencing, 443f, 444, 447, 448f, 657
  genomics, bioinformatics, proteomics, and, 9
Genome-wide association studies, 427, 433
Genomic data in Scientific Skills Exercise, 657
Genomic imprinting, 310f, 311, 372
Genomics, 9, 86, 87, 443–444, 583
  cell signaling, cancer, and, 391, 392f–393f, 394
  contributions of, 88f
  cracking the genetic code, 341f
  DNA sequence analysis, 8f, 9
  genetic material in, 7f–8f
  polyadenylation signal sequence, 345f
  simple sequence DNA, 452
  taxonomy and, 12f, 13
  transcription termination, 344
  transposable elements, 450f–451f, 452
Genotypes, 274
  DNA transformation and, 315
  gene expression as link between phenotypes and, 335–336
  genetic variation and, 487, 488f
  heterozygote advantage and, 500
  phenotypes vs., 274f, 282f, 283
  proteins and, 335
  relative fitness of, 497
Gentoo penguin, 14f
Genus (genera), 12, 554, 555f
Geochemical cycles, 1249f–1252f
Geoemyda spengleri, 736f
Geographical barriers, allopatric speciation and, 511f–512f, 513
Geographic species distributions. See Species distributions
Geologic record, 530f, 532f–533f
Geospiza fortis, 486f–487f, 1217f
Geospiza fuliginosa, 1217f
Germ cells
  fate maps and, 1058f–1059f
  human, 258, 1029f
  telomerase and telomeres in, 329
Germination
  gibberellins in, 851f
  light-sensitive, 871
  phytochromes in, 856, 857f
  seedling development after, 829, 830f
  strigolactones in, 855
Germ layers, 680, 1049, 1050f–1051f
Gestation, 1034, 1035f–1037f
Ghost crabs, 709f
Ghrelin, 917f
Giant panda (Ailuropoda melanoleuca), 448t
Giant sequoia tree, 774f
Giant squid, 702
Giardia intestinalis, 598f, 600
Gibberellins, 846t, 850, 851f
Gibbons, 89, 746f–747f
Gibbs, J. Willard, 147
Gibbs free energy, 147. See also Free-energy change
Gigantism, 1009, 1010f
Gills, 921f, 940f–941f
  annelid, 703
  arthropod, 707
  of axolotls, 921f

crustacean, 709
  of fishes, 728f
  mollusc, 699f
  osmoregulation by, 979f
Gill slits, 720f
Ginkgos, 642f
Giraffes, 931
GLABRA-2 gene, 778f
Glaciation
  ecological succession after, 1230f–1231f
  seedless vascular plants and, 633
Glacier Bay, Alaska, 1230f–1231f
Glands, endocrine system. See Endocrine glands
Glans, 1025f, 1026
Glanville fritillaries (Melitaea cinxia), 1206f, 1207
Glass lizard, 553f–554f
Glass sponge, 146f
Glaucomys volans, 517
Glaucus atlanticus, 686f
Glia (glial cells), 879f, 1069f, 1076f–1077f, 1090f
Glioblastoma, 250, 447
Global biogeochemical cycles, 1249
Global carrying capacity, human population, 1209, 1210f, 1211
Global change. See also Climate change
  atmospheric ozone depletion in, 1283f, 1284
  environmental toxins in, 1275, 1276f–1277f
  greenhouse gases and climate change in, 1278f–1283f
  human impacts on, 1265, 1266f, 1274, 1275f–1283f, 1284
  nutrient enrichment in, 1274, 1275f
  as threat to biodiversity, 1265, 1266f
Global climate change. See Climate change
Global climate patterns, 1166f, 1167
Global cooling, 633
Global ecology, 1165f. See also Ecology
  aquatic biomes in, 1177f–1182f
  of biosphere. See Biosphere
  effects of mass extinctions on, 542f
  evolution and, 1187f
  global climate in, 1166f, 1167
  global human population size issues in, 1209, 1210f, 1211
  human environmental impacts in. See Human environmental impacts
  importance of mycorrhizae to, 818
  ocean acidification in, 53f, 54
  species distributions in, 1164, 1170f, 1178, 1183f–1186f
  terrestrial biomes in, 1171f–1176f, 1189
Global energy budget, 1241
Global extinctions, 1262
Global human population, 1207f–1209f
Global net primary production, 1241, 1242f
Global temperatures, extinction rates and, 541f, 542
Global warming. See Climate change
  as ecosystem interaction, 11
  fossil fuel burning and, 11
Globigerina, 599f
Globin genes, human, 453f, 455, 456f, 458
Globular proteins, 78
Glomeromycetes, 663
Glomerulus, 987f
Glottis, 906f
Glucagon, 916f, 917, 1004f
Glucocorticoids, 1004f, 1012f, 1013, 1018
Glucose
  blood regulation of, 10f
  enzymatic catalysis and, 153, 155, 157
  as fuel for cellular respiration, 165–167
  glucocorticoids and metabolism of, 1012f, 1013
  homeostasis of, 915, 916f, 917
  insulin regulation of, 1004f
  as monosaccharide, 68f–69f
  oxidation of, to pyruvate by glycolysis, 170f–171f
  in photosynthesis, 41f, 190f, 206, 207f
  in positive gene regulation, 369f, 370
  in signal transduction pathways, 216
  transport of, 132, 136
  in treating diarrhea, 139
Glucose 6-phosphatase, 157
Glutamate, 233, 1081t, 1101f, 1120, 1121f
Glutamic acid, 77f, 152f, 341f
Glutamine, 77f, 152f
Glyceraldehyde, 68f
Glyceraldehyde 3-phosphate (G3P), 201, 202f, 206, 207f

Glycerol phosphate, 63f
Glycine, 63f, 77f, 1081t
Glycocalyx, 97f
Glycogen, 71
  in cell signaling, 216f, 226, 227f
  in glucose metabolism, 915, 916f
  in muscle contraction, 1128
  as storage polysaccharide, 70f, 71
Glycogen phosphorylase, 216–217
Glycolipids, 131
Glycolysis, 168
  ATP yield in, 177f
  evolutionary significance of, 182
  fermentation and, 180f
  oxidation of glucose to pyruvate by, 170f–171f
  as stage of cellular respiration, 168, 169f
Glycolytic fibers, 1130–1131
Glycoproteins, 105, 131
  in animal morphogenesis, 1056
  in cellular membranes, 127f, 130f, 131
  in extracellular matrix, 118, 119f
  genetic engineering of, using fungi, 671
  rough ER and, 105
  viruses and, 400f, 405f, 407f
Glycosidic linkages, 69
Glyoxysomes, 112
Glyphosate, 838
Gnathostomes, 725
  derived characters of, 725f, 726
  fossil, 726f
  ray-finned fishes and lobe-fins as osteichthyans, 728f–729f, 730
  sharks and rays as chondrichthyans, 726, 727f
  tetrapods as, 730
Gnetophytes, 642f, 647
Gnetum, 642f
Goats, 436f
Goatsbeard plants (Tragopogon miscellus), 514, 515f
Goiter, 29, 1009
Golden Rice, 838
Goldenrod (Solidago canadensis), 812
Golden spiny mouse, 1216f
Golgi apparatus, 100f, 105, 106f, 107, 109f, 131f, 208f, 405f
Gonadotropin-releasing hormone (GnRH), 1014, 1030, 1031f–1032f, 1033
Gonadotropins, 1008, 1014, 1015f, 1030, 1031f–1032f, 1033
Gonads, 237, 258, 1004f, 1014, 1015f, 1023
Gonorrhea, 576, 584f, 1039
Gonsalves, Dennis, 757
Gonyosoma oxycephala, 1053f
Goodall, Jane, 17f
Gorillas, 87, 89, 746f–747f
Gormley, Andrew, 1191f
Gout, 983
GPCRs (G protein-coupled receptors), 217f–219f, 218, 220, 224f, 1124
GPP. See Gross primary production
G protein-coupled receptors (GPCRs), 217f–219f, 218, 220, 224f, 1124
G proteins, 218f, 224f
Graded potentials, 1072, 1073f, 1078
Gradients, solute, 989, 990f, 991
Gradual model of speciation, 520f, 521
Grafting, plant, 835–836
Gram (unit), 50
Gram, Hans Christian, 574
Gram-negative bacteria, 574, 575f, 584f
Gram-positive bacteria, 574, 575f, 584f
Gram stain technique, 574, 575f
Grant, Peter and Rosemary, 21, 487
Granum, 110, 111f, 189f
Grapefruit, 645f
Grapes, 851f
Grapes of Wrath, The (book), 807f
Graphs in Scientific Skills Exercises
  bar, 23, 179, 376, 483, 590, 629, 700, 762, 864, 1186, 1217
  comparing two variables on a common x-axis of, 972
  estimating quantitative data from, 538
  histograms, 250, 283, 938
  interpreting changes in slope in, 1049
  line, 33, 157, 264, 411, 972, 1049, 1095, 1136, 1186, 1279
  with log scales, 1136

pie charts, 892
  scatter plots, 54, 136, 205, 513, 751, 1217
  sequence logos, 351
Grass, phototropism in coleoptiles of, 847*f*, 848
Grasshoppers, 464*f*, 710*f*, 712*f*, 905*f*, 923*f*, 942*f*, 1132*f*
Grasslands, 783, 1175*f*
Grassy stunt virus, 1263
Graves' disease, 1010
Gravitational motion, free energy and, 148*f*
Gravitropism, **861***f*
Gravity
  axis formation and, 1060
  blood pressure and, 931*f*–932*f*, 951
  locomotion and, 1135
  mechanoreceptors for sensing, 1112, 1115*f*
  plant responses to, 861*f*
Gray matter, **1087***f*
Gray tree frogs (*Hyla versicolor*), 500*f*, 514
Great auk, 1264–1265
Great Barrier Reef, 1282
Greater bilby, 743*f*
Greater prairie chickens (*Tympanuchus cupido*), 495*f*, 496, 1267*f*
Great Salt Lake, 142, 585
Great tits (*Parus major*), 740*f*
Green algae, 101*f*, 596*f*, 599*f*, **610***f*–611*f*, 614, 617, 618, 619*f*, 621, 663, 668*f*–669*f*
Green fluorescent protein (GFP), 342*f*
Greenhouse effect, **1278***f*
Greenhouse gases, 11, 48, 53*f*, 1170, 1278*f*–1283*f*
Greening, plant, 843*f*–844*f*, 845
Green manure, 817
Green parrot snake, 1218*f*
Griffith, Frederick, 315*f*
Grizzly bears (*Ursus arctos*), 510*f*, 1268*f*, 1269, 1272–1273
Grolar bears, 510*f*
Gross primary production (GPP), **1241**, 1242*f*
Ground meristem, 766
Ground squirrels, 893, 1156, 1158, 1159*f*, 1193*t*, 1194*f*, 1195, 1213
Ground tissue system, plant, 758, **763***f*
Groups, control and experimental, 20, 21*f*
Growth. *See also* Plant growth
  as cell division function, 235*f*, 776, 777*f*
  heterochrony and differential rates of, 544*f*–545*f*
  hormonal regulation of, 1004*f*, 1009, 1010*f*
  plant and animal, 894*f*
  as property of life, 3*f*
  vegetative, 830
Growth, population. *See* Population growth
Growth factors, **247**
  in cell cycle control system, 247, 248*f*
  cell fate and, 1062–1064
  in cell-signaling nuclear responses, 226*f*
  induction and, 383
  as local regulators in cell signaling, 215*f*, 216
Growth hormone (GH), 1004*f*, 1008*f*, **1009**–1010
Growth rings, tree, 773*f*
Grus species, 1144
GTP (guanosine triphosphate), 172*f*–173*f*, 218*f*, 352*f*–353*f*
GTPase, 229
Guanine, 84, 85*f*, 86, 317*f*, 318, 341*f*
Guano, 983*f*
Guanosine triphosphate (GTP), 172*f*–173*f*, 218*f*, 352*f*–353*f*
Guard cells, **763***f*, 770, 771*f*, 797*f*, 798
Guichon Creek, 1274*f*
Gulf of Carpentaria, 605*f*
Gulf of Mexico dead zone, 1275*f*
Gulf Stream, 1168*f*
Gulls, 898*f*, 1151*f*
Guppies (*Poecilia reticulata*), 483, 1152, 1153*f*, 1187*f*
Gurdon, John, 428, 429*f*, 432
Gustation, **1123***f*–1124*f*
Gutenberg, Johannes, 24
Guttation, **793***f*
Gymnosperms, **623***f*
  evolution of, 641*f*
  evolution of seeds in, 639
  gametophyte-sporophyte relationships in, 637*f*, 638
  life cycle of pine and, 640*f*, 641
  ovules and production of seeds in, 638*f*
  phylogeny of, 622*t*, 635, 641, 642*f*–643*f*
Gymnothorax dovii, 729*f*, 1214*f*

Gynandromorph, 313
Gyres, ocean, 1168*f*

H

H1N1 virus, 410–411, 1234
H5N1 virus, 410, 1234, 1235*f*
Habitat
  carrying capacity of, 1197, 1198*f*–1199*f*, 1198*t*, 1200
  critical, 1269*f*–1270*f*
  destruction of, in tropical rain forests, 651*f*, 652
  fragmented, 1263, 1264*f*, 1270, 1271*f*
  island, *1232*, 1233*f*
  loss of, as threat to biodiversity, 1263, 1264*f*
  nitrogenous wastes and, 982*f*–983*f*
  of red-cockaded woodpecker, 1269*f*–1270*f*
  sympatric speciation and differentiation of, 515–516
Habitat isolation, 508*f*
Hadean eon, 530*f*, 532*f*–533*f*
Haemophilus influenzae, 448*t*
Hagfishes, 719*f*, **723***f*
Haikouella, 724*f*
Hair, mammalian, 741, 920
Hair cells, 1112, 1113*f*–**1114***f*
Hair color, polygenic inheritance, 281
Hairy-cap moss, 626*f*
Hakea purpurea, 783
Haldane, J. B. S., 526, 1158
Half-life, **32**, **529**, 530*f*
Hall, Jeffrey, 883
Hallucigenia, 706*f*
Halobacterium, 585
Hamilton, William, 1157–1158
Hamilton's rule, **1158***f*–1159*f*
Hammerhead shark, 890*f*
Hamsters, 893*f*, 1095
Hansen's disease, 65
Haplo-diploid sex determination system, 298*f*
Haploid cells, **257**, 258*f*, 259
Haplotypes, 466
Haptophyte, 597*f*
Hardy-Weinberg equation, 490–493
Hardy-Weinberg equilibrium, 489, **490***f*–491*f*, 492*t*, 493
Hares, 1205*f*
Haustoria, 656*f*
Hawaiian Islands, 513, 543*f*, 544
Hawaiian silversword plants, 543*f*, 558, 1183
Hawkmoth, 825*f*, 869*f*, 1218*f*, 1219
Hazel, 824*f*
Hazel dormouse (*Muscardinus avellanarius*), 893*f*
Hazelnut, 645*f*
Head end, 680*f*
Head, insect, 710*f*
Head structure morphogen, 386*f*–387*f*
Hearing, 1112*f*–1116*f*
Heart attacks, 418*f*, 436, **937***f*, 938–939
Heartbeat rhythm, 928*f*
Heartburn, 908
Heart disease, 217, 418*f*, 428, 433, 436. *See also* Cardiovascular diseases
Heart murmurs, **928**
Heart rate, **927**, 928*f*
Hearts, **923**
  atrial natriuretic peptide hormone released by, 996*f*
  cardiac cycle of, 927*f*–928*f*, 930, 951
  in circulatory systems, 923*f*–924*f*
  effects of adrenal hormones on, 1012
  insect, 710*f*
  location of, for human embryo, 1043*f*
  mammalian, in cardiovascular systems, 926*f*–928*f*
  mollusc, 699*f*
  regulation of rhythmic beating of, 928*f*
Heartwood, 774*f*
Heat, **46**, 143–**144**
  as by-product of cellular respiration, 178
  carbon dioxide and, 11
  diffusion and, 132
  metabolic rate and loss of, 890*f*
  plant response to stress of, 865
  temperature vs., 46
  thermophiles and, 585*f*, 586
  thermoreceptors and, 1111*f*
Heat exchange adaptations, animal, 885*f*–889*f*

Heat of vaporization, **47**
Heat-shock proteins, **865**
Heavy chains, **958***f*–961*f*
Heavy metals, bioremediation of, 1253, 1255*f*
Hector's dolphins, 1191*f*, 1192
Hedgehog growth factor, 1063–1064
Height, polygenic inheritance, 281
Heimlich maneuver, 906
HeLa cancer cells, 248–249, 252
Helianthus species, 521*f*
Helical viruses, 400*f*
Helicases, **323***f*, 326*f*, 327*t*
Helicobacter pylori, 584*f*, 912, 913*f*
Helium, 30*f*, 35
Helper T cells, **963**, 964*f*
Hemagglutinin gene, 410–411
Heme group, 175
Hemichordata, 689*f*
Hemings, Sally, 437
Hemiptera, 712*f*
Hemispheres, brain, 1093*f*, 1098, 1106
Hemizygous organisms, 299
Hemochromatosis, 920
Hemocoel, **681***f*, 697, 707, 717
Hemocyanin, 947
Hemocytes, 953
Hemoglobin, **935**
  α-globin and β-globin gene families and, 453*f*, 455, 456*f*
  in circulation and gas exchange, 947*f*–948*f*
  cooperativity as allosteric regulation in, 160–161
  dissociation curves of, 948*f*, 951
  in erythrocytes, 935
  as measure of evolution, 87, 89
  polypeptides in, 337
  as protein, 76*f*
  protein quaternary structure and, 81*f*
  sickle-cell disease and, 82*f*, 286*f*, 287, 500–501, 502*f*–503*f*, 505
Hemolymph, 681, 707, **923***f*, 981, 985*f*
Hemophilia, **300**, 433, 936
Hemorrhagic fever, 409
Henslow, John, 472
Hepatic portal veins, **910**
Hepatitis B virus, 409, 974
Hepatophyta (liverworts), 620*f*, **623**, 626*f*
HER2 breast cancer, 220, 250, 392*f*–393*f*
Herbicide resistance, 438
Herbicides
  auxin in, 849
  transgenic, 837–838
Herbivores, **901**
  alimentary canals of, 912*f*
  animals as, 673*f*
  as biotic factors limiting species distributions, 1184*f*
  dentition and diet in, 911*f*, 912
  energetic hypothesis and biomass of, 1225*f*
  evolutionary links between plants and, 648, 649*f*
  insects as, 713
  mutualistic digestive adaptations of, 913, 914*f*
  plant defenses against, 867, 868*f*–869*f*
Herbivory, 551, **867**, 868*f*–869*f*, **1219***f*, 1220, 1256*f*
Herceptin, 220, 250, 393*f*
Hereditary factors, genes as, 269–270, 294*f*. *See also* Genes
Hereditary nonpolyposis colon cancer (HNPCC), 394
Hereditary variation, **255**. *See also* Genetic variation
Heredity, **255**, 269–270. *See also* Inheritance; Mendelian inheritance
Hermaphrodites, **691**, 705
Hermaphroditism, **1020**, 1042
Heroin, 40, 78, 1103*f*
Herpes simplex viruses, 973*t*
Herpesviruses, 405, 409, 973*t*
Herring gull, 898*f*
Hershey, Alfred, 316*f*, 317
Heterochromatin, 330*f*, **332**, 371, 381
Heterochrony, **544***f*–545*f*
Heterocysts (heterocytes), **582***f*
Heterokaryon mycelia, **658**
Heteromorphic generations, **604**
Heterosporous species, **631**, 638
Heterotrophs, **188**, 581*t*, 594, 655, 673–674, 889, 894*f*, 1240*f*
Heterozygote, **274**
Heterozygote advantage, **500**, 501*f*–503*f*

Heterozygote protection, 488
Heterozygous organisms, **274**
Hexapoda, 710. *See also* Insects
Hexoses, 68*f*
Hfr cells, 580*f*, 581
Hibernation, 178, **893***f*, 897
Hierarchical classification, 554, 555*f*
High-density lipoproteins (HDLs), **937**–938
Highly conserved genes, 460
High-throughput DNA technology, 9, 415*f*–417*f*, 443*f*, 444, 447
Hindbrain, 1085, **1091***f*–1092*f*
Hinge joints, 1134*f*
Hippocampus, 1095*f*, 1100, 1101*f*
Hippopotamus, 88*f*
Hirudin, 704
Hirudinea, 703
Histamine, 955*f*, **956**, 971
Histidine, 77*f*
Histograms in Scientific Skills Exercises, 250, 283, 938
Histone acetylation, **371***f*
Histone modifications, 371*f*, 372, 391
Histones, **330***f*
HIV (human immunodeficiency virus), **406**, **973**. *See also* AIDS
    AIDS and, 398*f*
    antiviral drugs and, 409
    applying molecular clock to origin of, 567, 568*f*
    attacks on immune system by, 973*f*, 974
    biotechnology in diagnosis of, 433
    CCR gene and, 435
    cell-surface proteins and blocking entry of, into cells, 130*f*
    as emerging virus, 409–410
    G protein-coupled receptors and, 217*f*
    host range of, 401
    rapid reproduction of, 489
    replicative cycle of, 407*f*
    as retrovirus, 406*t*
HIV-1 M strain, 567, 568*f*
Hoary marmot (*Marmota caligata*), 164*f*
Hodgkin's disease, 971, 1263*f*
Hoekstra, Hopi, 20, 21*f*, 22, 26
Holdfasts, **602**, 603*f*
Holoblastic cleavage, 1048
Holothuroidea, 715*f*
Homarus americanus, 979
Homeoboxes, **463***f*–464*f*, 675
Homeodomains, 463
Homeostasis, **881**. *See also* Thermoregulation
    of blood calcium levels, 1011*f*
    as feedback regulation of animal internal environment, 881*f*–883*f*
    of glucose, 915, 916*f*, 917, 1004*f*, 1012*f*, 1013
    hormonal regulation of kidneys for, 994*f*–996*f*
    of human breathing, 946*f*
    of macaques, 897
    of marine iguana, 998
    osmoregulation for, 977*f*, 978
    peripheral nervous system in, 1088–1089
    thyroid, 1008, 1009*f*, 1010
Homeotherms, 884
Homeotic genes, **385***f*, 463*f*–464*f*, **545***f*–546*f*, 547. *See also Hox* genes
Hominins, **748**
    Australopiths, 749, 750*f*
    bipedalism in, 750*f*
    derived characters of, 748
    earliest, 748*f*–749*f*
    early *Homo* genus, 750–751, 752*f*
    Homo sapiens, 753*f*–754*f*
    Neanderthals, 752*f*–753*f*
    tool use in, 750
Homo erectus, 752–753
Homo ergaster, 752*f*
Homo floresiensis, 754
Homogenization, 96*f*
Homo genus, 746*f*, 750–751, 752*f*–754*f*
Homo habilis, 750–752
Homologies, **479***f*–481*f*, **558***f*–559*f*
Homologous chromosomes (homologs), **256**
    alleles in, 259
    in chromosomal basis of Mendelian inheritance, 294*f*, 295
    human, 256*f*–257*f*
    in meiosis, 260
    in mitosis vs. in meiosis, 262, 263*f*, 264

Homologous genes, 565*f*, 566
Homologous structures, **479***f*–481*f*
Homo naledi, 753*f*, 754
Homo sapiens, 554, 665*t*. *See also* Humans
Homosporous species, **631**, 638
Homozygote, **274**
Homozygous organisms, **274**
Honeybees, 661, 824*f*, 887, 1142*f*, 1143, 1146, 1147*f*, 1156
Honeypot ant, 485
Hook, flagellum, 576*f*, 577
Hooke, Robert, 94
Hookworms, 705
Hopping, 1138
Horizontal cells, 1118*f*, 1120
Horizontal gene transfer, 568, **569***f*–570*f*, 579*f*–580*f*, 581, 583, 588, 689*f*
Horizontal transmission, viral, 412
Hormonal proteins, 76*f*
Hormone cascade pathways, 1008, 1009*f*, 1010, 1014
Hormones, **215**, **846**, **1000**. *See also* Animal hormones; Plant hormones
    animal vs. plant, 215–216, 894*f*. *See also* Animal hormones
    coordinate control of genes by, 375–376
    environment and, 1018
    in fight-or-flight responses, 212
    as intracellular chemical signals, 220, 221*f*
    kidney regulation by, 994*f*–996*f*
    in long-distance cell signaling, 215*f*, 216
    specificity of, 227, 228*f*
Hornworts, **623**, 626*f*. *See also* Bryophytes
Horowitz, Norman, 336*f*–338*f*
Horses, 548, 549*f*
Horseshoe crabs, 707, 708*f*
Horsetails, 630*f*–632*f*, 633, 635
Horvitz, Robert, 1058–1059
Hosken, David, 1024*f*
Host cells, endosymbiont, 534, 535*f*
Host ranges, viral, **401**
Hosts, parasite, **1220**, 1234
Hosts, symbiont, **587**
Hot spots, biodiversity, 1272*f*
Hot springs, 585*f*, 586
Hox genes. *See also* Homeotic genes
    as animal development genes, 463*f*–464*f*, 675
    arthropod body plan and, 706*f*, 707
    cell fate and, 1064
    jaws and, 725*f*, 726
    lancelet, tunicate, and vertebrate, 722*f*
    in macroevolution of development, 545*f*–546*f*, 547
    origin of, 677
    in plants, 778*f*
    in treehoppers, 466
HTLV-1 virus, 394
Hubbard Brook Experimental Forest, 1252*f*, 1266*f*
Human body
    bacteria in, 587, 588*f*
    brain in nervous system of, 1085*f*, 1091*f*–1096*f*. *See also* Brains; Nervous systems
    circadian rhythms in thermoregulation of, 882, 883*f*
    digestive systems. *See* Digestive systems
    ears of, 1113*f*
    endocrine glands and hormones of, 1004*f*, 1007*f*–1008*f*. *See also* Animal hormones; Endocrine systems; Nervous systems
    evolution of human eye, 547, 548*f*
    excretory systems of, 985, 986*f*–987*f*
    eyes of, 1118*f*–1119*f*
    glucose homeostasis in, 915, 916*f*
    heterochrony and differential growth rates in skulls of, 544*f*
    hypothalamus in thermoregulation of, 888, 889*f*
    locomotion by interaction of muscles and skeletons of, 1132*f*
    lymphatic system of, 956*f*
    mechanoreceptors in skin of, 1110*f*
    metabolic rates of, 891
    osmoregulation of, 980
    overnourishment, obesity, and, 917*f*, 918
    regulation of growth of, 1009, 1010*f*
    skeleton of, 1134*f*
    two-solute model of kidney function, 989, 990*f*, 991
Human chorionic gonadotropin (hCG), 969, 1035
Human embryonic development. *See also* Animal development

brains in, 1092*f*
cilia and cell fate in, 1064*f*
conception, pregnancy, and birth in, 1034, 1035*f*–1037*f*
embryo image, 1043*f*
gastrulation in, 1052, 1053*f*
maternal immune tolerance in, 1037
neuron competition in, 1099–1100
Human environmental impacts
    on biodiversity, 1260*f*, 1261
    biodiversity crisis, 1260*f*–1262*f*. *See also* Biodiversity
    biome disturbances, 1172
    climate change, 1278*f*–1283*f*
    community disturbances, 1228, 1231*f*
    global change, 1265, 1266*f*, 1274, 1275*f*–1283*f*, 1284
    habitat loss and fragmentation, 1263, 1264*f*
    introduced species, 1264*f*
    melting of Arctic sea ice, 48*f*
    ocean acidification, 53*f*, 54
    overharvesting, 1264, 1265*f*
    spread of pathogens, 1234, 1235*f*
    threats to biodiversity from, 1263, 1264*f*–1266*f*
Human genetics
    dominantly inherited disorders in, 287*f*
    genetic testing and counseling in, 287–288, 289*f*, 290, 293
    molecular tags and karyotypes of chromosomes in, 332*f*
    multifactorial disorders in, 287
    pedigree analysis in, 284*f*, 285
    recessively inherited disorders in, 285*f*–286*f*, 287
    skin pigmentation and, 281, 282*f*
Human genome
    α-globin and β-globin gene families in, 453*f*, 455, 456f
    comparing genomes of other species with, 454*f*–455*f*, 459, 460*f*–461*f*, 462
    complete sequence for, 443*f*
    evolution of, 524, 566
    function of *FOXP2* gene in, 461, 462*f*
    gene density of fungal vs., 665*t*
    microarray chips containing, 448*f*
    sequencing of, 443*f*, 444, 447, 448*f*
    size, number of genes, gene density, and noncoding DNA of, 448*t*, 449
    types of DNA sequences in, 450*f*
Human Genome Project, 87, 347, **443**–444, 449
Human growth hormone (HGH), 418*f*, 419, 436, 1009–1010
Human immunodeficiency virus (HIV). *See* HIV
Human nutrition. *See also* Animal nutrition
    assessing nutritional needs for, 902*f*
    bacteria in, 587
    dietary deficiencies in, 902*f*
    essential nutrients for, 899*f*, 900*t*–901*t*
    transgenic crops and, 438–439
Human papillomavirus (HPV), 974*f*
Human population
    density-dependent population regulation of, by diseases, 1204*f*
    global carrying capacity for, 1209, 1210*f*, 1211
    growth of, 1207*f*–1210*f*, 1211, 1213
    survivorship curves for, 1194*f*
Human reproduction. *See also* Animal reproduction
    conception, embryonic development, and birth in, 1034, 1035*f*–1037*f*
    contraception and abortion in, 1037, 1038*f*, 1039
    endocrine disruptors and, 1015
    female reproductive organs of, 1026, 1027*f*
    gametogenesis in, 1027, 1028*f*–1029*f*
    hormonal regulation of, 1030, 1031*f*–1032*f*, 1033–1034
    male reproductive organs of, 1025*f*, 1026
    maternal immune tolerance of embryo and fetus in, 1037
    reproductive technologies and, 1039*f*, 1040
    sexual response in, 1034
Humans (*Homo sapiens*). *See also* Human body; Human embryonic development; Human environmental impacts; Human genetics; Human genome; Human nutrition; Human population; Human reproduction
    analyzing polypeptide sequence data for monkeys and, 87, 89
    as anthropoid apes, 746–747

apoptosis of white blood cells of, 229f
biodiversity and welfare of, 1262, 1263f
biological species concept and, 507f
blood pH of, 52f, 53
blood pressure of, 931f
brown algae as food for, 603
catabolism and diets of, 182f, 183
chromosomes of, 102, 235, 236f, 237, 256f–257f, 258
cilia in windpipes of, 13f
circulatory systems of. *See* Cardiovascular systems
cloning of, 430
derived characters of, 748
determining gene function by analyzing genomes of, 427f, 433–434
digestive system of. *See* Digestive systems
DNA microarray assays on tissue of, 426f
DNA sequencing of genome of, 415f, 416
ecosystem interaction with, 11
essential elements and trace elements for, 29t
ethical issues on silencing gene expression in, 427
evolution of culture in, 1159f, 1160
fossils of, 753f, 754
gene flow with Neanderthals and, 752f–753f
genomics and proteomics in study of, 86, 87f
glycoproteins and blood types of, 131
as *Homo sapiens*, 748f, 749
importance of insects to, 713
importance of seed plants to welfare of, 651f, 652
inactivated olfactory receptor genes of, 489
lymphatic systems of, 933f
origin of, 533f, 537
overlap of Neanderthals and modern, 33
plastic waste and, 1277
practical uses of fungi for, 670, 671f
primate phylogenetic tree and, 746f
prokaryotic impacts on, 587, 588f–591f
red algae as food for, 609f, 610
reducing hunger and malnutrition in, with transgenic crops, 837, 838f
regulation of breathing in, 946f
regulation of molecular interactions in, 10f, 11
relationship of Neanderthals and, 752f–753f
sex chromosomes of, 298f, 299
skulls of chimpanzees and, 558
small intestine surface area for, 695f
sustainable development in Costa Rica and living conditions of, 1284, *1285f*
transgenic crops and health of, 839
urban, 1274
water gain and loss in, 998
Humboldt Current, 1168f
Hummingbird hawkmoth, 712f
Hummingbirds, 6, 522, 740f, 825f, 869f
Humoral immune response, 957f, **963**, 964f–967f
Humpback whales, 903f
Humus, **806**–807
Hundred Heartbeat Club, 1262f
Hunger, transgenic crops and reducing human, 837, 838f
Huntington's disease, **287**, 431–433
Hurricane Katrina, 1172
Hutton, James, 469f, 471
Hybrid breakdown, 509f
Hybridization, **271**, 518, 837, 839–841
Hybrid orbitals, 39f, 40
Hybrids, **508**
bears, 510f
DNA in, 524
reproductive barriers and sterility of, 509f
speciation rates and, 521f, 522
sterility of, 522–523
Hybrid zones, **516**, 517f–519f, 520
Hydrangea, 282f
Hydras, 255f, 687f, 691f–693f, 904f, 1086f
Hydration shells, **49**
Hydrocarbons, **60**, 61f, 671f
Hydrocarbon tail, chlorophyll, 194f
Hydrochloric acid, 51, 907f, 908
Hydrogen
covalent bonding and, 36f–37f
dinitrophenol and, 186
electronegativity of, 45
electrons of, 35
as essential element, 29t, 64
in organic compounds, 59f
oxidation of organic molecules containing, 166–167

in plant composition, 809
in saturated and unsaturated fats, 73f, 74
Hydrogen bonds, **39**
in DNA structure, 86f
floating of ice and, 48f
in water molecules, 44, 45f–46f
as weak chemical bonds, 38, 39f
Hydrogen ions, **51**, 52f, 53
Hydrogenosomes, 600f
Hydrogen peroxide, 112
Hydrogen sulfide gas, 540
Hydrolagus colliei, 727f
Hydrolysis, **67**
of ATP, 151f, 152
disassembling of polymers to monomers by, 67f
enzymatic, 904, 908f
by lysosomes, 107f, 108
Hydrolytic enzymes, fungal, 655
Hydronium ions, **51**, 52f, 53
Hydrophilic substances, **49**–50, 127f
amino acids as, 76, 77f
Hydrophobic interaction, **81**f
Hydrophobic substances, **49**–50, 127f
amino acids as, 76, 77f
Hydroponic culture, **810**f
Hydrostatic skeletons, **1133**f
Hydrothermal vents, **526**f, 585, 587, 592, 914f, **1182**f
Hydroxide ions, **51**, 52f, 53
Hydroxyl group, 63f
Hydrozoans, 692f–693f
Hyla versicolor, 500f, 514
Hylonomus, 735f
Hymen, 1026
Hymenoptera, 712f
Hypercholesterolemia, 139
Hypermastigote, 614f
Hyperosmotic solutions, 977–978, 991
Hyperpolarization, **1072**f–1073f
Hypersensitive response, plant, **866**, 867f
Hypertension, **939**
Hypertonic solutions, **134**f
Hyphae, fungal, 654, 655f–**656**f, 818
Hypoblast, 1052f–1053f
Hypocotyl, **829**f–830f
Hypoosmotic solutions, 977–978
Hypothalamus, **888**, 1004f, **1007**, **1093**f
in homeostasis, 1088–1089
in human brain, 1093f
in kidney regulation, 994f–995f
in neuroendocrine signaling, 1007f–1008f
in regulation of mammalian reproduction, 1030, 1031f–1032f
in stress responses, 1012f
suprachiasmatic nucleus (SCN) in, 893f, 1015, 1094–1095
in thermoregulation, 888, 889f
Hypotheses, **17**. *See also* Inquiry Figures; Scientific Skills Exercises
forming and testing of, in science, 17, 18f–19f
phylogenetic trees as, 564f, 565
theories and, 21–22, 483–484
Hypothyroidism, 1010
Hypotonic solutions, **134**f
Hyraccoidea, 745f
Hyracotherium, 548, 549f

I

Ibex, 1266
Ibuprofen, 62f, 1001, 1013–1014, 1112
Ice
floating of, on liquid water, 48f
fossils in, 529f
on Mars, 50f
as solid water, 44
Icosahedral viruses, 400f, 401, 408, 412
Identical DNA sequences, 453f
Identical twins, 1036, 1061, 1162
Idioblasts, 868f
IgE antibodies, 970f
Iguanas, 888, 998
Ileum, 909
Imaging, brain, 1085f, 1096f
Imaging techniques, fetal testing, 289
Imatinib, 435
Imbibition, **829**
Immigration, **1192**, 1196, 1197f, 1206f, 1207, 1256f
island equilibrium model and, 1232, 1233f

Immune defenses, plant, 866
Immune response
overview of, 957f
pathogen trigger of, 952f
primary and secondary, 962, 963f
trypanosome evasion of, 600–601
Immune systems, **953**
adaptations of pathogens to evade, 957, 971–972, 973f, 973t, 974, 976
adaptive immunity defense against infection in, 963, 964f–969f, 970
apoptosis in, 233
diseases of, 229
disruptions of, 970f–974f, 973t
HIV/AIDS and, 398f, 406. *See also* AIDS; HIV
immune rejection by, 969–970
immunization and, 968f, 976
innate immunity in, 952f–957f
leukocytes in, 934f–935f
lymphatic systems and, 933f
maternal immune tolerance of embryo and fetus during pregnancy, 1037
membrane carbohydrate cell-cell recognition in, 130f–131f
pathogen recognition in, 952f, 953, 957f–963f
in plants, 866
prostaglandins in, 1001
stem cells and, 431
trematode camouflage and, 696
Immunity
active and passive, 968
maternal immune tolerance of embryo and fetus during pregnancy, 1037
Immunization, **968**f, 976
Immunodeficiency diseases, 971, 972f, 973
Immunoglobulin (Ig), **958**f–961f, 966
Immunological memory, 960, 962, 963f
Impala, 212f, 216–217
Imprinting, **1144**, 1145f, 1152f–1153f
Inactivation, cell-signaling, 229
Inborn errors of metabolism, 336
Inborn immunodeficiency, 971
Inclusive fitness, **1157**, 1158f–1159f
Incomplete dominance, **279**f
Incomplete flowers, **824**
Incomplete metamorphosis, **711**, 712f
Independent assortment, law of, 274–**275**, 276f, 296, 297f, 302
Independent assortment of chromosomes, 265f
Independent variables, **21**, 513
Indeterminate cleavage, **681**, 682f
Indeterminate growth, **766**f–767f
Indian corn, 451f
Indian Ocean Subtropical Gyre, 1168f
Indian pipe, 819f
Indian rice, 1263
Indoleacetic acid (IAA), 848. *See also* Auxin
Indolebutyric acid (IBA), 849
Indonesia, 1254f
Induced fit, 155f, **156**
Induced pluripotent stem (iPS) cells, 431, 432f
Inducers, **368**f, 369
Inducible enzymes, 368f, 369
Inducible operons, 368f, 369
Induction, 382f, **383**, **1054**
Inductive reasoning, **17**
Inductive signals, cell fate determination and pattern formation by, 1061, 1062f–1063f, 1064
Industrialization, human population growth and, 1207f
Industrial toxins, 1275, 1276f
Inert elements, 35
Infant mortality, 1208–1209, 1284, 1285f
Infection
adaptive immunity defense against, 963, 964f–969f, 970
bacterial, 218f
cytotoxic T cell response to, 966f, 967
fungal, 670
inflammation in, 956–957
Infection thread, bacterial, 816f, 817
Inferences, 1030
Infertility, 1039f
Inflammation, 937f, 938–939, 1013–1014, 1111–1112
systemic and chronic, 956–957
Inflammatory response, **955**f–956f
Inflorescences, **824**

Influenza viruses
    antibody proteins and, 79*f*
    antigenic variation of, 973
    in density-dependent population regulation, 1204*f*
    as emerging viruses, 410
    immune system recognition and response to, 952
    oseltamivir for, 414
    structure of, 400*f*
    as zoonotic, 1234, 1235*f*
Information
    alleles and, 293
    B and T cells and DNA, 976
    biotechnology and, 441
    chromosomes and, 268, 313
    DNA structure and inheritance, 334, 364
    in eukaryotic cells, 93*f*
    F_1 hybrids, 524
    genetic material as, 6, 7*f*–8*f*
    genome and nervous system, 1106
    genomics, 8–9
    heritable, 1162
    in mitosis, 252
    phylogeny reconstruction with, 572
    as theme of biology, 2, 3
Information flow, intercellular, 1000*f*–1001*f*
Information processing
    cerebral cortex in, 1096, 1097*f*
    ion movements and gradients and, 993*f*
    neuron, 1067–1068, 1069*f*
    problem solving and, 1146, 1147*f*
Infrared receptors, 1111*f*
Ingestion, **902**, 903*f*
Ingroups, **561***f*
Inhalation, 944, 945*f*
Inheritance. *See also* Mendelian inheritance; Sexual
      life cycles
    of cancer predisposition, 394
    chromosomal basis of. *See* Chromosomal basis of
      inheritance
    of chromosomes and genes, 255
    Darwinian theory on, 475*f*, 476
    C. Darwin on, 14*f*–16*f*
    DNA structure and, 334, 364
    epigenetic, 372*f*, 373
    expression and transmission of genetic
      information in, 7*f*–8*f*
    genetics as study of, 255. *See also* Genetics
    genetic variation and, 487
    genomic imprinting and, 310*f*, 311
    Lamarck's theory on, 471
    learning and, 1162
    molecular basis of. *See* Molecular basis of
      inheritance
    of organelle genes, 311*f*
    of X-linked genes, 299*f*, 300
Inheritance of acquired characteristics principle, 471
Inherited variation, **255**
Inhibin, 1031*f*
Inhibiting hormones, 1004*f*, 1008
Inhibition, 158–161*f*
Inhibitors, 160*f*
Inhibitory postsynaptic potential (IPSP), **1078**
Initials, cell, 766
Initiation factors, 352*f*
Initiation stage
    regulation of transcription, 373*f*–378*f*, 379
    regulation of translation, 378–379
    transcription, 343*f*
    translation, 352*f*
Inland mouse (*Peromyscus polionotus*), 20*f*–21*f*
Innate behavior, **1143**
Innate immunity, **953**
    antimicrobial peptides and proteins in, 953–954,
      955*f*, 956–957
    barrier defenses of, 953–954
    cellular innate defenses of, 954, 955*f*
    inflammatory response of, 955*f*–956*f*
    invertebrate, 953*f*–954*f*
    molecular recognition by, 953
    overview of, 957*f*
    pathogen evasion of, 957
    pathogen recognition in, 952*f*
    systemic and chronic inflammation in, 956–957
    vertebrate, 954*f*–956*f*, 957
Inner cell mass, **1052**, 1053*f*
Inner ear, **1113***f*
Innocence Project, 24*f*, 437*f*

Inorganic components, topsoil, 806, 807*f*
Inositol triphosphate (IP₃), **225***f*
Inquiry, scientific, **16**. *See also* Inquiry Figures;
      Research Method Figures; Scientific inquiry
Inquiry Figures . *See* list, xxii–xxiii
Insecticide resistance, 496, 518
Insects
    alimentary canals in, 905*f*
    anatomy and features of, 710*f*–711*f*
    antiviral defense of, 954*f*
    axis formation in, 1060
    body plans of, 545*f*, 546
    camouflage of, 468*f*, 476*f*
    cleavage in, 1048
    CO₂ absorption and, 1245
    compound eyes of, 1117*f*, 1118
    CRISPR-Cas9 and gene editing of, 427
    defense mechanisms of, 28*f*
    evolution by natural selection in, due to food
      source changes, 477*f*, 478
    exoskeletons of, 1133
    eyes of, 894*f*
    flower pollination by, 649*f*, 822*f*–825*f*
    in fossil record, 529*f*
    gamete production and delivery in, 1024*f*
    gas exchange of, 951
    herbivory evolution in, 551
    Hox genes in, 463*f*–464*f*
    importance of, for humans, 713
    innate immunity in, 953*f*–954*f*
    insecticide resistance in, 496
    malaria and, 606*f*
    Malpighian tubules of, 985*f*
    mechanoreceptors and hearing in, 1112*f*
    nervous systems of, 1086*f*
    neuroendocrine coordination in, 1006*f*, 1007
    nonheritable variation in, 488*f*
    open circulatory systems of, 923*f*
    organogenesis in, 1055*f*
    parasitic, 1220
    phylogeny and diversity of, 689*f*, 708, 709*f*, 712*f*,
      713
    plant defenses against, 867, 868*f*–869*f*
    pollination by, 522, 641*f*
    sex chromosomes of, 298*f*
    sex determination of, 298*f*
    taste and smell in, 1123
    thermoregulation in, 887*f*
    tracheal systems for gas exchange in, 941, 942*f*
    wildfires and, 1244*f*
    zoopagomycete infection of, 662*f*
Insertions (mutations), 358*f*, **359**
In situ hybridization, **423**, 424*f*
Instability, free-energy change and, 147, 148*f*
Insulation, animal thermoregulation and, 885
Insulin, **916**, 1004*f*
    in diabetes mellitus, 916–917
    exocytosis and, 139
    in glucose homeostasis, 916*f*
    in glucose metabolism, 10*f*, 216
    neonatal diabetes and, 359
    as polypeptide, 1001*f*
    production of, by biotechnology, 436
    as protein, 76*f*
    in regulation of appetite and consumption, 917*f*
    rough ER and, 105
    stem cells and, 431
Insulin-dependent diabetes, 917
Insulin-like growth factor gene (*Igf2*), 310*f*, 311
Insulin-like growth factors (IGFs), 1009
Integral proteins, **129**
Integration
    cellular, 121*f*
    sensory, 1068*f*, 1108*f*, 1109
Integrins, **119***f*, 129
Integument, **638***f*, 826
Integumentary systems, **885**
Intelligence, 1106
Interactions
    Cambrian explosion and feedback regulation,
      685
    climate change and positive feedbacks, 1189
    diatoms and, 617
    in eukaryotic cells, 93*f*
    feedback mechanisms, 397
    between hormones and environment, 1018
    interspecific. *See* Interspecific interactions

across kingdoms and domains, 813*f*
    mycorrhizal, 657
    within organisms, 10*f*
    between organisms and physical environment,
      10*f*–11*f*
    oxygen transport, 951
    of pancreatic cell, 142
    physical environment and evolution, 485
    phytochromes and shoot growth, 871
    of plants with other kingdoms, 648, 649*f*
    prokaryotic ecological, 586*f*, 587
    species, 1214*f*–1221*f*, 1237
    as theme of biology, 2, 3, 9
Intercalated disks, 879*f*, 1131
Intercellular communication. *See* Cell signaling
Intercellular information flow, 1000*f*–1001*f*
Intercellular joining function, membrane protein,
    130*f*
Interdisciplinary research teams, 9
Interferons, **957**
Intermediate disturbance hypothesis,
    1228, 1229*f*
Intermediate filaments, 100*f*–101*f*, 102, 113*t*, **117**
Intermediate sexual (intersex) characteristics, 299
Intermembrane space, 110
Internal environments, animal, 874, 875*f*, 876,
    881*f*–883*f*
Internal factors, cell cycle control system, 246,
    247*f*–248*f*
Internal fertilization, 1022–1023, 1151. *See also*
    Fertilization, reproductive
International Union for Conservation of Nature and
    Natural Resources (IUCN), 1261
Internet resources, genome-sequence, 444,
    445*f*, 446
Interneurons, 1068, 1069*f*, 1088*f*
Internodes, **761**
Interphase, 234, **237***f*–238*f*, 252
Interphase nucleus transcription factories, 377, 378*f*
Interpret the Data, 21*f*, 23, 29*t*, 33, 47*f*, 50, 54, 58, 89,
    99, 136, 157, 158*f*, 179, 186, 194*f*, 205, 264,
    501*f*, 505, 540*f*, 562*f*, 567*f*, 592, 635, 717,
    773*f*, 804, 951, 963*f*, 998, 1018, 1021*f*, 1083,
    1102*f*, 1115*f*, 1138, 1149*f*, 1171*f*, 1189, 1205*f*,
    1213, 1217, 1243*t*, 1246*f*. *See also* Scientific
    Skills Exercises
Intersex characteristics, 299
Intersexual selection, **499***f*–500*f*, 1152
Interspecific competition, 1215, 1216*f*–1217*f*, 1237
Interspecific interactions, 1214*f*, **1215**
    character displacement and, 1216,
      1217*f*, 1237
    competition, 1215, 1216*f*–1217*f*, 1237
    exploitation, 1217, 1218*f*–1219*f*, 1220
    genomics and proteomics in, 88*f*
    positive interactions, 1220*f*–1221*f*
Interspecific mating, 507*f*. *See also* Mating
Interstitial fluid, **875***f*, 923, 932, 933*f*, 956*f*, 981,
    988–989, 990*f*, 991
Intertidal zones, **1181***f*
Intestinal bacteria, 912, 913*f*
Intestines, 587, 695*f*
Intracellular digestion, 904
Intracellular receptors, cell-signaling, 220,
    221*f*, 1018
Intracellular recording, 1072*f*
Intrasexual selection, **499***f*, 1152
Intrauterine devices (IUDs), 1038*f*
Intrinsic (physiological) factors, density-dependent
    population regulation by, 1204*f*
Intrinsic rate of increase, **1197**
Introduced species, 818, 1223, **1264***f*, 1287
Introns, 345*f*–347*f*, **346**, 449, 488*f*
Invagination, 1050*f*
Inversions, chromosome, **307**, 308*f*
Invertebrates, **683**, 687
    action potential conduction speed in, 1076
    chordates, 719*f*–722*f*
    cnidarians, 687*f*, 691*f*–693*f*
    deuterostomes, echinoderms, and chordates,
      689*f*, 713, 714*f*–715*f*
    digestive systems of, 717
    ecdysozoans and arthropods, 688*f*–689*f*, 705*f*–
      712*f*, 713. *See also* Arthropods; Ecdysozoans
    gamete production and delivery in, 1024*f*
    hydrostatic skeletons of, 1133*f*
    innate immunity in, 953*f*–954*f*

lophotrochozoans, 687f–688f, 694, 695f–704f, 705. *See also* Lophotrochozoans
mechanoreceptors in, to sense gravity, 1112
nervous systems of, 1086f, 1087
neuroendocrine signaling in, 1006f, 1007
organogenesis in, 1055f
osmoregulation in, 979, 980f
parental care in, 1023f
phylogeny and diversity of, 687f–689f
sponges, 687f, 690f, 691
In vitro culturing, angiosperm, 836
In vitro fertilization (IVF), 436, 1039f
In vitro mutagenesis, **426**
Iodine, 29
Iodine deficiencies, 1009
Ion channel proteins, 1076
Ion channel receptors, 220f
Ion channels, **135**, **1070**
    action potentials and gated, 1072f–1077f
    in mechanoreceptors, 1110
    neuron resting potential and, 1070f–1071f
Ionic bonds, **37**, 38f
Ionic compounds (salts), **38f**. *See also* Salts
Ion movements and gradients, in life processes, 993f
Ionotropic receptors, 1077, 1078f, 1079–1080
Ion pumps, 137, 138f, 1070f–1071f
Ions, **37**, 38f, 934f, 993f, 1070t
IP₃ (inositol triphosphate), **225f**
Iridium, 541
Iris, **1118f**
Iron
    as limiting nutrient in aquatic biomes, 1243t
    plant deficiency in, 810f
Iron overload, 920
Iron oxide, 534
Irrigation, 142, 808, 821
Island equilibrium model, 1232, 1233f
Island species, 483
Isolated systems, 145, 149f
Isoleucine, 77f
Isoleucine synthesis, 161f
Isomers, **61f–62f**
Isomorphic generations, **604**
Isoosmotic solutions, 977–978
Isopods, 709
Isotonic solutions, **134f**
Isotopes, **31**, 32f, 33, 190
Italy, age-structure pyramid for, 1209f
Iteroparity, **1201f**
Ivanowsky, Dmitri, 399
Ivory, 1265f

J

J. *See* Joule
Jackson, Rob, 1223f
Jacob, François, 366, 547
Jacoby, Gordon C., 773f
Japan, 1208
Japanese canopy plant (*Paris japonica*), 448t, 449
Japanese snails, 522
Jasmine, 855
Jasmonates, 846t, **855**
Jawfish, 1151f
Jawless vertebrates, 723f–725f
Jaws
    mammalian, 530, 531f–533f
    snake, 498, 499f
    vertebrate, 718, 723, 725f, 726
Jefferson, Thomas, 437
Jejunum, 909
Jellies (jellyfish), 687f, 691f–692f, 922f
Jellyfish gene, 342f
Jenner, Edward, 968
Jirtle, Randy, 372
Joints, human, 1134f
Joly, John, 794
Jost, Alfred, 1030
Joule (J), **46**, 890
J-shaped exponential growth curve, 1197f
Jumping genes. *See* Transposable elements
June solstice, 1167f
Juniper, 643f
Junk DNA. *See* Noncoding DNA
Juvenile hormone (JH), 1006f, 1007
Juxtaglomerular apparatus (JGA), **995**, 996f
Juxtamedullary nephrons, **986f**, 991

K

Kangaroo rat (*Dipodomys merriami*), 741f, 998
Kangaroos, 743, 744f, 1135f, 1153f, 1178
Kaposi's sarcoma herpesvirus, 974
Kartagener's syndrome, 1064
Karyogamy, **658f**, 662f
Karyotypes, **256f**, 289f, 308f, 332f
Katydids, 712f
Kaufman, D. W., 22–23
kcal. *See* Kilocalorie
Kelp perch (*Brachyistius frenatus*), 1203f
Kelps, 603, 804, 1189
Keratin, 76f, 337, 735, 920
Kestrels, 1201f
Ketone compounds, 63f
Ketoses, 68f
Keystone species, **1226**, 1270
Kidneys, **985**
    adaptations of vertebrate, to diverse environments, 741, 991f–993f
    evolution of, 991, 998
    homeostatic regulation of, 994f–996f
    hormonal regulation of, 1007f, 1008
    human cystic kidney disease, 1064
    osmoregulation by, in aquatic animals, 979f
    processing of blood filtrate in nephrons of, 987, 988f, 989
    solute gradients and water conservation by, 989, 990f, 991
    structure of, in mammalian excretory systems, 985, 986f–987f
Killifish, 483
Kilocalorie (kcal), **46**, 890
Kimberella, 535f, 536, 694
Kinases, 219f, 220. *See also* Protein kinases; Receptor tyrosine kinases
Kinetic energy, **46**, **144**, 145f, 163
Kinetochore microtubules, 238f, 240f–241f, 244f, 252
Kinetochores, 238f, **240f–241f**
Kinetoplastids, **600**, 601f
Kinetoplasts, 600
King, Mary-Claire, 394
King, Thomas, 428
Kingdoms, taxonomy, 12f, **554**, 555f, 568
King penguins (*Aptenodytes patagonicus*), 740f, 884f, 1192f
Kin selection, 1157, **1158f–1159f**
Kissimmee River Restoration Project, 1254f
Klinefelter syndrome, 309, 1040
Knee-jerk reflex, 1088f
KNOTTED-1 gene, 778f
Koalas, 743, 912f
Kombu, 603
Komodo dragon (*Varanus komodoensis*), 1042
Korarchaeota clade, 586
Krebs, Hans, 172
Krebs cycle. *See* Citric acid cycle
Krill, 710f
K-selection, **1202**
Kudzu, 1264f
Kuru, 412–413

L

Labeling, GM products, 438–439
Labia majora, **1026**, 1027f
Labia minora, **1026**, 1027f
Labor, childbirth and, 1036, 1037f
Lacks, Henrietta, 248–249
lac operon, 368f, 369
Lactase, 70
Lactate, 180f, 181
Lactation, 1036
Lacteals, 909f–**910f**
Lactic acid fermentation, **180f**, 181
Lactose, 69–70, 368f, 369
Laetoli footprints, 750f
Lagging strand, DNA, **324**, 325f
Lagomorpha, 745f
Lake Erie water snake (*Nerodia sipedon*), 496, 497f
Lakes, 1177f–1179f, 1275, 1276f
    primary production in, 1242–1243
Lake Vesijärvi, 1227f
Lake Victoria, 515f, 519f, 520
Lamarck, Jean-Baptiste de, 469f, 471
Laminaria, 603f, 604

Lampreys, 719f, **723f**, 724, 1091f
Lamp shells, 688f, 698f, 699
Lancelets, 689f, 715, 719f, **720**, 721f–722f
Land
    colonization of, 536, 537f
    global human use of, 1209
    locomotion on, 1135f
    subsidence of, 808f
Land plants. *See* Plants
Landscape ecology, **1165f**. *See also* Ecology
    biodiversity hot spots in, 1272f
    landscape fragmentation and edges in, 1270, 1271f
    movement corridors in, 1271f, 1272
    philosophy of nature reserves in, 1272–1273
    urban ecology and, 1273, 1274f
    zoned reserves in, 1273f
Landscapes, **1165f**
Land snails, 700f, 702f
Language
    brain function and, 1098f
    FOXP2 gene and, 461, 462f
lap alleles, 505
Larch, 643f
Large intestine, **910**, 911f–912f
Largemouth bass, 881f
Large-scale disturbances, 1229f
Large tree finch, 519
Larva, **674f**, 711f
Larval dispersal, 538
Larynx, 906f, **942**, 943f
La Sagrada Familia towers, 146f
Late-acting dominant lethal alleles, 293
Latency, viral, 973t
Lateral geniculate nuclei, 1121f
Lateral inhibition, 1120
Lateralization, **1098**
Lateral line system, 725–726, 728f, **1116f**
Lateral meristems, **766f**
Lateral roots, **759f**, 769f
Latitude
    animal size and, 897
    sunlight intensity and, 1166f
Latitudinal gradients, community diversity and, 1232f
Law of conservation of mass, **1239**
Law of independent assortment, 274–**275**, 276f, 296, 297f, 302
Law of segregation, 271f–275f, **272**–273, 296, 297f
Laws of probability, 276, 277f, 278
Laws of thermodynamics, 143, 145f, 146
Laysan albatross (*Phoebastria immutabilis*), 1277f
L-dopa, 1104
Leaching, 806
Leading strand, DNA, **324**, 325f
Leaf area index, 786f, 787
Leaf-cutter ants, 668f, 813f
Leaf primordia, 769f, **770**, 771f
Leafy liverworts, 626f
Learned behaviors, 1147
Learning, **1144**
    associative, 1146f
    cognition, problem solving, and, 1146, 1147f
    development of learned behaviors, 1147
    inheritance and, 1162
    neuronal plasticity and, 1100, 1101f
    sleep and, 1094
    social, 1147f–1148f
Leaves (leaf), **630**, 758, **761**
    abscisic acid in abscission of, 852
    anatomy of, in C₄ plants, 204f
    auxin in pattern formation of, 849
    ecosystem interaction of, 10f, 11
    effects of transpiration on wilting and temperature of, 798
    ethylene in abscission of, 854f
    evolution of, in vascular plants, 630, 631f
    green color of, 193f
    Hox genes in formation of, 777, 778f
    leaf area index and arrangements of, 786f, 787
    monocot vs. eudicot, 649f
    photosynthesis in, 189f, 190
    structure of, 761f–762f
    tissue organization of, 770, 771f
Leber's congenital amaurosis (LCA), 1122
Leber's hereditary optic neuropathy, 311
Leeches, 703f, 704, 1086f

Leeuwenhoek, Anton van, 1209
Left atrium, 926f–927f
Left ventricle, 926f–927f
Leghemoblobin, 816
Legless lizards, 553f–554f
Legumes, 815f–816f, 817
Lemurs, 746f
Length, carbon skeleton, 60f
Lens, **1118f**, 1138
Lenski, Richard, 578f
Lenticels, **774**
Leopards, 554, 555f
Lepidopterans, 712f
Lepidosaurs, **736**, 737f, 738
Leprosy, 584f
Leptin, 917f, 918
Leptogenys distinguenda, 1001f
Lepus americanus, 501f, 1205f
Lettuce seed germination, 857f
Leucine, 77f
Leukemia, 309f, 394, 434, 1263f
Leukocytes (white blood cells), 229f, 878f, 934f–**935f**
Lewis, Edward B., 385–386
Lewis dot structures, 36, 37f, 43, 65
Leydig cells, 1031f
Lichens, 663, **668f**–669f, 672, 813f, 1256f–1257f
    bioremediation using, 1253, 1255f
Life. *See also* Animals; Organisms; Plants
    abiotic synthesis of organic molecules as origin
        of, 57f, 58
    biological molecules of, 66. *See also* Biological
        molecules
    biology as scientific study of, 3. *See also* Biology;
        Science
    carbon in organic compounds of, 64
    cell division as fundamental to, 235
    cells as fundamental units of, 5f, 6f, 93f. *See also*
        Cells; Eukaryotic cells; Prokaryotic cells
    cellular respiration and energy for, 164f, 165.
        See also Cellular respiration
    classification of diversity of, 12f–13f
    colonization of land by, 536, 537f
    conditions on early Earth and origin of, 525f–528f
    diversity of. *See* Biodiversity
    domains of, 12f–13f
    effects of speciation and extinction on, 537f–543f,
        544
    as emergent property, 125
    essential elements and trace elements for, 29t
    evolution of, as theme of biology, 11. *See also*
        Evolution
    extension of, 233
    fossil record as history of, 529f. *See also* Fossil record
    geologic record and, 532f–533f
    history of, as limitation of natural selection, 504
    importance of water for. *See* Water
    levels of biological organization of, 4f–5f
    order as property of, 146f, 147
    origin of multicellular, 533f, 535f–536f
    origin of single-celled, 532f–535f
    photosynthesis as process that feeds, 188f. *See also*
        Photosynthesis
    phylogenies as evolutionary history of, 554f–
        556f. *See also* Phylogenies
    possible evolution of, on planets with water, 50f
    properties of, 3f
    silicon-based, 65
    tree of, 15f–16f. *See also* Tree of life
    unifying biological themes in, 3f–11f
    unity in diversity of, 13f, 26, 469, 473
    viruses and characteristics of, 398f, 399, 408
    web of, 570f
Life cycles, **256**. *See also* Sexual life cycles
    of angiosperms, 646f, 647, 653, 826, 827f, 828
    of apicomplexan *Plasmodium*, 606f
    of blood fluke *Schistosoma mansoni*, 696f
    of brown alga *Laminaria*, 603f, 604
    of cellular slime mold *Dictyostelium*, 613f
    of ciliate *Paramecium caudatum*, 607f
    developmental events in, 1044
    of *Drosophila melanogaster* (fruit fly), 385f
    of fern as seedless vascular plant, 630f
    of fungi, 658f–659f, 662f, 664f, 666f
    of green algal chlorophyte *Chlamydomonas*, 611f
    of humans, 257f, 258
    of hydrozoan *Obelia*, 693f
    of moss, 624f

of pine tree and gymnosperms, 640f, 641
    reproduction and, 256
Life expectancy at birth, 1208–1209, 1284, 1285f
Life histories, population, **1200**, 1201f–1202f
Life tables, population, **1193t**
Ligaments, **878f**
Ligand binding, 217
Ligand-gated ion channels, 220f, **1077**, 1078f,
    1079–1080
Ligands, **217**, 223
Light chains, 958f–961f
Light-detecting organs, 1117f
Light detector, euglenid, 601f
Light energy
    absorption of, determining primary production
        with, 1242f
    bioluminescence as, 143
    in energy flow and chemical cycling, 9f, 164f,
        165, 1240f
    excitation of chlorophyll by, 195f
    global energy budget and, 1241
    in photosynthesis, 187f, 206, 207f. *See also* Light
        reactions
    primary production in aquatic ecosystems and
        limitations of, 1242
    properties of, 192f
    sunlight as, 143, 145, 150
Light energy, plant responses to
    biological clocks and circadian rhythms in, 857,
        858f, 859
    blue-light photoreceptors in, 855, 856f
    de-etiolation signal transduction pathway for,
        843f–844f, 845
    germination and, 871
    photomorphogenesis and action spectrum of, 855
    photoperiodism and seasonal responses in,
        859f–860f
    phototropism and, 847f–848f
    phytochromes as photoreceptors in, 856, 857f
    plant shoot architecture and, 785, 786f, 787
    stomatal opening and closing as, 797–798
Light-harvesting complexes, **196f**
Light microscope (LM), **94**
Light microscopy (LM), 94f–95f
Light reactions, **191**. *See also* Photosynthesis
    chemiosmosis of, in chloroplasts vs. in
        mitochondria, 199f–200f, 201
    cyclic electron flow in, 198f, 199
    determination of absorption spectrum for, 193f
    excitation of chlorophyll by light energy in, 195f
    linear electron flow in, 197f–198f
    most effective wavelengths for, 194f
    nature of sunlight and, 192f
    photosynthetic pigments as light receptors in,
        192, 193f–194f, 195
    photosystems of, 195, 196f–198f, 199
    as stage of photosynthesis, 191f, 192, 206, 207f
Light responses, rod cell, 1121f
Lignin, **629**–630, 665, **764f**, 783
Likens, Gene, 1252f
Lily, 650f
Limbic system, 1095f, 1096
Limbs
    genes for formation of, 545f, 546
    as homologous structures, 479f, 480
    tetrapod, 718, 730, 731f
    vertebrate formation of, 1062, 1063f, 1064
Limiting nutrients, **1242**, 1243f–1244f, 1243t
Limnetic zone, **1179f**
Limp cells, 790, 791f, 797f
Limpets, 547f–548f
LINE-1 retrotransposons, 452
Lineage-based mechanisms, 777–778
Linear electron flow, **197f**–198f
Line graphs in Scientific Skills Exercises, 33, 157, 264,
    411, 972, 1049, 1095, 1136, 1186, 1279
Linkage groups, 306f
Linkage maps, 305f–306f
Linked genes, **301**
    genetic recombination and, 302, 303f
    identifying, 304
    inheritance of, 301f–302f
    mapping of, 305f–306f
    natural selection and genetic variation from
        recombination of, 305
Linker DNA, 330f
Linnaean classification, 554, 555f

Linnaeus, Carolus, 470, 554
Lionfish, 729f
Lions, 912f
Lipid bilayers. *See* Phospholipid bilayers
Lipid rafts, 128
Lipids, 66, **72**
    in cellular membranes, 102, 110, 127f, 128
    evolution of differences in cellular membrane
        composition of, 129
    fats as, 72, 73f, 74
    phospholipids as, 74f, 75
    in plasma membranes, 98f, 99
    smooth ER synthesis of, 104–105
    steroids as, 75f
    Tay-Sachs disease and, 280
Lipid-soluble hormones, 1001f–1003f
Lipopolysaccharides, 574, 976
Literacy rate, Costa Rican, 1285
Litter decomposition, 1248f
Litter size, 1200, 1201f–1202f
Littoral zone, **1179f**
Liver, **909**
    in digestion, 909–910
    in energy storage, 915, 916f
    lowering plasma LDL levels by inactivating
        enzyme of, 938
Liverworts, 620f, **623**, 626f. *See also* Bryophytes
Living fossils, 633, 708f
Lizards, 553f–554f, 735, 737f, 738, 925, 1021f, 1154f,
    1216f, 1260f
    climate change and, 11f
Loams, **806**
Lobe-fins, 718, **728**, 729f, 730
Lobes, brain, 1096, 1097f, 1098
Lobotomy, 1098
Lobsters, 707f, 708–709, 979
Local biogeochemical cycles, 1249
Local cell signaling, 215f, 216
Local extinctions, 1262
Local inflammatory response, 955f–956f
Local regulators, **1000f**, 1001
Lock-and-key specificity, viral, 401
Locomotion, 993f, 1133f, **1135f**, 1136, 1138
Locomotor play, 1159
Locus, gene, **255**, 272
Locusts, 1213
Lodgepole pines, 1229f
Logarithms, natural, 639
Loggerhead turtles, 1194, 1195f
Logistic equation in Scientific Skills Exercise, 1200
Logistic population growth, 1197, 1198f–1199f, **1198t**,
    1200
Logos, DNA sequence, 351
Log scales in Scientific Skills Exercise, 1136
Lokiarchaeotes, 586
Long-day plants, **859f**
Long-distance cell signaling, 215f, 216
Long Island Sound, 505
Long-night plants, 859f, 860
Long noncoding RNAs (lncRNAs), 380–**381**
Long-term memory, **1100**, 1101f
Long-term potentiation (LTP), **1101f**
Looped domains, DNA, 331f, 332
Loop of Henle, **987f**–990f, 998
Loose connective tissue, 878f
Lophophores, 683f, **684**, 694, 698f, 699
Lophotrochozoans, **683f**, 684
    annelids, 703f–704f, 705
    characteristics of, 694
    ectoprocts and brachiopods, 698f, 699
    flatworms, 694, 695f–697f
    molluscs, 699f–702f
    phylogeny of, 687f–688f
    rotifers, 697f, 698
Loricifera, 688f
Lorises, 746f
"Lost City" vent field, 527f
Lovelock, James, 1259
Low-density lipoproteins (LDLs), 139, 141, **937**–938
Loxodonta africana, 474f, 1197f, 1265f
LSD, 1081
Luminal A and luminal B breast cancer, 393f
Luna moths (*Actias luna*), 43
Lung cancer, 447
Lung cells, newt, 7f, 238f–239f
Lung disease, 951
Lungfishes, 480f, 719f, 729

Lungs, 728, **942**
    gas exchange in, 924f, 925, 942, 943f–946f
    ventilation of, 944, 945f–946f
Lupines, 1255
Lupus, 971
Luteal phase, 1032f, 1033
Luteinizing hormone (LH), 1004f, 1008f, **1030**,
    1031f–1032f, 1033
lux genes, 397
Lycophytes, **622**, 622t, 631, 632f, 633, 635
Lyell, Charles, 469f, 471–473
Lyme disease, 584f, 588f, 1234, 1235f
Lymph, **933f**, 956f
Lymphatic systems, 910, **933f**, 956f
Lymph nodes, **933f**, 956f
Lymphocytes, 934f–935f, **958f**. *See also* B cells; T cells
Lymphoid stem cells, 935f
Lymph vessels, 933f, 956f
Lynx, 1205f
Lyon, Mary, 300f
Lysine, 77f
Lysogenic cycle, **402**, 403f
Lysosomal storage diseases, 108
Lysosomes, 100f, **107**f–109f
Lysozymes, 49f, 79f, 456, 457f, **953f**, 954
Lytic cycle, **402**f, 414

M

Macaques (*Macaca fuscata*), 897
MacArthur, Robert, 1232, 1233f
MacLeod, Colin, 315
Macroevolution, **507**, **526**. *See also* Evolution
    adaptive radiations in, 542, 543f, 544
    colonization of land in, 536, 537f
    of development, 544f–546f, 547
    early Earth conditions for origin of life in,
      526f–528f
    fossil evidence for, 525f, 526, 528, 529f–531f, 530
    gene flow, genetic drift, and natural selection in,
      551
    geologic record of key events in, 532f–533f
    mass extinctions in, 540f–542f
    novelties and trends in, 547f–549f
    origin of multicellular organisms in, 533f, 535f–536f
    origin of single-celled organisms in, 532f–535
    plate tectonics and, 538f–539f, 540
    speciation and extinction rates of organisms in,
      537f–543f, 544
Macromolecules, 66f–**67**f, 526f–527f, 802
    in lysosomes, 107f, 108
    plasmodesmata and, 846
Macronuclei, ciliate, 606, 607f
Macronutrients, plant, **810**, 811t
Macrophages, 107f, 121f, **878**f, **955f**, 956
Macular degeneration, age-related, 432
Madagascar orchids, 825f
Mad cow disease, 83, 412
Madreporite, sea star, 714f
MADS-box genes, 545, 778f
Maggot flies, 508f, 515–516, 903f
Magnesium, 29t, 810
Magnetic field, Earth's, 1111f, 1141
Magnification, 94
Magnificent frigatebird (*Fregata magnificens*), 1139f
Magnolia tree, 650f
Magnoliids, **649**, 650f
Maiden veil fungus, 665f
Maize (corn), 451f, 651
    alleles in, 293
    artificial selection of, 836f
    health of transgenic Bt, 839
    phylogeny of, 557
    precocious germination in, 852f
    proteins in, 899
    seeds, 829f–830f
Major depressive disorder, **1102**
Major histocompatibility complex (MHC) molecule,
    959f–960f, 963, 964f, 969–970
Make Connections Figures. *See* list, xxii
Malacidins, 589f
Malacosoma americanum, 897
Malaria, 286f, 287, 501, 503f, 599f, 606f, 614, 713
Malaysia, 1232f
Malaysian orchid mantis (*Hymenopus coronatus*),
    468f, 469
Male gametophytes, angiosperm, 827f
Males

competition between, for mates, 1153f, 1154
female mate choice and, 1151, 1152f–1153f, 1154
hormonal regulation of reproductive systems of,
    1031f
hormones of, 999f, 1014, 1015f
parental care by, 1151f
reproductive anatomy of human, 1025f, 1026
sex chromosomes of, 298f, 299
sexual competition between, 499f–500f
spermatogenesis in human, 1027, 1028f, 1031f,
    1042
Malignant tumors, **249**f
Malnutrition, 837, 838f, 901–902
Malpighian tubules, 710f, **985**f
Malthus, Thomas, 469f, 475
Maltose, 69f
Mammals, **741**
    adaptations of kidneys of, 991f, 992
    adaptive radiation of, 542, 543f
    amniotic eggs of, 734, 735f
    axis formation in, 1060
    bats as, 15f
    brains in nervous systems of, 1091f. *See also*
      Nervous systems
    breathing in, 925, 945f–946f
    cardiovascular systems of. *See* Cardiovascular
      systems
    cellular respiration in hibernating, 178
    circadian clocks in hibernating, 893f
    circulation in, 924f, 925. *See also* Cardiovascular
      systems
    comparing genomes of, 454f–455f, 459,
      460f–462f
    control of circadian rhythms in, 1094
    convergent evolution of, 481f, 744f
    derived characters of, 741f
    digestive systems of. *See* Digestive systems
    early evolution of, 741, 742f
    endangered or threatened, 1261, 1262f, 1288
    eutherians and primates as placental, 744,
      745f–747f
    evolution of, 679
    evolution of melanocyte-stimulating hormone
      in, 1016
    extraembryonic membranes of, 1053f, 1054
    fertilization in, 1046f
    gas exchange adaptations of, 947f–949f
    genomic imprinting in, 310f, 311
    glia in brains of, 1069f
    hearts of, 926f–928f
    hominins and humans as, 748f–754f
    homologous structures in, 479f
    hormonal regulation of reproduction in, 1030,
      1031f–1032f, 1033–1034. *See also* Animal
      reproduction; Human reproduction
    inactivation of X-linked genes in female, 300f
    ion concentrations inside and outside of neurons
      of, 1070t
    kidneys in excretory systems of, 985, 986f–987f.
      See also Excretory systems; Kidneys
    marsupials, 743f–744f
    mechanoreceptors for hearing and equilibrium
      in, 1112f–1116f
    modeling neurons of, 1071f
    molecular clock for, 567f
    monotremes, 742, 743f
    nitrogenous wastes of, 982f
    opiate receptors in brains of, 1082
    organ systems of, 876f
    origination of cetaceans as terrestrial, 481f–482f
    origin of, 530, 531f–533f
    osmoregulation in, 980–981
    phylogeny of, 719f, 745f
    reproductive cloning of, 430f–431f
    respiratory systems of, 942, 943f–946f
    sex chromosomes of, 298f, 299
    taste in, 1123f–1124f
    thermoregulation in, 881f
Mammary glands, 392f, 741, 1005f–1008f, 1016,
    1026–1027
Mammoths, 421
Manatee, 572, 1219f
Mandibles, 709, 710f
Mangold, Hilda, 1062f
Mantises, 468f
Mantle, **699**f
Mantle cavity, **699**f

Maple tree leaves, 762
Mapping
    of brain activity, 1096f
    linkage, 305f–306f
Map units, **305**f
Maquis, 1174f
Maraviroc, 217, 409
Marchantia, 620f, 626f
March equinox, 1167f
Marine animals
    adaptations of kidneys of, 992, 993f
    mass extinctions and, 541–542
    osmoregulation in, 978, 979f
Marine benthic zones, **1182**f
Marine biomes, 1177f, 1178, 1224f
    primary production in, 1242, 1243f, 1243t
Marine birds, 982f
Marine reserves, 1273f
Marine snails, 538
Marine worm, 940f
Mark-recapture method, **1191**f
Marmota caligata, 164f
Mars, 50f
Marshall, Barry, 912
Marshes, 1186
Marsh gas, 585
Marsupials, 481f, 540, 543f, **743**f–745f
Marsupium, 743f
Mass, 31
    conservation of, 1239–1240
    trophic levels and, 1240f, 1241
Mass extinctions, **540**f–542f, 1287
    current sixth, 702f
    of dinosaurs, 736
    evolution and, 653
    life on earth and, 525
    speciation and, 522
    tropical deforestation and potential, 651f, 652
Mass number, **31**
Mast cells, 955f, 956, 970f
Master regulatory genes, 545f–546f, 547. *See also*
    Homeotic genes; *Hox* genes
Masting, 869f
Mate choice, 499f–500f, 515f, 519, 1151, 1152f–1153f, 1154
Mate choice copying, **1152**, 1153f
Mate recognition, 508f
Maternal age, Down syndrome and, 308–309
Maternal alleles, 372
Maternal chromosomes, 265
Maternal effect genes, **386**
Maternal inheritance, 311
Maternal mRNAs, 387f
Mating. *See also* Reproduction
    animal reproduction and, 1020f
    cell signaling in yeast, 213f, 214
    clumped dispersion and, 1192
    earthworm, 705
    external fertilization and, 1023
    genetic disorders from human, 285–286
    Hardy-Weinberg equilibrium and random, 492t
    human, 1034
    human sexual arousal and, 1081–1082
    hybrid zones and, 516, 517f
    insect, 711
    interspecific, and hybrids, 507f
    of pea plants, 270f, 271
    reproductive barriers to, 507, 508f–509f, 510
    reproductive cycles and, 1021f
    sexual selection and, 499f–500f
Mating behavior. *See also* Courtship rituals
    applying game theory to, 1154f
    mating systems,1149–1150, 1151f
    sexual selection and mate choice in, 1151,
      1152f–1153f, 1154
Mating systems, 1150, 1151f
Matorral, 1174f
Matter, 2, 3, 9, **29**–30. *See also* Energy and matter
Matteuccia, 632f
Maturation-promoting factor (MPF), **245**, 246f
Mauna Loa monitoring station, 1278f, 1279
Maungatautari restoration project, 1254f
Maximum likelihood, **563**
Maximum parsimony, **562**, 563f, 564
Mayer, Adolf, 399
Maze experiments, 1146, 1147f
McCarty, Maclyn, 315
McClintock, Barbara, 451f, 459

M checkpoint, 245f–247f
Meadowlarks, 507f
Meadow voles, 1155f
Measles, 401, 409
Measles virus, 968f
Mechanical defense, prey, 1218f
Mechanical isolation, 508f
Mechanical signaling, 119
Mechanical stimuli, plant responses to, 861, 862f
Mechanical stress, plant responses to, 853f
Mechanical work, 150, 152f
Mechanoreceptors, **1110**f, 1112f–1116f
Mediator proteins, 374, 375f
Medical records, 448
Medicine. *See also* Drugs; Pharmaceutical products
    antibodies as tools in, 969f
    application of systems biology to, 447, 448f
    biotechnology in, 433, 434f–436f
    blocking HIV entry into cells in, 130f
    fungi in, 670
    genomics and proteomics in, 88f
    medical leeches in, 703f, 704
    plant-derived medicines in, 651t, 652
    radioactive tracers in, 31, 32f
    stem cells in, 431–432
    treatments of nervous system disorders in, 1102
Mediterranean climate, 1168
Medulla oblongata, 946f, **1093**f
Medusa, **691**f, 692
Medusozoans, 692f–693f
Megapascal (MPa), **789**
Megaphylls, **631**f, 635
Megasporangia, 638f
Megaspores, **631**, 638f, **826**, 827f
Megasporocytes, 826
Meiosis, **258**
    in animal cells, 260f–261f
    crossing over and synapsis during, 260f, 262f
    DNA changes of yeast cells in, 264
    errors in, 306, 307f–309f
    gamete formation by, in sexual life cycles, 258
    genetic variation from gene alteration during, 489
    genome evolution and errors in, 455f–457f
    human gametogenesis and, 1027
    in human life cycle, 254, 257f, 258
    mitosis vs., 262, 263f, 264
    production of gametes by, 237
    stages of, 259f–262f
    in varieties of sexual life cycles, 258f, 259
Meiosis I, **259**f–260f, 262f–263f, 264
Meiosis II, **259**f, 261f, 263f, 264
Melanerpes formicivorus, 1162
Melanin, 337
Melanocyte-stimulating hormone (MSH), 1004f, 1008f, **1016**, 1018
Melatonin, 883f, 1004f, **1015**
Melitaea cinxia, 1206f, 1207
Membrane attack complex, 966f
Membrane potentials, 137, 138f, **1069**, 1070f–1071f.
    See also Action potentials, neuron; Resting potentials, neuron
Membrane proteins, 105, 127f–130f, 152f
Membranes, amniotic egg extraembryonic, 734, 735f
Membranes, cellular. *See* Cellular membranes
Memory
    emotion and, 1096
    formation of, 1085
    neuronal plasticity and, 1100, 1101f
    sleep and, 1094
Memory cells, **962**
Mendel, Gregor
    experimental, quantitative approach of, 270f, 271
    genes as hereditary factors of, 294f
    law of independent assortment of, 274–275, 276f
    law of segregation of, 271f–275f
    particulate model of inheritance of, 487
Mendelian inheritance
    chromosomal basis of, 294f, 295
    environmental impacts on phenotypes and, 282f
    evolution of gene concept from, 361
    exceptions to, 310f–311f
    extending, for multiple genes, 281f–282f
    extending, for single gene, 278, 279f–280f
    genetic variation and, 487
    human patterns of inheritance and, 284f–289f, 290

integrating, with emergent properties, 282–283
    law of independent assortment of, 274–275, 276f
    law of segregation of, 271f–275f
    laws of probability governing, 276, 277f, 278
    limitations of, 278
    making histograms and analyzing distribution patterns for, 283
    G. Mendel's experimental quantitative approach, 270f, 271
Menopause, **1033**
Menstrual cycle, **1032**f, 1033–1034
Menstrual flow phase, 1032f, 1033
Menstruation, **1032**f
Meristems, 766f–767f, 830, 849
Meroblastic cleavage, 1048
Merozoites, 606f
Meselson, Matthew, 322f
Mesenchyme cells, 1055
Mesoderm, **680**, 1050f–**1051**f
Mesoglea, 691f
Mesohyl, **690**f
Mesonychoteuthis hamiltoni, 702
Mesophyll, **189**f, 203, 204f, 206f, **771**f
Mesozoic era, 530f, 533f, 539f, 679
Messenger RNA (mRNA), **339**. *See also* RNA
    alteration of ends of, 345f
    in analyzing gene expression, 423, 424f–428f
    bicoid, 387f
    in breast cancer, 392f
    cell fractionation and, 125
    effects of miRNAs and siRNAs on, 380f
    in gene expression, 8f, 84f
    genetic code and, 340f, 341
    maternal, 387f
    mutations affecting, 357f–358f, 359
    in plant cells, 208f
    polyribosomes and, 355f
    regulation of degradation of, 379
    synthesis of, in cell signaling, 226f
    synthesis of, in eukaryotic cell nucleus, 102
    synthesis of, in transcription, 342, 343f–344f
    in transcription and translation, 337, 339f
    in translation, 347f–350f, 352f
    viral, 401f, 402, 405f, 406t
Metabolic defects, 336f–338f
Metabolic pathways, **144**. *See also* Metabolism
    evolution of, 163
    metabolic defects in, 336f–338f
    regulation of bacterial, 366f
Metabolic rates, **890**f–893f
Metabolism, **144**. *See also* Bioenergetics
    ATP energy coupling of exergonic and endergonic reactions in, 150, 151f–153f
    bioenergetics and metabolic rates of animal, 890f–893f
    catabolism and, 182f, 183
    effects of adrenal hormones on, 1012f, 1013–1014
    enzymatic catalysis of reactions in. *See* Enzymatic catalysis
    evolution of hormones regulating, 1016
    forms of energy for, 144, 145f
    free-energy change, equilibrium, and, 147, 148f–150f
    graphing reactions of, 157
    laws of thermodynamics and, 145f–146f, 147
    metabolic pathways of, 144
    nitrogenous wastes and, 983
    osmoregulation and, 980
    prokaryotic, 581t, 582f, 587
    protocell, 527, 528f
    radioactive tracers in research on, 31–32
    regulation of cellular respiration and, 183f, 184
    role of enzymes as catalysts in, 75, 76f
    thermogenesis in animal, 887f–888f
    thyroid regulation of, 1009, 1015f
Metabotropic receptors, 1080
Metagenomics, **444**, 583
Metamorphosis, **674**
    amphibian, 732f
    frog, 1015f
    insect, 711f, 1006f, 1007
    lancelet, 720
    tunicate, 721
Metanephridium, 699f, 704f, **985**f
Metaphase (mitosis), **237**, 239f, 243f, 252, 263f, 331f
Metaphase chromosomes, 331f–332f
Metaphase I, 260f, 263f

Metaphase II, 261f
Metaphase plate, 239f–**240**f, 262
Metapopulations, **1206**f, 1207
Metastasis, **249**f, 447–448
Metazoans (Metazoa), 682, 683f, 684
Meteorites, 527
Methamphetamine, 62
Methane
    carbon and bonds in, 59f
    combustion of, as redox reaction, 166f
    covalent bonding in, 37f
    molecular shape of, 39f, 40
Methanogens, **585**–586
Methanosarcina barkeri, 448t
Methicillin, 478f
Methicillin-resistant *Staphylococcus aureus* (MRSA), 214, 478f, 574, 589f, 590
Methionine, 77f, 341f
Methods, research. *See* Research Method Figures
Methylated compounds, 63f
Methylation, DNA, **371**, 372f, 373, 391, 430
Methyl group, 63f
Methyl jasmonate, 855
Methylsalicylic acid, 867f
Mexico, 1208
MHC (major histocompatibility complex) molecule, **959**f–960f, 963, 964f, 969–970
Mice. *See* Mouse
Microarray analysis, 392f
Microarray chips, human genome, 448f
Microbial diversity, 1223f
Microbiomes, **912**, 913f
Microcephaly, 409
Microclimate, **1169**–1170
Microevolution, **487**, **507**. *See also* Evolution
    adaptive evolution by natural selection in, 497, 498f–503f, 504
    alteration of allele frequencies by natural selection, genetic drift, and gene flow in, 493, 494f–497f
    genetic variation and, 487f–488f, 489
    populations as smallest units of, 486f–487f
    of sickle-cell disease, 503f
    speciation as conceptual bridge between macroevolution and, 507, 523. *See also* Macroevolution
    using Hardy-Weinberg equation to test, 489, 490f–491f, 492t, 493
Microfibrils
    cellulose, 777f
    in plant cell walls, 118
    in structural polysaccharides, 70f
Microfilaments, **115**
    animal cell, 100f
    in animal cytokinesis, 241, 242f
    cytoskeleton structure and function and, 113t
    in morphogenesis, 1056f
    plant cell, 101f
    structure and function of, 115, 116f–117f
Microglia, 1090f
Micronuclei, ciliate, 606, 607f
Micronutrients, plant, **810**, 811t
Microphylls, **631**f
Microplastics, 1163, **1277**
Micropyles, **647**, 826
MicroRNAs (miRNAs), **380**f, 391, 678
Microscopy, 94f–95f, 96
Microsporangia, 638f, 826
Microspores, **631**, 638f, **826**, 827f
Microsporidians, 660f–**661**f
Microsporocytes, 826
Microtubule-organizing center, 240
Microtubules, **114**
    in animal cell, 100f
    in cell division, 238f–241f, 244f
    centrosomes, centrioles, and, 114f
    cilia, flagella, and, 114, 115f–116f
    cytoskeleton structure and function and, 113t
    in plant cell, 101f
    structure and function of, 114f–116f
Microvilli, 98–99, 100f, 116f, 695f, **909**f, 910
Midbrain, 1085, **1091**f–1092f
Middle ear, **1113**f
Middle lamella, **118**
Midrib, 761
Mifepristone (RU486), 1039
Migration, **1140**

electromagnetic receptors and, 1111*f*
  as fixed action pattern, 1140, 1141*f*
  genetic variation in patterns of, 1156, 1157*f*
  movement corridors and, 1271*f*, 1272
Milk, mammalian, 741–742, 743*f*, 1005*f*–1008*f*, 1016
Milk duct, 392*f*
Milkweed, 645*f*, 653
Miller, Stanley, 57*f*, 58, 526*f*
Millipedes, 708, 709*f*
Mimic octopus, 1219*f*
Mimicry
  endorphin, 1081
  molecular, 40*f*, 78
  in prey and predator adaptations, 1218*f*–1219*f*, 1237
  as prey defensive adaptation, 868*f*
Mimivirus, 408
Mimosa pudica, 802, 862*f*
Mimulus species, 522*f*
Mineralized dental elements, 724*f*, 725
Mineralized organic matter, 529*f*
Mineralocorticoids, 1004*f*, 1012*f*, **1014**
Minerals, 899*f*, **900**, 901*t*
  deficiencies of, in plants, 810, 811*t*, 812
  mineralocorticoids and metabolism of, 1014
  mycorrhizae and plant deficiencies of, 818
  root architecture and acquisition of, 787
  roots interaction with, 10*f*, 11
  transpiration of, from roots to shoots via xylem, 792, 793*f*–795*f*, 796
  vascular plant transport of, 787, 788*f*–791*f*
Mines, restoration of, 1253*f*
Miniaturization, gametophyte, 637*f*, 638
Minimal medium, 336*f*–338*f*
Minimum viable population (MVP), **1268**
Minnows, 1143*f*, 1276
Mirounga angustirostris, 999*f*, 1267
Miscarriages, 306, 1039
Misfolding, protein, 83
Mismatch repairs, **327**
Missense mutations, **357**, 358*f*
Mississippi River, 1275*f*
Mistletoe, 819*f*
Mitchell, Peter, 177
Mites, 707, 708*f*
Mitochondria, **109**
  animal cell, 100*f*
  animal hibernation and, 178
  in apoptosis, 230*f*, 231
  ATP synthase and, 186
  chemical energy conversion by, 109–110, 111*f*
  chemiosmosis in, 175, 176*f*, 177, 199*f*–200*f*, 201
  dinitrophenol and, 186
  electron transport chains in, 168–169
  endosymbiotic origin of, 534, 535*f*
  enzymes in, 161*f*
  evolutionary origins of, 109, 110*f*
  fungal cell, 100*f*
  inheritance of genes of, 311
  origin of, in endosymbiosis, 595, 596*f*–597*f*
  plant cell, 101*f*, 209*f*
  protist, 594
  pyruvate in, 171*f*
  using cell fractionation to study, 97
Mitochondria DNA (mtDNA)
  evolutionary rate of, 565
  species identity in, 557*f*
Mitochondrial matrix, **110**
Mitochondrial myopathy, 311
Mitosis, **236**. *See also* Cell cycle; Cell division
  in animal cells, 238*f*–239*f*
  in chromatin packing, 332*f*
  in daughter cells, 234
  evolution of, 243, 244*f*
  in human life cycle, 257*f*, 258
  information in, 252
  meiosis vs., 262, 263*f*, 264
  nuclear envelope during, 252
  origin of term for, 237
  in plant cells, 241, 242*f*–243*f*
  in spermatogenesis, 1042
  in varieties of sexual life cycles, 258*f*, 259
Mitosomes, 600*f*
Mitotic (M) phase, **237***f*
Mitotic spindles, **240***f*–241*f*
Mixotrophs, **594**
Mobile genetic elements, evolution of viruses and, 408
Mobile River basin, 1264

Model organisms, **22**, **1044**. *See also Drosophila melanogaster; Escherichia coli* bacteria
  Arabidopsis thaliana, 22, 774, 776*f*, 783
  bread mold. *See Neurospora crassa*
  Caenorhabditis elegans, 22, 705
  in developmental biology, 1044
  for DNA research, 316*f*, 317
  mouse (*Mus musculus*), 22. *See also* Mouse
  Neurospora crassa, 664*f*, 665*t*
  scientific cooperation and, 22–23
  for T. Morgan's experiments, 295*f*
Models
  atomic, 30*f*
  community disturbance, 1228, *1229f*
  of covalent bonds, 36, 37*f*
  electron orbital, 35*f*
  exponential population growth, 1196, 1197*f*, 1207*f*, 1208
  island equilibrium, 1232, 1233*f*
  logistic population growth, 1197, 1198*f*–1199*f*, 1198*t*, 1200
  molecular-shape, 39*f*
  optimal foraging, 1149*f*
  process of science, 19*f*
  testing hypotheses with quantitative in Scientific Skills Exercise, 1150
Modified leaves, 762*f*
Modified roots, 760*f*
Modified stems, 761*f*
Mojica, Francisco, 253
Molarity, **50**
Molar mass, 50
Molar ratios in Scientific Skills Exercise, 58
Mold model organism. *See Neurospora crassa*
Molds, **658***f*–659*f*, 662*f*, 663
Molecular basis of inheritance
  chromatin packing of DNA and proteins in eukaryotic chromosomes, 330*f*–332*f*
  discovery of double helix structure of DNA in, 314*f*, 317, 318*f*–320*f*
  DNA replication and repair in. *See* DNA replication
  evidence for DNA as genetic material in, 315*f*–317*f*, 318
  evolution of gene concept from, 361
Molecular biology
  Arabidopsis thaliana as model organism for, 774, 776*f*
  determination of microbial diversity using, 1223*f*
  importance of viruses to, 400
  measures of evolution in, 87, 89
  mutants in, 845
  of plant development, 775*f*–780*f*
Molecular clocks, **566**, 567*f*–568*f*, 659*f*, 660
Molecular formulas, 36, 37*f*, 59*f*
Molecular genetics, in ecological forensics, 1265*f*
Molecular genetic variation, 487, 488*f*
Molecular homologies, 479–480, 558, 559*f*
Molecular identification tags, 107
Molecular-level herbivore defenses, plant, 868*f*
Molecular mass, **50**
Molecular recognition, immune system, 953, 957*f*
Molecular systematics, 563*f*, 583, 682, 683*f*, 684. *See also* Cladistics; Systematics; Taxonomy
Molecular tags, 332*f*
Molecules, 5*f*, **36**. *See also* Compounds
  biological. *See* Biological molecules
  chemical bonds and formation of. *See* Chemical bonds
  as level of biological organization, 5*f*
  organic. *See* Organic compounds
  origin of self-replicating, 526*f*–528*f*
  regulation of interactions of, 10*f*
  shape and function of, 39*f*–40*f*, 59*f*
  structure of DNA and RNA, 7*f*–8*f*
Moles, 1107*f*–1108*f*
Moles (animal), 558*f*, 559
Moles (mol), **50**, 58
Molluscs, 688*f*
  bivalves, 699, 701*f*–702*f*
  body plan of, 699*f*
  cephalopods, 699, 701*f*, 702
  chitons, 699*f*
  in Ediacaran period, 535*f*, 536
  eye complexity in, 547*f*–548*f*
  gastropods, 542, 699, 700*f*–701*f*, 702
  nervous systems of, 1086*f*

protecting freshwater and terrestrial, from extinction, 702*f*
Moloch horridus, 737*f*
Molothrus ater, 1271, 1287
Molting (ecdysis), 683–684, **705**, 707, 894*f*
Monarch butterflies (*Danaus plexippus*), 839, 1146*f*
Monera kingdom, 568
Monilophytes, **622**, 622*t*
Monkey flowers, 522*f*, 834
Monkeys, 56*f*, 87, 89, 746*f*–747*f*, 1122*f*. *See also* Chimpanzees
Monocilia, 1064
Monoclonal antibodies, **969**
Monocots, **649***f*–650*f*, 761, 769*f*–770*f*, 829*f*–830*f*
Monocytes, 934*f*–935*f*
Monod, Jacques, 366
Monogamous mating, 1024, **1149**–1150, 1151*f*
Monoglycerides, 910*f*
Monohybrid crosses, **274**
Monohybrids, **274**
Monomers, **67***f*
Monophyletic groups, **560***f*
Monosaccharides, **68***f*–69*f*
Monosodium glutamate, 1123
Monosomic cells, **307**
Monosomy X, 309
Monotremes, 543*f*, **742**, 743*f*, 745*f*
Monozygotic twins, 1036, 1061
Monterey County, California, 614*f*
Montmorillonite clay, 527, 528*f*
Montreal Protocol, 1284
Moon jelly, 922*f*
Moose, 1205*f*
Moray eel (*Gymnothorax dovii*), 729*f*, 1214*f*
Morels, **663***f*, 670
Morgan, Thomas Hunt, 295*f*–296*f*, 301*f*–303*f*, 315
Mormon tea, 642*f*
Morning-after birth control pills, 1038*f*
Morning sickness, 65, 1036
Morphine, 40*f*, 78
Morphogenesis, 381–**382**, **1049**. *See also* Embryonic development; Pattern formation
  apoptosis in, 1056–1057
  cytoskeletons in, 1056*f*–1057*f*
  developmental adaptations of amniotes in, 1053*f*, 1054
  gastrulation in, 1044, 1049, 1050*f*–1053*f*
  organogenesis in, 1044, 1049, 1054*f*–1055*f*
  plant development and, 776–777, 778*f*
Morphogen gradients, 386
Morphogens, **386***f*–387*f*
Morphological homologies, 558
Morphological isolation, 508*f*
Morphological species concept, **510**
Morphology
  animal phylogeny and, 682, 683*f*, 684
  fungal, 655*f*–656*f*
  macroevolution of, 544*f*–546*f*, 547
  species concepts and, 507
Mortality rates. *See* Deaths
Morton, Michael, 24*f*
Mosaicism, 300*f*
Mosquitoes, 409, 412*f*, 485, 496, 503*f*, 518, 606*f*, 712*f*, 713
Mosquitofish (*Gambusia hubbsi*), 511*f*, 512
Mosquito larva, 342*f*
Mosses, 621*f*, **623**, 624*f*–627*f*, 628, 635, 637*f*. *See also* Bryophytes
Mossy leaf-tailed gecko, 26
Moths, 43, 476*f*, 551, 712*f*, 717, 824*f*, 1001, 1006*f*, 1007, 1110, 1111*f*
Motile cilia, 1064
Motility, prokaryotic, 576*f*, 577
Motor, flagellum, 576*f*, 577
Motor areas, cerebral cortex, 1096
Motor cortex, 1097*f*, 1098
Motor neurons, 1068, 1069*f*, **1088***f*, 1128, 1129*f*–1130*f*
Motor output stage, 1068*f*
Motor proteins, 76*f*, **113***f*, 115, 116*f*, 123*f*, 152*f*, 241*f*, 252
Motor systems, **1088**
  cardiac and smooth muscle in, 1131–1132
  muscle function in, 1125, 1126*f*–1131*f*, 1132
  sensory systems and, 1107*f*–1108*f*. *See also* Sensory systems
  skeletal muscle contraction in, 1127*f*–1131*f*
  skeletal systems and locomotion in, 1132*f*–1135*f*, 1136

Motor unit, **1130**f
Mountain lions, 1149
Mountain pine beetles (*Dendroctonus ponderosae*), 1244f, 1245, 1280f
Mountains, 1168, 1169f
Mount Kilimanjaro, 1189
Mount St. Helens, 636f, 637
Mouse (mice), 22
    agouti gene and, 372f
    appetite regulation in, 918
    brains of, 1086f
    comparing human genome with genome of, 454, 455f, 460f–462f
    complete genome sequence for, 454, 455f
    density-dependent population regulation of, 1204f
    energy budgets of, 892
    FOXP2 gene evolution in, 461, 462f
    genomic imprinting of insulin-like growth factor gene of, 310f, 311
    homeotic genes in, 463f
    as model organisms. *See Mus musculus*
    modes of natural selection in, 498f
    osmotic homeostasis in desert, 981
    paw development of, 231f
    transfer of genetic trait between bacterial strains in, 315f
Mouse (*Peromyscus polionotus*), camouflage case studies with, 2f, 20f–21f, 23
Mouse model organism, 22
Mouth formation, 1050f
Movement, prokaryotic, 576f, 577
Movement corridors, **1271**f, 1272
MPF (maturation-promoting factor), **245**, 246f
mPGES-1 gene, 376
MRSA, 214, 478f
Mucoromycetes, 660f, **663**f
Mucous cells, 907f
Mucus, **906**, 907f, 954
Mucus escalator, 943
Mule deer, 871, 1149
Mules, 509f
Muller, Hermann, 360
Müllerian mimicry, 1218f, **1219**
Multicellular asexual reproduction, 255f
Multicellular organisms, 5f, 93, 533f, 535f–536f, 675f–676f
Multienzyme complexes, 161
Multifactorial characters, **282**f
Multifactorial disorders, human, 287
Multigene families, 452, **453**f
Multiple fruits, **831**f
Multiple myeloma, 65
Multiple sclerosis, 971
Multiplication rule, **277**f, 278
Multiprotein complexes, 174f, 175
Mumps, 409
Murchison meteorite, 527
Muscardinus avellanarius, 893f
Muscle
    cardiac and smooth, 1131–1132
    contraction of, 1126, 1127f, 1128
    regulation of contraction of, 1128, 1129f–1130f
    skeletal. *See* Skeletal muscle
Muscle cells, 117f, 383, 384f, 673, 915
Muscle fibers, 879f, 1130, 1131t
Muscle tissue, **879**f
Muscular dystrophy, 299, 433
Mushrooms, 655, 665f, 665f–667f, 670. *See also* Fungi
Mus musculus (mouse), 22. *See also* Mouse
Mussels, 505, 701f, 702f, 1264
Mustard plant model organism. *See Arabidopsis thaliana*
Mutagens, **360**
Mutant phenotypes, 295f–296f
Mutants
    designing experiments using genetic, 1095
    interpreting data from experiments with genetic, 918
    in molecular biology, 845
    nutritional, in gene-enzyme relationship experiment, 336f–338f
Mutations, **357**f
    in alterations of chromosome structure, 454f–455f
    in aquaporins causing diabetes insipidus, 995f
    in cancer development, 391f, 394

cancer genes and, 388, 389f
cellular slime mold, 613
CRISPR-Cas9 and gene editing for, 426, 434–435
of developmental genes, 545f–546f
in duplication and divergence of gene-sized regions, 455f–457f
in duplication of entire chromosome sets, 454
effects of, during cell division, 389f–390f
as embryonic lethals, 386
as errors in proofreading, 327
evolution and rate of, 334
evolution of enzymes by, 159f
in exon duplication and exon shuffling, 456, 457f
in flowering, 779f–780f
gene editing to correct, 360, 361f
genetic variation from, 305f–306f
genome evolution and, 454f–457f, 458–459
Hardy-Weinberg equilibrium and, 492t
of ion channel protein genes, 1076
in mitochondrial DNA, 311
molecular clock speed and, 566, 567f
mutagens as cause of, 360
natural selection and, 328
nucleotide-pair substitutions, insertions, and deletions, 357, 358f, 359
phenotypes and, 295f–296f, 306f
plants and creation of, in molecular biology, 776
point mutations, 357f–358f, 359
in prokaryotes, 573, 578–579
as source of alleles, 265
as sources of genetic variation, 488–489
transposable elements and, 459
of viruses, 409, 414
Mutualism, **587**, **1220**, 1256f–1257f
    bacterial, 587f
    in flower pollination, 825f
    fungal, 655, 661f, 667f–669f. *See also* Mycorrhizae
    as interspecific interaction, 1220f, 1221
    across kingdoms and domains, 813f
    mycorrhizae as plant-fungi, 787, 817, 818f
    nutrient limitations and, 1244
    plant-bacteria, 813f–816f, 817
    in vertebrate digestive systems, 912, 913f–914f
Myasthenia gravis, 1129
Mycelium (mycelia), 655f–659f, **656**, 662f
Mycetozoans, 612f–613f
Mycobacterium tuberculosis, 589, 957
Mycorrhizae, **656**, **787**, **817**
    basidiomycetes in, 665
    biological augmentation using, 1255
    in colonization of land by plants, 536, 537f
    evolution of, 660
    genomic analysis of interactions of, 657
    as mutualism, 1220–1221
    nutrient limitations and, 1244
    plant nutrition and, 817, 818f
    plant roots and, 760
    as root-fungi mutualism, 787
    specialized hyphae in, 656f, 657
    strigolactones and, 855
    terrestrial plants and, 621
Mycorrhizal associations, 760
Mycosis, **670**
Myelination, 1076f–1077f
Myelin sheath, **1076**f–1077f
Myeloid stem cells, 935f
Myllokunmingia fengjiaoa, 718f, 724
Myoblasts, 383, 384f
Myocardial infarctions, 937f, 938–939
MyoD activator, 374f
myoD gene, 383, 384f
Myofibrils, **1126**f
Myoglobin, **949**, **1130**, 1131t
Myopathy, 311
Myosin, 76f, **117**, 241, 384f, 879f
Myosin filaments, 1125, 1126f–1127f
Myotonia, 1034, 1076
Myriapods, **707**–708, 709f
Myrmecocystus, 485
Mytilus edulis, 505
Myxini (hagfishes), 719f, 723f
Myxobacteria, 213f
Myxococcus xanthus, 213f

N

NAD+ (nicotinamide adenine dinucleotide), 167f–168f, 172f–173f, 178, 180f, 181

NADH, 172f–173f, 178, 180f, 181
NADP+ (nicotinamide adenine dinucleotide phosphate), **191**f, 192, 206, 207f
NADPH, **191**f, 192, 197f–198f, 206, 207f
Naked mole rats, 1157f, 1158
Naloxone, 1082
Nanopores, 416
Nasal glands, marine bird, 982f
Nash, John, 1154
National Cancer Institute, 447
National Center for Biotechnology Information (NCBI), 444, 445f
National Institutes of Health (NIH), 444, 447
National Library of Medicine (NLM), 444
National Medal of Science, 944
Native Americans, 565
Natural family planning, 1037, 1038f
Natural killer (NK) cells, **955**
Natural logarithms in Scientific Skills Exercise, 639
Natural plastics, 590, 591f
Natural range expansions, 1183, 1184f
Natural selection, **14**, **473**. *See also* Adaptations; Evolution
    adaptations and, 472, 473f
    adaptive evolution and, 494
    Darwinian theory of descent with modification by, 14f–16f, 469–470, 473f–476f, 483–484
    of developmental genes, 545
    directional, disruptive, and stabilizing selection in, 497, 498f
    of ecological niches, 1215, 1216f
    evolution of drug resistance by, 478f
    evolution of enzymes by, 159f
    in evolution of life history traits, 1200, 1201f–1202f
    genetic variation for, from genetic recombination, 305
    Hardy-Weinberg equilibrium and, 492t
    insect evolution by, due to food source changes, 477f, 478
    key role of, in adaptive evolution, 498, 499f
    limitations of, in creating perfect organisms, 501f, 504
    in macroevolution, 551
    mutations and, 328
    predator role in, 26
    relative fitness and, 497
    of ribozymes, 528
    sexual reproduction, genetic variation, and, 266, 267f
    species selection as, 548–549
Natural vs. supernatural explanations, 18
Nature reserves
    philosophy of, 1272–1273
    zoned reserves and, 1273f
Nature vs. nurture, 282f, 762
Nautilus, chambered, 701f, 702
ncRNAs, 379, 380f, 381
Neanderthals (*Homo neanderthalensis*), 33, 88f, 461, 752f–753f
Near vision, 1122f
Nectar guides, 824f
Nectaries, 813f
Nectarine, 645f
Negative and positive correlations, 834
Negative feedback, **882**, **1005**
    in density-dependent population growth, 1203
    in endocrine system feedback regulation, 1005–1006
    in feedback regulation, 10f
    in homeostasis, 882
Negative gene regulation, bacterial, 369
Negative gravitropism, 861
Negative pressure breathing, **945**f, 946
Neisseria gonorrhoeae, 576, 584f
Nematocysts, **691**, 692f
Nematode model organism. *See Caenorhabditis elegans*
Nematodes, 689f, 705f, 706, 1058f–1059f, 1133f
Nemertea, 688f
Nemoria arizonaria, 488f
Neodenticula seminae (diatom), 1170
Neon, 35f
Neonatal diabetes, 359
Neoproterozoic era, 530f, 676f–677f
Neornithes, 739
NEP. *See* Net ecosystem production
Nephrons, **986**f

Bowman's capsule in, 987f
evolution adaptations of, 991f–992f
processing of blood filtrate to urine by, 987, 988f, 989
solute gradients and water conservation in, 989, 990f, 991
structure of mammalian kidneys and, 986f–987f
Nereimyra punctata, 703f
Neritic zones, **1182**f
Nernst equation, 1071f
Nerodia sipedon, 496, 497f
Nerve cells, 673
Nerve cord, 710f, 720f, 722f
Nerve gas, 1081
Nerve nets, 692, 1086f
Nerves, **1086**
Nervous systems, **880**, **1000**. See also Central nervous system; Peripheral nervous system
cerebral cortex control of voluntary movement and cognitive functions in, 1096, 1097f–1099f
disorders of, 1102f–1104f
endocrine system coordination with, 1000f, 1001, 1006f–1008f
faulty apoptosis in diseases of human, 231
genome and, 1106
Huntington's disease and, 287
long-distance cell signaling in, 215f, 216
memory, learning, and changes of synaptic connections in, 1099, 1100f–1101f
neurons and cell signaling in, 880f. See also Neurons
organization of, 1086f, 1087
regulation of blood pressure by, 931
regulation of digestion by, 915
regulation of heart rhythm by, 928
regulation of human breathing by, 946f
regulation of skeletal muscle contraction by, 1128, 1129f, 1130
research methods for studying brains and, 1085f
synaptic and neuroendocrine signaling of, 1000f, 1001. See also Neuroendocrine signaling
vertebrate, 1087f–1090f
vertebrate brain in, 1091f–1096f
Nervous tissue, **879**f
Nests
of birds and dinosaurs, 564f, 565
red-cockaded woodpecker, 1269f–1270f
Net ecosystem production (NEP), **1242**
climate change effects on, 1244f, 1245
Net primary production (NPP), **1241**, 1242f, 1259
climate change effects on, 1244f, 1245
in terrestrial ecosystems, 1243f–1244f, 1245
Neural crest, **722**, 723f, 1054f, **1055**
Neural pathways, 1121f
Neural plate, 1054f
Neural tube birth defects, human, 902f
Neural tubes, 1054f, **1055**
Neuraminidase, 410, 414
Neurodegenerative diseases, 413
Neuroendocrine signaling. See also Endocrine systems; Nervous systems
coordination of endocrine and nervous systems in, 1000f, 1001, 1006f–1008f
endocrine glands and hormones for, 1004f
feedback regulation in of, 1005–1006
invertebrate, 1006f, 1007
simple pathway of, 1005f
vertebrate, 1007f–1008f
Neurofibrillary tangles, 1104f
Neurohormones, 1000f, **1001**, 1005f–1007f
Neuromuscular junctions, 1080
Neuronal plasticity, **1100**f
Neurons, **879**f, **1068**
action potentials of, as signals conducted by axons, 1072f–1077f
in cell signaling by animal nervous systems, 880f
chemical and electrical signals of, 1067f, 1068
communication between cells and, at synapses, 1077, 1078f–1080f, 1081t, 1082
exocytosis and, 139
in human eye, 1118f
ion pumps, ion channels, and resting potential of, 1070f–1071f
in nervous systems, 1085f, 1086, 1088f, 1089. See also Nervous systems

in neuroendocrine signaling, 1000f, 1001
olfactory, 1124, 1125f
organization of, 1084
plasticity of, in memory and learning, 1099, 1100f–1101f
in sensory reception, 1108f–1109f
structure and function of, in information transfer, 1068f–1069f
Neuropathy, 311
Neuropeptides, **1081**t
Neurospora crassa (bread mold), 336f–338f, 664f, 665t
Neurotransmitters, **1001**, **1068**
as chemical messengers of neurons, 1068
exocytosis and, 139
mechanisms of terminating, 1080f
synaptic signaling by, 1077, 1078f, 1080
types of, 1080–1082, 1081t
Neurulation, 1054f, 1055
Neutralization, 965f
Neutral mutations, 567
Neutral variation, **489**
Neutrons, **30**f, 31
Neutrophils, 934f–935f, **955**f
Newborn screening, 289–290
New Caledonian crows, 1099
New Guinea, 742
Newton, Isaac, 22
Newts, 7f, 238f–239f
New World monkeys, 746f–747f
New York City, Catskill Mountain purchase by, 1263
New Zealand, restoration projects in, 1254f
Next-generation DNA sequencing, 415f–417f, 443f
Nicotinamide adenine dinucleotide (NAD$^+$), **167**f–168f, 172f–173f, 178, 180f, 181
Nicotinamide adenine dinucleotide phosphate (NADP$^+$), **191**f, 192, 206, 207f
Nicotine, 1080, 1103f, 1220
Night length, flowering and, 859f, 860
Nirenberg, Marshall, 341
Nitrates, 1275f
deforestation effects on, 1252f
in nitrogen cycle, 1251f
Nitric oxide (NO), 220, 224, 931, **1001**, 1026, 1081t, 1082
Nitrification, 815f
Nitrifying bacteria, 815f, 1251f
Nitrite, in nitrogen cycle, 1251f
Nitrogen
algal blooms and, 1242, 1243f
bacteria in plant acquisition of, 814f–816f, 817
bryophyte reduction of leaching of, from soil, 627f
as essential element, 29t, 64
as limiting nutrient, 1242, 1243f, 1243t, 1244
nutrient enrichment and pollution by, 1275f
in organic compounds, 59f
prokaryotic chemical recycling of, 586
soil, in ecological succession, 1230, 1231f
soil fertilization and, 805, 808
Nitrogen cycle, **815**f, 821, 1251f, 1257f
Nitrogen fixation, **582**, 586, **815**f
bacterial, 814f–816f, 817, 821, 1244
biological augmentation using, 1255
bryophyte, 626f, 627
forest floor moss and, 635
lichens and, 668–669
in nitrogen cycle, 1251f
prokaryotic, 582, 592
Nitrogenous bases, 84, 85f, 317f, 318. See also Nucleotides
Nitrogenous wastes, 977–978, 982f–983f
Nitrogen oxide emissions, 1266
NMDA receptors, 1101f
NO (nitric oxide), 220, 224, 931, **1001**, 1026, 1081t, 1082
Nobel Prize winners
B. Marshall and R. Warren, 912
B. McClintock, 451
C. Nüsslein-Volhard, E. Wieschaus, and E. Lewis, 386
for discovery of ncRNAs, 380
for discovery of Toll receptor in insects, 954
E. Sutherland, 216
F. Jacob, 547
F. Sanger, 416
G. Beadle and E. Tatum, 337
H. zur Hausen, 974

J. Gurdon and S. Yamanaka, 430, 432
J. Hall, M. Rosbash, and M. Young, 883
J. Watson, F. Crick, and M. Wilkins, 320
N. Tinbergen, 1140
P. Mitchell, 177
R. Axel and L. Buck, 1124
S. Brenner, R. Horvitz, and J. Sulston, 1058–1059
S. Prusiner, 413
Nociceptors, **1111**–1112
Nodes, lymph, 933f
Nodes, plant stem, **761**
Nodes of Ranvier, **1076**f–1077f
Nodules, **816**f, 817
Nomarski microscopy, 95f
Nonbreeding adults, territoriality and, 1204f
Noncoding DNA, 449, 450f–451f, 452
Noncoding RNAs (ncRNAs), 379, 380f, 381
Noncompetitive inhibitors, **158**, 159f
Nondisjunction, **307**f, 309
Nonequilibrium model, community, **1228**
Nonheritable variation, 488f
Nonidentical DNA sequences, 453f
Non-insulin-dependent diabetes, 917
Nonkinetochore microtubules, 238f, 240f, 241, 252
Non-native species, 1264f
Nonpolar covalent bonds, **37**
Nonpolar side chains, 76, 77f
Nonrenewable resources, human population size and, 1210f, 1211
Nonsense mutations, **358**f, 359
Nonshivering thermogenesis, 887f
Nonsister chromatids, 257f, 262, 264, 266f
Nonspontaneous processes, 146
Nonsteroidal anti-inflammatory drugs (NSAIDs), 1013–1014
Nonsteroid hormones, 376
Nontemplate DNA strand, 340f
Nonvascular plants. See Bryophytes
Norepinephrine (noradrenaline), 1004f, **1012**f–1013f, 1081t
Nori, 609f, 610
Normal range, homeostatic, 882
North America, 743–744
North Atlantic Subtropical Gyre, 1168f
Northern coniferous forests, **1175**f
Northern red maple tree leaves, 762
Northern white rhinoceros (*Ceratotherium simum cottoni*), 1266
North Pacific Subtropical Gyre, 1168f
Nosema ceranae, 661
Notation system, gene, 295
No-till agriculture, **809**
Notochords, **720**f, 1054f, 1055
Novelties, evolutionary, 547, 548f
NPP. See Net primary production
N-terminus, 78f, 352, 371f
Nucifraga columbiana, 1146
Nuclear envelopes, 100f–101f, **102**, 103f, 109f, 252, 332, 339f, 355
Nucleariids, 614, **659**
Nuclear lamina, **102**, 103f, 117, 332
Nuclear magnetic resonance (NMR) spectroscopy, 83
Nuclear matrix, 102, 326, 332
Nuclear responses, cell-signaling, 226f
Nuclear transplantation, animal cloning and, 428, 429f–430f
Nucleases, **327**
Nucleic acid hybridization, **416**
Nucleic acid probes, **423**, 433
Nucleic acids, **84**. See also DNA; RNA
components of, 84, 85f
digestion of, 908f
genes, nucleotides, and, 84f. See also Genes
as genetic material, 315. See also DNA
as macromolecules, 66
as nucleotide polymers, 85
roles of, in gene expression, 84f. See also Gene expression
separating, with gel electrophoresis, 420f
structures of molecules of, 86f
viral, 400, 401f–407f, 406t
Nucleoids, **97**f, **577**
Nucleolus, 100f–101f, **102**, 103f
Nucleomorphs, 597f
Nucleosides, 84, 85f
Nucleoside triphosphates, 324
Nucleosomes, **330**f

Nucleotide excision repairs, **327**, 328*f*
Nucleotide-pair insertions and deletions, 358*f*
Nucleotide-pair substitutions, **357**, 358*f*
Nucleotides, **84**. *See also* Nucleic acids
    coding and noncoding, 345*f*–347*f*
    as components of nucleic acids, 84, 85*f*
    in DNA-sequencing techniques, 415*f*–417*f*
    DNA vs. RNA, 337
    evolutionary significance of altered DNA, 328
    in genetic code, 7*f*–8*f*, 340*f*–341*f*
    genomics and proteomics in study of, 86, 87*f*
    mutations as base-pair substitutions, insertions,
        and deletions of, 357, 358*f*, 359
    ratios of, 317*f*, 318
    in telomeres, 328, 329*f*
    variability of, in genetic variation, 488*f*
Nucleus, atomic, **30***f*, 31
Nucleus, cell, **102**
    animal cell, 100*f*
    cell division of, 236, 243, 244*f*. *See also* Cell cycle;
        Cell division
    cell-signaling responses in, 226*f*
    ciliate types of, 606, 607*f*
    DNA in eukaryotic cell, 97, 102, 103*f*
    fungal cell, 100*f*
    hormone receptors in, 1002*f*–1003*f*
    plant cell, 101*f*
    regulation of gene expression and architecture of,
        377, 378*f*
    reproductive cloning by transplantation of
        eukaryotic, 428, 429*f*–430*f*
Nucleus accumbens, 1096*f*, 1103*f*
Nudibranchs, 700*f*, 1020*f*
Number, offspring, 1200, 1201*f*–1202*f*
Nursing, 1005*f*–1008*f*, 1036
Nurture vs. nature, 282*f*, 762
Nüsslein-Volhard, Christiane, 386, 387*f*
Nutrient cycling. *See also* Energy flow and chemical
        cycling
    biogeochemical cycles, 1249*f*–1252*f*
    decomposition and nutrient cycling rates in,
        1248*f*, 1249
    in Hubbard Brook Experimental Forest, 1252*f*
Nutrient enrichment, global, 1274, 1275*f*
Nutrients
    cycling of. *See* Energy flow and chemical cycling
    enrichment experiments with, 1242, 1243*f*, 1243*t*
    essential, 899*f*, 900*t*–901*t*, 901*f*
    limiting, 1242, 1243*f*–1244*f*, 1243*t*
    plant and animal absorption of, 895*f*
    prokaryotic recycling of, 586*f*
    small intestine and, 125
    uptake of, 898
Nutrition, **899**. *See also* Animal nutrition; Plant
        nutrition
    essential elements and trace elements for, 29*t*
    Eukarya kingdoms and, 12*f*, 13
    fungal, 655–656
    prokaryotic, 581*t*, 582*f*
    protist, 594
Nymphs, 711

O

Oak Ridge National Laboratory, bioremediation of,
        1255*f*
Oak trees, 614*f*, 650*f*, 1201*f*
Obelia, 692*f*–693*f*
Obesity, 372*f*, 917*f*, 918, 1018
ob gene, 918
Object play, 1159
Obligate aerobes, **582**
Obligate anaerobes, **181**, **582**
Observations
    of evolutionary change, 477*f*–478*f*
    scientific, 17*f*, 19*f*
    in Scientific Skills Exercise, 812
Occam's razor, 562
Occipital lobe, 1085, 1097*f*
Ocean acidification, 53*f*, 54–55
Ocean currents, climate and, 1168*f*–1169*f*
Oceanic pelagic zone, **1181***f*
Oceans
    acidification of, 53*f*, 54
    climate and currents of, 1168*f*–1169*f*
    marine benthic zones of, 1182*f*
    as marine biome, 1177*f*, 1178
    moderation of climate by, 47*f*

    pelagic zones of, 1181*f*
    plastic waste in, 1277*f*
    primary production in, 1242, *1243f*, 1243*t*
    tides of, 1181*f*
    trawling of, as community disturbance, 1231*f*
Ocelli, 1117*f*
Ocelloid, 594*f*
Ochotona princeps, 1280*f*
Ocotillo, 799*f*
Octopus, 688*f*, 699, 701*f*, 702, 1219*f*
Odocoileus virginianus, 1271
Odorant receptors (ORs), 1124, 1125*f*
Odorants, **1123**–1124, 1125*f*
Offspring
    life history traits in survival of, 1200, 1201*f*–1202*f*
    survival of, 1023*f*
Oil. *See also* Fossil fuels
    conodonts and, 725
Oil spills, 591*f*. *See also* Fossil fuels
Okazaki fragments, **324**, 325*f*
Old World monkeys, 746*f*–747*f*
Olden, Kenneth, 27
Oleander, 799*f*
Olfaction, **1123**–1124, 1125*f*
Olfactory bulb, 1091*f*, 1095*f*, 1124, 1125*f*
Olfactory receptor genes, human, 489, 566
Oligochaetes, 703
Oligodendrocytes, **1076***f*–1077*f*, 1090*f*
Oligotrophic lakes, **1179***f*
Omasum, 914*f*
Ommatidia, **1117***f*
Omnivores, **901**, 911*f*, 912
Oncogenes, **388**, 389*f*, 391*f*
Oncorhynchus keta, 89
Oncorhynchus kisutch, 89, 1200
Oncorhynchus nerka, 978*f*
One gene–one enzyme hypothesis, 336–337, 338*f*,
        665
One gene–one polypeptide hypothesis, 337
One gene–one protein hypothesis, 337
One-shot reproduction, 1200
Onion, 252
On the Origin of Species by Means of Natural Selection
        (Darwin, C.), 14*f*, 469, 473, 483–484, 487
Onychophorans, 689*f*
Oocytes, **1026**, 1027*f*, 1029*f*
Oogenesis, **1027**, 1029*f*
Oogonia, **1029***f*
Oomycetes, 604*f*
Oparin, A. I., 526
Oparin-Haldane hypothesis, 526
Open circulatory systems, 699*f*, **707**, **923***f*
Open systems, 145, 150*f*
Operant conditioning, 1146
Operators, **366**, 367*f*
Operculum, **728***f*
Operon model, 366
Operons, **367**
    basic concept of, 366, 367*f*
    inducible, 368*f*, 369
    positive gene regulation and, 369*f*, 370
    repressible, 367*f*, 368–369
Ophisaurus apodus, 553*f*–554*f*
Ophiuroidea, 714*f*, 715
Ophrys speculum, 822*f*
Opiates, 40*f*, 78, 1082
Opioids, 1103*f*
Opisthokonts, **613**–614, **659**
Opium poppy, 868*f*
Opossums, 743
Opposable thumb, **744**
Opposite phyllotaxy, 786
Opsin, **1119***f*, 1122
Opsonization, 965*f*
Optic chiasm, 1121*f*
Optic disk, 1118*f*
Optic nerves, 1121*f*
Optic neuropathy, 311
Optimal conditions, enzymatic catalysis, 157, 158*f*
Optimal foraging model, **1149***f*
Oral cavity, **905**, 906*f*, 908*f*
Orangutans, 746*f*–747*f*, 750
Orbitals, electron, **35***f*, 36, 39*f*, 40
Orchid mantis, 468*f*, 469
Orchids, 650*f*, 822*f*, 825*f*
Order, as property of life, 3*f*, 146*f*, 147
Orders, taxonomy, **554**, 555*f*

Ordovician period global climate change, 629
Organelles, **5***f*, **94**
    as enzyme locations, 161*f*
    of eukaryotic cells, 97*f*–101*f*
    inheritance of genes in, 311*f*
    as level of biological organization, 5*f*
    lysosomal digestion of, 107*f*, 108
    plastids in plant cells, 111
    of prokaryotic cells, 97*f*, 98
    using electron microscopy to study, 94
Organic acid, 63*f*
Organic chemistry, **57***f*, 58
Organic compounds. *See also* Biological molecules
    abiotic synthesis of, 57*f*, 58, 526*f*–527*f*
    ATP as, 64
    bonding of carbon atoms in, 58, 59*f*, 60
    carbon in, as backbone of life, 56, 64
    chemical functional groups and, 62*f*–63*f*
    diversity of, 60*f*–62*f*
    organic chemistry as study of, 57*f*, 58
    in plant cells, 209*f*
    working with moles and molar ratios of, 58
Organic fertilizers, 808, 983
Organic phosphate, 63*f*
Organismal cloning
    of animals, 428, 429*f*–430*f*
    of animal stem cells, 428, 430*f*–432*f*
    of plants, 428
Organismal ecology, **1165***f*. *See also* Ecology;
        Organisms
Organismal-level herbivore defenses, plant, 869*f*
Organisms, **4***f*. *See also* Animals; Fungi; Life; Plants
    acidic and basic conditions affecting, 51, 52*f*–53*f*,
        54
    adaptations of, to environments, 468*f*. *See also*
        Adaptations
    Cambrian explosion in numbers of, 536*f*
    carbon in organic compounds of, 56, 64
    cells as fundamental units of, 6*f*, 93*f*. *See also* Cells
    climate change effects on, 1280*f*
    cloning of. *See* Organismal cloning
    differential gene expression in multicellular. *See*
        Differential gene expression
    DNA in development of, 7*f*–8*f*
    ecology as interactions between environment
        and. *See* Ecology
    ecosystem interactions of, 10*f*–11*f*
    effects of continental drift on, 539–540
    effects of sickle-cell disease on, 503*f*
    effects of speciation and extinctions on diversity
        of, 537*f*
    elemental ratio in, 43
    environment interactions with, 1165, 1178,
        1183*f*–1186*f*
    genes shared between, 26
    genomics, bioinformatics, and proteomics in
        study of genomes of, 9
    geographic distributions of, 539–540
    imperfection of, and evolution, 505
    inherited DNA and development of, 7*f*–8*f*
    interactions of, as theme in biology, 10*f*–11*f*
    as level of biological organization, 4*f*
    model. *See* Model organisms
    multicellular, 5*f*
    as open systems, 145
    origin of mammalian, 530, 531*f*, 532
    origin of multicellular, 533*f*, 535*f*–536*f*
    origin of single-celled, 532*f*–535*f*
    possible effects of transgenic crops on nontarget,
        839
    regulation of molecular interactions within, 10*f*
    single-celled, 5*f*
    in topsoil, 807
    transgenic. *See* Transgenic animals; Transgenic
        plants
Organization
    amino acid structures and, 91
    apoptosis in, 233
    of cell, 6*f*
    chemical waste and, 43
    digestive tracts of invertebrates, 717
    DNA structure and inheritance, 334
    emergent properties of, 4–6
    enantiomer effectiveness and, 65
    flowers as emergent property, 841
    functionality of forms, 551
    of fungi relationships, 672

gastrula emergent properties, 1066
of hair and keratin, 920
of lens of eye, 1138
levels of biological, 4f–5f
life as emergent property, 125
lignin evolution, 783
in loop of Henle, 998
of neurons, 1084
oxidative phosphorylation and, 186
photosynthesis and shoot architecture, 804
in plant life cycle, 653
sickle-cell disease and, 505
structure and function in, 6
as theme of biology, 2, 3
viral structure and function, 414
of walking and breathing, 756
water as versatile solvent and, 55
Organizer of Spemann and Mangold, 1062f
Organ-level herbivore defenses, plant, 868f
Organ of Corti, **1113f**
Organogenesis, **1036f**, 1044, **1049**, 1054f–1055f
Organs, 5f, **759**, **876t**
digestive. See Digestive systems
embryonic germ layers and, 1051f
endocrine system, 1003, 1004f. See also Endocrine
glands
excretory, 985, 986f–987f
eyes and light-detecting, 1117f–1119f
floral, 823f–824f
human reproductive, 1025f–1027f
immune system rejection of transplanted,
969–970
as level of biological organization, 5f
organogenesis of, 1044, 1049, 1054f–1055f
plant roots, stems, and leaves as, 759f–762f
reverse positioning of human, in situs inversus,
1064f
smooth muscle in vertebrate, 1131–1132
Organ systems, **876**
internal exchange surfaces of, 875f
mammalian, 876t
Orgasm, 1034
Orientation
honeybees, 1142f
leaf, 787
migratory, 1156, 1157f
plant cell expansion, 777f
Origin-of-life studies, 57f, 58
Origin of Species, The (Darwin, C.), 14f, 469, 473,
483–484, 487
Origins of replication, **242**, 243f, **322**, 323f
Ornithine, 338f
Orthologous genes, **565f**, 566
Orthoptera, 712f
Oryza sativa, 438, 509f, 651, 817, 838, 850–851, 1263
Osculum, **690f**
Oseltamivir (Tamiflu), 414
Osmoconformers, **978**
Osmolarity, **978**, 989, 990f, 991, 998
Osmoreceptors, 1110
Osmoregulation, **134**, **978**
challenges and mechanisms of, 978f–980f, 981
energetics of, 980
excretion and. See Excretory systems
homeostasis by, 977f, 978
ion movement and gradients and, 993f
osmosis and, 134f–135f
osmosis and osmolarity in, 978
salinity and, 1185f
transport epithelia in, 981, 982f
Osmoregulators, **978f**
Osmosis, **134**, **788**, 977
diffusion of water by, across plant plasma
membranes, 788, 789f–791f
effects of, on water balance, 133f–135f
osmolarity and, 978
Osmotic pressure, blood, 933f
Osteichthyans, **728f**–729f, 730. See also Fishes
Osteoblasts, 878f, 1134
Osteoclasts, 1134
Osteons, 878f
Ostriches, 992f
Otoliths, 1115f
Otters, 1189
Ouabain, 1084
Outer ear, **1113f**
Outgroups, **561f**, 572

Ovalbumin, 76f
Oval window, **1113f**
Ovarian cancer, 447
Ovarian cycle, **1032f**, 1033
Ovaries, angiosperm, **644f**–646f, 647, **823f**–824f,
830f–831f. See also Fruits
Ovaries, human, 257f, 258, 1004f, 1014, 1015f, **1026**,
1027f
Overgrazing, 1231
Overharvesting, 702f, 727–728, 1264, 1265f
Overnourishment, 917–918
Overproduction, offspring, 475f, 476
Oviducts, **1026**, 1027f
Oviparous organisms, **727**
Oviraptor dinosaurs, 564f, 565
Ovoviviparous organisms, **727**
Ovulate cones, 640f, 641
Ovulation, **1021**, 1035f
Ovules, 636, **638f**, **823f**
Owl-mouse predation study, 23
Owls, 920, 1218
Oxidation, 112, 165, **166f**
Oxidative fibers, 1130–1131
Oxidative phosphorylation, 168, **169f**, 174f–177f, 186
Oxidizing agents, **166f**
Oxygen
atmospheric, in animal evolution, 677
catabolic pathways and, 165
in circulation and gas exchange, 947f–949f
covalent bonding and, 36, 37f
development of photosynthesis and atmospheric,
532f–534f
diffusion of, across capillary walls, 932
in double circulation, 924f, 925
in ecosystem interaction, 10f, 11
electronegativity of, 37, 45
as essential element, 29t, 64
in gas exchange, 921, 939t, 940
in human breathing, 946
plants production of, 619
in mammalian circulation, 926f
metabolic rate and, 890f
in net ecosystem production, 1242
in organic compounds, 59f
in oxidation, 166f
Permian mass extinction and low levels of, 540
in photosynthesis, 41f
in plant cells, 209f
in plant composition, 809
as product of photosynthesis, 188, 206, 207f
role of, in prokaryotic metabolism, 582
species distributions and availability of, 1185
thyroid hormone and consumption of, 179
transport of, 951
Oxytocin, 1004f–**1005f**, 1007f, 1037f
Oystercatcher, 1161
Oyster drills, 542
Oysters, 699, 701f, 1020
Ozone depletion, 1283f, 1284

P
p21 gene, 390
p53 gene, **389**, 390f–391f
P680 chlorophyll *a*, 196, 197f
P700 chlorophyll *a*, 196, 197f, 198
Pacemaker, heart, 928f
Pacific Island land snails, 702f
Pacific salmon (*Oncorhynchus* species), 89
Pac-man mechanism, 241f
Paedomorphosis, 544, **545f**, 732
Paedophryne swiftorum, 1164f
Paine, Robert, 1226f
Pain receptors, **1111**–1112
Pair bonding, 1144, 1155f
Pakicetus, 482f
Paleoanthropology, **748**
Paleogene period, 538
Paleontology, 13f, 88f, **470f**, 471
Paleozoic era, 530f, 533f, 539f, 677f, 678
Palisade mesophyll, 771f
Pallium, avian, 1099f
Palumbi, S. R., 557f
Pampas, 1175f
PAMP-triggered immunity, 866
Pan-Cancer Atlas, 448
Pancreas, **909**, 1004f
cellular interactions of, 142

in digestion, 909
exocytosis and, 139
in glucose homeostasis, 916f
rough ER and, 105
Pancrustaceans, **708**, 709f–712f, 713
Pandemics, **410**
Pandoravirus, 408
Pangaea, **482**–483, **539f**, 551
Pan genus, 746f. See also Chimpanzees
Panthera pardus, 554, 555f
Panthera tigris altaica, 1288
Panther chameleons (*Furcifer pardalis*), 735f
Paper, 651
Papaya, 757
Paper wasps, 712f
Papillae, 1124f
Papillomaviruses, 394
Pappochelys, 737
Papua New Guinea, 1164f
Parabasalids, **600f**
Parabronchi, 944, 945f
Parachutes, seed and fruit, 832f
Paracrine signaling, 215f, **1000f**, 1001
Parahippus, 548, 549f
Parakeets, 1136
Paralogous genes, 565f, **566**
Paramecium, 13f, 93f, 134, 135f, 607f, 617, 1199f, 1215
Paraphyletic groups, **560f**
Parapodia, 703
Parasites, 584f, **587**, **1220**
acanthocephalans, 698
animals as, 673f
antigenic variation and, 972
apicomplexans, 605, 606f
arachnids, 708f
cercozoans, 609
entamoebas, 613
flatworms, 696f–697f
fungi as, 655, 660, 663, 665, 669f–670f
insects as, 713
lampreys as, 723f, 724
nematodes, 705f, 706
plants as, 818, 819f, 820
protist, 598f, 600f, 614
in zoonotic diseases, 1234
Parasitism, **587**, **1220**
Parasitoid wasps, 869f
Parasympathetic division, peripheral nervous system,
928, 1088f–**1089f**
Parathion, 158
Parathyroid glands, 1004f, **1011f**
Parathyroid hormone (PTH), 1004f, **1011f**
Parenchyma cells, **764f**, 769f, 770
Parental alleles, genomic imprinting and, 310f, 311
Parental care, 1023f, 1150, 1151f, 1200, 1201f–1202f
Parental types, **302**
Parietal cells, 907f
Parietal lobe, 1085, 1097f
Paris Agreement, 1283
Paris japonica, 448t, 449, 843
Parkinson's disease, 65, 83, 231, 361, 413, 431–432,
671, 1081, **1104**
Parsimony, 562, 563f, 564
Parsnip, 757
Parthenogenesis, **697**–698, **1020**, 1021f
Partial pressure, **939**
Particulate model of inheritance, 487
Parus major, 740f
Passeriformes, 740f
Passive immunity, **968**
Passive transport, 126, **133**
active transport vs., 137f
diffusion as, 132, 133f
facilitated diffusion as, 135f, 136
interpreting scatter plots on glucose uptake as
facilitated, 136
in plant cells, 209f
of water across plant plasma membranes, 788,
789f–791f
water balance and osmosis as, 133f–135f
Patella vulgata, 547f–548f
Paternal alleles, 372
Paternal chromosomes, 265
Paternity
certainty of, 1150, 1151f
tests for, 437
Pathogen-associated molecular patterns (PAMPs),
866

Pathogenicity trait, bacterial, 315f
Pathogens, **587**, **953**, **1234**
    adaptations of, in immune system evasion, 957,
        971–972, 973f, 973t, 974, 976
    alteration of community structure by, 1234, 1235f
    bacterial, 315f, 583f–584f, 585, 587, 588f
    B cells and antibodies as responses to
        extracellular, 964, 965f–966f
    cytotoxic T cell response to cells infected by, 966f,
        967
    fungi as, 655, 663, 669f–670f
    immune system recognition and response to,
        952f, 953, 957f–958f
    innate immunity evasion by, 957
    plant defenses against, 866, 867f
    prokaryotic, 587, 588f
    viruses as, 398f–400f, 401, 408, 409f–413f
Pathways
    nervous system, 880
Pathways, sensory, 1108f–1109f
Pattern
    evolutionary, 470, 483
    taxonomy based on, 470
Pattern formation, **384**, 385f–387f, **777**, 778f, 849,
    1062, 1063f–1064f. See also Morphogenesis
Paulinella chromatophora, 608f, 609
Pauling, Linus, 317–318
Pavlov, Ivan, 1146
Pax-6 gene, 423
Pc, 198f, 199
PCSK9 enzyme, 938
PDGF (platelet-derived growth factor), 247, 248f
Pea aphids (*Acyrthosiphon pisum*), 570
Peacocks, 499f
Pea fruit, 831f
Pea plants, G. Mendel's experiments with, 269f–276f
Pearl mussels, 702f
Peat, **627**f, 628, 633–634, 1249, 1259
Pectin, 118
Pediastrum, 610f
Pedigrees, **284**f, 285
Pedipalps, 708
Pelagic zone, **1177**f
Pellicle, 601f
Peltigera, 813f
Penguins, 14f, 739, 740f, 873f, 874f, 884f, 892, 1192f
Penicillin, 159, 478, 574–575, 670
Penicillium, 658f
Penis, 1025f, **1026**, 1034, 1081–1082
Pentoses, 68f
PEP carboxylase, **203**, 204f
Pepsin, 158f, **907**f
Pepsinogen, **907**f
Peptide bonds, 75, **78**f, 353f
Peptides
    antimicrobial, 953–954, 955f, 956–957
Peptidoglycan, **574**, 575f
Per capita birth rate, 1196–1198
Per capita death rate, 1196–1198
Per capita rate of increase, 1196–1198
Perception, **1109**, 1121f
Perch, 979f, 1203f
Perennials, 766
Pericarp, 830–831
Pericycle, **768**, 769f
Periderm, **763**, 774
Perilymph, 1115f
Periodic table of elements, 34f
Peripheral nervous system (PNS), **1069**, 1081, **1086**
    central nervous system and, 1086, 1087f–1090f
    structure and function of vertebrate, 1088f–1089f
Peripheral proteins, **129**
Peripheral vision, 1122f, 1123
Perissodactyla, 745f
Perissodus microlepis, 500, 501f
Peristalsis, **906**, **1133**f, 1135
Peristome, **625**
Peritubular capillaries, **987**f
Periwinkles, 1263f
Permafrost, 1176f
Permease, 368f
Permian mass extinction, 540f, 541
Peromyscus polionotus, 2f, 20f–21f, 23
Peroxisomes, 100f–101f, **112**f
Personal genome analysis, 433
Personalized medicine, 88f, **433**–434
Pertussis, 218f

Pesticides, 1275, 1276f
    DDT, 158
    transgenic, 837–838
Pest resistance, plant, 418f, 438
Petals, **644**f, **823**f
Petioles, **761**
Petromyzon marinus, 724
Petromyzontida (lampreys), 719f, 723f, 724
PET scanners, 31, 32f
Pévet, Paul, 893f
Pfu polymerase, 421
P (parental) generations, 271f
P granules, 1059f
pH, **52**
    acid precipitation and, 1265, 1266f
    adjusting soil, 808–809
    buffers and, 52–53
    enzymatic catalysis and, 157, 158f
    and hemoglobin dissociation, 948f
    of human cerebrospinal fluid, 946
    pH scale and, 51, 52f
    protein denaturation and, 82f, 83
    species distributions and soil, 1186
    sucrose uptake by plant cells and, 142
PHA (polyhydroxyalkanoate), 590, 591f
Phage λ, 403, 404f
Phages, **315**, **401**
    capsids of, 400f, 401
    Cas9 protein and, 360, 361f
    defense against, 404f
    in DNA research, 315f–316f, 317
    in fighting infection, 872
    prophages and temperate, 402, 403f
    replicative cycles of, 402f–404f
    in transduction, 579f
    virulent, 402f
Phagocytic cells, 955f, 956
Phagocytosis, **107**, **140**f, **953**f
    cellular integration of, 121f
    as endocytosis, 139, 140f
    immune systems and, 953f, 955f, 956, 965f
    lysosomes and, 107f
Phalacrocorax harrisi, 506f, 507, 511
Phanerozoic eon, 530f, 533f, 539f
Pharmaceutical products
    biotechnology in production of, 435, 436f.
        See also Drugs; Medicine
    as environmental toxins, 1276, 1277f
    fungal, 670
    species and genetic diversity and, 1262, 1263f
Pharmacogenetics, 434
Pharyngeal clefts, **720**f
Pharyngeal slits, **720**f–721f
Pharyngolepis, 725f
Pharynx, 696f, **906**f, 908f
Phase changes, plant development, 778, **780**f
Phase-contrast microscopy, 95f
Phenobarbital, 104
Phenolics, 868f
Phenotypes, **274**
    DNA transformation and, 315
    dominant alleles and, 279–280
    gene concept and, 362
    gene expression as link between genotypes and,
        335–336
    genetic mapping and, 305f
    genetic variation and, 487f–488f
    genotypes vs., 274f, 282f, 283
    impact of environment on, 282f
    making histograms and analyzing distribution
        patterns of, 283
    mutant, 295f–296f, 328
    proteins and, 335
    relative fitness and, 497
Phenylalanine, 77f, 289–290, 341, 492–493
Phenylketonuria (PKU), 289–290, 492–493
Phenylthiocarbamide (PTC) pedigree analysis case,
    284f, 285
Pheromones, **658**, 670, **1001**f, 1023, **1142**, 1143f
Philadelphia chromosome, 309f
Philippine eagle, 1262f
Phloem, **629**, **763**, **785**
    primary growth and, 769f
    resource acquisition and, 785f
    sugar-conducting cells of, 765f
    sugar transport from sources to sinks via, 799,
        800f–801f

symplastic communication through, 802
vascular plant, 629–630
in vascular tissue systems, 763
Phloem sap, **799**, 800f–801f
Phoca hispida, 44f
Phoebastria immutabilis, 1277f
Phoenicopterus ruber, 14f, 740f
Phosphatases, 223
Phosphate group, 63f, 84, 85f
Phosphates, in phosphorus cycle, 1251f
Phosphodiesterase, 229, 233
Phosphofructokinase, 183f, 184, 186
Phospholipid bilayers, 74f, 75, 98f, 99, 102, 110, 127f,
    132
Phospholipids, **74**
    in cellular membranes, 102, 110, 127f, 132
    Golgi apparatus and, 106
    movement of, in cellular membranes, 128f, 129
    in plasma membranes, 98f, 99
    structure of, 74f, 75
Phosphorus
    as essential element, 29t, 64
    as limiting nutrient, 1242, 1243f, 1243t, 1244
    plant deficiency in, 818
    soil fertilization and, 805, 808
Phosphorus cycle, 1251f
Phosphorylated intermediates, **151**, 152f
Phosphorylation
    in cell-signal responses, 226f
    in cell-signal transduction, 222f, 223
    in light reactions of photosynthesis, 191f
Phosphorylation cascade, **222**f, 226f
Photic zone, **1177**f
Photoautotrophs, 188f, 581t, 594
Photoheterotrophs, 581t
Photomorphogenesis, **855**
Photons, **192**f, 195f
Photoperiodism, **859**f–860f
Photophosphorylation, **191**
Photoprotection, 195, 199
Photopsins, 1122
Photoreceptors, **1117**f–1119f
Photorespiration, **203**, 211
Photosynthates, 759
Photosynthesis, **188**
    alternative mechanisms of carbon fixation in,
        203, 204f–206f
    in aquatic ecosystems, 1242
    Calvin cycle of, 201, 202f
    in carbon cycle, 1250f
    cellular respiration vs., 190. See also Cellular
        respiration
    cercozoan, 608f, 609
    chemical reactions in, 41f
    chloroplasts in, 5f, 109–110, 111f
    climate change and, 211
    conversion of light energy to chemical energy of
        food by, 189f–191f
    cyanobacteria and, 584f
    determining rate of, with satellites, 1242f
    development of, and atmospheric oxygen, 533,
        534f
    in energy flow and chemical cycling, 9f, 164–165,
        211, 1240f
    evolution of adaptations for resource acquisition
        and, 784f–786f, 787
    as gas exchange, 895f
    importance of, 188f, 206, 207f
    lichens in, 188f–669f
    light reactions of. See Light reactions
    maximizing surface area for, 695f
    as plant nutritional mode, 894f
    principles of, 163
    prokaryotic, 577f
    protist, 594, 599f–600f, 614, 615f
    red and green algae, 609f–610f
    scale of, 122f
    shoot architecture and, 804
    stramenopile, 602
    sunlight availability and, 1185, 1186f
    two stages of, 191f, 192
    in vascular plants, 631
    water loss compromise with, 787
    in working plant cells, 209f
    zonation of aquatic biomes and, 1177f–1178f
Photosystem I (PS I), **196**, 197f–198f
Photosystem II (PS II), **196**, 197f–198f

Photosystems, 195, **196**f–198f, 199
Phototrophs, 581t
Phototropin, 856f
Phototropism, **847**f–848f
pH scale, 51, 52f
Phycoerythrin, 609
Phyla, taxonomy and, **554**, 555f
    angiosperm, 648f
    bryophyte, 626f
    gymnosperm, 642f–643f
    plant, 622t
Phyllotaxy, 786f, 849
Phylogenetic bracketing, 564f, 572
Phylogenetic trees, **555**. See also Evolutionary trees;
    Phylogenies
    analyzing sequence-based, to understand viral
      evolution, 411
    of animals, 683f
    application of, 557f
    of bilaterians, 717
    of chordates, 719f
    cladistics in construction of, 559, 560f–561f
    of eukaryotes, 612f
    as hypotheses, 564f, 565
    linking classification and phylogeny with,
      555f–556f
    of mammals, 745f
    maximum parsimony and maximum likelihood
      in, 562, 563f, 564
    of primates, 746f
    of prokaryotes, 583f
    proportional branch lengths in, 561, 562f
    of protists, 598f
    of tetrapods, 731f
    tree of life and, 15f–16f
    visualizing, 556f
Phylogenies, **554**
    of amniotes, 734f
    of angiosperms, 648f–650f, 653
    of animals, 682, 683f, 684
    of chordates, 719f
    constructing phylogenetic trees for, 559, 560f–
      564f, 565. See also Phylogenetic trees
    documentation of, in genomes, 565f, 566
    of eukaryotes, 612f
    of fungi, 660f–667f
    gymnosperm, 641, 642f–643f
    as hypotheses, 564f, 565
    inferring, from morphological and molecular
      data, 558f–559f
    investigating tree of life with, 553f–554f, 568,
      569f–570f
    of maize, 557
    of mammals, 745f
    molecular clocks and evolutionary time in, 566,
      567f–568f
    nitrogenous wastes and, 982f–983f
    of plants, 623f
    practical applications of, 557f
    of primates, 746f
    of prokaryotes, 583f
    systematics and, 554
    taxonomy and evolutionary relationships in, 554,
      555f–556f, 568, 569f
    of tetrapods, 731f
Physical reconstruction, ecosystem, 1253f–1254f
Physiological thermostats, 888, 889f
Physiology, 873f–**874**
Phytoalexins, 866
Phytochemicals, 195
Phytochromes, **856**
    in circadian rhythms, 858–859
    in plant signal transduction pathways, 844f
    in seed germination, 856, 857f
    in shade avoidance in plants, 857, 871
Phytophthora species, 604, 614f, 1234
Phytoplankton, 584f. See also Plankton
    in biomass pyramids, 1247f, 1248
    in carbon cycle, 1250f
    dinoflagellate, 605f
    green algae, 610
    nitrogen pollution and blooms of, 1275f
    primary production by, 1242, 1243f, 1243t
    seasonality and, 1167
Phytoremediation, **809**
Picoides borealis, 1269f–1270f

Pie charts in Scientific Skills Exercise, 892
Pied flycatchers (Ficedula hypoleuca), 519
Pigmentation
    DNA methylation and, 372f
    gene expression and, 335f, 336
    plant, 311f
    of Siamese cats, 364
    skin, 281, 282f
Pigments
    as photosynthetic light receptors, 192, 193f–194f,
      195
    in photosystems, 196f–200f, 201
    respiratory, 947f–948f, 949
    visual, 1122
Pikas, 1280f
Pili, **576**, 580f, 581
Pill bugs, 709
Pilobolus, 663f
Pimephales promelas, 1276
Pineal gland, 1004f, 1007f, **1015**, 1093f
Pineapple, 206f, 831f
Pine trees, 640f, 643f, 1229f, 1244f, 1245, 1280f
Pin flower, 835f
Pinocytosis, **140**f
Pinworms, 705
Piping plover (Charadrius melodus)., 1190f
Pisaster ochraceus, 1226f
Pistils, **644**f, **823**f
Pitch, 1114
Pitcher plants, 819f
Pitch pine canker, 669
Pith, **763**, 770f, 783
Pithovirus sibericum, 408
Pituitary dwarfism, 1010
Pituitary gland, **1007**, 1030, 1031f–1032f, 1093f
    in human endocrine system, 1004f
    in kidney regulation, 994f–995f
    in neuroendocrine signaling, 1007f–1008f
Pit vipers, 737f, 1111f
Pivot joints, 1134f
Piwi-interacting RNAs (piRNAs), 380–381
PKU (phenylketonuria), 289–290, 492–493
Placenta, **742**, **1035**f
Placental mammals. See Eutherians (placental
    mammals)
Placental transfer cells, 620f
Placoderm fossil, 533f
Placoderms, **726**f
Placozoans, 687f
Plains, 1175f
Planarians, 687f, **694**, 695f–696f. See also Flatworms
Planes, plant cell division, 776–777
Planets, possible evolution of life on other, 50f
Plankton, 608, 710f. See also Phytoplankton
Plantae kingdom, **12**f, 568, 599f, 609, 619f, 621
Plant cells. See also Eukaryotic cells
    cell fate in pattern formation, 777, 778f
    cell signaling in, 215f
    cellular activities of, 208f–209f
    cell walls of, 118f
    chloroplasts in, 109, 110f–111f
    common types of, 763, 764f–765f
    cytoplasmic streaming in, 117f
    division and expansion of, in growth, 776, 777f
    as eukaryotic cells, 101f
    gene expression and control of differentiation
      of, 778f
    mitosis and cytokinesis in, 241, 242f–243f
    photosynthesis in. See Photosynthesis
    plant cloning from single, 428. See also Transgenic
      plants
    plasmodesmata as cell junctions in, 119f, 120
    sucrose uptake by, 142
Plant development
    adaptations of, 894f
    auxin in, 849
    cell division and cell expansion in growth and,
      776, 777f
    gene expression and control of cell differentiation
      in, 778f
    genetic control of flowering in, 779f–780f
    growth, morphogenesis, and cell differentiation
      in, 775f, 776
    model organisms in the study of, 774, 776f
    morphogenesis and pattern formation in, 777, 778f
    phase changes in, 779, 780f
Plant growth

adaptations of, 894f
    cell division and cell expansion in, 776, 777f
    meristem generation of cells for, 766f–767f
    plant development and, 775–776
    primary, 768f–771f
    regulators of. See Plant hormones
    secondary, 772f–775f
Plant hormones (plant growth regulators), **846**t.
      See also Hormones
    abscisic acid, 846t, 852f, 863
    animal hormones vs., 216, 894f
    auxin, 846t, 847f–849f, 850, 854f, 861
    brassinosteroids, 845, 846t, 854–855
    cytokinins, 846t, 850f
    in de-etiolation (greening) responses, 845
    ethylene, 846t, 852, 853f–854f, 864f
    florigen, 860f
    gibberellins, 846t, 850, 851f
    jasmonates, 846t, 855
    in long-distance cell signaling, 216
    overview of, 846t
    as plant growth regulators, 846–847
    strigolactones, 846t, 850, 855
Plant nutrition
    essential elements for, 809, 810f, 811t
    mutualisms for, 812, 813f–819f, 820
    nutritional modes in, 894f
    photosynthesis and modes of, 188f
    soil as complex ecosystem for, 805f–809f
    unusual adaptations for, 806, 818, 819f, 820
    vascular plant acquisition of water and minerals,
      787
    vascular plant transport of water and minerals,
      787, 788f–791f
Plant responses
    to attacks by pathogens and herbivores, 866,
      867f–869f
    to environmental stresses, 862–863, 864f, 865
    to gravity, 861f
    to light, 855, 856f–860f
    to mechanical stimuli, 861, 862f
    plant hormones and, 845, 846t, 847f–854f, 855
    signal-transduction pathways linking signal
      reception to, 843f–844f, 845
Plants. See also Angiosperms; Plant development;
      Plant growth; Plant hormones; Plant
      nutrition; Plant responses; Plant structure;
      Seed plants
    adaptations of, that reduce terrestrial nutrient
      limitations, 1244
    adaptations of, to toxic elements, 30f
    adaptive radiations of, 543f, 544
    alternation of generations in life cycles of, 258f,
      620f
    in Archaeplastida supergroup, 599f, 609
    bioremediation using, 1253, 1255f
    cells of, 6f, 208f–209f. See also Plant cells
    cellulose as structural polysaccharide for, 70f–71f,
      72
    chemical communication by, 845–846
    climate change and, 11, 1279, 1280f–1281f, 1282
    cloning of, 428
    community stability and diversity of, 1223f
    crop. See Crop plants
    defensive adaptations of, 1219–1220
    derived traits of, 620f–621f
    diseases of. See Diseases, plant
    in domain Eukarya, 12f
    in ecosystem interaction, 10f, 11
    elements in composition of, 809
    endangered or threatened, 1261
    essential elements and trace elements for, 29
    evolutionary mystery of flowering, for C. Darwin,
      477
    evolution from green algae, 617, 618, 619f, 621
    fossil of, 529f
    fungal mutualisms with, 665, 667f, 668, 813f.
      See also Mycorrhizae
    fungal pathogens of, 663, 669f
    gametophyte-sporophyte relationships in, 637f,
      638
    genetic engineering of transgenic, 418f, 438
    genomics and proteomics in study of, 88f
    global climate change and species distributions
      of, 1170, 1170f
    habitat loss and, 1263–1264
    immune response in, 866, 867f

Plants (continued)
importance of insects to, 713
inheritance of organelle genes in, 311f
introduced, 818
kingdom and domain mutualisms with, 813f
land colonization by, 536, 537f
life challenges and solutions for, 894f–895f
nematode parasites of, 705
nonvascular, 623, 624f–627f, 628
nutrition of, 12f, 13
origin, diversification, and phylogeny of, 621, 622f–623f, 622t
pathogens of, 1234
as photoautotrophs, 188f
photosynthesis of. *See* Photosynthesis
polyploidy in, 307
protists as pathogens of, 614f
reproduction of. *See* Angiosperm reproduction
resource acquisition for vascular, 784f–786f, 787
salinity and species distributions of, 1186
starch as storage polysaccharide for, 70f, 71
sympatric speciation in, 514f–515f
terrestrial adaptations of, 619f, 621
vascular. *See* Transport in vascular plants; Vascular plants
water balance of cells of, 134f, 135
water transport in, 46f
Plant structure
cells in, 763, 764f–765f
diversity in, 759
hierarchy of organs, tissues, and cells in, 758f–765f
meristem generation of cells for growth of, 766f–767f
plant development and, 775f–780f
primary growth of roots and shoots of, 768f–771f
secondary growth of stems and roots in woody plants, 772f–775f
Plaque, arterial, 937f, 939
Plasma, 934f, 935
Plasma cells, 961, 962f, 970f
Plasma membranes, 98, 109f. *See also* Cellular membranes
animal cell, 100f
chemiosmosis in prokaryotic, 175, 176f, 177
electron transport chains in prokaryotic, 168
of eukaryotic cells, 98f, 99
hormone receptors in, 1003f
ion gradients and transport of ions across, 993f
microfilaments in, 116f
movement across plant cell, 209f
nuclear envelopes as, 102
plant cell, 101f
of prokaryotic cell, 97f
receptor proteins in, 217f–220f
short-distance and long-distance transport across, 787, 788f–791f
Plasmids, 418, 577
in antibiotic resistance of bacteria, 581
in bacterial conjugation, 580f, 581
evolution of viruses and, 408
prokaryotic, 577f
as recombinant DNA, 418f, 419–420
Plasmodesmata, 119
as cell junctions in plants, 119f, 120
macromolecules and, 846
in plant cells, 101f
in plant local cell signaling, 215f
in symplastic communication, 802f
Plasmodial slime molds, 612f
Plasmodium (protist), 599f, 606f, 614, 713, 953
Plasmogamy, 658f
Plasmolysis, 135, 790
Plasticity, neuronal, 1100f
Plastics, natural, 590, 591f
Plastic waste, 1163, 1277f
Plastids, 111
endosymbiotic origin of, 534, 535f
eukaryotic endosymbiosis and evolution of, 596f–597f
Plastocyanin (Pc), 198f, 199
Plastoquinone (Pq), 198f
Plateau phase, sexual, 1034
Platelet-derived growth factor (PDGF), 247, 248f
Platelets, 10, 878f, 934f–935f
Plate tectonics, 538f–539f, 540
Platyhelminthes (flatworms). *See* Flatworms

Platypus, 742, 1110–1111
Play, 1159
Pleasure, brain activity and, 1103f
Pleiotropy, 280
Pleurozium schreberi, 635
Plumule, 829
Pluripotent cells, 431
Pneumatophores, 760f
Pneumocystis jirovecii, 973–974
Pneumonia, 315f, 579
Poaching, elephant, 1265f
Podium, sea star, 714f
Poecile species, 517f
Poecilia reticulata (guppies), 483, 1152, 1153f, 1187f
Poikilotherms, 884
Point mutations, 357f–358f, 359, 388, 389f, 488–489
Poison dart frog, 1218f
Pokeweed, 830f
Polar bears (*Ursus maritimus*), 11, 510f
Polar covalent bonds, 37f, 45
Polarity, 877f
Polar molecules, 44–45
Polar side chains, 76, 77f
Polar transport, auxin, 848f, 849
Polio, 408–409, 968f, 976
Pollen cones, 640f, 641
Pollen grains, 636, 638f, 646f, 647, 649f, 826, 827f, 841
Pollen tubes, 825f, 826, 827f
Pollination, 638, 825
angiosperm cross-pollination, 646f, 647
asexual reproduction vs., 833
coevolution of flowers and pollinators in, 825f
cross-pollination in breeding plants, 837
flowers and angiosperm, 644f
flower shape and insect, 649f
genetic engineering of flowers to force self-pollination, 839–840
by insects, 641f
insects in, 713
mechanisms for preventing self-, 834, 835f
mechanisms of, 824f–825f
G. Mendel's techniques of, 270f–271f
seed plant, 638f
Pollinators
coevolution of flowers and, 825f
reproductive isolation and choice of, 522f
Pollution
biomanipulation and, 1227f
coral reefs and, 693
ecosystem services and, 1263
molluscs and water, 702–703
nutrient, 1274, 1275f
ocean acidification and, 53f, 54
plastic, 1163, 12771f
prokaryotes and bioremediation of, 591f
toxins, 1275, 1276f–1277f
Polyadenylation signal sequence, 345f
Polyandry, 1149, 1151f
Poly-A tail, 345f
Polychaetes, 703
Polychaos dubium, 449
Polychlorinated biphenyls (PCBs), 1275, 1276f
Polydactyly, 280, 292
Polygamous mating, 1149–1150, 1151f
Polygenic inheritance, 281, 282f, 287
Polygyny, 1149–1150, 1151f
Polymerase chain reaction (PCR), 420
amplification of DNA using, 420, 421f, 422
bacteria for, 589
diagnosing diseases with, 433–434
in estimating reproductive rates, 1194, 1195f
extreme thermophile archaea in, 585, 1263
in forensic science, 436, 437f
in genomic analysis of fetuses, 1040
in prokaryotic analysis, 583
in RT-PCR analysis, 424, 425f
in sequencing human digestive system microbiome, 912, 913f
in vitro amplification of DNA using, 422f
Polymerases, 401
Polymerization, protein, 240
Polymers, 67f
Polymorphisms, 427, 433
Polynucleotides, 84, 85f
Polypeptide backbone, 78f
Polypeptides, 75, 1001f

amino acid monomers of, 75, 76f–77f
as amino acid polymers, 78f
analyzing sequence data of, in Scientific Skills Exercise, 89
mutations affecting, 357f–358f, 359
one gene–one polypeptide hypothesis on, 337
proteins as composed of, 78f
synthesis of, in translation, 337, 339f
synthesizing multiple, with polyribosomes in translation, 355f
targeting, to specific locations, 354f, 355
translation stages in synthesis of, 350f–353f
Polyphemus moth (*Antheraea polyphemus*), 1001
Polyphyletic groups, 560f
Polyplacophora (chitons), 699f
Polyploidy, 307, 514f–515f
Polyps, 391f, 691f, 1019f
Polyribosomes (polysomes), 355f
Polysaccharides, 70f–72f, 106
Polyspermy, 1045
Polytomies, 569f
Polytrichum, 626f–627f
Pongo species, 746f
Pons, 946, 1093f
Poplar trees, 774, 838
Population conservation
conflicted demands in, 1270f
critical habitat in, 1269f–1270f
extinction risks in small populations, 1266, 1267f–1268f, 1269
Population cycles, 1205f, 1206
Population dynamics, 1191f, 1192, 1205f–1206f, 1207, 1256f. *See also* Population growth
Population ecology, 1165f. *See also* Ecology
demography of population vital statistics in, 1193t, 1194f–1195f
determining population size using mark-recapture method in, 1191f
human population in, 1207f–1210f, 1211, 1213
population density and dispersion in, 1191f–1192f, 1193, 1202, 1203f
population dynamics in, 1191f, 1192, 1205f–1206f, 1207
population growth models in, 1196, 1197f–1199f, 1198t, 1200. *See also* Population growth
population growth regulation in, 1202, 1203f–1206f, 1207
as study of populations in environments, 1190f, 1191. *See also* Populations
Population growth. *See also* Populations
density-dependent regulation of, 1202, 1203f–1206f, 1207, 1213
ecological impact of, 1213
exponential model of, 1196, 1197f, 1207f, 1208
of human population, 1207f–1210f, 1211, 1213
logistic model of, 1197, 1198f–1199f, 1198t, 1200
population dynamics and, 1191f, 1192, 1205f–1206f, 1207
using logistic equation to model, 1200
Population-level herbivore defenses, plant, 869f
Populations, 4f, 490, 1165f, 1191
climate change effects on, 1281f
C. Darwin on natural selection and, 14f–16f
demographics of, 1193t, 1194f–1195f
density and dispersion of, 1191f–1192f, 1193, 1202, 1203f, 1256f
determining size of, using mark-recapture method, 1191f
diffusion of, of molecules, 132
dynamics of, 1191f, 1192, 1205f–1206f, 1207, 1256f
effective size for, 1268
evolution of genetic variation in, 266, 267f
extinction risks in small, 1266, 1267f–1268f, 1269
gene pools of, 490f
genetic diversity of, 1261f
growth of. *See* Population growth
Hardy-Weinberg equilibrium and size of, 492t
human, 1207f–1210f, 1211, 1213
as level of biological organization, 4f
life histories of, 1200, 1201f–1202f
minimum viable size for, 1268
natural selection and evolution of, 475f, 476
population ecology as study of, 1190f, 1191. *See also* Population ecology
using Hardy-Weinberg principle to test microevolution in, 490, 491f, 492–493

Populus tremuloides, 791f, 833f
Porcupine, 1218f
Pore complexes, 102, 103f
Pores, nuclear, 102, 103f, 123f
Pores, plasmodesmatal, 802f
Porifera (sponges). *See* Sponges
Porphyra, 609f, 610
Porphyrin ring, 194f
Porpoises, 481f–482f
Portal vessels, 1008f
Positional information, **384**, **1062**, 1063f–1064f
Position-based mechanisms, plant, 778
Positive and negative correlations, 834
Positive feedback, **882**, **1006**
    in climate change, 1189
    in endocrine system feedback regulation, 1006
    in feedback regulation, 10
    in homeostasis, 882
Positive gene regulation, bacterial, 369f, 370
Positive gravitropism, 861f
Positive interactions, **1220**f–1221f
Positive pressure breathing, **944**
Positron-emission tomography (PET), 31, 32f, 1096
Possum, 743f
Posterior pituitary gland, 1004f, **1007**f, 1008
Postganglionic neurons, 1089
Postsynaptic cells, 1068f
Postsynaptic neurons, 1068f, 1079f, 1101f
Postsynaptic potentials, 1077–1078, 1079f
Post-translational protein modifications, 354
Postzygotic barriers, 509f, **510**
Potassium, 29t, 805, 808
Potassium ion channel protein, 135
Potassium ions, 797f, 798, 1070t, 1071f
Potato blight, 867
Potatoes, 651, 843f–844f
Potential, cell developmental, 1060, 1061f
Potential energy, **32**, **144**, 145f, 163
Potential range, 1184
Pq, 198f
Prairie chickens, 495f, 496
Prairies, 1175f, 1264
Prairie voles, 1155f
Precapillary sphincters, 932f
Precipitation
    climate change and, 11
    climographs of, 1171f
    evaporation and, 47
    global patterns of, 1166f
    mountains, rain shadow, and, 1169f
    primary production in terrestrial biomes and, 1243f, 1244, 1244f
    in water cycle, 1250f
Precocious germination, 852f
Predation, **1217**, 1256f
    allopatric speciation and, 511f, 512
    camouflage and, 468f
    camouflage case studies and, 20f–21f, 23
    density-dependent population regulation by, 1204f
    genetic variation in, 1155, 1156f
    as interspecific interaction, 1217, 1218f–1219f
    population cycles and, 1205f, 1206
    as risk, 1149
    top-down model of trophic control and, 1227f
Predators
    adaptations of, 1217–1218, 1219f, 1237
    animals as, 673f
    as biotic factors limiting species distributions, 1184f
    cephalopods as, 701f, 702
    evolution of animal, 677f
    insects as, 713
    mass extinctions and, 542
    natural selection and role of, 26
    plant recruitment of, as herbivore defense, 869f
Predictions, scientific, 17, 18f, 19f, 493
Predisposition, cancer and inherited, 394
Preganglionic neurons, 1089
Pregnancy, human, **1034**
    conception and trimesters of, 1035f–1037f
    detecting disorders during, 1039–1040
    detecting human, 969
    DNA methylation and, 372f
    ectopic, 1039
    endocrine disruptors and, 1015
    prevention of, 1037, 1038f, 1039

Pre-mRNA, 339f, 345f–346f, 373f
Prepuce, 1025f, 1026
Pressure
    hearing and, 1114
    receptors for, 1110f
    root, 793f, 794
    water potential and, 788–790
Pressure-flow hypothesis, 800, 801f
Pressure potential, **789**
Presynaptic cells, 1068f
Presynaptic neurons, 1068f, 1079, 1080f, 1101f
Prey
    adaptations of native predators to introduced, 1217
    defensive adaptations of, 1218f–1219f, 1237
    genetic variation in selection of, 1155, 1156f
Prezygotic barriers, 508f–509f, **510**
Priapula, 688f
Primary cell walls, **118**
Primary cilium, 114
Primary consumers, **1240**f
    trophic efficiency of, 1246, 1247f, 1248
Primary electron acceptors, **196**f
Primary growth, plant, **766**, 783
    meristem generation of cells for, 766f–767f
    of roots, 768f–769f
    of woody stems, 772f, 775f
Primary immune response, **962**, **963**f
Primary meristems, **766**
Primary motor cortex, 1097f
Primary oocytes, **1029**f
Primary producers, **1240**f, 1257f
Primary production, **1241**
    in aquatic ecosystems, 1242, 1243f, 1243t
    in arctic tundra ecosystem, 1238f, 1244, 1256f–1257f
    climate change effects on, 1244f, 1245
    determining, with satellites, 1242f
    ecosystem energy budgets in, 1241, 1242f
    global, 1241, 1242f
    gross and net, 1241, 1242f
    in terrestrial ecosystems, 1243f–1244f, 1245
Primary roots, 759
Primary somatosensory cortex, 1097f
Primary structure, protein, **80**f
Primary succession, **1229**, 1230f–1231f
Primary transcripts, **339**
Primary visual cortex, 1121f
Primases, 323f, **324**, 326f, 327t
Primates. *See also* Humans
    cloning of, 430
    derived characters of, 744, 746
    HIV in, 567
    living, 746f–747f
    in mammalian phylogeny, 745f
    phylogenetic tree of, 746f
Primers, 323f, **324**
Primitive streak, 1052f
Primordial germ cells, 1029f
Principle of conservation of energy, 145f, 146
Printing press, 24
Prions, **412**, 413f
Probability
    laws of, 276, 277f, 278
    principle of maximum likelihood and, 563
Problem solving, **1146**, 1147f
Problem-Solving Exercises. *See* list, xxi
Proboscidea, 745f
Proboscis, 688f, 712f, 825f
Procambium, 766
Process, evolutionary, 470, 483
Prochlorococcus, 552
Producers, **9**, 188f, **614**, 615f
Production efficiency, **1246**f
Products, **41**, 651
Progesterone, 1004f, **1014**, 1015f, **1030**, 1032f
Progesterone receptor (PR), 392f–393f
Progestins, 1014, 1015f, 1038
Proglottids, 697f
Programmed cell death. *See* Apoptosis
Projections, body surface area and, 695f
Prokaryotes
    adaptive abilities of, 573f, 574
    archaea as, 585f, 586
    bacteria as, 583, 584f, 585
    beneficial and harmful impacts of, on humans, 587, 588f–591f

    bioremediation using, 1255f
    cell signaling in, 213f, 214
    cells of. *See* Prokaryotic cells
    chemiosmosis in, 175, 176f, 177
    as decomposers, 188
    ecological roles of, in biosphere, 586f–587f
    electron transport chains in, 168
    as endosymbionts, 534, 535f
    evolution of glycolysis in, 182
    genetic diversity in, 578f–580f, 581
    genome sizes and number of genes for, 448t, 449
    horizontal gene transfer in genomes of, 569, 570f
    hydrothermal vents, energy, and, 592
    land colonization by, 536
    nutritional and metabolic adaptations of, 581t, 582f
    origin of, 532f–534f
    photosynthesis in, 188f
    phylogeny of, 583f
    structural and functional adaptations of, 574f–577f, 578
    taxonomy of, 568, 569f
Prokaryotic cells, **6**, **97**. *See also* Cells
    cell-surface structures of, 574, 575f–576f
    DNA replication in, 322f–327f
    eukaryotic cells vs., 6f, 97f–98f, 99. *See also* Eukaryotic cells
    evolution of cell division of, 243, 244f
    programming of, by viral DNA, 315f–316f, 317
    regulation of transcription in, 366f–369f, 370
    structure of, 97f
    transcription and translation in, 337, 339f, 342, 343f–344f, 355f
Prolactin, 1004f, **1008**f, 1016
Prolactin-releasing hormone, 1008
Proliferative phase, 1032f, 1033
Proline, 77f, 341–342
Prometaphase, **237**, 238f, 243f, 252, 331f
Promiscuous mating, 1149
Promoters, **342**, 343f
Proof, hypotheses and, 18
Proofreading, DNA, 327, 328f
Propanal, 63f
Properties, chemical, 34f, 35
Properties, emergent. *See* Emergent properties
Prophages, **403**
Prophase (mitosis), **237**, 238f, 243f, 252, 263f, 331f
Prophase I, 260f, 262f–263f, 266f
Prophase II, 261f
Propithecus verreauxi, 746f
Prop roots, 760f
Prostaglandins, **1001**, 1037f, 1111–1112
Prostate cancer, 397
Prostate cells, 397
Prostate glands, 1025f, **1026**
Prosthetic groups, 175
Proteases, **907**
Proteasomes, 379
Protected areas, 1272f–1273f
Protective secretions, 898
Protein Data Bank, 444
Protein interaction networks, 446, 447f
Protein kinase A, 223, 224f
Protein kinases, **222**f–224f, 228f, 229
Protein phosphatases, **223**
Proteins, **75**. *See also* Amino acids
    amino acids of, 67, 75, 76f–78f. *See also* Amino acids
    antibiotics and prokaryotic synthesis of, 577
    antifreeze, 888
    in bacterial binary fission, 242, 243f
    biotechnology in production of, 435, 436f
    blood plasma, 934f, 935
    cadherin in choanoflagellates and animals, 676f
    in cell-signaling nuclear responses, 226f
    cell-signaling receptor, 217f–220f
    in cellular membranes, 127f–130f
    cellulose-synthesizing, 619f
    as composed of polypeptides, 78f. *See also* Polypeptides
    Conserved Domain Database (CDD) of structures of, 445f
    denaturation and renaturation of, 82f, 83
    digestion and absorption of, 908f–909f
    DNA-binding, 91
    in DNA replication, 326f, 327t
    DNA vs., as genetic material, 315, 316f, 317
    domains of, 347f

Proteins (continued)
in egg yolks, 91
in electron transport chains, 174f, 175
enzymes as, 75, 76f, 153, 159f
essential amino acids in, 899
evolution and divergence of, 91
evolution of genes for, with novel functions, 456, 457f
expressing cloned genes for, 423
facilitated diffusion and, 135f, 136
florigen as, 860f
folding and post-translational modifications of, 354
folding of, in cells, 83
four levels of structure of, 79, 80f–81f
as fuel for catabolism, 182f
functions of types of, 66f, 76f
gene expression and, 7f–8f
genotypes and, 335
human dietary deficiencies in, 901
identifying genes coding for, 445f, 446
innate immune response and, 957
as macromolecules, 66
as measures of evolution, 87, 89
mediator, 374, 375f
motor. See Motor proteins
nonenzyme, 337
nucleic acids in gene expression and synthesis of, 84f
packing of DNA and, into chromosomes, 330f–332f
phenotypes and, 335
phosphorylation of, in cell-signal transduction, 222f, 223
in photosystems, 196f–200f, 201
plant antifreeze, 865
plant heat-shock, 865
post-translational modification of, in plant responses, 845
prions as infectious, 412, 413f
producing, with gene cloning, 418f, 419
in prokaryotic flagella, 576f, 577
protein interactions with, 221
proteomics in study of, 9, 86, 87f, 446
regulation of eukaryotic processing and degradation of, 378–379
regulatory, 378–379, 383, 384f
in replication of DNA telomeres, 328
repressor, 367f
scaffolding, 228f, 229
separating, with gel electrophoresis, 420f
structures of, 78, 79f
synthesis of. See Protein synthesis
synthesis of, in plant cells, 208f
systems biology in studying networks of, 446, 447f
transcription factors, 343f, 344
translation elongation factors, 352, 353f
translation initiation factors, 352f
translation release factors, 352, 353f
transport. See Transport proteins
in transport and mechanical work, 152f
using data on, to test hypotheses on horizontal gene transfer, 570
viral movement, 802f
viruses and, 400f, 402f–403f
visualizing, 79f
water-soluble, 49f
Protein sequences, 444, 445f, 446
in Scientific Skills Exercise, 570
Protein synthesis. See also Gene expression
evidence for, in study of metabolic defects, 336f–338f
gene concept and, 361–362
genetic code in, 340f–342f
mutations during, 357f–358f, 359
by ribosomes, 102, 103f, 104
summary of, 356f
via transcription, RNA processing, and translation, 337, 339f. See also RNA processing; Transcription; Translation
Proteobacteria, 584f
Proteoglycans, 118, 119f
Proteomes, 9, 446
Proteomics, 9, 86, 87f–88f, 446
Proterozoic eon, 530f, 533f
Prothoracicotropic hormone (PTTH), 1006f, 1007

Prothrombin, 936f
Protista, kingdom, 568, 594
Protists, 12f, 594
contractile vacuoles of, 108
in domain Eukarya, 12f, 13
ecological roles of symbiotic and photosynthetic, 614f–615f
endosymbiosis in evolution of, 594–595, 596f–597f, 608f, 609
excavates, 597, 598f, 600
origin of, 593
origin of fungi in, 659f, 660
photosynthesis in, 188f
phylogeny of, 597, 598f–599f
plants, 599f, 609. See also Plants
red and green algae, 599f, 609f–611f
SAR clade, 599f, 601f–608f, 609
sexual life cycles of, 258f, 259
as single-celled eukaryotes, 594
structural and functional diversity of, 594f
unikonts, 599f, 611, 612f–613f, 614
Protocell, 526, 527, 528f
Protoderm, 766
Protonema, 624f, 625
Protonephridia, 694, 984f, 985
Proton gradients, 176f, 177
Proton-motive force, 176f, 177
Proton pumps, 138f, 788, 789f, 848
Protons, 28, 30f, 31
Proto-oncogenes, 388, 389f, 391f
Protoplasts, 789
Protostome development, 681, 682f, 685
Proviruses, 406, 407f
Proximal control elements, 374
Proximal tubule, 987f–988f
Proximate causation, 1140
Prozac, 1081, 1103
Prusiner, Stanley, 413
Pseudemys nelsoni, 884f
Pseudocoelomates, 681
Pseudogenes, 450, 480
Pseudomonas aeruginosa, 582f
Pseudemys hermannsburgensis, 981
Pseudopodia, 117, 140f, 599f, 607, 608f, 690
Pseudostratified columnar epithelium, 877f
Psilotum, 632f, 633
P site (peptidyl-tRNA binding site), 350f, 352f–353f
Psychedelic rock gecko (Cnemaspis psychedelica), 1260f
Psychoactive drugs, 1081
PTC pedigree analysis case, 284f, 285
Pteraspis, 725f
Pterois volitans, 729f
Pteropus mariannus, 1262f
Pterosaurs, 736, 1135
Puberty, human, 1030
PubMed, 26
Puffballs, 665f
Puffer fish, 813f
Pulmocutaneous circuit, 924f, 925
Pulmonary circuit, 924f–926f
Pulp, fruit, 831
Pulse, 930
Punctuated equilibria, 520f, 521
Puncture vine, 832f
Pundamilia species, 515f, 519f, 520
Punnett squares, 273f, 274, 281f–282f, 292–293
Pupa, 711f
Pupil, 1118f
Pure elements, 37
Purines, 84, 85f, 320f
Purple sulfur bacteria, 188f
Pus, 956
Puszta, 1175f
Pygmy date palm, 650f
Pyramids
age-structure, 1209f
ecological, 1246, 1247f, 1248
Pyrenean ibex (Capra pyrenaica pyrenaica), 1266
Pyrimidines, 84, 85f, 320f
Pyrococcus furiosus, 421, 585, 589
Pyrophorus nyctophanus, 143f
Pyruvate
ATP yield in oxidation of, 177f
in fermentation, 181f, 182
oxidation of, as stage of cellular respiration, 168, 169f

oxidation of, to acetyl CoA, 171f–172f
oxidation of glucose to, by glycolysis, 170f–171f
Python molurus bivittatus, 887, 888f
Pythons, 887, 888f, 892
PYY hormone, 917f

Q
Q (ubiquinone), 175
Qinling golden snub-nosed monkeys, 56f
Qualitative data, 17f
Quantitative approach, G. Mendel's, 270f, 271
Quantitative characters, 281, 282f
Quantitative data, 17f. See also Scientific Skills Exercises
Quantitative RT-PCR (qRT-PCR), 424–425
Quaternary structure, protein, 81f
Quillworts, 631, 632f, 633
Quorum sensing, 213–214

R
RAAS (renin-angiotensin-aldosterone system), 995, 996f
Rabbits, 914
Radial canal, sea star, 714f
Radial cleavage, 681, 682f
Radial glia, 1090
Radial symmetry, 644f, 649, 679, 680f
Radiation, 885f
alterations of chromosome structure by, 307
animal heat exchange and, 885f
cancer and, 388, 394
as cancer treatment, 249
DNA damage from, 328
mutagenic, 360
Radicle, 829f–830f
Radioactive isotope decay curve in Scientific Skills Exercise, 33
Radioactive isotopes, 31–33, 316f, 317, 529, 530f
Radioactive tracers, 31, 32f
Radiolarians, 608f
Radiometric dating, 32–33, 529, 530f
Radula, 699f
Raft spider, 45f
Rain, acidic, 55
Rain shadow, 1169f
Random dispersion, 1192f, 1193
Random fertilization, 266, 305
Random mating, 492t
Random mutations. See Mutations
Randomness, entropy and, 146
Range expansions, species, 1183, 1184f
Ranges, actual and potential, 1184
Rangifer tarandus, 490f, 1021, 1256f–1257f, 1281f
Raphides, 868f
Rapid eye movements (REMs), 1094
ras gene, 389, 390f–391f
Raspberry fruit, 831f
Ras protein, 389, 390f
Rates, speciation, 520f–521f, 522
Ratfish, 719f, 726, 727f
Ratites, 739, 740f
Rattlesnakes, 1084, 1111f
Ravens, 1147
Ray-finned fishes, 719f, 728f–729f, 1091f
Rays, 719f, 726, 727f
Reabsorption, 984f
Reactants, 41
Reaction-center complexes, 196f
Reading frame, 341, 358f
Realized niches, 1215, 1216f
Reasoning
deductive, 18
inductive, 17
Receptacle, flower, 823f
Reception, sensory, 1108f, 1109
Reception stage, cell-signaling, 216
cell-surface transmembrane receptors in, 217f–220f
intracellular receptors in, 220, 221f
ligands, ligand binding, and receptor proteins in, 217
overview of, 217
in plant signal transduction pathways, 844f
plasma membrane proteins as receptors in, 217f–220f
Receptive fields, 1120–1121
Receptor-mediated endocytosis, 140f

Receptor potential, **1109**
Receptor proteins, 76*f*
Receptors
   antigen, 958*f*–961*f*
   dendrites as, 1068
   glutamate, 1101*f*
   hormone, 1000*f*–1004*f*, 1018
   opiate, 1082
   sensory, 1108*f*–1111*f*, 1112
   somatosensory, 1097*f*, 1098
   Toll-like, 955*f*, 976
Receptor tyrosine kinases (RTKs), **219***f*, 220, 222, 250
Recessive alleles, **272**, 273*f*–276*f*, 280, 285*f*–286*f*, 287, 293, 500. *See also* Alleles
Recessively inherited human disorders, 285*f*–286*f*, 287
Recessive traits, 271*f*, 272*t*, 285*f*, 299*f*, 300
Reciprocal altruism, 1158–**1159**
Reciprocal eco-evolutionary effects, 1187*f*
Recombinant bacteria, 418*f*
Recombinant chromosomes, **266***f*
Recombinant DNA, **418**
   in DNA cloning and gene cloning, 418*f*, 419
   ethical issues on, 438–439
   hirudin from leeches and, 704
   using restriction enzymes to make, 419*f*–420*f*
Recombinants (recombinant types), **302**
Recombinase, 960, 961*f*
Recombination. *See also* Recombinant DNA
   linkage maps based on frequency of, 305*f*–306*f*
   of linked genes, 302, 303*f*
   natural selection and genetic variation from, 305
   of unlinked genes, 302*f*
Recombination frequencies, 305*f*–306*f*, 313
Reconstruction, ecosystem, 1253*f*–1254*f*
Recovery, seed plants and, 636*f*
Recruitment, motor neuron, 1130
Recruitment of predatory animals, plant, 869*f*
Rectum, **911**
Red alga (*Chondracanthus harveyanus*), 1222
Red algae, 596*f*, 599*f*, **609***f*, 610
Red-bellied black snakes, 1217
Red blood cells. *See* Erythrocytes
Red-cockaded woodpecker (*Picoides borealis*), 1269*f*–1270*f*
Red-green color blindness, 299*f*
Red light, plant responses to, 856, 857*f*, 860*f*
Red lionfish (*Pterois volitans*), 729*f*
Red mangrove, 852*f*
Red maple tree leaves, 762
Red-necked phalaropes, 1151*f*
Redox (oxidation-reduction) reactions, **165**, 166*f*–168*f*, 190–191
Red-tailed hawk, 14*f*
Red-tailed racer snake (*Gonyosoma oxycephala*), 1053*f*
Red tide, 605*f*
Reduced hybrid fertility, 509*f*
Reduced hybrid viability, 509*f*
Reducing agents, **166***f*
Reduction, 165, 166*f*, 201, 202*f*
Reductionism, 4
Redundancy, genetic code, 341, 357, 489
Redwood trees, 255*f*
Reefs, ectoproct, 698
Reference genome, **443**
Reflexes, 906*f*, **1087**, 1088*f*
Refractory period, **1075**
Regeneration, 705, 713–714, 1020
Regenerative medicine, 432
Regional adaptive radiations, 543*f*, 544
Regional conservation, 1270, 1271*f*–1274*f*
Regional human population patterns, 1208
Regression lines in Scientific Skills Exercises, 54, 205, 751
Regulation. *See also* Osmoregulation
   of animal digestion, energy storage, and appetite, 914, 915*f*–917*f*, 918
   by animal endocrine and nervous systems, 880*f*
   of biological clocks, 1094–1095
   of biological rhythms, 1015
   of blood pressure, 930–931
   of cell-signaling responses, 226–227, 228*f*, 229
   of cellular respiration via feedback mechanisms, 183*f*, 184
   of cleavage, 1049
   by endocrine glands, 1011*f*–1015*f*, 1016
   of enzymatic catalysis, 159, 160*f*–161*f*

extracellular matrix role in, 119
feedback, of endocrine systems, 1005–1006
feedback in, of molecular interactions, 10*f*
of gene expression. *See* Gene regulation
   of growth, 1004*f*, 1009, 1010*f*
   of heart rhythm, 928*f*
   homeostatic feedback, 881*f*–883*f*
   hormonal, of mammalian sexual reproduction, 1030, 1031*f*–1032*f*, 1033–1034
   hormone cascade pathways in, 1008, 1009*f*, 1010, 1014
   of human breathing, 946*f*
   kidney, 994*f*–996*f*
   of muscle contraction, 1128, 1129*f*, 1130
   of neurons in sensory reception, 1108*f*
   plant and animal, 894*f*
   of population growth, 1202, 1203*f*–1206*f*, 1207
   as property of life, 3*f*
   thyroid, 1008, 1009*f*, 1010
Regulator animals, **881***f*
Regulatory genes, **367***f*
Regulatory proteins, 378–379, 383, 384*f*, 1128, 1129*f*, 1130
Regulatory T cells, 971
Reinforcement, hybrid zone, **518**, 519*f*
Rejection, immune system, 969–970
Relatedness, altruism and, 1158*f*–1159*f*
Relative abundance, species, **1222**. *See also* Species diversity
Relative fitness, **497**
Relay molecules, 216*f*, 217
Relay proteins, 228*f*, 229
Release factors, 352, 353*f*
Release stage, phage lytic cycle, 402*f*
Releasing hormones, 1004*f*, 1008
Renal cortex, **986***f*
Renal medulla, **986***f*
Renal pelvis, **986***f*
Renaturation, protein, 82*f*, 83
Renin-angiotensin-aldosterone system (RAAS), **995**, 996*f*
Repair, DNA, 327, 328*f*
Repetitive DNA, **450***f*
Replication fork, DNA, **323***f*
Replicative cycles, viral
   of animal viruses, 404, 405*f*, 406*t*, 407*f*
   general features of, 401*f*, 402
   of phages, 402*f*–404*f*
Repolarization, 1075
Repressible enzymes, 368*f*, 369
Repressible operons, 367*f*, 368–369
Repressors, **367***f*, 374
Reproduction. *See also* Life cycles; Mating; Sexual life cycles
   animal behavior evolution and, 1148, 1149*f*–1154*f*, 1161
   bacterial rate of, 592
   as cell division function, 235*f*
   crustacean, 709
   C. Darwin on natural selection and, 14*f*–16*f*
   delayed human, in population growth, 1208
   density-dependent population regulation by rates of, 1204*f*
   effective population size and, 1268*f*, 1269
   evolution and differential success in, 266, 267*f*
   fungal, 657, 658*f*–659*f*
   genetic disorders from human, 285–286
   heterochrony and differential reproductive development, 544–545*f*
   insect, 711
   iteroparity vs. semelparity in, 1200, 1201*f*
   life history traits in, 1200, 1201*f*–1202*f*
   overproduction of offspring and natural selection, 475*f*, 476
   plant and animal, 895*f*. *See also* Angiosperm reproduction; Animal reproduction
   prokaryotic binary fission, 577–578
   as property of life, 3*f*
   protist, 594
   protocell, 527, 528*f*
   rapid, as source of genetic variation in viruses, 489
   rapid prokaryotic, and mutation, 573, 578*f*, 579
   rates of, 1194, 1195*f*, 1204*f*, 1208, 1213
   sexual, as source of genetic variation, 489
   sexual vs. asexual, 255*f*
   of triploid organisms, 268
Reproduction, plant. *See* Angiosperm reproduction

Reproductive barriers
   in hybrid zones, 518, 519*f*, 520
   reproductive isolation and types of, 506–507, 508*f*–509*f*, 510
Reproductive cloning, 430*f*–431*f*
Reproductive cycles, animal, 1021*f*
Reproductive cycles, human, 1032*f*, 1033–1034
Reproductive isolation, **508**
   allopatric speciation and, 511*f*–512*f*, 513
   hybrid zones and, 516, 517*f*–519*f*, 520
   pollinator choice and, 522*f*
   reproductive barriers in, 506–507, 508*f*–509*f*, 510
   sexual selection and, 515*f*
Reproductive leaves, 762*f*
Reproductive organs, human
   female, 1026, 1027*f*
   gametogenesis and, 1027, 1028*f*–1029*f*
   male, 1025*f*, 1026
Reproductive rates, 1194, 1195*f*, 1204*f*, 1208, 1213
Reproductive success, 1024, 1042
Reproductive technologies, 1039*f*, 1040
Reptiles, 719*f*, **735**
   adaptations of kidneys of, 992
   amniotic eggs of, 734, 735*f*
   birds, 738*f*–740*f*
   breathing in, 925
   characteristics of, 735*f*, 736
   crocodilians, 737*f*, 738
   evolution of, 679
   extraembryonic membranes of, 1053*f*
   hearing and equilibrium in, 1116
   lepidosaurs, 737*f*, 738
   in Mesozoic era, 533
   origin and evolutionary radiation of, 736
   thermoregulation in, 887, 888*f*
   turtles, 737*f*
Research Method Figures. *See* list, xxiii
   absorption spectrum determination, , 193*f*
Reservoirs
   carbon, 1250*f*
   nitrogen, 1251*f*
   nutrient, 1249*f*
   phosphorus, 1251*f*
   water, 1250*f*
Residual volume, **945–946**
Resolution, 94
Resolution phase, sexual, 1034
Resource acquisition, vascular plant, 784*f*–786*f*, 787. *See also* Transport in vascular plants; Vascular plants
Resource competition, density-dependent population regulation by, 1204*f*
Resource partitioning, **1215**, 1216*f*
Respiration, cellular. *See* Cellular respiration
Respiration, plant and animal, 895*f*
Respiratory diseases, human, 1204*f*
Respiratory distress syndrome (RDS), 943, 944*f*
Respiratory media, 939*t*, 940
Respiratory pigments, **947***f*–948*f*, 949
Respiratory surfaces, 940
Respiratory systems. *See also* Gas exchange
   adaptations of, 947*f*–949*f*
   breathing to ventilate lungs in, 944, 945*f*–946*f*
   effects of adrenal hormones on, 1012
   internal exchange surfaces and, 875*f*
   lungs and components of, 942, 943*f*
   respiratory distress syndrome of, 944*f*
Response
   homeostatic, **882**
   immune, 952*f*, 957*f*, 962, 963*f*
   inflammatory, 955*f*–956*f*
Response pathways, endocrine system, 1000*f*–1004*f*
Response stage, cell-signaling, **217**
   cell-signaling specificity and coordination of responses in, 227, 228*f*
   increasing signaling efficiency in, 228*f*, 229
   nuclear and cytoplasmic responses in, 226*f*–227*f*
   overview of, 216*f*, 217
   regulation of responses in, 226–227, 228*f*, 229
   signal amplification in, 226–227
   signal termination in, 229
Response to environment, 3*f*, 894*f*. *See also* Environment
Rest-and-digest responses, 1089
Resting potentials, neuron, **1069**
   formation of, 1069, 1070*f*
   modeling of, 1071*f*

INDEX   **I-42**

Resting potential state, action potential, 1074f, 1075
Restoration ecology, 1253f
    biological augmentation in, 1255
    biomanipulation of trophic levels in, 1227f
    bioremediation in, 1253, 1255f
    worldwide projects of, 1254f
Restriction enzymes, **404**, **419**
    making recombinant DNA plasmids using, 419f–420f
    polymerase chain reaction and, in gene cloning, 421, 422f
Restriction fragments, **419**–420
Restriction sites, **419**f
Resurrection
    of extinct species, 1266
Reticular fibers, 878f
Reticular formation, 1094f
Reticulum, 914f
Retina, **1118**f, 1122, 1138
Retinal, **1119**f–1120f
Retinitis pigmentosa, 495
Retrotransposons, **451**f
Retroviral vectors, gene therapy using, 434f
Retroviruses, **406**t, 407f
Reverse transcriptase, **406**, 407f, 425f
Reverse transcriptase-polymerase chain reaction (RT-PCR), **424**, 424, 425f, 433
Reward system, brain, 1103f
Reward vs. risk, foraging behavior and, 1149
R groups, amino acid, 75, 76f–77f
Rhabditophora, 694, 695f–697f, 698
Rhagoletis mendax, 508f
Rhagoletis pomonella, 508f, 515–516
Rhesus monkeys, 89
Rheumatoid arthritis, 971f
Rhinoceros, 1199f, 1266
Rhizarians, 599f, **606**–607, 608f, 609
Rhizobacteria, **814**f–816f, 817
Rhizobium bacteria, 592, 814f–816f, 817
Rhizoids, **625**
Rhizomes, 761f
Rhizopus stolonifer, 662f, 663
Rhizosphere, **814**f
Rhodophytes, 609f, 610
Rhodopsin, **1119**f–1121f
Ribbon model, 79f
Ribbon worms, 688f
Ribonucleic acid. *See* RNA
Ribose, **85**
    in ATP, 150, 151f
    as monosaccharide, 68f
    in nucleic acids, 85f
Ribosomal RNA (rRNA), **349**, 350f
    in eukaryotic cell nucleus, 102
    evolutionary rate of, 565
    gene family of, 453f
    interpreting comparisons of sequences of, 595
    in molecular systematics, 583
    self-splicing in production of, 346
Ribosomes, **102**, **339**
    animal cell, 100f
    identifying binding sites on, with sequence logos, 351
    plant cell, 101f
    in plant cells, 208f
    polyribosomes and, **355**f
    prokaryotic, 577
    in prokaryotic and eukaryotic cells, 97f
    in protein synthesis, 102, 103f, 104
    protein synthesis in, 84
    in RNA virus replicative cycle, 405f
    rough ER and, 105
    structure of, 103f, 349, 350f
    in translation, 337, 339f, 349, 350f, 352f–353f
Ribozymes, 153, **346**, **528**
Ribulose, 68f
Rice (*Oryza sativa*), 438, 509f, 651, 817, 838, 850–851, 1263
Rieseberg, Loren, 521f
Right atrium, 926f–927f
Right ventricle, 926f–927f
Ringed seals (*Phoca hispida*), 44f
Ring structures
    of carbon skeletons, 60f
    of cellulose-synthesizing proteins, 619f
    of glucose, 69f
Ringworm, 670

Rising phase, action potential, 1074f, 1075
Risk factors, cardiovascular disease, 938
Risk vs. reward, foraging behavior and, 1149
Riverine wetlands, 1179f
River otters, 881f
Rivers, 1180f
RNA (ribonucleic acid), **84**. *See also* Messenger RNA; Nucleic acids; RNA processing; Transfer RNA
    in circadian clock genes during hibernation, 893f
    components of, 84, 85f
    development of self-replicating, 528
    in DNA replication, 323f, 324
    elongation of strands of, 344f
    in eukaryotic cells, 102
    gene density in genomes and, 449
    in gene expression, 8f
    in insect infection, 954f
    interpreting sequences of, 595
    noncoding, in gene regulation, 379, 380f, 381
    in plant cells, 208f
    post-transcriptional modification of, 345f–347f
    in regulation of cleavage, 1049
    ribosomal RNA gene family, 453f
    roles of, in gene expression, 84f
    sequencing of, 426f
    structure of, 86f
    synthesis of, in transcription, 337, 339f, 342, 343f–344f
    viral. *See* RNA viruses
RNA enzymes. *See* Ribozymes
RNA interference (RNAi), **380**, **427**
RNA polymerases, **342**, 343f–344f, 373f, 374
RNA processing, **345**
    alteration of mRNA ends in, 345f
    eukaryotic transcription and, 373f, 374
    evolutionary importance of introns in, 346, 347f
    in gene expression and protein synthesis, 337, 339f–340f
    regulation of eukaryotic, 378f
    ribozymes in, 346
    split genes and RNA splicing in, 345f–347f
    summary of eukaryotic, 356f
RNA sequencing (RNA-seq), **425**f, 426, 448
RNA splicing, **345**f–347f, 378f
RNA transcript, 343f–344f
RNA viruses
    classes of, 406t
    emerging viral disease and, 409–410
    HIV as, 406, 407f. *See also* HIV
    replicative cycles of, 401f, 402, 404, 405f
    retroviruses as, 406t, 407f
    structure of, 400f, 401
Robin, European, 14f
Rochman, Chelsea, 1163
Rock python, 903f
Rocks
    dating of, 529, 530f
    nonvascular plants and weathering of, in Ordovician period climate change, 629
    species distributions and, 1186
    weathering of, in phosphorus cycle, 1251f
Rod, ATP synthase, 175f
Rodents, 745f, 914
Rods (photoreceptor), **1119**f–1121f, 1122–1123, 1138
Rod-shaped prokaryotes, 97f, 574f
Rolling circle replication, 580f
Root caps, **768**f
Rooted phylogenetic trees, 556f, **557**
Root hairs, 758, 759f–**760**f, 778f, 792
Root pressure, 793f, 794
Roots, **630**, 758–**759**
    apical meristems of plant, 621f
    architecture of, and acquisition of water and minerals, 787
    bacterial mutualism with, 592
    compositions of bacterial communities of, 814f
    drought responses of, 863–864
    evolution of, in vascular plants, 63
    fungal mycorrhizae and, 787
    gravitropism in, 861f
    monocot vs. eudicot, 649f
    mycorrhizae and plant, 656f, 657
    nitrogen-fixing bacteria and legume, 815f–816f, 817
    primary growth of, 768f–769f
    rhizoids vs., 624
    secondary growth of, 772f–774f
    seedling, 830f

soil texture and, 806
    structure of, 759f–760f
    transport of water and minerals from, to shoots via xylem, 792, 793f–795f, 796
Root systems, **759**f–760f
Rosbash, Michael, 883
Rosy periwinkle (*Catharanthus roseus*), 1263f
Rotifers, 687f, 697f, 698
Rotor, ATP synthase, 175f
Rough ER, 100f–101f, 104, 105f, 109f
Round dance, honeybee, 1142f
Round window, **1114**, 1115f
Roundworms, 689f, 705f, 706, 1058f–1059f, 1133f
Rous, Peyton, 394
Rozella allomycis, 661f
R plasmids, **581**
r-selection, **1202**
RTKs (receptor tyrosine kinases), **219**f, 220, 222, 250
RU486 (mifepristone), 1039
Rubella, 409
RuBP carboxylase, **201**, 202f
Ruffed grouse (*Bonasa umbellus*), 1271
Rule of multiplication, 491
Rumen, 914f
Ruminants, 914f
Running, 1135–1136
Runoff
    experimental, and nutrient cycling, 1252f
    in nitrogen cycle, 1251f
    in water cycle, 1250f
Rupicapra rupicapra, 901f
Rusts, 665
Rusty-patched bumblebee (*Bombus affinis*), 1170f
Ryther, John, 1243f

S
Saccharomyces cerevisiae (yeast)
    budding of, 659f
    cell signaling in, 213f, 214
    DNA changes in cells of, in meiosis, 264
    genome size and number of genes of, 448t, 449
    human uses of, 670–671
    protein interaction network of, 447f
    toxic wastes of, in winemaking, 1204f
Saccule, 1115f
Sac fungi, 663f–664f, 665t
Safety issues
    on biotechnology, 438–439
    on plant biotechnology and genetic engineering, 839–840
Saguaro cactus (*Carnegiea gigantea*), 1178, 1183f
Sahara Desert, 525f, 526
Sahelanthropus tchadensis, 748f, 749
Salamanders, 509f, 513, 544, 545f, 719f, 731, 732f, 921f, 1066, 1086f
Salicin, 651
Salicylic acid, **867**
Salinity
    extreme halophiles and, 573f, 585
    lap alleles and, 505
    osmosis, water balance, and, 134
    plant responses to, 142, 865
    soil salinization and, 808
    species distributions and, 1185f, 1186
Saliva, 906
Salivary glands, **905**, 906f
Salmon, 89, 438, 978f, 979, 1200
Salmonella species, 588
Salmo salar, 89
Saltatory conduction, **1076**f–1077f
Salt marshes, 1186, 1221f, 1247
Salt marsh hay (*Spartina patens*), 1186
Salts (ionic compounds), **38**f
    in blood plasma, 934f
    diffusion of, across capillary walls, 932, 933f
    ion gradients and transport of, 993f
    osmoregulation of, 977f–982f
Salt stress, plant responses to, 865
Salt water, 977f, 979f, 982f
Salty tastes, 233, 1123f–1124f
Sampling techniques, population, 1191f
San Andreas fault, 539
Sand dollars, 235f, 689f, 715
Sandhill crane (*Grus canadensis*), 1144
Sandy inland mouse (*Pseudomys hermannsburgensis*), 981

Sanger, Frederick, 416
Sapwood, 774f
SAR clade, 599f, **601**f–**608**f, 609
Sarcomeres, 879f, **1126**f
Sarcoplasmic reticulum (SR), **1128**f–1129f
Sarcopterygii (lobe-fins), 728, 729f, 730
Sargasso Sea, 1242, 1243t
Sarin, 158, 1080
Satellites, determining primary production with, 1242f
Satiety center, 917f
Saturated enzymes, 156
Saturated fats, 73f, 74, 128f
Saturated fatty acids, **73**f, 74
Savannas, 750, **1174**f
Savory tastes, 1123f–1124f
Sawfish, 1020
Scaffolding proteins, **228**f, 229
Scala naturae (scale of nature), Aristotle's, 470
Scale bar in Scientific Skills Exercise, 99
Scale-eating fish (*Perissodus microlepis*), 500, 501f
Scales
    fish, 728
    reptile, 735
Scallops, 701f
Scanning electron microscope (SEM), **94**, 95f, 96
Scanning electron microscopy (SEM), 95f
Scarlet fever, 403
Scatter plots in Scientific Skills Exercises, 54, 136, 205, 513, 751, 1217
Sceloporus species, climate change and, 11f
Schindleria brevipinguis, 719
Schistosoma mansoni, 696f
Schistosomiasis, 696f
Schizophrenia, 372f, **1102**f
Schmidt-Nielsen, Knut, 1136
Schwann cells, **1076**f–1077f, 1090f
Science, **16**
    biology as study of life by, 3. *See also* Biology
    cooperative approach and model organisms in, 22–23, 24f. *See also* Model organisms
    diverse viewpoints in, 24
    inquiry process of, 17, 18f–21f, 22. *See also* Case studies; Inquiry Figures
    medical. *See* Medicine
    methods. *See* Research Method Figures
    skills. *See* Scientific Skills Exercises
    society, technology, and, 23, 24f
Scientific inquiry, 16–20. *See also* Case studies; Inquiry Figures; Research Method Figures; Scientific Skills Exercises
Scientific method, 17, 18f–21f, 22. *See also* Case studies; Inquiry Figures; Research Method Figures; Scientific Skills Exercises
Scientific notation in Scientific Skills Exercise, 1082
Scientific Skills Exercises. *See list*, xxi
Scion, **835**–836
Sclera, 1118f
Sclereids, 764f
Sclerenchyma cells, **764**f, 770
Sclerenchyma tissue, 868f
Scolex, 697f
Scorpions, 707, 708f
Scrapie, 412
Scr gene, 545
Scrotum, **1025**f
Scrub jays, 1099, 1162
Scutellum, 829f
Scyphozoans, 687f, 691f–692f
Sea anemones, 692f–693f
Seabirds, 1238f, 1244
Sea cucumbers, 715f
Sea daisies, 713, 714f
Sea horse, 729f
Sea lampreys, 723f, 724
Sea lettuce, 610f
Sea lilies, 715
Seals, 44f, 874f, 949f
Sea otters, 1189
Sea slugs, 700f, 1020f
Seasonality, 1167f
Seasonal turnover, lake, 1178f
Sea spiders, 707–708
Sea squirts, 721f
Sea stars, 689f, 713, 714f, 898f, 940f, 1086f, 1192f, 1226f
Sea urchin (*Centrostephanus rodgersii*), 1281f

Sea urchins, 509f, 689f, 713, 715f, 1044f–1047f, 1184f–1185f, 1189
Sea wasps, 692f
Seawater, 977f, 979f, 982f
Seaweed, 599f, 602, 603f, 604, 609f, 610, 1184f
Secondary cell walls, **118**
Secondary consumers, **1240**f
    trophic efficiency of, 1246, 1247f, 1248
Secondary endosymbiosis, **596**f–597f
Secondary growth, plant, **766**, 783
    cork cambium and periderm production in, 774
    evolution of, 774
    meristem generation of cells for, 766f–767f
    vascular cambium and secondary vascular tissue for, 772f–775f
Secondary immune response, **962**, 963f
Secondary oocytes, **1029**f
Secondary production, **1246**
    in arctic tundra ecosystem, 1256f–1257f
    production efficiency in, 1246f
    in salt marsh ecosystem, 1247
    trophic efficiency and ecological pyramids in, 1246, 1247f, 1248
Secondary structure, protein, **80**f
Secondary succession, 1230–**1231**
Secondary vascular tissue, 772f–774f
Second law of thermodynamics, 143, 145f, **146**, 1239
Second messengers, **223**, **844**
    calcium ions, inositol trisphosphate, and diacylglycerol as, 224, 225f
    cyclic AMP as, 223f–224f
    in hormone pathways, 1002, 1003f
    in plant signal transduction, 844f, 845
Secretin, 915f, 1005f
Secretion, excretory system, **984**f
Secretions
    animal hormone, 1001f–1004f
    cell signaling and. *See* Animal hormones; Neurohormones
    digestive, 898, 907f–908f, 909–910, 915f
    protective, 898
Secretory phase, 1032f, 1033
Secretory proteins, 105
Secretory systems, prokaryotic, 577
Secretory tubules, marine bird, 982f
Sedentarians, 703f–704f, 705
Sedimentary strata, 470f, 529f
Seed coat, **828**f–829f
Seedless vascular plants, **622**
    gametophyte-sporophyte relationships in, 637f, 638
    importance of, 633f, 634
    life cycles of, 630f
    origin and traits of, 628f–631f
    phylogeny of, 622t, 631, 632f, 633
Seedling development, 829, 830f
Seed plants. *See also* Angiosperms; Gymnosperms; Plants
    adaptations of, 653
    angiosperms, 637f, 638, 644f–650f. *See also* Angiosperms
    gymnosperms, 637f–643f
    importance of, to Earth's ecosystems, 636f
    importance of, to human welfare, 651f, 652
    phylogeny of, 623f
    seeds and pollen grains as terrestrial adaptations of, 637f–638f, 639
    threats to biodiversity of, 651f, 652
Seeds, **622**–623, 636–**637**, **828**
    abscisic acid in dormancy of, 852
    development of, 826, 827f–829f
    dispersal of, 832f
    dormancy of, 639
    evolutionary advantage of, 639
    fruits and angiosperm, 645f
    gibberellins in germination of, 851f
    glyoxysomes in, 112
    phytochromes in germination of, 856, 857f
    plant seedlings development from, 828f–830f
    strigolactones in germination of, 855
Segmentation, 1066
Segmented worms, 688f
Segregation, law of, 271f–275f, 296, 297f
Seizures, 1076
Selective breeding, 474, 475f, 841
Selective permeability, **131**, 132f, 1071f
Self-assembly, protocell, 527, 528f

Self-assembly stage, phage lytic cycle, 402f
Self-fertilization, 834, 835f, 840–841, 1042
Self-incompatibility, **834**, 835f, 840–841
Selfing, 834, 835f, 840–841
Self-pollination, 271f
Self-pruning, 786
Self-replicating molecules, 526, 527f–528f
Self-splicing, 346
Self-thinning, 801
Self-tolerance, 961
Semelparity, **1200**, 1201f
Semen, **1026**, 1034
Semicircular canals, **1113**f, 1115f, 1116
Semiconservative model, DNA replication, **321**f–322f
Semilunar valves, **927**f
Seminal vesicles, 1025f, **1026**
Seminiferous tubules, **1025**f
Senescence, **853**–854
Senile dementia, 83
Sensitive period, **1144**, 1145f
Sensitive plants (*Mimosa pudica*), 802, 862f
Sensors, homeostatic, **882**f
Sensory adaptation, **1109**–1110
Sensory areas, cerebral cortex, 1096, 1097f–1098f
Sensory input, 1068f, 1096, 1097f–1098f
Sensory neurons, 1068, 1069f, **1088**f
Sensory pathways, 1108f–1109f
Sensory reception, **1108**f, 1109
Sensory receptors, **1108**f–1111f, 1112
Sensory systems
    animal and plant, 894f
    arthropod, 707f
    mechanoreceptors for hearing and equilibrium in, 1112f–1116f
    motor systems and, 1107f–1108f. *See also* Motor systems
    muscle function and, 1125, 1126f–1131f, 1132
    sensory receptors in, 1108f–1111f, 1112
    shark, 727
    skeletal systems, locomotion, and, 1132f–1135f, 1136
    snake, 738
    taste and smell receptors in, 1123f–1125f
    visual receptors in, 1117f–1122f, 1123
Sensory transduction, **1109**, 1120f
Sepals, **644**f, **823**f
Separase, 240
Separate electron orbitals model, 35f
Septa, **656**f
September equinox, 1167f
Septic shock, 957, 976
Sequence-based phylogenetic tree in Scientific Skills Exercise, 411
Sequence logo in Scientific Skills Exercise, 351
Sequencing by synthesis, 415f–417f. *See also* DNA sequencing
Sequoiadendron giganteum, 774f
Sequoia trees, 643f, 774f
Serial endosymbiosis, **534**, 535f
Serine, 77f, 222
Serotonin, 1081t
Serpentine plant communities, 30f
Sertoli cells, 1031f
Serum, 934
Seta, **625**
Set point, homeostatic, **882**f
Severe combined immunodeficiency (SCID), 434f, 971
Sewage treatment, 586, 591
Sex
    chromosomal basis of, 298f, 299
    genomic imprinting and, 310f, 311
Sex attractant, 43
Sex chromosomes, **257**
    aneuploidy of, 309
    human, 256f–257f, 258
    patterns of inheritance of, 298f–300f
Sex determination, 1030
Sex hormones
    endocrine disruptors and, 1015
    as environmental toxins, 1276
    functional groups and, 62f
    in regulation of mammalian reproduction, 1030, 1031f–1032f, 1033–1034
    in sex determination, 1030
    smooth ER synthesis of, 104–105
    as steroids, 75f, 1014, 1015f

Sex-linked genes, 298f–300f, **299**
Sex pili, 576, 580f, 581
Sex reversal, 1020
Sexual dimorphism, **499f**, 752, 1149–1150, 1151f
Sexual intercourse, human, 1034, 1081–1082
Sexual life cycles. *See also* Life cycles
    alternation of fertilization and meiosis in, 256f–258f, 259
    angiosperm, 826, 827f, 828
    asexual reproduction vs., 255f
    chromosomal basis of Mendelian inheritance in, 294f, 295. *See also* Chromosomal basis of inheritance
    evolution from genetic variation produced in, 265f–267f
    genetics of heredity and genetic variation in, 254–255
    graphing DNA changes in meiosis of, 264
    human chromosome sets in, 256f–257f, 258
    inheritance of chromosomes and genes in, 255
    karyotypes of chromosomes in, 256f
    mitosis vs. meiosis in, 262, 263f, 264
    protist, 594
    stages of meiosis in, 259f–262f
    varieties of, 258f, 259
Sexually transmitted diseases (STDs), 600f, 1038–1039. *See also* AIDS; HIV
Sexually transmitted infections (STIs), 1038–1039
Sexual reproduction, **255**. *See also* Sexual life cycles
    in angiosperms. *See* Angiosperm reproduction
    in animals, 674f, 675. *See also* Animal reproduction
    asexual reproduction vs., 255f, 833–834
    in bryophytes, 623, 624f–627f
    flowers and angiosperm, 644f
    in fungi, 658f
    inheritance in, 255
    microevolution from sexual selection in, 499f–500f
    as source of genetic variation, 489
    switch to, 268
Sexual reproduction, animal, **1020**. *See also* Animal reproduction; Human reproduction
    asexual reproduction vs., 1021, 1022f
    as evolutionary enigma, 1021, 1022f
    fertilization mechanisms in, 1022f–1024f
    reproductive cycles in, 1021f
    variations in patterns of, 1020f
Sexual response, human, 1034
Sexual selection, **499f**–500f, 515f, 1151, 1152f–1153f, 1154
S-genes, 835
Shade avoidance, plant, 857
Shaffer, Mark, 1268
Shannon diversity index (*H*), **1222**, 1223f
Shapes
    cell, 116–117
    enzyme, 155f, 156, 160f–161f
    insect pollinators and flower, 649f
    molecular, 39f–40f
    morphogenesis and cell, 1056f–1057f
    of organic compounds, 59f
    prokaryotic, 574f
Shared ancestral characters, **560**, 561f
Shared derived characters, **560**, 561f
Shark Bay, 532f
Sharks, 442f, 719f, 726, 727f, 890f, 979, 1091f
Shelf fungi, 665f
Shell drilling adaptation, 542
Shells, electron, 32f, **34**, 35f, 36
Shewanella oneidensis, 1255f
Shivering thermogenesis, 887, 888f
Shoots, 848f
    apical meristems of plant, 621f
    light capture and architecture of, 785, 786f, 787
    photosynthesis and architecture of, 804
    phytochromes and, 871
    primary growth of, 769f–771f
    transport of water and minerals from roots to, via xylem, 792, 793f–795f, 796
Shoot systems, **759f**
Short-day plants, **859f**
Short tandem repeats (STRs), **436**, 437f, **452**, 462
Short-term memory, **1100**–1101
Shrews, 1234, 1235f
Shrimp, 464f, 512f, 513, 708–709
Siamese cats, 364

Siberian tiger (*Panthera tigris altaica*), 1288
Sickle-cell disease, **82f**, **286**, **935**
    abnormal hemoglobin in, 935
    CRISPR-Cas9 system for, 360, 361f
    emergent properties of, 505
    evolution of, 503f
    genetic diagnosis of, 433
    heterozygote advantage in, 501
    inheritance of, 293
    pleiotropy and, 280
    point mutations and, 357f, 488, 502f
    protein primary structure changes and, 82f
    as recessively inherited, 286f, 287
    scientific cooperation and, 22–23
Sickle-cell trait, 286
Side-blotched lizards, 1154f
Side chains, amino acid, 75, 76f–77f
Sidedness, cellular membrane, 131f
Sieve plates, **765f**, 799–800
Sieve-tube elements, **765f**
Sieve tubes, 799, 800f–801f
Signaling molecules, 215f, 216. *See also* Animal hormones; Neurohormones; Neurotransmitters
Signaling molecules, endocrine system, 1000f–1004f. *See also* Animal hormones
Signal peptides, **354f**
Signal-recognition particles (SRPs), **354f**
Signals, animal, **1141f**–1143f
Signal transduction, **1002**
    in hormone pathways, 1002, 1003f
    membrane proteins and, 130f
Signal transduction pathways, **216**
    cancer and interference with normal, 389f–390f, 391
    in cell signaling, 212, 216f, 221–222, 233
    coordinate control of, 376
    differential gene expression and induction in, 382f, 383
    evolution of, 213
    linking signal reception to plant responses, 843f–844f, 845
    neurotransmitters and, 1080
    second messengers in, 223f–225f
    in sensory amplification, 1109–1110
    in visual sensory transduction, 1120f
    in yeast cell mating, 214
Sign language, 1106
Sign stimulus, **1140f**
Silencing, gene expression, 427
Silencing, transcription, 374
Silent mutations, **357**, 358f
Silent Spring (Carson), 1276
Silicon-based life, 65
Silk, 80f, 708, 951
Silk moth, 1006f, 1007
Silkworm moths, 1110, 1111f
Silverfish, 712f
Silversword plants, 543f, 558, 1183
Simple columnar epithelium, 877f
Simple endocrine pathway, 1004, 1005f
Simple fruits, **831f**
Simple leaves, 761f
Simple neuroendocrine pathway, 1005f
Simple sequence DNA, **452**
Simple squamous epithelium, 877f
Single bonds, **36**, 37f
Single-celled organisms, 5f, 93, 532f–535f, 594. *See also* Protists
Single circulation, **924f**, 925
Single-lens eyes, **1118f**–1119
Single nucleotide polymorphisms (SNPs), **427f**, 428, 433–434, 441, 461–462, 466
Single-strand binding proteins, **323f**, 327t
Single-stranded DNA (ssDNA) viruses, 406t
Single-stranded RNA (ssRNA) viruses, 405f, 406t
Sinoatrial (SA) node, **928f**
Sirenia, 745f
Sister chromatid cohesion, 236, 259
Sister chromatids, **236f**, 237, 257f, 260f–263f, 264, 266f
Sister species, 512f
Sister taxa, **555**, **556f**
Situs inversus, 1064f
Sixth mass extinction, 541f, 542
Size
    brain, 756, 1091f

carrying capacity and population, 1197, 1198f–1199f, 1198t, 1200
changes in population, 1196
determining population, using mark-recapture method, 1191f
ecosystem, 1239f
evolution of axon, 1076f–1077f
extinction risks in small population, 1266, 1267f–1268f, 1269
of genomes, 448t, 449
of global human population, 1207f–1210f, 1211, 1213
Hardy-Weinberg equilibrium and population, 492t
of *Homo floresiensis*, 754
of hormones, 216
of insects, 951
latitude and animal, 897
life history traits and, 1200, 1201f–1202f
locomotion costs and, 1136
metabolic rate and animal body, 891f
of plasmodesmatal pores, 802
prokaryotic cell, 573–574
of prokaryotic vs. eukaryotic cells, 6f, 98
protist, 593f
of skeletons, 1134
Skeletal muscle, **879f**, **1126**. *See also* Muscle
    in human breathing, 945f
    locomotion from contraction of, in skeletal systems, 1132f–1135f, 1136
    muscle fibers of, 1130, 1131t
    regulation of contraction of, 1128f–1130f
    sliding-filament model of contraction of, 1126, 1127f, 1128
    structure of, 1125, 1126f
Skeletal systems
    bones and joints of human, 1134f
    cartilage, 723–726
    energy costs of locomotion by, 1136
    locomotion from interaction of muscles and skeletons in, 1132f, 1135f, 1136
    origins of, 725
    types of, 1132, 1133f–1134f
Skeletons, carbon, 60f–61f
Skills, scientific. *See* Scientific Skills Exercises
Skin
    cancer of, 328, 1284
    coloration, 1016, 1018
    gas exchange and, 924f, 925
    heat exchange and, in animals, 885
    mechanoreceptors in human, 1110f
    pigmentation of human, 281, 282f
Skinner, B. F., 1146
Skulls, human vs. chimpanzee, 544f, 558
Skunks, 508f, 1218f
Sleep, ACTH secretion during, 1014
Sleep, brain functions and, 1094f
Sleeping sickness, 65, 600, 601f, 713, 972
Sleep movements, plant, 857, 858f
Sliding-filament model, **1126**, 1127f, 1128
Slime, hagfish, 723f
Slime bacteria, 213f
Slime layer, 575
Slime molds, 612f–613f
Slope in Scientific Skills Exercises, 157, 1049
Slow block to polyspermy, 1045
Slow-twitch fibers, **1131**
Slugs, 699–700, 1155, 1156f
Small interfering RNAs (siRNAs), **380f**
Small intestine, 695f, **908**
    absorption in, 909f–910f
    adaptations of, 912f
    cells of, 125
    digestion in, 908f, 909
Small populations, conservation of, 1266, 1267f–1268f, 1269
Smallpox, 408–409, 968
Smithells, Richard, 902f
Smooth ER, 100f–101f, 104, 105f, 109f
Smooth muscle, **879f**, **1131**–1132
Smuts, 665
Snails, 508f, 522, 538, 688f, 699, 700f, 702f, 1067f–1068
Snakes, 498, 499f, 719f, 735, 737f, 738, 925, 968, 1053f, 1111f, 1217, 1218f, 1264
    gene flow in Lake Erie water snake, 496, 497f
    prey selection of, 1155, 1156f

Snapdragons, 279f
Snapping shrimp, 512f, 513
Snook, Rhonda, 1024f
Snow geese, 1256f–1257f
Snow pea, 650f
Snowshoe hares (*Lepus americanus*), 501f, 1205f
Soapberry bugs, 477f, 478, 1187
Social behavior, 1159f, 1160
Social learning, **1147**f–1148f, 1153
Social play, 1159
Society
    plant biotechnology and, 840
    population age structure and, 1208
Society, science and, 19f, 23, 24f
Sociobiology, 1159–1160
Sociobiology: The New Synthesis (book), 1159
Sockeye salmon (*Oncorhynchus nerka*), 978f
Sodium, 29t
Sodium chloride
    elimination of excess, by marine birds, 982f
    human diets and, 901
    kidney processing of, 988f–990f, 991
    osmoregulation of, 992, 993f
    plant responses to excessive, 865
    as table salt, 29f, 38f, 49f
    in treating diarrhea, 139
Sodium ions, 1070t, 1071f
Sodium-potassium pump, **137**, **1069**–1070
    as active transport, 136, 137f
    neuron resting potential and, 1070f–1071f
    ouabain and, 1084
Software, bioinformatics, 444, 445f–448f
Soil
    acid precipitation and, 821
    antibiotics from bacteria in, 590
    bacteria in, 584f, 586f, 814f–816f, 817
    bryophyte reduction of nitrogen leaching from, 627f
    determining diversity of bacteria in, 1223f
    fossils in, 529f
    fungi in, 656f
    limiting nutrients in, 1244
    plant resource acquisition from, 787
    plant response to excessive salt in, 142, 865
    species distributions and, 1186
    sustainable agriculture and conservation of, 807f–809f
    texture and composition of, 806f–807f
Soil conservation, 807f–809f
Soil horizons, **806**f
Soil worm model organism. *See Caenorhabditis elegans*
Solar cycle, 1282
Solar energy
    absorption of, determining primary production with, 1242f
    in energy flow and chemical cycling, 9f, 1240f
    global energy budget and, 1241
    in photosynthesis, 41f, 187f
    primary production in aquatic ecosystems and limitations of, 1242
Solidago canadensis, 812
Solute potential, **789**, 804
Solute potential equation, 804
Solutes, **49**
    animal osmoregulation of, 981, 982f
    diffusion of, 132, 133f
    diffusion of, across capillary walls, 932, 933f
    effects of, on water potential, 789–790
    ion gradients and transport of, 993f
    kidney gradients of, 989, 990f, 991
    osmoregulation of, 977f–982f
    short-distance transport of, across plant plasma membranes, 788, 789f
    solvents and, 49f, 50
    transport of, from roots to shoots via xylem, 792, 793f–795f, 796
    transport of, from sources to sinks via phloem, 799, 800f–801f
Solutions, 49f, 50
Solvents, **49**f, 50
Somatic cells, **236**, **255**
Somatosensory cortex, 1097f, 1098
Somites, 720, 1054f, **1055**
Songs, bird learning of, 1147
Sonic hedgehog growth factor, 1063–1064
Soredia, **668**, 669f
Sorghum, 813f

Sori, **631**
Soufriere Hills volcano, 551
Sound receptors, 1110, 1112f–1116f
Sour tastes, 1123f–1124f
South Africa, restoration projects in, 1254f
South America, 743–744
South American vampire bats, 991f, 992
South Atlantic Subtropical Gyre, 1168f
Southern flying squirrel (*Glaucomys volans*), 517
Southern stingray (*Dasyatis americana*), 727f
South Pacific Subtropical Gyre, 1168f
Soybeans, 211, 493, 815f–816f, 817, 1015
Space-filling models, 36, 37f, 39f, 59f, 74f, 79f, 319f
Spain, 573f
Spanish flu pandemic, 410
Spartina patens, 1186
Spatial learning, 1144, **1145**f
Spatial summation, 1079f
Spawning, 1022–1023
Speciation, **507**
    allopatric, 511f–512f, 513
    allopatric vs. sympatric, 511f, 516
    as conceptual bridge between microevolution and macroevolution, 507, 523
    Darwinian theory of, 473, 474f–476f
    genetics of, 522f, 523
    of humans, 524
    hybrid zones and reproductive isolation in, 516, 517f–519f, 520
    as origin of species in Darwinian evolution, 506f, 507
    orthologous genes and, 566
    reproductive isolation and biological species concept in, 507f–510f
    species selection and differential, 548–549
    sympatric, 513, 514f–515f, 516
    time course of, 520f–521f, 522
Speciation clock, 522
Species, **507**
    biological concept of, 507f–510f
    classification of, 12f–13f, 470, 554, 555f
    communities of. *See* Communities
    comparing developmental processes of, 463f–464f
    comparing genomes of, 459, 460f–462f
    complete genome sequences for, 448t
    cross-species gene expression in evolution of, 423
    C. Darwin's theory of origin and evolution of, 14f–16f, 473, 474f–476f. *See also* Evolution; Speciation
    discovery of new, 1164f, 1185, 1260f
    distributions of. *See* Species distributions
    diversity of. *See* Species diversity
    edge, 1271
    endangered or threatened, 1261, 1262f, 1272f, 1288
    extinction of mollusc, 702f
    foundation, 1214, 1226
    fusion of, 519f, 520
    genome size, number of genes, gene density, and noncoding DNA for, 448t, 449
    geographic distribution of, 482f, 483
    homologous genes in, 565f, 566
    interactions between. *See* Interspecific interactions
    interactions of, 1214f–1221f, 1237
    introduced, 1223, 1264f, 1287
    keystone, 1226f, 1270
    large-impact, 1225, 1226f
    loss of amphibian, 733
    metagenomics and genome sequencing of groups of, 444
    morphological and ecological concepts of, 507, 508f–509f, 510
    number of named, 1261
    phylogenies as history of, for species, 554f–556f. *See also* Phylogenies
    populations of. *See* Populations
    resurrection of extinct, 1266
    in tree of life, 15f–16f
    tropical deforestation and extinctions of, 651f, 652
    using phylogenetic gene trees to identify, of whale meat, 557f
Species-area curve, **1232**f
Species distributions
    abiotic factors in, 1178, 1183f–1186f

    in aquatic biomes, 1178
    biotic factors in, 1178, 1183f–1184f
    climate and, 1164, 1170f
    determinants of, 1164f
    dispersal in, 1183, 1184f
    evolution and, 1183, 1189
    natural range expansion, 1183, 1184f
Species diversity, **1222**f
    benefits of, to humans, 1262, 1263f
    biodiversity crisis and, 1260f–1262f
    biogeographical factors affecting, 1231, 1232f–1233f
    bottom-up and top-down controls in, 1226, 1227f
    climate change effects on, 1279
    community stability and, 1223f
    disturbances influencing, 1228, 1229f–1231f
    human impacts on, 1231f
    large-impact species in, 1225, 1226f
    as level of biodiversity, 1261f–1262f
    protection of, 1260
    resurrection of, 1266
    species richness and relative abundance in, 1222f
    sustainable development and, 1284, 1285f, 1286
    threats to, 1260f, 1261, 1263, 1264f–1266f
    trophic structure and, 1223, 1224f–1227f. *See also* Trophic structure
Species richness, **1222**
    biogeographic factors affecting, 1231, 1232f–1233f
    species-area curves of, 1232f
    in species diversity, 1222f. *See also* Species diversity
Species selection, 548–549
Species transplants, 1184
Specific heat, **46**, 47f
Specificity
    cell-signaling, 227, 228f
    enzyme substrate, 155f, 156
    polymerase chain reaction, 421, 422f
    viral, 401
Specific transcription factors, 374f–375f, 845
Spectrophotometer, **193**f
Speech
    brain function and, 1098f
    FOXP2 gene and, 461, 462f
Spemann, Hans, 1060, 1061f–1062f
Sperm, **1020**
    biased usage of, in female fruit flies, 1024f
    chromosomes in human, 236–237
    conception and, 1035f
    in fertilization, 1022f–1024f, 1044f, 1045
    flagellated, in plants, 619
    human spermatogenesis and, 1027, 1028f, 1031f, 1042
    mammalian sex determination and, 298f
    seed plant. *See* Pollen grains
Spermathecae, 711, **1024**f
Spermatids, 1028f
Spermatocytes, 1028f
Spermatogenesis, **1027**, 1028f, 1031f, 1042
Spermatogonia, **1028**f
Spermicidal foam or jelly, 1038f
Spermophilus parryii, 893
Sphagnum moss, **627**f, 628
S phase, **237**f
Sphenodon punctatus, 737f
Spherical prokaryotes, 574f
Sphincters, 906f, **907**
    precapillary, 932f
Sphygmomanometer, 931f
Spiders, 45f, 707, 708f, 951
Spikemosses, 631, 632f, 633
Spilogale species, 508f
Spina bifida, 1055
Spinal cord, 1086f–1089f
Spines, 762f
Spines, sea star, 714f
Spinnerets, 708
Spiny-headed worms, 687f
Spiny mouse, 1216f
Spiral cleavage, **681**, 682f
Spiral phyllotaxy, 786f
Spiral prokaryotes, 574f
Spiral valve, shark, 727
Spirobranchus giganteus, 703f
Spirochetes, 584f
Spirodela oligorrhiza, 101f

Spliceosomes, **346***f*
Split-brain effect, 1098
Sponges
    phylogeny of, 682, 683*f*, 684, 687*f*, 690*f*, 691
Spongin, 690
Spongocoel, **690***f*
Spongy mesophyll, 771*f*
Spontaneous abortions, 306, 1039
Spontaneous change, 148*f*
Spontaneous mutations, 360, 388
Spontaneous processes, **146**
Sporangia, **621***f*, 625, 631, 640*f*, 641, 662*f*
Spores, **620***f*, **657**
    in alternation of generations, 258*f*
    brown algae, 603*f*, 604
    cell signaling and bacterial, 213*f*
    fossilized plant, 622*f*
    fungal, 654, 655*f*, 657, 658*f*–659*f*, 662*f*–663*f*
    meiosis and production of, 270*n*
    of plants, 620*f*–621*f*
    seeds vs., 639
    variations of, in vascular plants, 631
Sporophylls, **631**
Sporophytes, **620***f*, 823
    in alternation of generations, 258*f*
    brown algae, 603*f*, 604
    of bryophytes, 625, 626*f*
    gametophyte relationships with, in plants, 637*f*, 638
    in pine tree life cycle, 640*f*, 641
    of plants, 620*f*–621*f*
    of seedless vascular plants, 628*f*–630*f*
Sporopollenin, **619**, 621*f*, 638
Sporozoites, 605, 606*f*
Spotted ratfish (*Hydrolagus colliei*), 727*f*
Spotted skunks (*Spilogale* species), 508*f*
Spruce, 1230*f*–1231*f*
Squamates, 737*f*, 738
Squamous epithelium, 877*f*
Squids, 688*f*, 699, 701*f*, 702, 1086*f*
Squirrel monkeys, 1122*f*
Squirrels, 832*f*
    Arctic ground, 893
    Belding's ground, 1156, 1158, 1159*f*, 1193*t*, 1194*f*, 1195, 1213
    flying, 481*f*, 517
Srb, Adrian, 337, 338*f*
SRY gene, 298–299
S-shaped logistic growth curve, 1198*f*–1199*f*
Stability, community, 1223*f*, 1228
Stability, equilibrium as, 147, 148*f*
Stability, hybrid zone, 519*f*, 520
Stability, population, 1205*f*, 1208
Stabilizing selection, **498***f*
Stable isotopes, 31
Staghorn coral, 1234
Staghorn fern, 819*f*
Stahl, Franklin, 322*f*
Stalk-eyed flies, 1152*f*
Stamens, 270*f*, 271, **644***f*, **823***f*
Staminate flowers, 834
Standard metabolic rate (SMR), **891**–892
Stanley, Wendell, 399
Staphylococcus aureus, 214*f*, 478*f*, 584*f*
Star anise, 650*f*, 653
Starches, **70**
    as fuel for catabolism, 182*f*
    as product of photosynthesis, 206, 207*f*
    as storage polysaccharides, 70*f*, 71
Starfish. *See* Sea stars
Starling, European, 1264
Star-nosed moles, 1107*f*–1108*f*
Start codons, 341*f*, 350
Start point, transcription, **343***f*
Statins, 938
Stationary cilia, 1064
Statistics, 17
Statocysts, **1112***f*
Statoliths, **861***f*, **1112***f*
Stator, ATP synthase, 175*f*
Steinbeck, John, 807*f*
Stele, **763**, 769*f*
Stem cells, **428**, **935**
    animal embryonic and adult, 430, 431*f*
    animal induced pluripotent, 431, 432*f*
    generation of differentiated cells from, 431*f*
    glia as, 1090*f*
    plant, 766

potential of, 428
in replacement of blood components, 935*f*, 936
in spermatogenesis, 1042
Stems, 758, **761**
    ethylene in triple response of, to mechanical stress, 853*f*
    gibberellins in elongation of, 851*f*
    monocot vs. eudicot, 649*f*
    primary and secondary growth of, 766*f*–767*f*
    primary growth of, 769*f*–771*f*, 775*f*
    secondary growth of, 772*f*–775*f*
    structure of, 761*f*
Stenohaline animals, 978
Stents, 937*f*
Steppes, 1175*f*
Sterility
    hybrid, 509*f*, 522–523
    transgenic plant, 839–840
Sterilization, human, 1038*f*, 1039
Steroid hormones, 75
    coordinate control of genes by, 375–376
    functional groups and, 62*f*
    as intracellular chemical signals, 220, 221*f*
    as lipids, 75*f*
    smooth ER synthesis of, 104–105
Steroids, **75**
    adrenal gland and, 1012*f*, 1013
    brassinosteroids as, 854–855
    as environmental toxins, 1276
    in human endocrine system, 1004*f*
    as lipid-soluble hormones, 1001*f*
    receptors for, 1003*f*
    sex hormones as, 1014, 1015*f*
Steward, F. C., 428
Stickleback fish (*Gasterosteus aculeatus*), 481, 546*f*, 547, 1140*f*
Sticky end, DNA, **420**
Stigma, **644***f*
Stigma, angiosperm, **823***f*
Stimulus, **882**
    environmental, 1140*f*–1141*f*
    homeostatic, 882
    sensory, 1108*f*–1109*f*
    in stimulus-response chains, 1141*f*, 1142
Stingrays, 726, 727*f*
Stink bugs, 712*f*
Stipes, **602**, 603*f*
Stock, **835**–836
Stolons, 761*f*
Stomach, **907**
    adaptations of, 912*f*
    bacteria in, 912, 913*f*
    digestion in, 907*f*–908*f*
    dynamics of, 908
Stomach ulcers, 584*f*
Stomata, **189**, **621**, 771*f*, 777*f*
    of CAM plants, 205, 206*f*
    of horsetail, 635
    ion gradients and, 993*f*
    in photosynthesis, 189*f*
    photosynthesis–water loss compromise, 787
    plant, 622
    regulation of transpiration by opening and closing of, 796, 797*f*–799*f*
    sporophyte, 625
    transpiration and, 203
Stop codons, 341*f*, 352, 353*f*
Storage leaves, 762*f*
Storage polysaccharides, 70*f*, 71
Storage proteins, 76*f*
Storage roots, 760*f*
Storms, 1172, 1228
    climate change and, 11
Stramenopiles, 599*f*, 602*f*–604*f*
Strangling aerial roots, 760*f*
Strasburger, Eduard, 794
Strata, **470***f*, 529*f*
Strathdee, Steffanie, 872
Stratified squamous epithelium, 877*f*
Strawberry poison dart frogs, 524
Streams, 1180*f*, 1274
Streptococcus, 575*f*, 584*f*, 691
Streptococcus pneumoniae, 315*f*, 579, 957
Streptococcus pyogenes, 404*f*
Stress
    adrenal gland response to, 1012*f*–1013*f*, 1014
    ethylene in plant responses to, 853*f*
    immune systems and, 971

response to, 1003
Stretch receptors, 1110
Striated muscle, 879*f*, 1126. *See also* Skeletal muscle
Striga, 855
Strigolactones, 846*t*, 850, **855**
Strobili, **631**
Strokes, **937***f*, 938–939
Stroke volume, **927**
Stroma, **110**, 111*f*, **189***f*, 191*f*, 192, 200*f*, 201
Stromatolites, 532*f*, **533**
Structural formulas, 36, 37*f*, 59*f*, 74*f*
Structural isomers, **61***f*
Structural polysaccharides, 70*f*–72*f*
Structural proteins, 76*f*
Structure, function and, 6
Strychnine, 1220
Studies. *See* Inquiry Figures
Sturnella magna, 507*f*
Sturnella neglecta, 507*f*
Sturtevant, Alfred H., 305*f*–306*f*
Styles, flower, **644***f*, **823***f*
Subatomic particles, 30*f*, 31
Suberin, 774
Subsidence, land, 808*f*
Substance P, 1081*t*
Substrate feeders, **903***f*
Substrate-level phosphorylation, 168, **169***f*
Substrates, **155***f*, 156, 160*f*–161*f*
Succulent Karoo restoration project, 1254*f*
Succulent plants, 205, 206*f*
Suckling, 1036
Sucrase, 153, 155
Sucrose
    as disaccharide, 69*f*
    enzymatic catalysis and, 153, 155
    molecular mass of, 50
    plant cell uptake of, 142
    as product of photosynthesis, 206, 207*f*
    transport of, in vascular plants, 799, 800*f*–801*f*
Sudden oak death (SOD), 614*f*, 867, 1234
Sugarcane, 206*f*
Sugar gliders, 481*f*
Sugar-phosphate backbone, DNA, 85*f*–86*f*, 317*f*–320*f*, 325*f*–326*f*
Sugars. *See also* Carbohydrates
    blood regulation, 10*f*
    as components of nucleic acids, 84, 85*f*
    conduction of, in plant cells, 765*f*
    monosaccharides and disaccharides, 68*f*–69*f*, 70
    polysaccharides, 70*f*–72*f*
    as products of photosynthesis, 191*f*, 192, 201, 202*f*, 206, 207*f*
    translocation of, from sources to sinks via phloem, 799, 800*f*–801*f*
Sugar sinks, **800**
Sugar sources, **800**
Suicide genes, 390
Sulfhydryl group, 63*f*
Sulfur, 29*t*, 64
Sulfur dioxide emissions, 1266
Sulston, John, 1058–1059
Sumner, Francis Bertody, 20
Sundews, 819*f*
Sunflowers, 521*f*, 894*f*
Sunlight
    absorption of, determining primary production with, 1242*f*
    aquatic biomes and, 1177*f*–1178*f*
    cancer and, 394
    DNA damage from, 328
    in energy flow and chemical cycling, 9*f*, 164–165, 1240f
    as energy for life, 143, 145, 150
    global energy budget and, 1241
    latitudinal variation in intensity of, 1166*f*
    in photosynthesis, 187*f*, 206, 207*f*
    primary production in aquatic ecosystems and limitations of, 1242
    properties of, 192*f*
    seasonal variations in, 1167*f*
    species distributions and availability of, 1185, 1186*f*
Supercontinent, 538, 539*f*
Supergroups, protist, 597*f*–599*f*
Superimposed electron orbitals model, 35*f*
Supernatural vs. natural explanations, 18
Super-resolution microscopy, 95*f*, 96
Superweeds, 438

Suprachiasmatic nucleus (SCN), 893*f*, 1015, **1094**–1095
Surface area
 calculating cell volume and, 99
 maximizing, by animals, 695*f*
Surface area-volume relationships, 98*f*
Surface tension, **45***f*, 46
Surfactants, **943**, 944*f*
Surroundings, systems and, 145
Survival
 adaptations, natural selection, and, 475*f*, 476
 animal behavior evolution and, 1148, 1149*f*–1154*f*
 gecko appearance and, 26
 life histories and, 1200, 1201*f*–1202*f*
Survivorship curves, **1193**, 1194*f*
Suspension feeding, 721*f*, 727, 903*f*
Suspensor cells, 828*f*
Sustainability, 1284
Sustainable agriculture, **808**
 in Costa Rica zoned reserves, 1273
 soil conservation in, 807*f*–809*f*
Sustainable development, **1284**, *1285f*, 1286
Sutherland, Earl W., 216*f*, 217, 223
Sutton, Walter S., 295
Swallowing reflex, 906*f*
Sweden, 1208
Sweet potatoes, 651, 837
Sweet tastes, 1123*f*–1124*f*
Swim bladders, **728***f*
Swimming, 1135
Swine flu, 410, 1234
Switchgrass, 838
Symbionts, **587**
Symbiosis, **587**, 1256*f*. *See also* Commensalism; Mutualism; Parasitism
 fungal, 662–663, 667*f*–669*f*
 protist, 614*f*
sym genes, 660
Symmetry
 animal body, 679, 680*f*
 body, 1059, 1060*f*, 1064*f*
 flower, 644*f*, 649*f*
Sympathetic division, peripheral nervous system, 928, 1088*f*–**1089***f*
Sympatric populations, character displacement in, 1216, 1217*f*
Sympatric speciation, **513**, 514*f*–515*f*, 516
Symplast, **787**, 788*f*, 800*f*
Symplastic communication, 801, 802*f*
Symplastic domains, 802
Symplastic route, 788*f*, 793*f*
Synapses, 1067–**1068**
 electrical and chemical, 1077, 1078*f*
 generation of postsynaptic potentials and, 1077–1078
 in memory and learning, 1099, 1100*f*–1101*f*
 modulated signaling at, 1080
 neurotransmitters and, 1077, 1078*f*, 1080*f*, 1081*t*, 1082
 in regulation of muscle contraction, 1128, 1129*f*, 1130
 scaffolding proteins and, 228*f*, 229
 summation of postsynaptic potentials and, 1078, 1079*f*
Synapsids, 531*f*, **741**, 742*f*
Synapsis, **262***f*
Synaptic cleft, 1077
Synaptic signaling, 215*f*, 216, 1000*f*, 1001
Synaptic terminals, 1068*f*
Synaptic vesicles, 1077
Synaptonemal complex, **262***f*
Syndermata, 687*f*
Syndromes, 308
Synthesis stage, phage lytic cycle, 402*f*
Synthetases, 349*f*
Syphilis, 584*f*
Systematics, **554**, 682, 683*f*, 684. *See also* Cladistics; Molecular systematics; Taxonomy
Systemic acquired resistance, **866**, 867*f*
Systemic circuit, 924*f*–926*f*
Systemic inflammation, 956–957
Systemic lupus erythematosus, 971
Systemic mycoses, 670
Systems
 thermodynamics and isolated vs. open, 145, 149*f*–150*f*
Systems approach, 425

Systems biology, **5**–6, **446**, 447*f*–448*f*
Systole, **927***f*
Systolic pressure, **930**, 931*f*
Szent-Györgyi, Albert, 900
Szostak, Jack, 528

T
T2 phage, 316*f*, 317
Table salt. *See* Sodium chloride
Tables in Scientific Skills Exercises, 58, 89, 157, 179, 205, 264, 304, 318, 458, 493, 513, 570, 590, 595, 629, 639, 657, 678, 751, 762, 790, 834, 918, 938, 972, 981, 1014, 1030, 1049, 1082, 1186, 1217, 1247, 1279
TACK supergroup, 586
Tactile communication, 1141*f*
Tadpoles, 381*f*, 732*f*
Taiga, 1175*f*
Tail, muscular post-anal, 720*f*
Tail end, 680*f*
Tails, histone, 330*f*, 371*f*
Takahe bird, 1254*f*
TAL protein, 334
Tamiflu, 414
Tamoxifen, 250, 393*f*
Tannins, 1220
Tapeworms, 687*f*, 696, 697*f*, 1220
Tapping, in fruit fly courtship, 1141*f*
Taproots, **759***f*–760*f*
Taproot systems, 759*f*, 760
Taq polymerase, 421, 1263
Tardigrades, 689*f*, 980*f*
Target cells, hormone, 1000*f*–1004*f*
Tarsiers, 746*f*
Tar spot fungus, 669*f*
Tastants, **1123***f*–1124*f*
Taste, 233, 1123*f*–1124*f*
Taste buds, **1124***f*
TATA boxes, 343*f*, **344**, 373
Tatum, Edward, 336*f*–338*f*
Tau protein, 1104
Taxis, **576**
Taxol, 249
Taxon, **554**, 555*f*–556*f*, 560
Taxonomy. *See also* Cladistics; Phylogenies; Systematics
 early schemes of, 470
 grouping of species in, 12*f*–13*f*
 mammalian, 745*f*
 phylogenies and, 554, 555*f*–556*f*
 possible plant kingdoms, 619*f*, 621
 ten phyla of extant plants, 622*t*
 three-domain system of, 12*f*, 13, 568, 569*f*
 tree of life of, 15*f*–16*f*
Tay-Sachs disease, **280**
 allele dominance and, 280
 fetal testing for, 288, 289*f*
 as lysosomal storage disease, 108
 recessive alleles in, 285
T cells, **958***f*
 antigen recognition by, 959*f*–960*f*
 clonal selection of, 962*f*, 963
 cytotoxic, 962, 966*f*, 967
 development of, 960, 961*f*–963*f*
 diversity of, 960, 961*f*
 DNA of, 976
 helper, 963, 964*f*
 immunological memory, 963*f*
 proliferation of, 961–962
 regulatory, 971
Tea, 651, 783
Teal, John, 1247
Technology, **23**, 24*f*
 DNA. *See* Biotechnology; DNA technology
 global carrying capacity and, 1211
 prokaryotes in research and, 589, 590*f*–591*f*
Tectonic plates, 538*f*–539*f*, 540
Teeth
 conodont mineralized dental elements, 724*f*, 725
 diet and adaptations of, 911*f*, 912
 mammalian, 530, 531*f*, 532, 741
 origins of, 725
Teixobactin, 478, 590
Telomerase, 329
Telomeres, **328**, 329*f*, 452
Telomeric DNA, 452
Telophase (mitosis), **237**, 239*f*, 243*f*, 252, 263*f*

Telophase I, 260*f*, 263*f*
Telophase II, 261*f*
Temperate broadleaf forests, **1176***f*
Temperate forests, decomposition in, 1248
Temperate grasslands, **1175***f*
Temperate phages, **403***f*
Temperature, **46**
 aquatic biomes and, 1177, 1178*f*
 climographs of, 1171*f*
 coefficients in Scientific Skills Exercise, 790
 correlation of atmospheric carbon dioxide with global, 1278*f*, 1279, 1282, 1283*f*
 decomposition rates and, 1248*f*
 effects of transpiration on leaf, 798
 enzymatic catalysis and, 157, 158*f*
 global, and extinction rates, 541*f*, 542
 heat vs., 46
 membrane proteins and, 128*f*
 moderation of, by water, 46, 47*f*–48*f*
 plant response to stress of, 865
 primary production in terrestrial biomes and, 1243*f*, 1244, 1244*f*, 1245
 protein denaturation and, 82*f*, 83
 regulation of body. *See* Thermoregulation
 species distributions and, 1164, 1185*f*
Templates, viral DNA and RNA, 405*f*, 406*t*
Template strands, DNA, 320, 321*f*–327*f*, **340***f*
Tempo, speciation, 520*f*–521*f*, 522
Temporal fenestra, 531*f*, 741
Temporal isolation, 508*f*
Temporal lobe, 1085, 1097*f*
Temporal summation, 1079*f*
Tendons, **878***f*, 1138
Tendrils, 762*f*
Tentacles, invertebrate, 691*f*, 701*f*, 702
Terminal cells, 828*f*
Termination, neurotransmission, 1080*f*
Termination codons, 341*f*
Termination stage
 transcription, 343*f*, 344
 translation, 352, 353*f*
Terminators, **342**, 343*f*
Termites, 143*f*, 614*f*, 655
Terpenoids, 868*f*
Terrestrial adaptations
 mycorrhizae as plant, 817
 seed plant, 637*f*–638*f*, 639
Terrestrial biomes
 animal osmoregulation in, 980–981
 biodiversity hot spots in, 1272*f*
 chaparral, 1174*f*
 climate and, 1171*f*
 climate change effects on, 1244*f*, 1245
 decomposition in, 1248*f*
 deserts, 1173*f*
 disturbance in, 1172
 food chains in, 1224*f*
 general features of, 1172*f*
 global distribution of, 1171*f*, 1189
 habitat loss in, 1263–1264
 locomotion in, 1135*f*
 northern coniferous forests, 1175*f*
 nutrient cycling in, 1250*f*–1251*f*
 plant adaptations to, 203
 primary production in, 1243*f*–1244*f*, 1245
 savannas, 1174*f*
 temperate broadleaf forests, 1176*f*
 temperate grasslands, 1175*f*
 tropical forests, 1173*f*
 tundra, 1176*f*, 1256*f*–1257*f*
Territoriality, **1192***f*, 1204*f*
Tertiary consumers, **1240***f*
 trophic efficiency of, 1246, 1247*f*, 1248
Tertiary structure, protein, **81***f*
Testcrosses, **274**, 275*f*, 301*f*–302*f*
Testes, 257*f*, 258, 1004*f*, 1014, 1015*f*, **1025***f*, 1031*f*
Testicles, 1025*f*
Testing, genetic. *See* Genetic testing
Testing, hypothesis, 17, 18*f*–19*f*
Testosterone, 62*f*, 397, 999*f*, **1014**, 1015*f*, **1030**, 1031*f*
Tests, foraminiferan, **608**
Test-tube cloning, 836*f*
Tetanus, **1130**
Tetrahymena, 346
Tetraploids, 514
Tetraploidy, 307

Tetrapods, **730**
  amniotes, 734f–735f
  amphibians, 731, 732f–733f
  derived characters of, 730
  evolution of, 678
  homologous characteristics of, 480f
  land colonization by, 537
  as lobe-fins, 730
  origin and phylogeny of, 730f–731f
  origin of, 530, 531f–533f
Tetrodotoxin, 813f
Texture, soil, 806f
Thalamus, **1093f**, 1095f, 1097
Thalassoma bifasciatum, 1020
Thalidomide, 65
Thalloid liverworts, 626f
Thaumarchaeota clade, 586
Themes of biology, 3f–16f
Theobroma cacao, 667f, 668
Theories, **21**–22, 483–484
Therapeutic cloning, 431
Therapsids, 531f
Thermal energy, **46**, 132, **144**
Thermocline, **1177**, 1178f
Thermodynamics, laws of, 143, 145f, 146, 1239
Thermogenesis, 887f–888f
Thermophiles, 441, 585f, 586
Thermoreceptors, **1111f**
Thermoregulation, **884**
  acclimatization in, 883f, 888
  aquatic animal feedback regulation in, 881f
  circadian rhythms in human, 882, 883f
  in endothermic and ectothermic animals, 884f
  heat loss and gain balance in, 885f–888f
  minimum metabolic rate and, 890–891
  nonliving example of, 881f–882f
  penguin form and function for, 873f
  physiological thermostats and fever in, 888, 889f
  variation in animal body temperatures and, 884–885
Thermostatic thermoregulation, 888, 889f
Thermus aquaticus, 421, 441, 1263
Theropods, **736**
Thick filaments, **1125**, 1126f–1127f, 1131–1132
Thigmomorphogenesis, **861**, 862f
Thigmotropism, **862f**
Thin filaments, **1125**, 1126f–1127f, 1131–1132
Thiol compounds, 63f
Thiomargarita namibiensis, 574, 584f
Third-generation sequencing, 416
Thirst, 1110
Thlaspi caerulescens, 809
Thompson seedless grapes, 851f
Thoracic cavity, 942, 945
Thorax, insect, 710f
Threatened species, **1261**, 1262f, 1272f
Three-parent babies, 311
Threonine, 77f, 222
Threshold, **1073**
Thrombin, 936f
Thrombus, **937**, 939
Thrum flower, 835f
Thucydides, 962
Thumbs, opposable, 744
Thunnus albacares, 729f
Thylakoid membranes, 189f, 190, 196f–200f, 201
Thylakoids, **110**, **189**
  ATP production and, 211
  in chloroplasts, 110, 111f
  light reactions in, 191f, 192
  as sites of photosynthesis in chloroplasts, 189f, 190
Thylakoid space, 189f
Thymine, 84, 85f, 86, 317f, 318, 337, 340f
Thymine dimers, 328
Thymus, **958**
Thyroid gland, 1004f, 1008, **1009f**, 1010
Thyroid hormone (T₃ and T₄), 1004f, **1008**, 1009f, 1010, 1015f
Thyroid hormones, 179, 220
Thyroid-stimulating hormone (TSH), 1004f, 1008f–1009f, 1010
Thyrotropin-releasing hormone (TRH), 1009f
Thyroxine (T₄), 1001f, 1003, 1008, 1009f, 1010, 1015f
Tiburon Mariposa lily (*Calochortus tiburonensis*), 30f
Ticks, 588f, 707–708, 1220, 1234, 1235f
Tidal volume, **945**

Tides, 1181f
Tiger moths, 717
Tigers, 293, 1288
Tiger swallowtail, 313
Tight junctions, 120f
Tiktaalik fossil, 533f, 730f, 731
Time
  phylogenetic tree branch lengths and, 561, 562f
  required for human cell division, 237
Tinbergen, Niko, 1140, 1145f
Tissue culture, plant, 836f
Tissue-level herbivore defenses, plant, 868f
Tissue plasminogen activator (TPA), 436, 457f
Tissues, 5f, **674**, **759**, **876**
  animal, 673–674, 876, 877f–879f
  animal body plan and, 673–674, 680–681
  culturing plant, 836f
  of human endocrine system, 1003, 1004f
  immune system rejection of transplanted, 969–970
  as level of biological organization, 5f
  plant, 759, 762, 763f
  proteins specific to, in cell differentiation, 383
  renewal of, as cell division function, 235f
  substitutes for, 897
  water and salt concentrations in, 977f, 978
Tissue-specific proteins, 383
Tissue systems, plant, **762**
  in leaves, 770, 771f
  in primary growth of roots, 768f–769f
  in primary growth of shoots, 769f–771f
  types of, 762, 763f
Tit-for-tat strategy, 1159
Tmesipteris, 632f, 633
Toadfish, 1131f
Toads, 516, 517f, 520, 732, 1217
Tobacco mosaic virus (TMV), 399f–400f, 412
Tobacco plant, 342f
Toll-like receptors (TLRs), **955f**, 976
Toll receptors
  in insects, 954
Tollund Man, 627f
Tomatoes, 645f, 849
Tongue, 906f, 1124f
Tonicity, **134f**–135f
Tools, hominin use of, 750
Tooth cavities, 213
Top-down control, 1226, **1227f**
Topoisomerases, **323f**, 327t
Topologically associated domains (TADs), 377, 378f
Topsoil, **806f**, 807f
Torpor, **892**
Tortoiseshell cats, 300f
Totipotent amoebocytes, 691
Totipotent cells, **428**, **1061**
Totipotent plants, **835**–836
Touch, plant response to, 861, 862f
Touch receptors, 1110f
Tourism, zoned reserves and, 1273, 1273f
Toxic waste
  bioremediation of, 1253, 1255f
  biotechnology in cleanup of, 418f, 437
  density-dependent population regulation by, 1204f
Toxins
  botulism, 1080
  as defensive adaptations, 1217–1220
  detoxification and, 104–105, 112
  dinoflagellate, 605f
  environmental, 1275, 1276f–1277f
  enzymatic catalysis and, 158–159
  evolution of tolerance to, 30f
  fungal, 669f
  soil, 809, 812
T phages, 400f–402f, 403
Trace elements, **29t**
Trace fossils, 529f
Tracers, radioactive, 31, 32f, 190
Trachea, 906f, **942**, 943f
Tracheal systems, insect, 707, 710f, **941**, 942f
Tracheids, **629**–630, **765f**
Trade-offs, 501
Trade-offs, life history, 1201f–1202f
Tragopogon species, 514, 515f
Traits, **270**. *See also* Alleles
  characters and, 270–271

  C. Darwin on natural selection and, 14f–16f
  dominant vs. recessive, 271f, 272t, 285f
  inheritance of, in Darwinian evolution, 475f, 476
  inheritance of X-linked genes and recessive, 299f, 300
  life history, 1200, 1201f–1202f
  noninheritance of acquired, 471f
  plant derived, 620f–621f
  seedless vascular plant, 628f–631f
Transacetylase, 368f
Transcription, **337**
  effects of ncRNAs on, 380–381
  eukaryotic gene regulation after, 377, 378f, 379
  in gene expression and protein synthesis, 7f–8f, 335–336, 337, 339f
  gene expression as, 371
  molecular components and stages of, 342, 343f
  regulation of, in plant responses, 845
  regulation of bacterial, 366f–369f, 370
  regulation of eukaryotic initiation of, 373f–378f, 379
  RNA processing after, 345f–347f
  scale of, 123f
  summary of eukaryotic, 356f
  synthesis of RNA transcript during, 343f–344f
  template strands in, 340f
Transcription factories, 377, 378f
Transcription factors, 343f, **344**
  in cell signaling, 221f, 226f
  in eukaryotic gene regulation, 374f–377f
Transcription initiation complex, 343f, **344**, 373f–375f
Transcription units, **342**, 343f
Transduction, genetic, **579f**
Transduction, sensory, 1109, 1114, 1115f, 1120f
Transduction stage, cell-signaling, **216**
  multistep pathways and signal amplification in, 221
  overview of, 216f, 217
  in plant signal transduction pathways, 844f, 845
  protein phosphorylation and dephosphorylation in, 222f, 223
  signal transduction pathways and, 221–222
  small molecules and ions as second messengers in, 223f–225f
  in yeast-cell mating, 214
trans face, Golgi apparatus, 106f, 107
trans fats, 61, **74**, 938
Transfer RNA (tRNA), **348**. *See also* RNA (ribonucleic acid)
  structure of, 86f, 348f
  in translation, 347f–350f, 352, 353f
Transformation, cancer and cellular, **249f**
Transformation, energy, 143f–146f, 147, 209f. *See also* Metabolism
Transformation, genetic, **315f**, **579**, 776
Transfusions, blood, 969–970
Transgender individuals, 299
Transgene escape issue, 839–840
Transgenes, **436f**, 776, **837**, 841
Transgenic animals, **436**
Transgenic plants, **837**. *See also* Genetically modified organisms
  biotechnology and genetic engineering of, 837, 838f
  issues about agricultural crops as, 438–439, 838–840
trans isomers, 61f
Transitional ER, 105f
Transition state, 154
Translation, **339**
  basic concept of, 347f
  building polypeptides in, 350f–353f
  completing and targeting functional proteins in, 354f, 355
  in eukaryotic cells, 356f
  in gene expression and protein synthesis, 335–337, 339f
  identifying ribosome binding sites with sequence logos in, 351
  molecular components of, 347f–350f
  post-translational protein modification in plant responses, 845
  regulation of eukaryotic initiation of, 378–379
  ribosomes in, 349, 350f
  scale of, 123f
  synthesizing multiple polypeptides with polyribosomes in, 355f

transfer RNA in, 347f–350f
Translation initiation complex, 352f
Translation initiation factors, 378–379
Translocation, **799**
    cancer gene, 388, 389f
    chromosome structure and, 307, 308f, 309f
    plant transport, 799, 800f–801f
    in translation, 352, 353f, 354–355
Transmembrane proteins, **129**f–130f, 217f–220f
Transmembrane route, 788f, 793f
Transmission, sensory, 1109f
Transmission electron microscope (TEM), **96**
Transmission electron microscopy (TEM), 95f
Transmission rate, disease, 1204f
Transpiration, **792**
    effects of, on plant wilting and leaf temperature, 798
    plant adaptations for reducing evaporative water loss by, 798, 799f
    regulation of, by opening and closing stomata, 796, 797f–799f
    in water and mineral transport from roots to shoots via xylem, 792, 793f–795f, 796
    in water cycle, 1250f
Transpirational pull, 794f–795f
Transplants
    immune system rejection of, 969–970
    species, 1184
Transport, plant and animal, 895f. *See also* Circulatory systems; Transport in vascular plants
Transport epithelia, **981**, 982f, 988f, 989
Transport function, membrane protein, 130f
Transport in vascular plants, 629–630. *See also* Vascular plants
    regulation of transpiration rate by stomata in, 796, 797f–799f
    resource acquisition adaptations and, 784f–786f, 787
    short-distance and long-distance mechanisms of, 787, 788f–791f
    sugar transport from sources to sinks via phloem in, 799, 800f–801f
    symplastic communication in, 801, 802f
    of water and minerals from roots to shoots via xylem, 792, 793f–795f, 796
    xylem and phloem in, 765f
Transport proteins, 76f, 126, **132**
    in active transport, 136, 137f
    aquaporins, 790–791
    cellular membrane selective permeability and, 132f
    as cotransporters, 138f, 139
    facilitated diffusion and, 135f, 136
    ion pumps and, 138
    plant solute transport and, 788, 789f
    water diffusion and, 790–791
Transport vesicles, **105**, 106f, 107, 109f, 139, 140f
Transport work, 150, 152f
Transposable elements, 450f–**451**f, 452, 459
Transposase, 451f
Transposition process, 450f–451f, 452, 459
Transposons, 380–381, 408, **451**f
Transthyretin protein, 80f–81f
Transverse (T) tubules, **1128**f–1129f
Trawling
    community disturbances by, 1231f
Tree frogs, 500f
Treehopper, 466
Tree of life
    Darwinian evolution and, 15f–16f, 474f. *See also* Evolutionary trees
    phylogenies in investigation of, 553f–554f, 568, 569f–570f
    three-domain taxonomy of, 568, 569f
Tree rings, 773f
Trees
    climate change effects on CO_2 absorption by, 1245
    ecosystem interaction with, 10f, 11
    petrified, 529f
    photosynthesis in, 187f
Tree trunks, 774f
Trematodes, 696f
Trends, evolutionary, 548, 549f
Treponema pallidum, 584f
Triacylglycerols, 72, 73f, 74
Trial-and-error learning, 1146f

Tricarboxylic acid cycle. *See* Citric acid cycle
Trichechus manatus, 572, 1219f
Trichinosis, 705f, 706
Trichomes, **763**f, 799f, 868f
Trichomonas vaginalis, 600f
Triglycerides, 72, 73f, 74, 910f
Triiodothyronine (T_3), 1008, 1009f, 1010
Trilobites, 706f
Trimesters, human pregnancy, 1035f–1037f
Trimethylamine oxide (TMAO), 979
Trioses, 68f
Triple response, plant, **853**f
Triplet code, **340**f. *See also* Genetic code
Triploblastic animals, **680**, 681f
Triploidy, 268, 307
Trisomic cells, **307**
Trisomy 21, 308f, 309. *See also* Down syndrome
Trisomy X, 309
Tristan da Cunha, 495
Triticum aestivum, 1204f
Triticum species, 514–515, *524*
Trochophore larva, 683f, **684**, 685, 694
Trophic efficiency, **1246**, 1247f, 1248
Trophic levels, **1224**, 1257f
    in ecosystem energy flow and chemical cycling, 1240f, 1241
    energy transfer between, 1246, 1247f, 1248
Trophic structure, **1223**
    bottom-up and top-down controls on, and biomanipulation of, 1226, 1227f
    energetic hypothesis on restriction of food chain length, 1225f
    food webs of food chains in, 1224f–1225f, 1237
    species with large impacts on, 1225, 1226f
Trophoblasts, **1035**, **1052**, 1053f
Tropical cone snails, 1067f–1068
Tropical dry forests, **1173**f, 1263
Tropical rain forests, **1173**f
    climate change and photosynthesis of, 211
    decomposition in, 1248
    deforestation of, as threat to biodiversity, 651f, 652, 1263
    fragmentation of, 1271f
    primary production in, 1244
Tropic hormones, 1008f
Tropic of Cancer, 1166f
Tropic of Capricorn, 1166f
Tropics, 1166f, 1232
Tropidolaemus wagleri, 737f
Tropisms, **847**
Tropomyosin, **1128**f–1129f
Troponin complex, **1128**f
Troponin T, 378f
Trout, 728f, 1265
trp operator, 367f
trp operon, 367f, 368–369
TRP (transient receptor potential) proteins, 1111, 1124
trp repressor, 367f
True-breeding organisms, **271**
True bugs, 712f
Truffles, **663**f, 670
Trypanosoma, 600, 601f, 713, 972
Trypsin, 158f
Tryptophan, 77f, 366f–367f, 1081
Tsetse flies, 713, 903f
Tuataras (*Sphenodon punctatus*), 735, 737f
Tubal ligation, 1038f, **1039**
Tubal pregnancies, 1039
Tube cells, 646f, 826
Tube feet, **713**, 714f
Tuberculosis, 584f, 589, 592, 957, 1204f
Tubers, 761f
Tubeworms, 914f
Tubulin, 114, 240–241
Tubulinids, 612
Tumbleweeds, 832f
Tumors, 435
Tumors, cancer, 249f, 250
Tumor-suppressor genes, **388**, 391f, 394
Tumor viruses, 394–395
Tuna, 729f, 874f, 1265f
Tundra, **1176**f, 1238f, 1244, 1256f–1257f
Tunicates, 715, 719f, **721**f, 722, 1058f
Turgid cells, 134f, **135**, **790**, 791f, 797f
Turgor movements, plant, 862f
Turgor pressure, 134f, 135, **789**
Turner syndrome, 309

Turnover, **1178**f
Turtles, 719f, 735, 736f, 737, 884f, 925, 1066, 1194, 1195f
Tutu, Desmond, 462
Twins, 1036, 1061, 1162
Twin studies, 1144
Tympanic membrane, 1112f–**1113**f
Tympanuchus cupido, 495f, 496, 1267f
Type 1 diabetes, 917
Type 2 diabetes, 372f, 917
Typha angustifolia, 1186
Typhoid fever, 588
Tyrannosaurus rex, 736, 874, 983
Tyrosinase, 337
Tyrosine, 77f, 222, 1009, 1081

U
Ubiquinone (Q), 175
Ubiquitin, 379
Ubx gene, 545f, 546, 706f
Ulcers, 584f, 912, 913f
Ultimate causation, 1140
Ultrasonic sound, 717
Ultrasound fetal testing, 289
Ultrasound imaging, 1039
Ultraviolet (UV) radiation
    cancer and, 394
    Chlamydomonas nivalis and, 211
    DNA damage from, 328
    elevation and, affecting species distributions, 1186f
    flowers and, 824f
    in insect vision, 1117–1118
    mutations from, 360
    ozone depletion and, 1283f, 1284
Ulva, 610f
Umami tastes, 233, 1123f–1124f
Undernourishment, 902
Undershoot phase, action potential, 1074f, 1075
Unger, Franz, 270
Ungulates, 481f–482f
Unicellular organisms, 532f–535f
Uniform dispersion, 1192f
Unikonts, 599f, **611**, 612f–613f, 614
Unisexual flowers, 824
United States, age-structure pyramid for, 1209f
Unity
    in biodiversity, 13f–14f, 26
    cellular structures and, 125
    evolution and, 469, 473
    within species, 510
Unlinked genes
    identifying, 304
    mapping, 305f–306f
    recombination of, 302f
Unpaired electrons, 36
Unsaturated fats, 73f, 74, 128f
Unsaturated fatty acids, **73**f, 74
Unselfish behavior, 1156f–1157f
Untranslated regions (UTRs), 345f, 378
Upright posture, hominin, 748
Upwellings, 1243
Uracil, 84, 85f, 86, 337, 340f–341f
Uranium
    bioremediation of, 1255f
    half-life of, 32
Uranium-238, 529, 530f
Urban ecology, 1273, **1274**f
Urea, 60, 979, 982f, **983**, 989, 990f, 991
Ureter, **986**f
Urethra, **986**f, **1025**f
Urey, Harold, 57, 526
Uric acid, 982f–**983**f
Urinary bladder, **986**f
Urine
    concentration of, 989, 990f, 991
    hyperosmotic, 991
    nephron processing of blood filtrate to, 987, 988f, 989
Urochordata (tunicates), 719f, 721f, 722
Urodela, 731, 732f
Ursus arctos, 510f, 1268f, 1269, 1272–1273
Ursus maritimus, 11, 510f
USA300 bacteria, 478f, 589f
Use and disuse principle, Lamarck's, 471
Uterine cycle, **1032**f, 1033
Uterus, 100f, **1026**, 1027f

Utricle, 1115*f*
Utricularia gibba, 448*t*, 449
Uyuni Salt Flat, 1185*f*

V
Vaccination, 968*f*, 976
Vaccines, **408**–409
 for amphibians, 733
Vacuolar sap, 777
Vacuoles, 100*f*–101*f*, **108***f*
Vagina, **1026**, 1027*f*, 1034
Vaginal pouch, 1038
Valence, **36**, **59***f*
Valence electrons, **35**–36
Valence shells, **35**
Valine, 77*f*
Valium, 1081
Vampire bats, 991*f*, 992
van der Waals interactions, **39**, 81*f*
van Leeuwenhoek, Antoni, 94
van Niel, C. B., 190
Varanus komodoensis, 1042
Variable (V) region, light and heavy chain, 958*f*–961*f*
Variables, **20**–21
 comparing two, 972
 identifying dependent and independent, 513
Variation, **255**
Variation, genetic. *See* Genetic variation
Variegation, 311*f*
Vasa recta, **987***f*
Vascular bundles, 763, 770*f*
Vascular cambium, **766***f*, 772*f*–774*f*, 849
Vascular cylinder, **763**
Vascular plants, **622**. *See also* Seedless vascular plants;
  Seed plants
 origin and traits of, 628*f*–631*f*
 phylogeny of, 622*t*
 resource acquisition for, 784*f*–786*f*, 787
 seedless. *See* Seedless vascular plants
 transport in. *See* Transport in vascular plants
Vascular rays, 773*f*
Vascular tissue, **622**, 785*f*
Vascular tissue system, plant, 758, **763***f*, 765*f*
Vascular transport, 536, 629–630
Vas deferens, **1025***f*
Vasectomy, 1038*f*, **1039**
Vasocongestion, 1034
Vasoconstriction, 885, **931**
Vasodilation, 885, **931**, 1001, 1026
Vasopressin. *See* Antidiuretic hormone
Vectors, zoonotic disease, **1234**, 1235*f*
Vegetable oil, 49–50, 74
Vegetarian diets, 899
Vegetal pole, **1048***f*
Vegetation, climate effects of, 1169*f*
Vegetative growth, 830
Vegetative propagation, **835**–836, 849
Vegetative reproduction, **833**–834
Veins, blood, **924***f*–926*f*, 929*f*–932*f*
Veins, leaf, 189*f*, **761***f*, 771*f*, 849
Veldts, 1175*f*
Velocity
 blood flow, 930*f*
Velvetleaf plants, 205
Velvet worms, 689*f*
Venae cavae, 926*f*, 930*f*
Venom, snail, 1067*f*
Venomous snakes, 738
Venter, Craig, 443*f*
Ventilation, **941**. *See also* Breathing
 of lungs, 944, 945*f*–946*f*
Ventral nerve cords
 earthworm, 704*f*
 planarian, 696*f*
Ventral side, 680*f*
Ventral tegmental area (VTA), 1103*f*
Ventricles, brain, 1087*f*
Ventricles, heart, **924***f*–927*f*
Venules, **924**, 929*f*–930*f*, 932*f*
Venus flower basket glass sponge, 146*f*
Venus flytrap (*Dionaea muscipula*), 802, 819*f*, 862
Vernalization, **860**
Verreaux's sifaka (*Propithecus verreauxi*), 746*f*
Vertebrae, 718
Vertebrates, **683**, **719**

action potential conduction speed in, 1076
adaptations of digestive systems of, 911*f*–914*f*.
  See also Digestive systems
amniotes and development of terrestrially
  adapted eggs in, 734*f*–740*f*. *See also* Amniotes
anatomical homologies in embryos of, 479*f*, 480
brains of, 1091*f*–1096*f*, 1099*f*. *See also* Nervous
  systems
cardiovascular systems of. *See* Cardiovascular
  systems
as chordates, 689*f*, 715, 719*f*, 722, 723*f*–725*f*.
  See also Chordates
circulatory systems of, 924*f*, 925. *See also*
  Cardiovascular systems
derived characters of, 722, 723*f*
energy budgets for terrestrial, 892
evolution of, 678
evolution of backbones and diversity vs. disparity
  in, 718*f*, 719
fossils and early evolution of, 724*f*–725*f*
gamete production and delivery in, 1024
gas exchange systems of. *See* Gas exchange
gnathostomes and development of jaws in,
  725*f*–729*f*, 730
hominins and humans, 748*f*–754*f*
innate immunity in, 954*f*–956*f*, 957
kidneys in excretory systems of, 985, 986*f*–987*f*,
  991*f*–993*f*. *See also* Excretory systems;
  Kidneys
limb formation in, 1062, 1063*f*, 1064
mammals, 741*f*–747*f*
mechanoreceptors for hearing and equilibrium
  in, 1112*f*–1116*f*. *See also* Sensory systems
nervous system of, 1086*f*–1090*f*. *See also* Nervous
  systems
neuroendocrine signaling in, 1007*f*–1008*f*
organogenesis in, 1055*f*
origins of bone and teeth in, 725
phylogeny of, 683*f*
small intestine surface area in, 695*f*
tetrapods and development of limbs in, 730*f*–733*f*
visual systems of, 1118*f*–1122*f*, 1123
Vertical layering, terrestrial biome, 1172
Vertical transmission, viral, 412
Vervet monkeys, 1148*f*
Vesicles, **104**
 abiotically produced, as protocells, 527, 528*f*
 in endomembrane system, 104
 in exocytosis and endocytosis, 139, 140*f*
 in plant cytokinesis, 241, 242*f*
 self-replicating RNA in, 528
 transport, 105, 106*f*, 107, 109*f*
Vessel elements, 648, **765***f*
Vessels, 765*f*
Vessels, blood. *See* Blood vessels
Vessels, lymph, 933*f*
Vestibular glands, 1027*f*
Vestigial structures, **479**
Viagra, 224, 1001, 1026, 1082
Vibrio cholerae, 224, 584*f*, 588
Viewpoints, science and diverse, 24
Villi, 695*f*, **909***f*, 910
Vinegar, 49–50
Viral envelopes, **400***f*, 401, 405*f*
Viral integration, 395
Viral movement proteins, 802*f*
Virtual plants, computer-generated, 776
Virulent phages, **402***f*
Viruses, **315**, **399**
 analyzing sequence-based phylogenetic trees to
  understand evolution of, 411
 antibodies as medical tool for detection of, 969*f*
 antigenic variation in, 972–973
 cancer-causing, 974*f*
 in cancer development, 394–395
 cellular RNAi pathway and, 380
 classes of animal, 406*t*
 climate change and, 412
 discovery of, 399*f*
 emerging, 409*f*–412*f*
 evolution of, 406, 408, 411
 features of replicative cycles of, 401*f*, 402
 host range of, 401
 immune system recognition of, 952*f*
 importance of, 400

infection of bacterial cells by, 315*f*–316*f*, 317
influenza. *See* Influenza viruses
insect defense against, 954*f*
latency of, 973*t*
mutations of, 409, 414
as pathogens, 398*f*–400*f*, 401, 408, 409*f*–413*f*
rapid reproduction of, as source of genetic
  variation, 489
replication of, 398
replicative cycles of animal, 404, 405*f*, 406*t*, 407*f*
replicative cycles of phages as, 402*f*–404*f*
structure and function of, 414
structure of, 399, 400*f*, 401
viral movement proteins of plant, 802
Visceral mass, **699***f*
Visible light, **192***f*
Visual communication, 1142*f*
Visual cortex, 1121*f*
Visualizing Figures
 biogeochemical cycles, 1249*f*
 DNA, 319*f*
 gastrulation, 1050*f*
 molecular machinery in cell, 122*f*–123*f*
 phylogenetic relationships, 556*f*
 primary and secondary growth, 767*f*
 proteins, 79*f*
 scale of geologic time, 532*f*–533*f*
Visual pigments, 1119*f*, 1122
Visual Skills Questions, 6*f*, 10*f*, 34*f*, 36, 45, 53*f*, 58,
  62, 64–65, 79*f*, 95*f*, 98*f*, 114*f*, 127*f*, 130*f*–133*f*,
  137*f*, 139, 140*f*, 150, 167*f*, 171, 174*f*, 177*f*,
  184, 186, 198*f*, 216*f*, 227*f*–228*f*, 246*f*, 252,
  257*f*–258*f*, 259, 280*f*–281*f*, 306, 320, 329,
  340*f*–341*f*, 348*f*, 355*f*, 367*f*, 374*f*, 377*f*, 401,
  408, 423, 428, 443*f*, 451*f*, 453*f*, 457*f*, 474*f*,
  480*f*, 482*f*, 494*f*, 504, 536*f*, 539*f*, 558, 564*f*,
  572, 576*f*, 579*f*, 583*f*, 596*f*, 603*f*, 611*f*, 613*f*,
  624*f*, 638*f*, 648*f*, 679, 683*f*, 722, 734*f*, 741,
  746*f*, 770*f*, 772*f*, 774*f*, 786*f*, 791*f*, 793*f*, 807*f*,
  815*f*–816*f*, 827*f*, 829*f*–830*f*, 853*f*, 860*f*, 906*f*,
  910*f*, 926*f*, 946*f*, 954*f*–955*f*, 962*f*, 967*f*, 996*f*,
  1003*f*, 1015*f*, 1024*f*, 1035*f*, 1037*f*, 1072*f*,
  1079*f*, 1096*f*–1097*f*, 1119*f*, 1126*f*, 1142*f*, 1147*f*,
  1188, 1195*f*, 1224*f*, 1240*f*, 1242*f*, 1252*f*, 1271*f*.
  See also Draw It Questions
Visual systems, 1117*f*–1123
Vital capacity, **945**
Vitamin A, deficiencies of, 838*f*
Vitamin D, 1003, 1011*f*
Vitamins, 158, **900**
 deficiencies of, 902*f*
 effects of supplementation of, on neural tube
  defects in, 902*f*
 as lipid-soluble hormones, 1003
 parathyroid hormone and, 1011*f*
 requirements for, 899*f*, 900*t*
Vitelline layer, 1045
Vitellogenin, 1003*f*
Vitreous humor, 1118*f*
Viviparous organisms, **727**
Vocal cords, 942, 943*f*
Vocal folds, 942, 943*f*
Vocalization, *FOXP2* gene and, 461, 462*f*
Vogt, Walther, 1058
Volcanic springs, 585*f*
Volcanism, 636*f*, 637
Volcanoes, 57–58, 526*f*, 527, 540, 551, 1282
Voles, 1155*f*
Voltage-gated ion channels, 220*f*, **1072***f*–1077*f*
Volume, 99, 1114
Volume-surface area relationships, 98*f*
Voluntary contraception, population growth and,
  1208
Volutidae family, 538
Volvox, 599*f*, 610*f*
von Frisch, Karl, 1142
von Humboldt, Alexander, 1232
Vulva, **1026**, 1027*f*

W
Waggle dance, honeybee, 1142*f*
Wagler's pit viper (*Tropidolaemus wagleri*), 737*f*
Waists, chromatid, 236
Walking, 756, 1135
Wallace, Alfred Russel, 469*f*, 473, 1232

Walled spores, 621*f*
Wandering albatross (*Diomedea exulans*), 977*f*
Warming, global. *See* Climate change
Warren, Robin, 912
WAS (Wiskott-Aldrich syndrome), 229
Washington, Earl, 437*f*
Wasps, 672, 712*f*, 822*f*, 869*f*, 1145*f*
Waste, toxic. *See* Toxic waste
Wastes, nitrogenous. *See* Nitrogenous wastes
Water
    acidic and basic conditions of, and living
        organisms, 51, 52*f*, 53
    acidification as threat to quality of, 53*f*, 54
    albatross drinking of salty, 977*f*
    biomanipulation and quality of, 1227*f*
    in blood plasma, 934*f*
    cellular regulation of, 142
    cohesion of molecules of, 45*f*–46*f*
    conduction of, in plant cells, 765*f*
    covalent bonding of, 37*f*, 45*f*
    evolution of life on planets with, 50*f*
    floating of ice on liquid, 44, 48*f*
    forms of, 44
    fruit and seed dispersal by, 832*f*
    gain and loss of, 998
    hydrogen bonds and, 39*f*
    imbibition of, by seeds, 829
    ions in, 38
    irrigation with, 808
    latitudinal gradients and evapotranspiration of,
        1232*f*
    as limiting factor on human population size, 1211
    maximizing body surface area and uptake of, 695*f*
    moderation of temperature by, 46, 47*f*–48*f*
    molecular shape of, 39*f*, 40
    molluscs and pollution of, 702–703
    photosynthesis and compromise with loss of, 787
    plant adaptations for reducing evaporative loss
        of, 798, 799*f*
    in plant composition, 809
    plant response to submergence in, 864*f*
    plant transport of, 46*f*
    properties of, 45*f*–50*f*
    regulation of transpiration and plant loss of, 796,
        797*f*–799*f*
    root architecture and acquisition of, 787
    roots interaction with, 10*f*, 11
    seed dispersal by, 645*f*
    as solvent of life, 49*f*, 50, 55
    species distributions and availability of, 1185
    splitting of, in photosynthesis, 190*f*
    transport of, across plant plasma membranes,
        788, 789*f*–791*f*
    transport of, from roots to shoots via xylem, 792,
        793*f*–795*f*, 796
    uptake of, 898
Water balance
    effects of osmosis on, 133*f*–135*f*
    hormonal regulation of, 994*f*–996*f*
    osmoregulation of, 977*f*–982*f*
Water bears, 689*f*
Water bodies, climate and, 1168*f*–1169*f*
Water bugs, 1023*f*
Water conservation
    kidney adaptations for, 991*f*–992*f*
    kidney role in, 989, 990*f*, 991
Water cycle, 1250*f*, 1259
Water cycling, in ecosystems, 1248*f*–1252*f*
Water fleas (*Daphnia pulex*), 448*t*, 1021, 1199*f*, 1200
Waterland, Robert, 372
Water lily, 650*f*, 653
"Watermelon snow," 211
Water potential, **788**, 789*f*–791*f*
Water scorpions, 708
Water snakes, 496, 497*f*
Water-soluble hormones, 1001*f*–1003*f*
Water-soluble vitamins, 900*t*
Water vascular system, **713**, 714*f*
Watson, James
    discovery of DNA molecular structure by, 4,
        23–24, 314*f*, 317, 318*f*–320*f*
    model of DNA replication by, 320, 321*f*
Wavelengths, electromagnetic, 192*f*–194*f*, 195

Wawona Sequoia, 774*f*
WD40 domains, 445*f*, 446
Weak acids, 51
Weather
    climate change and, 11
    population fluctuations and, 1205*f*. *See also*
        Climate
Weathering, rock, 629
Web of life, 570*f*
Weddell seals, 949*f*
Weeds, transgene escape and, 839
Weevils, 712*f*
Welwitschia, 642*f*
Wernicke, Karl, 1098
Wernicke's area, 1098
Westemeier, Ronald, 1267*f*
Western garter snakes, 1155, 1156*f*
Western gulls, 1151*f*
West Indian manatee (*Trichechus manatus*),
    572, 1219*f*
West Nile virus, 401, 409
Wetlands, **1179***f*, 1262
    restoration of, 1254*f*
Whales, 88*f*, 481*f*–482*f*, 525*f*, 526, 557*f*, 710, 719,
    903*f*, 1111*f*, 1265
Wheat (*Triticum aestivum*), 514–515, 524, 595, 651,
    1204*f*
Whisk ferns, 631, 632*f*, 633
White, John, 24*f*
White-band disease, 1234
White blood cells. *See* Leukocytes
White-crowned sparrows, 1147
White-footed mice, 1144*t*
White matter, **1087***f*
White rhinoceros, 1199*f*, 1266
White-tailed deer (*Odocoileus virginianus*), 1271
Whole-genome shotgun approach, DNA sequencing,
    443*f*, 444, 447, 448*f*, 452
Whooping cough, 218*f*
Whooping crane (*Grus americana*), 1144
Whorled phyllotaxy, 786
Widow's peak pedigree analysis case, 284*f*, 285
Wieschaus, Eric, 386
Wildfires, 1172, 1228, 1229*f*, *1244f*
Wild mustard, artificial selection and, 475*f*
Wild tobacco plants, 869*f*
Wild types, **295***f*
Wilkins, Maurice, 317–318, 320
Wilson, E. O., 1159, 1232, 1233*f*, 1262*f*, 1263
Wilting, **790**, 798, 804, 852, 863
Wind
    climate change and, 11
    dispersal of mosses by, 627
    flower pollination by, 824*f*
    fruit and seed dispersal by, 832*f*
    global patterns of, 1166*f*
    seed dispersal by, 645*f*
Windpipes, 13*f*
Winged fruits and seeds, 832*f*
Wings
    bat, as evolutionary adaptation, 15*f*
    bird, 738*f*, 739
    evolution of, 679
    flight muscles and, 1136
    insect, 710, 711*f*–712*f*
    muscle contraction and insect, 1132
    pterosaur, 736
    seed, 645*f*
Wireframe models, 79*f*
Wiskott-Aldrich syndrome (WAS), 229
Witchweed, 855
WNT4 gene, 299
Wobble, **349**
Wollemi pine, 643*f*
Wolves, 529*f*, 1205*f*, 1257*f*, 1270
Wood, 118, 651, 783
Wood ants, 28*f*
Wood lice, 709
Woolly mammoths, 421
Work, cellular, 150, 151*f*–153*f*
World Trade Center attack, 437
Worldwide adaptive radiations, 542, 543*f*
Worms, 905*f*, 1133*f*

X
Xanthomonas, 334
X chromosomes, 256*f*–257*f*, 298*f*–300*f*, 381, 1030,
    1040
Xenarthra, 745*f*
Xeroderma pigmentosum (XP), 328
Xerophytes, **798**, 799*f*
X-linked genes, **299***f*–300*f*
X-O sex determination system, 298*f*
X-ray crystallography, **83***f*, 96, 217, 317, 318*f*
X-rays
    cancer and, 388
    mutations from, 360
Xylem, **629**, **763**, **785**
    primary growth and, 769*f*
    resource acquisition and, 785*f*
    transport of water and minerals from roots to
        shoots via, 792, 793*f*–795*f*, 796
    vascular plant, 629–630
    in vascular tissue systems, 763
    water-conducting cells of, 765*f*
Xylem sap, **792**, 793*f*–795*f*, 796
X-Y sex determination system, 298*f*, 313

Y
Yamanaka, Shinya, 432*f*
Yangtze River dolphin, 1262*f*
Yarrow (*Achillea lanulosa*), 1189
Y chromosomes, 256*f*–257*f*, 298*f*, 299, 313, 1030
Yeast cells, 100*f*
Yeast infections, 670
Yeasts, **655**
    alcohol fermentation by, 180*f*
    asexual reproduction in, 659*f*
    cell division in, 244*f*
    cell signaling in, 213*f*, 214
    expressing cloned eukaryotic genes in, 422–423
    fungi as, 655, 670
    human uses of, 670–671
    model organism. *See Saccharomyces cerevisiae*
Yellowfin tuna (*Thunnus albacares*), 729*f*
Yellow jackets, 1218*f*, 1219
Yellow-legged frogs, 669, 670*f*
Yellowstone National Park, 441, 1270
    extreme thermophiles in, 585*f*, 586
    forest fire disturbance of, 1228, 1229*f*
    grizzly bear population in, 1268*f*, 1269,
        1272–1273
    Taq polymerase from bacterium in, 1263
Y-linked genes, 299
Yolk, 91, **1048***f*
Yolk sac, 735*f*, 1053*f*, 1054
Young, Michael, 883
Yucca, 824*f*

Z
Zambia, age-structure pyramid for, 1209*f*
Zambia, elephant poaching in, 1265*f*
Zea mays (corn). *See* Corn; Maize
Zeatin, 850
Zebra finches, 1152*f*
Zebrafish model organism, 22
Zebra mussels, 1264
Zero population growth, 1208, 1211
Zika virus, 409*f*
Zona pellucida, **1046***f*
Zonation, aquatic, 1177*f*–1178*f*
Zoned reserves, **1273***f*
Zone of cell division, 768*f*
Zone of differentiation, 768*f*
Zone of elongation, 768*f*
Zone of polarizing activity (ZPA), **1062**, *1063f*
Zoonotic pathogens, **1234**, 1235*f*
Zoopagomycetes, 660*f*, **662***f*, 663
Zooplankton, in biomass pyramids, 1247*f*, 1248
Zoospores, 603*f*, **661***f*
zur Hausen, Harald, 974
Z-W sex determination system, 298*f*
Zygentoma (silverfish), 712*f*
Zygnema, 619*f*
Zygomycetes, 660
Zygosporangium, 662*f*, **663**
Zygotes, 257*f*, **258**, **1020**, 1034, 1035*f*

The Metric System

Metric Prefixes:
10^{15} = peta (P) 10^{6} = mega (M) 10^{-2} = centi (c) 10^{-9} = nano (n)
10^{12} = tera (T) 10^{3} = kilo (k) 10^{-3} = milli (m) 10^{-12} = pico (p)
10^{9} = giga (G) 10^{-1} = deci (d) 10^{-6} = micro (µ) 10^{-15} = femto (f)

Measurement	Unit and Abbreviation	Metric Equivalent	Metric-to-English Conversion Factor	English-to-Metric Conversion Factor
Length	1 kilometer (km)	= 1,000 (10^3) meters	1 km = 0.62 mile	1 mile = 1.61 km
	1 meter (m)	= 100 (10^2) centimeters = 1,000 millimeters	1 m = 1.09 yards 1 m = 3.28 feet 1 m = 39.37 inches	1 yard = 0.914 m 1 foot = 0.305 m
	1 centimeter (cm)	= 0.01 (10^{-2}) meter	1 cm = 0.394 inch	1 foot = 30.5 cm 1 inch = 2.54 cm
	1 millimeter (mm)	= 0.001 (10^{-3}) meter	1 mm = 0.039 inch	
	1 micrometer (µm) (formerly micron, µ)	= 10^{-6} meter (10^{-3} mm)		
	1 nanometer (nm) (formerly millimicron, mµ)	= 10^{-9} meter (10^{-3} µm)		
	1 angstrom (Å)	= 10^{-10} meter (10^{-4} µm)		
Area	1 hectare (ha)	= 10,000 square meters	1 ha = 2.47 acres	1 acre = 0.405 ha
	1 square meter (m²)	= 10,000 square centimeters	1 m² = 1.196 square yards 1 m² = 10.764 square feet	1 square yard = 0.8361 m² 1 square foot = 0.0929 m²
	1 square centimeter (cm²)	= 100 square millimeters	1 cm² = 0.155 square inch	1 square inch = 6.4516 cm²
Mass	1 metric ton (t)	= 1,000 kilograms	1 t = 1.103 tons	1 ton = 0.907 t
	1 kilogram (kg)	= 1,000 grams	1 kg = 2.205 pounds	1 pound = 0.4536 kg
	1 gram (g)	= 1,000 milligrams	1 g = 0.0353 ounce 1 g = 15.432 grains	1 ounce = 28.35 g
	1 milligram (mg)	= 10^{-3} gram	1 mg = approx. 0.015 grain	
	1 microgram (µg or mcg)	= 10^{-6} gram		
Volume (solids)	1 cubic meter (m³)	= 1,000,000 cubic centimeters	1 m³ = 1.308 cubic yards 1 m³ = 35.315 cubic feet	1 cubic yard = 0.7646 m³ 1 cubic foot = 0.0283 m³
	1 cubic centimeter (cm³ or cc)	= 10^{-6} cubic meter	1 cm³ = 0.061 cubic inch	1 cubic inch = 16.387 cm³
	1 cubic millimeter (mm³)	= 10^{-9} cubic meter = 10^{-3} cubic centimeter		
Volume (liquids and gases)	1 kiloliter (kL or kl)	= 1,000 liters	1 kL = 264.17 gallons	
	1 liter (L or l)	= 1,000 milliliters	1 L = 0.264 gallon 1 L = 1.057 quarts	1 gallon = 3.785 L 1 quart = 0.946 L
	1 milliliter (mL or ml)	= 10^{-3} liter = 1 cubic centimeter (cc)	1 mL = 0.034 fluid ounce 1 mL = approx. ¼ teaspoon 1 mL = approx. 15–16 drops (gtt.)	1 quart = 946 mL 1 pint = 473 mL 1 fluid ounce = 29.57 mL 1 teaspoon = approx. 5 mL
	1 microliter (µL or µl)	= 10^{-6} liter (10^{-3} milliliter)		
Pressure	1 megapascal (MPa)	= 1,000 kilopascals	1 MPa = 10 bars	1 bar = 0.1 MPa
	1 kilopascal (kPa)	= 1,000 pascals	1 kPa = 0.01 bar	1 bar = 100 kPa
	1 pascal (Pa)	= 1 newton/m² (N/m²)	1 Pa = 1.0×10^{-5} bar	1 bar = 1.0×10^{5} Pa
Time	1 second (s or sec)	= ¹⁄₆₀ minute		
	1 millisecond (ms or msec)	= 10^{-3} second		
Temperature	Degrees Celsius (°C) (0 K [Kelvin] = −273.15°C)		°F = ⁹⁄₅°C + 32	°C = ⁵⁄₉ (°F − 32)